Fracture Mechanics of Ceramics

Volume 2

Microstructure, Materials, and Applications

Fracture Mechanics of Ceramics

Volume 2
Microstructure, Materials, and Applications

Edited by R. C. Bradt

Department of Materials Science
Ceramics Science Section
Pennsylvania State University
University Park, Pennsylvania

D. P. H. Hasselman

Ceramics Research Laboratory
Materials Research Center
Lehigh University
Bethlehem, Pennsylvania

and F. F. Lange

Research and Development Center
Westinghouse Electric Corporation
Pittsburgh, Pennsylvania

PLENUM PRESS • NEW YORK-LONDON

Library of Congress Cataloging in Publication Data

Symposium on the Fracture Mechanics of Ceramics,
  Pennsylvania State University, 1973.
  Fracture mechanics of ceramics.

  Includes bibliographical references.
  CONTENTS: v. 1. Concepts, flaws, and fractography. —v. 2. Microstructure,
materials, and applications.
  1. Ceramics—Fracture—Congresses. I. Bradt, Richard Carl, 1938-   ed. II.
Hasselman, D. P. H., 1931-   ed. III. Lange, F. F., 1939-   IV. Pennsylvania
State University. V. Title.
TA430.S97   1973   620.1'4   73-20399
ISBN 0-306-37592-3 (v. 2)

Second half of the proceedings of a symposium on the
Fracture Mechanics of Ceramics, held at Pennsylvania State University,
University Park, Pennsylvania, July 11-13, 1973

© 1974 Plenum Press, New York
A Division of Plenum Publishing Corporation
227 West 17th Street, New York, N.Y. 10011

United Kingdom edition published by Plenum Press, London
A Division of Plenum Publishing Company, Ltd.
Davis House (4th Floor), 8 Scrubs Lane, Harlesden, London, NW10 6SE, England

PREFACE

These volumes constitute the Proceedings of a Symposium on the Fracture Mechanics of Ceramics, held at the Pennsylvania State University, University Park, Pennsylvania, July 11, 12, and 13, 1973.

The theme of the symposium focussed on the mechanical behavior of brittle ceramics in terms of the characteristics of cracks. The 52 contributed papers by 87 authors, present an overview of the current understanding of the theory and application of fracture mechanics to brittle ceramics.

The program chairmen gratefully acknowledge the financial assistance for the Symposium provided by the Office of Naval Research, the College of Earth and Mineral Sciences of the Pennsylvania State University, the Materials Research Center of Lehigh University, Bethlehem, Pennsylvana and Westinghouse Research Laboratories, Pittsburgh, Pennsylvania.

Special appreciation is extended to the expert organization provided by the J. Orvis Keller Conference Center of the Pennsylvania State Conference Center of the Pennsylvania State University. In particular, Mrs. Patricia Ewing should be acknowledged for the excellent program organization and planning. Dean Harold J. O'Brien, who was featured as the after-dinner speaker and who presented a most stimulating talk on the communication between people, also contributed to the success of the meeting.

Finally, we also wish to thank our joint secretaries for the patience and help in bringing these Proceedings to press.

University Park   R. C. Bradt

Bethlehem   D. P. H. Hasselman

Pittsburgh, Pennsylvania   F. F. Lange

July, 1973

v

CONTENTS OF VOLUME 2

V.   SUBCRITICAL CRACK GROWTH

VI.   ENGINEERING, SCIENTIFIC, AND DESIGN APPLICATIONS

CONTENTS OF VOLUME 1

III.   FRACTOGRAPHY

IV. MICROSTRUCTURAL EFFECTS

EFFECTS OF MICROSTRUCTURE ON THE MECHANICAL PROPERTIES OF CERAMICS

R. W. Davidge

Materials Development Division, Atomic Energy Research

Establishment, Harwell, OX11 ORA, U.K.

ABSTRACT

This review considers the influence of microstructure on the strength of ceramics in terms of the key variables of fracture mechanics - a critical flaw size and a surface energy. Attention is focused on four types of material: single crystals, simple polycrystals, multiphase materials, and ceramic based composites. The relationship between flaw size and microstructure is well established and the flaw size is related to the size of pores or grains. The microstructural factors controlling surface energy are generally not understood. For ceramic composites it is not certain that the fracture mechanics approach is of direct applicability; however, the work of fracture can be explained by consideration of microstructural effects.

1. INTRODUCTION

There are two main groups of parameter that affect the mechanical properties of ceramics: material properties (including microstructure) and environmental conditions. Table I lists these. Although this review is concerned with the former it is obviously essential to ensure that the latter are kept constant when making general comparisons. To keep the scope of the discussion within reasonable bounds it is proposed to concentrate mainly on properties derived under ambient conditions of atmosphere and temperature using simple bend tests (equivalent to uniaxial tension). This excludes highly important areas such as the effects of multiaxial stressing, temperature, strain rate and time dependent effects, and statistical variations of strength, many of which are covered in other parts of this conference.

447

Material Parameters	Environmental Parameters
Composition	Temperature
Crystal structure	Atmosphere
Conventional microstructure: Pore size; grain size; geometrical phase distribution; cracks; surface condition	Strain rate
	Static or cyclic fatigue
	State of stress, e.g., uniaxial or biaxial.
Internal strains, related to differences in expansion coefficient and elastic constants	
Specimen shape and size.	

Table I. Some parameters affecting the mechanical properties of ceramics.

The key mechanical property is strength (σ). A group of properties - fracture toughness, effective surface energy and work of fracture - is also important and in some cases are related directly to strength. Defining γ_i as an effective surface energy for the initiation of fracture from an inherent flaw of size C, then[1]

$$\sigma = \frac{1}{Y}\left(\frac{2E\gamma_i}{C}\right)^{\frac{1}{2}} \qquad (1)$$

where Y is a geometrical constant and E Young's modulus. $(2E\gamma_i)^{\frac{1}{2}}$ is equal to the stress intensity factor K_{Ic} but for current purposes, when examining microstructural effects, it is more convenient to consider γ_i. Also related to γ_i is the work of fracture γ_f, defined as the energy required to generate unit area of fracture face. In general $\gamma_f > \gamma_i$ but when γ_f is measured in a test involving slow controlled crack growth (rather than a partly catastrophic crack growth) $\gamma_f = \gamma_i$ [2].

Because ceramics are brittle catastrophic failure is common once the fracture stress has been reached. However, this is not inevitable and it is instructive to consider what degree of toughness would prevent this behaviour. Consider for example a beam of material 10 x 10 x 100 mm. supported at the ends and deformed in three-point bending. Table II considers typical materials with strengths of 10^8 and 10^9 N m^{-2} with Young's moduli 4.10^{10} and 4.10^{11} N m^{-2}. When raised to the fracture stress the elastic energy in the bar is $\sigma^2/18E$ multiplied by the specimen volume [3] and Table II shows the works of fracture required to prevent complete fracture of the bar. Fig. 1 demonstrates the ranges of

σ Strength $N\ m^{-2}$	E Young's Modulus $N\ m^{-2}$	$10^{-5}\sigma^2/18E$ Elastic Energy J	γ_f Work of Fracture to Prevent Failure $J\ m^{-2}$
10^9	4.10^{11}	1.3	7000
10^8	4.10^{11}	0.013	70
10^8	4.10^{10}	0.13	700

Table II. Elastic energy in beams 10x10x100 mm, deformed to the fracture stress in three-point bending, and works of fracture required to prevent catastropic failures in the beams.

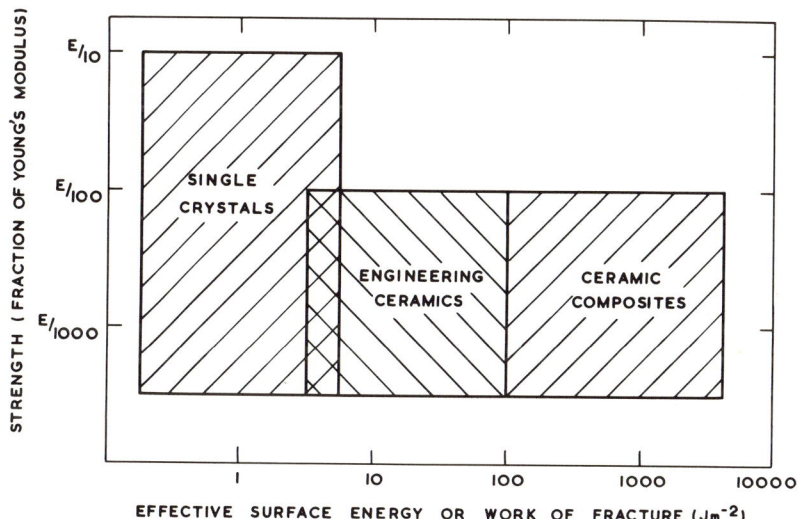

Fig. 1. Typical strength and surface energy values for ceramics.

strength (expressed as a fraction of modulus) and work of fracture values typically observed for various ceramics. Comparison of Table II and Fig. 1 indicates that there is little hope in making unreinforced ceramics adequately tough and the trend towards increased strengths exacerbates the situation. However, with reinforced ceramics work of fracture values of the right order are found and here there is hope of obtaining materials that are both strong and tough.

Four types of material will thus be discussed, with increasing complexity of microstructure: single crystals, simple polycrystals, multiphase materials, and ceramic based composites. The aim is to enquire how well the fracture mechanics approach to strength, essentially Eqn. (1), can be used to generate an understanding of strength in terms of microstructural variables. For individual materials from a single source the agreement between experiment and theory is usually good (1). It will be shown however that in comparing similar materials from different sources, or the results of different workers, many of the underlying principles are far from clear. This is partly because of the large number of factors involved, Table I, and partly because most materials are insufficiently characterised to enable general conclusions to be drawn.

An advantage of the fracture mechanics approach is that it allows strength to be discussed in terms of two key variables: γ_i and C. It is intuitively obvious to expect simple relationships to exist between flaw size and grain size or pore size, but in the past there has been too little emphasis on the factors controlling surface energy. It has been assumed that there is no variation of γ_i with microstructure, or any variation ignored, and this has led to the all-embracing type of review where general grain size strength relationships have been sought (4,5). This practice is dangerous and there is no reason to suppose, except in certain circumstances, that such relationships do exist.

2. SINGLE CRYSTALS

The study of single crystals plays an important historical role in the development of the fracture mechanics of ceramics. Workers in the early 60's measured the surface energy of a range of crystals, particularly of NaCl structures, (6,7) using double cantilever beams with large pre-formed sharp cracks. The general conclusion was that, if care was taken to suppress plastic deformation at the crack tip, γ_i was equal to the thermodynamic surface energy γ_o, as calculated from fundamental data on the strength of the atomic bonds. If plastic deformation did occur then an additional term γ_p appeared and $\gamma_i = \gamma_o + \gamma_p$. Further terms can be added to account for shear processes near cleavage steps and other surface irregularities, although Lange and Lambe (8) showed that in MgO relatively little energy is expended in cleavage step formation.

More recently, Wiederhorn (9) measured surface energies of 7.3 and 6.0 J m^{-2} for the $(10\bar{1}0)$ and $(\bar{1}012)$ planes of sapphire. Some surface irregularities were observed but there was no evidence of plastic deformation. The values obtained were higher than those expected from thermodynamic arguments and gave estimates of the theoretical strength of sapphire about three times greater than experimental values. Attempts to induce fracture on (0001) planes

gave estimates of $\gamma_i > 40$ J m^{-2}. It is likely then that the lack
of a preferred cleavage plane in sapphire leads to imperfect cleavage surfaces and high values of cleavage surface energy. Similar
arguments may apply to other crystals with poorly defined cleavage
planes but further work is required to elucidate the fundamental
mechanisms.

3. POLYCRYSTALS

To unravel the basic relationships between strength and microstructure it is necessary first to understand the separate effects
of structure on surface energy and flaw size.

3.1 Surface Energy

In view of the above situation for single crystals it is not
surprising that there has been little headway in arriving at a
quantitative understanding of the surface energy of polycrystals.
Nevertheless in large (200 μm) grain sized MgO Evans (10) calculated the contribution of γ_p to the surface energy, from etch pit
observations, and concluded that this was the major effect (γ_p at
least 9 J m^{-2} cf. γ_i 14 J m^{-2}). This conclusion is probably of
general validity because the high stresses near the tip of a propagating crack will exceed the flow stress for a narrow zone adjacent
to the fracture face. There is X-ray (11) and electron microscope
(12) evidence for such plastic flow in polycrystalline alumina.

For a given material, γ_i can be measured reliably and repro-
ducibly by techniques involving a number of specimen geometries.
Meredith and Pratt, (13) for example, measured γ_i for three 95%
polycrystalline aluminas using three different analytical techniques
- notched bend, double cantilever and compact tension - and a wide
range of initial crack sizes; for each material very consistent
results were obtained.

Problems arise, however, when it is required to study systematically changes in selected variables. In principle, this could
be achieved in a pure single phase polycrystal where grain size
could be varied by suitable heat treatments. But in real materials
even small amounts of impurity present will be distributed differ-
entially between grain interiors and grain boundaries. Thus the
chemical and physical nature of the grain boundaries will change
with grain size.

There are a large number of data in the literature for γ_i but
it is most useful to consider those for families of materials of
the same chemical origin. Attempts to correlate data for similar
materials of different origin usually produces little positive

Material	Grain Size (μm)	Porosity (%)	γ_i (J m^{-2})	Reference
Al$_2$O$_3$	10	0	18	Gutshall and Gross (17)
	30	0	27	
	45	0	46	
Al$_2$O$_3$	3	5	20	Evans and Tappin (23)
	30	5	20	
	100	5	10	
	3	20	16	
	3	50	10	
Al$_2$O$_3$ (95%)	25	5	47	Perry and Davidge (20)
	35	6	42	
	45	6	36	
Al$_2$O$_3$ (95%)	50	5	52	Perry and Davidge (20)
	50	6	46	
	90	6	110	
MgO	20	0	4	Evans and Davidge (19)
	50	0	5	
	200	0	5	
UO$_2$	8	3	8	Evans and Davidge (21)
	25	3	5	

Table III. Effects of Grain Size and Porosity on γ_i.

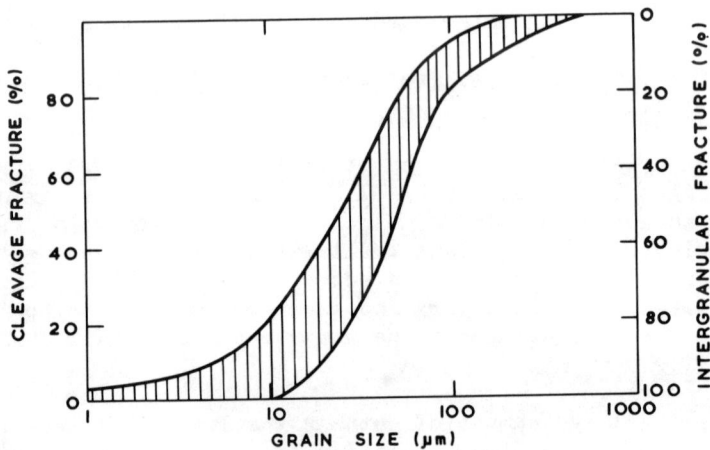

Fig. 2. Mode of failure of MgO as function of grain size (18).

conclusion: very detailed studies of twenty commercial aluminas
(14-16) led to no obvious correlations between surface energy and
strength, and a range of microstructural parameters.   Some self-
consistent batches of data are given in Table III.   Gross and
Gutshall (17) obtained an increase in γ_i with increase in the grain
size of Lucalox alumina and this was linked with an increase in the
amount of transgranular fracture.   This is a commonly quoted effect
and Fig. 2 shows how the mode of fracture of a batch of MgO changes
from predominantly intergranular fracture at small grain size to
predominantly transgranular fracture at large grain size (18).
However, Evans and Davidge (19) found very little grain size dependence of γ_i for MgO.   For 95% aluminas, Perry and Davidge (20)
observed an increase in γ_i with grain size for one material and a
decrease in the other.   In UO_2, material of the higher grain size
is associated with the smaller fracture energy; this was attributed tentatively by Evans and Davidge (21) to a marked grain size
dependence of the distribution of impurity.

Probably the most comprehensive and self-consistent results
have been obtained recently by Simpson (22) using notched beams of
Al_2O_3.   Material with a range of porosity and grain size was produced.   An important (but fortuitous) feature of these materials
is that fracture was almost entirely intercrystalline for all microstructures, so that one key variable, the mode of fracture, could
be eliminated.   This constancy of fracture mode was partly due to
the situation of the porosity at grain boundaries. γ_i falls
markedly with increase in porosity as expected and this in agreement

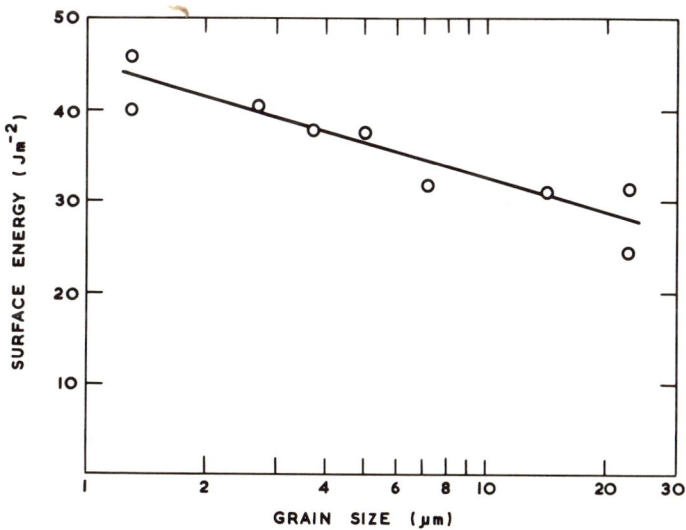

Fig. 3.   Variation in γ_i with grain size in Al_2O_3 (22).

with the data of Evans and Tappin (23) for similar aluminas. γ_i
falls also with increase in grain size, Fig. 3, and Simpson suggested that this may be due to the larger residual stresses present
in the large grain size material due to anisotropic thermal contractions on cooling after fabrication.   (In fact it is the
residual strain energy, not the residual stress, that increases
with grain size, but this does not affect the argument).

In view of the numerous factors affecting the grain size
dependence of γ_i no general conclusion can be drawn, and it seems
unlikely, in the current state of knowledge, that estimates of
surface energy can be made for a material based solely on microstructural information.   Instead it is best to regard γ_i as an
empirically determined parameter.   However, the main factors controlling the γ_i grain size dependence appear clear and thus data
can often be rationalised with hindsight as follows.

Firstly, the grain boundary fracture energy γ_{gb} should be
less than the cleavage energy because of several effects: atomic
bonding at grain boundaries is imperfect; impurities or grain
anisotropy can lead to concentrations of strain energy at the
boundaries; porosity if present is often situated at grain boundaries.   Secondly, γ_{gb} should be less at large grain size than at
small grain size because the concentrations of impurity, porosity
and strain energy are all greater at large grain size.   Thus in
materials where there is a transition from inter to transcrystal-
line fracture with increase in grain size γ_i should also increase,
and in materials without a transition γ_i should decrease.   In
general this seems to be observed.   This analysis is consistent
with γ_p being the major contributing effect to γ_i.   Fracture at
grain boundaries occurs at lower stresses ($\gamma_{gb} < \gamma_o$) than for
cleavage fracture and this will affect in sympathy the γ_p term.

An additional complicating factor, concerning the test geometry, has been noted very recently by Simpson (24).   Aluminas
were used of 3 and 20 μm grain size: the materials used in the
previous study (22), Lucalox and Coors AD 999.   Beams notched
with a 0.15 mm wide slitting wheel, and double cantilever beams
with sharp starting cracks were prepared, with the results in
Table IV.   The key observation is that the double cantilever
measurements are significantly greater than those from notched
beams for 20 μm grain size Al_2O_3 but very similar for 3 μm grain
size materials.   There is no satisfactory explanation for this
anomaly, but Simpson suggests that the biaxial stress state at the
crack tip in the double cantilever geometry may be important if
the crack deviates from the median plane.   Simpson notes that the
anomaly is absent in both silicon carbides and graphite, as it was
in the results of Meredith and Pratt (13) mentioned above on 95%
aluminas.   Hopefully, the effects in large grain size Al_2O_3 are
an isolated case but this needs checking.

Material	Grain Size μm	Fracture Energy J m^{-2}	
		Notched Beam	Double Cantilever
Simpson	20	21	39
"	3	23	21
Lucalox	20	20	32
Coors AD999	3	20	19

Table IV. Grain size/fracture energy/test
method data for Al_2O_3 (24).

3.2 Flaw Size and Strength

Although γ_i can only be determined by direct measurement, and
the links with microstructure are unclear at present, there is a
very closely identified relation between flaw size and microstructure.
The important microstructural features are grain size and pore size.
Some idealised structures are shown in Fig. 4, with some equivalent
micrographs.   This subject has been reviewed thoroughly (1,25,26)
and the main conclusion is that the flaw size approximates to the
largest microstructural feature, i.e., the largest grain size in
structures A and B, the largest pore in structure D and the sum of
the largest grain and largest pores in structure C.   Using these
values for the flaw size and measured γ_i values there is excellent
agreement between calculated (via Eqn. (1)) and observed strengths.

The state of the specimen surface is of some relevance.   Most
specimens are tested with machined surfaces.   Normal machining
operations tend to produce grain size cracks in the surface region
because cracks can propagate through the first grain with a low
energy requirement.   However, specimens with polished surfaces
tend to be stronger, and those with heavily damaged surfaces,
weaker.

Fig. 5 demonstrates the effect of polishing on the strength of
MgO polycrystals with a range of grain size (19).   Machined specimens exhibit strength values expected from the presence of grain
sized cracks.   (Note that γ_i is independent of grain size for this
material and hence the straight line relationship).   Polished
specimens are appreciably stronger.   Here flaws are generated by a
plastic flow mechanism related to grain size, and the strength at
infinite grain size extrapolates to the flow stress of a single
crystal.   The flaw generation mechanism involves the pile up of
dislocations at grain boundaries leading to the production of a
grain sized crack.   Thus the flaw initiation stress controls
strength.   MgO is unusual in that limited plastic flow occurs

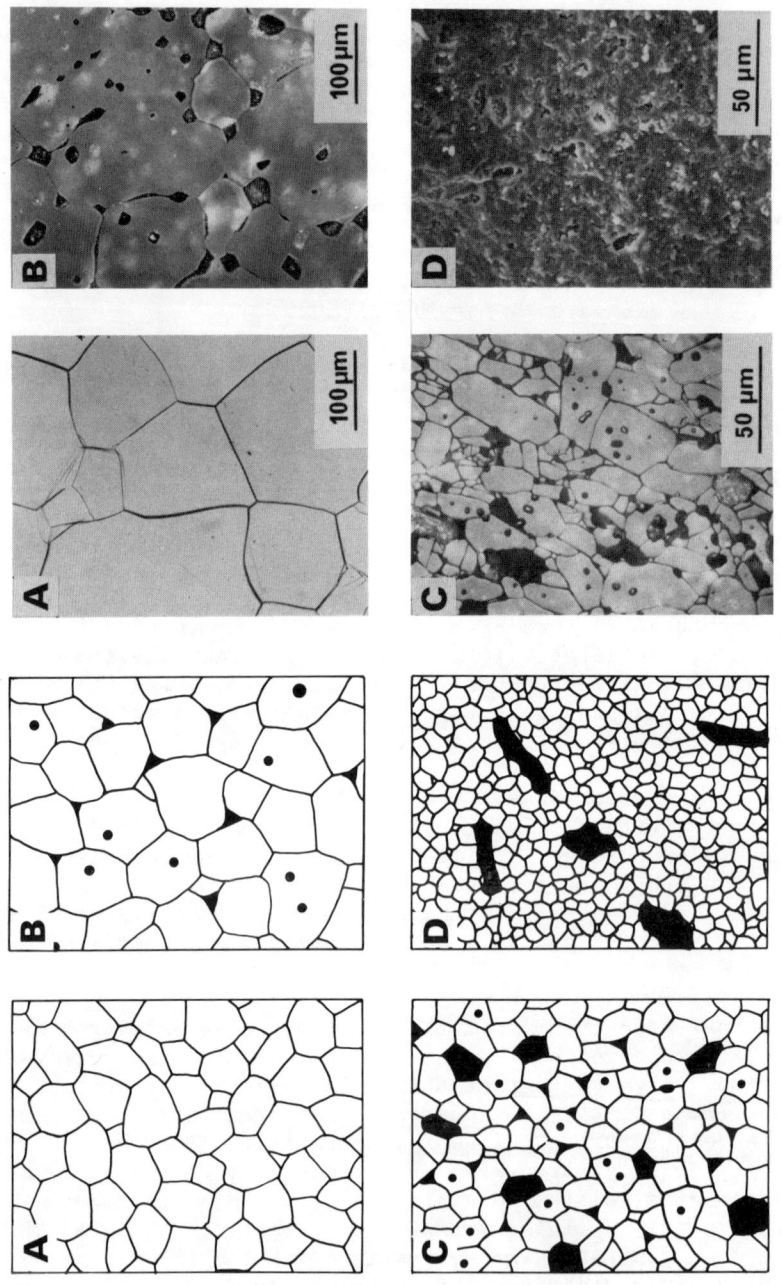

Fig. 4. Idealised and actual microstructures of ceramics. A-fully dense (MgO); B-porous with small pores (MgO); C-porous with pores and grains of similar size (Al$_2$O$_3$); D-porous with pores larger than grains (Si$_3$N$_4$) (25).

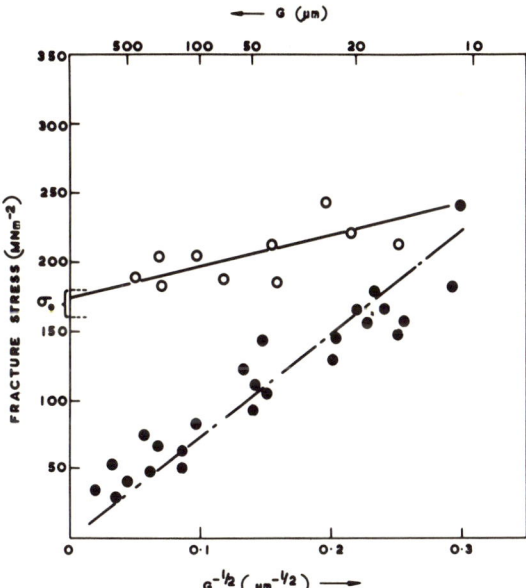

Fig. 5. Fracture strength of fully dense MgO as function of grain size. ● - machined surface; o - polished surface (19).

readily at ambient temperature but the same principle holds for other ceramics at higher temperature where an equivalence between flow stress and fracture stress is often observed (1).

The influence of machining or cutting operations on surface flaws is complex. Although flaws tend to be limited to a surface layer a grain deep this is not so for heavily abrasive conditions or in material of very fine grain size. A convenient way of investigating these effects is to measure the strengths of samples after single diamond indentations of varying severity. Wachtman (27) showed that the strength of Al_2O_3 of grain size of 2 and 5 μm was reduced for indentation loads greater than respectively 50 and 100 g. Below these loads strength was unaffected. For higher loads strength fell steadily with increase in indentation load. Fig. 6 gives strength values on a statistical basis for pyrolytic SiC subjected to indentation loads of 300-700 g (28). Both the mean strength and the scatter in strength fall with increase in indentation load.

Finally, a comment on the measurement of grain size is pertinent. Most workers quote only a mean grain size but as shown above it is the largest grain size that is important. Furthermore it is unlikely that materials of similar mean grain size show the same variation in grain size and thus comparisons between supposedly similar materials are dangerous. Aboav and Langdon measured the distribution of grain size in MgO (26) Fig. 7 gives their data

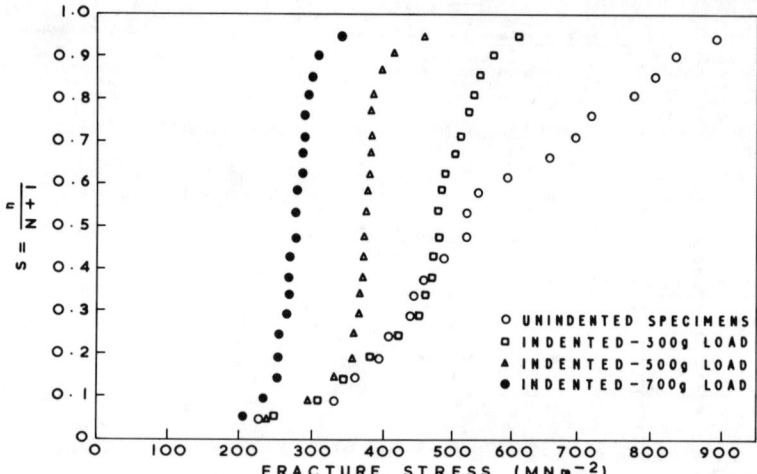

Fig. 6. Strength of pyrolytic SiC after
indentation at loads 300-700 g (28).

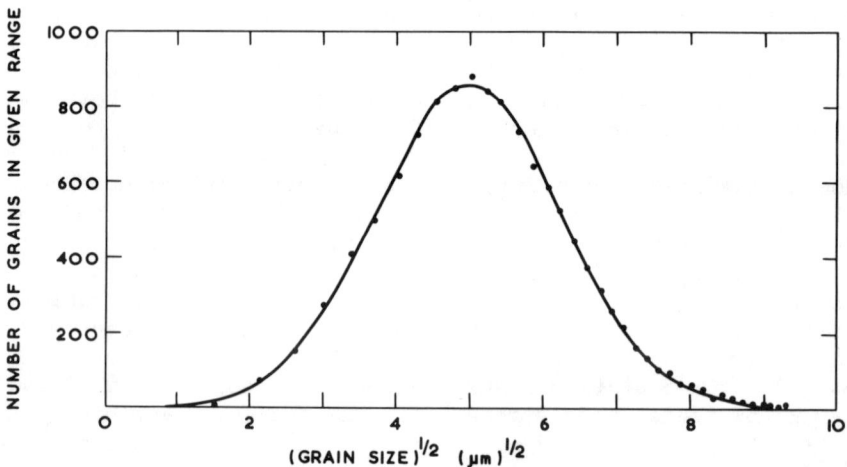

Fig. 7. Distribution of grain size (mean 25 µm) in MgO (29).

for MgO of mean grain size 25 µm based on measurements of 12000
grains.   The number of grains z having diameter x within a given
range is

$$z = 855 \exp\left[-0.314 \, (x^{\frac{1}{2}}-5.02)^2\right] \qquad (2)$$

There is a good general fit between the equation and the data points

but this is not good in the important large grain size region.   It
is recommended therefore that maximum grain sizes be determined by
direct measurement.

4. MULTIPHASE MATERIALS

In this section the effects of microstructure and physical
properties of the phases, on the surface energy, flaw size, and
strength of multiphase ceramics is considered.   Materials of
commercial significance include the wide range of pottery bodies
which, after firing, generally contain crystalline phases in a
predominantly glassy matrix.   A classic case is electrical porcelain containing quartz filler particles.   It is well known that if
the quartz particles are greater than a certain size then, because
the particles shrink more than the matrix during cooling, cracks
form round the particles.   Fig. 8 gives examples.   Finely ground
quartz is thus used to improve strength.   A large number of model
systems of this type have been studied, mostly ceramic particles
in a glassy matrix, and much of recent data has been reviewed in
detail by Lange (30).

To understand the microstructural effects it is necessary
first to introduce some elementary theory.   A spherical particle
in an infinite matrix is subjected to a pressure P.

$$P = \frac{\Delta \alpha \; \Delta T}{(1+\nu_m)/2E_m + (1-2\nu_p)/E_p} \tag{3}$$

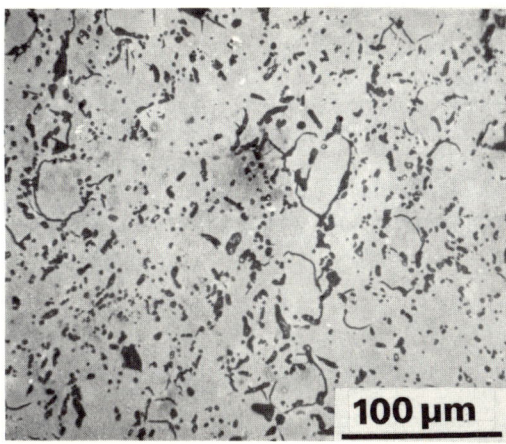

Fig. 8. Circumferential cracks around
quartz particles in electrical porcelain.

where $\Delta \alpha$ is the difference in expansion coefficients of the phases, ΔT the cooling range over which the matrix plasticity is negligible, and ν_m, ν_p, E_m, E_p the Poisson's ratio and Young's modulus of matrix and particle (31,32). When the matrix shrinks more than the particle ($\alpha_m > \alpha_p$), the particle is under compression: the radial and tangential stresses in the matrix are PR^3/r^3 (compression) and $PR^3/2r^3$ (tension) respectively, where R is the particle radius and r the distance from the particle centre. (The signs are reversed if $\alpha_m < \alpha_p$). When $\alpha_m > \alpha_p$, cracks tend to be generated that radiate from the particles and form a crazed network: when $\alpha_m < \alpha_p$ localised circumferential cracks tend to form round the particles. Systems of the former type are of little practical interest because of the associated poor strength.

Returning to the effects of particle size, the stresses set up in and around particles are independent of particle size. However, as commented above for electrical porcelain, for any given system cracks are observed only around particles greater than a specific size. Davidge and Green (33) rationalised this on an energy balance criterion. The strain energy in a particle matrix unit is proportional to the cube of the particle size whereas the energy to form the observed (approximately hemispherical) cracks is proportional to the square of the particle size. It was shown that cracks should form for particles of size greater than a critical radius R_c where

$$R_c = 8\gamma_s / \left\{ P^2 \left[(1+\nu_m)/E_m + 2(1-2\nu_p)/E_p \right] \right\} \qquad (4)$$

where γ_s is the surface energy.

Two important microstructural aspects of second phase particles are the generation of cracks and the influence on the fracture path. Fig. 9 shows structures of two ThO_2/glass composites, with $\alpha_m = \alpha_p$ and $\alpha_m < \alpha_p$, in the vicinity of a fracture path. When $\alpha_m = \alpha_p$ the fracture plane passes indiscriminately through matrix and particle; there are no cracks round the particles. When $\alpha_m < \alpha_p$ the fracture path is clearly deflected around the particles and tends to link up existing cracks.

Davidge and Green (33) measured the strengths of various glasses containing 10 v/o monosized ThO_2 spheres 50-700 μm diameter. In systems where $\alpha_m < \alpha_p$ and where cracking was observed around the particles the observed strengths agreed well with those calculated from Eqn. (1) on the assumption that the surface energy was the same as plain glass, 4 J m^{-2}. This is reasonable since fracture is solely in the glass matrix.

The situation is less clear when fracture occurs through both matrix and particles and where higher particle concentrations are

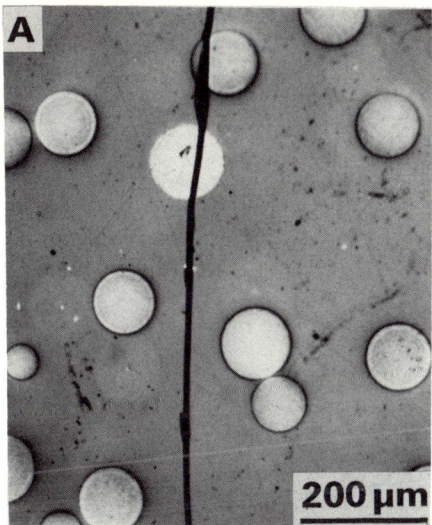 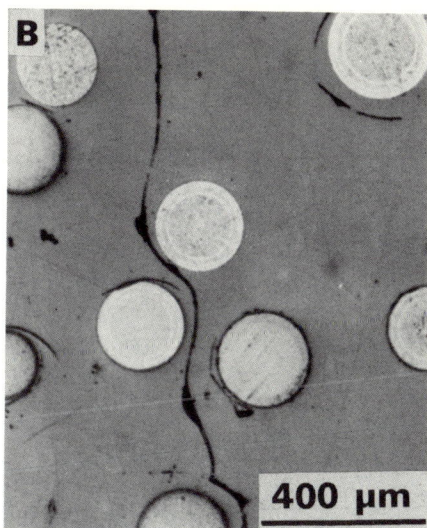

Fig. 9.   Fracture paths in ThO_2 sphere/glass composites.   A - $\alpha_m = \alpha_p$;   B - $\alpha_m < \alpha_p$ (33).

used.   Hasselman and Fulrath, (34) and Lange, (35) studied glass/ alumina composites where $\alpha_m = \alpha_p$.   Hasselman and Fulrath (34) suggested that the flaw size could be limited to the interparticle spacing at high particle concentrations.   From the slope of the strength interparticle spacing graph they calculated a surface energy of 4 J m^{-2}.   This value is certainly too low as indicated by subsequent work of Lange (35).   Data in Table V demonstrate clearly that in this system the surface energy may be up about five times the value for glass.   Surface energy increases with both increase in particle size and in volume function.   Lange (35) related the surface energy qualitatively to the microstructure using a hypothetical model involving the line energy of the crack front with the added assumption that the larger particles are more effective in pinning the crack front than the smaller particles. Finally it should be noted that there is little systematic variation with calculated flaw size and either particle size or volume fraction.

Even in relatively simple model systems of multiphase ceramics only certain qualitative aspects of the effects of microstructure are understood and there is very little quantitative understanding. Data on the more complex systems of practical interest must therefore remain on an empirical basis.

Alumina Particle Size μm	Volume Fraction	Fracture Energy J m^{-2}	Calculated Flaw Size μm
3.5	0.10	10.6	42
	0.25	12.7	40
	0.40	12.8	33
11	0.10	10.4	43
	0.25	13.8	35
	0.40	19.8	45
44	0.10	15.6	148
	0.25	22.7	177
	0.40	29.9	106

Table V. Fracture energy and calculated flaw sizes in Al_2O_3/glass composites (35).

5. CERAMIC BASED FIBRE COMPOSITES

The bulk of the experimental and theoretical work on fibre composites has been on metal or plastic matrix systems: only in the last few years have ceramic based materials been developed. Some of the early data on ceramics, mainly with metal wire reinforcement, have been reviewed by Bowen (36). A significant factor with ceramic matrix composites is that the failure strain of the matrix is generally less than that of the fibre. The reverse is true for metallic and plastic matrices which are ductile or of high compliance respectively.

To illustrate the basic principles, emphasis will be placed on two systems of aligned unidirectional composites developed at Harwell: carbon fibre reinforced glass (CRG) (37-40) and cement (CRC) (41). The fabrication processes are similar. A continuous tow of carbon fibre (∼10,000 fibres, 8 μm dia.) is passed through a slurry of particulate matrix material in an appropriate liquid (organic liquid plus binder or water). The impregnated tow is then, for example, wound on to a former. The cement materials are allowed to cure normally, possibly after further compaction, and the glass materials subjected to a hot pressing stage (∼1200°C) to densify the matrix. An important aspect of the process is that a very uniform dispersion of fibres is obtained, Fig. 10.

Table VI (42) illustrates the marked improvements in strength, stiffness, and particularly work of fracture, that result from the incorporation of fibres, and Fig. 11 gives strain curves for the glass samples. The Young's modulus of the composites E_c is

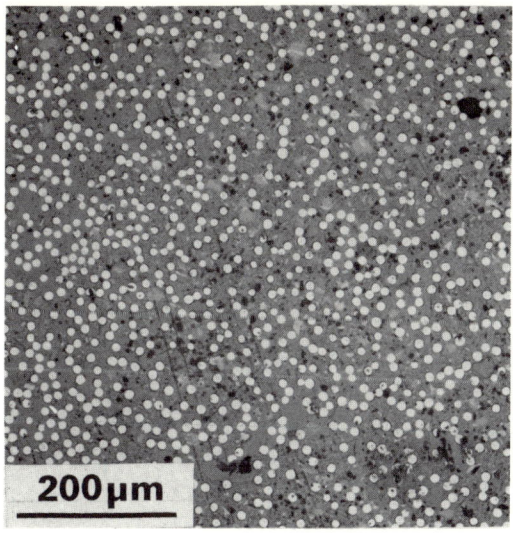

Fig. 10. Section through CRC
normal to fibres (10^V/o) (41).

	Young's Modulus GN m^{-2}	Bend Strength MN m^{-2}	Work of Fracture kJ m^{-2}
Unreinforced Pyrex	60	100	0.004
Pyrex composite (50^V/o type I carbon fibre)	193	700	5
Unreinforced cement paste	13	10	0.02
Cement composite (9^V/o type II carbon fibre)	34	170	7.5

Table VI. Mechanical property data for
unreinforced and reinforced ceramics.

understood in terms of the approximate mixtures law and

$$E_c = E_f V_f + E_m V_m \qquad (5)$$

where V is the volume fraction and the subscripts refer to fibre
and matrix. Fig. 12 shows that this relationship agrees well for
CRG after correcting for the small amount of residual porosity (38).

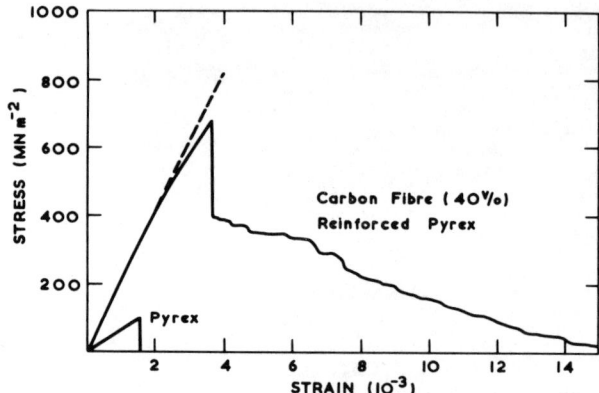

Fig. 11. Stress strain data for Pyrex
and CRG tested in bending (39).

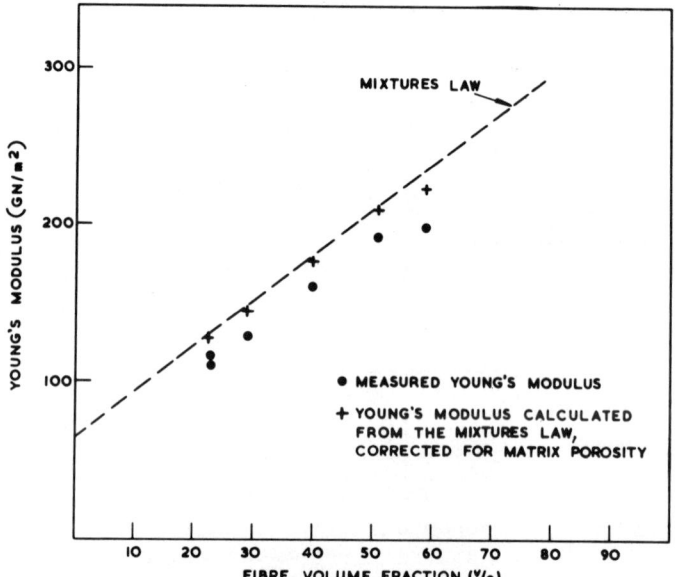

Fig. 12. Young's moduli vs volume
fraction of fibre in CRG (38).

A consequence of the low fracture strain of the ceramic matrix
is a deviation in the stress strain curve at stresses well below the
fracture stress.   In Fig. 11 this bendover occurs at 340 MN m^{-2}
(half the fracture stress).   The bendover has been identified with
the generation of fine scale matrix cracks in the regions of maximum

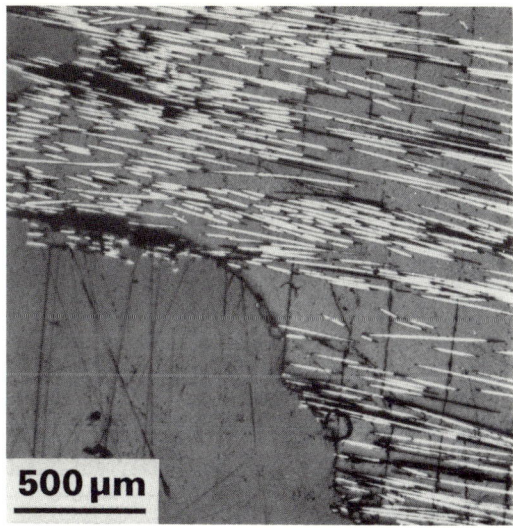

Fig. 13. Matrix cracking on tensile face of bend specimen of CRG (38).

Fig. 14. Fracture face of CRG (40ᵛ/o) (40).

tensile stress and these occur only when the bendover stress has been exceeded (38). Fig. 13 gives examples of cracking in CRG on the polished tensile face of a bend specimen, after failure. Similar stress strain and cracking behaviour is found in CRC. It

is interesting that the matrix cracking stress is considerably in
excess of that expected of the matrix itself, thus the presence of
the fibres must inhibit the initiation of cracks.   Cooper and
Sillwood (43) showed that the mean crack spacing t is given by

$$t = \frac{3}{4} \left(\frac{1 - V_f}{V_f} \right) \frac{\sigma_m r}{\tau} \tag{6}$$

where r is the fibre radius and τ the shear strength of the fibre/
matrix interface.   Large fibre volume fractions and small fibre
radii thus give small matrix crack spacings.   Agreement between
this analysis and experimental data for CRG is good (44).   Depending on the value of t, the strength of the matrix may exceed its
inherent strength.   This is the case for the CRG materials (large
V_f, small r).   The energy balance considerations for the generation of matrix cracks are limited to regions of specimen thickness
~t so that the strength of the matrix in the composite is enhanced
for small t (40).

Phillips (37) conducted a fracture mechanics analysis for CRG
and measured both γ_i and γ_f.   For $40^v/o$ CRG, $\gamma_i \sim 0.8$ kJ m^{-2} and
$\gamma_f \sim 3$ kJ m^{-2}.   To relate these values to microstructural features
is difficult.   The total fracture process involves successively:

(a) fracture of the matrix into blocks at the matrix
    cracking stress;

(b) fracture of fibres near the ultimate fracture stress;
    and

(c) pull out of the fibres under falling stress.

Fig. 14 shows a fracture face.

Fracture of the matrix, as discussed above, has a small energy
requirement.   Phillips (37) proposed that γ_i was related to fracture of the fibres and calculated the energy requirements.   These
are the fracture energy of the fibres themselves plus the loss in
elastic energy over a characteristic length of fibres near the
broken ends.   This gave an estimate of $\gamma_i \sim 0.25$ kJ m^{-2} - rather
lower than the experimental value. γ_f should be the sum of the
terms (a), (b) and (c) with the latter (fibre pull out) predomin-
ating at ~2 kJ m^{-2}.

In contrast to unreinforced ceramics, the fracture energy of
composites is reasonably understood in terms of microstructural
parameters.   On the other hand the concept of a critical flaw size
is difficult to visualise in composites which have a heavily
cracked matrix prior to failure.

REFERENCES

1. R. W. Davidge and A. G. Evans, Mater. Sci. Eng. 6 (1970) 281.
2. R. W. Davidge and G. Tappin, J. Mater. Sci. 3 (1968) 165.
3. R. W. Davidge and D. C. Phillips, J. Mater. Sci. 7 (1972) 1308.
4. R. W. Rice, Proc. Brit. Ceram. Soc. 20 (1972) 205.
5. S. C. Carniglia, J. Amer. Ceram. Soc. 55 (1972) 243.
6. J. J. Gilman, J. Appl. Phys. 31 (1960) 2208.
7. A. R. C. Westwood and T. T. Hitch, J. Appl. Phys. 34 (1963) 3085.
8. F. F. Lange and K. A. D. Lambe, Phil. Mag. 18 (1968) 129.
9. S. M. Wiederhorn, J. Amer. Ceram. Soc. 52 (1969) 485.
10. A. G. Evans, Phil. Mag. 22 (1970) 841.
11. R. W. Guard and P. C. Romo, J. Amer. Ceram. Soc. 48 (1967) 7.
12. J. Congleton and N. J. Petch, Acta Met. 14 (1966) 1179.
13. H. Meridith and P. L. Pratt, Paper presented at Third International Fracture Conference, Munich (1973).
14. D. B. Binns and P. Popper, Proc. Brit. Ceram. Soc. 6 (1966) 71.
15. F. J. P. Clarke, H. G. Tattersall and G. Tappin, Proc. Brit. Ceram. Soc. 6 (1966) 163.
16. R. W. Davidge and G. Tappin, Trans. Brit. Ceram. Soc. 66 (1967) 405.
17. P. L. Gutshall and G. E. Gross, Eng. Frac. Mech. 1 (1969) 463.
18. R. W. Rice, Proc. Brit. Ceram. Soc. 20 (1972) 329.
19. A. G. Evans and R. W. Davidge, Phil. Mag. 20 (1969) 373.
20. J. R. Perry and R. W. Davidge, Ceramurgica, in press (1973).
21. A. G. Evans and R. W. Davidge, J. Nucl. Mater. 33 (1969) 249.
22. L. A. Simpson, J. Amer. Ceram. Soc. 56 (1973) 7.
23. A. G. Evans and G. Tappin, Proc. Brit. Ceram. Soc. 20 (1972) 275.
24. L. A. Simpson, J. Amer. Ceram. Soc. 56 (1973) in press.
25. R. W. Davidge, Proc. Brit. Ceram. Soc. 20 (1972) 364.
26. R. W. Davidge, Sci. Ceram. 6 (1973) in press.
27. J. B. Wachtman, Proc. Int. Conf. Mech. Behaviour Mater. (Soc. Mater. Sci. Japan) 4 (1972) 432.
28. A. G. Evans, C. Padgett and R. W. Davidge, J. Amer. Ceram. Soc. 56 (1973) 36.
29. D. A. Aboav and T. G. Langdon, Metallography 1 (1969) 333.
30. F. F. Lange, 'Fracture and Fatigue of Composites', Ed. L. J. Broutman and R. H. Crock, Academic Press, (1973) in press.
31. D. Weyl, Ber. Deut. Keram. Ges. 36 (1959) 319.
32. J. Selsing, J. Amer. Ceram. Soc. 44 (1961) 419.
33. R. W. Davidge and T. J. Green, J. Mater. Sci. 3 (1968) 629.
34. D. P. Hasselman and R. M. Fulrath, J. Amer. Ceram. Soc. 49 (1966) 68.
35. F. F. Lange, J. Amer. Ceram. Soc. 54 (1971) 614.
36. D. H. Bowen, Fibre Sci. Tech. 1 (1968) 85.
37. R. A. J. Sambell, A. Briggs, D. C. Phillips and D. H. Bowen, J. Mater. Sci. 7 (1972) 676.

38. D. C. Phillips, R. A. J. Sambell and D. H. Bowen, J. Mater. Sci. 7 (1972) 1454.

39. D. H. Bowen, D. C. Phillips, R. A. J. Sambell and A. Briggs, Proc. Int. Conf. Mech. Behaviour Mater. (Soc. Mater. Sci. Japan) 5 (1972) 123.

40. D. C. Phillips, J. Mater. Sci. 7 (1972) 1175.

41. A. Briggs, private communication.

42. R. W. Davidge, Bull. Phys. Soc. 23 (1972) 649.

43. G. A. Cooper and J. M. Sillwood, J. Mater. Sci. 7 (1972) 325.

44. J. Aveston, G. A. Cooper and A. Kelly, Proc. Conf. 'Properties of Fibre Composites' National Physical Laboratory, Teddington, U.K. (1971).

FRACTURE MECHANICS OF CONCRETE

Naus, Dan J.

U. S. Army, Construction Engineering Research Laboratory, Champaign, Illinois

Batson, Gordon B.

Clarkson College, Potsdam, New York

Lott, James L.

Illinois Institute of Technology, Chicago, Illinois

INTRODUCTION

Concrete* is produced by mixing water, hydraulic cement,
sand, and coarse aggregate. Normal concrete has a low tensile
strength (approximately one-eighth to one-tenth its compressive
strength) and exhibits limited ductility, especially in tension.
Concrete can thus be classfied as a brittle material containing
many inherent flaws in the form of air voids and microcracks.

Fracture of concrete is a complex process due to its heterogeneity. Fracture may occur through the cement paste, through
the aggregate, through the cement paste-aggregate bond, or any
combination of the above. The size of the constituents, particularly the coarse aggregate, and the flaws, such as air voids,
present problems in defining a fracture mechanism. The variation
in constituent size may be one of the factors why fracture mechanics was not applied to concretes to investigate crack growth and
propagation until the 1960's when research indicated that the
fracture mechanics approach was applicable.[1,2,3,4]

---

* Cement paste consists of portland cement and water. The addition of a fine aggregate to cement paste produces a mortar; and,
the addition of coarse aggregate particles to mortar produces a
concrete.

RELEVANT FRACTURE MECHANICS CONCEPTS

The conditions for fracture or rupture of a solid can be stated in terms of an energy criteria based on the Law of Conservation of Energy. It assumes that the energy necessary for crack extension is provided by the energy released by the system during crack extension. An energy relationship of the following form can be written for a stressed body containing a crack,

$$\frac{dWe}{dA} + \frac{dU}{dA} \geq \frac{dS}{dA} + \frac{dWp}{dA} \tag{1}$$

where

$\frac{dWe}{dA}$ = work done by external forces per unit increase in crack area,

$\frac{dU}{dA}$ = elastic strain energy released per unit extension of crack area,*

$\frac{dS}{dA}$ = increase in the elastic energy of the body from the surface energy per unit extension of crack area,

$\frac{dWp}{dA}$ = irrecoverable work per unit extension of crack area.

The familar Griffith criteria for an elliptical crack of length 2a in a plane stress field involves only the second and third terms of Equation 1. The Griffith criteria for the stress at the onset of rapid crack propagation is

$$\frac{\pi \sigma a^2}{E} = 2T \tag{2}$$

or

$$\sigma = (\frac{2TE}{\pi a})^{1/2} \tag{3}$$

where

σ = remotely applied stress perpendicular to crack
E = modulus of elasticity
T = surface energy
a = half length of crack.

The Griffith approach can be extended to ductile materials to account for the plastic deformation which occurs near the crack tip by modifying Equation 2 to include the irrecoverable work

---

* The sign of this term depends on whether the physical conditions are for a "fixed grip" condition or for a "constant load" condition.

$$\frac{\pi\sigma^2 a}{E} = 2T + \frac{\partial W_p}{\partial a} \cdot \qquad (4)$$

The term $\partial W_p/\partial a$ for concrete arises from the microcracking that develops ahead of the crack tip. The left side of Equation 4 is the elastic energy release rate denoted by G. The right side of Equation 4 is the work required to extend the crack. The value of G at onset of crack propagation, G_c, is the critical strain energy release rate and is assumed to be a material constant independent of shape of the body and its loading; whereas, G is primarily a function of the loading and geometry of the body.

The critical strain energy release rate is given by:

$$G_c = \frac{\pi\sigma^2 a_c}{E} \text{ plane stress} \qquad (5)$$

or

$$G_c = \frac{\pi\sigma^2 a_c}{E} (1-\nu^2) \text{ plane stress} \qquad (6)$$

where
    ν = Poisson's ratio
    a_c = critical crack length

Substitution of typical values of modulus of elasticity (3 x 10^6 psi), G_c (0.05 in-lb/in^2), and tensile strength (500 psi) for plain concrete into Equation 5 indicates that the critical crack length in concrete is about 0.2 inches. Flaws of this magnitude are inherent in concrete and suggests that the tensile strength of concrete depends on the size of the flaw rather than the maximum cohesion of the material.*

Irwin introduced a parameter called the stress intensity factor, K, which relates the stress at the tip of a crack to the specimen geometry and loading. The stress intensity factor, K, is proportional to the square root of G

$$K^2 = EG \text{ (plane stress)} \qquad (7)$$

or

$$K^2 = EG/(1-\nu^2) \text{ (plane stress)} \qquad (8)$$

Cracking of concrete resulting from external loads, forces, and/or shrinkage forces will extend an existing crack slowly until it reaches a critical size at which point rapid crack propagation occurs. The value of G_c or K_c is used as a measure of the resistance of the concrete to cracking.

---

* The theoretical cohesive strength of concrete is two to three orders of magnitude larger than measured tensile strengths.

Of the three possible modes of crack extension, the most commonly reported value for concrete has been Mode I (opening displacement by symmetric loading). Recently Mode II (sliding of the crack surfaces by forces in the plane of the crack) has been considered in the extension of inclined cracks in a compressive stress field. Mode III cracks extension by forces perpendicular to the plane containing the crack surface has not been considered in concrete research.

APPLICATIONS OF FRACTURE MECHANICS TO CONCRETE MATERIALS

Kaplan[1] was the first to apply fracture mechanics concepts to concrete material systems. Analytical and direct experimental approaches were used to evaluate critical strain energy release rates, G_c, associated with rapid crack propagation. The analytical technique used stress analysis procedures to derive a mathematical relationship for G

$$G = (1-\nu^2) \, \sigma_n^2 h \, f(a/d) \quad (0.1h \le a \le 0.5h) \tag{9}$$

$$f(a/d) = \frac{\pi a}{d} (1 - \frac{a}{d})^3$$

where

  ν = Poisson's Ratio
  σ_n = nominal bending stress at root of notch
  h = net depth of beam at notch
  a = notch depth
  d = overall depth of beam.

The direct experimental determination of G was by a method suggested by Irwin and Kies and involved the use of the change of the specimen rigidity (spring constant*) with crack extension to measure the release of elastic energy, G. If

$$F = k\delta$$

where F is applied force per unit width of beam, k is the spring constant, and δ is the displacement of the applied load, then,

$$G = 1/2 \, F^2 \, \frac{d}{dc} \left(\frac{1}{M} \right) \tag{11}$$

where d/dc (1/M) is determined graphically from the particular beam load-deflection behavior. Variables investigated by Kaplan included material (mortar or concrete), specimen geometry (3" x 4" x 16" with notch depths of 0.0", 0.5", 1.0", or 1.5", and 6" x

---

* The particular spring constant for a notched beam is related to the load-deformation behavior of the beam in flexure.

6" x 20" beams with notch depths of 0.0", 1.0", or 2.0"), and
loading (center-point or third-point). Evaluating G_c assuming no
slow crack growth, Kaplan found that: beams with different notch
depths gave similar G_c values; G_c for third-point loading was 15%
lower than G_c for center-point loading; G_c values for the 3" x 4"
16" beams were 38% less than G_c for the 6" x 6" x 20" beams; and,
the experimental method gave G_c values which were 21% different
from the analytical values. It was concluded from the investigation that the Griffith concept of strain energy release rate was
applicable to concrete.

Glucklich[2] modified the Griffith-Irwin theory to introduce
a non-linear relationship to represent the energy requirements
to propagate a crack in a heterogeneous material such as concrete.
Such an alteration is necessary because each phase of the material
through which the crack propagates has distinct G_c values. In
compression it was proposed that the energy requirement curve will
be a series of stepped lines and in tension it will be parabolas.
Two types of crack growth in concrete were discussed: an initial
stage in which the release of strain energy with slow crack
extension is so low that any sudden increase in energy requirement, such as encountering an aggregate, will stabilize the propagating crack; and a final stage where the energy release rate
with rapid crack propagation is of such magnitude that any energy
demand encountered will be supplied so that an unstable situation
results (fracture).

Glucklich[5] investigated the growth of macrocracks in 2" x
4" x 42" mortar beams subjected to sinusoidal loadings. The
crack length during loading was estimated using a compliance
technique. The crack length, a, increased rapidly during the
early load cycles and continued to increase at a decreasing rate
until failure at 3,675,800 cycles. The value of crack growth
rate (da/dN) was 8μ inches per cycle during the first 500 cycles
and averaged only 0.09μ inches per cycle during the last
2,675,800 cycles. This represents two-orders of magnitude
decrease in the crack growth rate during the life of the fatigue
test. The crack growth rate, da/dN, can be represented as a
function of the square root of the crack length, $a^{1/2}$.*

---

* Glucklich's data indicates that the crack growth rate can be
represented approximately as da/dN $\simeq c(a^{1/2})^{-n}$. The effect of a
negative n value has been obtained for notched concrete beams in
terms of beam compliance.[6] A propagating macrocrack in a mortar or
concrete beam subjected to repeated loadings thus becomes less
critical relative to crack growth rate which suggests the formation of a microcracing zone of ever-increasing size in the tensile
region around the macrocrack. The microcracking region blunts
the effect of the macrocrack and retards crack growth. Micro-

Lott and Kesler[3] conducted a study to develop a hypothesis for propagation of cracks in plain concrete and to compare the hypothesis to results of an experimental investigation of crack propagation in several mortars and concretes. It was suggested that the critical stress intensity factor for plain concrete was derived from the stress intensity factor of the cement paste and the crack arresting mechanism developed by the heterogeneity of the concrete. Since the critical stress intensity factor of the cement paste was a material constant, variations in the critical stress intensity factor for the concrete were reflected through the arresting function. The effects of several concrete parameters (water-cement ratio, sand-cement ratio, and gravel-cement ratio) on the "pseudo" fracture toughness of concrete was determined by the fabrication and testing of 4" x 4" by 12" mortar and concrete beams with 0.0", 0.5", 1.0", or 1.5" flaws. The beams were tested in flexure and the critical stress intensity factor, neglecting slow crack growth, was evaluated from

$$K = \frac{6M}{Wd^2} \left[\frac{2d}{\pi} h(a/d) \right]^{1/2} \tag{12}$$

where $h(a/d) = 10.08 \ (a/d)^2 - 1.225 \ (a/d) + 0.1917$

   M = the applied bending moment
   W = specimen width
   a = notch depth
   d = depth of beam.

For the range of variables investigated it was found that: the critical stress intensity factor was independent of the water-cement ratio for mortars and concretes where the aggregate percentages remained constant; the critical stress intensity factor was independent of fine aggregate percentage for the mortars with the same water-cement ratio; the critical stress intensity factor varied directly with coarse aggregate content for concretes with the same water-cement ratio and fine aggregate content; and the critical stress intensity factor for concrete was approximately 20.0% greater for a concrete than for a mortar with the same water-cement ratio and fine aggregate content.

---

cracks have been found in the region of maximum cyclic tensile stress of unnotched beams for some combinations of stress range, load frequency, and number of load cycles[7]. In some cases the static strength of the beams subjected to repeated load was increased above the strength of the beams not subjected to cyclic loads and it was suggested that the observed microcracks would blunt the natural flaws in the concrete to increase the fracture load corresponding to static flexural strength.

Naus and Lott[8] conducted a study to determine the effects of
several concrete parameters on the fracture toughness of concrete.
Paste and mortar prisms 2" x 2" x 14" with flaw depths of 0.0",
0.25", 0.50", and 1.0", and, mortar and concrete specimens 4" x
4" x 12" with flaw depths of 0.0", 0.5", 1.0", and 1.5" were fabricated. The beams were tested in flexure and the effective fracture
toughness at the onset of rapid crack propagation, neglecting
slow crack growth, was evaluated from

$$K = Y \frac{6Ma^{1/2}}{Bw^2} \qquad (13)$$

$$Y = 1.99 - 2.47 \, (a/w) + 12.97 \, (a/w)^2 - 23.17 \, (a/w)^3$$
$$+ 24.80 \, (a/w)^4$$

where
    a = cast flaw depth
    w = specimen depth
    M = applied bending moment
    B = specimen width.

It was found that the effective fracture toughness was not significantly effected by fine aggregate content, air content, or
water-cement ratio for the range of parameters encountered in a
typical mix design; however, the effective fracture toughness of
concrete increased significantly as the fineness modulus of the
aggregate increased and also as the amount of coarse aggregate
increased. A rigid-plastic cracked strip model based on Rice's
rigid-plastic strip model[9] was also developed to approximate the
inelastic phenomenon in the region of the crack tip. The model
may be used to provide an estimate of the length of the inelastic
zone (microcracking region), ω, in the region of the crack tip
if the elastic stress intensity factor, K, for the given load and
geometry, and the threshold stress for microcracking (σ_{th}) are
known

$$\omega = \alpha \left[\frac{K}{\sigma_{th}} \right]^2 \qquad (14)$$

where α is a constant depending upon the stress distribution
along the inelastic region.

Naus[10] investigated the applicability of the concepts of
linear-elastic fracture mechanics to portland cement paste, mortar,
and concrete material systems. Plate specimen 2" x 12" x W", where
W varied from 18" to 36", were fabricated containing pre-cast
through-the-thickness flaws. The specimens were loaded at the
level of the flaw (wedge loaded specimen) and the stress intensity
factor at the onset of rapid crack propagation was evaluated
from

$$K = \frac{P}{B\sqrt{\pi a}} \frac{1}{\left[\frac{W}{2\pi a} \sin \frac{2\pi a}{W}\right]^{1/2}} \tag{15}$$

where

P = applied load
B = specimen thickness
a = one-half the total flaw length
W = specimen width

The results obtained indicated that: the concepts of linear-elastic fracture mechanics were not directly applicable to the cement pastes, mortars, and concretes for the specimen geometry; other analytical techniques such as the net section stress did not appear applicable as a failure criterion; and there is a localized region near the flaw tip in which large strains are apparent. The model developed in Reference 9 was correlated with the experimental results obtained.

Moavenzadeh and Kuguel[11] utilized notched beams 1" x 1" x 12" to investigate the fracture of cement pastes, mortars, and concretes. The true facture work of concrete was determined to be slightly lower than for cement paste because the cracks showed a preference for propagation through the paste-aggregate interface which generally was of lower bond strength (energy requirement) than the matrix. The extent of cracking in concrete was found to be considerably more extensive than in the pastes and mortars, and thus the energy requirements to fracture a concrete beam are higher than required to fracture a paste or mortar beams even though the surface energies of paste, mortar, and concrete are similar.

Romualdi and Batson[4] investigated the fracture of concrete in tension and proposed that the strength of concrete in tension is related to the largest flaw concept rather than maximum cohesive forces since typical flaws in concrete are on the order of magnitude of the critical flaw site for fracture. Remotely loaded plate specimens 2.5" x 24" x 32" containing through-the-thickness flaws ranging from 2" to 12" long were fabricated and tested. The critical strain energy release rate was evaluated from

$$G_c = \frac{\pi\sigma^2 a}{E} (1-\nu^2) \tag{16}$$

where σ is the remotely applied stress. No corrections were made for slow crack growth and it was found that in general the G_c values decreased with a decrease in flaw size.

Brown[12] used notched beams 1.5" x 1.5" by 10", and double cantilever beam (DCB) specimens 2.0" by 4.0" by 14" whose web was

contoured to provide a fracture toughness independent of crack
length. The stress intensity factor for the flexure specimens
was evaluated from

$$K_c = 3Pa^{1/2} (\ell_1 - \ell_2) \, \gamma / 2Wh^2 \tag{17}$$

where
$$\gamma = 1.99 - 2.47 \, (a/h) + 12.97 \, (a/h)^2 - 23.17 \, (a/h)^3$$
$$\qquad + 24.8 \, (a/h)^4$$

ℓ_1 = major span
ℓ_2 = minor span
W = beam width
h = beam depth.

The stress intensity factor for the double cantilever beam specimens was evaluated from

$$K_c^2 = 12 \, P_c^2 \, a^2 \, \xi / bWh^3 \tag{18}$$

where
$$\xi = 1. + 1.32 \, (h/a) + 0.532 \, (h/a)^2$$

h = DCB height
b = DCB width
W = web width

Equation 10 can be modified to produce a DCB where K is independent
of the crack length by conforming the web width to

$$W = a^2 \xi \, k \tag{19}$$

thus

$$K_c^2 = \frac{12 \, P_c^2 k}{bh^3} \tag{20}$$

The results obtained indicate that: the fracture toughness of
cement paste is approximately constant as the crack grows; the
fracture toughness of mortar increases with crack growth; and, for
paste and mortar the initiation of crack growth takes place at
a lower stress intensity factor than is needed to keep the crack
running.

Welch and Haisman[13] investigated the tensile and compressive
failure of concrete and the effects on the fracture toughness
if assumptions are made in the analysis relative to length of flaw
at failure (whether it is determined by a compliance technique or
a correction factor), and the modulus of elasticity. It was concluded from the investigation that: the fracture toughness values
are very dependent on assumptions; the energy concepts of fracture
mechanics may be applied to crack propagation and failure mechanisms of concrete; and G_c (or K_c) is a strength determining
property which increases with the quality of the cement paste,

increased solid volume fraction of cement, improved aggregate properties, and reduced flaws prior to loading.

Forzani[14] determined the K_{Ic} value of mortars using a wedge open loading (WOL) specimen. The specimens were 10" x 10" x 4" with various notch depends. The specimens were pulled in tension and the compliance of the specimen as a function of the crack-depth ratio was obtained by fitting a least square polynominal to the test data. The elastic energy release rate, G, was computed from the compliance data substituting test values and specimen geometry values into

$$G = 1/2 \left(\frac{P}{B}\right)^2 \left(\frac{dc}{da}\right) \tag{21}$$

where
  P = load at crack extension
  B = thickness of specimen
  c = compliance
  a = crack depth.

The stress intensity factor was also determined for test data by evaluating

$$K = \frac{Pa^{1/2}}{BW} Y \tag{22}$$

where
  P = load at crack extension
  B = thickness of specimen
  a = crack depth
  W = distance from line of loading to end of crack
  Y = compliance for particular crack-depth ratio (a/W).

The two measurements of the cracking resistance of mortar obtained were related by the expression

$$K^2 = EG \tag{23}$$

The computed values of G increased and K decreased with increasing values of crack depth ratio, a/W.

SUMMARY AND CONCLUSIONS

1. Fracture of concrete is a complex process due to its heterogeneity and the occurrance of microcracking.

2. The Griffith criteria is applicable to concrete materials, however, the irrecoverable work resulting from microcracking (analogous to yielding in metallic materials) must be included.

TABLE I
SUMMARY OF TYPES OF SPECIMENS AND RANGE OF
K_C VALUES OBTAINED FOR INVESTIGATION
OF CONCRETE MATERIAL SYSTEMS

Investigator	Type Specimen	Mat'l System	Experimental Data, K_C ksi $\sqrt{}$ in.
Kaplan[1]		Mortar	0.58 to 0.71
		Concrete	0.50 to 0.84
		Mortar	0.61 to 0.81
		Concrete	0.52 to 1.05
Lott and Kesler[3]		Mortar	0.27 to 0.30
		Concrete	0.31 to 0.37
Naus and Lott[8]		Paste	0.28 to 0.41
		Mortar	0.19 to 0.52
		Concrete	0.34 to 0.70

TABLE I
(con't.)

Investigator	Type Specimen	Mat'l System	Experimental Data, K_c ksi $\sqrt{\text{in.}}$
Naus[10]		Paste	0.07 to 0.24
		Mortar	0.27 to 1.15
		Concrete	0.31 to 1.30
Romualdi and Batson[4]		Mortar	0.31 to 0.47
Moavenzadeh and Kuguel[11]		Paste	0.12 to 0.15
		Mortar	0.12 to 0.14
		Concrete	0.21 to 0.23
Brown[12]		Mortar	0.27 to 0.45
		Concrete	0.42 to 0.62
		Mortar	0.36 to 0.41
		Concrete	0.59 to 0.73
Forzani[14] and Batson[15]		Mortar	0.59 to 0.96

3. Numerous investigators have used various types of specimens (Table I) to evaluate the resistance of concrete materials
to crack propagation, however, in most cases the investigations
have yielded relative results which exhibited trends rather than
valid fracture toughness values.

4. A unique fracture toughness for cement paste, mortar, or
concrete does not exist; i.e., values of fracture toughness
obtained are valid only for the particular concrete material
system investigated.

5. In general, the "effective" fracture toughness of concrete
is higher than the value for a mortar which in turn is higher than
for a paste; i.e., the addition of aggregate particles to the
matrix increases the energy requirements for fracture.

6. Stable crack growth in concrete has been observed in
fatigue tests of concrete. Limited test data indicates that the
crack growth rate is inversely proportional to the square root of
the crack length. It is suggested that fracture mechanics be
applied to the fatigue of concrete to develop a method for predicting the crack growth rate in concrete structures. Such a
model would provide a means for prediction of the useful life of
structures such as pavements. Also the effects of adverse environments such as salt water, distilled water and sulfate liquors, on
the cracking and fatigue strength of concrete can be investigated
by determining the change in the stress intensity factor with time
of exposure to the adverse environments by cyclic and static
loading of WOL or another appropriate type test specimen.[15]

REFERENCES

1. M. F. Kaplan, Proc. American Concrete Institute, 58, 591
(1961).

2. J. Glucklich, Theoretical and Applied Mechanics Report No.
217, University of Illinois, Champaign-Urbana, Illinois, (1962).

3. J. L. Lott, and C. E. Kesler, Theoretical and Applied
Mechanics Report No. 648, Univerity of Illinois, Champaign-Urbana,
Illinois, (1964).

4. J. P. Romualdi, and G. B. Batson, Jour. of Engineering
Mechanics Div., ASCE, EM3, 147 (1963).

5. J. Glucklich, Proc. 1st Int. Conf. on Fracture, Japan; 2,
1343, (1965).

6. J. Lloyd, J. L. Lott, and C. E. Kesler, Theoretical and Applied Mechanics Report No. 668, University of Illinois, Champaign-Urbana, Illinois, (1966).

7. A. Yoshimoto, S. Ogivo, and M. Kawakami, Proc. American Concrete Institute, 69, 233, (1972).

8. D. J. Naus, and J. L. Lott, Theoretical and Applied Mechanics Report No. 314, University of Illinois, Champaign-Urbana, Illinois, (1968).

9. J. R. Rice, Proc. 1st Int. Conf. Fracture, Japan, 1, 283, (1965).

10. D. J. Naus, Technical Manuscript M-42, Construction Engineering Research Laboratory, Champaign, Illinois, (1973).

11. F. Moavenzadeh, and R. Kuguel, Jour. of Materials, JMLSA, 4, 497, (1969).

12. J. H. Brown, Mag. of Concrete Research, 24, 185, (1972).

13. G. B. Welch and B. Haisman, Materiaux et Constructions, 2, 171, (1969).

14. A. Forzani, Master's Thesis, Clarkson College, Potsdam, New York, (1972).

15. G. B. Batson, Final Report NSF Grant 4120, (1972).

FRACTURE OF DIRECTIONALLY SOLIDIFIED CaO·ZrO$_2$-ZrO$_2$ EUTECTIC

C. O. Hulse and J. A. Batt

United Aircraft Research Laboratories

East Hartford, Connecticut

ABSTRACT

The fracture behavior of the CaO·ZrO$_2$-ZrO$_2$ eutectic (M.P. 2300°C) directionally solidified at 10 cm/hr was studied using the work-of-fracture technique and detailed examinations of the fracture surfaces. Below 1000°C, the lamellar structure appeared to have only a minor effect on the path of fracture when the direction of crack propagation was exactly perpendicular to the lamellae. There were usually appreciable effects for all other orientations. Above 1000°C, the eutectic behaved in the classic composite manner wherein a relatively strong minor phase (ZrO$_2$) reinforces a weaker ductile matrix phase. The energy required to propagate a slow moving crack through this microstructure increased by a factor of 28 from room temperatures to 1500°C (4.4 to 114 x 10^4 ergs/cm^2).

INTRODUCTION

The temperature requirements for materials used in gas turbine engines have continually increased until they are now so close to the melting temperature of nickel and cobalt-base superalloys that significant further materials development appears doubtful. A reasonable response to this development is to reexamine more completely than ever before the possibilities of completely different types of materials. Pure oxide ceramics have a number of properties which make them of interest: high melting points, good resistance

to corrosion by liquids and gases, and high potential strength to
weight ratios. The major limitations which currently restrict
their use are low tensile strength and an inability to plastically
relieve internal stress concentrations. As a result, ceramics are
brittle, heat shock sensitive and do not exhibit reproducible design strength values. The current interest and the potential
improvements in efficiency which could be attained through the use
of ceramics in gas turbine engines presents a challenge to ceramic
scientists to understand the fracture mechanisms which operate in
these materials and then to use this knowledge to develop the
strongest and toughest ceramics possible.

Although there have been relatively few studies of the toughness of ceramics, many investigations of strengthening mechanisms
in ceramics can be summarized by a few general requirements: (1)
extremely high density, (2) smallest possible microstructure,
(3) highest possible purity, (4) high bond strength (high melting
points and high elastic modulus), and (5) minimum possible internal
stress due to differences in thermal expansions. These requirements have been met by hot-pressing different single-phase oxides
to yield ceramics with excellent mechanical strengths and even good
optical transparency.

Unfortunately, some important problems are associated with the
fabrication of ceramics for high temperature strength applications
using the hot-pressing approach. Extremely fine, reactive charge
powders must be employed in order to hot press at low enough temperatures to achieve both theoretical density and a fine grain size.
These fine, reactive powders tend to pick up gases which cannot
generally be removed without destroying the reactivity so essential
to successful hot pressing. When many transparent hot-pressed
ceramic materials are subsequently heated to high temperatures, the
gases contained within the piece tend to form voids and pieces may
actually explode.[1]

Another difficulty with hot-pressed materials, perhaps more
serious, is that fine grain size material necessarily contains many
grain boundaries. These boundaries are not thermodynamically stable
and at high temperatures grain boundary area tends to be reduced by
grain growth. Thus, the advantage of the original fine grain size
is significantly reduced. Even if grain growth were avoided, the

----

[1]R. Rice, Meeting Am. Ceram. Soc. (April 1972)

presence of grain boundaries is detrimental to strength at high
temperatures. The major mechanisms of creep and fracture at high
temperatures are associated with enhanced diffusion along grain
boundaries and grain boundary sliding and separation.[2,3] Grain
boundaries are perhaps a more serious defect in ionic structures
than metals because of the greater complexity in crystal structure
brought about by valence balance requirements.

There are a number of circumstances which could produce improvements in the mechanical properties of ceramic materials with
a directional eutectic microstructure over that available from hot-
pressed ceramics. An immediate possibility is that at high temperatures the minor phase will have the high strength characteristic
of a material in whisker form and that this phase will directionally
reinforce the somewhat ductile ceramic matrix in a typical composite manner. The reality of this possibility has been demonstrated
in numerous metallic eutectic systems. Similar Petch type equations
relate strength to the spacing between phases in a eutectic[4] just
as they relate the strength to grain size in hot-pressed ceramics.

An important advantage of unidirectionally solidified eutectics for high temperature strength applications is that their
microstructure is extremely stable, practically to the melting
point.[5] This stability results from the fact that their microstructures are produced directly from the molten state under
conditions of thermodynamic equilibrium. If grain boundaries are
present in these microstructures they are relatively few in number
and generally parallel to the axis of primary reinforcement.

This paper reports on the fracture behavior of the directionally solidified CaO·ZrO₂-ZrO₂ eutectic using the work-to-fracture
technique. This eutectic melts at 2300°C and both phases are cubic.[6]
The coefficients of thermal expansion at the two phases are identical[7] and therefore any possible effects due to this variable are

[2]J. Stavrolakis and F. Norton, J. Am. Ceram. Soc., 33, 263 (1950).
[3]A. Paladino and R. Coble, J. Am. Ceram. Soc., 46, 133 (1963).
[4]E. Thompson, E. Kraft, and F. George, Tech. Rept. J910803-4,
United Aircraft Research Laboratories (1970).
[5]B. Bayles, J. Ford, and M. Salkind, Trans. AIME, 239, 844 (1967).
[6]P. Duwez, F. Odell, and F. Brown, Jr., J. Am. Ceram. Soc., 35,
109 (1952).
[7]J. Lynch, C. Ruderer, and W. Duckworth, Eng. Prop. Selected Ceram.
Mats., Pub. J. Am. Ceram. Soc. (1966).

removed. This eutectic is of special interest because it has been reported that $SrO \cdot ZrO_2$ and solid solutions of a small amount of $CaO \cdot ZrO_2$ in $SrO \cdot ZrO_2$ can be plastically deformed at room temperatures by a twinning process.[8,9]

EXPERIMENTAL PROCEDURE

Directionally solidified ingots of the $CaO \cdot ZrO_2-ZrO_2$ eutectic were prepared using the floating molten zone facility shown diagramatically in Figure 1. The sintered charge rod was traversed at 10 cm/hr through a small furnace which consisted of a carbon ring susceptor surrounded by ZrO_2 fiber insulation and a two turn R.F. coil operating at 550 KC. After one preliminary pass in vacuum at about 1900°C to sinter and degas the charge rod, two molten zone passes were made in an argon atmosphere.

The charge rods were prepared from fine powders of ZrO_2 and CaO supplied by Wah Chang, Albany and Research Organic/Inorganic Corp., with purities reported as 99.99+ and 99.999 percent, respectively. All weighings of CaO powders were made in argon to avoid complications due to hydration. The batch powders were hand mixed in air and hydrostatically pressed into bars at 10,000 psi without the use of a binder. The final preparation step was hand sanding into straight charge rods 5 inches long and 1/2 inch in diameter.

Work-to-fracture samples were prepared from directionally solidified ingots which had previously been annealed for 2 hrs at 1600°C in air. These samples were typically 0.25 in. x 0.25 in. x 2.0 in. with the direction of solidification parallel to the longest dimension. Two diagonal cuts were made with a 0.011 in. thick diamond saw to remove all material at the center of the 2 inch dimension except for a triangular web whose apex was inside the notch at the center of the sample. The sample configuration was similar to that used by Coppola et al[10,11] and Simpson.[12] No study was made of the possible effects of different notch depths.

[8]J. Tinklepaugh, Twelfth Sagamore Army Mat. Res. Conf. (1965).

[9]J. Funk, Graduate Thesis, Alfred Univ. (1964).

[10]J. Coppola and R. Bradt, J. Am. Ceram. Soc., 55, 455 (1972).

[11]J. Coppola and R. Bradt, Ceramic Bull., 51, 847 (1972).

[12]L. Simpson, J. Am. Ceram. Soc., 56, 7 (1973).

VERTICAL SECTION VIEW THROUGH CENTER OF CYLINDRICAL FURNACE

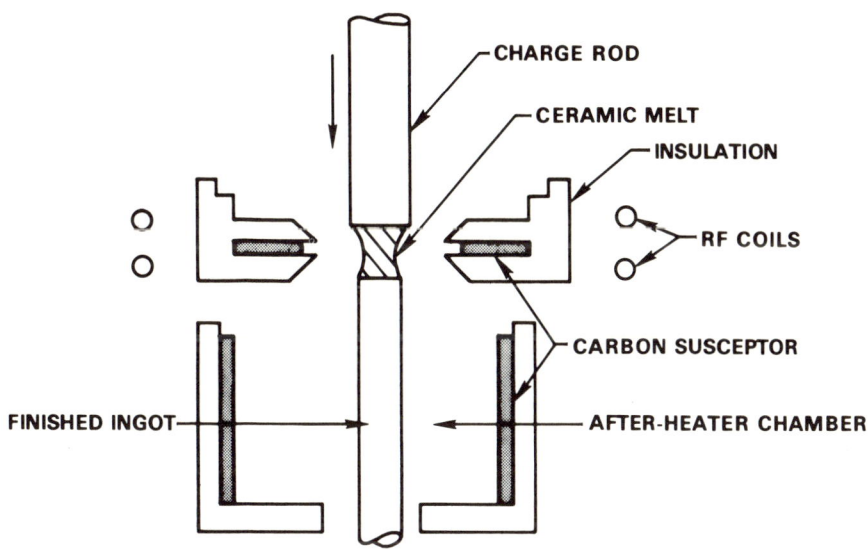

Fig. 1. Schematic of Directional Solidification Equipment for Ceramic Oxide Eutectics

The samples were broken with relatively stable fracture propagation using a 4-screw, 60,000 lb Tinius Olsen universal testing machine with the crosshead moving at .0025 in./min. Elevated temperature testing was done in air using alumina fixtures and solid alumina loading rams 2 inches in diameter. Fracture energies are reported per unit area of crack which is equal to twice the fracture surface energies.

RESULTS AND DISCUSSION

The CaO·ZrO$_2$-ZrO$_2$ eutectic is lamellar with the maximum spacing between the lamellae being approximately 3μ when solidified at 10 cm/hr. The volume fraction of ZrO$_2$ as determined from optical photographs was 32 v/o which may be compared with 41 v/o calculated from the published equilibrium phase diagram.[6]

Despite considerable effort, eutectic ingots sufficient for the preparation of 0.25 x 0.25 inch square work-to-fracture bars

could not be prepared with uniform eutectic microstructures throughout. One reason for this was that there was always some loss of
CaO during the zone melting which made it difficult to hold the
ingots to exact composition. The batch composition finally selected
consisted of 24 w/o CaO with the rest ZrO_2.

Near the surface of the ingots and inward typically for about
1/8 inch, relatively perfect lamellar eutectic microstructures were
always obtained. Figure 2 shows an example of this microstructure
in longitudinal section wherein the lamellae are parallel to the
direction of solidification. Extensive areas of apparent twins
are observed in the $CaO \cdot ZrO_2$ matrix phase, apparently resulting
from the stresses generated during mechanical polishing. Near the
centers of the ingots, the liquid-solid thermal gradients were not
sufficiently high to produce comparable microstructures. Figure 3
shows a transverse view of this microstructure which contains
eutectic colonies and dendrites. At least half of the fracture
process in work-to-fracture samples prepared from these ingots and
reported here, was through imperfect microstructures similar to
those shown in Figure 3.

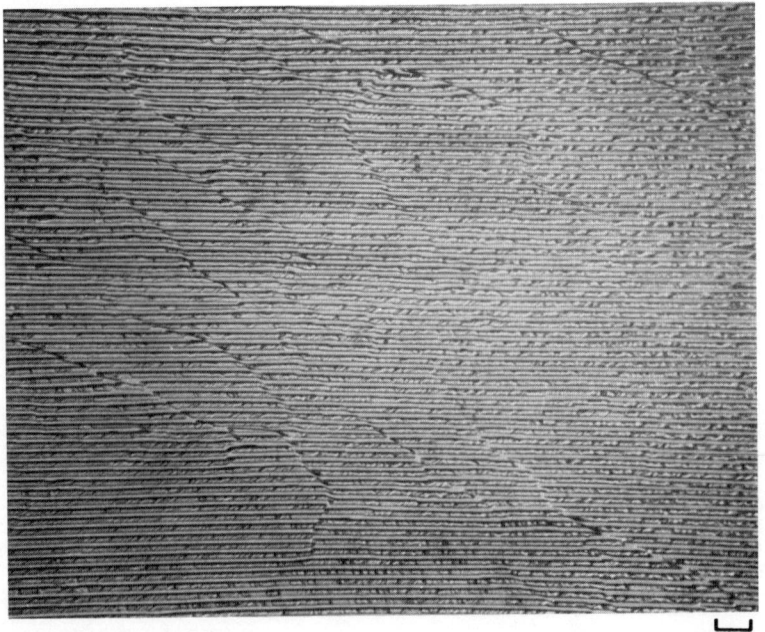

10 μ

Fig. 2. Longitudinal Section of $CaO \cdot ZrO_2 - ZrO_2$ Eutectic (Solidified
        10 cm/hr)

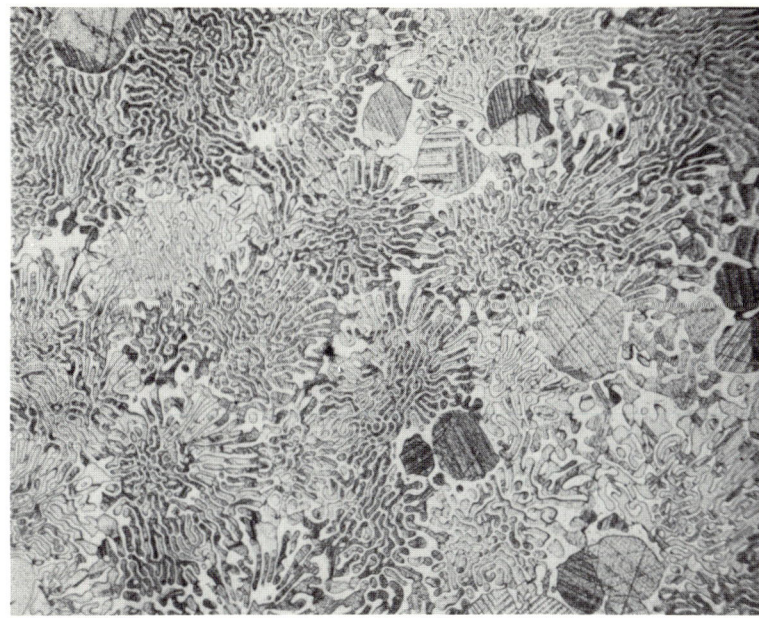

Fig. 3. Transverse View of CaO·ZrO₂-ZrO₂ Eutectic Microstructure
        Near Center of Ingot

Stable crack growth could be obtained in work-to-fracture
experiments at room temperatures but the details on the fracture
face were difficult to interpret because of the irregular nature
of the crack faces and the nonuniform microstructures. There was
no evidence of delamination between the two phases.

The effect of the eutectic microstructure on fracture was
orientation dependent. In case 1, in which the direction of fracture was perpendicular to the plane of the lamellae, there was no
observable interaction. In case 2, however, in which the normal
to the lamellae is rotated in the plane of the fracture surface,
there was noticeable interaction between the fracture path and the
microstructure. In case 3, in which the normal to the lamellae is
rotated out of the plane of fracture there was also interaction as
shown in Figure 4.

The fracture surfaces and the work-to-fracture energy at 1000°C
were similar to those obtained at room temperature. Figure 5 shows
how the fracture path for case 2 can be affected by the eutectic
microstructure at 1000°C.

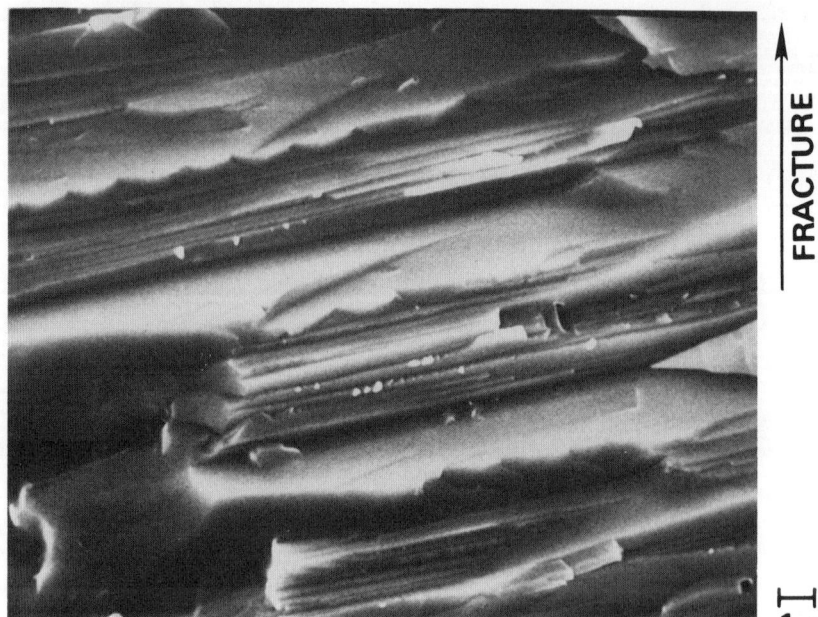

Fig. 4. Fracture Surface of CaO·ZrO$_2$-ZrO$_2$ Eutectic at Room
Temperature

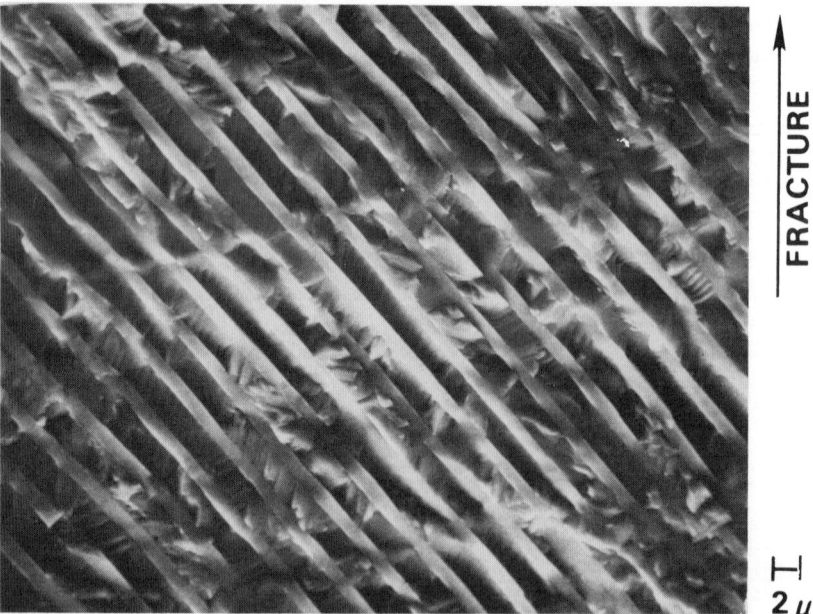

Fig. 5. Fracture Surface of CaO·ZrO$_2$-ZrO$_2$ Eutectic at 1000°C

In tests at even higher temperatures, much larger work-to-fracture energies were obtained as shown in Figure 6. Figure 7 shows force/deflection curves at different temperatures which indicate not only that there is a large increase in deformation at the higher temperatures but also that the forces required for fracture can be considerably greater than those required at lower temperatures. These results suggest that the strength of this eutectic at 1500°C is larger than that at room temperatures.

Examination of the fracture surfaces obtained at 1500°C, as shown in Figure 8, support the view that at these temperatures the eutectic behaves in a typical composite manner with the more ductile CaO·ZrO₂ matrix being reinforced by the stronger ZrO₂ phase. Energy is absorbed in deforming the more ductile calcium zirconate phase and in the shear processes involved in pulling the ZrO₂ phase out of this matrix. Figure 9 shows the partially extracted ZrO₂ phase in more detail. The ZrO₂ phase when removed from the matrix has begun to spheroidize by surface diffusion processes. A similar

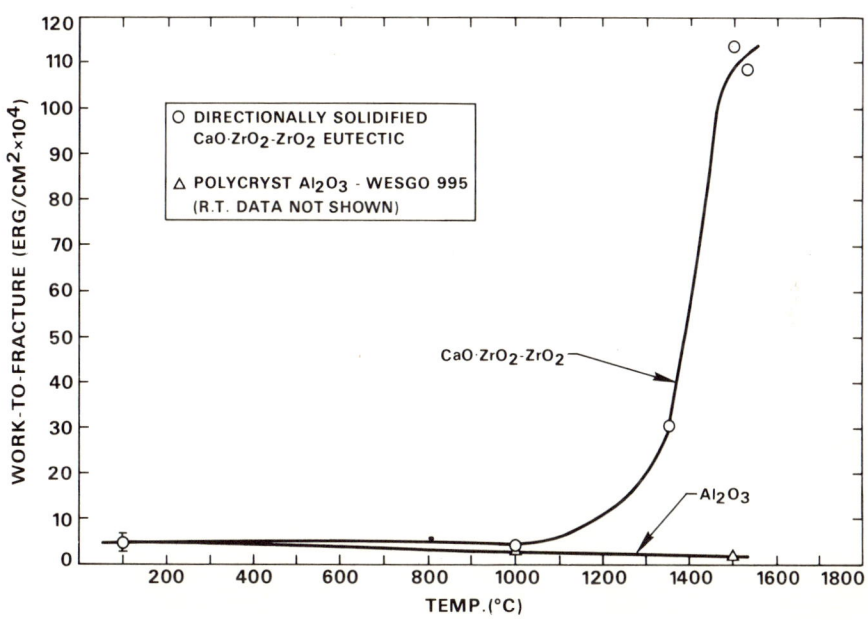

Fig. 6. Work-to-Fracture of CaO·ZrO₂-ZrO₂ Eutectic Directionally Solidified at 10 cm/hr and Polycrystalline Al₂O₃ Versus Temperature

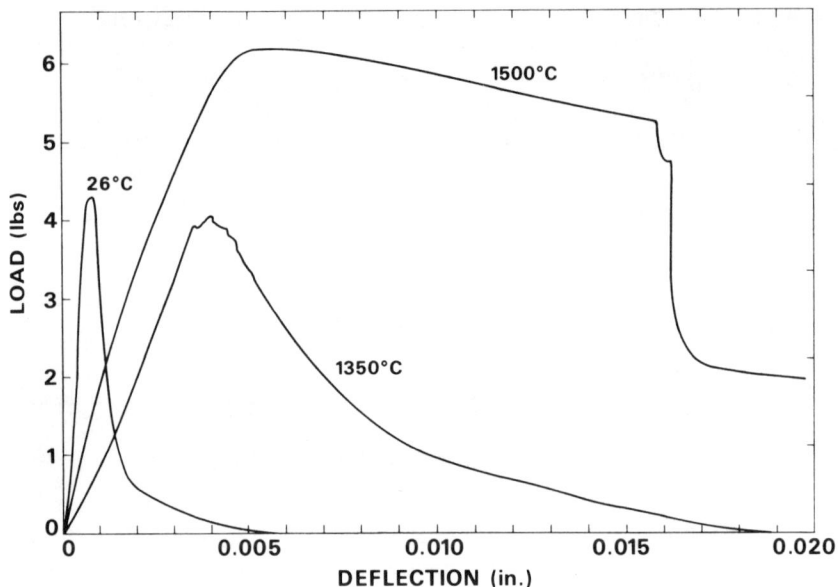

Fig. 7. Work-to-Fracture Force/Deflection Curves for CaO·ZrO$_2$-ZrO$_2$
Eutectic at Several Temperatures

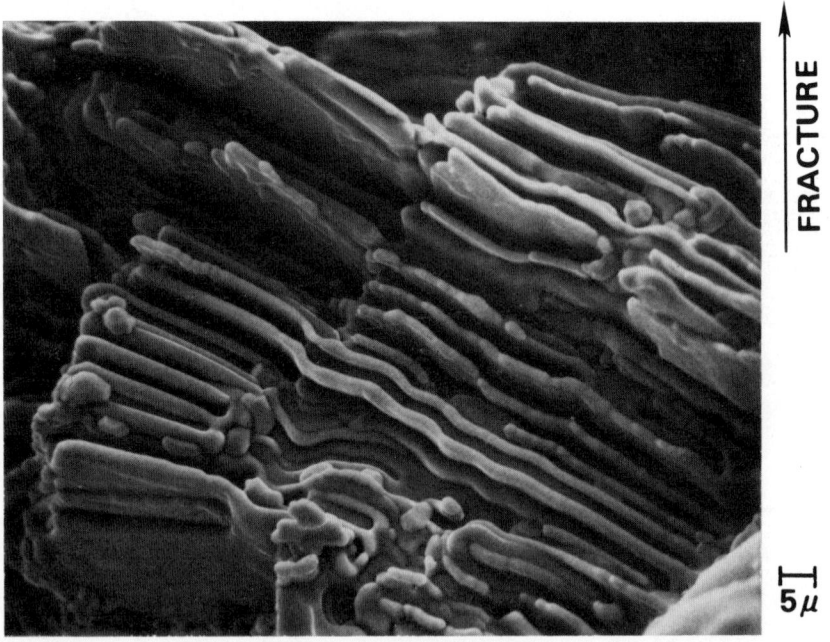

Fig. 8. Fracture Surface of CaO·ZrO$_2$-ZrO$_2$ Eutectic at 1500°C

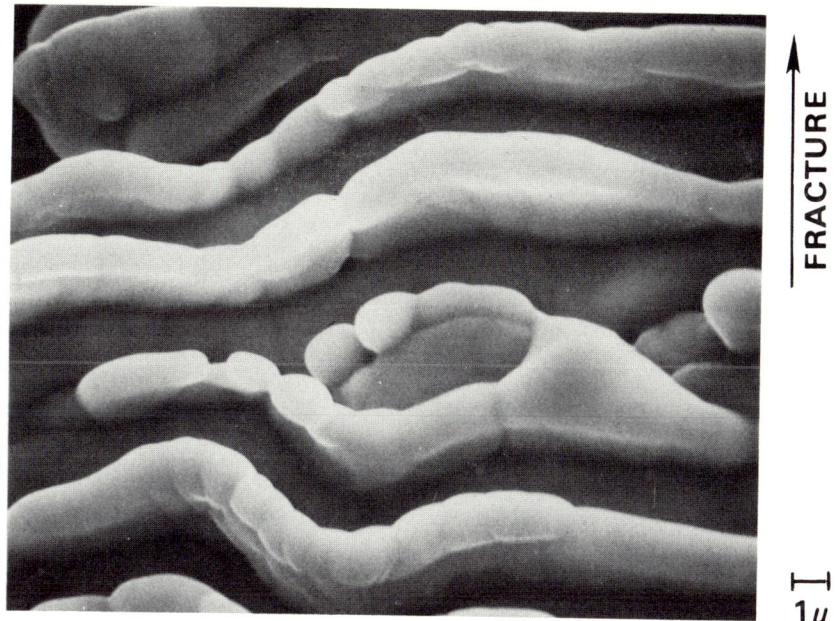

Fig. 9. Fracture Surface of CaO·ZrO₂-ZrO₂ Eutectic at 1500°C
    (30° Tilt)

even more extreme example of this process was observed in the
Al₂O₃-ZrO₂ (Y₂O₃) eutectic at similar temperatures.[13]

Figure 10 shows a polished section through a sample only
partially fractured at 1630°C. The fracture path is discontinuous
and is associated primarily with growth faults in the lamellar
structure. Examples can also be seen of cracks which were opened
in the calcium zirconate matrix being stopped by the ZrO₂ lamellae.

SUMMARY AND CONCLUSIONS

The propagation of slow cracks through the CaO·ZrO₂-ZrO₂
eutectic directionally solidified at 10 cm/hr was studied using
the work-to-fracture technique. The fracture direction was always
perpendicular to the direction of solidification. This eutectic
is lamellar with CaO·ZrO₂ as the matrix phase. At 1000°C and

----
[13]C. Hulse, Meeting Am. Ceram. Soc. (April 1973).

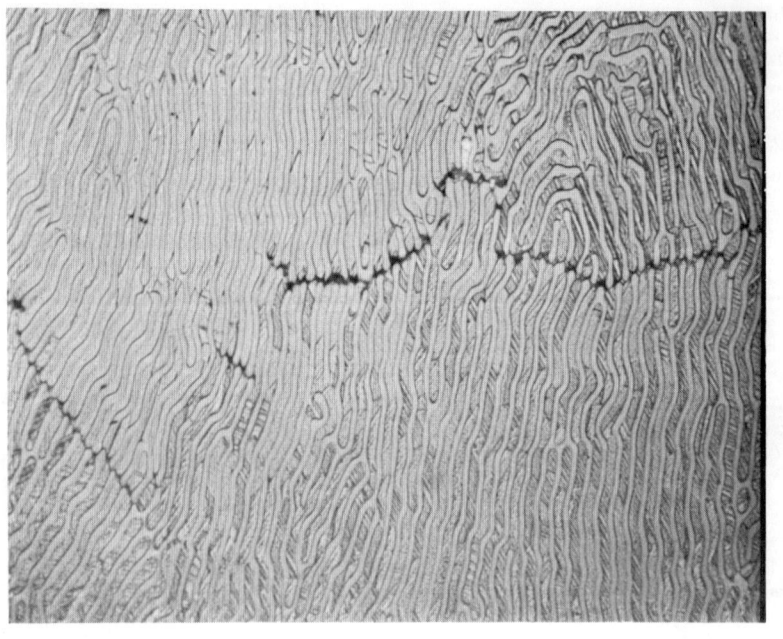

Fig. 10. Edge View of Fracture in $CaO \cdot ZrO_2 - ZrO_2$ Eutectic at 1630°C

below, there was no interaction with the microstructure when the crack direction was perpendicular to the plane of the lamellae. Interaction with the microstructure did occur when the fracture direction was not normal to the lamellae.

Although the $CaO \cdot ZrO_2$ phase is apparently somewhat ductile at room temperatures, this did not provide a mechanism for major energy absorption below 1000°C. The fracture energy at 1500°C, however, was 28 times greater than at room temperature. At these temperatures the eutectic behaves in a classic composite manner in which the relatively ductile calcium zirconate matrix phase is reinforced by a stronger ZrO_2 lamellar phase.

ACKNOWLEDGEMENTS

The authors are pleased to acknowledge that this work was supported by the Office of Naval Research. They would also like to thank Mr. L. Jackman for the SEM photographs and Mr. E. Ahlberg, Mr. K. Palmer and Mr. P. Molinari for conducting the mechanical testing.

FRACTURE ENERGY AND THERMAL SHOCK RESISTANCE OF MICA GLASS-CERAMICS

Kenneth Chyung

Research and Development Laboratory
Corning Glass Works
Corning, New York 14830

INTRODUCTION

Recent development of mica glass-ceramics provides a family of materials whose microstructures consist of highly interlocked two-dimensional mica flakes in a brittle matrix of glass.[1-3] Some of the most interesting properties of mica glass-ceramics include a high degree of machinability, high values of fracture surface energy, and a wide range of strength.

Mica flakes are characteristically easily cleavable, yet very strong in the direction parallel to the cleavage plane.[4,5] Incorporation of mica flakes in a brittle glass matrix, therefore, provides strength-controlling flaws as well as fracture paths. Highly interlocked mica flakes cause fracture deflections, formation of secondary cracks, and often fracture blunting, resulting in high values of fracture surface energy.[2]

The importance of fracture energy is well recognized in thermal shock resistance.[6,7] It is the purpose of this work to study the thermal shock behavior of a mica glass-ceramic in terms of fracture energy.

EXPERIMENTAL STUDIES

A glass composition (Code 9654) approaching a trisilicic fluorophlogopite ($KMg_3AlSi_3O_{10}F_2$) was melted at 1400°C for 4 hours in a platinum crucible and poured into 1/2"-1" thick slabs. A typical crystallization treatment involves heating to and holding for 4 hours each at 800°C and 900°C-1050°C. The volume

495

percent crystallinity after the crystallization treatment is approximately 60-65.

The effective surface energy γ_F was measured from load-deflection curves of notched bars in bending.[8-10] Specimens (1/4 in. x 1/4 in. x 2 in.) were tested in 3-pt bending with 1.5 in. span. Details of testing were reported earlier.[2]

Thermal shock behavior was studied by measuring flexural strengths after plunging samples into water at 25°C from elevated temperatures after holding at least 15 min. Strengths were measured in a 4-pt bending fixture having 1-1/2-in. major span and 3/8-in. minor span. A deflection rate of 0.005-0.01 cm/min was used. Samples were nominally 3/16 in.-diam x 2-in. long, and were tested in fine-ground surface condition.

Elastic moduli were measured sonically from 0.1 in. x 1.0 in. x 4.0 in. samples. Both the replica electron microscope and scanning electron microscope were used to study the microstructures and fracture morphology.

RESULTS AND DISCUSSION

As-cooled glass slabs were given various thermal treatments in order to obtain a wide range of microstructures. Table I lists five sets of samples with heat-treating temperature range from 900°C to 1040°C. Sample A consists mainly of norbergite ($Mg_2SiO_4 \cdot MgF_2$), mullite ($3Al_2O_3 \cdot 2SiO_2$), and a very small amount of mica crystals, after heat treating at 900°C for 4 hours. All other sets of samples contain 60-65 vol % of mica crystals with very small amounts of secondary phases in a glass matrix. Also shown in Table I are flexural strengths, elastic moduli, average coefficients of thermal expansion, and their densities.

The decreasing elastic modulus with increasing heat treatment temperature can be explained partly by the decrease in the secondary phases of high modulus (mainly mullite and norbergite), especially for samples A through C. The main cause for the decrease in the modulus for samples C through E appears to be due to the increase in the size of mica flakes, which in effect act as cracks due to the easy cleavage.

Except for sample A, the flexural strength decreases also with increasing temperature of heat treatment, and appears to be inversely proportional to the first power of the flake diameter, as shown in Figure 1. The relation $\sigma \propto d^{-1}$ suggests that the boundary shear stress due to the thermal expansion mismatch between the mica flake and residual glass controls the strength. Figure 1 also shows the plot of compressive strength vs. the

TABLE I

Sample	Heat Treat. °C/Hr.	MOR σ_o (psi)	Elastic Modulus E(10^6 psi)	Poisson's Ratio ν	Coefficient Thermal Exp. x 10^{-7}/°C (25-600°C)	Density (gr/cm^3)
A	800/4 900/4	10,600 ± 800	13.34	0.24	92	2.614
B	800/4 950/4	12,500 ± 1,400	10.74	0.27	78	2.612
C	800/4 975/4	10,700 ± 500	9.86	0.25	78	2.603
D	800/4 1040/6	7,500 ± 800	5.96	0.14	74	2.574
E	800/4* 1000/4	3,900 ± 400	3.78	0.10	71	2.518

* Thermal history of these as-cooled glass samples was such that subsequent thermal treatments resulted in very coarse grained microstructures and hence very low strengths.

flake diameter. Now the slope is approximately -0.5, rather than
-1.0 for the flexural strength.

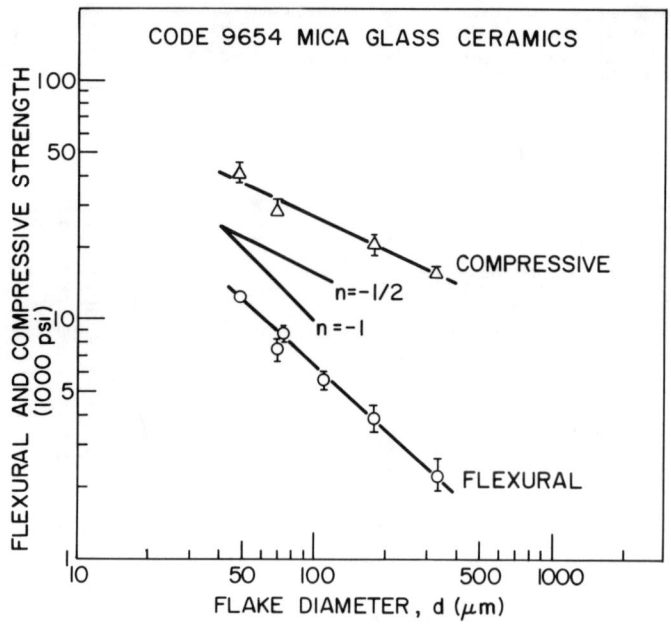

<u>Fig. 1</u>   Flexural and compressive strength vs. mica flake dia-
             meter. Note the flexural and compressive strengths are
             proportional to d^{-1} and $d^{-1/2}$, respectively.

        Figures 2a and 2b show typical microstructures for samples A
and B, respectively. Sample A shows submicron crystals of mullite
and norbergite which constitute approximately 40 vol %, whereas
sample B contains highly interlocked mica flakes ($\sim$ 50 μ in
diameter) in the matrix of residual glass, and the crystallinity
is about 60-65 vol %. Further increase in the ceramming temperature coarsens the mica flakes greatly and reduces the secondary
phases.

        Because of the very minor amount of mica crystals, sample A
would be very similar to ordinary glasses or glass-ceramics in
terms of fracture energy. As the mica crystallinity and crystal
size increase, the mica glass-ceramics become machinable, and the
fracture energy values sharply increase to 3-4 x 10^4 ergs/cm^2,
as will be shown later.

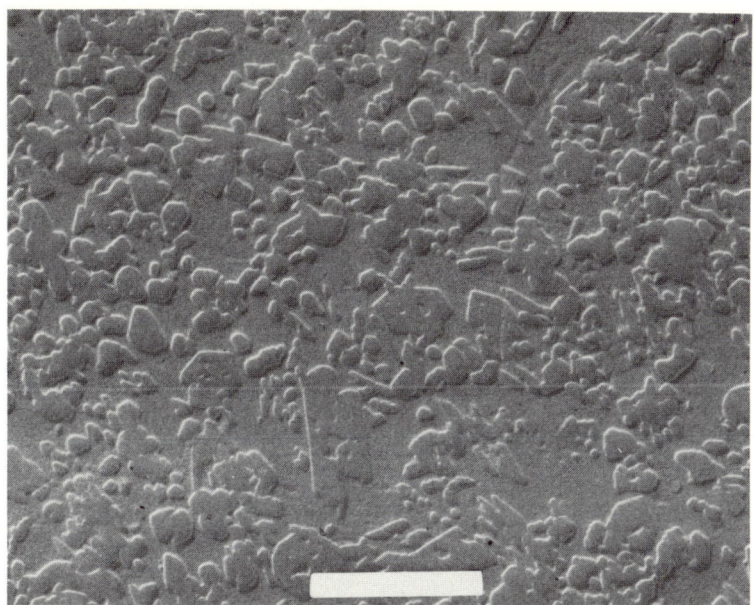

Fig. 2a   Typical microstructure of sample A showing submicron
           crystals of norbergite and mullite in the glass matrix.
           (bar = 1 μm)

Fig. 2b   Typical microstructure of sample B. Note the highly
           interlocking mica flakes. (bar = 1 μm)

Thermal Shock Tests

Figures 3-7 show variations of flexural strength as a function of severity of quenching. Generally, they are consistent with the unified theory of thermal shock resistance proposed by Hasselman.[6,7,11] There are generally four regions of strength; region I of more or less constant strength level similar to the original strength prior to the nucleation of cracks, region II of catastropic strength decrease accompanying the crack propagation, region III of constant strength over which subcritical cracks are stable, and finally a further decrease occurs in region IV, accompanying the deepening of subcritical cracks.

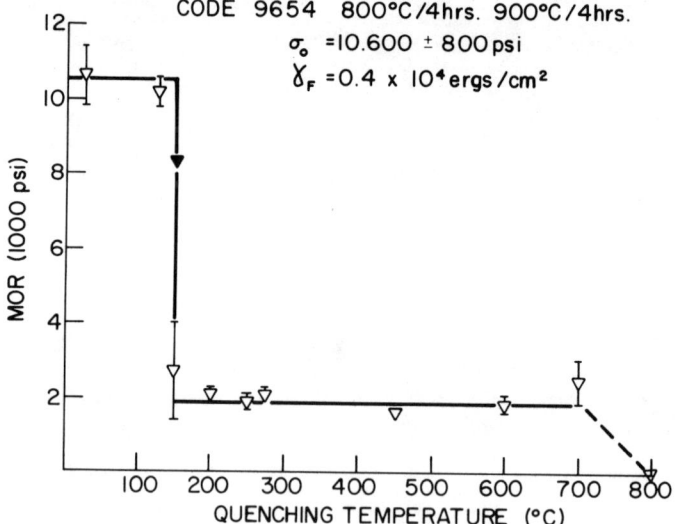

CODE 9654   800°C/4hrs. 900°C/4hrs.
σ_o = 10.600 ± 800 psi
γ_F = 0.4 x 10^4 ergs/cm^2

Fig. 3   Changes in fracture strength vs. quenching temperature for sample A.

Sample A is not substantially different from ordinary glasses in terms of strength and fracture energy. Figure 3 gives typical results, showing very low critical quenching temperature difference about 125°C. The sharp strength decrease at T_c from 10,600 psi to around 1,900 psi is a loss of about 82%.

Sample B contains relatively fine, yet highly interlocked, mica crystals (see Fig. 2b), and shows substantially higher fracture surface energy of 2.7 x 10^4 ergs/cm^2, compared to 0.4 x 10^4 ergs/cm^2 for sample A, as shown in Table II. The critical quenching temperature difference ΔT_c = 225°C is approximately

100°C higher than that of sample A (Fig. 4). The strength changes from 13,000 psi to about 6,000 psi (= -54%) at T_c. The subcritical cracks are stable to a much higher temperature of 800°C before a further decrease occurs.

With higher heat-treating temperatures, the mica flake size increases and the strength decreases. The fracture surface energy, however, tends to increase as the flake size increase, at least in the beginning.

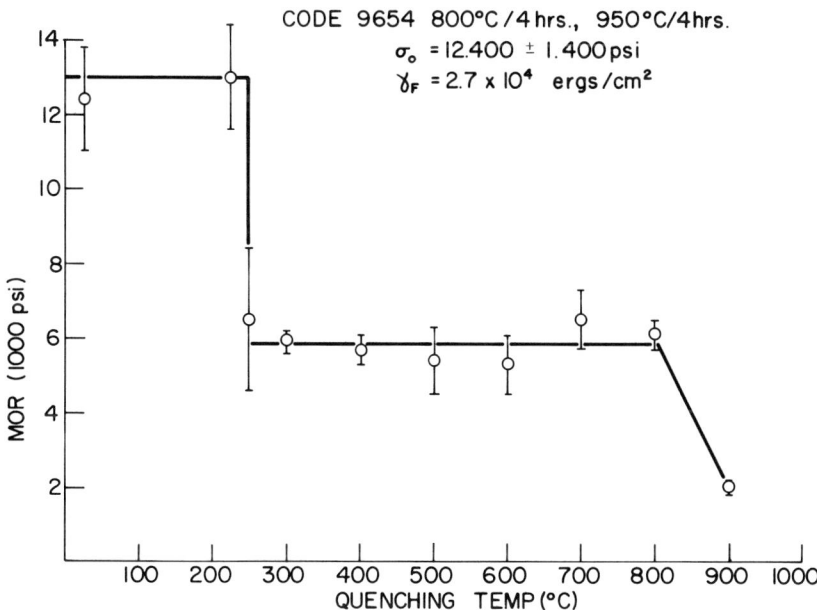

Fig. 4   Changes in fracture strength vs. quenching temperature for sample B.

Sample C shown in Fig. 5, obtained after 800°C/4 hr and 975°C/4 hr treatment, has a flexural strength of 10,700 psi and a fracture surface energy of 3.2 x 10^4 ergs/cm^2. The critical quenching temperature difference increased slightly to 250°C. Region II represents a strength change of -35%, followed by a further gradual decrease. This type of behavior has been observed earlier by Coppola and Bradt in their SiC samples.[11]

Sample D shown in Fig. 6, heat-treated at 800°C/4 hr and 1040°C/6 hr, shows a flexural strength of 7,500 psi and a fracture energy value of 4.0 x 10^4 ergs/cm^2 (the highest value measured for Code 9654). The critical quenching temperature

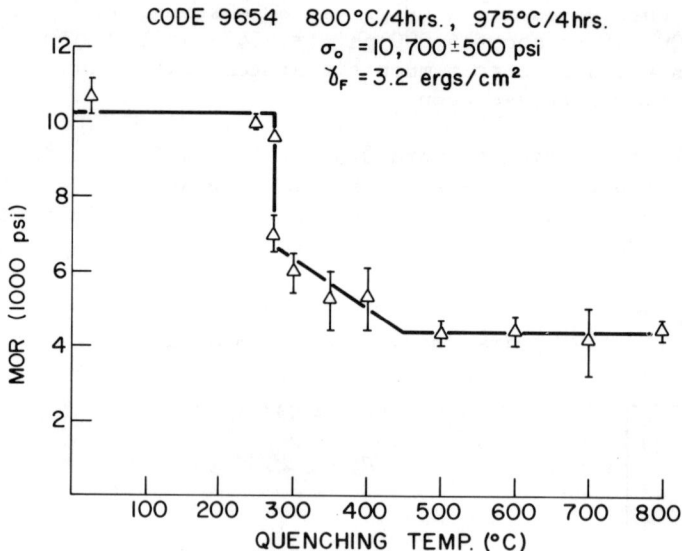

Fig. 5   Changes in fracture strength vs. quenching temperature
        for sample C.

difference ΔT_c is greatly increased to 350°C and the strength
change at T_c is -22%, which is followed by a further gradual
strength decrease. Above approximately 500°C, the strength remains constant at 4,200 psi to 800°C, and then increases to
5,500 psi at 900°C. The strength increase at 900°C might be
similar to the tempering effect observed earlier in mullite.[13]

Sample E contains very coarse mica crystals with diameters
in the range of 180 μ and, consequently, has a very low strength
of only 4,000 psi. The fracture energy decreases slightly to
3.5×10^4 ergs/cm^2, mainly because of the low strength. As
shown in Fig. 7, there is no loss in strength at least up to a
quenching temperature of 900°C. It may be considered that the
stored elastic energy is so small compared to the fracture energy
that no catastropic fracture propagation occurs. Indeed, the
fact that a material with a coefficient of thermal expansion of
71×10^{-7}/°C has such high thermal shock resistance is phenomenal.

Table II lists thermal-shock-resistance parameters $R =$
$\sigma_0(1-\nu)/\alpha E$, $R''' = E/\sigma_0^2(1-\nu)$, $R'''' = \gamma_F R'''$, and $R_{st} = (\gamma_F/\alpha^2 E)^{1/2}$
where E = Young's modulus, ν = Poisson's ratio, σ_0 = fracture
strength, γ_F = fracture surface energy, and α = coefficient of
thermal expansion. The parameter R represents the maximum

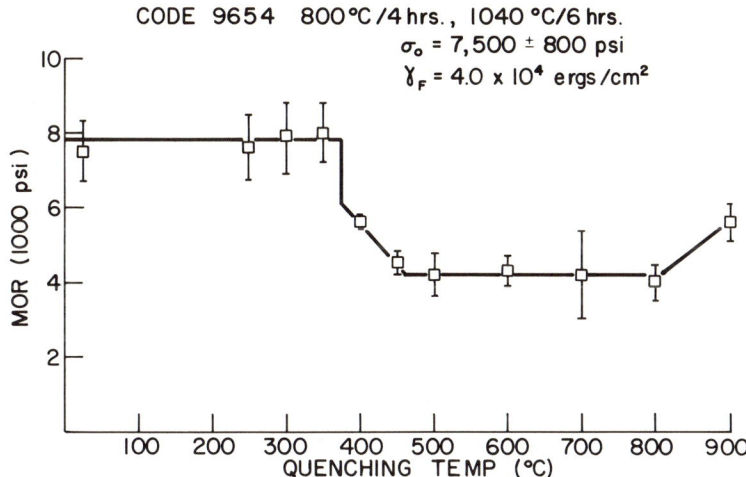

Fig. 6   Changes in fracture strength vs. quenching temperature
         for sample D.

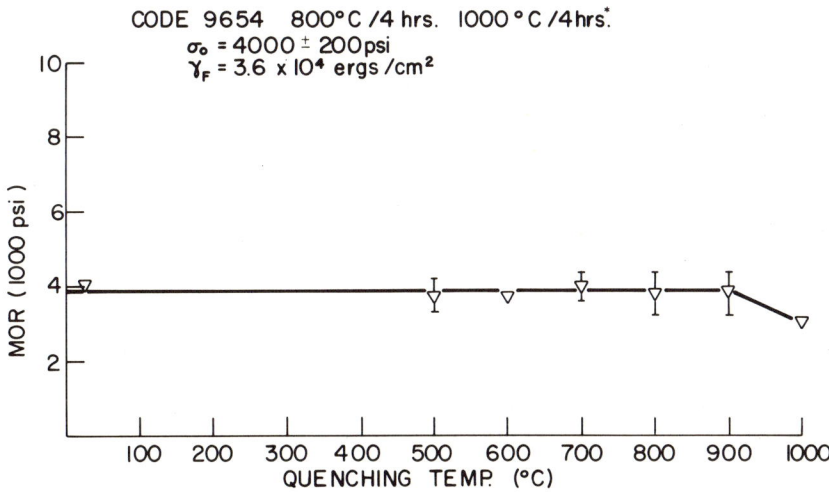

Fig. 7   Changes in fracture strength vs. quenching temperature
         for sample E.

quenching temperature difference a material can sustain without crack nucleation. Comparing the values of R to the actual observed quenching temperature differences, no correlation appears to exist. Especially for sample E, the parameter R decreases sharply from that of sample D, and the observed ΔT_c increases tremendously to 900°C. If the parameters R and R_{st} are equivalent except for a constant and the geometric quantities,[12] it is not evident from Table II. However, the crack stability parameter R_{st} appears consistent with the increase in ΔT_c. Qualitatively at least, the increase in ΔT_c can be explained in terms of R_{st}. If so, the resistance parameters governing region I appear to be directly related to the fracture energy, quite contrary to the maximum tensile stress criterion.

The thermal-stress-resistance parameters governing region II are R''' and R''''. The factor R''' represents the minimum available elastic energy at fracture, and R'''' indicates the minimum in the extent of crack propagation on fracture initiation.[13] Table II shows that R''' is not consistent with the amount of strength change at the critical quenching temperature, especially for sample A. Since R''' is inversely proportional to the stored elastic energy available for fracture propagation, the higher the value of R''', the lower the damage should be. According to the values of R''', sample A should have a lower damage than sample B, contrary to the observed strength change at T_c.

The parameter R'''', on the other hand, is consistent with the amount of damage. Further, the amount of variation in R'''' best describes the magnitude of strength change at T_c for all the samples studied. Perhaps it is not a coincidence that the quenching temperature difference is also well represented by R''''.

On the basis of these results, R'''' appears to be the best indicator of thermal shock behavior for mica glass-ceramic materials. The importance of γ_F in describing thermal shock behavior can not be overemphasized.

For regions III and IV, the governing parameter R_{st} is described as the maximum allowable temperature difference to propagate long cracks under severe thermal condition.[13] From Figs. 3-7, a moderate increase in the transition temperature from region III to IV occurs in agreement with R_{st} values.

Formation of mica crystals--randomly distributed two-dimensional flakes--in a glass matrix greatly increases the fracture surface energy. This is due to the nature of easily cleavable and flexible crystals with extremely high strength in the direction parallel to the basal plane. Consequently, fracture propagates mainly along the cleavage planes and undergoes

TABLE II

Sample	Fracture Surface Energy (10^4 ergs/cm^2)	$R(°C)$	$\Delta T_C(°C)$ Observed	Strength Change At T_C (%)	R''' $(10^{-6} \text{ cm}^2/\text{dyne})$	R'''' (10^{-2} cm)	R_{st} $(\text{cm}^{1/2} \text{ °C})$
A	0.4	65 ± 5	125	-82	2.27	0.9	7.2
B	2.7	109 ± 12	225	-54	1.39	3.6	24.5
C	3.2	104 ± 5	250	-35	1.67	5.3	27.8
D	4.0	146 ± 16	350	-22	1.78	7.1	42.2
E	3.5	111 ± 12	> 900	0	5.60	19.6	51.7

series of crack branching and deflections due to the highly interlocked mica flakes.

Crack blunting was often observed.[2] According to Cook and
Gordon, the propagation of elastic cracks in brittle solids containing planar regions of weakness can be blunted by the formation
of a T-shaped junction when the primary crack merges into the
secondary crack nucleated along the planar region of weakness
(i.e., mica crystals in this case) under the tensile stress concentration.[2,14] A series of crack branching, deflections, and
blunting would eventually exhaust the energy released during
crack propagation, and is well known to provide an energy-
absorbing mechanism, thereby increasing the fracture surface
energy.

In a similar microstructure of hypereutectic ZrC-C alloys,
Rossi and Carnahan[15] concluded that an energy-absorbing mechanism
was provided by the primary graphite flakes. The cracks were
observed to follow the graphite flakes and to undergo blunting
after a series of crack deflections. The thermal-shock resistance
was explained in terms of the stored-elastic-energy criterion.

SUMMARY

It has been shown that the crystallization of mica flakes in
a brittle glass matrix can increase the fracture surface energy
by an order of magnitude greater than an ordinary glass or other
fine-grained glass-ceramics (e.g. Code 9608 β-spodumene ss). The
large increase in the fracture energy is due to the crack
branching, deflection, and blunting that are direct results of
the highly interlocked, easily cleavable, but very strong, mica
flakes.

The high values of fracture energy increases greatly the
thermal-shock resistance of mica glass-ceramics, which appears to
be consistent with the Hasselman's theory based on the stored-
elastic-energy criterion, rather than on the maximum-tensile-
stress criterion. The thermal-stress-resistance parameter R'''',
which takes into account the fracture surface energy, appears to
best describe thermal shock behavior of mica glass-ceramics, not
only in terms of the thermal shock damage but also the quenching
temperature difference.

It is the microstructure of mica glass-ceramics which can be
characterized as having a uniform distribution of easily cleavable,
yet very strong, two-dimensional mica flakes in a brittle matrix
that increases the fracture surface energy.

ACKNOWLEDGEMENT

The author would like to express his deep gratitude to Drs. G. H. Beall and D. G. Grossman for their many helpful discussions. Special thanks are due to Prof. D. P. H. Hasselman for his encouragement and many valuable suggestions. Technical assistance provided by Mr. T. R. Kennedy is gratefully acknowledged.

REFERENCES

1. G. H. Beall, Symposium on Nucleation and Crystallization - Revisited, Ed. by L. L. Hench and S. W. Freiman (Amer. Ceram. Soc., 1972), p. 251.

2. C. K. Chyung, G. H. Beall and D. G. Grossman, Electron Microscopy and Structure of Materials, Ed. by G. Thomas, R. M. Fulrath, and R. M. Fisher, p. 1167.

3. D. G. Grossman, J. Amer. Ceram. Soc., Vol. 55, 446 (1972).

4. A. J. Bailey, Second International Congress of Surface Activity III, Solid-Liquid Interface (Butterworth Scientific Publ. 1957), p. 406.

5. H. R. Shell and K. H. Ivey, Fluorine Micas, (U.S. Bureau of Mines Bull. 647, 1969), p. 99.

6. D. P. H. Hasselman, J. Amer. Ceram. Soc., Vol. 46, 534 (1963).

7. D. P. H. Hasselman, Ibid., Vol. 52, 600 (1969).

8. R. W. Davidge and G. Tappin, J. Mat. Sci., Vol. 3, 165 (1968).

9. H. G. Tattersall and G. Tappin, J. Mat. Sci., Vol. 1, 296 (1966).

10. J. Nakayama, J. Am. Ceram. Soc., Vol. 48, 583 (1965).

11. J. A. Coppola and R. C. Bradt, J. Amer. Ceram. Soc., Vol. 56, 214 (1973).

12. D. P. H. Hasselman, J. Amer. Ceram. Soc., Vol. 54, 219 (1971).

13. D. P. H. Hasselman, "Thermal Stress Resistance Parameters for Brittle Refractory Ceramics: A Compendium", J. Amer. Ceram. Soc., Vol. 49, 1033 (1970).

14. J. Cook and J. E. Gordon, Proc. Roy. Soc., (London), A282, 508 (1964).

15. R. C. Rossi and R. D. Carnahan, Ceramic Microstructures Their Analysis, Significance, and Production, (John Wiley, 1968), p. 620.

FRACTURE OF POLYCRYSTALLINE GRAPHITE

C. A. Andersson and E. I. Salkovitz

Westinghouse Research Laboratories

and Office of Naval Research

INTRODUCTION

The fracture toughness parameters (the critical crack extension force, GI_c, and the critical stress intensity factor, KI_c) are considered to be material constants for given conditions of temperature and strain rate. As such, the fracture toughness values of two or more materials are often compared to ascertain the relative merits of these materials in a structural component. This use is similar to comparing yield strengths of materials to determine the relative resistances to plastic deformation. More precisely, however, with a knowledge of GI_c or KI_c along with the elastic modulus, the defect size, and the relative geometries of the defect and the structure, it is possible to predict the failure stress of that structure. The fracture toughness approach, therefore, constitutes a useful failure criterion for a wide range of materials.

There are certain classes of materials for which these direct fracture toughness approaches must be modified in order for them to be used as fracture criteria. In the particular case to be considered, polycrystalline graphites, extensive cracking occurs during deformation. Thus, the failure mechanism is similar to the familiar fatigue crack mechanism. To demonstrate by previewing some of the experimental results, failure stresses are plotted against critical stress intensity factor values in Fig. 1.

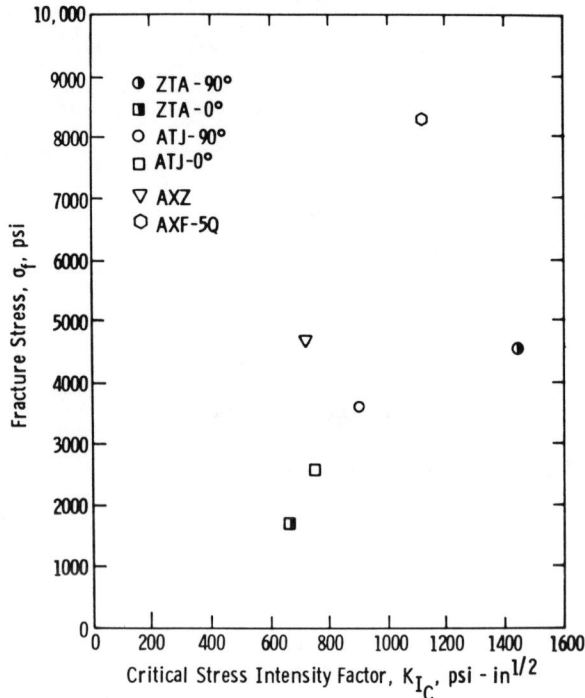

Figure 1.   A comparison of strengths with critical stress intensity
factors of polcrystalline graphites.

Although the AXZ and ZTA-90° materials have equivalent strengths,
the former has half the K_{IC} value of the latter. Also, the AXF-5Q
has an extremely high strength for its K_{IC}. Therfore, a comparison
of K_{IC} values does not necessarily allow comparative strength
predictions.

Before proceeding, a brief description of the constitution
and failure mechanism of artificial polycrystalline graphites is in

order. These materials are generally produced by mixing graphite
and/or carbon flour particles with a hydrocarbon binder, pressure
compacting and heating to temperature which ultimately exceeds
2500°C. During the initial stages of heating, the binders are
converted to carbon and later are graphitized to varying degrees
depending on the binder used. The flour particles used are generally, though not always, highly oriented and have easy planes of
fracture. The common compacting operations (uniaxial pressing or
extrusion) tend to align the particles giving the bulk material a
preferred orientation. The densities of the final products are
also less than theoretical ranging between 60 and 90 percent.

The strengths of polycrystalline graphites have been shown
to be dependent on the extent of preferred orientation and stress
axis directions,[1-3] the densities,[2-6] and flour particle
sizes.[6] The preferred orientations, stress directions and
densities effect the strain energies. The crack sizes are related
to the particle sizes.

The failure process has been determined by microscopic
observation during stress appliction.[6,7,8] As the stress is
increased above some low value, individual flour particles cleave
normal to the tensile axis. This is an easy process in single
crystalline graphite since the crack nuclei exist prior to deformation and do not have to be created.[9]

At high stress levels, still more particles cleave, but the
cracks also extend and/or link up through material which is not as
suitably oriented. The stress intensity, K_I, increases due to both
an increase in the stress and an increase in crack lengths, until
the critical value of fracture instability is reached in one. Any
failure criterion for graphite must accommodate the crack growth
with deformation, as well as account for the physical property
effects on the strength.

The equation for the failure of a body containing a disk
shaped crack normal to a large tensile stress field is:[10]

$$\sigma_f = \left[\frac{\pi E G_{I_c}}{4(1-\nu^2)c}\right]^{1/2} \tag{1}$$

where σ_f is the failure stress, E is the modulus of elasticity,
ν is Poisson's ratio and c is the crack radius. A fracture criterion was arrived at by determining the effects of the physical
properties of density, particle size and degree of preferred
orientation on each of the parameters. Since the crack size
increases with applied stress and its magnitude is unknown, the
equation was converted to one of proportionality.

EXPERIMENTAL

Materials

Four grades of graphite were chosen for the study: ZTA,* ATJ,*
AXZ** and AXF-5Q.** They were selected for their high strengths
relative to other graphites and to enable the determination of the
effects of density, particle size and degree of preferred orientation on the mechanical properties. The effects of stress axis
direction were determined by cutting test specimens from ZTA at 0°,
30°, 60° and 90° from the pressing direction (the axis of symmetry)
and at 0° and 90° from ATJ. The AXZ and AXF-5Q graphites are
isotropic.

Procedures

The fracture toughness parameters of ZTA and ATJ were determined from the center-notched specimens shown in Fig. 2. The AXZ
and AXF-5Q specimens were similar but proportionally 5/8 the size
due to material limitations. A natural crack was introduced
into the specimen and strain gauges were applied at sufficiently
distant locations. The specimens were loaded to failure, and K_{Ic},
E and G_{Ic} were determined on each specimen.[11]

The ZTA and ATJ tensile specimens in Fig. 3 were designed to
be geometrically similar to the fracture toughness specimens.
The specimen width in the gauge section was reduced to ensure
against loading pin hole breakage. Again, the AXZ and AXF-5Q
specimens were 5/8 the size. Resistance strain gauges were also
applied and these specimens were loaded to failure. Fracture
stress, σ_f, E, and ν were evaluated.

Densities were determined from the buoyancies of coated
specimens in water. Average particle sizes were measured on
polished specimens with a translating stage microscope.

Preferred orientation was obtained by determining the
distribution of basal planes of the hexagonal layered structure
as a function of the angle from the axis of symmetry. One
millimeter thick plates were machined from the gauge lengths of the

---

 * Produced by the Carbon Products Division, Union Carbide
   Corporation.
** Produced by Poco Graphite, Inc., Union Oil Company of
   California.

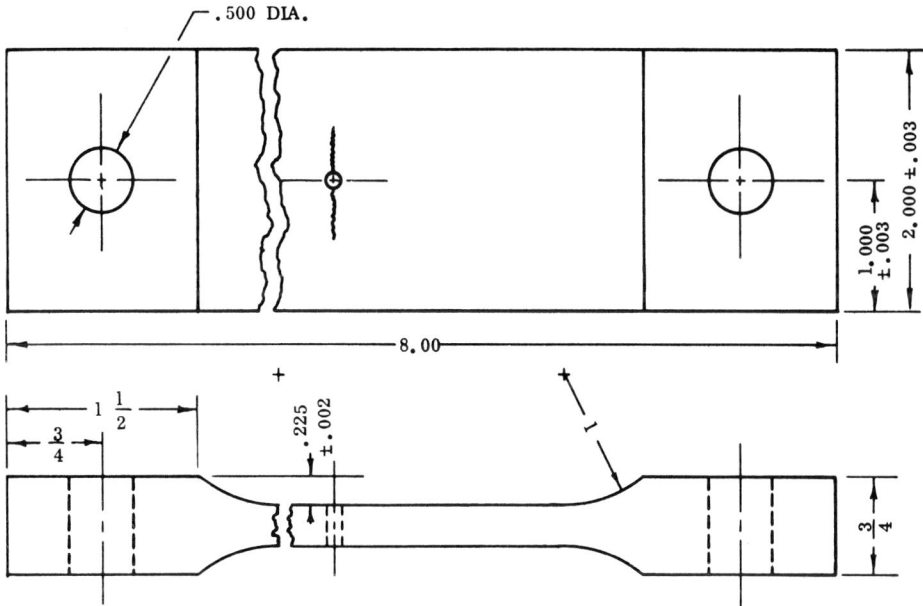

Figure 2.   The center-notched fracture toughness specimen used to
test ATJ and ZTA graphites.

broken tensile specimens. These were placed in an x-ray diffraction
unit at the Bragg angle. The specimens were then rotated about an
axis normal to the pressing direction and the relative intensities
of [0002] planes were determined.

RESULTS

The average values for fracture toughness parameters, K_{Ic} and
G_{Ic} are given in Table 1. The average tensile test results (tensile
strength, σ_f, elastic modulus calculated for theoretical density,
E_o, and Poisson's ratio, ν, as well as the average particle sizes,
d, and the volume fractions of porosity, f_p are presented in
Table 2.

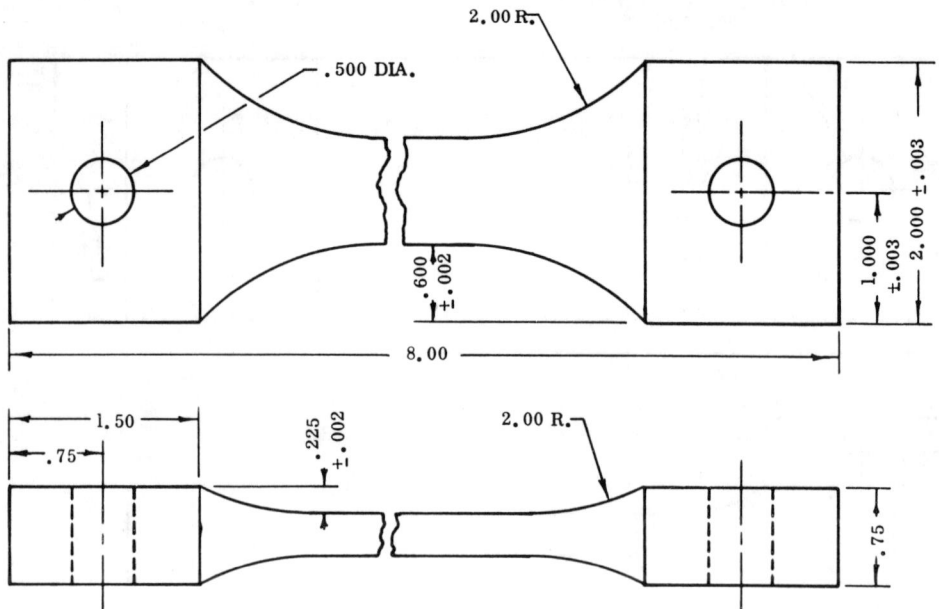

Figure 3.   The tensile specimen used for testing ATJ and ZTA
            graphites.

TABLE 1.   AVERAGE FRACTURE TOUGHNESS PARAMETERS FOR
           SEVERAL GRADES OF POLYCRYSTALLINE GRAPHITES

Specimen Type	Number of Specimens	K_{I_c} psi - in.$^{1/2}$	G_{I_c} psi - in.
ZTA-0°	5	660	.660
ZTA-90°	6	1440	.760
ATJ-0°	6	750	.580
ATJ-90°	6	905	.580
AXF-5Q	3	1120	.590
AXZ	8	720	.360

TABLE 2.   AVERAGE RESULTS OF TENSILE TESTS, PHYSICAL PROPERTY MEASUREMENTS
AND FRACTURE CRITERION CALCULATIONS

Specimen Type	No. of Spec.	σ_f psi	E_o x 10^6 psi	ν	d, x 10^{-3} in	f_p	c in	$\bar{n}$	F.C., x 10^4 psi
ZTA-90°	5	4530	5.28	.13	4.4	.158	.062	1.14	2.58
ZTA-60°	4	3290	3.53	.12	4.4	.158	.080	1.47	1.98
ZTA-30°	8	1740	1.62	.07	4.4	.158	.143	1.96	1.24
ZTA-0°	7	1640	1.46	.06	4.4	.158	.146	2.27	1.14
ATJ-90°	7	3490	4.06	.12	4.4	.271	.056	1.47	1.44
ATJ-0°	6	2570	2.67	.10	4.4	.271	.059	1.72	1.12
AXF-5Q	5	8270	4.39	.21	.5	.235	.011	1.54	5.34
AXZ	6	4630	4.02	.23	.5	.332	.016	1.54	3.37

The specimen type designations indicate not only the material but
the angle of the applied stress from the axis of symmetry (pressing
direction). The average stress-strain curves are shown in Fig. 4.
Finally the relative intensities of [0002] plane normals are shown
in Fig. 5 for ZTA. The ATJ distribution is similar, though not as
anisotropic.

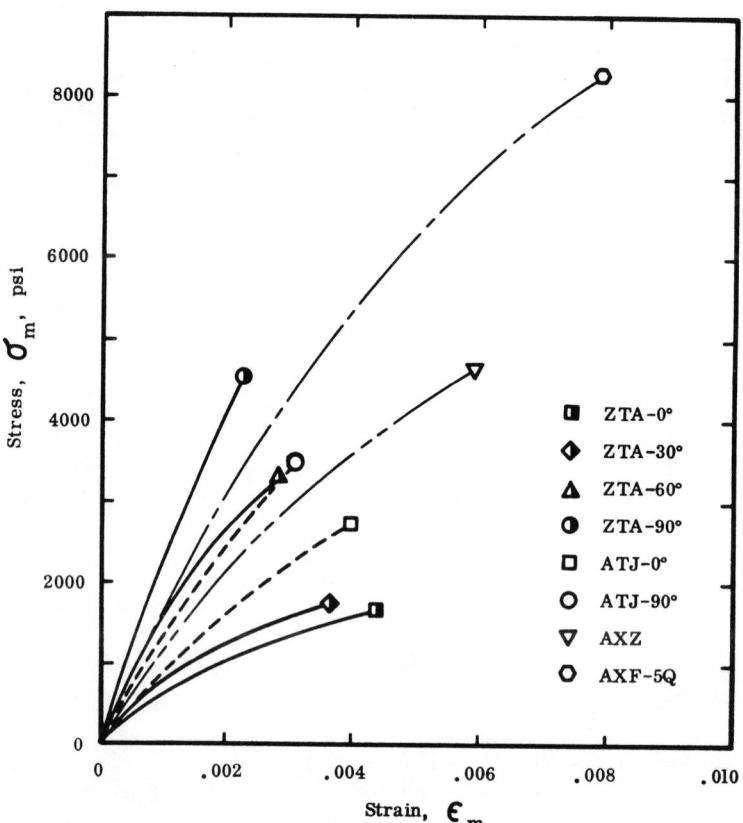

Figure 4.   The average stress-strain curves of four commercial
            graphite grades.

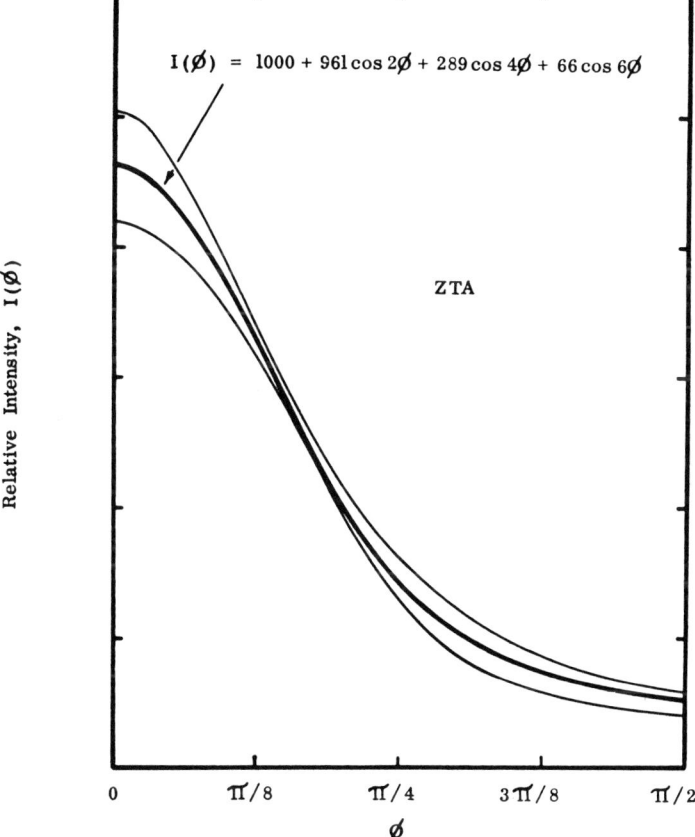

Figure 5. The relative intensity of 0002 plane normals as a
function of the angle ϕ from the pressing direction.
ZTA graphite.

DISCUSSION

The effects of the physical properties of density, particle
size, preferred orientation and stress direction on the mechanical
properties of elastic modulus, the critical crack extension force
and the crack size will be assessed in the following. A criterion
for the fracture of polycrystalline graphites will be suggested.

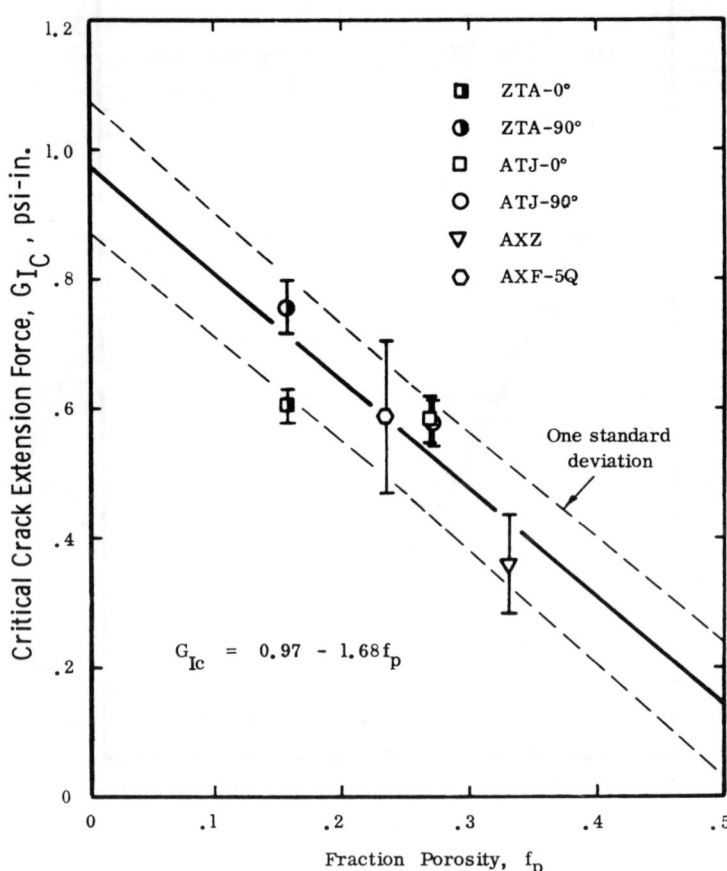

Figure 6.   The critical crack extension force (critical energy release rate or fracture surface energy), G_{Ic}, as a function of the fraction porosity, f_p, for several graphite grades.

Fracture Toughness

Figure 6 shows that G_{Ic} decreases with increasing pore fraction. The G_{Ic} can be considered to be the rate of change of strain energy per increase in crack area at failure. Since the actual surface area of a given size crack decreases linearly with pore fraction, G_{Ic} would be expected to also be a linear function, and

$$G_{Ic} = \overset{\circ}{G_{Ic}} - k_f f_p \qquad (2)$$

where $\overset{\circ}{G_{Ic}}$ is the critical crack extension force for the fully dense material and k_f is the rate of decrease with porosity, f_p. The value of $\overset{\circ}{G_{Ic}}$ was found to be .97 psi-in (.05 psi-in standard deviation) and k_f was 1.68 psi-in.

A small stress axis direction dependence of G_{Ic} also exists. There is a 15 percent difference in G_{Ic} for the orthogonal directs of ZTA. The graphite flour particles of ZTA are not spherical and the longer axes tend to align normal to the pressing direction. This causes the surface roughness differences which are reflected in the G_{Ic} values. Particle sizes do not seem to have any great effect on G_{Ic} either.

Elastic Modulus

Since the elastic properties of single crystal graphite are highly anisotropic, the elastic properties of the polcrystalline material are strongly effected by the degree of preferred orientation and the stress axis direction. As with all materials, porosity also has a major effect. Although details will not be given here, a method of calculating the elastic moduli in any direction of the polycrystalline material from the single crystal elastic constants and the distributions of basal planes has been developed. Therefore, the moduli of the fully dense material can be obtained with reasonable accuracy from the x-ray diffraction data.

Although various relationships between porosity and elastic moduli have been proposed, the exponential one has been shown to have theoretical foundation.[8] Briefly, this is accomplished by calculating the length change in a finite body under constant stress which contains a single void. Dividing by the original length and the stress, a new apparent elastic compliance is determined. Successively adding independent pores allows the calculation of new apparent compliances which increase in magnitude exponentially in the typical manner of this type of compounding process. Finally by inserting pore fraction, f_p, for numbers of pores, the relative sizes of the pores with respect to the bulk dimension cancel out and

$$A = A_o \exp (k_p f_p) \qquad (3)$$

where A is the apparent compliance of a material whose fully dense compliance is A_O, and k_p is a function of Poisson's ratio only. Correspondingly, the equation for the apparent elastic modulus is:

$$E = E_O \exp(-k_p f_p). \tag{4}$$

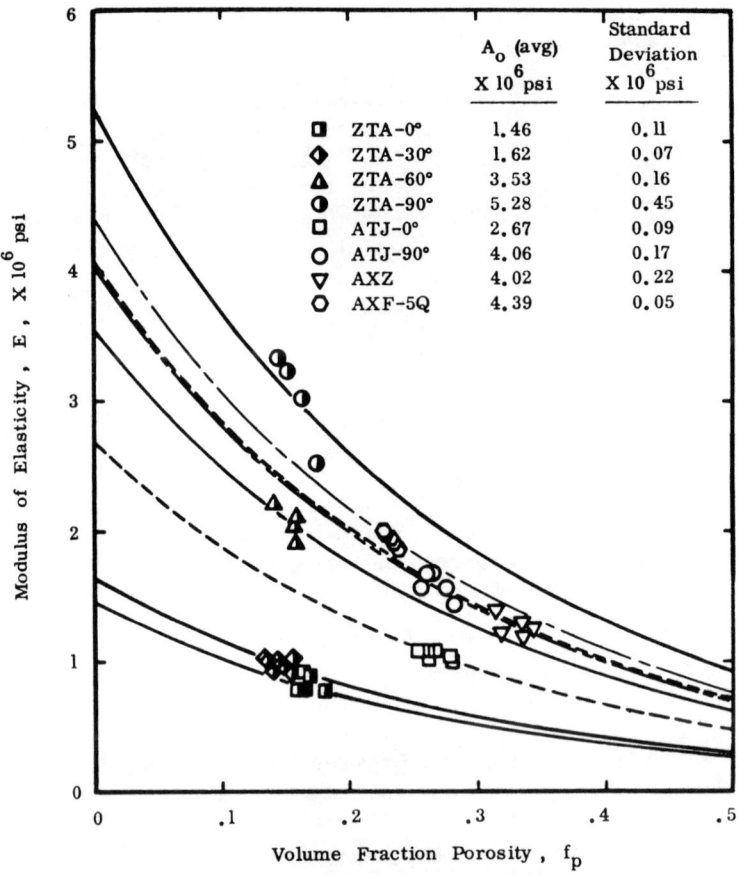

Figure 7. The effect of the volume fraction porosity on the elastic moduli of several graphite grades and orientations.

Since insufficient data was collected in this study to determine k_p, a literature value of 3.5 [11,12] was used. In the present case E_O was calculated and the extrapolation is shown in Fig. 7.

Critical Crack Size

Assuming a disk shaped crack, the radii of the critical sized cracks, c, were calculated and appear in Table 2. These values indicate that the crack size that a polycrystalline graphite can tolerate before failing is primarily a function of particle size, elastic modulus, crack extension force, and porosity. More subtly, it is also effected by the preferred alignment of basal planes normal to the stress axis.

Large cracks sometimes exist in graphite materials prior to deformation and these cracks probably account for the extremely weak structures occasionally encountered. Since this is not the general case, a procedure to handle the deformation cracks is necessary for the fracture criterion. This has been accomplished by past authors by noting that cracks are composed of an incremental number of grains or particles and that there exists a proportionality between grain size and crack size.

There is also a finite probability that two or more grains are oriented such that a common crack can occur through both. This probability varies with the degree of preferred orientation, and the stress axis direction. The average number of adjacently cracked grains was determined by the following method.

The relative densities of c-axes (cleavage plane normals) per unit solid conical angle, κ, as functions of the rotation, α, of the stress axes from the axes of symmetry are shown in Fig. 8 for ZTA as an example. These were calculated from the measured planar x-ray distributions using standard methods.[13] Similar distributions were calculated for the other graphites. The probability that a basal plane will cleave is approximated by $\cos^2 \kappa$. The probability that a single particle will cleave is then:

$$P_1 = \frac{\int_0^{\pi/2} L(\kappa,\alpha) \sin \kappa \cos^2 \kappa \, d\kappa}{\int_0^{\pi/2} L(\kappa,\alpha) \sin \kappa \, d\kappa} \tag{5}$$

Assuming independent events, the probability that n adjacent particles will cleave is:

$$P_n = P_1^n \tag{6}$$

Finally, the average number of adjacently cracked particles, $\bar{n}$, is

$$\bar{n} = \frac{\sum_{n=1}^{\infty} n P_1^n}{\sum_{n=1}^{\infty} P_1^n} \tag{7}$$

Calculated values of $\bar{n}$ are given in Table 2 and the P_n as functions, of n are shown in Fig. 9 for the various grades and orientations.

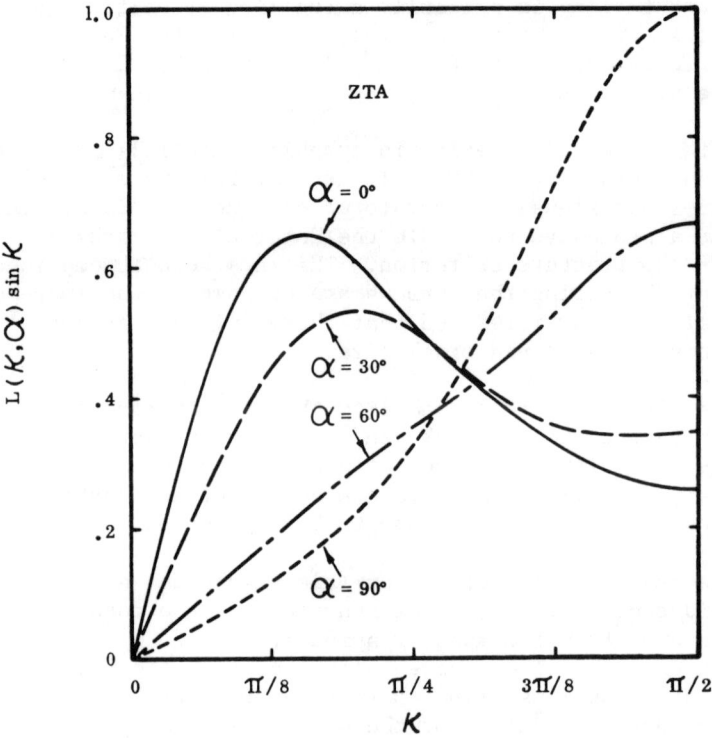

Figure 8.   The relative density of 0002 plane normals per unit solid conical angle, κ, from an axis that has been rotated by an angle α from the pressing direction. ZTA graphite.

These results are qualitatively confirmed by the microscopic studies of fracture. The samples of ZTA oriented with numerous basal planes normal to the tensile stress showed cracks early in deformation which went through several particles. Those samples orthogonally oriented showed few extended cracks until later in the deformation where the stresses were large enough for the cracks to circumvent misoriented particles.

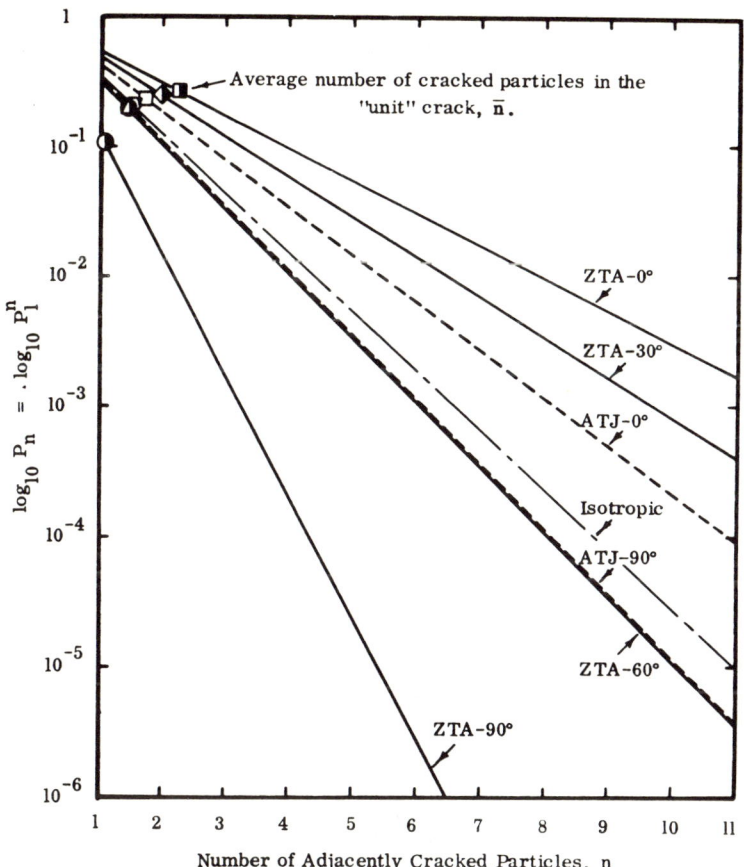

Figure 9. The probability, P_n, that n graphite flour particles in adjacent positions are cracked.

A comparison of the data in Fig. 9 with that of the G_{Ic} data indicate that the final fracture of graphite probably occurs through particle boundaries, around misoriented material. Although the probabilities of cracks extending through adjacent particles of ZTA-90° become very remote rather rapidly when compared to ZTA-0°, there is little difference in the critical crack intensity forces.

The critical crack area can be considered to consist of pores, as well as linked cleaved particles of which an average of $\bar{n}$ are adjacent to one another. The pores are pertinent since the higher the porosity, the fewer particles need be fractured for a given crack size. Assuming a disk shaped crack and spherical particles:

$$\pi \; c^2 \; \propto \; \frac{\pi d^2}{4} \; \frac{\overline{n}}{(1-f_p)} \tag{8}$$

or

$$c \; \propto \; \frac{d}{2} \; \left(\frac{\overline{n}}{1-f_p}\right)^{1/2} \tag{9}$$

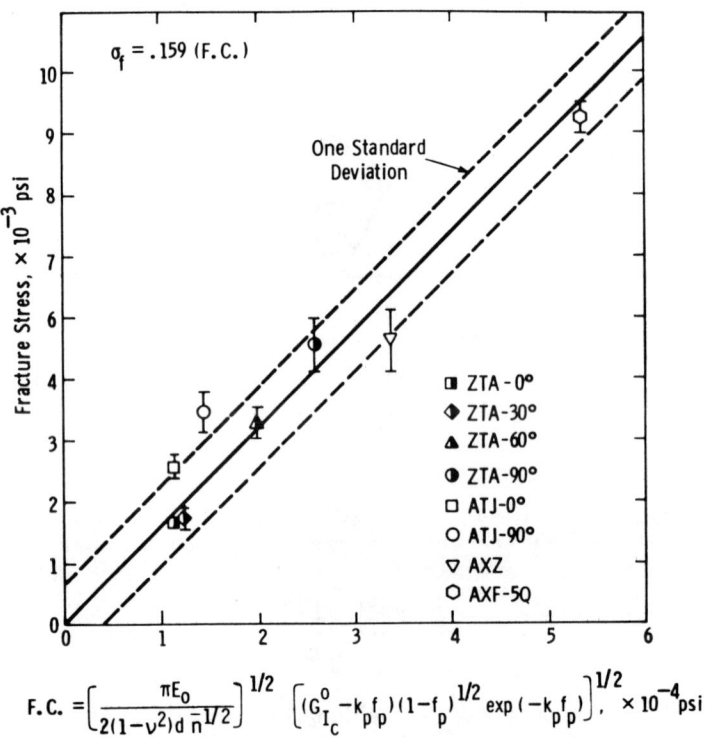

$$\text{F.C.} = \left[\frac{\pi E_0}{2(1-\nu^2)d \, \overline{n}^{1/2}}\right]^{1/2} \left[(G^0_{I_c} - k_p f_p)(1-f_p)^{1/2} \exp(-k_p f_p)\right]^{1/2}, \; \times 10^{-4} \, \text{psi}$$

Figure 10.   Fracture stresses, σ_f, versus the calculated fracture criteria, F.C., of several polycrystalline graphites.

Fracture Criterion

The fracture criterion is obtained by substituting Eqs. 2, 4 and 9 into Eq. 1 and determining the proportionality constant empirically. The results, F.C., of the calculations of the data in Table 3 are plotted versus fracture stress, σ_f, in Fig. 10. The equation relating the fracture stress to the physical properties is:

$$\sigma_f = .159 \left[\frac{\pi E_o}{2(1-\nu^2) d \, \bar{n}^{1/2}} \right]^{1/2} [(\overset{o}{G}_{I_c} - k_f f_p)(1-f_p)^{1/2} \exp{(-k_p f_p)}]^{1/2} \tag{10}$$

where $\overset{o}{G}_{I_c}$ is .97 psi-in, k_f = 1.68 psi-in, and k_p = 3.5. The standard deviation from this equation is 630 psi.

CONCLUSIONS

1. The fracture mechanism of artificial polcrystalline graphite was identified. On increasing load application, first individual particles fracture, then these cracks extend and/or link up through particle boundaries until one is sufficiently large to become unstable.

2. Based on the observed fracture mechanism and the equation for the fracture stress of a material containing a disk shaped crack, a failure criterion was proposed. The criterion was arrived at by determining the effects of density, particle size, preferred orientation and stress axis direction on each of the parameters of the equation. Since the crack sizes are deformation dependent, the Sneddon equation was transformed to a proportionality and the proportionality coefficient was empirically determined.

REFERENCES

1. E. J. Seldin, Carbon, 4, 177 (1966).

2. H. E. Martens, L. D. Jaffe and J. E. Jepsand, Proc. Third Conf. on Carbon, 529. Pergamon Press, New York (1959).

3. H. H. W. Losty and J. S. Orchard, Proc. Fifth Conf. on Carbon, 2, 519. The Macmillan Co., New York (1963).

4. I. B. Mason, Proc. Fifth Conf. on Carbon, 597. The Macmillan Co., New York (1963).

5. J. M. Hutcheon and M. S. T. Price, Proc. Fourth Conf. on Carbon, 645. Pergamon Press, New York (1960).

6. R. H. Knibbs, J. Nucl. Mater., 24, 174 (1967).

7. O. D. Slagle, J. Amer. Ceram. Soc., 50, 495 (1967).

8. C. A. Andersson, Ph.D. Thesis, Univ. of Pittsburgh, 1970.

9. D. E. Soule and C. W. Nezbeda, J. Appl. Physics, 8, 5122 (1952).

10. I. N. Sneddon, Proc. Royal Soc. London, 187, 229 (1946).

11. J. E. Srawley and W. F. Brown, Jr., Fracture Toughness Testing and Its Applications, STP 381, 133. ASTM, Philadelphia (1965).

12. J. R. Cost, K. R. Janowski and R. C. Rossi, Phil. Mag., 17, 851 (1968).

13. R. M. Spriggs, J. Amer. Ceram. Soc., 44, 628 (1961).

14. G. E. Bacon, J. Appl. Chem., 6, 477 (1956).

FRACTURE OF WÜSTITE AND WÜSTITE-Fe_3O_4-5v/o Fe VERSUS GRAIN SIZE[†]

M. I. Mendelson[*] and M. E. Fine

Northwestern University, Dept. of Materials Science
Evanston, Ill. 60201

In rapidly cooled wüstite the average fracture stress increased from about 64 to 100 MN/m^2 while the grain size decreased from 88 to 9μm. The fracture energy (7.5 J/m^2) was grain size independent. This was explained in terms of the theories for Griffith-Orowan flaw extension and plasticity-induced flaw nucleation assuming the intensive properties of the grain boundaries remained constant with change in grain size. In aged wüstite, the fracture energy increased from 23 to about 40 J/m^2 with decrease in grain size from 96 to 8μm. This change was discussed in terms of crack front pinning and secondary cracking. The fracture stress decreased with grain size to a peak value of 187 MN/m^2 at 24μm. The decrease was attributed to the increase in fracture energy and decrease in flaw size.

I. INTRODUCTION

The fracture energy, fracture strength and fracture stability of wüstite (>99.5% dense) were enhanced by eutectoid decomposition whereby a continuous network of α-Fe formed at the grain boundaries of the parent wüstite.[1] In the previous study the grain size was fixed at 23μm and the amount of α-Fe was varied from approximately 1 to 9.5 v/o by controlling the extent of the wüstite decomposition.

The objective of the present paper was to investigate the fracture behavior vs. grain size (8 to 96μm) for fixed vol.% α-Fe (1 and 5 v/o). Two types of microstructures were studied: (1)

† Sponsored by U. S. Air Force Office of Scientific Research, Office of Aerospace Research, Grant No. AF-AFOSR-68-1457.

* Western Electric Co., P. O. Box 900, Princeton, N. J. 08540

rapidly cooled (unaged) wüstite consisting of about 1 v/o α-Fe globules in a $Fe_{0.94}O$ matrix, and (2) aged wüstite decomposed at $480^\circ C$, consisting of 5 v/o α-Fe at the grain boundaries. When rapidly cooled wüstite of eutectoid composition was isothermally aged at $480^\circ C$, it decomposed forming 10.8 v/o α-Fe at completion by the reaction[1-5]

$$Fe_{0.94}O \rightarrow 0.19\ \alpha\text{-Fe} + 0.25\ Fe_3O_4 \ . \tag{1}$$

The fracture initiation energy was measured in pre-cracked double-cantilever beam (DCB) specimens, and the fracture stress was measured in 3-point bending.

II. EXPERIMENTAL PROCEDURE

Polycrystalline wüstite* disks were prepared by pressure sintering mixtures of Fe and Fe_2O_3 powders to give $Fe/O \cong 0.95$, as previously described.[1] The pressure-temperature-time schedules for hot-pressing were varied to give the different grain sizes shown in Table I. After pressure sintering, the specimens were furnace cooled to between $570-600^\circ C$ and then "rapidly" cooled retaining metastable wüstite.[1] The rapidly cooled wüstite samples were heat treated at $480^\circ C$ in argon for selected times, depending upon their grain size, to obtain about 5 v/o α-Fe as shown in Table II.

The DCB specimens were machined from the hot-pressed disks to a nominal size of 10mm X 30mm X 2.5mm with side grooves 0.5mm thick X 0.8mm deep. The specimen thickness accommodated at least 9 grains across the fracture section for the largest grain size.

---

* Chemical analysis in at. ppm: 100 Aℓ; 100 Cr; <1 Cu; 50 Mn; 75 Ni; 30 Pb; 1000 Si; 30 Sn; 35 Zn; B, Co, Mo, Ti, V not detected.

TABLE I. Hot-Pressing Procedures for Producing Various Grain Sizes

Plunger Material	Pressure, Temperature, Time Schedule	Grain Size (μm)
Inconel	1000 psi, $600-750^\circ C$, 1h 10,000 psi, $750-800^\circ C$, $\frac{1}{2}$h 10,000 psi at $800^\circ C$, 2h	8-12
Inconel	1000 psi, $600-750^\circ C$, 1h 10,000 psi, $750-850^\circ C$, 1h 10,000 psi at $850^\circ C$, 1h	23-25
Graphite	1000 psi, $600-750^\circ C$, $1\frac{1}{2}$h 7000 psi, $750-850^\circ C$, $1\frac{1}{2}$h 7000 psi at 850 to $930^\circ C$, $\frac{1}{2}$h	45-60
Graphite	1000 psi, $600-750^\circ C$, 1h 6000 psi, $750-950^\circ C$, 1h 6000 psi at $950^\circ C$, 1h	82-96

TABLE II. 480°C Aging Times to Obtain 5 v/o Fe

Grain Size (μm)	Aging Time (hours)	Total vol.% α-Fe
10	48	5.1 ± .4
23	68	5.0 ± .4
33	76	~5.0
54	96	5.0 ⊥ .5
69	512	5.1 ± .5
88	672	5.2 ± .4

The specimens were pre-cracked prior to testing. All tests were conducted in an Instron machine at a rate of 0.85μm/s at room temperature in air. The load was applied perpendicular to the fracture plane through steel pins which fit into holes machined in the specimens.[1]

The analytical expression for the fracture initiation energy, γ_i, of a DCB specimen has been reported for a crack length of > 1.5 times the beam height.[6] In specimens of known elastic modulus and dimensions, γ_i can be determined by measuring the crack length and critical fracture load. The crack length was measured from the surface marking of the arrested pre-crack using a traveling microscope. The critical fracture load was taken to be the first major load instability on the load-deflection test curve when the curve was essentially linear up to the critical load. In some cases the test curve was non-linear up to the critical load, and then a 2% offset criterion was used.[7]

The complete fracture of each DCB specimen yielded two rough beams. These were machined into 30mm × 2.5mm × 2.5mm beams, polished with 1μm diamond, and loaded in 3-point bending with steel rollers in air at room temperature at a rate of 0.85μm/s. The percentage of intergranular fracture was measured using a systematic point count[8] on the image of the fracture surface given by a scanning electron microscope. The average grain size, $\bar{D}$, was determined from the average intercept length[9] times a factor of 1.56.[10] The sampling error was 10%. The total volume fraction of α-Fe was measured metallographically using a systematic point count[8], which yielded about 10% error. The Young's modulus, E, of the two microstructures was reported.[1]

III. RESULTS

In rapidly cooled wüstite the microstructure consisted of 1 v/o α-Fe globules dispersed in a eutectoid $Fe_{0.94}$ matrix[1], as shown in Fig. 1. Both the α-Fe globule size and mean-free spacing increased with increasing grain size. After aging the rapidly

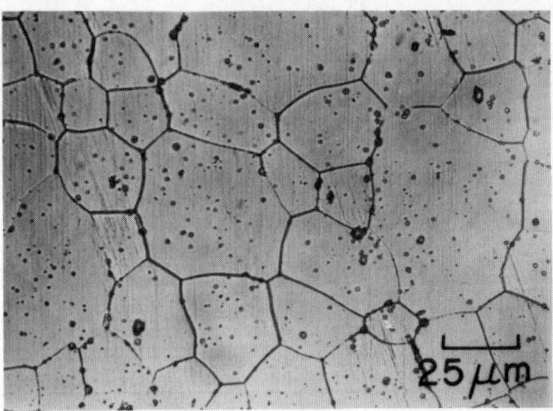

Fig. 1. Optical micrograph of rapidly cooled wüstite, 23μm grain size. Specimen was etched with 40% HCl-ethanol for 5 s.

cooled wüstite at $480°C$ for various times (Table II), α-Fe and adjacent Fe_3O_4 layers formed at the wüstite grain boundaries, as shown in Fig. 2 for two different grain sizes. The α-Fe boundary network sandwiched between two Fe_3O_4 layers is shown more clearly in Fig. 2B. After long time aging some α-Fe and Fe_3O_4 precipitates also formed in the wüstite grain interiors. In the present specimens the decomposition, Eq.(1), is approximately 46% complete, giving about 5 v/o total α-Fe and 41 v/o Fe_3O_4. The reported total vol.% α-Fe was only about 5% greater than the grain boundary vol.% α-Fe which may indicate that some of the globules dissolved.

The fracture initiation energy, γ_i, for rapidly cooled wüstite as a function of grain size is shown in Fig. 3. The least squares

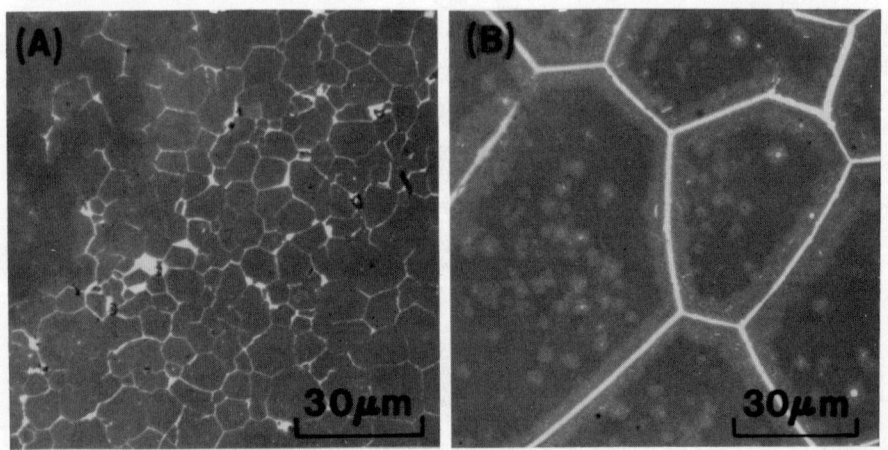

Fig. 2. Optical micrographs of aged wüstite (5 v/o α-Fe) with variable grain size: (A) 10μm, (B) 88μm. Specimens were not etched.

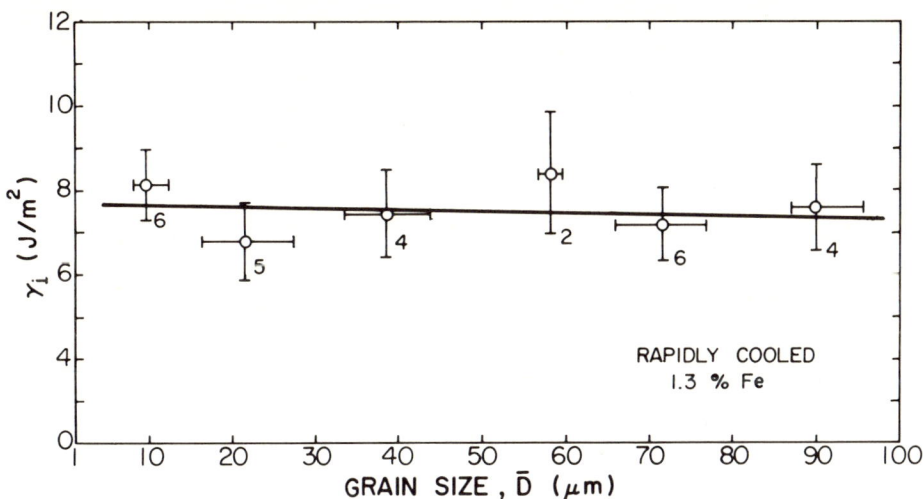

Fig. 3. Fracture energy vs. grain size for rapidly cooled wüstite.
Each datum point is the mean value for the indicated sample size.
Vertical and horizontal bars are 95% confidence limits and grain
size range, respectively.

line indicates a slight decrease with increasing grain size; however, an analysis of variance of the data showed that the regression line was statistically insignificant at 95% confidence.
Hence, within the precision of the data γ_i was independent of grain
size. The mean γ_i and standard error were 7.5 ± 1.0 J/m^2. The
fracture energy for aged wüstite (5 v/o α-Fe) is plotted as a function of grain size in Fig. 4. Here the average γ_i decreased from
approximately 40 to 23 J/m^2 as the grain size increased from 8 to
96μm. These values are 3 to 5 times greater than those for the
rapidly cooled wüstite.

The fracture stress, σ_f, of the rapidly cooled wüstite decreased with increasing grain size as shown in Fig. 5, where the
average σ_f decreased from about 100 MN/m^2 at 9μm to 64 MN/m^2 at
88μm. Using a "best-fit" least squares line through the data[11],
gave a weak grain size dependence, $\sigma_f = 133(\bar{D})^{-0.156}$. The least
squares line was significant at 95% confidence and the standard
error was 8 MN/m^2. The results of σ_f vs. $\bar{D}$ for the aged wüstite
(5 v/o α-Fe) are also shown in Fig. 5. The average fracture
stress decreased somewhat with increasing grain size (from 24 to
90μm) and reached a peak value of 187 MN/m^2 at 24μm. The peak
value was more than twice that for rapidly cooled wüstite of the
same grain size. At other grain sizes the average σ_f was about
20-50% greater for the aged wüstite.

The percentages of intergranular fracture for both rapidly

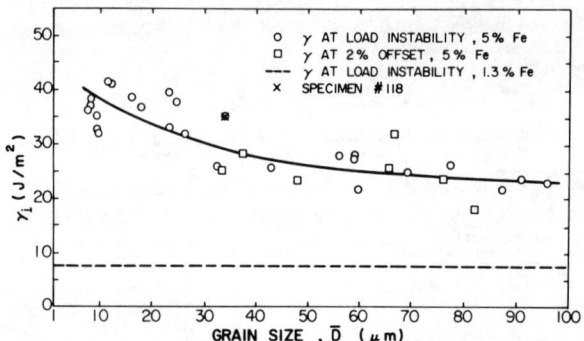

Fig. 4. Fracture energy vs. grain size for aged wüstite and for rapidly cooled wüstite (dotted line from Fig. 3).

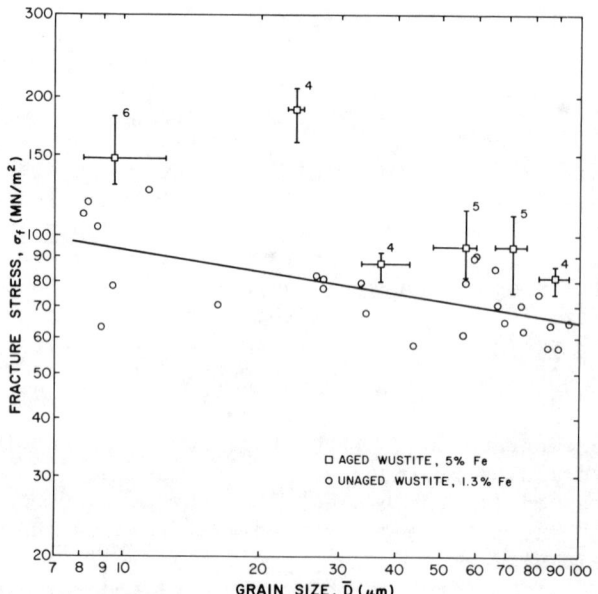

Fig. 5. Log-log plot of fracture stress vs. grain size for aged and rapidly cooled (unaged) wüstite. Each square datum point is the mean value for the indicated sample size and range of scatter.

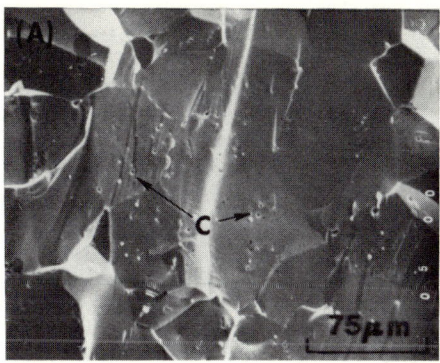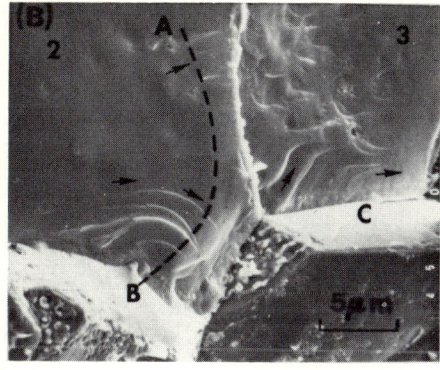

Fig. 6. SEM micrographs of fracture surfaces for (A) rapidly cooled wüstite, 86μm grain size, with cleavage steps indicated by arrows at C, and (B) aged wüstite, 69μm grain size. Arrows show local crack direction.

cooled and aged wüstite were approximately independent of grain size. The mean intergranular fracture percentages and standard errors were 20 ± 7% and 77 ± 14% for the rapidly cooled and aged wüstite, respectively. Scanning electron micrographs of the fracture surfaces are shown in Fig. 6.

IV. DISCUSSION

A. Rapidly Cooled Wüstite

The increase in σ_f with decreasing grain size (Fig. 5) and the grain size invariance of γ_i (Fig. 3) mean that the inherent flaw length, C, increases with grain size. Since $\sigma_f \sim (\bar{D})^{-0.156}$ following Knudson[11], $C \sim (\bar{D})^{0.312}$ assuming fracture initiated from inherent flaws.[12] Such a flaw length-grain size relationship has never been observed in polycrystalline ceramics[13] and no explanation exists which gives $\sigma_f \sim (\bar{D})^{-0.156}$.

Fracture of polycrystalline ceramics at room temperature is thought to occur by initiation from inherent flaws[12] and by plastic flow.[14] If fracture is initiated from inherent flaws the fracture stress, σ_I, is governed by the Griffith-Orowan relationship

$$\sigma_I = \sqrt{2E\gamma_i/\pi C} \quad . \tag{2}$$

If fracture is initiated by dislocation pile-up against a barrier, e.g. grain boundaries[15], the applied stress for crack nucleation, σ_{II}, has been suggested to follow a relation[14]

$$\sigma_{II} = \sigma_o + kL^{-1/2} \tag{3}$$

where σ_o is the applied stress for dislocation motion, k is a term depending upon the fracture energy for localized crack nucleation, and L is the pile-up length.

Carniglia[16] related the previous theories to a two-stage fracture process by assuming $C \cong \bar{D}$ in Eq.(2) and $L \cong \bar{D}$ in Eq.(3), while Rice[17] related Eq.(3) to single-stage fracture. To assist in the interpretation of the data, the results in Fig. 5 are plotted as σ_f vs. $(\bar{D})^{-1/2}$ as shown in Fig. 7. The data were fitted with a single least squares line[17] which was significant at 95% confidence, and the standard error was 15 MN/m^2. The result was approximately statistically the same when the data were fitted with a two-branched curve following Eqs.(2) and (3).[16] The dot-dashed line was constructed assuming Eq.(2) where $C = \bar{D}$, $\gamma_i = 7.5$ J/m^2 (Fig. 3) and $E = 1.28 \times 10^{11}$ N/m^2.[1] Equation (2) is reasonably close to the actual σ_f data for large grain sizes suggesting Griffith-Orowan fracture. However, additional data at larger grain sizes are necessary to verify this. If the fracture at small grain sizes is plastically-induced, then Eq.(3) may be extrapolated to infinite grain size and the intercept compared with the yield stress in a single crystal. The σ_o intercept from Fig. 7 is 54 MN/m^2. To the writers' knowledge, single crystal yield stress data on wüstite are not available; however, data exist on MgO which has the same crystal structure and similar ductility as wüstite. The friction stress for dislocation multipli-

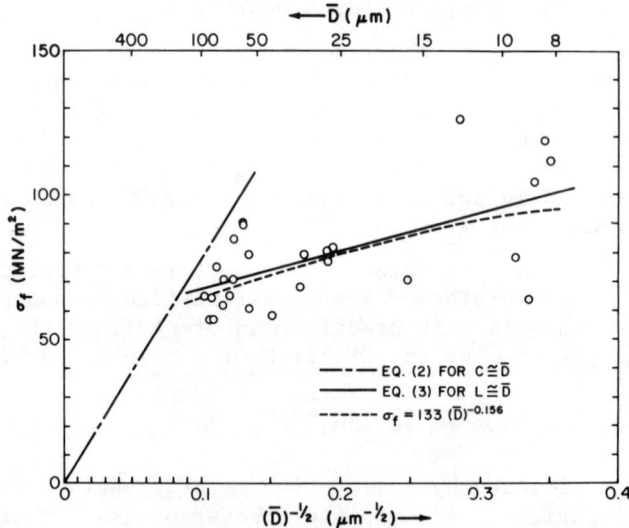

Fig. 7. Fracture stress vs. $(\bar{D})^{-1/2}$ for rapidly cooled wüstite data from Fig. 5.

cation in MgO bicrystals[18] and the yield stress in MgO single crystals[19] has been reported to be 50 and 55 MN/m^2, respectively, in reasonable agreement with the present extrapolated value. However, whether the σ_f data follows a Carniglia[16] two-branched curve, a Rice[17] single curve or a Knudson[11] $(\bar{D})^{-0.156}$ curve cannot be decided in view of the large scatter in the data.

The effective fracture energy may be thought to consist of additive component parts.[20] All of the terms are small except the plastic work term which accounted for about 70% of γ_i.[1,20] Wüstite must be plastic because dislocation motion was observed in the transmission electron microscope.[21] Following a dislocation decorating technique[22], the present fracture surfaces were etched. The fracture surfaces showed wavy slip traces and a polygonized dislocation structure. Such structures were previously observed after high temperature deformation of polycrystalline wüstite[23] and probably occurred, in the present study, during hot-pressing. Dislocations due to fracture were not clearly observed because the dislocation structure prior to fracture masked the fracture surface.

Since γ_i and the % intergranular fracture (20%) are independent of grain size, the characteristics of the grain boundaries as they affect γ_i must not change with change in grain size. Two characteristics which may not change appreciably with grain size are (1) the distribution of impurities and (2) the distribution of α-Fe globules between the grain interior and grain boundaries. Nonstoichiometric $Fe_{0.94}O$ has more vacant cation sites to accommodate impurities than stoichiometric ceramics. Of the detected impurities in the present specimens and their known binary phase diagrams with wüstite[24] only Si was observed not to form a solid solution with wüstite. Also, the dislocation substructure in wüstite can act as sites for impurity segregation within the grains.[25]

In stoichiometric oxides impurity segregation often occurs inhomogeneously at grain boundaries. The change in grain size by thermal treatment may also be accompanied by a redistribution of impurities at the boundaries, and this may be responsible for the change in γ_i with grain size as reported for several stoichiometric oxides.[26-29]

It should also be pointed out that the distribution of α-Fe globules between the bulk grains and boundaries did not change with grain size. The ratio of the number of boundary globules (per unit area) to the total number of globules (per unit area) was measured on both fracture surfaces and polished sections. In both, this ratio was about $\frac{1}{3}$ and was grain size independent. This is also consistent with both γ_i and % intergranular fracture being grain size independent, since the weak α-Fe/wüstite interfaces may provide a preferred fracture path. These weak interfaces may nucleate microcracks which link-up to the major crack and thus determine the fracture path.[30]

B. Aged Wüstite (5 v/o α-Fe)

In aged wüstite with α-Fe at the grain boundaries γ_i clearly decreases with increasing grain size. Two possibilities which could account for this decrease are (1) crack front pinning by the α-Fe boundary layer, and (2) secondary cracking at the α-Fe/Fe$_3$O$_4$ interface.

Lange[31] proposed that the crack front bows between pinning positions and breaks away when the bowed crack front diameter equals the spacing between pinning positions. The relation between γ_i and the average mean-free spacing, $\bar{d}$, between pinning positions is

$$\gamma_i = \gamma_o + T/\bar{d} \tag{4}$$

where γ_o is the energy to create a new surface in the matrix, and T is the line energy of the crack front. The α-Fe boundary network can pin the crack front at grain edges and corners, while bowing can occur at the α-Fe/Fe$_3$O$_4$ interfaces on grain surfaces and in the oxide grains.[1] Since about 80% of the fracture occurred preferentially at α-Fe/Fe$_3$O$_4$ interfaces, a non-random inter-network spacing was calculated using $\bar{d} \cong 0.97 \bar{D}$.[20] The data in Fig. 4 was replotted as γ_i vs. $1/\bar{d}$ in Fig. 8. The fracture energy clearly increases with $1/\bar{d}$ up to about $0.09\mu m^{-1}$, and beyond this the fracture energy decreases. Although there is too much data scatter to be certain of a linear relation between γ_i and $1/\bar{d}$, the data was "force-fitted" to Eq.(4) with a least squares line. The slope T and intercept γ_o with the 95% confidence limits were $(26.2 \pm 6.4) \times 10^{-5}$ J/m^2 and $\gamma_o = 21.3 \pm 2.4$ J/m^2, respectively. The value for T is in the range of that measured for glass.[32] Data on γ_o are not available, although a value of 21 J/m^2 is not unreasonable.[1]

The SEM micrograph in Fig. 6B indicates that crack front bowing has occurred in grains 2 and 3. The coalescence of multiple cleavage steps determines the direction of the advancing (orthogonal) crack front. The curved cleavage steps indicate that the crack front is changing morphology possibly due to being pinned near positions B and C on the α-Fe boundary layer. The dotted arc along AB is orthogonal to the cleavage steps and is consistent with a bowed crack front being pinned near position B.

The negative deviation in γ_i from Lange's theory for $1/\bar{d} > 0.09\mu m^{-1}$ may be attributed to the α-Fe layers being too thin to effectively pin the crack front when the grain size is too small. The pinning effectiveness has been reported to increase for increasing size of the pinning positions, and Eq.(4) was revised to

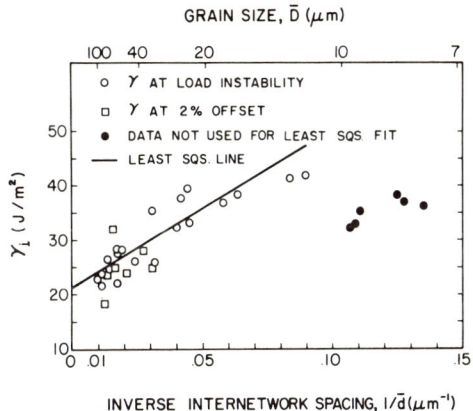

Fig. 8. Fracture energy vs. inverse inter-network spacing for aged wüstite data from Fig. 4.

weight the pinning term in proportion to the pinning particle size.[32]

Secondary cracking was previously reported at the α-Fe/Fe$_3$O$_4$ boundary interfaces.[1,20] A greater number of secondary cracks are expected with smaller grain sizes, and this may give an increase in γ_i with decreasing grain size. A model was formulated which predicts that γ_i will be inversely proportioned to grain size[20], as is approximately observed in Fig. 8. It is presently impossible to determine the relative importance of crack pinning and secondary cracking for causing the observed decrease in γ_i with increasing grain size.

The increase in fracture stress with decreasing grain size can partially be explained by the increase in γ_i with decreasing grain size. However, the inherent flaw size is expected to be related to the spacing of the α-Fe layers. Assuming that fracture is initiated from penny-shaped flaws, the flaw radius, C, was determined following Sack.[33] By using the average values of γ_i (Fig. 4) and σ_f (Fig. 5) for the different grain sizes and E = 1.38×10^{11} N/m^2[1], C was calculated to be 75, 61, 68, 83, 22 and 40μm for grain sizes 89, 72, 56, 37, 24 and 9μm, respectively. Except for the 9 and 37μm grain sizes, the results indicate that C $\cong$ D, i.e. the α-Fe interlayer spacing limits the flaw radius.

Lange[31,32] suggested that the increase in σ_f with decrease in grain size which we observe is due to an increase in γ_i by dispersion toughening. Hasselman and Fulrath[34] suggested that dispersion strengthening may occur by a reduction in C with γ_i assumed to be constant. The present results indicate that both theories apply.

As previously discussed, below about 11μm grain size the pinning effectiveness of the α-Fe layer is low which gives a lower γ_i. Also, a number of flaws at weak α-Fe/Fe_3O_4 interfaces may link-up to give a longer effective flaw length. These effects may explain the low σ_f for 9μm grain size. The abnormally low fracture stress at 37μm probably means that for some unknown reason the flaw size was unusually large, $C \sim 2\bar{D}$.

V. CONCLUSIONS

In rapidly cooled wüstite the fracture energy and % intergranular fracture were independent of grain size at 7.5 J/m^2 and $\approx$ 20%, respectively. This was attributed to the distributions of impurities and α-Fe globules between the grain interior and grain boundaries remaining constant with changing grain size. The fracture stress showed a weak dependence on grain size. The data were discussed in terms of Griffith-Orowan flaw extension and plastically-initiated crack nucleation. In aged wüstite the % intergranular fracture was $\approx$ 80% and independent of grain size. The fracture energy increased from 23 to about 40 J/m^2 as the grain size decreased from 96 to 8μm. This was attributed to crack front pinning and secondary cracking. The fracture stress increased slightly with decreasing grain size to a peak value of 187 MN/m^2 at 24μm. This was attributed to the observed increase in fracture energy and to a reduction in flaw size with decrease in α-Fe interlayer spacing resulting from a decrease in grain size.

ACKNOWLEDGEMENTS

The authors are pleased to acknowledge Mr. William Kobes for preparing the fracture strength specimens and performing the measurements and Drs. D. L. Johnson and S. M. Wiederhorn for many helpful discussions.

1. M. I. Mendelson, M. E. Fine, "Enhancement of Fracture Properties of Wüstite by Precipitation", submitted to J. Am. Ceram. Soc.
2. W. A. Fischer, A. Hoffman, R. Shimada, Arch. Eisenhüttenw 27, 521 (1956); W. A. Fischer, A. Hoffman, ibid 30, 15 (1959).
3. R. Collongues, Publ. Scient. et Tech. du Minist. de l'Air, No. 324, 1 (1957).

4. J. Manenc, T. Herai, J. Benard, 5th Intl. Sym. Reactivity of
    Solids, Munich, 1964; Publ. by Elsevier Publ. Co., Holland,
    1965, p. 432.
5. B. Ilschner, E. Mlitzke, Acta Met. 13, 855 (1965).
6. S. M. Wiederhorn, A. M. Shorb, R. L. Moses, J. Appl. Phys. 39,
    1569 (1968).
7. W. F. Brown, J. E. Srawley, ASTM Spec. Tech. Publ. No. 410,
    1967.
8. J. E. Hilliard, J. W. Cahn, Trans. AIME 221, 344 (1961).
9. C. S. Smith, L. Guttman, Trans. AIME 197, 81 (1953).
10. M. I. Mendelson, J. Am. Ceram. Soc. 52, 443 (1969).
11. F. P. Knudson, J. Am. Ceram. Soc. 42, 376 (1959).
12. A. A. Griffith, Phil. Trans. Roy. Soc. (London) 221A, 163 (1920);
    E. Orowan, Rept. Progr. Phys. 12, 186 (1948-49).
13. R. W. Davidge, A. G. Evans, Mater. Sci. and Eng. 6, 281 (1970).
14. A. N. Stroh, Advances in Physics 6, 418 (1957); C. Zener, Trans.
    ASM A40, 3 (1948).
15. T. L. Johnston, R. S. Stokes, C. H. Li, Phil. Mag. 7, 23 (1962).
16. S. C. Carniglia, J. Am. Ceram. Soc. 55, 243 (1972); ibid 48,
    580 (1965).
17. R. W. Rice, paper given at 73rd Annual Meeting of Amer. Ceram.
    Soc., Chicago, Ill., April 28, 1971 (Basic Science Div., No.
    47-B-71); abstract in Am. Ceram. Soc. Bull. 50, 374 (1971).
18. R. C. Ku, T. L. Johnston, Phil. Mag. 9, 231 (1964).
19. R. J. Stokes, T. L. Johnston, C. H. Li, Trans. AIME 215, 437
    (1959).
20. M. I. Mendelson, Ph.D. Thesis, Northwestern University, 1973.
21. G. P. Wirtz, F. B. Koch, private communication.
22. J. Manenc. G. Vagnard, J. Benard, C. R. Acad. Sci. Paris 254,
    1777 (1962).
23. B. Ilschner, B. Reppich, E. Riecke, Faraday Soc. Discuss. 38,
    244 (1964).
24. E. M. Levin, C. R. Robbins, H. F. McCurdie, Phase Diagrams for
    Ceramists, Amer. Ceram. Soc., Columbus, Ohio, 1964, 1969 Suppl.
25. A. H. Cottrell, Dislocations and Plastic Flow in Crystals,
    Oxford Univ. Press, London, 1953, p. 134.
26. A. G. Evans, R. W. Davidge, J. Nucl. Mater. 33, 249 (1969).
27. F. J. P. Clarke, H. G. Tattersall, G. Tappin, Proc. Brit.
    Ceram. Soc. 6, 163 (1966).
28. P. L. Gutshall, G. E. Gross, J. Eng. Frac. Mech. 1, 463 (1969).
29. L. A. Simpson, J. Am. Ceram. Soc. 56, 7 (1973).
30. U. Lindberg, Acta Met. 17, 521 (1969).
31. F. F. Lange, Phil. Mag. 22, 983 (1970).
32. F. F. Lange, J. Am. Ceram. Soc. 54, 614 (1971).
33. R. A. Sack, Proc. Phys. Soc. 58, 729 (1946).
34. D. P. H. Hasselman, R. M. Fulrath, J. Am. Ceram. Soc. 49, 68
    (1966).

MICROSTRUCTURAL DEVELOPMENT AND FRACTURE TOUGHNESS OF A CALCIA

PARTIALLY STABILIZED ZIRCONIA

D. J. Green, P. S. Nicholson and J. D. Embury

Department of Metallurgy and Materials Science
McMaster University
Hamilton, Ontario L8S 4L7
Canada

ABSTRACT

The effect of microcrack formation on the brittle fracture of ceramics is discussed and a model presented to explain the effect of such a process zone on crack stability. The transient phenomena of stable crack growth is explained in terms of a microcrack zone of increasing size.

Evidence supporting this model is presented for the case of a calcia partially-stabilized zirconia which can form microcracks at the grain boundaries.

INTRODUCTION

The phenomena of crack stability in brittle materials has been reported by several investigators. This increase in resistance to crack propagation is generally caused by irreversible processes in the vicinity of the crack tip. This process zone tends to take three related forms, i.e., the formation of discrete cracks ahead of the main crack; an increase in the tortuous nature of the crack due to secondary cracking processes; or a combination of these two mechanisms.

In ceramics, the production of microcracks is usually related to the presence of internal stress. This stress may be caused by such phenomena as anisotropic thermal expansion in single phase materials or stress concentrations produced by second phase particles associated with phase transformations or differing thermal expansion coefficients in polyphase materials. In brittle

541

polymers, these process zones are evidenced by the formation of a
crazed zone ahead of the crack. This crazing and its subsequent
effect on crack stability has been observed by Berry[1] and by
Van den Boogart[2] for P.M.M.A. and Bevis and Hull[3] have discussed
the distribution of these crazes in polystyrene for the case of
a stationary crack.

The mechanism of increased crack resistance resulting from
heterogeneity has been qualitatively discussed by Kaplan[4]. He
described the "plasticity" of concrete as being due to the formation of many discrete cracks and the arresting nature of the aggregate particles. Similar observations have also been made for
hardened cement paste [5], graphite[6,7], filled rubber[8] and mica
glass ceramics[9]. The "debonding zone" at notch tips in glass-
fibre composites observed by Beaumont and Philips[10] could be yet
another related mechanism of the formation of a decohesed process
zone.

The study of these process zones and their relation to crack
stability and fracture toughness remains largely unexplored and
it may be that study of these various phenomena could lead to a
basic understanding of controlling the fracture processes in
brittle materials. In calcia partially-stabilized zirconia (CaO-
PSZ), microstructures can be developed to enhance the production
of a process zone at a crack tip. The results of the investigation
of this material is reported here.

THE INFLUENCE OF MICROCRACK FORMATION ON FRACTURE

Recognition of the basic, inherent factors controlling the
brittle behaviour of ceramic materials has given rise to a variety
of attempts to increase their toughness by exploiting the interaction between a growing crack and second phase particles or
fibres embedded in the ceramic matrix. However, consideration
must also be given not only to methods of increasing the effective
energy expenditure at the crack tip but also to the nature of the
crack tip itself in complex commercial ceramics.

In the Griffith[11] formulation of the critical condition for
fracture we can write:

$$- \partial U/\partial a \geqslant \partial W/\partial a \tag{1}$$

where U and W are the stored elastic energy and the irreversible
work to fracture, respectively, and 2a is the crack length.

The rate of release of stored elastic energy for the case of
an elliptical crack normal to a tensile stress σ and under fixed
grip conditions or crack driving force G is given by:

$$G = - \left(\frac{\partial U}{\partial a} \right)_a = - \frac{2 \pi a \sigma^2}{E} \qquad (2)$$

The driving force increases with increasing crack length and thus in order to stabilize a crack, the resistance process, namely the work to fracture, must also increase with increasing crack length. The general condition which can lead to crack stability in terms of the crack extension and resistance forces, $\partial U / \partial a$ and $\partial W / \partial a$, have been considered recently by Glucklich[12].

From the viewpoint of ceramic materials the situation is illustrated in Figure I. In the absence of plastic deformation and other irreversible processes, the energy demand is independent of crack length. Thus for a given crack length the material is simply loaded until the energy release rate exceeds the energy demand whence brittle failure ensues.

However, if we envisage some alternative dissipation process such as the formation of microcracks ahead of the main crack at a stress σ_{MC} which is less than the bulk fracture stress, then a situation arises in which for a limited growth period the energy demand increases with crack length. The concentrated elastic stress, σ_{yy}, at a distance, r, ahead of the crack is given by $\sigma_{app} (a/2r)^{1/2}$ in the crack plane, where σ_{app} is the level of the

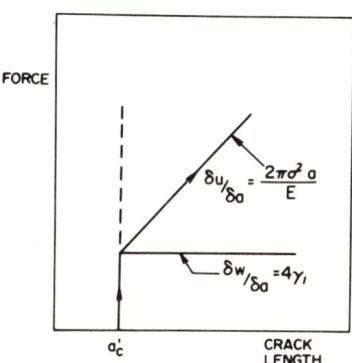

Fig. I   Representation of crack instability for ideal brittle materials in uniaxial tension.

applied stress. At some point this critical stress ahead of the crack reaches σ_{MC} and an array of microcracks form. The size of this process zone, r_{crit}, is of the order of:

$$r_{crit} = \frac{1}{2} \frac{\sigma^2 a}{\sigma_{MC}^2} \qquad (3)$$

Hence, on increasing the applied stress, the extent of the zone containing microcracks increases (alternatively the density of the microcracks within the zone may increase). Furthermore, during stable crack growth, the process zone might be expected to increase in size. The energy needed to form the microcracks is there-fore not available for growth of the main crack. This concept is analogous to the growth of a plastic zone in a ductile material and leads to a period of stable crack growth.

It should be noted that this is a transient phenomena in which the crack undergoes a period of stable crack growth and at some point the Griffith's inequality is fulfilled and catastrophic failure occurs but with a greater effective surface energy than predicted by the Griffith formulation. An example of this type of behaviour occurs in calcia partially-stabilized zirconia (PSZ) which can be heat-treated in such a way as to form microcracks at the grain boundaries.

SAMPLE FABRICATION

Test bars were made from micronised ZrO_2* (99.7%) and reagent grade calcia; all specimens contained 3.4 wt.% calcia. Bars 4 x 1 x 1/2 inches were pressed isostatically at 20,000 psi and fired for five hours at 1850°C. After firing, the specimens were slow-cooled to 1300°C and then fast-cooled to room temperature.

EXPERIMENTAL PROCEDURE

The fracture surface energy and work of fracture parameters were determined in four-point bending on an Instron testing machine** using lucalox knife edges and self-aligning alumina columns to trans-mit the load. The sample size was 2 1/2 x 7/16 x 1/10 inches with major and minor spans of two and one inch, respectively. The dimensions and the loading arrangements used minimised specimen

---

* Tizon Chemical Corp. (Flemington, New Jersey)
** Instron Corp. (Canton, Massachusetts)

indentation and friction at the supports and allowed pure bending within the minor span[13]. The samples were diamond machined and polished from the as-fabricated bars. A sharp, central notch was introduced into the sample using an ultrasonic impact grinder and the depth measured with a travelling microscope. The crack length was taken to be the notch depth plus the size of the observed machine flaws.

The stress intensity determination was carried for a series of notch depths at a cross-head speed of 0.002 inches/min. Six samples were fractured at each notch depth.

The occurrence of crack stability in this material was utilised to observe the load-deflection characteristics for samples which had been unloaded after partial crack propagation. These load-unload tests were carried out using a 1/2" Instron strain gauge extensometer* attached to symmetrical grips on either side of sample notches.

The work of fracture values were determined from the area under the load-deflection curves using a planimeter and dividing by twice the cross-sectional area of the fracture surface. These results were also calculated for different notch depths. This technique tends to be useful since no value of Young's modulus is needed. However, its dependence on machine modulus makes its application more comparative than absolute. Tensile tests were also carried out on a sample of the machined bars, with due regard to the grip and alignment problems, to determine the form of the stress-strain curves.

The approximate value of Young's modulus for the material was determined from the bend tests using elastic flexure theory[14]. The value was then confirmed by a uniaxial compression test utilising the Instron strain-gauge extensometer. An attempt was made to determine Young's modulus by an ultrasonic technique but the difficulty of measuring the attenuated wave due to the presence of the microcracks made the results difficult to interpret.

The fracture surfaces and machined edges of the samples were studied in a scanning electron microscope (SEM)** in the usual fashion. The morphology of the crack was studied using partially broken samples vacuum-set in hard epoxy resin.

---

*   Instron Corp. (Canton, Massachusetts)
**   Cambridge Stereoscan (Cambridge, England)

RESULTS AND DISCUSSION

Microstructural Development

The detailed development of the microstructure in calcia PSZ has been discussed elsewhere[15]. The grain structure was determined to be bimodal with small grains of pure monoclinic zirconia (15 μ) dispersed along the grain boundaries of the larger grains (80 μ) as shown in Figure 2. During the initial stages of cooling, after the sintering in the single phase cubic region at $1850°C^{16}$, the system becomes supersaturated with pure zirconia and this phase precipitates on cooling. The initial precipitation will occur at the grain boundaries and it is believed that the diffusion processes involved are enhanced by the presence of a liquid phase in the microstructure at these temperatures[17]. This liquid phase is associated with the silica impurity level in the source zirconia and accounts for the abnormal size of the grain boundary precipitate.

The identity of the grain boundary phase was established as pure zirconia using X-ray dispersive analysis to show these grains contained no calcia. This is compatible with the very low solubility of calcia in pure zirconia.

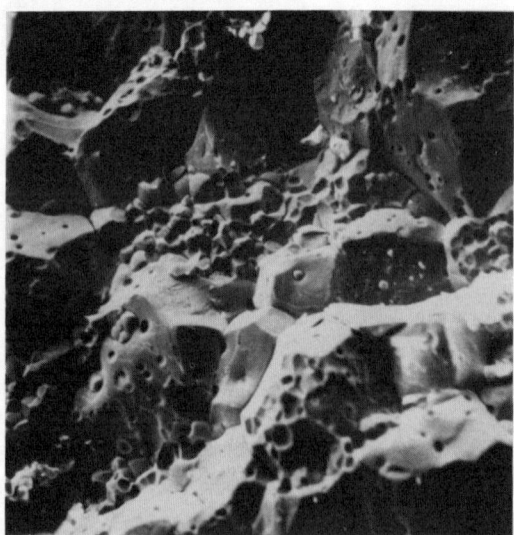

Fig. 2 Fractograph of calcia PSZ showing bimodal grain structure (225X).

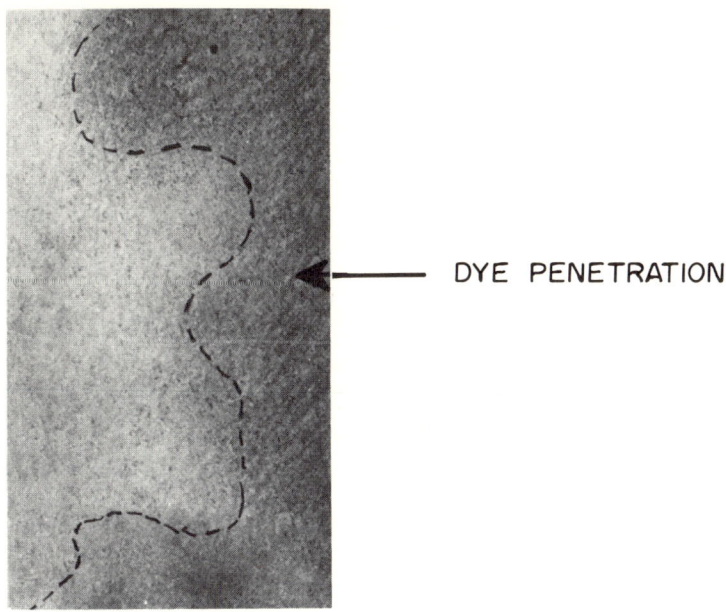

DYE PENETRATION

Fig. 3 Application of dye penetrant (X3).

It is postulated that the grain boundary ZrO_2 precipitate, on transformation, causes decohesion and weakening of the grain boundaries in the material. The martensitic phase transformation of the ZrO_2 from the tetragonal to the monoclinic modification on cooling involves a 3% volume expansion[18]. This decohesion process promotes the subsequent production of microcracks in the material during fracture and results in a material of very low elastic modulus (1.5×10^6 psi) and strength, but a high density (99.5%). The evidence of this decohesion is established by the easy dye penetration shown in Figure 3.

Mechanical Behaviour

The load-deflection characteristics of calcia PSZ were determined in four-point bending and tension and the results are shown in Figures 4 and 5. Controlled crack propagation was observed initially for both test configurations preceded by a non-linear elastic region. This stable propagation was observed to persist in the bend tests at long crack lengths and is thought to be associated with geometric crack stability induced by the bending configuration.

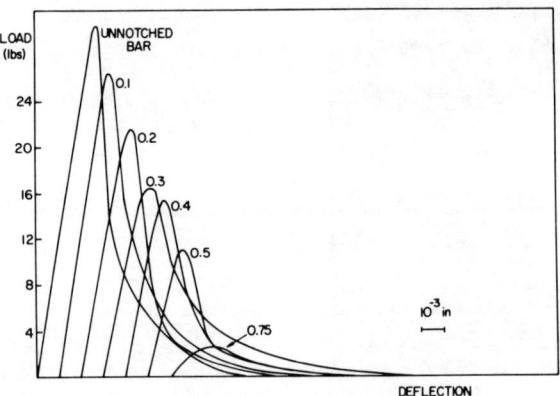

Fig. 4 Typical load deflection curves for different notch depth/spec-
imen width (bend test).

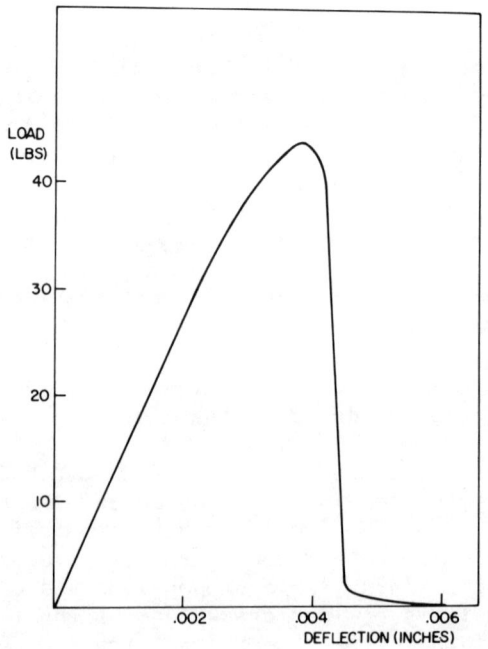

Fig. 5 Tensile test for unnotched bars.

As discussed earlier, crack stability can be caused by an increase in the crack resistance force ($\partial W/\partial a$). A similar effect can arise if there is a decrease of the crack driving force ($\partial U/\partial a$) and this is termed geometric stability. Clausing[19] has quantitatively discussed this effect for a variety of testing configurations and has shown that geometric stability can occur in the bend test at relatively long crack lengths especially in brittle materials with low elastic moduli.

The initial crack stability and non-linear elastic behaviour is thought to be associated with the ability of the material to form microcracks. At a critical load, the closed grain boundary microcracks open or extend and new microcracks form at the tip of the critical flaw. This effect will produce an envelope of decreasing elastic moduli, causing the observed curvature in the elastic section of the load-deflection curves. Further evidence of these microcracks can be inferred from the compression tests, which were carried out on both the fully and partially stabilized materials (Figure 6). The PSZ shows marked hysteresis, caused by crack friction and the lower modulus on unloading. This behaviour is very similar to that of rock samples under compression[20]. The fully-stabilized single phase material, however, showed little hysteresis effect. Similar hysteresis is also seen in the bend test, where an extensometer has been placed across the notch. The

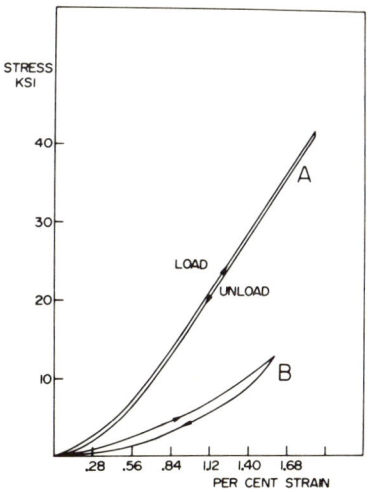

Fig. 6 Compression tests for the fully (A) and partially stabilized (B) zirconia materials.

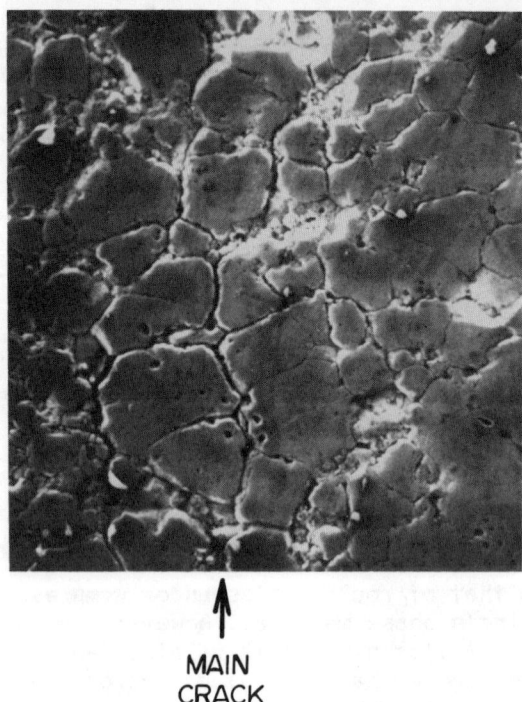

MAIN
CRACK

Fig. 7   Scanning electron micrograph of crack tip zone (160X).

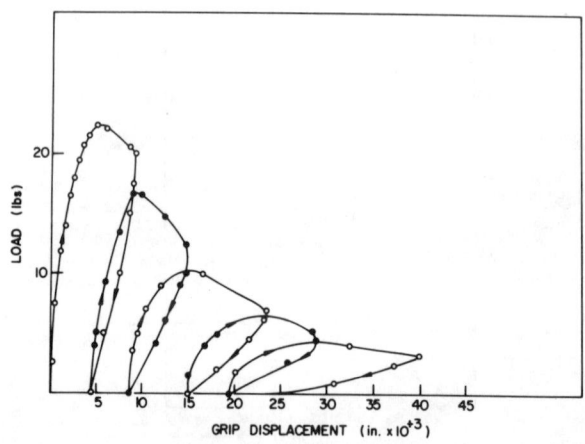

Fig. 8   Load-reload hysteresis.

energy associated with this hysteresis must again be utilised in the extension of inherent microcracks and crack surface friction on relaxation.

The initial crack stability is thought to be caused by the production of a microcrack "zone" and an increase in the zone size. The presence of a microcrack zone at the crack tip is shown in Figure 7. The tortuous nature and secondary cracking associated with the fracture is also evident in this figure. Direct evidence of the increasing zone size has proved difficult to obtain but an increase in the elastic non-linearity and hysteresis with increasing crack length (Figure 8) supports the model indirectly. This figure shows the results of the extensometer test where crack propagation was allowed. The elastic hysteresis and non-linearity is evident and furthermore permanent strain was associated with the crack growth due to the irreversible work needed to produce the microcrack zone. It should be emphasized that the microcrack zone is confined initially to the vicinity of the crack tip. In Figure 8, the modulus of the first re-load is similar to that of the initial load and it is therefore unlikely that the crack stability is simply caused by the low elastic modulus[21]. This is further substantiated by Clausing[19] who shows that crack stability is unlikely in the bend test at short crack lengths.

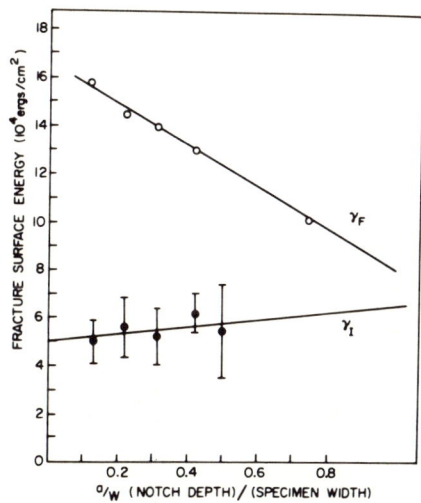

Fig. 9   Fracture surface energy parameters as a function of notch depth.

Fracture energy measurements made by Green et al[22] on this material produced relatively high values when compared to other commercial ceramics. The effective surface energy (γ_I) as measured by the standard single edge crack configuration was found to be 6×10^4 ergs/cm^2 and the work-of-fracture[2,3] (γ_F) was found to be 8×10^4 ergs/cm^2 when extrapolated to long crack lengths. These results are summarised in Figure 9. It is thought that the presence of a microcrack zone would have a greater effect on fracture toughness if the microcracks were induced by stress rather than by opening of decohesed boundaries since these also reduce the material's strength. The fracture energy measurements were also shown to be compatible with fracture from the flaws present in the material after fabrication and machining. The difference in the two fracture surface energy parameters is thought to reflect the difficulty of crack propagation in calcia PSZ.

CONCLUSIONS

The occurrence of stability and its related microstructures has been discussed for various brittle materials. A model is presented to show the effect of microcrack formation on the nature of the crack tip and its role in crack stabilization. Evidence supporting the model is presented for a calcia PSZ, which can form microcracks on the grain boundaries on application of stress. These microcracks form from the decohesed and weak grain boundaries which are present in the material after fabrication. Crack stability was observed to occur initially in both tension and bend tests. Furthermore, the material showed a non-linear elastic region preceding fracture. Both these effects are caused by the presence of the microcrack zone at the crack tip which is established by the observation of mechanical hysteresis and microscopy. Finally, the fracture surface energy measurements were summarised to show the difficulty of crack propagation in this material.

ACKNOWLEDGMENTS

The work was sponsored by the Atomic Energy of Canada Limited.

REFERENCES

1.   J. P. Berry, Fracture Processes in Polymeric Solids (Ed. B. Rosen), p. 195, Interscience, New York (1964).
2.   A. Van den Boogart, Proc. Conf. on Physical Basis of Yield and Fracture, p. 167. Oxford Inst. of Phys. and Physical Soc., Conf. Series, No. 1.

3. M. Bevis and D. Hull, J. Mat. Sci., 5 (1970) 983.
4. M. F. Kaplan, Paper DI, Proc. Int. Conf. On Structure of Concrete, London (1965).
5. J. Glucklich, T. and A.M. Rep. No. 622, Univ. Illinois (1962).
6. R. H. Knibbs, J. Nuc. Matls., 24 (1967) 174.
7. F. J. P. Clarke and R. S. Wilks, Proc. Conf. Nucl. App. Non-Fissionable Ceramics, Am. Nuc. Soc., (Ed. A. Baltax and J. H. Handwerk) (1965), p. 57.
8. E. H. Andrews and A. Walsh, Proc. Phys. Soc., 72 (1958) 42.
9. C. K. Chyung, G. H. Beall and D. G. Grossman, Electron Microscopy and Structure of Materials (Ed. G. Thomas et al), p. 1167, Univ. of California Press, Berkeley (1972).
10. P. W. R. Beaumont and D. C. Phillips, J. Mat. Sci., 7 (1972) 682.
11. A. A. Griffiths, Trans. Roy. Soc., London, A221 (1920) 163.
12. J. Glucklich, Eng. Fract. Mech., 3 (3) (1971) 333.
13. W. F. Brown, Jr., and J. E. Srawley, A.S.T.M., S.T.P. No. 410, (1967) 9 - 16.
14. S. P. Timoshenko and J. M. Gere, Mechanics of Materials, Van Nostrand-Reinhold Co., New York, p. 518 (1972).
15. D. J. Green and P. S. Nicholson, submitted to J. Am. Cer. Soc., June (1973).
16. R. C. Garvie, High Temperature Oxides, Vol. 2, p. 117-66, Ed. A. Alper, Academic Press, New York (1970).
17. J. F. Shackelford, P. S. Nicholson and W. W. Smeltzer, presented at the 75th Annual Meeting of the American Ceramic Society, Cincinnati (1973). Paper 15-B-73.
18. R. N. Patil and E. C. Subbarao, J. Appl. Cryst., 2 (1969) 281-288.
19. D. P. Clausing, Int. J. Fract. Mech., 5 (3) (1969) 211.
20. J. C. Jaeger and N. G. W. Cook, Fundamentals of Rock Mechanics, Methuen (1969) p. 310.
21. J. P. Berry, J. Mech. Phys. Solids, 8 (1960) 194-216.
22. D. J. Green, P. S. Nicholson and J. D. Embury, submitted to J. Am. Cer. Soc., April (1973).
23. H. G. Tattersall and G. Tappin, J. Mat. Sci., 1 (1966) 296.

DETERMINATION OF K_{Ic}-FACTORS WITH DIAMOND-SAW-CUTS IN CERAMIC MATERIALS

R. F. Pabst

Max-Planck-Institut für Metallforschung

Institut für Sondermetalle, Stuttgart, Germany

ABSTRACT

When determining K_{Ic}-factors in brittle materials like opaque Al_2O_3, Si_3N_4 or SiC it is very difficult to generate a plane crack front of a certain depth and measure the crack length accurately. To overcome these difficulties the data were obtained from diamond saw cuts. A critical examination is performed to determine whether this method will give useful K_{Ic}-factors. The data obtained with diamond saw cuts were independent of the normalized notch length, the specific test, the specimen volume and the specimen dimension. However, the data are dependent upon the width and shape of the ground notch and the specific structure of the base of the notch. A "true" K_{Ic}-factor can therefore be obtained from theory using a Neuber-formalism of stress concentration factors and from extrapolated experimental data. Because extrapolating is considered as physically unrealistic there must exist a small but finite range γ of linear dimension and therefore a finite notch radius ρ'. The relative values of K_{Ic} indicated as K_{Ic}^{+} can also be used for determining structure parameters and the influence of various media.

INTRODUCTION

With linear fracture mechanics and the experience we have from metals there is hope to get material's constants in ceramic material where the data show a low standard deviation. In brittle materials the low standard deviation will be the condition of getting valuable data as a function of structural parameters.

Fracture mechanics in form of K_{Ic}-factors is valid only for a mathematically sharp crack (notch radius $\rho \approx 0$). Contrary to metals or transparent brittle materials such as glass it is much more difficult in materials like opaque Al_2O_3, Si_3N_4 or SiC to generate a plane crack front of a specific depth and to measure the crack length accurately. When creating a sharp crack, for instance pressing a hardened steel blade in to a saw notch, there is moreover a danger of crack branching because of the high delivered energy, especially if the load mechanism is not stiff enough. One has therefore a large zone to dissipate energy and therefore large K_{Ic}-factors. The measure of the position of the crack front by a dye penetrant will influence the K_{Ic}-factor measurement due to the nature of the penetrant.

To overcome these difficulties one can use diamond saw cuts to simulate cracks as done previously in metals and in ceramics[1]. The problem is to perform a critical examination that will determine whether this method will give useful K_{Ic}-factors.

EXPERIMENTAL

Experiments were made with various Al_2O_3 qualities (Figure 1). The data are compared with Si_3N_4, SiC and graphite specimens. Notches of different width were introduced with diamond saws of 0.1; 0.2; 0.3; 0.4; 0.5 mm width of the blade.

The K_{Ic}-factors determined with saw cuts are indicated as K_{Ic}^+.

Most of the specimens were tested by four-point loading with varying span and varying span ratios. As to be seen in Figure 2 the A S T M specifications gained with metals are used. The finiteness of the specimen is considered in form of the boundary collocation method.

Material		Porosity %	$\bar{d}$ [μm]	E [kp/mm²] ×10⁴	K_{Ic}^* [kp/mm³/²]	[MN/m³/²]	Y_I [erg/cm²] ×10⁴
Al₂O₃	H: 96% MgO Mg-Silikat	7.4	20	3.2	14,4 ± 1%	4.6	3.4
	S: 97% SiO₂	7.6	10	3.5	14,2 ± 5%	4.5	3.2
	D: Lucalox	0.7	50	4.0	13,0 ± 4%	4.1	2.2
	B: 99,7% MgO	4.0	20	3.8	11,9 ± 3%	3.8	1.8
	P₂: off. Por. 11,9%	17.0	–	2.8	12,0 ± 7%	3.8	2.1
	P₁: off. Por. 41,5%	42.3	–	1.7	4,0 ± 8%	1.3	0.6
Si₃N₄		ρ[g/cm³] 2.32	–	1.4	8,2 ± 4%	2.6	2.4
Si C		2.84	–	3.2	10,6 ± 8%	3.4	1.7
Graphite		1.71	–	0.1	2,5 ± 2%	0.7	3.1
Saw-width 0,2 mm, Medium: Silicon oil							

Fig. 1. K_{Ic}^+-factor measured with saw width 0.2 mm and fracture energy Y_I as a function of grain diameter $\bar{d}$, porosity and impurity.

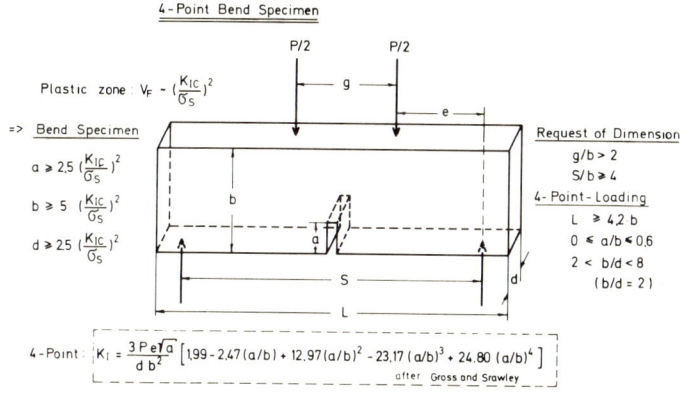

Fig. 2. Four-point-bend specimen with specified dimensions because of pin friction, pure bending and "plastic" zone (for metals).

For comparison with four-point loading, three-point loading, and the DCB test were carried out. The following test parameters were studied: normalized notch depth, specimen dimension, specimen volume, width of the base of the

notch. The roughness of the base of the notch can be varied by using diamond saws of different diamond particle
sizes and by annealing.

The Al_2O_3 specimen of quality S were tested in various media such as: silicon oil, water (20°C), air (50-
60 %), xylol, cyclohexane, methanol (Figure 3). A special apparatus has build which gives a good infiltration and diminishes the water content at the notch base.

RESULTS AND DISCUSSION

The experiments have shown that there exists no specific dependence of normalized saw cut length. This is
independent of the saw cut width used (Figure 4). The
data for the small specimen type A (3.0 x 1.2 x 26 mm)
are systematically a little bit lower than that for specimen B (7.5 x 2.0 x 60 mm) (3-4 %). This is not a result of different volume of different span length but an
incorrect measurement of thickness "d" because of a small
conic form. With this very small specimen A the danger of
incorrect measurement is much greater than with the larger type B.

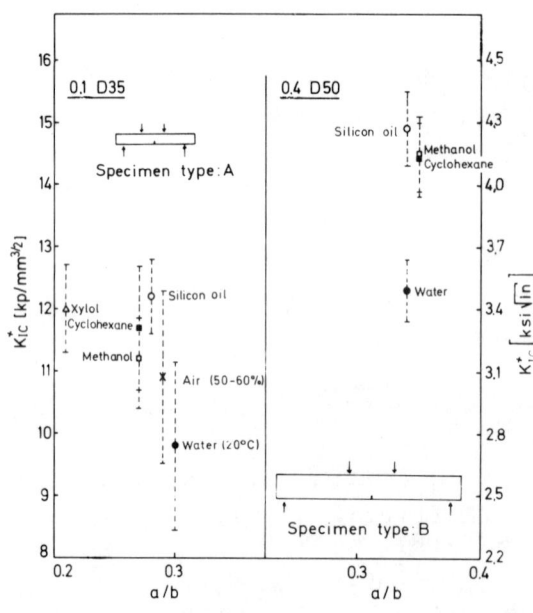

Fig. 3. K_{Ic}^+ as a function of various media
and saw width (0.1 mm
and 0.4 mm). Standard
deviation of small spe-
cimen A is much larger.

It may be that because of the small dimensions of A and
therefore due to a relative larger roughness and a larger inhomogenity of the base of the notch the measured
saw cut length will be too low and the standard devia-

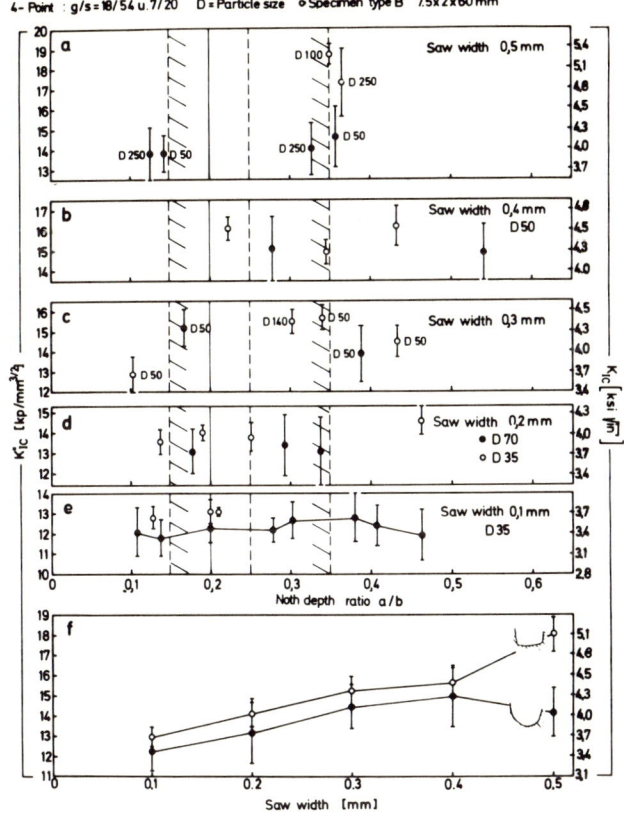

Material: Al$_2$O$_3$: S Medium: Silicon oil, 22 °C ● Specimen type A 3 x 1,2 x 26 mm
4- Point : g/s = 18/54 u. 7/20 D = Particle size ○ Specimen type B 7.5 x 2 x 60 mm

Fig. 4. K$_{Ic}^{+}$ as a function of normalized notch depth a/b, saw width, saw particle diameter D in μm and form of the notch base (with 0.5 mm). Dotted lines show limits and preferred measuring point a/b =0.2 for metals.

tion of 10 % will be greater then with the B specimen which is 5 %.

Figure 5 shows, that there is no influence of specimen volume in the range of at least a factor of 40 and no influence of span width and different span ratios. Four-point loading, three-point loading and the DCB test give the same values.

With metals there is reported[2] an influence of volume of the ground notch ρ^2 d (ρ = radius of the ground notch, d = thickness of the specimen). Therefore in Figure 6 K$_{Ic}^{+}$ is plotted against different b/d ratios (b = width of the specimen). There is no influence of "d" to a great extend independent of saw width. The dotted boundary lines indicate the limit measured with metals because of pin friction and pure bending. This limit is not decisive for ceramics in the measured range.

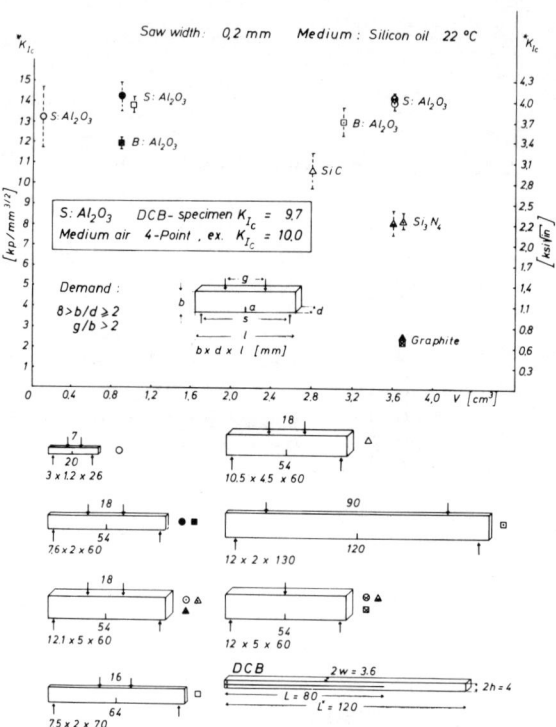

Fig. 5. K_{Ic}^{+} as a function of specimen volume, span width, span ratio, four-point-bending, three-point-bending, DCB-test and different materials

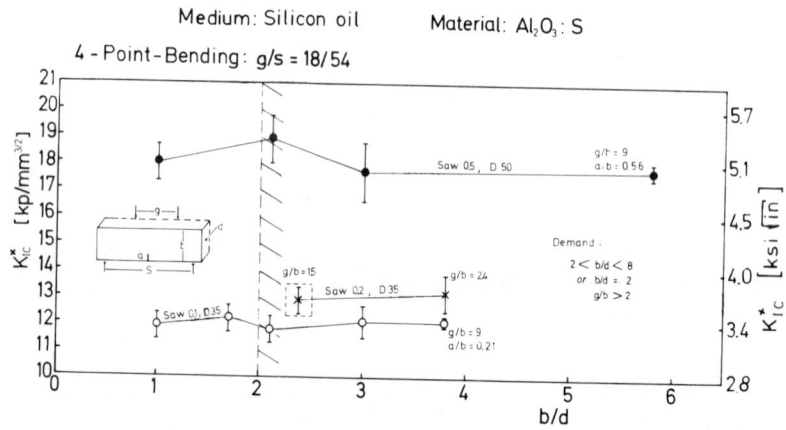

Fig. 6. Influence of volume of the ground notch $\rho^2 d$ (different b/d ratios with constant ρ)on K_{Ic}^{+}. Dotted lines indicate the limit of measurement because of pin friction and pure bending (for metals)

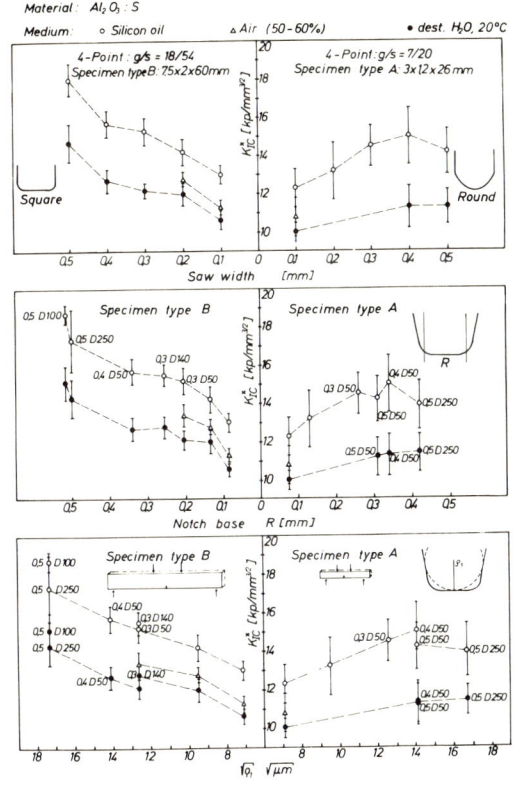

Fig. 7. K_{Ic}^+ as a function of saw width
(mean values), notch
base R and $\wp_1$, saw
particle diameter D
in μm and various
media.

Now, however, the K_{Ic}-factor is strongly influenced
by the width of the saw cuts and the form of the ground
notch. This is more than generally expected (Figure 7).
In Figure 7 (first part) K_{Ic}^+ is plotted against the saw
width where the measuring points are mean values regardless of form and structure of the ground notch. If there
is a more rounded notch base, particularly with the saw
width of 0.5 mm, one has a considerable lower K_{Ic}^+-factor.
This may be corrected by plotting K_{Ic}^+ against R or $\wp_1$.

The roughness of the notch base can be changed by
using diamond saws of different diamond particle diameter D in μm. The influence of different D is only measured with the largest particle diameter of D = 250 μm
(Figure 7).

The roughness can be changed to a greater degree by
annealing. The heat treatment rounds off or heals the
small Griffith cracks introduced by the diamond particle.

The effect of healing is to be seen in Figure 8. For a square notch base there exists a linear function between K_{Ic}^+ and the width of the diamond saw. With heat treatment of 1280°C there is no effect except with the round notch (0.5 mm saw width). At 1450°C one has a pronounced change of K_{Ic}^+ which increases with increasing saw width. The standard deviation is larger with the annealed specimens and increases with increasing saw width.

This may be because of the increasing inhomogenity at the base of the notch. The difference between annealed and not annealed specimens becomes small with a saw width of 0.1 mm. It can be an argument that there is no large difference between a saw cut width of nearly 0.1 mm and a sharp crack in alumina.

Because of the high notch sensivity in alumina and the linear relation between K_{Ic}^+- and the saw width, a "true" K_{Ic}-factor of a mathematically sharp crack can be obtained by extrapolation of saw width = 0. In this case, the linear curves of both the specimens with and without Griffith cracks at the root of the notch must intercept at saw width = 0, that is on the K_{Ic}^+-axis. This is full-filled exactly (Figure 8). As to be seen, the difference between the saw width of 0.1 mm and the extrapolation point is only 9 %.

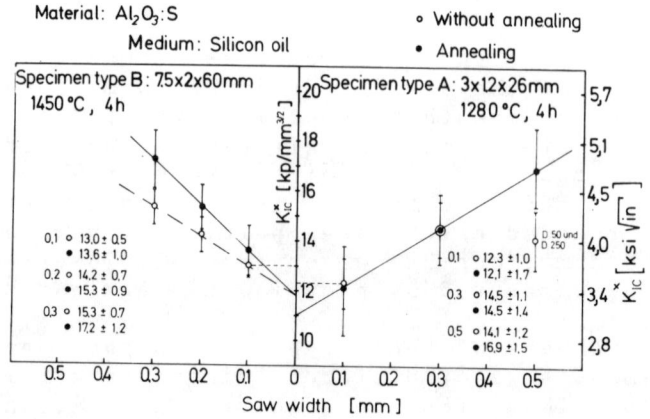

Fig. 8. Annealing of Griffith cracks at the notch base(reported values are mean values). "True" K_{Ic}-factor is obtained by extrapolation and intersection at saw width = 0.

The same can be found from theory using a Neuber-formalism of stress-concentration factor, using the formula below[3],[4]

$$K_{IC} = \lim_{\rho \to 0} \frac{1}{2} K_{tB} \, \sigma_N \sqrt{\pi \rho} \cdot k_b \qquad (1)$$

$$K_{IC}(\eta) = \lim_{\rho' \to 2\eta} \frac{1}{2} K_{tB} \, \sigma_N \sqrt{\pi \rho'} \cdot k_b \qquad (2)$$

$$\rho' = \rho + 2\eta \;\; (w = 0) \qquad (3)$$

Kt_B = stress concentration factor for pure bending

σ_N = nominal stress

k_b = correction because of finiteness of the specimen

η = finite element because of structure

ρ' = finite radius

The question is, will it be correct to extrapolate ($\rho \approx 0$), for instance, because of blunting of the crack tip as is the case with metals. Blunting with this extremely brittle material, if it occurs at all, is very small (computed ρ_B = 0.1 μm). Therefore, one may extrapolate. Because of the finite structure of grains, grain boundaries and pores in the material a formalism of continuum mechanics is physically unrealistic for $\rho \approx 0$. There must exist a small but finite range η of linear dimension and consequently a finite radius ρ'. Up to this η the structure is regarded as a continuum. If ρ' is determined, ρ' will be 2η after formula (3). Here the included angle of thread of the notch w must be zero. This is approximately the case with the saw cuts used here. The K_{Ic}-factor as a boundary value defined in formula (1) is therefore not to be taken at $\rho' \approx 0$ but at $\rho' \approx 2\eta$.

The question is how large this finite η element may be.

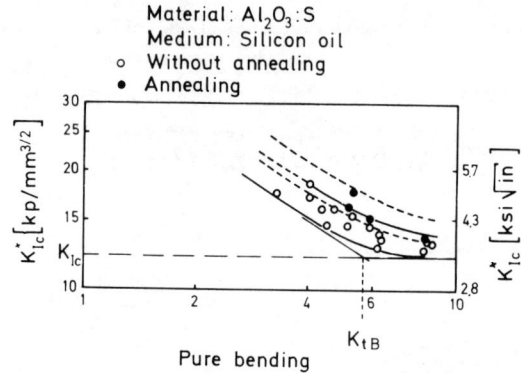

Pure bending

Fig. 9. K_{Ic}^{+} versus K_{tB} (stress concentration factor computed by Neuber formalism). Intersection of tangents defines K_{Ic}-value where continuum mechanics is valid.

In Figure 9 the K_{Ic}^{+}-factor is plotted versus the stress concentration factor K_{tB} for pure bending by Neuber formalism[3] (which can be very complicated as function of deepness and form of the notch). If we use not only the mean value measured with saw cut width 0.1 mm we find in this region a break in the curve leading to a constant K_{Ic}-factor designated by the intersection of the two different tangents of the curve.

This intersection indicates the point where continuum mechanics is valid and where the K_{Ic}^{+} has a minimum value. That means that this is the true K_{Ic}-factor for the special structure measured. Below this point measured K_{Ic}-factors will be too low.

In brittle alumina ceramics (average grain diameter $\bar{d}$: 10 μm) with notch analysis one can evaluate from this intersection $\rho' \approx 50$ μm and therefore there must exist a finite element of $\eta \approx 25$ μm.

This means that for our measurements with diamond saw cuts a true K_{Ic}-factor can be measured more exactly with a saw cut width of 0.1 mm than with the extrapolated value of a mathematically sharp crack, which will be too low. Nevertheless, there is no large difference between the two values (only 9 %).

No matter what the true K_{Ic}-factor is, the relative values of K_{Ic}^{+} measured with 0.2 mm saw width give information that is dependent upon structure parameters and various media (Figures 1, 3). There is a very low standard deviation which is a function of the structure.

The influence of grain diameter on the K$_{Ic}^{+}$-factor is clearly shown by replica pictures.

CONCLUSION

The K$_{Ic}$-factor is strongly influenced by the width of the saw cuts, which is more than generally expected.

Nevertheless, with diamond saw cuts, we have a useful instrument in getting K$_{Ic}$-factors if we use fracture mechanics formalism in connection with the formalism of notch analysis. We may even say, that with this connection, we have additional information about the structure of the material.

A great advantage lies in research technique. You may apply saw cuts of a certain depth to a lot of specimens simultaneously and in a very short time. Especially you are able to generate a plane crack front over the whole thickness of the specimen and measure the saw cut length exactly.

REFERENCES

1.   Davidge, R.W., Tappin, G.,
     Proc. Brith. Ceram. Soc. No. 15, 1970

2.   Weiss, V.,
     Notch Analysis of Fracture, Fracture Vol. III ed.
     by H. Liebowitz, Academic Press, New York and London, 1971

3.   Neuber, H.,
     Kernspannungslehre, Springer-Verlag, Berlin-Göttingen-Heidelberg, 1958

4.   Irwin, G.R.,
     Proc., First Symposium on Naval Structural Mechanics,
     Pergamon Press, London and New York, N.Y. 1960

MICROSTRUCTURAL CONSIDERATIONS FOR THE APPLICATION OF FRACTURE MECHANICS TECHNIQUES

Leonard A. Simpson

Atomic Energy of Canada Limited, Whiteshell Nuclear

Research Establishment, Pinawa, Manitoba, Canada

INTRODUCTION

Early attempts to understand the effect of microstructure on strength in ceramics have been mainly concerned with direct correlations between strength and grain size or porosity. The inverse square root dependence of strength on flaw size predicted by the Griffith equation has prompted many workers to plot strength against the inverse square root of grain size on the assumption that the flaw size is proportional to the grain size. Failure of the data to fit such a plot is often interpreted in terms of a two stage process involving a microplastic crack nucleation process at finer grain sizes[1,2]. Until recently, workers have neglected the fact that fracture energy as well as flaw size can depend on microstructure. This could provide an alternative explanation for the failure of strength data to fit a plot against the inverse square root of grain size for some ceramics.

Recently, a number of publications have appeared, concerned with the fracture energy-grain size relationship[3-7]. However, in spite of these, no clear trends have yet been established as many of these papers contain conflicting results. Part of the reason for the conflicts may be that several different techniques of fracture energy measurement have been used.

In the ceramics field, two fundamentally different approaches to fracture mechanics measurement are now commonly used, the total-work-of-fracture method (TWOF) and the energy balance technique.

The details of specimen type and analysis are described in other
papers in this conference. This paper will describe how the applicability of these methods can depend on the microstructure of the
test material using examples from work at WNRE.

TOTAL-WORK-OF-FRACTURE METHOD

The accuracy of this technique (often referred to as the
Tattersall-Tappin[8] or Nakayama[9] method) is highly dependent on the
degree of stability of crack propagation. In an earlier paper the
author described difficulties in achieving stable crack propagation
in strong, dense microstructures where there are few features capable
of blunting or arresting a moving crack[7]. Admittedly, good agreement
has been reported between this method and an energy balance technique
on some materials. However, there are equally many examples of non-
agreement, even in the same publications[7,10]. The use of such a
method must be considered unreliable unless used in conjunction with
an energy balance technique, especially if the results are to be
used for critical crack size estimates in the design of a structure.

On the other hand, for non-homogeneous microstructures such as
porous ceramics or refractories, the TWOF method can be useful.
Here cracks are frequently blunted as they traverse the microstructure, continually stopping, restarting and changing direction
with accompanying energy losses. A total energy measurement for
these materials may have more significance than an initiation
energy measurement especially in fixed displacement loading modes
such as thermal shock situations. This method is also frequently
used with composite materials in the absence of a well defined
analytical theory of crack propagation in these highly anisotropic
and non-homogeneous microstructures[11,12].

ENERGY BALANCE METHOD

The two most commonly used energy balance techniques for
fracture energy measurement in ceramics are the single-edge-
notched beam (SENB) and the double cantilever beam (DCB) methods.
Analytical expressions for determining fracture energy are available
for both specimen configurations[13,14]. Each method possesses
advantages and disadvantages, the relative importance of which
will depend to some extent on the microstructure of the test material.

The advantages of SENB over DCB are:

1. Specimens are of simple geometry and are easier and cheaper to
    fabricate (assuming precracking is not necessary).

2. The specimens are compact, minimizing material consumption.

The advantages of DCB over SENB are:

1. DCB's are easy to precrack.

2. One specimen can yield several determinations if proper care
    is exercised.

Although, on the basis of the above comparisons, the SENB
might appear to be the most practical technique from the point of
view of specimen fabrication, difficulties in precracking can
often nullify this advantage.

The mathematical solutions for the SENB and DCB specimens
describe zero volume cracks, that is, cracks of zero width. In
the DCB specimen such cracks can easily be introduced by driving
a wedge into the crack plane. A runaway situation can be avoided
by putting the crack plane into compression a short distance ahead
of the wedge. In the SENB specimen, production of a zero volume
crack is difficult because of the short length of the crack plane.
In practice, a notch of finite width is cut and a starter crack
induced at the notch root of sufficient length that the combination
behaves like a zero volume crack.

Microstructure and Notch Root Cracks in SENB Specimens

The American Society for Testing Materials has published
specifications for the relationship between root crack length and
notch size which essentially demand that the length of the root
crack be at least twice the notch width[15]. With metals this is
often satisfied by using large specimens and by inducing root cracks
by a fatigue process. However, root cracks are more difficult to
introduce in ceramics which, combined with the necessity to use
small specimens, often prevents satisfaction of the ASTM criterion.
Recently, the author has reported that a root crack length of 0.5
times the root radius should be sufficient for a valid test[16].

In some ceramics, such as alumina, root cracking is induced by
the machining damage during the sawing of the notch. Davidge and
Tappin[17] demonstrated for coarse grained alumina that these cracks
were sufficient to yield results in agreement with those for
specimens containing real cracks.

The results from a study of fine grained (3 μm), cold pressed
and sintered alumina are shown in Figure 1[16]. This is a comparison
of fracture energies determined from SENB specimens with as-cut
notches of two widths and pre-cracked DCB specimens. The SENB
results are also plotted against notch depth. The results for
the narrow notch are about 15% higher than those for DCB determinations for relative notch depths greater than 0.2.

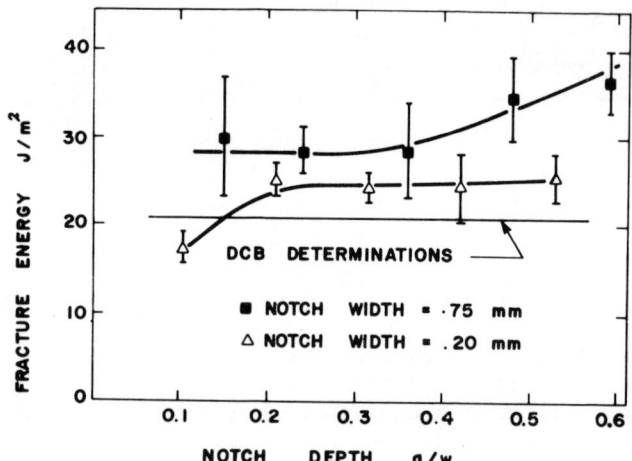

FIG. 1

Apparent Fracture Energy vs Ratio of Notch Depth, a, to Specimen
Depth, w, for SENB Specimens made from 3 μm Al_2O_3.

The wider notches yielded considerably larger fracture energies
which varied somewhat with notch depth. The differences between
the SENB results and those from DCB's were attributed to insufficient root crack depths in the as-cut notchs. Except for a small
but noticeable effect of saw blade width, the depth of root cracking
was controlled by the resistance of the microstructure to machining
damage. Thus it is not possible to specify a universal maximum
width for as-cut notches in ceramics and the presence of sufficient
root cracking to ensure a valid test should be demonstrated for each
microstructure being subjected to a fracture energy determination.

The importance of the root crack is demonstrated by some results
of as-notched SENB* specimens on two silicon carbides which are
compared in Table 1 with Mathew's[18] pre-cracked DCB determinations
on the same material.

---

* Unless otherwise indicated, all SENB specimens in this paper
  had notch widths = 0.2 mm, a/w = 0.4, and w = 6.4 mm.

TABLE 1

FRACTURE ENERGIES OF SILICON CARBIDES USING SINGLE EDGE
NOTCHED BEAM AND DOUBLE CANTILEVER BEAM SPECIMENS

| MATERIAL | GRAIN SIZE μm | FRACTURE ENERGY J/m^2 | | FLAW DEPTH FROM MOR (μm) |
		SENB	DCB	
Springfields*	20 μm	25.6 ± 3.0	15.4 ± 1.0	35
KT**	100 μm	15.2 ± 1.7	22.3 ± 2.9	100

* UKAEA, Springfields Works, Salwick, Lancs., England.
** Carborundum Co., Buffalo Ave., Niagara Falls, N.Y. 14302, USA.

As the results stand, they predict two opposite grain size
effects depending on which testing method is considered. A partial
explanation for this contradiction may be found by examining the
microstructures in Figure 2. The Springfields material has a fine
grained uniform microstructure, close to theoretical density, while
the KT microstructure is non-uniform with some very large grains
and pockets of free silicon. Extensive examinations of the notch
roots of the Springfields material indicated that the maximum
depth of through-the-thickness root cracking was between 15 and
30 μm. Flaw depths due to machining damage were also estimated
from modulus of rupture determinations (MOR) on as-ground bars
and the DCB fracture energy assuming failure was initiated from a
through-the-thickness surface crack. These estimated root crack
depths are too short to yield a valid test using 0.2 mm wide notches[16].
Thus the SENB values for Springfields material are probably considerable overestimates and the DCB test would appear to be a more
reliable approach for this microstructure.

The low SENB result relative to the DCB measurements for KT
material is more difficult to explain. The calculation of fracture
energy did not include the root crack length as part of the total
starting crack. If this were sizeable, the correction would
increase the SENB value in Table 1. The microstructures in Figure
2 show numerous grains of 200 μm in length as well as equally
large pockets of free silicon. It is therefore conceivable that
root cracking due to machining damage could penetrate at least
to this depth. The much larger flaw size estimated from modulus
of rupture measurements in Table 1 confirms the greater susceptibility of the material to machining damage.

The correction for a 200 μm root crack (about 20%) would not be

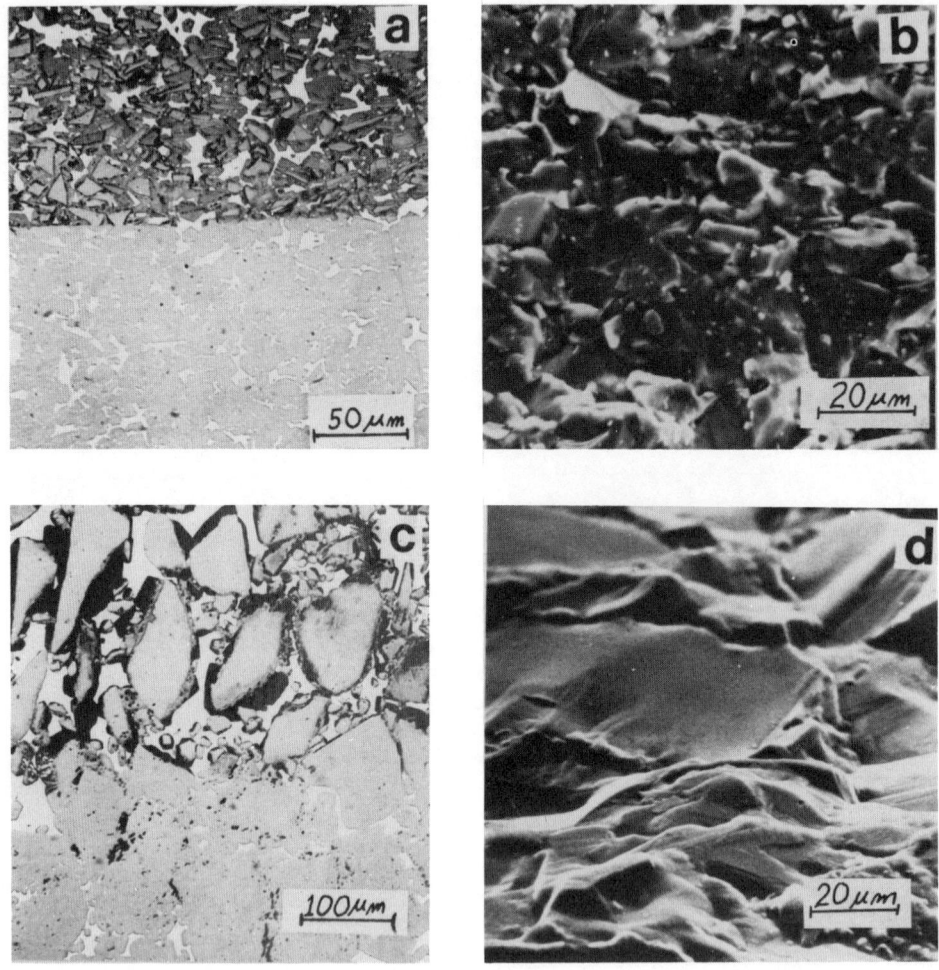

FIG. 2

(a) Microstructure of Springfields SiC (upper half etched)
(b) Fracture surface of Springfields SiC
(c) Microstructure of KT SiC (upper half etched)
(d) Fracture surface of KT SiC
(Figures 2a and 2c provided by R.B. Mathews[18]).

sufficient to raise the SENB values to those obtained from DCB
specimens (Figure 3). Root cracks deeper than 200 μm are unlikely
as dye penetrant examination gave no indication of their presence.
Hence there appears to still be a discrepancy between the DCB and
SENB results on large grained SiC. This type of discrepancy has
also been observed in large grained alumina and is discussed in
the next section.

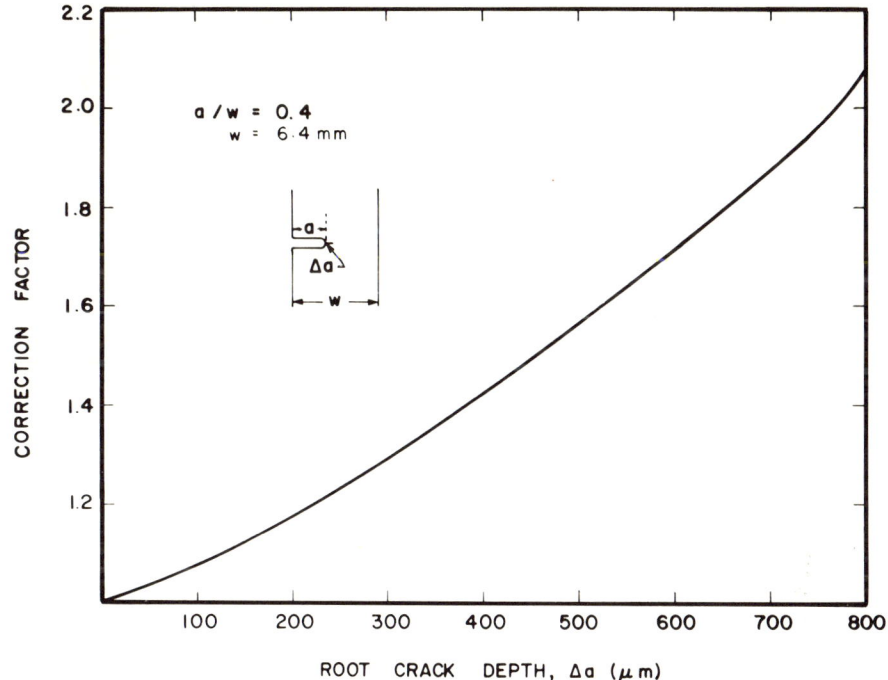

FIG. 3
Correction Factor to be Applied to SENB Fracture Energy Measurements to Correct for Neglect of Finite Depth of Root Crack, Δa.

Fracture Energy Measurements and Grain Size

A review of the literature on the grain size dependence of
fracture energy of alumina reveals a conflict of results in that
some workers report a considerable increase of fracture energy with
grain size[3] while others report no effect or a slight decrease[6,7].
A significant feature of these studies was that the workers reporting
an increase in fracture energy with grain size used DCB specimens
while the others used the SENB.

This conflict prompted a detailed comparison of the two methods
at WNRE using cold pressed and sintered alumina of two grain sizes,
3 μm and 20 μm. Several slabs of each material were fabricated from
Alcoa XA16[†] powder using techniques described elsewhere[7]. (The 3 μm
material was a different batch from that described in Figure 1.)
DCB and SENB specimens were machined from these slabs and the DCB's

---

[†] Aluminum Company of America, Pittsburgh, PA., USA.

were precracked. In addition, the used DCB arms were recut into
SENB specimens. The results from the latter were identical to
those for SENB specimens cut from the other slabs fabricated under
the same conditions. Two commercial aluminas were also tested in
a similar manner, a Coor's material of 3 μm grain size and a G.E.
Lucalox of 35 μm grain size. The results are summarized in Table 2.

TABLE 2

FRACTURE ENERGIES OF ALUMINA USING SINGLE EDGE NOTCHED
BEAM AND DOUBLE CANTILEVER BEAM SPECIMENS

MATERIAL	GRAIN SIZE μm	FRACTURE ENERGY J/m^2	
		SENB*	DCB
Al_2O_3 #1	20	21.1 ± 2.0	38.8 ± 3.0
Al_2O_3 #2	3	22.5 ± 3.0	20.8 ± 1.1
LUCALOX**	35	20.3 ± 1.1	32.1 ± 4.0
COORS AD999***	3	19.7 ± 1.8	19.05

*   Includes specimens made from DCB arms.
**   General Electric Co., 24400 Highland Rd.,
      Richmond Heights, Ohio, USA.
*** Coors Porcelain Co., 600 North St., Golden
      Colo. 80401, USA.

The two methods applied to the 3 μm alumina are in good
agreement. However the DCB tests yielded consistently higher values
for the large grained materials. By themselves, the DCB determinations indicate a strong grain size dependence of fracture energy
while little effect is indicated by the SENB specimens.

An increase in fracture energy with grain size has been
attributed to a transition from intergranular to transgranular
fracture at large grain sizes[3]. This is based on an argument that
transgranular fracture requires more energy than intergranular
fracture but the basis for this argument is neither rigorous nor
convincing. In this work no transition in fracture mode was
observed for the WNRE alumina in spite of the increase in DCB
fracture energy with grain size (Figures 4a and 4c). In fact,
the Lucalox (Figure 4d) contains more intergranular fracture
than the 20 μm WNRE alumina yet the DCB results indicate a lower
fracture energy. The fracture surface features resulting from

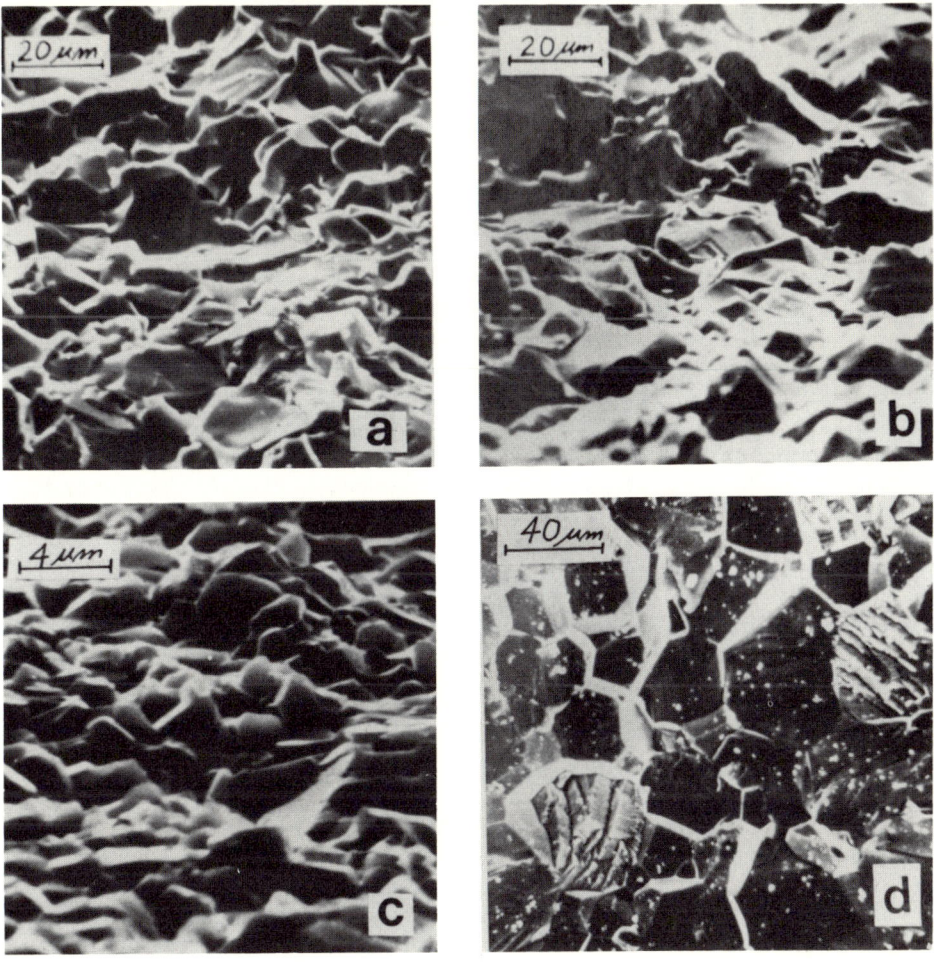

FIG. 4

Typical Fracture Surfaces of:
(a)   20 μm Al$_2$O$_3$ after DCB test
(b)   20 μm Al$_2$O$_3$ after SENB test
(c)   3 μm Al$_2$O$_3$ after DCB test
(d)   35 μm Lucalox after DCB test.

the two different testing methods on 20 μm material are indis-
tinguishable (Figures 4a and 4b). This rules out an explanation
of the discrepancy between techniques based on differences in
fracture mode.

It is unlikely that the lower SENB values from 20 μm material

are due to deep cracks at the notch root. Reference to Figure 3
reveals that through-the-thickness root cracks 750 μm deep would
have to have been present to account for the discrepancy. Dye
penetrant tests on specimens did not reveal any cracking beyond
100 μm.

Thus one is forced to consider the possibility that the fracture
energy may be a function of testing technique. One explanation may
involve the differences in stress states in the two specimen types
near the crack tip and how they effect cracks lying out of the
symmetry plane. The fracture surfaces in Figure 4 show that
propagation in the symmetry plane is the exception rather than
the rule because the crack prefers to follow the grain boundaries.
The stress state near the crack tip of the DCB specimen differs
from the SENB in that it is strongly biaxial[17]. Not only is
there the splitting force which acts to open cracks in the
symmetry plane but also there is a tension parallel to this
plane which tends to drive cracks at right angles to the symmetry
plane resulting in the breaking off of one of the beams. This
is why it is often necessary to side-groove DCB specimens to
confine the crack propagation to the symmetry plane. Because
in polycrystalline ceramics, the crack is frequently diverted from
the symmetry plane what is needed is an analysis of criticality
conditions for cracks oriented at an angle to this plane under
the stress state of the DCB specimen.

SUMMARY

1. Microstructure must be considered in the selection and
application of a method of fracture energy measurement and
necessary steps taken to ensure valid testing.

2. The degree of root cracking in as-cut SENB specimens
depends on microstructure and hence each microstructure should
be carefully examined to ensure that sufficient cracking exists
for the given notch width.

3. The disagreement in the grain size dependence of fracture
energies for alumina appears to be related to a fundamental
difference between two testing techniques the degree of which is
itself microstructure dependent.

4. A transition in fracture mode from intergranular to
transgranular does not accompany the increase in fracture energy
with grain size determined by the DCB technique for the cold
pressed and sintered alumina studied here.

REFERENCES

1. S.C. Carniglia. J. Amer. Ceram. Soc. 55, 243 (1972).
2. R.W. Rice. Proc. Brit. Ceram. Soc. 20, 205 (1972).
3. P.L. Gutshall, G.E. Gross. Eng. Fract. Mech. 1, 463 (1969).
4. F.J.P. Clark, A.G. Tattersall, G. Tappin. Proc. Brit. Ceram. Soc. 6, 163 (1966).
5. A.G. Evans, R.W. Davidge. J. Nucl. Mater. 33, 249 (1969).
6. A.G. Evans, G. Tappin. Proc. Brit. Ceram. Soc. 20, 275 (1972).
7. L.A. Simpson. J. Amer. Ceram. Soc. 56, 7 (1973).
8. H.G. Tattersall, G. Tappin. J. Mater. Sci. 1, 296 (1966).
9. J. Nakayama. J. Amer. Ceram. Soc. 48, 583 (1965).
10. J.A. Coppola, R.C. Bradt. J. Amer. Ceram. Soc. 55, 455 (1972).
11. L.A. Simpson, A. Wasylyshyn. J. Amer. Ceram. Soc. 54, 56 (1971).
12. D.C. Phillips. J. Mater. Sci. 1, 1175 (1972).
13. W.F. Brown, Jr. and J.E. Srawley. Amer. Soc. Test Mater., Spec. Tech. Publ. No. 410, 1966.
14. S.M. Wiederhorn, A.M. Shorb, R.L. Moses. J. Appl. Phys. 39, 1569 (1968).
15. Annual Book of ASTM Standards, Amer. Soc. Test Mater., Philadelphia (1971) p. 919.
16. L.A. Simpson. "Use of the Notched Beam Test for Evaluation of Fracture Energies of Ceramics." Presented at the Annual Meeting, American Ceramic Society, Cincinnati, Ohio, May 1, 1973 (Paper No. 31-B-73).
17. R.W. Davidge, G. Tappin. Proc. Brit. Ceram. Soc. 15, 47 (1970).
18. R.B. Mathews. To be published in J. Can. Ceram. Soc. 41 (1973).
19. M.J. Manjoine. J. Basic Eng. 87, 293 (1965).

STRENGTH AND MICROSTRUCTURE OF DENSE,

HOT-PRESSED SILICON CARBIDE

Svante Prochazka and R.J. Charles

General Electric Company

Corporate Research and Development

ABSTRACT

The microstructural and strength characteristics of dense SiC, hot-pressed from submicron SiC powders with small additions of boron, are shown to be strongly dependent on the control exercised over oxygen content during pressing. Uniform, equiaxed grain structures of SiC are obtained if sufficient oxygen is present to allow detectable particles of SiO_2, whereas uniform, elongated grain structures of β SiC are obtained if excess carbon is utilized to reduce oxygen content to undetectable limits. At intermediate oxygen levels, free silicon was formed and large, tabular α-SiC grains in a fine-grained β-SiC matrix were obtained.

The flexural strength of the two materials with uniform grain size was about 80,000 psi in as-ground conditions and was nearly independent of temperature up to 1400°C and 1500°C, respectively. The conclusion that strength was limited by surface defects formed on machining was supported by the observation of a 25% strength increase of polished test bars. At 1600°C, the strength was 44,000 psi and 64,000 psi.

The specimens containing large tabular α-SiC crystals showed an average strength of 38,000 psi with single values down to 20,000 psi. Examination of fracture surfaces revealed that the large grains invariably initiated rupture. The size of the fracture-initiating grains was correlated to fracture stress and yielded a relationship of the form of the Petch equation. The weakening mechanism of these large grains is attributed to stress concentration from elastic anisotropy modified by stresses due to thermal expansion mismatch and stresses resulting from the relaxation of strains imposed during fabrication.

579

INTRODUCTION

Refinement of grain size is an effective procedure for developing high strength in ceramics and has been extensively utilized for both commercial and laboratory materials.   There is, however, a lack of fundamental understanding of the relationship between grain size and strength of non-yielding polycrystalline solids and the subject remains in considerable controversy.

If it is assumed that a direct relationship exists between grain size and the strength-limiting flaw then, as predicted by Orowan,[1] an inverse proportionality of applied fracture stress and the square root of grain size is expected. Carniglia[2] and Rice[3] have made extensive review of existing experimental data and show that the above relationship is rarely obeyed and that ceramics generally show a more complex behavior. By including qualitative effects due to internal stresses, more satisfactory descriptions of experimental data are often obtained and, in many cases, the data can be adequately represented by the Petch[4] form of the Orowan relation. This procedure introduces a basic uncertainty, however, in that the origin of these internal stresses and their quantitative relationships to microstructure, processing parameters, and the fracture process are difficult to assess.

Strength-grain size dependence studies are inherently difficult since the grain size in ceramics cannot be varied as an independent parameter. In ceramic processing to achieve variations in grain size, there are always accompanying changes of other microstructural and chemical characteristics such as porosity level, pore size and pore size distribution, grain bonding, impurity segregation, etc., which complicate the results. Final fabrication procedures may also play a role as has been recently illustrated by Evans and Tappin[5] who have shown that the strength of a sintered porous alumina was almost entirely dominated by surface machining flaws and that grain size had only a secondary effect.

In the work to be described, an unusual microstructure was developed in a pressure-sintered form of silicon carbide. This microstructure consisted of a dense, fine-grained matrix which contained large, well-isolated tabular crystals which resulted from exaggerated grain growth. Such a microstructure offered a unique possibility to relate strength to the size of the single grain in which fracture of the ceramic was initiated. The purpose of this contribution is to describe processing resulting in this microstructure and the effect of the large crystals on fracture behavior.

PREPARATION OF MATERIALS

Thick disks of high-density SiC, up to 2" in diameter, were prepared by hot-pressing two types of SiC powders at 1950°C and

0.68 kbars. The first powder was a submicron size fraction of a commercial SiC (Type E277 of the Norton Company) to which 1% elemental boron was added as a densification aid. The other powder was prepared in the laboratory by carbon reduction of a mixture of an amorphous silica and a boron compound. In some cases, additions of 1% carbon were made to the second powder before hot-pressing to suppress grain growth.[6]

Specimens for metallography were prepared by diamond cutting and polishing and grain boundaries were revealed by Murakami's boiling reagent or by thermal etching for 20 minutes at 1550°C in argon. Thin sections were prepared for optical and electron transmission microscopy and crystal structures were identified by x-ray diffraction of individual grains.

In general, test bars for strength measurements were cut and machined with a 220 grit diamond wheel to dimensions of 0.25 cm by 0.25 cm and 2.5 cm long (0.1" x 0.1" x 1"). While the machining marks on the faces were not removed, the two longitudinal edges to be subjected to bending tensile stresses were smoothly champhered. Specimens with obvious surface defects were rejected. The machined surfaces showed no variations attributable to the type of material. Typical surface features were 10-15 micron-sized depressions formed by removal of groups of grains during during cutting and grinding. The micrograph in Fig.1 shows the high density and uniform distribution of these depressions in the machined surface.

Fig.1 Surface morphology of dense, fine-grained silicon carbide as machined with 220 mesh diamond wheel (Spec type B).   SEM.   1050X

MICROSTRUCTURE AND PHASE COMPOSITION

Three different types of microstructures were observed in the high-density billets. ($>99\%$ of theoretical density taken as 3.214 g/cc.) Analyses have shown a correlation of the microstructure and the presence of some secondary phases with the initial and final oxygen content. The characteristics of the three SiC types are listed in Table I and discussed below in detail.

TABLE I

CHARACTERISTICS OF HOT-PRESSED SiC

Specimen Type	A	B	C
Hot-Pressing Conditions	1950°C 10 kpsi 30 min.	Same	Same
Density, g/cc	3.19	3.20	3.20
Grain size, μ	3.2	3.8 (matrix)	3.0
Largest grains, μ	10	>500	25
Phases detected	β SiC α SiC SiO_2	β SiC α SiC Si + unknown	β SiC C
Boron, %	0.88	0.40	0.46
Oxygen, %	0.54	0.1	<0.006
Total metal Impurities, % (Fe, Al, Ti, Zr, V)	<0.4	--	<0.2

A. Microstructure Type A: 1-3 microns equiaxed grains

This morphology is obtained on hot-pressing of E277 powders with relatively large contents of oxygen, typically 5%, and is shown in Fig.2. Only a fraction of the initial oxygen is retained in the SiC pressing (Table I) and forms a silica phase distributed along SiC grain edges, as revealed by selective etching in HF, Fig.3. It is likely that this phase, a viscous melt at the forming conditions, conditions, controlled boundary migration and gave rise to a grain shape typical for relatively slow, normal grain growth.

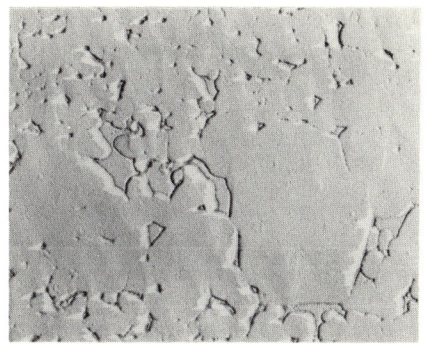

Fig.2 Fine-grained silicon
hot-pressed from a powder
containing 5% oxygen. Type
A – Table I. Double-stage
replica of a thermally
etched surface.   10,000X

Fig.3 Silica phase in the
specimen shown in Fig.2
revealed by HF etching.
Double stage replica.
                    10,000X

This microstructure, termed Type A, was composed almost entirely
of β -SiC. No other SiC phase was detected; however, in addition to
small amounts of SiO_2, the microstructures showed a small number of
2-5 μm, shiny grains which were presumed to be traces of elemental
silicon.

B. Microstructure Type B: Large tabular crystals in a 3-5 μ
    grain size matrix.

This type of microstructure was obtained by hot-pressing
of SiC powders prepared by SiO_2 reduction with carbon. The resultant
SiC powder was leached with HF, before pressing, and typically showed
an order of magnitude lower oxygen ($\simeq 0.5\%$) level than the preceding
powder. There is a marked variation in the morphology, size, and
density of the idiomorphic grains with some up to several hundred
microns longs. [Figs.4(a) + 4(b)] Another micrograph in Fig.5
reveals the tabular morphology. The exaggerated grain growth in SiC
is a rapid process and ultimately transforms all the matrix into the
tabular grains. Such an advanced stage of recrystallization is
shown in Fig.6. X-ray analysis and electron diffraction identified
the matrix as prevailingly β -SiC and the tabular grains as α -SiC(6H),
implying that the exaggerated grain growth involves a phase transformation. Beside these two phases, a small amount of elemental
silicon was always found as shiny grains in as-polished sections.
A dark-brown, optically isotropic phase was found trapped inside the
pores engulfed in the large α -SiC grains. It was not identified;
however, microprobe analysis showed it to have a variable Si/C ratio
and to contain traces of Fe, V, and Al.

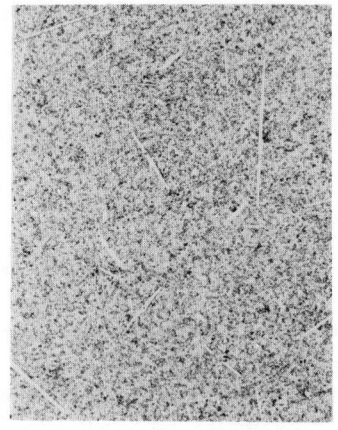 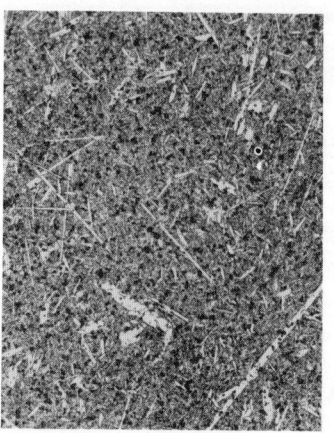

Fig.4 Growth of idiomorphic crystals on hot-pressing of a silicon
carbide powder with low oxygen content. Microstructure type B.
Thermal etch. a and b corresponds to different powders sintered
at the same conditions.   100X

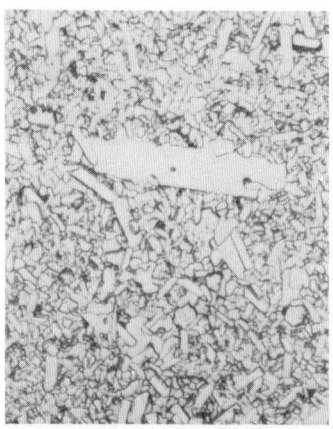 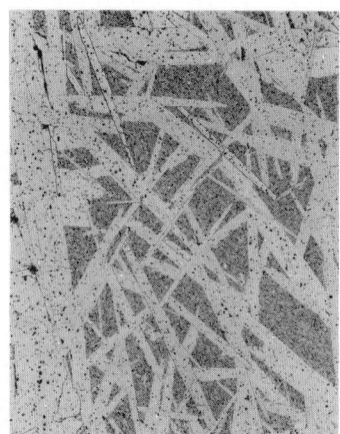

Fig. 5 Tabular morphology of   Fig.6 Advanced stage of re-
  the idiomorphic grain grown   crystallication of a dense,
  on hot-pressing in silicon   fine-grained β-SiC matrix on
  carbide. Chemical etch.   extended hot-pressing. Thermal
                      500X   etch.   50X

C. Microstructure Type C: 3-5μ elongated grains

     A small addition of carbon, typically 1%, to the synthesized
powders suppressed the exaggerated grain growth shown in the Type B

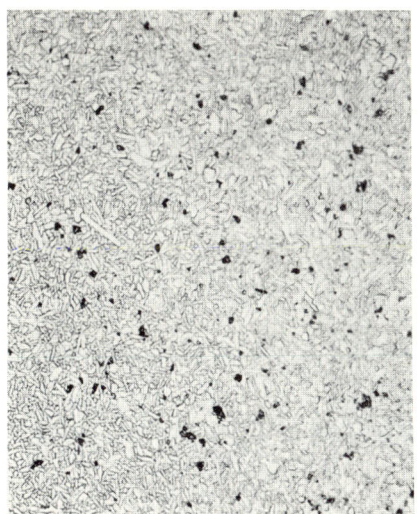

Fig.7 Silicon carbide hot-pressed with 1%
excess carbon. Exaggerated grain growth
shown in Figs.4-6 was suppressed. Thermal
etch.   625X

material and yielded microstructures composed of uniformly fine,
elongated grains as shown in Fig.7. The dark regions in this micrograph are carbon grains resulting from inhomogenieties in carbon
distribution. Type C silicon carbide was β-phase and exhibited no
secondary phases other than a small amount of residual carbon.
Chemical analyses showed an extremely low oxygen content (<0.006%)
as shown in Table I.

INTERNAL STRESS ANALYSIS

The large tabular grains in B-type microstructures have stresses
associated with them easily detectable as photoelastic patterns in
thin sections by circularly polarized, transmitted light. They are
localized at grain boundaries or grain-matrix boundaries in regions
typically 20 microns thick and appear to fall off rapidly away from
boundaries, as predicted by Mar and Scott.[7]

There are three possible sources of the internal stresses in the
present system: a) volume change on phase transformation; b) thermal
expansion anisotropy; c) stresses generated by relaxation of strains
imposed on hot-pressing.

(a) The density change on β-αSiC transition is, according to
x-ray data of deMesquita,[8] very small, about 0.1%. Moreover,

the transformation occurs at a high temperature where
stresses would be probably relaxed by yielding.

(b) The maximum thermal expansion stresses, σ_{th}, developed in
the B-type SiC, can be estimated treating the α SiC grains
as inclusions in a homogeneous matrix. If the values of
the elastic moduli and Poisson's ratios are considered
equal, then the expression given by Selsing[9] simplifies to:

$$\sigma_{th} = \frac{2E\Delta T\Delta\alpha}{3(1-\nu)} \tag{1}$$

where E = elastic modulus

ν = Poisson's ratio

$\Delta\alpha$ = thermal expansion differential

ΔT = temperature above which stresses are removed by
yielding.

X-ray thermal expansion data for β and α SiC by Taylor and
Jones[10] are limited to 1250°C. An extrapolation was performed in order to obtain the coefficients over the full
temperature range to 1700°C and then $\Delta\alpha \cdot \Delta T$ obtained by
graphical integration. The temperature limit of yielding
(1700°C) was taken somewhat arbitrarily but was based on
creep measurements and sintering experiments under pressure.
Using the above, the normal boundary stresses, σ_{th}, for E
= 6.12 x 10^7 psi, ν = 0.19 calculates to be 9500 psi. Tangential stresses, equal in magnitude, would also develop in
the inclusion.

(c) The third component of the internal stress arises from the
relaxation of strains imposed by the forming pressure during
fabrication. As the elastic moduli vary with crystallographic directions, a difference in stress appears at the
grain boundaries during pressing and is relaxed by yielding.
When the outer pressure is removed (on cooling through
1700°C) the strain differences result in the re-introduction
of stresses. An estimate of their magnitude can be obtained
from the following.

The fabrication pressure, P, is magnified at the boundary
of an inclusion by an elastic stress concentration factor,
F, which according to Hasselman,[11] achieves a maximum
value of A = E_2/E_1, where E_2 and E_1 are, respectively,
the largest and smallest elastic moduli of the system.
In pressing, a stress differential, P - FP, would be
relaxed by yielding and may appear as a boundary stress
if the outside loading is removed below the temperature

of yielding. Thus, a residual stress, σ_r, may be introduced during fabrication equal to

$$\sigma_r = P(1 - F) \tag{2}$$

which, for the maximum case where ΔE is the moduli difference, would be approximately equal to $(\Delta E/E_{average})P$.

Hasselman[11] further analyzed stresses generated by elastic anisotropy of a two-dimensional elliptical inclusion and gave an analytical solution for the stress concentration factor F. It increases with the grain shape ratio r (the ratio of the major and minor axes) and asymptotically reaches a value close to the anisotropy factor, A. For a stress applied parallel to the minor axis, 96% of the limiting value is obtained when r = 10. In other words, elastic stress concentrations are essentially independent of the shape parameter beyond a certain limit ($r \sim 5$).

Elastic anisotropy factors for the various forms of SiC may be calculated from elastic constant data given by Toplygo[12] and Arlt and Schodder[13] and formulae by Nye[14] and Voigt.[15] For the present case, the ratio of $E_{\alpha-SiC}$ (maximum value) to $E_{\beta-SiC}$ (average value) was taken as the desired room temperature anisotropy factor A and amounted to 1.35. The specific data is summarized in Table II.

For an applied forming pressure of 10,000 psi and for this anisotropy factor, the maximum residual stress, σ_r, due to the presence of α and β silicon carbide phases in a structure, calculates to be 3500 psi.

RESULTS OF STRENGTH MEASUREMENTS

The results to 1500°C of fluxural strength measurements, performed in three point bending with a span length of 0.625" at a loading rate of 0.002"/min. are summarized for the three materials in Table III. As previously indicated, the characteristics of the materials are given in Table I. Additional three-point tests at 1600°C were performed but in a different test device using specimens 0.05" x 0.1" x 1.0" on an 0.75" span. The strength vs temperature results are plotted in Figure 8.

The two fine-grained SiC (A and C) retain a strength of about 80,000 psi up to 1400° and 1500°C, respectively. The earlier fall-off in strength of Type A is attributed to its content of silica. The oxygen-free material, Type C, shows an increase in strength from 71,000 to 84,000 psi over the same range. A strength increase

TABLE II

ELASTIC PROPERTIES OF SiC

CONSTANT	UNIT	VALUE	REFERENCE COMMENT
C_{11}		3.52×10^{12}	(12)
C_{12}	$\dfrac{\text{dyne}}{\text{cm}^2}$	1.40×10^{12}	(13)
C_{44}		2.33×10^{12}	
C'_{11}/C_{11}	--	1.63	
$\bar{E}$	$\dfrac{\text{dyne}}{\text{cm}^2}$	4.18×10^{12}	calculated
$\bar{E}$	psi	6.14×10^7	calculated
$\dfrac{E\ max}{\bar{E}}$	--	1.35	
$\bar{E}$	psi	6.65×10^7	this work, 99.8% dense
C_{11}	$\dfrac{\text{dyne}}{\text{cm}^2}$	5.00×10^{12}	(22)
C_{12}		0.92×10^{12}	
C_{13}		0.0×10^{12}	
C_{33}		5.64×10^{12}	
C_{44}		1.68×10^{12}	
C_{66}		2.04×10^{12}	
C'_{33}/C_{33}	--	1.68	
$\bar{E}$	$\dfrac{\text{dyne}}{\text{cm}^2}$	4.86×10^{12}	calculated
$\bar{E}$	psi	7.14×10^7	calculated
$\bar{E}$	psi	6.5×10^7	
$\dfrac{E\ max}{\bar{E}}$		1.31	

TABLE III

FLEXURAL STRENGTH OF HOT-PRESSED SiC, kpsi

| Temperature | Specimen Type | | | |
	SiC-A*	SiC-B*	SiC-C*	SiC-A (polished)
Room Temp	82.0 (12) ± 6.6	39.0 (5) ± 8.8	71.9 (10) ± 9.2	103.2 (4) ± 7.6
1300°C	78.9 (7) ± 7.4	38.7 (5) ± 9.6	80.0 (5) ± 7.2	-- --
1400°C	78.2 (7) ± 4.0	37.5 (5) ± 11.9	81.9 (4) ± 4.8	-- --
1500°C	67.8 (7) ± 3.4	36.6(5) ± 5.9	84.1 (5) ± 6.6	64.0 (4) ± 4.4
1600°C	44.0 (4) ± 4.8	--	63.7 (6) ± 9.6	--

*See Table I for characteristics.

Note: Number of specimens tested given in perentheses.

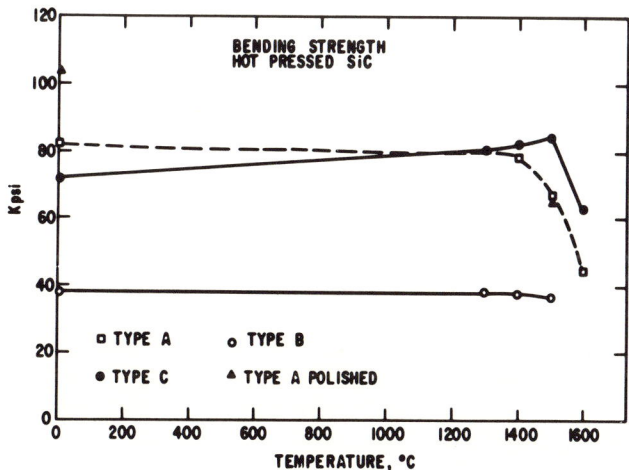

Fig.8 Three point bending strength of dense hot-pressed silicon
   carbide.

with temperature in SiC was also reported by Alliegro[16] and Gulden.[17] In the present case, however, it may not be a real phenomenon; it may be related to the yielding of alumina support pins which is sufficiently large at temperatures above 1300°C to improve aligning and reduce twist. The noticeable decrease in scatter above 1300°C is also attributed to this effect.

Types A and C specimens show small differences in strength; both have their maximum grain size well below the estimated flaw size necessary to initiate fracture and were fabricated and machined by the same procedure. It is concluded that their strength was limited by similar stress concentrators, probably surface damage introduced by machining. This is supported by the increase of room temperature strength by 25% measured on polished and lapped test bars of spec. A, Table III.

The specimens of the B type SiC show an average strength about 38,000 psi independent of temperature and single values down to 20,000 psi. They were prepared by the same procedures as the former two specimen series and the only marked difference is the presence of the large tabular graims to which is attributed to the drastic reduction of strength.

The fracture surfaces of Type B specimens were investigated and, in the majority of cases, fracture initiation was clearly localized at the large crystals intersecting the tensile face of the specimens. Table IV lists the strength results in relation to the size of these grains and Figs. 9(a,b,c) show examples of the fracture surfaces including those with the smallest and the largest grains. In two cases, there was fracture initiated at large interior grains and in four cases, the direction of fracture propagation was uncertain. The data in Table IV gives the largest dimension of the crystals to which fracture was attributed. At this point, it is of value to note that the use of the 3-point bend test tends to select a wider range of sizes of fracture initiating crystals than a 4-point test and thus is preferred for this type of investigation.

The crystals were easily observable in the fracture surfaces using a microscope with an extra-long focus objective. These measurements were, however, somewhat ambiguous in that two fragments of one test piece usually did not yield the same length. Therefore, two measurements are given in Table IV whenever this difference was larger than 20%. Comparing matching pairs of specimen fragments showed that fracture propagated by cleavage through the bulk of the large crystals and avoided the path along the crystal-matrix boundary.

DISCUSSION

To account for the marked weakening of the Type B material relative to Types A and C, one must conclude that stress concentrations

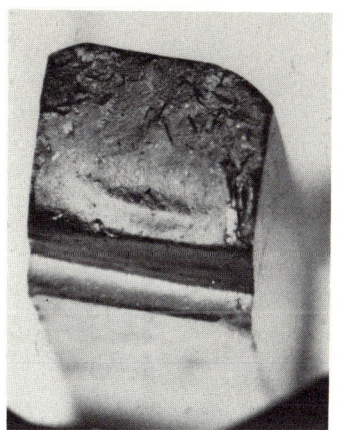

Fig.9   Fracture surfaces of test specimens of material having B-
    type microstructure showing idiomorphic grains which initiated
    fracture.   17X

are associated with the large, isolated crystals in this material.
These grains play a role in the fracture process as if they were
large cracks; yet one must assume that the only cracks present in
the specimen tensile regions would be those arising from machining,
and such cracks would be of the same small size as those initiating
fracture in the strong materials. It is useful, therefore, to
attempt to describe, in stages, the fracture process occurring in
these Type B materials.

    From the foregoing treatment of residual stresses, the effec-
tive stress, σ_{eff}, acting on a flaw in the stress field of a large

TABLE IV

Strength and Size of Fracture Initiating Grains in Specimens of Type B

Temp.°C	Strength kpsi	Maximum Dimension of Grain Initiating Fracture, μ	
Room Temp.	50.4	230	
Room Temp.	36.6	495	310
Room Temp.	44.0	315	
Room Temp.	40.2	290	
Room Temp.	23.8	990	
1300	34.2	415	
1300	17.4**	1680	1090
1300	47.0	155	
1300	52.0	175	63
1300	32.0	670	400
1400	53.5	145	
1400	31.8	930	740
1400	25.9	1250	
1400	24.1**	1300	800
1400	48.0	250*	
1500	33.8	730	390
1500	36.8	640	500
1500	34.2**	400	
1500	18.7	1140*	

---

*Fracture initiated at internal grain.

**Values corrected for fracture off center.

crystal in a fine-grained matrix under an applied external stress, σ_a, would be

$$\sigma_{eff} = F(\sigma_a \pm \sigma_i) \tag{3}$$

where F is the elastic stress concentration factor and $\pm\sigma_i$ are the maximum tensile or compressive values of local internal stress. Let us assume the large crystals intersect the surface at which points machining cracks of depth c have been introduced. The condition for propagation of such a crack in a surface crystal would be

$$\sigma_{eff}\Big|_{critical} = K \left[\frac{E\gamma_{fs}}{c}\right]^{\frac{1}{2}} \tag{4}$$

where K = a constant related to the crack geometry $[\sqrt{\pi}/2(1-\nu^2)$ for a penny-shaped crack[18]] and γ_{fs} is the fracture energy of the large crystal in a plane of easy cleavage.

From Eqs. 3 and 4, for the critical conditions for crack propagation at minimum applied stress, i.e., the tensile component of σ_i is operative,

$$\sigma_{eff}\Big|_{critical} = K\left[\frac{E\gamma_{fs}}{c}\right]^{\frac{1}{2}} - F(\sigma_{ac} + \sigma_i)$$

or

$$\sigma_{ac} = \frac{K}{F}\left[\frac{E\gamma_{fs}}{c}\right]^{\frac{1}{2}} - \sigma_i \ . \tag{5}$$

The fracture stress of the matrix of the above material may be given by

$$\sigma_f = K\left[\frac{E\gamma_{fm}}{c}\right]^{\frac{1}{2}} \tag{6}$$

and may be approximated by the values for the materials (Type A) which do not contain large grains. Thus, from Eqs. 5 and 6.

$$\sigma_{ac} = \frac{1}{F}\left[\frac{\gamma_{fs}}{\gamma_{fm}}\right]^{\frac{1}{2}} \sigma_f - \sigma_i \ . \tag{7}$$

Although values of the cleavage fracture energy, γ_{fs}, are not available, the ratio $\gamma_{fs}/\gamma_{fm} \simeq \frac{1}{4}$ may be substantiated by the work of Weiderhorn.[19]

The value of σ_{ac} is estimated to be, therefore,

$$\sigma_{ac} \simeq 0.37\sigma_f - \sigma_i \ . \tag{8}$$

σ_{ac} is the minimum stress which will propagate a crack through the large grain in the tensile surface in terms of the fracture stress in the absence of tabular grains. The estimate in Eq.8 holds, provided the aspect ratio, r, of the large grain is sufficient for F to achieve its maximum value and the minor axis of the grain is parallel to the specimen surface. Taking σ_f = 75,000 psi (the average strength in the absence of large grains) σ_{ac} calculates to 15 kpsi and is thus in agreement with the lowest strengths observed in specimens of Type B material. The remarkably low value of stress at which local damage occurs is a consequence of low cleavage and of unfavorable and significant internal stress.

The crack extending through the grain will progressively relieve the local residual tensile stresses and, on reaching the crystal-matrix boundary, will experience an abrupt increase of fracture energy from γ_{fs} to γ_{fm}.

If all residual stresses are relieved at this point, the crack may be arrested or continue to propagate to failure, depending on whether or not

$$\sigma_{ac} \begin{array}{c}> \\ <\end{array} K \left[\frac{E\gamma_{fm}}{\ell}\right]^{\frac{1}{2}} \tag{9}$$

where ℓ is now the length of the crack propagated through the large grain. In the first case, when arrested, further propagation through the matrix to failure takes place when the applied stress is increased above σ_{ac} to the value

$$\sigma_a = K \left[\frac{E\gamma_{fm}}{\ell}\right]^{\frac{1}{2}} . \tag{10}$$

Thus, the observed fracture stress σ_a becomes related to $\ell^{-\frac{1}{2}}$. As ℓ increases, the fracture stress decreases until the second condition is met; i.e., the crack is not arrested at the crystal-matrix boundary and propagates to failure at σ_{ac}. This occurs for all ℓ larger than a particular ℓ_c which can be evaluated from Eqs. 5 and 9.

$$\ell_c = \frac{K^2 F^2 E\gamma_{fm} c}{K^2 E\gamma_{fs} - 2KF\sigma_i (E\gamma_{fs} c)^{\frac{1}{2}} + \sigma_i^2 F^2 c} \tag{11}$$

All specimens having crystals larger than ℓ_c in the stressed part of the tensile surface will fracture at a constant stress σ_{ac}, which is independent of ℓ.

The above discussion, relating to Eqs. 10 and 11, is predicated on the assumption that all residual stresses are relieved when the initial crack reaches the grain-matrix boundary. This condition is, however, not the most likely case. Polycrystalline materials with elastic anisotropy can generate residual stresses because they are multiply-connected structures and, as indicated in Eq. 3, are stable in the absence of outside forces due to a precise balance of both tensile (+) and compressive (-) components of internal stress. It is true that a running crack tends to deviate and follows directions where the tensile residual stress components are maximized. A crack which is, however, constrained to propagate in a given direction must encounter regions of both tensile and compressive residual stress. In the present case, a crack in a large surface grain initiates due to a tensile component of residual stress and is constrained to follow a cleavage plane until it reaches the grain-matrix

boundary. At this point, and particularly if the probability is high
that other sub-surface tabular crystals are nearby (see Figures 4a
and b), it is most likely that the crack tip will encounter a local
region of residual compressive stress. Equation 10 should, therefore be altered to account for such stress. This may simply be done
by introducing a term $\alpha\sigma_i$, where α is a constant which would not exceed
one and would probably not be smaller than 1/2. Equation 12 gives the
modified result for Eq. 10 which, as may be noted, is of the Petch
form.

$$\sigma_a = K \left[\frac{E\gamma_{fm}}{\ell}\right]^{\frac{1}{2}} + \alpha\sigma_i \ . \tag{12}$$

On a plot of σ_a versus $\ell^{-\frac{1}{2}}$, we note that introduction of the
constant term $\alpha\sigma_i$ does not influence the slope, $K(E\gamma_{fm})^{\frac{1}{2}}$, of the
straight lines represented by Either Eq. 10 or 12.

Figure 10 plots the strength data of Table III in terms of the
square root of the lengths of the fracture-initiating grains in the
Type B material. A least-squares fit of a straight line through
these points yields a slope from which is calculated a matrix fracture energy, γ_{fm}, of 23,800 ergs/cm^2. This value is in good agreement with fracture energy values for SiC measured by cantilever beam
techniques (26,000 ergs/cm^2 by McLaren et al.[20] and 1900-30,000 ergs/
cm^2 by Coppola and Bradt[21]).

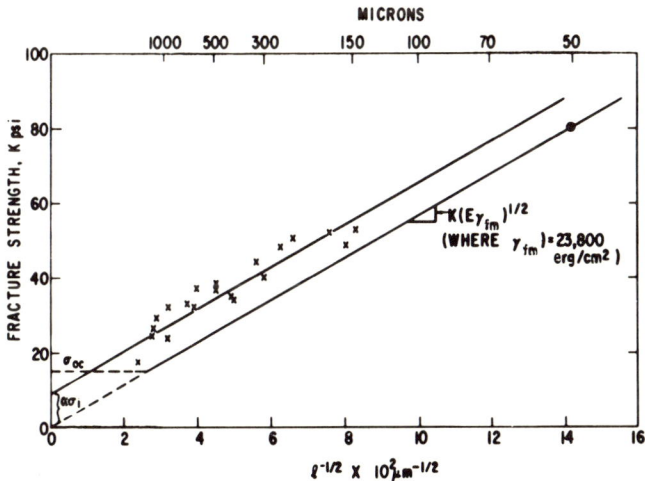

Fig.10   Strength of β-SiC containing isolated and strength determining grains of α-SiC. E is the Young's modulus of SiC, γ_{fm} the
fracture energy, K a geometrical constant and ℓ is the length of
the α-SiC grain initiating fracture.

The plot also shows the calculated value of σ_{ac} which represents the level of stress above which the initial cracks are propagated but arrested at the grain-matrix boundary. Since σ_i has been estimated at ± 13 kpsi, the ordinate intercept of the above line through the data points gives a value for α of $9/13 \simeq 0.7$. As indicated previously, this is a reasonable value and tends to confirm the importance of considering both the tensile and compressive components of residual stresses in analyzing fracture behavior of brittle materials. In the present case for Type B material, it indicates that once σ_{ac} is exceeded and a crack propagates in a surface grain, a compressive residual stress assists in strengthening the material.

Also included in Figure 10 is a point corresponding to a fracture stress of 75 kpsi. Since this value results from the materials (Types A and C) which are fine-grained and do not contain α-SiC tabular crystals, this point is placed on the lower curve representing anticipated fracture of these kinds of materials. The defect size corresponding to this point is about 50μ in length. Although some uncertainty exists relative to the origin of such a strength-controlling defect, we tentatively attribute it to surface damage arising from the original machining with a 220-grit diamond wheel. Such a large defect size indicates that marked improvements in the strength of fine-grained, hot-pressed SiC should be attainable with improved surface finishing.

The above experimentation and analysis of fracture behavior relates to a material which is inhomogeneous and, in fact, two-phase. The question arises as to whether or not the conclusions may be generalized to apply to other polycrystalline materials which are single phase. Examination of the grain size-strength data for many materials with elastic anisotropy indicates that the majority of materials studied[2,3] show a positive ordinate intercept on plots of strength vs the square root of grain size for specimen sizes much larger than the grain size. This intercept may most easily be interpreted as resulting from some compressive residual stress affecting the fracture process. The foregoing analysis suggests that a polycrystalline material may develop spatially random regions of residual stress (necessarily both tensile and compressive) and that these regions of residual stress and that these regions will be separated at distances approximately equal to the grain size. More importantly, for grain sizes smaller than some critical size, ℓ_c, the effect of only the compressive residual stress components can be reflected in a strength-(grain size)$^{-\frac{1}{2}}$ plot and it is these components that determine the ordinate intercept. Thus, it is unnecessary to attribute a positive intercept in every case to yielding in terms of the mechanism proposed by Petch[4] or to the effect of a uniform compressive surface layer on a specimen which is balanced by a uniform subsurface tensile stress. The model also indicates that for grain sizes greater than ℓ_c, fracture occurs at a very low stress, σ_{ac}, and by essentially a different mechanism. In this case, a fracture, once initiated at a surface, passes unimpeded through the sample. This is in contrast to the previous case where continued fracture propagation requires a stress considerably greater than that necessary to temporarily grow a surface defect to a depth about equal to a grain dimension.

The above discussion, in line with experience, emphasizes the need to develop uniform, fine-grain size in ceramics if high strengths are desired. It also draws attention to another important and desirable effect of fine-grain size which is less obvious. In general, the processing of anisotropic materials to fine-grain size results in a more equiaxed grain morphology than if exaggerated grain growth is allowed. The stress concentration factor, F, for such morphologies can be much reduced and thus the effect of residual stresses can also be advantageously reduced.

CONCLUSIONS

Several SiC materials have been examined with respect to their fracture behavior and it is concluded that in order to achieve high-strengh, these materials must be processed to fully suppress exaggerated grain growth of related SiC phases. This is of particular importance if surface damage cannot be prevented, since cracks in large surface grains are highly sensitive to residual stresses and can propagate at very low applied stresses.

It is further concluded that both the compressive and tensile components of residual stresses must be given equal emphasis in analyzing the step-wise process by which failure proceeds in poly-crystalline materials with elastic anisotropy.

ACKNOWLEDGMENTS

We gratefully acknowledge the assistance of W. Dondalski and C. Bobik for the preparation and testing of the materials in this work. The research reported herein was conducted under NASC Contract No. N000019-72-C-0129 with Mr. I. Machlin as the responsible monitor.

REFERENCES

1. E. Orowan, Reports on Progress in Physics, 12, 185 (1948).
2. S. C. Carniglia, J. Am. Cer. Soc. 55, 243 (1972).
3. R. W. Rice, Proc. Brit. Cer. Soc. 20, 205 (1972).
4. N. J. Petch, J. Iron Steel Inst., London, 174, 25 (1953).

5. A. G. Evans, G. Tappin, Proc. Brit. Cer. Soc. 20, 275 (1972).

6. S. Prochazka, Rept. SRD-72-171, Dec. 1972.

7. H.Y.B. Mar, W. D. Scott, J. Am. Ceram. Soc. 53, 555 (1970).

8. A. H. Gomes DeMesquita, Acta. Cryst. 23, 610 (1967).

9. J. Selsing, J. Am. Cer. Soc., 44, 419 (1961).

10. A. Taylor, R. M. Jones in "Silicon Carbide," J. R. O'Connor and J. Smiltens, Eds., 1960.

11. D.P.H. Hasselman, "Anisotropy of Single Crystals," W. Vahldiek, Ed., 1968.

12. K. B. Toplygo, Soviet Phys.-Solid State 2, 2367 (1961).

13. G. Arlt, G. R. Schodder, J. Ac. Soc. Am. 37, 384 (1965).

14. J. F. Nye, "Physical Properties of Crystals," 1957.

15. H. B. Huntington, Solid State Physics, 7, 213 (1958).

16. R. A. Alliegro, L. B. Coffin. J. R. Tinklepaugh, J. Am. Cer. Soc., 39, 386 (1956).

17. T. D. Gulden, J. Am. Cer. Soc., 52, 591 (1969).

18. R. A. Sack, Proc. Phys. Soc. (London), 58, 729 (1946).

19. S. M. Wiederhorn in "Ultrafine-Grain Ceramics," J. J. Burke, N. L. Reed, V. Weiss, Eds., 1970, p. 317.

20. J. R. McLaren, G. Tappin, R. W. Davidge, Proc. Brit. Cer. Soc., 20 (1972).

21. J. A. Coppola, R. C. Bradt, J. Am. Cer. Soc., 55, 455 (1972).

22. R. D. Carnahan, J. Am. Cer. Soc., 51, 223 (1968).

CRITERIA FOR CRACK EXTENSION AND ARREST IN RESIDUAL, LOCALIZED
STRESS FIELDS ASSOCIATED WITH SECOND PHASE PARTICLES

F. F. Lange

Westinghouse Research Laboratories

Pittsburgh, Pennsylvania   15235

ABSTRACT

An energy balance approach is taken to determine the
conditions in which crack extension will occur within the highly-
localized, residual stress field associated with second phase
particles embedded within a matrix phase. The unique result of
this analysis is that the particle size R is one of the factors
that governs the criterion of crack instability. The maximum
stress (σ), the elastic properties of the two phases, the fracture
energy of the cracked phase, and the size of the pre-existing crack
are the other factors. It is shown that for a given material,
crack extension will not occur unless $\sigma^2 R \geq$ a constant, regardless
of the pre-existing crack size. Once this condition is satisfied,
the size of the pre-existing crack will then govern the stress
required to extend the crack. It is also shown that crack arrest
will occur once the above conditions are satisfied and that the
length of the stable crack depends on the particle size. The
implications of these results are briefly discussed in relation
to observed phenomena.

INTRODUCTION

Second phase particles are commonly found in ceramic
materials. Many 'classical' and several 'new' ceramic materials
can be classified as particulate composites, i.e., a dispersed
particulate phase within either a glassy or a polcrystalline
matrix. 'Single-phase' ceramics also contain second phase

particulates which are accidently incorporated during powder preparation and fabrication.

Because the thermal expansion of the particle phase is usually different from that of the matrix phase, highly localized stresses arise as the body is cooled from its fabrication temperature; thus, the second phase particles are usually considered as precursors to cracks that form either during cooling or during subsequent stressing.[1] The criteria which governs the formation of these unwanted cracks are therefore important.

Observations made by Binns [2] and later confirmed by Davidge and Green [3] are significant relative to understanding these criteria. They showed that during the cooling of several different particulate composites, cracks only formed adjacent to the larger particles and not the smaller particles. This size effect appeared to be inconsistent with the usually accepted Griffith fracture equation [4,5] which would indicate that since the maximum tensile stress is independent of particle size [6,7] cracks should be observed adjacent to all particles, regardless of their size. Davidge and Green recognized this inconsistency and pointed out that since the stresses associated with the particles are highly localized, only a limited amount of stored strain energy is associated with a particle and therefore available for crack extension. They showed that the residual stored strain energy (U_{SE}) depends on the size of the particle cubed, whereas the energy required to form the crack (U_S) can be related to the size of the particle squared. By hypothesizing that the same relative crack, e.g., a hemispherical crack, would form around each particle and that

$$U_{SE} \geq U_S, \tag{1}$$

they concluded that cracks will only form around particles greater than a critical size, R_c. Although this conclusion was consistent with the general experimental observations, it can be shown that Eq. (1) can also be satisfied for any particle size as long as the crack size, which also governs the magnitude of U_S, is sufficiently small. That is, Eq. (1) predicts that smaller cracks (relative to the particle size) should be observed adjacent to smaller particles. Since this is not observed, their concept is not consistent and it is therefore, incomplete.

The object of this article is to present a general analysis for crack extension in the stress fields associated with a spherical particle embedded within a homogeneous matrix material. Its basic form is similar to that first proposed by Griffith [4] in which the energy of the body is examined as a function of crack

length to obtain the conditions for which crack instability
(extension) and stability (arrest) will occur.

STRAIN ENERGY ASSOCIATED WITH PARTICLES

Due to the difference in the thermal expansion of the
particle and matrix phases, stresses arise within and around the
particle as the body is cooled from its fabrication temperature.
For the case of a single, spherical particle of radius R within an
infinite matrix, a uniform isostatic stress σ arises within the
particle and radial and tangential stresses of $-\sigma R^3/r^3$ and
$\sigma R^3/2r^3$, respectively, arise within the surrounding matrix.[6,7]
The stress σ depends on the differential thermal expansion
coefficients of the two phase ($\alpha_p - \alpha_m = \Delta\alpha$), the elastic constants
of the two phases ($E_{m,p}$, $\nu_{m,p}$) and the temperature change (ΔT):

$$\sigma = \frac{\Delta\alpha\Delta T}{k} \tag{2}$$

where $k = (1+\nu_m)/2E_m + (1-2\nu_p)/E_p$. (Subscripts m and p denote
matrix and particle, respectively). σ is a tensile stress (defined
as positive) where $\alpha_p > \alpha_m$ and a compressive stress (negative)
where $\alpha_m > \alpha_p$. It should be noted that the maximum stress σ is
independent of particle size.

The method of calculating the strain energy within the particle
(U_p) and within the matrix (U_m) has been reported by Davidge and
Green.[3] Their results are:

$$U_p = 2\pi \frac{\sigma^2(1-2\nu_p)}{E_p} R^3 \tag{3}$$

and

$$U_m = \pi \frac{\sigma^2(1+\nu_m)}{E_m} R^3. \tag{4}$$

Thus, the total stored strain energy ($U_p + U_m$) associated with the
particle is

$$U^\circ_{SE} = 2\pi k \sigma^2 R^3. \tag{5}$$

CONDITIONS FOR CRACK EXTENSION AND ARREST

It will be assumed that a pre-existing crack, which is both
specified by a single dimensional parameter (c) and much smaller
than the particle radius (c << R), is associated with the particle
and favorably oriented on a plane perpendicular to the tensile

stresses.* Although it is not necessary for this argument, its most probable location would be at the particle-matrix interface.

If the pre-existing crack is allowed to extend into the surrounding stress field, the initial stored strain energy U_{SE}^{o} will be reduced due to the diminished stress field. The strain energy associated with the particle for a given normalized crack length $\mu = c/R$ can be expressed as

$$U_{SE} = U_{SE}^{o} \, f(\mu),\qquad (6)$$

where $f(\mu)$ is a dimensionless function and by definition, $1 \geq f(\mu) \geq 0$.

The energy required to create the new crack surfaces is

$$U_S = \gamma A,\qquad (7)$$

where γ = the fracture energy of the phase into which the crack is extending and A = the surface area of the crack. U_S can be expressed as a multiple of the particles surface area $(4\pi R^2)$ as

$$U_S = 4\pi\gamma R^2 \, g(\mu)\qquad (8)$$

where $g(\mu)$ is a dimensionless function of the normalized crack length. Again, by definition, $g(\mu) \geq 0$.

Neglecting other energy terms which are assumed not to change during crack extension, the total energy associated with the system for a given relative crack size is

$$U_T = U_{SE} + U_S = 2\pi k \, \sigma^2 R^3 \, f(\mu) + 4\pi\gamma R^2 \, g(\mu)\qquad (9)$$

Using the thermodynamic criterion introduced by Griffith,[4] crack extension will only occur when it is accompanied by a free energy change ≤ 0, i.e., for an unstable extension of the normalized crack considered here

$$\frac{\partial U_T}{\partial \mu} = 2\pi k \, \sigma^2 R^3 \, f'(\mu) + 4\pi\gamma \, R^2 \, g'(\mu) \leq 0\qquad (10)$$

This condition must correspond to a maximum in the U_T vs μ curve where

$$\frac{\partial^2 U_T}{\partial \mu^2} = 2\pi k \, \sigma^2 R^3 \, f''(\mu) + 4\pi\gamma \, R^2 \, g''(\mu) < 0\qquad (11)$$

---

*The tensile stress distribution depends on the relative values of α_m and α_p, which determines the fracture plane during crack extension.

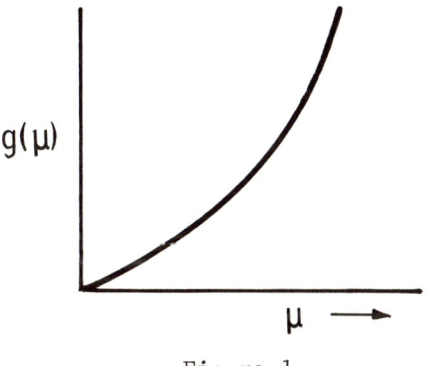

Figure 1

Likewise, crack arrest will occur when $\partial U_T/\partial\mu = 0$ and $\partial^2 U/\partial\mu^2 > 0$. This corresponds to a minimum position in the U_T vs μ curve. A third condition may also exist where the U_T vs μ curve exhibits neither a maximum nor minimum point, but only a single inflection. This arises when $\partial U_T/\partial\mu = 0$ and $\partial^2 U_T/\partial\mu^2 = 0$. $f'(\mu)$, $g'(\mu)$ and $f''(\mu)$, $g''(\mu)$ are the first and second derivatives of $f(\mu)$ and $g(\mu)$ with respect to μ.

Up to this point, no assumptions have been made concerning the specific functional forms of either $f(\mu)$ or $g(\mu)$. It is beyond the scope of this article to derive their explicit forms. This would require specific knowledge of the crack front shape and knowledge of the stress redistribution during crack extension.* The purpose of the following paragraphs will be to indicate the general forms of these functions in order to draw important conclusions from the previous analysis.

The function $g(\mu)$ is the fractional increase in the surface area. If it is assumed that the geometry of the crack's periphery does not change in such a way as to decrease the rate of area formation during crack extension, the function $g(\mu)$ will not possess an inflection point, viz., $g'(\mu) > 0$ and $g''(\mu) \geq 0$. This is a reasonable assumption based on known crack geometries. An example of such a function is shown in Fig. 1.

----

*For particles in tension ($\alpha_p > \alpha_m$) cracks can occur either within the particle or within the matrix. For particles in compression ($\alpha_m > \alpha_p$), cracks should only occur within the matrix.

The function $f(\mu)$ is defined as the fractional release of
stored strain energy during crack extension. This function must
possess two properties. First, by definition, $f'(\mu)$ is always
< 0, i.e., at no point during crack extension can the stored strain
energy be regained. Second, since the magnitude of the stress
decreases during crack extension, very little of the strain energy
will be relieved once the crack becomes large ($\mu >> 1$). Thus
$f(\mu) \to f_\infty$ and $f' \to 0$ when $\mu >> 1$. Figure 2a illustrates the case
where the crack is coincident with the particle-matrix interface
(one of the specific cases when $\alpha_p > \alpha_m$). In this case, when the
crack completely encompasses the particle, all of the strain energy
is relieved and $f_\infty = 0$. For the case illustrated in Fig. 2b,
where the crack extends in a radial direction (the case where
$\alpha_m > \alpha_p$), the total strain energy can never be completely relieved
and thus, $f_\infty > 0$.

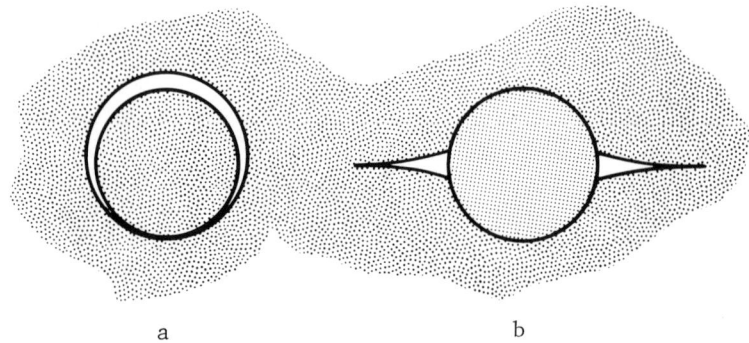

a   b

Figure 2

Although not implicit by definition, several strong reasons
can be given that $f(\mu)$ must possess at least one inflection point.
First, in analyzing Eqs. (10) and (11), it can be seen that $f''(\mu)$
must be < 0 over some range of μ in order to have crack instability.
Otherwise, a pre-existing crack would always extend in a stable
manner as the body is cooled. Since cracks are not observed
adjacent to particles in composite systems where $\Delta\alpha$ is small,
and only adjacent to large particles when $\Delta\alpha$ is large, [2,3] crack
extension must occur in an unstable manner after the stress reaches
some defined value. This means that $f(\mu)$ must contain an inflection
point. The second reason is related to crack propagation in a
uniform stress field. For penny-shaped cracks in uniform tension,
the magnitude of the strain energy release rate increases with

increasing crack size.[5] In the highly localized stress fields
associated with second phase particles, the strain energy release
rate, which is directly proportional to f(μ), will depend on both
the crack size and the stress gradient. Thus, it follows that the
magnitude of the strain energy release rate will be small for both
small and large cracks and exhibit a maximum value for some
intermediate crack size. This means that f(μ) will possess
an inflection at some intermediate value of μ. Without proof of
these arguments, it will be hypothesized that f(μ) does possess a
single inflection similar to the form shown in Fig. 3.

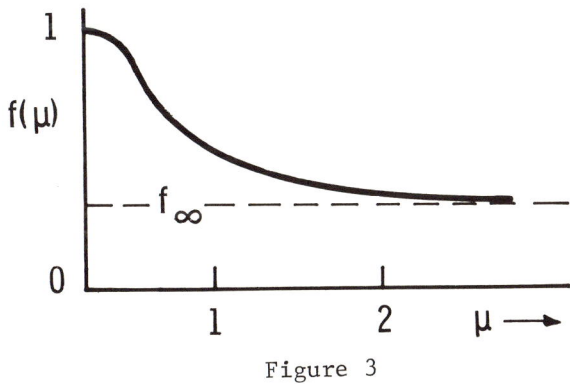

Figure 3

Based on this hypothesis, the conditions for crack extension
and arrest will now be examined. The first and most important
property of the function U_T is that it can possess a single
inflection point without exhibiting either a maximum or a minimum,
i.e., when $\partial U_T/\partial \mu = 0$ and $\partial^2 U_T/\partial \mu^2 = 0$, as shown in Fig. 4. This
is the limiting condition where crack extension will not occur
regardless of the size of the pre-existing crack. By rearranging
Eq. (10) and (11), an important relation between the maximum
stress and the particle size is obtained:

$$\sigma^2 R = \frac{2\gamma}{K} H(\mu_0) \tag{12}$$

where $H(\mu_0) = - g'(\mu_0)/f'(\mu_0) = - g''(\mu_0)/f''(\mu_0)$ and μ_0 defines
the inflection point. Since the right hand side of this relation
is a constant for a given particle-matrix phase pair, the first
condition that must be satisfied is that

$$\sigma^2 R \geq \text{constant.} \tag{13}$$

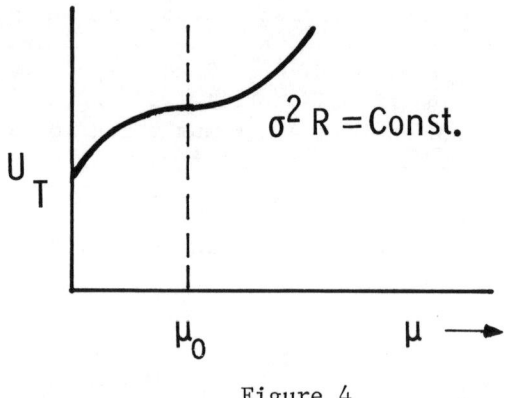

Figure 4

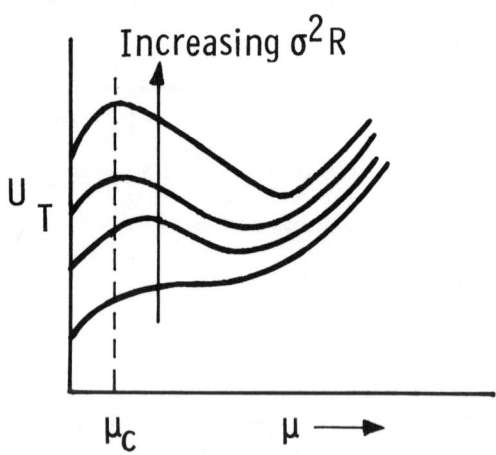

Figure 5

Once this is satisfied, U_T will possess a maximum and a minimum
point corresponding to values of μ where crack extension and crack
arrest will occur. By examining Eq. (9), the maximum and minimum
will shift to lower and higher values of μ, respectively as the
product $\sigma^2 R$ is increased. Thus, once Eq. (13) is satisfied, the
size of the pre-exist crack, μ_c, will determine the value of $\sigma^2 R$
where crack instability will occur and it will also govern the
final length of the stable crack. This is schematically
illustrated in Fig. 5.

Thus, by hypothesizing that the fractional strain energy release rate function, $f(\mu)$, possess an inflection point, it has been shown that two conditions are required for crack extension. First, the product $\sigma^2 R$ must be greater than a given value which depends on the properties of the two phases and the mode of crack propagation. Second, once this condition is satisfied crack extension will occur when the maximum in the U_T vs μ curve is shifted by increasing the value of $\sigma^2 R$ to the value μ_c defined by the size of the pre-existing crack. At this value, the crack will extend until its normalized size reaches the minimum position in the U_T vs μ curve. Between these two values of μ, the condition for unstable crack extension $(\partial U_T/\partial \mu \leq 0)$ is maintained.

DISCUSSION

The results of the previous analysis will now be applied to a body containing a widely spaced dispersion of second phase particles. During cooling, the same stress, σ, will arise within each particle regardless of its size (see Eq. (2)). Once the condition that $\sigma^2 R >$ constant is met for the largest particle, a crack will form upon subsequent cooling when the stress is increased to the critical value necessary to extend the pre-existing crack. Likewise, cracks may form around the next largest particle and subsequently smaller particles as the body is cooled to room temperature. Thus, at a given temperature, cracks will only be observed adjacent to the larger particles despite the fact that an unusually high stress may exist within and around the smaller particles which did not satisfy the conditions for crack instability. If it is assumed that the same pre-existing, normalized crack length μ_0 exists adjacent to each particle, a critical particle size R_c exists below which cracks are not observed:

$$R_c = \frac{\text{constant}}{\sigma^2} .$$

Since one would expect a more statistical crack size distribution, some of the particles larger than R_c, which have normalized crack sizes smaller than μ_0, will not form cracks. Thus, when the body is cooled to room temperature cracks will only be observed adjacent to larger particles. Cracks may or may not be observed adjacent to particles which are $\sim R_c$. This is in direct agreement with the observations of Davidge and Green.[3]

The subsequent stressing of such a body can be viewed as follows. Neglecting the inhomogeneous stress fields that arise around each particle due to the difference in the elastic properties of the two phases, the externally applied stress (σ_a) can be super-positioned onto the residual stresses. Thus, Eq. (13) can be re-written as:

$$(\sigma + \sigma_a)^2 R > \text{constant} \tag{14}$$

to show that particles that have not already cracked will form stable cracks once this condition is met.* Thus, during stressing, larger particles will form stable cracks first. Subsequently cracks will form around smaller particles as the applied stress is increased until fracture. Although many cracks will exist within the body prior to fracture, fracture will originate from the largest particle, which, as shown in Fig. 5, produces the largest, stable crack. This analysis suggested that the size distribution of second phase particles can be obtained by monitoring the acoustic emission of the body during loading. It also indicates that acoustic emission may be a useful tool in proof testing.

Other implications can also be drawn from the analysis. First, it suggests that second phase particles are not necessarily crack precursors as long as the particle size is very small. This means that strong particulate composites can be fabricated regardless of their differential thermal expansion. This has already been shown to be a fact for the Si_3N_4-SiC [8] and the $SiC-Al_2O_3$ [9] composite systems.

Second, although the analysis was presented to explain crack extension adjacent to second phase particles, it can also be applied to other highly localized stress fields as for example, Hertzian contact stresses. For this specific example, a result,[10] similar to Eq. (13) predicts the load-indentor size relation known as Auerbach's Law (see e.g., Ref. (11)).

In conclusion, further effort in this area might best be directed towards determining the explicit function $f(\mu)$ and $g(\mu)$ in order that this theory might be more quantitative and of greater engineering use.

ACKNOWLEDGMENT

This work was supported by the Office of Naval Research, Contract No. N00014-68-C-0323.

---

*The sign of σ_a is important. If it is opposite to that of σ, a portion of the residual strain energy will be relieved by the applied stress and the total stress $(\sigma + \sigma_a)$ is reduced.

REFERENCES

1. F. F. Lange, "Strength Behavior of Brittle Matrix, Particulate Composites", Fracture and Fatigue of Composites, Ed. by L. J. Broutman and R. H. Knock, Academic Press (in press).

2. D. B. Binns, "Some Physical Properties of Two-Phase Crystal-Glass Solids", Science of Ceramics, Ed. by G. H. Steward, Vol. 1, pp. 315-335, Academic Press (1962).

3. R. W. Davidge and T. J. Green, "The Strength of Two-Phase Ceramic/Glass Materials", J. Mat. Sci. 3, 629 (1968).

4. A. A. Griffith, "The Phenomena of Rupture and Flow in Solids", Phil. Trans. Roy. Soc. Lond. 221A, 163 (1920).

5. R. A. Sack, "Extension of Griffith's Theory of Rupture to Three Dimensions", Proc. Phys. Soc. Lond., 58, 729 (1946).

6. V. D. Weyl, "Uber den Einfluss innerer Spannunger auf das Gefuge and die mechanische Festigkeit des Prozellans", Ber. dent. Keram. Ges. 36, 319 (1959).

7. J. Selsing, "Internal Stresses in Ceramics", J. Am. Ceram. Soc., 44, 419 (1961).

8. F. F. Lange, "The Si_3N_4-SiC Composite System: Effect of Microstructure on Strength", J. Am. Ceram. Soc. (in press), (1973).

9. F. F. Lange, "The Strength and Fracture Mechanics of SiC Hot-Pressed with Al_2O_3", ONR Technical Report No. 10, Contract No. N00014-68-C-0323, (1973).

10. F. F. Lange, "Criterion for Crack Extension in Hertzian Stress Fields", unpublished.

11. J. P. A. Tillet, "Fracture of Glass by Spherical Indentors", Proc. Phys. Soc. Lond., B69, 47 (1956).

V. SUBCRITICAL CRACK GROWTH

SUBCRITICAL CRACK GROWTH IN CERAMICS

S. M. Wiederhorn

Institute for Materials Research
National Bureau of Standards
Washington, D. C. 20234

ABSTRACT

Subcritical crack growth that causes delayed failure is
discussed in terms of fracture mechanics concepts. Techniques
of characterizing subcritical crack growth are presented and
the available crack growth data are discussed with particular
emphasis on fracture mechanisms. Finally a design technique is
presented to predict useful component lifetime from crack growth
data after proof testing.

Key Words: Delayed failure, fracture, crack growth, ceramics,
            proof testing.

1. INTRODUCTION

Brittle fracture of ceramic materials is often preceded by
subcritical crack growth and this results in a time dependence of
strength,known as delayed failure. Because delayed failure can
occur without warning weeks or even months after the first application of load, this type of failure must be understood to avoid
structural failure of ceramic components. One means of gaining
this understanding is through fracture mechanics,
to characterize the subcritical crack growth that causes the
failure. Once the growth has been characterized the mechanisms
that control crack growth may be understood and design criteria
can be developed to prevent delayed failure.

Fracture mechanics is of value in characterizing subcritical
crack growth because the crack tip stresses that cause crack growth

are directly proportional to the stress intensity factor, K_I [1].
Therefore, the crack motion can be related to the crack tip
stresses by expressing the crack velocity as a function of the
stress intensity factor. This fundamental relationship between
the crack tip stress and the stress intensity factor accounts
for the unique correlations that are found experimentally between
the crack velocity and the stress intensity factor [2]. As noted
earlier by Johnson and Paris [2], the stress intensity factor
is the controlling mechanical parameter for crack extension.

This paper presents a general review of subcritical crack
growth in ceramic materials and shows how these data may be
used for failure prediction. The experimental techniques used
to obtain crack growth data are described; crack growth data
are discussed with particular regard to fracture mechanisms
and, a method of using crack growth data to develop design criteria
is presented. Throughout the paper, the use of fracture mechanics
to obtain crack growth data and to develop failure criteria
is emphasized.

2.   EXPERIMENTAL TECHNIQUES FOR OBTAINING CRACK PROPAGATION DATA

Crack propagation data may be obtained either by direct
or indirect methods. The direct methods require crack velocities
to be measured as a function of stress intensity factor on fracture
mechanics type specimens. These specimens contain macroscopic
cracks allowing accurate measurement of the crack velocity and
stress intensity factor. Parameters describing crack growth
are determined directly from the crack growth data. The indirect
methods of obtaining crack growth data require crack propagation
parameters to be inferred from strength measurements on specimens
that simulate structural components. Both methods have their
advantages. Small details of fracture behavior are easily observed
by direct techniques.   These details are often missed when
indirect methods are used because only the average crack behavior
(for the total failure time) is measured. Since indirect methods
use specimens that closely simulate structural parts, however,
crack behavior is the same in the test specimens and structural
parts. Consequently, an assumption of common behavior for large
and small cracks is not necessary. It is advised therefore
that data from both direct and indirect techniques should be
obtained for unequivocal interpretation of experimental data
and for certainty of structural design.

2.1 Direct Methods of Obtaining Crack Velocity Data

Crack propagation in ceramic materials has been studied on
a number of different specimen geometries. These include the

double cantilever specimen, the edge cracked tensile specimen,
the center cracked tensile specimen and the double torsion specimen.
Of these, the double torsion and double cantilever specimens
are the geometries most frequently used on ceramic materials
because stable crack growth in these specimens facilitates data
acquisition.

Two principal methods of loading are used to obtain
crack growth data. In one method, the displacement of the loading
points is fixed and the crack velocity is measured from the load
relaxation as the crack grows [3]. In the second method, a constant
load is applied to the specimen and the crack velocity is measured
by direct optical observation of the crack tip. The constant
load technique is insensitive to small changes in temperature;
therefore, crack velocity measurements can be made over long
periods of time, without errors resulting from temperature changes.
In contrast, fixed displacement techniques are sensitive to
temperature changes which cause variations in the displacement
due to thermal expansion of the loading instrument. Consequently,
low velocity studies by constant displacement techniques require
equipment with long term temperature stability. The main advantage
of the constant displacement technique is that measurements can
be made without direct observation of the crack tip. This permits
crack velocity measurements on opaque materials and in hostile
environments where visibility is not possible.

2.1.1 The Double Cantilever Beam Technique

The double cantilever beam technique (see figure 1) was first
used by Gilman [4] to measure the energy for fracture of a

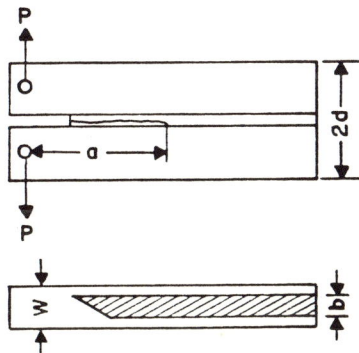

Figure 1. Double cantilever beam specimen configuration.
Cross-hatched area represents the uncracked
portion of the specimen between the guiding
notches.

variety of crystalline materials. Gilman obtained the energy
for fracture by approximating a solution for the double cantilever
beam specimen. Gilman's solution, which assumes that the elastic
energy of the specimen results solely from bending of the beams,
is adequate for cracks that are long with respect to the specimen
height (see figure 1). However, a more complete solution for
the double cantilever beam configuration [5,6] demonstrates.a
significant contribution to the elastic energy from shear stresses
in the beams and from stresses in regions of the specimen past
the crack tip. The stress intensity factor, K_I, and the displacement of the loading points, y, is given by the following
equations [6,7]:

$$K_I = (Pa/w^{1/2}b^{1/2}d^{3/2})(3.467+2.315d/a) \qquad (1)$$

$$y = 4Pa^3/Ewd^3 + 7.92Pa^2/Ewd^2 + 2.65Pa/Gwd \qquad (2)$$

where E and G are Young's modulus and the shear modulus respectively,
and the other terms are defined in figure 1.

Fixed loading conditions are usually used for crack velocity
studies with the double cantilever beam technique [7]. The crack
position is monitored optically and the crack velocity is calculated
from the time necessary for the crack to move a measured length
increment. These increments must be small in relation to the
crack length in order to avoid large changes in the stress
intensity factor. Crack length increments of 0.3 mm are usual
for cracks 2 to 4 cm long, which results in a crack velocity
error of approximately 3 percent, and an error in the stress
intensity factor due to crack growth of approximately 1
percent.

A modification of the double cantilever technique has been
proposed by Freiman, Mulville and Mark [8] who applied a constant
moment to the crack arms of the double cantilever rather than a
constant load. Since the stress distribution in the bent beam
is constant for an applied moment, the stress intensity factor
for this type of specimen depends only on the applied moment and
not on the crack length. Specimens having this property are
referred to as constant K_I specimens.

2.1.2 The Double Torsion Technique

The double torsion technique uses the same type of specimen
as the double cantilever technique, however, torsional loading

is used to propagate the crack, rather than tensile loading [9,10].
Load application is by four or three-point bending at the cracked
end of the specimen (see figure 2). A partial solution is available

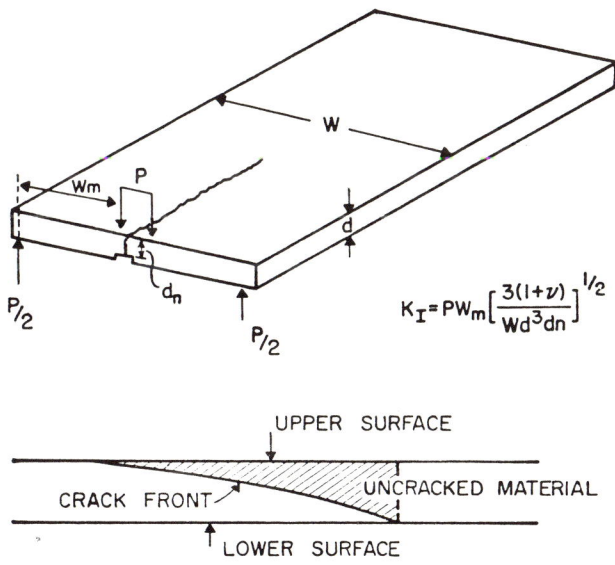

$$K_I = PW_m \left[\frac{3(1+\nu)}{Wd^3dn} \right]^{1/2}$$

UPPER SURFACE

CRACK FRONT   UNCRACKED MATERIAL

LOWER SURFACE

Figure 2.   The double torsion specimen. The crack front
            curvature is depicted in the lower portion of the
            figure.

for this crack geometry. McKinney and Smith [11] have recently
shown that this solution gives an accurate prediction of the
displacement of the loading points as a function of crack length.
Under a constant applied torsion, the displacement of the loading
points is linearly proportional to the crack length and the
applied load P [9].

$$y = P(Ba + C) \tag{3}$$

where B and C are constants.

Since $K_I = \sqrt{EG}$ and $G = (P^2/2t)(\partial\lambda/\partial L)$ [1]—where λ is the
specimen compliance (Ba + C) and G is the strain energy release
rate — it follows that the stress intensity factor for the
double torsion configuration depends only on the applied load
and not on the crack length. The double torsion configuration
is thus a constant K_I specimen. The shape of the crack is
depicted in figure 2. Despite this curved crack front, crack
velocity data obtained with these specimens are identical to

data obtained with the double cantilever beam specimens [9].
The relationship between the stress intensity factor and
the applied load P is

$$K_I = PW_m\left[\frac{3(1+\nu)}{Wd^3d_n}\right]^{1/2} \tag{4}$$

where ν is Poisson's ratio, and the other symbols are given
in figure 2.

The double torsion technique was first devised by Outwater
[12] who used a constant load to study crack growth in glass
and other materials. The crack growth rate was monitored
optically with a traveling microscope and the data was presented
as the strain energy release rate, G, versus the logarithm
of the crack velocity. This technique has been used in the
same way by other investigators [13], but has no advantage
over the double cantilever beam technique under these conditions.
By contrast, Evans [9] has shown that when a constant displacement
or a constant displacement rate is used to propagate the crack,
crack growth can be measured directly from the load, P, or
the load relaxation rate $\dot{P}$. In this way both the crack velocity
and the stress intensity factor can be measured from the load.
This method provides a simple way of monitoring crack growth
when direct visual observation of the crack tip is not possible.

The theory behind the fixed displacement or constant
displacement rate methods of measuring crack growth is not
complex [9,10]. From equation (3) the time derivative of the
displacement can be obtained:

$$\dot{y} = \dot{P}(Ba+C) + PB v \tag{5}$$

For a fixed displacement, $\dot{y} = 0$, the crack velocity, v, is
given by

$$v = -\dot{P}(Ba+C)/PB \tag{6}$$

For constant displacement $(Ba+C) = P_f(Ba_f+C)/P$, where P_f
and a_f are simultaneous measurements of load and crack length
at any time after the displacement has been fixed.   Therefore,
substituting for the specimen compliance, $(Ba+C)$ in equation
(6), and recognizing that $C<<B$, the following equation is
obtained for the crack velocity:

$$v = -\dot{P}P_f a_f/P^2 \tag{7}$$

To use the above equation: the specimen is quickly loaded until crack propagation occurs; the displacement is fixed by turning off the test machine; and the time rate of load change is measured from the strip chart recorder of the test machine. After the experiment has been completed, the crack length is measured to determine a_f. P_f is then the final load on the specimen. The stress intensity factor is then obtained from the strip chart recorder using equation (4).*

The double torsion specimen may also be used by applying a constant rate of displacement to the test specimen [9]. When the rate of machine displacement is just compensated by the specimen compliance change due to crack growth, then the load becomes constant and the following equation for crack velocity is obtained from equation (5):

$$v = \dot{y}/PB \tag{8}$$

In this way the crack velocity is measured from the constant load, P, and the rate of motion, $\dot{y}$, of the cross head of the testing machine. The constant B is determined analytically [10]:

$$B = 3W_m^2 a/Wt^3 G \tag{9}$$

Equation (8) may also be used by fixing the load, P, and measuring the displacement rate $\dot{y}$. $\dot{y}$ will be constant if crack growth is not accompanied by gross plastic flow [14]. A limitation of the constant level and constant displacement rate techniques is that few determinations can be made on each specimen, because the crack growth that occurs during each measurement is extensive.

The constant rate of displacement method can be combined to good effect with the constant displacement method to eliminate the crack length in quation (7). If the test machine is stopped after a constant load is obtained, then the

---

* The proportionality constant between K_I and P in equation (4) may be obtained experimentally by rapidly loading the specimen after the run has been completed [9]. The rapid fracture load, P_{IC}, divided into an independently measured value of K_{IC} then gives the proportionality constant, $D = K_{IC}/P_{IC}$. A comparison of D determined by this method and by equation (4) shows that the two methods give equivalent values.

specimen compliance at the instant the machine is stopped
is calculated by equating equations (6) and (8). In this manner,
$Ba_i + C = y/P_i = (Ba+C)P/P_i$, where a_i, P_i and P_i represent arrest
values for these parameters. Substituting for the compliance, (Ba+C),
in equation (6), the following equation is obtained for the crack
velocity after the machine displacement is fixed:

$$v = (\dot{P}/\dot{P}_i)(P_i/P)^2 v_i \qquad (10)$$

where v_i is the initial crack velocity determined from equation
(8).

2.2 Indirect Methods of Obtaining Crack Velocity Data

Two types of strength measurement have been used to obtain
crack velocity data. In one, a constant load is applied to the
specimen and the time to failure is determined as a function
of load. In the second, specimens are loaded at a constant strain
rate and the load is determined as a function of the strain rate.
To obtain crack growth information from these measurements, a
general crack propagation equation is required to relate the crack
velocity to the stress intensity factor. Strength data are then
used to determine the constants of the crack propagation equation.
For most ceramic materials the crack velocity can be expressed
as a power function of the applied stress intensity factor,
giving the following crack propagation equation [15]:

$$v = AK_I^n \qquad (11)$$

where A and n are the constants to be evaluated from strength
data.

2.2.1 Failure Under Constant Load [16]

The time to failure under a constant load can be derived
from the definition of crack velocity, v = da/dt, and the relationship between stress, σ, flaw size, a, and stress intensity
factor [9], K_I,

$$K_I = \sigma Y \sqrt{a} \qquad (12)$$

(Y is a constant that depends on the crack geometry [17].)
Substituting equation (12) into the definition of crack velocity

and using equation (11), one obtains the time to failure under a constant load [16]:

$$t = 2K_{Ii}^{2-n}/A\sigma_a^2 Y^2 (n-2) \tag{13}$$

K_{Ii} is the initial stress intensity factor at the most serious flaw in the ceramic and σ_a is the applied stress. Since the stress intensity factor and the applied stress are linearly related (equation (12)), the initial stress intensity factor is given by the following equation,

$$K_{Ii} = (\sigma_a/\sigma_{IC})K_{IC}, \tag{14}$$

where σ_{IC} is the critical fracture stress in an inert environment. By substituting equation (14) into (13) the following equation is obtained for failure time [16]:

$$t = 2\sigma_a^{-n}(K_{IC}/\sigma_{IC})^{2-n}/AY^2 (n-2). \tag{15}$$

n is conventionally determined from a logarithmic plot of failure time, t, versus applied load, σ_a; K_{IC} and σ_{IC} are evaluated from a second set of strength measurements and the constant A of equation (11) can then be calculated from equation (15).

Since mean or median values of σ_{IC} and t are usually used in equation (15), this method of determining the crack propagation constants, n and A, does not account for the strength distribution of the specimens tested. This may lead to erroneous values of n and A if, by some chance, the strength distribution at one load differs from the distribution at a different load. Recently, Evans and Wiederhorn [16] developed a method of handling constant load data which takes proper account of the statistical distribution of strength. Since failure probability is related to strength or time to failure, Evans and Wiederhorn note that strength data obtained at different load levels (or loading rates) can be compared only at equal failure probability. The failure probability is given by simple order statistics; $p_q(t) = q/Q+1$; where $p_q(t)$ is the cumulative failure probability for specimen q from an ordered set of Q specimens. Designating the order of the specimen in the group by q, the following equation is obtained directly from equation (15):

$$t_{q1}/t_{q2} = (\sigma_{a2}/\sigma_{a1})^n (\sigma_{ICq2}/\sigma_{ICq1})^{n-2} \qquad (16)$$

where σ_{a1} and σ_{a2} correspond to the two applied stress levels and σ_{ICq2} and σ_{ICq1} are the critical fracture strengths of the specimens. If the strength distributions of the specimens are the same at the two applied stress levels, then $\sigma_{ICq2} = \sigma_{ICq1}$ and a logarithmic plot of the time to failure at one stress level versus the time to failure at a second stress level should generate a straight line of slope unity [16].

$$\log t_{q1} = \log t_{q2} + n \log (\sigma_2/\sigma_1) \qquad (17)$$

If the slope of the data differs from unity, then the strength distribution at the two load levels are not identical and the data cannot be compared. The position of the straight line given by equation (17) can be used to calculate n.

Once the value of n has been determined, the constant A can be obtained from equation (15), by determining the distribution of critical strengths, σ_{IC}, and ordering them in the same way as the failure times. A logarithmic plot of the failure time at a stress σ_a versus the strength σ_a gives a straight line with slope of n-2.

$$\log t_q = (n-2)\log \sigma_{ICq} + \log[2\sigma_a^{-n}K_{IC}^{2-n}/AY^2(n-2)] \qquad (18)$$

The slope of this equation should confirm the value of n determined from equation (17) for validity and the position of the equation should give the value of the constant A, since σ_a, n, K_{IC} and Y have already been determined.

2.2.2 Constant Strain Rate Experiments [16]

The constant strain rate method of evaluating crack growth parameters was studied first by Charles [18], and later by Evans [15]. The technique is based on the fact that crack growth causes a strain rate dependence of the strength. Therefore, by measuring strength at various strain rates one may obtain the crack growth parameters.

The derivations given in this section, suggested to the author by Evans, are valid for ceramic materials containing

surface cracks, sufficiently small so that the Young's modulus
of the specimen is unaffected by the crack growth.* This assumption
is valid for many ceramic materials. It is also assumed that the
crack velocity can be expressed as a power function of the stress
intensity factor (equation (11)). This latter assumption is
necessary to separate variables that will appear in the
derivation.

At a constant rate of strain, the loading rate is directly
proportional to the strain rate, and the differential increase
in stress is proportional to the differential increase
in time, $d\sigma = \dot{\sigma}dt$, where $\dot{\sigma}$ is the rate of loading and is load
independent. The total derivative of the stress with respect to
crack length is $d\sigma/da = \dot{\sigma}/v$ and since the crack velocity is a power
function of the stress intensity factor,

$$d\sigma/da = \dot{\sigma}/AK_I^n = \dot{\sigma}/A\sigma^n Y^n a^{n/2} \qquad (19)$$

By separating the variables of this equation, the following
integrated equation for the fracture stress, σ, as a function of
strain rate, $\dot{\varepsilon}$, is obtained [15]:

$$\sigma^{n+1} = 2(n+1)E\dot{\varepsilon}\sigma_{IC}^{n-2}/AY^2 K_{IC}^{n-2}(n-2) \qquad (20)$$

where $\dot{\sigma} = E\dot{\varepsilon}$. It follows from equation (20) that a logarithmic
plot of strength versus strain rate should give a straight line
with a slope of $1/(n+1)$. Thus, the value of n is obtained from
strain rate data. As in the case of the constant load data, the
mean or median fracture is usually used at any given strain
rate. Once n has been determined, the constant A may be
evaluated from equation (20) provided values of K_{IC} and σ_{IC}
are available for the ceramic. Again this technique does not
account for the statistical distribution of strengths in the specimens tested, and is therefore liable to error if by some chance
the distribution of strengths at one strain rate differs from
those at another.

As in the case of the static loading experiments, the strength
distribution can be taken into account by ordering the specimens
that have been tested [16]. The ratio of strengths of specimens

---

* A more complete discussion of the subject is given in reference
  15.

tested at one strain rate to those tested at a second strain rate
is given by

$$\sigma_{q1}/\sigma_{q2} = (\dot{\epsilon}_1/\dot{\epsilon}_2)^{1/n+1} (\sigma_{ICq1}/\sigma_{ICq2})^{n-2/n+1} \tag{21}$$

where q gives the ordered position of the datum point. If the
strength distributions of the specimens at each strain rate are
identical then the rapid loading strength is identical for any
given position in the order, $\sigma_{ICq1} = \sigma_{ICq2}$, so that equation
(21) reduces to the following:

$$\sigma_{q1}/\sigma_{q2} = (\dot{\epsilon}_1/\dot{\epsilon}_2)^{1/n+1} \tag{22}$$

Therefore, a logarithmic plot of the strength at one strain rate
versus the strength at a second strain rate gives a straight
line of slope unity provided the distributions of strength within
each group is the same. The value of n is obtained from the
position of the line described by equation (22). Once n has been
determined the constant A may be evaluated from a second set of
experiments in which σ_{ICq} are determined. A logarithmic plot of
σ_q versus σ_{ICq} then gives a straight line of slope $(n-2)/(n+1)$,
as shown by the following equation [16]:

$$\log \sigma_q = [(n-2)/(n+1)] \log \sigma_{ICq}$$

$$+ [1/(n+1)]\log[2E\dot{\epsilon}(n+1)/AY^2 K_{IC}^{n-2}(n-2)] \tag{23}$$

Since n, Y, K_{IC}, E and $\dot{\epsilon}$ are known, the position of the line given
by equation (23) can be used to evaluate A.

3. REVIEW OF CRACK PROPAGATION DATA

Over the past eight years, crack propagation data has been
collected on a variety of ceramic materials including glasses,
porcelains, high alumina ceramics, and silicon nitride. These
data have been obtained in vacuum, in corrosive environments
and at high temperatures, and have been used to understand the
fracture process and to predict the lifetime of ceramic parts
under stress. In this section, the available crack propagation
data will be summarized and discussed with reference to crack
propagation mechanisms.

3.1 Glass

The greatest amount of crack propagation data on any ceramic
material has been collected on glass, because glass is homogeneous
and transparent permitting crack growth to be easily monitored.
Crack velocity studies in nitrogen gas [19,20] containing various
percentages of water vapor showed that there are three characteristic regions of crack growth in glass. At low values of the
stress intensity factor, region I (see figure 3), the crack growth

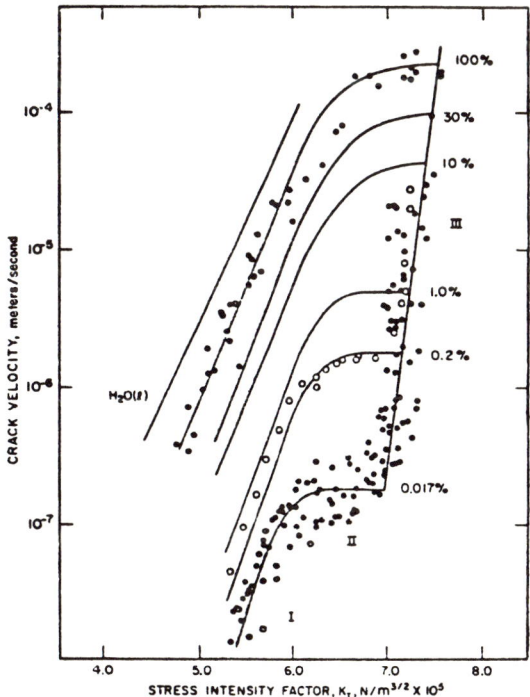

Figure 3.   Crack propagation in nitrogen, soda lime
            silicate glass. The percent relative
            humidities are given on the right hand
            side of the diagram. After Wiederhorn,
            reference 19.

is exponentially dependent on the stress intensity factor and is
also dependent on the amount of water in the environment. Crack
growth in region I is attributable to a stress enhanced chemical
reaction between water and the glass [21,22]; the rate of growth
is reaction rate controlled. In the plateau region, region II,
the crack velocity is nearly independent of the applied stress
intensity factor, but is strongly dependent on the amount of
water in the environment. Crack growth in region II is also due

to a stress enhanced chemical reaction between water and glass, but now the rate of growth is determined by the rate of transport of water to the crack tip. In region III, the crack velocity is again exponentially dependent on the applied stress intensity factor, but is not dependent on water in the environment. Instead, crack growth is controlled by the chemical composition and structure of the glass. Crack propagation studies in vacuum [23,24] have shown that region III does not occur for all glasses. Crack propagation in silica, 96 percent silica (Vycor), and Pyrex becomes so rapid at a critical stress intensity factor, that failure appears to be instantaneous, and, as a result, crack velocities cannot be measured.

Crack propagation in region I depends on temperature as well as load and environment. In water, the crack velocity in glass increases as the temperature is increased [7]. A fit of the crack propagation data to an equation derived from reaction rate theory gives activation energy values that range from 20 to 30 Kcal/mole. This range of activation energies is consistent with the chemical reaction mechanism for fracture proposed by Charles and Hillig [20,21]. Although alkali ion diffusion in glass falls in approximately the same activation energy range, a diffusion mechanism for fracture is not likely since slow crack growth is observed in glasses such as silica that contain low concentrations of alkali ions (∿1 PPM for silica).

Crack propagation studies in aqueous environments of acids, bases, and salts indicate that crack growth in region I depends on the pH of the test solution [25] (figure 4). The slope of

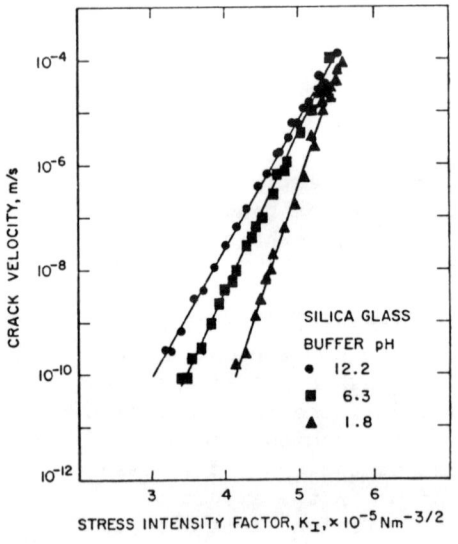

Figure 4. Effect of pH on crack propagation in glass. The slope of the lines decrease as the pH increases. After Wiederhorn and Johnson, reference 25.

crack propagation curves in low pH solutions (acidic solutions)
is about twice that in high pH solutions (basic solutions). This
change in slope is probably related to similar slope variations
obtained from strength studies [26] in which glasses containing
highly basic constituents such as alkali ions were shown to have
slopes that were less by a factor of two than glasses such as
silica that contained no basic constituents. The components
of the glass reacting with water at the crack tip probably controls
the pH of the solution at the crack tip. In silica glass, which
contains no mobile alkali ions, this reaction creates an acidic
environment, while in soda lime silicate glass, ion exchange
between the sodium ions in the glass and the hydrogen ions in
the water creates a highly basic solution at the crack tip.
As demonstrated by measurements of the pH of glass-water
slurries [27], this crack tip solution can range from a pH as
low as 4.5 for fused silica to as high as 12.0 for soda lime
silicate glass. These studies demonstrate that the composition
of the glass and the reaction of the glass with water have a
strong influence on crack growth.

The effect of n-alcohols on crack propagation has been
studied by Freiman [28], and by Evans and Wiederhorn [29], who
have shown that crack growth in alcohol depends on the presence
of dissolved water in the alcohols. Freiman showed that in normal
alcohols the fracture behavior of soda lime silicate glass is
essentially the same as it is in nitrogen gas. Trimodal crack
propagation curves are obtained and the crack propagation behavior
in regions I and II is controlled by the relative humidity of the
alcohol solutions (figure 5). The same crack propagation curve is
obtained in region I for water saturated nitrogen gas as for
saturated alcohol solutions [28,29]. In region II, crack propagation is controlled by the transport of water through the alcohol

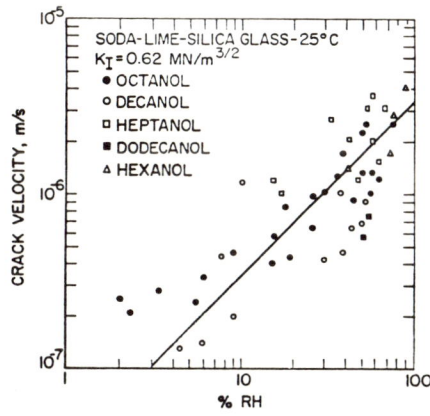

Figure 5. Crack propa-
gation in alcohol at a
stress intensity factor
of 6.2 MN/m$^{3/2}$. After
Freiman, reference 28.

solution, while in region III crack propagation is independent
of water in the alcohol solution. In region III, however, there
is also a small dependence of crack propagation on the chain
length of the alcohol.

Crack propagation studies in vacuum [20,23,24,30], region III,
have demonstrated that slow crack growth in glass is possible even
in the absence of a corrosive environment (figure 6). Crack growth

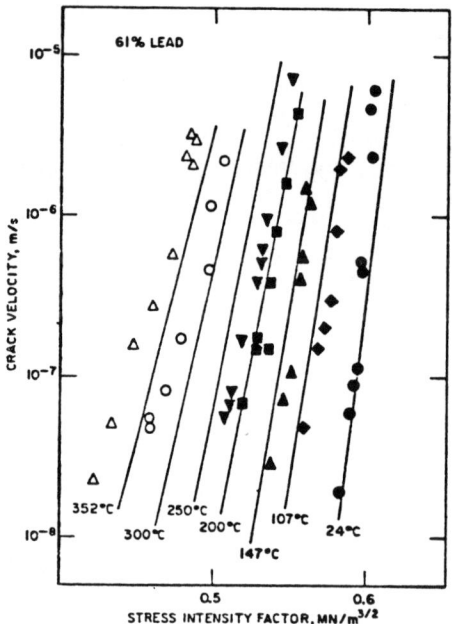

Figure 6.   Crack propagation in a 61 percent lead glass
            in vacuum, 10^{-5} Torr. After Wiederhorn,
            Johnson, Diness and Heuer, reference 24.

depends strongly on the composition of the glass, and for some
glass compositions does not occur in the absence of water [23,24].
When crack growth does occur, it is dependent on temperature,
suggesting an activated process for crack growth. Activation
energies estimated from the crack growth data range from 70 to
200 Kcal/mol, which is considerably higher than the activation
energy for alkali ion diffusion [31], 15 to 30 Kcal/mol, suggesting
that the motion of alkali ions does not control crack propagation.
The possibility that a chemical reaction controls crack growth
can be eliminated by similar arguments, since activation energies
for chemical reactions normally range from about 15 to 60
Kcal/mol [32].

Crack growth in vacuum can be related to the elastic properties of the glass [24]. Glasses having normal elastic

properties (the elastic constants decrease with increasing
temperature, and increase with increasing pressure) exhibit
subcritical crack growth in vacuum, while glasses (Vycor, silica,
Pyrex) having anomalous elastic properties exhibit no subcritical
crack growth, but instead fracture suddenly at a critical value of the
intensity factor. Wiederhorn et al [24] suggest that the growth
of cracks in vacuum is determined by the atomic configuration of
the crack tip. If the crack has a wide cohesive region as
described by Barenblatt [33], then crack growth will not occur.
Conversely, a narrow cohesive region leads to crack growth.
A kinetic theory describing crack growth in these terms has been
presented recently by Thomson and his co-workers [34,35].

Crack propagation data obtained by indirect means has been
recently reviewed by Doremus [36]. Constant loading rate data
and constant stress data indicate large variations in the slope
of the crack propagation curve, depending on the glass composition,
surface preparation, and chemical composition of the test
environment [7,26,37-42]. Ritter [26] reports good agreement
on a variety of glasses between slopes obtained from crack propagation studies and constant loading rate studies. Good agreement between slopes determined by direct and indirect means is
also obtained from constant load studies on silica fibers [7]
and on soda lime silicate glass, in water [7] (figure 7) and in

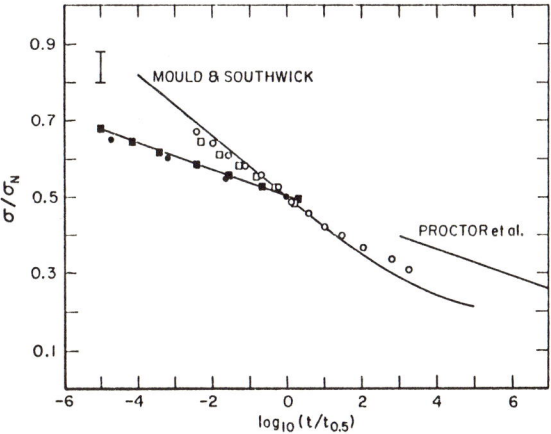

Figure 7.   Universal fatigue curves for soda lime silicate
            glass (open circles and squares) and silica
            glass (closed circles and squares). After
            Wiederhorn and Bolz, reference 7.

acid-base solutions [25]. Constant load studies on Pyrex
(abraded and chemically polished) however, do not agree with crack
propagation studies on the same glass [41]. Similarly, constant

strength data from abraded silica does not agree with data from
crack propagation studies [36,41]. These differences might be
explained by differences in crack tip chemistry for the two types
of studies [25]. For example, if the pH of the crack tip differed
for crack propagation studies by the direct and indirect methods
then differences in slopes might be expected from the two types
of techniques. Because estimates of failure time are based on
crack growth data (section 4), it is of practical importance to
resolve these differences in fracture behavior.

Crack propagation studies can be used to cast some light on
the possible existence of a static fatigue limit. The static
fatigue limit is the load or stress below which failure will not
occur regardless of how long a load is left on the component.
This limit has been estimated to be 20 to 30 percent of the
strength of glass in liquid nitrogen [21,43]. The fatigue limit
can be estimated from crack propagation studies by determining
the applied stress intensity factor at which crack motion stops.
Recent studies by Wiederhorn and Johnson [44] (figure 8) at

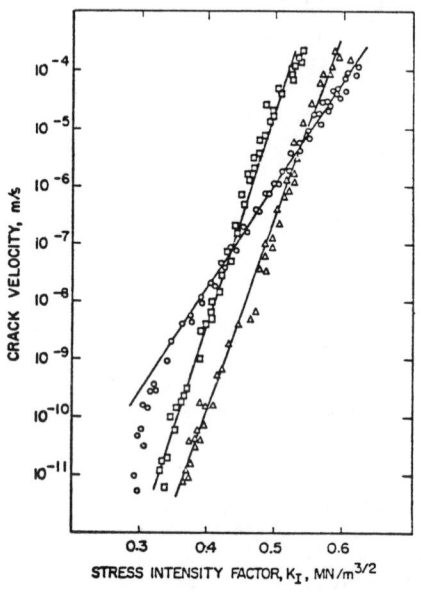

Figure 8. Crack propagation in
air (50 percent relative humidity).
Δ silica glass, □ borosilicate
glass (Pyrex), o soda lime
silicate glass. After Wiederhorn
and Johnson, reference 44.

crack velocities as low as 10^{-12} m/s do not indicate a fatigue
limit in water for silica, Pyrex or soda lime silicate glass,
although severe bending of the curve for soda lime silicate does
suggest an approach to a fatigue limit for this glass. However,
a fatigue limit has been observed for Pyrex tested in an acidic
environment [25], which appears to have caused the crack tip to
round off resulting in an abrupt halt in crack motion at a
relatively high stress intensity factor. The fact that a static

fatigue limit is not usually observed in glass may indicate that
this limit is at a lower stress intensity factor than has yet
been studied. For practical purposes, the absence of a fatigue
limit means that design methods will have to be based on crack
motion to zero load.

3.2 Porcelain

The strength of porcelain also depends on the water content of
the environment, a result that is hardly surprising since porcelain
consists largely of a continuous glass matrix. The effect of
water on the strength of porcelain suspension insulators was
first studied by Taylor [45] in 1939. Although no mention was
made of water causing the strength loss, Taylor showed that the
average loss in strength over a three-year period was greater
than 50 percent. In 1946, Preston and his co-workers [46,47]
showed that the same type of static fatigue curve in air was
obtained for porcelain and glass (figure 9). More recently,

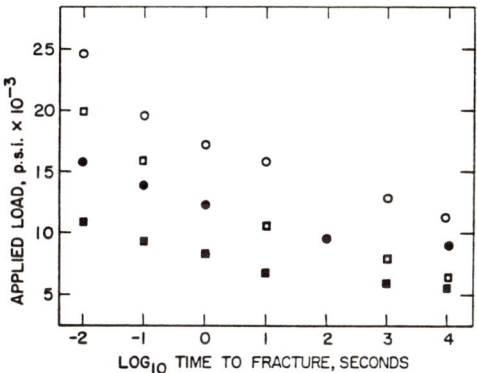

Figure 9.   Static fatigue of glass and porcelain.
o Pyrex, □ soda lime silicate glass, ● porcelain
A, ■ porcelain C. After Glathart and Preston,
reference 47.

Creyke [48] demonstrated a loading rate dependence of strength.
The value of n, 25, calculated from Creyke's data, was about
the same as that obtained for many glasses [26]. Evans and Linzer
[49] have recently shown that this value of n is consistent with
slopes obtained from direct measurements of crack propagation using

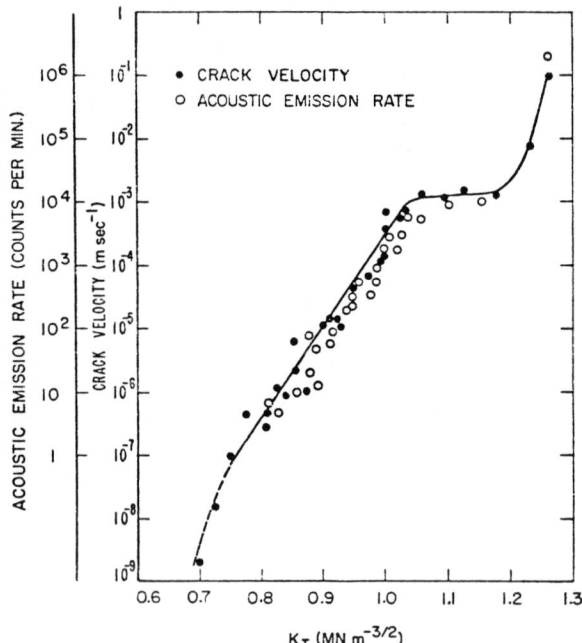

Figure 10.   Crack velocity and acoustic emission rate in
             porcelain as a function of applied stress
             intensity factor. After Evans and Linzer,
             reference 49.

double torsion specimens (figure 10). Evans and Linzer obtained
trimodal crack propagation curves for porcelain. Because the study
was conducted in water, the plateau region could not be explained
by the water transport. The solution at the crack tip should
always have had sufficient water for a chemical reaction. The
occurrence of a plateau region in porcelain still awaits a good
explanation.

An interesting use of the double torsion technique was
demonstrated by Evans and Linzer [49] on porcelain. Not only
was crack propagation data obtained on an opaque material, but
the crack propagation data was related to the rate of acoustic
emission from this material. The data obtained by these authors
are shown in figure 10 where it can be seen that the rate of
acoustic emission exhibits the same functional dependence on
stress intensity factor as does the crack velocity. This result
suggests that acoustic emission may be used as part of a failure
prediction scheme for porcelain ceramics. This type of data
demonstrates the power of the double torsion technique as a tool
for obtaining fracture mechanics data.

3.3 Aluminum Oxide

Delayed failure in aluminum oxide was first demonstrated by Roberts and Watt [50] on polycrystalline material and by Wachtman and Maxwell [51] on single crystal material (sapphire). Pearson [52] showed that delayed failure could be eliminated by heating the alumina to 350°C and testing it in a vacuum. He concluded from this study that delayed failure was caused by atmospheric attack of water on Griffith flaws in the stressed material. Charles and Shaw [53] showed that delayed failure in alumina was temperature dependent, and attributed failure to a stress enhanced chemical attack on the alumina, which resulted in crack growth and failure. These conclusions are supported by crack growth studies on alumina which show the crack growth behavior of alumina to be similar to glass in many ways. Fracture studies on (10$\bar{1}$2) planes of sapphire [54] give the same type of crack propagation curves as glass tested in nitrogen gas (compare figures 11 and 3). In both materials, trimodal curves are obtained,

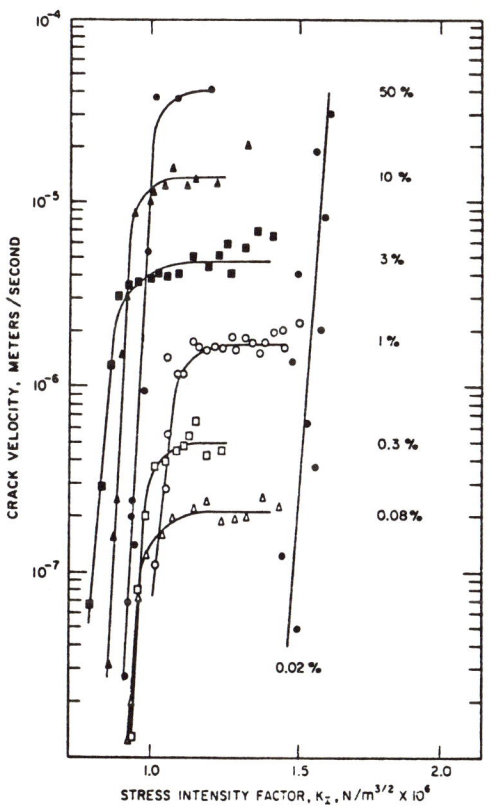

Figure 11. Crack propagation in sapphire, (10$\bar{1}$2) plane. Percent relative humidity is given on the right hand side of the diagram. After Wiederhorn, reference 54.

the position of each curve depending on the amount of water in the nitrogen. In polycrystalline aluminium oxide trimodal crack propagation curves were obtained in toluene and in air (50 percent

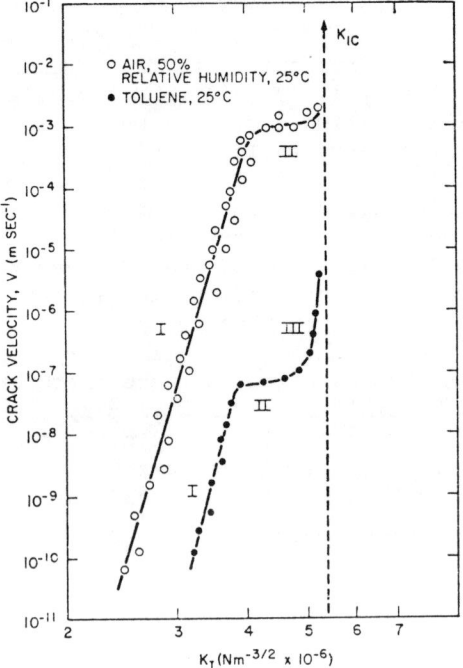

Figure 12. Crack propagation in aluminum oxide.
After Evans, reference 9.

relative humidity) by Evans [9] (see figure 12). The mechanisms
of crack propagation in these environments are probably the
same as for glass, crack propagation being caused by the water
in air or toluene. Evans demonstrated excellent agreement
between the crack propagation and strength data, supporting the
suggestion that crack growth controls strength.

3.4 Hot Pressed Silicon Nitride

Because of its excellent thermal shock properties, hot pressed
silicon nitride shows promise in applications at high temperatures
for gas turbines. Although high strengths are maintained to 1300°C,
a loading rate dependence of strength is observed at temperatures
greater than 1200°C [55]. This loading rate dependence of strength
has been attributed to crack propagation from small flaws in the
surface of the silicon nitride [55]. Since the strength and the
critical stress intensity factor of silicon nitride are relatively
independent of the test environment (air, water, etc.), subcritical
crack growth is primarily environment independent. Crack growth in
silicon nitride therefore differs from that in other ceramics, resulting from processes such as creep or viscous flow rather than

chemical reaction with the environment [14].

Subcritical crack growth in silicon nitride has been confirmed
recently by Evans and Wiederhorn [14] at temperatures as high as
1400°C (figure 13).   Two distinct crack growth behaviors were observed.

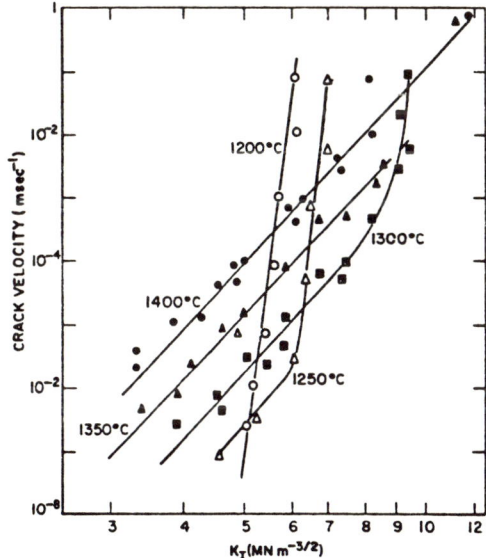

Figure 13.   Crack propagation in hot pressed silicon
             nitride. After Evans and Wiederhorn, reference
             14.

For a given temperature, crack growth at low velocities, region B,
is characterized by a constant slope n ≃ 10 and by the fact that
crack propagation becomes easier as the temperature is increased.
A least squares fit of the data to $v = v_0 \exp(-\Delta H + bK_I)/RT$ gives
an activation energy of approximately 220 Kcal/mol.   This
value is much larger than the activation energies typically
observed for ion diffusion in glass or for chemical reactions;
therefore, fracture cannot be attributed to these processes.
The mechanism of crack growth in region B most likely results
from viscous flow, or creep which typically have activation
energies of the magnitude estimated from the crack growth
measurements.

The second region of crack growth in silicon nitride, region
A, occurs at relatively high velocities for any given temperature.
The crack propagation curve has a much greater slope, n ≃ 50,
than in region B, and the stress intensity factor for crack
propagation increases as the temperature is increased.   This
increase in K_I for crack propagation causes an increase
in the critical stress intensity factor, K_{IC}, at the higher

temperatures (see figure 14). A mechanism for crack growth in
region B has not yet been proposed.

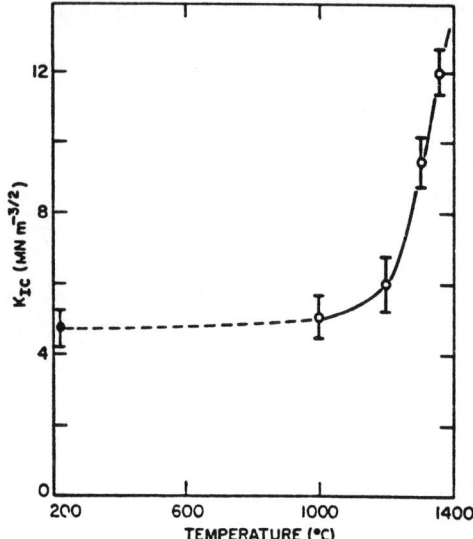

Figure 14. Critical stress intensity factor of silicon
            nitride as a function of temperature. After
            Evans and Wiederhorn, reference 14.

4. THE USE OF SUBCRITICAL CRACK GROWTH
DATA FOR DESIGN PURPOSES

This section presents a technique that can be used to predict
a component's useful lifetime [16]. Crack propagation data combined
with proof testing are used to develop design diagrams that
predict a useful material lifetime and thus aid the engineer in
selecting materials for structural use. The method of design
is embodied in equation (13) which gives the time to failure of
a ceramic component under constant load. Once the crack propagation data necessary for equation (13) has been obtained, then
the failure time is determined only by K_{Ii}, the initial stress
factor at the most serious flaw in the ceramic. Thus, a prediction of failure time depends only on the ability to characterize
K_{Ii}. Since K_{Ii} is related to flaw size, the characterization of
K_{Ii} is equivalent to characterizing the most serious flaw in the
structural component.

Classically, non-destructive techniques have been used to
measure flaw sizes in structural materials [56]. These techniques
are capable of determining flaws as small as a millimeter.
Unfortunately flaws of this size are much larger than can be
tolerated in ceramic components which must maintain useful mechani-

cal properties.   Therefore other methods of establishing the
maximum flaw size must be used for ceramic materials.

Proof testing offers one way of characterizing the maximum
flaw size in a ceramic component.   By proof testing, we mean
that before a ceramic component is placed in service, a load
greater than that expected in service is first applied to
the component.   This assures us that during the proof test the
stress at the most serious flaw is greater than will be
experienced in service.   The proof test must simulate the service
stresses so that the most serious flaw does in fact experience a
high stress level.   In combination with crack propagation data,
proof testing does give the information necessary for design
purposes.

As noted by Tiffany and Masters [57], the proof test
characterizes the largest size flaw present in the tested
component.   During the proof test, the stress intensity
factor at the most serious flaw in the component, K_{IP}, has to be
less than the critical stress intensity factor, K_{IC}, otherwise
the ceramic component would have failed during the proof test.
Thus, the initial flaw size, a_i, is limited to a value given
by the following relationship:

$$K_{IC} > K_{IP} = \sigma_p Y\sqrt{a_i} \tag{24}$$

where σ_p is the proof stress.   When the ceramic component is
placed in service, the flaw size, a_i, will be the same
as it was at the end of the proof test.   Therefore, K_{Ii}, the
stress intensity factor at the most serious flaw at the beginning
of service is given by the following relationship:

$$K_{Ii} = \sigma_a Y\sqrt{a_i} \tag{25}$$

where σ_a is the service stress.

Dividing equation (24) by (25) the following relationship
is obtained for the stress intensity factor at the most serious
flaw at the onset of service:

$$K_{Ii} \leq (\sigma_a/\sigma_p)K_{IC} \tag{26}$$

Substituting this relationship into equation (13) the following
equation is obtained for the minimum time to failure [16,58]:

$$t_{min} = 2\sigma_a^{-2} (K_{IC}\sigma_a/\sigma_p)^{2-n}/AY^2 (n-2) \qquad (27)$$

The main feature of this equation is that the minimum time to failure after proof testing is inversely proportional to the service stress squared and directly proportional to some function of the ratio of the proof test stress to the service stress.

$$t_{min} = \sigma_a^{-2} \cdot f(\sigma_p/\sigma_a) \qquad (28)$$

This function of proportionality is determined from crack propagation data.

The fact that the minimum time to failure can be expressed by the relationship given in equation (28) permits one to present crack propagation data in the terms of proof test diagrams [16]. By plotting $\log(t_{min}\sigma_a^2)$ versus $\log(\sigma_p/\sigma_a)$ a curve is obtained that gives a prediction of the failure time once the service stress and proof test ratio are known (see figure 15). Diagrams of this

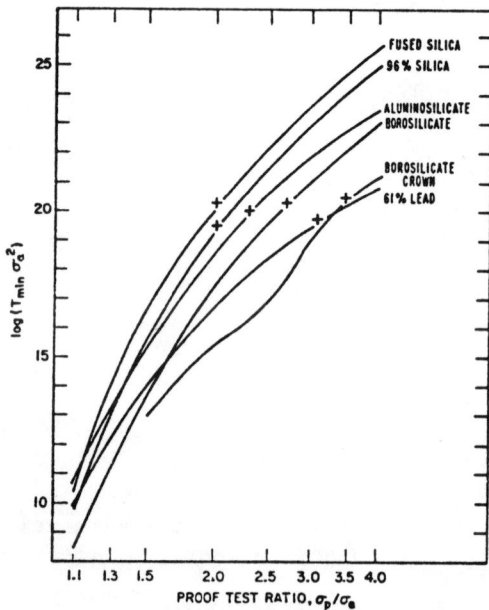

Figure 15.   Proof test diagram for glass. The crosses mark the limit of crack propagation data. Curves were extrapolated to higher proof test ratios using an empirical fit to the data. After Wiederhorn, Evans and Roberts, reference 23.

type are useful because crack propagation data for several
different materials can be presented on a single diagram. A more
useful proof test diagram for a single material is obtained by
plotting $\log(t_{min})$ versus $\log \sigma_a$ (figure 16). The data then appears

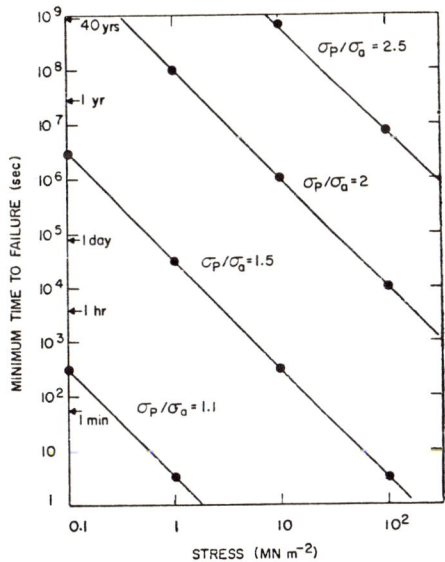

Figure 16. Proof test diagram for porcelain. After
Evans and Linzer, reference 49.

as a series of straight lines with a slope of -2, each line representing a different proof stress ratio. From this type of diagram,
one can directly read the time to failure as a function of applied
stress and proof test ratio. Each line on the second type of
diagram is represented by a point on the first type of diagram.

   The use of the second type of diagram is illustrated in
figure 16 for porcelain [49]. Porcelain suspension insulators
should have a life expectancy of about 40 years, during which
time stresses of the order of 2000 psi (~ 13 MN/m^2) have to be
supported. Referring to the proof test diagram, a porcelain
insulator will support the load for 40 years provided a proof
test of slightly more than 2.5 times the service load is first
applied to the insulator before putting it into service.

   This proof testing technique has been studied in the laboratory on glass by Evans and Wiederhorn [16] and on porcelain by
Evans and Linzer [49]. Evidence for the applicability of equation
(27) to glass is presented in figure 17. Using a proof test ratio
of 1.35, two types of proof tests were conducted on glass micro-

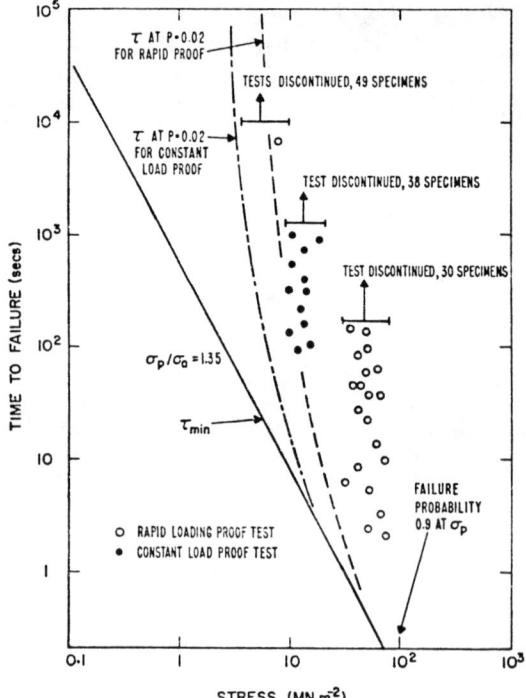

Figure 17.   A comparison of failure times after proof
             testing for ground soda-lime glass in water,
             with the predicted minimum failure time for
             the soda-lime glass/water system (full line),
             and the times in excess of the minimum for a
             probability of 0.02. After Evans and Wiederhorn,
             reference 16.

scope slides. In one, the open circles, specimens were rapidly
loaded and then unloaded so that the specimen was exposed to the
proof test for only a short time (< 0.1 sec). In the second proof
test, a constant load was applied to the specimen for five minutes
after which the load on the specimens was reduced to the service
stress load. In both cases, failure occurred at times greater than
predicted by equation (27). At the lower service stress levels,
the predictions of equation (27) were extremely conservative.
Failure occurred in times several orders of magnitude greater
than predicted by the equation.

    Evans and Wiederhorn [16] also discussed weakening of the
specimens during proof testing as a result of crack growth. They
showed that after proof testing, the distribution of strengths
was better than the initial distribution (strengths were higher

at all probability levels) provided m, the Weibull parameter
characterizing the initial stress distribution, was less than
(n-2), m < n-2.

The proof test method has been expanded to cyclic loading
by Evans and Fuller [59]. Their analysis enables one to predict
crack propagation rates during cyclic loading from crack growth
parameters determined from static load experiments. To a good
approximation, the failure time under a cyclic load, t_c, is
directly proportional to the failure time under a static load,
t_s,

$$t_c = g^{-1} t_s. \tag{29}$$

g^{-1}, a proportionality factor that depends on the type of cyclic
load, has been evaluated for square wave loading, sinusoidal loading
and saw-tooth type of loading. Values of g^{-1} are available in
chart form as a function of the exponent n and the ratio of the
load amplitude and the average cyclic load (figure 18). This

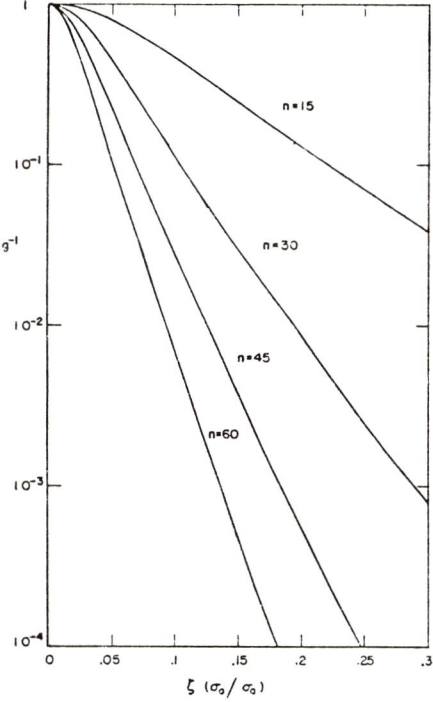

Figure 18.   Curves to estimate g^{-1} for sinusoidal load.
             After Evans and Fuller, reference 59.

analysis permits a comparison of predicted times to failure for
cyclic and static load conditions, assuming that crack growth
results from the same causes for both types of loading. Results
at room temperatures on porcelain, alumina and glass suggest
there is no enhanced effect of cyclic loading on slow growth.
This conclusion was verified in a series of measurements on slow
crack growth rates under static and cyclic conditions [59]. The
result is consistent with the idea that ceramic materials do not
exhibit significant fatigue due to cyclic loading.

When failure does not endanger life or property, the failure
criteria given by equation (28) can be relaxed to permit
a certain fraction of failures. Here, economics must be
used to decide the relative merits of replacing broken parts
rather than designing the original part to be failure free. A
method of design that permits a given fraction of failures has been
described by Evans and Wiederhorn [15]. The method of design is
not based on proof testing, but instead combines Weibull statistics
of strength with crack growth data to develop design diagrams.
If proof testing has been used then the change in the flaw distribution due to proof testing can also be incorporated into the method
of design. The dashed lines in figure 17 were obtained by this
method for a 2 percent probability of failure for both rapid and
constant load proof testing. The failure data presented in
figure 17 fall much closer to these lines than they do to the
solid line giving the minimum time to failure.

5. SUMMARY

Methods of obtaining crack growth data are described and
their uses are discussed. Data can be collected either by direct
or indirect methods. The direct methods require crack growth to
be monitored by fracture mechanics techniques, while the indirect
methods infer crack growth data from strength measurements.
Agreement between both types of data is essential for confidence
in design and for interpreting mechanisms of crack growth.

There are several direct methods of measuring crack growth,
among which the double cantilever beam technique and the double
torsion technique are the most useful. The double cantilever
technique has the advantage of great long term load stability so
that very low crack velocities, $\sim 10^{-12}$ m/s, can be measured.
The double torsion technique can be used with fixed displacement
loading so that crack velocities can be made without viewing the
crack tip directly. Thus crack velocity studies can be conducted
on opaque materials and in hostile environments where visibility
is not possible.

Crack propagation techniques have been used to collect data

on glass, porcelain, aluminum oxide, and silicon nitride. This
data has been useful in understanding mechanisms of crack growth
and in developing failure criteria for ceramic materials. In this
paper we describe a promising method of design that uses crack
growth data with proof testing to estimate failure times. The
method may be applied either to cyclic or static loading conditions.
This method emphasizes the practical importance of crack growth
data because the only experimental input necessary for predicting
failure times is data on crack growth.

REFERENCES

1. P. C. Paris and G. C. Sih, ASTM Special Tech. Publ. No.
    381 (1965).

2. H. H. Johnson and P. C. Paris, Eng. Fract. Mech. $\underline{1}$, 3 (1968).

3. A. G. Evans, this volume.

4. J. J. Gilman, J. Appl. Phys. $\underline{31}$, 2208 (1960).

5. J. E. Srawley and B. Gross, Mat. Res. Stds. $\underline{7}$, 155 (1967).

6. S. M. Wiederhorn, A. M. Shorb, and R. L. Moses, J. Appl.
    Phys. $\underline{39}$, 1569 (1968).

7. S. M. Wiederhorn and L. H. Bolz, J. Am. Ceram. Soc. $\underline{53}$,
    543 (1970).

8. S. W. Freiman, D. R. Mulville and P. W. Mark, Report of NRL
    Progress, Feb. 1972, p. 36.

9. A. G. Evans, J. Mater. Sci. $\underline{7}$, 1137 (1972).

10. D. P. Williams and A. G. Evans, J. of Testing and Evaluation,
    August 1973, to be published.

11. K. R. McKinney and H. L. Smith, J. Am. Ceram. Soc. $\underline{56}$, 30
    (1973).

12. J. O. Outwater and D. J. Jerry, "On the Fracture Energy of
    Glass," Interim Rept., Contract NONR-3219(ONX), University
    of Vermont, Burlington, Vt, August 1966.

13. J. A. Kies and A. B. J. Clark, "Fracture-1969" (Ed. P. L.
    Pratt), pp. 483-491, Chapman and Hall, Ltd., London (1969).

14. A. G. Evans and S. M. Wiederhorn, J. Mat. Sci., to be
    published.

15. A. G. Evans, Int. J. Fract. Mech., to be published.

16. A. G. Evans and S. M. Wiederhorn, NBS Report No. NBSIR 73-147, Int. J. Fract. Mechanics.

17. W. F. Brown and J. E. Srawley, ASTM Special Tech. Publ. No. 410 (1966).

18. R. J. Charles, J. Appl. Phys. 29, 1657 (1958).

19. S. M. Wiederhorn, J. Am. Ceram. Soc. 50, 407 (1967).

20. K. Schönert, H. Umhauer and W. Hlemm, "Fracture-1969," (Ed. P. L. Pratt), pp.474-482 , Chapman and Hall, Ltd., London (1969).

21. R. J. Charles and W. B. Hillig, pp. 511-527 in "Symposium on Mechanical Strength of Glass and Ways of Improving It," Florence, Italy, September 25-29, 1961. Union Scientifique Continentale du Verre, Charleroi, Belgium, 1962.

22. W. B. Hillig and R. J. Charles, "High Strength Materials," (Ed. V. F. Zackey), pp. 682-705, John Wiley and Sons, Inc., New York (1965).

23. S. M. Wiederhorn, A. G. Evans, and D. L. Roberts, this volume.

24. S. M. Wiederhorn, H. Johnson, A. Diness and A. H. Heuer, J. Am. Ceram. Soc., to be published.

25. S. M. Wiederhorn and H. Johnson, J. Am. Ceram. Soc. 56, 192 (1973).

26. J. E. Ritter and C. L. Sherburne, J. Am. Ceram. Soc. 54, 601 (1971).

27. S. M. Wiederhorn, J. Am. Ceram. Soc. 55, 81 (1972).

28. S. W. Freiman, J. Am. Ceram. Soc., to be published.

29. A. G. Evans and S. M. Wiederhorn, J. Am. Ceram. Soc., to be published.

30. V. P. Pukh, S. A. Laterner, and V. N. Ingal, Soviet Physics-Solid State 12, 881 (1970).

31. R. H. Doremus, "Modern Aspects of the Vitreous State-Vol. 2," p. 1, Butterworths, London (1962).

32. S. Glasstone, "Textbook of Physical Chemistry," D. Van Nostrand, New York (1940).

33. G. Barenblatt, Adv. Appl. Mech. 7, 55 (1962).

34. R. Thomson, C. Hsieh and R. Rana, J. Appl. Phys. 42, 3154 (1971).

35. C. Hsieh and R. Thomson, J. Appl. Phys. 44, 2051 (1973).

36. R. H. Doremus, "Corrosion Fatigue: Chemistry, Mechanics and Microstructure," pp. 743-748, National Association of Corrosion Engineers, Houston, Texas (1972).

37. R. J. Charles, J. Appl. Phys. 29, 1549 (1958).

38. B. A. Proctor, I. Whitney and J. W. Johnson, Proc. Roy. Soc., 297A, 534 (1967).

39. R. E. Mould and R. D. Southwick, J. Am. Ceram. Soc. 42, 589 (1959).

40. F. W. Preston, T. C. Baker and J. L. Glathart, J. Appl. Phys. 17, 162 (1946).

41. J. E. Ritter, Jr. and J. Manthuruthil, "Glass Technology," to be published, 1973.

42. J. E. Burke, R. H. Doremus, W. B. Hillig and A. M. Turkalo, "Ceramics in Severe Environments," (Editors, W. W. Kriegel and Hayne Palmour III) Plenum Press, New York (1971).

43. E. B. Shand, J. Am. Ceram. Soc. 37, 52 (1954).

44. S. M. Wiederhorn and H. Johnson, unpublished data.

45. J. J. Taylor, "Conférence Internationale des Grands Réseaux Electrique a Haute Tension," June 29 to July 8, 1939, Tome II 2nd Section, Rapport No. 228.

46. T. C. Baker and F. W. Preston, J. Appl. Phys. 17, 170 (1946).

47. J. L. Glathart and F. W. Preston, J. Appl. Phys. 17, 189 (1946).

48. W. E. C. Creyke, Trans. Brit. Ceram. Soc. 67, 339 (1968).

49. A. G. Evans and M. Linzer, J. Am. Ceram. Soc., October 1973, to be published.

50. J. P. Roberts and W. Watt, Ceram. Glass 10, 53 (1952).

51. J. B. Wachtman, Jr. and L. H. Maxwell, J. Am. Ceram. Soc.
     37, 291 (1954).

52. S. Pearson, Proc. Phys. Soc. (London) 69(B), 1293 (1956).

53. R. J. Charles and R. R. Shaw, "Delayed Failure of Poly-
     crystalline and Single-Crystal Alumina," General Electric
     Co., Report No. 62-RL-3081M, July 1962.

54. S. M. Wiederhorn, Int. J. Fract. Mech. 4, 171 (1968).

55. F. F. Lange, J. Am. Ceram. Soc., to be published.

56. W. J. McGonnagle, "Nondestructive Testing," Gordon and Breach,
     New York, 2nd Edition (1966).

57. C. F. Tiffany and J. N. Masters, ASTM Spec. Tech. Publ. No.
     381, pp. 249-277 (1965).

58. S. M. Wiederhorn, J. Am. Ceram. Soc. 56, 227 (1973).

59. A. G. Evans and E. R. Fuller, Met. Trans., to be published.

THE PROPAGATION OF CRACKS BY DIFFUSION

R. Dutton

Atomic Energy of Canada Limited, Whiteshell Nuclear

Research Establishment, Pinawa, Manitoba, Canada

INTRODUCTION

In the recent literature in various fields of research, an interest has developed in several aspects of mass transport to the tips of Griffith-type cracks by diffusional processes. This phenomenon is, for example, of importance to creep rupture behaviour where fracture may result from the growth of grain boundary cracks by diffusion of vacancies to the crack tip. Similar growth processes can also be responsible for static fatigue observed in many ceramic materials. Diffusion of chemically active species to crack tips can also control stress corrosion cracking.

Crussard and Friedel (1) were perhaps the first people to discuss the role of diffusion in creep rupture. They considered, qualitatively, the net result of several competing vacancy fluxes due to several different driving forces around an elliptical crack lying normal to a tensile stress. They argued that as regions at the crack tip were under tension, while regions at the upper and lower surfaces of the crack were under compression, the resulting vacancy flux would result in spheroidisation. Alternatively, as the strain energy is greater at the crack tip, a flux of vacancies will be generated in the opposite direction, causing crack growth. The latter process might be thought to dominate but they suggest that under the influence of the surface energy, the high strain energy field would be removed by spheroidisation. Crussard and Friedel conclude that diffusion of vacancies from the lattice to the crack tip must compensate this to maintain a sharp crack. The net effect is that vacancies arrive at the crack tip to cause crack growth.

A quantitative theory of diffusional controlled creep rupture was developed by Hull and Rimmer (2). They considered the growth of spherical grain boundary cavities by a process of grain boundary diffusion. Their model ignored the small strain energy term associated with the cavity and is therefore inapplicable to the growth of elongated Griffith-type cracks. However, McLean (3) asserted that the criterion for the stability of an elongated crack is the same as that for the spherical sphere although he expected the strain field at the crack tip to promote a high supersaturation of vacancies resulting in the enhancement of the flux of vacancies to the crack tip.

Liu (4) has considered the diffusion of point defects to a Griffith-type crack, lying normal to a tensile stress, by considering the first order elastic interaction (size effect) of the point defect with the hydrostatic component of the stress gradient at the crack tip. From this treatment, he derived expressions for the steady state concentration gradient of point defects (interstitial solute atoms in the case considered by him) in the vicinity of the crack tip. Using this analysis, he developed a kinetic model for stress corrosion cracking based on the enhanced reaction of solute atoms at the crack tip resulting in the formation of a low strength corrosion product.

A corollary of Liu's model is that vacancies are always repelled from the crack tip whereas interstitials are always attracted to it. However, Heald et al (5) have shown that if the second order in-homogeneity effect is considered, the crack tip always attracts vacancies but repels interstitials. This contradicts the stress corrosion cracking model developed by Liu for the case where crack propagation is controlled by the diffusion of interstitial solute atoms to the crack tip. That this second order elastic interaction must not be ignored when considering the diffusion of point defects to a crack tip, was first pointed out by Crussard (6). However, Crussard suggested that the interaction of a point defect with the shear component of the stress at the crack tip is predominant. He pointed out that this causes vacancies always to diffuse towards the crack tip, in agreement with Heald et al.

Recently, Heald and Williams (7) have modified their earlier model of creep fracture based on the growth of a wedge crack by grain boundary sliding, to include a contribution to crack growth by diffusion. The diffusion kinetics are introduced via their analysis (5) of the elastic interaction of a vacancy with the stress field of the crack tip. This model has been criticized by Dutton and Stevens (8) on the basis that diffusion of vacancies to the crack considered by Heald and Williams, is thermodynamically unfavourable.

The thermodynamic stability of cracks has been recently considered by Stevens and Dutton (9) who also developed a model of the high temperature static fatigue of ceramic materials by the propagation of cracks solely through diffusional processes. The model is based on a calculation of the total chemical potential of atoms at the crack tip (including strain energy and surface energy terms) and predicts that vacancies will be either attracted to or repelled from a crack tip according to the magnitude of the applied stress. We shall first review the important aspects of this model and will outline some recent improvements we have made which allow the inclusion of terms in the diffusional equations which account for the elastic interaction of the vacancies with the stress gradient at the crack tip.

THERMODYNAMIC STABILITY OF CRACKS

It is assumed that brittle materials always contain inherent flaws so that there is no nucleation event to be studied. Also, it is assumed that the cracks are of pure Griffith-type, i.e. have no volume except that produced by elastic distortion. In practice, the crack must have some intrinsic volume otherwise it would disappear on removal of the stress, allowing the material to realize its full theoretical strength on subsequent testing. However, the theory, as originally presented by Griffith, is applicable with good accuracy to the typical non-ideal crack (9) under complex stress fields. In addition, various crack geometries can be accommodated by the theory by a suitable choice of solutions to the elastic equation. Thus, the extension of the theory to the three-dimensional case, e.g. penny-shaped crack, alters the equations by a numerical factor only such that the theory presented in this paper is applicable to many practical situations with minor numerical modifications. In fact, the generality of the approach has been shown by Dutton and Stevens (8) who demonstrated that the thermodynamic stability of any crack can be calculated if only the local stress at the tip of the crack is known. Elasticity theory enables these crack tip stresses to be calculated in many circumstances.

Figure 1 shows how changes in crack shape and volume can be effected by mass transport (surface, bulk and vapour phase diffusion). The driving force for these processes will depend on the thermodynamic stability of the crack and has its origin in the free energy change of the system. Let such a system comprise a flaw-free plate specimen stressed by a freely suspended weight. If a crack of length 2c is now introduced normal to the stress, the total Helmholtz free energy per unit thickness of plate can be written as:

$$F = F_o - \sigma v - \frac{\pi\sigma^2 c^2 (1 - \nu^2)}{Y} + 4c\gamma \qquad [1]$$

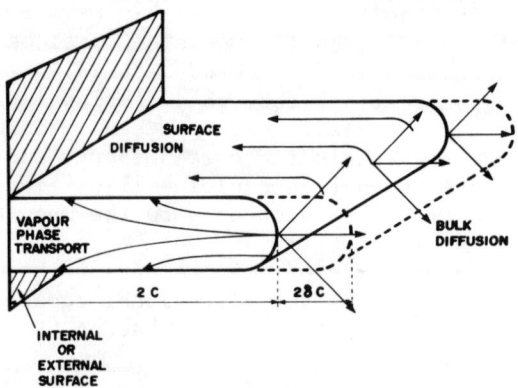

Figure 1. Schematic diagram of a crack
propagating by bulk diffusion, surface
diffusion and vapour phase transport.

where F_o is an arbitrary constant, σ is the applied stress, v is the
volume of any material which may be removed to form the crack, ν
is Poisson's ratio (we are considering the plane strain situation),
Y is Young's modulus and γ is the specific surface energy of the
material. The σv term represents the maximum work done in depositing
any atoms removed from the crack at some sink in the material, e.g.
at a grain boundary normal to the stress. This term is significant
in the case of small circular cavities as produced during creep
fracture (2) where the strain energy term is small. However, in
the present case of large thin cracks, for which the latter term
is large, the σv term can be ignored. The third term represents
the net change in the strain energy of the system and the last
term represents the increase in surface energy. It is important
to note that the value of the surface energy in this term is not
the effective fracture energy which is appropriate to the normal
instantaneous brittle fracture dealt with in classical fracture
mechanics. This latter quantity comprises the thermodynamic
surface energy, γ, together with other terms associated with
surface irregularity, crack branching and other irreversible work
done due to dislocation generation and phonon emission during
catastrophic bond rupture. In contrast, we are concerned here with
the surface energy pertinent to slow crack propagation by diffusion
which maintains the crack path normal to the tensile stress and
promotes smooth fracture surfaces. Thus, the initial crack length
defined in equation [2] below is much less than that for ordinary
brittle fracture and therefore a crack just greater than this length
is thermodynamically unstable with respect to growth by diffusion
but stable with regard to catastrophic failure. A similar statement,
but based on a fundamentally different argument, could be made from
the recent publication of Hasselman et al (10).

The critical crack length for crack growth is defined by $\partial F/\partial c = 0$ and is given by

$$c_c = \frac{\gamma}{B\sigma^2} \tag{2}$$

where B is a material constant equal to $\pi(1 - \nu^2)/2Y$.

The driving force for mass transport is usually expressed in terms of a chemical potential gradient. In order to find this gradient, the chemical potential of atoms at the crack tip must first be determined. To simplify the growth geometry, it is assumed that the crack tip terminates at one end at a free surface or grain boundary so that the addition of δn molecules to the crack tip results in a decrease in crack length of $2\delta c$ (see Figure 1). From Equation [1] and using $\delta c = (\Omega/4\ell)\delta n$ where Ω is the atomic volume and ℓ is the crack tip radius, the chemical potential of atoms, μ, at the crack tip is given by

$$\mu = \frac{\partial \Delta F}{\partial n} = \frac{\Omega}{\ell}(B\sigma^2 c - \gamma) \tag{3}$$

When the crack length is greater than c_c and μ is positive with respect to the surrounding material involved in the diffusion process, atoms will tend to migrate away from, or vacancies migrate to, the crack tip, resulting in crack growth and sharpening of the crack tip. Alternatively, if μ is negative with respect to the surrounding material, the crack will shrink and spheroidise as vacancies diffuse away from the crack tip. It should be emphasised that this conclusion is independent of any consideration of the interaction of vacancies, or any other point defect involved in the mass transport process, with the stress field of the crack. In the steady state situation, when the local concentrations of point defects are constant, the sign of the free energy change of the system during crack growth or shrinkage and thus whether the crack grows or shrinks, is determined by the sign of the chemical potential of atoms at the crack tip to that at other sources in the material. The interaction of point defects with the stress field will therefore only affect the kinetics of growth or shrinkage and not which of these actually occurs. A comparison with dislocation climb may be helpful here. The direction of climb depends only on the sign of the externally applied stress (which determines the chemical potential at the dislocation line) while the stress field of the dislocation and its interaction with point defects only affects the kinetics of the climb process.

Our model of crack stability resolves the uncertainty of the net effect due to the surface energy and strain energy terms expressed by Crussard and Friedel (1), by considering the total force for crack growth by mass transport processes. Thus, when μ is positive with respect to the surrounding material, surface diffusion and bulk diffusion both contribute to crack propagation and crack sharpening. Spheroidisation will occur only when μ is negative.

The value of σ when μ equals zero defines a stress value below which a crack will not grow, i.e. this represents the static fatigue limit in ceramic materials. This has been discussed elsewhere by Dutton and Rogowski (11).

Using the expression for the chemical potential of atoms at the crack tip, Stevens and Dutton (9) have developed some simple kinetic models for crack growth. These, together with a model for the growth of a grain boundary crack by grain boundary diffusion, will be briefly reviewed below.

CRACK PROPAGATION BY GRAIN BOUNDARY DIFFUSION

Following the creep rupture model developed by Hull and Rimmer (2), we will calculate the linear gradient of chemical potential $\partial\mu/\partial x$ assuming the value given by equation [3] at the crack tip and the value of $-\sigma\Omega$ at a distance L from the crack tip, where L is approximately half the grain diameter. (Note that $\partial\mu/\partial x$ is negative when the x axis lies in the grain boundary). Thus, the potential gradient is given by

$$\frac{\partial\mu}{\partial x} = \frac{\Omega}{L\ell} \left(\gamma_{gb} - B\sigma^2 c - \sigma\ell\right) \qquad [4]$$

The flux of atoms J_{gb} from the crack tip to the grain boundary, per unit thickness of crack is then

$$J_{gb} = -\frac{D_{gb}\partial z}{kT\Omega} \frac{\partial\mu}{\partial x} \qquad [5]$$

where D_{gb} is the grain boundary diffusion coefficient, ∂z is the grain boundary width, k is Boltzmann's constant and T is the absolute temperature. The rate of crack growth $\partial c/\partial t$ is given by

$$\frac{\partial c}{\partial t} = \frac{\Omega}{4\ell} \frac{\partial n}{\partial t} = \frac{\Omega}{4\ell} J_{gb} \qquad [6]$$

and whence, finally,

$$\frac{\partial c}{\partial t} = \frac{\Omega D_{gb}\delta z}{4L\ell^2 kT} \left(B\sigma^2 c - \gamma_{gb} + \sigma\ell\right) \qquad [7]$$

CRACK PROPAGATION BY LATTICE DIFFUSION

An alternative method of crack growth is the absorption of vacancies at the crack tip from the surrounding lattice by a process of bulk diffusion. To derive the rate of diffusion, cylindrical geometry is assumed along the length of this crack tip. The appropriate steady state diffusion equation is:

$$\frac{\partial}{\partial r}\left(\frac{r\partial C}{\partial r}\right) = 0 \qquad [8]$$

where r is the radial distance from the crack tip and C is the concentration of vacancies. The boundary conditions are (a) at the surface of the crack tip ($r = \ell$), the vacancy concentration (C_ℓ) is fixed in the usual way, by the value of the chemical potential at that location such that:

$$C_\ell = C_o \exp(-\mu/kT) \qquad [9]$$

(b) at a distance $r = L$, far removed from the crack tip (L is approximately equal to half the grain diameter) the vacancy concentration is at the equilibrium value, C_o. The solution to equation [8] which satisfies these boundary conditions is

$$C - C_o = (C_\ell - C_o) \ln\left(\frac{L}{r}\right)/\ln\left(\frac{L}{\ell}\right) \qquad [10]$$

Note that this solution is insensitive to the exact values of L and ℓ and hence the choice of boundary conditions does not substantially alter the result. The net vacancy flux J_b ($= - \partial n/\partial t$) per unit thickness of crack is given by

$$J_b = - 2\pi r D_v \frac{\partial}{\partial r}(C - C_o) = \frac{2\pi(C_\ell - C_o)D_v}{\ln(L/\ell)} \qquad [11]$$

where D_v is the vacancy diffusion coefficient and is related to the self diffusion coefficient, D, by $D = \Omega C_o D_v$. Whence, using the identity shown in equation [6], the crack growth rate is given by

$$\frac{\partial c}{\partial t} = \frac{\pi D}{2\ell \ln\left(\frac{L}{\ell}\right)}\left[1 - \exp\left\{\frac{\Omega}{kT\ell}\left(\gamma_{gb} - B\sigma^2 c\right)\right\}\right] \qquad [12]$$

CRACK PROPAGATION BY SURFACE DIFFUSION

We will now calculate the rate of propagation of a crack by the removal of atoms from the crack tip followed by their deposition on the flat surface of the crack by surface diffusion. We approximate the steady state distribution of chemical potential by a constant gradient. The chemical potential at the crack tip is given by equation [3] while that at the flat surface, a distance c away, is zero giving a gradient (negative along the x axis) of

$$\frac{\partial \mu}{\partial x} = \frac{\Omega}{\ell} \left(\frac{\gamma_{gb}}{c} - B\sigma^2 \right) \tag{13}$$

The flux of atoms, J_s, is given by

$$J_s = - \frac{\partial \mu}{\partial x} \frac{N_a D_s}{kT} \tag{14}$$

where N_a is the number of atoms per unit area and D_s is the surface diffusion coefficient. The rate of removal of atoms, $\partial n/\partial t$, from the crack tip is equal to $2J_s$ since there are two surfaces along which diffusion takes place. Thus, the rate of crack growth is given by

$$\frac{\partial c}{\partial t} = \frac{\Omega^2 ND_s}{2\ell^2 kT} \left(B\sigma^2 - \frac{\gamma_{gb}}{c} \right) \tag{15}$$

CRACK PROPAGATION BY VAPOUR PHASE TRANSPORT

A fourth mass transport process is that of evaporation of atoms at the crack tip followed by their condensation on the stress free flat surface. The relationship between the vapour pressure at the crack tip, p', and that at the flat surface, p, is given by the Gibbs-Thompson formula:

$$kT \ln\left(\frac{p'}{p}\right) = \mu \tag{16}$$

and p is expressed through the Clausius-Clapeyron equation as $p = p_o \exp(-H/RT)$ where p_o is a constant and H is the latent heat of evaporation. The rate of emission of atoms from a surface in equilibrium with its vapour at pressure p is given by the kinetic theory of gases as $p/(2\pi mkT)^{1/2}$ where m is the atomic mass. Thus the flux of atoms, J_v, leaving the crack tip is given by

$$J_v = \frac{p' - p}{(2\pi mkT)^{1/2}} \tag{17}$$

and, the rate of crack growth $(= J_V \Omega/2)$ is given by

$$\frac{\partial c}{\partial t} = \frac{\Omega p}{(8\pi mkT)^{1/2}} \left[\exp\left\{ \frac{\Omega}{kT\ell} (B\sigma^2 c - \gamma_{gb}) \right\} - 1 \right] \qquad [19]$$

THE EFFECT OF THE STRESS GRADIENT

As pointed out in section 2, the criterion of crack stability is true regardless of the elastic interaction of point defects with the stress gradients at the crack tip. These interactions only become important when considering the kinetics of crack growth or shrinkage by bulk diffusion (lattice or grain boundary). There is therefore a need for a rigorous kinetic treatment of crack growth which should incorporate these effects with the chemical potential at the crack tip included as a boundary condition. This solution has not been presented in the literature for the crack tip situation but a similar problem has recently been considered by Nix et al (12) for the case of diffusion of vacancies to a climbing edge dislocation where the effect of the dislocation stress field was included. We have adopted a similar approach in order to attempt a complete calculation of the velocity of a crack propagating by lattice diffusion. As this will be published in full elsewhere, only the essence of the calculation will be presented here.

The vacancy flux (J_V) equation can be written generally as

$$J_V = \frac{C_V^* D_V}{kT} \frac{d(\mu_A - \mu_V)}{dr} \qquad [20]$$

where μ_A and μ_V are the chemical potentials of atoms and vacancies respectively, $\mu_A - \mu_V$ being the general driving force for diffusion. The expression for μ given by equation [3] is the boundary value of the driving force at the crack tip. r is the radial distance in the cylindrical geometry considered, C_V^* is the number of vacancies per unit volume and is given by $C_V N$ where N is the number of lattice sites per unit volume. Equation [20] can be written in terms of vacancy concentration as

$$J_V = - ND_V \left(\frac{dC_V}{dr} + \frac{C_V}{kT} \frac{d\phi}{dr} \right) \qquad [21]$$

where $\phi(r)$ is the interaction energy of a vacancy and is given by

$$\phi(r) = p_c(r)\Delta V_V + \text{second order terms} \qquad [22]$$

where p_c is the stress due to the presence of the crack tip and ΔV_V is the relaxation volume of the vacancy. Note that we have refined the formulation of the chemical potentials to include, in addition to the first order size effect $p_c\Delta V_V$, the possibility of other elastic interactions. However, for simplicity, we have omitted entropy and electrical interactions. If the form of these interactions were known, they could be included in $\phi(r)$ in a straightforward manner. In addition, we have assumed that the effect of stress on D_V can be ignored in comparison with the effect of stress on $\mu_A - \mu_V$ or C_V.

To determine the crack velocity we require the steady state value of J_V and this is obtained by putting $\nabla \cdot J_V = 0$. The diffusion equation is then solved subject to the same boundary conditions as given in section 4.

Using the expressions for the stress field at a crack tip, the two dominant elastic interaction energies for the vacancy can be written down. These are, the size effect and the inhomogeneity effect. The size effect occurs because the vacancy can change its volume when subjected to a hydrostatic stress. The inhomogeneity interaction is a second order effect which results from the fact that the local elastic constants associated with the vacancy differ from the corresponding elastic constants of the matrix. For a vacancy the size effect interaction energy is given by (4)

$$\phi_s(r) = -\frac{8\pi(1 + \nu)\varepsilon\, r_o^3 K}{3\sqrt{2\pi}} \cdot \frac{\cos\theta/2}{r^{1/2}} \qquad [23]$$

While the inhomogeneity interaction is written (5)

$$\phi_i(r) = -\frac{(1 - \nu^2)\, r_o^3\, K^2}{3G} \frac{\cos^2\theta/2}{r} \qquad [24]$$

$$\left\{1 + \frac{15}{2(1 + \nu)(7 - 5\nu)}\sin^2(\theta/2)\cdot\cos^2(3\theta/2)\right\}$$

where G is the shear modulus and K is the stress intensity factor. In these two equations the vacancy has been formed in a continuum by placing a sphere of radius r_o into a hole of radius $r_o(1+\varepsilon)$ and then forcing the surfaces of the two spheres together into perfect contact. This is an alternate way of expressing the relaxation volume of a vacancy, ΔV_V.

Following the procedure of Nix et al (12) we can obtain the flux of vacancies to the crack tip and hence, through equation [6], derive the crack velocity as:

$$\frac{dc}{dt} \simeq \frac{\pi D}{2 \ell F(\ell)} \ (1 - e^{\mu/kT}) \qquad\qquad [25]$$

This equation is very similar to the previous equation when stress field interactions were ignored (equation [12]). The effect of the interactions enters into equation [68] in a straightforward way via the expression $F(\ell)$ which is a complex function of ℓ, L, $\phi(r)$ etc. (readers are referred to our complete paper on this work, which is to be published, for the exact form of $F(\ell)$). Thus, the final result of this calculation is that the velocity of a crack growing by the absorption of vacancies is modified by the term $\ln(L/\ell)/F(\ell)$. We have evaluated $F(\ell)$ for the case of alumina, which was considered previously (9): it is found that for small crack sizes, the effect is to retard the crack velocity by about 5%. However, as the crack grows, the attractive inhomogeneity interaction (equation [24]), which increases with K^2 (as opposed to the repulsive size effect interaction which depends linearly on K), gradually becomes predominant. The result is that the crack velocity is enhanced to a maximum of about 25% when the crack length is approaching the critical length for catastrophic brittle failure.

The main conclusion of this work is therefore that the modifying effect of the stress gradient at the crack tip is relatively minor. This means that for most cases, the simpler equation [12] can be used with good accuracy.

REFERENCES

1. C. Crussard and J. Friedel, Creep and Fracture of Metals at High Temperatures, H.M. Stationary Office, London, 1956.
2. D. Hull and D.E. Rimmer. Phil. Mag. 4, 673 (1959).
3. D. McLean. Rep. Prog. Phys. 29, 1 (1966).
4. H.W. Liu. Trans. ASME, J. Basic Eng. 92, 633 (1970).
5. P.T. Heald, J.A. Williams and R.P. Harrison. Scripta Met. 5, 543 (1971).
6. C. Crussard. Acta Met. 5, 475 (1957).
7. P.T. Heald and J.A. Williams. Phil. Mag. 24, 1215 (1971).
8. R. Dutton and R.N. Stevens. Scripta Met. 6, 969 (1972).
9. R.N. Stevens and R. Dutton. Mats. Sci. Eng. 8, 220 (1971).
10. D.P.H. Hasselman, J.A. Coppola, D.A. Krohn and R.C. Bradt. Mat. Res. Bull. 7, 769 (1972).
11. R. Dutton and A.J. Rogowski. J. Can. Ceram. Soc. 41, 53 (1972).
12. W.D. Nix, R. Gasca Neri and J.P. Hirth. Phil. Mag. 23, 1339 (1971).

SLOW CRACK GROWTH IN POLYCRYSTALLINE CERAMICS

S. W. Freiman, K. R. McKinney and H. L. Smith

Naval Research Laboratory

Washington, D. C. 20375

INTRODUCTION

It is now well known that slow crack growth occurs in glasses at stress levels well below that which will cause immediate fractures. Work performed on polycrystalline ceramics[1,2] has shown that many undergo delayed failure in a manner similar to glasses and some slow crack growth studies have been reported[3-5]. Evans[6] recently showed a correlation between slow crack propagation in an alumina and its delayed failure characteristics. He also showed that the time-to-fail of the alumina under a given stress could be determined through knowledge of crack velocity as a function of the stress intensity factor, K_I.

It will be shown in this paper that slow crack growth is a common phenomena in ceramics. Examples of this behavior in PZT, barium titanate, and several aluminas of various purities and grain sizes are presented and a relationship between slow crack growth behavior and microstructure is shown. The use of these fracture mechanics data for prediction of strengths will be discussed. Finally, we will discuss the use of acoustic emission measurements for analyzing crack propagation in ceramics.

EXPERIMENTAL PROCEDURE

Crack propagation data for this study were obtained for several test methods. These included the double-torsion test under both constant load[3] and constant deflection[6] and a double-cantilever type test in which a bending moment rather than an end

load is applied to the specimen[7]. Specimen sizes ranged from
(12" x 3" x 0.125") in the double torsion tests to (2" x 0.5" x
0.04") in the double cantilever test. Since all of these tests
have been described elsewhere no details will be given here.
Changes in crack length were determined either by direct observation with a traveling microscope or computed through measurement
of the compliance change in the specimen[3]. Results on the same
materials obtained in different tests agreed very well.

Acoustic emission (AE) measurements were taken simultaneously
with crack propagation in some materials using commercially available equipment manufactured by Nortec Inc. The double torsion
technique described by Evans[6] was used to acquire the crack growth
data in order to obtain a wide spectrum of crack velocities in a
relatively short time. The load applied to the specimen was
recorded on a strip chart recorder along with the emission data
allowing a direct correlation to be made at a given time.

With this equipment each AE burst above the selected threshold is counted as an event. In this manner, one can relate the
count frequency to some other active parameter such as stress
intensity or velocity. It is not known what role the strength
(energy) of the event plays so that some caution must be taken
when attempting to correlate AE to some fracturing process. In
addition, because the number of emissions is dependent on specimen
size, the type of equipment used, and the instrument settings,
comparison of results between different investigators is not very
meaningful.

Examination of the fracture surfaces of the specimens was performed by replicating with C-Pt-Pd replicas and examining in an
electron microscope.

RESULTS

Slow Crack Growth in Alumina

The results of slow crack growth measurements in various
commercial aluminas are presented in Figure 1. The steepness of
the curves indicates the extreme sensitivity of crack velocity to
changes in stress intensity. It was observed that crack growth in
all of the aluminas was somewhat discontinuous, i.e. the crack
would spurt ahead for a short distance, then slow down or stop.
The velocities plotted in Figure 1 represent an average taken over
a distance large enough to include a number of these oscillations.
The tendence for discontinuous growth was greater at the lower
average crack velocities. A critical stress intensity factor can
be obtained from Figure 1 by taking K_{Ic} as that value of K_I needed

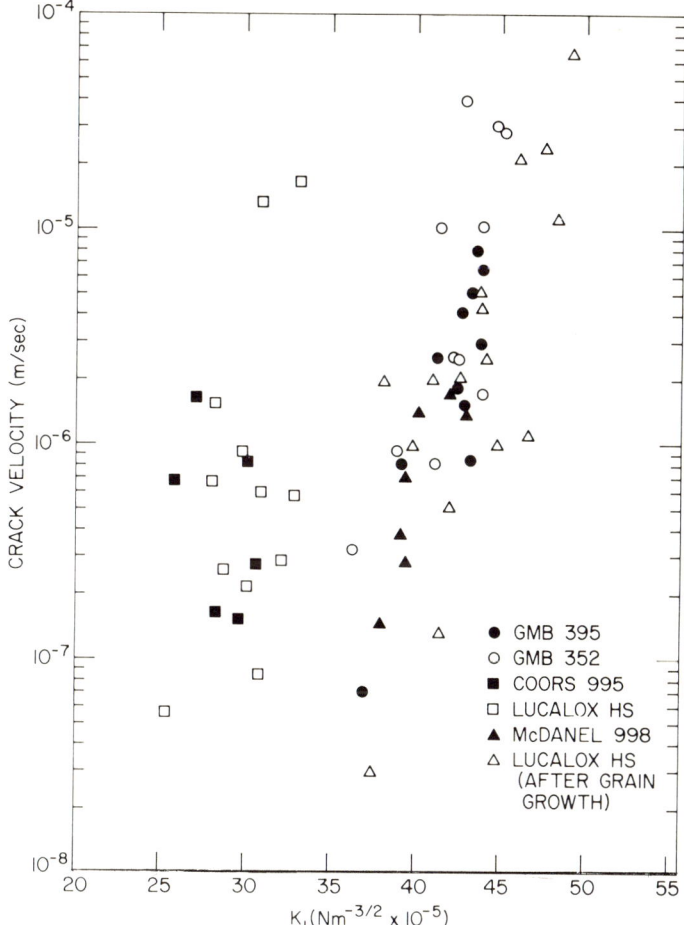

Figure 1. Crack velocity as a function of stress intensity factor for several commercial aluminas. Data were taken in air at 25°C and 40% RH.

to drive the crack at a velocity of 10^{-4}m/sec. This velocity was chosen because critical fracture energies calculated in this way ($\gamma_c = K_{Ic}^2/2E$) are in good agreement with measurements made by the double cantilever technique for glass[8] and alumina[9]. These data are presented in Table I. The aluminas fall into two groups in terms of their resistance to crack growth, the primary difference between the materials in each group being the average grain size of the alumina (Table II). Even though similar observations have been made by Gutshall and Gross in alumina[10] and Clarke et al. in MgO[11], it was still somewhat surprising that the larger grain size

TABLE I

Fracture Characteristics of Ceramics

Material	% Al_2O_3	Source	Flexural Strength Ksi	Elastic Modulus psix10^{-6}	K_{Ic} Nm-3/2 X 10^{-5}	γ_c J/m^2
GMB 395	95	Gladding-McBean	37.5	40	49	44
GMB 352	99.3	"	52.5	50	47	32
AD-85	85	Coors Porcelain Co.	43.0	33	30	20
ADS-995	99.5	"	68.0	52	37	19
Lucalox	99.9	General Electric Co.	40.0	57	41	22
Lucalox-HS	99.9	"	50.0	57	37	18
Lucalox-HS (after grain growth)	99.9	"	----	57	50	32
McDanel 998	99.8	McDanel Refr. Porcelain Co.	55.0	50	49	35
AlSiMag 614	96	American Lava Corp.	48.0	44	32	17
PZT 5800	---	Channel Industries	11.3	10.9	6.5*	2.8
$BaTiO_3$	---	"	14.7	17.8	10.5	4.5

*Measured in H_2O

---

materials have the greater resistance to crack propagation. Simpson[12] measured a decrease in γ_c with an increase in grain size in alumina. As shown in Table I, the data for AD 85 and AlSiMag 614 fall into the fine grain size group as they should. Slow crack growth data for these will be shown later in conjunction with acoustic emission measurements. No effect of the purity of the alumina can be noted, i.e. AD 85 has the same resistance to cracking as Lucalox-HS. Typical electron micrographs of the fracture surfaces of several aluminas are presented in Figures 2 through 5. Fracture in these materials varies from almost completely intergranular in the Lucalox alumina (Figs. 2 and 3) to about 20% intergranular in the Ad 85 (Fig. 4). In the other materials, transgranular failure generally took place through large grains, intergranular failure being limited to the smaller grain size regions. No effect of crack velocity on the relative proportion of intergranular to transgranular areas could be noted in a given

Figure 2. Fracture surface of Lucalox-HS. Note presence of
second phase particles on several grain boundaries.

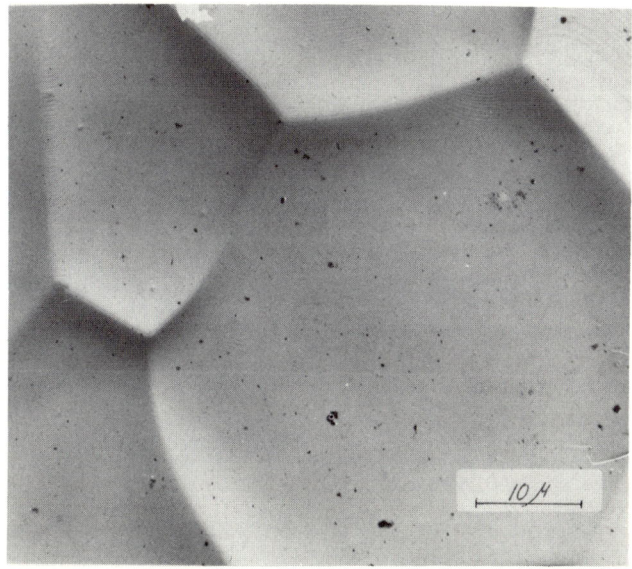

Figure 3. Fracture surface of Lucalox-HS after grain growth.
Little evidence was found of precipitates seen in Figure 2.

Figure 4.   Primarily transgranular type fracture in AD 85.

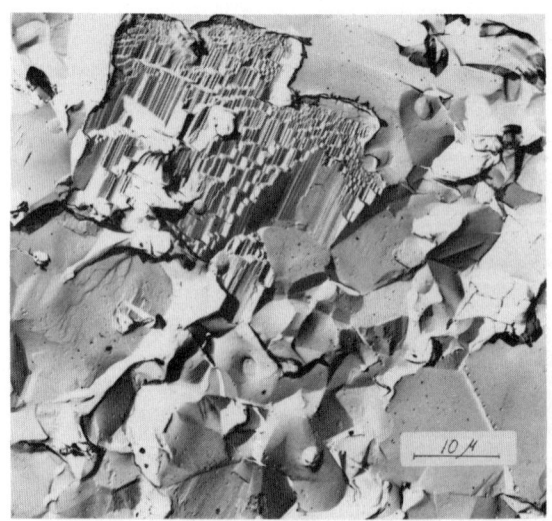

Figure 5.   Fracture surface of GMB 352 showing both intergranular
and transgranular failure.   Note porosity within many of the grains.

material. Some evidence of second phase particles present in grain
boundaries can be seen in the Lucalox-HS (Fig. 2). It was noted
that the amount of grain boundary precipitates was decreased
noticeably after grain growth in the Lucalox (Fig. 3). Fairly extensive porosity was noted within the grains of the GMB 352 (Fig.
5) with lesser amounts being present in several of the other
materials. Taken as a whole these observations indicate that
while porosity or second phase may have some effect on fracture
resistance, the most important parameter is grain size.

Flaw Size Determinations

Using the values of γ_c in Table I plus the flexural strength,
σ, and elastic modulus, E, of the materials which were supplied by
the manufacturer, Eq. 2 can be used to determine the critical size
of a penny-shaped flaw in the surface[13].

$$\sigma = \frac{1}{1.12}\sqrt{\frac{\pi E \gamma_c}{2a}} \qquad (1)$$

$$\text{or} \quad a = \frac{1.25 E \gamma_c}{\sigma^2} \qquad (2)$$

where a is the depth of the flaw necessary to produce failure.
The results of these calculations are given in Table II. Because
these materials have undergone differing surface treatments and
have differing volume fractions and distributions of porosity,
some care must be taken in analyzing these results. It is evident,
however, that there is no obvious correlation between flaw size
and grain size as one might expect from strength-grain-size relationships. This fact suggests that one should be very careful
in using fracture mechanics data obtained for macroscopic cracks
in polycrystalline bodies to predict the behavior of microscopic
flaws.

Wiederhorn[14] suggested that in those materials where flaw
sizes should be limited to the grain size, a value of γ_c obtained
from single crystals should be used in the Griffith equation. If
one calculates flaw sizes for the aluminas using a γ_c determined
for the (10$\bar{1}$1) plane in sapphire (6 J/m^2)[15] one obtains values
very close to the grain sizes of the larger grained materials (Table II). Sedlacek et al[16] found similar results for aluminas with
grain sizes from 10 to 32μ. The strength of those materials with
grain sizes of 20μ or greater appear in this case to be governed
by a different mechanism than those having grain sizes of 10μ or
less. This difference in behavior may account for the discontinuity in the strength vs (grain size)$^{-\frac{1}{2}}$ curve reported by
Stokes[17] for alumina, which occurs at about a 25μ grain size. This
type of behavior was discussed by Rice[18] for a number of ceramic
systems.

TABLE II

Comparison of Flaw and Grain Sizes in
Polycrystalline Aluminas

Material	Calculated a(μ) using poly-crystalline γ_c	Calculated a(μ) using single crystal γ_c	Grain Size*(μ) Range	Avg
GMB 395	220	30	10-50	20
GMB 352	105	20	10-50	20
AD-85	62	19	3-10	7
ADS-995	38	12	3-5	4
Lucalox	140	38	30-40	35
Lucalox HS	74	25	6-10	8
McDanel 998	104	18	5-30	20
AlSiMag 614	59	21	2-12	4

*Grain sizes were determined from both optical observation of
fracture surfaces as well as from replicas. Range should be
taken as approximate since some extreme grains may have been
missed.

———————————

It is theorized that the greater resistance to crack growth
exhibited by the larger grain-size material is due to the fact
that the path of least resistance is more tortuous than in the
finer grain ceramics. Additional energy is required to divert the
crack around obstacles such as larger grains. Evidence for such
diversions can be seen by observing the crack during propagation,
as well as by the greater roughness of the fracture surfaces of
the larger grain size aluminas. In addition, for a microscopic
surface flaw, one should use a micromechanics analysis to determine the stress intensity at the crack tip. Parameters such as
porosity, grain boundary impurities, and residual stresses which,
on the average, have a small affect on a macro-crack may be significant in determining whether a micro-crack will propagate or not.

Some data is available in the literature in which σ, E and
γ_c have been determined for different microstructures in the same
material. These include Si_3N_4[19], SiC[20], alumina[21], and a
$3BaO \cdot 5SiO_2$ glass-ceramic[22]. Equation (1) which relates these
parameters is shown plotted as σ vs $\sqrt{E\gamma_c}$ in Figure 6. The interesting point regarding these graphs is that a reasonably good

straight line fit can be made of the data for each of the materials,
indicating that the flaw size is constant irrespective of microstructure. Those data points for alumina that deviate significantly
from the curve can possibly be explained by the extensive grain
boundary porosity in these specimens. The fact that the curves for
Si_3N_4 and alumina do not pass through the origin may be due to internal stresses in the material as discussed by Rice[23] and Clarke[24].
The flaw sizes determined from the slopes of these curves are
presented in Table III. Since the calculated flaw sizes are much
larger than the grain size, use of polycrystalline fracture energies in the analysis appears to be correct. Because of the different types of tests used to determine γ, including the work of
fracture method as well as the notched beam and double cantilever

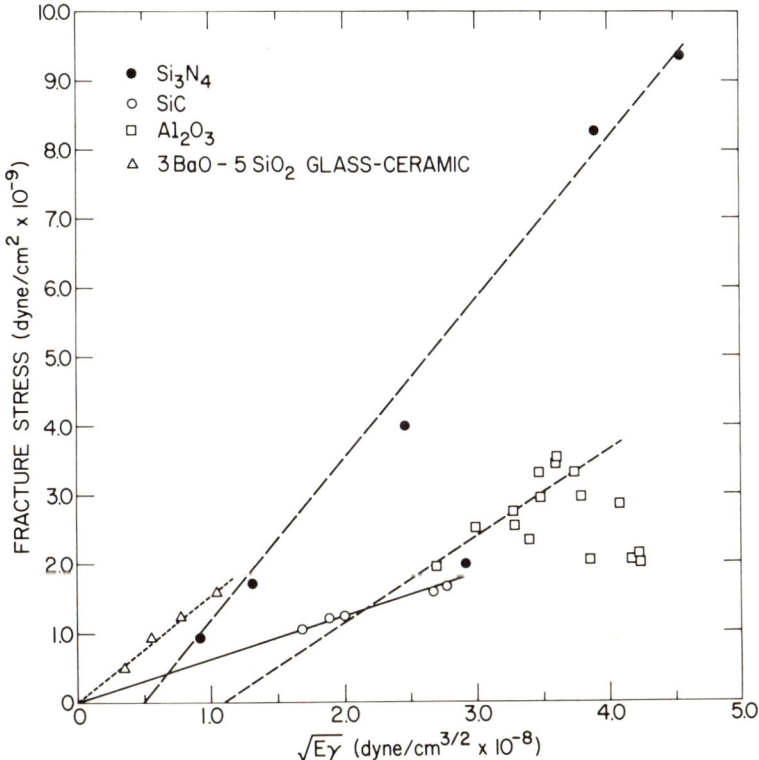

Figure 6. Plots of fracture stress versus $\sqrt{E\gamma_c}$ for a number of
ceramics.

TABLE III

Flaw Sizes Calculated From Slopes
of Curves in Figure 6

Material	Avg. Grain Size (μ)	Calculated Flaw Size (μ)
Si_3N_4 [19]	.5 - 1.5	23
SiC [20]	*	325
Alumina [21]	4 - 50	80
$3BaO \cdot 5SiO_2$ [22]	3	54

*An average grain size was difficult to determine since both fine grained (2μ) and coarse grained (250μ) material was present in the same body.

———————

techniques, there may be some question whether the true γ_c was measured in each case. Coppola and Bradt[20] showed, however, that for SiC all of these tests gave somewhat the same values for γ_c. The results of Figure 6 imply that the size of the flaws in a given ceramic is determined by surface treatment and that their strength depends primarily on resistance to propagation of these flaws. Proof of this hypothesis must include actual measurements of flaw sizes on fracture surfaces and correlation with calculated values.

Crack Growth in PZT

The relationship between crack velocity and K_I for PZT is shown in Figure 7 for both moist air and liquid water environments. It can be seen that the presence of liquid water at the crack tip increases by an order of magnitude the crack velocity at a given K_I. Work reported by Sedlacek et al[25] as well as studies carried out in this laboratory show that PZT undergoes delayed failure in the presence of water. The scatter in our results was too great, however, to allow us to make a prediction of times to fail from the fracture mechanics data.

Fracture occurred transgranularly almost exclusively throughout the bodies, although as seen in Figure 8 there were some areas of intergranular failure. A calculation of the critical flaw size in PZT was made using Eq. 2. The value for σ was determined by the authors using a 3-point bend test on the halves of the double cantilever specimens remaining after the crack propagation tests. A value of $\underline{a}$ of 44μ was calculated, which is much larger than the grain size of about 5μ.

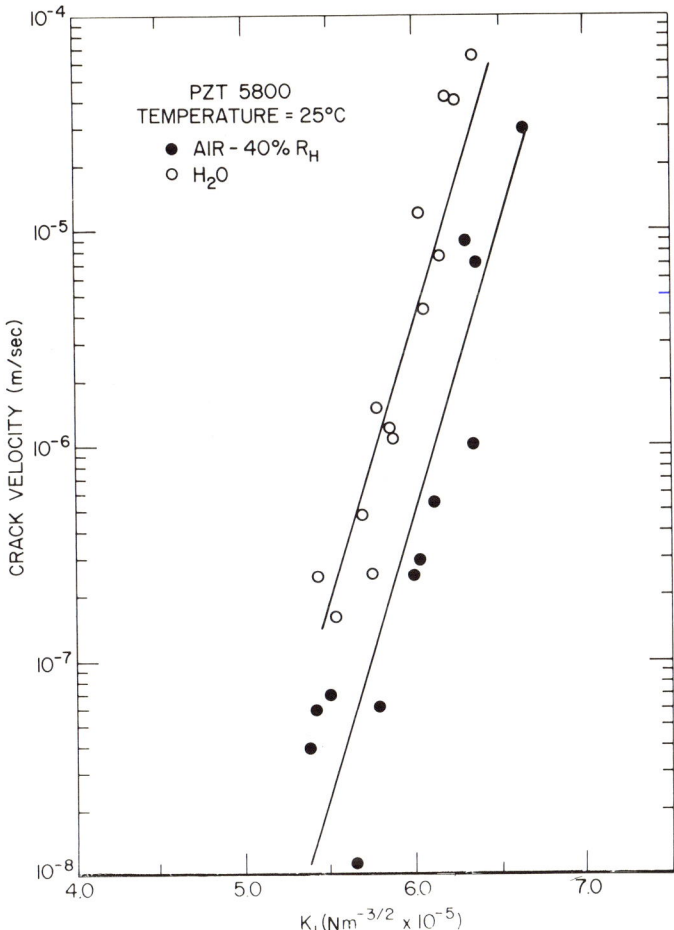

Figure 7. Crack velocity as a function of K_I for PZT in air and water.

Crack Growth in Barium Titanate

The typical transgranular fracture of barium titanate is seen in Figure 9. The herringbone pattern seen in various areas may be due to crack interaction with the domain walls. No obvious differences in the microstructure of the fracture surface between the poled and unpoled specimens could be noted.

Plots of crack velocity vs K_I for barium titanate in both a

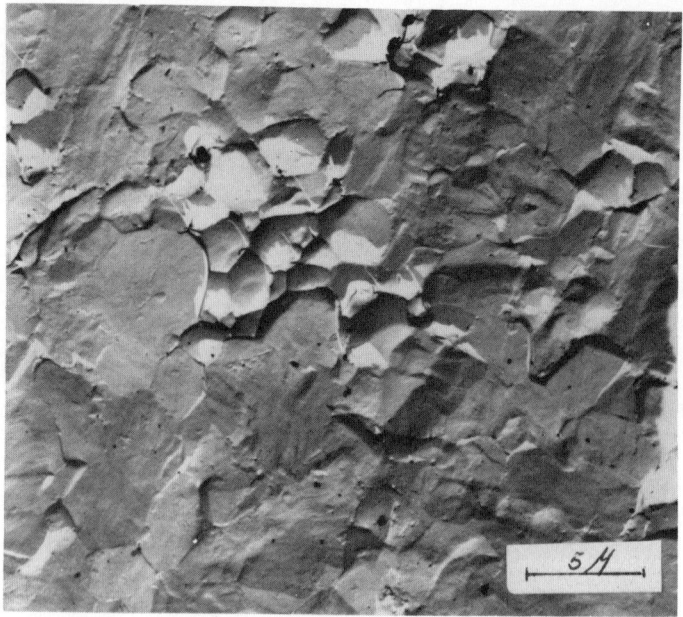

Figure 8.   Typical fracture surface of PZT.

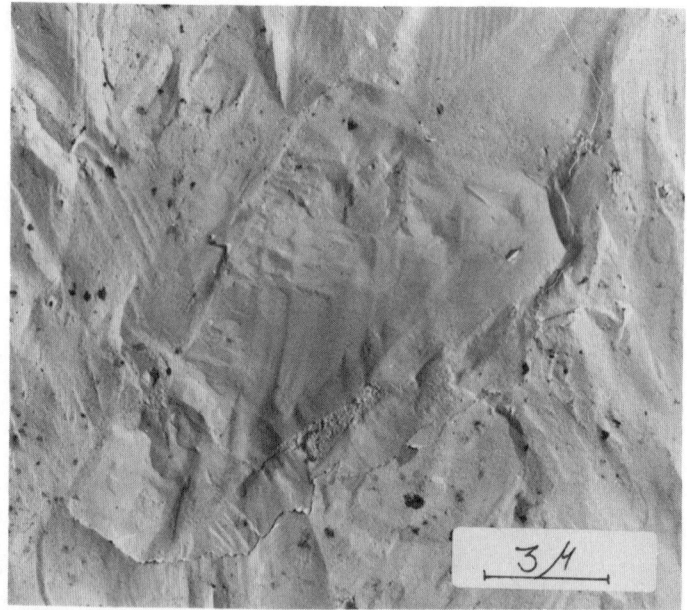

Figure 9.   Typical transgranular type fracture in barium titanate.

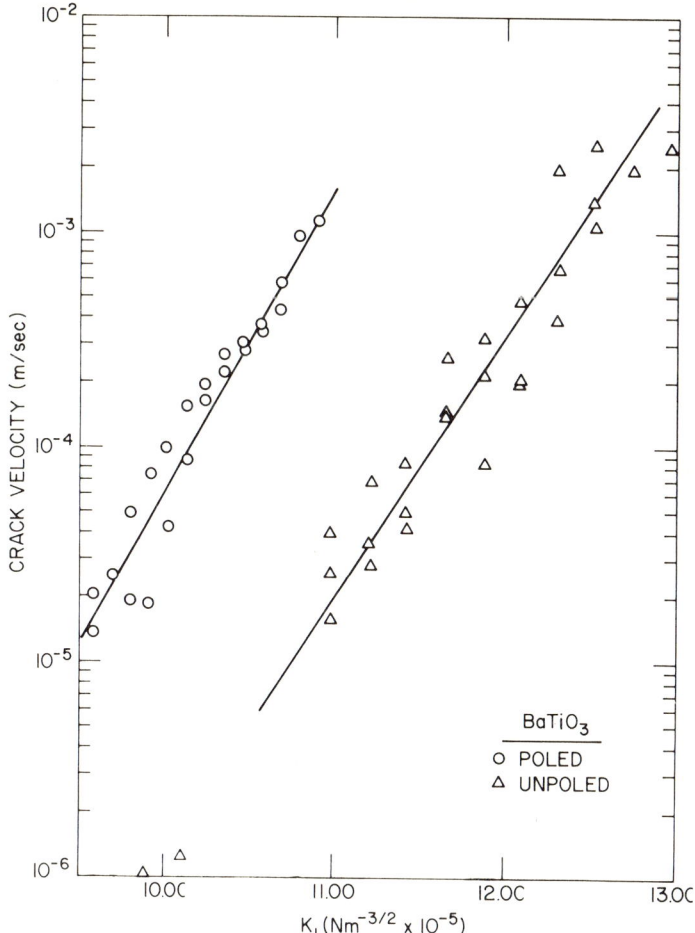

Figure 10. Crack velocity as a function of K_I for poled and unpoled barium titanate.

poled as well as an unpoled condition are shown in Figure 10. The ordering of the structure by the electrical poling parallel to the thickness of the specimen produced greater than an order of magnitude increase in crack velocity at the same K_I. One reason for this increase could be the strains set up in the material by the poling, although some effect of domain orientation may occur.

The fracture strength of the poled barium titanate was measured using a ball and ring test. Because this is a biaxial

test, the strengths are somewhat lower than would be measured in
a flexural test. Calculation of the critical flaw size in this
material from this data yielded a value of $\underline{a}$ of 66μ, which is much
larger than the 5μ grain size and is also larger than the largest
pores (≈30μ) observed in limited sampling.

Acoustic Emission Studies

Acoustic emission measurements were taken in conjunction with
the fracture mechanics data for AD 85 and AlSiMag 614 aluminas.
The data for both crack velocity and acoustic emssion rate in counts
per second as a function of stress intensity are presented in
Figure 11. The fracture mechanics data fall in the same V-K$_I$
region as the other finer grain size materials discussed previously.

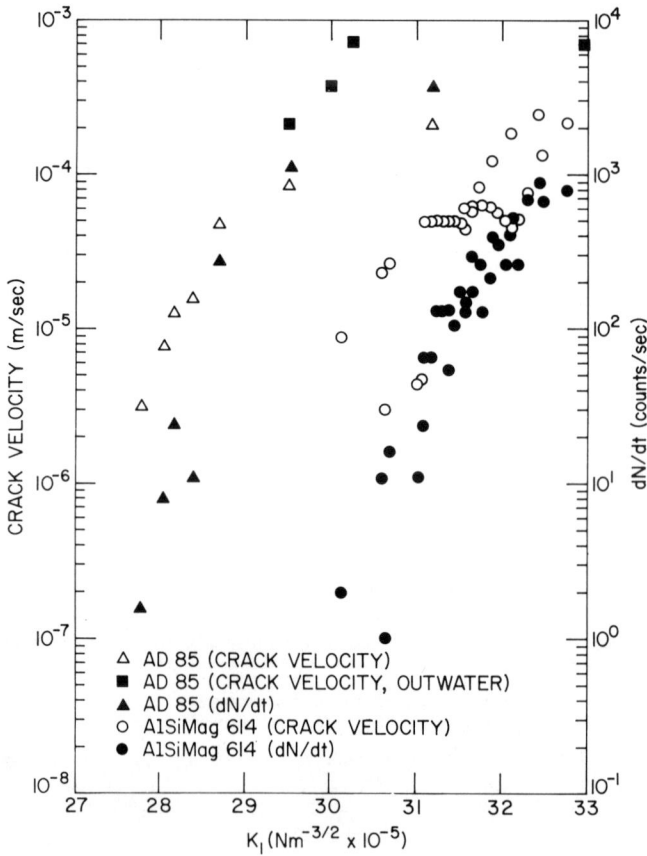

Figure 11. Both crack velocity and acoustic emission rate plotted
as a function of K$_I$ for AD 85 and AlSiMag 614.

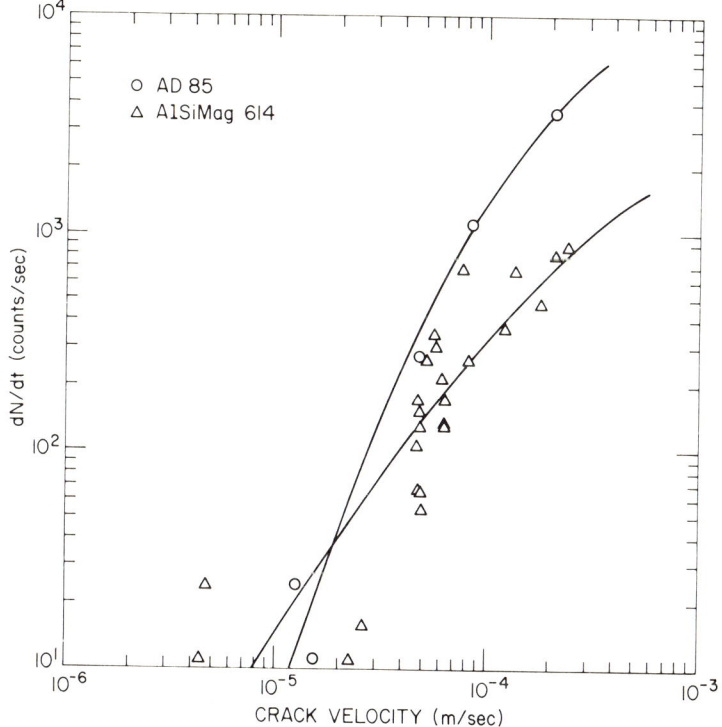

Figure 12.   Plot of acoustic emission rate as a function of crack velocity for AD 85 and AlSiMag 614.

Several data points obtained in a double torsion test under constant load are presented for the AD 85 alumina to demonstrate the equivalence of this test to that employing constant deflection.

The relationship between crack velocity and acoustic emission rate was analyzed by assuming that the number of emissions is related only to the amount of crack extension, therefore:

$$\frac{dN}{dt} = \frac{dN}{da}\frac{da}{dt} = \frac{dN}{da} V \qquad\qquad (3)$$

This expression is plotted in Figure 12 as log dN/dt vs log V. If Eq. (3) is correct, these plots should be straight lines of slope equal to one. While the plots for both materials do not differ greatly from a straight line at the smaller velocities, their slopes are greater than one suggesting that:

$$\frac{dN}{da} = f(V) \qquad\qquad (4)$$

Further work is necessary before any firm relationship between acoustic emission rates and crack velocities can be established.

SUMMARY AND CONCLUSIONS

Slow crack growth was shown to take place in aluminas of varying purity and grain sizes, PZT, and barium titanate. Crack velocity was shown to be a strong function of the stress intensity factor at the crack tip. Grain size appeared to be the primary factor in determining crack propagation rates in the alumina, with the large grain material offering more resistance to growth. There was no observable effect of crack velocity on the fracture topography.

Critical values of stress intensity factor were determined from the slow crack growth data. These values of K_{Ic} were then used to calculate critical flaw sizes based on measured flexural strengths and elastic moduli. These calculated flaw sizes were always much larger than the grain sizes. When the value of the fracture surface energy for the ($10\bar{1}1$) plane of sapphire was used to calculate $\underline{a}$, values were obtained which agreed very well with the average grain diameter in those materials where the grain size was 20μ and greater. These results suggested that in finer grain size materials where flaws may extend over a number of grains, use of polycrystalline fracture energies yields reasonable results, while in those in which flaws are limited to a single grain, the use of single crystal values for γ_c is more correct. It was theorized that this kind of behavior may account for the discontinuity found in plots of strength versus (grain size)$^{-\frac{1}{2}}$ for many ceramics.

Crack velocities as a function of stress intensity factor were also determined for PZT and barium titanate. The distinct effect of water on crack velocity in PZT could easily be seen. Crack velocities in barium titanate were shown to be strongly dependent on whether the material was in a poled or unpoled condition. The higher crack velocities at a given K_I in the poled barium titanate were theorized to be due to strains in the material.

Finally, it was shown that there is a direct correlation between crack velocity and acoustic emission rate in alumina. Analysis of the experimental relationship between these parameters suggested that the number of emissions per unit increase in crack length, dN/da, was also a function of crack velocity.

ACKNOWLEDGEMENTS

The authors thank Mr. G. P. Kendrick for preparing the specimens, Mr. P. C. Miller for his help in obtaining the crack propagation data, Mr. H. H. Chaskelis for his aid in taking the acoustic emission data, and Mrs. J. N. Robinson for the electron micrographs.

We would also like to thank Dr. G. Scott of the General Electric
Co., who carried out the grain growth heat treatments on the
Lucalox specimens. Mr. R. W. Rice contributed much in the way of
helpful discussions during the course of this work.

REFERENCES

1. H. P. Kirchner and R. E. Walker, Mater. Sci. Eng., 8 301
   (1971).

2. B. K. Sarkar and T. G. J. Glinn, Trans. Brit. Ceram. Soc.,
   69, 199 (1970).

3. K. R. McKinney and H. L. Smith, J. Am. Ceram. Soc., 56, 30
   (1973).

4. M. H. Leipold, Jr. and A. H. Kelkar, Bull. Am. Ceram. Soc.,
   52, 344 (1973).

5. A. G. Evans and S. M. Wiederhorn, Bull. Am. Ceram. Soc.,
   52, 427 (1973).

6. A. G. Evans, J. Materials Sci. 7 1137 (1972).

7. S. W. Freiman, D. R. Mulville and P. W. Mast, accepted for
   publication in J. Materials Sci.

8. S. M. Weiderhorn, J. Am. Ceram. Soc., 52, 99 (1969).

9. G. D. Swanson, J. Am. Ceram. Soc., 55, 48 (1972).

10. P. L. Gutshall and G. E. Gross, Eng. Fract. Mech. 1, 463
    (1969).

11. F. J. P. Clarke, H. G. Tattersall, and G. Tappin, Proc. Brit.
    Ceram. Soc., 6, 163 (1966).

12. L. A. Simpson, J. Am. Ceram. Soc., 56, 7 (1973).

13. G. R. Irwin and P. C. Paris, Fracture 3 ed. by H. Liebowitz
    (Academic Press) (1971).

14. S. M. Wiederhorn, Mechanical and Thermal Properties of
    Ceramics, NBS Special Publication 303, 217 (1969).

15. S. M. Wiederhorn, J. Am. Ceram. Soc., 52, 485 (1969).

16. R. D. Sedlacek, F. A. Halden, and P. J. Jorgensen, The Science of Ceramic Machining and Surface Finishing, NBS Special Publication 348, 89 (1972).

17. R. J. Stokes, Ceramic Microstructures, ed. by R. M. Fulrath and J. A. Pask (John Wiley and Sons), 379 (1968).

18. R. Rice, This Volume.

19. J. A. Coppola, R. C. Bradt, D. W. Richerson, and R. A. Alliegro, Bull. Am. Ceram. Soc., 51, 847 (1972).

20. J. A. Coppola and R. C. Bradt, J. Am. Ceram. Soc., 55, 455 (1972).

21. D. B. Burns and P. Popper, Proc. Brit. Ceram. Soc., 6, 77 (1966).

22. S. W. Freiman, G. Y. Onoda, Jr. and A. G. Pincus, submitted for publication to J. Am. Ceram. Soc.

23. R. W. Rice, Proc. Brit. Ceram. Soc., 20, 205 (1972).

24. F.J.P. Clarke, Acta Met. 12, 139 (1964).

25. R. Sedlacek and V. Salmon, Final Technical Report Con. No. N0024-70-C-1224, Naval Ship Systems Command (1971).

PRECIPITATION STRENGTHENING IN NON-STOICHIOMETRIC Mg-Al SPINEL

G. K. Bansal[*] and A. H. Heuer

Department of Metallurgy and Materials Science

Case Western Reserve University, Cleveland, Ohio 44106

ABSTRACT

Non-stoichiometric Mg-Al spinel (MgO : 3.5 Al_2O_3), when decomposed at 850° for different times, form several metastable intermediate precipitates prior to the formation of the equilibrium corrundum phase. Knoop hardness and four point bending strengths have been measured at room temperature and show two maxima as a function of ageing time. Electron microscopic studies indicated that the first (short time) peak is associated with a pre-precipitation stage (Guinier-Preston zones?) and the second with the precipitation of the intermediate phases.

In addition, K (stress intensity factor) vs. V (velocity of crack propagation) diagrams have been determined using the double torsion technique for both as-grown and aged crystals. Samples aged to the pre-precipitate stage exhibited an increase of ~70% of K_{IC} compared to the single phase material.

[*]Now with Battelle Memorial Institute, Columbus Laboratories, Columbus, Ohio.

INTRODUCTION

It has been well established that precipitate particles increase the strength and hardness of metallic alloys [1,2]. On the other hand, only a limited number of experimental investigations of precipitation strengthening of ceramics have been made [3,4]. Of these ceramic systems, non-stoichiometric spinel is one of the more interesting, as it has found industrial application*.

Saalfeld and Jagodzinski [5], while studying the exsolution of Mg-Al spinels supersaturated with Al_2O_3, described three stages of exsolution, corresponding to the formation of pre-precipitates (also called G-P zones), the formation of an intermediate metastable structure, and the final exsolution of α-Al_2O_3. The effects of the precipitation reactions on the mechanical properties have been examined by several workers. Mangin and Forestier [6] found an increase in Knoop hardness of MgO : 3.5 Al_2O_3 spinel with time of annealing at 900°C. Eppler [7] observed that water-clear spinels became translucent upon ageing, which was accompanied by considerable increase in "corrosion hardness"; after heat treatment at 1000° for 12 hours, the hardness reached a maximum. Grabmaier and Falckenberg [8] measured the fracture load of flame-fusion grown Mg-Al spinels and found that the strength was influenced by the temperature of heat treatment.

During the present research, the precipitation sequence in MgO : 3.5 Al_2O_3 spinel has been followed using transmission electron microscopy and the effects of the various precipitate substructures on hardness, bend strength and slow crack growth have been determined.

MATERIAL AND METHOD

Single crystal boules of non-stoichiometric magnesium aluminate spinel (Al_2O_3/MgO = 3.5/1), grown by the Verneuil process, were obtained from Hrand Djevahirdjian Ltd., Monthey, Switzerland. Four point bending specimens, 1.0" x 0.11" x 0.05", were machine ground with faces parallel to (110), (1$\bar{1}$0), and (001) spinel planes; no further surface preparation was employed. Specimens were aged at 850°C for different lengths of time, quenched in air, and then tested at room temperature. The

---

*Spinels used for watch bearings are ground to shape in the quenched single phase state, but are then heat treated to cause precipitation hardening, before being put into service [4].

hardness values were obtained on mechanically polished and
optically featureless (001) surface, along $\langle 100 \rangle$ and $\langle 110 \rangle$
directions, using a Knoop indentor and a 300 gram load.

Specimens for transmission electron microscopy studies were
prepared by ion-bombardment from specimens used for mechanical
properties measurements. Electron transparent foils were coated
with a thin layer of carbon and were examined using a HU-650B
(CWRU) electron microscope. In addition, fracture surfaces of
the bend specimens were examined by scanning electron microscopy.

K (stress intensity factor) vs. V (velocity of crack
propagation) plots were obtained using the double torsion technique, a method suggested by Outwater and developed by Kies and
Clark [9] and Evans [10]. Rectangular plates, 2" x 0.5" x 0.05",
with faces parallel to $\{100\}$ (spinel) planes, were used for this
purpose. Specimens were aged at 850°C for different lengths of
time and were quenched in air. All the specimens were side
grooved ($\sim$0.025" deep) along the bottom face to guide the crack
along the specimen axis. The bars were further notched on one
end and were pre-cracked by pushing a knife edge into the notch
[11]. (The special problems involved in pre-cracking aged
specimens are discussed in the Appendix.) The load was applied
to the pre-cracked specimen at a cross head speed of ~ 0.2" min^{-1}
up to a pre-determined load; the cross head was arrested and the
load was allowed to relax. The crack velocities and K_I values
were obtained directly from the rate of load relaxation at
constant displacement, using the relationships discussed by
Evans [10] and McKinney and Smith [12].

RESULTS

The room temperature bending strengths and hardness along
$\langle 110 \rangle$ are shown in Figure 1. Both properties show two ageing
peaks, one at $\sim$ 15 minutes and the other at $\sim$ 25 hours. The
hardness values along $\langle 100 \rangle$ were always greater than along $\langle 110 \rangle$
but the variation with ageing times were similar for both
orientations.

Crack propagation data for the as-grown (i.e. single phase)
spinel are shown in Figure 2, along with data taken from the
literature for glass [10,13]. Spinel is much stronger than
glass; its strength is very similar to that of sapphire along
the (10$\bar{1}$1) rhombohedral plane [13]. The as-grown specimens
tested in toluene showed three regimes of crack-growth, behavior found
previously for glass and sapphire [10,13] and indicating that static
fatigue will play a role in the mechanical properties of spinel.
Three stage V-K plots may perhaps be typical of all oxide ceramics.

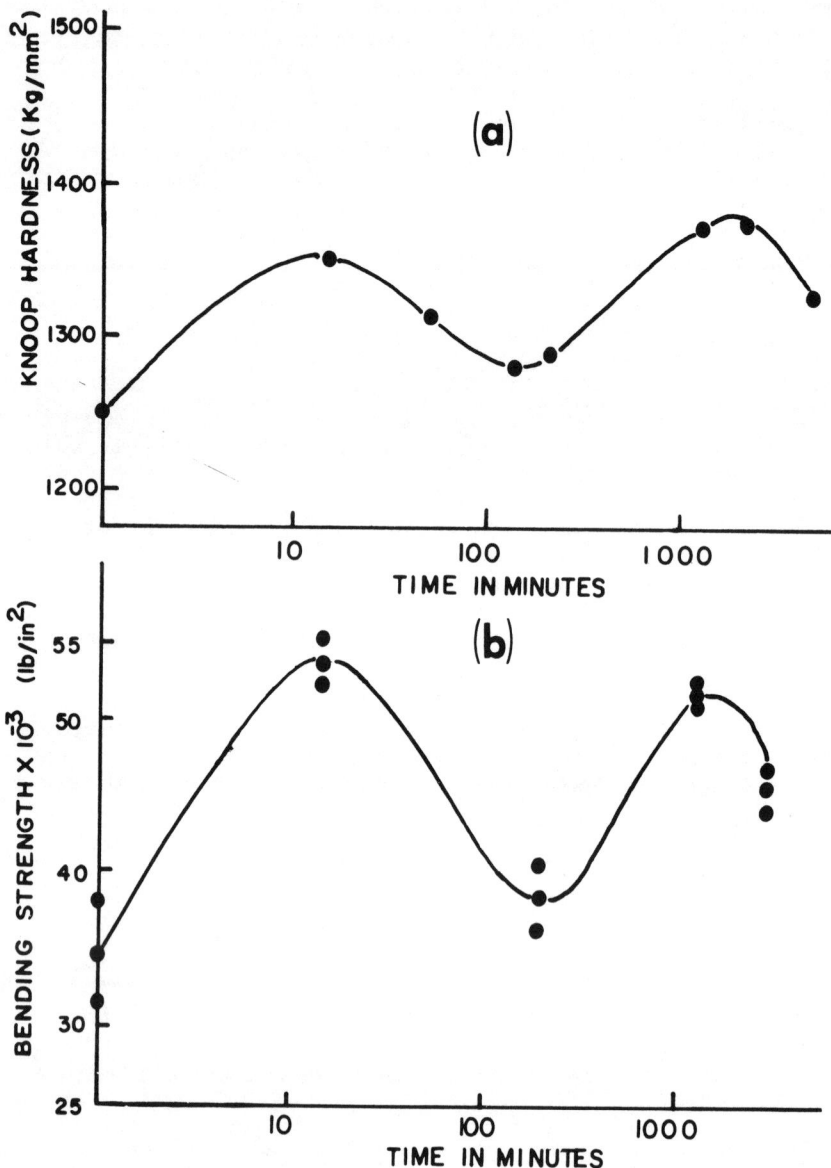

FIG. 1
Variation of Knoop hardness (a) and bend strength (b) with ageing
time at 850°C.

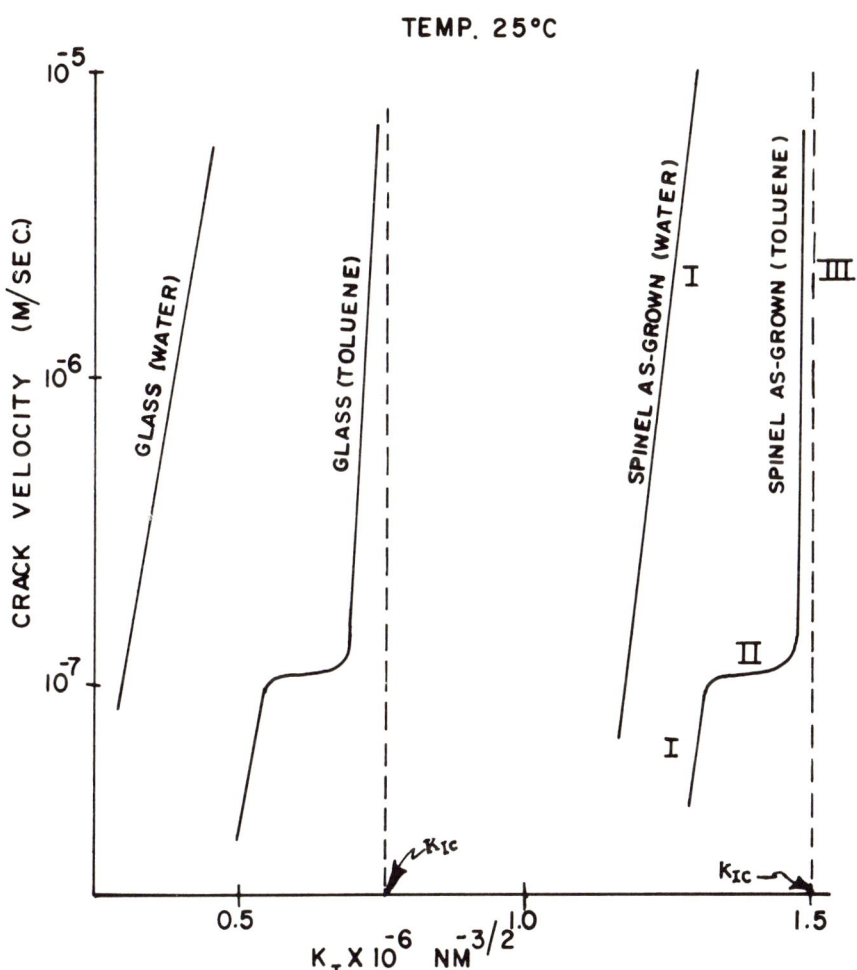

FIG. 2
V-K curves for glass [10] and as-grown spinel.

Region II, where the crack propagation is controlled by
diffusion in the corrosive environment (water) and is insensitive
to the structure of the test material for a given corrosive
species, occurs at a crack velocity of $\sim 10^{-7}$ meters/sec, which
is similar to the behavior of glass/toluene and alumina/toluene
systems. Aged crystals were tested only in distilled water at
25°C; stage I of the crack propagation for these specimens are
shown in Figure 3.

Microstructures which result at various stages of precipitation were examined using high voltage transmission electron
microscopy (HVEM). Only the salient features will be discussed
here, as a detailed analysis of the entire precipitation sequence
will be published elsewhere [14]. Microstructures corresponding
to the two peaks in Figures 1a and b are shown in Figures 4 and 5.
A very fine structure (less than $\sim 40\overset{\circ}{A}$), similar to spherical
Guinier--Preston (G-P) zones [15,16] in metals, results after $\sim$15
minutes of annealing at 850°C. (G-P zones are believed to have a
similar structure to and are coherent with the matrix.) As ageing
continues, these zones coarsen (G-P precipitates?), as shown in
Figure 4(b).

After longer ageing times, two monoclinic metastable intermediate phases form [17]. One of these (Type I) has an acicular
morphology, as shown in Figure 5, while the second intermediate
phase (Type II) has a lathe-type morphology (this latter precipitate phase is not shown here). Both phases preferentially nucleate
on dislocations and subgrain boundaries (Fig. 5); the abundance
of Type I precipitates is generally greater than Type II, because
its nucleation is easier. At longer ageing times, presumably both
of these intermediate phases are replaced by the stable corundum
phase. Figure 6 shows that the Type I metastable acicular needles
are still present, along with coarse platelets of the equilibrium
corrundum phase, in crystals aged for 10 days at 850°C. (Note
that this last microstructure corresponds to the over-aged
condition in Fig. 1).

DISCUSSION

Figure 1 shows that heat treatment of Mg-Al spinel at 850°C
caused the bending strength to increase by 60% and the Knoop
hardness to increase by $\sim$10%. The similarity of the hardness
and the bend strength trends with ageing times, as well as the
V-K plot (Fig. 3), all indicate that the improved properties
realized upon ageing can be attributed to the precipitation
sequence shown in Figs. 4-6, and not to any other possible effect
of the heat treatment. Specifically, the increase in bending
strength is due to the increasing difficulty of crack propagation

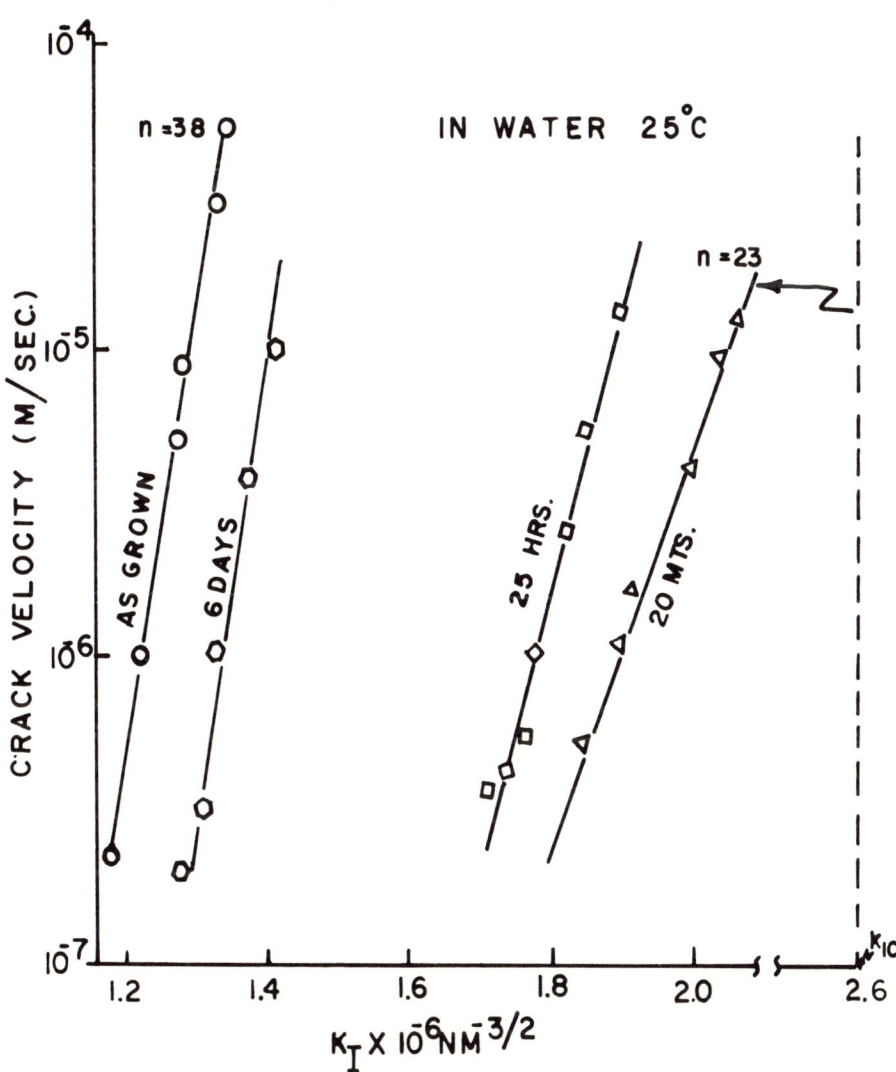

FIG. 3
V-K curves for as-grown and aged spinel crystals treated in
distilled water at room temperature.

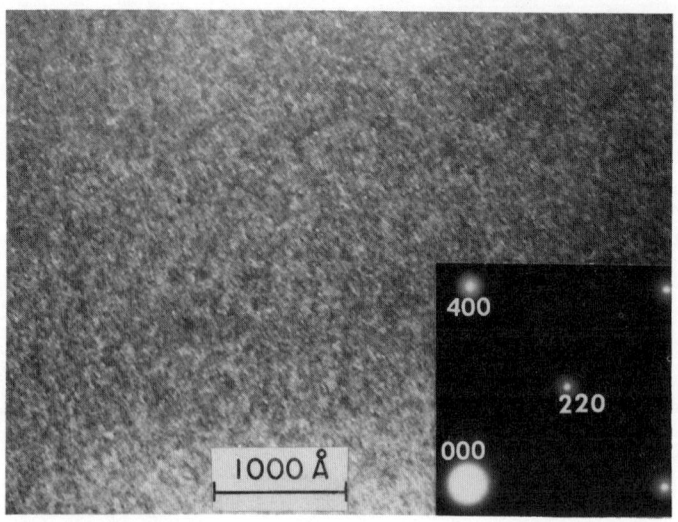

FIG. 4a.
Electron micrograph of very fine (~40Å) preprecipitates (G-P zones)
in non-stoichiometric spinels aged for 15 min. at 850°C. These
pre-precipitates give rise to diffuse halos around the matrix
diffraction spots. Bright field electron micrograph, 650 kV.

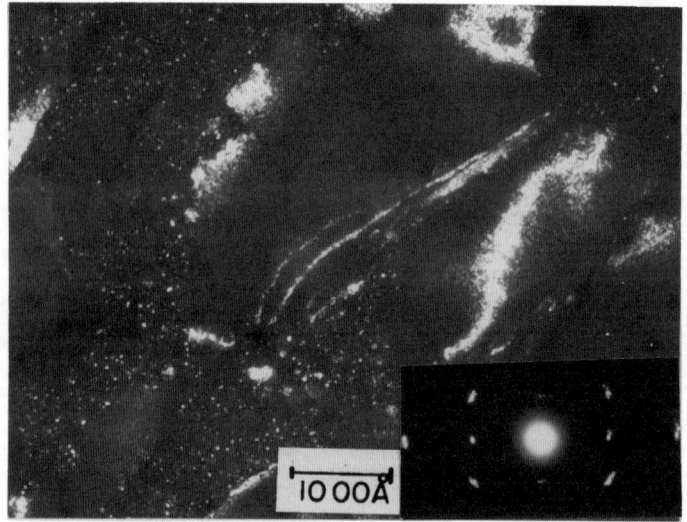

FIG. 4b
Electron micrograph showing G.P. precipitates, which give rise to
characteristic diffraction spots close to the matrix reflections.
Aged 3 hrs. at 850°C. Dark field electron micrograph.

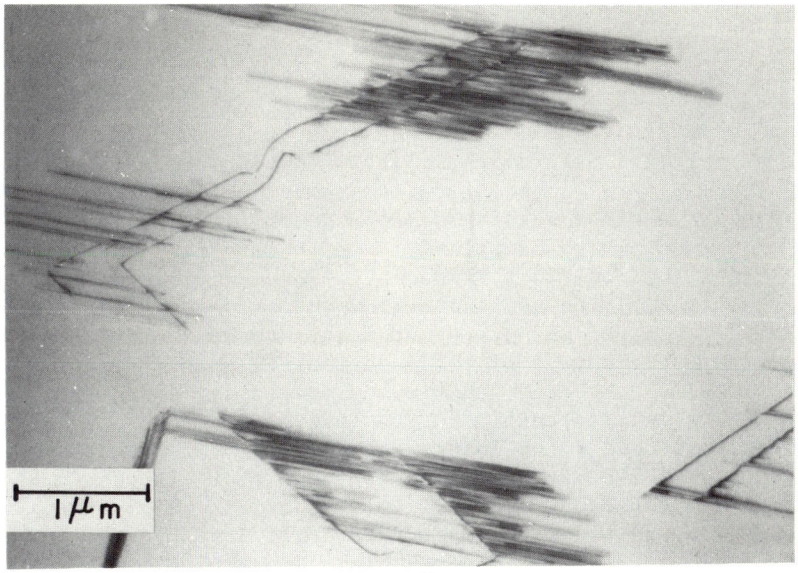

FIG. 5

Electron micrograph of Type I acicular metastable intermediate
precipitates, which have nucleated on dislocations. Aged 25 hours.
Bright field electron micrograph, 650 kV.

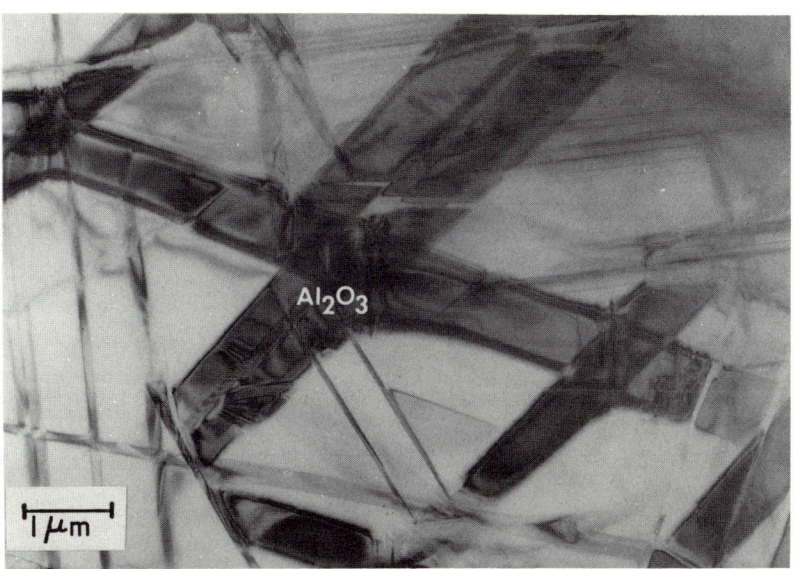

FIG. 6

Electron micrograph of sample aged 10 days, showing Type I
precipitate and platelets of the stable α-Al$_2$O$_3$ phase.

in the two-phase material compared to single phase (quenched)
spinel (Fig. 3), and not to any possible diminution of machining
damage as a result of thermal etching. (Based on analogy with
the thermal etching of alumina [18], temperatures in excess of
~1300°C would be needed before any noticeable healing of surface
flaws would occur in spinel.) Before discussing the bend strength
and crack propagation data further, it is appropriate to discuss
the microhardness data. There is little doubt that a microhardness indent requires massive plastic deformation, even in a
material with such a high brittle-ductile transformation temperature as spinel (cf. the work of Hockey [19] on microhardness
indents in alumina). The correlation of the hardness increases
with the development of the pre-precipitate structure (Fig. 4a)
and the maximum development of the intermediate phases (Fig. 5)
is strikingly similar to the behavior of age-hardened aluminum
alloys, and is almost certainly due to interaction of the
precipitates (presumed to be coherent and semi-coherent,
respectively (see below)) with dislocations. (This is the
subject of a separate study.) Similarly, the decay after the
first peak is caused by coarsening of the G.P. zones (Fig. 4(b))
while the decay after the second peak is associated with the
formation of the incoherent α-Al_2O_3 phase (Fig. 6). It is also
interesting that the hardness at the first (short time) peak is
less than that of the second peak (Fig. 1a), similar to the trend
found for metallic alloys [20]. At first sight, the similar
trends shown by the hardness and strength in Fig. 1 is remarkable,
in that the bend strength of spinel at room temperature is not
believed to involve much (if any) plastic deformation. However,
it has recently been proposed by Lange [21] and Evans [22] that
a propagating (primary) crack bows out between precipitates (just
as a dislocation bows out between obstacles) forming secondary
cracks in the matrix. Thus, a pseudo-line tension (similar to that
possessed by dislocations) exists for cracks, which can cause
considerable strengthening of multi-phase materials. To test this
theory, fracture surfaces of the bend bars were examined using
scanning electron microscopy and the results are shown in Figure 7.

The fracture surfaces of the single phase material (Fig. 7a)
and the "over-aged" sample (Fig. 7d) were much smoother and showed
less evidence for secondary crack formation than did samples aged
to give the maximum strength (Figs. 7b and 7c), in qualitative
accord with Lange and Evans'model. The crack propagation data
of Fig. 3 also supports this notion, in that K_{IC} for the sample
aged for 15 minutes is the highest of any tested; the sample of
the second ageing peak also has a much higher K_{IC} than either the
as-received or the over-aged samples.

Although the electron microscopy data is not yet fully in
hand, the microhardness and crack propagation data suggest that
the pre-precipitatesand the intermediate phases exhibit a degree

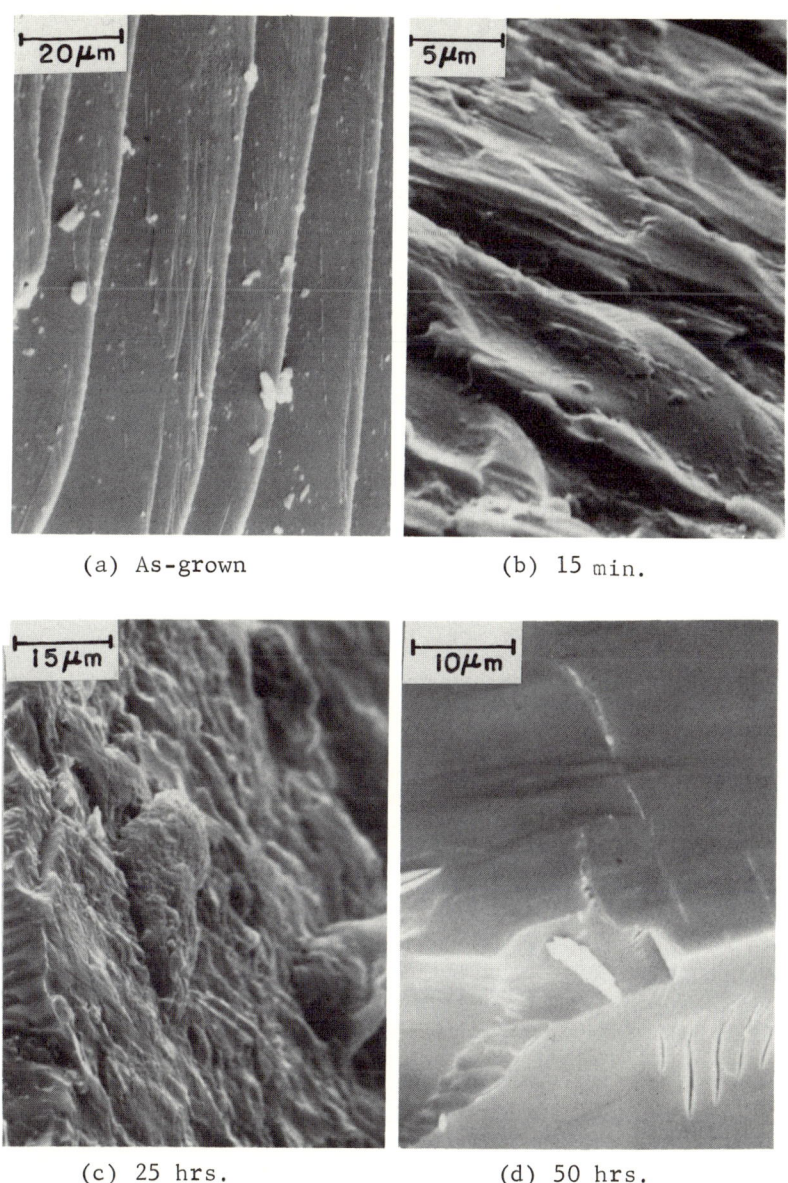

(a) As-grown   (b) 15 min.

(c) 25 hrs.   (d) 50 hrs.

FIG. 7

Scanning electron micrographs showing fracture surfaces of as-grown and aged crystals. Note the rough nature of the fracture surfaces in (b) and (c).

of coherency not possessed by the stable corundum phase. In addition, such fully or partially-coherent particles are much more effective in contributing to the pseudo-line tension of a crack than are incoherent precipitates. Finally, it is interesting that under the ageing conditions leading to the strongest samples (15 minutes), the lowest slope in region I was found (see Fig. 3), suggesting that the presence of coherent particles also improves the static fatigue properties.

CONCLUSION

Ageing of non-stoichiometric Mg-Al spinels leads to "age-hardening" phenomenon which is very similar to that observed in many binary metallic systems. The strongest materials are obtained by ageing spinel for a very short time, to give rise to very fine pre-precipitates (G.P. zones). A second ageing peak is observed and correlated with the maximum development of metastable intermediate phases. The increase in microhardness of aged crystals is attributed to the interaction of dislocations with the precipitate particles. The improvements in the strength are due to a crack-particle interaction, which appears to increase with the coherency between the precipitates and the matrix.

APPENDIX
PRE-CRACKING OF THE AGED CRYSTALS

As-grown and over-aged samples were pre-cracked successfully by the technique described by Davidge and Tappin [11] but considerable difficulties were experienced in attempting to pre-crack the stronger, correctly-aged samples by this technique. Much higher loads (by a factor of three) were required for crack initiation, which caused the crack to propagate at high velocities. It was thus difficult to arrest the crack at any desired position. Efforts were made to pre-crack the specimens in the as-grown state prior to ageing; however, the subsequent air-quenching caused complete cracking of the specimen along the long axis. Therefore, the following technique was developed for these specimens.

A slide groove (~0.025" deep) was first made on one face of the aged crystal to guide the crack and the specimen was then mounted horizontally in a waffering machine with the grooved face down. A short machined notch was then placed in the specimen using a 6 mil diamond saw blade. Before completing this operation, the specimen was lightly thrust against the saw blade; this produced a crack with a sharp front. With a little practice, the length of the crack could be controlled by the force with which the specimen was pushed against the blade. The translucent nature

of the aged crystals facilitated locating the crack position
without any external means. However, opaque and stronger
materials might also be pre-cracked by this technique, if the
crack front is monitored by some optical means.

ACKNOWLEDGMENT

The authors wish to thank our colleague I. Alvarez for
his help in conducting the V-K experiments. Thanks are also due
to T.E. Mitchell and G.L. Nord for their fruitful suggestions and
constructive criticisms. This research was sponsored by the
Office of Naval Research under Contract No. N00014-67-A-0404-003,
NR 032-508.

REFERENCES

1. R.B. Nicholson, G. Thomas and J. Nutting, Acta Metallurgica,
    8, 172 (1960).

2. D. Dew-Hughes and W.D. Robertson, Acta Metallurgica, 8, 156
    (1960).

3. G. W. Groves, "Precipitation Strengthening of Ceramics" in
    Strengthening Methods in Crystals, edited by A. Kelley and
    R.B. Nicholson, Halsted Wiley Press (1971).

4. M.E. Fine, Amer. Ceram. Soc. Bull., 51 [6], 510 (1972).

5. Von H. Saalfeld and H. Jagodzinski, Zeitschrift fur
    Kristallographie, 109, 87 (1957).

6. A. Mangin and H. Forestier, Academie des Sciences, Seance
    Du 9 Avril (1956).

7. W. Fr. Eppler, Zeitschrift fur Angewandte Mineralogie, Band IV,
    4, 345 (1943).

8. Grabmaier and Falckenberg, J. Amer. Ceram. Soc. 52 [12], 648
    (1969).

9. J.A. Kies and A.B.J. Clark, Proc. Br. Cer. Soc. 15, 59 (1970).

10. A.G. Evans, J. Mat. Sci. 7, 1137 (1972).

11. R.W. Davidge and G. Tappin, Proc. Second International Conference
    on Fracture, Brighton 1969, paper 42.

12. K.R. McKinney and H.L. Smith, J. Amer. Ceram. Soc. $\underline{56}$, [1] 30 (1973).

13. S.M. Wiederhorn, in Mechanical and Thermal Properties of Ceramics, NBS Special Publication 303, edited by J.B. Wachtman, 1969.

14. G.K. Bansal and A.H. Heuer, Abstract Submitted to Amer. Ceram. Soc. Meeting, Cincinnati, Ohio, May 1973.

15. A. Guinier, Nature, $\underline{142}$, 569, Sept. 24, 1938.

16. G.D. Preston, Nature, $\underline{142}$, 570, Sept. 24, 1938.

17. M.H. Lewis, Phil. Mag. VIII, $\underline{20}$, 985 (1969).

18. M.J. Noone and A.H. Heuer, NBS Special Publication 348, 218 (1972).

19. B.J. Hockey, J. Amer. Ceram. Soc. $\underline{54}$ [5], 223 (1971).

20. P.G. Shewmon, Transformations in Metals, McGraw-Hill Book Co. pp. 286-321 (1969).

21. F.F. Lange, Phil. Mag., $\underline{22}$, 983 (1970).

22. A.G. Evans, Phil. Mag. VIII, $\underline{26}$, 1327 (1972).

FATIGUE FRACTURE OF AN ALUMINA CERAMIC AT SEVERAL TEMPERATURES

C.P. Chen and W.J. Knapp

School of Engineering and Applied Science

University of California, Los Angeles, California 90024

ABSTRACT

Specimens of a polycrystalline alumina ceramic (96% Al_2O_3) were subjected to static and cyclic loading to fracture at several temperatures in a dry argon atmosphere containing less than one part per million each of water vapor and oxygen. A set of specimens also was tested in humid air (50% relative humidity) at room temperature under static loading as a reference. All the specimens were notched, and were loaded in four-point bending fixtures so as to produce 94% of the nominal fracture stress in static loading and a peak stress of 94% of the nominal fracture stress in cyclic loading.

The resistance to both static and cyclic fatigue of this material decreased with increasing temperature from 25°C to 216°C and then increased as the temperature was raised to 359°C. At room temperature, the resistance of the material to fatigue under cyclic loading was lower than to fatigue under static loading. This weakening effect due to cyclic loading decreased with increasing temperature. Above 216°C the cyclic fatigue effect was less significant. The fatigue life of specimens tested in humid air under static loading was drastically lower than that for specimens tested in dry argon.

The dominant mechanism of fatigue of the alumina ceramic of this study, at temperatures ranging from 25°C to 216°C, appeared to be a thermally-activated corrosion reaction with water vapor. The apparent activation energy for fatigue of this material is 8.8 Kcal/mole for static loading and 6.3 Kcal/mole for cyclic loading.

INTRODUCTION

Ceramics are increasingly considered for applications in which the material would be subjected to steady loading, or repeated loading, in tension for prolonged periods. Consequently, the use of ceramic materials for structural elements and mechanical components has raised questions concerning their fatigue behavior. The phenomenon of time delay to fracture when a ceramic is subjected to a steady stress is commonly termed static fatigue, or, if it involves delayed fracture under repeated stress, cyclic fatigue.

In spite of prior investigations[1-6] on fatigue fracture of polycrystalline ceramics, the mechanisms causing static and cyclic fatigue fracture require more investigation. Most of these previous studies involving cyclic or static loading have been carried out at room temperature and under ordinary atmospheric conditions.

Attempts have been made to compare cyclic loading with static loading by Sedlacek and Halden,[7] who studied the fatigue behavior of an alumina under conditions of constant and cyclic tensile stress with the use of a ring specimen loaded hydraulically. They indicated that the resistance to cyclic stresses is somewhat lower than that for static stresses. A quantitative comparison of the fatigue behavior of a polycrystalline alumina under various loading conditions has been reported by Krohn and Hasselman.[8] Their results demonstrate that a cyclic fatigue mechanism exists in Al_2O_3 which depends upon the amplitude and frequency of cyclic stress. Glasses have been extensively studied, which are susceptible to delayed fracture in the presence of moisture. (Distinctive review papers in this subject are by Charles,[9] Hillig,[10] Mould,[11] Ernsberger[12] and Widerhorn[13].) The impairment of strength of glass is proposed to be a stress-enhanced chemical reaction between glass and water vapor in the environment, to form one of the many hydrated states of the compound. The role of cyclic stress on the fatigue behavior of glass has been investigated by Gurney and Pearson[14] and Proctor et al.[15] and their results suggest no cyclic loading effect.

The purpose of this investigation was to study the fatigue fracture of a polycrystalline alumina ceramic under static and cyclic loadings at several temperatures. A direct comparison of static fatigue and cyclic fatigue at several temperatures was expected to throw more light on the mechanisms of fatigue of ceramic materials. Since the existence of another mechanism which causes crack propagation, in addition to stress corrosion, was to be considered, most of the experiments were conducted in a dry argon atmosphere to minimize the stress corrosion effect.

Experimental Procedure

1) Material. The material used in the study was a polycrystalline alumina ceramic,[*] comprised of 96% Al_2O_3 and 4% additives (clay and talc). The test specimens were obtained from the manufacturer in the form of bars with a machined square cross-section approximately 0.124 inches on each side, and 4 inches in length. A microscopic study of the ceramic material showed the presence of a well-distributed glassy phase at grain-boundaries. Specimens for fatigue testing were notched in the laboratory with a diamond saw. The height and width of each specimen were measured individually with a micrometer and the notch was measured with a projecting comparator.[**] These dimensions were used in the calculation of bending stress. Certain physical properties of the specimens were determined, and the strength values for un-notched specimens and notched specimens were obtained by loading in air in four-point bending, using an Instron testing machine[†] with a cross head speed of 0.02 in/min. The fracture stress of an un-notched specimen was determined to be σ_f=40.62 × 10^3 psi, (standard deviation = 5.16 × 10^3 psi) and the nominal fracture stress of a notched specimen was σ_n-25.20 × 10^3 psi (standard deviation = 1.37 × 10^3 psi), both calculated from elementary strength theory.

To minimize the stress corrosion effect due to moisture absorbed on the surface of the specimen during notching, all test specimens were heated in air to 225°C in a drying oven. After holding at this temperature for overnight in the oven, they were then removed to a dessicator for slow cooling, and later were transferred to the glove box.

2) Apparatus. All fatigue tests were performed in a glove box system,[††] which provided an isolated working area with an atmosphere of dry argon. Moisture and oxygen were continually removed from the box area by a dessicator and oxygen furnace, respectively, minimizing these constituents to less than one part per million of each.

Each test specimen was loaded in four-point bending. The test apparatus, as shown in Figure 1, consisted essentially of a frame for supporting the loading system, lead loading weights, an adjustable weight holder, a spring connected to the weight holder, a furnace, and a moving table which oscillated vertically. The testing temperature was controlled by an electric furnace which was lowered to surround the specimen and fixture. For a static

---

[*]AD-96, H.F. Coors Company, Inglewood, California.
[**]Manufactured by Bausch & Lomb Optical Co., Serial No. VD5350.
[†]Type TT-B-1 and TT-C-L, Instron Corp., Canton, Mass.
[††]Dry-Lab-Dry-Train, Vacuum/Atmosphere Corp., Model No. HE-43-6 and HE-93-BIN.

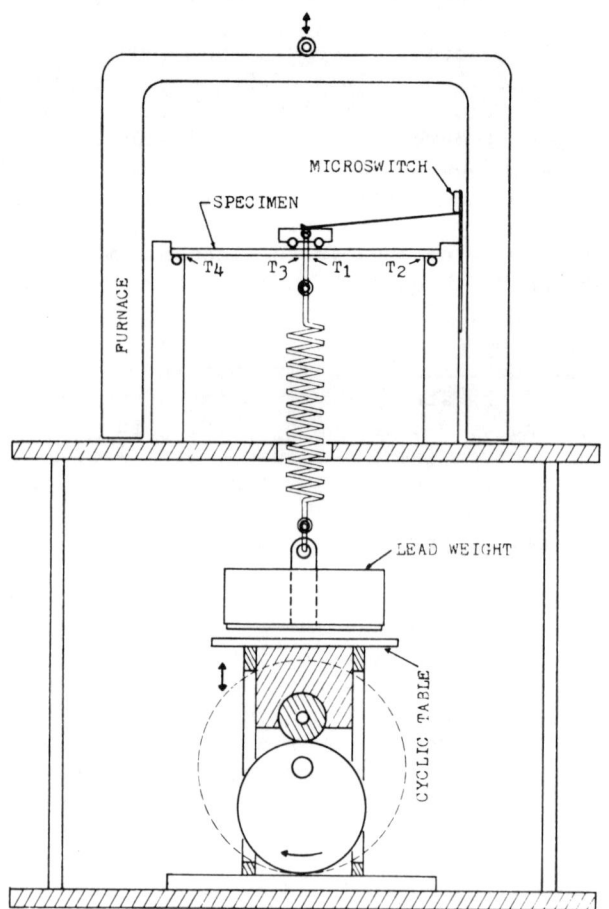

Figure 1.   Schematic Diagram of the Test Apparatus. T_1- T_4:
Positions of Thermocouples

fatigue test, the fatigue life time was recorded with a timer
which was connected to a microswitch. For a cyclic test, the
number of cyclics was registered with a ratchet counter. Two
individual loading systems of the type described above were built
inside the controlled atmosphere glove box, and permitted the
simultaneous loading of two specimens. In each loading system,
two sets of thermocouples were attached to each specimen in the
positions shown in Figure 1.

3) Fatigue Tests. Fatigue testing was initiated by arranging
the two loading systems, in the controlled-atmosphere glove box,
so as to develop an identical fatigue stress level of 23.7×10^3
psi (94% of the nominal fracture stress) in each notched specimen.
After the specimen in the furnace reached the required temperature,

a static fatigue test was started by applying the full load,
through the spring, by lowering the moving table to its minimum
position. This condition of loading was maintained until fracture
occurred, or until a decision was made to terminate the loading.
In addition to the fatigue testing in dry argon, for comparison
fifteen specimens were tested under static loading in a humid air
atmosphere (40% - 50% relative humidity), using otherwise identical
experimental conditions.

To study the effects of repeated-loading, as compared with
those of static loading, cyclic fatigue tests were performed under
identical maximum loading conditions as in the static test. Figure
1 illustrates schematically the cyclic loading apparatus, in which
the required weight was attached to the center loading fixture
through a spring. The time at peak loading (complete suspension
of weight) was held nearly constant at 1.5 seconds for cyclic
tests. This was accomplished for different loading levels by
adjusting the length of the weight holder with a pre-calibrated
load-extension curve of the spring. A typical load-time profile
for cyclic loading is shown in Figure 2.

Five temperature levels (room temperature, 113°C, 216°C,
304°C and 359°C) were used to perform the static and cyclic
fatigue tests in a dry argon atmosphere. Subsequently, a set of
tests under static loading was conducted in air (at 40% to 50%
relative humidity) at room temperature.

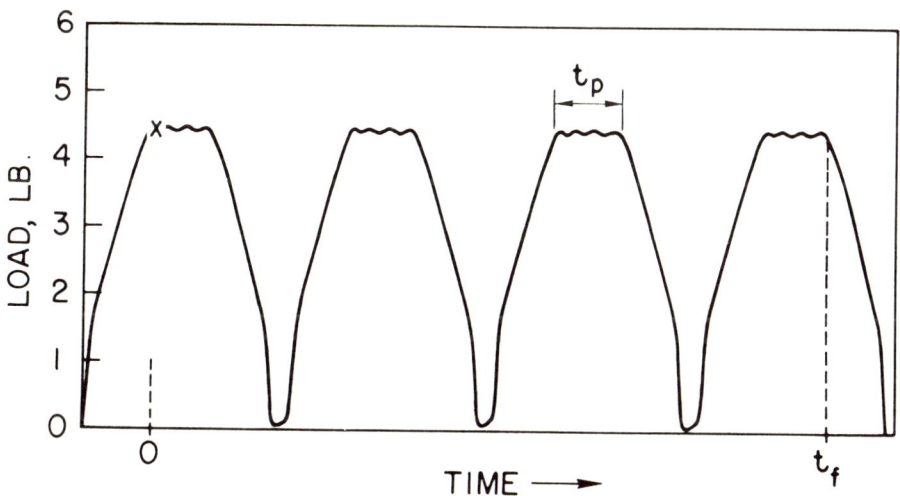

Figure 2. A Typical Load-Time Profile for Cyclic Fatigue Loading.
Cyclic Fatigue Life is $0-t_f$. Time to Fracture at Cyclic Peak
Loading = $\Sigma \ t_p$. Where t_p = 1.5 sec.

Experimental Results

The experimental data show that fatigue life is an extremely variable characteristic which requires some processing to make it tractable. The conventional method of representing the observed quantities by the arithmetic mean of the fatigue life and its standard deviation can not give a meaningful correlation. However, the cumulative fracture probability G may be expressed, using the Weibull distribution equation,[18] in terms of the logarithm of the fatigue life time[19] in the following form

$$G = 1 - \exp\left[- \int_A \left(\frac{L_t - L_u}{L_o} \right)^m dA \right] \qquad (1)$$

since the specimen was loaded in four-point bending, the surface flaw distribution governs the fracture of the specimen. Where

L_t = the logarithm of time to failure

L_u = lower bound of the logarithm of time to fracture

L_o = scale parameter

dA = surface element under stress

m = Weibull Modulus.

In this testing, since the surface area under bending moment stress is constant in all the specimens, Equation (1) can be expressed as

$$G = 1 - \exp\left[- \left(\frac{L_t - L_u}{L_o} \right)^m A \right] \qquad . \qquad (2)$$

A detailed discussion of this analysis is presented in Reference 16.

The cyclic fatigue life time at various temperatures is represented in two ways: (a) the total time to fracture at peak cyclic loading, (b) the total life time to fracture under cyclic loading. (See Figure 2) It was desired to determine meaningful representative cyclic endurance values for comparison with static endurance values. The stress level applied on the specimen at the time under peak loading of cyclic fatigue is selected to be identical to those of the corresponding static fatigue tests. A direct comparison of the total time to fracture at peak cyclic loading with the life time under static loading may shed light on the mechanism of cyclic fatigue fracture.

The times to fracture, under static loading producing 94% of the nominal fracture stress, are plotted against failure probability in Figure 3 for the tests conducted at several temperatures in argon, and for the test conducted at room temperature in air.

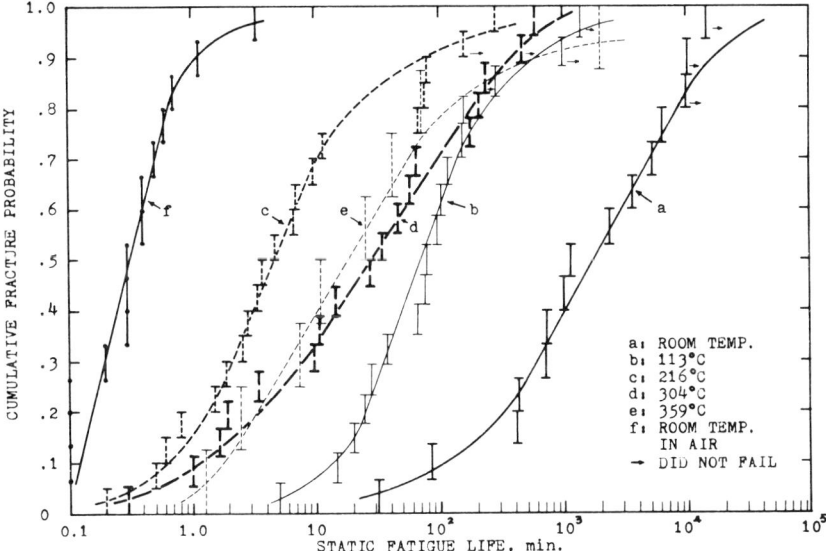

Figure 3.   Time to Fracture under Static Loading vs.
Fracture Probability

In Figure 3 the fracture probability is shown in each 1/N
interval, where N is the sample size. A G value for any observation is calculated by i/(1+N). [Reference 20]

Similarly, the total times to fracture at the peak cyclic
loading of various temperatures in argon are plotted against
fracture probability in Figure 4. Figures 5 to 7 show the comparisons of cyclic fatigue life with static fatigue life at
various temperatures. The cyclic fatigue life times shown in
Figures 5 to 7 were represented in two ways; one curve shows the
total time to fracture under cyclic loading and the other curve
shows the total time to fracture at peak cyclic loading (as
defined in Figure 2). At each temperature, a direct comparison
of the total time to fracture at peak cyclic loading with the life
time under static loading may provide an insight regarding cyclic
fatigue effects.

Discussion

1) Atmosphere. One of the most dramatic influences on static
fatigue fracture of the alumina ceramic was shown by the testing
atmosphere. The results from a set of 15 specimens tested in air
(50% relative humidity) may be compared with those from a set of
15 specimens tested in the dry argon atmosphere. As shown in
Figure 3, the fatigue life times for those specimens tested in dry

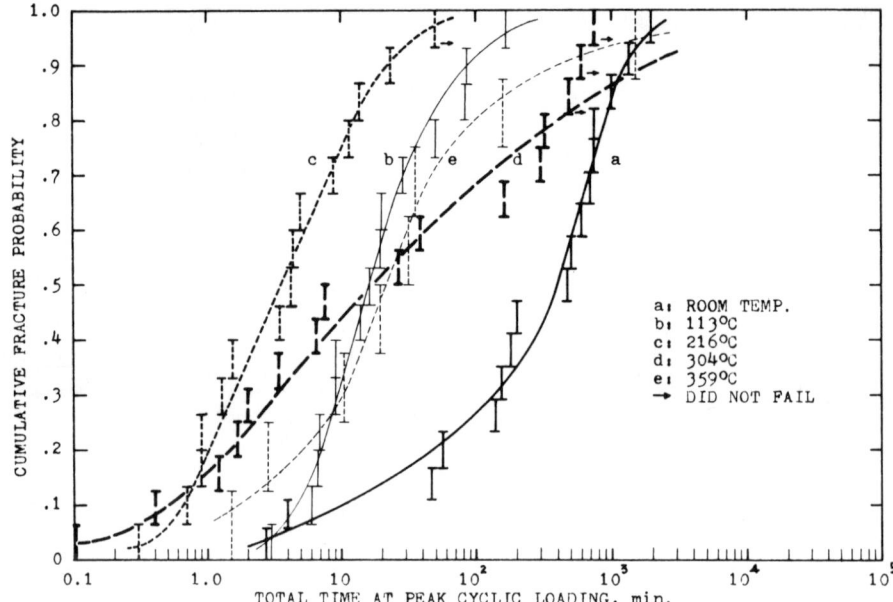

Figure 4.   Total Time to Fracture at Peak Cyclic Loading vs.
Fracture Probability

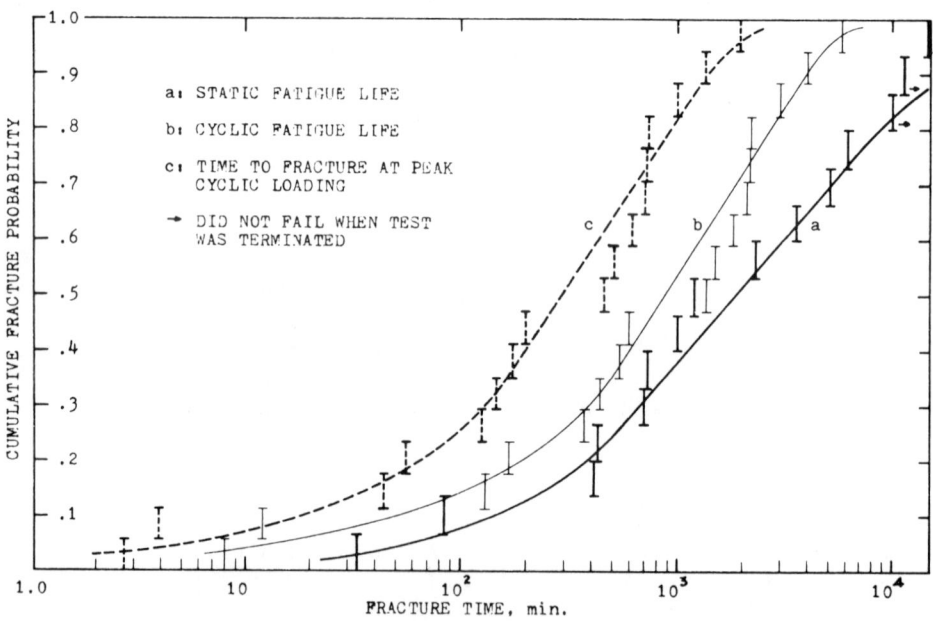

Figure 5.   Static and Cyclic Fatigue Tests at Room Temperature.
Cumulative Fracture Probability vs. Fracture Time

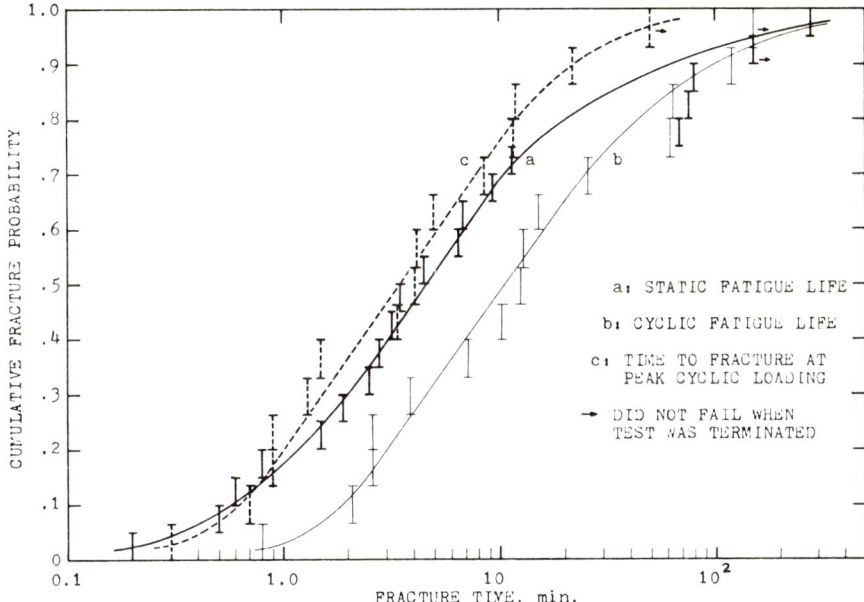

Figure 6.   Static and Cyclic Fatigue Tests at 216°C. Cumulative
Fracture Probability vs. Fracture Time

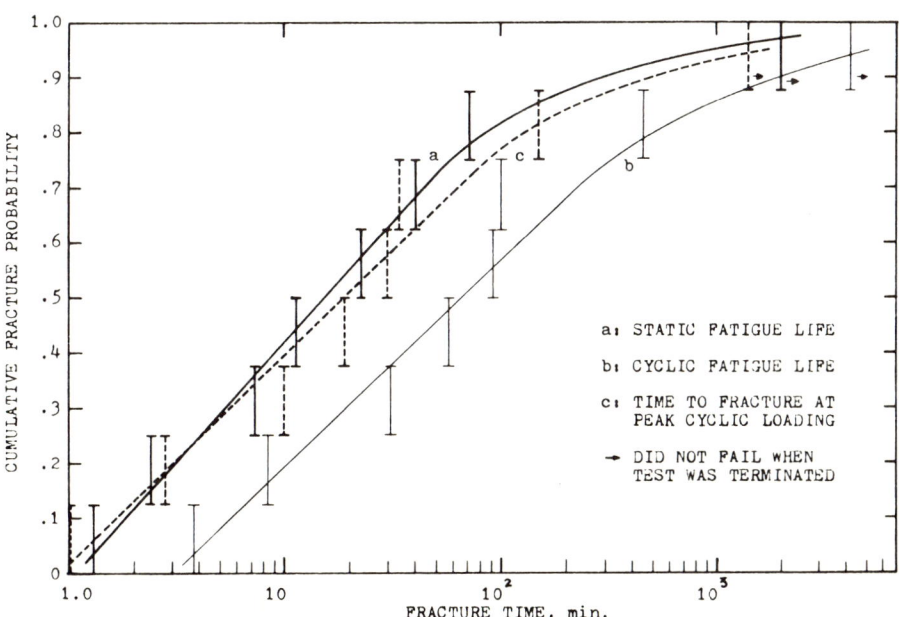

Figure 7.   Static and Cyclic Fatigue Tests at 359°C. Cumulative
Fracture Probability vs. Fracture Time

argon at room temperature have a range of time to fracture from 33
minutes to 14,500 minutes (did not fracture), and the mean (equivalent) fatigue life is about 1,400 minutes. The fatigue life
times for those specimens tested in air at 40% to 50% relative
humidity have a range from 0.1 minute to 3.3 minutes, with a mean
(equivalent) fatigue life of only 0.3 minutes. It is well known
that static fatigue fracture of ceramic materials is due to stress
corrosion resulting from a chemical interaction between the
stressed ceramic material and water vapor in the atmosphere.
Therefore, even for those fatigue tests conducted in a dry argon
atmosphere, the water vapor content in the atmosphere is very
important. In the atmosphere-controlled system of the glove box,
three furnaces were operated continuously to remove oxygen, nitrogen and moisture from the argon, and it was reported by the system
manufacturer that the oxygen and moisture contents of the circulating atmosphere were reduced to a few parts per million. A simple
test to check the oxygen and moisture contents in the atmosphere
was used by energizing a 25 watt light bulb with the glass envelope removed in the glove box; if the atmosphere is in the area
of 1 or 2 parts per million of oxygen and moisture, the light will
burn for several hours, since prior experience has shown that the
filament of the light bulb will soon burn out when exposed to
more than five p.p.m. of oxygen and moisture. The light bulb test
was used from time to time during these fatigue experiments, and
in each case the light bulb with its glass envelope removed
stayed illuminated for a week inside the glove box, indicating
that the oxygen and moisture content in the testing atmosphere
was something less than 1 or 2 parts per million.

2) Temperature. One of the major objectives of this investigation is to study the influence of temperature on the static and
cyclic fatigue fracture of an alumina ceramic in a controlled dry
environment. Very limited data from previous studies are available. On the other hand, the temperature dependence of the nominal strength of polycrystalline alumina ceramics has received much
consideration.[21,22] Most of these studies were concerned with
the strength of alumina ceramics of high purity. One such result
has been reported by Davidge and Tappin[23] on the temperature
dependence of fracture stress for a 95% alumina ceramic, indicating
that its strength experiences a small decrease with increasing
temperature up to about 500°C, then an increase in strength rising
to a very sharp peak at about 850°C. The peak strength, according
to these authors, is associated with viscous flow in the glassy
silicate phase, since this peak is not observed in pure polycrystalline alumina tests. The temperature region of interest in the
present work is from room temperature (25°C) to 359°C. In this
temperature range, the strength of most alumina ceramics normally
exhibits a small decrease with increasing temperature. For the
tests of the present work, it can be seen (Figures 8 and 9) that

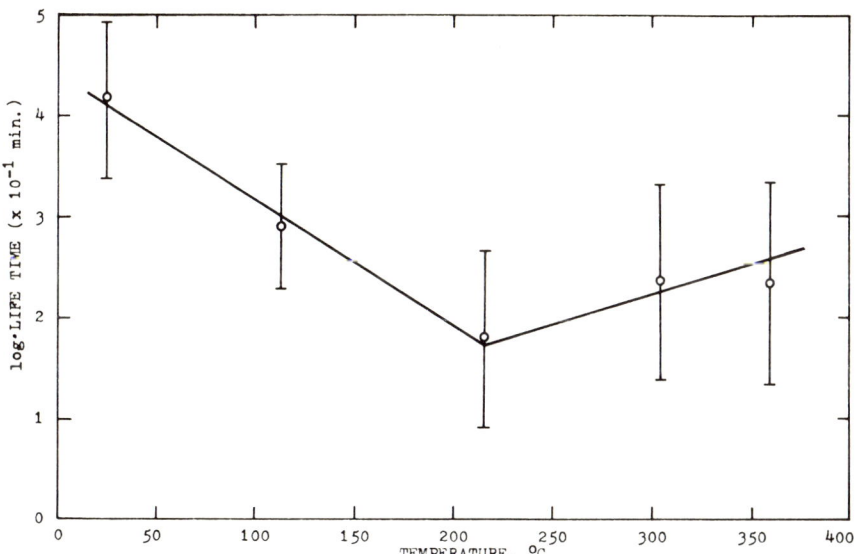

Figure 8. Time to Fracture under Static Loading in Argon. Points
and Bars Indicate Mean log·Fatigue Life ($\times 10^{-1}$min.) $\pm$ 1 Standard
Deviation of L_t

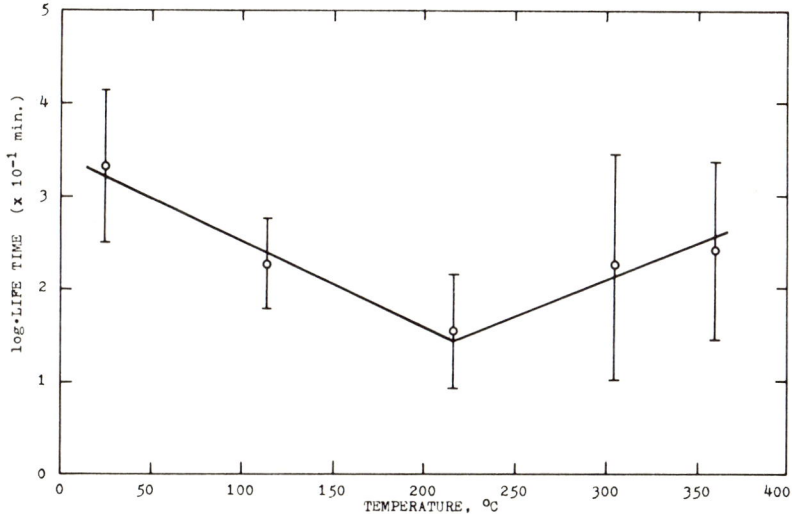

Figure 9. Time to Fracture under Peak Cyclic Loading in Argon.
Points and Bars Indicate Mean log·Fatigue Life ($\times 10^{-1}$min.) $\pm$ 1
Standard Deviation of L_t

the time to fracture under both static and cyclic loading tests
decreases several orders of magnitude from room temperature to
about 216°C. Beyond this temperature, the time to fracture
increases again. This phenomenon was first observed for the
static fatigue of glass.[24] It is an interesting coincidence that
the minimum strength of glass is observed at about 200°C, and that
it has been interpreted in terms of a stress-corrosive attack of
water vapor on the silica network. For sintered glass-free
alumina ceramics,[26] this minimum strength was observed at higher
temperatures (about 600°C).

Scanning electron microscope examinations of fracture surfaces revealed that the fracture occurs mainly as an intergranular
process as seen in Figure 10. Transgranular fracture on the fracture surface was often observed, as shown in Figure 11, and was
considered to be due to the stress concentration at grain boundary
triplets where pores normally exist.

It is quite evident that excessively large pores, significant
porosity and inclusions did influence the fatigue strength.

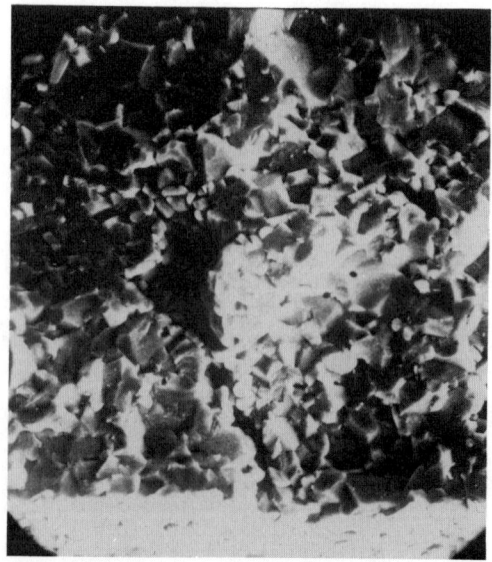

Type of Loading: Cyclic
Temperature: 304°C
Fatigue Life: 0.1 minute
Cumulative Fracture Probability: 5.9%

Figure 10.   Intergranular Fracture Near the Notch.   An Excessively
Large Pore is Observed

Type of Loading: Static
Temperature: 216°C
Fatigue Life: 79.8 min.
Cumulative Fracture Probability: 86%

Figure 11. Transgranular Fracture

Examination of the fracture surfaces of the specimen at each
temperature and at each loading condition, reveals this individually and collectively. In addition, more than four percent of
a glassy silicate phase was observed in the grain boundaries of
the material of this investigation. The strength of the grain
boundary material was recognized to be weaker than the strength of
grains. Therefore, it may be concluded that stress corrosion of
the grain boundary glassy phase is the controlling factor for
fatigue fracture of the alumina ceramic. Furthermore, Davidge and
Tappin[23] verified that the glassy phase in a 95% alumina ceramic
did play an important role in the fracture of the ceramic.

    The temperature dependence of mean fatigue life time under
both static loading and peak cyclic loading is shown in Figure 12.
An apparent activation energy for static fatigue was calculated,
using the plot of Figure 12, to be 8.8 kcals/mole, and a similar
calculation for cyclic fatigue gives 6.3 kcals/mole. We may compare the above results with those for the static fatigue of a
soda-lime glass in air as studied by Charles,[27] where the activation energy was calculated to be 18.8 kcals/mole. Later, Charles
and Shaw[4] determined the activation energy for static fatigue of

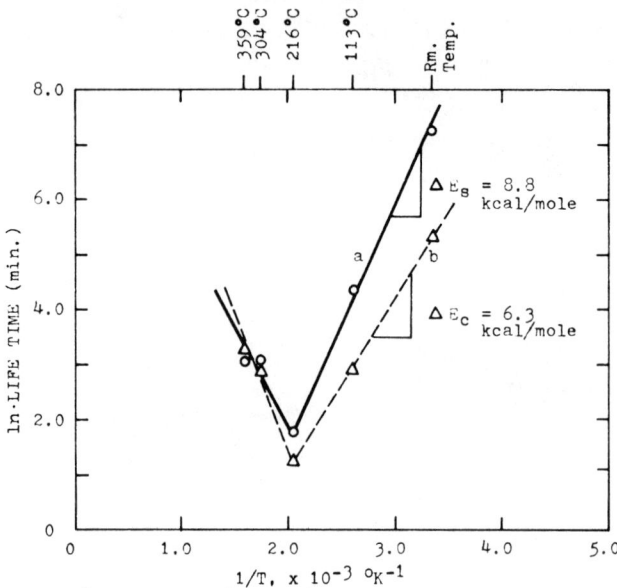

Curve a: Equivalent Mean Fatigue Life
Time (min.) at Static Loading
Curve b: Equivalent Mean Fatigue Life
Time (min.) at Peak Cyclic Loading

Figure 12.   Temperature Dependence of Fatigue Life Time under
Static Loading and Peak Cyclic Loading

single crystal sapphire to be 14 kcals per mole. For the polycrystalline alumina ceramic of this study, the apparent activation
energy values for static fatigue of 8.8 kcals per mole, and for
cyclic fatigue of 6.3 kcals per mole, seem reasonable. It can be
concluded that the fatigue fracture of the alumina ceramic of this
study, under either static or cyclic loading, is a thermally-
activated process up to 216°C, above which additional mechanisms
are operative.

Conclusions

(1) The observed results are consistent with those of previous investigators, and indicate that the static fatigue of a
polycrystalline ceramic results from stress corrosion with water
vapor in the atmosphere.

(2) The dominant mechanism of fatigue of the alumina ceramic
of this study, at temperatures ranging from 25°C to 216°C, appears
to be a thermally-activated corrosion reaction with water vapor.

The apparent activation energy for static fatigue of this material is 8.8 kcal/mole and for cyclic fatigue is 6.3 kcal/mole.

(3) A minimum resistance to fatigue loading (both static and cyclic loading) of this material was shown at about 216°C. It is predicted that glass-free sintered alumina materials should possess a minimum fatigue strength temperature of about 600°C.

(4) At room temperature, the resistance of the polycrystal-line alumina ceramic to cyclic loading is lower than to static loading. This weakening effect due to cyclic loading decreases with increasing temperature. Above 216°C a cyclic fatigue effect does not appear.

(7) The fatigue behavior of the polycrystalline alumina ceramic at each temperature can be written as a Weibull distribution function in terms of the logarithm of the fatigue life time. The probability of fracture is expressed as

$$G = 1 - \exp\left[- \int_A \left(\frac{L_t - L_u}{L_o} \right)^m dA \right]$$

The logarithm of the time to fracture L_t of each loading at each temperature has been shown to be normal distribution.

ACKNOWLEDGMENTS

Appreciation is expressed for the valuable advice and assistance provided by Professor Peter Kurtz, as well as for the suggestions of Professors George H. Sines and J.D. Mackenzie. The preparation of thin sections by Professor John M. Christie was particularly helpful.

REFERENCES

1. Williams, L.S., "Stress-endurance of sintered alumina," Trans. Brit. Ceram. Soc., 55: 287-312, 1956.

2. Mizushima, J.S. and W.J. Knapp, "Behavior of a ceramic under cyclic loading," Ceramic News, 5: 26-29, 1956.

3. Pearson, S., "Delayed fracture of sintered alumina," Proc. Phys. Soc., (London), 69: 1293-1296, 1956.

4. Charles, R.J. and R.R. Shaw, "Delayed failure of polycrystal-line and single-crystal alumina," General Electric Research Laboratory Report, No. 62-RL-3081 M. July 1962.

5. Williams, L.S., "Fatigue and ceramics," Paper No. 18 in
    <u>Mechanical Properties of Engineering Ceramics</u>, W.W. Kriegel
    and H. Palmour III, (eds.) Interscience, New York, 1961,
    p. 245-302.

6. Acquaviva, S.J., "A method for fatigue testing of ceramic
    materials," <u>Review of Scientific Instruments</u>, 42[12]: 1858-
    1859, 1971.

7. Sedlacek, R. and F.A. Halden, "Static and cyclic fatigue of
    alumina," <u>Structural Ceramics and Testing of Brittle
    Materials</u>, S.J. Acquaviva and S.A. Bortz, (eds.) Proc. Semi-
    nar at IIT, March 1967, p. 211-220.

8. Krohn, D.A. and D.P.H. Hasselman, "Static and cyclic fatigue
    behavior of a polycrystalline alumina," <u>J. Am. Ceram. Soc.</u>,
    55[4]: 208-211, 1972.

9. Charles, R.J., "A review of glass strength," <u>Progress in
    Ceramic Science</u>, Vol. 1, J.E. Burke, (ed.), Pergamon Press,
    New York, 1961, p. 1-38.

10. Hillig, W.B., "Sources of weakness and the ultimate strength
    of brittle amorphous solids," <u>Modern Aspects of the Vitreous
    State</u>, Vol. 2, J.D. Mackenzie, (ed.) Butterworth and Co.,
    London, 1962, p. 152-194.

11. Mould, R.E., "The strength of inorganic glasses," <u>Fundamental
    Phenomena in the Material Sciences</u>, Vol. 4: Fracture of
    Metals, Polymers and Glasses, L.J. Bonis, J.J. Duga and
    J.J. Gilman, (eds.) Plenum Press, New York, 1967, p. 119-149.

12. Ernsberger, F.M., "Strength of glasses," Proc. of the <u>Eighth
    International Congress on Glass</u>, 1968, p. 123-139.

13. Weiderhorn, S.M., "Environmental stress corrosion cracking of
    glass," <u>National Bur. Std. Report</u>, No. 10565, April 1971.

14. Gurney, C. and S. Pearson, "Fatigue of mineral glass under
    static and cyclic loading," <u>Proc. Roy. Soc.</u>, (London), 192:
    537-544, 1948.

15. Proctor, B.A., I. Whitney and J.W. Johnson, "Strength of fused
    silica," <u>Proc. Roy. Soc.</u>, 297A: 534-557, 1967.

16. Chen, C.P., "Fatigue fracture of a polycrystalline alumina
    ceramic at several temperatures," Ph.D. dissertation in Engi-
    neering, University of California, Los Angeles, March 1972.

17. Neuber, H., _Theory of notch stresses_, translation published by J.W. Edwards Co., Ann Arbor, Michigan, 1946.

18. Weibull, W., "Statistical theory of strength of materials," _Ingenioersvetenskapsakad._, Handl., No. 151, 44, (1939).

19. Leichter, H.L. and E.Y. Robinson, "Fatigue behavior of a high-density graphite and general design correlation," _J. Am. Ceram. Soc._, 53: 197-204, 1970.

20. Mandel, J., _The statistical analysis of experimental data_, Interscience Publishers, New York, 1964, Chapter 5.

21. Gitzen, W.H., "Alumina Ceramics," _Air Force Materials Laboratory, Research and Technology Division, Air Force Systems Command, Technical Report_, No. AFML-TR-66-13, Wright-Patterson Air Force Base, Ohio, January 1966.

22. Spriggs, R.M., J.B. Mitchell and T. Vasilos, "Mechanical properties of pure, dense aluminum oxide as a function of temperature and grain size," _J. Am. Ceram. Soc._, 47: 323-327, 1964.

23. Davidge, R.W. and G. Tappin, "The effects of temperature and environment on the strength of two polycrystalline aluminas," _Proc. Brit. Ceram. Soc._, 15: 47-60, 1970.

24. Vonnegut, B. and J.L. Glathart, "The effect of temperature on the strength and fatigue of glass rods," _J. Appl. Phys._, 17: 1082-1085, 1946.

25. Mould, R.E., "The strength and static fatigue of glass," _Glastech. Ber. Sounderband V_, International Glass Congress, 32K: III/18, 1959.

26. Schwartz, B., "Thermal stress failure of pure refractory oxides," _J. Am. Ceram. Soc._, 35: 325-333, 1952.

27. Charles, R.J., "Static fatigue of glass, II," _J. Appl. Phys._, 29: 1554-1560, 1958.

28. Gurney, C., "Delayed fracture in glass," _Pros. Phys. Soc._, (London) 59: 169-185, 1947.

29. Rosenwasser, S.N., "Static and Dynamic Fatigue of an Alumina Ceramic," Master of Science Thesis in Engineering, University of California, Los Angeles, 1967.

STRESS-CORROSION CRACKING IN POLYCRYSTALLINE MgO

W. H. Rhodes, R. M. Cannon, Jr., and T. Vasilos

AVCO SYSTEMS DIVISION

Lowell, Massachusetts 01851

ABSTRACT

Stress-corrosion cracking (SCC) was studied in four grades of $99^+\%$ dense MgO with the major variables grain size and purity. Testing consisted of four-point bend and static fatigue tests in H_2O, DMF, and DMSO-DMF solutions. In an H_2O environment the highest purity material gave the slowest strength loss. SCC in low purity grades was judged to be controlled by a chemical interaction of OH^- with a $(CaNaSiAl)O_x$ or LiF grain boundary phase. The highest purity grade tested ($99.98^+\%$ MgO) may not have a discrete grain boundary phase, so the low corrosion rate and high static fatigue limit of $\simeq 0.83\,\sigma_D$ may be characteristic of an intrinsic process. The possibility exists that this was caused by a shift in mechanism, but a passive film model consistent with the data is proposed. When H_2O is present, chemical corrosion is believed to have faster SCC kinetics than possible competing processes. This was demonstrated for the second purest ($99.92^+\%$ MgO) specimens. Testing in DMSO + 10% DMF and DMF gave a sufficient separation of the data to conclude that the Rebinder effect was also operative. Thus, under certain conditions, SCC can result from a dislocation model of crack nucleation. Stress intensity factors, K, were calculated based on the conclusion that the Griffith model was operative. The calculated K-V (velocity) diagram was thought to be qualitatively correct in showing that the second purest sintered grade of MgO had the best overall behavior in terms of dry strength and static fatigue in a H_2O environment. However, the static fatigue performance and nearly identical K_{ISCC} for the highest purity material suggested that the high purity hot pressed grade showed the most promise from a materials development viewpoint.

709

I. INTRODUCTION

Crack propagation by stress corrosion is a frequently encountered phenomena in the structural application of materials. Allowable long time loads in ceramics are dictated then not only by the
dynamic strength, standard deviation, and application of the Weibull
parameters to the volume under load, but knowledge of the static
fatigue limit (stress below which failure does not occur at infinite
times) for the material and perhaps grade of material under question.
Among ceramics materials, silicate glasses have been most extensively
studied (Refs. 1 through 4) with the general conclusion that a chemical corrosion model applies although some question remains (Ref. 5)
as to whether the Charles and Hillig (Ref. 6) stress dependent chemical reaction between water and a pre-existing flaw theory, the Cox
(Ref. 7) atomistic weakening theory, or the ion exchange theory
(Ref. 8) applies. Alumina is the crystalline ceramic most extensively studied (Refs. 9 through 11) with investigators initially
proposing fatigue resulting from chemical processes (Ref. 12).
Alumina was thought to behave in a completely brittle manner, but
more recently (Refs. 9 and 13) it has been suggested based on static
and cyclic fatigue results that dislocation motion and crack tip
lattice defect creation are likely causes of failure.

Few static fatigue tests have been performed on MgO, which is
surprising because it has served as a model system for the examination of mechanical phenomena in ceramics. Charles (Ref. 14) performed dynamic 240° C compressive tests on single crystal MgO in
saturated H_2O vapor and dry N_2 with resulting failure stresses of
8 Kpsi and 26.6 Kpsi, respectively. Considering the known hydration
behavior of MgO, he speculated that a chemical stress corrosion
model similar to glass applied. Janowski and Rossi (Ref. 15) noted
loss of strength in polycrystalline MgO by water vapor attack with
a similar interpretation. In contrast, Rice (Ref. 16) interpreted
a loss of dynamic strength for both single and polycrystalline MgO
tested in H_2O to be a result of enhanced dislocation mobility by
the Rebinder effect. A number of models have been proposed to
explain the Rebinder effect (Ref. 17), which basically applies to
absorption-induced reductions in hardness or enhanced dislocation
mobilities. Westwood, et al (Ref. 17) measured slight increases in
dislocation mobility for the H_2O environment compared to ambient,
and further enhancements by a factor of seven in dimethyl formamide
(DMF), a high dipole moment organic molecule. They proposed that
the observed Rebinder effect was caused by chemisorption-induced
bond bending which altered the electronic core structure of near
surface dislocations and point defects and, consequently the resistance of the lattice to dislocation glide. Shockey and Groves
(Ref. 18) concluded that the increased fracture surface energy in
H_2O as measured by the double cantilever method resulted from increased surface roughness because of a change in the fracture plane

perhaps imposed by a chemically altered surface layer. Measurements in DMF failed to show any enhanced toughening caused by higher dislocation mobilities.

Clear evidence has been generated for a dislocation mechanism of crack nucleation in single crystal MgO (Refs. 19, 20). Debates continue, however, between investigators (Refs. 16 and 21 through 24) on the mechanism of dynamic fracture under ambient conditions in polycrystalline MgO. On one hand, fracture can be caused by the fulfillment of the Griffith criteria on the elastic propagation of existing flaws, or alternatively mobile dislocations can interact with other defects (Ref. 18) or structural features (Refs. 20 and 24) to nucleate a critical crack. The answer probably lies in between with the mechanism dependent on the surface condition and lattice hardness, e.g., elastic extension of cracks for machined surfaces and Stroh model (Refs. 25, 26) dislocation initiated fracture on chemically polished surfaces (Ref. 24). Further, Rice (Ref. 16) has shown that increased strengths are observed for specimens which experienced slow annealing which apparently distilled off impurities. The mechanical behavior of polycrystalline MgO is obviously very complex and not subject to simple analysis, but the vast background available offers an opportunity to study phenomena that may occur in many systems.

A major goal of this study was to define the level of stress-corrosion cracking, SCC, for dense polycrystalline MgO not only to indicate the severity of the problem from a design viewpoint, but to determine if such measurements could further elucidate the mechanism of crack nucleation and/or propagation in polycrystalline MgO. A second major objective was to determine which material property such as microstructure or chemistry controlled the level of SCC.

II.   EXPERIMENTAL

A.   Materials

Testing was conducted on four grades of MgO spanning the range of commercially available electronic quality materials plus a higher purity grade developed specifically for the purposes of these experiments. Preliminary testing suggested that a high purity material might be essential to interpreting results. The latter material was vacuum hot pressed using MgO powder that was rotary calcined from 99.9998% $MgCO_3$. Table I describes the general character of the four materials tested. Grade II material was the highest quality sintered MgO commercially available and also the second purest grade. Grade III was ambient atmosphere hot pressed. Grade IV was optical grade MgO produced with a 0.3% LiF densification aid;

a post pressing anneal of 60 hours at 1000° C reduced the LiF content to 0.1%. In general, one large billet of each grade was employed for the program to eliminate the problem of sample-to-sample
inhomogeneity. The problem of within-billet homogeneity was
approached by dividing the billets into zones, and spreading specimens from each zone among the testing environments and conditions.

TABLE I

DESCRIPTION OF MATERIALS

Grade	Method of Fabrication	Density % Theoretical	Grain Size μ m	Purity %	Major Impurities (in ppm W)
I	Vacuum Hot Press	99.6	46	99.98+	*100 Fe, 40 S, 30 Ca 20 Al, 20F
II	Sintered**	99.3	43	99.92	300 Si, 300 Na, 70 Al, 50 Ca, 50 Fe
III	Ambient Hot Press	99.8	26	99.40	1500 Ca, 500 Fe, 750 Si, 100 Na, 150 Al
IV	Vacuum Hot Press and Annealed	100	30	99.60	1000 Na, 750 Si, 800 Ca, 500 Li, 500 F

*Mass Spectroscopy - other grades by emission spectroscopy.
**Honeywell

All specimens were tested with machined surfaces where the
final operation was grinding with a 400 grit diamond wheel parallel
to the long axis of the bar. The use of machined surfaces having
an abundance of surface flaws was thought desirable to ensure a
surface fracture origin. The question of whether machining gave
a work hardened layer preventing surface nucleated fracture (Ref.
16) was considered in the fractographic analysis. A 1/64-inch
radius was machined on the two tensile edges to reduce corner
stress concentrations.

The chemistry of Grade I was examined in detail by spark source
mass spectroscopy and emission spectroscopy. The values reported
in Table I are spark source numbers on the billet employed in the
testing program. The starting powder was analyzed by both techniques and the reported spark source numbers were 2 to 10 times
higher than the emission spectrograph values, e.g., 30 ppm Ca by
spark source spectroscopy and 3 ppm Ca by emission spectroscopy.
This discrepancy in analytical techniques underestimates the difference in Grades I and II. Grade I is probably 10 to 30 times purer

than Grade II instead of the factor of four indicated in Table I. The analytical comparison of the fabricated piece and powder for Grade I revealed that Fe was the only element introduced bringing the concentration from 30 ppmw to 100 ppmw. The analyses for the remaining grades are typical emission spectroscopy results not specifically of the specimens tested.

The microstructures for the four materials are illustrated in Figure 1. All of the structures are equiaxed. The grain sizes reported are the linear intercepts corrected for geometry by the factor 1.5. The strengths for each grade were normalized relative to the dynamic dry strength of that billet, making it possible to test materials of different grain size, etc. In general, the porosity was located at a grain surface site. Grade IV had 82% total transmission confirming its low porosity, while Grade III was transparent but to a lesser degree. The structure of the grain boundaries were examined by electron microscopy techniques on fracture surfaces. (See Figure 2.) The fractographs for Grades II, III, and IV show an apparent 0.1 to 0.2 μm width which is interpreted as being evidence for a grain boundary phase. Electron diffraction was performed on numerous "pull-off" particles on the Grade III replicas. Many of these were MgO as expected, but one pattern indexed as $Na_6Al_4Si_4O_{17}$ which is taken as further evidence for discrete grain boundary phases. Grade I, the highest purity material was examined by both replica and scanning electron microscopy and, although the boundaries appeared clean, it was unclear if they were completely free of a second phase. Secondary x-ray emission mapping was not productive, but integrated counting on a fracture surface demonstrated a high Ca concentration relative to that expected based on the bulk analysis. This was taken as evidence for Ca segregation at grain boundaries in Grade I material.

B.   Test Technique

Base line dynamic four-point bend strengths on bars 0.1 by 0.2 by 1.75 inches with a 1.5 inch outer and 0.5 inch inner span were established for the four grades by testing in corrosion-free environments. It is generally agreed that testing in liquid N_2 provides one such environment (Ref. 11 and 14). Because a mechanical model of fracture initiation was also being considered, it was important to consider the fact that the $\{110\}<110>$ yield stress increases by a factor of two (Ref. 27) from 23° C to -196° C. Thus, an alternate base line strength, testing technique consisted of heating the sample to 900° C in argon, holding for 1 hour, cooling to 23° C, and loading the specimen without altering the environment. Base line tests on Grade IV gave 209 MN/m^2 in liquid N_2 and 210 MN/m^2 in argon. The strengths were judged equivalent; however, base line strengths for the other grades were established by the bake-out argon test at a strain rate of 8×10^{-5} sec^{-1} and are designated, σ_D .

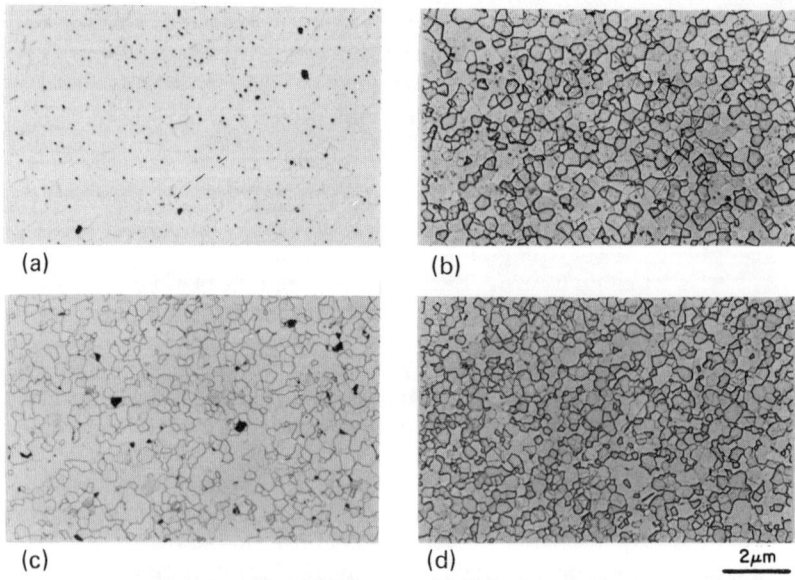

Figure 1 ETCHED MICROSTRUCTURE OF (a) GRADE I, (b) GRADE II, (c) GRADE III, AND (d) GRADE IV MgO

Figure 2 ELECTRON MICROGRAPH OF FRACTURE SURFACES; (a) GRADE I, (b) GRADE II, (c) GRADE III, AND (d) GRADE IV MgO

Long term tests were conducted in a lever arm test frame equipped with a microswitch to record time to failure at a given stress. A brass four-point bend test fixture accommodating bars 0.070 by 0.125 by 1.25 inches with a 1.00 inch outer and 0.30 inch inner span was equipped for holding liquids. Tungsten (outer) and alumina (inner) knife edges employed a thin mylar sheet to give separation from the test bar. This precaution prevented chemical interactions even in the two to three week tests. Distilled H_2O was employed as a test environment that was judged highly corrosive by the reaction;

$$MgO + H_2O \longrightarrow Mg(OH)_2 \qquad\qquad (1)$$

The effect of increasing dislocation mobility was tested by employing the environments Westwood, et al., (Ref. 17) used in their classic MgO dislocation mobility experiments. Dimethyl formamide (DMF), either pure or as 1 M DMF should enhance dislocation mobility by a factor of 6 to 7 over H_2O (both were employed). A third testing environment used was dimethyl sulphoxide (DMSO) plus 10 percent DMF which depresses dislocation mobility over pure DMF to about the same level as H_2O; thus, in combination with the other environments, provided a separation of chemical and mechanical effects. Each sample was immersed for 1 hour prior to loading to ensure complete absorption and equilibration of surface and near surface charge effects. The loading and timing devices were constructed such that data was judged accurate for ≥ 2 sec.

III. RESULTS

A. Dynamic Tests

Dynamic bend strength measurements first raised the question of the importance and mechanism of stress-corrosion in MgO. A billet of Grade III material having a 6.5 μm grain size was tested in air and argon after a 900° C - 1 hour outgas (Ref. 28). An average of four 8×10^{-5} sec^{-1} strain rate tests gave strengths of 298 MN/m^2 and 296 MN/m^2 for the two environments, respectively. This was in marked contrast to similar tests on Al_2O_3[28] where strength reductions of 20% were experienced for ambient tests.

Limited dynamic testing was performed on the billets for this study. Not only was it important to establish the "corrosion-free" strength, but the effect of the various liquid test environments on dynamic strength was determined. The average of 3 to 6 strength measurements for each condition is reported in Table II.

The ambient tests on Grade IV are lower than the argon test values in the direction expected from stress corrosion. Testing

TABLE II

DYNAMIC STRENGTH OF MgO SPECIMENS IN VARIOUS ENVIRONMENTS

	Four-Point Bend Strength, MN/m^2					
	Argon- After Anneal, σ_D	Liquid N_2	Ambient	H_2O	DMF	DMSO – 10% DMF
Grade I 46 μm G.S.	149 ± 10.7					
Grade II 43 μm G.S.	192 ± 26.2			172	174	187
Grade III 26 μm G.S.	167 ± 20.7					
Grade IV 30 μm G.S.	210 ± 9.6	209	183			

on Grade II was performed in liquid H_2O and, although lower strengths were measured compared with the dry strength, they are within the standard deviation. Thus, on the basis of dynamic tests, the evidence for the stress corrosion by H_2O appeared to vary depending on the grade of MgO. Similar tests on other ceramics, notably Al_2O_3 and glass, would have clearly demonstrated the loss of strength caused by stress-corrosion.

The tests of Grade II in DMF and DMSO-10% DMF gave results within the scatter of the dry strengths. Thus, no firm conclusions can be drawn concerning dynamic crack initiation and propagation. This is in contrast to the work of Westwood and Latanision (Ref. 29) where drilling rate in MgO and CaF_2 was affected by n-alcohol environments, and the results interpreted to be the result of the Rebinder effect. The drilling experiments suggest that the processes involved are rapid and should be complete in the time scale of a bend test. One interpretation of the dynamic bend tests is that slow crack growth or dislocation crack nucleation effects are absent as flaws of sufficient size are already present. This would be consistent with the results of Shockey and Groves (Ref. 18) where DMF did not alter the fracture surface energy of MgO. Alternatively, the strength reduction in H_2O and DMF could be the order of the expected effect as a reduction in yield stress or increase in dislocation mobility would not have a 1:1 effect on bend strength in a polycrystalline body by a Stroh model.

The σ_D values appear consistent with previous literature values on MgO produced by similar processes and with similar microstructures. Leipold and Nielsen (Ref. 30) and Rice (Ref. 16) noted lower strengths for the highest purity samples after normalizing for microstructure effects. This was attributed to higher dislocation mobility resulting from reduced drag effects which promoted early dislocation nucleated fractures. The Grade II sintered specimens might be expected to have slightly higher strengths than hot pressed specimens of equivalent microstructure caused by the expected reduction in residual gases or products thought to weaken bonding in some hot pressed products. This also explains the higher strength of Grade IV compared with Grades I and III, as Grade IV was annealed in a slow heating rate cycle that has been shown by Rice (Ref. 16) to reduce anion concentration and increase strength.

B. Static Tests

Dead load tests in distilled H_2O are plotted in Figure 3 for Grades I and II. Due to the large scatter which is typical of stress-corrosion results, the data for Grades III and IV are plotted in Figure 4. Figure 4 also shows the straight lines for the best fit to the data of Figure 3. Points highlighted with an arrow indicate a test discontinued without failure. Such points were used in drawing the indicated lines with a strong weighting factor if the point was above the general fit of the data and lesser weight if the point was below the line. Grades II and III exhibit different slopes, but the closeness and scatter of the data require a qualification of the distinction between these sets of data. It is clear, however, that the delayed failure curves of Grade(s) I, (II and III), and Grade IV are markedly different both in slope and relative effect. It is possible that the data lies on different portions of a general fatigue curve, but the differences in slope imply that different kinetics are involved. This could mean one type of process, say for example, chemical corrosion was proceeding at different rates through different phases or pseudo phases of different chemistry. The term pseudo phase is meant to include grain boundaries with different concentrations and/or species of segregated impurities. A second explanation for the different slopes centers on different fundamental phenomena controlling delayed failure of one group compared to the next, e.g., chemical stress enhanced corrosion in one and dislocation glide induced crack growth in another. Third, a different initial flaw size could account for these observations. Further discussion of this and the apparent differences in the fatigue limit (the applied stress below which no failure of the specimen can occur) is deferred.

Both DMF and one molar DMF were employed for the data reported in Figure 5. Westwood, et al., (Ref. 17) reported both media to

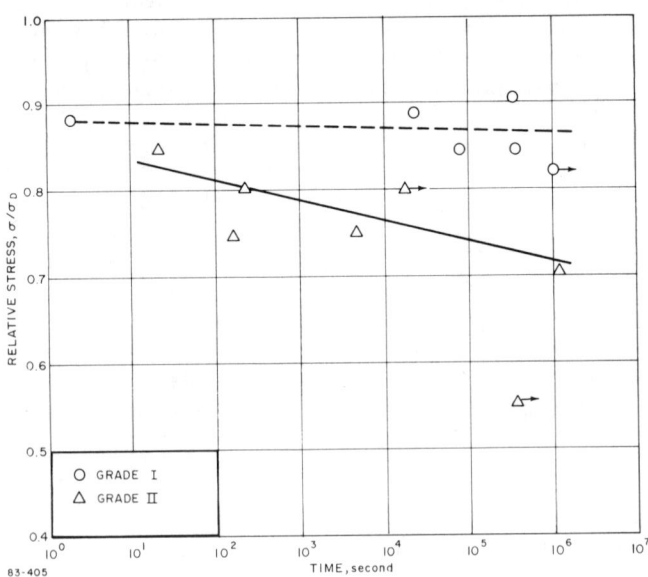

Figure 3 DELAYED FAILURE FOR GRADES I AND II MgO in H_2O

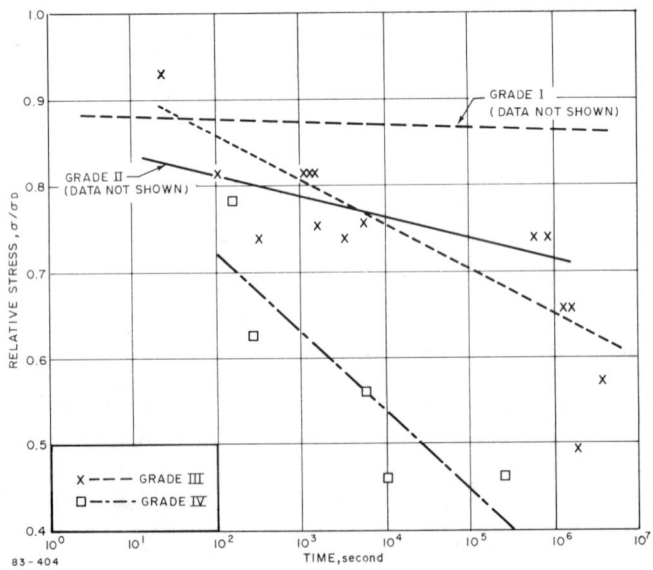

Figure 4 DELAYED FAILURE FOR GRADES III AND IV ALONG WITH
LINE REPRESENTING GRADES I AND II MgO IN H_2O

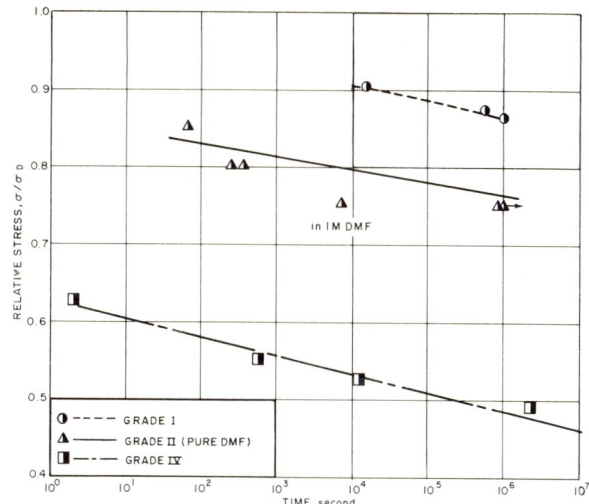

Figure 5 DELAYED FAILURE IN 1M DMF

have approximately equal effect on the enhanced dislocation mobility provided a 4000 second soak was employed. The tests reported were held under the liquid environment for 1 hour prior to loading. One advantage to pure DMF environment was its relative freedom from OH^-, which could provide competitive chemical corrosion. The different grades of MgO again exhibited distinctly different behavior in terms of σ/σ_D intercepts, but unlike the behavior in H_2O the slopes appear about equal. This implies that a similar mechanism may be responsible for the onset of failure in each grade.

SCC results in DMSO-10% DMF environment on Grade I material are compared with data for the other environments in Figure 6. The data for this environment exhibited little scatter. However, due to the scatter of data in the other environments, it does not appear possible to draw conclusions regarding distinguishable behavior from one environment to the next for Grade I.

The data for Grade II, plotted in Figure 7, exhibits greater separation. The DMF data falls midway between failure data in H_2O and DMSO-10% DMF. Three orders of magnitude in time separate the expected failure times in H_2O and DMSO-10% DMF, and approximately one order of magnitude separate the DMF and the DMSO-10% DMF data. The significance of the separation for the latter two environments may be argued, but the authors attach moderate significance to all three sets of data.

Grade IV was tested in more than one environment and a comparison for the delayed failure in H_2O and 1M DMF is shown in Figure 8. This short time data in 1M DMF is quite interesting. Attempts to load above $\sigma/\sigma_D = 0.627$ in this environment resulted in immediate failure, thus the effect of 1M DMF is the reverse of that found for Grade II.

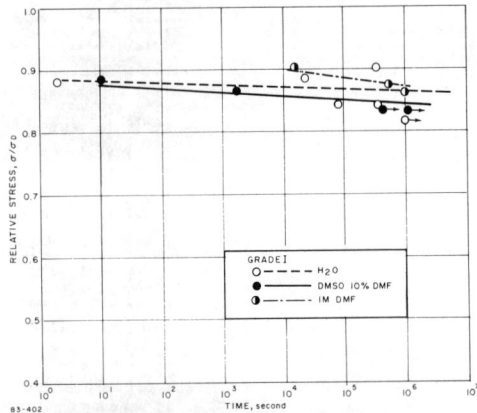

Figure 6 DELAYED FAILURE OF GRADE I IN H_2O,
1M DMF AND DMSO · 10% DMF

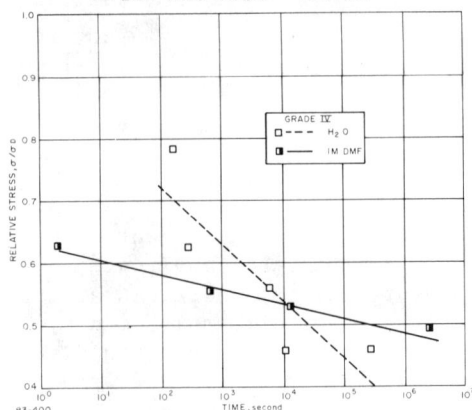

Figure 7 DELAYED FAILURE OF GRADE II IN H_2O,
DMF AND DMSO · 10% DMF

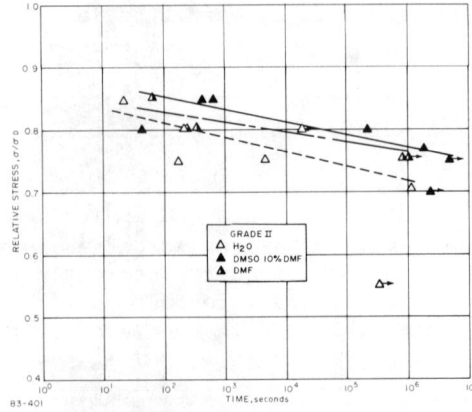

Figure 8 DELAYED FAILURE OF GRADE IV IN H_2O
AND 1M DMF

C. Fractography

Replica and scanning electron micrographic techniques were employed to examine the fracture surfaces near the tensile edge. Grades II, III, and IV exhibited predominantly intergranular fracture surfaces, while Grade I was about 50% transgranular. Figure 9a illustrates a series of flat bottom etch pits found on a grain face in a Group II H_2O tested specimen which fractured after 1.2×10^6 sec. Another grain face of this specimen (closer to the tensile surface) also showed etch pits, but with apparently random crystallographic orientations and flat bottoms, Figure 9b. The etch pit lines are interpreted as arrest points for the crack front. The randomness of the pattern on Figure 9b and the apparent flat bottoms argue against their being associated with dislocations. Thus, the interpretation is that chemical corrosion generates the pits with increased activity at regions of stress concentration. A Group IV specimen also fractured in H_2O after 6.8×10^3 sec to failure exhibited a corroded grain face and a semicircular pattern, Figure 10. This semicircular pattern was also interpreted as being caused by intermittent crack movement associated with chemical corrosion.

A zone near the tensile surface on a fracture face of a Group I specimen tested in 1M DMF was examined by scanning microscopy techniques. The fracture origin was not located nor were any features located that could differentiate the mechanism of crack nucleation or growth. Figure 11 illustrates one region about 20 μm from the tensile surface where the cleavage fracture tracings reveal that the crack front was moving away (in direction of arrow) from the tensile surface. Because this distance is essentially one grain in from the surface, it was concluded that a surface crack caused the failure. Furthermore, the crack had probably reached critical dimension and was traveling at high velocity by the time the cleavage fracture tracings were created.

An extensive survey by scanning microscopy was also performed on Group I specimens tested in H_2O and DMSO-10% DMF. No positive identifications of fracture origins were found in either case. This in part spoke for the uniformity of the material and freedom from flaws much larger than the grain size. Figure 12 illustrates a zone on a DMSO-10% DMF fracture surface adjacent to the tensile surface. The saw-toothed cleavage fracture is unusual in polycrystalline materials and may have a similar origin to that observed by Shockey and Groves (Ref. 18) in single crystal MgO. They attributed this phenomena to chemical corrosion which caused alternate cleavage planes to become favorable. In the case of a polycrystalline material, the crack front is influenced by the energy for propagation through adjacent grains as well, so it is not clear that the same reasoning can be applied. In fact, several grains near the saw-toothed grain have undergone transgranular fracture,

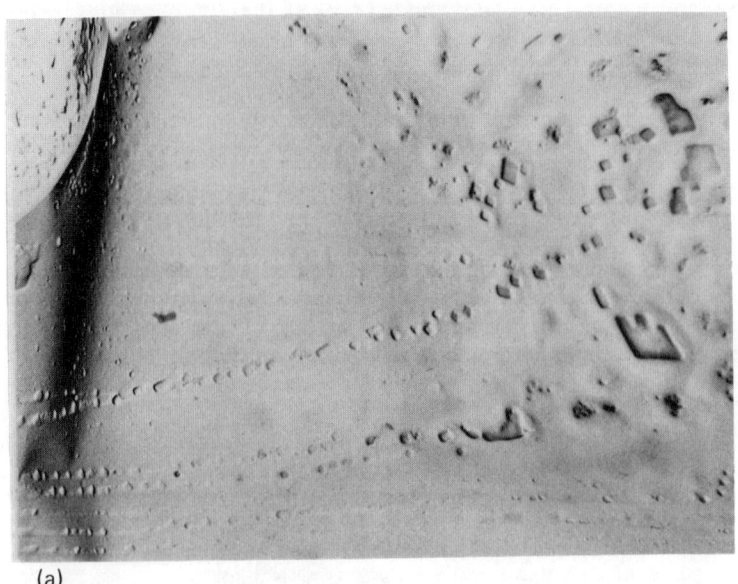

(a)

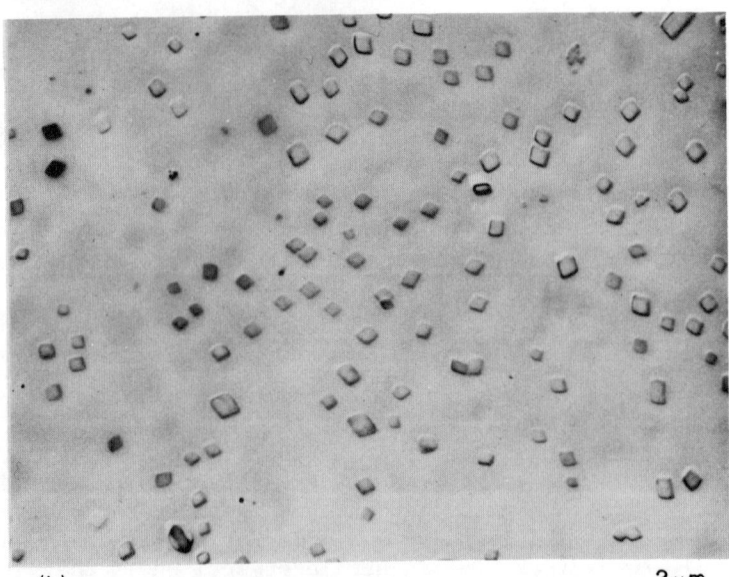

(b)

2 μm

Figure 9 FRACTOGRAPH OF GROUP IV, 1.2 x 10^6 SEC., H_2O ENVIRONMENT
SPECIMENS EXHIBITING (a) PITS ASSOCIATED WITH CRACK ARREST
LINES, AND (b) RANDOM FLAT BOTTOM PITS

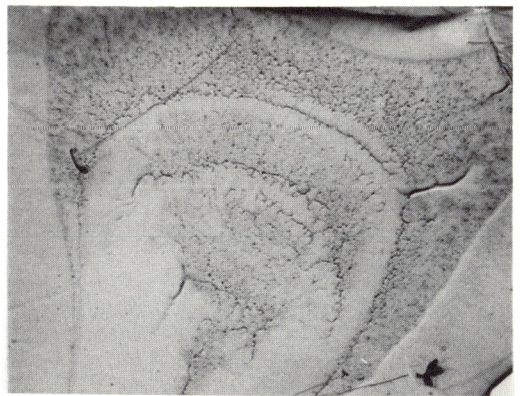

2 μm

Figure 10 FRACTOGRAPH OF GROUP IV, 6.8 x 10^3 SEC., H$_2$O
ENVIRONMENT SPECIMEN CORRODED GRAIN FACE
AND ASSOCIATED CRACK ARREST PATTERN

10 μm

Figure 11 FRACTOGRAPH OF ZONE ADJACENT
TO TENSILE EDGE OF GROUP I, 1 x 10^6 SEC.,
1M DMF ENVIRONMENT SPECIMEN
SHOWING DIRECTION OF CRACK
PROPAGATION (ARROW)

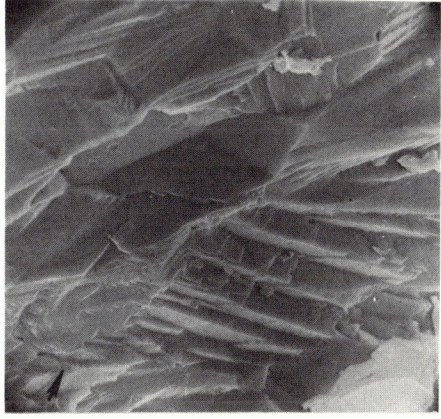

10 μm

Figure 12 FRACTOGRAPH OF ZONE ADJACENT
TO TENSILE EDGE OF GROUP I, 1.8 x 10^3 SEC.,
DMSO-10% DMF ENVIRONMENT SPECIMEN
SHOWING THREE TYPES OF FRACTURE
AND DIRECTION OF CRACK
PROPAGATION

a process that probably requires more energy than the saw-toothed higher surface area cleavage fracture. Also note that one grain fractured intergranularly, which undoubtedly requires the least energy of the three types illustrated in this one micrograph. The fracture tracings indicate that the direction of crack growth (in direction of arrow) was away from the tensile surface as expected for growth of a surface crack.

IV. DISCUSSION

Testing in DMF and DMF-10% DMSO was aimed toward further elucidation of the crack growth mechanism, but the question of whether or not the environment could penetrate to the crack tip in the time of the experiment was first addressed. An estimate of the time required for these molecules to diffuse to the crack tip was made by assuming semi-infinite plate diffusion (no radial loss of diffusing species), a $D = 3 \times 10^{-6}$ cm^2/sec which is about the lower limit of self-diffusion of large molecules in liquids (Ref. 35), a diffusion distance of one grain length (45 μm maximum), and 50% saturation to be effective. These conservative conditions are satisfied in approximately 5 seconds, which is a short time compared with the presoak and total time to fracture for all experiments. From a molecular size viewpoint, the longest chain length was calculated to be 5.1 Å and 5.6 Å for DMF and DMSO, respectively. Thus, it appears reasonable from both a size and time consideration that these special environments were present at the tips of surface cracks for these experiments. The negative slopes shown in Figures 5 through 7, as well as the fracture studies, were taken as evidence for surface initiated crack growth.

One concern with the use of machined bend specimen surfaces was the possibility that the surface would be work hardened as discussed by Rice (Ref. 16) to the point where subsurface fracture origins would control failure. In this case, fatigue response to the environment would probably not be observed because the environment would not have access to the crack tip. Because pronounced negative slopes were observed particularly on the fatigue curves for Grades II, III, and IV MgO, it was concluded that surface fracture origins were controlling. The extremely flat curves for Grade I required more extensive consideration of this question. A sample from each testing environment was examined extensively by scanning electron microscopy. For the cases illustrated, Figures 11 and 12 demonstrate that although the origin of fracture could not be found, the crack front was traveling away from the surface within one grain depth from the surface. Similar observations were noted for specimens tested in the H_2O environment, and dynamically. This, combined with the negative slopes on the fatigue curve was taken as evidence for surface fracture origins. Rebinder

effects of over 30 μm (the grain size) would be required to explain
the negative slopes of the fatigue curves and subsurface dislocation
nucleation events resulting indirectly from a work hardened surface
layer. Rebinder effects in the order of 10 μm have been discussed,
(Ref. 17) and, although more extensive effects may be possible in
single crystals, it does not seem likely that these effects could
extend two or more layers deep in a 30 μm grain size material.

Figure 4 illustrates marked differences in SCC resistance in
H_2O for the four grades of MgO. It would appear that the origin
for these differences lies in either microstructural or chemical
differences. The specimens were all impermeable because they were
> 99.3% density, and the observed phenomena should not be a con-
sequence of density differences. This view is substantiated by the
results of Janowski and Rossi (Ref. 15) who found evidence for internal attack by H_2O with a corresponding loss of strength for hot
pressed MgO having > 1.5% porosity, but not for specimens < 0.58%
porosity. Further, Grades I and II were the most resistant grades
and had the highest porosity. The program did not cover a wide
range of grain sizes. Grades I and II were essentially the same
grain size and were a factor of 1.6 larger than Grades III and IV.
This resulted in their being ~30% less grain boundary area in Grades
I and II than Grades III and IV. If corrosion followed grain boundaries, the finer grain size material would possess the greatest
slopes. In a qualitative sense, this behavior is followed. However, several factors lead to the conclusion that grain size differences alone do not explain the observed behavior. First, a
considerable difference in fatigue behavior was measured between
Grades I and II having the same grain size and similarly Grades III
and IV exhibit marked fatigue differences with similar grain sizes.
Second, Grades II and III show somewhat similar fatigue behavior
but possess the maximum difference in grain size. Thus, it was
concluded that the differences in fatigue behavior in H_2O are not
strictly a consequence of grain size differences.

The third material property considered to explain the H_2O
fatigue curves was chemical composition. Particularly noteworthy
was the very low fatigue slope for Grade I, the purest material
followed by Grade II, the next purest grade. The fatigue behavior
of Grades III and IV is reversed from the expected behavior based
strictly on purity. The electron microscopy of Figure 2 and findings of others (Refs. 16 and 31) demonstrate that some impurities
segregate at grain boundaries and in fact form discrete second
phases. One would expect that the composition of the grain boundary
phase would depend on the specific species involved depending on
its solid solubility and concentration. Ca^{+2}, Na^{+1}, Si^{+4}, and Al^{+3},
for example, have extremely low solubilities. It is postulated
that stress-corrosion resistance in H_2O is controlled by stress
enhanced chemical attack along a grain boundary crack, the material

free of a grain boundary phase, or showing very little phase would exhibit the greatest stress-corrosion resistance. If several grades had about the same level of total impurity, but basically had a grain boundary phase, different stress-corrosion rates could be observed depending on the rate of stress enhanced chemical attack on the particular grain boundary phase in question. This would explain the apparent reversal in corrosion resistance for Grades III and IV. Weiderhorn and Bolz (Ref. 32) have shown that different glass compositions have different values of K_{SCC}. A dislocation model for crack nucleation and growth would predict that the purest material would exhibit the least lattice resistance to dislocation flow. This effect would predict that Grade I would show the least stress-corrosion resistance rather than the greatest resistance as observed. This, combined with the correlation of behavior with purity and the fractographic evidence for corrosion associated with crack arrest lines, lead to the conclusion that SCC in a H_2O environment is caused by a chemical corrosion mechanism with impurity phases in the grain boundaries being the principal point of attack.

Most glass (Refs. 1 and 3 through 6) and crystalline oxide (Refs. 9 through 13) materials exhibit static fatigue limits, σ_L, of $0.2\ \sigma_D < \sigma_L < 0.6\ \sigma_D$; thus, the apparent fatigue limits shown in Figure 4 warrant discussion. MgO Grades III and IV appear to approach fatigue limits in the expected range. The establishment of a fatigue limit cannot be exact because of the scatter and difficulty in collecting data $> 10^6$ sec (11.6 days). However, Grade I and possibly Grade II is clearly outside the expected range. One possible explanation is that in high purity MgO, free of grain boundary phases, stress enhanced corrosion builds a layer of coherent corrosion product, and that product achieved a semi-stable geometry which extended very slowly under the conditions of these tests. The formation of $Mg(OH)_2$, which is known (Ref. 33) to have some matching coherent planes with MgO, would seem a likely corrosion product in an H_2O environment. Crack extension may become limited by the rate of OH^- diffusion through this layer. This model would be termed the "passive film" model by terminology common to explaining stress-corrosion in metals (Ref. 34). In this model, crack growth rates increase each time the film breaks by any one of several mechanisms, e.g., thermal cycling, mechanical cycling, or lattice strains caused by misfit between the reactant and product. The very flat fatigue curve for Group I specimens might be a result of very little breakage of the film. The explanation offered for the less pure grades is that stress corrosion was controlled by corrosion of the grain boundary phase. From the available evidence, this phase may be a silicate glass phase or a LiF phase in the case of Group IV; thus, fatigue behavior more in line with that found for glass could be observed. A second model to explain the low slope for Group I specimens is that the mechanism shifted for this group of specimens to an internal

dislocation nucleated fracture, for example. It can only be stated
that evidence for the latter mechanism was sought by scanning electron microscopy but not found, and the passive film model seems
self-consistent and the best explanation at this time.

The data for Grade I (Figure 6) shows sufficient scatter and
similarity of behavior from one media to the next that it would be
difficult to draw firm conclusions between a chemical and mechanical
corrosion mechanism. The high σ/σ_D is one of the causes of this
scatter, which is related to the distribution of strengths around
σ_D. Greater separation in the data for the three environments exists
for Grade II material (Figure 7). The three orders of magnitude
time separation between limits appears to be significant. There
is one order of magnitude difference in time between the lines for
DMF and DMSO + 10% DMF; thus, with the scatter in strength, the
significance of this separation may be questioned. The relative
order seems reasonable from a dislocation mobility viewpoint, however. The DMF environment, which should give the highest dislocation mobility, gave the shortest times to failure. This would
follow the predicted behavior for the Stroh model where dislocations
pile up at a grain boundary. Under these conditions, greater dislocation mobility leads to more rapid crack nucleation. Actually,
any crack nucleation model requiring the movement of mobile dislocations as an integral part of fracture would be similarly affected.

Examination of the H_2O and 1 M DMF environment data for Group
IV specimens is instructive (Figure 8). The short-time high-load
behavior indicated that 1 M DMF greatly enhanced the time to failure.
At longer times (> 10^4 sec), the curves converged and perhaps should
be drawn as one curve with a static fatigue limit of $\sigma/\sigma_D \simeq 0.46$.
It may be noteworthy that the 1M DMF environment also contains OH^-
leading to the possibility of competitive kinetic processes. Is
it possible that at high loads, dislocation enhanced crack nucleation proceed, but at lower loads chemical corrosion by OH^- proceeds
at a rate that surpasses dislocation nucleated fracture? This
indeed seems unlikely as Figure 5 shows the various grades to behave
quite differently, and there was no apparent reason for dislocation
processes to operate at the different relative σ_D for Grades II and
IV, for example. The higher purity of Grade II should result in
greater not less dislocation mobility than Grade IV. The finer grain
size of Grade IV does not explain the operation of an inherent MgO
dislocation mechanism at lower relative stresses. Westwood and
Latanision (Ref. 36) found that high dipole moment complexes also
affect the hardness and drilling rate in soda-lime glass by an
absorption induced change in near surface flow behavior. It was
previously stated that the H_2O environment behavior of Grade IV
material was controlled by stress enhanced chemical corrosion of
OH^- at the grain boundary phase. Thus, it appears reasonable to
suggest that the behavior in 1M DMF is a result of the influence

of this environment on the grain boundary phase in a manner analogous to that suggested by Westwood, et al. (Ref. 36).

It was concluded that interpretation of the H_2O environment tests based on an extension of a pre-existing crack was valid. The conditions for fracture stress, σ, which must be satisfied as derived by Griffith are

$$\sigma = \left(\frac{2E\gamma}{\pi C}\right)^{1/2} \tag{2}$$

where E is Young's modulus, γ is the surface energy, and C is the flaw size. It is recognized that γ should include a geometrical factor to account for the inclination of the crack path to the stress direction, a term for the energy absorbed by dislocation motion associated with the moving crack (this does not necessarily imply dislocation nucleated fracture), a term for subsidiary crack-ing, and a term for cleavage step formation. Evans (Ref. 37) con-cluded that the dislocation motion term was dominant, accounting for about 9 of the measured 14 J/m^2 in 200 μm grain size MgO.

Using notched bars and slit cracks in 20 μm and 50 μm grain size MgO Evans and Davidge (Ref. 25) measured $\gamma \simeq 4$ J/m^2 for 50 μm cracks and increasing to a plateau of $\gamma \simeq 14$ J/m^2 for 400 μm cracks. In the absence of macroscopic flaws, C is thought to be between G/2 and 2G where G is the grain size. Using 2G, the σ_D data of Table II, γ was evaluated by Equation (2) with the following results:

Grade	γ, J/m^2
I	10.3
II	16.1
III	7.43
IV	13.3

These values agree reasonably well with those of Evans and Davidge (Ref. 25) although γ for Grade II is large for an 86 μm crack and γ for Grade IV is large for a 60 μm crack. This may imply that C was underestimated for these two grades of material.

Because stress intensity factors, K, for polycrystalline MgO have not been determined directly, it may be of some value to report calculated values based on a Griffith crack where

$$K = \sigma(\pi C)^{1/2} \tag{3}$$

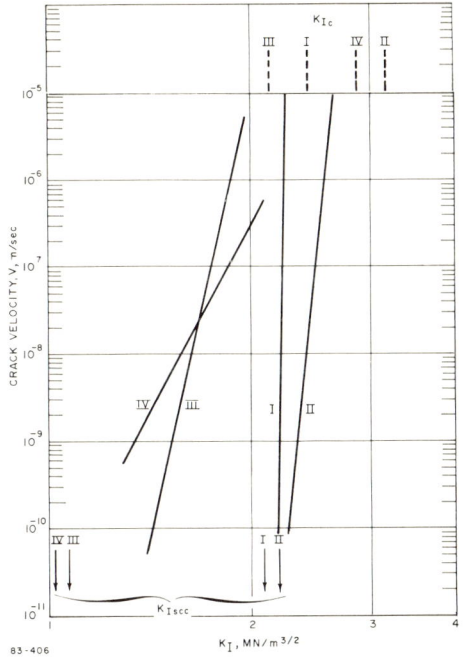

K_{Ic}

III I IV II

IV III I II

IV III I II

K_{Iscc}

K_I, MN/m$^{3/2}$

83-406

**Figure 13 STRESS INTENSITY FACTORS
CALCUALTED FROM FAILURE DATA**

The fast fracture value K_{IC} was calculated from σ_D in Table II. For consistency and lacking direct measurement of the flaw size, C was assumed to be 2G. The stress controlled cracking (SCC) limit K_{ISCC} for an H_2O environment was obtained from a best estimate based on the data of Figures 3 and 4. Stress intensity versus crack velocity relations were also calculated based on these data and are plotted in Figure 13.

Weiderhorn and Bolz (Ref. 32) have used directly measured K-V curves to predict static fatigue curves for several silicate glass compositions. These calculations agreed qualitatively but not quantitatively with the directly measured fatigue curves of Mould and Southwick (Ref. 1). Thus, it is probably unreasonable to expect Figure 13 to be more than a qualitative representation of a K-V curve for the four grades of polycrystalline MgO in a H_2O environment. It is interesting to note that higher K_{IC} values were calculated for Grades II and IV than the high purity material, Grade I. However, only Grade II exceeded Grade I and by a very small margin for K_{ISCC}. This, of course, is because of the very low slope to the K-V curve (or σ/σ_D - t curve) thought to be a result of the high purity and freedom from a grain boundary phase in Grade I. From a design application veiwpoint, Grade II material is to be preferred over Grade I because of the consistently higher K. However, from a materials development viewpoint, it would be advantageous to produce a material with higher σ_D using powder of Grade I quality. One obvious avenue of approach would be to reduce the grain size of the product made with this powder.

V. CONCLUSIONS

The rate of stress-corrosion cracking of polycrystalline MgO in H_2O is strongly influenced by the chemical purity of the body with the highest purity material giving the slowest corrosion rates. SCC in lower purity grades is probably controlled by chemical interaction of OH$^-$ at the crack tip which is thought to lie in a

(CaSiNaAl)O_x or LiF grain boundary phase, explaining the different corrosion rates for the various grades of material. Static fatigue limits of 0.35 σ_D to 0.5 σ_D are consistent with this conclusion. The highest purity grade tested (99.98$^+$% MgO) may not have a discrete grain boundary phase, so the low stress-corrosion rates and high static fatigue limit of $\simeq$ 0.83 σ_D may be characteristic of intrinsic SCC in polycrystalline MgO.

When H_2O is present, chemical corrosion is believed to have faster SCC kinetics than possible competing processes. This was demonstrated most clearly for Group II specimens. Testing in DMF and DMSO plus DMF mixtures gave evidence for the operation of the Rebinder effect which is interpreted as resulting from enhanced SCC by a Stroh dislocation model. This condition applies only in the absence of H_2O in the environment and on moderately high purity material ($>$ 99.92% MgO) where the influence of a discrete grain boundary phase does not dominate. In lower purity material, 1 M DMF may affect the grain boundary phase in a manner analogous to the effect of high alcohols on hardness and drilling rates in silicate glass.

Calculated stress intensity factors qualitatively show that Grade II, a sintered large grain size material, exhibits the best overall behavior in terms of dry strength and static fatigue in an H_2O environment. However, the static fatigue performance and nearly identical K_{ISCC} for Grades I and II suggested that the Grade I, high purity material, would have an advantage if a higher dynamic strength version were produced.

ACKNOWLEDGMENTS

The authors wish to gratefully acknowledge the support of the Office of Naval Research, Contract N00014-70-C-0138 and Naval Air Systems, Contracts N00019-70-C-0171, N00019-69-C-0198, and N00019-68-C-0108. Helpful discussions were held with B.J. Wuensch, and S.K. Dutta. J.A. Centorino, C.L. Houck, and B.W. MacAllister are acknowledged for their able experimental assistance.

IV.   REFERENCES

1.   R.E. Mould and R.D. Southwick, "Strength and Static Fatigue of Abraded Glass Under Controlled Ambient Conditions, II", *J. Am. Ceram. Soc.*, 42, 582 (1959).

2.   S.M. Wiederhorn, "A Chemical Interpretation of Static Fatigue", *J. Am. Ceram. Soc.*, 55, 81 (1972).

3.   J.E. Ritter and C.L. Sherburne, "Dynamic Static Fatigue of
     Silicate Glasses," J. Am. Ceram. Soc., 54, 601 (1971).

4.   C. Gurney and S. Pearson, "The Effect of the Surrounding
     Atmosphere on the Delayed Fracture of Glass," Proc. Phys. Soc.,
     62, 469 (August 1949).

5.   J.E. Ritter and J. Manthuruthil, "Static Fatigue of Silicate
     Glasses," School of Eng., Univ. of Mass. Report No. UM-72-5
     (May 1972).

6.   W.B. Hillig and R.J. Charles, in High Strength Materials, Ed.
     V.F. Zackay, John Wiley & Sons, Inc., New York, (1965) p. 682.

7.   S.M. Cox, "Glass Strength on Ion Mobility," Phys. Chem. Glasses,
     10, 226 (1969).

8.   A.G. Metcalfe, M.E. Gulden, and G.K. Schmitz, "Spontaneous
     Cracking of Glass Filaments," Glass Tech., 12, 15 (1971).

9.   L.S. Williams, "Stress-Endurance of Sintered Alumina," Trans.
     Brit. Ceram. Soc., 55 287 (1956).

10. S. Pearson, "Delayed Fracture of Sintered Alumina," Proc. Phys.
     Soc., 69B, 1293 (1956).

11. J.E. Burke, R.H. Doremus, W.B. Hillig, and A.M. Turkalo,
     "Static Fatigue in Glasses and Alumina," Ceramics in Severe
     Environments, 5, Eds. W. Wurth Kriegel and Hayne Palmour III,
     p. 435.

12. R.J. Charles and R.R. Shaw, "Delayed Failure of Polycrystalline
     and Single-Crystal Alumina," General Electric Res. Lab Report
     No. 62-RL-3081M, (July 1962).

13. D.A. Krohn and D.P.H. Hasselman, "Static and Cyclic Fatigue
     Behavior of a Polycrystalline Alumina," J. Am. Ceram. Soc.,
     55, 208 (1972).

14. R.J. Charles, "The Strength of Silicate Glasses and Some
     Crystalline Oxides," in Fracture, Ed. Averbach, et al., John
     Wiley & Sons, Inc., New York, (1959) p. 225.

15. K.R. Janowski and R.C. Rossi, "Mechanical Degradation of MgO
     by Water Vapor," J. Am. Ceram. Soc., 51 (8) 453 (August 21,
     1968).

16. R.W. Rice, "Strength and Fracture of Hot-Pressed MgO," Proc.
     Brit. Ceram. Soc., 20, 329 (1972).

17. a. A.R.C. Westwood, D.L. Goldheim, and R.G. Lye, "Rebinder Effects in MgO," Phil. Mag., 16, 505, (1967).

    b. A.R.C. Westwood, D.L. Goldheim and R.G. Lye, "Further Observations on Rebinder Effects in MgO," ibid, 17, 951 (1968).

18. D.A. Shockey and G.W. Groves, "Origin of Water-Induced Toughening in MgO Crystals," J. Am. Ceram. Soc., 52, (2), 82 (February 21, 1969).

19. A.E. Gorum, E.R. Parker, and J.A. Pask, "Effect of Surface Conditions on Room Temperature Ductility of Ionic Crystals," J. Am. Ceram. Soc., 41, 161 (1958).

20. R.J. Stokes, T.L. Johnston, C.H. Li, "Effect of Surface Condition on the Initiation of Plastic Flow in Magnesium Oxide," Trans. of Metall. Soc. of AIME, 215, 437 (June 1959).

21. F.J.P. Clarke, R.A.J. Sambell, and H.G. Tattersall, "Cracking at Grain Boundaries Due to Dislocation Pile-up," Phil. Mag., 1, 393 (1962).

22. S.C. Carniglia, "Grain Boundary and Surface Influence on Mechanical Behavior of Refractory Oxides - Experimental and Deductive Evidence," Mater. Sci. Res., 3, 425 (1966).

23. T. Vasilos, J.B. Mitchell, and R.M. Spriggs, "Mechanical Properties of Pure, Dense Magnesium Oxide as a Function of Temperature and Grain Size," J. Am. Ceram. Soc., 47, 606 (1964).

24. W.B. Harrison, "Influence of Surface Condition on the Strength of Polycrystalline MgO," J. Am. Ceram. Soc., 47, (11), 574 (November 21, 1964).

25. A.G. Evans and R.W. Davidge, "The Strength and Fracture of Fully Dense Polycrystalline Magnesium Oxide," Phil. Mag., 20, (164), 373 (Aug. 1969).

26. A.N. Stroh, Proc. R. Soc., 223, 404 (1954).

27. C.O. Hulse and J. Pask, "Mechanical Properties of Magnesium Single Crystals in Compression," J. Am. Ceram. Soc., 43, 375 (1960).

28. W.H. Rhodes, D.J. Sellers, R.M. Cannon, and A.H. Heuer, "Microstructure Studies of Polycrystalline Refractory Oxides," Contract N00019-67-C-0336, Summary Report (May 1968).

29. A.R.C. Westwood and D.L. Goldheim, "Mechanism for Environmental Control of Drilling in MgO and CaF$_2$ Monocrystals," J. Am. Ceram. Soc., 53, (3), 142 (March 21, 1970).

30. M.H. Leipold and T.H. Nielsen, "The Mechanical Behavior of Tantalum Carbide and Magnesium Oxide," NASA Report 32-1201 (December 1967).

31. M.H. Leipold and T.H. Nielsen, "Hot-Pressed High-Purity Polycrystalline MgO," Bull. Am. Ceram. Soc., 45, 281 (1966).

32. S.M. Wiederhorn and L.H. Bolz, "Stress-Corrosion and Static Fatigue of Glass," J. Am. Ceram. Soc., 53, 543 (1970).

33. R.S. Gordon and W.D. Kingery, "Thermal Decomposition of Brucite: I, Electron and Optical Microscope Studies," J. Am. Ceram. Soc., 49, 654, (1966).

34. E.N. Pugh, J.A.S. Green, and A.J. Sedriks, "Current Understanding of Stress-Corrosion Phenomena," Martin-Marietta Corp. RIAS Tech. Report 69-3, (March 1969).

35. R.C.L. Bosworth, Transport Processes in Applied Chemistry, John Wiley & Sons, Inc., New York, (1956), p. 216.

36. A.R.C. Westwood and R.M. Latanision, "Environment-Sensitive Machining Behavior of Nonmetals," Chapter in The Science of Ceramic Machining and Surface Finishing, NBS Special Publication 348, Eds. Schneider & Rice (May 1972).

37. A.G. Evans, "Energies for Crack Propagation in Polycrystalline MgO," Phil. Mag., 22, 841 (1970).

EFFECT OF POLYMERIC COATINGS ON STRENGTH OF SODA-LIME GLASS

J. E. Ritter, Jr.

Mechanical Engineering Department

University of Massachusetts, Amherst, Mass.   01002

The effectiveness of polymeric coatings (acrylic, epoxy, and silicone) in preventing stress corrosion was determined by measuring the loading rate sensitivity of bend strength for coated, abraded and acid-polished soda-lime glass. The coatings were found to limit the availability of water to the glass surface and, thereby, cause an increase in the strength of abraded glass. The coatings also significantly improved the mechanical abrasion resistance of the acid-polished glass. However, the coatings did not eliminate the stress corrosion reaction between water and the glass surface and, hence, did not affect the sensitivity of glass strength to loading rate. During our testing program, it became evident that significant friction can arise during the bending test due to the supports "digging" into the coatings. Therefore, the differential equations governing beam deflection in four-point bending were solved to include frictional forces between the test samples and the supports. The results of the bending analysis showed that frictional forces become significant at large deflections and can cause an increase in the observed strength.

I.   INTRODUCTION

The detrimental effect of moisture on the strength of glass is easily demonstrated by the fact that higher strength values are always obtained for glass tested in dry air than in moist air. Under extreme conditions, such as in vacuum or in liquid nitrogen, where surface water is either absent or made chemically inactive, the strength is not only high, but also the dependence of strength on time of loading tends to disappear. Thus, complete and permanent absence of surface-adsorbed water is highly desirable. It is possible that this can be achieved by applying a coating to the glass which can replace the surface-adsorbed water and is imperme-able to ambient moisture. It has previously been shown that silane and epoxy coatings can retard the adsorption of moisture somewhat since short term strength values of coated E-glass fibers are slightly higher than for uncoated fibers; however, after long term exposure to high humidity or in water immersion, both coated and uncoated fibers decrease to similar strengths[1,2]. The purpose of this study was to explore in detail the effectiveness of polymeric coatings in preventing the detrimental attack of moisture on glass.

Ritter[3] has shown that the failure strength of glass (σ_f) is proportional to the rate of loading (β) as follows:

$$\sigma_f = k\beta^{\frac{1}{n+1}} \tag{1}$$

where k and n are constants. The strength constant k is a measure of the strength level of the glass being tested and has been shown to be dependent on the relative humidity of the ambient environment and the surface condition of the glass[3]. The constant n is a mea-sure of the stress corrosion susceptibility of a glass and has been shown to be dependent on the chemical composition of the glass[4]. Therefore, it was thought that the effectiveness of polymeric coat-ings in inhibiting the stress corrosion reaction between water and the glass surface could be determined by measuring the sensitivity of fracture strength to loading rate of coated glass and comparing the resulting constants k and n to those obtained for uncoated glass. In addition, it became evident in preliminary experiments that in the bend test of samples with relatively soft coatings significant friction could exist between the beam and the supports which could affect the measurement of strength. Therefore, the nonlinear bending analysis of Vrooman and Ritter[5] was extended to account for frictional forces at the support points.

II. EXPERIMENTAL PROCEDURE

Soda-lime laboratory glass rods (Kimble, Standard Flint R-6 glass) of 3 mm diameter were cut to 5-in. lengths and annealed at 500°C for one hour. The specimens were then given one of two standardized surface treatments. One group of specimens was abraded with a standard blast of No. 240 SiC grit. The grit-blast apparatus rotated the specimens at 250 RPM to give uniform damage around the circumference and nitrogen compressed at 7 psi supplied a 5-S blast. After abrasion, the samples were allowed to age in distilled water for approximately 24 hours to normalize the strength. Another group of specimens were acid-polished in an aqueous solu-

tion by volume of 15% hydrofluoric and 15% sulfuric acid until
approximately 0.025 in. was removed from the diameter, which is
adequate to give maximum strength. After the surface treatment,
the samples were rinsed with acetone, dried, and the diameter of
each rod measured at one end with a micrometer. The samples were
then stored in a desiccator until they were coated.

The polymeric coatings chosen for evaluation were acrylic,
epoxy, and silicone resins since all are commercially available
and are reported to form water-resistant, durable coatings. The
acrylic coatings were prepared from Lucite Transparent Molding
Powder L-149[+]. The powder was mixed in the ratio of one part powder to four parts ethyl acetate and the mixture was heated to about
150°F and stirred until all the powder was dissolved. The glass
samples were then dip coated in the acrylic solution and allowed
to dry in air overnight. The epoxy consisted of Conapoxy PA-122
and Conacure EA-011[≠] mixed in a weight ratio of 100 to 7 with 15%
Conap S-1 Thinner. The samples were dip coated and the coating
was cured at 60°C for three hours. The silicone coating was Pan
Shield[*] which could be sprayed onto the glass samples, allowed to
dry for one hour, and cured at 60°C for three hours. All three
coatings were well-adhered and continuous with a thickness of about
0.003 in.

After coating, groups of 20 samples were broken in 4-point
bending at seven different loading rates on an Instron testing
machine. The bending apparatus had inner and outer spans of 0.813
and 2.808 in., respectively. The supports were ball bearings
0.750 in. in diameter, and the inner rollers were fitted with brass
sleeves having a peripheral radial groove with a minimum diameter
of 0.780 in. This groove enabled a specimen to sit stably on the
center rollers while the outside rollers were brought into contact
with it. Both the inner and outer rollers were pinned to eliminate
friction between the beam and its support.

The samples were wetted with distilled water immediately
before the test. For brief loadings, one dipping was sufficient
to wet the samples for the duration of the test; for longer times,
the samples were rinsed periodically with distilled water during
the test. The fracture stress was calculated from the well-known
simple bending formula.

[+] Fisher Scientific Co., Pittsburgh, Pa.

[≠] Conap Corp. Allegany, N.Y.

[*] Dow Corning, Midland, Michigan.

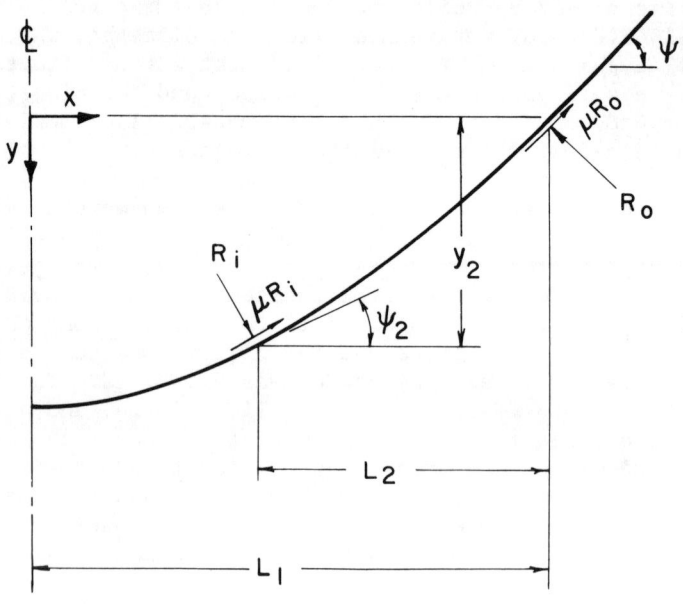

Figure 1.   Deflected position of one-half a thin beam in four-
            point bending.

III.   BENDING ANALYSIS

Figure 1 shows one-half of a symmetrically loaded thin beam
in four-point bending. The distance in the x-direction is measured from the center of the beam and in the y-direction from the
point of contact at the outer support. From Figure 1 the forces
at the supports consist of the normal reactions R_o and R_i and the
tangential frictional forces μR_o and μR_i, where $\mu = \tan \alpha$ is the
coefficient of friction between the beam and the supports. Since
the frictional forces do not act in the neutral plane of the beam,
a moment will be set up at each support; however, these moments
will be quite small for thin beams and consequently have been neglected in the analysis. The lateral extension or compression of
the beam due to the reactive forces is also neglected. Further,
it can be shown that the corrections due to the shortening of the
moment arms L_1 and L_2 resulting from a shifting in the point of
tangency at the support rollers are negligible for rollers of diameter small compared with the length of the beam[5].

Since the resultant vertical force at each support is equal to one-half the applied load on the beam, P, we have:

$$P = R_o \cos \psi_1 + \mu R_o \sin \psi_1 \tag{2}$$

and

$$P = R_i \cos \psi_2 - \mu R_i \sin \psi_2 \tag{3}$$

hence

$$R_o = \frac{P}{\cos \psi_1 + \mu \sin \psi_1} \tag{4}$$

and

$$R_i = \frac{P}{\cos \psi_2 - \mu \sin \psi_2} \tag{5}$$

The resultant horizontal forces at the outer and inner supports are respectively:

$$H_o = \frac{P(\sin \psi_1 - \mu \cos \psi_1)}{\cos \psi_1 + \mu \sin \psi_1} = P \tan (\psi_1 - \alpha) \tag{6}$$

and

$$H_i = \frac{P(\sin \psi_2 + \mu \cos \psi_2)}{\cos \psi_2 - \mu \sin \psi_2} = P \tan (\psi_2 + \alpha) \tag{7}$$

The bending moments at a distance x from the center of the beam are therefore:

$$L_1 - L_2 \leq x \leq L_1$$

$$M = P[(L_1 - x) + (y) \tan (\psi_1 - \alpha)] \tag{8}$$

$$0 \leq x \leq L_1 - L_2$$

$$M = P[(L_1 - x) + (y) \tan(\psi_1 - \alpha) - (L_1 - L_2 - x)$$

$$- (y - y_2) \tan(\psi_2 + \alpha) \tag{9}$$

The bending moment equations can be substituted into the beam deflection equation:

$$\frac{d^2y/dx^2}{[1 + (dy/dx)^2]^{3/2}} = \frac{M}{EI} \tag{10}$$

where E = elastic modulus and I = moment of inertia of the beam, to obtain the following expressions for curvature:

$$L_1 - L_2 \leq x \leq L_1$$

$$\frac{d^2y}{dx^2} = \frac{1}{EI} [1 + (dy/dx)^2]^{3/2} P[L_1 - x + (y) \tan(\psi_1 - \alpha)] \tag{11}$$

$$0 \leq x \leq L_1 - L_2$$

$$\frac{d^2y}{dx^2} = \frac{1}{EI} [1 + (dy/dx)^2]^{3/2} \, P[L_2 + (y) \, \tan(\psi_1 - \alpha)$$
$$- (y - y_2) \, \tan(\psi_2 + \alpha) \tag{12}$$

The nonlinear differential equations (11) and (12) can be readily solved using numerically techniques as described previously[5]. Figure 2 is the result of such calculations where the applied load is plotted vs. the angle at the outer support, ψ_1. Also shown in Figure 2 is the simple beam stress, σ, which was calculated from

$$\sigma = \frac{P \, L_2 \, C}{I} \tag{13}$$

where C = distance from neutral axis to the outer fibers of the beam.

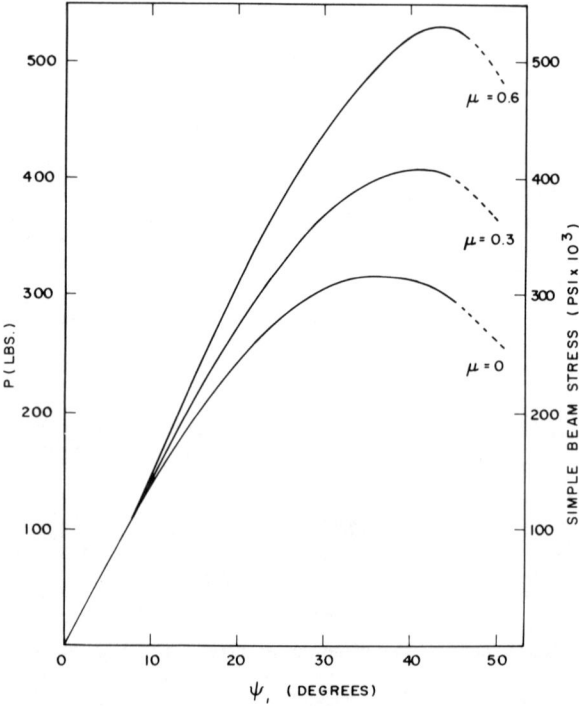

Figure 2. One-half the applied load, P vs. the angle at the outer support ψ_1. Bending apparatus had dimensions: $L_1 = 1.5$ in. and $L_2 = 0.75$ in. Beam is a borosilicate glass rod with a diameter of 0.197 in. and $E = 9.1 \times 10^6$ psi.

The curves in Figure 2 were calculated using the dimensions of the bending apparatus and test samples of Teeg, et al.[6],[7] for reasons that will be discussed later in the paper. The maxima in the curves of Figure 2 occur because the moments due to the horizontal load components that tend to buckle the beam become greater than the moments from the vertical load components. If a load greater than the maximum is applied, an instability occurs which for brittle, high strength materials can result in premature fracture.[5]

From Figure 2 it is seen that frictional forces do not become significant until large deflections (ψ_1 greater 10°) and that the applied load is resisted by the frictional forces so that to obtain the same beam deflection, the load must be increased to offset the effect of friction. Therefore, from a practical point of view, the existence of frictional forces at the supports could cause an erroneous increase in strength since the applied load must be increased to achieve the same degree of bending in the beam. Experimental proof of this seemingly "strengthening" effect will be discussed later in this paper.

By plotting the maximum applied load P_m, corresponding to the bending instability, as a function of the coefficient of friction,

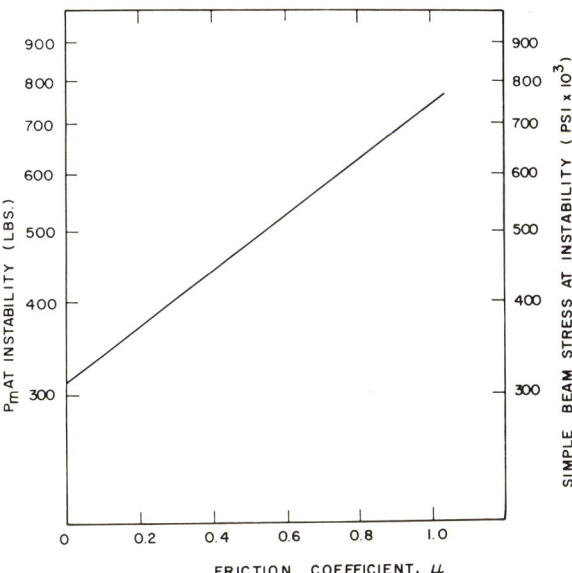

Figure 3.   One-half the applied load at the bending instability, P_m, vs. coefficient of friction, μ, for the bending apparatus and beam given in Figure 2.

it was found that a simple relationship exists between P_m and μ, namely, that the logarithm of P_m is proportional to μ. This is shown in Figure 3 where the calculations were made for the bending apparatus and beam given in Figure 2. The corresponding simple beam stresses are also included in Figure 3.

IV. RESULTS AND DISCUSSION

In Figure 4 the loading rate sensitivity data for abraded, coated and uncoated soda-lime glass are plotted with a least square straight line drawn through each set of data points. The loading rate in Figure 1 is given as the cross-head speed of the Instron testing machine; stress rate (psi/min) can be computed by multiplying the cross-head speed by the factor 3×10^5 and strain rate (in/in/min) by multiplying by 2.9×10^{-2}. From equation (1) it is seen that the constants n and k can be evaluated from the slope and intercept of a log-log plot of failure stress vs. loading rate such as Figure 4. For convenience the intercept, k, of the log-log plot was taken as the stress value obtained from the least square analysis for a cross-head speed of 0.5 cm/min. Table I summarizes the results from the loading rate sensitivity data.

Since the coefficient of variation for the abraded samples was on the average about 11% and for the acid-polished samples 23%, it can be seen from Table I that the acrylic, epoxy, and silicone coatings were all effective in significantly increasing the short-term strength of abraded glass; however, none of these polymeric

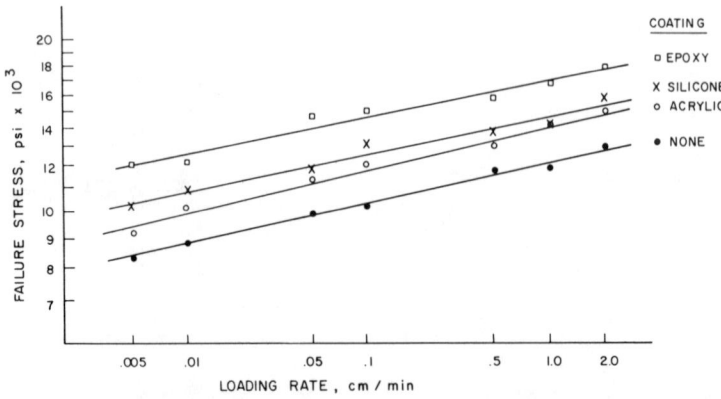

Figure 4.   Effect of loading rate on room temperature strength of abraded soda-lime glass tested wet with a. epoxy coating; b. silicone coating; c. acrylic coating; d. no coating.

Table I. Strength Constants (k) and Stress Corrosion Suscepti-
          bility Constants (n) for Polymeric-Coated, Soda-Lime Glass

Surface Condition	Coating	k (psi x 10^3)	n
Abraded	Acrylic	13.5	12.0
	Epoxy	15.8	16.9
	Silicone	14.0	14.4
	None	11.7	13.0 - 16.0
Acid-Polished	Acrylic	344.2	17.4
	Epoxy	349.6	15.1
	Silicone	324.4	13.8
	None	317.3	13.0 - 16.0

coatings increased significantly the strength of acid-polished
glass. These results are similar to ones obtained previously
where abraded glass tested dry in air was about 10% stronger than
tested wet with water but acid-polished samples showed no difference in strengths between these two test conditions[3]. Thus, it
is believed that these polymeric coatings do limit the availability
of water to the glass surface, presumably by acting as a diffusion
barrier, and, hence, cause an increase in strength for abraded
glass. Since the strength of acid-polished glass is not as sensitive within experimental error to the availability of water, these
polymeric coatings are not as effective in increasing the strength;
however, it should be noted that all three coatings greatly improved the mechanical abrasion resistance of the acid-polished
glass and the coated samples could be freely handled with no detrimental strength effect.

From Table I it is also seen that the acrylic, epoxy, and
silicone coatings have little, if any, effect on the stress corrosion susceptibility of either abraded or acid-polished soda-lime
glass. These results agree with previous findings that the loading rate sensitivity of glass is not affected by having the glass
tested wet or dry[3]. Thus, it is concluded that the polymeric
coatings do not alter the basic reaction between water and the
glass surface and that to minimize sensitivity to loading rate, a
glass with a low stress corrosion susceptibility, such as fused

quartz (n = 37.8) or borosilicate (n = 27.4)[4], should be used.

In contrast to the above discussed results, Teeg, et al.[6,7] found that soft coatings increased significantly the strength of acid-polished borosilicate glass and also increased the sensitivity of strength to loading rate. Since these researchers observed similar strength increases when the glass was only coated at the support edge areas, they believed that this increase was due to the effectiveness of the coatings in preventing damage to the glass by the supports of the bend test apparatus. Since their bend apparatus utilized knife edge supports which tended to "dig" into the coatings and in some cases cause an actual "tearing" of the coating, the present author suspected that friction between the coated samples and the supports was the cause of their contradictory strength effects. This suspicion was furthered when the present author (using the nonlinear bending analysis of Vrooman and Ritter[5]) calculated that the maximum strength, corresponding to the simple beam stress at the bending instability, that Teeg, et al. could measure for 5 mm (0.197 in.) diameter borosilicate glass with their bend apparatus was 310,000 psi (see Figure 2 where μ = 0). This corresponded very well to their observed maximum strength of 300,000 psi for 5 mm, uncoated, acid-polished borosilicate glass. Therefore, it was postulated by this author that the strengths observed by Teeg, et al. in excess of 300,000 psi for the coated glass must be due to frictional effects.

Table II summarizes the average and maximum strengths recorded by Teeg, et al[6,7] for 5 mm acid-polished borosilicate glass having various coatings. By assuming that the maximum strength observed

Table II.   Strengths measured by Teeg, et al.[6,7] for acid-polished borosilicate glass. The estimated friction coefficient is based on the results given in Figure 3. The coating labeled ASI is an arsenic-sulfur-iodine coating.

Coating	Average Strength (psi)	Maximum Strength (psi)	Estimated Friction Coefficient
None	200,000	300,000	0.0
Paraffin	330,000	510,000	0.55
Acrylic	270,000	390,000	0.28
Acrolyoid	285,000	470,000	0.48
Epoxy	355,000	535,000	0.62
ASI	450,000	680,000	0.89

for each glass-coating combination was due to the bending stability,
an estimate of the coefficient of friction due to the coating
could be obtained from Figure 3 and are included in Table II.
These calculated friction coefficients are quite reasonable and
are of the same order of magnitude (0.2 to 1.0) as measured for a
steel slider (1.2 mm radius) sliding on a variety of plastic
materials[8].

Since Teeg, et al.[6,7] observed the greatest strength increases
with the arsenic-sulfur-iodine (ASI) coating, a variety of other
types of 5 mm acid-polished glasses were coated with ASI and their
strength measured. By assuming a friction coefficient of 0.89
(Table II) for the ASI coatings and the appropriate elastic modulus
for the given glass, the maximum strengths corresponding to the
bending instability could be calculated from curves such as in
Figure 2 and compared to the maximum strengths actually observed
by Teeg, et al.[6,7]. These results are summarized in Table III and
the agreement between the observed and calculated maximum strengths
is almost perfect. It should be noted that the differences in the
calculated maximum strengths in Table III are just proportional to
the elastic moduli of the glasses since this is the only parameter
in the bending equations that would vary from glass to glass for a
given test set-up.

Therefore, it is concluded that the coatings studied by Teeg,
et al.[6,7] did not provide any unique mechanical protection for the
glass during testing; but instead, his strength measurements were
just a reflection of the frictional forces existing between the
knife edge supports and the coated glass. This frictional explanation explains a number of their other results such as the increased

Table III. Observed (ref. 6) and calculated maximum strengths for
            various ASI coated, acid-polished glasses assuming a
            friction coefficient of 0.89.

ASI Coated Glass	Observed Maximum Strength (psi)	Calculated Maximum Strength (psi)
7740 (Borosilicate)	680,000	680,000
R-6 (Soda-Lime)	750,000	750,000
0120 (Potash Soda Lead)	650,000	640,000
1720 (Alumino Silicate)	1,000,000	950,000

sensitivity of strength to loading rate for the coated glass compared to the uncoated glass since an increase in the loading rate
would give the coating less time to adjust to the stresses at the
supports and, hence, cause an increase in the friction at the
support which in turn causes a greater increase in strength than
would be expected from just stress corrosion effects. The presence
of friction at the supports also explains why the increase in
strength was less for ASI coatings that had a high iodine content
since increasing the iodine content gave a noted less viscous
coating.

V.   CONCLUSIONS

1. Polymeric coatings (acrylic, epoxy, and silicone) do not effect
   the stress corrosion susceptibility of abraded and acid-
   polished soda-lime glass.

2. Polymeric coatings do limit the availability of water to the
   glass surface and, consequently, increase the short term
   strength of abraded glass. Also, these coatings significantly
   improve the mechanical abrasion resistance of acid-polished
   glass.

3. A numerical solution to the problem of large deflection bending
   of a thin beam loaded and supported at two points that includes
   the possibility of frictional forces at the supports was obtained. The very high strengths (up to a million psi) of
   coated, acid-polished glass observed by Teeg, et al.[6,7] is
   shown to probably be a result of friction at the supports and
   not, as thought by these researchers, to a unique protection
   from mechanical damage afforded by the coating to the glass
   during testing.

ACKNOWLEDGMENTS

This research was supported by the Office of Naval Research.
The author is particularly grateful to Messrs. J. Manthuruthil,
J. Kessler, and A. Avitable for their assistance in the experimen-
tal work.

REFERENCES

1. W.H. Otto, Tech. Report AD-629 370, 1965.

2. S. Freske, W.H. Otto, R.A. Long, Tech. Report AD-645-880, 1966.

3. J.E. Ritter, Jr., J. Appl. Phys., 40, 340 (1969).

4. J.E. Ritter, Jr., and C.L. Sherburne, J. Am. Ceram. Soc., 54, 601 (1971).

5. D.L. Vrooman, and J.E. Ritter, Jr., Am. Ceram. Soc. Bull., 49, 789 (1970).

6. R.O. Teeg, R.W. Hallman, J.S. Anderson, Final Report of Contract N00019-68-C-0380, 20 May 1969, TRN 900-011.

7. R.O. Teeg, Final Report of Contract N00019-69-C-0212, 20 December 1969, TRN-900-011.

8. K.V. Shooter and D. Taber, Proc. Phys. Soc., B65, 661 (1952).

THE GRIFFITH CRITERION AND THE REVERSIBLE AND IRREVERSIBLE

FRACTURE OF BRITTLE MATERIALS

D. P. H. Hasselman*, D. A. Krohn*, R. C. Bradt** and

J. A. Coppola*

 * Lehigh University, Bethlehem, Pa.

 ** Pennsylvania State University, University Park, Pa.

ABSTRACT

On the basis of the concepts of global and local instability
of mechanical systems, it is concluded that the Griffith criterion
represents a necessary, but not sufficient condition for catastrophic fracture. Instead, a brittle material with a microcrack
will have two values of critical fracture stress. The lower critical fracture stress which corresponds to the Griffith criterion for
reversible fracture, represents the minimum stress level for slow
crack growth by a reversible thermally activated process, i.e the
fatigue limit. The higher critical fracture stress corresponds to
the stress required for irreversible catastrophic fracture, which
requires an energy expenditure in excess of the surface free energy of the new crack surfaces even in the absence of energy dissipative processes, such as plastic flow at the crack tip. Over
the stress range intermediate of the two critical fracture stresses,
the crack exhibits slow crack growth, without requiring the presence of stress-corrosion reactions. It is concluded that the surface free energy of a material cannot be measured by a fracture
experiment.

INTRODUCTION

The low tensile strength of brittle solids was attributed by
Griffith to the presence of microcracks (1). In the formulation
of his theory Griffith suggested that under conditions of thermodynamic reversibility, fracture will occur at a stress level when
on crack extension the decrease in elastic energy of the stress

field of the crack (strain energy release rate) equals or exceeds
the surface free energy of the new crack surfaces.

Theoretical analyses (2,3,4) of the stresses at the tip of a
crack have shown that under the stress condition of the Griffith
criterion, the stress at the crack tip approximates the value of the
interatomic cohesive stress, i.e., the theoretical strength of the
material. This led to the conclusion that the energy condition for
crack instability as suggested by Griffith also represents the sufficient condition for catastrophic fracture. The good agreement
with theoretical and experimental values of surface free energy determined by cleavage experiments (5,6,7) on single crystals appeared
to provide further evidence for the validity of this conclusion.

Numerous other studies (8-16), however, have shown that for
many brittle materials the apparent surface energy (γ_f) required
to propagate a crack in a brittle material greatly exceeds the
surface free energy (γ_s). This discrepancy was attributed to surface roughness effects and to energy dissipative processes such as
plastic flow at the crack tip. Brittle materials also exhibit static and cyclic fatigue behavior, attributable to the slow (subcritical) growth of microcracks (17-20). If the energy condition
is both necessary and sufficient for catastrophic fracture, the
Griffith theory does not appear to predict such sub-critical crack
growth (21). To account for this behavior a number of fatigue
theories based on stress-corrosion reactions, mass-transport processes or viscoelastic behavior at the crack tip were proposed (22-
28).

It is the purpose of this paper to suggest a new physical interpretation of the Griffith criterion, which automatically provides for the existence of a range of stress over which a crack
will exhibit slow crack growth without having to rely on a stress-
corrosion mechanism. In addition, high values for surface fracture
energy are also explained without having to resort to plastic flow
or other energy dissipative processes at the crack tip.

DISCUSSION

The discussion centers around the criteria of global and
local instability of mechanical structures. A structure is considered unstable globally, if sufficient potential energy is
released from the system as a whole to supply the surface free energy of the new fracture surfaces resulting from failure at a single
site. Under these conditions, potential energy is converted into
surface energy without energy loss, i.e., failure is reversible.
A structure is considered unstable locally, if at any site within
the structure the stress level equals the failure stress (i.e.,
the strength) of the material. The energy and stress conditions
for these two criteria will be compared. Throughout the discussion

the material is assumed to be entirely brittle, and to exhibit linear stress-strain behavior up to its value of strength taken equal to the interatomic cohesive stress (σ_c).

Consider a structure with arbitrary geometry and stress distribution and dimensions well in excess of the interatomic distance. It is generally agreed that the potential energy density of a material stressed to its value of cohesive strength, is such that sufficient potential energy is available within a volume of material of the order of a few interatomic distances to supply the surface energy of the new fracture surfaces. As a result, under conditions of global instability which requires potential energy released from the structure as a whole, the potential energy density anywhere within the structure must be less than the value of energy density required for the stress to equal the cohesive stress. As a consequence, the value of stress (σ') applied to the structure required for global instability is insufficient to satisfy the condition for local instability.

For the same structure and stress distribution, local instability anywhere within the structure can be achieved by applying a stress (σ'') which exceeds the stress (σ') for global instability by the ratio of cohesive stress to maximum stress at global instability. Under these conditions the potential energy released on failure will exceed the energy released for global instability, i.e., the surface free energy of the new fracture surfaces. The excess energy which is not recoverable for useful work is lost in the form of acoustic and vibrational energy and/or heat, as suggested previously (29). As a result, failure under conditions of local instability is irreversible.*

Although over the range of applied stress $\sigma' < \sigma < \sigma''$ catastrophic failure cannot occur since the stress condition for local instability is not satisfied, the structure nevertheless is unstable in a thermodynamic sense. As a result, deformation and eventual failure still can occur as the result of a stress-enhanced, thermally activated process and the structure will exhibit stress-time-dependent

---

*These conclusions are easily demonstrated quantitatively for a simple mechanical model consisting of a uniaxially stressed fiber of length L>>a, the interatomic distance. For global instability, the potential energy per unit crossectional area and interatomic distance ~$2\gamma_s a/L$, where γ_s is the surface free energy. For L>a, the energy density is below the value of $2\gamma_s$ required for the stress to equal the cohesive stress (σ_c). The total energy expended for failure at local instability (i.e. $\sigma'' = \sigma_c$) is of the order ~$2\gamma_s L/a$. As a result, only if L = a will the stress conditions for local and global instability be identical, with a net energy expenditure of $2\gamma_s$.

failure characteristics. The stress (σ') for global instability
represents the minimum stress level for which this type of failure
can be operative. Failure at the stress (σ'') for local instability will be instantaneous.

The above conclusions were obtained without restriction on
geometry or stress distribution and should also be valid for a
brittle material containing a crack. Generally, the sizes of
cracks and the dimensions of the stress field of the crack from
which the elastic energy is obtained on fracture, are well in excess of the interatomic distance. A crack is unstable globally if
the strain energy released from the stress field of the crack is
converted reversibly into surface free energy. The Griffith criterion, in fact, represents the condition of global instability
of a crack. Local instability of a crack implies that the maximum
value of stress (which occurs at the crack tip) equals the cohesive
stress (σ_c). In terms of the conclusions reached above, the stress
condition for global instability of a crack cannot simultaneously
represent the condition for local instability. This latter conclusion disagrees with the generally accepted opinion that the
Griffith criterion represents both the necessary and also sufficient condition for catastrophic fracture, which in effect is say-
ing that for a crack the stress conditions for global and local
instability are identical.

The concept of potential energy distribution, invoked earlier,
can be used to show that the Griffith criterion cannot constitute
both the necessary and sufficient condition for catastrophic fracture. If the stress at the crack tip were to equal the cohesive
stress, no need would exist for the energy released from the regions of the stress field away from the crack tip. This is in direct defiance of the condition of reversibility of the Griffith
criterion. As a result, the potential energy density at the crack
tip for the value of applied stress defined by the Griffith theory,
must correspond to a value of stress at the crack which is less
than the cohesive stress. The theoretical analyses (2,3,4) which
concluded otherwise must have been based on erroneous assumptions
for the stress distribution and geometry of the crack tip and did
not include the necessary thermodynamic constraints. In fact, the
statement that the Griffith criterion represents the necessary as
well as sufficient condition for catastrophic fracture implies
that on loading potential energy is created at the crack tip which
is not supplied by the applied load.

As a direct consequence of the insufficiency of the Griffith
criterion for catastrophic fracture, a brittle material with a
crack must have two critical values of applied stress, rather than
only one. The lower value of critical stress (σ') corresponds to

the stress defined by the Griffith criterion. The higher value of critical stress (σ'') corresponds to a value of applied stress for which the stress at the crack tip equals the cohesive stress. Fracture at σ'' occurs irreversibly with an amount of energy released from the stress field of the crack in excess of the surface free energy. As a result no need exists to invoke surface roughness or energy dissipative processes at the crack tip to explain high values of surface fracture energy. Values in excess of the surface free energy simply occur as the consequence of the Griffith criterion's being insufficient for catastrophic fracture.

Over the range of applied stress, $\sigma'<\sigma<\sigma''$, catastrophic fracture cannot take place since the crack tip stress is below the value of cohesive stress. Nevertheless, the crack is unstable from a thermodynamic point of view and can change its dimensions and geometry under the influence of a thermally activated process such as diffusion (25,27). The rate of the process is expected to be most rapid at the site of maximum stress, namely the crack tip, which manifests itself in an increase in crack length and corresponding increase in stress at the crack tip. Failure will occur when the crack tip stress reaches the value of cohesive stress. In short, the material exhibits fatigue behavior typical of brittle materials, without requiring the presence of a stress-corrosive environment. The stress (σ') represents the lowest stress level at which such fatigue can occur. As a result, the Griffith criterion, rather than representing the instantaneous fracture stress of an entirely brittle material, in fact represents the lowest stress level required for slow crack growth, i.e. the fatigue limit. Support for this conclusion is provided by experimental data for the minimum stress levels required for slow crack growth in soda-lime-glass (30) and single crystal sapphire (31) in moist air which correspond to surface free energies of 370 and 320 ergs.cm^{-2}, resp. In terms of the present discussion, the role of "stress-corrosive" environments is to decrease the level of the stress (σ') as a result of a decrease in surface energy. For a given level of stress this leads to greatly accelerated rates of crack growth. At the stress (σ'') for local instability, no thermally activated crack growth is required for catastrophic fracture, failure occurring instantaneously.

Finally, as an additional consequence of the irreversibility of catastrophic fracture, it must be concluded that the surface free energy of a brittle material cannot be measured by means of a fracture experiment. Studies (5,6,7) with conclusions to the contrary will require re-examination. In principle, surface free energy can be measured by determining the minimum stress level for subcritical (slow) crack growth. At low temperatures such experiments are expected to be time-consuming.

Acknowledgments

This study resulted from the cooperation between research programs supported by the Pennsylvania Science and Engineering Foundation and the Army Research Office - Durham under Grant: DA - AROD - 31 - 124 - 73 - G45.

References

1. A. A. Griffith, Proc. Roy. Soc. (London) 221A, 163 (1920).
2. E. Orowan, Repts. Prog. Phys., XII, 185 (1948).
3. G. R. Irwin, J. A. Kies and H. L. Smith, Proc. ASTM, 58, 640 (1958).
4. J. L. Sanders, Jr., Trans., ASME 27, 352 (1960).
5. J. J. Gilman, J. Appl. Phys. 31, 2008 (1960).
6. A. R. C. Westwood and D. L. Goldheim, J. Appl. Phys, 34, 335 (1963).
7. Y. P. Gupta and A. T. Santhanam, Acta Met. 17, 419 (1969).
8. J. Nakayama, J. Am. Ceram. Soc. 48, 583 (1965).
9. R. W. Davidge and G. Tappin, J. Mat. Sci., 3, 165 (1968).
10. A. G. Evans and R. W. Davidge, Phil. Mag. 30, 373 (1969).
11. A. G. Evans and R. W. Davidge, J. Mat. Sci. 5, 314 (1970).
12. J. Congleton and N. J. Petch, Acta Met. 14, 1179 (1966).
13. F. J. P. Clarke, H. G. Tattersall and G. Tappin, Proc. Brit. Ceram. Soc. 6, 163 (1966).
14. J. Congleton, N. J. Petch and S. A. Shiels, Phil. Mag. 19, 795 (1969).
15. J. A. Coppola and R. C. Bradt, J. Am. Ceram. Soc., 55, 455 (1972).
16. A. G. Evans, Phil. Mag., 22, 84 (1970).
17. R. J. Charles, J. Appl. Phys. 29, 1549 (1958).
18. J. C. V. Runsey and A. L. Roberts, Proc. Brit. Ceram. Soc. 7, 233 (1967).
19. J. E. Burke, et al., pp 435-48 in Ceramics in Severe Environments, Ed. by W. W. Kriegal and H. Palmour III, Plenum Press, N. Y. (1971).
20. D. A. Krohn and D. P. H. Hasselman, J. Am. Ceram. Soc., 55, 208 (1972).
21. S. M. Cox, Phys. Chem. of Glasses 10, 226 (1969).
22. D. A. Stuart and O. L. Anderson, J. Am. Ceram. Soc. 36, 416 (1953).
23. W. B. Hillig and R. J. Charles, Chptr. 17 in High-Strength Materials, John Wiley, N. Y. (1965).
24. S. N. Zhurkov, Int. J. Fr. Mech. 1, 311 (1965).
25. D. P. H. Hasselman, Chptr. 14 in Ultra-Fine Grain Ceramics, Ed. J. J. Burke et al., Syracuse University Press (1970).
26. K. Schönert, et al., Chptr. 41 in Fracture, Ed. P. L. Pratt, et al., Chapman and Hall (1969).

27. R. N. Stevens and R. Dutton, Mat. Sc. and Eng. $\underline{8}$, 220 (1971).
28. W. G. Knaus, Int. J. of Fracture Mech. $\underline{6}$, 7 (1970).
29. D. P. H. Hasselman, J. A. Coppola, D. A. Krohn and R. C. Bradt, Mat. Res. Bull. 1, 769 (1972).
30. S. M. Wiederhorn, J. Amer. Ceram. Soc. $\underline{50}$, 407 (1967).
31. S. M. Wiederhorn, in Mechanical and Thermal Properties of Ceramics, NBS Spec. Pub. 303.

VI. ENGINEERING, SCIENTIFIC, AND DESIGN APPLICATIONS

THERMAL SHOCK RESISTANCE OF CERAMIC MATERIALS

Junn Nakayama

Research Laboratory, Asahi Glass Company, Limited

Kanagawa-ku, Yokohama 221, Japan

ABSTRACT

Concepts and theories as well as typical experimental results on thermal shock resistance of ceramic materials are reviewed with an emphasis on the Hasselman theory. On the basis of the strain energy concept, in the latter half of this paper, a new thermal shock test method for refractory firebricks is introduced.
A rod specimen with a square cross-section is heated by radiation on its opposite two surfaces. After slow cooling, the specimen is cut into two pieces at the central plane parallel to the exposed faces followed by strength measurement for each piece by bending. Strength behavior of six brands of firebrick against varying radiation temperature is compared with thermal shock fracture resistance and thermal shock damage resistance parameters as well as thermal shock crack stability parameter of the materials tested. The results show that the present method is useful for comparing both fracture resistance and damage resistance of materials.

1. INTRODUCTION

It is widely recognized that the thermal stress fracture of ceramic materials is one of the most fatal and probably everlasting problems when they are used as structural components or containers under severe thermal environments. Up to the present many efforts have been exerted to describe theoretically thermal shock resistance in terms of physical properties of materials. Many test methods also have been developed to compare the relative resistance and to obtain information on the behavior of materials under various types of thermal shock occuring in practical applications.

759

Fundamental theories of thermal shock resistance developed today appear to have succeeded in principle in providing useful means in the form of thermal shock resistance parameters[1] so that one can compare materials and select favorite ones from candidate materials for particular applications, or one can make use of the parameters as a guide to improve thermal shock resistance of particular materials of interest. The parameters as represented with pertinent physical properties are conveniently classified into two groups, namely, "thermal stress fracture resistance parameters" (hereafter referred to as fracture resistance parameters) and "thermal stress damage resistance parameters" (hereafter referred to as damage resistance parameters). The fracture resistance parameters apply to crack initiation problems, while the damage resistance parameters apply to crack propagation problems.

On the other hand, a large variety of thermal shock test methods has been developed to date. Most of them are designed to compare the relative resistance in terms of relative effect of cyclic thermal shock on the physical condition or physical properties. For instance, for a given number of thermal shock cycles, weight loss[2], reduction in strength[3] or the change in Young's modulus of elasticity[4], may be determined. The number of cycles required to produce a given percentage loss of weight[5] or to produce a crack of a given width[6] may also be determined. The effects of single quenching on internal friction[7], Young's modulus[7] and strength[7-13] have also been investigated.

This paper first makes a review of the present state of theories along with some examples of experimental support which appeared in the literature. Emphasis will be put on the theory which is based on fracture mechanics, which has become a powerful tool not only for interpreting results of various thermal shock tests but also for improving thermal shock resistance of materials in various applications. In the latter half of this paper, a thermal chock test method with one dimensional radiation heating for firebrick is presented and the results for six brands of commercial firebrick are discussed.

II. GENERAL REVIEW

Fracture Resistance

Most theories of thermal stress fracture have been based upon the stress-strength criterion. In these treatments, the temperature distribution in a specimen subjected to a specified thermal shock environment is calculated first, and then the resultant thermal stress is compared with the strength of the material. By setting the maximum thermal stress level equal to the tensile strength, the allowable maximum thermal conditions are obtained. For instance,

Table I.   Summary of Thermal Stress Fracture Resistance Parameters[1].

Literature designation	Fracture resistance parameter
R	$S(1-\nu)/\alpha E$
R'	$S(1-\nu)k/\alpha E = R \cdot k$
R''	$S(1-\nu)k/\alpha Ec\rho = R'/c\rho$
Rcr	$S(1-\nu)/\alpha\eta$
R'cr	$S(1-\nu)k/\alpha\eta = Rcr \cdot k$
R_{rad}	$(S(1-\nu)k/\alpha E\varepsilon)^{1/4} = (R'/\varepsilon)^{1/4}$
R_{transp}	$(S(1-\nu)k/\alpha E\varepsilon(1-F\lambda_0))^{1/4}$

For notations, see Ref. 1.

the temperature difference in the specimen[14], the temperature
difference to which the specimen can be subjected in convective
heat transfer[15,16], the steady state heat flow in a body[17], the
temperature difference or the steady state heat flow with thermal
stress relaxation by creep[18], the rate of change of surface temer-
ature for a flat plate[16], the radiation temperature to which small
or thin bodies at low initial temperature can be subjected[19,20],
the temperature gradient in a body with spherical cavity[21], the
temperature gradient in a flat plate with crack perpendicular to
heat flow[22] are calculated. For each of these thermal environments,
the allowable maximum thermal conditions can be expressed as a
product, in general, of a parameter, size factor and heat transfer
factor. The parameters as composed of pertinent physical properties
are called fracture resistance parameters. Analytical expressions
for the fracture resistance parameters along with the designations
used in the literature are listed in Table I as compiled by
Hasselman[1]. These parameters should be applicable to such ceramic
bodies as dinnerware, crucibles, spark plug insulators, rocket
nozzles, etc., in which initiation of thermal stress fracture must
be avoided. In order to attain an increase in the fracture resistance, high values of strength, thermal conductivity and thermal
diffusivity, and low values of coefficient of thermal expansion,
Young's modulus, Poisson's ratio, emissivity and viscosity are
required.

Other approaches to thermal stress fracture problems are those
based on Weibull's statistical theory of fracture[23-25]. This type
of approach, which predicts the effect of body size on strength,
should be taken into account in combination with fracture resistance
parameters for design problems involving scaling up.

Besides these theoretical approaches, efforts are also made to
minimize thermal stresses arising in structural bricks by changing
their geometrical designs with the aid of computer calculation[26].

Damage Resistance

In the industry of refractory bricks, materials are compared under conditions in which all would fail. The results of the test indicate the relative degree of damage rather than the critical levels of allowable thermal severities. With the purpose of describing resistance to damage due to thermal stress, Hasselman[27] introduced the concept of damage resistance to thermal shock with a simplified model of a sphere subjected to thermal shock by heating. It was assumed that the area over which the crack will propagate is directly proportional to the elastic energy stored at the time of fracture and inversely proportional to the fracture surface energy required to propagate a crack. The damage resistance parameters, R''' and R'''' as defined by Hasselman, are given by:

$$R''' = E/S^2(1 - \nu) \tag{1}$$

$$R'''' = EG/S^2(1 - \nu) \tag{2}$$

where E is Young's modulus, S tensile strength, ν Poisson's ratio, and G fracture surface energy. Since the parameter R''' is proportional to the inverse of elastic energy stored in the body, it applies to comparison of materials having similar value of fracture surface energy. The parameter R'''' is proportional to the inverse of crack area propagated by the thermal shock.

Relying also on a strain energy criterion, Clarke, Tattersall and Tappin[28] presented the "toughness parameter" defined by:

$$T = -\log(S^2/12EG)^3 \tag{3}$$

The parameter T, a logarithm of the inverse of total number of fragments resulting from thermal stress fracture initiation, has essentially the same physical meaning as that of the damage resistance parameter R''''.

High values of these damage resistance parameters require high values of Young's modulus and fracture surface energy, and low values of tensile strength. It should be noted that requirements for high Young's modulus and low tensile strength are directly converse to the requirements of low Young's modulus and high strength to avoid the initiation of thermal stress fracture.

The damage resistance parameters, R''', R'''' and T, can be used to compare the relative degree of damage expected for materials under severe thermal shock rather than the critical levels of allowable thermal conditions. Experimental support for the damage resistance parameter R'''' was reported by Nakayama and Ishizuka[29]. They compared the theoretical parameters, R, R', R''' and R'''',

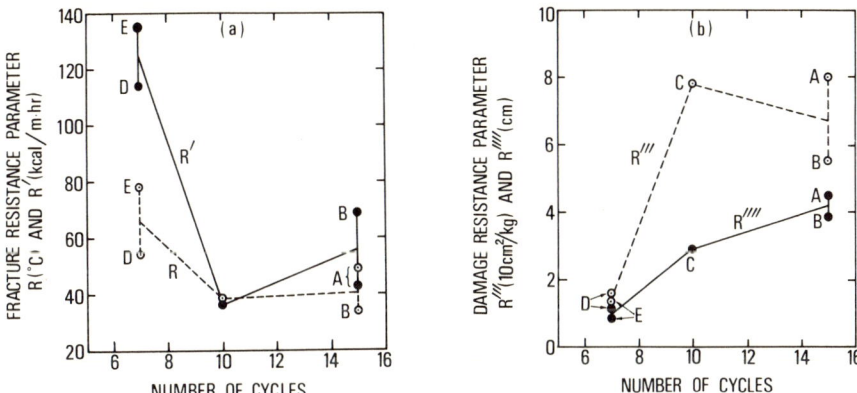

Fig. 1. Correlation between thermal shock resistance determined by an industrial test and various resistance parameters calculated for five brands of commercial firebrick. (Ref. 29).

for five kinds of refractory bricks with the practical spalling resistance determined with a modified DIN-test method in which the spalling resistance is represented as the number of cycles of heating and cooling required to produce a given percentage loss of specimen. The result is shown in Fig. 1. Good correlation is seen for the damage resistance parameter R'''' which involves the term of fracture surface energy (G).

Unified Theory of Fracture Initiation and Crack Propagation

The theories of the damage resistance parameters as originally developed are strictly applicable only to a dynamic crack propagation when temperature differences in the specimens are just sufficient for fracture initiation. For situations where the temperature difference exceeds the minimum required to initiate thermal stress fracture, and for a possible presence of a quasistatic crack propagation, Hasselman[30,31] presented a theory using a fracture mechanical approach which followed Berry's analysis[32] of crack propagation under constant deformation. The mechanical model on which the theory is based is a three dimensional solid body[30] or two dimensional thin flat plate[31] uniformly cooled through a temperature difference, with the external surfaces or edges rigidly restrained to induce a uniform state of triaxial or uniaxial tensile stress, respectively. On the basis of interaction of strain energy and surface energy, an expression for the stability of latent cracks in the body or plate is derived. For a given temperature difference, a region of unstable crack length exists where cracks will grow until they reach a point where strain energy released is not enough to supply the increase in surface energy. The minimum tem-

perature difference required to initiate propagation of the cracks
with the shortest length in the unstable region gives an expression
essentially identical to the fracture resistance parameter, R, given
in Table I. From the expression for the extent of crack propagation
after the initiation of dynamic fracture, the damage resistance
parameter R'''', as given in Equation (2), is derived. The expression for the minimum temperature difference required to repropagate
long cracks gives the following parameters for crack stability:

$$R_{st} = (G/\alpha^2 E_o)^{1/2} \tag{4}$$

$$R'_{st} = (k^2 G/\alpha^2 E_o)^{1/2} \tag{5}$$

where α is the coefficient of thermal expansion, E_o Young's modulus
of crack-free material, and k thermal conductivity. Equation (4) is
essentially identical with that obtained by Goodier and Florence[22]
for the critical temperature gradient required to initiate the propagation of a crack in a plate with linear temperature gradient.

As a practical application of the theory, the expected relative
changes in crack length, and corresponding changes in strength, of
a brittle material subjected to thermal shock by quenching is estimated as a function of increasing temperature difference ΔT.
Figures 2(a) and (b) schematically show, respectively, the expected
crack length and the corresponding strength as a function of ΔT for
those materials which have initial cracks of a length C_o smaller than
the critical crack length C_m given by:

$$C_m = (\frac{80(1-\nu^2)}{9(1-2\nu)}N)^{-1/3} \quad \text{(for solid body)} \tag{6}$$

$$C_m = (6\pi N)^{-1/2} \quad \text{(for thin flat plate)} \tag{7}$$

Crack length C_o and strength S_o remain unchanged until a critical
temperature difference ΔT_c is reached. At ΔT_c, cracks start to
propagate dynamically to a final length C_c, and strength decreases
discontinuously from S_o to a final value of S_c. With further increase in ΔT, crack length and strength do not change until another
critical temperature $\Delta T'_c$ is reached, and then crack length increases
and strength decreases gradually. The final crack length reached
from the initial crack length C_o in the dynamic process is given,
for $C_o \ll C_m$, by:

$$C_c = (8(1-\nu^2)NC_o/3(1-2\nu)^{-1/2} \quad \text{(for solid body)} \tag{8}$$

$$C_c = (4\pi NC_o)^{-1} \quad \text{(for thin flat plate)} \tag{9}$$

Figure 2(c) shows a typical discontinuous curve presented by
Hasselman for the strength behavior of polycrystalline alumina rods
subjected to water quenching. The discontinuous characteristics

were reported for alumina by Davidge and Tappin[7], Ainsworth and
Moor[8], and Gupta[9], for silicon carbide by Coppola[10], for an aluminosilicate ceramic by Gebauer, Krohn and Hasselman[11], for porcelain
by Kato and Okuda[33], and for a refractory brick by Nakayama[34].

For those materials for which initial crack length is larger
than the critical crack length given by Equation (6) or (7), a continuous characteristic curve is expected to exist, as shown in Figs.

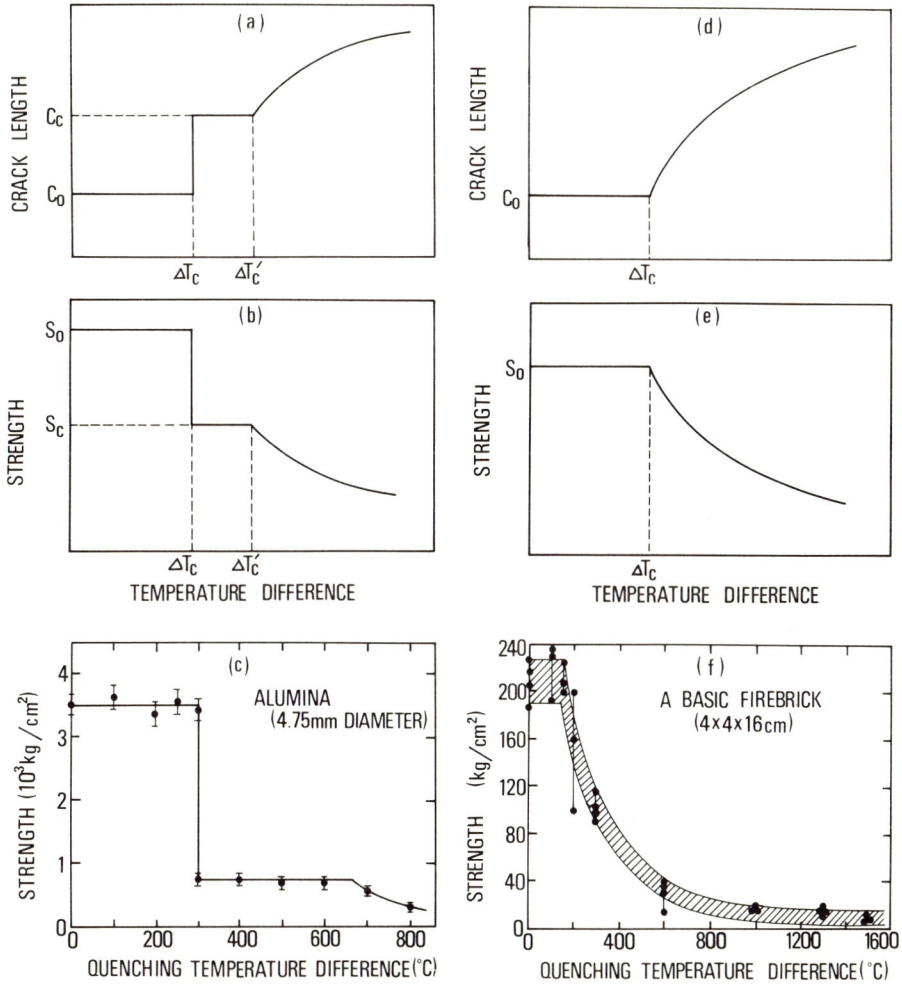

Fig. 2. Variation of crack length and strength with increasing
temperature difference predicated by the theory. (a) and (b) for
small initial crack length, and (d) and (e) for large initial
crack length. (c) and (f) are quenching test results, respectively,
for alumina rods and for a refractory brick specimen. (Ref. 30 for
(a), (b), (d) and (e), Ref. 9 for (c) and Ref. 34 for (f)).

2(d) and (e). Crack extension does not occur and strength remains
unchanged until ΔT reaches ΔT_C as is the case for short initial
cracks. As ΔT exceeds ΔT_C, cracks begin to propagate quasistati-
cally resulting in a gradual decrease in strength. Figure 2(f)
shows strength after water quenching as a function of temperature
difference for a basic refractory brick[34]. The continuously decreasing curves were observed for porous porcelain bodies by Kato
and Okuda[33], for cermet materials by Tacvorian[35], and for large
grain sized alumina by Gupta[12].

     As stated in the foregoing, which type of fracture character-
istic would occur, discontinuous type or continuous type, depends
upon the initial crack length compared with a critical crack length
given by Equation (6) or (7). Since the critical crack length is
determined, as the equations indicate, mainly by the number of
cracks per unit volume or per unit area, and since the number of
cracks may vary with the type of thermal shock or the mode of distribution of thermal stress, most materials may exhibit both types
of characteristic curves. For instance, for a basic refractory
brick, for which a continuous curve was shown in Fig. 2(f) with a
quenching test, Nakayama[34] reported also a discontinuous curve shown
in Fig. 3 as a result of single radiation heat shock followed by
slow cooling.

     Presence of another type of characteristic curve than those
demonstrated in Fig. 2 was reported by Coppola[10] for silicon carbide
rectangular bars subjected to water quenching as shown in Fig. 4.
The discontinuous drop in strength was followed directly by a gradual
decrease in strength without a constant strength level. He speculated that the high value of fracture surface energy might tend
to mask the region of constant strength level.

     Coppola, Krohn and Hasselman[36] presented an expression for the
fractional loss in strength of circular rods subjected to a water

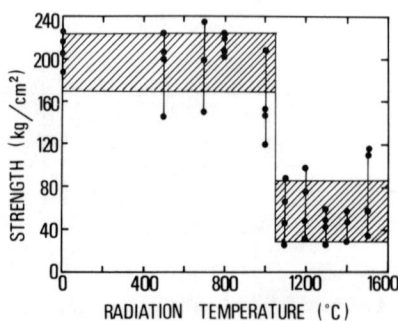

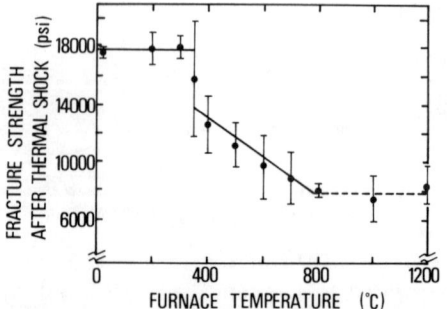

Fig. 3.   Strength of a brick
after radiation heat shock.
(Ref. 34).

Fig. 4.   Strength of a silicon
carbide after water quenching.
(Ref. 10).

quenching in the form:

$$F_s = 1-S_c/S_o = 1-(8E^3G^3N/0.57(1-\nu^2)^2bS_o^6)^{1/4} \qquad (10)$$

where N is the resultant surface crack density and b is rod radius.
From this equation, they derived the "strength-loss resistance parameter", Rstr, as given by:

$$Rstr = (E^3G^3N/S_o^6(1-\nu^2)^2)^{1/4} \qquad (11)$$

Comparing this with Equation (2), and putting $(1-\nu) = (1-\nu^2) = 1$ and
N = constant, one can find out that Rstr is proportional to $(R'''')^{3/4}$
for a given size of specimens.

For more detailed review of thermal shock problems and suggestions for improving the thermal shock resistance of materials,
readers are referred to the literature presented by Rossi and
Hasselman[37].

III. SINGLE THERMAL SHOCK BY RADIATION HEATING

Introductory Remarks

In most industrial tests for refractory bricks, thermal cycling
of combined heating and cooling has been adopted probably because
of two reasons: that thermal cycling is encountered in most practical applications of refractory bricks, and that the cycling in the
test may magnify the effect of thermal shock, although the analysis
of the effects of cycling on the strength behavior involves very
complicated problems to be solved in future. However, in industrial
use of refractory bricks, there are cases where large blocks suffer
fatal damage due to severe single heating resulting in the formation
of large plane crack parallel to the hot face. This type of fracture
was investigated by Kienov[38] with the stress-strength approach.
In view of the fracture mechanical approach reviewed in the previous
section, N number of cracks per unit volume for this type of fracture
is considered to be as small as 1/(volume of a brick) for bricks
with nearly cubic geometries or 1/(minimum cross-sectional area of
a brick)$^{3/2}$ for bricks with relatively columnar geometries, and,
as a result, the critical crack length C_m as defined by Equation
(6) is very large. The mode of fracture for these situations, in
the light of the Hasselman theory, would always be a dynamic one,
i.e., a crack initiated in the body propagates dynamically, even if
the length and density of the latent initial cracks are considerable.
Material selection should, therefore, be based on the comparison
of the damage resistance parameter R''''.

In order to confirm the validity of R'''' with analytical tests
and to establish an adequate method for comparing materials to be

used in these situations, a series of thermal shock tests was performed by making use of single thermal shock by radiation heating.

Experimental Procedure

Specimen preparation. The following six brands of commercial
firebrick, namely, A, B, C, D, E and F, were tested.

> A-brick: Hard burned, dense type aluminosilicate brick.
> B-brick: High alumina, clay bonded refractory brick.
> C-brick: Dense type aluminosilicate brick.
> D-brick: High alumina, spalling resistant refractory brick.
> E-brick: High magnesia, direct bonded basic brick.
> F-brick: Chamotte brick fired to SK-34.

Specimens were sawed with a diamond cutting wheel from large blocks
of standard sizes. The specimen size for thermal shock test was
2 by 2 by 7 cm. Only for F-brick, another size of 4 by 4 by 10 cm
was added to examine size effect. They were finished with 100 mesh
silicon carbide abrasives. For fracture surface energy measurement,
specimens with the dimensions of 1.5 by 1.5 by 6.0 cm were prepared.

Measurement of physical properties. Strength was measured by
transverse three-point bending with a span length of 5 cm and a
loading speed of 0.03 cm per minute at the center. The strength was
calculated with the conventional formula:

$$S = 3LW/2BD^2 \qquad\qquad\qquad (12)$$

where L is span length, W is fracture load, B and D are width and
thickness of a specimen, respectively. Young's modulus was determined by measurement of sonic wave velocity using original large
bricks. Poisson's ratio was assumed to be 0.25 throughout the
study. Thermal expansion was measured with a silica glass dilatometer from room temperature to 1000°C, but the expansion from
room temperature to 500°C was taken into account. Thermal conductivity was determined by hollow cylinder method[39] with specimens 4
and 0.5 cm in outer and inner diameters, respectively, and 4 cm in
height. Fracture surface energy was measured by work of fracture
method[40] with a span length of 5 cm and with loading speed of
0.015 cm per minute at the center of specimen.

Thermal shock test method. Two pieces of specimens were tested
at a time. They were sandwiched side by side with thermal insulation blocks having the same geometry as that of specimens, and
the whole assembly was tied with platinum wires to form a heat
shock test unit as shown in Fig. 5(a). The unit was inserted
quickly into an electric furnace which had been kept at a given
high temperature. The inner sides of the furnace were 15 by 15 cm

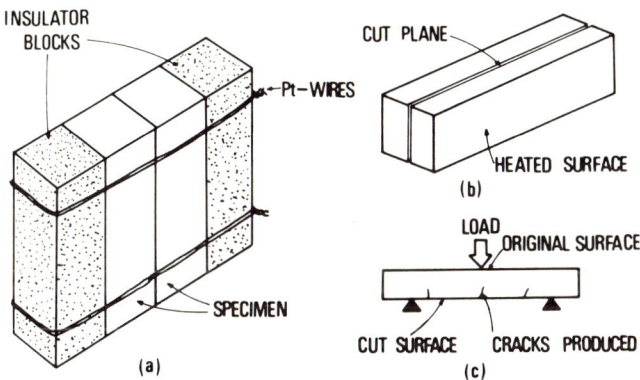

Fig. 5.   Schematic illustrations of (a) a thermal shock specimen
unit which is heated on both sides by radiation, (b) cutting direction in a specimen after thermal shock, and (c) strength measurement
after cutting.

in vertical cross-section and 30 cm in horizontal depth.   Each of the
two vertical side walls was installed with six silicon carbide heating elements, so that both sides of the specimen unit were exposed
directly to the main heat source of radiation.   The furnace temperature was controlled through a thermocouple placed at a position
near the specimen unit.   The specimen unit was kept in the furnace
for a given time depending upon the specimen size; about 2 and 8
minutes for specimens with cross-section of 2 and 4 cm square,
respectively.   These values were determined with much allowance by
computer calculation for the maximum tensile stress at the center
of a specimen to be reached.   The specimen unit was then taken out
of the furnace and cooled to room temperature in the annealing powder
sil-o-cell.   After cooling, each specimen was sawed with a 0.04 cm
thick diamond wheel, parallel to the heated surfaces as shown in
Fig. 5(b).   Bending strength of each of the "twin" samples was
measured in such a manner as shown in Fig. 5(c).   Original strength
was also measured following the cutting procedure.   The calculated
strength was corrected by a factor, $(1-2\Delta L/L)$, where ΔL is the distance between the crack and the center of the bending span.

Results and Discussion

Physical properties of each brick are listed in Table II.
Thermal shock resistance parameters, R', R'''', and R'_{st}, and critical stress intensity factor, K_{IC}, calculated from the physical properties are summarized in Table III.   Thermal shock test results,

Table II.   Physical Properties of Bricks Tested

Brick	S_0 (at 25°C) dyne/cm^2 (10^6)	E (at 25°C) dyne/cm^2 (10^{11})	G (at 25°C) erg/cm^2 (10^4)	α (at 25-500°C) 1/deg (10^{-5})	k (at 200-300°C) cal/cm·sec·deg (10^{-2})
A	258 (38)*	7.42	4.01	1.55	0.70
B	200 (17)	5.57	4.86	0.35	0.30
C	142 (27)	3.19	4.47	1.55	0.30
D	160 (22)	4.76	3.91	0.35	0.30
E	220 (32)	9.13	4.96	1.26	2.20
F	48 (4)	1.04	4.12	0.85	0.25

* Numbers in parentheses are standard deviations.

Table III.   Thermal Shock Resistance Parameters

Parameter	R'	R''''	$R'st$	K_{IC}
Equation	*	(2)	(5)	(15)
Brick	cal/cm·sec (10^{-3})	cm (10^{-2})	cal/cm$^{1/2}$·sec (10^{-3})	dyne/cm$^{3/2}$ (10^6)
A	118 (17)**	59 (18)**	105	244
B	230 (20)	90 (15)	253	233
C	65 (12)	94 (36)	72	169
D	217 (30)	96 (26)	245	193
E	315 (46)	125 (36)	407	301
F	102 (8)	248 (41)	185	93

* Equation for R' is given in Table I.
** Numbers in parentheses are standard deviations.

Table IV. Critical Temperature and Strength Retained
After Thermal Shock of T_{rc} and 1500°C Radiations.

Brick	T_{cr} °C	S_c dyne/cm^2 (10^6)	$S(1500°C)$ dyne/cm^2 (10^6)	S_c/S_0
A	950	34 (14)*	0	0.13
B	1050	107 (9)	21 (9)*	0.54
C	850 ±50**	78 (23)	26 (9)	0.55
D	1050	102 (12)	13 (5)	0.65
E	1100	186 (9)	108 (3)	0.85
F	950 ±50	44 (5)	18 (4)	-

* Numbers in parentheses are standard deviations.
** Limit of uncertainty.

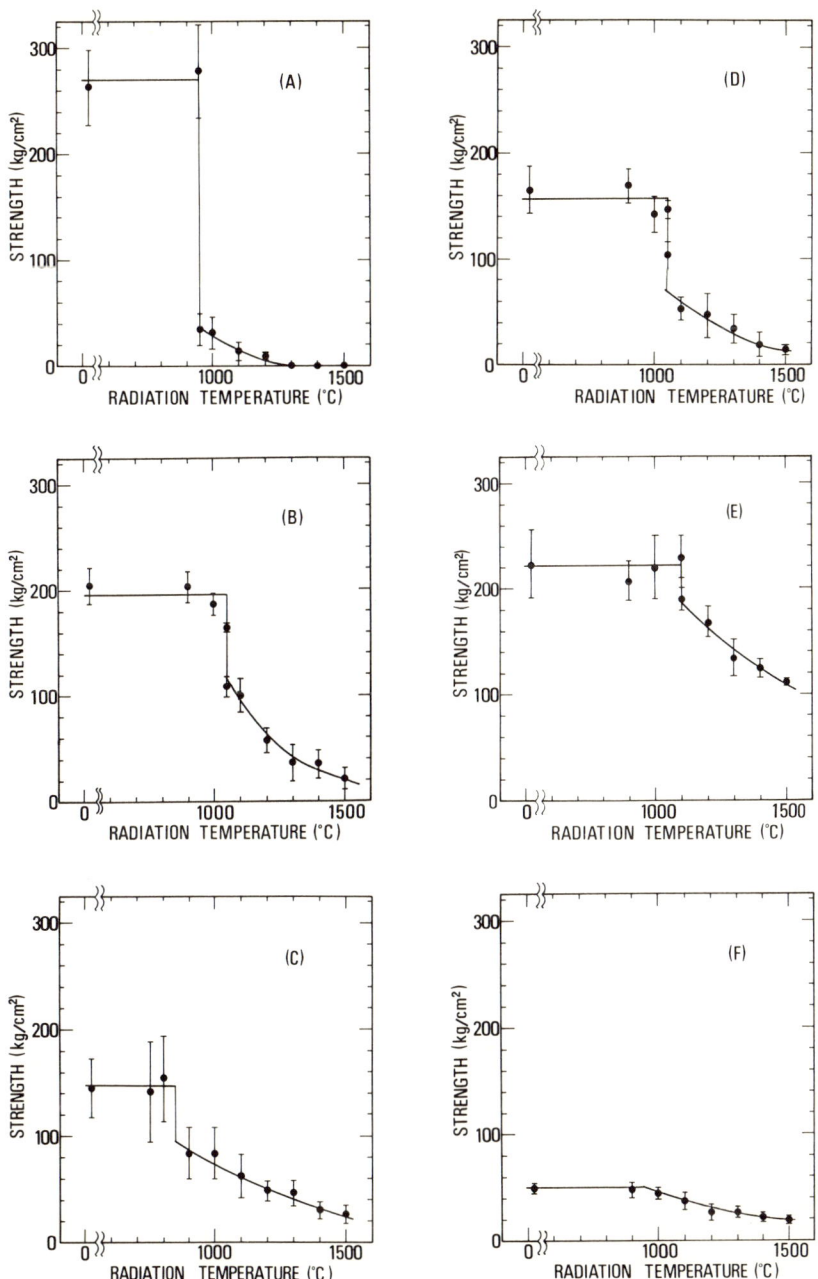

Fig. 6.   Strength variation of specimens subjected to radiation heating as a function of radiation temperature.   The capital letters shown in parentheses correspond to the brands of firebrick.

as expressed in the form of characteristic curves, are given in Fig.
6. Critical radiation temperatures T_{rc}, at which strength begins
to decrease, and strength values, S_c and $S(1500°C)$, retained after
thermal shock with radiation temperatures of T_{rc} and 1500°C, and
strength ratio S_c/S_o are listed in Table IV.

For each of those bricks other than F-brick, discontinuous
decrease in strength at a critical radiation temperature was observed. For F-brick, however, a typical continuous characeristic
curve was obtained.

The strength behavior of those materials which showed a discontinuous decrease at T_{rc} is, however, defferent from either of
those predicted by the Hasselman theory, as were illustrated in Fig.
2. It rather resembles that reported by Coppola[10], as was shown in
Fig. 4, in which the strength in the region $\Delta T > \Delta T_c$ decreased
continuously with increasing temperature difference. This behavior
in the present results may be explained by the increase of elastic
energy stored at the time of initiation of fracture with increasing radiation temperature.

Figure 7(a) illustrates computer calculated stress distributions* in a specimen of D-brick at the time when the maximum
temsile stress at the center reached a tentative value of 250 kg/cm^2
for various radiation temperatures. The corresponding elastic energy stored in the specimen per unit axial length of 1 cm, calculated
by elementary theory of elasticity, is shown in Fig. 7(b) as a
function of radiation temperature. The higher radiation temperature
gives a larger amount of elastic energy to the specimen, resulting
in the formation of a longer crack by the dynamic crack propagation
process. Accordingly, the fracture process in the range immediately
following the discontinuous decrease is considered to be purely
dynamic. This mechanism of gradual decrease in strength with a
dynamic process is supported by the result of observation of fracture mode of A-brick specimens subjected to the radiation temperatures of 1300, 1400 and 1500°C. These specimens, as plotted at
zero strength in Fig. 6(A), showed an instantaneous splitting
fracture with a noise emission, forming a plane crack perpendicular
to the long axis.

Contrary to quench tests, in which N number of cracks per unit
area varies from one material to another as reported by Coppola,
Krohn and Hasselman[36], the present method of radiation heating is
expected to produce only one crack in a central region of a specimen, at least at critical radiation temperature.

---

* For simplicity, only the principal stress in the direction parallel to the axis of the specimen was taken into account. Emissivity
was assumed to be 0.8.

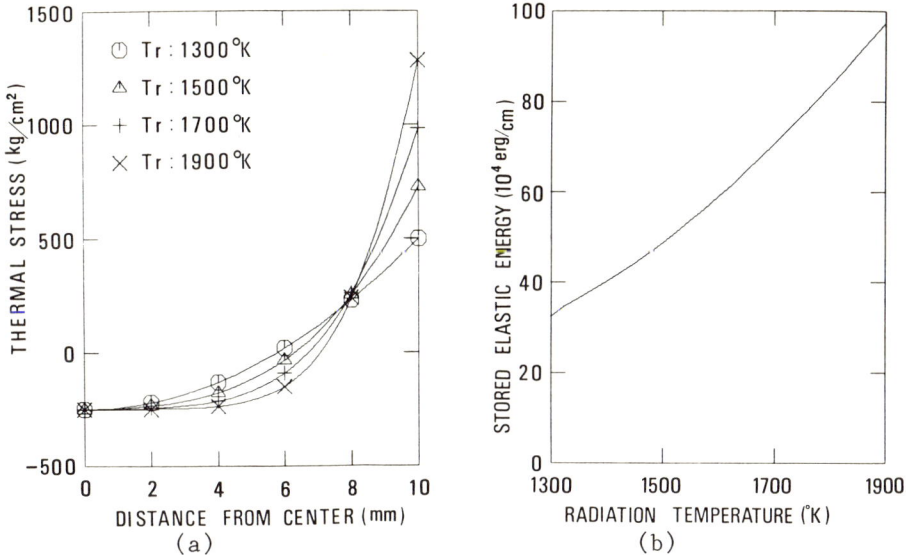

Fig. 7.   (a)Axial stress distributions in a specimen at fracture
for various radiation temperatures, and (b)elastic energy stored in
unit axial length as a function of radiation temperature.

The extent of crack propagation in specimens which suffered
dynamic cracking was estimated from the bending strength and the
critical stress intensity factor.   Since the temperature distribution in the center cross-section of a specimen is one dimensional,
stress distribution before fracture may also be approximated to
be one dimensional.   Accordingly, an assumption was made that the
shape of a crack propagated under the one dimensional stress distribution was rectangular, B cm wide and C cm deep in a bending specimen, although an analysis of the shape of crack front in bending
fracture of glass specimens made by Lange[41] showed elliptical shapes.

For a rectangular bar with a plane crack on a single edge subjected to three point bending, the critical stress intensity factor
K_{IC}, as given by Brown and Srawley[42], is:

$$K_{IC} = (3LWC^{1/2}/2BD^2) \cdot F(C/D, L/D) \tag{13}$$

where C is the depth of a crack, and F(C/D, L/D) is a non-dimension-
al numerical factor determined as a function of C/D and L/D.
Combining this equation with Equation (12), the ratio of K_{IC} to
bending strength (S) may be written in the form:

$$K_{IC}/S = C^{1/2} \cdot F \tag{14}$$

On the other hand, K_{IC} is related to fracture surface energy (G) and Young's modulus (E) by[10]:

$$K_{IC} = (2EG)^{1/2} \qquad\qquad (15)$$

From the measured values of K_{IC} (or E and G) and S, final crack length in a specimen after thermal shock may be determined. Figure 8 presents curves for obtaining the length of a crack from the value of K_{IC}/S for the cases of D = 0.73, 0.98 and 1.48 cm and L = 5 cm.

The effective lengths of initial crack and final crack remaining in bending specimen for the radiation temperature of T_{rc} and 1500°C are summarized in Table V. Since the area over which a crack propagates is proportional to the final crack length, and since the resistance parameter R'''' is defined as a quantity proportional to the inverse of the area of a crack propagated at the critical radiation temperature, a close correlation is expected to exist between R'''' and the inverse of a final crack length reached by the radiation of T_{rc}. From the point of view of material selection for practical uses, final crack length produced by a common radiation temperature should better be compared rather than C_c. Correlation of the inverse of crack lengths, C_o, C_c and $C(1500°C)$, with R'''' is illustrated in Fig. 9(a). Good correlations of R'''', not only with C_c but also with $C(1500°C)$, are observed. A clear, but negative, correlation between reciprocal C_o and R'''' is also apparent. The smaller values of C_o resulted in the larger values of C_c, as expected from Equation (8) or (9). The unexpectedly low

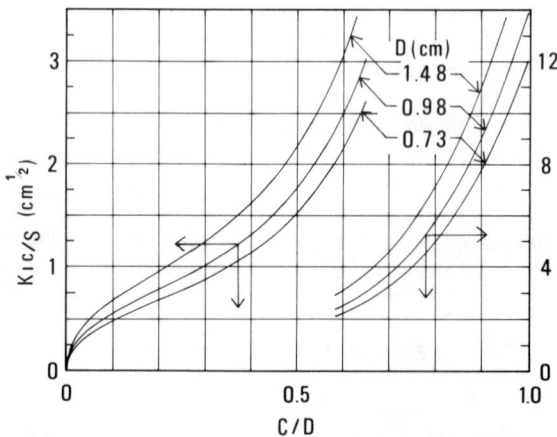

Fig. 8. K_{IC}/S (= $C^{1/2} \cdot F$) vs. C/D curves for obtaining crack length C from values of K_{IC}, S and D. The curve with D = 0.98 corresponds to the thermal shock specimen of 2.0 cm square cross-section.

Table V. Estimated Crack Length in Bending Specimen
Before and After Thermal Shock. (cm)

Brick	C_O	C_C	$C(1500°C)$
A	0.27	0.84	1.00
B	0.35	0.55	0.93
C	0.36	0.55	0.81
D	0.36	0.51	0.99
E	0.41	0.50	0.62
F	0.52	-	0.76

value of $C(1500°C)^{-1}$ for F-brick, which was shown in parentheses
in Fig. 9(a), might be attributed to a possible damage of surface
layer of thermal shock specimen due to the high compressive stress
induced by the severe radiation heating of 1500°C.

In view of Rstr, the strength-loss resistance parameter, as
given in Equation (11), the ratio S_c/S_0 was compared with R''''.
The result is shown in Fig. 9(b). An excellent correlation was
found to exist.

The critical radiation temperature at which strength begins to
decrease can be compared with the fracture resistance parameter,
R' or R_{rad}, given in Table I. Because of the absence of data on
emissivity of the bricks tested, only R' parameters were compared
with the observed values of T_{rc}. Results are shown in Fig. 10.
In spite of uncertainty for emissivity values, a good positive
correlation between R' and T_{rc} is observed.

No significant correlation was observed between the crack
stability parameter R'_{st} and length of cracks propagated by thermal
shock.

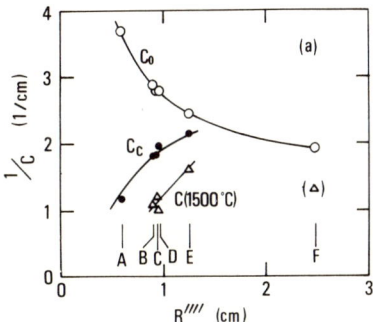 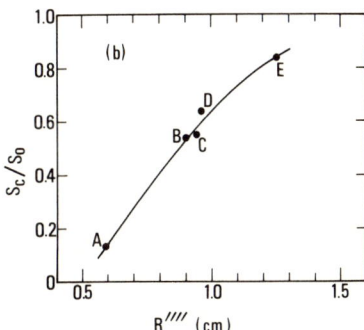

Fig. 9. Comparison between test results and damage resistance
parameter R''''. (a)Reciprocal crack length vs. R'''', and
(b)strength ratio S_c/S_0 vs. R''''.

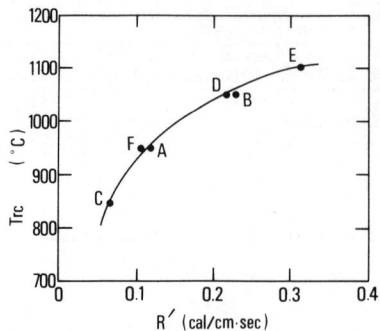

Fig. 10. Critical radiation
temperature T_{rc} vs. R'.

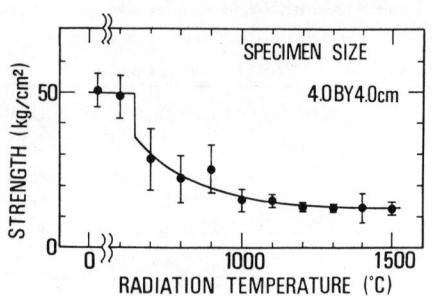

Fig. 11. Discontinuous curve obtained with large specimens of
F-brick.

From the foregoing discussion, it is expected that dynamic
fracture will be realized even for F-brick, which exhibited a continuous fracture curve as illustrated in Fig. 6(F) with specimens
of 2 cm cross-section, by the use of specimens with larger cross-
section. By using specimens with 4 cm square cross-section and 10
cm in length, the expected discontinuous decrease in strength was
obtained as illustrated in Fig. 10.

Thus, the present method of thermal shock testing was found to
be useful for comparing materials such as firebricks in views of
both R' and R'''' parameters. However, application of the method
to those materials, which have much larger R' or much smaller R'''',
is considered to be inadequate. For instance, these parameters for
sintered alumina materials, in rough estimation, are, $R' \approx 3.0$ and
$R'''' \approx 0.02$, both in c.g.s. units, compared with those of E-brick
in the present study of $R' \approx 0.3$ and $R'''' \approx 1$. A simple calculation
based on the parameter $R_{rad}(= R'/\varepsilon)^{1/4}$ given in Table I indicates
that about one order of magnitude larger value of R' for alumina
requires an impractical critical radiation temperature of more than
2000°C, if specimens with 2 cm square cross-section are used.
Even if this temperature is realized, the fracture will undoubtedly
be catastrophic because of the very small value of R''''. Although
the use of specimens with much smaller cross-section may solve the
difficulty of the catastrophic fracture, it would require far higher
radiation temperature to initiate fracture. Difficulty in the cutting procedure also becomes serious.

IV. SUMMARY

Concepts, theories and typical experimental results on thermal shock resistance of ceramic materials were briefly reviewed. On the basis of the strain energy concept, a new thermal shock test method for refractory firebricks was introduced. Six brands of commercial refractory brick were tested by the method. It was found to be useful in comparing these materials in view of R' or R_{rad} from T_{rc} and R'''' from C_c or S_c/S_o.

ACKNOWLEDGMENTS

The writer is indebted to T. Ono, K. Tsuji and Y. Iwabuchi for performing the thermal shock test, to Y. Furuse for programming the temperature, stress and energy equations and the representation of calculation curves on an electronic computer, and to T. Sawamoto for discussing the present technique of thermal shock testing.

REFERENCES

1. D.P.H.Hasselman, Bull. Am. Ceram. Soc., 49 (12) 1033-37 (1970).
2. Standard Method for Basic Procedure in Panel Spalling Test for
        Refractory Brick, ASTM Designation C38-58, pp.293-99 in 1958
        Book of ASTM Standards, Part 5.
3. a) W.R.Morgan, J. Am. Ceram. Soc., 14 (12) 913-23 (1931).
    b) C.W.Parmelee and A.E.R.Westman, J. Am. Ceram. Soc., 11
        (12) 884-95 (1928).
4. M.Ueda, Taikabutsu (Refractories), 15 (73) 87-102 (1963).
5. German Standard for Structural Refractory Materials, DIN Desig-
        nation 1068, 1931.
6. Russian Standard for Chemical Stoneware GOST 473-53; Ceram.
        Abstr., 1955, June, p.107h.
7. R.W.Davidge and G.Tappin, Trans. Brit. Ceram. Soc., 66 (8)
        405-22 (1967).
8. J.H.Ainsworth and R.E.Moore, J. Am. Ceram. Soc., 52 (11)
        628-29 (1969).
9. D.P.H.Hasselman, ibid., 53 (9) 490-95 (1970).
10. J.A.Coppola, Investigation of the Fracture Surface Energy,
        Fracture Strength and Thermal Shock Behavior of Polycrystal-
        line Materials, A Thesis in Ceramic Science, The Pennsyl-
        vania state University, 1971.
11. J.Gebauer, D.A.Krohn and D.P.H.Hasselman, J. Am. Ceram.
        Soc., 55 (4) 198-201 (1972).
12. T.K.Gupta, ibid., 55 (5) 249-53 (1972).
13. J.A.Coppola, D.A.Krohn and D.P.H.Hasselman, ibid., 55
        (9) 481 (1972).
14. W.D.Kingery, ibid., 38 (1) 3-15 (1955).
15. W.D.Kingery, Introduction to Ceramics; p.636. John Wiley &
        Sons, Inc., New York, 1959.
16. W.R.Buessem, Sprechsaal F. Keram-Glas-Email, 93 (6) 137-41
        (1960).

17. R.L.Coble and W.D.Kingery, J. Am. Ceram. Soc., $\underline{38}$ (1) 33-37 (1955).

18. D.P.H.Hasselman, ibid., $\underline{50}$ (9) 454-57 (1967).

19. D.P.H.Hasselman, ibid., $\overline{46}$ (5) 229-34 (1963).

20. D.P.H.Hasselman, ibid., $\overline{49}$ (2) 103-04 (1966).

21. A.L.Florence and J.N.Goodire, J. Appl. Mech., $\underline{26}$ 293-94 (1959).

22. J.N.Goodier and A.L.Florence; pp562-68 in Proceedings XIth International Congress of Applied Mechanics, Munich 1964.

23. W.Weibull, Ing. Vetenskaps Akad. Hand., $\underline{151}$ 1939, 45pp; Ceram. Abstr., $\underline{19}$ (3) 78 (1940).

24. S.S.Manson and R.W.Smith, J. Am. Ceram. Soc., $\underline{38}$ (1) 18-27 (1955).

25. R.N.Newman, ibid., $\underline{55}$ (9) 464-69 (1972).

26. S.Saito, A.Nohara and Y.Hayakawa, Taikabutsu (Refractories), $\underline{24}$ (179) 545-51 (1972).

27. D.P.H.Hasselman, J. Am. Ceram. Soc., $\underline{46}$ (11) 535-40 (1963).

28. F.J.P.Clarke, H.G.Tattersall and G.Tappin, Proc. Brit. Ceram. Soc., (6) 163-72 (1966).

29. J.Nakayama and M.Ishizuka, Bull. Am. Ceram. Soc., $\underline{45}$ (7) 666-69 (1966).

30. D.P.H.Hasselman, J. Am. Ceram. Soc., $\underline{52}$ (11) 600-04 (1969).

31. a) D.P.H.Hasselman, Int. Journ. Fracture Mech., $\underline{7}$ (2) 157-61 (1971).
    b) D.P.H.Hasselman, pp89-103 in Proceedings of a Conference on Ceramics in Severe Thermal Environments. Edited by H. Palmour III and W.W.Kriegel. Plenum Press, New York, 1971.

32. J.P.Berry, J. Mech. Phys. Solids, $\underline{8}$ 207-16 (1960).

33. S.Kato and H.Okuda, Nagoya Kogyo Gijutsu Shikensho Hokoku, $\underline{8}$ (5) 37-43 (1959).

34. J.Nakayama, Inorganic Materials Science; pp282-309,M.Kunugi (Editor), Seibundo Shinko-sha, Tokyo, 1972.

35. M.S.Tacvorian, Soc. Franc. Ceram. Bull., 29 20-40 (1955).

36. J.A.Coppola, D.A.Krohn and D.P.H.Hasselman, J. Am. Ceram. Soc., $\underline{55}$ (9) 481 (1972).

37. R.C.Rossi and D.P.H.Hasselman, Aerospace Report No. TR-0172 (2250-40)-6, April 1972.

38. S.Keinov, Ber. Deut. Keram. Ges., $\underline{47}$ (7) 426-30 (1970).

39. W.D.Kingery, Property Measurements at High Temperature; pp 104-106. John Wiley & Sons, Inc., New York, 1959.

40. a) J.Nakayama, J. Am. Ceram. Soc., $\underline{48}$ (11) 583-87 (1965).
    b) H.G.Tattersall and G.Tappin, J. Matls. Sci., $\underline{1}$ 269-301 (1966).

41. F.F.Lange, Office of Naval Research, Technical Report No. 7, N-00014-68-C-0323, March 1972.

42. W.F.Brown and J.E.Srawley, ASTM Special Technical Publication No. 410, 13-14 (1966).

CRACK PROPAGATION IN A THERMALLY STRESSED CERAMIC REACTOR FUEL*

G. G. Trantina** and J. T. A. Roberts

Materials Science Division, Argonne National Laboratory

Argonne, Illinois   60439

Ceramic reactor fuels are inherently brittle, and therefore extensive cracking occurs from thermal stresses generated during reactor transients (e.g., reactor startup, shutdown, or overpower transients). The fuel cross section in Fig. 1[1] shows the type of flaws that are present in the as-fabricated fuel and the extensive cracking that occurs during irradiation. Fuel cracking influences fuel swelling and fission-gas release. Also, the loss of fuel integrity that results from cracking can have a deleterious effect on the cladding stress and on "fuel motion" during an accident situation. For these reasons, an analysis of fuel cracking is an essential part of the computer models[2,3] of fuel-element behavior that are being developed to predict normal and off-normal (or accident) performance. The analytical study of cracking presented in this paper was undertaken to (1) improve the treatment of fuel cracking in the mechanical models of fuel behavior,[2,3] and (2) provide insight into possible ways of controlling fuel-cracking behavior through manipulation of material parameters or flaw densities.

The main feature of the CRACK subroutine in the LIFE-II code[2] is its ability to relax the large thermoelastic stresses generated in the oxide fuel during transient operations in-reactor. In an isotropic body, the effect of cracks must be described by assuming

---

* Work performed under the auspices of the U.S. Atomic Energy Commission.

** Present address: General Electric Company, Corporate Research and Development, Schenectady, N. Y.   12301.

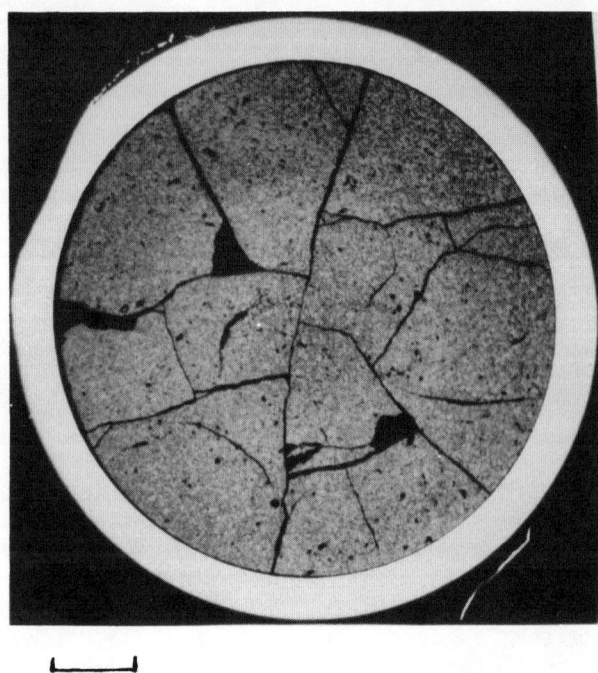

·04″

Fig. 1. Appearance of a Fuel Pellet after Irradiation at Low
         Power to Low Burnup.

an "effective" elastic modulus or compliance that is reduced in
proportion to the number of cracks. The incremental strain or
"swelling" increase due to cracking must also be consistent with
the assumption of an isotropic body. Hence, the cracks have no
real identity, and their lengths and positions in the fuel are
arbitrary. The next stage in the development of a cracking model
is to provide a realistic description of a "crack morphology,"
inasmuch as the crack pattern will influence the aforementioned
aspects of fuel-element performance. We have used a combination
of experimental results and analysis to establish the conditions
necessary for crack propagation to occur. A finite-element stress
analysis and fracture-mechanics concepts were used to obtain
strain energy-release rates G and stress-intensity factors K for
radial cracks in a thermally stressed cylinder. These values
were then compared with experimentally measured critical values[4]
to predict crack initiation.

     The study was composed of two parts: first, the propagation
of a single crack under thermal loading and the effect of a
central void on its propagation behavior were considered; second,

an analysis of the effect of the number of regularly distributed, equal-length, initial flaws (or cracks) on their propagation characteristics was made.

FINITE-ELEMENT ANALYSIS

For the finite-element stress analysis of a thermally stressed cylinder, six-node linear strain triangles were used. The plane-strain, linear elastic analysis provides the displacements and, hence, the stresses and strains in the r-θ plane of a cylinder. After the appropriate arrangement of the elements (grid) was chosen, the proper elastic properties and temperatures were assigned to each element. An approximate solution to the heat-flow equation that provides the radial temperature distribution in a mixed-oxide fuel[5] was used to determine the nodal temperatures, which were, in turn, converted to a linear temperature variation within the element. The temperature-dependent coefficient of thermal expansion α for the fuel[2] can be determined from the nodal temperature. The thermal strains were then introduced into the cylinder as "initial" boundary conditions. The modulus of elasticity E and Poisson's ratio ν were corrected for temperature and porosity[2] for each element by using the temperature at the centroid of the element and the porosity of the fuel.

The strain energy-release rate G can be defined as the change in stored elastic strain energy of the body with a change in crack area. The strain energy of the body can be determined at two crack lengths with a finite-element analysis; G was then calculated from

$$G = -\frac{dU}{dA} = -\frac{U_2 - U_1}{A_2 - A_1} , \qquad (1)$$

where, for unit thickness, $A_2 - A_1 = 10^{-10}$ in. in the present analysis.

An alternate method of determining G was used as a check of the direct-energy approach. With the displacement method, K was obtained by evaluating the plane-strain crack-tip displacement equation along the crack line

$$v = \frac{4K}{E} \sqrt{\frac{r}{2\pi}} (1 - \nu^2) , \qquad (2)$$

where v is the displacement perpendicular to the crack line, and r is the distance to the crack tip. At each crack-line node, K was calculated, and the results were extrapolated to the crack tip to obtain the desired stress-intensity factor. Inasmuch as

$$K^2(1 - \nu^2) = GE \qquad\qquad (3)$$

for plane strain, the results can be used to check the results of the direct-energy method. The displacement method is less desirable because it requires the accurate determination of the displacements of the crack-line nodes as well as an extrapolation.

As an example of the generation of the grid system, Fig. 2a shows the cross section of a fuel pellet with a central void and four uniformly spaced, equal-length, radial cracks. Since two planes of symmetry exist, only the section shown in Fig. 2b requires analysis. The boundary conditions allow no displacement

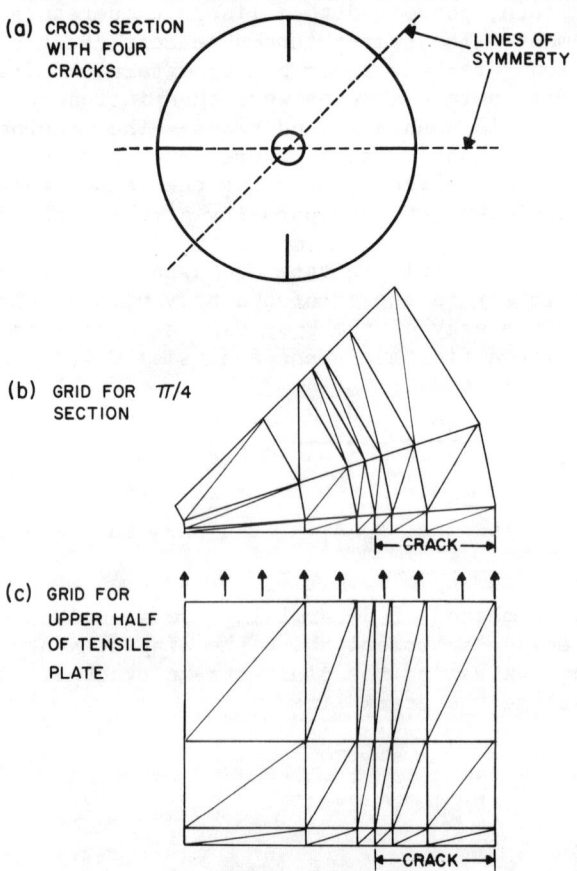

Fig. 2.   Finite-element Grid Patterns.   Coarse grids are shown as examples.

perpendicular to the symmetry planes, except for the crack line,
which is completely free. The coarse grid shown as an example
was generated by first locating nodes at a constant distance
above, to the right, and to the left of the crack tip. One
radial and three circumferential lines were constructed, which
form two almost equal squares at the crack tip. Other radial
and circumferential lines were formed by an arithmetic progression,
thus forming quadrilaterals that are divided into triangles, which
result in a grid with smaller elements near the crack tip.

To check the procedure for determining the strain energy-
release rate, a single-edge-notch tension (σ) plate with a similar
grid was considered (see Fig. 2c). For a crack length a, which
is much less than the width of the plate, the stress-intensity
factor is given by[6]

$$K = 1.12\sigma \sqrt{\pi a} . \tag{4}$$

The G values determined by the direct-energy method were converted
to stress-intensity factors by using Eq. 3. For crack lengths of
1 and 3 mils and configurations similar to the wedge shape of
Fig. 2b for 4-18 cracks, the error was always less than 9%, and
the K's determined by the finite-element method were always lower.

RESULTS AND DISCUSSION

The first case considered was a single radial crack (Fig. 3).
In a conventional fuel pellet, crack-initiating flaws are of

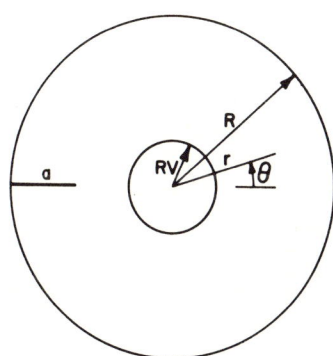

Fig. 3. Radially Cracked, Thermally Stressed Cylinder.

random size and distribution, and, generally, a single crack
originates first from the largest flaw. The strain energy-release
rate G for a radially cracked cylinder is shown in Fig. 4. The
dimensionless strain energy-release-rate parameter was chosen to
normalize the results with respect to the loading condition σ,
which is related to EαΔT, where ΔT is the difference in temperature
of the inside and outside of the fuel. A constant fuel-surface
temperature of 450°C and a fuel radius of 100 mils were used for
this study. If both sides of the modified Griffith equation

$$G = \frac{\sigma^2 a}{E} Y \tag{5}$$

are divided by R (where Y is a geometric factor, and R is the
fuel radius) and if EαΔT is substituted for σ, the following
strain energy-release-rate parameter is obtained:

$$\frac{G}{RE(\alpha\Delta T)^2} = \frac{a}{R}(Y) . \tag{6}$$

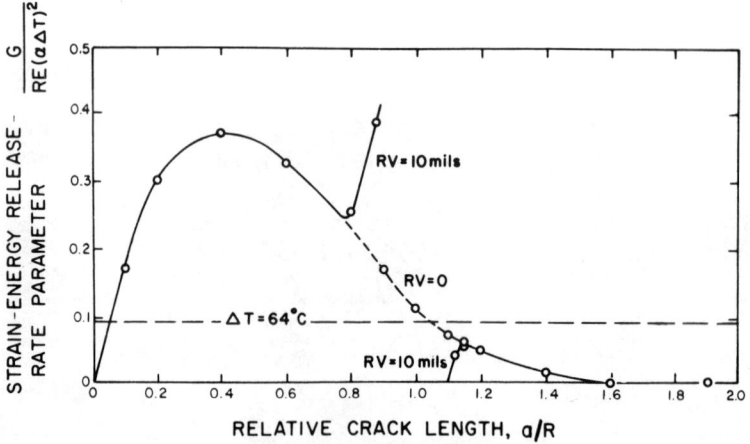

Fig. 4.   Strain Energy-release Rate for a Thermally Stressed
          Cylinder with one Radial Crack.

The peaked nature of the curve in Fig. 4 can be attributed to the fact that the crack length is increasing but the stress field is decreasing. For the initial portion of the curve, the increasing crack length apparently has more influence than the decreasing stress field. The opposite is true for the decreasing portion of the curve. The horizontal line represents a critical value of G for UO_2[4] at a ΔT of 64°C. Thus, if an initial radial surface flaw with a length of about 5 mils ($\sim$125 µm) were present, it would extend under the loading produced by the ΔT of 64°C. Fabrication flaws of this size are typical, as can be seen in Fig. 1. Figure 5 shows the stress distribution for that thermal loading before cracking and after one crack has propagated to the center of the cylinder.

In the central portion of the fuel, G varies significantly with the central void size. For a fuel without a central void, a continuous decrease in G occurs (Fig. 4). Thus, the inherent 5-mil flaw that extended at a ΔT of 64°C would not arrest until it extended to at least 3 mils past the center. Dynamic effects could cause the crack to extend much further. If the crack did arrest 3 mils past the center, a subsequent increase in ΔT would cause the crack to extend further. In either case, the crack should extend to a point where the remaining ligament has little influence on the overall stress field. Thus, for subsequent cracking considerations, the fuel can be considered to be divided longitudinally into two equal parts. Figure 1 shows this type of crack, which probably initiated in the upper right-hand portion of the fuel and extended, splitting the fuel diagonally into two semicircular segments.

Additional crack progression can be analyzed, first of all, by considering the semicircular segment. Large tensile stresses in the central portion of the fuel that remain after the first crack has split the fuel, and which increase as ΔT increases, indicate a further division of the semicircular segments. Consideration of further cracking must allow for the possibility of circumferential cracking inasmuch as radial tensile stresses have increased continuously up to this point. Incomplete results have indicated the possibility of at least one circumferential crack, followed by several more radial cracks, during a normal reactor startup thermal history. The resultant crack pattern is similar to those observed in irradiated fuel.

For a central void of 10 mils, G increases to a peak, decreases, and increases again as the crack propagation is influenced by the central void (Fig. 4). This final increase can be attributed to the tendency of the cracked cylinder to open to the extent that tensile stresses are produced near the inner surface (Fig. 5). Once the crack has propagated to the central

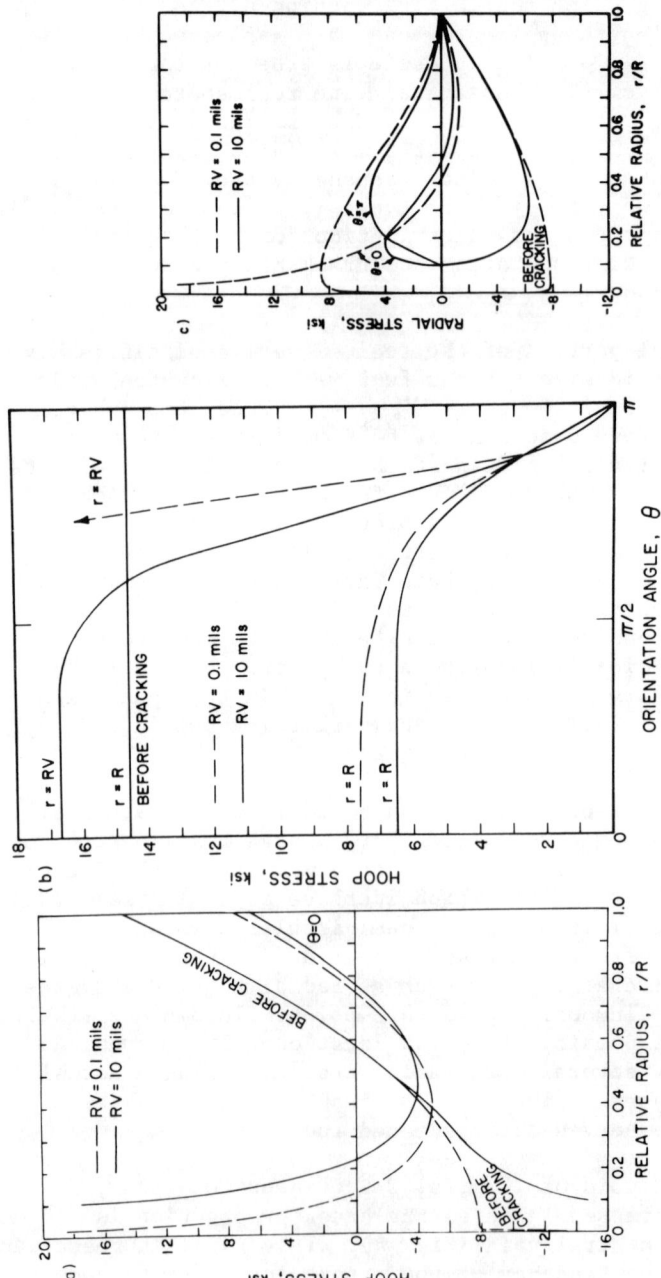

Fig. 5.   Stress Distribution before Cracking and with One Crack to the Center of the Cylinder.

void, reinitiation must be considered. Since the hoop stress is
more than 2-1/2 times larger on the inner surface than on the
outer surface, the inner surface was considered. Although the
maximum hoop stress on the inner surface was almost constant
over a wide range of θ (Fig. 5b), for simplicity, $\theta = 0$ was chosen.
As shown in Fig. 4, the strain energy-release rate for the re-
initiated crack increases to a peak (below the $\Delta T = 64°C$ line) and
then decreases to a small value. Thus, if the 5-mil inherent radial
flaw extends at a ΔT of 64°C and if the excess energy is dissipated
by behavior such as microcracking (secondary cracking), the crack
will arrest at the central void and will not reinitiate, regardless
of the flaw size. However, a maximum stress criterion would
predict reinitiation, inasmuch as the hoop stress before cracking
(at $r = R$) is 14.6 ksi and after cracking (at $r = RV$, $\theta = 0$) is
16.7 ksi (Fig. 5a). This discrepancy can be resolved by considering
the stress gradient; 5 mils from the surface, the hoop stress
before cracking (at $r = 0.95R$) is 12.5 ksi and after cracking
(at $r = 0.15R$, $\theta = 0$) is 6.5 ksi (Fig. 5a).

A more detailed investigation of the stress distribution as
the single crack propagates to a central void indicates that no
additional cracking would occur, as long as ΔT remains constant,
because large tensile stresses are present only in the crack-tip
region. Cracks would not initiate from the outer surface since
the hoop stress at the surface decreases as the crack extends.
When the crack is only 2 mils from a central void of 10-mil
radius or 10 mils from a central void of 0.1 mil radius, the hoop
stress is zero at the inner surface 180° from the crack line.
However, when the crack has propagated to the central void, large
tensile hoop stresses are present at the inner surface (Fig. 5).

The second part of this study considered the problem of
inhibiting or controlling cracking by fabricating a fuel with a
number of equal-size surface flaws that would not extend under
reactor thermal-loading conditions, and/or by fabricating a fuel
with a high toughness value (G_c). The strain energy-release
rates were determined for N equally spaced radial cracks all with
the same length (Fig. 6). The crack length at which the
individual curves break away from the common curve indicates the
crack length at which one crack would begin to influence the
strain energy-release rate of the adjacent cracks. Therefore,
the larger the number of cracks, the shorter the crack length at
which these cracks would begin to interact.

For the determination of the crack-extension behavior, the
critical strain energy-release rate G_c must be obtained
experimentally. This is a material parameter that is sensitive
to microstructure (grain size porosity). Evans and Davidge[4]
have determined G_c for UO_2 as a function of temperature for two

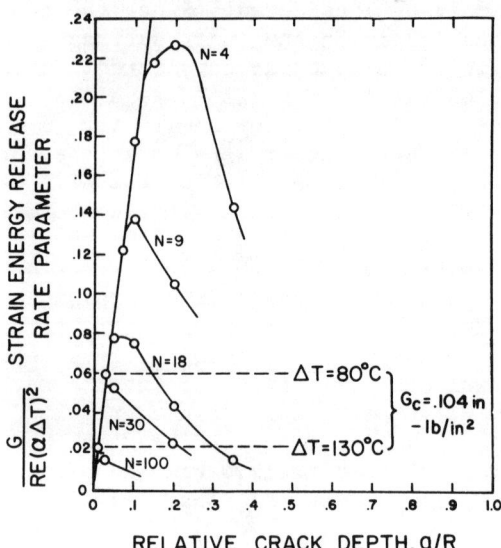

Fig. 6.   Strain Energy-release Rate as a Function of Crack Length
           for N Equally spaced, Equal-length Radial Cracks.

grain sizes. In Fig. 6, two horizontal dashed lines have been
constructed that represent the critical strain energy-release-rate
parameter (G_c = 0.104 in.-lb/in.^{2} for a grain size of 8 μm) at two
loadings (ΔT = 80 and 130°C). The data indicate that, to prevent
crack initiation for a ΔT of 80°C, any noninteracting crack must
be less than 3 mils, or 30 equally spaced, equal-length cracks of
any crack length must be present. For ΔT = 130°C, any noninteracting crack must be less than 1 mil, or 100 equally spaced,
equal-length cracks of any crack length must be present. Therefore,
on the basis of these calculations, a large number of initial
surface flaws must be introduced into the fuel pellet to avoid
cracking under the ΔT's of ∿1000°C that can exist across the
pellet radius. However, as the ΔT increases, the absolute
temperature increases, and the effect of plastic flow must be
included. It might be possible to raise the ΔT for cracking
sufficiently to introduce plastic stress-relaxation effects that

would inhibit cracking. In this case, the heating rate would be a critical parameter inasmuch as the brittle-to-ductile transition temperature increases with an increase in elastic strain rate (thermal loading rate).[7] For example, for a normal reactor startup condition, the fuel temperature would have to be 1400°C before plastic stress-relaxation would be favorable. However, for an accident transient, a temperature of 1800°C would be required.[7]

In addition to flaw density, the characteristic G_c is an important fuel parameter that should be as high as possible if cracking is to be inhibited; G_c increases with an increase in temperature.[4] Also, several studies have shown that ceramic microstructure, such as grain size and porosity, influences G_c, but large changes (e.g., in alumina a grain-size reduction from 100 to 3 μm or a porosity decrease from 50 to 5 vol %,[8] and in UO_2 a grain-size reduction from 25 to 8 μm[4]) result in no more than approximately a factor of two increase in G_c.

It must therefore be concluded that fuel cracking cannot be prevented under the extreme thermal stress environment seen in-reactor, but that a large number of surface flaws, in addition to as high a G_c as possible, could prevent cracking sufficiently for fuel temperatures to rise to the point where plastic flow could inhibit crack propagation. A large number of small cracks in the fuel would be an acceptable structure in that structural integrity is still retained.

SUMMARY AND CONCLUSIONS

A finite-element stress analysis of a thermally stressed cylinder, with material parameters adjusted for temperature and porosity, was performed. A direct-energy approach was used to obtain strain energy-release rates from the finite-element results, and a displacement method was used to obtain stress-intensity factors. These values were then compared with experimentally measured critical values to predict crack initiation. Based on the analysis of the propagation of a single radial crack and of a number of regularly distributed, equal-size cracks that originate from the surface, the following conclusions can be made:

(1) In a conventional, thermally stressed, solid fuel pellet, a radial crack initiated from a typical inherent radial flaw would propagate beyond the center of the fuel pellet. Further cracking considerations indicate additional radial cracks as well as circumferential cracks.

(2) The presence of a central void or annulus stabilizes the crack at the hole.

(3) The severe thermal environment in-reactor makes crack
prevention impossible even if the surface flaw density and the
fracture toughness are maximized. However, this approach will
cause the ΔT necessary for crack propagation to be increased,
allowing the possibility of thermal stress relaxation by plastic
flow to inhibit crack propagation. The influence of plasticity
on crack propagation was outside the scope of the present
analysis.

ACKNOWLEDGMENTS

The writers would like to acknowledge interesting discussions
with J. C. Voglewede and A. A. Solomon.

REFERENCES

1. J. Lambert, Private Communication, Argonne National Laboratory,
    (April 1973).

2. V. Z. Jankus and R. W. Weeks, "LIFE-II-A Computer Analysis of
    Fast-reactor Fuel-element Behavior as a Function of Reactor
    Operating History," Nucl. Eng. Design 18(1), 83-96 (1972).

3. A. Watanabe and A. Judd, "A Computer Code for Predicting the
    Behavior of Oxide Fuel in Accidents," Trans. Am. Nucl. Soc.
    14, 733 (1971).

4. A. G. Evans and R. W. Davidge, "The Strength and Fracture of
    Stoichiometric Polycrystalline UO_2," J. Nucl. Mater. 33,
    249-260 (1969).

5. R. O. Meyer and B. J. Buescher, "A Simple Method of Calculating
    the Radial Temperature Distribution in a Mixed-oxide Fuel
    Element," Nucl. Tech. 14, 153-156 (1972).

6. P. C. Paris and G. Sih, "Stress Analysis of Cracks," Fracture
    Toughness Testing and Its Applications, ASTM-STP 381, 30-83
    (1965).

7. J. T. A. Roberts and B. J. Wrona, "Deformation and Fracture of
    UO_2-20 wt % PuO_2," ANL-7945 (June 1972).

8. A. G. Evans and G. Tappin, "Effect of Microstructure on the
    Stress to Propagate Inherent Flaws," Proc. Brit. Ceram. Soc.
    20, 275-297 (1972).

DETERMINATION OF FRACTURE TOUGHNESS PARAMETERS

FOR TUNGSTEN CARBIDE-COBALT ALLOYS*

R. C. Lueth**

Carboloy Systems Department

General Electric Company

ABSTRACT

The fracture toughness parameters for nine tungsten carbide cobalt alloys has been determined. These alloys range in cobalt content from 3 to 15 wt. % and from 1.5 to 8 microns in tungsten carbide grain size. A wedge loaded double cantilever beam approach was used to obtain the G_{IC} and K_{IC} values for these high modulus low toughness materials. This was done in order to obtain the necessary displacement control such that a crack situation could be realized.

The fracture toughness of tungsten carbide colbalt alloys has been found to depend strongly on binder film thickness such that generally the thicker the binder film, the higher the fracture toughness. The strength in bending of tungsten carbide cobalt alloys has been found to depend on the yield strength, nominal flaw size, and fracture toughness of the material.

---

*This paper is based on a thesis submitted by R. C. Lueth in partial fulfillment of the requirements for the degree of Doctor of Philosophy of Metallurgy, College of Engineering, Michigan State University. The work was conducted in the Hard Metals Research Section of the Carboloy (R) Systems Department, General Electric Company.

**Presently Research Metallurgist, Carboloy Systems Department, Detroit, Michigan.

(R) Trademark of the General Electric Company.

INTRODUCTION

The successful application of tungsten carbide cobalt alloys in almost all cases depends on the ability of these alloys to resist both wear and fracture in service. In many respects wear is more easily studied and predicted since it occurs progressively. The wear experienced in many mining applications, for example, correlates fairly well with data obtained on the abrasion test as given in the Cemented Carbide Producers Association Procedure (# 112). The ability of the material to resist fracture in use, however, is more difficult to predict.

The present methods generally used to predict the fracture resistance of tungsten carbide cobalt alloys include compression testing, transverse rupture testing, and charpy impact testing. In order for a material to operate in the field it must have the necessary compressive strength to resist crushing. This resistance to crushing can be reasonably well predicted by the ASTM Procedure for Compression Testing (E9-61) with modifications proposed by Lueth and Hale [1]. The ability of the material to withstand tensile loadings such as those produced by contact stresses or bending loads in use is, however, not well predicted by transverse rupture strength or by data on charpy impact testing. Data from the above tests do not correlate well with field performance.

An actual cemented carbide insert, component or test specimen contains surface defects. These defects are due to the preparation of this surface for testing or use, or incurred during use. In the case of the tungsten carbide-cobalt alloys prepared for transverse rupture testing and charpy impact testing, the defects may consist of the damage resulting from surface grinding. The fracture of these relatively brittle materials is greatly affected by these defects. In fact, it is the size and shape of these defects coupled with the materials ability to resist the growth of these defects that is the major factor which controls the values measured for transverse rupture testing. Also this ability to resist the extension of defects coupled with other material parameters such as modulus of elasticity controls the values obtained for fracture energy in the charpy impact test [2]. In the case of those alloys which fracture at stress levels above that necessary for yielding, in transverse rupture testing the initial extension of the defects will be a slow growth through plastic action or other stable mechanism such as grain cracking. When the defect reaches a certain size rapid extension will occur. For alloys which fail at stresses below yield, the defects will extend catastrophically at a certain stress level which is dependent on the defect size and shape, and the toughness of the specific carbide alloy. It is this rapid fracture and the conditions which promote it that are the subjects of this paper.

The need to study such behavior is brought out clearly in the application of cemented carbides in mining and metalworking tools, and in various component applications. The surface of the cemented carbide in mining tools, for example, frequently shows evidence of many relatively large surface defects incurred during usage. Terms such as "alligatoring" [3] have been used to describe some of these surface cracks or defects. These cracks have been observed by the author to extend up to 1/4" deep in 1/2" mining insert before failure occurred. These cracks can be ascribed to thermal cracking, fatigue, high contact stress, etc. With these defects present, the failure of the material now largely depends on the materials inherent ability to resist the propagation of these cracks. The parameters which measure this ability of a material to resist catastrophic propagation of a crack result from the application of linear elastic fracture mechanics to the problem. The two specific parameters of interest are the critical strain energy release rate G_{IC} and the critical stress intensity parameter K_{IC}. This paper discusses an experimental method for obtaining these two parameters for cemented carbide.

DISCUSSION

The subject of linear elastic fracture mechanics is extensively documented (4,5,6,7) and will not be described here except for a brief statement of principles. Linear elastic fracture mechanics concerns itself with the circumstances under which existing cracks or defects will extend at high speed in a material under stress. Only the linear elastic case will be discussed here since WC-Co alloys are nearly linearly elastic up to the failure stress in tension. Consider a material under load with a defect located somewhere in the body. Fracture mechanics deals with the energies involved when the defect extends. If the defect extends a small amount, new surface is created and thus some energy dU_S is used. Also the strain energy in the body and the energy associated with the boundary forces will change dU_e. The energy associated with the extended crack is $U = U_{orig} - dU_S + dU_e$. Consider how this energy difference $U - U_{orig} = -dU_S + dU_e$ changes as the crack increases a small amount. $\frac{\Delta U}{\Delta a} = -\frac{dU_S}{da} + \frac{dU_e}{da}$. If this quantity is positive the crack will propagate since the energy released as the crack grows is greater than that absorbed. If the quantity is negative, the crack will not move as the necessary energy cannot be supplied. If, however, this quantity is zero, a condition of metastable equilibrim exists such that the energy available per unit crack extention is equal to that necessary for extension. Then $\frac{dU_S}{da} = \frac{dU_e}{da}$.

Through proper analysis $\frac{dU_e}{da}$ can be calculated from the exterior boundary forces and displacements and the inverse spring constant of the system and its change with respect to crack length (see

Appendix I). Since at the point of metastable equilibrium $\frac{dUs}{da}$ = $\frac{dUe}{da}$ the energy per unit area necessary to propagate a defect can be calculated. This is known as the critical strain energy release rate (G_{IC}). The stress intensity parameter (K_{IC}) is derived from a linear elastic solution of the stress field around the tip of a crack (6). For a plane strain condition $K_{IC}^2 = \frac{GE}{1-v^2}$ where E is the elastic modulus of the material and v is Poisson's ratio. K_{IC} is a measure of the stress which can be safely exerted at the tip of a crack in a body. K_{IC} is a material parameter as is G_{IC}.

<u>EXPERIMENTAL</u>

Equipment and Technique

In order to evaluate the fracture toughness parameters a test specimen must be chosen such that the necessary experimental meas-urements may be obtained to allow a calculation of energy changes as the crack length increases. A wedge loaded double cantilever beam technique (hereafter referred to as a W.L.D.C.B. test) was chosen to evaluate K_{IC} and G_{IC} for WC-Co alloys. This test was chosen because of the unique problems presented by the high modulus, high strength, and low toughness exhibited by these alloys. In or-der to measure the fracture toughness parameters a natural crack must be started and stopped at least once. In a high modulus mat-erial a small displacement at the crack opening involves a relat-ively large energy input, this necessitates that displacement must be accurately controlled such that the crack propagation does not become catastrophic. Also, in order to control the crack, a test must be chosen such that as the crack increases in length the en-ergy available for propagation decreases, thus inducing a stable crack situation. Both of these criteria are met by WLDCB test. The mechanics of this type test or technique are give in detail in ref. (8). The specimen itself consists of a 3" x 1/2" x 1/8" bar with a 3/4" deep slot approximately .030" wide cut into one end such that the wedge could be easily introduced (see Fig. 1). Since the maximum tensile stress and the maximum distortional strain energy positions are not on the central beam axis of a WLDCB specimen, a groove 1/16" deep x .030" wide was also cut the length of the sample to insure that the crack would not wander off center.

In order to evaluate G_{IC} and K_{IC}, consider the energy balance of the system. (See Figure 2).

During an increment of crack extension which creates new sur-face dA, the work done by the loading force is Pd where is the displacement of the force in its own direction. The stored ener-gy V is always positive and contributes -dV during crack extension.

Figure 1

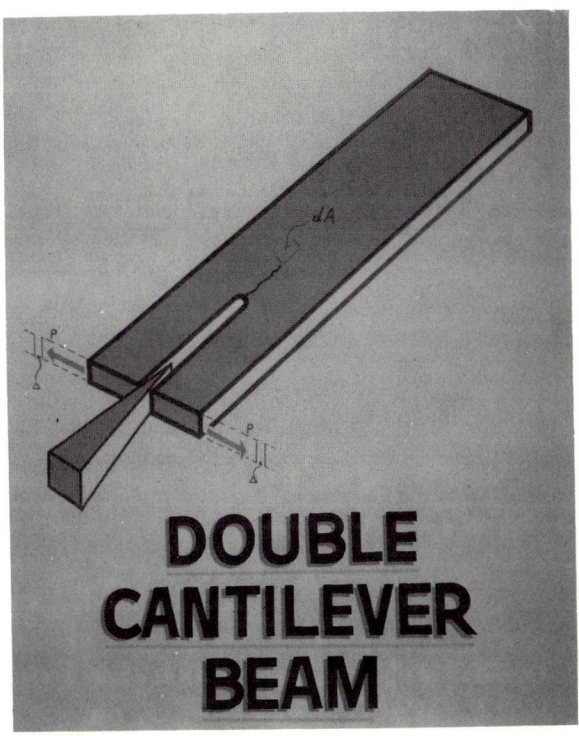

Figure 2

Therefore: $G = P \dfrac{d\Delta}{da} = \dfrac{dv}{da}$.   Load is related to displacement by $\Delta = \lambda p$ where λ is the inverse spring constant. The strain energy can be written as the work done during loading at constant crack length:

$$V = \frac{P\Delta}{2} \quad \frac{P^2}{2} \qquad \text{where } \lambda = \text{compliance} = \frac{\Delta}{P}$$

$$G = P \frac{\partial (\lambda p)}{\partial a} - \frac{\partial \left(\frac{\lambda p^2}{2}\right)}{da}$$

$$G = \frac{P^2}{2} \frac{\partial \lambda}{\partial a} = \frac{\Delta^2}{2 \lambda^2 b} \frac{\partial \lambda}{\partial a}, \qquad$$ where b is the thickness of the sample and a is the crack length.

Thus if we know the compliance, its change with respect to length ($\lambda, \frac{\partial \lambda}{\partial a}$) and the displacement at any given crack length, G can be calculated, and since at the point of incipient crack growth $G \to G_{IC}$ then G_{IC} can be calculated. The method of obtaining λ and $\frac{\partial \lambda}{\partial a}$ at any crack length as given in Appendix I. Since the modulus of the various alloys of WC-Co alloys is not the same, the effect of modulus on λ and $\frac{\partial \lambda}{\partial a}$ is also discussed in Appendix I.

Experimental Plan

Nine alloys were used in this study with various cobalt contents and grain sizes. The nominal grain sizes and cobalt contents of these alloys are listed below:

Alloy No.	Cobalt Content wt. %	Nominal Grain Size
1	3	1.5
2	3	3
3	6	7.8
4	9	1.5
5	9	3
6	9	7.8
7	15	1.5
8	15	3
9	15	7.8

These alloys were sintered at appropiate temperatures, and the carbon contents were in the median range for the alloy, that is, close to stoichiometric values for WC. All sample preparation was accomplished by grinding with a diamond grinding wheel of 100 grit, and all grinding marks were parallel to the longitudinal axis at the specimen. The ungrooved side of the specimen was polished prior to testing to aid in observing the length of the crack.

Figure   3

     Testing was done in a specially designed fixture, (see Figure
3) which allowed the advance of a carbide wedge (via a micrometer
screw) into the slot provided in the sample. The arms of the
double cantilever beam could then be displaced and this displacement measured via attached transducers. The length of the crack
was measured through the use of a traveling stage (to which the
total fixture was attached) and a microscope mounted above the
sample (see Figure 4). The data gathered then consisted of arm
displacement and crack length.

Test Procedure

     The test samples were placed in the test fixture and the arms
of the double cantilever beam displaced until a natural crack was
initiated and arrested. The length of this pre-crack varied from
1/2" to 2" depending on the fracture toughness of the material.
The total length of the crack was then measured (including the
original diamond wheel notch) and the wedge withdrawn to close the
natural crack. The displacement transducers were then zeroed prior
to beginning the test. The arms of the WLDCB were displaced until
the crack began to move, data were then taken on total crack length

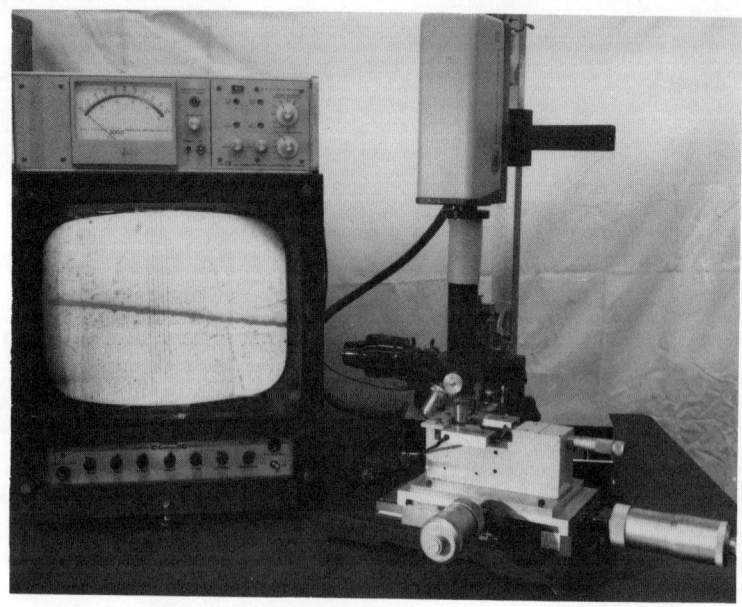

Figure   4

and arm displacement. When ten readings had been taken in this manner the wedge was then withdrawn and the procedure repeated. This was done to eliminate any cumulative error which might result if all readings were taken in succession. Usually four to six sets of ten readings were taken on each sample. After completion of the test the fractured sample was examined to ascertain whether the crack had run off center into the thicker material on either side of the slot. If this had occurred, those data points taken from the affected area were considered invalid and discarded. The consequences of a crack running off center would be to have the crack traversing a double thickness area (1/8" thick) and to make the arms of the beam unequal in thickness, thus negating the compliance calibration. The arm displacement, crack length and modulus of elasticity data were converted to G_{IC} and K_{IC} values using a computer program based on the compliance calibration and calculations in Appendix I.

Samples were also prepared to obtain other pertinent physical property data such as transverse rupture strength, compressive

strength, yield, and modulus of elasticity. The transverse rupture data were obtained from 1/4" to 1/4" bars broken in three
point bending on 5/8" centers. The samples were ground using a
100 grit diamond wheel with the grinding marks running parrallel
to the longitudinal axis. The compression samples were prepared
in a similar manner and were 1/4" x 1/4" x 1" with the ends ground
parallel to better than .003". The test was run with 1 mil steel
shims top and bottom of the sample (see Reference #1). The compression samples were also instrumented with two strain gages on
opposing sides and the strain data used in the modulus of elasticity calculations were the average of these two gages.

All of these physical property data along with binder film
thickness are compiled in Table I.

RESULTS AND DISCUSSION

The critical strain energy release rate G_{IC} for cemented carbides bears a near linear relationship to binder film thicknesses
with some leveling out at the higher values of this parameter (see
Figure 5). The compressive strength of these alloys continues to
increase as the binder content goes down even though the fracture
toughness values goes down.

The reason for this is that any surface flaws, as well as
flaws perpendicular to the longitudinal axis of the specimen, will
tend to close on compression and thus their effect will be minimal.
The fracture toughness then is not consequential in the compression
strength of this material unless a flaw of large dimensions and
critical orientation is present.

The transverse rupture strength of these alloys is determined
by the yield strength, the size of flaws present on the surface,
and the strain energy release rate. For the low binder materials
the transverse rupture strength is below or close to the yield (3%
1.5 and 3 micron grain size, 9% 1.5 micron grain size alloys and 6%
8 micron grain size alloys) and the materials will fail when the
strain energy release rate for crack propagation becomes critical
for the flaw sizes which are present. The surface flaws on the 3%
1.5 micron grain size and 9% 1.5 micron grain size materials which
result from grinding should be approximately the same size and as
a result the transverse rupture strength of the 9% alloy is about
33% higher due to the 33% higher strain energy release rate (the
modulus of these two materials being fairly close). The 3% alloys
with 3 micron grain size should have a larger inherent flaw size
(assuming that the inherent flaws from grinding, etc., are proport-

PHYSICAL PROPERTY DATA

Wt. % Co	Grain Size in microns	Transverse Rupture Strength psi	Compressive Yield psi .002% offset	Ultimate Compressive Strength psi	Critical GI Strain Energy Release Rate in lbs/in2	Critical Stress Intensity Parameter psi in KI	Hardness Rockwell A	Modulus of Elasticity 106 psi	Binder Film Thickness in microns
15	1.5	430,000	158,000	662,000	1.9	12,500	89.0	83.1	.29
	2.5-3	420,000	102,000	556,000	3.25	15,000	87.0	72.4	.65
	7-8	388,000	78,000	458,000	6.5	22,000	85.0	71.9	1.4
9	1.5	340,000	221,000	709,000	.88	8,400	90.5	81	.16
	2.5-3	306,000	212,000	631,000	1.7	11,600	88.9	79.2	.38
	7-8	340,000	142,000	515,000	4.78	19,000	87.3	76.0	1.05
3	1.5	196,000	420,000	860,000	.58	6,500	92.8	95.5	.09
	2.5-3	250,000	280,000	693,000	.98	9,600	91.3	94.7	.14
6	7-8	298,000	190,000	566,000	1.98	13,000	88.6	84.5	

TABLE I

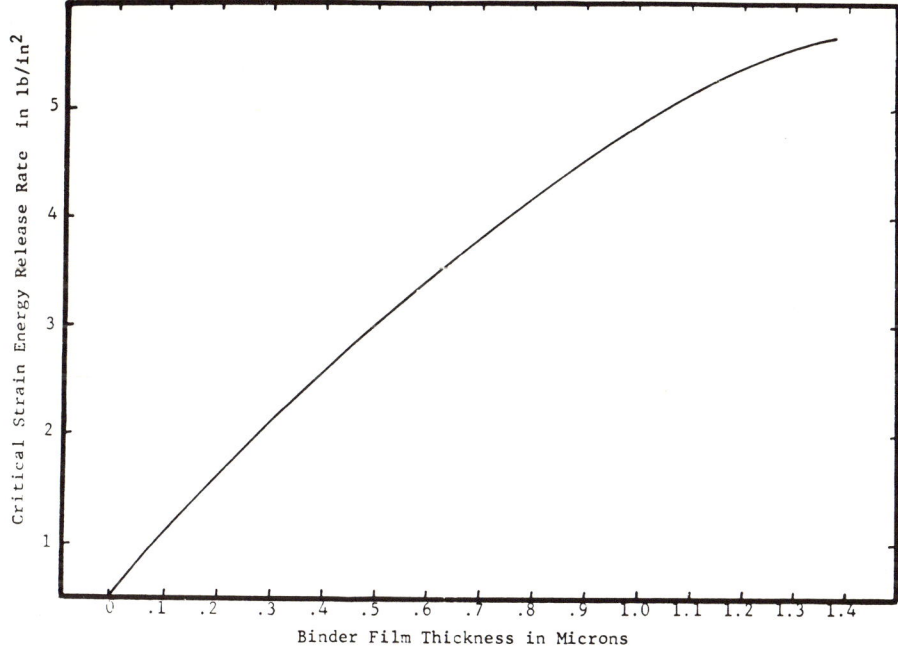

Figure   5

ional to the grain size of a given alloy) than the 9% 1.5 micron
grain size alloy, and since they have approximately the same critical strain energy release rates, the 3% alloy is less strong in
the transverse rupture sense. The 6% Co 8 micron grain size alloy
would have the great flaw size of these alloys but also the greatest critical strain energy release rate, and as a result the transverse rupture strength is in the same range as the other alloys in
this group.

For the alloys whose transverse rupture strength is below the
yield strength the critical flaw size with respect to its critical
strain release rate must be present; that is, no crack growth is
necessary as the binder has not yielded. Thus an increase in binder yield for these alloys would not effect an increase in transverse
rupture strength.

In the materials which show a transverse rupture strength above the yield strength, the transverse rupture strength is controlled by different mechanisms. For these alloys the strain energy release rate is such that the critical flaw sizes are not present initially. As these materials are loaded, fracture on a microscale leads eventually to a critical size flaw and total failure then ensues. This mechanism operates in the large grained higher cobalt alloys, especially where the critical strain release rate is high and the grain strength is low. In materials where the grain strength is sufficient to withstand the stresses imposed on them (15% 1.5 micron grain size) the binder itself must fail in a normal ductile manner until the critical flaw size is reached, then catastrophic failure will occur. For alloys of medium grain size all of these mechanisms can work, depending on the specific cobalt content and grain size. For this type of alloy, strengthening the binder should raise the transverse rupture strength and lower the critical strain energy release rate.

CONCLUSIONS

The tests which are commonly used to predict the performance of WC-Co alloys in the field with respect to fracture are, compressive strength, transverse rupture strength, and charpy impact testing. The last two relate to the tensile failure and "toughness" of the material respectively, and these have been shown by experience and other work (2) to be of limited value in the prediction of field breakage. With fracture toughness data, such as presented here, the failure of WC-Co alloys in the field can be studied as a multifaceted problem. The final fracture will occur when the critical strain energy release rate for a given material is exceeded. Just when the above will happen is a function of both nominal stress level in the area of the effective flaw and the size and shape of the flaw. The nominal stress level in a carbide part under given conditions will be relatively the same regardless of the grade of carbide used. The fracture problem then reduces to a study of the ability of different carbides to resist surface damage, and to resist the growth of this damage to a critical size flaw. If all of these factors are well understood, namely the fracture toughness, and the ability of a material to resist surface damage and growth of this surface damage, then a better prediction of field performance should be achieved.

APPENDIX I

Compliance Calibration

The strain energy release rate G is equal to

$$G = \frac{\Delta^2}{2\lambda^2 b} \frac{\partial \lambda}{\partial(a)}$$

where b is the width of the fractured surface.

In order to determine the strain energy release rate of any given material we must first determine the compliance   of the material in the configuration of the test. $\int$ (Compliance = $\frac{\Delta}{p}$)

This is essentially the inverse spring constant of the system.

The accuracy of the compliance can depend only on the accuracy of the measurment and since the compliance changes very little with crack length, the correctness and sensitivity of the measuring system is paramount.

LOAD MEASURING SYSTEM

To measure the load on the sample a transducer was built which consisted of tungsten carbide cobalt wedge with a groove machined along the midline of the wedge (see Figure 6).

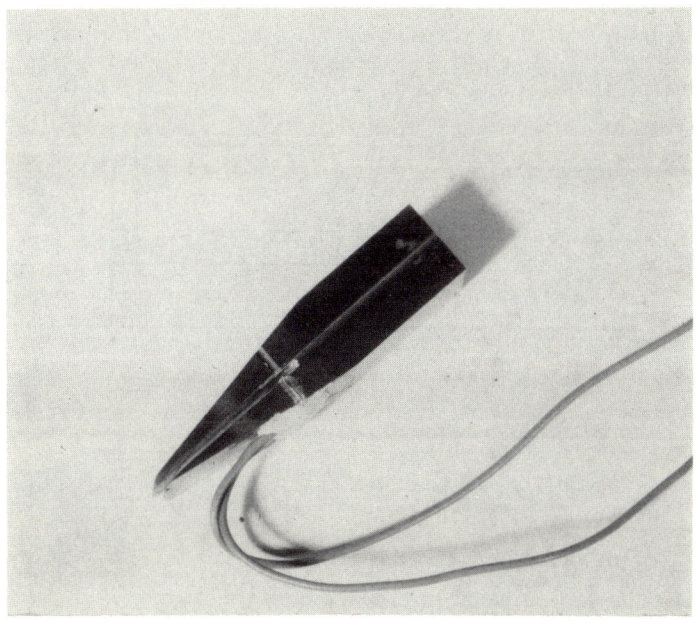

Figure   6

Strain gages were placed at the bottom of the groove. These gages produced a signal when the wedge was loaded at the tip. Since the wedge was displaced into the machined portion of the specimen to produce increasing displacements, and thus loads, the point of contact of the specimen on the wedge transducer would change; therefore the wedge must be calibrated along its usable portion. This was accomplished via a fixture which allowed the wedge to be loaded by an exterior mechanism at any point on the usable part of the wedge (see Figure 7.).

This total apparatus was then placed on an Instron load cell and data on load versus output from the wedge was taken as a function of loading position. The wedge strain gage signal was read through a BAM 1 amplifier and a Barber Coleman recorder. The output was found to be linear with respect to load at any given point of application on the wedge. The slope of the load versus wedge output data were plotted against position on the wedge and this was also found to be linear. The load could be read to within + .25 lbs. over a range of 50 lbs., or about + .5% error. From the above information, the known starting point, and the displacement read off the micrometer, the load could be determined on the specimen during compliance calibration in the test fixture.

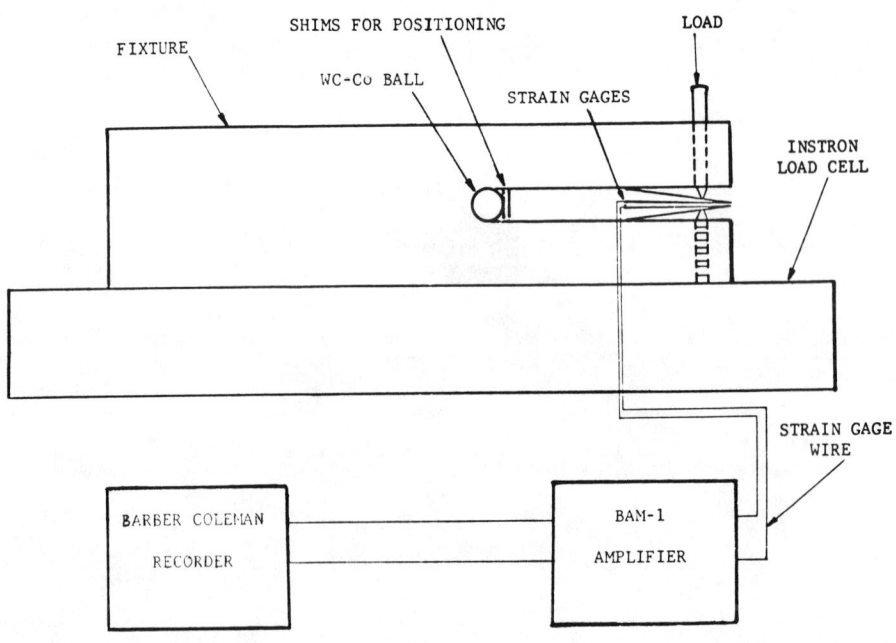

Figure   7

DISPLACEMENT MEASURING SYSTEM

The displacement was measured via a daytronic variable induct-
ance transducer and read on a daytronic transducer amplifier meter
(see Figure 4). This system has a summing capability for the two
transducers such that lateral displacement of the specimen, if any,
will not be recorded as displacement. This system allows measurement of five micro inches; however, readings were taken routinely
of .0002", thus error here could only be attributed to positioning
of probe and initial calibration of the instrument, an error of less
than .5% is expected.

MEASUREMENT OF DATA AND TREATMENT

Load versus displacement data were taken and several samples
were taken at crack length increments of 1" over the length of the
specimens from 1" to 3". All these data were then plotted for each
crack length and the compliance determined. The deflections were
found to be linear with load as required. The compliance data from
all samples and crack lengths were then plotted on a master curve,
all of these data correlated quite well. A regression analysis was
performed on these data and equation of the form $Y = AX^B$ gave the
best fit with an index of determination of .996. This result agrees
well with theory. The displacement using the regular theory is:

$$y = \frac{FL^3}{3EI} \quad \text{or} \quad \frac{\Delta}{p} = \frac{L^3}{3EI}$$

From the regression analysis $\frac{\Delta}{p} = 90.8 \times 10^{-6} L^{2.64}$. To
obtain $\frac{\partial \Delta}{\partial a}$ the first derivative was taken of this equation in
the normal manner. This procedure eliminated the need to deter-
mine experimentally the slope of the master plot at any point.
With this information G for this material may be determined with
only displacement and crack length measurements.

ACKNOWLEDGEMENT

The author wishes to express his gratitude to Dr. A. J. Smith,
Thesis Advisor at Michigan State University, and to Dr. E. W.
Goliber, of the Carboloy Systems Department, General Electric
Company, for their assistance and encouragement. The author also
wishes to express his appreciation to W. A. Powell, Carboloy
Systems Department, for his advice on the formulation of the de-
sired alloys, and to G. F. Paluch, also of the Carboloy Systems
Department, who performed most of the experimental work.

BIBLIOGRAPHY

1. Lueth, R. C. and Hale, T. E., "Compressive Strength of
    Cemented Carbide - Failure Mechanics and Testing Methods",
    Materials Research and Standards, M. T. R. S. A., Volume 10,
    Number 2, page 23.

2. Lueth, R. C., "An Analysis of Charpy Impact Testing as Applied
    to Cemented Carbides," A. S. T. M. Symposium on Instrumental
    Impact Testing, June 26 and June 27, 1973.

3. Atkinson, E. F., "Button Bit Failure and Preventive Mainten-
    ance", Pit and Quarry, March 1972, page 98.

4. Fracture Toughness Testing and its applications, A. S. T. M.,
    S. T. P. 381, American Society for Testing and Materials, 1965.

5. Fracture Toughness, A. S. T. M., S. T. P. 514 J, American
    Society for Testing and Materials, 1972.

6. Stress Analysis and Growth of Cracks, A. S. T. M., S. T. P.
    513, American Society for Testing and Materials, 1972.

7. Plain Strain Crack Toughness Testing of High Strength Metalic
    Materials, A. S. T. M. 410, American Society for Testing and
    Materials, 1966.

8. Lueth, R. C., "A Study of the Strength of Tungsten Carbide
    Cobalt from a Fracture Mechanics Viewpoint", PH. D. dissertation
    Michigan State University, June 1972.

FRACTURE RESISTANCE OF METAL-INFILTRATED POROUS CERAMICS

R. A. Queeney and R. L. Turner

Department of Engineering Mechanics

The Pennsylvania State University

INTRODUCTION

The development of infiltrated porous ceramics as specialty structural materials is an area of technological promise in the engineering of materials. The attractive features of selected ceramics-e.g. hardness, oxidation resistance-can be exploited without the necessity of achieving full density in the solid. The direct production of complex geometries, through the cold-pressing and sintering of powders, or related techniques, is complemented by relatively facile machining. Within certain limits, the strength levels of the solids can approach that of the fully dense product when the porous elements are fully infiltrated by the proper infiltrant.

Porous ceramic materials suffer from a degradation of most mechanical strength measures[1], as, in fact, do most metals[2]. Although economies may be realized in the production of brittle, porous solids, their lowered strength and fracture resistance renders them poor candidates for mechanical structural application. Their degraded strength levels, as compared to the fully dense forms of the same solid, are attributed to two features of their porosity-a reduction in actual cross section load bearing area and the generation of high local stress states near the pores[1]. The stress states near pores, and changes in these states upon filling the pores, is one of the topics developed in the present work. In addition, the possible effects of plastic flow within the pores are explored as they pertain to fracture resistance. Rational choices of ceramic matrix-pore infiltrant couples are indicated by these studies and their application in an experimental study.

STRESS STATE ANALYSIS

An infiltrated porous body possesses an extremely complex
microstructure. From the macrostructural stress analysis viewpoint,
it resembles a highly randomized network composite, with the ceramic
matrix reinforced by a three-dimensional infiltrant skeleton. The
skeletal structure of the infiltrant arises from interconnected,
not isolated, pores if infiltration is, in fact, successful. Rather
than attempt a continuum mechanical analysis of these intractable
geometries, one might examine more simplified analysis problems,
without undue loss in predictive accuracy. In particular, the
probing of stress states around isolated voids and inclusions will
provide ample foundation for the engineering of matrix/filler
couples.

Analysis of isolated pores and inclusions is well-established
in the literature[3,4,5]. These analyses are three-dimensional in
character: since the plane of fracture will cut through the infiltrant network, a two-dimensional analysis will be more useful, and
numerical calculations less tedious. The two dimensional analysis
is an extension of the work of Inglis[6] to include elliptical inclusions as well as elliptical voids[7]. Figure 1 illustrates the geometries and loadings in question.

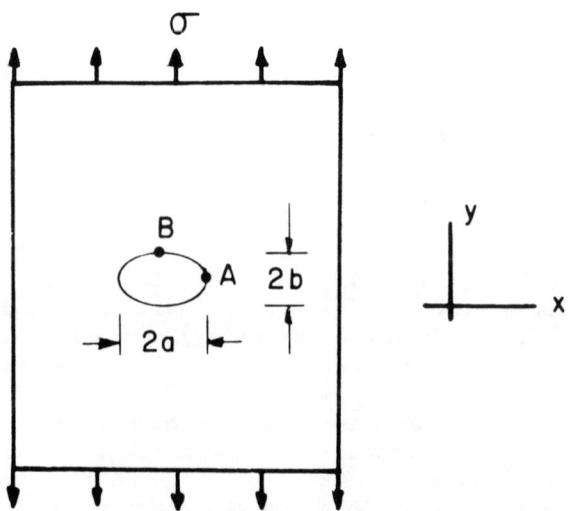

Figure 1:   Elliptical void/inclusion imbedded in
            tensile stress field

The latter results of Donnell[7] indicate that the stress concentrations at the point A in Figure 1 are reduced by any filler material inserted into the void. If the void filler or infiltrant
is stiffer than the matrix, a reinforcement is produced, with
stresses there less than the nominal net section stress σ. Detailed
calculations for matrix/infiltrant couples investigated experimentally are given below. The application of this analysis[7] to predict the elastic response of infiltrated ceramics has been reported[8].

Were the pore infiltrant of the same degree of brittleness as
the ceramic matrix, the above elastic considerations, more fully developed, would probably be sufficient to provide theoretical predictions of fracture strengths. However, if the infiltrant is a
ductile metal or metallic alloy, plasticity of the infiltrant at
fracture may prove significant or even dominant. Stress states
near sharp flat flaws have been analyzed by Irwin[9] who found that
the stresses near the flaw were governed by a constant K_I (the
stress intensity factor) dependent on structural geometry and
boundary conditions, and that these stresses displayed an invariant
spatial distribution. Furthermore, K_I (the subscript refers to
mode I, or opening mode, loading of the flaw) is a measure of
fracture resistance, and at the onset of unstable flaw extension
is referred to as the fracture toughness K_{IC}.

Using the Irwin stress field equations[7], and applying an
appropriate yield criterion, such as the distortion energy theory,
the elastic plastic interface, ahead of, and in the plane of,
the crack, r_p, can be defined to be[10];

$$r_p = \frac{K_I^2}{2\pi\sigma_o^2} (1-2\nu)^2$$

Equation(1) pertains to the plane strain case, and σ_o is the
phenomenological yield stress and ν the poisson's ratio of the
solid. Here, the elastic-plastic interface refers to yielding
in the ductile infiltrant, where the stresses are not those of
the net sections[7] or the homogeneous media assumed by Irwin[9]. However, if r_p, from eq. (1), is comparable to the interpore spacing
on the fracture plane, then the plastic response of the filler will
be significant. In particular, the lower its yield stress σ_o, the
greater the size of the plastic zone and the higher will be the
composite's resistance to fracture.

EXPERIMENTAL STUDIES

The above analytical considerations concerning the response of

metal infiltrated ceramic bodies were tested by mechanically infil-
trating (via pressure) molten 2024 aluminum alloy and an aluminum
bronze (10% Al, 2.5% Fe, 5% Ni, 82.5% Cu) alloy into a commercial
graphite (Grade 2020, Stackpole Carbon Co., St. Marys, Pa.). The
aluminum alloy was subsequently solution treated at 494°C for one
hour and furnace cooled (-TO state), or quenched and room temperature aged (-T4 state) or artificially aged at 190°C for 12 minutes
(-T6 state). Pertinent properties and responses of the metal alloys
are given in Table I below. Physical properties and mechanical
response measures determined from smooth-sided test specimens have
been previously reported[11].

	2024 Al			Bronze
	-TO	-T4	-T6	
Young's Modulus, psi	----- 10.6x10^6 ---------			16.2x10^6
Poisson's Ratio	-------- 0.33 ------------			0.33
0.2% Offset Yield, σ_o psi.	11,000	42,000	52,000	75,000
Ultimate stress, σ_u, psi.	27,000	64,000	65,000	118,000
% elongation	19	18	10	15
Modulus of toughness, T, psi.	3,600	9,500	5,850	14,550

Table I: Properties of Metallic Infiltrant Alloys

A typical infiltrated microstructure is shown in Figure 2. An
average pore diameter of 2.4×10^{-4} inches was determined via the
line intercept method: the directly measured a/b ratio was 3.5.
Interpore spacing was found to be 1.2×10^{-3} inches.

Three-point loading bend tests were performed on notched beam
specimens, and fracture toughness K_{IC} calculated from the critical
load value P_c and the following stress intensity expression[12]:

$$\frac{K_I BW}{3PLa^{1/2}} = 1.96 - 2.75\left(\frac{a}{w}\right) + 13.66\left(\frac{a}{w}\right)^2 - 23.98\left(\frac{a}{w}\right)^3 + 25.22\left(\frac{a}{w}\right)^4 \qquad (2)$$

For the specimens employed, B=W=0.25 inches (square cross-section)
and the beam span L was 2.0 inches, the crack depth a was 0.062
inches. Four specimens were tested for each material microstate.

Figure 3 indicates the results of the fracture toughness
testing. The bronze infiltrated graphite exhibited a three-fold
increase in fracture resistance, whereas the aluminum infiltrated
material exhibited a four-fold toughness increase. Calculated
changes in cross-sectional area from pore size and spacing data
would only account for about a 5.0% increase in actual load-bearing

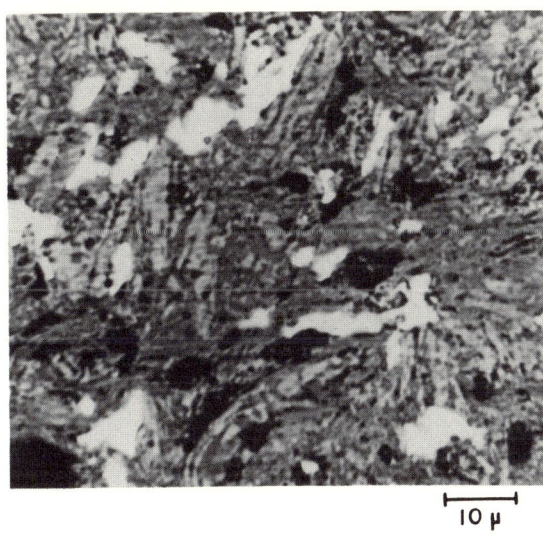

Figure 2:   Bronze Infiltrated Graphite Microstructure

cross section. Other factors apparently govern changes in fracture
resistance, as discussed below.

Table II summarizes the changes in elastic stress states
caused by filling the pores with infiltrants stiffer than the
ceramic matrix. For the shape ratio of the pores noted in the
present study ($\frac{a}{b} \cong 3.5$) equatorial stress concentration factors
(point A in Fig. 1) are reduced from 8.0 to 0.045 for the aluminum
infiltrant, and to a negative value (compressive stress) of -0.046
for the bronze infiltrated material. Both infiltrants, of course,
act as reinforcements. Thus, if the concentrated stress states
associated with the pores extend only to a distance of a pore radius
from the pore periphery[7], then about 10% of the fracture plane, for
the present material, is not only relieved of these high local
stresses, but actually experiences lower than net-section stresses.
If the relief of these high local stress states were the only factor
acting to alter the fracture resistance of the graphite, then the
stiffer bronze would produce the most toughening. The test results
indicate that this is not so, leading to the conclusion that other
effects, such as plasticity considerations, must be examined.

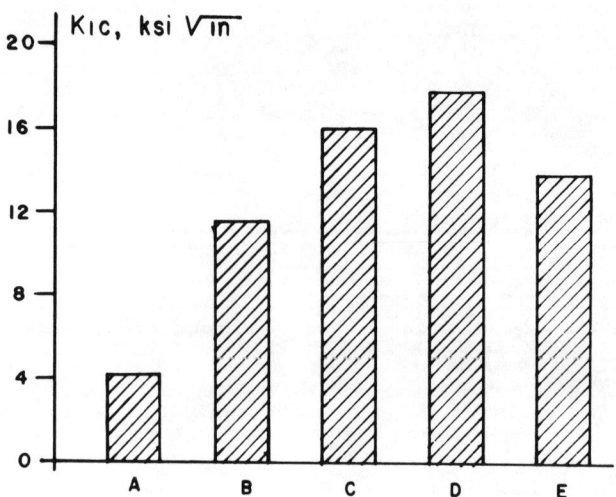

Figure 3:   Fracture Toughness K_{IC} of; A. Pure graphite;
B. Bronze infiltrated; 2024 Al infiltrated in C. -T0;
C. -T4; E. -T6 heat treatment states.

The plastic zone size, as calculated from eq. (1), indicates
that yielding can be induced in the 2024-T0 infiltrant 0.5×10^{-3}
inches ahead of the flaw tip.   Similar calculations for the bronze-
infiltrated graphite predict the elastic plastic interface for the
latter two orders of magnitude nearer the flaw tip.   This would
appear to account for the difference observed in the K_{IC} values
for the aluminum vs. bronze infiltrated material.   However, the
2024-T4 infiltrated material has a higher K_{IC} measure than the -T0
heat treatment material, even though it possesses a greater yield
stress.   The -T4 state has a calculated elastic-plastic zone size
much closer to that of the bronze.   There is, as commented upon
earlier, some inaccuracy in applying eq. (1) directly to the stress
states in the pores without taking into account the perturbation
in local stresses caused by infiltration reinforcement.   Further,
the use of wrought alloy mechanical data, Table I, is questionable,
since alloy microstructural effects (e.g., grain size) may change
yield stress values for the metals.

Examination of the data in Figure 3 indicates that the variation in mechanical properties, achieved through heat-treatment of

Equatorial Normal Stress (σ_y) Concentration Factor K (point A, Figure 1)

Shape Ratio a/b	Void	Aluminum	Bronze	Rigid
1.0	3.0	0.18	0.21	
1.4	3.9	0.15	0.15	-0.03
2.0	5.0	0.11	0.08	-0.06
3.5	8.0	0.09	-0.046	-0.15
5.0	11.0	0.006	-0.15	-0.25

Polar Normal Stress (σ_y) Concentration Factor K (point B, Figure 1)

0.1	0	3.85	4.57	6.56
0.3	0	2.28	2.46	2.84
1.0	0	1.41	1.49	1.50
2.0	0	1.19	1.20	1.22
3.46	0	1.09	1.10	1.10
5.0	0	1.05	1.05	1.05

Table II: Stress Concentration Factors K[7] for Selected Sites around Pores and Inclusions, including Inclusions filled with Completely-Rigid (E→∞) Filler.

a given alloy composition also play a role in determining absolute fracture resistance. The differences in K_{IC} measured for 2024-T0, -T4, and -T6 infiltrants do not correlate either with strength or ductility measures alone, as the latter two change monotonically with heat treatment. Figure 4 shows the fibrous nature of the fracture of the ductile infiltrant phase. The work to fracture un-notched specimens or ductile metals, per unit volume, in tensile loading, is specified by the modulus of toughness, T, where:

$$T = \frac{(\sigma_o + \sigma_u)}{2} \varepsilon_f \qquad (4)$$

Here, ε_f is fracture strain. The calculated T values for the various heat-treated aluminum infiltrated graphite correlate well with the observed changes in K_{IC} for the same material. However, the modulus of toughness is higher for the bronze infiltrated graphite than for any of those graphites containing aluminum, so that this measure alone is not sufficient to explain absolute K_{IC} enhancement.

Finally, the writers feel that chemical interaction effects may also be of some importance, and may, ultimately, lead to understanding the absolute values of K_{IC} (or other strength measures) achieved in a given matrix/infiltrant couple. For the present study, significant quantities of Al_4C_3 were detected, using x-ray

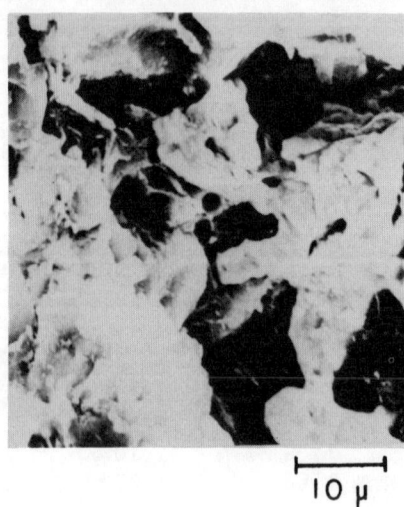

Figure 4:   Scanning electron fractograph of bronze infiltrated
            graphite

diffraction in the 2024 Al infiltrated graphite. Similar reaction
products were not detected for the bronze composite, even though
the alloy contains some aluminum. In view of the slight differences in stress concentration factor between the metals used and an
hypothetical rigid filler, it does not seem likely that the stiffen-
ing afforded by an Al_4C_3 layer would account for the higher K_{IC} of
the aluminum-graphite composite. Instead, there must be a more
complete bonding of the matrix to the infiltrant. The toughening
effects of such bonds, and the exact extent of reaction product
formation, remain to be characterized.

                            SUMMARY

    In summary, there appear to be several contributions to the
toughening of porous ceramics when they are metal-infiltrated.
These contributions to enhancing fracture resistance include:

a. amelioration of high local stress states near voids upon
    their becoming filled with infiltrant. If the Young's
    modulus for the infiltrant exceeds that of the matrix,

the filled pores may actually become reinforcements.

b. Replacement of non-load bearing regions with material
that can flow plastically and absorb strain energy released by an extending flaw.

c. The heat treatment of treatable alloys to maximize the
alloy network's resistance to flaw extension.

d. Achievement of better matrix/infiltrant bonding through
the creation of mutually compatible reaction products at
the phase interface.

More thorough investigations are needed on chosen matrix/in-
filtrant couples to ascertain which of the foregoing considerations
are of prime concern in achieving optimized strength and fracture
resistance.

REFERENCES

1. S.C. Carniglia, J. Am. Ceram. Soc., 55, 610 (1972).
2. G.A. Clarke and R.A. Queeney, Int. J. Pow. Met., 8, 81 (1972).
3. J.N.Goodier, J. Appl. Mech., Trans. ASME, 55, 39 (1933).
4. M.A. Sadowsky and E. Sternberg, J. Appl. Mech., 17, 149 (1949).
5. R.H. Edwards, J. Appl. Mech., 19, 19 (1951).
6. C.E. Inglis, Trans. Inst. Naval Arch., 60, 219 (1913).
7. L.H. Donnell in Theodore von Karman Anniversary Volume,
    California Institute of Technology, Pasadena, Calif., 293-309
    (1941).
8. D.P.H. Hasselman, J. Gebauer, and J.A. Manson, J. Am. Ceram.
    Soc., 55, 588 (1972).
9. G.R. Irwin, J. Appl. Mech., 24, 361 (1957).
10. F.A. McClintock and G.R. Irwin, in Fracture Toughness Testing
    and Its Applications, ASTM Spec. Tech. Pub. No. 381, 84-113
    (1964).
11. J.C. Conway and A.J. Shaler, Am. Ceram. Soc. Bull., 50, 656
    (1971).
12. W.F. Brown, Jr., and J.E. Srawley in Plane Strain Crack Tough-
    ness Testing of High Strength Metallic Materials, ASTM Spec.
    Tech. Pub. No. 410, 1-65 (1966).

FRACTURE AND STRAIN ENERGY RELEASE IN PRESTRESSED GLASS AND GLASS-CERAMICS

E. K. Beauchamp

Sandia Laboratories, Albuquerque, New Mexico 87115

INTRODUCTION

The engineering strength of glasses and glass-ceramics has increased dramatically with the development of ion-exchange techniques which generate very high compressive surface stresses. As a result of the relatively low elastic moduli of these materials, the strain energy stored in the compressive layers can be quite high. For example, the energy in a 1-inch-diameter, 1/8-inch-thick disc of an experimental Corning Glass Works material was found to be greater than 14 inch-pounds.[1]

Not only are the energy densities comparable to those in electrical capacitor storage, but as a storage device an ion-exchanged membrane is highly stable. The stress relaxation (by viscous flow or diffusion) that would be required to reduce the stored energy requires exceedingly long times at room temperature.

A consequence of these very high compressive stresses is that, with a moderately thick ion-exchange layer on a plate, quite high internal tensile stresses are generated. As a result, when the plate is broken, very rapid disintegration (dicing) of the plate occurs and the fragments separate violently. This characteristic has earned this class of materials the generic name "frangibles."

Frangibles have been employed in a variety of applications which exploit this tendency to dice and release strain energy. In most of these applications, such as the Maverick missile dome cover[2] and the Astro-Dart housing,[3] the released energy is used simply to separate the particles and remove them from an area. However, it is also possible to couple a transducer to a frangible plate and use the fracture of the plate to trigger a reaction.

This paper describes efforts to characterize commercially available frangibles in terms of stored strain energy and energy released in dicing. It also describes a model for the generation of a secondary fracture, which is essential to the dicing process.

STRAIN ENERGY RELEASE

Stress Measurement in Opaque Materials

A preliminary step in the determination of how much strain energy is released from the frangibles in dicing is the measurement of strain energy initially stored in a membrane. The energy calculation is simple and straightforward once the stress profile is known. The problem attending measurement of the stress profile in these materials, however, is that many of the glass-ceramics are opaque. Even for the transparent ion-exchanged materials, stress birefringence measurements are not so simple as for tempered glasses. Not only is the stress birefringence constant in the ion-exchanged layer different from that in the base glass, but the refractive index is also changed and it becomes difficult to pass a light ray along a path of constant stress.

A procedure developed in this program[4] for measuring stresses in the opaque frangibles consists basically of etching away thin layers of the surfaces of a plate and observing the change in strain, $\Delta\epsilon$. The plot of $\Delta\epsilon$, or the equivalent length change, $\Delta\ell$, as a function of the thickness change is differentiated (usually by obtaining a polynomial fit and taking its first derivative), and the stress is calculated from the equation

$$\sigma(z) = \frac{E}{(1-\nu)\,\ell_o} \left(-\, z\frac{d\ell}{dz} + \int_z^b d\ell \right) ,$$

where z is the distance from the central plane of the sample plate, E is Young's modulus, and ν is Poisson's ratio (assumed to be the same for the base material and the exchanged layer), ℓ_o is the initial gauge length, $d\ell/dz$ is the slope of the length change vs. thickness change curve, and $\int_z^b d\ell$ is the total length change produced in etching the plate from its original thickness (2b) to 2z.

In practice, there is a problem, with this procedure, of providing a well-defined gauge length on a sample so that the length change can be measured with a dilatometer. In most cases, it has been proved convenient to introduce conical indentations on opposite edges of a plate by using an S.S. White Airbrasive drill. These indentations were protected with beeswax while the rest of the plate was etched. The indentations were then located on sapphire balls which had been bonded permanently to the dilatometer platens for the length measurements. To obtain the desired precision in length measurements, it was necessary to provide very good temperature control in the vicinity of the dilatometer and to use reference bars.

One of the dividends of careful stress measurements on these materials is additional insight into the ion-exchange process and the mechanism for stress generation. As an example, Sandia has been working with Corning Glass Works to develop a stressed glass-ceramic similar to Code 9611 but with improved stability. Compressive surface stresses of over 150,000 psi are generated by ion-exchanging the base high-expansion magnesium aluminosilicate glass-ceramic to produce a low-expansion surface.[5] It had been assumed, prior to the stress measurements, that only a single phase change was involved in the ion-exchange of the new material. However, as the stress profile in Fig. 1 shows, there are two steps

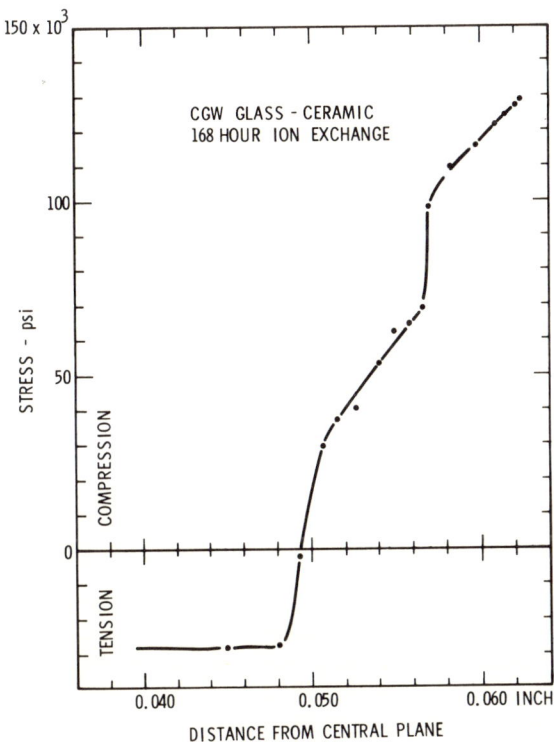

Fig. 1 – Ion-Exchange Stress Profile for a Corning Glass Works
          Experimental Magnesium Aluminosilicate Glass-Ceramic.

in the compressive stress region. That fact suggests two different
expansion layers; i.e., two separate phase changes. Confirmation
of the existence of these two layers was subsequently obtained
through an x-ray diffraction study and electron microprobe measurements.[6]

Computer Calculations

One method of estimating the energy released in the dicing
process is to calculate the residual strain energy in the fragments.
For these calculations it was convenient to assume that the fragments approximated a right-circular cylinder. That assumption permitted the use of the SAAS computer codes for finite element analysis. The results of the calculations for a 1/8-inch-thick plate
of 9611 are shown in Fig. 2.[7] The upper dotted curve is the

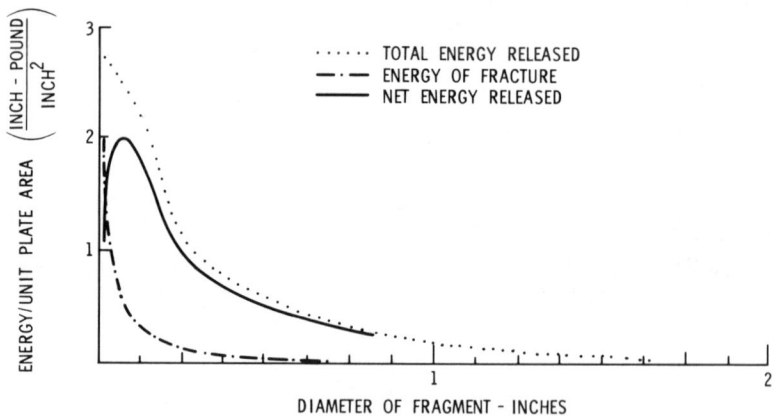

Fig. 2 - Calculated Strain Energy Release vs. Dicing Fragment Size for Corning Glass Works Code 9611.

difference between the initial strain energy (per unit plate area)
and the residual strain energy; i.e., the total energy released.
The lower solid curve shows the released energy less that required
to create the new fracture surface in dicing the plate to that size.
The value of the fracture surface energy ($\gamma = 8.6 \times 10^{-2}$ inch-pound/
inch^2) used in this calculation is probably an overestimate. For
this value of γ, the maximum energy available to separate particles
or actuate a transducer takes place when the fragment diameter is
approximately equal to the plate thickness (1/8 inch). For that
particle size, which is the size produced when 1/8-inch-thick 9611
sheets were diced, the calculations show that approximately 75 percent of the original strain energy of the plate is released.

Ballistic Energy Measurements

The first attempt in this program to experimentally determine
the energy released in dicing involved a direct measurement of the
ballistic energy of fragments after separation.[7] In these experiments, a flat plate of the frangible material was mounted on a
pedestal 4 feet above a series of cylindrical ring section collectors as shown in Fig. 3. The plate was fractured by drilling a

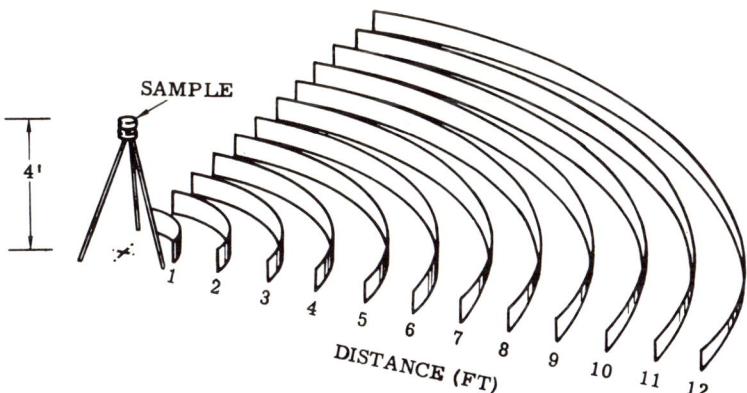

SAMPLE

4'

DISTANCE (FT)

1 2 3 4 5 6 7 8 9 10 11 12

Fig. 3 - Sketch of Dicing Fragment Ballistic Energy Test Setup.

hole in its center. The fragments ejected into each collector
section were then swept up and weighed.

There is an aspect of fracture in the frangible plates which
makes it possible to determine the velocity (and, because the mass
is known, the kinetic energy) of the fragments in a collector ring
from the radius of the ring. The stresses in the plate are parallel to the plane of the plate (except at the edges). Therefore,
the velocity vectors of the fragments at separation should be in
the plane of the plate. This was confirmed by high-speed photography
of a fractured plate. As a result, for a plate originally mounted
in the horizontal plane, the range of a fragment (i.e., the radius
of the collector ring it enters) can be used to calculate the initial velocity of the fragment. The total released kinetic energy
is then determined by summing the energy for fragments ejected
into all collectors in the array.

In general, the values for released energy determined by this
ballistic technique were 30-35 percent of the initial strain energy,
or less than one-half the calculated release energies. Loss of
energy due to viscous drag on the fragments by the air could account
for only a small part of the difference.

Calorimetric Measurement of Released Energy

If all the strain energy in a plate of glass with an average energy density of 0.5 Joule/cc could be converted to heat, the plate temperature would increase approximately 0.3°C. Accurate measurement of a temperature excursion of this magnitude is well within the capabilities of modern calorimetry. However, there are some special problems which complicate the measurement of energy released in dicing. First, the dicing process must be initiated without introducing an error into the measurement. Second, the calorimeter cup must be strong enough to contain the fragments during separation. The extra mass, and increased thermal capacity, of the cup required for this containment reduces the sensitivity of the instrument.

The calorimeter used for dicing energy measurements[1] was of isoperibol design, consisting of an evacuated chamber in a thermo-stated bath. The calorimeter cup containing the sample disc was suspended within the chamber on nylon lines. A second, identical cup and disc were suspended below the first. To measure the temperature, a pair of thermistors was bonded to each cup and the four were connected to form a bridge circuit, with the thermistors on the lower cup serving as references. Also bonded to each of the calorimeter cups was a heating coil. Calibration of the calorimeter was accomplished by introducing a specified amount of electrical energy into the coil on the sample cup and observing the temperature change.

Dicing was initiated by means of a tungsten carbide tipped probe which entered the top of the vacuum chamber through a metal bellows. It was essential to initiate fracture without introducing substantial additional strain energy into the sample. This was accomplished by etching a cavity in the center of the circular disc sample through the compressive stress region. Fracture could be initiated with a force of only a few hundred grams when the probe was located in the cavity. Temperature changes produced by the energy generated in pressing the probe into the sample and by thermal transport along the probe gave errors in measurements of approximately 4 percent.

The results of measurements on Corning Glass Works 0319 glass and on the experimental magnesium aluminosilicate glass-ceramic are shown in Table I. In each case, the calorimeter measurement was slightly less than 80 percent of the original (calculated) stored energy. This agrees with the predictions from the computer calculations of strain energy release.

TABLE I

Strain Energy in 1.0-Inch-Diameter, 1/8-Inch-Thick Disc

	Calorimeter Measurement (inch-pounds)	Calculated from Stress (inch-pounds)	Percent Released
0319 Glass	1.36	1.75	78%
CGW Experimental	1.34		
Glass-Ceramic	11.4	14.41	79%
(Ion-Exchanged 48 Hr)	11.2		

MECHANISM FOR FRACTURE BRANCHING

Although fracture mechanics indicates that fracture branching
in brittle materials should occur when the strain energy release
rate is twice that required for generation of a single fracture,[8]
it suggests no mechanism for branching. The strain energy release
rate requirement is thus a necessary, but not sufficient, condition
for branching.

It is, at least intuitively, obvious that branching cannot
occur at the atomic level; i.e., simultaneous rupture of bonds on
each side of atoms at the crack tip is highly unlikely. To produce a second crack, a second source is required.

Congleton and Petch[9] have examined the problem of crack branching and have proposed a mechanism in which existing flaws can be
activated by the locally high stress field of a passing (primary)
crack to produce secondary fracture. This concept is intriguing.
It suggests that in the application of frangibles, the fragment
size might be modified by changing the distribution and size of
the existing flaws.

In an attempt to determine whether some trace of these flaws
might exist and in the hope of gaining some insight into how secondary fracture progresses, the fracture surfaces of fragments from
a diced plate of thermally tempered window glass were examined
directly by SEM and through replicas by transmission electron
microscopy. Acloque[10] had pointed out that secondary fractures
begin as feather-shaped cracks which curve away from the primary
fracture plane. The final, secondary fracture surface which sepa-
rates fragments does not contain the source. Within a dicing
fragment, many of these feather-shaped cracks can be seen. Most

of these extend only a short distance from the primary plane and
do not result in complete secondary fracture. It was at the base
of these cracks on the primary that the examination was centered.

    With the electron microscope it was found that at the tip of
the feathers (where the secondary fracture was initiated) the
fracture surface contained features like that shown in Fig. 4.

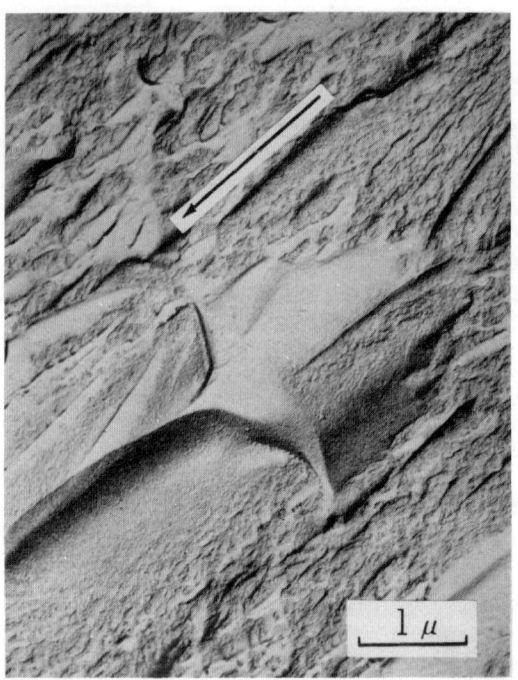

Fig. 4 - Secondary Fracture Source. TEM of Fracture Surface
        Replica. The arrow indicates the propagation direction
        of the primary fracture.

The interesting point about these features is that the fracture
surface which eventually forms the feather is apparently a continuous extension of the primary fracture surface. It strongly suggests
that the secondary fractures originate not in existing flaws, as
Congleton and Petch[9] had suggested, but in flaws produced by the
passage of the primary fracture.

    A model for flaw generation by the primary fracture is shown
in Fig. 5.[11] The first step in the generation (Fig. 5a) is deviation of a short segment of the primary fracture front out of the
primary plane. This deviation can be produced by an inhomogeneity

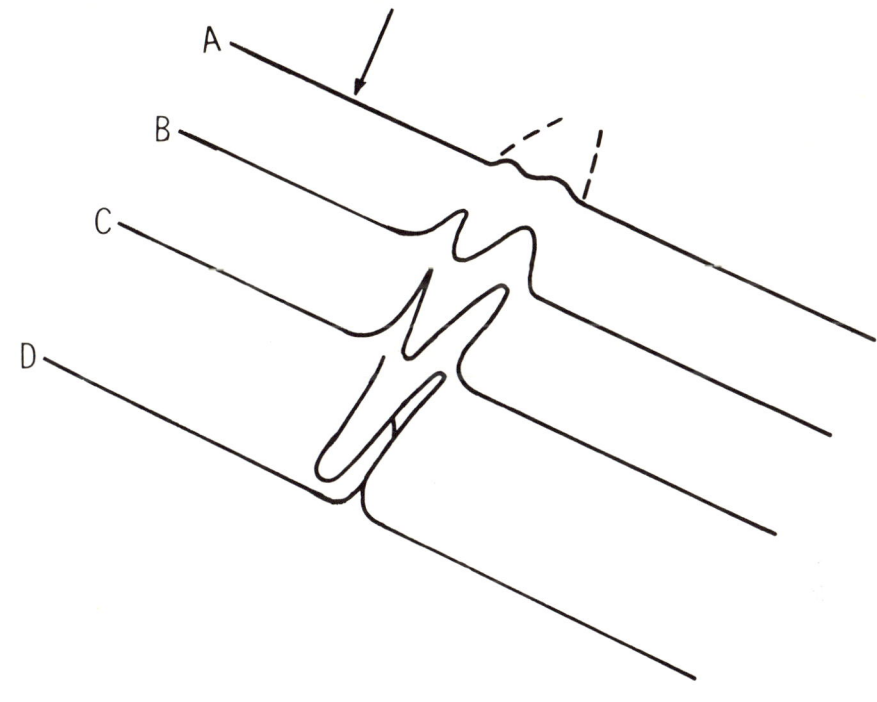

Fig. 5 - Model of Secondary Fracture Initiation.

in the material or, as in Wallner lines, through interaction with
a stress wave generated by the rapidly moving primary and reflected
back to the primary. (Secondary fracture occurs only when the
strain energy release rate is high enough to produce high fracture
velocity. The high strain energy release rate also insures that
enough energy is available for the extra surface produced in the
deviation out of the primary.)

        The next step in the development of the flaw is isolation
(Fig. 5b). As the deviated segment moves forward, the section
parallel to the primary moves at about the same velocity as the
primary, while the sections at an angle to the primary are retarded.
The result is a "tongue" of fracture running parallel to the primary but separated from it. This growth of a segment of fracture

out of the primary plane has been described by Frechette[12] in a
discussion of the development of "hackle."

Now the two separated portions of the fracture on the primary
plane grow together and eventually coalesce into a single fracture
front. In the stress field of this primary front the flaw grows
(Fig. 5d) and turns away from the primary plane. Initially, its
growth is determined by the magnitude and direction of the primary
stress field. Whether or not it is able to grow in the static
stress field (i.e., the original plate stress) and form a secondary
depends on the size it attains while under the influence of the
primary fracture stress field.

It is not obvious that the tongue-shaped flaw shown in Fig. 4
will produce the feature shown in Fig. 5. To understand the difference, it should be noted that in order to separate a plate on the
primary fracture plane it is necessary to generate additional fracture surface. There are still connected regions in the vicinity of
the tongue in Fig. 5d. To separate the fragments on the primary,
it would be necessary to generate additional fracture extending the
sides of the tongue down to the primary and extending the primary
up to intersect the tongue.

Although the fracture surface in Fig. 4 is from a thermally
tempered soda lime glass plate, similar features have been found
on dice fragments from ion-exchanged Chemcor$^{(R)}$ glasses and from
fine-grained prestressed glass-ceramics. It appears that the generation of secondary fracture sources during the passage of the
primary fracture may be a general phenomenon in amorphous brittle
materials.

ACKNOWLEDGMENT

The research was jointly supported by the U. S. Atomic Energy
Commission and the U. S. Air Force. The author is grateful for the
contributions of his co-workers, including, in particular, G.
Treadwell, R. H. Altherr and H. P. Stephens.

REFERENCES

1. H. P. Stephens and E. K. Beauchamp, Sandia Laboratories Rep.
    SC-DR-72 0819 (1972).
2. DMS Market Intelligence Rep., Missiles/Spacecraft, Maverick,
    p. 2 (Nov. 1972).
3. R. Barnhart, Technology Week 20 (9), 16 (1967).
4. E. K. Beauchamp and R. H. Altherr, J. Am. Cer. Soc. 54, 103
    (1971).
5. G. H. Beall, Advances in Nucleation and Crystallization of
    Glasses, The American Ceramic Society, Columbus, Ohio (1971).

6. E. K. Beauchamp, Sandia Laboratories Rep. SLA-73 0233 (1973).
7. E. K. Beauchamp and R. H. Altherr, Sandia Laboratories Rep. SC-RR-67 526 (1967).
8. J. W. Johnson and D. G. Holloway, Phil. Mag. 14, 721 (1966).
9. J. Congleton and N. J. Petch, Phil. Mag. 16, 749 (1967).
10. P. Acloque, Proceedings of the Fourth International Glass Congress, Paris, France, p. 1, (1956).
11. E. K. Beauchamp, Sandia Laboratories Rep. SC-RR-70 766 (1970).
12. V. D. Frechette, Proc. Brit. Cer. Soc. 5, 97 (1965).

A FRACTURE MECHANICS STUDY OF THE SKYLAB WINDOWS

S. M. Wiederhorn, A. G. Evans and D. E. Roberts

Institute for Materials Research
National Bureau of Standards
Washington, D. C. 20234

ABSTRACT

Design criteria based on fracture mechanics concepts are
developed for spacecraft windows. Critical stress intensity
factor data and crack velocity data are used for lifetime
predictions and for the development of acceptance tests for the
eight candidate glass compositions for the Skylab. Design charts
are presented which give the minimum time to failure of the
Skylab windows as a function of service stress and proof test
stress. Surface adsorbed water is shown to be detrimental to the
strength of spacecraft windows, even after the spacecraft has left
the earth's atmosphere, because of the slow rate of evaporation
of water from surface cracks.

1. INTRODUCTION

Since glass is known to exhibit a time delay to failure in
tensile loading [1], an understanding of the long term tensile
strength of glass is considered by NASA to be essential for the
design of space-craft windows. As has been demonstrated in a
number of recent studies, this understanding can be obtained by
applying fracture mechanics to the candidate glasses for spacecraft windows [2,3]. Fracture mechanics can be used to predict
structural lifetime and to develop acceptance tests for glass
subjected to load. In this paper a summary is presented of
fracture mechanics data obtained for the Skylab module window
glass. A method of using these data for lifetime predictions is
illustrated by developing design charts that give the minimum
time to failure of the Skylab windows as a function of service

stress and proof test stress. This approach was used to support
the design of the Skylab module windows.

2. EXPERIMENTAL PROCEDURE

Fracture mechanics data for the design of spacecraft windows
were collected on the eight glass compositions given in Table 1.

TABLE 1. Glass Compositions, % by Weight

Glass	SiO_2	Al_2O_3	B_2O_3	Na_2O	K_2O	MgO	CaO	Others	Annealing Temp.
Fused Silica	99.9								1050°C
96% Silica I	96	0.3	3						910°C
96% Silica II	96.5	0.5	3						910°C
Aluminosilicate	57	15	5			7	10	6% BaO	710°C
Borosilicate	81	2	13	4					565°C
Borosilicate Crown I	69		11	10	7		0.2	2% BaO, 1% CeO	559°C
Borosilicate Crown II	70		11	10	7		0.2	2% BaO	563°C
61% Lead Glass	35			4				61% PbO	417°C

With the exception of the fused silica and 96% silica I and II,
all of the glasses were annealed for 15 minutes at the annealing
temperature and then slowly cooled to room temperature. After
annealing, crack velocity measurements were made on these glasses
as a function of stress intensity factor, both in air (100%
relative humidity) and in vacuum (10^{-5} Torr). In addition, critical
stress intensity factor measurements were made in vacuum and in
dry nitrogen gas (r.h. < 0.02%). The double cantilever beam (DCB)
technique and the edge-cracked, three point bending technique
were used to obtain critical stress intensity factor data. Tests
were conducted both with and without a 30 minute preheat at 300°C
to evaluate the effect of adsorbed water on the critical stress
intensity factor. The experimental techniques and the vacuum
equipment for these studies are fully described in references
3-4.

For crack velocity studies in 100% relative humidity, a dead
weight loading system was used to apply the load to the double
cantilever specimens. The equipment consisted of a pan balance
and a semi-enclosed environmental chamber that was maintained
at 100% relative humidity by water in the bottom of the chamber.
Velocity measurements ranged from 10^{-3} to 10^{-11} m/s. Specimens
were introduced cautiously into the environmental chamber to avoid

water condensation at the crack tip. Condensation always occurred
if the slides were slightly cooler than the water used to control
the environment. To avoid condensation, slides were introduced
into a chamber with an initial water temperature approximately
10 degrees less than room temperature. The specimen chamber
was then enclosed and measurements were made after the water had
reached room temperature. In this way, condensation was avoided
at the crack tip during the initial stages of the experiment.
Despite these precautions, crack tip condensation was unavoidable
for many of the glass compositions for cracks moving at low
velocities $<10^{-10}$ m/s (which required runs of the order of a
week to accomplish).

3. PRESENTATION OF RESULTS

3.1 Critical Stress Intensity Factor Measurements

The results of the critical stress intensity factor measurements in vacuum are given in Table 2. The number of measurements
for each determination is presented along with the average and
standard deviation of each determination.

TABLE 2. Critical Stress Intensity Factors, K_{IC}, $MN/m^{3/2}$

Glass	Vacuum, DCB No Preheat	Vacuum, DCB Preheat 300°C. for 30 Minutes	Vacuum, 3 Point Bend No Preheat	Dry N_2 Gas <0.02% RH
Silica	0.741 ± 0.025 (6)	0.729 ± 0.023 (3)	0.753 ± 0.030 (4)	0.758 (1)
96% Silica, I	0.700 ± 0.024 (4)	0.711 ± 0.009 (3)	0.709 ± 0.040 (3)	- -
96% Silica, II	0.715 ± 0.007 (3)	0.699 ± 0.012 (4)	0.746 ± 0.009 (4)	- -
Aluminosilicate	0.846 ± 0.023 (3)	0.874 ± 0.011 (4)	0.836 ± 0.032 (5)	- -
Borosilicate	0.760 ± 0.007 (6)	0.770 ± 0.012 (3)	0.777 ± 0.032 (5)	0.764 ± 0.008 (5)
Borosilicate Crown, I	0.862 ± 0.032 (4)	0.927 ± 0.010 (4)	0.842 ± 0.007 (4)	- -
Borosilicate Crown, II	0.886 ± 0.001 (3)	0.879 ± 0.034 (3)	0.904 ± 0.014 (3)	- -
61% Lead	0.624 ± 0.009 (3)	0.625 ± 0.008 (3)	0.643 ± 0.009 (3)	- -

NOTE: The uncertainty given in the table is the standard deviation; the number of specimens
for each determination is given in brackets.

The critical stress intensity factors of Table 2 ranged from
approximately 6.2 to 9.3 x 10^5 $N/m^{3/2}$, depending on glass composition, and were essentially independent of the prior heat
treatment. Within statistical scatter, the same values of
the critical stress intensity factor were obtained using the double
cantilever and the edge-cracked, three-point bend specimens.

Finally, the same values for the critical stress intensity factor
are obtained for borosilicate glass tested in vacuum and in dry
nitrogen, a result that conflicts with an earlier one by Linger
and Hollaway [5].

3.2 Crack Velocity Studies

Crack velocity data for crack propagation in air (100% r.h.)
are presented in figures 1 and 2. The data is plotted as the

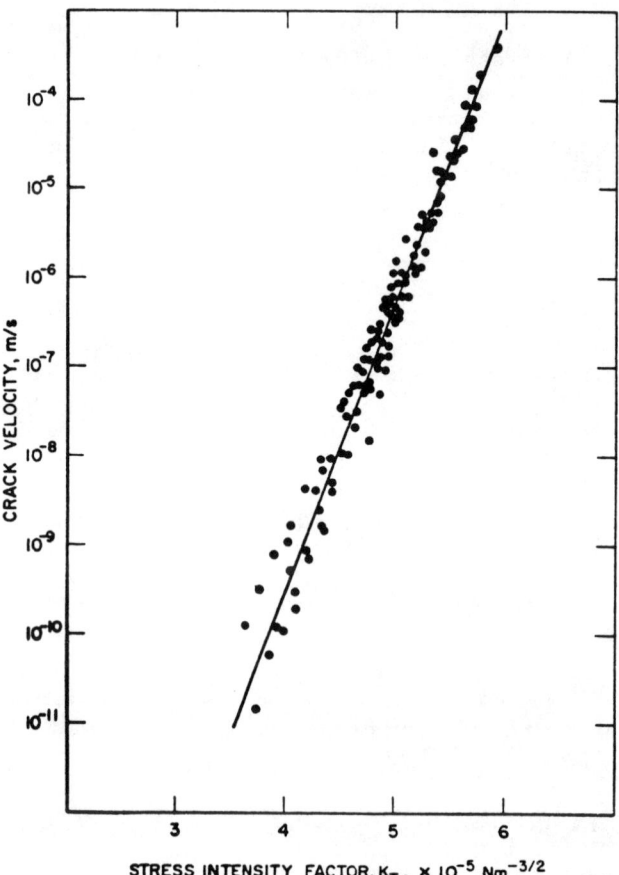

Figure 1. Crack propagation data for fused silica. Data
obtained in water-saturated air using the double cantilever
beam technique.

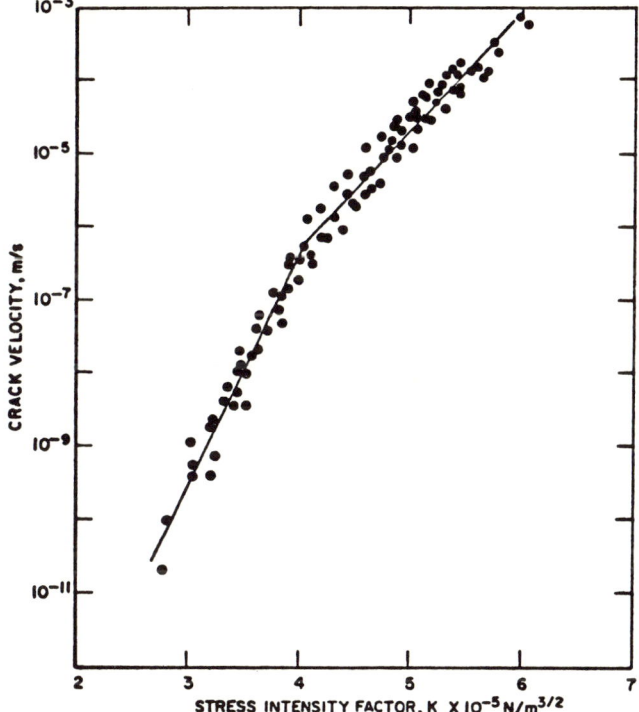

Figure 2.   Crack propagation data for borosilicate crown glass
II.   Data obtained in water-saturated air using the double
cantilever beam technique.

logarithm of the crack velocity versus the applied stress intensity
factor. Data plotted in this manner appear as straight lines
over the entire range of velocities for four of the eight glasses
investigated. For two of the glasses, some curvature occurs
at the low velocity portion of the curve, suggesting the existence
of a stress corrosion limit. For the borosilicate crown glasses,
curvature occurs over the entire range. The crack velocity data
were fitted by the method of least squares to the equation
K_I = a+bℓnv minimizing the error along the stress intensity factor
axis. The average slope and intercept of these data are presented
in Table 3. The data for the borosilicate crown glasses were
fitted to two linear components.

TABLE 3.   Least Squares Fit of Crack Velocity Data to
$K_I = a + b \ln v$; $MN/m^{3/2}$

Glass	a $MN/m^{3/2} \times 10^{-1}$	σ_a	b $MN/m^{3/2} \times 10^{-2}$	σ_b
Silica	6.931	0.034	1.342	0.022
96% Silica, I	6.490	0.048	1.388	0.029
96% Silica, II	6.507	0.034	1.298	0.022
High Alumina	7.950	0.041	1.797	0.027
Borosilicate	6.431	0.028	1.414	0.018
61% Lead	5.717	0.023	1.557	0.017
Borosilicate Crown I	7.946	0.13	2.53	0.12
	5.826	0.35	1.177	0.19
Borosilicate Crown II	7.872	0.11	2.674	0.107
	5.976	0.17	1.348	0.087

The data was fitted to two straight lines for the borosilicate crown
glasses. The upper set of numbers were obtained for $v > 10^{-6}$ m/s,
the lower set for $v < 10^{-7}$ m/s.

Finally, it is noted that crack velocity data taken from
crack tips containing condensate agreed within experimental error
with data obtained from specimens that appeared to have dry crack
tips. Thus, the presence of condensate at crack tips had little
influence on the positions or slopes of crack propagation curves
obtained in water-saturated air.

Data for crack propagation in vacuum are presented in figure 3. The data also appear as straight lines over the range of variables. However, the slopes of these curves are steeper than those obtained in water environments. The positions of the curves depend on the temperature of the study, the curves shifting to lower stress intensity factors as the temperature is increased. Attempts at slow crack propagation in vacuum were unsuccessful for the borosilicate glass, the 96 percent silica glasses, and the silica glass. Failure was always very sudden and slow crack propagation was not obtained. Crack propagation in vacuum and the effect of temperature on crack propagation is discussed in reference [4].

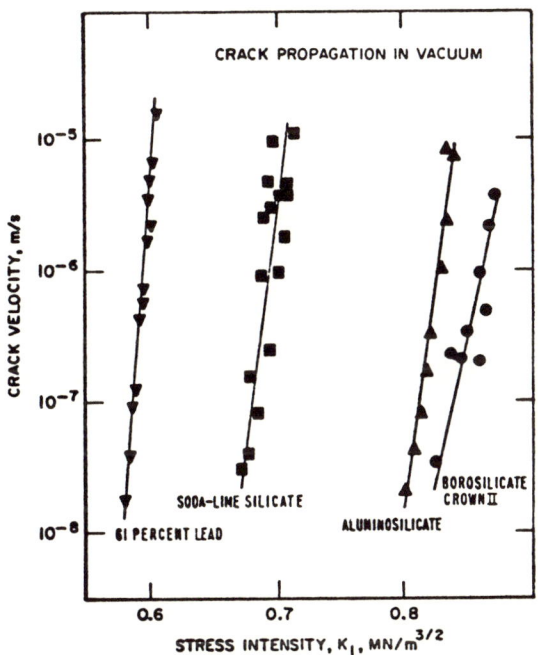

Figure 3. Crack propagation in a vacuum of 10^{-5} Torr. Data was obtained using the double cantilever cleavage technique. The data for the soda lime silicate glass was taken from reference [4].

3.3 Crack Tip Condensation

Some experiments were conducted to elucidate the characteristics of glass after condensate had formed at the crack tip.
Fused silica specimens with condensate at the crack tip were
transferred to an environmental chamber in which the percent
relative humidity could be controlled. The crack tip was observed
optically as the environment was changed from 100% relative
humidity to less than 0.02%. Condensate at the crack tip required
approximately an hour to evaporate, and a residue was left after
the condensate had evaporated. This residue left the portion of
the crack surface near the crack tip non-reflecting, suggesting
that it consisted of a silicate gel. Condensate reappeared after
introducing water-saturated nitrogen gas into the chamber. The
disappearance of the reflecting quality of the glass surface is
attributed to an increase in index of refraction of the gel as
water evaporated. The occurrence of condensate at crack tips and
the long period required for evaporation of the condensate suggest
that water will play a role in determining the fracture of glass
even after the spacecraft has left the earth's atmosphere. As
long as a condensate is at the crack tip, the crack will behave
as if it were exposed to a moist environment.

4. DISCUSSION OF RESULTS

The crack velocity data presented in this report are being
used to develop design criteria for spacecraft windows [6].
The data for fused silica has been used to develop acceptance
tests for the windows for the Skylab module. A method of estimating the time to failure, described earlier by Wiederhorn [7],
and Evans and Wiederhorn [8], is illustrated here for the eight
glasses studied. The method assumes that failure results from the
growth of preexisting cracks and that the kinetics of crack growth
can be obtained by fracture mechanics techniques of the type used
in this paper. The total time to failure, t, for a stressed
window containing a small surface crack is given by the following
equation [7-9]:

$$t = (2/\sigma_a^2 Y^2) \int_{K_{Ii}}^{K_{IC}} (K_I/v)\,dK_I \tag{1}$$

where σ_a is the applied stress, Y is a constant that depends on
initial crack geometry, K_{IC} is the critical stress intensity
factor, and K_{Ii} is the stress intensity at the most serious flaw
in the glass surface when the load is first applied to the windows.

Equation 1 can be evaluated numerically or analytically if K_I is known as a function of v and the result can be expressed as a function only of the initial stress intensity K_{Ii}. The initial stress intensity factor may be estimated by proof testing the glass windows.

Proof testing establishes an upper bound for maximum flaw size, a_i, present in the glass window at the time of the proof test [10]. If a glass window survives the proof test, we know that the stress intensity factor, K_{IP}, at the most serious flaw in the glass surface has not exceeded the critical stress intensity, K_{IC};

$$K_{IP} = \sigma_p Y \sqrt{a_i} < K_{IC} \qquad (2)$$

Between the time that the proof test is completed and the time at which the glass window is put into service, the most serious flaw does not change shape or size since the glass window is free of stress. When the glass is placed into service, therefore, $K_{Ii} = \sigma_a Y \sqrt{a_i}$. Substituting for $Y \sqrt{a_i}$ in equation (2), the following estimate of the initial stress intensity factor, K_{Ii}, is obtained from the proof test.

$$K_{Ii} \leq (\sigma_a / \sigma_p) K_{IC} \qquad (3)$$

If this value of K_{Ii} is substituted as the lower limit of the integral of equation (1), an estimate for the minimum time to failure, T_{min}, under an applied load σ_a is obtained. Thus, a prediction of the minimum time to failure of the spacecraft windows can be obtained from the fracture mechanics data summarized in Table III and proof testing.

This approach can be used to develop design diagrams which establish proof testing levels that guarantee the minimum survival time at a given load. By plotting log ($T_{min} \sigma_a^2$) vs log (σ_p / σ_a), a survival diagram for six of the glasses studied in this paper* is

---

*The other two, 96% silica II and borosilicate crown glass I, were similar in composition to two glasses presented in this diagram and so gave redundant data.

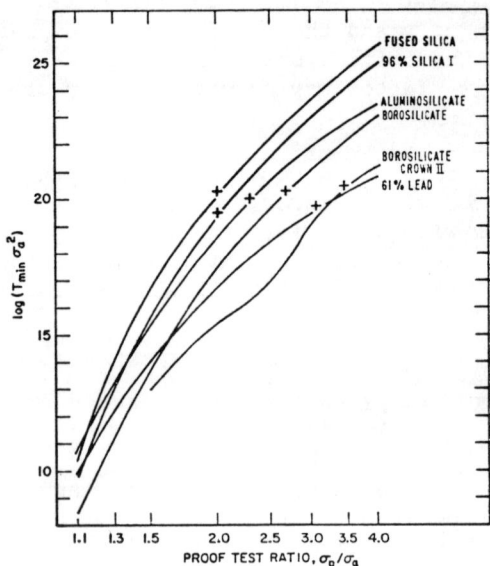

Figure 4. Proof test diagram showing six of the eight glasses
studied. The crosses indicate the limit of the crack growth
data. Curves are extrapolated to higher proof test ratios using
an empirical fit to the crack propagation data (see table III).

obtained (Figure 4). The discontinuity for the borosilicate
crown glass shown in figure 4 is due to the curvature of the K_I,V
data shown in figure 2. Figure 4 can be used to evaluate any one
of the design parameters, T_{min}, σ_a or σ_p/σ_a, when the other two
are known. For the design of spacecraft windows, the applied stress
level and the minimum permissible time to failure are known, and
hence the requisite σ_p/σ_a can be determined.

A second more descriptive type of design diagram may be obtained
from equation (1). If the logarithm of the minimum time to failure,
$\log(T_{min})$ is plotted as a function of the logarithm of the applied
stress, $\log(\sigma_a)$, then a straight line of slope minus 2 is obtained
for each value of σ_p/σ_a. Two such diagrams are illustrated in
figures 5 and 6 for the fused silica and the borosilicate crown II

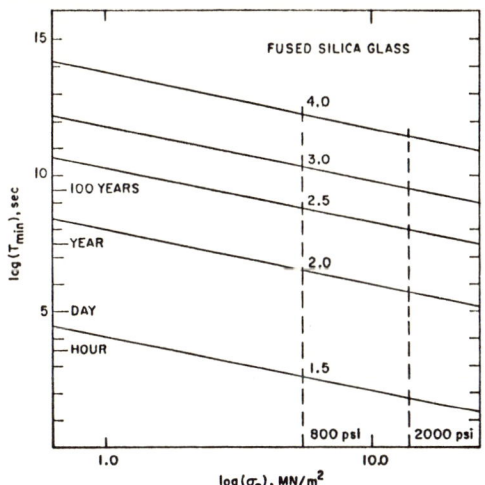

Figure 5. Proof test diagram for fused silica; minimum time to failure, T_{min}, given as a function of service stress, σ_a, and proof test ratio σ_p/σ_a.

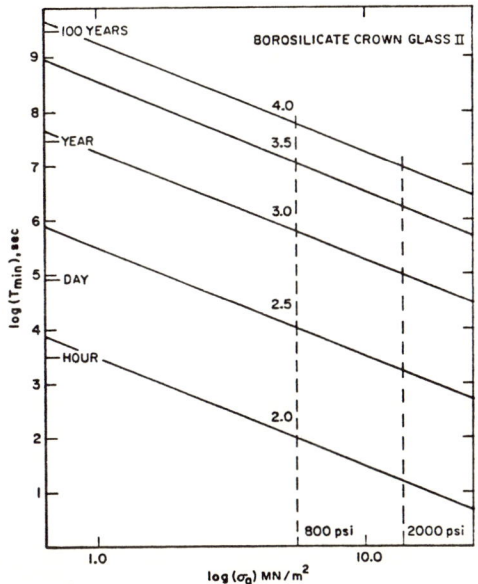

Figure 6. Proof test diagram for borosilicate crown glass II; minimum time to failure, T_{min}, given as a function of service stress, σ_a, and proof test ratio, σ_p/σ_a.

glass respectively. This type of diagram is related to the earlier
one since each point on the first type of diagram transforms into a
line on the second type. The second type of diagram has the
advantage of displaying each of the design parameters independently,
while the first diagram is more compact and can be used to present
more data.

The use of these diagrams can be illustrated for the case of
the fused silica used in the Skylab modules. The maximum stress
expected in the largest silica window is approximately 2,000 psi
($\sim$14MN/m^2). This stress is to be supported for 24 hours after blast-
off. Since the proof stress for this window is three times the
expected operating stress, a lifetime of 100 years in saturated
air at an operating stress of 2000 psi is estimated from figure 5.*
Similar considerations for the borosilicate crown glass give a
lifetime estimate of only one day, figure 6. This difference in
lifetime estimate demonstrates the enormous effect of glass composition on performance. After the first 24 hours, these estimates
of lifetime will be inordinately conservative because water will
have evaporated from the window, and the operating stress reduced
to 800 psi (5.6MN/m^2). Nevertheless, safe design practices dictate
that lifetime estimates be based on the most severe condition
expected, and this will occur during the first 24 hours of the
mission.

In conclusion, this paper describes a method of predicting
the lifetime of brittle structural materials that normally fail
by the growth of small preexisting cracks. The method uses
fracture mechanics data in combination with proof-testing for
lifetime predictions. Although this method has been applied to
the development of an acceptance test for the windows of the
Skylab module, it is applicable to any brittle material provided
failure occurs by crack growth. The method, therefore, provides
a rational basis for design with brittle materials.

ACKNOWLEDGMENT

The authors are grateful for the partial support of the
National Aeronautics and Space Administration in this work.

---

*We assume here that the crack tips in the silica glass contain
condensate, and that, because of the slow rate of water evaporation from these cracks, they behave as if they were exposed to
100 percent relative humidity air.

REFERENCES

1. R. J. Charles, pp. 1-38 in *Progress in Ceramic Science*,
    Vol. 1. Edited by J. E. Burke, Pergamon Press Inc., Elmsford,
    N. Y. (1961).

2. S. M. Wiederhorn, J. Am. Ceram. Soc. 50, 407 (1967).

3. S. M. Wiederhorn and L. H. Bolz, J. Am. Ceram. Soc. 53,
    543 (1970).

4. S. M. Wiederhorn, H. Johnson, A. M. Diness, and A. H. Heuer,
    to be published in J. Am. Ceram. Soc.

5. K. R. Linger and D. G. Hollaway, Phil. Mag. 18, 1269 (1963).

6. J. M. Davidson and V. Harrell, Private Communication.

7. S. M. Wiederhorn, J. Am. Ceram. Soc. 56, 227 (1973).

8. A. G. Evans and S. M. Wiederhorn, NBS report No. NBSIR 73-147,
    March, 1973.

9. A. G. Evans, J. Mat. Sci. 7, 1137 (1972).

10. C. F. Tifany and J. N. Masters, pp. 249-77 in *Fracture
    Toughness Testing and Its Applications*. ASTM Spec. Tech.
    Publ. No. 381. American Society for Testing and Materials,
    Philadelphia, Pa. (1965).

THE APPLICABILITY OF FRACTURE MECHANICS TECHNOLOGY TO PORCELAIN

CERAMICS

W. G. Clark, Jr. and W. A. Logsdon

Westinghouse Research Laboratories

Pittsburgh, Pennsylvania

ABSTRACT

An experimental investigation was conducted to evaluate the applicability of existing linear elastic fracture mechanics technology to multiphase electrical porcelain ceramics. The effect of test specimen geometry, crack length and notch radius on the nominal fracture toughness of electrical porcelain ceramic was evaluated and a K_{Ic} testing procedure developed. Results show that fracture mechanics technology can be used to develop realistic component design, material selection and nondestructive inspection criteria for ceramic materials. A hypothetical example problem is included and the application of fracture mechanics technology to a ceramic structure is demonstrated.

INTRODUCTION

Linear elastic fracture mechanics technology provides a unique quantitative approach to the evaluation of the effect of defects on the fracture properties of structural materials. The essence of the approach is to relate the magnitude of the localized elastic crack tip stresses required to cause failure to the applied nominal stress on the structure, the material properties, and the defect size.[1-3] Once the appropriate fracture mechanics parameters and material properties have been established for a given loading configuration, it is then possible to compute either the maximum allowable stress or defect size which will not result in failure of the structure. This information can then be directly incorporated into the development of realistic component design, material selection, and nondestructive inspection criteria.

The basic concepts underlying linear elastic fracture mechanics technology apply primarily to brittle plane strain loading conditions (limited crack tip plasticity). As a result, it would appear that this technology is directly applicable to the design of relatively brittle ceramic structures. However, to date, little progress has been made in applying the concepts of fracture mechanics to ceramic materials. The use of fracture mechanics in the area of ceramics has been limited primarily because of the experimental problems associated with generating the required material properties data. Consequently, sufficient data have not yet been developed which demonstrate that existing fracture mechanics technology applies to the design of ceramic materials. Because of the multiphase nature of ceramic materials, there are some questions as to whether fracture mechanics concepts apply. Specifically, fracture mechanics is essentially a contiuum mechanics concept and little work has been done to evaluate this technology as it applies to multiphase materials--especially where the defect sizes of concern are on the order of the same size as the individual phase particles.

Because of the increasing use of ceramic materials for critical structural applications, it is desirable to have available a design procedure which permits a realistic evaluation of the inherent brittle fracture properties of these materials. Existing fracture mechanics concepts appear to offer the technology required to develop a practical approach to the design and evaluation of ceramic structures. In view of the potential advantages of a fracture mechanics approach to design, a test program was under-taken to develop a practical experimental approach to the determination of fracture mechanics properties (K_{Ic} fracture toughness) for electrical porcelain ceramics and in turn, to evaluate the applicability of fracture mechanics technology to this multiphase ceramic material.

Three point bend bars and compact tension fracture toughness specimens were used to evaluate the effect of specimen geometry, crack length and notch radius on the nominal fracture toughness of electrical porcelain ceramic. In addition, a technique for controlled precracking of relatively large (1 in. thick) ceramic fracture toughness specimens was developed and room temperature K_{Ic} fracture toughness data generated for electrical porcelain ceramic.

These data are used to demonstrate the application of fracture mechanics technology to the design of a ceramic pressure vessel.

CONCLUSIONS

1. Existing linear elastic fracture mechanics technology is directly applicable to the evaluation of multiphase electrical porcelain ceramics.

2. Fracture toughness test specimen geometry and precrack length has no effect on the K_{Ic} fracture toughness of electrical porcelain ceramic.

3. The room temperature K_{Ic} fracture toughness of electrical porcelain ceramic is approximately 1.0 ksi$\sqrt{\text{in}}$.

4. The nominal toughness of electrical porcelain ceramic was found to be a function of defect acuity. Specifically, a linear relationship exists between nominal toughness measurements, K_Q and the square root of the notch radius, ρ. This relationship can be expressed as

$$K_Q = 0.952 + 4.47 \ (\rho)^{1/2}$$

5. A test specimen notch radius of less than 0.01 in. is required to determine the limiting value of toughness, K_{Ic}, for electrical porcelain ceramic.

6. A technique for controlled precracking of relatively large (one inch thick) electrical porcelain ceramic fracture toughness test specimens was developed.

7. The respective order of the firing and machining phases of test specimen manufacture has no influence on the K_{Ic} fracture toughness of electrical porcelain ceramic.

MATERIAL

The material evaluated in this investigation included 1 in. x 2 in. x 11 in. bars of wet process, electrical porcelain ceramic. The nominal composition (weight percent) of this material is presented below.*

Material	Weight Percent
Ball Clay	37
Kaolin	16
Feldspar	28
Flint (S_iO_2)	19

*A ceramic material of this or a similar composition is typically referred to as an electrical porcelain and no more specific designation is available.

The test bars were extruded and then subjected to a "cone 11" firing in a production tunnel kiln. The firing temperature was 2280°F. All bars were evaluated in the first fire, unglazed condition.

Typical room temperature mechanical properties are summarized below.[4]

Property

Apparent Density	0.087 lb/in.^{3} (2.41 gm/cm^3)
Flexural Strength (3 point loading)	11,300 psi
Charpy Impact Strength	1.30 ± 0.12 ft-lbs/in.^{2}
Modulus of Elasticity	8.5 x 10^6 psi

Figure 1 illustrates the typical microstructure of the electrical porcelain ceramic studied here.

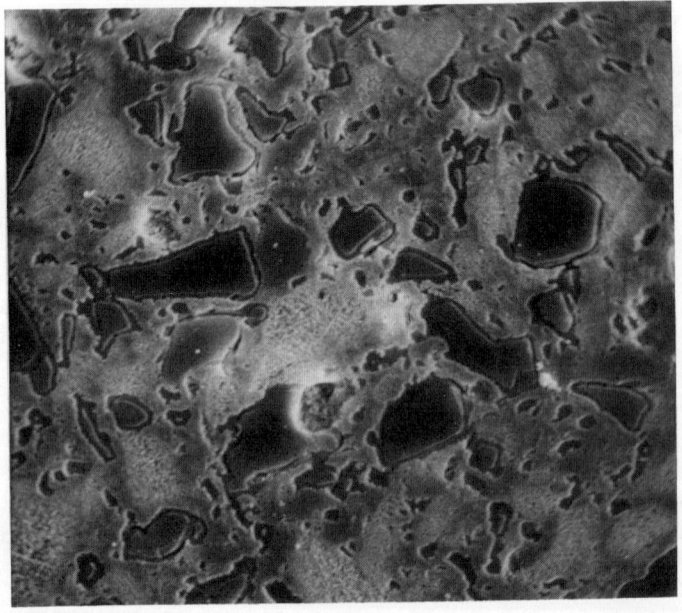

Fig. 1. An etched micrograph of electrical porcelain ceramic at a magnification of 760. (Quartz particles in a glassy matrix)

EXPERIMENTAL PROCEDURE

Three test specimen configurations were involved in this study. These specimens and their pertinent dimensions are shown in Fig. 2.

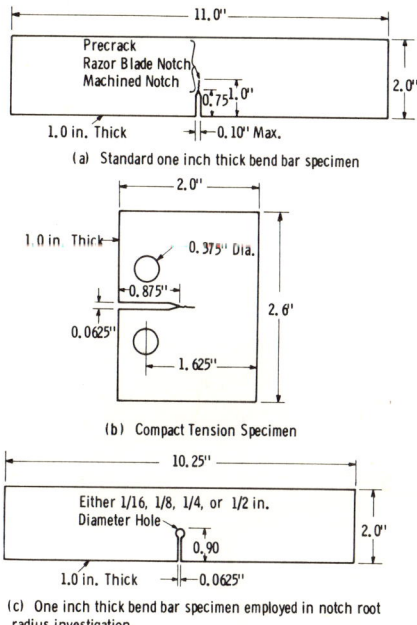

Fig. 2. Test specimens used to evaluate the fracture toughness of electrical porcelain ceramic.

The 1" x 2" x 11" bend bar specimen was used to evaluate the influence of notch preparation on subsequent toughness measurements and also to develop a controlled precracking technique for the brittle ceramic under investigation.

Several potential precracking techniques were evaluated. Attempts were made to precrack sharp notched bend bars by fatigue loading, thermal shock and wedge loading. Only the wedge loading procedure produced satisfactory results. This precracking technique involves driving a steel wedge into the machined slot until a crack develops and grows to the required length. The wedge is then removed immediately. Figure 3 shows the fixture used to precrack the ceramic specimens involved in this study. A torque of approximately 20 in.-lb was required to develop a precrack with this fixture.

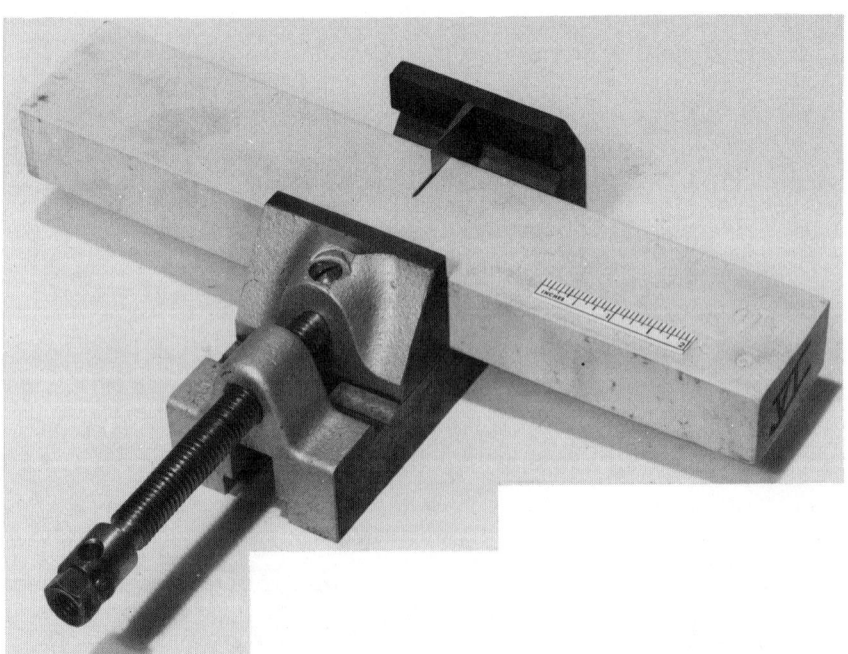

Fig. 3. Wedge fixture arrangement used to precrack ceramic bend bar and compact tension specimens.

Three point bend fracture toughness tests were conducted with precracked specimens, as machined sharp notch specimens ($\rho \approx 0.01$ in.) and notched specimens in which the machined notch was further sharpened by making a single pass across the notch root with a razor blade ($\rho \approx 0.001$ in.). The effect of notching the test specimens before and after firing was also evaluated. In those cases where the bars were notched prior to firing, extensive warping occurred as the result of firing and it was necessary to surface grind the specimens to get flat-parallel test surfaces. The notches were not re-machined.

The 1-inch thick compact tension specimens were used to evaluate the effect of precrack length on fracture toughness and also, when compared to the results of the bend bar tests, to evaluate the effect of specimen geometry. The compact tension specimens were made from the broken halves of the bend bars and precracked in exactly the same manner as the bend bars. Figure 4 shows three compact tension specimens precracked to various lengths. In order to clearly identify the extent of precracking, the specimens were subjected to a dye penetrant inspection prior to testing. The dye was allowed to dry for 1 week before testing in order to prevent any influence of the penetrant environment on the subsequent toughness values. Figure 5 shows the fracture appearance of the same specimens illustrated in Figure 4.

The blunt notched three point bend toughness specimens were used to extend the study of the effect of notch preparation and in particular, notch radius on subsequent fracture measurements.

All tests involved in this investigation were conducted at room temperature.

EXPERIMENTAL RESULTS

The fracture toughness test results generated with the precracked, razor blade notched and "as machined" notched bend bars and the precracked compact tension specimens are summarized in Table 1. The fracture toughness test results for the blunt notched bend bar specimens are summarized in Table 2.

The nominal fracture toughness values, K_Q, presented in Tables 1 and 2 were determined in accordance with the ASTM specification for the fracture toughness testing of metals E399-72.[5] Specifically, the K_Q values for the three point bend specimens and the compact tension specimens were computed from the respective stress intensity expressions given below.

3 pt Bend Bar Specimen
$$K_Q = \frac{3Y\,P_Q\,L\,\sqrt{a}}{BW^2}$$

Compact Tension Specimen
$$K_Q = \frac{Y\,P_Q\,\sqrt{a}}{BW}$$

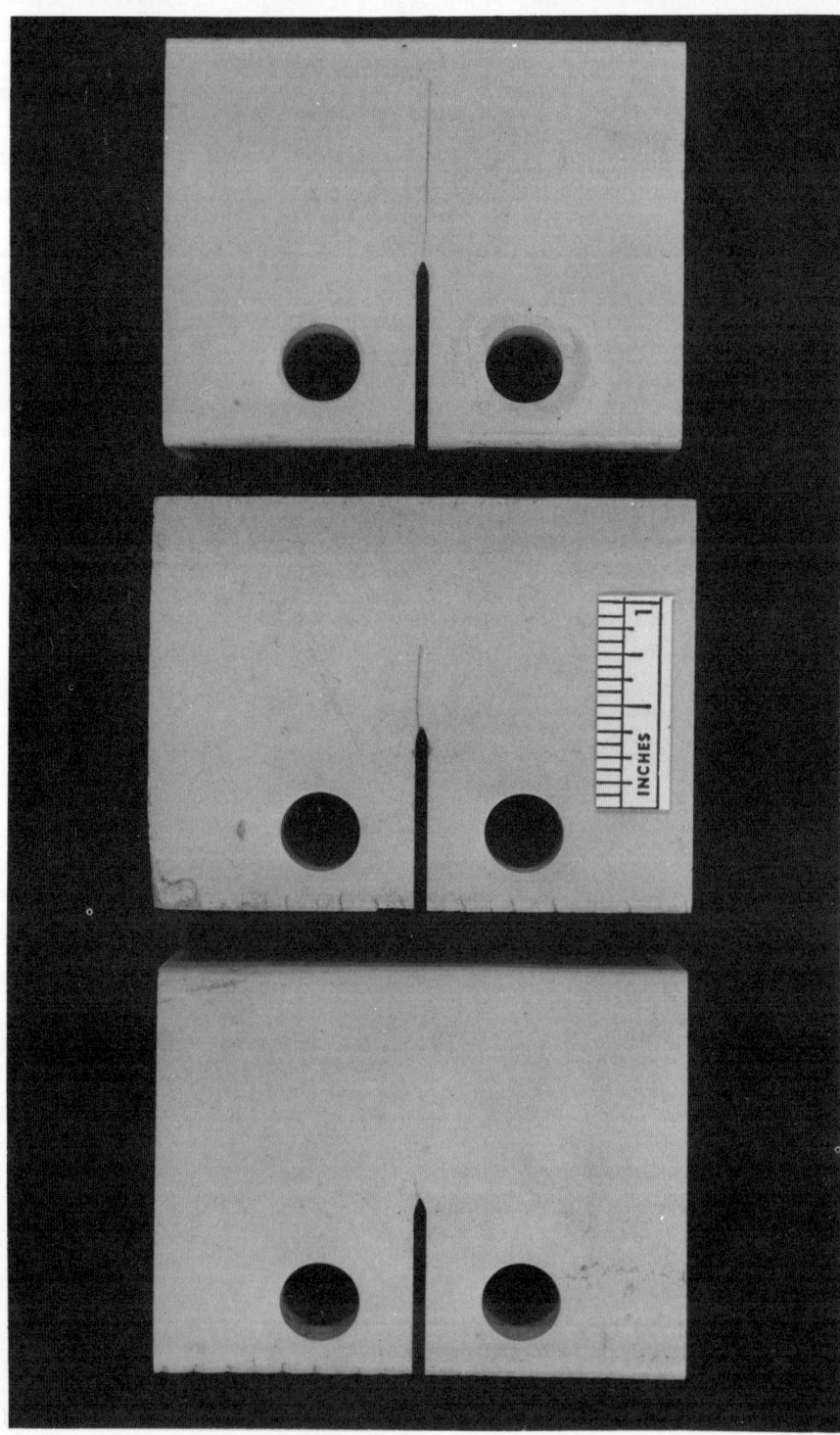

Fig. 4. Three precracked electrical porcelain ceramic compact tension specimens.

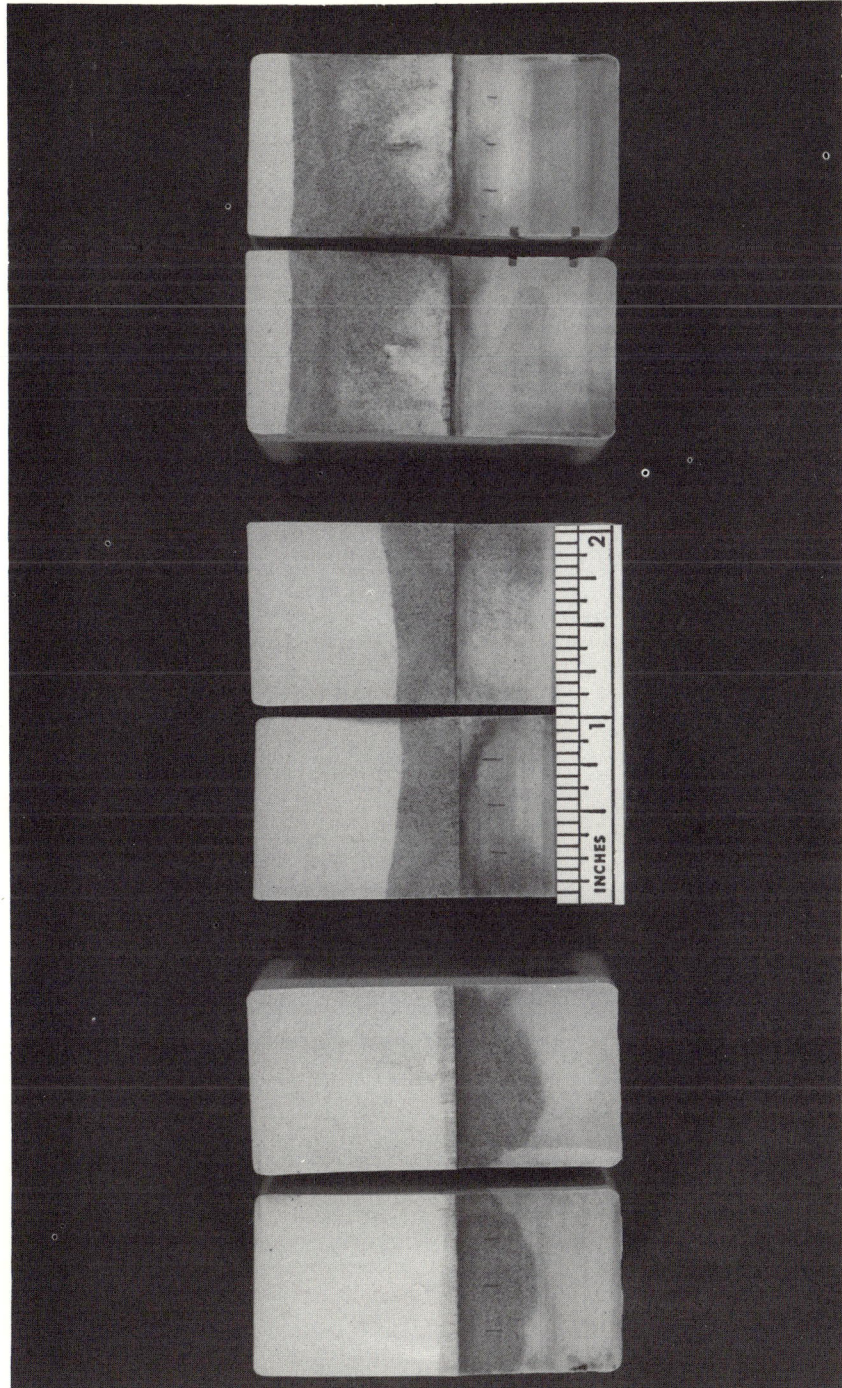

Fig. 5. Fracture surfaces of three compact tension specimens. The darker areas of the fracture surfaces represent the specimen precrack.

TABLE 1

ELECTRICAL PORCELAIN CERAMIC FRACTURE TOUGHNESS DATA

Specimen Identity	Specimen Type	Crack Length (in.)	K_Q* (ksi√in.)	Notch Description	Notch Root Radius (ρ) (in.)	Comments
1	Bend	0.78	1.02	Machined	ρ < 0.01	Specimens Machined, Then Fired, Warped,
2	Bend	0.80	1.02	Machined	ρ < 0.01	Specimens Face
3	Bend	0.84	1.09	Razor Blade	ρ = 0.001	Ground Again
4	Bend	0.80	1.04	Razor Blade	ρ = 0.001	
5	Bend	0.82	1.07	Razor Blade	ρ = 0.001	Specimens Fired,
7	Bend	0.80	1.10	Razor Blade	ρ = 0.001	Then Machined
8	Bend	0.80	1.03	Machined	ρ < 0.01	
9	Bend	0.80	0.94	Machined	ρ < 0.01	
II	Bend	1.11	1.14	Precracked	ρ ⋘ 0.001	
III	Bend	1.34	1.07	Precracked	ρ ⋘ 0.001	Specimens Fired,
VI	Bend	0.91	1.34	Machined	ρ = 0.01	Then Machined
VII	Bend	0.91	1.38	Machined	ρ = 0.01	
IX	Bend	0.91	1.29	Machined	ρ = 0.01	
2	Compact Tension	0.58	1.07	Precracked	ρ ⋘ 0.001	Specimens Fired,
3	Compact Tension	1.36	0.93	Precracked	ρ ⋘ 0.001	Then Machined
4	Compact Tension	0.84	0.93	Precracked	ρ ⋘ 0.001	

*When ρ < 0.01, $K_Q \approx K_{Ic}$.

TABLE 2

ELECTRICAL PORCELAIN CERAMIC FRACTURE TOUGHNESS DATA

Specimen Identity	Specimen Type	Crack Length (in.)	K_Q (ksi$\sqrt{\text{in.}}$)	Notch Description	Notch Root Radius (ρ) (in.)	Comments
5A	Bend	0.90	2.38	Machined	ρ = 0.125	
6A	Bend	0.90	2.92	Machined	ρ = 0.250	
7A	Bend	0.90	2.14	Machined	ρ = 0.125	
8A	Bend	0.90	2.10	Machined	ρ = 0.0625	
9A	Bend	0.90	1.98	Machined	ρ = 0.0625	
10A	Bend	0.90	1.92	Machined	ρ = 0.03125	
12A	Bend	0.90	2.20	Machined	ρ = 0.0625	
13A	Bend	0.90	3.02	Machined	ρ = 0.125	
14A	Bend	0.90	1.94	Machined	ρ = 0.03125	All Specimens
18A	Bend	0.90	3.45	Machined	ρ = 0.250	Fired, Then
19A	Bend	0.90	1.68	Machined	ρ = 0.03125	Machined
20A	Bend	0.90	2.10	Machined	ρ = 0.125	
21A	Bend	0.90	2.31	Machined	ρ = 0.0625	
22A	Bend	0.90	2.06	Machined	ρ = 0.03125	
23A	Bend	0.90	1.75	Machined	ρ = 0.03125	
24A	Bend	0.90	2.97	Machined	ρ = 0.125	
26A	Bend	0.90	3.24	Machined	ρ = 0.250	
27A	Bend	0.90	2.98	Machined	ρ = 0.250	
28A	Bend	0.90	1.94	Machined	ρ = 0.0625	

where P_Q is the applied load at failure determined from the load-deflection curve, W is the specimen width, B is the specimen thickness, "a" is the crack length (measured from the centerline of loading for the compact tension specimen), Y is a compliance constant dependent upon crack length and L is one-half the span employed in the bend bar test.[6]

In all the fracture toughness tests conducted in this study, the resulting load-deflection curves were esssentially straight lines to failure. Thus, the load corresponding to specimen failure was used for the value of P_Q. In those tests where the total crack length was long enough to yield an a/W ratio substantially greater than 0.6, a deep flaw K_I correction factor was used to compute the K_Q values.[7]

Examination of the fracture toughness data generated with the sharp notched 3 point bend specimens (Table 1) indicates that there is little influence of notch preparation on the K_Q value when the notch radius is less than 0.010 in. The precracked, razor blade notched, and sharpest machined notch specimens yield K_Q values ranging from 0.94 to 1.14 ksi$\sqrt{\text{in}}$. Note, however, for machine notched specimens with a notch radius of approximately 0.010 in., the values of K_Q were substantially higher, ranging from 1.29 to 1.38 ksi$\sqrt{\text{in}}$. These results indicate that in order to measure the lower limiting value of nominal toughness, K_{Ic}, for this ceramic material a notch radius of less than 0.01 in. is required.

The data presented in Table 1 also show that in the presence of a sharp notch ($\rho < 0.01$ in.) the crack length and test specimen geometry does not have a significant influence on the fracture toughness of electrical porcelain ceramic. Note that for the bend bar specimen, crack lengths ranging from 0.78 in. to 1.34 in. yield similar K_Q values. The absence of a crack length effect on K_Q is also apparent from the compact tension test results.

Comparison of the range of toughness values developed with the sharp notched ($\rho < 0.01$ in.) bend bars (K_Q = 0.94 to 1.14 ksi$\overline{\sqrt{\text{in}}}$.) clearly indicates that the toughness measurements are independent of test specimen geometry. Note also that the K_Q values generated in this study are not dependent on whether the bars were notched in the green state and then fired or notched after firing.

The toughness data generated with the blunt notched bend bars (Table 2) clearly demonstrate the effect of notch root radius on the nominal fracture toughness. Figure 6 presents the nominal fracture toughness, K_Q, as a function of notch root radius, ρ. The data from the sharp notched specimens (Table 1) are also included.

Figure 7 presents the average value of K_Q as well as the associated scatter band for various notch root radii as a function

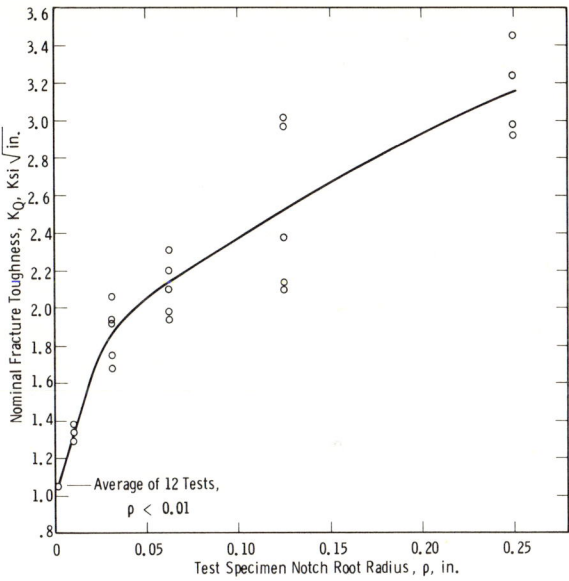

Fig. 6. Nominal fracture toughness versus ρ for electrical porcelain ceramic bend bar specimens.

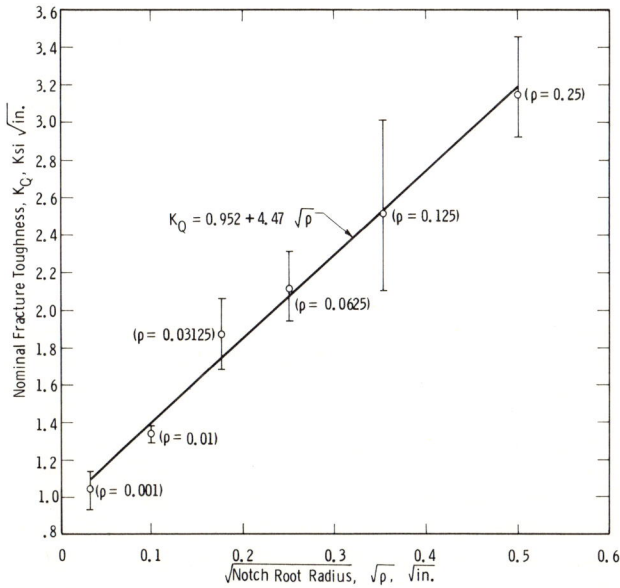

Fig. 7. Nominal fracture toughness versus ρ for electrical porcelain ceramic bend bar specimens.

of the square root of the specific notch root radius. Note that a
linear relationship exists between nominal toughness and the square
root of ρ. This linear relationship can be described by the
following expression:

$$K_Q = 0.952 + 4.47 \ (\rho)^{1/2}$$

DISCUSSION

A basic requirement concerning the applicability of linear
elastic fracture mechanics technology to a given material is that
brittle plane-strain (limited plasticity) conditions prevail up to
the point of failure. A related requirement is that within
prescribed limits, a material's inherent fracture toughness be
independent of structural geometry and crack length. Under these
conditions, the K_{Ic} concept provides an accurate failure criteria.
The results of this investigation clearly demonstrate that elastic
failure conditions prevail and test specimen geometry and crack
length do not effect the fracture toughness measurements for
electrical porcelain ceramic. In addition, the small amount of
scatter in the fracture toughness data generated with the precracked
or sharp notched ($\rho < 0.01$ in.) specimen indicates that the multiphase nature of this ceramic material (Table 1) has little
influence on the inherent fracture toughness. In view of these
observations, we can conclude that existing linear elastic fracture
mechanics technology is directly applicable to the design and
evaluation of electrical porcelain ceramic structures. The room
temperature K_{Ic} fracture toughness of electrical porcelain ceramic
is approximately 1.0 ksi$\sqrt{\text{in}}$.

The effect of notch radius on the nominal fracture toughness
as illustrated in Figs. 6 and 7 is somewhat similar to data presented
previously for high strength steels.[8] One significant difference,
however, is the fact that for high strength steels, the effect of
notch root radius on apparent toughness can be much more significant
than that observed for electrical porcelain ceramic. For the
ceramic, increasing the notch root radius from a natural precrack
to a 0.01 in. radius produced a change in K_Q from about 1.0 ksi$\sqrt{\text{in}}$.
to 1.3 ksi$\sqrt{\text{in}}$. A similar change in notch root radius for an H-11
steel has been shown to increase the K_Q values by a factor of more
than 4.[8]

In view of the strong dependence of toughness measurements on
the notch root radius that can be encountered for metals, it has
been necessary to develop a specific fatigue precracking technique
for the K_{Ic} testing of metals.[5] However, based on the results of
this study, it appears that a precracked test specimen may not be

necessary for the determination of K_{Ic} values for brittle ceramic
materials. Instead, it may be possible to conduct fracture
toughness tests with sharp machine notched specimens and to
determine a meaningful limiting value of K_{Ic} by extrapolating the
results to a natural crack on the basis of a curve such as that
shown in Fig. 7. Obviously, this technique for computing K_{Ic}
values must be explored further, particularly for other ceramic
materials, however, in view of the difficulty associated with
controlled precracking of brittle materials any procedure which
would eliminate the need for precracking is worth exploring.

The wedge loading procedure used in this study to precrack
ceramic fracture toughness specimens is a delicate operation. The
force required to precrack the specimen is just slightly lower than
that required to cause failure. As a result, the yield of the
precracking procedure is only about 50 percent. In addition, the
positioning of the steel wedge is critical since a small misalignment
produces an angular crack front which is difficult to analyze.
Unlike a fatigue precracked metal toughness specimen, the extent
of the precrack developed in a ceramic material cannot be
distinguished from the fast fracture portion of the failure.
Therefore, it is necessary to mark the extent of the precrack prior
to toughness testing so that it can be accurately measured after
the test. The dye penetrant procedure used in this study to mark
the extent of the precrack appears satisfactory, however, the dye
must be allowed to dry thoroughly to prevent any possible influence
of the dye itself on the subsequent fracture behavior.

EXAMPLE PROBLEM

In order to demonstrate the use of fracture mechanics concepts
as they apply to a ceramic structure, a hypothetical problem
involving a ceramic pressure vessel is presented below. The
pertinent considerations and computations are described in detail.

Suppose we are required to establish the fracture potential as
well as meaningful nondestructive inspection requirements for an
electrical porcelain ceramic pressure vessel. Let us assume the
pressure vessel is a 20 in. outside diameter cylinder with a 2 in.
wall thickness and that the maximum internal pressure likely to be
encountered in service is 400 psig. In addition, let us assume all
loading takes place at room temperature in a nonhostile environment.

To attack this problem from the fracture mechanics point of
view, we must first establish the maximum nominal stresses acting
on the structure. For the purposes of this example, we will limit
our considerations to the cylindrical body of the pressure vessel.
The largest stresses on the component would be the circumferential

hoop stresses. If the outside diameter of the vessel is subjected
to atmospheric pressure, the circumferential hoop stresses acting
on the inside diameter of the pressure vessel can be computed in
accordance with conventional thick-wall pressure vessel design
concepts as:

$$\sigma = \frac{P_i(r_o^2 + r_i^2)}{r_o^2 - r_i^2}$$

where σ is the applied nominal circumferential hoop stress, P_i is
the internal pressure, r_i is the inside radius and r_o is the
outside radius. Solving the above equation for σ when $P_i = 400$ psi,
$r_i = 8$ in. and $r_o = 10$ in. yields a maximum circumferential hoop
stress of 1822 psi.

Next, we must obtain the pertinent fracture mechanics material
properties representative of the material and loading condition of
interest. Recall, the lower limiting value of nominal toughness,
K_{Ic}, for an electrical porcelain ceramic material equals 1.0 ksi$\sqrt{in}$.
Therefore, this value of fracture toughness will be employed in our
fracture mechanics analysis. In addition, electrical porcelain
ceramic does not appear to fatigue. Consequently, no fatigue crack
growth rate data is available for this material. Therefore, the
primary aim of our fracture mechanics analysis will be to utilize
our K_{Ic} fracture toughness data to calculate the critical flaw
size required to cause failure of the pressure vessel in one cycle
of loading.

The last piece of information required for our fracture
mechanics analysis is an appropriate stress intensity expression
which relates applied stress, material properties, and defect size
to the combination of structural configuration, flaw orientation
and flaw geometry of interest. In order to select the appropriate
stress intensity expression, we must first estimate the most critical
type defect likely to be encountered in the pressure vessel.
Surface defects are considered more severe than internal defects.[9]
Therefore, we will limit our analysis to surface cracks located
on the inside diameter (region of maximum circumferential hoop
stress) of the electrical porcelain ceramic pressure vessel.
Figure 8 presents a schematic illustration of the structural
configuration and flaw orientation being considered. The pertinent
stress intensity expression for this configuration (a surface
crack oriented with the major plane of the crack perpendicular to
a uniform tension stress field) is

$$K_I^2 = \frac{1.21 \, a \, \pi \, \sigma^2}{Q} \tag{1}$$

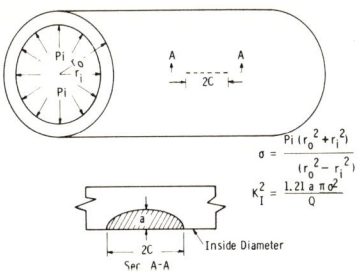

Fig. 8. Structural configuration and flaw geometry for example problem.

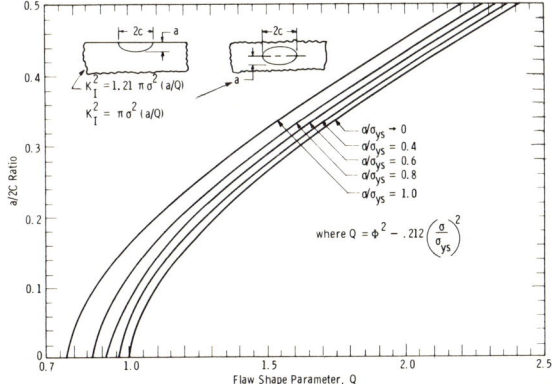

Fig. 9. Flaw shape parameter curves for surface and internal cracks.

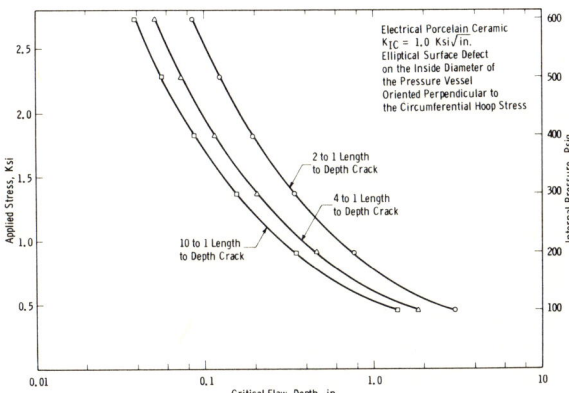

Fig. 10. Critical flaw size analysis for electrical porcelain ceramic pressure vessel.

where K_I = nominal stress intensity factor, $psi\sqrt{in.}$
     a = surface crack depth, in.
     σ = applied nominal stress, psi
     Q = flaw shape parameter (from Fig. 9)

The flaw shape parameter (Q) permits us to evaluate the severity of defects with different shapes by accounting for surface or internal defects with various length to depth ratios. To cover the range from long shallow flaws to short deep flaws, we will consider surface cracks with three length to depth ratios; 10 to 1, 4 to 1, and 2 to 1, where the corresponding flaw shape parameters (Q) are 1.10, 1.45 and 2.42, respectively (from Fig. 9).

We now have all the information required to calculate the critical flaw sizes necessary to cause failure of the electrical porcelain ceramic pressure vessel in one cycle of loading. If we rearrange the terms in Eq. (1) and set K_I equal to K_{Ic} and "a" equal to the critical flaw size, a_{cr}, we obtain the following critical flaw size expression

$$a_{cr} = \frac{K_{Ic}^2 \, Q}{1.21 \, \pi \, \sigma^2} \qquad\qquad (2)$$

By substituting the appropriate values of K_{Ic} and Q into this expression, we can readily solve for the critical flaw sizes (a_{cr}) which correspond to various applied stresses (σ). Figure 10 illustrates critical flaw size curves for each of the three surface defect geometries considered in our fracture mechanics analysis over the stress range of 456 psi to 2733 psi.

Based on our maximum expected circumferential hoop stress of 1822 psi, the critical flaw sizes corresponding to the three surface defect geometries are presented below.

Flaw Length to Depth Ratio	Q	Critical Flaw Size Depth	Length
10 to 1	1.10	0.087 in.	0.87 in.
4 to 1	1.45	0.115 in.	0.46 in.
2 to 1	2.42	0.192 in.	0.38 in.

These are the critical flaw sizes of a given geometry necessary to cause failure of the pressure vessel in one cycle of loading when the pressure vessel is subjected to its maximum anticipated internal pressure of 400 psi, corresponding to a maximum circumferential hoop stress of 1822 psi. Obviously, associated nondestructive inspection requirements must insure that no defects larger than these particular flaw sizes are present in the electrical porcelain ceramic pressure vessel.

REFERENCES

1. Irwin, G. R., "Fracture Mechanics," Structural Mechanics, Pergamon Press, New York and London, 1960.

2. Irwin, G. R., Kraft, J. M., Paris, P. C. and Wells, A. A., "Basic Aspects of Crack Growth and Fracture," U. S. Naval Research Laboratories, Report 6598, November 1967.

3. "Fracture Toughness Testing and Its Applications," ASTM STP 381, April 1965.

4. Daniels, W. H., Unpublished Westinghouse data.

5. "Standard Method of Test for Plane-Strain Fracture Toughness of Metallic Materials," E399-72, ASTM Standards, Part 31, July 1972.

6. Brown, W. F., Jr. and Srawley, J. E., "Plane Strain Crack Toughness Testing of High Strength Metallic Materials," ASTM STP 410, January 1967.

7. Wilson, W. K., "Stress Intensity Factors for Deep Cracks in Bending and Compact Tension Specimens," Westinghouse Scientific Paper 70-1E7-FMPWR-P4, June 1970.

8. ASTM Special Committee, 3rd Report, "Fracture Testing of High-Strength Sheet Materials," Materials Research and Standards," November 1961, p 877.

9. Clark, W. G., Jr., "Fracture Mechanics and Nondestructive Testing of Brittle Materials," Journal of Engineering for Industry, February 1972, pp. 291-298.

APPLICATION OF FRACTURE MECHANICS TO GRAPHITE UNDER COMPLEX

STRESS CONDITIONS*

G. T. Yahr and R. S. Valachovic[†]

Oak Ridge National Laboratory

Oak Ridge, Tennessee 37830

ABSTRACT

The purpose of this study was to examine the applicability of
linear-elastic fracture mechanics to graphite under multiaxial
stress conditions. Earlier studies have demonstrated the applicability of fracture mechanics to graphite under a limited range of
biaxial states of stress where one principal stress was tensile and
the other was compressive.[1] In this paper, the study was extended
to consider other combined stress states.

The specimens used in the present study were thick-walled
graphite cylinders with flat heads which were internally pressurized. Two series of specimens were used. The first series had
complete circumferential notches machined diagonally into the head-
cylinder juncture region, while the second series was unnotched.

The methods of linear-elastic fracture mechanics and a finite-
element analysis were used to predict pressures to cause fracture
for both notched and unnotched specimens. In the latter case, it
was necessary to postulate a "size" for a naturally occurring flaw.
These predicted pressures were in good agreement with measured

*Research sponsored by the National Aeronautics and Space
Administration under interagency agreement 40-182-69, NASA-SNSO-C
Order SNC-79, and performed at the Oak Ridge National Laboratory
operated by Union Carbide Corporation for the U.S. Atomic Energy
Commission.

†Present address is ACF Industries, Inc., St. Charles,
Missouri 63301.

results for the vessels with machined notches. The unnotched vessels, however, withstood a greater pressure than was predicted, but
the agreement is considered good because the flaw sizes used were
inferred from tests on other specimens and hence were very approximate. The results from this study indicate that fracture mechanics
can be a valuable technique for predicting fracture of graphite
under complex stress states, with the method being especially useful where regions of high stress concentration exist.

INTRODUCTION

We begin with a brief review of our prior studies in the series on linear-elastic fracture mechanics. We measured the fracture toughness[2] of ATJ* and AXM† graphites with several different
specimen types including notched beams, compact tension specimens,
double cantilever beams, and circumferentially notched tensile
specimens. From these tests, it was shown that fracture-mechanics
concepts can be applied to these specimens with gross flaws, and
specimens and methods for use in determining fracture toughness
values were identified. At the same time, extension of fracture-
mechanics concepts to graphite components where the flaws are the
relatively small naturally occurring "flaws" was examined with
positive results.

To illustrate the latter, beams without machined notches, but
which were otherwise identical to the notched beams used for fracture toughness determinations, were tested. The effective crack
depths were calculated, based on the bending moment at fracture and
the K_{Ic} values determined from the beams with notches. The calculated crack depths for naturally occurring "flaws" in ATJ and AXM
graphites were 0.010 in. and 0.003 in. respectively. These depths
are compatible with dimensions of observed microstructural features
in both materials.

Applicability of fracture-mechanics concepts and methods for
predicting fracture of graphite under complex stress conditions was
then studied through tests on splitting tensile specimens of AXM
and ATJ graphites.[1] The splitting tensile specimens were cylindrical or elliptical disks that were loaded in compression along the

---

*ATJ graphite is a molded graphite that has a maximum grain
size of 0.006 in. It has been produced by Union Carbide Corporation, Carbon Products Division, for a number of years, and its mechanical behavior has been well characterized. It was supplied in
14-in.-diam by 15-in.-long billets.

†AXM graphite is an extremely fine grain, isotropic graphite
manufactured by POCO Graphite, Inc. This material was supplied in
2-in.-diam by 12-in.-long cylinders.

major axis. Nearly uniform biaxial stress states exist in the central portions of circular or elliptical disks loaded by lateral compressive loads. The maximum tensile stress occurs normal to the loading direction at the center of the disk and is the maximum principal stress at that point. The minimum principal stress at the center is compressive. The ratio of the absolute values of the principal stresses is dependent on the shape of the disk; it ranged from 3.0 to 6.9 in the specimens tested. Slots in the central region that were aligned with the loading direction caused premature failures that were predicted very accurately using fracture-mechanics methods.

Since these tests demonstrated that fracture-mechanics methods can be used for predicting the fracture of graphite under biaxial stress conditions where one principal stress is tensile and the other is compressive, other combined stress states were considered. This paper describes results from an investigation of fracture under the complex stress conditions that exist in an internally pressurized thick-walled cylindrical vessel with a flat head.

SPECIMENS AND TEST APPARATUS

Thirty thick-walled vessel specimens were tested: fifteen each of AXM and ATJ graphites. The longitudinal axes of the ATJ vessels were oriented in the across-grain direction. The configuration of the thick-walled vessels is shown in Fig. 1. Details of

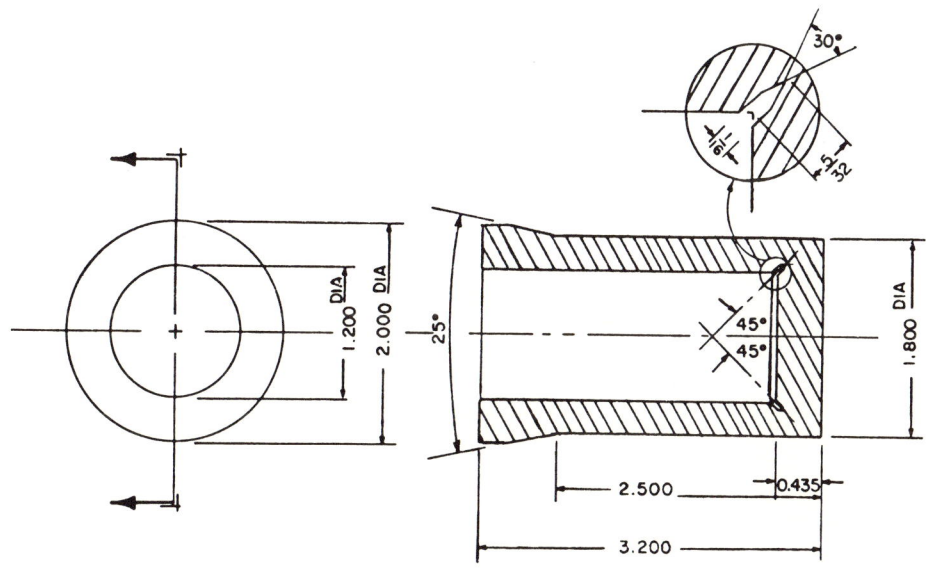

Fig. 1. Thick-walled vessel with notch.

the circumferential notch are shown in the inset. Some of the
specimens were unnotched; in these cases there was a square corner
at the intersection of the head and the cylindrical wall.

The rig for testing the vessels is illustrated in Fig. 2.
Hydraulic oil was used to pressurize the specimen, and a latex rubber liner prevented the oil from entering the pores in the graphite
and also acted as a gasket to seal the vessel.

THEORETICAL ANALYSIS

Finite-Element Analysis

The finite-element method[3,4] was used for a theoretical analysis of the thick-walled vessel specimens. Since the vessels are
axisymmetric, a two-dimensional analysis is sufficient for determining the stresses in each specimen. The axisymmetric finite-
element idealization constructed for the unnotched thick-walled
vessels consisted of 784 nodal points and 716 discrete elements;
the finite-element idealization of the notched vessel was similar.
An internal pressure of 100 psi on the entire inner surface of the
vessel was assumed in the calculations. In the notched vessels,
the pressure was considered to extend into the notch but not to act
at the very tip. Even though silicone mold-release grease was put
into the notch to transfer the pressure, it was felt that such
transfer would not be effective all the way to the tip because the
grease has a high viscosity. Analyses showed an appreciable

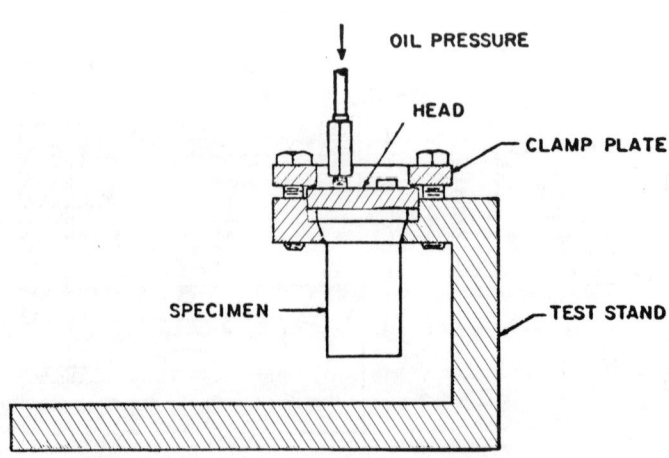

Fig. 2. Test rig for thick-walled vessels.

difference in results, depending on whether or not there was pressure in the notch.

The uniaxial tensile properties of the two graphites were determined for use in the analyses. Neither of the two graphites exhibited any purely elastic behavior. The stress-strain curves were all nonlinear, even at extremely low stress levels, and the strain ratios* decreased as the stress increased. Since the method used for analyzing the thick-walled vessel specimens was based on linear elasticity, a linear approximation to the stress-strain curve and a constant value of strain ratio were required. The secant modulus and strain ratio at a selected stress level were chosen for use in the analyses. In this way, the nonlinear behavior of the graphite was accounted for, in an approximate fashion, so that linear-elastic analyses could be used. These values are a better approximation of the nonlinear behavior than the tangent modulus and initial strain ratios.

The average strength of three AXM tensile specimens was 7300 psi, and their average strain-to-failure was 0.57%. The secant modulus at a stress of 1000 psi was 1.60×10^6 psi, and the strain ratio at 1000 psi was 0.27. The strength, strain-to-failure, and strain ratio of AXM graphite are all markedly high.

Since ATJ graphite is transversely isotropic, it was necessary to conduct two sets of tensile tests on this material to determine the response to with-grain and across-grain loading. Three across-grain ATJ tensile specimens gave average strength and strain-to-failure values of 2400 psi and 0.41%, respectively. The average secant modulus and average ratio of transverse strain to axial strain at a stress of 500 psi were 0.83×10^6 psi and 0.08, respectively. Four with-grain specimens of ATJ graphite were tested. The average strength was 3500 psi and the average strain-to-failure was 0.36%. The average secant modulus at a stress of 500 psi was 1.39×10^6 psi. The transverse strain varies around the circumference of a with-grain ATJ graphite specimen because of anisotropy. Therefore the transverse strain was measured in both the with-grain and across-grain direction. The ratio of transverse strain in the across-grain direction to axial strain was 0.14 at 500 psi stress. The ratio of transverse strain in the with-grain direction to axial strain was 0.09 at that same stress.

The properties determined from the tensile tests were used in the finite-element analyses of the AXM and ATJ thick-walled vessels. The analyses showed that the maximum stress in the unnotched vessels was at the intersection of the head and wall. For the notched vessels, it was at the tip of the notch.

---

*Strain ratio is the negative of the ratio of the strain normal to the direction of loading to the strain in the loading direction.

The stress state in the element with the largest principal
stress, along with the orientation of the maximum principal stress,
is given in Table 1 for each of the thick-walled pressure vessels.
The maximum principal stress is always in the meridional plane; the
angle given is the angle between the direction of the maximum principal stress and the longitudinal axis of the cylinder. Because of
the large gradient in stress through this area of the vessel's cross
section, the calculated stress values are strongly dependent on the
sizes of the discrete elements. But the calculated values do give
an indication of the triaxiality of the stress state at the point
of fracture initiation. The elements at the intersection were made
smaller for the AXM graphite than for the ATJ graphite because of
the smaller natural flaw size for AXM graphite. Therefore the calculated stresses in the unnotched AXM vessel were higher than in
the unnotched ATJ vessels because the element size was smaller in
the former.

Fracture Mechanics

The procedure used for determining the relationship among
strain-energy release rate, crack length, and pressure was developed by Watwood[5] and by Mowbray.[6] It is basically an analog to the
compliance method of experimentally determining values of the
strain-energy release rate at onset of rapid crack propagation.
The finite-element method was used to compute the strain energy
stored in several vessels which are identical except for differences
in crack lengths. This was done by sequentially removing two elements at the notch tip or wall-to-head intersection. With the
strain energy U known as a function of crack area A, Irwin's[7] definition of the strain-energy release rate

Table 1. Stress states at point of maximum stress
in thick-walled pressure vessels

Graphite	Notched	Maximum meridional stress (psi)	Minimum meridional stress (psi)	Orientation of maximum meridional stress[a] (deg)	Circum- ferential stress (psi)
AXM	No	1328	241	38	443
	Yes	1287	690	25	513
ATJ	No	690	99	44	73
	Yes	1232	529	34	105

[a]The angle between the direction of the maximum principal
stress and the longitudinal axis of the cylinder.

$$G = -\partial U/\partial A \qquad (1)$$

was used to find the strain-energy release rate for a particular
vessel with a given crack length. Since the strain energy in an
elastic body is proportional to the square of the pressure p, the
relation between the pressure and the strain-energy release rate is

$$p = \sqrt{G/F} , \qquad (2)$$

where F is a constant of proportionality for a particular graphite
and crack length that is found from the finite-element analyses.
The values of F for the notch depths of interest are given in Table
2. A value of F was determined for a notch the depth of the natural flaw size for each graphite as well as for the machined notch.

These values may be used to predict the pressures necessary
to break the vessels, if the value of G_{Ic} for the material is known.
The values of G_{Ic} are listed in Table 3.

Table 2.   Calculated values of F for
thick-walled pressure vessels

Graphite	Notch depth (in.)	F (in.^{3}/lb)
AXM	0.003	0.83×10^{-6}
	0.156	1.99×10^{-6}
ATJ	0.010	1.55×10^{-6}
	0.156	2.94×10^{-6}

Table 3. G_{Ic} values for ATJ and AXM

Graphite	Loading direction	Cracking direction	G_{Ic} (in.-lb/in.^{2})
AXM	(Isotropic)	(Isotropic)	0.81
ATJ	With grain	Across grain	0.48
	Across grain	With grain	0.57

At this point we were confronted with a difficulty in applying fracture mechanics to the ATJ vessels. Since the cracks are not aligned with the principal material axes, we did not know the value of G_{Ic} to use to predict the pressure that will fracture the vessels. In these studies, we assumed that the values of G_{Ic} for cracks aligned with the principal axes of the material are the maximum and minimum values of G_{Ic} and that the value of G_{Ic} varies smoothly between the two extremes at intermediate orientations. Although data do not exist for establishing the validity of this assumption, it was used here to provide a comparison between calculated and experimental results.

The form of the variation was assumed to be

$$G_{Ic}^{\theta} = G_{Ic}^{r} - (G_{Ic}^{r} - G_{Ic}^{z}) \sin \theta , \qquad (3)$$

where θ is the angle that the crack makes with the axis of the cylinder, and G_{Ic}^{z} and G_{Ic}^{r} are the values of G_{Ic} in the direction of the axis of the vessel and in the radial direction respectively. G_{Ic}^{r} is the value for with-grain loading. The value to be used for θ is the complement of the orientation of the maximum meridional stress given in Table 1. The calculated values for G_{Ic}^{θ} are 0.54 in.-lb/in.^{2} for the unnotched ATJ vessels and 0.55 in.-lb/in.^{2} for the notched vessels.

TESTS

AXM Graphite

Fracture data for the AXM thick-walled vessels are given in Table 4. The predicted failure-pressure values were calculated by substituting the appropriate F values from Table 2 and a G_{Ic} value of 0.81 in.-lb/in.^{2} into Eq. (2). All of the unnotched vessels failed at higher pressures than predicted. The average maximum pressure is 21% higher than predicted. In the case of the notched vessels, the predicted failure pressure is in good agreement with the experimental results. The average maximum pressure is only 3% higher than predicted.

Strain gages were mounted in the center of the head and on the outside of the wall, 17/32 in. from the head and oriented in the axial direction, to monitor crack initiation. These gages did not give any indication of crack growth prior to complete fracture in the unnotched AXM vessels. The gages did indicate that the cracks grew some in the notched AXM vessels prior to complete fracture.

Table 4. Data from AXM thick-walled vessels

Nominal notch depth (in.)	Predicted failure pressure (psi)	Specimen No.	Maximum pressure (psi)
0	990	3	1120
		4	1190
		5	1180
		6	1210
		7	1430
		8	1190
		9	1080
		Average	1200
5/32	640	1	680
		2	690
		10	610
		11	680
		12	670
		13	640
		14	680
		15	680
		Average	660

All the unnotched AXM vessels failed at the intersection of the head and wall. All the heads were broken off, with the separation being associated with one circumferential crack. All the notched AXM vessels failed at the tip of the notch. Again the heads were all separated from the rest of the vessel by a circumferential crack.

ATJ Graphite

The ATJ graphite specimens were geometrically identical to the AXM graphite specimens and were tested in the same manner. The ATJ vessels all failed in the same manner as the AXM vessels: the heads broke off with a circumferential crack. Data from the ATJ thick-walled vessels are given in Table 5. The predicted failure pressures were calculated with Eq. (2), using the F values given in Table 2 and G_{Ic} values of 0.54 in.-lb/in.^{2} for the unnotched specimens and of 0.55 in.-lb/in.^{2} for the notched vessels. The average pressure at fracture for the unnotched vessels is 19% higher than predicted and for the notched vessels it is 7% lower than predicted. Strain data from the crack-detection gages indicated crack growth prior to fracture in both the notched and unnotched ATJ vessels.

Table 5. Data from ATJ thick-walled vessels

Nominal notch depth (in.)	Predicted failure pressure (psi)	Specimen No.	Maximum pressure (psi)
0	590	4	700
		5	710
		6	690
		7	700
		8	680
		9	690
		10	720
		Average	700
5/32	430	1	420
		2	440
		3	410
		11	380
		12	340
		13	400
		14	410
		15	420
		Average	400

CONCLUSIONS

The failure pressures for the notched thick-walled vessels were in good agreement with predictions based on fracture mechanics. The unnotched vessels were stronger than predicted, but the agreement there is reasonably good considering the approximation of flaw size used in the calculation. The tests on ATJ graphite specimens illustrate the complexities involved in applying fracture mechanics to anisotropic materials under general states of stress. In this particular case, there is a need for additional basic data, that is, fracture toughness values for cracks that are not aligned with the principal axes of the material are required.

When the results of these tests and the earlier tests on splitting tensile specimens are considered, it is concluded that fracture-mechanics methods can be used to predict fracture of these graphites under general stress conditions at room temperature. These studies also demonstrate that the finite-element method can be used effectively in predicting failure. This demonstration is of importance because of the versatility of the finite-element method in handling varied and complex geometries.

ACKNOWLEDGMENTS

The authors gratefully acknowledge the assistance of E. H. Guinn in instrumenting and testing the specimens; the assistance of H. A. MacColl in reducing the data and performing the finite-element analyses; and the support, encouragement, and helpful suggestions received from W. L. Greenstreet.

REFERENCES

1. G. T. Yahr, R. S. Valachovic, and W. L. Greenstreet, Graphite Structures for Nuclear Reactors, 7-9 March 1972 Conference (Institution of Mechanical Engineers, London, 1972), p. 29.

2. G. T. Yahr, R. S. Valachovic, and W. L. Greenstreet, Summary of Papers, Tenth Biennial Conference on Carbon, p. 223 (1971).

3. O. C. Zienkiewicz and Y. K. Cheung, The Finite Element Method in Structural and Continuum Mechanics, McGraw-Hill, London, 1967.

4. E. L. Wilson, AIAA Journal 3, 2269 (1965).

5. V. B. Watwood, Jr., Nuclear Engineering and Design 11, 323 (1969).

6. D. F. Mowbray, Engineering Fracture Mechanics 2, 173 (1970).

7. G. R. Irwin, "Fracture Mechanics," p. 557 in Structural Mechanics, edited by J. N. Goodier and N. J. Hoff, Pergamon Press, New York, 1960.

ENGINEERING DESIGN AND THE PROBABILITY OF FATIGUE FAILURE OF CERAMIC MATERIAL

B. J. S. Wilkins

Atomic Energy of Canada Limited

Whiteshell Nuclear Research Establishment

INTRODUCTION

Ceramic materials can exhibit a large variability in instantaneous fracture stress. Hence, their engineering design requires an approach that deals with probability of failure at a given applied stress. The variability in instantaneous fracture stress (σ_i) is attributed to the distributions of flaw size and flaws throughout the material, at which brittle fracture may be initiated.[1] The flaw that needs the smallest applied stress (σ_A) to propagate determines the value of σ_i. Since, σ_i depends on the distribution of flaws, it must depend also on the material volume and the manner of loading. Weibull[2] has taken this into account in his familiar empirical relationship,

$$P_i = 1 - \exp\{-KV[\sigma_i/\sigma_o]^m\} \tag{1}$$

where P_i is the cumulative probability of a given instantaneous fracture stress, σ_o and m are distribution constants, K is a constant related to manner of loading and V is a constant related to volume of material under load.

Engineering design should also take into account the phenomenon of fatigue. Here, failure probabilities increase with service and have to be estimated as a function of service time. Static (delayed fracture) and dynamic fatigue (varying stress, mostly cyclic) data are commonly expressed as applied fatigue stress (σ_A) versus time-to-failure (t_r) or stress cycles to failure (N). Data in this form generally show large scatter and are often of limited value to both the engineer and theoretician. They need instead to

relate fatigue life (t_r, N) to σ_i (which reflects the maximum flaw
size), and to σ_A.

Fatigue data for brittle materials are expressed better as a
family of curves of homologous fatigue stress ratio (σ_H) versus t_r
or N for a range of σ_A values, as in Fig. 1.[3,4,5,6] For any specimen, σ_H is the ratio of the fatigue stress (σ_A) applied to it and
the instantaneous fracture stress (σ_i) in the same mode of loading,
i.e., $\sigma_H = \sigma_A/\sigma_i$. This eliminates effects due to size and mode of
stressing, i.e., the V and K terms in eqn. (1). To ensure no fatigue failures occur at a given σ_A and life time, σ_H must not exceed an initial critical value for all specimens. Strictly, information such as in Fig. 1, applies only to specimens from which it
was obtained. However, if the material is reasonably consistent,
the information can be used to predict fatigue behaviour of any
specimen made from the same material, under any conditions of loading.[3,7]

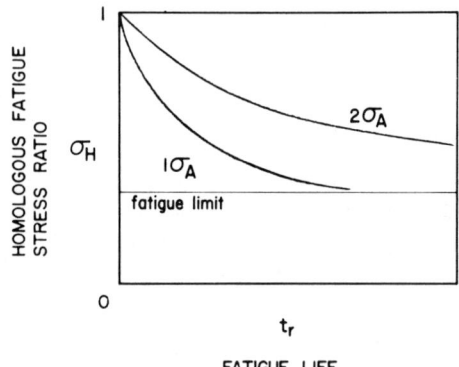

FATIGUE LIFE

Fig. 1 Homologous Fatigue Stress Ratio (σ_H) Versus Fatigue Life
        for Applied Stresses $1\sigma_A$ and $2\sigma_A$

To obtain fatigue data, in terms of σ_H, instantaneous fracture
stress (σ_i) and fatigue life (t_r, N) have to be measured on the
same specimens. Since, this is not possible, σ_i has to be estimated, statistically from additional specimens from the same population that are not fatigue tested. For the specimens that are
fatigued, ranked estimates of σ_H are determined from the applied
stress (σ_A) and the values of σ_i that correspond to $P_i = 1/(A + 1)$,
$2/(A + 1)$......$A/(A + 1)$, where A is the number of specimens fatigue
tested. The values of fatigue life found experimentally, for a
given σ_A, are ranked and paired with the ranked σ_i (σ_H) values.
This assumes the weakest specimen has the shortest fatigue life and
that the origin of fracture is the same for both fatigue and instantaneous failure, i.e., the propagation of Griffith-type flaws.
It is unnecessary to measure fatigue lives that are impractical,

i.e., very short or long, because all specimens contribute to the
estimation of σ_i values.

The information generated, gives for any fatigue life and σ_A
a most likely estimate of a critical value of σ_H, which if exceeded
will result in fatigue failure. Depending on experimental errors
the critical σ_H value determined, at any fatigue life and σ_A, will
be distributed between 1 and a value that corresponds to a fatigue
limit.   This complicates the estimation of cumulative probability
of fatigue failure. Essentially, the probability of a specimen
being acted on by a given σ_H has to be combined with the probability
of fatigue failure at that σ_H value.[6] However, fatigue results for
graphites show the critical σ_H value, for a given fatigue life and
σ_A, is not widely distributed and in effect has only one value.[5,6,7]

This simplifies the estimation of fatigue failure probability.
The σ_H, fatigue life, σ_A relationship depends on ranked best estimates of σ_i values. Each σ_i, and hence σ_H value can be established
between limits for a given confidence level. If engineering design
requires a probability of fatigue failure of 10^{-x}, the limits of
σ_H of interest are those at the $(1 - 10^{-x})$ confidence level.   These
may be calculated from one-sided tolerance limits for a normal distribution. The procedure is described in detail by Natrella[8] and
Ownens.[9] Fig. 2 shows the form of the confidence limits for a given
σ_H, fatigue life, σ_A relationship. The confidence limits shown are
conservative for they ignore the fact that $\sigma_H = 1$ when t_r (N) = 0.
For conservative design, at all values of t_r or N, the critical
value of σ_H is set equal to the maximum in the lower confidence
limit. The maximum occurs at the mean value of σ_i.[8] This critical
value of σ_H is indicated in Fig. 2 by the heavy broken horizontal
line.

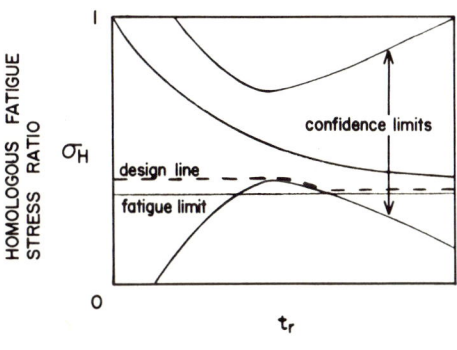

FATIGUE LIFE

Fig. 2   Homologous Fatigue Stress Ratio (σ_H) Versus Fatigue Life
for a Given Applied Stress (σ_A) with Confidence Limits.
Design Line Is Based on Lower Confidence Limit.

CRITICAL CRACK SIZE AND FRACTURE MECHANICS FOR STATIC FATIGUE

Fatigue data expressed as a family of curves of σ_H versus t_r over a range of σ_A, as in Fig. 1, is consistent with two critical crack sizes, one for slow propagation by the fatigue process and a large one for instantaneous propagation.[10] In terms of stress intensity at the crack tips (K), the critical crack sizes correspond to critical values of K, i.e., $K_{fat\ c}$ and K_c, respectively. Also,

$$K = B\sigma_A L^{1/2} \tag{2}$$

where B is a constant and L is crack size and

$$L = C\sigma_i^{-2} \tag{3}$$

where C is a constant. From eqns. (2) and (3),

$$K = D(\sigma_A/\sigma_i) = D\sigma_H \tag{4}$$

where D is a constant. Hence, K depends on σ_H and in Fig. 1, for all values of σ_A, K_c is where $\sigma_H = 1$, $t_r = 0$ and $K_{fat\ c}$ is where σ_H = fatigue limit, $t_r = \infty$.

For any value of σ_H between 1 and the fatigue limit the value of t_r depends on σ_A. This follows because the fatigue crack growth distance (Y) (from its initial size to that corresponding to K_c) depends on σ_A and is given by

$$Y = E(\sigma_A^{-2} - \sigma_i^{-2}) \tag{5}$$

where E is a constant. Initially, at a fixed σ_H, between 1 and the fatigue limit, for two values of σ_A, say $1\sigma_A$ and $2\sigma_A$, there are two fatigue cracks, not equal in size, but with the same stress intensity (K) at their tips. Both cracks grow slowly to reach their critical size for instantaneous propagation. These critical sizes are not the same, but are where the stress intensity has reached the K_c value. If crack velocity during fatigue growth is the same function of stress intensity (K), at the crack tip, for both cracks, the initial and final growth velocities are the same for both cracks. The fatigue growth distances (Y) are not equal and hence their t_r values also differ.

Fatigue life (t_r) as a function of σ_A, at a given σ_H, can be anomalous depending on the density and size distribution of cracks in the material. In hypothetical single-crack material, at a fixed σ_H, between 1 and a fatigue limit, t_r decreases as σ_A increases. In multi-crack material, as variability increases, t_r can increase with increasing σ_A, due to interference between the growing crack and other cracks in its path.[11]

THE CRACK VELOCITY, STRESS INTENSITY RELATIONSHIP
FOR STATIC FATIGUE

A popular empirical expression, relating crack velocity (V) to stress intensity is

$$V = \frac{dL}{dt} = FK^n \tag{6}$$

where F and n are constants. The information contained in the σ_H, t_r, σ_A relationship can be used to estimate a value for n. Consider Fig. 3, which is a plot of σ_H versus t_r at applied stress σ_A. The two values of σ_H shown, $\sigma_{H(x)}$ and $\sigma_{H(y)}$, correspond to two different specimens whose instantaneous fracture stress (σ_i) are σ_x and σ_y. They fail at times t_x and t_y, under stress σ_A. From eqn. (6) we have

$$\int_{L(\sigma_x)}^{L(\sigma_A)} K^{-n}dL = F \int_0^{t_x} dt = Ft_x \tag{7}$$

where $L(\sigma_x)$ and $L(\sigma_A)$ are the size of the crack responsible for fatigue failure at $t_r = 0$ (when $\sigma_i = \sigma_x$) and at $t_r = t_x$ (when $\sigma_i = \sigma_A$). Similarly,

$$\int_{L(\sigma_y)}^{L(\sigma_A)} K^{-n}dL = Ft_y \tag{8}$$

From eqns. (2), (3) and (7) we have,

$$\frac{1 - (\sigma_x/\sigma_A)^{(n-2)}}{1 - (\sigma_y/\sigma_A)^{(n-2)}} = \frac{t_x}{t_y} \tag{9}$$

where, $n \neq 2$. If n is large and positive, say >10, eqn. (9) further reduces to,

$$n = 2 + \frac{\ln(t_x/t_y)}{\ln[\sigma_{H(y)}/\sigma_{H(x)}]} \tag{10}$$

Fig. 4 shows the static fatigue behaviour of POCO* (moulded) graphite rods, at room temperature in air. It is a plot of $\ln\sigma_H$ versus $\ln t_r$ for the applied stress (σ_A), which corresponds to the stress that breaks instantaneously 18% of the virgin population of rods. The slope of the line, i.e., $d\ln\sigma_H/d\ln t_r$ is $1/n$. The value of n, given by the line drawn through the experimental points,

---

*POCO Graphite Inc., 1200 Jupiter Road, Garland, Texas, 75070, USA.

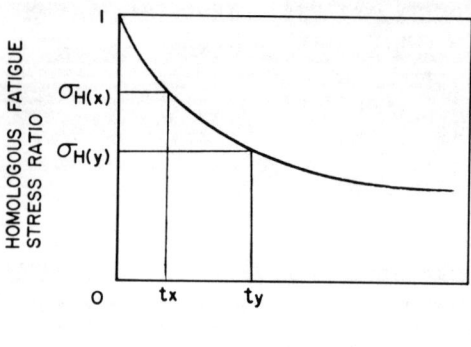

Fig. 3 Homologous Fatigue Stress Ratio (σ_H) Versus Fatigue Life
        for a Given Applied Stress (σ_A).

in Fig. 4, is ∿850. The reason for the large value of n is that the
data, in Fig. 4, are from specimens that failed. Hence, the value
of n estimated reflects an average value between K initial ($>K_{fat\ c}$)
and K final (K_c).

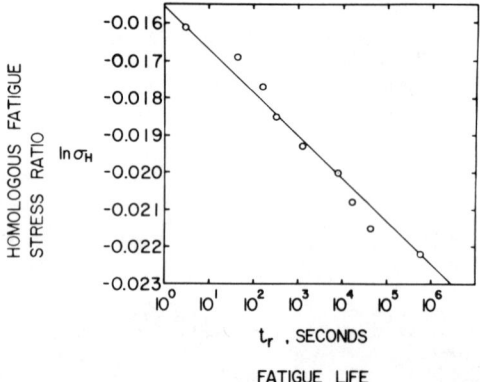

Fig. 4 Static Fatigue Behaviour of POCO Graphite Rods, at Room
        Temperature, in Air. The Value of n Is Given by the Recip-
        rocal of the Slope of the Line.

        Clearly, in eqn. (6) n is not constant and approaches ∞ as σ_H
approaches 1 or the fatigue limit, i.e., as the stress intensity at
the fatigue crack tip approaches K_c or $K_{fat\ c}$. The shape of $\ln\sigma_H$,
$\ln t_r$ curves are consistent with the fatigue crack velocity, stress
intensity relationship shown in Fig. 5. Further, if n can be found

in terms only of fatigue crack velocity and the correpsonding stress intensity at the crack tip, the family of $\ln\sigma_H$, $\ln t_r$ curves (over a range of σ_A values) must be parallel at all values of σ_H.

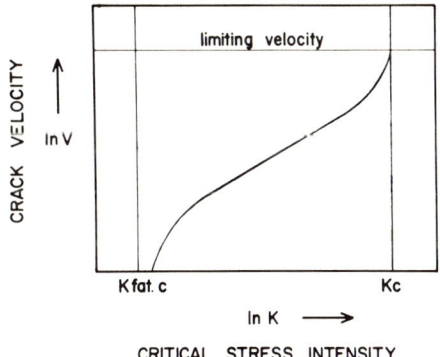

Fig. 5   Crack Velocity Versus Critical Stress Intensity at Crack Tip.

REFERENCES

1. A. A. Griffith, Phil. Trans, Roy. Soc. London, Sect. A, 221 [4] 163-98 (1920).

2. W. A. Weibull, J. Appl. Mech. 18 [3] 293-97 (1951).

3. H. Leichter and E. Y. Robinson, Tech. Rept. UCRL-50337, 1967.

4. B. J. S. Wilkins and A. R. Reich, Rept. AECL-3958, 1972.

5. B. J. S. Wilkins, J. Amer. Ceram. Soc. 54 [12] 593-95 (1971).

6. B. J. S. Wilkins, J. Mat., 7 [2] 251-6 (1972).

7. B. J. S. Wilkins and A. R. Reich, Rept. AECL 4216, 1972.

8. M. G. Natrella, Experimental Statistics, National Bureau of Standards Handbook 91; pp. 2-14, August 1963.

9. D. B. Ownens, Rept. SCR-13 Sandia Corporation, April 1958.

10. R. N. Stevens and R. Dutton, Mater. Sci. Eng., 8, 220-34 (1971).

11. B. J. S. Wilkins and B. F. Jones, The Influence of Flaw Density and Flaw Size Distribution on the Static and Dynamic Fatigue Behaviour of Graphite, to be published J. Mater. Sci., (1973).

FRACTURE MECHANICS OF BINARY SODIUM SILICATE GLASSES

C. R. Kennedy, R. C. Bradt, and G. E. Rindone

Ceramic Science Section, Material Sciences Department

The Pennsylvania State University, University Park, Pa.

INTRODUCTION

The generally accepted explanation for the discrepancy between the theoretical strengths of 10^6 psi and the commonly observed strengths of 10^4 psi for glasses lies in the flaw-stress concentration theory originally proposed by Griffith[1] nearly fifty years ago. The often modified equation may be expressed as:

$$\sigma_f = k\sqrt{\frac{E\gamma}{c}} = \frac{k}{\sqrt{2c}}K_{Ic} \qquad (1)$$

where: σ_f = fracture strength, E = elastic modulus, γ = surface energy, c = length of the idealized flaw, K_{Ic} = fracture toughness, and k = a constant ($0.8 \rightarrow 1.3$). Ernsberger[2] has reviewed several alternatives to the Griffith theory. Cox[3] and Poncelot[4] both propose that flaws are not inherent but appear as the direct result of the application of stress. Cox's theory is dynamic in nature, assuming that due to the variation in vibrational energies there are a certain fraction of Si-O bonds broken at any given time. If a critical number of neighboring bonds are broken, a fracture occurs. Watanabe and Moriya[5] relate the strength of glass to the size of its constituent "microphases", where the Griffith flaw is related to the interface between the microphases, or perhaps their size and/or spacing. Marsh[6,7] proposes that a flow process governs the strength of brittle glasses. He cites permanent micro-indentations left during microhardness tests as tangible evidence of flow, and reports that the strength of several types of flaw free glasses are equal to their yield stresses as calculated from microhardnesses. Consequently, factors other than those in equation (1) may be pertinent to the strength of glass.

Several studies have reported measurements of the fracture mechanics variables which are important in the strength of glass[8,9,10];
however, these, for the most part, were performed on a wide variety
of commercially available glasses. This paper deals with the
pertinent strength parameters over a systematic range of glass-
forming compositions in the binary glass system Na_2O-SiO_2, and
examines them in view of equation (1) and some of the other proposed theories.

EXPERIMENTAL

Glasses of compositions from $Na_2O \cdot 2SiO_2$ (65.9 w/o Na_2O) to
$Na_2O \cdot 5SiO_2$ (82.9 w/o Na_2O) were prepared from Mallinckrodt Reagent
Grade Na_2CO_3 and potters' flint (99.9% SiO_2). Batches of 200 grams
each were melted in air in a platinum crucible at 1400°C for 8
hours in an electric furnace. During firing, the melts were stirred three times with a fused silica rod to aid homogenization.
Fused silica plate for fracture surface energy measurements was
obtained from Corning Glass Works. It was prepared by the $SiCl_4$
flame hydrolysis method. Densities of the glasses were determined
by Archimede's technique using kerosene.

For strength measurements, rods were drawn using the apparatus
of Caporali [11] in an open furnace held at 950°C. One rod, about
80 inches in length, was obtained from each draw, about 10 minutes
in duration. This rod was then cut into three inch lengths for
testing. Care was taken to handle only the ends of these rods,
which were stored vertically in a plastic sample holder in a
vacuum desiccator until testing. The rods varied from 0.070-0.100
inches in diameter with roundness limited to $\pm$ 0.001 inches, and
taper to 0.004 inches over a three inch length. Strengths of
these were determined under liquid nitrogen in a four point bend
test over a 1.2 inch span with equidistant 0.4 inch knife edge
separations. A model TTD Instron testing machine at a crosshead
speed of 0.05 in/min was utilized.

Strengths were determined in the pristine as-drawn condition
and also after abrasion. Since fragmentation occurred upon testing, the diameters of the pristine rods were measured before
testing by averaging the diameters of each end of the rod. In
order to abrade the rods uniformly, Caporali's [11] apparatus was
used. A constant amount ($\approx$0.045 g) of 600 grit SiC was blown onto
the rod by dry nitrogen gas at 7.0 psi over a ten second period.
The rod was rotated at 150 rpm during the abrasion. Although the
abrasion treatment may appear slightly arbitrary, it was reproducible and yielded a flawed surface with a reduced level of
strength. The diameters of the abraded rods were measured at the
point of the break after testing, since the two pieces were left
intact.

Samples for modulus and fracture surface energy measurements were prepared by casting the Na_2O-SiO_2 glasses (at 1400°C) into graphite molds and annealing from 530°C at the rate of 1°C/min. For elastic modulus measurements, cast and annealed samples were ground into 5" x 1/2" x 1/4" bars using 120, 240, 320, and 400 grit SiC. The elastic modulus was determined at liquid nitrogen temperature by the resonance technique discussed by Spinner and Teft [12] with a Nametre Acoustic Spectrometer. The equation used to determine the modulus was:

$$E = 0.9464 \frac{df^2 l}{At^2} T \qquad (2)$$

where E = elastic modulus (psi), d = density, l = length, t = thickness, f = resonant frequency, A = constant (386.09 in/sec^2), and T = correction factor.

The fracture surface energy was measured by both the double cantilever beam method (DCB) and by the notched beam technique (NBT). Double cantilever beam samples were prepared by grinding cast specimens to a 400 grit SiC finish. Final sample dimensions were approximately 3" x 1" x 1/4". Notches to guide the crack were sawed along the midplane with a 0.010" x 4" lapidary diamond saw. A single atomically sharp crack was initiated and propagated by repeated controlled thermal shock until the tip of the crack lay at least 1.5 (t) from either end of the sample, where (t) is the thickness of one beam. The double cantilever beams were tested under liquid nitrogen on the Instron at a crosshead speed of 0.02 in/min. To avoid thermal shock, the samples were held just above the level of the liquid nitrogen for about 30 minutes before immersion. The equation [13] used for calculation of the fracture surface energy was:

$$\gamma_f = \frac{F^2 L^2}{2Ewbt^3} [2.38 \left(\frac{t}{L}\right) + 3.46]^2 \qquad (3)$$

where: F = force, L = length of the crack, E = elastic modulus, w = width of the web, b = width, and t = thickness of one beam. Notched beam samples were cut from the elastic modulus bars to give specimens 1" x 1/2" x 1/4". A notch was sawed halfway through the specimen using a 0.010" x 4" diamond blade lapidary saw with oil as a coolant. The notched beam samples were tested under liquid nitrogen in three point bend on the Instron at a 0.75 inch span at a crosshead speed of 0.05 in/min. The equation [14], used to evaluate the notched beam fracture surface energy was:

$$\gamma_f = \frac{9F^2L^2C}{8EW^2D^4} [A_o + A_1 \left(\frac{C}{D}\right) + A_2 \left(\frac{C}{D}\right)^2 + A_3 \left(\frac{C}{D}\right)^3 + A_4 \left(\frac{C}{D}\right)^4]^2 \qquad (4)$$

where: F = force, L = span, C = notch depth, E = elastic modulus,
D = depth, W = width, A_o = 1.90 + 0.0075X, A_1 = -3.39 + 0.08X,
A_2 = 15.40 - 0.2175X, A_3 = -26.24 + 0.2815X, A_4 = 26.38 - 0.145X,
and X = L/D.

Microhardness measurements were taken on cast samples which
were polished to a one micron finish. Care was taken to protect
the samples from moisture attack. A Leitz Miniload Hardness Tester
with a 100 gram load and a Knoop indentor was utilized.

RESULTS AND DISCUSSION

Densities of the Na_2O-SiO_2 glasses (Figure 1) decrease linearly with increasing silica content, in excellent agreement with the
data of Winks and Turner [15]. The Young's modulus (Figure 2)
increases slightly with increasing silica content, in agreement
with results reported by Phillips [16], and those of Gehlhoff and
Thomas [17].

Strengths, in both the pristine and abraded conditions
(Figures 3 and 4) decrease with increasing silica content over the
65 to 85 mole percent region, similar to the results of Gehlhoff
and Thomas [17]. This trend is opposite to that exhibited by the
elastic modulus. Clearly the strength of the Na_2O-SiO_2 glasses is
not a function the elastic modulus alone [18].

The fracture surface energy, as measured by the double cantilever beam technique (Figure 5), decreases approximately linearly
with increasing silica content, with values between 7,000 and 3,700
ergs/cm^2. These magnitudes are consistent with those reported by
Wiederhorn [8] and others [9,10], and are about an order of magnitude
above thermodynamic surface free energies cited from surface tension
measurements [19]. The difference probably reflects irreversible
processes occuring at the crack tip, such as flow processes, densification, or the presence of non-recoverable elastic strain energy.
The fracture surface energy, as measured by the notched beam technique (Figure 6) exhibits a similar decrease with increasing silica
content; but, for any given composition, is approximately three
times greater than the double cantilever beam value. These higher
values result from the use of a sawed notch to approximate an
atomically sharp crack.

The fracture surface energy of the Na_2O-SiO_2 glasses does not
linearly follow the modulus, as reported by Wiederhorn [8] for several
of a variety of commercial glasses. In fact, it exhibits an inverse
relationship. Since Wiederhorn also noted a number of exceptions
to the linear E vs. γ_f behavior, it is probable that the tentative
agreement which he observed was fortuitous.

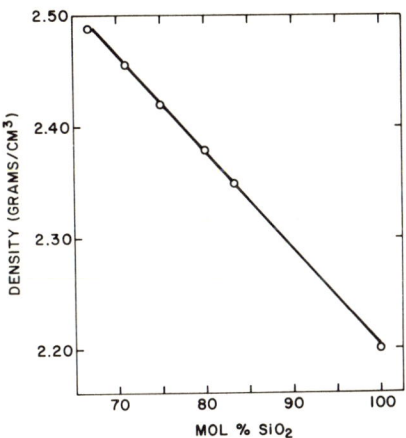

Figure 1. Density of the sodium silicate glasses.

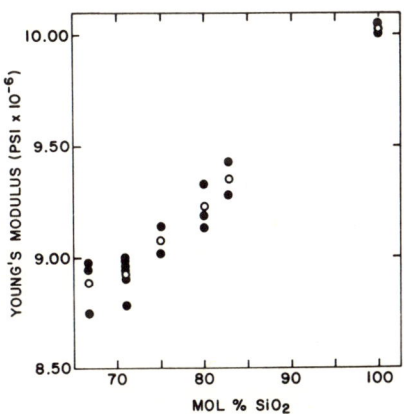

Figure 2. Elastic modulus of the sodium silicate glasses.

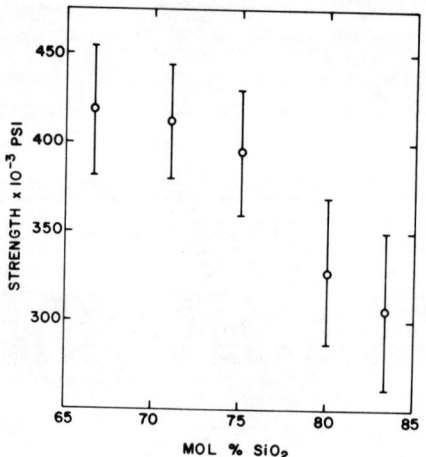

Figure 3.   The pristine strengths of the sodium silicate glasses.

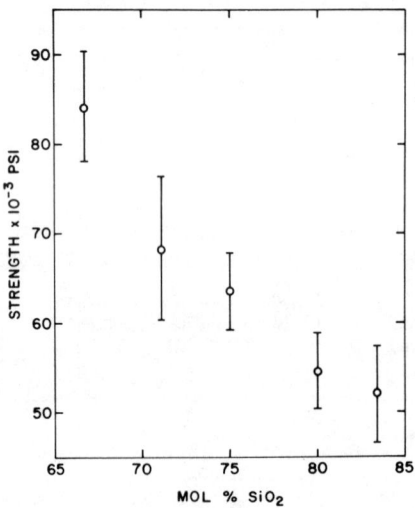

Figure 4.   The abraded strengths of the sodium silicate glasses.

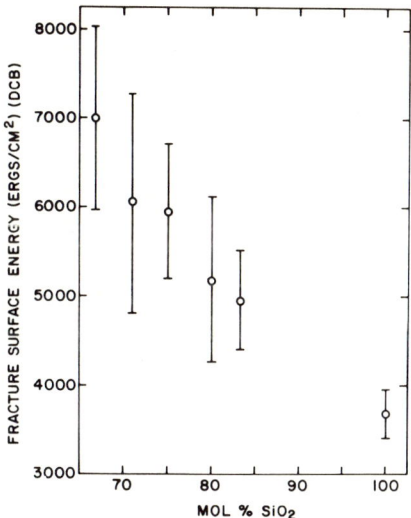

Figure 5.   The fracture surface energies of the sodium silicate
            glasses, measured by the double cantilever beam method.

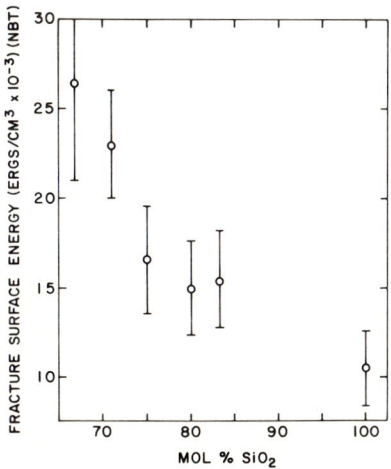

Figure 6.   The fracture surface energies of the sodium silicate
            glasses, measured by the notched beam technique.

Figure 7 illustrates the behavior of the strength versus the fracture toughness, K_{Ic} or $\sqrt{2E\gamma_f}$, based on the double cantilever beam data. If the flaw size is constant, the Griffith relationship predicts a straight line intercepting zero, with a slope equal to $k/\sqrt{2c}$. Between 65% and 85%, the strength does vary linearly with K_{Ic}, but the intercept is not zero. Calculations of the flaw size yields 70Å for the pristine glasses, and 900Å for the abraded glasses. However, the extremely high strengths reported [20] for fused silica suggest that the nature of the flaw in fused silica differs from that in the sodium-silicate compositional range reported in this study. This lends credence to the microphase-flaw concept advanced by Watanabe and Moriya [5] and supported by others [21]. Frieman [22] has also observed negative intercepts (in Si_3N_4 and Al_2O_3), as well as zero intercepts (in SiC and $BaO-SiO_2$ glass ceramics), but it is not clear at present what a negative intercept signifies. Similar plots of strength vs. K_{Ic}, determined using the notched beam technique, results in intercepts very near zero; however, this is considered fortuitous.

The Knoop microhardness (Figure 8) increases with increasing silica content, as previously noted by Gehlhoff and Thomas [17]. Since this trend is opposite to that of the strength, it would appear that Marsh's [6,7] theory cannot be applied successfully to these glasses. However, Marsh's claim that permanent indentations produced by microhardness measurements are an indication of flow in glass, along with Wiederhorn's [8] suggestion of deformation in the vicinity of the crack tip, indicate that the decrease in fracture surface energy with increasing silica content in these glasses may be related to the decrease in flow in the vicinity of the crack tip. This conceptual approach must be applied with some reservation since Neely and MacKenzie [23] have shown that the predominant processes of permanent deformation in microhardness indentations in fused silica is densification rather than flow. Just how this applies to these sodium silicate glasses which increase in hardness as they decrease in density (Figures 1 and 8) is puzzling. However, Neely and MacKenzie do not completely eliminate the presence of flow and pile-up of glass around the indentation; rather they limit the height of the pile-up to about 300Å. Since Wiederhorn [8] has estimated that the size of the zone undergoing flow in glass need only be 10Å to 100Å, it is reasonable to suggest that micro-flow processes may be approximated in trend by the microhardness, and possibly control the fracture surface energy, which in turn controls the strength of these sodium silicate glasses when the flaw character remains unchanged.

Preliminary results of parallel studies on potassium silicate glasses indicate the same type of compositional and mechanical trends reported here for sodium silicate glasses.

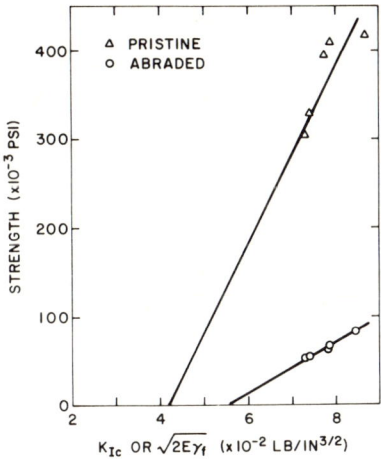

Figure 7. The strength vs. fracture toughness of the sodium silicate glasses. Note the linear behavior and the non-zero
intercept.

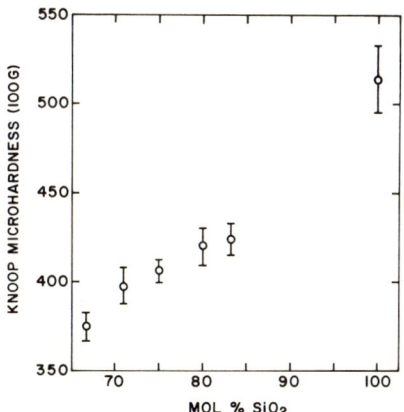

Figure 8. The Knoop, 100 gm., microhardness of the sodium
silicate glasses.

SUMMARY AND CONCLUSIONS

The pertinent parameters for examining the strength behavior of binary sodium silicate glasses have been measured. In addition some measurements were made on fused silica.

The strength was found to vary linearly with K_{Ic}, as predicted by the Griffith theory, although a non-zero intercept is as yet unexplained. The observed decrease in fracture surface energy with increasing silica content may be related to the decrease in microflow processes at the crack tip.

ACKNOWLEDGEMENT

The authors would like to acknowledge the support of the Applied Research Laboratory of The Pennsylvania State University for the support of this study, and the Corning Glass Works for supply of the fused silica.

1. Griffith, A.A., Phil. Trans. Roy. Soc., A221, 163 (1921).

2. Ernsberger, F. M., Eighth In. Cong. on Glass, 123 (1968).

3. Cox, S. M., J. Soc. Glass Tech., 32, 127 (1945).

4. Poncelot, E. F., Glass Ind., 43, 612 (1962).

5. Watanabe, M., and Moriya, T., Rev. Elect. Comm. Lab., 9, 50 (1961).

6. Marsh, D. M., Proc. Roy. Soc., A282, 240 (1964).

7. Marsh, D. M., Proc. Roy. Soc., A279, 240 (1964).

8. Wiederhorn, S. M., J. Amer. Cer. Soc., 52, 99 (1969).

9. Linger, K. R., and Holloway, D. G., Phil. Mag., 18, 1269 (1968).

10. Davidge, R. W., and Tappin, G., J. Mater. Sci., 3, 165 (1968).

11. Caporali, R. V., Ph.D. Thesis, Pennsylvania State University, (1969).

12. Spinner, S., and Tefft, W. E., Proc. ASTM 61, 1221 (1961).

13. Srawley, J. E., and Gross, B., Mat. Res. and Stand., 7, 155 (1967).

14. Brown, W. F., and Srawley, J. E., ASTM Spec. Tech. Pub. No.
     410, 13 (1966).

15. Winks, F., and Turner, W. E. S., J. Soc. Glass Tech., $\underline{4}$, 3
     (1920).

16. Phillips, C. J., Glass Tech., $\underline{5}$, 216 (1964).

17. Gehlhoff, G., and Thomas, M., Z. Tech. Physik., $\underline{7}$, 105 (1926).

18. Phillips, C. J., Amer. Scient. 53, 20 (1965).

19. King, F. B., Trans. Soc. Glass Tech. $\underline{35}$, 241 (1951).

20. Morely, V. G., Andrews, P. A., and Whitney, I., Phys. Chem.
     Glasses, $\underline{5}$, 1 (1964).

21. Rindone, G. E., Sproull, J. F., Kennedy, C. R., and Bradt, R.C.,
     Second Cairo Solid State Conference, 21-26, April, 1973.

22. Frieman, S. W., private communications.

23. Neely, J. E., and MacKenzie, J. D., J. Mat. Sci., $\underline{3}$, 603
     (1968).

FAILURE PREDICTION IN BRITTLE MATERIALS USING FRACTURE MECHANICS AND ACOUSTIC EMISSION

A. S. Tetelman[*] and A. G. Evans[†]

[*]Materials Dept.
School of Engineering
U.C.L.A.
Los Angeles, Calif.

[†]Inorganic Materials Div.
National Bureau of Stds.
Washington, D.C.

ABSTRACT

Acoustic emission testing has found many uses in recent years and new applications are constantly being uncovered. The interesting areas for application lie in failure prediction and the characterization of the microscopic processes of yielding and fracture, and the macroscopic processes of slow crack growth and onset of fast crack propagation. This paper has described these applications and emphasized that the total counts and count rate depend on the energy released per event and the density of events per unit of deformation. Where models are presented, they should be regarded as first order approximations which await confirmation.

1. INTRODUCTION

During the last several years the field of acoustic emission testing has grown very rapidly.[1,2] This growth has been stimulated by (1) the realization that acoustic emission can be used to characterize and monitor the processes of flow and fracture that occur during mechanical testing, and (2) the availability of commercial testing equipment at modest cost. Acoustic emission testing has been utilized in several ways:[2] (a) as a research tool to help understand various processes of flow and fracture; (b) as a monitor of slow crack growth in structural components; (c) as a monitor of specific phenomena, such as weld cracking or susceptibility to stress corrosion cracking; and (d) as a part of a large scale, computerized system designed to monitor large structures against premature failure, particularly during proof testing.

The growth of acoustic emission testing has paralleled our increased understanding of plastic flow and fracture in structural materials. Acoustic emission testing (AET) is the only method of nondestructive inspection whose signal output can be directly correlated with the crack tip stress intensity factor K, the sole parameter that describes both the rate of slow crack growth in fatigue and/or reactive environments and the conditions for the onset of unstable crack propagation. This correlation provides the principal driving force for the development of acoustic emission testing in the last few years, and its application to failure prevention in large structural systems.

This paper reviews some of the principal correlations between AET and the processes of flow and fracture in structural materials. Most studies have been conducted on metals and composites, although recently there has been application to ceramics. We begin with a short review of the principal aspects of fracture mechanics and show how various parameters can be used to predict the conditions for structural failure. We then discuss the nature of the information that can be gained from AET, and show how this correlates with fracture mechanics parameters. Finally, we discuss some applications where AET offers great promise as a failure prevention tool.

2.   PRINCIPLES OF FRACTURE MECHANICS

2.1.   The Critical Condition for Fracture

Most structural materials contain flaws or cracks that are introduced during fabrication or service. Under various combinations of static and alternating loads and reactive environments, these flaws grow slowly and stably. Unstable fracture occurs in a structure when a flaw has developed to a critical size that is a function of both the applied (or residual) stresses acting on the structure and the toughness of the material. The principles of linear elastic fracture mechanics can be used to describe these functional relationships for unstable fracture that occurs at nominal stresses below the general yield stress.

For simplicity, we consider a plate of thickness B and width W that contains a sharp, through-crack of length 2a and tip radius ρ, (Fig 1). The plate is subjected to a uniform gross section tensile stress, σ, which is less than the gross section yield strength, σ_Y. It is then possible to describe the state of stress in the vicinity of the crack tip by a parameter K, known as the stress intensity factor.

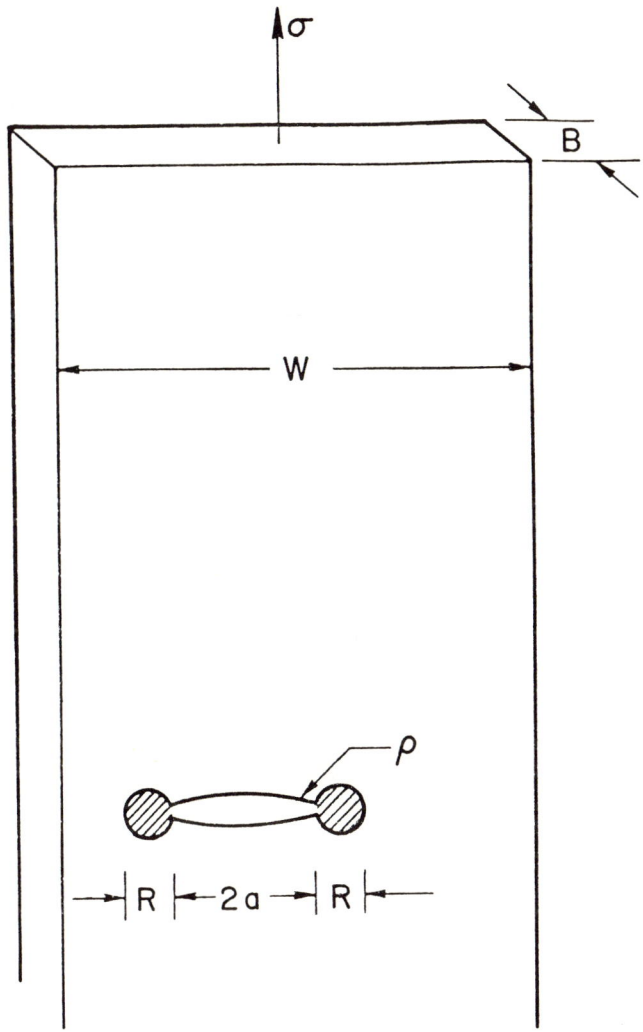

Figure 1.   The formation of a plastic zone in a cracked plate.

K is defined by the relation[3]

$$K = \frac{\sigma_{MAX}}{2} \sqrt{\pi\rho}$$

(1)

$$\lim \rho \to 0$$

where (σ_{MAX}) is the maximum stress at the tip of the crack in the limit as the root radius $\rho \to 0$, assuming the material remains perfectly elastic. For example, for an elliptical center crack of length 2a that extends all the way through the thickness

$$\sigma_{MAX} = \sigma(1 + 2\sqrt{a/\rho})$$

(2)

so

$$K = \sigma\sqrt{\pi a}$$

(3)

    Thus, K is a direct function of the applied stress and the crack length. For other crack shapes and loading modes,

$$K = \sigma\sqrt{\pi a}\ f(g)$$

(4)

where f(g) is a geometrical parameter of the order of 1.0-2.0 that can be determined analytically for a given crack and structural shape.

    Near the crack tip, at some point (r,θ), the local stress σ_{ij} and displacements u are of the form[4]

$$\sigma_{ij} = \frac{Kf(\theta)}{\sqrt{2\pi r}}$$

(5)

$$u = \frac{K\sqrt{r/2\pi}\ f(\theta)}{\mu}$$

(6)

where μ is the shear modulus and ν is Poisson's ratio. The work done G in opening the crack may be obtained by integrating the product of the stress and displacement such that[4]

$$G = \frac{(1-\nu^2)K^2}{E}$$

(7)

where E is Young's modulus.

    Several criteria have been proposed to describe the condition for unstable fracture:

2.1.1. <u>Energy Approach</u>. According to this approach, unstable fracture occurs when the work done at the crack tip G reaches a critical value, G_C; i.e., when K reaches a critical value, K_C. For the simple case of a wide plate containing a center crack of length 2a, f(g) = 1 and eqn (4) indicates that the fracture strength, σ_F, is

$$\sigma_F = \frac{K_C}{\sqrt{\pi a}} \equiv \left(\frac{EG}{\pi a (1-\nu^2)} \right)^{1/2} \tag{8}$$

which is the same relation originally proposed by Griffith if we set $G_C = 2\gamma_i$, where γ_i is the energy required to produce unit area of fracture surface. For metals, G_C is several orders of magnitude larger than the thermodynamic surface energy, γ_o, due to the large amount of plastic work that is expended prior to the onset of un- stable crack propagation. However, in ceramics, G_C is found to vary between $\sim 2\gamma_o$ and $100\gamma_o$, depending on the microstructure and temperature.

2.1.2. <u>Crack Opening Displacement Approach</u>. This method pre- dicts that fracture will occur at the crack tip when the material there is displaced sufficiently, such that atomic bonds are broken across the fracture plane. For completely brittle solids, this implies that the critical crack opening displacement, $(COD)_C$, will be of the order of the lattice spacing, a_o. Since the maximum stress at the crack tip will be the cohesive strength $\sigma_C \approx E/10$, the work done at the tip, G_C, is

$$G_C = \sigma_C (COD)_C = Ea_o/10 \tag{9}$$

which is approximately twice the thermodynamic surface energy $\gamma_o \approx Ea_o/20$. Thus the critical (COD) method and the critical work methods are <u>approximately</u> equivalent.

When plastic deformation occurs at the crack tip (at the yield stress σ_Y) the (COD) is a function of the size of the plastically deformed region, R,[5,6] which is in turn a function of K.[7]

$$(COD) = \beta\sigma_Y R/E, \tag{10}$$

$$(2<\beta<4)$$

$$R \approx \frac{1}{3\pi} \left(\frac{K}{\sigma_Y} \right)^2. \tag{11}$$

Thus,

$$(COD) \approx \frac{4}{3\pi E} \sigma_Y \left(\frac{K}{\sigma_Y}\right)^2 \tag{12}$$

Fracture occurs at a critical value of (COD)--where $K = K_C$--and hence,

$$K_C = \left[\frac{3\pi}{4} E\sigma_Y (COD)_C\right]^{1/2} \tag{13}$$

When localized plastic flow triggers the unstable fracture, the critical (COD) will be of the order of the slip band spacing, $10^4 a_O$, and the local stresses in the deforming grain near the crack tip will be equivalent to the dislocation flow stress, $\sim 10^{-4}$ E. Then, $G_C \approx Ea_O$ or about ten times that for completely brittle fracture. This situation applies for several polycrystalline ceramic materials prior to the onset of the ductile/brittle transition. When large scale plastic flow occurs in metals, $(COD)_C$ can be as high as $10^6 a_O$, and with $\sigma_Y \sim 10^{-3}$ E, G_C is another three orders of magnitude larger. The toughness of a material is thus primarily controlled by the amount and distribution of plastic strain which can be accommodated prior to the onset of unstable fracture.

2.2.   Sub-critical Crack Propagation

The lifetime of a structural component t_f is the time required for a crack to nucleate (by fatigue, reactive environments, etc.) and grow slowly and stably out to the critical size a_c.

The rate of slow crack growth is a unique function of the stress intensity factor K. For environmentally induced cracking, it has recently been established that three stages exist (Fig. 2):[8] (1) The crack growth rate increases very rapidly with K above the threshold value K_O; (2) a plateau region occurs where da/dt is essentially independent of K or varies slowly with K; and (3) a region of either combined environmental and mechanical cracking or environmentally independent crack propagation occurs prior to rapid failure at K_C. Similar observations have been made for fatigue loading,[9,10] although the crack growth per cycle does not level off completely in region II, but does show a lower dependence on K than in either regions I or III.

Various analytic expressions have been proposed to compute structural lifetime. In the simplest model $K_O = 0$ and the three stages of crack growth are described by one approximate equation

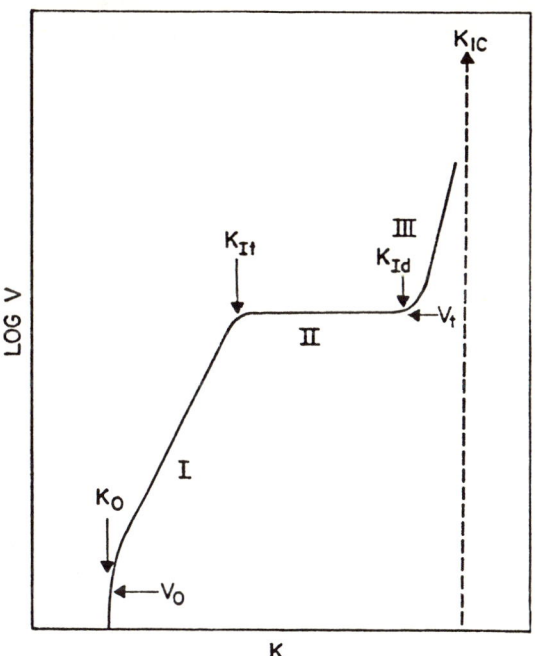

Figure 2.   The relation between crack growth rate, V, and stress intensity factor, K, for environmentally induced slow crack growth.

$$\frac{da}{dt} = AK^n \qquad (14)$$

If we take $n = 4$ as an example, then

$$\frac{da}{dt} = AK^4 = A\sigma^4\pi^2a^2 \qquad (15)$$

The lifetime may now be computed from the time required for an initial flaw of size a_i to grow out to critical size a_C.   From eqn (4),

$$a_C = \frac{1}{\pi}\left(\frac{K_C}{\sigma}\right)^2 \qquad (16)$$

for $f(g) = 1$.   Then,

$$\int_{a_i}^{a_C} \frac{da}{a^2} = A\sigma^4\pi^2 \int_{o}^{t_f} dt \qquad (17)$$

$$\frac{1}{a_i} - \frac{1}{a_C} = A\sigma^4 \pi^2 t_f \tag{18}$$

Substituting for a_C, we have

$$t_f = \frac{1}{A\sigma^4 \pi^2} \left[\frac{1}{a_i} - \frac{\pi\sigma^2}{K_C^2} \right] \tag{19}$$

This equation is often written in terms of the <u>initial</u> stress intensity factor of the flaw,

$$K_i = \sigma \sqrt{\pi a_i} \tag{20}$$

as,

$$t_f = \frac{1}{A\sigma^2 \pi} \left[\frac{1}{K_i^2} - \frac{1}{K_C^2} \right] \tag{21}$$

Consequently, it is possible to predict the lifetime of a cracked part if the initial stress intensity factor can be determined and the other parameters are known. It is in this realm that acoustic emission testing offers its greatest advantages, as discussed below.

2.3. Failure Prediction and Proof Testing

One of the most important applications of fracture mechanics methods lies in the ability to predict the optimum schedule for conducting periodic overstress (proof) tests to insure against failure during service.[11] The method can be used for ceramic materials provided the da/dt vs. K curve is known,[12] and provided that the structure can be periodically stressed to a level $\alpha\sigma_W$ where $\alpha > 1$ is called the proof test factor and σ_W is the peak service stress.

Suppose that a structure is stressed to a level

$$\sigma_P = \alpha\sigma_W \tag{22}$$

and failure does <u>not</u> occur. Then the flaw size that exists in the part, a_i, must be smaller than a certain value given by

$$a_i \leqslant \frac{K_C^2}{\pi (\alpha \sigma_W)^2} \tag{23}$$

so that the maximum initial stress intensity under service conditions is

$$K_i = \sigma_W \sqrt{\pi a_i} \tag{24}$$

or

$$(K_i)^2 = \sigma_W^2 \pi a_i \leqslant \frac{K_C^2}{\alpha^2} \tag{25}$$

Then,

$$\frac{K_i}{K_C} \leqslant \frac{1}{\alpha} \tag{26}$$

From da/dt vs. K curves it is possible therefore to determine structural lifetime, obtained by combining eqns (21) and (26); this gives for the case n = 4

$$t_f = \frac{(\alpha^2 - 1)}{A K_C^2 \sigma_W^2 \pi} \tag{27}$$

Knowing the growth constant A and K_C and σ_W, we can determine the proof factor α that must be used to guarantee a particular service life. If the structure fails in the proof test at σ_p, or perhaps before σ_p is achieved, then it is obvious that the structure would not have fulfilled its service requirement.

It is not always required to guarantee complete service life by one proof test. Suppose that a component is required to reliably function for one thousand hours. One could either proof to the requisite α in one test, or one could guarantee the first 500 hours by proof tests to a lower α and then guarantee the final 500 hours by repeating the procedure midway through the life of the structure.

One of the limiting factors to the successful use of the periodic proof test in metals is that σ_p must be smaller than the general yield strength σ_Y, otherwise large-scale deformations will occur and the structure may not be useful even if it survives the proof test. In ceramics, the ratio of working stress to gross yield strength is very high and large α values are possible.

Furthermore, K_C values are smaller so that it is possible to "proof out" smaller flaws. Consequently, periodic proof testing should be an extremely attractive method of "nondestructively" inspecting critical ceramic structures where 100% reliability is required, and where several loadings can be duplicated in test.

One of the principal uses of acoustic emission testing in metals is the monitoring of proof tests on large vessels to detect sites of premature failure and allow repair before continuing with the test. Most ceramic structures are not easily repaired and acoustic emission testing would then have little benefit, except to warn the operator of the impending failure so that the test could be stopped and the part examined in detail.

3.   PRINCIPLES OF ACOUSTIC EMISSION TESTING

Acoustic emissions are the stress waves spontaneously gener- ated within the volume of a material that is being deformed or fractured. In the most common commercial systems used today the emissions are detected by coupling a PZT sensor directly to the specimen or structure under load. The transducer signals are pre- amplified and filtered before being inserted into a secondary amplifier and filter. The amplified and filtered signals are then fed into a counter which counts the number of times the signal exceeds a certain threshold level for triggering the counter.

As discussed by Harris et al.,[13] the signals consist of damped sinusoids with a frequency corresponding approximately to the resonant frequency of the transducer, as shown schematically in Fig. 3.   A given acoustic emission event within the structure does

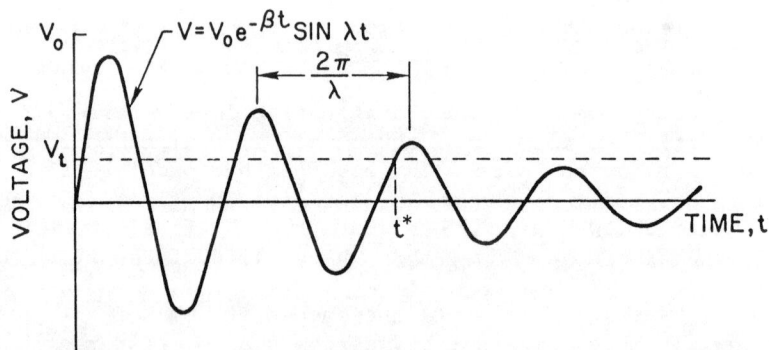

Figure 3.   Schematic of a damped sinusoid signal showing threshold voltage and the multiple counts from a single event within the test piece.[13]

not produce a single count, but rather produces several counts
associated with the number of times the signal crosses the threshold V_t in "ringing down" to a voltage below the trigger level.
Three counts would result from the signal shown in Fig. 3. Larger
events produce more signals in ringing down to a voltage below the
trigger level. Consequently, the number of counts is a function of
the energy released in the event, ΔE_g.

Harris et al.,[13] have proposed that the initial voltage output
from the transducer V_o is proportional to the square root of the
energy released during a given deformation process.

$$V_o = \Psi \sqrt{\Delta E_g} \tag{28}$$

where Ψ is a proportionality factor. They have also shown that the
number of "counts" η associated with the damped sinusoid is

$$\eta = \frac{f}{\Delta} \ln\left(\frac{V_o}{V_t}\right) \tag{29}$$

where f = linear frequency, Δ = damping constant, and V_t is the
threshold voltage. Combining eqns (28) and (29) gives

$$\eta = \frac{f}{\Delta} \ln\left[\frac{\Psi \sqrt{\Delta E_g}}{V_t}\right] \tag{30}$$

Acoustic emissions are produced by a variety of processes.
The two that are of principal concern here are plastic deformation
and cracking--both microcracking and macrocrack growth. The emissions that occur at the tip of a macrocrack result from plastic
deformation, in the plastic zone, and/or crack growth (and therefore can be used to characterize the K factor of the crack).

The relaxation process (ΔE_g) that produces acoustic emission
counts is a function of a deformation variable, i.e., stress or
strain in a tensile test conducted on flaw free materials, or K in
a fracture mechanics test on a cracked specimen. Let us denote
this variable as x, such that

$$\eta = \eta(x) \tag{31}$$

the number of events ϕ that occur during deformation or fracture
is rarely a constant, but is also a function of the deformation
state. Thus,

$$\phi = \phi(x) \tag{32}$$

Let $N(x)$ be the total number of counts that will have occurred after a material has reached a deformation state x. $N(x)$ will be a function both of the number of counts per event and the number of events that will have occurred.

$$N(x) = \int \eta(x)\, d\phi = \int \eta(x)\left(\frac{d\phi}{dx}\right) dx \tag{33}$$

then

$$dN = \eta(x)\left(\frac{d\phi}{dx}\right) dx \tag{34}$$

and thus the count rate dN/dt is

$$\frac{dN}{dt} = \eta(x)\left(\frac{d\phi}{dx}\right)\left(\frac{dx}{dt}\right) \tag{35}$$

For example, the count rate measured during the growth of a crack due to stress corrosion cracking is

$$\frac{dN}{dt} = \left(\frac{dN}{dA'}\right)\left(\frac{dA'}{dt}\right) \tag{36}$$

where A' is the area swept out by the growing crack and dA'/dt is the areal rate of crack growth. For the linear growth of a through-thickness crack A' is proportional to the crack length, a, so

$$\frac{dN}{dt} = \left(\frac{dN}{da}\right)\left(\frac{da}{dt}\right) \tag{37}$$

It has been noted that when hydrogen cracking occurs in cathodically charged 4340 steel, (da/dt) is proportional to K and dN/dA' is proportional to K^4. Consequently, dN/dt is proportional to K^5 and small changes in K lead to rapid increases in count rate.[14] For porcelain dN/dt during moisture-assisted slow crack growth is also related exclusively to K,[15] although the K dependence of dN/dA' is less than that for the 4340 steel ($<K^2$). It is often possible therefore to associate a critical count rate with the critical K value at failure (K_C). dN/dt measurements can then be used directly to predict the onset of structural failure. For example, the count rate for four cathodically charged 4340 bolts at failure (Fig. 4) approached the critical value associated with K_C measured in calibration tests on precracked specimens. (It is noted that it is not necessary to determine total counts to predict the onset of failure; in fact, in most applications, dN/dt provides a more sensitive measure of K).

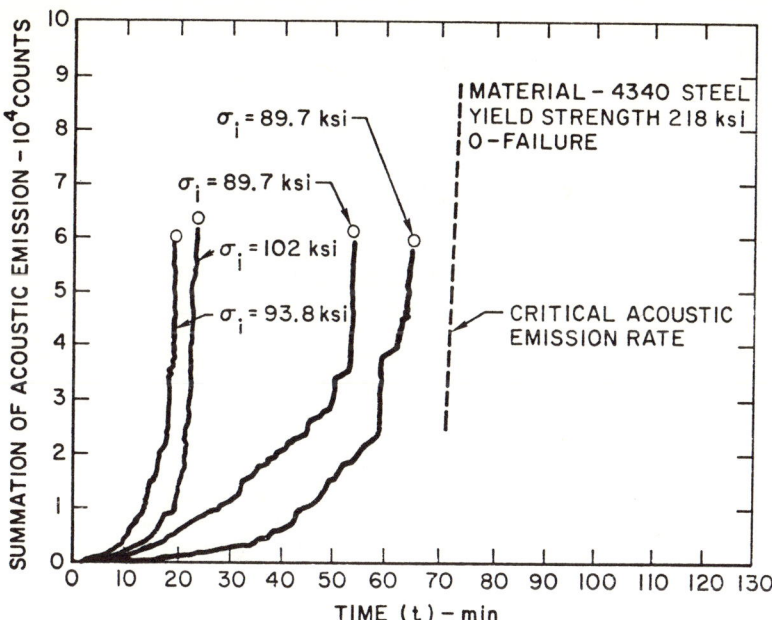

Figure 4. Summation of acoustic emission for cathodically charged and torqued bolts of 4340 steel. Failure predicted to occur at the critical count rate shown by the dotted line.[14]

4. MICROSCOPIC PROCESSES LEADING TO ACOUSTIC EMISSION

Acoustic emissions result from various processes of deformation and fracture which release elastic energy. This energy is transmitted to a transducer mounted on the specimen or structural component. In order to further develop AET methods, it is necessary to relate microscopic processes of flow and fracture to $\eta(x)$ and $\phi(x)$. Equation (30) shows that $\eta(x)$ is a function of the energy released per event, ΔE_g. Consequently, it is necessary to determine relationships between ΔE_g and deformation processes that are functions of σ, ε, K, etc. As a first approximation, we have

$$\Delta E_g = (w)(v)\lambda \tag{38}$$

where w is the strain energy released per event per unit volume, v is the volume being relaxed and λ is a parameter that measures the attenuation and dissipation of sonic waves prior to their arrival at the transducer. Little is known about this parameter and for purposes of subsequent discussion we will take $\lambda = 1$ although this is certainly a gross assumption.

4.1. Fracture

The energy per unit volume released during brittle fracture at a stress, σ, is

$$w = \frac{\sigma^2}{2E} \tag{39}$$

When cleavage occurs in one grain alone (i.e., a microcrack is formed but is blocked by grain boundaries) the stress will be relaxed over a volume of the order of d^3 (where d is the grain diameter), so that

$$\Delta E_g \cong \left(\frac{\sigma^2}{2E}\right) d^3 \tag{40}$$

Note that although stress is relaxed over the entire cleaved grain, the specimen as a whole exhibits only a miniscule load drop, since the adjacent grains immediately pick up the load that had been carried by the failed grain. The event is thus too small to be detected on the conventional stress-strain curve but may be detected by AET.

Similar concepts apply in the case of uniaxially loaded fiber composites and particulate composites. Assuming a simple isostrain model where the strain in the fiber and the composite is identical, and the fiber remains elastic until fracture, then

$$\Delta E_g = \left(\frac{\varepsilon^2 E}{2}\right) d^2 \left(\frac{\ell_c}{4}\right) \tag{41}$$

where d is the fiber diameter and $\ell_c/2$ is the length of fiber over which relaxation followed by partial reloading occurs.[13]

For the particulate composite where, for example, the strain in the particle and matrix are due primarily to thermal expansion mismatch, $\Delta\alpha$, the strain energy at a temperature ΔT below the fabrication temperature is[17]

$$E_g = 4\pi R^3 (\Delta\alpha \, \Delta T)^2 / k \tag{42}$$

where k is a constant that depends on the elastic constants of the particle and matrix and R is the particle radius. This strain energy is released when a certain stress is applied--due to circumferential matrix cracking---and eqn (42) gives a value for ΔE_g, to a good first approximation.

The energy release due to the extension of a pre-existing crack is obtained from the strain energy E_g of an infinite body containing a small central through crack of length $2a$[18]

$$E_g = \frac{\pi\sigma^2 a^2}{E} + \text{constant} \tag{43}$$

The strain energy release, for fixed grip conditions, during an increment Δa in crack length is thus, for unit thickness [15]

$$\Delta E_g \approx \frac{2\pi\bar{\sigma}^2 a\Delta a}{E} \equiv \frac{2K^2\Delta a}{E} \tag{44}$$

where $\bar{\sigma}$ is the average stress during the increment ($\sim\sigma$ for small increments).

4.2. Plastic Deformation

The yield drop in one grain due to dislocation source operation at a stress σ produces an energy release

$$\Delta E_g = \left(\frac{\sigma^2 - \sigma_i^2}{2E}\right) d^3 \tag{45}$$

where σ_i is the lattice friction stress. From elementary dislocation theory, a yield drop $(\sigma - \sigma_i)$ in a single grain produces a pile up of n dislocations, Burgers vector b, against the grain boundary;

$$(\sigma - \sigma_i) \approx \left(\frac{\bar{n}b}{d}\right) E \tag{46}$$

Hence, the strain energy release becomes simply

$$\Delta E_g = \frac{\bar{n}b}{2} d^2 (\sigma + \sigma_i) \tag{47}$$

5. ACOUSTIC EMISSION TESTING

5.1. Fracture Without Slow Crack Growth

In the absence of slow crack growth several events which occur prior to fracture can lead to acoustic emissions and these may be useful as failure indicators. These are particle or fiber fracture in composites, grain fracture and plastic deformation. Each of these processes are considered below.

5.1.1. Composites. In most composites, the particles (or fibers) exhibit a distribution of strengths such that the number of broken particles increases with increasing stress. The cumulative probability of particle fracture, P, is generally given, to a good approximation, by an extreme value function of the Weibull type, such that

$$P = 1 - \exp\left[-B\left(\frac{\sigma}{\sigma_o}\right)^m\right] \tag{48}$$

where m is a distribution constant and σ_o is a normalizing stress constant. The total number of broken particles ℓ at a stress, σ, is thus

$$\ell = (L+1)\left(1 - \exp\left[-B\left(\frac{\sigma}{\sigma_o}\right)^m\right]\right) \tag{49}$$

where L is the number of available particles. For $\ell \ll L$ (the general case), this reduces to

$$\ell = L\ B\left(\frac{\sigma}{\sigma_o}\right)^m \tag{50}$$

The number of acoustic emission counts per event is next obtained by substituting the appropriate value for ΔE_g per event into eqn (30). For the fiber composites this gives

$$\eta(\varepsilon) = \frac{f}{\Delta}\ \ell n\left[\frac{\Psi}{V_t}\ \frac{\varepsilon d}{2}\left(\frac{\ell_c E}{2}\right)^{1/2}\right] \equiv C\ \ell n\left(\frac{\varepsilon}{\varepsilon_o}\right) \tag{51}$$

where C is a system constant and ε_o is the threshold strain for activation of the transducer: for the particulate composite,

$$\eta(\Delta T) = \frac{f}{\Delta}\ \ell n\left[\frac{\Psi}{V_t}\ \Delta\alpha\Delta T\left(\frac{4\pi}{k}\right)^{1/2}\ R^{3/2}\right] \tag{52}$$

The total number of counts is now obtained by substituting η and ℓ --expressed in terms of strain--into eqn (33) to give

$$N(\varepsilon) = C'\int_{\varepsilon_o}^{\varepsilon}\frac{\exp[-(\varepsilon/\varepsilon_o)^m B]}{\varepsilon^{1-m}}\ \ell n\left(\frac{\varepsilon}{\varepsilon_o}\right)d\varepsilon \tag{53}$$

for the fiber composites (where C' is a constant). When η is a
weak function of the variable, N can be obtained to a good approximation by simply multiplying the number of counts per event by the
number of broken particles, e.g. for the particulate composite

$$N(\Delta T) = LC^* \exp\left(\frac{\Delta T}{\Delta T_o}\right)^m \ln\left(\frac{\Delta T}{\Delta T_o}\right) \tag{54}$$

where C^* is a constant and ΔT_o is the threshold temperature differential.

Finally, if particle fracture is time dependent (due to stress
corrosion for example) the time dependence can be incorporated,[15]
and the number of counts obtained from a particulate composite
(where the stresses are now primarily mechanical in origin) at a
stress σ is

$$N(\sigma,t) \approx C^{**}\sigma^m t^{m/n} \ln\left(\frac{\sigma}{\sigma_o}\right) \tag{55}$$

where C^{**} is a constant and n is the slow crack growth exponent in
eqn (14).

The predicted dependence of N on σ has been verified for a
Aℓ- AℓNi$_3$ fiber composite[13] (Fig. 5) and the dependence of N on σ
and t has been verified for a glass-quartz particulate composite
(porcelain)[15] as shown in Fig. 6.

Acoustic emission testing can thus be utilized to monitor the
breaking of particles or fibers prior to the onset of failure.
Alternatively (AET) provides a simple technique for determining
$\ell(\sigma)$ without recourse to tedious metallography.

5.1.2. Single Phase Polycrystals. Acoustic emission is
expected in single phase polycrystals, prior to ultimate fracture,
in the regime where fracture is dislocation initiated, e.g. polycrystalline MgO, since grain fracture leads to the formation of
non-propagating microcracks. In low carbon steel, for instance,
the number of cracks varies with stress as[19]

$$\ell = (\sigma/\sigma_o)^9 \tag{56}$$

giving,

$$\frac{d\ell}{d\sigma} = \frac{9\sigma^8}{\sigma_o^9} \tag{57}$$

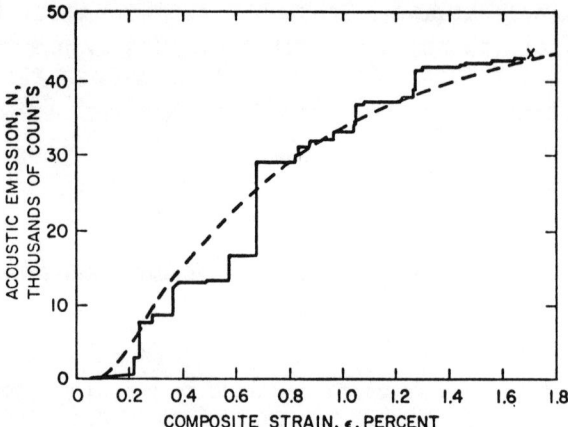

Figure 5.   Results of theoretical calculations of summation of acoustic emission as a function of strain compared with experimental results for an Al/AlNi$_3$ fiber composite.[11]

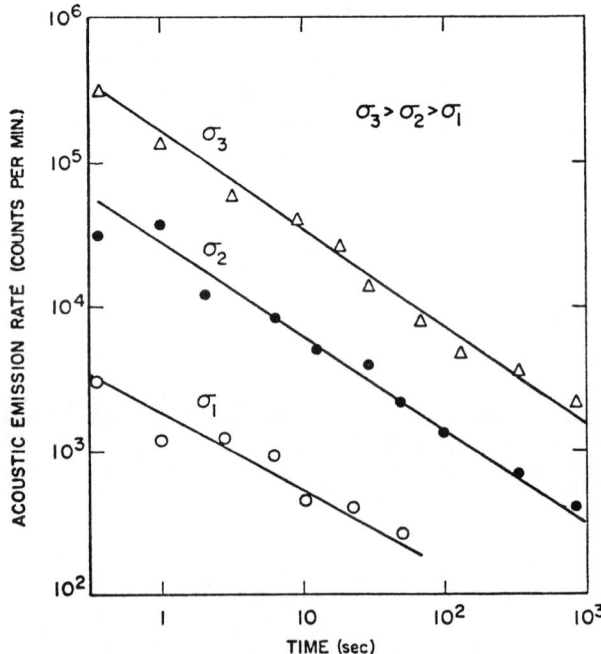

Figure 6.   The variation of acoustic emission rate with time for porcelain at various stress levels. The stress separation and time dependence give values for m and n which are used to verify the analytical description.[15]

where σ_O is the stress level at which one microcrack per unit volume has formed. The total number of acoustic emission counts thus becomes

$$N(\sigma) = \frac{d^3}{2E\sigma_1^9}\left[\sigma^9(\ln \sigma/\sigma_1 - 1/9) + \sigma_o^9/9\right] \qquad (58)$$

where σ_1 is the threshold stress. The stress range between σ^1 and the fracture stress is relatively small (fig 7) so that no acoustic emission will be observed until just prior to failure. This is in marked contrast to the $A\ell\text{-}A\ell Ni_3$ and the porcelain where emissions were observed well before failure.

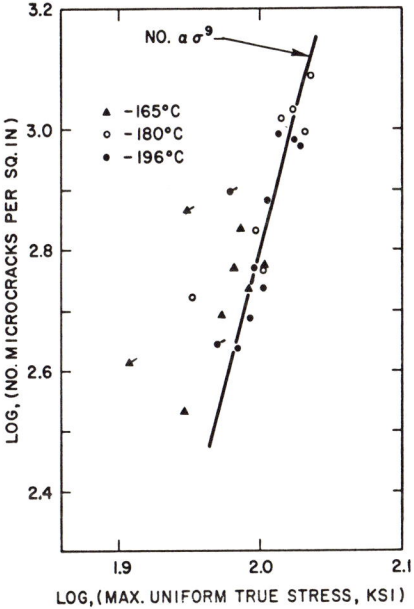

Figure 7. The stress dependence of microcrack formation in low-carbon steel tension specimens.[19]

5.1.3.   <u>Yielding and Plastic Deformation</u>.   Various studies of
yielding and plastic deformation have been conducted during the past
several years.[20,21,22]   The results of these tests indicate that[21]
(1) considerable acoustic emission activity occurs before the
yield strength σ_y is reached (Fig. 8); (2) dN/dt goes through a

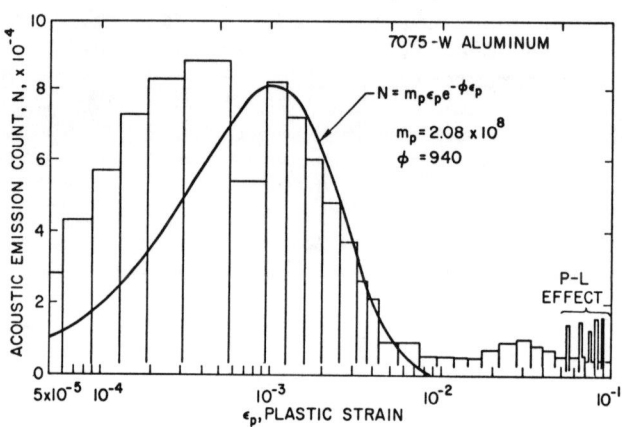

Figure 8.   Fit of dislocation mobility model to acoustic emission
data for 7075 aluminum.[22]

maximum at or close to σ_y; (3) the count rate decays rapidly after
the yield strength is reached; (4) not all materials are noisy--
generally, (dN/dt) is high for "brittle" materials (e.g. beryllium)
and low for soft, ductile materials (copper, stainless steel, etc.);
(5) in a given material, (dN/dt) increases with yield strength at
a given strain level; and (6) coarse-grained materials emit more
sound than fine-grained materials.

We believe that these acoustic emission bursts result from
the operation of pinned dislocation sources (or the creation and
operation of fresh sources)[23] and the generation of glide band
packets containing n dislocations each.   As shown in eqn (47) the
energy released by each source operation is proportional to the
lattice friction stress σ_i (in accord with observation (5)) and
the grain size (in accord with observation (6)).   The number of
dislocations in the pile up, n, and the source operation stress, σ,
are also important.   These are inter-related parameters which
are larger for brittle materials, in which unstable plastic flow

events can occur, than for ductile materials, which yield more homogeneously and which strain harden more rapidly (consistent with observation (4)). The increase in acoustic emission in pressure vessel steels after irradiation[24] probably results from both an increase in σ_i and an increase in the number of dislocations $\bar{n}$ generated by unstable microplastic flow events.

Observations (1-3) can be accounted for by assuming that the density of sources of glide bands (dislocation packets) follows an extreme value distribution of the Weibull type, then

$$\ell(\varepsilon) = g[1- \exp(-h(\varepsilon-\varepsilon_o)^j)] \qquad (59)$$

$$\frac{d\ell(\varepsilon)}{d\varepsilon} = \frac{g\,\exp[-h(\varepsilon-\varepsilon_o)^j]}{(\varepsilon-\varepsilon_o)^{(1-j)}}\,(jd) \qquad (60)$$

where $\ell(\varepsilon)$ is the number of sources that will have operated after a strain $\varepsilon > \varepsilon_o$ and g, h and j are constants; $(d\ell/d\varepsilon)$ is the rate of source operation. It is of course possible that ℓ is an independent function of stress σ rather than strain ε and this point needs experimental confirmation. This model predicts that the deviations from linearity before σ_Y result from a series of discrete load drops (Fig. 9) which are generally too small to be detected in the

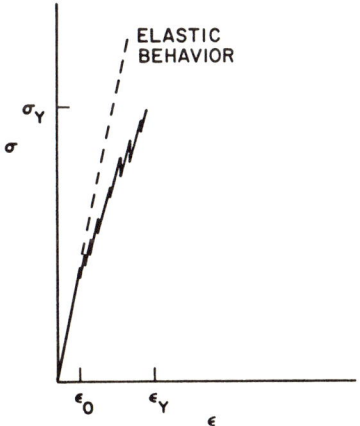

Figure 9. Microyielding and deviations from elastic behavior shown as a series of microyield drops.

conventional tensile test. The model is in agreement with the
fact that slip band sources have been observed to operate well
before the yield point is reached,[25] and that at the yield point
most sources will be operating. There is absolutely no reason
a priori to expect that fresh sources cannot operate after yield.
However, as the "easy" sources are used up, slip will tend to occur
by the motion of dislocations already generated and the acoustic
emission rate drops off rapidly for $\varepsilon > \varepsilon_Y$. Note that this form of
acoustic emission, like that of fiber cracking, is irreversible;
once a source has been used, it cannot operate again.

    5.1.4. <u>Failure Indication</u>. A schematic diagram of the varia-
tion of dN/dt with deformation variable (σ, ε or K) is used (Fig. 10)

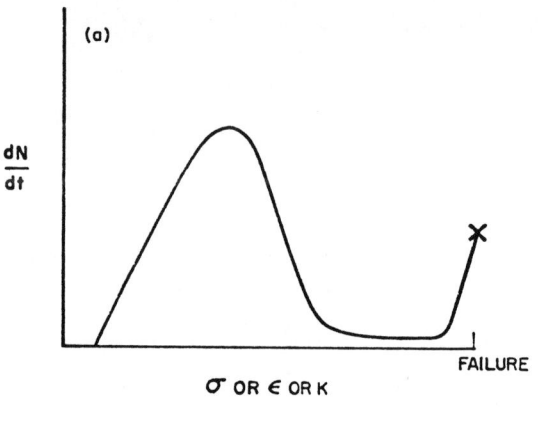

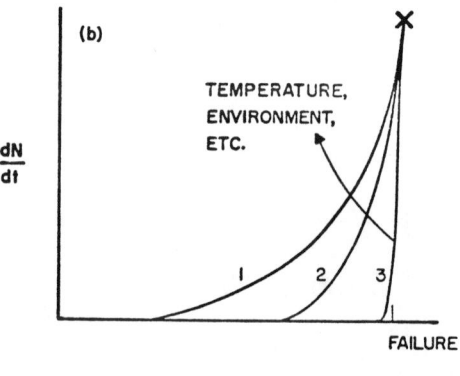

Figure 10. A schematic showing the variation of acoustic emission
rate with deformation variable σ, ε or K, (a) for plastic deformation
and (b) for various fracture processes where (1) represents say porcelain, (2) the optimum behavior, and (3) low carbon steel.

to discuss the systems in which acoustic emission may be used as an
effective failure indicator. The case in Fig.10a represents ductile
materials where the first high count rate results from yield and the
second from fracture. It is relatively difficult to predict failure
in these materials because dN/dt generally varies slowly with ε until just prior to fracture and there is no distinct cut-off rate
that may be used as a failure indicator. The case in Fig. 10b
represents the composites, grain fracture and (as shown later) slow
crack growth. The effectiveness of acoustic emission for failure
prediction for this case depends on the deformation range where
significant counts are obtained. The range should not be too small,
as for low carbon steel, or the structure cannot be unloaded rapidly
enough prior to failure. An excessively large range is also undesirable, as for porcelain, because it is difficult to set an
effective cut-off rate which relates fairly closely to fracture im-
minence. A deformation range for acoustic emission commencing at
about 70% of the deformation at fracture is optimum.

5.2.   Fracture with Prior Slow Crack Growth

When slow crack growth occurs, the acoustic emission rate may
be uniquely related to K, as discussed in section 3. In this case,
acoustic emission is clearly a very effective tool for failure prediction. It is of interest, therefore, to examine the various
contributions to acoustic emission that originate from the deformation zone at a crack tip.

Acoustic emission can occur from three entirely separate
processes near the crack tip--the growth of the plastic zone R,
microcracking in the process zone near the crack tip and the continuous extension of the primary crack itself. Each of these
processes may show a separate dependence on K. As discussed above,
the acoustic emission that occurs during plastic strain will mainly
come from grains that are just undergoing yield so that most of the
activity comes from a thin ring of material at the plastic-elastic
interface (Fig. 11). The volume (area in the two dimensional case
shown here) of the ring undergoing yielding at the plastic-elastic
interface (and hence the source density) is proportional to R^2.
In the simplest model[20] it is assumed that $\eta(\varepsilon)$ is constant for all
sources operating in the ring and hence N is proportional to the
source density (i.e., to R^2). From eqn (11) we then have

$$N(K) \approx A'' \left(\frac{K}{\sigma_Y} \right)^M \qquad (61)$$

where A" and M are constants. Experimental values between 4 and 8
have been observed for M and some recent data by Nakamura[1] can

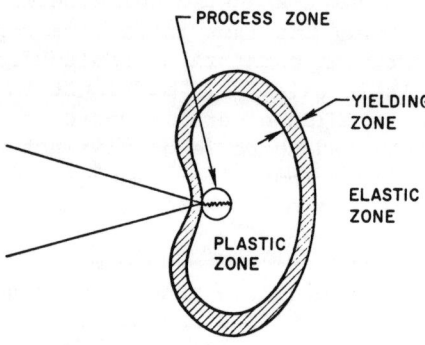

Figure 11.   Zones producing acoustic emission near crack tip.

account for these variations. The low level acoustic emission
associated with the plastic zone is generally difficult to detect
in high yield strength materials. It is most easily detected in
low strength materials (K/σ_Y large) which show unstable glide
band formation (e.g. low carbon steel, irradiated alloy steels,
etc.).

The acoustic emission that results from grain cracking and
particle or fiber cracking can most easily be investigated in
fiber composites. We have investigated the Boron/epoxy system[16]
using a ben l specimen designed such that the specimen thickness
increases as the crack propagates. In this "decreasing K" type
specimen, it is possible to control the crack propagation process
such that increased load must be applied to break each succeeding
layer of fibers. Independent measurements of specimen compliance
and electrical resistance can be used to determine the number of
broken fibers and relate this number to the acoustic emission
counts. The results showed that the total number of counts was
proportional to the number of broken fibers. This suggests that,
on average, each fiber fracture contributes equally to the total
number of acoustic counts which, if correct, infers that the
fibers break at a constant average strain $\bar{\varepsilon}_b$ (eqn 41). A model of
acoustic emission consistent with this result is equivalent to the
plastic zone model and considers that the microcracking (of the
fibers, grains or particles) occurs at the perimeter of the process
zone as K increases, i.e. from eqn (5)

$$\bar{\sigma}_b = \frac{Kf(\theta)}{\sqrt{2\pi r_o}} \tag{62}$$

where $\bar{\sigma}_b$ is the average stress for cracking and r_o is the radius
of the process zone. The number of counts per cracking event is
thus constant and the total number of counts obtained depends only
on the number of fracture events in the process zone. If the
density of microcracks is q, then the number of fractures Q per
unit thickness of material for an increment Δa in crack length
at constant K is; $Q \approx 2r_o q \Delta a$. Thus, from eqn (33), $dN/dA' \, \alpha \, K^2$
($\equiv G$).* Acoustic emission data obtained on a boron-epoxy system
(Fig. 12) is consistent with this relation.

---

*This result is only strictly applicable to grain or particle frac-
ture. For aligned fiber composites, if the fibers only break once
in the process zone, then the number of broken fibers during an
increment in crack length $\Delta a \, (\gg \, r_o)$ is independent of K. The
observation of Fig. 12 can then only be explained by invoking a
strain dependent fiber fracture condition so that η is K dependent
and hence $dN/dA' \, \alpha \, \ell n \, K$.

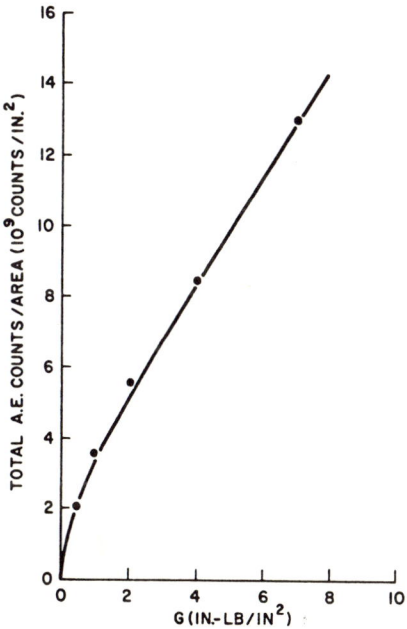

Figure 12. Variation of dN/dA' with strain energy release rate,
G, for a boron/epoxy composite.[16]

The acoustic emission due to crack length increments is obtained from eqns (47) and (30).

$$\eta(K) = C'' \ln [C'''K\sqrt{\Delta a}] \tag{63}$$

where C'' and C''' are constants. For constant crack growth rates (and hence constant K) the number of events for a total increment, a, is

$$\phi = \frac{a}{\Delta a} \tag{64}$$

The acoustic emission rate corresponding to this crack growth rate, da/dt, is thus;[15]

$$\frac{dN}{dt} = \frac{(da/dt)}{\Delta a} \; C'' \ln [C'''K\sqrt{\Delta a}] \tag{65}$$

The acoustic emission rate derived from increments in the primary crack length is thus expected to depend primarily on the crack growth rate, with only a small (logarithmic) K dependence once the threshold K for acoustic emission is effectively exceeded. In terms of the number of counts per unit area, eqn (65) becomes

$$\frac{dN}{dA'} \; \alpha \; \ln (K\sqrt{\Delta a}) \tag{66}$$

The K dependence of dN/dA' can often be used to identify the primary contribution to acoustic emission during crack propagation. An interesting example is the data obtained for porcelain (Fig. 13). The acoustic emission obtained in this material cannot be due to crack tip plasticity (because there is no plastic relaxation), but there are possible contributions from increments in the primary crack and the cracking of quartz particles in the "process" zone. The data show that the K dependence of dN/dA is very small and certainly $< K^2$. This suggests that the principal contribution to the acoustic emission comes from increments in the primary crack.

A knowledge of the origin of the acoustic emission is not needed, of course, for the practical application to failure prediction. Once it has been established that dN/dt is uniquely related to K (as it is for the two cases described here, 4340 steel and porcelain), then it is merely required to measure the acoustic emission rate associated with crack propagation to predict the remaining lifetime t_r of the component from[12]

$$t_r = \frac{2}{\pi \sigma^2 f^2(g)} \int_{K_i}^{K_C} \frac{K}{da/dt} \, dK \qquad (67)$$

where K_i is the stress intensity factor determined from the acoustic emission rate.

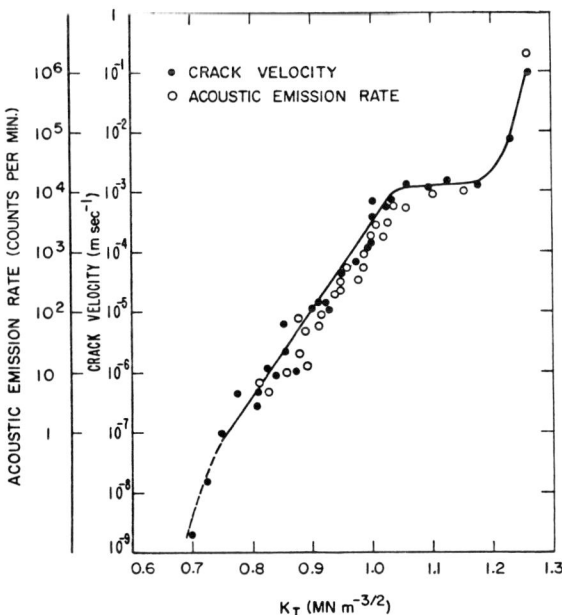

Figure 13. The variation of acoustic emission rate and crack growth rate with stress intensity factor for porcelain.[15]

There are clearly many practical difficulties associated with isolating the acoustic emission from crack growth in real applications, but it is not intended to discuss these in this paper. It is merely pointed out that recent developments enable crack growth emissions to be identified in many cases. The type of behavior which is best suited to the adaptation of acoustic emission for detection of slow crack growth has already been discussed in section 5.1.4. with reference to Fig. 10.

Finally, it is noted that acoustic emission can be used in

other applications and two of particular interest are its use as a
stress corrosion susceptibility monitor and its application to the
detection of thermal shock cracks. The extreme sensitivity of
(AET) to slow crack growth processes in reactive environments
means that susceptibility to stress corrosion cracking can be
quickly and easily determined by placing pre-cracked, pre-loaded
specimens of questionable microstructure or composition in known
solutions and measuring dN/dt. As shown in fig 14, the rapid
emission from material A indicated that it is considerably more
sensitive to SCC than B or C in this solution.

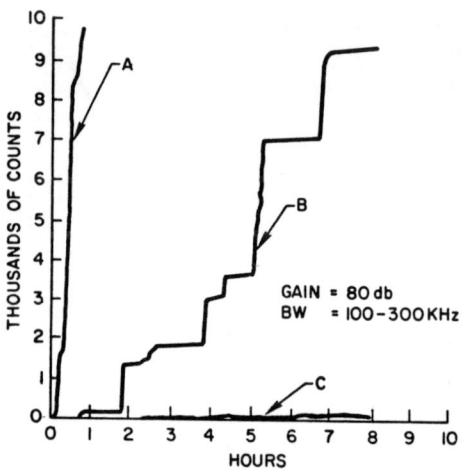

Figure 14.   Summation of acoustic emission for three different heat
treated specimens of an aluminum alloy for 3 percent salt solution.[26]

Thermal shock cracks tend to arrest in a component before final
fracture (due to the fixed grip deformation) and often, the cracks
that form are difficult to detect with conventional nondestructive
techniques.   It is very simple, however, to monitor acoustic emission during the thermal shock cycle to detect the onset of crack
propagation.

6.   CONCLUSIONS

    1.   The paper describes how fracture mechanics may be used in
conjunction with acoustic emission and proof testing to predict
failure in brittle structural components.

2. The optimum conditions for the effective use of acoustic emission are discussed in terms of the relation between the acoustic emission rate and the deformation variable (stress, strain, stress intensity factor).

3. The origins of acoustic emission during deformation and fracture are discussed and models which relate the magnitude of the emission to the type of event are presented.

4. The functional relationships for the acoustic emission are used to relate the acoustic emission observed in various systems--brittle composites (fibrous and particulate) brittle single phase polycrystals, ductile polycrystals--to the events which contribute to the emission. Hence, showing how acoustic emission can be used in studies of failure processes.

5. Other uses for acoustic emission are also discussed; notably, its use as a monitor for stress corrosion susceptibility and thermal shock cracking.

REFERENCES

1. Acoustic Emission Testing, ASTM, Philadelphia, STP505, (1972).

2. Materials Research & Standards, Vol. 11, No. 3, (1968).

3. A. S. Tetelman and A. J. McEvily, Fracture of Structural Materials, Wiley, New York (1967).

4. P. C. Paris and G. Sih, Fracture Toughness Testing, ASTM, Philadelphia, STP381 (1965), p. 30.

5. J. F. Knott and A. H. Cottrell, J.I.S.I., 201, 249 (1963).

6. T. R. Wilshaw, C. A. Rau, and A. S. Tetelman, Engineering Fracture Mechanics, 1, 191 (1968).

7. F. A. McClintock and G. R. Irwin, Fracture Toughness Testing, ASTM, Philadelphia, STP381 (1965), p. 84.

8. S. M. Wiederhorn, J. Am. Ceram. Soc., 50, 407 (1967).

9. D. P. Williams, Intl. J. Frac. Mech., to be published.

10. A. J. McEvily et al., presented at Fourth National Symp. on Fracture Mechanics, Carnegie-Mellon University, (1970).

11. C. F. Tiffany and J. N. Masters, Fracture Toughness Testing, ASTM, Philadelphia, STP381 (1965), p. 249.

12. A. G. Evans and S. M. Wiederhorn, NBSIR 73-147, March 1973; Intl. J. Frac. Mech., in press.

13. D. Harris, A. S. Tetelman, and F. Darwish, Acoustic Emission Testing, ASTM, Philadelphia, STP 505 (1972), p. 238.

14. H. Dunegan and A. S. Tetelman, Engineering Fracture Mechanics, 2, 387 (1971).

15. A. G. Evans and M. Linzer, J. Am. Ceram. Soc., Oct. 1973.

16. J. Fitz-Randolph, et al., J. Composite Materials, 5, 542 (1971).

17. R. W. Davidge and T. J. Green, J. Mat. Sci. 3, 629 (1968).

18. J. P. Berry, J. Mech. Phys. Solid, 8, 194 (1960).

19. L. Kaechele and A. S. Tetelman, Acta Met. 17, 463 (1969).

20. H. L. Dunegan, D. Harris, C. Tatro, Eng. Fracture Mechanics, 1, 105 (1968).

21. H. L. Dunegan and A. T. Green, Mat. Res. & Stds. 11, No. 3, p. 21. (1968).

22. W. Gerberich and W. Reuter, Final Rept. to ONR, Contract No. 00014-66-60340, (1969).

23. D. James and S. Carpenter, J. Appl. Phys. 42, 4685 (1971).

24. D. Ireland et al., Conn. Yankee Reactor Vessel Surveillance Program Report, Battelle, Columbus, Ohio (1970).

25. J. Suits and J. R. Low, Jr., Acta Met. 5, 285 (1957).

26. A. S. Tetelman and H. L. Dunegan, Research/Development, May 1971, p. 21.

CONTRIBUTORS

Co-Chairmen

R. C. Bradt, Associate Professor, Department of Materials
Science, Ceramics Science Section, Pennsylvania State
University, University Park, Pennsylvania

D. P. H. Hasselman, Director, Ceramics Research Laboratory,
and Associate Professor of Metallurgy and Materials
Science, Lehigh University, Bethlehem, Pennsylvania

F. F. Lange, Senior Research Scientist, Research and
Development, Westinghouse Corporation, Pittsburgh,
Pennsylvania

Conference Session Chairmen

W. R. Buessum, Pennsylvania State University, University Park,
Pennsylvania

R. J. Charles, General Electric Corporation, Schenectady,
New York

F. Erdogan, Lehigh University, Bethelehem, Pennsylvania

V. D. Frechette, Alfred University, Alfred, New York

R. F. Pabst, Max Planck Institute, Stuttgart, Germany

E. T. Wessel, Westinghouse Research Laboratory, Pittsburgh,
Pennsylvania

Arrangements

Patricia Ewing, Coordinator, J. Orvis Keller Conference
Center, Pennsylvania State University, University Park,
Pennsylvania

AUTHORS

G. A. Alers, Science Center, Rockwell International, Thousand Oaks, California

C. A. Andersson, Senior Scientist, Materials Research Laboratories, Westinghouse Research Laboratories, Pittsburgh, Pennsylvania

G. K. Bansal, Department of Metallurgy and Materials Science, Case Western Reserve University, Cleveland, Ohio; now with Battelle Memorial Institute, Columbus Laboratories, Columbus, Ohio

Gordon B. Batson, Department of the Army, Corps of Engineers, Champaign, Illinois

J. A. Batt, United Aircraft Research Laboratories, East Hartford, Connecticut

H. R. Baumgartner, Senior Research Engineer, Norton Company, Worcester, Massachusetts

E. K. Beauchamp, Supervisor, Ceramics Development, Sandia Laboratories, Albuquerque, New Mexico

R. M. Cannon, Jr., AVCO Systems Division, Lowell, Massachusetts

C. P. Chen, Apparatus Engineer, Southern California Edison, Rosemead, California

J. L. Chermant, Groupe de Cristallographie et Chimie du Solide, Universite de Caen, Caen, France

K. Chyung, Research and Development Laboratory, Corning Glass Works, Corning, New York

William G. Clark, Jr., Fellow Engineer, Westinghouse Research, Pittsburgh, Pennsylvania

John A. Coppola, Senior Engineer, New Products Branch, Carborundun Company, Niagara Falls, New York

R. W. Davidge, Materials Development Division, Atomic Energy Research Establishment, Harwell, Didcot, Berkshire, England

A. Deschanvres, Groupe de Cristallographie et Chimie du Solide, Universite de Caen, Caen, France

R. Dutton, Atomic Energy of Canada, Limited, Whiteshell Nuclear Research Establishment, Pinawa, Manitoba, Canada

J. D. Embury, Department of Metallurgy and Materials Science, McMaster University, Hamilton, Ontario, Canada

F. M. Ernsberger, Senior Scientist, PPG Industries, Pittsburgh, Pennsylvania

A. G. Evans, National Bureau of Standards, Washington, D. C.

M. E. Fine, Department of Materials Science, Northwestern University, Evanston, Illinois

Iain Finnie, Mechanical Engineering Department, University of California, Berkeley, California

Stephen W. Freiman, Ceramic Engineering, Naval Research Laboratory, Washington, D. C.

Edwin R. Fuller, Jr., Research Physicist, Inorganic Material Division, National Bureau of Standards, Washington, D. C.

Otto R. Gericke, Chief, Nondestructing Testing, Army Materials and Mechanical Research Center, Watertown, Massachusetts

L. J. Graham, Science Center, Rockwell International, Thousand Oaks, California

David J. Green, Graduate Student, Department of Metallurgy and Materials Science, McMaster University, Hamilton, Ontario, Canada

Robert M. Gruver, Technical Director, Ceramic Finishing Company, State College, Pennsylvania

J. M. Harris, Science Center, Rockwell International, Thousand Oaks, California

A. H. Heuer, Department of Metallurgy and Materials Science, Case Western Reserve University, Cleveland, Ohio

M. R. Hoover, Jr., Materials Research Laboratory, Pennsylvania State University, University Park, Pennsylvania

C. O. Hulse, Research Scientist, United Aircraft Corporation, East Hartford, Connecticut

Shin-Ichi Hyodo, Applied Physics Department, University of Tokyo, Bukyo-ku, Tokyo, Japan

A. Iost, Groupe de Cristallographie et Chimie du Solide, Universite de Caen, Caen, France

Om Johari, Illinois Institute of Technology, Chicago, Illinois

G. G. Johnson, Jr., Materials Research Laboratory, Pennsylvania State University, University Park, Pennsylvania

C. R. Kennedy, Ceramic Science Section, Material Sciences Department, Pennsylvania State University, University Park, Pennsylvania

M. Kimura, Department of Applied Physics, Faculty of Engineering, University of Tokyo, Bunkyo-yu, Tokyo, Japan

H. P. Kirchner, Ceramic Finishing Company, State College, Pennsylvania

W. J. Knapp, School of Engineering and Applied Science, University of California, Los Angeles, California

D. A. Krohn, Lehigh University, Bethelehm, Pennsylvania

W. A. Logsdon, Westinghouse Research Laboratories, Pittsburgh, Pennsylvania

J. L. Lott, Illinois Institute of Technology, Chicago, Illinois

R. C. Lueth, Carboloy Systems Department, General Electric Company, Detroit, Michigan

H. L. Marcus, Science Center, Rockwell International, Thousand Oaks, California

Charles W. Marschall, Associate Chief, Deformation and Fracture Research, Battelle Columbus Laboratories, Columbus, Ohio

Frank A. McClintock, Mechanical Engineering Department, Massachusetts Institute of Technology, Cambridge, Massachusetts

K. R. McKinney, Naval Research Laboratory, Washington, D. C.

H. A. McKinstry, Materials Research Laboratory, Pennsylvania State University, University Park, Pennsylvania

R. L. Mehan, Space Sciences Laboratory, General Electric Company, King of Prussia, Pennsylvania

Mel I. Mendelson, Member Research Staff, Ceramic Processing Unit, Western Electric, Princeton, New Jersey

Junn Nakayama, Asahi Glass Research Laboratory, Kanagawa-ku, Yokohama, Japan

D. J. Naus, United States Army Construction Engineering Research Laboratory, Champaign, Illinois

Patrick S. Nicholson, Associate Professor, McMaster University, Hamilton, Ontario, Canada

Michael J. Noone, Ceramist, Space Science Laboratory, General Electric, Philadelphia, Pennsylvania

N. M. Parikh, Illinois Institute of Technology, Chicago, Illinois

Svante Prochazka, Research and Development Center, General Electric Corporation, Schenectady, New York

Richard A. Queeney, Associate Professor, Pennsylvania State University, University Park, Pennsylvania

William H. Rhodes, Section Chief, Ceramics Research and Development, Avco Corporation, Lowell, Massachusetts

Roy W. Rice, Supervisor, Chemistry Department, Naval Research Laboratory, Washington, D. C.

D. W. Richerson, Norton Company, Worcester, Massachusetts

G. E. Rindone, Ceramic Science Section, Material Sciences Department, Pennsylvania State University, University Park, Pennsylvania

John E. Ritter, Jr., Associate Professor, Mechanical Engineering Department, University of Massachusetts, Amherst, Massachusetts

D. E. Roberts, Institute for Materials Research, National Bureau of Standards, Washington, D. C.

J. T. A. Roberts, Materials Science Division, Argonne National Laboratory, Argonne, Illinois

A. Rudnick, Head, Mechanical Department, Standards Institute of Israel, Tel Aviv, Israel

E. I. Salkovitz, Office of Naval Research, Washington, D. C.

J. M. Samuels, Industrial Engineering Department, Pennsylvania
    State University, University Park, Pennsylvania

John J. Schuldies, Research Engineer, Turbine Research, Ford Motor
    Company, Dearborn, Michigan

G. P. Sendeckyj, Department of the Air Force, Air Force Flight
    Dynamics Laboratory, Wright Patterson Air Force Base, Ohio

L. A. Simpson, Research Officer, Atomic Energy of Canada, Limited,
    Whiteshell Nuclear Research Establishment, Pinawa, Manitoba,
    Canada

H. L. Smith, Naval Research Laboratory, Washington, D. C.

B. Steverding, Department of the Army, Development and Engineering
    Laboratory, United States Army Missile Command, Redstone
    Arsenal, Alabama

Frederick Szalkowski, Member, Technical Staff, North American
    Rockwell, Thousand Oaks, California

L. Tarhay, Materials Research Laboratory, Pennsylvania State
    University, University Park, Pennsylvania

A. S. Tetelman, Materials Department, School of Engineering,
    University of California, Los Angeles, California

Robb M. Thomson, Senior Research Scientist, Inorganic Materials
    Department, National Bureau of Standards, Washington, D. C.

G. G. Trantina, Materials Science Division, Argonne National
    Laboratory, Argonne, Illinois; now with General Electric
    Company, Corporate Research and Development, Schenectady,
    New York

R. L. Turner, Department of Engineering Mechanics, Pennsylvania
    State University, University Park, Pennsylvania

S. Vaidyanathan, Department of Mechanical Engineering, University
    of California, Berkeley, California

R. S. Valachovic, Oak Ridge National Laboratory, Oak Ridge,
    Tennessee; now with AFC Industries, Incorporated, St.
    Charles, Missouri

T. Vasilos, AVCO Systems Division, Lowell, Massachusetts

J. B. Wachtman, Jr., National Bureau of Standards, Washington, D. C.

E. W. White, Materials Research Department, Pennsylvania State University University Park, Pennsylvania

S. M. Wiederhorn, National Bureau of Standards, Washington, D. C.

B. J. S. Wilkins, Atomic Energy of Canada, Limited, Whiteshell Nuclear Research Laboratory, Pinawa, Manitoba, Canada

G. T. Yahr, Engineer, Oak Ridge National Laboratory, Oak Ridge, Tennessee

SUBJECT INDEX

Pages 1-443 are found in Volume 1, pages 445-924 in Volume 2.

Thure von Uexküll

Psychosomatische Medizin

Herausgegeben von

Rolf Adler · Jörg Michael Herrmann
Karl Köhle · Othmar W. Schonecke
Thure von Uexküll · Wolfgang Wesiack

4., neubearbeitete und erweiterte Auflage
Mit 206 Abbildungen im Text und auf Tafeln
sowie 117 Tabellen

Urban & Schwarzenberg · München–Wien–Baltimore

Lektorat: Dr. Rainer Broll
Redaktion: Sabine Burgdorf, Dr. Susanne Hinke, Dr. Brigitte Zakaria

CIP-Titelaufnahme der Deutschen Bibliothek

Psychosomatische Medizin / Thure von Uexküll. Hrsg. von Rolf Adler . . . – 4., neubearb. u. erw. Aufl. –
München ; Wien ; Baltimore : Urban u. Schwarzenberg, 1990
  ISBN 3-541-08844-3
NE: Uexküll, Thure von [Mitverf.]; Adler, Rolf [Hrsg.]

1. Auflage 1979   ISBN 3-541-08841-9
2. Auflage 1981   ISBN 3-541-08842-7
3. Auflage 1986   ISBN 3-541-08843-5

93 92 91
5 4 3 2 1

Satz und Druck: C. H. Beck, Nördlingen. Printed in Germany.

© Urban & Schwarzenberg 1990

ISBN 3-541-08844-3

Vorwort zur vierten Auflage

Die rasche Entwicklung auf einzelnen Gebieten der Psychosomatischen Medizin und der Medizin überhaupt hat neue Einsichten gebracht und neue Akzente gesetzt. Probleme, wie AIDS, verlangen eine Darstellung der psychosomatischen Aspekte. Aus diesem Grunde ergab sich bereits nach vier Jahren die Notwendigkeit einer neuen Auflage dieses Werkes. Dabei hat sich die Richtigkeit des Grundkonzeptes des Buches weiterhin bestätigt. Auch die Gliederung konnte beibehalten werden.

Die meisten Kapitel sind von den bisherigen Autoren überarbeitet worden. Für einige wurden neue Autoren gefunden. Darüber hinaus wurde eine Anzahl neuer Themen aufgenommen:

In Teil I (Wissenschaftstheorie) wurde das Grundsatzkapitel (Nr. 1) durch ein Kapitel über sozialmedizinische Aspekte ergänzt.

In Teil II (Allgemeine Psychosomatik) wurden ein Kapitel über die Klassifikationssysteme ICD-9/DSM III sowie ein Kapitel über körperorientierte Psychotherapie aufgenommen. Das Kapitel 9 der dritten Auflage, mit dem Bericht der klassischen Arbeit Aders über die Konditionierbarkeit des Immunsystems, wurde durch ein Kapitel „Psychoimmunologie" ersetzt, da dieses Gebiet inzwischen zu einem eigenen Forschungszweig geworden ist. Das Kapitel „Epidemiologie" wurde von einem neuen Autor den speziellen Problemen der Psychosomatischen Medizin angepaßt. Das Kapitel über den vertrauensärztlichen Dienst berichtet über die inzwischen veränderten Aufgaben dieser Institution und die damit gegebenen Möglichkeiten für eine Berücksichtigung psychosomatischer Fragen in der Begutachtung.

Teil III (Spezielle Psychosomatik) bringt vier neue Kapitel: „Bulimie", „Schlafstörungen", „AIDS" und – als besondere Akzentsetzung – „Psychiatrie".

Das Kapitel 51 der dritten Auflage über das Krebsproblem sollte in Anbetracht des außerordentlich angewachsenen Schrifttums, sowohl über Fragen der Pathogenese als auch über Möglichkeiten der Therapie, auf zwei Autoren verteilt werden. Leider konnte C.B. Bahnson sein Kapitel über Familientherapie bei Krebskranken aufgrund äußerer Umstände nicht rechtzeitig fertigstellen. Wir hoffen, daß es in der nächsten Auflage erscheinen wird. Daher enthält diese Auflage nur das Kapitel über die Bedeutung psychischer und sozialer Faktoren für die Entstehung und den Verlauf maligner Erkrankungen von C. Hürny.

Für das Kapitel „Pädiatrie" fanden wir Autoren (D.B. Bürgin und B. Rost), die bereit waren, im Rahmen der entwicklungspsychologischen Aspekte die Bedeutung der neuen Selbst-Psychologie darzustellen, und damit ein Thema abzudecken, das für die Theorienbildung der Psychosomatischen Medizin von grundsätzlicher Bedeutung ist.

Für das Kapitel 58 der dritten Auflage „Funktionelle Störungen in der HNO-Heilkunde" wurde ein neuer Autor (J. Sopko) gefunden. Das Thema wird, von einem zweiten Autor (H. Bauer), durch ein Kapitel „Phoniologische Störungen" ergänzt.

In Teil V (Aus- und Fortbildung) wurde das Kapitel „Fort- und Weiterbildung" im Hinblick auf die Einführung einer „Psychosomatischen Grundversorgung" erweitert.

Wir waren wieder – wie wir meinen, mit Erfolg – bemüht, die integrierende Funktion des theoretischen Grundkonzepts in der wachsenden Vielfalt der Einzelbeiträge zum Ausdruck zu bringen.

Trotz des erweiterten Inhalts gelang es dem Verlag, das bisherige Format des Buches beizubehalten.

Wir danken den Autoren für ihre Bereitschaft zur Diskussion ihrer Beiträge mit den Herausgebern und für ihre Disziplin bei der Einhaltung der Termine. Dem Verlag danken wir besonders für sein großzügiges Entgegenkommen bei der Vorbereitung dieser vierten Auflage.

Freiburg,
April 1990

Im Namen der Mitherausgeber
Thure von Uexküll

Vorwort zur ersten Auflage

Wer dieses Lehrbuch in die Hand nimmt, möchte – ehe er sich entschließt, es zu lesen, darüber informiert sein, was ihn erwartet. Vor allem möchte er wissen, was unter „Psychosomatischer Medizin" verstanden, und was nicht darunter verstanden wird. Nicht darunter verstanden wird eine Disziplin, die der Meinung ist, eine begrenzte Anzahl von Krankheiten als „psychosomatisch" etikettieren zu können. In diesem Buch wird vielmehr die Auffassung vertreten, daß psychosoziale Einflüsse auf Entstehung, Verlauf und Endzustände von Krankheiten ebenso wichtige und legitime Probleme für die Heilkunde aufwerfen wie die Einflüsse physikalischer, chemischer oder mikrobiologischer Faktoren. Psychosomatische Medizin unter einem – wie man es auch nennen kann – psychobiologischen Aspekt – hat in Deutschland eine lange Geschichte, die in diesem Jahrhundert auf Ludolf Krehl, Gustav v. Bergmann und Viktor v. Weizsäcker zurückgeht, und die ihren Schwerpunkt in der Inneren Medizin hatte. Mit der Aufsplitterung dieses Faches in eine Reihe von Subdisziplinen ist eine neue Situation entstanden. Eine Reaktion darauf besteht in dem Rat, Psychosomatische Medizin solle sich auf einen ihrer methodischen Ansätze zurückziehen, sei er nun psychoanalytischer oder lerntheoretischer Provenienz, und sich wie die anderen medizinischen Disziplinen zu einem Spezialfach verengen. Eine andere Antwort auf diese Situation verlangt von dem psychosomatisch tätigen Arzt, daß er zu den bisherigen Aufgaben und Problemen auch das Problem einer Integration spezialistischer Ansätze sieht und sich diesem Problem stellt. Die Konsequenz ist eine Öffnung der Psychosomatischen Medizin zu neuen Fragestellungen, neuen Konzepten und neuen Methoden. Diese Antwort auf die Herausforderung der heutigen Situation wird in dem Lehrbuch vertreten. Sie verlangt von dem psychosomatisch tätigen Arzt, daß er bereit ist, sich mit dem Problem der Integration bei Diagnose und Therapie jedes einzelnen Kranken und auf theoretischer Ebene mit neuen Konzepten und deren Erprobung auseinanderzusetzen. Sie verlangt darüber hinaus das Suchen nach neuen Formen der kollegialen Zusammenarbeit, sowohl in der Krankenversorgung wie in der Forschung.

Der Plan zu diesem Lehrbuch und die ersten Anfänge zu seiner Verwirklichung wurden in einer Gruppe von Ärzten gefaßt, die im Rahmen einer Abteilung für Innere Medizin und Psychosomatik zehn Jahre an einer modernen, hochspezialisierten internistischen Universitätsklinik gearbeitet haben. Ich möchte an dieser Stelle allen Mitarbeitern danken, welche die Belastungen einer doppelten Weiterbildung in Innerer Medizin und Psychotherapie, die inneren und äußeren Spannungen und die Schwierigkeiten der Identitätsfindung ertragen haben, denen sie in diesen Jahren ausgesetzt waren. Ihre Mitarbeit an zahlreichen Kapiteln dieses Lehrbuchs ist auch ein Zeugnis ihres Entwicklungsprozesses. Darüber hinaus möchte ich allen anderen Autoren für ihre Bereitwilligkeit danken, ihre Beiträge uns anzuvertrauen. Das gilt besonders für die drei amerikanischen Autoren, deren Beiträge uns in eindrucksvoller Weise zeigen, was Psychosomatische Medizin durch eine Öffnung zu enger Grenzen gewinnen kann. Mein Dank gilt auch Ursula Lodders, Gertrud Müller, Klara Paulus, Waltraud Pfister, Heidrun Richter und Renate Senft für die vorbildliche Mitarbeit bei der Abschrift von Protokollen und Manuskripten. Schließlich gilt mein Dank nicht zuletzt dem Verleger, vor allem für die Geduld, die er mit uns gehabt hat.

Freiburg, Frühjahr 1979   *Thure von Uexküll*

Inhaltsverzeichnis

Inhaltsverzeichnis

Inhaltsverzeichnis

Inhaltsverzeichnis

Inhaltsverzeichnis

Inhaltsverzeichnis

Verzeichnis der Herausgeber und Verfasser

Herausgeber

Prof. Dr. Thure von Uexküll
Sonnhalde 15
7800 Freiburg

Prof. Dr. Rolf Adler
C.L. Lory-Haus
Med. Abteilung
Inselspital
CH-3010 Bern

Prof. Dr. Jörg Michael Herrmann
Klinik für Rehabilitation Glotterbad
der LVA Württemberg
7804 Glottertal

Prof. Dr. Karl Köhle
Institut für Psychosomatik und
Psychotherapie der Universität
Joseph-Stelzmann-Str. 9
5000 Köln 41

Priv.-Doz. Dr. Othmar W. Schonecke
Institut für Psychosomatik und
Psychotherapie der Universität
Joseph-Stelzmann-Str. 9
5000 Köln 41

Prof. Dr. Wolfgang Wesiack
Institut für Med. Psychologie und
Psychotherapie der Universität
Sonnenburgstr. 16
A-6020 Innsbruck

Verfasser

Prof. Dr. Hans Hermann Bauer
Poliklinik für Phoniatrie und
Pädaudiologie
Kardinal-von-Galen-Ring 10
4400 Münster

Dipl.-Psych. Sophinette Becker
Abt. für Sexualwissenschaft
Klinikum der Universität
Theodor-Stern-Kai 7
6000 Frankfurt 70

Priv.-Doz. Dr. Wolfgang Beischer
Medizinische Klinik II
Bürgerhospital
Tunzhoferstr. 14–16
7000 Stuttgart 1

Dipl.-Psych. Dr. Manfred Beutel
Institut für Medizinische Psychologie
und Psychotherapie
Technische Universität
Langerstr. 3
8000 München 80

Dr. Wilfried Biebl
Universitätsklinik für Psychiatrie
Anichstr. 35
A-6020 Innsbruck

Dr. Friederike Bischof
Wielandstr. 73
7900 Ulm

Priv.-Doz.-Dr. Claus Bischoff
Psychosomat. Fachklinik
Kurbrunnenstr. 12
6702 Bad Dürkheim

Prof. Dr. Dr. Klaus Bosse
Universitäts-Hautklinik
von-Siebold-Str. 3
3400 Göttingen

Prof. Dr. Dieter Bürgin
Psychiatrische Universitäts-Poliklinik
für Kinder und Jugendliche
Schaffhauser Rheinweg 55
CH-4058 Basel

Dipl.-Psych. Dr. Ulrich Clement
Psychosomat. Klinik der Universität
Thibautstr. 2
6900 Heidelberg

Prof. Dr. Bernhard Dahme
Psycholog. Institut III der Universität
von-Melle-Park 5
2000 Hamburg 13

Dr. Hans-Joachim Demmel
Lietzenburger Str. 51
1000 Berlin 30

Prof. Dr. Peter Diederichs
Südendstr. 3
1000 Berlin 41

Dipl.-Psych. Dr. Dirk Fehlenberg
Landeskrankenhaus Weißenau
7980 Ravensburg-Weißenau

Prof. Dr. Horst-Lorenz Fehm
Med. Klinik und Poliklinik
der Universität
Abt. Innere Medizin I
Oberer Eselsberg
Postfach 3880
7900 Ulm

Prof. Dr. Hubert Feiereis
Klinik für Psychosomatik und
Psychotherapie der Universität
Ratzeburger Allee 160
2400 Lübeck

Dr. Ekkehard Gaus
Städt. Krankenanstalten
Psychosomat. Abteilung
Hirschlandstr. 97
7300 Esslingen

Dr. Werner Geigges
Klinik für Rehabilitation Glotterbad der
LVA Württemberg
7804 Glottertal

Dr. Rüdiger Großpietzsch
Ubbo-Emmius-Str. 2
2960 Aurich

Dr. Ernst-Albrecht Günthert
Leopoldstr. 58
8000 München 40

Dipl.-Psych. Willi Hemmeler
C.L. Lory-Haus
Med. Abteilung
Inselspital
CH-3010 Bern

Marianne Holzamer
Landeskrankenhaus Weißenau
7980 Ravensburg-Weißenau

Dr. Christoph Hürny
C.L. Lory-Haus
Med. Abteilung
Inselspital
CH-3010 Bern

Dr. Peter Joraschky
Psychiatrische Universitätsklinik
und Poliklinik
Schwabachanlage 6
8520 Erlangen

Dr. Sybille Klosterhalfen
Institut für Med. Psychologie
der Universität
Moorenstr. 5
4000 Düsseldorf 1

Priv.-Doz. Dr. Wolfgang Klosterhalfen
Institut für Med. Psychologie
der Universität
Moorenstr. 5
4000 Düsseldorf 1

Prof. Dr. Uwe Koch
Psycholog. Institut der Universität
Belfortstr. 16
7800 Freiburg

Prof. Dr. Hugo M. Krott
Gänsbühl 2
7980 Ravensburg

Prof. Dr. Bernhard Kubanek
Med. Klinik und Poliklinik
der Universität
Abt. Transfusionsmedizin
Oberer Eselsberg 10
7900 Ulm

Dr. Mechthilde Kütemeyer
St. Agatha-Krankenhaus
Psychosomat. Abteilung
Feldgärtenstr. 97
5000 Köln 60

Prof. Dr. Friedrich Lamprecht
Psychosomat. Fachklinik
Dr. Schröder-Weg 12
7542 Schömberg

Dr. Wolfgang Langosch
Rehabilitationszentrum für Herz-
und Kreislaufkranke
Südring 15
7812 Bad Krozingen

Prof. Dr. Kurt Loewit
Institut für Med. Biologie und Genetik
Abt. für Fortpflanzungsbiologie und
Sexualmedizin
Schöpfstr. 41
A-6020 Innsbruck

Prof. Dr. Reinhard Lohmann
Hüttenrothstr. 9
3524 Immenhausen 2

Dipl.-Psych. Rolf Manz
Zentralinstitut f. Seelische Gesundheit
J5
Postfach 5970
6800 Mannheim 1

Dipl.-Psych. Christine Muck-Weich
Institut für Psychosomatik und
Psychotherapie der Universität
Joseph-Stelzmann-Str. 9
5000 Köln 41

Prof. Dr. Hans Müller-Braunschweig
Volpertstriesch 4
6301 Wettenberg-Launsbach

Dr. Fritz A. Muthny
Psycholog. Institut der Universität
Belfortstr. 16
7800 Freiburg

Prof. Dr. Dr. Peter Novak
Universität Ulm
Abt. Med. Soziologie
Am Hochsträß 8
7900 Ulm

Dr. Gerhard H. Paar
Gelderland-Klinik
Fachklinik für Psychotherapie und
Psychosomatik
Clemensstraße
4170 Geldern

Prof. Dr. Hannes G. Pauli
Holligenstr. 93
CH-3008 Bern

Merita J. Poremba
Melanchthonstr. 16
7400 Tübingen

Dipl.-Psych. Christa Probst-Geigges
Klinik für Rehabilitation Glotterbad
des LVA Württemberg
7804 Glottertal

Prof. Dr. Volker Pudel
Ernährungspsycholog. Forschungsstelle
der Universität
von-Siebold-Str. 5
3400 Göttingen

Prof. Dr. Michael von Rad
Städt. Krankenhaus Bogenhausen
Abt. für Psychosomat. Medizin und
Psychotherapie
Englschalkinger Str. 77
8000 München 81
und
Institut und Poliklinik für Psycho-
somat. Medizin, Psychotherapie und
Med. Psychologie
Technische Universität
Langerstr. 3
8000 München 80

Prof. Dr. Hartmut Radebold
Gesamthochschule Kassel
Interdisziplinäre Arbeitsgruppe für
Angewandte Soziale Gerontologie
Mönchebergstr. 19
3500 Kassel

Dr. Andreas Radvila
Med. Poliklinik
Inselspital
CH-3010 Bern

Prof. Dr. Dr. Hans-Heinrich Raspe
Institut für Sozialmedizin d. Universität
Sophienstr. 2
2400 Lübeck

Dr. Michael Rassek
Märchenweg 11
3500 Kassel

Dr. Michaela E. Rauch
Berliner Str. 35
7980 Ravensburg

Prof. Dr. Dietmar Richter
Kreiskrankenhaus
Gynäkolog. Abteilung
7880 Bad Säckingen

Dr. Rainer Richter
II. Med. Universitätsklinik u. Poliklinik
Martinistr. 52
2000 Hamburg 20

Dr. Barbara Rost
Psychiatr. Universitäts-Poliklinik
für Kinder und Jugendliche
Schaffhauser Rheinweg 55
CH-4058 Basel

Dr. Lore Schacht
Panoramastr. 6/2
7830 Emmendingen 13-Windenreute

Priv.-Doz. Dr. Nikolaus Schäfer
Kreiskrankenhaus
Innere Abteilung
Am Kälblesrainweg 1
7080 Aalen

Priv.-Doz. Dr. Thomas H. Schmidt
Med. Hochschule Hannover
Abt. Epidemiologie und Sozialmedizin
Konstanty-Gutschow-Str. 8
3000 Hannover 61

Dr. Doris Schmitt
Droste-Hülshoff-Str. 22
4000 Düsseldorf 30

Priv.-Doz. Dr. Dietrich Schneider-
Helmert
Med. Centrum Mariastein
CH-4115 Mariastein/Basel

Prof. Dr. Wolfram Schüffel
Zentrum für Innere Medizin
der Universität
Abt. Psychosomatik
Baldingerstraße
3550 Marburg

Dr. Ulrich Schultz
Institut für Psychosomatik und
Psychotherapie der Universität
Joseph-Stelzmann-Str. 9
5000 Köln 41

Dr. Wolfgang Schultz-Zehden
Mehringdamm 40
1000 Berlin 61

Dr. Wolfgang Senf
Psychosomat. Universitätsklinik
Thibautstr. 2
6900 Heidelberg

Dipl.-Psych. Claudia Simons
Universität Ulm
Abt. Med. Psychologie
Am Hochsträß 8
7900 Ulm

Dipl.-Sozialpäd. Beate Sollors-Mossler
Zentralinstitut f. Seelische Gesundheit
J5
Postfach 5970
6800 Mannheim 1

Priv.-Doz. Dr. Joseph Sopko
Universitätsklinik und Poliklinik
für IINO
Kantonsspital Basel
Petersgraben 4
CH-4031 Basel

Prof. Dr. Manfred Stauber
I. Universitäts-Frauenklinik
Maistr. 11
8000 München 2

Dipl.-Psych. Wolfgang Stiels
Klinik für Rehabilitation Glotterbad der
LVA Württemberg
7804 Glottertal

Prof. Dr. Albert J. Stunkard
Hospital of the University of
Pennsylvania
205 Piersol Building
Philadelphia, PA 19104, USA

Dipl.-Psych. Dr. Walter Thomas
Institut für Psychosomatik und
Psychotherapie der Universität
Joseph-Stelzmann-Str. 9
5000 Köln 41

Priv.-Doz. Dr. Harald Traue
Universität Ulm
Abt. Med. Psychologie
Am Hochsträß 8
7900 Ulm

Priv.-Doz. Dr. Dr. Wolfgang Tress
Zentralinstitut f. Seelische Gesundheit
J 5
Postfach 5970
6800 Mannheim 1

Dr. Daniel Vasella
Med. Universitätsklinik
C.L. Lory-Haus
Inselspital
CH-3010 Bern

Prof. Dr. Karlheinz Voigt
Institut für Normale und Pathologische
Physiologie der Universität
Deutschhausstr. 1–2
3550 Marburg

Prof. Dr. Herbert Weiner
University of California
Neuropsychiatric Hospital
Center for the Health Sciences
760 Estwood Plaza
Los Angeles, CA 90024, USA

Prof. Dr. Michael Wirsching
Zentrum für Psychosomat. Medizin
der Universität
Friedrichstr. 28
6300 Gießen

Verzeichnis der Herausgeber und Verfasser

Prof. Dr. Helmuth Zenz
Universität Ulm
Abt. Med. Psychologie
Am Hochsträß 8
7900 Ulm

Jutta Zenz
Weiterbildungsbeauftragte des
Pflegesektors, Zentrum für
Innere Medizin
der Universität
Steinhövelstr. 9
7900 Ulm

Prof. Dr. Siegfried Zepf
Klinisches Institut für Psychotherapie
und Psychosomatik der Universität
Moorenstr. 5
4000 Düsseldorf 1

Einleitung

Thure von Uexküll

Ein Lehrbuch muß sich an seinen Adressaten und seinem Gegenstand orientieren. Beides bietet im Fall der Psychosomatischen Medizin Schwierigkeiten, die von Anfang an deutlich gemacht werden müssen.

Beginnen wir mit den Adressaten: Das Buch wendet sich in erster Linie an Ärzte und Medizinstudenten, darüber hinaus aber auch an alle, die sich für die Frage interessieren, welche Rolle die individuell erlebte Wirklichkeit eines Menschen für dessen Gesundsein oder Kranksein spielt, wenn wir unter dieser Wirklichkeit nicht nur die hell erleuchtete Bühne verstehen, auf der die bewußten Auseinandersetzungen mit den Mitmenschen und Gegenständen unserer Umgebung stattfinden, sondern auch den Raum hinter den Kulissen, in dem Stimmungen und Gefühle für die wechselnde Beleuchtung des Vordergrundes sorgen. Damit sind alle angesprochen, die der Beruf mit Problemen menschlichen Verhaltens und Reagierens konfrontiert, also auch Psychologen, Pädagogen, Soziologen, Sozialarbeiter und andere.

Derartig verschiedene Adressaten unterscheiden sich nicht nur durch ihre Interessen, sondern auch durch ihre Vorkenntnisse, Vorerfahrungen und Vorurteile. Entsprechend verschieden wird ihr Urteil ausfallen. So werden Medizinstudenten, die für das Examen arbeiten, in dem Buch manches zu ausführlich, zu speziell oder zu abstrakt dargestellt finden. Sie stoßen erst später auf die Probleme und theoretischen Hintergründe der Gegenstände, die im Examen geprüft werden. Immerhin – seit Psychologie, Soziologie und Psychosomatische Medizin an den Medizinischen Fakultäten der Bundesrepublik gelehrt und geprüft werden, ist der Inhalt dieses Buches für die heutigen Medizinstudenten kein so fremdes Land mehr, wie es noch vor wenigen Jahren der Fall war.

Ärzten wird je nach Ausbildung, Fachrichtung und Spezialinteresse manches zu vereinfacht oder zu kurz abgehandelt erscheinen. Sie werden eine Reihe wichtiger Themen vermissen. Anderen wieder wird vieles überflüssig und kompliziert vorkommen. Ähnlich verschiedenartig wird es Lesern aus anderen Berufen mit anderen Erfahrungshintergründen ergehen.

Jedoch in einem Punkt werden die meisten Adressaten einander ähnlich sein: Sie üben einen Beruf aus, in dem Menschen Menschen anvertraut sind; aber sie wurden nicht dafür ausgebildet, andere Menschen und deren Motivationen zu verstehen und sich Rechenschaft zu geben, welche Auswirkungen ihr Verhalten auf die Motivationen, das Verhalten und das Befinden der ihnen anvertrauten Menschen hat. Die ärztliche Ausbildung ist dafür nur ein Beispiel unter vielen: Sie vermittelt kognitive Werkzeuge, um Krankheiten zu erkennen und zu behandeln; d.h., die Studenten lernen gedankliche Schemata (Krankheitsbilder), nach denen Krankheitszeichen (Symptome) erfaßt und für diagnostische Zwecke eingeordnet werden. Sie üben in Kursen für Inspektion, Auskultation, Perkussion und Palpation ihre Augen, Ohren und Hände, diese Krankheitszeichen zu sehen, zu hören und zu tasten. Sie haben aber kaum Gelegenheit, gedankliche Schemata zu erwerben und einzuüben, mit deren Hilfe sie lernen, auch ihre Gefühle bei der Begegnung mit kranken Menschen zu registrieren und als Reaktionen auf Gedanken, Wünsche, Gefühle und Motivationen dieser Menschen zu deuten und zu ordnen. „Empathie", wie man diese Fähigkeit zum gefühlsmäßigen Erfassen anderer Menschen nennt, bleibt für die meisten ein magischer Begriff.

Psychotherapeuten haben schon seit langem Situationen beschrieben, in denen zwischen ihnen und ihren Patienten keine Kommunikation gefühlsmäßiger Art zustande kommt. Neuere Theorien – sie werden in Kap. 5 des Teils II genauer dargestellt – deuten solche Erfahrungen als Unfähigkeit der Patienten, Gefühle zu entwickeln und mitzuteilen. Man hat dafür den Begriff „Alexithymie" geprägt (von den griechischen Wörtern legein = lesen und thymos = Gefühl) und eine spezifische Art, in klischeehaften, technischen Vorstellungen zu denken, „pensée opératoire" genannt. An dieser Theorie ist noch das meiste umstritten: Wir wissen noch nicht, ob derartige Erfahrungen Persönlichkeitsmerkmale bestimmter Patienten widerspiegeln – oder deren Reaktionen auf den Arzt – oder ein Unvermögen des Arztes, die Sprache des Patienten richtig zu deuten. An der Existenz des Phänomens ist jedoch kein Zweifel möglich: Es gibt zwischenmenschliche Situationen, in denen kein gefühlsmäßiger Kontakt zustande kommt und in denen sich Verständigung – wenn sie überhaupt erreicht wird – auf die oberflächlichsten Themen beschränkt. Es unterliegt auch keinem Zweifel, daß menschliche Kommunikation gelernt und geübt werden muß, und daß sie – wie alles, was wir gelernt haben – auch wieder verlernt werden kann. Unsere traditionelle ärztliche Ausbildung birgt daher die Gefahr, daß Medizinstudenten während ihres Studiums das meiste von dem verlernen, was sie an natürlichen Fähigkeiten zu einem empathischen Verstehen ihrer Mitmenschen mitgebracht hatten; daß sie in wenigen Jahren zu „emotionalen Analphabeten" werden, die über kranke Menschen nur noch in technischen Modellen denken und sich im Umgang mit ihnen nur an technischen Modellen orientieren. Das Ergebnis ist ein Gesundheitssystem, in dem nicht Ärzte alexithymen Pa-

1

tienten gegenübersitzen, sondern in dem Patienten ihre Sorgen und inneren Nöte emotionalen Analphabeten schildern müssen.

Eine derartige Kritik ist in dieser oder jener Form schon oft geäußert worden. Man hat auch darauf hingewiesen, daß die medizinische Ausbildung nur den spezifischen Umgangsstil unserer Industriekultur und ihre Überschätzung technischer Faktoren widerspiegelt, und daß ihre Tendenz, nur technische Errungenschaften als Fortschritt anzuerkennen, zu einer Vernachlässigung der gefühlsmäßigen und emotionalen Seiten des Lebens geführt hat.

Klagen dieser Art haben nicht viel an den Verhältnissen zu ändern vermocht, aber es scheint ein ermutigendes Zeichen zu sein, daß sie sich immer weniger leicht als bloßer Kulturpessimismus abtun lassen. Der überraschende Zuwachs an Interesse und Sympathie, den Psychosomatische Medizin in den letzten Jahren in der Öffentlichkeit und bei der jungen Generation gefunden hat, läßt hoffen, daß allen Widerständen und Hindernissen zum Trotz ein Prozeß der Besinnung und Neuorientierung eingesetzt hat.

Aber hier beginnen auch die Schwierigkeiten, mit denen der Gegenstand eines Lehrbuchs für Psychosomatische Medizin behaftet ist: „Die Einführung des Menschen als Subjekt in die Medizin", um ein Wort Viktor von Weizsäckers zu zitieren, wird in einer alexithymen Kultur ein sehr „langer Marsch" sein. Die Notwendigkeit, immer wieder einen fundamentalen Widerspruch in der modernen Heilkunde deutlich zu machen, schafft Widerstände, die es der Psychosomatischen Medizin nicht erleichtern, die Mittel zu entwickeln, die notwendig sind, um diesen Widerspruch zu überwinden.

Voraussetzung jeder wirksamen Hilfe für Kranke ist eine genaue Analyse ihrer Krankheitsbilder; aber erst die Synthese der analytisch gewonnenen Resultate zeigt dem Arzt den Weg zum Helfen. Die moderne Heilkunde hat die analytischen Methoden zur Gewinnung von Teildiagnosen immer weiter verfeinert. Sie hat sich auf diesem Weg in immer weitere Spezialdisziplinen aufgefächert. Aber sie hat die Frage, wie die Synthese der Teildiagnosen zur Gesamtdiagnose gewonnen wird, ebenso vernachlässigt wie die Frage, auf welche Weise die Spezialdisziplinen für die Behandlung von Kranken organisatorisch integriert werden können. Dieses Dilemma ist ein zentrales Thema der Psychosomatischen Medizin. Sie muß einerseits auf einer genauen Analyse der Krankheitsbilder ihrer Patienten mit den modernsten Methoden der somatischen und der psychologischen Medizin bestehen. Aber sie ist kein einfaches additives Nebeneinander somatologischer und psychologischer Methoden. Ihr Ausgangspunkt und Ziel ist vielmehr das Bemühen um deren Synthese. Aber sobald sie diese zentrale Aufgabe in Angriff nimmt, wird sie auf allen Ebenen zu einem Ärgernis:

Auf der Ebene der Theorien- und Konzeptbildung stellt sich heraus, daß somatische und psychologische Medizin auf zwei heterogenen Grundmodellen basieren, die sich nicht gegenseitig ergänzen, sondern gegenseitig ausschließen. Die somatische Medizin arbeitet mit dem allgemeinen Modell eines anatomischen Körpers, in dem sich Organe und Organsysteme aufgrund komplizierter biochemischer Stoffwechselvorgänge erhalten. Dieses Modell gibt den allgemeinen Rahmen, in dem der Arzt spezielle Störungen, z.B. einen Diabetes mellitus, eine Gefäßsklerose, ein Emphysem usw. suchen, feststellen und für therapeutische Eingriffe chirurgischer oder medikamentöser Art räumlich lokalisieren kann. Die psychologische Medizin verwendet das ganz andere Modell eines „psychischen Apparates", der im Verlauf einer biographischen Geschichte entstanden ist. Auch dieses Modell gibt einen allgemeinen Rahmen, in dem der Arzt spezielle Störungen suchen, feststellen und für sein therapeutisches Vorgehen „lokalisieren" kann. Aber die speziellen Störungen heißen jetzt Konflikt, Konversion, Regression usw., und Lokalisation bedeutet jetzt Orientierung in der biographischen Entwicklung eines Kranken. Beide Modelle erfassen verschiedenartige Ausschnitte aus dem Gesamtgeschehen eines menschlichen Schicksals. Aber dieses Gesamtgeschehen läßt sich nicht durch eine Addition der beiden Ausschnitte rekonstruieren.

Hier wird das theoretische Ärgernis zu einem praktischen: Ohne ein Bild von dem Gesamtgeschehen ist der Arzt außerstande, das Gewicht der Störungen, die er in den beiden Ausschnitten vorfindet, zu ermitteln. Nur wenn es ihm gelingt, die verschiedenen speziellen Diagnosen, die bei einem Kranken gestellt wurden, als Teildiagnosen in eine Gesamtdiagnose einzuordnen, kann er abschätzen, wie ein Diabetes mellitus, eine Gefäßsklerose, ein Emphysem und eine neurotische Störung sich bei dem individuellen Patienten gegenseitig beeinflussen und welches Gewicht sie für diesen und sein Kranksein haben. Erst aufgrund einer solchen Gesamtdiagnose kann er einen rationalen Therapieplan aufstellen. Aber weder die somatische noch die psychologische Medizin besitzen ein Modell oder ein Schema für die Synthesen zu einer umfassenden Gesamtdiagnose.

An dieser Stelle wird aus dem praktischen Ärgernis ein berufspolitisches: Sowohl die somatische wie die psychologische Medizin begnügen sich mit Teildiagnosen und verleugnen die Notwendigkeit von Gesamtdiagnosen, die über ihren jeweiligen Bezugsrahmen hinausgehen. Sie rationalisieren ihre Verleugnung mit großen Worten. Die Somatiker zitieren Naunyn: „Medizin wird Naturwissenschaft sein – oder sie wird nicht sein."* Die Psychoanalytiker berufen sich auf Viktor von Weizsäckers Ausspruch: „Die

---

* Mit dem (falsch zitierten) Ausspruch setzt sich Naunyn mit der These auseinander, die Medizin sei nicht nur Wissenschaft, sondern auch Kunst. Dagegen wendet er sich und schreibt: „Die Medizin wird Wissenschaft sein, oder sie wird nicht sein". Wie wenig Naunyn Wissenschaft mit Naturwissenschaft gleichsetzte, geht aus einem anderen Ausspruch hervor: „Es ist richtig, wenn man sagt, das 19. Jahrhundert habe die Entwicklung der Medizin zu einer Naturwissenschaft gebracht; ... eine Naturwissenschaft aber ist sie darum auch im 19. Jahrhundert nicht geworden und wird sie auch schwerlich jemals werden ... Dazu sitzt ihr die Humanität zu tief im Blute" (Naunyn, 1906b).
Wir verdanken diesen Hinweis Prof. Dr. E. Seidler, Institut für Geschichte der Medizin, Freiburg.

Psychosomatische Medizin wird eine tiefenpsychologische sein – oder sie wird nicht sein" und beanspruchen damit die Psychosomatische Medizin für sich. Aber weder der Ausspruch Naunyns noch der Viktor von Weizsäckers enthält die ganze Wahrheit. Medizin wird nicht Naturwissenschaft oder Tiefenpsychologie sein, sondern für beide die Synthese finden müssen.

Es gibt keinen Gegensatz zwischen Medizin und Psychosomatischer Medizin. Es gibt aber ein psychosomatisches Problem der Medizin, das von beiden Seiten verleugnet wird, wenn sich deren Vertreter wie Wissenschaftler verhalten, von denen Feigel (1958) gesagt hat: „Robust strukturierte Wissenschaftler neigen heute dazu, das Leib-Seele-Problem in die Rumpelkammer der spekulativen Metaphysik zu verbannen. Nachdem sie vielleicht – aber mit zweifelhaftem Erfolg – versucht haben, sich selbst mit dem Rätsel herumzuschlagen, ziehen sie es vor, sich die Sache leicht zu machen: Sie überlassen es den Philosophen, sich daran die Zähne auszubeißen, oder sie erklären geradewegs, das Ganze sei ein Scheinproblem und nicht wert darüber nachzudenken ...; in Wahrheit handelt es sich (aber) um einen Komplex verwickelter Rätsel, von denen einige empirischer, andere wissenschaftstheoretischer, einige syntaktischer, andere semantischer und wieder andere pragmatischer Art sind."

Der Gegenstand eines Lehrbuchs der Psychosomatischen Medizin wird also notwendigerweise unbequem sein. Er konfrontiert die Ärzte mit der Tatsache, daß ein Grundproblem ihres Berufes nicht dadurch aus der Welt geschafft werden kann, daß man es ignoriert.

Aber das ist der geringere Teil der Schwierigkeiten, mit denen der Gegenstand für ein solches Lehrbuch belastet ist. Viel schwerer wiegt die Tatsache, daß der Weg zur Synthese lang und steinig ist. Das bedeutet für ein Lehrbuch, daß Psychosomatische Medizin nicht als ein einheitliches, in sich abgeschlossenes Systemgebäude dargestellt werden kann. Mit der Aufgabe, den Menschen als psychophysische Einheit zu erfassen, steht sie vor einer Schwierigkeit, welche den Fächern der Medizin, die im Zuge der arbeitsteiligen Differenzierung unserer Industriekultur der allgemeinen Tendenz zur Spezialisierung und Subspezialisierung folgen, erspart bleibt. Diese Fächer können ihre Gebiete als methodisch und begrifflich einheitliche Systeme darstellen. Psychosomatische Medizin hat – auch wo sie im Rahmen der einzelnen Fächer, etwa der Inneren Medizin, der Gynäkologie usw., mit den speziellen Problemen dieser Disziplin befaßt ist – immer die Aufgabe, diese Probleme unter dem Aspekt der Einheit des erkrankten Menschen und der Veränderung dieser Einheit zu sehen. Sie hat auch dort – und gerade dort – eine integrative Aufgabe.

Das verlangt von dem psychosomatisch tätigen Arzt nicht nur die Bereitschaft zu ständiger kritischer Analyse des Menschenbildes, von dem die traditionellen Fächer ausgehen, sondern genauso die Bereitschaft zu ständiger kritischer Analyse und Revision seiner eigenen Voraussetzungen. Er muß bereit sein, das Bild, das er sich vom Menschen und damit auch von sich selbst gemacht hat, zu verändern, sobald sich die Notwendigkeit ergibt. Unter diesem Aspekt ist die Vorläufigkeit und Unabgeschlossenheit der Psychosomatischen Medizin kein bloßer Ausdruck einer Unfertigkeit ihrer Entwürfe, Konzepte und Methoden, sondern auch ein Niederschlag der prinzipiellen Unabgeschlossenheit und Offenheit des Menschen, der sich mit dem Bild, in dem er sich selbst erfährt, auch selbst verändert.

Es lag den Herausgebern des Lehrbuchs daran, diese Offenheit, die sich auch in einer Vielfalt der Aspekte und Ansätze äußert, nicht zugunsten einer einheitlichen, aber notwendigerweise einseitigen Darstellung zu verdecken. Ein Lehrbuch, das einen sich entwickelnden Gegenstand darzustellen versucht, darf die Vorläufigkeit nicht scheuen und muß auch heterogene Standpunkte zu Wort kommen lassen.

Die kurze Geschichte der Psychosomatischen Medizin unterstreicht die Notwendigkeit einer solchen Haltung: In den USA wurde Psychosomatische Medizin lange Zeit nahezu mit Psychoanalyse identifiziert. Das brachte zunächst eine große Bereicherung. Dann aber führte der einseitige Versuch, Krankheit ausschließlich psychodynamisch zu deuten, dazu, daß Psychosomatische Medizin für zwei Jahrzehnte in einen Dornröschenschlaf verfiel, ja – wie Lipowski (1977) schreibt – fast ausgelöscht wurde. Nachdem sie sich dann aber, auch unter dem Eindruck der Arbeiten von H. G. Wolff und seiner Mitarbeiter, entschloß, einen multifaktoriellen Ansatz zu akzeptieren und neben den psychoanalytischen Konzepten auch psychophysiologische und epidemiologische Modelle sowie die Methoden der Streß-Forschung zu berücksichtigen, mit anderen Worten Krankheit als psychobiologisches Problem (Weiner, 1977) zu betrachten, erlebte sie in den siebziger Jahren ein geradezu spektakuläres Comeback (Lipowski). Heute „findet sie sich mitten im Hauptstrom der Gedanken- und Ideenbildung zeitgenössischer Medizin, in der Fragen nach der relativen Bedeutung psychischer, biologischer und sozialer Faktoren für Entstehung, Verlauf und Endzustände somatischer und psychiatrischer Krankheiten eine beherrschende Rolle spielen."

So hat in diesem Lehrbuch das Menschenbild der Psychoanalyse als noch immer eines der differenziertesten Modelle der menschlichen Psyche zwar einen bevorzugten Platz; aber von ihm führt keine tragbare Brücke zu dem Modell eines menschlichen Körpers, dessen subtile Modelle für biologische Mechanismen die Möglichkeiten moderner somatischer Diagnostik und Therapie eröffnen. Schon aus diesem Grund ist es notwendig zu zeigen, daß neben Freud auch Pawlow und Cannon grundlegende Beiträge für psychosomatische Modelle beigesteuert haben. Es ist weiter erforderlich, andere – vor allem verhaltenstheoretische – Modelle darzustellen, obgleich auch von ihnen eine Brücke zu den somatischen Modellen noch nicht genau sichtbar ist. Aber auch neuere Konzepte wie das der Alexithymie oder der Pensée opératoire, die alte Vorstellungen Freuds wieder aufgreifen und vertiefen, gehören in das Bild der modernen Psychosomatischen Medizin.

Das Buch wendet sich also an Leser, die nicht nur an Fakten, sondern auch an den Problemen interessiert sind, die hinter den Fakten stehen. Es wünscht sich Leser, die bereit sind, sich mit diesen Problemen auseinanderzusetzen, weil sie nicht nur die Psychosomatische Medizin, sondern die moderne Heilkunde überhaupt und damit jeden Arzt angehen.

Wir haben uns bemüht, die sonst kaum überschaubare Landschaft dieser Probleme zu gliedern:

Teil I bringt eine Auseinandersetzung mit wissenschaftstheoretischen Fragen der Psychosomatik. In ihm wird versucht, einen allgemeinen Bezugsrahmen zu entwickeln, oder anspruchsvoller ausgedrückt, eine allgemeine Theorie der Medizin. Uns ist bewußt, daß dies ein sehr kühner Versuch ist. Wir glauben aber, daß man Psychosomatische Medizin nicht darstellen kann, ohne diesen Versuch zu wagen. Wir glauben auch, daß der von uns entworfene Rahmen trotz unvermeidbarer Vorläufigkeit, Subjektivität und Korrekturbedürftigkeit einen brauchbaren Ansatz darstellt und die Möglichkeit bietet, die verschiedenen Konzepte Psychosomatischer Medizin zu ordnen und zueinander in Beziehung zu setzen. Wir haben auch in einzelnen der speziellen Kapitel versucht, diese Beziehungen aufzuzeigen und so die Brauchbarkeit des theoretischen Rahmens zu erproben. Bei den meisten Kapiteln, die ja aus den Federn verschiedener Autoren stammen, bleibt dies jedoch weitgehend eine Aufgabe für den interessierten Leser.

In **Teil II** „Psychosomatische Konzepte und Theorien" werden in drei Abschnitten die verschiedenen epidemiologischen und pathogenetischen Konzepte, die diagnostischen und therapeutischen Verfahren sowie die Organisationsformen psychosomatischer Krankenversorgung abgehandelt.

In **Teil III** „Psychosomatik einzelner Erkrankungen und in verschiedenen Fachgebieten" werden Störungen von Funktionsabläufen und spezielle internistische Krankheitsbilder dargestellt; daneben kommen andere klinische Fächer zu Wort. Hier zeigt sich die Zunahme des Interesses für die psychosomatische Betrachtungsweise in den letzten Jahren am eindrucksvollsten. Waren in den ersten beiden Auflagen des Lehrbuchs neben der Inneren Medizin nur die Gynäkologie und Geriatrie vertreten, so findet der Leser jetzt Kapitel über Neurologie, Pädiatrie, Dermatologie, Urologie, Augen-, Hals-Nasen-Ohren-Krankheiten und Zahnheilkunde sowie über Arbeit und Krankheit. Diese Kapitel machen deutlich, daß die Integration der Psychosomatischen Medizin jedem klinischen Fach Gewinn bringt. Es ist zu wünschen, daß das Beispiel der Autoren mehr Kollegen ihrer Fächer von dieser Tatsache überzeugt, als es bisher der Fall ist.

Die einzelnen Kapitel zeigen auch den unterschiedlichen Entwicklungsstand der Psychosomatischen Medizin auf den verschiedenen Gebieten, der das unterschiedliche Interesse der Ärzte an psychosozialen Faktoren in den verschiedenen Fächern und bei verschiedenen Krankheitsbildern widerspiegelt.

In **Teil IV** sind psychosomatische Probleme bei Schwerkranken abgehandelt. Diese fünf Kapitel zeigen eindrucksvoll, daß die Situation der Kranken in den Kliniken und in der Praxis an eine qualifizierte ärztliche Betreuung Anforderungen stellt, denen die moderne, immer arbeitsteiligere Medizin allein nicht gewachsen ist. Auch die Vorstellung, die diagnostischen und therapeutischen Probleme der Kranken durch Herbeiziehung konsiliarischer Experten lösen zu können, ist nur dann realistisch, wenn der erstbehandelnde verantwortliche Arzt in der Lage ist, die Probleme zu erkennen. Das aber ist in vielen Fällen schon mit der Fähigkeit identisch, sie auch zu lösen.

Überall stoßen wir auf Aufgaben, denen nur ein Arzt wirklich gewachsen ist, der gelernt hat, psychosomatisch zu denken und zu handeln.

Teil V besteht aus drei Kapiteln, in denen Fragen der Aus- und Weiterbildung behandelt werden. Er schließt mit einem Kapitel über wissenschaftstheoretische und berufspolitische Fragen, die durch die Forderung nach einer Integration der Psychosomatischen Medizin in die moderne Heilkunde aufgeworfen werden, und das Gedanken zum Problem der ärztlichen Verantwortung bringt.

Wir sind nicht der Meinung, daß in dem Lehrbuch „die" Psychosomatische Medizin dargestellt worden ist. Wir nehmen bewußt die Unvollständigkeit in Kauf, die darin begründet liegt, *daß Psychosomatische Medizin kein Fach ist, sondern in jedem einzelnen Fach dessen allgemeine Probleme zum Gegenstand hat.* Dieses Lehrbuch unterscheidet sich damit in seiner Absicht und in seinem Rahmen von anderen in deutscher Sprache erschienenen Büchern über dieses Thema. Es ist kein Kompendium und auch nicht nur mit dem Blick auf die praktische Anwendung geschrieben. Indem es auch die Probleme der Psychosomatischen Medizin und deren Position innerhalb der modernen Heilkunde sowie die sich daraus ergebenden Aufgaben darstellt, will es eine Lücke in dem System unserer modernen Gesundheitsversorgung ausfüllen. Auch das Wagnis, eine allgemeine Theorie der Heilkunde zu entwerfen, muß unter diesem Aspekt gesehen werden. *Dieser Versuch wird von einer Situation herausgefordert, in der die Medizin ohne Konzept für eine Integration der Gefahr einer hemmungslosen Aufsplitterung in immer neue Spezialdisziplinen ausgesetzt ist.*

Zu der Vorgeschichte des Lehrbuchs ist noch zu sagen, daß es ursprünglich aus Vorlesungen entstanden ist, die in Fortbildungsseminaren und Diskussionen mit den Mitarbeitern einer Abteilung für Innere Medizin und Psychosomatik erweitert und verändert wurden, einer Abteilung, die zehn Jahre lang in ein Department für Innere Medizin der seinerzeit neu gegründeten Universität Ulm integriert war. Daraus ist ein Vielautorenbuch mit einem Kern aus den Beiträgen der Mitarbeiter dieser Abteilung entstanden, die in den zehn Jahren die Dringlichkeit, aber auch die theoretischen und praktischen Schwierigkeiten erfahren haben, ein Modell für die Durchführung einer psychosomatischen Betreuung organisch Kranker zu entwickeln.

4

Teil I: Wissenschaftstheorie

1 Wissenschaftstheorie und Psychosomatische Medizin, ein bio-psycho-soziales Modell

Thure von Uexküll und *Wolfgang Wesiack*

1.1 Begründung für eine theoretische Einführung in ein Lehrbuch der Psychosomatischen Medizin

Medizinische Lehrbücher verzichten gewöhnlich auf eine theoretische Einführung. Sie kommen gleich zur „Sache". Ein Arzt, der sich über Infektionskrankheiten, Unfallchirurgie oder Vergiftungen informiert, braucht keine Einführung in die Theorie, auf der die betreffenden Lehrbücher aufbauen. So entsteht die Meinung, das Problem einer Theorie der Medizin würde entweder gar nicht existieren oder habe mit der Sache, die medizinische Lehrbücher vermitteln, nichts zu tun.

In Wahrheit können Ärzte und Medizinstudenten aber diese Lehrbücher nur deswegen ohne theoretische Einführung verstehen, weil sie während der ersten Semester ihrer medizinischen Ausbildung die Theorie erlernt haben, die dort vorausgesetzt wird. Wenn der Medizinstudent nach dem vorklinischen Studienabschnitt mit kranken Menschen in Berührung kommt, weiß er bereits, was „die Sache der Medizin" ist. Er hat während des Studiums in Physik, Chemie, Anatomie, Biochemie und Physiologie die Theorie erlernt, nach der er sich den Aufbau des menschlichen Körpers und die komplizierten Mechanismen, die in seinem Inneren ablaufen, vorzustellen hat. Soweit hier noch theoretische Probleme der Medizin existieren, gehören sie zu den Aufgaben der sogenannten Grundlagenwissenschaften, der Molekularbiologie, der Genetik, der Immunbiologie usw., die an der Aufklärung immer subtilerer biologischer Mechanismen arbeiten.

Auf diese Weise lernen Ärzte schon als Medizinstudenten, ein Modell auf den menschlichen Körper zu übertragen, das die Physik zur Lösung technischer Probleme entwickelt hat und das in der zweiten Hälfte des 19. Jahrhunderts seinen Siegeszug durch die Welt antrat: das Modell der Maschine.

Die Faszination dieses Modells für Ärzte beruht auf seiner Fähigkeit, ein räumliches Ordnungsschema bereitzustellen, von dem sich einfache Handlungsanweisungen für manuelle Eingriffe in den menschlichen Körper ableiten lassen. So wurde der nach dem Maschinenmodell gedeutete Körper „zur Sache der Medizin".

Von Ferber (1971) hat auf den inneren Zusammenhang dieser Lehre mit dem Entstehen der Industriekultur hingewiesen. Dieser Aspekt läßt uns besser verstehen, wie es möglich war, daß der Grundsatz der räumlichen Orientierung für manuelle Eingriffe im Verlauf der stürmischen Entwicklung zu dem hochkomplexen Theoriegebäude der modernen Medizin konsequent durchgehalten wurde. Die zunehmende Verfeinerung der Möglichkeiten für direkte Eingriffe der menschlichen Hand durch technische Apparaturen und für indirekte Eingriffe durch Pharmaka erzwang eine fortschreitende Differenzierung dieses Körpermodells, das dann umgekehrt wieder die Verfeinerung der Technik für Eingriffe vorantrieb. So entstand das imponierende Gebäude der modernen Medizin, das den menschlichen Körper nach dem Modell einer hochkomplexen physikalisch-chemischen Maschine interpretiert. Krankheit ist nach diesem Modell eine räumlich lokalisierbare Störung in einem technischen Betrieb, der zwar eine sehr komplexe, aber aufgrund des technischen Vorbilds doch überschaubare Struktur besitzt. Von diesem allgemeinen Modell lassen sich Diagnosen für konkrete Krankheiten als spezielle Spielregeln für den Umgang mit Kurzschlüssen, Rohrbrüchen, Transportproblemen oder ähnlichen technischen Fragen ableiten. Wie ein Techniker auf der Basis eines Schaltplans den Betriebsschaden eines Autos, eines Fernsehers oder Computers lokalisieren und danach die Reparatur durchführen kann, so kann der Arzt eine Krankheit, die als Betriebsschaden im menschlichen Körper – als Klappenfehler im Herzen, als Geschwür im Magen oder als Enzymdefekt in einem Gewebe oder Transportsystem – lokalisiert wurde, mit gezielten technischen Eingriffen (chirurgischer oder medikamentöser Art) reparieren.

Damit geriet der einfache Tatbestand, daß die „Sache" der Medizin immer gemeinsame Angelegenheit eines Kranken und eines Arztes ist, mehr und mehr in Vergessenheit, und mit ihr die noch bis tief in das 19. Jahrhundert hineinwirkende, teils vorwissenschaftliche, teils sozialepidemiologisch fundierte Erfahrung über Zusammenhänge zwischen Lebenssituationen von Individuen, insbesondere ihrer sozialen Lage, und der Entstehung spezifischer Krankheiten (Siegrist, 1975). Die Möglichkeit der Lokalisie-

5

rung von Krankheitsursachen im Körper machte es scheinbar überflüssig, nach psychischen oder sozialen Ursachen zu suchen.

So entstand die Vorstellung, daß psychische oder sozial ausgelöste Störungen weder „wirkliche" Krankheiten seien, noch zu „wirklichen" Krankheiten führen könnten. Störungen auf psychischer oder sozialer Grundlage würden neben den „wirklichen" Krankheiten, die in der Inneren Medizin, der Chirurgie und den anderen somatischen Fächern gelehrt werden, bestenfalls eine Sondergruppe von Beschwerdebildern darstellen, für deren Behandlung wieder eine neue Spezialdisziplin zuständig sei.

Wie sehr diese Vorstellung an der Wirklichkeit vorbeigeht, erfahren Studenten und Ärzte, sobald sie mit Patienten konfrontiert sind. Hier stellen sie fest, daß Magenbeschwerden, Herzsymptome und andere somatische Erscheinungen psychische und soziale Determinanten haben, und daß auf der anderen Seite seelische Störungen wie ein Delir oder Stimmungsschwankungen und deren soziale Auswirkungen somatisch bedingt sein können. Sie erfahren, daß der Arzt ständig vor der Frage steht, ob und wie weit Symptome eines Kranken oder der Verlauf einer Krankheit durch physische, psychische oder soziale Determinanten oder durch eine Kombination aus allen dreien bedingt sind, daß er immer wieder entscheiden muß, ob und welche biochemische, physikalische oder psychologische Methode für die Diagnostik und Therapie eingesetzt werden muß. Dies soll an einem exemplarischen Krankheitsfall, auf den wir uns auch später wiederholt beziehen werden, erläutert werden.

1.1.1 Ein exemplarischer Krankheitsfall

Das Sprechzimmer betritt eine 52jährige Frau und berichtet, daß sie in den letzten drei Wochen zweimal nachts Anfälle von akuter Atemnot bekommen habe. Die Luft sei ihr weggeblieben, und sie habe gemeint, sterben zu müssen. Auf die Bitte des Arztes, die Umstände zu schildern, unter denen die Atemnotanfälle aufgetreten seien, berichtet sie unter tiefem Seufzen, daß sie mit einem Ausländer in schlechter Ehe verheiratet sei, der sie vernachlässige und oft nächtelang wegbliebe. Die so bedrohlich empfundenen Atemnotanfälle seien aufgetreten, als ihr ältester, achtzehnjähriger Sohn ihr erklärt habe, er wolle sich nun von der Familie trennen und wegziehen. Nachdem sie dies alles in recht vorwurfsvollem Ton vorgebracht hat, bricht sie in Tränen aus.

Während des Berichtes der Patientin ändert sich die Stimmungslage des Arztes. Beim Eintreten nahm er eine kleine, adipöse – sie wog, wie sich später herausstellte, bei 161 cm Größe 108 kg – und kurzatmige Frau mit etwas zyanotischen Lippen wahr, die auf ihn zunächst einen „schmuddeligen" und unsympathischen Eindruck machte, obwohl sie, wie er später bemerkte, keineswegs ungepflegt war. Diese ablehnende Stimmung des Arztes, die der erste Eindruck hervorgerufen hatte, wandelte sich während des Berichtes der Patientin in wohlwollendes Interesse und Hilfsbereitschaft.

Die weitere Untersuchung der Patientin ergab Anzeichen einer durch Adipositas und eine leichte Hypertonie bedingten Herzinsuffizienz mit Linkshypertrophie des Herzens sowie eine leichte Erhöhung der Blutfette.

Dieser „banale Alltagsfall" aus der Sprechstunde wirft bei etwas genauerem Hinsehen bereits eine Fülle von Problemen auf. Ohne einen Anspruch auf Vollständigkeit zu erheben, wollen wir einige, die uns besonders wichtig erscheinen, aufzählen:
1. Warum kommt die Patientin gerade jetzt zum Arzt?
2. Warum sucht sie diesen und keinen anderen Arzt auf?
3. Woran leidet sie?
4. Wird der Arzt ihr Leiden erkennen?
5. Wird er ihr helfen können?

Ad 1: Warum kommt die Patientin gerade jetzt zum Arzt?
Sie hat doch offenbar schon lange eine Adipositas und die dadurch bedingte Herzinsuffizienz. Die unerfreuliche Ehesituation besteht ebenfalls schon seit einiger Zeit. Ob die leichte Hypertonie und Hyperlipidämie älteren oder jüngeren Datums sind, läßt sich, da keine früheren Untersuchungsergebnisse vorliegen, nicht entscheiden. Der Grund ihres Kommens ist wohl darin zu sehen, daß sie die beiden Atemnotanfälle, die möglicherweise etwas mit dem drohenden Auszug ihres Sohnes zu tun haben, sehr beunruhigen. Jetzt erst fürchtet sie herzkrank zu sein und womöglich sterben zu müssen. Deshalb sucht sie wohl jetzt den Arzt auf.

Ad 2: Warum sucht sie diesen und keinen anderen Arzt auf?
Wir wissen es nicht. Wir können nur vermuten, daß sie – aus welchen Gründen auch immer – gerade von diesem Arzt erwartet, er werde ihr Leiden richtig erkennen und ihr auch helfen können.

Ad 3: Woran leidet sie?
Der Versuch, diese Frage zu beantworten, wirft sofort eine Reihe weiterer Fragen auf. Daß sie eine Adipositas, eine Herzinsuffizienz und eine leichte Hypertonie hat, ist offensichtlich. Aber wie steht es mit den nächtlichen Atemnotanfällen? Welche Rolle spielen dabei physiologische und welche psychische Faktoren? Die Enttäuschung und Verzweiflung über den bevorstehenden Auszug des Sohnes hat offensichtlich etwas damit zu tun, aber was? Wie wirkt sich Enttäuschung und Verzweiflung auf die Hämodynamik aus? Wie kam es zur schlechten Ehe, zur Adipositas und der wohl daraus folgenden Herzinsuffizienz? Fragen über Fragen, die sich nicht kurzschlüssig beantworten lassen!

Ad 4: Wird der Arzt ihr Leiden erkennen?
Diese Frage ist mit den unter 3. erörterten auf das engste verknüpft und wirft zudem eine Reihe weiterer Fragen auf; wir wollen zwei herausgreifen. Erstens, wie läuft der Erkenntnisprozeß im allgemeinen und der ärztlich-diagnostische Prozeß im besonderen ab? Weiter unten werden wir noch darauf zurückkom-

men. Zweitens, es steht wohl außer Frage, daß das diagnostische Urteil des Arztes sehr stark von seinen theoretischen Konzepten und „Vor-Urteilen" abhängen wird. Ein reiner Somatiker würde wahrscheinlich nur die Adipositas und die Hyperlipidämie registrieren. Er würde vermutlich die Arzt-Patient-Interaktion so gestalten, daß die Patientin kaum Gelegenheit hätte, von ihren familiären Schwierigkeiten zu berichten und ihre Verzweiflung auszudrücken. Umgekehrt würde sich ein behandelnder Psychologe sehr eingehend für letzteres interessieren, dabei aber möglicherweise die Herzinsuffizienz und damit verbundene Gefahren übersehen.

Ad 5: Wird der Arzt der Patientin helfen können? Obwohl diese Frage zweifellos die wichtigste ist – denn was nützt alle Diagnostik, wenn sie nicht zu einer zufriedenstellenden Therapie führt – so können wir sie doch erst dann befriedigend beantworten, wenn wir vorher auf die unter 3. und 4. gestellten Fragen eine Antwort gefunden haben. Dies ist aber noch keineswegs der Fall.

Wir sehen, daß schon ein so banaler Alltagsfall eine Fülle von schwerwiegenden Fragen aufwirft, selbst dann, wenn wir uns nur auf die allerwichtigsten beschränken. Wie sieht es erst aus, wenn der Arzt mit schwierigen „Problemfällen" konfrontiert ist?

Probleme dieser Art müssen in einem Lehrbuch der Psychosomatischen Medizin abgehandelt werden. Psychosomatische Medizin kann aber noch nicht auf eine Theorie der Heilkunde zurückverweisen, die somatische, psychische und soziale Faktoren in einen Zusammenhang bringt, weil die heutige Medizin eine derartige Theorie noch nicht besitzt. Es existieren nur verschiedene, zum Teil sich widersprechende Theorien.

Wir müssen uns daher den Weg zu einer umfassenden Theorie der Heilkunde im Rahmen einer Einführung in das vorliegende Lehrbuch selbst suchen. Wir halten es für notwendig, den Leser zu bitten, uns auf diesem Weg zu begleiten; denn das Konzept für ein umfassendes Ordnungsschema, das wir auf diesem Wege erarbeiten wollen, kann zunächst nicht mehr als ein hypothetisches Modell geben. Es soll zwar in den folgenden Kapiteln genauer ausgeführt und an konkreten Fragestellungen untersucht werden, bewähren kann es sich aber nur, wenn der Leser das Konzept immer wieder für den eigenen Gebrauch als Orientierungshilfe erprobt.

1.1.2 Die Psychosomatische Medizin in historischer Perspektive

In seiner Geschichte der Psychosomatischen Medizin schreibt der spanische Medizinhistoriker Pedro Lain Entralgo (1950): „Die Heilkunde war zu allen Zeiten in der einen oder anderen Art psychosomatisch, und sie mußte es auch immer sein; nicht so jedoch die Pathologie." Er meint damit, daß der psychosomatische Ansatz ärztlichen Handelns „so alt ist wie die Heilkunde selbst". Eine ausgearbeitete Krankheitslehre, also eine psychosomatische Pathologie, aber ist nach Lain Entralgo erst eine Errungenschaft der jüngsten Zeit.

In einem Aufsatz über die Geschichte der Psychosomatischen Medizin wirft Weiner (1984) die interessante Frage auf, worauf es wohl zurückzuführen sei, daß nur eine Medizin für geschädigte Organe, Gewebe und Zellen zur offiziellen Medizin der westlichen Welt wurde und nicht eine umfassende Medizin des kranken Menschen.

Er nimmt Descartes gegen den Vorwurf in Schutz, der Urheber unseres medizinischen Dualismus zu sein, weil er mit seiner Metaphysik einer res extensa und einer res cogitans eine unüberbrückbare Kluft zwischen einem körperlichen und einem geistigen Sein aufgerissen habe. Er meint, Descartes habe ganz andere Vorstellungen gehabt, sobald es um medizinische Probleme ging, und zitiert als Beweis einen Satz aus der Sechsten Meditation, in der es heißt:

„Die Natur lehrt mich durch die Erfahrung von Schmerz, Hunger, Durst usw …, daß ich in meinem Körper nicht wie der Kapitän in einem Schiff wohne, sondern daß ich innig mit ihm vereint, sozusagen mit ihm vermischt bin, so daß ich mit ihm zusammen eine Einheit zu bilden scheine."

Weiner vertritt die These, der Dualismus sei von Ärzten verschuldet, die von Galen über Morgagni und Virchow bis heute die Erklärung für Krankheitssymptome in anatomischen Strukturen von Leichen suchten, nicht aber in veränderten Lebensfunktionen. Er meint, es habe diese beiden Zugänge zu den Problemen kranker Menschen schon seit Beginn der Medizin gegeben, und schon unter den Ärzten im Altertum hätten sich Anhänger einer Lehre der Strukturen und einer Lehre der Funktionen gegenübergestanden. Die eigentliche Kluft liege daher gar nicht zwischen Körper und Seele, sondern zwischen Ärzten, die kranke Menschen und deren Lebensfunktionen untersuchen und behandeln, auf der einen Seite und Medizinern auf der anderen, welche Leichen auf den Tischen der Pathologie sezieren und die dort erhobenen Befunde auf Lebende übertragen. Pathologen, sagt er, könnten Strukturveränderungen an toten Organen, Geweben und Zellen, aber nicht deren Funktionen im Leben beschreiben. William Harvey hätte es schwer gehabt, den Blutkreislauf an einer Leiche zu demonstrieren.

Es wäre in der Tat eine lohnende Aufgabe, die Geschichte des Dualismus in der Medizin zu schreiben. Dabei würde sich herausstellen, daß die Gewichte zwischen den einander bekämpfenden Lehren zu verschiedenen Zeiten verschieden verteilt waren, ja, daß eine wirkliche Polarisierung wohl erst ein relativ spätes Ereignis war. Sicher ist nur, daß die Anhänger der Strukturlehre erst im 17. Jahrhundert ihre Position auszubauen und ihre Theorie konsequent zu formulieren vermochten. Damals war es der Physik gelungen, eine in sich geschlossene Lehre der mechanischen Kräfte zu entwickeln und den Begriff der Kausalität von den ihm noch anhaftenden metaphysischen Vorstellungen zu befreien (Tsouyopulos, 1979; Th. von Uexküll, 1984).

Erst im Gefolge davon entstand das Paradigma der Maschine als Erklärungsmodell für Lebensvorgänge, das vorher den sogenannten „Iatromechanikern" nur relativ vage vorgeschwebt hatte. Das Erklärungsmodell der Maschine hat für Ärzte nicht nur die besondere Anziehungskraft klarer und einfacher Deutungs- und Handlungsanweisungen, mit deren Hilfe sich Krankheiten als Betriebsstörungen infolge eines Maschinenschadens deuten lassen; das Modell hat auch den Vorteil, immer modern zu sein, denn sobald die Technik eine neue, noch kompliziertere und noch leistungsfähigere Maschine erfindet, kann die Medizin ihr Bild des Maschinenmenschen entsprechend verfeinern, ohne das Grundprinzip preisgeben zu müssen. Die gestörte Struktur, die der Pathologe in den Organen Verstorbener findet, gilt immer als Ursache der Krankheit, an welcher der Patient verstorben ist. Es stört den Glauben an dieses Prinzip auch kaum, daß der Pathologe bei der Sektion eines Verstorbenen in vielen Fällen keine Strukturveränderungen findet, die dessen Tod erklären können.

Man übersieht, daß sich die Medizin ihr psychophysisches Problem selbst geschaffen hat, indem sie die Symptome von Kranken auf Befunde reduziert, die Anatomen und Pathologen an Leichen erheben. Der bekannte Ausspruch Rudolph Virchows, er habe schon viele Leichen seziert, ohne je eine Seele anzutreffen, illustriert, wie wir uns in der selbstgelegten Falle gefangen haben.

1.1.3 Die Konsequenzen des Maschinenparadigmas

Wir wollen von der Feststellung ausgehen, daß der Körperbegriff der modernen Medizin trotz der großen Erfolge, die er zunächst gebracht hat, an zwei entscheidenden Punkten versagt:
- Er reicht nicht aus, um Körpervorgänge als spezifische Lebensphänomene zu beschreiben. Entscheidende Begriffe, die dazu notwendig wären, kommen in der Terminologie der Physik und Chemie nicht vor: Autonomie oder Selbstreduplikation enthalten den Begriff „Selbst", der für eine adäquate Beschreibung lebender Systeme von zentraler Bedeutung ist. Die Begriffe müssen geklärt werden, wenn man die Phänomene verstehen will, die wir mit ihnen bezeichnen. Das gleiche gilt für so grundlegende Termini wie Reiz und Reaktion, von den Begriffen Subjekt und Objekt gar nicht zu reden.
- Zu den Folgen der Verkürzung der Lebensphänomene im Körperbereich durch das Maschinenparadigma gehört die Unmöglichkeit, psychische und soziale Faktoren mit Körpervorgängen in Verbindung zu bringen. So entstand die absurde Aufspaltung der heutigen Medizin in Ärzte und Kliniken für Körper ohne Seelen auf der einen Seite und in Therapeuten und Neurosekliniken für Seelen ohne Körper auf der anderen. Die staatliche Festschreibung dieses absurden Zustandes durch Verordnungen und Gesetzesvorschriften macht rasche Fortschritte.

Viktor von Weizsäcker (1949, 1955) hatte noch geglaubt, mit der Einführung der Psychoanalyse in die Medizin den Menschen als Subjekt in die Heilkunde einführen zu können. Seine Hoffnung hat sich nicht erfüllt. Das Auseinanderbrechen der Medizin in zwei heterogene Bereiche ließ sich allein dadurch nicht aufhalten.

Versuchen wir diese Patt-Situation zu analysieren, so ist sie dadurch gekennzeichnet, daß die somatische Medizin, das heißt Physiologie, Biochemie, Anatomie und Pathologie, behauptet „zu wissen", was „Körper" sei, und daß die psychologische Medizin, das heißt Psychologie, Psychoanalyse usw., vorgibt zu wissen, was „Seele" sei, daß aber keine sich oder die andere Seite danach fragt, woher sie dieses Wissen haben.

In dieser Situation müssen sich „Somatiker" und „Psychiker" für eine von zwei Alternativen entscheiden: Entweder sie unterstellen, psychisches Sein sei prinzipiell anders als das physisch-materielle; dann sind sie Dualisten und haben sich alle Schwierigkeiten und Aporien eingehandelt, mit denen unsere dualistische Medizin heute zu kämpfen hat. – Oder sie unterstellen psychisches und physisches Sein sei von prinzipiell gleicher Natur. Dann sind sie Monisten, müssen sich aber jetzt entscheiden, ob dieses im Prinzip überall gleichartige Sein physischer oder psychischer Natur sei. Mit anderen Worten, sie müssen sich entscheiden, ob sie materialistische oder spiritualistische Monisten sein wollen. Alle drei Positionen führen in Schwierigkeiten, die zwar primär erkenntnistheoretischer Art sind, die sich aber in der Praxis auf Schritt und Tritt als Fallgruben und Sackgassen auswirken; denn keine führt zu einer praktikablen Vorstellung, wie ein Modell aussehen könnte, in dem Physisches, Psychisches und Soziales miteinander in Verbindung stehen.

Um die Frage stellen zu können, wie eine Einheit, „ein größerer Organismus", aussehen könnte, innerhalb dessen sich Körper und Seele als interdependente und interaktive Organe (Weiss und English, 1943) definieren lassen, kann man nicht von dem Körperbegriff der somatischen und dem Seelenbegriff der psychologischen Medizin ausgehen und dann versuchen, die beiden zu addieren. Man muß anders vorgehen und als erstes die Frage stellen, was wir als Ärzte, mit lebenden Menschen umgehen, unter lebenden Einheiten verstehen, um dann in einem zweiten und dritten Schritt zu fragen, was im Rahmen dieser Einheit die Termini Seele und Körper bedeuten. Das ist – so kann man einwenden – eine abenteuerliche Zumutung: Ärzte des ausgehenden 20. Jahrhunderts, die über die subtilsten Vorgänge in Zellen und Organen des Körpers informiert sind, sollen sich die Frage stellen, was das eigentlich sei, was sie als Körper bezeichnen? Ebenso werden es die Psychologen als eine Zumutung empfinden, daß sie, die viel genauer über die Zusammenhänge des Psychischen informiert sind als jede Generation vor ihnen, sich fragen sollen, was sie eigentlich mit diesen Zusammenhängen beschreiben?

Aber genau das ist es, was Psychosomatische Medizin von uns verlangt, wenn wir ihren Auftrag ernst nehmen. Sie verlangt, daß wir eine Gegenposition zu jedem Objektivismus beziehen, der ein psychisches und/oder physisches Sein unreflektiert voraussetzt und damit eine Theorie, die alle unsere Vorstellungen von Realität bestimmt.

Erinnern wir uns an unseren exemplarischen Krankheitsfall. Es ist unmöglich, die Symptomatik der Patientin entweder nur auf ein somatisches oder aber ausschließlich auf ein psychologisches Paradigma zu beziehen. Beide greifen zu kurz; sie sind zu reduktionistisch. Sie lassen sich aber auch nicht, ohne den Phänomenen Gewalt anzutun, einfach addieren. Hier wird deutlich, daß wir einen neuen Ansatz suchen müssen, um Psychosomatische Medizin praktizieren zu können.

1.1.4 Das Problem der Erkenntnis und der Theorienbildung

Im Anschluß an die Schilderung des exemplarischen Krankheitsfalls (S. 6) haben wir fünf daraus sich ergebende Fragenkomplexe abgeleitet. Einer davon lautete: Wird der Arzt das Leiden der Patientin erkennen? Damit ist das Problem der Erkenntnis angesprochen, dem wir, wie sich zeigt, nicht ausweichen können.

Zunächst scheinen uns einige Grundüberlegungen und Begriffsbestimmungen notwendig zu sein. Als Wissen wollen wir die Summe unserer Erkenntnisse definieren, die der Überprüfung durch die Erfahrung standhalten. Es steht gewissermaßen zwischen dem Glauben und der Meinung, wobei ersterer Überzeugungen (subjektive Gewißheiten) enthält, die das Individuum zur Bewältigung seines Lebens benötigt, ohne sie in der Umwelt überprüfen zu können, während Meinungen (in der Wissenschaft sprechen wir von Hypothesen) relativ leicht durch korrigierende Erfahrungen veränderbar sind. Wird an Meinungen starr wie an unveränderbaren Glaubenssätzen festgehalten, dann sprechen wir von Dogmen oder von Wahnvorstellungen.

Unter **Wissenschaft** wollen wir die methodisch-systematische Erweiterung unseres Wissens im Zusammenhang mit einer Verbreiterung der empirischen Basis und der Ausarbeitung einer Theorie verstehen. Die **Theorie** ist ein sich nicht widersprechendes System von Aussagen, das die empirischen Daten (die Basissätze) ordnet und uns ermöglicht, über unsere Erfahrungen nachzudenken und zu sprechen.

Nur wenn man den Wissenschaftsbegriff so weit und allgemein faßt und ihn nicht an eine bestimmte Methode bindet, kann man die wissenschaftlichen Bemühungen der verschiedensten wissenschaftlichen Disziplinen unter einer Definition zusammenfassen. Dann wird auch die provozierende Frage, ob die Psychoanalyse eine Wissenschaft sei, gegenstandslos, denn nach dieser allgemeinen Definition kann natürlich gar kein Zweifel darüber bestehen, daß Psychoanalyse eine Wissenschaft ist.

Empirie und Theorie sind immer aufeinander angewiesen – nach dem bekannten Ausspruch Immanuel Kants, demzufolge Begriffe ohne Anschauung leer, Anschauungen ohne Begriffe aber blind sind. Dieses untrennbare Aufeinanderangewiesensein von Empirie und Theorie, von Anschauung und Begriff führt uns zur ersten Schwachstelle der Wissensgewinnung, unabhängig davon, ob es sich dabei um den vorwissenschaftlichen oder wissenschaftlich-systematischen Wissenserwerb handelt. Wir meinen die empirische Datengewinnung. Da der Mensch keine „Tabula rasa" ist, die einfach Eindrücke sammelt, sondern ein Wesen mit beschränkten Wahrnehmungsorganen, mit einer Geschichte, mit einem komplizierten Sozialisationsprozeß und sehr subjektiven Erfahrungen, sind uns objektive empirische Ausgangsdaten, wie sie von Empirismus und Positivismus als Ausgangsbasis unseres Wissens gefordert wurden, nirgends unverfälscht zugänglich. Alle unsere Wahrnehmungen sind immer schon durch ein Vorwissen, eine wie immer geartete „Theorie" mitgestaltet, so daß Einstein mit Recht darauf hinweisen konnte, daß unsere Theorien darüber bestimmen, was wir sehen und beschreiben. Ludwig Wittgenstein (1967) formulierte es noch schärfer: „Die Grenzen meiner Sprache sind die Grenzen meiner Welt."

Ist, so gesehen, bereits die Datengewinnung äußerst problematisch und stets mehr oder weniger „subjektiv verfälscht", so wird diese erste Unsicherheit des Wissenserwerbs noch durch eine zweite ganz erheblich verstärkt. Alle Daten sind nämlich mehr oder weniger unterschiedlich interpretierbar.

Wissen kann auf zweierlei Arten erworben werden: zum einen durch eigene handelnde Erfahrung, zum anderen durch Übernehmen und Aneignen des tradierten Wissens, das heißt der Erfahrungen, die andere Menschen, manchmal ganze Generationen, vor uns gemacht haben. Da wir immer nur einen begrenzten Ausschnitt von Erfahrungen selbst machen können, sind wir, insbesondere in der Wissenschaft, aber auch sonst, auf das tradierte Wissen angewiesen. Um zu verdeutlichen, wie Wissen durch handelnde Erfahrung erworben wird, wollen wir das Modell des Wissenserwerbs als Handlung kurz umreißen.

Um die Handlung, unverfälscht und unreduziert auf irgendwelche Teilaspekte, zum Modell für unsere Deutungen zu machen, müssen wir zunächst fragen, was sie als eigenständiges Phänomen darstellt. Ursprünglich und auf die allgemeinste Form gebracht ist **Handlung: Umgang mit der Welt.** An diesem Umgang sind wir in irgendeiner Weise beteiligt.

Wenn wir das Gesamtgeschehen einer Handlung analysieren, lassen sich darin verschiedene Phasen oder Etappen unterscheiden.

Nehmen wir als Beispiel, wie ich einen Baum ersteige und einen Apfel pflücke:
- „Ich sehe etwas, zum Beispiel Farben und Formen, die durch eine gleichzeitig einsetzende Deutung als Baum vor einer Mauer mit einem Apfel im Geäst interpretiert werden.
- Apfel, Baum und Mauer geben mir Handlungs-

anweisungen, die Mauer als Stütze und den Baum als Leiter zu benützen und den Apfel zu ergreifen.
– Sobald ich versuche, diese Anweisungen auszuführen, stellt sich heraus, ob die Deutung richtig war. Es könnte ja sein, daß die Mauer nachgibt, der Stamm oder der Apfel faul ist" (Th. v. Uexküll, 1963, S. 93).

> Eine **Handlung** läuft also stets nach folgendem Schema ab:
> – Ein Ausschnitt der mich umgebenden Welt wird gedeutet.
> – Das Gedeutete gibt mir bestimmte Handlungsanweisungen.
> – Im Umgang mit der Welt erfolgt eine Prüfung, ob die Deutung und die Handlungsanweisungen richtig waren.

Nach diesem Grundschema entsteht jede menschliche Erfahrung, und zwar sowohl vorwissenschaftlich als auch im Bereich der empirischen Wissenschaften (s. Anmerkung 1, S. 36).

Ohne auf Einzelheiten eingehen zu können, wollen wir doch darauf hinweisen, daß dieses Grundschema nicht nur eine kognitive, sondern auch eine emotionale Bedeutung hat, die sogar für die Psychosomatische Medizin von besonderer Wichtigkeit ist. Die Frage, ob eine Handlungsanweisung, die sich aus der hypothetischen Deutung einer Situation ergibt, deren Probleme lösen kann oder nicht, kann für das Überleben des Betreffenden entscheidend sein. Wir werden daher in diesem Zusammenhang von einem **„pragmatischen Realitätsprinzip"** sprechen und aufzeigen, daß eine ungelöste Problemsituation bei entsprechender Dringlichkeit eine Alarmreaktion mit allen psychischen und somatischen Begleiterscheinungen auslöst. In unserem Beispiel des Apfelpflückens würde der Anblick des Apfels im Geäst eines Baumes bei einem Verhungernden eine Aktivität auslösen, die mit Hoffnung auf Errettung vor dem Hungertod einhergeht. Würde sich die Handlungsanweisung dann als nicht praktikabel erweisen, könnte die Stimmung in Verzweiflung und schließlich in Rückzug und Apathie umschlagen, wie das etwa im Rahmen einer Nausea der Fall sein kann (Th. v. Uexküll, 1952). Erweist sich die Handlungsanweisung dagegen als praktikabel, bedeutet das einen Zuwachs an Vitalität. Verhaltenspsychologisch handelt es sich um Bestrafung oder Belohnung.

Papousek (1975) macht darauf aufmerksam, daß die Grundform dieses Handlungsschemas schon angeboren ist. Er schreibt:

„Situationen, die dem Kind Probleme stellen, für deren Lösung es über keine Programme verfügt, werden schon im frühesten Kindesalter mit einer Alarmreaktion beantwortet ... Wenn der Säugling die richtige Lösung in einer Problemsituation nicht findet, steigert er zunächst sein Bemühen, aber seine Reaktionen verlieren bald an Koordination ... Dies kann soweit gehen, daß eine Überlastung des Organismus droht. Hier können wir beim Säugling eine plötzliche Verhaltensänderung beobachten, die an die Pawlowsche Schutzhemmung oder den biologischen Totstellreflex erinnert. Der Säugling bleibt bewegungslos liegen mit konvergenzlos starrenden Augen und geht zur Schlafatmung über."

Es handelt sich um den Umschlag in einen Zustand, in dem die Umwelt ausgelöscht und der Säugling nur noch Körper ist, ein Vorgang, von dem wir noch sprechen werden.

Neben dem „pragmatischen Realitätsprinzip" mit seinen beiden Reaktionsmustern Aktivierung oder Rückzug müssen wir ein **„kommunikatives Realitätsprinzip"** annehmen (1979), das ebenfalls sehr frühe, möglicherweise auch schon angeborene Vorstufen hat. Wir werden es als das Gefühl eines Echos beschreiben, das alle unsere Deutungen und Verhaltensreaktionen bei relevanten Personen unserer Mitwelt haben, und das uns die Sicherheit gibt, nie ganz einsam und isoliert zu sein. Winnicott (1973) hat beschrieben, wie in der frühen Kindheit die Anwesenheit der Mutter dem Kind diese Sicherheit gibt, die notwendig ist, damit es sich mit sich selbst beschäftigen kann. Offenbar braucht in diesem Stadium das Gefühl des Echos noch eine sichtbare oder hörbare Unterstützung. Der Verlust des kommunikativen Realitätsgefühls spielt bei bestimmten depressiven Zuständen eine Rolle. Wir werden die Vermutung äußern, daß es in Krankengeschichten, in denen Patienten sich isoliert und von der Wirklichkeit getrennt erleben, wie das bei chronischer Krankheit, aber auch in der Vorgeschichte von Krebskranken geschildert wird, eine Rolle spielen könnte.

Fassen wir die Überlegungen dieses Abschnittes zusammen: Wissen entsteht durch handelnde Erfahrung und hilft uns, Programme zu entwickeln, mit deren Hilfe wir unsere Lebensaufgaben mehr oder weniger gut bewältigen können.

> Sowohl die vorwissenschaftliche als auch die systematisch und methodisch herbeigeführte wissenschaftliche **Wissensbildung** verläuft immer über die folgenden drei Stufen:
> 1. Wahrnehmung (Datensammlung);
> 2. Deuten (Interpretieren) des Wahrgenommenen als etwas Bestimmtes (als ein Objekt unseres „Interesses"), das uns Handlungsanweisungen für unser weiteres Verhalten und Vorgehen gibt;
> 3. Realitätsprüfung.

Die Datensammlung und der Deutungsvorgang sind Schwachstellen jeder Wissenschaftstheorie. Die Schwachstellen sind bei der psychoanalytischen Methode besonders offenkundig, denn die Daten sind die kaum exakt beschreibbaren Assoziationen und Verhaltensweisen des Analysanden und die Beziehung zwischen Analytiker und Analysand, die wiederum einem zwar nicht beliebigen, aber immerhin mehr- bzw. vielschichtigen Deutungsprozeß unterworfen werden. Diese Schwachstellen sind aber auch bei allen anderen wissenschaftlichen Methoden nachzuweisen. Sie treten dort nur nicht so offenkundig hervor.

Die beiden Schwachstellen des Wissenserwerbs,

die Datensammlung und die Interpretation der gesammelten Daten, machen alle unsere diagnostischen
Bemühungen so problematisch, unvollkommen und
unabgeschlossen. Mit diesen Unsicherheiten müssen
wir leben, und es wäre ein verhängnisvoller Irrtum –
er wird immer wieder begangen – zu meinen, wir
könnten diese Unsicherheiten durch immer neuere
und „exaktere" Untersuchungen beheben. Zunächst
können wir durch zusätzliche Untersuchungen und
zusätzliche Datensammlungen zwar den Grad der
Unsicherheit verringern. Von einem gewissen Punkt
an – er ist von Fall zu Fall sehr schwer zu bestimmen
– tragen jedoch neue Daten nicht mehr zur Klärung,
sondern im Gegenteil nur zur weiteren Verwirrung
der Situation bei.

1.2 Die Notwendigkeit, einen neuen Ansatz zu finden

Wir sind in der Überzeugung aufgewachsen, die Geschichte der Wissenschaft sei, unbeschadet einiger
Umwege und vorübergehender Rückschläge, eine
Einbahnstraße des Fortschritts. Nach dieser Überzeugung ist die Gegenwart klüger als die Vergangenheit
und die Zukunft ein sicherer Hafen für heute noch
ungelöste Fragen. Rückblick in die Vergangenheit ist
nur für Geschichtsprofessoren und Festredner interessant, die darstellen wollen, wie weit wir es gebracht
haben.

Anders sieht das Bild aus, wenn wir die Geschichte
der Wissenschaft als abenteuerlichen Weg auffassen,
der von einer Herausforderung zu einer Antwort
führt, welche wieder neue Herausforderungen
schafft. In diesem Bild ist die Gegenwart ein Arsenal
von Antworten auf in der Vergangenheit mitverschul-
dete Herausforderungen und die Zukunft die Summe
der Herausforderungen, die durch unsere gegenwärtigen Antworten konstelliert sind.

Diese Auffassung wird durch die Untersuchungen
Thomas Kuhns (1973) über die Struktur wissenschaftlicher Revolutionen gestützt. Für ihn ist die
Antwort, die eine Wissenschaft auf eine Herausforderung gibt, mit dem Begriff **„Paradigma"** zu fassen. Er
versteht darunter ein „anerkanntes Modell" oder
„Schema", das bei der Lösung eines bisher ungelösten
Problems erfolgreich war und nun als Beispiel – Paradigma – dient, um auch verwandte Probleme lösen zu
helfen. Dabei geht das Beispiel in Lehrbücher, Vorlesungen und Anleitungen zu Experimenten ein, durch
deren Studium die Adepten ihr Fach lernen. Das Paradigma gewinnt damit auch eine soziale Funktion:
Es begründet eine Gemeinschaft von Wissenschaftlern, die den Anwendungsbereich des Paradigmas erweitern und seine Regeln präzisieren. Damit entsteht
das, was Kuhn die „normale Wissenschaft" nennt. Sie
beschäftigt sich – wie er sagt – mit „Aufräumarbeiten"
der Möglichkeiten, die ein Paradigma zunächst übrig-
gelassen hat. „Aufräumarbeiten", sagt er, „sind das, was
die meisten Wissenschaftler während ihrer ganzen
Laufbahn beschäftigt", ein Unternehmen, das sich bei

näherer Betrachtung als der Versuch erweist, „die Natur in die vorgeformte Schublade zu pressen, die das
Paradigma bereitstellt".

Ein Paradigma ist demnach eine Art Welterklä-
rungsprinzip für ein begrenztes Universum. Es spezifiziert nicht nur, wie Kuhn betont, „die Entitäten, die
das (von dem Paradigma umrissene) Universum bevölkern, sondern auch, welche es nicht enthält"; oder
genauer gesagt, nicht enthalten darf. Anders ausgedrückt: In dem Universum, das von einem Paradigma
geschaffen und begrenzt wird, herrscht das Gesetz
von Christian Morgenstern, nach dem nicht sein
kann, was nicht sein darf.

Eine wissenschaftliche Revolution – und mit ihr der
Anfang eines Weges der Wissenschaft in eine neue
Richtung – kann erst stattfinden, wenn ein neues Paradigma neben dem alten aufgetaucht ist – und wenn
es ihm gelingt, den Widerstand der Normalwissenschaft zu überwinden. Dieser Kampf zwischen verschiedenen Paradigmen ist daher für die Thematik
der Wissenschaftsentwicklung von zentralem Interesse.

Für unser Thema sind damit zwei wichtige Ansätze
gewonnen:
– Die Geschichte der Medizin der letzten 200 Jahre
  läßt sich als ein Weg beschreiben, auf dem ein Paradigma für den Körper dadurch zu einer Herausforderung für die Ärzte wurde, daß es die Existenz
  körperlicher Entitäten in einer Weise definierte,
  welche die Existenz seelischer Entitäten verbot.
  Die Antwort auf diese Herausforderung war ein Paradigma für den Umgang mit den verbotenen seelischen Entitäten, das nun seinerseits dadurch zu einer Herausforderung für die Ärzte geworden ist,
  daß es die seelischen Entitäten in einem eigenen
  Universum ansiedelt, in dem Entitäten der nach
  dem Körperparadigma definierten Natur nicht existieren dürfen.
– Der andere Ansatz ist die Einsicht, daß es nicht
  möglich ist, Beobachter, Beobachtung und das die
  Beobachtung leitende Modell (das Paradigma) zu
  trennen. Für unser spezielles Problem heißt das –
  in Worten Balints –, daß der Arzt, sein Patient und
  die Krankheit in einer Theorie und Praxis umgrei-
  fenden Einheit verschränkt sind.

Hier zeigt sich, und das ist für unsere Überlegungen
entscheidend, der Beginn eines Weges zu einer Zwei-
Personen-Medizin, um das Wort Balints von einer
Zwei-Personen-Psychologie abzuwandeln.

Dieser Exkurs soll plausibel machen, warum wir
von der Vorstellung Abschied nehmen müssen, es gäbe so etwas wie einen Körper und eine körperliche
Welt neben einer Psyche und einer psychischen Wirklichkeit als zwei verschiedene Seinsbereiche im Sinne
einer „res extensa" und einer „res cogitans", wie Descartes sie genannt hat.

Alles, was wir über einen Körper und über eine
Psyche wissen, wissen wir aufgrund von Interpretationen, die wir bestimmten Phänomenen gegeben haben. Und die Probleme, die dabei entstehen, sind
durch unsere Interpretationen geschaffen. Im einzelnen sah das folgendermaßen aus:

Die Definition der rätselhaften Naturerscheinung des lebenden Körpers als Maschine und seiner Krankheiten als Maschinenschäden dekretierte nicht nur, daß Entitäten, welche das Universum der Medizin bevölkern, mit den bloßen oder durch technische Instrumente verlängerten Händen greifbar und manipulierbar sein müssen; sie dekretierte auch, daß es Entitäten wie Gedanken, Gefühle oder Triebe, die sich weder mit bloßen noch mit technisch bewaffneten Händen greifen lassen, nicht geben darf. In einer Maschine gibt es nirgendwo Angriffspunkte für Gedanken, Gefühle oder Triebe. Wenn die Maschinendefinition für den Körper zutrifft, kann es so etwas wie eine Seele nur als „Gespenst in einer Maschine" geben.

In einem Berufsfeld, in dem aber nicht nur körperliche Defekte, sondern auch Schmerzen, Ängste, Zwänge usw. – also seelische Entitäten – operationalisierbare Definitionen erzwingen, bedeuten diese Definitionen eine Herausforderung. Das Werk Sigmund Freuds und die Entwicklung der Psychoanalyse waren eine Antwort darauf.

Die Ideengeschichte der Begriffe Körper und Seele in der modernen Medizin führt damit von dem Maschinen-Paradigma für den Körper zu dem Paradigma des „psychischen Apparats" für die Seele. Wenn wir auch in diesem Fall von Paradigma sprechen, so tun wir das mit Bedacht; denn Freuds Methode, mit Kranken zu sprechen, statt sie mit direkten oder indirekten Eingriffen der Hand, also mit physischen Mitteln zu behandeln, löste bis dahin ungelöste Probleme des Umgangs mit seelisch Kranken. Seine Methode ist als Beispiel in Lehrbücher, Vorlesungen und Anweisungen für Psychodiagnostik und Psychotherapie eingegangen, durch deren Studium die Adepten ihr Fach lernen. Auf diese Weise entstanden die Psychoanalyse, ganz im Sinne Kuhns, als „Normalwissenschaft" und eine auf das Paradigma eingeschworene Gemeinschaft von „Normalwissenschaftlern".

In ihrem Universum gilt die Regel, daß nicht sein kann, was nicht sein darf, genau wie in dem Universum des Physischen: Ebenso wie das Maschinen-Paradigma die Existenz seelischer Entitäten verbietet, so verbietet das Paradigma des psychischen Apparats die Existenz physischer Entitäten.

Aber – und damit wird die Geschichte spannend – mit dem Freudschen Paradigma hat die Medizin eine neue Dimension gewonnen, in der die scheinbar selbstverständlichen, eindeutigen und unanfechtbaren Definitionen für Körper und körperliche Entitäten fragwürdig werden. Wir sind plötzlich vor die Notwendigkeit gestellt, unsere wissenschaftlichen Definitionen für den Körper neu zu überdenken und die Wissenschaft kritisch zu prüfen, die sich Biologie nennt, aber in Wahrheit eine Subdisziplin der Physik geworden ist.

Damit wird ein Nachholbedarf der Medizin sichtbar, die vor etwa 100 Jahren „beschloß, Naturwissenschaft zu werden", und die diesen Beschluß noch heute als Befreiung von scholastischen und romantischen Reminiszenzen der Naturphilosophie feiert. Aber die Naturwissenschaften haben inzwischen die Voraussetzungen radikal revidiert, von denen sie im 19. Jahrhundert ausgegangen waren, ohne daß die Medizin daraus die erforderlichen Konsequenzen gezogen hat. So ist die Medizin im 20. Jahrhundert eine Naturwissenschaft des 19. Jahrhunderts geblieben.

Auf die Frage, was sich in den Naturwissenschaften verändert hat, und warum diese Veränderung für die Medizin so bedeutungsvoll ist, läßt sich die Antwort in der kurzen Formel zusammenfassen: Die Naturwissenschaften haben die Bedeutung des Beobachterproblems entdeckt. Glaubten sie früher, ihre Aufgabe sei, eine objektive Realität zu enthüllen, wie sie unabhängig von unseren Vorstellungen aussieht, so stellte sich jetzt heraus, daß diese Zielsetzung falsch war. Eine in diesem Sinne objektive Realität ist uns verschlossen. Einstein hat das eindrucksvoll formuliert:

„Physikalische Konzepte", schreibt er 1938, „sind freie Schöpfungen des menschlichen Geistes, und, wenn es auch so aussehen mag, nicht eindeutig von der äußeren Welt determiniert. In unserem Bemühen, die Realität zu verstehen, gleichen wir in gewisser Weise einem Menschen, der den **Mechanismus einer verschlossenen Uhr** zu verstehen sucht. Er sieht das Zifferblatt und die Bewegungen der Zeiger, er hört ihr Ticken, aber er hat keine Möglichkeit, sie zu öffnen. Wenn er scharfsinnig ist, kann er sich das Bild eines Mechanismus ausdenken, das imstande ist, alles das zu erklären, was er beobachten kann. Aber er darf niemals sicher sein, daß sein Bild das einzige ist, das seine Beobachtungen zu erklären vermag. Er wird nie in die Lage kommen, sein Bild mit dem wirklichen Mechanismus zu vergleichen. Ja, er kann sich nicht einmal ausdenken, was für einen Sinn ein solcher Vergleich haben könnte" (A. Einstein und L. Infeld, 1938).

Wir müssen den Glauben aufgeben, eine Realität aufdecken zu können, die von unserem Beobachtungsvorgang unabhängig ist. Der Beobachter kann sich nur – wenn er intelligent ist – ein Bild von dem machen, was er beobachtet, aber in diesem Bild ist er immer selbst mit seinen Fragestellungen, seinen Zielsetzungen und seiner Methode enthalten. Die Beobachter können sich lediglich über Fragestellung, Zielsetzung und Methodik einigen. Dann wird der einzelne Beobachter austauschbar. Das ist alles. Sie bekommen auf diese Weise ein intersubjektiv übereinstimmendes Bild, das sie gemeinsam korrigieren und für ihre Zielsetzungen verbessern können. Das ist die Aufgabe der Institution, die wir Wissenschaft nennen.

Wenn wir diese Einsicht für die Medizin fruchtbar machen wollen, müssen wir das Beispiel Einsteins auf die Situation übertragen, in welcher der Arzt einem Kranken und die Kranke einem Arzt gegenübersteht und in der beide versuchen, sich ein Bild von dem anderen zu machen. Wir verstehen dann die Forderung nach einer Zwei-Personen-Medizin, deren Elemente der Arzt, sein Patient und die Krankheit sind. Die Krankheit entspricht dem Bild des „Mechanismus", den sich der Arzt ausdenken muß, um all das zu erklären, was er an dem Patienten beobachten kann. Auch der Kranke macht sich ein Bild von der Krankheit und von dem Arzt und dessen für ihn sonst ganz unverständlichen Aktivitäten. Beide müssen ein Minimum an Übereinstimmung ihrer Bilder erzielen, wenn eine gemeinsame Wirklichkeit zustande kommen soll, in der Kommunikation und Kooperation möglich werden. In Kapitel 69 wird an einer Krankengeschichte gezeigt, wie in dem heutigen Medizinbetrieb die Interaktion zwischen Kranken und Ärzten abläuft, und die Frage diskutiert, welche praktischen Konsequenzen sich ergeben, wenn die Medizin die Wandlung in den Voraussetzungen zur Kenntnis nehmen würde, die Einstein mit seinem Gleichnis von der verschlossenen Uhr dargestellt hat. Hier wollen wir lediglich betonen, daß dieses Problem für jede Heilkunde, nicht nur für die unserer Industriekultur, von Bedeutung ist.

Kleinman (1973, 1980), der als Arzt und Kulturanthropologe die Interaktionen zwischen Patienten und Vertretern westlicher und anderer medizinischer Systeme untersucht hat, beschreibt, wie verschieden die Erklärungsmodelle für Krankheiten sein können, die Patienten und die Vertreter der Gesundheitsberufe für die diversen Leiden haben können. Er betont, daß der Erfolg der Verhandlungen zwischen beiden Parteien über ein gemeinsames Erklärungsmodell für die Resultate medizinischer Maßnahmen von großer Bedeutung ist.

„Der Effekt von Gesundheitsmaßnahmen (health care outcomes: compliance, Patientenzufriedenheit usw.) ist direkt von dem Grad der Unterschiedlichkeit der Erklärungsmodelle bei Patienten und deren Ärzten und der Effektivität klinischer Kommunikation abhängig. Ist die Differenz groß, läßt sich ein unbefriedigendes Resultat vorhersagen."

Er erklärt damit auch die Anziehungskraft paramedizinischer Heilmethoden für Kranke:

„In jeder Gesellschaft beinhaltet Volksmedizin verglichen mit professioneller ärztlicher Praxis geringere soziale und kulturelle Unterschiede zwischen Patienten und Heilbeflissenen. Die Differenzen sind kleiner ... und die Resultate der Gesundheitsmaßnahmen sind daher besser" (Kleinman, 1980, S. 114).

Jetzt sehen wir, warum die „Sache der Medizin" eine gemeinsame Angelegenheit von Kranken und Ärzten ist, und daß die Krankheit als Sache der Medizin einen völlig neuen Aspekt gewinnt, wenn man den objektivistischen Ansatz überwindet. Aus einer Sache, die Anatomen und Pathologen als neutrale Beobachter beschreiben, ist eine gemeinsame Sache geworden, die, wie die Handlung eines Dramas, nach für die Akteure verbindlichen Spielregeln in der Interaktion zwischen Menschen entstehen muß. Je nach dem Stück, in das die Akteure verwickelt sind, ändern sich die Rollen der Teilnehmer und die Bilder, die sie sich gegenseitig voneinander machen. Die Textbücher, in denen die Rollen der verschiedenen Stücke festgeschrieben sind, nennen wir „Diagnosen".

Kehren wir kurz zum Beispiel unserer Patientin (S. 6) zurück. Schlüpfen wir in die Rolle des somatisch interpretierenden Arztes, dann finden wir im Textbuch die Diagnose: Adipositas, Herzinsuffizienz und Hypertonie. Umgekehrt finden wir im Textbuch des Psychologen bei der gleichen Patientin die Diagnose: Eheschwierigkeiten, Depression und akute Trennungsproblematik. Wieder stellen wir fest, daß die beiden Diagnosen in keinen Zusammenhang gebracht werden können, der die Symptomatik – die nächtlichen Anfälle von Atemnot und Todesangst – erklären kann. Ja, wir verstehen nicht einmal, wie ein solcher Zusammenhang aussehen sollte. Ein psychosomatisches Erklärungsmodell muß daher vor allem anderen ein Bild entwerfen, in dem der Zusammenhang zwischen der somatischen und der psychologischen Diagnose sichtbar und plausibel wird.

1.3 Systemtheorie und das Phänomen der Emergenz

Ein Ansatz, der diese Forderung verwirklicht, findet sich in der Systemtheorie. Sie hat zur Beschreibung der Naturzusammenhänge das Konzept einer hierarchischen Ordnung entwickelt (Bertalanffy, 1968), in der einfachere Systeme (z.B. Zellen) als Elemente oder Subsysteme in komplexere Systeme (z.B. Organe) integriert sind, die dann wieder als Elemente oder Subsysteme in noch komplexeren Systemen (z.B. Organismen) zusammengefaßt werden. So kann sie bis hinauf zu sozialen Systemen eine Hierarchie entwerfen, in der sich verschiedene Integrationsebenen oder -stufen unterscheiden lassen. Damit ergibt sich die Möglichkeit, Physik, Biologie, Psychologie und Soziologie diesen Integrationsebenen oder -stufen zuzuordnen.

Mit der Betonung der Begriffe „Integration" und „Ebene" oder „Stufe" rückt die Systemtheorie eine Tatsache in den Mittelpunkt, die man bisher nicht zur Kenntnis genommen hatte, obwohl sie schon Ende des vorigen Jahrhunderts durch v. Ehrenfels, den Begründer der Gestalttheorie, klar formuliert worden war, nämlich die Tatsache, daß ein Ganzes (ein System) mehr ist als die Summe der Teile (der Elemente oder Subsysteme). Anders formuliert heißt das, daß mit der Bildung eines Systems neue Eigenschaften auftreten, d.h. Eigenschaften, die es auf der Ebene der Subsysteme oder Elemente nicht gibt. Dieses Phänomen hat man als **„Emergenz"** bezeichnet (Popper, 1977; Medawar und Medawar, 1977). Der Nobelpreisträger Medawar beschreibt es folgendermaßen:

„Wenn wir in der oben skizzierten Hierarchie (der Wissenschaften) aufsteigen, finden wir, daß der empirische Reichtum und Informationsgehalt der Wissenschaften zunehmend größer wird ... teilweise weil jede Wissenschaft die Theoreme der Wissenschaft unter ihr enthält, teilweise weil die Restriktionen, welche fortschreitend mögliche Interaktionen zwischen den Elementen begrenzen, auf jeder höheren Ebene Ideen und Konzepte hervorbringen, die spezifisch für diese Ebene sind. Das sind die ‚emergenten Eigenschaften'."

Ein Beispiel für solche Restriktionen ist die Beschränkung, die den Bewegungsmöglichkeiten der einzelnen Muskeln auferlegt wird, wenn sie eine koordinierte Bewegung, zum Beispiel Greifen oder Gehen, zustande bringen sollen. Die „neuen Ideen der höheren Ebene" sind hier die neuromuskulären Programme für Greifen und Gehen. Sie lassen sich nicht aus einer Addition der Bewegungsmöglichkeiten der einzelnen beteiligten Muskeln ableiten, sondern enthalten „Pläne", welche diese Möglichkeiten limitieren und einander zuordnen, zum Beispiel als Agonisten und Antagonisten.

Für unser Problem heißt das folgendes: Mit dem Übergang von einfacheren zu komplexeren Systemebenen, also vor allem mit dem Übergang von der Ebene anorganischer Stoffe zu der biologischer Systeme, dann wieder mit dem Übergang zu psychischen Systemen und schließlich zu sozialen Systemen treten sprunghaft neue Phänomene auf. Sie zwingen uns, jedesmal eine neue wissenschaftliche Disziplin zu entwickeln, deren Terminologie in der Lage ist, die neuen Phänomene adäquat zu beschreiben.

> Die Sprachen der verschiedenen Wissenschaften (Physik, Biologie, Psychologie und Soziologie), die auf diese Weise entstehen, lassen sich nicht ohne weiteres ineinander übersetzen, und vor allem kann keine auf die Sprache der Wissenschaft für die einfachere Systemebene reduziert werden.

Damit wird das Problem sichtbar, wie man sich die Verbindung zwischen diesen heterogenen Systemebenen oder -stufen vorstellen soll. Um es lösen zu können, müssen wir uns zunächst die Unterschiede der Phänomene auf den verschiedenen Integrationsebenen konkret vor Augen führen. Das heißt, wir müssen uns klarmachen, wodurch sich biologische Systeme von anorganischen Zusammenhängen, dann psychische Systeme von biologischen und schließlich soziale Systeme von den vorhergehenden unterscheiden. Diese Fragen hatte man bisher nicht ernstgenommen, weil man überzeugt war, es sei nur eine Frage der Zeit, bis die Naturwissenschaften alle komplexeren Phänomene auf physikalische Zusammenhänge reduziert haben würden. Man wollte nicht sehen, daß ein solcher Reduktionismus zu einer Denaturierung der Phänomene führt, da ihre spezifischen Eigenschaften ja bei einer Zurückführung auf die elementarere Ebene verlorengehen. Halten wir also zunächst fest, daß ein Erklärungsmodell, das somatische und psychische Diagnosen in einen Zusammenhang bringen will, verschiedene Integrationsebenen im Sinne der Systemtheorie darstellen muß.

1.3.1 Die emergenten Eigenschaften der biologischen Systemebene

Mit dem Übergang von der Ebene physikalischer und chemischer Prozesse zu der Stufe biologischer Systeme treten Gebilde auf, welche die neue Eigenschaft besitzen, Zentren eigener Aktivität oder, wie Bertalanffy (1968) es formuliert hat, primär aktive Systeme zu sein. Das Neue, das uns auf dieser Ebene entgegentritt, und das dann auf allen höheren Ebenen in differenzierterer und reicherer Form wieder angetroffen wird, kennzeichnet das Wort „eigen" oder „selbst". Der Mikrobiologe Csaba (1984) schreibt, die Fähigkeit zwischen **selbst** und **nicht-selbst** zu unterscheiden

sei von fundamentaler Bedeutung in der lebenden Welt. Er fährt dann fort:

„Lebende Organismen oder bestimmte Teile von ihnen haben die Fähigkeit, zwischen Stoffen zu unterscheiden, die identisch mit oder komplementär zu dem ‚selbst' sind und solchen, die so fremd oder unähnlich sind, daß eine Interaktion mit ihnen verderblich oder unmöglich sein würde. Diese Differenzierung zwischen selbst und fremd, komplementär und unähnlich findet auf allen Ebenen des Lebendigen statt, von der molekularen bis zu hoch organisierten Systemen."

Was damit gemeint ist, soll etwas näher ausgeführt werden. Die Feststellung, daß biologische Systeme primär aktive Einheiten sind, macht zunächst eine Neufassung der klassischen Reflextheorie notwendig, welche die Beziehung zwischen einem lebenden System und seiner Umgebung nach dem Modell von Reiz und Reaktion beschreibt. Das Ungenügen dieses Modells, das die physikalischen Zusammenhänge von Ursache (Reiz) und Wirkung (Reaktion) auf biologische Vorgänge übertragen will, beruht auf der Vorstellung, der **Reiz** sei ein unabhängig von dem Organismus existierendes Ereignis, das dessen Verhalten (als kausale Folge) hervorbringen würde. Wenn der Organismus und seine Organe aber primär aktive Systeme sind, kann ein Vorgang der Umgebung dort kein Geschehen bewirken (wie er es in einem ruhenden Gebilde könnte), sondern lediglich das Verhalten des bereits aktiven Systems modifizieren.

Für die **Reaktion** des biologischen Systems ist also nicht nur der äußere Vorgang (der Reiz) entscheidend, sondern ebenso dessen innerer Zustand (die Reaktionsbereitschaft), den man zum Beispiel mit Hilfe des kybernetischen Modells als Abweichung von einem Sollwert oder als gestörtes homöostatisches Gleichgewicht beschreiben kann. So wird ein Organismus nicht primär durch Reize, sondern erst durch ein Bedürfnis nach Nahrung, nach einem Geschlechtspartner, nach Wärme usw. veranlaßt, auf Reize zu reagieren. Ohne das Bedürfnis würde der Reiz gar nicht existieren, aber ohne den Reiz könnte die Reaktion, die zur Befriedigung des Bedürfnisses führen soll, nicht zustande kommen.

Auf diesem Zusammenhang beruht die Notwendigkeit, zur Beschreibung selbst einfachster biologischer Vorgänge, die linearen Ursache-Wirkungs-(Reiz-Reaktions-)Modelle durch kreisförmige Modelle zu ersetzen, deren einfachstes das **kybernetische Modell des Regelkreises** darstellt. Bereits dieses Modell beschreibt – und hier wird deutlich, was die Termini „eigen" und „selbst" besagen – wie der (durch einen Sollwert repräsentierte) Zustand eines Systems dessen Umgebung deutet: Nur das, was für den Zustand des Systems (zur Befriedigung eines Bedürfnisses, zur Rückführung eines Istwertes zum Sollwert o. ä.) von Bedeutung ist, existiert für das System, alles andere, was in der Umgebung sonst noch vorhanden sein mag, existiert für das System nicht.

Die so erreichte relative Unabhängigkeit von der Umgebung bezeichnen wir als **„Autonomie"** (von den griechischen Worten autos = selbst und nomos = Gesetz), durch die sich lebende Systeme von unbelebten Gebilden unterscheiden. Wir haben betont, daß damit der Begriff eines „selbst" auftaucht, auf den man nicht mehr verzichten kann, wenn man Lebensvorgänge, welcher Stufe sie auch angehören mögen, adäquat beschreiben will. Unter diesem Gesichtspunkt kann man die Eigenschaft „selbst" als eine Fähigkeit oder ein Verhalten beschreiben, das alles Erreichbare in drei Kategorien einteilt: „selbst", „nicht-selbst" und „nicht-existent". Die Kategorie „selbst" kennzeichnet das System, „nicht-selbst" dessen Umgebung, soweit sie für das System von Bedeutung ist; alles übrige fällt unter die Kategorie „nicht-existent" (s. Anmerkung 2, S. 36).

Die Differenzierung und Verwandlung dieser Kategorien auf den komplexeren Stufen des Lebendigen ist ein wichtiges und interessantes Thema. Hier sei nur angedeutet, daß diese Kategorien von einer bestimmten Organisationsstufe an von spezifischen Schutzsystemen, zunächst dem Immunsystem, auf einer noch komplexeren von dem Schmerzsystem und schließlich auf der Ebene sozialer Systeme von einem System wahrgenommen werden, das Angst hervorbringt.

Da das Phänomen eines „selbst" auf allen Ebenen des Lebendigen in Erscheinung tritt, ist es notwendig darzustellen, wodurch es sich auf den verschiedenen Stufen unterscheidet. Damit erhalten wir einen Zugang zu Problemen, die zum Nachteil einer klaren Begriffsbildung in der Psychosomatischen Medizin lange nicht verstanden waren: Wir sehen dann, daß die Begriffe **Psyche** und **Soma** nicht den Unterschied zwischen zwei Seinsweisen, sondern zwei Systemebenen, die **animalische** und die **vegetative** Systemebene beschreiben.

J.v. Uexküll (1940) hat diesen Unterschied durch eine anschauliche Formulierung deutlich gemacht: Er sagt, daß Systeme der animalischen Ebene **„Umwelten"** aufbauen, während Systeme der vegetativen Stufe, also Einzeller und Pflanzen, nur eine **„Wohnhülle"** besitzen. Was ist damit gemeint?

Bertalanffy (1968) beschreibt, wie nach der Umwelttheorie J.v. Uexkülls jedes Lebewesen aus dem großen Kuchen der Realität eine Scheibe herausschneidet, die seinen rezeptorischen und effektorischen Einrichtungen entspricht. Er erläutert das an dem Beispiel eines einzelligen Organismus, der wie das Pantoffeltierchen (Paramecium) für alle möglichen Reize chemischer, thermischer oder taktiler Art nur über eine einzige Antwort, die Fluchtreaktion, verfügt. Für Paramecium besteht die Welt daher nur aus zwei Eigenschaften, den feindlichen (nichtselbst), vor denen es flieht und den freundlichen, bei denen es bleibt und seine Nahrung findet, wobei es durch Aufnahme der Nahrung aus „nicht-selbst" „selbst" macht, so wie es umgekehrt durch Ausstoßen von nichtverwendungsfähigen Bestandteilen „selbst" wieder in „nicht-selbst" verwandelt. Diese primitive „Welt" genügt vollständig, um das Lebewesen, das keine spezifischen Sinnesorgane besitzt, durch alle Hindernisse und Gefahren sicher in die Bereiche optimaler Lebensbedingungen zu führen. Die vielen Dinge,

die der Beobachter in der Umgebung von Paramecium wahrnimmt, Algen, andere Infusorien, kleine Crustaceen, mechanische und chemische Hindernisse usw. existieren für Paramecium samt und sonders nicht. Sie fallen unter die Kategorie „nicht-existent".

Wir haben hier den Begriff „Welt" in Anführungsstriche gesetzt, denn diese „Welt" existiert für Paramecium nur in Veränderungen des Zustandes der Rezeptoren, die sich in der Oberflächenmembran des einzelligen Lebewesens verteilt finden. Auch mehrzellige Lebewesen und Pflanzen stehen mit ihrer Umgebung nur durch die Rezeptoren in den Zellmembranen ihrer oberflächlichen Zellschichten in Verbindung, und alle Reaktionen, mit denen vegetative Systeme auf Einwirkungen der Umgebung antworten können, sind Reaktionen auf Veränderungen in diesen Zellschichten. Diesen Sachverhalt bringt die Formulierung zum Ausdruck, daß Pflanzen, oder allgemeiner Gebilde der vegetativen Systemebene, nicht in „Umwelten", sondern in „Wohnhüllen" leben. Selbst, Nicht-Selbst und Nichtexistenz sind auf dieser Stufe noch weitgehend undifferenziert.

1.3.2 Die emergenten Eigenschaften animalischer Systeme

Der Zusammenhang zwischen vegetativen Systemen und dem Begriff „Körper" oder „Soma" wird deutlich, wenn wir im Unterschied zu den Phänomenen dieser Stufe Systeme der animalischen Systemebene beschreiben, und wenn wir hinzufügen, daß der Terminus „animalisch", wie Bateson (1982) feststellt, von „anima", dem lateinischen Wort für Seele kommt. Dieser Begriff bezeichnet eine Fähigkeit, die J. v. Uexküll (1920) den Tieren im Unterschied zu den Pflanzen zuerkennt: **die Fähigkeit, eine „Umwelt" zu bilden.**

Der Unterschied läßt sich am einfachsten durch die Feststellung deutlich machen, daß man das Verhalten von vegetativen Systemen mit dem Modell des Regelkreises beschreiben kann. Die in den Zellmembranen verteilten Rezeptoren entsprechen den Fühlern von Regelkreisen, welche die Umgebung nach Sollwerten messen. Um das Verhalten von Tieren zu beschreiben, die spezifische Sinnes- und Bewegungsorgane sowie ein diese Organe verbindendes Nervensystem besitzen, genügt das einfache Modell des Regelkreises nicht mehr.

J. v. Uexküll hat dafür das Modell des **Funktionskreises** entwickelt (1920), und damit das Fundament für die moderne Verhaltensforschung gelegt, deren physiologische Aspekte dann von N. Tinbergen und K. Lorenz weiterentwickelt worden sind. Dabei sind aber die spezifisch subjektiven oder rezeptorischen Aspekte, die **„Merkwelten"**, wie J. v. Uexküll sie definiert hat, nicht berücksichtigt worden. Diese Seite der Umwelttheorie und ihre erkenntnistheoretischen Konsequenzen werden erst heute wieder neu entdeckt.

Was beschreibt also das Modell des Funktionskreises (siehe Abb. 1–1, S. 17)? Es unterscheidet sich von dem des Regelkreises durch die Berücksichtigung spezifischer rezeptorischer und effektorischer Organe sowie deren Verbindung durch ein Nervensystem auf der Seite des Subjekts und den dieser Organisation entsprechenden Phänomenen auf der Seite des Objekts. Es beschreibt, wie **rezeptorische Eindrücke (Merkzeichen) effektorische Antworten** auslösen, die (als **Wirkzeichen**) zu einer Veränderung der rezeptorischen Vorgänge (der Merkzeichen) führen. Dadurch ist eine ständige Rückkopplung zwischen rezeptorischen und effektorischen Aktivitäten (Merken und Wirken) gegeben, die – und hier begegnen wir dem Neuen – auf etwas projiziert wird, das außerhalb des Organismus stattfindet. J. v. Uexküll spricht davon, daß die rezeptorischen Eindrücke (die Merkzeichen) als Eigenschaften **(Merkmale)** von etwas, das außerhalb des Organismus entsteht – d. h. als Eigenschaften der „Objekte" einer subjektiven Umwelt – „hinausverlegt" werden. Es entsteht also außerhalb des Organismus eine Sphäre, in der (bisher für den Organismus nicht existierende) Vorgänge der Umgebung als „Objekte" (dessen Nicht-Selbst) in Erscheinung treten können. Diese Sphäre, die sich nach den Worten J. v. Uexkülls wie eine feste, aber für den außenstehenden Betrachter unsichtbare Hülle um den Körper des Subjektes legt, ist das emergente neue Phänomen der animalischen Stufe. Sie läßt sich als **psychisches System** oder Organ beschreiben, das für den Körper vitale Aufgaben übernimmt. Diese Aufgaben bestehen einmal in einem Schutz, den die **Umwelthülle** leistet, zum anderen in der Fähigkeit, aus der neutralen (für das Lebewesen nicht existenten) Umgebung Phänomene aufleuchten zu lassen, die für dessen Ernährung und Fortpflanzung eine vitale Bedeutung haben.

J. v. Uexküll beschreibt die „Anatomie" des Organs Umwelthülle, die sich wie eine zweite Haut um den Körper jedes Lebewesens der animalischen Stufe – und dazu gehören auch wir Menschen – legt, sehr anschaulich. Er schreibt 1936:

„Jeder Mensch, der in der freien Natur um sich schaut, befindet sich in der Mitte eines runden Eilands, das von der blauen Himmelskuppel überdacht ist. Das ist die ihm zugewiesene anschauliche Welt, die alles für ihn Sichtbare enthält. Und dieses Sichtbare ist entsprechend der Bedeutung, die es für sein Leben hat, angeordnet. Alles, was nah ist und unmittelbar auf den Menschen einwirken kann, steht in voller Größe da; das Ferne und daher ungefährlichere ist klein. Die Bewegungen der fernen Dinge können ihm unsichtbar bleiben, während die Bewegungen der nahen Dinge ihn aufschrecken. ... Dinge, die sich dem Menschen unsichtbar nähern, weil sie durch andere Gegenstände verdeckt sind, verraten sich seinem Ohr durch Geräusche oder seiner Nase als Geruch und, wenn sie ganz nahe herangekommen sind, durch den Tastsinn.

Die Nähe ist durch einen immer dichter werdenden Schutzwall der Sinne ausgezeichnet. Tastsinn, Geruchssinn, Gehörsinn und Sehsinn umgeben den Menschen wie vier Hüllen eines nach außen hin immer dünner werdenden Gewandes."

Umwelt als psychisches System

Mit den Umwelten treten also auf der „animalischen" Ebene sprunghaft (emergent) neue Phänomene auf. Wir können sie als „psychische Systeme" bezeichnen

und ihren Unterschied zu den Phänomenen der vegetativen Stufe dadurch deutlich machen, daß wir diese „somatische Systeme" nennen.

> Die biologische Funktion **psychischer Systeme** ist eine differenziertere Herstellung der Beziehungen zwischen Organismus und Umgebung als auf der vegetativen, **somatischen Stufe,** auf der sie in noch primitiver Form von den Rezeptoren der äußeren Zellschicht (als Wohnhülle) übernommen wird.

Unter theoretischem Aspekt ist es wichtig, daß die Umwelttheorie ein Konzept entwickelt hat, das weder den Organismus noch die Umgebung voraussetzt, sondern von der Beziehung zwischen beiden ausgeht (Lazarus, 1971). Für die Umwelttheorie ist die Umgebung weder die physikalisch-chemische Außenwelt noch die Biosphäre – und der Organismus weder mechanisch noch psychologisch definierbar. Umgebung und Organismus lassen sich vielmehr erst aufgrund der Beziehungen definieren, die zwischen ihnen bestehen, und die mit Hilfe des Modells des Funktionskreises beschrieben werden können, in dem „Merken" und „Wirken" ineinandergreifen und ihren Gegenstand ständig neu definieren.

> Umwelt und Organismus bilden zusammen ein dynamisch sich entwickelndes Ganzes, d. h. ein System (s. Anmerkung 4, S. 37).

Wir sahen, daß in einer Umwelt von den neutralen Vorgängen und Gegenständen der Umgebung nur ein mehr oder weniger enger und veränderter Ausschnitt existiert, der – wie wir sagten – jedes Lebewesen der animalischen Stufe als feste, aber für den außenstehenden Betrachter unsichtbare Hülle umgibt. Der außenstehende Betrachter kann diese Hülle jedoch aufgrund der Kenntnisse der Merk- und Wirkorgane sowie der spezifischen Bedürfnisse des Lebewesens als Einheit einer Merk- und Wirkwelt rekonstruieren, indem er das Verhalten des Lebewesens als Ablauf eines Funktionskreises analysiert, der aus einem Merk- und einem Wirksektor besteht. Nach diesem Modell heften zum Beispiel die Sinnesorgane eines hungrigen Lebewesens bestimmten, bis dahin neutralen (für das Lebewesen nicht existenten) Faktoren der Umgebung mit bestimmten taktilen, optischen und olfaktorischen Merkmalen die Bedeutung „Nahrung" als eine Art Etikett an. Damit taucht ein Nahrungsobjekt in der Umwelt des Lebewesens auf, das dem Merksektor entspricht. Das Etikett Nahrung löst ein Verhalten, das heißt eine Aktivität der Wirkorgane aus, das dem Nahrungsobjekt (durch ergreifen, zubeißen und hinunterschlucken) „Wirkmale" erteilt und dadurch das Merkmal (das Bedeutungsetikett) subjektiv (durch Sättigung) oder objektiv (durch Verschlingen) auslöscht. Das entspricht dem Wirksektor. Mit ihm ist der Funktionskreis abgelaufen und das Verhalten des Lebewesens kommt zur Ruhe, bis ein neuer Funktionskreis in Gang kommt (Abb. 1–1).

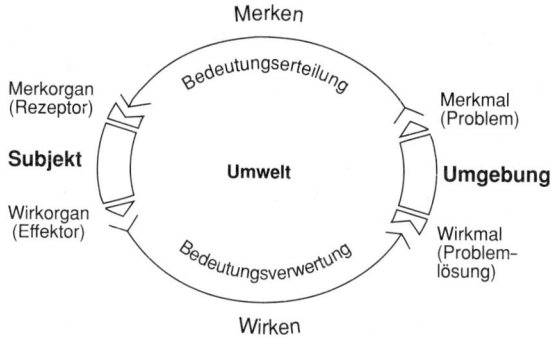

Abb. 1–1. Der Funktionskreis (modifiziert nach J. v. Uexküll, 1936). Das Lebewesen (Subjekt) prägt seiner Umgebung durch Merken ein Merkmal auf und definiert bzw. erschafft damit ein Objekt. Dadurch wird ein Verhalten in Gang gesetzt, das dem Objekt ein Wirkmal aufprägt, welches das Merkmal auslöscht oder verändert. Merken entspricht als Bedeutungserteilung der Strukturierung der Umgebung als Problem, das durch Wirken als Bedeutungsverwertung gelöst werden muß.

Man hat J. v. Uexküll den Vorwurf gemacht, er habe eine neue und eigenwillige Terminologie erfunden, die das Verständnis seiner Arbeiten erschweren würde. Dazu ist jedoch zu sagen, daß man etwas Neues nur beschreiben kann, wenn man eine neue Sprache entwickelt. Die von J. v. Uexküll geschaffenen Begriffe machen Zusammenhänge sichtbar, die ohne sie nicht oder nicht in dieser Weise gesehen werden könnten. Die Begriffe **„Merken"** und **„Wirken"** bezeichnen Grundfunktionen lebender Systeme, die noch nichts mit einem bewußten Wahrnehmen oder Wollen zu tun haben, aber deren Voraussetzungen schaffen. Der Begriff **„Bedeutungserteilung"** bezeichnet die Umwandlung neutraler Umgebungsfaktoren in Bestandteile der subjektiven Umwelt und der Begriff **„Bedeutungsverwertung"** bezeichnet deren Assimilation. Schließlich bekommen auch **„Subjekt"** und **„Objekt"** als biologische Phänomene einen neuen Inhalt: Sie bezeichnen Elemente einer Einheit, in der ein primär aktives System mit der Eigenschaft „selbst" eine bestimmte Beziehung zu seiner Umgebung aufbaut.

Für J. v. Uexküll findet der „Sprung" von einer körperlichen zu einer psychischen Welt nicht an der Grenze zwischen anorganischen Prozessen und biologischen Systemen, sondern an der Grenze zwischen vegetativen und animalischen Systemen statt. Insofern stimmt er mit der Auffassung Freuds, Piagets und Batesons überein, für die der Ursprung der menschlichen Psyche ebenfalls im biologischen Bereich liegt. Das darf aber nicht den Blick für die Tatsache verstellen, daß mit dem Auftreten animalischen Lebens ein Sprung von einer einfacheren (vegetativen) auf eine komplexere (animalische) Systemebene stattgefunden hat.

Auf dieser komplexeren Stufe bildet das Subsystem Organismus oder Körper mit dem Subsystem Umwelt oder Psyche zusammen ein System, in dem beide als Elemente oder Subsysteme integriert sind. Dabei handelt es sich nicht um eine Addition, sondern um schöpferische Verwandlung der integrierten Teile.

Die erkenntnistheoretischen Konsequenzen

Wir sagten, die Naturwissenschaften hätten mit der Entdeckung des Beobachterproblems eine neue Dimension erobert, und Biologie und Medizin müßten daraus die notwendigen Konsequenzen ziehen. Dafür ist das Modell des Funktionskreises hilfreich; denn es beschreibt Subjekt und Objekt als Elemente eines Systems, in dem sie sich gegenseitig definieren. Das bedeutet für den Beobachter die Notwendigkeit, sich selbst als Subjekt und damit als Zentrum seiner Aktivität beziehungsweise als kreativen Pol einer Subjekt-Objekt-Interaktion zu reflektieren, in der er einem Gegenpol Merk- und Wirkmale aufprägt, welche diesen als Objekt definieren. Er erkennt sich damit als Element in dem pragmatischen System einer Handlung (S. 10). In diesem System entscheidet das Entgegenkommen oder der Widerstand des Gegenpols, welche der dem Subjekt verfügbaren Merk- und Wirkmale realisiert werden können, d. h., mit welchen Eigenschaften der Gegenpol als Objekt in der Umwelt des Beobachters in Erscheinung treten kann.

Den gleichen erkenntnistheoretischen Ansatz finden wir bei Piaget (1975), dessen zentrale Begriffe „Assimilation" und „Akkommodation" polare Beziehungen zwischen Subjekt und Objekt bezeichnen.

> **„Assimilation"** steht für eine Beziehung, in welcher der Subjekt-Pol dominiert und der Objekt-Pol sich dessen kreativen Aktivitäten fügt. Demgegenüber bezeichnet der Begriff **„Akkommodation"** eine Beziehung, in welcher der Objekt-Pol das Übergewicht hat, und das Subjekt seine kreativen Schemata modifizieren muß, damit sie von dem Gegenpol toleriert werden.

In unserem Zusammenhang ist es ferner wichtig zu sehen, daß Winnicott (1973) den gleichen Ansatz gewählt hat. Für ihn ist der Raum zwischen Säugling und Mutter das primäre Spannungsfeld zwischen Subjekt und Objekt und der Schauplatz, auf dem sich die Kreativität des Kindes entwickelt – oder im ungünstigen Fall nicht oder krankhaft entwickelt. Er betont die Bedeutung, welche der „Objekt-Pol" Mutter als Prototyp der späteren Umgebung für das Kind und dessen Entwicklung als Subjekt hat. Die „genügend gute Mutter" muß von einer „genügend guten Umgebung" abgelöst werden, wenn die psychische Entwicklung des Kindes keinen Schaden nehmen soll.

Piaget und Winnicott beschreiben zwei verschiedene Seiten der psychischen Entwicklung des Menschen, deren Zusammenhang noch keineswegs durchschaut ist: die kognitive und die affektive Seite (Ciompi, 1982).

1.3.3 Die emergenten Phänomene der Stufe des Humanen

Wir haben beschrieben, wie mit den Umwelten der Tiere emergent psychische Systeme auftauchen. Wie

17

entwickeln sich diese Systeme einer Tierpsychologie zu Systemen, mit denen sich die Psychologie des Menschen befaßt? Hier stoßen wir auf die Schwierigkeit, daß Phänomene, die einer individuellen und einer sozialen Ebene angehören, so eng ineinandergreifen, daß es kaum möglich ist, die Anteile auseinanderzuhalten, ja, daß genau gesehen die sozialen den individuellen vorangehen. Das soziale System der symbiotischen Einheit zwischen Säugling und Mutter existiert früher als das Individuum, das sich erst in einer „zweiten Geburt" (Mahler, 1980) aus der Symbiose lösen muß. Im Grunde ist bereits der Begriff „Symbiose" problematisch, weil er dazu verleitet zwei Individuen vorauszusetzen, die dann eine Verbindung eingehen. In Wahrheit handelt es sich aber um eine Verbindung, die der Individualität des Kindes vorausgeht. Um das zu unterstreichen, sagte Winnicott (1940*): „There is no such thing as an infant" (s. Anmerkung 3, S. 37 und Kap. 6).

Wir wollen von der Feststellung ausgehen, daß die Fähigkeit der Beobachtung und der Aufbau der Umgebung im Sinne einer objektiven Außenwelt Phänomene sind, die zum ersten Mal mit dem Menschen auftauchen. Hier hätten wir also emergente Phänomene der Ebene des Humanen vor uns. Im Kern handelt es sich um das, was wir mit dem Terminus **„Gegenstand"** bezeichnen, und das Winnicott (1973) sehr präzise das **„objektive Objekt"** nennt. Er definiert damit den Unterschied zu den Objekten auf der Stufe des animalischen Lebens, die als „subjektive Objekte" keine Konstanz besitzen, und zu denen die Subjekte keine Distanz halten können. Die subjektiven Objekte der Tiere und kleinen Kinder lösen sich in Nichts auf, sobald sie aus ihrer Umwelt verschwinden, und Tiere und kleine Kinder bleiben als Subjekte an ihre Objekte gekettet, solange diese in ihrer Umwelt gegenwärtig sind.

Den Beweis für die überraschend klingende Behauptung, daß Tiere niemals mit Gegenständen in Beziehung treten, führt J. v. Uexküll anhand eines einfachen Beispiels (1940):

„Gesetzt den Fall", schreibt er, „ich werde auf der Landstraße von einem wütenden Hund angebellt; um ihn loszuwerden, hebe ich einen Chausseestein auf und verjage den Angreifer mit einem geschickten Wurf – dann wird niemand, der den Vorgang beobachtete und nachher den Stein aufhob, daran zweifeln, daß es derselbe Gegenstand „Stein" war, der anfangs auf der Straße lag und nachher dem Hunde nachgeworfen wurde.

Weder die Form noch die Schwere, noch die sonstigen physikalischen und chemischen Eigenschaften des Steins haben sich geändert. Seine Farbe, seine Härte, seine Kristallbildung sind die gleichen geblieben – und doch hat sich eine grundsätzliche Wandlung an ihm vollzogen: Er hat seine *Bedeutung* gewechselt.

Solange der Stein der Landstraße eingegliedert war, diente er dem Fuß des Wanderers als Unterstützung. Seine Bedeutung lag in seiner Teilnahme an der Leistung des Weges. Er hatte, wie wir uns ausdrücken, einen „Wegton" (eine „Wegqualität").

Das änderte sich von Grund auf, als ich den Stein aufhob, um ihn nach dem Hunde zu werfen. Der Stein wurde zum Wurfgeschoß – eine neue Bedeutung wurde ihm aufgeprägt. Er erhielt einen „Wurfton" (eine „Geschoßqualität").

Der Stein, der als *beziehungsloser Gegenstand in der Hand des Beobachters liegt*, wandelt sich in einen Bedeutungsträger, sobald er in Beziehung zu einem Subjekt tritt. Da kein Tier jemals als Beobachter auftritt, darf man behaupten, daß kein Tier jemals zu einem „Gegenstand" in Beziehung tritt. Durch die Beziehung allein wandelt sich der Gegenstand in den Träger einer Bedeutung, die ihm von einem Subjekt aufgeprägt wird" (und d.h. zu einem Objekt) (J. v. Uexküll, 1970, S. 107/108).

Das Beispiel beschreibt nicht nur den Unterschied zwischen dem Objekt, das als beziehungsloser Gegenstand in der Hand des Beobachters liegt, dem „Gegenstand" oder „objektiven Objekt" und dem „subjektiven" Objekt in der Umwelt des Hundes. Es beschreibt auch, wie der Mensch in zwei verschiedenen Welten lebt: einer Beobachterwelt, die von Gegenständen erfüllt ist, und einer Welt der Affekte, in der Objekte auftauchen, die keine Konstanz haben, und zu denen er keine Distanz halten kann. Der wütende Hund brachte es fertig, daß der unbeteiligte Beobachter in eine dramatische Umwelt versetzt wurde, in der ein vorher objektives Objekt, der Gegenstand „Stein", plötzlich zu einem subjektiven Objekt, dem Wurfgeschoß „Stein" wurde. In beiden Fällen war der Stein eine gemeinsame „Sache". Für den Beobachter, der die physikalischen, chemischen und kristallographischen Eigenschaften des Steins untersuchte, war der Stein gemeinsame Sache einer Gruppe von Physikern, Chemikern und Geologen, deren Zielsetzungen, Fragestellungen und Methoden sich der Beobachter zu eigen gemacht hatte. Für den von dem Hund Angegriffenen wurde der Stein zur gemeinsamen Sache von ihm und dem Hund, die sich für den Menschen nur deswegen nicht wieder in Nichts auflöste, weil er als Gegenstand konserviert werden konnte. Erst auf der Stufe des Humanen tritt die Fähigkeit auf, Phänomenen als Beobachter (ohne affektives Interesse) gegenüberzutreten, und ihnen eine von jeder subjektiven Einschätzung unabhängige Existenz und Dauerhaftigkeit zu unterstellen. Die Eigenschaften „Objektivität" und „Gegenstandhaftigkeit" sind emergente Phänomene der humanen Integrationsstufe. Gleichzeitig bleiben die Phänomene aber „subjektive Objekte", d.h. „Eigentum" des Subjekts, dessen „animalische" Kreativität sie als sinnliche Wahrnehmung erzeugt.

Damit löst sich eine Paradoxie, mit der wir ständig konfrontiert sind: Die Gegenstände der menschlichen Welt sind auf der einen Seite Objekte unserer subjektiven Umwelten. Die Sonne, die ich am Himmel sehe, ist Eigentum meiner Umwelt. Unter diesem Aspekt gibt es so viele Sonnen, wie es Menschen gibt, deren Augen eine sichtbare Welt entstehen lassen. Aber die Sonne ist auf der anderen Seite auch Gegenstand einer Welt des Wissens, in der es für alle Menschen, die dieses Wissen teilen, nur eine Sonne gibt, die jeder entsprechend dem Standpunkt, von dem aus er die Sonne betrachtet, auf seine subjektive Weise wahrnimmt.

Piaget hat mit seinen Untersuchungen über die Entwicklung der Intelligenz beim Kinde Licht in den

---

* Zitiert nach L. Schacht, Kap. 6.1 dieses Lehrbuches.

rätselhaften Vorgang gebracht, den man als Umwandlung einer natürlichen Umwelt in eine individuelle Wirklichkeit bezeichnen kann. Er spricht von einer **„kopernikanischen Wende"** und versteht darunter eine Veränderung des Verhaltens des Kindes zu sich selbst und zu seiner Umgebung, die in der gesamten Natur nur beim Menschen beobachtet wird und die selbst bei den uns am nächsten verwandten Primaten nur in Andeutung zu finden ist: Mit etwa zwei Jahren beginnt das Kind seine Umwelt, die es bis zu diesem Zeitpunkt ähnlich wie alle höheren Lebewesen nach sensomotorischen Programmen oder Schemata aufgebaut hat, die seine Sinneseindrücke (seine Merkwelt) seinen Bewegungen (seiner Wirkwelt) zuordnen, nach einem neuen Prinzip zu organisieren. Dieses neue Prinzip ist sein Vorstellungsvermögen, mit dem außer einer Außenwelt eine Innenwelt der Phantasie entsteht, in der das Kind Objekte und Vorgänge reproduzieren kann, die aus seiner Wahrnehmung verschwunden sind.

„Von diesem Augenblick an", sagt Piaget, „kehrt das Kind seine anfängliche Welt ganz um, deren bewegte Bilder (bisher) auf eine, ihrer selbst unbewußten Aktivität zentriert waren, und formt sie zu einer festen Welt von koordinierten Objekten, die den eigenen Körper als Element miteinschließt".

Jetzt können auch Schäden, die das Kind durch seine Umgebung erlitten hat, festgehalten und zu Gegenständen umgeformt werden, vor denen man sich schützen kann, und mit denen man umzugehen lernt. Es beginnt der lange Prozeß des Aufbaus einer Welt des Wissens, für den die Sprache und damit die Gesellschaft, in die das Kind hineingeboren wurde, eine zunehmende Rolle spielen. So entsteht aus Umwelt eine **„Wirklichkeit"**, in der neue Formen des Wirkens aufgrund gemeinschaftlicher Programme zum beherrschenden Prinzip werden und in der die Objekte, die aus dem Horizont der Wahrnehmung verschwunden sind, aufbewahrt und zu permanenten Gegenständen umgeformt werden (Th. v. Uexküll, 1984).

Diese Wirklichkeit ist der individuelle Besitz jedes Einzelnen, und bleibt zeit seines Lebens dessen feste, für den außenstehenden Betrachter unsichtbare Hülle oder zweite Haut. Aber diese Haut wird jetzt nicht mehr allein von Programmen biologischer Bedürfnisse, sondern auch von den Begriffen der Sprache geprägt, welche die Konzepte und Vorstellungen für den Umgang mit der Umgebung vermitteln, die von der Sprachgemeinschaft einer bestimmten Kultur entwickelt worden sind. Berger und Luckmann (1969) sprechen von der „gesellschaftlichen Konstruktion der Wirklichkeit", die dem Einzelnen dann als „Allerweltswirklichkeit" oder dem Wissenschaftler als abstrakte Berufswirklichkeit entgegentritt. Hier stoßen wir überall auf „gemeinsame Sachen" für Zielsetzungen und Methoden gesellschaftlicher Gruppen, deren Spielregeln wir internalisiert haben.

Mit dieser Verankerung der Spielregeln einer gesellschaftlichen Gruppe, mit der wir uns identifizieren, in unserer Vorstellungswelt, haben wir eine komplexere Ebene der Integration erreicht als die Ebene animalischer Umwelten.

1.3.4 Biologische, psychologische und soziale Phänomene

Wir wollen die Ergebnisse unserer Analyse der emergenten Eigenschaften zusammenfassen, die wir als neuen Ansatz gewonnen haben. Ausgangspunkt war die Feststellung, daß mit der Einführung eines Begriffs für „Seele" durch die Psychoanalyse in die Medizin deren traditioneller, nach dem Maschinenmodell gedeuteter Begriff für „Körper" fragwürdig geworden war:

– Die Notwendigkeit, das Phänomen der Emergenz ernst zu nehmen, das durch die Systemtheorie in den Mittelpunkt unserer Aufmerksamkeit gerückt wird, bedeutet das Ende des Reduktionismus: Die Phänomene der einfacheren Systemebene bilden zwar eine notwendige, aber nicht eine hinreichende Voraussetzung für das Zustandekommen der Phänomene komplexerer Stufen. Diese lassen sich nicht auf die Phänomene der einfacheren Stufe zurückführen.

– Die Analyse der emergenten Eigenschaften auf den drei Stufen oder Ebenen des vegetativen, des animalischen und des humanen Lebens führte von dem Prinzip der „Wohnhülle" zu dem Prinzip der „Umwelt" und von dort zu dem Prinzip des „Gegenstandes" und der gegenständlichen oder objektiven Wirklichkeit. Diese Prinzipien bedeuten jeweils andere Subjekt-Objekt-Interaktionen mit entsprechend verschiedenen Definitionen der Begriffe „Subjekt" und „Objekt".

– Mit der Feststellung der Verschiedenartigkeit der Eigenschaften, die auf jeder Stufe emergent auftreten, und der Einsicht, daß soziale Vorgänge nicht auf psychologische und diese nicht auf biologische zurückgeführt werden können, wird das Problem konkret, wie wir uns die Verbindungen zwischen den verschiedenen Systemebenen vorstellen sollen. Die Lösung dieses Problems ist eine Aufgabe, vor die jedes psychosomatische Modell gestellt ist.

Wir verstehen jetzt die Problematik, vor welcher der Arzt steht, dem die Patientin von ihren nächtlichen Anfällen erzählt, etwas besser: Wir sehen, daß sich ihr Leben auf verschiedenen Ebenen abspielt, die offenbar miteinander in Verbindung stehen: auf der somatischen Ebene lassen sich die Symptome als Adipositas, Hypertonie und leichte kardiale Dekompensation deuten. Auf dieser Ebene sind das Herz, der Kreislauf und die Lungen der Patientin anatomisch definierte Organe, die nach physiologischen Gesetzen zusammenarbeiten. Nach diesen Gesetzen kann eine Überlastung des Herzens durch Adipositas und Hochdruck zu Lungenstauung führen, die unter Umständen in Form eines Asthma cardiale auch als Anfälle nächtlicher Atemnot in Erscheinung treten kann. Auf dieser Ebene der Diagnostik spielen der Ehemann und der Sohn der Patientin keine Rolle für die Erkrankung der Frau. Sie haben keine Bedeutung für Organreaktionen, die nur mit der O_2- und CO_2-Spannung des Blutes, mit Atem- und Kreislaufreflexen zu tun haben.

Was der Arzt auf der somatischen Ebene untersuchen und berücksichtigen muß, sind im Sinne des obigen Beispiels (S. 18) „Gegenstände", das heißt objektive Objekte und Vorgänge, die in der Sprache der Physik und Chemie beschrieben werden können. Sie bleiben in der Hand des Arztes, wie der Stein in der Hand des Beobachters, Objekte, deren Bedeutung sich in der Zuordnung zu Organreaktionen erschöpft.

Auf der Ebene der **individuellen Wirklichkeit,** die von der Patientin erlebt und emotional gedeutet und beantwortet wird, tauchen statt dessen der Mann und der Sohn nicht nur als neutrale objektive Objekte auf, sondern als **Bedeutungsträger** und das heißt als subjektive Objekte, die für das Erleben und Verhalten der Kranken eine entscheidende Rolle spielen. Auf dieser Ebene läßt sich das Seufzen als unbewußter Appell an die Mitwelt, eine Atemnot als Hyperventilation im Rahmen einer Bereitstellung für Flucht oder Kampf verstehen.

Darüber hinaus stellt sich aber die Frage, ob und wieweit Vorgänge, die sich in der individuellen Wirklichkeit der Patientin auf einer Erlebens- und einer Verhaltensebene abspielen, die Vorgänge auf der somatischen Ebene, d.h. die Herz- und Kreislaufaktionen, die Reaktionen der Lungen, des Endokriniums usw. im Sinne psychosomatischer Abläufe beeinflussen. Daß derartige Beeinflussungen stattfinden, wissen wir nicht nur aus der praktischen Erfahrung, sondern auch aufgrund exakter experimenteller Beobachtungen. Aber wir können uns vorläufig nicht vorstellen, wie solche Einflüsse vor sich gehen sollen. Hier stehen wir also vor der Frage nach einem Modell, das in der Lage ist, solche Einflüsse, die zwei oder sogar drei verschiedene Systemebenen verbinden, abzubilden.

1.3.5 Das Konzept des „lebenden Systems"

Wir können diesen Paragraphen nicht abschließen, ohne den Begriff des „lebenden Systems" genauer besprochen und definiert zu haben. Das ist auch notwendig, um einem Mißverständnis vorzubeugen, das sich aus der eben gemachten Feststellung ergeben könnte: Wir sagten, die Phänomene, die der Arzt auf der somatischen Ebene beobachtet, seien „Gegenstände", die in der Sprache der Physik und Chemie beschrieben werden könnten. Das ist richtig, aber auch zugleich falsch. Es ist richtig, weil der Beobachter Vorgänge, die im Organismus ablaufen, in Form physikalischer und chemischer Prozesse registrieren und messend erfassen muß, um sie manipulieren zu können. Es ist aber falsch, wenn man daraus den Schluß zieht, mit der Aufdeckung der physikalischen und chemischen Zusammenhänge das Rätsel gelöst zu haben. Ein solches Urteil beweist nur, daß man das eigentliche Problem gar nicht gesehen hat.

Das eigentliche Problem besteht nicht darin, wie die Vorgänge im Organismus dem Beobachter erscheinen, und was er mit den Vorgängen macht, sondern darin, was sie für den Organismus bedeuten, und was er aus und mit den Vorgängen macht. Solange

man dieses Problem nicht sieht, bleibt einem auch das Phänomen der Emergenz verschlossen, das ja darin besteht, daß Atome und Moleküle, die in Zellorganellen und Zellen integriert sind, Eigenschaften an den Tag legen, die im Bereich der Atome und Moleküle nicht existieren. Viktor von Weizsäcker formuliert diese Feststellung folgendermaßen:

„Wenn wir physikalische Denkweisen auf Psychisches und Biologisches anwenden, kommen wir zu Sätzen, die an sich nicht falsch sind, bei denen aber immer auch das Gegenteil richtig ist … Der Gegensatz der beiden Sachverhalte ist unaufhebbar; aber beide zusammen machen erst das wirkliche Leben aus … Der Widerspruch muß daher von der Wissenschaft nicht durch Erklärungen beseitigt, sondern gesetzt werden, sobald sie sich dem Phänomen des Psychischen oder Lebenden nähert" (Ges. Werke, Bd. I, Kap. IV, 1986).

Weizsäckers Ziel war die Einführung des Menschen als Subjekt in die Heilkunde, und er hat selten so deutlich wie in diesen Sätzen formuliert, daß diese Einführung nie gelingen kann, wenn man nicht gleichzeitig die Zelle als Subjekt in die Biologie einführt.

Weizsäckers Programm wurde von den meisten seiner Zeitgenossen als Verzichterklärung auf ein rationales Verstehen biologischer und psychischer Zusammenhänge mißverstanden. In Wirklichkeit ging es ihm um die Überwindung eines zu eng gefaßten Begriffs für Rationalität. Heute läßt sich das „Antilogische", um dessen Anerkennung als Kriterium des Lebens er sich bemüht hat, leichter definieren, als es noch vor 30 Jahren möglich war: Die Theorie des lebenden Systems und die Lehre von den Zeichen, die Semiotik, stellen begriffliche Werkzeuge bereit, die Viktor von Weizsäcker nicht zur Verfügung standen.

Ich will mit zwei Beispielen illustrieren, wie die Theorie des lebenden Systems und die Zeichenlehre „den unaufhebbaren Gegensatz der Sachverhalte" nicht wegerklären, sondern zur Maxime erheben:

- Eine, wenn nicht **die** Grundformel der Systemtheorie sagt: Ein lebendes System erzeugt (als „autopoietisches System" im Sinne Maturanas, 1982) seine Teile selbst. Gleichzeitig ist das System ein Erzeugnis seiner Teile.
- Eine Grundformel der Bio-Semiotik lautet: Lebende Systeme reagieren nicht mechanisch auf physikalische und chemische Ursachen, sondern antworten auf Zeichen. Gleichzeitig brauchen Zeichen physikalische und chemische Ursachen als Vehikel für Bedeutung bzw. Nachrichten.

Beide Formeln bringen die „antilogischen" Beziehungen zwischen den Teilen und dem Ganzen – als System und als Zeichenprozeß – zum Ausdruck, die nach den Worten Weizsäckers für Psychisches oder Lebendes konstitutiv sind.

Ich will fünf Punkte herausheben, die für den Begriff des lebenden Systems wichtig sind und seine Verbindung zu dem Begriff des Zeichens deutlich machen:

1. Die Formel „autopoietisches System" besagt, daß sich lebende Systeme auf jeder Integrationsstufe selbst erzeugen. Maturana und Varela (1988) erläutern diesen Vorgang an den Beziehungen des Stoffwechsels zwischen Zellinnerem und Zellmembran:

„Auf der einen Seite sehen wir ein dynamisches Netzwerk von Transformationen (Stoffwechselvorgängen), das seine eigenen Bestandteile erzeugt und die Bedingung der Möglichkeit eines Randes (der Membran) ist. Auf der anderen Seite sehen wir einen Rand (eine Membran), der die Bedingung der Möglichkeit des Operierens eines Netzwerks von Transformationen (Stoffwechselvorgängen) ist, welche das Netzwerk als Einheit erzeugt."

2. Auf der animalischen und humanen Integrationsebene hat Piaget (1936) die Selbsterzeugung lebender Systeme als „Assimilation", d.h. Transformation von Teilen der Umgebung in Elemente einer subjektiven Umwelt (im Sinne J. v. Uexkülls) bzw. einer individuellen Wirklichkeit beschrieben. Damit zeichnet sich der zweite Punkt ab, der für die Definition des Begriffs „lebendes System" wichtig ist. Er macht die erkenntnistheoretische Bedeutung dieses Begriffs, aber auch seine enge Verbindung zu dem Zeichenbegriff deutlich. Bateson (1985) hat das folgendermaßen formuliert:

„Jetzt fangen wir an, einige erkenntnistheoretische Trugschlüsse der abendländischen Zivilisation zu sehen. Übereinstimmend mit dem allgemeinen gedanklichen Klima in England um die Mitte des neunzehnten Jahrhunderts entwickelte Darwin eine Theorie der natürlichen Selektion und der Evolution, in der die Einheit des Überlebens entweder der Stammbaum oder die Spezies, Subspezies oder etwas dieser Art war. Heute ist es aber ziemlich offenkundig, daß dies nicht die Einheit des Überlebens in der realen biologischen Welt ist. Die Einheit des Überlebens besteht aus **Umwelt plus Organismus**. Wir lernen durch bittere Erfahrung, daß der Organismus, der seine Umwelt zerstört, sich selbst zerstört."

3. Lebende Systeme sind keine statischen Gebilde. Sie verändern sich (im Verlauf ihrer ständigen Autopoiese), aber bleiben trotz – oder vielmehr nur wegen – dieses ständigen Sich-selbst-Veränderns sie selbst. Sie sind historische bzw. Zeit-Gestalten. Das Sich-selbst-Erzeugen autopoietischer Systeme (Bertalanffy sprach von „primär aktiven Systemen") läßt sich nicht nach dem Modell linearer Ursache-Wirkungs-Ketten der Mechanik, sondern nur mit kreisförmigen Modellen beschreiben. Solche Modelle sind der Funktionskreis (J. v. Uexküll, 1920), der Gestaltkreis (V. v. Weizsäcker, 1933), die sensomotorische Zirkulärreaktion (J. Piaget, 1936) und der Regelkreis (N. Wiener, 1943). Wir werden das Modell des „Situationskreises" entwickeln, um auf der humanen Integrationsebene den kognitiven Vorgang zu beschreiben, mit dem der Mensch seine individuelle Wirklichkeit hervorbringt.

4. Lebende Systeme verkehren mit ihrer Umgebung durch Zeichen. Sie kodieren physikalische und chemische Prozesse, die ihre Rezeptoren verändern, zu Zeichen, die sie über die Bedeutung der physikalischen und chemischen Vorgänge für ihre biologischen, psychischen oder sozialen Bedürfnisse unterrichten. Darauf beruht ihre Autonomie.

5. Lebende Systeme sind „Einheiten", aber keine „Einerlei-heiten". Sie haben eine hierarchisch gegliederte Struktur, in der sich die verschiedenen Integrationsebenen unterscheiden lassen, die wir beschrieben haben. Maturana und Varela (1988) führen diesen Gedanken für die biologische Integrationsebene, und ihr Verhältnis zu den physikalischen und chemischen Elementen, aus denen sie sich aufbaut, genauer aus. Sie zeigen auf diese Weise, wie die neuen Eigenschaften des Biologischen „emergent" und unableitbar von den Eigenschaften der Elemente entstehen:

„So spezifizieren die autopoietischen Einheiten die **biologische Phänomenologie** als die ihnen eigene Phänomenologie mit Charakteristika, die von denen der physikalischen Phänomenologie verschieden sind. Dies ist nicht etwa so, weil die autopoietischen Einheiten irgendeinem Aspekt der physikalischen Phänomenologie widersprechen; da sie molekulare Komponenten haben, müssen sie auch die gesamte physikalische Gesetzlichkeit erfüllen. Vielmehr hängen die Phänomene, die autopoietische Einheiten mit ihrem Operieren erzeugen, von der Organisation der Einheit ab und von der Art, wie diese verwirklicht wird, und nicht von den physikalischen Eigenschaften ihrer Bestandteile, welche nur den Raum ihrer Existenz bestimmen."

Jakob von Uexküll hat das Phänomen an einem einfachen Beispiel erläutert. Er führt sinngemäß aus: Das Problem sei nicht, wie aus Holz und Eisen ein Rad entstehe, sondern wie aus Rädern, Deichsel und Polstern ein Gebilde zustande kommen könne, das die vorher unbekannte Eigenschaft an den Tag lege, Menschen zu transportieren. Das emergente Neue ist ein Erzeugnis der Komposition der Teile und nicht der Teile als solcher. Daher betont J. v. Uexküll, daß Biologie „Kompositionslehre der Natur" sein müsse (1980).

1.4 Die Neuformulierung der Aufgabe eines Modells der Psychosomatischen Medizin

Wir haben in den früheren Auflagen (1979 und 1981) zwei Problemkreise definiert, die ein Modell der Psychosomatischen Medizin lösen muß:
1. Die Beziehungen zwischen Organismus und Umgebung.
2. Die Beziehungen zwischen biologischen, psychischen und sozialen Vorgängen.

Eine Lösung des ersten Problems sehen wir in dem Modell des Funktionskreises bzw. den Modifikationen, die dieses Modell auf den verschiedenen Integrationsebenen (als Regelkreis, Funktionskreis und, wie wir noch sehen werden, als Situationskreis) erfährt. Die Lösung des zweiten Problems besteht in einer Antwort auf die Frage nach den Beziehungen zwischen den Vorgängen, die sich auf verschiedenen Integrationsebenen abspielen. Wir wollen versuchen auf diese Frage eine Antwort zu geben, die sich, soweit wie möglich, an konkreten Beispielen orientiert.

1.4.1 Das Triebkonzept Freuds

Das erste psychosomatische Modell, das dem Unterschied zwischen verschiedenen Systemebenen Rechnung trägt und auf die Frage nach ihrer Verbindung eine Antwort zu geben versucht, stammt von Freud. Es ist das erste psychosomatische Modell, das diesen Namen verdient. Es handelt sich nicht, wie man irrtümlicherweise meinte, um sein Konversionskonzept, sondern um den **Triebbegriff,** den er bereits 1895 in dem Entwurf einer Psychologie konzipiert hat (Freud, 1975). Damals schrieb er, das psychische System sei den endogenen Erregungsquantitäten schutzlos ausgeliefert, und darin liege die Triebfeder der psychischen Mechanismen:

„Was wir von den endogenen Reizen wissen, läßt sich in der Annahme ausdrücken, daß sie interzellulärer Natur sind, kontinuierlich entstehen und nur periodisch zu psychischen Reizen werden."

An diesem Entwurf sind zwei Punkte hervorzuheben:
- Die Unterscheidung zwischen interzellulären Vorgängen, die mit Begriffen der Physik und Chemie beschrieben werden können, auf der einen Seite, und Vorgängen in einem psychischen System auf der anderen Seite, für deren Beschreibung andere Begriffe erforderlich sind. Eine der großen Leistungen Freuds ist die Entwicklung einer Sprache, die in der Lage ist, diese andersartigen Phänomene zu beschreiben.
- Die Feststellung, daß zwischen dem System interzellulärer Chemismen, in denen Freud die Ursachen der „endogenen Erregungsquantitäten" vermutet, und dem System psychischer Vorgänge periodisch, das heißt nur unter bestimmten Bedingungen, Transformationen stattfinden.

Damit sind zwei große Forschungsthemen formuliert: einmal das erkenntnis- und wissenschaftstheoretische Problem des Unterschiedes und der Beziehungen zwischen einem System interzellulärer Chemismen und einem System psychischer Vorgänge; zum anderen die Frage nach den Bedingungen für das Zustandekommen von Transformationen interzellulärer Chemismen in psychische Reize und umgekehrt. Freud hat an der Weiterentwicklung dieser beiden Themen keinen Anteil genommen, aber sein Entwurf blieb dafür die Basis.

Betrachten wir zunächst das erste Thema.

In dem Triebmodell wird einmal ein hypothetisches **somato-psychisches** Geschehen beschrieben: Bestimmte chemische Stoffe in den Körperflüssigkeiten sollen bei entsprechender Konzentration in einen psychischen Antrieb transformiert werden.

1895 kannte noch niemand derartige Stoffe, unter denen sich Freud irgendwelche Sexualstoffe vorstellte. Der Begriff des Hormons wurde erst 1902 von Bayliss und Starling formuliert.

Zum anderen wird ein **psycho-somatisches** Geschehen postuliert: Der psychische Antrieb soll ein Verhalten in Gang setzen, das (z. B. über sexuelle Befriedigung) wieder auf den interzellulären Chemismus zurückwirkt und durch Herabsetzung der Konzentration dieser Stoffe die Quelle für den psychischen Drang zum Versiegen bringt.

Mit der Vorstellung einer somato-psychischen Triebquelle und einer psycho-somatischen Triebabfuhr wird die Skizze für einen Regelkreis mit negativer Rückkoppelung entworfen, der zwei heterogene Phänomenebenen verbindet: ein System körperlicher Vorgänge, die durch chemische Reize gesteuert werden, und ein System psychischer Abläufe (Abb.1–2).

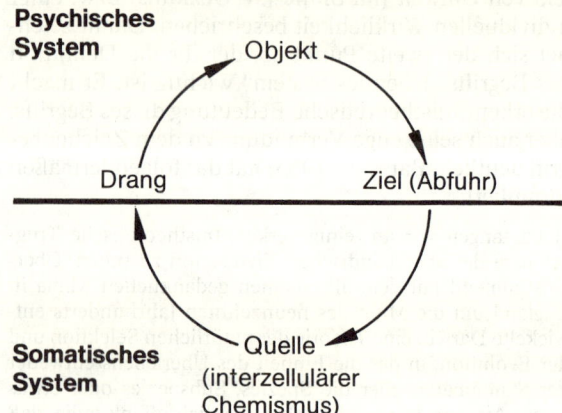

Psychisches System — Objekt

Drang — Ziel (Abfuhr)

Somatisches System — Quelle (interzellulärer Chemismus)

Abb. 1–2. Freuds Triebkonzept als Regelkreis, der ein System körperlicher Prozesse mit einem System psychischer Vorgänge verknüpft. 1915 beschrieb Freud (in „Triebe und Triebschicksale") vier Kriterien für den Trieb, die diese Skizze präzisieren: Die Triebquelle, den Triebdrang, das Triebziel und das Triebobjekt. Die Triebquelle würde dem interzellulären Chemismus bzw. den durch diesen ausgelösten Reaktionen physiologischer Systeme entsprechen. Wenn diese Systeme die durch den Chemismus verursachte Homöostasestörung nicht innerhalb des Organismus kompensieren können, wird ein psychischer Drang ausgelöst, der ein Verhalten (das Triebziel) in Gang setzt, das unter Inanspruchnahme der Kompensationsmöglichkeiten der Umgebung (des Triebobjekts) das innerkörperliche Problem löst.

Ein solches Modell bleibt jedoch nebelhaft, solange die erkenntnis- und wissenschaftstheoretische Aufgabe nicht gelöst ist, die Verschiedenartigkeit der beiden Bereiche und ihre Beziehungen zu definieren. Dafür haben erst, wie wir es beschrieben haben, die Systemtheorie und – wie wir gleich darstellen werden – die moderne Zeichenlehre die notwendigen Voraussetzungen geschaffen (s. Anmerkung 4, S. 37).

In den vorhergehenden Abschnitten wurde dargestellt, wie die Systemtheorie die verschiedenen Ebenen des Freudschen Modells genauer zu definieren erlaubt, und wie sich damit die Frage nach der Verbindung zwischen diesen verschiedenen Ebenen konkret stellt.

1.4.2 Das Konzept der Zeichenlehre und der Beitrag Pawlows

Ein Konzept, das auf diese Frage eine Antwort gibt, stammt aus der modernen Zeichenlehre. Sie kennt das Problem der Kluft zwischen verschiedenen Zeichensystemen, zum Beispiel die zunächst unüberbrückbare Grenze zwischen zwei Menschen, die verschiedene Sprachen sprechen. Diese Grenze kann nur überwunden werden, wenn beide gelernt haben, die Sprache des anderen in ihre eigene Sprache zu übersetzen. Der Vorgang der Übersetzung ist daher zeichentheoretisch von besonderem Interesse. Jakobson (1971) hat dafür ein Konzept entwickelt, das zwischen **Interpretationen** (Übersetzungen in ein und derselben Sprache), zwischensprachlichen **Übersetzungen** (von einer Sprache in eine andere) und **Transmutationen** (Übersetzungen von einem nichtsprachlichen in ein sprachliches Zeichensystem oder von einem nichtsprachlichen in ein anderes nichtsprachliches Zeichensystem) unterscheidet.

Unter diesem Aspekt läßt sich die Notwendigkeit, für jede Integrationsebene der Systemhierarchie eine eigene Sprache zu entwickeln, durch die Annahme interpretieren, daß auf jeder dieser Ebenen spezifische Zeichensysteme die Verbindung zwischen den Elementen der dort angesiedelten Systeme aufrechterhalten. Die Sprachen, welche von der Biologie, der Psychologie und der Soziologie entwickelt werden, um die Phänomene ihrer Integrationsebene zu beschreiben, würden dann Transmutationen entsprechen, mit deren Hilfe averbale Zeichenprozesse der verschiedenen Integrationsebenen in Wortsprachen übersetzt werden.

So vermitteln nach unseren heutigen Vorstellungen auf der Ebene der Zellen Zeichen des genetischen Codes den Informationsaustausch zwischen Zellelementen. „The information is not merely being transcribed and translated but is operating as instructions – if you want to put it in a fancy jargon, as ‚algorithms‘. The DNA makes RNA and the RNA makes a protein and the protein then does something to its surroundings, which result in the production of more varieties of molecules than before" (Waddington, 1968). Auf der Ebene des Organismus vermitteln Hormone und Nervenaktionsströme den Informationsaustausch zwischen Organen, und auf der nächst komplexeren Ebene vermitteln psychische Prozesse die Verbindung zwischen Organismus und Umgebung.

Das Problem, wie man sich die Verbindungen zwischen diesen verschiedenen Ebenen vorstellen soll, das man systemtheoretisch als die Frage formulieren kann, wie **„Aufwärts"**- und **„Abwärts"-Effekte** in einem hierarchischen System (Popper, 1977) zustande kommen, läßt sich dann mit der Annahme lösen, daß die Zeichensysteme der verschiedenen Ebenen durch Übersetzungen verbunden werden.

Im Rahmen einer solchen Betrachtungsweise ist das sogenannte psychophysische Problem nur eines unter anderen. Freuds Annahme einer periodisch erfolgenden Transformation interzellulärer Chemismen in ein psychisch erlebtes Drängen bedeutet unter diesem Aspekt, daß Übersetzungen zwischen interzellulären Chemismen und psychischen Zeichen erfolgen, daß sie aber nur unter bestimmten Bedingungen zustande kommen.

Diese in dem Modellentwurf Freuds nur sehr allgemein formulierten Hypothesen, die zeichentheoretisch formuliert zunächst sehr abstrakt und theoretisch klingen, haben durch die Versuche Pawlows eine empirische Bestätigung und Konkretisierung gefunden, die für das Triebkonzept und eine psychosomatische Theorienbildung von kaum zu überschätzender Bedeutung sind: Der Vorgang, den Pawlow als **Bildung bedingter Reflexe** beschrieben hat, ist nichts anderes als eine Übersetzung von Nachrichten aus einem psychischen Zeichensystem in Nachrichten eines Systems somatischer Zeichen und umgekehrt. Mit dieser Übersetzung kommt eine Verbindung zwischen der psychischen und der somatischen Ebene zustande, die ohne diese Übersetzung nicht existiert.

> Dabei finden **„Bedeutungskoppelungen"** statt: Zeichen, welche im Körper Nachrichten über die Bedeutung einer Organreaktion für andere Organe übertragen, werden mit Zeichen zusammengekoppelt, die den Organismus über die Bedeutung von Vorgängen in seiner Umgebung informieren.

Pawlow hat gezeigt, daß solche Bedeutungskoppelungen oder Übersetzungen nur „periodisch", das heißt nur in Situationen zustande kommen, in denen bestimmte Bedingungen erfüllt sind, und daß die Dauerhaftigkeit solcher Koppelungen von diesen Bedingungen abhängt. In dem bekannten Beispiel begannen seine Versuchshunde auf Geräusche hin, die der Labordiener im Nebenraum bei der Zubereitung des Futters verursachte, Speichel und Magensaft zu sezernieren. Das war bisher nur während der Fütterung geschehen, wenn Geschmacks-, Geruchs- und Berührungsreize nach einem angeborenen Code (als „unbedingter Reflex") in nervale Zeichen für die Aktivierung der Speichel- und Magendrüsen übersetzt wurden. Jetzt waren bisher neutrale akustische Phänomene, „akustisches Rauschen", als Zeichen für Vorgänge von Bedeutung in der Umgebung mit den nervalen Zeichen zusammengekoppelt worden. Solche Koppelungen kamen jedoch nur zustande, wenn die Hunde gesund und hungrig waren. Bei kranken oder gesättigten Hunden blieb die Konditionierung aus (s. Anmerkung 5, S. 38).

Diese Versuche zeigen zweierlei:

1. Die von Freud angenommene Periodizität beruht auf einem „organischen Entgegenkommen" oder einer „vulnerablen Phase" als Bedingung für das Zustandekommen solcher Koppelungen oder Übersetzungen. Anders formuliert: es muß eine entsprechende „Stimmung" vorhanden sein.

2. Wir dürfen nicht nur mit einer, für alle Individuen konformen Physiologie rechnen. Es gibt auch individuelle Physiologien für bestimmte Organe und Funktionen (Adler und v. Uexküll, 1987). Die Speichel- und Magendrüsen der Pawlowschen Hunde

hatten seit ihrer Konditionierung eine andere Physiologie als die Speichel- und Magendrüsen nichtkonditionierter Hunde. Die individuellen Physiologien können nur biographisch, das heißt nur aufgrund der Geschichte verstanden werden, welche über die Situationen informiert, in denen Bedeutungskoppelungen stattgefunden oder nicht stattgefunden haben. Wir wissen heute, daß auch das **Immunsystem** zu den Organen zählt, deren Reaktionen eine, nur biographisch verstehbare, individuelle Physiologie aufweisen (s. Kap. 11).

Für das Problem, das der geschilderte Krankheitsfall uns aufgibt, bedeutet das folgendes: Die somatischen Zeichensysteme endokriner und nervaler Art, welche die Herz-, Kreislauf- und Lungenfunktionen der Patientin regulieren, sind mit psychischen Zeichensystemen gekoppelt, welche die Patientin über – für sie vital wichtige – Aspekte ihrer Umwelt informieren. Der Ehemann und der Sohn sind Bedeutungsträger, die entscheiden, welchen Aspekt die Umwelt hat und welche somatischen Bereitstellungen der Organe jeweils erforderlich sind.

Bedeutungskoppelungen, in denen Bezugspersonen eine derartige Wichtigkeit erlangen, sind meist schon in der frühen Kindheit geknüpft worden. Psychoanalytisch formuliert, handelt es sich dann um Übertragung von Gefühlen, die ursprünglich der Mutter oder dem Vater gegolten haben, auf andere Personen.

Um solche Vorgänge, die eine Verbindung zwischen humanen, animalischen und vegetativ-somatischen Integrationsstufen betreffen, anschaulich und vorstellbar zu machen, müssen wir das Modell des Funktionskreises erweitern.

1.5 Das Modell des Situationskreises

Wir haben betont, daß auf den verschiedenen Stufen des Organischen (N. Hartmann, 1969), die das Konzept der Systemtheorie beschreibt, emergent neue Phänomene auftreten. Die neuen Phänomene lassen sich jedoch als Differenzierungen der Phänomene der einfacheren Ebenen beschreiben. Die Begriffe Subjekt und Objekt, Merken und Wirken, Selbst und Nicht-Selbst, die Phänomene beschreiben, die auf der einfachsten Ebene mit den ersten primär aktiven Systemen auftreten, müssen für jede Stufe neu definiert werden, um diesen Zusammenhang zwischen den einfacheren und den differenzierteren Phänomenen sichtbar zu machen.

Das gilt nicht nur für die Begriffe zur Beschreibung der Phänomene, sondern auch für die Modelle, die wir zur Analyse der Vorgänge auf den verschiedenen Ebenen heranziehen müssen: Das Modell des Regelkreises, das Vorgänge auf der vegetativen Ebene beschreiben kann, mußte zu dem Modell des Funktionskreises erweitert werden, um Vorgänge auf der animalischen Ebene beschreiben zu können, und für die Ebene des Humanen benötigen wir, wie bereits

angedeutet wurde, ein noch komplexeres Modell, den **Situationskreis,** von dem wir jetzt sprechen wollen.

Dabei müssen wir jedoch den Gedanken festhalten, daß die **Geschichtlichkeit** auf jeder höheren Stufe des Lebens eine zunehmende Rolle spielt. Auf der Stufe des vegetativen Daseins spielen Zeit und Geschichtlichkeit eine verhältnismäßig geringe Rolle. Einzellige Organismen sind praktisch unsterblich. Auf der Stufe des animalischen Lebens begegnet uns Geschichtlichkeit nicht nur in der Form der Lebenszyklen, die den einzelnen Lebewesen zugewiesen sind, sondern auch in Form einer individuellen Entwicklung angeborener Möglichkeiten, wie es das Beispiel der Konditionierung zeigte. Beim Menschen gewinnt die Geschichtlichkeit seiner ontogenetischen Entwicklung eine zentrale Bedeutung. Wir werden bei der Darstellung des Situationskreismodells daher diesem Gedanken Rechnung tragen müssen.

1.5.1 Primärprozeß und Funktionskreis

Wenn wir die im Funktionskreismodell dargestellten Systemzusammenhänge für die Medizin fruchtbar machen wollen, müssen wir bei sehr viel komplexeren Phänomenen als denen des tierischen Verhaltens nach analogen Zusammenhängen suchen. Dazu ist eine terminologische Vorbemerkung nötig: Wir haben darauf hingewiesen, daß zwischen den Phänomenen der verschiedenen Systemebenen bei allen Unterschieden grundsätzlich Ähnlichkeiten, Homomorphien (Bertalanffy, 1968), gefunden werden, die in den zu ihrer Beschreibung verwendeten Begriffen zum Ausdruck kommen müssen. So bezeichnen die Begriffe „Merken" und „Wirken" zunächst nur die rezeptorische und effektorische Aktivität lebender Systeme und haben noch nichts mit bewußtem Wahrnehmen und Wollen zu tun. Es hängt von der Organisation der Rezeptoren und des Nervensystems ab, wie vielfältig die Reize sind, die als „Merkmale" in die Umwelt projiziert werden, und von der Organisation komplizierter Rückmeldungen, ob sich ein Lebewesen über sein Merken und Wirken Rechenschaft geben, d.h., ob es sein Merken und Wirken „merken" kann. Das **Reafferenzprinzip** (v. Holst und Mittelstädt, 1950) beschreibt eine solche Differenzierung. Merken und Wirken werden von Programmen geleitet, die ursprünglich aus der Quelle biologischer Bedürfnisse stammen, wie sie vor allem bei den immer wiederkehrenden Homöostasestörungen im Körper bei Durst, Hunger, Müdigkeit usw. auftreten. Diese Programme integrieren, wie das Beispiel der Bedeutungskoppelungen bei den Pawlowschen Hunden zeigte, Sensationen, die aus dem Körper stammen, und Eindrücke der Sinnesorgane aus der Umgebung für die jeweiligen Bedürfnisse des Lebewesens zu dessen spezifischen Problemsituationen. Sie sorgen für deren Lösung durch ein adäquates, das heißt von erprobten Programmen gesteuertes Verhalten. Das stellt das Modell des Funktionskreises als kreisförmigen Ablauf dar, der bei Piaget als **„sensomotorische Zirkulärreaktion"** beschrieben wird.

Wenn wir diese beiden Modelle vergleichen, so entspricht Merken und Wirken nach angeborenen Programmen weitgehend dem, was Piaget als „Assimilation" bezeichnet, während er die Bildung neuer Programme als „Akkommodation" beschreibt. Beides, Programmbildung und programmgesteuertes Merken und Wirken, entspricht dem, was in der angelsächsischen Literatur als „coping behaviour" bezeichnet wird. Die praktische Bedeutung dieser Modelle für die Medizin wird sichtbar, wenn wir uns klarmachen, daß einer ungelösten Problemsituation „Maladaptation" und „Streß" entspricht, während Problemlösung nach neuen Programmen „Adaptation" bedeutet.

Wenn wir nach diesen terminologischen Vorbemerkungen beim Menschen nach Vorgängen suchen, die sich nach dem Modell des Funktionskreises beschreiben lassen, stoßen wir auf folgenden Zusammenhang: Das automatische Reagieren, mit dem Tiere Problemsituationen nach angeborenen Programmen lösen, entspricht weitgehend einem Verhalten, das die Psychoanalyse als **„Primärprozeß"** beschreibt. Dabei soll ökonomisch-dynamisch gesehen psychische Energie ohne jene Hindernisse abströmen, die der psychische Apparat des erwachsenen Menschen aufgebaut hat. Unter diesem Aspekt entspricht der Vorgang, den der Funktionskreis beschreibt, weitgehend dem, was Freud vorschwebte, als er davon sprach, daß in der Frühphase menschlicher Entwicklung „orales", „anales" und „genitales" Triebverhalten primärprozeßhaft realisiert wird. Wir können uns vorstellen, daß beim menschlichen Säugling, ähnlich wie beim Tier, vom ersten Tag an orale Funktionskreise der Nahrungsaufnahme und anale Funktionskreise der Ausscheidung nach angeborenen Programmen ablaufen, und daß der genitale Funktionskreis ähnlich wie beim Tier erst später strukturiert wird.

Auf diese Weise wird die Übereinstimmung des Verhaltens menschlicher Säuglinge mit dem Verhalten von Tieren, was Zwanghaftigkeit, Unaufschiebbarkeit und Unbelehrbarkeit angeht, deutlich. Es wird darüber hinaus sichtbar, wie physiologische Prozesse im Körper („interzelluläre Chemismen") und sensomotorisches Verhalten zur Umgebung noch eng gekoppelt zusammengehören. Wenn irgendwo, dann wird hier die psychophysische Einheit, wie sie das Triebmodell Freuds beschrieben hat, sichtbar. In ihr werden physiologische Homöostasevorgänge in ein psychisches Drängen übersetzt, das dann Umgebungsfaktoren durch Bedeutungserteilung in Umweltobjekte verwandelt, deren Bedeutungsverwertung als Nahrung oder Ausscheidungsprodukte wieder auf den Körper zurückwirkt.

> Entwicklungspsychologisch entspricht das Verhalten des Säuglings während der ersten 2–3 Lebensmonate weitgehend dem Schema des Funktionskreises.

Säuglinge reagieren auf Außenweltreize noch **primärprozeßhaft.** Psychoanalytiker sprechen von einer **undifferenzierten** Phase (A. Freud, 1969; H. Hartmann, 1970 u.a.), von einer Phase des **Autismus** (M. S. Mahler et al., 1980) oder einer **objektlosen** Phase (R. A. Spitz, 1965, 1972).

Diese Kennzeichnung macht einen Unterschied zwischen dem Funktionskreisschema deutlich, nach dem das Verhalten menschlicher Säuglinge abläuft, und dem Funktionskreismodell, nach dem tierisches Verhalten gedeutet werden kann: Tiere sind bald nach der Geburt autark. Sie können dann ihre Problemsituationen durch ihr Merken und Wirken, das heißt durch ihre sensomotorischen Aktivitäten selbst lösen. Der menschliche Säugling ist während der ersten Lebensmonate auf eine spezifisch menschliche Umgebung angewiesen, die ihm die Lösung seiner biologischen Probleme zu einem großen Teil abnimmt. Das bedeutet, daß biologische Funktionskreise der Nahrungsaufnahme und der Ausscheidung beim Menschen zwar von Geburt an nach angeborenen Programmen verlaufen, aber – im Unterschied zu den Tieren – vom ersten Tag an **mit Anforderungen der Gesellschaft** konfrontiert sind. Die lebensnotwendige Ergänzung kindlicher Funktionskreise durch die Mutter, die, ohne es zu wollen oder zu wissen, als „Agent der Gesellschaft" fungiert, zwingt den Säugling zu einer Veränderung und Differenzierung primär biologischer Verhaltensformen oder zu einer „Sozialisation biologischer Funktionskreise".

Diese **„primäre Sozialisation"** des Kindes, die für die Nahrungsaufnahme im ersten Lebensjahr – dem „extrauterinen Frühjahr" (Portmann, 1969) – stattfindet, spielt sich in einer engen Wechselbeziehung zwischen Mutter und Kind ab. Erst in deren Verlauf wird die Mutter für das Kind aus einer anonymen bedürfnisbefriedigenden Umwelt zu einer geliebten Person, einem Objekt, wie es die Psychoanalyse präzise formuliert. Das eng verschränkte Sich-Entwickeln biologischer Reifungsprozesse der Sensomotorik mit psychischen Lernvorgängen, zum Beispiel die Unterscheidung zwischen Innen und Außen und später auch das Erlernen der Sprache, sind eng an die Aktivitäten der Mutter oder der sie vertretenden Pflegepersonen gebunden. In dieser Verschränkung erfolgt die erste Modifikation und Sozialisation angeborenen, triebhaften – zunächst oralen, später analen – Verhaltens. Dieser Entwicklungsprozeß, der vom biologischen Funktionskreis schließlich zum Situationskreis des erwachsenen Menschen führt – was darunter zu verstehen ist, werden wir gleich darstellen – läßt sich heute vor allem durch die Untersuchungen von R. A. Spitz (1972), Jean Piaget (1975), M. S. Mahler und Mitarbeitern (1980) sowie von M. Balint (1966), D. W. Winnicott (1973), H. Kohut (1979) und anderen in großen Zügen rekonstruieren.

1.5.2 Terminologische Probleme

Dabei gibt es jedoch terminologische Probleme. Wir haben schon darauf hingewiesen, wie problematisch manche Begriffe sind, mit denen wir frühe Phasen der psychophysischen Entwicklung beschreiben. Wir sagten, Begriffe wie „Symbiose", „Dyade" oder „Zweierbeziehung" setzten zwei Individuen und damit etwas voraus, das in dem von den Begriffen bezeichneten Zeitraum noch gar nicht existiert. Die Begriffe sind aber in unsere Terminologie eingeführt, und es wäre unrealistisch zu erwarten, daß sie wieder verschwinden. Vor allem, haben wir vorläufig keine besseren, die uns umständliche und wahrscheinlich wenig präzise Umschreibungen ersparen. Wir müssen sie also beibehalten, aber uns klarmachen, daß wir mit ihnen Zustände bezeichnen, die gerade noch keine Zweierbeziehung sind, sondern eher einer Beziehung zwischen einem Individuum und einem seiner Organe entsprechen (s. Anmerkung 3, S. 37).

Ein anderes Problem wirft der Begriff des „undifferenzierten Zustands" auf, der den Anfang der psychophysischen Entwicklung bezeichnen soll. Auch hier verführt der Begriff zu falschen Vorstellungen: Der Embryo ist kein undifferenziertes Gebilde oder gar eine amorphe Masse. Er ist ein außerordentlich differenziertes System, das jedoch in eine völlig andere Umgebung integriert ist als nach der Geburt, und das mit dieser anderen Umgebung eine organische Einheit bildet. Diese andere Umgebung und die Integration der differenzierten Funktionen des Embryos mit dem Plazentarkreislauf, dem Fruchtwasser, der konstanten Außentemperatur usw. zu einer lebenden Einheit muß man sich genau vergegenwärtigen und dann mit der neuen Einheit vergleichen, die sich nach der Geburt zwischen dem Kind und der Mutter bilden muß, wenn der Säugling am Leben bleiben und sich entwickeln soll. Dann wird deutlich, daß es sich nicht um eine „Differenzierung" aus einem undifferenzierten, quasi amorphen Zustand, sondern um eine Neuintegration durch Verknüpfung vieler Funktionen des kindlichen Organismus mit einer anderen Umgebung handelt, die an die Stelle der intrauterinen Umgebung treten muß. Diese Neuintegration ist zu einem überwiegenden Teil die Aufgabe der Mutter. Sie muß die neuen Verknüpfungen ermöglichen und die Programme für den Umgang mit der neuen Umwelt einüben. Die **Mutter** ist für das Neugeborene – und darin liegt ihre entscheidende Bedeutung für die Frühentwicklung – der **Prototyp der späteren Umwelt** des Kindes.

Die Programme, welche die Verknüpfung mit der neuen Umgebung durch Riechen, Schmecken, Tasten, Hören, Saugen usw. herstellen, müssen im Gehirn des Kleinkindes gespeichert werden. Sie sind die Grundlage für die späteren Programme des Umgangs zwischen dem Menschen und seiner Umgebung. Damit die frühen Programme, die dann allerdings im Vergleich zu denen des Heranwachsenden relativ primitiv und undifferenziert sind, rechtzeitig und mit der erforderlichen Plastizität entstehen und eingeübt werden können, ist eine – um den Winnicottschen Ausdruck zu gebrauchen – **„genügend gute Mutter"** unerläßlich. Mit ihr bildet dann der Organismus des Säuglings eine Einheit, die sich im Verlauf der ersten Wochen und Monate differenziert.

Störungen in dieser dynamischen Einheit, wie sie durch zeitweise Trennungen von der Mutter unumgänglich sind, bilden die ersten Problemsituationen, für die vorübergehend keine Lösungen zur Verfügung

stehen. Sie legen damit das frühe Muster für Streßerfahrungen, und es kommt alles darauf an, daß diese so dosiert werden, daß sie Anstöße zur Entwicklung und nicht zu Störungen und Retardierungen bilden.

Der früheste Funktionskreis knüpft als „symbiotischer Funktionskreis" die Fäden zwischen dem Neugeborenen und der entstehenden Umwelt, die zunächst der Körper der Mutter ist. Tierversuche, vor allem die von H. Weiner und Mitarbeitern (Ackerman, 1981; Ackerman et al., 1978; Hofer, 1976 und 1982; Hofer und Weiner, 1972) durchgeführten sehr genauen Beobachtungen der frühen Beziehungen zwischen kleinen Ratten und ihren Müttern, haben Aufschluß über die unglaublich differenzierten und genau abgestimmten Muster gebracht, die realisiert werden müssen, wenn diese Fäden geknüpft werden sollen. Sie zeigen, wie der Geruch der mütterlichen Brust die kleinen Ratten anzieht, wie die mütterliche Milch den Herzrhythmus der Kleinen, die Frequenz der Fütterung ihren Schlafrhythmus, die Nestwärme ihre Temperaturregulation steuert. Die von der Mutter ausgehenden Stimuli sind Voraussetzung für das rechtzeitige Einsetzen bestimmter endokriner und enzymatischer Reaktionen im Organismus und vor allem auch im Gehirn der Neugeborenen. Die Beobachtungen Winnicotts zeigen, wie bedeutsam diese Gegenseitigkeiten für die Verhältnisse zwischen menschlichen Müttern und ihren Neugeborenen sind.

Damit zeichnet sich ein erster Akt in der Frühentwicklung ab, der, wie gesagt, als Neuinszenierung und Neuintegration bezeichnet werden muß, und der das dramatische Geschehen einleitet, von dem dann Anna Freud (1969) und Margret Mahler (1980) spätere Szenen untersucht und beschrieben haben.

1.5.3 Der symbiotische Funktionskreis

Wir haben aus diesem dramatischen Geschehen einen Aspekt herausgegriffen und wollen jetzt seine Konsequenzen verfolgen: Nicht nur tierexperimentelle (Harlow und Zimmerman, 1959; Henry und Stephens, 1977; Weiner, 1981; Ackermann et al., 1978), auch entwicklungspsychologische Untersuchungen an Neugeborenen (Hofer, 1981; Lichtenberg, 1983; 1984 u.a.) zeigen, daß die Vorstellung eines **primären Narzißmus,** nach dem das Neugeborene wie ein Ei in der Schale nur mit sich selbst in Interaktion steht, ebenso wie die Vorstellung eines „natürlichen Autismus" (M. S. Mahler) zugunsten einer anderen Vorstellung aufgegeben werden muß. Danach ist das Kind (und bereits der Embryo) von Anfang an mit komplexen Verhaltensmustern oder deren Vorläufern für eine soziale Interaktion mit der Mutter ausgerüstet, denen auf seiten der Mutter ebenfalls z.T. angeborene Antwortbereitschaften gewissermaßen „kontrapunktisch" (d.h. wie Rolle und Gegenrolle) entsprechen.

Wissenschaftstheoretisch heißt das zweierlei:
Das soziale System ist früher als das Individuum.
Die Einheit oder das Ganze ist früher als die Teile.

Für die Psychosomatische Medizin bedeutet das eine hochinteressante Konstellation: einen Funktionskreis, der zwei Subjekte umfaßt, von denen das eine die Umwelt des anderen bildet. In diesem „symbiotischen Funktionskreis" wird das Verhalten des einen Subjekts unmittelbar in physiologische Vorgänge im Organismus des anderen übersetzt. Schematisch läßt sich das wie in Abbildung 1–3 darstellen.

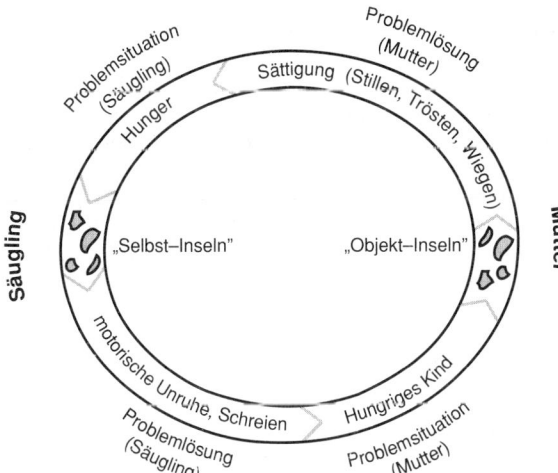

Abb. 1–3. Der symbiotische Funktionskreis. In dem symbiotischen Funktionskreis einer Zweierbeziehung oder Dyade (wir erinnern uns an die Problematik dieser Begriffe) erlebt der hungrige Säugling seine Umgebung als Problemsituation, die nach angeborenen Programmen gedeutet und mit motorischer Unruhe beantwortet wird. Die Problemsituation des Säuglings kann nur durch die Mutter gelöst werden, die ihrerseits den unruhigen Säugling als ihre Problemsituation (ihre Umwelt) erlebt und diese mit Trösten, Liebkosen, Wiegen und Stillen beantwortet (Verhaltensweisen, die zum Teil nach angeborenen, zum Teil nach erlernten Programmen ablaufen). Damit löst sie gleichzeitig mit ihren eigenen Problemen die Probleme des Säuglings.

Das bedeutet, daß Kinder in dieser symbiotischen Phase ihrer Entwicklung durch bestimmte Menschen (die Mutter oder ihre Stellvertreter) außerordentlich vulnerabel und von ihnen in einer kaum vorstellbaren Weise abhängig sind. Das Modell des symbiotischen Funktionskreises illustriert aber nicht nur das Ausgeliefertsein des Säuglings an die Mutter als Agentin der Gesellschaft, es macht auch nachvollziehbar, daß diese Konstellation dem Neugeborenen etwas völlig Neues eröffnet: Die Anfänge einer Welt als Quelle künftiger Kreativität und Basis für die Wirklichkeit, die es um sich herum aufbauen muß, um autark zu werden.* Im symbiotischen Kreisgeschehen lernt der Säugling, sein angeborenes Zeichensystem der Signale aus seinem Körperinneren und seinen Sinnesorganen in ein neues, soziales Zeichensystem zu übersetzen. Dabei werden strikt priva-

---

* In den Kapiteln 41.3 „Essentielle Hypertonie", 42 „Asthma bronchiale" und 43 „Ulcus duodeni" werden die Konsequenzen dieser Zusammenhänge für unser Krankheitsmodell genauer dargestellt.

te Sensationen, Gefühle und Sinneszeichen, die der Säugling spürt, an Zeichen gekoppelt, die aussagen, was diese privaten Signale für die Gesellschaft (repräsentiert durch die Mutter) bedeuten, daß zum Beispiel eine besondere Art Unlust Hunger, eine andere Schmerzen und eine dritte Durst bedeutet usw.

Erst im Laufe dieses Differenzierungsprozesses wird aus dem bedürfnisbefriedigenden Medium, das noch, wie Balint sagt, in „primärer Liebe" wie die umgebende Luft gebraucht wird, ein geliebtes Objekt, das aktiv umworben, verlassen und wieder aufgesucht werden kann.

Dabei müssen die Programme gelernt und internalisiert werden, mit denen schließlich eine autonome Ich-Instanz und eine Umwelt entstehen.

> Nach den heutigen Vorstellungen haben Störungen in diesem frühen Differenzierungsprozeß psychosomatische Dispositionen zu späteren Erkrankungen zur Folge.

Tierversuche wie die Henrys an Mäusen, Harlows an Affen und Weiners an jungen Ratten haben diese Vorstellungen in eindrucksvoller Weise untermauert.

1.5.4 Die Entstehung von Wirklichkeit aus Umwelt

Trotzdem fehlen uns heute noch wichtige Elemente, um die Ergebnisse dieser Tierversuche mit frühen Störungen der psychischen Entwicklung beim Menschen, wie sie zum Beispiel als Ich-Defekte oder „Selbst-Pathologie" (Kohut, 1979) beschrieben werden, in klare Beziehungen zu setzen. Der Grund dafür ist, wie bereits erwähnt, die Tatsache, daß die psychische Entwicklung beim Menschen – jedenfalls vom zweiten Lebensjahr an – anders verläuft als bei Tieren, selbst bei den ihm am nächsten stehenden Anthropoiden. Dabei spielt die Entstehung des Vorstellungsvermögens eine entscheidende Rolle, das, wie Piaget (1975) überzeugend nachweist, etwa von diesem Zeitpunkt an den Aufbau der Umwelt des Kindes zunehmend bestimmt. Jetzt beginnt die Trennung in die Innenwelt, einer zunehmend von der unmittelbaren Fesselung an die biologischen Triebe gelösten Phantasie, und in eine nach den Programmen oder Schemata unserer Sensomotorik gedeuteten, direkt oder indirekt durch unsere Motorik beherrschbaren Außenwelt kausal geordneter Gegenstände (Th. v. Uexküll, 1984).

Diese Entwicklungen lassen vermuten, daß die Entstehung unserer dualistischen Vorstellungen von einer physisch-materiellen und einer psychisch-immateriellen Welt als Gegenbilder unserer Motorik und unserer Phantasie im Sinne Piagets genetisch in dem Kontinuum unserer Ontogenese zurückverfolgt werden kann. Sie machen verständlich, daß dabei zwei völlig verschiedenartige Körpervorstellungen entstehen: Die Vorstellung eines anatomischen Körpers als Gegenstand der Sensomotorik, dessen Innen und Außen als **räumliche Dimension** immer wieder zu verhängnisvollen Verwechslungen mit dem Innen und Außen einer **semantischen Dimension** führt, in der wir von einem Körper als Repräsentanz innenweltlicher Erlebnisse sprechen.

Das Ergebnis dieser Differenzierung der zunächst nach dem Modell des Funktionskreises, dann nach dem Modell des symbiotischen Funktionskreises gedeuteten Einheit, läßt sich für den erwachsenen Menschen mit Hilfe des Situationskreismodells darstellen (Abb. 1–4).

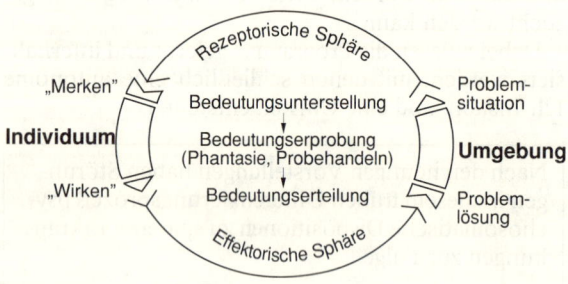

Abb. 1–4. Der Situationskreis unterscheidet sich von dem Funktionskreis durch eine obligatorische Zwischenschaltung der Vorstellung, in der Programme für Bedeutungserteilung (**"Merken"**) und Bedeutungsverwertung (**"Wirken"**) zunächst probeweise als Bedeutungsunterstellung und Bedeutungserprobung durchgespielt werden können, ehe das Ich sie für die Sensomotorik freigibt. Dabei wird in der Vorstellung die Situation (die dem Merkmal bzw. der Problemsituation im Funktionskreis entspricht) quasi experimentell vorstrukturiert: d.h. Bedeutungserteilung erfolgt zunächst als (hypothetische) Bedeutungsunterstellung, deren Konsequenzen (in der Phantasie durch "Probehandeln") abgetastet werden.

Das Modell beschreibt, wie der Automatismus der Primärvorgänge schließlich mit Hilfe der zahlreichen Schranken und Kanäle, die der psychische Apparat inzwischen als Programmreservoir und innere Bühne zur Erprobung der Programme aufgebaut hat, zunehmend durch Sekundärvorgänge ersetzt wird. Aus der biologischen Umwelt, die nach dem Funktionskreismodell aufgrund angeborener Programme aufgebaut wurde, ist eine individuelle Wirklichkeit geworden, die jeder einzelne nach Programmen aufbauen muß, die er im Laufe seiner individuellen Lebensgeschichte in seiner Kultur erworben hat. Damit entsteht wieder jene feste, aber für den außenstehenden Beobachter unsichtbare Schale, die den Körper des einzelnen umgibt und vor dem unmittelbaren Einwirken des Verhaltens seiner Objekte auf seine körperlichen Vorgänge abschirmt.

1.6 Der Körper als Organ der psychosomatischen Einheit

Damit ist das Stichwort gefallen: Was sollen wir im Rahmen dieser Vorstellung unter dem Begriff "Körper" verstehen? Die bisherigen Ausführungen haben das psychosomatische Modell als **Geschichte** beschrieben, in der, beginnend mit der Geburt, Funktionskrei-

se über Zwischenstufen schließlich in Situationskreise mit individuellen, kulturspezifischen Wirklichkeiten verwandelt werden. In diesem Modell läßt sich Seelisches, "der psychische Apparat", als ein Organ beschreiben, das eine Außenwelt für die Sensomotorik als äußere Bühne und eine Innenwelt der Phantasie als innere Bühne aufbaut, auf der Programme zur Lösung von Problemen der äußeren Bühne durchgespielt werden. Der Körper tritt im Rahmen dieser Entwicklung in zwei Formen auf: als Gegenstand der Außenwelt, der wie alle physischen Gegenstände manipuliert werden kann – und als Repräsentanz unseres Körpererlebens in der Innenwelt. In beiden Formen ist Körper lediglich ein Produkt der psychischen Tätigkeit. In dieser Form läßt sich der Körper nicht als ergänzendes, d.h. interdependentes und interaktives Organ (im Sinne von Weiss und English, 1943), das mit dem "Organ Psyche" in ständigem Nachrichtenaustausch steht, in das Modell der Zeitgestalt dieser Geschichte einfügen. Dafür genügt es also nicht, die Entwicklungsgeschichte mit der Geburt beginnen zu lassen. Wir müssen auch die Zeit davor, d.h. die Keimentwicklung im Uterus hinzudenken.

In dieser Phase unterscheidet sich der Zustand der heranwachsenden Frucht von dem Zustand des Neugeborenen, wie wir beschrieben haben, dadurch, daß für sie die Probleme der Beschaffung und Einverleibung von Nachschub für den Energie- und Baustoffwechsel, und das heißt auch für den Ausgleich der Homöostasestörungen im Körper, noch nicht existieren. Alle diese Probleme werden im Inneren des Körpers mit Reserven gelöst, welche der mütterliche Organismus in reichem Maße zur Verfügung stellt.

Nach der Formulierung Helmuth Plessners (1976) wäre die Phase vor der Geburt dadurch gekennzeichnet, daß der Embryo zwar "Körper" ist, aber seinen Körper noch nicht "hat", weder als Werkzeug, um sich in einer Außenwelt (die es für ihn noch nicht gibt) zu behaupten, noch als Gegenstand, mit dem man umzugehen lernen muß. In diesem Zustand eines reinen "Körper-Seins" gibt es noch keine Umgebung, die eine Bedeutung hat, welche die Fernsinne durch Merken und Wirken in eine Umwelt transponieren. Es besteht noch eine Organisationsform, die sich als semantisch geschlossenes System beschreiben läßt. Sie entspricht der Organisationsform von Pflanzen, die über keine Sinnes- und Bewegungsorgane verfügen, mit denen sie ein Umwelt-Kompartiment außerhalb ihres Körpers aufbauen könnten. Vorgänge der Umgebung werden nur als Veränderungen der Rezeptoren ihrer oberflächlichen Zellschicht erfahren. Der Embryo hat daher, wenn wir die Formulierung, die J. v. Uexküll für Pflanzen angewendet hat, übernehmen, nur eine "Wohnhülle", aber noch keine Umwelt.

Um sich eine Vorstellung davon zu bilden, was "Wohnhülle" bezeichnet, braucht man sich nur einen Zustand zu denken, in dem man blind, taub und ohne Geruch nur auf den Tast-, Temperatur- und Vibrationssinn der Haut angewiesen ist. Alle Interaktionen mit der Umgebung müßten sich dann auf der Oberfläche der Haut (der Wohnhülle) abspielen. Wir verstehen dann, daß unsere Haut auch für den gesunden

Menschen ständig eine Wohnhülle aufbaut, die aber als Zentrum seiner Umwelt bzw. individuellen Wirklichkeit nur unter besonderen Umständen (z. B. bei einer Hauterkrankung) als Wohnhülle erlebt wird.

Für den Embryo kommt als entscheidend hinzu, daß er – wie eine Pflanze – seine Wohnhülle nicht als Reservoir für vital benötigte Reserven erlebt, da der Nachschub ja durch die Nabelschnur gedeckt wird.

Nach der Geburt wird das Kind zum ersten Mal mit Problemen des Nachschubs für die Reserven in seinem Körperinneren konfrontiert. Jetzt entsteht für den Säugling etwas Neues, etwas, das es vorher nie gab: Eine Welt, wenn auch zunächst nur als erste primitive Vorstufe für den Aufbau einer Umwelt, jener festen, für den außenstehenden Beobachter unsichtbaren Schale, die sich als neues Kompartiment um das Kompartiment „Körper" bilden wird. Von diesem Augenblick an besteht das vorher semantisch geschlossene System eines bloßen Körpers aus zwei Subsystemen, die als interdependente und interaktive Organe eine neue Organisationsform repräsentieren.

Aber sowohl die einfachere wie die komplexere Organisationsform bleiben als zwei Zustandsmöglichkeiten während des ganzen Lebens erhalten: Im Schlaf, in der Nausea, im Schock oder im Koma zieht sich der Organismus auf die Organisationsform des semantisch geschlossenen Systems, den **histiotropen** Zustand im Sinne von W. R. Hess (1949, 1956, 1957) oder die Rückzug-Konservierungs-Reaktion (G. Engel und Schmale, 1972) zurück.

Jetzt ist alle Libido von den Objekten abgezogen. Sie „gehen uns nichts mehr an". (Sie fallen – wie Piaget es formuliert – in das „affektive Nichts".) Wir sind wieder „nur Körper". Im **ergotropen** Zustand sind dagegen unsere Aktivitäten nach außen gerichtet, um Umgebung unserer Umwelt bzw. unserer individuellen Wirklichkeit einzuverleiben. Jetzt herrscht die Organisationsform des „offenen Systems".

> **Histiotrope** Reaktionen des Rückzugs und **ergotrope** Reaktionen der aktiven Bewältigung eines Außens sind daher zwei Grundformen psychosomatischer Abläufe.

Auf dem Hintergrund des Modells, das sich damit abzuzeichnen beginnt, lassen sich die Begriffe „Körper" und „Seele" oder Soma und Psyche mit konkreten Vorstellungen verbinden, die sich allerdings nicht mehr mit der dualistischen Annahme eines materiellen und eines psychischen „Seins" decken. Statt dessen können wir verstehen, warum somatische Medizin, die den Körper als Hierarchie von Subsystemen eines, von der Haut begrenzten, semantisch geschlossenen Systems beschreibt, durch eine psychosoziale Medizin ergänzt werden muß, welche die individuellen Wirklichkeiten der Menschen erforscht, die den Körper als zweites Kompartiment umgeben. Wir können aber auch verstehen, daß eine bloße Addition einer somatischen und einer psychosozialen Medizin keine Psychosomatische Medizin ergibt, weil die Interaktionen und Interdependenzen zwischen Soma

und Psyche als Organe einer Einheit nicht erfaßt werden, jener Einheit von Körper und Seele, die das Prinzip bildet, welches jeden Arzt bei seinem Umgang mit kranken Menschen leiten muß.

1.7 Die praktische Anwendung des Situationskreismodells

Kehren wir abschließend wieder zu unserer Patientin (S. 6) zurück. Wir hatten festgestellt, daß wir ihre Problematik weder mit dem somatisch-naturwissenschaftlichen noch mit dem psychologisch-psychoanalytischen Paradigma ganz erfassen können. Im ersten Fall haben wir einen Körper ohne Seele, im zweiten eine Seele ohne Körper vor uns. Versuchen wir beide Theoriesysteme additiv zu verbinden, dann zwingt uns der Dualismus zu entscheiden, ob ihre Beschwerden psychischer oder aber organischer „Natur" seien. Auch bezüglich der Adipositas, der Hypertonie und der Hyperlipidämie stellt sich auf der Basis des Dualismus die Frage: Sind diese Befunde „organisch" oder „psychisch" bedingt? Eine Entscheidung für die eine oder aber die andere Seite beinhaltet bereits Handlungsanweisungen für die einzuschlagende Therapie. Wir sind im dualistischen Vorurteil gefangen (s. Anmerkung 6, S. 38)!

Da unsere Theorien und Vorurteile unsere Handlungen leiten – dies gilt sowohl für das Leben schlechthin als auch für die Heilkunde – wird die Interaktion zwischen Arzt und Patient zwangsläufig von der Theorie des Arztes bestimmt sein. Ist der Arzt Anhänger der Maschinentheorie, dann werden ihn nur objektivierbare Befunde interessieren. Das Erleben der Patientin, ihre psychosoziale Problematik wird zu einer „quantité négligeable". Ist er aber Anhänger der psychoanalytischen Theorie, dann wird er zwangsläufig sein Interesse so gut wie ausschließlich auf die psychodynamischen Vorgänge richten und bestenfalls den Körperbefund von einem „Körperarzt" abklären lassen. Wenn wir uns nochmals die Sprechstundensituation mit unserer Patientin vergegenwärtigen, dann wird klar, daß beide Alternativen mit hoher Wahrscheinlichkeit nicht zum Erfolg führen werden.

Wenn der Arzt versucht, beide Theoriesysteme additiv oder simultan zu berücksichtigen, wird er der Problematik der Patientin zwar eher gerecht, bleibt aber trotzdem in den Aporien des Leib-Seele-Dualismus gefangen und sieht sich, wie oben angedeutet, bei jedem Symptom gezwungen zu entscheiden, ob es nun „psychisch" oder „organisch" bedingt ist. Bei den nächtlichen Atemnotanfällen der Patientin wird ihm dies alles andere als leicht fallen. Arbeitet der Arzt jedoch mit dem von uns dargestellten Situationskreiskonzept, dann vermag er weitgehend den Fallstricken des Dualismus zu entgehen und der Patientin als Person zu begegnen.

Die Interaktion zwischen Arzt und Patient läßt sich in Anlehnung an das Situationskreiskonzept schematisch folgendermaßen darstellen:

> Problemsituation – Bedeutungserteilung – Bedeutungsverwertung.

Durch die Bedeutungsverwertung entsteht eine neue Problemsituation, die wiederum den gleichen Kreisprozeß durchläuft usw. So entstehen Kreisprozesse, die sich einerseits spiralenförmig vom Allgemeinen zum Detail und zurück, sowie „abwärts" und „aufwärts" bewegen. Entsprechend der früher beschriebenen Hierarchie der Systemebenen, finden wir jetzt eine Hierarchie der theoretischen Konzepte und therapeutischen Maßnahmen. Der Fokus der diagnostischen und therapeutischen Bemühungen wird immer wieder wechseln, bleibt aber immer auf die Gesamtsituation der Patientin bezogen. So läßt sich Reduktionismus vermeiden, ohne die Detailproblematik zu vernachlässigen.

In der Interaktion mit der Patientin sieht das aus der Sicht des Arztes sehr vereinfacht so aus:

Problemsituation I: Das Sprechzimmer betritt erstmals eine etwas kurzatmige, adipöse Patientin mit zyanotischen Lippen und berichtet klagend-anklagend von ihren nächtlichen Atemnotanfällen und ihrer Trennungsproblematik. Der Arzt registriert bei sich Gefühle der Ablehnung und Impulse, die Patientin zurückzuweisen.

Bedeutungserteilung I: Der Arzt vermutet, daß die nächtlichen Atemnotanfälle etwas mit der Trennungsproblematik der Patientin zu tun haben, vermag aber noch nicht zu entscheiden, inwieweit nicht auch Asthma-cardiale-Zustände mit Sauerstoffmangel eine Rolle spielen. Die Gefühle der Ablehnung deutet er (im Sinne Balints) als Gegenübertragungsreaktionen auf die psychosoziale Situation der Patientin, die sich von allen abgelehnt und zurückgewiesen fühlt.

Bedeutungsverwertung I: Es muß zunächst geklärt werden, welchen Einfluß die Trennungsproblematik auf den Gesamtzustand, insbesondere auch auf das Kreislaufsystem hat („Abwärtsbewegung"). Somit wird Bedeutungsverwertung I zur **Problemsituation II.**

Bedeutungserteilung II: Die Untersuchung des Kreislaufsystems ergibt Befunde (mäßige Linkshypertrophie des Herzens, mäßige Hypertonie, Herzinsuffizienz durch Adipositas), die eine nicht allzu hochgradige chronische Belastung des Kreislaufsystems wahrscheinlich, eine akute Gefährdung aber eher unwahrscheinlich machen.

Bedeutungsverwertung II: Maßnahmen zur Behandlung der Herzinsuffizienz und Fokussierung der Aufmerksamkeit auf die Trennungsproblematik. Damit wird Bedeutungsverwertung II zur **Problemsituation III.** Diese besteht darin zu klären, welche Bedeutung im Leben dieser Patientin Zurückweisungen und Trennungen spielen und früher schon gespielt haben.

Bedeutungserteilung III: Es wird erkennbar, daß es sich hier um einen Menschen handelt, der schon in der Primärfamilie abgelehnt wurde, von dort in eine Ehe flüchtete, in der sich die Patientin wiederum abgelehnt fühlte, reaktiv depressiv wurde und ersatz-

weise vermehrt zu essen begann, wodurch sich ein Circulus vitiosus von Gewichtszunahme, verstärkter Ablehnung und vermehrter Nahrungsaufnahme entwickelte. In diesem Circulus vitiosus können wir sehr gut die Bedeutungskoppelungen zwischen den verschiedenen Integrationsebenen und die durch sie gebahnten **„Abwärts"-** und **„Aufwärtsbewegungen"** von der psychosozialen Systemebene zur körperlichen und von hier wieder zurück beobachten. Weil sich die Patientin abgelehnt fühlt, beginnt sie vermehrt zu essen und wird adipös. Durch diese körperliche Veränderung stößt sie nun vermehrt auf Ablehnung, wodurch sie verstärkt depressiv wird und noch mehr Nahrung zu sich nimmt usw.

Bei der Patientin können wir aber noch eine weitere „Abwärts"- und „Aufwärtsbewegung" feststellen. Durch die Gewichtszunahme kam es allmählich zu einer relativen Herzinsuffizienz mit Linkshypertrophie des Herzens und leichter Hypertonie, sowie zu eventuellen Asthma-cardiale-Zuständen. Jetzt ist die Patientin „herzkrank". Damit ist für sie eine neue Erlebnisstufe erreicht, denn wenn sie herzkrank ist, dann muß im naturwissenschaftlich-reduktionistischen Krankheitsverständnis vor allem das Organ Herz beziehungsweise das Kreislaufsystem behandelt werden.

Wir erkennen an diesem Beispiel die Aufgaben des mit dem Situationskreiskonzept arbeitenden Arztes. Er muß zunächst mit der Patientin und unter Zuhilfenahme der notwendigen technischen Hilfsmittel alle **„Abwärtsbewegungen"** bis ins Detail nachvollziehen, um die erhobenen Befunde vor dem Hintergrund der Gesamtsituation der Patientin zu interpretieren, wohlwissend, daß letzte und endgültige Sicherheiten des Urteils nie erreichbar sind.

Dann aber – und dies sind die entscheidenden (psycho)therapeutischen Schritte – muß der Arzt in einer **„Aufwärtsbewegung"** der Patientin den Zusammenhang der erhobenen Befunde mit der Gesamtsituation erklären. Jetzt kann er erfolgversprechende therapeutische Empfehlungen geben. Die notwendige Gewichtsreduktion und Unterstützung des Kreislaufsystems werden aber nur von Erfolg begleitet sein, wenn auch die psychosoziale Situation der Patientin berücksichtigt und nach Möglichkeit therapeutisch modifiziert wird. Wie schwer das einerseits oft ist und wie erfolgversprechend es andererseits aber auch sein kann, weiß jeder erfahrene Arzt.

Da die Begegnung zwischen Arzt und Patient sozial festgelegten Programmen folgt, sind die Prozesse der Bedeutungserteilung und Bedeutungsverwertung nur in gewissen Grenzen frei. Darüber hinaus sind beide, wie alle Lebewesen, an vorgegebene angeborene oder erworbene Wahrnehmungs- und Verhaltensstrukturen gebunden, die beim Menschen ihre individuelle Geschichte haben. Die spezifische Problemsituation des Arztes besteht darin, bei den Patienten charakteristische Merkmale dieser Wahrnehmungs- und Verhaltensstrukturen, die sich in organischen Strukturveränderungen, in Funktionsstörungen, aber auch in Verhaltens- und Beziehungsstörungen manifestieren können, zu erkennen, um sie dann nach Möglichkeit so zu modifizieren, daß sie lebens- und situationsgerechter werden. Den ersten Teil dieses Prozesses, nämlich das Erkennen, bezeichnen wir als diagnostischen, den zweiten Teil, das Modifizieren, als therapeutischen Teil dieses Prozesses. Da aber Bedeutungserteilung und Bedeutungsverwertung im Situationskreisprozeß stets untrennbar miteinander verbunden sind, lassen sich auch im ärztlichen Bereich Diagnose und Therapie nur künstlich voneinander trennen. Um diese Einheit zu betonen, nennen wir diesen Prozeß den **„diagnostisch-therapeutischen Zirkel"**.

Hier kann man einwenden, daß diese Einheit von Diagnostik und Therapeutik in der Realität der Krankenversorgung so gut wie immer, manchmal sogar institutionell, getrennt ist. Man müsse – so wird immer behauptet – erst die Diagnose gestellt haben, ehe man mit der Therapie beginnen könne. Dieser Einwand ist richtig und falsch zugleich und muß deshalb weiter differenziert werden. Er ist insofern richtig, als der Bedeutungsverwertung, also dem „Handeln", stets die Bedeutungserteilung, also das „Erkennen", zeitlich vorausgehen muß. So gesehen haben tatsächlich die „Götter die Diagnose vor die Therapie gesetzt" (Volhard, 1982). Er ist aber falsch, wenn man aus dieser zeitlichen Sukzession einzelner Schritte folgert, man könne unabhängig voneinander erst das „Erkennen" (die Diagnose) zu Ende bringen, ehe man mit dem „Handeln" (der Therapie) beginnen dürfe. Das ist schon deshalb nicht möglich, weil es kein „Erkennen" ohne „Handeln" gibt – selbst wenn sich das Handeln nur in Form von „Zuwendung" des Arztes zum Patienten äußert.

Hier müssen wir, um nicht mißverstanden zu werden, einfügen, daß der Terminus „Diagnose" zwei keineswegs deckungsgleiche Bedeutungen hat. Zunächst einmal bedeutet „Diagnose" die „Summe der Erkenntnis" über diesen einen konkreten Patienten. Diese erweiterte Diagnose ist nie abgeschlossen und wird durch jede Arzt-Patient-Interaktion vertieft und vervollständigt.

Gemeinhin verstehen wir jedoch unter „Diagnose" etwas viel engeres, nämlich die „Zuordnung" eines konkreten Krankheitsgeschehens zum Klassifikationssystem unserer Nosologie. Dementsprechend könnten wir unsere Patientin den „Diagnosen" „reaktive Depression", „Adipositas" und „Herzinsuffizienz" zuordnen. Diese „Diagnosen" im engeren Sinne sind Integrations- und Ordnungsbegriffe und dürfen nicht mit der Diagnose im Sinne der „Summe der Erkenntnis" über diese Patientin verwechselt werden.

Durch die Arbeit mit dem Situationskreiskonzept ergeben sich einerseits für die Forschung neue Fragestellungen, andererseits ändert sich auch der Interaktionsprozeß zwischen Arzt und Patient. Durch Überwindung des Leib-Seele-Dualismus und durch Aufheben der prinzipiellen Trennung zwischen Diagnostik und Therapeutik wird es möglich, den Patienten als ganze Person zu untersuchen und zu behandeln. Wird der Patient bei Anwendung des Maschinenmodells und der objektivierenden naturwissenschaftli-

chen Methode zwangsläufig zum naturwissenschaftlichen Objekt des Arztes, mit allen damit verbundenen nachteiligen Folgen, so wird es mit Hilfe des Situationskreiskonzeptes möglich, den Patienten ganzheitlich zu betrachten, was natürlich nicht ausschließt, sich je nach Notwendigkeit mit Detailproblemen zu befassen, jetzt jedoch immer bezogen auf die Gesamtperson und die Gesamtsituation.

1.8 Schlußbetrachtung: Medizin als kulturgebundenes System und die soziale Dimension von Krankheit

Zwei Gedanken, die bei unseren Überlegungen eine Rolle gespielt haben, müssen wir vor Abschluß unserer Darstellung noch einmal aufgreifen und in einem größeren Zusammenhang betrachten:
– Die Tatsache, daß für die Interaktion zwischen Patient und Arzt sozial festgelegte Grenzen existieren,
– daß Theorien und Modelle handlungsleitende Funktionen – allgemein im menschlichen Leben und speziell in den Wissenschaften – haben, und daß diese Beziehung zwischen Theorie und Praxis auch für die Medizin Gültigkeit hat.

Die **Ethnologie** (Lévi-Strauss, 1967; Pflanz, 1962) kann erklären, warum diese beiden Feststellungen zusammengehören: Sie klärt uns darüber auf, daß nicht nur Sprache, Religion, Verwandtschaftsbeziehungen und andere Formen, die menschliches Zusammenleben möglich machen und ordnen, Erzeugnisse der Kultur sind, in der die Menschen leben, sondern daß auch Medizin das Erzeugnis einer bestimmten Kultur ist.

Der Vergleich unserer westlichen, technologisch orientierten Medizin mit der Medizin einer anderen Kultur zeigt, daß die Theorien und Modelle, welche das diagnostische und therapeutische Verhalten der Ärzte und ihre Auffassungen von Gesundheit und Krankheit bestimmen, kulturelle Erzeugnisse, d.h. von der Kultur vorgeschrieben sind. Jedes Urteil über den Wert oder die Effektivität einer Medizin muß daher in die Irre führen, wenn es unreflektiert den Maßstab unserer westlichen Medizin zugrunde legt.

Diese Feststellung ist für uns nicht leicht zu akzeptieren. Wir machen uns selten klar, wie sehr wir durch unsere technische Zivilisation geprägt sind. Ihr Glaube an den Fortschritt durch Wissenschaft und Technik, die ihre Erzeugnisse sind, hat unsere Überzeugung von der Rückständigkeit jeder anderen Medizin geprägt. Nach unserer Vorstellung kann die Medizin jeder anderen Kultur im besten Fall Museumswert besitzen.

Diese westliche Arroganz hat durch die Untersuchungen Thomas Kuhns über die Struktur wissenschaftlicher Revolutionen (1973) einen ersten Stoß bekommen. Sie zeigen uns, daß die Geschichte der Wissenschaften keinem kontinuierlichen Weg folgt, sondern daß sie von einem Paradigma zum anderen immer wieder Neuanfänge mit dem Zwang zu Neuinterpretationen des vorliegenden Wissens erkennen läßt. Die Betrachtung der Medizin als kulturgebundenes, von einer Kultur erzeugtes System bedeutet den Durchbruch zu einer Meta-Position, von der aus wir die westliche, technologisch orientierte Medizin mit ihren Theorien, Erfolgen und Problemen auf einer Ebene mit der Medizin anderer Kulturen sehen, und nun deren Theorien, Erfolge und Probleme mit denen der westlichen Medizin unter neuen Gesichtspunkten vergleichen müssen:

Jetzt geht es nicht mehr um die Frage, ob vielleicht diese oder jene Therapiemethode einer fremden Kultur für die Übernahme in das Arsenal unserer westlichen Methoden geeignet ist. Das kann als Nebenprodukt anfallen, und muß von Fall zu Fall (wie z.B. bei der Akupunktur) geprüft werden. Der Wert einer Betrachtung verschiedener Medizin-Formen im Rahmen eines Kulturvergleichs liegt auf einer anderen Ebene: Er eröffnet uns nicht nur die Möglichkeit, die Kulturabhängigkeit bestimmter Theorien und Vorstellungen zu sehen, sondern auch die Kulturabhängigkeit der Begriffe und Vorstellungen, die Menschen von Gesundheit und Krankheit entwickeln. Er kann uns zeigen, daß die Aufgaben, die verschiedene Kulturen an ihre Medizin stellen, von Fall zu Fall sehr verschieden sein können. Das macht deutlich, warum das Urteil, ob eine Medizin fortgeschritten oder rückständig ist, von den Zielen her beantwortet werden muß, die eine Kultur ihrer Medizin vorgeschrieben hat. Jetzt muß sich auch unsere westliche Medizin die Frage gefallen lassen, wie diese Ziele in der Industriekultur aussehen, was sie unter Gesundheit und Krankheit versteht, und wie die Aufgaben aussehen, welche der Medizin unter dem Aspekt dieser Ziele gestellt werden.

Eine Erhellung der Vorstellungen von Gesundheit und Krankheit, die hinter den Theorien und Methoden der Medizin verschiedener Völker stehen, kann wichtige Fragen der Anthropologie beantworten. Eine dieser Fragen hat unmittelbare Bedeutung für die Praxis jeder Medizin, auch – und nicht zuletzt – für unsere westliche, technologisch orientierte Heilkunde:

> Wenn jede Kultur ihre Medizin erzeugt, erzeugt nicht vielleicht auch jede Kultur ihre Krankheiten?

Ethnomedizinische Untersuchungen über sogenannte „kulturgebundene Syndrome" lassen an der Richtigkeit dieser Vermutung keinen Zweifel: Untersuchungen über „kulturgebundene Syndrome" im Iran (B.J. Good, 1977), in Korea (D. Sich, 1979) und in Taiwan (A. Kleinman, 1980) machen aber noch ein weiteres Faktum deutlich: „Kulturgebundene Krankheiten" lassen sich nur durch die Medizin ihrer Kultur erfolgreich behandeln.

Unter **„kulturgebundenen Syndromen"** wird eine Reihe laienhafter Bezeichnungen für Beschwerdebilder zusammengefaßt, die sich nicht in das nosologische System der westlichen Medizin einordnen lassen (Good, 1977; Kleinman, 1980; Sich, 1979). Nach Albers (1988) ist die Diskussion über diesen Begriff noch nicht abgeschlossen. Er meint aber, daß die Definition nach dem heutigen Stand der Meinungsbildung folgendermaßen lauten könne:

Es handelt sich „um Kombinationen körperlicher, seelischer und anderer (astrologischer, geophysikalischer) Zeichen, die von medizinischen „Laien" kulturspezifisch wahrgenommen werden, und die aufgrund von Sozialisationserfahrungen des einzelnen mit kulturtypischen, aber gering standardisierten Begriffen, Erklärungsmodellen, Gefühlen, Werten, Erwartungen, Meinungen, Interpretationen und Normen verknüpft sind, weshalb nur therapeutische Verfahren der gleichen Kultur Erfolg in der Behandlung dieser Syndrome haben."

Damit stehen wir vor einem Problem, das von Jahr zu Jahr größeres Gewicht erlangt: Was geschieht mit den Kranken, die an diesen Syndromen leiden, in Ländern, deren kulturspezifische Heilkunde durch die westliche, technologisch orientierte Medizin abgelöst wird? Die bisher vorliegenden Erfahrungen enthüllen ein trübes Bild: Die Beschwerdebilder, mit denen diese Patienten zu den in den USA oder in Europa ausgebildeten Ärzten kommen, passen nicht in deren nosologischen Katalog. Sie lassen sich nur gewaltsam den in diesem Katalog enthaltenen Diagnosen zuordnen, mit dem Erfolg, daß Somatisierungen und Chronifizierungen der Beschwerden, Fehlbehandlungen und iatrogene Krankheiten durch unnötige und schädliche diagnostische und therapeutische Eingriffe die Probleme der medizinischen Versorgung in den Ländern der Dritten Welt zusätzlich erschweren.

Diese Erfahrungen werfen aber auch ein spannendes Problem auf, das uns selbst betrifft: Die Konsequenzen, welche die Einführung der westlichen, technologischen Medizin in den Ländern der Dritten Welt hat, erinnern in verblüffender Weise an die Verhältnisse, die uns in unserer Industriekultur nur zu geläufig sind. Hier begegnen wir ja ebenfalls einer sehr großen Gruppe von Patienten, deren Beschwerdebilder auch nicht in das nosologische System der westlichen, technologischen Medizin passen. Dies sind die Patienten mit sogenannten **„funktionellen Syndromen"**, d.h. Kranke, die ohne organische Ursachen krank sind. Epidemiologische Untersuchungen in verschiedenen Industrienationen haben ergeben, daß 30–60% aller Kranken, die einen Arzt oder ein Krankenhaus aufsuchen, zu dieser Gruppe gehören. Weiner (1988) stellt fest, daß in den USA die jährlichen Kosten für die medizinischen Maßnahmen an dieser Gruppe von Kranken die entsprechenden Kosten der Durchschnittsbevölkerung um den Faktor 9 übertreffen. Er schreibt, diese Patienten seien eine schwere Belastung für sich selbst, für ihre Familien und die Gesellschaft. Sie seien ständig in Gefahr Opfer falscher Diagnosen, nicht indizierter chirurgischer Eingriffe und ständig wiederholter nutzloser diagnostischer Maßnahmen und auf diesem Wege iatrogener Krankheiten zu werden. Sie würden ihre Ärzte verwirren und frustrieren.

Untersuchungen aus der Bundesrepublik haben ergeben, daß es im Durchschnitt 7–11 Jahre dauert, ehe bei diesen Patienten die richtige Diagnose gestellt wird (Reimer et al., 1979). Wenn solche Patienten dann nach einer Odyssee durch die verschiedensten Institutionen der Organmedizin in eine psychosomatisch orientierte Behandlung gelangen, ist bei vielen von ihnen im Verlauf ihrer langjährigen Patientenkarriere eine Abhängigkeit von medizinischer Betreuung entstanden, die an die Abhängigkeit Süchtiger von ihrer Droge erinnert. Damit meinen wir nicht die „Artefakt-Patienten" (s. Kap. 55), deren Problematik spezieller Natur ist (Plassmann, 1987).

Es drängt sich die Hypothese auf, daß wir es bei den Patienten mit funktionellen Krankheitsbildern mit den „kulturgebundenen Syndromen" unserer Industriekultur zu tun haben, ja, daß „kulturgebundene Syndrome" ganz allgemein die Reaktion bestimmter Menschen auf das sein könnten, was Freud als das „Unbehagen in der Kultur" bezeichnet hat.

Lennart Levi (1971) hat von einem „schlecht passenden Schuh" gesprochen, um ein derartiges Verhältnis zwischen einem Individuum und seiner psychosozialen Umwelt zu illustrieren.

Gegen diese Hypothese kann man folgenden Einwand erheben: Die Ethnomedizin nimmt an, daß jede Kultur für die von ihr erzeugten Krankheiten die passende Medizin hervorbringen würde. Die Medizin der Industriekultur erweist sich aber als unfähig, die Beschwerdebilder der Patienten, die an funktionellen Syndromen leiden, erfolgreich zu behandeln. Damit scheint doch die Annahme widerlegt zu sein, daß es sich bei diesen Beschwerdebildern um die „kulturgebundenen Syndrome" unserer Industriekultur handeln könnte. Aber dieses Argument vergißt, daß die moderne technologische Medizin in Europa und in den USA vor weniger als hundert Jahren begonnen hat, die Medizin einer noch weitgehend landwirtschaftlich orientierten Kultur zu verdrängen. Damals geschah in Europa und den USA, was heute in den Ländern der Dritten Welt geschieht. Nur das Tempo des kulturellen Wandels ist verschieden.

Weder in Europa, noch in den USA, noch in den Ländern der Dritten Welt wurden und werden die traditionellen Heilmethoden durch die technologische Medizin völlig ausgelöscht. Sie werden „nur", aber offenbar mit dem gleichen Erfolg für die Gesundheit der betroffenen Bevölkerung, in den Hintergrund oder den Untergrund gedrängt. Der große (und wachsende) Zulauf, den Naturheilverfahren, Homöopathie und zahllose Formen einer Volksmedizin auch unter „aufgeklärten" Patienten in den Industrieländern haben, beweist das Bestehen anderer Heilmethoden neben der modernen biotechnischen Medizin. Es bestätigt die Feststellung Kleinmans (1980) über die komplexe Struktur jedes Gesundheitssystems (health care system), in dem sich ein professioneller Sektor einer offiziell anerkannten, privilegierten Medizin mit einem volksmedizinischen Sektor überschneide, der wieder aus verschiedenen Sektoren einer individuellen, einer familiären, einer sozialgruppenspezifischen und einer gemeindegebundenen Medizin bestehe.

Bezüglich der Einstellung zu dem Problem der Beziehung zwischen Gesellschaft und Krankheit stehen sich gewöhnlich zwei unversöhnliche Gruppen gegenüber: Die eine sieht die Wurzel der Krankheit allein in der Person des Kranken. Sie macht eine angeborene Konstitution, eine erworbene Disposition oder ein zufälliges Schicksal dafür verantwortlich, daß ein Mensch vor den Anforderungen seiner Gesellschaft versagt und krank wird. Krankheit ist für diese Gruppe letzten Endes Ausdruck, wenn auch nicht der Schuld, so doch des biologischen oder psychischen Ungenügens des Kranken. Es ist noch nicht lange her, da sprachen Ärzte von „rassischer Minderwertigkeit" der Kranken und träumten den gefährlichen Traum der Herstellbarkeit einer Gesellschaft ohne Krankheit durch Selektion und Züchtung.

Die andere Gruppe macht die Gesellschaft für die Krankheiten ihrer Mitglieder verantwortlich und wirft den Ärzten vor, mit ihren Behandlungsverfahren das Ziel zu verfolgen, die Patienten an eine kranke Gesellschaft anzupassen. Sie würden dadurch die notwendigen Veränderungen des gesellschaftlichen Systems verhindern.

Beide Positionen begehen, wie Schäfer (1975) es formuliert, den Fehler der „Häresie": Sie geben Teilwahrheiten für die ganze Wahrheit aus. Die Medizin hat in jeder Kultur sowohl systemerhaltende wie systemkritische Aufgaben. Aber diese Aufgaben lassen sich nur richtig sehen, wenn die Medizin ihre partikularistische Betrachtungsweise überwindet und sich zu einer systemischen Sicht durchringt. Dann zeigt sich, daß die beiden Aufgaben einander nicht widersprechen, sondern sich gegenseitig ergänzen. Damit werden die Probleme zwar komplizierter, aber wirklichkeitsnäher.

Lebende Systeme sind autopoietische (Maturana, 1982), dynamisch wachsende und im Verlauf ihres Wachstums sich verändernde Gebilde. Sie haben eine historische Dimension. Eine systemische Betrachtungsweise sieht den Menschen in einem Beziehungsgeflecht zu seiner sozialen Umwelt, das sich im Laufe seiner Biographie bildet und verändert. Dabei spielen die kreativen Potenzen des einzelnen und die fördernden und hemmenden Einflüsse der Umgebung eine nur in ihrer gegenseitigen Bezogenheit verständliche Rolle.

Das Modell des Situationskreises erlaubt uns das Bild eines autopoietischen Systems der Beziehungen zwischen einem Menschen und seiner Umgebung zu entwerfen, in dem sich dieses Wachstum verfolgen läßt. Das Bild beschreibt, wie einerseits die kreativen Potenzen des einzelnen immer wieder (autopoietisch) Umgebung in individuelle Wirklichkeit transponieren, und wie andererseits seine kreativen Potenzen durch das Entgegenkommen und die Widerstände der Umgebung entwickelt, geprägt und umgeprägt werden. In diesem Prozeß einer Assimilation der Umgebung und einer Akkommodation der kreativen Potenzen des einzelnen (Piaget, 1936) spielt die Umgebung nicht nur eine passive Rolle. Sie hat die wichtige Aufgabe, die von der Biologie mit dem Begriff der „Nische" umschrieben wird.

Unter einer Nische versteht die Biologie einen Ausschnitt aus der Umgebung eines Lebewesens, der die Ressourcen bereithält, die sein Überleben ermöglichen, und deren Gefahren seinen Möglichkeiten, mit ihnen umzugehen, angepaßt sind. Da es einerseits von dem Lebewesen abhängt, was unter „Ressourcen" und „Gefahren" zu verstehen ist, es andererseits aber Sache der Umgebung ist, was an konkreten Möglichkeiten vorliegt, definieren sich Lebewesen und Nische immer gegenseitig.

Winnicott (1973) hat den Begriff der „genügend guten" Umgebung geprägt, um deren Rolle für das Zustandekommen der lebensnotwendigen Beziehungen zwischen dem Menschen (und jedem Lebewesen) und seiner Umgebung zu charakterisieren. Die soziale Dimension von Gesundheit und Krankheit wird in dem Augenblick sichtbar, in dem wir das Entsprechungsverhältnis (die „kontrapunktische Zuordnung" nach J. v. Uexküll, 1940) der kreativen Potenzen des einzelnen und die „Nischenpotenzen" der sozialen Umgebung ins Auge fassen, d.h., wenn wir beides unter dem Gesichtspunkt von Elementen sehen, denen mit ihrer Integration in ein System Eigenschaften zuwachsen (oder im Falle der Krankheit nicht zuwachsen), die sie zu Elementen des Systems werden lassen.

> „Kultur" kann unter diesem Gesichtspunkt als ein System von Regeln verstanden werden, die in einer Gesellschaft die Nischenfunktion der sozialen Umgebung und die kreativen Potenzen ihrer Mitglieder aufeinander abstimmen und das subtile Gleichgewicht zwischen beiden aufrechterhalten. Für eine solche Betrachtungsweise wird es verständlich, daß jede Kultur ihre Medizin entwickeln muß, deren Aufgabe es ist, über dieses Gleichgewicht zu wachen.

Die kulturgebundenen Syndrome können als Symptome dafür aufgefaßt werden, daß dieses Gleichgewicht gestört ist.

Byron J. Good (1979) entwickelt den Vorschlag, semantische Felder zu entwerfen, in denen sich die Bedeutungsbeziehungen zwischen den Beschwerden, über die ein Patient klagt, und bestimmten als Bedrückung, Einschränkung, Enttäuschung oder unerfüllbare Forderung erlebten Aspekten seiner sozialen Umgebung analysieren lassen. Er betont, daß der Name einer Krankheit nicht einfach als Bezeichnung einer charakteristischen Gruppe von Symptomen oder physiologischen Parametern aufgefaßt werden dürfe. Der Name einer Krankheit sei vielmehr ein Ausdruck der Funktion, einen Komplex von Symbolen, Gefühlen und Bedrückungen, die tief in die Struktur einer Gemeinschaft und ihrer Kultur integriert sind, zu einem potenten Bild zu verknüpfen. Good zeigt, wie sich von dem Bild, das der Name der Krankheit (für ein kulturgebundenes Syndrom) bei den Mitgliedern einer bestimmten Kultur entwirft, das Bild der sozialen Umgebung und der Störungen ihrer „Nischenfunktion" gewinnen läßt. Er schließt mit der Forderung, „Krankheit als soziohistorisches und kul-

turelles Phänomen aufzufassen, als verschlungenes und vielschichtiges Netzwerk von sozialen, persönlichen und organischen Kontexten – sozusagen von der Gesellschaft zur Zelle – in dem der Arzt an bestimmten Punkten diagnostisch und therapeutisch interveniert."

In Kapitel 31 werden wir ein Modell entwickeln, das diese Forderung erfüllt. Es greift auf eine Untersuchung von Christian und Haas (1949) zurück, die nachweist, daß Gesundheit, als Gefühl vollkommener Autonomie, d.h. des freien Verfügen-Könnens über die eigenen Leistungen, ein ständiges, unbemerktes Entgegenkommen der Umgebung voraussetzt. Die eigenen Leistungen kommen nur zustande, wenn sie durch die passenden Gegenleistungen der (anorganischen, biotischen und/oder sozialen) Umgebung ergänzt werden.

V. v. Weizsäcker (1955) hat betont, daß Gesundheit kein Kapital ist, das man aufzehren kann. Sie muß in einem Prozeß der „Salutogenese" (Antonowsky, 1988) ständig erzeugt werden. Wird sie nicht mehr erzeugt, ist der Mensch bereits krank. Dieser Prozeß bedarf in jedem Augenblick einer Umgebung, die unsere Leistungen durch die passenden Gegenleistungen und unsere Rollen durch passende Gegenrollen ergänzt (Th. v. Uexküll, 1981).

Wenn wir uns diesen Zusammenhang klarmachen, wird deutlich, was wir unter systemischer Struktur und unter der Beziehung zwischen Element (oder Subsystem) und System zu verstehen haben. Wir können uns dann auch vorstellen, wie in einem hierarchisch (aus Integrationsebenen) aufgebauten System „Aufwärts"- und „Abwärts"-Effekte zustande kommen. Die Metaphern von dem „Unbehagen in der Kultur" oder dem „schlecht passenden Schuh" für eine ungenügende Nischenfunktion der Umgebung füllen sich mit konkretem Inhalt, wenn man das Angewiesensein jeder Leistung auf die passende Gegenleistung vor Augen hat.

1.9 Zusammenfassung

Ausgehend von dem Ungenügen des Maschinenmodells für den menschlichen Körper haben wir zunächst auf die Problematik hingewiesen, die das psychoanalytische Modell der Seele in der Medizin aufgeworfen hat. Neben dem „Be-handeln" war damit das „Sprechen" mit dem Kranken eine wichtige Aufgabe des Arztes geworden, die sowohl eine diagnostische wie eine therapeutische Funktion zum Inhalt hat. Nach dem neuen Paradigma, das damit in die Medizin eingeführt wurde, hat Sprache die Aufgabe, im Laufe der Ontogenese den Aufbau einer individuellen Wirklichkeit zu leisten. Wenn dieser Aufbau gestört ist, entstehen Defekte dieser Wirklichkeit als Basis für Krankheiten. Das psychotherapeutische Gespräch hat die Aufgabe, die biographische Situation, in welcher der Bildungsschaden entstand, wieder in das Erleben des Patienten zurückzurufen und neu zu bearbeiten.

Aber der Körper und der dort entstandene Schaden bleiben aus dem Modell ausgespart. Es kann lediglich erklären, warum die Schutzfunktion der individuellen Wirklichkeit als „zweite Haut" an einer bestimmten Stelle oder in einer bestimmten Situation versagt. Aber wie und warum dadurch ein körperlicher Schaden, ein Ulcus duodeni, eine Colitis ulcerosa oder ein Hochdruck entstehen, kann von diesem Modell aus nicht einmal gefragt werden (Weiner, 1977).

Ein psychosomatisches Modell hat daher die Aufgabe, ein Modell für den Körper zu entwickeln, das dem Modell für die Seele entspricht, und dem Arzt die Möglichkeit gibt, Vorgänge, die sich auf der physiologischen, der psychologischen und der sozialen Ebene zutragen, in Verbindung zu bringen. Dafür wurde das Konzept der Systemtheorie herangezogen, nach dem die Vorgänge verschiedener Integrationsebenen einer Hierarchie von Systemen und Suprasystemen angehören.

G. Engel (1977, 1982) hat auf die Bedeutung dieses Modells für die Orientierung des Arztes bei den vielen und verschiedenartigen Problemen des Kranken, aber auch für die Ausbildung des Arztes hingewiesen.

Auf jeder der verschiedenen Integrationsebenen gelten andere Zeichensysteme: Auf der physiologischen Ebene „verständigen" sich Zellen, Organe und Organsysteme mit biochemischen und/oder elektrophysiologischen Zeichen. Unter ihnen lassen sich wieder verschiedene spezielle Zeichensysteme, wie das endokrine, das immunologische und das nervale, unterscheiden. Alle diese Zeichensysteme sind „endosemiotisch", das heißt, sie spielen sich innerhalb des Organismus ab.

Auf der psychischen Ebene gibt es wieder spezifische und unter sich differenzierte Zeichensysteme. Zu ihnen gehören Sensationen und Gefühle, die wir aus unserem Körper und/oder über die Sinnesorgane aus der Umgebung empfangen. Mit Hilfe dieser Zeichen baut das Subjekt seine subjektive Umwelt auf, einen Vorgang, den wir als „Seelentätigkeit" auf der animalischen Ebene beschrieben haben. Nach den Untersuchungen Piagets gilt das auch für die frühen Entwicklungsstadien des Kindes, das seine Umwelt aufgrund sensomotorischer Programme oder Schemata aufbaut, in denen Körper und Umwelt noch als ungetrennte Einheit erlebt werden. Etwa vom zweiten Lebensjahr an beginnt dann aber eine **Sozialisation der Umweltbildung,** die sich nicht mehr allein der averbalen Interaktion mit der mütterlichen Umgebung bedient, sondern zunehmend sprachliche Konzepte verwendet, und bei der nach Piaget das Vorstellungsvermögen eine führende Rolle übernimmt. Diesem verdanken wir die Fähigkeit, konstante Objekte zu bilden und zu den Objekten (als Gegenständen) Distanz zu gewinnen. Damit beginnt eine Differenzierung in eine Innenwelt und eine Außenwelt.

Hier wird die enge Interaktion zwischen der psychischen und der sozialen Ebene deutlich, die dazu führt, daß die individuelle Wirklichkeit des einzelnen zunehmend aufgrund sozial entwickelter Programme aufgebaut wird. Aufgabe der Interaktion zwischen verschiedenen menschlichen Individuen, die alle in ihren spezifischen individuellen Wirklichkeiten leben, ist immer wieder der Aufbau gemeinsamer, d.h. sozialer Wirklichkeiten. Dafür kann die Interaktion zwischen Arzt und Patient als exemplarisches Beispiel dienen.

Analysen der Situation, in der Arzt und Patient einander begegnen, und der Funktion, die das Gespräch dabei hat, zeigen einen Ablauf, den man in Analogie zu der Interaktion zwischen dem Kleinkind und der Mutter oder den anderen Familienmitgliedern deuten kann. In dieser Situation lernt das Kleinkind, seine individuelle Wirklichkeit aufzubauen, wobei es von der Mutter und den übrigen Familienmitgliedern unterstützt werden muß. In Analogie dazu läßt sich vorstellen, daß der Kranke, der allein nicht in der Lage ist, die notwendige Neuorientierung und Korrektur einer deformierten oder verfremdeten Wirklichkeit zu finden, von dem Arzt unterstützt werden muß. Dabei spielen die verschiedenen Wünsche und Ziele sowie die verschiedenen Ängste der beiden Partner als Hindernisse und Widerstände eine wechselnde Rolle. Das Gespräch läßt sich so als Handlungssystem mit verschiedenen Akteuren beschreiben, in dem die Beiträge der Partner Teilhandlungen darstellen, die das Gelingen des Gesprächs ermöglichen oder verhindern.

Das bio-psycho-soziale Modell, das auf diese Weise entsteht, respektiert die Eigenständigkeit der Phänomene auf jeder der verschiedenen Systemebenen. Statt reduktionistisch die komplexen Phänomene von den einfachen physiologischen oder biochemischen ableiten zu wollen, zeigt es durch den Rückgriff auf die Zeichenlehre und den dort geläufigen Begriff der „Übersetzung" Verbindungen zwischen den Systemebenen auf, die Bedeutungskoppelungen entsprechen. Damit werden somatopsychische „Aufwärts-Effekte" und psychosomatische „Abwärts-Effekte" sichtbar, wobei zwischen angeborenen und erworbenen Bedeutungskoppelungen unterschieden werden muß. Sie entsprechen den unbedingten und den bedingten Reflexen Pawlows. Sein Modell der Konditionierung hat zum ersten Mal gezeigt, daß und wie Bedeutungskoppelungen entstehen.

In diesem Zusammenhang ist der dynamische Aspekt des so konzipierten Modells zu betonen. Er besagt, daß die individuelle Wirklichkeit jedes Menschen eine historische Dimension hat, was bedeutet, daß ihr Aufbau und Umbau immer wieder bei der „Stunde 0" beginnt. Darunter ist jener Zustand zu verstehen, der vor der Berührung mit Außenweltreizen besteht, und der als Bereitschaft oder Bereitstellung zu Aktivitäten oder zum Rückzug beschrieben werden kann. Dieser Zustand bildet die Grundlage, auf der auch Übersetzungen (Bedeutungskoppelungen) möglich oder verhindert werden. Er hat unter diesem Aspekt eine entscheidend wichtige regulatorische Funktion.

Diese Bereitschaften lassen sich als verschiedene Formen von „Gestimmtsein" beschreiben, worunter Ordnungszustände innerhalb des Organismus zu verstehen sind, die aus dem Aufeinander-Abgestimmtsein verschiedener Teilfunktionen (z. B. des Gastrointestinaltraktes, des Kreislaufs und der Muskulatur) im Sinne von Bereitstellungszuständen zu bestimmten Auseinandersetzungen mit der Umgebung bestehen. Stimmungen haben ihr psychisches Äquivalent, das dem entspricht, was die Verhaltensforscher unter „Appetenz" oder die Angelsachsen unter „state" verstehen. So ist der Rückzug, das Schrumpfen der individuellen Wirklichkeit im Rahmen einer Nausea ebenso eine Stimmung, wie die vigilante Aufmerksamkeit im Rahmen einer Aktivierung mit ihrer entsprechenden Wirklichkeitserfahrung Ausdruck einer Stimmung ist. In beiden Fällen entspricht dem psychischen Gestimmtsein ein Zusammenstimmen somatischer Funktionen. Auf der Basis solcher, die psychosomatischen Einheiten repräsentierender Stimmungen können dann Motive entstehen, die spezifische Objekte in der individuellen Wirklichkeit herausheben, die aber auch Bedeutungskoppelungen zur Herstellung neuer Programme begünstigen oder verhindern.

Es gibt spezifische „Stimmungssignale", zu denen zum Beispiel Phänomene gehören, die Ekel, Schreck oder Angst auslösen und die biologisch die Bedeutung haben, bestimmte Bereitstellungen hervorzurufen oder zu unterhalten.

Zum Abschluß wollen wir im Rahmen unseres theoretischen Konzeptes drei Begriffe neu definieren, die für die Medizin eine zentrale Rolle spielen: Krankheit, Diagnose und Symptom.

Unter **Krankheit** verstehen wir eine ungelöste Problemsituation oder deren Folgen auf einer, mehreren oder allen Ebenen des hierarchischen Systems.

Diagnose ist eine Spielregel für den Umgang des Arztes mit dem Patienten im Dienst der gemeinsamen Sache – Krankheit zu heilen oder zu lindern. Sie schreibt dem Arzt und dem Kranken spezifische Rollen für eine gemeinsame Handlung vor, die auf der physiologischen, psychischen und sozialen Ebene verwirklicht werden muß.

Symptome sind Zeichen, die über den Zustand lebender Systeme informieren (Th. v. Uexküll, 1984). Der Gang, die Haltung und der Gesichtsausdruck eines Patienten informieren den Arzt über den Zustand, in dem sich ein Patient befindet. Symptome unterscheiden sich von anderen Zeichen dadurch, daß der Zeichengeber mit ihnen keine Intention zur Information eines Empfängers verbindet, wie das im Rahmen einer Kommunikation der Fall ist.

Im Rahmen unseres Modells der hierarchisch gegliederten Integrationsebenen gewinnen Symptome darüber hinaus eine prinzipielle Bedeutung: Sie informieren die komplexere Systemebene über den Zustand der Systeme der einfacheren Ebene, das heißt, sie werden dort in Zeichen übersetzt, die „Appell um Unterstützung und Hilfe" bedeuten oder die besagen, daß eine solche Hilfe nicht erforderlich ist. Auf diese Weise sind Symptome die Grundlage für „Aufwärts- und Abwärts-Effekte" in komplexen lebenden Systemen oder mit anderen Worten Auslöser für somato-psychische und psychosomatische Effekte.

Anmerkungen

Anmerkung 1: Das Handlungsmodell entspricht gewissermaßen dem Bild eines Situationskreises (S. 28) von innen. Wir werden noch ausführen, wie der Situationskreis aus dem Funktionskreis entsteht. Das Modell des Funktionskreises (S. 27) wurde von J. v. Uexküll erstmals 1920 entwickelt, um die Beziehungen zwischen einem Tiersubjekt und seinen Objekten darzustellen. Diese Beziehungen werden als ein Ineinandergreifen von „Merken" und „Wirken" beschrieben, wobei Merken den rezeptorischen und Wirken den effektorischen Anteil bezeichnet.

Um diesen Zusammenhang deutlich zu machen, heißt es (1936):

„Bildlich gesprochen greift jedes Tiersubjekt mit zwei Gliedern einer Zange sein Objekt an – einem Merk- und einem Wirkgliede. Mit dem einen Glied erteilt es dem Objekt ein Merkmal und mit dem anderen ein Wirkmal. Dadurch werden bestimmte Eigenschaften des Objekts zu Merkmalträgern und andere zu Wirkmalträgern. Da alle Eigenschaften eines Objekts durch den Bau des Objekts miteinander verbunden sind, müssen die vom Wirkmal getroffenen Eigenschaften durch das Objekt hindurch ihren Einfluß auf die das Merkmal tragenden Eigenschaften ausüben und auch auf dieses verändernd einwirken. Dies drückt man am besten kurz so aus: *das Wirkmal löscht das Merkmal aus*" (J. v. Uexküll und Kriszat, 1936).

Das Auslöschen des Merkmals durch das Wirkmal kann subjektiv, zum Beispiel durch Sättigung, oder objektiv, zum

Beispiel durch Verschlingen des Objekts, erfolgen. In jedem Fall kommt dadurch der Funktionskreis zur Ruhe und kann jetzt durch ein neues Merkmal wieder in Tätigkeit versetzt werden.

Dieses Modell bildet also unser Handlungsmodell in einfachster Form ab. Wir wollen schon hier anmerken, daß damit ein Grundschema für ein dynamisches Systemmodell (vgl. Anm. 4) entworfen ist: Es erlaubt uns ein System nicht als statisches Gebilde, sondern als ein sich vollziehendes Geschehen aufzufassen.

Anmerkung 2: Der außenstehende Beobachter kann diese Fähigkeit bei lebenden Systemen nur indirekt erschließen. Winnicott zeigt, wie die eigene Erfahrung oder besser die Erfahrung des eigenen Selbst eine zentrale Bedeutung für die psychische Entwicklung des Menschen hat (s. Kap. 6.1: Früheste Kindheitsentwicklung und ihre Störungen aus der Sicht Winnicotts).

Von besonderem Interesse ist darüber hinaus die erkenntnistheoretische Bedeutung der Entdeckung des Selbst. Bei Winnicott ist sie deutlich ausgesprochen, wenngleich sie nicht sein eigentliches Interesse berührt. Für ihn ist die Erfahrung des Selbst identisch mit Sein. Damit ist angedeutet, daß wir mit dieser Einsicht am Anfang einer neuen Ontologie stehen.

Dieser Eindruck wird zur Gewißheit, wenn wir die neuen Entwicklungen der Systemtheorie, vor allem die Arbeiten

Maturanas (1969, 1975, 1978, 1980) und Varelas (1974, 1979, 1981) über autopoietische und selbstreferentielle Systeme verfolgen (Roth und Schwegler, 1981).

Anmerkung 3: In Wahrheit sind die Zusammenhänge noch verwickelter: Auf der biologisch-vegetativen Ebene ist das Kind schon von dem ersten Augenblick seiner Existenz an, das heißt schon seit der Befruchtung der Eizelle, ein von der Mutter genetisch unterschiedenes Individuum. Auf der animalischen Ebene, die sich erst nach der Geburt entwickelt, besteht dagegen die soziale Einheit der Mutter-Kind-Dyade vor der Individualität des Kindes.

Das wird in den Vorstellungen deutlich, die Winnicott, aber auch Piaget von den frühesten Stadien der psychischen Entwicklung des Kindes entworfen haben.

Für Winnicott (1973) ist das Kind zunächst die Mutter, das heißt die Mutter wird als Teil des kindlichen Selbst erfahren. Diesen Zustand einer Subjekt-Objekt-Identität nennt Winnicott „Objekt-Beziehung" und unterscheidet davon die „Objekt-Verwendung", die erst in einem späteren Stadium, nach einer kritischen Entwicklungsphase, erreicht wird.

Er macht den Psychoanalytikern den Vorwurf: „Sie haben die Subjekt-Objekt-Identität, die ganz am Anfang der Fähigkeit steht, zu sein, und auf die ich hier aufmerksam mache, ... außer acht gelassen" (S. 95).

Piaget betont ebenfalls die Subjekt-Objekt-Identität, ohne sie allerdings so zu nennen, in den frühesten Entwicklungsphasen des Kindes. So heißt es zum Beispiel bei der Interpretation einer Handlung des acht Monate alten Sohnes:

„Das Kind betrachtet also den Effekt B (der in Wahrheit von dem Vater gewissermaßen in Abstimmung mit dem kindlichen Verhalten hervorgebracht war) als eines der zahlreichen Phänomene, die seine eigene Handlung verlängern und nicht etwa als das Produkt eines von dieser Handlung unabhängigen Prozesses" (1975, S. 2, 238).

Anmerkung 4: Wenn wir in den vorangehenden Abschnitten von „der Systemtheorie" sprechen, so soll damit nicht der Eindruck erweckt werden, es gäbe bereits eine abgeschlossene Lehre mit festen, allgemein anerkannten Begriffsbestimmungen. Systemtheorie ist eine sehr junge Wissenschaft. Sie befindet sich noch in voller Entwicklung. Es gibt daher verschiedene Ansätze und Konzepte, die sich zum Teil sogar zu widersprechen scheinen.

Dasselbe gilt bis zu einem gewissen Punkt für die moderne Zeichenlehre, die Semiotik, von der wir noch sprechen werden. Auch bei ihr können die verschiedenen Ansätze und Konzepte den Eindruck mangelnder Einheitlichkeit erwecken. Für Oehler (1984) ist das aber nur das „systemische Problem einer Grundlagenwissenschaft", deren Aufgabe sich je nach den Ausgangspositionen der Einzelwissenschaften unter verschiedenen Aspekten darstellen würde.

„Der Eindruck (fehlender Einheit) verweist auf ein systemisches Problem der Semiotik als Wissenschaft. Die moderne Semiotik befindet sich trotz ihrer bereits über einhundertjährigen Geschichte immer noch in der formativen Phase. Der Hinweis darauf ist kein Einwand, sondern spricht für die Größe und Bedeutung der Aufgabe der Semiotik als Grundlagenwissenschaft."

Die Parallelität der Probleme, die sich der Systemtheorie und der Zeichenlehre, der Semiotik, stellen, ist Ausdruck der Tatsache, daß in beiden der Paradigmawechsel als unaufhaltsamer Wandel unserer Einstellung zur Wirklichkeit unter verschiedenen, aber sich ergänzenden Aspekten sichtbar wird. In beiden geht es letzten Endes um das Bemühen, etwas sehr einfaches, zugleich aber sehr rätselhaftes, das unsere bisherige naturwissenschaftliche Betrachtungsweise übersehen hat, in den Blick zu bekommen: die Tatsache, daß ein Ganzes „mehr" ist als die Summe seiner Teile. Die Frage, was dieses „mehr" sei, das Begriffe wie Ganzes, Ganzheit oder Einheit nur annäherungsweise beschreiben, soll durch den Systembegriff und den Begriff des Zeichens genauer beantwortet werden. Das macht verständlich, daß die Definition der Begriffe System und Zeichen in den beiden Wissenschaften im Zentrum der Diskussion steht.

Dabei schält sich als Kern sowohl in der Systemtheorie wie in der Semiotik eine zweifache Einsicht heraus:
- daß der Beobachter nicht von den beobachteten Phänomenen geschieden werden kann, und
- daß die bisherigen, an statischen und strukturellen Vorstellungen orientierten Konzepte zur Deutung der Phänomene zugunsten dynamischer Modelle aufgegeben werden müssen. Systeme müssen als Geschehnisse und Zeichen als Prozesse aufgefaßt werden, an denen wir beteiligt sind.

Diesem Wandel stehen die traditionellen, an dem dualistischen Paradigma orientierten Denkformen im Wege. Sie wollen Systeme und Zeichen als statische Gegebenheiten mit bleibenden Strukturen beschreiben, und den Beobachter als an dem Beobachtungsprozeß unbeteiligte Instanz von den beobachteten Phänomenen trennen. Beide sollen – das ist der Kern der dualistischen Voraussetzung – zwei verschiedenen Seinsbereichen angehören.

Eng verknüpft mit der Alternative zwischen einer dynamischen und einer statischen Betrachtungsweise ist die Frage, ob wir in den Naturerscheinungen Kontinuität oder Diskontinuität voraussetzen. Diese Frage hat eine lange Geschichte, die zuerst bei den Vorsokratikern in dem Gegensatz zwischen dem atomistischen Weltbild des Demokrit und dem dynamischen Ansatz des panta rhei des Heraklit faßbar wird. Wir haben darauf hingewiesen, daß das Ringen um diese beiden Paradigmen in der Medizin als Streit um die Priorität der Struktur oder der Funktion zum Ausdruck kommt.

Seine scharfsinnigste und folgenreichste Formulierung fand das dualistische Paradigma, nach dem der Beobachter von den beobachteten Phänomenen getrennt und für deren Veränderungen Kontinuität vorausgesetzt werden muß, in dem Infinitesimalkalkül Newtons und Leibnitz's. Ihm liegt die Annahme zugrunde, daß Veränderungen nur kontinuierlich über beliebig viele Zwischenstufen erfolgen können, und daß der Beobachter an diesen Veränderungen keinen Anteil hat. Damit wird die Möglichkeit diskontinuierlicher, sprunghafter Veränderungen ausgeschlossen (Dell und Goolishian, 1981), und die Beteiligung des Beobachtungsvorgangs geleugnet.

Dieses Paradigma hat uns dreihundert Jahre lang die Möglichkeit gegeben, die Natur mit beispiellosem Erfolg für unsere Ziele zu manipulieren. Heute werden die Konsequenzen einer Einstellung, welche den Beobachter als unbeteiligte Instanz einer Natur als Objekt seiner Beobachtung gegenüberstellt, in den ökologischen Katastrophen sichtbar, die sich immer deutlicher als unabwendbare Gefahren erweisen. Unter diesem Gesichtspunkt gewinnt der Paradigmawandel, der sich in unserer Naturbetrachtung vollzieht, eine geradezu existentielle Bedeutung für das Leben und Überleben der Menschheit.

Wir haben als Beispiel für den Wandel, der sich in den Voraussetzungen der Naturwissenschaft vollzogen hat, das Gleichnis von der verschlossenen Uhr zitiert, das Einstein gebracht hat. Die Unhaltbarkeit des dualistischen Ansatzes zeigte sich der Physik auch in der Unvereinbarkeit des Korpuskel- und Wellenkonzepts, in denen die Lehren des Demokrit und des Heraklit in modernem Gewand wiederauftauchen. Das Dogma der Kontinuität in den Naturvorgängen war schon vorher mit der Entdeckung des Wirkungsquants durch Planck gefallen.

In der Systemtheorie zeigt sich der Wandel unserer Voraussetzungen in der Entwicklung von Modellen, die Systeme als in Entwicklung begriffene Gebilde auffassen. Die einzelnen Phasen dieser Entwicklung lassen sich nur in Momentaufnahmen als Gleichgewichtszustände beschreiben, die schon im nächsten Augenblick in einen anderen Gleichgewichtszustand übergegangen sind. Ferner ist für diese Modelle charakteristisch, daß die Übergänge nicht kontinuierlich, sondern sprunghaft erfolgen. Dafür sind die Vorstellungen, die Prigogine (1978, 1982) über Ordnung durch Fluktuation entwickelt hat, von besonderer Bedeutung.

Mit diesen Konzepten stellen sich grundsätzliche erkenntnistheoretische Probleme. Was sollen wir zum Beispiel unter den Begriffen „Dauer" und „Identität" bei Gebilden verstehen, die sich ständig verändern? Dabei wird deutlich, daß diese Probleme bereits bei der Betrachtung eines jeden lebenden Gebildes auftauchen, so daß man sich nur wundern kann, wie stark die erkenntnistheoretischen Brillen die Wirklichkeit verändert haben, daß wir diese Diskrepanz zwischen unseren Begriffen und den beobachteten Phänomenen nicht bemerkt haben.

Ein Baum, ein Vogel oder ein Mensch sind für uns Wesen, die sowohl Dauer wie Identität aufweisen, und zwar nicht obgleich, sondern weil sie sich ständig verändern. Eine Eichel hat mit dem Baum, ein Ei mit dem Vogel, ein Embryo mit dem Menschen nicht die geringste Ähnlichkeit, und doch haben wir keinen Zweifel, daß es sich in allen Fällen um ein sich identische und andauernde Gebilde handelt. Unsere tägliche Wahrnehmung löst diese Probleme spielend. Nur kennen wir vorläufig die Spielregeln nicht, nach denen ihr das gelingt. Diesen Spielregeln versuchen die neuen Konzepte der Systemtheorie auf die Spur zu kommen.

Von dort aus wird einsichtig, daß die Konzepte der Umwelt und des Funktionskreises die Forderungen nach dynamischen, nicht an Strukturen, sondern an Funktionen orientierten Modellen und das Kriterium des Sich-Entwickelns weitgehend erfüllen. Umwelt ist eine nach den Regeln des Funktionskreises sich entwickelnde Handlung, in der sich Subjekt und Objekt als Elemente dieser Handlung ständig neu definieren. Der Begriff der Handlung ist im Sinne eines Dramas zu verstehen, in dem sich eine Geschichte verwirklicht, die durch das Aufeinander-Angewiesensein der Rollen und Gegenrollen ihrer Akteure, durch ihren Anfang und ihr Ende (wie ein Baum, ein Vogel oder ein Mensch) eine sich verwirklichende Einheit – ein System – bildet (Th. v. Uexküll, 1983).

Einblick in derartige sich verwirklichende Geschichten (Systeme) kann kein unbeteiligter Beobachter gewinnen. Für ihn bleiben sie verschlossene Uhren. Einblick gewinnt man nur, wenn man sich an ihnen beteiligt. Dann aber hört die Uhr auf, noch eine Uhr zu sein und löst sich in viele einzelne Handlungen oder Dramen auf, in denen Teilnehmer mit Teilen der Uhr, mit dem Zifferblatt, mit den Zeigern, mit Zahnrädern, Schrauben usw. in Interaktionen verstrickt sind.

Uhren, Bäume, Vögel und Menschen sind als „Gegenstände" Abstraktionen, in denen unsere Vorstellung das ständig sich Verwandelnde in Symbole einschließt, die ihm Dauer verleihen. Hier ist einer der Berührungspunkte zwischen der Systemtheorie und der Lehre von den Zeichen. Hier liegt auch eine Möglichkeit die Paradoxie des „Dinges an sich" aufzulösen, die Kant den Philosophen hinterlassen hat.

Anmerkung 5: Die Bereitschaft oder das „Organische Entgegenkommen" für Bedeutungskoppelungen entspricht einem Zustand, den Th. v. Uexküll (1952, 1963) als „Stimmung" bezeichnet und in Anlehnung an W. B. Cannon (1975) als „Bereitstellung" gedeutet hat. Der Begriff „Stim-

mung" soll den Gleichklang oder das Aufeinander-Abgestimmtsein körperlicher Funktionen und seelischer Erlebnisbereitschaften zum Ausdruck bringen, der unserem aktiven Verhalten, unseren Handlungen vorausgeht. Danach sind Stimmungen Ordnungsprinzipien beziehungsweise Integrationsschemata, die in engem Zusammenhang mit den Affekten und Emotionen stehen.

Der Zusammenhang mit dem Handlungsmodell wird darin gesehen, daß Stimmungen den Boden vorbereiten, auf dem Motive entstehen können, welche – wie in dem Beispiel des Apfelpflückens gezeigt wurde (S. 9) – die Umwelt für den Verlauf einer Handlung (durch Bedeutungserteilung) vorstrukturieren. Handlungen werden neurophysiologisch mit dem animalischen Nervensystem in Zusammenhang gebracht, während Stimmungen dem vegetativen Nervensystem unterstehen.

In der Darstellung der Unterschiede zwischen Stimmungen und Handlungen sowie des Zusammenhangs zwischen beiden heißt es:

„... die Welt, die wir im Rahmen von Stimmungen erleben, unterscheidet sich von der Welt, die wir im Bann von Motiven erfahren ... Stimmungen geben uns keine Handlungsanweisungen. Sie geben nur Anweisungen für Bereitstellungen. Was draußen entsteht, ist nur Bühne, auf der Handlungen sich abspielen können, ja, auf der alles zur Handlung drängt. Aber die Handlung ist nur vorbereitet, das Stichwort, das sie in Gang setzt und das nur von einem Motiv kommen kann, steht noch aus. Es ist noch „alles" gefährlich, verheißend, feindlich, gleichgültig oder ekelhaft, aber das konzentriert sich noch nicht auf diesen oder jenen Gegenstand. Es gibt nur einen gemeinsamen Ton, auf den unser Körper, unser Ich und unsere Welt abgestimmt sind" (1963, S. 177).

Wir haben das Handlungsmodell im Zusammenhang mit dem Problem der vorwissenschaftlichen und wissenschaftlichen Erfahrung dargestellt und betont, daß es das Verfahren beschreibt, nach dem wir (empirisches) Wissen erwerben. Der enge Zusammenhang mit den Stimmungen macht darauf aufmerksam, wie eng die kognitiven Funktionen mit dem Affektiven und Emotionalen zusammenhängen.

Anmerkung 6: L. L. Weed (1969) hat auf die Tatsache aufmerksam gemacht, daß die meisten Krankenblätter, auf denen in den Kliniken mit viel Mühe und Zeitaufwand die Dokumentation der Krankheitsverläufe vorgenommen wird, praktisch wertlos sind. Als Grund nennt er ungelöste Schwierigkeiten, die medizinisch relevanten Probleme der Kranken zu identifizieren und einander zuzuordnen. Er hat daher eine problemorientierte Patientendokumentation vorgeschlagen, die zweifellos einen wichtigen Fortschritt bringt, aber das Problem des Ordnungsprinzips für eine derartige Dokumentation nicht löst.

Hinter diesen Schwierigkeiten steht das Fehlen einer Theorie, die imstande ist zwei Dinge zu leisten:
- Ereignisse in der Interaktion eines Kranken mit seinem Körper, mit sich und mit seiner Umgebung (auch den Ärzten) als medizinisch relevante Probleme zu identifizieren.
- Diese Probleme aufgrund ihrer Bedeutung für das Krankheitsgeschehen zu ordnen und zu einander in Beziehung zu setzen.

Wir haben schon erwähnt, daß G. Engel (1977, 1982) auf die Wichtigkeit eines systemtheoretischen Ansatzes für diese Aufgabe hingewiesen hat.

Das Beispiel der Interaktion zwischen dem Arzt und der Patientin, die über die nächtlichen Anfälle von Atemnot und Todesangst klagt, zeigt, wie das Modell des Situationskreises die praktische Orientierung für das diagnostische und therapeutische Vorgehen des Arztes übernehmen kann.

2 Sozialmedizinische, medizinsoziologische und soziosomatische Aspekte zur Entstehung und Erhaltung von Gesundheit und Krankheit

Hannes G. Pauli

In Kapitel 1 wurde ein biopsychosoziales Modell der Medizin vorgestellt. Ausgangspunkt dazu war eine ökologische, d.h. die Lebewesen und ihre Umgebung[1] umfassende Biologie, wie sie Jakob von Uexküll zu Beginn dieses Jahrhunderts entwickelt hat. Thure von Uexküll hat diese mit den Instrumenten der Systemtheorie und Semiotik vertieft und in den Humanbereich hinein erweitert. Diese Erweiterung geschah unter Einbezug einer neueren Entwicklung in den Sozial- und Kulturwissenschaften. Auf konzeptioneller und empirischer Grundlage ist damit eine Humanmedizin (Th. von Uexküll und Wesiack, 1988) entstanden, deren Ansätze sich in der bisherigen Medizingeschichte bis etwa zur Mitte dieses Jahrhunderts lediglich auf der philosophischen Ebene manifestiert haben. Diese Entwicklung wird auch als „zweite medizinische Revolution" (Foss und Rothenberg, 1987) – nach der ersten naturwissenschaftlichen – bezeichnet. Was sind die Konsequenzen für ärztliches Handeln, die Organisation der Gesundheitsdienste und für die ärztliche Ausbildung?

Im Vordergrund steht ein gewaltiger Vorsprung an humanwissenschaftlicher Erkenntnis bzw. Theorie gegenüber einem, entsprechend den vorwiegend technischen Entwicklungspotenzen unserer Industriegesellschaft ins Gigantische gewachsenen, als biotechnisch zu bezeichnenden, ärztlichen Handlungsbereich. So ist der Begriff Psychosomatik auf der Erkenntnisebene einerseits historisch gefestigt, andererseits auf der Handlungsebene kaum umgesetzte Zielvorstellung geblieben: Seit der Beschreibung psychisch konditionierter somatischer Reflexe durch Pawlow (1953) zu Beginn dieses Jahrhunderts haben einerseits unzählige experimentelle, klinische und epidemiologische Studien die Existenz psycho-somatischer Verbindungsstrukturen erhärtet, andererseits ist der ärztliche Handlungsbereich noch immer nach nosologischen, meist monokausal gedeuteten Krankheitsbegriffen (World Health Organization, 1977, 1978) organisiert. Es wird weiterhin von einem Leib-Seele-Dualismus ausgegangen. Psychosomatik ist im Denken vieler Ärzte noch immer mit der Aura einer Glaubensrichtung umgeben. Sie hat demzufolge auf der Ebene ärztlichen Handelns weitgehend den Stellenwert eines uneingelösten Versprechens beibehalten, obwohl die Theorieentwicklung und die Sammlung empirischer Daten in Übereinstimmung mit dem erwähnten Modell, das nicht nur die biotische und psychische, sondern auch die soziale und kulturelle Ebene umfaßt, weit über dieses Vorurteil hinausweist. Es ist für die ärztliche Praxis offensichtlich immer noch notwendig, von sozial und kulturell geprägten Denk- und Organisationsformen auszugehen, die sich an spezifischen Krankheiten bzw. Fächern orientieren. Sie kommen auch in diesem Buch zum Ausdruck. Ihre Restrukturierung sollte allerdings eine realistische Zukunftsperspektive darstellen.

2.1 Zwei grundsätzliche Makel

Es soll in diesem Kapitel versucht werden, unabhängig von diesen etablierten Kategorien der ärztlichen Wissenschaft, den heutigen Kenntnisstand bezüglich der Umstände, die für die Entstehung und Erhaltung von Gesundheit und Krankheit von Bedeutung sind, zu skizzieren. Es wird daraus hervorgehen, daß diese Umstände in der heutigen Industriegesellschaft vorwiegend dem soziokulturellen Bereich zuzurechnen sind. Gegenüber den gängigen Konzepten ärztlicher Wissenschaft müssen dabei vor allem zwei Vorbehalte gemacht werden:

1. Ein systemtheoretisches Modell, wie es in Kapitel 1 entwickelt worden ist, muß sich von einem im 19. Jahrhundert etablierten medizinisch-naturwissenschaftlichen **Kausalitätsbegriff** absetzen. Dieser basiert auf der in Kapitel 1 beschriebenen mechanistischen Vorstellung des Lebendigen. Im deutschen Sprachraum hatte der Begründer der modernen Physiologie, Johannes Müller (1801–1856), mit seinem Konzept einer „spezifischen Sinnesenergie" den Grundstein zu einer Biologie gelegt, welche die Interaktion zwischen dem Organismus und seiner Umgebung nicht als linear und mechanistisch verstand, sondern im Sinne einer Eigenaktivität, die nicht direkt und physikalisch auf die von außen einwirken-

---

1 Die Begriffe „Umwelt" und „Umgebung" werden hier im Sinne Jakob von Uexkülls (1920) verwendet. Danach meint „Umwelt" die von den Rezeptoren eines lebenden Systems aufgebaute subjektive Welt, die dessen Körper wie eine feste, aber für den außenstehenden Beobachter unsichtbare Hülle umgibt. „Umgebung" meint demgegenüber die Summe der neutralen, einem wissenschaftlichen Registrieren zugänglichen „objektiven" Fakten.

den Faktoren zurückzuführen ist. Seine Schüler Emil Du Bois-Reymond, Ernst von Brücke, Hermann von Helmholtz und Carl Ludwig setzten dieser Vorstellung (sie kann als Vision eines Systemmodells betrachtet werden), die sie als „vitalistisch" und damit unwissenschaftlich erachteten, ein auf physikalische Gesetze reduzierbares Aktions-Reaktions-Modell entgegen (Th. von Uexküll und Wesiack, 1988). Die Verlockung, das Leben physikalisch zu erklären und damit nennbare und angehbare Ursachen von Gesundheitsstörungen zu identifizieren, blieb seither bestehen.

Auch die Einführung der Autopsie als Grundlage der pathologischen Anatomie zu Beginn des 19. Jahrhunderts in Frankreich hat mit der Möglichkeit einer dinghaften Visualisierung von Läsionen dieses materialistische Modell gestützt (Cassell, 1979). Die Perspektive, Krankheiten nach stabilen, sichtbaren und beschreibbaren Phänomenen an der Leiche kategorisieren zu können, versprach die Befreiung von den inkohärenten und inkonsistenten Nosologien des 18. Jahrhunderts (Foucault, 1973). Die spektakuläre Entwicklung von Physiologie, Biochemie und Mikrobiologie hat in der Folge die an der Leiche vorgefundenen Phänomene zunehmend auch erklärbar gemacht. Die daraus abgeleitete Verbindung: physische (mechanische) Ursache → somatische Läsion → Diagnose → Krankheit → Therapie → Nicht-Krankheit, hat damit eine scheinbar offensichtliche Rationalität gewonnen. Der naturwissenschaftlich (vor der Begründung einer „neuen" Physik und Biologie) entstandene Begriff Krankheit wird damit zu einem Konstrukt, das mit der erwähnten Sequenz die Vorstellung einer linearen (= mechanischen) Kausalität erweckt (bestimmte Ursachen führen zu bestimmten Krankheiten). Außerdem wird damit ursächliche Qualität ausschließlich dem somatischen, vom psychischen und vor allem vom sozialen abgesetzten Bereich zugestanden. Ätiologie wird damit identisch mit mechanischen Ursachen, womit gleichzeitig die Ausgangslage für die nachfolgende gewaltige Technikentwicklung in der Medizin gegeben war. Kausalität beschränkt sich auf numerisch-quantitativ faßbare Phänomene; qualitative Daten und Einsichten im psychischen und sozialen Bereich werden, wenn überhaupt, mit abschwächenden Termini wie „Risiko" bzw. „statistisch-probabilistisch korrelierend" ausgestattet. Diese Unterscheidung hat im Rahmen des gängigen wissenschaftlichen Paradigmas zu einer kaum hinterfragten Einteilung in „harte" (für mechanische Beziehungen brauchbare) und „weiche" (... unbrauchbare) Daten und schließlich entsprechend selektionierten Erkenntnissen geführt. Dem ist hinzuzufügen, daß mit den Erkenntnissen einer modernen Physik auch die Verknüpfung „harter" Daten als statistisch-probabilistisch bezeichnet werden muß.

Der fundamentale Irrtum dieses pseudorationalen Denkgebäudes ergibt sich aus der scheinbar trivialen Tatsache, daß die Leiche nicht identisch ist mit dem Individuum, dem Träger einer auf diese Weise nicht definierten Störung der Gesundheit. Nicht nur der Mensch, der lebende Organismus generell unterschei-

det sich von der im biotechnischen Denken als Modell betrachteten Leiche durch seine **Autonomie** bzw. **Autopoiese.** Mit diesen Begriffen hat eine moderne Biologie die Vorstellung von Johannes Müller einer „spezifischen Sinnesenergie" wieder aufgenommen und erweitert. Lebewesen sind fähig, „ihre eigene Gesetzlichkeit bzw. das ihnen Eigene zu spezifizieren" (Maturana und Varela, 1987). Die Leiche reagiert auf einen physikalischen oder chemischen Reiz in einer sehr spezifischen, mechanischen und auf den Ort der Einwirkung begrenzten Weise. Der lebende Organismus reagiert auf solche und auf psychologische Reize individuell-variabel und als Ganzes mittels einer veränderten (adaptierten) Selbstorganisation bzw. Autopoiese. Die Rolle des Stimulus von außen ist zu relativieren, Maturana und Varela (1987) haben sie mit dem Ausdruck „Perturbation" qualifiziert. Auf der human-psychologischen Ebene drückt sich die Autopoiese u.a. in der autonomen Perzeption und Verarbeitung von Gesundheit und Krankheit durch die davon Betroffenen aus. Die eminente Bedeutung dieser individuell-subjektiven Prozesse für das ärztliche Handeln ist eines der Hauptanliegen dieses Buches. Die Vorstellung von Krankheit als Ausdruck einer physikalischen Interaktion zwischen Umgebung und Organismus greift zu kurz.

2. Der zweite grundsätzliche Vorbehalt gegenüber den Grundbegriffen der etablierten medizinischen Wissenschaften betrifft die „Entstehung" von Gesundheit. Der Begriff Entstehung erweckt in diesem Zusammenhang eine für traditionelles ärztliches Denken ungewohnte Vorstellung. Gesundheit wird eher als gegebene statistische Norm betrachtet, die durch eine (sehr wohl „entstehende") Krankheit abgebaut wird. Die Betrachtung des für den lebenden Organismus konzipierten Funktionskreises (vgl. Abb. 1–3, Kap. 1) sowie des für das menschliche Individuum spezifizierten Situationskreises (vgl. Abb. 1–4, Kap. 1) läßt aber die Vorstellung einer „Entstehung" von Gesundheit als durchaus sinnvoll, ja selbstverständlich erscheinen. Der tierische bzw. menschliche Organismus interagiert mit seiner Umgebung im Sinne deren **Nutzung** sowie des **Überlebens,** bzw. einer Vermeidung, einer Abwehr schädlicher Entwicklungen. Die Alltagsbegriffe Nutzung und Überleben müssen hier allerdings epistemologisch präzisiert werden. Nutzung ist im Sinne von **Assimilation** zu verstehen, die dem Aufbau lebendiger Struktur aus den Elementen einer für das Lebewesen „natürlichen" Umgebung dient. Überleben meint **Akkommodation** der Sollwerte eines lebendigen Systems angesichts veränderter Umgebungsbedingungen. Erst wenn die Möglichkeiten der Akkommodation erschöpft sind, kommt es zu Phänomenen des Schadens bzw. des Mangels. Daraus ergibt sich eine quasi axiomatische Logik der Entstehung und der Existenz lebender Strukturen: Sie müssen aus der Umgebung aufgebaut werden (Assimilation). Da sich diese Umgebung laufend verändert, dienen Akkommodationsprozesse von Anfang an der Erhaltung dieser Struktur. Störung und Zerstörung (Pathogenese) müssen dann als Konsequenzen einer Überforderung vitaler Prozesse interpretiert werden. Das Funktions- und das Situationskreismodell beschreiben Nutzung und Überleben, die Entstehung und Erhaltung von **Leben,** was in der medizinischen Perspektive als Entstehung und Erhaltung von Gesundheit bezeichnet werden kann.

Aaron Antonovsky (1987) hat, als Soziologe von außerhalb des medizinischen Kulturkreises kommend, die Einschränkung der tradierten ärztlichen Sichtweise auf das Konzept der „Pathogenese", d. h. auf die Einwirkung und Auswirkung von Schäden auf den menschlichen Organismus, kritisiert. Für das wissenschaftliche Brachland, in dem es um die Entstehung und Erhaltung von **Gesundheit** geht, mußte er zunächst einen Begriff – „Salutogenese" – schaffen. Empirische Daten in diesem Bereich sind zwar vorhanden, den ärztlich-wissenschaftlichen Stellenwert, der ihnen bisher zuteil geworden ist, kann man jedoch vernachlässigen, während sich das Denkgebäude der Pathogenese auf unüberblickbare Ausmaße ausgedehnt hat. Antonovsky hat die zentrale Eigenschaft der „salutogenen" Potenz lebender Systeme mit dem Begriff „Kohärenz-Sinn" („sense of coherence"; Antonovsky, 1987) umschrieben. Damit ist die Fähigkeit gemeint, aus einer Umgebung, die sich in Richtung von Entropie (Chaos) bewegt, diejenigen Elemente nutzbar zu machen („herauszusaugen"), die dem Aufbau der eigenen (negentropen) Struktur dienen, und diejenigen Elemente zu meiden, welche die Entropie verstärken würden. Antonovskys salutogenetisches Modell deckt sich weitgehend mit Jakob von Uexkülls und Thure von Uexkülls Modellen des Funktions- und Situationskreises. In diesen Modellen wird der „sense of coherence" Antonovskys als die kreative Potenz eines autopoietischen Systems beschrieben, die Teile der Umgebung durch Bedeutungserteilung und Bedeutungsverwertung assimiliert, d. h. dem sich selbst erzeugenden und selbst erhaltenden System eingliedert. Diese kreative Potenz ist sowohl auf der biologischen wie auf der psychischen und der sozialen Integrationsstufe aktiv. Ein Beispiel, wie diese Aktivität objektiv beobachtet und subjektiv erlebt werden kann, haben Christian und Haas (1949) mit ihren Versuchen über „Bipersonalität" vorgelegt. In ihnen wird gezeigt, daß bei einer gemeinsamen Arbeit zu zweit subjektiv das Gefühl der Autonomie nur in den Augenblicken entsteht, in denen sich objektiv Leistung und Gegenleistung entsprechen.

In der Abbildung 2–1 sind exemplarisch Funktions- und Situationskreise in ein Modell eingefügt, welches das biopsychosoziale System des Individuums und seiner Umgebung veranschaulicht. Die graphischen Komponenten (biotische, psychische, soziokulturelle Ebene) entsprechen verschiedenen Integrationsstufen des hierarchisch aufgebauten Systems. Die Verbindungswege (= Vernetzung) zwischen ihnen durch „Aufwärts"- und „Abwärts"-Effekte sind durch vertikale Pfeile links in der Graphik symbolisiert. Die Existenz dieser Vernetzung läßt sich mit der Perzeption und Erfahrung betroffener Individuen einerseits und mit wissenschaftlichen Daten andererseits verdeutli-

chen. Norman Cousins (1976, 1979) hat über den von ihm selbst vorangetriebenen Heilungsprozeß seines als unheilbar betrachteten Leidens (Morbus Bechterew) 1976 berichtet. Seither mehren sich in der Literatur Darstellungen von miteinander verflochtenen perzipierten Phänomenen auf der somatischen, psychischen und sozialen Ebene bei Veränderungen des Gesundheitszustandes. Ein Beispiel dafür findet sich im Buch des Neurologen und Schriftstellers Oliver Sacks (1989), dessen vielschichtige Darstellung der Ereignisse nach der somatischen und psycho-neuralen „Abkoppelung" seines verunfallten Beines mit multiplen Unterbrüchen anhand der in Abbildung 2–1 dargestellten horizontalen und vertikalen Verbindungen erklärt werden kann. Solche systemischen Sichtweisen werden in einer sich wissenschaftlich gebenden Literatur abgelehnt oder lächerlich gemacht

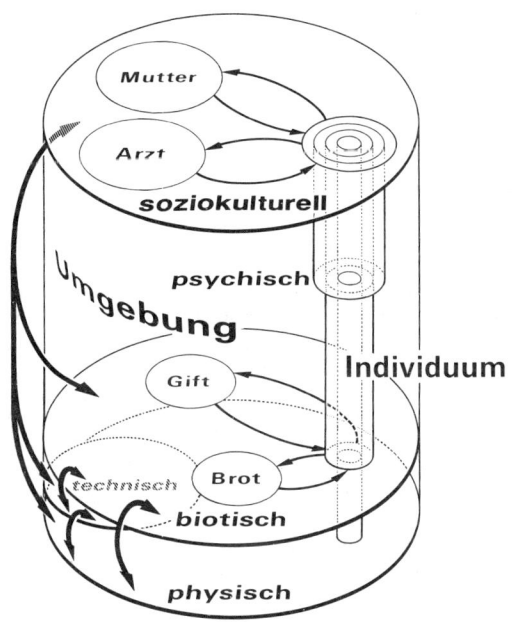

Abb. 2–1. Graphische Darstellung des Individuum-Umgebungs-Systems. Das Individuum – als Teil der Umgebung – läßt sich auf vier Ebenen von „physisch" bis „soziokulturell" projizieren bzw. wissenschaftlich beschreiben. Die Eigenschaften auf den „unteren" Ebenen finden sich auf den „oberen" Ebenen als in komplexere Phänomene integrierte Elemente. Von „unten" nach „oben" kommt es zur Ausprägung von neuen, durch die Verhältnisse „unten" nicht erklärbaren Eigenschaften (Phänomen der Emergenz), was graphisch durch erweiterte Abbildungsflächen angedeutet ist. Auf der biotischen und soziokulturellen Ebene sind je zwei Beispiele für Funktions- bzw. Situationskreisen (vgl. Abb. 1.1, 1.4, Kap. 1) eingetragen. Der Person der Mutter bzw. des Arztes, dem Gegenstand Brot bzw. Gift wird eine Bedeutung erteilt und diese Bedeutung wird verwertet (Bereitschaft zur mütterlichen oder ärztlichen Zuwendung, Nahrungsaufnahme, Vermeidung von oder Schutz vor Vergiftung). Die Graphik stellt die Verhältnisse vereinfacht dar: Situationskreise können mehrere Ebenen („aufwärts" und „abwärts") beeinflussen (z.B. der Vorgang des Stillens bzw. die gemeinschaftliche Nahrungsbeschaffung mit ihren biotischen und soziokulturellen Anteilen); diese Vernetzung ist mittels der Pfeile links angedeutet (aus H. G. Pauli, 1989).

(Angell, 1985). Eine derartige Ablehnung wäre aber nur durch Bestätigung einer „Nullhypothese" zu rechtfertigen, d. h. aufgrund der Unmöglichkeit, die in der anekdotischen Literatur geschilderte Vernetzung im einzelnen nachzuweisen. Es mehren sich aber Studien, die diesen Nachweis – siehe nachfolgendes Beispiel – mittels biologischer und epidemiologischer Methoden durchaus erbringen. Anstelle einer Ausgrenzung perzipierter Gesundheitsphänomene in den außerwissenschaftlichen Bereich stellt sich vielmehr die Aufgabe, diese durch eine Erweiterung der klinischen Methodik, etwa im Sinne einer „klinischen Ontologie" (Sacks, 1989), zu validieren.

Johannes Siegrist et al. (1980, 1988) haben die Bedeutung der Situation am Arbeitsplatz untersucht. In retrospektiven Studien konnten sie bestimmte Streßsituationen beschreiben, die mit einer hohen Inzidenz von Myokardinfarkt verbunden waren. Diese Situationen können zusammenfassend als langfristig überbeanspruchend und durch die Betroffenen nicht oder kaum beeinflußbar beschrieben werden. In einer prospektiven Studie haben die gleichen Autoren unter derartigen Belastungen Blutlipidwerte erhoben, die mit einem erhöhten Myokardinfarktrisiko zu verbinden sind. James Henry (1982) hat zwei Verbindungswege beschrieben, über die sich soziale Situationen somatisch auswirken. In einer Streßsituation, deren Bewältigung oder Abwendung die betroffene Person für möglich hält, wird die hypothalamisch-adrenomedulläre neuroendokrine Achse aktiviert. Erweist sich die soziale Streßsituation für das betroffene Individuum als unabwendbar bzw. unbeeinflußbar, kommt es zur hypothalamisch-hypophysär-adrenokortikalen Aktivierung. Im ersteren Fall scheint dies mit einer sympathischen („fight or flight"), im zweiten Fall mit einer trophotropen Stimmung („conservation-withdrawal"; Engel, 1972) des autonomen Nervensystems einherzugehen. Während eine akute Aktivierung dieser Achsen als physiologischer Regelmechanismus zu verstehen ist, kann es heute als gesichert gelten, daß eine langfristig wiederholte Beanspruchung dieser Mechanismen über einen gegebenen Bereich hinaus mit pathologischen Veränderungen an den Zielorganen einhergeht (Henry, 1982). Das sympathisch-parasympathische Reaktionssystem, das Walter Hess (1948) vor über 40 Jahren aufgrund von Tierexperimenten beschrieben hat, läßt sich damit als Verbindungselement zwischen bestimmten sozialen Situationen und der Entwicklung von somatischen Veränderungen und schließlich von Läsionen interpretieren.

Zwischen dem System Individuum (Organismus und individuelle Wirklichkeit) und dem Suprasystem Umgebung (im obigen Beispiel die soziale Situation am Arbeitsplatz) besteht somit eine zirkuläre Verbindung. Das Individuum erteilt der speziellen Situation in der Umgebung eine bestimmte Bedeutung[2] und versucht, diese Bedeutung zu „verwerten", d. h. auf die Situation einzuwirken. Diese zirkuläre Verbindung Individuum–Umgebung läßt sich anhand des Schemas des Situationskreises (vgl. Abb. 1–4, Kap. 1) darstellen. Es sei hier die Hypothese aufgestellt, daß eine für die Existenz des Individuums „brauchbare" Bedeutungserteilung und eine entsprechende Bedeutungsverwertung mit Lebensvorgängen einhergeht, die wir als „gesund" zu bezeichnen geneigt sind, und daß eine derartige Situation weitgehend von einem positiven Kohärenz-Gefühl (Antonovsky, 1987) begleitet wird.

Stressoren werden nicht nur erfolgreich aufgefangen, sie dienen unter Umständen sogar der Erhaltung oder der Verbesserung der Gesundheit.

Streß in der Arbeitssituation bei erhaltener Einflußmöglichkeit durch davon Betroffene (z. B. der Streß des unabhängigen Managers) läßt sich dann nicht mit „krankhaften" Prozessen korrelieren; im Gegensatz dazu ist unkontrollierbarer Streß mit biochemischen und strukturellen Schäden (Lipidstoffwechsel, Schädigung der Kreislauforgane) verbunden. Im letzteren Fall ist der zirkuläre Fluß im Situationskreis an mindestens einer Stelle (Bedeutungsverwertung: Unbeeinflußbarkeit der Arbeitssituation) blockiert.

Unter dieser Perspektive wird Gesundheit ein integraler Teil des biopsychosozialen Modells und kann nicht mehr als unerforschtes oder sogar unerklärbares Residuum betrachtet werden, wie dies im gängigen medizinischen Denksystem geschieht.

2.2 Die systemische Sichtweise

Es ist nun klargeworden, daß uns eine systemtheoretische und semiotische Betrachtungsweise vor dem irreführenden Schluß bewahren muß, die Begriffe „biotisch", „psychisch" und „soziokulturell" würden voneinander unabhängige Ebenen beschrieben. Die Einsicht in die Vernetzung durch Aufwärts- und Abwärts-Effekte macht für die Medizin den Nachholbedarf im soziokulturellen Bereich unübersehbar. Ohne auf diese für eine Forschungsplanung und -politik zentrale Frage einzugehen, soll nun eine systemische Beschreibung des Feldes versucht werden, in dem Gesundheit und Krankheit entstehen. Dieses Feld deckt sich, wie wir gesehen haben, weitgehend mit demjenigen, in dem Phänomene des Lebens ermöglicht oder behindert bzw. verunmöglicht werden. Man ist versucht, in Analogie zu der Situation in der Evolution der Erdgeschichte, in der erstmals Lebensphänomene auftraten, von einer „Ursuppe" (**„Ur-Sache"**) zu sprechen, aus der durch Organisation aus Entropie Strukturen (= Negentropie) entstehen bzw. wiederum abgebaut werden. Was entsteht bzw. beeinträchtigt wird, drückt sich als **Gesundheitsphänomen** aus, das entweder durch Forscher/Beobachter oder durch betroffene Individuen oder Gruppen wahrgenommen und beschrieben wird. Damit verbunden sind **Gesundheitsfolgen**, die sich wiederum für Betroffene „subjektiv", für außenstehende Betrachter „objektiv" darstellen. In Abbildung 2–2 (revidiert nach Pauli, 1986) ist diese sequentielle Sicht des „Gesundheitsfeldes" graphisch dargestellt und dessen Abbildung durch wissenschaftliche Fachbereiche angedeutet. Von spezieller Bedeutung im Zusammenhang mit dieser Übersicht sind die Felder „Gesundheits-Ursachen und -Folgen", während der mittlere Anteil („Phänomene") als traditionell ärztlich-wissenschaft-

---

2 Bedeutungserteilung muß hier im weitesten Sinne verstanden werden: alles was am und im Individuum irgendwelche Prozesse in Gang setzt, sei es auf bewußter oder unbewußter Ebene.

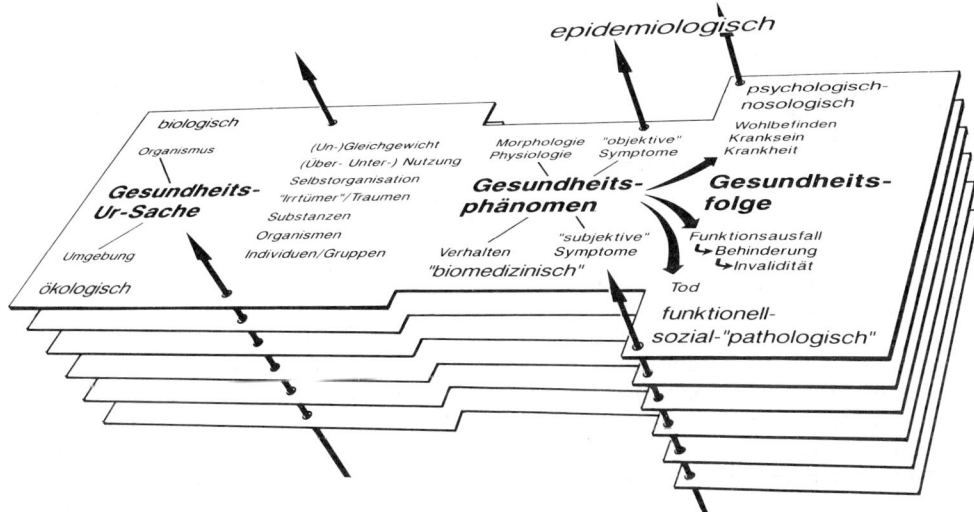

Abb. 2–2. Die Entstehung, die Phänomene und die Folgen von Gesundheit bzw. Gesundheitsstörungen sowie die wissenschaftlichen Dimensionen, auf denen diese Bereiche analysiert werden. Die einzelne Ebene symbolisiert das „Gesundheitsfeld" eines einzigen Individuums. Die Juxtaposition mehrerer individueller Ebenen schafft die Möglichkeit einer (durch Pfeile angedeuteten) „epidemiologischen" Zusammenfassung von Ur-Sachen, Phänomenen und Folgen. Dabei kommen im allgemeinen statistische Methoden zur Anwendung.

liches Arbeitsgebiet weniger einer Erläuterung bedarf. In den folgenden Abschnitten wird auf die drei Teilbereiche in Abbildung 2–2 im einzelnen eingegangen.

2.2.1 Die „Ur-Sache" von Gesundheit und Krankheit

„Gesundheit" wird in diesem Zusammenhang als bipolarer Begriff verwendet, der sowohl „gute" als auch „gestörte" Gesundheit umfaßt. Den im Bereich „Ur-Sache" beschreibbaren Prozessen und Sachverhalten brauchen keine im traditionell-naturwissenschaftlichen Sinne kausalen Funktionen zuzukommen. Sie bilden vielmehr ein Substrat, aus dem heraus Gesundheitsphänomene (d.h. Lebensphänomene inklusive ihrer Beeinträchtigungen) im systemischen Sinn entstehen (emergieren).

Gleichgewichte – Ungleichgewichte

Der Begriff Gleichgewicht darf hier keineswegs im physikalischen Sinn verstanden werden. Das klassische biologisch-physiologische Konzept in diesem Zusammenhang ist die **Homöostase.** Für Organismen bzw. Individuen charakteristische labile Gleichgewichte lassen sich auf biologischer, psychologischer sowie soziokultureller Ebene beschreiben. So werden beispielsweise intra-/extrazelluläre Gradienten von Elektrolyt- und Eiweißkonzentrationen in den Körperflüssigkeiten durch energetische Prozesse aufrechterhalten. Im psychologischen Bereich lassen sich periodische und ausgleichende Phasen von Motivation bzw. Erregung und Relaxation beschreiben. Auf der soziokulturellen Ebene mag ein Gleichgewicht im Wechsel zwischen sozialen Außenkontakten und der Kommunikation in Kleingruppen wie Familien und Personengruppen am Arbeitsort als Bei-

spiel dienen. Dem sozialen Gleichgewichtsbereich kommt in der heutigen Gesellschaft ein noch nie dagewesenes Gewicht zu.

Vor allem die Technologieentwicklung verändert die menschliche Umgebung mit einer Geschwindigkeit, für die in der bisherigen Evolution von Lebewesen, inklusive des Menschen, keine Beispiele bekannt sind. Die gleichzeitige beträchtliche Zunahme der mittleren menschlichen Lebensdauer hat dazu geführt, daß die Gesamtzahl von Veränderungen während eines einzigen Lebensalters zusätzlich erhöht ist (s. Abb. 2–3). Diese Situation birgt andererseits ein Potential an Negentropie durch die Möglichkeit kreativ genutzten Ungleichgewichts. Der in der gesamten Menschheitsentwicklung für die Gesundheit zentrale Umstand einer genügenden bzw. ungenügenden Ernährung wird mindestens in den Industriegesellschaften durch das Phänomen der soziokulturellen Adaptation/Fehladaptation bzw. Akkommodation/Fehlakkommodation verdrängt (McKeown, 1982).

(Über-, Unter-)Nutzung

Das Leben bzw. das Überleben jedes Organismus ist von einem bestimmten Ausmaß an Belastung bzw. Nutzung abhängig. Ein völliges Ausbleiben dieser Belastung beispielsweise im Zustand der Schwerelosigkeit oder der sensorischen Deprivation führt zum Verlust von Strukturen bis hin zum Tode. Diesem Phänomen der salutogenen Nutzung – Piaget (1975) spricht von einem notwendigen Maß an Akkommodation – ist dasjenige der Unter- oder Übernutzung im pathogenen Sinn entgegenzusetzen.

Selbstorganisation

Auch Selbstorganisation/Autopoiese stellt ein existentielles Grundphänomen lebender Systeme dar

(Jantsch, 1982; Maturana und Varela, 1987). Selbstorganisation bedeutet einerseits Grundlage der Autonomie von Organismen und Individuen. Autonomiefähigkeit schließt andererseits die Möglichkeit zur Isolation ein, die auf biotischer, psychischer und soziokultureller Ebene das Überleben gefährden kann.

„Irrtümer"/Traumen

Unter „Irrtümern" – in semiotischer Sicht als „Übersetzungsfehler" zu bezeichnen – sind hier Anomalien gemeint, die im Innern des Organismus (z.B. genetische strukturelle bzw. biochemische Veränderungen) auftreten und denen im wesentlichen pathogene Bedeutung zukommt. Analog dazu umschreiben Traumen (im weitesten Sinn) Schädigungen durch Einwirkung der physisch-biotischen oder soziokulturellen Umgebung auf das Individuum, die ebenfalls als pathogen zu bezeichnen sind. Die letztere Ebene hat in der Industriegesellschaft an Bedeutung gewonnen. Die Ausschaltung solcher Traumen mittels Herstellung von „künstlichen Nischen" für den Menschen ist zur primären Aufgabe dieser Gesellschaft geworden. So ermöglichen beispielsweise energiekonsumierende Transportsysteme die Existenz in modernen großen Agglomerationen. Andererseits produziert ebendiese soziokulturelle Umgebung mit ihren Nischen physische, gesundheitsaktive Elemente im Sinne der „Umweltverschmutzung", so daß sich die Gewichtung wiederum in die pathogene Richtung sowie auf die physisch-biotische Ebene verschieben könnte. Im oben erwähnten Beispiel wirkt sich die durch Transportsysteme verursachte Umgebungsverschmutzung in diesem Sinne aus. Am extremen Ende dieses Spektrums von Einwirkungen ist auf jeden Fall die endgültige Bedrohung menschlicher Gesundheit, der atomare Holocaust, zu lokalisieren.

Substanzen

Leben ist vom Stoffumsatz, insbesondere der Assimilation von Substanzen, abhängig und gleichzeitig durch Substanzen bedroht (vgl. Beispiele der Abb. 2–1). Substanzen können sowohl salutogene als auch pathogene Eigenschaften aufweisen (z.B. essentielle Spurenelemente, die in hohen Konzentrationen toxisch wirken).

Organismen

Mikroorganismen (z.B. die für den Stoffumsatz unabdingbare Darmflora bis hin zu den Erregern von akut tödlichen Infektionskrankheiten), Pflanzen und Tiere können in gleicher Weise einerseits eine salutogene, andererseits eine pathogene bzw. lebensbedrohende Rolle spielen.

Individuen/Gruppen

Soziale Beziehungen von der Bipersonalität (Christian, 1949) über Bezugsgruppen und Gemeinden bis hin zur nationalen Identität sind zweifellos für die Gesundheit des Individuums von ausschlaggebender Bedeutung. Im Vordergrund stehen in den Industriegesellschaften einerseits die Lebenspartnerschaft bzw. die Familie, andererseits die Arbeitsumgebung. Die Forschung hat sich im letzteren Bereich vor allem mit seinen männlich-professionellen Anteilen befaßt, während für den Existenzbereich der Frau, insbesondere der „Hausfrau", ein erhebliches Erkenntnisdefizit besteht.

Die Bedeutung des umgebenden „sozialen Netzes" für die Gesundheit kommt u.a. in einer kalifornischen Studie zum Ausdruck, in der die Sterblichkeit in einer Zufallsstichprobe von fast 7000 Personen während neun Jahren registriert wurde, nachdem und währenddem deren soziale Beziehungen eingehend analysiert worden waren. Die relative alterskorrigierte Sterblichkeit der sozial am meisten isolierten Gruppe gegenüber derjenigen, die sozial am besten integriert war, betrug 2,3 bzw. 2,8 für Männer respektive Frauen. Diese Korrelation zwischen sozialem Netz und Lebenserwartung ließ sich nachweisen unabhängig vom subjektiven Gesundheitszustand (zu Beginn der Studie), vom sozioökonomischen Status, von gesundheitsfördernden oder -beeinträchtigenden Lebensweisen, vom Körpergewicht und vom Ausmaß, in dem Leistungen der Gesundheitsversorgung in Anspruch genommen wurden (Berkman und Syme, 1979).

2.2.2 Gesundheitsphänomene

Gesundheit als Phänomen (s. Abb. 2–2), wiederum definiert entlang eines Spektrums zwischen „gut" und „gestört", wird auf verschiedenen Erkenntnisebenen erfaßt (Pauli, 1983). In der biotechnischen Sichtweise hat die Analyse von Strukturen und Funktionen (Morphologie, Physiologie bzw. Pathologie und Pathophysiologie) Priorität. Im professionellen Bereich (mit Ausnahme der Psychiatrie) ist die Beurteilung von Gesundheit durch Beobachtung auf der Verhaltensebene im Vergleich dazu randständig. Klinisch stehen objektive Symptome („signs") aufgrund von direkten Beobachtungen bzw. durch Erfassung unter Zwischenschaltung von Instrumenten einerseits und subjektive Symptome durch Vermittlung bzw. Aussagen des vom Gesundheitsphänomen betroffenen Individuums im Vordergrund. Das letztere hat als Mitglied einer biotechnisch orientierten Gesellschaft die Tendenz, seine Befindlichkeit entsprechend biotechnisch zu beschreiben („Medikalisation", „Somatisation", s.u.). Trotzdem wird dies durch Ärzte kaum als individuell und kulturell determiniertes „Kranksein" wahrgenommen.

2.2.3 Gesundheitsfolge

Hier muß unterschieden werden, auf wen sich die Folge bezieht. Die Unterschiede zwischen professioneller und subjektiver Perzeption können sehr bedeutsam sein, bis hin zur Kommunikationsunmöglichkeit. Für die Betroffenen reicht das Spektrum vom Wohlbefinden zum Kranksein; in der professionellen Sichtweise steht Krankheit und deren Kategorisierung (World Health Organization, 1977, 1978) im Vordergrund. Es ist bemerkenswert, wie stark sich die professionell-nosologische Kategorisierung und Differenzierung entwickelt hat, währenddem Kategorien der Behinderungen der menschlichen Existenz und Lebensqualität (World Health Organization, 1980) wenig Beachtung gefunden haben.

2.3 Der Gesundheits-Gesamtbereich: Perzeption und wissenschaftliche Abbildung

Der Gesamtüberblick in Abbildung 2–2 verdeutlicht das oben angedeutete Auseinanderklaffen der subjektiven Perzeption durch die Betroffenen einerseits und der professionell-wissenschaftlichen Repräsentation des Gesundheitsbereiches andererseits, wie sie für unsere Kultur charakteristisch ist. Betroffene haben vergleichsweise begrenzte rationale Erkenntnismöglichkeiten im „Ur-Sachen"-Bereich; für sie steht der „Folgen"-Bereich im Vordergrund. Die westliche biotechnische Medizin hat das Schwergewicht in den vermeintlich empirisch/„objektiv" faßbaren Phänomenbereich gelegt. Im „ursächlichen Bereich" ist sie stark materiell orientiert („Irrtümer"/Traumen, Substanzen, Organismen), während die angeführten Prozeßbegriffe (Gleichgewichts- und Nutzungskonzepte, Selbstorganisation) eher nichtwestlichen Medizinkulturen (Foster, 1986) bzw. neueren Erkenntnissen in Biologie und Sozialwissenschaften entstammen. Im „Folgen"-Bereich schließlich stehen für die biotechnische Sichtweise einerseits wiederum eine materielle – mit dem in der Folge der Autopsie-Methode entstandenen zentralen Fachbereich Pathologie – und davon abgeleitet eine phänomenologisch klassifizierende Orientierung („Nosologie", World Health Organization, 1977, 1978) im Vordergrund. Das subjektiv-professionelle Auseinanderklaffen läßt sich auch so formulieren: Was als „wissenschaftliche Repräsentation" gelten will, ist nicht sehr wissenschaftlich, weil dabei grundlegende Daten ausgeklammert werden. Diese Ausklammerung betrifft sowohl die subjektive Perzeption als auch, insbesondere im Bereich der „Ur-Sachen", systemische (z. B. Un-/Gleichgewichte) und soziokulturelle Phänomene.

2.4 Folgerungen: Soziale Probleme in der Medizin unserer Zeit

Die „Landschaft Gesundheit und Krankheit" läßt sich ohne Einsicht in die Natur des „abbildenden Mediums" (d. h. der Natur und Kultur sowohl der betroffenen Individuen als auch der Praktiker und Forscher im Gesundheitsbereich) nur unvollständig bzw. einseitig darstellen. So lassen sich in der Folge zwei Problemfelder unterscheiden, von denen das erste eher den sozialen Dimensionen von Gesundheit und Krankheit, das zweite eher den sozialen Umständen der Gesundheitsversorgung zuzurechnen ist, die aber in mancher Hinsicht miteinander verflochten sind.

2.4.1 Die soziale Determination von Gesundheit und Krankheit

Die tiefgreifenden Veränderungen von Morbidität und Mortalität während des zu Ende gehenden Jahrhunderts mit dem drastischen Rückgang der Infektionskrankheiten einerseits und der beschleunigten Veränderung der menschlichen Umgebung andererseits haben die sozialen Gesundheitsdeterminanten gegenüber physisch-somatischen in den Vordergrund treten lassen. Auch individuelle Gesundheits- und Krankheitszustände lassen sich häufig nur im Zusammenhang mit soziokulturellen, u.a. biographischen, ethnischen und ökonomischen Umständen verstehen. Dies wird besonders deutlich anhand der Situation von Individuen mit sog. funktionellen Syndromen, bei denen keine ihr Kranksein erklärenden organischen Veränderungen festzustellen sind. Diese Syndrome erweisen sich als stark abhängig von den soziokulturellen Umständen, unter denen sie auftreten, insbesondere vom Vorhandensein oder vom Fehlen einer für das betroffene Individuum „tragfähigen" sozialen Umgebung, vor allem im Bereich der Familie und des Arbeitsortes (Weiner, 1987). Das heutige biotechnische Versorgungssystem selbst spielt bei der Ausprägung von funktionellen Syndromen und der Auswirkung auf sie eine wesentliche Rolle. So kommt es in diesem Zusammenhang zu den Phänomenen der „Somatisierung" sowie – seltener – der „Psychologisierung" (Th. von Uexküll, 1988) d. h., das betroffene Individuum präsentiert sein Kranksein in der „Sprache" des ihm zur Verfügung stehenden Versorgungssystems, in unserer Kultur überwiegend derjenigen der somatisch orientierten biotechnischen Medizin sowie neuerdings zunehmend der Psychotherapie (s. Ausführungen zum Kausalitätsbegriff in Abschnitt 2.1).

Die Zusammenhänge zwischen sozialen Umständen und erhaltener bzw. gestörter Gesundheit der beteiligten Individuen beschränken sich wie oben beschrieben nicht auf die Manifestationen der funktionellen Syndrome. Seit der klassischen Studie von Berkman und Syme (1979) zur Bedeutung sozialer Netze für die Gesundheit haben eine große Zahl weiterer sozialepidemiologischer Studien ein Ausmaß dieses Chancen- und Risikofaktors aufgezeigt, das sich höchstens mit demjenigen für den Gesundheitsfaktor Rauchen vergleichen läßt (House, 1989).

Unter Berücksichtigung der in Abschnitt „Gleichgewichte – Ungleichgewichte" beschriebenen Akzeleration der Veränderungen menschlicher Lebensbedingungen (s. Abb. 2–3) ist schließlich zu bedenken, daß sich auch die Perzeption und das Bewußtsein von Gesundheit bzw. ihrer Störungen in gleicher Weise dramatisch verändern. Während einige der wichtig-

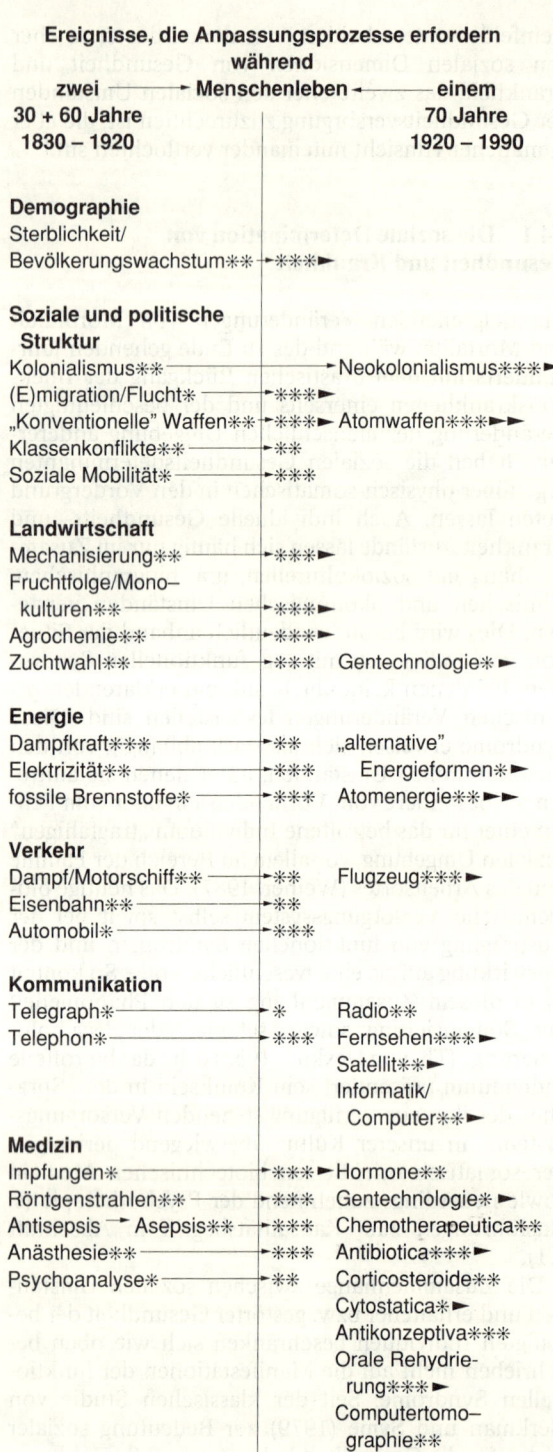

Ereignisse, die Anpassungsprozesse erfordern während

zwei ⟶ Menschenleben ⟵ einem

| 30 + 60 Jahre | | 70 Jahre |
| 1830 – 1920 | | 1920 – 1990 |

Demographie
Sterblichkeit/
Bevölkerungswachstum** ⊢***►

Soziale und politische Struktur
Kolonialismus** ⟶ Neokolonialismus*** ►
(E)migration/Flucht* ⟶ ***
„Konventionelle" Waffen** ⊢*** Atomwaffen*** ►►
Klassenkonflikte** ⟶ **
Soziale Mobilität* ⟶ ***

Landwirtschaft
Mechanisierung** ⟶ *** ►
Fruchtfolge/Mono–
kulturen** ⟶ *** ►
Agrochemie** ⟶ *** ►
Zuchtwahl** ⟶ *** Gentechnologie* ►

Energie
Dampfkraft*** ⟶ ** „alternative"
Elektrizität** ⟶ *** Energieformen* ►
fossile Brennstoffe* ⟶ *** Atomenergie** ►►

Verkehr
Dampf/Motorschiff** ⟶ ** Flugzeug*** ►
Eisenbahn** ⟶ **
Automobil* ⟶ ***

Kommunikation
Telegraph* ⟶ * Radio**
Telephon* ⟶ *** Fernsehen*** ►►
Satellit** ►
Informatik/
Computer** ►

Medizin
Impfungen* ⟶ *** ►Hormone**
Röntgenstrahlen** ⟶ *** Gentechnologie* ►
Antisepsis ⟶ Asepsis** ⊢*** Chemotherapeutica**
Anästhesie** ⟶ *** Antibiotica*** ►
Psychoanalyse* ⟶ ** Corticosteroide**
Cytostatica* ►
Antikonzeptiva***
Orale Rehydrie-
rung*** ►
Computertomo-
graphie**
Kernspinresonanz* ►

Abb. 2–3. Ereignisse, die sich während der vergangenen 160 Jahre in Industriegesellschaften stark (✳ bis ✳✳✳ = geschätztes zunehmendes Ausmaß) auf die Gesundheit ausgewirkt haben. Es werden zwei Zeitperioden à 90 bzw. 70 Jahre betrachtet. In der ersten stieg die Lebenserwartung von rund 30 auf 60 Jahre, in der zweiten auf 70 Jahre, so daß sich während der ersten Periode im Mittel **zwei** Generationen, in der zweiten Periode **eine** einzige Generation an die sich beschleunigt verändernden Verhältnisse anpassen mußten. In der linken Kolonne sind somit die Ereignisse eingetragen, die in zwei Generationen, in der rechten diejenigen, welche in nur einer bewältigt werden mußten. Mit ► sind Ereignisse bezeichnet, die zukünftig wahrscheinlich zusätzliche Auswirkungen haben werden. Es werden vorwiegend materielle/technische Ereignisse aufgeführt, da sich das Ausmaß von Veränderungen auf den Ebenen von sozialen Interaktionen, Einstellungen und Werthaltungen kaum abschätzen läßt und diese in den Industriegesellschaften stark von materiellen bzw. technischen Entwicklungen abhängen. Die Auswahl und Quantifizierung dieser Liste erfolgte intuitiv und subjektiv. Es ist anzunehmen, daß andere Beobachter andere Termini und andere Gewichtungen wählen würden, jedoch in der Gesamtbewertung zum analogen Schluß einer gewaltigen Zunahme von Veränderungen pro Lebensalter kommen würden.

züglich Schlüsse zu. So wurde im Rahmen von Befragungen größerer Bevölkerungsgruppen in den zwanziger Jahren im Mittel über 0,82 Episoden schwerer, akuter und behindernder Gesundheitsstörungen pro Person und Jahr berichtet; in einer vergleichbaren Untersuchung 60 Jahre später wurden 2,12 derartige Episoden angegeben. Auch bezüglich der Dauer dieser Episoden wurde im gleichen Zeitraum ein Anstieg von 16 auf 19 Tage festgestellt (Shorter, 1985). Während des gleichen Zeitraums hat offenbar der Anteil von Personen abgenommen, die auf Befragung keinerlei Beeinträchtigung der Befindlichkeit angeben. Abgesehen von methodischen Problemen lassen derartige Erhebungen unterschiedliche Interpretationen zu. Generell ergibt sich jedoch die Annahme einer zunehmenden Empfindlichkeit gegenüber Befindlichkeitsstörungen bzw. einer Tendenz, derartige Störungen als Kranksein einzustufen. Einheitlich kommt eine derartige Tendenz zum Ausdruck in der zunehmenden Beanspruchung sämtlicher Gesundheitsdienste in denjenigen Industrieländern, in denen diese zur Verfügung stehen (Weltgesundheitsorganisation, 1983).

2.4.2 An die heutigen Umstände von Gesundheit, Kranksein und Krankheit angepaßte Gesundheitsdienste?

Die unter 2.4.1 beschriebene „Gesundheitsparadoxie" (Barsky, 1988), der historische Rückgang von akuten, lebensbedrohenden Zuständen bei gleichzeitiger Zunahme von chronischen und schwer faßbaren Befindlichkeitsstörungen wirft Fragen auf nach der Rolle der gleichzeitig vermehrt beanspruchten und kostspieligen Gesundheitsdienste. Gesundheitsdienste spielen einerseits eine bedeutsame Rolle in dieser Evolution, und sie haben sich andererseits den sich

sten Krankheits- und Todesursachen (mit Ausnahme der Krebssterblichkeit) in den Industriegesellschaften während der vergangenen 20–50 Jahre eine rückgängige Tendenz aufwiesen und die Lebenserwartung nochmals angestiegen ist, hat sich das gesundheitliche Befinden gegenläufig entwickelt. Es ist zwar methodisch schwierig, über einen Zeitraum von mehreren Jahrzehnten vergleichbare Daten zu gewinnen, einige nordamerikanische Studien lassen aber diesbe-

daraus ergebenden Erwartungen und Anforderungen zu stellen. Die Interaktion zwischen der Natur des Angebotes der Gesundheitsdienste und der Art, in der sich das Gesundheitsbefinden der Bevölkerung ausdrückt, im Begriff der „Medikalisierung" zusammengefaßt, ist zu einem zentralen Gesundheits- und Gesundheitsversorgungsproblem geworden (von Ferber, 1988). Sie findet ihren Ausdruck in der Vermarktung der biotechnischen Angebote (inklusive der Erzeugung und Förderung entsprechender Bedürfnisse und Erwartungen) im Sinne eines sich selbst erhaltenden, ja positiv rückkoppelnden Prozesses. Ein gewisses Mißverhältnis zwischen diesem Angebot und den Erwartungen auf qualitativer Ebene kommt außerdem in der zunehmenden Beanspruchung von sog. alternativen Methoden zum Ausdruck, selbst dort, wo orthodoxe Verfahren in genügendem Ausmaß zur Verfügung stehen. Einer Fortsetzung dieser Entwicklung stehen ethische, gesundheits- und finanzpolitische Argumente entgegen. Ansätze einer entsprechenden Wende müssen auf konzeptioneller und struktureller Ebene sowie im Bereich von Bildung und Ausbildung von Angehörigen der Gesundheitsberufe und von „Laien" gesucht werden (Pauli, 1989).

An die konzeptionelle Ebene braucht im Zusammenhang dieses Buches (vor allem Kapitel 1) und dieses Kapitels lediglich nochmals erinnert zu werden. Die Einsicht in die dichte Vernetzung im Gesundheitsbereich zwischen biologischen, psychologischen, soziokulturellen und ökologischen Determinanten und Resultanten muß zu einer Ausweitung vor allem der professionell-fachspezifischen Denkweise auf diesem Gebiet führen. Niemand ist „Laie" bezüglich seiner Gesundheit. Im Gesundheitsberufsbereich wird Spezialisierung weiterhin als Dienstleistung, quasi Zulieferung an die Ebene der Grundver-

sorgung, ihre Bedeutung beibehalten. Diese selbst muß, mit Ausnahme von Spezialsituationen, durch Generalisten mit einer interdisziplinären Ausbildung und Arbeitsweise erbracht werden. Die interprofessionelle Kooperation im Gesundheitsbereich muß in einer sinnvollen Weise Berufsangehörige im Bereich der Pflege, der Physio- und Psychotherapie miteinbeziehen. In der heutigen Situation richtet sich diese Herausforderung vor allem an den ärztlichen Beruf, der im Bereich der Berufssoziologie geradezu zum Prototyp für die Begriffe „Expertise" und „Dominanz" geworden ist. In der medizinischen Wissenschaft müssen die heute noch marginalen sozial- und geisteswissenschaftlichen, die ökologischen, ökonomischen und ethnologischen als ebenso bedeutende Spezialitäten wie die biotechnischen mitintegriert werden (Pauli, 1983). Praktizierende Ärztinnen und Ärzte benötigen nicht weniger, sondern mehr Wissenschaft. Insbesondere muß die unerläßliche soziale Erfahrung, die der immer wieder implorierte „gute alte Hausarzt" im Rahmen einer vergleichsweise stabilen tradierten Gesellschaft handlungsimmanent aus direkter Anschauung erwerben konnte, zu einem beträchtlichen Anteil wissenschaftlich ergänzt werden (Eisenberg, 1988). Die medizinische Aus- und Weiterbildung (vgl. Kap. 68) schließlich hat die Vermittlung entsprechender Inhalte und vor allem die Sozialisation der Ärzte in diesem Sinne zu gewährleisten. Studierende und Praktizierende in der Medizin müssen gezielt für die Teamarbeit ausgebildet werden. Eine Mehrzahl medizinischer Fakultäten ist für diese Aufgabe denkbar ungeeignet geworden. Sie werden die Grundausbildung zunehmend in die dezentralisierten Bereiche der extra-institutionellen, „gemeindeorientierten" Versorgung zu verlegen haben (Association for American Medical Colleges, 1984).

Teil II: Psychosomatische Konzepte und Theorien
Verständniskonzepte und Epidemiologie

3 Psychosomatische Medizin als eine Sozialwissenschaft

Siegfried Zepf

3.1 Vorbemerkung

Individuum und Gesellschaft existieren in der Realität immer in einer engen Verflechtung und sind nur durch wechselseitige Abstraktionen voneinander abzulösen. Dies gilt sowohl für das gesunde wie auch für das kranke Individuum. Die gesellschaftlichen Verhältnisse, in denen die Individuen leben, haben somit sowohl eine protektive, vor Erkrankungen schützende Funktion wie auch einen Stellenwert in der Ätiopathogenese körperlicher Störungen. Daß Krankheiten sozial bedingt sein können, ist heute auch eine Annahme geworden, die in vielfältigen Publikationen vertreten wurde (z.B. Alexander, 1951; Cassel, 1970; Christian, 1952; Cannon, 1939; Dodge und Martin, 1970; Halliday, 1948; Hinkle, 1961; Lipowski, 1972; Mitscherlich, 1966, 1967; v. Uexküll, 1963; Wolff, 1950, 1953). Schaefer und Heinemann (1975) beschreiben die gesellschaftliche Einflußnahme im Hinblick auf die Art der Reaktion des Individuums als „technosomatische", physikochemische Wirkungen, wobei keine psychische Beteiligung vorliegt, als „soziosomatische Wirkungen", bei denen Verhaltensänderungen aufgrund gesellschaftlicher Zwänge und Normen – sog. „Risikofaktoren" – ohne emotionale Beteiligung erfolgen, und als „überwiegend emotionale Wirkungen", die Levy (1971) unter dem Begriff des „psychosozialen Streß" zusammenfaßte. Auf sie richtete sich die Aufmerksamkeit der psychosomatischen Medizin besonders. Sozialempirische Untersuchungen wurden vor allem unter diesem Aspekt interpretiert (z.B. Dunbar, 1938, 1948; Halliday, 1948; Jores, 1961, 1962, 1973; Strotzka, 1965; v. Weizsäcker, 1930).

Obwohl es in der psychosomatischen Medizin sehr verschiedene theoretische Konzepte gibt, die sich nicht zuletzt im anthropologischen Vorverständnis unterscheiden, beschränkt sich die folgende Darstellung auf jene Unternehmungen, in denen versucht wurde, eine soziale Bedingtheit oder Mitverursachung psychisch vermittelter Körperstörungen unter Beibehaltung einer Sichtweise theoretisch zu erfassen, die wenigstens in Ansätzen der Psychoanalyse verpflichtet war. Der Grund dieser Begrenzung liegt nicht nur darin, daß andere Konzepte und Problembereiche in anderen Kapiteln zur Darstellung kommen. In der Auffassung von Soziologen (z.B. Parsons, 1961) bietet die Psychoanalyse die einzige Theorie über die menschliche Persönlichkeit, welche auf dem Niveau soziologischer Theorien über soziale Beziehungen liegt und deshalb auch mit ihnen in Verbindung gebracht werden kann.

Auf eine Kurzformel gebracht entsteht in psychoanalytischer Auffassung Krankheit im Gefolge eines sinnvollen Mißverständnisses. Der Patient versteht seine gegenwärtigen Lebensumstände nicht mehr so, wie sie sind, sondern er mißversteht sie, verzerrt sie subjektiv aus Gründen, die in seiner Vergangenheit liegen und die ihm nicht bewußt sind. Diese Gründe werden als frühkindliche Konflikte verstanden, welche durch gestörte zwischenmenschliche Beziehungen hervorgebracht wurden und die für das Kind in der damaligen Situation nicht mehr lösbar waren. Deshalb blieben sie unbewußt oder wurden sie ins Unbewußte verdrängt. An diese unbewußten Konflikte appelliert nun eine aktuelle Situation und aktualisiert unter Angstentwicklung die früheren, traumatischen Erlebnisse. Mit der Bildung eines Symptoms sucht der Patient diese Angst aufs neue zu bewältigen, wobei die Bedingungen, die letztlich darüber entscheiden, zu welcher Art von Erkrankung es kommt, ebenfalls von der individuellen Lebensgeschichte mitverantwortet werden. Der psychoanalytische Ansatz erlaubt damit, auch das psychosomatische Symptom als eine sinnvolle Verhaltensstrategie zu verstehen, in welche Gesellschaft in doppelter Weise eingeht: einmal in der primären, frühkindlichen Sozialisation und zum anderen in Gestalt der aktuellen sozialen Wirklichkeit der sekundären Sozialisation, in der das Symptom entsteht. Da auch die Familie, in der die primäre Sozialisation stattfindet, nicht außerhalb, sondern innerhalb gesellschaftlicher Verhältnisse steht[1], kann freilich eine psychoanalyti-

---

[1] Claessens (1962, S. 154) beispielsweise schreibt: Die Familie ist „im Prozeß der Sozialisation, dem der Enkulturation und der Vermittlung gesellschaftlich erwünschter, positionsabhängiger Verhaltensweisen ebenso Agentin der Kultur-Gesellschaft, legt also dem Kind nicht nur die ‚Chance' nahe, Kultur und gesellschaftliche Verhaltensweisen zu übernehmen, sondern wirkt praktisch unnachsichtig auf solche Übernahmen hin" und muß „im Hinblick auf die Aufgabe gegenüber dem Nachwuchs, Werte und Normen des Verhaltens zu tradieren, (...) wegen ihrer Prägekraft als optimales Medium angesehen werden."

sche Erklärung psychosomatischer Körperstörungen allein nicht genügen.

Die Begrenzung auf Psychoanalyse zieht eine weitere Einschränkung auf jene körperlichen Störungen nach sich, in deren Ursachenbündel psychische und soziale Prozesse im Vorverständnis der Psychoanalyse – aus welchen Gründen auch immer – eine wesentliche Rolle spielen. Diese Einschränkung ist problematisch. Wie die Erfahrung vermuten läßt, sind auch bei anderen körperlichen Erkrankungen psychische und soziale Prozesse nicht nur im Umgang mit der Körperstörung, sondern auch in der Ätiopathogenese von Bedeutung. Dort ist allerdings der Zusammenhang zwischen körperlichem und seelischem Geschehen noch weniger durchschaut als bei den sog. „psychosomatischen Erkrankungen im engeren Sinne" (z.B. Colitis ulcerosa, Asthma bronchiale, Ulcus duodeni etc.), so daß sich auch bei diesen Erkrankungen die Rolle sozialer Gegebenheiten noch weit ungenauer bestimmen ließ und läßt. Auch das „Situationskreismodell", das für eine Analyse des Gesamtzusammenhanges den theoretischen Bezugsrahmen abgeben könnte, ist heute noch nicht auf dem Stand, der erlauben würde, das jeweils besondere Zusammenspiel und Gewicht von gesellschaftlichen, psychischen und sozialen Prozessen in der Genese bestimmter Körperstörungen genau zu verfolgen. Dazu wäre zuvor in diesem Modell die psychoanalytische wie auch die naturwissenschaftliche und soziologische Begrifflichkeit insgesamt und systematisch zu verbinden. Diese Arbeit ist zwar in Angriff genommen. Solange aber die Verbindungen noch lückenhaft sind, solange ist es ebenso problematisch, psychoanalytische Einsichten in bestimmte Körperstörungen auf „Körperstörungen überhaupt" zu übertragen, wie es schwierig sein dürfte, für jegliche Körperstörung eine psychosomatische und soziale Genese theoretisch zu begründen. Jedenfalls konnten bisher empirisch beobachtbare Zusammenhänge in entsprechenden Arbeiten nur sehr vage theoretisch begriffen werden.

Aber auch wenn das Thema so eingegrenzt ist, eine weitere Beschneidung ist dennoch unvermeidlich. Die verschiedenen Überlegungen können hier nicht in der ihnen gebührenden Breite, sondern allenfalls in Auszügen vorgestellt werden. Zu Wort kommen werden die Untersuchungen und Überlegungen von Laurence E. Hinkle, Talcott Parsons, James Halliday, Alexander Mitscherlich und Alfred Lorenzer, anhand derer sich die sozialwissenschaftliche Dimension der psychoanalytischen Psychosomatik allerdings nicht so sehr in „Lösungen", sondern eher als „Problemgeschichte" entfalten wird.

3.2 Der „subjektive Faktor" in der Sozialempirie

Empirische Unterlagen über die Bedeutung, die in der Krankheitsgenese dem subjektiven Faktor zukommen kann, liegen seit einiger Zeit vor. Exemplarisch sind hier insbesondere die Untersuchungen der Arbeitsgruppe um Hinkle. Hinkle und Mitarbeiter gingen von der Beobachtung aus, daß Krankheitshäufungen über die Bevölkerung nicht gleichmäßig verteilt sind, sondern daß sie sich auf bestimmte Personengruppen konzentrieren (Hinkle und Wolff, 1957, 1958). In mehreren Untersuchungen wurde von ihnen gezeigt, daß die Gruppe der öfter erkrankenden Individuen über viele Jahre hinweg konstant bleibt. Dabei stellte sich heraus, daß die Gruppen der öfter Erkrankenden und der „gesunden" Individuen nicht durch einheitliche Parameter abzugrenzen waren. Auch in der Art der Erkrankung unterschied sich die „most healthy" nicht von der „least healthy" Gruppe, die Erkrankungen traten dort nur öfter auf. Gesunde wie auch relativ häufig kranke Individuen fanden sich in einer Vielzahl ethnischer Gruppen sowie in allen Gesellschaftsschichten. Individuen, die Armut, psychische Härte, Deprivation, Trennung ihrer familiären Bindungen, Statusverluste, Situationen der Ungewißheit und vielerlei zwischenmenschliche Konflikte erlebt hatten, konnten durchaus durchgängig einen relativ guten Gesundheitszustand aufweisen, während sich bei Individuen, die nichts dergleichen erfahren hatten und die augenscheinlich unter den besten sozialen Bedingungen lebten, wiederholt Krankheiten unterschiedlichster Art beobachten ließen.

Zur Erklärung dieser Sachlage wurde ein subjektives Moment eingeführt. Der Einfluß der sozialen und zwischenmenschlichen Umwelt auf den Gesundheitszustand gründete offensichtlich in der Art, wie diese Umwelt vom betroffenen Individuum eingeschätzt wird. Für seine Reaktion auf die soziale Umwelt ist es nicht wichtig, wie andere sie einschätzen. Das Individuum reagiert auf die soziale Umwelt aufgrund seiner eigenen Einschätzung (Hinkle, 1961).[2] Im Urteil von Hinkle und Mitarbeitern erfolgen diese subjektiven Einschätzungen nicht nur im Rahmen bewußter und unbewußter psychischer Prozesse. Sie können auch auf körperlicher Ebene stattfinden, schrieb Hinkle (1961)[3] und gab damit eine Deutung, die sich problemlos in das Situationskreismodell einfügen läßt.

In verschiedenen Arbeiten wurde versucht, eine Antwort auf die Frage zu formulieren, wie die Krankheitshäufung in bestimmten Zeiten bei bestimmten Personen möglicherweise mit ihrer jeweiligen Lebenssituation zusammenhängen könnte. Unter diesem Aspekt wurden u.a. Ungarn untersucht, die aus politischen Gründen aus ihrem Heimatland in die USA geflüchtet waren (z.B. Hinkle und Wolff, 1958; Hinkle et al., 1958). Von der Geburt an wurden die biographischen Daten dieser Personen erhoben (sozialer und kultureller Hintergrund, Entwicklung, Familie etc.). Ferner wurden ihre gewöhnlichen Verhaltensmuster (usual reaction patterns) und ihre persönlichen Charakteristika durch

---

2 „(...) for it is this to which he reacts, and not the social environment as someone else evaluates it" (Hinkle, 1961).

3 „His evaluation does not take place entirely within the central nervous system (...); the ‚evaluation' of horse serum as an antigen takes place in the cells of the blood vessels and the reticulo-endothelial system, and the evaluation of the ‚meaning' of the genetic material within a spermatozoon takes place within the ovum, with exquisite precision" (Hinkle, 1961).

Beobachtungen eingeschätzt und mittels psychologischer Testverfahren bestimmt. Zugleich wurden die medizinischen Daten ihrer Krankheitsgeschichten und ihres jeweiligen Gesundheitszustandes erhoben.

Die vielfältigen biographischen Daten über ein Individuum wurden drei unabhängigen Ratern vorgelegt, die – ohne Kenntnis der medizinischen Daten – auf ihre Grundlage für jedes Jahr auf einer 5-Punkte-Skala einzuschätzen hatten, inwieweit ein Individuum seine jeweilige gesamte Lebenssituation als „highly satisfactory" oder als „highly unsatisfactory" interpretierte. Es stellte sich heraus, daß in den Jahren, in denen nach Ansicht der Rater die Individuen ihre Lebenssituation als unbefriedigend wahrnahmen, eine statistisch signifikante Krankheitshäufung beobachtet werden konnte. Nach Ansicht der Autoren waren für diesen Zusammenhang nicht so sehr die „objektiven" Lebensumstände entscheidend, sondern vielmehr deren subjektive Deutung. Es stellte sich heraus, daß die mehr kranken Individuen in jeder der untersuchten Gruppen ihre Lebenssituation als bedrohlich, konflikterzeugend, überfordernd oder als zu Entbehrungen zwingend (depriving) erlebten. Obwohl die gesünderen Individuen in vielen Fällen in einem sozialen Milieu lebten, das sich von dem der kranken Individuen objektiv nicht unterschied, wurde dieses Milieu von ihnen dennoch viel positiver eingeschätzt (Hinkle, 1961).

Da die verschiedenen Krankheiten praktisch über die ganze Palette menschlicher Erkrankungen verteilt waren und sich keine positive Korrelation zwischen Störungen auf psychischer Ebene (in der Stimmung, im Denken und Verhalten) und bestimmten körperlichen Erkrankungen fand, sahen Hinkle (1961) sowie Hinkle und Wolff (1958) auch keinen legitimen Grund für die Eingrenzung jener körperlichen Erkrankungen in einer bestimmten Krankheitseinheit, welche die psychoanalytische Psychosomatik als sog. „psychosomatische Erkrankungen im engeren Sinne" betrachtete.

Angesichts der inzwischen vielfach diagnostizierten psychischen Unauffälligkeit, der Verhaltensnormalität von Patienten, die an einer der „klassischen" psychosomatischen Erkrankungen leiden (Brede, 1972; Zepf, 1976, 1981) kann heute freilich mit dem Argument einer mangelnden Korrelation von psychischen und bestimmten somatischen Störungen allein die Annahme spezifischer psychosomatischer Erkrankungen nicht zurückgewiesen werden. In den Überlegungen dieser Arbeitsgruppe fehlt auch eine theoretische Klärung der Fragen, aufgrund welcher inneren und äußeren Bedingungen diese Individuen in die Lebensumstände kamen, in denen sie waren, warum diese Lebensumstände so eingeschätzt wurden, wie dies offensichtlich der Fall war, und wie über diese Einschätzung soziale Umstände zu körperlichen Erkrankungen führten. Obwohl die Einschätzung der Lebenssituation als bedrohend, konfliktproduzierend etc. signalisierte, daß die Individuen affektiv beteiligt waren, die körperliche Erkrankung damit psychisch vermittelt war und der subjektive Faktor somit selbst gesellschaftlichen Einflüssen unterliegen mußte, wurde die Möglichkeit nicht bedacht, daß sowohl die subjektiven Einschätzungen wie auch die körperliche Symptomwahl selbst gesellschaftlichen Einflüssen unterliegen konnten. Zwischenzeitlich hat Brede (1971) darauf aufmerksam gemacht, daß sich

in der psychosomatischen Symptombildung möglicherweise die Übernahme des gesellschaftlichen und in Gestalt der Krankenrolle in bestimmter Weise sanktionierten „naturwissenschaftlichen" Krankheitsbegriffs darstellt und die psychische und soziale Unauffälligkeit psychosomatisch Kranker sinnvoll ergänzt.

Die Untersuchungen der Arbeitsgruppe um Hinkle sind zweifelsohne in der Lage, den Objektivismus vieler sozialepidemiologischer Arbeiten zu relativieren, die auf Schichtzusammenhänge zielen. Wie in den meisten sozialempirischen Untersuchungen wurde jedoch auch in ihren Arbeiten Gesellschaft auf einen Einfluß durch empirisch auffindbare soziale Faktoren reduziert. Autoren wie Strotzka (1965, 1969; Strotzka und Grundmiller, 1972) sahen beispielsweise in der entfremdeten Arbeit eine Mitursache somatischer Gesundheitsstörungen, die bei den Betroffenen über die psychische Verarbeitung oder über ein konstitutionelles Entgegenkommen wirksam werden kann. Die Kategorie der „entfremdeten Arbeit" wurde dabei allerdings aus ihrem soziologischen Bezugsrahmen herausgelöst, in dem sie einen analytischen Charakter hat. Sie wurde auf einen bloß empirischen Inhalt bezogen, dessen gesellschaftliche Genese und gesellschaftlich-systematischer Stellenwert dann außerhalb der Frageperspektive bleibt.

3.3 Vermittlungsversuche von Individuum und Gesellschaft

Es liegen bisher nur wenige theoretische Arbeiten vor, in denen versucht wurde, Daten aus den verschiedenen Wissenschaftsbereichen, die in der Genese psychosomatischer Erkrankungen eine Rolle spielen und die mit differenten Methoden erfaßt werden – psychoanalytisch, mit natur- und sozialwissenschaftlichen Verfahren – so zu vermitteln, daß einzelwissenschaftliche Einsichten unverkürzt ihre Geltung behielten. Das Verhältnis von Individuum und Gesellschaft wurde dabei sehr kontrovers ausgelegt.

Freud selbst sah von Anfang an die Gesellschaft in Abhängigkeit vom Individuum. Nicht nur die Sozialpsychologie, sondern auch die Soziologie reduzierte sich für ihn auf „angewandte Psychologie", auf Psychoanalyse. „Auch die Soziologie, die vom Verhalten der Menschen in der Gesellschaft handelt, kann nichts anderes sein, als angewandte Psychologie. Strenggenommen gibt es nur zwei Wissenschaften, Psychologie, reine und angewandte, und Naturkunde" (Freud, 1933). Die meisten Sozialwissenschaftler dagegen sahen im Gefolge von Durkheim (1922) und vor allem unter dem Einfluß der Theorie von Parsons (1964, 1968) die Menschwerdung des Individuums als gesellschaftlich bedingt an. In den sozialpsychologischen Unternehmungen wurden dabei meist nur Teile aus der psychoanalytischen Konzeption der menschlichen Subjektivität herausgesondert (z.B. Parsons, 1968a; Riesman, 1963; Marcuse, 1965), die zudem oft „privat" und unter Verlust wesentlicher Inhalte so ausgelegt wurden, daß sie mit den soziologischen respektive sozialpsychologischen Konzepten kompatibel waren. Eine detaillierte Übersicht über den Umgang mit der Psychoanalyse in der amerikanischen Sozialpsychologie gab Hinkle (1951). In den psychoanalytischen Erklärungsansätzen wiederum wurden meist theoretische Kategorien der Soziologie (z. B. „Familie", „Arbeit" etc.) aus dem systematischen Zusammenhang herausgelöst, in dem sie standen und nur empirisch genommen. Einige Psychoanalytiker, die sog. „Kulturisten" wie Horney (1951) und Sullivan (1947), kamen der Soziologie und der Sozialpsychologie gleichsam einen Schritt entgegen, indem sie diesen Wissenschaften ein psychoanalytisches Konzept anboten, welches – wie z. B. Adorno (1955) und Marcuse (1965) kritisch anmerkten – um die psychoanalytische Trieblehre verkürzt war. Politisch engagierte Psychoanalytiker wie Bernfeld, Fenichel und Reich suchten unter Wahrung der psychoanalytischen Trieblehre eine Vermittlung von Psychoanalyse und historisch-materialistischer Gesellschaftswissenschaft. Sie scheiterten am Anspruch, die in dieser Gesellschaftskonzeption als innerlich-widersprüchlich gefaßte dialektische Beziehung von Individuum und Gesellschaft auf theoretischer Ebene einlösen zu müssen. Psychoanalytische Triebtheorie und die Lehre von der Marxschen politischen Ökonomie mit ihrem Monopolanspruch auf die Erklärung gesellschaftlicher Strukturen ließen sich beim damaligen Stand, der sowohl durch wechselseitige Verständnisbarrieren wie durch sterile Kämpfe zweier Orthodoxien gekennzeichnet war, nur in eine äußerliche Beziehung setzen. Dokumentiert ist diese Kontroverse bei Gente (1972) und bei Dahmer (1982).

Es blieben im wesentlichen wechselseitige Vorwürfe. Der Vorwurf eines „Psychologismus" aus soziologischer Sicht, in dem die fundamentale Wirkung der Gesellschaft auf die Bedürfnisstrukturen des Individuums verleugnet wurde, korrespondierte dem eines „Soziologismus", welcher mit der biologischen Basis, den Trieben, auch das menschliche Individuum als Subjekt abschaffen würde. Zugleich aber führte der Kombinationsversuch der am historischen Materialismus interessierten Psychoanalytiker doch zu einigen bedeutsamen Einsichten, die von Dahmer (1968) summarisch dargestellt wurden. Ihnen ist nicht nur eine Historisierung psychoanalytischer Kategorien (vor allem des Lust- und des Realitätsprinzips) und die heute geläufige Vorstellung, daß die Familie als Sozialisationsinstanz gesellschaftskonforme Charakterstrukturen ausbildet, zu verdanken, sondern auch eine relative Gewichtung psychologischer und sozialökonomischer Erklärungen: „Je mehr das geschichtliche Handeln von Menschen und Menschengruppen durch Erkenntnis motiviert ist, um so weniger braucht der Historiker auf psychologische Erklärungen zurückzugreifen (…). Je weniger das Handeln aber der Einsicht in die Wirklichkeit entspringt, ja dieser Einsicht widerspricht, um so mehr ist es notwendig, die irrationalen, zwangsmäßig die Menschen bestimmenden Mächte psychologisch aufzudecken (…)" (Horkheimer, 1932).

Die psychosomatische, der Psychoanalyse verpflichtete Medizin konnte sich in ihren Versuchen nicht auf Lösungsformeln stützen, in denen das Verhältnis Individuum–Gesellschaft konsensfähig durchschaut war. Sie mußte (und muß) sich eine Antwort auf diese Frage selbst erarbeiten. Dabei ist zumindest fraglich, ob eine Reduzierung der Gesellschaft auf nicht weiter hinterfrag- und erklärbare soziale „Faktoren", aufs „Milieu", als Antwort genügen kann. Soziale Faktoren sind zwar Bedingung menschlicher Entwicklung und Erkrankung, sie werden aber zugleich auch im sozialen Handeln der gesellschaftlich aufeinander bezogenen menschlichen Individuen hergestellt. Soziales Handeln aber unterliegt wesentlich sozialen Gesetzen. Der theoretische Begründungszusammenhang psychosomatischer Erkrankungen kann dann auch nicht auf die sozialen Gesetze verzichten, die dem gesellschaftlichen Handeln der Individuen zugrunde liegen.

In diesen Versuchen sah (und sieht) sich die psychosomatische Medizin darüber hinaus mit dem Problem konfrontiert, daß der körperliche Organismus des menschlichen Individuums in den verschiedenen wissenschaftlichen Disziplinen sehr unterschiedlich konzeptualisiert war und ist. In der Psychoanalyse wird der körperliche Organismus zugleich als energetisches Reservoir der Triebbedürfnisse und als Terrain der Symptombildung beschrieben, in der „klassischen", naturwissenschaftlich orientierten Humanmedizin gilt er als ein Naturobjekt, welches von psychischen und gesellschaftlich-zwischenmenschlichen Verhältnissen weitgehend unbeeinflußt ist und im wesentlichen durch natürlich-objektive, physiologische und pathophysiologische Prozesse charakterisiert ist, während er in der Soziologie die Voraussetzung dafür ist, überhaupt handeln zu können und als taugliches Handlungsinstrument und leibliches Ausdrucksmittel in Erscheinung tritt, welches den Handlungssinn erläutert.

3.3.1 Individualpsychologie aus sozialpsychologischer Sicht (Talcott Parsons)

Im Bestreben, die psychosomatischen Erkrankungen einer sozialwissenschaftlichen Fragestellung zugänglich zu machen, hat sich insbesondere Parsons darum bemüht, psychoanalytische Gesichtspunkte zur Geltung zu bringen. Zu einem menschlichen Wesen wird in seiner Sicht das Individuum dadurch, daß mit der Geburt die Gesellschaft die psychodynamischen Prozesse im Individuum in Gang setzt, in denen sich die psychischen Funktionen herausbilden, welche Voraussetzung für die Fähigkeit zu sozialem Handeln sind. Gegen den Utilitarismus mit seinem auf Hobbes und Locke gegründeten Menschenbild versucht Parsons (1959) in einer Theorie des sozialen Handelns nachzuweisen, daß der Mensch nicht von Natur aus selbstsüchtig, aggressiv und böse, sondern vielmehr ganz auf soziales Handeln hin angelegt sei. Anstelle der Triebe sieht er die maßgeblichen Bestimmungsmomente des Handelns in kollektiven Werten, die von allen geteilt werden. Selbstinteressen und Kollektivinteressen sieht er nicht durch die menschliche Natur, sondern soziokulturell begründet. Dem als unstrukturiert gedachten Individuum wird eine Gesellschaft gegenübergestellt, die aufgrund einer kollektiven Rationalität funktioniert und deren kulturelle Werte und Normen sich in Rollen darstellen.

Parsons konzipiert sein sozioanalytisches Strukturmodell des Handelns nach Art einer Kontrollhierarchie. Er unterscheidet darin in absteigender Reihenfolge das kulturelle, das soziale, das Persönlichkeits- oder psychologische System und den „behavioral organism". Private Motivation ist immer schon gesellschaftlich strukturiert. Diese Strukturierungsprozesse beginnen mit den kindlichen Objektbeziehungen, in denen normativ bestimmte Rollen übernommen werden. Diese primären Strukturierungsprozesse verfolgt Parsons in den psychoanalytischen Begriffen der Identifikation, der Objektbesetzung, der Verinnerlichung und der Introjektion. Durch die Identifikation mit einem besetzten Objekt erfolgt die Internalisierung von dessen Rolle und damit der gesellschaftlichen Normen und Funktionen, die mittels der jeweiligen Rolle unmittelbar oder mittelbar verbunden sind.

Der „behavioral organism" wurde von Parsons u.a. auch deshalb in die „hierarchy of control" eingeführt, um die psychosomatischen Störungen in seiner Handlungstheorie fassen zu können (vgl. Brede, 1979). Der „behavioral organism" wird durch das psychische Persönlichkeitssystem kontrolliert. In dieser Auffassung findet die Annahme ihren Niederschlag, daß die Individuen ihren Körper, seine Reaktionen und reifenden biologischen Funktionskreise, soziokulturell geltenden Interpretationen unterwerfen, wodurch er in soziales Handeln einbezogen wird. Allerdings knüpft Parsons (1968b) nur die Gesundheit an diese Bedingung, nicht jedoch die psychosomatische Erkrankung. Diese wird von Parsons darauf zurückgeführt, daß sich der Organismus hier der psychosozialen Disposition nicht fügt. Er widersetzt sich einer Kontrolle durch die übergeordneten Systeme, weil seine „organischen Interessen" (Parsons, 1959) dort nicht zureichend repräsentiert sind. Deshalb setzt der körperliche Organismus diese Interessen selbsttätig durch, woraus dann ein funktionsstörender Einfluß „sowohl auf die Kontrolle der Mechanismen des äußeren Verhaltens als auch auf viszerale Prozesse" resultiert (Parsons, 1968b).

Brede (1979) macht darauf aufmerksam, daß damit die Grenze soziologischer Argumentationen überschritten wird. Während sich der gesunde Körper im Verhalten des Organismus als Quelle ungerichteter Energie auflöst und so als Bestandteil sozialer Handlungen analysiert werden kann, fällt er als kranker Organismus zumindest partiell aus dem sozialen Zusammenhang heraus. Daß der Organismus biologische Forderungen stellt, kann in der Handlungstheorie von Parsons nicht mehr verfolgt werden. Bei Parsons kommt in der Krankheit ein gesellschaftlich nicht mehr erfaßbares Eigeninteresse des körperlichen Organismus zum Vorschein, wodurch zwar noch der Zugriff der naturwissenschaftlichen Medizin, aber nicht mehr derjenige der Psychoanalyse theoretisch begründet werden kann. Im Verständnis der psychoanalytischen Psychosomatik sind psychosomatische Störungen auf soziokulturell konstellierte, unbewußte psychodynamische Konflikte zurückzuführen. Durch deren Verarbeitung wird das organische Substrat krankhaft verändert. Neben einer pathologischen Veränderung körperlicher Funktionsabläufe liegt das entscheidende Kriterium für diese Erkrankungen im Nachweis eines sinnvollen Zusammenhangs der körperlichen Störung mit einem psychischen Konflikt. Parsons löst die psychosomatischen Erkrankungen aus dieser theoretischen Verankerung. Mit ihm kann nurmehr eine psychische und sozialpsychologische Untersuchung des individuellen oder kollektiven Umgangs mit Erkrankungen begründet werden.

Im Konzept von Parsons läßt sich die psychosomatische Erkrankung unter Wahrung psychoanalytischer Grundannahmen nicht mehr auf ihre Soziogenese hin befragen. Ebensowenig kann dieses Konzept als gelungener Vermittlungsversuch von Psychoanalyse und Gesellschaftstheorie angesehen werden. Die Parsonssche Konzeption ist kontrapunktisch zu der Einschränkung des sozialpsychologischen Gesichtspunkts auf „angewandte Psychologie" durch Freud angelegt. Freud löste die Soziologie in der Psychoanalyse auf, bei Parsons verschwindet die Psychoanalyse in der Soziologie. Für Parsons (1961) ist auch nicht die „gesamte Psychoanalyse (...) von gleichem Belang". Vielmehr ist ein „kompliziertes Verfahren gegenseitiger Adaptation und Selektion notwendig (...), bevor eine wirkliche Anpassung" von Soziologie und Psychoanalyse zustande kommen kann (Parsons, 1961). Bergmann (1967) konnte zeigen, daß diese Selektion auf eine Revision der Freudschen Theorie hinausläuft. In seinem Urteil wird die Psychoanalyse um ihr kritisches Potential ausgehöhlt, nämlich um die Fähigkeit, jene Konfliktdimensionen angeben zu können, die in der gesellschaftlichen Auseinandersetzung mit der inneren, biologischen Natur des Menschen entstehen. Die Differenzen zwischen Individuum und Gesellschaft werden eingeebnet und die bestehende Realität wird als eine harmonistische Verschränkung beider expliziert, bei der die „Einordnung des Individuums in das soziale System perfekt und endgültig" ist (Bergmann, 1967). Eine Störung des Verhältnisses zwischen Individuum und Gesellschaft geht hier zu Lasten einer schlechten Anpassung des einzelnen. Der Mensch erscheint in dieser Konzeption als eine Art „Knetmännchen", das sich in die Konturen der Gesellschaft einfügen soll und einfügt.

Diese Konzeption findet sich auch in den sozialpsychologischen Theorien der sog. Ostblockländer, für die der Ansatz von Hiebsch und Vorwerg paradigmatisch ist. Hiebsch und Vorwerg (1971) gehen von weitgehend unspezialisierten Bedürfnissen aus, welche „in einem Komplex von dynamischen (triebmäßigen) und selbstregulativen (instinktmäßigen) Mechanismen angeboren" sind und durch Lernen entsprechend den gesellschaftlichen Normen komplettiert werden müssen. Lernen wird vorgestellt als eine Verhaltensprägung durch Verstärkung und Hemmung. In Frage gestellt wird allerdings, ob eine „direkte Prägung" durch frühkindliche „Aufzuchtstrategien" überhaupt stattfinden kann (Hiebsch, 1971). Sie sehen Verhaltensprägung eher als Resultat einer permanenten Wirkung der elterlichen Erziehung, „die vermutlich nach denselben Prinzipien wie die Aufzucht betrieben wird". Die Annahme, daß es sich bei den frühkindlichen Aufzuchtstrategien nicht um „langdauernde Prägungen" mit „spätzündender" Wirkung handelt, sondern um die „permanente Wirkung der elterlichen Erziehung", wird fortgeführt mit dem Konzept der „Sozialrolle". Diese wird definiert als ein „in sich zusammenhängender Komplex der einer Gruppe zugeordneten und für sie geltenden Verhaltensnormen und -anforderungen, die von den Mitgliedern mit einer gewissen Verbindlichkeit erlebt werden" (Hiebsch, 1971). Diese sozialen Rollen prägen dann die Person „in dem Maße, in dem diese sich mit ihnen auseinandersetzt und sie aktiv – mehr oder weniger adäquat – erfüllt" (Hiebsch, 1971). Was mit den Verhaltensweisen geschieht, die im Verlaufe der kindlichen Entwicklung durch neue Verhaltensweisen abgelöst und ersetzt werden, wird nicht verfolgt. Da Verhaltensprägung voraussetzt, daß der Mensch „immer schon Verhaltensteile in die Welt (entläßt S. Z.) und (...) nachträglich die Reaktion der Welt in Form einer Verstärkung oder in der des Ausbleibens einer Verstärkung (erfährt S. Z.)" (Cornell, 1969), werden damit die Subjektivierungsprozesse zu einem Prozeß der Beschneidung und Zusammensetzung herausgeschnittener Verhaltensweisen, in welchem sich die amputierten Teile in Nichts auflösen. Das Subjekt wird so zu einem bloßen Knotenpunkt verschiedener sozialer Rollen. Die psychosomatische Erkrankung wie auch eine Divergenz zwischen Individuum und der sich in sozialen Rollen darstellenden Gesellschaft werden hier zum Signal einer mißlungenen Vergesellschaftung, deren Gründe jedoch in dieser konzeptuellen Fassung nicht mehr in der Gesellschaft, sondern nurmehr im Individuum aufgesucht werden können.

3.3.2 Sozialpsychologie aus individualpsychologischer Sicht

Sozialpsychologische Zusammenhänge, die in der Ätiopathogenese psychosomatischer Erkrankungen eine Rolle spielen, wurden auch vereinzelt aus individualpsychologischer Sicht untersucht. Dabei wurden allerdings Individuum und Gesellschaft mehr punktuell als systematisch zueinander in Beziehung gesetzt. Die begrifflichen Differenzen verschiedener theoretischer Bezugssysteme wurden nicht aufgearbeitet.

James Halliday

Halliday führt die psychosomatischen Erkrankungen auf eine „desintegrierte" Gesellschaft zurück. Eine Gesellschaft ist integriert und sozial gesund, wenn sie in der Lage ist, den Stand ihrer sozialen Güter zu behalten oder zu erhöhen. Ihre Mitglieder reflektieren diese soziale Gesundheit dadurch, daß sie emotional integriert, d.h. psychologisch gesund sind. Wenn jedoch die psychologischen Bindungen der Mitglieder schwächer werden, dann verliert die Gesellschaft ihre Kohärenz. Solch eine Gesellschaft kann nach Halliday (1948)[4] als desintegriert, als sozial krank, als kranke Gesellschaft beschrieben werden. Ihre Mitglieder reflektieren diese soziale Krankheit dadurch, daß sie emotional desintegriert, d.h. psychologisch

krank sind. Krank wird beispielsweise eine Gesellschaft, in welcher die Rationalität des industriellen Produktionsprozesses sowohl in der Primär- wie in der Sekundärsozialisation quersteht zu den natürlich angelegten Lebensrhythmen („innate biological working rhythm"), wodurch den einzelnen ontogenetischen Entwicklungsphasen des Individuums der „social circle" fehlt, der ihnen entspricht (Halliday, 1948).

Das Individuum erkrankt, wenn Veränderungen bisher eingespielter sozialer Muster („social patterns") bestimmter Gruppen stattfinden, die in der gesellschaftlichen Güterproduktion umschriebene und auf das Wohlergehen aller gerichtete Teilfunktionen erfüllen, und nicht durch entsprechende Änderungen der sozialen Muster der anderen Gruppierungen wieder so ins gesamtgesellschaftliche Gleichgewicht gebracht werden, wie es die menschliche Natur vorschreibt. Mit großer Intensität bemüht sich Halliday um den Nachweis einer Zunahme psychosomatischer Erkrankungen im Zusammenhang mit dem gesellschaftlichen Krisenzustand nach dem Ersten Weltkrieg. Zu psychosomatischen Erkrankungen disponiert ihm dabei eine Primärsozialisation, die besonders in der prägenitalen Phase dem biologischen Rhythmus einer natürlichen Entwicklung („natural unfolding") nicht entspricht und zu einer zwangsneurotischen Abwehrstruktur führt. Der dadurch erreichbare hohe Grad an Anpassung an die Umwelt geht auf Kosten einer Abtrennung von dem inneren emotionalen Leben. Zerfallen nun in ökonomischen Krisensituationen die gesellschaftlichen Werte, dann bricht diese Abwehrformation zusammen. Der Organismus verliert seine Abwehr gegen den physiologischen Ausdruck besonderer infantiler emotionaler Zustände.[5] Die daraus resultierende physiologische Dysfunktion ist für Halliday der somatische Ausdruck einer bestimmten emotionalen Konstellation, wobei die körperlichen Beschädigungen als Nebenprodukte gelten, die keine direkte emotionale Signifikanz haben (Halliday, 1948).

Im Unterschied zu einem hysterischen Symptom gründet der Sinn einer psychosomatischen Erkrankung nicht im Bezugsrahmen einer individuellen Lebensgeschichte (Halliday, 1948)[6], sondern vielmehr

---

4 „A group which is able to produce and also to reproduce (i.e., maintain or increase its social goods) is attractive and integrated (i.e., is socially healthy); and its members reflect its social health by being emotionally integrated (i.e., psychologically healthy). If, however, the psychological bonds of a community become weakened (...) the group looses its coherence, becomes repellent, suffers dispearsal, and ceases to be able to fulfill its particular social function; that is, it no longer produces ‚social goods' but ‚social evils'. Such a group may be described as desintegrated (i.e., socially unhealthy) or as sick community or sick society, and its members reflect its social ill-health by being emotionally desintegrated (i.e., psychologically unhealthy)" (Halliday, 1948; Sperrungen aufgehoben).

5 Der Organismus wird „‚defenseless' against the physiological expression of the particular infantile emotional states as a counterbalance to which the overwhelmed defense had originally been divised" (Halliday, 1948).

6 „The feature of purpose that appears in hysterical illness is not seen in psychosomatic affections precipitated by the breakdown of pregenitally founded defenses" (Halliday, 1948).

darin, daß sie dem Individuum erlaubt, aus der gefährlichen Arbeitsumwelt zu entfliehen (Halliday, 1943).[7] Die psychosomatische Erkrankung wird so zu einem ersatzweisen Handlungsziel, mit dem sich der Organismus des Individuums einer destruktiven sozialen Desintegration aus biologisch begründeten Selbsterhaltungstendenzen entzieht. In Hallidays Auffassung hat die psychosomatische Erkrankung keinen subjektiven, sondern einen „objektiven" Sinn. Er erwächst aus der allgemeinen Diskrepanz, die zwischen dem faktischen und dem von der menschlichen Biologie geforderten natürlichen Zusammenspiel von Individuum und Gesellschaft besteht.

Obwohl Halliday die zu einer psychosomatischen Erkrankung disponierenden Bedingungen in der Primärsozialisation ansiedelt, wird die krankheitsauslösende Situation nicht mehr in einem subjektiven Bezugsrahmen verfolgt. Damit entzieht er diese Erkrankungen dem psychoanalytischen Zugriff in einer Weise, die der Freudschen Ausgrenzung der Angstneurose analog ist. Zugleich wird von ihm bereits die vorgesellschaftliche menschliche Natur subjektiviert, d.h. mit Absichten ausgestattet, die auf soziales Handeln hin angelegt sind und denen die gesellschaftliche Organisation entsprechen muß. Zur Frage wird dann, wie die menschlichen Individuen überhaupt dazu kommen, die einzelnen gesellschaftlichen Gruppierungen, die sich auch in Hallidays Konzept selbst herstellen, so zu gestalten, daß sie ihrer inhärenten Natur widersprechen. Offen bleibt auch die Frage, wie der Zusammenbruch von Abwehr – einem psychologischen Konzept – physiologische Dysfunktionen freisetzt.[8] Trotz dieser und anderer ungelöster Probleme, die durch die teleologische Fassung der menschlichen

---

7 „(...) a blind purpose, namely that of escaping from a dangerous working environment" (Halliday, 1943).
8 Dies gilt auch für die Konzeption von Flanders Dunbar (1938, 1948). Sie stellt dem Individuum eine Gesellschaft gegenüber, welche einerseits die angeborene Tendenz zur Selbstentwicklung fördert und sie andererseits dadurch einschränkt, daß sie die Übernahme kulturell bereitgestellter, sich in Persönlichkeitsprofilen darstellender Verhaltensmuster erzwingt. Je mehr das Individuum sich selbst in diese Verhaltensmuster integriert, desto mehr wird die Befriedigung individueller und biologischer Bedürfnisse verunmöglicht. Die psychosomatische Erkrankung wird begriffen als Resultat der Versuche des selbstinteressierten Organismus, sich vom Druck, den die Kultur auf das Individuum ausübt, dadurch zu entlasten, daß er seine blockierten Handlungsantriebe über selbst gesuchte somatische Kanäle abführt. Angesichts der antagonistischen Entgegensetzung von Individuum und Gesellschaft kann nun im Konzept von Dunbar Vergesellschaftung nicht mehr zugleich als Subjektivierung, als Vermenschlichung, sondern nurmehr dazu kontradiktorisch als Anpassung aufgefaßt werden (Habermas, 1983). Wie bei Halliday bleibt auch bei Dunbar unklar, warum sich die Individuen eine Kultur schaffen, unter der sie nur unter mehr oder weniger großem Verlust ihrer Spontaneität und der Befriedigung ihrer natürlichen Bedürfnisse leben können. Da die Gesellschaft die Entwicklung der nach eigenen Gesetzmäßigkeiten ablaufenden „inneren", biologischen Triebnatur der Individuen nur provoziert, aber inhaltlich nicht gestaltet, macht es der Tatbestand der psychosomatischen Erkrankungen erforderlich, daß der biologische Organismus selbst als intentional, als „beseelt", d.h. als Körpersubjekt gedacht werden muß. Dabei bleibt unklar, wie die dem biologischen Organismus innewohnenden Absichten physiologische Prozesse im Krankheitsgeschehen in pathophysiologische wenden.

Natur partiell verdeckt sind, kann jedoch nicht übersehen werden, daß Halliday die psychosomatischen Erkrankungen – wie Freud u.a. die Angstneurose – als ein Argument nimmt, das er kritisch gegen die Gesellschaft wendet, die sie hervorbringt, und in dem sich für ihn zugleich eine Erkenntnis verbirgt, welche eine weitere Desintegration der Gesellschaft verhindern kann.

Alexander Mitscherlich

Auch nach Mitscherlich verbirgt sich in der Existenz psychosomatischer Erkrankungen eine Chance für die Gesellschaft, zu ihrer eigenen Pathologie vorzudringen. Psychosomatische Medizin sei eine „Sozialmedizin in einem völlig veränderten Sinn. Sie wird im Einzelfall, wie bruchstückhaft auch immer, die krankheitserregenden Lebensbedingungen der Gesellschaft zu erkennen versuchen. Ein solch neuer sozialmedizinischer Aspekt bedeutet aber, daß die Gesellschaft hier in die Lage versetzt wird, etwas über sich selbst zu erfahren, und zwar gerade das, wofür sie sonst kein Wahrnehmungsorgan besitzt, was sie aus ihrer gegenwärtigen Bewußtseinslage noch nicht überschauen und also noch nicht zu korrigieren vermag" (Mitscherlich, 1966). In der Genese psychosomatischer Erkrankungen spielt eine Störung des „psychosomatischen Simultangeschehens" eine entscheidende Rolle, das Mitscherlich zur Grundvoraussetzung des menschlichen Lebens überhaupt macht. Auch „Triebäußerungen sind ihrer Natur nach psychosomatisch. Die Erregung, die sie bewirken, ist eine nervöse und eine erfahrene, erlebte Unruhe" (Mitscherlich, 1967). Triebe sind eine primäre Verhaltensorganisation mit intentionalem Charakter, der durch „die **wechselseitige Repräsentanz** von Leiblichem in Seelischem und von Seelischem in Leiblichem" (Mitscherlich, 1972) garantiert ist. Werden in der Sozialisation Triebspannungen verdrängt, dann kann das psychosomatische Simultangeschehen bei einer Triebhandlung nicht mehr bewußt verlaufen. Damit wird zwar das natürlich festgelegte Zusammenspiel von psychischen und somatischen Prozessen nicht aufgebrochen. Es wird jedoch unbewußt vollzogen. Die Triebspannungen dauern unbewußt fort, führen bei dem einen zu einer psychoneurotischen Verarbeitung und können bei einem anderen zu einer „autoplastischen" Veränderung von Organleistungen, zur psychosomatischen Störung führen. Möglicherweise werden diese autoplastischen Veränderungen über widersprüchliche Affektkorrelate vermittelt. Bewußte Triebspannungen können sich nach außen wenden und an Objekten zur Entspannung kommen; unbewußte Triebspannungen verlaufen autoplastisch und finden lediglich eine partielle Abfuhr im Symptom. Unbewußt wird dabei der Körper zugleich als ein „Ausdrucksmittel" eingesetzt: „Die Konversion eines seelisch-geistigen Inhaltes (...) in einen körperlichen Vorgang, z.B. eine Geste, das Herzklopfen der Aufregung, das Erröten der Scham, die Beschleunigung der Darmbewegung in der Angst, ihre Lahmlegung in der Depression, diese und unzählige andere Möglichkeiten des Ausdrucks sind in der Mimik vorgegeben (...). Die Pantomime kommt meist auf anderem Wege zustande, und zwar durch ein gestörtes Verhältnis zwischen Bewußtsein und Unbewußtem" (Mitscherlich, 1948).

Der Prozeß, der zu einer psychosomatischen Erkrankung führt und der von Mitscherlich an die Bedingung einer neurotischen Fehlhaltung gebunden wird, ist damit noch nicht abgeschlossen. Die unter „permanentem Verdrängungsaufwand deformierten Organvorgänge können in einen irreversiblen Zustand übergehen. Ist eine solche Zerreißung der höheren Organisationseinheit, also eine Zerreißung des psychosomatischen Simultangeschehens eingetreten, so liegt eine **Defektautonomie** in der Regulation körperlicher Prozesse vor" (Mitscherlich, 1967). Mitscherlich beschreibt die daraus resultierende Chronifizierung psychosomatischer Symptome als Resultat einer **„zweiphasigen Verdrängung oder Abwehr"**, wobei unter der Bedingung, daß die psychischen Mittel der Konfliktbewältigung nicht ausreichen, „in einer zweiten Phase die Verschiebung in die Dynamik **körperlicher** Abwehrvorgänge" erfolgt. Das Individuum regrediert zur Konfliktlösung auf „organische Abläufe, deren Steuerung als biologische Mitgift gesichert ist", „auf die ‚biologische Intelligenz'" dann, „wenn die ‚höhere Intelligenz' der psychischen Instanzen versagt hat". Entscheidend für diesen Vorgang sind für ihn konstitutionell verstärkte oder abgeschwächte Organfunktionen, die unter diesem Gesichtspunkt in eine Schlüsselposition geraten. Das dann ablaufende Geschehen dient „nur noch biologischen Selbsterhaltungsprozessen, die partikulär sind und einzig dem einzelnen Organ oder Organsystem, nicht mehr dem übergeordneten Organismus nützen". (Alle Zitate aus: Mitscherlich, 1967)

Da es die Gesellschaft ist, die letztlich die Verhaltensprägung steuert, über Triebregulation und Triebverzicht das faktische Verhalten gestaltet, wird sie von Mitscherlich für die Genese psychosomatischer Erkrankungen verantwortlich gemacht. Dies geschieht nach Mitscherlich dadurch, daß die Gesellschaft eine unbewußt verlaufende kollektive Einstimmung auf bestimmte Verhaltensstrukturen erzwingt. In seiner Auffassung ist der Mensch vor allem „homo socialis", ein „zoon politikon", das primär auf mitmenschliches Leben angewiesen ist. Auch die Anlagefaktoren, denen Mitscherlich für die Genese psychosomatischer Erkrankungen zweifelsohne eine wesentliche Funktion zuweist, werden seiner Ansicht nach nicht aus sich selbst heraus wirksam. Sie bekommen „erst unter den sozial erworbenen Verhaltensweisen (...) einen spezifischen Stellenwert im psychosomatischen Gesamtgeschehen einer Person" (Mitscherlich, 1967).

Als globale Bedingung für psychosomatische Störungen diagnostiziert Mitscherlich (1966) einen Prozeß, in welchem „der Anspruch der Gesellschaft so terroristisch in das Individuum hinein vorgetragen wird, daß Abweichungen von Geboten und Verhaltensnormen permanente, intensive Angst erwecken und damit die spontane Rückäußerung des Individu-

ums auf die gesellschaftlichen Zustände gelähmt erscheint". In seinem Konzept kann man sich dies so vorstellen, daß in Krisensituationen der Sekundärsozialisation dann auf eine psychoneurotische Symptombildung verzichtet wird, wenn sie als Abweichung von den gesellschaftlichen Verhaltensnormen gilt. Diese Abweichungen können dann vermieden werden, wenn Konflikte unter Einbeziehung des gesellschaftlich etablierten, naturwissenschaftlichen Verständnisses von Krankheiten körperlich ausgetragen werden (Brede, 1972). Die psychosomatische Symptombildung erscheint so als eine Ich-Leistung, in der mit der Erkrankung noch die Übereinstimmung mit gesellschaftlichen Verhaltensnormen gewahrt wird.

Obwohl diese These zum Verständnis der Wahl der Symptomstätte beitragen kann, so läßt sie doch offen, wie der von Mitscherlich angenommene Prozeß einer gesellschaftlichen Beeinflussung körperlicher Funktionsabläufe genau verläuft. Problematisch ist ferner, daß die Sozialisationsbedingungen ungeklärt bleiben, die in seinem Modell für eine „zweiphasige Verdrängung" anzuschuldigen sind. Er sieht die spezifischen Ursachen für die Chronifizierung einer psychosomatischen Organkrankheit „in der Störung des psychobiologischen Reifens durch versagende oder verwöhnende Fixierung an emotionelle Notlagen, Konflikte und deren Bewältigungsversuche in den frühen Entwicklungsstufen des Individuums" (Mitscherlich, 1967). Ob diese Bedingungen spezifischen Charakter haben, ist fraglich; sie liegen gewiß auch anderen psychisch vermittelten Erkrankungen zugrunde. Darüber hinaus sprengt seine Auslegung des Verdrängungsbegriffes die psychoanalytische Metapsychologie entscheidend. Metapsychologisch ist die Verdrängung konzipiert als ein Vorgang innerhalb eines bestimmten Begriffssystems, welches sich auf den psychischen Apparat bezieht. Dessen neurophysiologisches Substrat ist bis heute unbestimmt geblieben. Freud (1900) warnte ausdrücklich vor „der Versuchung (...), die psychische Lokalität etwa anatomisch zu bestimmen". Weil das neurophysiologische Substrat des psychischen Apparates weder geortet noch mit der metapsychologischen Begrifflichkeit der Psychoanalyse vermittelt wurde, deshalb ist das Mitscherliche Konzept einer „zweiphasigen Verdrängung" eher eine metaphorische Beschreibung als eine theoretische Erklärung des psychosomatischen Krankheitsprozesses.

In Frage zu stellen ist auch das Gesellschaftsverständnis von Mitscherlich, das ganz in jener Tradition entfaltet wird, die von Freud (1921) mit seiner Arbeit „Massenpsychologie und Ich-Analyse" begründet wurde. Darin wurde der Versuch unternommen, die Unterwerfung einer Masse unter eine Führerfigur psychoanalytisch zu erklären. Freud begründete diese Erscheinung mit einer sado-masochistischen Bindung der Individuen an einen idealisierten Führer, der in die Funktion eines passager aufgelösten individuellen Über-Ichs eintritt und über den sich dann die einzelnen in ihrem Ideal-Ich miteinander identifizieren. Das auch von Mitscherlich angewandte gesellschaftsanalytische Verfahren einer „angewandten Psychologie" setzt aber die Annahme voraus, daß die Masse als eine Art „Makrosubjekt" gedacht werden kann und daß deren reale Interaktionen denen eines intrapsychischen Geschehens homolog sind. Diese Annahme ist zumindest fragwürdig. Habermas ist der Ansicht, daß sich die Dynamik von Gesellschaftssystemen nicht aus den aggregierten Perspektiven vieler einzelner Lebensgeschichten erklären läßt und fügt hinzu: „Weil Märkte oder Verwaltungen in die Lebenswelt eines Subjektes anders eingreifen als Personen (und Ereignisse, die Personen zugerechnet werden), weil sie in der anonymen Gestalt von Systemimperativen auf Handlungszusammenhänge einwirken, können ihre deformierenden Einflüsse auch nicht wie die Einwirkungen eines charismatischen Führers in Begriffen der Massenpsychologie erfaßt werden" (Habermas, 1983). So erklärt Mitscherlich beispielsweise die Einbeziehung von Kindern in den Arbeitsprozeß aus einer besonderen Bewußtseinslage, die daran hindern würde, sich mit den Ausgebeuteten zu identifizieren. „Die bürgerlichen Unternehmer identifizierten sich untereinander, eine Einfühlung in das Arbeiterkind war durch diese Identifizierung ausgeschlossen" (Mitscherlich, 1966). Deshalb zentrierte er auch die Frageperspektive seiner analytischen Sozialpsychologie vor allem auf den Prozeß der individuellen Bewußtseinsbildung. Daß diese wechselseitigen Identifizierungen unter Ausschluß des Arbeiterkindes zugleich auch immer das Resultat ökonomischer Handlungszwänge sind, kann in diesem analytischen Verfahren nurmehr unterstellt, aber theoretisch nicht mehr verfolgt werden.

3.4 Die körperliche Erkrankung – ein biologischer, psychoanalytischer und soziologischer Gegenstand

Die dargestellten Konzepte machen insbesondere zwei Problemzonen kenntlich.

1. Durchgängig wird die Hypothese vertreten, daß sich in der Bildung psychosomatischer Erkrankungen die Interessen des körperlichen Organismus selbständig durchsetzen, weil sie in der bestehenden gesellschaftlichen Organisationsform nicht angemessen berücksichtigt werden. Letztlich wird die psychosomatische Erkrankung zurückgeführt auf eine biologische Natur, die schon intentional handeln kann, mithin schon subjektiviert ist, noch ehe sie mit der Gesellschaft in Berührung kam. Diese vorgesellschaftliche Subjektivierung der menschlichen Natur durch die psychosomatische Medizin läßt sich begreifen als Reaktion auf die Reduzierung des menschlichen Subjekts auf ein bloßes Naturobjekt durch die vorherrschende Medizin. In ihrem Krankheitskonzept wird die gesellschaftliche Reduzierung menschlicher Subjekte auf entpersönlichte Objekte im gesellschaftlichen Verwertungsprozeß des Kapitals blind reproduziert – Krankheit bricht wie eine kapitalistische Krise mit naturhafter Gewalt über die Individuen herein und ist aus dem subjektiven Bezugsrahmen der Indi-

viduen herausgelöst. In der psychosomatischen Medizin dagegen erscheint durch die Subjektivierung der menschlichen Natur, die sich unter den gegebenen gesellschaftlichen Bedingungen nurmehr in einem Krankheitsprozeß durchsetzen kann, die gesellschaftliche Zerstörung menschlicher Subjektivität noch in biologischem Gewande. Diese Erscheinungsform ist freilich eine Mystifikation, weil sich die Überdehnung des Subjektbegriffs auf vorgesellschaftliche Natur verbietet. Im psychoanalytischen Verständnis jedenfalls wird das Individuum zum Subjekt erst im Prozeß seiner Vergesellschaftung und zwar über die Bildung einer psychischen Repräsentanzwelt, in deren Bezugssystem dann dem Stellenwert, d.h. dem Sinn einer psychosomatischen Erkrankung im psychoanalytischen Verfahren nachgespürt wird. Wenn aber die Subjektivierung des Individuums, seine Vermenschlichung, an seine Vergesellschaftung gebunden ist, dann wird in den vorgetragenen Konzepten der Sinn einer Erkrankung nicht mehr vom Subjekt und auch nicht mehr von der Gesellschaft, sondern nurmehr von einer vorgesellschaftlichen Natur gestiftet. Die psychosomatische Erkrankung ist damit am entscheidenden Punkt dem psychoanalytischen Verständnis entzogen.

Ungelöstes Kernproblem blieb, daß das menschliche Individuum im Prozeß seiner Vergesellschaftung sowohl in psychologischer wie auch in körperlicher Hinsicht ein Subjekt wird – auch die Organisation körperlicher Funktionsabläufe entwickelt sich in Abhängigkeit von der besonderen Lebenspraxis des Individuums –, als welches es psychodynamischen und von der Psychoanalyse angebbaren Prozessen unterliegt; zugleich hat es zum anderen einen Körper als Objekt, der Gesetzen unterworfen ist, die mit den naturwissenschaftlichen Verfahren der Humanmedizin zu erforschen sind. Die Überdehnung des Subjektbegriffs auf ein vorgesellschaftliches Körperobjekt verdeckt zwar dieses Problem, löst es jedoch nicht. Auf theoretischer Ebene stellt es sich vielmehr als Problem einer metatheoretischen Vermittlung von psychoanalytischen und naturwissenschaftlich gewonnenen Einsichten in das menschliche Individuum.

2. Es gelang nicht, das Zusammenspiel von Individuum und Gesellschaft logisch widerspruchsfrei zu begründen, und die gesellschaftliche Einflußnahme und Gestaltung menschlichen Verhaltens wurde reduziert auf ein sich wechselseitig konstellierendes Verhalten von Personen oder Personengruppen. Die gesellschaftlichen Verhältnisse, in denen die Personen oder Personengruppen zueinander stehen, wurden nicht mehr verfolgt. Diese Verhältnisse werden zwar auch von den menschlichen Individuen selbst hergestellt, indem sie bewußt und über Werkzeuge im Akt der Kooperation auf ihre Umwelt einwirken und ihre Lebensmittel selbst produzieren. Aus diesen Verhältnissen aber ergeben sich zugleich Handlungszwänge, die menschliches Verhalten bestimmen. Sie werden dadurch zum genuinen Gegenstand der Soziologie, daß sie als gesellschaftliche Verhältnisse sozialen Gesetzen unterliegen werden. In soziologischer Auffassung verdankt sich diesen gesellschaftlichen Verhältnissen die Vermenschlichung des Individuums. Gesellschaft wurde jedoch im wesentlichen als „Gemeinschaft" (vgl. Dahmer, 1982) begriffen, als eine Art instinktgebundene, animalische Zusammenbindung einzelner Individuen, die nur unter diesem Vorverständnis im Verfahren einer analytischen Sozialpsychologie zureichend aufgeschlüsselt werden konnte.

Versteht man jedoch Gesellschaft nicht als Gemeinschaft, sondern als ein Ensemble von Verhältnissen (ökonomische, kulturelle etc.), das sich nach eigenen Gesetzen ausgestaltet und entwickelt, dann erfordert eine Einbeziehung der sozialpsychologischen Dimension psychosomatischer Erkrankungen nicht nur eine metatheoretische Vermittlung von psychoanalytischer Metapsychologie und menschlicher Naturwissenschaft, sondern ebenso eine Anvermittlung soziologischer Theorien.

Beide Problembereiche weisen dann zurück auf das eingangs von v. Uexküll skizzierte Problem einer systematischen Vermittlung der Theorien der verschiedenen Bereiche. Durch ein bloß punktuelles In-Beziehung-Setzen einzelner Kategorien kann das empirisch beobachtbare Zusammenspiel von somatischen, psychischen und sozialen Prozessen, die hier in eine psychosomatische Erkrankung einmünden, nicht auf den Status begriffener Zusammenhänge angehoben werden. Die kategorialen Inhalte sind abhängig vom jeweiligen theoretischen Bezugssystem definiert, dem sie angehören und in dem sie eine Erklärungsfunktion haben (Althusser, 1972). Kategorien haben einmal einen empirischen Bezug und stehen zum anderen in einem theoretischen. Verwendet man z. B. die Kategorie der „Arbeit" im Zuge einer Bedingungsanalyse subjektiver Bildungsprozesse ohne vorherige metatheoretische Vermittlung der verschiedenen Theoriebereiche, dann stehen für die Verwendung dieser Kategorie prinzipiell drei Möglichkeiten offen. Man kann dieser Kategorie einmal den spezifischen Inhalt belassen, den sie innerhalb einer Gesellschaftstheorie hat. Dann aber läßt sie sich der psychoanalytischen und naturwissenschaftlichen Konzeption von Krankheit bloß additiv anfügen. Die Beziehungen zwischen ihr und den anderen, psychoanalytischen und naturwissenschaftlichen Kategorien bleiben unklar. Man kann sie weiter umgangssprachlich, empiristisch und/oder entsprechend dem jeweiligen Vorverständnis des Untersuchers definieren, wobei sie dann ihren systematischen Stellenwert und ihre Erklärungspotenz verlieren würde. Schließlich kann man sie dem psychoanalytischen oder naturwissenschaftlichen Theoriegebäude über den Menschen einverleiben und sie entweder aus dieser oder jener Sicht definieren, wobei sie allerdings wiederum ihres spezifischen theoretischen Inhalts verlustig ginge.

3.4.1 Ansätze zu einer metatheoretischen Vermittlung der Humanwissenschaften

Die Vermittlungsarbeit sollte dort ansetzen, wo sie in der Praxis beginnt, nämlich in der Primärsozialisation. Unter Verweis auf Lazarus (1971) und Jakob v. Uexküll (1936) hat v. Uexküll in den einleitenden Kapiteln eine Überlegung ausgeführt, welche für eine Untersuchung dieses Bereiches von prinzipieller Bedeutung ist. Bei einer Analyse des Zusammenspiels von Organismen und ihrer Umgebung ist nicht vom Organismus oder von der Umgebung, sondern von der Beziehung zwischen beiden auszugehen. Wenn man also menschliche Erkrankungen auf dem Hintergrund des Verhältnisses von Individuum und Gesellschaft thematisiert, dann kann dabei nicht das Individuum zum Ausgangspunkt genommen werden, um es dann etwa über Triebe oder Bedürfnisse und deren Befriedigung in Beziehung zur Gesellschaft zu setzen. Die Beziehung zwischen Individuum und Gesellschaft ist grundlegend, Triebe oder Bedürfnisse erwachsen aus ihr und sind von ihr abzuleiten. Implizit liegt diese Überlegung auch einigen neueren psychoanalytischen Schriften (z. B. Kernberg, 1981) zugrunde. Durchgängig und konsequent wird dieser Gesichtspunkt jedoch insbesondere in den Arbeiten von Lorenzer vertreten, in denen er die „Theorie der Interaktionsformen" entwickelt und die er zugleich als Bausteine auf dem Weg zu einer psychoanalytischen Metatheorie ansieht, über die sich psychoanalytische wie soziologische Einsichten in das menschliche „Wesen" zueinander in Beziehung setzen lassen.

„Theorie der Interaktionsformen" (Alfred Lorenzer)

In diesem Versuch einer Vermittlung von Psychoanalyse und Gesellschaftswissenschaft werden zunächst die Zuständigkeitsbereiche beider Wissenschaften abgegrenzt. Psychoanalyse wird ausschließlich als eine Analyse der Person, ihrer subjektiven Struktur verstanden, welche die objektiven Bedingungen, denen sich die subjektiven Bildungsprozesse verdanken, nicht erfassen kann. Die objektiven, gesellschaftlichen Bedingungen sind in soziologischer Analyse einzuholen. Die Notwendigkeit einer Vermittlung gründet in der Tatsache, daß der Sozialisationsprozeß selbst kein eigenes analytisches Verfahren hat, so daß er in Vermittlung beider Perspektiven gegenläufig durchdrungen werden muß. Beide Verfahren werden, im Begriff der „Interaktionsform" aufeinander bezogen, zu den Elementen subjektiver Strukturbildung. Sie begreift Lorenzer als Niederschläge realer Interaktionen in den Individuen. Zugleich sind aber Interaktionsformen nichts anderes „als konkrete Darstellungen, als Inszenierungen im Rahmen sozialer Organisationen und Institutionen, (…) als Realisierung von sozialen Verkehrsformen" (Lorenzer, 1974), als gesellschaftliches „Produktionsmittel" menschlicher Subjektivität.

Die Grundlegung der subjektiven Struktur eines Individuums wird durch Interaktionsformen besorgt. Sie sind das Beziehungsresultat von Mutter und Kind, das sich in wiederholten Einigungssituationen zwischen kindlichem Körperbedarf und mütterlichen Interaktionsangeboten einstellt, welche durch ein befriedigendes Wechselverhältnis gekennzeichnet sind. Interaktionsformen sind Lorenzer weder eine „äußere", dem Kind auferlegte Realität, noch sind sie eine innere, „apriorische" Verhaltensformel. Sie sind das Produkt der Auseinandersetzung zwischen kindlicher Natur und mütterlichen Interaktionsangeboten in der konkreten Praxis der Mutter-Kind-Dyade. Da die Mutter-Kind-Dyade zugleich im Feld der mütterlichen Sekundärsozialisation steht, die Praxis des mütterlichen Pols somit immer Teil der gesellschaftlichen Gesamtpraxis ist, deshalb wird die gesellschaftliche Struktur dieser Gesamtpraxis – gebrochen durch den sozialen Ort, den die Mutter darin konkret einnimmt und gebrochen über die sozialen Verkehrsformen, die sich als Interaktionsformen in der bisherigen Sozialisation der Mutter in ihr niedergeschlagen haben – über die mütterlichen Interaktionsangebote an das Kind vermittelt. Die subjektiven Strukturen wie die Persönlichkeitsinstanzen Es, Ich und Über-Ich können so begriffen werden als Produkte eines „praktisch-dialektischen Prozeß, der – kontrapunktisch zur großen Auseinandersetzung des Menschen mit der äußeren Natur – Auseinandersetzung mit der inneren Natur (des Kindes) ist" (Lorenzer, 1974).

Ausgangsbasis dieses Prozesses ist die intrauterine Einheit von Mutter und Kind, die mit der Geburt aufgehoben wird. Die kontinuierliche Bedarfsstillung wird nun ersetzt durch die Praxis eingeübter Interaktionsformen. Der kindliche Organismus wird zunächst in undifferenzierte Spannungszustände versetzt, die sich in unkontrollierten, ganzheitlichen Körperreaktionen äußern. Indem die Mutter auf diese undifferenzierte „organismische Entladung eines noch unprofilierten Körperbedarfs" (Lorenzer, 1973) in relativ konstanter Weise mit einem bestimmten Verhalten reagiert, das Entspannung herbeiführt, qualifiziert sie diesen Körper**bedarf** zu spezifischen (Trieb-)**Bedürfnissen** des Säuglings nach den besonderen Gegebenheiten, die aus dem mütterlichen Verhalten resultieren.

Zentraler Punkt ist, daß die Subjektivität des Neugeborenen nicht in ahistorischen Bedürfnissen, nicht in bereits vorgesellschaftlich präformierten Inhalten festgemacht wird. Das Neugeborene benötigt zwar aus Gründen der Lebenserhaltung z.B. bestimmte Nahrungsmittel, eine bestimmte Flüssigkeitszufuhr, Sauerstoff etc. Diese objektiv feststellbaren körperlichen Bedürfnisse sind jedoch nicht schon subjektiv, für das Neugeborene selbst, bereits als „bestimmte" Bedürfnisse existent. Lorenzer geht davon aus, daß ein körperlicher Zustand, in dem ein objektiver Mangel herrscht, erst inhaltlich als ein „bestimmter" Mangel definiert werden muß, damit er vom Neugeborenen subjektiv nicht bloß als ein diffuser Spannungszustand empfunden, sondern als Mangel auch erlebt werden kann. Um zu einer subjektiven Tatsache zu werden, muß der Mangelzustand in Beziehung gesetzt werden können zu Aktionen oder Gegenständen, die ihn beheben.

In dieser Sicht erfährt auch die kindliche Biologie eine gesellschaftliche Formbestimmung. Lorenzer erläutert diesen Sachverhalt unter dem Begriff des „Interaktionsengramms". Gemeint ist damit das somatische, zentralnervöse Substrat einer Interaktionsform (Lorenzer, 1972). Diese neurophysiologischen Bausteine subjektiver Strukturbildung sind bedingt-reflektorische Zusammenschaltungen von bestimmten intero- und exterozeptiven Reizen und motorischen Impulsen, von bestimmten unbedingt-reflektorisch ablaufenden Körperprozessen, die das somatische Substrat dieser Reize und Impulse abgeben. Im Eingehen auf die mütterlichen Interaktionsangebote werden aus der Fülle der unbedingt-reflektorisch ablaufenden Körperprozesse bestimmte Prozesse dadurch ausgesondert, daß sie in eine bedingt-reflektorische Zusammenschaltung gebracht werden, die engrammatisch fixiert wird. Diese engrammatische Fixierung ist identisch mit der Niederschrift von Funktions- und Situationskreisen in den „Programmen" von v. Uexküll oder mit den „funktionellen Systemen" Anochins (1967), die in der Lebenspraxis entstehen und in welchen die notwendigen Bedingungen eines bestimmten Interagierens gespeichert sind.

Der engrammatische Niederschlag der Interaktionsformen im Individuum erweist sich als das zentrale Strukturmoment, welches die Art und Weise bestimmt, in der im Individuum endokrinologische, biochemische und sonstige Körperprozesse zusammenwirken.

In der „Theorie der Interaktionsformen" wird ihr Zusammenspiel geregelt durch die sozialen Gesetze der gesellschaftlichen Existenz des Individuums. Die mütterliche Biologie, mit der sich der kindliche Organismus intrauterin in Wechselwirkung befindet, kann so als eine bereits sozialisierte, besonders strukturierte Biologie verstanden werden, so daß auch bereits der fetale Organismus eine gesellschaftliche Formbestimmung erfährt.

Für eine Vermittlung von natur- und sozialwissenschaftlichen Erkenntnissen ergibt sich daraus eine bedeutsame Konsequenz. Das Verhältnis von biologischen und sozialen Prozessen stellt sich dar als eine Beziehung zwischen den allgemeinen biologischen Gesetzen, die auch für das Tier Gültigkeit haben, und den gesellschaftlichen Gesetzen, welche für die menschliche Existenz spezifisch sind. Diese spezifischen Gesetze definieren die besonderen Bedingungen, unter denen im Menschen die allgemeinen biologischen Gesetze wirksam werden können, und bestimmen dadurch den Effekt ihrer Wirkung in spezifischer Weise. Anders ausgedrückt: Die biologischen Gesetze werden im Menschen durch seine soziale Existenz „dialektisch" aufgehoben.

Die zunächst vorsprachlichen und unbewußten Interaktionsformen entwickeln sich weiter zu „symbolischen" und bewußten. Unter dem Stichwort „Einführungssituation von Sprache" beschreibt Lorenzer (1974) diese Entwicklung als zweites Moment seines Konzeptes. Sprache und Bewußtsein werden in einem Prozeß der Sprachvermittlung erworben, der sich über das Vorsprechen der Mutter realisiert und in welchem über die von der Mutter prädizierten Interaktionsformen die innere Triebnatur des Kindes in die Sprache eingeholt wird.[9] In die vorbestehende sprachliche Semantik einer Gesellschaft wird die Praxis individueller Lebensgeschichten in Gestalt von Beziehungssituationen eingebracht, über welche sich die subjektiven, „privaten" Bedeutungen sprachlicher Gebilde konstituieren. Als Symbole verweisen Wortbedeutungen zurück auf erfahrene Interaktion, wodurch das Sprachsymbol zum wichtigsten Mittel wird für eine reflexive Verfügbarkeit über die individuelle Lebensgeschichte.

Allerdings nur idealiter, wie Lorenzer (1972) anhand der „Klischeebildung" am Beispiel der Pferdephobie des kleinen Hans darlegt. Im Gefolge eines neurotogenen Konflikts wird hier eine bestimmte Beziehungslage von Hans und seinem Vater, eine bestimmte Interaktionsform, aus dem insgesamt für Hans bewußt und widersprüchlich gewordenen Beziehungsgefüge zum Vater ausgegliedert. Die verpönte Interaktionsform wird „desymbolisiert". Durch den Verlust ihres Sprachsymbols entzieht sie sich von nun an der bewußten Reflexion. Zum Klischee geworden wird die desymbolisierte Interaktionsform nun jenen symbolischen Interaktionsformen hinzuaddiert, die das bisherige Beziehungsgefüge von Hans und Pferd kennzeichnen. Die desymbolisierte Interaktionsform wird nun mit dem Pferd interagiert. In das bewußt verfügbare Symbol „Pferd" wurde ein unbewußter Bedeutungsanteil aufgenommen, der früher als bewußte Bedeutung dem Symbol „Vater" angehört hat. Die verpönte Interaktionsform wurde zwar desymbolisiert, aber gleichsam „hinter dem Rücken" des Subjekts unter falschem Namen wieder in dessen Symbolsystem eingebracht.

Dieses Beispiel macht auch deutlich, wie die psychoanalytischen Begriffe der „Verdrängung" und „Verschiebung" in den Begriffen der „Theorie der Interaktionsformen" aufgehoben sind.

Über symbolische Interaktionsformen wird jene individuelle Wirklichkeit hergestellt und als Teil in eine gemeinsame, gesellschaftliche Wirklichkeit eingebracht, die v. Uexküll als eine dem außenstehenden Beobachter unsichtbare Hülle beschreibt, die das menschliche Individuum umgibt. Das „pragmatische Realitätskriterium" gründet in einer Übereinstimmung von Interaktionsform und realer Interaktion, das „kommunikative Realitätskriterium" setzt voraus, daß die Deutung einer Sachlage, die der einzelne aufgrund seines individuellen Gefüges von symbolischen Interaktionsformen durchführt, zumindest der Möglichkeit nach auch in dem Gefüge symbolischer Interaktionsformen der anderen angelegt ist. Das Klischee entspricht den zu „Stereotypen" erstarrten Programmen, welche die individuelle, private Wirklichkeit aus der gemeinsamen Wirklichkeit ausschließen. Sie kann nurmehr „falsch" verstanden werden.

Die sich in Gestalt einer Psychosomatose, Neurose oder Psychose darstellenden Brechungen des Subjekts sind in der Auffassung Lorenzers (1974) „die Folge objektiver gesellschaftlicher Widersprüche", die Folge widersprüchlicher Spannungen zwischen den gesellschaftlichen Produktivkräften und den Produktionsverhältnissen, der „Gesamtheit der Bedingungen und Formen der Aneignung und der Kontrolle der Produktivkräfte und des Sozialprodukts" (Godelier, 1973).

Wie sich nun die Genese der unter dem Stichwort

„Alexithymie" beschriebenen psychischen Eigentümlichkeiten psychosomatisch Kranker („pensée opératoire", Verhaltensnormalität, narzißtische Bedürftigkeit in Form von Selbst-Objekt-Beziehungen, Objektverlust als symptomauslösende Ursache) in diesem Entwurf darstellt, wird in Kapitel 5 skizziert. Sie verdanken sich wie die psychosomatische Symptombildung einer „restriktiven Praxis" (Zepf, 1981) in der Primärsozialisation, in welcher auch der menschliche Körper nur mangelhaft sozialisiert wird.[10]

Auch im Konzept Lorenzers kann allerdings beim gegenwärtigen Stand der Theoriebildung der Zusammenhang von Individuum und Gesellschaft weder für die psychosomatischen, noch für andere psychische Erkrankungen begrifflich genau und in einer materialen Analyse lückenlos verfolgt werden. Ein solches Unterfangen ist an eine weitere und detaillierte metatheoretische Vermittlung der Begrifflichkeit der Theorie der Interaktionsformen mit den naturwissenschaftlichen und den analytischen Kategorien der Gesellschaftswissenschaften (z. B. „Gebrauchswert", „Tauschwert", „Kapital", „Arbeit" etc.) gebunden. Die Untersuchung gliedert sich dann in mehrere Fragebereiche. Die Erwachsenen, welche in der Primärsozialisation die Subjektivierung des kindlichen Individuums durchführen, sind in gesellschaftlichen Verhältnissen postiert, die eine bestimmte organisatorische Struktur aufweisen. Die gesellschaftlichen Bedingungen, unter denen sich die Sekundärsozialisation der Erwachsenen vollzieht, sind in soziologischer Analyse einzuholen. In einer Analyse der subjektiven Struktur der Erwachsenen ist zu klären, wie sich im Zusammenspiel mit ihrer Primärsozialisation die aktuellen gesellschaftlichen Bedingungen, unter denen sie leben, in die Primärsozialisation der Kinder hinein durchsetzen. Dann kann die Frage verfolgt werden, wie sich unter diesen, durch die subjektiven Strukturen der Erwachsenen gebrochenen Bedingungen das kindliche Individuum als ein Subjekt konstituiert, das in seinem späteren Leben psychosomatisch erkranken kann. Die äußeren Konfliktbedingungen der Auslösesituation, die in einer subjektiven Strukturanalyse des erkrankenden Subjekts ausgemacht werden können, sind dann wiederum in einer soziologischen Analyse auf ihre gesellschaftliche Genese hin zu befragen.

Die psychosomatische Struktur in der Rollentheorie

Als Sozialwissenschaft hat die psychoanalytisch orientierte Psychosomatik ihren Gegenstand nicht nur struktur- und bedingungsanalytisch, sondern

---

9 S. dazu Kap. 16
10 Die Bedingungen einer „psychosomatischen Genese" und die daraus resultierenden psychischen Eigentümlichkeiten können prinzipiell bei jeder körperlichen Erkrankung vorliegen. Die kumulierte klinische Erfahrung und die bisherigen Untersuchungen haben jedoch gezeigt, daß sie insbesondere bei Patienten mit Herzangstsyndrom, extrinsischem Asthma bronchiale, essentieller Hypertonie, Ulcus duodeni, Colitis ulcerosa und Morbus Crohn vorhanden zu sein scheinen (z. B. Liedtke, 1987; Zepf, 1986).

ebenso auch unter der Frage nach der gesellschaftlichen Funktion einer Primärsozialisation zu erkunden, die zu einer zu psychosomatischen Erkrankungen neigenden Persönlichkeitsstruktur führt. Auch wenn eine genaue Antwort auf diese Frage ebenfalls mit dem noch in weiten Stücken ungelösten Vermittlungsproblem konfrontiert ist, so kann sie unter Verwendung rollentheoretischer Kategorien dennoch – wenn auch relativ abstrakt – zumindest angedeutet werden.

Unter einer sozialen Rolle wird ein Komplex normativer Erwartungen verstanden, die von einer Gruppe gebildet werden und die sich im Interesse eines Funktionierens des gesamtgesellschaftlichen Lebenszusammenhanges an das Verhalten eines Trägers von Positionen in Interaktionssituationen richten. Soziale Rollen sind zu verstehen als eine Bündelung bestimmter, gesellschaftlich notwendiger sozialer Verkehrs- und d.h. Interaktionsformen, deren Einhaltung von den Trägern bestimmter Positionen erwartet wird. Normativ wird das Verhalten durch die wechselseitigen Verhaltenserwartungen der Akteure als geregelt angesehen. Die Aktionen des einen orientieren sich an den Verhaltenserwartungen des jeweils anderen, wobei das Einhalten der Rollennormen über Gratifikationen und Sanktionen einer wechselseitigen Kontrolle unterliegt.

Das problematische Verhältnis von Individuum und Gesellschaft läßt sich in der Rollentheorie in der Beziehung des gesellschaftlichen, in der Primärsozialisation hergestellten Subjekts zu den sozialen Rollen, welche ihm in der Sekundärsozialisation gegenübertreten, auf Dimensionen erfassen, die insbesondere von Habermas (1968) entwickelt wurden. Diese Dimensionen sind die der „Repressivität" – dem Ausmaß der Differenz zwischen in der Primärsozialisation hergestellten Bedürfnissen und in der Sekundärsozialisation erlaubten Bedürfnisbefriedigungen, der „Rigidität" – dem Verhältnis von gesellschaftlicher Rollendefinition und gewährtem subjektivem Interpretationsspielraum, und der „Internalisierung" – der Art und dem Ausmaß der Verhaltenskontrolle. Die in der primären Sozialisation erworbenen Grundqualifikationen lassen sich dann daran messen:

- ob das Subjekt der Rollenambivalenz gewachsen ist, ob die Wechselseitigkeit der Erwartungen im offenen Rollenkonflikt verletzt oder ob diese Wechselseitigkeit nur vorgespiegelt und zwanghaft aufrechterhalten wird;
- ob eine kontrollierte Selbstdarstellung durch ein angemessenes Verhältnis von Rollenübernahme und subjektivem Rollenentwurf gelingt oder ob überwiegend Rollendefinitionen übernommen werden;
- ob sich das Subjekt relativ autonom verhalten und gut verinnerlichte Normen reflexiv anwenden kann, oder ob es aufgrund von Konditionierungen auferlegte Normen reaktiv anwendet.

Unter diesen Fragen gewinnen die unter dem Titel „Alexithymie" beschriebenen psychischen Eigentümlichkeiten (s. Kap. 5) psychosomatisch erkrankter Patienten einen besonderen Stellenwert. Der mehr zeichenhaft-öffentliche als private Charakter ihrer Sprache verweist auf Handlungs- und Interpretationsentwürfe, die sich als besondere Erscheinungsform der in der Sprache einer Gesellschaft objektivierten Handlungs- und Deutungszusammenhänge in den Bedeutungssystemen ihrer relevanten Bezugspersonen niedergeschlagen haben. Aus Gründen einer narzißtischen Bedürftigkeit müssen sie vom psychosomatisch erkrankenden Patienten übernommen werden. Bei einem Mangel an symbolischen Interaktionsformen gerät dadurch auch die inhaltliche Bestimmung des eigenen Erlebens und Verhaltens unter die Herrschaft fremder Handlungs- und Interpretationssysteme. Auf der bewußten Ebene gilt für den psychosomatisch Kranken nicht „weil es andere von mir erwarten, muß ich mich so und so verhalten, obwohl ich das nicht will", sondern „ich verhalte mich so und so, weil ich mich so verhalten will". Das eigene Verhalten wird nicht unter Druck eines äußeren Zwanges, sondern quasi von „innen" heraus auf die Verhaltenserwartungen abgestimmt.

Das Problem, welches sich aus einer Diskrepanz von Rollendefinition und Rolleninterpretation ergeben könnte, existiert damit für den psychosomatisch Kranken ebensowenig wie das Problem, welches dann erwachsen könnte, wenn einer gesellschaftlich institutionalisierten Wertorientierung kein in der Primärsozialisation internalisierter Wert entspricht. Im Modell von Habermas gilt für den psychosomatisch Kranken, daß er auf Kosten der Bildung eines körperlichen Symptoms die Wechselseitigkeit der Verhaltenserwartungen unter Vorspiegelung einer faktisch nicht vorhandenen Entsprechung der Befriedigungen zwanghaft aufrechterhält, überwiegend Rollendefinitionen übernimmt und aufgrund von Konditionierungen auferlegte Normen reaktiv anwendet. In diesem Modell erscheint der psychosomatisch Erkrankte als ein bloß verdinglichtes Strukturmoment sozialer Beziehungen. Die Primärsozialisation führt hier zu einer normenkonformen Adaptation an vorgegebene Rollensysteme und verhindert die Entwicklung einer Rollendistanz mit der Möglichkeit, soziale Normen auch reflektieren und damit problematisieren zu können. Das psychosomatisch erkrankende Individuum gewinnt keine „personale Identität" (Goffman, 1970). Es kann ein Gefühl der Identität nicht mehr aus seiner individuellen Biographie ableiten, sondern nurmehr aus der Quersumme gesellschaftlich vorgegebener Rollensysteme destillieren.

Eine ausschließlich rollentheoretische Fassung der „psychosomatischen Struktur" scheint allerdings im Hinblick auf die Einwände problematisch, die von der Position des französischen Existentialismus wie auch insbesondere von der des historischen Materialismus formuliert wurden. Von der einen Position wurde vorgebracht, daß sich der Mensch in der Verkörperung gesellschaftlicher Sozialrollen seiner eigentlichen Natur entäußern und entfremden würde. Von seiten einer geschichtsmaterialistischen Gesellschaftsauffassung lautet die Kritik, daß die Rollentheorie ein Konzept sei, das die Erscheinung für das

Wesen nimmt, der Oberfläche des Selbstverständnisses dieser Gesellschaft verhaftet bleibt und deren wahre Bewegungsgesetze, letztlich den Widerspruch zwischen Produktivkräften und Produktionsverhältnissen, nicht mehr erfassen kann. Gottschalch et al. (1971) vertreten die Ansicht, das unter Verwendung rollentheoretischer Kategorien die gegenwärtige Gesellschaft als eine Leistungsgesellschaft erscheint, in der Rollenspiel als Folge einer Verteilung von Machtpositionen verstanden werden muß, die sich individuellen Fähigkeiten verdanken und die nicht Folge des Besitzes oder Nicht-Besitzes von Produktionsmitteln sind. Die Rollentheorie wäre nicht von einem Zuschnitt, der es gestatten würde, die verschiedenen Zwischenglieder aufzuspüren, über die sich ökonomisch bedingte Interessen in Rollenerwartungen vermitteln, noch könnte mit ihr Aufschluß gewonnen werden, weshalb sich menschliche Subjekte trotz des Gegensatzes, in dem sie sich mit der Gesellschaft befinden, erwartungsgemäß verhalten (Furth, 1971; Haug, 1972). In dieser Sicht wird dann zur Frage, ob das gesellschaftsfunktionale Verhalten psychosomatisch Kranker dem gesellschaftlichen Gesamtinteresse oder eher den Partialinteressen bestimmter gesellschaftlicher Gruppierungen nützt.

Zumindest läßt die Rollentheorie die Frage offen, wodurch letztlich die Sozialrollen inhaltlich definiert werden. Diese Definitionen sind sicherlich von den gesellschaftlich aufeinander bezogenen Individuen zu verantworten. Identifiziert man Gesellschaft nicht mit Gemeinschaft, dann liegt der inhaltlichen Ausgestaltung sozialer Rollen die Struktur des gesellschaftlichen Systems selbst zugrunde, in das die Menschen hineingeboren werden und das – über sie vermittelt – in sozialen Rollen zur Darstellung kommt.

3.5 Schlußbemerkung

Wenn menschliche Gesellschaft, Psychologie und Biologie drei Wissenschaftsbereiche markieren, die sich im Verfahren wie im Gegenstand qualitativ voneinander unterscheiden, dann steht für die psychoanalytische Psychosomatik das Vermittlungsproblem zentral. Da sich die Kategorien verschiedener Wissenschaftssysteme nicht unmittelbar ineinander übersetzen lassen, können bei der Analyse der Zusammenhänge, in denen sich die psychosomatischen Erkrankungen befinden, begriffliche Verkürzungen nur über eine angemessene Vermittlung vermieden werden. Gleichgültig, von welcher ontologischen Position aus man sich die Gesellschaft vorstellt, die psychoanalytische Psychosomatik ist in jedem Fall auf eine allgemeine Theorie des Subjekts angewiesen. Die besonderen Zusammenhänge von somatischen, psychischen und sozialen Prozessen können jedenfalls erst dann theoretisch genau verfolgt werden, wenn im Subjekt der allgemeine Vermittlungszusammenhang seiner verschiedenen Seiten begriffen ist. Dies ist der Weg einzelwissenschaftlicher Erkenntnisbildung. Er kann von der psychoanalytisch orientierten psychosomatischen Medizin nicht abgekürzt werden, wenn sie die Entwicklung ihres Gegenstandes als das begreifen möchte, was er ihrem Vorverständnis nach auch ist: eine gesellschaftlich bedingte (oder zumindest mitbedingte), psychisch vermittelte Störung körperlicher Funktionsabläufe. Auch als eine Sozialwissenschaft, welche die Genese körperlicher Krankheiten begreifen will, ist die psychoanalytische Psychosomatik „im Kern" auf eine interdisziplinäre Zusammenarbeit angewiesen.

4 Epidemiologie in der Psychosomatischen Medizin

Wolfgang Tress, Rolf Manz und *Beate Sollors-Mossler*

Epidemiologie strebt – auch im Dienste der psychosomatischen Medizin – auf der Grundlage von Operationalisierungen zählend und messend nach Aussagen über Häufigkeiten von krankheitsbezogenen Daten und ihre wissenschaftlich-gesetzmäßigen (nomothetischen) Zusammenhänge. Zwangsläufig gerät sie damit in Kontrast (vgl. Tress, 1988) zu einer psychosomatischen Medizin, welche sich zur Aufgabe setzt, intentional-kommunikatives Verhalten in den Verschlüsselungen gestörter physiologischer Funktionsweisen aufzuspüren und in die Sinnhorizonte individueller Biographie einzubringen oder gar, noch weiter greifend, der psychophysischen Ganzheit des Menschen gerecht zu werden. Epidemiologie zergliedert die hier angedeutete hermeneutische Operation nach klinisch-pragmatischen Gesichtspunkten. Sie definiert ein- und ausschließende Merkmale nosologischer oder syndromatologischer Einheiten („Fälle von…"), um ihr statistisch-mathematisches Procedere überhaupt erst in Gang setzen zu können.

Deshalb ist der quantifizierende, kausalanalytische Weg auch in unserem Kapitel der einzig gangbare, um zu Häufigkeiten, Verteilungen, Verläufen und sozialen Bedingungsanalysen psychosomatischer Störungsbilder zu gelangen. Wir bewegen uns dabei im Bereich zwischen leib-seelischer Individualbetrachtung und festgeschriebenen nosologischen Entitäten auf einige jener klinischen Syndrome zu, die der spezielle Teil des Lehrbuches behandelt und die wir meist als funktionelle bzw. psychovegetative (ohne Organbefund) und psychosomatische Erkrankungen im engeren Sinne (mit Organbefund) bezeichnen. Für solche Leidenszustände dürfen wir mit hoher Sicherheit auch triftige psychosoziale, biographisch geformte, also psychogene Bedingungskonstellationen voraussetzen. Psychogene Syndrome begegnen uns in variierender Intensität auf unterschiedlichen Dimensionen klinischer Phänomenologie, auf der körperlichen, der innerseelischen wie auf der des (oft zwischenmenschlichen) Verhaltens. Die psychosomatische Medizin im Unterschied zur Klinik der Psychoneurosen und der psychiatrischen Psychopathologie bemüht sich um körperliche Syndrome (auch) psychosozialer Ätiologie. Hierbei erzeugt der geläufige Sachverhalt, daß neben ausgeprägten körperlichen auch innerseelische und/oder zwischenmenschliche Auffälligkeiten vorliegen, nicht selten unfruchtbare Diskussionen (vgl. Degkwitz, 1981) um die nosologische Extension, um die klinische Weite unserer Begriffe von psychosomatischen gegenüber psychoneurotischen oder charakterneurotischen und erst recht primär organischen Erkrankungen. Gerade diesbezüglich aber vermag die operationalisierende Analyse vorteilhaft sich verschiedenen Konzepten anzupassen, etwa in den unten referierten Arbeiten häufig der ICD (WHO, 8. Revision), deren jeweiligen Nutzen in der Konfrontation mit den Phänomenen sie dann sowohl theoretisch-begrifflich wie auch versorgungspraktisch der kritischen Diskussion anheimstellt.

4.1 Grundlagen

Die Epidemiologie handelt von der qualitativen und quantitativen Verteilung gesundheits- und krankheitsrelevanter Merkmale im zeitlichen Quer- und Längsschnitt. Solche Merkmale gehören sowohl der biologischen, der innerseelischen, aber auch der sozialen Befundebene an. Ebenso beschäftigen den Epidemiologen kausale Bedingungsfelder wie die Abhängigkeiten und Auswirkungen gesundheits- und krankheitsrelevanter Merkmale auf allen drei genannten Ebenen. Nach der Komplexität des Ansatzes und des erhofften Erkenntnisgewinnes gestaffelt, beginnt Epidemiologie **deskriptiv** mit der Beschreibung von Häufigkeiten einschlägiger Syndrome innerhalb definierter Zeiträume (Prävalenz) sowie ihres Neuauftretens zwischen vorgegebenen Zeitpunkten (Inzidenz), beides zumeist aufgeschlüsselt nach demographischen Variablen (Alter, Geschlecht, Sozialstatus, Stadt/Land etc.). Die **analytische Epidemiologie** ermittelt regelhafte Zusammenhänge und Abfolgen verschiedener Einzelvariablen, um derart zumindest wahrscheinliche Teilursachen von Krankheiten und ihren Verläufen zu erfassen. Solche prüft dann die **experimentelle Epidemiologie,** etwa im Rahmen von Präventiv- oder Interventionsstudien, auf ihre Stichhaltigkeit.

Zu diesem Zweck sucht der Epidemiologe zumeist in repräsentativen Stichproben (nur selten steht die zu beforschende Grundgesamtheit in toto zur Verfügung) nach „Fällen von…", was ohne abwertende Konnotation die jeweiligen Träger definierter Merkmale (z. B. Indikatoren eines erblichen Risikos oder einer gesundheitlich riskanten Lebensweise bis zu klar definierten Krankheitsbildern) meint. Während die **Falldefinition** festlegt, wie sich ein gesuchtes Merkmal inhaltlich bestimmt, gibt die **Fallidentifikation** diagnostische Methoden vor, die in der epidemiologischen Forschungspraxis Fälle von Nicht-Fällen verläßlich (reliabel, valide, spezifisch und selektiv) unterscheiden. Zumeist handelt es sich dabei um standardisierte Interviews oder psychometrische Testverfahren.

Epidemiologische Forschungsprojekte begegnen uns entweder als **administrative Studien** oder als **Feldstudien.** Erstere betreiben die Sekundäranalyse von Daten, die sich in den Dokumentationen von Behörden und Institutionen finden (Todesursachen-Statistiken, meldepflichtige Erkrankungen, Statistiken der Krankenkassen und Krankenhäuser), und hängen damit von der Inanspruchnahme der dokumentierenden Versorgungseinrichtung ab. Ihr epidemiologischer Wert ist deshalb im Kontext psychogener, d.h. auch psychosomatischer Erkrankungen recht gering, ganz im Gegensatz zu Aussagen über unmittelbar bedrohliche Zustände wie etwa delirante Bilder oder bestimmte Malignome, die nahezu vollständig den Institutionen zur Kenntnis gelangen. – Hingegen haben Feldstudien, in denen sich der Epidemiologe direkt an die Bevölkerung wendet, eine Chance, zur wahren Häufigkeit von Inzidenz und Prävalenz psychogener Krankheiten vorzudringen, sofern sie die Probleme der Stichprobenauswahl sowie der selektiven Verweigerung bewältigen. – Administrative Erhebungen und Feldstudien kombinieren sich in **Reihenuntersuchungen** an vorselektierten Personengruppen, etwa im Rahmen von gewerberechtlich vorgeschriebenen Vorsorgeuntersuchungen. – Schließlich unterscheiden wir epidemiologische **Querschnittuntersuchungen** (Punktprävalenz) von **Longitudinaluntersuchungen** (Periodenprävalenz und Inzidenz). Longitudinalstudien untersuchen in aufeinanderfolgenden zeitlichen Querschnitten dieselben Individuen mehrfach nach, oder sie ermitteln im Laufe der Zeit an immer wieder anderen repräsentativen Stichproben aus derselben Grundpopulation die Konstanz bzw. Veränderungen in deren Morbidität.

4.2 Grundprobleme der Epidemiologie psychogener Erkrankungen

Als **psychogen** bezeichnen wir solche Störungen und Krankheiten, die vorrangig unter dem Einfluß der psychosozialen Biographie eines Individuums einschließlich der aktuellen Lebensumstände und ihrer inneren Verarbeitung stehen. Der Begriff der psychogenen Erkrankung kam in dem Maße in Gebrauch, in dem die Fachsprache den Neurosebegriff auf die Psychoneurosen (ICD 300.0–300.9)[1] einengte. Neben ihnen zählen zu den psychogenen Erkrankungen die Charakterneurosen, einschließlich der sexuellen Störungen und der Süchte (ICD 301–304), sowie die psychovegetativ-funktionellen und die psychosomatischen (psychogenen körperlichen) Erkrankungen im engeren Sinne mit pathologischem Organbefund (ICD 305.0–305.9; 306.4; 306.5; 306.8) (vgl. Tab. 4–1).

Dieses Spektrum, das viele der in diesem Lehrbuch behandelten Krankheitsbilder nicht einmal enthält

(z.B. chronische Polyarthritis, Diabetes mellitus, Infektionskrankheiten etc.), stellt den Epidemiologen vor spezielle Probleme, die in der Vergangenheit nicht immer mit der wünschenswerten Klarheit erkannt wurden und von daher die Aussagekraft älterer Untersuchungen deutlich einschränken. Folgende, stichwortartig aufgeführte Charakteristika psychogener Erkrankungen bereiten dem Epidemiologen Schwierigkeiten:

– Stufenlose Variation der Ausprägungsschwere erfordert komplexe Vorentscheidungen zur Fallidentifikation.
– Beginn und Dauer der Störung liegen oft nicht eindeutig fest.
– Die beobachteten Verläufe variieren in starkem Maße (kurzzeitige vs. langfristige Episoden, schleichender vs. akuter Beginn).
– Meist bestehen mehrere Symptome auf körperlicher, seelischer und sozialer Ebene nebeneinander, wobei die Akzentuierungen erheblich wechseln, von Person zu Person wie auch im Krankheitsverlauf. Eine wirkliche Monosymptomatik indessen darf als Rarität gelten. Die damit angedeutete Vielfalt der Erscheinungsbilder erschwert die diagnostische Festlegung erheblich.
– Im Gegensatz zu vornehmlich somatischen Erkrankungen zeigen psychogen Kranke extreme Unterschiede in ihrem Inanspruchnahmeverhalten. Sie beschäftigen nicht nur die „Psycho-Disziplinen", sondern nahezu alle Sparten der Medizin, paramedizinische Anbieter, diverse Beratungsdienste, Kirchen, Sozialämter, gelegentlich Gerichte. Häufig nehmen sie überhaupt keine Hilfe in Anspruch.

In der Konsequenz muß der Epidemiologe seine diagnostischen Zielgruppen präzisieren und spezifische Fallfindungsstrategien auf sie anwenden. So taugen beispielsweise psychiatrische Explorationen wenig, wenn auch psychovegetative Störungen zu den angezielten Symptomgruppen gehören. Ferner sind eingebürgerte Fragebögen zur Selbstbeurteilung schon als Screeningverfahren weitgehend ungeeignet, während halbstrukturierte, offen gestaltete Explorationen durch erfahrene Kliniker die Erhebung der wahren Prävalenz solide fundieren. Die Qualität jeglicher Studien über psychogene Erkrankungen steht und fällt mit der klinisch-psychotherapeutischen Kompetenz des Feldforschers. Trainierte Laien etwa, wie in vielen großen Studien eingesetzt, bleiben grundsätzlich als Diagnostiker insuffizient. Schließlich sind klare Vorgaben zu Prävalenzzeiträumen, von der Punkt- bis zur lebenslangen Prävalenz, für die Falldefinition unerläßlich. Zunächst interessiert meist der Querschnitt, die Punktprävalenz, von der aus man retrospektiv Krankheitsbeginn, Dauer und Verlauf wenigstens in grober Annäherung beschreiben kann. Ferner dienen Instrumente zur Einschätzung

---

1 Obwohl die 10. Revision der International Classification of Diseases unmittelbar vor ihrer Einführung steht, orientieren wir uns in diesem Kapitel noch an der ICD-8 (Degkwitz et al., 1975), auf welche die meisten der aktuellen epidemiologischen Arbeiten bezogen sind.

Tabelle 4–1. Psychogene Erkrankungen im Spiegel der International Classification of Diseases (ICD; WHO, 8. Rev.).

Neurosen, Persönlichkeitsstörungen (Psychopathien) und andere nicht-psychotische psychische Störungen

300	Neurosen		304	Medikamentenabhängigkeit (Sucht und Miß- brauch)
.0	Angstneurose		.0	Opium, Opium-Alkaloide und deren Derivate
.1	Hysterische Neurose		.1	Synthetische Analgetika mit morphinähnlicher Wirkung
.2	Phobie		.2	Barbiturate
.3	Zwangsneurose		.3	Andere Schlafmittel und Sedativa oder Tranquili- zer
.4	Depressive Neurose		.4	Cocain
.5	Neurasthenie		.5	Haschisch, Marihuana (Cannabis sativa)
.6	Neurotisches Depersonalisationssyndrom		.6	Andere Stimulantien
.7	Hypochondrische Neurose		.7	Halluzinogene
.8	Andere Neurosen		.8	Andere Medikamente (und kombinierte)
.9	Nicht näher bezeichnete Neurosen		.9	Nicht näher bezeichnete Medikamente

300 Neurosen
.0 Angstneurose
.1 Hysterische Neurose
.2 Phobie
.3 Zwangsneurose
.4 Depressive Neurose
.5 Neurasthenie
.6 Neurotisches Depersonalisationssyndrom
.7 Hypochondrische Neurose
.8 Andere Neurosen
.9 Nicht näher bezeichnete Neurosen

301 Persönlichkeitsstörungen
(Psychopathien, Charakterneurosen)
.0 Paranoide Persönlichkeit
.1 Cyclothyme (thymopathische) Persönlichkeit
.2 Schizoide Persönlichkeit
.3 Erregbare Persönlichkeit
.4 Anankastische Persönlichkeit
.5 Hysterische Persönlichkeit
.6 Asthenische Persönlichkeit
.7 Antisoziale Persönlichkeit
.8 Andere Persönlichkeitsstörungen
.9 Nicht näher bezeichnete Persönlichkeitsstörun- gen

302 Sexuelle Verhaltensabweichungen
(„sexuelle Perversionen")
.01 Homosexualität
.1 Fetischismus
.2 Pädophilie
.3 Transvestitismus
.4 Exhibitionismus
.8 Andere sexuelle Verhaltensabweichungen
.9 Nicht näher bezeichnete sexuelle Verhaltensab- weichungen

303 Alkoholismus mit Ausnahme der Alkoholpsycho- sen
.0 Episodischer Alkoholmißbrauch
.1 Gewohnheitsmäßiger Alkoholmißbrauch
.2 Chronischer Alkoholmißbrauch (Trunksucht)
.9 Andere und nicht näher bezeichnete Formen des Alkoholismus

304 Medikamentenabhängigkeit (Sucht und Miß- brauch)
.0 Opium, Opium-Alkaloide und deren Derivate
.1 Synthetische Analgetika mit morphinähnlicher Wirkung
.2 Barbiturate
.3 Andere Schlafmittel und Sedativa oder Tranquili- zer
.4 Cocain
.5 Haschisch, Marihuana (Cannabis sativa)
.6 Andere Stimulantien
.7 Halluzinogene
.8 Andere Medikamente (und kombinierte)
.9 Nicht näher bezeichnete Medikamente

305 Psychosomatische Störungen (körperliche Stö- rungen wahrscheinlich psychischen Ursprungs)
.0 Haut
.1 Muskulatur und Skelettsystem
.2 Atmungsorgane
.3 Herz- und Kreislaufsystem
.4 Blut- und Lymphsystem
.5 Magen-Darm-Trakt
.6 Urogenitalsystem
.7 Endokrines System
.8 Sinnesorgane
.9 Andere psychosomatische Störungen

306 Besondere Symptome, die nicht anderweitig klas- sifiziert werden können
.0 Stammeln und Stottern
.1 Spezielle Lernstörungen
.2 Tick
.3 Andere psychomotorische Störungen
.4 Schlafstörungen
.5 Eßstörungen
.6 Enuresis
.7 Enkopresis
.8 Kopfschmerzen
.9 Andere Symptome

307 Vorübergehende kurzfristige psychische Auffällig- keiten, die mit situativen Belastungen im Zusam- menhang stehen

des Schweregrades des Krankheitsbildes dem klini- schen Bezug, wobei häufig ein operationalisierter cut- off-point die epidemiologischen „Fälle" von den „Nicht-Fällen" trennt.

4.3 Die Epidemiologie psychosomatischer Störungen im Spiegel der Literatur

Bis in die Mitte der 70er Jahre können wir den Er- kenntnisstand der deskriptiven Epidemiologie psy- chosomatischer Krankheitsbilder, ganz zu schweigen von den komplexeren Varianten, auch bei gutwilliger Betrachtung nur sehr bescheiden nennen. Oberfläch- lich gesehen liegt der Hauptgrund dafür in der Tatsa- che, daß frühere Autoren speziell die funktionellen psychovegetativen Syndrome unter dem Oberbegriff der Neurosen subsumierten. Ihnen das klinisch oder theoretisch als Mangel an Reflexion anzulasten, wäre voreilig. Vielmehr sehen wir darin eher ein Zeichen für die nach wie vor nur willkürlich und unbefriedi- gend gelöste Schwierigkeit, einen individuellen Pa- tienten entweder der Kategorie der Psychoneurosen oder der psychosomatischen Syndrome ohne erhebli- chen Entscheidungszwang zuzuordnen.

Zu erwähnen sind hier die Midtown-Manhattan- Studie I und II (Srole et al., 1962; Langner und Mi- chael, 1963; Srole, 1975; Srole und Fischer, 1986; zur Kritik vgl. Robins, 1986), die Sterling-County-Studie (Leighton et al., 1963) oder die Studie von Hagnell (1969, 1970) an der gesamten Bevölkerung zweier südschwedischer Gemeinden. In Abhängigkeit von der Weite der jeweils angelegten Forschungskriterien wird unter Einschluß der zumeist funktionellen psychosomatischen Syndrome eine Prävalenz neurotischer Störungen zwischen 13% und 80% berichtet. Immerhin kommt Pflanz (1962) in einer Übersicht über 54 Studien aus westlichen Ländern zu dem Schluß, daß etwa ein Drittel der Patienten, die um ärztliche Hilfe nachsuchen, an psychovegetativen Störungen leidet. Neugebauer und Dohrenwend (1980) stellten 24 bedeutende Feldstudien aus Nordamerika und Europa aus der Zeit nach 1949 zusammen, wiederum ohne funktionell-psychosomatische Störungen gesondert zu betrachten. Für Neurosen wird dabei ein Median der wahren Prävalenz von 9,4% in der Gesamtbevölkerung und für Persönlichkeitsstörungen von 4,8% mitgeteilt, freilich mit Maximalschwankungen von unter 1% bis über 60%. – Schepank (1986, S. 10) nennt als Ursache für diese extremen Schwankungsbreiten „1. fehlende Differenzierung nach verschiedenen Kennwerten in der Gesamtauflistung: Inzidenzraten ergeben sehr niedrige Werte, lebenslange Prävalenzraten sehr hohe; 2. unterschiedlich strenge Falldefinitionen in den verschiedenen Forschungsprojekten, sowohl den Schweregrad der Störung als auch die Art der in die Untersuchung einbezogenen Symptome/Störungen/Krankheiten betreffend; 3. Fallidentifikations-Instrumente von unterschiedlicher Sensitivität; 4. unterschiedliche Kompetenz der Untersucher und schließlich 5. unterschiedliche Untersuchungs-Designs (bezüglich Probandengewinnung, Fokussierung der Fragestellung, Altersgruppen).“

Für die Belange einer psychosomatischen Epidemiologie sind exemplarische Erörterungen (Helmchen und Rüger, 1980) zu dem alltäglichen Dilemma relevant, das entsteht, wenn individuelle Patienten unter einer einzigen Rubrik traditioneller Klassifikationen (hier der ICD) zugeordnet werden sollen. Als effiziente Lösung hierfür bietet sich bislang lediglich das multiaxiale Diagnostizieren an, etwa nach den Dimensionen Symptomatik, Ätiologie, Persönlichkeitsstruktur und Charakteristik des zeitlichen Verlaufs. Nach wie vor fehlt es noch an der probeweisen Realisierung einer solchen psychiatrisch bereits bewährten Lösung auch für psychosomatische Erkrankungen, wenngleich das Verfahren der dimensionalen Beeinträchtigungsschwere nach Schepank (vgl. Tress, 1987) dem schon recht weit entgegenkommt.

In einem ungewöhnlich ausgefeilten Tabellenwerk unternahm Schepank (1986) den Versuch, die Prävalenz- und gegebenenfalls Inzidenzdaten sowohl für psychosomatische Erkrankungen im engeren Sinne wie für funktionelle psychosomatische Beschwerden zusammenzustellen. Wir geben hiervon eine vereinfachte Version wieder (vgl. Tab. 4–2).

Die hochkomplexe Originalauflistung dokumentiert den bestenfalls uneinheitlichen, häufig aber undifferenzierten Gebrauch der Begriffe Prävalenz und Inzidenz in der Literatur. So verbleiben alle Versuche, zu allgemeinen Aussagen über die wahre Prävalenz psychosomatischer Störungen in der Gesamtbe-

Tabelle 4–2. Vereinfachte Übersicht von Prävalenzangaben psychosomatischer Störungen in westlichen Industriestaaten (verändert nach Schepank, 1986).

Psychosomatische Erkrankungen i.e.S.	Prävalenz*		Bezugspopulation
Asthma bronchiale**	0,5–1,0%	P?	„Bevölkerung"
Ulcus ventriculi**	1,0%	1JP	alle Männer
Adipositas	35,0%	Pktp	Männer 40 J.
	40,0%	Pktp	Frauen 40 J.
Anorexia nervosa	2,0%	lIP	(Studentinnen 18 J.)
Morbus Crohn	0,009–0,03%	1JP	„Bevölkerung"
Colitis ulcerosa	0,1%	bP	(intern. beh. Pat.)
Arthritis (rheumatoide)	0,3–3,0%	P?	„Bevölkerung"

Funktionelle psychosomatische Beschwerden	Prävalenz*		Bezugspopulation
Funktionelle Syndrome	25,5–40,0%	bP	(Inanspr. med. Polikliniken u. Praxen)
Kopfschmerzen	20,0%	P?	„Bevölkerung"
Obstipation	15,0%	P?	„Bevölkerung"
Herzneurose, Herz-Kreislauf-Beschwerden	6,5%	bP	(Pat. aus nervenärztl. Praxis)
Funktionelle gastrointestinale Beschwerden	7,4–8,8%	bP	(Pat. aus med. Polikliniken)
Essentielle Hypertonie	8–19,0%	P	(20–80j. Erw. in USA)
	ca. 10,0%	P	(alle Einw. d. BRD)

* Spezifizierung der jeweiligen Prävalenzangaben
** Diese Zahlen konnten noch nicht die heute bekannte psychosomatische Heterogenie dieser Krankheitsbilder berücksichtigen.
bP   „behandelte" Prävalenz (= Inanspruchnahme-Klientel)
P?   unbestimmte Häufigkeitsangabe
1JP 1-Jahres-Prävalenz
lIP lebenslange Prävalenz
Pktp Punktprävalenz
P   undifferenzierter Prävalenzbegriff

völkerung zu gelangen, im Stadium der Mutmaßungen.

Wenn die Literatur zur Prävalenz psychosomatischer Syndrome nichts Verbindliches aussagen kann, so um so weniger zu einzelnen Aspekten der Demographie, zur Verteilung psychosomatischer Erkrankungen nach Alter und sozialer Schicht; lediglich geschlechtsspezifische Mitteilungen nehmen sich hiervon aus. So berichten Dohrenwend und Dohrenwend (1969) sowie Neugebauer et al. (1980) von sehr unterschiedlichen Angaben über Altersgipfel neurotischer (und damit implizit psychosomatischer) Syndrome vor oder nach dem 40. Lebensjahr. Hinsichtlich der mit dem Alter ansteigenden Somatisierungsraten bleibt es nach Hönmann (1986) unklar, ob hier die Tendenz zur Somatisierung tatsächlich ansteigt, oder „ob die im Alter gehäuft auftretenden Beeinträchtigungen es dem Probanden ermöglichen, das mehr psychoneurotisch/psychovegetative Beschwerdeangebot aufgeben zu können". Lediglich hinsichtlich der Geschlechtsvariable kommt die allenthalben so divergente Literatur zu der einheitlichen Feststellung, daß psychosomatische Störungen beim weiblichen Geschlecht signifikant häufiger anzutreffen sind. Dies gilt sowohl für Prävalenz- wie auch für Inzidenzraten (vgl. Dohrenwend und Dohrenwend, 1976, sowie Neugebauer et al., 1980). Bezüglich der sozialen Schichten schließlich setzte sich seit Freedman und Hollingshead (1956) die Auffassung durch, psychosomatische Syndrome häuften sich in den unteren sozialen Schichten. Diese Annahme jedoch basiert auf einer Inanspruchnahme-Klientel, deren Repräsentativität angesichts unübersichtlicher Selektionsmechanismen erheblich zu bezweifeln ist. Zusätzlich ist zu bedenken, daß, anders als Alter und Geschlecht, das soziologische Konzept der sozialen Schicht umfangreiche Vorentscheidungen hinsichtlich der Theoriebildung und Operationalisierung verlangt, wobei unterschiedliche Ansätze zu unterschiedlichen Ergebnissen führen müssen.

Die 70er Jahre brachten ausgehend von Europa mit Schwerpunkten in England und der Bundesrepublik eine Neubesinnung auf die Epidemiologie als Grundlagendisziplin für den Gesamtbereich seelischer Gesundheit und Krankheit (vgl. Dilling, 1983). Dem entsprachen erhebliche theoretische Anstrengungen, Grundbegriffe und Standards des Fachs zu klären (Pflanz, 1973; Cooper und Morgan, 1977; Häfner, 1978; Wing, 1981) und verbindlich festzulegen, die dann mit hohem Einsatz etwa in Gestalt des Sonderforschungsbereichs 116 „Psychiatrische Epidemiologie" in der Bundesrepublik ins Werk gesetzt wurden. Von dieser Entwicklung profitierte auch unsere epidemiologische Kenntnis psychosomatischer Beeinträchtigungen. Davon sei nun ohne Anspruch auf Vollständigkeit berichtet.

Dilling und Weyerer stellten 1984 die Ergebnisse einer zwischen 1975 und 1977 durchgeführten repräsentativen Felduntersuchung an 1536 Einwohnern ländlicher bzw. kleinstädtischer Gemeinden in Oberbayern (psychiatrische Assistenzärzte als Interviewer; Fallidentifikation: Goldberg-Cooper-Interview, 1970, Beschwerdelisten nach v. Zerssen, 1976) zusammen. Man diagnostizierte gemäß ICD-8 in einem einstündigen Untersuchungsgespräch aktuelle (Punktprävalenz), dementielle und psychotische Erkrankungen (ICD 290–299), Psychoneurosen (ICD 300), psychosomatische Syndrome (ICD 305, 306.4, 306.5, 306.8), Persönlichkeitsstörungen (ICD 301, 302) sowie Süchte (ICD 303, 304). Die Hauptdiagnose einer neurotischen oder psychosomatischen Erkrankung (wobei die Untersucher hier deutliche Unsicherheiten berichten) wurde am häufigsten gestellt, und zwar für

26,9% der erwachsenen Bevölkerung. Mehrheitlich handelte es sich mit 20,5% dabei um Psychoneurosen, und hiervon zur Hälfte um depressive Syndrome. Psychosomatische Störungen (ICD 305, 306.4, 306.8), angeführt von Beeinträchtigung des Magen-Darm-Traktes und des Herz-Kreislauf-Systems, waren mit 5,6% (Bevölkerungsanteil) unter den Hauptdiagnosen vertreten und Persönlichkeitsstörungen (ICD 301, 302, 303, 304) mit 7,3%. Wie immer wieder in der Literatur berichtet, überwogen die Frauen eindeutig bei den neurotischen und psychosomatischen Diagnosen, die Männer dagegen bei Alkoholproblemen. Dem Kriterium der psychiatrisch-psychotherapeutischen Behandlungsbedürftigkeit, wie in jener Arbeitsgruppe gehandhabt (vgl. Tress, 1985), entsprachen für die Psychoneurosen 9,4% der Gesamtbevölkerung, für die psychosomatischen Störungen 1,8% Bevölkerungsanteil und 2,5% unter der Rubrik der Persönlichkeitsstörungen, vornehmlich des Alkoholismus. – Auch bei dieser überwiegend von psychiatrischen Assistenzärzten, die sich in der Diagnostik psychosomatischer und neurotischer Erkrankungen nicht sonderlich kompetent fühlten, durchgeführten Studie sind die Angaben zu den psychogenen Erkrankungen mit Skepsis zu betrachten. Die Quote der psychosomatischen Syndrome dürfte insbesondere deshalb zu niedrig ausgefallen sein, weil diesbezüglich das diagnostische Hauptinstrument, das Goldberg-Cooper-Interview, im Vergleich zu den übrigen Passagen recht grobmaschig gehalten ist.

1984 legte das National Institute of Mental Health (NIMH) der USA erste Ergebnisse seines Epidemiologic Catchment Area Program (Regier et al., 1984; Robins et al., 1984; Myers et al., 1984) aus bis dahin drei amerikanischen Großstädten vor. Darin hatten hochtrainierte Laienrater mit Hilfe des sog. Diagnostic Interview Schedule (DIS) 9543 Personen befragt und anhand dieses Manuals gegebenenfalls einer Kategorie des Diagnostic and Statistical Manual of Mental Disorders, 3. Aufl. (DSM-III, American Psychiatric Association, 1980, deutsch: Köhler und Saß, 1984) zugeordnet. Für sämtliche psychiatrisch relevanten Diagnosen zusammengenommen erbrachte die ungewöhnlich aufwendige Untersuchung eine 6-Monats-Prävalenz zwischen 16% und 24% und eine lebenslange Prävalenz zwischen 30% und 40% für alle psychiatrischen Erkrankungen insgesamt. Die Rangreihe der Diagnosen wird angeführt von Alkoholabusus und Persönlichkeitsstörungen bei Männern sowie von Phobien und depressiven Episoden bei Frauen. Lediglich im Fall der funktionell-vegetativen und der psychosomatischen Störungen im engeren Sinne blieb die Erhebung hinter aller klinischer Erfahrung unrealistisch zurück: Die Diagnose „somatization" wurde an nur 0,1% der gesamten Stichprobe vergeben!

Dieses Artefakt macht das Epidemiologic Catchment Area Program für alle Belange der psychosomatischen Medizin leider gegenstandslos und wirft erhebliche Zweifel an der diesbezüglichen Brauchbarkeit des DSM-III auf. – Darüber darf der Psychosomatiker deshalb nicht zur Tagesordnung übergehen, weil das DSM-III, ein multiaxiales diagnostisches Instrument, das die Unterscheidung von Neurosen und Psychosen aufgegeben hat und damit dem Geltungsanspruch der biologischen Psychiatrie (Stichwort: „Panikattacke") kräftigen Vorschub leistet, weltweit immer stärkere Anerkennung als das führende Klassifikationssystem für sämtliche seelischen Erkrankungen findet. Verschwinden aber erst einmal die Begriffe wie „Neurose" oder „psychosomatische Erkrankung", gerät auch der Gegenstand bald in Vergessenheit.

Allerdings zeigt die „Somatization Study" in Manchester (Bridges und Goldberg, 1985; Goldberg und Bridges, 1988) an den Patienten von 15 General Practitioners, daß auch DSM-III-orientierte Erhebungen bei entsprechenden Vorgaben die tatsächlichen Verhältnisse adäquat abbilden können. Am Ausgangspunkt stand die These, Somatisierung sei ein basaler Mechanismus der Spezies Mensch, auf psychosozialen Streß im Erleben und in zwischenmenschlichen Kommunikationen zu reagieren, und etwa in Entwicklungsländern der am weitesten verbreitete Ausdruck seelischer Erkrankung. Dazu führte die Manchester-Studie folgende Operationalisierung ein: Somatisierung liegt vor, wenn ein Patient somatische Beschwerden ausschließlich körperlich attribuiert, für ihn aber gleichzeitig eine psychiatrische DSM-III-Diagnose gilt und die zugrundeliegende affektive Störung zusammen mit dem körperlichen Beschwerdebild voraussichtlich positiv auf eine psychopharmakologische (Psychotherapie wurde nicht in Betracht gezogen) Behandlung anspricht. Auf dieser Grundlage konnte etwa ein Drittel der Patienten gemäß DSM-III als psychiatrisch krank diagnostiziert werden, allerdings nur 5% als „rein psychiatrisch", d.h. frei von einer gleichzeitig vorliegenden körperlichen Störung. Die Gruppe der somatisierenden psychiatrisch Kranken wurde von den General Practitioners nur zur Hälfte, die rein psychiatrischen Fälle aber zu 95% erkannt. Hausärzte neigen also dazu, seelische Störungen dann zu übersehen, wenn sie sich auch körperlich manifestieren. Offenbar ziehen viele Allgemeinärzte ein seelisches Krankheitsgeschehen erst in Betracht, wenn der leidende Patient keine körperlichen Symptome anbietet. – Ferner gelang es in der Studie trotz intensiven Bemühens nicht, die somatisierenden von den eher psychologisierenden Patienten anhand ihres psychosozialen Umfeldes oder ihrer Biographien zu differenzieren. – Die Häufigkeitsangaben der Manchester-Studie, einer der jüngsten epidemiologischen Untersuchungen an der Klientel von Allgemeinärzten, konvergieren eindrucksvoll mit Praxisstudien aus Österreich (Strotzka et al., 1969) sowie aus Oberbayern (Dilling et al., 1978) und Mannheim (Zintl-Wiegand et al., 1978). Stets wird von reichlich einem Drittel unter den Patienten der Allgemeinärzte berichtet, welches mit psychogenen, vorwiegend vegetativ-funktionellen Beschwerden zur Behandlung kommt.

Das Konzept der Somatisierung und seine klinische Anwendung greift Lipowski (1988) auf. Diverse von ihm referierte Mitteilungen, meist US-amerikanischer

Herkunft, zur Inanspruchnahme-Prävalenz bestätigen die eben genannten Zahlen. Zumeist stehen depressive oder Angstkrankheiten im Hintergrund. Andererseits betont Lipowski in aller wünschenswerten Klarheit, daß eine Vielzahl somatisierender Patienten körperliche Klagen als Eröffnungszug (opening gambit) der Konsultation ihres Allgemeinarztes einsetzen und damit Somatisierung auch ein soziales Kunstprodukt der gegenseitigen Rollenerwartungen von Ärzten und Patienten aneinander, besonders in Entwicklungsländern und in unteren sozialen Schichten, darstellt.

4.4 Das Mannheimer Kohortenprojekt

4.4.1 Ein Überblick

Seit 1979 erforscht die Gruppe um Schepank (1987; Schepank und Tress, 1987) Häufigkeiten und Bedingungen psychogener Erkrankungen in der Stadtbevölkerung, wobei die ICD-Ziffern 300 bis 307 (8. Rev.) zugrunde liegen (vgl. Tab. 4–1): Es ist unseres Wissens das einzige Projekt, welches, von Psychotherapeuten/Psychosomatikern durchgeführt, sich ausschließlich den psychogenen und folglich mit geschärftem klinischem Problembewußtsein auch den psychosomatischen Syndromen zuwendet.

Das Projekt untersuchte in seiner ersten Querschnitterhebung von 1979 bis 1983 (A-Studie) und in seiner zweiten Erhebung von 1983 bis 1985 (B-Studie) zweimal dieselben 600, zufällig ausgelesenen Mannheimer Bürger, um die Verbreitung und den Verlauf psychogener Erkrankungen für die psychotherapeutisch besonders relevante Altersgruppe der 25- bis 50jährigen aufzuklären. Man entschied sich für die Untersuchung von je 200 Personen der Geburtsjahrgänge 1935, 1945 und 1955 (Kohortendesign), um so den eventuellen Einfluß unterschiedlicher zeitgeschichtlicher und sozialer Entwicklungsbedingungen auf die Entstehung psychogener Krankheitsbilder offenzulegen. Die Interviewer mit fortgeschrittener psychotherapeutisch-psychosomatischer Qualifikation suchten die Probanden zu Hause auf. Im Mittelpunkt des dreistündigen Untersuchungsgesprächs stand eine halbstandardisierte, strukturierte biographische Anamnese. In der Vielzahl gegenwärtiger und anamnestischer Befunde zur Morbidität, zur Entwicklung, zu frühkindlichen Lebensbedingungen, zu derzeitigen Aspekten des Berufs-, Familien- und Freizeitlebens gingen demographische, psychologische, psychiatrische und psychoanalytische Fragenkomplexe ineinander über. Dazwischen waren zahlreiche Fragebögen eingeschoben. Der persönliche Einsatz aller Mitarbeiter hielt die Verweigererquote mit 23% vergleichsweise niedrig. Substantielle Hinweise (vgl. Schepank, 1990) legen nahe, daß selektive Verzerrungen unserer Ergebnisse dadurch sehr gering blieben.

Wie eingangs erläutert, kommt der Falldefinition für die Epidemiologie psychogener Erkrankungen eine kardinale Bedeutung zu. Sie war hier dreifach gefaßt:

– Zeitliches Kriterium: Im Sinne der Punktprävalenz waren die psychogenen Syndrome für die zurückliegenden 7 Tage einzustufen (1-Jahres-, 3-Jahres- und lebenslange Prävalenzen wurden ebenfalls notiert).
– Qualitatives Kriterium: Die psychogenen Syndrome hatten einer der ICD-Ziffern 300 bis 307 (8. Rev.) zu entsprechen.
– Quantitatives Kriterium: Nach Vergleichsstudien an Patienten der psychosomatisch-psychoanalytischen Ambulanz und Bettenstation mußte der für dieses Projekt eigens modifizierte Beeinträchtigungsschwere-Score für psychogene Erkrankungen (BSS) nach Schepank (vgl. Tress, 1987; Manz, 1987) mindestens 5 Punkte bzw. das Goldberg-Cooper-Interview (1970) 20 Punkte ausweisen.

Letzteres Verfahren diente vornehmlich der Vergleichbarkeit mit anderen epidemiologischen Projekten, weist allerdings erhebliche Schwächen hinsichtlich psychosomatischer Syndrome auf (vgl. Manz, 1987). Der Beeinträchtigungsschwere-Score (BSS) indessen gewichtet als konzeptorientiertes, komplexes Rating primär psychogene (nicht organisch bedingte) Symptome nach ihrer **körperlichen,** ihrer **psychischen** und ihrer **sozialkommunikativen** Schwere der Beeinträchtigung, jeweils zwischen den Punktwerten 0 und 4 (Ankerbeispiele vgl. Schepank, 1987, S. 319ff.). Diese drei Einzelratings summieren sich zu einem Gesamtwert von maximal 12 Punkten auf (vgl. Abb. 4–1).

Probanden mit mehr als 5 Punkten gelten als Fälle. Ein solcher cut-off-point ergibt sich aus dem Vergleich mit Patienten der psychosomatischen Ambulanz und der Klinik. Es wäre aber eine Fehleinschätzung zu meinen, Probanden mit einer Belastungsschwere von „nur" 3 oder 4 Summenpunkten seien ohne psychogene Auffälligkeiten. Vielmehr liegen auch hier klinisch prägnante Syndrome vor, die angesichts der Risiken zukünftiger Progredienz unter präventiven Gesichtspunkten erhöhte Beachtung verdienen. Gemäß dieser Definition und bezogen auf die zurückliegenden 7 Tage (Punktprävalenz) waren 26% der untersuchten Mannheimer Bevölkerung als Fälle von psychogener Erkrankung einzustufen. Noch einmal fast ebenso viele Probanden (weitere 24%) wiesen klinisch identifizierbare psychogene Syndrome auf (ICD-Diagnose), ohne deshalb schon als Fälle zu rangieren.

Die 26% Fälle verteilen sich gemäß ihrer Erstdiagnose so:

Psychoneurosen (ICD 300):	7,2%
Persönlichkeitsstörungen (ICD 301):	5,7%
Süchte (ICD 303, 304):	1,5%
Funktionelle und psychosomatische Störungen (ICD 305, 306):	11,6%
	26,0%

Für jeden männlichen fanden wir knapp zwei weibliche Fälle: einer Fallrate von 34% der Frauen standen nur 18% der Männer gegenüber. Während die Män-

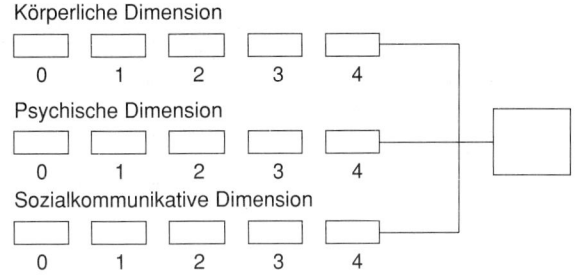

Abb. 4–1. Der Beeinträchtigungsschwere-Score

ner in ihren Hauptdiagnosen bei Persönlichkeitsstörungen (einschließlich des Alkoholmißbrauchs) anteilmäßig überwiegen, besetzen die Frauen vornehmlich den Bereich der psychoneurotischen und der psychosomatischen bzw. funktionellen Syndrome. – Wie in anderen Untersuchungen sind hier unter den Fällen die Angehörigen der Unterschicht überproportional vertreten. – Nicht bestätigt hat sich die Erwartung, deutlich unterschiedliche Fallraten für die drei Jahrgangskohorten und damit Hinweise auf soziohistorisch-politische Einflüsse des frühkindlichen Kollektivschicksals auf die Häufigkeit psychogener Erkrankungen zu finden.

Auf der Symptomebene von Beschwerdefragebögen (nicht zu verwechseln mit der epidemiologischen Falleigenschaft!) berichten 96% aller 600 Probanden irgendwelche psychogenen Beeinträchtigungen während der vergangenen 7 Tage. Das jeweilige Leitsymptom war in 23% eine psychoneurotische, in 22% eine charakterneurotische, in ca. 48% aber eine psychosomatische Manifestation! Dazu gehören in abfallender Folge: Kopfschmerzen, allgemeine innere Unruhe, Oberbauchbeschwerden, Schlafstörungen und Ermüdungserscheinungen. Die Untergruppe der „Fälle" im Sinne unserer Definition gab als psychosomatische Symptome vornehmlich Herzklopfen und Herzschmerzen an. Betrachten wir die Geschlechter für sich, dann tendieren Frauen im allgemeinen zu psychosomatischen Beschwerden wie Alibidinie, Appetitstörungen, Kopfschmerzen und funktionellen gynäkologischen Beschwerden, Männer hingegen zu Übelkeit und Potenzstörungen. Begrenzen wir uns wiederum auf die männlichen und weiblichen Fälle, so berichten uns Frauen von Ermüdung und Erschöpfung (43% der Fälle), Kopfschmerzen (41%), Schlafstörungen (40%), Männer indessen von Oberbauchbeschwerden (39%), Schlafstörungen (38%) sowie Unterbauchbeschwerden (21%). Die Punktprävalenz prägnanter psychosomatischer Krankheitsbilder im engeren Sinne, differenziert nach Geschlechtern, ist in Tabelle 4–3 aufgelistet.

Die angeführten Zahlen gelten für die Ersterhebung (A-Studie), unterscheiden sich aber nicht bedeutsam von jenen des 2. Querschnitts (B-Studie) und korrespondieren, soweit nach epidemiologischen Kriterien vergleichbar, weitgehend mit der Literatur (vgl. Tab. 4–2). – Einschränkend sei jedoch die Tatsache nicht übersehen, daß psychosomatische Erkrankungen im engeren Sinne (vgl. Tab. 4–2) bevölkerungsstatistisch

Tabelle 4–3. Durchschnittliche Häufigkeit (Punktprävalenz) psychosomatischer Krankheitsbilder im engeren Sinne in der erwachsenen Allgemeinbevölkerung auf der Basis subjektiver Angaben. Gemittelte Zahlen aus 2 Untersuchungen im Abstand von 3 Jahren (280 Männer, 248 Frauen, 25.–48. Lebensjahr).

	Männer	Frauen
Anorexia nervosa	./.	0,6%
Asthma bronchiale	0,3%	1,0%
Magenulkus	2,3%	1,8%
Morbus Crohn	0,3%	./.
Colitis ulcerosa	0,3%	0,6%
rheumatische Arthritis	1,2%	1,2%
Hyperthyreose	./.	0,6%
psychogen mitbedingte		
Dermatosen	1,5%	1,2%
Adipositas	1,5%	3,2%
Migräne	2,1%	6,0%

zu den seltenen Ereignissen gehören. Deshalb kann ihre wahre Prävalenz zuverlässig nur anhand einer deutlich größeren Stichprobe als der unseren geschätzt werden. Erst bei einem Stichprobenumfang von etwa 2000 Männern und ebenso vielen Frauen können Prävalenzwerte den wahren epidemiologischen Tatbestand verläßlich wiedergeben.

4.4.2 Zur vergleichenden Epidemiologie psychovegetativ-funktioneller Störungen

Der Beeinträchtigungsschwere-Score (BSS) für psychogene Erkrankungen enthält u.a. die Dimension der subjektiven und objektiven psychogenen körperlichen Beeinträchtigung, einzustufen zwischen 0 und 4 Punkten (vgl. Tress, 1987). Entsprechend liegen für jeden Probanden des Mannheimer Kohortenprojektes mittlerweile zwei Beurteilungen im 3-Jahres-Abstand vor. Sie ergeben zusammen mit den Subskalen der psychogenen Beeinträchtigung des seelischen Erlebens sowie des sozialkommunikativen Verhaltens den Gesamtscore der Beeinträchtigungsschwere. Das erlaubt uns, unabhängig von nosologischen Klassifikationen (hier der ICD-8) die Verbreitung und den 3-Jahres-Verlauf psychovegetativ-funktioneller Beeinträchtigungen auch in der Verquickung mit psycho- und charakterneurotischen Störungen auf den beiden anderen klinischen Beeinträchtigungsskalen des BSS zu studieren. Zugleich erfahren wir auch, welche Beeinträchtigungsprofile diagnostisch wie etikettiert wurden. Einzelheiten dieser detaillierten Untersuchungen sind bei Reister et al. (1990) nachzulesen. Wir referieren hier einige der Hauptergebnisse.

Entsprechend ihrem klinischen Profil, wie es der BSS abbildet, wurden drei Probandengruppen definiert:

Gruppe I: **Psychovegetative Prägnanzsyndrome** (incl. psychosomatische Syndrome im engeren Sinne).
Diese Gruppe leidet vornehmlich, ausgewiesen durch den körperlichen BSS-Subscore und die klinischen Hauptsymptome, unter psychovegetativen Beeinträchtigungen, während psychosoziale Beschwerden auf den beiden anderen Subscores auch quantitativ eine nachgeordnete Rolle spielen.

Gruppe II: **Psychovegetativ-psychosoziale Mischsyndrome.**
Auch hier sind nach den unter Gruppe I genannten Kriterien psychogene körperliche Symptome eines zumindest merklichen Beeinträchtigungsgrades eindeutig vorhanden (körperlicher Subscore) und tauchen auch unter den Hauptsymptomen des Patienten auf, ebenso stark sind aber entsprechend den übrigen beiden Subskalen psychosoziale Störungen zugelassen. Diese Gruppe repräsentiert solche Patienten, die auf allen Symptomebenen psychogener Pathologie merklich tangiert sind. Die nosologischen Kategorien, unter denen wir im ärztlichen Alltag diese Patienten wiederfinden, spiegeln deshalb auch die fachliche Grundorientierung des ärztlichen Diagnostikers (internistisch, neurologisch, gynäkologisch, orthopädisch, psychiatrisch etc.) wider.

Gruppe III: **Psychosoziale Prägnanzsyndrome.**
Bei ausgeprägten charakterneurotischen und seelischen Beeinträchtigungen tauchen unter den ersten drei Hauptsymptomen dieser Patienten keine körperlichen auf, so daß, selbst wenn deutlich vorhanden, sie im klinischen Gesamtbild zurücktreten und allenfalls das Gewicht einer erheblichen körperlichen Mitreaktion bei klinisch vorwiegend psychosozialen Krankheitsbildern haben.

Gruppe IV: **Klinisch Gesunde.**
Diese Menschen werden von ihren bunt ausgeprägten psychogenen Symptomen kaum beeinträchtigt. Sie gehören damit zu jenen rund 50% der Bevölkerung, die wir klinisch-pragmatisch als frei von psychogenen Syndromen einstufen dürfen.

Die Kenndaten der Gruppen I bis IV (vgl. Tab. 4–4), gemittelt aus zwei Erhebungsquerschnitten im Abstand von drei Jahren, grenzen die wahren Werte für die Bevölkerung einer mittleren westdeutschen Großstadt mit weitgehender Sicherheit ein.

Der von allen Untersuchern gefundene **Überhang der Frauen** gegenüber den Männern bei psychogenen Störungen rührt besonders von ihrem prozentual höheren Anteil an der psychovegetativ-psychosozial beeinträchtigten Mischgruppe (II) und in geringerem Maße von der psychovegetativ-psychosomatischen Prägnanzgruppe her. Überhaupt bilden psychovegetativ-psychosomatisch eindeutig belastete Probanden (Gr. I, II) die Mehrheit aller psychogenen Auffälligkeiten, psychoneurotisch-charakterneurotische Prägnanztypen (Gr. III) machen hingegen nur 10% der Bevölkerung aus. Mit Ausnahme der psychosozial auffälligen Frauen (Gr. III) und der männlichen psychovegetativ-psychosozialen Mischgruppe (II) tendieren die belasteten Gruppen I bis III eher zu den unteren **sozialen Schichten,** ein durchaus bekannter, wenn auch nach unseren Zahlen keineswegs linear durchgängiger Befund. – Die Daten zum **Zivilstand** drücken bei den Männern vielleicht ein Bedürfnis

Tabelle 4–4.   Klinische Typen psychogener Erkrankungen im Vergleich zu Gesunden: geschlechtsdifferenzierte repräsentative Kenndaten, gemittelt aus 2 Erhebungen im 3-Jahres-Abstand.

	Gruppe I psychovegetativ- psychosomatische Prägnanzsyndrome		Gruppe II psychovegetativ- psychosoziale Mischsyndrome		Gruppe III psychosoziale Prägnanzsyndrome		Gruppe IV klinisch Gesunde	
	Männer	Frauen	Männer	Frauen	Männer	Frauen	Männer	Frauen
Kennlinien des Beeinträchtigungs-schwere-Scores	kö / psy / VH		kö / psy / VH		kö / psy / VH		kö / psy / VH	
Anteil an der Allgemeinbevölkerung (σ, φ)	15–20%	20–25%	10–15%	ca. 20%	ca. 10%	ca. 10%	ca. 60%	45–50%
soziale Schicht								
unt. Unterschicht	0– 5%	5– 7%	5–10%	ca. 10%	5–10%	0– 5%	0– 5%	ca. 5%
obere Unterschicht	ca. 40%	40–45%	45–50%	35–40%	ca. 30%	20–25%	25–30%	ca. 30%
unt. Mittelschicht	ca. 35%	35–40%	20–25%	35–40%	25–30%	40–45%	30–35%	ca. 45%
obere Mittelschicht u. Oberschicht	ca. 20%	ca. 5%	ca. 25%	5–10%	ca. 15%	ca. 25%	25–30%	15–20%
Zivilstand								
ledig	ca. 15%	ca. 15%	25–30%	ca. 15%	ca. 50%	15–20%	ca. 25%	15–20%
verheiratet	ca. 80%	70–75%	ca. 65%	ca. 70%	35–40%	ca. 70%	65–70%	70–75%
geschieden (getrennt)	ca. 0%	0– 5%	0– 5%	ca. 5%	ca. 0–5%	ca. 0%	ca. 0%	ca. 5%
verwitwet	ca. 0–5%	ca. 10%	ca. 5%	10–15%	5–10%	ca. 15%	ca. 10%	ca. 5%
Geburtsjahr								
1935	ca. 30%	ca. 35%	ca. 40%	ca. 35%	ca. 20%	20–25%	ca. 35%	30–35%
1945	35–40%	ca. 35%	ca. 40%	40–45%	35–40%	ca. 45%	ca. 30%	30–35%
1955	30–35%	ca. 30%	ca. 20%	ca. 20%	40–45%	ca. 35%	35–40%	ca. 35%
psychogene Beeinträchtigungsschwere im Durchschnitt								
BSS-Gesamtscore (0–12)	4,0	4,4	5,6	6,1	5,1	5,4	2,5	2,9
körperlich (0–4)	2,3	2,4	2,2	2,3	1,2	1,5	0,8	0,9
psychisch (0–4)	0,9	1,0	1,6	1,9	1,9	2,1	0,8	1,1
Verhalten (0–4)	0,8	1,0	1,8	1,9	2,0	1,8	0,9	0,9
Hauptdiagnosen								
keine	ca. 25%	15–20%	ca. 5%	0– 5%	ca. 25%	ca. 20%	ca. 65%	ca. 60%
nur psychosomatische (ICD 305, 306)	50–55%	55–60%	35–40%	30–35%	5–10%	ca. 5%	ca. 15%	ca. 15%
gemischte	10–15%	ca. 10%	ca. 30%	25–30%	ca. 10%	ca. 5%	ca. 5%	ca. 5%
nur psychosoziale (ICD 300–304)	10–15%	10–15%	ca. 30%	35–40%	55–60%	65–70%	ca. 15%	ca. 20%
5 häufigste Leitsymptome (Mehrfachnennungen bis zu max. 3/Proband möglich)	inn. Unruhe 30–35%	Kopf-schm. 40–45%	Nikotin/ Alkohol 30–35%	Kopf-schm. 30–35%	Depress. ca. 50%	Depress. 75–80%	Nikotin/ Alkohol ca. 35%	inn. Unr. ca. 25%
	Ober-bauch ca. 25%	Beweg.-App. 20–25%	inn. Unr. 25–30%	Depress. 25–30%	Kontakt-stör. 35–40%	Partner-probleme ca. 50%	inn. Unr. ca. 30%	Kopf-schm. 20–25%
	Kopf-schm. 15–20%	Ober-bauch ca. 20%	Herz-beschw. 20–25%	inn. Unr. ca. 25%	Zwänge 30–35%	Ängste 35–40%	Ober-bauch 15–20%	Depress. ca. 15%
	Herz-beschw. 15–20%	Müdigk. ca. 20%	Ober-bauch 20–25%	Partner-Probleme ca. 20%	Ängste ca. 30%	Zwänge 30–35%	Beweg.-App. 15–20%	Müdigk. ca. 15%
	Schlaf-stör. 15–20%	inn. Unr. 15–20%	Ängste ca. 20%	Nikotin/ Alkohol 15–20%	Partner-Probleme ca. 30%	Kon-zentr. ca. 20%	Kopf-schm. ca. 15%	Beweg.-App. ca. 15%

	Gruppe I psychovegetativ-psychosomatische Prägnanzsyndrome		Gruppe II psychovegetativ-psychosoziale Mischsyndrome		Gruppe III psychosoziale Prägnanzsyndrome		Gruppe IV klinisch Gesunde	
	Männer	Frauen	Männer	Frauen	Männer	Frauen	Männer	Frauen
Alkoholkonsum/d	35–40g	ca. 15g	ca. 50g	15–20g	ca. 40g	ca. 35g	30–35g	15–20g
Anzahl der Medikamente/d	0,9	1,4	0,8	1,5	0,7	0,8	0,5	0,6
Anz. d. Arztbesuche während d. vergangenen 12 Mon.								
keine	10–15%	ca. 5%	ca. 15%	ca. 10%	25–30%	ca. 15%	20–25%	5–10%
bis 2mal	35–40%	ca. 25%	30–35%	25–30%	35–40%	15–20%	35–40%	35–40%
3–4mal	ca. 15%	20–25%	ca. 20%	15–20%	ca. 10%	ca. 15%	ca. 15%	ca. 20%
und öfter	35–40%	45–50%	30–35%	ca. 45%	20–25%	ca. 50%	ca. 25%	35–40%
Dauer der **Krankschreibung** während d. vergang. 12 Mon.								
keine	ca. 45%	ca. 40%	ca. 55%	35–40%	ca. 55%	ca. 50%	ca. 60%	ca. 55%
bis zu 2 Wo.	30–35%	ca. 30%	ca. 25%	25–30%	30–35%	ca. 25%	20–25%	ca. 25%
2–12 Wo.	ca. 20%	ca. 20%	10–15%	ca. 25%	ca. 10%	15–20%	10–15%	ca. 15%
länger	0– 5%	5–10%	5–10%	5–10%	ca. 5%	5–10%	0– 5%	0– 5%
psychiatrisch-psychotherap. Inanspruchnahme (lebenslänglich)								
psych. Konsult.	5–10%	ca. 10%	ca. 10%	ca. 20%	5–10%	ca. 35%	ca. 5%	ca. 15%
amb. Ther. bis 3 Mon.	0– 5%	ca. 55%	0– 5%	ca. 10%	ca. 5%	10–15%	0–5%	ca. 5%
stat. Therapie	–	–	–	–	–	ca. 5%	–	ca. 5%
	–	–	ca. 5%	0– 5%	–	5–10%	–	0– 5%

nach verläßlichen Objektbezügen in der vornehmlich psychovegetativ belasteten Gruppe I aus. – Auch die **Altersverteilung** bleibt ohne geschlechtsspezifische Schwankungen recht homogen, abgesehen von relativ wenigen psychosozialen Prägnanzsyndromen im Jahrgang 1935 und ebenso wenigen Mischsyndromen im Jahrgang 1955. Ferner imponiert ein leichtes Überwiegen derselben Gruppen II und III für den Geburtsjahrgang 1945. – Die durchschnittliche **Schwere** der psychogenen Beeinträchtigung im Vergleich der vier Gruppen auf den drei Unterskalen des BSS liegt in der Konsequenz unserer Operationalisierung. Die psychovegetative Prägnanzgruppe I hebt sich von den klinisch Gesunden ausschließlich in der körperlichen Subskala ab und zeigt somit unter den pathologischen Gruppen die leichtesten Beeinträchtigungen. Indessen fallen die Belastungen der Gruppen II und III mit deutlichem Anstieg auf den beiden Subskalen des psychischen Erlebens und des Verhaltens stärker aus. Wenngleich sie das klinische Bild im Einzelfall nicht unbedingt beherrschen, sind psychovegetative Beschwerden im Spektrum der psychogenen Syndrome mithin ubiquitär. Ferner manifestieren sich schwere psychogene Krankheitsbilder unabhängig von unserer klassifikatorischen Festlegung gleichermaßen im körperlichen Befinden, im seelischen Erleben, in der zwischenmenschlichen Interaktion wie

auf der Verhaltensebene generell. – Hinsichtlich der **Hauptdiagnosen** erhielten immerhin 10–15% der psychovegetativen Prägnanzgruppe I keine einschlägige Hauptdiagnose. Andererseits stoßen wir bemerkenswert oft auch in allen anderen Gruppen auf psychosomatische Hauptdiagnosen. Folglich unterscheiden klinische Diagnosen als solche nicht verläßlich zwischen überwiegend psychoneurotisch und überwiegend psychovegetativ-psychosomatisch alterierten Patienten. Sie geben vielmehr aufgrund weiter klinischer Überschneidungen der fachlich-theoretischen Perspektivität des Diagnostikers beträchtlichen Spielraum, beispielsweise für die Zuordnung eines Patienten mit psychogenem Mischsyndrom (Gr. II) zur Kategorie psychovegetativer Leiden (ICD 305, 306) oder jener der Persönlichkeitsstörungen (ICD 301). – Die häufigsten **Leitsymptome** der Gruppen I bis IV bedürfen nur insofern des Kommentars, als sie klinisch unsere Gruppeneinteilung validieren. Bemerkenswert ist allerdings, daß die Beschwerden der gesunden Vergleichsgruppe sich mehr im Körperlichen und weniger im Seelischen häufen. – Zwischen Alkoholkonsum (Selbstmedikation) und der Anzahl der eingenommenen Medikamente zeichnet sich mit aller Vorsicht ein gegenläufiger Zyklus ab, insofern als bei vorwiegend körperlich akzentuierten Syndromen das medizinische Therapeutikum, bei seelischen

Syndromen die Droge prävaliert. Zudem mag dem gesteigerten Tablettengebrauch der Frauen der erhöhte Alkoholkonsum von Männern mit psychogenen körperlichen Syndromen entsprechen. – Der Gang zum Arzt erscheint besonders für Männer mit ausgeprägten leib-seelischen Beschwerden eine Option, während psychosozial belastete Männer die Vorstellung beim Arzt im Vergleich zu Gesunden sogar eher vermeiden. Psychosozial auffällige Frauen nehmen die Ärzteschaft hingegen durchaus in Anspruch. Ähnlich verhält es sich mit der Dauer der Krankschreibung innerhalb eines Jahres. – Schließlich die lebenslange (!) Inanspruchnahme des psychiatrisch-psychotherapeutischen Behandlungsnetzes: Hier halten sich Männer durchgängig und unabhängig von ihren wie auch immer klinisch geprägten psychogenen Syndromen hartnäckig zurück. Frauen sind aufgeschlossener: 20–35% wenden sich besonders unter dem Eindruck psychosozialer Belastung (Gr. II und III) wenigstens einmal im Leben dem psychiatrisch-psychotherapeutischen Fachangebot zu. Vermutlich wegen des nach wie vor mangelhaft ausgebauten psychotherapeutischen Versorgungsnetzes kommen dabei vorwiegend psychiatrische Konsultationen zustande, die nur sehr selten in ambulante oder stationäre Psychotherapien einmünden. Das unterstreicht die hohe Verantwortung der allgemeinen Ärzteschaft für die psychotherapeutisch-psychosomatische Versorgung der Bevölkerung und zugleich die Bedeutung der 1987 in die neugefaßten Psychotherapierichtlinien aufgenommenen „psychosomatischen Grundversorgung".

4.4.3 Die zeitliche Variabilität aller psychogenen Prägnanzsyndrome

Unsere wiederholte Untersuchung von 528 Probanden im Abstand von 3 Jahren erlaubt schließlich den Blick auf einen umgrenzten Verlaufsausschnitt psychogener und damit auch „psychosomatischer" Syndrome im Vergleich zu Gesunden für beide Geschlechter (Abb. 4–2 und 4–3).

Vor allem ist von einem Querschnitt in der Zeit zum nächsten die relative Stabilität der einzelnen Subgruppen in der Gesamtbevölkerung zu betonen (linke vs. rechte Säulen), stärker bei den Männern (Abb. 4–2) als bei den Frauen (Abb. 4–3). Dennoch herrschen unter den vier Subgruppen während der drei Jahre trotz der zugrundegelegten 1-Jahres-Prävalenz starke Wechselbewegungen. Am stabilsten scheint bei Männern und Frauen die Gruppe der Gesunden, während für die pathologischen Syndromprofile (Gr. I–III) die zeitliche Variation vor der Konstanz bei weitem (etwa im Verhältnis 3:1) überwiegt, ohne daß besondere Richtungstendenzen deutlich würden.[2] Die Rede von den psychosomatischen, den psychoneurotischen und den persönlichkeitsgestörten Patienten wird damit genauso fragwürdig wie die mancherorts darauf aufgebauten Abgrenzungen ärztlicher Berufsbilder bis in klinisch-institutionelle Strukturen hinein. Statt dessen empfehlen uns die

erörterten Daten, vom **übergeordneten Begriff der psychogenen Erkrankungen** auszugehen, also von seelisch bedingten (neurotischen im weiteren Sinne) Störungsbildern, die uns einmal mehr im leib-seelischen, ein andermal mehr im seelisch-erlebnismäßigen und sehr häufig gleichzeitig auch im zwischenmenschlichen Verhalten begegnen, wobei der Einzelfall im Laufe der Zeit mit einiger Wahrscheinlichkeit seine klinische Akzentuierung mehrfach ändert. Das legt die Vorstellung einer klinischen Übergangsreihe für psychogene Erkrankungen mit zeitstabilen Prägnanzgruppen sowohl am leib-seelischen als auch am psychoneurotisch-persönlichkeitsstrukturellen Endpol nahe, während psychovegetativ beeinträchtigte Patienten den Übergangsbereich in wechselnder Lokalisierung dicht bevölkern. Ihre sich über die Zeit wandelnden Verlaufsgestalten respektieren keine Markierungen unserer gängigen nosologischen Taxonomie und stellen darauf abgestimmte Berufsbilder in Frage.

Vielmehr unterstützt der epidemiologische Befund die verstärkte Psychologisierung der allgemeinen Ärzteschaft, die in Aus- und Weiterbildung für die Besonderheiten psychogener Krankheiten zu sensibilisieren und mit der hier adäquaten personalen Grundeinstellung, Diagnostik und Basistherapie jenseits spezieller, schulisch ausgerichteter Technologien vertraut zu machen ist. Ferner sprechen unsere epidemiologischen Erkenntnisse gegen zeitgenössische Bestrebungen, die Versorgung psychogen Kranker aus der Verantwortung der Ärzte zu entlassen und ganz in die Hände von Psychologen zu legen. Gerade die bei allen psychogen Kranken so wesentlich mitbetroffene Manifestationsebene des Körpers käme dabei zu kurz. Psychotherapie bleibt vielmehr immer auch eine Aufgabe des Arztes und kann durch Angebote der klinischen Psychologie wesentlich ergänzt, aber nicht ersetzt werden. Oft hat nur der niedergelassene Arzt die Chance, psychogene Syndrome in statu nascendi zu erkennen und zu behandeln. Freilich bedarf er hierzu einer psychosomatisch-psychotherapeutischen Grundhaltung, die wiederum nicht an eine spezialisierte Fachgruppe abgetreten werden kann.

---

2 Der in den Abbildungen 4–2 und 4–3 dargestellte deutliche individuelle Syndromwandel sowohl bei Männern als auch bei Frauen über einen Zeitraum von drei Jahren gibt u.a. zu methodischen Fragen Anlaß, etwa nach der Güte der einschlägigen Meßinstrumente. Deren Validität und Reliabilität wurde hinreichend bestätigt: So liegen beispielsweise die Reliabilitätskoeffizienten der in diesem Zusammenhang interessierenden Subskalen der körperlichen und psychischen Beeinträchtigung sowie des BSS-Gesamtwertes zwischen .70 und .93 (vgl. Manz, 1987). Während sich für die Männer bezüglich der Randverteilungen in Abb. 4–2 und 4–3, d.h. bezüglich der prozentualen Verteilung der Prägnanzsyndrome zum A- bzw. B-Meßzeitpunkt keine Differenzen ergeben, zeigt sich für die Frauen eine deutliche Zunahme in der Gruppe der klinisch Gesunden. Dies geht vorwiegend zu Lasten der psychovegetativ-psychosozialen Mischsyndrome sowie der psychosozialen Prägnanzsyndrome. Nur für diese Abnahme der Geschlechtsdifferenz dürften methodische Ursachen, nämlich veränderte Beurteilungsstandards der Interviewer im Zuge wachsender Projekterfahrung, verantwortlich sein. Dies allerdings verändert nur sehr wenig am Gesamtbefund eines ausgesprochen häufigen Syndromwandels auf der Ebene des einzelnen Probanden.

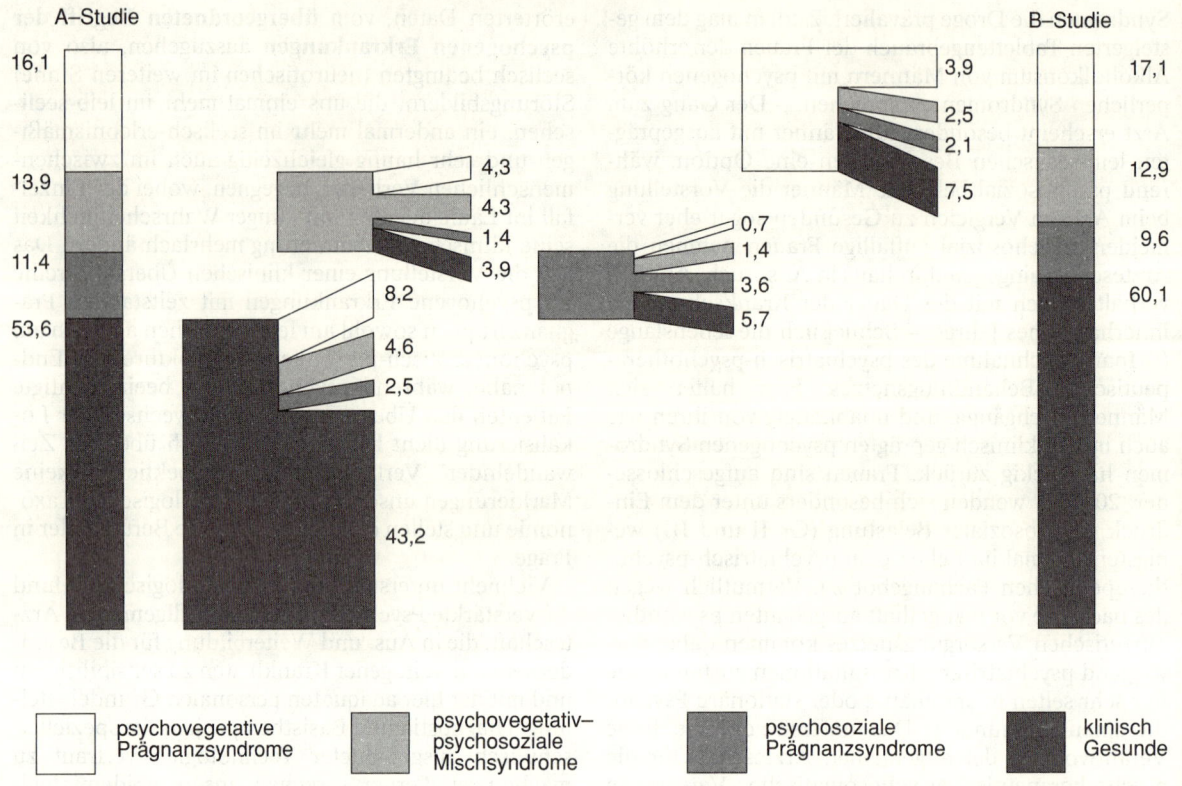

Abb. 4–2. Die Variabilität psychogener Prägnanzsyndrome im Verlauf von drei Jahren (1-Jahres-Prävalenz; Angaben in Prozent der Stichprobe; n = 280 Männer).

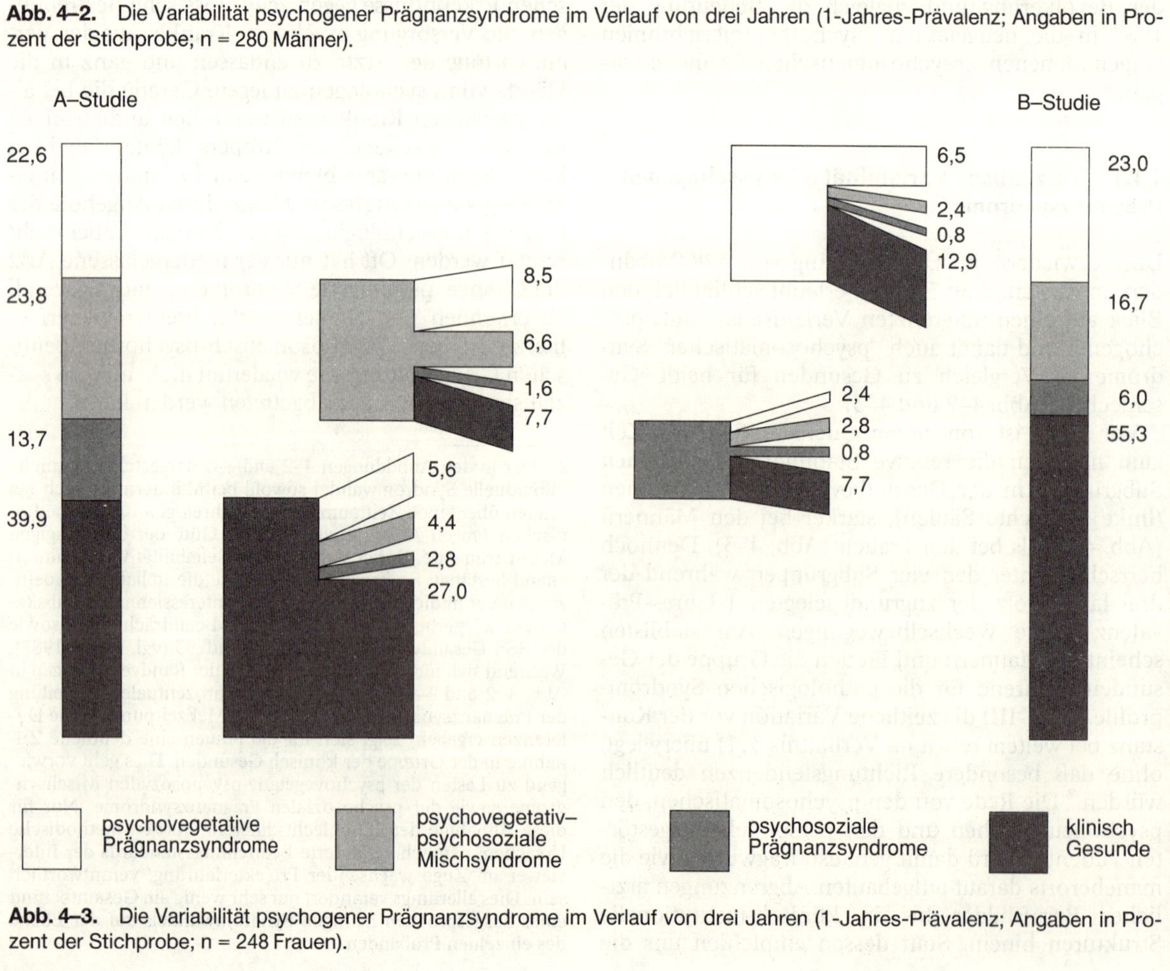

Abb. 4–3. Die Variabilität psychogener Prägnanzsyndrome im Verlauf von drei Jahren (1-Jahres-Prävalenz; Angaben in Prozent der Stichprobe; n = 248 Frauen).

5 Psychoanalytische Konzepte psychosomatischer Symptom- und Strukturbildung

Michael von Rad und *Siegfried Zepf*

Im Selbstverständnis der Psychoanalyse wurde die psychosomatische Symptombildung über lange Zeit neurosenpsychologisch verstanden und erklärt. In neuerer Zeit haben sich Bestrebungen Gehör verschafft, die der psychosomatischen Symptomentwicklung den Charakter einer nosologischen Krankheitseinheit zuweisen. Unbeschadet der Verschiedenheit dieser Erklärungsansätze stützen sich beide Perspektiven in ihrer Sichtweise des psychosomatischen Symptoms auf zwei Grundcharakteristika: Einerseits liegt eine Störung physiologischer Funktionskreise (häufig mit einer Organdestruktion verbunden) vor, die andererseits Züge eines intentional-kommunikativen Verhaltens aufweist, welches in verschlüsselter Form sinnvoll in die individuelle Lebensgeschichte eingebettet ist. Der Terminus „psychosomatisch" krank oder „psychosomatische Krankheit" wird hier beibehalten, obwohl er zweifellos in sehr vielen verschiedenen Sinnzusammenhängen benützt wird, unscharf und in mancher Hinsicht auch unrichtig ist. Er hat sich aber gegenüber alternativen Begriffen wie „psychogene Erstmanifestation organischer Krankheiten" nicht nur wegen seiner größeren Kürze durchgesetzt, sondern weil er durch die sprachliche Verschmelzung gleichzeitig die unlösliche Verbindung psychogener, individuell biographischer Faktoren mit dem Auftreten somatischer Veränderungen kenntlich macht. Aus dieser Doppelbestimmung ergibt sich der Anspruch, dem die zu referierenden Konzepte sich unter heutiger Sicht stellen müssen:

Unter welchen physiologischen und psychosozialen Bedingungen entwickelt ein Individuum mit einer bestimmten biologischen und psychisch-konstitutionellen Ausstattung welche Körperstörung, die unter welchen Bedingungen rasch abklingt, krisenhaft verläuft oder chronifiziert?

Für die Symptombildung der sog. klassischen psychosomatischen Erkrankungen im engeren Sinne ist dabei festzuhalten, daß sie zwar als körperliche Erscheinungen in den Bereich der naturwissenschaftlichen Humanmedizin gehören – ihrem Wesen nach jedoch in zwischenmenschlichen Beziehungsstörungen der individuellen Lebensgeschichte gründen, die in hermeneutischen Verfahren der Psychoanalyse eingeholt und aufgearbeitet werden können.

Es hat an Konzepten nicht gefehlt, in denen versucht wurde, unter unterschiedlicher Akzentuierung verschiedener psychoanalytischer Aspekte des Gesamtgeschehens – insbesondere der trieb-, ich- und objektpsychologischen – der körperlichen Natur des Symptoms Rechnung zu tragen und einige dieser Fragen einer Lösung näher zu bringen. Obwohl Freud selbst (in einem Brief an Viktor von Weizsäcker) die Ansicht äußerte, „daß wir den Sprung vom Seelischen ins Körperliche doch nie mitmachen können", hat er dennoch zwei Konzepte vorgelegt, mit denen er versuchte, diese Grenzlinie zu überschreiten und beide Bereiche zu vermitteln. Diese beiden Modelle – das der Konversion und das der Angstneurose – wurden Ausgangspunkt der weiteren theoretischen Entwicklung.

5.1 Das Konversionsmodell (Freud)

Im Verständnis von Freud (1932) wird durch die Konversion eine unlustbereitende Vorstellung dadurch unschädlich gemacht, daß ihre „Erregungssumme" ins Körperliche umgesetzt wird. Für das Konversionssymptom selbst hat Freud sechs Charakteristika herausgearbeitet. Es entsteht, wenn 1. ein Triebwunsch mit inneren oder äußeren Normen in Konflikt gerät und so zu einer unerträglichen Vorstellung führt, die 2. deshalb aus dem Bewußtsein verdrängt wird, und 3. der Konflikt genital-sexueller (ödipaler) Natur ist. Wird 4. dieser Triebwunsch reaktualisiert und kann die bisherige Verdrängung nicht mehr aufrechterhalten werden, erfolgt eine Konversion, eine Verschiebung der psychischen Energie (Libido) von der Besetzung seelischer zur Besetzung somatischer Prozesse. Das führt 5. zu einem körperlichen Symptom, das den zugrundeliegenden Triebwunsch wie dessen Verbot in einem Kompromiß verschlüsselt symbolisch zum Ausdruck bringt. D.h.: Die somatischen Veränderungen sind in Sprache übersetzbar und damit zu verstehen. Das Symptom bindet die psychische Energie und hält die unerträgliche Vorstellung unbewußt, erfordert aber auch zusätzliche Aufmerksamkeit und führt sekundär zu verstärkter libidinöser Besetzung – es hat also sowohl Befriedigungs- als auch Bestrafungscharakter. Freud hat ferner 6. immer daran festgehalten, daß ein „somatisches Entgegenkommen" postuliert werden muß, ein körperlicher Faktor, der für die „Organwahl" bedeutsam ist und den man sich in seiner Entstehung von einer genetischen Disposition über eine aktuelle Überbeanspruchung bis hin zu frühkindlichen Prägungen im Rahmen besonderer Körpererfahrungen vorstellen kann. Auch wenn im Gefolge gesellschaftlicher Entwicklungen solche psychosomatischen Symptome seltener geworden sind, so ist doch bis heute die theoretische und therapeutische Validität dieses Modells – auch in der strengen Form – z.B. bei bestimmten Patienten mit funktionellen Lähmungen, Dysbasien, aber auch Sensibilitäts- oder Stimmstörungen unbestritten.

Bald aber wurde deutlich, daß damit viele körperliche Symptombildungen, bei deren Entstehung offensichtlich psychosoziale Faktoren eine gravierende Rolle spielen, nicht befriedigend zu erklären waren. So wurde, um das Konversionsmodell als Erklärungsparadigma zu erhalten und zu erweitern, die Bindung an einen genital-sexuellen Konflikt aufgegeben: Fenichels Begriff der „prägenitalen Konversion" erweiterte den Bereich relevanter Konflikte auch auf Störungen frühester, infantiler Bedürfnisse, die sich um Wünsche nach Nähe, Wärme, Versorgung usw. zentrieren. Auch wenn später Rangell (1959) und andere Versuche unternahmen, den Konversionsbegriff wieder einzugrenzen, so fand doch im wesentlichen eine fast grenzenlose Ausweitung dieses Modells auf praktisch alle möglichen Konflikte statt, in deren Zusammenhang auch körperliche Störungen auftraten, so daß sich bald die Frage stellte, ob damit nicht lediglich ein Begriff aufrechterhalten wurde, ohne daß der Kern des Freudschen Vorschlages – die symbolische Darstellung des Konfliktes und die triebdynamische Entlastung durch das Symptom – noch nachweisbar war. (Hierher gehören auch manche „Sünden" der psychoanalytischen Psychosomatik, wenn etwa alle möglichen körperlichen Vorgänge pseudo-„symbolisch" interpretiert wurden – z.B. das „Weinen" der Bronchien beim Asthma usw. – waghalsige Spekulationen zumindest, die zu Recht von der Medizin übelgenommen und von nicht wenigen Patienten erlitten wurden.)

5.2 Psychosomatik als Psychoanalyse des „Es" (Groddeck)

Fast gleichzeitig mit der Begründung der Psychoanalyse durch Freud entwickelte Georg Groddeck einen Ansatz, welcher Eigenständigkeit für sich beanspruchen kann, obwohl er dem Konversionsmodell Freuds strukturell ähnlich ist. Er versuchte mit ihm den psychosomatischen Ursprung und die psychotherapeutische Beeinflußbarkeit aller Erkrankungen theoretisch zu begründen. Gegenüber den mehr theoretischen, metapsychologisch-systematischen Intentionen Freuds war Groddeck stark geprägt von den Erfahrungen des praktischen Arztes, den Problemen allgemeinärztlicher Therapie.

> Zunächst soll eine kurze Krankengeschichte seine Konzeption erläutern. Groddeck berichtet über eine Dame, die über heftige Kopfschmerzen klagt. Sie habe im Zug zwischen zwei offenen Fenstern gesessen; da-

> von sei es gekommen. Groddeck fragt, wozu man den Kopf hat. Sie antwortet, zum Denken. Groddeck fährt fort: „Und wenn Ihre Kopfschmerzen nun immer und immer schlimmer werden, was wird dann aus dem Denken?" „Man kann nicht mehr denken", antwortet die Patientin. Also, so Groddeck, „haben Sie den Kopfschmerz, weil Sie etwas nicht denken sollen". Groddeck fragt nun, welches denn dieser „unerträgliche, gewissermaßen verbotene Gedanke" sein könnte. Er müsse wohl mit der Zugluft in Verbindung stehen. Nach einem Zögern berichtet die Patientin, daß ihre Mutter an einer Bronchitis gestorben sei und der Arzt gemeint habe, sie müsse sich erkältet haben; da sie lange bettlägerig war, könne sie die Erkältung nur durch Zugluft bekommen haben. Sie habe sich schwere Vorwürfe gemacht, am Tod der Mutter schuldig zu sein, da sie tatsächlich die Mutter kurz vor Beginn der Todeskrankheit „beim Lüften des Zimmers der Zugluft zwischen zwei offenen Fenstern ausgesetzt hatte. Übrigens ist heute der Todestag meiner Mutter. Meine Kopfschmerzen sind weg". Groddeck fügt hinzu, die Dame sei Begleiterin einer seiner Patientinnen, die große Angst vor Erkältungen habe; deshalb habe sie die Abwesenheit dieser Freundin benutzt, um das gemeinsame Wohnzimmer gründlich zu lüften.

Kopfschmerzen, so erläutert Groddeck diesen Fall, „auch wenn sie durch schwere organische Leiden bedingt sind", kämen nur dann zustande, „wenn ein unerträglicher, zu irgendeiner Zeit sehr wichtiger Gedanke nicht gedacht, verdrängt wird: hier war es die Idee, die Mutter getötet zu haben". Daß besonders oft Kopfschmerzen „als Feier der Mutter" gewählt werden, liege daran, daß die Schädelhöhle von dem Es als Mutterleib aufgefaßt wird und die Gedanken als Kinder. Die Tatsache, daß Menschen im Anschluß und in Verbindung mit dem Aufenthalt im Zug erkranken, sei unbestreitbar. Dies liege am „Zusammentreffen physikalischer Faktoren mit unbewußten Gedankengängen, die vom Es geleitet werden. Jede Erkältung setzt voraus, daß eine drohende Erhitzung vom Unbewußten durch zu starke physikalische Kälte bekämpft wird".

Groddecks zentrale These lautet: Wir leben gar nicht, wie wir immer zu meinen glauben, aktiv und selbstentscheidend – wir werden gelebt. Alle unsere Äußerungen und auch die der Natur gehen für Groddeck zurück auf eine Art kreative Urkraft, ein primum movens, das allerdings nicht als realer Anfang einer kausalen Kette angesehen wird, welches aber gleichwohl vor allem Individuellen da ist. Groddeck nennt diese formende und gestaltende Urkraft das „Es". Dieses „Es umfaßt bewußt und unbewußt Ich und Triebe, Körper und Seele, Physiologisches und Psychologisches; dem Es gegenüber gibt es keine Grenzen zwischen Physiologischem und Psychischem. Beides sind Äußerungen des Es, Erscheinungsformen" (Groddeck, 1950/51).

Damit wird deutlich, daß im Entwurf Groddecks „Es" und „Ich" nicht – wie bei Freud – als topische Instanzen begriffen werden. Für Groddeck ist das Ich nur eine der unendlichen Äußerungsformen des Es, wie der Körper oder die Seele, Organisches und Anorganisches. Das Es spricht und stellt sich dar in jeder Geste und Gestalt, es formt Körper und Seele und schafft – wenn der Mensch dem Es nicht angemessen lebt – auch Krankheit und Symptom. Ist aber der Körper, die Krankheit, das Symptom erst einmal als Symbol in den Blick gekommen, dann gibt es auch etwas zu verstehen, dann ist die körperliche Erkrankung nicht mehr ein rein naturgesetzlich ablaufendes, sinnloses, der psychotherapeutischen Beeinflußbarkeit enthobenes Geschehen. In Anlehnung an Freud, den er zeitlebens bewunderte und dabei doch mit großer Entschiedenheit seine Eigenständigkeit bewahrte, unterscheidet Groddeck analog der Traumtheorie einen manifesten und latenten Inhalt des Symptoms, für welches er auch die frühkindliche Determiniertheit sowie die Unterwerfung unter den Wiederholungszwang in Anspruch nimmt. Das Symptom wird verstanden als Mahnung „bis hierher und nicht weiter", als eine Art „Schutzhaft", deren Bedingungen und Sinnhaftigkeit zu ergründen seien. Nicht nur psychoanalytische, sondern jedwede therapeutische Maßnahmen sollen dann dazu genutzt werden, „dem kranken und eigenwilligen Es zu beweisen, daß es mit seinen gesunden Ausdrucksformen wieder zurechtkommen kann" (Groddeck, 1966).

Als erster hat Groddeck die Bedeutung und den Einfluß unbewußter Prozesse auf die Entstehung körperlicher Erkrankungen erkannt. Sein Versuch fußt theoretisch auf einem monistischen Entwurf der Person oder, in Groddecks Sprache, dessen, was die Person lebt: des Es (in unserem Fallbeispiel wird das Zusammentreffen physikalischer Faktoren mit unbewußten Gedankengängen vom Es geleitet). Groddeck behandelt sozusagen einen Menschen noch vor der karthesianischen Spaltung in Körper und Seele. Der durch die Ausweitung der Psychotherapie auf alle körperlichen Erkrankungen erzielte Gewinn wird bezahlt mit einer Ausweitung der Begriffe, die dadurch unbestimmbar werden. Wenn das Es alles macht, die Krankheit und die Gesundheit, dann ist es auch nie eindeutig zu fassen – es wird zu einem unbegrenzten Spielraum der Phantasie und Projektion des Untersuchers. (Warum werden im Fallbeispiel die Schädelhöhle als Mutterleib und die Gedanken als Kinder aufgefaßt? Oder umgekehrt: Warum bekam die Patientin keine Bronchitis, die aufgrund der Zugluft und des Konfliktes mit der Mutter symbolisch ebensogut verständlich wäre?) Für Groddeck gibt es kein „somatisches Entgegenkommen" – überhaupt hat der Körper als Spielart des Es keine Eigenständigkeit. Er wird auf eine bloße Erscheinungsform des Es reduziert, welches jenseits des Menschen liegt und in dessen konkreter Einmaligkeit sich dieses „Es" bloß materialisiert. Dieses „Es", welches den Menschen lebt, wird transzendental, metaphysisch-ahistorisch begriffen. Damit ist aber auch letztlich die Sinnhaftigkeit leiblicher Bedeutungen der lebensgeschichtlichen Prägung einer entwicklungsgeschichtlichen Entschlüsselung entzogen (in unserem Fallbeispiel: Wenn jede Erkältung damit zusammenhängt, „eine unangenehme Hitze seiner Seele loszuwerden" – oder Kopfschmerzen nur dann zustande kommen, wenn ein unerträglicher Gedanke nicht gedacht wird, dann haben wir es mit zeitlosen und überindividuellen Gültigkeiten zu tun) (Groddeck, 1966).

Es liegt auf der Hand, daß das ausgeprägt Spekulative an Groddecks Krankheitsverständnis auch gefährliche Wirkungen gezeitigt hat; vielleicht liegt darin auch der Grund, warum sein Entwurf sich bei den nichtärztlichen Psychotherapeuten größerer Beliebtheit erfreut.

Hellsichtig haben Groddeck und Freud ihre Gemeinsamkeiten wie ihre unüberbrückbaren Differenzen in ihren ersten Briefen festgehalten. Freud schließt den seinen mit folgendem Passus: „Es scheint mir ebenso mutwillig, die Natur durchweg zu beseelen, wie sie radikal zu entgeistern. Lassen wir ihr doch ihre großartige Mannigfaltigkeit, die vom Unbelebten zum organisch Belebten, vom körperlich Lebenden zum Seelischen aufsteigt. Gewiß ist das Unbewußte die richtige Vermittlung zwischen dem Körperlichen und dem Seelischen, vielleicht das langentbehrte ‚missing link'. Aber weil wir das endlich gesehen haben, sollen wir darum nichts anderes mehr sehen können? Ich fürchte, Sie (...) haben die monistische Neigung, alle die schönen Differenzen in der Natur gegen die Lockungen der Einheit geringzuschätzen. Werden Sie damit die Differenzen los?" (Groddeck und Freud, 1974).

Weitere Überlegungen zur (Psycho-)Genese psychosomatischer Erkrankungen wurden von verschiedenen Psychoanalytikern vorgetragen, im deutschen Sprachraum beispielsweise von Harald Schultz-Hencke und Werner Schwidder. Eine Ausnahme macht Alexander Mitscherlich, der ein für die psychosomatischen Erkrankungen spezifisches Konzept einer „zweiphasigen Verdrängung bzw. Abwehr" entworfen hat, das eine gewisse Sonderstellung einnimmt (Mitscherlich, 1967). Mitscherlich nimmt an, daß beim psychosomatisch Erkrankenden die neurotische Abwehr insuffizient wird, und er deshalb veranlaßt ist, ins „Somatische" zu verdrängen. Dabei läßt er allerdings die Bedingungen offen, die für diese Insuffizienz verantwortlich sind, und sprengt mit seiner Auslegung des Verdrängungsbegriffs den konzeptionellen Rahmen der psychoanalytischen Metapsychologie. Verdrängung ist als Vorgang innerhalb eines bestimmten theoretischen Bezugssystems konzipiert, dem des psychischen Apparates. Schon Freud hatte ja ausdrücklich vor der Versuchung gewarnt, die „psychische Lokalität etwa anatomisch zu bestimmen" (Freud, 1900). Solange das neurophysiologische Substrat des psychischen Apparates mit der metapsychologischen Begrifflichkeit nicht vermittelt ist, solange bedeutet die These einer Verdrängung ins Körperliche keinen Zuwachs an Erkenntnis. Es erscheint fraglich, ob die Kategorien, die zur Erklärung der Funktionsweisen des psychischen Apparates entwickelt wurden, in gleicher Weise auch auf körperliche Prozesse anwendbar sind und übertragen werden dürfen (Schneider, 1973). Eine genauere Darstellung findet sich im Kapitel 3, in dem auch die Konzeption von Flanders Dunbar erörtert wird.

5.3 Die Angstneurose (Freud)

Etwa parallel zum Konversionskonzept beobachtete und beschrieb Freud 1895 Körpersymptome wie Schwitzen, Schwindel, Durchfall als **Äquivalente** eines Angstanfalles und grenzte sie scharf von den Konversionssymptomen ab. Auch hier nahm Freud einen sexuellen Ursprung an, der jedoch nicht in einer konflikthaften Fehlentwicklung der Sexualität, sondern in einer Störung des aktuellen Sexuallebens lag. Ihre Ätiologie wurde nicht psychisch, sondern somatisch-toxisch (als Folge nicht abgeführter sexueller Spannungen) verstanden. Ihr entscheidendes Charakteristikum ist, daß hier nicht aufgrund eines seelischen Konfliktes mit Hilfe psychischer Verdrängungsarbeit – also einer Ich-Leistung – ein somatisches Symptom entsteht, sondern daß bei ihnen eine seelische Verarbeitung gar nicht zustande kommt und statt dessen die somatisch gedachte Erregung direkt in ein Körpersymptom überführt wird. Er beschreibt die Dynamik der in der damaligen Terminologie „Angstneurose" genannten Symptomgenese im Unterschied zu Konversionsmechanismen, die er der Hysterie zuordnete, 1895 folgendermaßen: „... So ergeben sich Gesichtspunkte, welche die Angstneurose geradezu als das somatische Seitenstück der Hysterie erscheinen lassen. Hier wie dort Anhäufung von Erregung ... Hier wie dort psychische Unzulänglichkeit, der zufolge abnorme somatische Vorgänge zustande kommen. Hier wie dort tritt an Stelle einer psychischen Verarbeitung eine Ablenkung der Erregung in das Somatische ein, der Unterschied liegt bloß darin, daß die Erregung, in deren Verschiebung sich die Neurose äußert, bei der Angstneurose eine rein somatische ..., bei der Hysterie eine psychische ... ist." Oder, 1917, unmißverständlich: „... Die Symptome der Aktualneurosen ... haben keinen Sinn, keine psychische Bedeutung." Es handele sich, sagt er an anderer Stelle, „um eine Entfremdung zwischen dem Psychischen und dem Somatischen" (Freud, 1895). Seiner Meinung nach bot die Aktualneurose für die Psychoanalyse keine Angriffspunkte und wurde deshalb außerhalb der Reichweite der Psychoanalyse angesiedelt.

5.4 Vegetative Neurose und Konfliktspezifität (Alexander)

Durch diese Entscheidung Freuds blieb längere Zeit verborgen, daß dieses Konzept bereits Erklärungsansätze der psychosomatischen Symptombildung in sich barg.

Wieder einbezogen in die psychoanalytische Krankheitslehre wurde eine Reihe körperlicher Erkrankungen, von denen Freud einige als Aktualneurosen ausgegliedert hatte, durch Franz Alexander, der in vieler Hinsicht die Fundamente einer modernen Psychosomatik gelegt hat. Unter der heute etwas irreführenden Bezeichnung „vegetative Neurose" respek-

tive „Organneurose" faßte er eine Gruppe von Krankheitsbildern (wie z. B. das Ulcus duodeni, die essentielle Hypertonie, das Asthma bronchiale, aber auch rein funktionelle Störungen ohne Organläsionen) zusammen und grenzte sie scharf von den Konversionssymptomen ab.

Nach seiner Ansicht entstehen auch die „vegetativen Neurosen" auf der Grundlage eines unbewußten Konfliktes im Zuge einer neurotischen Fehlentwicklung, die dazu führt, daß eine auf Außenobjekte gerichtete Handlung unterlassen wird. Die emotionale Spannung kann so nicht abgeführt werden, während die sie begleitenden vegetativen Veränderungen persistieren. In einem zweiten Schritt könne es dann zu Gewebsveränderungen und irreversiblen organischen Erkrankungen kommen. Der physiologischen Reaktion wird hier also keinerlei Ausdruckscharakter zugemessen. „Hier sind die körperlichen Symptome nicht ersatzweise Ausdruck einer verdrängten Emotion, sondern sie sind normale physiologische Begleiterscheinungen der Emotion ... Sie entlasten die verdrängte Wut nicht, sondern sie begleiten sie. Sie sind Anpassungsvorgänge des Organismus ... Der erhöhte Blutdruck oder die Zuckermobilisierung entlasten die Wut nicht im geringsten; diese Symptome treten nicht auf an Stelle der emotionalen Spannung; sie begleiten einfach die Emotion Wut; sie sind ein untrennbarer Bestandteil des Gesamtphänomens, das wir Wut nennen" (Alexander, 1978).

Zentral beschäftigte sich Alexander ferner mit der Frage der emotionalen Spezifität der vegetativen Neurosen. Er vermutete, daß die Entgleisung vegetativer Funktionen mit emotionalen Zuständen in bestimmten Konfliktsituationen in einem unmittelbaren Zusammenhang steht als mit Persönlichkeitstypologien vergangener Zeiten. Für eine Erkrankung wie beispielsweise das Ulcus duodeni war für ihn nicht ein bestimmter Persönlichkeitstyp das gemeinsame Charakteristikum, sondern vielmehr eine typische unbewußte Konfliktsituation, die sich bei sehr unterschiedlichen Persönlichkeiten entwickeln und dann für das Leben dieser Menschen eine beherrschende Bedeutung gewinnen kann. So kann, um ein Beispiel Alexanders aufzugreifen, in Situationen, die von dem späteren Kranken als Verlust fürsorglicher Zuwendung und Versorgung erlebt werden, der unbewußte Wunsch, geliebt zu werden, sich auf das Bedürfnis gefüttert zu werden verschieben. Das unter dem Einfluß des Über-Ichs verdrängte Verlangen nach emotionaler Zuwendung mobilisiert dann die Magensekretion. Die Bereitschaft, gerade in diesen Konflikt zu geraten und die Unfähigkeit, ihn auf adäquate Weise zu lösen, wird von ihm als ein Persönlichkeitsfaktor angesehen.

Bleibt die Befriedigung von Geborgenheits- und Abhängigkeitswünschen aus, dann manifestieren sich die vegetativen Reaktionen in Fehlfunktionen, die in der Auffassung Alexanders das Ergebnis einer gesteigerten parasympathischen Erregung sind. Werden die Ausdrucksmöglichkeiten von Konkurrenz- und Aggressionshaltungen im Verhalten gehemmt, dann ist ein Dauererregungszustand des sympathisch-adrenergen Systems die Folge. Diese vegetativen Symptome sind dann das Resultat der nicht abgeführten sympathischen Erregung, die andauert, weil der Vollzug einer adäquaten Kampf- bzw. Fluchtreaktion nicht stattfinden konnte. Er veranschaulicht dies am Kranken mit essentieller Hypertonie: Seiner Ansicht nach handelt es sich bei diesen Patienten um aggressionsgehemmte, beherrschte Menschen, die weder psychisch noch physisch in der Lage sind, ihre aggressiven Regungen angemessen abzuführen.

Für jede vegetative Neurose suchte Alexander ein dynamisches Grundschema zu entwerfen. Beispielsweise führt bei den Ulkus-Patienten die Versagung oral-rezeptiver Wünsche zu oral-aggressiven Reaktionen, die wiederum Schuldgefühle und Ängste auslösen. Dies wird durch ein ausgeprägtes Leistungsstreben überkompensiert, welches jedoch wieder die unbewußten oralen Abhängigkeitsbestrebungen verstärkt.

Ein weiterer Einwand betrifft die Konflikttypologien, die sich bei genauerem Hinsehen nicht so sehr von den früheren Persönlichkeitstypologien unterscheiden. Denn Alexander formulierte seine spezifischen Konfliktkonstellationen als spezifisch für alle Patienten einer nosologischen Gruppe (z.B. Ulcus duodeni) – eine Annahme, die auf dem Hintergrund der neueren psychosomatischen Forschung fraglich geworden ist. Von einer anderen Position her bezweifelt auch Grinker die „haarespalterischen" Unterschiede bestimmter Persönlichkeits-/Konflikttypologien, in denen er nichts anderes zu erkennen vermochte, „als die gleichen Verkettungen von Abhängigkeit, Versagung und Feindseligkeit, die in monotoner Wiederholung bei allen Menschen vorkommen" (Grinker, 1961).

Auf Kritik traf auch eine wesentliche Annahme im Konzept Alexanders, daß nämlich „Jeder emotionale Zustand ... sein eigenes physiologisches Syndrom" (Alexander, 1951, S. 44) aufweist, und die in seinem Konzept relativ zentral steht. Diese Annahme ergab sich nicht zuletzt aus seiner grundsätzlichen Bestimmung des Verhältnisses von psychischen und physiologischen Prozessen. Alexander reduzierte die psychischen Phänomene auf die „subjektive Seite gewisser physiologischer (Hirn-)Prozesse" (Alexander, 1951, S. 28) und betrachtete sie als deren Spiegelungen. „Während die Physiologie die Funktionen des Zentralnervensystems in Begriffen von Raum und Zeit angeht, findet die Psychologie ihren Zugang dazu in Begriffen einer Reihe von subjektiven Phänomenen, die die subjektiven Spiegelungen (Reflexionen) physiologischer Prozesse sind" (Alexander, 1951, S. 18, vgl. auch S. 28f., S. 32). Psychische und physiologische Prozesse werden damit in ein gnostisches Verhältnis gesetzt. In objektiver Hinsicht sind physiologische Prozesse dann bereits selbst die Emotionen, die durch die „Spiegelung" auch zu einer subjektiven Tatsache werden.

Angesichts vielfältiger empirischer Befunde hat sich die Hypothese – jede emotionale Konstellation hat ihr eigenes physiologisches Muster – nicht halten lassen (z.B. Schachter und Singer, 1962). Die vege-

tativen Reaktionsmuster scheinen begrenzt, eher einförmig und wenig plastisch und sind auf die verschiedensten Reize hin relativ monoton auslösbar. Darüber hinaus scheint die Einbeziehung des somatischen Affektkorrelats („Organwahl") auch durch die individuelle Biographie mitbestimmt. Experimentalpsychologische Untersuchungen (z.B. Lacey und Lacey, 1958) haben deutlich gemacht, daß verschiedene Individuen in identischen Situationen zwar mit den gleichen psychologischen Affekten reagieren können, diese aber mit ganz unterschiedlichen somatischen Abläufen gekoppelt sind.

So wie die Emotionen auf psychischer Ebene als Spiegelungen objektiv ablaufender (durch „emotionale Reize" (Alexander, 1951, S. 35) ausgelöster) Hirnprozesse entstehen, so sind seiner Meinung nach psychische Inhalte überhaupt als subjektive Spiegelungen der durch Reize in Gang gesetzten Hirnprozesse anzusehen. „Wenn wir von Psychogenie sprechen, so denken wir dabei an physiologische Prozesse, die aus zentralen Erregungsabläufen im Nervensystem bestehen, und die mit psychologischen Methoden untersucht werden können, weil sie subjektiv in der Form von Emotionen, Ideen oder Wünschen wahrgenommen werden" (Alexander, 1951, S. 32). Die Umwelt wird hier als ein Reizgefüge gesehen, welches zu körperlichen Veränderungen oder Empfindungen führt, die dann wahrgenommen werden und zu psychischen Inhalten führen können. Damit rückt Alexander in die Nähe der Position von James-Lange[1] und wird mit seiner Konzeption auch empfindlich für die Kritik, die aus verschiedenen Richtungen an dieser erkenntnistheoretischen Position geübt wurde (z.B. Rubinstein, 1968; für eine zusammenfassende kritische Darstellung des Alexanderschen Konzepts vgl. Inderfrey, 1986).

Trotz dieser Kritik sind insbesondere zwei Aspekte aus Alexanders Werk von bleibendem Wert. Dies betrifft einmal die Präzisierung und Eingrenzung des Freudschen Konversionsbegriffs für den Bereich der klinischen Psychosomatik. Zum anderen wird in seinem Konzept die relative Autonomie physiologischer Prozesse berücksichtigt, ohne daß diese damit zugleich aus der je individuellen Lebensgeschichte entlassen wären. Auf diese Weise wurde Franz Alexander zum Begründer der modernen psychosomatischen Medizin.

5.5 Das De- und Re-Somatisierungsmodell (Schur)

Schur, Freuds „Leibarzt" in langen Jahren der Krankheit und des Sterbens, stützt sich zwar auf Beobach-

---

[1] James beschrieb diese Theorie so: „Meine Theorie (...) ist die, daß die körperlichen Veränderungen direkt auf die Wahrnehmung der erregenden Tatsache folgen, und daß das Bewußtsein vom Eintritt eben dieser Veränderungen die Gemütsbewegung ist" (James, 1909, S. 376). Freud hielt von dieser Theorie nicht sehr viel. Er merkte an, daß „die James-Langesche Theorie (...) für uns Psychoanalytiker geradezu unverständlich und undiskutierbar" (Freud, 1917, GW XI, S. 411) ist.

tungen bei der psychoanalytischen Behandlung von Neurodermitis-Patienten, hält aber sein Konzept für prinzipiell anwendbar auf alle „psychosomatischen" Störungen (Schur, 1974, S. 355). Sein Ausgangspunkt ist die Ich-Psychologie, sein Ziel ist die psychologische Erfassung von menschlichen Entwicklungs- und Reifungsvorgängen unter den Bedingungen sowohl innerer (libidinöser und aggressiver Impulse) als auch äußerer (z. B. umweltbedingter) Anforderungen.

Schur beschreibt die Entwicklungs- und Reifungsvorgänge beim Säugling und Kind als einen fortlaufenden Prozeß der „De-Somatisierung" respektive der Neutralisierung libidinöser und aggressiver Energien: Während das Neugeborene aufgrund seiner noch unentwickelten, nicht ausdifferenzierten (psychischen und somatischen) Strukturen auf Störungen seiner Homöostase immer auch körperlich unkoordiniert, unbewußt-primärprozeßhaft reagiert, führe die Reifung zunehmend zur „Entfaltung des Sekundärprozeßdenkens als wesentlicher Komponente der Bildung des Ichs" (Schur, 1974, S. 340). Dabei laufen verschiedene Prozesse parallel. Aus dem undifferenzierten Stadium entwickelt sich neben dem zentralen Nervensystem die Fähigkeit zu koordinierter Muskelaktion; die Ausbildung der Wahrnehmungsfunktionen erlaubt progredient eine Überprüfung der Realität (und damit auch möglicher Gefahrenquellen); die wachsende Mobilität, die zunächst im Dienst der Spannungsabfuhr stand, unterstützt die Fähigkeit zur Realitätsprüfung; mit der Entwicklung des Gedächtnisses und der Möglichkeit zu planen, ergeben sich erste Ansätze, die sowohl die Vorwegnahme von Gefahren als auch einen Handlungs- und Triebaufschub ermöglichen (Schur, 1953, S. 70). Aus der undifferenzierten Matrix strukturiert sich mehr und mehr das Ich heraus. Vegetative Abfuhrprozesse treten mehr und mehr in den Hintergrund.

Die Reifungsentwicklung des gesunden Kindes erscheint – bedeutend mehr als beim zeitlebens stärker instinktgebundenen Tier – als ein Prozeß der De-Somatisierung. Die Entwicklungstendenz liegt in der koordinierten Integration unkoordinierter somatischer Prozesse, in der Reduzierung vegetativer Entladungserscheinungen und in der Ersetzung von Handlungen durch Gedanken. Das wünschenswerte Entwicklungsergebnis ist die Beherrschung von Reizen mit einem Minimum an Energieverbrauch. Eine Reaktionsweise, in der „das Ich im ganzen Verlauf der Reaktion mit Sekundärvorgängen operiert und durchwegs neutralisierte Energie verwendet", entspricht nach Schur am ehesten der Vorstellung von der Reaktion eines „normalen Ichs" (Schur, 1974, S. 342). Jede Störung dieses Entwicklungsprozesses stellt dann eine Gefahr für die Ökonomie des leibseelischen Gleichgewichtes dar (Schur, 1974, S. 340).

Gleichzeitig mit und als Folge dieses Reifungsprozesses entwickelt sich das Ich und damit die Fähigkeit, auf innere oder äußere Konflikte oder Gefahren mit frei verfügbarer, sog. „neutralisierter" psychischer Energie zu reagieren. Schur stützt sich bei diesen Überlegungen auf das Konzept der Neutralisierung von Triebenergien, wie es von Hartmann und Mitar-

beitern (1949) und Kris (1962) entwickelt worden war, sowie auf die Annahme von Rappaport (1956), der eine Entsprechung zwischen dem Neutralisierungsgrad psychischer Energie und dem Grad des Vorherrschens sekundärprozeßhafter Denkweisen annimmt. Diesem Konzept liegt zugrunde, daß das Ich sich auf zwei verschiedene Arten von gebundener Energie stützen kann, nämlich einerseits auf neutralisierte Energie, andererseits auf Libido und Aggressionen in nicht neutralisierter Form. Das Ausmaß der Fähigkeit, libidinöse und aggressive Triebregungen in neutralisierte, dem Individuum frei verfügbare Energien zu überführen, wird als ein Hinweis auf die Stärke und Unabhängigkeit der Ich-Funktionen angesehen. „Die Fähigkeit des Ichs, Triebenergien zu neutralisieren und diese neutralisierte Energie in seinen Reaktionen zu nutzen, stellt also ein weiteres Ergebnis der Ich-Reifung dar" (Schur, 1974, S. 342). Das bedeutet mehr Elastizität, mehr Spielraum, um z. B. durch vorwegnehmendes Planen oder aber durch Zurückstellung aktueller Triebimpulse mit bestimmten Gefahrensituationen fertig zu werden. Die Entwicklung solcher frei verfügbarer Energien ist gebunden an die Ausbildung einer frühkindlichen Homöostase, die von der adäquaten Versorgung durch die Umwelt, z. B. die Mutter, abhängig ist – sie ist also „objektvermittelt". Während für den Säugling zunächst der Hunger die Gefahr darstellt, so verschiebt diese sich im Laufe der Entwicklung mehr und mehr z. B. auf die Abwesenheit der Mutter, die die Sättigung garantiert, später auf den Verlust der Liebe überhaupt usw. Eine innere Gefahr hat sich also in eine äußere verwandelt und ist damit auch einer Bewältigung zugänglicher geworden (Schur, 1953, S. 71).

Schurs zentrale Hypothese besagt nun, „daß zwischen dem Vermögen des Ichs, auf Sekundärprozeß-Niveau zu operieren und Triebenergien zu neutralisieren, und der De-Somatisierung von Reaktionsweisen eine wechselseitige Abhängigkeit bestehen muß. Damit verknüpft ist eine weitere Annahme, nämlich daß umgekehrt die Re-Somatisierung von Reaktionen mit einem Vorherrschen primärprozeßhafter Denkweisen unter Verwendung deneutralisierter Energieformen einhergeht" (Schur, 1974, S. 342). Der oben beschriebene natürliche Reifungsvorgang der De-Somatisierung ist unter bestimmten Bedingungen umkehrbar. Wenn z. B. eine äußere oder innere Gefahr nicht mehr mit den frei verfügbaren, „neutralisierten" Energien bewältigt werden kann, so kommt es unter dem Druck der dadurch entstehenden Angst zu einer Regression in somatische Reaktionen. Anlaß dieses Prozesses, der entsprechend „Re-Somatisierung" genannt wird, ist ein Rückgriff des Ichs auf frühere Verhaltensmuster primärprozeßhafter Art, wenn eine Gefahren- oder Angstsituation nicht mehr mit psychischen Mitteln, d. h. mit frei verfügbarer Energie, bewältigt werden kann. Schur nennt dies eine „physiologische Regression" und meint damit eine „Regression zu den Vorläufern von Denkprozessen, Affekten, Trieb- und Abwehrhandlungen, die aber hier ausschließlich auf somatischer Ebene zum Ausdruck kommen" (Schur, 1974, S. 367).

Als notwendige Bedingung einer physiologischen Regression wird eine „Ich-Schwäche" angesehen. Ich-Schwäche bedeutet in diesem Zusammenhang, daß bereits in den frühesten Phasen des De-Somatisierungsprozesses Störungen aufgetreten waren, die Fixierungen an bestimmte Organsysteme und damit auch eine allgemeine Einschränkung der Fähigkeit des Ichs, Konflikte mit psychischen Mitteln zu lösen, zur Folge haben. Sie stellt sich insbesondere dar in einer nicht ausreichenden Neutralisierungsfunktion des Ichs (Schur, 1974, S. 345). Hier sind neben erblichen Determinanten frühkindliche Erkrankungen, bestimmte Interaktionsstile in der Familie, aber auch gravierende äußere Traumatisierungen zu nennen. Um diese unterschiedlichen Faktoren in ihrer Bedeutung für die allgemeine wie auch die spezielle Ätiologie einer Störung zu fassen, führt Schur den Begriff der „Gesamtkonstellation (total condition)" (Schur, 1974, S. 387) ein, der zum Ausdruck bringen soll, daß man es jeweils mit „Konstellationen von Kausalfaktoren" zu tun hat, „mit bestimmten Typen der Trieb- und Ich-Ausstattung und -Entwicklung, mit einem bestimmten anlage- und entwicklungsbedingten Zustand der Organe und Organsysteme, wobei diese Faktoren wiederum in Wechselwirkung stehen mit bestimmten Umwelteinflüssen" (Schur, 1974, S. 385 f.).

Liegt nun eine derartige Ich-Schwäche vor, dann ist die Folge, daß in einer Gefahren- oder Angstsituation besonders aufgrund einer unbewußten, regressiven Fehleinschätzung (wenn z.B. eine äußere Situation als Wiederholung eines frühkindlichen Konfliktes wahrgenommen wird) das Ich nicht in der Lage ist, mit psychischen Mitteln zu reagieren. Es kommt dann zu einer physiologischen Regression auf die Primärprozeßebene mit somatischen und vegetativen Reaktionsweisen. In Abänderung des Freudschen Ausdrucks der „Organsprache" nennt Schur dies eine „Organhandlung", womit er ausdrückt, daß auf dieser Ebene nicht mehr „gesprochen", sondern durch das Symptom „handelnd" ein Konflikt ausgetragen wird. Art und Ausmaß der Re-Somatisierung werden als abhängig gesehen von der Entwicklungsphase, in die die Traumatisierung und Fixierung an körperliche Abläufe, also die frühe Störung der De-Somatisierung eingesetzt hatte.

Hier ergibt sich auch eine Verbindung zum umstrittenen Problem der Spezifität. Im Gegensatz zu Alexander formuliert Schur: „Bei meinen eigenen Untersuchungen fand ich in keinem Fall so etwas wie eine Spezifität hinsichtlich des Persönlichkeitstyps, des Konfliktes, der Abwehr gegen sexuelle Triebregungen bestimmter Phasen oder auch der Aggression und der gegen sie gerichteten Abwehr" (Schur, 1974, S. 383). Spezifität besteht für ihn nur „in einer Bereitschaft zu bestimmten spezifischen Organreaktionen" (Schur, 1974, S. 386). Es geht also hier nicht um eine sog. Spezifität des Konfliktes, der die auslösende Situation einer Erkrankung bestimmt.

Gerade darin hat das Schursche De- und Re-Somatisierungsmodell Bedeutung für die Psychosomatik gewonnen. Es wurde vorstellbar, daß auch eine allgemeine (nicht spezifische) Belastungs- oder Gefahren-/Angstsituation („Streß") unter bestimmten Bedingungen zu einer psychosomatischen Regression führen und dennoch eine spezifische Organreaktion auslösen kann.

Probleme des Re-Somatisierungsmodells (vgl. Hartkamp, 1986) liegen allerdings u.a. darin, daß es sich auf eine Reihe zum Teil umstrittener ichpsychologischer Hypothesen stützt, z.B. auf das Konzept „neutralisierter" frei verfügbarer Ich-Energien und auf die Annahme primärautonomer Wurzeln einzelner Ich-Funktionen (Schur, 1974, S. 340). Das Ich wird einerseits als Produkt der Entfaltung des Sekundärprozeßdenkens und der damit einhergehenden Neutralisierung libidinöser und aggressiver Energien vorgestellt, gleichzeitig aber wird diese Neutralisierung als eine Ich-Funktion angesehen (Schur, 1974, S. 340, 342, 345, 355).

Problematisch ist ferner eine unzulässige Ausweitung der Anwendungsbereiche psychoanalytischer Termini. Begriffe wie Ich, Primärvorgang, Neutralisierung, Regression stehen in der Psychoanalyse in einem bestimmten theoretischen Bezugsrahmen, dem der psychoanalytischen Metapsychologie des psychischen Apparates, und haben darin eine Erklärungspotenz. Der menschliche Körper ist jedoch in seinen naturgesetzlich ablaufenden physiologischen und pathophysiologischen Prozessen in der psychoanalytischen Metapsychologie nicht enthalten. Damit kann aber auch mit Hilfe der Annahme einer „physiologischen Regression" die enge Verklammerung von Primärvorgang, Deneutralisierung und Re-Somatisierung nicht über den Status eines nur korrelativen Zusammenhanges angehoben werden.

Das gleiche gilt auch für den umgekehrten Zusammenhang von Sekundärvorgang/Neutralisierung und De-Somatisierung. Durch die enge Verknüpfung von Normalität und De-Somatisierung (Schur, 1974, S. 342, 349) erscheinen affektive Körperreaktionen letztendlich nur als eine Art Symptom. Zur Frage wird auch, wie eine Störung des De-Somatisierungsprozesses sowie ein auf der Ebene des Primärvorganges funktionierendes Ich mit der vielfach beschriebenen Verhaltensnormalität und dem „operativen Denken" psychosomatisch Erkrankender, die eine De-Somatisierung und eine sekundärprozeßhafte Funktionsweise des Ichs offensichtlich schon voraussetzen, in Einklang gebracht werden können.

5.6 Eine psychoanalytische Theorie der psychosomatischen Erkrankungen (Engel und Schmale)

Die amerikanischen Autoren Engel und Schmale verknüpfen in ihrem Ansatz verschiedene bereits vorliegende Konzepte mit eigenen Beiträgen zu einem interessanten Entwurf auf psychoanalytischer Grundlage (Engel und Schmale, 1978). Ihnen geht es um die Präzisierung der Rolle seelischer Faktoren bei der Genese somatischer Symptombildungen (hier insbesondere um die Klärung des Stellenwerts der „Konversion" im Rahmen der Entstehung einer körperlichen Krankheit) sowie um die Bedeutung der Situation beim Ausbruch der Erkrankung. Sie sind der Meinung, daß die Einschränkung des Konversionsmechanismus auf das sensomotorische System, wie es insbesondere von Alexander vertreten wurde (Alexander, 1951), inzwischen weitgehend durchbrochen sei. Ihrer Ansicht nach ist die Konversion „ein psychologisches Konzept; es ist durch neuroanatomische Begriffe nicht zu definieren oder abzugrenzen, auch wenn Funktionen und Struktur des Nervensystems sekundär als biologische Folge der Konversion von dem Organismus einbezogen sein können. Die Körperpartien oder Systeme, die sich zur Konversion herleihen, sind nicht durch ihre willkürliche oder autonome Innervation gekennzeichnet, sondern durch die Fähigkeit, eine seelische Repräsentanz übernehmen zu können, ein Vorgang, der Innervationen, Wahrnehmungen mittels Distanzrezeptoren und Phantasie umfaßt" (Engel und Schmale, 1978, S. 247). Am Beispiel etwa der Hauterscheinungen im Rahmen einer Neurodermitis vertreten sie die These, daß der Zeitpunkt und Ort der Läsion, jedoch nicht die Läsion selbst die Kriterien einer Konversion erfüllen (Engel und Schmale, 1978, S. 247).

Es ist für das Konversionskonzept der Autoren wichtig, daß sie scharf zwischen dem Akt der Konversion und den sekundären Folgeerscheinungen (der Art der Läsion) unterscheiden. Nervensystem bzw. Willkürmotorik seien nur als biologische Folge einbezogen, und zwar nicht aufgrund ihrer willkürlichen Innervation, sondern aufgrund der Fähigkeit, eine seelische Repräsentanz übernehmen zu können. Im Rahmen etwa objektgerichteter Aktivitäten, „wenn in der Beziehung zu einem Objekt ein Körperteil zum Zwecke der Entladung, Ausdruck oder Mitteilung mobilisiert wird, können die beteiligten physiologischen wie pathophysiologischen Prozesse sehr wohl auch eine seelische Repräsentanz erlangen und dadurch nicht nur der späteren Reaktivierung durch symbolische Reize, sondern auch der Beteiligung an primär symbolischen Ausdrucksformen fähig werden" (Engel und Schmale, 1978, S. 247 f.). Sie erläutern dies am Beispiel des Erbrechens, das einmal als Folge verdorbener Nahrungsmittel, das andere Mal als Folge einer Angst, verdorbene Nahrungsmittel aufgenommen zu haben, und schließlich drittens als Konversionssymptom auftreten kann. Dabei kann sich die Konversion am Brechreiz abspielen – beispielsweise dann, wenn die Nahrungsaufnahme im Erleben des Individuums eine unbewußte Bedeutung angenommen hat (Engel und Schmale, 1978, S. 248) – oder auch an der körperlichen Funktion, dem Erbrechen selbst. In Generalisierung dieser Überlegungen notieren sie, daß es einmal über eine phantasierte Schädigung der Körperoberfläche allein – analog hypnotisch induzierter Hautsymptome – oder aufgrund einer früheren Läsion der Haut in diesem Bereich zu einer Aktivierung neurosekretorischer Aktivitäten in diesen Bereichen kommen kann, die dann zu einer körperlichen Läsion führen können. Zum anderen kann

sich die Konversion an den spezifischen Innervationen selbst abspielen, wobei dann „durch Erotisierung der örtlichen Schutzsysteme die einer Konversion zugrundeliegende unbewußte Phantasie sich auf einen Körperteil konzentrieren kann, der zur Abwehr gegen Verletzung innerviert ist und nun reagiert, als wenn die Verletzung wirklich stattgefunden hätte" (Engel und Schmale, 1978, S. 249). Die dann einsetzende Läsion selbst und die sich aus ihr ergebenden Symptome haben weder eine primär symbolische Bedeutung noch dienen sie der Abwehr. Engel und Schmale machen dies nochmals am Beispiel der Hyperventilation deutlich. Die Atmung wird hier in einen unbewußten Bedeutungszusammenhang einbezogen, der „zu exzessiver Atmung führt, während die Symptome der Parästhesie, Schwindel und Tetanie, selbst nicht Konversionen sind, sondern die Folgen der respiratorischen Alkalose, die als Komplikation der Hyperventilation auftritt" (Engel und Schmale, 1978, S. 250).

Wie der Ort der Schädigung auf Konversionsgrundlage symbolisch determiniert ist, so sind die Autoren auch der Meinung, daß Patienten mit Erkrankungen des gleichen Systems (z.B. Colitis ulcerosa und Enteritis regionalis Crohn) sich in der Dynamik eher gleichen als z.B. Patienten mit Pneumonie oder Hepatitis (Engel und Schmale, 1978, S. 251). Insgesamt sei deshalb die (prägenitale) Konversion als „Glied in der Kette der Ereignisse beteiligt" anzusehen, obwohl die Autoren eine primär psychogene Auslösung solcher Organstörungen eher für unwahrscheinlich halten (Engel und Schmale, 1978, S. 250). Um dies behaupten zu können, müsse man nachweisen, daß primär die Symbolisierung, dann die Konversion und dann die Läsion da war, und nicht umgekehrt primär eine Läsion, die dann sekundär eine seelische Repräsentanz erwarb und schließlich wieder zu sekundären Symbolisierungen Anlaß gab. Die Autoren ziehen deshalb den von Engel vorgeschlagenen Begriff der somatopsychisch-psychosomatischen Störung vor, um die Entwicklungsdynamik deutlich zu machen.

„Somatopsychisch-psychosomatisch" nennen sie solche Erkrankungen, „bei denen die prädisponierenden biologischen Faktoren nicht nur schon von Geburt oder früher Säuglingszeit an vorhanden sind, sondern auch direkt oder indirekt an der Entwicklung des psychischen Apparates beteiligt sind. Dies bedeutet nicht unbedingt das Primat des biologischen Faktors; es könnte auch eine gemeinsame, undifferenzierte Matrix vorliegen. Es bedeutet jedoch, daß an irgendeinem Punkt das fragliche somatische System anfängt, auf die psychische Entwicklung einen spezifischen Einfluß auszuüben" (Engel und Schmale, 1978, S. 252).

In Erweiterung des Alexanderschen Modells der Spezifität sind sie der Meinung, daß für die Entstehung einer psychosomatischen Störung eine Verbindung unspezifischer und spezifischer Faktoren wahrscheinlich ist. In diesem Zusammenhang wird insbesondere die Bedeutung eines (realen oder phantasierten) Objektverlustes mit den daraus folgenden Gefühlen der Hilflosigkeit und Hoffnungslosigkeit, dem Komplex des „giving up – given up", für das Entstehen körperlicher Erkrankungen herausgearbeitet. Sie schließen sich dabei dem Konsens an, der über verschiedene Ansätze hinweg besteht, daß die Situation bei Krankheitsausbruch in der Regel durch ein Versagen der psychischen Abwehrmechanismen gekennzeichnet ist. Im Vorfeld des Erkrankungsbeginns wurden nun häufig affektive Zustände von Verzweiflung, Depression, allgemeinem Rückzug beobachtet. In diesem Zusammenhang differenzieren die Autoren das Gefühl der Hilflosigkeit von einem Gefühl der Hoffnungslosigkeit. „Hilflosigkeit meint einen Verlust an Ich-Autonomie, verbunden mit einem Gefühl von Entbehrung wegen des Verlustes von Befriedigung, die von einem außerhalb des Selbst vorhandenen Objekt ersehnt wird. Hoffnungslosigkeit dagegen spiegelt einen Autonomieverlust mit einem Gefühl von Verzweiflung wider, das aus dem Gewahrwerden der Unfähigkeit des Selbst, sich die gewünschte Befriedigung zu verschaffen, herrührt" (Engel und Schmale, 1978, S. 258f.). Entwicklungsgenetisch wird das Gefühl der Hilflosigkeit der oralen Phase zugeordnet, während die Hoffnungslosigkeit stärker internalisierte Befriedigungsquellen voraussetzt und nach Ansicht der Autoren der phallischen Phase nähersteht. Typisch für den Komplex des „giving up – given up" ist das Gefühl des Nicht-mehr-intakt-Seins. Es finden sich oft labile und aufgegebene Objektbeziehungen, ein verändertes Erleben der Umwelt mit der Folge, daß Erfahrungen aus der Vergangenheit nicht mehr für die Zukunft verwendet werden können und daß der Zusammenhang von Vergangenheit und Zukunft subjektiv unterbrochen ist. Eine Reihe eindrucksvoller Untersuchungen unterstützen die These der Autoren, daß die Gefühle der Hilf- und Hoffnungslosigkeit vor dem Krankheitsbeginn für die Auslösung verschiedenster körperlicher Erkrankungen von Bedeutung sind.

Der Komplex des „giving up – given up" steht jedoch nach Ansicht der Autoren in keiner direkten Kausalbeziehung zum Auftreten einer somatischen Erkrankung. „Er ist weder als notwendige noch als hinreichende Bedingung, sondern nur als Beitrag zum Auftreten einer somatischen Erkrankung anzusehen und auch das nur dann, wenn die prädisponierenden Faktoren vorhanden sind" (Engel und Schmale, 1978, S. 263).

Die Theorie von Engel und Schmale versucht eine Verknüpfung unspezifischer und spezifischer Bereitschaften des Individuums mit unspezifischen und spezifischen Konflikten. Es ist ein Entwurf, der nicht nur die psychologischen Bedingungen, sondern auch die angeborenen oder krankheitsbedingt erworbenen körperlichen Faktoren in ihrer Bedeutung zu erfassen sucht und somit auf eine differenzierte Berücksichtigung somatischer, psychischer und sozialer Faktoren bei der Entstehung von Krankheiten verweist. Allerdings enthält auch ihre Theorie, in welcher die Konversion zwar von einer phasenspezifischen Zuordnung gelöst, aber dennoch im Einklang mit der psychoanalytischen Metapsychologie als ein „psychologisches Konzept" verstanden wird – also als die Verschiebung libidinöser oder aggressiver Besetzungen

von szenischen Objektrepräsentanzen auf die psychischen Repräsentanzen des Körpers – einige ungelöste Probleme. Sie kristallisieren sich im wesentlichen um die Frage: Unter welchen Bedingungen erhalten diese Repräsentanzen eine unbewußte Bedeutung und unter welchen Bedingungen reagiert das Individuum darauf mit pathophysiologischen Mechanismen?

Engel und Schmale machen drei Faktoren kenntlich, die dabei eine Rolle spielen sollen: ein somatisches Entgegenkommen (Engel und Schmale, 1978, S. 251), reale, drohende oder symbolische psychische Objektverluste (Engel und Schmale, 1978, S. 261), die freilich als nicht-spezifisch angesehen werden, sowie ein Versagen der Abwehrmechanismen (Engel und Schmale, 1978, S. 254, 261). Das somatische Entgegenkommen wird von ihnen gedacht als ein – wie auch immer – begründeter Krankheitsprozeß, dessen Ablauf strukturell dem abzuwehrenden psychischen Inhalt entgegenkommt, psychische Repräsentanz gewinnt und sich so als eine Ersatzvorstellung anbietet. Diese Überlegung macht jedoch nicht einsichtig, warum ein Individuum bestimmte libidinöse oder aggressive Impulse zum Zwecke der Abwehr auf diese Vorstellung und nicht auf Ersatzvorstellungen aus dem Bereich seiner Objektrepräsentanzen verlagern soll, die sich dazu ebenfalls anbieten können. Offen bleibt auch, warum das Individuum auf die unbewußte Bedeutung beispielsweise eines Körperteils mit pathophysiologischen Mechanismen reagiert und diese nicht psychologisch verarbeitet.

Ferner gewinnt der Zusammenhang zwischen einem psychischen Objektverlust und einem Auftreten des Komplexes „giving up – given up" keine ausreichende Transparenz. Auf theoretischer Ebene wird der Objektverlust nicht als ein Moment ausgewiesen, welches beispielsweise beim psychosomatisch Erkrankenden aufgrund einer bestimmten psychischen Entwicklung die Abwehrmechanismen versagen läßt. Das Argument eines Versagens der Abwehrmechanismen ist ferner nur dann hilfreich für die Klärung, wie es im Zuge einer Konversion zur Bildung eines körperlichen Symptoms kommt, wenn man zugleich annimmt, daß hiermit die reiferen Abwehrmechanismen gemeint sind und die Konversion eine primitivere Form der Abwehr darstellt. Eine ausführliche Kritik dieses Konzepts findet sich bei Hartkamp (1986a).

5.7 Alexithymie

Obwohl die eigentliche Erstbeschreibung durch Ruesch (1948) inzwischen mehr als 40 Jahre zurückliegt, hat das zumeist unter dem Namen „Alexithymie" (Sifneos, 1973) oder „pensée opératoire" (Marty und de M'Uzan, 1978) diskutierte Verhalten vieler Patienten erst in den letzten Jahren eine wirkliche Aufmerksamkeit gefunden. Dabei waren Einzelaspekte der Kernsymptomatik alexithymen Verhaltens – die Phantasieschwäche, die geringe Fähigkeit zum verbalen Gefühlsausdruck sowie die soziale Überangepaßtheit – schon sehr viel früher beobachtet worden.

Schon Ferenczi (1924) erwähnt „einen Menschentypus, der sich in der Analyse wie im Leben besonders phantasiearm, wenn auch nicht phantasielos, gebärdet, Menschen, an denen die eindrucksvollsten Erlebnisse spurlos vorüberzugehen scheinen. Solche sind imstande, in der Erinnerung Situationen zu reproduzieren, die nach unserer Schätzung in jedem Menschen notwendigerweise heftige Affekte der Angst, der Rache, der erotischen Erregung usw. und die zur Affektabfuhr erforderlichen Handlungen, Wallungen, Phantasien oder zumindest äußerliche oder innerliche Ausdrucksbewegung hätten erwecken müssen, ohne auch nur die Spur solcher Reaktionen zu fühlen oder zu äußern". Er empfahl bereits damals bei diesen Patienten eine aktivere psychotherapeutische Technik im Umgang mit Assoziationen und nahm so in vielem den späteren Theorie- und Therapiestreit vorweg. Sehr ähnliche Beschreibungen finden sich – jeweils unter verschiedenen Termini – auch bei Zilboorg (1933) und bei Fenichel, der 1945 den „emotionally frigide type" folgendermaßen kennzeichnet: „Certain persons of this type avoid becoming aware of their insufficiencies by proving to themselves that they are ‚efficient' …; ‚fleeing' to reality from their feared fantasies but to a dead and lifeless reality …".

Waren solche Beschreibungen bislang unsystematisch und eher kasuistisch-impressionistisch, so kommt Jürgen Ruesch (1948) das Verdienst zu, die typischen Merkmale „alexithymen" Verhaltens unter dem Begriff „infantile personality" nicht nur zusammenfassend beschrieben, sondern auch erstmals in einen Zusammenhang mit sog. „psychosomatischen" Patienten gebracht zu haben. In der Struktur der „infantile personality" und dem daraus resultierenden Verhalten sah er das „Kernproblem" psychosomatischer Patienten und eröffnete damit eine Diskussion, die bis heute kontrovers und leidenschaftlich geführt wird.

Das zunehmende Interesse an diesem Thema hat aber auch zu wachsenden Unklarheiten darüber geführt, was denn als Alexithymie anzusehen, wie sie beobachtet und begrifflich gefaßt, verstanden, erklärt und möglicherweise behandelt werden kann. Zwar läßt sich hinsichtlich der Kernsymptomatik alexithymen Verhaltens (s. u.) bei den verschiedensten Autoren aus den verschiedensten Ländern immerhin eine erhebliche Übereinstimmung feststellen, doch finden sich kontroverse Standpunkte, wenn die Bedeutung dieses Phänomens in ihrem klinischen Kontext und insbesondere ihre ätiologische Zuordnung diskutiert werden. Eine Reihe solcher grundsätzlicher Fragen, die sich auf die sehr verschiedenen Beobachtungs- und Erklärungsansätze der Alexithymie beziehen, sollen hier vorangestellt werden, um eine Orientierung zu ermöglichen.

Nimmt man die Alexithymie als ein Ensemble bestimmter Verhaltensweisen, so ist zu klären, ob es sich dabei um ein Persönlichkeitsmerkmal oder um eine situativ bestimmte Verhaltensweise – einen „trait" oder einen „state" – handelt. Man kann sogar fragen, ob es sich dabei um ein Phänomen sui generis handelt oder um ein künstliches Produkt der Untersuchungssituation. Nimmt man es als ein zu Recht abgrenzbares Phänomen, so bleibt zu entscheiden, ob es sich dabei um einen Mangel – ein Defizit – „normaler" Re-

aktionsweisen oder vielmehr um eine vielleicht unangemessene, aber produktive, kreative Leistung des Individuums handelt. Die gleiche Frage auf anderer Ebene stellt sich, wenn man zu klären versucht, inwieweit Alexithymie als Entwicklungsdefizit oder als Abwehrformation, als Struktur oder als Symptom zu verstehen ist. Darin liegt ein weiteres Grundproblem, inwieweit nämlich alexithymes Verhalten als primäres Persönlichkeitsmerkmal – vielleicht vererbt oder im Zuge des Sozialisationsprozesses erworbenes Verarbeitungsmuster – anzusehen ist. Sehen wir uns mit einem unspezifischen, mehr oder minder allen Menschen eigenen Verhaltensmuster konfrontiert, oder ist die Alexithymie spezifisch zumindest für bestimmte Patientengruppen (wie z. B. „psychosomatische" oder Unterschichtpatienten)? Das gleiche Dilemma begleitet uns, wenn wir nach ätiologischen Faktoren fragen: Finden sich genetische, körperliche, entwicklungsbiologische, psychologische oder soziologische Ursachen, die allein oder multifaktoriell gemeinsam verantwortlich zu machen sind? Dies führt schließlich zu dem besonders umstrittenen Fragenkomplex, ob nämlich alexithyme Patienten einer tiefenpsychologisch aufdeckenden Psychotherapie zugänglich sind oder besser mit anderen (z. B. stützenden, übenden oder suggestiven) psychotherapeutischen Verfahren behandelt werden können und sollen.

5.7.1 Klinische Befunde

Wir wollen hier alexithymes Verhalten hinsichtlich seiner affektiven, seiner kognitiven und seiner Selbst- bzw. Objektbeziehungsstruktur genauer darstellen.

Affektive Struktur

Die Beobachtung, daß viele psychosomatische Patienten ihre Gefühle kaum oder gar nicht in Worte fassen können, gab den hier beschriebenen Phänomenen ihren Namen: Alexithymie bezeichnet den „thymos", für den es keine Worte gibt. Oft findet sich eine Art äußerer „Reisebericht", eine minuziöse Beschreibung der realen Umstände, die Schilderung körperlicher Sensationen oder bestimmter Handlungen ohne die Erwähnung ihrer gefühlsmäßigen Relevanz. Deshalb wurde vorgeschlagen, zwischen dem erlebnishaft-bewußten Ausdruck von Gefühlen („feeling") und deren physiologisch-körperlichen Begleiterscheinungen („emotions"), die bei diesen Patienten allein bewußtseinsfähig sein sollen, zu unterscheiden (Sifneos, 1975). Die in diesem Zusammenhang entstehende Frage, inwieweit solche Patienten Gefühle überhaupt erleben oder sie nur nicht benennen können, wird kontrovers diskutiert. Es scheint jedoch so zu sein, daß zumeist undifferenzierte Empfindungen zum Ausdruck kommen, die eher die allgemeine Qualität von Wohlbehagen bzw. Mißbehagen aufweisen, als etwa spezifische Ängste oder Aggressionen. Verschiedene Gefühlsqualitäten können oft nicht unterschieden werden (Warnes, 1979; Krystal, 1979; Lefebvre, 1980). Solche Patienten „wissen" ihre Gefühle nicht, sind in dieser Hinsicht „farbenblind", schließen höchstens indirekt aus der Reaktion anderer auf ihre eigene Verfassung und wirken in ihrem Verhalten oft hölzern-steif.

Kognitive Struktur

Eng verknüpft mit der Schwierigkeit, Gefühle zu äußern, ist eine bestimmte Struktur des Denkens, Handelns und Sprechens, die oft als operatives Denken („pensée opératoire" im engeren Sinn) beschrieben wurde (Marty und de M'Uzan, 1978; Marty et al., 1963). Damit ist eine funktionale, ganz auf die konkret faßbare Realität bezogene Denk- und Erlebnisweise gemeint, die äußere Gewohnheiten wie innere Zustände fast völlig auf ihr mechanistisch-instrumentelles Gefüge im Rahmen objektiver und allgemeiner Bezugssysteme reduziert. Solche Patienten werden als phantasiearm, farblos-trocken, unkreativ, unlebendig und wenig einfallsreich geschildert, obwohl sie beruflich oft außerordentlich erfolgreich und hinsichtlich ihrer Intelligenz sicher nicht eingeschränkt sind. Sie scheinen überhaupt wenig Beziehung zu ihrem „Innenleben" zu haben, vertreten eine pragmatische, handlungsorientierte Einstellung, so daß McDougall (1980) in diesem Zusammenhang von „acting disorders" spricht. Solche Patienten können sich – wenn überhaupt – nur selten an Träume erinnern und, etwa im Rahmen von Psychotherapien, durch assoziative Einfälle Zugang zu deren Bedeutung finden. Diese Phantasiearmut ist besonders an der Art und Struktur der Sprache aufgefallen und untersucht worden, die als dürr und eingeengt, devitalisiert und schablonenhaft, oft am nebensächlichen Detail haftend bezeichnet wird (Ruesch, 1948; Shands, 1975; Marty und de M'Uzan, 1978; Nemiah et al., 1976; Zepf, 1976; v. Rad und Lolas, 1978). Es ist der geringe oder ganz fehlende Symbolgehalt der Sprache, ihre Armut an privaten, konnotativen Bedeutungen, der sie eher als Handlungsfragment denn als Produkt einer psychischen Verarbeitung von Ereignissen erscheinen läßt. Aussagen und Wortinhalte werden kaum in einer persönlich angeeigneten Weise benützt, sie werden vielmehr in einer schablonenhaft-vagen, allgemeingültigen Form eingesetzt, so wie man und möglichst jeder andere in ähnlicher Situation etwas ausdrücken kann und würde (Shands, 1958; v. Rad und Lolas, 1978). Die Art dieses Denkens und Sprechens ist deshalb als ahistorisch-geschichtslos, ohne Bezug auf Vergangenes oder Zukünftiges, ganz dem funktionalen Aspekt des Hier und Jetzt verpflichtet, beschrieben worden. Dies sollte allerdings nicht verwechselt werden mit einer allgemeinen Restriktion sprachlicher Fähigkeiten, die z.B. mit einer Unterschichtzugehörigkeit in Verbindung gebracht werden kann – ein Befund, den Rost (1981) als „Pseudoalexithymie" bezeichnet hat.

Struktur der Selbst- und der Objektbeziehungen

In der bisherigen Beschreibung sind implizit bereits einige typische Wesensmerkmale der Selbst- und Ob-jektbeziehungsstruktur enthalten. Es ist von einer „disorder of individuation" gesprochen worden, da es solchen Menschen nicht möglich sei, den Begriff „Ich" in einem emotional bedeutungsgefüllten Kontext zu benutzen (Shands, 1958, 1975). In ihren zwischenmenschlichen Beziehungen sind sie symbiotisch eng an einen Partner – ihre Schlüsselperson – gebunden, mit dessen Hilfe sie ihr fehlendes Identitätsgefühl und ihre mangelnde Autonomie über ein System äußerer Absicherungen zu stabilisieren versuchen. Ein Defizit an Selbstwertgefühl und innerer Unabhängigkeit macht sie extrem abhängig von der Harmonie mit und der Zuwendung von der Schlüsselperson – eine ständig potentiell bedrohte Lebenssituation, die vielleicht erklärt, warum solche Menschen in hohem Maße verletzlich und krankheitsgefährdet sind, wenn es zur Trennung kommt. Umgekehrt wird einsichtig, daß unter günstigen Außenbedingungen eine symbiotische Partnerbeziehung lebenslang stabil bleiben kann. Für die symbiotischen Beziehungen psychosomatisch Kranker wurde der Terminus „réduplication projective" (Marty et al., 1963; Stephanos, 1973) eingeführt. Er nimmt Bezug darauf, daß solche Individuen stereotyp den anderen projektiv wie eine Art Doppelgänger ihrer selbst sehen und damit versuchen, ihre symbiotische Ungeschiedenheit aufrechtzuerhalten. Die mit diesem Terminus gefaßte Annahme ist in letzter Zeit kritisiert worden (Zepf und Gattig, 1986; Ullrich, 1988). Es scheint eher so zu sein, daß die psychosomatisch Kranken die Repräsentanz ihrer Schlüsselperson in sich selbst verdoppeln.

Ein weiterer wesentlicher Aspekt ihrer Abhängigkeit und Angewiesenheit auf einen Partner, der das eigene leib-seelische Gleichgewicht garantiert, äußert sich auch in einer ausgeprägten Unsicherheit beim Durchsetzen eigener Wertvorstellungen, in einem hohen Maß sozialer Konformität. Sie sind widerspruchsarm, neigen zum „goldenen Mittelweg", orientieren sich an dem, was „man" tut und zeigen in ihrer unauffälligen Angepaßtheit ein Verhalten, das sozial zumeist erwünscht und mit den Begriffen „pseudonormal" und „übernormal" beschrieben worden ist (McDougall, 1974; Brede, 1972; Zepf, 1976a). In dieser pseudonormalen Unauffälligkeit könnte ein Grund liegen, warum diese Merkmale so lange wenig Aufmerksamkeit fanden. In der Beziehung zu anderen Menschen, etwa einem Untersucher oder Psychotherapeuten gegenüber, macht sich deshalb leicht eine gewisse Leere und Langeweile breit, da die Gesprächspartner (wie auch andere Personen oder Gegenstände) lediglich in ihrer Faktizität, kaum jedoch mit affektiver Beteiligung oder in ihrer Beziehung zum Patienten deutlich werden. Das gleiche gilt auch für ihre Beziehung zum eigenen Körper, der wie etwas Fremdes erlebt und in seinen Störungen und Behinderungen im Rahmen von Erkrankungen oft mit einer Art stoischer Unbeteiligtheit ertragen wird. Diese Beziehungsleere, die „relation blanche" (Marty et al., 1963; de M'Uzan, 1977; Schneider, 1973) stellt eine spezifische Schwierigkeit in der Untersuchungs- und Behandlungssituation dar, die beim Arzt oft das Gefühl erweckt, er müsse diese Leere mit eigenen

Phantasien oder Aktivitäten, mit einer „energetischen Zufuhr" ausgleichen (Marty und de M'Uzan, 1978). So scheinen Psychotherapeuten, zumindest anfangs beim Erstgespräch oder im Rahmen von Behandlungen, bei Alexithymen häufiger zu intervenieren als bei neurotisch Kranken (Overbeck, 1977; v. Rad und Lolas, 1978; Balzer, 1981). Andererseits wird dieser Befund auch immer wieder so diskutiert, ob die „Alexithymie" nicht eher auf eine „Gegenübertragungs"-Reaktion des Untersuchers in einer bestimmten Situation zurückzuführen ist, als eine Verhaltensweise des Patienten darstellt (Cremerius, 1977, 1979; Wolff, 1977).

Insgesamt gibt es einen breiten Konsens im Schrifttum über die Phänomenologie und klinisch-deskriptiv beschreibbaren Merkmale alexithymen Verhaltens. In mehr oder minder typischer Ausprägung wurden sie beobachtet: bei Patienten mit Magen-Darm-Erkrankungen, insbesondere Colitis ulcerosa (Jackson, 1977; Freyberger, 1977; Zepf, 1976a, 1981; Fava und Pavan, 1976/77; de M'Uzan et al., 1958; Zepf und Tschirsch, 1987), bei chronischer Pankreatitis (Nakai et al., 1979), bei Asthmatikern (Zepf, 1976a, 1981; Kleiger und Jones, 1980), Hautkranken (Marty, 1958, 1969; Lefebvre et al., 1980), Koronarpatienten (Dongier, 1974; Defourmy et al., 1976/77), bei Fettsüchtigen (Waysfeld et al., 1977), Kopfschmerzpatienten (Timsit et al., 1975), Schwangerschaftsgestosen (Berger et al., 1976/77), bei rheumatisch Kranken (Shands, 1975), Patienten mit sog. „myofacial pain dysfunction" (Heiberg et al., 1978) oder „unexplained physical complaints" (Flannery, 1978) sowie bei manchen gesunden Zwillingen (Heiberg und Heiberg, 1978). Als „alexithym" interpretierbare Merkmale fanden sich auch bei heterogenen Gruppen psychosomatisch Kranker im Vergleich mit Gesunden (Zepf, 1976a, 1981), Neurotikern (v. Rad et al., 1979a, b; Tress, 1979; Taylor et al., 1981) und organisch Kranken (Tress, 1979). Bei diesen Untersuchungen wurden neben rein deskriptiven zumeist psychometrische und andere objektivierende Verfahren angewandt.

Demgegenüber ist die Frage der Ätiologie und der Pathogenese alexithymen Verhaltens Gegenstand einer sehr kontrovers geführten Diskussion. Auch über die Zusammenhänge zwischen alexithymen Auffälligkeiten und psychosomatischen Erkrankungen gibt es sehr unterschiedliche Vorstellungen. Im folgenden werden vor allem die psychodynamischen Beiträge zu diesem Themenkomplex diskutiert, während zahlreiche genetische (z.B. Heiberg und Heiberg, 1977, 1978) sowie neurophysiologische und neuroanatomische Ansätze (z.B. Nemiah, 1975, 1977; Hoppe, 1988; Ten Houten et al., 1985 a–d) unberücksichtigt bleiben. Zunächst werden die psychodynamischen Beiträge zu Einzelaspekten der Alexithymie wiedergegeben und anschließend die theoretischen Systematisierungsbemühungen skizziert. Dabei ist darauf hinzuweisen, daß aufgrund des empirischen Forschungsstandes gegenwärtig weder die eine noch die andere theoretische Konzeption als hinreichend verifiziert oder falsifiziert gelten kann. Eine kritische

Würdigung der empirischen Forschung findet sich bei v. Rad (1983) und bei Gerhards (1988).

5.7.2 Modellvorstellungen innerhalb der psychoanalytischen Strukturtheorie

Nachdem sich gezeigt hatte, daß die klassischen psychosomatischen Konzepte alexithyme Verhaltensweisen nur wenig erklären können, rückten neue Überlegungen in den Vordergrund. Gestützt auf Erfahrungen aus zum Teil langjährigen Psychoanalysen wurden die alexithymen Besonderheiten im Rahmen der neueren Entwicklungs- (Mahler, 1968) und Objektbeziehungspsychologie (Jacobson, 1964) sowie der Narzißmustheorie (Kohut, 1971, 1973; Kernberg, 1975, 1976) diskutiert. Entscheidend hierfür war die Erfahrung, daß sehr viele körperliche Symptome im Rahmen sog. typischer psychosomatischer Erkrankungen symbolisch ausdruckslos, ohne primären privaten Bedeutungsgehalt sind, auch wenn sie im Laufe einer individuellen Lebensgeschichte (sekundär) spezifische Bedeutung erlangen können.

Ausgangspunkt der meisten theoretischen Ansätze war der empirische Tatbestand, daß in der Primärsozialisation psychosomatisch Erkrankender mehrheitlich ein mütterliches Verhalten dominiert, das entweder als „überfürsorglich" oder als „offen oder verdeckt zurückweisend" bezeichnet worden ist (Zepf, 1976a; Liedtke, 1987), wobei oft beide (scheinbar widersprüchlichen) Verhaltensweisen gemeinsam nachweisbar sind. Es sind die eigenen ungelösten, besonders narzißtischen Konflikte dieser Mütter (und Väter), die auf das Kind übertragen, an ihm und mit seinem Körper abgehandelt werden (z.B. Sperling, 1949; Zepf, 1976a). Solche Mütter brauchen ihr Kind „wie eine Droge" (McDougall, 1980), überwachen es ängstlich-mißtrauisch mit steter Präsenz und entfalten dabei eine so intensive Kontrolle, als wäre es „ein Teil (...) des eigenen Körpers" (Sperling, 1949; McDougall, 1980). Das Kind wird zu einem „Selbst-Objekt" (Kohut, 1971) einer selbst symbiotisch abhängigen und trennungsunfähigen primären Bezugsperson. Oft stützt der Vater unbewußt die symbiotische Mutter-Kind-Verschränkung, um sich selbst dieser bedrohlichen Umklammerung zu entziehen. Werden Spannungen oder Unlustgefühle beim Kind deutlich, so versucht die Mutter diese durch beruhigende Manipulation zu beseitigen. Diese Spannungsbewältigung bleibt an ihre Realpräsenz als Reizschutz gebunden, so daß vom Kind keine eigenen Bewältigungsstrategien entwickelt werden können. In diesem Zusammenhang fiel auf, daß psychosomatische Patienten in ihrer Kindheit häufig kein Übergangsobjekt hatten, das als Grundlage späterer Phantasietätigkeit betrachtet wird (Winnicott, 1976; Rost, 1981; Eicke, 1973; Gaddini, 1974, 1977; Hoppe, 1964; M. Mitscherlich, 1977), welches ihnen eine allererste Unabhängigkeit von der Mutter erlaubt hätte.

In dieser Beziehungslage entstehen keine von Objekten abgegrenzten Selbstrepräsentanzen, deren Entwicklung in dieser frühen Phase unlösbar mit körperlichen Sensationen und deren Wahrnehmungen verknüpft ist. Da die Mütter dieser Kinder in hohem Maße auf die körperlichen Signale und Störungen ihrer Kinder reagieren, sie kontrollieren und manipulieren, wird der Körper als Teil der Selbstrepräsentanz nicht oder nur verzerrt eine seelische Vertretung und Bedeutung gewinnen können – er bleibt etwas Fremdes, nicht zum Selbst Gehörendes, das unter der Kontrolle der Schlüsselperson (Mutter) verbleibt. Das „Körperschema" ist pathologisch. Körperfunktionen werden so zum unmittelbaren Ausdruck für die Befindlichkeit im Umgang mit der Bezugsperson, sie spiegeln den symbiotischen Umgang des Primärobjektes, das oft ausschließlich auf solche konkreten Körperfunktionen in der Interaktion reagiert. Dies kann sich einmal in einer fast vollständigen Mißachtung, einer mangelnden Fürsorge und gelegentlich vital-gefährlichen Fehlwahrnehmungen von körperlichen Störungen äußern, die in ihrer Bedeutung und Signalfunktion nicht erkannt, differenziert und adäquat bearbeitet werden können. Ein solcher gefährlicher Mangel an Selbstfürsorge ist bei vielen psychosomatischen Patienten zu beobachten (Krystal, 1979; McDougall, 1980). Zum anderen zeigt sich auch oft eine überfürsorgliche, liebevoll-besorgte Überwachung des (körperlichen) Symptoms, das dann wie ein lebensnotwendiger Partner umsorgt, gepflegt und aufmerksam in seinen feinsten Befindlichkeiten beobachtet wird. In diesen beiden Grundverhaltensmustern, die sich auch abwechseln oder mischen können, läßt sich unschwer die Spiegelung der oben genannten Grundformen mütterlicher Verhaltensweisen bei psychosomatischen Patienten wiedererkennen: einmal die verdeckt oder offen zurückweisende, zum anderen die überfürsorglich-besorgte Haltung gegenüber dem Kind. Die Art, den Körper zu leben, entspricht der Art, nach dem Muster der primären Bezugspersonen mit Gefühlen umzugehen. In einem solchen Beziehungsfeld, in dem jegliche Trennung vermieden wird und ein stummes Reiz-Reaktionsmuster auf der Basis gegenseitiger optimaler Angleichung das Ziel bildet, ist es nicht überraschend, daß allgemeine Gefühlsvektoren, wie etwa Wohlbehagen und Mißbehagen, das Terrain bestimmen und somit differenzierte Gefühle kaum sichtbar werden.

Damit ergibt sich, daß die Fixierung in einer derartigen Mutter-Kind-Symbiose, die weit über die physiologisch notwendige Phase hinaus aufrechterhalten wird, die Entwicklung symbolischer Repräsentationen realer Gegebenheiten verhindert, z. B. die Entwicklung der Sprache anstelle von Handlungen, der Phantasien und der Internalisierung von Objekten. Da keine guten internalisierten Objektrepräsentanzen vorliegen, bleiben solche Menschen in einer suchtähnlichen Weise abhängig von der konkreten Verfügbarkeit realer äußerer (Ersatz-)Objekte, die sie anstelle symbolischer Repräsentanzen in der äußeren Welt in Dienst zu nehmen versuchen (McDougall, 1974). Dies macht auch die beschriebene Pseudonormalität verständlich, die zum Ausdruck bringt, daß sich diese Patienten an die Verhaltenserwartungen ihrer relevanten Beziehungspersonen übermäßig an-

passen. Einen ähnlichen Sachverhalt beschreibt Krystal (1974, 1978, 1979) in einem anderen begrifflichen Kontext. Bei Alexithymen seien die normalen Reifungsschritte der „emotions" (physiologische Vorgänge im Sinne von Sifneos, 1975) so gestört, daß es weder zur Affektdistanzierung, noch zu ihrer Verbalisierung und De-Somatisierung komme. Trifft dies zu, ist die Entwicklung „primär arretiert" (Ruesch, 1948), dann scheidet das Modell der „Re-Somatisierung" (Schur, 1974; A. Mitscherlich, 1967) zur Erklärung psychosomatischer Symptombildungen aus. Denn wenn ein Affekt gar nicht erst „desomatisiert" wurde, dann kann er auch nicht als Produkt einer „Re-Somatisierung" im Gefolge einer psychisch nicht erträglichen Konfliktsituation begriffen werden.

Uneinheitlich sind bisher die Antworten auf die Frage, ob und wenn ja, welche Abwehrmechanismen in der Genese einer psychosomatischen Erkrankung eine Rolle spielen. War eine am Neurosenmodell orientierte psychoanalytische Psychosomatik bislang weitgehend davon ausgegangen, daß eine Abwehr analog der Verdrängungsvorgänge bei neurotischen Symptomen angenommen werden kann, so wurden solche Überlegungen inzwischen – angesichts alexithymer Verhaltensweisen – im wesentlichen als fragwürdig angesehen. Statt dessen wurde oft zur Charakterisierung typischer psychosomatischer Abwehrvorgänge von Verleugnung („denial") gesprochen. Dieser Begriff bezeichnet jedoch ursprünglich die Vermeidung der Wahrnehmung äußerer Realität und nicht die Ausschaltung innerer Ereignisse (Nemiah, 1973, 1975). Soweit die körperlichen Symptome keinen Symbolcharakter haben, können hier nur sehr wenig ausdifferenzierte, globale und entwicklungsgeschichtlich frühe Abwehrmechanismen, wie sie sonst nur bei psychotischen und Borderline-Patienten vorkommen, wirksam geworden sein. Dabei handelt es sich im wesentlichen um die sog. Verwerfung (foreclosure), Spaltung (splitting), Projektion und projektive Identifikation. Durch ein „splitting" kommt es beispielsweise „zu keiner stabilen ganzheitlichen unbewußten Objektrepräsentanz, zu keiner Synthese zwischen negativen und positiven Aspekten des Objektes. Beim Überwiegen der ersteren und bei der noch mangelnden Fähigkeit des Ichs, sie ins Unbewußte zu verdrängen, werden die negativen von den positiven Qualitäten des Objektes ferngehalten und paranoid auf andere Partner verschoben …" (Benedetti, 1980). Die Selbst- und Objektanteile, die nicht abgespalten oder mit Hilfe der Verwerfung abgewiesen und annulliert werden, können so „gut" und idealisiert bleiben und die Fiktion einer harmonischen, ambivalenzlosen Einheit aufrechterhalten.

Zu Abwehroperationen dieser Art kommt es, wenn ein labiles Gleichgewicht entweder durch eine allzu große Nähe der primären Bezugsperson oder aber durch einen realen oder phantasierten Objektverlust bedroht ist. Die Nähe birgt die Gefahr einer Verschmelzung auf Kosten der rudimentären Ich-Identität bis hin zur psychotischen Dekompensation in sich, wie es sich in psychotherapeutischen Behandlungen psychosomatischer Patienten beobachten läßt

(Benedetti, 1980; Kütemeyer, 1953; McDougall, 1974; Meng, 1934). Häufiger jedoch führt eine Trennung zur Desintegration, weil mit dem Objekt unwiederbringlich Selbstanteile verlorengehen. Es bleibt dann nur noch hilflose Verzweiflung und ohnmächtige Wut, die nicht durch die psychische Bewältigung mit Hilfe von Trauerarbeit oder neuen Identifizierungen aufgefangen werden können.

Es sind die beiden Gefahrenpole des „Eins-sein" oder „Nichts-sein" („oneness oder noneness"), die das stabile narzißtische Gleichgewicht vieler psychosomatisch Kranker begrenzen und die mühsame und permanente Bemühung um einen optimalen Nähe-Abstand zur Schlüsselperson verständlich machen (Lefebvre, 1980). In solchen Gefahrsituationen wird bei psychosomatisch Kranken „der eigene biologische Organismus zum Projektionsschirm für ihre pathogenen Objektrepräsentanzen" (De Boor, 1976), sei es, daß er unbewußt als Teil der idealisierten oder der gehaßten Mutter angehört.

5.7.3 Spezifische Erklärungskonzepte

Eine Sonderstellung nehmen unter den psychoanalytischen Modellen die linguistisch beeinflußten Überlegungen der französischen Schule und der Entwurf von Zepf ein.

David et al. begannen etwa um 1960 eine Konzeption der psychosomatischen Erkrankung zu entwickeln, die sich von den bisherigen theoretischen Vorstellungen unterschied. Entsprechend ihrer Auffassung, daß es sich bei den psychosomatischen Erkrankungen um eine nosologische Einheit handelt (z.B. De Boor und A. Mitscherlich, 1973), die auch die Annahme einer besonderen psychosomatischen Struktur als Resultat einer besonderen Sozialisationspraxis impliziert, suchten die französischen Autoren – wie auch Zepf – dem Rechnung zu tragen. In Anlehnung an Ruesch (1948) beschrieben sie eine spezifische psychosomatische Persönlichkeitsstruktur, die sie von einer neurotischen, psychotischen, genitalen oder perversen Struktur abgrenzten (Fain, 1966; Fain und Marty, 1964, 1965; Marty, 1974; Marty und De M'Uzan, 1978; Marty et al., 1963; De M'Uzan und David, 1960).

Drei Merkmale wurden von ihnen als pathognomonisch für eine psychosomatische Struktur angesehen:
„La pensée opératoire": Mit diesem Begriff wird ein Denken beschrieben, das sich stets am Aktuellen und Konkreten orientiert. Es kann abstrakt und intellektuell sein, bleibt aber pragmatisch und instrumentell und ist in jedem Fall abgekoppelt von den innerpsychischen Objektrepräsentanzen. Diese Art des Denkens bestimmt die Sprache des psychosomatisch Kranken. Personen und Ereignisse werden ohne Beziehung zum Subjekt, gleichsam wie abphotographiert, geschildert.
„Réduplication projective": Gemeint ist damit, daß der psychosomatisch Kranke seine Objekte immer nur nach dem Muster seines eigenen, undifferenzierten Selbstbildes wahrnimmt.
Wie dieses, so ist auch das dritte Merkmal einer psychosomatischen Struktur, die „inhibition fantasmatique du base",

ein Mangel an unbewußten Phantasien, mit dem operativen Denken verbunden. Wenn diese Patienten phantasieren, dann sind die Phantasien entweder in der Struktur undifferenziert oder aber sie bestehen in einer einfachen Reproduktion faktisch erlebter Situationen. Später wurde noch als weiteres Merkmal eine soziale Überangepaßtheit hinzugenommen (McDougall, 1974; Stephanos und Auhagen, 1979).

In einer anders gearteten Terminologie rechnet Zepf (1976, 1976a, 1981) psychosomatisch Kranken ferner mehrheitlich folgende psychische Eigentümlichkeiten zu: Undifferenzierte unbewußte Phantasien, narzißtische Objektbeziehungen (Selbst-Objekte im Sinne von Kohut, 1973), ein labiles oder reduziertes Selbstwertgefühl sowie fehlende aggressive Verhaltensweisen. Die französischen Autoren wie auch Zepf nehmen praktisch dieselben empirischen Befunde zum Ausgangspunkt ihrer theoretischen Überlegungen, in denen sie jedoch in entscheidenden Punkten differieren.

„La pensée opératoire": Die französische psychosomatische Schule

Der theoretische Ansatz der französischen Schule ging von dem Versuch aus, die von ihnen beschriebenen psychischen Eigentümlichkeiten psychosomatisch Kranker im Rahmen psychoanalytischer Modelle einheitlich und systematisch zu interpretieren (De M'Uzan, 1977). Ihren Überlegungen lag dabei zunächst das topographische Modell Freuds zugrunde, welches den psychischen Apparat in die Systeme Bewußtsein, Vorbewußtes und Unbewußtes gliedert. Das operative Denken wird als Resultat einer Diskontinuität von Bewußtem und Unbewußtem verstanden, die durch das Fehlen des Vorbewußten bedingt sei. Es erscheine aufgrund seiner Bindungen an Kausalität, Kontinuität und Realität als Modalität des Sekundärvorganges, habe sich aber nicht über die Bearbeitung von Phantasien aus dem Primärvorgang ausgeformt und sei deshalb unfähig zur symbolischen Repräsentation. Worte würden Handlungen und Sachen lediglich verdoppeln, wobei die Distanz zwischen Bezeichnendem (signifiant) und Bezeichnetem (signifié) gleichsam aufgehoben sei (Marty und De M'Uzan, 1963). Als Ursache wird eine übertriebene Reizabschirmung durch eine „mère calmante" wie auch eine das Kind zurückweisende Mutter angeführt, die sich dazu „antithetisch" verhält (Fain, 1971; Fain und Kreisler, 1970; McDougall, 1974). In beiden Fällen werde die Entwicklung autoerotischer Aktivitäten unterdrückt und das Stadium der halluzinatorischen Wunscherfüllung nicht erreicht. Das Es könne sich somit nicht von seinem somatischen Ursprung lösen und keine psychischen Repräsentanzen für triebhafte Ziele bilden (Fain, 1971; Fain und Marty, 1965).

Dieser erste Begründungsversuch der „pensée opératoire" wirft Fragen auf. So wird das operative Denken einerseits als logisch, kausal und realitätsbezogen bezeichnet und als Modalität des Sekundärvorganges hingestellt; andererseits aber wird festgestellt, daß im operativen Denken die Distanz zwischen Bezeichnendem und Bezeichnetem aufgehoben sei. In dieser Hinsicht funktioniert das operative Denken damit analog dem Primärvorgang, der Logik, Kausalität und

Realitätsorientiertheit ausschließt. Offen bleibt auch, inwiefern ganz widersprüchliche Erfahrungen der Mutter gleichermaßen eine mangelhafte Verbindung von libidinösen Aktivitäten und Sprache bewirken sollen. Die Annahme einer spezifischen psychosomatischen Struktur wird ferner dadurch problematisch, daß sie innerhalb des topographischen Modells nicht von andersartigen Strukturen abgegrenzt wird. Ferner wird innerhalb des Bezugsrahmens der topographischen Theorie immer wieder mit strukturtheoretisch definierten Begriffen argumentiert, ohne daß die in vielerlei Hinsicht prinzipiell unterschiedlichen und sich widersprechenden Theorien (Arlow und Brenner, 1976) metatheoretisch vermittelt worden wären. Ihre Gesamtkonzeption besteht damit aus einer Reihe von unverbundenen Hypothesen.

Konsequenterweise nahmen die französischen Autoren 1976 Abstand vom topographischen Modell Freuds. Sie erweiterten ihren Bezugsrahmen, um so auch die Genese der „primären Fixierungsmechanismen" als Prämissen psychosomatischer Symptombildungen theoretisch einordnen zu können (Marty, 1976). Die Fixierung auf primäre physiologische Mechanismen erfolgt nach Marty (1973) bereits intrauterin durch eine pathologische humorale Interaktion zwischen Fetus und mütterlichem Organismus. An dem Konzept einer Diskontinuität von psychischen Abläufen und körperlichen Triebprozessen wird festgehalten, welches nun auch auf das Fehlen eines sinnvollen Zusammenhangs zwischen psychischem Konflikt und psychosomatischem Körpersymptom bezogen wird (z.B. Stephanos und Auhagen, 1979). Diskontinuität und primäre Fixierungsmechanismen werden jetzt als Resultat eines pathologischen Zusammenspiels von Lebens- und Todestrieb verstanden (Marty, 1976). Es wird angenommen, daß alle psychologischen und physiologischen Entwicklungsprozesse durch die Wechselwirkung dieser beiden Triebe bestimmt werden, wobei sich der „Eros" in „evolutionären" Strukturbildungen und der „Thanatos" in „antievolutionären", desorganisierenden Veränderungen materialisieren sollen. Unter der zeitweiligen Herrschaft des Todestriebes würden im Säuglingsalter pathologische Dysfunktionen den Lebenstrieb zu Reorganisationsprozessen mit einer Stärkung der Libido und des psychischen Apparates aktivieren. Als psychosomatische Fixierungsmechanismen blieben jedoch die Dysfunktionen erhalten. In Situationen, in denen der Lebenstrieb geschwächt sei und/oder keine adäquaten psychischen Verarbeitungsmechanismen verfügbar wären, drohe ihre Reaktivierung und eine pathologische Desorganisation, die aber bei einem erneuten Erstarken des Lebenstriebes, z.B. durch Zuwendung eines Objektes, wieder überwunden werden könne.

Bedingt durch eine konstitutionelle Schwäche des Lebenstriebes (Marty, 1973, 1976), welche nicht durch mütterliche Zuwendungen kompensiert würde (Stephanos und Auhagen, 1979), komme es nun bei psychosomatisch Erkrankenden zu einer biophysischen Reifungsstörung mit mangelhafter Strukturierung des psychischen Apparates, dem Fehlen des „gu-

ten" inneren Objekts und der Unfähigkeit zur Symbolisierung. Daraus resultiere das operationale Denken. In Krisensituationen bestehe die Tendenz zur Regression auf die Fixierungsstellen mit der Gefahr eines psychischen Zusammenbruchs, deren Folge dann das psychosomatische Symptom sei.

An dieser Neuformulierung des Konzeptes bleibt problematisch, daß die Lehre vom „Eros" und „Thanatos" dem topographischen und dem Strukturmodell der Psychoanalyse ohne metatheoretische Vermittlung nur angefügt wird. Es ist ferner zweifelhaft, ob sich für eine Reduktion der menschlichen Natur auf diese beiden Grundtriebe – Freud (1969) selbst bezeichnete sie als „spekulativ" – eine wissenschaftliche Begründung finden läßt (vgl. Braun, 1979; Brun, 1953, 1954; Fenichel, 1935; Jones, 1962; Reich, 1970). In der Biologie ließ sich bisher keine Beobachtung finden, welche die Annahme eines Todestriebes zwingend notwendig machen würde.

Restriktive Praxis und defizitäre Symbolbildung

In diesem Konzept von Zepf (1976, 1986) wird die psychosomatische Strukturbildung in der von Lorenzer entwickelten „Theorie der Interaktionsformen" entfaltet, der zufolge das menschliche Individuum dadurch Geschichte gewinnt und zu einem Subjekt wird, daß sich die Beziehungen, in die es involviert war, als sog. „Interaktionsformen" im Individuum niederschlagen. Durch die Art und Weise, in der Mutter und Neugeborenes miteinander umgehen, werden die zunächst undifferenzierten Körperspannungen zu bestimmten Triebbedürfnissen in Interaktionsformen qualifiziert. Neurophysiologisch kann man sich diesen Prozeß so vorstellen, daß die lebenspraktisch bestimmten Interaktionsformen in Gestalt einer Zusammenschaltung der daran beteiligten somatischen Abläufe als sog. „Interaktionsengramme" gespeichert werden. Die Lebenspraxis führt somit nicht nur auf psychologischer, sondern auch auf somatischer Ebene zur Gründung der Subjektivität „dieses" Kindes. Mit der Einführung von Sprache entwickeln sich die Interaktionsformen zu „symbolischen" und damit zu bewußten. Der Symbolbegriff bezieht sich hier immer auf eine bestimmte Kombination von Sprache und Interaktionsformen, wobei die Interaktionsformen die konnotativen, privaten Bedeutungen der „öffentlichen" Sprache konstituieren, die ihre denotativen Bedeutungen über eine bloß „gezeigte" Praxis, in der das Kind nicht an Interaktionsprozessen teilhat (Zepf, 1976, 1976a), sowie über metasprachliche Definitionen erworben hat. Als allgemeine Bedingungen für die Entwicklung dieser symbolischen Interaktionsformen gelten hier die Einführung vielfältiger Prädikatoren sowie ein breites und konstantes mütterliches Interaktionsangebot.

Auch Zepfs Ausgangspunkt ist das überfürsorgliche und/oder offen zurückweisende Verhalten der Mütter psychosomatisch Kranker. Dieses Verhalten wird als Folge einer ungelösten narzißtischen Problematik der Mutter verstanden, wobei sich im faktischen mütterlichen Verhalten einmal die eine, ein andermal die andere Seite dieses Konflikts offener darstellen und die jeweils dazu gegenläufige Tendenz unterschwellig wirken kann. Durch dieses widersprüchliche Verhalten wird sich ein breites, in sich differenziertes und konstantes mütterliches Interaktionsangebot, welches zur Einübung vielfältiger und selektiver Interaktionsformen notwendig wäre, kaum einstellen können. Es herrscht vielmehr eine „restriktive Sozialisationspraxis" (Zepf, 1976) vor, in der eine Aufgliederung des kindlichen Triebbedarfs in spezifische und selektive identifizierbare Triebbedürfnisse nur mangelhaft gelingt. Weil sich die kindlichen Spannungszustände nicht genügend in Interaktionsformen ausdifferenzieren können, lassen sie sich auch nicht in die sprachliche Semantik einbinden. Die sich herstellenden Interaktionsformen sind undifferenziert, bleiben vorsprachlich und unbewußt. Da sich in der „Theorie der Interaktionsformen" das Beziehungsgefüge von Mutter und Kind erst auf der Höhe symbolischer Interaktionsformen in ein kindliches und mütterliches Subjekt auseinanderdifferenziert, wird auch die Genese der als „symbiotisch" oder „narzißtisch" beschriebenen Objektbeziehungen psychosomatisch Kranker auf eine restriktive Sozialisationspraxis zurückgeführt.

Die sinnlich-emotionale, innere Leere psychosomatisch Kranker wird darauf zurückgeführt, daß die Beziehungen zu den Objekten eingeschränkt sind, wodurch die eingeführten sprachlichen Prädikatoren sich mehrheitlich auch nur auf Objekte beziehen können, deren Beziehungsaspekt rudimentär ist. In einer restriktiven Praxis bleibt die eingeführte Sprache vorwiegend in einem bloß denotativen Bezug.

Die Sprache gewinnt hier den Charakter von „primären" Zeichen, die Zepf von den „sekundären" Zeichen abhebt, die Lorenzer als Produkt einer neurotogenen Sozialisation beschreibt. „Sekundäre" Zeichen sind das Resultat von Abwehroperationen, in denen eine vorher bestehende Beziehung von Sprache und Interaktionsformen sekundär aufgespalten wird. Bei „primären" Zeichen ist eine derartige Beziehung nicht zustande gekommen, so daß hier die Sprache durch eine konnotative Armut, durch einen Mangel an lebensgeschichtlicher Körpererfahrung ausgezeichnet ist.

Geht man davon aus, daß in einer restriktiven Sozialisationspraxis auch das Es nur ungenügend in Interaktionsformen strukturiert wird, dann erklären sich daraus auch die unstrukturierten unbewußten Phantasien psychosomatisch Kranker, die nur in rudimentärer Beziehung zur Sprache stehen (vgl. De Boor, 1964). Die soziale Überangepaßtheit wird durch die Annahme verständlich gemacht, daß im Gefolge einer mangelhaften Differenzierung auch die Spannungszustände undifferenziert und unlustvoll bleiben. Sie entstehen als subjektives Korrelat von Trennungserfahrungen und lassen sich dadurch vermeiden, daß die „primäre Ungeschiedenheit" (primary confusion; Sandler und Joffe, 1967) mit der Mutterfigur wieder hergestellt wird. Aufgrund dieser narzißtischen Bedürftigkeit vermeiden psychosomatisch Kranke aggressive Verhaltensweisen und entwickeln

ihre instrumentellen (Ich-)Funktionen vor allem im Hinblick darauf, sich selbst in Übereinstimmung mit den Erwartungen der Mutter oder ihrer späteren Ersatzfigur („key figure"; Engel, 1955) zu bringen und zu halten.

Auch in diesem Konzept hat der „Objektverlust", d.h. eine phantasierte, drohende oder reale Auflösung der narzißtischen Beziehung zum mütterlichen Ersatzobjekt, einen zentralen Stellenwert für die Bildung eines psychosomatischen Symptoms. Die Auflösung dieser Beziehung bedeutet für den psychosomatisch Erkrankenden eine Wiederbelebung jener früheren Situationen, in denen es im Gefolge einer Trennung zu unlustvollen Spannungszuständen gekommen war. Das psychosomatische Symptom wird dann verstanden als Resultat der erneuten Versuche, die reaktualisierten, undifferenzierten Körperspannungen zu bewältigen. Zepf greift dabei auf eine Überlegung von Schur (1974) zurück und nimmt an, daß die aktuelle Situation nicht nur regressiv eingeschätzt, sondern auch regressiv beantwortet wird. Engel und Schmale (1978) haben das gleiche Verfahren als Konversion beschrieben. Damit ist gemeint, daß die in dieser Situation aktualisierten unbewußten Interaktionsformen unter der Bedingung, daß keine Ersatzobjekte mehr zur Verfügung stehen, im Zuge einer Verschiebung den Repräsentanzen äußerer, dinglicher Gegebenheiten (z. B. der Luft, dem Allergen oder dem Dosieraerosol beim Asthmapatienten) oder eines Körperorgans zugeschlagen werden. Im Erleben des Patienten gewinnen sie dadurch eine unbewußte Bedeutung. Dieser unbewußte Bedeutungszuwachs kann z. B. verständlich machen, warum psychosomatische Patienten in der Schilderung ihres Körpers oft wesentlich mehr emotional engagiert sind als in der Schilderung ihrer Objektbeziehungen.

Auf diesen unbewußten Bedeutungszuwachs erfolgt nun eine Reaktion mit pathophysiologischen Mechanismen. Dies wird damit begründet, daß beim psychosomatisch Kranken aufgrund einer restriktiven Praxis auch die somatischen Funktionsabläufe nur mangelhaft strukturiert wurden. Da körperliche Funktionsabläufe über die Bildung von „Interaktionsengrammen" zueinander in Beziehung gesetzt und entsprechend der Lebenspraxis aufeinander abgestimmt werden, bleibt unter der Bedingung einer restriktiven primären Sozialisationspraxis die Struktur ihres Zusammenspiels auf jenem undifferenzierten Status, den z. B. Grinker (1953) als „physiologischen Infantilismus" beschrieb. Dieses nur mangelhaft sozialisierte, relativ unorganisierte Zusammenspiel von Körperprozessen wird unter der drohenden oder realen Auflösung einer narzißtischen Objektbeziehung aktuell. Somatisch „wildwüchsig" oder „archaisch" wird nun auf den unbewußten Bedeutungszuwachs reagiert. Mit der körperlichen Reaktion zieht sich der Patient auf jene physiologische Ebene der Interaktion zurück, auf der der mütterliche und der kindliche Organismus in noch ungeschiedener Einheit miteinander verbunden waren. Dadurch werden die Objektbeziehungen auf den Status von biologischen Reiz-Reaktions-Zusammenhängen zurückgenommen.

Auch dieser Entwurf läßt Fragen offen. So werden zwar die allgemeinen Bedingungen kenntlich gemacht, welche über die Wahl des Organismus als Stätte der Symptombildung entscheiden. Die besonderen Gegebenheiten jedoch (somatisches Entgegenkommen?), die für die jeweils spezifische Körpersymptomatik verantwortlich zu machen sind, werden nicht expliziert. Auch das Problem der sog. „Mischstrukturen", bei denen zugleich eine neurotische und psychosomatische Erkrankung nachweisbar ist, wird in diesem Konzept nicht zureichend gelöst. So ist beispielsweise der Vorgang der „Desymbolisierung", der bei neurotischen Erkrankungen in diesem konzeptionellen Rahmen im Vordergrund steht, bisher noch nicht unter den Bedingungen einer restriktiven primären Sozialisationspraxis untersucht worden.

5.7.4 Zusammenfassung

In den letzten Jahren gab es zunehmend Versuche, das Alexithymiekonzept empirisch zu validieren und zu spezifizieren. Dabei gab es kaum neue Fragestellungen, wohl aber neue Untersuchungsmethoden und -instrumente. Gleichwohl blieben die Befunde inkonsistent. Dies liegt einmal daran, daß sehr verschiedene Testverfahren verwendet wurden, die in methodologischer Hinsicht problematisch sind und deren Ergebnisse sich nicht vergleichen lassen (Taylor, 1984; Lesser, 1985; Bagby et al., 1988; Gerhards, 1988). Zum anderen gründet die Inkonsistenz der erhobenen Befunde auch darin, daß die theoretischen Bezugsrahmen der Untersuchungen kaum expliziert wurden, so daß auch unklar blieb, auf welche Art von Gefühlen – Affekte, Stimmungen, Orientierungs-, kognitiv-situative oder Persönlichkeitsgefühle – sich die erhobenen Befunde beziehen. Ein Gefühlsbegriff wie etwa „Furcht" bezieht sich nicht auf empirisch homogene, sondern auf heterogene Gefühle, deren qualitative Spezifik sich aus dem jeweiligen theoretischen Bezugsrahmen ergibt. Wenn man sich etwa vor einem herannahenden Raubtier fürchtet, dann ist die dabei erlebte Furcht in der Theorie von Heller (1981) ein Affekt, weil sie an die wirkliche Anwesenheit eines auslösenden Reizes gebunden ist; fürchtet man sich auf dieses oder jenes Unternehmen einzulassen, dann handelt es sich um ein kognitiv-situatives Gefühl, weil das, weswegen und wem gegenüber man fühlt, zum Gefühl selbst gehört, und ist man ein allgemein furchtsamer Mensch, dann handelt es sich bei dieser Furcht um ein Persönlichkeitsgefühl.

Mittels objektivierender Verfahren konnte bisher jedenfalls nicht nachgewiesen werden, daß Defizite in der Verarbeitung von Gefühlen generell für psychosomatisch Kranke (im engeren Sinne) spezifisch sind. Aus der mangelhaften Befundlage läßt sich jedoch nicht der Schluß ziehen, daß das Alexithymiekonzept widerlegt sei. Gravierender erscheint die Frage, ob nicht der Versuch, Alexithymie als Persönlichkeitskomponente psychometrisch zu erfassen, der Verdinglichung eines Phänomens entspricht, welches in Wirklichkeit nur Ausdruck einer sozialen Beziehung ist und sich auch nur in sozialen Beziehungen darstellen kann. Wiederholt wurde alexithymes Verhalten als Artefakt der Untersuchungsbedingungen gedeutet. Dabei wurden situative Effekte hervorgehoben, u.a. Aspekte der Gegenübertragung des Untersuchenden (Cremerius, 1977; Taylor, 1977; Wolff, 1977; Ahrens, 1988). Es wurde auch auf eine Begünstigung des operativen Denkens durch die Normen technokratischer Gesellschaften hingewiesen (De M'Uzan, 1977). Aus heutiger Sicht beschreibt der Begriff der Alexithymie ein bestimmtes kommunikatives Verhalten, welches sich vermutlich auf der Grundlage von hereditären und physiologischen, vor allem über lebenspraktische Gegebenheiten und insbesondere in dyadischen Situationen manifestiert.

5.8 Die anthropologische Medizin Viktor von Weizsäckers

Die abschließend zu besprechenden Überlegungen Viktor von Weizsäckers zu einer anthropologischen Medizin stellen insofern einen Sonderfall dar, als es sich bei seinem Konzept nicht um einen psychoanalytischen Entwurf somatischer Symptombildungen im engeren Sinne handelt. Viktor von Weizsäcker, zunächst als Physiologe und Philosoph und erst später dann neurologisch-internistisch tätig, wurde zu einem der Begründer der Psychosomatischen Medizin in Deutschland. Seine Begegnung und Auseinandersetzung mit Sigmund Freud prägten seinen eigenständigen und bis heute nicht ausgeschöpften philosophischen Entwurf weitreichend. Wir beschränken uns hier auf einen kleinen Ausschnitt seines Werkes, nämlich die Skizzierung einiger weniger Grundelemente dessen, was als „anthropologische Medizin" oder als „Einführung des Subjektes in die Medizin" (und die Biologie) mit seinem Namen verknüpft ist.

Gestützt auf seine Gestaltkreis-Lehre (v. Weizsäcker, 1973), entwickelte er eine Anthropologie, die den Menschen zu verstehen sucht in seiner Abhängigkeit „von einem Grund, der selbst nicht Gegenstand werden kann" (v. Weizsäcker, 1973). Es ist ein Mensch, der als Teil der Natur geprägt ist durch seine ontische und pathische Existenz – Modalitäten, die an dem für Weizsäcker zentralen Begriff der „Krise" verdeutlicht werden. Einerseits ist für ihn das Subjekt in seiner Existenz ständig durch Krisen (die sich medizinisch z.B. in Form von Krankheiten äußern können) bedroht – andererseits hat es auch die Möglichkeit, über sich selbst hinauszufinden. Die Krise markiert also einen Wendepunkt, ja sie ist geradezu „Vermittlerin biologischen Geschehens" (Schneemann, 1967). Weizsäcker faßt die Struktur solcher krisenhaften Erkrankungen, die oft den Charakter der Stellvertretung für einen ungelösten Konflikt haben, folgendermaßen zusammen: „Eine Situation ist gegeben, eine Tendenz kommt auf, eine Spannung steigt an, eine Krise spitzt sich zu, ein Einbruch der Krankheit erfolgt, und mit ihr, nach ihr ist die Entscheidung da; eine neue Situation ist geschaffen und kommt zur Ruhe; Gewinne und Verluste sind jetzt zu übersehen, das Ganze ist wie eine historische Einheit: Wendung, kritische Unterbrechung, Wandlung" (v. Weizsäcker, 1935).

Jede Krankheit ist für v. Weizsäcker nicht nur objektives Schicksal, sondern hat ihren ganz persönlichen Sinn in der Geschichte und dem Lebensentwurf des einzelnen. „Das Problem des Menschen in der Medizin – oder speziell in dieser neuen Art Medizin – ist, daß er, der Mensch, seine Krankheit, die als Teil seiner ganzen Biographie zu verstehen ist, nicht nur hat, sondern auch macht. Daß er die Krankheit, die Ausdrucksgebärde, die Sprache seines Körpers produziert, wie er jede andere Ausdrucksgebärde und jedes andere Sprechen formt. Noch verstehen wir diese Sprache nicht ganz, aber wir kommen immer näher an sie heran" (v. Weizsäcker, 1953).

Von Weizsäckers Subjektbegriff nimmt Bezug auf einen Menschen, der auch in seiner Krankheit gleichermaßen bestimmt ist durch seine Geschichtlichkeit, seine Sozialität und seine Finalität (v. Rad, 1974). Das hat Konsequenzen z.B. für sein Krankheitsverständnis: „Die Krankheit liegt jetzt zwischen den Menschen, ist eines ihrer Verhältnisse und ihrer Begegnungsarten. Hier beginnt anthropologische Medizin" (v. Weizsäcker, 1947). Dabei beschreibt er das Verhältnis von Körper und Seele als komplementären „Umgang" gegenseitiger Darstellung und Erläuterung. „Wir hörten, daß die Darstellungsfunktion gegenseitig ist: der Leib stellt die Seele, die Seele den Leib dar. Das Wichtigste in diesem Wechselspiel ist, daß sie einander vertreten" (v. Weizsäcker, 1948). Lange vor dem Entstehen einer Sozialmedizin oder Medizinischen Soziologie hat v. Weizsäcker die soziale Dimension der Krankheit immer wieder kritisch artikuliert und versucht, sie theoretisch zu begründen (v. Weizsäcker, 1955). Seine medizinische Anthropologie schließt konsequent ihrem Ansatz folgend die Finalität, den Entwurfcharakter des Lebens und die Sinnfrage mit ein.

Hier insbesondere liegt auch seine Kritik an der Psychoanalyse und einer Psychosomatischen Medizin, die nur „psychische Gesichtspunkte" additiv den unverändert beizubehaltenden somatologisch-physikalischen Denkweisen ergänzend an die Seite zu stellen versucht. „Die Psychosomatische Medizin fragt meistens: Was ist dieser Mensch? Die anthropologische aber: Was wird dieser Mensch? Aber freilich wird das Pathische das Ontische nie los …" (v. Weizsäcker und Wyss, 1957). An anderer Stelle fragt er pointiert: „Warum immer nur Schuldgefühle und nie Schuld?" (v. Weizsäcker, 1949).

Gerade auch die Hartnäckigkeit, mit der v. Weizsäcker die Dimension der Wert- und Sinnfragen im Zusammenhang mit Krankheit immer wieder herausgestellt hat, hat verhindert, daß sein bedeutender Entwurf in die Medizin integriert wurde. Das hängt wohl auch damit zusammen, daß der Begriff der Krankheit für v. Weizsäcker immer auch unlösbar verknüpft war mit der Beziehung des Kranken zu seinen Mitmenschen, seinem Platz in der Gesellschaft. Krankheit ist für ihn stets so etwas wie ein Herausfallen aus der Wahrhaftigkeit; ein Riß in der Verwirklichung des „richtigen" Lebens, das möglich wäre; ein Bruch in der Solidarität der Gegenseitigkeit, in die die Menschen gestellt sind. Krankheit ist damit nicht mehr einzugrenzen und zu isolieren auf gestörte Organfunktionen, auf rein objektive Bedingungen, z.B. auf irgendwelche Krankheitserreger. Sie liegt für ihn „zwischen den Menschen", also auch zwischen Arzt und Patient, und stellt somit ebenso unerbittlich die Frage nach der Wahrhaftigkeit, der Solidarität und der Verantwortung des Arztes. Dieser soziale Krankheitsbegriff bezieht die ganz persönliche Sinnfrage des menschlichen Lebens und Leidens bis hin zur religiösen Erfahrung mit ein, wodurch für ihn die „recht verstandene Psychosomatische Medizin (…)" einen umstürzenden Charakter" (v. Weizsäcker, 1949/50) gewinnt. Diese Herausforderung an das Selbstverständnis der Medizin und ihrer Vertreter blieb in der Einschätzung von De Boor und Mitscherlich jedoch „die Parole einer Revolution, die wir analytischen Psychosomatiker in den vergangenen 23 Jahren nicht zustande gebracht haben" (De Boor und A. Mitscherlich, 1973).

Viktor von Weizsäcker hat wie nur wenige andere die psychosomatische Zugangsweise nicht als ein Spezialfach „Psychosomatische Medizin", sondern als revolutionäre Änderung der gesamten Medizin und Biologie verstanden. Dabei bleibt kritisch anzumerken, daß sein faszinierender Entwurf sich oft sehr schillernder und im ärztlichen Alltag schwer konkretisierbarer Begriffe bedient, die der Sache nicht immer förderlich waren. Wenn der späte Viktor von Weizsäcker etwa sich zu der These hinreißen ließ, „daß jede Erkrankung unbewußte Schuld enthält …" (v. Weizsäcker, 1956), dann rückt sein Krankheitsbegriff in eine gefährliche Nähe zu alten theologischen Überlegungen, in denen Krankheit als Schuld interpretiert wurde. Dennoch hat er gerade auch mit der Frage nach der Finalität des Menschen und seiner Krankheit wichtiges Neuland erschlossen, das von Medizin und Psychoanalyse gleichermaßen bislang gemieden wird. Eine entscheidende Frage Viktor von Weizsäckers greifen De Boor und A. Mitscherlich (1973) erneut auf und drängen auf Präzisierung: „Körper und Seele gehen miteinander um, hat Viktor von Weizsäcker formuliert, aber mit welcher Syntax?"

6 Die früheste Kindheitsentwicklung und ihre Störungen aus der Sicht Winnicotts

Lore Schacht

Thure von Uexküll

„A word like ‚self' naturally knows more than we do; it uses us, and can command us." *D. W. Winnicott* (* Anmerkung 1)

Vorbemerkung

Die vorangehenden Kapitel machten deutlich, wie groß das Interesse der Psychosomatischen Medizin an den frühesten Perioden der kindlichen Entwicklung und deren Störungen ist. Der in vielen Untersuchungen auftauchende Hinweis auf frühe Störungen in den Beziehungen zur Mutter als disponierendem Faktor zu den verschiedensten körperlichen Erkrankungen kann ohne einen Einblick in die vielfältigen und verschiedenartigen Aufgaben, die diese Beziehungen für die körperliche und seelische Entwicklung des Kindes haben, gar nicht verstanden werden.

Die besondere Schwierigkeit, mit der die Erforschung der frühen Entwicklungsphasen und die Darstellung ihrer Ergebnisse zu kämpfen hat, liegt darin begründet, daß unsere vorsprachliche Existenz nur in Annäherung und oft genug nur metaphorisch mit den Mitteln der Sprache zu fassen ist. Die Gefahr einer adultomorphen Mißdeutung der kindlichen Welt wird um so größer, je frühere Phasen unserer Entwicklung gedeutet werden müssen.

Schließlich tauchen mit der Hinwendung zu den frühen Vorstufen unserer kognitiven Entwicklung auch erkenntnistheoretische Probleme von großem Gewicht auf. Sie zu sehen und zu formulieren, erfordert eine philosophische Betrachtungsweise und ein hohes Maß an Unvoreingenommenheit – vor allem aber eine neue Sprache, die in der Lage ist, die neuen Einsichten zu beschreiben. Das alles können wir bei Winnicott lernen.

In Kapitel 16 wird näher auf die Bedeutung der von Winnicott entscheidend beeinflußten neuen Entwicklungen in der Theorie für die psychoanalytischen Therapieverfahren eingegangen.

6.1 Vorwort

Der englische Titel des ersten Sammelwerkes, „Through Paediatrics to Psycho-Analysis" (1958), umgrenzt besser als seine deutsche Übersetzung „Von der Kinderheilkunde zur Psychoanalyse" in nur wenigen Worten den Bereich klinischer Tätigkeit des Pädiaters und Psychoanalytikers D. W. Winnicott (1896–1971), wirkt er doch im Originaltext dynamischer und läßt

somit mehr das Hin und Her zwischen Winnicotts Aufmerksamkeit an der therapeutischen Arbeit mit Kindern und der mit Erwachsenen – eine die andere fortwährend stimulierend – ahnen.

Schon in seinen frühen Artikeln ist das Prisma seiner Beobachtungsfähigkeit und seiner Interpretation des Umganges zwischen dem heranreifenden Subjekt und dessen Umwelt als eben das gleiche erkennbar, das in dem weiteren Werk erst explizit dargelegt wird. So verweist der klassische Aufsatz „Die Beobachtung von Säuglingen in einer vorgegebenen Situation" (1941) nicht nur auf sein Interesse an der klinischen Situation, d.h. am Setting, sondern unterstreicht darüber hinaus sein Gespür für die Interdependenz zwischen der Mutter-Kind-Beziehung und der möglichen Erfahrung des Spielens – beschrieben in dem Spiel des Säuglings mit dem Spatel. Dies ist die Erfahrung, die Winnicott in dem späteren Konzept vom Übergangsobjekt im „intermediären Raum" ansiedeln wird. Nicht zuletzt werden aber schon damals Hinweise auf Winnicotts Idee von der fördernden Umwelt gegeben.

Der Kreis seiner originellen klinischen Forschung, die in der jahrzehntelangen Tätigkeit am Paddington Green Children's Hospital ihren Ausgang genommen und später Bereicherung ganz besonders in der Behandlung erwachsener Borderline-Patienten gefunden hatte, schließt mit der Beschreibung des Schnörkelspiels während der „therapeutischen Arbeit mit Kindern" (1973). Hier wird die Erfahrung des „intermediären Raumes" sowohl beim Therapeuten als auch beim Kind als Voraussetzung für die geglückte therapeutische Begegnung demonstriert.

In dem folgenden Essay wird der Versuch unternommen, einer der weiteren gebündelten Linien des Winnicottschen Denkens zu folgen, die, durch seine so eigene Betrachtungsweise entstanden, sporadisch über sein Werk verteilt auftaucht.

6.2 Der Begriff des Selbst

Das Selbst ist für Winnicott immer das werdende Selbst, verankert in der frühen Beziehung zwischen dem Kleinkind und der Mutter und unlösbar in seinem Schicksal, ob bzw. wie es sich entwickeln kann,

---

* (1960c). Übersetzung siehe Anmerkungen am Ende des Kapitels.

mit ihr verbunden. Das Wort Selbst führt direkt auf den zentralen Bereich seines klinischen Denkens und Verstehens zu, den er um 1940 vor der British Psychoanalytic Society in dem beinahe dramatischen Ausspruch anklingen ließ: „There is no such thing as a baby." (* Anm. 2)

Später gestand er sich rückblendend ein, daß er alarmiert war, als er sich diese Worte aussprechen hörte, mit denen er zum Ausdruck bringen wollte, daß, so man ihm ein Baby zeige, man ihm gewiß auch jemanden zeigen werde, der für das Baby sorgt, wenigstens aber „a pram with someone's eyes glued to it" (wenigstens aber einen Kinderwagen, an dem jemand mit Augen und Ohren hängt). Er fügte hinzu, daß er jetzt ruhiger sagen würde:

„That before object relationships the state of affairs is this: that the unit is not the individual, the unit is an environment-individual set-up. The centre of gravity of the being does not start off in the individual. It is in the total set-up. By good-enough child care, technique, holding and general management the shell becomes gradually taken over and the kernel (which has looked all the time like a human baby to us) can begin to be an individual" (1952a). (* Anm. 3)

Ausreichend gute oder nicht ausreichend gute „child care" oder Management ist, so könnte man sagen, ein Grundthema Winnicotts, das in immer neuen Variationen auftaucht und von dem seine klinischen Hypothesen ihren Ausgang nehmen.

Schon in einer seiner frühen Arbeiten heißt es:

„The mental health of the human being is laid down in infancy by the mother who provides an environment in which complex but essential processes in the infant's self can become completed" (1948). (* Anm. 4)

Von 1949 an taucht der Begriff Selbst in den meisten Aufsätzen auf und steht dabei bis zu den späteren Artikeln in einem sich stets erweiternden Zusammenhang. Winnicott gibt indessen vorläufig keine ausführliche Beschreibung dessen, was er unter seinem Konzept vom Selbst versteht. Es scheint ihm nichts daran zu liegen. Er läßt den Begriff gleichsam für weitere klinische Bedeutungen offen. Dieser auf Vorläufigkeit eingestellte Umgang mit einem Begriff ist nicht ungewöhnlich für sein Vorgehen und läßt daran denken, wie Winnicott andere Begriffe handhabe. So meinte er im Hinblick auf „personalisation":

„Nevertheless, it is possible for me to take the word …, which I have used in another context, and to see how it becomes illustrated in detailed clinical work in child psychiatry and psychoanalysis …" (1972). (* Anm. 5)

Der Begriff Selbst wird für Winnicott durch detaillierte klinische Arbeit illustriert. Deshalb ist es kaum erstaunlich, daß Winnicott bei dem Versuch einer Definition, den er schließlich machte, die wesentlichen klinischen Konzepte aus den verschiedenen Phasen seiner Arbeit mit einbezog. Seine französische Übersetzerin, Mme. Jeannine Kalmanovitch, mit der Übertragung seines Aufsatzes „Basis for Self in Body" befaßt, hatte auf Schwierigkeiten bei der Übertragung des Begriffes „Selbst" hingewiesen. Winnicott antwortete ihr in einem Brief vom 19. 1. 1971 ausführlich. Der Brief traf bei ihr nach seinem Tode ein.

Ich möchte diesen Briefauszug, soweit er in der Nouvelle Revue de Psychanalyse 1971 (in Englisch!) abgedruckt ist, ausführlich wiedergeben. Es ist dies die letzte, aber auch umfassendste Äußerung Winnicotts zum Begriff Selbst, wie er ihn zu der Zeit sah, die zugleich eine unmittelbare Begegnung mit den klinischen Konzepten zuläßt, denen Winnicott im Hinblick auf das Selbst Bedeutung zumißt. Zudem zeigt sich Winnicott hier in seinem abwägenden und differenzierten, zögernden und sich immer in Frage stellenden Denken, das auf Änderung vorbereitet bleibt: „Certainly I might want to alter it".

„In regard to this article, the main thing has to do with the word self. I did wonder if I could write something out about this word, but of course as soon as I came to do it I found that there is much uncertainty even in my own mind about my own meaning. I found I had written the following:

For me the self, which is not the ego, is the person who is me, who is only me, who has a totality based on the operation of the maturational process. At the time the self has parts, and in fact is constituted of these parts. These parts agglutinate from a direction interior-exterior in the course of the operation of the maturational process, aided as it must be (maximally at the beginning) by the human environment which holds and handles and in a live way facilitates. The self finds itself naturally placed in the body, but it may in certain circumstances become dissociated from the body or the body from it. The self essentially recognizes itself in the eyes and facial expression of the mother and in the mirror which can come to represent the mother's face. Eventually the self arrives at a significant relationship between the child and the sum of the identifications which (after enough incorporation and introjection of mental representations) become organised in the shape of an internal psychic living reality. The relationship between the boy or girl with his or her own internal psychic organisation becomes modified according to the expectations that are displayed by the father and mother and those who have become significant in the external life of the individual. It is the self and the life of the self that alone makes sense of action or of living from the point of view of the individual who has grown so far and who is continuing to grow from dependence and immaturity towards independence, and the capacity to identify with mature love objects without loss of individual identity.

You may find this unhelpful but at any rate it seemed to me to be a valuable thing to do to try to write it down. Certainly I might want to alter it.

You of course are left with the same problem that you had at the beginning, which is how to translate the self without using the same word that you would use to translate the ego. Let me try to be more helpful. I think that the user of the term self is on a different platform from the user of the term ego. The first platform has to do with life and living in a direct way; the second, where the word le moi is used, the speaker or writer is more detached, less involved, perhaps clearer because of being able to use all that there is of the intellectual approach …" (1971a). (* Anm. 6)

Diese komprimierte und vielschichtige, ja multidimensionale Definition läßt erkennen, daß man kaum eine Idee vom Selbst bei Winnicott gewinnen kann, ohne eine Reihe seiner wichtigsten theoretischen Konzepte zu berücksichtigen und sie in einen Zusammenhang zu stellen, dessen Ausgangspunkt die frühe

Beziehung zwischen dem Kleinkind und der Mutter ist.

Winnicott weist hier auf den unterschiedlichen Gebrauch der Begriffe Selbst und Ich hin. Seine hilfreiche Erklärung, die er dazu aus der Distanz des Rückblickes auf sein Werk gegeben hat, entspricht im wesentlichen dem Vorgehen Winnicotts in seinen Arbeiten. Sie läßt zugleich vermuten, daß es ihm nicht von vornehmlicher Wichtigkeit war, eine genaue Anwendung der Begriffe zu erreichen, sondern daß es ihm vielmehr darum ging, mit dem Begriff des Selbst etwas von der Erfahrung des Lebendigseins, etwas vom Leben in einer sehr direkten Weise festzuhalten. Mir scheint, daß bei Winnicott der Begriff Selbst zumeist an den Begriff der Erfahrung gebunden ist. Mit dem Begriff des Selbst versucht Winnicott die Bedeutung der Erfahrungen des Individuums vom frühesten Anfang an innerhalb des „environment-individual set-up" festzuhalten.

Ich möchte hier eine Passage, in der Winnicott Ich und Selbst voneinander abzugrenzen versucht, einer solchen gegenüberstellen, in der der Begriff Selbst nur verständlich erscheint, weil er in Verbindung mit dem Wort Erfahrung auftritt, weil eben der Begriff Selbst Winnicott dazu verhilft, den menschlichen Säugling als ein erlebendes (experiencing) Individuum in der Beziehung zu seiner Umwelt zu sehen:

„The term ego can be used to describe that part of the growing human personality that tends, under suitable conditions, to become integrated into a unit ... It will be seen that the ego offers itself for study long before the word self has relevance. The word self arrives after the child has begun to use the intellect to look at what others see or feel or hear and what they conceive of when they meet this infant body ..." (1962a). (* Anm. 7)

Im Gegensatz dazu spricht Winnicott an anderer Stelle von einem äußerst frühen Beginn des Selbst, und hier taucht zugleich das Wort „experiencing" auf:

„Certainly before birth it can be said of the psyche that there is a personal going-along, a continuity of experiencing. This continuity, which could be called the beginnings of the self, is periodically interrupted by phases of reaction to impingement. The self begins to include memories of limited phases in which reaction to impingement disturbs the continuity" (1949). (* Anm. 8)

Dies schließt nicht aus, daß auch die Begriffe Ich und Selbst in der Verbindung mit dem Wort „experiencing" verwandt werden, wie z. B. im folgenden Text:

„Ego here implies a summation of experience. The individual self starts as a summation of resting experience ..." (1956). (* Anm. 9)

Wenn ich der Frage nach Winnicotts Gebrauch des Begriffes Selbst nachgehe, so werde ich dies entlang der imaginären Linie des sich entwickelnden Selbst tun, wie er sie in seiner Definition 1971 gezeichnet hat. Das Denken Winnicotts läßt Linien erkennen, die von seinen frühen Arbeiten bis zu seinen späten Schriften reichen, und die sich dann leichter erkennen lassen, wenn man seine Aufsätze in ihrer chronologischen Abfolge liest. Es verhält sich oftmals mit einem Gedanken so, daß er kurz auftaucht, dann später wieder aufgegriffen wird, um einige neue Überlegungen oder klinische Beobachtungen erweitert. Manchmal greift Winnicott die Linie in der Weise wieder auf, daß er Bezug darauf nimmt, wann er zuvor schon den selben Gedanken gebracht hat – dann wiederum kann es geschehen, daß Winnicott die Verbindung zu dem vorherigen Gebrauch des Begriffes nicht herstellt. Dies kann zu Unklarheiten führen, weiß man als Leser doch nicht, ob ihm die Beziehung, die sich zu sehen anbietet, nicht mehr wichtig erscheint oder ob er voraussetzt, daß der Leser diese Brücke zu früheren Ausführungen selber herstellt.

Das eine Thema, das niemals fallengelassen, sondern stets weiter differenziert wird, ist das der Mutter-Kind-Beziehung, wenngleich auch hier Aspekte dieser Beziehung, wie sie in früheren Arbeiten beschrieben worden waren, in den späteren nicht mehr aufgegriffen werden.

Winnicott selber hat auf eine andere Eigenart seines Arbeitsstiles hingewiesen, die mir hier erwähnenswert erscheint. So beginnt er seinen Vortrag „Primitive Emotional Development" wie folgt:

„I shall not first give an historical survey and show the development of my ideas from the theories of others, because my mind does not work that way. What happens is that I gather this and that, here and there, settle down to clinical experience, form my own theories and then, last of all, interest myself in looking to see where I stole what. Perhaps this is as good a method as any" (1945). (* Anm. 10)

Neben dem besonderen Umgang mit den Begriffen, auf den ich schon hingewiesen hatte, wird sich insbesondere in Winnicotts Umgang mit dem Begriff des Selbst zeigen, daß er sich nicht oder wenig daran interessiert zeigt, was andere Autoren darunter verstehen, sondern daß er für sich und unabhängig eine Vorstellung davon sucht. Für die Ausrichtung seiner klinischen Tätigkeit war die Tatsache entscheidend, daß Winnicott in den frühen 20er Jahren pädiatrischer Facharzt war, als er erstmals mit der Psychoanalyse in Kontakt kam. Er führt aus, weshalb er sich, wenn er als Psychoanalytiker spricht, auch zugleich als Pädiater mitteilen möchte, als

„a paediatrician who is in the habit of thinking of the developing child and indeed of the developing infant. For the paediatrician there is a continuity of development of the individual ... The aim in child care is not only to produce a healthy child but also to allow of the ultimate development of the healthy adult" (1952b). (* Anm. 11)

Es wird sich zeigen, wie dieser Blick, der auf die anhaltende menschliche Entwicklung ausgerichtet ist, seine Idee vom Selbst formt, so wie er dies in der oben angeführten Definition schon in wenigen Worten zum Ausdruck bringt: „... and who has grown so far and who is continuing to grow from dependence and immaturity towards independence." Von hier aus ergibt sich auch eine Beziehung zu dem Optimismus in seinem klinischen Vorgehen, in das Winnicott

---

* Übersetzungen siehe jeweils Anmerkungen am Ende des Kapitels.

immer wieder so überraschend Einblick gewährt. „A belief in human nature and in the developmental process exists in the analyst if work is to be done" (1954). (* Anm. 12)

Winnicott, der sich immer als Pädiater verstanden hat, hat seine klinisch-theoretischen Beiträge aus der wechselseitigen Anregung bezogen, die sich aus seiner Arbeit mit Müttern und Säuglingen einerseits und mit schizoiden oder Borderline-Patienten andererseits ergab.

In diesem Sinne schrieb er (1967), daß wir die Antwort über frühe Vorgänge zwischen Baby und Mutter von den psychoanalytischen Patienten erhalten, die zu frühen Phänomenen zurückreichen können und die verbalisieren können „(when they feel they can do so) without insulting the delicacy of what is preverbal, unverbalized, and unverbalizable except perhaps in poetry". (* Anm. 13)

Ein Jahr später griff Winnicott die Frage der Wechselseitigkeit abwägend wieder auf. Er betonte, daß wir von Müttern und Babies für die Übertragungssituation mit schizoiden Patienten lernen und umgekehrt, daß wir von schizoiden Patienten lernen können, wie wir Mütter und ihre Babies sehen können und damit einen klareren Blick dafür gewinnen, was sich dort abspielt. Er schließt mit der Überlegung, daß unser Wissen über die Bedürfnisse von Patienten in psychotischen Phasen ganz wesentlich von Müttern und Babies stammt (1968).

Ich habe die Absicht, möglichst Winnicott selber sprechen zu lassen und bedeutsame Passagen und die manchmal hier und da verstreuten Bemerkungen den Linien im Werk Winnicotts folgend zusammenzutragen und zusammenzufügen, ohne eine falsche Systematisierung vorzutäuschen. Mir scheint nämlich, daß ich dem Leser etwas Kostbares vorenthalten würde, ließe ich ihn nicht die lebendige und so ganz und gar persönliche Sprache Winnicotts mithören, dem Hin und Her folgen, dem Zaudern und Innehalten, das dann überraschend recht bestimmten Formulierungen Platz macht, die Anspruch darauf erheben, gehört zu werden.

Als Winnicott später einmal auf die Schriften zurückblickte, die die Entwicklung seines Denkens markierten, gelangte er bis zu seinem ersten Buch „Clinical Notes on Disorders in Childhood" (aus dem Jahre 1931, 1972b, S. 3). Aus diesem frühen Buch stammt ebenfalls ein Aufsatz, in dem erstmals der Begriff Selbst auftaucht. Er taucht dort so wie selbstverständlich und wie nebenbei auf, und doch schon eine Tonart verratend, die unmißverständlich die von Winnicott ist. Er beschreibt ein Mädchen, zwei Jahre und 6 Monate alt, und zwar in folgenden Worten:

„Instead of being her own contented self she now gets quickly tired of things, losing interest in one toy after another" (1931). (* Anm. 14)

Zwischen diesem „instead of being her own contented self" 1931, in Verbindung mit der Unfähigkeit zu spielen, und der Formulierung in der obigen Definition 1971: „. . . It is the self and the life of the self that alone makes sense of action and of living . . ." erstreckt sich das Werk Winnicotts.

6.3 The Unit Self

Winnicott unterscheidet zwei Aspekte des Selbst:
- Das Selbst, das sich in der interpersonalen Kommunikation erfährt und aus ihr Lebendigkeit bezieht, das aus der Erfahrung erwächst, in seiner ersten Impulsivität, in seiner spontanen Geste beantwortet zu sein. Die Entwicklungslinie führt von der Beantwortung der frühesten spontanen Geste durch die ausreichend-gute Mutter über die Erfahrung der Illusion auf der Basis einer gemeinsamen Lebenserfahrung zwischen Kleinkind und Mutter zur Erfahrung des intermediären Raumes, zum Erleben in der kulturellen Erfahrung.
- Das nichtkommunizierende Selbst, das im Falle von Gesundheit primär nicht kommuniziert, und das er auch das zentrale Selbst nennt.

„Certainly before birth it can be said of the psyche (apart from the soma) that there is a personal going-along, a continuity of experiencing. This continuity, which could be called the beginnings of the self, is periodically interrupted by phases of reaction to impingement. The self begins to include memories of limited phases in which reaction to impingement disturbs the continuity (1949).

Here it may be observed that the infant that is disturbed by being forced to react is disturbed out of a state of ‚being'. This state of being can obtain only under certain conditions. When reacting, an infant is not ‚being' (1949).

It may be pointed out that the most important thing is the trauma represented by the need to react. Reacting at this stage of human development means a temporary loss of identity. This gives an extreme sense of insecurity, and lays the basis for an expectation of further examples of loss of continuity of self, and even a congenital (but not inherited) hopelessness in respect of the attainment of a personal life" (1949). (* Anm. 15)

In einem anderen Zusammenhang spricht Winnicott von einer „continuity of being" (1960a), die nicht unterbrochen werden darf, so eine gesunde Entwicklung möglich sein soll. Voraussetzung dafür ist eine perfekte Umgebung. Eine solche perfekte Umgebung paßt sich aktiv den Bedürfnissen des neugeformten Psycho-Somas an. Die Umgebung, die sich nicht in dieser Weise anpaßt, wird zum Übergriff, auf den das Kleinkind reagieren muß.

„The first ego organization comes from the experience of threats of annihilation which do not lead to annihilation and from which, repeatedly, there is recovery" (1956). (* Anm. 16)

Auf der Basis solcher Erfahrungen entwickelt sich die Fähigkeit zum Vertrauen auf Wiederherstellung. Das Ich beziehungsweise das Selbst – die Begriffe werden austauschbar verwandt – entsteht aus der Summation von Erfahrung. Gelingt es der Mutter nicht, sich den Bedürfnissen des Säuglings anzupassen, so führt dies zu einer „annihilation of the infant's sense of self".

Winnicott unterscheidet zwischen zwei extremen Fällen: Babies, die aufgrund geglückter Erfahrungen einen Glauben an Zuverlässigkeit entwickeln können, und Babies, die das Versagen der Umwelt in solch hohem Maße erfahren mußten, daß eine

Erholung davon nicht mehr möglich ist. Von diesen nimmt Winnicott an, daß sie die Erfahrung von unvorstellbarer oder archaischer Angst in sich tragen:

„They know what it is to be in a state of acute confusion or the agony of disintegration. They know what it is like to be dropped, to fall forever, or to become split into psychosomatic disunion. – In other words, they have experienced trauma, and their personalities have to be built round the organization of defences following trauma …" (1970). (* Anm. 17)

Winnicott findet immer wieder neue Ausdrucksweisen, um zu betonen, daß der Reifungsprozeß, den das Individuum erlebt, nur insofern eine Vorwärtsbewegung zustande bringt, als eine fördernde Umwelt vorhanden ist. Dabei schreitet das Individuum von absoluter Abhängigkeit (es besteht keine Möglichkeit, etwas von der mütterlichen Fürsorge zu wissen) zu relativer (der Säugling kann das Bedürfnis nach den Einzelheiten der mütterlichen Fürsorge bemerken) hin zur Unabhängigkeit (Anhäufung von Erinnerungen an Fürsorge, Projektion, persönliche Bedürfnisse und Introjektion von Einzelheiten der Fürsorge, die zu Vertrauen führen und, um intellektuelles Verstehen bereichert, bewirken können, daß der Säugling die Möglichkeit entwickeln kann, ohne Fürsorge auszukommen).

Winnicott klassifiziert die Entwicklung des Kindes als Integration, Personalisation und „object-relating".[1] Der mütterlichen Umwelt kommt dabei ein eigenes Wachstum zu, das in Abhängigkeit von den sich stets verändernden Bedürfnissen des Säuglings steht, und das Winnicott als primäre Mütterlichkeit definiert. Die Mutter, die in der Lage ist, diesen Zustand erhöhter Sensitivität zu entwickeln,

„provides a setting for the infant's constitution to begin to make itself evident, for the development tendencies to start to unfold, and for the infant to experience spontaneous movement and become the owner of the sensations that are appropriate to this early phase of life" (1956). (* Anm. 18)

Winnicott, der die Bedeutung der stets sich den Bedürfnissen des Kindes anpassenden Aufmerksamkeit der Mutter hervorhebt, hat dies selten in so anschaulicher Weise getan, mit so viel Bemühen, die Behutsamkeit seiner Schilderung der unaufdringlichen mütterlichen Präsenz anzugleichen, wie in der folgenden Passage:

„Let us attempt to study the mother's job. If the infant is to be able to start to develop into a being, and to start to find the world to know, to start to come together and to cohere, then the following things about a mother stand out as vitally important: she exists, continues to exist, lives, smells, breathes, her heart beats. She is there to be sensed in all possible ways. She loves in a physical way, provides contact, a body temperature, movement, and quiet according to the baby's needs. She provides opportunity for the baby to make the transition between the quiet and the excited state, not suddenly coming at the child with a feed and demanding a response. She provides suitable food at suitable times. At first she lets the infant dominate, being willing (as the child is so nearly a part of herself) to hold herself in readiness to respond. Gradually she introduces the external shared world, carefully grading this according to the child's needs

which vary from day to day and hour to hour. She protects the baby from coincidences and shocks (the door banging as the baby goes to the breast) trying to keep the physical and emotional situation simple enough for the infant to be able to understand, and yet rich enough according to the infant's growing capacity. She provides continuity. By believing in the infant as a human being in its own right she does not hurry his development and so enables him to catch hold of time, to get the feeling of an internal personal going-along. For the mother the child is a whole human being from the start, and this enables her to tolerate his lack of integration and his weak sense of living-in-the-body" (1948). (* Anm. 19)

Ich werde die einzelnen Abschnitte der frühen Entwicklung hin zur Einheit Selbst skizzieren, wie sie Winnicott unterscheidet, mit der Einschränkung, daß, wie er einmal sagt, „the beginning is a summation of beginnings" (1962a). (* Anm. 20)

Aus der primären Unintegration entwickelt sich allmählich Integration. Zu Beginn ist der Säugling durch eine Anzahl von Motilitätsphasen und Sinneswahrnehmungen bestimmt. Winnicott stellt hier die Beziehung zum Konzept des primären Narzißmus her. Ein Weg zurück zur Unintegration ist dann möglich, wenn sich das Kind durch die Mutter gehalten fühlt, Unintegration und Reintegration können dann, ohne daß Angst eintritt, zusammen stattfinden. „Relaxation for an infant means not feeling a need to integrate, the mother's ego-supportive function being taken for granted" (1962a) (* Anm. 21). Winnicott stellt sich vor, daß es lange Zeitphasen gibt, in denen es für das Baby nichts ausmacht, ob es im Gesicht seiner Mutter lebt, in seinem eigenen Körper oder ob es aus vielen Einzelstücken besteht, wenn es durch das Gehaltenwerden von Zeit zu Zeit die Erfahrung machen kann, daß es sich wieder zusammenfügt und etwas fühlen kann. Die Auflösung dieser Integration zu einer Einheit Selbst jedoch ist Desintegration, und Desintegration ist schmerzlich. Das Ergebnis der gesunden Entwicklung in dieser Phase ist, daß das Kind einen „unit status" erreicht, daß es eine Person wird, ein Individuum um seiner selbst willen. Zu dieser Phase gehört als besondere Form der mütterlichen Pflege das Halten, das in dem vorangegangenen längeren Zitat so eindrucksvoll von Winnicott beschrieben worden ist.

Ebenso bedeutsam wie Integration ist der Beginn einer engen Beziehung zwischen Psyche und Soma, „the feeling that one's person is in one's body" (1945) oder die psychosomatische Existenz (1960a) oder das, was er das Innewohnen im Soma nennt. Schließlich wählt er den Begriff der Personalisierung, um damit zum Ausdruck zu bringen, daß es sich um ein „achievement in health" (1972) handelt.

„The basis for the indwelling is a linkage of motor and sensory and functional experiences with the infant's new state of being a person. As a further development there comes into existence what might be called a limiting membrane,

---

[1] Meine frühere Übersetzung dieses schwierigen Begriffes war „Beziehung zu einem Objekt haben" (Schacht, 1973); in der Kindler-Ausgabe 1974 lautet die Übersetzung „Kontaktaufnahme mit dem Objekt".
* Übersetzungen siehe jeweils Anmerkungen am Ende des Kapitels.

which to some extent (in health) is equated with the surface of the skin, and has a position between the infant's ‚me' and his ‚not me'. So the infant comes to have an inside and an outside, and a body-scheme" (1960a). (* Anm. 22)

An anderer Stelle sagt Winnicott, daß bei einer ausreichend guten mütterlichen Fürsorge

„the centre of gravity of being in the environment-individual set-up can afford to lodge in the centre, in the kernel rather than in the shell. The human being now developing an entity from the centre can become localized in the baby's body and so begin to create an external world at the same time as acquiring a limiting membrane and an inside" (1952a). (* Anm. 23)

Der Personalisierung entspricht auf der Seite der mütterlichen Pflege das Handhaben. Damit meint Winnicott einen großen Teil der physischen Pflege, die dem Kind zuteil wird, wie Anfassen, Baden, und wodurch dem Kind die Möglichkeit, eine psychosomatische Existenz zu erreichen, vermittelt wird (1977).

Die Mutter, die aufgrund ihrer Identifikation mit dem Kind während und nach der Schwangerschaft sich an die Stelle des Kindes versetzen kann, vermag sich auf die Bedürfnisse des Kindes einzustellen, sie zu beantworten. Diese Bedürfnisse, die, wie Winnicott sagt, erst körperliche Bedürfnisse sind und allmählich in Ich-Bedürfnisse umgewandelt werden, insofern, als eine Psychologie aus der imaginativen Verarbeitung physischer Erfahrungen erwächst (1956), werden entweder beantwortet oder nicht beantwortet. Der Effekt jedoch dessen, ob sie beantwortet werden oder nicht, ist ein ganz anderer, als wenn ein Es-Impuls auf Befriedigung oder Frustration stößt. Anders ausgedrückt vermittelt die Mutter, die sich in ihrem Halten und Handhaben sensibel auf das Kind einstellt, ihm mehr als die Befriedigung eines Triebbedürfnisses. „One might say, that the mother makes the baby's weak ego into a strong one, because she is there, reinforcing every-thing, like power-assisted steering on a motor-bus" (1962b) (* Anm. 24). Für diese besondere Beziehung zwischen Mutter und Säugling wählt Winnicott den Begriff der ego-relatedness (Ich-Beziehung) und hebt hervor, daß die Mutter sich von ihr erholt (1956). Im Rahmen dieser Ich-Beziehung zwischen Mutter und Kind können Es-Beziehungen das Ich stärken. Auf die traumatisierende Diskrepanz zwischen Ich-Bedürfnissen und Es-Bedürfnissen werde ich im Zusammenhang mit dem Konzept vom Falschen Selbst ausführlicher eingehen.

In seiner obigen Definition von 1971 sagte Winnicott, daß sich das Selbst ganz wesentlich in den Augen und im Gesichtsausdruck der Mutter erkennt und in dem Spiegel, der das Gesicht der Mutter vertreten kann. Hier ist von dem die Rede, was Winnicott die Spiegel-Funktion der Mutter nennt: eine gesunde Entwicklung vorausgesetzt, sieht sich das Kind im Gesicht der Mutter. Winnicott stellt sich vor, wie in diesem Falle der visuelle Austausch zwischen Baby und Mutter, dieser Zwei-Wege-Prozeß abläuft. Das Baby schaut, schaut in das Gesicht der Mutter und findet sich darin, d.h., die Mutter hat das Baby angeschaut, und ihr Gesicht gibt nun wieder, was sie

sieht. Man könnte sagen, daß sie in ihrem Anschauen dem Bild, das sie vom Gesicht des Babys gewinnt, Raum in ihrem Gesicht gibt. „What she looks like is related to what she sees there" (1967) (* Anm. 25). Dies ist der Beginn eines bedeutsamen Austausches mit der Welt, in dem Selbst-Bereicherung mit der Entdeckung dessen abwechselt, was die Welt der Dinge für das Kind bereithält. Den historischen Prozeß in der individuellen Entwicklung, der sich auf das Gesehenwerden gründet, hält Winnicott in folgenden Zeilen fest:

„When I look I am seen, so I exist.
I can now afford to look and see.
I now look creatively and what I apperceive I also perceive." (* Anm. 26)

Eine Beziehung zwischen den vorangegangenen Phasen und den Begriffen Selbst und unit self ergibt sich in dem Begriff des „imaginative self". Winnicott meint damit, daß der lebendige Körper mit seinen Begrenzungen, mit seiner Innenseite und seiner Außenseite von dem Individuum als Kern seines imaginativen Selbst gefühlt wird. Hierher gehört auch die Definition, die Winnicott von der inneren psychischen Realität gibt:

„Let us look ... at inner psychic reality, the personal property of each individual in so far as a degree of mature integration has been reached which includes the establishment of a unit self, with the implied existence of an inside and an outside, and a limiting membrane, there again there is to be seen a fixity that belongs to inheritance, to the personality organization, and to environmental factors introjected and to personal factors projected" (1971b, S. 8). (* Anm. 27)

6.4 Der Anfang des Object-Relating

Winnicott hat sein ganzes Augenmerk auf diese frühesten Phasen in der Entwicklung des Kindes gerichtet, in denen das Kind zunächst noch verschmolzen mit der Mutter sich durch eine Reihe von komplizierten Mechanismen aus diesem Zustand herausbewegt und infolgedessen allmählich Beziehung zu Objekten aufnimmt, die außerhalb des Selbst liegen, die nicht Teil des Selbst sind. Es ist der Reifungsprozeß, der das Baby dazu drängt, Objektbeziehungen zu erreichen. Die Fähigkeit jedoch, Beziehung zu Objekten zu haben, kann nur dann wachsen, wenn sich die Umwelt aktiv auf das Kind einstellt. Hinsichtlich des Object-Relating besteht für den Reifungsprozeß dieselbe Abhängigkeit von der fördernden Umwelt wie für die Integration und Personalisation. Als die wesentliche Aufgabe der fördernden Umwelt in dieser Phase bezeichnet Winnicott das Präsentieren des Objektes („object-Presenting"). Der Darstellung dessen, was die Mutter zum Gelingen dieser notwendigen Entwicklungsschritte beiträgt, versucht Winnicott durch eine immer weitere Auffächerung ihrer Funktionen gerecht zu werden. So geht er in seinen klinischen und theoretischen Überlegungen in vielfachen Versionen auf die sich überlagernden Prozesse ein, die sich in dieser Phase zwischen Mutter und Kind abspielen, in der sich das Kind auf die Entdeckung der Nicht-Ich-Welt zubewegt.

Das Präsentieren des Objektes besagt, daß die Mutter dem Kind das Objekt in einer solchen Weise präsentiert, daß das Kind den Eindruck bekommen kann, das Objekt selbst geschaffen zu haben. Das Kind hat eine vage Erwartung, die einem unformulierten Bedürfnis entspringt. Die sich adaptierende Mutter präsentiert nun dem Baby ein Objekt oder eine Handhabung, die den Bedürfnissen des Babys entspricht. So geschieht es, daß das Baby just das zu brauchen beginnt, was die Mutter für das Kind bereithält oder ihm präsentiert. Auf diese Weise gewinnt das Baby allmählich das Vertrauen, die aktuelle Welt schaffen zu können. Die sich in dieser Weise anpassende Mutter gewährt dem Kind eine kurze Periode, in der für das Kind Omnipotenz erlebbar wird. Mit dieser „initialen Erfahrung von Omnipotenz" (1968) meint Winnicott mehr als magische Kontrolle, schließt sie doch „den kreativen Aspekt von Erfahrung" ein. Das Kind, das in dieser Weise die Erfahrung von Omnipotenz machen kann, schafft und wiedererschafft das Objekt. Das Objekt wird geschaffen, nicht gefunden (created, not found). Allerdings muß das Objekt gefunden werden, um geschaffen zu werden, ein Paradox, auf das Winnicott aufmerksam macht.

Winnicott führt hier den Begriff der Illusion an und sagt, daß die Mutter, die dem Kind die Illusion gewährt, daß das, was dort ist, vom Baby geschaffen ist, damit die Voraussetzung dafür vermittelt, daß die Realität vom Baby als etwas gesehen wird, über das man Illusionen haben kann. Im diesem Sinne spricht er vom Baby und der Brust der Mutter als zwei Phänomenen, die nicht in Beziehung miteinander treten können, ehe nicht Mutter und Kind miteinander eine Erfahrung gelebt haben. Die Mutter ist diejenige, die die Situation schaffen muß, in der, wenn alles gutgeht, das Kind einen ersten Kontakt mit einem äußeren Objekt macht, einem Objekt, das vom Standpunkt des Kindes aus außerhalb des Selbst liegt.

„I think of the process as if two lines came from opposite directions, liable to come near each other. If they overlap, there is a moment of illusion – a bit of experience which the infant can take as either his hallucination *or* a thing belonging to external reality" (1945). (* Anm. 28)

Indem die Mutter dem Kind die Illusion gewährt, daß das, was dort ist, vom Baby geschaffen ist, vermittelt sie mehr als Triebbefriedigung. In der Wiederholung dieser Erfahrung lernt das Baby die Illusion zu gebrauchen, ohne die kein Kontakt zwischen Psyche und Umwelt möglich wird. Setzt man nun an die Stelle des Wortes Illusion das Wort Daumen oder Tuchzipfel, so sagt Winnicott, gelangt man zu dem, was er das Übergangsobjekt nennt (1952b).

„From the observer's point of view there may seem to be object-relating in the primary merged state, but it has to be remembered that at the beginning the object is a ‚subjective object'. I have used this term subjective object to allow a discrepancy between what is observed and what is being experienced by the baby" (1971b, S. 10). (* Anm. 29)

Winnicott nimmt an, daß die Mutter, die in glücklicher Weise ihrem Kind die Fähigkeit zur Illusion

vermitteln kann, mit ihrer späteren Aufgabe, nämlich der graduellen Desillusionierung, keine Schwierigkeiten hat. Die Phase, die Illusion und Desillusion umfaßt, in der das Kind in die Erfahrung der Omnipotenz eintritt bis hin zum Verzicht auf die Omnipotenz, bis das Kind die Omnipotenz als Lebensraum aufgibt, ist die Phase, in der sich für das Kind die Natur des Objektes, mit dem es kommuniziert, vom zunächst subjektiv wahrgenommenen zum objektiv wahrgenommenen wandelt.

Das Konzept, das die Prozesse auf seiten der Mutter festzuhalten trachtet, ist das der „good-enough mother", die zunächst eine fast vollständige Anpassung an die Bedürfnisse des Kindes vornimmt und im Laufe der Zeit sich zunehmend weniger anpaßt in Abstimmung an die wachsende Fähigkeit des Kindes, mit ihrem Versagen fertig zu werden. Winnicott weist hier auf den entscheidenden Unterschied zwischen den Wegen hin, auf denen Mutter und Baby zu dieser Gegenseitigkeit von Beziehung gelangen: Die Mutter ist selber einmal ein Baby gewesen, in ihr ist das Erfahrungsgut, das ihren Weg von der Abhängigkeit zur Unabhängigkeit umfaßt, sie hat vielleicht schon gespielt, wie es ist, ein Baby zu sein, während für das Baby alles „eine erste Erfahrung" ist.

„In describing communication between baby and mother, then, there is the essential dichotomy – the mother can shrink to infantile modes of experience, but the baby cannot blow up to adult sophistication" (1968). (* Anm. 30)

6.5 Das Konzept des Wahren und Falschen Selbst

Hinsichtlich der Frage der Ätiologie treffen sich in seinem Aufsatz 1960 „Ego Distortion in Terms of True und False Self" zwei Linien der Definition.

In seinen früheren Arbeiten, in denen Winnicott über die Entstehung des Falschen Selbst gesprochen hatte (das Konzept wurde zum ersten Mal 1949 in dem Aufsatz „Mind and its Relation to the Psycho-Soma" gebraucht), hatte Winnicott hinsichtlich der Genese den Begriff „impingement" angeführt.

„In the early development of the human being the environment that behaves well enough enables *personal growth to take place*. The self processes then may continue active, in an unbroken line of living growth. If the environment behaves not well enough, then the individual is engaged in reactions to impingement, and the self processes are interrupted. If the state of affairs reaches a quantitative limit the core of the self cannot make new processes unless and until the environment failure situation is corrected … With the true self there develops a false self built on a defence-compliance basis, the acceptance of reaction to impingement. The development of a false self is one of *the most successful defence organizations* designed for the protection of the true self's core, and its existence results in the sense of futility" (1954). (* Anm. 31)

Wie bereits dargelegt, impliziert der Begriff „impingement" (Übergriff) bei Winnicott die Möglichkeit einer

---

* Übersetzungen siehe jeweils Anmerkungen am Ende des Kapitels.

sehr frühen Beeinträchtigung der Entwicklung. Nachdem er von einem Übergriff durch die Umwelt vor der Geburt gesprochen hatte, erscheint es nur konsequent, wenn er (im Jahre 1949) von der Möglichkeit einer falschen oder ungesunden Vorwärtsbewegung in der emotionalen Entwicklung vor der Geburt spricht.

Als eine besondere Form der Traumatisierung beschreibt er in seinem Aufsatz 1960 Es-Erregungen, für die das Ich noch nicht reif genug ist, indem er auf den Begriff der Es-Bedürfnisse und Ich-Bedürfnisse zurückgreift:

„It must be emphasized that in referring to the meeting of the infant's needs I am not referring to the satisfaction of instincts. In the area that I am examining the instincts are not yet clearly defined as internal to the infant. The instincts can be as much external as can a clap of thunder or a hit. The infant's ego is building up strength and in consequence is getting towards a state in which id-demands will be felt as part of the self, and not as environmental. When this development occurs, then id-satisfaction becomes a very important strengthener of the ego, or of the True Self, but id-excitements can be traumatic when the ego is not yet able to include them, and not yet able to contain the risks involved and the frustrations experienced up to the point when id-satisfaction becomes a fact" (1960b). (* Anm. 32)

Dazu vermittelt Winnicott ein anschauliches Beispiel in einem anderen Aufsatz im Jahre 1960, der in enger Verbindung mit dem über das Konzept vom Wahren und Falschen Selbst gesehen werden kann:

„Example: a baby is feeding at the breast and obtains satisfaction. This fact by itself does not indicate whether he is having an ego-syntonic id-experience or, on the contrary, is suffering the trauma of seduction, a threat to personal ego continuity, a threat by an id-experience which is not ego-syntonic and with which the ego is not equipped to deal" (1960a). (* Anm. 33)

Indem Winnicott auf diese besondere Form der Störung der Kontinuität der Lebenslinie hingewiesen hat, hat er die Linie der Beschreibung, zentriert um den Begriff des „impingement", fortgesetzt.

Dem fügt er nun einen neuen Akzent hinzu, oder, wie ich sagen würde, eine neue Linie der Beschreibung. Er zentriert seine weiteren Ausführungen zur Ätiologie um den Begriff der „good-enough mother" (hinreichend gute Mutter), die dem Kind die Erfahrung der Omnipotenz vermittelt, mit der im Einklang das Kind die Beantwortung seiner spontanen Geste erleben kann. Als Folge dessen, daß es der hinreichend guten Mutter immer wieder gelingt, die spontane Geste oder die sensorische Halluzination des Kindes zu beantworten, wird das Wahre Selbst zur lebendigen Realität.

Für Winnicott hat es wenig Sinn, die Idee des Wahren Selbst zu formulieren, außer um das Konzept vom Falschen Selbst zu verstehen. Das Konzept vom Falschen Selbst muß gewissermaßen durch die Idee vom Wahren Selbst ausbalanciert werden. Mit diesem Vorbehalt sei nun kurz von dem Ursprung des Wahren Selbst die Rede, wie ihn sich Winnicott vorstellt.

Das Wahre Selbst – ich passe mich hier der Schreibweise Winnicotts in seinem Aufsatz von 1960 an, in welchem er die Begriffe Wahres Selbst und Falsches Selbst durch die Großschreibung hervorhebt – stammt aus der Lebendigkeit des Körpers, vom Schlagen des Herzens, vom Atmen, von der Aktion der Körperfunktionen. Winnicott geht hier zu der frühen Phase des Reifungsprozesses zurück, die mit den Begriffen „Unintegration – Integration" gekennzeichnet ist, und in der die haltende Funktion der Mutter den verschiedenen Empfindungselementen des Kindes Kohäsion verleiht. In dieser frühen Phase nun findet das Kind Ausdruck in einer spontanen Geste, es tut es wiederholt.

Das Wahre Selbst ist die Quelle dieser Geste, oder umgekehrt formuliert, die Geste weist auf die Existenz eines potentiellen Wahren Selbst hin. Winnicott stellt so ausdrücklich eine Beziehung zwischen der spontanen Geste des Kindes und dem her, was er Wahres Selbst nennt. Er nennt in dieser frühen Phase das Wahre Selbst eine theoretische Position, von der die spontane Geste und der persönliche Impuls ausgehen. Die spontane Geste ist das Wahre Selbst in Aktion.

Ihm liegt daran, diese Phase als eine sehr frühe zu verdeutlichen, in welcher er den Ursprung der spontanen Geste vermutet, indem er sie von der späteren Phase abhebt, in welcher man bereits von einer inneren, psychischen Realität sprechen kann. In der frühen Phase, in der das Kind die spontane Geste zeigt, ist es von entscheidender Bedeutung, das Verhalten der Mutter in Betracht zu ziehen. In dieser frühen Phase ist die Abhängigkeit wirklich gegeben und fast absolut. Winnicott unterscheidet zwei extreme Möglichkeiten: entweder die Mutter ist eine hinreichend gute Mutter, oder sie ist eine nicht hinreichend gute Mutter (good-enough mother oder not-good-enough mother). Ich hatte zuvor im Zusammenhang mit dem Begriff „object-presenting" gesagt, daß die hinreichend gute Mutter dem Kind die Erfahrung der Omnipotenz vermittelt, indem sie ihm das Objekt in dem Moment präsentiert, in dem das Kind ein vages Bedürfnis nach dem Objekt verspürt. Das Kind beginnt, an eine äußere Realität zu glauben, die sein Gefühl von Omnipotenz nicht stört.

Die zusätzliche Betonung, die Winnicott jetzt im Hinblick auf sein Konzept Wahres Selbst und Falsches Selbst macht, ist die, daß er den Begriff der spontanen Geste oder der sensorischen Halluzination beim Kind hervorhebt und die Notwendigkeit unterstreicht, daß die Geste von der Mutter beantwortet wird. Es ist ein essentieller Teil seiner Theorie, daß das Wahre Selbst nur eine lebende Realität wird, wenn es der Mutter wiederholt gelungen ist, der spontanen Geste des Kindes oder seiner sensorischen Halluzination zu begegnen. Das Wahre Selbst beginnt nun dadurch zu leben, daß die Mutter dem schwachen Ich des Kindes dadurch Stärke vermittelt hat, daß sie seine omnipotenten Ansprüche erfüllt hat (1960b). Das Wahre Selbst besitzt nun eine Spontaneität, die beantwortet worden ist, und – einen weiteren Kreis des Erlebens andeutend – fügt Winnicott hinzu, daß die Spontaneität in Beziehung zu den Ereignissen der Welt getreten ist.

Die nicht hinreichend gute Mutter ist unfähig, die Omnipotenz des Säuglings zu erfüllen, und so verpaßt sie wieder und wieder die Gesten des Säuglings. Statt dessen drängt sie dem Kind ihre eigene Geste auf. Damit ist das Kind gezwungen, sich der Geste der Mutter anzupassen. Diese Anpassung des Säuglings an die Mutter, die einer Umkehrung der Verhältnisse entspricht, wie sie für die Beziehung zwischen Baby und hinreichend guter Mutter postuliert werden, nennt Winnicott Gefügigkeit. Dies führt zu der frühesten Stufe des Falschen Selbst. Das kleine Kind wird verführt, sich anzupassen, oder es wird zur Gefügigkeit verführt.

Im Sinne des Object-Relating heißt es hier:

„The process that leads to the capacity for symbol-usage does not get started (or else it becomes broken up, with corresponding withdrawal on the part of the infant from advantages gained)" (1960b). (* Anm. 34)

In der klinischen Arbeit unterscheidet Winnicott Organisationen des Falschen Selbst, die von dem Extrem reichen, in dem das sich anpassende Selbst als real betrachtet wird und das Wahre Selbst versteckt bleibt, bis hin zu dem natürlichen Zustand, in dem das Falsche Selbst gekennzeichnet ist durch die höfliche Haltung, daß man „sein Herz nicht auf der Zunge trägt".

Als einen besonderen Fall beschreibt er den, daß es bei einem Individuum mit hohem intellektuellem Potential zu einer Dissoziation zwischen intellektueller Aktivität und psychosomatischer Existenz kommen kann, mit der Tendenz, daß der Intellekt der Ort des Falschen Selbst wird. Mit dem Begriff der Dissoziation schließt er an eine Definition an, die er in seinem Aufsatz „Primitive Emotional Development" (1945) macht:

„Out of the problem of unintegration comes another, that of dissociation ... According to my view there grows out of unintegration a series of what are then called Dissociations, which arise owing to integration being incomplete or partial." (* Anm. 35)

Winnicott wird den Begriff der Dissoziation noch verwenden, um andere Formen der Veränderung der Persönlichkeitsstruktur zu kennzeichnen, auf die ich jedoch hier nicht eingehen kann, wie Delinquenz oder, wie in der obigen Definition von 1971 angedeutet, die Gegengeschlechts-Dissoziation.

Ich vermute, daß das theoretische Argument, das die Frage in den Mittelpunkt stellt, ob die Mutter die spontane Geste, den spontanen Impuls oder die sensorische Halluzination aufgreift oder nicht, eine Verschiebung des Akzentes oder eine Veränderung des Akzentes im Denken Winnicotts anzeigt. Winnicott, der früher einmal gesagt hatte, daß es eine heikle Sache sei, ob oder ob es nicht zwischen Mutter und Säugling „klickt" (1948), beschreibt folgendermaßen, was eintreten kann, falls die Mutter die spontane Geste des Kindes hat aufgreifen können:

„The infant can now begin to enjoy the *illusion* of omnipotent creating and controlling, and then gradually come to recognize the illusory element, the fact of playing and

imagining. Here is the basis for the symbol which at first is *both*, the infant's spontaneity or hallucination, *and also* the external object created and ultimately *cathected*" (1960b). (* Anm. 36)

Hier hat Winnicott sein Konzept vom Wahren und Falschen Selbst ausdrücklich mit dem des Übergangsobjektes verbunden, ein Schritt, der in seinem früheren Aufsatz „Psychosis and Child Care" (1952b) vorbereitet schien. Hier wird die Beziehung zwischen Lebendig-Sein, Lebendig-Werden, Wahrem Selbst und Kreativität in dem Bereich der frühen geglückten Wechselseitigkeit zwischen Baby und Mutter explizit formuliert, ein Gedanke, um den Winnicott in seinen späten Aufsätzen immer wieder kreisen wird. Im Gegensatz dazu hatte sich die Formulierung, die sich um die Frage des Übergriffes zentriert hatte, auf die Vorstellung der Beeinträchtigung, des Störens oder Nichtstörens einer Entwicklung beschränkt.

Schließlich greift Winnicotts Beschreibung des klinischen Erscheinungsbildes bei verschiedener Ausprägung der Organisation des Falschen Selbst beide Gedankenlinien auf:

„In the healthy individual who has a compliant aspect of the self but who exists and who is a creative and spontaneous being, there is at the same time a capacity for the use of symbols. In other words health here is closely bound up with the capacity of the individual to live in an area that is intermediate between the dream and the reality, that which is called cultural life.

By contrast where there is a high degree of split between the True Self and the False Self which hides the True Self, there is found a poor capacity for using symbols, and a poverty of cultural living. Instead of cultural pursuits one observes in such persons extreme restlessness, an inability to concentrate, and a need to collect impingements from external reality so that the living-time of the individual can be filled by reactions to these impingements" (1960b). (* Anm. 37)

Wenn Winnicott sagt, daß es wenig Sinn hat, die Idee des Wahren Selbst als solche zu verfolgen, wenn nicht zu dem alleinigen Zweck, das Falsche Selbst zu verstehen, so gewinnt man den Eindruck, daß er unterstreichen möchte, daß das Wahre Selbst außerhalb des Konzeptes vom Falschen Selbst seinen Sinn verliert. Tatsächlich taucht der Begriff in seinen späten Aufsätzen 1971, in denen Winnicott so oft vom Selbst spricht, so gut wie nicht mehr auf.

Es ergibt sich jedoch eine Beziehung zu seinem Aufsatz „The Theory of the Parent-Infant-Relationship", ebenfalls aus dem Jahre 1960, der in der Nähe des Konzeptes vom Wahren und Falschen Selbst zu sehen ist. Hier stellt Winnicott eine Beziehung zwischen dem Wahren Selbst und dem zentralen Selbst her, ein Gedanke, den er 1963 in der Idee vom nichtkommunizierenden Selbst aufgreifen wird.

Ich werde nicht den Versuch unternehmen, das Konzept des Wahren Selbst und Falschen Selbst in Zusammenhang mit der psychoanalytischen Theorie zu diskutieren (Morse, 1972), nicht nur, weil es zu Mißverständnissen führen könnte, sondern nicht

---

* Übersetzungen siehe jeweils Anmerkungen am Ende des Kapitels.

zuletzt auch deshalb, weil es für Winnicott kein Anliegen zu sein schien. Sein Versuch, das Wahre Selbst im Hinblick auf die psychoanalytische Theorie einzuordnen, bleibt knapp: „It is closely linked with the idea of the Primary Process ... and is, at the beginning, essentially not reactive to external stimuli, but primary" (1960b) (* Anm. 38). Winnicott schloß seinen Artikel mit der Bemerkung, daß seines Erachtens sein Konzept des Wahren und Falschen Selbst keine wesentliche Veränderung der grundlegenden Theorie erforderlich macht.

6.6 Die Beziehung zu objektiv wahrgenommenen Objekten

Das Übergangsobjekt oder die Übergangsphänomene weisen auf die frühen Phasen des Gebrauches von Illusion hin, „without which there is no meaning for the human being in the idea of a relationship with an object that is perceived by others as external to that being" (1951) (* Anm. 39). Das Übergangsobjekt hat eine Funktion der Nicht-Ich-Welt, oder, wie Winnicott sagt, man kann am Gebrauch des Übergangsobjektes, das er auch den ersten „Nicht-Ich"-Besitz nennt, die Fähigkeit des Kindes studieren, das Übergangsobjekt als ein Nicht-Ich-Objekt zu erkennen. Der Übergang von dem Beziehung-Haben zu subjektiven Objekten zum Beziehung-Haben zu objektiv wahrgenommenen Objekten stellt eine Entwicklung dar, die die Fähigkeit voraussetzt, auf die Erfahrung der Omnipotenz zu verzichten und das objektiv wahrgenommene Objekt zu entdecken. Diese Reise, die das Kind bewältigen muß, wird von Winnicott in seinem theoretisch bedeutsamen Aufsatz „The Use of an Object and Relating through Identifications" 1969 als das vielleicht schwierigste Ding bezeichnet, das das Kind im Laufe seiner Entwicklung zu bewältigen hat. Winnicott versucht dem durch den neuen Ausdruck vom Gebrauch des Objektes (use of an object) gerecht zu werden.

6.7 Das Leben des Selbst

Wenn Winnicott in der oben erwähnten Definition von 1971 schrieb, „Es ist das Selbst und das Leben des Selbst, das allein der Aktion oder dem Leben vom Gesichtspunkt des Individuums aus Sinn verleiht", so ist damit all das impliziert, was er auf dem Hintergrund des intermediären Raumes über das Spielen gesagt hat.

„It is in playing and only in playing that the individual child or adult is able to be creative and to use the whole personality, and it is only in being creative that the individual discovers the self" (1971b, Kap. 4). (* Anm. 40)

Winnicott wird schließlich von dem potentiellen Raum zwischen Baby und Mutter und von dem dritten Bereich, dem Bereich der kulturellen Erfahrung, der ein Abkömmling des Spielens ist, sprechen.

Wie ich schon zu Beginn sagte, hat Winnicott niemals einen solchen umfassenden Versuch gemacht, zu dem Begriff Selbst, wie er ihn verwandte, Stellung zu nehmen, wie in der obigen Definition 1971. Der erste, längere Teil seiner Ausführungen hier meint den Entwicklungsprozeß, die frühen Phasen des Selbst, während er gegen Ende vom Leben des Selbst spricht. Die Frage nach dem Leben des Selbst entspricht dem Anliegen, das ihn zunehmend in einigen seiner späten Aufsätze beschäftigt, die in dem Buch „Playing and Reality" (Vom Spiel zur Kreativität) zusammengefaßt sind, nämlich „was es mit dem Leben auf sich hat" oder, formuliert als Titel eines Artikels, „The Place where we Live". Es ist nicht mehr nur die Frage, wie das Baby das Individuum verpassen kann – ich beziehe mich hier auf einen Ausdruck Winnicotts, als er meinte, daß Babies, die eine nicht hinreichend gute Pflege erhalten, „sich nicht erfüllen, auch nicht als Babies" (1968) –, sondern wie sich das Individuum in seiner Kreativität gewinnen kann, wie es sich „spielend" erfahren kann als Selbst, in der Begegnung mit der Umwelt, aber auch mit sich selbst.

Winnicott stellt das Spielen als ein Subjekt als solches in den Raum und in die Zeit. Spielen ist eine Erfahrung, die an das Raum-Zeit-Kontinuum gebunden ist. Es ist eine basale Form des Lebens, es ist Begegnung, es ist ein universales Phänomen. Spielen, so sagt Winnicott,

„facilitates growth and therefore health, playing leads into group relationships; playing can be a form of communication in psychotherapy, and lastly, psychoanalysis has been developed as a highly specialized from a playing in the service of communication with oneself and others" (1971b, Kap. 3). (* Anm. 41)

Winnicott ist fasziniert vom Phänomen Spielen. Er sagt von sich, daß er schon immer in seiner konsultativen Technik ein großes Interesse für das Spiel empfunden habe, das sich auf der Basis von Vertrauen zwischen Baby und Mutter entwickeln kann.

„I have always known Freud's description of the game with the cotton reel and have always been stimulated by it to make detailed observations on infant play" (1941). (* Anm. 42)

Früher hatte Winnicott einmal mit dem Hinweis auf das erste Spiel des Babys an der Brust der Mutter gesagt, daß ohne die Chance des Spiels Baby und Mutter füreinander Fremde bleiben (1948). Nun verlegt er das Spielen in den potentiellen Raum zwischen Baby und Mutter. Damit meint er die „hypothetical area that exists (but cannot exist) between the baby and the object (mother or part mother) during the phase of the repudiation of the object as not-me, that is at the end of being merged in with the object" (1971b, Kap. 8). Die Trennung von Baby und Mutter, die durch den Gebrauch des Übergangsobjektes zugleich auch initiiert wird, wird vermieden, indem der potentielle Raum mit dem Gebrauch von Symbolen, mit kreativem Spielen angefüllt wird, was schließlich zum kulturellen Leben führt. Winnicott, der die Idee des Kreierens unabhängig von dem sehen will, was Kunst ist, sondern als etwas, was universal ist, das dazugehört, wenn von Lebendigsein die Rede ist, greift auf den Begriff des kreativen Impulses zurück, den er zusammen mit dem der spontanen Geste in Zusammenhang mit dem Konzept vom Wahren und Falschen Selbst verwandt hatte. Wenn auch keine Erklärung für den kreativen Impuls zu finden ist, so sind doch die Gründe zu studieren, weshalb kreatives Leben verloren werden kann oder weshalb das Gefühl, daß das Leben real und lebenswert ist, einem Menschen verlorengehen kann: Die Unfähigkeit des Babys, in der beschriebenen Weise die Verläßlichkeit der Mutter zu erfahren, führt zu einer Einschränkung des Spielbereiches. Der „potentielle Raum" hat keine Bedeutung, und der kreative Gebrauch von Objekten ist unsicher oder fehlt ganz. In früheren Worten Winnicotts würde es heißen, daß die kreative Geste nicht beantwortet worden ist. Die Entwicklung, die sich daraus ergeben kann, führt zurück zum Konzept des Falschen Selbst. Winnicott drückt sich selbst so aus, daß das Falsche Selbst das Wahre Selbst versteckt, das die Fähigkeit zum kreativen Gebrauch von Objekten hat. An sein Konzept vom Wahren und Falschen Selbst anschließend, werden jetzt zwei extreme Möglichkeiten in der Kommunikation des Babys und der Mutter gegenübergestellt und mit Begriffen von schöpferischer Haltung und Anpassung zusammengefaßt.

„On the basis of seeing and reaching to the world creatively the baby can become able to comply without losing face. When the pattern is the other way round and compliance dominates, then we think of ill-health and we see a bad basis for the development of the individual" (1968). (* Anm. 43)

Für die Entwicklung der Fähigkeit zu Objektbeziehungen, im Laufe derer das Kind von der Beziehung zu subjektiven zu der zu objektiv wahrnehmbaren Objekten überwechselt, hatte Winnicott den Begriff des „living with" diskutiert. „The term ‚living with‘ implies object relationships, and the emergence of the infant from the state of being merged with the mother, or his perception of objects as external to the self" (1960a). Dieser Begriff ist der Vorläufer der idealen Sequenz von Beziehungen, die im Laufe des Entwicklungsprozesses zu dem hinführen, was Winnicott das Zusammenspielen in einer Beziehung nennt (1971b, Kap. 3).

Bereichert um den Begriff des kreativen Impulses und um die Betonung der Zuverlässigkeit der Mutter, die den „potentiellen Raum" zur möglichen Erfahrung macht, setzt Winnicott fort, was er hinsichtlich der Entstehung der Spontaneität des Wahren Selbst, das Leben zu haben beginnt, angedeutet hatte. Geändert, so könnte man sagen, hat sich das Vorzeichen. Während es ihm mit dem Konzept vom Wahren und Falschen Selbst primär darum ging, die Entstehung des Falschen Selbst zu verstehen, und er infolgedessen die Idee vom Wahren Selbst einführte und den theoretisch notwendigen Schritt machte, das Wahre Selbst zu definieren, geht es Winnicott in den späteren Aufsätzen darum, die Idee von der Erfahrung der Lebendigkeit zu formulieren. Er widmet sich zunehmend der Frage, was es mit dem Leben auf sich hat. Ähnlich wie er zuvor gesagt hatte, daß es die Patienten seien, die die Idee vom Falschen Selbst nahelegen oder aufkommen lassen, so führt er auch jetzt wiederum die Patienten an, die zu der Frage nach dem Sinn des Selbst anregen. Indessen, es wird deutlich, daß sich die Frage von den Patienten löst, daß sie unabhängig von der Psychopathologie Raum einnimmt und Zeit.

Das Wort Selbst taucht in den letzten Aufsätzen 1971 in solchen Formulierungen auf wie „sense of self" oder „lack of sense of self" oder „searching for self" oder in so eng damit verbundenen Ausdrücken wie „sense of existing as a person" oder „creative experiences" oder „creative living". In der folgenden Sequenz hat Winnicott die Begriffe Erfahrung, Spielen, Kreativität und Sinn vom Selbst zueinander in Beziehung gesetzt:

„a) Relaxation in conditions of trust based on experience;
b) creative, physical, and mental activity manifested in play;
c) the summation of these experiences forming the basis for a sense of self" (1971b, Kap. 4). (* Anm. 44)

Wenn man die Definition, die Winnicott vom Selbst gegeben hat, auf dem Hintergrund seines Werkes liest, so wird einem bewußt, daß sie in ihrer Dynamik dem zu entsprechen scheint, was Winnicott zu benennen versucht, nämlich die ständige Entfaltung des Selbst. Zunächst geht es ihm darum, die ungeheure Fragilität dieser Entfaltung des Selbst in den frühesten Phasen hervorzuheben und diese Entfaltung auf das engste mit der Beziehung zwischen Baby und Mutter zu verbinden. Das Selbst erwacht in der Gegenseitigkeit der Beziehung, in der Begegnung, in der Aufeinanderfolge und Überlagerung der Erfahrung, gehalten, gehandhabt, gesehen und in der kreativen Geste beantwortet zu werden. In Anschluß an seine oben zitierte Definition spricht Winnicott von der „ersten Plattform der Betrachtung", die mit Leben und Lebendigkeit zu tun hat. Es ist dort, wo er sich zumeist aufhält. Pontalis (1975) hat darauf aufmerksam gemacht, daß Winnicott gerade in seinem letzten Buch immer wieder Verlaufsformen benutzt, wodurch er den Prozeß des Geschehens, das Anhalten eines Prozesses oder der Bewegung festzuhalten trachtet, wie z.B. playing, being, experiencing. Die Unverwechselbarkeit der Sprache Winnicotts und ihre Unmittelbarkeit kommt dadurch zustande, daß sie das Miterleben eines Prozesses möglich machen will. Es gibt aber einen weiteren Grund. Winnicott mußte vermutlich eine ganz eigene Sprache entwickeln, um die Intimität dieser sehr frühen Erfahrungen zu beschreiben oder, wie er sich ausdrückte, diese „sacred area".

„It is as if a work of art were being subjected to an analytic process. Can one be sure that the capacity to appreciate the work of art fully will not be destroyed by the searchlight that is played upon the picture?" (1968) (* Anm. 45)

Es liegt Winnicott daran, immer wieder an das Werden und an das Gewordensein zu erinnern, an die frühen Phasen des Reifungsprozesses, in denen von Anfang an die Chance der persönlichen Entwicklung

---

* Übersetzungen siehe jeweils Anmerkungen am Ende des Kapitels.

neben der Gefahr besteht des Sich-Verfehlens. Der Begriff des Selbst wird, wie sich in der Definition Winnicotts zeigt, zu einer Verdichtung seines theoretischen und klinischen Anliegens, das Person-Sein aus der frühen Mutter-Kind-Beziehung sich entfalten zu sehen und zu zeigen, daß es sich auch weiterhin in der Ausdehnung des potentiellen Raumes zwischen Baby und Mutter bewegt und realisiert.

6.8   Das nichtkommunizierende Selbst

Es mag einem in den Sinn kommen zu fragen, was Winnicott in seiner Definition vom Selbst (1971) nicht aufgegriffen hat von dem, was er früher schon über das Selbst gesagt hatte. Winnicott hat 1963 einen Aufsatz geschrieben, den er mit einer sehr persönlichen Bemerkung einleitete: „Während ich ... von keinem festgelegten Punkt ausging, kam ich zu meiner Überraschung bald dazu, das Recht zur Nicht-Kommunikation zu verteidigen. Dies war ein Protest aus meinem Innersten gegen die erschreckende Phantasie, unendlich ausgenützt zu werden". Der Aufsatz heißt „Communicating and not Communicating Leading to a Study of Certain Opposites". In Anlehnung an den Gedanken, den er vorher (1960a) in der Frage nach einem zentralen Selbst und dabei noch in Verbindung mit dem Begriff des Wahren Selbst geäußert hatte, vertritt Winnicott hier die Auffassung, daß in jedem Menschen, im Zentrum einer jeden Person ein geheimes Selbst ist, ein „incommunicado element", das für immer schweigt. Er nimmt an, daß es auch im Falle von Gesundheit einen Kern der Persönlichkeit gibt, der niemals mit objektiv wahrgenommenen Objekten kommuniziert, und weiter, daß die einzelne Person darum weiß, daß niemals mit diesem Kern kommuniziert werden darf, ebenso wie er für die äußere Realität unerreichbar sein muß. Winnicott nimmt an, daß beim gesunden Individuum ein Bedürfnis besteht, das dem Wahren Selbst der Persönlichkeit mit einer Organisation von einem Falschen Selbst entspricht, oder, wie er kurz sagt, der gespaltenen Persönlichkeit, nämlich ein Bedürfnis mit subjektiven Objekten umzugehen.

„Although healthy persons communicate and enjoy communicating, the other fact is equally true, that *each individual is an isolate, permanently non-communicating, permanently unknown, in fact unfound*" (1963) (* Anm. 46)

Hier besteht eine Beziehung zu dem, was Winnicott zuvor über die Fähigkeit, allein zu sein, geschrieben hatte, von der er sagte, daß sie entweder ein höchst verfeinertes Phänomen ist, das sich nach der Herstellung der Dreierbeziehung entwickelt, oder daß sie ein Phänomen der frühen Kindheit ist, das deshalb von Interesse sei, weil es die Grundlage für die differenzierte Form des Alleinseins darstellt.

Bei einer normalen Entwicklung nimmt Winnicott drei Linien der Kommunikation an: die Kommunikation, die für immer eine schweigende ist (aktiv im Gegensatz zur reaktiv schweigenden Kommunikation beim Wahren Selbst in der gespaltenen Persönlich-

keit), weiter die explizite Kommunikation, die indirekt ist, Freude bereitet und unter anderem den Gebrauch der Sprache einschließt, und schließlich die Kommunikation, die dem intermediären Raum verhaftet ist, zum Spielen führt und weiter zur kulturellen Erfahrung, eine Kommunikation, die Winnicott einen höchst wertvollen Kompromiß nennt.

Die Nicht-Kommunikation ist nicht averbal: „it is like music of the spheres, absolutely personal. It belongs to being alive. And in health, it is out of this that communication naturally arises" (1963). Der Künstler ist für Winnicott der Mensch, der in ganz besonderer Weise einem Dilemma ausgesetzt ist – da er sich zwei gleichzeitig bestehenden Tendenzen ausgesetzt sieht – dem Bedürfnis, das ein sehr drängendes ist, sich mitzuteilen, und dem noch drängenderen Bedürfnis, nicht gefunden zu werden, d.h. das Zentrum seiner Person zu schützen und schweigend mit subjektiven Objekten zu kommunizieren.

Winnicott streift auch in diesem Artikel nur kurz die Möglichkeit, seine theoretische Annahme mit der psychoanalytischen Theorie in Verbindung zu diskutieren. Die Frage, ob die hier beschriebene schweigende Kommunikation mit dem Konzept des primären Narzißmus zu tun hat, wird von ihm aufgeworfen und unbeantwortet gelassen. Statt dessen geht er auf die klinische Bedeutung seiner Hypothese ein. Klinisch hat für Winnicott die Annahme vom Individuum als einem isolierten Wesen Relevanz für das Studium der Kindheit, der Adoleszenz und der Psychose. Ich möchte mit einem Auszug aus seinen Gedanken über die Adoleszenz schließen, der klinische Erfahrung und persönliche Aussage in einer für Winnicott ungewöhnlichen Weise verbindet:

„Jungen und Mädchen in der Adoleszenz lassen sich auf mancherlei Weise beschreiben, und eine davon betrifft den Jugendlichen als Isolierten. Die Bewahrung der persönlichen Isoliertheit ist ein Teil der Suche nach Identität und nach dem Erwerb einer persönlichen Kommunikationstechnik, die nicht zu einer Vergewaltigung des zentralen Selbst führt. Dies kann ein Grund sein, warum Jugendliche im allgemeinen die psychoanalytische Behandlung scheuen, obwohl sie sich für psychoanalytische Theorien interessieren. Sie haben das Gefühl, sie würden durch die Psychoanalyse vergewaltigt, nicht sexuell, aber geistig. In der Praxis kann es der Analytiker vermeiden, derartige Ängste des Jugendlichen zu bestätigen, aber er muß darauf gefaßt sein, ganz und gar auf die Probe gestellt zu werden, und bereit, indirekte Kommunikation anzuwenden und einfache Nicht-Kommunikation zu erkennen.

In der Adoleszenz, wenn das Individuum pubertären Veränderungen unterworfen ist und noch nicht ganz bereit ist, ein Mitglied der Erwachsenengesellschaft zu werden, verstärkt sich die Abwehr gegen das Gefundenwerden, das heißt gegen das Gefundenwerden, bevor man da ist, um gefunden zu werden. Was wirklich personal ist und sich real anfühlt, muß um jeden Preis verteidigt werden, selbst dann, wenn dies eine vorübergehende Blindheit gegen den Wert des Kompromisses bedeutet. Jugendliche bilden eher ‚Haufen' als Gruppen, und dadurch, daß sie gleich aussehen, betonen sie die fundamentale Einsamkeit jedes Individuums" (1963).

---

* Übersetzungen siehe jeweils Anmerkungen am Ende des Kapitels.

Anmerkungen

Anmerkung 1 (S. 93): „Ein Wort wie ‚Selbst' weiß mehr als wir, es gebraucht uns und vermag uns zu beherrschen." (Übersetzung von L. Schacht)

Anmerkung 2 (S. 94): „So ein Ding wie ein Baby gibt es gar nicht." (Übersetzung von L. Schacht)

Anmerkung 3 (S. 94): „Daß, bevor Objektbeziehungen bestehen, sich die Sachlage folgendermaßen darstellt: Die Einheit ist nicht das Individuum, die Einheit ist ein Gefüge aus Umwelt und Individuum. Der Schwerpunkt des Seins geht nicht vom Individuum aus. Er liegt im Gesamtgefüge: Durch genügend gute Kinderpflege, Technik, genügend gutes Halten und genügend gute Versorgung wird die Schale allmählich übernommen, und der Kern (der für uns die ganze Zeit wie ein menschliches Baby ausgesehen hat) kann anfangen, ein Individuum zu sein." (Übersetzung aus Winnicott, 1976)

Anmerkung 4 (S. 94): „Die geistige Gesundheit des menschlichen Wesens wird in der frühen Kindheit durch die Mutter festgelegt, die eine Umwelt vermittelt, die komplexe, aber wesentliche Prozesse im Selbst des Kleinkindes vervollständigt werden können." (Übersetzung von L. Schacht)

Anmerkung 5 (S. 94): „Nichtsdestoweniger, es ist mir möglich, das Wort zu nehmen ..., das ich in einem anderen Zusammenhang gebraucht habe, und zu sehen, wie es durch detaillierte klinische Arbeit in der Kinderpsychiatrie und -psychoanalyse veranschaulicht wird." (Übersetzung von L. Schacht)

Anmerkung 6 (S. 94): „Im Bezug auf diesen Artikel hat die Hauptsache mit dem Wort ‚Selbst' zu tun. Ich habe mich gefragt, ob ich etwas über dieses Wort niederschreiben könnte, aber natürlich, kaum hatte ich mich daran gemacht, fand ich, daß auch in meiner Vorstellung reichlich Ungewißheit dazu besteht, welche Bedeutung ich ihm beimesse. Ich fand, daß ich das Folgende geschrieben hatte:

Für mich ist das Selbst, das nicht das Ich ist, die Person, die sie selbst ist, die ausschließlich sie selbst ist, die in ihrer Ganzheit auf der Operation des Reifungsprozesses basiert. Zur gleichen Zeit besitzt das Selbst verschiedene Teile, ja, es setzt sich in der Tat aus diesen Teilen zusammen. Diese Teile agglutinieren im Verlaufe der Operation des Reifungsprozesses in der Richtung, die von innen nach außen geht, unterstützt, wie dies der Fall sein muß (maximal zu Beginn), durch die menschliche Umgebung, die hält, die handhabt und in lebendiger Weise begünstigend wirkt. Das Selbst findet sich natürlich im Körper angesiedelt, doch mag es unter Umständen vom Körper oder umgekehrt der Körper von ihm dissoziiert werden. Das Selbst erkennt sich wesentlich in den Augen und im Gesichtsausdruck der Mutter wieder und im Spiegel, der das Gesicht der Mutter vertreten kann. Schließlich gelangt das Selbst an einer signifikanten Beziehung zwischen dem Kind und der Summe von Identifikationen, die (nach genügender Inkorporation und Introjektion von mentalen Repräsentanzen) in der Gestalt einer inneren psychischen lebendigen Realität organisiert werden, in Erscheinung. Die Beziehung zwischen dem Jungen oder dem Mädchen mit seiner oder ihrer eigenen inneren psychischen Organisation wird entsprechend den Erwartungen modifiziert, die von seiten des Vaters oder der Mutter oder von denen ausgehen, die im äußeren Leben des Individuums bedeutend geworden sind. Es ist das Selbst und das Leben des Selbst, das allein der Handlung oder dem Leben vom Standpunkt des Individuums aus Sinn verleiht, das bisher gewachsen ist und das fortfährt von der Abhängigkeit und Unreife in Richtung Unabhängigkeit und in Richtung der Fähigkeit zu wachsen, sich mit reifen Liebesobjekten zu identifizieren, ohne die individuelle Identität aufzugeben.

Sie mögen das nicht hilfreich finden, doch auf jeden Fall schien es mir eine wertvolle Sache zu sein, dies versuchsweise niederzuschreiben. Wahrscheinlich mag ich es ändern wollen.

Sie stehen natürlich allein da mit demselben Problem, das Sie zu Beginn hatten, das darin liegt, wie ‚Selbst' zu übersetzen ist, ohne dasselbe Wort zu wählen, das Sie auch für die Übersetzung von ‚Ich' wählen würden. Lassen Sie mich versuchen hilfreicher zu sein. Ich nehme an, daß derjenige, der den Begriff ‚Selbst' verwendet, sich auf einer anderen Ebene befindet als derjenige, der den Begriff ‚Ich' verwendet. Die erste Ebene hat mit dem Leben und mit leben in direkter Weise zu tun, auf der zweiten Ebene, wo das Wort ‚Ich' (le moi) gebraucht wird, ist derjenige, der spricht oder schreibt, distanzierter, weniger involviert, vielleicht klarer wegen der Fähigkeit, all das gebrauchen zu können, was den intellektuellen Zugang umfaßt." (Übersetzung von L. Schacht)

Anmerkung 7 (S. 95): „Man kann den Ausdruck ‚Ich' gebrauchen, um jenen Teil der wachsenden menschlichen Persönlichkeit zu bezeichnen, der dazu neigt, sich unter geeigneten Bedingungen zu einer Einheit zu integrieren ... Wir werden sehen, daß das Ich sich der Untersuchung darbietet, lange bevor das Wort ‚Selbst' relevant ist. Das Wort ‚Selbst' wird sinnvoll, wenn das Kind angefangen hat, seinen Intellekt zu benützen, um das anzuschauen, was andere sehen oder fühlen oder hören und was sie begreifen, wenn sie diesem Säuglingskörper begegnen." (Übersetzung aus Winnicott, 1974)

Anmerkung 8 (S. 95): „Gewiß kann vor der Geburt über die Psyche gesagt werden, daß es ein persönliches Vorwärtsgelangen gibt, eine Kontinuität der Erfahrung. Diese Kontinuität, die auch der Beginn des Selbst genannt werden könnte, wird periodisch durch solche Phasen unterbrochen, die Reaktionen auf Übergriffe beinhalten. Das Selbst beginnt Erinnerungen von begrenzten Phasen einzuschließen, in denen die Reaktion auf einen Übergriff die Kontinuität stört." (Übersetzung von L. Schacht)

Anmerkung 9 (S. 95): „Ich bedeutet hier eine Summe von Erfahrung. Das Individuelle Selbst beginnt als eine Summation von Erfahrungen der Ruhe." (Übersetzung von L. Schacht)

Anmerkung 10 (S. 95): „Ich werde nicht damit beginnen, einen historischen Überblick zu geben und zu zeigen, wie sich meine Ideen aus den Theorien anderer entwickelt haben, denn meine Gedanken gehen andere Wege. Ich nehme dies und das hier und dort auf, und dann, zu allerletzt, schaue ich interessiert nach, um herauszubekommen, wo ich was gestohlen habe. Vielleicht ist diese Methode nicht schlechter als irgendeine andere." (Übersetzung aus Winnicott, 1976)

Anmerkung 11 (S. 95): „... weil ich als Pädiater sprechen möchte, der gewohnheitsmäßig an das sich entwickelnde Kind und an den sich entwickelnden Säugling denkt. Für den Pädiater gibt es eine Kontinuität in der Entwicklung des Individuums ... Das Ziel der Kinderpflege besteht nicht nur darin, ein gesundes Kind hervorzubringen, sondern auch darin, die Entwicklung eines gesunden Erwachsenen zu ermöglichen." (Übersetzung aus Winnicott, 1976)

Anmerkung 12 (S. 96): „Wenn überhaupt analytische Arbeit geleistet werden soll, muß der Analytiker an die menschliche Natur und an den Entwicklungsprozeß glauben." (Übersetzung aus Winnicott, 1976)

Anmerkung 13 (S. 96): „(wenn sie spüren, daß sie dazu in der Lage sind) ohne das Delikate an dem zu verunglimpfen, was präverbal, nicht in Worte gefaßt oder faßbar ist, es sei denn in der Dichtung." (Übersetzung von L. Schacht)

Anmerkung 14 (S. 96): „Anstatt ihr ganz eigenes Selbst zu sein, wird sie jetzt schnell der Dinge überdrüssig und verliert jedes Interesse an einem nach dem anderen Spielzeug." (Übersetzung von L. Schacht)

Anmerkung 15 (S. 96): „Gewiß kann vor der Geburt … (s. Anm. 8)

Hier mag beobachtet werden, daß der Säugling zur Reaktion gezwungen ist, aus einem Zustand des ‚Seins' herausgedrängt wird. Dieser Zustand des Seins kann nur unter bestimmten Bedingungen zustande kommen. Wenn er reagieren muß, befindet sich der Säugling nicht im Zustand des ‚Seins'. Es mag hervorgehoben werden, daß am bedeutendsten das Trauma ist, das die Notwendigkeit zu reagieren darstellt. Reagieren kommt in dieser Phase der menschlichen Entwicklung einem vorübergehenden Verlust an Identität gleich. Dadurch wird ein extremes Gefühl von Unsicherheit bewirkt und die Basis für die Erwartung weiterer Beispiele von Verlust der Kontinuität des Selbst, ja sogar eine angeborene (wenn nicht ererbte) Hoffnungslosigkeit im Hinblick auf das Erlangen eines persönlichen Lebens." (Übersetzung von L. Schacht)

Anmerkung 16 (S. 96): „Die erste Ich-Organisation entsteht aus dem Erleben drohender Vernichtung, zu der es jedoch nicht kommt, auf die immer eine Wiederherstellung folgt." (Übersetzung aus Winnicott, 1976)

Anmerkung 17 (S. 97): „Sie wissen, was es heißt, im Zustand akuter Konfusion oder in der Todesangst der Desintegration zu sein. Sie wissen, was es heißt, fallen gelassen zu werden, für immer zu fallen oder in die psychosomatische Uneinigkeit gespalten zu werden. Mit anderen Worten, sie haben Trauma erfahren, und ihre Persönlichkeiten müssen um die Organisierung von Abwehrformen, die auf das Trauma gefolgt war, aufgebaut werden." (Übersetzung von L. Schacht)

Anmerkung 18 (S. 97): „… bietet die Voraussetzung, daß die kindliche Konstitution hervortreten kann, daß die Entwicklungsmöglichkeiten beginnen können, sich zu entfalten, und daß das Kind spontan Bewegung erleben und die diesen frühen Lebensphasen angemessenen Empfindungen haben kann." (Übersetzung aus Winnicott, 1976)

Anmerkung 19 (S. 97): „Lassen Sie uns versuchen, den Job der Mutter zu studieren. Wenn der Säugling sich zu einem Wesen entwickeln soll und die Welt zu erfassen beginnen soll, wenn er je innerlich zusammenwachsen und kohärent werden soll, dann sind folgende Gegebenheiten auf seiten der Mutter lebensnotwendig: sie muß existieren, fortfahren zu existieren, sie muß leben, riechen, atmen, und ihr Herz muß schlagen. Sie muß in jedweder Weise spürbar werden. Sie liebt in einer körperlichen Weise, vermittelt Berührung, Wärme, Bewegung und Ruhe entsprechend den Bedürfnissen des Babys. Sie stellt für das Baby die Gelegenheit dar, einen Übergang zwischen dem Ruhezustand und dem der Erregung zu durchstehen – indem sie das Baby nicht plötzlich mit Nahrung überfällt oder mit der Forderung nach einer Erwiderung auf sie. Sie stellt angemessene Nahrung zur angemessenen Zeit zur Verfügung. Zu Beginn läßt sie den Säugling bestimmen, während sie bereit ist, sich für eine Antwort zur Verfügung zu stellen (da das Kind noch zu sehr ein Teil ihrer selbst ist). Allmählich führt sie die gemeinsame Umwelt ein, indem sie diesen Vorgang vorsichtig abstuft entsprechend den von Tag zu Tag und Stunde zu Stunde wechselnden Bedürfnissen des Kindes. Sie schützt das Baby vor Zwischenfällen und Schocks (Türenknallen während das Baby gestillt wird), indem sie die körperliche und gefühlsmäßige Situation einfach genug hält, so daß der Säugling diese verstehen kann und dennoch reich genug im Einklang mit der wachsenden Fähigkeit des Kindes. Sie stellt Kontinuität zur Verfügung. Indem sie an das Kind als an ein eigenständiges, um seiner selbst willen existierendes Wesen

glaubt, überstürzt sie diese Entwicklung nicht und versetzt den Säugling somit in die Lage, Zeitgefühl zu entwickeln und das Gefühl eines inneren persönliches Wachstums. Für die Mutter ist das Kind von Anfang an ein ganzheitliches menschliches Wesen, das sie auch, und das befähigt sie, seinen Mangel an Integration und sein schwaches Im-Körper-Leben zu ertragen." (Übersetzung von L. Schacht)

Anmerkung 20 (S. 97): „… der Beginn ist eine Summation von Anfängen." (Übersetzung von L. Schacht)

Anmerkung 21 (S. 97): „Entspannung bedeutet für den Säugling, daß er nicht das Gefühl haben muß, Integration leisten zu müssen, da die ichstützende Funktion der Mutter gewährleistet ist." (Übersetzung von L. Schacht)

Anmerkung 22 (S. 97): „Die Grundlage für dieses Innewohnen ist eine Verknüpfung motorischer und sensorischer und funktionaler Erfahrungen mit dem neuen Zustand des Säuglings, eine Person zu sein. Als weitere Entwicklung entsteht etwas, das man als begrenzende Membran bezeichnen könnte, die (im gesunden Zustand) in gewissem Maß mit der Oberfläche der Haut gleichzusetzen ist und ihre Stellung zwischen dem Ich und dem Nicht-Ich des Säuglings einnimmt. So kommt der Säugling dazu, ein Innen und ein Außen und ein Körperschema zu haben." (Übersetzung aus Winnicott, 1974)

Anmerkung 23 (S. 98): „… kann der Schwerpunkt des Seins im Mittelpunkt des Gefüges aus Umwelt und Individuum liegen, im Kern und nicht in der Schale. Der Mensch, der sich nun als Wesen vom Mittelpunkt her entwickelt, kann seinen Ort im Körper des Kindes finden, er kann beginnen, eine äußere Welt zu erschaffen und zugleich eine begrenzende Membran und ein Inneres zu erwerben." (Übersetzung aus Winnicott, 1976)

Anmerkung 24 (S. 98): „Man könnte sagen, die Mutter mache aus dem schwachen Ich des Babys ein starkes, weil sie da ist und alles verstärkt, wie die Servolenkung bei einem Autobus." (Übersetzung aus Winnicott, 1974)

Anmerkung 25 (S. 98): „Wie sie ausschaut, hängt davon ab, was sie selbst erblickt." (Übersetzung von L. Schacht)

Anmerkung 26 (S. 98): „Wenn ich schaue, werde ich gesehen, also existiere ich. Ich kann es mir jetzt leisten zu schauen und zu sehen. Ich schaue nun in kreativer Weise und was ich erfasse, nehme ich auch wahr." (Übersetzung von L. Schacht)

Anmerkung 27 (S. 98): „Betrachten wir … die innere psychische Realität, den individuellen Besitz jedes einzelnen, soweit er in seiner Entwicklung eine reife Integration und den Aufbau eines einheitlichen Selbst mit Innenwelt, Außenwelt und Grenzschicht erreicht hat. Auch hier ergeben sich aus dem Erbgut und dem Persönlichkeitsaufbau, aus introjizierten Anteilen der Umwelt und projizierten Anteilen der Persönlichkeit fest vorgegebene Voraussetzungen." (Übersetzung aus Winnicott, 1973)

Anmerkung 28 (S. 99): „Ich stelle mir den Prozeß so vor, als ob sich aus entgegengesetzten Richtungen zwei Linien einander näherten, die sich wahrscheinlich berühren werden. Wenn sie sich schneiden, entsteht ein Augenblick der Illusion – ein Stückchen Erfahrung, das der Säugling entweder als seine eigene Halluzination oder als ein Ding nehmen kann, das zur äußeren Realität gehört." (Übersetzung aus Winnicott, 1976)

Anmerkung 29 (S. 99): „Vom Standpunkt des Beobachters aus scheint eine ‚Beziehung zu einem Objekt haben' im Sinne eines primären Fusionszustandes vorzuliegen, aber man muß sich daran erinnern, daß das Objekt zu Beginn ein subjektives Objekt ist. Ich habe diesen Begriff subjektives Objekt deshalb gewählt, um die Diskrepanz zwischen dem, was beobachtet, und dem, was vom Baby erfahren werden kann, hervorzuheben." (Übersetzung von L. Schacht)

Anmerkung 30 (S. 99): „Wenn man die Kommunikation zwischen Baby und Mutter beschreibt, dann besteht da die essentielle Dichotomie. Die Mutter kann zu den kindlichen Formen des Erlebens zusammenschrumpfen, aber das Baby kann sich nicht zur erwachsenen Aufgeklärtheit aufplustern." (Übersetzung von L. Schacht)

Anmerkung 31 (S. 99): „In der Frühentwicklung des Menschen ermöglicht die Umwelt, die sich gut genug verhält (die eine ausreichend gute aktive Anpassung zuwege bringt), persönliches Wachstum. Die Prozesse des Selbst können dann aktiv weiterlaufen, in einer ununterbrochenen Abfolge lebendiger Weiterentwicklungen. Wenn die Umwelt sich nicht gut genug verhält, wird das Individuum zu Reaktionen auf Übergriffe veranlaßt, und die Prozesse des Selbst werden unterbrochen. Wenn dieser Zustand eine gewisse quantitative Grenze erreicht, wird der Kern des Selbst geschützt; es tritt eine Stockung ein, das Selbst kann keine weiteren Fortschritte machen, bis die verfehlte Umweltsituation auf die Weise wieder in Ordnung gebracht wird, die ich beschrieben habe. Während das wahre Selbst geschützt wird, entwickelt sich ein falsches Selbst, das auf der Grundlage von Abwehr und Gefügigkeit, auf der Annahme der Reaktion auf Übergriffe aufgebaut ist. Die Entwicklung eines falschen Selbst ist eine der erfolgreichsten Abwehrorganisationen, die den Kern des wahren Selbst schützen soll, und ihr Vorhandensein ruft das Gefühl der Vergeblichkeit hervor." (Übersetzung aus Winnicott, 1976)

Anmerkung 32 (S. 100): „Es muß betont werden, daß ich, wenn ich von der Erfüllung der Bedürfnisse des Säuglings spreche, nicht Triebbefriedigung meine. In dem Bereich, den ich untersuche, sind die Triebe für den Säugling noch nicht klar als etwas Inneres definiert. Die Triebe können ebenso außen sein wie ein Donnerhall oder ein Schlag. Das Ich des Säuglings baut Stärke auf und erlangt infolgedessen allmählich einen Zustand, in dem Es-Forderungen als Teil des Selbst empfunden werden und nicht als etwas, das aus der Umwelt kommt. Wenn diese Entwicklung eintritt, wird die Es-Befriedigung zu einer sehr wichtigen Verstärkung des Ichs oder des wahren Selbst; aber Es-Erregungen können traumatisch sein, wenn das Ich noch nicht fähig ist, sie einzubeziehen, und die damit verbundenen Risiken und die Frustrationen, die erlebt werden bis zu dem Zeitpunkt, in dem die Es-Befriedigung zur Tatsache wird, zu ertragen." (Übersetzung aus Winnicott, 1974)

Anmerkung 33 (S. 100): „Ein Beispiel: Ein Baby trinkt an der Brust und erlangt Befriedigung. Dieser Umstand an sich zeigt noch nicht, ob es ein ichsyntones Es-Erlebnis hat oder ob es, im Gegenteil, das Trauma einer Verführung erleidet, eine Bedrohung der persönlichen Ich-Kontinuität oder eine Bedrohung durch ein nicht ichsyntones Es-Erlebnis, mit dem das Ich nicht fertigzuwerden imstande ist." (Übersetzung aus Winnicott, 1974)

Anmerkung 34 (S. 101): „Der Prozeß, der zur Fähigkeit des Symbolgebrauchs führt, kommt nicht in Gang (oder er wird unterbrochen), womit ein Sich-Zurückziehen des Säuglings von bereits gewonnenen Vorteilen einhergeht." (Übersetzung aus Winnicott, 1974)

Anmerkung 35 (S. 101): „Aus dem Problem der Unintegriertheit erwächst ein anderes, das der Dissoziationen ... Meiner Meinung nach erwächst aus der Unintegriertheit eine Reihe von Erscheinungen, die man dann Dissoziationen nennt, weil die Integration unvollständig oder bruchstückhaft geblieben ist." (Übersetzung aus Winnicott, 1976)

Anmerkung 36 (S. 101): „Der Säugling kann jetzt anfangen, die Illusion des omnipotenten Erschaffens und Lenkens zu genießen; dann kann er allmählich das illusorische Element erkennen lernen, die Tatsache, daß er spielt und phantasiert. Hier ist die Grundlage für das Symbol, das zunächst sowohl die Spontaneität oder Halluzination des Säuglings ist, als auch das geschaffene und schließlich besetzte äußere Objekt." (Übersetzung aus Winnicott, 1974)

Anmerkung 37 (S. 101): „Das gesunde Individuum, dessen Selbst einen gefügigen Aspekt hat, das aber existiert und ein kreatives und spontanes Lebewesen ist, hat zugleich eine Fähigkeit, Symbole zu gebrauchen. Mit anderen Worten, Gesundheit ist hier eng verbunden mit der Fähigkeit des Individuums, in einem Bereich zu leben, der zwischen Traum und Realität liegt, dem Bereich, den man das ‚kulturelle Leben' nennt. Im Gegensatz dazu findet sich, wo ein hoher Grad der Spaltung zwischen dem wahren Selbst und dem falschen Selbst besteht (das das wahre Selbst verbirgt), eine schlechte Fähigkeit des Symbolgebrauchs und eine Verarmung des kulturellen Lebens. An Stelle kultureller Aktivitäten beobachtet man bei solchen Menschen äußerste Ruhelosigkeit, Konzentrationsunfähigkeit und ein Bedürfnis, aus der äußeren Realität störende Einflüsse auf sich zu ziehen, so daß die Lebenszeit des Individuums mit Reaktionen auf diese Störungen ausgefüllt werden kann." (Übersetzung aus Winnicott, 1974)

Anmerkung 38 (S. 102): „Es ist eng verknüpft mit der Vorstellung vom Primärvorgang und ist am Anfang im wesentlichen nicht reaktiv gegenüber äußeren Reizen, sondern primär." (Übersetzung aus Winnicott, 1974)

Anmerkung 39 (S. 102): „... ohne sie hat die Vorstellung von einer Beziehung zu einem Objekt, das von anderen als etwas außerhalb des betreffenden Menschen Liegendes wahrgenommen wird, keine Bedeutung für diesen Menschen." (Übersetzung aus Winnicott, 1976)

Anmerkung 40 (S. 102): „Gerade im Spielen und nur im Spielen kann das Kind und der Erwachsene sich kreativ entfalten und seine ganze Persönlichkeit einsetzen, und nur in der kreativen Entfaltung kann das Individuum sich selbst entdecken." (Übersetzung aus Winnicott, 1973)

Anmerkung 41 (S. 102): „... Spielen ermöglicht Reifung und damit Gesundheit, es führt zu Gruppenbeziehungen; es kann eine Form der Kommunikation in der Psychotherapie sein; und schließlich hat sich Psychoanalyse als eine hochdifferenzierte Art des Spielens im Dienste der Kommunikation des Patienten mit sich selbst und anderen entwickelt." (Übersetzung aus Winnicott, 1973)

Anmerkung 42 (S. 102): „... Obwohl ich Freuds Beschreibung des Spiels mit der Garnrolle schon lange kenne und obwohl sie mich immer schon zur eingehenden Beobachtung des Spiels kleiner Kinder angeregt hat ..." (Übersetzung von L. Schacht)

Anmerkung 43 (S. 103): „Auf der Basis des Sehens und des kreativen Nach-der-Welt-Ausreichens kann das Baby in die Lage versetzt werden, sich anzupassen, ohne das Gesicht zu verlieren. Wenn das Muster anders herum verläuft und die Willfährigkeit überwiegt, dann denken wir an einen krankhaften Zustand und sehen eine schlechte Voraussetzung für die Entwicklung des Individuums." (Übersetzung von L. Schacht)

Anmerkung 44 (S. 103): „a) Entspannung, die sich unter der Voraussetzung vollzieht, daß der Patient aufgrund von Erfahrung genügend Vertrauen entwickelt hat; b) schöpferische, körperliche und geistige Aktivität, die sich im Spiel manifestiert; c) die Zusammenfassung dieser Erfahrungen, die die Grundlage für ein Selbstgefühl abgibt." (Übersetzung aus Winnicott, 1973)

Anmerkung 45 (S. 103): „Es ist, als wenn ein Kunstwerk einem analytischen Prozeß unterzogen würde. Kann man sicher sein, daß die Fähigkeit, das Kunstwerk wertzuschätzen, nicht durch das Suchlicht zerstört werden wird, das auf das Bild geworfen wird?" (Übersetzung von L. Schacht)

Anmerkung 46 (S. 104): „Wenn auch gesunde Menschen kommunizieren und es genießen, so ist doch die andere Tatsache ebenso wahr, daß jedes Individuum ein isoliertes Wesen ist, ständig nichtkommunizierend, ständig unbekannt, tatsächlich ungefunden." (Übersetzung von L. Schacht)

7 Lernpsychologische Grundlagen für die Psychosomatische Medizin

Othmar W. Schonecke

7.1 Zum Verhältnis von Lernpsychologie und Psychosomatischer Medizin

In den letzten Jahren hat sich der Einfluß lerntheoretisch orientierter Konzepte in der Psychosomatischen Medizin zunehmend verstärkt. Dies gilt sowohl für die Erforschung pathogener Bedingungen von Erkrankungen mit vornehmlich organischer Symptomatik, die für die Psychosomatische Medizin von besonderem Interesse sind, als auch für Möglichkeiten der Behandlung derartiger Störungen.

Die Psychosomatische Medizin begegnet diesen Problemen auf verschiedene Weise. So fordert sie, die Beziehung zwischen Arzt und Patient müsse in der Medizin ganz allgemein den Bedürfnissen der Patienten angemessener sein. Der Patient dürfe nicht nur als Träger einer Krankheit aufgefaßt werden, sondern als eine Person, die erkrankt ist und die Krankheit und ihre Folgen verarbeiten muß. Diese Auffassung wird meistens „ganzheitlich" genannt, sie dividiere den Patienten nicht in verschiedene Bereiche von Leib und Seele auseinander.

Tatsächlich ist die genannte Ganzheitlichkeit nur selten anzutreffen. In der psychosomatischen Therapie von Patienten sind selten bis nie Maßnahmen vorhanden, die auf körperliche Vorgänge zielen. Diese werden dann „somatisch" arbeitenden Kollegen überlassen. „Solche Erfahrungen lassen sich weitgehend auf den inneren Widerspruch unserer dualistischen Medizin zurückführen, d.h. auf die Tatsache, daß wir nicht eine, sondern zwei sich einander nicht etwa ergänzende, sondern sich bereits im Prinzip gegenseitig ausschließende Heilkunden haben: eine Medizin für Körper ohne Seelen mit hoch spezialisierten Organdisziplinen und dazugehörigen Spezialkliniken auf der einen Seite, und eine Medizin für Seelen ohne Körper, ebenfalls mit Spezialdisziplinen und dazugehörigen Neurosekrankenhäusern, auf der anderen" (v. Uexküll, 1981).

Der dualistische Standpunkt, wie er von v. Uexküll für die Krankenversorgung aufgezeigt worden ist, beherrschte weitgehend auch die Erforschung pathogener Bedingungen sog. „psychosomatischer Erkrankungen". Bis vor einiger Zeit versuchte man zu erfassen, ob es für bestimmte Krankheitsbilder bestimmte Dispositionen in Form von spezifischen Persönlichkeitsmerkmalen oder Merkmalskonfigurationen gebe. Auch bei derartigen Versuchen, dem Problem der Erforschung pathogener Bedingungen näherzukommen, kann eigentlich nicht davon die Rede sein, daß ein ganzheitlicher Standpunkt bzw. Ausgangspunkt bestanden hat, denn die konzeptuell und methodisch hergestellte Beziehung zwischen psychischen Bedingungen und körperlichen Krankheitsfolgen war höchstens korrelativ.

Neben der Geltung der Psychosomatischen Medizin für den gesamten Bereich der Medizin, d.h. neben der Forderung, Medizin müsse im Grunde stets psychosomatische Medizin sein, hat sie sich ebenfalls als eine Spezialdisziplin etabliert, als eine Disziplin mit einem bestimmten Gegenstandsbereich und mit bestimmten Grundlagenwissenschaften.

Schonecke (1972) hat zu zeigen versucht, daß Psychosomatische Medizin verstanden werden kann als Disziplin, deren Gegenstand unter anderem in der empirischen Klärung des als Methoden- und Sprachproblem aufgefaßten Leib-Seele-Problems besteht. Dieser Auffassung zufolge ist ein wesentliches Element des Leib-Seele-Problems ein Sprachproblem, insofern als es derzeit kein theoretisches Sprachsystem gibt, das die Unterscheidung in Leib und Seele aufheben würde und darüber hinaus alle relevanten Gegenstandsbereiche erfassen würde. Ein derartiges Sprachsystem, eine Metasprache, erhielte ihren empirischen Bezug, ihre „empirische Sinnhaftigkeit", durch die Verknüpfung mit einer Beobachtungsmethode, die Psychisches und Physisches methodisch aufeinander bezogen erfaßt. So könnte die Psychosomatische Medizin es ermöglichen, verschiedene Wissenschaftsbereiche im Hinblick auf die entsprechenden Sachverhalte zu verknüpfen, und damit deren jeweiligen Geltungsbereich zu erweitern (vgl. Kap. 8 zum Begriff der Emotion).

Diese Auffassung hat eine weitere Voraussetzung. Sie geht nicht von einer Aufteilung in Natur- und Geisteswissenschaften aus, sondern davon, daß es wissenschaftliches Vorgehen gibt, das bestimmten Regeln folgt, oder aber ein Vorgehen, das diesen Regeln nicht folgt und damit nicht als wissenschaftlich (aber keineswegs als wertlos) zu gelten hat. Das bedeutet, daß sich das Vorgehen im Bereich der Psychologie nicht grundsätzlich von dem in der Physik unterscheidet. Es muß ein bestimmtes logisches Verhältnis zwischen theoretischen Aussagen bestehen, sowie zwischen diesen und solchen Aussagen, die Methoden der Beobachtung beschreiben. Die in solchen Aussagen verwendeten Begriffe müssen eindeutig sein, d.h. nicht mehrere Interpretationsmöglichkeiten zulassen. So wird der Vorgang der Beobachtung (keineswegs immer des Experiments) so präzise definiert sein müssen, daß er wiederholbar ist. Durch diese Definition wird auch das Beobachtete mit definiert, da die Aussagen über Beobachtungsmethoden Teil des theoretischen Aussagesystems über das Beobachtete sind. Dies bedeutet nicht, daß der Vorgang des Verstehens in der Psychologie keinen Platz hätte. Er hat schon darum einen besonderen Platz, weil das „Sich-verstanden-Fühlen" einer Person einen Einfluß auf diese haben kann.

Tatsächlich entsprechen psychologische Modelle, auf die die Psychosomatische Medizin als Teil ihrer Grundlagen zurückgreift, den oben genannten Anforderungen nicht immer. Betrachtet man den Gesamtbereich psychologischer Erkenntnisse, so fällt auf, daß dieser in sehr verschiedene Wissensbereiche aufgeteilt werden kann. Dies gilt einmal im Hinblick auf formale Aspekte, wie auch im Hinblick auf ein Merkmal, das „Nähe zum Leben" genannt werden könnte oder, für den vorliegenden Zusammenhang, klinische Relevanz. Beide Merkmale scheinen nicht unabhängig voneinander zu sein. Es gibt Bereiche, die diese Relevanz nicht unmittelbar aufweisen, und Psychologen ist gelegentlich der Vorwurf gemacht worden, sie würden Sachverhalte mit großer Akribie untersuchen, die vollständig uninteressant seien. Andererseits sind klinisch orientierte psychologische Modelle im Hinblick auf ihre wissenschaftliche Überprüfbarkeit, aufgrund zu allgemeiner Annahmen, deren Verknüpfung mit Daten der Beobachtung in einem methodischen Sinne nicht gegeben ist, häufig fragwürdig.

Von vielen Vertretern der Psychoanalyse wird diese als verstehende, hermeneutische, als Geisteswissenschaft verstanden, deren Zugang zur Erfassung psychischer Phänomene grundsätzlich vom Zugang der Naturwissenschaften verschieden sei. Auf diese Weise wird ein Dualismus eingeführt zwischen Methoden der Naturwissenschaften und denen einer verstehenden Psychologie. Geht man nun davon aus, daß sich die Methode am Gegenstand orientiert, so ergibt sich daraus, daß der Gegenstand der Psychologie grundsätzlich von dem der Naturwissenschaften verschieden ist. Auf diese Weise wird ein Dualismus zwischen Psychischem und Physischem eingeführt, der dem oben für die Praxis geschilderten dann wohl nicht zufällig entspricht.

7.2 Der lerntheoretische Ansatz

Die Lernpsychologie hat einen Ausgangspunkt u.a. in den Arbeiten Pawlows zum Vorgang des klassischen Konditionierens. In der angelsächsischen Literatur wird daher häufig auch der Begriff „Pavlovian conditioning" verwendet. Die Versuche, die Pawlow in diesem Zusammenhang durchgeführt hat, bezeichnete er als „psychische Versuche", da sie sich mit dem Einfluß äußerer Reize auf körperliche Reaktionen beschäftigten (Pawlow, 1953 III). Etwa zur selben Zeit eta-

blierte sich in den USA eine psychologische Denk- und Methodenrichtung, verbunden u.a. mit den Namen Thorndike und Watson, die sich den Namen „Behaviorism" (Behaviorismus) gab. Dieser wissenschaftstheoretisch sehr strikte Ansatz, der mit Skinner theoretische Formulierungen über interne, nicht beobachtbare Zustände von Organismen für überflüssig hielt, hat für einen längeren Zeitraum psychologisches Denken in den USA beeinflußt.

Es ist jedoch nicht sinnvoll, lerntheoretische Ansätze mit dem Behaviorismus gleichzusetzen, er stellt lediglich eine Strömung dar, die lerntheoretisch orientiert ist. Für die Lernpsychologie wesentlich ist allerdings, daß das Interesse zunächst auf äußere Bedingungen des Verhaltens gerichtet ist. Damit ist ein wesentlicher Unterschied zu den tiefenpsychologischen Modellen vorhanden. Letztere versuchen, Verhalten hauptsächlich aus internen, der direkten Beobachtung nicht zugänglichen Bedingungen, psychischen Strukturen, zu erklären. Aber auch in der Lernpsychologie werden Konstrukte verwendet, die sich auf interne Bedingungen beziehen, wie Triebstärke (Hull, 1952). Dennoch ist beispielsweise Fahrenberg (1979) der Meinung, daß die Lernpsychologie aufgrund ihrer Methoden und ihres Interesses an direkt beobachtbaren Sachverhalten, die Verhalten beeinflussen, letztlich in einem engeren Zusammenhang mit der Physiologie steht als mit einer Psychologie, für die im Begriff des „Psychischen die beiden sehr verschiedenen Phänomenbereiche Verhalten und Erleben zusammengefaßt sind" (Fahrenberg, 1979).

Betrachtet man die Konzepte der Lernpsychologie, so scheint es, daß eine entsprechende Trennung zwischen Psychischem und Physischem nicht vorhanden ist und schon gar nicht methodisch gefordert wird. Dies gilt sowohl für die Anfänge der Lernpsychologie, z.B. die Ergebnisse der Arbeiten Pawlows oder Thorndikes, als auch für neuere Entwicklungen in besonderem Maße (Rescorla, Mackintosh, Grant usw.). Es muß allerdings eingeräumt werden, daß Vertreter der Lernpsychologie häufig dazu neigen, ihre Arbeiten in einem anderen Licht zu betrachten. So lehnt Skinner (1950) theoretische Formulierungen über nicht beobachtbare Vorgänge ab und fordert, sich auf mit Methoden der Psychologie beobachtbare Sachverhalte in einer „funktionalen Analyse" zu beschränken. Für ihn sind Aussagen, die physiologische Komponenten für den Vorgang des Lernens mitberücksichtigen, theoretische Aussagen, da physiologische Prozesse im Hinblick auf das Lernen interne, mit Methoden der Lernpsychologie nicht erfaßbare Vorgänge sind.

7.2.1 Funktionale Analyse

Lerntheoretische Konzepte beziehen sich ganz allgemein auf die Beeinflussung von „Lebensvorgängen". Diese können sowohl in physiologischen Reaktionen bestehen, als auch in recht komplexen Verhaltensweisen der Einwirkung auf die Umgebung. Dabei wird paradigmatisch und methodisch kein Unterschied gemacht, ob eine physiologische Reaktion auf einen Reiz konditioniert wird oder eine Verhaltenssequenz.

Für diesen Sachverhalt ist der funktionale Charakter lerntheoretischer Modelle von Bedeutung, d. h. sie beschreiben Regelmäßigkeiten bestimmter Lebensvorgänge, nicht deren Inhalt. Der jeweilige Inhalt des Lernens ist vom individuellen Organismus und von den Ereignissen, auf die er in seiner Umgebung trifft, abhängig. Da die jeweiligen Inhalte dessen, was gelernt wird, nicht Teil der lernpsychologischen Modelle sind, gehen auch die von ihnen abgeleiteten Therapieformen der Verhaltenstherapie am Einzelfall orientiert vor, um jeweils zu ermitteln, was in der individuellen Lerngeschichte gelernt worden ist. Das bedeutet allerdings nicht, daß es nicht eine Fülle empirischer Ergebnisse gibt, die sich auf die Auseinandersetzung von bestimmten Organismen mit bestimmten Ereignissen beziehen und die ebenfalls zum Bereich der Lernpsychologie gehören. Diese Ergebnisse wurden jedoch mit der methodischen Anwendung lerntheoretischer Prinzipien gewonnen, und ihr Inhalt läßt sich daher über die Elemente der funktionalen Analyse eindeutig auf Beobachtbares beziehen. Damit kommt der funktionalen Analyse in einem zweiten Schritt auch eine methodische Dimension zu.

Wird ein neutraler Reiz mit einem unkonditionierten Reiz nach bestimmten Regeln (s. u.) zusammen dargeboten, so erhält der dann konditionierte Reiz die Qualität eines Signals und damit eine bestimmte, für den Organismus spezifische Bedeutung. Bezieht man diesen Sachverhalt auf sehr komplexe Situationen, etwa solche des sozialen Lernens, das funktional nach Gesichtspunkten des Konditionierens betrachtet werden kann, so wird deutlich, daß die individuelle Lerngeschichte eines Menschen auch soziale Bedeutungen bedingt, also Bedeutungen, die andere Menschen und deren Verhaltensweisen für ein Individuum haben. Bedeutungen, die im Verlauf der Lerngeschichte eines Individuums erworben worden sind, sind also subjektiv, in einem gewissen Sinne einmalig, auch wenn angenommen wird, daß sie nach beschreibbaren, allgemeinen Prinzipien des Lernens erworben worden sind.

In der funktionalen Analyse werden Verhaltensweisen betrachtet, unabhängig davon, ob sie nach gängiger Klassifikation als physisch oder psychisch eingestuft würden.

7.2.2 Assoziation und Bedeutung

Ein Grundgedanke, der dem Modell des klassischen Konditionierens zugrunde liegt, besagt, daß eine Verbindung zwischen Reizen stattfindet, wenn sie in zeitlicher Nähe auftreten, es findet eine Assoziation zwischen beiden statt. Tritt mit einem unkonditionierten Reiz (UCS) in bestimmten Grenzen „gleichzeitig" ein für die unkonditionierte Reaktion (UCR) neutraler Reiz auf, d. h. ein Reiz, der die UCR nicht hervorruft, so wird durch diesen Reiz nach einigen Wiederholun-

gen eine Reaktion hervorgerufen, die der UCR zumindest ähnlich ist, die konditionierte Reaktion (CR). Der neutrale Reiz ist zu einem konditionierten Reiz (CS) geworden. Im Hinblick auf den CS wird vom Organismus eine CR erworben, der Vorgang wird Akquisition genannt.

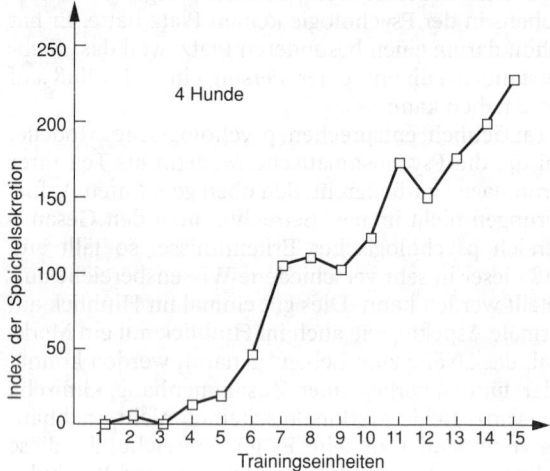

Abb 7–1. Verlauf der Akquisition von konditioniertem Speichelfluß (aus Hilgard und Bower, 1966).

Dabei besteht die „Gleichzeitigkeit" am besten darin, daß der CS vor dem UCS beginnt und beide Reize sich zumindest teilweise zeitlich überlappen. Aufgrund dieser Tatsache wird der CS auch als „Signalreiz" bezeichnet, er kündigt gleichsam an, daß der UCS auftreten wird. Bei näherer Betrachtung der konditionierten Reaktion stellt sich heraus, daß sich diese von der unkonditionierten unterscheidet, es ist nicht dieselbe Reaktion, ist ihr allerdings gewöhnlich ähnlich. In einigen Fällen ist dies jedoch nicht der Fall. So besteht die unkonditionierte Reaktion auf einen elektrischen Reiz unter anderem in einer Erhöhung der Herzfrequenz, die auf einen mit einem Schmerzreiz konditionierten Reiz unter anderem in einer Verringerung der Herzfrequenz (Obrist et al., 1965; vgl. auch Kap. 8).

Pawlow ging dabei von neurophysiologischen Vorstellungen der Ausbreitung von Erregung aus, indem er annahm, daß die Erregung, die von einem Reiz ausgeht, sich mit der, die von einem gleichzeitig dargebotenen ausgeht, verbindet. „So ist die zeitweise nervöse Verbindung das universellste physiologische Phänomen, sowohl in der Welt der Tiere als auch in unserer eigenen. Gleichzeitig ist es ein psychologisches Phänomen – das, was die Psychologen Assoziation nennen, egal ob sie von einer Kombination von irgendwelchen Aktionen stammt, oder von Eindrücken, Buchstaben, Worten und Gedanken" (Pawlow, 1934, zit. nach Hilgard und Bower, 1966).

So besteht im Falle des Lernens im Sinne des klassischen Konditionierens eine Verknüpfung von Reizen, die dem Verhalten vorausgehen oder gleichzeitig mit ihm auftreten. Dabei wird die Verknüpfung zwischen einem Reiz, dem unkonditionierten Reiz (UCS = unconditioned stimulus), und einer Reaktion, der unkonditionierten Reaktion (UCR = unconditioned response), als angeborene Verknüpfung vorausgesetzt. Dies bedeutet, daß der UCS jedesmal bei seinem Auftreten zu einer UCR führt.

Bedenkt man, daß durch die gleichzeitige Darbietung des CS mit dem UCS eine Verbindung zwischen beiden Reizen sowie eine zwischen dem CS und der CR hergestellt wird, betrachtet man weiterhin den Vorgang der Akquisition, bei dem z.B. die Reaktionsamplitude der CR mit der Anzahl der Versuchsdurchgänge ansteigt, so wird deutlich, daß die gleichzeitige Darbietung der beiden Reize die genannten Verbindungen verstärkt. Infolgedessen bezeichnet man die gelegentliche gemeinsame Darbietung vom UCS mit dem CS, auch nachdem die Konditionierung stattgefunden hat, als Verstärkung. Der Begriff der Verstärkung im Rahmen des klassischen Konditionierens ist nicht zu verwechseln mit dem Begriff der Verstärkung im Zusammenhang des operanten Lernens (s.u.).

Unterbleibt die Verstärkung, so werden die genannten Verbindungen zwischen dem CS und der CR sowie zwischen dem UCS und dem CS abgeschwächt. Dieser Vorgang wird Extinktion oder Löschung genannt. Unter Umständen, wenn die Verbindung gar nicht mehr verstärkt wird, verschwindet die CR vollständig.

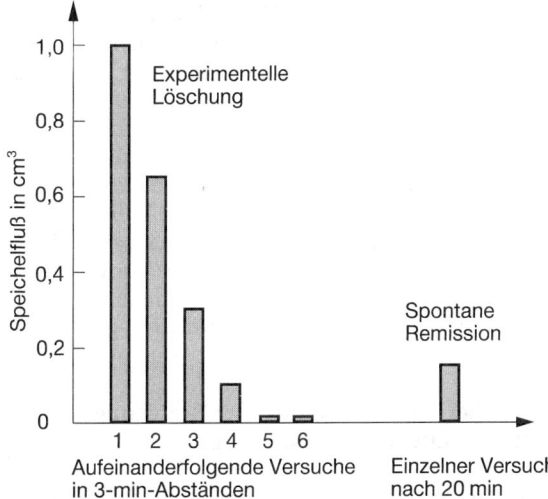

Abb. 7–2. Verlauf der Extinktion einer Speichelsekretion sowie das spontane Wiederauftreten der Reaktion (spontaneous recovery) (aus Hilgard und Bower, 1966).

Betrachtet man den Verlauf der Löschung, so zeigt sich, daß nach einigen Durchgängen die CR nicht mehr auftritt, nach einigen weiteren jedoch wieder. Dieser Vorgang wird „Spontanremission" (spontaneous recovery) genannt. Zu beachten ist dieses Phänomen beispielsweise bei Behandlungen, bei denen eine Verhaltensweise gelöscht werden soll, z.B. eine Angstreaktion. Es besteht dann die Tendenz, daß diese Reaktion nach einiger Zeit „von selbst" wieder auftritt, was häufig von Patienten mit Beunruhigung aufgenommen wird.

Der Vorgang der Löschung ist außerordentlich wichtig, wenn man sich vergegenwärtigt, daß er eine

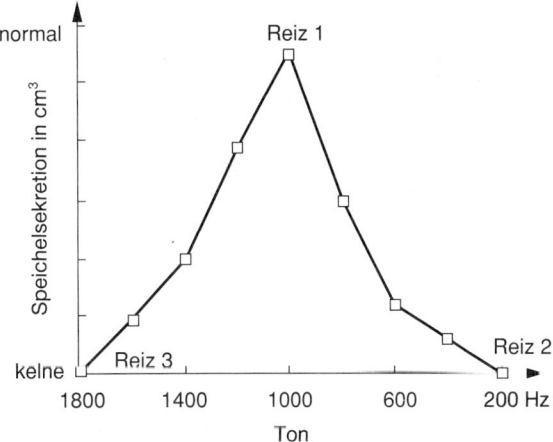

Abb. 7–3. Generalisationsgradient. Mit zunehmender Unähnlichkeit des Reizes nimmt die Reaktionsamplitude ab (aus Hilgard und Bower, 1966).

Anpassung an veränderte Umgebungsbedingungen darstellt. Hat sich ein Tier daran gewöhnt, daß bestimmte Reize beispielsweise die Verfügbarkeit von Nahrung ankündigen, und ändert sich dieser Sachverhalt, d.h. der ehemals ankündigende Reiz kündigt keine Nahrung an, so wäre es für das Tier ungünstig, wenn es diesen Reizen weiter nachginge. Die Tendenz des Tieres, auf diese Reize mit Suchverhalten usw. zu antworten, muß gelöscht werden, will es überleben.

Es läßt sich nun auch feststellen, daß nicht nur der CS eine CR hervorruft, sondern auch andere, ihm ähnliche Reize. Mit abnehmender Ähnlichkeit verringert sich ebenfalls die „Ähnlichkeit" der hervorgerufenen Reaktion mit der CR. So nimmt die Frequenz des Auftretens der Reaktion ab, die Amplitude wird geringer, die Latenz nimmt zu, d.h. die Zeit, die vergeht von der Darbietung des Reizes bis zum Auftreten der Reaktion. Dieser Sachverhalt wird Generalisierung genannt. Die Beziehung zwischen den oben genannten Merkmalen der Reaktion und der Ähnlichkeit des Reizes heißt „Generalisationsgradient".

Der Vorgang der Generalisierung ist beispielsweise wichtig bei der „Ausbreitung" von Angst in einer zunehmenden Anzahl von Situationen. So kommt es oft vor, daß Patienten, die etwa in einem Stau auf der Autobahn Angst erlebt haben, zuerst bestimmte Strecken nicht mehr befahren können, schließlich gar nicht mehr auf der Autobahn, dann zu bestimmten Zeiten nicht mehr in der Stadt oder auf der Landstraße, dann unter Umständen gar nicht mehr Auto fahren können usw. Auf vergleichbare Weise treten auch häufig soziale Ängste in immer mehr Situationen auf, die dann vermieden werden. Wichtig ist dabei, daß es in diesen Fällen von Generalisierung um „Ähnlichkeit" von Situationen geht, die im Hinblick auf ein bestimmtes Merkmal der Situation ähnlich sind. Im Falle des Autofahrens geht es meist darum, daß die Autobahn oder sonstige Straßen im Stau nicht verlassen werden können, d.h. die Situation nicht kontrolliert werden kann. So ist eine kritische Situation dann oft das Warten in einer Schlange an der Kasse eines Supermarktes.

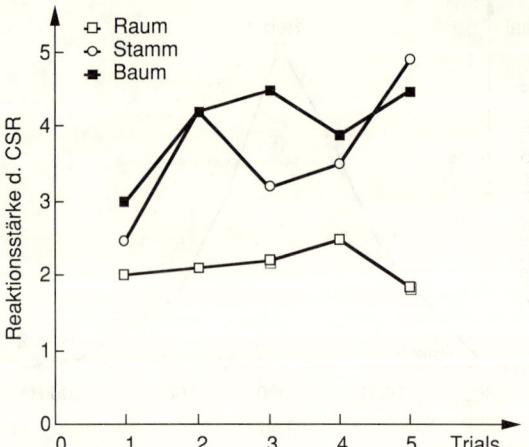

Abb. 7–4. Semantische Generalisierung einer elektrodermalen Reaktion in Abhängigkeit von der Bedeutungs- oder Lautähnlichkeit (aus Malzman, 1968).

Besteht der CS in einem Wort, so läßt sich feststellen, ob eine Generalisierung auf laut- oder bedeutungsähnliche Wörter stattfindet (Razran, 1949, zit. nach Foppa, 1968). Eine Generalisierung anhand der Bedeutungsähnlichkeit wird semantische Generalisierung genannt. Interessanterweise zeigen Kinder bis etwa zum Alter von knapp 10 Jahren eher eine Generalisierung anhand der Lautähnlichkeit, danach anhand der Bedeutungsähnlichkeit (Riess, 1946, zit. nach Foppa, 1968).

Der Vorgang des Diskriminationslernens beruht prinzipiell auf der differentiellen Verstärkung spezifischer Reize. Er bewirkt, daß die Reizgeneralisierung im Hinblick auf bestimmte Reize nicht stattfindet. Es wird dabei ein bestimmter Reiz stets mit dem UCS zusammen dargeboten, ein weiterer dagegen stets ohne den UCS: Der Organismus lernt dabei „diskriminativ", daß der eine Reiz den UCS ankündigt, der andere Reiz jedoch keinen Zusammenhang mit diesem aufweist. Dadurch verläuft der Generalisierungsgradient anders, als dies aufgrund der Reizähnlichkeit zu er-

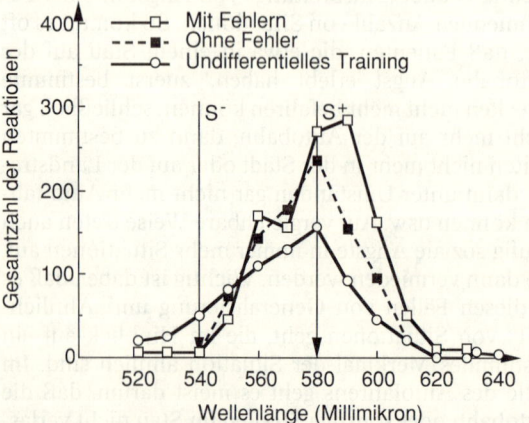

Abb. 7–5. Diskriminationslernen: Unterschiedliche Generalisationsgradienten, einmal ohne Diskriminationslernen o———o und zweimal mit unterschiedlichem Diskriminationslernen ■- - - - -■ □———□ (aus Hilgard und Bower, 1966).

warten wäre. Dabei ist wichtig, daß der Unterschied zwischen den Gruppen nicht nur darin besteht, daß die Gruppe mit Diskriminationslernen bei S^- im Gegensatz zur Gruppe ohne Diskriminationslernen keine Reaktionen zeigt, sondern auch, daß sie beim S^+ eine größere Anzahl von Reaktionen zeigt.

Das Diskriminationslernen spielt in vielen Situationen „natürlichen" Verhaltens, d.h. außerhalb der Laboratorien, eine ganz wesentliche Rolle, ebenso wie Vorgänge der Generalisierung oder der Konditionierung höherer Ordnung. In Situationen werden ständig neue Reizkonfigurationen vorhanden sein, deren untereinander bestehende Beziehungen das Verhalten beeinflussen und damit auch ständig modifizieren. Sich in vergleichbaren (ähnlichen) Situationen vergleichbar zu verhalten (Generalisation) oder etwa bei der zusätzlichen Anwesenheit einer anderen Person verschieden zu verhalten (Diskrimination), setzt eine wenn auch endliche so doch fast unüberschaubare Menge von Einzellernleistungen voraus.

Pawlow (1953 III, 1) merkt an, daß das „Psychische" am Vorgang des Lernens darin besteht, daß vom Organismus eine Beziehung hergestellt wird zwischen Reizen, die für die (in seinen Versuchen physiologischen) Reaktionen „unwesentlich" sind. Für eine unkonditionierte Reaktion ist nur der unkonditionierte Reiz „wesentlich", er ist starr mit der Reaktion verbunden. Ein akustischer Reiz beispielsweise ist zunächst für den Speichelfluß unwesentlich. Macht ein Organismus jedoch eine bestimmte Erfahrung, so daß der akustische Reiz die Bedeutung erhält, ein Signal für Futter zu sein, so wird er für diesen Organismus und dessen Reaktion des Speichelflusses wesentlich. Das „Psychische" im Sinne Pawlows besteht demnach darin, daß ein Reiz eine Bedeutung erhält, die er vorher nicht hatte, zusätzlich zu der, die er schon haben mag. Es besteht damit im Vorgang des Lernens selbst, d.h. das Psychische ist in der Funktionalität des Lernens und nicht in dem, was gelernt wird, gegeben.

7.2.3 Lernen als Änderung der Beziehung zwischen Organismus und Umgebung

Es sei in diesem Zusammenhang auf das theoretische Konzept des „Funktionskreises" verwiesen (vgl. Kap. 1), in dem das Erteilen einer Bedeutung für Außenreize ein wesentliches Element darstellt. In diesem Modell wird allerdings nicht unterschieden zwischen unkonditionierter und konditionierter „Bedeutung", beide werden als subjektiv in Abhängigkeit von den rezeptorischen Fähigkeiten des Organismus betrachtet, also nur das, was die Sinne vermitteln, ist für einen Organismus Umgebung, seine „subjektive" Umwelt.

Wesentlich für das Modell des Funktionskreises wie für Modelle des Lernens ist der Ausgangspunkt, der Vorgänge des Lebens eines Organismus in der Beziehung zur Umgebung versteht und nicht Umgebung und Organismus isoliert betrachtet. So betrachtet die funktionale Analyse des Verhaltens die Beziehung zwischen dem Organismus und seiner Umgebung, die im Falle des Lernens beinhaltet, daß Regelmäßigkeiten (Kontingenzen, Beziehungen zwischen Ereignissen) in der Umgebung für den Organismus insofern bedeutsame werden, als er diese erfassen kann und in seinem Verhalten „berücksichtigt" und entsprechend sein Verhalten ändert. Diese Fähigkeit zur Anpassung bzw. „Erfahrung" setzt voraus, daß der Organismus – als Teil seiner Umwelt – Assoziationen bilden kann.

Lernen bewirkt eine mehr oder weniger stabile Änderung der Beziehung des Organismus zu seiner Umgebung. Es bewirkt keine Änderung der Reize, geändert wird lediglich ihre Beziehung zum Organismus, also ihre Bedeutung. Im lerntheoretischen Schrifttum wird häufig die Formulierung verwendet, ein Reiz erwerbe durch Lernen die „Fähigkeit", eine Reaktion auszulösen. Diese Formulierung mag der Tatsache entsprechen, daß in den meisten lernpsychologischen Untersuchungen das Verhalten die abhängige und die Reize die unabhängigen Variablen darstellen, sie ist jedoch ansonsten irreführend. Lernfähigkeit ist ein Merkmal von Organismen und nicht von Reizen.

Deutlich wird dies in Untersuchungen zum latenten Lernen. Dabei lernt beispielsweise ein Tier in einem Labyrinth zu einer Zielbox zu laufen. Es macht an Entscheidungspunkten eine meßbare Anzahl von Fehlern. In der Zielbox wird es zunächst nicht belohnt. Wird es dann schließlich doch dort belohnt, so sinkt die Fehlerquote an den Entscheidungspunkten in mitunter nur einem Durchgang auf das Maß derjenigen Tiere, die dort von Anfang an belohnt wurden. Das Tier hat also bereits in den vorherigen Durchgängen etwas gelernt, auch ohne daß eine Belohnung wirksam geworden wäre (vgl. 7.2.11).

7.2.4 Lernen als „Distanz" zum Reflex und das „Realitätsprinzip"

Der Reiz erlangt also keine Fähigkeit, sondern eine Bedeutung für einen individuellen Organismus, der sich dadurch in einer für ihn veränderten Umgebung wiederfindet. Lernen betrifft immer die Beziehung zwischen einem Organismus und seiner Umgebung. Das „Psychische" des Vorgangs besteht darin, daß das Verhalten seine „Reflexhaftigkeit" verliert. Deshalb ist es sinnvoll von einem unkonditionierten Reflex zu sprechen, da er fest mit dem unkonditionierten Reiz verbunden ist, irreführend ist es jedoch, von einem konditionierten Reflex zu sprechen statt sinnvollerweise von konditioniertem Verhalten.

Wird auf einen CS ein weiterer zunächst neutraler Reiz konditioniert, so wird dieser Vorgang als Konditionieren höherer Ordnung bezeichnet (second order conditioning). In der folgenden Abbildung sind die verschiedenen Phasen des Konditionierens höherer Ordnung dargestellt. Zunächst wird eine Abwehrreaktion auf einen Tonreiz konditioniert, der alleine dargeboten die Reaktion hervorruft (Konditionieren 1. Ordnung). Dann wird ein Lichtreiz zusammen mit dem Tonreiz dargeboten, der dann ebenfalls die Abwehrreaktion hervorruft (Konditionieren 2. Ordnung). Schließlich wird mit dem Lichtreiz ein taktiler Reiz zusammen dargeboten, der dann ebenfalls die Abwehrreaktion hervorruft (Konditionieren 3. Ordnung).

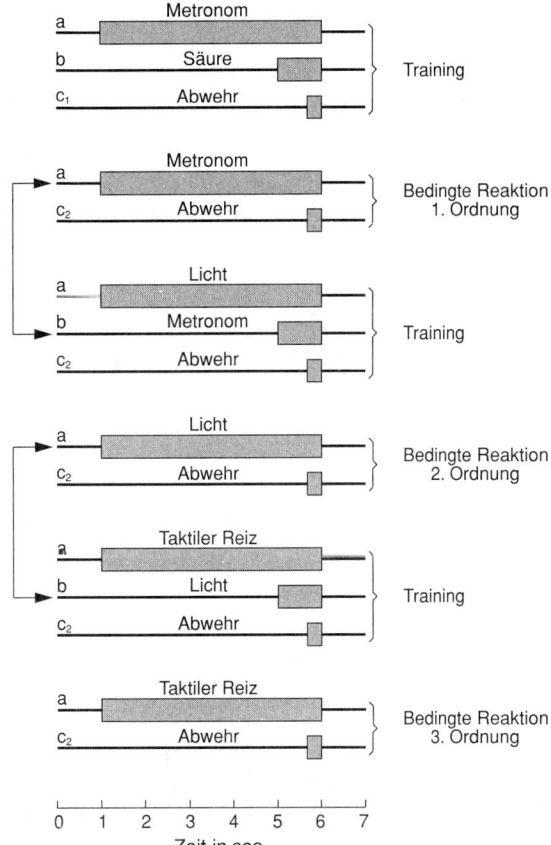

Abb. 7–6. Bedingte Reaktionen höherer Ordnung (aus Angermaier und Peters, 1973).

Auf diese Weise lassen sich sehr mittelbare Verbindungen zwischen Reizen mit einem unkonditionierten Reiz denken, ein Vorgang, der sicherlich sehr wichtig ist, versucht man die Komplexität menschlicher Verhaltensweisen zu erklären. Es wird dabei ebenfalls deutlich, daß der Vorgang des Konditionierens bzw. dessen Resultat, die Reiz-Reiz-Verbindung (UCS-CS) und/oder die Reiz-Reaktions-Verbindung (UCR-CS), als sinnvolle Grundeinheit von Verhalten angenommen wird, d. h. daß sich auch komplexeres Verhalten aus derartigen Verhaltenseinheiten zusammensetzt.

Beim sensorischen Vorkonditionieren (sensory preconditioning), das auch als assoziatives Konditionieren bezeichnet wird (Foppa, 1968), werden im ersten Versuchsteil zwei oder mehrere Reize gemeinsam dargeboten (CS1 und CS2). Im zweiten Versuchsteil wird einer der beiden Reize, z. B. CS1, mit einem UCS gemeinsam dargeboten und so eine CR auf CS1 ausgebildet. Im letzten Versuchsteil wird schließlich geprüft, inwieweit auf CS2 mit einer CR geantwortet wird. Durch die im ersten Versuchsteil hergestellte Assoziation zwischen CS1 und CS2 wird auch auf CS2 mit einer CR geantwortet, obwohl nie eine gemeinsame Darbietung mit dem UCS stattgefunden hat und auch bei der Bildung der Assoziation zwischen CS1 und CS2 noch keine Beziehung zwischen CS1 und dem UCS vorhanden war.

Besteht die Reiz-Reiz-Verbindung nicht in der „Gleichzeitigkeit" des Auftretens der Reize in dem Sinne, daß sie sich zumindest teilweise zeitlich überlappen, sondern nur in einer zeitlichen Nähe insofern, als der zu konditionierende Reiz dem UCS vorausgeht, so findet eine Spurenkonditionierung (trace conditioning) statt. In der folgenden Abbildung wird deutlich, daß die Reaktion (Speichelfluß) deutlich geringer ausfällt.

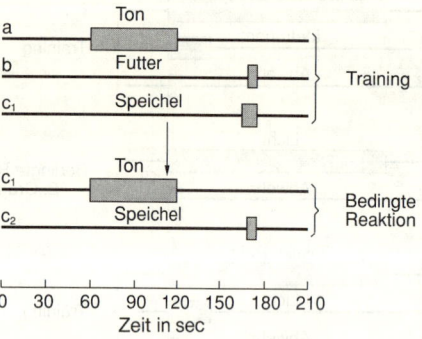

Abb. 7–7. Spurenkonditionierung (aus Angermaier und Peters, 1973).

Geht der CS dem UCS nicht voraus, sondern folgt ihm, wobei keine zeitliche Überlappung stattfindet, so handelt es sich um eine sog. Rückwärtskonditionierung (backward conditioning). Es ist umstritten, ob dieser Vorgang tatsächlich wirksam ist, Razran (1971) bezweifelt ihn jedoch nicht. So gibt es eine Möglichkeit zur Behandlung der Enuresis, die im wesentlichen darin besteht, daß in einer Bettunterlage eine stromleitende Anordnung von Drähten angebracht ist, die bewirkt, daß durch Feuchtigkeit ein Kontakt geschlossen wird und eine Klingel ertönt, die das Kind weckt. Der Ton wäre demnach zunächst als CS zu betrachten, der dem UCS der Blasenfüllung und der UCR der Blasenentleerung folgt. Trotzdem ist diese Anlage recht wirksam. Sie ist auch anders interpretierbar, indem die Blasenfüllung als CS für den Klingelton als UCS (Weckreiz) und die UCR, das Aufwachen, betrachtet wird.

Das Reflexhafte eines konditionierten Verhaltens wird wohl darum mitunter angenommen, weil der Vorgang des Lernens von den Kontingenzen der Reize in der Umgebung abhängig ist, die im Experiment manipuliert werden. Wäre das Lernen unabhängig davon, wäre es sinnlos. Aber gerade weil es nicht sinnlos ist, weil es dem Organismus angemessenere Einflußmöglichkeiten auf seine Umgebung ermöglicht, handelt es sich beim Resultat des Lernens – einem außerhalb der Laboratorien niemals abgeschlossenen Vorgang – nicht um Reflexe, sondern um Verhalten, das im Sinne eines „Realitätsprinzips" an den Regelmäßigkeiten der Umgebung, an der Information, die die Umgebung enthält, orientiert ist.

Es ist also ganz wesentlich, daß gelernte Verhaltensweisen oder Reaktionen die ausschließliche Bindung des Organismus an Reflexe relativieren. Die Freiheitsgrade im Hinblick auf die Auseinandersetzung mit der Umwelt werden dadurch erweitert bzw. kommen dadurch erst zustande.

Für Pawlow bestand in der Sprachfunktion der wesentlichste Unterschied zwischen Tieren und Menschen. „Wenn unsere Empfindungen und Vorstellungen, die sich auf die Außenwelt beziehen, für uns die ersten und dabei konkreten Signale der Wirklichkeit sind, so bildet die Sprache, und in erster Linie speziell die kinästhetischen Reize, die von den Sprachorganen der Hirnrinde übermittelt werden, eine zweite Ordnung von Signalen, die Signale der Signale. Sie stellen selbst eine Abstraktion von der Wirklichkeit dar und gestatten die Verallgemeinerung, die unser übriges, speziell menschliches, höheres Denken bildet, das zuerst die allgemeine menschliche Erfahrung und schließlich die Wissenschaft begründet hat" (Pawlow, 1953, zit. nach Foppa, 1968).

Wichtig dabei ist, daß im zweiten Signalsystem dieselben Regeln des Lernens gelten wie im ersten, allerdings gestattet die Sprache für den Erwerb von Verhalten neue Möglichkeiten über das Verwenden von Instruktionen. Neue Verhaltensweisen können durch die sprachliche Formulierung von Regeln erfaßt und mitgeteilt werden, die z. B. das Verhältnis von CS und CR betreffen („wenn der Reiz S auftritt, drücke auf den Knopf" (CR-Äquivalent)).

7.2.5 Lernen als Orientierung an Beziehungen in der Umgebung

Im Modell des Funktionskreises wird als ein wesentliches Element eine Art von Erkenntnisfunktion angenommen, die im „Merken" die Umwelt als „merkbare" für den individuellen Organismus definiert. Dies geschieht in Abhängigkeit von einer „Merkfähigkeit" (hier in diesem ganz speziellen Sinne) des Organismus. Diese Vorstellung, die in einer gewissen Nähe zur Philosophie Kants zu stehen scheint, macht jedoch keine Unterscheidung zwischen bedingtem und unbedingtem „Programm". Auch der unbedingte Reiz wird „gemerkt" und erhält durch das Merken eine bestimmte Bedeutung im Hinblick auf das unbedingte Programm, mit dem er fest verbunden ist. Könnte der Organismus ihn nicht wahrnehmen, könnte er auch nicht darauf reagieren. Wahrnehmen und Merken bedeuten hier nicht dasselbe. Wahrnehmen und „Wahrnehmen-Können" beziehen sich auf Grundeigenschaften eines Organismus, nämlich die Möglichkeiten der Informationsaufnahme und -verarbeitung. Sie sind durch die Eigenschaften des Organismus begrenzt. So können beispielsweise elektromagnetische Wellen nur eines bestimmten Frequenzbandes wahrgenommen werden. Merken ist ebenfalls begrenzt durch die Bedeutung, die prinzipiell wahrnehmbare Reize für einen Organismus besitzen. Bedeutungslose Reize werden nicht gemerkt. Konditionieren wird hierbei aufgefaßt als Verbindung zweier Integrationsebenen, der „innerkörperlichen" mit der „psychischen", als Verbindung zweier Ebenen von Bedeutung, als „Bedeutungskoppelung".

Der Unterschied zwischen einem unbedingten Reflex und bedingtem Verhalten soll hier als sehr wesentlicher betont werden. Lernen ist damit das „Lernen von Beziehungen zwischen Ereignissen in der Umgebung, die außerhalb der Kontrolle des Organismus geschehen" (Rescorla, 1978). Lernen besteht demnach in einer „sinnvollen" Bedeutungserteilung für im Sinne Pawlows „unwesentliche Reize", die Neues schafft, nämlich die Integration von verschiedenen Informationen. Sie setzt eine ganz besondere „Merkfähigkeit" des lernenden Organismus voraus, das „Merken" von Beziehungen zwischen Ereignissen, deren logischer oder „probabilistischer" (s. u.) Verknüpfung. Diese Fähigkeit, die sowohl als Grundlage für das klassische Konditionieren als auch für das instrumentelle Lernen gilt, ist die Voraussetzung für die Umstrukturierung der subjektiven Umwelt. Der lernende Organismus findet sich in einer durch das Lernen für ihn veränderten Umwelt wieder.

7.2.6 Lernen und Anpassung

Pawlow hat häufig die Anpassungsqualität des Lernens hervorgehoben. Der Organismus ist durch das Lernen in der Lage, Reize, die im Hinblick auf das (physiologische) System „distanter" sind, zu beantworten. „Es besteht kein Zweifel, daß wir die Tatsache einer weiteren Anpassung vor uns haben. Im gegebenen Fall lenkt eine derart feine Verbindung auf Distanz, wie die Verbindung der charakteristischen Schrittgeräusche eines bestimmten Menschen, der dem Tier gewöhnlich die Nahrung bringt, mit der Funktion der Speicheldrüse, sicher nur wegen ihrer Feinheit und nicht wegen einer besonderen physiologischen Wichtigkeit die Aufmerksamkeit auf sich" (Pawlow, 1953 III/1).

Pawlow betont weiterhin, daß der Organismus durch das Lernen „vorbeugend" auf seine Umgebung reagieren kann. Lernen ist „zukunftsorientiert", ein Merkmal, das häufig mit dem Begriff des Handelns in Verbindung gebracht wird. Dieser Aspekt wird um so deutlicher, wenn daran gedacht wird, daß Lernen ebenfalls durch die Konsequenzen, die auf ein Verhalten folgen, beeinflußt wird.

Hat ein Organismus gelernt, daß auf ein bestimmtes Verhalten eine verstärkende Konsequenz (Belohnung) folgt, so könnte angenommen werden, daß die Erhöhung der Frequenz des Verhaltens erfolgt, damit die Konsequenz eintritt. Hier wäre wieder eine Zukunftsorientierung bzw. eine Zielgerichtetheit des Verhaltens deutlich. Beobachtbar ist diese nicht, sondern nur die Zunahme der Auftretenshäufigkeit eines bestimmten Verhaltens oder dessen Stabilisierung.

7.2.7 Lernen und Erwartung – Zielgerichtetheit

Besteht beim klassischen Konditionieren die kritische Bedingung für das Lernen in der Beziehung zwi-

schen Reizen, die einer Reaktion zeitlich vorausgehen, so besteht sie im Falle des operanten oder instrumentellen Lernens in der Beziehung von Reizen, die zeitlich auf das Verhalten folgen. Auch im Modell des operanten Lernens ist eine Reaktion auf einen Reiz oder eine spontane Reaktion, für die kein direkt auslösender Reiz beobachtet werden kann, die grundlegende Verhaltenseinheit.

Vor allem Skinner hat das Modell des operanten Lernens geprägt, und die folgende Darstellung orientiert sich weitgehend an seinen Vorstellungen.

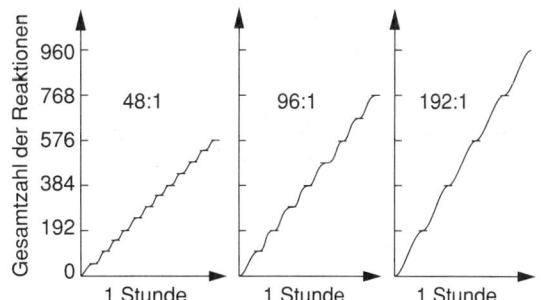

Abb. 7–8. Kumulativkurve von Reaktionen, die mit einem Verhaltens-Verstärkerplan bekräftigt wurden, mit verschiedenen Verhältnissen zwischen Anzahl der Reaktionen und dem Auftreten des Verstärkers. Die höchste Reaktionsrate zeigt sich beim niedrigsten Verstärkungsverhältnis (aus Hilgard und Bower, 1966).

Von Skinner (1953) wurde spontan „geäußertes" (emitted) Verhalten als operant bezeichnet, von Reizen ausgelöstes Verhalten wird als respondent bezeichnet. Dabei wird als spontane Äußerung von Verhalten angenommen, daß auslösende Bedingungen nicht direkt beobachtbar sind, und zwar mit den Methoden der Verhaltenspsychologie. Physiologische Reize, bzw. deren Annahme als ein Verhalten auslösend, wird von Skinner als theoretische, und damit unzulässige Erklärung verworfen, zumindest für den Bereich einer Verhaltenspsychologie.

Vorgänge des klassischen Konditionierens bezeichnet Skinner als Konditionieren vom „Typ S", solche des operanten Konditionierens als „Typ R".

Beim klassischen Konditionieren geht es gleichsam um die Erweiterung des „Verhaltensrepertoires" eines Organismus, indem neue Beziehungen zwischen Reizen und Reaktionen gebildet werden. Wenn man bedenkt, daß eine konditionierte Reaktion nicht identisch ist mit der unkonditionierten, zu der eine Beziehung besteht, so werden durch den Vorgang des klassischen Konditionierens auch neue Reaktionen erworben. Beim operanten Lernen ist dies zunächst prinzipiell nicht der Fall. Hierbei wird lediglich die Wahrscheinlichkeit für das Auftreten einer im Repertoire eines Organismus vorhandenen Reaktion verändert oder beeinflußt.

Ein Reiz hat für einen Organismus die Wirkung der Verstärkung (reinforcement), wenn er die Auftretenswahrscheinlichkeit eines zeitlich vor ihm stattfindenden Verhaltens des Organismus erhöht. Für Skinner gilt diese rein operationale Definition, er bezieht sich nicht darauf, wie z.B. Hull (1952), daß ein Verstärker einen Trieb reduziert und darum verstärkend wirkt, was etwa durch den Begriff der Belohnung nahegelegt wird. In seinen Untersuchungen hat er allerdings z.B. durch Nahrungsentzug einen Zustand der Deprivation hergestellt, bzw. manipuliert, um die Abhängigkeit des Verstärkungswertes von Futterpillen von der Dauer der Deprivation zu untersuchen. Er verzichtet jedoch auf das theoretische Konstrukt „Trieb", da es redundant sei, wenn man es operational als Deprivation definiert, sonst jedoch wissenschaftlich unzulässig sei.

Als positive Verstärker werden Reize bezeichnet, deren Vorkommen die Auftretenswahrscheinlichkeit von Verhalten erhöht, bei der negativen Verstärkung wird sie durch die Beendigung von Reizen erhöht. Somit ist die negative Verstärkung nicht zu verwechseln mit Strafreizen, deren Vorkommen die Auftretenswahrscheinlichkeit verringert. Diese Verringerung ist jedoch nur vorübergehend, die Reaktion wird nicht aus dem Verhaltensrepertoire gelöscht, sondern wird wieder auftreten. Der Vorgang der operanten Extinktion kann durch Strafreize, die emotional wirken, sogar behindert werden (Skinner, 1938, zit. nach Hilgard und Bower, 1966).

Als Verhaltensmaß für die Wirkung der Verstärkung, bzw. des operanten Konditionierens, wird die Reaktionsrate (response rate) benutzt. Im Bereich des operanten Lernens ist die Registrierung von Kumulativkurven üblich. Jedes Ansteigen der Reaktionsrate resultiert in einer größeren Steigung der Kurve. Im allgemeinen wird die Gabe eines Verstärkers durch einen kurzen senkrechten oder waagerechten Strich durch die Kurve angezeigt. Auf diese einfache Weise ist es möglich, den Lerneffekt sehr übersichtlich darzustellen.

Durch die Verbindung („Assoziation") eines Reizes mit einem primären oder sonst wirksamen Verstärker erwirbt dieser Reiz verstärkende Wirkung. Er wird sie verlieren, wird er nicht gelegentlich von einem primären Verstärker gefolgt, vergleichbar der Löschung beim klassischen Konditionieren. Ähnlich gelten hierfür auch die Gesetze der Generalisierung im Sinne des klassischen Konditionierens. Geld ist wohl das beste Beispiel für einen sekundären Verstärker.

Bei der Behandlung autistischer Kinder beispielsweise war es notwendig, „normale" Verstärker wie Zuwendung in Form von Streicheln durch die Koppelung mit der Gabe von Bonbons als sekundären Verstärkern zu etablieren, um sie später einsetzen zu können.

Die operante Löschung bzw. Extinktion besteht in der Wegnahme des verstärkenden Reizes. Dadurch wird die Verhaltensrate reduziert. Die Resistenz gegenüber der Löschung wird als ein Maß für die Wirksamkeit der Verstärkung benutzt (operant strength). Dabei kann sich zeigen, daß bereits eine einzige Verstärkung eine gewisse Extinktionsresistenz erzeugt.

Skinner unterschied zwei Arten der Diskrimination. In einem Fall wird die Verstärkung gegeben, wenn ein Verhalten, z.B. Hebeldruck in Gegenwart eines Reizes (z.B. Licht) stattfindet. Das Verhalten wird bei Abwesenheit des Reizes (Licht) nicht verstärkt. Der Lichtreiz ist in diesem Falle ein diskriminativer Reiz, die Diskrimination heißt „Reizdiskrimination".

Bei der Reaktionsdiskrimination wird dagegen nur eine Reaktion einer bestimmten Art verstärkt, z.B. wenn der Hebeldruck mit der linken Vorderpfote durchgeführt wird.

Unter Verstärkungsplänen (schedules of reinforcement) versteht man die Art der Kontingenz, mit die die Verstärkung auf ein Verhalten folgt. Es lassen sich prinzipiell kontinuierliche und intermittierende (intermittent) Verstärkung unterscheiden. Bei der kontinuierlichen Verstärkung wird jedes Verhalten verstärkt, bei der intermittierenden nur eine bestimmte Anzahl von Reaktionen. Dies kann in Abhängigkeit von einem Zeitintervall, das verstrichen sein muß, geschehen oder aber in Abhängigkeit von der Anzahl der aufgetretenen Reaktionen.

Bei Verstärkung nach festen Quotenplänen (fixed ratio (FR)) wird nach einer bestimmten Anzahl von Reaktionen eine Verstärkung gegeben, z.B. wird nach 100 Reaktionen oder nach 10 Reaktionen verstärkt. Abgekürzt schreibt man FR 100 oder FR 10. In derartigen Programmen wird meist mit relativ niedrigen Quoten begonnen und dann zu höheren Quoten übergegangen. Dieser Vorgang heißt „Ausschleichen der Verstärkung" (fading out of reinforcement).

Bei der variablen Quotenverstärkung (variable ratio (VR)) wird nach einer durchschnittlichen Anzahl von Reaktionen eine Verstärkung gegeben, z.B. VR50: nach durchschnittlich 50 Reaktionen. Dem Verstärkungsplan wird eine Zufallsfolge von Raten zugrunde gelegt, deren Mittelwert in diesem Falle 50 wäre. Die Grenzen dieser Zufallsfolge haben einen Einfluß auf das Verhalten und sind willkürlich. Man kann in diesem Beispiel bei 0 beginnen und bis 100 gehen oder aber den Bereich begrenzen von 45–55.

Bei der Verstärkung nach einem festen Intervallplan (fixed interval (FI)) wird die erste Reaktion nach einem festen Zeitintervall, z.B. nach 1 Minute (FI1) verstärkt. Bei der Verstärkung nach einem variablen Intervallplan ist wiederum das durchschnittliche Intervall angegeben, z.B. VI10: nach durchschnittlich 10 Minuten wird die erste Reaktion verstärkt. Die Regeln für die Erstellung der Zeitintervalle sind vergleichbar denen bei der variablen Quotenverstärkung.

Jeder dieser Verstärkungspläne hat bestimmte Charakteristika des resultierenden Verhaltens zur Folge. Bei Quotenplänen kommt es zu recht hohen Reaktionsraten, bei Intervallplänen je nach der Dauer des Intervalls zu stabilen, aber niedrigen Raten, wobei nach jeder Reaktion eine Pause eintritt.

Es gibt darüber hinaus eine ganze Reihe sog. gemischter Verstärkungspläne, in denen zwei oder mehrere Pläne alternierend oder in einer bestimmten Reihenfolge oder Abhängigkeit wirksam sind. Jeder dieser Pläne bewirkt eigene Charakteristika des Verhaltens.

Durch intermittierende Pläne etablierte Verhaltensraten sind schwerer löschbar als solche, die durch kontinuierliche Verstärkerpläne aufgebaut worden sind. Dies ist beispielsweise bei Erziehungsmaßnahmen wichtig. Läßt man ein unerwünschtes Verhalten, das für das Kind belohnende Konsequenzen hat, gelegentlich zu, so wird es intermittierend verstärkt und ist dann sehr schwer zu löschen, einer von mehreren Gründen für den Vorteil „konsequenter" Erziehung.

Unter Verhaltensformung (shaping) versteht man das approximative und schrittweise Annähern an eine Zielreaktion, die ursprünglich nicht im Repertoire des Verhaltens vorhanden ist. Zunächst wird eine vorkommende, der Zielreaktion möglichst ähnliche Reaktion verstärkt. Da diese Reaktion nicht immer genau die gleiche ist, sondern es kleine Abweichungen gibt, wird es zu Reaktionen kommen, die der erwünschten noch ähnlicher sind. Diese werden selektiv weiter verstärkt usw., bis die erwünschte Reaktion schließlich erreicht ist. Dieser Vorgang spielt eine wichtige Rolle im alltäglichen Leben, und Skinner (1953) vergleicht ihn mit der Tätigkeit eines Bildhauers, der das Ausgangsmaterial in kleinen Schritten in einen Zielzustand überführt. Durch diesen Vorgang lerne ein Kind zu stehen, zu laufen, nach Objekten zu greifen usw.

Der Begriff der Zielgerichtetheit würde den nicht unproblematischen Sachverhalt erklären, daß eine Konsequenz, im Prinzip also ein äußerer Reiz wirksam ist, bevor er tatsächlich vorhanden ist. Zum Zeitpunkt des instrumentellen Verhaltens, z.B. dem Drücken auf einen Hebel, ist der auf das Verhalten folgende, es belohnende Reiz materiell ja nicht vorhanden. Wie kann er dennoch als Konsequenz wirksam sein, wenn man den Organismus als lediglich passiv reagierend versteht, d.h., wie kann ein Organismus auf einen noch nicht vorhandenen Reiz reagieren?

Von seiten „kognitiver" Lerntheoretiker (z.B. Tolman, 1932) wurde angenommen, daß der Organismus die Konsequenz seines Verhaltens „erwartet". Diese Annahme wird z.B. durch das Phänomen der überschießenden (overshooting) Reaktion oder durch den „Depressionseffekt" (undershooting) gestützt. Diese Effekte gehen von der Tatsache aus, daß die Größe der Belohnung einen Einfluß hat auf Merkmale eines Verhaltens, beispielsweise die Laufgeschwindigkeit in einem Gang, der zu einer Zielbox führt. Crespi (1942) konnte zeigen, daß eine Ratte, die mit einer geringen Belohnung trainiert wurde, den Laufgang entlangzulaufen, die Laufgeschwindigkeit steigerte, wenn sie mit einer größeren Belohnung verstärkt wurde. Das „overshooting"-Phänomen besteht nun darin, daß sie noch schneller läuft als ein Tier, das von Anfang an mit dieser größeren Belohnung verstärkt worden war. Wurde nun umgekehrt dieses Tier so belohnt wie das erste, so verringerte es seine Laufgeschwindigkeit unter die des ersten Tieres.

Die Interpretation dieses letzten Effektes als „frustrative Nicht-Belohnung" (Amsel, 1962) setzt voraus, daß das Tier eine bestimmte Erwartung im Hinblick auf das Ergebnis seines Verhaltens hat. Es wird vom Tier offenbar eine gespeicherte Verbindung (Assoziation) hergestellt zwischen dem Verhalten und dessen

Konsequenz. Darüber hinaus legen diese Ergebnisse nahe, daß das Tier Abweichungen von dieser Erwartung feststellen kann, es kann vergleichen.

Lernen und Vorhersage

Schon Pawlow betrachtete den CS als Signal, der den UCS ankündigt. Noch deutlicher wird dieser Sachverhalt in den Untersuchungen zum sog. „Blockierungseffekt" (blocking). Dieses zuerst von Kamin (1968) systematisch untersuchte Phänomen bezieht sich auf folgenden Sachverhalt: Eine Gruppe von Versuchstieren erhält in einem Versuchsabschnitt eine Darbietung von einem UCS zusammen mit einem zusammengesetzten Reiz, der die Elemente AB enthält. Ein Teil dieser Gruppe hatte vorher Lerndurchgänge, in denen nur das Element A zusammen mit dem UCS dargeboten worden war, der andere Teil hatte diese Erfahrung nicht. Das hier interessierende Ergebnis dieser Prozedur besteht darin, daß in dem Fall, in dem A vorher den UCS angekündigt hat, eine bedeutend geringere Konditionierung vom Element B erfolgt, als wenn keine Konditionierung von A der von AB vorausgegangen ist. Die vorherige Konditionierung von A „blockiert" die von B. Dies läßt sich dadurch nachprüfen, indem man B alleine darbietet, nachdem die Konditionierung von AB erfolgt ist.

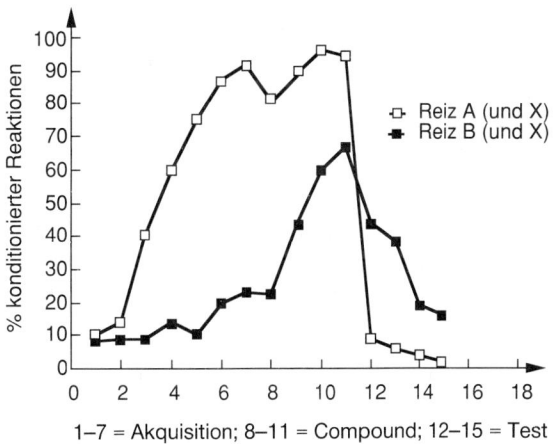

Abb. 7–9. Durchschnittliche Lidschlußreaktion während der drei Phasen eines „Blocking"-Experiments. Während der Akquisitionsphase wird einer Gruppe der Reiz A zusammen mit dem UCS dargeboten, in der „Compound"-Phase werden beide Reize A und B zusammen mit dem UCS beiden Gruppen dargeboten. In der Testphase wird nur der Reiz B beiden Gruppen dargeboten. Es zeigt sich in der Stärke der CR auf den Reiz B in der Testphase, daß die vorherige Darbietung des Reizes A in dieser Gruppe die Assoziation von Reiz B mit dem UCS blockiert (aus Wagner, 1978).

Rescorla (1980) hat dieses Paradigma in experimentellen Strategien zum Konditionieren zweiter Ordnung verwendet, um nachzuweisen, daß es letztlich der Informationsgehalt des CS im Hinblick auf das Auftreten des UCS ist, der die Stärke der Konditionierung moduliert. Die Stärke hängt ab vom Ausmaß, in dem beide Reize korreliert sind. Ist beim Blockierungsphänomen A zusammen mit dem UCS vorher dargeboten worden, so enthält er bereits die Information für das Auftreten des UCS. Der Reiz B fügt dieser in den darauffolgenden Durchgängen, in denen AB zusammen mit dem UCS dargeboten werden, keine neuen Informationen hinzu, was die Konditionierung von B blockiert, da B redundant ist, und somit die „Aufmerksamkeit" beeinflußt (Mackintosh, 1975, 1978).

Wagner, Logan, Haberlandt und Price (1968) (zit. nach Mackintosh, 1977) konnten darüber hinaus zeigen, daß nach einer Konditionierung eines CS, der den UCS mit einer Wahrscheinlichkeit von 50 Prozent voraussagte, diese Konditionierung gelöscht wurde, wenn ein anderer Reiz später den UCS genauer voraussagte. Somit ist die wesentliche Bedingung für den Vorgang des Konditionierens nicht nur die zeitliche Nähe, sondern der prädiktive Wert des CS für das Auftreten des UCS.

7.2.8 Lernen und qualitative Beziehungen zwischen Reizen

Daß die zeitliche Kontiguität zwischen CS und UCS nicht der allein wirksame Faktor sein kann, belegen auch die Untersuchungen zur Geschmacksaversion (Garcia und Koelling, 1966; Garcia et al., 1966; Revusky und Bedarf, 1967). Hierbei kommt es zu einer Assoziation zwischen einer bestimmten Geschmacksqualität und dem Auftreten von Nausea, auch wenn Zeiträume bis zu Stunden zwischen dem Auftreten beider Reize liegen. Eine Voraussetzung für dieses Phänomen besteht allerdings in der „Ungewohntheit" der Geschmacksqualität, „eine vergiftete Nahrung muß neu sein, sonst wäre die Ratte vermutlich schon tot" (Revusky und Bedarf, 1967). Im Sinne von Rescorla könnte auch argumentiert werden, daß durch das Gewohntsein der Geschmacksqualität in einem Zusammenhang, der keine Nausea produziert, der Vorhersagewert für die Nausea sehr gering ist.

Bei der Konditionierung der Geschmacksaversion ist ein weiterer Sachverhalt wichtig, der von Garcia und Koelling (1966) als „cue to consequence" bezeichnet wurde. Nicht jeder Reiz eignet sich als CS für die Konditionierung der Geschmacksaversion. So bildet sich keine Assoziation im Hinblick auf akustische oder visuelle Reize, wenn sie zusammen mit der Aufnahme einer Geschmackslösung dargeboten wurden. Umgekehrt war bei diesen Reizen eine Assoziation mit peripher applizierten Schocks möglich, die dagegen mit Geschmacksreizen nicht zustande kam. Im einen Fall wird die CS-UCS-Assoziation durch ein interozeptives Ereignis (Nausea) verstärkt, im anderen Fall durch ein exterozeptives (Schock). Es scheint sich darin eine biologische Prädisposition von Organismen zu zeigen, die den Gesetzen des Lernens Grenzen setzt. „Wenn ein Organismus in ein Experiment zum klassischen Konditionieren gebracht wird, können die verschiedenen CS mehr oder weniger wahrnehmbar sein für einen Organismus, und die

UCS mehr oder weniger starke Reaktionen hervorrufen, aber auch die CS und UCS können mehr oder weniger assoziierbar sein. Das Tier kann mehr oder weniger durch die Evolution der Art vorbereitet (prepared) sein, einen gegebenen CS und UCS oder eine gegebene Reaktion und einen Verstärker zu assoziieren" (Seligman, 1972). Von Seligman und Hager (1972) wurden diese Befunde also als Zeichen eines speziellen Lernsystems interpretiert, das die üblichen Prinzipien des klassischen Konditionierens in Frage stellt.

Ob qualitative Beziehungen zwischen CS und UCS die Bildung von Assoziationen fördern oder hemmen, kann in einem speziellen Paradigma, das von Rescorla und Furrow (1977) als „doppeltes Dissoziationsexperiment" bezeichnet wurde, untersucht werden. Dabei werden verschiedenen Gruppen von Individuen qualitativ zusammengehörige CS1 und UCS1 oder CS2 und UCS2 dargeboten. Anderen Gruppen werden dagegen nicht zusammengehörige (z.B. CS1 und UCS2) gemeinsam dargeboten. Die Ergebnisse zeigen, daß qualitative Ähnlichkeit oder qualitative Beziehung die Bildung von Assoziationen fördert. Dies gilt also nicht nur für spezielle biologisch relevante Systeme, sondern scheint ein generelles Prinzip zu sein, das allerdings in unterschiedlichem Zusammenhang auch ein unterschiedliches Gewicht hat.

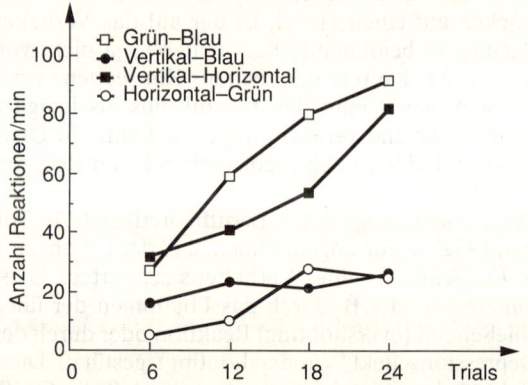

Abb. 7–10. Konditionierung zweiter Ordnung als Funktion der Ähnlichkeit der beiden konditionierten Reize CS1 und CS2. Gehören beide Reize zu einer Dimension (Farbe oder Lage), so zeigt sich eine stärkere konditionierte Reaktion (aus Mackintosh, 1977).

7.2.9 Lernen als natürliche Rationalität

Revusky (1971) konnte darüber hinaus zeigen, daß die Kontiguität durchaus als ein die Konditionierung unterstützendes Element auch in derartigen Untersuchungen erhalten bleibt. Gibt man nämlich in größerer zeitlicher Nähe zur Nausea den Tieren eine weitere noch unbekannte Substanz, so wird die Geschmacksaversion auf diese konditioniert. „Die Organismen scheinen sich auf komplexe, ja fast rationale Weise zu verhalten. Sie schreiben das Auftreten eines Verstärkers seiner wahrscheinlichsten vorauslaufenden Ursache zu und assoziieren ihn auf Kosten entlegenerer Ereignisse selektiv mit Ereignissen, die ihm in engerem Abstand vorausgehen, sowie mit Ereignissen, die durch eine Sequenz von Durchgängen hindurch auf Kosten nicht ganz so gut korrelierter Ereignisse besser mit ihm korreliert sind" (Mackintosh, 1977). In der folgenden Abbildung wird die Abhängigkeit der Konditionierungsstärke von der Prädiktivität des CS dargestellt. Diese ist dabei von zwei Faktoren abhängig:

– davon wie hoch die Wahrscheinlichkeit ist, daß mit dem CS auch der UCS auftritt (Verlauf der Kurven) und

– davon wie hoch die Wahrscheinlichkeit ist, daß der UCS auch ohne den CS auftritt (Typ der Kurve).

Beide Faktoren zusammen bedingen dann die tatsächliche Konditionierungsstärke. Wird also der UCS niemals ohne den CS dargeboten (offenes Viereck, obere Kurve), so ergibt das bei einer Wahrscheinlichkeit von .4 für das gemeinsame Auftreten des CS mit dem UCS die höchste Reaktionsstärke. Wird der CS genausohäufig mit dem UCS dargeboten, der UCS aber genausooft ohne den CS, so ist die Reaktionsstärke sehr gering (geschlossener Kreis), da der CS keinen prädiktiven Wert besitzt.

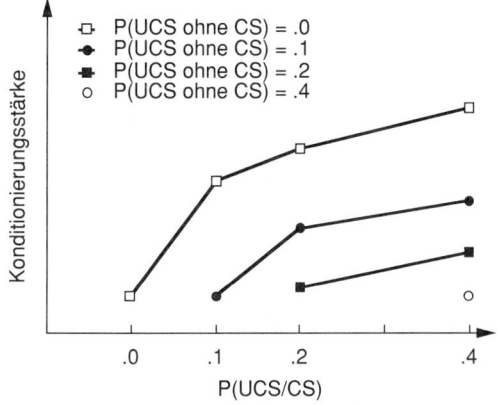

Abb. 7–11. Abhängigkeit der Konditionierung von der Wahrscheinlichkeit des Auftretens des UCS während des CS (Verlauf der Kurven) und des zusätzlichen Auftretens des UCS ohne CS (verschiedene Kurven) (aus Rescorla, 1988).

Der Vorgang des Konditionierens scheint, auch in einfachen Formen, Regeln zu folgen, die auch im Hinblick auf wissenschaftliche Theorien gelten. Diese müssen einen Vorhersagewert besitzen für zukünftige Ereignisse und sollten nur die mindestens notwendige Anzahl von Elementen enthalten, also möglichst ökonomisch sein („Ockham's razor"). Auch im „blocking"-Phänomen zeigt sich diese Ökonomie, es wird keine überflüssige Assoziation gebildet zwischen einem Reizelement und dem UCS, das den Vorhersagewert eines Ereignisses (UCS) nicht erhöht. Des weiteren ist ein Merkmal wissenschaftlicher Hypothesen, daß sie nur vorläufig gültig sind. Wenn eine Hypothese bzw. ein theoretisches Modell einen besseren prädiktiven Wert besitzt, löst sie das frühere Modell ab. Die Untersuchungen von Wagner et al. (1967) zeigen

etwas ganz entsprechendes beim Konditionieren. Ist ein Reiz mit einem UCS aufgrund eines bestimmten prädiktiven Wertes assoziiert, so wird diese Verbindung zugunsten eines neuen Reizes gelöst, wenn der neue Reiz das zukünftige Ereignis, den UCS, besser vorhersagt.

Diese Entsprechung ist vermutlich nicht zufällig. Das Bilden von Assoziationen, wie es beim Konditionieren stattfindet, setzt, wie bereits oben ausgeführt, eine bestimmte Art der „Erkenntnisfähigkeit" voraus, eine Erkenntnisfähigkeit für Beziehungen zwischen Ereignissen. Diese Fähigkeit könnte als eine sehr grundlegende aufgefaßt werden, die bei unbedingtem Reagieren nicht notwendig ist, sondern erst beim Lernen wirksam wird. Lernen beinhaltet danach immer einen „Erkenntnisgewinn", der bewirkt, daß der Organismus sich anhand von „Hypothesen", also damit anhand von Vorhersagen verhält. Es ist sinnvoll, anzunehmen, daß die Methode des Erkenntnisgewinns, der ja nicht unabhängig von der Umwelt stattfindet, durch die Beziehung zur Umwelt mitbedingt ist und somit an verschiedenen Orten anzutreffen ist, an denen es um diesen Erkenntnisgewinn geht, also auch bei der Bildung wissenschaftlicher Theorien. Das Grundlegende am Lernen wäre demnach auch nicht ein konditionierter Reflex, sondern das Wirksamwerden dieser Fähigkeit von Organismen in Beziehung zu ihrer Umgebung. Lernen ist somit immer ein „kognitiver" Prozeß, ein solcher findet nicht erst statt, wenn z.B. sprachliche Prozesse beteiligt sind.

7.2.10 Lernen und Abstraktion

Bei komplexeren Lernleistungen treten deutliche Speziesunterschiede zutage, wie beispielsweise beim Zuordnungslernen. Hierbei wird die Beziehung zwischen zwei Reizen zum kritischen Element, von dem abhängt, ob eine Belohnung erfolgt oder nicht. So wird ein Standardreiz zusammen mit zwei anderen Reizen dargeboten. Die richtige Reaktion des Tieres besteht dann darin, daß der kleinere bzw. größere im Vergleich zum Standardreiz für die Belohnung gewählt werden muß. Die Tiere können lernen, diese Regel zu „abstrahieren". Es wird ja nicht auf den Reiz als solchen, sondern auf das Verhältnis, das er zu einem anderen hat, reagiert.

Noch deutlicher wird dies, wenn von einer gelernten Situation die Regel auf eine Situation mit vollständig neuen Reizen, in denen jedoch die Regel enthalten ist, übertragen werden muß. Hier zeigen sich deutliche Speziesunterschiede. Dies ist eine Aufgabe, die sehr leicht von Primaten gelöst wird, Tauben jedoch einige Schwierigkeiten bereitet. Ähnliche Unterschiede treten auf, wenn der Standardreiz vor den kritischen Auswahlreizen dargeboten wird und zur Zeit der Wahlaufgabe nicht mehr wahrnehmbar ist. Hier können Primaten ebenfalls bei Verzögerungen der Auswahl gegenüber der Darbietung des Standardreizes von mehreren Minuten korrekt reagieren. Hier spielt offenbar auch die Dauer von Gedächtnis eine Rolle.

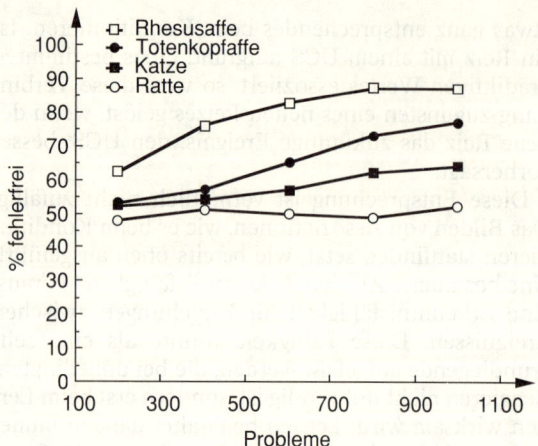

Abb. 7–12. Die Leistung verschiedener Tiergruppen bei einer Reihe von visuellen Diskriminationsproblemen (aus Mackintosh, 1977).

So ist für Premack (1978) die Übertragung von Erlerntem eine operationale Definitionsmöglichkeit von „Abstraktheit", d. h. die Übertragung erlernter Beziehungen zwischen Reizen, die eine Regel darstellen. Spence (1937) hatte das von Koehler (1929) als „Transposition" bezeichnete Phänomen eher als Ergebnis „absoluten" Lernens und nicht als das von Beziehungslernen aufgefaßt. Er sah es als das Resultat von Diskriminationslernen, wobei die Erregungsstärke eines Reizes, auf der Dimension von Grautönen, von erfahrener Nichtverstärkung (Hemmung) und Verstärkung (Erregung) abhängt. Dabei wird der Nettobetrag der Erregungsstärke eines neuen Reizes abhängen von der Generalisation der früheren Erfahrung auf diesen Reiz, die zusammengesetzt ist aus spezifischer Erregung und Hemmung.

Abstraktionsfähigkeit: Lernen ist nicht immer das gleiche

Premack (1978) ist der Ansicht, daß bei verschiedenen Tierarten beide Prozesse – absolutes und Beziehungslernen – von unterschiedlichem Gewicht sind. Dies sei an folgendem Beispiel zum sog. „Umlernen" (reversal learning) verdeutlicht.

Hat z. B. eine Ratte im Sinne des Diskriminationslernens gelernt, daß in einem T-Labyrinth das Futter in der schwarzen und nicht in der weißen Seite zu finden ist, so benötigt das Tier eine bestimmte Anzahl von Lerndurchgängen, um umzulernen, daß ab einem bestimmten Zeitpunkt das Futter nunmehr in der weißen Seite zu finden ist. Menschen lernen derartiges sehr viel schneller. Dies ist jedoch nicht das Ergebnis einer größeren Lerngeschwindigkeit beim Menschen, sondern es liegt daran, daß Menschen in derartigen Situationen ein vermittelndes Konzept bilden (hell/dunkel), das die Funktion einer Regel hat und in der neuen Situation beibehalten werden kann („reversal shift"-Aufgabe). In folgender Situation lernen jedoch Tiere schneller als Menschen: Der diskriminative Stimulus beinhaltet neben einer kritischen Dimension (Helligkeit: schwarz/weiß) eine weitere

Dimension (Größe: groß/klein), die jedoch in der ersten Lernphase zufällig variiert wird, also für die Diskrimination irrelevant ist. In der zweiten Lernphase jedoch ist diese zweite Dimension die wichtige, d. h. der richtige Reiz ist immer der große und nicht der kleine. Es besteht kein Zusammenhang mehr zur ersten relevanten Dimension (Helligkeit) („non-reversal shift"-Aufgabe). Diese Aufgabe lernen Tiere und Kinder bis etwa zum 6. Lebensjahr schneller als erwachsene Menschen (Kendler und Kendler, 1970).

Dieser Sachverhalt läßt sich folgendermaßen erklären: Im ersten Fall, der „reversal shift"-Aufgabe, lernen Tiere absolut, während Menschen nach dem 6. Lebensjahr ein Konzept bilden, das auch in der zweiten Lernphase beibehalten werden kann. Im Falle der „non-reversal shift"-Aufgabe muß diese Regel jedoch zuerst gelöscht werden, damit ein neues vermittelndes Konzept (Größe) gebildet werden kann. Damit wird die Akquisition der korrekten Reaktion gegenüber dem absoluten Lernen verzögert. Die folgende Abbildung zeigt die Lernleistung von vier- und siebenjährigen Kindern in Abhängigkeit von der Benennung der Dimension. Bezieht sich die Benennung auf die irrelevante Dimension, ist die Lernleistung der älteren Kinder besonders schlecht.

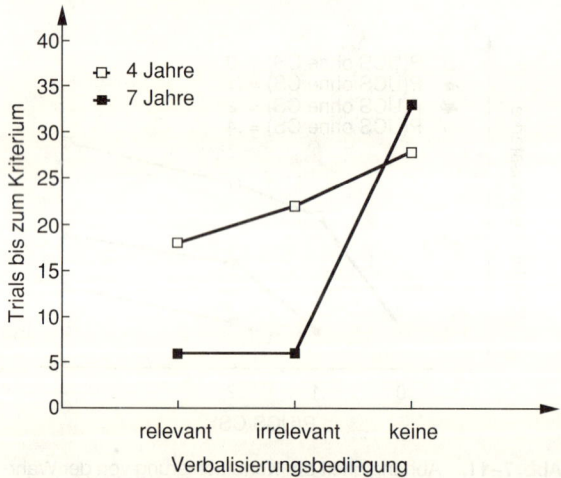

Abb. 7–13. Abhängigkeit der Leistung in Reversal-shift-Aufgaben vom Alter (aus Razran, 1971).

An diesem Beispiel wird deutlich, daß die Kapazität für die Abstraktion und Anwendung von Regeln nicht nur zwischen den Arten unterschiedlich, sondern beim Menschen von der Entwicklung abhängig ist.

7.2.11 Interne Bedingungen des Verhaltens

Das vorangehende Beispiel zeigt einen weiteren wichtigen Zusammenhang. Bei Ratten und bei Menschen wurde ein gleiches Experiment vorgenommen, das Verhalten war jeweils verschieden. Diese Unterschiedlichkeit des Verhaltens läßt sich somit nicht aus den unterschiedlichen Reizgegebenheiten erklären, die Varianz stammt aus Merkmalen der verschiedenen Organismen, also aus der direkten Beobachtung nicht zugänglichen Sachverhalten. Somit beziehen sich lerntheoretische Erklärungen keineswegs nur auf die direkte Beobachtung zugängliche Sachverhalte, sondern auch auf interne, nicht direkt beobachtbare Faktoren. Dasselbe gilt für wesentliche Elemente etwa der Theorie Hulls. Die darin verwendeten Begriffe wie „Trieb" (drive) oder „Anreiz" (incentive) beziehen sich ebenfalls auf interne Zustände von Organismen (Hull, 1943).

Tolman (1932) legte großes Gewicht auf die Tatsache, daß es einen Unterschied zwischen Lernen und Verhalten gibt. Dieser Unterschied ist wichtig, betrachtet man Lernen, wie es hier geschieht, als eine in irgendeiner Form „erkennende" Auseinandersetzung von Organismen mit ihrer Umgebung. Dies gerät häufig aus dem Blickfeld, wenn beispielsweise davon gesprochen wird, ein Reiz erwerbe die Fähigkeit, bei einem Organismus irgendeine Reaktion hervorzurufen. Dieser Sprachgebrauch kommt vermutlich daher, daß in der Mehrzahl von Experimenten der Effekt des Lernens durch das Auftreten bestimmter Reaktionen erfaßt wird, und daß dabei das Verhalten des Organismus als abhängige, die Reize als unabhängige Variablen betrachtet werden. Tatsächlich lernen nicht die Reize, sondern der Organismus wird durch das Lernen ein anderer.

Man kann dies deutlich am Beispiel des sog. „latenten Lernens" zeigen. Hierbei werden zwei Gruppen von Tieren gebildet, von denen die Tiere der einen Gruppe von Anfang an nach dem Durchlaufen eines T-Labyrinths in einer der beiden Seiten durch Futter belohnt werden, die Tiere der anderen Gruppe jedoch nicht. Nach etwa zehn Versuchsdurchgängen ist die Leistung der stets in einer Zielbox verstärkten Tiere deutlich besser als die der Tiere, die keinerlei Verstärkung erhalten hatten; dies ist nach den Prinzipien des operanten Lernens auch nicht anders zu erwarten. Nicht ohne weiteres zu erwarten ist hingegen die Veränderung des Verhaltens der Tiere, die nicht verstärkt worden waren, wenn man beginnt, sie nun für die richtigen Reaktionen, also die Wahl einer Seite am Entscheidungspunkt, zu belohnen. Ihre Leistung wird schlagartig besser.

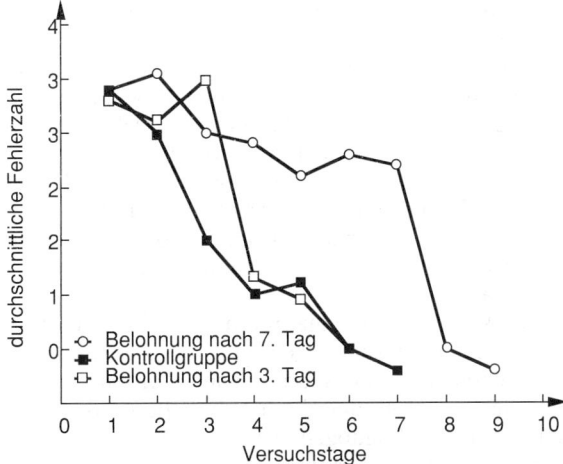

Abb. 7–14. Lernleistung nach unterschiedlichem Beginn von Belohnung (aus Foppa, 1968).

Diese Tiere haben offensichtlich etwas über das T-Labyrinth gelernt, das im unverstärkten Verhalten bis dahin nicht sichtbar geworden ist, aber die anschließend vorhandene Leistung vermittelt (Tolman und Honzik, 1930).

Allerdings äußert das Tier in dieser Situation auch in den Durchgängen, in denen es nicht belohnt wurde, ein Verhalten. Es geht ja beim operanten Lernen um die Änderung der Wahrscheinlichkeit des Auftretens eines Verhaltens. Daß, auch ohne daß es zu beobachtbarem Verhalten kommt, etwas gelernt wird, sich also der Organismus „latent" verändert bzw. Reize eine Bedeutung erhalten, zeigt sich beim sog. „sensorischen Vorkonditionieren". Dabei werden zwei Reize häufig miteinander dargeboten. Anschließend wird einer der Reize auf einen UCS konditioniert. Man kann dann zeigen, daß auch der andere Reiz eine bestimmte assoziative Verbindung mit dem UCS besitzt, und zwar durch seine Verbindung mit dem ersten CS. Daran zeigt sich deutlich, daß sich beim Lernen der Organismus verändert und nicht der Reiz, wie der oben erwähnte Sprachgebrauch nahelegen könnte.

Tolman war der Auffassung, daß im Verlauf des Lernens ein inneres Abbild, eine „kognitive Karte" der relevanten Aspekte der Umgebung angelegt wird, an der sich der Organismus dann orientiert. Zumindest scheint ein Speichervorgang im Hinblick auf die räumliche Anordnung bestimmter Elemente der Umgebung stattzufinden, dessen Ergebnis die Leistung dann später erleichtert. Auch hierbei sollte man sich klarmachen, daß die Nutzung der gespeicherten Struktur zu einem späteren Zeitpunkt im eigentlichen Sinne hypothetischen Charakter hat, nämlich die Hypothese beinhaltet, daß sich diese Struktur in der Zeit erhält und in Zukunft nicht ändert und damit für zukünftiges Verhalten valide ist. Dies ist also hier ganz ähnlich wie beim klassischen Konditionieren, bei dem zeitliche Ereignisabfolgen gleichsam erwartet und darum assoziiert werden.

Diese internen Zustände haben einen wesentlichen Einfluß auf die Beziehung des Organismus zu seiner Umgebung, z.B. darauf, ob ein bestimmter Reiz positiv bekräftigend wirkt oder nicht. Ein sattes Tier kann mit der Gabe von Futterpillen schlecht verstärkt werden. Sind nun die Beziehungen zwischen Organismen und ihrer Umgebung wesentlich von Zuständen und Merkmalen der Organismen selbst abhängig, so können Organismen nicht als passiv auf die Umgebung reagierend verstanden werden, sie verhalten sich in bezug auf ihre Umgebung aktiv.

Neugier – Organismen verhalten sich aktiv

Von Popper (1977) wurde die Theorie des Konditionierens u. a. in dem Punkt kritisiert, daß der Organismus in dieser Theorie als passiv reagierend verstanden werde. Aus diesem Grund lehne er den Begriff des Reflexes ab. „Pawlows Interpretation sieht den Hund als einen passiven Mechanismus, während meine Interpretation dem Hund ein aktives (wenn auch zweifellos unbewußtes) Interesse an seiner Umgebung zuerkennt, einen exploratorischen Instinkt" (Popper, 1977). Ein derartiger Instinkt wurde allerdings durchaus auch von seiten der Lerntheorie gesehen, allerdings unterschiedlich benannt. So nahm Thorndike (1913, zit. nach Hunt, 1971) Bezug auf einen „Neugiertrieb" (curiosity drive), Harlow (1950) benutzte den Begriff „Explorationstrieb" (exploratory drive) und wie Hunt (1971) den der „intrinsischen Motivation" (intrinsic motivation).

Für diesen Zusammenhang sind die Untersuchungen zur sensorischen Verstärkung wichtig. Unter dem Begriff „sensorische Verstärkung" (sensory reinforcement) wird verstanden, daß die Gabe von sensorischen Reizen, die lediglich in der gegebenen Situation den Informationsgehalt vergrößern, einen verstärkenden Effekt hat (Kish, 1966). Von Berlyne (1950) wurde beispielsweise das Ausmaß des Neugierverhaltens bei Tieren untersucht, indem gemessen wurde, in welchem Umfang sie sich neuen Reizen näherten, daran rochen usw., d.h. Aktivität zeigten.

Neuigkeit ist eine wesentliche Bedingung für die Auslösung der Orientierungsreaktion, und von Miller, Galanter und Pribram (1960) wurde der Inkongruenz zwischen einem Sollwert und einem Istwert ein motivationaler Charakter zugemessen. Nimmt man eine Bedingung wie Neugier als eine Verhalten bedingende Größe an, so ist es zweifelhaft, ob ein dadurch bedingtes Verhalten im Sinne der Wiederherstellung einer Homöostase verstanden werden kann. Popper geht davon aus, daß der Organismus Hypothesen im Hinblick auf seine Umgebung entwirft und diese gezielt durch sein Verhalten testet.

Die oben dargestellten Ergebnisse zum klassischen Konditionieren ließen sich durchaus so interpretieren, daß derjenige Reiz, der die beste Hypothese über das Auftreten des UCS beinhaltet, als CS vom Organismus mit dem UCS verknüpft wird, und es wurde eine gewisse Entsprechung der beim Konditionieren gültigen Regelmäßigkeiten zu Regeln wissenschaftlichen Vorgehens gefunden.

7.2.12 Sprache

Bekanntlich hat Pawlow das sog. „zweite Signalsystem", die Sprachfunktion, als wesentlich menschliche Fähigkeit angesehen. Über den Erwerb dieser Fähigkeit gibt es eine weitreichende Kontroverse. Es geht dabei um die Frage, ob Sprache ausschließlich nach Regeln von Reiz-Reaktions-Verknüpfungen erworben wird, oder ob der Erwerb hierarchische Regeln beinhaltet (z.B. Arbib, 1969; Suppes, 1969).

Daß Sprache bei menschlichem Lernen einen ganz wesentlichen Einfluß hat, haben zahlreiche Untersuchungen gezeigt. Hier soll das Phänomen der „semantischen Generalisierung" als Beispiel kurz erläutert werden. Konditioniert man beispielsweise eine Änderung der Hautleitfähigkeit (PGR) auf ein bestimmtes Wort, so zeigt sich ein Generalisierungseffekt im PGR bei Kindern bis etwa zum 18. Lebensjahr hauptsächlich auf lautähnliche Wörter, d.h. die physikalische

Ähnlichkeit des Reizes ist ausschlaggebend. Bei älteren Kindern und Erwachsenen hingegen tritt der Generalisierungseffekt bei Wörtern auf, die mit der sprachlichen Bedeutung des ersten Wortes zusammenhängen. Dabei kommt es interessanterweise bei Kindern von etwa 10 Jahren zur stärksten Generalisierung bei Bedeutungsgegensätzen, später, ab etwa dem 14. Lebensjahr bei Bedeutungsähnlichkeiten (Grant, 1972).

Vorgänge des Lernens, d.h. solche, die den Prinzipien des Lernens folgen, spielen auch in sehr komplexen menschlichen Lebenssituationen eine wesentliche Rolle, auch wenn derartige Vorgänge nicht darauf reduzierbar sind. Die Ergebnisse der sozialen Lerntheorien sind dafür ein wichtiges Beispiel.

7.2.13 Soziales Lernen

Soziale Lerntheorien gehen prinzipiell davon aus, daß die Klasse von Reizen, die Verhalten kontrollieren, um solche erweitert werden muß, die von anderen Personen ausgehen. Dies bedeutet nicht, daß damit die funktionalen Gesetze etwa des operanten Konditionierens keine Gültigkeit mehr besitzen. Dieser Ansatz trägt zunächst eher der Tatsache Rechnung, daß eine Vielzahl sozialer Verhaltensweisen verstärkende Wirkung besitzen. Dies gilt für eine ganze Reihe „normaler" Verhaltensweisen, deren verstärkender Charakter nicht unbedingt unmittelbar erkenntlich ist.

Wichtig bei dieser Betrachtungsweise ist die Tatsache, daß soziales Verhalten wechselseitig das Verhalten des jeweiligen oder der jeweiligen Interaktionspartner kontrolliert. In den oben referierten Modellen, deren Lebensnähe nicht unmittelbar einsichtig ist, wurde meist von in hohem Maße isolierten und standardisierten Laborbedingungen ausgegangen, die die Erfassung möglichst „reinen" Verhaltens ermöglichen sollten, sozusagen die Grundeinheit von Verhalten. Eine Ratte in einer sog. Skinner-Box, deren Zweck es war, neben präzisen und möglichst automatisierten Steuerungs- und Registrierbedingungen eine weitgehende Ausschaltung von störenden Situationseinflüssen zu gewährleisten, konnte wohl kaum im Sinne einer wechselseitigen Verhaltensbeeinflussung das Verhalten des Versuchsleiters kontrollieren. Betrachtet man die Interaktion zwischen zwei Menschen, so verfügen beide prinzipiell über vergleichbare Möglichkeiten sozialen Verhaltens und damit auch über die entsprechenden Kontrollmöglichkeiten in dem Maße, in dem dies Verhalten verstärkende Wirkung hat.

Das Modell des sozialen Lernens ist in vielen Studien und Untersuchungen auf die Beeinflussung kindlichen Verhaltens angewandt worden. Es hat sich dabei gezeigt, daß die funktionelle Analyse verstärkender Bedingungen, wie die Betrachtung der Wirkung von Verstärkungsplänen, die Berücksichtigung zeitlicher Verhältnisse von Kontingenzen, in derartigen Plänen von großer Wichtigkeit ist. Dies soll an einem Beispiel von Gewirtz (1977) verdeutlicht werden, in dem ein eigentlich sensationelles Ergebnis einer Studie von Bell und Ainsworth (1972) analysiert wird.

Das Schreien von Kindern ist nicht immer das gleiche – Beispiel einer funktionalen Analyse sozialen Verhaltens

Bell und Ainsworth (1972) hatten in einer Studie gefunden, daß Kinder weniger „Schreiverhalten" zeigen, wenn Mütter sich jedesmal den Kindern zuwenden, wenn sie schreien. Dies Ergebnis scheint den Prinzipien des operanten Lernens zu widersprechen, insofern, als Zuwendung eine Verstärkung darstellt und somit das Schreiverhalten, da es verstärkt wird, in seiner Frequenz zunehmen müßte; derartig behandelte Kinder müßten also mehr schreien. Dem gegenüber steht ein Ergebnis von Etzel und Gewirtz (1967), daß sog. „operantes" Schreien seltener auftritt, wenn Bezugspersonen nicht darauf reagieren. Operantes Schreien wird dabei definiert als nicht durch aversive Reize, wie Hunger oder Durst, ausgelöstes Schreien, sondern ist Schreien, das gleichsam einem Ziel dient, z.B. Zuwendung zu erreichen. Nicht-operantes Schreien wird als „nicht konditioniert", „ausgelöst" oder „expressiv" bezeichnet.

Die Analyse dieser Ergebnisse richtet sich auf folgende Merkmale des Sachverhalts:

– Das Schreien der Kinder ist kein einheitliches Geschehen, es kann kurz oder lang, laut oder weniger laut geschrien werden usw.
– Das Verhalten der Mütter, die nicht auf das Schreien der Kinder reagierten, ist ebensowenig als einheitlich zu erwarten. So ist anzunehmen, daß sie auf bestimmte Arten zu schreien intermittierend reagierten und so recht stabile Frequenzen des Auftretens des Schreiens bewirkten. Aus den Prinzipien des operanten Lernens (s.o.) ist ableitbar, daß bei intermittierender Verstärkung relativ wenige Verstärker ausreichen, um hohe Reaktionsraten zu erreichen.
– Ganz wesentlich für eine angemessene Interpretation ist die Berücksichtigung der „Latenz" der mütterlichen Reaktionen. Aus verschiedenen Untersuchungen ist bekannt, daß gerade bei kleinen Kindern die Verstärkung durch soziale Zuwendung recht schnell auf das zu verstärkende Verhalten folgen muß (2–3 Sek.). Es läßt sich nun denken, daß die auf das Schreien reagierenden Mütter wenigstens zum Teil nach dieser kritischen Latenz auf die Kinder reagiert haben und damit nicht das Schreien, sondern das auf das Schreien folgende Verhalten verstärkt haben, das möglicherweise mit dem Schreien inkompatibel war. Damit würde das Schreien in der Frequenz reduziert.

Diese Überlegungen machen deutlich, daß man nicht einfach das Verhalten der Mütter und der Kinder als Gesamtheit in Beziehung bringen kann, sondern es im Hinblick auf seine Funktionalität im Sinne des operanten Lernens betrachten muß.

7.2.14 Wirkung der „Kenntnis" von Verhaltenskontingenzen

Im Falle des klassischen Konditionierens der Lidschlußreaktion im Humanexperiment läßt sich zeigen, daß die Kenntnis der Konditionierungskontingenz zu einer Form der Akquisition der Lidschlußreaktion führt, die sich unterscheidet von der Akquisition ohne Kenntnis der Kontingenz (awareness of contingency). Es wird dabei zwischen „wahr konditionierten" und „willkürlichen" Reaktionen unterschieden (V-form conditioning für voluntary und C-form conditioning für classical) (Spence und Taylor, 1951, zit. nach E. Saltz, 1973).

Rotter (1954, 1972) hat den Gedanken betont, daß Personen im Hinblick auf ihr Verhalten bestimmte verstärkende Konsequenzen erwarten. Er definiert Erwartung als „die Wahrscheinlichkeit, die eine Person dem Auftreten einer bestimmten Verstärkung nach einem bestimmten Verhalten in einer bestimmten Situation zumißt" (Rotter, 1972). Diese Einschätzung der Wahrscheinlichkeit des Auftretens der Verstärkung ist unabhängig von dessen Verstärkungswert. Dies setzt voraus, daß eine Person Kenntnis von der wirksamen Kontingenz der Verstärkung hat (awareness). De Nike (1964) führte eine Untersuchung durch, in der Personen durch einen sozialen verbalen Verstärker dafür belohnt wurden, wenn sie in einer Aufgabe, in der Wörter benannt werden sollten, bestimmte Hauptwörter nannten. Die Rate der Nennung dieser Wörter stieg in dem Augenblick sprunghaft an, in dem den Probanden die Verstärkungskontingenz klarwurde. Die Probanden, die die Kontingenz nicht erkannten, verhielten sich wie die Probanden der Kontrollgruppe, die zufällig verstärkt worden waren.

Um die Abhängigkeit der Lernleistung von der Einsicht in die Kontingenzen des Verhaltens zu ermitteln, sind unter Umständen wie bei der Verhaltensformung sehr lange Untersuchungen notwendig, bis etwa eine Ratte gelernt hat, zu tun, was man von ihr will. Men-

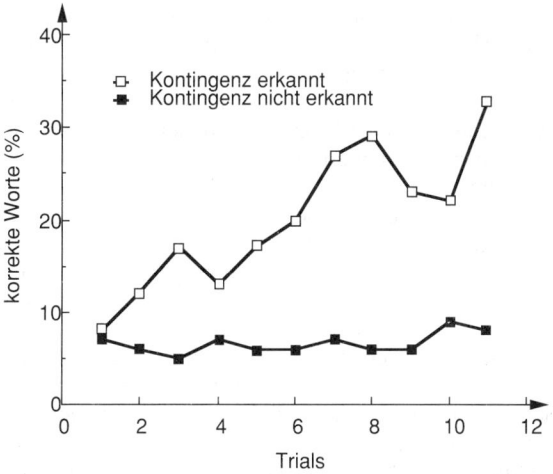

Abb. 7–15. Prozentsatz korrekter Worte, deren Nennung verstärkt wurde, in Abhängigkeit von der Einsicht in die Kontingenz (aus Spielberger und DeNike, 1966).

schen kann man mittels einer Instruktion mitteilen, was man von ihnen will, was beim Tier nicht möglich ist. In verhaltenstherapeutischen Zusammenhängen wird man also Instruktionen benutzen, um einen bestimmten Verhaltenseffekt zu erzielen. Mit der Instruktion erhält eine Person dann eine Einsicht in das, was von ihr erwartet wird. Die folgende Abbildung zeigt, daß eine derartige Einsicht zwar wirksam, gleichsam „der erste Schritt zur Besserung" ist, aber alleine, ohne daß eine Person „etwas davon hat", nicht ausreicht, um einen stabilen Effekt zu erzielen, die Kurve fällt ab, um dann bei Einführung der zusätzlichen Belohnung wieder steil anzusteigen.

Lernen durch Beobachtung eines anderen

Der Begriff „Beobachtungslernen" (vicarious learning (stellvertretendes Lernen) oder observational learning) wird von vielen Autoren austauschbar benutzt. Ebenso werden die Begriffe Imitation oder Modellernen verwendet. Man versteht darunter den Sachverhalt, daß einer Person die Gelegenheit gegeben wird, eine andere zu beobachten, die einer bestimmten Lernkontingenz ausgesetzt ist. Dies gilt sowohl im Sinne des Paradigmas des klassischen als auch des operanten Lernens. So können Verhaltensweisen geändert werden, indem ein „Modell" beobachtet wird, das vergleichbares Verhalten ausübt, z.B. Annäherung an einen angstauslösenden Reiz. Ebenso kann emotionales Verhalten erworben werden, ohne daß z.B. die Aversivität eines Reizes selbst erfahren wurde.

Daß die Beobachtung in diesem Sinne auch bei Tieren wirksam ist, konnte Miller (1967, zit. nach Bandura, 1969) zeigen. Er trainierte Rhesusaffen in einem signalisierten Vermeidungsversuch, in dem ein Lichtreiz einen Schock signalisierte, der dann durch das Drücken eines Hebels vermieden werden konnte. In einem zweiten Versuchsabschnitt wurden die Tiere paarweise angeordnet und zwar so, daß ein Tier den Lichtreiz wahrnehmen konnte, aber keinen Reaktionshebel hatte, also die Schocks nicht vermeiden konnte. Das andere Tier verfügte über den Hebel, konnte aber den die Schocks signalisierenden Lichtreiz nicht sehen. Die Tiere konnten sich gegenseitig beobachten. Es zeigte sich, daß das Tier, das zwar reagieren, jedoch den ankündigenden Reiz nicht wahrnehmen konnte, dennoch in der Lage war, die Schocks zu vermeiden, indem es das andere Tier beobachtete und offensichtlich an dessen Verhalten die Furchtmerkmale erkennen konnte.

In diesem Zusammenhang wird von vielen Autoren angenommen, daß die Beobachtung der Kontingenz bei einem Modell „Einsicht" vermittelt, die einen Vorgang der Übung überflüssig macht. Dies ist in dem oben zitierten Tierversuch von Miller sicher nicht der Fall. Das Tier, das durch das Drücken des Hebels die Schocks vermeiden konnte, hatte bereits vorher dieses Verhalten als Vermeidungsreaktion gelernt. Der Versuch zeigt aber, daß das Verhalten eines anderen Tieres als Hinweisreiz (cue) dienen kann, sogar ohne daß die äußere Verhaltenskonsequenz beobachtbar wäre. Tatsächlich hat das Tier ja die Schocks auch für

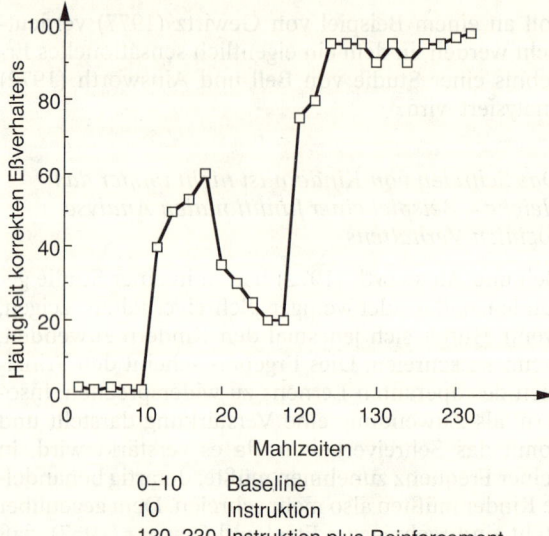

Abb. 7–16. Korrektes Eßverhalten nach Instruktion allein (10) und Instruktion mit zusätzlicher Belohnung (120–230) (aus Ayllon und Azrin, 1964).

sich selbst vermieden, d.h. das Verhalten wurde weiter durch „äußere" Konsequenzen verstärkt bzw. aufrechterhalten.

Bandura (1969) sieht den Vorgang des Beobachtungslernens als einen kognitiven Vorgang. „Eine stellvertretende (vicarious) Verstärkung vermittelt nicht nur Informationen die wahrscheinlichen Verstärkungskontingenzen betreffend, Wissen über die Arten der Situationen, in denen das Verhalten angemessen ist, sowie eine Darbietung von Belohnungen, die aktivierende Merkmale enthalten, sie schließt darüber hinaus einen affektiven Ausdruck des Modells ein, das den belohnenden oder bestrafenden unterzogen ist" (Bandura, 1969). Das genannte Wissen über die Kontingenz der Verstärkung oder Bestrafung ersetzt dieser Auffassung zufolge die Notwendigkeit der Übung. Die folgende Abbildung zeigt die Wirkung der Beobachtung eines oder mehrerer Modelle bei der Annäherung an ein Angstobjekt. In der Kontrollgruppe stand kein Modell zur Verfügung.

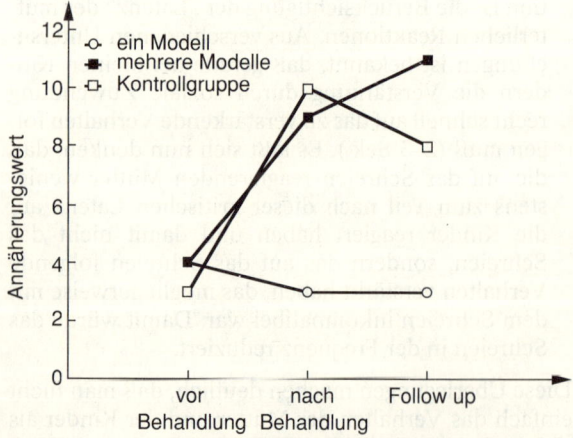

Abb. 7–17. Durchschnittliches Annäherungsverhalten von Kindern an einen Hund nach der Beobachtung von einem oder mehreren Modellen (aus Bandura und Menlove, 1968).

Gewirtz (1977) hat die kognitive Interpretation des Beobachtungslernens als nicht stringent kritisiert. Er wendet ein, „daß imitative Reaktionen einfach instrumentelle Reaktionen sind, die den Hinweisreizen zugeordnet werden, die von den verschiedenen Reaktionen der Demonstrationsmodelle geliefert werden. Derartige imitative Reaktionen sind konditional, wie jene beim gewöhnlichen Zuordnungslernen, und bilden für das Kind eine funktionale Klasse von Zuordnungsreaktionen" (Gewirtz, 1977). Geht man jedoch wie in dieser Darstellung des Lernens und seiner Bedeutung für Vorgänge des Lebens von der Annahme aus, daß Lernen sinnvoll erst in der Beziehung eines Organismus, oder im vorliegenden Zusammenhang besser Individuums, zu seiner Umgebung gesehen werden kann, so zeigt sich, daß Lernen immer ein kognitiver Vorgang ist. Dies gilt erst recht, wenn die Bildung von Erfahrung und die „Erteilung von Bedeutung" im Sinne Pawlows von den konkreten physikalischen Ereignissen in einem sehr weitreichenden Sinne „distant" ist.

7.3 Die Rolle des Lernens in der Genese organischer Störungen

Die Lernpsychologie hat in den letzten Jahren erheblich dazu beigetragen, das Verständnis für Krankheitsprozesse und deren Entstehung zu erweitern. Eine der dafür wesentlichen Voraussetzungen besteht in der Annahme, daß organische Prozesse mittelbar oder unmittelbar durch Vorgänge des Lernens beeinflußt werden. Es ist dargestellt worden, daß lerntheoretische Modelle prinzipiell keine Unterscheidung zwischen Verhaltensvorgängen machen, die man gemeinhin als psychisch klassifizieren würde, und solchen, die als somatisch eingestuft würden. Bereits im klassischen Versuch Pawlows wurde das Auftreten einer organischen Reaktion durch für sie „unwesentliche" Reize hervorgerufen.

Die Orientierungsreaktion (OR) (Sokolow, 1963) ist die Reaktion eines Organismus auf neue, unerwartete Reize. Sie besteht in einer unspezifischen Desynchronisation im gesamten Cortex, einem Ansteigen der Hautleitfähigkeit, einer biphasischen Reaktion der Herzfrequenz (unmittelbar nach dem Reiz kommt es zu einer kurzen Erniedrigung der Herzfrequenz, dann zu einem Ansteigen), einer Erhöhung der Muskelaktivität, einer Erhöhung der Atemfrequenz, einer Vasodilatation im Kopfbereich sowie Vasokonstriktion in den Fingern. Es kommt zu einer Hinwendung des Organismus zur Reizquelle, etwa durch eine Kopfdrehung. Das Auftreten der OR ist abhängig vom Neuigkeitsgrad des Reizes. Wird er einige Male wiederholt dargeboten, so verschwindet die OR, sie habituiert.

Der Vorgang der Habituation besteht darin, daß ein „Organismus lernt, daß eine Reaktion unnötig geworden ist und diese Reaktion einstellt. Dadurch wird das informationsverarbeitende Nervensystem für neue Reaktionen frei, da immer nur eine begrenzte, kleine Zahl von Reaktionen zur selben Zeit ablaufen kann" (Birbaumer, 1975). Der Vorgang der Habituation ist kein assoziativer wie das Konditionieren und betrifft nur angeborene, also unkonditionierte Reaktionen (Razran, 1971) wie die Orientierungsreaktion, die sehr schnell habituiert, zumindest unter normalen Umständen. Bei bestimmten Personen ist die Habituation verlangsamt. Solche Personen zeichnen sich in Fragebögen zur Erfassung von Merkmalen der Persönlichkeit durch ein recht hohes Maß an Neurotizismus aus und neigen zu Angstreaktionen (Lader und Wing, 1966). Es ist wichtig, den Vorgang der Habituation von dem der Extinktion, der konditionierte Reaktionen betrifft, zu unterscheiden.

Die OR könnte als Reaktion auf Neuigkeit bezeichnet werden, die Habituation als Antwort auf das Fehlen von Information oder Neuigkeit. Untersucht man die Habituation auf einen recht komplexen Reiz, der durch mehrere Dimensionen beschreibbar wäre, und hat eine Habituation stattgefunden, so kommt es zur Dishabituation, wenn eine Dimension des Reizes geändert oder ein Reiz hinzugefügt wird (Vinogradova und Lindsley, 1963, zit. nach Razran, 1971).

Unter Sensibilisierung kann ein der Habituation entgegengesetzter Vorgang verstanden werden. Besteht die Habituation in der Abschwächung einer Reaktion bei wiederholter Stimulierung mit einem gleichbleibenden Reiz, so besteht die Sensibilisierung in einer Verstärkung der Reaktion. Die Reaktionsamplitude nimmt zu, die Latenzzeit wird kürzer.

Es treten unter denselben Reizbedingungen ebenfalls neue Reaktionen auf, die ursprünglich nicht mit dem Reiz verknüpft waren. Dieser Vorgang wird Pseudokonditionierung genannt.

Harris und Brady (1974) klassifizierten drei Typen des Lernens autonomer bzw. physiologischer Reaktionen, die für das Verständnis der Entstehung von Krankheitsprozessen nützlich sind:
– klassisches Konditionieren autonomer Reaktionen,
– gleichzeitiges Konditionieren autonomer Reaktionen,
– operantes Konditionieren autonomer Reaktionen.
Wichtig ist eine weitere Kategorie, das
– interozeptive Konditionieren.

7.3.1 Klassisches Konditionieren autonomer Reaktionen

Hierzu würde der Versuch Pawlows zählen, in dem eine physiologische Reaktion auf einen neutralen Reiz konditioniert wurde. Hierbei handelte es sich jedoch um eine isolierte Reaktion, wichtiger im vorliegenden Zusammenhang dürfte der Sachverhalt sein, daß auch ein ganzes Muster autonomer Reaktionen auf neutrale Reize konditionierbar ist, bzw. ein Ablauf derartiger Reaktionen. In der Pathogenese funktioneller Herz-Kreislauf-Störungen findet sich häufig der Sachverhalt, daß bestimmte Reize, oder besser Situationen, recht zuverlässig die Beschwerden auslösen. Der Vorgang der Reizgeneralisierung führt dann weiterhin dazu, daß in einer zunehmend größeren Anzahl von Situationen die Beschwerden auftreten.

Von gegenwärtig noch nicht absehbarer Bedeutung sind die Studien zur klassischen Konditionierung immunbiologischer und immunsuppressiver Reaktionen (vgl. Kap. 11). Ausgehend von Studien zum Erwerb von Geschmacksaversionen fanden Ader und Cohen (1975), daß ein Teil der Tiere, die mit Cyclophosphamid, einer Nausea produzierenden Substanz behandelt worden waren, in der Phase der Löschung der Geschmacksaversion starben. Dabei war interessant, daß die Tiere gestorben waren, die die größte Menge einer Saccharinlösung als CS erhalten hatten. Da Cyclophosphamid eine starke immunsuppressive Wirkung hat, schlossen die Autoren daraus, daß die Tiere möglicherweise aufgrund einer Abwehrschwäche an einem Infekt gestorben waren, die durch den Vorgang der Konditionierung von Saccharin (CS) mit Cyclophosphamid (UCS) verstärkt und verlängert worden war. Die Tiere starben nicht nach der Injektion des Endoxans, sondern in der Zeit, in der die Saccharinlösung ohne UCS dargeboten worden war.

In den folgenden Jahren führten Ader und Cohen sowie andere Autoren eine Reihe von Untersuchungen durch, mehr oder weniger mit dem Ziel, eine „reine" konditionierte Immunsuppression nachzuweisen. Trotz vieler methodischer Verbesserungen, die dem Ausschluß möglicher Fehlerquellen dienten, konnte in vielen Untersuchungen ein stabil reproduzierbarer, jedoch immer recht geringer immunsuppressiver Effekt durch das Konditionieren nachgewiesen werden. Dafür könnte die Tatsache ausschlaggebend sein, daß „zirkulierende Antikörper das Resultat einer komplexen Kette von Ereignissen sind, und der Einfluß des Konditionierens auf ein oder mehrere frühe Ereignisse stattfindet, die für die Produktion von Antikörpern wichtig sind" (Ader und Cohen, 1981).

Eine weitere Überlegung zur Erklärung des stets geringfügigen Ausmaßes der konditionierten Immunsuppression geht davon aus, daß durch das verwendete Paradigma zwei gegenläufige Vorgänge ausgelöst werden. Die Applikation von Antigenen stelle für die Produktion von Antikörpern einen UCS dar. Zunächst werde dann ein neutraler mit einem immunsuppressiv wirksamen Reiz konditioniert, der dann für die Immunsuppression ein konditionierter Reiz wird. In der eigentlichen Testphase der Wirkung der Konditionierung wird der CS für Immunsuppression zusammen mit dem UCS für eine Immunreaktion dargeboten. Die Ergebnisse bestehen entsprechend in einer Verminderung der Immunreaktion, nicht in ihrer totalen Unterdrückung.

Wie bei den Untersuchungen zur Geschmacksaversion ist auch hierbei ein wichtiger Sachverhalt, daß nicht alle Reize gleichermaßen als CS tauglich sind, d.h. nicht jeder CS mit dem UCS zu assoziieren ist. Von Garcia und Koelling (1966) wurde dieser Sachverhalt als „cue to consequence" bezeichnet. Eine Geschmacksaversion war entsprechend nicht auf akustische oder visuelle Reize konditionierbar, sondern nur

auf Geschmacksreize (s.o.). Auch bei der konditionierten Immunsuppression scheinen „Reize, die proximale Rezeptoren im Vergleich zu distalen Rezeptoren beeinflussen, den Vorgang der Konditionierung zu fördern" (Ader, 1981).

Die Ergebnisse zur konditionierten Immunsuppression sind in einer weiteren Hinsicht wichtig, sie legen neben den Untersuchungen zur Beeinflussung von Immunreaktionen durch „Streß", Untersuchungen, die oft dem Paradigma des gleichzeitigen (concurrent) Konditionierens (s.u.) entsprechen, den Einfluß zentralnervöser Vorgänge auf das Abwehrsystem nahe. Es ist dies ein weiterer Hinweis dafür, daß Verhalten den ganzen Organismus betrifft und beeinflußt, und der ganze Organismus von der Beziehung zu seiner Umgebung betroffen ist.

7.3.2 Gleichzeitiges Konditionieren autonomer Reaktionen

Beim gleichzeitigen autonomen Konditionieren (concurrent autonomic conditioning) wird ein willkürliches Verhalten durch Konditionieren beeinflußt. Mit diesem Vorgang gehen erfaßbare physiologische Veränderungen einher, die in einer systematischen Beziehung dazu stehen.

Betrachtet man das experimentelle Vorgehen in Untersuchungen, die sich am Prinzip des gleichzeitigen Konditionierens orientieren, so zeigt sich, daß die Art der Ergebnisse sehr stark vom methodischen Vorgehen beeinflußt wird. Mit Hilfe des „yoked control designs", bei dem die Erfahrung, die ein Tier in der experimentellen Situation macht (Anzahl, Stärke usw. von Schocks beispielsweise), von der Reaktion eines anderen Tieres abhängt und nicht vom Verhalten des „yoked" Tieres, läßt sich ein Verhalten erreichen, das als depressiv interpretiert wurde (Seligman, 1975). Es kam zu einer Ulzeration der Magenschleimhaut (Weiss, 1972) oder zu erheblichen kardiovaskulären Effekten, zum Teil bis zum Tod (Corley, Mauk und Shiel, 1975). Das, was in der entsprechenden experimentellen Situation gemessen wird, hängt vom Interesse des Untersuchers ab, entsprechend auch die Ergebnisse und deren Interpretation. In derartigen Untersuchungen zeigt sich, daß ein Organismus als ganzer auf eine entsprechende Situation reagiert, also „gleichzeitig" mit einer Fülle von Verhaltenskomponenten.

Es existiert eine Fülle von Untersuchungen, in denen mit bestimmten Versuchsplänen z.T. erhebliche Effekte auf organische Abläufe erzielt werden. Hierzu gehört die konditionierte emotionale Reaktion (conditioned emotional response, CER) (Estes und Skinner, 1941), die einen erheblichen Einfluß auf physiologische Funktionen und hormonelle Prozesse hat (Brady, 1975). Im Kapitel 8 wird eine Reihe derartiger Untersuchungen ausführlich dargestellt.

In vielen dieser Untersuchungen werden Individuen, meist Tiere, einem bestimmten Paradigma ausgesetzt, zum Teil auch in längerfristigen Versuchsplänen, und es wird festgestellt, welche Wirkung diese

Situation auf den Organismus hat. Es stellt sich in diesem Zusammenhang die Frage, inwieweit in der realen Lebenssituation, wenn vergleichbare Bedingungen längerfristig wirksam sind, habituelle Verhaltensmuster entstehen, die ihrerseits unter Umständen schädlich auf den Organismus wirken. Ein Beispiel hierfür wäre das sog. „coronary prone behavior" (Typ-A-Verhalten), das durch eine Reihe von habituellen Merkmalen geprägt ist. In einer Vielzahl von Untersuchungen konnte gezeigt werden, daß Personen, die bestimmte Merkmale (Feindseligkeit, Arbeit unter Zeitdruck und eine kompetitive Einstellung) dieses Verhaltensmusters aufweisen, ein höheres Risiko für die Entwicklung einer koronaren Herzerkrankung haben (z.B. Rosenman und Chesney, 1981).

Die bisherigen Ergebnisse scheinen dafür zu sprechen, „daß Typ-A-Personen keine konstitutionelle Hyperreaktivität besitzen, sondern verstärkte adrenerge Reaktionen aufgrund ihrer Wahrnehmung der meisten sozialen Stressoren (milieu stressors) als herausragende Herausforderungen" (Rosenman, 1983), d.h. aufgrund der Bedeutung, die bestimmte Situationen für sie besitzen und auf die sie entsprechend reagieren. Versuche, meist mit Hilfe von Techniken der Entspannung, dies Verhaltensmuster zu ändern, zeigten, daß die verstärkten adrenergen Reaktionen durch derartige Techniken reduziert werden können (Benson, 1983). Darüber hinaus ist erfolgreich versucht worden, die Ausprägung der oben genannten Merkmale des Typ-A-Verhaltensmusters zu reduzieren und damit eine Reduzierung des Risikos für eine koronare Herzerkrankung zu erreichen (Rosenman und Friedman, 1977). Es ist jedoch darauf hingewiesen worden, daß es schwierig ist, einem Patienten klarzumachen, eine Art des Verhaltens aufzugeben, das „ein integraler Faktor in moderner Berufskarriere ist, d.h. als eine Stärke wahrgenommen wird, die belohnt wird und nicht mit psychologischem Mißbefinden (distreß) einhergeht" (Rosenman und Chesney, 1981).

Es läßt sich nun denken, daß das Typ-A-Verhalten das Resultat früherer Erfahrungen darstellt, indem diese Art von Verhalten belohnt, d.h. operant verstärkt wurde. Im Sinne der oben angestellten Betrachtungsweise des Lernens, in der Lernen in der Beziehung eines Individuums zu seiner Umgebung gesehen wird, in der die Bedeutung der Umgebung für das Individuum von dessen Bedürfnissen und Erfahrungen, seiner Lerngeschichte abhängt, ist es ganz wesentlich, daß die Personen, von denen hier die Rede ist, ihre Umgebung auf eine für sie typische Weise interpretieren. Diese Interpretation ist keineswegs unabhängig von der Art der Umgebung, bzw. den Kontingenzen, Beziehungen, die in der Umgebung vorhanden sind und waren und die die Erfahrung bedingt haben. So zeigt sich, daß andere Personen in denselben experimentellen Situationen, in denen versucht wird, das spezifische Erleben von Typ-A-Personen zu erfassen, ganz anders darauf reagieren, sie offensichtlich nicht oder weniger als Herausforderung erleben.

7.3.3 Instrumentelles autonomes Konditionieren

Unter instrumentellem autonomem Konditionieren versteht man den Vorgang, daß eine autonome Reaktion durch die Konsequenzen ihres Auftretens oder Unterbleibens oder ihrer Veränderung kontrolliert wird. Aufgrund der größeren Komplexität der Lernsituation wurde das operante oder instrumentelle Lernen stets als höhere Form des Lernens angesehen. So wurde angenommen, daß autonome Funktionen als „niedere" Prozesse lediglich durch Vorgänge des klassischen Konditionierens beeinflußbar seien. Diese Annahme folgte der funktionalen Aufteilung des Nervensystems in einen niederen viszeralen oder autonomen und in einen höheren zerebrospinalen Teil.

Man glaubte daher, daß operantes Lernen lediglich willkürliches Verhalten beeinflussen könne, klassisches Konditionieren dagegen vegetative oder viszerale Funktionen (vgl. Kimmel, 1974). N.E. Miller bezweifelte jedoch die Notwendigkeit, zwei verschiedene Arten des Lernens anzunehmen. „Alles, was für den vorliegenden Zweck nötig ist, ist eine Theorie der Verstärkung im weitesten Sinne des Wortes. Man könnte ... wie Skinner (1938) und Mowrer (1947) annehmen, es gäbe zwei Arten der Verstärkung: Triebreduktion für somatische und Kontiguität für autonome Reaktionen (responses). ... Alle Gewohnheiten scheinen exakt denselben Gesetzen zu gehorchen, d.h. dem Gradienten der Verstärkung, der Generalisierung, Löschung und Spontanremission usw. ... Dies scheint anwendbar zu sein auf Gewohnheiten (habits), die instrumentelle Reaktionen beinhalten, unter der Kontrolle des somatischen Nervensystems, als auch solche, die emotionale oder viszerale Reaktionen, unter der Kontrolle des autonomen Nervensystems, beinhalten" (Dollard und Miller, 1950).

Um zu zeigen, daß diese Annahme zutreffend ist, mußte der Nachweis erbracht werden, daß autonome Reaktionen, wie etwa die Herzfrequenz oder der Blutdruck, direkt operant zu beeinflussen sind. Bei einem derartigen Nachweis ist es wesentlich, daß der

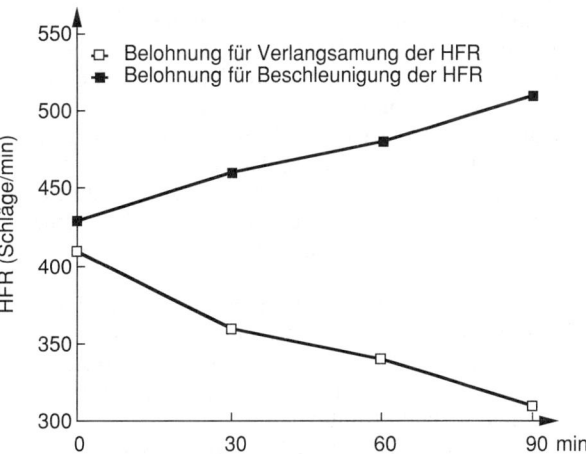

Abb. 7–18. Herzfrequenzveränderungen von kurarisierten Ratten, die für Anstiege oder Verringerungen der Herzfrequenz mit Elektrostimulation belohnt worden waren (aus Di-Cara, 1971).

vermittelnde Einfluß willkürlicher Reaktionen ausgeschlossen ist. Untersucht man etwa die Modifizierbarkeit der Herzfrequenz durch operantes Konditionieren, so ist es notwendig, bekannte willkürliche Beeinflussungsmöglichkeiten wie z. B. die Atmung auszuschließen. Ohne diesen Ausschluß könnte lediglich eine Willkürreaktion beeinflußt werden und die damit autonome Reaktion nur mittelbar, wie dies beim „gleichzeitigen Konditionieren autonomer Funktionen" (s. o.) der Fall ist. Aus diesem Grunde wurden in den zahlreichen Untersuchungen von Miller und DiCara die Versuchstiere kurarisiert.

DiCara und Miller (1968) untersuchten beispielsweise die Möglichkeit, bei kurarisierten Ratten den systolischen Blutdruck zu beeinflussen. Die eine Hälfte der Tiere wurde durch Vermeidung eines milden elektrischen Schocks dafür belohnt, den Blutdruck zu senken, die andere Hälfte dafür, ihn zu erhöhen. Um den absoluten Einfluß der Schocks zu kontrollieren, wurden jeweils zwei Kontrollgruppen gebildet, die in einem „yoked control design" identisch wie die Versuchstiere behandelt wurden, mit dem einzigen Unterschied, daß die Schocks in keinerlei Beziehung zum Verhalten des Blutdrucks standen.

Die Ergebnisse zeigten erwartungsgemäß einen Einfluß der operanten Prozedur auf den Blutdruck, interessanterweise ohne daß eine Korrelation (0.08) zu Veränderungen der Herzfrequenz bestanden hätte. Diese Art von Arbeiten war hauptsächlich theoretisch und methodisch motiviert gewesen. In den Labors von Miller und DiCara zeigten sich recht stabile Effekte in Replikationsstudien. In anderen Laboratorien ließ sich auch stets ein derartiger Effekt nachweisen, nur war er meist sehr viel geringer. Es hat ausgiebige Diskussionen über den Einfluß der Technik der Kurarisierung gegeben, wobei u. a. angenommen wurde, daß die Art der künstlichen Beatmung der Tiere ihrerseits unter Umständen einen Einfluß auf die Ergebnisse von DiCara gehabt haben könnte. Es scheint jedoch festzustehen, daß autonome Reaktionen direkt operant beeinflußbar sind, auch wenn über das Ausmaß dieses Effekts Uneinigkeit bestehen mag.

Die Übertragung dieser fast ausschließlich aus dem tierexperimentellen Bereich stammenden Befunde auf den Menschen ist sicher problematisch. Hier muß stets an die Möglichkeit einer kognitiven Vermittlung gedacht werden. Katkin und Murray (1968) schlagen daher vor, bei Anwendung derartiger Methoden bei Menschen nicht von operantem Lernen, sondern von Kontrollernen zu sprechen. Brener (1974) schlägt den Begriff der „willkürlichen Kontrolle" (voluntary control) vor.

Für den vorliegenden Zusammenhang sind derartige Ergebnisse von einiger Bedeutung. Sie eröffnen die konzeptuelle Möglichkeit, körperliche Prozesse im Sinne einer Ersatzhandlung zu verstehen, vor allem in einem sozialen Kontext.

Cahoon und Turner (1972) führen als einfaches Beispiel für die direkte operante Kontrolle autonomer Funktionen das psychogene Fainting an. Es trete auf, damit sozial unangenehme Situationen vermieden werden können. Die Vermeidung der Situation

würde dabei die dem Fainting zugrundeliegenden physiologischen Prozesse verstärken, wodurch die Auftretenswahrscheinlichkeit des Faintings erhöht würde. Auf vergleichbare Weise lassen sich zahlreiche Möglichkeiten denken, in denen etwa die Reaktionen von Bezugspersonen eher dazu führen, daß Funktionsänderungen stabilisiert werden, anstatt, wie meist intendiert, zu verschwinden.

7.3.4 Interozeptives Konditionieren

Razran (1961) zeigte anhand einer Fülle sowjetischer Studien, daß interozeptive Stimuli, ohne daß sie bewußt wahrgenommen werden (conditioning without awareness), als konditionierte Reize dienen können. Eine Vielzahl der Studien benutzte Magensonden, die mit verschieden temperiertem Wasser gefüllt werden konnten. Dabei konnten dann Temperaturunterschiede, die von den Versuchspersonen nicht bewußt unterschieden werden konnten, als konditionierter Reiz benutzt werden. Dabei ist es unwesentlich, daß der Vorgang des Konditionierens länger dauert, also die Akquisitionskurve flacher verläuft.

Studien zum Phänomen der Wahrnehmungsabwehr belegen deutlich, daß vor einer bewußten Wahrnehmung Prozesse der Informationsverarbeitung ablaufen, die die notwendige Darbietungszeit, die zu einer bewußten Wahrnehmung des Reizes führt, beeinflussen (vgl. zusammenfassend Erdelyi, 1974). Ebenso zeigt der sog. „Subception"-Effekt, der darin besteht, daß es zu Reaktionen im Hautwiderstand kommt, bevor ein aversiver Reiz bewußt wahrgenommen wird, daß nichtbewußte Wahrnehmungen einen Einfluß, in diesem Falle auf eine physiologische Reaktion, haben.

Wichtig sind in diesem Zusammenhang Untersuchungen, die von Hefferline (1973) zusammenfassend referiert werden. In einer von ihm 1950 durchgeführten Untersuchung trainierte er eine Albinoratte, durch das Niedergedrückthalten eines Hebels helles Licht zu vermeiden. Das Versuchstier hielt den Hebel bis zu 45 Minuten niedergedrückt. Danach rannte es wild im Käfig umher und drückte erneut den Hebel herunter. Nach einer Löschungsphase konnte beobachtet werden, daß das Tier, wenn es zufällig den Hebel berührte, diesen erneut herunterdrückte und festhielt. Es war unfähig, den Hebel loszulassen. In einer Folgestudie wurde die Versuchsanordnung derart geändert, daß festgestellt werden konnte, auf welche Art der Hebel gedrückt gehalten wurde. Der Weg des Hebels war genau im Hinblick auf kleinste Bewegungen des Loslassens registrierbar. So konnte festgestellt werden, daß das Versuchstier auch nach der Löschungsphase den Hebel für kleine Bewegungen losließ, dann aber wieder niederdrückte (Winnick, 1956).

Die Hypothese, daß die propriozeptiven Reize der Bewegung des Loslassens ihrerseits als Reiz für eine emotionale Reaktion dienten, wurde in einem weiteren Experiment, in dem die Atmung registriert wurde, überprüft. Es zeigte sich, daß die Atemfrequenz während des Loslassens, vor allem kurz vor dem erneuten Drücken des Hebels, stark anstieg (Eldridge, 1954). Daß derartige „subliminale Körperwahrnehmungen" auch eine diskriminative Funktion haben können, zeigt eine Untersuchung von Hefferline und Perera (1963). Sie registrierten elektromyographisch kleinste Bewegungen eines Daumens und ließen diese so schnell wie möglich von einem Ton folgen. Die Probanden hatten die Aufgabe, so schnell wie möglich nach jedem Ton mit dem Zeigefinger auf einen Knopf zu drücken. War dieser Zusammenhang gut gelernt, wurde die Intensität des Tons langsam reduziert, bis kein Ton mehr präsentiert wurde. Die Probanden berichteten, daß die Töne immer schwerer zu entdecken waren, aber nicht, daß sie keinen Ton mehr hörten. Darüber hinaus zeigte es sich, daß 72% der Veränderungen im Elektromyogramm des Daumens vom Drücken auf den Knopf gefolgt wurden. Die Probanden gaben an, daß sie auf den Knopf drückten, weil sie den Ton gehört hätten.

Für die Frage des dabei stattfindenden Wahrnehmungsvorgangs, obwohl es sich um einen solchen nicht handelt, zumindest im Hinblick auf die nur vermeintlich wahrgenommenen Töne, ist es notwendig einen Faktor einzuführen, der in klassischen psychophysischen Wahrnehmungsversuchen als „Raten" bezeichnet wurde. Im Rahmen der „signal detection theory" wird dieser Faktor als „response bias" bezeichnet. Darunter ist ein „verfälschender" Einfluß, der unabhängig von der Signalintensität ist, zu verstehen.

Eine Untersuchung von Davis (1952) zeigt deutlich einen solchen Einfluß. Er präsentierte Versuchspersonen zwei aufeinanderfolgende Töne vollständig gleicher Qualität und Intensität. Die Probanden hatten die Aufgabe, mit der einen Hand auf einen Knopf zu drücken, wenn der erste Ton als lauter empfunden wurde, mit der anderen Hand, wenn der zweite als lauter empfunden wurde. Tatsächlich waren ja beide Töne gleichlaut. Eine Entscheidung muß in diesem Falle von anderen als von Wahrnehmungsfaktoren abhängen.

Die Ergebnisse zeigen, daß jeweils mit der Hand der Knopf gedrückt wurde, deren Muskelaktionspotentiale kurz vor dem zweiten Ton höher waren. Damit zeigt sich deutlich ein Einfluß eines Faktors, der vollständig unabhängig vom wahrgenommenen Signal war. Hefferline (1973) nimmt in diesem Zusammenhang Bezug auf das Modell der „signal detection theory" und interpretiert das eben geschilderte Ergebnis als „response bias".

Die eben dargestellten Untersuchungen zeigen, daß Körpervorgänge als diskriminative Reize dienen können, ohne daß sie bewußt wahrgenommen werden. Sie können so Verhalten steuern und beeinflussen. In neueren Studien aus dem Bereich der Biofeedback-Forschung (Katkin et al., 1981; Clemens, 1979; Ashton et al., 1979) wird ebenfalls deutlich, daß die Wahrnehmung von Körpervorgängen erlernt werden kann bzw. durch Biofeedback verbessert wird (Brener und Jones, 1974). Hierbei handelt es sich jedoch auch um bewußte Wahrnehmungen meistens der Herzfrequenz, obwohl in einigen der Arbeiten deutlich wird, daß wiederum ein „Rate"-Faktor eine wesentliche Rolle spielt. In diesen Untersuchungen geht es hauptsächlich darum, zu ermitteln, inwieweit eine bewußte Wahrnehmung der Herzfrequenz möglich ist. Es wird im allgemeinen dabei so vorgegangen, daß Probanden entscheiden müssen, ob eine rhythmische Abfolge von Tönen oder sonstigen Reizen mit den eigenen Herzaktionen synchron verläuft (McFarland, 1975; Brener, 1977). Ein anderer methodischer Ansatz erfordert vom Probanden, zwischen zwei rhythmischen Abfolgen von Tönen zu unterscheiden und zu entscheiden, welche der eigenen Herzfrequenz entspricht. Dabei wurden von verschiedenen Autoren methodische Besonderheiten eingeführt, um sicherzugehen, daß die Probanden, etwa durch Atemmanöver, die eigene Herzfrequenz nicht beeinflussen und damit Entscheidungshilfen haben. Darüber hinaus haben diese Untersuchungen, wenigstens zu einem wesentlichen Teil, gezeigt, daß eine Wahrnehmung der Herzfrequenz dieser Art durch Biofeedback insofern beeinflußt werden kann, als die Trefferrate erhöht werden kann (Katkin et al., 1981).

7.4 Therapeutische Anwendungen

Über Anwendungsmöglichkeiten lernpsychologischer Prinzipien in der Psychosomatischen Medizin informiert Kapitel 19. Dort werden sehr ausführlich die einzelnen Strategien sowie deren Anwendung bei spezifischen Krankheitsbildern berichtet. Insofern muß an dieser Stelle nicht weiter hierauf eingegangen werden. Hier soll lediglich auf einige spezielle Besonderheiten der Anwendung von Verhaltenstherapie hingewiesen werden.

Die Verhaltenstherapie hatte ihren Ausgangspunkt in der Behandlung der Angst durch Wolpe (1958). Er entwickelte ein Verfahren, das später die Bezeichnung „systematische Desensibilisierung" erhielt. Er ging davon aus, es müsse ein der Angst entgegengesetzter Zustand hergestellt werden und der Patient, wenn er sich in diesem Zustand befindet, mit mäßigen angstauslösenden Reizen konfrontiert werden. Wolpe (1958) ging davon aus, daß Angst ein über das sympathische Nervensystem vermittelter Zustand sei, so daß der damit inkompatible Zustand parasympathisch vermittelt sein müsse. So kam er zur Anwendung der „progressiven Relaxation" nach Jacobson (1938).

Wichtig im vorliegenden Zusammenhang ist dabei, daß Angst als „autonomes Reaktionsmuster, das charakteristischerweise Teil der Reaktion (response) eines Organismus auf schädliche Reize ist" (Wolpe, 1958), angesehen wird. Wird Angst so definiert, wird die Antwort eines Organismus auf seine, in diesem Falle schädliche Umgebung „ganzheitlich" betrachtet. Die auf dieser Basis entwickelte Therapie ist in einem gewissen Sinne eine „psychosomatische" Therapie, da sie körperliche Reaktionsanteile als wesentlich miteinbezieht. Entsprechend werden gerade auch im Bereich sog. „psychosomatischer" Störungen häufig Methoden der Entspannung angewandt.

Die im Kapitel 19 „Verhaltenstherapie" dargestellten Interventionsformen entsprechen ebenfalls weitgehend diesem „ganzheitlichen" Ansatz, auch wenn kognitive Therapiekomponenten betont werden. Für Ansätze, die auf dem Prinzip des operanten Konditionierens autonomer Reaktionen beruhen, wie die Me-

thoden des Biofeedback, gilt dies in besonderem Maße. Aber auch wenn es um die Änderungen von Einstellungen geht, werden diese, wie oben am Beispiel des Typ-A-Verhaltens verdeutlicht wurde, konzeptuell als intervenierende Variable gesehen, als Teil des Verhaltens und gleichzeitig als Verhaltensbedingung, indem Verhalten nicht als ein lineares Geschehen, sondern als dynamischer Prozeß gesehen wird.

8 Psychophysiologie

Othmar W. Schonecke und *Jörg Michael Herrmann*

8.1 Einleitung

Im vorliegenden Kapitel wird der Ansatz der Psychophysiologie in seiner Bedeutung für die Psychosomatische Medizin, bzw. Medizin ganz allgemein, dargestellt. In der Einleitung wird in Anlehnung an das medizintheoretische Modell von v. Uexküll Psychophysiologie als notwendige Konsequenz der in der Medizin aufgegebenen „ganzheitlichen" Betrachtungsweise herausgestellt. Der unbestreitbare und wertvolle Fortschritt in der Medizin basiert zumindest teilweise auf Vorgängen der Isolierung. Der einen Sachverhalt Erkennende ist von seinem Gegenstand, einem Teil der Natur, im Sinne einer „objektiven" Methode isoliert, obwohl er selbst Teil der Natur ist, ebenso wie der Vorgang der Erkenntnis. Weiterhin wird der natürliche Gegenstand der Erkenntnis zur wissenschaftlichen Analyse in immer kleinere Untereinheiten isoliert, was den Fortschritt der Erkenntnis einerseits fördert, aber auch zu oft dazu führt, daß das Ganze aus dem Blickfeld gerät. In der Medizin wird der Körper eines Menschen auf diese Weise zu sehr als Körper im physikalischen Sinne verstanden und weniger als Organismus, dessen Eigenschaften ohne die Betrachtung seines „In-einer-Umgebung-Leben" keinen Sinn ergeben (Schonecke, 1988). Das bedeutet nicht, daß das „isolierende" Erkennen sinnlos ist, es ist vielmehr die Voraussetzung für die Erkenntnis des Ganzen. In der Immunologie hat die molekularbiologische Forschungsmethodik einen extrem wertvollen Fortschritt der Erkenntnis erbracht, aber auch dazu geführt, daß die Vertreter dieses Fortschritts das Immunsystem als vom Organismus isoliert, als autonom betrachtet haben, trotz zunehmend vorliegender Evidenz, daß dies nicht so ist. Gleichzeitig hat dieser Fortschritt die Möglichkeit geschaffen, die Verbindung des Immunsystems mit dem endokrinen und Nervensystem zu zeigen (vgl. Kap. 11).

Die Konsequenz aus diesen Überlegungen kann nun nicht sein, nur das Ganze zu betrachten und die naturwissenschaftliche Methodik aufzugeben, sondern sie besteht darin, die Grenzen fachspezifischer Methodik zu überschreiten, unter Beibehaltung einer wissenschaftlichen oder auch Denkmethodik, wenn es ein zu erklärendes Phänomen erfordert. Psychophysiologie versucht nun gerade, dies zu tun, d.h. die Methodik der Physiologie mit der der Psychologie zu verbinden, um beispielsweise ein Phänomen wie Emotion hinreichend zu erklären. Hieraus ergeben sich besondere methodische Schwierigkeiten, die im folgenden nur angedeutet werden können, vor allem im Abschnitt über Aktivierungsmessung. Diese Schwierigkeiten ergeben sich vor allem auch aus der Tatsache, daß in der Psychophysiologie das Objekt der Erkenntnis ein in einer bestimmten Umgebung lebendes Subjekt ist, dessen „objektivierbare" Interaktionen mit einer z.B. experimentellen Umgebung untersucht werden. Die sich daraus ergebenden „Unschärfen" des Erkennens machen den umfangreichen Gebrauch statistischer Analyseverfahren notwendig.

In allen drei Abschnitten des Kapitels, Aktivierung, Streß als Sonderform von Aktivierung und schließlich Emotion, wird dieser Aspekt deutlich, wobei die im Bereich der Psychophysiologie vorliegenden Ergebnisse im Hinblick auf ihre, so verstandene, „medizintheoretische" Relevanz dargestellt werden, aber auch im Hinblick auf ihre unmittelbare Brauchbarkeit für medizinisch-inhaltliche Fragen, etwa zur Erklärung pathogenetischer Aspekte von Erkrankungen.

8.1.1 Wissenschaftshistorische Überlegungen

Sowohl „Psychophysiologie" als auch „Psychosomatik" beziehen sich entsprechend den begrifflichen Elementen, die sie in der Bezeichnung für ihren Inhalt enthalten, auf „Körperliches" und „Psychisches". Man kann der Meinung sein, Psychophysiologie habe den Stellenwert einer Grundlagendisziplin, die sich mehr auf Forschung bezieht, während Psychosomatik sich mehr auf die medizinische Anwendung richtet. Vergleichbares kommt etwa im Titel eines Aufsatzes von Fahrenberg (1979) zum Ausdruck: „Das Komplementaritätsprinzip in der psychophysiologischen Forschung und psychosomatischen Medizin". Th. v. Uexküll (1979) stellt dazu die folgenden drei Fragen und versucht, diese auf „dem Hintergrund einer wissenschaftsgeschichtlichen Überlegung" zu beantworten:

– Sind „Physiologie" und „Somatik" identisch?
– Verstehen Psychophysiologie und Psychosomatik unter „Psyche" das gleiche?
– Was verstehen schließlich beide Disziplinen unter „Beziehung" zwischen Seelischem und Körperlichem?

Von Uexküll verweist darauf, daß der Begriff „Physis" ursprünglich soviel wie Natur bedeutet hat, also bedeutend weiter gefaßt war als der Begriff „Soma", der sich nur auf „Körper" bezog. Mit dem Fortschritt der Naturwissenschaften schränkte sich die Physiologie immer mehr auf die physikalischen Prozesse im Körper ein und wurde zu einer „Körper-Physik". So sei die Psychologie zunächst Teil einer „ganzheitlichen" Physiologie als der „Lehre von den belebten Naturerscheinungen" gewesen, dann aber im Laufe der genannten Entwicklung als eigene Disziplin entstanden, jedoch „nicht ohne Verfremdung der Teilstücke", da Leben nicht zusammengesetzt sei wie Kochsalz aus Natrium und Chlor.

Der Zerfall einer ganzheitlichen Betrachtungsweise von Natur wird dann am „Werdegang" des Begriffs der „psychischen Energie" verdeutlicht, die bei Johannes Müller (1840) als „Sinnesenergie", eine für das Bewußtsein durch den Zustand eines Sinnesnerven vermittelte Einheit „äußerer" Ursachen mit der spezifischen Qualität der Sinne darstellt und so eine Empfindungsqualität bewirkt. Nach dem genannten Zerfall einer ganzheitlichen Betrachtungsweise wird Wahrnehmung das Thema einer Psycho-Physik, die versucht, Wahrnehmung als Folge eines physikalischen Vorgangs objektiv zu erfassen, d.h. mit physikalisch-quantitativen Methoden. In den Augen der Physiologie sei dieser Versuch gescheitert. Psychische Energie als Begriff der Psychoanalyse verweist auf Triebenergie und gelte als „Markenzeichen" einer unwissenschaftlichen Einstellung.

So sei auch Pawlow auf dem Standpunkt eines „äußeren Beobachters" bei seinen „psychischen Versuchen" geblieben. „Sowohl die Methoden und die Verhältnisse unseres Experimentierens, als auch die Planung der einzelnen Aufgaben, die Bearbeitung des Materials und schließlich seine Systematisierung, alles das bleibt im Bereich der Tatsachen, der Begriffe und der Terminologie der Physiologie des Nervensystems." So wird ein äußerer von einem inneren, subjektiven Anteil beim Phänomen des Lernens abgetrennt. Lernen wird als Teil einer höheren Nerventätigkeit angesehen, die die Verbindung des Organismus zur Außenwelt herstellt (vgl. Kap. 7).

W. B. Cannons Theorie der Emotion könne dagegen als „psychosomatisches" Modell angesehen werden, da durch die Einbeziehung von Gefühlsqualitäten eine subjektive Wirklichkeit Teil dieses Modells wird. Dies gilt ebenso für das Modell der Bereitstellung (emergency states), in dem der Organismus in der Auseinandersetzung mit der Umgebung gesehen wird, als Gesamtorganismus auf eine Gesamtsituation antwortend. Insofern als das Verhalten als Reaktion im Hinblick auf die Umgebung gesehen wird, läßt sich daraus eine Ordnung im Sinne eines gemeinsamen „teleonormen Nenners" aufzeigen. Emotionen sind daher keine das Verhalten störenden Prozesse, sondern „organisierte Reaktionsmuster, deren Programme man erforschen kann, und die Vorbereitungen für Handlungen beziehungsweise Bereitstellungen zuwege bringen, in denen jeweils eine bestimmte Umweltsituation vorweggenommen wird" (v. Uexküll, 1979). Wichtig dabei sei die Subjektivität der in der Emotion resultierenden Interpretation der Umgebung.

Das Dilemma der unvermittelten Wissenschaftssysteme der Physiologie und Psychologie, das darin besteht, daß keines der Systeme Sachverhalte des jeweiligen anderen Bereichs wissenschaftlich zu Begriff bringen kann, läßt sich nach Th. v. Uexküll dadurch lösen, daß man ein „Modell für die Einheit eines Systems konstruiert, in dem physische und psychische Elemente im Rahmen von Aufgaben miteinander in Beziehung stehen, die sie für einander und für das System erfüllen" (v. Uexküll, 1979). Ein solches Modell enthielte Begriffe, die es gestatten würden, physiologische und psychologische Begriffe gleichsam zu übersetzen (vgl. Schonecke, 1972).

8.1.2 Der Situationskreis und das Leib-Seele-Problem

Th. v. Uexküll zeigt anschließend auf, wie das Modell des Situationskreises als systemtheoretisches Konzept die oben genannten Schwierigkeiten überwinden könnte. Wesentlich für eine systemtheoretische Betrachtungsweise ist, daß die jeweilige Ebene der Betrachtung relativ bleibt, d.h. als ein Element einer Analyse einer höheren Ebene dient, insofern ist es eine hierarchische Betrachtungsweise. So läßt sich ein Körper als ein geschlossenes System denken, solange sein Gleichgewicht autonom durch innere Prozesse aufrechterhalten werden kann. Ist dies nicht mehr der Fall, d.h. werden äußere Quellen benötigt, so entsteht ein Bedürfnis, indem die Umgebung im Hinblick auf dieses Bedürfnis bedeutungsvoll wird, – das System ist nicht mehr geschlossen. Durch das Bedürfnis wird die Umgebung als individuelle Wirklichkeit in das Gesamtsystem mit einbezogen. Dieser Sachverhalt impliziert einen Bedeutungssprung und kann nicht als ein Kontinuum angesehen werden, da Programme für innerkörperliche Vorgänge des zunächst geschlossenen Systems in Beziehung gesetzt werden mit Programmen, die „Umgebung in individuelle Wirklichkeit transponieren". Damit ist für die Betrachtung „Psyche" als Systemanteil integriert.

Konditionieren sei bei dieser Betrachtungsweise aufzufassen als Entstehen von Programmen, die eine Bedeutungskoppelung zweier Integrationsebenen herstellen, der physiologischen und der psychologischen Integrationsebene. Derartige Programme der Bedeutungskoppelung, abhängig von den Erfahrungen eines Organismus, stellen den Organismus zu einem gegebenen Zeitpunkt neben einem kausalen in einen „historischen" Zusammenhang, da die individuelle Wirklichkeit u.a. von diesen Erfahrungen abhängig ist. Aus diesem Grunde ist für die psychophysiologische Forschung zu erwarten, daß psychophysische Reaktionen eine starke individuelle Spezifität aufweisen.

Programm bedeutet dabei stets die Integration verschiedener Einzelelemente zu einer „Gestalt", die dabei jedoch nicht als ein „nur subjektives Phänomen der Wahrnehmung" aufgefaßt wird. In einer Sinnesempfindung sind Nachrichten über die Umgebung und den Zustand verschiedener körperlicher Zustände und Prozesse zu einem „einheitlichen Phänomen verschmolzen".

Programme lassen sich einerseits unterscheiden nach ihrer Dringlichkeit, womit der Intensitätsaspekt angesprochen wird, zum anderen nach „Offenheit" und „Geschlossenheit", womit die Flexibilität möglicher Änderungen des Ablaufs bezeichnet wird. Geschlossene Programme lassen sich durch Erfahrung nicht modifizieren, sie laufen, einmal ausgelöst, starr ab. Lediglich ihre Einbeziehung in verschiedene andere, dann komplexere Programme ist änderbar, nicht ihr Ablauf. Offene Programme dagegen können durch Erfahrung, durch Lernen geändert werden, es sind lernfähige Programme. In der menschlichen Entwicklung etwa findet eine Zunahme offener Programme statt, ein Sachverhalt, der von der Psychoanalyse mit dem Begriff des „Sekundärprozesses" gekennzeichnet wird.

Geht man davon aus, daß Programme strukturell im Gehirn gespeichert sind, wobei deutlich wird, daß das Speichersubstrat nicht mit dem Gespeicherten identisch ist, so ist es auch notwendig anzunehmen, daß dieses Substrat durch Erfahrung geändert wird, d.h., Vorgänge auf der psychischen Integrationsebene ändern und beeinflussen die physiologische. Somit sind beide Integrationsebenen von einander abhängig. Erfahrungen ändern im weitesten Sinne auch das ZNS, dieses beeinflußt die später geänderte Interpretation der Umgebung.

Für die Psychosomatische Medizin sei es wesentlich, „den kranken Menschen unter dem Aspekt eines offenen Systems" zu betrachten und nicht, wie die klassische Medizin, als geschlossenes System, im Sinne einer „komplizierten anatomisch-biochemischen Maschinerie". Krankheit, auf diese Weise betrachtet, ist ein „persönliches Schicksal" in einem historischen Zusammenhang der individuellen Lebensgeschichte, bedingt auch durch die Art der Verknüpfung beider Integrationsebenen. Die Auswirkungen derartiger Verknüpfungen sowie der Einfluß der Umgebungsbedingungen auf die Verknüpfung stellen den Gegenstandsbereich der Psychophysiologie dar, die mit ihren Methoden versucht, dem die Bereiche bzw. Integrationsebenen übergreifenden Aspekt gerecht zu werden.

Betrachtet man Psychophysiologie als eine Forschungsrichtung bzw. als einen Inhaltsbereich, dessen Gegenstand sich auf das Zusammenwirken von Prozessen physiologischer, behavioraler und erlebnismäßiger Art bezieht, so ist es zumindest naheliegend, dieses Zusammenwirken als wesentlich für die Entstehung von Krankheiten oder ihre Verarbeitung anzunehmen. Dies gilt um so mehr, wenn davon ausgegangen wird, wie dies im Bereich der Psychosomatischen Medizin getan wird, daß für Krankheiten, wenn auch in unterschiedlichem Maße, psychologisch beschreibbare Bedingungen eine wesentliche Rolle spielen. „Psychosomatische Medizin muß versuchen zu erklären, wie psychosoziale Reize übersetzt werden in akute oder chronische Veränderungen von Struktur und physiologischen und biochemischen Funktionen. ... Um es noch bündiger zu formulieren, wir verstehen einfach nicht, wie unmaterielle, symbolische Ereignisse – wie psychologische Reaktionen

auf Lebenserfahrungen und Lebensereignisse – „übersetzt" werden in materielle Änderungen – wie die Ausschüttung von Hypophysenhormonen, anhaltende Blutdruckerhöhungen, Veränderungen von Immunprozessen, autonom neurale Entladungen oder die Induktion von Enzymen oder Viren" (Weiner, 1977).

Unterschiedliche Konzepte einer „Psychogenese" sind kaum geeignet, Fragen dieser Art zu klären, zumal sie meist vage und mißverständlich verwendet werden. Sie beschreiben nicht den von Weiner angesprochenen Prozeß, sondern konstatieren höchstens dessen Resultat. In der als psychophysiologisch im weitesten Sinne zu bezeichnenden Forschung wird versucht, eben jene Prozesse und ihre Bedingungen zu erfassen und zu klären, von denen in der Psychosomatischen Medizin angenommen wird, daß sie für die Entstehung von Krankheiten wesentlich sind.

Im Bereich der Psychophysiologie sind mehrere Ansätze zu nennen, mit denen versucht wird, diese Probleme einer Lösung näher zu bringen. So wird versucht, Prozesse einer möglicherweise speziellen Reaktivität bei Patienten mit verschiedenen Störungen zu untersuchen. Es gibt für die hier interessierenden Krankheitsbilder inzwischen zahlreiche Untersuchungen, die in den jeweiligen Kapiteln dargestellt werden. Im Abschnitt „Emotion und Verhalten" sind Arbeiten dargestellt (z.B. Weiss zur Pathogenese des Ulcus ventriculi), in denen versucht wird, bestimmte Merkmale belastender Lebensbedingungen experimentell herzustellen, in der Annahme einer pathogenen Relevanz dieser Merkmale. In mehreren dieser Untersuchungen hatte sich beispielsweise gezeigt, daß die Dauer der Belastung eine wesentliche Rolle spielt (Brady, 1975), oder die Schwierigkeit, eine Reaktion zu erlernen, um aversive Ereignisse zu vermeiden bei der Anwendung sog. „Sidman Avoidance Schedules" (Forsyth und Harris, 1970). Diesem Ansatz, wie den im folgenden kurz darzustellenden, am Begriff Streß orientierten Ansätzen, ist die Annahme gemeinsam, daß bestimmte oder weniger spezifische Bedingungen zu Reaktionen des Organismus führen, die langfristig schädigend wirken und Krankheiten verursachen können. Dabei sind die eher lernpsychologisch orientierten Ansätze mehr an spezifischen Paradigmen interessiert, mit denen schädigende Reaktionen produziert werden können. Ebenso ist die Erforschung der belastenden Lebensereignisse zunächst an diesen Ereignissen interessiert gewesen, also an der belastenden Bedingung, und erst später an der Art der Prozesse, die die Wirkung vermitteln. Gemeinsam ist diesen Ansätzen jedoch, daß sie nicht davon ausgehen, daß spezifische Bedingungen zu spezifischen Erkrankungen führen. Mit dem Paradigma des „Yoked Control Designs" lassen sich ulzerative Veränderungen der Magenschleimhaut produzieren, aber auch kardiovaskuläre Veränderungen, die sogar zum Tode führen können.

Die im folgenden dargestellten psychophysiologischen Modelle und Forschungsergebnisse stellen lediglich einen exemplarischen Ausschnitt dar, von dem angenommen wird, daß er für die Psychosomatische Medizin Relevanz besitzt. Diese Relevanz muß nicht unbedingt darin bestehen, daß Ergebnisse oder Konzepte auf vermutete pathogene Prozesse oder therapeutische Anwendungsmöglichkeiten bezogen werden können, sondern kann auch dadurch gegeben sein, daß eher theoretische Fragen, wie sie in der Einleitung als sehr grundsätzliche für die Psychosomatik dargestellt wurden, einer Klärung näher gebracht werden. Dies gilt etwa für die Ergebnisse zum Problem der Spezifität, nicht nur im Sinne der Auffindung möglicher spezifischer pathogener Prozesse, sondern ebenso im Sinne der Integration verschiedener Ebenen eines Organismus in der Beziehung zu seiner Umgebung. Der Schwierigkeit derartiger Probleme entspricht die Komplexität und Aufwendigkeit des methodischen Vorgehens beim Versuch, einer Lösung dieser Probleme näher zu kommen. Die Bedeutung, die die Psychophysiologie für die Psychosomatische Medizin, nicht nur im Sinne einer Grundlagenforschung besitzt, rechtfertigt es jedoch in hohem Maße, diese Schwierigkeiten auf sich zu nehmen.

8.2 Aktivierung

Für den vorliegenden Zusammenhang ist das Konzept der Aktivierung grundlegend wichtig, da es diejenigen Aspekte psychischer Prozesse bezeichnet, die über die rein psychologische Betrachtungsweise hinausgehen und körperliche Prozesse miteinbeziehen. Man kann emotionale Phänomene unter rein psychologischen Gesichtspunkten betrachten, und im weiteren Verlauf dieses Kapitels werden derartige Modelle auch vorgestellt, um dem Gesamtphänomen von Emotion erklärend gerecht zu werden, muß die Grenze psychologischer Betrachtungsweisen jedoch überschritten werden, da das Phänomen der Emotion auch die Erlebnisebene körperlicher Vorgänge beinhaltet und von diesen bestimmt wird. Das Konzept der Aktivierung bezieht sich nun genau auf den Aspekt von Emotion, der körperliche und psychische Phänomene aufeinander bezieht, indem psychische Phänomene eine körperliche Bedeutung gewinnen oder umgekehrt. Dies ist beispielsweise dann der Fall, wenn ein Außenreiz ein Gefahrensignal darstellt, das Gefühle der Angst auslöst und gleichzeitig den Körper aktiviert, damit der Organismus sich durch Flucht dem Reiz entziehen kann.

Diese Betrachtungsweise beinhaltet, daß psychologische Modelle um physiologische Modelle erweitert werden und umgekehrt. Darin liegt eine besondere Schwierigkeit, da die jeweiligen Modelle für sich in einer physiologischen bzw. psychologischen Wissenschaftssprache beschrieben werden und nun mit einer anderen Sprache in Berührung kommen, wobei es nicht möglich ist, eine Sprache auf die andere zu reduzieren oder in die andere zu übersetzen. Es gibt jedoch auch keine Metasprache, die beide Wissenssysteme integriert beschreiben würde. In der Psychophysiologie führt dies neben dem Aufwand an Meßmethodik aus zwei Wissensbereichen auch zu einem Aufwand an methodologischen und statistischen

Konzepten, mit denen beide Bereiche aufeinander bezogen werden. Dies wird besonders an dem grundlegenden Konzept der Aktivierung deutlich, mit dem sich das „Grenzüberschreitende" der Psychophysiologie verbindet.

„Psychophysische Aktivierungsprozesse begleiten alle menschlichen Lebensäußerungen. Es sind ebenso universelle Funktionen wie die Informationsverarbeitung und das Lernen" (Fahrenberg et al., 1979). Die Allgemeinheit des Begriffs der Aktivierung beinhaltet jedoch auch Schwierigkeiten der Abgrenzung. Von Fahrenberg (1979) wurde ebenfalls darauf hingewiesen, daß in diesem Begriff im Grunde Begriffe wie Emotion oder auch Streß mit enthalten sind, bzw. diese bei einer genaueren Betrachtung darin aufgehen. Entsprechend groß ist die Vielfalt von Modellen, die auf unterschiedliche Art versuchen, den Vorgang von Aktivierung zu beschreiben. „Aktivierung ist einer der wichtigsten, aber unschärfsten Begriffe der Psychophysiologie. Häufig ist in einem engeren Sinne mit Aktivierung (Aktivation) nur die Intensitätsdimension psychophysischer Prozesse gemeint, doch bleiben die anderen Aspekte dann implizit oder sind theoretisch vernachlässigt. Dieses Gebiet zeichnet sich durch ungewöhnlich viele theoretische Ansätze bei sehr heterogener und oft sehr einseitiger und schmaler empirischer Basis aus. Außerdem sind die Operationalisierungen und die Messung inter- und intraindividueller Unterschiede von Aktivierung beziehungsweise Aktiviertheit beim Menschen methodologisch durchweg noch unbefriedigend" (Fahrenberg et al., 1979).

Die Begriffe „Aktivierung" (Aktivation) und „Streß" werden häufig, aber nicht immer austauschbar verwendet. „Arousal" wird meist von Aktivierung abgegrenzt und einem verschiedenen neurophysiologischen Prozeß zugeordnet. Geht man von der eigentlichen Bedeutung des Wortes Aktivierung aus, so wird damit der Umstand gekennzeichnet, daß ein Organismus aktiv ist, wenn er aktiviert ist, zumindest normalerweise. Eingangs wurde darauf verwiesen, daß ein Organismus als ein geschlossenes System gedacht werden kann, solange ein Gleichgewicht durch innere Prozesse aufrechterhalten werden kann. Ist dies nicht mehr der Fall, so entsteht ein Bedürfnis, das den Organismus im Hinblick auf seine Umgebung aktiv werden läßt. Das Ausmaß der Aktivierung oder – als Zustand – Aktiviertheit ist abhängig von der Dringlichkeit oder Intensität des Bedürfnisses und der damit verbundenen Relevanz und Bedeutung einer Situation. Der Begriff Aktivierung würde demnach auf das Ausmaß hinweisen, in dem ein Organismus durch irgendwelche inneren oder äußeren Ereignisse „aus der Ruhe gebracht" wird. Aus dieser Bestimmung ergibt sich einerseits eine gewisse Nähe zum Begriff der Motivation (Bedürfnis), zum anderen aber auch die Möglichkeit, Aktivierung oder Aktiviertheit als ein eindimensionales Konzept aufzufassen, wie dies auch im Sinne einer Intensitätsdimension geschehen ist (z. B. Duffy, 1972). Danach ließe sich Aktivierung nur auf das Mehr oder Weniger des Aktivseins eines Organismus beziehen.

Dieser Auffassung widerspricht, daß anhand vorliegender Ergebnisse Aktivierung nicht als ein solcher eindimensionaler Prozeß nur im Hinblick auf Intensität aufgefaßt werden kann, sondern Spezifität in verschiedener Hinsicht beinhaltet, z.B. im Hinblick auf die Merkmale von Reizen, aber auch auf die Merkmale von Individuen. Bezieht man nun Aktivierung auf Aktivität, so beinhaltet das keineswegs, daß jede Aktivität identische psychophysische Muster aufweisen muß, Aktivierung also auch keineswegs ein eindimensionaler Prozeß ist.

Es lassen sich verschiedene Ansätze im Bereich psychophysiologischer Aktivierungsforschung unterscheiden. Eine Reihe von Modellen, die eher neurophysiologisch orientiert sind, also nicht im eigentlichen Sinne als psychophysiologisch zu bezeichnen sind, legen ihren Schwerpunkt auf die zentrale Vermittlung der Prozesse von Aktivierung oder Emotionen (z.B. Pribram und McGuinness, 1975; Pribram, 1980). Diese Ansätze werden im Kapitel 9 ausführlich dargestellt.

Andere Ansätze legen den Schwerpunkt auf die aktivierenden Bedingungen, wie die lerntheoretisch orientierten (vgl. Kap. 9) oder klassifikatorischen Ansätze (z.B. Janke, 1974). Hierzu zählen vor allem auch diejenigen Ansätze, die im Sinne der Stimulusspezifität versuchen, die spezifische Wirkung bestimmter aktivierender Bedingungen zu erfassen.

Ein weiterer Ansatz stellt Prozesse der internen Verarbeitung von aktivierenden oder belastenden Bedingungen in den Vordergrund (z.B. Lazarus et al., 1980). Hierbei wird angenommen, daß diese intervenierenden Prozesse auch maßgeblich für die Art der beteiligten physiologischen Prozesse sind (vgl. Abschn. „Emotion als Gefühlszustand").

Ein weiterer Ansatz ließe sich als typologisch bezeichnen oder als Ansatz zu einer differentiellen Aktivierungsforschung (vgl. hierzu Myrtek, 1980). Hierbei geht es um die Frage, ob Personen anhand bestimmter Merkmale, z.B. der Persönlichkeit, aber auch des Musters von physiologischen Aktivierungsreaktionen klassifizierbar sind. Dabei wird angenommen, daß derartige Merkmale eine bestimmbare zeitliche Invarianz aufweisen. Hierzu zählen die klassischen Konzepte etwa der Sympathikotonie-Vagotonie (Eppinger und Hess, 1910; Wenger, 1941, 1948) oder der vegetativen Labilität (Eysenck, 1967). Von Mandler und Kremen (1958) wurde beispielsweise versucht, Personen mit Hilfe des „Autonomic Perception Questionnaire" (APQ) im Hinblick auf das Ausmaß der Wahrnehmung autonomer Funktionsänderungen einzuteilen. Es entstand dann die Frage, ob Personen, die angaben, häufige und starke vegetative Reaktionen zu spüren, die also einen hohen APQ-Wert hatten, in einem Aktivierungsexperiment auch tatsächlich stärkere Reaktionen zeigen als solche Personen, die eine geringe Körperwahrnehmung angaben, also einen niedrigen APQ-Wert hatten. Hierbei zeigte sich allerdings, daß zumindest in Wiederholungsversuchen ein Zusammenhang nicht reliabel nachgewiesen werden konnte (Mandler und Kremen, 1958).

Schließlich ist ein hauptsächlich methodisch orientierter Ansatz zu nennen, der sich um die Standardisierung von Methoden der Erfassung von Aktivierung, also um eine Operationalisierung der Aktivierungskonzepte im Sinne einer „Aktivierungsdiagnostik" bemüht (Fahrenberg et al., 1979; Dahme und Richter, 1980). Ziel dieses Ansatzes ist es also zunächst nicht, Modelle interner Aktivierungsvermittlung zu ermitteln, neuropsychologischer oder kognitionspsychologischer Art, sondern in Anlehnung an Meßkonzepte der Testtheorie die Güte von Meßverfahren zur Erfassung von Aktivierung zu verbessern. Daß auch dieses Vorgehen Grundannahmen über das zu Erfassende enthält, ist selbstverständlich.

8.2.1 Probleme der Erfassung von Aktivierung

Psychophysiologische Forschung betrachtet vor allem unter methodischem Aspekt die physiologischen Funktionen als abhängige Variablen, abhängig von Bedingungen, die mit Methoden der Psychologie herstellbar und beschreibbar sind. Dies bedeutet, daß man psychologisch möglichst exakt beschreibbare Bedingungen schafft, die auf einen Organismus einwirken. Wichtig dabei ist, daß diese Bedingungen nicht willkürlich gewählt werden, sondern letztlich konkreter Teil eines theoretischen Systems sind und dabei auf die Fragestellungen bezogen sind, die man untersuchen will. Dies wird in vielen Fällen nur annäherungsweise der Fall sein. Demnach wäre die Meßebene die der Erfassung physiologischer Funktionen als abhängige Variable. In vielen Untersuchungen zur Aktivierung wird versucht, den Einfluß psychologisch beschreibbarer Bedingungen auf physiologische Funktionen zu erfassen. Aus dieser Vorgehensweise läßt sich leicht ein „einseitig gerichtetes" kausales Modell ableiten, in dem angenommen wird, die eine oder andere psychische Bedingung wirke auf die physiologischen Funktionen und verändere sie. Gleichzeitig wird davon ausgegangen, daß sich die Intensität einer Emotion oder das Ausmaß der Aktiviertheit in der Größe der Veränderungen der physiologischen Funktionen zeigt, so daß sich die Intensität von Emotion gleichsam physiologisch bestimmen lasse und damit auch die Relevanz, die eine Situation für einen Organismus besitzt. Diese eingeengte Interpretationsweise ergibt sich jedoch nicht notwendigerweise aus der Art des methodischen Vorgehens, wenn man in Betracht zieht, daß Emotion ein ganzheitlicher Vorgang ist und einen Organismus als Ganzes betrifft. Hinzu kommt, daß die Korrelation verschiedener Ebenen z.B. von Emotion sehr unterschiedlich, oft recht gering ist. So kann ein Patient nach einer Behandlung kaum noch Angst in einer ehemals angstauslösenden Situation verspüren, aber dennoch mit deutlichen Änderungen physiologischer Parameter, etwa der Herzfrequenz, darauf reagieren.

Faßt man Aktivierung als ein psychophysisches Geschehen auf, das den gesamten Organismus auf zum Teil unterschiedliche Weise betrifft, so ergeben sich daraus eine Reihe methodischer Anforderungen. Prinzipiell reichen die Meßmethoden der Physiologie, d.h. die Messung physiologischer Funktionen nicht aus, sondern sie müssen durch Methoden der Psychologie erweitert werden. Je nach Fragestellung wird es notwendig sein, physiologische Methoden und solche zur Erfassung erlebnismäßiger Vorgänge oder solche des Verhaltens zu verwenden.

Aktivierung als theoretischer Begriff ist nicht direkt erfaßbar, sondern kann durch bestimmte Daten (Indikatoren) erschlossen werden. Eine Untersuchung zur Erfassung von Aktivierung muß also derartig gestaltet werden, daß die Untersuchungs- und Meßbedingungen mit dem theoretischen Modell der Aktivierung verknüpft sind. Sowohl für die Herstellung von Untersuchungsbedingungen, deren Einfluß auf einen Organismus untersucht werden soll, als auch für die Erfassung dieses Einflusses auf Erleben, Verhalten und physiologische Vorgänge sollten also Operationalisierungen vorliegen, die etwa die Wiederholbarkeit einer Untersuchung, aber vor allem eine möglichst stringente Interpretation der Ergebnisse ermöglichen. Dies ist abhängig davon, wie stringent eine Versuchssituation aus einem theoretischen System abgeleitet ist, impliziert also oft eine Reihe von Schwierigkeiten.

So ist es beispielsweise nicht immer leicht, eine „Belastung" optimal für viele Probanden zu standardisieren. Benutzt man etwa eine Rechenaufgabe einer bestimmten Schwierigkeit, so ist bereits die Schwierigkeit aufgrund unterschiedlicher Rechenfähigkeit nicht für alle Probanden gleich. Darüber hinaus ist vermutlich die Motivation zur Lösung von Rechenaufgaben bei den einzelnen Probanden verschieden. Liegt eine ausgeprägte Leistungsmotivation vor, so könnte die Belastung stärker wirken. Es ist in diesem Beispiel prinzipiell an eine Interaktion zu denken zwischen den die Ergebnisvarianz bestimmenden Faktoren Motivation und Leistungsfähigkeit, wobei die Motivation von weiteren Bedingungen der Untersuchung beeinflußt sein kann, so ist etwa die Wettbewerbssituation wichtig für die Reagibilität von Personen mit ausgeprägtem Typ-A-Verhalten. Benutzt man eine solche Bedingung, um etwa die Reagibilität von Personen auf Belastung zu erfassen, so ergibt sich aus diesen Erwägungen die Notwendigkeit, möglichst viele Bedingungen von Belastung herzustellen, um die spezifischen Einflüsse einer speziellen Belastung (Stimulusspezifität – SSR, Motivationsspezifität – MSR) im Hinblick auf die Fragestellung erfassen und kontrollieren zu können. Die Verwendung mehrerer Belastungen würde es jedoch gestatten, diesen für die aktivierende Bedingung spezifischen Anteil der Varianz zu bestimmen.

Wenn bedacht wird, daß die Kovariation zwischen Veränderungen von Werten innerhalb einer der genannten Meßebenen häufig sehr gering ist, so wird unmittelbar deutlich, daß die Intensität einer Emotion nicht nur durch die Messung der Veränderung etwa der Herzfrequenz erfaßbar ist, also einer einzelnen Funktion der physiologischen Meßebene, sondern daß eine möglichst große Anzahl von Variablen auf möglichst vielen Ebenen erfaßt werden muß.

Die Erhebung von Daten aus verschiedenen Berei-

chen und Funktionssystemen ist auch darum notwendig, weil die Beziehung zwischen verschiedenen Funktionssystemen nicht nur nicht eng, sondern auch nicht notwendigerweise linear ist. So ist der Fall denkbar, daß eine Person angibt, sie empfinde ein bestimmtes Gefühl in großer Intensität, ohne daß es zu größeren Änderungen physiologischer Funktionen oder zu nennenswerten Änderungen des Verhaltens kommt. Dies kann einerseits abhängen von der Art der Emotion, oder aber von einem nicht linearen Zusammenhang zwischen dem Erleben einer Emotion und den entsprechenden Änderungen physiologischer Funktionen. Dieser Zusammenhang kann bei unterschiedlichen Emotionen im Hinblick auf physiologische oder endokrine Funktionen jeweils verschieden sein. Daraus ergibt sich die Forderung nach einer multivariaten Untersuchungsstrategie, die eine Beziehung der Daten aus verschiedenen Funktionssystemen untereinander und auf ein Konstrukt, z.B. einer bestimmten Emotion, ermöglicht.

Für den Bereich der hauptsächlich klinisch orientierten Psychophysiologie wurde von Richter und Mitarbeitern (1984) darüber hinaus die Forderung nach einer „Funktionsdiagnostik" aufgestellt, worunter die Autoren die „bevorzugte Untersuchung derjenigen physiologischen Funktionen, die bei der jeweiligen psychosomatischen Erkrankung beeinträchtigt sind", verstehen. So hatten schon Malmo und Shagass (1959) eine Beziehung zwischen Reaktivität von physiologischen Funktionen und auf diese beziehbaren Beschwerden gefunden. Richter und Mitarbeiter (1984) schlagen folgende Untersuchungsbedingungen für die Funktionsdiagnostik vor: experimentell kontrollierte Reaktionsmuster auf emotionale Belastungen, symptomatische Veränderungen im therapeutischen Setting, interozeptive Wahrnehmung der beeinträchtigten Funktion und deren Veränderung, Langzeitverlauf der beeinträchtigten Funktion in natürlichen Lebensbedingungen und operante Kontrolle der symptomatischen Funktion. Die Autoren sehen vor allem einen Nutzen im Hinblick auf die „Aktivierungsdiagnostik" für den Einzelfall, was jedoch unter dem Gesichtspunkt der Ökonomie fragwürdig erscheint.

Die Forderung nach der Messung physiologischer Funktionen und „natürlichen Lebensbedingungen" impliziert die Annahme, daß unter den „künstlichen" Bedingungen von Laboruntersuchungen z.B. keine für eine Erkrankung typischen Reaktionen auftreten, da im Labor lebensechte Anforderungen und Belastungen nicht herstellbar seien.

Es ist sicher zutreffend, daß der Herstellung experimenteller Bedingungen im Labor im Vergleich zu Feldstudien recht enge Grenzen gesetzt sind. Dies muß jedoch keineswegs bedeuten, daß Laborbefunde nicht auch für bestimmte Personen oder Personengruppen typisch bzw. spezifisch und für die Erforschung pathogener Mechanismen relevant sein können. Dies wäre nur dann der Fall, wenn bestimmte, für eine Personengruppe spezifische Aktivierungsmuster, die z.B. als pathogenetisch bedeutsam erscheinen, nur unter sehr wenigen, im Labor nicht herstellbaren Bedingungen auftreten. Tatsächlich aber scheint es häufig so zu sein, daß belastende Situationen durch bestimmte Merkmale der Situation belastend bzw. aktivierend wirken, etwa durch den Mangel an Kontrollmöglichkeiten beim Vermeiden aversiver Bedingungen (Thompson, 1981) oder durch Wettbewerbscharakter usw. Laborsituationen können derartige Merkmale durchaus aufweisen, allerdings im Vergleich zur „echten" Lebensbedingung nur in abgeschwächter Form. Tatsächlich zeigen zahlreiche Laboruntersuchungen, stellt man Situationen mit bestimmten Merkmalen her, daß es zu „krankheitsspezifisch" relevanten Reaktionen kommt. Deren Wert wird nicht dadurch verringert, daß sie unter natürlichen Bedingungen oft stärker ausfallen würden. Dies ist allerdings dann anders, wenn sich mit dem Zuwachs an Reaktionsintensität auch deren Muster ändert und erst unter der Bedingung großer Intensität für eine Patientengruppe z.B. spezifische Reaktionsmuster auftreten würden. Die Klärung derartiger Fragen allein rechtfertigt bereits die Durchführung von Feldstudien, in denen physiologische Funktionen unter natürlichen Bedingungen gemessen werden.

Im Vergleich zu Laboruntersuchungen weisen Untersuchungen in der natürlichen Lebenssituation eine Reihe zusätzlicher Schwierigkeiten auf. Die gesamte Meßelektronik muß soweit miniaturisiert sein, daß sie für den Probanden „tragbar" ist. Durch die Entwicklung im Bereich der Mikroelektronik ist dies heute weitgehend zu realisieren. Weiterhin müssen die zum Teil sehr komplizierten Geräte auf irgendeine Weise vom Probanden bedient werden, was zu Fehlern führen kann. Elektroden müssen so angebracht werden, daß sie für längere Zeiträume einen sicheren Sitz aufweisen, so daß sie auch nicht durch Bewegungen verändert werden. Es entstehen Probleme, die durch die Bewegung der Probanden gegebenen Artefakte bei der Auswertung zu kontrollieren usw.

Eines der wesentlichsten Probleme besteht jedoch in der Schwierigkeit, die physiologischen Meßdaten mit den Ereignissen und Situationen in Beziehung zu setzen, die während der Messung, etwa eines Tages, auftreten. Angaben der Personen sind dabei oft unzuverlässig. Sind sie es nicht, d.h. werden auch die Situationen und ihre Bedeutungen möglichst genau erfaßt, z.B. durch das Beantworten einiger Fragen auf einem Bogen, so ist bereits dadurch die Situation verändert, indem über die Situation und ihre Valenz nachgedacht wird, was vielleicht sonst nicht geschehen wäre, so daß die Messung eigentlich gar nicht unter „natürlichen" Bedingungen erfolgt. Dennoch ist es von Wert, wenn bei einer über einen Tag dauernden Messung etwa des Blutdrucks eines Hypertonikers festgestellt werden kann, was gerade geschah oder was gedacht oder empfunden wurde, wenn der Blutdruck besonders hoch oder auch niedrig gewesen ist, auch wenn man niemals sicher sein kann, daß es nicht noch relevantere Situationen geben könnte, in denen verschiedene Muster der Aktivierung auftreten könnten.

Aus diesen Überlegungen kann als Ergebnis festgehalten werden, daß es in jeder Hinsicht „ideale Untersuchungsbedingungen" nicht gibt, es wird stets notwendig sein, irgendwelche Einschränkungen in Kauf zu nehmen. Von der jeweiligen Fragestellung wird es dann abhängen, welche Einschränkungen gerade tolerierbar sind.

Will man die aktivierende Wirkung einer bestimmten Situation erfassen, so ist es notwendig, den Grad der Aktivierung mit einem Bezugspunkt zu vergleichen. Im allgemeinen wird eine Ruhesituation verwendet, die der aktivierenden Situation vorausgeht. Das Ausmaß von Aktiviertheit kann dann definiert werden als der Unterschied des Befindens, Verhaltens und verschiedener physiologischer, endokriner und immunologischer Funktionen zwischen der Ruhe- und der Aktivierungssituation, also als die Differenz zwischen den Werten, die diese Funktionen in den verschiedenen Situationen einnehmen. Wilder (1931) hat darauf hingewiesen, daß diese Differenz abhängig ist vom Ruhe- bzw. Ausgangswert. Er formulierte das sog. „Ausgangswertgesetz", das besagt, daß die Differenz zwischen Ruhe- und Belastungswert um so geringer sein wird, je höher die Aktiviertheit bereits bei der Ruhesituation ist, je höher also auch die Werte in dieser Situation sind. Andererseits wird der Differenzwert um so höher sein können, je niedriger der Ausgangswert ist, d.h. die Korrelation des Differenz- oder Reaktionswertes mit dem Ausgangswert ist negativ. Diese Auffassung ist verschiedentlich kritisiert worden (Johnson und Lubin, 1972; Myrtek, 1980). So hat Myrtek die Ausgangswertabhängigkeit von 64 Reaktivitätsmaßen überprüft und dabei nur in 3 Fällen (5%) eine entsprechende Abhängigkeit der Reaktivitätswerte nachweisen können, bei 66% dieser Werte bestand eine positive Korrelation zwischen Ausgangs- und Reaktivitätswert, also genau umgekehrt, wie dies vom Ausgangswertgesetz postuliert wird.

Von Lacey (1956) wurde ein Korrekturverfahren der Ausgangswertabhängigkeit vorgeschlagen, der sog. „autonomic lability score" (ALS), der in der standardisierten Abweichung des bei einer Person gemessenen Belastungswertes vom erwarteten Belastungswert besteht. Der erwartete Belastungswert entspricht dem mittleren Belastungswert derjenigen Personen, die denselben Ausgangswert besitzen (vgl. Myrtek et al., 1977; Myrtek, 1980). Sinnvoll ist jedoch die Anwendung des ALS nur in Fällen, in denen eine echte Ausgangswertabhängigkeit besteht, was, wie Myrtek gezeigt hat, nur selten der Fall ist.

Neben der Abhängigkeit vom Ausgangswert ist die Größe eines Reaktivitätswertes weiterhin abhängig von der „Bandbreite" der individuell möglichen und tatsächlich in einer Untersuchung vorhandenen Variabilität der Funktionen, deren Reaktionen untersucht werden sollen. Will man etwa die relative Wirkung einer bestimmten Belastung ermitteln, so können zwei Personen denselben Ausgangswert besitzen, jedoch reagiert die eine Person viel geringer auf die Belastung als die zweite. Dies kann daran liegen, daß die Belastung bei dieser Person weniger wirksam ist, jedoch auch daran, daß diese Person insgesamt weniger „reagibel" ist, so daß die aktivierende Wirkung einer Belastung bei dieser Person unterschätzt würde. Lefave und Neufeld (1980) schlagen daher eine Korrektur von Meßwerten einer Belastung am Bereich der Meßwerte (Maximum – Minimum) der gesamten Untersuchung vor.

8.2.2 Klassische Konzepte von Aktivierung

Duffy (1972) definiert Aktivierung folgendermaßen: „Ein Individuum, das heißt ein Organismus als Ganzer, ist manchmal erregt, manchmal entspannt und manchmal in einem der vielen dazwischen möglichen Zustände. Diese offensichtlichen Zustände legen das Konzept von Aktivierung oder Arousal nahe, das versucht, die Physiologie dieser Zustände zu beschreiben, sowie ihre Ursachen und Effekte."

Mit dem Begriff der Aktivierung ist nicht immer auch offen beobachtbare Aktivität bezeichnet, sondern oft das Bereitstellen von Energie für eine solche Aktivität. Dabei kann es durchaus der Fall sein, daß es trotz der Bereitstellung nicht zu einer nach außen gerichteten Aktivität kommt, etwa weil externe oder interne Bedingungen zu einer Hemmung der angestrebten Aktivität führen. Von Uexkülls (1965) Konzept der Bereitstellung als pathogene Bedingung in der Ätiologie der Hypertonie ist so zu verstehen.

Cannon (1931) betonte ebenfalls den Gedanken einer Bereitstellung von energetischen Reserven des Körpers für plötzlich notwendig werdende Aktivität. In seinem Modell der „emergency states" der Emotion ist Aktivierung eng verknüpft mit der Erregung des sympathischen Nervensystems bei gleichzeitiger Inhibition des parasympathischen Systems und wird dem aktivierenden Effekt von Adrenalin gleichgesetzt.

Malmo (1962) steht mit seinem Aktivierungsmodell der Hullschen Theorie des Lernens nahe. In dieser Theorie Hulls spielt das Triebniveau eines sich verhaltenden Organismus eine wesentliche Rolle. „Wenn ich den Begriff ‚Aktivierung' benutze, verweise ich damit auf eine Intensitätsdimension. Arousal wird oft als Begriff vertauschbar mit Aktivierung benutzt; und der Begriff des Triebniveaus (Hulls D) ist ein sehr ähnliches Konzept. Zum Beispiel ist ein schläfriges Subjekt wenig, ein erregtes Subjekt hoch aktiviert." Aktivierung wird als Kontinuum angenommen von tiefem Schlaf bis hin zu höchster Erregung und werde kontrolliert durch die Aktivität des „Ascending Reticular Arousing System" (ARAS).

Lindsley (1957) stützt sich in seinen Arbeiten zur EEG-Desynchronisation als Indikator für Aktiviertheit vor allen Dingen auf Untersuchungen des ARAS. So fand er, daß Läsionen in diesem System das Aktiviertheitsmuster im EEG verhinderten und das Bild einer Verhaltenslethargie und Somnolenz produzierten. Umgekehrt konnte durch Stimulation im ARAS ein Aktiviertheitsmuster im EEG hervorgerufen werden.

Nimmt man den Vorstellungen Hulls entsprechend für ein derartiges System eine Reaktionshierarchie an, so kann eine Übererregung von außen dazu führen, daß auch irrelevante Reaktionen erleichtert werden,

so daß es zu einer „response competition" kommt. Aus derartigen Überlegungen ergibt sich die Annahme einer umgekehrt U-förmigen Beziehung zwischen dem Grad von Aktiviertheit und Leistung. Aus der folgenden Abbildung 8–1 wird deutlich, daß die Leistung zunächst mit steigender Aktiviertheit zunimmt, dann jedoch nach Überschreiten eines optimalen Niveaus von Aktiviertheit wieder abnimmt.

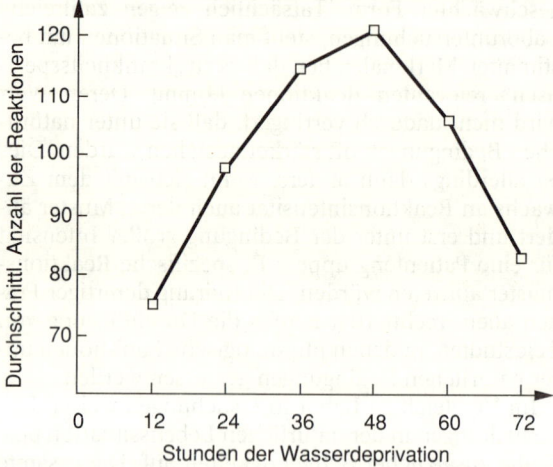

Abb. 8–1. Daten von Belanger und Feldman, die die Beziehung zwischen Wasserdeprivation und Leistung in einer Skinner-Box bei 7 Ratten zeigen (aus Malmo, 1962).

Malmo (1962) setzt allerdings Triebniveau – operational definiert als Deprivationsdauer -- nicht gleich mit Aktiviertheit. „Der wichtige Punkt, der hier sichtbar wird, besteht darin, daß das höhere Aktivierungsniveau das kombinierte Produkt der Reize, ihrer Anforderungscharakteristik und der Schlafdeprivation darstellt. Ohne eine solche Stimulation würde das Subjekt sicherlich einschlafen, und wir wissen von unseren Schlafstudien, daß das physiologische Niveau sehr schnell absinkt, wenn man zu schlafen beginnt. Es ist daher offensichtlich, daß in der Abwesenheit von Aufgaben die physiologischen Indikatoren nach einer 60stündigen Schlafdeprivation niedrigere und nicht höhere Aktivierungswerte haben würden im Vergleich zu einer Ruhebedingung" (Malmo, 1962).

8.2.3 Der Spezifitätsaspekt von Aktivierung

In den im vorangegangenen dargestellten Modellen wird Aktiviertheit als ein eindimensionaler Zustand im Sinne von Erregungsintensität aufgefaßt, der auf stattfindende Prozesse einen Einfluß ausübt. Malmo faßte Aktiviertheit als Produkt der Einflußfaktoren Deprivation, Reiz und dessen Anforderungscharakteristik auf, unterschied jedoch nicht spezifische Aktivierungsmuster etwa in Abhängigkeit von extremen Reizbedingungen und deren Anforderungscharakter. Die meist nur geringe Kovariation zwischen verschiedenen Funktionsgrößen legt die Vermutung nahe, daß physiologische Teilsysteme unter Umständen unabhängig oder nur mäßig abhängig voneinander, d.h.

inhomogen, auf verschiedene Bedingungen reagieren. Der Spezifitätsaspekt von Aktivierung betont die spezifischen Einflüsse der Faktoren: Reiz (oder Stimulus), Individuum und Motivation. Dies bedeutet, daß spezifische Stimulusgegebenheiten spezifische Aktivierungsmuster hervorrufen, bzw., daß ein erfaßtes Aktivierungsmuster zu einem bestimmten Anteil für die aktivierende Bedingung spezifisch ist (SSR = Stimulus-spezifische Reaktion). Davon unterscheidbar ist eine bei einer Person erfaßte Aktivierungsreaktion zu einem bestimmten Anteil für diese Person spezifisch, also unabhängig von der aktivierenden Bedingung (ISR = Individual-spezifische Reaktion). Nochmals davon unabhängig ist die aktuelle Motivationslage einer Person, die die Reaktion auf eine aktivierende Bedingung beeinflußt. Zu einem anderen Zeitpunkt könnte die Motivation einer Person, etwa zur Lösung einer Rechenaufgabe, verschieden sein, so daß eine erfaßte Reaktion auf eine spezifische Situation bei einer spezifischen Person nochmals für die aktuelle Motivationslage spezifisch ist (MSR = Motivations-spezifische Reaktion). Aus der Wirksamkeit dieser Einflußgrößen auf Aktivierungsreaktionen ergeben sich bestimmte Notwendigkeiten für deren Erfassung. Auf eine kurze Formel gebracht besagen sie, daß möglichst viele Funktionen bei möglichst vielen Personen unter möglichst vielen Bedingungen zu möglichst vielen Zeitpunkten erfaßt werden müssen. Faßt man Personen, Zeitpunkte und Bedingungen als Einflußfaktoren auf, so ergibt sich daraus die Notwendigkeit für mehrfaktorielle Versuchspläne. Spezifitäten lassen sich ermitteln durch korrelationsstatistische, univariate oder multivariate varianzanalytische Verfahren.

In letzter Zeit haben Fahrenberg und Mitarbeiter (1979; vgl. auch Walschburger, 1976; Myrtek, 1980) ausführlich zu methodischen Fragen der Aktivierungsforschung Stellung genommen und den Einfluß von Spezifitäten betont. Dabei wird allerdings nicht bestritten, daß Emotion als ein „Simultangeschehen von erlebter und introspektiv-verbal mitteilbarer Befindensänderung mit bestimmten Ausdrucks- und Verhaltensweisen sowie zentralnervöser und vegetativ-endokriner Funktionsanregung gilt" (Fahrenberg, 1980). So habe sich in den vorliegenden Untersuchungen durchaus bestätigt, daß sich unter Belastung die verschiedensten Parameter erwartungsgemäß geändert haben im Sinne eines „allgemeinen Belastungs-Beanspruchungs-Effekts". Nicht aufrechterhalten werden können jedoch Annahmen eines eindimensionalen Aktivierungsprozesses, der beinhalten würde, daß in verschiedenen Funktionsbereichen psychischer und körperlicher Art homogene Prozesse der Aktivierung stattfinden, so daß ein vorhersagbarer Zusammenhang zwischen diesen Systemen bestünde. „Die individuellen Aktivierungsprozesse sind so verschieden, daß weder aus Selbsteinstufungen der erlebten Aktivierung die physiologischen Aktivierungsvariablen, noch aus einzelnen physiologischen Aktivierungsvariablen die Veränderungswerte in anderen Aktivierungsvariablen systematisch und substantiell, das heißt praktisch relevant, vorhergesagt werden

können" (Fahrenberg, 1980). Daraus läßt sich u.a. unter einem methodischen Gesichtspunkt schließen, daß „einfache Funktionsprüfungen" zur Erfassung etwa der „Reagibilität" von Personen mit einfachen univariaten Versuchsplänen unzureichend sind. Das bedeutet andererseits nicht, daß nicht „symptomatische Funktionsänderungen", etwa bei Patienten mit bestimmten Funktionsstörungen, auf ihre Veränderbarkeit hin untersucht werden könnten, allerdings mit der Einschränkung, daß Ergebnisse dieser Art nur auf diese Funktionen hin interpretiert werden dürfen, also keine Aussagen über eine allgemeine Reagibilität zulassen.

Stimulusspezifität (SSR)

Von Lacey (1962, 1974) und Mitarbeitern wurde ein eindimensionales Konzept der Aktivierung kritisiert. Neben dem Grad der Aktivierung in einer Situation, aufgefaßt als intervenierende Intensitätsvariable, wurde von ihm dem Aspekt der Gerichtetheit von Verhalten besondere Bedeutung zugemessen. Dabei wurde zwischen interner und externer Informationsverarbeitung unterschieden, so daß je nach dem Anforderungscharakter, den eine Situation für ein Individuum besitzt, ein unterschiedliches Aktivierungsmuster auftritt.

Lacey untersuchte seine Hypothese, indem er Probanden verschiedene Aufgaben stellte, bei denen ein Aufgabentyp eine vermehrte Aufmerksamkeit für Außenreize erforderte, etwa indem stroboskopische Lichtreize beobachtet (Lacey et al., 1963) oder ein „Signalton" von 513 Hz in einer Serie von 500-Hz-Tönen entdeckt werden sollte (Lacey und Lacey, 1974 b). Diesem Typ von Aufgaben wurden Aufgaben eines zweiten Typs gegenübergestellt, von denen angenommen wurde, daß sie eine interne Informationsverarbeitung beinhalten, wie z.B. das Lösen von Rechenaufgaben. Die folgende Abbildung 8–2 zeigt, daß bei der Aufgabe, in der ein Signalton entdeckt werden sollte, ein Abfall der Herzfrequenz stattfindet,

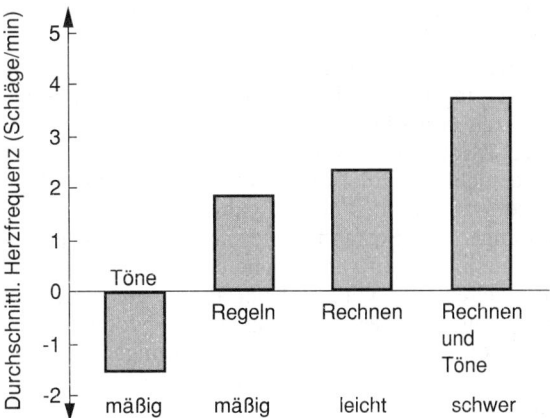

Abb. 8–2. Durchschnittliche Herzfrequenzveränderung von einminütigen Perioden erhöhter Aufmerksamkeit im Gegensatz zu Aufgabenperioden mit Problemlösen. Die Schwierigkeitseinstufungen der Aufgaben sind unterhalb der Blöcke vermerkt (aus Lacey und Lacey, 1974).

dagegen bei den Aufgaben, die eine interne Informationsverarbeitung erfordern, die Herzfrequenz ansteigt.

In vielen Untersuchungen dieser Art zeigte sich eine vergleichbar unterschiedliche Reaktion der Herzfrequenz bei den beiden Typen von Aufgaben. Wesentlich dabei ist die Tatsache, daß andere Indikatoren von Aktivierung, wie die Veränderung der Hautleitfähigkeit, bei Aufgaben beider Typen nicht unterschiedlich beeinflußt wurden, so daß sich ein verschiedenes Muster der physiologischen Reaktion ergibt.

Viele Untersuchungen wurden von Lacey so durchgeführt, daß der Aufgabentyp vorher angekündigt wurde, mit einem Hinweis auf die Art der erwarteten Reaktion. Daraus wird abgeleitet, daß die Beurteilung einer Situation im Hinblick darauf, ob sie Informationsaufnahme oder -verarbeitung erfordert, für die Veränderungen der Herzfrequenz und des Blutdrucks wichtig ist. Diese Veränderungen werden als instrumentell interpretiert, weil sie der Erreichung eines Ziels in Abhängigkeit von der Beurteilung der Situation dienen und nicht als Folge der Art der Informationsverarbeitung angesehen werden.

Als Voraussetzung für diesen instrumentellen Mechanismus wird den afferenten Bahnen, ausgehend von den Barorezeptoren, vor allem des Karotissinus und des Aortenbogens, eine negative Feedback-Wirkung zugeschrieben, die in Abhängigkeit von den Blutdruckverhältnissen eine die Aktivität des Sympathikus hemmende Wirkung ausübt. Diese gehe über den homöostatischen Effekt innerhalb des Kreislaufsystems hinaus und beeinflusse daher auch höhere Prozesse des Zentralnervensystems. Durch diesen Mechanismus stelle sich der Organismus auf die Notwendigkeit einer Situation ein und erreiche eine Verbesserung der erforderlichen Aufmerksamkeitsleistung. Durch die Erhöhung des Blutdrucks erreiche der Organismus eine „Stimulus-Barriere", so daß die interne Informationsverarbeitung ungestörter ablaufen könne und nicht durch irrelevante Außenreize behindert werde. So konnten auch Lacey und Lacey (1974) zeigen, daß für das Muster der Reaktion der Typ der Aufgabe von größerem Einfluß war als die Schwierigkeit der Aufgabe.

Lacey und Mitarbeiter zitierten verschiedene Untersuchungen, die eine Beziehung zwischen Reaktionszeit und Blutdruckschwankungen während des EKG-Zyklus herstellen. Dabei wurde die Hypothese untersucht, daß zu Zeiten höherer Blutdruckverhältnisse während des EKG-Zyklus die Reaktionszeit verlängert sei. So fanden Heymans und Neil (1958) wie auch Callaway und Layne (1964) kürzere Reaktionszeiten kurz vor der Kontraktion des Herzmuskels, während des QRS-Komplexes jedoch längere Reaktionszeiten. Derartige Untersuchungen scheinen die Annahme des von Lacey postulierten Einflusses zu stützen.

Das Modell Laceys ist verschiedentlich kritisiert worden, z. B. von Elliott (1972, 1974a, b). Er kritisiert hauptsächlich zwei Punkte. Die Kritik richtet sich gegen die relativ unscharfe Definition der Aufgabentypen (Aufnahme versus Verarbeitung von Information,

„intake versus rejection"). „Das Fehlen einer unabhängigen, klaren Definition von dem, was eine ‚intake'-Aufgabe und was eine ‚rejection'-Aufgabe ist, macht es schwer, Vorhersagen zu treffen, obwohl der Posthoc-Gebrauch der Intake-Rejection-Dimension, wie bei allen Theorien, alarmierend einfach ist" (Elliott, 1974b). So bestehe beispielsweise eine weitere Schwierigkeit darin, daß in vielen Situationen, die als Rejection-Aufgaben deklariert würden, Verbalisierungen notwendig seien, deren Einfluß auf die Herzfrequenz von dem interner Informationsverarbeitung beim Problemlösen nur schwer zu trennen sei. Entsprechende Schwierigkeiten träten bei vielen Untersuchungen auf, die nicht als direkte Tests der Hypothese Laceys durchgeführt worden seien, aber post hoc von Lacey im Sinne seines Modells interpretiert wurden.

Die Kontroverse um das Phänomen der Abhängigkeit von Veränderungen der Herzfrequenz vom Zeitpunkt der Darbietungen eines Stimulus innerhalb eines Herzzyklus („cycle time effect") ist noch immer theoretisch interessant, zumal ebenfalls ein Einfluß auf Veränderungen der Herzfrequenz durch den Zeitpunkt einer Reaktion innerhalb des Herzzyklus festgestellt worden ist („vagal inhibition") (Jennings und Wood, 1977; Jennings et al., 1987). Diese Untersuchungen stützen die Auffassung Laceys eines vagalen Einflusses äußerer Reize, der sich in einer Verlangsamung der Herzfrequenz auswirkt, ebenso wie die Erwartung eines signifikanten Reizes sowie eines die Herzfrequenz steigernden Einflusses von Reaktionen. Die gegenwärtige Kontroverse bezieht sich hauptsächlich auf die Frage, ob Stimulus- und Reaktionseffekte innerhalb des Herzzyklus auf einen oder verschiedene Mechanismen zurückzuführen sind. Von Barry (1987) wurde argumentiert, es handele sich dabei um eine einfache „Verzögerungsfunktion", d. h., je später im Zyklus das Ereignis auftritt, desto kürzer ist seine Wirkzeit innerhalb des Zyklus, so daß die Annahme einer Interaktion zweier Prozesse überflüssig ist.

Lacey hatte in seinen Untersuchungen zur Situationsstereotypie gefunden, daß eine verstärkte Stimulation der Barorezeptoren zentral-inhibitorisch wirkt und dadurch eine Barriere gegenüber der Verarbeitung von Außenreizen bedingt. Diese Ergebnisse bezogen sich zunächst auf einen Aufgabentyp, der von Probanden verlangte, bei „internem Problemlösen" Außenreize als störende Ablenkung zu unterdrücken.

Dworkin (1979) zeigte, daß Ratten, die durch Laufen eine aversive Stimulation vermeiden konnten, weniger liefen, wenn durch Phenylephrin der Blutdruck erhöht wurde. Vergleichstiere mit denervierten Barorezeptoren zeigten diesen Effekt nicht. Der Autor zieht daraus den Schluß, daß Hypertonie auch als operant bedingtes Verhalten angesehen werden kann. So hatten schon Richter-Heinrich et al. (1982) bei Hypertonikern andere Schwellenwerte für Außenreize im Vergleich zu Normotonikern gefunden. Dies hat die Hypothese bekräftigt, daß eine chronische Erhöhung des Blutdrucks und damit eine verstärkte Aktivität der Barorezeptoren belohnenden

Charakter haben könnten, indem aversive Reize dadurch weniger deutlich wahrgenommen würden. Unter Bedingungen chronisch aversiver Belastung durch Schmerz oder sonstige Belastungen könnte durch eine Erhöhung des Blutdrucks das Ausmaß der Aversivität verringert werden.

Erste Untersuchungen, die dieser Hypothese nachgingen, waren bestätigend, es ergaben sich jedoch darüber hinaus Hinweise dafür, daß die Modulation der Schwelle für Schmerzreize durch die Aktivität der Barorezeptoren ihrerseits abhängig ist vom tonischen Niveau des Blutdrucks (Larbig et al., 1985). Diese Vermutung wurde von Elbert et al. (1988) bestätigt. Sie manipulierten die Aktivität der Barorezeptoren durch Druckveränderungen einer Halsmanschette. Bei Probanden mit normotonen (bis 120 mmHg systolisch) oder labil hypertonen Blutdruckwerten (130–160 mmHg systolisch) wurde neben der Herzfrequenz, der Pulsamplitude und der Pulslaufzeit auch das EEG abgeleitet. Unter verschiedenen Druckverhältnissen der Halsmanschette und damit verschiedener Aktivität der Barorezeptoren wurden am Unterarm Schmerzreize durch elektrische Schocks gesetzt und die Schmerzschwelle durch Abbruch der Schmerzreize durch die Probanden erfaßt. In einer Vorphase der Untersuchung wurde die Schockintensität ermittelt, bei der die Probanden die Schocks ohne irgendeine Manipulation abgebrochen hatten. Die Ergebnisse zeigten, daß es bei erhöhter Aktivität der Barorezeptoren nur bei den labilen Hypertonikern zu einer Erhöhung der Schmerzschwelle kam, bei den normotonen Probanden kam es dagegen zu einer Verringerung (Abb. 8–3).

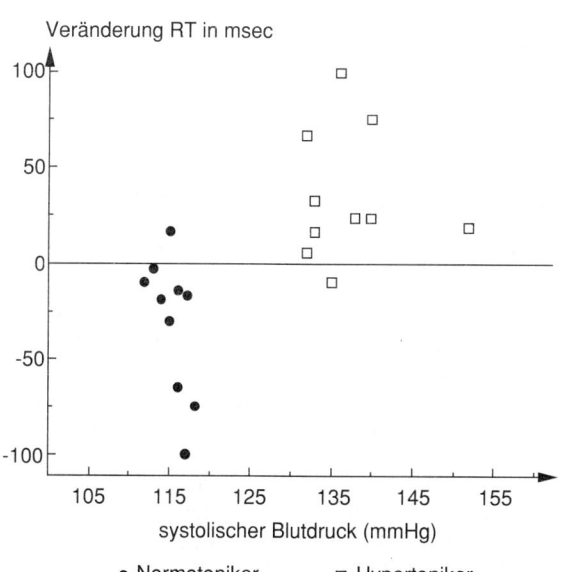

Abb. 8–3. Veränderungen der Reaktionszeit (RT) zur Beendigung des Schmerzreizes durch Stimulierung der Barorezeptoren bei Normotonikern und Hypertonikern (aus Elbert et al., 1988).

Dabei gab es keine Unterschiede der Reaktivität der gemessenen Kreislaufparameter, so daß angenommen werden kann, daß die Stimulierung bei beiden Grup-

pen vergleichbare afferente Effekte auf das Kreislaufzentrum hatte.

Lacey hatte bei seinen Untersuchungen auch festgestellt, daß der situationsspezifische Effekt seiner Aufgabentypen bei einer Reihe von Probanden nicht auftrat, die etwa bei der Aufgabe, einen Signalton zu entdecken, nicht mit Verringerung, sondern Anstiegen der Herzfrequenz reagierten (Lacey und Lacey, 1974). Er bezeichnete diese Probanden als „accelerators". Vergleichbare Ergebnisse hatten auch Williams und Mitarbeiter (1975) gefunden. Sie stellten fest, daß ein Teil der Probanden bereits unter Ruhebedingungen ein höheres Ausgangsniveau der Herzfrequenz hatte. Bei diesen Probanden kam es in Abhängigkeit vom Ausgangswert nicht zu einem Abfall, sondern zu einem Anstieg der Herzfrequenz, der um so größer war, je höher das Ausgangsniveau gewesen war. Dieses Verhalten der Herzfrequenz ist dem Ausgangswertgesetz von Wilder (1931) genau entgegengesetzt. Wie die folgende Abbildung 8–4 zeigt, trat dieser Effekt für die Unterarmdurchblutung besonders deutlich auf.

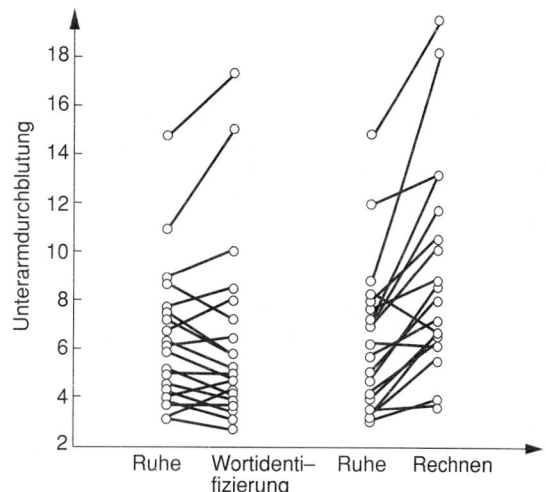

Abb. 8–4. Unterarmdurchblutung von 19 Versuchspersonen während Basiswertperioden, Wortidentifikations- und Rechenaufgaben (aus Williams et al., 1975).

Die Autoren interpretieren diese Ergebnisse dahingehend, daß diejenigen Probanden, die bereits ein hohes Ausgangsniveau unter Ruhebedingungen besitzen, habituell die Informationsaufnahme inhibieren, auch wenn die Situation dies erfordern würde.

Von Obrist und Mitarbeitern wurde das Spezifitätsmodell der „kardio-somatischen Koppelung" entwickelt. Sie gehen dabei von einer biologischen Basis aus, davon, daß die Steigerung der Herzfrequenz „einer der hauptsächlichsten Faktoren ist, um das Herzminutenvolumen, das heißt die Menge von Blut, die für die Muskulatur verfügbar ist, zu ändern, wobei die Herzfrequenz in direkter Beziehung zum Herzminutenvolumen steht" (Obrist et al., 1970). In einer Reihe von Untersuchungen versuchte Obrist nachzuweisen, daß die Verbindung zwischen Veränderungen der Herzfrequenz und körperlicher Aktivität auch für Situationen gilt, die paradigmatisch nicht auf körperliche Aktivität bezogen sind, in denen man etwa am Effekt eines bestimmten Konditionierungsparadigmas interessiert ist. So faßt er das Ergebnis zweier solcher Studien folgendermaßen zusammen: „In beiden Experimenten also ist die kardiale Veränderung eine Funktion dessen, was das Tier körperlich getan hat, und nicht eine Funktion der Signal-Schock- oder Signal-Futter-Kontingenz" (Obrist et al., 1970).

Er stellt seine Auffassung der Laceys gegenüber, der angenommen hatte, daß z.B. eine Senkung der Herzfrequenz als instrumenteller Mechanismus im Sinne der Erleichterung der Aufmerksamkeit für äußere Reize aufzufassen ist. In einem Reaktionszeitexperiment stellt er fest, daß die Blockierung phasischer Veränderungen der Herzfrequenz durch Atropin keinen negativen Einfluß auf die Reaktionsleistung hatte, wie dies nach der Hypothese Laceys zu erwarten gewesen wäre.

Wichtig sind diese Überlegungen auch deshalb, weil in vielen Untersuchungen Veränderungen der Herzfrequenz – als ein einfach zu erfassendes Maß – benutzt werden, um sympathisch vermittelte Erregung zu erfassen. In vielen Situationen jedoch scheinen Veränderungen der Herzfrequenz nicht unter sympathischer, sondern unter vagaler Kontrolle stattzufinden. In derartigen Situationen verändern sich körperliche Aktivität und die Herzfrequenz parallel. Die Veränderung der Herzfrequenz ist dabei nicht als Folge von Anforderungen des Stoffwechsels zu betrachten, sondern als Folge einer zentralen Steuerung, die körperliche Aktivität und Herzfrequenz gleichermaßen betrifft, und als Folge einer somatomotorischen Rückmeldung (Obrist, 1981).

Obrist (1976) zitiert zwei Untersuchungen, aus denen hervorgeht, daß Veränderungen der Herzfrequenz bei intendierter Muskelarbeit auch ohne tatsächliche Arbeit, also auch ohne Stoffwechselanforderung, stattfinden. „In zwei Arbeiten wurden menschliche Probanden instruiert, die Muskulatur in einem Arm anzuspannen. Aber in beiden Fällen war die Muskulatur paralysiert, entweder durch die lokale Anwendung von Curare (Freyschuss, 1970), oder als Folge einer Erkrankung (Dworkin et al., 1975), z.B. Polio. In beiden Untersuchungen kam es zu Anstiegen der Herzfrequenz."

Die kardio-somatische Koppelung ist für Situationen gültig, in denen eine vagale Kontrolle vorherrscht, also auch geringere sympathische Erregung zudecken kann. In derartigen Situationen ist jedoch ein sympathischer Einfluß etwa auf das Gefäßsystem und die Schweißsekretion nachweisbar. Davon zu unterscheiden sind Situationen, in denen ein deutlicher sympathischer Einfluß auf die Herztätigkeit stattfindet. Es sind dies Situationen, in denen sich ein Organismus aktiv mit einer Situation auseinandersetzen muß („active coping").

In einer weiteren Untersuchung wurde daher die Applikation eines Schocks von der Leistung in einem Reaktionszeitversuch abhängig gemacht. Acht Sekunden nach einem Signal wurde ein Signal, auf das reagiert werden mußte, dargeboten. Weitere acht Sekunden später erfolgte ein Schock, falls er durch die

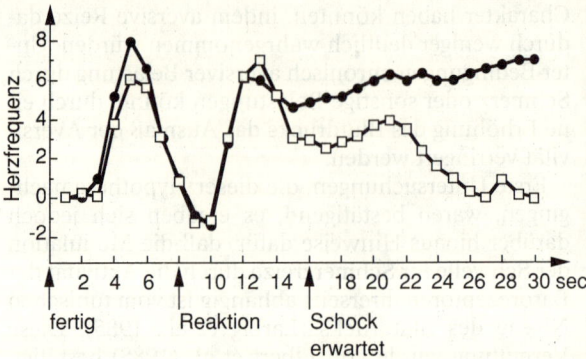

Abb. 8–5. Veränderungen der Herzfrequenz pro Minute mit –□– und ohne –●– pharmakologische Blockade des sympathischen Einflusses (aus Obrist, 1974).

Reaktionsleistung nicht vermieden werden konnte. Bei jedem dritten Probanden wurde der sympathische Einfluß durch eine Infusion von 4 mg Propranolol blockiert. Die Abbildungen 8–5 und 8–6 zeigen den Verlauf der Herzfrequenz und der allgemeinen körperlichen Aktivität (Positionsveränderungen usw.) für Versuchsdurchgänge, in denen der Schock erfolgreich vermieden werden konnte.

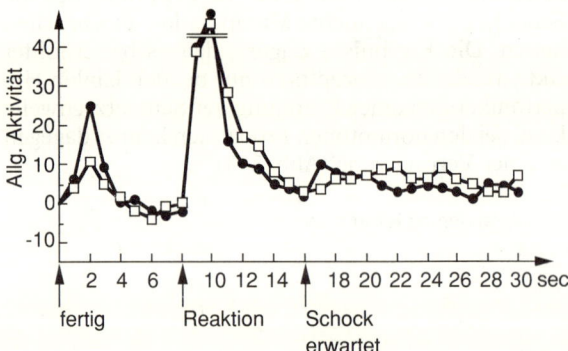

Abb. 8–6. Veränderungen pro Sekunde der allgemeinen Aktivität mit –□– und ohne –●– pharmakologische Blockade des sympathischen Einflusses (aus Obrist, 1974).

Aus den Abbildungen wird deutlich, daß bis zu dem Zeitpunkt, an dem der Schock erwartet wird, der Verlauf der Herzfrequenz unter vagalem Einfluß steht und starke Schwankungen aufweist. Bis zu diesem Zeitpunkt kovariieren körperliche Aktivität und Herzfrequenz. Danach bleibt bei denjenigen Probanden mit erhaltenem sympathischem Einfluß die Herzfrequenz erhöht, trotz abnehmender körperlicher Aktivität, bei den Probanden mit sympathischer Blockade fällt die Herzfrequenz entsprechend ab, d.h., von diesem Zeitpunkt an steht die Herzfrequenz normalerweise unter hauptsächlich sympathischem Einfluß.

In weiteren Untersuchungen wurde die Dimension „aktive" versus „passive" Bewältigung als Bedingung manipuliert, mit dem Ergebnis, daß das Ausmaß der aktiven Einstellung für den sympathischen Einfluß ausschlaggebend ist (Elliott, 1969). Glaubten Probanden beispielsweise, sie könnten das Auftreten der aversiven Ereignisse beeinflussen, ohne daß dies tatsächlich der Fall war, so zeigten sie im Verlauf der Untersuchung ein erhöhtes Niveau der kardiovaskulären Aktivität, im Gegensatz zu denjenigen Probanden, die dies nicht glaubten (Houston, 1973; Malcuit, 1973).

Weitere Einflüsse, vor allem auf das tonische Niveau der kardiovaskulären Aktivität, wurden nachgewiesen für die Neuigkeit eines aversiven Ereignisses, wobei diese bei einer aktiven Auseinandersetzung deren Effekt erhöht (Light und Obrist, 1980). Für das Ausmaß der Motivation (involvement), sich mit der Situation auseinanderzusetzen, gilt dasselbe.

Individuelle Spezifität (ISR)

Neben dem Konzept der Stimulusspezifität ist das der individuellen Spezifität von Wichtigkeit. Für das Problem der Spezifität psychosomatischer Störungen kommt dem Konzept der ISR eine besondere Bedeutung zu, denn es besagt, daß physiologische Reaktionsmuster für ein Individuum, im Sinne der individuellen Konsistenz, spezifisch sind. Eine Person reagiert auf aktivierende Reize hauptsächlich mit bestimmten (z.B. kardiovaskulären) Funktionsänderungen möglicherweise intensiver als andere vergleichbare Personen. Entsprechend fanden Malmo und Shagass (1959), daß die Reaktionen von psychiatrischen Patienten in den Funktionssystemen am größten waren, die den bei ihnen vorhandenen Beschwerden zugeordnet werden konnten. Hierbei zeigt sich, daß das Konzept der individuellen Spezifität unter Umständen für die Frage einer psychosomatischen Spezifität wichtig sein könnte, nachdem Versuche, diese durch persönlichkeitsmetrische Verfahren zu erfassen, bisher fehlgeschlagen sind. Es ließe sich jedoch denken, daß Personen eine spezifische Disposition aufweisen, mit bestimmten Organsystemen auf Belastungen besonders stark zu reagieren, was dann unter Umständen schädigende Wirkung haben und zu spezifischen Erkrankungen führen könnte.

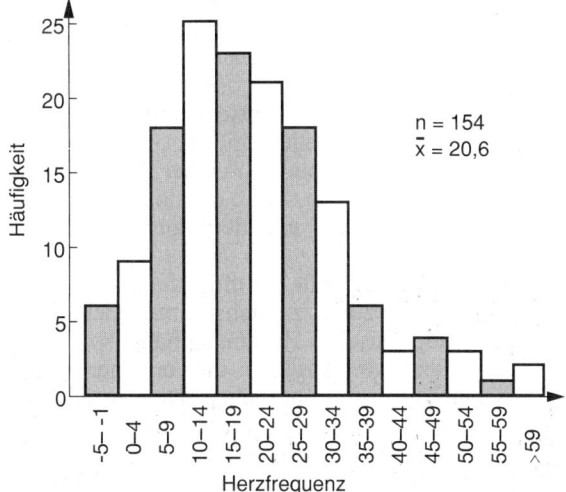

Abb. 8–7. Verteilung der Veränderung der Herzfrequenzwerte zwischen Baseline und den ersten beiden Minuten der Schockvermeidung (aus Obrist, 1981).

Obrist (1981) fand in seinen Untersuchungen zum Teil erhebliche Unterschiede in der Reaktivität der Probanden, wie die Abbildung 8–7 für die Unterschiede der Herzfrequenz zwischen Ruhewerten und Schockvermeidung zeigt.

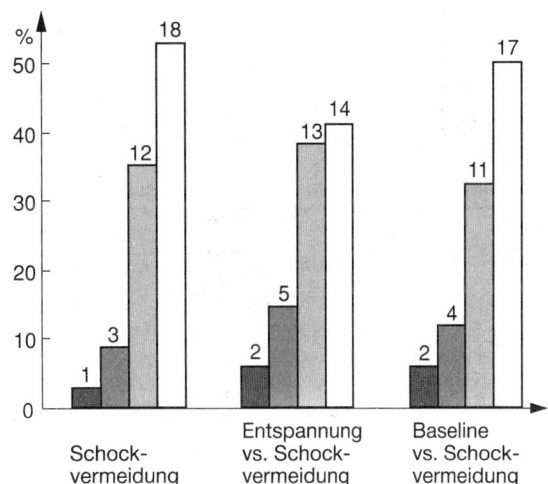

Abb. 8–8. Anzahl der Elternteile mit Hypertonie für jedes Quartil als Funktion der durchschnittlichen Herzfrequenz während Baseline, dem Beginn der Schockvermeidung sowie des durchschnittlichen Unterschieds zwischen jeder Baseline und der Schockvermeidung (aus Obrist, 1981).

Diese Unterschiede zeigten sich auch für andere Versuchsbedingungen, wie den „Cold-Pressure-Test". Dabei war allerdings ein Sachverhalt wichtig: die Unterschiede zur Ruhebedingung, die den jeweiligen Belastungen vorausging, waren nur für die Schockvermeidung individuell verschieden. Sehr deutlich wurden sie aber im Vergleich zu einer „Entspannungssituation". Hier zeigten sich auch deutliche Unterschiede im Niveau der beiden Ruhesituationen, etwa bezüglich der Herzfrequenz bei verschiedenen Probanden.

Ähnlich verhielt es sich bei genetisch im Hinblick auf eine Hypertonie unterschiedlich belasteten Person. Wie die Abbildungen 8–8 und 8–9 zeigen, ist die Reaktivität um so höher, je größer die entsprechende Belastung ist.

Es gibt inzwischen eine ganze Reihe von Studien, die zeigen, daß bei Probanden eine höhere kardiovaskuläre Reaktivität vorliegt, die entweder hypertone Eltern haben oder aber bereits unter Ruhebedingungen einen höheren Blutdruck aufweisen. Dies gilt auch für adoleszente Probanden, wie Manuck und Proietti (1982) zeigen konnten. Sie verglichen die Reaktivität von insgesamt 36 Probanden, von denen 18 normotone und 18 hypertone Eltern hatten. Die Anstiege von Herzfrequenz und Blutdruck waren unter verschiedenen kognitiven Belastungsbedingungen bei den Probanden mit hypertonen Eltern deutlich größer als bei denjenigen mit normotonen Eltern. Drummond (1985) konnte dabei zeigen, daß diese Unterschiede stabil sind, da er vergleichbare Belastungssituationen an drei verschiedenen Tagen wiederholte, wobei die Unterschiede bestehenblieben.

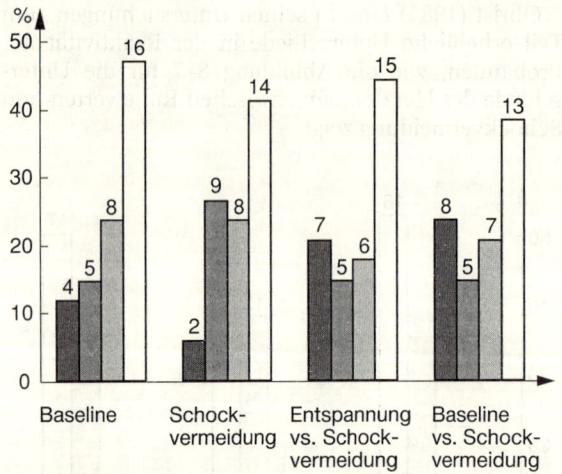

Abb. 8–9. Anzahl der Elternteile mit Hypertonie für jedes Quartil als Funktion des durchschnittlichen systolischen Blutdrucks während Baseline, dem Beginn der Schockvermeidung sowie des durchschnittlichen Unterschieds zwischen jeder Baseline und der Schockvermeidung (aus Obrist, 1981).

Dies galt ebenfalls für die Reaktion auf Orthostasebelastung. Keine Unterschiede gab es jedoch in der subjektiv erlebten Belastung zwischen den Gruppen. Lovallo, Pincomb und Wilson (1986 a, b) konnten dies für Komponenten der Blutdruckregulation (HMV, peripherer Widerstand, Herzfrequenz und Kontraktilität) sowie Cortisol und Noradrenalin nachweisen, wobei sie verschiedene Belastungstypen verwendeten und für Personen mit einer hohen Reaktivität der Herzfrequenz beim Cold-Pressure-Test bei den übrigen Belastungen eine konsistente Reaktivität der kardiovaskulären Parameter fanden. Die verschiedenen Arten der Belastung wirkten auf dieses Muster der Reaktivität steigernd im Falle der Notwendigkeit, aktiv einen aversiven Reiz zu vermeiden. Es gab in dieser Untersuchung keinen Unterschied von Typ-A- im Vergleich zu Typ-B-Personen, von denen angenommen wird, daß sie ein erhöhtes Risiko für das Auftreten eines Herzinfarkts haben.

McCann und Matthews (1988) untersuchten ebenfalls den Einfluß des Vorliegens von Hypertonie bei den Eltern auf die kardiovaskuläre Reaktivität bei Probanden. Die Ergebnisse zeigen ebenfalls, daß Probanden mit hypertonen Eltern einen höheren diastolischen Blutdruck während aller Belastungen hatten. Dieser Effekt war jedoch bei Typ-A-Personen besonders ausgeprägt. Außerdem hatten während isometrischer Belastung Personen mit hohen Werten auf einer Aggressionsskala die höchsten Blutdruckwerte. Filipowski et al. (1988) spezifizierten Aggression in nach innen und nach außen gerichtete Aggression. Die Ergebnisse einer Aktivierungsstudie zeigen, daß Probanden mit niedriger nach innen gerichteter Aggression in allen Aufgaben die höchste Reaktivität des systolischen Blutdrucks und der Herzfrequenz hatten, aber unter mentaler Belastung die niedrigsten Reaktionen des diastolischen Blutdrucks.

Matthews und Stoney (1988) konnten den genetischen Einfluß auf die kardiovaskuläre Reaktivität

noch weiter bestimmen. Dabei zeigte sich, daß die Reaktivität des diastolischen Blutdrucks zwischen Eltern und Kindern in multivariaten Ähnlichkeitsanalysen in keiner Belastungssituation ähnlich war. Nur bei isometrischer Belastung gab es eine Ähnlichkeit in den Reaktionen des systolischen Blutdrucks, während das Vorliegen elterlicher Hypertonie zu ähnlichen Reaktionen zwischen Eltern und ihren Kindern unabhängig vom Alter der Kinder führte. Dies bedeutet, daß die familiäre Belastung durch Hypertonie eine Ähnlichkeit kardiovaskulärer Reaktionen bedingt, nicht aber die familiäre Beziehung für sich.

Andere Untersucher wie Ditto (1988) oder Cumes-Raxner und Price (1988) konnten zeigen, daß auch unter Ruhebedingungen deutliche Unterschiede zwischen durch elterliche Hypertonie belasteten und Vergleichspersonen bestehen, die Unterschiede sogar oft größer sind als die der Reaktivität unter Belastungsbedingungen. Dabei gab es kaum Unterschiede in Persönlichkeitsmaßen, lediglich die Korrelation des systolischen Blutdrucks mit Neurotizismus war bei Hypertonikern höher (Sims et al., 1988).

Die genannten Untersuchungen sprechen also für das Vorliegen einer teilweise genetisch bedingten Disposition, die als spezifisch kardiovaskuläre Überreaktivität in einer Gruppe von Personen im Sinne einer Symptomspezifität vorliegt. Mit dieser spezifischen Aktivierungsreaktivität sind jedoch keine besonderen, psychologisch erfaßbaren Merkmale verbunden, zumindest nicht solche, die mit persönlichkeitsmetrischen Verfahren erfaßbar sind.

Die hier referierten Studien beziehen sich zunächst auf eine Spezifität physiologischer Reaktionen auf externe psychische Belastung. In den Untersuchungen von Elbert et al. wird dies jedoch um eine Komponente psychischer Reaktion auf Schmerz oder sonstige Aversivität erweitert, so daß eine psychophysische Disposition angenommen wird, die sich sowohl auf Merkmale psychischer als auch auf solche physischer Reaktionen bezieht.

Möglicherweise besteht hierbei eine Beziehung zu Ergebnissen einerseits aus der Angstforschung, zum anderen zu kognitiven Modellen der Persönlichkeit. Das Konzept „Repression-Sensitization" beschreibt, ähnlich wie das Modell der modulierten oder unmodulierten Angstkontrolle (Epstein und Fenz, 1965), einerseits einen habituellen Stil im Umgang mit gefahrenrelevanter Information in einer Situation, zum anderen unterschiedlich starke physiologische Reaktionen beim Umgang mit derartiger Information (Byrne, 1964).

Bezeichnen die Begriffe der individuellen Spezifität oder Stimulusspezifität zunächst einen nicht unbedingt psychophysischen Zusammenhang, so existiert doch eine Reihe von Ansätzen, die von einer psychophysischen Disposition ausgehen. Dies bezieht sich auf die Annahme derartiger Modelle, daß es eine Entsprechung psychischer und physischer Prozesse gibt, wie sie etwa im Konzept der „emotionalen Labilität" angenommen wird. Dies ist eine Erweiterung der Annahme, daß etwa Hypertoniker im Vergleich zu normotonen Personen eine möglicherweise vererbte

Tendenz haben, insgesamt mit stärkeren kardiovaskulären Reaktionen auf Umweltreize zu antworten. Diese Annahme beinhaltet keine Aussagen über Merkmale psychischer Prozesse oder Merkmale der Persönlichkeit, die mit denen physiologischer Prozesse einhergehen oder kovariieren.

8.2.4 Psychophysische Aktivierungsdisposition

Im vorangegangenen Abschnitt ging es um Merkmale physiologischer Aktivierung, mit denen sich Personen oder Gruppen von Personen unterscheiden lassen. Diese Frage ist für die Erklärung bestimmter Krankheitsbilder, z.B. der essentiellen Hypertonie oder des Herzinfarkts, von großer Bedeutung. Stellt man nun bei Personen derartige Merkmale fest und versucht, das Auftreten einer Erkrankung damit zumindest teilweise zu erklären, so impliziert dies die Annahme, daß diese Merkmale schon vor Ausbildung der Erkrankung bestanden haben, die Personen also durch diese Merkmale eine Disposition aufweisen, die die Erkrankung bei ihnen wahrscheinlicher macht als bei Personen ohne diese Disposition.

Der ursprüngliche Spezifitätsansatz der Psychosomatik bezog sich auf psychologisch beschreibbare Merkmale beispielsweise der Persönlichkeit, die Personen mit spezifischen „psychosomatischen Erkrankungen" besitzen sollen. Derartige Annahmen gingen von der kumulativen klinischen Erfahrung im Umgang mit diesen Patienten aus, ließen sich jedoch mit den verfügbaren Methoden empirischer Überprüfung, also persönlichkeitsmetrischen Verfahren, nicht bestätigen. Dies bedeutet nicht, daß derartige Annahmen unbedingt unzutreffend sein müssen, es bedeutet nur, daß sie beim gegenwärtigen Methodenstand nicht zu bestätigen sind, was auch daran liegen kann, daß die verfügbaren Methoden für die Überprüfung dieser Fragestellung nicht tauglich sind.

Andererseits enthalten persönlichkeitstheoretische Ansätze nicht selten auch Aussagen über physiologische oder andere somatische Prozesse. So wird von Personen mit ausgeprägter „emotionaler Labilität" angenommen, sie seien auch im Hinblick auf ihre vegetative Regulation labil. Für die Psychosomatik sind derartige Modelle von Bedeutung, da sie zumindest die Kovarianz psychischer und physischer Funktionsabläufe für Personen als zeitlich relativ invariant beschreiben. Es wäre daher nun denkbar, daß eine derartige Kovarianz zur Beschreibung einer möglichen psychosomatischen Spezifität tauglicher sein könnte.

Im folgenden sollen daher einige dieser persönlichkeitstheoretischen Ansätze insoweit dargestellt werden, als es dabei um Annahmen geht, die das Verhältnis somatischer und psychischer Prozesse betreffen.

Grundlage für Eysencks Modell einer psychophysiologischen Persönlichkeitstheorie (1967, 1975) ist die Unterscheidung zwischen Aktivierung und Arousal, sowie die zwischen den Persönlichkeitsdimensionen Introversion/Extraversion und Neurotizismus. Er schlägt vor, den Begriff Aktivierung lediglich für diejenigen Aktivitäten zu verwenden, die durch viszerale Strukturen, den Hippokampus, die Amygdala, den Gyrus cinguli, das Septum und den Hypothalamus vermittelt werden. Derartige Aktivierung beziehe sich auf emotionale Reaktivität oder Erregbarkeit. Unter Arousal versteht er lediglich kortikale Erregung, die ausschließlich durch das aufsteigende aktivierende Retikularsystem (ARAS) vermittelt wird. Zwischen beiden Erregungssystemen besteht eine Interdependenz insofern, als Aktivierung stets zu Arousal führt, wohingegen kortikales Arousal häufig durch eine Stimulation hervorgerufen wird, die nicht notwendigerweise Aktivierung beinhaltet. So kann kortikales Arousal „einerseits durch sensorische Stimulation oder durch Problemlöseaktivitäten des Gehirns hervorgerufen werden, ohne notwendigerweise viszerale Strukturen überhaupt zu involvieren. Andererseits kann kortikales Arousal auch hervorgerufen werden durch Emotionen, wobei in diesem Fall die Formatio reticularis involviert ist durch auf- und absteigende Bahnen, die eine Verbindung zum Hypothalamus herstellen" (Eysenck, 1975). Die Persönlichkeitsdimension Extraversion/Introversion beruht diesem Modell zufolge auf verschiedenen Schwellenwerten für kortikales Arousal, die ihrerseits durch verschiedene Schwellenwerte im ARAS bedingt sind. Introvertierte hätten demzufolge eine niedrigere Reizschwelle als Extravertierte. Introvertierte ertrügen eine sensorische Deprivation besser als Extravertierte, die ihrerseits eine höhere Schmerztoleranz aufweisen sollen.

Die folgende Abbildung 8–10 zeigt das hypothetische Modell, wie es von Eysenck postuliert wird. Auf der Abszisse sind verschiedene Stufen sensorischer Stimulation aufgetragen, auf der Ordinate verschiedene Grade der Angenehmheit oder Unangenehmheit.

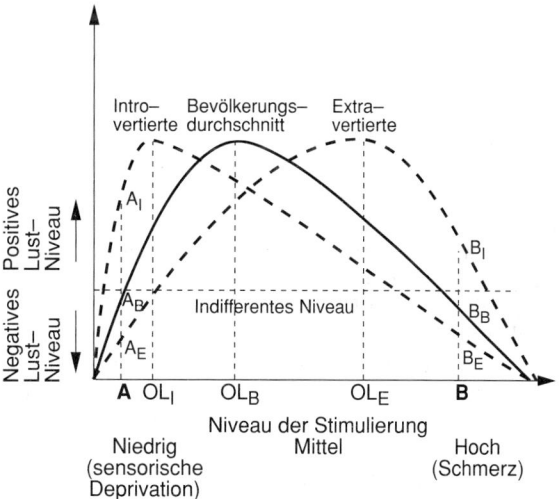

Abb. 8–10. Lust-/Unlustniveau als Funktion des Stimulusniveaus für verschiedene Personengruppen (aus Eysenck, 1975).

Die Dimension Neurotizismus schließlich sei verknüpft mit einer größeren Labilität des autonomen Nervensystems. Neurotische Personen reagierten stärker und variabler auf Stimulation, auch sei bei ihnen die Rückbildung der Erregung verzögert. In dem durch diese beiden Dimensionen definierten Raum lassen sich verschiedene Störungstypen definieren. So seien dysthyme Patienten gekennzeichnet durch eine hohes Maß an Neurotizismus und Introversion, hysterische Patienten als neurotisch und extravertiert. Extraversion und Introversion ohne Neurotizismus seien Varianten einer normalen Persönlichkeit. Delius und Fahrenberg (1966) sehen im Merkmal der Dysthymie eine Disposition für die Entstehung funktioneller Störungen (vgl. Kap. 32).

Die Vorstellungen Eysencks sind verschiedentlich aus methodischen, aber auch anderen Erwägungen kritisiert worden. Zum Teil wurden Untersuchungen zur Stützung seines Modells nur mit wenigen Probanden durchgeführt, die Ergebnisse entsprechend überinterpretiert. So merkt Myrtek (1980) an: „In der ersten Arbeit (Eysenck und Eysenck, 1967) wurde bei nur 12 Mädchen (je 6 Intro- und Extravertierte) die Speichelsekretion auf 4 Tropfen Zitronensaft gemessen. Zur anspruchsvollen Interpretation der Ergebnisse wird Pawlows Gesetz der transmarginalen Hemmung bemüht." So blieben derartige Ergebnisse kontrovers und konnten häufig von anderen Autoren nicht bestätigt werden.

Gray (1971, 1972, 1973) versuchte zwischen den von ihm ermittelten drei emotionalen Systemen: Annäherung, Verhaltensinhibition und Kampf/Flucht (vgl. Abschn. 8.4), eine Beziehung zu den von Eysenck definierten Persönlichkeitsfaktoren Extraversion/Introversion und Neurotizismus herzustellen. Durch Rotation dieser Faktoren um 45° ergeben sich seinem Ansatz zufolge zwei neue Faktoren: Angst und Impulsivität. Von Gray wurde dieses zunächst grundlegende Modell weiter elaboriert. Danach unterscheidet er dann ein System für Annäherungsverhalten oder Aktivierung (behavioral activation system, BAS), ein Hemmsystem (behavioral inhibition system, BIS), ein Bestrafungssystem („flight/fight system") und ein „Arousal-System". Letzteres ist gleichzusetzen mit dem „Erregungsmechanismus" in der Abbildung des Entscheidungsmechanismus. Dieses Arousal-System repräsentiert somit eine Intensitätsdimension und ist mit dem Retikularsystem verknüpft, wobei relativ unklar bleibt, wie die Verknüpfung mit den anderen Systemen zu denken ist.

Für die weiteren Überlegungen Grays, das Verhältnis dieser Systeme zu psychobiologischen Persönlichkeitsdimensionen betreffend, sind jedoch nur drei der Systeme wichtig geblieben, das „behavioral activation system" (BAS), das inhibitorische System (BIS) und das „flight/fight system".

Wie aus der folgenden Abbildung 8–11 ersichtlich ist, reicht die Dimension Angst vom Quadranten stabiler Extraversion bis zu dem neurotischer Introversion und soll durch Instrumente, wie etwa die Manifest Anxiety Scale gemessen werden können. Die Dimension Impulsivität reicht vom Quadranten stabiler Introversion zum Quadranten neurotischer Extraversion und ist meßbar durch die Skalen behavioraler Extraversion oder Impulsivität von Eysenck.

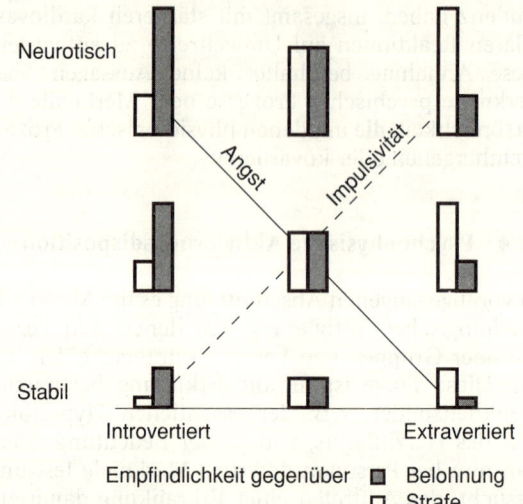

Abb. 8–11. Angenommene Beziehung zwischen a) Empfindlichkeit gegenüber Belohnungs- und Bestrafungssignalen und b) den Dimensionen Introversion/Extraversion und Neurotizismus. Die Dimensionen Angst und Impulsivität (Diagonalen) repräsentieren die größten Zunahmen der Empfindlichkeit gegenüber Belohnungs- und Bestrafungssignalen (aus Gray, 1972).

Fowles (1980) bezieht sich in seinen Überlegungen hauptsächlich auf die drei Grayschen Systeme BAS, BIS und Arousal-System. Im Unterschied zu Gray versucht er die Wirkung dieser Systeme auf physiologische Funktionen zu beschreiben, im wesentlichen auf die Herzfrequenz und die elektrodermale Aktivität. Dabei verknüpft er das behavioral activation system (BAS) mit kardiovaskulärer Aktivität, vornehmlich dem Parameter der Herzfrequenz. Betrachtet man dies unter dem Aspekt der Parametrisierung, so ist es anhand der oben dargestellten Untersuchungen von Obrist, in denen mit Hilfe der Blockierung sympathischer Einflüsse auf die Herzfrequenz nachgewiesen werden konnte, möglich, daß die Schwankungen der Herzfrequenz in einem bestimmten Bereich von Aktivierung von vagalen Einflüssen abhängig sind. Andererseits ist durch die Unterscheidung zwischen „aktivem" und „passivem Coping" im Modell von Obrist die Möglichkeit gegeben, das BAS mit aktivem Coping in Verbindung zu bringen, so daß, wenn auch nicht auf der Ebene der Parametrisierung, so doch auf einer konzeptuellen Ebene die Arbeiten von Obrist den Annahmen von Fowles nicht widersprechen. In verschiedenen Arbeiten konnte gezeigt werden, daß Steigerungen der Herzfrequenz durch belohnende Einflüsse bedingt werden, d. h. entweder durch erfolgreiches Vermeiden aversiver Reize oder durch Herstellen positiver oder belohnender Reize.

Die Sachlage ist viel weniger klar bei der Verbindung von Fluktuationen der elektrodermalen Aktivität (EDA) und dem behavioral inhibition system (BIS). Es gibt zwar Untersuchungen, die zeigen, daß die Fluktuationen der EDA von der Kontrollierbarkeit von Schocks abhängig waren, im Gegensatz zur Herzfrequenz, die durch diese Bedingung nicht beeinflußt wurde, jedoch gibt es auch Untersuchungen, die einen „nicht-inhibitorischen" Effekt anderer Reizbedingungen auf die EDA deutlich machen. In den Untersuchungen von Andresen (1987) ließ sich diese Zuordnung peripher-physiologischer Parameter zum behavioral activation system oder behavioral inhibition system nicht auffinden.

Fowles (1988) hat versucht, sein Modell auf die Erklärung psychopathologischer Zustände anzuwenden. Grundannahme ist wieder die Unterscheidung zwischen einem appetitiven und einem aversiven Motivationssystem, die antagonistisch wirken, das dritte Erregungssystem spielt hierbei keine bzw. nur eine untergeordnete Rolle. Das appetitive System wirkt als Aktivierung und resultiert in Annäherungsverhalten (BAS). Es wirkt auch bei negativer Verstärkung (hope and relief). Relief ist Resultat eines konditionierten Reizes als Sicherheitssignal. Das aversive Motivationssystem hemmt Aktivierung (BIS) in Anwesenheit von Reizen, die aversive Konsequenzen signalisieren. Passive Vermeidung hemmt Annäherungsverhalten, das durch Belohnung motiviert wird, und produziert Angst, ebenfalls ist „frustrative non-reward" Teil des BIS. Aus diesen Komponenten läßt sich eine mehrdimensionale Matrix generieren mit folgenden Einflüssen: Typ der Motivation (BIS oder BAS), Temperamentstärke der Motivation, Umgebungsbedingungen und Regulation des Motivationssystems durch die Umgebung.

Angstneigung als zeitlich überdauerndes Merkmal wird dabei einer starken Ausprägung des BIS zugeschrieben, aktuelle Angst- oder Panikattacken seien das Resultat einer fehlenden Regulierung durch externe Reize. Die in verschiedenen Untersuchungen deutlich geringere Beeinflussung durch aversive Reize bei der Psychopathie wird durch eine geringe Ausprägung des BIS erklärt. Depression wird durch eine Überaktivierung des BIS als Reaktion auf Unkontrollierbarkeit bedingt, bei verringerter appetitiver Motivation durch das BAS, das ja auch als Belohnungssystem bezeichnet wird. Führe man den Begriff der Erwartung von Belohnung ein, so sei dieses Modell kompatibel mit den kognitiven Theorien der Depression, der erlernten Hilflosigkeit. Fowles wendete sein Modell weiter zur Erklärung der Manie, der Sucht und der Schizophrenie an.

Um den Einfluß von Belohnung auf die Herzfrequenz zu untersuchen, wurden Studien durchgeführt, in denen Belohnung systematisch variiert wurde. Die Studien zeigten, daß mit zunehmender Belohnung auch die Herzfrequenz zunahm, bei 100%igem Erfolg. Aversive Motivation durch Manipulation von Mißerfolg zeigte keinen Einfluß auf die Herzfrequenz, allerdings auf die elektrodermale Aktivität. Die Gruppe mit dem höchsten Mißerfolg hatte die höchsten GSR-Reaktionen. Allerdings gelang eine Replikation dieser Ergebnisse nicht.

Myrtek (zusammenfassend 1980) hat mit einem sehr hohen methodischen Aufwand unter anderem Konzepte einer psychophysischen Reaktivität überprüft. Bei rund 700 Studenten und Patienten wurde eine Fülle physiologischer, biochemischer, anthropometrischer und psychologischer Daten erfaßt. Es kann hier nicht im einzelnen auf die, auch in methodischer Hinsicht, sehr interessanten Ergebnisse eingegangen werden. Für die Konzepte der Sympathikotonie-Vagotonie sowie psychovegetativer Labilität konnte in dieser Studie keine stichhaltige Bestätigung gefunden werden. Es gibt Hinweise auf individualspezifische Muster: „Individuen, die zum Beispiel im Reaktionszeit-Versuch mit dem Herzminutenvolumen stärker reagieren, zeigen diese Tendenz auch im Cold-Pressure-Test und beim Zahlenreihen-Versuch. Man kann jedoch nicht von einer allgemeinen Hyperreaktivität sprechen, wie die niedrigen Korrelationen ... zwischen verschiedenen Variablen bei verschiedenen Funktionsprüfungen und zwischen verschiedenen Variablen innerhalb derselben Funktionsprüfung anzeigen" (Myrtek, 1980).

Ebenso ließen sich keine Beziehungen zwischen Merkmalen der Persönlichkeit, etwa emotionaler Labilität, und objektiven vegetativen Funktionsprüfungen nachweisen, so daß das Konzept einer psychophysischen Reaktivität experimentell nicht gestützt werden konnte. „Die individuellen Aktivierungsprozesse entwickeln sich funktional wahrscheinlich nicht ‚unabhängig' von solchen Persönlichkeitsmerkmalen, die aufgrund empirischer Reliabilitäts- und Stabilitäts-Studien durchaus als habituelle Kennzeichen eines Individuums gelten können, doch entziehen sich solche Zusammenhänge – zumindest in diesem Datensatz – der systematischen Darstellung durch lineare Korrelationen und multiple Regressionen. Aus den Korrelationsmatrizen und kanonischen Analysen läßt sich auch der Umkehrschluß stützen: Jene habituellen Merkmale sind aus den hauptsächlichen Aktivierungsvariablen nicht vorherzusagen" (Fahrenberg, 1980). Dieses Ergebnis ist vor allem für das Verständnis funktioneller Störungen von Bedeutung, denn es impliziert, daß kein nennenswerter Zusammenhang besteht zwischen den Angaben über Körpervorgänge (Klagen über Beschwerden) und den Körpervorgängen selbst. In diesem Zusammenhang ist an die Untersuchungen von Mandler und Kremen (1958) zu erinnern, die keinen wiederholbaren Zusammenhang gefunden hatten zwischen der Wahrnehmung körperlicher Reaktivität und tatsächlicher Reaktivität. Im Gegenteil, Stalmann et al. (1988) fanden, daß Patienten mit Mitralklappenprolaps, Agoraphobie oder „Herzphobie" im 24-Stunden-EKG Abweichungen ihrer Herztätigkeit (Extrasystolen oder sonstige Arrhythmien) im Vergleich zu gesunden Kontrollpersonen weniger genau wahrnahmen. Im Labor konnten derartige Unterschiede nicht nachgewiesen werden.

Schonecke (1987) konnte in einer Aktivierungsstudie bei Patienten mit funktionellen Herz-KreislaufStörungen im Vergeich zu gesunden Kontrollpersonen nachweisen, daß die Patienten im Vergleich zu den Kontrollpersonen sowohl eine erhöhte kardiovaskuläre Reaktivität aufwiesen, als auch auf persönlichkeitsmetrischen Skalen, die Angstneigung, Neurotizismus und Depression erfassen, deutlich höhere Werte erreichten. Die physiologische Reaktivität kann dabei durchaus im Sinne einer Symptomspezifität interpretiert werden. Für den vorliegenden Zusammenhang bedeutsam erscheinen diejenigen Ergebnisse, die sich auf den Zusammenhang zwischen physiologischen und psychologischen Merkmalen beziehen. Dabei wurden zunächst für jede Variablengruppe (physiologisch, psychologisch) getrennt Faktorenanalysen errechnet, deren Ergebnisse dann in eine Faktorenanalyse zweiter Ordnung eingingen. Dabei ergaben sich vier Faktoren 2. Ordnung, die jeweils sowohl mit den psychologischen als auch den physiologischen Faktoren einen Zusammenhang aufwiesen. Anhand zwei dieser Faktoren ließen sich Patienten und Kontrollpersonen diskriminanzanalytisch sehr gut zuordnen. Dabei war der eine Faktor bestimmt durch Reaktivität des kardiovaskulären Systems und Merkmale wie Neurotizismus, Angstneigung, Beschwerdehäufigkeit und Depression, der zweite durch starke elektrodermale und muskuläre Aktivität (M. frontalis) sowie Aggressivität, Erregbarkeit, mangelnde Gelassenheit, Dominanzstreben und geringere Beachtung sozialer Normen, Merkmale, die eine gewissen Nähe zu Grays „Impulsivität"-Dimension aufweisen.

Diese Sekundärfaktoren können als Konstrukte aufgefaßt werden, die eine Beziehung zwischen psychischen und körperlichen Merkmalen enthalten und im Hinblick auf einen klinischen Sachverhalt sinnvoll sind. In dieser Untersuchung zeigte sich auch, daß ein Zusammenhang zwischen physiologischen und psychologischen Merkmalen bei verschiedenen Personen verschieden sein kann. So gab es eine positive Beziehung zwischen kardiovaskulärer Reaktivität und einem Merkmal „Erregbarkeit" nur bei den Patienten, nicht jedoch bei den gesunden Personen. Es zeigte sich ebenfalls, daß ein Konstrukt, das auf der Ebene psychologischer Daten allein nicht erfaßbar ist (repressive vs. sensitive Angstabwehr; Byrne, 1964), mit Hilfe der psychophysiologischen Methode aufzeigbar ist. So hatten nur die Patienten mit einer stärkeren Reaktivität kardiovaskulärer Parameter höhere Werte auf Fragebogenskalen, die insgesamt eine Vermeidung von Belastung und weniger erlebte Angst erfassen.

8.3 Streß

Es ist bereits darauf hingewiesen worden, daß die Begriffe Streß, Aktivierung und mitunter Emotion fast austauschbar verwendet werden. Selye (1981) definiert Streß in einem sehr weiten Sinne: „Streß ist die unspezifische Reaktion des Körpers auf irgendeine Anforderung." Er räumt zwar ein, daß die genannten Anforderungen spezifisch sind, „alle diese Anforderungen jedoch haben eines gemeinsam, sie erhöhen die Notwendigkeit für eine Wiederanpassung (readjustment), für die Leistung adaptiver Funktionen, die Normalität wieder herstellen." Er unterscheidet weiter zwischen „Eustreß" und „Distreß", zwischen angenehmem oder heilsamem Streß und solchem, der unange-

nehm ist und zur Krankheit führen kann. Als Konsequenz der Allgemeinheit dieser Definition folgert er, „Streß ist nicht etwas, das vermieden werden muß. Tatsächlich kann er per definitionem nicht vermieden werden. ... Komplette Freiheit von Streß ist Tod."

Es ist fraglich, ob eine so breite Definition eines Sachverhalts einen Nutzen hat, denn wenn jeder Lebensvorgang als Streß(-Reaktion) aufzufassen ist, fügt die Verwendung des Begriffes nichts Wesentliches zu irgendeiner Erklärung hinzu. Dennoch erfreut sich der Begriff großer Beliebtheit und wird im alltagssprachlichen Bereich häufig verwendet, hier allerdings in der Bedeutung, die von Selye dem Begriff Distreß zuerkannt wird. Unter Streß wird dann jedoch meist nicht eine Reaktion, sondern eine belastende Bedingung verstanden, die im wissenschaftlichen Gebrauch eher als Stressor bezeichnet wird, andererseits wird häufig dann auch von Streßreaktion gesprochen.

Die relative Verwirrung im Hinblick auf das, was durch den Begriff Streß bezeichnet wird, rührt möglicherweise daher, daß mit diesem Begriff weder die belastende Bedingung noch die Antwort eines Organismus darauf jeweils allein bzw. isoliert bezeichnet wird, sondern Streß den Aspekt der Beziehung zwischen beiden bezeichnet, er bezeichnet die Einheit zwischen Organismus und Umgebung in einer Situation bzw. als Situation, insofern als die Umgebung für den Organismus bedeutsam ist.

Aktivierung wurde definiert als Ergebnis der Notwendigkeit für den Organismus, sich mit Bedingungen der Umgebung auseinanderzusetzen. Eine derartige Definition ließe sich auch für den Begriff Streß verwenden, zumindest beinhaltet Streß diese Notwendigkeit. Betrachtet man Streß als einen „Sonderfall" von Aktivierung (Schandry, 1981), der belastende oder schädigende Umgebungsbedingungen enthält, so treffen die unter dem Begriff Aktivierung dargestellten Merkmale des Prozesses oder des Zustands auch auf Streß zu. Die schädigend oder belastend aktivierende Wirkung einer Bedingung ist abhängig von ihrer Art, Dauer und Intensität. Es ist, wenn überhaupt, nur sinnvoll, den Begriff Streß zur Bezeichnung einer Organismus-Umgebung-Interaktion zu verwenden, die für den Organismus entweder schädigend oder belastend ist.

Es gibt verschiedene Ansätze, mit denen versucht wird, das Phänomen Streß näher zu spezifizieren. Von Selye (1956, 1971, 1981) wurde Streß als biologisches Phänomen betrachtet, ähnlich wie von Henry und Stephens (1977), Ely und Henry (1980), jedoch mit einem Schwerpunkt auf der Erforschung allgemeiner körperlicher Anpassungs- und Reaktionsmuster. Henry und Mitarbeiter gehen von einem eher ethologisch orientierten Standpunkt aus und betrachten Streß ebenfalls als ein biologisches Phänomen, das jedoch bei sozial lebenden Organismen auch durch soziale Verhältnisse bedingt sein kann. In beiden Ansätzen jedoch geht es darum, die körperlichen Prozesse, die für die Streßreaktionen wesentlich sind, in einem System zu erfassen, indem sie die Interaktion zwischen Umgebungsbedingungen, Prozessen des Zentralnervensystems, endokrinen Funktionen und dem beobachtbaren Verhalten zu systematisieren versuchen. Diese Modelle enthalten differenzierte Vorstellungen über die Rolle gegenseitiger Interdependenzen der beteiligten körperlichen Teilsysteme. Mit hormonellen Reaktionsanteilen haben sich Mason (1972, 1975) und Frankenhaeuser (1971, 1975a, b) beschäftigt (vgl. Abschn. 8.4.2). Bei diesen Autoren liegt der Schwerpunkt ihres Interesses auf den physiologisch-endokrinen Reaktionsanteilen (vgl. Kap. 10).

Andere Autoren sind eher an der Art der belastenden Bedingungen, den „Stressoren", interessiert. In diesem Zusammenhang ist vor allem der Ansatz zu nennen, der den Beginn von Erkrankungen im Zusammenhang mit belastenden Lebensereignissen sieht (Hinkle und Wolff, 1957; Hinkle, 1961; Holmes und Rahe, 1967). Dieser Ansatz steht dem klinischen Konzept von G. Engel (1972) und Schmale (1969) des „giving up" nahe, das zwar eine Art der Reaktion eines Individuums beschreibt, aber eine Reaktion auf Bedingungen, die das Individuum vor ein oder mehrere ihm unlösbar erscheinende Probleme stellt, so daß es aufgibt zu versuchen, sie zu lösen. Beide Autoren sehen einen derartigen Vorgang in einem lebensgeschichtlichen Zusammenhang und sind der Meinung, daß der Krankheitsbeginn ebenfalls in Zusammenhang mit den belastenden Lebensereignissen stehen kann.

Andere Ansätze sehen die Art der „Bewältigung" (coping) von Belastungen als wesentliche Bedingung für Streß. Ob eine Bedingung beispielsweise schädigend ist, hängt von der Art und Weise des Umgangs damit ab. Die Rolle psychologischer Prozesse der Informationsverarbeitung wurde vor allem von der Arbeitsgruppe um Lazarus (1971, 1975, 1980) untersucht. Die Wichtigkeit derartiger Prozesse wird aber auch zunehmend in anderen Ansätzen gesehen, z.B. im Bereich der Erforschung belastender Lebensereignisse (Siegrist und Dittmann, 1981).

Wie bereits oben dargestellt, faßt Selye den Streßbegriff sehr weit, so daß Streß fast alle Lebensprozesse begleitet. Obwohl er unter Streß einerseits die Reaktion eines Organismus auf irgendwelche Anforderungen versteht, spricht er dennoch von Streßreaktionen, die auf die Anforderungen hin entweder lokal (LAS – lokales Adaptations-Syndrom) oder generell (GAS – generelles Adaptations-Syndrom) erfolgen. Für den vorliegenden Zusammenhang ist hauptsächlich das GAS von Bedeutung. Es besteht aus mehreren Phasen:

– Alarmreaktion
– Widerstandsphase
– Phase der Erschöpfung.

Die Alarmreaktion tritt auf, wenn der Organismus mit irgendwelchen Umständen konfrontiert wird, an die er nicht „angepaßt" (adaptiert) ist. Die Reaktion beginnt mit der Schockphase, der ersten unmittelbaren Reaktion auf die belastende Bedingung. Es kommt zu körperlichen Reaktionen wie Tachykardie, verringertem Muskeltonus und Blutdruck sowie einem Abfall der Körpertemperatur. In der Gegenschockphase kommt es zur Gegenregulation gegen diese körperlichen Reaktionen. Die Sekretion von Nebennierenrindenhormonen wird verstärkt.

In der Widerstandsphase kommt es zur Anpassung an die belastende Bedingung und zu einem Verschwinden der Symptome. Jedoch ist der Widerstand gegen andere Belastungen verringert. Dauert die Belastung jedoch zu lange oder ist sie zu stark, so kommt es zur Erschöpfung. Auch ist der direkte Übergang von Alarmreaktion in die Phase der Erschöpfung möglich.

Selye sieht in Cannons „Notfallreaktion", die mit erhöhter Sekretion von Adrenalin unter anderem zu Anstiegen von Blutdruck und Herzfrequenz führt, eine mögliche Phase der Alarmreaktion als Abwehr gegen eine äußere Bedrohung (Selye, 1981). Wichtig an Selyes Auffassung ist, daß er den Organismus in der Auseinandersetzung mit der Umgebung als veränderbar ansieht. Dies nicht nur auf einer organischen Ebene, obwohl er Streß immer als „biologischen Streß" ansieht. „Ich habe die Bezeichnung ‚generelles Adaptations-Syndrom' gewählt, um eine Klasse von Symptomen zu beschreiben, die in einer Vielzahl von Situationen konsistent beobachtet werden können. Es muß ein gemeinsamer Nenner existieren; ich habe ihn ‚biologischen Streß' genannt" (Selye, 1981).

Der Zustand eines Organismus ist in jedem Augenblick von der Einwirkung verschiedenster Belastungen bedingt, an die er sich anpassen muß. Ihr Effekt ist kumulativ, d.h., Auseinandersetzung mit einer Belastung verändert die Möglichkeiten der Anpassung an eine andere Belastung. Streß sei zwar nicht zu vermeiden, aber man könne unnötigen Streß vermeiden oder verhindern, daß neutrale Ereignisse zu Stressoren würden, empfiehlt Selye. „Wir müssen lernen ‚overstress' zu erkennen, wenn wir die Grenzen unserer Anpassungsfähigkeit überschritten haben; oder ‚understress', wenn wir unter einem Mangel an Selbstverwirklichung leiden (physische Immobilität, Langeweile, sensorische Deprivation)" (Selye, 1981). In dieser Formulierung wird deutlich, daß Selye den Streßbegriff nicht auf organische Bedingungen von Belastung beschränkt, obwohl seine Beispiele dies mitunter vermuten lassen.

Henry und seine Mitarbeiter gehen von einem Standpunkt aus, der für Organismen, welche in sozialen Gruppen leben, biologische und soziale Bedingungen des Lebens zugrunde legt. „Soziale Systeme von Säugetieren, einschließlich Affen und Nagetiere, können im Kontext dreier gemeinsamer Spezialisierungen von Rollen und Verhalten betrachtet werden.

1. Alle besitzen eine solche Hierarchie, in der einige Individuen dominant, andere untergeordnet sind.

2. Alle besitzen eine funktionelle Unterteilung in männliche und weibliche Individuen mit unterschiedlichen biologischen Rollen, die dennoch in das Gebilde einer einheitlichen Gruppe verwoben sind.

3. Alle Individuen entwickeln sich von einer hilflosen, verwundbaren Kindheit durch Reifung, Stärke und Aktivität hin zu Abhängigkeit und Tod" (Henry und Stephens, 1977).

In zahlreichen Untersuchungen mit Mäusen, „die in zwei Jahren das Äquivalent einer vollen menschlichen Lebensspanne durchlaufen", fanden sie, daß für die Fähigkeit, stabile soziale Hierarchien zu bilden oder in ihnen zu leben, die ersten frühen Lebenserfahrungen von Wichtigkeit sind. Diese können gestört werden durch soziale Isolierung oder durch ein gestörtes Verhalten der Muttertiere. Derartig gestörte Tiere sind später nicht fähig, in sozialen Gruppen zu leben. Diese bilden für sie eine ständige Belastung. „Als Konsequenz zeigen sie wiederholtes Arousal des sympathischen Nervensystems und des Nebennierenmarks (Henry, 1971) und der Nebennierenrinde (Henry und Stephens, 1977b). So sind die notwendigen Bedingungen zur Induzierung chronischer Erkrankungen erfüllt" (Ely und Henry, 1980).

Dominante Tiere in sozialen Verbänden zeigen eine höhere Aktivität der „Sympathikus-Nebennierenmark-Achse", submissive Tiere eine höhere Aktivität der „Hypophysen-Nebennierenrinden-Achse", wobei die Aktivität dieser Systeme dem Ausmaß der indizierten sozialen Belastung entspricht. So neigen dominante Tiere unter sozialer Belastung zur Ausbildung einer Hypertonie als Anzeichen bzw. Folge einer chronisch erhöhten sympathischen Erregung, wie die folgende Abbildung 8–12 zeigt.

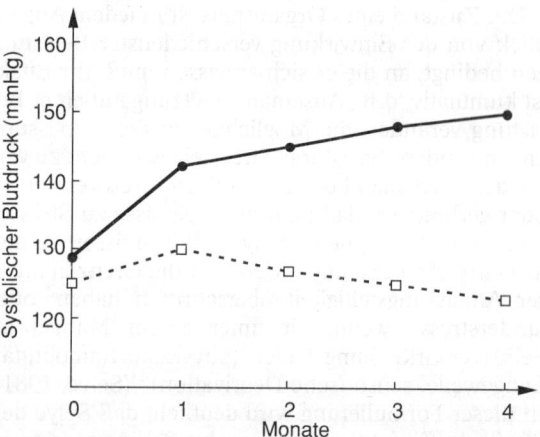

Abb. 8–12. Blutdruckreaktionen dominanter männlicher Tiere –●– im Vergleich zu submissiven Tieren –□–, die in einer Kolonie sozial interagieren (aus Ely und Henry, 1980).

Die Autoren sehen eine Beziehung zum Konzept der erlernten Hilflosigkeit von Seligman (1975), in dem die Möglichkeit einer aktiven und erfolgreichen Bewältigung von Umgebungssituationen, hier sozialer Natur, einhergeht mit sympathisch-adrenerger Erregung. Bleibt der Erfolg jedoch aus, bzw. besteht keine Möglichkeit, die Umgebung erfolgreich zu beeinflussen, so kommt es zu erhöhter Nebennierenrindenaktivität und Hilflosigkeit. Solange eine Situation von Unsicherheit und Neuheit gekennzeichnet ist, also noch nicht entschieden ist, ob eine Bewältigung erfolgreich sein wird, solange sind beide Systeme aktiv.

In zahlreichen Untersuchungen wurde versucht, den belastenden Charakter bestimmter Situationen zu ermitteln. Dies geschieht im allgemeinen dadurch,

daß in einer Laborsituation die entsprechenden Bedingungen hergestellt werden und ihre aktivierende Wirkung ermittelt wird. Aus der aktivierenden Wirkung kann jedoch noch nicht unbedingt auf die Erlebnisvalenz (lästig/unangenehm) oder eine schädigende Wirkung geschlossen werden. Tatsächlich implizieren fast alle Ansätze, daß belastende Bedingungen erst dann eine schädigende Wirkung besitzen, wenn sie eine bestimmte Dauer oder Häufigkeit haben. Dies wird im Modell von Selye oder in dem von Henry und Mitarbeitern ebenfalls angenommen. Dennoch ist es von Interesse zu ermitteln, ob bestimmte Situationen in hohem Maße aktivierend und belastend sind. Dies gilt vor allem für berufsbezogene Belastungen, die die jeweiligen Personen ja nur schwer vermeiden können. Als Beispiel hierfür soll eine Untersuchung von Becker-Carus (1981) dienen, der bei einem Showmaster die Herzfrequenz in einer Probe- und in der Auftrittssituation einer Live-Sendung gemessen hat. Die folgende Abbildung 8–13 zeigt den Verlauf der Herzfrequenz für beide Situationen.

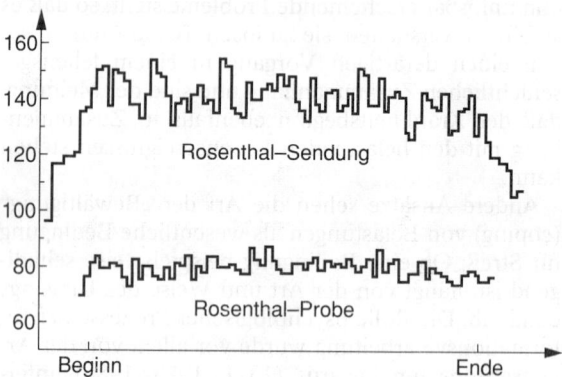

Abb. 8–13. Streßbelastung bei öffentlichen Auftritten gemessen an der Herzfrequenz. Obere Linie: Live-Sendung (Rosenthal), untere Linie: Probe derselben Sendung (aus Becker-Carus, 1981).

8.3.1 Belastende Lebensereignisse (life stress)

Außer in direkt traumatischen Situationen werden Bedingungen erst dann schädigend wirken, wenn sie eine bestimmte Dauer und Intensität aufweisen. Geht man davon aus, daß Streß beinhaltende Situationen eine Anforderung an einen Organismus zur Anpassung stellen, so ist dies in Lebenssituationen der Fall, die eine mehr oder weniger starke Änderung der Lebensumstände mit sich bringen, an die sich eine Person anpassen muß. Hinkle und Wolff (1957) kamen in einer epidemiologischen Untersuchung zur Verteilung von Krankheiten in einer Stichprobe von fast 3000 Angestellten einer amerikanischen Telefongesellschaft zu dem Ergebnis, daß die Krankheiten sich nicht auf alle Personen nach der statistisch zu erwartenden Häufigkeit verteilten. Das Auftreten von Krankheiten war bei denjenigen Personen gehäuft, die starken sozialen Belastungen ausgesetzt waren. Auch bei der Betrachtung dieser Personen zeigte sich, daß Krankheiten hauptsächlich zu Zeiten auftraten, in denen die Personen den genannten Belastungen ausgesetzt waren. Mit der Anzahl der Krankheitsepisoden stieg die Anzahl der betroffenen Organsysteme sowie die Anzahl der Stimmungs- und Verhaltensstörungen.

Holmes und Rahe (1967) entwickelten einen Fragebogen, der 43 Lebensereignisse enthielt, das „Schedule of Recent Experiences" (SRE). Da die belastende Bedeutung der einzelnen Ereignisse verschieden war, und damit auch die Notwendigkeit zur Anpassung, wurde eine Skala zur Einschätzung der für jedes Ereignis notwendigen Anpassungsleistung, die „Social Readjustment Rating Scale" (SRRS) entwickelt. Diese Skala ordnet jedem Ereignis ein bestimmtes Gewicht entsprechend der notwendigen Anpassungsleistung zu. Das Ergebnis dieses Fragebogens besteht dann in der sog. Life Change Unit (LCU), d.h. dem Gewicht eines Ereignisses, multipliziert mit der Häufigkeit seines Auftretens. Mit diesem Instrument konnten die Autoren in vielen Untersuchungen einen Zusammenhang zwischen hoher Lebensbelastung und dem gehäuften Auftreten von Erkrankungen aufzeigen. In einer prospektiven Studie bei Angehörigen der US-Marine konnte der prädiktive Wert des LCU-Wertes gezeigt werden, wie die folgende Abbildung 8–14 zeigt (Rahe et al., 1970).

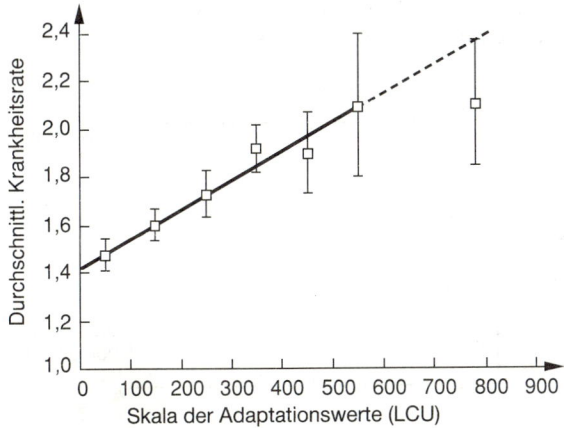

Abb. 8–14. Beziehung der Durchschnittszahl der Krankheiten zum mittleren Gesamt-LCU-Wert während des Einsatzes zur See (Rahe et al., 1970).

Trotz der zum Teil eindrucksvollen Ergebnisse dieses zunächst eindimensionalen Ansatzes ergibt sich eine Reihe von Fragen, die im Prinzip denjenigen gleichen, die sich im Zusammenhang mit einem eindimensionalen Konzept der Aktivierung gestellt hatten, es sind Fragen nach dem möglichen Einfluß von Spezifitäten. In der oben erwähnten Untersuchung von Hinkle und Wolff (1957) hatte es sich gezeigt, daß bestimmte Personen häufiger belastet und entsprechend häufiger erkrankt waren. Damit stellt sich die Frage nach einer besonderen psychosozialen Vulnerabilität (Hinkle, 1974), die ihrerseits wiederum mehrere Faktoren enthalten kann, wie Konfliktanfälligkeit, Bewältigungsmöglichkeiten u.ä. (Siegrist und Dittmann,

1981). Johnson und Sarason (1979) geben eine Übersicht über sog. Moderatorvariablen, die den Zusammenhang zwischen belastenden Lebensereignissen und Krankheit beeinflussen. Sie nennen im einzelnen: „soziale Unterstützung", „Locus of Control", „Perceived Control", „Sensation Seeking" und „Level of Arousability". Die Autoren zitieren verschiedene Studien, in denen gezeigt werden konnte, daß diese Variablen durchaus den Zusammenhang zwischen LCU-Veränderungen und Krankheit beeinflussen. Von Dohrenwend und Dohrenwend (1978) wurde daneben als wichtiger Faktor die individuelle Erfahrung einer Person im Umgang mit Belastungen herausgestellt. Rahe (1978) betont den Einfluß von Verarbeitungsmechanismen wie Verleugnung oder Verdrängung, denen in diesem Zusammenhang eine schützende Funktion zuerkannt wird.

Letztlich stellt sich in diesem Zusammenhang wiederum das „Spezifitätsproblem" der Psychosomatik, indem gefragt werden kann, ob spezifische Ereignisse bei spezifischen Personen mit einer spezifischen Disposition über spezifische vermittelnde Prozesse zu spezifischen Erkrankungen führen. In der Herzinfarktforschung scheinen sich möglicherweise gewisse Spezifitäten anzudeuten, indem bei bestimmten Personen eine Disposition vorliegt (Typ-A-Persönlichkeit), die u.a. bewirkt, daß sie auf bestimmte Situationen (Leistung im Zusammenhang mit sozialem Wettbewerb) mit bestimmten (sympathisch-adrenergen) Prozessen und Verhalten (coronary prone behavior) antworten, was zu einer erhöhten Wahrscheinlichkeit des Auftretens einer bestimmten Erkrankung (Herzinfarkt) führt (vgl. z.B. Dembrowski, 1979).

Ein weiteres Problem, das nicht unabhängig vom Problem der Spezifität gesehen werden kann, besteht in der Vermittlung von belastenden Ereignissen, d.h., über welche Mechanismen üben derartige Ereignisse ihre schädigende Wirkung auf den Organismus aus. Die allgemeinen Streßmodelle versuchen, psychophysiologische oder psychoendokrine Mechanismen aufzuzeigen, die als Vermittlung der schädigenden Wirkung bestimmter Umgebungsbedingungen (Stressoren) auf den Organismus angesehen werden. So geht auch der Ansatz der Lebensereignisforschung davon aus, daß diese Ereignisse über Mechanismen auf den Organismus wirken, die in Labor- oder Feldstudien als Streßreaktionen ermittelt wurden.

Für den Bereich der Koronarerkrankung gibt es in der Zwischenzeit eine Fülle von Untersuchungen, in denen eine im weitesten Sinne kardiovaskuläre Übererregbarkeit von Personen mit einem hohen Risiko für eine koronare Herzerkrankung festgestellt wurde. Dabei ist wichtig, daß sowohl auf der Ebene der Stimuli oder Stressoren als auch auf der Ebene der Reaktionen diese Spezifitäten gefunden wurden. So fanden Steptoe und Ross (1981), daß Personen mit hoher kardiovaskulärer Reaktivität, gemessen an Änderungen der Herzfrequenz und der Pulswellengeschwindigkeit, in Parametern der Atmung (Atemtiefe und Frequenz) und der Hautleitfähigkeit nicht mehr reagierten als Personen mit vergleichsweise niedriger kardiovaskulärer Reaktivität. Allerdings zeigte sich in dieser Studie kein Zusammenhang mit Werten auf der „Jenkins Activity Scale" (JAS), mit der das Typ-A-Verhalten erfaßt worden war. Dembrowski und Mitarbeiter (1979) hatten entsprechend gefunden, daß die JAS für die Erfassung des Anteils des Typ-A-Verhaltens, das mit erhöhter kardiovaskulärer Reaktivität einhergeht, weniger geeignet ist als ein semistrukturiertes Interview.

Die Autoren führen den fehlenden Zusammenhang zwischen den Werten auf der JAS mit kardiovaskulärer Reaktivität u.a. darauf zurück, daß die Belastungssituation einen zu geringen Grad an „Anforderung" (challenge) aufwies. So hat sich in einer Studie von Gastorf (1981) gezeigt, daß nicht nur das Ausmaß der Aufgabenschwierigkeit bei Typ-A-Personen zu einer stärkeren kardiovaskulären Reaktivität führt, sondern daß es auch durch die Erwartung einer hohen Schwierigkeit bei objektiver Einfachheit von Aufgaben zu der erhöhten Reaktivität kommt. Goldband (1980) konnte zeigen, daß Merkmale einer Belastungssituation, die für das Typ-A-Verhalten Relevanz besitzen, wie Zeitdruck oder Wettbewerbscharakter, für die erhöhte Reaktivität von Typ-A-Personen wesentlich sind und ebenfalls nicht der Schwierigkeitscharakter der Aufgabe. Nur unter dieser Bedingung unterschieden sie sich von Personen des Typs B. Die Ergebnisse von Dembrowski und Mitarbeitern (1979) zeigen auch, daß bei Vorliegen der entsprechenden Verhaltensmerkmale bereits eine geringfügige Relevanz der Situation im Sinne von „challenge" bei Typ-A-Personen zu deutlich höheren physiologischen Reaktionen führt im Vergleich zu Typ-B-Personen.

Neuere Untersuchungen (Perini und Bühler, 1984; Shekelle et al., 1984; Appels et al., 1984) machen es jedoch zweifelhaft, daß das Typ-A-Verhalten, so wie es zunächst global als unabhängiger Risikofaktor definiert worden war, tatsächlich einen Vorhersagewert für das Auftreten von koronaren Herzerkrankungen besitzt. Es scheint vielmehr, daß lediglich einige Komponenten des Verhaltens wie „hostility" (Feindseligkeit) prädiktive Potenz für die Vorhersage der KHK besitzen (vgl. Kap 41.1).

Eine ausführliche Darstellung von Forschungsergebnissen aus diesem Bereich findet sich in Kapitel 41.1. Für den vorliegenden Zusammenhang ist es jedoch wesentlich, darauf hinzuweisen, daß für eine bestimmte Gruppe von Personen spezifische Verhaltensmerkmale typisch sind, und sich diese Personen anhand dieser Merkmale des Verhaltens identifizieren lassen. Weiterhin neigen diese Personen in bestimmten Situationen zu spezifischen physiologischen Reaktionen. So reagieren sie mit stärkeren Änderungen kardiovaskulärer Funktionen im Vergleich zu anderen Personen, aber auch im Vergleich zu anderen eigenen physiologischen Funktionen. Von der erhöhten kardiovaskulären Reaktivität wird weiterhin angenommen, daß sie die Umgebungseinflüsse umsetzt in eine schädigende Wirkung auf den Organismus, der schließlich koronar erkrankt.

Weiterhin ist festzuhalten, daß es nicht die „objektiven" Merkmale von Anforderungen sind, die diese an-

genommenen pathogenen Mechanismen in Gang setzen, sondern die Bewertung der Situation durch das Individuum. Wird diese als Herausforderung oder Wettbewerbssituation bewertet, dann kommt es zu den entsprechenden Reaktionen, d.h., die individuelle Wirklichkeit, die sich aus der Interaktion zwischen Individuum und Umgebung ergibt (vgl. Abschn. 8.1.2), beinhaltet dann Reaktionen, die möglicherweise der Situation nicht angemessen sind und damit pathogen sein können. Die erhöhte Anstrengung und Erregung führt nicht immer zu erhöhter Leistung (Gastorf, 1981).

Darüber hinaus spielt für die aktivierende und möglicherweise schädigende Wirkung einer Belastung nicht nur ihre „objektive" Schwierigkeit eine Rolle, sondern auch die Möglichkeiten des Individuums zu ihrer Bewältigung oder ihrer Kontrolle. Karasek und Russel (1982) schlagen ein zweidimensionales Modell zur Wirkung belastender Situationen auf einen Organismus vor, mit den Dimensionen „Stressor" (stark/schwach) und „Kontrolle" (niedrig/hoch). Einen hohen Belastungswert haben nur die Situationen mit starken Stressoren und niedriger Kontrollmöglichkeit in bezug auf die Situation. Situationen, die starke Stressoren enthalten, aber durch das Individuum kontrolliert werden können, führen zu einer Erhöhung der eigenen Kapazität. Die Autoren versuchen, ihr Modell mit Modellen der Aktivierung (z.B. Hebb) oder dem Streßmodell von Selye zu verbinden, wie die folgende Abbildung 8–15 zeigt.

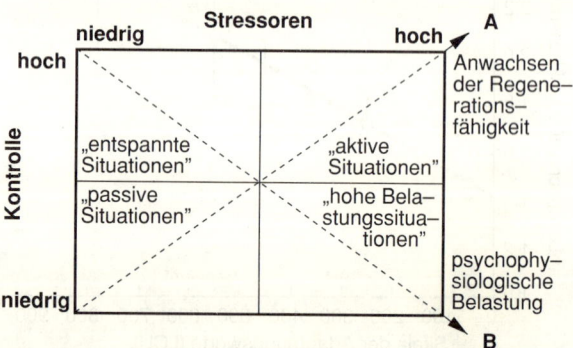

Abb. 8–15. Kategorisierung von Umweltsituationen und physiologischen Reaktionen (aus Karasek et al., 1982).

Das Beispiel des Typ-A-Verhaltens und seine Beziehung zum erhöhten Risiko für eine koronare Herzerkrankung wurde gewählt, um pathogene psychophysiologische Mechanismen zu erläutern, weil es zum einen eine Vielzahl von Untersuchungen auf diesem Gebiet gibt, zum anderen eine Verbindung zu Spezifitätskonzepten der allgemeinen Aktivierungstheorie herstellbar ist. Es ist sicher noch zu früh, um eine tatsächliche Gültigkeit der in diesem Modell implizierten Spezifitätsannahmen anzunehmen. Das Modell ist jedoch auch aus einem weiteren Grund interessant, es ist in einem methodischen Sinne spezifisch. Im allgemeinen wurde mit bestimmten vorhandenen Methoden zur Erfassung von Merkmalen der Persönlichkeit versucht, spezifische Merkmale von Personen mit bestimmten Erkrankungen aufzufinden, von denen dann angenommen werden sollte, daß sie ursächlich in der Pathogenese der entsprechenden Erkrankung wirksam seien. Diese Forschungsrichtung hat ihren Ursprung in psychoanalytischen Spezifitätsannahmen, die sich auf spezifische Entwicklungsdefekte und daraus resultierende Merkmale und Defekte der dann unter Umständen erkrankten Persönlichkeit beziehen. Davon unterscheidet sich der Ansatz des Typ-A-Verhaltens vor allem dadurch, daß aus der klinischen Beobachtung spezifische Methoden entwickelt wurden, um das klinisch Beobachtete zu erfassen, und nicht auf vorhandene Methoden, etwa zur Messung der Persönlichkeit, zurückgegriffen wurde, deren Validität in diesem Zusammenhang fragwürdig gewesen wäre. Dies bedeutet nicht, daß die Operationalisierung des Typ-A-Verhaltens gegenwärtig als endgültig gelöst anzusehen wäre (Matthews, 1982, 1983), es bedeutet jedoch, daß der Ansatz auch in methodischer Hinsicht Spezifitäten berücksichtigt. Es ist vom theoretischen Standpunkt abhängig, ob man annimmt, es handele sich bei Personen mit dem Typ-A-Verhalten um spezifische Persönlichkeiten, deren persönliche Spezifität sich u.a. im Typ-A-Verhalten zeigt. Von Vertretern dieses Ansatzes wird oft betont, daß es sich bei Typ-A-Verhalten nicht um Zeichen eines oder mehrerer Merkmale der Persönlichkeit (trait) handele (Matthews, 1982). Dieser Aussage liegt eine Merkmalsdefinition zugrunde, die ein Persönlichkeitsmerkmal als global betrachtet, d.h. seine Wirkung auf Verhalten nicht in Abhängigkeit von bestimmten Situationen sieht. Vom Typ-A-Verhalten wird hingegen angenommen, daß es von ganz bestimmten situationellen Bedingungen abhängt, wie es im moderneren Persönlichkeitskonzept des sog. „Interaktionismus" ohnehin auch betrachtet wird (vgl. Kap. 14 und 19).

8.4 Emotion

8.4.1 Emotion als Gefühlszustand

Emotion als Gefühlszustand betrifft das Erleben eines Zustands. Dabei beinhaltet „Zustand" ein einheitliches Phänomen, das die Beziehung eines Organismus zu seiner Umgebung erlebbar enthält. Diese Beziehung ist geprägt von für den Organismus bedeutsamen Merkmalen der Umgebung, die etwa „Anforderungscharakter" haben können, von Reaktionen auf die Umgebung, deren Bewertung usw. Im folgenden Abschnitt werden diejenigen Konzepte dargestellt, die von diesem Aspekt von Emotion ausgehen. Diese Konzepte haben ihrerseits verschiedene Schwerpunkte. Kognitiv orientierte Konzepte gehen zum Teil kaum auf physiologische Reaktionsanteile ein, sie setzen diese voraus. Zum Teil wird die Veränderung physiologischer Funktionen und deren Wahrnehmung als wesentliche Bedingung für das Entstehen von Emotionen angesehen, wobei die Spezifität von „Gefühlen" zum einen als abhängig von spezifischen physiologischen Reaktionen angesehen, zum

anderen als durch Merkmale der Umgebung bedingt betrachtet wird.

Fahrenberg (1979) unterscheidet zwischen Stimmungen, Gefühlen, Emotionen und Affekten. Stimmungen bezeichnen danach „relativ überdauernde Qualitäten, welche das persönliche Erleben färben, entweder als leibbezogenes Befinden ... oder als mehr atmosphärische Qualitäten ... Wenn sich aus diesen eher diffusen Gestimmtheiten unter dem Einfluß bestimmter Ereignisse und Reize aktuelle Regungen herausdifferenzieren, bezeichnet man sie als Gefühle. Introspektiv lassen sie sich als stärker umrissene, gerichtete und aktualisierte Erlebnisqualitäten beschreiben". Werden diese weiter verstärkt, handele es sich um Emotionen oder, bei einer weiteren Steigerung der Erlebnisintensität, „welche die ganze Person ergreift, erschüttert und ausrichtet", um Affekte. Hierbei wird eine Typisierung an einer Dimension vorgenommen, die eine Kombination von Intensität und Zielgerichtetheit enthält. Sie bezieht sich damit auf zentrale Elemente verschiedener Emotionstheorien, in denen nicht immer eine derartige Bemühung um Definitionen enthalten ist, in denen aber ebenfalls die erlebnisbezogene Betrachtungsweise deutlich wird.

8.4.2 Valenz

Von vielen Emotionstheorien wird die Valenz als ein wichtiges Element von Emotion angesehen. Im allgemeinen wird zwischen positiver und negativer Valenz unterschieden oder entsprechend der lerntheoretischen Unterscheidung zwischen Annäherung und Vermeidung oder der Art der Verstärkung (positiv/negativ).

Theoretisch erfordert dieser Gedanke einen Entscheidungsmechanismus, der negative von positiven Reizen unterscheiden kann. In jedem Fall müssen Merkmalskonfigurationen verglichen und aufgrund des Vergleichs Entscheidungen getroffen werden. Dieser Vorgang ist in den verschiedensten Differenzierungen denkbar, wobei eine Differenzierung vor allem durch den Einfluß von Erfahrungen angenommen werden kann. Gray (1971, 1973) hat ein Modell für Annäherung und Vermeidung vorgestellt, in dem auch ein Entscheidungsmechanismus enthalten ist. Dieser entspricht einem „Entweder/oder-Schalter", wie er aus der Schaltalgebra bekannt ist. Das Modell beruht auf der Annahme zweier exklusiver Valenzen. Es wird darüber hinaus ein „Vergleicher" eingeführt, der tatsächliche mit erwarteten Konsequenzen eines Verhaltens vergleicht und so unmittelbar über eine Rückkoppelung auf den Entscheidungsmechanismus einwirkt, wie die folgende Abbildung 8–16 zeigt. Diese Rückkoppelung erst läßt das System lernfähig werden. Verhalten wird auf diese Weise als ein dynamisch geregeltes Geschehen aufgefaßt, wobei deutlich wird, daß das Prinzip des operanten Lernens keineswegs, wie häufig irrtümlicherweise angenommen wird, eine Linearität von Erklärungen impliziert.

Andresen (1987) hat sich sehr ausführlich mit dem Valenzkonzept in der Emotionsforschung, aber auch

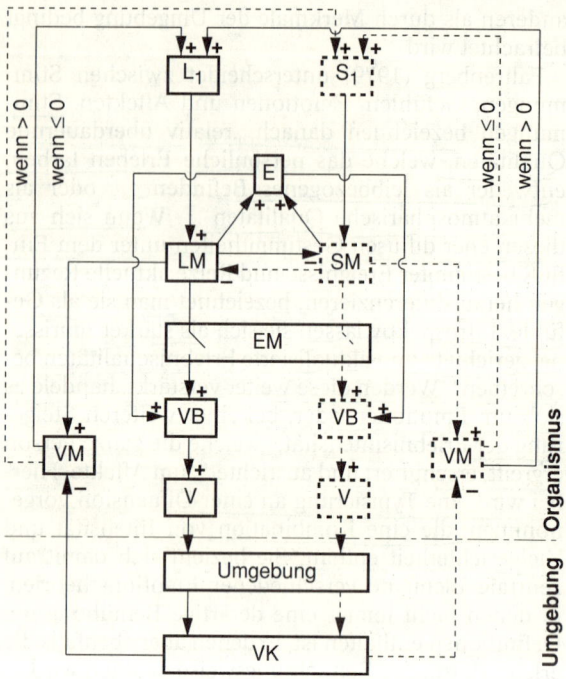

Abb. 8–16. Vollständige Darstellung des Gray-Smithschen Modells. L_1 und S_1 = Input in den Belohnungsmechanismus (LM) bzw. Bestrafungsmechanismus (SM). EM = Entscheidungsmechanismus. E = Erregungsmechanismus. VB = Verhaltensbefehl „Annäherung" (auf der Belohnungsseite) bzw. „Stop" (auf der Bestrafungsseite). V = beobachtetes motorisches Verhalten. VK = Konsequenzen (Lohn oder Strafe) des auftretenden Verhaltens. VM = Vergleichermechanismen, die die tatsächlichen mit den erwarteten Folgen vergleichen und von denen ein entsprechender Belohnungs- bzw. Bestrafungsinput ausgeht.

der „differentiellen Psychophysiologie" auseinandergesetzt. Mit dem Begriff „valenzkonträre Aktivierungsdimension" bezeichnet er den Umstand, daß „viele Gefühlskategorien auf einer Ordnungsdimension der emotionalen Valenz bipolar korreliert bzw. geladen sind". Dies zeigt sich nicht nur in bipolaren Emotionsbezeichnungen wie Freude/Trauer, sondern hat auch einen Stellenwert in Konzepten habitueller Reaktionstendenzen, anhand derer Personen unterscheidbar sind, etwa im Modell Grays und Fowles oder Eysencks. In diesen Modellen spielt die Annahme der Bestimmtheit von Verhalten durch den unterschiedlich ausgeprägten Einfluß von Annäherungs- oder Vermeidungssystemen (Gray, 1973; Fowles, 1980) oder eine unterschiedlich ausgeprägte Empfindlichkeit gegenüber bestrafenden oder belohnenden Ereignissen in der Umgebung (Eysenck, 1975) eine wesentliche theoretische Rolle. In einer eigenen, sehr aufwendigen Untersuchung konnte Andresen den Einfluß einer bipolaren Valenzdimension nicht nur auf der Ebene subjektiven Befindens, sondern auch auf der peripher-physiologischer Reaktivität nachweisen. Dabei waren die Valenzdimensionen, ausgehend vom Modell Izards, durch „Angst" und „Reizsuche" (information seeking) definiert. Untersuchungssituationen ließen sich anhand dieser Dimension gruppieren und deren Einfluß war der bedeut-

samste auf die physiologische Reaktivität. Vor allem war in dieser Untersuchung der Einfluß individuell invarianter Reaktionsprofile sehr gering ausgeprägt, so daß der Schluß gezogen wird, es ergebe sich „die Unmöglichkeit, diagnostische Aussagen über individuelle multivariate Reaktivitäten zwischen den Situationen zu generalisieren bzw. vorherzusagen".

8.4.3 Kognitive Komponenten von Emotion

Im Modell Grays waren bereits die Elemente der Erwartung und auch der Bewertung enthalten. Bei Lazarus und Mitarbeitern (1971, 1975, 1980) wurde der Begriff der Bewertung oder Einschätzung zum zentralen Element ihrer Emotionstheorie. „Das Konzept, das benutzt wird, um die bewertenden Prozesse zu bezeichnen, die für eine kognitive Theorie der Emotionen zentral sind, ist kognitive Einschätzung (‚appraisal‘, Bewertung) ... Ganz deutlich formuliert besagt eine kognitiv orientierte Theorie, daß jede emotionale Qualität und Intensität – Angst, Schuld, Eifersucht, Liebe, Freude, oder was auch immer – erzeugt und geleitet wird durch ihr eigenes Muster von Einschätzung" (Lazarus et al., 1980).

Mit diesem Begriff von „Appraisal" knüpfen die Autoren zunächst an das Exzitationsmodell von Arnold (1967) an. Arnold hatte versucht, spezifische Regionen im limbischen System als wesentlich für diese Funktionen nachzuweisen. Neben der Appraisal-Funktion hatte Arnold vor allem auf die eines affektiven Gedächtnisses als Funktionskreis, vom hippokampischen System zu den vorderen thalamischen Nuclei und zum Gyrus cinguli reichend, hingewiesen, sowie auf eine Imaginationsfunktion als Funktionskreis vom limbischen System über die Amygdala und den Thalamus zu den kortikalen Assoziationsfeldern (vgl. Kap. 9).

Die bewertende Funktion hat bei Lazarus zunächst eine vergleichbare Bedeutung wie bei Gray der Entscheidungsmechanismus. Sie entscheidet zwischen positiven und negativen Bedeutungen von Ereignissen für einen Organismus. Die Bewertungsfunktion wird „primär" genannt. Es gibt dabei drei Möglichkeiten eines Ergebnisses: ein Ergebnis kann im Hinblick auf das Wohlergehen irrelevant, positiv oder negativ (stressful) sein. Ein negatives Ereignis kann wiederum auf drei verschiedene Weisen negativ sein, schädlich, wenn es bereits eingetreten ist, bedrohlich, wenn es antizipiert wird, und als Herausforderung, wenn es Möglichkeiten sowohl positiver als auch negativer Art enthält. Mit dem Konzept der sekundären Bewertung wird dieser Rahmen sehr bedeutsam erweitert. Die sekundäre Bewertung richtet sich auf die Möglichkeiten, die einem Organismus zur Verfügung stehen, mit einer z.B. bedrohlichen Situation umgehen zu können.

In einem, wenn auch sehr eingeschränkten Sinne ist eine sekundäre Bewertung auch im Modell von Gray enthalten. Der Kampf-Flucht-Mechanismus beantwortet aversive Reize einer bestimmten Art. Dabei ist dieser Mechanismus als einheitlich zu verstehen, also nicht, daß es bei bestimmten Reizen zu Flucht-, bei anderen zu Kampfverhalten kommt. Ob es zu Kampf oder Flucht kommt, hängt dagegen von zusätzlicher, in der Situation enthaltener Information und von deren Bewertung ab.

Der Begriff der sekundären Bewertung ist bei Lazarus jedoch bedeutend weitreichender. Sie resultiert nicht in der Wahl von zwei Verhaltensalternativen, sondern sie determiniert auch die Art der Emotion. So wird die Furcht einer Person größer sein, wenn sie im Zuge der sekundären Bewertung keine Möglichkeit findet, der Bedrohung adäquat zu begegnen. Lazarus mißt der sekundären Bewertung inzwischen eine größere Bedeutung zu als dem belastenden Ereignis selbst (Lazarus et al., 1980).

Hat eine Auseinandersetzung mit der Umgebung stattgefunden, so werden wiederum deren Ergebnis und die Situation als Ganzes erneut bewertet (Wiederbewertung, „reappraisal"). Diese Wiederbewertung kann in zwei Arten stattfinden, einmal als Informationsaufnahme aus der Umgebung, zum anderen aber auch als Beurteilung der Bedeutung, die die Situation beinhaltet. Darin wird eine intrapsychische Bewältigung (coping) gesehen, die z.B. in Verleugnung oder Ablenkung bestehen kann. Diese Art der Wiederbewertung wird als defensiv bezeichnet. Die Art der Bewertung kann also einen Bewältigungsversuch beinhalten, im Gegensatz zu Aktionen, die auf eine Veränderung oder Beeinflussung der Situation abzielen.

Die Einschätzung der eigenen Möglichkeiten, sich in einer Situation angemessen zu verhalten, ist in neuen Situationen aufgrund der mangelnden Erfahrung entsprechend erschwert. In einer Reihe von Studien haben Epstein und Fenz den Einfluß von Erfahrung auf die Angstbewältigung bei Fallschirmspringern untersucht. Dabei hat es sich gezeigt, daß eine adaptive Angstkontrolle erst gelernt wird, d.h., daß sich auch das physiologische Muster der Aktivierung vor einem Absprung in Abhängigkeit vom Neuheitsgrad der Situation ändert. Es wurde deutlich, daß sich die Verarbeitung von Reizen, die einen Hinweischarakter auf den Angstinhalt besitzen, im Sinne einer erhöhten Hinwendung verstärkt, so daß Angst in kleinen Quantitäten verarbeitet wird (Epstein und Fenz, 1965; Epstein, 1967).

Die Neuheit einer Situation, bzw. die Menge verfügbarer Information, ist also ein wichtiges Element für die Art der Emotion in Abhängigkeit von der sekundären Bewertung, vor allem auch im Hinblick auf die Bewältigung einer bedeutsamen Situation. Der Prozeß der Bewertung ist ein kontinuierlicher Vorgang, der emotionales Erleben beeinflußt und zu ständigen Änderungen der Art der erlebten Emotion führt. Koriat und Mitarbeiter (1972) versuchten nicht nur, den Einfluß emotionaler Kontrolle im Sinne einer Reduktion von Streß durch emotionale Distanzierung (detachment), sondern auch der verstärkten Anteilnahme (involvement) zu erfassen. Die Probanden sahen Filme von Arbeitsunfällen, bei denen z.B. Gliedmaßen abgetrennt wurden oder die zum Tode einer Person führten. Bei zwei Durchgängen wurden keine besonderen Instruktionen gegeben, beim drit-

ten Durchgang wurde die eine Hälfte der Probanden dahingehend instruiert, sich von der emotionalen Bedeutung des Films zu distanzieren, die andere Hälfte, sich verstärkt zuzuwenden. Beide Gruppen ließen sich anhand der Einschätzung der Stärke des Streßinhalts der gesehenen Filme, aber auch am Ausmaß der Veränderungen der Herzfrequenz unterscheiden. Dabei wurden verschiedene Strategien der emotionalen Zu- oder Abwendung berichtet. Personen, die instruiert worden waren, sich verstärkt zuzuwenden, berichteten, sie hätten sich vorgestellt, selbst von dem Unfall betroffen zu sein, während die Abwendung meist in der Vorstellung bestand, das Geschehen sei nur gestellt und filmisch übertrieben.

Diese internen Coping-Strategien, die als Methoden der Informationsverarbeitung bezeichnet werden, sind für das Modell von Lazarus am wichtigsten. Direkte, auf eine Situation folgende Verhaltensaktionen, zum Teil angeborene Reaktionen, gelten als starr und lassen nur wenige Freiheitsgrade zu. „Nichtsdestoweniger verschiebt sich mit dem phylogenetischen Fortschritt von einfacheren Lebewesen bis hin zum Menschen die Regulation der Emotionen, weg von festen Reaktionen auf äußere Auslöser und hin zu Informationsverarbeitung und kognitiver Kontrolle. Regulation wird Selbst-Regulation, womit eine zunehmend geringere Abhängigkeit von angeborenen, fest eingebauten Mechanismen und eine zunehmende Abhängigkeit von Lernen und Kognitionen verbunden ist" (Lazarus, 1975).

Die von Lazarus angesprochenen Strategien der Informationsverarbeitung kommen einigen Abwehrmechanismen nahe, die von der psychoanalytischen Theorie beschrieben werden. Aus dem Bereich der experimentellen Gedächtnis- und Wahrnehmungspsychologie sowie der kognitiven Psychologie sind empirisch gut gesicherte Modelle menschlicher Informationsverarbeitung bekannt, die als theoretischer Hintergrund der Vorstellungen von Lazarus angesehen werden können (Broadbent, 1958, 1971; Miller et al., 1960; Neisser, 1967; Schroder et al., 1967; Lindsay und Norman, 1972).

Mit der Einbeziehung sog. „interner" Prozesse in theoretische Modelle ist zwar ein Verständnisgewinn für das Phänomen Emotion verbunden, andererseits entstehen aber auch erhebliche methodische Schwierigkeiten, etwa der experimentellen Unterscheidung der beiden Stufen der Bewertung. Dennoch scheint der Verständnisgewinn, bzw. die Unvollständigkeit von Erklärungsversuchen, die derartige Prozesse ausklammern, die Einführung derartiger Erklärungselemente zu rechtfertigen.

Die Verleugnung ist eine häufig untersuchte Strategie des Umgangs mit Belastung. Wolff (1964) untersuchte die Eltern von Kindern, die an Leukämie erkrankt waren. Diejenigen Eltern, die den Ernst der Erkrankung ihrer Kinder verleugneten, hatten eine niedrigere 17-Hydroxycorticosteroid-Ausscheidung im Urin als diejenigen Eltern, die die Realität anerkannten. In einer prospektiven Studie bei 367 Patienten mit Herzinfarkt wurde von Havik und Maeland (1988) während des Klinikaufenthalts auf drei verschiedene Weisen „Verleugnung" erfaßt und in einer Nachuntersuchung nach drei und fünf Jahren mit Gesundheit und Befinden in Beziehung gesetzt. Ein hohes Maß an „denial of illness" (Verleugnung der Krankheit und der damit verbundenen Behandlungsnotwendigkeiten) war mit weniger Problemen bei der Arbeit, Sexualität, physischer Aktivität und geringerer Mortalität verbunden; „denial of impact" (Verleugnung der mit den Krankheitsepisoden verbundenen Angst) mit besserem emotionalem Befinden, aber höherer Mortalität. „Suppression" (das bewußte Vermeiden krankheitsbezogener Gedanken) hatte keinen Einfluß auf die Überlebenszeit.

Fenz (1975) untersuchte 16 unerfahrene Fallschirmspringer zu mehreren Zeitpunkten vor dem ersten Absprung. Es wurden Testreize verwendet, die ähnlich wie die des Thematischen Apperzeptionstests (TAT) waren, jedoch inhaltlich auf Fallschirmspringen bezogen. Die Ergebnisse zeigen, daß eine Verleugnung von Furcht mit einer geringeren Reaktivität, gemessen an Veränderungen der Hautleitfähigkeit, einherging. Diese Unterschiede zeigten sich nur am Tag des ersten Absprungs, nicht jedoch an einem früheren Tag, der zeitlich vom Absprung, also der Belastung entfernt war.

8.4.4 Spezifität emotionalen Verhaltens

Die im groben mögliche Einteilung in positive und negative Emotionen kommt der Aufteilung in positive, die Auftretenswahrscheinlichkeit von Verhalten vergrößernde Verhaltenskonsequenzen und negative nahe, die diese Wahrscheinlichkeit verringern. So definiert Gray Emotionen als „jene (hypothetischen) Zustände des konzeptuellen Nervensystems, die durch verstärkende Ereignisse hervorgerufen wurden oder durch Reize, die in der vorangegangenen Erfahrung des Subjekts von solchen verstärkenden Reizen gefolgt wurden" (Gray, 1972). Man brauche dann nur noch die Klassen unterscheidbarer verstärkender Ereignisse aufzuzählen und hätte damit gleichzeitig die Anzahl operational unterscheidbarer Emotionen gefunden. So würde die Klassifizierung der auf das Verhalten einwirkenden Umgebungsbedingungen einer der emotionalen Grundstimmungen entsprechen und umgekehrt.

Gray weist in diesem Zusammenhang darauf hin, daß die Gegenwart von positiven unkonditionierten Ereignissen eigentlich nicht in ein entsprechendes Emotionsschema paßt. Während bei der Darbietung eines konditionierten positiven Reizes ein mehr oder weniger einheitliches Annäherungsverhalten stattfindet – Mowrer (1960) beschrieb den entsprechenden Gefühlszustand als Hoffnung –, sei dies bei unkonditionierten positiven Reizen nicht der Fall. Derartige Reize beendeten eher interne Zustände, genauer Triebzustände, als sie zu induzieren. Die weiteren Felder des obigen Schemas lassen sich zum Teil zusammenlegen. So entsprechen sich etwa unkonditionierte Strafreize und die Beendigung von unkonditionierten positiven Reizen. Beide rufen zum Teil aggres-

sives oder Fluchtverhalten hervor. Zu trennen sind im Hinblick auf emotionales Verhalten konditionierte und unkonditionierte negative Reize. So ruft beim Menschen ein unkonditionierter elektrischer Reiz eine Herzfrequenzsteigerung, ein mit diesem Reiz konditionierter Reiz einen Abfall der Herzfrequenz hervor (Obrist et al., 1965). Ebenfalls auf der Seite negativer Ereignisse läßt sich ein Unterschied zwischen konditionierten und unkonditionierten Reizen in der Wirkung auf Verhalten aufzeigen. Konditionierte negative Reize führen zu einer Verhaltenshemmung, zu passiver Vermeidung und Löschung von Verhalten, unkonditionierte negative Reize zu Flucht- oder Angriffsverhalten. So lassen sich drei übergreifende emotionale Systeme unterscheiden:

– Annäherungsverhalten (BAS – behavioral approach system),
– Verhaltensinhibition (BIS – behavioral inhibition system),
– Kampf – Flucht.

Brady (1975) zieht für seine Klassifikation von Emotionen die „Zwei-Prozeß-Theorie" des Lernens nach Rescorla und Solomon (1967) heran und unterscheidet die Paradigmen des klassischen und des operanten Konditionierens. Unkonditionierte, aber auch konditionierte Reaktionen auf einen Reiz werden als „respondent" zusammengefaßt, instrumentelles Verhalten im weitesten Sinne als „operant". Dies bezieht sich sowohl auf interne, also z.B. autonome Reaktionen, wie auch auf externe, also direkt beobachtbare Verhaltensweisen. So lassen sich interne Respondents von internen Operants unterscheiden, ebenso externe Respondents von externen Operants. Wie Gray unterscheidet er dann zwischen Reaktionen auf positive und solchen auf negative Reize, bzw. solchen Verhaltenskonsequenzen. Positive Reizwirksamkeit wird als „appetitiv", negative als „aversiv" bezeichnet.

Die Frage nach der Spezifität von Emotionen wurde auf sehr verschiedene Weise beantwortet. Im Abschnitt „Emotion und Verhalten" werden Ansätze referiert, in denen versucht wurde, mit Hilfe lerntheoretischer Paradigmen spezifische Emotionen anhand des Verhaltens von Organismen zu klassifizieren. Geht man davon aus, daß Lernen die „Koppelung von Bedeutungen" beinhaltet, d.h., daß Verhalten immer auch Bedeutung enthält, so ist es denkbar, aus dem Verhalten auf Bedeutung und damit auch auf Erleben zu schließen. Andererseits hat es sich gezeigt, daß das Ergebnis dieser Art von Klassifikation, da sie sich hauptsächlich an funktionalen Bedingungen des Lernens orientierte und nicht so sehr am Inhalt des Gelernten, sehr grob war und der Vielfalt, die etwa emotionales Erleben bei Menschen beinhaltet, keineswegs entsprach. Es hatte sich allerdings auch gezeigt, daß mit Hilfe dieses Ansatzes Bedingungen realisiert werden konnten (Hilflosigkeit), die vom klinischen Standpunkt aus als gültig angesehen wurden.

Sowohl James (1884) als auch Lange (1885) gehen davon aus, daß emotionalem Erleben spezifische körperliche Prozesse zugrunde liegen, deren Wahrnehmung Emotionen verursacht. Es wird angenommen, daß zuerst ein bestimmter Reiz die Sinnesorgane stimuliert, dann die Erregung über afferente Bahnen zur Großhirnrinde gelangt, und schließlich von dort periphere Organe stimuliert werden. Diese Stimulierung wird der Großhirnrinde über afferente Bahnen rückgemeldet, was zu emotionalem Erleben führt. Wichtig an diesem Ansatz ist die Annahme spezifischer körperlicher Prozesse, die Emotion ist mehr oder weniger als Epiphänomen dieser spezifischen Prozesse anzusehen. Von James stammt der Satz: „Man weint nicht, weil man traurig ist, sondern man ist traurig, weil man weint."

Die Grundannahmen dieses Modells sind vielfach kritisiert worden, u. a. von Cannon (1927, 1931), vor allem die Annahme, daß emotionales Erleben auf der Rückmeldung oder Wahrnehmung der Veränderung physiologischer Funktionen beruhen soll. Andererseits ist immer wieder versucht worden, für verschiedene Emotionen verschiedene Muster entweder von physiologischer Erregung oder von hormonellen Reaktionen aufzufinden. Die Annahme, daß eine wie auch immer geartete Wahrnehmung körperlicher Erregung für das Zustandekommen von emotionalem Erleben eine wesentliche Bedingung ist, auch wenn die Annahme spezifischer körperlicher Erregung abgelehnt wird (Schachter und Singer, 1962), ist wesentlicher Bestandteil vieler Emotionstheorien geblieben.

Für neuere Emotionstheorien ist die Annahme kognitiver Faktoren für die Erklärung spezifischer Emotionen wesentlich. So gingen Schachter und Singer (1962) davon aus, daß Emotion zum einen auf der Wahrnehmung körperlicher Aktivierung beruht, zum anderen die Art der Emotion durch Merkmale der Situation und deren kognitive Verarbeitung bedingt ist. Sie nahmen an, daß eine gleiche physiologische Aktivierung in Abhängigkeit vom kognitiven Aspekt einer Situation einmal als Freude, ein anderes Mal als Zorn oder Aggression erlebt werden kann. Die Autoren injizierten Versuchspersonen Adrenalin und informierten die Probanden über den Effekt der Injektion auf unterschiedliche Weise, indem ein Teil der Probanden zutreffend über die aktivierende Wirkung des Adrenalins aufgeklärt wurde, ein Teil der Probanden jedoch darüber uninformiert blieb. Einer Vergleichsgruppe wurde als Plazebo eine Kochsalzlösung injiziert. Nach der Injektion kam eine in den Versuch eingeweihte Person in den Raum und verhielt sich entsprechend dem Versuchsplan in einem Fall ärgerlich, im anderen Fall euphorisch. Das Ausmaß der Information über den Effekt der Injektion hatte einen wesentlichen Einfluß auf die erlebte Emotion. So war z. B. das Ausmaß der erlebten Euphorie in der uninformierten Gruppe etwa doppelt so hoch wie in der Gruppe, die über den aktivierenden Effekt der Injektion aufgeklärt worden war.

Schachter hatte bereits früher (1959) angenommen, daß die Wahrnehmung einer physiologischen Aktivierung zu einem Erklärungsbedürfnis führen würde. Ist, wie im vorliegenden Fall, durch die Instruktion eine Erklärung für die erlebte Aktivierung vorhanden, so braucht keine neue gesucht zu werden. Ist sie je-

doch nicht vorhanden, so wirkt die erlebte Aktivierung, indem sie das aus der Bewertung der Situation resultierende Erleben verstärkt. Die Annahme Schachters und Singers, daß Emotion das Resultat des Wirksamwerdens der zwei Faktoren „Kognition" und „Aktiviertheit" darstellt, ist kritisiert worden (Erdmann und Janke, 1978; Marshall und Zimbardo, 1979; Maslach, 1979). Erdmann und Janke kritisierten zum einen die Methode, mit der von Schachter und Singer emotionales Erleben gemessen wurde, als zu wenig differenziert. Zum zweiten wurden nur die Emotionen Ärger und Euphorie erfaßt, zudem gab es keine neutrale Kontrollbedingung und einen Mangel an signifikanten Ergebnissen, vor allem der Unterschiede zwischen der Plazebobedingung und der kritischen Bedingung (Adrenalin und fehlende Information). So könne nach Erdmann und Janke nicht davon ausgegangen werden, daß die Untersuchung von Schachter und Singer gezeigt habe, daß das Ausmaß emotionaler Veränderung mit dem von Aktiviertheit variiert. Ähnlich fanden Marshall und Zimbardo (1979), daß mit steigender Dosierung von Adrenalin lediglich negative Emotionen und nicht Euphorie zunahmen. Dem entgegneten Schachter und Singer (1979), daß sie keinen linearen Zusammenhang zwischen der Dosierung von Adrenalin und erlebter Emotion annähmen, sondern daß die „emotionale Plastizität", d. h. Beeinflussung durch kognitive Faktoren, nur bei einer „milden" Dosierung möglich sei. Die Untersuchung von Erdmann und Janke allerdings zeigt ein der ursprünglichen Untersuchung vergleichbares Resultat für die Bedingung „Euphorie", nicht jedoch für Angst.

Diese Kontroverse zeigt, daß zumindest mit den angewandten Strategien der Manipulation des kognitiven Faktors für das Erleben spezifischer Emotionen der von Schachter und Singer angenommene Effekt nicht generell erzeugt werden kann. Allerdings sind auch die Kritiker dieses Modells darin einig, daß kognitive Faktoren, z. B. solche der Bewertung, für emotionales Erleben eine wichtige Bedingung darstellen.

Stemmler (1984) versuchte spezifische physiologische Emotionsmuster aufzuzeigen. Bei diesem Versuch ist bemerkenswert, daß neben der Messung einer Vielzahl physiologischer Funktionen auch die Ebene des Erlebens sehr differenziert erfaßt wurde. Neben einer freien Befindensäußerung, Selbsteinstufung von Gefühlen nach einer Befindlichkeitsliste sowie einer Befragung im Interview wurden die sprachlichen Äußerungen von Probanden einer Sprachinhaltsanalyse unterzogen.

Auf diese Weise ließ sich zeigen, daß die Mehrzahl der Befindensäußerungen mit den experimentellen Bedingungen zur Emotionsinduktion von Freude, Angst und Ärger übereinstimmten, allerdings nicht ausnahmslos bei allen Probanden.

In der univariaten statistischen Analyse zur Bestimmung des Profils der physiologischen Variablen, die die Emotionsbedingungen unterscheiden, zeigte sich, daß 41% dieser Variablen zwischen den Emotionsbedingungen statistisch signifikant unterschieden, im Sinne verschiedener physiologischer Emotionsprofile. In die Analyse gingen Messungen in einer sog. Referenzphase ein, die zeitlich auf die Induktion der Emotion folgte. In dieser Phase traten keine irgendwie gearteten Anforderungen an die Probanden auf, da es sonst schwierig gewesen wäre, die „reine" Emotion von der Anforderung der Situation zu trennen. Damit folgte Stemmlers Interpretation von Emotion der „jahrhundertalten und auch in zeitgenössischen Emotionstheorien geäußerten Vorstellung, daß Emotionen mit ‚Passivität' (Stop-Mechanismus bei Pribram) verknüpft seien" (Stemmler, 1984). Physiologische Reaktionen, die ohne erkennbare Ursache in der Umgebung auftreten, würden eher mit Emotion verknüpft als solche, die durch die Umgebung und ihre Anforderungen begründet seien. Aus diesen Überlegungen ergab sich die Art der Messung in der Referenzphase.

Auch im multivariaten Diskriminanzraum, der durch die Reaktionen auf die drei Emotionsinduktionen definiert wurde, bestätigten sich die Profilunterschiede zunächst. Andererseits hatte sich bereits bei den univariaten Analysen gezeigt, daß auf der Ebene der Befindlichkeit auch in Antizipationsphasen beispielsweise Angst erlebt worden war, die jedoch mit physiologischen Mustern verbunden gewesen war, die sich erheblich von denen der Angst-Referenzphase unterschieden. Infolgedessen wurden multivariate Diskriminanzanalysen berechnet, in die alle 52 Untersuchungsphasen mit eingingen.

Im Diskriminanzraum aller 52 Untersuchungsbedingungen, einer „physiologischen Landkarte" der Emotionen, ließ sich ein Unterschied zwischen den Bedingungen Ärger und Freude nicht mehr nachweisen, was jedoch aus den entsprechenden Befindlichkeitsmaßen nicht mehr erklärbar war.

Stemmler erklärte diesen Befund mit der Annahme, daß physiologische Muster eine Person-Umwelt-Interaktion widerspiegeln, bei der eine emotionale Reaktion nur ein Teil einer bestimmten Interaktion sein könnte. „Aus der Lage eines Musters auf der ‚physiologischen Landkarte' wird die Person-Umwelt-Interaktion, momentan und aktuell stattfindend oder antizipiert, sichtbar, soweit sie sich auf die physiologische Landkarte projiziert" (Stemmler, 1984).

Diese Interpretation psychophysiologischer Erlebnismuster (nicht Emotionsmuster im Sinne der zitierten Arbeit) als Beziehung zwischen Person und Umwelt definiert Emotion als unter Umständen daraus resultierende Bedeutung, die eine Person in einer Situation erlebt (vgl. Abschn. 8.1). Es ist dabei sicherlich anhand dieser Ergebnisse zu überlegen, ob die „Isolierung reiner Emotionen" möglich und auch sinnvoll ist.

Izard und Buechler (1980) gehen von angeborenen Grundemotionen aus, die durch Information in der Umgebung ausgelöst werden. Sie nehmen an, daß bewußtes Erleben auch stets emotionales Erleben ist, so daß Situationsinformation nicht schlechthin Emotionen bewirkt, sondern Emotionen ändert. Ebenso können Gedanken oder propriozeptive Impulse Emotionen ändern bzw. beeinflussen. Sie „ändern das Niveau oder Muster elektrochemischer Aktivität im Nervensystem". Diese Änderung ruft einen bestimmten angeboren determinierten Gesichtsausdruck hervor, der als sensorische Rückmeldung das subjektive Erleben der Emotion bedingt. Ist die Emotion auf diese Weise einmal hervorgerufen, so „sind hormonelle, kardiovaskuläre, auf die Atmung bezogene und andere Lebenssysteme in die Verstärkung und Regulation der Emotion eingeschlossen". So haben die Autoren versucht, über die Einschätzung des Gesichtsausdrucks Grundemotionen aufzufinden. Sie kamen zu zehn Grundemotionen, die als Interesse, Freude, Überraschung, Traurigkeit, Ärger, Abscheu, Verachtung, Furcht, Scham und Schuld erlebt werden. Tatsächlich erlebte Emotionen stellen meist eine Mischung oder schnelle Abfolge verschiedener dieser Grundemotionen dar.

Plutchik (1977, 1980) geht von in der Evolution entwickelten „primären Anpassungsreaktionen" aus, die auf allen phylogenetischen Stufen aufzufinden seien, wie „Schutzreaktionen, Zerstörungsreaktionen und solche der Fortpflanzung". Kognitive Prozesse hätten sich in der Evolution, im Dienste dieser mit Emotionen verbundenen Anpassungsreaktionen, entwickelt. Sie hätten hauptsächlich die Funktion, die Umgebung zu bewerten und Vorhersagen über zukünftige Ereignisse zu ermöglichen.

„Da eine Emotion eine komplexe Abfolge von Reaktionen ist, kann man verschiedene Aspekte einer solchen Sequenz mit verschiedenen Begriffen beschreiben" (Plutchik, 1980), man könne die Reizcharakteristik beschreiben, die zur Emotion führt, oder das Gefühl, Verhalten usw. Man könne darüber hinaus Emotion nach zwei grundlegenden Dimensionen ordnen, Ähnlichkeit und Gegensatz. Daraus ergebe sich eine „analoge" Kreisstruktur, wobei die meisten Emotionen am Rand des Kreises anzuordnen seien. Plutchik geht darüber hinaus davon aus, daß die „State-Trait" (Zustand-Merkmal)-Unterscheidung nicht nur für Angst gelte, sondern für jede Emotion. Daraus lasse sich eine Beschreibung von Persönlichkeit ableiten, anhand des Vorherrschens verschiedener Emotionen. Den faktorenanalytisch ermittelten acht Grundemotionen entsprechen dann bestimmte Typen der Persönlichkeit, denen auch diagnostische Klassifikationen zugeordnet werden können sollen (vgl. Tab. 8–1).

Tabelle 8–1. Klassifizierungsschema für auf Verhalten einwirkende Ereignisse (Erklärung siehe Text)

	Unkonditioniert		Konditioniert	
Vorhanden +	+ (+ E)	+ (– E)	+ (+ E)	+ (– E)
Nicht vorhanden –	– (+ E)	– (– E)	– (+ E)	– (– E)
Beendet				

In diesem Klassifikationsschema wird nicht direkt Bezug genommen auf physiologische oder neurophysiologische Prozesse, die der Spezifität zugrunde liegen könnten, wie dies etwa bei Gray (1972) der Fall war. In einer Studie von Ax (1953) wurde der Versuch unternommen, die Emotionen Furcht und Ärger anhand verschiedener körperlicher Reaktionsmuster zu differenzieren. Die Ergebnisse versucht Ax in Beziehung zu setzen zur Wirkung von Adrenalin für Furcht und Noradrenalin für Ärger. So waren beispielsweise die Anstiege des diastolischen Blutdrucks, Verringerungen der Herzfrequenz, Anstiege der Muskelpotentiale sowie die Anzahl der Hautwiderstandsänderungen bei Ärger größer als bei Furcht, bei der es zum Anstieg der Herzfrequenz, Anstieg des Niveaus der Hautleitfähigkeit, der Anzahl der Muskelpotentiale und Anstieg der Atemfrequenz kam. Ax weist darauf hin, daß die Unterschiede zwischen den beiden Emotionen sich nicht nur in den Amplituden der Meßwerte zeigten, sondern daß die Richtung der Reaktionen bei fast allen Variablen verschieden war, so daß es sich um „echte" Unterschiede handelt.

8.4.5 Vermittlung von emotionalem Verhalten

Es existiert eine Reihe von Untersuchungen, deren Ergebnisse die Klassifizierung der Emotionen nach Gray stützen. So fanden Olds und Olds (1965), daß elektrische Selbststimulation im medialen Vorderhirnbündel, im lateralen Hypothalamus und in rostralen Punkten der Septumregion zu einer Verstärkung der Selbststimulation führen. Dagegen führt die Reizung von medialen Septumkernen, von Punkten im medialen Frontalkortex und im Hippokampus zu einer Verringerung der Selbststimulation. Läsionen in diesen Gebieten führen nicht zu einer Beeinträchtigung von passivem Vermeidungsverhalten, sondern auch zu einer Löschung von Annäherungsverhalten, wenn es nicht mehr zu Bekräftigung kommt.

Strukturen in der Amygdala, dem medialen Hypothalamus und dem zentralen Grau des Mittelhirns sind wesentlich für die Vermittlung des Kampf-Flucht-Systems. De Molina und Hunsperger (1962) fanden, daß es bei elektrischer Stimulation in diesen Regionen zu defensiver Aggression (Kampf) oder in Abhängigkeit von Umgebungsbedingungen zu Flucht kommt (Ulrich, 1967). Adams und Flynn (1966) konnten entsprechend zeigen, daß auf Beutetiere bezogene Aggression nicht durch eine elektrische Stimulation in diesen Regionen ausgelöst werden konnte, sondern durch Stimulation im lateralen Hypothalamus. Eine derartige Stimulation hatte darüber hinaus eine belohnende Wirkung, während eine Stimulation, die defensive Aggression auslöst, eine aversive Wirkung hatte.

Die Entscheidung darüber, ob in einer Situation Kampf oder Flucht erfolgt, wird nach Olds im medialen Hypothalamus vermittelt. Es wird angenommen, daß diese Region die aus den drei Systemen auftretende Erregung moduliert oder hemmt. „Wir nehmen an, daß der ventromediale Kern des Hypothalamus über die zentrale Substanz des Mittelhirns eine tonische (d.h. eine normalerweise vorhandene) Hemmwirkung ausübt, und daß jene Substanz letztlich darüber entscheidet, ob Flucht oder defensiver Angriff erfolgt. Wir nehmen weiterhin an, daß jene Hemmwirkung ihrerseits entweder von der Amygdala gehemmt

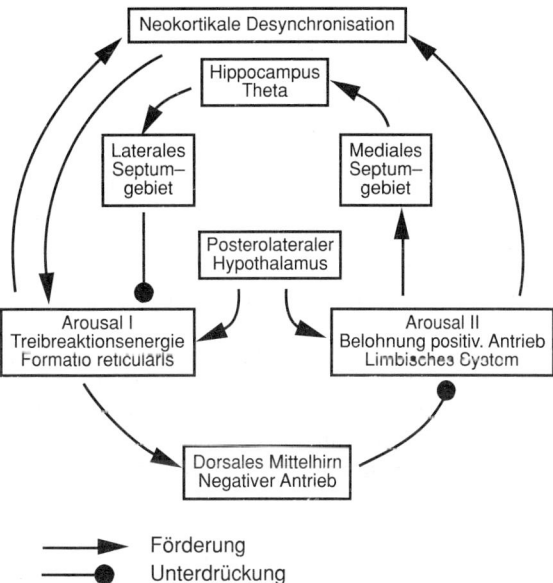

Abb. 8–17. Schema der Beziehungen, die der „two-arousal"-Hypothese zugrunde liegen (aus Routtenberg, 1968).

(dann entsteht Kampf-Flucht-Verhalten) oder durch Einflüsse aus dem septohippokampischen passiven Vermeidungssystem verstärkt wird. Diese letzteren Einflüsse bilden ein Gegengewicht zur Reaktion der Amygdala auf neue oder erschreckende Stimuli, wobei das daraus resultierende Verhalten (passive Vermeidung oder Kampf – Flucht) davon abhängt, welcher Input in den ventromedialen Hypothalamus bei einer bestimmten Kombination von Umweltbedingungen sich als der stärkste erweist. Dieses Schema erklärt nicht nur die Interaktionen zwischen den Läsionen im Septum, in der Amygdala und im Hypothalamus, sondern macht auch begreiflich, warum bei Gefahr Erstarren und Kampf-Flucht-Verhalten gleichzeitig intensiviert werden können, um dann rasch einander abzuwechseln" (Gray, 1973).

Routtenberg (1968) stellte ein „two arousal"-Modell der Aktivierung vor, das einerseits bestimmte Übereinstimmung mit dem Modell Eysencks der Trennung zwischen Aktivierung und Arousal aufweist, zum anderen ebenfalls, wie das Modell Grays, sich an den Untersuchungen von Olds und Olds (1962, 1965) orientiert. Wie das Arousal-System Eysencks ist bei Routtenberg das Arousal-System I bezogen auf das Aktivierungssystem der Formatio reticularis, wobei eine hohe Aktivität dieses Systems in einem Desynchronisationseffekt im Hippokampus sowie im Kortex besteht. Das Arousal-System II entspricht dem Aktivierungssystem Eysencks und ist auf das limbische System bezogen. Es kontrolliert die Verarbeitung triebbezogener Reize und wird teilweise als „reward"-System bezeichnet. „Belohnung oder positive Triebreize (incentive stimuli) erregen hauptsächlich das Arousal-System II, was normalerweise zu Konsequenzen führt, die hier reinforcement genannt werden. So wird reinforcement primär nicht als belohnend angesehen, sondern eher als die Konsequenz

159

eines positiven Anreizes, was zur verstärkten Wahrscheinlichkeit des Wiederauftretens einer Reaktion führt" (Routtenberg, 1968).

Die Abbildung 8–17 gibt eine Übersicht über die Interaktion beider Arousal-Systeme, wobei prinzipiell anzunehmen ist, daß sie einen gegenseitig hemmenden Einfluß aufeinander ausüben. So kann eine Verhaltensbekräftigung etwa auch dadurch zustande kommen, daß die Aktivität des Arousal-Systems I unterdrückt wird.

Ebenso ist anzumerken, daß eine milde Aktivierung des Arousal-Systems I Thetawellen im Hippokampus produzieren und entsprechend einen Belohnungswert haben kann. Vor allem Berlyne (1968) hat gezeigt, daß der Belohnungswert eines Reizes abhängig ist vom Aktivierungsanstieg oder -abfall, der mit ihm verbunden ist. So kann auch Triebinduktion belohnend wirken. Bekannt ist diese Tatsache vor allem im Hinblick auf den Grad der Neuigkeit von Umgebungsreizen, wobei ein mittleres Maß an Neuigkeit belohnend wirken kann.

8.4.6 Die konditionierte emotionale Reaktion

Emotion und Verhaltensbeeinflussung

Die häufig geäußerte Auffassung, Emotion störe Verhalten (z. B. Lang et al., 1972), rührt vermutlich daher, daß sich eine große Anzahl von Untersuchungen, ausgehend von Estes und Skinner (1941), mit dem Phänomen der Verhaltensunterdrückung (behavioral suppression oder behavioral freezing; Hoffmann, 1966) beschäftigt haben. Für Estes und Skinner war das Ausmaß der Verhaltensunterdrückung ein Maß für die Quantität einer Emotion, von Angst oder einer konditionierten emotionalen Reaktion (CER). Wie in vielen lernpsychologisch orientierten Untersuchungen, vor allem solchen mit Tieren, wird ein bestimmtes experimentelles Paradigma entworfen, um aus dem Verhalten der Tiere ein „Erleben" abzuleiten, bzw. dieses zu erfassen. Ähnlich werden Veränderungen physiologischer Funktionen herangezogen, um etwa Aktiviertheit zu bestimmen. Ob derartige Maße tatsächlich geeignet sind, um Emotionen hinreichend oder gar vollständig zu erfassen oder auch zu klassifizieren, ist fraglich. Dennoch haben derartige Untersuchungspläne für den vorliegenden Zusammenhang Bedeutung, da sie ganz offensichtlich, wie die Ergebnisse zeigen, den gesamten Organismus betreffen, und es zu erheblichen organischen Veränderungen kommen kann (vgl. Kap. 7).

Um den Effekt einer Verhaltensunterdrückung herzustellen, bestehen grundsätzlich zwei Möglichkeiten: Es wird ein bestimmtes Verhalten, z. B. das Drücken eines Hebels, durch eine positive Bekräftigung aufrechterhalten. Die Darbietung eines klassisch konditionierten aversiven Reizes, eines Signalreizes, der von einem Schock gefolgt wird, führt zur Verhaltensunterdrückung. Die Darbietung des Signalreizes wird dem Verstärkerplan „überlagert" (superimposed). Wesentlich dabei ist, daß ein aversiver Respondent mit einem positiven Operant interagiert. Umgekehrt wäre es denkbar, daß es zu einer Interaktion zwischen einem aversiven Operant und einem positiven Respondent kommt. Im Klassifikationsschema von Brady wurde Emotion entsprechend stets als Interaktion zwischen Respondents und Operants definiert. Stimmen die Vorzeichen, d. h. die Valenz beider Reize, nicht überein, kommt es zum Effekt der Verhaltensunterdrückung, stimmen sie überein, müßte ein verhaltenfördernder (facilitating) Effekt resultieren. Obwohl die Verhaltensunterdrückung am häufigsten untersucht worden ist, gibt es auch Arbeiten, die den verhaltenfördernden Effekt nachweisen konnten. So konnte Sidman (1960) diesen Effekt durch Überlagerung von Signal-Schock-Verbindungen mit einem Verhalten des Hebeldrückens, das durch einen Sidman-Vermeidungsplan aufrechterhalten wurde, hervorrufen. In diesem Falle stimmten also die negativen Vorzeichen beider Reize überein.

Sidman (1966) beschreibt den Vermeidungsplan folgendermaßen: „Zwei sich wiederholende Zeitgeber programmieren die Schocks. Wenn das Tier den Hebel nicht drückt, wird das Zeitintervall zwischen den Schocks – Schock-Schock-Intervall genannt – durch den ersten Zeitgeber definiert. Das Tier kann durch das Drücken des Hebels den Schock verzögern. Der Betrag der Zeit, um den das Tier durch das Drücken des Hebels den Schock verzögern kann, wird Response-Schock-Intervall genannt und wird durch den zweiten Zeitgeber programmiert. Die beiden Zeitgeber arbeiten niemals simultan. Jeder Schock startet von neuem das Schock-Schock-Intervall, drückt das Tier das erste Mal den Hebel nach einem Schock, beendet es damit das Schock-Schock-Intervall und startet den Response-Schock-Zeitgeber … Kein exterozeptiver Reiz warnt das Tier vor einem Schock. Die Schockdauer ist auf den Bruchteil einer Sekunde begrenzt, so daß das Tier den Schock nicht beenden kann."

Das belastende Moment dieses Planes liegt vermutlich in der Tatsache, daß die Tiere keinen äußeren Anhaltspunkt für das zur Schockvermeidung erforderliche Verhalten haben und entsprechend lernen müssen, Zeitintervalle abzuschätzen, wobei der erste Erfolg einer Vermeidung das dann geschätzte Intervall außer Kraft setzt. Fügt man ein Warnsignal in den Versuchsplan ein, so entfällt der belastende Charakter des Versuchsplanes.

Der Sidman-Vermeidungsplan (Sidman-avoidance, free operant avoidance, free paced avoidance) ist in diesem Zusammenhang ein sehr häufig verwendeter Verstärkungsplan. Er erzeugt aufgrund seiner Komplexität recht große Belastungen, wie in einer Untersuchung von Forsyth und Harris (1970) deutlich wird. Mußten Rhesusaffen anhand eines derartigen Planes eine Vermeidungsreaktion erlernen, so kam es während der gesamten Versuchsdauer, auch außerhalb der eigentlichen Lernphasen, zu einer anhaltenden Erhöhung des arteriellen Blutdrucks. Bei anderen, weniger komplexen Versuchsplänen traten Blutdruckerhöhungen lediglich am 1. Tag des Trainings im ersten Versuchsdurchgang auf, später nicht mehr.

Dieser Befund ist darum wichtig, weil er zeigt, daß durch Situationen, in denen das Erlernen eines Verhaltens durch komplexe Umgebungsbedingungen erschwert wird, langanhaltende Blutdruckerhöhungen erzeugt werden, die über die belastende Situation hinaus anhalten (vgl. Kap. 41.3).

Der Einfluß konditionierter emotionaler Reaktionen auf physiologische und endokrine Funktionen

Die im folgenden behandelten Untersuchungen sind für die Psychosomatische Medizin von besonderem Interesse, obwohl es sich bei den hier dargestellten Untersuchungen ausnahmslos um Tierversuche handelt. Sie stellen einen experimentellen Ansatz dar, pathogene Bedingungen zu operationalisieren. Es gelang dabei, Bedingungen herzustellen, deren Übertragbarkeit auf Lebensverhältnisse von Patienten durchaus möglich ist.

So kommentiert Engel (1972) die Arbeiten von Weiss zur Pathogenese des Ulkus folgendermaßen: „Ihre Untersuchungen sind für mich aufregend, weil Sie erfolgreich etwas getan haben, was vorher niemand erreicht hat. Sie haben einen Versuchsplan gefunden, der dem sehr nahe kommt, was wir klinisch vorfinden. Nach unseren Erfahrungen besteht das einzige und wichtigste Ereignis, das mit dem Beginn der Krankheit korreliert war, darin, daß, bevor die Krankheit manifest wird, der Patient eine Periode des ‚giving up' durchmacht mit Affekten, die Schmale als Hilflosigkeit oder Hoffnungslosigkeit definiert hat (Schmale, 1958, 1969; Sweeney et al., 1970). Sie haben im Sinne eines Laborexperiments das operationalisiert, was der Kliniker ‚giving up' nennt, weil giving up bedeutet, daß eine Person das Gefühl hat, keine Lösungsmöglichkeit zu besitzen, oder diese tatsächlich nicht besitzt, und nichts tun kann. Egal, was sie tut oder denkt (dies kann intrapsychisch geschehen oder als direkte Reaktion auf die Umgebung), sie erhält keine Erfolgsrückmeldung." In dieser Übereinstimmung klinischer Beobachtungsdaten und ihrer Konzeptualisierung mit experimentell operationalisierbaren Bedingungen kommt nicht nur der Wert dieser Untersuchungen zum Ausdruck, sondern auch der des klinisch gewonnenen Konzepts der Hoffnungslosigkeit von Engel und Schmale.

In vielen Untersuchungen hat sich gezeigt, daß die Technik der Überlagerung, wie sie von Estes und Skinner (1941) zur Erzeugung einer CER benutzt worden war, zu besonderen Veränderungen auch im endokrinen und physiologischen Bereich führt. Mason und Mitarbeiter (1966) untersuchten den Zusammenhang zwischen 17-OH-CS-Plasmakonzentration und Verhaltensunterdrückung. Abbildung 8–18 zeigt die entsprechenden Veränderungen der 17-OH-CS-Plasmakonzentration und des Hebeldrückens. Abbildung 8–19 stellt diesen Einfluß normalem Vermeidungslernen gegenüber, Abbildung 8–20 schließlich zeigt die Wirkung einzelner, die Lernleistung erschwerender Komponenten auf die Plasmakonzentration von Adrenalin und Noradrenalin.

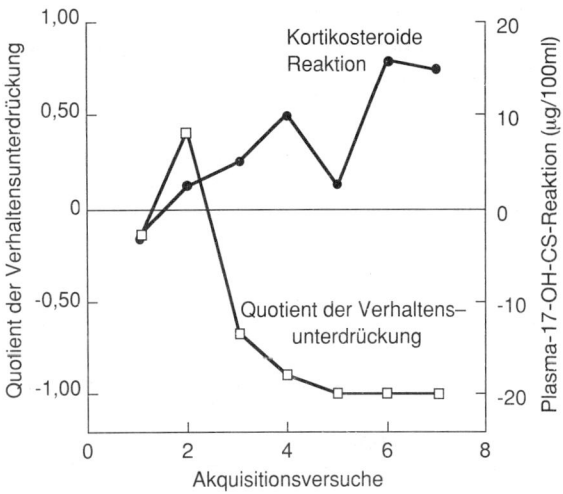

Abb. 8–18. Beziehung zwischen emotionalem Konditionieren und Veränderungen der 17-OH-CS-Plasmakonzentration (aus Brady, 1975).

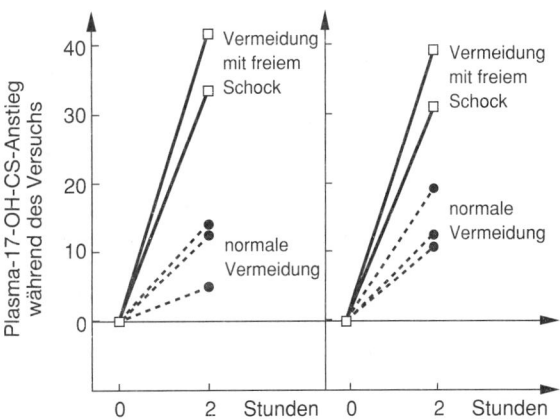

Abb. 8–19. 17-OH-CS-Plasmakonzentration während „normalen" Vermeidungsversuchen ohne exterozeptives „Warnsignal" (––●) und während Vermeidungsversuchen mit zufälligen Schocks (——□) (aus Brady, 1975).

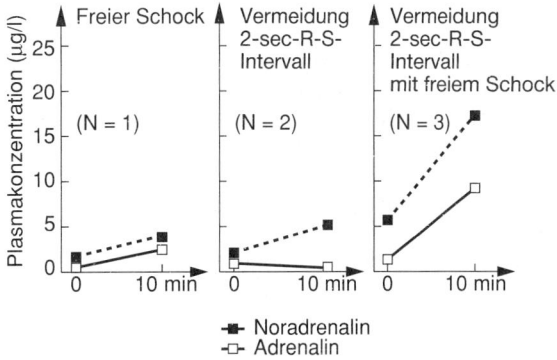

Abb. 8–20. Adrenalin- und Noradrenalin-Plasmakonzentration während eines Zufalls-Schock-Versuchs vor einem Vermeidungstraining (linker Teil), während eines Leistungsversuchs nach einem Vermeidungstraining ohne Warnsignal (mittlerer Teil) und während eines Leistungsversuchs mit Zufalls-Schocks (rechter Teil) (aus Brady, 1975).

Im Kapitel 10 werden derartige Befunde ausführlich dargestellt, so daß hier nicht weiter darauf eingegangen werden muß.

In mehreren der geschilderten Untersuchungen bestand eine wesentliche Bedingung darin, daß die Schocks in zufälligen Zeitabständen erfolgten und daher für die Versuchstiere nicht vorhersagbar waren. Dies stellt offensichtlich eine besondere Belastung dar. Die folgende klassische Untersuchung von Brady (1958) und Brady und Mitarbeitern (1958) scheint dem zu widersprechen. Die Autoren hatten in mehreren Untersuchungen jeweils zwei Affen paarweise einem „Yoked-Control-Design" (Jochkontrolle) unterworfen. Die Versuchstiere waren in Primatenstühlen weitgehend immobilisiert. Eines der beiden Tiere, das „Exekutivtier", konnte durch Hebeldrücken Schocks vermeiden, das andere Tier konnte dies nicht und wurde identisch wie das Exekutivtier behandelt, erhielt also dieselbe Anzahl von Schocks derselben Dauer, Intensität usw. Die Tiere waren diesem Untersuchungsplan für 6–7 Wochen unterworfen, wobei 6stündige Perioden der geschilderten Anordnung mit ebenso langen schockfreien Perioden abwechselten. Nach 3–4 Wochen hatten die Exekutivtiere ausnahmslos gastrointestinale Läsionen entwickelt, die anderen Tiere nicht, obwohl sie das Auftreten der Schocks weder vorhersagen noch beeinflussen konnten. Bei kürzeren Trainingsphasen kam es nicht zu den Läsionen.

Die Autoren interpretierten ihre Ergebnisse dahingehend, daß nicht die Tatsache, aversiven Ereignissen ausgesetzt zu sein, allein schädigend wirke, sondern das ständige Bemühen, diese zu beeinflussen, also ein ständiges und langanhaltendes Aktiviertsein.

Seligman (1975) wie auch Weiss (1972a, b) weisen jedoch darauf hin, daß eine bestimmte methodische Besonderheit der Experimente von Brady zu diesem Ergebnis geführt hat. Brady hatte die Zuordnung seiner Versuchstiere zu den beiden experimentellen Gruppen nach einem Maß der „Emotionalität" vorgenommen, das darin bestand, daß alle Tiere zunächst der „Exekutivbedingung" unterworfen wurden, und diejenigen Tiere, die als erste begannen, auf den Hebel zu drücken, in die Exekutivgruppe kamen, die restlichen in die „Yoked-Gruppe". Es hat sich gezeigt, daß diese Art der Zuordnung der Tiere die Ergebnisse wesentlich beeinflußt hat.

Weiss wiederholte den Versuch in exakt derselben Weise und kam zu den gleichen Ergebnissen, wie die folgende Abbildung 8–21 zeigt.

Nach den oben referierten Ergebnissen hätte eigentlich die Bedingung der Unbeeinflußbarkeit der Schocks besonders belastend sein müssen, die Tiere der Yoked-Gruppe hätten also die Läsionen aufweisen müssen und nicht die Exekutivtiere. Es sei in diesem Zusammenhang daran erinnert, daß der belastende Effekt des Sidman-Vermeidungsplanes dadurch erheblich verringert werden kann, daß ein Warnsignal zur zeitlichen Orientierung gegeben wird. Dadurch wird die Vorhersagbarkeit der Schocks erhöht und damit auch die Voraussetzung für ihre Kontrollierbarkeit.

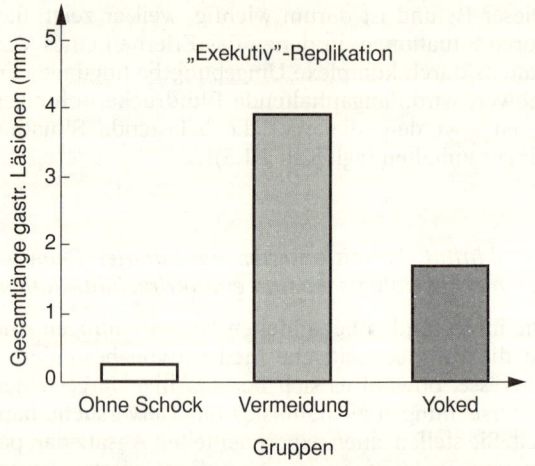

Abb. 8–21. Durchschnittliche Gesamtlänge gastrischer Läsionen bei Versuchstieren, die der gleichen Gruppenzuordnung unterworfen waren wie im Experiment von Brady et al. 1958 (aus Weiss, 1972).

Weiss (1970) wiederholte die Versuchsanordnung der ursprünglichen Untersuchung von Brady, allerdings mit Ratten und dem wesentlichen Unterschied, daß die Exekutivtiere ein Warnsignal erhielten, so daß die Schocks vorhergesagt und leichter kontrolliert werden konnten. Die Zuordnung der Tiere zu den beiden Gruppen erfolgte randomisiert und nicht entsprechend eines Wertes von „Emotionalität". Auch hierbei waren die Exekutivtiere zu ständiger Aktivität gezwungen, jedoch mit erheblich reduzierter Unsicherheit. Die Abbildung 8–22 zeigt deutliche Unterschiede zur Versuchsanordnung, wie sie von Brady verwendet wurde. Jetzt zeigen die Tiere in der Yoked-Gruppe deutlich mehr Läsionen als die Tiere, die die Schocks kontrollieren konnten.

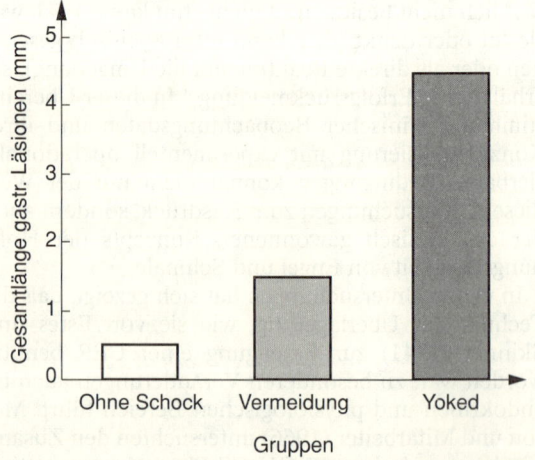

Abb. 8–22. Durchschnittliche Gesamtlänge gastrischer Läsionen für die Gruppen ohne Schock, Vermeidung möglich, Yoked-Control. Die Läsionen wurden bei jedem Tier (in mm) gemessen und die Gesamtlänge für jedes Tier errechnet. Die Graphik zeigt die Durchschnittswerte für jede Gruppe. Wie sich zeigt, kommt es auch bei der Gruppe ohne Schocks zu geringfügiger Ulzeration. Dies ist auf die Versuchsanordnung mit teilweiser Immobilisierung, Futter- und Wasserdeprivation zurückzuführen (aus Weiss, 1970).

Caul und Mitarbeiter (1972) untersuchten den Einfluß abgestufter Vorhersagbarkeit von Schocks, ohne daß die Schocks in irgendeiner Weise kontrolliert werden konnten. Hierbei zeigte sich, daß die Bedingung der Vorhersagbarkeit die Bildung von gastrischen Läsionen beeinflußte.

Corley und Mitarbeiter (1975) fanden ebenfalls eine erheblich größere Belastung durch eine Yoked-Control-Bedingung im Vergleich zu einer Vermeidungsbedingung, auch bei Verwendung eines Sidman-Vermeidungsplanes. Diese Autoren untersuchten den Einfluß der Belastung auf das kardiovaskuläre System. Bei dieser Untersuchung wird besonders deutlich, wie wichtig die Dauer einer derartigen Untersuchung ist. Ursprünglich war geplant, das Experiment für jeweils ein Paar von Versuchstieren zu beenden, wenn das Exekutivtier aufhörte, den Hebel zu drücken. Durch das Auftreten besonderer Schwächeeffekte (Bradykardie) bei den Yoked-Control-Tieren mußte der Versuch jedoch vorzeitig beendet werden, die folgende Abbildung 8–23 faßt die Ergebnisse für die Herzfrequenz zusammen.

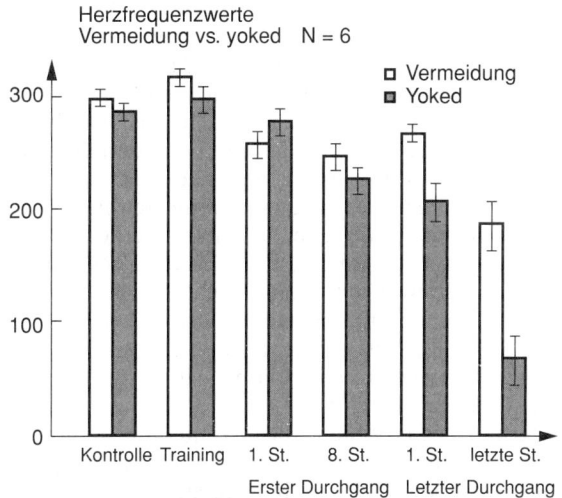

Abb. 8–23. Durchschnittliche Herzfrequenzwerte mit Standardabweichungen von 6 Vermeidungs- und 6 Yoked-Control-Affen zu verschiedenen Zeitpunkten des Versuchs (aus Corley et al., 1975).

Wie in dieser Untersuchung deutlich wird, spielt die Dauer einer Belastung neben anderen Bedingungen eine wesentliche Rolle. Seligman und Groves (1970) fanden, daß ein einmaliges Erlebnis von unkontrollierbarem Schock zumindest bei Hunden seinen negativen Effekt auf effektives Fluchtverhalten nach einiger Zeit wieder verliert. Sind die Tiere dieser Erfahrung jedoch für längere Zeit ausgesetzt gewesen, so kann es mehrere Wochen dauern, bis die Tiere wieder normales Fluchtverhalten zeigen. Bei Ratten reichen jedoch bedeutend kürzere Erfahrungszeiträume der Unkontrollierbarkeit von Schocks, um diesen Effekt zu erzielen (Seligman et al., 1974), so daß auch hier deutlich wird, daß die Merkmale eines Organismus die Wirksamkeit derartiger belastender Bedingungen wesentlich mitbeeinflussen. Dies hatte sich

schon bei den Untersuchungen von Brady gezeigt, indem die Selektion der Tiere für die experimentellen Gruppen nach ihrer „Emotionalität" den Einfluß der Versuchsbedingungen verdeckt hatte.

Darüber hinaus hatte sich bei den Langzeituntersuchungen herausgestellt, daß die Zeit vor den Vermeidungsphasen von Veränderungen physiologischer Parameter begleitet war. Brady (1975) hatte ähnlich wie Anderson und Brady (1973) einen stetigen Anstieg des systolischen Blutdrucks gefunden, und weniger ausgeprägt auch des diastolischen. Dabei ist bemerkenswert, daß die Herzfrequenz zunächst leicht erhöht war, dann jedoch, trotz weiter ansteigenden Blutdrucks, nach etwa 9 Stunden abfiel. Die Dauer der entsprechenden Phasen vor dem Vermeidungstraining betrug dabei 15 Stunden.

Bei diesem Befund stellt sich die Frage nach der Regulation des Blutdrucks an den verschiedenen Zeitpunkten der Phase vor der Vermeidung, da Herzfrequenz und Blutdruck zu verschiedenen Zeitpunkten verschieden korreliert sind. Dies könnte bedeuten, daß zu Beginn der Phase der erhöhte Blutdruck durch ein erhöhtes Herzminutenvolumen, später aber durch einen erhöhten peripheren Widerstand zustande kommt. Interessant ist dies im Hinblick auf Überlegungen zur Pathogenese der essentiellen Hypertonie. Zwei verschiedene pathogene Mechanismen, von denen angenommen wird, daß sie im zeitlichen Verlauf der Erkrankung unterschiedliches Gewicht besit-

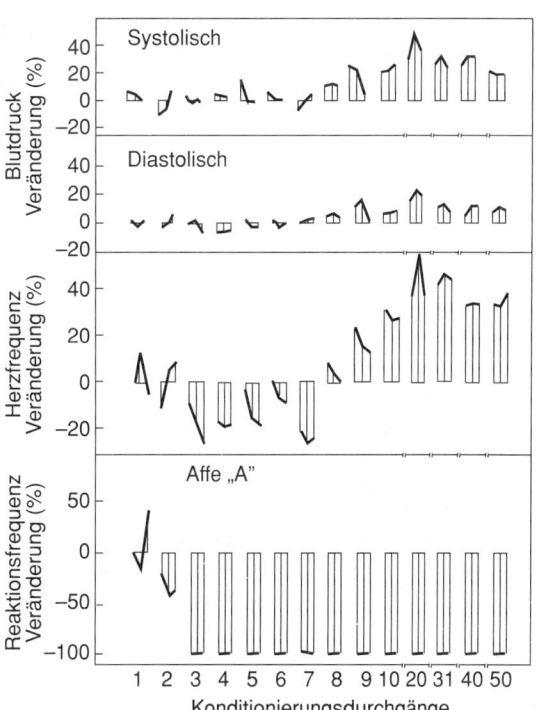

Abb. 8–24. Minütliche Veränderungen von Blutdruck, Herzfrequenz und der Rate des Hebeldrückens eines Affen bei aufeinanderfolgenden 3minütigen Klick-Schock-Versuchen während der Akquisition einer konditionierten emotionalen Reaktion. Die Null-Punkte zeigen Kontrollwerte, die vom 3-Minuten-Intervall berechnet wurden, das dem Klick vorausging (aus Brady, 1975).

zen, treten möglicherweise unter bestimmten experimentellen Bedingungen kurz hintereinander auf.

Aus den Untersuchungen mit relativ langer Dauer wird weiter die weitgehende Unabhängigkeit behavioraler und physiologischer Parameter deutlich. In einer Reihe von Untersuchungen zur Verhaltensunterdrückung fanden Brady und Mitarbeiter (1969), daß bereits nach drei Versuchsdurchgängen eine für den Rest der Untersuchung stabile Verhaltensunterdrückung erreicht worden war. Die Veränderung kardiovaskulärer Größen hatte einen ganz anderen, biphasischen Verlauf. Die Abbildung 8–24 zeigt, daß es zunächst zu einem starken Abfall der Herzfrequenz bei relativ unveränderten Blutdruckwerten kommt und etwa nach dem 8. Versuchsdurchgang zu einer Erhöhung von Herzfrequenz und Blutdruck.

8.4.7 Angst

Eines der wohl am häufigsten im Bereich der Psychophysiologie untersuchten Phänomene ist das der Angst. Dies hat sicherlich viele Gründe, es ist ein Grundgefühl organismischer Existenz, sein Erleben beinhaltet in der Regel die Wahrnehmung körperlicher Erregung, und der Angst wird eine zentrale Bedeutung im klinischen Bereich zuerkannt, als wesentliche Bedingung psychischer Störungen. Viele der bisher referierten Ansätze implizieren mehr oder weniger deutlich eine Konzeption des Phänomens Angst als auf Bedrohung erfolgende Emotion extrem negativer Valenz, die eine extreme Intensität annehmen kann und dann gelegentlich als Panik bezeichnet wird. Ein wesentlicher Teil der kognitiven Emotionsmodelle bezieht sich eigentlich auf den Prozeß des Umgangs mit Bedrohung, d.h. zumindest potentiell angstinduzierenden Situationen, wobei Angst sicherlich nicht die einzige Reaktionsmöglichkeit auf Bedrohung darstellt. Angst ist ebenfalls diejenige Emotion, von der am plausibelsten gesagt werden kann, daß sie stattfindendes Verhalten stört. Die konditionierte emotionale Reaktion, deren Parameter die Verhaltensunterdrückung ist, d.h. das Ausmaß, in dem stattfindendes Verhalten durch einen konditionierten aversiven Reiz gestört wird, ist im wesentlichen ein Operationalisierungsversuch, experimentell herstellbare Angst im Verhalten zu quantifizieren. Ist Angst eine mögliche Antwort auf Bedrohung, so ist sie abhängig von der Einschätzung der Möglichkeiten, und wohl auch den tatsächlichen Möglichkeiten, mit der Bedrohung umzugehen, d.h. abhängig von der Kompetenz einer Person (vgl. Lazarus et al., 1980). Möglicherweise ist dies ein Grund dafür, daß Angst oder emotionales Reagieren oder emotionaler Ausdruck ganz allgemein, wenn auch kulturell unterschiedlich ausgeprägt, oft unterdrückt wird. So fand Ekman (1971), daß Japaner und Amerikaner beim Betrachten von Streßfilmen ihr emotionales Mitreagieren mimisch zum Ausdruck brachten, japanische Probanden jedoch deutlich weniger, wenn eine zweite Person mit im Raum anwesend war.

Angst wird in der Psychologie unter zwei Aspekten behandelt, einmal als psychischer Prozeß der Auseinandersetzung mit Bedrohung, zum anderen aber auch im Sinne eines Merkmals einer habituellen Art der Auseinandersetzung mit Bedrohung, anhand dessen man Personen unterscheiden kann, dem Merkmal „Angstneigung", „Ängstlichkeit" oder „Angstbereitschaft". Beide Aspekte spielen in der Psychophysiologie eine Rolle, Angst als psychophysisches Geschehen und Ängstlichkeit als Merkmal einer differentiellen Psychophysiologie, beispielsweise bei der Frage, ob ängstliche Personen physiologisch reagibler sind als nicht-ängstliche. Es wurde bereits oben darauf hingewiesen, daß beispielsweise das Konzept der emotionalen Labilität in diesem Sinne auch eine Labilität der Regulation physiologischer Prozesse beinhaltet.

Angst als psychophysischer Prozeß

Angst ist eine Reaktion von Organismen auf Bedrohung oder Gefahr, die zunächst als nicht zu bewältigen eingeschätzt wird. Es kommt zu einer allgemeinen sympathisch vermittelten Erregung (Aktivierung) mit Anstieg der Herzfrequenz, des Blutdrucks, Erhöhung der Hautleitfähigkeit aufgrund erhöhter Schweißsekretion, Erhöhung des Muskeltonus und so unter Umständen zur Inhibition motorischer Reaktionen, Anstieg der Atemfrequenz bis zur Hyperventilation, jedoch ohne daß die Veränderungen der CO_2-Konzentration im Blut damit kovariieren (Suess, 1980). Die Veränderung dieser Parameter bietet auch die Möglichkeit, Angst bei Patienten zu erkennen. Eine beschleunigte Atmung, beim Händedruck feuchte Hände, die unter Umständen leicht zittern, eine veränderte höhere und vielleicht etwas heisere Stimme und Schweißperlen auf der Stirn, die nicht durch die Umgebungstemperatur erklärbar sind, sprechen für das Vorliegen von Angst. Auch eine gewisse Starre aufgrund des erhöhten Muskeltonus und reduzierte Aufmerksamkeit im Hinblick auf die Umgebung (weniger Blickbewegungen) sind beobachtbar.

Es kommt weiterhin zur Ausschüttung von Katecholaminen, wobei die Anstiege der Konzentration von Adrenalin gegenüber der von Noradrenalin bei Angst im Vergleich zu Zuständen hoher Aggression überwiegen sollen (Funkenstein, 1955; Frankenhaeuser und Rissler, 1970). Diese Unterschiede scheinen davon abhängig zu sein, ob eine Person bei derartigen Reizen die Möglichkeit zur Bewältigung hat. Dabei kommt es dann wohl eher zu Anstiegen von Noradrenalin. Trotz derartiger Befunde ist es bisher nicht überzeugend gelungen, ein spezifisches physiologisches Reaktionsmuster für Angstreaktionen zu definieren.

Von Easterbrook (1959) wurde die Hypothese aufgestellt, daß es bei großen Ausmaßen von Erregung zu einer veränderten bzw. zu einer eingeschränkten Aufmerksamkeit kommt, so daß die kognitiven Leistungen in Situationen mit großer Angst eingeschränkt sind (z.B. Wirkung von Examensangst in Prüfungssituationen). Dabei spielt das Ausmaß von Besorgtheit um die zu erbringende Leistung eine Rolle, da ein Teil der informationsverarbeitenden Kapazität durch Gedanken an die Leistung besetzt ist und auf diese Weise nicht für die Lösung von Aufgaben zur Verfügung steht. Ebenso ist es wahrscheinlich, daß ein Teil der Aufmerksamkeit durch die mit der Angst verbundene körperliche Erregung und die dadurch vorhandenen körperlichen Empfindungen in Anspruch genommen wird. Dies gilt vor allem, wenn die körperliche Erregung mit der Erwartung von Mißerfolg erklärt wird, d. h. als Anzeichen von Mißerfolg interpretiert wird (Douglas und Anisman, 1975).

Neben der körperlichen Reaktion kommt es zu einer intensiven Tendenz, dem bedrohenden oder angstauslösenden Reiz zu entkommen. Zunächst handelt es sich dabei um eine Fluchtreaktion, die zur Folge hat, daß die Erregung nachläßt und damit die negative Emotion. Durch diesen Vorgang der negativen Verstärkung wird auch die Tendenz verstärkt, den Reiz zu vermeiden, wodurch die Vermeidungsreaktion weiter verstärkt wird. Als Zwei-Stufen-Theorie der Angstentstehung wurde von Mowrer (1956) ein Modell vorgestellt, das in der ersten Stufe das klassische Konditionieren eines neutralen auf einen aversiven Reiz annimmt und in der zweiten Stufe die eben beschriebene negative Verstärkung. Ein Organismus kann auf diese Weise kaum die Erfahrung machen, daß ein konditionierter Reiz harmlos ist, da er sich ihm nicht aussetzt. Als Modell zur Erklärung persistierender, unrealistischer Ängste ist dies Modell sicher nicht ausreichend, enthält jedoch wesentliche, zu jeder Erklärung notwendige Elemente. So enthalten fast alle therapeutischen Ansätze in irgendeiner Weise die Konfrontation mit den angstauslösenden Reizen, verhindern also die Vermeidungsreaktion.

Das Inkubationsmodell der Angstentstehung nach Eysenck (1968) versucht ebenfalls, den Sachverhalt zu erklären, daß Angstreaktionen sich relativ schnell gleichsam verselbständigen, die Reaktion also trotz ausbleibender externer, negativer Konsequenzen nicht gelöscht wird. Der Grundgedanke in diesem Modell besteht darin, daß die physiologischen Anteile der konditionierten Angstreaktion in sich stark aversiv erlebt werden, so daß die sie auslösenden Reize nicht als harmlos erlebt werden, da sie zu den aversiven Körperreizen führen, so daß die konditionierte Reaktion durch diese Anteile eher noch verstärkt wird. Es handelt sich dabei also um eine Angst vor der Angst, bzw. vor der Angsterregung, ein Sachverhalt, der in vielen Aussagen von Patienten enthalten ist.

Die von Seligman (1971) als „Preparedness-Hypothese" bezeichnete Auffassung versucht ebenfalls zu erklären, warum Angstreaktionen unter Umständen sehr schnell gelernt werden und in hohem Maße löschungsresistent sind. Seiner Auffassung nach treffen diese Merkmale von Angstreaktionen nur auf solche, die durch bestimmte Reize ausgelöst wurden, zu. Gerade bei Tierphobien erscheint die Annahme plausibel, daß in der Evolution Angstreaktionen auf gefährliche Reize durch andere Tiere ein gewisser Überlebenswert zukommt, so daß Tierphobien aufgrund einer Art phylogenetischen Gedächtnisses schneller zu-

stande kommen und auch schwerer zu löschen sind; Organismen sind also vorbereitet (prepared), auf diese Weise auf Reize, die von Tieren ausgehen, zu reagieren. In zahlreichen Untersuchungen wurde die Hypothese psychophysiologisch untersucht und konnte hauptsächlich im Hinblick auf die Löschungsresistenz bestätigt werden, jedoch weniger für die Schnelligkeit der Akquisition (McNally, 1987). In derartigen Untersuchungen wird im allgemeinen so vorgegangen, daß die zur Diskussion stehenden Reize mit einem aversiven Reiz, einem elektrischen Schock als Schmerzreiz kontingent dargeboten, also konditioniert werden. Die abhängigen Variablen sind dann meist Veränderungen der elektrodermalen Aktivität oder der Herzfrequenz. Im zweiten Teil der Untersuchung werden dann die unterschiedlichen konditionierten Reize (z. B. Schlange, Spinne, Gewehr) ohne die aversiven Reize dargeboten (Extinktion oder Löschung) und die Schnelligkeit der Löschung wird durch die Stärke der physiologischen Reaktion erfaßt.

Johnsen und Hugdahl (1988) konnten zeigen, daß auch für kulturelle und spezifisch menschliche Gefahrenreize (ein Bild eines auf den Betrachter gerichteten Gewehres) als konditionierten Reizen für eine Angstreaktion eine höhere Löschungsresistenz besteht. Dies ist nicht der Fall, wenn das Gewehr zur Seite gerichtet ist. Dimberg und Öhman (1983) konnten für den emotionalen Gesichtsausdruck von menschlichen Gesichtern einen ähnlichen Zusammenhang ermitteln. Zornige (angry) Gesichter als konditionierte Reize resultierten nur dann in einer höheren Löschungsresistenz für Angstreaktionen, wenn sie auf die Probanden gerichtet waren, sie also ansahen. Die Autoren interpretieren dies ebenfalls in einem phylogenetischen Sinn, da von zornigen Personen eine Gefahr ausgeht.

Angstbereitschaft

Im vorangegangenen wurde Angst kurz als ein allgemeines Phänomen menschlicher oder tierischer Existenz beschrieben. Es ist nun weiterhin möglich, Individuen im Hinblick auf ihre Bereitschaft und das Ausmaß zu unterscheiden, mit dem sie mit Angst z. B. auf Bedrohung reagieren. Dies bedeutet, daß eine Person in einer für eine andere Person mäßig oder gar nicht bedrohlich erscheinenden Situation bereits mit Angst reagiert. Man kann Personen mit Hilfe von Fragebögen unterscheiden, d. h. Personen können verbal über das Ausmaß ihrer Ängstlichkeit Auskunft geben. Dabei wird angenommen, daß dieses Merkmal (Angstneigung, Angstbereitschaft oder Ängstlichkeit) zeitlich relativ invariant ist, sich also nicht schnell ändert. Es stellt sich nun die Frage nach dem Zusammenhang dieses Merkmals mit anderen psychologisch oder psychophysiologisch erfaßbaren Merkmalen.

Es gibt nun verschiedene Ansätze, mit denen versucht wird, Angstbereitschaft als psychologisches Merkmal zu konzeptualisieren. Wie bereits dargestellt, stützt sich Eysenck in seinen Annahmen zu Dimensionen der Persönlichkeit auf die Unterscheidung zwischen kortikalem Arousal und autonomer oder vegetativer Aktivierung. Ein hohes Ausmaß vegetativer Aktivierung kennzeichnet seiner Auffassung zufolge Personen mit einer hohen emotionalen Labilität, während ein hohes Maß an kortikalem Arousal Personen mit hoher Introversion kennzeichnet. Personen mit einer hohen Angstbereitschaft sind diesem Modell zufolge durch ein hohes Maß an emotionaler Labilität und Introversion gekennzeichnet. Wie ebenfalls bereits erwähnt, wurden diese Faktoren von Gray um 45 Grad rotiert, wodurch sich die neuen Faktoren „Impulsivität" (extravertiert/emotional labil) und „Angst" (introvertiert/emotional labil) ergeben. Die so im System Eysencks als „angstbereit" definierten Personen zeichnen sich also durch eine hohes kortikales Arousal aus, sind leichter konditionierbar und neigen schneller zu vegetativer Aktivierung bei einer labileren autonomen Regulation.

In vielen Untersuchungen wurde das Zusammenspiel zwischen kortikalem Arousal und autonomer Aktivierung bei Personen mit einer hohen Angstbereitschaft durch die Bestimmung des Habituationsverlaufs erfaßt. Die Habituation ist verzögert bei hoher autonomer Erregung bzw. bei einem hohen emotionalen Wert eines Reizes. So verglich beispielsweise Lader (1967) Patienten mit verschiedenen Arten von Angststörungen im Hinblick auf deren Habituationsgeschwindigkeit. Er fand dabei, daß diejenigen Personen, die am ehesten durch Introversion und hohe emotionale Labilität gekennzeichnet waren, die längsten Habituationsverläufe aufwiesen, was als Bestätigung der Einordnung in das Schema von Eysenck gewertet werden kann. In vielen anderen Untersuchungen wurde lediglich die Reaktivität physiologischer Parameter auf verschiedene Belastungen erfaßt, wobei sich oft bestätigte, daß Personen mit hohen Werten auf Angstskalen in derartigen Untersuchungen stärker vegetativ reagierten und daß die Erregung sich langsamer zurückbildete. Bond et al. (1974) zeigten darüber hinaus, daß sich Patienten mit Angststörungen nicht so sehr in Situationen, die einen deutlich aktivierenden Einfluß hatten, von normalen Kontrollpersonen unterschieden, sondern daß die Unterschiede zwischen beiden Gruppen in sehr milde aktivierenden Situationen am größten waren. Dies bedeutet, daß die Personen mit Angststörungen bereits durch sehr gering aktivierende oder belastende Reize stärker als Kontrollpersonen aktiviert wurden, was dem Konzept der erhöhten Angstbereitschaft zumindest im Sinne der Plausibilität entspricht. Andererseits muß in diesem Zusammenhang auf die, wenn überhaupt vorhandene, meist sehr geringe Kovarianz zwischen auf Fragebogenebene erfaßten und physiologisch gemessenen Daten zur emotionalen Labilität hingewiesen werden.

Andere Ansätze beziehen sich auf die sog. „Angstverarbeitung" und versuchen kognitive Prozesse zu beschreiben, die einen Umgang mit bedrohender oder gefahrenrelevanter Information in der Umgebung beinhalten. Diese Ansätze wurden im wesentlichen bereits dargestellt, gemeint sind hier die Modelle von Epstein und Fenz, Byrne und Lazarus. Allerdings

sind diese Konzepte weniger geeignet, auf der Ebene von Fragebögen Personen mit einer hohen von solchen mit einer geringen Angstbereitschaft zu unterscheiden, wie sich vor allem am Konzept „Repression – Sensitization" gezeigt hat.

8.5 Psychophysiologische Behandlungsmöglichkeiten

Im Kapitel 7 wurde kurz das Prinzip der operanten oder beim Menschen besser willkürlichen Kontrolle autonomer physiologischer Funktionen erläutert. Geht man davon aus, daß bestimmte körperliche Erkrankungen durch Störungen bestimmter physiologischer Funktionen bedingt sind, so liegt es nahe, Prinzipien und Methoden anzuwenden, die es gestatten, eben diese Funktionen zu beeinflussen. Studien zur willkürlichen Beeinflussung autonomer Funktionen haben gezeigt, daß dies durchaus möglich ist. Zunächst, nachdem es vor allem in tierexperimentellen Untersuchungen der Arbeitsgruppe um Miller und DiCara (1967, 1969, 1970) gelungen war, autonome Funktionen, vor allem unter der Zielsetzung einer theoretischen Klärung des Konditionierens, mit Methoden des operanten Lernens zu beeinflussen, kam es in den Jahren bis ca. 1978 zu einer Fülle von Untersuchungen, in denen versucht wurde, diese Ergebnisse auch klinisch als Behandlungsmöglichkeit zu nutzen. Es zeigte sich jedoch, daß die Höhe der jeweils erzielten Effekte wie in den im folgenden kurz referierten Arbeiten relativ gering war, häufig zu gering, um als adäquate Behandlung dienen zu können.

So versuchten Benson und Mitarbeiter (1971) bei sieben Patienten mit essentieller Hypertonie mittels operanten Konditionierens den Blutdruck zu senken. Senkungen des Blutdrucks, die einem bestimmten Kriterium genügten, wurden mit Geld belohnt. Die Ergebnisse der Untersuchung zeigen einen deutlichen Einfluß dieser Prozedur auf die Höhe des Blutdrucks,

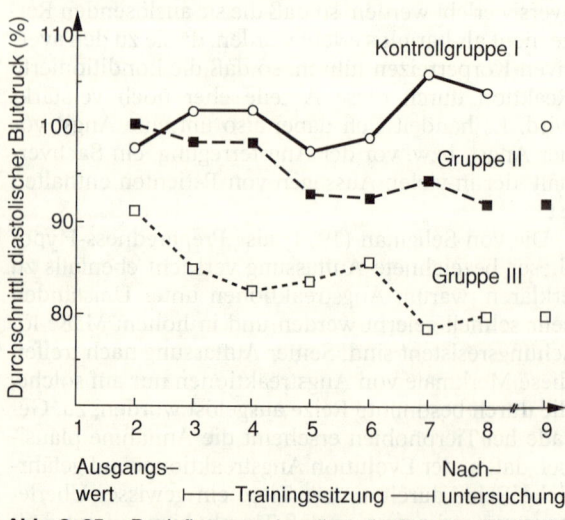

Abb. 8–25. Beeinflussung des diastolischen Blutdrucks (s. Abb. 8–12) (aus Elder et al., 1973).

es konnte eine durchschnittliche Senkung des Blutdrucks von 16,5 mm Hg (systolisch) erreicht werden. Der Erfolg der Konditionierung war jedoch sehr unterschiedlich und reichte von einer Senkung um 3,5 mm Hg bis zu einer von 33,8 mm Hg. Elder und Mitarbeiter (1973) fanden, wie die Abbildung 8–25 zeigt, daß die Rückmeldung des diastolischen Blutdrucks allein (Gruppe II), wie häufig angenommen wurde, keinen Einfluß hatte, daß aber eine „Belohnung" in Form eines je nach Erfolg abgestuften Lobs durchaus einen Einfluß ausübte (Gruppe III).

Für die Anwendung derartiger Methoden als Behandlung „psychophysiologischer Störungen" stellen sich jedoch eine Reihe von Problemen. Zum einen sind die erzielten Ergebnisse anhand eines klinischen Maßstabs zu gering. Zum zweiten sind sie bei verschiedenen Patienten sehr verschieden hoch, d.h., einige Patienten ließen sich auf diese Weise durchaus behandeln. Eine Senkung des systolischen Blutdrucks um 55 mm Hg kann durchaus genügend sein, eine von 9 mm Hg bei einem anderen Patienten derselben Studie ist wohl eher zu gering. Ein weiteres Problem besteht in der Übertragung aus einer Laborsituation auf die Alltagssituation. Diese Übertragung muß beinhalten, daß der Patient es lernt, mit den streßbeinhaltenden Situationen besser umzugehen.

So wurde versucht, verschiedene Methoden der Entspannung, wie transzendentale Meditation (Benson et al., 1974) oder Yoga (Patel, 1973, 1975), anzuwenden. Der Erfolg dieser Bemühungen reichte als alleinige Behandlungsstrategie in diesen Fällen essentieller Hypertonie nicht aus, führte jedoch zu einer Reduktion des notwendigen Medikamentenkonsums der Patienten (Patel, 1975).

Erinnert man sich an die Merkmale des Typ-A-Verhaltens und die Ergebnisse verschiedener Studien, so wird einsichtig, daß die Beeinflussung einer oder mehrerer physiologischer Funktionen (Birbaumer, 1977) zur effektiven Behebung etwa einer kardiovaskulären Überreaktivität nicht ausreichend sein kann. Diese kommt, wie gezeigt wurde, in bestimmten Situationen zustande, und zwar aufgrund eines Anforderungscharakters, der nicht unabhängig von der Interpretation und damit der Bewertung und Bedeutung der Situation gesehen werden kann. Erinnert man sich weiterhin an die Ansätze zur „Emotion als Gefühlszustand", so wird einsichtig, daß ganz wesentlich für das Zustandekommen von „Risikoreaktionen" kognitive Prozesse sind. Auch wenn es durchaus eine Reihe methodischer Probleme bei der Erfassung derartiger Prozesse gibt, so bedeutet dies nicht, daß sie keinen Einfluß ausüben könnten. Im Bereich der Verhaltenstherapie ist die Notwendigkeit der Berücksichtigung dieser Prozesse anerkannt (Mahoney, 1974) und inzwischen zur Regel geworden (vgl. Kap. 19). Entsprechend werden auch im Bereich einer im weitesten Sinne psychophysiologischen Behandlung derartige Prinzipien mitberücksichtigt.

Vor allem gibt es Ansätze, die versuchen, ein breites Spektrum von Methoden auf ein breites Spektrum von Problemen anzuwenden. So entwickelte Patel (1983) ein Therapieprogramm zur Behandlung der essentiellen Hypertonie, das für Therapie, aber auch Prävention des Herzinfarkts Anwendung finden kann. Dieses Programm enthält eine „kognitive Restrukturierung", mit der dem Patienten der Zusammenhang zwischen seinen körperlichen Reaktionen und seinem Erleben bestimmter Situationen verdeutlicht wird. Durch „Atemübungen" wird ihm gezeigt, wie rhythmisches Atmen zu einer gewissen Beruhigung führen kann, die in tiefer „Muskelrelaxation" weiter geübt wird. Mit Hilfe von „Meditation" wird versucht, einen mental wachen Zustand zu üben, der ebenfalls eine wache Beruhigung enthält. Das „Biofeedback" dient auch weitgehend der Herstellung eines entspannten Zustands, ist also zunächst nicht nur auf den Blutdruck gerichtet, sondern es wird entweder der Hautwiderstand oder das Elektromyogramm rückgemeldet, allerdings dann auch der Blutdruck am Ende einer Sitzung, um den Patienten den Erfolg ihrer Bemühungen zu zeigen und sie damit weiterhin zu motivieren. Das „Streßmanagement" dient hauptsächlich dazu, die Erfolge der Übungen auf Alltagssituationen zu übertragen, indem Regeln für den Umgang mit bestimmten Situationen erarbeitet werden.

Richter-Heinrich et al. (1988) wandten verhaltenstherapeutische Methoden ebenfalls bei Patienten mit essentieller Hypertonie an. Diese zielten einerseits auf die Beeinflussung der bei diesen Patienten bekannten kardiovaskulären Reaktivität, zum anderen auf die Veränderung des Umgangs mit Belastung, wie er mit Hilfe des Streßverarbeitungsbogens (Janke, 1985) erfaßt wurde. Interessant dabei war, daß nur 20% der 75 untersuchten und dann behandelten Patienten eine aktive Auseinandersetzung, 61,4% dagegen einen eher passiven Umgang mit Belastung angaben. Die Behandlung beinhaltete ein durch Biofeedback unterstütztes Entspannungstraining mit täglichen Übungen im häuslichen Bereich, tägliches Selbstmessen des Blutdrucks und ein aus mehreren Komponenten bestehendes Streßmanagement-Training. Wie in allen derartigen Untersuchungen waren die Ergebnisse interindividuell unterschiedlich. Es ließen sich jedoch Merkmale von denjenigen Patienten ermitteln, die schlecht oder gar nicht auf die Therapie ansprachen. Sie hatten eher eine genetische Belastung durch Hypertonie bei den Eltern, sie reagierten in einem Aktivierungsversuch auch nach der Behandlung mit Anstiegen kardiovaskulärer Parameter, vor allem des diastolischen Blutdrucks. Sie hatten ein höheres Angstniveau, waren nervöser, aggressiver und hatten mehr Dominanzstreben (Freiburger Persönlichkeitsinventar; Manifest Anxiety Scale). Im Jenkins Activity Survey hatten sie ebenfalls höhere Werte, also ein ausgeprägtes Typ-A-Verhalten. Sie hatten auch nach der Therapie einen evasiveren Umgang mit Belastung, so daß das Streßmanagement-Training keine Veränderung erbracht hatte. Offensichtlich wirkt sich die genetische Belastung eher ungünstig auf die Wirksamkeit verhaltenstherapeutischer oder psychophysiologischer Behandlungen aus, möglicherweise auch aufgrund eines verschiedenen Gefäßstatus. Bei fast der Hälfte einer behandelten Gruppe mit labiler Hypertonie konnte im Anschluß an die Behandlung auf die Medikation verzichtet werden.

8.6 Ausblick

Die für den Bereich der Psychophysiologie ausschnitthafte Darstellung verschiedener theoretischer und auf den Bereich der Psychosomatischen Medizin angewandter Modelle und Methoden hat gezeigt, daß die Ergebnisse aus diesem Inhaltsbereich für die Psychosomatische Medizin zu wesentlichen Erkenntnissen geführt haben. Das Verständnis pathogener Prozesse ist detailreicher und der Sache angemessener geworden, auch wenn viele Fragen noch nicht beantwortet worden sind. Mit der Zunahme des Interesses an psychophysiologischer Forschung wuchs die Einsicht in die Notwendigkeit bestimmter methodischer Anforderungen, die der Komplexität der zu behandelnden Probleme auch im Bereich der Psychosomatischen Medizin entsprechen. Für den Bereich psychophysiologischer Forschung wäre es wünschenswert, wenn methodische Ansätze entwickelt würden, die den kognitiven oder Erlebensbereich, mehr als dies bisher geschehen ist, integrieren. So hat sich bei der Darstellung der verschiedenen Konzepte gezeigt, daß eine gewisse methodische Kluft zwischen den kognitiv orientierten und etwa den behavioral orientierten Ansätzen besteht. Es wird hier nicht davon ausgegangen, daß diese Kluft prinzipiell bestehen muß, da Erleben einer wissenschaftlichen Erfassung nicht zugänglich sei, sondern eher daher rührt, daß die damit verbundenen Probleme einerseits erheblich größer sind, und zum anderen eine gewisse Vorliebe der Forschenden für technisch-methodisch Realisierbares vorherrscht. Andererseits hat sich am Beispiel des Typ-A-Verhaltens gezeigt, wie notwendig eine Be-

rücksichtigung des Erlebens für das Verständnis von Krankheitsprozessen, aber auch für Möglichkeiten der Behandlung ist.

Ebenso ist es zumindest denkbar, daß eine weitere Verfeinerung einer Spezifitäten berücksichtigenden multivariaten psychophysiologischen Forschung, die multivariat eben auch in dem Sinne ist, daß sie die Ebene des Erlebens mehr miteinbezieht, der weiterhin bestehenden Frage der psychosomatischen Spezifität näher kommt. Diese nachzuweisen ist bis heute möglicherweise eher aufgrund mangelnder Methoden nicht gelungen, was nichts über ihre Gültigkeit aussagt. So wird es notwendig sein, gerade wegen der Wichtigkeit der Psychophysiologie für die Psychosomatische Medizin, neben klinisch orientierter Forschung auch in verstärktem Maße methodisch orientierte Grundlagenforschung zu betreiben. Viele der klinisch orientierten Studien haben bislang aus methodischen Gründen eher heuristischen Wert. Dies könnte daran liegen, daß im klinischen Bereich trotz vieler Bemühungen und der tatsächlichen Notwendigkeit interdisziplinäre Forschung eher die Ausnahme als die Regel ist. Betrachtet man den die Bereiche von Physiologie und Psychologie verbindenden Charakter von Psychophysiologie und den die Bereiche Medizin und Psychologie verbindenden Charakter der Psychosomatik, so wird hierfür die Notwendigkeit der Interdisziplinarität wissenschaftlicher Handlung unmittelbar einsichtig. Dies widerspricht in keiner Weise der Forderung nach einer ganzheitlichen Betrachtungsweise von Lebensvorgängen in der Medizin, von der eingangs die Rede war, sondern entspricht ihr ausdrücklich. Die Forderung nach einer multivariaten Forschungsstrategie setzt sich also gleichsam fort in der nach einer multidisziplinären Strategie wissenschaftlichen Handelns, da diese erst die Ganzheitlichkeit der Betrachtung und der Handlung ermöglicht.

9 Neurophysiologische Grundlagen des emotionalen Verhaltens

Hugo M. Krott, Merita J. Poremba und *Michaela E. Rauch*

9.1 Geschichtliches

Sämtliche Theorien über Seele und Gemüt wurden noch bis ins zweite nachchristliche Jahrhundert von Philosophen erdacht und keineswegs von Medizinern geprägt. Demokrit und Platon definierten Geist und Seele als Prinzip des abstrakten Denkens, ohne von ihrem Landsmann Hippokrates Kenntnis zu nehmen, der bereits um 400 v. Chr. die Medizin zu einer eigenständigen Naturwissenschaft erhoben und dabei die Seele als von einem Körper abhängig beschrieben hatte. Es war Galen, der zum ersten Mal das Gehirn als Organ der Seele bezeichnete. Der Kirchenvater Augustinus verlegte seelische Vorgänge getrennt nach Vorstellung, Vernunft und Gedächtnis in die vordere, mittlere und hintere Schädelgrube, der Philosoph Descartes alle zusammen in die Zirbeldrüse (1649). Darüber mokierte sich Voltaire so nachhaltig, daß zwei Jahrhunderte lang keine weitere Theorie über Gefühle, deren Entstehung oder Sitz auftauchte. Anfang des 19. Jahrhunderts eröffnete die Lehre der Phrenologie des Franz-Josef Gall noch einmal wilden Spekulationen über Wesen und Sitz der Emotionen Tür und Tor, bis Claude Bernard aufgrund seiner klinischen Beobachtungen den Begriff Lebensvitalität einführte und diese im Hirnstamm ansiedelte. Es war ein Chirurg und Anthropologe, nämlich der wegen seiner Aphasieforschung bekannte Broca (1878), der als erster die überragende Bedeutung der wallartig um den Hirnstamm gelegenen Zellgruppen für emotionales Verhalten erkannte und diese anatomisch zusammenhängenden Strukturen als großen limbischen Lappen bezeichnete (grand lobe limbique).

Parallel zu dieser Entwicklung, aber unabhängig voneinander vertraten James (1884) in London und Lange (1885) in Kopenhagen die Ansicht, daß Emotionen durch Veränderungen der Körperperipherie zustande kämen. Beide Forscher erklärten Gefühle als Produkt peripherer Reizverarbeitung, beispielsweise Angst als Folge von Schwankungen des Blutdruckes oder der Körpertemperatur, Wut als Ergebnis veränderter vegetativer Reaktionen. Die James-Lange-Theorie – heute nur noch von historischem Interesse – basierte u.a. auf der Beobachtung, daß Querschnittsgelähmte eine qualitative Veränderung ihres Gefühls bemerken. Paraplegiker erleben Unangenehmes weniger intensiv, Angst weniger bedrohlich als zu gesunden Zeiten und reagieren auf Provokation weniger heftig, als es ihrem früheren Naturell entsprochen hätte. – Physiologisch erklärte sich dieses Verhalten dadurch, daß der aktivierende und modulierende Einfluß der Peripherie auf das Gehirn fortfällt.

Die medizinische Forschung der letzten 50 Jahre ist weniger durch philosophische Theorien als durch Beschreibungen von Morphologie und Funktion gekennzeichnet. Durch Reiz- und Ausschaltungsversuche an Tieren und in therapeutischer Anwendung vereinzelt auch am Menschen war es möglich, neue Kenntnisse über den Aufbau des Gehirns zu gewinnen und ein Konzept von seinem Funktionieren aufzustellen, wenn auch vorerst nur mosaikhaft. Dabei erwiesen sich die histologischen und physiologischen Untersuchungsergebnisse in dem von Broca abgegrenzten limbischen Lappen als so weittragend und in sich kohärent, daß McLean (1949) den heutigen Begriff des Limbischen Systems prägte.

9.2 Anatomie des Limbischen Systems und des Hypothalamus

Unter dem Begriff „Limbisches System" fassen wir anatomisch paarig um den Hirnstamm und den III. Ventrikel gelagerte Neuronengruppen mit ihren Faserverbindungen zusammen. Im Unterschied zur Definition von Broca zählt der olfaktorische Apparat (das Rhinenzephalon) nicht mehr zum Limbischen System im engeren Sinne. Im einzelnen verstehen wir darunter folgende morphologische Strukturen: Hippokampus (Ammonshorn oder Seepferdchen), Gyrus hippocampi, Nucleus amygdalae (Mandelkern), Gyrus cinguli, Inselanteile, Septum, Isthmus, Area paraolfactoria und die orbitale Gegend des Frontalhirns. Der Hippokampus ist über Fimbrien, die sich als Fornix bündeln, mit den Corpora mamillaria des Hypothalamus verbunden, der seinerseits über den Gyrus cinguli mit dem Limbischen System zu einem Regelkreis verknüpft ist. Diese fest vernetzte Neuronenkette innerhalb des Limbischen Systems wird nach dem Erstbeschreiber Papez-Kreis genannt (1937). Daneben existieren noch zwei weitere Verschaltungen: ein innerer Ring aus dem phylogenetisch älteren Archikortex mit Verbindungen zwischen Hippokampus, Indusium griseum und Area septalis und ein äußerer bzw. jüngerer aus Gyrus cinguli und

Abb. 9–1. Kerngebiete und Bahnverbindungen des Limbischen Systems (nach Krieg, 1975).

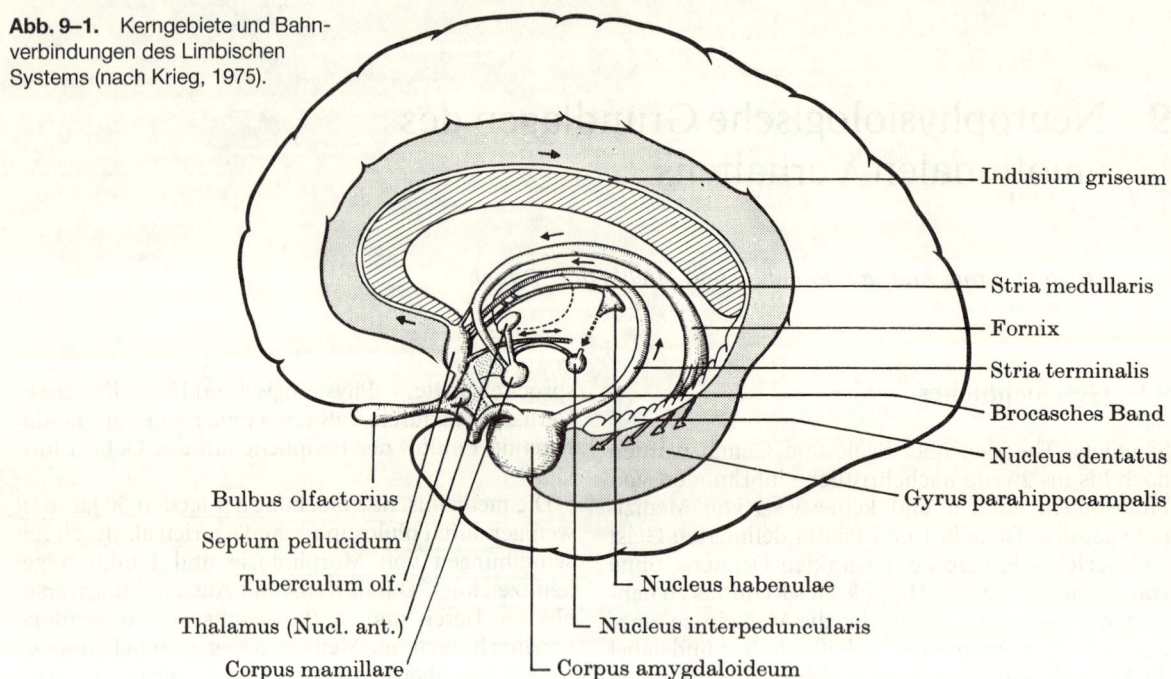

Nucleus amygdalae. – Mit der Formatio reticularis ist das Limbische System über das mediale Vorderhirnbündel, mit dem Kortex über im einzelnen nicht näher bekannte Nervenbahnen verschaltet. – Efferenzen des Limbischen Systems ziehen in das Zwischen- und Mittelhirn und üben eine Kontrolle über das zentrale Höhlengrau mit den lebenswichtigen vegetativen Zentren und über den neuroendokrinen Anteil der Hypophyse aus (Abb. 9–1) (Akert und Hummel, 1968; Ganong, 1977; Hassler, 1964; Noback und Demarest, 1981).

Der Hypothalamus zählt anatomisch nicht zum Limbischen System, hängt aber funktionell eng mit diesem zusammen. Topographisch handelt es sich um einen unterhalb des Thalamus gelegenen Teil des Zwischenhirns, der aus Seitenwand und Boden des III. Ventrikels, Area praeoptica, Corpora mamillaria, Tuber cinereum, Infundibulum mit Lamina terminalis und Neurohypophyse besteht. Neben den oben skizzierten Verbindungen erhält der Hypothalamus In-

formationen vom Thalamus selbst über den Fornix, der zu allen hypothalamischen Kernen zieht und in den Corpora mamillaria endet, sowie von der olfaktorischen Gegend über das mediale Vorderhirnbündel, das gleichzeitig Informationen des Limbischen Systems enthält. Vom Kortex wird der Hypothalamus via Thalamus, vom Rückenmark via Formatio reticularis und Fasciculus longitudinalis beeinflußt. – Efferent ist der Hypothalamus mit dem Limbischen System, mit Thalamus und Kortex, mit Formatio reticularis und Hirnnervenkernen verbunden (Abb. 9–2 bis 9–4).

9.3 Limbisches System

Tierexperimentelle Befunde und klinische Beobachtungen ergeben heute ein relativ abgerundetes Bild vom Limbischen System (Valenstein, 1973). Es gilt

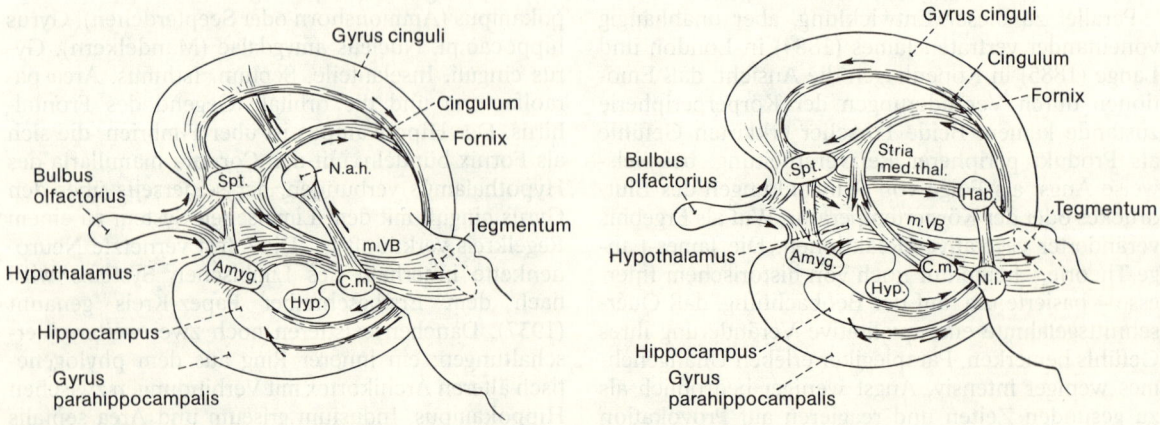

Abb. 9–2 und 9–3. Regelkreis im Limbischen System (nach Patton et al., 1976).

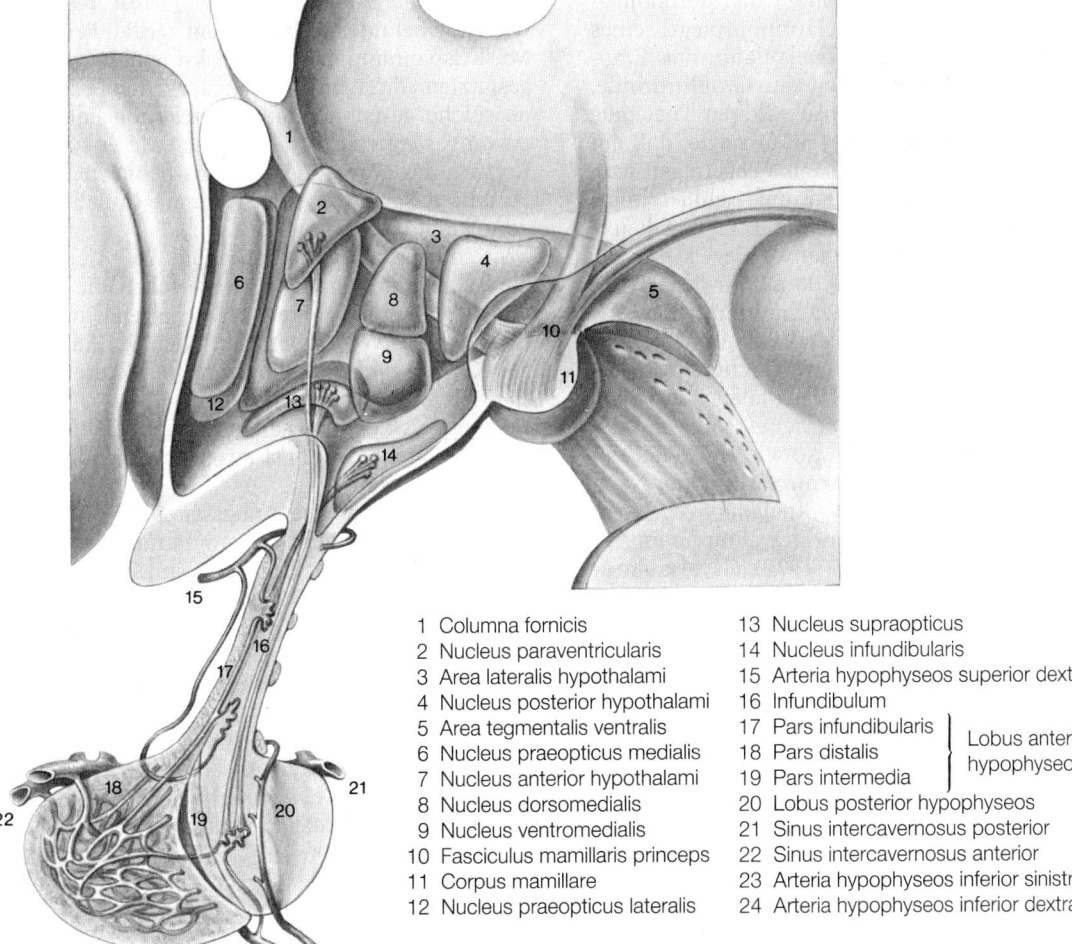

1 Columna fornicis
2 Nucleus paraventricularis
3 Area lateralis hypothalami
4 Nucleus posterior hypothalami
5 Area tegmentalis ventralis
6 Nucleus praeopticus medialis
7 Nucleus anterior hypothalami
8 Nucleus dorsomedialis
9 Nucleus ventromedialis
10 Fasciculus mamillaris princeps
11 Corpus mamillare
12 Nucleus praeopticus lateralis

13 Nucleus supraopticus
14 Nucleus infundibularis
15 Arteria hypophyseos superior dextra
16 Infundibulum
17 Pars infundibularis ⎤
18 Pars distalis ⎥ Lobus anterior
19 Pars intermedia ⎦ hypophyseos
20 Lobus posterior hypophyseos
21 Sinus intercavernosus posterior
22 Sinus intercavernosus anterior
23 Arteria hypophyseos inferior sinistra
24 Arteria hypophyseos inferior dextra

Abb. 9–4. Hypothalamische Kerne und Beziehung zwischen Hypothalamus und Hypophyse (aus Nieuwenhuys et al., 1980).

unumstritten als wichtigste Integrationszentrale für jede Ausdrucksform animalischen Verhaltens. Das System ist weniger der Ort, in dem Emotionen primär entstehen, als vielmehr die Schaltstelle, in der Emotionen ein komplexes Verhaltensmuster hervorrufen und zielgerichtetes Handeln ermöglichen. Dabei sind jeweils bestimmte Neuronengruppen für ein jeweils bestimmtes Verhalten verantwortlich. Das Großhirn wählt individuell wie willkürlich eine solche Zellpopulation an, die daraufhin artspezifisch mit vorprogrammiertem Verhaltensmuster reagiert. Dabei hat der Kortex einen anregenden oder hemmenden Modulationseffekt. – Im Tierversuch wird der modulierende Einfluß des Großhirns durch die Art der Versuchsanordnung imitiert: Elektrostimulation eines Zellverbandes führt prinzipiell zur Auslösung und/ oder Verstärkung eines Verhaltens, die neurochirurgische Entfernung zu dessen Abschwächung bis Beseitigung (Pfaffmann, 1969).

9.3.1 Thalamische und hypothalamische Ausschaltungsversuche

Cannon (Cannon und Britton, 1926; Cannon, 1929) versuchte als erster, umschriebene Hirnareale von Tieren experimentell zu erregen, um dadurch ein bestimmtes Verhalten auszulösen. Es handelte sich dabei um Ausschaltungsversuche. Durch chirurgische Entfernung eines umschriebenen Hirnareals sollte eine bestimmte Funktion ausgelöscht werden, wobei man anschließend die Reaktion des Versuchstieres, wenn auch nicht als direkte Folge der Ausschaltung, so doch als hiervon beeinflußt bis abhängig bezeichnete. Der Versuchshund von Cannon reagierte trotz operativer Entfernung des sensorischen und motorischen Kortex noch auf artspezifische Reize. Das Thalamus-Tier zeigte beispielsweise auf Vorhalten einer Katze ausgeprägte vegetative Reaktionen wie Pilomotorerektion, sympathische Pupillenerweiterung und vermehrten Speichelfluß, sowie eine Angriffsstellung mit Knurren, Zähneblecken und Beißen. Währenddessen stieg der Adrenalin-Spiegel im Blut des Versuchstieres auf Werte gereizter, aber nichtoperierter

Vergleichstiere an. Cannon sah in den autonomen Reaktionen und dem erhöhten Hormonspiegel seines dekortizierten Tieres einen Beweis dafür, daß Emotionen nicht von der Intaktheit der Großhirnrinde, sondern eher vom phylogenetisch älteren Thalamus abhängen. Sein Schüler Bard (1934) zeigte, daß ein Tier, dem sogar noch der Thalamus abgetragen war, ebenso artspezifisch emotional auf Schlüsselreize reagierte und den eine allgemeine Erregung kennzeichnenden Anstieg des Adrenalin-Spiegels aufwies. Cannon und Bard vermuteten daraufhin den Hypothalamus als Entstehungsort der Emotionen bzw. als Schaltstelle zwischen Sinneseindruck und Verhaltensreaktion. – Spätere Versuche an nichtdekortizierten Tieren zeigten jedoch, daß das thalamo-hypothalamische System nicht autark arbeitet, sondern stets vom Kortex kontrolliert wird, „so wie ein Reiter ein Pferd beherrscht", wie es McLean ausdrückte. Nach diesem Modell werden im Hypothalamus vegetative Reaktionen, in der Großhirnrinde Empfindungen ausgelöst, was Arnold später (1970) als exzitatorisches Gefühlsmodell bezeichnete.

9.3.2 Versuche am Nucleus amygdalae

Seit den dreißiger Jahren führten Neurophysiologen im Limbischen System von Tieren erfolgreich Reiz- und/oder Ausschaltungsversuche durch. Der entscheidende Unterschied gegenüber früheren Experimenten, beispielsweise den dekortizierten Hunden von Cannon und Bard, bestand in der differenzierteren Versuchsanordnung: die Tiere wurden unter nahezu reproduzierbaren Standardbedingungen ohne grobe Zerstörung von Zellgewebe in definierten Hirnstrukturen gereizt. Sie waren während der Versuche bei Bewußtsein und konnten im komplexen Ablauf ihrer Reaktionen beobachtet werden.

Elektrische Reizung des Nucleus amygdalae führt im Einzelexperiment zu jeweils reproduzierbaren Ergebnissen. Die verschiedenen Forscher machten jedoch trotz gleicher oder zumindest vergleichbarer Reizbedingungen im Nucleus amygdalae erheblich voneinander abweichende Beobachtungen. Offenbar sind in dieser Neuronenanhäufung Strukturen für unterschiedliches (bis gegensätzliches) Verhalten angelegt. Es ist bisher nicht gelungen, aus diesen homogenen Neuronenpopulationen histologisch oder histochemisch unterschiedliche Zellstrukturen oder verschiedene Transmitter zu differenzieren und diese einem bestimmten Reaktionsmuster zuzuordnen. Reizt man Affen im Nucleus amygdalae über implantierte Drahtelektroden, so können die Tiere folgende Reaktionen zeigen:

- Freßverhalten: Die Tiere beginnen zu schnüffeln, sich das Maul zu lecken, Freßbewegungen zu imitieren, zu beißen, zu schlucken und gelegentlich auch sich zu erbrechen.
- Hemmung eines angefangenen Freßverhaltens: Die Tiere hören auf zu fressen und zu trinken, obwohl sie nach der Versuchsanordnung Hunger und Durst haben müßten.

- Angespannte Aufmerksamkeit (arrest reaction): Die Tiere befinden sich in einem Zustand erhöhter Muskelspannung mit leicht gekrümmtem Rücken, gespitzten Ohren und zurückgezogenen Lefzen. Eine solche gespannte Stillhaltereaktion entspricht einem gesteigerten Wachzustand, wie er sonst bei Reizung der Formatio reticularis beobachtet wird. Funktionell handelt es sich um ein Initialstadium von Flucht oder Kampf.
- Wut- oder Aggressionsverhalten: Die Tiere nehmen eine Sprunghaltung ein, blecken die Zähne und beginnen zu schreien. Sie zeigen eine hohe Erregung und vegetative Reaktionen wie Aufstellen der Körper- und Schwanzhaare, vermehrten Speichelfluß und erweiterte Pupillen.
- Störung des Sozialverhaltens: Innerhalb einer Affenhorde brach die sonst ausgewogene Sozialordnung zusammen.
- Sezernierung von Verdauungssäften: Diese werden unabhängig von der Nahrungsaufnahme ausgeschüttet und können im Verdauungstrakt Erosionen und peptische Geschwüre hervorrufen (ein analoger Mechanismus für die Entstehung peptischer Ulzera wird unter aggressionsauslösenden Situationen auch beim Menschen angenommen; Gray, 1971).

Nach Reizende, also nach Abschalten des Stromes, beruhigten sich die Tiere innerhalb von Sekunden, beendeten die jeweils experimentell ausgelösten Verhaltensweisen und kehrten „zur Tagesordnung" zurück. Länger anhaltende Stimulation oder auch häufige Wiederholung einer einzelnen Reizsituation führten jedoch bei den Versuchstieren regelmäßig zu einer Veränderung ihres Verhaltens: wird beispielsweise eine von zwei Katzen im Nucleus amygdalae elektrisch stimuliert, kommt es regelmäßig zum Kampf zwischen den Tieren. Nach Abschalten des Reizes kehrt die Katze zunächst zu ihrem friedlichen Ruheverhalten zurück. Nach mehrfacher Stimulation behält sie jedoch ihre Kampfeslust bei. Es wird vermutet, daß sich die Erregung im Nucleus amygdalae über die äußere Ringschaltung bis zum Hippokampus ausbreitet, dessen synchrone Nachentladungen neurophysiologisches „Lernen" bewirken sollen. Andererseits kann es sich bei dem Verhalten auch um den neuropsychologischen Konditionierungsvorgang eines bedingten Reflexes handeln (Pribram, 1971).

Kaada (1967) beobachtete an Affen und Katzen, daß elektrische Reizung des Nucleus amygdalae je nach Lage der Elektrodenspitze an bestimmten Neuronengruppen unterschiedliches Verhalten auslösen konnte: an einer Stelle kam es reproduzierbar zu einer Furchtreaktion mit Zusammenkauern und Zurückweichen, abwehrenden Bewegungen der Pfoten und besorgtem Gesicht, bei wiederholter Reizung zur Flucht, an einer anderen Stelle löste die Elektrostimulation Wut mit entsprechendem Gebahren wie Krümmen des Rückens, Aufrichten des Schwanzes und Spreizen der Haare, Zähnezeigen und Beißen aus. Bei Dauerreiz kam es regelmäßig zu entsprechendem Angriffsverhalten.

Delgado (1963, 1967, 1971) erweiterte die oben beschriebene Versuchsanordnung dahingehend, daß er bei seinen Affen nicht nur eine einzelne Eigenschaft, beispielsweise Furcht oder Flucht, beobachtete, sondern ein komplexes Verhalten beschrieb. Er entdeckte während langdauernder Reizversuche, daß Elektrostimulation des Nucleus amygdalae sogar eine Änderung sozialer Verhaltensstrukturen bewirken kann. Eine von allen Mitgliedern einer Affenhorde akzeptierte Rangordnung wurde instabil. Dabei war es gleichgültig, ob man isoliert den Leitaffen, nur die Horde oder randomisiert beide reizte. Die Affen kämpften während der Reizung mit den nächsthöheren Chargen und stießen im Gegensatz zu ihrem früheren Verhalten die schwachen Mitglieder aus der Horde aus. Am deutlichsten waren die situativen Veränderungen, wenn nur der Leitaffe gereizt wurde. Dieser verhielt sich dabei unnatürlich zahm und zeigte eine Abnahme seiner früheren aggressiven Potenz, was in der Folge schnell zu einem Respektverlust bei der Herde führte. Untergebene Affen drangen in sein Revier ein und machten ihm das Futter streitig. Schaltete man den elektrischen Stimulus ab, wurde der Leitaffe sofort wieder aggressiv, und die Affen kehrten binnen kurzem zu ihren früheren Umgangsformen zurück. – Häufige Wiederholung der Reizsituation führte jedoch dazu, daß der Reizaffe offenbar „lernte", sein verändertes Benehmen zu akzeptieren und schließlich beizubehalten. In der Folge reagierte er auch ohne Reizung zahm, und es kam zu dem oben beschriebenen respektlosen Verhalten seiner untergebenen Affen. – Delgado folgerte, daß der Nucleus amygdalae für ausgewogenes Sozialverhalten innerhalb einer Gruppe verantwortlich sei. Trotz Kenntnis der Versuche von Kaada, nach denen im Nucleus amygdalae gegenteiliges Verhalten etwa gleich häufig angelegt ist, nahm er an, daß Reizung dieser Neuronenpopulation im Nucleus amygdalae stets eine Fehlbeurteilung der jeweiligen Situation zur Folge hat und damit eine Störung des sozialen Gleichgewichtes bewirkt. Delgado vermutete, daß eine ähnliche Versuchssituation beim Menschen zu neurotischen und psychosomatischen Störungen führt. – Nach anderen Autoren (Ganong, 1977; Hassler, 1964; Clemente und Chase, 1973; Grossmann, 1964) löst elektrische Reizung des Nucleus amygdalae beim Menschen je nach Lage der Elektrodenspitze entweder Furcht oder Wut aus.

Im Unterschied zu den oben beschriebenen tierexperimentellen Ergebnissen nach elektrischer Reizung kommt es bei elektrischer Koagulation oder operativer Entfernung des Nucleus amygdalae regelmäßig zu einem Verlust sowohl der Furcht- als auch der Wutreaktion. Die Tiere werden entgegen ihrem ursprünglichen Naturell zahm und zeigen auch auf artspezifische Reize weder Fluchtreflexe (beispielsweise Katze vor Hund) noch Angriffstendenzen (beispielsweise Katze gegenüber Maus). So konnte nach stereotaktischer Ausschaltung des Nucleus amygdalae ein Puma als Maskottchen in einem Labor gehalten werden.

Auf dem Boden dieser Beobachtungen und in Kenntnis der Tatsache, daß Patienten mit Temporallappenepilepsien nahezu regelmäßig ein sanftes Naturell und eine herabgesetzte Sexualität aufweisen, haben in den letzten Jahren japanische Neurochirurgen (Ganong, 1977) psychochirurgische Eingriffe durchgeführt. Bei Patienten mit therapieresistenter Agitiertheit und Aggression wurden bilaterale Reizungen mit anschließender Elektrokoagulation oder operativer Entfernung des Nucleus amygdalae vorgenommen. Während der Reizung der Neuronenpopulation waren die Patienten stark erregt und zeigten gesteigerte vegetative Reaktionen, nach der Elektrokoagulation waren sie angstfrei und gefügig, ohne daß von den Japanern eine meßbare Gedächtnis- oder Sexualstörung beschrieben wurde.

9.3.3 Hippokampus

Die Neurone des Hippokampus besitzen im Vergleich zu allen übrigen Hirnzellen eine einzigartige Fähigkeit: sie können über den Initialreiz hinaus salvenartige Spikes entladen (Liberson und Akert, 1955; Penfield und Milner, 1958). Offenbar ermöglicht die Netzgitterstruktur der Hippokampusneurone und ihre ringartige Verschaltung zu einem geschlossenen System eine Erregung, die sich in einer Art Kippschwingung aufschaukelt und in benachbarte Ringstrukturen des Limbischen Systems fortgeleitet wird. Solche Erregungssalven bleiben offenbar auf das System selbst beschränkt, da zwar „Hippokampusanfälle" mit bizarren Verhaltensmustern beobachtet werden, nicht dagegen generalisierte zerebrale Krampfanfälle mit Bewußtlosigkeit. – Beim Déjà-vu-Erlebnis einer Temporallappenepilepsie handelt es sich wahrscheinlich um einen auf den Hippokampus oder seine direkte Umgebung beschränkten epileptischen Minianfall mit unwillkürlicher Erlebnisreproduktion früherer Sinneseindrücke.

Der Hippokampus ist sowohl für sexuelle Aktivität als auch für verbale Kommunikation verantwortlich. – Elektrostimulation seiner Zellverbände löst bei Tieren im allgemeinen ein Balzverhalten aus, beim Männchen regelmäßig eine Erektion. Daneben verändert sich je nach Ausgangslage die auch bei Tieren vorhandene Körpersprache. So wechselt während der Stimulation der Gesichtsausdruck ständig und reicht von wütend bis apathisch, von ratlos bis aufmerksam. Aus ungerichtetem Grimassieren wird arteigenes Eß- und Trinkverhalten. – Lubar und Peracchio (1965) beobachteten, daß Elektrostimulation an Hippokampuszellen zwei stereotype Verhaltensmuster erzeugte: einmal ergriffen die Katzen ausschließlich die Flucht, während der alternativ zum Fluchtreflex sonst vorhandene Totstellreflex nicht mehr auftrat. Zum anderen kam es bei mehr in Richtung Gyrus cinguli liegender Reizelektrode, sonst aber analoger Stimulationstechnik, zum gegenteiligen Effekt: die Tiere zeigten ausschließlich den Totstellreflex, während das alternative Fluchtverhalten eliminiert schien. Auch Durchtrennung der Faserverbindungen von Hypothalamus und Gyrus cinguli führte regelmäßig zum Verlust der Fluchtreaktion. Die Autoren interpretierten dieses Verhalten dahingehend, daß Flucht- und Totstellreflex situationsangepaßte Reaktionen seien, die letztlich der Arterhaltung dienten.

Elektrische Reizung des Hippokampus löst beim Menschen zunächst eine Änderung von Atemrhythmus und Herzschlag aus, während längere Reizung eine sexuelle Appetenz, beim Mann eine Erektion bewirkt.

Eine im Rahmen einer neurochirurgischen Operation durchgeführte einseitige Entfernung des Gyrus hippocampus führt beim Menschen zu einer zeitlichen und örtlichen Desorientierung, zu einer Störung der Merk- und Konzentrationsfähigkeit sowie zur emotionalen Veröddung. Bei bilateraler Entfernung kommt es zu einem Verlust des Kurzzeitgedächtnisses (s. u.). Chronische Schädigung des Ammonshorns mit partiellem Zelluntergang, beispielsweise nach Kohlenmonoxidvergiftung, ist gefolgt von einem Korsakow-Syndrom: das Kurzzeitgedächtnis und das Zeitgefühl sind gestört, und es kommt zu einer Abnahme der intellektuellen Fähigkeiten, während das Bewußtsein und das Altgedächtnis erhalten bleiben. Patienten mit Korsakow-Psychose neigen zur „kompensatorischen" Konfabulation. Eine ihnen soeben gestellte Frage haben sie schneller vergessen als beantwortet. Da sie selbst diesen Zustand nicht mehr realisieren, antworten sie konfabulatorisch, was dem Syndrom die Bezeichnung Psychose eingebracht hat.

Das Gedächtnis, also die Fähigkeit, Sinneswahrnehmungen oder seelische Vorgänge im Gehirn zu speichern und die Information willentlich wieder abzurufen, ist an die Intaktheit des Limbischen Systems, insbesondere des Hippokampus und Nucleus amygdalae, gebunden. Diese beiden Nervenknoten sind als einzige in der Lage, einen soeben erlebten Sachverhalt für einige Sekunden bis zu 5 (maximal 10) Minuten zu speichern. Während dieser Zeit entscheiden ihre Neurone (auf welche Weise, ist nicht bekannt), ob das unmittelbar Erlebte (beim Menschen auch innerlich Erlebte, also Erdachte) als mnestische Spur in das Gehirn aufgenommen oder „vergessen" werden soll. Noback und Demarest (1981) haben hierfür das Bild einer Tonbandaufzeichnung geprägt: entweder wird die Information auf Band gespeichert und für eine spätere Verwendung bereitgehalten oder das Band wird gelöscht. Hippokampus und Nucleus amygdalae selbst gelten dabei nicht als der eigentliche Gedächtnisspeicher (Eccles, 1977/78). Offenbar fungiert das Gehirn in toto als Engrammatrize, ohne daß die Neuronenverbindung zwischen den zwei Kernen des basalen Temporallappens und den Speicherstellen bekannt ist.

Ein 35jähriger Mann wies nach einer wegen eines Tumors erfolgten Entfernung des linken Temporallappens deutliche Störungen des Kurzzeitgedächtnisses auf. Der Patient war fähig, einem Gespräch von 5 bis 10 Minuten Länge über ein geläufiges Thema zu folgen, konnte jedoch anschließend keine Einzelheit der Unterhaltung wiedergeben. Nach Unterbrechung von 10 Minuten wußte der Patient nicht mehr, daß er und der Arzt bereits ein Gespräch ähnlichen Inhalts geführt hatten. Der Patient wurde jedoch während solcher Gespräche zunehmend unsicher und unruhig, wobei er den Grund nicht nennen konnte. Nach einer Gesprächspause von etwa 30 Minuten war der Patient wieder ruhig, und die Unterhaltung ließ sich in der oben beschriebenen Weise wiederholen, ohne daß der Patient eine Erinnerung an die vorausgegangenen Szenen hatte.

Bei Untersuchungen 6 und 12 Monate nach der Operation bestand noch eine deutliche Schwäche des Kurzzeitgedächtnisses, vor allem aber lag ein ausgesprochen devotes, klebriges Benehmen vor.

Gloor (1967) und Milner (1970) fanden bei Ratten ein gestörtes Lernverhalten, sofern Hippokampus und Nucleus amygdalae operativ entfernt worden waren. Die Tiere konnten unangenehme Situationen nicht mehr als solche erkennen, bzw. nicht mehr lernen, solche zu vermeiden. So waren Ratten nicht mehr in der Lage, zwischen einem neutralen Weg zu einem Freßtopf und einem mit elektrisch geladenem Gitter zu unterscheiden, das ihnen unangenehme Schläge an den Pfoten erteilte. – Nichtoperierten Versuchstieren gelang die Unterscheidung dagegen bereits nach ein bis maximal drei Versuchen.

9.3.4 Versuche nach Klüver und Bucy

Klüver und Bucy (1939, 1958) entdeckten, daß die operative Entfernung des basalen Temporallappens bei den auf diese Weise operierten Tieren eine spezifische Veränderung ihres Verhaltens hervorrief. Postoperativ wiesen die Tiere folgende, vorher nicht beobachtete Verhaltensmerkmale auf:

– Ausgeprägte orale Tendenz. Affen versuchten nahezu alle, auch nicht eßbare Gegenstände ins Maul zu stecken und zu fressen. Das sonst vorhandene Sättigungsgefühl, das frei lebende Tiere nie mehr Nahrung zu sich nehmen läßt als physiologisch notwendig, schien nicht mehr vorzuliegen.

– Optische Agnosie. Sie verhielten sich situativ inadäquat. So erkannten sie beispielsweise eine ihnen sonst geläufige Gefahrenquelle nicht mehr.

– Keine arteigenen Furchtreaktionen. Affen spielten beispielsweise mit sonst von ihnen gefürchteten Schlangen. Eine artspezifische Vermeidungsreaktion (avoidance reaction) war offensichtlich nicht mehr vorhanden.

– Ausgeprägte Hypersexualität. Die männlichen Tiere waren ständig hinter den weiblichen her und versuchten auch mit gleichgeschlechtlichen oder artfremden Tieren, ja sogar mit Gegenständen zu kopulieren.

Neurohistologische Untersuchungen ergaben, daß es sich bei dem nach den Erstautoren benannten Klüver-Bucy-Syndrom im wesentlichen um die Folge einer Läsion des Gyrus hippocampus und der ihm benachbarten Zellverbände und Fasersysteme des basalen Temporallappens handelt. – Die Trias optische Agnosie, Hyperphagie und Hypersexualität kommt auch beim Menschen nach Teilläsion des basalen Schläfenlappens einschließlich des Nucleus amygdalae und Hippokampus vor und wird analog auch als Klüver-Bucy-Syndrom bezeichnet. – Es findet sich pathoätiologisch u.a. als Folge einer Enzephalitis, Encephalomyelitis disseminata, nach manchen Vergiftungen, beispielsweise mit Kohlenmonoxid, sowie bei Hirntumoren.

Ein 45jähriger Mann wies als Folge einer Hirnkontusion ein organisches Psychosyndrom mit ausgeprägten Verhaltensstörungen auf. Der Mann aß wahllos und ohne Sättigungsgefühl alles, dessen er habhaft werden konnte. Während der stationären Behandlung machte er sowohl Krankenschwestern als auch -pflegern eindeutige Anträge. Sein übriges Verhalten war situationsgerecht und geordnet, insbesondere lag kein Frontalhirnsyndrom vor. – Der zentrale und periphere Neurostatus war normal, das EEG lediglich allgemeinverändert, während das kranielle Computertomogramm eine traumatisch bedingte Verquellung zeigte. – 6 Wochen nach dem Unfallereignis war der Patient annähernd beschwerdefrei. – 4 Monate später hatte er 10 kg zugenommen, klagte über ein ständiges Hungergefühl, wobei es ihm ausschließlich auf die Menge und nicht auf die Qualität der Speisen ankam. Die Ehefrau berichtete, daß sich die Nachbarschaft über die Anzüglichkeiten ihres Mannes beklage, er den Angestellten nachstelle und mehrfach wöchentlich mit dem Auto halbtägige Touren mit unbekanntem Ziel unternehme. Darauf angesprochen, erklärt der Patient, daß er die Bordelle der umliegenden Großstädte aufsuche und zusätzlich häufig masturbiere. – Der zentrale Neurostatus war auch bei der Nachuntersuchung normal, das EEG weitgehend unauffällig. Neben der Hyperphagie und Hypersexualität lagen nach testpsychologischer Untersuchung noch eine Störung des Neuzeitgedächtnisses und eine partielle visuelle Agnosie vor. Die computertomographische Kontrolle zeigte eine deutliche Atrophie der basalen Temporalanteile bei sonst unauffälligem Befund. – Diagnostisch handelte es sich um ein traumatisch bedingtes Klüver-Bucy-Syndrom mit zusätzlich visueller Agnosie infolge Läsion der Projektionsbahnen zur Area 20.

9.3.5 Lust- und Bestrafungszentren

Olds (Olds und Milner, 1954; Olds, 1958, 1961, 1966) und Milner (1970) entdeckten während Stimulationsversuchen an Ratten, daß ein Tier mit einer im Gehirn implantierten Elektrode bemüht war, den elektrischen Impuls möglichst oft zu erhalten. Die Spitze der Reizelektrode lag dabei im Septum des Limbischen Systems und nicht, wie eigentlich beabsichtigt, in der Formatio reticularis. Die Forscher nahmen das Zufallsergebnis als Ausgangsbasis für weitere Versuche. Das Neue daran waren der Reizort im Septum des Limbischen Systems und die Technik der Selbststimulation. 1956 veröffentlichte Olds seine mittlerweile legendären Rattenversuche.

Die Forscher folgerten aus ihren Beobachtungen, daß ein Reiz im Septum des Limbischen Systems offensichtlich eine „belohnende" Wirkung auf das Tier ausüben müsse. Olds konditionierte Ratten, mit den Pfoten kleine Kontakthebel herunterzudrücken, wo-

bei die Tiere über die im Septum implantierten Drahtelektroden jeweils einen elektrischen Schlag erhielten. Die Ratten stimulierten sich ununterbrochen und immer schneller und erreichten Reizfrequenzen bis 7000mal in der Stunde. Sie zogen die Selbststimulation jedem anderen Reiz, auch jedem sexuellen Kontakt oder jeder Art von Nahrungsaufnahme, vor und führten sie gelegentlich bis zur totalen Erschöpfung durch. – Auch unter erschwerten Versuchsbedingungen, beispielsweise nach Hungern oder Dursten oder wenn vor dem Hebelwerk zur Selbststimulation ein elektrisches Gitter angebracht war, behielten die Tiere das oben beschriebene Verhalten bis zur Erschöpfung bei.

Olds beobachtete, daß Ratten das Hebelwerk zur Selbststimulation auch dann noch minutenlang bedienten, wenn der Strom bereits abgeschaltet war, sie also keinen Stromstoß mehr erhielten. Die Physiologen erklären dieses Phänomen damit, daß eine im Limbischen System ausgelöste hypersynchrone Erregung als Kippschwingung weiterläuft, und es bei den zirkulären Verschaltungen und geschlossenen Neuronenkreisen zu ständigen hypersynchronen, den Initialreiz minutenlang überdauernden Nachentladungen kommt.

Lag die Elektrodenspitze jedoch in den basalen Anteilen des dorsalen Septums – ähnlich der Versuchsanordnung von Lubar und Peracchio – so fiel der Reizeffekt gegenteilig aus: die Ratten vermieden es streng, den Reizhebel auch nur zu berühren. Offenbar übt das Selbstelektrisieren an dieser Stelle des Limbischen Systems eine „bestrafende" Wirkung aus.

Lag die Elektrodenspitze zwischen einem Reizort mit hoher Selbstreizungs- und hoher Vermeidungsquote, so bewegte sich die Häufigkeit der Selbststimulation im Zufallsbereich.

Olds vermutete, daß Ratten mit hoher Selbstreizungsrate den elektrischen Impuls als angenehm empfinden und bezeichnete das entsprechende Areal im Septum als Lustzentrum. Analog nahm er an, daß Ratten, die die Selbststimulation vermeiden, den elektrischen Impuls als unangenehm empfinden und nannte das Areal in den basalen Anteilen des Septums Bestrafungszentrum.

Systematische Untersuchungen der Septalregion des Limbischen Systems bestätigten die Oldschen Lust- und Unlustzentren. Gut 30%, hauptsächlich obere Septumanteile lösen eine Lustreaktion mit hoher Reizfrequenz und -dauer aus, nur etwa 5%, hauptsächlich untere Septumanteile eine Bestrafungsreaktion. Der Rest, also der weitaus größte Teil des Septums, weist bezüglich der Selbststimulation ein indifferentes Verhalten auf, hat also weder eine belohnende noch bestrafende Wirkung. – Es ist weder Olds noch späteren Untersuchern bisher gelungen, die Neurone des Limbischen Systems nach ihren unterschiedlichen Wirkungen einzuteilen: neurohistologisch und neurochemisch unterscheiden sich die Zellen des Lust- und Unlustzentrums nicht voneinander. Beide arbeiten mit dem Neurotransmitter Acetylcholin. Olds nahm an, daß das Gefühl Lust bzw. die Belohnung oder das Gefühl Unlust bzw. die Bestrafung weniger von einem bestimmten Zelltyp als vielmehr vom jeweiligen Erregungszustand der Neuronenpopulation abhinge.

Verabreichung des männlichen Sexualhormons Testosteron steigerte bei Ratten die intrakranielle Stimulationsrate, während sie durch das weibliche Sexualhormon Östradiol gesenkt wurde. In diesem Verhalten bestand zwischen männlichen und weiblichen Versuchstieren kein meßbarer Unterschied.

Zwischenzeitlich sind nahezu alle Strukturen des Limbischen Systems mit modifizierten Reizversuchen untersucht worden. Die Ergebnisse wurden in topographische Gehirnkarten eingezeichnet. Als Orte mit hoher Selbstreizungsquote gelten in fallender Reihenfolge: Septum, Hippokampus, Gyrus cinguli und gelegentlich auch der Nucleus amygdalae.

Gallistel fand 1973, daß eine durch Abschalten des Stromes erzwungene Reizpause den „Circulus vitiosus" der Selbststimulation unterbricht, sofern sie etwa 10 Minuten dauert. Die konditionierte wie konditionierende Reiz-Antwort-Kette, wonach der Reiz die Motivation für einen neuen Stromstoß darstellt, wird offenbar nur in den Hippokampusstrukturen verarbeitet, aber nicht im Langzeitgedächtnis gespeichert. Entsprechend müssen die Tiere 10 Minuten nach Unterbrechung der Selbststimulation wieder neu konditioniert werden. Liegt die Pause unter 10 Minuten, unterbrechen die Tiere trotz abgeschalteter Reizelektrode ihre Reizversuche nicht. – Neurophysiologisch wurde dieses Phänomen mit hypersynchronen Nachentladungen der gereizten Neuronenpopulationen innerhalb bestimmter Regelkreise des Limbischen Systems bzw. als Kippschwingung erklärt. Gallistel gelang es, in der Septalregion des Limbischen Systems neurophysiologisch (nicht dagegen histologisch) zwei Arten von Neuronen zu identifizieren. Die einen hatten eine lange Refraktärzeit von 1 ms (mit Refraktärzeit ist hier die Zeit zwischen zwei Neuronenentladungen gemeint), während die anderen mit 0,6 ms eine deutlich kürzere Periode absoluter Unerregbarkeit aufwiesen. Die Neurone mit der langen Abklingquote nannte Gallistel Triebneurone, die mit der kurzen Zeitkonstante Verstärkerneurone. Während, wie oben beschrieben, das Soma der beiden Neuronentypen keine Unterschiede aufwies, zeigten die Neuriten einen unterschiedlichen Querdurchmesser und damit verschieden schnelle Impulslaufzeiten. Die aus der Faserdicke und der Isolierschicht errechnete Leitungsgeschwindigkeit beträgt für die Neuriten der Triebneurone 1 bis 2 m/sec, für die der Verstärkerneurone 5 bis 15 m/sec. Die Triebneurone reagieren auf mechanische und elektrische Stimulation mit hypersynchroner oder salvenartiger Nachentladung bis zu einigen Minuten und weisen das Phänomen der synaptischen Summation auf. Bei der zeitlichen Summation führt der repetitive Charakter der durch den Reiz ausgelösten Aktionspotentiale an den Synapsen zwar jeweils zu einem unterschwelligen Effekt, der aber durch Summierung eine überschwellige Wirkung und damit eine Erregung der Zelle auslöst. Der räumlichen Summation liegt ein Konvergenzverhalten zugrunde, durch das mehrere Afferenzen, die jede für sich an der Zelle ohne Effekt bleiben würde, zusammen einen überschwelligen Reiz darstellen und die Zelle erregen.

9.3.6 Septum und Limbisches System

Heath (1972) und DeFrance (1976) haben beobachtet, daß beim Menschen elektrische Stimulation der Septalgegend angenehme Gefühle, insgesamt eine gehobene Stimmungslage und einen vermehrten Redefluß hervorruft. Heath berichtete von einem Mann, der bei Stimulation kranialer Septumanteile einen Orgasmus bekam. Während dieses Ereignisses konnte er über die implantierten Elektroden hypersynchrone Entladungen mit Spike-Wave-Komplexen registrieren, die qualitativ denen eines epileptischen Anfalles gleichen, quantitativ jedoch wesentlich geringere Amplituden und Sequenzen besitzen. Im Unterschied zum generalisierten zerebralen Anfall bleiben diese hypersynchronen Entladungen auf das Septum und die septumnahen Areale des Limbischen Systems begrenzt. – Bei einer weiblichen Versuchsperson, bei der die Elektrostimulation des Septums ebenfalls einen Orgasmus auslöste, wiesen die registrierten hypersynchronen Potentiale im Vergleich zu denen des Mannes deutlich niedrigere Amplituden auf.

Injektion von Acetylcholin in das Septum einer weiblichen Versuchsperson löste ebenfalls einen Orgasmus mit hypersynchronen Potentialen und Spike-Wave-Mustern aus. Im direkten Anschluß daran wurde im Serum ein hoher Spiegel von Oxytocin und mehreren Peptiden gefunden (vgl. Kap. 10). Heath machte vor allem letztere für das Phänomen verantwortlich, daß ein Orgasmus Wohlbefinden hervorruft und Schmerzen überdeckt. Er vermutete, daß diese während des Orgasmus freigesetzten Stoffe eine endorphinähnliche Wirkung auf die Lustzentren von Septum und Hypothalamus ausüben.

Routtenberg (1968) beschrieb, daß im medialen Vorderhirnbündel gereizte Ratten sich wie die Versuchstiere von Olds ununterbrochen selbst stimulieren, also offenbar den Reiz als angenehm und erstrebenswert empfinden. – Eine vergleichbare Elektrostimulation im dorsalen Mittelhirn, hier insbesondere in den periventrikulären Kernen, führte bei den Versuchstieren zu einer ausgesprochenen Aversivreaktion. Routtenberg folgerte, daß diese Areale des medialen Vorderhirnbündels die Formatio reticularis hemmen, wonach es zu einer Verschiebung des inneren Gleichgewichtes (Homöostase) von der Empfindung Lust zu der der Unlust käme. Während Olds sich auf eine anatomische Beschreibung der Lust- und Unlustzentren beschränkt hatte, interpretierte sie Routtenberg mehr unter funktionellen Gesichtspunkten und sprach vom Phänomen des Lust- bzw. Unlustprinzips. Nach seiner Interpretation werden Motivation und Handeln weitgehend durch den jeweiligen Erregungszustand der antagonistisch arbeitenden Neuronenpopulationen des Limbischen Systems bestimmt.

9.4 Hypothalamus

Diese neurophysiologische Funktionseinheit zählt neuroanatomisch und damit strenggenommen nicht zum Limbischen System, ist aber neurophysiologisch nicht von diesem zu trennen. Welche wichtige Funktion der Hypothalamus ausübt, wird in der von McLean geprägten Bezeichnung „visceral brain" deutlich.

Die Erforschung des Hypothalamus verdanken wir vor allem Ranson (1934) und W. R. Hess (1954). Die Forscher stimulierten hauptsächlich bei Katzen über implantierte Drahtelektroden verschiedene hypothalamische Kerngebiete und lösten dabei unterschiedliche Verhaltensmuster aus. Im einzelnen beobachteten sie vegetative, endokrine und motorische Wirkungen. – Dabei war das unter experimentellen Reizbedingungen ausgelöste Verhalten der Versuchstiere nicht von dem spontanen Benehmen vergleichbarer, aber nichtoperierter Kontrolltiere zu unterscheiden. Hess fand heraus, daß Reizung des Hypothalamus nicht nur eine einzelne Eigenschaft auslöst, beispielsweise Wut, sondern einen in sich geordneten Verhaltensablauf hervorruft. Die Tiere zeigten als vegetative Reizantwort nahezu regelmäßig einen Anstieg des Blutdrucks, ein Aufstellen der Körperhaare, eine vermehrte Schweißbildung und eine Erweiterung der Pupillen, als endokrine Reaktion einen Anstieg der Releasing-Faktoren (Näheres vgl. Kap. 10) und als motorische Wirkung ein Krümmen des Rückens mit gesteigertem Muskeltonus, Aufrichten des Schwanzes und Strecken der Krallen sowie Kratz- und Beißversuche. Dabei ist es Hess erstmals gelungen, die verschiedenen hypothalamischen Kerne entsprechend ihren verschiedenen Verhaltensmustern in funktionelle Gruppen zu gliedern. – Reizung des lateralen Hypothalamus ruft Wut hervor, Läsion dieser Gegend Sanftmut. – Cannon nannte diesen wie Wut aussehenden motorischen Ablauf mit gespannter Haltung, Kratzen und Beißen Scheinwut (shamrage), da er zwar alle äußeren Verhaltensweisen der Wut feststellen konnte, die Wut selbst aber nicht. Er zog diese Folgerung aus dem Verhalten der Tiere, die nach Abschalten des elektrischen Reizes ihre Wut schlagartig einstellten und – vorausgesetzt, daß es sich jeweils um Einzelreizungen und nicht um längere Reizserien handelte – sofort wieder zur allgemeinen Tagesordnung übergingen. Die Tiere ließen jedenfalls weder ein positives noch negatives Nachklingen der Erregung erkennen, wie dies bei der „echten" Emotion Wut der Fall ist. Ein weiteres Argument, die hypothalamisch ausgelöste Wut als shamrage zu bezeichnen, bestand in der Beobachtung, daß sich die Tiere vor lauter Aggressivität selbst zu beißen begannen, was unter nichtexperimentellen Bedingungen in der Natur nicht beobachtet wird. Die Scheinwut ist an den intakten Hypothalamus gebunden, da sie nach dessen Entfernung nicht mehr ausgelöst werden kann. – Elektrische Einzelreizung des lateralen Hypothalamus und des medialen Vorderhirnbündels

nahe dem Fütterungszentrum ruft bei dem Versuchstier dagegen offensichtlich angenehme Gefühle hervor. Jedenfalls waren die Versuchstiere bemüht, möglichst oft einen elektrischen Schock zu bekommen. Analog der Interpretation von Olds könnte man dieses Areal als hypothalamisches Belohnungs- bzw. Lustzentrum bezeichnen. Demgegenüber führt elektrische Reizung des medialen Hypothalamus zu Unlustgefühlen, da die Tiere streng den Reiz zu vermeiden suchten. Man könnte diese medialen Areale als Unlust- bzw. Bestrafungszentrum bezeichnen. – Die ventrale Neuronengruppe des Hypothalamus mit den Nuclei supraopticus und paraventricularis dient der Flüssigkeitsregulation. Reizung dieser Zellkerne mit Stromstößen oder mit Spuren des Neurotransmitters Noradrenalin führt dazu, daß die Tiere keine Flüssigkeit mehr zu sich nehmen, während das Freßverhalten ungestört bleibt. Bringt man Acetylcholin in dieselbe Gegend des ventralen Hypothalamus, so wollen die Tiere nicht mehr fressen, während sie jedoch weiter trinken (DeWied und Gipsen, 1977; Epstein, 1971; Reichlin, 1978). Eine neurohistologische Differenzierung der ventralen Zellpopulation nach ihrer jeweiligen Verhaltensweise ist bisher nicht gelungen. Gleiche oder sehr ähnliche Symptome lassen sich bei ein und demselben Versuchstier aus jeweils mehreren Reizorten sowohl des vorderen als auch des seitlichen Hypothalamus auslösen. Die Flüssigkeits- und Nahrungsregulation ist offensichtlich an mehreren Stellen des ventralen Hypothalamus lokalisiert, was als Sicherung der Natur bei einem so lebenswichtigen Verhalten interpretiert wird. – Die zentrale Kerngruppe des Hypothalamus ist für die Körpertemperatur und Schweißsekretion verantwortlich. Ihre Zerstörung führt zum Diabetes insipidus und zu Fieber. – Die dorsale Kerngruppe des Hypothalamus mit den Nuclei posterior und mamillaris regelt Appetit und Sättigung sowie die Reflexe der Nahrungsaufnahme. Läsionen in diesem Gebiet führen zu einem Verlust der auch bei Tieren appetitgesteuerten Nahrungsaufnahme. Die Tiere gehen an Abmagerung ein. In seltenen Fällen kommt es allerdings zu einem paradoxen Verhalten, nämlich zur extremen Freßlust mit entsprechender Verfettung. Offenbar dienen die ventrale und dorsale Kerngruppe dem Gleichgewicht zwischen Hunger und Durst bzw. Sättigung und Trinken. – Teitelbaum (1971) reizte die noch weiter seitlich des lateralen Hypothalamus gelegenen Hirnstrukturen. Sofern der Kortex nicht lädiert war, konnte er von diesen lateralen Arealen, die neuroanatomisch nicht mehr dem Hypothalamus zugerechnet werden, die aber mit ihm vielfältige, nicht näher definierte Faserverbindungen aufweisen, dieselben Phänomene und Verhaltensweisen wie bei direkter Hypothalamusreizung auslösen. Er folgerte, daß der Kortex den Funktionsausfall des Hypothalamus kompensieren kann. War die Funktion der Hirnrinde dagegen gestört, sei es nach chirurgischer Unterschneidung oder chemischer Betäubung, konnte

Teitelbaum das Verhalten der Versuchstiere weder durch Stimulation intakter Hypothalamusanteile noch durch Reizung ihrer Umgebung beeinflussen, was eine übergeordnete Steuerung des hypothalamischen Systems durch die Hirnrinde beweist.

Als Beispiel für das funktionelle Wechsel- und Zusammenspiel von Kortex, Hypothalamus und Limbischem System soll die Wutreaktion der Katze dienen (Haymaker et al., 1969). Bei diesem Tier zeigt sich Wut durch Krümmen des Rückens, Sträuben der Schwanzhaare, Vorstrecken der Krallen, Fauchen und Zischen, Kratzen und Beißen sowie durch eine Pupillenerweiterung. Diese Reaktionen können von mehreren Stellen des hypothalamischen Systems ausgelöst werden durch:

- Reizung des lateralen Hypothalamus,
- Reizung der den lateralen Hypothalamus umgebenden Hirnareale,
- Reizung der den lateralen Hypothalamus umgebenden Hirnareale bei gleichzeitiger Zerstörung des lateralen Hypothalamus, aber intakter Großhirnrinde,
- Reizung des ventro-medialen Hypothalamus,
- Reizung des Septums im Limbischen System,
- Reizung des Nucleus amygdalae und des Gyrus cinguli,
- Zerstörung ventro-medialer Hypothalamuskerne bei gleichzeitiger Läsion des Nucleus amygdalae.

Wie oben ausgeführt, führt die alleinige Zerstörung des Nucleus amygdalae zu einer sonst in der Natur nicht beobachteten Sanftheit des betreffenden Tieres (s. o. Puma als Maskottchen im Labor), während Reizung Wut auslöst. Diese kann wiederum durch eine ipsilaterale stereotaktische Koagulation lateraler Hypothalamusanteile beseitigt werden, was die oben erwähnte unnatürliche Sanftheit hervorruft.

Die Tierexperimente im Hypothalamus führen zu dem Schluß, daß das visceral brain mehr ein Regelkreisverhalten zeigt, als daß die einzelnen Funktionen streng topographisch isoliert in einzelnen Zellpopulationen abgespeichert sind. Der Hypothalamus dient so allgemein der eigenen Homöostase und der des ganzen Limbischen Systems, also der Aufrechterhaltung des inneren emotionalen und vitalen Gleichgewichtes.

9.5 Zusammenfassung

Wir besitzen heute von den Funktionen des Limbischen Systems und des Hypothalamus ein relativ abgerundetes Bild. Das erstere gilt morphologisch und funktionell als wichtigste Integrationszentrale für alle Ausdrucksformen animalischen Verhaltens, der letztere als Steuerzentrum der vegetativen Funktionen. Das Limbische System selbst erzeugt dabei keine Emotionen, setzt aber Emotionen in komplexe Verhaltensmuster um, die zielgerichtetes Handeln ermöglichen.

Neurophysiologische Untersuchungen an Menschen und Tieren zeigen, daß Reizung oder Ausschal-

tung verschiedener Zellpopulationen im Limbischen System und Hypothalamus jeweils unterschiedliches, aber reproduzierbares Verhalten bewirkt. Umgekehrt besitzt auch jede Verhaltensäußerung eine spezifische Repräsentation in umschriebenen Strukturen des Limbischen Systems. Ganz allgemein obliegt dem System die Regulierung des affektiven Benehmens, Speicherung von Gedächtnisinhalten und Regulierung des Sexualverhaltens. Die damit verbundenen vegetativen Reaktionen steuert der Hypothalamus, der zwar anatomisch nicht zum Limbischen System zählt, funktionell aber eng mit diesem verknüpft ist. Zwischen beiden Systemen bestehen sowohl abhängige als auch sich gegenseitig modulierende Wechselbeziehungen. Die dem limbischen und hypothalamischen System innewohnende Codierung von Verhaltensprogrammen ermöglicht ein jeweils situationsadäquates, den vielfältigen Erfordernissen angepaßtes Verhalten. Zu diesem Zweck kann sowohl ein weitgehend variables als auch vorgeformtes Verhaltensrepertoire in Gang gesetzt werden. – Als Beispiel soll der Geschlechtsakt dienen: Das Limbische System steuert beim Menschen den Wunsch zur körperlichen Vereinigung, das Werbeverhalten mit einem geordneten Ablauf der einzelnen Schritte und auch den Akt selbst. – Demnach sind im Limbischen System sowohl emotionale Reaktionen als auch individuelle Grundmuster menschlichen Verhaltens angelegt.

Im Tierexperiment führt Elektrostimulation bestimmter Zellpopulationen des Limbischen Systems zu definierten Verhaltensweisen. – Der Nucleus amygdalae steuert u. a. die Nahrungsaufnahme, das Aufmerksamkeitsniveau sowie das Wut- und Aggressionsverhalten. Durch diese Funktionen greift er sogar in komplexe Verhaltensweisen ein, die die soziale Integration und die Rangordnung in der Gruppe ausmachen. – Bei agitierten Patienten konnte durch Entfernung des Nucleus amygdalae Angstfreiheit und Beruhigung ohne Beeinträchtigung anderer Eigenschaften erreicht werden. – Reizung des Hippokampus steuert im Tierversuch das Balzverhalten, analog wird beim Menschen die sexuelle Appetenz moduliert. – Im Hippokampus liegt die Integrationsstelle für das Kurzzeitgedächtnis. Hier wird die Entscheidung getroffen, ob ein Gedanke oder Erlebnis im Langzeitgedächtnis gespeichert oder vergessen werden soll. – Zerstörung der basalen Temporalanteile einschließlich Nucleus amygdalae und Hippokampus führt bei Tier und Mensch zum Klüver-Bucy-Syndrom mit der Trias optische Agnosie, Hyperphagie und Hypersexualität.

Olds und Milner entdeckten bei Stimulation im Septum des Limbischen Systems sog. Lust- und Bestrafungszentren. Nach der Quote der Selbststimulation vermuteten sie bei hoher Reizfrequenz eine Lust-, bei niedriger Rate eine Unlustreaktion. – Analog gelang Heath am Menschen durch Septumstimulation die Auslösung eines Orgasmus, wobei er Hirnpotentiale ableitete, die einem zerebralen Minianfall entsprechen und die u. a. im Hypothalamus beobachtet werden. Im Auftreten dieser Spikes im Hippokampus liegt die Erklärung des Déjà-vu-Erlebnisses bei

Temporallappenepilepsie. – Bei chronischer, über ein bestimmtes Zeitmaß hinausgehender Reizung kam es im Tierversuch zu einer Fortdauer der durch die Reizung hervorgerufenen Verhaltensweisen. Eine solche Störung des Limbischen Regelkreises führte in der Folge zu einer Änderung des Gleichgewichtes. Dies hat zur Folge, daß Reiz und Reizantwort nicht mehr adäquat sind. Von diesem Modell wird angenommen, daß eine psychosomatische Erkrankung durch eine Störung der Homöostase entstehen kann.

Es ist jedoch bisher nicht gelungen, neurotisches Verhalten einer Störung oder Läsion anatomisch definierter Zellpopulationen im Limbischen System oder Hypothalamus zuzuordnen. In Umkehrung kann bis heute auch nur angenommen werden, daß funktionellen Störungen oder pathologischen Verhaltensweisen eine Desintegration von Zellverbänden aus dem kortikalen, limbischen und vegetativen System zugrunde liegt. Alle in diese Richtung zielenden Versuche und Ergebnisse gehen nicht über die Arbeitshypothese hinaus, daß eine Störung des Limbischen Systems zu einer Störung der Homöostase mit Entgleisung affektiver Vorgänge und bei Einbe-

ziehung des Hypothalamus auch vegetativer Funktionen führt. Welche Affekte dabei entgleisen, läßt sich in vivo nicht wie im Tierversuch voraussehen. Emotionale Störungen führen nach dieser Vorstellung zur Dysfunktion limbisch-hypothalamischer Strukturen und in der Folge zu Funktionsstörungen einzelner Organe, bzw. zu den bisher als psychosomatisch bezeichneten Erkrankungen.

Ähnliche Überlegungen werden in der Neuropharmakologie in praktische Therapie umgesetzt. Mit Psychopharmaka wird in den Stoffwechsel der Neurotransmitter Acetylcholin, Serotonin und Dopamin eingegriffen, beispielsweise mit Butyrophenon. Hierauf basiert der therapeutische Ansatz, psychosomatische Erkrankungen außer mit Psychotherapie auch mit psychotropen Medikamenten zu behandeln.

Die großzügigste Interpretation des Limbischen Systems stammt von dem Londoner Physiologen H. J. Campbell, der Lust im weitesten Sinne mit Aktivierung von Teilen des Limbischen Systems gleichsetzt. Damit liegen die Ansichten von Descartes und Campbell über Seele und Emotionen gar nicht so weit auseinander.

10 Psychoneuroendokrinologie

Karl H. Voigt und *Horst L. Fehm*

In diesem Kapitel ist der Versuch gemacht worden, die Grundlagen der modernen Psychoendokrinologie als einen Zweig der Neurobiologie so darzustellen, daß die auch klinisch relevanten Bezüge von Hormonen und psychosomatischen Veränderungen, wie sie im Kapitel 46 (Klinische Psychoneuroendokrinologie) beschrieben sind, besser verständlich werden. Wegen der schnellen Entwicklung auf diesem Gebiet der neurobiologischen Forschung (Krieger, 1983; Voigt und Fehm, 1983a) wurde es notwendig, den Beitrag für diese Auflage vollkommen neu zu gestalten. Insbesondere ist dies durch die Einbeziehung der kürzlich charakterisierten Neuropeptide in den Komplex der Psychoendokrinologie bedingt worden (Guillemin, 1978; Snyder, 1980; Wuttke et al., 1980; Cooper und Martin, 1982).

Die Psychoneuroendokrinologie ist ein interdisziplinäres Fach, wobei die wesentlichen Beiträge von Endokrinologen, Neurologen und Neurophysiologen sowie von Psychiatern und Psychologen kommen. Damit ist diese klinische und naturwissenschaftliche Fachrichtung an der Schnittstelle der Betrachtung von psychosomatischen und somatopsychischen Vorgängen angesiedelt. Ein bekanntes Beispiel stellt die multiple endokrine Reaktion (Aktivierung der Hypophysen-Nebennieren-Achse und der Katecholamine) auf starke und psychische Belastungen („Streß") dar. (Die psychischen Veränderungen, die durch jede deutliche Hormonstörung bei endokrinen Erkrankungen hervorgerufen werden, sind im Kapitel 46 beschrieben.)

Bei näherer Analyse der psychosomatischen und somatopsychischen Abläufe wird auch die Problematik einer solchen Kausalitätszuweisung offensichtlich. So werden nach bestimmten Stimuli sowohl die entsprechenden Verhaltensformen induziert, als auch die adäquaten physiologischen Reaktionen realisiert. Diese enge Verknüpfung wird auch durch die Bedeu-

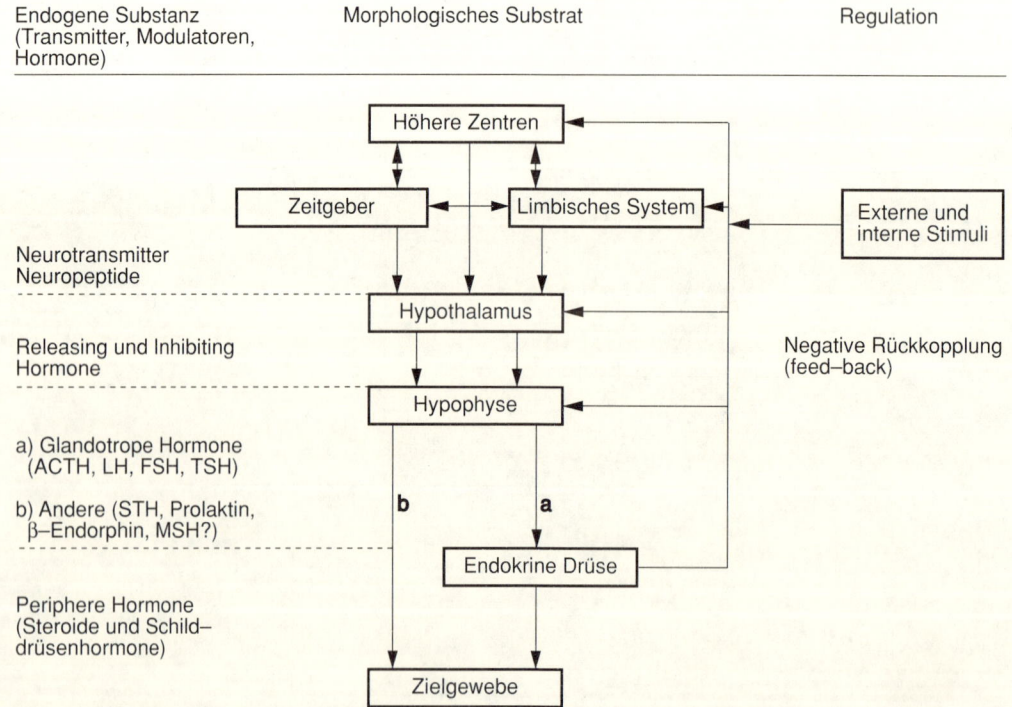

Abb. 10–1. Ein allgemein für neuroendokrine Regelkreise gültiges Schema. Der Terminus „Zeitgeber" soll die verschiedenen rhythmischen Zentren (zirkadian, episodisch, zyklisch) vertreten. Nicht berücksichtigt ist die efferente nervale Steuerung einiger peripherer endokriner Organe, z. B. der Nebennierenrinde (nach Voigt und Fehm, 1983a).

tung der gleichen Neuropeptidsubstanzen (z.B. Angiotensin II und Vasopressin) für psychische (z.B. Trinkverhalten) und periphere physiologische Vorgänge (Homöostase) illustriert (Kap. 10.2.1).

10.1 Grundlagen der Psychoneuroendokrinologie

10.1.1 Formen der Kommunikation in neuroendokrinen Systemen

Hormone und Neuropeptide werden in endokrinen Organen (Hypophyse, Nebennieren, Schilddrüse, Magen-Darm-Trakt) und in Nervenzellen (ZNS und peripheres NS) produziert. Die bekanntesten neuroendokrinen Regelkreise schließen verschiedene Kerngebiete des Hypothalamus, die Hypophyse und periphere Erfolgsorgane oder Target-Zellen ein (Abb.

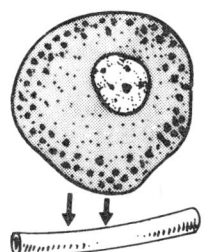 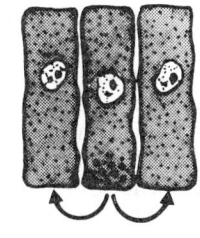

a) endokrin
Adenohypophyse
Gastrointestinaltrakt

b) parakrin
Gastrointestinaltrakt

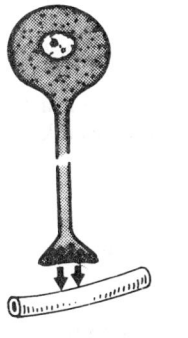 

c) Neurosekretion
Hypothalamus
Nebennierenmark
Neurohypophyse

d) Neuromodulator (synaptisch)
ZNS, Autonomes Nervensystem

Abb. 10–2. Formen der Signalübertragung durch Peptide. In d) können die (postsynaptischen) Rezeptoren an Nervenzellen, an Drüsenzellen oder an der glatten Muskelzelle (Gefäße und intestinale Hohlorgane) liegen. Statt ,Neuromodulator' könnte auch ,Neurotransmitter' stehen (nach Voigt und Fehm, 1983a).

10–1). Um die Formen der Signalübermittlung durch diese Substanzen angemessen beschreiben zu können, muß das Konzept der Hormonwirkungen erweitert werden: Die Hormone/Neurotransmitter können nicht nur über eine Distanz nach ihrer Sekretion in das Blut (Definition der Inkretion) wirken (Abb. 10–1), sondern auch über andere Mechanismen, die eine synaptische Übertragung mit einschließen (Abb. 10–2 a–d):

a) Die klassische endokrine Zelle gibt ihr Produkt in den Blutkreislauf ab. Peptidhormone werden in Sekretgranula gespeichert und durch Exozytose sezerniert, während Steroidhormone nicht gespeichert werden und die Hormonzelle per diffusionem verlassen. Diese Form ist anzutreffen bei allen endokrinen Drüsen und den endokrinen Zellen des Magen-Darm-Traktes.

b) Viele endokrine Zellen im Magen-Darm-Trakt sezernieren ihr Peptid auch zur Regulation der Nachbarzellen: Parakrine Sekretion.

c) Unter Neurosekretion versteht man die Abgabe von Peptiden, die in Nervenzellen gebildet worden sind, in das Blut. Die bekanntesten Beispiele stellen die neurohypophysären Hormone Vasopressin und Oxytocin dar, die in magnozellulären Kernen des Hypothalamus (Ncl. supraopticus und Ncl. paraventricularis) gebildet werden. In den Axonen der peptidergen Neurone werden sie dann zum Hypophysenhinterlappen, der Neurohypophyse, transportiert und hier schließlich in das Blut sezerniert (Scharrer und Scharrer, 1940). Auch die hypothalamischen Releasing- (freisetzende) und Inhibiting- (hemmende) Hormone für die Regulation des Hypophysenvorderlappens werden nach dem Prinzip der Neurosekretion in den hypophysären Portalkreislauf abgegeben (Blackwell und Guillemin, 1973).

d) Der vierte Typ der Signalübertragung durch Peptide entspricht im wesentlichen den Verhältnissen, wie sie von der Wirkung der Neurotransmitter bekannt sind. Die in Neuronen produzierten Peptide werden Neuropeptide genannt, sie werden häufig zur Abgrenzung gegenüber den klassischen Transmittersubstanzen wie z.B. Noradrenalin, Acetylcholin und Dopamin als Neuromodulatoren bezeichnet. Man kann jedoch heute davon ausgehen, daß die Neuropeptide, ebenso wie die Neurotransmittersubstanzen, über prä- und postsynaptische Rezeptoren ihre Wirkung entfalten und so eine Abgrenzung zwischen Neurotransmittern und Neuromodulatoren nicht weiter hilfreich erscheint (s. TiNs, 1983). Im peripheren Nervensystem werden die Neuropeptide zusammen mit den klassischen Überträgerstoffen Acetylcholin und Noradrenalin aus Nervenendigungen abgegeben, die in engem Kontakt zu ihren Effektorzellen liegen. Diese Erfolgszellen (Drüsenepithel, glatte Muskulatur der Blutgefäße und des Magen-Darm-Traktes, Herzmuskulatur) besitzen spezifische Rezeptoren, die dann jene „vegetativen" Effekte vermitteln, die man bisher lediglich dem sympathischen und parasympathischen Nervensystem zuordnen konnte (Burnstock und Hökfelt, 1979; Brown und Fisher, 1984).

10.1.2 Biosynthese

Eine kurze Darstellung der Prinzipien der Biosynthese von Neuropeptiden ist wegen der funktionellen Bedeutung der Vielfalt von Signalsubstanzen, durch die wahrscheinlich eine bessere Abstimmung und Harmonisierung von spezifischen Verhaltensweisen ermöglicht werden kann, unumgänglich. Jedes Neuropeptid ist als Signalsubstanz durch eine spezifische molekulare Form (shape) charakterisiert. Seine Struktur paßt genau in entsprechende Bindungsstellen an der Membran der Effektorzelle (= Rezeptoren). Durch Variation ihrer Anordnung von Aminosäureketten (= Sequenz), die das strukturelle Element von Peptiden darstellen, ist die Biosynthese einer fast unendlich großen Zahl von Signalsubstanzen möglich („Neurobiologists looking into the brain opened a box that released not all the evils of the world but a seemingly never-ending stream of peptides"; Marx, 1979). Die Neuropeptide werden im endoplasmatischen Retikulum als Pre-Pro-Moleküle synthetisiert (Translation von mRNA), in den Golgi-Apparat eingeschleust, wo sie ihre Signalsequenz verlieren und in Sekretgranula verpackt werden (Habener, 1981). Diese Granula werden dann zur Zellmembran transportiert, wobei die Peptide „reifen". In dieser Phase werden durch spezifische Enzyme die bioaktiven Endprodukte abgespalten und teilweise noch modifiziert, sog. „posttranslational processing" (Mains und Eipper, 1983). Auf diese Weise entstehen aus einem Pro-Peptid mehrere aktive Substanzen, die jeweils unterschiedliche Wirkungen entfalten können. Die bekanntesten Pro-Peptid-Systeme sind diejenigen, von denen die endogenen Opiate (= Opioide) stammen (Abb. 10–3). Aus dem Vorläufermolekül Pro-opio-melanocortin (POMC) entstehen so in einem bestimmten molaren Verhältnis das Hormon ACTH und das Peptid β-LPH, von dem das bekannte Opioid β-Endorphin abgespalten wird. Daneben können auch einige MSH-ähnliche Neuropeptide freigesetzt werden. Nach einem Stimulationsreiz werden dann diese verschiedenen, in identischen Sekretgranula enthaltenen Peptide durch Exozytose gemeinsam sezerniert.

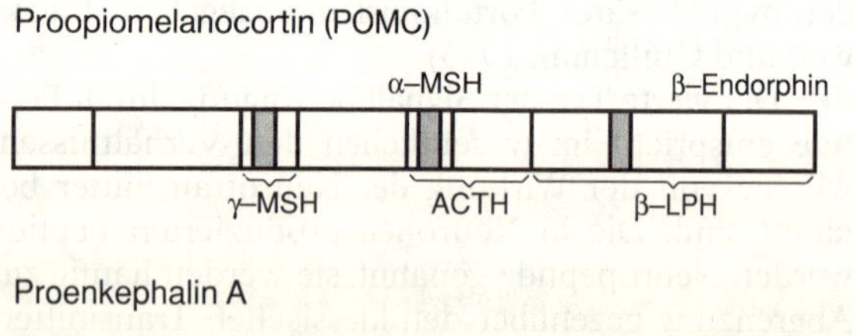

Welche funktionell bedeutsamen Phänomene ergeben sich nun aus der Vielfalt der Peptid-Biosynthese?

– Eine Nervenendigung bedient sich nicht nur **eines** chemischen Signals, sondern einer ganzen Peptidfamilie (aus einem Pro-Hormon) oder sogar mehrerer Pro-Hormon-Systeme. Es ist damit zu rechnen, daß die Steuerung physiologischer und pathologischer Prozesse prinzipiell durch mehrere Substanzen realisiert wird. Schon dieser Grund allein könnte das Scheitern monokausaler Betrachtungen psychischer Vorgänge (z.B. Dopamin-Hypothese der Schizophrenie) erklären.

– Die Signalstoffe entfalten ihre Wirkung über eine Interaktion mit verschiedenen spezifischen Rezeptoren. Diese Rezeptoren können von Neuropeptiden wahrscheinlich auch in einer größeren Distanz als den synaptischen Spalt erreicht werden (parasynaptische Übertragung). Dadurch werden sowohl eine Harmonisierung komplexer Vorgänge als auch eine größere Variabilität bei der Realisierung von funktionellen Abläufen gewährleistet.

– Da die bisher bekannten Neurotransmittersubstanzen (Acetylcholin, Noradrenalin, Adrenalin, GABA, Dopamin und Serotonin) deutlich weniger als die Hälfte aller Synapsen im ZNS versorgen, ist womöglich der größere andere Teil von Neuropeptiden belegt. Das könnte auch von besonderer therapeutischer Bedeutung sein, da alle heute verwendeten Psychopharmaka relativ unspezifisch über Neurotransmitter-Synapsen wirken, und so nur einen kleinen Ausschnitt des ZNS beeinflussen können.

Proopiomelanocortin (POMC)

Proenkephalin A

Proenkephalin B

ACTH 4–9   Met–Enkephalin   Leu–Enkephalin

Abb. 10–3. Schematische Darstellung der drei Vorläufersysteme für endogene Opioide. Jeweils kommen spezifische kurze Aminosäuresequenzen als das typische Peptid mehrfach vor (siehe die Balken). Diese charakteristischen Peptide können in verschiedener Länge vergrößert werden und bilden so eine ganze Familie von spezifischen Peptiden. Jeweils am linken Ende (N-Terminus) ist die für die Regulation der Biosynthese wichtige Signalsequenz (nach Voigt und Fehm, 1983c).

10.1.3 Allgemeine Regulationsmuster und biologische Rhythmen

Die durch neuere Erkenntnisse immer komplizierter werdenden unterschiedlichen Regelkreise für wichtige Lebensvorgänge sind am allgemeinen Beispiel einer neuroendokrinen Regelstrecke in Abbildung 10–1 schematisch dargestellt. Auf der linken Seite sind die Typen der verschiedenen Überträgersubstanzen aufgetragen, die auch für diagnostische und therapeutische Eingriffe in diesem System wichtig sind. Das Zentralnervensystem bedient sich der Neurotransmitter und der Neuropeptide für die synaptische Übertragung (s.a. Abb. 10–2), die dann an den hypothalamischen Neuronen die Produktion und Abgabe der hypophyseotropen Hormone (Releasing- und Inhibiting-Faktoren) in die Portalgefäße steuern. So wird deutlich, daß alle Substanzen, die in den Neurotransmitter-Stoffwechsel eingreifen (Neuropharmaka und Psychopharmaka) auch einen Effekt auf die Sekretion von Hypophysenhormonen haben (s. 10.1.4). Entscheidend für die Regulation der Hormonsekretion ist auch die Intaktheit der ZNS-Strukturen, die endogene Rhythmen steuern. Sie sind hier mit dem Begriff „Zeitgeber" berücksichtigt worden. Dazu sind die diurnalen (oder zirkadianen; z.B. 24-Stunden-Periodik von Cortisol), die vom Schlaf-Wach-Rhythmus gesteuerten (z.B. Wachstumshormon) und die über längere Funktionskreise (z.B. Menstruationszyklus) wirkenden Regelkreise zu zählen.

Endogene Reize aus der Peripherie (Hormonkonzentration, Osmolalität, Blutvolumen, Aminosäurezusammensetzung etc.) und aus dem „psychischen Apparat" (Emotionen, Aufmerksamkeit, Schmerz, „Streß" etc.) sowie externe Stimuli der Umwelt bilden den Faktorenkomplex, der in einem kompliziert abgestimmten Steuerungsprozeß die Hormonsekretion moduliert. Während das Phänomen der negativen Rückkopplungssteuerung (feed-back) der peripher sezernierten Hormone auf jene zentralen Strukturen, durch die ihre Sekretion reguliert wird (Hypothalamus, Hypophyse), seit langem allgemein bekannt ist, wird das Zusammenspiel von Neurotransmittersubstanzen und deren Beeinflussung durch häufig benutzte Psychopharmaka und Neuropharmaka auf die Hormonsekretion weniger beachtet (s. Tab. 10–2). Neben der humoralen Steuerung der peripheren endokrinen Gewebe ist wahrscheinlich eine zusätzliche nervale Regulation durch das autonome NS sehr bedeutsam.

Der wichtigste und bisher am besten untersuchte Biorhythmus ist die 24-Stunden-Periodik oder der zirkadiane Rhythmus. Die zirkadiane Rhythmik zeigt sich in charakteristischen Hormonsekretionsprofilen und ist relativ konstant. Periodische Außenfaktoren (z.B. Tageslicht) synchronisieren unter natürlichen Bedingungen die 24-Stunden-Rhythmik, die ohne solche Einflüsse sich auf eine endogene Periodik von etwa 25 Stunden einstellt (Aschoff, 1978; Weitzman et al., 1975).

Cortisol und andere Nebennierenrinden-Steroide zeigen eine deutliche zirkadiane Periodik mit einem prominenten Sekretionsschub in den frühen Morgenstunden und einer nahezu sekretionsstummen Zeit während des späten Abends. Diese Rhythmik scheint relativ unabhängig vom Schlafzustand und von der Hell/Dunkel-Periodik (Weitzman et al., 1975) zu sein, da nach Schlafentzug und bei willkürlicher Schlafverschiebung diese Rhythmik für längere Zeit erhalten bleiben kann. Eine aufgehobene Cortisol-Tagesrhythmik ist typisch für das Cushing-Syndrom (vgl. Kap. 46) und bei einem Teil endogen depressiver Patienten. Eine zirkadiane Periodik des Hypophysenhormons ACTH in Parallelität zu Cortisol ist jedoch nicht mit Sicherheit nachgewiesen worden (Fehm et al., 1984b).

Die Sekretion von Wachstumshormon (STH) ist eng an die Schlafperiodik gekoppelt: Vor allem während der Einschlafphase in den Schlafstadien III und IV wird STH sezerniert. Während der wachen Phase am Tage sind auch als „Streß" wirkende Situationen in der Lage, eine zusätzliche STH-Sekretion zu stimulieren.

Prolaktin ist wie STH vom Schlafrhythmus abhängig, hat jedoch sein Sekretionsmaximum in der zweiten Nachthälfte. Ebenso kann es am Tage bei besonderen Anforderungen („Streß") verstärkt sezerniert werden.

Die Gonadotropine LH und FSH zeigen beim Erwachsenen nur eine geringe Tagesrhythmik. Im Schnitt ist das LH etwas höher während des REM-Schlafes. Sekretionsphasen von LH können zu jeder Tages- und Nachtzeit gefunden werden. Typisch andere Verhältnisse des LH-Sekretionsmusters finden sich jedoch während der Pubertät bei Knaben und Mädchen (s. 10.5.1). Andere Hormone, wie TSH, FSH und Vasopressin, werden offenbar nicht nach einem zirkadianen Muster sezerniert.

Die Rhythmik der Hormonsekretion wird wahrscheinlich in einigen Systemen zur Frequenzmodulationssteuerung benutzt. Bisher ist dieses Prinzip bei der Induktion von Ovulationen bei Frauen nachgewiesen worden. Bei Frauen mit hypothalamischer Amenorrhoe konnten Ovulationen nur dann ausgelöst werden, wenn das hypothalamische Releasing-Hormon Gn-RH in einer bestimmten Periodik („pulsatil") mit automatischen Pumpen injiziert wurde, nicht jedoch nach konstanten Infusionen (Leyendecker et al., 1983).

10.1.4 Neuropeptide, Hormone, Neurotransmitter

Um mit der verwirrenden Vielfalt der endogenen Substanzen, die als Signalstoffe verwendet werden, besser arbeiten zu können, haben wir sie nach funktionellen und topographischen Gesichtspunkten eingeteilt: A) Neuropeptide, B) Hormone (im engeren Sinne) und C) „klassische" Neurotransmittersubstanzen (Tab. 10–1). Es ist damit zu rechnen, daß die Neuropeptide (A) in der Zukunft eine besondere Bedeutung bei der neurobiologischen Erklärung auch von menschlichem Verhalten erlangen werden. Vielleicht werden auch deren synthetische Analoga und Antagonisten dann vermehrt in der Klinik für die Diagnostik und Therapie angewendet werden können. Tabelle 10–1 soll als kleine Einführung in ein schnell wachsendes Forschungsgebiet dienen.

Neuropeptide werden von Neuronen produziert und können jeweils mehrere Funktionen ausüben, von denen momentan jedoch nur einige bekannt sind. Ein möglicherweise generelles Prinzip ist der duale Charakter ihres Wirkprofiles: in der Körperperipherie als Regulatoren von homöostatischen Prozessen und im ZNS als Signalüberträger spezifischer Verhaltensmuster (s.a.10.2). Das könnte sie zu wichtigen Integratoren von Körper und Gehirn qualifizieren (Iversen, 1981).

Die unter B eingeteilten Hormone sind bisher nur als Produkte endokriner Zellen bekannt. Die peripheren Hormone der Schilddrüse, die Steroidhormone der Nebennieren und der Gonaden können die Blut-Hirn-Schranke überwinden und üben eine Reihe von Wirkungen im ZNS aus (s.a. 10.3), die bei endokrinen Erkrankungen zu charakteristischen psychischen Veränderungen führen können (vgl. Kap. 46).

Die Hormonsekretion wird auch durch die „klassischen" Neurotransmittersubstanzen gesteuert. Daher sind Eingriffe in deren Stoffwechsel durch Psychopharmaka und Neuropharmaka auch an entsprechenden Sekretionsmustern von Hypophysenhormonen abzulesen (Tab. 10–2). Ja, sehr häufig (z.B. bei Neuroleptika) ist der entsprechende Hormonspiegel (hier Prolaktin) einer Verlaufsbeobachtung besser zugänglich als die Messungen von Neurotransmittern. Sie werden daher auch als „Marker" für Psychopharmaka-Wirkungen benutzt.

Die Beschreibung des Profils endokriner Wirkungen und Nebenwirkungen einzelner Substanzen ist von größerem wissenschaftlichem Interesse, weil dadurch Aussagen über die Bedeutung einzelner Neurotransmittersysteme für bestimmte endokrine Funktionen möglich werden. Ebenso können diese Befunde zur weiteren Aufklärung des Wirkungsmechanismus der verschiedenen neuropharmakologischen Substanzen beitragen. In Tabelle 10–2 ist versucht worden, für einzelne Neuropharmaka die Wirkung auf die Partialfunktionen des Hypophysenvorderlappens tabellarisch darzustellen. Es ist zu erkennen, daß Wachstumshormon und Prolaktin besonders empfindlich auf neuropharmakologische Manipulationen reagieren. Die Aussagen der Tabelle müssen jedoch auch mit Vorbehalt betrachtet werden, da einzelne Neuropharmaka in den meisten Fällen nicht konsistent die Hormonsekretion beeinflussen, sondern nur einzelne Aspekte unter bestimmten Voraussetzungen. So soll Cyproheptadin die Sekretion des Wachstumshormons im Rahmen eines Funktionstestes (Insulin-induzierte Hypoglykämie) blockieren, jedoch den Anstieg von Wachstumshormon während des Schlafs fördern. Bromocriptin wird bei Patienten mit Prolaktinomen der Hypophyse therapeutisch eingesetzt.

10.2 Neuropeptide/Hormone und motivationales Verhalten

Im Kontext dieses Kapitels über mögliche Zusammenhänge zwischen psychologischen Phänomenen und der Regulation von Neuropeptiden/Hormonen bieten sich insbesondere solche Verhaltensmuster an, die eine eher klare Verbindung von Stimulus und Handlung zeigen (Voigt und Fehm, 1987). Diese Verhaltensarten können in den Bereich der homöostatischen Motivationsmechanismen eingeordnet werden und umfassen die Komplexe Appetit und Sättigung, Durst, Temperaturregulation und auch nichthomöostatische Prozesse wie Schmerz, Flucht und Vermeidungsverhalten. Wir möchten im folgenden einige

Tabelle 10–1. Aufstellung der heute bekannten endogenen Signalsubstanzen im nervalen und endokrinen System. Die Gruppe A (Neuropeptide) wird wahrscheinlich in der nächsten Zeit noch deutlich vergrößert werden.

A. Peptide, die im Nervengewebe nachgewiesen worden sind und wahrscheinlich auch dort produziert werden

1. Hypophysen-Hormone
   POMC-Peptide (ACTH, MSH, β-Endorphin)
   STH
   Prolaktin
2. Hypothalamische Peptide
   Vasopressin
   Oxytocin
   TRH (Thyreotropin-Releasing-Hormon)
   Gn-RH (Gonadotropin-Releasing-Hormon)
   CRF (Corticotropin-Releasing-Faktor)
   GH-RH (Wachstumshormon-Releasing-Hormon)
   Somatostatin
   Pro-Enkephalin-B-Peptid
3. Darmpeptide
   Neurotensin
   Pro-Enkephalin-A-Peptid
   VIP (Vasoaktives intestinales Peptid)
   CCK-8 (Cholecystokinin-Oktapeptid)
   Substanz P
   Bombesin
   Insulin
   Glukagon
   Pankreatisches Polypeptid
   Neuropeptid Y
   Sekretin
4. Andere
   ANP (Atriales natriuretisches Peptid)
   Bradykinin
   Angiotensin II
   Carnosin, Homocarnosin
   Schlafpeptid(e)
   Hydra-Kopf-Aktivator
   (CGRP) Calcitonin gen-related peptide

B. Hormone, von endokrinen Zellen produziert

1. Hypophyse
   STH (Wachstumshormon)
   POMC-Peptide
   PRL (Prolaktin)
   TSH (Thyreoidea-stimulierendes Hormon)
   LH (Luteotropes Hormon)
   FSH (Follikel-stimulierendes Hormon)
2. Endokrine Drüsen
   Schilddrüsenhormone (T₃ und T₄)
   Glukokortikoide
   Mineralokortikoide
   Östrogene, Gestagene, Androgene

C. Neurotransmitter, von zentralen und peripheren Neuronen produziert

Acetylcholin
Adrenalin
Noradrenalin (NA)
GABA (γ-Aminobuttersäure)
Serotonin
Dopamin (DA)
Glycin

Tabelle 10–2. Beeinflussung der Hormonsekretion durch Neuro- und Psychopharmaka (nach Fehm und Voigt, 1983).

Substanz	Wirkungsmechanismus	STH	PRL	LH/FSH	TSH	Cortisol
L-DOPA	Substrat für DA- und NA-Bildung	↑	↓	–	(↓)	–
L-Tryptophan	Substrat für Serotonin-Bildung	↑	↑	–	↓	↑
Bromocriptin	Dopamin-Rezeptor-Agonist	↑	↓	–	–	–
Clonidin	alpha$_2$-Rezeptor-Agonist	↑	–	–	–	–
Isoproterenol	beta-Rezeptor-Agonist	↓	–	?	?	
Haloperidol	Dopamin-Rezeptor-Antagonist	↓	↑	↓	–	–
Phentolamin	alpha-Rezeptor-Antagonist	↓	–	–	–	–
Propranolol	beta-Rezeptor-Antagonist	↑	–	–	–	(↑)
Cyproheptadin	Serotonin-Rezeptor-Antagonist	↓↑	↓	–	–	↓
Reserpin	entleert Aminspeicher	↓	↑	(↓)		↑
Imipramin	Re-uptake-Hemmer	↓	↑	–	–	–
Amphetamin	bewirkt Aminfreisetzung	↑	↑	?	?	↑

dieser Verhaltensqualitäten unter der besonderen Berücksichtigung der Beteiligung bestimmter Neuropeptidsysteme betrachten, ohne auf die Problematik der Konstrukte Motivation, Emotion, Trieb, Aktivierung und Belohnung näher einzugehen. Dagegen soll der Komplex der menschlichen Sexualität in einem eigenen Kapitel (s. 10.3) behandelt werden.

Die Tabelle 10–3 zeigt eine Auflistung der bisher bekannten Beteiligung von Neuropeptiden an einigen psychischen Prozessen, die auch für eine Reihe von psychosomatischen Vorgängen bedeutsam sein könnten.

10.2.1 Trinkverhalten

Das Trinkverhalten wird reguliert durch eine Feedback-Kontrolle der Gewebeosmolalität und des Blutvolumens. Die Osmolalität wird vor allem durch Sensoren in Hypothalamuszellen gemessen und das Blutvolumen durch Barorezeptoren im venösen System bestimmt (Ganten et al., 1978; Epstein, 1978). Das Oktapeptid Angiotensin II des Gehirns ist hierbei die entscheidende Signalsubstanz für die Einleitung des Trinkverhaltens. Aber auch das in der Körperperipherie produzierte Angiotensin II ist am Trinkverhalten beteiligt über die Vermittlung von spezifischen Rezeptoren, die an einigen besonderen Kontaktstellen der Blutzirkulation des Gehirns, an denen keine Blut-Hirn-Schranke existiert (Subfornikal-Organ, Organum vasculorum der Stria terminalis und der Area praeoptica des Hypothalamus), nachgewiesen wurden (Lang et al., 1983). Daneben wird durch Angiotensin auch die Abgabe von Vasopressin stimuliert. Auf diese Weise kann die Homöostase des Blutvolumens und der Osmolalität realisiert werden: Aldosteron (ein Mineralokortikoid, dessen Sekretion von Angiotensin stimuliert wird) sichert die Na-Retention und Vasopressin die Rückresorption von Wasser in den Nierentubuli. Daneben modulieren Angiotensin II und in gewissen Gefäßabschnitten auch Vasopressin die Durchblutung durch Vasokonstriktion (und -dilatation).

Zusätzlich zur Aufnahme von Flüssigkeit in Form von Wasser scheint bei Säugetieren auch ein spezifi-

Tabelle 10–3. Bedeutung von Neuropeptiden und Hormonen bei der Regulation einiger homöostatischer und nicht-homöostatischer Verhaltensmuster.

Verhaltensqualität	beteiligte Hormone
Trinken	Angiotensin II
	ANP
	Vasopressin
Essen	Opioide
	CCK-8
	Bombesin
	(TRH, CRF)
Temperatur	TRH
	VP
Schmerz	Endorphine
	Enkephaline
	Dynorphine
	Substanz P
	(CCK)
Sexualverhalten	Sexualsteroide
	Gn-RH
Aggression	Testosteron
Aufmerksamkeit, Lernen, Gedächtnis	Vasopressin
	Oxytocin
	ACTH/MSH
Affekt (Depression, Euphorie)	ACTH
	Glukokortikoide
	TRH

scher Appetit auf Salz zu bestehen. Dieses Phänomen wird ebenfalls von Angiotensin II und Aldosteron reguliert (Epstein, 1978). Der starke Salzdurst von Patienten mit einigen Formen von maligner Hypertonie oder mit Herzfehlern scheint auch auf erhöhten Spiegeln von Angiotensin II im Plasma zu beruhen (Denton, 1982). Hieran ist auch das kürzlich entdeckte Hormon des Herzens, das atriale natriuretische Peptid (ANP) beteiligt, das sowohl die Sekretion von Aldosteron und dessen Funktion an der Niere, als auch im ZNS das Trinkverhalten beeinflußt (de Bold, 1985).

Die multiplen Wirkungen von Angiotensin II im ZNS und in der Peripherie stellen das bisher am besten untersuchte Modell zur Funktion von Neuropeptiden dar, sowohl auf der Ebene des Verhaltens als auch in seiner entsprechenden in der Peripherie liegenden homöostatischen Realisierung.

10.2.2 Appetit und Sättigung[1]

In Strukturen des Hypothalamus werden schon seit längerer Zeit die Regulationszentren vermutet, die über Appetit und Sättigung das Eßverhalten regulieren. Umfangreiche Versuche mit verschiedenen Läsionen im lateralen und medialen Hypothalamus bei der Ratte haben eine große Komplexität dieser Systeme gezeigt, die viele kortikale (Verarbeitung von Umgebungsfaktoren) und subkortikale (metabolische, hormonale, neurogene, thermische) Komponenten mit einschließen (Epstein, 1971; Teitelbaum, 1971; Morley, 1980).

Ein mediales Hypothalamuszentrum soll die Sättigung steuern, wobei Serotonin stimulierend und Adrenalin inhibierend wirkt. Dazu reziprok ist die Funktion des lateralen Hypothalamus mit vorwiegend dopaminerger Innervierung. Aber auch einige Neuropeptide sind wesentlich an der Regulierung des Eßverhaltens beteiligt:

– So können endogene Opioide nach zentraler Gabe die Nahrungsaufnahme stimulieren (Grandison und Guidotti, 1977). Entsprechend ist bei Ratten mit experimentell induzierter Fettsucht der Opioid-Spiegel erhöht, bei hungernden Tieren erniedrigt (Gambert et al., 1980).

– Cholecystokinin-8 (CCK-8), ein Oktapeptid-Fragment von Cholecystokinin, das in fast allen Teilen des ZNS gefunden worden ist und als einziges Neuropeptid in hoher Konzentration im Cortex vorkommt, soll zu einem Sättigungsverhalten führen (Dockray, 1982). Das gastrointestinale Hormon Cholecystokinin ist als ein Verdauungshormon an der regelrechten Funktion der Gallenblase beteiligt. Der Sättigungseffekt nach Gabe von CCK-8 scheint sowohl peripher über den Vagus und noch unbekannte Afferenzen aus dem Magen-Darm-Trakt, als auch durch eine direkte Wirkung am Sättigungszentrum vermittelt zu werden. Beim Menschen ist CCK-8 auch schon zur Behandlung der Adipositas angewendet worden.

– Andere zentrale Neuropeptide, die ebenso wie Cholecystokinin in der Körperperipherie Verdauungsvorgänge steuern, wie Bombesin und Insulin, aber auch die hypothalamischen Peptide TRH und CRF, können zu einer experimentellen Sättigung führen (Morley, 1980). Die geschilderten Verhältnisse scheinen analoge Vorgänge zum Trinkverhalten zu sein und bieten so weitere Beispiele für eine Körper und Gehirn integrierende Funktion der Neuropeptide.

10.2.3 Temperaturregulation

Es gibt eine Reihe von Hinweisen darauf, daß die Körpertemperatur von verschiedenen Neuropeptiden reguliert wird (Yehuda und Kastin, 1980). Vor allem ist ein zentraler hyperthermischer Effekt von TRH bekannt (Brown et al., 1977). Diese TRH-induzierte Hyperthermie wird jedoch nicht durch seine Hormonfunktion, also nicht über TSH und die Schilddrüsenhormone, vermittelt. Da dieser Effekt mit Indometacin zu blockieren ist, kann man einen Wirkungsmechanismus mit einer Beteiligung des Prostaglandin-Systems vermuten, das ja auch für die Wirkung der altbekannten antipyretischen Therapie mit Aspirin wesentlich ist. Mit β-Endorphin und α-MSH kann dagegen eine Hypothermie hervorgerufen werden. Eine besonders deutliche fiebersenkende Wirkung wird dem Vasopressin zugeschrieben (Veale et al., 1981), das als endogenes Antipyretikum auch Fieberkonvulsionen bei Kindern induzieren kann. Kürzlich sind einige Lymphokine, vor allem Interleukin-I und der Tumor-Necrosis-Factor als fieberinduzierende endogene Substanzen charakterisiert worden (Blatteis, 1989). Da Interleukin-I auch die ACTH-Nebennieren-Achse stimuliert, ist der ZNS-Effekt von Signalsubstanzen ein erstes, experimentell überprüfbares Beispiel für die Interaktion des Immunsystems mit dem endokrinen System und dem ZNS (vgl. Kap. 11).

10.2.4 Schmerz[2]

In den letzten Jahren ist mit der Entdeckung von endogenen Substanzen, die eine Modulation der Schmerzaufnahme, seiner Weiterleitung und seiner Empfindung bewirken können, die Schmerzforschung in ein neues Stadium getreten. Wir kennen heute bereits über ein Dutzend vom Körper hergestellte Peptide, die in die Schmerzkontrolle involviert sind, die bekanntesten sind die Opioide.

Nachdem im Gehirn Rezeptoren für Morphinalkaloide entdeckt wurden (Goldstein et al., 1971; Pert und Snyder, 1973; Terenius, 1974), sind in rascher Folge drei Opioid-Systeme charakterisiert worden (Höllt, 1983; Bloom, 1983) (Abb. 10–2). Die daraus entstehenden Peptide werden Opioide genannt. Sie besitzen eine Reihe von Partialfunktionen des wirksamsten und seit Jahrtausenden benutzten Schmerzmittels, dem Alkaloid Morphium. Die Grundstruktur aller Opioide wird von den Pentapeptiden Met[5]-Enkephalin und Leu[5]-Enkephalin gebildet, die dann mit unterschiedlichen Aminosäureketten verlängert werden. Enkephalin besteht aus der Aminosäuresequenz Tyr-Gly-Gly-Phe-Met (oder Leu), wobei das Tyrosin essentiell für die Bindung an Opiatrezeptoren ist. Da die Opioid-Systeme wahrscheinlich das am besten bekannte vielschichtige Prinzip der endogenen Regulation von wichtigen Lebensfunktionen darstellen, und wir für die anderen Verhaltensqualitäten mit einer ähnlichen Neuropeptid-Organisation rechnen können, ist den Opioiden hier eine etwas ausführlichere Darstellung eingeräumt worden. Darüber hinaus verleitet das breite Wirkungsspektrum der Opioide (und der Morphinalkaloide) zu der interessanten

---

1 S. auch Kapitel 38.1 „Adipositas".
2 S. auch Kapitel 36 „Schmerz".

Spekulation, daß auch negative Empfindungen des Menschen, die nicht durch einen körperlichen Schmerz verursacht, wohl aber als Schmerz gespürt und in der deutschen Sprache auch mit „Schmerz" bezeichnet werden, wie Heimweh, Weltschmerz etc., durch ähnliche Neuropeptide im Nervensystem beeinflußt werden könnten.

Im folgenden werden die beteiligten Neuropeptid-Systeme vorgestellt:

– **Pro-opio-melanocortin-System:** Die Hypophysenhormone ACTH, MSH und β-Endorphin werden durch Enzyme aus einem sehr viel größeren gemeinsamen Vorläufermolekül, dem Pro-opio-melanocortin (POMC), abgespalten. Dieser Vorläufer wird im Hypophysenvorderlappen, im Nucleus arcuatus des Hypothalamus und in verschiedenen Geweben der Körperperipherie synthetisiert. Sein opioides Hauptprodukt, β-Endorphin, führt bei zentraler Gabe zu einer starken, langanhaltenden Analgesie und kann bei Versuchstieren auch ein Katalepsie-Äquivalent hervorrufen (Snyder, 1977).

– **Enkephalin-A-System:** Das Pro-Enkephalin-A-Molekül, das bisher aus dem Nebennierenmark isoliert worden ist und auch im ZNS vorkommt, enthält vier Kopien von Met-Enkephalin und eine von Leu-Enkephalin sowie zwei weitere Met-Enkephalin-Opioide, die C-terminal verlängert sind.

– **Enkephalin-B-System:** Kürzlich ist ein drittes endogenes Opioid-System entdeckt worden, im Gewebe des Hypothalamus und des Hypophysenhinterlappens. Alle Opioide dieses Systems enthalten die Aminosäuresequenz von Leu-Enkephalin und zusätzliche hochspezifische Verlängerungen der Peptidkette. Die einzelnen bioaktiven Fragmente

werden auch als Dynorphine und α-, β-Neoendorphine bezeichnet.

Es ist heute allgemein akzeptiert, daß die endogenen Opiate ihre Wirkung durch Bindung an verschiedene Rezeptoren entfalten, die dann unterschiedliche Partialfunktionen vermitteln (Mansour et al., 1988): Der μ-Rezeptortyp scheint hauptsächlich für die körpereigene Analgesie, die δ-Rezeptoren für Emotionen und einige Formen von ZNS-Krämpfen und die k-Rezeptoren vielleicht für die Induktion von Halluzinationen und auch für die Sedierung durch Opioide verantwortlich zu sein. Diese Rezeptortypen werden wahrscheinlich noch Unterklassen bilden, die die große Vielfalt der durch Opioide beeinflußten Funktionen ermöglichen. Die verschiedenen Opioide haben eine unterschiedlich starke Affinität zu den einzelnen Rezeptortypen und können daraufhin einem bestimmten Funktionsprofil zugeordnet werden.

Die Bedeutung der körpereigenen Schmerzmodulation (Watkins und Mayer, 1982; Voigt und Fehm, 1983b) wird vielleicht auch dadurch illustriert, daß neben den beschriebenen Opioid-Systemen zusätzlich noch mehrere andere körpereigene Peptid-Signalstoffe an dieser auch evolutionär sehr bedeutsamen Regulation der Schmerzempfindung beteiligt sind.

– **Substanz P:** Substanz P ist ein Undeka-Peptid, das im autonomen Nervensystem und im ZNS weit verbreitet ist. Es wird heute als der Transmitter für viele sensorische Systeme angesehen, vor allem gilt Substanz P als Überträger der Schmerzsignale in den afferenten Schmerzfasern (s.a. Tab. 10–4). Entleerung der Nervenendigungen von Substanz P durch Capsicain, jenem spezifischen Stoff, der für die „Schärfe" in der Pfefferschote verantwortlich ist,

Tabelle 10–4. Zuordnung der verschiedenen endogenen schmerzmodulierenden Substanzen zu den morphologischen Ebenen der Schmerzbahnen. Rechts sind die pharmakologischen Substanzen an der Stelle ihrer pharmakodynamischen Angriffspunkte aufgelistet (nach Voigt und Fehm, 1983c).

Lokalisation	Funktion	Endogene Substanz Bezeichnung	Art	Pharmakologische, analgetische Substanz
Peripherie (Gewebezerstörung)	analgetisch, Stimulation der Nozizeptoren	Bradykinin	Peptid	
		Histamin	Neurotransm.	
		Serotonin	Neurotransm.	
		Prostaglandine (PGE$_2$, PGF$_2$α)	Fettsäuren	Glukokortikoide und Aspirin-ähnliche Medikamente
Afferente Fasern (Aδ, C) von Nozizeptoren	sensorische Übertragung des Schmerzes	Substanz P (SP)	Neuropeptid	Capsicain ? SP-Antagonisten
ZNS („Schmerz-Zentren" und efferente Bahnen)	analgetisch	Serotonin	Neurotransm.	
		Noradrenalin	Neurotransm.	
		Neurotensin	Neuropeptid	
		Bradykinin	Neuropeptid	
		β-Endorphin	Opioid	Opiat-Alkaloide und
		Enkephaline	Opioid	synthetische Opiate
		Dynorphine	Opioid	Ethylketozyklazozine

führt nach kurzem initialem Schmerz (durch Freisetzung von Substanz P) zu einer Analgesie (Marx, 1979; Nagy, 1982).
– Zusätzlich scheinen auch Neurotensin, Bradykinin, TRH, CCK und Somatostatin an verschiedenen Ebenen der endogenen Schmerzkontrolle einzugreifen.

Nach der Vorstellung, daß an der Schmerzkontrolle Neuropeptide beteiligt sind, sollen im folgenden die für die Klinik interessanten funktionellen Mechanismen der körpereigenen Schmerzmodulation beschrieben werden. Dabei hat insbesondere der Einsatz eines Rezeptorantagonisten für Morphinabkömmlinge und auch für Opioide, das Naloxon, die physiologische Relevanz der endogenen Schmerzkontrollsysteme aufgezeigt. Bei Patienten mit schweren chronischen Schmerzen neurogener, aber nicht bei jenen psychogener Ursache, war der Endorphingehalt im Liquor deutlich erniedrigt (Terenius, 1978). Eine Verminderung von Substanz P in den Rückenmark-Hinterhörnern und in der Medulla wurde bei Patienten mit einer familiären kongenitalen Schmerzunempfindlichkeit (Rilay-Day-Syndrom) gefunden (Pearson et al., 1982). Es ist auch möglich, durch Manipulationen beim Menschen die endogene Schmerzkontrolle durch Opioide zu beeinflussen. Ein Beispiel stellen bestimmte Formen der Akupunktur-Analgesie dar, bei denen sowohl ein Anstieg von Opioiden als auch eine Blockade des Effektes durch Naloxon nachgewiesen worden sind (Han und Terenius, 1982). Es sollte jedoch betont werden, daß dieser Mechanismus nur einen Teilaspekt der Akupunktur-Analgesie aufzeigt. Sehr interessant sind Befunde über eine Placebo-Analgesie bei Patienten nach Zahnextraktion (Levine et al., 1978). Es konnte gezeigt werden, daß bei den als „placebo-responders" klassifizierten Patienten mit Naloxon die Analgesie ausgelöscht werden konnte. Auch die Ergebnisse von Tierversuchen, bei denen eine Streß- oder Fuß-Schock-Analgesie konditioniert werden konnte, stützen eine ähnliche Erklärung des Mechanismus von Placebowirkungen (Watkins und Mayer, 1982). Vielleicht werden sich noch andere sog. Placebowirkungen auf eine Aktivierung endogener Neuropeptidsysteme zurückführen lassen.

Es ist damit zu rechnen, daß in der Zukunft Präparate entwickelt werden, die nur bestimmte Partialfunktionen der Opioide enthalten, und so vielleicht analgetische Peptide zur Verfügung stehen werden, die eine gleiche Potenz wie heute verwendete Alkaloide haben, jedoch keine Toleranz und Sucht entwickeln. In Tabelle 10–4 sind die endogenen Substanzen der Schmerzphysiologie aufgelistet und den heute benutzten Analgetikagruppen an ihrer pharmakodynamisch entscheidenden Position gegenübergestellt.

10.3   Hormone und Sexualität

Von allen menschlichen Verhaltensweisen wird die Sexualität am ehesten in Zusammenhang gesehen mit der Wirkung von Hormonen. Aber gerade auf diesem sehr komplexen Gebiet sind die psychosozialen Einflüsse in der menschlichen Gesellschaft so groß, daß die in zahlreichen Tierversuchen beobachteten und vielleicht auch für das menschliche Verhalten relevanten Beziehungen zwischen Hormonveränderungen und Sexualität sich keinesfalls so offensichtlich darstellen. Wir möchten aus diesem Komplex solche Themen vorstellen, bei denen bestimmte Relationen zwischen Hormonmustern und menschlichem Verhalten nachgewiesen worden sind. Dazu zählen die sexuelle Differenzierung in der Entwicklung, die Abläufe der Pubertät, einige Aspekte des Sexualverhaltens und die Beeinflussung einiger Qualitäten emotionalen Verhaltens durch Sexualhormone.

10.3.1   Die sexuelle Differenzierung

Neben dem genetischen Geschlecht, das durch die Sexchromosomen determiniert ist, dem gonadalen, bestimmt durch Testes oder Ovarien, und dem somatischen, erkennbar an den sekundären Sexualmerkmalen, gibt es offensichtlich auch eine sexuelle Differenzierung des Gehirns (Goy und McEwen, 1980). Dieser Geschlechtsdimorphismus ist abhängig von der Anwesenheit prägender Sexualsteroide in bestimmten kritischen Differenzierungsphasen des Hypothalamus. Bei Versuchstieren sind sogar deutliche geschlechtsspezifische morphologische Veränderungen in den sog. Sexualzentren des rostralen Hypothalamus nachgewiesen worden (Gorski et al., 1978).

Viele Tierversuche, die auch solche an Primaten mit einschließen, haben bestätigt, daß in bestimmten Entwicklungsphasen des Hypothalamus, der „critical period", das männliche Sexualsteroid Testosteron eine Differenzierung der zentralen Regelstrukturen bewirkt, die dann bei den adulten Tieren ein typisch männliches neuroendokrines System bedingen und die auch für „männliches" Sexualverhalten verantwortlich sein sollen (McLusky und Naftolin, 1981). Der zugrundeliegende Mechanismus ist auf molekularbiologischer Ebene schon eingehend untersucht worden. Testosteron wird enzymatisch im Hirngewebe zu Östrogen umgewandelt, und dieses Östrogen beeinflußt dann nach Bindung an einen spezifischen Rezeptor genetisches Material im Zellkern, durch das die nachfolgenden Differenzierungen induziert werden (McEwen, 1981). Es ist also kurioserweise das weibliche Sexualhormon, das in der molekularen Endstrecke eine „Vermännlichung" des Hypothalamus bewirkt. Das in größeren Mengen zirkulierende Östrogen selbst kann während der Schwangerschaft die Blut-Hirn-Schranke nicht überwinden, da es an Plasmaeiweiße (vor allem α-Fetoprotein) gebunden ist. Diese Erkenntnisse haben zu der Spekulation geführt, daß womöglich die beschriebene Sexualdifferenzierung auch beim Menschen bedeutsam sein könnte und so einige Störungen des Sexualverhaltens, vor allem die männliche Homosexualität, mit einem Mangel an Testosteron in der kritischen Phase in ursächlichen Zusammenhang gebracht werden könnten (Dörner, 1981).

Diese Hypothese hat verständlicherweise zu heftigen wissenschaftlichen Kontroversen geführt (Goy und McEwen, 1980; Naftolin, 1981; Ehrhardt und Meyer-Bahlburg, 1981; Wuttke und Horowski, 1981). Unter diesem Aspekt betrachtet sind Untersuchungen des geschlechtsspezifischen Verhaltens bei Patienten mit einem adrenogenitalen Syndrom (AGS) besonders bedeutsam. Beim AGS liegt eine verminderte Fähigkeit zur Cortisolproduktion vor, die durch Enzymdefekte in der Nebenniere verursacht wird. Dadurch kommt es regulativ zu einer vermehrten ACTH-Sekretion, durch die eine pathologische Produktion von Nebennierenrindenhormonen stimuliert wird, auch von Androgenen. Die pränatale Exposition mit Androgenen führt zu einer gewissen Maskulinisierung der äußeren Genitalien. Der wichtigen Frage, ob auch eine „Vermännlichung" des Verhaltens bei diesen Patientinnen vorliegt, ist vor allem A. Ehrhardt in großangelegten Untersuchungen nachgegangen (Ehrhardt und Meyer-Bahlburg, 1981). Sie hat das psychosexuelle Verhalten dieser Patientinnen (deren biochemischer Defekt mit Cortisolgabe und deren genitale Fehlbildungen chirurgisch in den ersten Wochen nach der Geburt korrigiert wurden) mit dem ihrer Geschwister verglichen. Es zeigte sich, daß die Geschlechtsidentität der Mädchen mit AGS auch weiblich war, also entsprechend ihrer „mädchenspezifischen" Erziehung. Jedoch waren einige Verhaltensweisen, die als typisch „knabenhaft" gelten, bei ihnen häufiger zu finden. Dazu gehören Herumtoben auf der Straße, Bevorzugung von Knaben als Spielkameraden bei nur geringer Beachtung von Puppen, Schwangerschafts- und Mutterspielen, kaum Interesse an Kleinkindern, ebensowenig an Schmuck, Makeup, Frisur und attraktiver Kleidung. Bei Knaben mit AGS sind dagegen keine besonderen Auffälligkeiten beobachtet worden. Das charakteristische Verhalten von Mädchen mit AGS war nicht nur kurzzeitig, sondern während der gesamten Kindheit nachweisbar. Eine ähnliche Beeinflussung des Verhaltens konnte auch bei Mädchen beobachtet werden, deren Mütter während der Schwangerschaft eine „maskulinisierende" Hormontherapie (als Nebeneffekt) erhielten. Das biologische Gegenexperiment zur Androgenisierung von Mädchen (genetisch weiblich) wird durch genetisch männliche Knaben dargestellt, die einen Rezeptordefekt für Androgene haben. Dieser Effekt einer verminderten Androgenwirkung führte dann zu einem eher typisch weiblichen Verhalten (Masica et al., 1971).

Die Befunde zur pränatalen Bedeutung der Androgene für Aspekte des späteren geschlechtsspezifischen Verhaltens zeigen (Goy und McEwen, 1980), daß unabhängig vom genetischen Geschlecht die Verfügbarkeit von Androgenen für spezifische Hirnstrukturen die Entwicklung eines männlichen Verhaltens bewirkt, ein Fehlen von Androgenen (oder intakter Rezeptoren dafür) dagegen weibliche Verhaltensformen verstärkt. Die postnatale Korrektur der Hormonspiegel kann diese Situation nicht beeinflussen. Dagegen bieten die Beobachtungen dieser Fälle keinen Hinweis auf eine Prädisposition zur Ausbildung eines homosexuellen Verhaltens (Goy und McEwen, 1980).

Im Gegensatz zu der geschilderten Modulation der Psychosexualität durch pränatale Hormonexposition gilt die Geschlechtsidentität (also das sexuelle Selbstverständnis) als ein Resultat des Lernens in den frühen Kinderjahren. Es ist abhängig von der geschlechtsspezifischen Erziehung und dem soziokulturellen Umfeld und scheint nicht vom biologischen Geschlecht bestimmt zu sein (Money, 1973). Wenn jedoch die Diskrepanz zwischen dem biologischen und dem Zuweisungs- und Erziehungsgeschlecht (Money, 1973; McLean, 1973) sehr offensichtlich wird, ergeben sich erhebliche Probleme für die Betroffenen. Solche Fälle können sich z. B. während der Pubertät bei männlichen Pseudohermaphroditen ereignen. In einer vielbeachteten (Lancet, 1979 a; Wilson, 1979) Untersuchungsreihe haben Imperato-McGinley und Mitarbeiter (1974, 1979) die Beobachtung veröffentlicht, daß Hormonkonzentrationen während der Pubertät eine weitgehende „Umpolung" der Geschlechtsidentität und des psychosexuellen Verhaltens induzieren können. Aufgrund eines phänotypisch weiblichen Aspektes der äußeren Genitalien, der durch einen partiellen Enzymdefekt (5α-Reduktase bewirkt die Umwandlung von Testosteron in die im Gewebe wirksame Form) bedingt war, sind 19 genotypisch männliche Neugeborene als Mädchen aufgezogen worden. Die in diesem Kulturgebiet (Dorf in Mittelamerika) deutlich geschlechtsspezifische Erziehung war bis etwa zum 10. Lebensjahr im Einklang mit der Geschlechtsidentität (als Mädchen) und auch der weiblichen Psychosexualität. Als mit Beginn der Pubertät der Einfluß von Testosteron offenbar wurde, begann der Prozeß einer mit dem biologischen Geschlecht übereinstimmenden „Vermännlichung" („... no longer feeling like girls, to feeling like men, and finally to conscious awareness that they were indeed men"; Imperato-McGinley et al., 1979). 17 von den Beobachteten nahmen die neue Geschlechtsidentität und 16 auch die männliche Psychosexualität an, einschließlich Ehe etc.

Wenn es auch etliche theoretische und experimentelle Einwände gegen die Interpretation einer hormoninduzierten Veränderung der Geschlechtsidentität gibt (Green, 1979), so sind diese Untersuchungen doch ein Hinweis auf die Notwendigkeit einer undogmatischen Betrachtung auch von vordergründig schon fixierten Gegebenheiten.

10.3.2 Pubertät

Nicht die Gonaden und nicht die sie steuernde Hypophyse sind für die Einleitung der Pubertät verantwortlich, sondern integrative Zentren des Nervensystems. So gilt als ein wesentliches Zeichen für das Einsetzen der Pubertät die episodische Sekretion des gonadotropen Hormones LH in Abhängigkeit von bestimmten Schlafstadien.

Die Pubertät stellt eine besondere Phase in der menschlichen Ontogenese dar, in der, gesteuert vom ZNS, zirkulierende Hormone nicht nur die sekundären Geschlechtsmerkmale deutlicher ausprägen und die Fertilität sicherstellen, sondern auch den sexuellen Dimorphismus an vielen anderen Organen und Funktionssystemen determinieren (z.B. in der Funktion von Leber, Nieren und Muskulatur) (Bardin und Catterall, 1981). Auch die geschlechtsspezifischen Unterschiede bei bestimmten sensorischen und kognitiven Leistungen werden dann stärker ausgeprägt (Wittig und Petersen, 1979).

Die neuroendokrine Regulation in der Kindheit vor der Pubertät ist gekennzeichnet durch eine niedrige, wahrscheinlich nichtepisodische Sekretion von LH und FSH und deren geringe Stimulierbarkeit durch exogenes Gn-RH, dem hypothalamischen Releasing-Hormon für die Gonadotropine. Man nimmt an, daß die Gonadotropinsekretion zusätzlich durch eine hohe Sensibilität der Sexualsteroidrezeptoren für die negative Rückkopplung im ZNS und der Hypophyse blockiert wird.

Als erstes Zeichen einer beginnenden Pubertät kommt es, wie eingangs erwähnt, zu einer Enthemmung der zentralen Gn-RH-Sekretion, die zunächst nur während des Schlafes einsetzt (Boyar et al., 1972). Dieses Phänomen tritt auch bei agonadalen Kindern auf, ist also eine Funktion des ZNS und nicht primär von Sexualsteroiden abhängig. Neuerdings scheinen interessante Resultate bei Kindern eine seit langem vermutete Bedeutung des Corpus pineale für das Einsetzen der Pubertät zu bestätigen (Kolata, 1984). Danach beginnt die episodische oder pulsatile Gn-RH-Sekretion auch in der Wachphase mit Sekretions-Episoden von 90 Minuten. Die Empfindlichkeit der hypophysären gonadotropen Zellen auf Gn-RH nimmt merkbar zu, und es werden mehr Sexualsteroide sezerniert. Diese wirken nun erst in einer deutlich höheren Konzentration hemmend im Rahmen der negativen Rückkopplungs-Regulation (s.a. Abb. 10–1), die Schwelle ist also erhöht. Zur vollständigen Ausbildung des komplizierten weiblichen Sexualzyklus ist noch die Reifung des Prinzips der positiven Rückkopplung erforderlich: Unmittelbar vor der Ovulation führt der Anstieg von Östrogen nicht zu einer Hemmung, sondern zu einer deutlichen Stimulation der Gonadotropine, wodurch dann die Ovulation ausgelöst wird.

Die peripuberale schlafabhängige LH-Sekretion wird inzwischen als ein verläßliches Maß für die Diagnostik von Störungen der normalen Pubertät, bei Hypothalamustumoren und bei der Anorexia nervosa betrachtet.

10.3.3 Sexualverhalten

Normales männliches und weibliches Sexualverhalten beim Menschen ist an die Integrität der Hypotha-

lamus-Hypophysen-Gonaden-Achse gebunden, jedoch lassen sich keinesfalls bestimmte Hormonmuster differenzierten sexuellen Verhaltensweisen zuordnen (Money und Ehrhardt, 1972; Bancroft, 1978). Eine solche direkte Beziehung ist von vielen Tierversuchen bekannt (Pfaff, 1980), das menschliche Sexualverhalten scheint sich aber von der Hormonsekretion „emanzipiert" zu haben (Symons, 1979).

Das weibliche Sexualverhalten hängt von der regelrechten Steuerung des weiblichen Sexualzyklus ab, in dem nicht nur das hypothalamische Releasing-Hormon Gn-RH und seine pulsatile Sekretion (Leyendecker et al., 1983), sondern auch die Gonadotropine LH und FSH, sowie Prolaktin und die Sexualsteroide – Östrogene und Gestagene – mit eingeschlossen sind. Darüber hinaus scheinen auch die in geringen Mengen produzierten Nebennieren-Androgene eine Rolle bei der Entwicklung der normalen weiblichen Libido zu spielen, wie Ergebnisse bei Patientinnen mit primärer und sekundärer Nebennierenrindeninsuffizienz zeigen (Waxenberg et al., 1959). Sorgfältige Untersuchungen einer großen Zahl verschiedener Aspekte weiblicher Sexualität in Abhängigkeit von den biologischen Veränderungen während des Menstruationszyklus haben die älteren Vermutungen einer gesteigerten sexuellen Erregbarkeit zum Zeitpunkt der Ovulation bestätigt (Adams et al., 1978). So könnten vielleicht auch einige der mit der Einnahme von Ovulationshemmern verbundenen Mißempfindungen eine Erklärung finden. Ebenso ist die in der Postmenopause beobachtete Veränderung der Sexualität neben vielen anderen Faktoren auch von einer reduzierten Sexualsteroidproduktion abhängig (Bancroft, 1978).

Das männliche Sexualverhalten ist an eine normale Testosteronproduktion gebunden. Das kann durch den Einfluß einer Androgensubstitution bei hypogonaden Männern beobachtet werden. Bei normaler Funktion der Testes sind jedoch keine Korrelationen zwischen der sexuellen Aktivität und aktuellen Veränderungen des Hormonspiegels nachweisbar (Lincoln, 1974). Die zirkulierenden Androgene (Bancroft und Skakkebaek, 1979) haben aber neben ihrer Sexualfunktion auch noch etliche andere für das Verhalten wichtige Eigenschaften, die mit motorischer Aktivität bis hin zu aggressivem Verhalten (s. 10.3.5) beschrieben werden. Die Behandlung von kriminellen Sexualtätern mit einem antiandrogenen Präparat (Neumann, 1971) ist nicht als eine Regulation einer überschießenden Testosteronproduktion anzusehen, sondern stellt quasi eine medikamentöse Kastration dar.

Neben den Wirkungen der männlichen und weiblichen Sexualsteroide werden auch den Neuropeptiden Gn-RH und Oxytocin Einflüsse auf die zentrale Steuerung des Sexualverhaltens zugeschrieben. Jedoch sind die bisher vorgelegten Resultate nicht eindeutig (Doering et al., 1977).

Zusammenfassend läßt sich sagen, daß bei endokrinen Störungen ein Einfluß von Hormonen auf das Sexualverhalten nachweisbar ist, jedoch bei einer normalen Regulation der Sexualsteroide keine deutlichen Hinweise auf spezifische Korrelationen zwischen Hormonmustern und der Sexualität bei Menschen bestehen (Bancroft, 1978).

10.3.4 Das prämenstruelle Syndrom

Ein gutes natürliches Experiment für die Beurteilung einer Korrelation zwischen einer spezifischen Hormonkonstellation und emotionalem Verhalten stellt der weibliche Menstruationszyklus dar. Collins und Mitarbeiter (1985) konnten nachweisen, daß in der Follikelphase und während der Ovulation eher positive Stimmungen vorherrschten und im Gegensatz depressive Verstimmungen und Nervosität, verbunden mit erhöhter „Streß-Anfälligkeit", häufiger in der lutealen Phase des Zyklus auftraten. Eine seit langem bekannte klinische Beobachtung ist eine negative Stimmungslage, die bei einigen Frauen mehrere Tage vor der Menstruation auftritt. Dieses, als prämenstruelles Syndrom bekannte Verhalten (Israel, 1938) umfaßt einen ganzen Symptomenkomplex, der Ruhelosigkeit, Nervosität, Spannungen, Depressionen und Suizidgedanken, Appetitlosigkeit, Schlaflosigkeit und viele andere psychosomatische Beschwerden mit einschließt (vgl. Kap. 52). Die zahlreichen Untersuchungen zur Abhängigkeit des prämenstruellen Syndroms von den zyklischen Hormonschwankungen haben sowohl einem erhöhten Östrogenspiegel als auch einer Verminderung von Progesteron oder einem Mißverhältnis dieser beiden Sexualsteroide eine pathogenetische Bedeutung zugemessen (Dennerstein und Abraham, 1982). Daneben ist auch eine vermehrte Prolaktinsekretion bei Patientinnen mit prämenstruellem Syndrom beobachtet worden. Es wurde konsequenterweise auch mit Medikamenten behandelt, die eine Prolaktinsekretion supprimieren (Horrobin et al., 1976; Halbreich und Kas, 1977).

Wenn man die Vielzahl der zu diesem klinisch relevanten Problem veröffentlichten Untersuchungen betrachtet, wird ganz offensichtlich, daß nicht die Hormonveränderungen allein das prämenstruelle Syndrom hervorrufen. Vielmehr liegt eine multifaktorielle Genese vor, an der auch das Persönlichkeitsbild der Patientin unter besonderer Berücksichtigung der Komponenten des Sexuallebens, der Partnerschaftsbeziehung und der Streßbewältigung sowie das soziokulturelle Umfeld gleichermaßen beteiligt sind (Abplanalb et al., 1979; Dennerstein et al., 1981). Dementsprechend muß sich auch die Therapie der prämenstruellen Symptomatik an dieser multifaktoriellen Genese orientieren. Dennerstein und Abraham (1982) fassen diese Situation wie folgt zusammen: „The large placebo effect and the consistent finding of associated psychological and sexual problems suggests a hypothesis of premenstrual tension that related biological, psychological and sexual factors".

10.3.5 Testosteron und Aggression

Das männliche Sexualhormon Testosteron ist in zahlreichen Studien mit Primaten als wesentlich für das Erreichen und Verteidigen des hierarchischen Status gefunden worden (Rose, 1975; Mazur, 1976), und es wird auch beim Menschen mit aggressivem Verhalten in Verbindung gebracht (Houser, 1979; Rubin et al., 1981). Jedoch sind auch hier die Korrelationen zwischen aggressivem, kämpferischem Verhalten und Testosteronkonzentration im Plasma nur selten eindeutig. Die Untersuchungen mit normalen Probanden ergaben keine konsistenten Korrelationen von verschiedenen Testparametern der Aggression und Feindschaft zu Testosteron (Rose, 1975).

Dagegen sind bei sehr aggressiven Gewalttätern meist erhöhte Testosteronwerte gemessen worden. Aber auch hier waren die Hormonkonzentrationen nicht in einer konstanten Beziehung zum aktuell gemessenen „Aggressions-Index" (Rose, 1975; Nieschlag, 1979). Wenn jedoch die Anamnese mitberücksichtigt worden ist, stellte sich heraus, daß jene wegen körperlicher Gewalt straffällig gewordenen Gefangenen höhere Testosteronwerte hatten, die schon in früher Jugend durch besonders brutale Gewalttaten auffielen (Kreuz und Rose, 1972). Also auch bei diesem Beispiel scheint ein besonderes Hormonprofil (nur?) zusammen mit einem Komplex von sozialen und genetischen Faktoren einen Einfluß auf spezifisches Verhalten zu haben.

Die Berichte über eine besondere Aggressivität bei Männern mit einem überzähligen Y-Chromosom (XYY-Karyotyp) sind nicht mit einem direkten Einfluß von vermehrter Testosteronsekretion zu erklären (Meyer-Bahlburg, 1974), da auch hier viele andere Faktoren (z.B. geistige Retardierung, Hospitalisierung etc.) berücksichtigt werden müssen.

Zusammenfassend läßt sich sagen, daß sich eine von Testosteron abhängige Aggressionssteigerung nur in sehr extremen Fällen vermuten läßt, aber für das normale menschliche Leben bisher nicht nachgewiesen worden ist.

Um so interessanter sind die Resultate von Mazur und Lamb (1980), die bei College-Studenten den Zusammenhang von Testosteron im Plasma und verschiedenen Arten einer Statusveränderung untersuchten. Sie fanden, daß nicht nur ein erfolgreiches Ergebnis an sich, sondern auch das Gefühl des „erkämpften" Gewinns für eine parallele Steigerung der Testosteronsekretion notwendig ist. Der Gewinn von 100 Dollar war nur nach einem kämpferischen Tennisspiel (für den Sieger), nicht jedoch nach einem zufälligen Gewinn des gleichen Geldbetrages bei einer Verlosung, mit erhöhtem Testosteron verbunden. Wenn sich diese Resultate replizieren lassen, können sie als ein deutliches Beispiel für die Bedeutung von diskreten Hormonveränderungen auch für das „normale" menschliche Verhalten angesehen werden.

Die exzessive illegale Einnahme von Anabolika, die immer auch androgene Nebenwirkungen (bei Frauen) haben, führte bei einem Drittel der untersuchten Sportler zu schweren affektiven Störungen (nach

DSM-III-R), nicht jedoch zu aggressivem Verhalten (Pepe und Katz, 1988).

10.4 Streß und Emotionen

Die endokrine Streßreaktion wird von den Psychophysiologen hauptsächlich als eine Stimulation der Nebennierenmarkhormone Adrenalin und Noradrenalin angesehen, wie es zuerst von Cannon (1911) beschrieben worden ist. Vertreter der „klassischen" Endokrinologie meinen mit Selye (1936), daß die Nebennierenrindenhormone (vor allem Cortisol) die entscheidende Rolle spielten, und Neuroendokrinologen (Blackwell und Guillemin, 1973) messen den Hypophysenhormonen (ACTH, β-Endorphin, STH und Prolaktin) eine größere Bedeutung zu.

Nach Mason (1968) kann die Sekretion von Cortisol beim Streß durch folgende Faktoren beeinflußt werden:
- Psychische Einflüsse (emotionale?) gehören zu den stärksten natürlichen Stimuli, die die Aktivität des Hypophysen-Nebennierenrinden-Systems beeinflussen.
- Die Empfindlichkeit des Plasma-Cortisolspiegels auch gegenüber relativ subtilen, alltäglichen psychischen Einflüssen legt den Schluß nahe, daß das Zentralnervensystem einen konstanten „tonischen Impuls" auf dieses endokrine System ausübt.
- Die Erhöhung des Cortisolspiegels bezieht sich nicht auf einen spezifischen emotionalen Zustand, sondern scheint ein Zeichen auch für unspezifische Erregungen zu sein.
- Die Faktoren Neuheit, Ungewißheit und Unvorhersagbarkeit bewirken besonders kräftige Anstiege von Cortisol.
- Auch intensive, desintegrierende emotionale Reaktionen mit Zusammenbruch der Verhaltensmuster gehen mit starken Erhöhungen des Cortisolspiegels einher.
- Die erheblichen individuellen Unterschiede der Reaktion des Hypophysen-Nebennierenrinden-Systems auf eine gegebene Situation sind eine stets beobachtete Erscheinung bei psychoendokrinologischen Untersuchungen.

Für die Psychosomatik ist die letzte Feststellung von besonderer Bedeutung, wonach die Kortikosteroidspiegel nicht nur ein Maß für die emotionale Erregtheit, sondern ebenso für entgegengesetzte psychische Abwehrmechanismen sein sollen. Dieser Aspekt wird durch eine Langzeitstudie an Eltern leukämischer Kinder eindrucksvoll illustriert (Friedman et al., 1963). Es zeigte sich, daß sich diese Eltern in Gruppen einteilen lassen mit anhaltend (über Monate und Jahre) „hoher", „mittlerer" und „niedriger" Steroidausscheidung. Bei manchen Eltern mit allgemein niedrigen Steroidwerten fielen diese an Tagen mit aufwühlenden Ereignissen noch weiter ab, während die Eltern mit hohen Spiegeln eher mit einem weiteren Anstieg reagierten. Diese Befunde konnten erst interpretiert werden, als man die Art und Wirksamkeit der psychischen Abwehrmechanismen bei jeder einzelnen Person beobachtete. Eltern, die stark zur Verleugnung und Verdrängung neigten, hatten meist niedrige 17-OHCS-Spiegel mit einer Tendenz zu weiterer Erniedrigung bei Zunahme der psychischen Abwehr. Greene und Mitarbeiter (1970) untersuchten STH und Cortisol bei Patienten, die sich einer Herzkatheteruntersuchung unterzogen. Dabei beurteilten zwei unabhängige Beobachter die Affektlage, die Erregtheit und das interpersonale Engagement der Patienten. Aufgrund dieser Daten und mit Hilfe retrospektiver Interviews wurden die Patienten in 4 Gruppen eingeteilt, nämlich in: ängstlich-zugewandt, ängstlich-nichtzugewandt, depressiv und ruhig. In der Gruppe der ruhigen wie der depressiven Patienten fand sich weder ein Anstieg des STH noch des Cortisols. Die ängstlich-nichtzugewandten Patienten zeigten einen signifikanten Anstieg beider Hormone, während die ängstlich-zugewandten Patienten nur einen Anstieg des Cortisols aufwiesen. (Bei experimentellen Streßsituationen hat sich gezeigt, daß die Beurteilung der Kontrollierbarkeit und der Vorhersehbarkeit einer Belastung die Hormonreaktionen wesentlich beeinflussen kann; Voigt et al., 1989.) Die Untersuchung zeigt damit auch, daß der durch einen psychischen Streß ausgelöste Anstieg von Wachstumshormon und von Cortisol durch verschiedene Mechanismen hervorgerufen werden muß, da beide Phänomene auch dissoziiert auftreten können (Armario et al., 1984).

Prolaktin reagiert mindestens ebenso empfindlich wie Wachstumshormon auf Streßreize wie chirurgische Eingriffe und körperliche Arbeit (Noel et al., 1972). Welche physiologische Bedeutung dem Anstieg von STH und Prolaktin im Rahmen der Streßreaktion zukommt, ist momentan noch unklar.

Überraschenderweise haben sorgfältige Messungen von ACTH im Plasma beim Menschen in vielen Fällen nicht die durch Streß induzierten Cortisolanstiege reflektieren können (Fehm et al., 1984 a, b). Damit ist die bisherige Vorstellung einer ausschließlich von ACTH regulierten Stimulation von Cortisol beim Menschen in Frage gestellt worden. Es muß vielmehr angenommen werden, daß einige Streß-induzierte Cortisolanstiege beim Menschen über ACTH-unabhängige (nervale?) Mechanismen vermittelt werden. Die neueren Entwicklungen auf den Gebieten der Neuroendokrinologie und Neuroimmunologie haben zu einer zunehmenden Kenntnis der biologischen Phänomene bei experimentellen Streßsituationen geführt. So werden über den Corticotropin-Releasing-Faktor (CRF) die endokrinen und die vegetativen (autonomes Nervensystem) Reaktionen gesteuert (Sapolsky et al., 1986; Voigt und Fehm-Wolfsdorf, 1989).

Nicht selten werden die Begriffe Streß und Emotion synonym verwendet (Edholm, 1978). In den meisten Artikeln sind die Hormonmuster bei unangenehmen Emotionen gemessen und dann auch als „Streß" bezeichnet worden (Frankenhaeuser, 1978). Wir möchten an dieser Stelle Überlegungen zur Beteiligung von Neuropeptiden (Voigt und Fehm, 1983 c) an der Generierung und Expression von Emotionen einfügen.

Trotz der teilweise kontroversen Emotionstheorien (vgl. Kap. 9) besteht doch der Konsens, daß bei emotionalen Prozessen sowohl zentrale (Thalamus, Limbisches System, Aktivierung, Belohnungsstrukturen) als auch periphere (Katecholamine, Nebennierenrindenhormone, autonomes Nervensystem) Mechanismen beteiligt sind. Es scheint ganz offensichtlich zu sein, daß alle bisher bekannten psychophysiologischen Phänomene von Emotionen auf eine wesentliche Mitbeteiligung von Neuropeptiden hindeuten:

- Neuropeptide als spezifische Signalsubstanzen kommen sowohl im ZNS als auch im autonomen Nervensystem und im gesamten Magen-Darm-Trakt vor.
- Die zentralen Neuropeptide haben (soweit bisher bekannt ist) zwei für emotionale Mechanismen wichtige Funktionen: Sie beeinflussen spezifisch wichtige motivationale Prozesse (s. 10.2) und daneben die zentrale Steuerung des sympathischen und parasympathischen Systems. Die Lokalisation der peptidergen Neurone im ZNS und deren Projektionen in das Limbische System und die vegetativen Zentren entspricht den für emotionale Prozesse wichtigen Hirnstrukturen.
- Die peripheren Phänomene der Emotionen (unabhängig von der kausalen Bedeutung, die man ihnen vom Blickwinkel der unterschiedlichen Theorien beimißt), wie Veränderung der Herzfrequenz und des Blutdrucks, Motilitätsänderungen der intestinalen Hohlorgane, Modulation des Muskeltonus und die Funktion von exokrinen und endokrinen Drüsen, werden durch Neuropeptide gesteuert, die wahrscheinlich ubiquitäre, aber hochspezifische Signalstoffe im vegetativen Nervensystem darstellen (neben Katecholaminen und Acetylcholin).
- Das gemeinsame Vorkommen mehrerer Neuropeptidsubstanzen (häufig zusammen mit Neurotransmittern) in einer Nervenendigung deutet darauf hin, daß generell auch eine bestimmte Kombination von Signalsubstanzen für einen spezifischen Aspekt einer Emotion essentiell ist und nicht nur eine Substanz (z.B. Adrenalin).
- Für psychosomatische Erkrankungen könnten die dargestellten Beziehungen zwischen Neuropeptiden und Emotionen gleichermaßen zutreffen und vielleicht nach Erweiterung unserer Kenntnis der für physiologische Mechanismen notwendigen endogenen Signalsubstanzen auch einen spezifischen therapeutischen Angriffspunkt darstellen.

10.5 Neuroendokrinologie bei psychischen Erkrankungen

Bei einigen psychischen Erkrankungen sind Messungen von Parametern der Neuroendokrinologie zu wichtigen Hilfsmitteln für die Diagnose und auch für prognostische Aussagen geworden (Sachar, 1975;

Beumont und Burrows, 1982). Allerdings ist das gestörte Hormonmuster im Plasma allein kein wesentliches Kriterium, noch weniger sind bei unserem gegenwärtigen Wissensstand hierdurch Aussagen zur Pathogenese möglich. Die Neuroendokrinologie von Patienten mit depressiven Störungen sind ausführlich in Kapitel 51 beschrieben. Das große Gebiet der anderen psychischen Veränderungen (Krankheiten?) ist noch schwerer einzugrenzen, und nur sehr selten sind diese Patienten auch unter endokrinologischen Aspekten untersucht worden. (Einige psychosomatisch interessante Hinweise mögen in Abschnitt 10.4 zu finden sein.)

10.5.1 Anorexia nervosa[3]

Die Hormonstörungen bei Anorexia nervosa sind auffällig und charakteristisch, jedoch können sie als Begleitsymptome der extremen Mangelernährung beurteilt werden (Pirke und Ploog, 1986). Die eigene pathogenetische Bedeutung ist umstritten (Beumont und Russell, 1982). Als Leitsymptom kann die Amenorrhoe gelten, die meist schon vor einem offensichtlichen Gewichtsverlust, jedoch etwa zur gleichen Zeit mit den pathologischen Eßgewohnheiten auftritt. Eine im vorderen Hypothalamus gelegene primäre Störung ist seit längerem in der Diskussion (Katz und Weiner, 1975; Russell, 1977). Dafür sprechen einige typische endokrinologische Befunde: Die LH-Spiegel sind erniedrigt und folgen dem infantilen oder peripuberalen Sekretionsmuster (s. 10.3.2), die pulsatile Induktion durch Gn-RH fehlt (Boyar et al., 1972). Die Ansprechbarkeit der LH-Sekretion auf exogene Gn-RH-Applikation ist jedoch im Normbereich (Wiegelman und Solbach, 1972). Auch die häufig beobachtete Störung der Temperaturregulation bei diesen Patientinnen ist zusammen mit einem gestörten Eßverhalten als Zeichen einer primären hypothalamischen Insuffizienz angesehen worden.

Neben der pathologischen Gonadotropinsekretion wird über mildere Unregelmäßigkeiten für alle anderen Hormone berichtet (Beumont und Russell, 1982), vor allem ist die Cortisolsekretion eher erhöht und der Dexamethason-Hemmtest fällt pathologisch aus (s. Tab. 10–5).

Die meisten Störungen der neuroendokrinen Regulation normalisieren sich parallel mit dem Anstieg des Körpergewichts (Pirke und Ploog, 1986). Es gibt einige Hinweise dafür, daß nicht nur ein bestimmtes Körpergewicht für das Einsetzen der Menarche und den normalen Ablauf des weiblichen Sexualzyklus entscheidend ist, sondern auch ein bestimmter Anteil an Fettgewebe ist offenbar notwendig. Vielleicht kann so auch die bei einigen Leistungssportlerinnen auftretende Amenorrhoe erklärt werden (Brooks et al., 1984).

---

3 S. auch Kapitel 38.2 „Anorexia nervosa".

Tabelle 10–5.   Gegenüberstellung von Laborparametern und klinischen Symptomen bei (einer Subpopulation von) endogener Depression, Morbus Cushing und Anorexia nervosa (nach Voigt et al., 1985). (DST = Dexamethason-Suppressionstest)

	Depression	Morbus Cushing	Anorexia nervosa
Neuroendokrinologie			
Cortisol	erhöht	erhöht	erhöht
zirkadiane Rhythmik	aufgehoben	aufgehoben	abgeflacht
DST	pathologisch	pathologisch	pathologisch
STH	verminderte		normal oder erhöht
TSH	Stimulierbarkeit	verminderte	normal
LH/FSH	?	Stimulierbarkeit	erniedrigt
PRL	erhöht?		normal?
CRF-Test	verminderte Stimulierbarkeit	erhöhte Stimulierbarkeit	?
Klinische Symptome	Depression	Affektive Störungen, meist depressiv	Emotionale Störungen, auch depressiv
	keine Zeichen von Hypercortisolismus, häufig Amenorrhoe	Hypercortisolismus mit Amenorrhoe	Magersucht mit Amenorrhoe

10.6   Zusammenfassung und Ausblick

Die Entwicklungen auf dem Gebiet der Endokrinologie und vor allem der Neurobiologie von Peptiden haben unsere Erkenntnisse über die Beteiligung von Signalstoffen (Hormone, Neurotransmitter und Neuropeptide) an psychischen Vorgängen und einigen pathologischen Störungen wesentlich erweitert. Es hat sich auf der anderen Seite jedoch gezeigt, daß einfache Korrelationen von endokrinen und psychologischen Parametern nicht zu erwarten sind, was an folgenden Gründen liegen könnte:

– Eine Verhaltensqualität wird mit einer ganzen Palette von Signalstoffen realisiert.
– Ein Signalstoff kann sehr viele Funktionen ausüben.
– Die Hormonprofile (im Plasma oder Liquor cerebrospinalis) unterliegen einer bestimmten Sekretionsdynamik, die nur durch engmaschige Messungen zu erfassen ist.

Die neuen Dimensionen einer neurobiologisch orientierten Forschung auf dem Gebiet des menschlichen Verhaltens sind bisher nur bei einigen Phänomenen (Schmerzkontrolle, Appetitverhalten, Aspekte der Sexualität) offenbar geworden, sie werden aber sicherlich in der Zukunft unser Verständnis der Pathogenese von psychosomatischen Erkrankungen erweitern helfen.

Abkürzungen

POMC	Pro-opio-melanocortin
ACTH	Adrenocorticotropes Hormon
mRNA	messenger-Ribonucleinsäure
Gly	Glycin
Leu	Leucin
Met	Methionin
Phe	Phenylalanin
β-LPH	Lipolytisches Hormon
GABA	γ-Aminobuttersäure
STH	Somatotropes Hormon
LH	Luteotropes Hormon
FSH	Follikel-stimulierendes Hormon
REM	Rapid-Eye-Movement (Phase des Schlaf-EEG)
Gn-RH	Gonadotropin-Releasing-Hormon
CCK-8	Cholecystokinin-Oktapeptid
TRH	TSH-Releasing-Hormon
AGS	Adrenogenitales Syndrom
17-OHCS	17-Hydroxycorticosteron
TSH	Thyreotropes Hormon
MSH	Melanozyten-stimulierendes Hormon
DST	Dexamethason-Suppressionstest

11 Psychoimmunologie

Wolfgang Klosterhalfen und *Sibylle Klosterhalfen*

Vorbemerkungen

Kaum ein anderes Teilgebiet der psychosomatischen Grundlagenforschung hat in den letzten Jahren so viel Beachtung gefunden wie die Psychoimmunologie (Synonym: Psychoneuroimmunologie). Zahlreiche Monographien und Kongreßberichte (Ader, 1981 a; Frederickson et al., 1986; Guillemin et al., 1985; Hellhammer et al., 1988; Jankovic et al., 1987; Locke et al., 1985; Spector et al., 1985), ein neues Journal (vgl. Ader et al., 1987 a), zwei umfangreiche Bibliographien (Locke, 1986; Locke und Hornig-Rohan, 1983) und Hunderte von Zeitschriftenartikeln, von denen viele in „Science" erschienen sind, zeugen von einem regelrechten Forschungsboom.

In der Psychoimmunologie geht es primär um die Effekte von psychologischen Faktoren und damit assoziierten Neurohormonen auf das Immunsystem und auf den Verlauf von Krankheiten, bei denen das Immunsystem abwehrende oder mediierende Funktionen hat.

Im ersten Teil dieses Kapitels gehen wir auf entsprechende Forschungsarbeiten ein. Im zweiten Teil diskutieren wir als ein Beispiel für neuroimmunologische Interaktionen, d. h. die wechselseitige Beeinflussung von Nervensystem und Immunsystem, das Phänomen der konditionierten Immunmodulation. Im Unterschied zu vielen anderen Psychoimmunologie-„Reviews" beschränken wir uns nicht auf eine selektive Darstellung „der schönsten Ergebnisse der Psychoimmunologie", sondern muten unseren Lesern auch viele nicht hypothesenkonforme Ergebnisse zu. Langfristig gesehen berechtigt die Psychoimmunologie zu Hoffnungen im Hinblick auf präventive oder therapeutische Maßnahmen; beim gegenwärtigen Stand des (Un-)Wissens sind entsprechende Interventionen aber noch nicht von den vorliegenden Forschungsergebnissen ableitbar (vgl. Schulz und Ferstl, 1989).

Auf eine Erläuterung immunologischer Grundbegriffe und -prinzipien haben wir verzichtet. Eine kurze, sehr empfehlenswerte Einführung in die Immunologie geben Staines et al. (1987). Ebenfalls sehr attraktiv gestaltet ist das ausführlichere Lehrbuch von Roitt et al. (1987).

11.1 Streß, Persönlichkeitsfaktoren und Immunität

11.1.1 Falldarstellung

Norman Cousins war zehn Jahre alt, als er im Jahre 1925 aufgrund einer Fehldiagnose sechs Monate in einem Sanatorium verbringen mußte. Dem Jungen fiel auf, daß sich in diesem Sanatorium zwei verschiedene Gruppen von Kindern gebildet hatten. Während die einen zuversichtlich waren, daß sie ihre Tuberkulose besiegen würden, hatten die anderen resigniert und sich auf eine langwierige, tödlich verlaufende Krankheit eingestellt. Norman gehörte zu den untereinander befreundeten Optimisten, die versuchten, Neuankömmlinge auf ihre Seite zu ziehen, „bevor die bleiche Brigade ans Werk ging". Das Kind konnte sich des Eindrucks nicht erwehren, daß die Chance, als geheilt entlassen zu werden, in seiner Gruppe erheblich größer war als bei den Pessimisten.

Mit einer bedrohlichen Diagnose wurde Norman Cousins erst wieder im Alter von 39 Jahren konfrontiert, als er mit Rücksicht auf seine Familie eine Erhöhung seiner Lebensversicherung beantragt hatte. Die ärztliche Diagnose lautete, er habe eine Verengung der Herzkranzgefäße und – bei völliger Schonung – höchstens noch eineinhalb Jahre zu leben. Als er nach Hause kam, rannten ihm seine Töchter entgegen, die es liebten, von ihm zur Begrüßung in die Luft geworfen zu werden. Cousins zögerte nur kurz. Dann warf er seine Töchter noch höher als je zuvor. Am nächsten Tag nahm er wie geplant aktiv an einem Tennisturnier teil. Den geliebten Sport aufzugeben, kam für ihn ebensowenig in Frage, wie seine erfolgreiche Arbeit als Herausgeber des „Saturday Review" zu beenden.

Zehn Jahre später reiste Cousins als Leiter einer amerikanischen Delegation nach Leningrad und Moskau. In Leningrad konnte er wegen der Hitze und des durch die geöffneten Fenster zu hörenden Baustellenverkehrs nachts nicht gut schlafen; beim Aufstehen war ihm etwas übel. Ein Empfang in Moskau verlief frustrierend, weil er aufgrund eines Mißverständnisses erst mit vierstündiger Verspätung bei seinen Gastgebern ankam. Beim Abflug geriet er auf dem Moskauer Flughafen in die Abgase einer startenden Verkehrsmaschine. Als er nach einem langen Flug in einer überfüllten Maschine schließlich wieder in den USA landete, fühlte er sich schlecht. Eine Woche später wurde er ins Krankenhaus eingeliefert. Er konnte sich kaum noch bewegen. Die Blutsenkungsrate stieg auf 115 mm/h. Es wurde eine systemische Bindegewebserkrankung (ankylosierende Spondylitis; Synonym: Bechterewsche Krankheit) diagnostiziert. Cousins erfuhr von seinem Arzt, mit dem er befreundet war, ein hinzugezogener Experte beurteile die Heilungschance mit 1:500.

Dem sich bis zu diesem Zeitpunkt eher angepaßt verhaltenden Patienten wurde nach dieser Hiobsbotschaft klar, daß er, um „der eine von 500" zu sein, mehr als nur ein passiver Beobachter sein müßte. Da er bei Selye (1956) gelesen hatte, daß negative Emotionen negative körperliche Folgen haben, fragte er sich, ob nicht positive Emotionen therapeutisch wirksam werden könnten. Er ließ sich einen Filmprojektor und lustige Filme bringen und machte die Entdeckung, daß Lachen eine anästhetische Wirkung hatte und er anschließend mindestens zwei Stunden lang gut schlafen konnte. Sein Lachen störte andere Patienten, aber Cousins war schon unabhängig davon zu der Auffassung gelangt, daß ein Krankenhaus für einen Menschen, der ernsthaft krank ist, kein geeigneter Ort ist. Er zog in ein Hotel und genoß es, nicht laufend geweckt zu werden („Bettenbau", Waschen, Medikamenteneinnahme, Blutabnahme, Untersuchungen, Mahlzeiten!). Außerdem war der Hotelaufenthalt wesentlich billiger und das Essen besser. Die Lachtherapie kombinierte er mit einer „Überdosis" Vitamin C (25 g pro Tag). Es gelang ihm, ohne entzündungshemmende Mittel und Schlaftabletten auszukommen. Sein körperlicher Zustand war nach einer Woche deutlich verbessert, und bereits nach vierzehn Tagen lernte er am Strand einer karibischen Insel wieder zu stehen, zu gehen und bald auch zu laufen.

Cousins erholte sich fast vollständig. Als Teile seiner ungewöhnlichen Krankengeschichte in der Öffentlichkeit bekannt wurden und er deswegen häufig Anfragen bekam, schrieb er für das „New England Journal of Medicine" einen ausführlichen Bericht (Cousins, 1976). Die Resonanz auf diesen Artikel war ebenfalls ungewöhnlich. Er erhielt über 3000 Zuschriften von Ärzten, zahlreiche Briefe und Anfragen von Kranken oder von deren Angehörigen (vgl. Cousins, 1979) und Spendengelder zur Förderung einschlägiger Forschungsarbeiten in Höhe von mehreren Millionen Dollar (G. F. Solomon, pers. Mitteilung, San Francisco, November 1984).

Waren für das Auftreten der Spondylitis die frustrierenden Ereignisse der Auslandsreise entscheidend? Cousins (1976, S. 1459) schreibt dazu:

„I found myself increasingly convinced . . . that the reason I was hit hard by the diesel and jet pollutants, whereas my wife was not, was that I had had a case of adrenal exhaustion, lowering my resistance."

Die in den folgenden Abschnitten referierten Ergebnisse der Psychoimmunologie zeigen, daß diese Auffassung im Prinzip keineswegs abwegig ist. Es ist in den letzten Jahren gezeigt worden, daß Lärm, Schlafentzug und verschiedene andere psychisch belastende Ereignisse Funktionen des Immunsystems beeinflussen können. Darüber hinaus gibt es Hinweise, daß der Verlauf von Krankheiten, für die das Immunsystem (potentiell) wichtig ist, von Persönlichkeitsfaktoren abhängt. In diesem Zusammenhang werden auch – bei Norman Cousins offensichtlich stark ausgeprägte – Eigenschaften wie „fighting spirit" und „hardiness" diskutiert. Über die Effekte psychotherapeutischer Interventionen auf Parameter des Immunsystems ist hingegen noch wenig bekannt. Nicht nur in dieser Hinsicht bleibt die hier referierte Fallgeschichte eine Herausforderung für die Psychoimmunologie.

11.1.2   Streßhormone und Immunität

Das Nervensystem besitzt durch das neuroendokrine System zahlreiche Möglichkeiten, das Immunsystem zu beeinflussen (Berczi, 1986; Berczi und Kovacs, 1987). Neurohormone können unmittelbar auf Lymphozyten und akzessorische Zellen einwirken, denn diese haben Rezeptoren für Kortikosteroide (Cake und Litwack, 1975; Homo-Delarche und Duval, 1987; Werb et al., 1978), Wachstumshormon (Arrenbrecht, 1974), Östrogen (Gillette und Gillette, 1979), Testosteron (Abraham und Bug, 1976), β-adrenerge Substanzen (Hollenberg und Cuatrecasas, 1974; Singh et al., 1979), Acetylcholin (Richman und Arnason, 1979; Strom et al., 1974) und β-Endorphin (Hazum et al., 1979).

Das größte Interesse haben bisher die Glukokortikosteroide gefunden (Übersicht bei Bach und Strom, 1985; Claman, 1987; Mantovani, 1985; Munck et al., 1984), deren entzündungshemmende Wirkung allgemein bekannt ist. In pharmakologischer Dosierung überwiegen suppressive Effekte, aber geringe Kortikoidmengen können auch die Abwehr fördern (z.B. Ambrose, 1964; Newson und Darrach, 1954; Sherman et al., 1973; Tuchinda et al., 1972). Bei kortikosteroidsensitiven Tieren (Mäuse, Ratten, Hamster, Kaninchen) führt eine Dauerbehandlung mit Steroiden zu einer Gewichtsabnahme des Thymus (Shewell und Long, 1956), wobei Thymuszellen zerstört werden (Claesson und Ropke, 1969; Cowan und Sorenson, 1964; Lundin und Schelin, 1969). In vitro werden Thymuszellen dieser Tiere durch Steroide in geringer Konzentration (10^{-7}M) geschädigt (Batra und Schrek, 1967; Burton et al., 1967). Meerschweinchen, Affen und Menschen sind vergleichsweise resistent; bei Kindern, die eine Woche lang täglich 1–2,5 mg/kg Prednison oral erhielten, kam es jedoch zu einer (reversiblen) Verkleinerung des Thymusschattens um 19–44% (Caffey und Silbey, 1960). Bei Gesunden (Yu et al., 1974) und Patienten (Fauci und Dale, 1974) führte schon eine einmalige Kortikoidgabe zu einer vorübergehenden Leukozytopenie, die offensichtlich nicht auf Lyse, sondern auf Redistribution der Leukozyten zurückgeht. In Tierexperimenten wurde außerdem die Infektionsresistenz durch Kortikoidgaben herabgesetzt (vgl. Monjan, 1981).

Bei den Katecholaminen sind die Auswirkungen auf immunologische Parameter uneinheitlich (vgl. z.B. Crary et al., 1983 a, b; Rosenthal und Holzer, 1921; Sanders und Munson, 1985; Tonnesen et al., 1984). Wie die Übersichten von Chang (1984), Fischer und Falke (1984), Heijnen und Ballieux (1986), Plotnikoff et al. (1986) und Wybran (1985) zeigen, gilt dies auch für Opioide.

Neuroanatomische bzw. histologische Befunde sprechen aber zusätzlich dafür, daß Noradrenalin, Adrenalin und eventuell noch weitere Neurotransmitter wesentlich an der Regulation des Immunsystems beteiligt sind. Die Innervation von Lymphknoten (Tonkoff, 1899) sowie des Thymus (Hammar, 1935) wurde schon früh beschrieben. Die meisten Veröffentlichungen zur Innervation lymphatischer

Gewebe sind jedoch erst in den letzten Jahren erschienen (Übersicht bei Bulloch, 1985; Felten et al., 1987). Giron et al. (1979) fanden noradrenerge Nervenfasern in zervikalen Lymphknoten von Ratten; diese Fasern waren nicht nur mit Blutgefäßen assoziiert, sondern fanden sich auch in medullären und internodularen Gebieten. Auch Thymus und Milz von Mäusen enthalten noradrenerge Nervenfasern, die überwiegend – aber nicht ausschließlich – Gefäßen folgen. Im Thymus liegen sowohl perivaskuläre als auch parenchymale Fasern nahe an Mastzellen, so daß eine sympathische Modulation der Histaminausschüttung möglich erscheint; in der Milz gibt es einzelne adrenerge Fasern, die die Gefäße verlassen und sich in lymphozytenreiche Gebiete der weißen Pulpa verzweigen (Reilly et al., 1979; Williams und Felten, 1981). Es ist daher denkbar, daß das vegetative Nervensystem nicht nur über eine Regulation der Gefäßdurchmesser in lymphatischen Organen, sondern auch neurohumoral auf das Immunsystem einwirkt. Ob eine solche Beeinflussung in vivo tatsächlich stattfindet, ist bisher nicht bekannt.

11.1.3 Stressoren und Immunität

Streßeffekte auf immunologische Parameter

Innerhalb der Psychoimmunologie befaßt sich eine größere Anzahl von Arbeiten mit den Effekten von Stressoren (z. B. Lärm, Bewegungsrestriktion, schmerzhafte elektrische Reize) auf immunologische Funktionen bei Ratten und Mäusen. Die Ergebnisse solcher Experimente sind nicht „auf den Menschen übertragbar", aber wegen der geringeren Kontrollmöglichkeiten bei nicht-experimentellen bzw. ethischen Bedenken bei experimentellen Humanstudien und der offensichtlich eher komplizierten Beziehungen zwischen Nervensystem und Immunsystem unter heuristischen Aspekten außerordentlich wichtig.

In vielen dieser Arbeiten wurden als nicht unumstrittener (vgl. Maier und Laudenslager, 1988) Indikator immunologischer „Abwehrbereitschaft" Lymphozytenproliferationsraten bestimmt. Dabei werden in vitro T-Zellen durch die pflanzlichen Antigene Phytohämagglutinin (PHA) oder Concanavalin A (Con A) oder B-Zellen durch Lipopolysaccharide (LPS) stimuliert; „pokeweed mitogen" (PWM) regt T- und B-Zellen zur Teilung an. Die Zellteilungsrate wird indirekt über den Einbau radioaktiv markierter Eiweiße ermittelt.

Unmittelbar nach der Streßexposition ist in diesen Experimenten meist eine deutliche Verminderung der Lymphozytenproliferationsrate festgestellt worden. Wie die folgenden Literaturbeispiele zeigen, kommt es aber bei immunologischen Streßeffekten (ähnlich wie bei pharmakologischen Effekten) auf die „Streßdosis", die Meßintervalle und die untersuchten immunologischen Variablen an.

Monjan und Collector (1977) setzten Mäuse an 0, 4, 20, 26, 34 oder 39 Tagen weißem Rauschen aus (100 db, 5 sec/min, 1 h täglich). Die Proliferation von B-Lymphozyten durch LPS und von T-Lymphozyten durch Con A war nach 4 und nach 20 Tagen erniedrigt, nach 26 und 34 Tagen erhöht und nach 39 Tagen im Vergleich zu Kontrolltieren unverändert. Zu diesem vielzitierten Experiment ist kritisch anzumerken, daß keine Inferenzstatistik vorliegt und pro Meßzeitpunkt jeweils nur zwei Versuchs- und zwei Kontrolltiere untersucht wurden.

In mehreren Arbeiten wurden Elektroschocks als Stressor eingesetzt. Keller et al. (1981) bestimmten die Anzahl verschiedener Zellen im peripheren Blut und die Lymphozytenproliferation durch PHA bei Ratten, die zuvor 20 Stunden lang sehr intensive oder mittelstarke elektrische Schocks (2 sec/min) erhalten hatten oder mäßig stark bewegungsrestringiert worden waren; eine vierte Gruppe wurde nicht aversiv stimuliert. Bei der Anzahl von Monozyten oder polymorphkernigen Leukozyten sowie beim Prozentsatz der T-Zellen gab es keine Gruppenunterschiede. Es kam jedoch zu einer streßinduzierten Lymphozytopenie. Die Lymphozyten der Streß-Gruppen ließen sich durch PHA sowohl im Gesamtblut als auch nach Isolation weniger gut stimulieren. In einer bemerkenswert systematischen Nachfolgeuntersuchung, in der außerdem Adrenalektomie als Faktor eingeführt wurde, konnten Keller et al. (1983) diese „dosisabhängigen" Streßeffekte bestätigen. Interessanterweise verhinderte Adrenalektomie zwar die streßinduzierte Lymphozytopenie, sie hatte aber relativ wenig Einfluß auf die Streßeffekte bei den PHA-Tests. Bei hypophysektomierten Ratten fielen die Streßeffekte auf die Stimulation von Lymphozyten im peripheren Blut sogar besonders stark aus (Keller et al., 1988), woraus die Autoren schließen, daß a) hypophysäre Mechanismen an einer Gegenregulation von Streßeffekten beteiligt sind und b) wahrscheinlich Katecholamine eine wesentliche Rolle bei der Vermittlung immunologischer Streßeffekte spielen.

Croiset et al. (1987) verglichen eine Kontrollgruppe von Ratten („home-cage") mit Tieren, die mehrfach auf eine hell erleuchtete Plattform gesetzt worden waren, von der sie Zugang zu einer dunklen Box hatten. Beim vierten von fünf Durchgängen erhielt ein Teil der Tiere in der Box einen „footshock" (0,9 mA, 2 sec). Überraschenderweise war gegenüber den Kontrolltieren die Lymphozytenproliferationsrate (Con A) der nicht-geschockten Versuchstiere stark erhöht. Dieser Effekt trat nicht bei den Tieren auf, die den Schock erhalten hatten und 5 Tage später erneut auf die Plattform gesetzt worden waren (sog. „passive avoidance"-Test). Die Autoren vermuten, daß nicht der Schock, sondern der „avoidance"-Test sich depressiv ausgewirkt hat.

In einer faktoriellen Versuchsanordnung variierten Lysle et al. (1987) die Anzahl von Schocks pro Sitzung (4, 8 oder 16 Schocks) sowie die Anzahl solcher Sitzungen (1, 3 oder 5 Tage); eine Kontrollgruppe erhielt keine Schocks. Nach 4 Schocks war die Proliferationsrate (Con A) praktisch noch unverändert; am Ende der dritten bzw. fünften Sitzung lag sie bei ca. 70%. Nach einer Sitzung mit 8 Schocks ging die Proliferationsrate aber auf ca. 20% zurück, wobei durch

Wiederholung der Sitzungen der Effekt nicht verstärkt wurde. Bei 16 Schocks pro Sitzung lagen die Werte unabhängig von der Anzahl der Sitzungen bei jeweils etwa 10%. In einem zweiten Experiment wurde geprüft, wie lange solche Streßeffekte anhalten. Bei Lymphozyten aus der Milz war schon 24 Stunden nach einer Sitzung mit 16 Schocks kein Effekt mehr nachzuweisen. Lymphozyten aus dem peripheren Blut sprachen dagegen noch 48 Stunden später weniger stark (ca. 35% der Kontrollwerte) auf das Mitogen an; nach 96 Stunden war die Reaktion aber wieder normal.

Funk und Jensen (1967) benutzten eine erheblich weniger aufwendige Methode, um Streßeffekte auf zelluläre Reaktionen zu quantifizieren. Sie beschallten (123 db) Mäuse vom 5. Tag vor bis zum 14. Tag nach der subkutanen Implantation eines Baumwolltupfers. Bei den beschallten Tieren wurden anschließend weniger Kapselbildung, weniger zelluläre Infiltrate und ein geringeres Gewicht des Implantats als bei der Kontrollgruppe festgestellt.

Shavit et al. (1984) leisteten einen interessanten Beitrag zur Frage der Reizspezifität, indem sie bei Ratten intermittierende Fußschocks (an 4 Tagen je 10 min) mit kontinuierlich applizierten Schocks (an 4 Tagen je 2 min) verglichen. Aus früheren Arbeiten der Forschergruppe war bekannt, daß die intermittierenden, aber nicht die kontinuierlichen Schocks die Freisetzung endogener Opioide bewirken (Lewis et al., 1980). Als abhängige Variable diente die Aktivität von „natural killer" (NK)-Zellen. (Killerzellen können ohne vorherige Aktivierung Tumorzellen und virusinfizierte Zellen zerstören; vgl. Reynolds und Ortaldo, 1987.) Im Unterschied zu den kontinuierlichen Schocks bewirkten die „opioiden" Schocks eine Reduzierung der NK-Zellaktivität. Dieser Effekt konnte auch durch die Gabe von 4 × 30 mg/kg Morphium hergestellt bzw. durch den Opiat-Antagonisten Naltrexon blockiert werden.

Cunnick et al. (1988) bestimmten bei Ratten sowohl die NK-Zellaktivität als auch die Stimulierbarkeit von T-Lymphozyten. Eine Sitzung mit 16 signalisierten Schocks wirkte sich auf beide Parameter hemmend aus. Da sich nur der Streßeffekt auf die NK-Zellaktivität durch Naltrexon blockieren ließ, ist anzunehmen, daß den Streßeffekten auf die Lymphozytenproliferation ein anderer neuroendokrinologischer Mechanismus zugrunde liegt. Glukokortikoide scheinen hier, wie schon berichtet (Keller et al., 1983), keine wesentliche Rolle zu spielen.

Häufig wird argumentiert, immunologische und andere Streßeffekte seien wesentlich von der Kontrollierbarkeit aversiver Ereignisse abhängig. Laudenslager et al. (1983) verglichen bei Ratten kontrollierbare mit unkontrollierbaren Schocks. Im Vergleich zu einer nicht-geschockten Kontrollgruppe trat nur unter der Bedingung „unkontrollierbare Schocks" eine Hemmung der T-Zell-Proliferationsrate auf. Diese Ergebnisse konnten jedoch weder von Gamzu et al. (1984) noch von Laudenslager und dessen Kollegen (vgl. Maier und Laudenslager, 1988) repliziert werden.

Zu den Effekten von Stressoren auf humorale Immunfunktionen liegen nur relativ wenige Tierexperimente vor. Die Antikörperbildung gegen Flagellin war bei neun Wochen alten Ratten, die in den ersten drei Lebenswochen täglich ein „handling" erhielten, erhöht (Solomon et al., 1968), nach „overcrowding" bei erwachsenen Ratten reduziert und nach Elektroschocks unverändert (Solomon, 1969). Einzeln aufgezogene Mäuse zeigten verminderte Antigen-Antikörper-Reaktionen (Edwards et al., 1980), wenn sie zwei Wochen lang gemeinsam mit anderen Tieren (16/Käfig) gehalten wurden.

Boranic et al. (1982) untersuchten den Einfluß von Bewegungsrestriktion auf die Anzahl der plaquebildenden Zellen in der Milz. Dazu wurden am Tag 0 Ratten intraperitoneal mit Schafserythrozyten stimuliert. An den Tagen 0 bis 3 wurden die Tiere mit Hilfe von Klebeband jeweils drei Stunden lang immobilisiert. Bei den immobilisierten Ratten lag die Zahl der plaquebildenden Zellen niedriger als in der nicht-restringierten Kontrollgruppe.

Laudenslager et al. (1988) berichten über einen bei Ratten noch nach zwei Monaten nachweisbaren Effekt von Schockprozeduren auf die Antikörperbildung gegen ein Neoantigen („keyhole limpet hemocyanin").

In den psychoimmunologischen Humanstudien (vgl. auch Schulz und Raedler, 1986) sind hauptsächlich prä- und quasiexperimentelle Untersuchungsanordnungen verwendet worden (vgl. Biondi und Pancheri, 1987), wobei berufliche und private Belastungen zu immunologischen Variablen in Beziehung gesetzt wurden. Besonders bekannt geworden sind Untersuchungen an Astronauten, an Witwen und Witwern und an studentischen Prüflingen.

In den Astronautenstudien wurden nach der Landung häufig verminderte Lymphozytenproliferationsraten festgestellt (vgl. Cogoli und Tschopp, 1985; Beisel und Talbot, 1987). Während des Fluges – die Astronauten nahmen sich am dritten Flugtag Blut ab und führten die Tests selbst durch – war die Proliferation sogar vollständig gehemmt (Bechler et al., 1986). Solche Ergebnisse sollten jedoch aus verschiedenen Gründen nicht zur Stützung psychoimmunologischer Hypothesen herangezogen werden. Es ist allgemein bekannt, daß Astronauten psychisch extrem stabil sind, und daher nicht verwunderlich, wenn Beisel und Talbot (1987, S. 198) schreiben:

„The reported changes in levels of stress hormones during space flight appear unlikely to affect the immune system."

Offensichtlich sind die während des Fluges und nach dem Flug feststellbaren Effekte auf die Mitogenstimulation eine Folge der Schwerelosigkeit: Bechler et al. (1986) ließen einem gesunden Spender Blut abnehmen. Ein Teil der daraus isolierten Lymphozyten wurde am Boden stimuliert, ein anderer Teil von Astronauten mitgenommen und während des Fluges untersucht. Im Unterschied zu den Bodenproben ließen sich die „Weltraumproben" praktisch nicht durch Con A stimulieren.

Bartrop et al. (1977) erhoben bei 26 Witwen und

Tabelle 11–1. Tod des Lebenspartners und immunologische Veränderungen

Autor(en)	Jahr	Meßzeitpunkte	Kontrollgruppe	immunol. Maße	Effekt
Bartrop et al.	1977	2 Wochen *post*	ja	Lymphozyten-	
		6 Wochen *post*		proliferation	↓
				Anzahl der	
				T- u. B-Zellen	=
				IgM, IgG, IgA	=
				Hauttests	=
Schleifer et al.	1983	einige Monate *prä*	ja	Lymphozyten-	
		1–2 Mon. *post*		proliferation	↓
		4–14 Mon. *post*		Anzahl der	
				T- u. B-Zellen	=
Monjan	1984	1 Monat *post*	ja	Lymphozyten-	
		3 Monate *post*		proliferation	=
Irwin et al.	1987	1–5 Mon. *post*	ja	NK-Zellaktivität	↓
		1–2 Mon. *prä*	nein	NK-Zellaktivität	=
		0–1 Mon. *post*			

einer nach Alter, Geschlecht und ethnischer Zugehörigkeit parallelisierten Kontrollgruppe (Klinikpersonal) immunologische Maße (vgl. Tab. 11–1). Die Blutproben wurden etwa zwei und sechs Wochen nach dem Tod des Ehepartners entnommen. Unter Verwendung von PHA war die Lymphozytentransformation bei den Witwen zu beiden Meßzeitpunkten, unter Con A nur beim zweiten Termin, relativ zur Kontrollgruppe erniedrigt; bei beiden Methoden lagen die 6-Wochen-Werte der Witwen niedriger als die 2-Wochen-Werte. Bei den folgenden Parametern ergaben sich jedoch keine Unterschiede: Anzahl der T- und B-Zellen, IgG (Immunglobulin G), IgA, IgM, α_2-Makroglobulin, verschiedene Autoantikörper und verzögerte Hypersensitivität (verschiedene Hauttests). Bemerkenswert erscheint, daß die Gruppen sich nicht bei Hormonen unterschieden, die häufig als Streßindikatoren verwendet werden: Thyroxin, Trijodthyronin, Cortisol, Prolaktin und Wachstumshormon.

Schleifer et al. (1983) erhielten bei Männern (N=15), deren Frauen Brustkrebs hatten, in einer prospektiven Studie ähnliche Ergebnisse: Im Vergleich zu Ausgangswerten, die wenige Monate vor dem Tod der Frauen erhoben wurden, war die Lymphozytentransformation der Witwer auf PHA und Con A 1 bis 2, jedoch nicht 4 bis 14 Monate post mortem erniedrigt; bei der Anzahl der peripheren T- und B-Lymphozyten gab es keine Unterschiede. Bei zeitlich parallel untersuchten Männern (N=13) mit gesunden Ehefrauen wurden keine Veränderungen in den Mitogentests registriert.

Monjan (1984) berichtet jedoch, er habe bei Witwern (N=17) keine veränderte Lymphozytenfunktionen (PHA, PWM) feststellen können, und bei Kontrollpersonen (N=17) seien die Werte merkwürdigerweise innerhalb von zwei Monaten abgefallen.

Irwin et al. (1987) stellten bei Witwen (N=10) eine gegenüber Kontrollpersonen (N=8) verminderte NK-Zellaktivität fest. In einer zweiten Studie ohne Kontrollgruppe wurde bei Witwen (N=6) nach dem Tod des Ehepartners keine wesentliche Veränderung der NK-Zellaktivität registriert.

Irwin et al. (1988) berichten offensichtlich über eine Erweiterung der Studie 1 von Irwin et al. (1987). Frauen (N=11) von Männern mit Lungenkrebs im Endstadium hatten ebenfalls gegenüber der Kontrollgruppe eine deutlich verminderte NK-Zellaktivität.

Die hier zitierten „bereavement"-Studien scheinen insgesamt dafür zu sprechen, daß die mit dem Verlust des Ehepartners verbundenen emotionalen Vorgänge (vgl. Stroebe und Stroebe, 1987) bzw. deren neuroendokrinologischen Korrelate sich hemmend auf die Lymphozytenproliferation und die NK-Zellaktivität auswirken können. Alternativ ist aber mit der Möglichkeit zu rechnen, daß mit dem Partnerverlust einhergehende Verhaltensänderungen immunologische Alterationen nach sich ziehen. Hier wäre z.B. an eine vermehrte Einnahme von Psychopharmaka (vgl. Descortes et al., 1985) oder Alkohol zu denken, sowie an eine Änderung der Eßgewohnheiten (vgl. auch Kiecolt-Glaser und Glaser, 1988). Natürlich können Änderungen der Lebensgewohnheiten auch zu wesentlichen Veränderungen im Spektrum der auf den Organismus einwirkenden Antigene und entsprechenden immunologischen Effekten führen.

In einer Reihe weiterer Streßstudien wurde Immunglobulin A (IgA) im Speichel bestimmt. Das sekretorische IgA dient der Infektabwehr (Brandtzaeg, 1970; Everhart et al., 1977; Hein, 1975; Yodfat und Silvian, 1977) und kann durch die Methode der radialen Immundiffusion relativ einfach und sehr genau quantifiziert werden (Kallestad, 1984; Mancini et al., 1965). Zwei Maße sind üblich: Konzentration und Sekretionsrate (Konzentration × Speichelvolumen/Zeit). Für die Konzentrationswerte ermittelten wir bei einem Zeitabstand von vier Tagen unterschiedliche (r = .24 bis .70), für die Sekretionsraten extrem hohe (r = .94 bis .98) Reliabilitätskoeffizienten (Ernst et al., 1987). Wie Tabelle 11–2 zeigt, sind die Ergebnisse dieser Studien heterogen. Da beim Speichel-IgA starke jahreszeitliche Schwankungen auftreten (Hendrickx, 1981), ist bei der Interpretation der in unterschiedlichen Monaten (September bis Juli) erhobenen Daten von Jemmott et al. (1983) Vorsicht ange-

Tabelle 11–2. Belastende Ereignisse und Immunglobulin A (IgA) im Speichel

Autoren	Jahr	Stressoren	Kontrollgruppe	Effekt auf IgA	
McClelland et al.	1980	„life events"	ja	Konzentration	↓
		„perceptual and learning tasks"	nein	Konzentration	↓
Jemmott et al.	1983	akad. Prüfungen	nein	Sekretionsrate	↓
Kiecolt-Glaser et al.	1984a	akad. Prüfungen	nein	„levels"	=
		„life events"	ja	„levels"	=
McClelland et al.	1985	akad. Prüfungen	nein	Konzentration	(↑)
Kubitz et al.	1986	„daily hassles"	ja	Konzentration	=
Ernst et al.	1987	akad. Prüfungen	nein	Konzentration	=
				Sekretionsrate	=
Klosterhalfen et al.	1987	Film über die Milgram-Experimente	ja	Konzentration	=
				Sekretionsrate	=
Jasnoski und Kugler	1987	„vigilance task" (1 h)	(ja)	Konzentration	↓
Stone et al.	1987	(„daily mood")	nein	siehe Text	
Jemmott und Magloire	1988	akad. Prüfungen	nein	Konzentration	↓

bracht. Stone et al. (1987) bezogen „daily mood" und IgA-Werte aufeinander. Die Antikörpertiter gegen Kaninchen-Albumin lagen an Tagen mit „high-positive mood" höher als an Tagen mit „high-negative mood". Ihre Daten für „low-positive mood" und „low-negative mood" sowie die Konzentrationen des gesamten IgA zeigen jedoch einen gegenläufigen Trend.

Gegenwärtig bieten die an Menschen durchgeführten psychoimmunologischen Streßstudien ein eher verwirrendes Bild. Es fehlt zwar nicht an Hinweisen, daß sich Streßreaktionen auch im Immunsystem manifestieren, hinreichend konsistente Ergebnisse traten bisher aber lediglich bei der Lymphozytenproliferationsrate auf, die in 11 von 12 der von Schulz und Raedler (1986) aufgelisteten entsprechenden Studien eine streßassoziierte Hemmung zeigt. Zu diesen Studien gehören auch Arbeiten von Palmblad et al. (1976, 1979a, b), in denen über verminderte Lymphozytenproliferationen sowie erhöhte Blutsenkungsraten nach Schlafentzug berichtet wird. Es handelt sich hier jedoch nicht um Experimente, sondern um unkontrollierte Verlaufsstudien. Erstaunlicherweise ist unseres Wissens bisher noch kein psychoimmunologisches Streßexperiment publiziert worden, bei dem Versuchspersonen nach dem Zufallsprinzip einer Streß- bzw. einer Kontrollgruppe zugeordnet wurden. Zu der wichtigen Frage, ob immunologische Funktionen durch Streß*reduktion* bzw. die Induktion angenehmer Gefühlszustände aktiviert werden können, liegen jedoch erste experimentelle Befunde vor.

Kiecolt-Glaser et al. (1985) haben Bewohner eines Altenheims, die bereit waren, an einer medizinisch-psychologischen Untersuchung teilzunehmen, nach Zufall auf Gruppen aufgeteilt: Mit Gruppe 1 wurde in 12 Sitzungen ein Entspannungstraining durchgeführt, Gruppe 2 erhielt 12mal Besuch von einem Studenten, Gruppe 3 diente als Kontrolle und hatte keine zusätzlichen Kontakte. Die Sitzungen bzw. Besuche verteilten sich über einen Monat und dauerten jeweils 45 Minuten. Als Folge der Relaxationssitzungen gab es nach einem Monat einen am Prozentsatz lysierter Zellen gemessenen deutlichen Anstieg der NK-Zellaktivität sowie niedrigere Antikörpertiter gegen Herpes-simplex-Viren. (Dieser niedrigere Antikörperspiegel wird aufgrund einer separaten Untersuchung als Zeichen einer verbesserten Abwehr interpretiert.) Bei den Mitogen-Stimulationstests (PHA und PWM) gab es keine gruppenspezifischen Effekte. Diese Arbeit ist wegweisend, denn langfristig wird die Psychoimmunologie zu der Frage Stellung nehmen müssen, ob mit Hilfe psychologischer Methoden Immunfunktionen und „immunologische" Krankheiten günstig beeinflußt werden können.

Dillon et al. (1985/86) zeigten Studierenden einen lustigen 30minütigen Videofilm (Richard Pryor Live) sowie einen gleichlangen Lehrfilm (The Thin Edge: Anxiety). Während sich beim Kontrollfilm die Konzentration von Speichel-IgA praktisch nicht veränderte, lag sie nach dem „humorous videotape" um durchschnittlich 16% über den Ausgangswerten.

Jasnoski und Kugler (1987) verglichen experimentell zwei Relaxationsbedingungen mit einer Vigilanzaufgabe (Dauer: jeweils 1 h). Unter den Entspannungsbedingungen kam es praktisch zu keiner Veränderung, unter der Aktivierungsbedingung zu einem leichten Abfall der Konzentration von IgA im Speichel.

Über Beziehungen zwischen Meditationsübungen und immunologischen Werten haben kürzlich D.E. Smith et al. (1989) und G.R. Smith et al. (1989) auf einer Tagung berichtet.

Streßeffekte bei Infektionskrankheiten, Krebs und Arthritis

Wie die Literaturübersicht von Plaut und Friedman (1981) zeigt, ist zur Frage von Streßeffekten auf das Auftreten und den Verlauf von **Infektionskrankheiten** bereits eine größere Zahl tierexperimenteller Arbeiten publiziert worden. (In der Bundesrepublik führte Schlewinski [1975, 1976, 1980] entsprechende Arbeiten durch.) Dabei fällt die erhebliche Variation der verwendeten Krankheitsmodelle auf. Nach Streßexposition traten häufig erhöhte, gelegentlich aber auch unveränderte oder erniedrigte Morbiditäts- und Mortalitätsraten auf. Daß überwiegend krankheitsfördernde Effekte festgestellt wurden, könnte daran liegen, daß der Zeitabstand zwischen Streßexposition und Krankheitsinduktion in den meisten Arbeiten kurz war (vgl. auch Rogers et al., 1979). „Die angeschnittenen Fragen erfordern noch viel Kleinarbeit, da die einzelnen Tiergattungen, darunter die Menschen, verschieden reagieren" (Haberland, 1926, S. 1393).

Jemmott und Locke (1984) haben in ihrer ausführlichen Literaturübersicht dargestellt, daß psychische Belastungen in der Regel die Entwicklung von Infektionskrankheiten beim Menschen begünstigen. Über entsprechende Befunde bei Herpes labialis haben Schmidt et al. (1985) in einer retrospektiven Studie berichtet. In der Woche vor einem erneuten Ausbruch der Krankheit traten „daily hassles, stressful life events, anxiety" gehäuft auf. Über eine neuere prospektive Studie berichten Graham et al. (1986). Personen, die zu Beginn oder während des sechsmonatigen Untersuchungszeitraums bei allen drei verwendeten Streß- bzw. Streßbewältigungsmaßen (Life Events Inventory; Daily Hassles Scale; General Health Questionnaire) überdurchschnittlich hohe Werte hatten, litten häufiger und insgesamt länger an Atemwegserkrankungen als Personen, die bei allen drei Maßen unterhalb des Medians lagen. (Hohe Werte im General Health Questionnaire reflektieren eine geringe Fähigkeit, „Streß" zu bewältigen.)

Streßeffekte auf die Entwicklung von **Krebs** hängen nach einer Literaturanalyse von Justice (1985) sowohl von zeitlichen Faktoren als auch von der Methode der Tumorinduktion ab. Bei viral induzierten Tumoren und kurzem Intervall zwischen Streßexposition und Induktion treten meist wachstumsfördernde, bei größerem Abstand aber protektive Streßeffekte auf; entgegengesetzte Effekte, also Hemmung bei akutem Streß und Stimulation bei länger zurückliegender Streßbelastung ergeben sich nach Justice hingegen, wenn der Tumor chemisch ausgelöst oder transplantiert wird. Aus der Fülle der Arbeiten sollen hier zwei – nicht in dieses Schema passende – Untersuchungen herausgegriffen werden. Sie sind besonders interessant, weil ihre Ergebnisse dafür sprechen, daß die Kontrollierbarkeit von Stressoren für das Tumorwachstum kritisch ist. Mäuse bzw. Ratten wurden kontrollierbaren oder physikalisch gleichen, aber unkontrollierbaren Schocks ausgesetzt, nachdem ihnen 24 Stunden zuvor Tumorzellen injiziert worden waren. Während im Vergleich zu einer ungestreßten Gruppe die Tumorentwicklung durch kontrollierbare Schocks nicht wesentlich beeinflußt wurde, führten unkontrollierbare Schocks zu einem schnellen Tumorwachstum und einer Verkürzung der Überlebenszeit (Sklar und Anisman, 1979) bzw. zu einer verminderten Abstoßungsrate (Visintainer et al., 1982). Wegen der Verschiedenartigkeit der Krankheiten und ihrer Wirte (vgl. Fox, 1981) bleibt es jedoch fraglich, ob Streßexperimente an Nagetieren mit künstlich indu

zierten Tumoren wesentlich zu einem besseren Verständnis der Pathogenese von Krebserkrankungen beim Menschen beitragen können.

Von der Annahme ausgehend, daß Zellen des Immunsystems die Entwicklung von (bestimmten) Tumoren verhindern oder zumindest behindern können (Burnet, 1970; Herberman und Holden, 1979; Levy et al., 1987; vgl. auch Schulz und Raedler, 1986), wird in der Psychoimmunologie auch die Bedeutung von psychisch belastenden Ereignissen für Krebserkrankungen beim Menschen diskutiert. Wie die methodenkritischen Übersichten von Temoshok und Heller (1984) sowie Scherg (1986) zeigen, ist offensichtlich bisher keine prospektive Streßstudie veröffentlicht worden. Die Ergebnisse retrospektiver Untersuchungen sind heterogen. Vom methodischen Ansatz her interessante retrospektive Studien haben Muslin et al. (1969), Greer und Morris (1975) und Schonfield (1975) durchgeführt (vgl. aber Schwarz und Geyer, 1984). Frauen, bei denen Verdacht auf Brustkrebs bestand, wurden vor der Biopsie nach belastenden Lebensereignissen befragt. Zwischen „Streß in den letzten Jahren" und Diagnose (gutartige versus bösartige Geschwulst) wurde kein Zusammenhang gefunden. Diese Ergebnisse sprechen gegen Streßeffekte auf das Wachstum schon vorhandener, aber noch nicht eindeutig diagnostizierter Tumoren (zwischen dem Auftreten der ersten malignen Zelle und der Diagnose eines Mammakarzinoms vergehen viele Jahre; vgl. von Fournier, 1982), lassen aber die Möglichkeit offen, daß Stressoren über mutationsfördernde Effekte (Fischman und Kelly, 1987) das Risiko, an Krebs zu erkranken, erhöhen. In der obengenannten Studie von Muslin et al. waren allerdings „early separation experiences" und Biopsieergebnisse ebenfalls nicht korreliert. Weniger, aber als stärker belastend bewertete „life events" hatten Frauen mit Krebsdiagnose in der Studie von Cooper et al. (1986).

Zu Streßeffekten bei Autoimmunkrankheiten liegen erst wenige tierexperimentelle Arbeiten vor. Neben Diabetesmodellen (Capponi et al., 1980; Carter et al., 1987; Huang et al., 1981) sind hauptsächlich experimentell induzierte Gelenkentzündungen untersucht worden. „Crowding" hatte fördernde (Amkraut et al., 1971) oder hemmende (Sofia, 1980) Effekte auf die Entwicklung einer Adjuvans-Arthritis, einer bei Ratten induzierbaren Krankheit, die unter morphologischen, klinischen, immunologischen und pharmakologischen Aspekten der rheumatoiden **Arthritis** ähnlich ist (vgl. Klosterhalfen, 1987). Divergierende Streßeffekte sind auch bei einem anderen Arthritismodell, der kollageninduzierten Arthritis, beschrieben worden (Rogers et al., 1980, 1983). In einer Serie eigener Experimente haben wir als Stressoren Lärm oder Bewegungsrestriktion verwendet und den Zeitabstand zwischen Streßexposition und Induktion der Adjuvans-Arthritis systematisch variiert (Klosterhalfen, 1987, 1988; Klosterhalfen und Klosterhalfen, 1988b). Wegen methodischer Probleme bei den obengenannten „crowding"-Experimenten und relativ schwacher und inkonsistenter Streßeffekte bei den von uns durchgeführten Experimenten sind wir der

Ansicht, daß Streßeinflüsse für die Entwicklung einer Adjuvans-Arthritis eher von geringer Bedeutung sind.

Bei der Gruppe der Autoimmunkrankheiten beziehen sich die bisher vorliegenden Streßstudien an Patienten überwiegend auf die rheumatoide Arthritis (Synonym: chronische Polyarthritis; Übersicht bei Anderson et al., 1985; Koehler, 1985; Köhler, 1989; Klosterhalfen, 1987; Krüskemper, 1985). In retrospektiv angelegten Untersuchungen stellten z.B. Meyerowitz et al. (1968) und Baker (1982) vor Ausbruch der Erkrankung eine Häufung psychisch belastender Ereignisse fest. Über negative Ergebnisse haben jedoch Lewis-Faning (1950) und Hendrie et al. (1971) berichtet. Erst in den letzten Jahren ist experimentell untersucht worden, ob der Verlauf der Arthritis durch psychologische Interventionen günstig beeinflußt werden kann. Die Ergebnisse waren leider durchweg enttäuschend (Kaplan und Kozin, 1981; Shearn und Fireman, 1985; Strauss et al., 1986).

11.1.4 Persönlichkeitsfaktoren und Immunität

Persönlichkeitsfaktoren und immunologische Parameter

Neben Untersuchungen an psychiatrischen Patienten (vgl. Müller et al., 1987; Solomon, 1981b; Stein et al., 1985) sind in einer Reihe von Arbeiten bei gesunden Probanden Beziehungen zwischen Persönlichkeitsfaktoren und Immunfunktionen untersucht worden. Entsprechende Korrelationen werden meist im Sinne einer Beeinflussung des Immunsystems durch das Nervensystem interpretiert. Umgekehrt läßt sich aber auch spekulieren, daß Monokine und Lymphokine psychotrope Effekte haben (zu neurotropen Effekten von Cytokinen vgl. Abschn. 11.2.9). Natürlich ist auch denkbar, daß dritte (z.B. genetische) Faktoren psychoimmunologische Zusammenhänge herstellen.

Obwohl sich die folgenden Arbeiten in der psychologischen Methodik wesentlich von den psychoimmunologischen Streßstudien unterscheiden, ist inhaltlich die Nähe zum Streßparadigma nicht zu übersehen.

Mit Hilfe des „Thematischen Apperzeptionstests" (Murray, 1943) selektierten McClelland et al. (1980) eine Gruppe von Studenten, denen u.a. überdurchschnittlich starke Machtbedürfnisse zugeschrieben wurden. Im Vergleich zu den restlichen Studenten war die Konzentration von IgA im Speichel bei diesen Studenten erniedrigt. Diese Ergebnisse konnten an Strafgefangenen repliziert werden (McClelland et al., 1982). Jemmott et al. (1983) verglichen in einer ähnlichen Arbeit zwei Gruppen von Studenten, die entweder durch ein „inhibited power motive syndrome" (IPS) oder durch ein „relaxed affiliative syndrome" (RAS, Bedürfnis nach sozialer Nähe und Wärme) charakterisiert waren. Die RAS-Gruppe hatte eine höhere Speichel-IgA-Sekretionsrate als die IPS-Gruppe.

Ursin et al. (1984) korrelierten bei Lehrerinnen Fragebogenergebnisse (u.a. Extraversion, Neurotizismus,

„internal locus of control") mit immunologischen Werten (IgM, IgG, IgA; Komplementfaktoren C3, C4; C1-Inhibitionsfaktor); die Untersuchung wurde ein und zwei Jahre später wiederholt. In allen drei Untersuchungen gab es signifikante negative Korrelationen zwischen Extraversion und C1-Inhibitionsfaktor sowie zwischen „internal locus of control" und IgA; C4 war bei allen Tests und Untersuchungen unkorreliert.

Eine Beziehung zwischen Einsamkeitsgefühlen (UCLA Loneliness Scale) und immunologischen Maßen fanden Kiecolt-Glaser et al. (1984b) bei neu aufgenommenen psychiatrischen Patienten, die weder Psychosen noch Alkohol- oder Drogenprobleme hatten und auch nicht schwachsinnig waren. Patienten mit „high loneliness" (Medianhalbierung) zeigten eine verminderte NK-Zellaktivität sowie schwächere Reaktionen auf PHA; bei einer Stimulation mit PWM traten aber keine Unterschiede auf. Studierende mit „high loneliness" hatten unter Prüfungsstreß geringere Antikörpertiter gegen das Epstein-Barr-Virus als ihre weniger einsamen Kollegen (Glaser et al., 1985). Zwischen Skalen des Minnesota Multiphasic Personality Inventory (MMPI; Hathaway und McKinley, 1967) und der NK-Zellaktivität fanden weder Kiecolt-Glaser et al. (1984a) noch Heisel et al. (1986) hohe Korrelationen. (Kiecolt-Glaser et al. machen keine näheren Angaben; bei Heisel et al. lag der höchste Wert bei rho = −.30.) In einer eigenen Untersuchung (Ernst et al., 1987) erhielten 120 Medizinstudenten die Kurzfassung des Freiburger Persönlichkeitsinventars (FPI-K; Fahrenberg et al., 1978) und den Streßverarbeitungsfragebogen (SVF; Janke et al., 1985). Die Untersuchung wurde eine Woche später an einer zweiten Stichprobe wiederholt. Die Ergebnisse waren im wesentlichen negativ, d.h. es gab praktisch keine reproduzierbaren Korrelationen zwischen FPI- oder SVF-Maßen einerseits und IgA-Werten andererseits.

Aus einer psychoimmunologischen Untersuchung an Patienten mit AIDS oder ARC („AIDS related complex") liegen erste Ergebnisse vor. Die Anzahl von Leukozyten, „Entzündungszellen" und Lymphozyten im peripheren Blut war (bis zu r = 0.38) mit psychologischen Variablen (z.B. Dysphorie, Angst, Hoffnungslosigkeit, Mangel an „hardiness") korreliert (Solomon und Temoshok, 1987). Über die Resultate aus einer zweiten Studie an ARC-Patienten schreiben Temoshok et al. (1988) in ihrem Vortragsmanuskript:

„In the subjects not taking antiviral or immune enhancing medication (N=60), impaired performance on Immediate Visual Memory and Delayed Visual Memory was strongly (r = 0.28 to 0.39) and significantly associated with absolute numbers of B cells, T cells, Cytotoxic Suppressor cells, T Helper cells, Virucidal cells, Cytotoxic cells, and large granular lymphocytes. In addition both the Relational concepts and the Logical grammatical relations sub-tests were significantly associated with overall T cell count, Cytotoxic Suppressor cells numbers, and for Logical grammatical relations only, Cytotoxic cell counts.

In terms of relationships with psychosocial variables, NK cell activity was strongly (r = 0.20 to 0.52) and significantly associated with higher scores on the UCLA Loneliness Scale, the Beck Hopelessness and Depression Scales, and all 6 subscales from the POMS (Profile of Mood States)."

Es ist unklar, wie diese ungewöhnlichen Ergebnisse zu interpretieren sind. Denkbar erscheint, daß es durch das Fortschreiten der Erkrankung einerseits zu psychischen Störungen und andererseits zu opportunistischen Infektionen kommt, die erhöhte Zellzahlen bewirken. Hinsichtlich der NK-Zellaktivität schreiben Temoshok et al. (1988): „One possibility to be evaluated in this sample is whether more p24 antigen might stimulate more NK cell activity, where p24 is a negative prognostic indicator of HIV progression and associated with distress directly or indirectly."

Persönlichkeitsfaktoren bei Infektionskrankheiten, Krebs und Arthritis

Einige prospektive Untersuchungen weisen auf die Bedeutung von Persönlichkeitsfaktoren für die Entwicklung von **Infektionskrankheiten** hin. Bei jungen Soldaten, die im Hinblick auf die angestrebte Militärlaufbahn ehrgeizig waren oder ehrgeizige Väter hatten und bereits Antikörper gegen das Epstein-Barr-Virus besaßen, war das Risiko, an Pfeiffer-Drüsenfieber (Mononucleosis infectiosa) zu erkranken, erhöht (Kasl et al., 1979). Canter (1972), der bei freiwilligen Versuchsteilnehmern eine fiebrige Infektion induzierte, fand bei Probanden, die zuvor als psychologisch vulnerabel eingestuft worden waren, vermehrt klinische Symptome. Totman und Kiff (1979) verwendeten Rhinoviren bei freiwilligen Versuchspersonen. In ihrer Studie zeigten Introvertierte stärkere Erkältungssymptome; Neurotizismus war in diesem Zusammenhang hingegen kein Prädiktor für die Stärke des Schnupfens.

Die Bedeutung von Persönlichkeitsfaktoren als Risikofaktoren für **Krebserkrankungen** wird seit etwa 20 Jahren (Bahnson und Bahnson, 1969; Stavraky et al., 1968) wissenschaftlich diskutiert (Übersicht bei Bahnson, 1980, 1981; Cooper, 1988; Fox, 1978, 1981, 1983; Levy, 1985; Scherg, 1986; Temoshok, 1987; Temoshok und Heller, 1984). In prospektiven Studien waren die folgenden Persönlichkeitsmerkmale mit erhöhten Krebsraten assoziiert: „substable personality" (Hagnell, 1966), ein geringer Grad an „closeness-to-parents" (Thomas, 1976) und „high scores for rationality and antiemotionality" (Grossarth-Maticek et al., 1985). Diese Korrelationen sind jedoch insofern mit Zurückhaltung zu beurteilen, als jeweils mehrere Variablen ins Rennen geschickt wurden. Umgekehrt sollte auch bei den negativen Ergebnissen von Cassileth et al. (1985) berücksichtigt werden, daß es sich um Patienten mit statistisch nur noch geringer Lebenserwartung handelte und nur zwei („hopelessness" und „adjustment to diagnosis") der sieben getesteten Variablen aus der psychosomatischen Krebsliteratur übernommen wurden. Graves et al. (1986) werteten Rorschach-Antworten von Medizinstudenten hinsichtlich der Art der beschriebenen sozialen Interaktionen aus. Nach 19–35 Jahren trat bei Personen mit einem „avoidant response pattern (distant, withdrawn, no scorable responses)" häufiger Krebs auf als bei Personen, die ein „flexible (well-adjusted) pattern" aufwiesen (4,1 : 1). Noch kaum zu beurteilen sind die

positiven Ergebnisse der kürzlich in vorläufiger Form publizierten Heidelberger Studie (Grossarth-Maticek, 1988; vgl. auch Eysenck, 1987). Leider sind bei dieser Studie die psychologischen Prädiktoren bisher nicht veröffentlicht worden; sie wurden erst nach dem Tod der in die Auswertung übernommenen Probanden bei einem zunächst nicht involvierten Kollegen (N. Bischoff, Zürich) hinterlegt.

In einigen „retrospektiven" Untersuchungen sind psychologische Datensätze, mit deren Hilfe z.B. Herzinfarktraten vorhergesagt werden sollten, im Hinblick auf Krebserkrankungen reanalysiert worden. Bei Probanden, die später an Krebs erkrankten, waren die Depressionswerte im MMPI erniedrigt (Dattore et al., 1980). In der Arbeit von Shekelle et al. (1981) waren jedoch relativ hohe MMPI-Depressionswerte mit einer Verdoppelung des Risikos verbunden, innerhalb der folgenden 17 Jahre an Krebs zu sterben (vgl. auch Persky et al., 1987). Diese Ergebnisse werden durch die kürzlich erschienene Arbeit von Kaplan und Reynolds (1988) „vervollständigt", in der keine Assoziation zwischen Depression und Krebsmorbidität oder -mortalität festgestellt wurde.

Shaffer et al. (1987) definierten 14 Persönlichkeitsmerkmale und berichten:

„The group characterized as ‚loners', who may well have suppressed their emotions, had the most unfavorable survival curve and was 16 times more likely to develop cancer than was the group characterized by acting out and emotional expression" (S. 41).

Fox et al. (1987) fanden bei Typ-A-Personen (ermittelt durch ein strukturiertes Interview) eine im Vergleich zum Typ B erhöhte Krebsmortalität von ca. 1,5:1. In einer prospektiven Studie stellten Cooper et al. (1986) jedoch bei Frauen, bei denen durch eine Biopsie Brustkrebs festgestellt wurde, ein weniger stark ausgeprägtes Typ-A-„Verhalten" (Bortner und Rosenman, 1967) fest.

In einer dritten Gruppe von Längsschnittuntersuchungen ist versucht worden, psychologische Prädiktoren des Krankheitsverlaufs zu finden. Als Verlaufskriterium diente in diesen „semiprospektiven" Studien meist das zeitliche Intervall zwischen der Krebsdiagnose und dem Tod der Patienten. Die „Prognose" war bei latenter Feindseligkeit ohne Verlust der emotionalen Kontrolle (Stavraky et al., 1968) bzw. bei kämpferischer Einstellung oder Leugnung (Greer et al., 1979) relativ günstig, bei Angst, Depression, Schuldgefühlen und Feindseligkeit (Derogatis et al., 1979), wenig Aggressivität und eher starker emotionaler Kontrolliertheit (Mastrovito et al., 1979) sowie bei wenig Auseinandersetzung mit der Krankheit (Rogentine et al., 1979) relativ ungünstig. Bei Frauen mit Brustkrebs waren „expressive activities at home or away from home, extroversion, low anger, low cognitive disturbance" Prädiktoren für relativ günstige Verläufe (Hislop et al., 1987).

Es muß gegenwärtig konstatiert werden, daß es trotz vieler Untersuchungen noch kaum gelungen ist, Persönlichkeitsmerkmale zu definieren, die in reproduzierbarer Weise zur Vorhersage des Auftretens und des Verlaufs von Krebserkrankungen beitragen. Entsprechende Bemühungen konzentrieren sich gegenwärtig besonders auf den „fighting spirit" bzw. den sogenannten Typ C („cancer-prone personality"; Greer und Watson, 1985; Morris und Greer, 1980; Temoshok und Fox, 1984).

Innerhalb der Gruppe der Autoimmunkrankheiten hat die rheumatoide **Arthritis** (RA) das meiste Interesse gefunden. Johnson et al. (1947) veröffentlichten eine Hypothese, der das Freudsche Konzept der Konversionshysterie (vgl. Freud, 1952) zugrunde liegt:

„Wir nehmen an, daß muskuläre Verspannungen und gesteigerter Muskeltonus, die durch verdrängte feindselige Antriebe verursacht sind, unter gewissen Bedingungen einen arthritischen Anfall auslösen können" (Alexander, 1951, S. 160 f.).

Ausgehend von Alexander, ist eine umfangreiche Literatur entstanden, in der u.a. mit testpsychologischen Methoden versucht wurde, die Bedeutung von Persönlichkeitsvariablen für Auftreten und Verlauf der RA zu untersuchen. So herrscht denn auch kein Mangel an Behauptungen über psychologische Merkmale von Rheumatikern (vgl. Solomon, 1981a). Nach einer Auswertung durch Moos und Solomon (1965) wurden RA-Patienten in den von Moos (1964) analysierten 80 Arbeiten insgesamt 140(!) „diskriminative" Persönlichkeitseigenschaften zugesprochen. Besonders häufig wurde das MMPI eingesetzt. Regelmäßig wurden erhöhte Depressionswerte (mindestens eine Standardabweichung über der Norm), häufig aber auch erhöhte Werte bei den Skalen „Hypochondrie" und „Hysterie" (z.B. Liang et al., 1984) oder „soziale Introversion" (z.B. Krüskemper und Zeidler, 1975) festgestellt. Wie z.B. Zeidler et al. (1978) diskutieren, kann jedoch nicht ausgeschlossen werden, daß solche Werte lediglich Folgen der Erkrankung sind. Spergel et al. (1978) halten die häufig postulierte „RA-Persönlichkeit" für einen psychodiagnostischen Mythos. Tatsächlich finden sich die oben angegebenen MMPI-Ergebnisse oder Hinweise auf „gehemmte Aggressivität" z.B. auch bei Patienten mit chronischem Lumbalsyndrom (Freeman et al., 1976; Kügler, 1980).

Eine psychoimmunologische Hypothese scheint in bezug auf die RA erstmals Moos (1964) formuliert zu haben:

„It is also possible that there is partial central nervous system control over various autoimmunological factors in the blood, such as the rheumatoid factors. These might, then, be affected by personality variables. This, of course, is a neglected general area of research of importance in all psychosomatic diseases" (Moos, 1964, S. 51).

Diese Hypothese wird inzwischen durch Arbeiten von Crown und Crown (1973) und Vollhardt et al. (1982) gestützt, in denen seropositive gegenüber seronegativen Patient(inn)en erhöhte Neurotizismuswerte hatten. Gardiner (1980) fand bei den genannten Untergruppen jedoch ähnliche Neurotizismuswerte. In einer kürzlich erschienenen Arbeit (Pow, 1987) war der Rheumafaktor mit „negative style of thinking" und „believe in powerful external forces" assoziiert.

Die Bedeutung psychologischer Faktoren für den Verlauf der RA haben McFarlane et al. (1987) untersucht. Überraschenderweise waren Depression und Neurotizismus Prädiktoren für einen relativ günstigen Verlauf (Beobachtungszeitraum: 3 Jahre). Nach außen gerichtete Feindseligkeit und die Leugnung der emotionalen Implikationen der Krankheit gingen mit schlechteren Verläufen einher.

Prospektive Untersuchungen, die von Gesunden ausgehen, liegen bisher nicht vor. Insgesamt muß festgestellt werden, daß sowohl die „psychomechanische" Hypothese Alexanders als auch die psychoimmunologische Hypothese von Moos nach etwa vier bzw. zwei Jahrzehnten wenig von ihrem hypothetischen Charakter eingebüßt haben.

11.2 Konditionierte Immunmodulation: ein Beispiel für neuroimmunologische Interaktionen

11.2.1 Konditionierte Immundepression

Vor ca. 15 Jahren formulierte Robert Ader (1974) in einem „letter to the editor" eine Hypothese, die ihn bzw. die Rochester-Gruppe sowie viele Forscher in unabhängigen Labors zu einer Überprüfung und Präzisierung veranlaßt hat. Ader hatte beobachtet, daß Ratten in einem Experiment zur gelernten Geschmacksaversion unerwartet starben, als sie eine Saccharinlösung, den konditionierten Stimulus (CS), statt Wasser zu trinken erhielten. Vorher war die Saccharinlösung (Sac) nicht mit LiCl, dem beim Geschmacksaversionslernen am häufigsten verwendeten unkonditionierten Stimulus (US), sondern mit Cyclophosphamid (Endoxan®) gepaart worden (vgl. Kap. 7). Je nach Dosierung bedingt Cyclophosphamid (CY) nicht nur Übelkeit und Erbrechen, was wahrscheinlich die Basis für das Geschmacksaversionslernen ist (vgl. Klosterhalfen und Klosterhalfen, 1985 b), sondern es wirkt auch stark immunsuppressiv. Daher spekulierte Ader, durch die Paarung von Sac und CY seien nicht nur die übelkeitserregenden Effekte von CY konditioniert worden, sondern auch die immunsupprimierende Wirkung. Infolgedessen hätte die Anfälligkeit der Ratten gegenüber Pathogenen sich stark erhöht, so daß letztere bei einigen Tieren letal wirkten.

In nachfolgenden Experimenten der Rochester-Gruppe wurde geprüft, ob tatsächlich die immunsupprimierende Wirkung von CY konditioniert werden kann. Es lag nahe, die Hypothese über die Konditionierbarkeit pharmakologisch induzierter Immundepression zu testen, indem ein immunologischer Parameter gemessen wurde. In ihrem ersten Experiment wählten Ader und Cohen (1975) die Menge von Antikörpern gegen Schaferythrozyten (Antigen) als abhängige Variable. Das von den Autoren verwendete Design wurde von vielen Forschern auf dem Gebiet der konditionierten Immunmodulation im wesentlichen übernommen und wird daher kurz dargestellt.

Tabelle 11–3. Design zur Überprüfung konditionierter immunpharmakologischer Effekte nach Ader und Cohen, 1975 (modifiziert)

	Akquisition		Test		
	Tag 0		Tag 3		Tag 9
Gruppe	CS	US	Antigen	CS	Messung
CS	Sac	CY	SRBC	Sac	Anti-
CSo	Sac	CY	SRBC	H$_2$O	körper-
NC	H$_2$O	CY	SRBC	Sac	titer

CS: konditionierter Stimulus; US: unkonditionierter Stimulus; Sac: Saccharinlösung; CY: Cyclophosphamid; SRBC: Schafserythrozyten (sheep red blood cells)

Insgesamt wurden sechs Gruppen von Ratten untersucht; für die Frage der Konditionierbarkeit immunpharmakologischer Effekte sind jedoch nur die folgenden drei kritisch: Gruppe CS, für die nach einmaliger Paarung von Sac und CY der CS unverstärkt präsentiert wird; Gruppe CSo, für die Sac und CY einmal gepaart werden, der CS aber nicht wieder präsentiert wird, und Gruppe NC, für die Sac und CY nicht gepaart, d.h. zeitlich getrennt präsentiert werden (vgl. Tab. 11–3).

Prozedural unterscheiden die Gruppen sich weder durch die CY-Injektion (hier 50 mg/kg) noch durch die Stimulation mit Schafserythrozyten, sondern durch die Paarung von Sac und CY (Gruppen CS und CSo versus NC) bzw. durch die unverstärkte Sac-Darbietung (Gruppen CS und NC versus CSo).

Es sei angemerkt, daß es methodisch sauberer und ökonomischer ist, Konditionierungseffekte zu überprüfen, indem nur zwei Gruppen miteinander verglichen werden. Diese sollten sich nicht durch die Häufigkeit von CS- und US-Präsentationen unterscheiden, sondern durch kontingente bzw. nicht kontingente Verabreichung der beiden Stimuli (wegen weiterer Kritikpunkte vgl. Klosterhalfen und Klosterhalfen, 1985, 1989).

Am Tag 3 zeigten die Tiere der Gruppe CS eine ausgeprägte Geschmacksaversion, d.h. die stark wasserdeprivierten Tiere vermieden es, Sac zu trinken, während die Tiere der Gruppen CSo und NC die für flüssigkeitsdeprivierte Ratten übliche Menge an Wasser bzw. Sac konsumierten. Am Tag 9 hatte die Gruppe CS einen signifikant niedrigeren Antikörpertiter als die beiden Gruppen CSo und NC.

Ader und Cohen (1975) interpretierten ihr Ergebnis, nämlich einen signifikant geringeren Antikörpertiter in der Gruppe CS als in den Gruppen CSo und NC, als Evidenz für eine „behavioral konditionierte Immunsuppression".

Mit „behavioral" wird hier nicht auf eine Verhaltensverstärker-Kontingenz rekurriert – obwohl natürlich die Vermeidung von Sac nach der Paarung mit CY eine instrumentelle Komponente enthält –, sondern auf Pawlowsches oder klassisches Konditionieren; aufgrund von Verhaltensänderungen wird auf den entsprechenden Lernvorgang geschlossen. Die auch in der traditionellen Immunologie verwendete Methode des Konditionierens hat dagegen mit einem Lernvorgang nichts zu tun.

Bereits in diesem ersten Experiment stellte sich die Frage, ob die relative Immunsuppression in der Gruppe CS nicht auf unspezifische Streßeffekte durch die gelernte Geschmacksaversion zurückzuführen ist. Ader und Cohen (1975, 1985) erschien dies jedoch nicht plausibel, da die Gruppen sich weder nach der unverstärkten CS-Präsentation in ihrem Plasma-Corticosteron-Spiegel (vgl. Kap. 10) unterschieden, noch in einem zweiten Experiment in ihren Antikörpertitern, nachdem die Geschmacksaversion gegenüber Sac mit LiCl (von dem anzunehmen ist, daß es in der verwendeten Dosierung keine immunologischen Effekte hat) anstatt mit CY induziert worden war. CS-bedingte verminderte Antikörpertiter gegen Schafserythrozyten nach Sac-CY-Paarung wurden auch von Rogers et al. (1976) und Wayner et al. (1978) gemessen. Daß Alterationen immunologischer Reaktionen nach CS-Darbietung nicht auf humorale Effekte eingeschränkt werden müssen, legen Experimente nahe, in denen zellmediierte Immunreaktionen durch ähnliche Konditionierungsprozeduren mit CY als US supprimiert (Ader und Cohen, 1981, 1985; Kusnecov et al., 1988) und auch stimuliert (Bovbjerg et al., 1987a) werden konnten. In dem vielzitierten Experiment von Ader und Cohen (1982) schließlich konnte die Überlebenszeit von Mäusen mit systemischem Lupus erythematodes dadurch verlängert werden, daß bei konditionierten Tieren zwischen Sac-CY-Paarungen die CS (Sac) unverstärkt gegeben wurde, während für nicht-konditionierte Tiere Sac und CY mit gleicher Häufigkeit, aber immer explizit ungepaart appliziert worden waren.

Die Resultate scheinen zu belegen, daß die Reaktivität des Immunsystems durch Lernvorgänge, die sich im ZNS abspielen, modifiziert werden kann. Wie der folgende Abschnitt zeigt, ist diese Sichtweise keineswegs neu.

11.2.2   Konditionierte Immunstimulation

In der psychosomatischen Literatur wird häufig eine Fallbeschreibung von MacKenzie (1886) zitiert, in der mit einer künstlichen Rose bei einer gegen Rosen allergischen Frau ein Heuschnupfenanfall provoziert wurde. Eine echte Rose, die sich die Patientin einige Tage später demonstrativ unter die Nase hielt, löste merkwürdigerweise aber keine allergische Reaktion aus. 1926 berichteten Metal'nikov und Chorine, daß nach wiederholter Paarung thermischer und taktiler Reize (CS) mit intraperitonealen Injektionen von Staphylokokken-Infiltrat (US) beim Meerschweinchen durch den CS allein eine unspezifische immunologische Reaktion, nämlich eine Erhöhung der Anzahl polymorphkerniger Leukozyten im Peritoneum, evoziert wurde. Im Gegensatz zu den weiter oben beschriebenen Experimenten wird hier also nicht eine immunpharmakologische Substanz, sondern ein Antigen als US angesehen. Aus heutiger Sicht sind die

Versuche Metal'nikovs und auch die vielen Arbeiten zur konditionierten Immunstimulation, die in den 30er und 50er Jahren in der Sowjetunion erschienen sind (vgl. Ader, 1981b), unzureichend kontrolliert und dokumentiert.

Neue Experimente zur konditionierten Immunstimulation (in denen die antigene Stimulation als US dient) scheint es nur wenige zu geben. Aus einer niederländischen Laborstudie gibt es Hinweise, daß ein CS bei Patienten asthmatische Symptome induzieren kann (Dekker et al., 1957). Interessant ist eine Arbeit von Russell et al. (1984), die dafür spricht, daß sich bei Meerschweinchen nach Paarung von Geruch (CS) mit einem die Histaminausschüttung stimulierenden Antigen (US) ein erhöhter Histaminspiegel im Plasma als konditionierte Reaktion (CR) hervorrufen läßt. In einem Replikationsversuch ergab sich, daß nur „gestreßte" Tiere einen erhöhten Histaminspiegel nach CS-Darbietung zeigten (Peeke et al., 1987). Häufig – aber unkritisch – zitiert wird noch ein Transplantationsexperiment von Gorczynski et al. (1982), bei dem die Transplantationsprozedur als CS und das Transplantat als US definiert wurden. Nach dreimaliger Paarung führte eine Scheintransplantation bei einigen der Versuchstiere zu einer Erhöhung der Anzahl von Vorläufern zytotoxischer T-Lymphozyten. Konditionierte und nicht-konditionierte Tiere wurden jedoch immunologisch unterschiedlich behandelt; bei der Konditionierungsgruppe wurde Haut von Mäusen eines anderen Stamms transplantiert, bei den Kontrolltieren jedoch eine bestimmte Menge allogener Lymphozyten injiziert.

11.2.3 Implikationen konditionierter Immunmodulation

Prinzipiell impliziert sowohl die Konditionierung mit Antigenen als US als auch die Konditionierung immunpharmakologischer Effekte, daß das ZNS Signale aus dem Immunsystem empfangen kann und das ZNS wieder auf das Immunsystem zurückwirkt. Der Grund, weshalb zunehmend mehr Labors Konditionierungseffekte auf immunologische Parameter untersuchen, liegt wohl u.a. darin, daß die klassische Konditionierung ein elegantes Modell ist, um neuroimmunologische Interaktionen, d.h. bidirektionale oder „feedback"-Mechanismen zu untersuchen (vgl. auch Ferstl und Müller-Ruchholz, 1987; Klosterhalfen und Klosterhalfen, 1987). Trotz vielfältiger Bemühungen war es bisher jedoch schwierig auszuschließen, daß immunologische Alterationen nach CS-Darbietung nicht das Resultat von Streßeffekten sind und damit lediglich auf einem „feedforeward"-Mechanismus (Bovbjerg et al., 1982) beruhen. Dies liegt daran, daß in den bisher veröffentlichten Experimenten zur konditionierten Immunmodulation (vgl. Ader und Cohen, 1985) die verwendeten Pharmaka bzw. verschiedene Aspekte der Konditionierungsprozedur als Stressor aufgefaßt werden können (z.B. Veldhuis und DeWied, 1985). Schon auf der semantischen Ebene wird dies deutlich, denn CY, das in weitaus den mei-

sten der entsprechenden Experimente als US diente, induziert reliabel eine Geschmacks*aversion*, wenn eine neu schmeckende Substanz, z.B. Sac, mit einer CY-Injektion gepaart wird. Auf physiologischer Ebene bewirkt eine CY-Injektion bei Ratten einen dramatischen Anstieg des Plasma-Corticosterons (Ader, 1976), der als Streßindikator angesehen wird und der eine Immundepression mediieren könnte (Riley, 1981; Riley et al., 1981). Daß die Erhöhung des Plasma-Corticosteron-Spiegels konditionierbar zu sein scheint (Ader, 1976; vgl. auch Ader und Cohen, 1986), steht nicht im Widerspruch zu der These, die Sac-evozierte Immundepression nach Sac-CY-Paarung sei eine Streßreaktion (Cunningham, 1985; Kelley und Dantzer, 1986), denn der CS muß nicht immunpharmakologische Effekte signalisieren, um eine ACTH-Ausschüttung und damit die adrenerge Reaktion zu evozieren: LiCl hat den gleichen unspezifischen Streßeffekt (Ader, 1976; Smotherman, 1985). Außerdem steht mit der „Streßhypothese" in Einklang, daß die Präsentation von CS's, die vorher mit Elektroschocks gepaart worden waren, in einer Depression von Immunfunktionen (Lymphozytenproliferationsrate nach Stimulation mit Con A bzw. PHA) resultieren kann (Lysle et al., 1988). Die „Streßhypothese" wird schließlich direkt durch den Befund gestützt, daß eine mit Hilfe von LiCl induzierte Geschmacksaversion hinreichend war, eine zellmediierte Immunreaktion bei Mäusen zu supprimieren (Kelley et al., 1984, 1985).

11.2.4 Ein alternativer Erklärungsansatz: die „Streßhypothese"

Wie die im folgenden referierten Experimente deutlich machen, gibt es zur konditionierten Immunmodulation eine ganze Reihe von Arbeiten, in denen versucht wurde, die generelle Gültigkeit der „Streßhypothese" zu überprüfen.

Hier wird zwar nicht versucht, „Streß" zu definieren (vgl. Kap. 8); es sollte jedoch nicht der Eindruck entstehen, daß unter „Streß" lediglich die adrenerge Reaktion auf Stressoren verstanden wird. Vielmehr schließen wir uns der Meinung der in diesem Zusammenhang zitierten Autoren an, die davon ausgehen, daß ein erhöhter Corticosteron-Spiegel ein Streßindikator ist. Der Umkehrschluß ist unseres Erachtens jedoch nicht möglich: weder ist ein Corticosteron-Spiegel innerhalb des Normbereichs gleichzusetzen mit „kein Streß" – denn andere (neuro)endokrine Reaktionen können ebenfalls als Streßindikator angesehen werden und unabhängig von Corticosteron variieren (vgl. Kap. 10) –, noch ist eine Immunmodulation von der adrenergen Reaktion abhängig (Keller et al., 1983). Daß im folgenden primär die adrenerge Reaktion als potentieller Mediator bei der Konditionierung immunpharmakologischer Effekte diskutiert wird, liegt daran, daß das Nebennierenrindenhormon bzw. ACTH das in den einschlägigen Experimenten bisher einzige gemessene Hormon ist.

Wie schon erwähnt, wurde eine humorale Immunreaktion von einer Geschmacksaversion gegenüber Sac, das mit LiCl gepaart worden war, nicht modifiziert, noch hatte exogenes Corticosteron anstelle der

CS-Exposition einen immunmodulierenden Effekt (Ader et al., 1979). Auch war eine LiCl-induzierte Geschmacksaversion (im Gegensatz zu CY) nicht effektiv, die Anzahl der Leukozyten im peripheren Blut bei Ratten zu reduzieren (Klosterhalfen und Klosterhalfen, 1987b). Bovbjerg et al. (1984) versuchten, eine konditionierte Geschmacksaversion von einer Immundepression zu dissoziieren und fanden eine CS-bedingte verminderte „graft-versus-host"-Reaktion, nachdem die CY-induzierte Geschmacksaversion gegenüber dem CS extingiert worden war. Klosterhalfen und Klosterhalfen (1987a) machten sich Ergebnisse von Smotherman (1985) zunutze, der gezeigt hat, daß Corticosteron nicht ansteigt, wenn Ratten nicht „gezwungen" werden, Sac (CS) zu trinken, d. h. wenn sie nicht wasserdepriviert sind, bzw. wenn konditionierte Tiere Wasser statt des CS erhalten (Ader, 1976). Sie verglichen daher die Effekte von CS-Expositionen in einem „one-bottle" (die Flasche enthält den CS) versus einem „two-bottle"-Test (eine Flasche enthält den CS, die andere Wasser) und fanden unabhängig von der Art der CS-Exposition eine geringere Anzahl von Leukozyten bei konditionierten im Vergleich zu nicht-konditionierten Tieren. Außer Ader und Cohen (1975) und Kelley et al. (1985) haben nur King et al. (1987) Corticosteron in einem Experiment zur konditionierten Immunmodulation direkt gemessen. Sie paarten Sac mit Antilymphozytenserum (ALS) und fanden nach CS-Präsentation keinen signifikanten Anstieg im Plasma-Corticosteron. Da konditionierte relativ zu unkonditionierten Tieren aber eine Immundepression zeigten, schlossen die Autoren, zwischen dem Corticosteron-Spiegel und dem konditionierten immundepressiven Effekt bestehe keine Beziehung. Insgesamt sprechen zwar diese Resultate dagegen, daß CS-bedingte immunologische Alterationen grundsätzlich auf die adrenerge Reaktion zurückzuführen sind. Wie bereits angemerkt, ist jedoch denkbar, daß beim Geschmacksaversionslernen auch andere „Streßhormone" freigesetzt werden, die, wie weiter unten diskutiert, für die Mediierung der Konditionierung immunpharmakologischer Effekte ebenfalls in Frage kommen.

11.2.5 Konditionierte immunpharmakologische Effekte ohne gelernte Geschmacksaversion

Die Tatsache, daß Tiere einen für sie neuen Geschmack vermeiden, nachdem dieser Geschmack (z. B. Sac) mit einem immunmodulierenden Pharmakon (vgl. auch Husband et al., 1986; Neveu et al., 1987) oder auch Antigen (Markovic et al., 1988) gepaart wurde, weist unserer Meinung nach darauf hin, daß die als US verwendeten Substanzen als „aversiv" bzw. als Stressor wahrgenommen werden. Mit dieser Annahme steht in Einklang, daß sogar ALS, das offenbar keine sensorischen Nebenwirkungen hat (Kusnecov et al., 1983), eine Erhöhung des Plasma-Corticosteron-Spiegels provoziert (King et al., 1987).

Auch wenn dies kein Beweis im strengen Sinne ist, könnte das **Fehlen** einer Geschmacksaversion darauf

hindeuten, daß die Applikation des entsprechenden US bzw. dessen Wirkung nicht als „aversiv" oder „stressend" perzipiert wird (vgl. auch Goudie, 1987). Nach Ader et al. (1987b) weist das Fehlen einer Geschmacksaversion unter bestimmten Bedingungen sogar darauf hin, daß der Geschmack mit **positiven** Effekten des US assoziiert wurde, dann nämlich, wenn z. B. CY therapeutisch wirkt, wie etwa bei Autoimmunkrankheiten. Die „Aversivität" ist u. a. sicher eine Frage der Dosis. Für CY ist auf der Basis der Dosierung eine Vorhersage über den Grad der „Aversivität", gemessen an der Stärke einer konditionierten Geschmacksaversion oder deren Extinktionsresistenz, besonders schwierig, was damit zusammenhängen könnte, daß CY viele unterschiedliche physiologische Effekte hat (Hengst und Kempf, 1984; Hill, 1975). Als immunmodulierendes Pharmakon wirkt es nicht nur supprimierend, sondern in geringer Dosierung immunstimulierend (Turk und Parker, 1982). Auf den ersten Blick entgegen den bekannten Gesetzmäßigkeiten des klassischen Konditionierens (vgl. Kap. 7) und des Geschmacksaversionslernens (vgl. z. B. Klosterhalfen und Klosterhalfen, 1985b) müßte der Annahme Aders et al. (1987b) zufolge CY in einer **geringen** Dosierung (wegen seiner immunstimulierenden Effekte) bei Tieren mit einer Autoimmunkrankheit ein potenter US für eine zu lernende Geschmacksaversion sein, während CY in mäßig hoher Dosierung bei solchen Tieren (wegen seiner therapeutischen Effekte) keine Geschmacksaversion induzieren sollte.

Diese Hypothese ist sehr interessant und erinnert an das „Homöostase-Konzept" (Cannon, 1939): Tiere scheinen z. B. lernen zu können, ihre Nahrungsaufnahme entsprechend ihrem „milieu interne" zu modifizieren (Garcia et al., 1974; Rozin und Kalat, 1971); für immunologische Alterationen wurde entsprechendes bisher jedoch nicht demonstriert. Ader et al. (1987b; vgl. auch Grota et al., 1987) sind der Hypothese nachgegangen und verglichen den Sac-Konsum von Mäusen mit (Lupus) und ohne genetisch bedingter Autoimmunerkrankung nach Paarung von Sac mit unterschiedlich hohen CY-Dosierungen. Bei der niedrigen Dosierung ergaben sich jedoch keinerlei Effekte: weder unterschieden die genetisch unterschiedlichen Stämme sich in ihrem Sac-Konsum, noch lernten die nicht-kranken Mäuse eine Geschmacksaversion (obwohl bei einem anderen Mäusestamm CY in dieser Dosierung zur Induktion einer deutlichen Geschmacksaversion hinreichend war; vgl. Bovbjerg et al., 1987a). In hoher Dosierung war CY offenbar für beide Stämme aversiv; beide vermieden Sac. Nur bei einer mittleren Dosierung zeigten die nicht-kranken Mäuse eine Aversion gegenüber Sac, nicht aber die erkrankten. Ader et al. (1987b) konzidieren, wegen der Konfundierung von genetischen und Verhaltensdifferenzen sei es schwierig, die Daten eindeutig zu interpretieren. Nichtsdestoweniger erscheint die Hypothese einer weiteren Überprüfung wert; die Konfundierung ließe sich leicht vermeiden, indem kranke Tiere differentiell konditioniert werden, um dann die Effekte vorher verstärkter bzw. explizit nicht-verstärkter Geschmacksreize in unabhängigen Gruppen zu untersuchen.

Entgegen der Annahme Aders et al. (1987b) verhalten sich Ratten mit Adjuvans-Arthritis, die auch als Autoimmunkrankheit angesehen wird (Chang et al., 1980), wenn CY als US verwendet wird. Sie entwikkeln zwar nach einer niedrigen und krankheitsverstärkenden Dosis von CY eine ausgeprägte Geschmacksaversion (Klosterhalfen und Klosterhalfen, 1985a); nach einer höheren, krankheitssupprimierenden Dosis ist die Differenz bezüglich des CS-Konsums zwischen konditionierten und nicht-konditionierten arthritischen Ratten jedoch noch deutlicher (Klosterhalfen und Klosterhalfen, 1983). Mit CY scheint es – zumindest bei Ratten (und möglicherweise auch beim Menschen; Bernstein, 1978) – besonders schwierig zu sein, **keine** Geschmacksaversion auszulösen. Solange eine Geschmacksaversion vorhanden ist, bleibt aber, wie oben aufgeführt, die Möglichkeit offen, eine Immunmodulation nach CS-Präsentation auf Streßeffekte bzw. lediglich auf einen „feedforward"-Mechanismus zurückzuführen. Aus dieser Perspektive ist es verwunderlich, daß manche Autoren (Cohen et al., 1979; Ghanta et al., 1987a; Jenkins et al., 1983) mit Hilfe von LiCl die Aversivität des CS verstärken. Abgesehen von der Interpretationsmöglichkeit im Sinne der „Streßhypothese" hat ein aversiver CS auch prozedural den Nachteil, daß das Tier durch seine Vermeidungsreaktion verhindert, dem CS lange exponiert zu werden; die von Bovbjerg et al. (1987b) gefundene negative Korrelation zwischen der Stärke einer konditionierten Geschmacksaversion und Immundepression nach CS-Exposition ist somit nicht überraschend. Gorczynski et al. (1984) fanden allerdings keine Korrelation zwischen dem Ausmaß einer Geschmacksaversion und dem einer konditionierten Immundepression.

Offenbar wäre es wünschenswert, ein Pharmakon mit definierter immunologischer Wirkung als US einzusetzen, das unabhängig vom Immunstatus keine Geschmacksaversion induziert. In eigenen Pilotstudien vermieden Ratten jedoch nach ein oder zwei Paarungen eines für die Tiere neuen Geschmacks (CS) mit einer von unterschiedlichen Dosen von beispielsweise Azathioprin, Levamisol (vgl. auch Husband et al., 1986) oder CY in Kombination mit Dexamethason (vgl. Revusky, 1985) den CS. Ob Poly I:C (das die Aktivität von NK-Zellen steigert) auch eine Geschmacksaversion induziert, ist unklar, da in den entsprechenden Experimenten keine Trinkdaten erhoben wurden (Dyck et al., 1986, 1987, 1989; Ghanta et al., 1987a). Cyclosporin (CyS) scheint aber die oben gewünschten Kriterien – zumindest unter bestimmten Bedingungen – zu erfüllen: wenn immunologisch nicht-stimulierten Ratten wiederholt eine geringe Menge CyS injiziert wird, trinken sie weiterhin von einer mit Cyclamat gesüßten Trinklösung, die ihnen unmittelbar vor jeder Injektion angeboten wurde (Hampel, 1986). CyS, das in zunehmendem Maße Azathioprin und andere Immunsuppressiva bei Organtransplantationen ablöst und auch bei der Behandlung von Autoimmunkrankheiten eingesetzt

wird, ist in hoher Dosierung toxisch (Magnus et al., 1985) und induziert dann auch eine Geschmacksaversion (Neveu et al., 1987); in niedriger Dosierung scheint es jedoch vornehmlich auf T-Lymphozyten zu wirken (wobei wahrscheinlich die Expression von Rezeptoren für Interleukin 2 und/oder dessen Produktion inhibiert wird; Shevach, 1985; White, 1982), aber keine (nicht-immunologischen) Nebenwirkungen zu haben. CyS supprimiert u. a. auch die Entwicklung von Adjuvans-Arthritis, und zwar für den Zeitraum, in dem das Pharmakon gegeben wird (Borel et al., 1976; Hampel, 1986; Kaibara et al., 1984). Wir untersuchten daher, ob die therapeutischen Effekte, die CyS auf die Entwicklung der experimentellen Autoimmunkrankheit hat, sich konditionieren lassen, ohne daß dabei eine Geschmacksaversion auftritt (Klosterhalfen und Klosterhalfen, 1988a). Dazu wurden arthritische Ratten differentiell konditioniert, indem für alle Tiere ein bestimmter Geschmack (süß oder sauer) siebenmal mit einer CyS-Injektion (US) gepaart wurde (CS$^+$), während der andere Geschmack (süß oder sauer) nie mit dem US gepaart wurde (CS$^-$). Danach erhielt die Hälfte der Tiere (Gruppe CS$^+$) unverstärkte CS$^+$-, die andere Hälfte (Gruppe CS$^-$) unverstärkte CS$^-$-Präsentationen. Es zeigte sich, daß gegenüber der Gruppe CS$^-$ die Gruppe CS$^+$ eine weniger stark ausgeprägte klinische Symptomatik der Arthritis (Pfotenschwellung) hatte. Bezüglich ihres CS$^+$- bzw. CS$^-$-Konsums unterschieden die Gruppen sich nicht signifikant; interessanterweise hatte sich aber ein Interaktionseffekt (Gruppen × Zeit) ergeben, der darauf beruhte, daß die Tiere der Gruppe CS$^-$ im Verlauf der unverstärkten CS$^-$-Darbietungen immer weniger tranken, während der Konsum der CS$^+$-Tiere konstant blieb. Spekulativ lassen sich die Ergebnisse zum Trinkverhalten wie folgt interpretieren: Die Tiere hatten während der Akquisition allmählich die negative Kontingenz zwischen CS$^-$ und CyS-Injektion gelernt (CS$^-$ signalisiert die **Abwesenheit** therapeutisch wirkender CyS-Injektionen) und zögerten deswegen, den CS$^-$ während des Tests zu konsumieren. Falls die Annahme richtig ist, dürfte sich ein Interaktionseffekt nicht ergeben, wenn die differentielle Konditionierung (Akquisition) **vor** Induktion der Arthritis liegt, d. h. wenn CyS noch keine therapeutischen Effekte auf die Autoimmunerkrankung haben kann.

Wir überprüften diese Hypothese und fanden tatsächlich weder einen Gruppenunterschied noch gruppenabhängige Verlaufsunterschiede beim Trinkverhalten während der unverstärkten CS-Darbietungen. Wie im ersten Experiment hatten jedoch die Tiere der Gruppe CS$^+$ signifikant weniger starke Pfotenschwellungen als die der Gruppe CS$^-$.

Unseres Wissens ist dies die erste Demonstration von konditionierten immunpharmakologischen Effekten bei gleichzeitiger (explizit registrierter) Abwesenheit einer Geschmacksaversion. Das Fehlen einer Geschmacksaversion muß nicht notwendig eine Assoziation zwischen Geschmacksreiz und **positiven** (therapeutischen) Effekten des US indizieren (vgl. Ader et al., 1987b), es legt jedoch nahe, daß das als

US verwendete Pharmakon **keine aversiven** Effekte hat oder „Streß" induziert (Klosterhalfen und Klosterhalfen, in Vorbereitung). Daher interpretieren wir die Ergebnisse als bisher beste mit Hilfe einer Konditionierungsprozedur gewonnene Stütze für den postulierten bidirektionalen Mechanismus zwischen Immunsystem und ZNS.

11.2.6 Konditionierte kompensatorische Effekte

Dyck et al. (1986, 1987, 1989) haben einen anderen Weg eingeschlagen, um Evidenz für „feedback"-Mechanismen zwischen Immunsystem und ZNS zu suchen. Sie fanden nach wiederholten Poly I:C-Injektionen bei Mäusen eine Toleranzentwicklung der NK-Zellaktivität und untersuchten, inwieweit dabei klassische Konditionierung eine Rolle spielt (vgl. z.B. Siegel, 1976). Wenn Stimuli, die vorher mit den Poly I:C-Injektionen gepaart wurden, im Intervall zwischen einer entwickelten Toleranz und einer Test-Injektion unverstärkt präsentiert (extingiert) wurden, wurde die Toleranz gegenüber dem immunstimulatorischen US aufgehoben; die Exposition von Stimuli, die vorher nicht mit dem US gepaart wurden, hob die Toleranz gegenüber Poly I:C jedoch nicht auf. Die Resultate liefern Evidenz dafür, daß die gegenüber dem Immunmodulator Poly I:C entwickelte Toleranz auf klassische Konditionierung zurückzuführen ist, und provozieren dazu, an ihre potentielle klinische Relevanz zu denken, zumal in der Klinik immunologisch wirksame Pharmaka meist in ritualisierter Weise und im gleichen Kontext appliziert werden, so daß es leicht zu einer Assoziation zwischen Ritual und/oder Kontext und Pharmakon kommen könnte. Wir geben jedoch zu bedenken, daß die Effekte bei der Toleranzentwicklung nicht immer vorhersagbar sind: Die Daten von Ghanta et al. (1987a) enthalten nach wiederholten Poly I:C-Injektionen in der gleichen Dosierung wie bei Dyck et al. (op. cit.) keinen Hinweis auf eine Toleranzentwicklung.

In letzter Zeit wurden außerdem zur konditionierten Immunmodulation Daten publiziert, die dafür sprechen, daß immunsuppressive CY-Effekte durch Konditionierung kompensiert werden können, d.h., der CS hat nach Paarung mit einer immunsupprimierenden CY-Dosis eine immunstimulierende Funktion (Krank und MacQueen, 1988; MacQueen und Siegel, 1989). Möglicherweise, aber dies müßte zuerst noch genauer erforscht werden, ist dabei die Modalität des CS kritisch: Geschmacksreize scheinen eher Reaktionen zu fördern, die in derselben Richtung liegen wie die direkt beobachtbaren Effekte von CY (vgl. MacQueen und Siegel, 1989), während Kontextstimuli eher kompensatorische Reaktionen auszulösen scheinen (Krank und MacQueen, 1988).

11.2.7 Konditionierungsversuche beim Menschen

Zwar wird durch die Möglichkeit, konditionierte immunpharmakologische Effekte bei experimentellen Autoimmunkrankheiten (Ader und Cohen, 1982, 1985; Klosterhalfen und Klosterhalfen, 1983, 1985a, 1988a) nachzuweisen, und ansatzweise auch bei einem Krebsmodell (Ghanta et al., 1987b), ebenfalls die klinische Relevanz der konditionierten Immunmodulation suggeriert, im Humanbereich sind entsprechende Hinweise jedoch eher spärlich: Ausgehend von Studien zum Einfluß von Hypnose (vgl. auch Bongartz, 1986) auf den Tuberkulin-Hauttest (Black et al., 1963), versuchten Smith und McDaniel (1983) beim Menschen eine DTHR (delayed type hypersensitivity response) durch Konditionierung zu modifizieren. Dazu wurde in monatlichen Intervallen fünfmal Tuberkulin mit einer roten Spritze in denselben Arm injiziert, als Kontrolle physiologische Kochsalzlösung mit einer grünen Spritze in den anderen Arm. Als bei der sechsten Injektion Verum und Placebo vertauscht wurden, löste die Kochsalzlösung keine DTHR aus, aber die Hautreaktion im vorher mit Placebo behandelten Arm war stark reduziert. Dieser Effekt konnte jedoch nicht repliziert werden (Smith, pers. Mitteilung, Bethesda, November 1984).

Hennig und Buske (1988) paarten viermal den Geschmack bzw. das Prickeln eines Brausebonbons mit einer NK-Zellaktivitäts-steigernden Adrenalin-Injektion (Tonnesen et al., 1984). Die fünfte Injektion bestand aus physiologischer Kochsalzlösung. In ihrer Diplomarbeit berichten die Autoren, die Placebo-Injektion ginge zwar mit einer gesteigerten NK-Zellaktivität einher, der Effekt sei jedoch nicht signifikant. Es bleibt abzuwarten, ob sich ein Konditionierungseffekt im Vergleich zu einer adäquaten Kontrollbedingung nachweisen läßt.

11.2.8 Zur Frage klinischer Anwendungsmöglichkeiten

Es ist wiederholt vorgeschlagen worden, die Konditionierbarkeit immunpharmakologischer Effekte klinisch anzuwenden (z.B. Ader, 1985a, b). Ein Vorteil wäre z.B., Pharmaka einzusparen, wenn anstelle des US ab und zu ein CS gegeben würde. (Konzeptuell läßt sich dies natürlich auch auf nicht-immunmodulierende Pharmaka generalisieren.) Dem ist das oben Dargestellte zusammenfassend wie folgt entgegenzuhalten:

- Die Richtung der konditionierten immunpharmakologischen Effekte ist selbst im gut kontrollierten Tierexperiment derzeit nicht mit Sicherheit vorherzusagen.
- Entsprechende Konditionierungseffekte sind bei Krankheitsmodellen zwar eher reliabel, aber klein. Die Erfahrungen beziehen sich im wesentlichen auf zwei Modelle von Autoimmunkrankheiten.
- Relevante Konditionierungseffekte sind im Humanbereich bisher nicht überzeugend nachgewiesen worden.

Außerdem ist zu bedenken, daß, wenn ein CS, der vorher mit der Applikation eines Immunsuppressivums etwa bei Organtransplantationen (vgl. z.B. Kap. 56) oder bei der chemotherapeutischen Krebsbehandlung (vgl. Kap. 50) gepaart wurde, anstelle des US verabreicht würde, er nach der Stimulussubstitutions-Theorie (Pavlov, 1927; Eikelboom und Stewart, 1982) u.a. auch die **unerwünschten Nebenwirkungen** des US provozieren sollte! Als potentielles Beispiel seien antizipatorische Nebenwirkungen unter Chemotherapie bei Krebspatienten zitiert (Carey und Burish, 1988; vgl. aber Klosterhalfen et al., 1989).

Bevor an eine Anwendung der Konditionierung immunpharmakologischer Effekte zu denken ist, erscheint es dringend erforderlich, zunächst zu erforschen, was den US und die unkonditionierte Reaktion (UR) des jeweiligen Pharmakons konstituiert (Klosterhalfen, 1989). Zwar wird die Verabreichung des Pharmakons prozedural als US definiert; welche Effekte aber nach Metabolisierung des Pharmakons als **effektiver** US auf den afferenten Arm des ZNS treffen (vgl. Eikelboom und Stewart, 1982), ist für die in den einschlägigen Konditionierungsexperimenten verwendeten Pharmaka nicht bekannt. Die UR ist nicht die unmittelbare Reaktion einer Zelle des Immunsystems, die sich etwa in der Expression bestimmter Rezeptoren äußert, sondern die Antwort des ZNS auf das Pharmakon als US. Da die üblicherweise verwendeten Pharmaka eine ganze Reihe von Effekten (meist an verschiedenen Organsystemen) haben, ist es wahrscheinlich, daß es auch eine Vielzahl von US's und damit UR's gibt, die jeweils ihre eigene Kinetik haben.

Falls die Gesetzmäßigkeiten des klassischen Konditionierens für immunmodulierende Pharmaka zutreffen – bisher spricht zumindest wenig gegen diese Hypothese (vgl. McCoy et al., 1986) – ist z.B. zu erwarten, daß CS's, die in kurzem Intervall vor dem effektiven US präsentiert werden, eher mit diesem assoziiert werden, als solche in längerem Intervall. Außerdem ist mit CS-US-Interaktionen zu rechnen, d.h. CS's einer bestimmten Modalität, Intensität und Dauer werden mit bestimmten US's leichter assoziiert werden als andere (vgl. z.B. Rescorla, 1988). Die von Krank und MacQueen (1988) gefundene CS-evozierte kompensatorische Immunreaktion nach Paarung von CY mit Kontextstimuli (nicht aber nach Paarung von CY mit einem gustatorischen Stimulus) könnte im Sinne eines solchen CS-US-Interaktionseffekts interpretiert werden. Auf der Basis der Kenntnis des effektiven US und seiner UR(s) ließen sich nicht nur gerichtete Hypothesen bezüglich der CR formulieren, sondern möglicherweise auch gewünschte Konditionierungseffekte gezielt vergrößern und unerwünschte vermeiden.

11.2.9 Zur Frage der Mechanismen

Prinzipiell kommen als UR's alle neurohormonellen Reaktionen in Frage. Für die konditionierte Immunmodulation sind vor allem solche Neurohormone und Neuropeptide interessant, von denen bereits bekannt ist, daß sie Effekte auf immunologische Vorgänge haben (z.B. Berczi und Kovacs, 1987; Plotnikoff et al., 1986). Es ist damit zu rechnen, daß weitere

entdeckt werden (vgl. Kap. 10). Sofern sie nicht peripher-physiologisch mediiert wurde, kann als UR auch eine Veränderung in der Körpertemperatur angesehen werden. Von einigen Substanzen ist die Richtung der Änderung bekannt: Poly I : C und Interleukin 1 sind beispielsweise pyrogen, während CyS die Körpertemperatur reduziert (Dantzer et al., 1987).

Zur Frage neuroimmunologischer Interaktionen haben Besedovsky et al. (Übersicht: Besedovsky und del Rey, 1986) eine Reihe außerordentlich interessanter Untersuchungen vorgelegt, die dafür sprechen, daß das Nervensystem auf Vorgänge im Immunsystem reagiert. Die Autoren registrierten nach Antigen-Gabe auf der Höhe der Immunreaktion eine gesteigerte neuronale Aktivität im ventromedialen Hypothalamus (Besedovsky et al., 1977). Sie wiesen außerdem nach, daß in vitro stimulierte Lymphozyten einen „glucocorticoid increasing factor (GIF)" produzieren, der in vivo via Hypophyse bei Ratten den Corticosteron-Spiegel im Blut erhöht und so möglicherweise überschießende immunologische Reaktionen verhindert (Besedovsky et al., 1985). Ferner zeigten sie, daß auch (subpyrogene) Dosen von Interleukin 1 eine starke ACTH- bzw. Corticosteron-Ausschüttung provozieren (Besedovsky et al., 1986; Berkenbosch et al., 1987; Besedovsky und del Rey, 1987; vgl. auch Lumpkin, 1987; Rettori et al., 1987; Uehara et al., 1987) und stimulierende Effekte auf den Noradrenalinmetabolismus im Gehirn haben (Kabiersch et al., 1988). Bei der von den „Immunohormonen" (Sorkin) „GIF" und Interleukin 1 induzierten ACTH-Ausschüttung handelt es sich um eine UR (vgl. Abb. 11–1). Es wäre interessant zu untersuchen, ob sich diese UR konditionieren läßt; erste Hinweise liefern die Ergebnisse von Bovbjerg (1988) und Dyck et al. (1988).

Tierexperimente mit der Zielsetzung, die UR's immunologisch wirksamer Substanzen aufzuspüren und zu prüfen, ob und unter welchen Bedingungen sich diese UR's konditionieren lassen, könnten zur Klärung der Mechanismen der konditionierten Immunmodulation entscheidend beitragen. Darüber hinaus könnten entsprechende Experimente einem besseren Verständnis neuroimmunologischer Interaktionen dienen. Es ist z.B. noch unbekannt, aus welchen Molekülen „GIF" besteht. Unter der Vorausset-

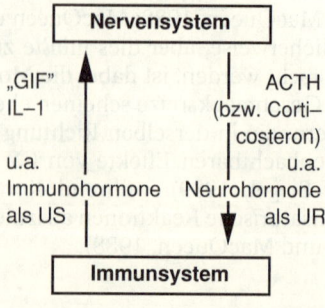

Abb. 11–1. Neuroimmunologische Interaktionen, bei denen Immunohormone als unkonditionierte Stimuli (US) und Neurohormone bzw. deren Effekte auf das Immunsystem als unkonditionierte Reaktionen (UR) aufgefaßt werden.

zung, daß sich aus dem Überstand stimulierter Lymphozyten (der „GIF" enthält) hinreichend große, d.h. biologisch wirksame Mengen unterschiedlicher Moleküle isolieren und dann Tieren als US injizieren lassen, könnte das Phänomen der konditionierten Geschmacksaversion dazu benutzt werden, Lymphokine (und weitere Monokine) mit neurotroper Wirkung zu identifizieren.

11.3 Zusammenfassende Einschätzung der Forschungsergebnisse

– In den meisten Arbeiten hatten Stressoren hemmende Effekte auf immunologische Funktionen. Immundepressive Streßeffekte sind durch Tierexperimente unter methodischen Aspekten besser belegt als durch Humanstudien; sie scheinen bei zellulären häufiger als bei humoralen Parametern aufzutreten.
– Bei Mensch und Tier wird die Entwicklung von Infektionskrankheiten wahrscheinlich durch Streßeinflüsse gefördert. In Tierexperimenten sind sowohl stimulierende als auch hemmende Effekte von Stressoren auf die Entwicklung von Tumoren festgestellt worden; ob beim Menschen belastende Lebensereignisse die Entwicklung von Krebs fördern, ist noch unklar. Streßeffekte auf die Entwicklung von Gelenkentzündungen sind bisher nur in einzelnen Tierexperimenten sowie einigen retrospektiven Humanstudien beobachtet worden; die Ergebnisse dieser Tierexperimente sind heterogen, Arthritispatienten berichten meist über eine Häufung von belastenden Ereignissen vor Ausbruch der Krankheit.
– In einer Reihe von Arbeiten ergaben sich Zusammenhänge zwischen unterschiedlichen Persönlichkeitsmerkmalen und immunologischen Parametern. Angesichts der Vielzahl der in diesen Arbeiten insgesamt untersuchten psychologischen und immunologischen Maße bleibt noch zu prüfen, inwieweit es sich um Zufallskorrelationen oder tatsächliche Zusammenhänge handelt.
– Beziehungen zwischen Persönlichkeitsmerkmalen einerseits und der Entwicklung von Infektionskrankheiten, Krebserkrankungen oder Gelenkentzündungen (rheumatoide Arthritis) andererseits sind noch nicht überzeugend dokumentiert worden. Auch hier fehlt es an Replikationsversuchen.
– Erste Experimente sprechen dafür, daß Immunfunktionen durch streßreduzierende Interventionen günstig beeinflußt werden können.
– Immunologische Vorgänge lassen sich beim Tier durch konditionierte Stimuli, die vorher mit der Verabreichung immunmodulierender Substanzen (unkonditionierte Stimuli) gepaart worden waren, beeinflussen (Phänomen der konditionierten Immunmodulation). Die Ergebnisse der Experimente zur konditionierten Immundepression sind bemerkenswert konsistent. Konditionierte immunmodulierende Effekte sind beim Menschen noch nicht überzeugend nachgewiesen worden.
– Die Mechanismen psychoimmunologischer Streß- und Konditionierungseffekte sind bisher nicht bekannt. Neuroanatomische, psychoneuroendokrinologische und immunpharmakologische Untersuchungen sprechen aber für eine – im einzelnen noch nicht hinreichend verstandene – neuroendokrinologische Mediierung, an der eine Vielzahl von Hormonen und Neuropeptiden sowie unterschiedlichen Zellen des Immunsystems beteiligt sind.
– Langfristig gesehen sind von der Psychoimmunologie (weiterhin) wesentliche Beiträge zur psychosomatischen Medizin zu erwarten. Die bisherigen Ergebnisse lassen es aussichtsreich erscheinen, Fragen der Entstehung, des Verlaufs, der Prävention und der Therapie von Infektionskrankheiten, Krebserkrankungen, Allergien und Autoimmunkrankheiten unter Verwendung psychologischer, neuroendokrinologischer und immunologischer Methoden zu untersuchen.

12 Anamneseerhebung

Rolf Adler

Die Verwirklichung einer biopsychosozialen Medizin beruht auf drei Voraussetzungen:
- Der Arzt muß entscheiden können, welche psychischen und sozialen anamnestischen Angaben im Zusammenhang mit den erhobenen somatischen Daten wichtig sind, und welche Bedeutung ihnen zukommt. Dies setzt medizinisch-psychologisches Wissen und ärztliche Erfahrung voraus.
- Er muß die Fähigkeit besitzen, diese psychischen und sozialen Daten überhaupt zu erheben. Dies verlangt eine bestimmte Technik der Anamneseerhebung.
- Der Arzt muß es verstehen, die Beziehung zum Patienten von Anfang an so zu gestalten, daß der Patient Vertrauen fassen kann, damit sich ein „Arbeitsbündnis" zwischen ihm und seinem Arzt errichten läßt. Dieses Arbeitsbündnis ermöglicht erst die optimale Anwendung und Wirksamkeit der Technik.

Dieses Kapitel befaßt sich mit der zweiten Voraussetzung, nämlich mit einer Technik der Anamneseerhebung, die es erlaubt, psychische, soziale und somatische Daten integriert zu erfassen. Der dritte Aspekt, der Aufbau eines Arbeitsbündnisses, kann praktisch von der zweiten Voraussetzung, der Technik, nicht getrennt werden. Die beste Technik ist unnütz, wenn der Aufbau eines Vertrauensverhältnisses nicht gelingt. Aus didaktischen Gründen wird die Technik hier gesondert behandelt. Die therapeutischen Aspekte, die das Arbeitsbündnis einschließen, werden in Kapitel 27 besprochen.

Anamnesen können auf verschiedene Arten erhoben werden. Alle gehen aber auf zwei Prinzipien zurück: das Prinzip der „**offenen**" und das der „**geschlossenen**" (determinierten) **Frage.** Die typische internistische Anamnese und auch die psychiatrische Exploration erfolgen mehr nach dem Prinzip der „geschlossenen" Frage. Sein Vorteil besteht in der vermeintlichen Zeitökonomie und der Übersichtlichkeit der Befragung, die es erlaubt, nach einem vorentworfenen Schema vorzugehen. Die Nachteile sind Informationen in Form von Antworten, die den in den Patienten hinein gefragten Vorstellungen des Arztes entsprechen und ein falsches Bild von seinen Überlegungen, Vorstellungen und Erlebnissen ergeben können. Das psychoanalytische Interview, welches die „offene Frage" bevorzugt, die dem Patienten einen breiten Spielraum gibt und ihm ermöglicht, das auszusprechen, was ihn beschäftigt, hilft dem Interviewer, psychische Vorgänge im Patienten und die Interaktion zwischen dem Patienten und ihm selbst zu beobachten. Es erlaubt aber nicht, diejenigen Informationen einzuholen, die neben der Beurteilung der Persönlichkeit des Patienten, seiner psychischen Entwicklung und seinen Konflikten, die Erfassung der somatischen Störung ermöglichen.

Der Abschnitt 12.1 stellt eine Technik der Anamneseerhebung dar, die es erlaubt, beide Anamneseformen zu integrieren. Er wendet sich vor allem an den Nichtpsychiater, obwohl grundsätzliche Aspekte der Technik nach meiner Meinung auch für das psychiatrische Interview gelten. Die dargelegte Anamnesetechnik wurde von Engel entwickelt, der praktizierender Internist und Psychoanalytiker ist. Dies muß vorausgeschickt werden, um klarzumachen, daß der Hintergrund für das Verständnis bedeutsamer psychologischer Daten in der psychoanalytischen Entwicklungslehre besteht, wie sie z. B. von Engel in „Psychisches Verhalten in Gesundheit und Krankheit" (1969) dargelegt wurde, und daß die Entwicklung dieser Anamnesetechnik sowohl auf Engels praktischer internistischer Tätigkeit als auch auf seiner psychoanalytischen Ausbildung beruht.

Diese Technik soll es Arzt und Student erleichtern, in **einem** Arbeitsgang die mehr unpersönlichen, objektiven Daten zusammen mit den mehr persönlich-subjektiven zu erheben, und zwar in Form einer Handlungsanweisung, die beide Aspekte harmonisch und zwanglos kombinieren läßt. Den meisten Vorschlägen für eine integrale Erfassung des Kranken und seines Leidens haften Mängel an. Häufig wird die Aufgabe des Somatikers nicht berücksichtigt, oder es erfolgt keine konkrete Anweisung, wie sich das „klinisch-objektive" und das „subjektiv-teilnehmende" Vorgehen kombinieren lassen. Die nachfolgende Darstellung zielt darauf hin, die biopsychosoziale Medizin dorthin zu tragen, wo der Arzt dem Patienten im ersten Kontakt begegnet. Sie möchte den Nachteil ausmerzen, daß ein Patient erst dann subjektiv-teilnehmend, d.h. Psychisches und Soziales einschließend, erfaßt wird, wenn der Somatiker keine oder nur ungenügende Befunde zur Erklärung der vorgebrachten Beschwerden erhoben hat, denn die diagnostischen und therapeutischen Um- und Irrwege dieses zweistufigen Vorgehens liegen auf der Hand. Dabei ist es klar, daß jedes Interview wieder anders verläuft, und daß ein sklavisches Befolgen der Interviewanleitung nicht das Ziel sein kann. Die aus didaktischen Gründen systematische Darlegung der Interviewtechnik darf nicht darüber hinwegtäuschen, daß ihr intellektuelles Verstehen noch keine erfolgreiche Handhabung bedeutet. Diese verlangt Kenntnisse der psychoanalytischen Entwicklungslehre, fundierte Kenntnisse in somatischer Medizin und Übung unter Anleitung.

Das Üben unter Anleitung muß neben der Interviewtechnik vor allem darauf hinarbeiten, den Interviewer seine eigenen Empfindungen und Gedanken, die sich während der Anamneseerhebung in ihm entwickeln, sorgfältig beobachten zu lassen. Nur so kann er lernen, den Einfluß seiner eigenen Reaktionen auf die Arzt-Patient-Beziehung wahrzunehmen und zu verstehen. (Beim Beispiel des unter Schritt 3 des Interviewschemas erwähnten Patienten mit zerebraler Insuffizienz reagiert der Arzt auf die Mühe des Patienten, die Anamnese geordnet zu schildern, häufig mit Verärgerung. Diese findet in der Krankengeschichte dann Ausdruck in Begriffen wie „unkooperativer Patient". Eine Selbstbefragung des Interviewers über sein Ärgergefühl hätte ihn nicht nur die Diagnose „zerebrale Insuffizienz" stellen lassen, sondern ihm auch erlaubt, die Schwierigkeiten des Patienten zu verstehen, die eigenen Affekte von Wut und Ungeduld aus der Beziehung zum Patienten herauszuhalten und das Interview ruhiger zu führen. Im Kommentar des Interviewers zum wörtlich wiedergegebenen Interview finden sich solche Selbstbeobachtungen auf S. 217–220.)

12.1 Interviewtechnik und -schema

Hier liegt die Erfahrung zugrunde, daß die Krankengeschichte, einschließlich der Bedeutung von psychischem Material und dessen Zusammenhang mit somatischen Daten, um so zuverlässiger und typischer für den jeweiligen Patienten wiedergegeben wird, je geschickter und behutsamer der Arzt das Interview führt und strukturiert und es dem Patienten ermöglicht, seine Angaben in seinen Worten, in seiner Reihenfolge und zu einem ihm möglichen Zeitpunkt zu machen.

Der Vorwurf übermäßigen Zeitaufwandes kann durch diese schriftliche Darlegung nicht entkräftet werden. Ich kann lediglich aus meiner Erfahrung sagen, daß der in dieser Technik Geübte nicht mehr Zeit benötigt, um eine zuverlässige Anamnese zu erheben, als andere Ärzte, und daß sich diese Technik mit entsprechender Modifikation auch beim Schwerkranken, in der Notfallsituation und beim Befragen von Angehörigen bewährt. Dauert ein Erstinterview dabei eine ganze Stunde, so handelt es sich gewöhnlich um Patienten, bei denen das Verpassen bedeutsamer psychischer und sozialer Faktoren bereits zu einem diagnostischen und therapeutischen Zeitauf-

wand geführt hat, der weit über die für das Erstinterview benötigte Zeit hinausgegangen ist.

Erster Schritt: Der Arzt begrüßt den Patienten und stellt sich vor.

Zweiter Schritt: Er bringt den bettlägerigen wie auch den ambulanten Patienten für das Interview in eine möglichst bequeme Lage. (Er achtet beispielsweise darauf, daß der Patient nicht durch mühsames aktives Heben des Kopfes in Augenkontakt mit dem Interviewer bleibt, sondern erleichtert dies durch stützende Anordnung des Kopfkissens. Auch erkundigt er sich beispielsweise, ob der Zeitpunkt für das Interview günstig ist, oder ob es für den Patienten in die Essens- oder Besuchszeit fällt.)

Die ersten beiden Schritte sollen dem Patienten von Beginn an das warme und wohlwollende Interesse des Arztes signalisieren. Fühlt sich der Patient seines Arztes nicht sicher, so hält er wesentliche Informationen zurück oder entstellt sie.

Diese beiden Schritte mögen für selbstverständlich und für nicht erwähnenswert gehalten werden. Die Beobachtung von Ärzten und Studenten zeigt aber immer wieder ihre Vernachlässigung mit nachteiliger Auswirkung auf den Interviewablauf.

Die Mutter eines 12jährigen, stark abgemagerten Jungen, der seit zwei Monaten an einer Eßstörung leidet, wird interviewt. Obwohl die Hospitalisation des Kindes als dringlich erscheint, lehnt es die Mutter, die selbst Krankenschwester war, ab, den Knaben für längere Zeit im Krankenhaus zu lassen. Bei der Besprechung des auf Tonband aufgenommenen Gesprächs zwischen Student und Mutter fällt dem Tutor die schon zu Beginn gespannt-ärgerliche Stimme der Mutter auf. Er fragt, ob dem Interviewer aufgefallen sei, daß er am Anfang des Gesprächs und auch in seinem Verlauf die Mutter nie nach ihren eigenen Gedanken und Empfindungen zu der Krankheit ihres Kindes gefragt habe. Jetzt teilt der Student mit, daß drei Monate vor Beginn der Eßstörung des Kindes beim Ehemann der Mutter eine Myelose festgestellt worden sei.

Es muß angenommen werden, daß das Übersehen des gespannten Zustandes der Mutter und das Vermeiden des Studenten, gleich zu Beginn darauf einzugehen und mit der Mutter die Einwirkung der schweren Erkrankung ihres Mannes zu besprechen, am Zusammenbruch der Beziehung zwischen ihm und der Mutter beteiligt waren.

Dritter Schritt: Der Patient wird mit einer „offenen" Frage (z.B. „Wie fühlen Sie sich?") angeregt, alle Beschwerden und den Grund für das Aufsuchen des Arztes mit seinen eigenen Worten zu schildern. Dieser Schritt versichert dem Patienten, daß er sich frei äußern darf. Er umfaßt die Hauptbeschwerden und -probleme, ihre wichtigsten zeitlichen Zusammenhänge, gibt einen Überblick über die derzeitigen Lebensumstände, die bedeutsamen Bezugspersonen und vermittelt einen Eindruck vom „Stil" des Patienten und seiner Persönlichkeit. Der Arzt kann daraus ableiten, ob er im weiteren Verlauf des Interviews seine Technik modifizieren muß.

Einen weitschweifigen Patienten bringt er durch strukturierende Fragen auf das eigentliche Thema zurück, einen ängstlich-unsicheren regt er zu spontaner Schilderung an, indem er beispielsweise den letzten Teil des vom Patienten gesprochenen Satzes wiederholt oder fragt: „Haben Sie sonst noch etwas bemerkt?" Er soll schon zu diesem Zeitpunkt erkennen, daß ein Patient mühsam nach Worten sucht, beim Bemühen, Daten zu erinnern, angestrengt die Brauen runzelt, sich zeitlich in Widersprüche verwickelt, Fragen lächelnd oder sarkastisch abweist und vermutlich das Bild der zerebralen Insuffizienz zeigt (Delir, Demenz), wie es bei hospitalisierten Patienten so häufig vorliegt und praktisch bei jeder schweren Störung vorkommt, die den Gehirnstoffwechsel indirekt oder direkt beeinträchtigt (Anämie, respiratorische Insuffizienz, Elektrolytstörung, Medikamentennebenwirkung usw.). Ein ausgedehntes Interview lohnt sich hier wegen der gestörten Gedächtnisfunktionen, der erlahmenden Aufmerksamkeit und Konzentrationsfähigkeit nicht, und das Befragen einer dritten Person ist angezeigt. Der Arzt soll zu diesem Zeitpunkt auch schon den pseudounabhängigen Patienten erkennen – wie er sich beispielsweise unter den an Myokardinfarkt Erkrankten häufig findet (s. Kap. 41.1) – und das Interview dessen Persönlichkeitszügen anpassen. Er muß wissen, daß ein solcher Patient seine Symptome bagatellisiert, seine Gesundheit betont, ängstlich reagiert, wenn er sich als hilflos und krank erkennen muß, und gereizt und verärgert antwortet, den Interviewer als lästigen Eindringling sogar zurückweist, wenn dieser auf die detaillierte Symptombeschreibung drängt. Hier muß der Interviewer die Symptome indirekt erfahren, indem er beispielsweise im Verlauf der Anamnese das Augenmerk darauf richtet, zu welchem Zeitpunkt der Patient in seinen üblichen Tätigkeiten eingeschränkt zu werden begann. Er sollte also nicht auf das Symptom lossteuern, das der Patient aus intrapsychischen Gründen bagatellisiert oder verleugnet.

Wird zu früh aktiv und detailliert gefragt, so gerät der Patient in passives Abwarten, das Interview führt zum „Ausfragen" und läuft Gefahr, diejenige Anamnese zu ergeben, die der Arzt in den Patienten hineinlegt und nicht mehr dessen eigene Krankengeschichte. Die Folgen sind diagnostische Irrtümer und eine von Beginn an gestörte Wechselbeziehung zwischen Arzt und Patient.

Vierter Schritt: Der Arzt erforscht das jetzige Leiden. Er erhellt jedes der bei Schritt 3 erwähnten Symptome nach:
– seinem zeitlichen Auftreten (a);
– seiner Qualität (b);
– seiner Intensität (c);
– der Lokalisation und eventuellen Ausstrahlung (d);
– dem Zusammenhang mit anderen Beschwerden (e);
– den Umständen, unter denen es auftritt (f);
– den Umständen, unter denen es sich intensiviert und mildert (g).

(a)–(g) sind die **„7 Dimensionen" des Symptoms.**

Beim **zeitlichen Auftreten (a)** achtet der Interviewer auf den Zeitpunkt des Beginns, die Dauer, die Reihenfolge, die Periodizität und freie Intervalle des Symptoms; bei der **Qualität (b)** auf den gewählten beschreibenden Ausdruck, der unter Umständen die Ätiologie eines Symptoms schon ein Stück weit verraten kann.

Die Bemerkung: „der Arm ist gelähmt, ich kann ihn nicht mehr heben", stellt bei einem mit Verdacht auf Herzinfarkt zugewiesenen Mann, der den in den linken Arm ausstrahlenden Schmerz beschreibt, einen Hinweis auf die Möglichkeit eines konversionsbedingten Schmerzes dar (vgl. Kap. 36.2.1).

Bei der Qualität darf man sich mit Ausdrücken wie „es war ein Bauchkrampf" nicht zufriedengeben. Der Patient wird aufgefordert zu beschreiben, was er dabei verspürt hat. Der Interviewer stellt dann fest, daß der eine Patient unter Bauchkrampf ein Völlegefühl versteht, ein anderer Blähungen und ein dritter eine Kolik. Wenn der Patient nicht imstande ist, seine Empfindungen genau zu beschreiben, kann man ihn fragen, ob er sie mit früher erlebten Empfindungen vergleichen könne. Zuletzt bietet man ihm verschiedene Möglichkeiten wie „ein innerhalb einiger Sekunden ansteigender Schmerz", „ein Gefühl, wie wenn man Stuhlgang haben müßte" usw. an, ohne dabei aber eines der Angebote überzubetonen. Denn gewisse suggestible Patienten neigen dazu, die vom Interviewer angebotenen Möglichkeiten zu übernehmen und zu bestätigen, so daß zuletzt nicht mehr eruierbar ist, ob der Patient wirklich das empfunden hat, was er jetzt angibt.

> Eine 38jährige ledige, bei ihrer Mutter wohnende Frau, die in ihrer Anamnese die verschiedensten Schmerzzustände vieler Körperregionen, die schwer einem bekannten Krankheitsbild zuzuordnen sind, und viele schmerzhafte, diagnostische und therapeutische Eingriffe mit zum Teil fraglicher Indikation beschreibt, schildert „Bauchkrämpfe". Sie ist aber unfähig, diese näher zu beschreiben. Sie übernimmt das Angebot des Interviewers eines Schmerzes „wie wenn Winde abgehen müßten" so unauffällig und bereitwillig, daß der unachtsame Interviewer irregeleitet wird. Erst die Beobachtung, daß sich dasselbe bei den verschiedensten Körperregionen und Organsystemen wiederholt, läßt ihn die Suggestibilität erkennen, die wahrscheinlich an einigen der schlecht indizierten Eingriffe in der Vergangenheit schuld war (eine beidseitige Nephropexie bei Wanderniere und drei Laparotomien mit Adhäsiolyse innerhalb von knapp zwei Jahren).

Die **Intensität (c)** betrifft den Stärkegrad, die Funktionseinbuße, das Volumen (z.B. Menge des erbrochenen Blutes) und die Auftretenshäufigkeit (z.B. Fieberschübe) eines Symptoms. Auch hier soll der Interviewer darauf achten, in welchem Zusammenhang das Symptom geschildert wird, beispielsweise welche mimischen Veränderungen und Gebärden es begleiten.

Das distanzierte, zweideutige Lächeln einer 28jährigen Frau während der Beschreibung unerträglich heftiger „Neuralgien", die von der Gegend des linken Ohres in die Schläfenregion, die Schulter und den Nacken ausstrahlen, erwecken beim Interviewer den Verdacht auf die für ein konversionsneurotisches Schmerzsyndrom typische „belle indifférence" (Kap. 36.2.1). Der Interviewer ist im weiteren Verlauf des Interviews nicht erstaunt zu erfahren, daß dem Schmerz ein Konflikt am Arbeitsplatz mit einer Vorgesetzten vorangegangen ist, und daß die Patientin eine Kindheit mit brutalen Züchtigungen (Schläge ins Gesicht) durch die sich schlecht vertragenden Eltern durchgemacht hat.

Bei der **Lokalisation (d)** achtet man darauf, ob eine Empfindung beispielsweise in der Tiefe oder oberflächlich liegt, und wohin sie ausstrahlt.

Ein 34jähriger Mann leidet seit einer Erkältung im Militärdienst vor acht Jahren an einer „Trigeminusneuralgie" des zweiten Astes rechts, die auf die verschiedensten Medikamente nicht angesprochen hat. Das Interview ergibt, daß die Lokalisation im Bereich der rechten Wange die rechte Nasenseite, die Oberlippe und das rechte Unterlid nicht einbezieht, was von einer klassischen Trigeminusneuralgie des zweiten Astes erwartet werden müßte.

Der **Zusammenhang mit anderen Beschwerden (e)** oder das Fehlen von Begleitsymptomen ist für das Verständnis ebenfalls bedeutsam.

Ein 32jähriger Mann mit einem gut eingestellten Diabetes mellitus klagt über ausgesprochene Müdigkeit. Dieses Symptom steht allein da und tritt jeweils kurz nach Arbeitsbeginn auf. Nächtlicher Schlaf sowie Ausruhen und Schlafen während des Tages ändern am Müdigkeitsgefühl nichts. Die Beschwerde Müdigkeit wird vom Patienten stark betont. Dies sind Merkmale, die auf eine psychogene Genese dieses Symptoms hinweisen (Engel, 1969). Der Interviewer erfährt weiter, daß sich der Patient am Arbeitsplatz in der Funktion als Vorgesetzter überfordert fühlt, kurz vor der Entlassung steht und sich mit seinem schon jahrelang bestehenden Diabetes keineswegs auseinandergesetzt hat.

Die **Umstände, unter denen sich ein Symptom intensiviert oder mildert (f),** sind für das Symptomverständnis sehr wichtig.

Ein 60jähriger Mann klagt über Brennen im linken Oberbauch, das bis in den linken Unterbauch und die Analgegend ausstrahlt. Das Symptom steigert sich während der Woche von Montag bis Freitag und klingt übers arbeitsfreie Wochenende wieder etwas ab.

Die körperliche Untersuchung und die Labortests ergeben bis auf die Zeichen einer mäßigen chronischen obstruktiven Bronchitis normale Befunde; insbesondere wird bei der klinischen Untersuchung der Wirbelsäule und des Abdomens keine Abnormität festgestellt.

Am Arbeitsplatz besteht eine ausgeprägte Konfliktsituation, auf die der Patient seit Monaten mit Hilflosig-

keit und Verzweiflung und mit dem Wunsch reagiert, sich nach 39 Dienstjahren vorzeitig pensionieren zu lassen. Diesem Wunsch steht ein starkes Streben nach Pflichterfüllung und mannhaftem Ertragen gegenüber.

Der zuweisende Arzt denkt an eine Depression. Der Interviewer stellt fest, daß der Patient beim Sitzen den Oberkörper leicht nach vorn und seitlich links neigt und berichtet, daß das Brennen beim Flachliegen abnimmt und sich bei Arbeiten mit erhobenem linkem Arm steigert. Der Interviewer denkt deswegen an eine organische Störung und vermutet, daß das Brennen ein sog. Substitutionssymptom bei einem hyposensitiven Mann sein könnte. Die Prüfung der Schmerzsensitivität mit dem Libman-Test (Libman, 1934; Adler und Lomazzi, 1973) erglbt Hyposensitivität. Die Vermutung liegt nahe, daß dieser Patient Brennen angibt, wo ein normo- oder hypersensitiver Mensch bei gleicher organischer Läsion Schmerz als Symptom angeben würde. Die Untersuchung wird ausgedehnt. Auf der Röntgenaufnahme der Brustwirbelsäule finden sich multiple Kompressionsfrakturen im Bereich der Wirbelkörper, die dem Dermatom, wo das Brennen empfunden wird, entsprechen.

Fünfter bis achter Schritt: Handelt es sich nicht um ein eng umschriebenes Symptom bei einem bis dahin ganz gesunden Individuum, was ja selten der Fall ist, dann kommt der Patient während seiner Schilderung spontan auf frühere Krankheiten (5. Schritt), die Gesundheit seiner Angehörigen (6. Schritt), seine persönliche Entwicklung (7. Schritt) und seine sozialen Lebensumstände (8. Schritt) zu sprechen. Der Arzt folgt den Assoziationen des Patienten zu den Schritten 5–8, während er das jetzige Leiden erforscht, und integriert sie dadurch in Schritt 4.

Die Berücksichtigung assoziativer Verknüpfungen, die mit Empathie und Intuition wahrgenommen werden, erlaubt, Zusammenhänge zu erkennen, die der direkten Befragung entgehen.

Eine ledige junge Frau wird hospitalisiert. Sie leidet seit einigen Monaten an Enge im Halsbereich und Atemnot. Sie erwähnt, daß ihre Mutter (6. Schritt) etwa vor einem halben Jahr an Herzversagen gestorben sei. Wegen der Erkrankung der Mutter habe sie widerwillig ihre Stelle im Ausland aufgegeben (7. und 8. Schritt). Vom Leiden der Mutter habe sie durch die Tonbandkorrespondenz mit ihr erfahren, auf dem die schwere Atmung der jetzt Verstorbenen sie so beeindruckt habe. Der Interviewer denkt aufgrund der zeitlichen Zusammenhänge, der Ähnlichkeit der Symptome von Mutter und Patientin sowie deren Einstellung gegenüber der Heimkehr an die Möglichkeit eines Konversionssymptoms.

Ein Mann Mitte Sechzig wird mit atemabhängigen Schmerzen im Bereich des linken unteren seitlichen Thorax ins Krankenhaus eingewiesen. Die körperliche Untersuchung ergibt eine Dämpfung links lateral basal, das Atemgeräusch ist dort abgeschwächt. Das Thoraxbild zeigt einen Zwerchfellhochstand links sowie pleurale Veränderungen. Obwohl Temperatur, BKS und EKG normal sind, und der Patient kein Blut gehustet hat, ist als wahrscheinlichste Diagnose eine Lungenembolie angenommen worden.

Der unvoreingenommene Interviewer erfährt über die jetzige Lebenslage und die Umstände, unter denen das Leiden aufgetreten ist (8. Schritt), zusätzlich, daß die Frau des Patienten (6. Schritt) genau eine Woche vor Krankheitsbeginn des Patienten im gleichen Spital an einer Lungenembolie verstorben ist. Ein daraufhin gesuchtes früheres Thoraxröntgenbild des Patienten wird gefunden und ergibt, daß sich die pleuralen Veränderungen schon vor mehr als einem Jahr feststellen ließen. Die Beschwerden der Frau haben also als Modell für die Symptomatik beim Patienten gedient.

Dieser Fall macht auch deutlich, daß zur Familienanamnese nicht nur die Blutsverwandten, sondern alle bedeutsamen Bezugspersonen gehören.

Neunter Schritt: Der Arzt forscht systematisch nach Symptomen der verschiedenen Organsysteme, von denen er die für das Leiden des Patienten bedeutsamen schon in den Schritten 3–8 erfahren haben sollte, und vervollständigt sie. Hier entspricht sein Vorgehen der traditionellen Anamnese, wobei er aber auch an dieser Stelle Suggestivfragen, Fragen, die die Antwort bereits enthalten, Doppelfragen sowie den Gebrauch von Worten, die der Kranke bis dahin nicht zur Beschreibung seiner Symptome verwendet hat, möglichst vermeidet.

Zehnter Schritt: Abschließend soll der Patient Gelegenheit erhalten, Fragen aufzuwerfen und noch nicht Besprochenes beizufügen. Erkundigt sich der Kranke nicht spontan nach den Vorstellungen, die sich der Arzt während des Interviews von Ursache und Behandlungsmöglichkeit des Leidens gebildet hat, so bringt sie der Interviewer ins Gespräch und ersucht den Kranken, sie zu beantworten (z.B. „Wie stellen Sie es sich vor, wie es zu Ihrer Krankheit kam?" und „Wie soll die Behandlung in Ihren Augen vor sich gehen?").

Die Beantwortung durch den Patienten bringt oft entscheidende Klarheit darüber, wie bewußtseinsnah Zusammenhänge zwischen psychischen Problemen und Symptomen dem Patienten sind, oder umgekehrt, wie stark dieser deren Erkennen noch von sich weisen muß. Daraus gewinnt der Arzt Anhaltspunkte, inwieweit er Widerstand und Abwehr – beide Begriffe sind im psychoanalytischen Sinn verstanden – des Kranken im Behandlungsplan zu berücksichtigen hat (Meerwein, 1969), und er vermag sich für die Erläuterung seiner diagnostischen und therapeutischen Pläne, die den Abschluß des Interviews bilden, patientgerechter einzublenden.

Ein Lehrer wird mit der Klage über Doppeltsehen eingewiesen. Er führt seine Störung auf eine Zeit voller Konflikte und Meinungsverschiedenheiten mit seinen Amtskollegen zurück, die sich über die Art der Einrichtung des Naturkundezimmers nicht zu einigen vermögen: Seine Augen würden so auseinanderweichen – bemerkt der Patient – wie die Auffassung seiner Arbeitskollegen von seiner eigenen Meinung. Es wird eine organische neurologische Störung diagnostiziert.

Dieser Fall illustriert den Versuch des Patienten, eine Kausalität zwischen Lebensumständen und Erkrankung herzustellen unter Benützung des psychischen Abwehrmechanismus der Rationalisierung. Aus ihm läßt sich die Faustregel ableiten, daß die Betonung von Konflikten und Lebensumständen als Ursache eines Symptoms den Interviewer mehr an eine organische Läsion denken lassen soll und umgekehrt.

12.2 Schwierigkeiten der Interviewtechnik

Eine Schwierigkeit liegt darin, daß die Information nicht einem vorgefaßten Plan eingeordnet und in einer logischen Reihenfolge erhoben werden kann, wie bei der traditionellen Technik. Läßt der Interviewer die Informationsfäden sich nach ihrer eigenen Logik (die psychodynamischen Prozessen des Patienten entspricht) zu einem Teppich weben, dann tritt aber ein Anamnese-„Muster" hervor, das den Vorzug hat, für den jeweiligen Patienten charakteristisch zu sein und somatische, psychische und soziale Faktoren in engem Zusammenhang wiederzugeben. Eine weitere Erschwerung kommt hinzu. Die Interviewtechnik zwingt den Interviewer, sich den psychischen Spannungen und Konflikten des Patienten zu stellen, was seine eigenen unerledigten Konflikte aktiviert (eine Schwierigkeit, aus der dem Interviewer im Verlaufe des Erlernens der Interviewtechnik die Chance erwächst, seine eigene Persönlichkeit ein Stück weit kennenzulernen; dies ist eine Voraussetzung für die biopsychosoziale ärztliche Tätigkeit).

Die Technik sollte anfänglich bei Patienten geübt werden, die nicht nach psychologischen Gesichtspunkten ausgewählt wurden. So entgeht der Interviewer der Gefahr, anzunehmen, daß psychologische Beobachtungen nur am psychiatrischen Patienten gemacht werden können, und es wird selbstverständlich, daß jeder körperlich Kranke auch psychische Phänomene – und meist auch Probleme – aufweist.

Der Interviewer, der die dargestellte Technik benützt, wird auf spezielle Fragen stoßen: Schweigepausen, Weinen, Feindseligkeit, Verwirrtheit, Einfluß schwerer körperlicher Erkrankung auf die Bedürfnisse des Patienten während des Interviews, persönliche an den Interviewer gerichtete Fragen usw. Diese speziellen Fragen zu diskutieren, würde den Rahmen dieses Kapitels sprengen. Näheres darüber findet der Leser im Buch von Adler und Hemmeler (1986).

Die in diesem Buch dargestellten Interviews verhelfen auch zur vermehrten Einsicht in die häufig vorkommenden psychischen Aspekte der vom Somatiker gesehenen Kranken:

- hypochondrische und konversionsneurotische Beschwerden (Kap. 36.2.2);
- anhaltende und unbeeinflußbare Schmerzzustände (Kap. 36.2.1);
- unspezifische Syndrome wie Nervosität, Müdigkeit und Erschöpfung;
- psychische Faktoren, die bei Auslösung, im Verlauf

und in der Heilungsphase organischer Leiden wirken (Kap. 41.1, 41.2, 41.3, 42, 43, 44.1, 44.2, 45.1, 45.2, 45.3);

– psychische Komponenten bei den früher als eigentliche „psychosomatische" Leiden bezeichneten Krankheiten (die heute besser unter den Begriff „somatopsychisch-psychosomatisch" eingereiht werden) (Kap. 41.1, 41.2, 41.3, 42, 43, 44.1, 44.2);

– psychische Probleme des Schwerkranken und des sterbenden Patienten (Kap. 62, 63, 64, 65, 66).

12.3 Beispiel

Herr H. S., geb. 1967 (Interview vom Dezember 1987)
Fragestellung des Hausarztes: Verdacht auf „verborgene psychosomatische Störungen" bei 20jährigem Mann mit unklaren Schmerzen im rechten Handgelenk, Diagnose? Therapie?

1 Die Begrüßung hat im Wartezimmer stattgefunden, zu Interviewbeginn haben sich Arzt (A) und Patient (P) bereits gesetzt.
(Dimensionen des Symptoms a–g und Schritte des Interviewers – gemäß Abschnitt 10.2 – am Rand, *Kommentar kursiv*.)
A: Ich habe ihr den Bericht des Hausarztes, den ich noch nicht gelesen habe, damit ich mir ein eigenes Bild machen kann. Ich werde ihn nach unserem Gespräch lesen.

Solche Klarstellungen tragen zum Aufbau des Arbeitsbündnisses bei. Der Patient wird als Partner ernstgenommen.

2 A: Sind Sie so bequem?
P: Ja, ja.
A: Und hören Sie normal, so daß ich normal laut sprechen kann?
P: Ja, ja.

3 A: Sagen Sie mir bitte, wie Sie sich fühlen.
P: Also im Moment arbeite ich nicht, ich habe starke Schmerzen beim Arbeiten im rechten Handgelenk.
A: mhm
P: Was sonst noch? ... einfach diese Schmerzen beim Arbeiten. Am Abend gehen sie zurück, auch am Sonntag. Sie sind am Morgen schwächer und gegen Abend stärker. Daran leide ich seit etwa einem Jahr, damals setzten die Schmerzen ein. Sie zwangen mich wiederholt zum Pausieren bis zu zwei Monaten.
A: Bis zu zwei Monaten.
P: Vom Mai an habe ich dann nicht mehr gearbeitet. Bin dann in den Wehrdienst gegangen, mußte ihn wegen des Handgelenks abbrechen, und habe dann bis anfangs Oktober nicht gearbeitet.
A: Und der Wehrdienst? ...

(d) P: Abgebrochen wegen dem Handgelenk.
A: Haben Sie außer den Problemen mit dem Handgelenk noch andere Beschwerden?

*Ergänzung der Landkarte**

P: Mit dem Knie habe ich etwas wegen des Fußballspielens ... es hat keinen Zusammenhang ...
A: Aber es scheint mir wichtig, alles zu erfahren ...

---

* Landkarte: Ein von G. L. Engel eingeführter Begriff für die Lokalisation der psychosomatischen Beschwerden.

P: Bei einem Zusammenstoß beim Fußballspiel wurde das Knie unterhalb der Kniescheibe verletzt ... jetzt muß ich deswegen aussetzen.
A: In welcher Liga spielen Sie?

Ich möchte den Stellenwert des Sports in seinem Leben erfahren und sein Verhalten bei der Ausübung.

P: 4. Liga, manchmal auch 1. (Amateurliga).
A: Bei welchem Verein?
P: Fußballclub U.
A: Andere Beschwerden außer Handgelenk und Knie?
P: Keine
A: Keine ... wenn sie Handgelenk und Knie nicht schmerzhaft spüren würden, arbeiteten Sie und würden Fußball spielen?
P: Ja.

4 (a) A: Wann waren Sie letztmals wirklich gesund?

Die Landkarte ist skizziert, das jetzige Leiden muß erfaßt werden.

P: Im Oktober vor einem Jahr.
A: Erzählen Sie mir bitte vom Augenblick an, wo Sie etwas verspürten, bis heute.

Diese offene Frage übergibt Initiative und Verantwortung dem Patienten.

(d) P: Anfang Oktober 1986 erlitt ich am linken Handgelenk eine Sehnenscheidenentzündung. Ich erhielt eine Schiene, natürlich belastete ich jetzt mehr rechts. Die Schiene half nichts. Nach Wiederaufnahme der Arbeit traten links und rechts Schmerzen auf. Man versuchte es mit Elektroden zu heilen, das brachte auch kein Ergebnis.

Der Patient ist ein einfacher, zutraulicher junger Mann.

Dann kamen die Weihnachtsferien ... es besserte sich ...

4 (a, f) *Die Schmerzen scheinen mit der Arbeit zusammenzuhängen und in der Zeit ohne Arbeit abzunehmen. Bei psychogenen Schmerzen, z. B. wegen eines Konflikts am Arbeitsplatz, würden die Zusammenhänge zwischen Schmerz und Arbeiten nicht so deutlich erwähnt.*

P: Nach den Ferien und der Wiederaufnahme der Arbeit traten die Schmerzen wieder auf, ich ging zum Arzt, es war Dr. Z., der mich zu einem Rheumatologen, Dr. A., sandte. Damals arbeitete ich noch, dann wieder zwei Monate nicht. Dann habe ich wieder mit einer Schiene versucht zu arbeiten, bis Mai, dann empfahl man mir mit dem Beruf aufzuhören, eine neue Lehre zu machen ...
A: Sie haben Maler gelernt.
P: Ja, dann bin ich in den Wehrdienst eingerückt. Ich wurde aufgefordert, mich beim Auftreten der Schmerzen zu melden. Dann wurde ich um ein Jahr zurückversetzt.
A: Wie lange nach Beginn war das?
P: Nach 17 Tagen. Dann kam ich ins Krankenhaus zur Untersuchung. Zu der Zeit war ich aber schmerzfrei, weil ich ja nicht arbeitete. Dann wurde ich zur Arbeit geschickt, damit man mich untersuchen konnte, während ich Schmerzen hatte.
A: Wann war das?
P: Anfang August. Dann ging ich arbeiten und wieder zur Untersuchung. Man sagte mir, ich hätte eine Schwäche der Handgelenke.

A: In welcher Abteilung waren Sie zur Untersuchung?
P: Handgelenk ... in einer Poliklinik.
A: Es gibt verschiedene ...
P: Die handchirurgische.
A: Und was sagten Ihnen die Ärzte?
P: Eine Schwäche des Handgelenks, mehr habe ich nicht erfahren.
A: Was wurde Ihnen empfohlen?
P: Nichts.
A: Haben Sie vom Hausarzt ein Ergebnis erfahren?
P: Nein.
A: Gearbeitet haben Sie jetzt nicht?
P: Doch ich arbeite.
A: In Ihrem Beruf?
P: Ja.
A: Und wie geht das?
P: Schlecht, ich bin vom Gerüst gefallen, ich mußte mich festhalten, kriegte Schmerzen, mußte loslassen und rutschte ab. So ist es gefährlich.
A: Können Sie mir diesen Vorfall genau schildern, ich verstehe ihn nur zum Teil. Sie mußten auf ein Gerüst steigen, sich halten, mit einer Hand, der kranken, während des Malens ...
P: und bin ausgerutscht mit der Hand und zwei Meter hinuntergefallen.
A: mh ... sind Sie Rechts- oder Linkshänder?
P: Rechts.
A: Wie kann das geschehen, daß man als Maler und Rechtshänder mit der Linken malt?
P: Mauer und Gerüst standen so, daß es nicht anders möglich war ...
A: und so kam es, daß Sie sich nicht festhalten konnten. Was war der Grund, daß Sie sich nicht festhalten konnten?
P: Es tat weh, so ließ ich los.

(e) A: Man könnte sich auch vorstellen, daß die Kraft nachließ, wie ist es mit der Kraft?
P: Schwächer als in der linken Hand.
A: Und wie ist es mit dem Gefühl in der Hand?
P: Das ist gut, das Gefühl habe ich.

Die Angaben klingen vom anatomischen und physiologischen Gesichtspunkt aus logisch. Ein sekundärer Gewinn (s. Kap. 27: Konversion) ist nicht ersichtlich.

4 (f, g) A: Und Sie haben gesagt, daß die Schmerzen in den
(a) Ferien, übers Wochenende und in der Nacht abnehmen, habe ich das richtig verstanden?
P: In den Ferien nehmen sie ab, am Morgen sind sie merklich geringer.

(g) A: Lindern andere Maßnahmen als Nichtarbeiten die Schmerzen?
P: Ich weiß nicht.
A: Man macht etwa die Erfahrung, daß man ein linderndes Moment feststellt und sich danach richtet.
P: ... Ich habe alles versucht, Tabletten, gar nichts hat geholfen.
A: Auch Tabletten nicht?
P: Nein.
A: Was für Tabletten haben Sie genommen?
P: Ich weiß es nicht mehr.
A: Sie haben noch von Elektroden gesprochen ...
P: Zwei Plättchen ...
A: Mit Drähten zu einem Apparat?

Ich denke an TENS (transkutane elektrische Neurostimulation). Die Effekte von Medikamenten und anderen Hilfsmitteln können aufschlußreich sein.

P: Das war in der Therapie.
A: Und wie häufig hat man behandelt?
P: 10 Minuten pro Tag.
A: Es war nicht ein Apparat, den man Ihnen mit nach Hause gegeben hat?
P: Nein.

(g) A: Haben Sie bemerkt, ob Sie mit der Lagerung des Armes oder der Hand etwas beeinflussen können?
P: Daß die Schmerzen schlimmer werden? ... Sie werden mit der Bewegung stärker.
A: Können Sie mir sagen, was sie schlimmer macht?
P: Wenn ich die Hand so nach hinten halte, so – und sie dann drehe (gebeugter Unterarm, ventral flektierte Hand und Supination).

Umschriebene Abhängigkeit von der Willkürmotorik spricht für organischen Faktor.

A: Gibt es noch eine andere Bewegung, die schmerzt?
P: Bewegen der Hand nach oben und unten, seitwärts weniger.
A: Gibt es eine Haltung, die mehr schmerzt?
P: Abwärts tut mehr weh.
A: Und höher lagern? Bringt das etwas?
P: Nein.
A: Hat die Temperatur einen Einfluß?
P: Nein.
A: Und die linke, wie ist sie jetzt?
P: Ich arbeite mit der rechten.

7 A: Und Sie haben eine Malerlehre gemacht und wann beendet?
P: 1986.
A: War die jetzige Stelle Ihre erste nach der Lehre?
P: Ich bin am gleichen Ort geblieben, wo ich die Lehre gemacht habe.
A: Warum?
P: Weil mir so die Zeit während des Wehrdienstes bezahlt wurde.
A: Hat es andere Gründe gegeben in dieser Stelle zu bleiben oder wegzugehen?
P: Es hat mir dort gefallen zu arbeiten.

8 A: Wie war die Lehrzeit dort, können Sie mir darüber erzählen?

Ich möchte herausfinden, ob am Arbeitsplatz Konflikte bestehen.

P: Normaler Maleralltag, Streichen, Schleifen, Spachteln, ich konnte alles machen ... und dann hat es mit der Sehnenscheidenentzündung angefangen.

Ich finde kein Material, das die Vermutung des Hausarztes stützt, der Patient habe seinen Beruf nicht gern.

4 (b) A: Beschreiben Sie mir bitte die Art der Schmerzen, die Sie verspürt haben!
P: ... stechend war es nicht, es ist einfach ...
A: Gibt es sonst Wörter, die passen würden?
P: (Pause)
A: Es gibt viele Wörter wie schneidend, reißend, brennend ...

Die Vorschläge dürfen nicht suggestiv sein.

P: Eher ein Brennen ...
A: Können Sie es mit einem früher erlebten Schmerz vergleichen?
P: Mit einer Muskelzerrung, es ist ähnlich.
A: Wie beim Fußballspiel?
P: Ja.

5 A: Welche Verletzungen haben Sie beim Fußball gehabt?

P: Zweimal leichte Dehnungen in der Leiste.
A: Sonst?
P: Das Knie, man mußte einen Teil der Kniescheibe entfernen.
A: Hatten Sie sonst noch Unfälle?

Menschen, die aus seelischen Gründen dazu neigen, Schmerz zu erleiden, verunfallen auffällig häufig.

P: Nein, nichts.
A: Treiben Sie neben dem Fußball sonst noch Sport?
P: Joggen.
A: Fahren Sie Auto ... wie fahren Sie?
P: Normal, nicht zu schnell, nicht zu langsam.
A: Haben Sie je einen Unfall gehabt?
P: Einmal beim Rückwärtsfahren.

Besonders Männer mit psychogenem Schmerz drücken Aggressionen in der Fahrweise aus.

A: Skifahren?
P: Sehr viel.
A: Wie fahren Sie dort?
P: Ich wedle.
A: Fahren Sie riskant?
P: Ich kenne meine Grenzen.
A: Motorradfahren, Deltasegeln, Tauchen?
P: Nein.
A: Und Sie sind aus U.
P: Ja.
A: Und wie war Ihre Gesundheit als Kind?
P: Gut.
A: Können Sie mir mehr darüber sagen?

Offene Frage, stimuliert den Patienten, aktiv am Gespräch teilzunehmen.

P: Ich habe nie etwas gehabt, bis auf das Knie.
A: Blinddarm, Mandeln?
P: Nein.
6,8 A: Sind Sie allein aufgewachsen oder haben Sie noch Geschwister?
P: Ich bin der Jüngste, habe einen Bruder. Der Bruder, er ist 5 Jahre älter.
A: Was macht der Bruder?
P: Er ist Schreiner.
A: Wie geht es dem Bruder?
P: Gut; die Mutter hat auch Sehnenscheidenentzündung gehabt, oft. Der Vater hat ebenfalls am Handgelenk, er mußte einen Knorpel operieren lassen.

Das „auch" zeigt, daß die Krankheiten der Eltern vom Patienten bewußt mit der eigenen in Beziehung gebracht werden. Damit fallen sie als Modell für eine Konversion weg (s. Kap. 30).

A: Wie alt ist Ihre Mutter?
P: 44.
6 A: Hat sie außer den Sehnenscheidenentzündungen sonst noch gesundheitliche Probleme?
P: Nein, gar nichts.
A: Und als Sie ein Kind waren, wie war damals die Gesundheit der Mutter?

Bei Menschen mit psychogenen Schmerzen findet man in ihrer Jugend häufig kränkelnde Eltern.

P: Gut.
A: Wie alt ist der Vater?
P: 48.
7 A: Erzählen Sie mir, wie es zu Hause während Ihrer Kindheit zugegangen ist.

Patienten mit psychogenen Schmerzen haben häufig eine belastende Kindheit durchgemacht mit Brutalität zwischen den Eltern, zum Kind, Verwöhnen des Kindes, nachdem es geschlagen worden ist, Nähe zu ihm nur in Zeiten von Krankheit, nach Unfall ... (s. Kap. 36).

P: Fußball gespielt, lieber als Aufgaben gemacht; es gibt nicht viel zu erzählen, eine normale Kindheit.
A: Wie haben die Eltern Sie erzogen?
P: Von der 7. Klasse an durfte ich bis 21 Uhr draußen sein.
A: Wie war die Strenge der Eltern?
P: Gerade gut, nicht zu streng und nicht zu wenig streng.
A: Hat es je Schläge gegeben?
P: Es kam vor.
A: Durch Vater oder Mutter?
P: Eine Ohrfeige vom Vater.
A: Sind Sie je mit einem Gegenstand geschlagen worden?
P: Nein.
A: Und die auslösende Situation?
P: Zu recht.
A: Können Sie mir ein Beispiel geben?
P: Eine Scheibe eingeschlagen.
A: Hatte es damit sein Bewenden oder kam der Vater einige Tage lang darauf zurück?
P: Nein.
A: Wie war die Mutter zu Ihnen?
P: Normal, wenn ich die Aufgaben erledigt hatte.
A: War sie streng?
P: Normal, nicht zu streng.
A: Und im Vergleich zu den Geschwistern?
P: Wir sind gleich behandelt worden.
A: Hat je ein Geschwister während Ihrer Kindheit einen schweren Unfall erlitten?

(Und damit Anlaß zu Schuldgefühlen gegeben.)

P: Nein, nichts.
A: Und wie ist die Stimmung zwischen den Eltern?
P: Gut.

Die Stimme klingt ruhig, spontan.

A: Gibt es Spannungen zwischen ihnen?
P: Das Normale.
6,8 A: Wer gehört sonst noch zu Ihren wichtigen Bezugspersonen?
P: Die Freundin, die Kollegen, die Großeltern.
A: Leben noch alle Großeltern?
P: Der Großvater mütterlicherseits lebt nicht mehr.
A: Wann starb er?
P: Dies ist schon lange her.
A: Haben Sie zu einem der Großeltern eine besonders nahe Beziehung gehabt?
P: Zu allen gleich, wobei die Eltern des Vaters in U., die der Mutter in W. leben.
A: Und Sie haben eine Freundin, Kollegen; hat es bei diesen eine Krankheit, einen Unfall gegeben?

Ich denke an Modelle; an Schuldgefühle.

P: Ein Kollege hat die Kreuzbänder gerissen.
7 A: Befreundet sind Sie seit?
P: Anderthalb Jahren.
A: Und wie finden Sie, geht es?
P: Gut.
A: Hat sie Krankheit oder Unfall gehabt?
P: Nein.

A: Haben Sie das Gefühl, es sei stabil mit der Freundin oder bricht es auseinander?

P: Es ist stabil.

A: Wenn ich Sie frage, wie Sie sich diese Schmerzen erklären, was würden Sie antworten?

P: ... weiß nicht ...

A: Gut, Sie sind nicht Doktor.

P: Keine Ahnung, ich werde von Doktor zu Doktor geschickt, Sie sind der siebente.

A: Einer hat gesagt, Sie hätten ein schwaches Handgelenk. Was haben die anderen gesagt?

P: Außer von dem des Krankenhauses habe ich von keinem etwas gehört.

4 (b) A: Haben Sie am Handgelenk je etwas gesehen?

P: Hier (am rechten) war es hie und da leicht geschwollen, aber nie war es rot.

Diese nüchterne (nicht farbig-dramatische) Beschreibung paßt zu einer organischen Schmerzentstehung.

10 A: Habe ich etwas zu fragen vergessen?

P: ... Ich glaube es nicht.

A: Mir kommt noch etwas in den Sinn: Wenn man von Umschulen spricht, woran hat man gedacht?

P: An noch nichts.

A: Wie wäre es für Sie?

P: Mühsam ... es würde mich etwas anscheißen.

A: Käme etwas in Frage?

P: Computer, das hat Zukunft.

A: Wäre das von Ihrer schulischen Begabung her eine Möglichkeit?

P: Schon.

7 A: Was waren Ihre Lieblingsfächer in der Schule?

P: Rechnen, ... Turnen.

A: Wie wäre es, wenn Sie beim Malen eine Schiene tragen würden?

P: Das kann ich nicht, ich kann damit nicht streichen.

A: Da braucht man das Handgelenk!? Also geht es jetzt um vieles, beim Verstehen, was man tun kann. Und Sie möchten bei diesem Beruf bleiben?

P: Wenn's geht, schon.

10 A: Haben Sie sonst noch Fragen? ... Ich habe noch die Frage, ob Ihnen außer den Handgelenken noch andere Gelenke je weh getan haben?

P: Das Knie wie gesagt, sonst keines.

9 A: Und Sie haben gesagt, es habe keine anderen Krankheitszeichen wie Fieber, Schüttelfrost ...

P: Nein, nichts.

A: Gut, ich schlage vor, den Bericht zu lesen, die Be-
richte der Handchirurgie zu bestellen, dann Ihrem Hausarzt Bericht zu geben, so daß er Sie einbestellen kann, um mit Ihnen über all das zu sprechen, von dem Sie zu recht gesagt haben, Sie hätten noch nichts erfahren. Ich werde 14 Tage benötigen, bis der Bericht fertig sein wird. Haben Sie noch eine Frage? ...

P: ... Nein.

A: Gut, dann schließen wir hier ab; sollte ich noch Ideen haben nach dem Lesen der Unterlagen, wegen denen ich Sie nochmals sehen möchte, werde ich mit Ihnen telefonieren.

Dieser junge Mann hat nach meiner Auffassung somatogene Schmerzen, das Interview hat keine Anhaltspunkte für psychische Faktoren ergeben. Ich bestelle ihn deshalb für die 8 Tage später anberaumte Schmerzsprechstunde ein. Der Handchirurg untersucht ihn, betrachtet die Röntgenbilder der Hand und gibt die folgende Beurteilung: Druckdolenz des distalen Radioulnargelenks ohne pathologische Luxierbarkeit oder Synovitis. Schmerz bei dorsalem Druck auf den Processus styloideus ulnae. Normale Bewegungsamplituden, keine interkarpale Instabilität, jedoch allgemein laxe Kapselbandführung der Handwurzelknochen. Bei Prüfung der sagittalen Schublade Gelenk in Ulnarduktion recht gut subluxierbar. Im Röntgen steiler Anstellwinkel der Radiusgelenkfläche, rechts im Scaphoulnarspalt diskrete Unschärfe. Er denkt an **Überlastungsarthropathie durch repetitive Handgelenksbewegungen bei allgemein mäßiger Bandlaxität.** Ein Schaden des „triangular fibrocartilage complex" könnte vorliegen. Schlägt Lokalanästhesie-Infiltration des distalen Radioulnargelenks und in zweiter Linie ein Handgelenksarthrogramm vor.

Verlauf: Elf Monate später berichtet der Hausarzt: Infiltrationen brachten jeweils eine leichte Besserung. Die Arbeit als Maler konnte wieder aufgenommen werden, erzeugte aber immer wieder die gleichen Schmerzen. Der Patient begann berufsbegleitend eine Handelsschule zu besuchen mit dem Ziel, einen Beruf in der elektronischen Datenverarbeitung zu erlernen. Psychisch ging es ihm in den vergangenen Monaten deutlich besser als im letzten Jahr vor dem Konsilium.

13 Psychologische Aspekte der körperlichen Untersuchung

Daniel Vasella

Während der körperlichen Untersuchung findet zwischen Patient und Arzt auf psychischer Ebene eine – teilweise unbewußte – Interaktion statt. Mit wenigen Ausnahmen (Morgan und Engel, 1977) wird diesem Teil der Körperuntersuchung selbst in spezialisierten Lehrbüchern (Anschütz, 1973; Dahmer, 1984; Fritze, 1983; Lodewick, 1981) wenig Beachtung geschenkt. In diesem Kapitel werden auf der Grundlage psychoanalytischer Konzepte Regeln zur Durchführung der körperlichen Untersuchung erläutert und Möglichkeiten der Handhabung häufig auftretender schwieriger Situationen aufgezeigt.

13.1 Übertragung, Gegenübertragung[1] und Arbeitsbündnis

Im Vertrauen auf fachliche Kompetenz und menschliches Verständnis begibt sich der Kranke oder von Krankheit Bedrohte in ärztliche Behandlung. Sowohl die Krankheit wie die ärztliche Behandlung sind meist mit dem Auftreten verschiedener Unlustgefühle wie Angst, Scham oder Schmerz verbunden. Hilfsbedürftigkeit und Abhängigkeit vom Arzt begünstigen eine Regression (Rückzug auf lebensgeschichtlich frühe Arten des Denkens, Fühlens und Handelns) (Laplanche und Pontalis, 1982) des Patienten. Auf der Matrix früherer Konflikte in der Beziehung zu Autoritätspersonen, oft den Eltern, können in dieser Situation bezüglich Intensität oder Qualität unangebrachte Gefühle und Verhaltensweisen auftreten, die auf einer unbewußt verzerrten Interpretation von Verhalten und Eigenschaften des Arztes beruhen (Übertragungsreaktion) (S. Freud, 1905a, 1912b). Andererseits kann der Arzt, beeinflußt durch die eigene Vergangenheit und ohne die Besonderheit der Situation zu erkennen, der Übertragung des Patienten entsprechende Affekte erleben und in die Beziehung einfließen lassen (Gegenübertragungsreaktion) (Heimann, 1964).

---

[1] Es wird zwischen Übertragungs- und Gegenübertragungsreaktionen einerseits und der Übertragungsneurose andererseits unterschieden. Bei der Übertragungsneurose, die sich während der psychoanalytischen Behandlung in Beziehung zum Analytiker entfaltet, nehmen die Übertragungs- und Gegenübertragungsmanifestationen, im Gegensatz zu den Übertragungsreaktionen, eine feste Organisation an (Neuauflage der klinischen Neurose, s. Freud, S.: Erinnern, Wiederholen und Durcharbeiten, Ges. Werke, Bd. X, 134–135, und Laplanche und Pontalis, 1982).

> Die 71jährige Frau K. wird notfallmäßig wegen einer transitorisch ischämischen Attacke mit linksseitiger Arm- und Beinparese eingewiesen.
> Während der Prüfung der Muskeleigenreflexe fährt sie die Assistenzärztin so ungehalten an, sie habe von der Untersucherei genug, daß diese die Aufnahme des Neurostatus abbricht und sich bei der Patientin entschuldigt.
> Anläßlich der Besprechung des Falles wurde der Ärztin bewußt, daß die Patientin in ihr trotz regelrechter Untersuchungstechnik heftige Schuldgefühle ausgelöst und sie deshalb die Untersuchung abgebrochen hatte. Dies machte eine korrekte Beurteilung des neurologischen Beschwerdebildes unmöglich.

Zur Stellung einer Diagnose und Einleitung einer Therapie muß der Arzt während der Anamnese oft intime Fragen stellen, einen vollständigen Körperstatus erheben und meist weiterreichende Untersuchungen und Eingriffe vornehmen. Die Fähigkeit des Patienten, aufgrund einer partiellen Identifikation mit dem Arzt und im Einvernehmen mit ihm die Notwendigkeit diagnostischer und später therapeutischer Schritte einzusehen und sich ihnen trotz der damit verbundenen Unlustgefühle zu unterziehen, wird Arbeitsbündnis genannt (Greenson, 1966; Greenson und Wexler, 1982).

> Nachdem sich die Ärztin ihrer durch die Interaktion ausgelösten, der Situation quantitativ unangebrachten Gefühle (Gegenübertragung) klarwurde, teilte sie der Patientin ihr Verständnis für das Belastende der Erkrankung und Untersuchung mit. Auch erläuterte sie ihr die Notwendigkeit der Untersuchung, räumte der Patientin aber die Möglichkeit ein, diese jederzeit zu unterbrechen.

Empathie, Klärung des Arbeitsbündnisses und Stützung der Autonomie ermöglichten der Patientin eine bessere Zusammenarbeit (Adler, 1981). Der weitere Verlauf zeigte, daß sie auf psychische Belastungen oft mit Trotz und Rückzug reagierte und ihr Verhalten auch bei mehreren Krankenschwestern Schuldgefühle auslöste.

Von der Gegenübertragung im engeren Sinn sollte das Auftreten von Affekten unterschieden werden, die ausschließlich eine Signal- oder Triebentladungsfunktion erfüllen.

Herr M., ein 24jähriger Heroinsüchtiger, wird von der Polizei in somnolentem Zustand auf die Notfallstation eingewiesen, nachdem er bei einem Ladendiebstahl ertappt wurde.

Er ist verwahrlost, stinkt, gibt keine Antwort. Seine Fingernägel sind schmutzig und gelb verfärbt, seine langen Haare verfilzt. Sein Gesicht ist zur Hälfte durch Verbrennungsnarben entstellt, die linke Kornea völlig weiß vernarbt. Beide Arme sind tätowiert. In seinen Hosentaschen findet man Methadon und eine Packung mit verschiedenen Benzodiazepinderivaten.

Bei der Betrachtung des Patienten spürt der Arzt eine Mischung von Abscheu und Mitleid, aber auch Wut, diesen Patienten untersuchen und behandeln zu müssen.

Unter Annahme einer Tablettenintoxikation wird zur Magenspülung ein Schlauch eingeführt, worauf der Patient erbricht und sich selbst wie auch den Arzt besudelt. Nach erfolgter Magenspülung wird der Patient von der Krankenschwester gereinigt. Er erholt sich innerhalb kurzer Zeit und bestätigt eine Tabletteneinnahme in nicht-suizidaler Absicht als Ursache der Bewußtseinsstörung. Als erstes verlangt er eine Zigarette, die ihm der Arzt im Behandlungsraum verweigert, worauf sich der Patient anzieht und seine Drogen zurückverlangt. Er bekommt das ärztlich verordnete Methadon, nicht aber die Benzodiazepine. Als er protestiert und meint, so könne er nicht schlafen, wird ihm ein Neuroleptikum angeboten, das er ablehnt.

Der Arzt ist im Zweifel, ob er dem Patienten nicht einfach alle Medikamente ausliefern soll, um ihn los zu sein, und ob er überhaupt das Recht besitze, die Benzodiazepine zurückzuhalten. Ekel und Abscheu ermöglichten als Signaleffekte, die tiefe Verachtung zu erkennen, die der Patient im Umgang mit sich selbst an den Tag legte.

Diese Erkenntnis ermöglichte es, ihn nicht nur anzunehmen, sondern auch Mitleid zu empfinden, das der Patient sich selbst gegenüber wahrscheinlich verloren hatte. In seinem Aussehen und Verhalten lag eine Provokation, die im Arzt Wut auslöste.

Das Zurückhalten der nicht absolut notwendigen Drogen war ohne Zweifel richtig. Der Arzt verweigerte damit die Beteiligung an der Selbstaufgabe des Patienten. Dies wurde möglich, nachdem sich der Arzt seiner Wut und seines Bedürfnisses, diesen Patienten mit der Wegnahme der Drogen zu bestrafen, bewußt wurde.

Eine 28jährige, attraktive, auffallend sorgfältig geschminkte Patientin wird wegen eines akuten Schubes einer Colitis ulcerosa eingewiesen.

Während der Untersuchung des Abdomens entblößt sie unnötigerweise und ohne Aufforderung die Brüste und die Schamgegend. Zugleich sieht sie den Arzt sehr intensiv an. In ihrem Verhalten liegt zugleich etwas Verführerisches und Schamloses. Während der Palpation zittert die Hand des Arztes unmerklich.

In den folgenden Gesprächen berichtet die Patientin über ihre Frigidität, die Schwierigkeiten in der Beziehung zu ihrem Freund und ihre Promiskuität auf der Suche nach einer befriedigenden Beziehung.

Das Auftreten von Affekten im Arzt wie Ekel, Ungeduld, Wut oder Erregung ist ein diagnostisch und letztlich therapeutisch nützliches Phänomen. Es soll aber nicht ohne Erkennen der Motive den Arzt zu unreflektiertem Handeln oder einer verzerrten Beurteilung der Patienten verleiten. So stellt zum Beispiel das Anknüpfen einer sexuellen Beziehung mit einer verführerischen Patientin einen nicht selten vorkommenden Mißbrauch der ärztlichen Rolle dar (Kardener et al., 1973).

Die Behauptung bestimmter Ärzte, eine solche Beziehung diene der Befreiung der Gefühle der Patientin, ist unhaltbar: Die Patientin handelt ja aus der Übertragung und nicht aus einem reifen Entschluß heraus, der Bedürfnisse und Realität harmonisch vereint. Zudem erzeugt der Arzt in ihr die Illusion, daß frühkindliche, idealisierende Wünsche doch noch in Erfüllung gehen, anstatt ihr bei der Trauerarbeit zu helfen, daß der Erwachsene gerade darauf verzichten muß. Schließlich soll die Therapie zur Führung der Bedürfnisse durch das „Ich" verhelfen und dient nicht der Entfesselung der Triebe.

In unserem Beispiel diente die Sexualität lediglich zur Abwehr depressiver Verstimmungszustände bei einer narzißtischen Störung (Miller, 1979). Einen ebenso großen Fehler stellt der Verzicht auf eine indizierte Rektal- oder Genitaluntersuchung dar, zum Beispiel aus Rücksicht auf das Schamgefühl des Patienten oder wegen eigener Scham oder Furcht vor sexueller Erregung. Meist ist der Fehler Folge einer Übertragungs- oder Gegenübertragungsreaktion des Arztes auf den Patienten.

13.2 Die körperliche Untersuchung als Ursache von psychischem Streß

Neben unmittelbar durch die körperliche Untersuchung ausgelösten Unlustgefühlen (Scham, Schmerz, Angst) können frühere Erfahrungen oder Phantasien zu Angst vor Schmerz, vor Entdeckung einer gefährlichen Krankheit, vor Regression mit Verlust von Autonomie und Selbstkontrolle (liegende Position, Nacktheit, Berührung), vor Penetration oder Durchbohrtwerden (Racheninspektion, Genital- und Rektaluntersuchung), vor Beschmutzen (Inkontinenz) oder vor Verschlungenwerden (drohender Selbstverlust durch Anonymität, Untersuchung mittels hochtechnisierter Apparate) führen. Dieser psychische Streß löst je nach Persönlichkeit oder psychischer Entwicklung normal adaptive, bei latenten Konflikten akut pathologische oder bei seelischen Erkrankungen chronisch pathologische Reaktionen aus.

13.3 Averbale Kommunikation

Emotionen und Einstellungen äußern sich, wie es die obigen Fallbeispiele bereits darlegen, meist unbewußt und damit weitgehend der Kontrolle entzogen, mittels der Körpersprache. In der menschlichen Entwicklung diente sie früher als die Sprache vorwiegend der Vermittlung von Lust- und Unlustgefühlen. Zugleich lösten Berührung, Bewegungen, Gerüche, Sehen und Gesehenwerden Lust oder Unlust aus. Beim Erwachsenen sind Entblößung, Berührung intimer Körperstellen und Unterschreiten einer kritischen räumlichen Distanz meist nur dem Geschlechtspartner, Eltern und Kindern sowie wenigen Berufsgruppen, zu denen Ärzte gehören, erlaubt (Argyle, 1979). Diese für den Patienten intimen Handlungen sind für die Arzt-Patient-Beziehung charakteristisch, sie lösen meist Unlustgefühle aus und fördern die Regression. Diese Handlungen enthalten zwar eine Information (z.B. die im eigenen Interesse vom Patient akzeptierte Dominanz des Arztes), doch ist ihr erstes Ziel nicht Kommunikation. Ihre Ausdrucksformen können von jenen unterschieden werden, die vorwiegend Vermittler von Affekten und Gedanken sind, vorwiegend eine Triebentladungsfunktion besitzen oder vorwiegend Ausdruck somatischer Reaktionen auf psychischen Streß darstellen.

Dem Arzt geben bereits Kleidung, Körperhaltung und Gang wesentliche Informationen über den sozialen Status, die Persönlichkeit, die momentanen Affekte und eventuell die Erkrankung des Patienten. Er wird den Patienten hinsichtlich Körperpflege, Haltung, Gesichtsausdruck und Gestik bewußt beobachten, ob dieser zum Beispiel steif daliegt, dauernd herumnestelt, die Augen geschlossen hält, starr an die Decke schaut, bei jeder Berührung zusammenzuckt, zittert, die Hände zu Fäusten ballt, den Kopf wegdreht, beim Sprechen die Hand vor den Mund hält, die Arme wie ein Säugling hochhält oder sie schützend über den Körper legt, oder ob er immer wieder die Decke bis zum Hals hochzieht. Meist erlaubt die Kenntnis der Bedeutung einzelner Gesten, die vom kulturellen und sozio-ökonomischen Hintergrund beeinflußt werden, und deren Abfolge und der dadurch im Beobachter ausgelösten Gefühle, Rückschlüsse auf die Vorgänge im Patienten zu ziehen. Wenn durch Erkrankung, Anamneseerhebung oder Untersuchung ein so großer psychischer Streß entsteht, daß dieser mittels psychischer Abwehrmechanismen nicht mehr bewältigt werden kann, treten als Folge somatische Reaktionen in den Vordergrund. Diese sind Vorbereitung oder Ausdruck somatischer Bewältigungsmechanismen. Auf eine Vielzahl verschiedener Affekte reagiert der Körper vornehmlich mittels zweier Muster: bei Angst, Wut oder ähnlichen Affekten mit dem Flucht-Kampf-Muster, bei Hilf- und Hoffnungslosigkeit mit dem Rückzug-Konservierungs-Muster. Die damit verbundenen körperlichen Reaktionen müssen um so mehr vom Arzt wahrgenommen werden, als die zugrundeliegenden Affekte dem Patienten möglicherweise nicht bewußt sind und mit körperlicher Erkrankung verwechselt werden können. Diffuse Angst löst Blässe, kalte und feuchte Extremitäten, Zittern, einen erhöhten Muskeltonus, unregelmäßige Atmung, Mundtrockenheit, erhöhten systolischen und erniedrigten diastolischen Blutdruck aus. Dagegen bewirken zielgerichtete Angst und Wut eine Hautrötung, warme und trockene Extremitäten, ebenfalls einen erhöhten Muskeltonus

(geballte Fäuste, zusammengepreßter Kiefer), eine verstärkte, regelmäßige Atmung und eine Erhöhung sowohl des systolischen wie des diastolischen Blutdruckes. Ein herabgesetzter Muskeltonus mit entsprechender Körperhaltung und Mimik sowie verminderter motorischer Aktivität spricht – oft mit Gefühlen von Hilf- und Hoffnungslosigkeit verbunden – für das Vorliegen eines Rückzug-Konservierungs-Musters (Engel, 1976; Engel, nicht publiziert).

13.4 Das Verhalten des Arztes während der körperlichen Untersuchung

Aufgrund des bisher Gesagten wissen wir, daß die während der Untersuchung im Patienten auftretenden Affekte in ihm mehr oder weniger bewußte Phantasien bezüglich der Persönlichkeit des Arztes, dessen fachlicher Kompetenz, therapeutischer Einstellung und Fähigkeit zur Empathie wecken oder modifizieren. Für den Arzt gilt es, die realitätsgerechten Erwartungen des Patienten zu erfüllen. Die körperliche Untersuchung soll rational, entsexualisiert und frei von Aggression vor sich gehen. Persönlichkeit, Erwartungen, Schamgefühl und Autonomie des Patienten sollen respektiert, Angst und Ärger möglichst vermieden werden.

Der Untersuchungsraum soll ein ungestörtes Entkleiden (Vorhang, Ablage für die Kleider) und Untersuchen erlauben. Weder soll der Arzt von der Untersuchung weggerufen werden, noch sollen Patienten ein und aus gehen. Es soll genügend Zeit zur Verfügung stehen, um eine ruhige und vollständige Untersuchung durchführen zu können, deren Zeitpunkt, außer in Notfällen, mit dem Patienten vereinbart werden soll (Rücksicht auf Besuchs- und Essenszeiten). Der Arzt soll möglichst neutral erscheinen, weder allzu modebewußt noch vernachlässigt. Saubere Kleidung und Körperpflege (Haare, Hände, Körpergeruch) sowie Pünktlichkeit drücken Achtung und Respekt vor dem Patienten aus. Zusammen mit einer exakten und vollständigen Untersuchung, diese spiegelt das Bemühen und Interesse des Arztes wider, fördert das Erscheinungsbild des Arztes die Vertrauensbildung, wie auch das Vermeiden von kumpelhaftem Benehmen, da der Patient in seinem Arzt einen kompetenten und verständnisvollen Fachmann und nicht einen Kameraden erwartet. Eine kurze Information des Patienten über den geplanten Untersuchungsablauf und sein Recht, diesen jederzeit bei Schmerzen oder Mißbehagen unterbrechen zu lassen, fördern die Mitarbeit und verringern seine Angst durch die Gewißheit, als mündiger Patient an diagnostischen und therapeutischen Entscheidungen mitwirken zu können. Es ist nützlich, zu erwähnen, daß nach der Untersuchung deren Resultate zusammenfassend erläutert werden und zur Beantwortung eventueller Fragen noch Zeit übrigbleibt. Um entspannt und konzentriert untersuchen zu können, soll es sich auch der Arzt am Krankenbett möglichst bequem machen (z.B. Höherstellen des Bettes). Zur Vermeidung von

Verspannung und des Gefühls, überrumpelt zu werden, empfiehlt es sich, die Untersuchung an der Hand (Pulspalpation, Inspektion) zu beginnen (Morris, 1971), einem Körperteil, dessen Exponiertheit und Berührtwerden der Patient gewohnt ist, nach Körperregionen vorzugehen und intime (Genital- und Rektaluntersuchung) oder schmerzende Untersuchungen am Schluß durchzuführen. Der Patient soll nie völlig nackt daliegen oder -stehen, was durch Planung des Untersuchungsablaufes vermieden werden kann. Während der Untersuchung des Thorax bedeckt eine Decke den restlichen Körper, danach soll der Patient sein Hemd wieder anziehen dürfen. Die Berührung mittels eines Instrumentes wird meist als weniger intim als die Berührung mit den Händen empfunden. Deshalb soll die Auskultation des Bauches vor der Palpation durchgeführt werden. Der Verdacht auf konversionsbedingte Schmerzen wird erhärtet, wenn anläßlich von Druck und plötzlicher Entlastung mittels des Stethoskopes kein, mittels der Hand aber ein starker Schmerz ausgelöst wird. Die Palpation soll nicht mit kalten oder feuchten Händen durchgeführt werden. Es ist unangebracht, über einen Befund Erstaunen oder Erschrecken zu zeigen, zu scherzen oder gar abschätzige Bemerkungen zu machen. Stellt man eine pathologische Veränderung fest, untersucht man oft länger oder ein zweites Mal. Aufmerksame Patienten bemerken dies und geben es durch einen Blick, eine Unruhe oder eine Frage zu verstehen. Man schuldet ihnen dann eine kurze Erklärung. Ist ein Patient stark geschwächt, soll zur Erleichterung des Patienten und zur Entlastung des Arztes, der sich besser auf die Befunderhebung konzentrieren kann, zum Aufsitzen oder Wenden eine Hilfsperson zugezogen werden. Nach der Untersuchung bringt man beim bettlägrigen Patienten alles wieder in die gewünschte Ordnung.

13.5 Körperliche Untersuchung und Persönlichkeitstypen

Die Kenntnis der verschiedenen Persönlichkeitstypen, von denen die wichtigsten erwähnt werden, und deren Verhaltensmuster bei psychischem Streß (z.B. während der Untersuchung) erlaubt es dem Arzt, gewisse Persönlichkeitszüge, die der Abwehr dienen, zu erkennen und sein Verhalten anzupassen (Engel, nicht publiziert; Kahana und Bibring, 1964).

Abhängige Persönlichkeiten benötigen besondere Aufmerksamkeit und Umsorgung sowie Verständnis des Arztes. Während der Untersuchung stellen sie oft Forderungen oder Bitten an den Arzt, die sie selbst erfüllen könnten (z.B. Höherstellen des Bettes, Aufsitzen usw.). Werden ihre Wünsche erfüllt, bleiben sie kompensiert, ansonsten reagieren sie oft mit Angst, Hilflosigkeit oder Ärger.

Pseudounabhängige Persönlichkeiten verachten Schwäche und Hilflosigkeit. Sie befürchten die Abhängigkeit, verhalten sich kontrollierend und geben sich unabhängig, doch ist ein unterschwelliges Gefühl von Unsicherheit spürbar. Gibt ihnen der Arzt klare Anweisungen, befolgen sie diese in ihrer Rolle als Patient meist gern.

Mütterlich-fürsorgliche Persönlichkeiten versuchen dem Arzt jegliche Mühe zu ersparen, die Lösung jeglicher drohender Schwierigkeit vorwegzunehmen und zu helfen, unbedacht der Tatsache, daß sie gerade dadurch die Untersuchung oft stören. Erkennt man dieses Verhalten als Abwehr, kann man die Hilfe annehmen oder dem Patienten erklären, daß Nichthelfen die größte Unterstützung darstellt.

Hysterische Persönlichkeiten im jungen Alter verhalten sich während der körperlichen Untersuchung offenherzig bis gar exhibitionistisch und verführerisch, während ältere Patienten oft besonders schamhaft sind, insbesondere da die Krankheit, neben dem Alter mit Verlust an Attraktivität, eine Einbuße an Selbstwertgefühl nach sich zieht. Eine Bedrohung der körperlichen Integrität tritt bereits anläßlich kleiner Eingriffe auf. Während Frauen versuchen, sie durch Flirten abzuwehren, reagieren Männer mit Aggressivität oder – besonders Ärztinnen gegenüber – mit betonter Männlichkeit.

Gehemmte Persönlichkeiten als Varianten der hysterischen Persönlichkeit sind verletzlich, beeindruckbar, ängstlich, unsicher und anklammernd. Männer halten sich oft für wenig männlich oder, zu Unrecht, für homosexuell. Die körperliche Untersuchung stellt eine Belastung dar. Frauen sind besonders schamhaft. Ihre Unsicherheit zeigt sich durch dauerndes Fragen, ob sie es auch recht machen. Nimmt der Arzt eine warmherzige, beschützende Haltung ein, nimmt die Ängstlichkeit ab.

Zwanghafte Persönlichkeiten sind darauf angewiesen, auch während der körperlichen Untersuchung die Kontrolle über die Situation und die eigenen Gefühle behalten zu können. Eine genaue Statuserhebung mit schrittweiser Orientierung über den Ablauf und ruhiges, von Rationalität geprägtes Verhalten, das den Patienten in Entscheidungen miteinbezieht, verhindern unnötiges Entstehen von Angst.

Masochistische Persönlichkeiten betonen das Ausmaß ihres Leidens in der Hoffnung, sich damit die Zuneigung des Arztes zu sichern. Bemühungen, dem Patienten die Untersuchung möglichst angenehm zu gestalten, bewirken eher eine Zunahme des Klagens, während die Anerkennung ihres Leidens beruhigt (vgl. Kap. 36).

Aktive, aggressive und paranoide Persönlichkeiten werden ungern dominiert. Paranoiker werfen dem Arzt nicht selten zu Unrecht eine grobe oder unfachmännische Untersuchung vor. Die Versicherung, daß man wertvolle Kritik schätze und sich bemühe, es möglichst gut zu machen, führt weiter als jede Auseinandersetzung.

Schizoide Persönlichkeiten sind verschlossen, passiv, eventuell mißtrauisch oder scheinen unbeteiligt. Man begegnet diesem Persönlichkeitstyp oft bei verwahrlosten Patienten, die in schlechtem Zustand hospitalisiert werden. Wegen leicht auftretender Interesselosigkeit ist der Arzt bei diesen Patienten in Gefahr, oberflächlich zu untersuchen.

Primitiv-magische Persönlichkeiten sehen im Arzt eher einen Zauberer als einen Wissenschaftler. Sie können der Untersuchung, einer Berührung, einem Blick oder gewissen Instrumenten leicht eine besondere Bedeutung zumessen.

Narzißtische Persönlichkeiten sind von sich eingenommen, fühlen sich mächtig, bewundernswert und besitzen wenig Empathie. Falls man ihren Wünschen nicht entspricht, reagieren sie explosiv-aggressiv. Für die Untersuchung ist nur der beste Arzt gerade gut genug.

13.6 Besondere Situationen anläßlich der körperlichen Untersuchung

Der Patient fängt zum Beispiel an zu weinen:
Man soll die Untersuchung unterbrechen, dem Patienten ein Taschentuch reichen, was ihn oft zu beruhigen vermag, und ihn nach der Ursache der Tränen fragen. Bei fachgerechter Untersuchungstechnik sind sie meist Ausdruck von Angst, Schmerz oder Verletzung, Trauer über einen phantasierten oder durchgemachten Objektverlust oder eines aktivierten Konflikts.

Eine 19jährige, etwas älter erscheinende Patientin wird wegen leichtgradiger Peritonitis mit Aszites und subfebrilen Temperaturen eingewiesen.

Ihre Krankengeschichte, die vor zwei Monaten mit Abdominalschmerzen begann, erzählt sie scheu, mit gesenktem Kopf, oft lächelnd und mit hoher, kindlicher Stimme. Während der sorgfältigen und vorsichtigen Untersuchung durch Oberarzt und Assistenten beginnt sie zu weinen. Auf die Frage, ob es denn so schlimm sei, antwortet sie ja, worauf der Oberarzt lächelnd meint, dies sei doch fast nicht möglich.

In der Krankengeschichte wird die Verdachtsdiagnose eines Fitz-Hugh-Curtis-Syndroms festgehalten, mit der Bemerkung, daß eine psychogene Komponente eine Rolle spiele. Da der gynäkologische Befund unauffällig ist, zweifelt man an der Diagnose und schlägt der Patientin eine Laparoskopie vor, welche diese ablehnt.

Zwei Tage später übernimmt ein anderer Arzt die Abteilung. Erneut weint die Patientin anläßlich der Untersuchung. Er ist vom spürbaren Kummer und der Verzweiflung ergriffen, unterbricht die Statuserhebung und teilt der Patientin mit, er habe das Gefühl, sie sei sehr traurig. Schluchzend meint diese, sie versuche zu lächeln und ihre Trauer zu verbergen, etwas anderes habe doch keinen Sinn. Sie befürchte, wegen der Erkrankung die Arbeitsstelle zu verlieren, wie dies nach einem Suizidversuch vor acht Monaten geschehen sei. In der Folge berichtet sie über die Gründe der damaligen Verzweiflungstat. Sie stamme als einziges Kind aus der ersten Ehe der Mutter. Ein 9 Jahre jüngerer Halbbruder werde in allem bevorzugt. Vor vier Monaten habe die Mutter sie in betrunkenem Zustand zusammengeschlagen, bis Nachbarn eingegriffen hätten. Als die Mutter der Patientin vor zwei Monaten an den Kopf warf, sie werde es immer bereuen, vor 20 Jahren keine Abtreibung durchgeführt zu haben, zog die Patientin von zu Hause aus und nahm sich eine kleine

Wohnung. Als sie nach dem Vater gefragt wird, errötet sie und meint, die Mutter habe ihr jeglichen Kontakt mit ihm verboten. Sie bejaht die Frage, ob ihr Vater im „Milieu" tätig sei und fügt hinzu, daß er vielleicht gar nicht ihr Vater sei. Kurz nach dem Suizidversuch habe sich ihre Tante das Leben genommen. Seither habe sie ein schlechtes Gewissen, weil sie während des letzten Telefongesprächs mit der Tante nur kurz mit ihr gesprochen habe. Zweieinhalb Monate vor der Hospitalisation habe ihr Freund sie verlassen, und sie habe seit dieser Zeit an den Wochenenden zusätzlich als Kellnerin gearbeitet, um nicht allein in der Wohnung zu sein.

Nachdem ihr gedeutet wurde, daß sie die Untersuchung und die geplante Laparoskopie als zusätzliche Quälerei und Eingriff in ihre Integrität erleben mußte, und sie über die diagnostischen Schwierigkeiten der Ärzte informiert wurde, ließ sie sich ruhig untersuchen und stimmte der Laparoskopie zu. Es fand sich eine seröse Peritonitis. Wenig später wurde serologisch ein signifikanter Chlamydientiter bestimmt und eine entsprechende antibiotische Therapie eingeleitet. Nach einem weiteren Gespräch entschloß sich die Patientin, die sich anläßlich dieser Pubertätskrise in einer ausgeprägt depressiven Verstimmung befand, zu einer Psychotherapie.

Dieses Fallbeispiel macht unter anderem deutlich, daß die Tränen weniger durch eine schmerzhafte Untersuchung als durch aktivierte Konflikte und die damit verbundene Trauer verursacht wurden, was alle beteiligten Ärzte richtig bemerkten.

Wesentlich ist jedoch, auf Hintergründe heftiger Emotionen, die während der Untersuchung auftreten, einzugehen und die zugrundeliegenden Motive zu verstehen. Die Beurteilung der Entstehung und Verarbeitung einer Erkrankung wird erschwert, falls die psychosozialen Daten anläßlich der Anamneseerhebung ungenügend berücksichtigt wurden (vgl. Kap. 12).

13.7 Forderung des Patienten nach nochmaliger Untersuchung oder weiterführender Abklärung

Die Forderung nach nochmaligen oder weiterreichenden Untersuchungen ist ein Zeichen von Mißtrauen und Unzufriedenheit bezüglich der bisherigen Resultate oder der Beurteilung und vorgeschlagenen Therapie des Arztes. Sie stellt ein Maß für die Bedrängnis und Angst bei Bestehen unverständlicher Körpersymptome dar. Aus naheliegenden Gründen tritt diese Situation vor allem bei Patienten mit somatischen Beschwerden psychischen Ursprungs auf. Ist sich der Arzt seiner Diagnose sicher, und hält er eine weitere Untersuchung unter Berücksichtigung der damit verbundenen Kosten und Risiken im Gegensatz zum Patienten nicht für indiziert, geht er fehl, diese trotzdem durchzuführen, in der Hoffnung, den Patienten zu beruhigen und um ihm zu beweisen, daß er an keiner organischen Krankheit leidet. Manche Ärzte fürchten, den Patienten zu verlieren oder versuchen, mittels einer möglichst vollständigen Abklärung jeglichem späteren Vorwurf mit dem damit verbundenen Prestigeverlust zu entgehen, d.h. sich selbst durch eine Untersuchung zu beruhigen. Diese Gefahr ist besonders dann groß, wenn der Patient durch sein Verhalten im Arzt Angst auslöst und dieser den Affekt nicht bewußt wahrnimmt. Wohl kann ein unauffälliger Befund den Patienten kurzdauernd erleichtern, doch zieht ein normales Resultat bei Weiterbestehen der Beschwerden auch eine Enttäuschung nach sich, hoffen doch diese Patienten im Gegensatz zu den geäußerten Befürchtungen meist unbewußt, es werde endlich eine organische Ursache für ihr Leiden festgestellt. Damit können sie dem Bewußtwerden ihrer Konflikte entgehen und dem Gefühl, als Simulanten angesehen zu werden, und einen vollen Krankheitsgewinn beziehen. Meist gelingt es durch Klärung und Deutung der Hintergründe des Wunsches nach erneuter oder weiterführender Untersuchung im Rahmen einer tragfähigen Beziehung, den Patienten vor unnötigen Abklärungen zu schützen. Eine Ausnahme stellen hypochondrische Patienten dar, die häufig nur mittels einer einfachen körperlichen Untersuchung neben stützender Gesprächstherapie für manchmal längere Zeit beruhigt werden können. Bei allen Patienten mit körperlichen Beschwerden psychischen Ursprungs wirken sich regelmäßige Konsultationen, unabhängig vom Bestehen oder Fehlen von Symptomen, günstig aus.

13.8 Körperliche Untersuchung und Psychotherapie

Als psychosomatisch tätiger Arzt kommt man regelmäßig in die Lage, Patienten, welche zur Abklärung oder zur Behandlung zugewiesen werden oder selbst den Arzt wegen körperlicher Beschwerden aufsuchen, zu untersuchen und bei ihnen später eine Psychotherapie durchzuführen. Die Untersuchung erlaubt, sich ein eigenes Urteil über die Erkrankung zu bilden. Der Patient erwartet sie und fühlt sich, falls darauf verzichtet wird, nicht ernstgenommen. Die somatische und psychische Behandlung in der Anfangsphase durch ein und denselben Arzt erlaubt, der Dichotomie in Körper und Psyche entgegenzuarbeiten. Diese Haltung ermöglicht die oft schwierige Herstellung eines psychotherapeutischen Arbeitsbündnisses. Die körperliche Untersuchung stellt unter diesen Umständen zwar einen Parameter (Eissler, 1953) dar, führt aber nur selten zu großen Übertragungsschwierigkeiten, wenn sie von seiten des Arztes regelrecht durchgeführt wird (Ruffler, 1953/54).

Anders verhält es sich, nachdem eine eigentliche Psychotherapie begonnen wurde und sich eine Übertragungsneurose möglichst ungestört entwickeln soll. In diesem Fall ist es ratsam, sich die Aufgabe der psychischen und somatischen Behandlung mit einem Kollegen zu teilen. Man erspart sich damit die Schwierigkeit, Symptome neu auftretender Erkrankungen als solche erkennen zu müssen und die entsprechende Abklärung und Behandlung durchzuführen. Letztere können zu unüberwindbaren Schwierigkeiten in der Übertragungsneurose des Patienten (z.B. körperliche Untersuchung als Inzesthandlung) führen und jegliche Aussichten auf einen Erfolg der Psychotherapie zunichte machen (S. Freud, 1916). Nach einer anfänglichen Besserung können körperliche Symptome auch als Folge eines verstärkten Widerstandes (S. Freud, 1926), der bearbeitet werden muß, auftreten. Besonders in diesem Fall ist Verständnis und Respekt für die Psychotherapie von seiten des Kollegen, welcher die somatische Betreuung gewährleistet, von großem Wert. Kann mit dem Patienten zu einem frühen Zeitpunkt die Indikation einer Psychotherapie besprochen werden und erkennt er deren Notwendigkeit, liegt die Überweisung an einen Psychotherapeuten nahe. Es ist wichtig, darauf zu achten, daß der Patient sie nicht als eine Abweisung erlebt und daß der Arzt ihm bei der Auswahl des Psychotherapeuten behilflich ist. Manchmal wünschen Patienten, deren Hemmungen besonders groß sind, daß man sie beim Psychotherapeuten anmelde, was am besten in ihrem Beisein geschieht. Wird die Überweisung nicht sorgfältig mit dem Patienten vorbereitet, mißlingt sie sehr häufig.

13.9 Die Untersuchung sterbender Patienten

Es ist bekannt, daß Ärzte im Zimmer unheilbar kranker Patienten häufig wenig Zeit verbringen. Dies hat je nach Arzt verschiedene Gründe, wie Angst vor Auseinandersetzung mit dem Tod, Verlust von Omnipotenzgefühlen, Kränkung der narzißtischen Befriedigung, heilen zu können, Scham oder allgemeines Desinteresse. Gewisse Ärzte bewältigen die eigene Angst vor dem Tod, indem sie dem Patienten immer neue Medikamente verordnen, unnötigerweise Opiate und Psychopharmaka verabreichen oder aber nützliche Medikamente vorenthalten, um den Sterbeprozeß zu beschleunigen. Andere Ärzte werden von einem Gefühl der Machtlosigkeit befallen und widmen sich unter Vernachlässigung des somatischen Leidens völlig der psychischen Betreuung. Da der Arzt durch den bevorstehenden Tod weniger als der betroffene Patient bedroht wird, gibt er in Kenntnis der Prognose die Hoffnung meistens früher auf. Die meisten Kranken besitzen die Fähigkeit, das Verhalten der Ärzte genau zu beobachten und bemerken rasch, wenn sie aufgegeben werden. Dies kann sich dadurch äußern, daß der Arzt den Patienten nicht mehr untersucht. Die gelegentliche, aber regelmäßige Untersuchung, die das Gespräch mit dem Patienten jedoch nicht ersetzt, ist nicht nur als ein Zeichen von Interesse und Bemühen um das körperliche Wohlergehen des Patienten indiziert, sondern es läßt sich aufgrund der erhobenen Befunde hie und da eine sinnvolle Anpassung der Therapie vornehmen. Die Untersuchung kann dem Patienten auch Anlaß zu Fragen über den Verlauf der Krankheit geben, d.h. einen wichtigen Bestandteil in der Krankheitsverarbeitung bilden (vgl. Kap. 65).

14 Testpsychologie

Walter Thomas und *Othmar W. Schonecke*

14.1 Begriffsbestimmung

Unter einem psychologischen Test versteht man nach Lienert (1969) ein „wissenschaftliches Routineverfahren zur Untersuchung eines oder mehrerer empirisch abgrenzbarer Persönlichkeitsmerkmale mit dem Ziel einer möglichst quantitativen Aussage über den relativen Grad der individuellen Merkmalsausprägung". Diese Definition begrenzt die Bedeutung des Wortes „Test", so daß ein psychologisches Untersuchungsverfahren nur dann Anspruch auf die Bezeichnung Test hat, wenn man bei der Testkonstruktion die folgenden Bedingungen berücksichtigt:

— Zunächst muß das Verfahren wissenschaftlich begründet sein. Tests sind also nicht irgendwelche beliebigen Untersuchungsmethoden, die sich einem Untersucher im Laufe der Zeit bei der Diagnosestellung als nützlich erwiesen haben.
— Einem Test kommt durch die Tatsache, daß man dieses Verfahren routinemäßig einsetzen kann, die Bedeutung eines Handwerkzeugs zu. Man provoziert unter standardisierten Bedingungen ein Verhalten, das zwar auch ohne Testanweisung der freien Beobachtung zugänglich ist, auf dessen Auftreten der Untersucher aber unter Umständen sehr lange warten kann.
— Nur dann, wenn man die relative Position einer untersuchten Person hinsichtlich des zu testenden Merkmals im Vergleich zu einer Referenzpopulation auf einer Skala bestimmen kann, sind Aussagen über ein „mehr oder weniger" möglich.
— Diejenigen Eigenschaften, Bereitschaften, Fähigkeiten oder Fertigkeiten, die ein Test untersuchen soll, sind der direkten Beobachtung nicht zugänglich. Sie müssen aber empirisch bestimmt worden sein und dürfen nicht auf vorwissenschaftlichen Annahmen beruhen.

Aus den bisherigen Darstellungen ergibt sich, daß Tests zwei Aspekte zum Inhalt haben, zum einen die Aufgabe oder Untersuchungssituation, die ein bestimmtes Verhalten auslösen soll, und zum anderen eine Anweisung, wie man von dem beobachteten Verhalten auf die diesem zugrundeliegenden Persönlichkeitsmerkmale schließen kann. Im folgenden Überblick sollen die Methoden beschrieben werden, mit denen Tests konstruiert werden, und wie man die Konstrukte, die Tests zu untersuchen vorgeben, zu erfassen sucht.

14.2 Geschichtlicher Überblick

Das Wort „Test" hat in der heutigen Umgangssprache mehrere Bedeutungen. So versteht man darunter nicht nur ein psychologisches Untersuchungsinstrument, sondern auch ein statistisches Prüfverfahren, z.B. t-Test, aber auch eine Untersuchung von Waren oder Dienstleistungen, man denke nur an die Stiftung Warentest. Das Wort Test übernahm man Anfang des 20. Jahrhunderts aus dem Englischen. Dort bedeutet es soviel wie Probe. Laut Dudens Herkunftswörterbuch (1963) läßt es sich auf das altfranzösische Wort „test" zurückführen, was soviel wie irdener Topf bzw. Tiegel für alchimistische Experimente bedeutet.

Zwar gibt es erst seit einigen Jahrzehnten eine wissenschaftlich begründete Testmethodik, jedoch haben schon immer Menschen versucht, Methoden oder Techniken zu entwickeln, mit denen sie feststellen wollten, wie gut ein einzelner für wichtige gesellschaftliche Funktionen geeignet ist. Diese Versuche lassen sich als die Vorläufer der modernen psychologischen Tests ansehen. In praktisch allen archaischen menschlichen Gesellschaften kann man heute noch Initiationsriten beobachten. Durch sie soll geprüft werden, ob der jugendliche Stammesgenosse den Anforderungen eines Erwachsenen genügt. Meist handelt es sich um Methoden, mit denen man Mut und Selbstbeherrschung, manchmal auch die Verstandeskraft prüft. In der geschichtlichen Überlieferung vieler Völker finden sich Hinweise darauf, daß man sich bemühte, Prüfungen zu entwickeln, mit denen man diejenigen Personen im voraus erfassen konnte, die in für die Allgemeinheit bedeutsamen oder gefahrvollen Situationen nicht versagen würden. Diese Prüfungen wurden nötig, weil sich entsprechende Fähigkeiten im Alltagsleben nur schwer oder gar nicht beobachten lassen.

In die Psychologie führte der Psychologe James McKeen Cattell das Wort Test 1890 durch das Buch „Mental tests and measurement" ein. Aber schon etwas früher begann der französische Forscher Galton (1883) sich für individuelle Unterschiede zu interessieren. Damit stellte er sich ebenso wie andere Forscher in den USA und Europa, die Testverfahren auszuarbeiten begannen, gegen die experimentelle Psychologie jener Zeit. Diese bemühte sich damals in ihren psychophysikalischen Laboratorien, die allen Menschen gemeinsamen Gesetze des Denkens, der Wahrnehmung oder des Handelns zu erforschen.

In dieser ersten Phase der Testpsychologie versuchte man, psychische Einzelfunktionen wie Gedächtnis, Konzentrationsfähigkeit, Reaktionsgeschwindigkeit oder Wahrnehmungsfunktionen zu prüfen. Erstaunlicherweise haben einige dieser ersten Testverfahren mit nur unwesentlichen Modifikationen bis zum heutigen Tage überdauert. Der jetzt noch gebräuchliche Pauli-Test nach Arnold (1970) zur Erfassung der Konzentration über einen längeren Zeitraum bei relativ einfach strukturierten, stereotypen Rechenaufgaben, geht beispielsweise auf einen Rechentest des deutschen Psychiaters Kraepelin (1895) zurück. Ebenso haben sich Techniken, die die Reaktionsgeschwindigkeit erfassen, bis auf die apparative Ausstattung nur unwesentlich geändert. Eine Reihe von Gedächtnistests läßt sich auf die Lückenprobe des Psychologen Ebbinghaus (1897) zurückführen.

Die nächste Phase der Testentwicklung zeichnet sich dadurch aus, daß man nun versucht, die hinter diesen Einzelfunktionen stehenden psychischen Konstrukte zu erfassen. Beispielhaft wird diese Entwicklungsgeschichte der Tests im folgenden an den Persönlichkeitsfragebogen dargestellt, die für die psychosomatische Praxis und Forschung von besonderer Bedeutung sind.

14.2.1 Zur Geschichte der Persönlichkeitsfragebogen

Persönlichkeitsfragebogen bestehen aus Listen mit Fragen und vorgegebenen Antwortmöglichkeiten, von denen der Proband sich für eine entscheiden soll. Es liegt somit eine gebundene Antwortform vor, der Proband kann nicht individuell oder beliebig antworten. Im Vordergrund der Entwicklung dieser Testart stand nach Angleitner (1976), der eine ausgezeichnete Einführung in die Konstruktionsprinzipien von Persönlichkeitsfragebogen sowie eine kritische Überprüfung und Bewertung aller bedeutenden deutschsprachigen Fragebogen gibt, die Diagnostik psychischer Fehlanpassung. Darüber hinaus wurden diese bevorzugt als Instrumente für schul- und berufspsychologische Eignungsuntersuchungen entwickelt.

Die Phase der intuitiven Testkonstruktion

Der erste wirkliche Persönlichkeitsfragebogen, zumindest für den Bereich des neurotischen Fehlverhaltens, ist das „Personal Data Sheet" von Woodworth aus dem Jahre 1917. Man entwickelte ihn, um in Massenerhebungen seelische Störungen bei Rekruten zu entdecken. Der Test bestand aus 116 mit „ja" oder „nein" zu beantwortenden Items und stellte ein abgekürztes psychiatrisches Interview dar. Probanden, die eine gewisse Anzahl dieser Items in Auswertungsrichtung beantworteten, wurden anschließend psychiatrisch interviewt.

Bei den ersten Fragebogen standen intuitive Konstruktionsprinzipien im Vordergrund: aufgrund von Expertenaussagen formulierte man Items, die das zu messende Konstrukt, in diesem Fall „emotionelle Fitness", erfassen sollten. Eine Nachprüfung der Gültigkeit der Expertenmeinung und damit der Validität der einzelnen Items entfiel. Zu Beginn der Testkonstruktion hatten Woodworth und seine Mitarbeiter eine Stichprobe von 200 Items zusammengestellt, die man einer Gruppe von Rekruten und College-Studenten zur Beantwortung vorlegte. Nachdem man zuvor die Verschlüsselungsrichtung der einzelnen Items festgelegt hatte, schied man jene Fragen aus, denen mehr als ein Viertel der Analysestichprobe zustimmte, da mit ihnen Symptome erfaßt wurden, die auch in unauffälligen Stichproben relativ häufig vorkamen.

Bei der intuitiven Testkonstruktion lassen sich nach Hase und Goldberg (1967) zwei Vorgehensweisen unterscheiden, nämlich eine rationale Strategie ohne direkten Bezug zu einer psychologischen Theorie, sowie eine mehr theoriegeleitete Vorgehensweise, bei der man die Testfragen aufgrund theoretisch fundierter Annahmen über das zu untersuchende Konstrukt formuliert. Für beide Strategien gilt, je exakter und ausführlicher man das zu untersuchende Merkmal definiert, desto eher kann nachgeprüft werden, inwieweit es verschiedenen Skalenkonstrukteuren gelingt, mit derselben Merkmalsdefinition gleichwertige Skalen zu erarbeiten. Durch diese Kontrollmöglichkeit relativiert sich der Nachteil der intuitiven Testkonstruktion, der in einer starken Abhängigkeit der Skala von der Erfahrenheit und Geschicklichkeit des Testkonstrukteurs besteht.

Während Woodworths „Personal Data Sheet" eine Einzelskala war, erwies es sich bald als nützlich, mehrere Persönlichkeitsskalen in einem Test zusammenzufassen. Für diese mehrdimensionalen Fragebogen, sog. Persönlichkeitsinventare, ist Bernreuters Personality Inventory „BPI" aus dem Jahre 1931 der bekannteste frühe Vertreter. Bei der Konstruktion dieses Fragebogens wurden, wie es bis heute noch üblich ist, Items und auch ganze Skalen aus bereits veröffentlichten Fragebogen übernommen.

Die Phase der externalen Testkonstruktion

Um die Mitte der dreißiger Jahre dieses Jahrhunderts ändern die Testkonstrukteure ihre Strategie, mit der sie geeignete Items für Fragebogentests auswählen. Die Auswahl stützt sich nun in der Regel nicht mehr auf intuitive Vorannahmen oder rationale Überlegungen, sondern gründet auf empirischen Analysen, indem man die Gültigkeit der einzelnen Testfragen an Kriterien- und Kontrollgruppen, z.B. Depressive und Gesunde, überprüft. Nur die Items werden in den endgültigen Test aufgenommen, in denen sich die beiden Gruppen hinsichtlich ihrer Beantwortungstendenzen deutlich unterscheiden. Voraussetzung für diese Art der Testkonstruktion ist zum einen ein großer Itempool, von dem man annimmt, daß er das zu untersuchende Merkmal erfaßt, und zum anderen die Existenz eines geeigneten Außenkriteriums. Ein entscheidender Unterschied zur intuitiven Teststrategie besteht darin, daß man für die einzelnen Items noch nicht festlegt, wie die Antworten auf die einzelnen Fragen in der zukünftigen Skala überhaupt bewertet werden sollen. Diese Festlegung erfolgt erst, nachdem die Items sowohl von der Kriteriengruppe, die das zu untersuchende Merkmal extrem ausgeprägt aufweisen soll, als auch von einer unauffälligen, vergleichbaren Kontrollgruppe beantwortet wurden. Von dieser Gruppe nimmt man an, daß sie sich auf dem entgegengesetzten Pol des zu messenden Merkmals einordnen läßt. Man wählt für den endgültigen Test nur diejenigen Items aus, die die beiden Gruppen bedeutsam trennen. Dann verschlüsselt man die einzelnen Fragen in derjenigen Richtung, die den von der Kriteriengruppe bevorzugten Antworten entspricht.

Mit diesem Wandel des Konstruktionsprinzips von Persönlichkeitsfragebogen ändert sich zugleich die Auffassung, wie die Antworten der getesteten Personen auf die einzelnen Testfragen zu interpretieren sind. Während man zuvor aus den jeweiligen Antworten unmittelbar auf ein Vorhandensein des erfragten Symptoms schloß, erkannte man nun, daß die Annahme, die Antworten der Probanden als Tatsachenberichte anzusehen, erhebliche methodische Schwierigkeiten mit sich brachte. Wenn beispielsweise jemand die Frage: Ich habe häufig Kopfschmerzen, mit „stimmt" beantwortet, so weiß man im Grunde genommen nicht, was der Betreffende sich unter häufig vorstellt und wie er den Begriff Kopfschmerzen interpretiert. Darüber hinaus stellt diese Art der Iteminterpretation einen hohen Anspruch an die Fähigkeit des Untersuchten, sich selbst zu beurteilen und seine Bereitschaft, bei der Untersuchung mitzuwirken. In der Phase der externalen Testkonstruktion glaubte man nun, diese Probleme als nicht mehr relevant ansehen zu müssen, da man die Items empirisch analysierte. Es interessierte nicht mehr, warum zwei Versuchspersonen eine Testfrage unterschiedlich beantworteten, entscheidend war nun, daß man die getesteten Personen anhand der Summe aller differenzierenden Items überzufällig richtig klassifizieren konnte.

Auf diese Art und Weise wurde u.a. der wohl bekannteste Persönlichkeitsfragebogen, das Minnesota Multiphasic Personality Inventory „MMPI", von Hathaway und McKinley (1943) entwickelt. Zielsetzung bei der Konstruktion des MMPI war es, eine bessere Zuordnung von Patienten zu den Kategorien der Kraepelinschen Typologie zu ermöglichen. Die Items entstammten psychiatrischen Lehrbüchern und Prüfungsformen, wurden aus anderen Tests übernommen und/oder aufgrund der klinischen Erfahrung der Autoren formuliert. Mit diesen Fragen beabsichtigte man, einen weiten Bereich menschlichen Verhaltens zu erfragen, nicht zuletzt, um aus diesem Itempool zukünftig weitere Skalen entwickeln zu können.

Diese Fragen mußten verschiedene Kriteriengruppen, die als repräsentativ für die zu bildenden Skalen angesehen wurden, sowie verschiedene Kontrollgruppen beantworten. Die Stichprobengrößen der Kriteriengruppen für die einzelnen zehn klinischen Skalen des MMPI waren erstaunlich klein, sie umfaßten im einzelnen 50 Hypochondriker, 50 Depressive, 50 Hysteriker, eine unbekannte Anzahl von Psychopathen, 13 männliche Homosexuelle, eine unbekannte Anzahl von Paranoikern, 20 Psychastheniker, 50

Schizophrene, 24 manisch Kranke sowie 50 Personen mit hohen und 50 Personen mit niedrigen Werten im Test „soziale Introversion". Zu jeder Skala wählte man diejenigen Items aus, in denen sich die Kriterien- und Kontrollgruppen bedeutsam unterschieden. Das Konstruktionsverfahren der MMPI-Skalen bedingt, daß sich mehrere Kriteriengruppen in gleichen Items von ihren Kontrollgruppen unterscheiden können. Folglich sind viele Items gleichzeitig in unterschiedlichen Skalen vertreten. Diese Tatsache führt zu artifiziellen Korrelationen der einzelnen Skalen, die gemeinsame Konstrukte vortäuschen.

Zusammenfassend läßt sich zu dieser Strategie sagen, daß der Iteminhalt im Grunde genommen unwichtig ist. Auch unsinnig erscheinende oder widersinnige, sog. „subtile" Items werden dann als brauchbar angesehen, solange sich in ihnen Kriterien- und Kontrollgruppe unterscheiden. Der Nachteil der externalen Strategie besteht zum einen darin, daß ein gewisser Prozentsatz der differenzierenden Itembeantwortungen rein zufällig auftritt und nicht sicher von den wirklich differenzierenden Items unterschieden werden kann. Zum anderen wurde das Problem der Gültigkeit der Expertenaussage nicht gelöst. Es wurde im Grunde nur von der Formulierung der Items auf die Definition der Kriteriengruppen verschoben.

Die Phase der internalen Testkonstruktion

Bei Nachanalysen von external konstruierten Fragebogentests traten „Schönheitsfehler" auf, indem einzelne Items oft wesentlich bedeutsamer mit einer fremden Skala als mit der eigenen korrelierten. Weiter fand man, daß einzelne Skalen eines Fragebogens oft in starker Beziehung zueinander standen. Dieser Effekt trat nicht nur bei solchen Skalen auf, die zum Teil gemeinsame Items enthielten, sondern auch bei vollkommen unterschiedlich zusammengesetzten. Bald vermutete man, daß sich Gesetzmäßigkeiten, die „eigentlichen" Persönlichkeitsdimensionen, hinter den Mustern, die sich in diesen Zusammenhängen andeuteten, verbargen.

Die Faktorenanalyse, deren Entwicklung fast parallel mit der der Persönlichkeitsfragebogen verläuft, findet jetzt als neue statistische Methode Einzug in die Psychologie. Bei Untersuchungen im psychologischen Bereich handelt es sich nicht um direkt meßbare Größen, wie z.B. Blutdruck, Körpertemperatur oder Gewicht. In den Reaktionen der Probanden auf bestimmte Teststimuli, wie z.B. Fragen, Aufgaben oder Lichtreize, werden individuelle Differenzen der zugrundeliegend gedachten hypothetischen Konstrukte erfaßbar. Solche sind z.B. Intelligenz, Gedächtnis oder im persönlichkeitspsychologischen Bereich Neurotizismus, Aggressivität oder Extraversion. In der psychologischen Forschung entwickelte man mathematische Methoden, die es erlauben, die den einzelnen Variablen zugrundeliegenden gemeinsamen Faktoren oder Dimensionen zu berechnen. Diese Methoden faßt man unter dem Oberbegriff „Faktorenanalyse" zusammen. Mit ihrer Hilfe konnte man die Vermutungen über die Persönlichkeitsdimensionen überprüfen, sowie die Beziehungen der einzelnen Items zueinander untersuchen und bei der Skalenbildung berücksichtigen.

Man war durch diese neuen Methoden auf einmal nicht nur unabhängig von geeigneten Kriteriengruppen, sondern auch von den theoretischen Annahmen über die Außenkriterien. Die internale Strategie postuliert, daß man das Merkmal dann optimal messen kann, wenn man die interne Struktur der Items berücksichtigt. Die Zugehörigkeit von Items zu einer Skala bestimmt der Testkonstrukteur, indem er ihre interne Konsistenz, signifikante Trennschärfen oder die Höhe und Richtung ihrer Ladungen auf einem Faktor berechnet. Ein bekanntes Beispiel für diese Art der Testkonstruktion ist das Sixteen Personality Factor Questionnaire „16-PF" von R. B. Cattell und Eber (1964).

Durch diese neue Methode wurde es relativ einfach, die einzelnen Dimensionen eines Fragebogens exakt zu bestimmen, indem man den Itempool einer Stichprobe von Probanden zur Beantwortung vorlegt. Nach einer Faktorenanalyse nimmt man von denjenigen Items, die jeweils einen Faktor bilden, an, daß sie einer gemeinsamen Skala angehören. Deren Benennung erfolgt meist nach Durchsicht der Texte der betreffenden Items, sie ist daher nur in seltenen Fällen unabhängig von der subjektiven Eindrucksbildung des Testkonstrukteurs. So gibt es Skalen unterschiedlicher Tests, die so hoch korrelieren, daß man sie fast als Paralleltests einsetzen kann. Nach Meinung der einzelnen Testautoren sollen diese Skalen jedoch Depression, Angst, Neurotizismus oder Beschwerden messen.

Vor allem nach der rapiden Entwicklung der elektronischen Datenverarbeitung und der Verfügbarkeit von Statistikprogrammen, bei denen der Forscher in einem der Umgangssprache ähnelnden „Dialekt" eine Fragestellung formulieren kann, kam es zu einem rapiden Ansteigen der veröffentlichten Fragebogen bzw. zu „Unter-" oder „Zusatzskalen" bestehender Tests. Für das MMPI nahm diese Tendenz inflationäre Ausmaße an, laut Dahlstrom (1975) gibt es für das MMPI mehr Zusatzskalen als überhaupt Items vorhanden sind.

Obwohl man nun in der Lage war, Fragebogen nach empirischen Konstruktionsprinzipien zu entwickeln, wurden sehr oft zusätzliche Kontrollskalen, die die Bezeichnung Lügenskalen, Korrekturskalen oder Offenheitsskalen erhielten, in die Persönlichkeitsinventare aufgenommen. Dies deutet darauf hin, daß man sich von der externalen Sichtweise, Itembeantwortungen als Annäherungen an den wahren Sachverhalt zu sehen, doch nicht vollständig löste. Man nahm Verfälschungstendenzen bei der Beantwortung von Fragebogen an und versuchte sie mit den Kontrollskalen zu erfassen, um von da aus auf die Gültigkeit der Beantwortung aller Fragen schließen zu können. Bei solchen Verfälschungstendenzen handelt es sich beispielsweise um die Gewohnheit, eine Frage eher mit ja als mit nein zu erwidern. Oft beantwortet der Getestete die einzelnen Fragen nicht wahrheitsgemäß, sondern so, daß er sich durch die Antworten in ein gutes Licht setzt, indem er sozial unerwünschte Verhaltensweisen oder Gedanken leugnet. Eine andere Verfälschungstendenz ist die mangelnde Motivation, diesen Fragebogen überhaupt auszufüllen. Bei der Auswertung ist es daher nützlich, wenn man Maße zur Verfügung hat, mit denen man feststellen kann, ob der Proband überhaupt hingesehen hat, als er die Fragen ankreuzte, ob er bereit oder fähig war, kleinere Fehler an sich selbst zu erkennen bzw. diese dem Testleiter oder sich selbst gegenüber zuzugeben, und wie groß seine Tendenz war, auf Fragen mit ja zu antworten.

Kritik an der klassischen Testtheorie

In den letzten Jahren wurde von quantitativ arbeitenden Forschern, so von Fischer (1974), Kritik an den Annahmen der klassischen Testtheorie geäußert. Als deren Folge entwickelte man allgemeinere, stochastische, testtheoretische Modelle, die jedoch keine kontinuierliche Weiterentwicklung der klassischen Testtheorie darstellen. In dieser kurzen Übersicht kann die probabilistische Testtheorie nicht adäquat dargestellt werden, zumal zur Zeit kaum Tests im Handel sind, die ausschließlich nach probabilistischen Modellen konstruiert wurden.

Die klassische Testtheorie hat zur Voraussetzung, daß ein quantitativer Rohwert, der sich als Summe aus „wahrem Wert" und Meßfehler darstellt, bereits vorliegt. Laut Fischer (1983) wird dabei übersehen, daß der Meßwert X willkürlich definiert, nicht aber empirisch begründet und hinterfragt wurde. So ist es z. B. sehr unwahrscheinlich, daß die Summation von Testpunktwerten zu einer Skala mit gleichen Abständen führt.

Des weiteren sind Veränderungsmessungen in Längsschnittstudien mit klassisch konstruierten Tests nicht unproblematisch und können leicht fehlinterpretiert werden. Der Meßbereich der manifesten Skala ist durch die Anzahl der Items des Tests begrenzt. Sehr schlechte Probanden können sich bei der zweiten Messung nur verbessern, sehr gute Probanden nur verschlechtern. Solche Ausgangswerteffekte, die zu falschen bzw. nichtinterpretierbaren Ergebnissen führen, sind bei Tests, die mit Hilfe probabilistischer Modelle konstruiert wurden, ausgeschlossen, weil die manifeste Testskala in eine nichtbeschränkte, latente Skala transformiert wird.

Der schwerwiegendste Einwand gegen die klassische Testtheorie besteht in der Abhängigkeit der Reliabilitätskoeffizienten von der Verteilung der wahren Werte in der jeweiligen Referenzpopulation. Die Retest-Reliabilität muß z. B. in Gruppen, die hinsichtlich des zu untersuchenden Merkmals relativ homogen zusammengesetzt sind, wesentlich niedriger ausfallen als in heterogenen Stichproben, weil in der homogenen Gruppe die Variabilität der Meßwertdifferenzen im Verhältnis zur Variabilität der durchschnittlichen Leistung wesentlich stärker ins Gewicht fällt als in der heterogenen Gruppe.

Diese Argumente gegen die klassische Testtheorie

sind theoretischer Art. Wie Fischer (1983) hervorhebt, ist es mit den Methoden der probabilistischen Testtheorie praktisch möglich, computerunterstützte, „adaptive Untersuchungen" durchzuführen, was mit den Methoden der klassischen Testtheorie nicht möglich ist. Die Reihenfolge der Testfragen, die dem Probanden vorgelegt werden, ist bei solch einem Test nicht mehr fest vorgegeben, dasjenige Item wird jeweils als nächstes dargeboten, welches den vermutlich größten Informationsbeitrag liefert. Bei solchen Untersuchungen richtet sich die Auswahl des nächsten Testitems also laufend an dem wahrscheinlichen Leistungsniveau der betreffenden Person aus, dessen Höhe aus dem Verlauf der bisherigen Untersuchung geschätzt wird. Bei falschen Antworten erhält die Versuchsperson daher leichtere Testfragen, bei richtigen Antworten werden schwierigere Items ausgewählt.

14.3 Klassifikation von Tests

Es fehlt an dieser Stelle der Platz, um einen erschöpfenden Überblick über die gebräuchlichsten psychologischen Tests zu geben. Statt dessen soll versucht werden, ein Ordnungsschema darzustellen, in das sich die unterschiedlichen Tests einordnen und bewerten lassen. Jedes Klassifikationssystem, das Tests in verschiedene Gruppen einteilt, ist letztlich willkürlich. Daher sollte man demjenigen Schema den Vorzug geben, das für den Anwender am nützlichsten ist. Die Typisierung von Tests kann zum einen aufgrund formaler und zum anderen aufgrund inhaltlicher Gesichtspunkte erfolgen und hat jeweils eine unterschiedliche Auswirkung auf die praktische Bedeutung.

Formale Klassifikationsgesichtspunkte sagen nichts über den eigentlichen Zweck des betreffenden Tests aus, sondern informieren über Art und Gestaltung der einzelnen Aufgaben, Anforderungen an das Untersuchungsmaterial, Auswertungsmodalitäten, Standardisierung oder Anforderungen an den zu Untersuchenden. Formale Klassifikationskriterien können für den Praktiker durchaus brauchbar sein, wenn er eine Untersuchung oder Versuchsreihe plant. So kann es wichtig sein, zu wissen, ob für eine Untersuchung ein kostspieliges Gerät nötig ist, welche Anforderungen an das Sprachverständnis des Untersuchten gestellt werden, oder ob es sich um einen standardisierten Test handelt usw. Lienert (1969) führt eine Reihe wichtiger, zumeist formaler Ordnungsschemata für psychologische Tests an, die zum größten Teil dichotome Typen bilden.

Anzumerken ist an dieser Stelle, daß die beliebte Aufgliederung in psychometrische Tests einerseits und projektive Verfahren andererseits keine sinnvolle Einteilung sein kann, da diese beiden Begriffe keine Gegensatzpaare sind, sondern unterschiedliche Aspekte psychologischer Tests hervorheben. Psychometrische Tests versuchen mit bestimmten Methoden quantitative Differenzen individueller Persönlich-

keitsmerkmale zu erfassen. Projektion beinhaltet im Zusammenhang mit psychologischen Tests keinen Abwehrmechanismus, sondern bestimmte Annahmen, wie man sich die Reaktionen des Probanden auf das Testmaterial erklärt. Demnach ist nicht jeder non-metrische Test automatisch ein projektiver Test, so wie es auch projektive Tests gibt, die metrischen Ansprüchen genügen.

Formale Ordnungsschemata geben keine Auskunft darüber, welche psychischen Bereiche oder Funktionen ein bestimmter Test erfassen will, dazu ist eine Klassifikation nach inhaltlichen Kriterien notwendig. Diese kann so gestaltet werden, daß man idealtypische Anwendungsbereiche für psychologische Tests wie klinische Tests, Schultests, Eignungstests usw. benennt, und die vorhandenen Untersuchungsverfahren in diese Klassen einordnet. Diese Einteilungsart ist unabhängig davon, ob es für die einzelnen Gebiete überhaupt schon irgendwelche bzw. genügende Testverfahren gibt. Einteilungen nach Anwendungsbereichen werden immer widersprüchlich sein, da es sehr oft möglich sein wird, ein und dasselbe Verfahren unterschiedlichen Gebieten zuzuordnen. Eine eindeutige, disjunktive Klassifikation ist durch diese Methode also nicht möglich.

Anstatt Tests in solch vorgegebene Anwendungsbereiche einzuordnen, ist es auch möglich, von den vorhandenen Tests auszugehen und aus diesen selbst die klassifikatorische Ordnung zu bilden. Bei dieser Art der Gruppenbildung kann man zwei Typisierungsmöglichkeiten unterscheiden, einmal eine Nebeneinanderstellung oder Aufzählung von gleichwertigen Bereichen wie Intelligenztests, Begabungstests, Persönlichkeitstests usw. Tests, die sich nicht einordnen lassen, bilden eine inhomogene Residuen-Gruppe. Die andere Möglichkeit ist die der hierarchischen Ordnung der inhaltlichen Klassen. So wählt Lienert (1969) als Oberbegriffe Intelligenztests, Leistungstests und Persönlichkeitstests. Innerhalb der Leistungstests unterscheidet er sodann zwischen motorischen, sensorischen und psychischen Leistungstests. Aber auch diese Art der Klassifikation schließt nicht aus, daß Tests uneindeutig zugeordnet werden können oder daß sich einzelne Bereiche überschneiden.

14.3.1 Das Klassifikationsmodell nach Brickenkamp

Im Laufe der Zeit sind eine Vielzahl von Klassifikationsmodellen entwickelt worden, die alle mehr oder weniger logische Unstimmigkeiten aufweisen. Sehr brauchbar ist das Brickenkampsche „Klassifikationsmodell psychologischer und pädagogischer Tests" (1975). Er hebt zunächst die Leistungstests von den Persönlichkeitstests ab, ohne damit eine Aufspaltung der Persönlichkeit in zwei relativ unabhängige Bereiche zu implizieren; er hält diese Unterscheidung aufgrund der fundamental unterschiedlichen Konstruktionsgesichtspunkte für gerechtfertigt. Der Persönlichkeitstestbereich wird in zwei Klassen weiter untergliedert, nämlich in psychometrische Persönlichkeitstests und in Persönlichkeitsentfaltungsverfahren. Diese letzte Gruppe wird wegen ihrer Heterogenität aufgrund formaler Merkmale klassifiziert, während die Einteilung der übrigen Tests anhand inhaltlicher Kriterien erfolgt.

Leistungstests

Entwicklungstests

Entwicklungstests sind alle diejenigen Verfahren, die primär im Dienste der Entwicklungsdiagnostik stehen. Sie haben somit zur Aufgabe, den aktuellen Entwicklungsstand eines Probanden zu erfassen. Zumeist entwickelte man diese Tests für Kinder und Jugendliche. Es sind jedoch Tests für alle menschlichen Entwicklungsphasen denkbar.

Intelligenztests

Da eine allgemeingültige Definition des Intelligenzmerkmals fehlt, werden unter diesem Oberbegriff sämtliche Verfahren zusammengefaßt, von denen die Testautoren angeben, daß sie in der Hauptsache intellektuelle Fähigkeiten zu erfassen beanspruchen.

Allgemeine Leistungstests

Wie Brickenkamp angibt, geht diese Kategorie auf Bartenwerfer zurück. Es handelt sich dabei um Tests, die Aufmerksamkeit, Konzentration, allgemeine Aktiviertheit, also das, was man unter anhaltender Konzentration bei geistiger Tempoarbeit versteht, erfassen.

Schultests

Die Gruppe der Schultests läßt sich weiter unterteilen, nämlich einmal in die Gruppe der Schulreife- und Schulfähigkeitstests und zum anderen in jene Methoden, die den augenblicklichen Leistungs- und Kenntnisstand der Lernenden während eines definierten Ausbildungsabschnittes in einem Lernfach erfassen sollen. Es handelt sich also um die sehr heterogene Gruppe der pädagogischen Tests, deren feinere Einteilungen bei Brickenkamp wiedergegeben sind.

Spezielle Funktionsprüfungen und Eignungstests

Diese Gruppe umfaßt noch wesentlich heterogenere Tests als die vorangehende Gruppe. Es geht um die Prüfung spezieller Funktionen wie Händigkeit, Psychomotorik, Geschicklichkeit oder technischen Verständnisses. Es sind Fertigkeiten, die nötig sind, um den Ansprüchen eines bestimmten Berufes nachkommen zu können. Diese Berufseignungstests unterscheiden sich notwendigerweise entsprechend den vielfältigen und unterschiedlich gestalteten Anforderungen der einzelnen Berufe stark voneinander.

Persönlichkeitstests

Psychometrische Persönlichkeitstests

Persönlichkeitsstrukturtests
Der Strukturbegriff ist in diesem Zusammenhang nicht derjenige, den bestimmte Persönlichkeitstheorien wie beispielsweise die Psychoanalyse benutzen. Er ist wesentlich allgemeiner gefaßt und bedeutet, daß es sich hier um eine Gruppe mehrdimensionaler Persönlichkeitstests handelt, denen jeweils eigene Ordnungsgesichtspunkte zugrunde liegen. Ihnen allen ist gemeinsam, daß sie mehrere Persönlichkeitsmerkmale aus dem „normalpsychologischen", d.h. dem nicht psychopathologischen Bereich quantifizierbar erfassen.

Einstellungs- und Interessentests
Mit Einstellungsskalen erfaßt man die Einstellung einzelner oder mehrerer Probanden zu Sachverhalten. Mit ähnlichen Methoden kann man die Meinung oder Vorurteile über bestimmte Gruppen oder Personen zu messen versuchen. Im Rahmen der Sozialpsychologie wurden sehr viele Einstellungsskalen entwickelt, von denen jedoch nur ein sehr geringer Teil als standardisierte Tests in den Handel gekommen ist.

Die Interessentests unterscheiden sich von den Einstellungstests darin, daß hier ein mehr intentionaler Objektbezug gemessen wird. Im Gegensatz zu manch anderen Persönlichkeitstests haben diese beiden Testarten zur Voraussetzung, daß der Proband über seine Vorlieben, Abneigungen und sonstigen Einstellungen in unmittelbarer Reaktion auf die Fragen Auskunft geben kann.

Klinische Tests
Tests aus der Gruppe der klinischen Tests sollen Anhaltspunkte für eine klinische Diagnosestellung geben, psychopathologische Erscheinungen erfassen und Hilfen für eine differentialdiagnostische Fragestellung anbieten. Sie sind jedoch nicht in der Lage, eine psychiatrische Diagnose zu ersetzen. Es ist verständlich, daß diese Verfahren ein besonderes Maß klinisch-psychologischer Erfahrung voraussetzen, ohne die eine Interpretation der Testergebnisse nicht möglich sein kann.

Persönlichkeitsentfaltungsverfahren

Der Projektionsbegriff wurde von Frank in die Testpsychologie eingeführt, er verstand darunter „Methoden, welche die Persönlichkeit dadurch untersuchen, daß sie die Versuchsperson einer Situation gegenüberstellen, auf welche sie entsprechend der Bedeutung reagiert, die diese Situation für sie besitzt ... Das Wesen eines projektiven Verfahrens liegt darin, daß es etwas hervorruft, was – auf verschiedene Art – Ausdruck der Eigenwelt, des Persönlichkeitsprozesses der Versuchsperson ist" (1948). Brickenkamp verzichtet in seiner Klassifizierung auf den Projektionsbegriff, da er zu mehrdeutig und problembefrachtet ist. Es scheint ihm sinnvoller zu sein, diese Art von Tests als Entfaltungsverfahren zu bezeichnen.

Während sich die bisherigen Persönlichkeitstests darauf beschränken, bestimmte, wohldefinierte Verhaltensmerkmale zu erfassen, ist der Anspruch der Persönlichkeitsentfaltungsverfahren ein anderer. Der Aufforderungscharakter der einzelnen Testaufgaben ist nicht mehr eindeutig, die formale Gliederung in Testaufgaben kann vollkommen aufgehoben sein, und es soll kein bestimmtes fest umschriebenes Verhalten provoziert werden. Statt dessen versucht man mehr oder weniger unbestimmte Verhaltensaspekte des Probanden zu provozieren, die der Diagnostiker nach heterogenen Konzepten meist qualitativ deutet. Aus diesem Grunde werden diese Verfahren auch häufig „Breitbanddiagnostika" genannt.

Diese kurze Einführung macht bereits deutlich, daß es sich bei diesen Tests um eine Ansammlung sehr unterschiedlicher Untersuchungsverfahren handelt, für die eine Klassifikation nach inhaltlichen Gesichtspunkten fast nicht möglich ist. Daher erfolgt diese aufgrund formaler Aspekte, die sich aus den unterschiedlichen Reaktionsweisen, die die Tests provozieren wollen, ergeben.

Formdeuteverfahren

Wie schon der Name sagt, ist den Formdeuteverfahren gemeinsam, daß die Probanden ein relativ unstrukturiertes, uneindeutiges Bildmaterial, wie z.B. Klecksbilder, deuten sollen. Diese Verfahren begründen sich aus der bekannten Tatsache, daß der Wahrnehmungsprozeß von Persönlichkeitseigenschaften gesteuert wird. Aufgabe des Diagnostikers ist es, diese Deutungen vorgegebenen Oberbegriffen zuzuordnen, zu signieren und zu „verrechnen", und sie dann zu interpretieren. Dazu haben Autoren unterschiedlicher „Schulen" heterogene, manchmal auch widersprüchliche und daher kaum vergleichbare Richtlinien zur Auswertung und Interpretation formuliert, die zumeist intuitiv-theoretisch ausgearbeitet worden sind. Der Rorschach-Test (1962) ist wohl das bekannteste Verfahren dieser Testklasse.

Verbal-thematische Verfahren

Die verbal-thematischen Verfahren konfrontieren den Probanden mit Reizen, die ihn zu einer thematischen Auseinandersetzung anregen sollen. Dazu gehören Verfahren, bei denen der Proband auf vorgegebene „problematische" Worte mit dem ersten Einfall antworten soll. Diese Assoziationstests gehen auf C. G. Jung (1906) zurück. Wird der Proband aufgefordert, unvollständige Sätze oder angefangene Geschichten zu ergänzen, so spricht man von Ergänzungsverfahren. Eine andere Gruppe von Tests bietet den Probanden Bilder dar, auf denen uneindeutige, meist soziale Situationen dargestellt sind, zu denen er eine möglichst spannende Geschichte erzählen soll. Der Thematische Apperzeptionstest (TAT) nach Murray (1943) soll als Beispiel für diese Erzählverfahren dienen. Auch hier gibt es, wie bei den Formdeuteverfahren, keine eindeutigen Auswertungs- und Interpretationsrichtlinien.

Zeichnerische und Gestaltungsverfahren

Zeichnerische Verfahren sollen den Probanden anregen, eine oder mehrere Zeichnungen zu produzieren, die etwas über ihn aussagen sollen. Dazu kann man ihm entweder ein Thema vorgeben, also einen Menschen, einen Baum oder seine Krankheit zu zeichnen bzw. zu malen, oder man läßt ihn vorgegebene „offene", abstrakte Formen zu thematisch freien, sinnvollen Gebilden vervollständigen, wie es im Wartegg-Zeichentest (1953) geschieht.

Die Gestaltungsverfahren setzen sich aus noch heterogeneren Tests zusammen, als dies schon bei den zeichnerischen Verfahren der Fall ist. Dem Probanden werden in den einzelnen Tests die unterschiedlichsten Materialien, wie Puppenstuben, Knetmasse oder bunte Plättchen, zur Verfügung gestellt, mit denen er irgend etwas gestalten soll. Typisch für diese Methoden ist, daß von den Probanden keine Leistung verlangt wird, es gibt keine richtige oder falsche Lösung, statt dessen soll man sich mit diesen Verfahren, die nicht selten als psychotherapeutische Hilfsmittel dienen, neue oder andere Ausdrucksmöglichkeiten erschließen.

14.4 Die Gütekriterien von Tests

Lienert (1969) fordert, daß ein guter Test Objektivität, Reliabilität und Validität als Hauptgütekriterien besitzen muß. Darüber hinaus wäre eine Normierung des Tests an einer Eichstichprobe wünschenswert. Nützlich ist es, wenn man die Testergebnisse anhand von Paralleltests oder validitätsähnlichen Tests vergleichen kann. Ein guter Test sollte darüber hinaus ökonomisch durchgeführt, ausgewertet und interpretiert werden können. Schließlich muß ein Test auch nützlich sein, das bedeutet, daß für die Untersuchung des Persönlichkeitsmerkmals einmal ein Bedürfnis vorliegt, und die Beschaffenheit des Tests zum anderen so ist, daß man ihn durch keinen anderen ersetzen kann.

14.4.1 Objektivität

Laut Wilde (1951) gilt ein Verfahren in der psychologischen Diagnostik dann als objektiv, wenn es von einer Reihe von Beurteilern in identischer Weise gedeutet oder bewertet werden kann. Lienert unterscheidet dabei drei verschiedene Aspekte der Objektivität.

Durchführungsobjektivität

Die Durchführungsobjektivität betrifft den „Grad der Unabhängigkeit der Testergebnisse durch zufällige oder systematische Verhaltensvariationen des Untersuchers während der Testdurchführung, die ihrerseits zu Verhaltensvariationen des Probanden führt und dessen Ergebnis beeinflußt" (Lienert, 1969). Die höchste Durchführungsobjektivität erhält man demnach, wenn die Untersuchungsinstruktionen möglichst genau, eindeutig und verständlich in schriftlicher Form festgelegt werden. Die Untersuchungssituation muß standardisiert werden, indem die Reihenfolge der einzelnen Aufgaben feststeht. Für eine ganze Reihe von Tests müssen aber noch weitere Standardisierungsbedingungen erfüllt sein. Bei manchen Wahrnehmungstests müssen gleiche Beleuchtungsverhältnisse vorliegen. Nachteilig wirkt sich aber auf diese hohe Durchführungsobjektivität aus, daß die soziale Interaktion zwischen Versuchsleiter und Proband stark eingeschränkt wird, was notwendigerweise im Widerspruch zu den Intentionen einer Reihe von Tests, insbesondere der Entfaltungsverfahren, steht.

Auswertungsobjektivität

Die Auswertungsobjektivität „betrifft die numerische oder kategoriale Auswertung des registrierten Testverhaltens nach vorgegebenen Regeln" (Lienert, 1969). Sie ist fast immer dann gegeben, wenn der Proband eine Reihe von Antwortmöglichkeiten vorgelegt bekommt, aus denen er lediglich die richtigen oder für ihn zutreffenden auszuwählen braucht, oder wenn ein elektronisches Gerät die Reaktionen der Probanden fehlerfrei registriert. Beispiele für diese Art Test sind Fragebogentests oder die meisten Leistungstests. Nachteilig macht sich jedoch bei diesen Testarten bemerkbar, daß manche Probanden die vorgeschlagenen Antwortmöglichkeiten für zu allgemein, zu grob oder für sie nicht zutreffend halten. Die Auswertungsobjektivität ist dann weniger gegeben, wenn es Aufgabe des Versuchsleiters ist, anhand der Antwort des Probanden zu entscheiden, ob diese im Sinne der Schlüsselrichtung zu bewerten ist. Besonders problematisch wird dies, wenn der Proband sich ungeschickt, umständlich oder wenig verständlich ausdrückt. Ungünstig wirkt es sich auf die Auswertungsobjektivität aus, wenn die Antworten der Versuchsperson völlig frei sein können, wie dies z.B. bei projektiven Tests der Fall ist, denn diese Art der Testauswertung bringt es mit sich, daß einzelne Auswerter die Antworten der Probanden unterschiedlich deuten können und verschiedenen Testkategorien zuordnen.

Interpretationsobjektivität

Die Interpretationsobjektivität „betrifft den Grad der Unabhängigkeit der Interpretation des Testergebnisses von der Person des interpretierenden Psychologen, der nicht mit dem Untersucher oder Auswerter identisch zu sein braucht" (Lienert, 1969).

Die Interpretationsobjektivität ist bei allen normierten Tests gegeben, die es erlauben, die relative Position der getesteten Person auf einer Skala anzugeben. Bei vielen Tests ist dies nicht der Fall, da die Angaben des Testautors dem Testleiter nur ungenaue oder widersprüchliche Hinweise zur Interpretation geben. Oft wird die Auswertung solcher Verfahren als „Kunst" hingestellt, zu deren Erlernung große Erfahrung und ein besonderes „Gespür" notwendig sind. Da aufgrund von Testergebnissen nicht selten Entscheidungen über Probanden getroffen werden, und die großen Künstler auch in der psychologischen Diagnostik nicht allzu zahlreich vertreten sein dürften, sollte man aus ethischen Gründen kritisch prüfen, ob der Einsatz solcher „Tests" gerechtfertigt ist.

14.4.2 Reliabilität

Unter der Reliabilität eines Tests, dieser Begriff wurde durch Spearman (1910) in die Testtheorie eingeführt, versteht man, wie genau dieses Verfahren ein bestimmtes Persönlichkeits- oder Verhaltensmerkmal mißt, und zwar unabhängig davon, um welches Merkmal es sich überhaupt handelt. Jede Messung und damit auch jede Testung kann trotz sorgfältigster und gewissenhaftester Durchführung niemals vollkommen fehlerfrei sein. Diese zufallsbestimmten, unsystematischen Fehlerwerte lassen sich in mehrere Komponenten aufteilen. Sie betreffen einmal die Ungenauigkeit des Meßinstrumentes selbst. Sodann spielen Umgebungsfaktoren, unter denen man die Messung durchführte, wie etwa Lärmpegel oder Beleuchtungsverhältnisse, eine Rolle. Temporäre Veränderungen der untersuchten Person selbst, wie Ablenkung, Ermüdung oder Desinteresse an der Untersuchung, können die Meßergebnisse ebenso verfälschen wie eine ungenaue Durchführung und Auswertung des Tests durch den Versuchsleiter.

Die Axiome der klassischen Testtheorie

Die klassische Testtheorie geht von zwei Grundbegriffen, dem wahren Wert und dem Meßfehler, aus. Würde man einen Probanden beliebig oft untersuchen können, so verteilten sich die einzelnen Meßergebnisse nach einer bestimmten Funktion. Der „wahre Wert" des Probanden wird als Erwartungswert über diese unabhängigen Meßwiederholungen definiert. Rechnerisch ist er der Mittelwert dieser unendlich großen Anzahl von Meßwiederholungen, da die Fehler genauso häufig zu erniedrigten wie zu erhöhten Messungen führen und sich im Mittel aufheben. Der Meßfehler wird als die Differenz zwischen dem beobachteten Wert und dem wahren Wert der Versuchsperson definiert.

Aus diesen Definitionen lassen sich die Grundannahmen der klassischen Testtheorie herleiten:
- Der durchschnittliche Meßfehler in jeder beliebigen Population oder Teilpopulation ist Null.
- Die Korrelation zwischen dem wahren Wert und dem Meßfehler strebt gegen Null, weil der unsystematisch streuende Meßfehler in keiner systematischen Beziehung zu dem wahren Wert stehen kann. Bei einem systematischen Fehler würde man dagegen, wie bei einem zu langen oder zu kurzen Metermaß, konstant entweder eine zu hohe oder zu niedrige Punktzahl erhalten.
- Ebenso haben die Fehlerwerte in zwei verschiedenen Messungen nichts miteinander zu tun, ihre Korrelation ist ebenfalls Null.

– Ungleich Null dagegen ist die Korrelation zwischen beobachteten Werten und Fehlerwerten, da definitionsgemäß der beobachtete Wert die Summe aus wahrem Wert und Fehler ist. Es läßt sich aber nachweisen, daß diese Korrelation um so stärker gegen Null geht, je geringere Unterschiede sich zwischen wahren und beobachteten Werten ergeben, je größer also die Meßgenauigkeit ist. Denn in diesem Falle wird der Anteil der Fehlervariabilität im Verhältnis zur Variabilität der wahren Werte immer unbedeutender.

In dieser letzten Grundannahme ist bereits der Begriff der Reliabilität enthalten, denn diese wird definiert als Verhältnis der wahren Varianz zur gemeinsamen Varianz, die sich aus wahrer und Fehlervarianz zusammensetzt. Der Begriff der Reliabilität beinhaltet nicht ein einheitliches Meßkonzept, sondern ist vielmehr ein Oberbegriff für eine Reihe von Konzepten, die jeweils bestimmte Aspekte der Meßgenauigkeit betreffen. Man unterscheidet vier Methoden der Reliabilitätsbestimmung, nämlich die Paralleltest-, Testwiederholungs- und Testhalbierungsmethode sowie die Konsistenzanalyse. Bevor auf die einzelnen Reliabilitätsmaße im einzelnen eingegangen wird, muß gesagt werden, daß in die Berechnung der Testwiederholungs- und Paralleltest-Reliabilität sowohl die Meßungenauigkeit des Tests selbst als auch situationsabhängige Fehler eingehen, wohingegen letztere bei der Testhalbierungs-Reliabilität und den Konsistenzmaßen zwar miterfaßt werden, als Konstanten die Berechnung jedoch nicht beeinflussen. Daher führen diese beiden Methoden zu günstigeren Reliabilitätskoeffizienten.

Paralleltest-Reliabilität

Die Methode der Reliabilitätsbestimmung durch parallele Tests ist für die klassische Testtheorie besonders bedeutsam, weil sich mit ihrer Hilfe nicht nur Angaben über beobachtbare Werte machen lassen, wie bei einer einzigen Messung auch, sondern weil man nun auch Berechnungen über die nicht beobachtbaren wahren Werte und die Meßfehler durchführen kann.

Definitionsgemäß liegen parallele Tests dann vor, wenn in beiden Tests zum einen die wahren Werte und zum anderen die Varianzen der Fehlerwerte gleich sind. Aus dieser Definition folgt, wenn man die oben geschilderten Grundannahmen der klassischen Testtheorie berücksichtigt, die mathematischen Ableitungen finden sich z.B. bei Lord und Novick (1968), daß die Korrelation zwischen zwei parallelen Tests gleich dem Verhältnis von wahrer Varianz zur beobachteten Varianz und damit gleich der Reliabilität ist.

Da sich die Korrelation zweier paralleler Tests berechnen läßt und die Varianz der beobachteten Werte bekannt ist, läßt sich der Meßfehler als sog. Standardmeßfehler ebenfalls bestimmen. Mit ihm steht ein Maß zur Verfügung, durch das man nach einer Testung angeben kann, innerhalb welcher Grenzen der wahre Wert eines Probanden wahrscheinlich liegt.

Es ist schwierig, wirklich parallele Tests zu konstruieren, zumal dann, wenn die einzelnen Testaufgaben einen gewissen Originalitätsgrad aufweisen, wie es z.B. bei Denkproblemen der Fall ist, wenn ein Transfer der Lösungsstrategie von dem einen auf den anderen Test relativ einfach ist oder wenn eine Testung bei der untersuchten Person einen Übungsfortschritt im Hinblick auf das Testergebnis mit sich bringt.

Testwiederholungs-Reliabilität

Die Testwiederholungs- oder Retest-Reliabilität eines Tests läßt sich berechnen, nachdem eine Probanden-Stichprobe diesen zweimal hintereinander bearbeitete, indem man beide Testreihen miteinander korreliert. Diese Methode wendet man innerhalb der Naturwissenschaften an, um die Exaktheit einer Messung zu vergrößern. In diesem Wissenschaftsbereich ist es im Gegensatz zur Psychologie zumeist möglich, einen Gegenstand oder ein Ereignis beliebig oft wiederholt zu messen. Bei psychologischen Messungen verändert man fast immer den Untersuchungsgegenstand allein durch die Tatsache der Messung, so daß die zweite Untersuchung zugleich den Einfluß der vorhergehenden miterfaßt. Bekannte Phänomene für diese Tatsache sind der Übungsfortschritt bei Determinationsaufgaben, Einsicht in die Lösungsstrategien bei Intelligenztests oder Ablenkung der Versuchsperson durch eine stereotype Testsituation bzw. eine zu lang dauernde Untersuchung. Zu den psychologischen Untersuchungen, bei denen sich Testwiederholungseffekte nur unbedeutend auswirken, gehören einfache Wahrnehmungsversuche und psychophysiologische Messungen. In der Praxis berechnet man die Testwiederholungs-Reliabilität oft bei Fragebogen- und Speed-Tests. Diese enthalten sehr viele leichte Aufgaben. Mit Absicht ist die Untersuchungsdauer so bemessen, daß es keinem Probanden gelingt, alle Aufgaben in der zur Verfügung stehenden Zeit zu lösen. Es kommt also besonders auf die Schnelligkeit der Aufgabenbearbeitung an. Für die meisten anderen psychologischen Tests, dies betrifft vor allen Dingen Leistungstests, ist es empfehlenswert, Testwiederholungsmethoden zur Reliabilitätsbestimmung zu vermeiden. Falls dies nicht möglich ist, sollte zumindest ein längeres Intervall zwischen zwei Untersuchungen liegen. Ansonsten wird eine Scheinreliabilität bestimmt, indem man nur zum Teil die Genauigkeit des Tests erfaßt und zum anderen das Gedächtnis der Probanden prüft.

Die Testhalbierungsmethode

Oft ist es nicht möglich, für einen Test eine Parallelform zu konstruieren, es ist aber nicht angebracht, die Reliabilität mittels der Testwiederholungsmethode zu bestimmen. Man kann dann der Analyse-Stichprobe den Test nur einmal vorlegen und anschließend diesen künstlich in zwei gleichwertige Teile zerlegen. Nachdem man die Rohwerte jeder Testhälfte miteinander korrelierte, korrigiert man diese Korrelation mittels einer geeigneten Formel für die volle Testlänge. Eine ideale Voraussetzung für die Durchführung der Testhalbierungsmethode ist dann gegeben, wenn der Test ein homogenes Merkmal mißt. Ein Test, der komplexere Merkmale wie z.B. Studieneignung, Intelligenz oder Therapiefähigkeit erfaßt, besteht naturgemäß aus vielen unterschiedlichen Items, so daß eine Halbierung in zwei gleichwertige Teiltests nicht ohne weiteres möglich ist. Zum mindesten müssen die beiden Testhälften die zu untersuchenden Dimensionen gleichsinnig erfassen, in beiden Testhälften muß sich also die gleiche Faktorenstruktur nachweisen lassen. Die verschiedenen Möglichkeiten, mit denen man Tests halbieren kann, werden bei Lienert (1969) dargestellt.

Konsistenzanalyse

Ausgehend von der Methode der Testhalbierung entwickelte man die Konsistenzanalyse, die der älteren Methode mit Einschränkungen überlegen ist und die man als deren Verallgemeinerung ansehen kann. Theoretisch lassen sich für einen Test so viele Halbierungskoeffizienten berechnen, wie Halbformen zu bilden möglich sind. Um die innere Konsistenz zu bestimmen, versucht man nun alle Informationen der Items, nämlich Aufgabenschwierigkeit und Trennschärfe, für eine Schätzung der Reliabilität auszunutzen. Kuder und Richardson (1937) sowie Hoyt (1941) entwickelten dazu eine Reihe von Formeln, die jeweils auf verschiedenen Annahmen aufbauen. Einschränkend muß gesagt werden, daß die Berechnung der inneren Konsistenz nur dann sinnvoll ist, wenn alle Items des Tests homogen sind, d.h., wenn der Test oder die Skala nur einen singulären Faktor erfaßt.

14.4.3 Validität

Unter Validität versteht man den Grad der Genauigkeit, mit dem ein Test das mißt, was er messen soll. Die Validität erlaubt also eine Aussage über den Zusammenhang zwischen dem Meßinstrument und dem Merkmal, welches der Test erfassen will, wohingegen die Reliabilität lediglich angibt, wie genau die Messung ist. Eine zufriedenstellende Meßgenauigkeit ist notwendige, nicht jedoch hinreichende Bedingung für eine befriedigende Validität eines Tests. Ist die Reliabilität niedrig, handelt es sich also um ein ungenaues Meßinstrument, dann sind auch keine validen Voraussagen möglich. Ist die Reliabilität gut, aber die Methode ungeeignet, das zu untersuchende Merkmal zu erfassen, dann ist die Validität des Verfahrens dennoch gering.

Auch bei der Validität kann man wie bei der Reliabilität verschiedene Aspekte berücksichtigen. Dies führte dazu, daß im Laufe der Zeit eine Vielzahl verwirrender Bezeichnungen entstanden. Darum empfahl die American Psychological Association 1954 die Validitätsmaße in vier Klassen zu unterteilen: Konstruktvalidität, Inhaltsvalidität, Übereinstimmungsvalidität und Vorhersagevalidität.

Merkmale, die Tests erfassen wollen, können psychologische Konstrukte sein, wie z.B. Intelligenz, bestimmte „Fähigkeiten" oder „Eigenschaften", die der direkten Beobachtung nicht zugänglich sind. Bei der „Konstruktvalidierung" versucht man einen Zusammenhang zwischen dem Testverhalten und den diesem Verhalten zugrundeliegend gedachten Konstrukten herzustellen, diese Art der Validierung hat also u.a. die theoretische Klärung dessen, was der Test überhaupt mißt, zum Ziele.

Die Frage der psychologischen Bedeutung des Testresultats ist für die drei anderen Validitätsmaße nur von nachgeordnetem Interesse. Diese prüfen, ob der Beobachtung zugängliche Merkmale, also Verhaltensweisen außerhalb der Testsituation, mit dem Testergebnis kovariieren. Bei solchen operational definierten äußeren Merkmalen, sog. Kriterien, kann es sich um das gleiche Verhalten außerhalb der Testsituation handeln. Durch einen „Repräsentationsschluß" (Michel, 1971) folgert man von der Verhaltensstichprobe, die der Test erfaßte, auf ein Gesamtverhalten außerhalb der Testsituation. Man nennt diese Art der Validierung inhaltliche Gültigkeit.

Der „Korrelationsschluß" beruht auf einem empirisch nachgewiesenen Zusammenhang zwischen dem Testverhalten und anders gearteten Verhaltensweisen außerhalb der Testsituation. Wenn diese Kriterien zur gleichen Zeit existieren, spricht man von „concurrent validity" oder Übereinstimmungsvalidität. Will man die Gültigkeit der Vorhersage hinsichtlich eines erst in der Zukunft beobachtbaren Kriteriums ermitteln, so heißt dieses Maß Vorhersagevalidität oder „predictive validity".

Konstruktvalidität

Das Hauptanliegen der Konstruktvalidierung ist die Erforschung der psychologischen Bedeutung der Testresultate. Theoriegeleitet überprüft sie, ob der Test wirklich das mißt, was er zu messen vorgibt. Eine solche Validierung ist methodisch wesentlich komplizierter als die Berechnung der Korrelation zu einem Außenkriterium, wie Schulerfolg oder Risikoverhalten, denn es gibt zunächst einmal keine Möglichkeit, die Testergebnisse unmittelbar mit psychischen Konstrukten wie Extraversion, Intelligenz oder Konzentrationsfähigkeit in Beziehung zu setzen. Wie Lienert (1969) feststellt, stehen hier Empirie und Theorie in enger wechselseitiger Beziehung. So kann das Testverfahren die Annahmen über das Konstrukt genauso beeinflussen, wie es möglich ist, daß die Einsicht in das Konstrukt eine Modifikation des Tests zur Folge hat.

Lienert führt in Anlehnung an Cronbach und Meehl (1955) sieben Methoden auf, die zwar nicht das Konstrukt selbst messen können, wohl aber in der Lage sind, den in Frage kommenden psychischen Aspekt von verschiedenen Seiten her zu betrachten und ihn dadurch gleichsam „einzukreisen", denn ein einheitliches Maß der Konstruktvalidität gibt es nicht. Zunächst untersucht man den Zusammenhang zwischen dem Test und einem geeigneten Außenkriterium. Durch Vergleiche mit Untersuchungsverfahren, die einen ähnlichen Validitätsanspruch haben, bestimmt man sodann die sog. konvergierende Validität. Darüber hinaus darf der zu untersuchende Test in nur geringer Beziehung zu Persönlichkeitsmaßen stehen, die er nicht messen soll. Die Korrelationen mit Tests, von denen bekannt ist, daß sie ganz andere Persönlichkeitsmerkmale erfassen, sollen also möglichst unbedeutend sein. Ist dies der Fall, so spricht man auch von diskriminierender Validierung.

Sehr ökonomisch kann man mit Hilfe von Faktorenanalysen eine Ordnung dieser Vielzahl von Korrelationsmaßen erhalten, wobei es notwendig ist, daß der zu validierende Test möglichst hoch auf jenem Faktor lädt, der das interessierende Konstrukt repräsentiert.

Eine andere Methode, die man bei der Konstruktvalidierung heranziehen kann, ist, zu überprüfen, ob sich unterschiedlich zusammengesetzte Stichproben so in ihrem Testverhalten voneinander unterscheiden, wie dies schon vorher aufgrund der Theorie gefordert wurde. Längsschnittstudien haben bei der Konstruktvalidierung zur Aufgabe, festzustellen, ob mit dem Test habituelle oder aktuelle Persönlichkeitsmerkmale erfaßt werden, und ob sich diese Ergebnisse mit den theoretischen Annahmen decken. Schließlich darf eine solche Validierung nicht nur den vollständigen Test betreffen, sondern muß auch die Einzelaufgaben inhaltlich validieren.

Inhaltliche Validität

Bei der inhaltlichen Validität ist das Verhalten in der Testsituation eine repräsentative Stichprobe eines Gesamtverhaltens, das der Test diagnostizieren soll. Es ist in diesem Falle also kein Symptom für andere Verhaltensweisen; daher lassen sich die Ergebnisse dieser Untersuchung verallgemeinern, indem man vom Testergebnis auf das zu untersuchende Verhalten schließt. Eine Grundvoraussetzung für diese Art der Validierung ist, daß man einerseits das zu erfassende Verhalten exakt begrifflich bestimmen kann und daß andererseits auch die Testaufgaben geeignet sind, um dieses zu untersuchen.

Aus dem Gesagten ergeben sich drei nicht selten zu beobachtende Fehler, die zu vermeiden sind, wenn man einen Test inhaltlich validieren will, nämlich einmal eine ungenaue bzw. zu weite Definition des Untersuchungszieles, sodann eine unausgewogene Verhaltensstichprobe und schließlich eine zu starke Verallgemeinerung des Untersuchungsergebnisses. Eine inhaltliche Validierung wird im allgemeinen bei Kenntnis- oder Schulleistungstests möglich sein, jedoch sind auch hier Fehler nicht ausgeschlossen.

Übereinstimmungs- und Vorhersagevalidität

Die kriterienbezogene Übereinstimmungs- und Vorhersagevalidität läßt sich bestimmen, indem man die Korrelation zwischen Test- und Kriterienpunktwert berechnet. Es braucht sich bei dem Zusammenhang, der sich auf diese Art und Weise aufzeigen läßt, lediglich um eine empirisch nachgewiesene Wenn-Dann-Beziehung zu handeln, es ist jedoch nicht erforderlich, den Zusammenhang zwischen beiden Phänomenen wissenschaftlich aufzuklären. Die Hauptschwierigkeit dieser Art der Validierung besteht darin, ein einwandfreies und geeignetes Außenkriterium zu finden, denn es ist offensichtlich, daß dieses selbst den Ansprüchen hinsichtlich Reliabilität und Validität genügen muß. Betrachtet man einen Validitätskoeffizienten für sich allein, so ist es von diesem Maß aus noch nicht möglich zu beurteilen, ob der Test für eine bestimmte Fragestellung praktisch brauchbar ist, denn jedes Maß, das sich auf ein Kriterium bezieht, ist eng und spezifisch. Es besagt zunächst nur, wie eng der Zusammenhang bei einer möglichst genau beschriebenen Stichprobe von Versuchspersonen unter festgelegten Versuchsbedingungen zwischen dem Testverhalten und dem Kriteriumsverhalten war. Es gibt daher nicht „die" Validität eines Tests; ein Test kann theoretisch beliebig viele Validitätsmaße haben, die sich zudem stark voneinander unterscheiden können. Wenn ein Test gegenüber unterschiedlichen Kriterien unterschiedliche Validitätsmaße aufweist, spricht man von der differentiellen Validität des Verfahrens. Der Anwender des Tests muß im Einzelfall entscheiden, ob der Test für die aktuelle Fragestellung der Untersuchung geeignet ist. Dies kann er nur, wenn die Angaben der Validitätsuntersuchungen möglichst exakt, ausführlich und eindeutig beschrieben worden sind.

Selbst beim besten Willen lassen sich nicht alle Bedingungen einer psychologischen Untersuchung eindeutig festlegen. Daher ist es sinnvoll, wenn andere Versuchsleiter, Auswerter sowie Beurteiler der Kriterienvariablen die Validitätsuntersuchung replizieren. Sind die Ergebnisse unabhängiger Gültigkeitsstudien vergleichbar, so spricht man von einer sog. Kreuzvalidität.

Zur Validitätsproblematik sagt Lienert (1969) abschließend: „Für eine praktisch-psychologische Diagnostik ist der Nachweis von Korrelationen zwischen Test und Kriterium mindestens so unentbehrlich wie die Aufklärung der psychologischen Faktoren, die hinter einem Test stehen. Auf die Kenntnis dessen, was hinter einem Test steht, kann vom pragmatischen Aspekt her unter Umständen verzichtet werden, nicht aber auf die kriterienbezogene Validität. Die nachgewiesene Korrelation eines Tests mit einem wohldefinierten Kriterium allein berechtigt den in der Praxis stehenden Diagnostiker, seine Methoden als brauchbar und anderen Verfahren als überlegen zu kennzeichnen."

14.5 Normen

Ein individuelles Testergebnis ist nicht aus sich heraus interpretierbar, sondern erst dann, wenn man es in Beziehung setzt zu den Testergebnissen einer vergleichbaren Stichprobe. Für einen guten Test müssen daher Normdaten zur Verfügung stehen, durch die man die relative Position des Probanden innerhalb der Verteilung der Testresultate einer Referenzpopulation bestimmen kann.

In Anlehnung an die technische Sprache wird die Prozedur zur Gewinnung von Normdaten auch „Eichung" genannt. Liegen für einen Test Normdaten vor, so gilt dieser als „geeicht". Wie Michel (1971) feststellt, kann dies zu Mißverständnissen führen, da es sich bei der Eichung psychologischer Tests um einen ganz anderen Vorgang als im Bereich der Physik oder der Technik handelt. Dort eicht man einen Gegenstand, indem man prüft, ob er mit einem festgelegten Maß, z. B. dem Ur-Meter übereinstimmt, ob also der Gegenstand richtig mißt. Bei der Eichung eines psychologischen Tests gewinnt man dagegen Vergleichsdaten an einer möglichst exakt definierten Stichprobe von Probanden. Michel weist weiter darauf hin, daß die „Eichung" eines Tests häufig überschätzt wird. Auch wenn ein Test an umfangreichen Stichproben geeicht worden ist, kann er trotzdem unbrauchbar sein, wenn die Reliabilität schlecht ist.

Allgemein gilt, daß sich nur für hochreliable Tests große Eichstichproben lohnen, und auch nur für solche Tests feinere Normen berechnet werden sollten, da ansonsten der Standardmeßfehler bei weitem die scheinbare Genauigkeit der Vergleichsskala übertrifft.

14.5.1   Organisation einer Eichstichprobe

Repräsentative Eichstichproben sollten mit den gleichen Methoden organisiert werden, die die Soziologie für Bevölkerungsumfragen ausgearbeitet hat. Testautoren haben in der Regel nicht die Möglichkeit, nach der Phase der Testkonstruktion mit einem erfahrenen und eingearbeiteten Mitarbeiterstab eine der Testintention entsprechende repräsentative Normgruppe zu untersuchen, in der zudem keine Angehörigen der Analysestichprobe vertreten sein dürfen. Daher übernahmen es renommierte Meinungsforschungsinstitute in den letzten Jahren, für einige gut konstruierte deutschsprachige Fragebogentests Eichstichproben zu erheben.

14.5.2   Klassifikation der Normmaße

Normmaßstäbe lassen sich in zwei Gruppen differenzieren, nämlich Äquivalentnormen einerseits und Variabilitätsnormen andererseits. Wie bereits bei der Darstellung des Intelligenzquotienten dargestellt, orientieren sich Äquivalentnormen an Mittelwerten, also an Lageparametern von Gruppen. So sollen einem IQ von 100 die durchschnittlichen Testergebnisse einer interessierenden Altersgruppe im Hamburg-Wechsler-Intelligenztest für Erwachsene entsprechen. Allen Variabilitätsmaßen dagegen ist gemeinsam, daß sie die Stellung eines Probanden innerhalb einer Häufigkeitsverteilung angeben, sie orientieren sich also an Streuungsparametern. Beispiele für Variabilitätsnormen sind die üblichen Standardwerte, T-

Werte oder Stanine-Werte von Fragebogen- oder Leistungstests, ein anderes Variabilitätsmaß sind Prozentrangnormen von Schulleistungstests.

Standardnormen

Standardnormen lassen sich nur dann sinnvoll berechnen, wenn sich die Rohwerte nach einer charakteristischen Glockenkurve verteilen. Die Streuungsparameter müssen mit anderen Worten die Bedingungen der Normalverteilung erfüllen. Dann ist es leicht, die Rohwerte in sog. z-Werte der Standard-Normalverteilung zu überführen. Diese hat als Mittelwert 0 und die Standardabweichung 1. Werte unter 0 geben demnach unter-, Werte über 0 überdurchschnittliche Testleistungen wieder. Da es aber unpraktisch und wenig anschaulich ist, wenn negative Dezimalbrüche ein Testergebnis ausdrücken, transformiert man die z-Werte so, daß sich der Mittelwert der Skala von 0 auf die Punktzahl 5, 50 oder 100 verschiebt und die Standardabweichung so gedehnt wird, daß sie im ersten Fall 2 und für die beiden übrigen Fälle 10 Punkte beträgt. Rechnerisch geschieht dies, indem man den individuellen z-Wert mit der gewünschten Standardabweichung multipliziert und als Konstante den entsprechenden neuen Mittelwert addiert. Die erste Skalenart wird C-Skala genannt. Ihr Bereich reicht von −1 bis +11. Diese Skala ist heutzutage kaum mehr gebräuchlich, wohl aber eine abgeleitete Form, bei der sich der Meßbereich auf 1 bis 9 beschränkt. Höhere oder niedrigere Werte, die sehr selten auftreten, werden den jeweiligen Endpunkten zugeordnet. Diese Skala nennt man Stanine-Skala; Stanine ist ein Kunstwort, das sich aus Standard und Nine zusammensetzt.

Skalen, die einen Mittelwert von 50 und eine Standardabweichung von 10 haben, werden T-Wert-Skalen genannt. Beträgt der Mittelwert 100 und die Standardabweichung 10, so spricht man von Z-, manchmal auch von Standardwerten.

Darüber hinaus gibt es vor allem bei Tests, die der Wechsler-Test-Familie angehören, noch einige seltener verwendete Transformationen, deren Gebrauch bei unerfahrenen Benutzern leicht zu Mißverständnissen führt. Es ist einmal die Wertpunktskala der einzelnen Untertests, mit einem Mittelwert von 10 und der Standardabweichung 3, und die IQ-Äquivalente.

Prozentrangnormen

Bei den oben genannten Transformationen handelt es sich um lineare Transformationen, die den Mittelwert der Skala verschieben und den Meßbereich entsprechend der gewünschten Standardabweichung spreizen. Diese Transformationsart hat jedoch die Normalverteilung der Rohdaten zur Voraussetzung. Bei Prozentrangnormen führt man eine Flächentransformation der Rohwerte, an deren Verteilungsform man zudem keine besonderen Anforderungen stellt, durch. Berechnet werden diese Normen, indem man feststellt, wieviel Prozent der untersuchten Probanden die gleiche oder eine niedrigere Punktzahl erreichen. Häufig bildet man aus Prozentrangnormen Quartilsnormen, die angeben, in welchem Viertel der Verteilung die Untersuchungsergebnisse eines Probanden einzuordnen sind.

Nachteilig wirkt sich bei Prozentrangskalen aus, daß diese gerade in demjenigen Bereich, der von den meisten Probanden abgedeckt wird, also zumeist im Mittelbereich, stark differenzieren, in den Randbereichen der Verteilung dagegen kaum noch, weil relativ wenige Probanden sehr niedrige oder sehr hohe Testergebnisse erzielen. So ist der Leistungsunterschied zwischen zwei Probanden, von denen einer in einem Intelligenztest den Prozentrang von 45% und der andere von 55% erreicht, wesentlich geringer als der von Probanden, die 98% bzw. 99% erzielen. Es handelt sich bei Prozentrangnormen um einen nicht-linearen Maßstab, der z.B. keine Mittelwertsberechnung erlaubt, denn die Abstände zwischen zwei benachbarten Punkten sind nicht in jedem Bereich der Skala gleichgroß.

Weitere nicht-lineare Transformationen

Auf McCall (1939) geht die Idee zurück, nicht-lineare Rohwertverteilungen so zu transformieren, daß die berechneten Standardwerte kaum noch von der Normalverteilung abweichen. Dazu bestimmt man zunächst den Prozentrang jedes Rohwerts. Jedem Prozentrang entspricht eine bestimmte Fläche unter der Standardnormalverteilung. Die dazugehörige Abszisse verkörpert den zu berechnenden z-Wert, den man nun mühelos in einen T-Wert oder einen anderen Maßstab transformieren kann. Die Berechnung der z-Werte aus Prozenträngen ist relativ mühsam, jedoch liegen diese Werte tabelliert z.B. bei Weber (1964) vor. Darüber hinaus ist die Berechnung mit Hilfe der elektronischen Datenverarbeitung sehr leicht, entsprechende Unterprogramme finden sich in sehr vielen Programmpaketen.

14.6 Anwendungsbereiche psychologischer Tests in der Psychosomatik

14.6.1 Anwendungsbereiche psychologischer Tests in der Forschung

Lange Zeit versuchte man mit Hilfe standardisierter psychologischer Tests Erkenntnisse über ausgewählte Krankengruppen zu gewinnen. Dabei hatte man zum Ziel, diejenigen psychischen Strukturen zu identifizieren, die an der Manifestation der Erkrankung beteiligt sind oder im Verlauf als Reaktion auf die Erkrankung oder deren Behandlung auftreten. Solche Untersuchungsansätze, die auf den ersten Blick plausibel erscheinen, sind in Wirklichkeit äußerst problematisch:

Die Tests wurden in der Regel an körperlich gesunden Probanden analysiert und geeicht. Dabei fragt ein relativ großer Anteil der Items von Persönlichkeitsfragebogen nach körperlichen Symptomen, die in somatisch unauffälligen Stichproben mit Dimensionen der emotionalen Labilität korreliert sind und daher in entsprechenden Skalen verrechnet werden. Diese Beschwerden können jedoch bei körperlich Kranken Folge der Erkrankung sein und dürfen nicht ohne weiteres zur Diagnostik emotionaler Probleme herangezogen werden. Bei hoher Meßgenauigkeit vieler Persönlichkeitstests ist deren Validität daher bei diesen Patienten eingeschränkt.

Bei Untersuchungen von chronisch körperlich Kranken mit dem MMPI zeigt sich z.B. durchgängig, daß, unabhängig vom Krankheitsbild, die ersten drei klinischen Skalen – man faßt sie unter dem Begriff „neurotische Trias" zusammen – bedeutsam erhöht sind. Der Schluß, daß bei chronisch Kranken als Folge der Erkrankung neurotische Störungen im Sinne der Testautoren auftreten, ist sehr gewagt.

Psychologische Tests, die in der psychosomatischen Forschung eingesetzt werden sollen, müssen demnach den zu Beginn des Kapitels beschriebenen Anforderungen Lienerts an psychologische Tests genügen. Darüber hinaus müssen sich mit diesen Tests die zu untersuchenden Merkmale bei den Kollektiven valide erfassen lassen. Dies stellt den Forscher vor die oft nur mit erheblichem Aufwand zu lösende Aufgabe, neue Untersuchungsinstrumente, die für seine Fragestellung adäquat sind, zu entwickeln.

Als Reaktion hierauf zeichnet sich zur Zeit der Trend ab, so wie am Anfang der Entwicklung psychodiagnostischer Testverfahren ad hoc intuitiv-konstruierte Items zu Skalen zusammenzustellen. Typisch hierfür ist die unüberschaubare Menge von Fragebogen zur Erfassung von Lebensqualität. Dabei ist das zu untersuchende Merkmal in der Regel nicht besser definiert als Woodworths „emotional fitness". Die Bogen haben in der Regel eine hohe „face validity", genügen jedoch in keiner Weise den Anforderungen, die ein psychologisches Untersuchungsinstrument zu erfüllen hat.

14.6.2 Anwendungsbereiche von Tests in der psychosomatischen Praxis

Auf viele Tests, die in der psychosomatischen Praxis durchgeführt werden, treffen die oben beschriebenen Einschränkungen zu, wenn mit ihnen körperlich Kranke untersucht werden sollen. Zur Lösung dieses Problems gibt es kein Patentrezept. Sinnvoll erscheint es, wenn sich der Untersucher sorgfältig ein Set einander ergänzender Untersuchungsinstrumente zusammenstellt und dieses nach Möglichkeit nicht verändert. So hat er die Möglichkeit, über einen längeren Zeitraum Erfahrungen mit den Instrumenten zu gewinnen. Bei neu untersuchten Patienten kann er deren Befunde aufgrund der bisherigen Erfahrung relativieren und sie in Beziehung zu anamnestischen und biographischen Daten setzen.

Am häufigsten führt man Tests in der Psychosomatik wohl im Rahmen einer Querschnittsdiagnostik zur Untersuchung bestimmter Persönlichkeitsmerkmale eines Probanden durch. Anhand von Normwerten kann der Testleiter nach solch einer Untersuchung die relative Position des Individuums auf einer Skala bestimmen. Werden mehrere Merkmale untersucht, so lassen sich in sog. Profilen die Intensitäten der einzelnen Ausprägungen miteinander vergleichen. Bei Fragestellungen, die eine Differentialdiagnose zum Ziel haben, stellt man darüber hinaus fest, ob und in welcher Ausprägung die geforderten Merkmale vorhanden sind.

Im Rahmen der Längsschnittdiagnostik untersucht man mit psychologischen Tests, ob sich innerhalb eines bestimmten Zeitraums, beispielsweise während einer Psychotherapie oder einer Rehabilitationsmaßnahme, das untersuchte Merkmal bei einem Individuum verändert hat.

14.7 Schlußbemerkung

Von seiten der neueren Testtheorie wurde der klassischen Testtheorie begründete Kritik entgegengebracht. Leider sind zur Zeit kaum Tests verfügbar, die ausschließlich nach den neuen Methoden konstruiert wurden, so daß es schwerfällt, deren praktische Nützlichkeit zu beurteilen. Diese geringe Verbreitung hat mehrere Ursachen. Einmal ist diese Theorie noch relativ jung und ihre Methodik wesentlich komplizierter zu verstehen als die der klassischen Testtheorie. Ihre Beherrschung setzt fundierte Statistikkenntnisse voraus, die Berechnungen sind überhaupt nur mit geeigneten Rechenprogrammen möglich. Diese sind bei

weitem noch nicht in dem Maße verbreitet und so benutzerfreundlich geschrieben, wie diejenigen, die zur klassischen Testkonstruktion benutzt werden können. Es ist jedoch nur eine Frage der Zeit, bis dieser Nachteil, den die neuere Testtheorie den älteren Methoden gegenüber hat, ausgeglichen ist. Aber nicht nur das Verständnis der komplizierten neuen Theorie hemmt die Entwicklung entsprechender Tests, es gibt auch praktische Probleme, denn die Testkonstruktion ist nur mit umfangreicheren Analysestichproben möglich. Gerade für klinische Tests kann es Jahre dauern, bis man so viele Probanden untersucht hat, daß die Stichprobe für eine Itemanalyse ausreicht.

Der Praktiker steht zur Zeit vor dem Problem, daß die verfügbaren Tests offensichtlich Nachteile aufweisen, und andere, neue Methoden noch nicht verfügbar sind. Andererseits kann er nicht auf Tests verzichten, wenn er sich nicht in subjektiven Spekulationen verlieren will. Gerade Fischer (1974) als herausragender, empirisch arbeitender Gegner der klassischen Testtheorie hebt hervor, daß er ein Ausweichen in eine vorwissenschaftliche Diagnostik nicht als adäquate Reaktion auf seine Kritik betrachtet. Rey (1977) vertritt die Meinung, daß so lange den klassisch konstruierten Tests der Vorzug gegeben werden soll, bis das probabilistische Testmodell den Beweis der uneingeschränkten praktischen Anwendbarkeit erbracht hat. Das Festhalten an den Methoden der klassischen Testtheorie sei aber nur berechtigt, wenn die Testkonstrukte präziser definiert werden und der Anwendungsbereich und damit der Validitätsbereich genauer als bisher angegeben wird.

15 ICD-9 und DSM-III
Eine kritische Stellungnahme zum Gebrauch der internationalen Diagnoseschlüssel

Jörg Michael Herrmann, Marianne Holzamer und *Wolfgang Stiels*

„Classification is the process by which a person reduces the complexity of phenomena by arranging them into categories according to some established criteria for one and more purposes" (Freedman et al., 1978).

Divergierende Definitionen (wie z.B. „Neurose"), unterschiedliche Krankheitskonzepte und verschiedenartige Traditionen in Psychiatrie, Psychologie und Psychotherapie, sowie die mangelnde Vergleichbarkeit wissenschaftlicher Arbeiten und Vorgehensweisen auf diesen Gebieten lassen die Notwendigkeit einer übergreifenden, gemeinsamen Klassifikation pathophysiologischer und besonders psychopathologischer Vorgänge und Zustände deutlich werden.

Dieser übergreifenden Klassifikation liegen die jeweiligen Denkmodelle zugrunde, so ist z.B. in den USA die psychosomatische Medizin lediglich eine Subdisziplin der Psychiatrie, obwohl Psychiatrie eher als Subdisziplin der Psychosomatik aufgefaßt werden sollte (vgl. Kap. 54). Dies beinhaltet ein anderes Paradigma, das im DSM nicht berücksichtigt wird.

Vielfältige Kompromisse, Überschneidungen, Unschärfen und Vernachlässigungen führen dann zwar zu einer Lösung dieser Probleme, rufen jedoch bei fast allen an der Kompromißbildung Beteiligten Unzufriedenheit und Kritik hervor.

Neben diesen Sachfragen nehmen jedoch auch gesundheits-, wissenschafts-, berufs- und gesellschaftspolitische Überlegungen und Entscheidungen Einfluß auf eine derartige Lösung.

Zwei der namhaftesten Versuche dieser Art, die ICD und das DSM, sollen hier exemplarisch und auszugsweise vorgestellt und diskutiert werden.

15.1 International Classification of Diseases (ICD)

Da die in der ICD aufgeführten psychiatrischen Krankheiten vor allem durch deskriptive Kriterien definiert werden, ist die genaue Beobachtung des Kranken und seiner Symptome Voraussetzung für die richtige Benutzung dieses Glossars.

„Der englische Ausdruck ‚to gloss over' oder ‚to gloze' (deutsche Übersetzung: beschönigen), der von der gleichen Wortwurzel wie Glossar abstammt, bezeichnet eine unehrliche Tätigkeit ... Unsicherheit und Tücken ... psychiatrischer Glossare ... werden durch die Spärlichkeit objektiver Daten erhöht, von denen Definition und Diagnose abhängen müssen. Ein psychiatrisches Glossar kann entweder auf klinischen Querschnittsbildern (Syndromen) oder auf dem klinischen Verlauf aufgebaut sein; es kann psychodynamisch, ätiologisch (genetisch) oder pathogenetisch ausgerichtet sein. Da jedoch Krankheiten in jedem Fall abstrakte Konzepte sind, nimmt es nicht wunder, daß sich die Krankheitskonstrukte, mit denen die Psychiater arbeiten, überlappen und undeutliche Begrenzungen haben. Fehlende Beobachterübereinstimmung ist hierfür von entwaffnender Evidenz; die Reliabilität ist zu niedrig, um wissenschaftlich zufriedenzustellen; Widersprüche können in einzelnen Fällen verringert, in anderen minimalisiert werden, in Abhängigkeit davon, ob sie von ungenauer Beobachtung, subjektiver Beurteilung oder von Diskrepanzen der benutzten nosologischen Systeme oder Fachausdrücke herrühren" (Lewis, 1974).

Die ICD ist in 17 Hauptkapitel, die mit römischen Ziffern bezeichnet sind, unterteilt. Diese Hauptkapitel sind nach unterschiedlichen, vor allem eklektischen Kriterien erarbeitet worden, so wurden z.B. die Infektionskrankheiten nach ätiologischen, die Atmungskrankheiten nach topographischen oder die Schwangerschaftskomplikationen nach situationsabhängigen Gesichtspunkten geordnet.

Die psychiatrischen Krankheiten wurden – bei zumeist nicht geklärter somatischer Ätiologie oder Pathogenese – durch Symptome, Verhaltensähnlichkeiten oder den Krankheitsverlauf definiert. Daher basiert das Glossar zum Hauptkapitel V „Psychiatrische Krankheiten" der ICD-9 (1980) hauptsächlich auf der Beschreibung von Symptombildern und Syndromen und nicht auf klaren Definitionen. Zwei Zusatzklassifikationen für die äußeren Ursachen bei Verletzungen und Vergiftungen („E-Schlüssel") und die Faktoren, die den Gesundheitszustand und die Inanspruchnahme der Gesundheitsdienste beeinflussen („V-Schlüssel"), erlauben eine genauere, insbesondere statistische Zuordnung psychiatrischer Krankheitsbilder.

Das Kapitel V, das die psychiatrischen Krankheiten enthält, wurde von 1965 bis 1972 durch eine internationale Expertengruppe der WHO erarbeitet, die jährlich ein spezielles Thema diskutierte:

1965 (London): Funktionelle Psychosen mit besonderer Berücksichtigung der Schizophrenie
1966 (Oslo): Borderline- und reaktive Psychosen
1967 (Paris): Psychiatrische Störungen im Kindesalter

1968 (Moskau): Psychische Störungen im Alter
1969 (Washington, D. C.): Oligophrenie
1970 (Basel): Neurotische und psychosomatische Störungen
1971 (Tokio): Persönlichkeitsstörungen und Medikamenten- und Drogenabhängigkeit
1972 (Genf): Zusammenfassung, Schlußfolgerungen, Empfehlungen und Vorschläge für zukünftige Forschung.
Seither wurde die ICD nicht mehr revidiert.

Von den Herausgebern der 9. Revision der ICD (ICD-9) werden folgende Ziele für die Benutzer angegeben:
– einheitlicher Gebrauch der wesentlichen diagnostischen Begriffe der Psychiatrie
– Verminderung internationaler Unterschiede zwischen den diagnostischen Konzepten bei der statistischen Dokumentation psychiatrischer Erkrankungen
– bessere internationale Verständigung in bezug auf klinische Arbeit und wissenschaftliche Forschung
– Verwendung für den Unterricht
Dabei wird vernachlässigt, daß dadurch in die Theorien- und Modellbildung und damit letztlich in die Beziehungsgestaltung und Therapie eingegriffen wird.

15.2 ICD-9 und psychosomatische Medizin

Für die psychosomatische Medizin sind die Kategorien für Neurosen (300 bis 300.9), Persönlichkeitsstörungen (301 bis 301.9), körperliche Funktionsstörungen psychischen Ursprungs (306 bis 306.9) sowie „anderweitig klassifizierte Erkrankungen, bei denen psychische Faktoren eine Rolle spielen (psychosomatischen Erkrankungen im engeren Sinne)" (316) von Bedeutung. Darüber hinaus führt die ICD-9 noch ein sich überschneidendes und heterogenes Sammelsurium anderer „nichtpsychotischer psychischer Störungen" an:
– sexuelle Verhaltensabweichungen und Störungen (302 bis 302.9)
– Alkoholabhängigkeit (303)
– Medikamentenabhängigkeit (304 bis 304.9)
– Drogen- und Medikamentenmißbrauch ohne Abhängigkeit (305 bis 305.9)
– spezielle, nicht anderweitig klassifizierbare Symptome oder Syndrome (307 bis 307.9)
– psychogene Reaktion (akute Belastungsreaktion) (308 bis 308.9)
– psychogene Reaktion (Anpassungsstörung) (309 bis 309.9)
– spezifische nichtpsychotische psychische Störungen nach Hirnschädigung (310 bis 310.9)
– anderweitig nicht klassifizierbare depressive Zustandsbilder (311)
– anderweitig nicht klassifizierbare Störungen des Sozialverhaltens (312 bis 312.9)
– spezifische emotionale Störungen des Kindes- und Jugendalters (313 bis 313.9)

– hyperkinetisches Syndrom des Kindesalters (314 bis 314.9) und
– umschriebene Entwicklungsrückstände (315 bis 315.9)

Trotz anderer Klassifikationskriterien bei den psychiatrischen Krankheiten werden die funktionellen Syndrome („körperliche Symptome oder Bilder physiologisch-funktioneller Störungen psychischen Ursprungs **ohne** Gewebsschädigung, die gewöhnlich durch das autonome Nervensystem vermittelt werden") nach groben und schlecht definierten topographischen Gesichtspunkten unterteilt:

306.0 Muskulatur und Skeletsystem
        dazugehöriger Begriff:
        psychogener Schiefhals
306.1 Atmungsorgane
        dazugehörige Begriffe:
        Atemnot
        (psychogener) Singultus
        Hyperventilation
        psychogener Husten
        Gähnen
306.2 Herz- und Kreislaufsystem
        dazugehörige Begriffe:
        Herzneurose
        Herz-Kreislauf-Neurose
        neurozirkulatorische Asthenie
        psychogene Herz-Kreislauf-Störung
306.3 Haut
        dazugehöriger Begriff:
        psychogener Pruritus
306.4 Magen-Darm-Trakt
        dazugehörige Begriffe:
        Aerophagie (Luftschlucken)
        psychogenes periodisches Erbrechen
306.5 Urogenitalsystem
        dazugehöriger Begriff:
        psychogene Dysmenorrhoe
306.6 Endokrines System
306.7 Sinnesorgane
306.8 Andere körperliche Funktionsstörungen psychischen Ursprungs
        dazugehöriger Begriff:
        Zähneknirschen
306.9 Nicht näher bezeichnete körperliche Funktionsstörungen psychischen Ursprungs
        dazugehörige Begriffe:
        nicht näher bezeichnete psychophysiologische Störung
        nicht näher bezeichnete psychosomatische Störung

Zum Teil werden dazugehörige Begriffe „als Synonyma" (z.B. 306.2) oder als eigene nosologische Entitäten (z.B. 306.1) aufgeführt. Überschneidungen (z.B. Angstneurose 300.0, Hyperventilation 306.1 oder Herzneurose 306.2) werden dabei in Kauf genommen (vgl. auch Kap. 32).

Die Borderline-Persönlichkeit findet sich innerhalb der Kategorien für Psychosen (290 bis 299) unter dem Begriff „Latente Schizophrenie" (295.5) als „Borderline-Schizophrenie" („dazugehöriger Begriff").

Als „Psychosomatische Erkrankungen im engeren Sinne" (316) werden psychische Erkrankungen oder Symptome **mit** Gewebsschädigung angesehen, wobei die „psychische Störung ... gewöhnlich leichteren Grades" ist, und „unspezifische psychische Symptome (Besorgtheit, Furcht, Konflikte usw.) ohne eine offensichtliche psychiatrische Erkrankung vorhanden sein" können.

Auf die multikausale Genese dieser Erkrankungen wird nicht eingegangen, sondern es wird lediglich zwischen psychisch und somatisch unterschieden, so daß beispielsweise das Asthma bronchiale mit 493.9, das „Psychogene Asthma" aber mit 316 **und** 493.9 kodiert wird.

Die „Anorexia nervosa" (307.1) wird unter „spezielle, nicht anderweitig klassifizierbare Symptome oder Syndrome" (307) eingruppiert und mit einigen unscharfen Symptomen definiert. Quantitative Angaben zum Gewichtsverlust oder der häufige Laxantien- und Diuretika-Abusus (vgl. Kap. 38.2) fehlen.

„Die gegenwärtigen psychiatrischen Debatten über Klassifikationssysteme, die vielen hypothetischen und unbestätigten Schemata psychodynamischer Mechanismen zugrunde liegen, und die Beschäftigung mit ätiologischen Schlußfolgerungen statt einer Beweisführung aufgrund objektiver Beobachtung, sind nosologische Aktivitäten, die einen manchmal an mittelalterliche Taxonomisten erinnern" (Feinstein, 1974).

15.3 DSM-III-R

Auch die Klassifikation psychischer Erkrankungen („Störungen") durch das DSM-III-R („Diagnostic and Statistical Manual of Mental Disorders", Revision der dritten Auflage der American Psychiatric Association, 1987) beruht auf einer klinisch unbefriedigenden Beschreibung dieser psychischen Störungen. Im Vordergrund dieses deskriptiven Vorgehens stehen diagnostische Kriterien, die aus Symptomen, zeitlichen und Verlaufsaspekten, Schweregraden und psychosozialen Merkmalen bestehen. Die über 200 DSM-III-Kategorien sollen mit Hilfe eines multiaxialen Systems beurteilt werden:

Achse I und II beinhalten alle psychischen Störungen, mit Achse III werden alle körperlichen Störungen erfaßt, Achse IV bestimmt den Schweregrad psychosozialer Belastungsfaktoren und mit Achse V erfolgt die Globalbeurteilung des psychosozialen Funktionsniveaus. Zwar haben diese diagnostischen Kategorien des DSM Eingang in die wissenschaftliche Literatur gefunden, eine weitergehende Zuordnung entsprechend den 5 Achsen erfolgte jedoch bisher zumeist nicht.

Besondere Probleme bietet die Übersetzung des DSM-III-R ins Deutsche, z. B. wird „mental disorder" durchgehend mit „psychische Störung", „mental illness" mit „psychische Erkrankung" oder „mood disorder" mit „affektive Störung" übersetzt. Der deutsche Terminus „schizophrener Schub" wird – wegen fehlender amerikanischer Entsprechung – durch „schizo-

phrene Episode" oder „schizophrene Phase" ersetzt. An der „zyklothymen Störung" im Sinne Kraepelins wird festgehalten.

Während noch in der ersten Auflage des DSM (1952) psychische Störungen als Reaktionen der Persönlichkeit auf psychische, soziale und biologische Faktoren dargestellt wurden, wurde dieser theoretische Bezugsrahmen mit dem DSM-II (1968) verlassen. Mit dem DSM-III (1980) und dem DSM-III-R (1987), von 26 Beratungsausschüssen mit mehr als 200 Mitgliedern erstellt, wurde ein rein deskriptiver Ansatz verwirklicht.

Im DSM-III-R entsprechen nun erstmals alle Kodierungsnummern der ICD-9. Für 1992 soll nicht nur die 4. Auflage des DSM, sondern auch eine mit dem DSM-IV in noch größerer Übereinstimmung stehende 10. Version der ICD (ICD-10) vorliegen.

Von der Arbeitsgruppe zur Revision des DSM-III wurden folgende Ziele für die Benutzer genannt (DSM-III-R, 1987):

– klinische Brauchbarkeit für therapeutische und administrative Entscheidungen in verschiedenen klinischen Arbeitsgebieten
– Reliabilität der diagnostischen Kriterien
– Akzeptanz für Kliniker und Forscher mit unterschiedlicher theoretischer Ausrichtung
– Brauchbarkeit als Ausbildungsgrundlage für Heilberufe
– Erhaltung der Kompatibilität mit der ICD-9
– soweit wie möglich Vermeidung neuer Terminologien und Konzepte, die mit der Tradition brechen
– Erreichen eines Konsensus über den Bedeutungsinhalt derjenigen notwendigen diagnostischen Begriffe, die zuvor unterschiedlich verwendet wurden, und ferner der Verzicht auf solche Begriffe, die ihren Nutzen überlebt haben
– Angleichung an die Ergebnisse von Forschungsstudien über die Validität diagnostischer Kategorien
– Brauchbarkeit zur Beschreibung von Probanden in Forschungsstudien
– während der Entwicklung von DSM-III-R Aufgeschlossenheit gegenüber Kritik anderer Kliniker und Forscher.

15.4 DSM-III-R und psychosomatische Medizin

Während in der ICD-9 die Neurosen (300 bis 309.9) noch als Kategorie aufgeführt werden, spart das DSM-III-R das Neurosenkonzept fast vollständig aus:

Die Angstneurose (300.0) erscheint als „nicht näher bezeichnete Angststörung", die Konversion wird den „Somatoformen Störungen" und die „Hysterische Neurose" den „Dissoziativen Störungen" zugerechnet und dann unterteilt in „Multiple Persönlichkeitsstörung" (300.14), „Psychogene Fugue" (300.13), „Psychogene Amnesie" (300.12), „Depersonalisationsstörung" (300.60) sowie „nicht näher bezeichnete dissoziative Störung" (300.15), insgesamt Begriffe, die in der ICD-9 nicht vorkommen.

Funktionelle Beschwerden wurden im DSM-III-R überhaupt nicht aufgenommen und die „Psychosomatischen Erkrankungen im engeren Sinne" (316) werden kurz auf 1½ Seiten als „körperlicher Zustand, bei dem psychische Faktoren eine Rolle spielen", abgehandelt. Als „gebräuchliche Beispiele" für diese Kategorie werden dann Adipositas, Spannungskopfschmerz, Migräne, Angina pectoris, Menorrhagie, Sakroiliakalschmerz, Neurodermitis, Akne, rheumatoide Arthritis, Asthma, Tachykardie, Arrhythmie, Ulcus ventriculi, Ulcus duodeni, Kardiospasmus, Pylorospasmus, Übelkeit und Erbrechen, regionale Enteritis, Colitis ulcerosa und Pollakisurie angeführt. Diese Auswahl wird nicht begründet und genügt dem im Vorwort betonten wissenschaftlichen Anspruch in keiner Weise.

Die „Eßstörungen" (u. a. Anorexia nervosa – 307.10 – und Bulimia nervosa – 307.15), bei denen im DSM-III, im Gegensatz zum DSM-III-R, noch H. Bruch und A. J. Stunkard verantwortlich mitarbeiteten, erscheinen unter „Störungen mit Beginn typischerweise im Kleinkindalter, Kindheit oder Adoleszenz".

Wie auch bei den anderen „psychischen Störungen" werden zwar Haupt- und Nebenmerkmale, Alter bei Beginn, Verlauf, Beeinträchtigungen, Komplikationen, prädisponierende Faktoren, Prävalenz, Geschlechtsverteilung, familiäre Häufung oder Differentialdiagnose entsprechend dem augenblicklichen wissenschaftlichen Forschungsstand beschrieben, jedoch werden theoretische Konzepte zu Ätiologie und Pathogenese, insbesondere psychodynamische Konzepte nicht erwähnt. Konsequenterweise wird auch die „Neurotische Depression" (300.4) zur „Dysthymen Störung (oder depressiven Neurose)", obwohl zu dem im angloamerikanischen Sprachbereich weitgefaßten Begriff der „Dysthymia" auch die funktionellen Syndrome gerechnet werden.

Die im DSM-III neu eingeführten Begriffe der „major depression" und „minor depression" (vgl. Kap. 51) lassen sich weder einigermaßen adäquat ins Deutsche übersetzen noch entsprechen sie traditionsreichen psychiatrischen Krankheitsbildern unseres Sprachraumes. Insofern kann z. B. die „major depression" im Deutschen folgende Bezeichnung tragen (Peters, 1984): Monopolare endogene Depression, bipolare endogene Depression, schizophrene Depression, körperlich begründbare Depression, Erschöpfungsdepression, neurotische Depression und depressive Erlebnisreaktion – allerdings nur dann, wenn die diagnostischen Kriterien erfüllt sind.

15.5 Kritik

Das deskriptive Konzept des DSM-III und DSM-III-R, das sich grundlegend von allen früheren psychiatrischen Klassifikationen, inklusive DSM-I und DSM-II, unterscheidet, ist das Produkt einer kleinen homogenen psychiatrischen Arbeitsgruppe in den USA, die sich selbst als „Neo-Kraepelinians" bezeichnen.

Der deutsche Psychiater Emil Kraepelin (1856 bis 1926) war nicht nur der Begründer des neuzeitlichen psychiatrischen Klassifikationssystems, sondern vertrat auch den Standpunkt, daß psychische Störungen den organischen Krankheiten ähnlich sind und insofern die für die Organmedizin gültigen Methoden und Perspektiven von der Psychiatrie übernommen werden müssen (Kraepelin, 1896).

Daher ist es nicht verwunderlich, daß in der Einleitung des DSM-III-R die diagnostische Hierarchiebildung durch folgendes Prinzip bestimmt wird: „Wenn eine organisch bedingte psychische Störung die Symptome verursacht, so steht diese Diagnose über jeder anderen Diagnose irgendeiner psychischen Störung mit den gleichen Symptomen (z. B. steht die organisch bedingte Angststörung vor der Panikstörung)".

Auf Kraepelin verweisend, fordern die Neo-Kraepelinianer eine fortschrittliche psychiatrische Klassifikation nach folgenden Gesichtspunkten (Young, 1988):
- Jede psychische Störung hat – wie jede andere Krankheit – eine bestimmte Ätiologie.
- Eine Klassifikation psychischer Störungen kann durch sorgfältige empirische Beobachtungen erreicht werden.
- Die Forschung wird beweisen, daß psychische Störungen eine organische oder biochemische Ursache haben.

Dabei wird impliziert, daß ein pathophysiologischer Prozeß für alle Krankheiten verantwortlich ist, bzw. daß es eine monokausale Erklärung für die meisten psychischen Störungen gibt. Andere theoretische Erklärungsmodelle, insbesondere die Psychoanalyse, werden abgelehnt.

Mit dem „DSM-III ist eine Richtung führend geworden, die man als ‚statistische Phänomenologie' oder nach einer bestimmten Methodik auch als ‚Exklusio – Inklusionismus' bezeichnen kann" (Peters, 1984).

Im deutschen Sprachraum blieb die in den USA heftig geführte Diskussion um das DSM-III relativ unbeachtet. Von den wenigen Kritikern wurde die Abkehr vom psychodynamischen Konzept der Neurosen problematisiert (Bluestone, 1985). Bluestone (1985) und Schuster et al. (1985) weisen jedoch auf gesundheitspolitische Entwicklungen hin, die eher Ökonomiegrundsätzen folgend eine Hinwendung zu nichtanalytischen Therapien begünstigten. Auch Hoffmann (1985) macht nicht das DSM-III für die Abschaffung des Neurosenbegriffs verantwortlich, sieht jedoch die neue Nomenklatur als Wegbereiter für eine Rücknahme von Kassenfinanzierung der Psychotherapie neurotischer Erkrankungen.

Die gesellschaftliche und vor allem ökonomische Verflechtung jedes Klassifikationsschemas kann nicht wegdiskutiert oder gar versachlicht werden, sondern bedarf der kritischen Reflexion sowohl auf der Basis der wissenschaftlichen Kriterien als auch der obengenannten wissenschaftspolitischen Ebene.

Diese kritische Auseinandersetzung mit den augenblicklichen psychiatrischen Konzepten wurde bisher nur von Young (1988) in Angriff genommen.

„Mit dem Paradigmawechsel wird eine völlig neue nicht-anatomische Systematik möglich" (v. Uexküll, 1988).

16 Psychoanalyse und psychoanalytisch orientierte Therapieverfahren

Wolfgang Wesiack

Die Psychoanalyse ist nicht nur für die Theorie, sondern ebenso für die therapeutische Praxis der Psychosomatischen Medizin ein Grundelement. Sie soll deshalb nachfolgend sowohl in den Grundzügen ihrer klassischen Form, als auch in ihren für die Psychosomatische Medizin wichtigsten Weiterentwicklungen dargestellt werden. Auf Vollständigkeit muß dabei natürlich verzichtet werden. Es kommt nur auf die Herausarbeitung der wichtigsten Gesichtspunkte an (vgl. auch Kap. 5).

Um Mißverständnisse zu vermeiden, muß auch darauf hingewiesen werden, daß man psychotherapeutische Techniken nicht aus Büchern, sondern nur durch geduldige Übung unter Anleitung und Supervision von Erfahrenen erlernen kann.

Wir werden so vorgehen, daß wir in diesem Kapitel zunächst versuchen wollen, einen zusammenfassenden Überblick über die von der Psychoanalyse ausgehenden und für die Psychosomatische Medizin bedeutsamen therapeutischen Techniken zu geben, um dann das nächste Kapitel dem diagnostisch-therapeutischen ärztlichen Gespräch zu widmen, das viele Elemente sowohl psychoanalytischer als auch nicht-psychoanalytischer Herkunft enthält.

16.1 Psychoanalyse

In seinem umfangreichen Werk hat uns Freud mehrere Definitionen dessen gegeben, was wir unter Psychoanalyse zu verstehen haben. Eine der klarsten ist am Anfang seines Aufsatzes „Psychoanalyse und Libidotheorie" zu finden. Sie lautet: „Psychoanalyse ist der Name 1. eines Verfahrens zur Untersuchung seelischer Vorgänge, welche sonst kaum zugänglich sind; 2. einer Behandlungsmethode neurotischer Störungen, die sich auf diese Untersuchung gründet; 3. einer Reihe von psychologischen, auf solchem Wege gewonnenen Einsichten, die allmählich zu einer neuen wissenschaftlichen Disziplin zusammenwachsen."[1] Über die Grundpfeiler der psychoanalytischen Theorie schreibt Freud in der gleichen Arbeit: „Die Annahme unbewußter seelischer Vorgänge, die Anerkennung der Lehre vom Widerstand und der Verdrängung, die Einschätzung der Sexualität und des Ödipus-Komplexes sind die Hauptinhalte der Psychoanalyse und die Grundlagen ihrer Theorie, und wer sie nicht alle gutzuheißen vermag, sollte sich nicht zu den Psychoanalytikern zählen."[2] Und wieder einige Seiten weiter heißt es: „Die Psychoanalyse ist kein System wie die philosophischen, das von einigen scharf definierten Grundbegriffen ausgeht, mit diesen das Weltganze zu erfassen sucht, und dann, einmal fertig gemacht, keinen Raum mehr hat für neue Funde und bessere Einsichten. Sie haftet vielmehr an den Tatsachen ihres Arbeitsgebietes, sucht die nächsten Probleme der Beobachtung zu lösen, tastet sich an der Erfahrung weiter, ist immer unfertig, immer bereit, ihre Lehren zurechtzurücken oder abzuändern. Sie verträgt es so gut wie die Physik oder die Chemie, daß ihre obersten Begriffe unklar, ihre Voraussetzungen vorläufige sind, und erwartet eine schärfere Bestimmung derselben von zukünftiger Arbeit."[3]

Viel wichtiger als die Theorie der Psychoanalyse, die sog. „Metapsychologie", die Freud selber in seiner „Selbstdarstellung" den „spekulativen Überbau der Psychoanalyse" genannt hat, „von dem jedes Stück ohne Schaden und Bedauern geopfert oder ausgetauscht werden kann, sobald eine Unzulänglichkeit erwiesen ist",[4] ist für unsere Betrachtungsweise die psychoanalytische Methode. Sie ist es, die erstmals einen neuen wissenschaftlichen Zugang zur Subjektivität des Menschen bahnte, der sich bisher zwangsläufig naturwissenschaftlicher Objektivierung verschloß und lediglich philosophischer Kontemplation und Spekulation zugänglich war.

Freud entwickelte seine Methode, nachdem er vorher bei Charcot in Paris und bei Bernheim in Nancy Erfahrungen mit hypnotischen Heilbehandlungen gesammelt hatte, und nachdem er die sog. kathartische Therapie Breuers kennengelernt und selbst an einer größeren Anzahl von Kranken erprobt hatte. Bei den Hypnosetherapien lernte er den Einfluß der ärztlichen Suggestionen, aber auch die Kraft der unbewußten Phantasien kennen, während ihm die kathartische Therapie Breuers zeigte, daß es mit ihrer Hilfe in Hypnose gelang, „seelische Vorgänge zu einem anderen als dem bisherigen Verlaufe zu bringen, der in die Symptombildung eingemündet hat".[5]

Breuer und Freud erklärten sich die therapeutische Wirksamkeit dieses Verfahrens mit der „Abfuhr" bzw.

---

1 Vgl. S. Freud (1923): Ges. W. Bd. XIII S. 211.
2 Vgl. S. Freud (1923): Ges. W. Bd. XIII S. 223.
3 Vgl. S. Freud (1923): Ges. W. Bd. XIII S. 229.
4 Vgl. S. Freud (1925): Ges. W. Bd. XIV S. 58.
5 Vgl. S. Freud (1904): Ges. W. Bd. V S. 4.

dem „Abreagieren" des bis dahin gleichsam „eingeklemmten" Affektes, der an unterdrückten seelischen Aktionen gehaftet hatte.[6] Freud verzichtete später auf Suggestion und Hypnose, „da das Hypnotisiertwerden, trotz aller Geschicklichkeit des Arztes, bekanntlich in der Willkür des Patienten liegt ..."[7], und ersetzte sie durch die Technik der freien Assoziation. „Die technische Grundregel, dieses Verfahren der ‚freien Assoziation', ist seither in der psychoanalytischen Arbeit festgehalten worden. Man leitet die Behandlung ein, indem man den Patienten auffordert, sich in die Lage eines aufmerksamen und leidenschaftslosen Selbstbeobachters zu versetzen, immer nur die Oberfläche seines Bewußtseins abzulesen und einerseits sich die vollste Aufrichtigkeit zur Pflicht zu machen, andererseits keinen Einfall von der Mitteilung auszuschließen, auch wenn man 1. ihn allzu unangenehm empfinden sollte, oder wenn man 2. urteilen müßte, er sei unsinnig, 3. allzu unwichtig, 4. gehöre nicht zu dem, was man suche. Es zeigt sich regelmäßig, daß gerade Einfälle, welche die letzterwähnten Ausstellungen hervorrufen, für die Auffindung des Vergessenen von besonderem Wert sind."[8]

Freud ließ dabei seine Patienten „ohne andersartige Beeinflussung eine bequeme Rückenlage auf einem Ruhebett einnehmen ... während er selbst, ihrem Anblick entzogen, auf einem Stuhle hinter ihnen"[9] Platz nahm.

„Die Erfahrung zeigte bald, daß der analysierende Arzt sich dabei am zweckmäßigsten verhalte, wenn er sich selbst bei gleichschwebender Aufmerksamkeit seiner eigenen unbewußten Geistestätigkeit überlasse, Nachdenken und Bildung bewußter Erwartungen möglichst vermeide, nichts von dem Gehörten sich besonders im Gedächtnis fixieren wolle, und solcher Art das Unbewußte des Patienten mit seinem eigenen Unbewußten auffange. Dann merkte man, wenn die Verhältnisse nicht allzu ungünstig waren, daß die Einfälle des Patienten sich gewissermaßen wie Anspielungen an ein bestimmtes Thema herantasteten, und brauchte selbst nur einen Schritt weiter zu wagen, um das ihm selbst Verborgene zu erraten und ihm mitteilen zu können. Gewiß war diese Deutungsarbeit nicht streng in Regeln zu fassen und ließ dem Takt und der Geschicklichkeit des Arztes einen großen Spielraum, allein wenn man Unparteilichkeit mit Übung verband, gelangte man in der Regel zu verläßlichen Resultaten, d.h. zu solchen, die sich durch Wiederholung in ähnlichen Fällen bestätigten."[10]

Mit Hilfe dieser Methode gelang es Freud, hinter dem „manifesten Inhalt" auch den „latenten", d.h. den „dynamisch unbewußten" Teil der Mitteilung des Patienten zu erfassen, und so kommt er zu den beiden „Grundpfeilern" der psychoanalytischen Technik, die darin bestehen, „daß mit dem Aufgeben der bewußten Zielvorstellungen die Herrschaft über den Vorstellungsablauf an verborgene Zielvorstellungen übergeht, und daß oberflächliche Assoziationen nur ein Verschiebungsersatz sind für unterdrückte tiefer gehende ..."[11]

Die konsequente Anwendung der psychoanalytischen Methode brachte nun Freud eine Reihe von fundamentalen Entdeckungen, wie etwa diese: Die „verborgenen Zielvorstellungen" der Patienten sind nicht nur ihnen selbst unbewußt, sie sind auch so gut wie immer triebbedingt.[12] Da sie mit der Zielvorstellung der Person, den bewußten „Ich"-Anteilen und den im „Über-Ich"[13] introjizierten Normen der Gesellschaft nicht vereinbar sind, werden sie abgewehrt[14] und damit nicht nur am Bewußtwerden, sondern an jeglicher Integration und harmonischen Verschmelzung mit der Gesamtpersönlichkeit gehindert. Die triebbedingten „verborgenen Zielvorstellungen" werden so gleichsam zu störenden Fremdkörpern innerhalb der psychischen Struktur der Patienten, konstituieren mit den abwehrenden Instanzen „Ich" und „Über-Ich" einen ungelösten intrapsychischen Konflikt und führen so zu den verschiedensten Symptomen.

Die sich daran anschließenden Entdeckungen Freuds waren folgende: Die „verborgenen Zielvorstellungen" der Patienten zeigten sich nicht nur in ihren Träumen, Fehlhandlungen und Symptomen, sondern bereits in allen ihren Äußerungen und Mitteilungen – sofern man nur darauf achtete. Freud erkannte, daß die neurotisch Kranken so gut wie immer sowohl die Person des Arztes als auch alle anderen wichtigen Beziehungspersonen, wie z.B. Familienangehörige, Vorgesetzte, Mitarbeiter und Untergebene, unbewußt in ihre neurotischen Konflikte einbeziehen, und nannte dies „die Übertragung".[15] In der Übertragung manifestieren sich also die alten ursprünglichen und infantilen Verhaltensmuster, die meist mit den Forderungen der Realität unvereinbar sind. Aufgabe der psychoanalytischen Therapie ist es, die Manifestationen der Übertragung möglichst auf den Arzt zu konzentrieren, weil sie auf dieser Ebene einer therapeutischen Bearbeitung und Auflösung am besten zugänglich sind. „Was sind die Übertragungen? Es sind

---

6 Vgl. S. Freud (1904): Ges. W. Bd. V S. 4.
7 Vgl. S. Freud (1904): Ges. W. Bd. V S. 5.
8 Vgl. S. Freud (1923): Ges. W. Bd. XIII S. 214/215.
9 Vgl. S. Freud (1904): Ges. W. Bd. V S. 5.
10 Vgl. S. Freud (1923): Ges. W. Bd. XIII S. 215.
11 Vgl. S. Freud (1900): Ges. W. Bd. II/III S. 536/537.
12 Der Triebbegriff und die Trieblehre haben für die psychoanalytische Betrachtungsweise eine zentrale Bedeutung. In der psychoanalytischen Theorie versteht man unter „Trieb" Kräfte, die ihren Ursprung in einer somatischen Triebquelle haben, sich durch ihren dranghaften Charakter und durch ihre Vorstellungs- und Affektrepräsentanzen psychisch repräsentieren und ihr Ziel in der Befriedigung an einem Objekt suchen. (Vgl. z.B. Loch: Die Krankheitslehre der Psychoanalyse, S. 17ff.)
13 In seiner zweiten Theorie des psychischen Apparates unterscheidet Freud drei psychische Instanzen: Das Es, das Ich und das Über-Ich. Das Es bildet das Hauptreservoir der psychischen Energie, der Triebe. Es ist unbewußt. Das Über-Ich wird durch Verinnerlichung der elterlichen Gebote und Verbote gebildet, während das Ich zwischen den triebbedingten Ansprüchen des Es, den Geboten des Über-Ich und den Forderungen der Realität vermitteln muß. (Vgl. z.B. Loch: Die Krankheitslehre der Psychoanalyse, S. 27ff.)
14 Der Abwehrbegriff ist ebenfalls ein Zentralbegriff der psychoanalytischen Theorie. Mit Hilfe verschiedener Abwehrmechanismen versucht sich das Ich der Triebansprüche des Es zu erwehren. (Vgl. Anna Freud: Das Ich und die Abwehrmechanismen und Loch: Die Krankheitslehre der Psychoanalyse, S. 38f.)
15 Vgl. auch Sandler et al., 1973.

Neuauflagen, Nachbildungen von den Regungen und Phantasien, die während des Vordringens der Analyse erweckt und bewußtgemacht werden sollen, mit einer für die Gattung charakteristischen Ersetzung einer früheren Person durch die Person des Arztes.“[16] Die sich in allen mitmenschlichen Beziehungen manifestierende Übertragung nannte Argelander später treffend „die Szene“ und meint damit die „situationsgerechte Darstellung einer unbewußten, infantilen Konfiguration – einer relativ stabilen, persönlichkeitsgebundenen Triebszene“.[17] Demnach ist „szenisches Verstehen“ eine der wichtigsten Aufgaben des Psychoanalytikers. Da sich „die Szene“ im Gegensatz zu anderen objektiven und subjektiven Informationen des Patienten, die oft erst nach längerer Zeit zu erhalten sind, meist „in Sekundenschnelle“ schon im ersten Sprechstundeninterview entfaltet, ist das „szenische Verstehen“ vor allem für die Sprechstundenpsychotherapie von überragender Bedeutung.[18]

Bei dem Versuch, die „verborgenen Zielvorstellungen“ der Patienten bewußtzumachen und den intrapsychischen Konflikt und die Übertragung durchzuarbeiten, stieß Freud auf das Phänomen des Widerstandes und stellte fest: „Der Widerstand in der (psychoanalytischen) Kur geht von denselben höheren Schichten und Systemen des Seelenlebens aus, die seinerzeit die Verdrängung durchgeführt haben.“[19]

Nach Greenson (1973) umfaßt die psychoanalytische Technik vier mehr oder weniger deutlich voneinander unterschiedene Verfahren: die Konfrontation, die Klärung, die Deutung und das Durcharbeiten. Die Konfrontation mit dem und die Klärung des Konfliktes des Patienten bereiten die Deutung vor, die das Herzstück und das Spezifische der Psychoanalyse darstellt. Durch die Deutung werden unbewußte Phänomene bzw. unbewußte Anteile der „Situation“ (im Sinne unseres Situationskreismodells, vgl. Kap. 1) bewußtgemacht. Das zeitraubendste Element der psychoanalytischen Therapie ist aber meist das Durcharbeiten, weil sich neue Einsichten im allgemeinen nicht sofort, sondern erst allmählich und unter Überwindung von Widerständen in Verhaltensänderungen umsetzen lassen.

Hier möchte ich einige Anmerkungen darüber machen, wie sich Theorie und Praxis der Psychoanalyse zwanglos in das in dem einführenden Kapitel entwickelte Situationskreismodell einfügen lassen:

Zunächst wird die Problemsituation des Patienten durch die Konfrontation mit dem Konflikt und seiner eventuellen weiteren Klärung etwas aufgehellt. Bei sehr bewußtseinsnahe gelegenen Konflikten genügen schon manchmal diese ersten beiden Schritte, die die Problemsituation klar herausarbeiten, um dem Patienten selbst die weitere Problemlösung zu ermöglichen und zu überlassen. Häufig gelingt dies dem Patienten selbst jedoch nicht, weil er unfähig ist, die unbewußten Anteile der Situation, meist die triebbedingten Programme, mit den Forderungen des eigenen Über-Ich, der Sozietät und der Außenwelt in Einklang zu bringen. Hier muß durch das Interpretationsangebot (= die Deutung) des Arztes das Programm-Repertoire und damit ein Stück individueller

Wirklichkeit des Patienten verändert und umstrukturiert werden.

Wieso wirken sich jedoch therapeutische Deutungen des Analytikers im günstigen Fall „mutativ“ auf das Erleben und damit auch früher oder später auf das Verhalten des Patienten aus? Wieso vermögen sie seine individuelle Wirklichkeit im Sinne des Situationskreismodells zu verändern? Die Antwort darauf scheint mir folgende zu sein: Freud war es durch Schaffung des typischen psychoanalytischen „Settings“ gelungen, eine therapeutische oder genauer gesagt eine Lebensatmosphäre zu schaffen, die in mancher Hinsicht die symbiotische Mutter-Kind-Beziehung wiederbelebt. Der den Blicken des entspannt liegenden Patienten entzogene Analytiker ist, wie seinerzeit die verstehende Mutter, einfach da; er kann gehört, aber nicht gesehen, wohl aber „erfühlt“ werden.

Die psychoanalytischen Deutungen sind im Gegensatz zu sonstigen Feststellungen des Arztes besonders dann wirksam, wenn es dem Analytiker gelingt, in der Regression auf die symbiotische Stufe (vgl. Kap. 1) die schützende Hülle der „Realität“ des Patienten zu durchstoßen und mit ihm gemeinsam in der partiell symbiotischen analytischen Dyade eine neue Wirklichkeit zu konstituieren. Im psychoanalytischen Durcharbeiten wird dann diese neue Wirklichkeit mit der Entwicklung neuer Programme Schritt für Schritt weiter ausgebaut und es werden dabei neue Erlebens- und Verhaltensweisen (= Programme) eingeübt. So vollzieht sich im psychoanalytischen Prozeß ein Stück Neuaufbau von Wirklichkeit.

Dies gilt, wenn auch in etwas abgeschwächter Form, für die psychoanalytisch orientierten Therapien überhaupt, wie auch für andere Formen der (Psycho-)Therapie. Auch hier sind die ärztlichen Deutungen und Suggestionen dann besonders wirksam (= mutativ), wenn es gelingt, wie z. B. mit der Flash-Technik (s. u.), die unsichtbare Hülle der „Realität“ des Patienten zu durchstoßen und zumindest für Augenblicke die symbiotische Ebene zu erreichen. In Kapitel 1 haben wir gezeigt, wie im symbiotischen Funktionskreis in der Interaktion zwischen Mutter und Kind „Wirklichkeit“ entsteht; gelingt es dem Arzt (bzw. dem Psychoanalytiker) mit seinem Patienten zumindest partiell auf diese Stufe zu regredieren, dann vollzieht sich hier ein ähnlicher kreativer Prozeß und es entsteht zwischen Arzt und Patient ein Stückchen „gemeinsamer Wirklichkeit“.

Nach diesen Ausführungen über das Situationskreiskonzept und die Freudsche psychoanalytische Methode, die erstmals einen wissenschaftlichen Zugang zur Subjektivität, also gewissermaßen zum Persönlichkeitskern der Patienten ermöglichte, müssen wir noch ein Ergebnis dieser Methode, nämlich die Theorie der Symptombildung näher ins Auge fassen.

---

16 Vgl. S. Freud (1904): Ges. W. Bd. V S. 275.
17 Vgl. A. Argelander: Die szenische Funktion des Ichs ... Psyche 24 (1970) S. 325.
18 Vgl. O. Goldschmidt, 1973.
19 Vgl. S. Freud (1920): Ges. W. Bd. XIII S. 17.

Bereits 1896 erkannte Freud, daß die Symptome bei den Psychoneurosen „... Kompromißbildungen zwischen den verdrängten und den verdrängenden Vorstellungen"[20] darstellen. 20 Jahre später schreibt er in den berühmten „Vorlesungen zur Einführung in die Psychoanalyse": „Von den neurotischen Symptomen wissen wir bereits, daß sie der Erfolg eines Konfliktes sind, der sich um eine neuere Art der Libidobefriedigung erhebt. Die beiden Kräfte, die sich entzweit haben, treffen im Symptom wieder zusammen, versöhnen sich gleichsam durch den Kompromiß der Symptombildung. Darum ist das Symptom auch so widerstandsfähig; es wird von den beiden Seiten her gehalten. Wir wissen auch, daß der eine der beiden Partner des Konfliktes die unbefriedigte, von der Realität abgewiesene Libido ist, die nun andere Wege zu ihrer Befriedigung suchen muß. Bleibt die Realität unerbittlich, auch wenn die Libido bereit ist, ein anderes Objekt an Stelle des versagten anzunehmen, so wird diese endlich genötigt sein, den Weg der Regression einzuschlagen und die Befriedigung in einer der bereits überwundenen Organisationen oder durch eines der früher aufgegebenen Objekte anzustreben. Auf den Weg der Regression wird die Libido durch die Fixierung gelockt, die sie an diesen Stellen ihrer Entwicklung zurückgelassen hat."[21]

In dem oben zitierten Freud-Text sind wir auf zwei weitere psychoanalytische Grundbegriffe, nämlich Libido und Regression, gestoßen, die noch kurz erklärt werden müssen. „Libido ist ein Ausdruck aus der Affektivitätslehre. Wir heißen so die als quantitative Größe betrachtete – wenn auch derzeit nicht meßbare – Energie solcher Triebe, welche mit all dem zu tun haben, was man als Liebe zusammenfassen kann."[22] Der Freudsche Begriff der Libido ist dadurch gekennzeichnet, daß er qualitativ auf den Sexualtrieb und seine verschiedenen Ausprägungen bezogen ist, darüber hinaus aber als quantifizierbare psychische Energie aufgefaßt wird. Nach Freud durchläuft die Libido in der psychosexuellen Entwicklung jedes Individuums die „orale",[23] die „anale"[24] und die „phallische"[25] Stufe, ehe sie in der Pubertät die reife „genitale" Stufe erreicht (Engel, 1970; Erikson, 1965). Jede dieser Entwicklungsstufen ist wiederum mit bestimmten „Objektbeziehungen" verknüpft, denn die Libido ist immer auf ein Objekt gerichtet. Dieses Objekt muß nicht immer eine wichtige Bezugsperson, sondern kann auch, je nach Organisations- und Reifungsstufe der Libido, lediglich ein Teil von ihr (etwa die Mutterbrust) sein. Wenn sich die Libido von den Objekten der Umwelt zurückzieht und sich dem eigenen Selbst, dem eigenen Körper oder Teilen desselben zuwendet, dann sprach Freud in Anlehnung an die griechische Mythologie, in der der Jüngling Narkyssos in sein eigenes Spiegelbild verliebt war, von Narzißmus.

Die verschiedene Stufen durchlaufende Entwicklung bzw. Organisation der Libido kann gestört werden, wodurch es zu Fixierungen auf den einzelnen Stufen kommen kann. Im Gegensatz zur Fixierung spricht man von Regression, wenn das Individuum eine bereits erreichte Entwicklungsstufe wieder auf-

gibt und auf eine bereits früher durchlaufene Stufe der psychosexuellen Entwicklung bzw. Objektbeziehung zurückfällt. Diese wenigstens summarische Kenntnis der psychoanalytischen Grundbegriffe ist nötig, falls man die psychoanalytische Theorie der Symptombildung verstehen will.

Obwohl Freud selbst Organkranke nie behandelt hat und allen Versuchen einiger seiner Schüler, die psychoanalytische Methode auch auf Organkranke anzuwenden, mit erheblicher Zurückhaltung begegnete, wurde er doch zum Initiator der modernen Psychosomatischen Medizin. Cremerius hat in einer Übersichtsarbeit „Freuds Konzept über die Entstehung psychogener Körpersymptome" (1957) dargestellt. Demnach hat Freud schon sehr früh sowohl beim Studium der Hysterie als auch bei der Angstneurose die Entstehung grundsätzlich verschiedener psychogener Körpersymptome beobachten können. Wie wir oben gehört haben, erkannte Freud, daß der neurotische Konflikt aus dem Gegensatz zwischen unbefriedigten Triebwünschen und abwehrenden Instanzen besteht, und daß man die Symptome am besten als „Kompromißbildungen zwischen den verdrängten und den verdrängenden Vorstellungen" auffassen könne. „Bei der Hysterie erfolgt die Unschädlichmachung der unverträglichen Vorstellung dadurch, daß deren Erregungssumme ins Körperliche umgesetzt wird, wofür ich den Namen der Konversion vorschlagen möchte."[26] Diese Umsetzung der „Erregungssumme ins Körperliche" konnte sich Freud offenbar nur mit Hilfe eines quantitativen Libidobegriffes vorstellen. Obwohl der Konversionsbegriff inzwischen durch Deutsch (1959), Rangell (1969) und andere eine erhebliche Erweiterung erfahren hat, ist er nach wie vor auch für die Psychosomatische Medizin ein Grundbegriff, mit dessen Hilfe wir einen Teil der psychosomatischen Symptombildung erklären können.

Völlig anders geartete Körpersymptome entdeckte Freud beim Studium der Angstneurose: Es sind dies vasomotorische Störungen, wie Tachykardie und Schwindelzustände, Störung der Atmung, Schweißausbrüche, Zittern und Schütteln, Heißhunger, Durchfälle und Parästhesien. Er schreibt, daß sich die Psyche so verhalte, „als projiziere sie die Erregung nach außen" und „das Nervensystem reagiert gegen eine innere Erregungsquelle wie in dem entsprechenden Affekt gegen eine analoge äußere".[27] Der Mecha-

---

20 Vgl. S. Freud (1896): Ges. W. Bd. I S. 387.
21 Vgl. S. Freud (1917): Ges. W. Bd. XI S. 373.
22 Vgl. S. Freud (1921): Ges. W. Bd. XIII S. 98.
23 Erste Stufe der Libidoentwicklung. Die Lustempfindungen sind vorwiegend um die Mundzone zentriert.
24 Zweite Stufe der Libidoentwicklung, in der die Vorgänge der Defäkation (Ausstoßen-Zurückhalten) von besonderer Bedeutung sind.
25 Auf dieser Entwicklungsstufe rückt das Genitale in den Mittelpunkt des Interesses. Nach der psychoanalytischen Theorie besteht die Unterscheidung der Geschlechter auf dieser Entwicklungsstufe in der Feststellung a) mit Phallus (männlich) oder b) ohne Phallus (kastriert bzw. weiblich).
26 Vgl. S. Freud (1894): Ges. W. Bd. I S. 63.
27 Vgl. S. Freud (1895): Ges. W. Bd. I S. 339.

nismus der körperlichen Symptomentstehung ist also bei der Hysterie und bei der Angstneurose nach Freud wesensverschieden. Bei der Hysterie entsteht das Symptom durch Konversion und es ist der Repräsentant eines ins Unbewußte verdrängten Erlebnisses. Bei der Angstneurose entsteht jedoch das Symptom durch Projektion der Angstquelle nach außen oder ist überhaupt nicht im Bewußtsein enthalten, und es ist lediglich das somatische Äquivalent eines psychischen Zustandes, nämlich der Angst.

Diese von Freud erstmals beschriebene Unterscheidung einerseits von Konversionssymptomen bei der Hysterie, die Kompromißbildungen eines intrapsychischen Konfliktes sind und somit verdrängte Triebwünsche repräsentieren bzw. ausdrücken, und andererseits körperlichen Begleitsymptomen bei der Angstneurose, die keine verdrängten Triebwünsche ausdrücken, sondern lediglich somatische Angstäquivalente sind, wurde später von vielen psychosomatischen Forschern übernommen. So unterschieden z.B. Fenichel (1945) Konversionssymptome und Organneurosen, Alexander (1951) Konversionsneurosen und vegetative Neurosen und v. Uexküll (1963) Ausdrucks- und Bereitstellungserkrankungen. Allen diesen Unterscheidungen liegt der Freudsche Gedanke zugrunde, daß viele Körpersymptome als Kompromißbildungen und Repräsentanten unterdrückter Triebwünsche aufgefaßt werden können, andere aber lediglich Begleitsymptome verschiedener Affekte sind und somit im Gegensatz zu den ersteren nichts repräsentieren bzw. ausdrücken, also auch nicht symbolisch interpretiert werden können.

Neben diesen beiden von Freud entwickelten und von der Psychosomatischen Medizin übernommenen Grundbegriffen, nämlich der Konversion und dem Affektäquivalent, wurde für manche Autoren auch Freuds Narzißmuskonzept zur wichtigen theoretischen Grundlage der Psychosomatischen Medizin. Bereits weiter oben haben wir festgestellt, daß Freud sich die „Libido" als prinzipiell quantifizierbare psychische Energie vorstellte, die das Individuum zu den Objekten aussendet, und er sprach von Narzißmus, wenn diese Libido von den Objekten abgezogen und auf das Individuum selbst zurückgezogen wurde. Psychosen und hypochondrische Zustände erklärte er libidotheoretisch so, daß bei diesen Krankheitszuständen die „Objektlibido" aufgegeben und in das Ich zurückgenommen wurde. So schreibt er z.B. in seiner Arbeit „Zur Einführung des Narzißmus": „Wir bilden so die Vorstellung einer ursprünglichen Libidobesetzung des Ichs, von der später an die Objekte abgegeben wird, die aber, im Grunde genommen, verbleibt und sich zu den Objektbesetzungen verhält wie der Körper eines Protoplasmatierchens zu den von ihm ausgeschickten Pseudopodien. Dieses Stück der Libidounterbringung mußte für unsere von den neurotischen Symptomen ausgehende Forschung zunächst verdeckt bleiben. Die Emanationen der Libido, die Objektbesetzungen, die ausgeschickt und wieder zurückgezogen werden können, wurden uns allen auffällig. Wir sehen auch im groben einen Gegensatz

zwischen der Ichlibido und der Objektlibido. Je mehr die eine verbraucht, desto mehr verarmt die andere. Als die höchste Entwicklungsphase, zu der es die letztere bringt, erscheint uns der Zustand der Verliebtheit, der sich uns wie ein Aufgeben der eigenen Persönlichkeit gegen die Objektbesetzung darstellt und seinen Gegensatz in der Phantasie (oder Selbstwahrnehmung) der Paranoiker vom Weltuntergang findet."[28]

Im Gegensatz zu den Psychoneurosen, die Freud auch „Übertragungsneurosen" nannte, weil bei ihnen die „Objektlibido" auf den Arzt übertragen und damit der neurotische Konflikt in der Analyse bearbeitet werden konnte, nannte Freud die Psychosen, psychotischen Reaktionen und Hypochondrien „narzißtische Neurosen", weil die Libido bei diesen Krankheitsbildern ganz auf das Individuum zurückgezogen ist. Er war deshalb auch überzeugt, daß diese Krankheitsbilder einer psychoanalytischen Behandlung grundsätzlich nicht zugänglich sind, weil ja das Prinzip der psychoanalytischen Therapie in der deutenden Bearbeitung von Übertragung und Widerstand besteht und zwangsläufig da nicht anwendbar ist, wo es keine Übertragung gibt. Die inzwischen gesammelten Erfahrungen bei der psychoanalytischen Behandlung von Psychosen haben allerdings gezeigt, daß auch Psychotiker „übertragen". Allerdings entnehmen sie die ihren „Übertragungen" zugrundeliegenden Verhaltensmuster sehr frühen Entwicklungsstadien, und es ist letztlich eine Frage der Definition, ob man diese Form der Beziehung zum Therapeuten noch Übertragung nennt oder nicht. Kein Zweifel aber kann darüber bestehen, daß die Beziehung, d.h. Übertragung des neurotisch Kranken zum Therapeuten eine andere ist als die des psychotisch Kranken und daß Freuds Unterscheidung in „Übertragungsneurosen" und „narzißtische Neurosen" jenseits von theoretischen und terminologischen Schwierigkeiten zwei wichtige unterschiedliche Sachverhalte kennzeichnet.

Meng machte bei der psychoanalytischen Behandlung von Patienten mit Magersucht, Tuberkulose, Diabetes und Gallenleiden schon sehr früh die Beobachtung, daß diese Patienten frühe Ich-Schädigungen, wie wir sie sonst nur bei Psychosen zu beobachten gewöhnt sind, und einen weitgehenden Rückzug der Objektlibido, entsprechend dem Freudschen Konzept der narzißtischen Neurose, aufweisen, und machte deshalb bereits 1934 den Vorschlag, diese Erkrankungen nicht als Organneurosen, sondern als „Organpsychosen" aufzufassen.

Die von Freud libidotheoretisch interpretierte Zweiteilung der psychischen Erkrankungen in „Übertragungsneurosen", die der psychoanalytischen Therapie zugänglich sind, und in „narzißtische Neurosen", die der psychoanalytischen Therapie nicht zugänglich sind, wurde in jüngster Zeit durch Balints Theorie der Grundstörung noch präzisiert.

---

28 Vgl. S. Freud (1914): Ges. W. Bd. X S. 141.

16.2   Die psychoanalytisch orientierten Psychotherapien

Das Ziel der Psychoanalyse, das mittels deutender Bearbeitung von Übertragung und Widerstand erreicht werden soll, hat Freud folgendermaßen definiert: „Die Psychoanalyse ist ein Werkzeug, welches dem Ich die fortschreitende Eroberung des Es ermöglichen soll"[29]: „Wo Es war, soll Ich werden."[30] Dieses weitgesteckte Ziel wird mittels der psychoanalytischen Behandlungsmethode zu erreichen gesucht.

Um die klassische Psychoanalyse von den psychoanalytisch orientierten Therapieverfahren, denen ebenfalls die Theorie der Psychoanalyse zugrunde liegt, abzugrenzen, hat man sich darauf geeinigt, von Psychoanalyse im klassischen Sinne nur dann zu sprechen, wenn die Therapie den Freudschen Vorschriften entsprechend im Liegen und zwar mindestens drei- bis viermal wöchentlich durchgeführt wird. Alle anderen von der Psychoanalyse abgeleiteten Therapieformen nennt man psychoanalytisch orientierte Therapieverfahren. Sie sind dadurch gekennzeichnet, daß sie an der Theorie der Psychoanalyse und an Freuds Forderung, „dem Ich die fortschreitende Eroberung des Es" zu ermöglichen, festhalten, das ursprüngliche psychoanalytische „Setting" der viermal wöchentlichen Behandlung auf der Couch aber modifiziert bzw. aufgegeben haben.

Wir werden uns hier nur auf eine kurze Charakterisierung und Beschreibung jener Methoden beschränken, die für den Psychosomatiker besonders bedeutsam sind und denen keine gesonderten Kapitel innerhalb dieses Lehrbuches gewidmet wurden. Ich meine die analytische Psychotherapie, die Fokaltherapie, die Flash-Technik und die analytisch orientierte Notfallpsychotherapie.

Die Methoden der Fokal-, Flash- und Notfallpsychotherapie spielen als Kurzpsychotherapie eine besondere Rolle. Da letztere aber auch aus anderen Elementen besteht, wird anschließend in einem gesonderten Kapitel noch auf die Sprechstundenpsychotherapie eingegangen werden.

16.2.1   Die analytische Psychotherapie

Die geringste Modifikation der klassischen psychoanalytischen Technik stellt die analytische Psychotherapie dar. Die Behandlung erfolgt in der Regel auf der Couch und nur in Ausnahmefällen im Sitzen. Die Stundenzahl ist jedoch auf ein bis drei, am häufigsten wohl auf zwei Wochenstunden reduziert. Die im Abschnitt 16.1 beschriebene Technik der Psychoanalyse wird im Grundsatz unverändert beibehalten, wobei allerdings zu berücksichtigen bleibt, daß die Verringerung der Behandlungsstunden pro Woche auch häufig eine geringere Regression des Patienten in der Behandlung zur Folge hat, d.h., die sog. „Übertragungsneurose" ist nicht so stark ausgebildet wie in der klassischen Psychoanalyse. Dadurch sieht sich der Therapeut in der analytischen Psychotherapie

manchmal genötigt – was bei richtiger Handhabung nicht unbedingt ein Nachteil sein muß – sich etwas aktiver zu verhalten als in der klassischen Psychoanalyse.

Meist sind es zeitliche und finanzielle, also außerhalb des eigentlichen Krankheitsgeschehens liegende Gründe, die zur Anwendung dieser Behandlungsmodifikation zwingen. Indiziert ist die analytische Psychotherapie bei allen der klassischen Psychoanalyse zugänglichen Erkrankungen, in erster Linie also bei den Psychoneurosen. Darüber hinaus wird sie auch bei psychosomatischen Erkrankungen, bei Charakterstörungen, Süchten, Perversionen und neurotisch-psychotischen Grenzfällen (sog. borderline cases) und vereinzelt auch bei Psychosen angewandt.

Wie die klassische Psychoanalyse, ist auch die analytische Psychotherapie dann indiziert, wenn vor allem eine psychische Strukturänderung des Patienten und nicht lediglich eine Symptombeseitigung intendiert wird. Je stärker die Strukturänderung des Patienten angestrebt wird, um so eher wird man sich für die klassische Psychoanalyse entscheiden. Ist die therapeutische Zielsetzung, was häufig der Fall sein wird, nur auf eine bessere Einsichtsfähigkeit und verbesserte Lebensbewältigung des Patienten ausgerichtet, dann ist die analytische Psychotherapie oft der klassischen Psychoanalyse sogar vorzuziehen.

16.2.2   Die Fokaltherapie

Der Wunsch, die psychoanalytische Behandlung abzukürzen, ist nahezu so alt wie die Psychoanalyse selbst. Bereits 1913 hat Freud in seiner Arbeit „Zur Einleitung der Behandlung" gesagt, „die Abkürzung der analytischen Kur bleibt ein berechtigter Wunsch . . .". Er hat aber gleich einschränkend hinzugefügt: „Es steht ihr leider ein sehr bedeutsames Moment entgegen, die Langsamkeit, mit der sich tiefgreifende seelische Veränderungen vollziehen, in letzter Linie wohl die ‚Zeitlosigkeit' unserer unbewußten Vorgänge."[31] Freuds ambivalente Einstellung zur Kurzpsychotherapie spiegelt sich bis heute in den Diskussionen der Psychoanalytiker wider. Der klinisch und praktisch arbeitende Psychosomatiker wird allerdings an der Notwendigkeit einer psychoanalytischen Kurzpsychotherapie nicht mehr zweifeln können.

Die ersten Versuche, die Psychoanalyse „aktiver" und damit kürzer zu gestalten, sind von S. Ferenczi (1939) und von W. Stekel (1938) unternommen worden. Ende der vierziger Jahre entwickelten dann Deutsch (1949) die „Sektor"- und Alexander (1949) die „Vektor-Therapie". Erst durch die Arbeitskreise um Bellak und Small (1972) und um Balint (1972) (vgl. auch Malan, 1965) wurde die analytische Kurzpsychotherapie und insbesondere die Fokaltherapie in ihrer heutigen Form geschaffen.

---

29   Vgl. S. Freud (1923): Ges. W. Bd. XIII S. 286.
30   Vgl. S. Freud (1932): Ges. W. Bd. XV S. 86.
31   Vgl. S. Freud (1913): Ges. W. Bd. VIII S. 462.

Nach D. Beck ist die Fokaltherapie im allgemeinen indiziert „bei relativ ichstarken Patienten mit gutem Behandlungsmotiv, bei denen sich ein umschriebenes Problem als Therapieziel finden läßt, und die mit ihrem Therapeuten und seinen Deutungen arbeiten können" (1974).

Sie wird im Sitzen, in der Regel einmal wöchentlich, durchgeführt und umfaßt einen Behandlungsumfang von 10 bis 30 Behandlungsstunden. Die therapeutischen Interventionen (= Deutungen + Durcharbeiten) werden fast ausschließlich auf das vorher als Fokus definierte Problem ausgerichtet, unter relativer Vernachlässigung aller anderen Probleme des Patienten.

Dies soll durch ein Beispiel, das auch gleichzeitig als Beispiel für eine Notfalltherapie gelten kann, illustriert werden:

Eine 45jährige unverheiratete Angestellte erkrankt an Symptomen einer depressiven Erschöpfung. Sie fühlt sich zunehmend beruflich und persönlich überfordert, wird zunächst völlig appetit- und teilnahmslos und psychisch „wie gelähmt". Todesängste stellen sich ein und steigern sich nachts zu panikartigen Zuständen. Wegen starker stenokardischer Beschwerden, die in den linken Arm ausstrahlen und von Schwindelzuständen begleitet sind, wird die Patientin ins Krankenhaus eingewiesen. Dort wird ein „nervöser Erschöpfungszustand mit hypotonen Kreislaufregulationsstörungen" diagnostiziert und ein psychischer Hintergrund vermutet.

Zum Erstinterview wird sie von ihrem um fast 30 Jahre älteren Freund gebracht und meint, etwas verlegen, als ob sie sich dessen schämen müßte, daß sie einen so totalen Zusammenbruch erlitten habe und am Ende ihrer Kräfte sei, daß ihr Leben vermutlich bald zu Ende gehen werde, obwohl die Krankenhausärzte an ihrem Herzen bisher noch keinen ernsten organischen Schaden feststellen konnten. Während des Gespräches ist sie offensichtlich bemüht, mit dem Arzt zusammenzuarbeiten, und macht im ganzen einen etwas zwanghaften Eindruck. Sie berichtet, daß sie als Tochter eines Großgrundbesitzers in den ehemals deutschen Ostgebieten aufgewachsen sei. Zur Mutter habe sie kein gutes Verhältnis gehabt, weil diese sie gegenüber dem älteren Bruder und der jüngeren Schwester „vernachlässigt" habe. Der Vater sei ein „fernes Ideal" gewesen, von dem sie aber „nicht genügend Sicherheit" bekommen habe. Die Ehe der Eltern sei nicht gut gewesen. Sie und ihre Geschwister seien „streng und gottesfürchtig" erzogen worden. Als sie 13 Jahre alt war, sei bei ihr eine Schieloperation durchgeführt worden, und ein Jahr darauf sei der Vater im Krieg gefallen. Nach dem Krieg habe sie zunächst mit der Mutter und beiden Geschwistern in ärmlichsten Verhältnissen gelebt. Die ganzen Anstrengungen der Familie waren darauf ausgerichtet, dem Bruder, dem „Star der Familie", ein Studium zu ermöglichen.

Sie habe sich dann in zäher Arbeit nach dem Abitur und nach einer kaufmännischen Lehre zur Abteilungsleiterin eines großen Unternehmens hochgearbeitet. Die Arbeit sei ihr ganzer Lebensinhalt gewesen. Seit sie vor eineinhalb Jahren durch Organisationsänderungen des Betriebes einen jüngeren Vollakademiker zum Vorgesetzten bekommen habe, fühle sie sich durch die Arbeit in zunehmendem Maße überfordert, habe aber in eiserner Disziplin ausgehalten.

Freunde und Freundinnen hatte sie in ihrer Jugend eigentlich kaum. Im Alter von 35 Jahren hatte sie ein erstes und nur flüchtiges sexuelles Erlebnis. Seit 8 Jahren hat sie nun mit ihrem ehemaligen, inzwischen pensionierten Chef eine intime Freundschaft. Obwohl seine Ehe „ganz zerrüttet" sei, belasten sie diese „ungeklärten Verhältnisse" ebenso wie ihre nach wie vor bestehenden sexuellen Hemmungen und das Alter ihres Freundes.

Über die mögliche Verursachung ihres „Zusammenbruchs" wurde gemeinsam herausgearbeitet:

- Die völlige Dekompensation ist erst nach einem glücklicherweise glimpflich verlaufenen Autounfall, den sie gemeinsam mit ihrem Freund hatte, und nach einer vom behandelnden Arzt verabreichten Spritze aufgetreten.
- Die „Überarbeitung" im Betrieb, verbunden mit dem neuen Vorgesetzten und der dadurch erfolgten Zurücksetzung.
- Das „ungeklärte" Dreiecksverhältnis mit starken Schuld- und Aggressionsgefühlen gegenüber der Ehefrau des Freundes, Vorwürfen gegen den Freund, daß er nicht längst „klare Verhältnisse" geschaffen habe, und eigenen Zweifeln, ob diese Bindung an einen so alten Mann für sie „das Richtige" sei.
- Eine allgemeine Identitäts- und Lebenskrise in der Lebensmitte mit der Frage, ob nicht der Lebenssinn bisher weitgehend verfehlt wurde.

Nach dieser hier natürlich nur in Stichworten und stark verkürzt wiedergegebenen „Situationsanalyse" erhob sich die Frage: Um welchen Problemkreis läßt sich das Krankheitsgeschehen dieser Patientin am zwanglosesten fokussieren? Welcher Fokus ist am ehesten geeignet, als gemeinsamer Nenner zu dienen und dabei gleichzeitig dazu beizutragen, den unbewußten (und damit vor allem pathogenen) Anteil der Situation zu erhellen?

Die zunächst nicht eindeutig ausgesprochene bzw. nur vorsichtig angedeutete Vermutung des Arztes, daß die Patientin jetzt wohl eine Neuauflage ihrer noch nie ganz durchgearbeiteten Ödipalproblematik erlebe, wird von ihr in der folgenden (dritten) Behandlungsstunde durch einen Traum „bestätigt", den sie überrascht und völlig unaufgefordert berichtet: Sie habe mit ihrem Vater sexuelle Beziehungen. Beide werden deshalb von ihrer Mutter gehaßt, die in einer mittelalterlichen Festung einen Aufstand gegen sie vorbereite.

Ausgehend von diesem Traumbild wird die Ödipalproblematik zum Fokus bestimmt und der Patientin zu zeigen versucht, wie sehr sie unerledigte infantile Konflikt- und Verhaltensmuster noch in der Gegenwart agiert. Die aktuellen Probleme und Konflikte konnten nachfolgend zwanglos in einen sinnvollen Zusammenhang mit den weitgehend unbewußten infantilen Problemen und damit mit dem Fokus gebracht werden. Es war aber sehr beeindruckend zu beobachten, wie sehr sich nach der Formulierung des Fokus durch den Arzt und der Annahme desselben durch die Patientin ihr Befinden gebessert hat.

Durch diese knappen Auszüge einer Behandlungsgeschichte habe ich versucht, das Prinzip der Fokaltherapie darzustellen, die auch bei der Behandlung von psychosomatisch Kranken dann angezeigt ist, wenn die Patienten über eine genügende Ich-Stärke (= Reife) verfügen und der Fokus klar definiert werden kann.

16.2.3 Die Flash-Technik

Als sich Michael Balint bemühte, im Rahmen seiner kurzpsychotherapeutischen Forschungen bei seinen Patienten jeweils einen Fokus zu bestimmen, machte er die Entdeckung, daß dies auf zweierlei verschiedenen Wegen möglich ist: Einmal auf dem klassisch psychoanalytischen Wege einer sog. „Detektivtechnik", die unter Beachtung von Übertragung und Widerstand den Fokus allmählich sich „herauskristallisieren" läßt, und dann durch ein „Einstimmen" (= tuning-in) des Arztes in die Problematik des Patienten, die dann zu einem blitzartigen „Aha-Erlebnis" führt, das von Balint und Mitarbeitern „Flash" genannt wurde. In der Einleitung zur deutschen Ausgabe des Buches von Enid Balint und J. S. Norell schreibt W. Loch über den Flash (S. 9): „Wird das im Flash erfahrene Reaktionsmuster einschließlich der ihm zugehörigen Gefühle nun in geeigneter Weise, in einer der Situation entsprechenden Weise formuliert, dann konstituieren sich für Patient und Arzt neue Einstellungen und Erwartungen. Es wird so eine neue Realität für das Erleben des Patienten geschaffen, was bedeutet, daß er die alten pathogenen Verhaltensmuster aufzugeben vermag" (Balint und Norell, 1975).

Mit der Flash-Technik wird also gleichsam die ganze Psychopathologie des Patienten einschließlich seiner Abwehrsysteme unterlaufen. Inwieweit dieser die durch das „Aha-Erlebnis" gewonnene neue Einsicht und Lebenserfahrung ohne ein intensives Durcharbeiten zu nutzen vermag, muß allerdings von Fall zu Fall offenbleiben.

Obwohl wir erst am Anfang der Erforschung der Flash-Phänomene stehen, und diese deshalb auch noch nicht planmäßig herbeiführen, sondern nur sich ereignen lassen können, läßt sich heute schon nach den Arbeiten von Balint und Mitarbeitern folgendes sagen:

– Mit dem, was wir heute „Flash" nennen, ist nicht so sehr die Neuschöpfung einer weiteren therapeutischen Technik gemeint, als vielmehr die Wiederentdeckung und systematische Erforschung eines schon immer von erfolgreichen Ärzten intuitiv geübten therapeutischen Verfahrens.
– Flashes vollziehen sich in Sekundenschnelle. Sie sind deshalb im Gegensatz zu allen anderen recht zeitaufwendigen psychotherapeutischen Verfahren in der allgemeinärztlichen Sprechstunde keine Fremdkörper.
– Flashes beziehen sich immer auf die gesamte „Situation" und bearbeiten nicht, wie die klassischen psychoanalytischen Deutungen, selektiv nur den unbewußten Anteil der Situation.
– Im Gegensatz zur klassischen psychoanalytischen Technik, die auf dem Umweg über die „Übertragungsneurose" heilt und den Patienten dadurch vorübergehend vermehrt abhängig macht und in-

fantilisiert, stellen geglückte Flashes für den Patienten eine Ich-Stärkung dar, ermöglichen ihm dadurch etwas mehr Freiheit und geben ihm so die Chance einer Neuorientierung. Sie sind deshalb durchaus auch bei Patienten mit einer Grundstörung im Sinne von Balint (= Alexithymie) mit Erfolg anwendbar und gewinnen so für die Psychosomatische Medizin besondere Bedeutung.

Nachfolgend soll auch die Flash-Technik an einem einfachen Beispiel illustriert werden:

> Eine 29jährige Frau, Mutter einer 3jährigen Tochter, die ich schon seit einiger Zeit kenne, weil sie an Angstzuständen und funktionellen Herzbeschwerden leidet, kommt eines Tages in einem so hochgradig verängstigten Zustand in die Sprechstunde, daß sie schon einen geradezu verstörten Eindruck macht. Auf meine Frage, wie es ihr denn gehe, bricht sie sofort in Tränen aus und berichtet, daß sie immer weinen müsse, wenn sie jemand nach ihrem Befinden frage. Sie müsse auch immer dann weinen, wenn sie ihre arme kleine Tochter ansehe. Vor einigen Tagen sei nun eine um zwei Jahre jüngere Arbeitskollegin an Brustkrebs verstorben und hinterlasse drei kleine unversorgte Kinder. Seither könne sie selber keinen klaren Gedanken mehr fassen und sei völlig verzweifelt. Auf mich macht sie dabei den Eindruck eines völlig verängstigten hilflosen kleinen Kindes (= szenische Information im Sinne von Argelander).[32]

Während ich die Worte der Patientin und die „Szene" auf mich wirken lasse, erinnere ich mich plötzlich daran, daß mir die Patientin beim Erheben der Vorgeschichte erzählt hat, daß sie selbst im Alter von etwa vier Jahren ihre Mutter durch vorzeitigen Tod verloren habe. Blitzartig schießt mir der Gedanke durch den Kopf (= Flash), die Frau erlebt jetzt die Verzweiflung und Angst wieder, die sie als Kind beim Tode ihrer Mutter erfahren hat. Sie ist überzeugt, selbst in naher Zukunft sterben zu müssen und glaubt, daß ihre arme kleine Tochter dann bald das gleiche Elend und die gleiche Verzweiflung wird durchmachen müssen, wie sie damals.

Während ich ihr das alles sage und diese Thematik mit ihr gemeinsam noch etwas vertiefe, beruhigt sie sich zusehends und fragt erstaunt: „Können Sie denn Gedanken lesen, Herr Doktor?!" (= Flash). Nach diesem kurzen Gespräch, das eine weiterhin kontrollierte völlige Beschwerdefreiheit („alles wie weggeblasen") zur Folge hatte, hatten wir wohl beide den Eindruck, einen zentralen Punkt ihres Krankheitsgeschehens getroffen und ein Stückchen positiver therapeutischer Arbeit geleistet zu haben.

Da blitzartige Erkenntnisse, die Karl Bühler „Aha-Erlebnisse" genannt hat, in jeder Analyse vorkommen, sei abschließend nochmals der Unterschied der Flash-Technik zur klassischen psychoanalytischen Deutungstechnik herausgearbeitet: Während letztere mit Hilfe der Methode des „Meisterdetektivs" „jeden Stein umdrehen" muß, wie Balint sich ausdrückte, um

---

32 Vgl. H. Argelander: Die szenische Funktion des Ichs ... Psyche 24 (1970).

dann in mühevoller Kleinarbeit Übertragung und Widerstand deutend zu bearbeiten, versucht die Flash-Technik durch ein „Sich-Einstimmen" (tuning-in) die Situation blitzartig zu erhellen. Die Deutungen, die der Arzt gibt, sind dann weder Übertragungsdeutungen noch gar sog. tiefe Deutungen, sondern beziehen sich ausschließlich auf die gegenwärtige Situation des Patienten, die ihm in der Rückspiegelung durch den Arzt verständlicher wird.

Der Flash-Technik, die mit dem „Sich-Einstimmen" arbeitet, auf das engste verwandt ist die von Loch beschriebene „Episoden-Technik". „Mit der Episoden-Technik ist gemeint, daß der Arzt ein plötzlich ihn in der Interaktion mit dem Patienten überfallendes Gefühl oder auch einen plötzlichen Einfall als Indikator für das momentane interaktionelle Problem des Patienten nimmt und sofort benutzt, um nach rascher Analyse dieses Gefühls oder Einfalls eine sinnvolle Intervention anzuschließen" (Loch, 1972, 1975).

Das „Einstimmen" in den Patienten, dieses „tuning-in", dieses blitzartige Erfassen der „Situation" läßt sich jedoch nicht rational erlernen, sondern nur im Umgang mit Patienten erfahren. Die lernende Verarbeitung dieser Erfahrungen geschieht am zweckmäßigsten in sog. Balint-Gruppen.[33]

Es kann nicht nachdrücklich genug darauf hingewiesen werden, daß man Flashes nicht einfach wie ein Medikament applizieren kann. Man kann nur den ermöglichenden Raum schaffen, warten bis sie sich ereignen und muß dann adäquat mit ihnen umgehen können. Insofern kann man auch nur mit Einschränkung von einer therapeutischen „Technik" sprechen. Flashes ereignen sich relativ selten.

16.2.4 Die analytisch orientierte Notfallpsychotherapie

Mit der Fokaltherapie und der Flash-Technik haben wir schon zwei therapeutische Prinzipien kennengelernt, die wir für die Notfallpsychotherapie benötigen. „Als Notfallpsychotherapie bezeichnen wir eine Kurzpsychotherapie in besonderen Dringlichkeits- und Krisensituationen" (Bellak und Small, 1972). Die drohende oder bereits eingetretene (psychische) Dekompensation oder Desintegration der Patienten ist ihre Domäne.

Bedient man sich zum besseren Verständnis der psychischen Vorgänge der psychoanalytischen Strukturtheorie mit ihren „Konstrukten"[34] „Es", „Ich" und „Über-Ich", dann kommt dem Ich eine vermittelnde, eine „synthetische" Funktion (Nunberg) zu. Es hat zwischen den Ansprüchen des Es, des Über-Ich und der äußeren Realität zu vermitteln. Mißlingt diese Synthese, dann kommt es zur Dekompensation und zur Desintegration.

So gesehen hätte die Notfallpsychotherapie zwei Aufgaben zu erfüllen: Sie muß einmal dem desintegrierten oder von Desintegration bedrohten Patienten Schutz und Anlehnungsmöglichkeit bieten, um ihn vor weiterer Desintegration zu bewahren, und sie muß zweitens die synthetischen Funktionen seines

„Ich" stärken, um ihm behilflich zu sein, zu einer besseren Integration zu finden.

Die erste Aufgabe erfüllt der Arzt dadurch, daß er für den desintegrierten Patienten zumindest zeitweilig die beschützende Mutterrolle einnimmt, ihm also gestattet, eine symbiotische Beziehung (vgl. Kap. 1) zum Arzt einzunehmen. Der Arzt muß für den Patienten einfach dasein, zumindest stets (telefonisch) erreichbar. Diese beschützende und stützende Funktion ist vom Arzt aus äußeren und inneren Gründen nicht immer leicht durchzuhalten. Äußerlich stören ihn die vielen anderen Aufgaben und Verpflichtungen, innerlich muß er erst seine eigene Abwehr und Angst überwinden, nämlich von dem sich anklammernden Patienten ganz in Beschlag genommen, „aufgefressen" und „ausgesaugt" zu werden, ehe er diese tragende und schützende Funktion dem in Not Geratenen gegenüber einnehmen kann. Hat man erst einmal erkannt, daß man einen von Desintegration bedrohten Patienten ebensowenig im Stich lassen kann wie einen Unfallverletzten, und daß die phantasierte Bedrohung durch das „Aufgefressen-" und „Ausgesaugtwerden" viel größer ist als die reale Gefahr, – das Angebot jederzeit erreichbar zu sein, schützt meist davor, zur Unzeit gestört zu werden –, dann wird man allmählich auch fähig, diese schwierige ärztliche Aufgabe zu erfüllen.

Zur schützenden und stützenden Funktion des Arztes gehört es auch manchmal, den Patienten allzu massiven pathogenen Einwirkungen seiner Umgebung zu entziehen, ihn also in den schützenden Bereich einer Klinik aufzunehmen.

Die Ich-Stärkung des von Desintegration bedrohten oder bereits desintegrierten Patienten kann durch mehrere Stufen bzw. Schritte erfolgen:

– Die eben beschriebene stützende und schützende Haltung führt bereits zu einer gewissen Ich-Stärkung des Patienten, was man sich mit Hilfe des libidotheoretischen Modells der Psychoanalyse so vorstellen kann, daß in der symbiotischen Phase der Zweieinheit von der Mutter zum Kind bzw. vom Arzt zum Patienten ein Zufluß einer Art von „psychischer Energie" stattfindet.
– Eng verbunden, vielleicht sogar identisch mit jener stützenden und schützenden (= symbiotischen) Funktion des Arztes ist seine bedingungslose Annahme des Patienten, der sich dadurch vom Arzt bestätigt und damit narzißtisch gestärkt fühlt.
– Eine weitere Stärkung der synthetischen Funktion des Ich wird durch die innere Spannungslösung, also durch die Katharsis der Affekte, erreicht. Man sollte deshalb den erregten, verzweifelten oder weinenden Patienten sich ruhig ausdrücken und im Rahmen des Möglichen sich auch abreagieren las-

---

[33] Unter Balint-Gruppen versteht man Fallbesprechungsseminare, in denen unter sachkundiger Leitung eines psychoanalytisch ausgebildeten Arztes die Sensibilität der Seminarteilnehmer für die Arzt-Patient-Interaktion und insbesondere ihre unbewußten Komponenten gesteigert wird.

[34] Der Terminus „Konstrukt" soll ausdrücken, daß es sich dabei nicht um „Sachen", sondern um wissenschaftliche Modelle im Sinne der Modelltheorie handelt.

sen und nicht versuchen, durch vorzeitiges Trösten oder sonstige Aktivitäten den kathartischen Prozeß zu unterbrechen. Anteilnehmendes und verständnisvolles Zuhören hilft dem Patienten im allgemeinen viel mehr als noch so gut gemeinte Aktivität!

- Einen weiteren Schritt der Ich-Stärkung finden wir in geglückten Flashes. Das im vorhergehenden (Flash-Technik-) Abschnitt gebrachte Beispiel der in Panik geratenen Patientin illustriert die Ich-stärkende und damit angstlösende Funktion eines geglückten „Aha-Erlebnisses".

- Man kann auch durch die Fokaltherapie eine entscheidende Ich-Stärkung erzielen, wie das Fallbeispiel im entsprechenden (Fokaltherapie-) Abschnitt zeigt, bei dem es sich ja auch wiederum um einen psychischen bzw. psychosomatischen Notfall handelt. Die Anwendung der Fokaltherapie setzt aber, wie dort bereits ausgeführt, eine gewisse Ich-Stärke voraus, ebenso wie die klare Definition eines Fokus. Ist der Patient völlig desintegriert, dann muß man sich zumindest zunächst mit den ersten vier hier erwähnten Stufen begnügen.

Dem psychosomatisch tätigen Arzt begegnet die akute Desintegration vor allem in drei Formen:
- Hochgradige, bis zur Panik sich steigernde Angstzustände.
- Der akute depressive Rückzug.
- Die akute psychosomatische Dekompensation in Form von akuter Verschlechterung des körperlichen Befindens.

Obwohl die oben erwähnten Maßnahmen der Ich-Stärkung bei allen Notfallpatienten Anwendung finden können und sollen, seien nachfolgend noch einige spezielle Maßnahmen, bezogen auf die oben erwähnten drei Erscheinungsformen der akuten Dekompensation, kurz skizziert:

Beim Angstpatienten hat die Angst nicht nur Signalfunktion, sondern gewinnt von einer gewissen Intensität an eine ausgesprochen desintegrative Kraft. Der Arzt wird daher bestrebt sein müssen, den Circulus vitiosus der Angst notfalls medikamentös zu durchbrechen, um dem Patienten wieder eine Besinnungs- und Wiederfindungspause zu gewähren, die dann allerdings psychotherapeutisch genutzt werden muß.

Beim depressiven Rückzug des Patienten besteht eine der größten Gefahren darin, daß er alle Aggressivität, die er nicht mehr nach außen zu richten vermag, gegen die eigene Person (z.B. Suizid) wendet. Hier hat sich gezeigt, daß eine gute und tragfähige Arzt-Patient-Beziehung nach wie vor die beste Suizidprophylaxe ist. Dies ist deshalb so wichtig, weil ja nicht jeder akut depressive und suizidgefährdete Patient in eine Klinik eingewiesen werden kann und soll. Hier wird der behandelnde Arzt bestrebt sein müssen, neben der Gabe entsprechender (antidepressiver) Medikamente dem Patienten behilflich zu sein, äußere Aggressionsobjekte zu finden, um seine innere Aggressionsspannung im Sinne der Katharsis etwas zu entlasten.

Bei der akuten psychosomatischen Dekompensation schließlich ist der Arzt zunächst einmal genötigt, die akute somatische Bedrohung (z.B. den Status asthmaticus, die Kolik, die Ulkusblutung, den Angina-pectoris-Anfall usw.) zu versorgen. Hier schafft eine gute und sachkundige somatische Versorgung jene stützende und schützende (symbiotische) Patient-Arzt-Beziehung, die dann ohne besondere Schwierigkeiten zu einer weiteren (psycho-)therapeutischen Interaktion ausgebaut werden kann. Gerade bei der akuten psychosomatischen Dekompensation zeigt sich, wie wichtig und vorteilhaft es ist, wenn die somatische und psychische Versorgung des Patienten in einer ärztlichen Hand liegt.

16.3 Anhang: Die neuen Narzißmustheorien und ihr Einfluß auf die psychoanalytische Therapie

Die Erforschung der frühen und frühesten psychischen Entwicklungsstufen des Menschen stößt auf große methodische Schwierigkeiten. Wenn wir die präzisen Beobachtungen der Interaktionen zwischen Mutter und Kind interpretieren, sind wir gezwungen diese Interpretationen aus der Sicht des Erwachsenen vorzunehmen. Das unmittelbare Erleben des Neugeborenen und des Kleinkindes ist uns nicht zugänglich.

Die Forschungen von Piaget (1974), R. Spitz (1967), M. Mahler und Mitarbeitern (1978) und anderen haben aber unser Wissen über diese frühen Entwicklungsphasen doch sehr erweitert und vertieft und damit nicht nur die Theorie der Psychoanalyse befruchtet. Obwohl man Abschließendes über die weitere Entwicklung auf diesem Forschungsfeld noch nicht sagen kann, erscheint ein Hinweis auf diese Entwicklungen in einem Lehrbuch der Psychosomatischen Medizin deshalb so wichtig, weil vieles dafür spricht, daß viele auch organisch sich manifestierende Erkrankungen ihre Wurzeln in dieser höchst vulnerablen Phase der menschlichen Entwicklung haben. Hinweisen möchte ich deshalb ergänzend auf die Ausführungen in Kapitel 1 „Wissenschaftstheorie und Psychosomatische Medizin..." sowie auf die Kapitel 5 „Psychoanalytische Konzepte psychosomatischer Symptom- und Strukturbildung" und 6 „Früheste Kindheitsentwicklung und ihre Störungen aus der Sicht Winnicotts".

In seinem Buch „Die Urform der Liebe und die Technik der Psychoanalyse" weist M. Balint (1965) darauf hin, daß die klassische Theorie der Psychoanalyse eigentlich eine „Ein-Körper-Psychologie" ist. Er führt aus: „Fast alle unsere Bezeichnungen und Begriffe stammen aus dem Studium pathologischer Formen und gehen kaum über die Region der Ein-Körper-Psychologie hinaus (Zwangsneurose, Melancholie, Schizophrenie). Deswegen kann sie nur eine grobe, annähernde Beschreibung dessen liefern, was in der psychoanalytischen Situation geschieht, die doch im wesentlichen eine Zwei-Personen-Situation ist" (S. 271).

Seit Balint – es war im Jahre 1950 – diese Zeilen schrieb, hat sich einiges verändert. Die psychoanalytische Entwicklungspsychologie hat ihr Forschungsinteresse in immer stärkerem Ausmaß (z. B. R. Spitz und M. Mahler) den frühen Objektbeziehungen zugewandt, so daß uns heute – insbesondere aus dem Blickwinkel der Psychosomatischen Medizin – die Störungen und Defekte in diesen frühen Stadien der Entwicklung noch viel bedeutsamer und folgenschwerer erscheinen als die späteren auf der Ödipalebene sich ereignenden Konflikte.

Balint (1970) spricht deshalb von „zwei Ebenen der analytischen Arbeit", die sich einerseits mit Hilfe der klassischen analytischen Technik – durch Deutung von Übertragung und Widerstand – auf der Ebene der Ödipalproblematik bewegt, andererseits aber in frühe Bereiche der symbiotischen Mutter-Kind-Beziehung vorzustoßen sucht. Er nennt diese Ebene die der „Grundstörung" und beschreibt sie wie folgt:

„Die Hauptmerkmale der Ebene der Grundstörung sind,
1. daß alle in ihr sich abspielenden Vorgänge zu einer ausschließlichen Zwei-Personen-Beziehung gehören – es gibt dabei keine dritte Person;
2. daß diese Zwei-Personen-Beziehung sehr eigenartig und gänzlich verschieden ist von den wohlbekannten menschlichen Beziehungen auf der ödipalen Stufe;
3. daß die auf dieser Ebene wirksame Dynamik nicht die Form eines Konfliktes hat, und
4. daß die Erwachsenensprache oft unbrauchbar und irreführend ist, wenn sie Vorgänge auf dieser Ebene beschreiben will, da die Worte nicht mehr ihre konventionelle Bedeutung haben."

Die Grundstörung entwickelt sich nach Balint auf einer sehr frühen Ebene der „Objektbeziehung", die er „primäre Liebe" genannt hat. Diese ist dadurch charakterisiert, daß die Mutter vom Säugling noch nicht als eigenständige Person, sondern als ein bedürfnisbefriedigendes Wesen wahrgenommen wird, das noch ein Teil des kindlichen Selbst ist.

Da jedoch die Entwicklung dieser frühen Objektbeziehung zwischen Mutter und Kind ein dynamischer Prozeß ist, entspricht der primären Liebe des Kindes auf der Seite der Mutter eine „primäre Mütterlichkeit", die Winnicott (1958) beschrieben hat. Winnicott versteht unter „primary maternal preoccupation" eine erhöhte Sensibilität der Mutter, die bereits während der Schwangerschaft einsetzt und sie in die Lage versetzt, sich auf die Bedürfnisse des Kindes optimal einzustellen und es zu „tragen". Winnicott spricht von der „holding function" der Mutter. In dieser frühen Phase der Objektbeziehung erlebt auch die Mutter, zumindest teilweise, ihr Kind als Teil ihrer selbst. Durch die „holding function" schafft die Mutter das „facilitating environment", den ermöglichenden Raum, in dem sich das Kind allmählich zurechtfinden und entwickeln kann.

Balint (1970) meint nun, daß Arzt und Patient manchmal auf diese Ebene regredieren müssen, der Arzt dem Patienten einen ermöglichenden Raum schaffen sollte, damit der Patient einen „Neubeginn" wagen kann. Er schreibt: „Wenn es dem Analytiker gelingt, auf die primitiven, unrealistischen Wünsche des Patienten auf die rechte Weise zu antworten, kann ihm geholfen werden, die bedrückende Ungleichheit zwischen sich und seinem Objekt zu verringern. Mit dem Schwinden dieser Ungleichheit kann auch die Abhängigkeit vom Primärobjekt, die der Patient in der Phase des „Neubeginns" wieder aufleben ließ, ebenfalls beträchtlich nachlassen oder sogar gänzlich aufhören. Wenn die Ungleichheit und die damit zusammenhängende Abhängigkeit reduziert werden, ist die Abwehr gegen sie nicht mehr nötig, der Haß kann weitgehend aufgegeben werden, und die aggressiven, destruktiven Impulse lassen nach." H. Thomä (1983) hat sich erst kürzlich mit der Problematik der „Grundstörung" und des „Neubeginns" im Sinne von Balint gründlich und kritisch auseinandergesetzt und kommt, im Gegensatz zu Balint, zu dem Schluß, daß der „Neubeginn" kein plötzliches und einmaliges Ereignis, sondern Teil eines kontinuierlichen, immer wiederkehrenden therapeutischen Prozesses sei, den man dem Durcharbeiten an die Seite zu stellen habe.

Von den neuen Narzißmustheorien wurden am bekanntesten die von Grunberger (1976), von Kohut (1973, 1979) und von Kernberg (1978).

Nach Grunberger hat der Narzißmus seinen Ursprung im pränatalen Leben. Hier erlebt jeder Mensch „eine Situation, aus der er auf traumatische Weise vertrieben wurde und die er sein Leben lang wiederzufinden versucht. Dieser fundamentale Wunsch ist die Basis unserer Narzißmus-Hypothese" (S. 22). Grunberger macht den Vorschlag, den Narzißmus als „autonomen Faktor im topischen Rahmen des Freudschen Systems" anzuerkennen und „in den Rang einer psychischen Instanz wie das Es, das Über-ich und das Ich zu erheben" (S. 128).

Die gelungene Synthese zwischen Narzißmus und Triebentwicklung, die „narzißtische Vollständigkeit", wird nach Grunberger im Unbewußten durch das Bild des „Phallus" repräsentiert, das keineswegs mit dem männlichen Sexualorgan, dem Penis, identisch ist. „Das phallische Bild drückt die Integrität in all ihren Erscheinungsformen aus und die Kastration all die Schwierigkeiten des Subjektes, sich im Zeichen dieser Integrität zu konstituieren" (S. 299), und an anderer Stelle heißt es: „Jede Triebbefriedigung oder Ich-Bereicherung des Kindes, die zur Steigerung seines Wertgefühles beiträgt und als solche bekräftigt wird, nimmt in seinem Unbewußten phallischen Charakter an, während umgekehrt das Fehlen von Bestätigung oder die Abwertung ohne anschließende narzißtische Kompensation als Kastration erlebt wird" (S. 209). Demnach gibt es nach Grunberger nicht nur eine sexuelle Kastration, sondern ebenso eine anale (den Verlust von Kot, materiellen Verlust oder Herrschaftsverlust), eine orale (die Entwöhnung) und schließlich die Geburt als „Urkastration". Grunbergers

Thesen werden in der Gegenwart keineswegs so heftig diskutiert wie die Publikationen zum Narzißmusproblem von Kohut (1973, 1979) und Kernberg (1978).

Die Ausgangspositionen von Kohut und Kernberg decken sich mit den oben beschriebenen von Balint. Sie stellten in Übereinstimmung mit vielen Psychotherapeuten fest, daß es eine große Zahl von Patienten gibt, die mit den Mitteln der klassischen Psychoanalyse nicht erfolgreich behandelt werden können. Das gemeinsame Kennzeichen dieser Patienten ist ein schwer gestörtes Selbstwertgefühl und pathologische Objektbeziehungen. Beide sind sich darüber im klaren, daß es nötig ist, zum Zweck der Behandlung dieser Patienten die klassische psychoanalytische Technik zu modifizieren. Der Hauptunterschied zwischen den beiden Autoren besteht darin, daß Kernberg bemüht ist, lediglich die Theorie der Psychoanalyse zu erweitern, aber in ihrem triebtheoretischen Rahmen zu verbleiben, während Kohut eine separate narzißtische Entwicklungslinie postuliert, die getrennt von der Entwicklung der Libido verläuft.

Kohut schreibt: „Warum, um die Frage in persönlichen Begriffen zu stellen, fühlte ich mich trotz meiner langandauernden Bindung an die Theorien der klassischen Psychoanalyse, die Wissenschaft, die ich während meines ganzen Berufslebens studierte und lehrte, warum fühlte ich mich trotz meiner tief verwurzelten konservativen Instinkte, die mir sagen, daß man in ein funktionierendes System nicht eingreifen sollte, gezwungen, eine Ausdehnung, eine Veränderung vorzuschlagen? Warum, um die Frage in den breiteren Rahmen zu stellen, dem sie angehört, braucht die Psychoanalyse nun zusätzlich zur klassischen Theorie und Technik eine Psychologie des Selbst und eine dieser entsprechende Technik? Sie braucht sie, sage ich, weil der Mensch sich verändert, wie die Welt, in der er lebt, sich verändert; sie braucht sie, denn wenn die Psychoanalyse die führende Kraft bei dem Versuch des Menschen bleiben soll, sich selbst zu verstehen, wenn sie tatsächlich lebendig bleiben will, dann muß sie mit neuen Einsichten reagieren, wenn sie mit neuen Daten und folglich mit neuen Aufgaben konfrontiert wird" (1979, S. 266).

Kohut sieht die Ursache der narzißtischen Störung in einer pathologischen Fixierung an das „archaische Größen-Selbst" und an die archaische „idealisierte Eltern-Imago". Im Laufe seiner frühen Entwicklung erfährt das Kind – und dies stellt eine große Frustration und Verunsicherung für es dar –, daß die Mutter ein von ihm unabhängiges Wesen ist, das nicht seiner magischen Kontrolle unterliegt. Um die dadurch hervorgerufene Verunsicherung zu kompensieren, kann es seine illusionären Omnipotenzgefühle entweder dem eigenen Selbst oder den elterlichen Objektrepräsentanzen zuschreiben.

Kohut schreibt: „Das Gleichgewicht des primären Narzißmus wird durch die unvermeidlichen Begrenzungen mütterlicher Fürsorge gestört, aber das Kind ersetzt die vorherige Vollkommenheit a) durch den Aufbau eines grandiosen und exhibitionistischen Bildes des Selbst: das Größen-Selbst; und b) indem es die vorherige Vollkommenheit einem bewunderten, allmächtigen (Übergangs-)Selbst-Objekt zuweist: der idealisierten Eltern-Imago."

Bei der Fixierung an das archaische Größen-Selbst kommt es nach Kohut zu einer „Persönlichkeitsspaltung", die den Patienten zwischen Größenideen und Minderwertigkeitsgefühlen, häufig verbunden mit depressiven Verstimmungen und hypochondrischen Befürchtungen, hin und her schwanken läßt.

Bei der Fixierung an die archaische Eltern-Imago mißlingt den Patienten die natürliche Entidealisierung der Eltern und der Aufbau eines soliden Selbstwertgefühles. Sie bleiben zeitlebens abhängig, sei es von Autoritätspersonen oder aber von Stoffen, an die sie suchtartig gebunden sind.

Während Kohut die Ursache des pathologischen Narzißmus in einer mehr oder weniger isolierten Entwicklungsstörung des Selbst sieht, die unabhängig von der libidinösen Triebentwicklung verläuft, läßt sich nach Kernberg die Entwicklung des pathologischen Narzißmus nicht von der libidinösen und aggressiven Triebentwicklung trennen. Nach Kernberg stellt das Größen-Selbst ein pathologisches Verschmelzungsprodukt von Anteilen des Real-Selbst, des Ideal-Selbst und der Ideal-Objekte dar. Das Real-Selbst beinhaltet nach Kernberg die Vorstellung „jemand besonderes zu sein"; das Ideal-Selbst umfaßt Größenphantasien und die Ideal-Objekte beinhalten Phantasien von grenzenlos liebenden und spendenden Elternfiguren.

Kernberg (1978) hat auch den Versuch unternommen, die Borderline-Persönlichkeitsstörungen eindeutig einerseits von den Neurosen und andererseits von den Psychosen abzugrenzen. Der Unterschied zwischen den narzißtischen und den Borderline-Persönlichkeitsstörungen besteht darin, daß die narzißtischen Persönlichkeitsstörungen über ein relativ kohärentes Selbst verfügen, während die Kohärenz des Selbst der Borderline-Patienten so mangelhaft ist, daß sie große Schwierigkeiten haben, zwischen Selbst- und Objektrepräsentanzen zu differenzieren, was dann entweder zu massiven Desintegrationsängsten oder aber infolge permanenter Vulnerabilität der „individuellen Wirklichkeit" des Patienten (vgl. Kap. 1) zu psychosomatischen Störungen führen kann.

Rohde-Dachser (1979) hat in einer Monographie über das Borderline-Syndrom die umfangreiche Literatur darüber kritisch zusammengefaßt. Wegen der großen Bedeutung für die Psychosomatische Medizin wollen wir anschließend, in Anlehnung an Mertens (1981), die nach Rohde-Dachser für das Borderline-Syndrom charakteristischen Krankheitszeichen zusammenfassen:

– Chronische, frei-flottierende Angst
  Die Angst, die häufig als allgegenwärtig erfahren wird, kann von Individuen mit einer Borderline-Störung vor allem dann eingesetzt werden, wenn andere bewußtseinsnahe, aber unvereinbare Affekte zugedeckt werden sollen.
– Multiple Phobien
  Hierzu gehören vor allem Phobien, welche die Körperlichkeit oder die leibliche Erscheinung betreffen (z. B. Errötungsphobie, Furcht vor öffentlichen Auftritten oder vor dem Angeschautwerden) und mit Beschämungsängsten verbunden sind.

– Zwangssymptome, die vorübergehend die Qualität unumstößlicher Gewißheit erhalten
Zwangsgedanken (z. B. hypochondrischen oder paranoiden Inhalts), die lange Zeit als Ich-fremd erfahren werden, können vorübergehend (wie beim psychotischen Individuum) Ich-synton werden, wobei sich die Realitätsprüfung nach einigen Stunden oder Tagen wieder einstellt.
– Multiple, bizarre Konversionssymptome
Hierunter fallen chronische oder auch massive monosymptomatische Konversionssymptome, Konversionssymptome mit der Tendenz zu Körperhalluzinationen oder mit bizarren Bewegungsabläufen.
– Dissoziative Reaktionen
Traum- oder Dämmerzustände, häufig schwere Depersonalisationserlebnisse werden vom Borderline-Patienten leicht übersehen, weil sie für ihn etwas sehr Vertrautes darstellen.
– Depression
Die Borderline-Depression stellt sich zumeist im Anschluß an den Zusammenbruch eines grandiosen Selbstbildes ein, manifestiert sich in ohnmächtiger Wut oder Gefühlen der Hilflosigkeit und löst gegenübertragungsmäßig wenig helferische Aktivitäten aus.
– Polymorph-perverse Sexualität
Das Vorliegen mehrerer perverser Züge (wie z. B. heterosexuelle und homosexuelle Promiskuität mit sadistischen Elementen) bei einer gleichzeitigen Instabilität von Beziehungen verweist – im Unterschied zu Individuen mit einer stabilen sexuellen Devianz bei konstanten Beziehungen – auf ein Borderline-Symptom.
– Vorübergehender Verlust der Impulskontrolle
Hierzu gehören z. B. episodische Freßsucht, Alkoholismus, Kleptomanie, Drogenabhängigkeit, die nach Beendigung der Impulsdurchbrüche als Ich-fremd erlebt werden.

16.4 Der Einfluß der neueren Narzißmustheorien auf die therapeutische Technik bei frühen Störungen

Als therapeutische Richtlinien gelten nach Rohde-Dachser (1979) folgende Hinweise, die sich von der klassischen psychoanalytischen Therapie durchaus erheblich unterscheiden:

– Variables, den jeweiligen Bedürfnissen des Patienten angepaßtes Setting
– Durchführung der Therapie in der Regel im Sitzen
– Steuerung der inhaltlichen Mitteilungen des Patienten in die Richtung eines verbesserten Realitätsbezuges anstelle der Aufforderung zur freien Assoziation
– Ausgiebige Information des Patienten über die Art seiner Krankheit, über den Sinn des jeweils gewählten therapeu-

tischen Settings und des technischen Vorgehens des Analytikers und über psychodynamische Zusammenhänge
– Verbesserung des Arbeitsbündnisses durch Forcierung der positiven Übertragung (z. B. dadurch, daß der Analytiker eindeutig für den Patienten Partei ergreift)
– Schnelles Unterbrechen von Schweigepausen
– Wiederkehrende verbale Bestätigungen, daß die Abstinenz des Analytikers keine Ablehnung des Patienten bedeute, und wiederkehrende verbale Versicherungen, daß der Analytiker die Integrität des Patienten respektiere
– Keine Interpretation der positiven Übertragung
– Aufspüren der abgespaltenen und außerhalb der Therapie agierten negativen Übertragung
– Sorgfältiges Aufspüren der am wenigsten konflikthaften Persönlichkeitsbereiche des Patienten und Konzentration der Deutungen zunächst auf diese Peripherie; Deutung des depressiven Materials in der Regel vor dem paranoiden Material, des Masochismus vor dem Sadismus
– Statt genetischer Deutungen überwiegend Deutungen, die den Realitätsbezug des Patienten verbessern, insbesondere Deutung der pathologischen Abwehrmechanismen in ihrer destruktiven Auswirkung auf diesen Realitätsbezug
– Freimütiges Mitteilen von Gegenübertragungsgefühlen, durch die der Analytiker für den Patienten als eigenständiges Individuum erlebbar wird; sofortige Richtigstellung der verzerrten, oft paranoid getönten Wahrnehmungen der Person des Analytikers (auch durch Beantwortung von Fragen); alsbaldiger Abbau der illusionären Erwartungen gegenüber dem Analytiker, die sich an die primitive Idealisierung knüpfen
– Kontrolle des Agierens des Patienten, gegebenenfalls durch strikte Grenzsetzungen oder auch durch eine vorübergehende Hospitalisierung
– Notfalls massive Konfrontation des Patienten mit hartnäckig verleugneten Inhalten, insbesondere mit verleugneten realen Gefahren
– Wiederkehrende Bestätigung der grundsätzlichen Liebesfähigkeit des Patienten (und seiner frühen Bezugspersonen); Deutung der Verzerrungen, in denen sich diese Liebesbedürfnisse manifestieren, und Aufzeigen befriedigender Möglichkeiten für die Verwirklichung dieser Bedürfnisse
– Entzerren der Bilder von den frühen Bezugspersonen („Entteufelung" und „Entidealisierung") zu realen Menschen mit Vorzügen und Schwächen
– Übersetzung des „Borderline-Dialogs" in wirkliche Kommunikation
– Herausarbeiten der unbewußten Identifikationsphantasie, nach der der Patient seine „Schicksalsneurose" gestaltet, mit dem Ziel, die Fremdbestimmung durch eine sichere eigene Identität zu ersetzen.

Die Hauptunterschiede der Borderline-Therapie bestehen gegenüber der klassischen Psychoanalyse in einem veränderten Setting, einer veränderten Deutungstechnik und dem Vermeiden tieferer Stadien der Regression.

17 Das ärztliche Gespräch – Versuch einer Strukturanalyse

Wolfgang Wesiack

17.1 Vorbemerkungen

Die Psychosomatische Medizin, die sich die Aufgabe gestellt hat, die gesamte Interaktion zwischen Arzt und Patient zu erfassen, und dabei vor allem die Aspekte der Beziehung und des emotionellen Erlebens berücksichtigt, ist in ganz besonderem Maße genötigt, die diagnostischen und therapeutischen Qualitäten des gesprochenen Wortes zu untersuchen und zu nützen. In Kapitel 1 haben wir im Anschluß an die Darstellung des Situationskreismodells den diagnostisch-therapeutischen Zirkel beschrieben und darauf hingewiesen, daß im ärztlichen Gespräch diagnostische und therapeutische Interventionen stets auf das engste miteinander verknüpft und kaum voneinander zu trennen sind. Im Kapitel über das so wichtige Erstinterview (vgl. Kap. 12) hat Adler die verschiedenen Funktionen und die Technik des Erstinterviews eingehend beschrieben, und im Kapitel 16 habe ich kurz die Psychoanalyse und die für die Psychosomatische Medizin wichtigsten psychoanalytisch orientierten Therapieformen dargelegt.

Jetzt wollen wir noch auf das ärztliche Gespräch als zentrales Kommunikationsmittel zwischen Arzt und Patient eingehen und es zunächst als Über- bzw. Sammelbegriff für alle zwischen Arzt und Patient gewechselten Worte verstehen. Eine so weite Definition des Terminus „ärztliches Gespräch" umfaßt dann auch das Erstinterview, die Psychoanalyse, die verschiedenen psychoanalytisch und nicht-psychoanalytisch orientierten Gesprächstherapieformen und reicht dann bis zu dem mehr oder weniger funktionsbezogenen Gespräch zwischen Arzt und Patient am Krankenbett oder in der ärztlichen Praxis. Aus diesem sehr weiten Feld des „ärztlichen Gesprächs" haben wir bereits verschiedene klar umgrenzbare Formen und Techniken ausgegliedert und gesondert beschrieben. Da aber „ärztliches Gespräch" nicht nur in Form dieser besonders beschriebenen Verfahren angewandt, sondern ununterbrochen in den Sprechstunden und am Krankenbett praktiziert wird, wollen wir unsere Aufmerksamkeit in diesem Kapitel ganz jenem bisher in Theorie und Praxis der Medizin stark vernachlässigten Bereich der Medizin zuwenden, um zu sehen, ob es sich nicht auch brauchbar strukturieren läßt.

Seit es eine wissenschaftliche Medizin gibt, ist diese nahezu ausschließlich damit beschäftigt, vorhandene diagnostische und therapeutische Techniken zu überprüfen, zu verbessern und neue zu entwickeln. In diesem Zusammenhang wurde das gesprochene Wort meist nur als notwendiges Hilfsmittel im Rahmen dieser Bemühungen gesehen, ohne daß ihm selbst besondere Aufmerksamkeit gewidmet worden wäre. Dies hatte zur Folge, daß das ärztliche Sprechstundengespräch lange Zeit keinen Platz in der medizinischen Theorie hatte, sondern der sog. ärztlichen Kunst zugeordnet wurde, worunter man eine Mischung mehr oder weniger verschwommener Vorstellungen aus den magischen und charismatischen Fähigkeiten des Arztes einerseits und seinem Einfühlungsvermögen, Taktgefühl und allgemeiner Lebenserfahrung andererseits verstand. Ohne diese Bereiche „ärztlicher Kunst" geringschätzen zu wollen, scheint es doch höchste Zeit zu sein, auch das ärztliche Sprechstundengespräch einer wissenschaftlichen Analyse zugänglich zu machen und es damit aus dem Bereich der mehr oder weniger unverbindlichen „ärztlichen Kunst" in den Bereich der lehr- und lernbaren ärztlichen Verhaltensweisen überzuführen.

Eine Analyse des ärztlichen Sprechstundengesprächs kann methodisch auf verschiedenen Wegen erreicht werden. Als praktisch brauchbares Gerüst erscheint mir vorläufig sowohl die informationstheoretische als auch die psychoanalytische Methode, die sich gegenseitig ergänzen und zusammen ein umfassendes Bild des ärztlichen Gesprächs ergeben, ausreichend zu sein. Ich will mich deshalb darauf beschränken.

Hier sei noch eine kurze Anmerkung über den Zeitfaktor eingefügt. Da in der Vergangenheit die theoretische Bedeutung des ärztlichen Gesprächs nicht erkannt oder mißachtet wurde, wurde auch im praktischen Vollzug der Gesprächskontakt zwischen Arzt und Patient immer mehr reduziert und durch, von der rein naturwissenschaftlichen Theorie her gesehen, „wichtigere" Maßnahmen ersetzt. So kommt es, daß verschiedenen Untersuchungen zufolge (Braun, 1965, 1970; Erdmann et al., 1974) der Gesprächskontakt zwischen Arzt und Patient sowohl in der ärztlichen Sprechstunde als auch in der Klinik nur auf wenige Minuten reduziert ist. Durch eine mehr psychosomatisch ausgerichtete Heilkunde und durch organisatorische und strukturelle Änderungen unserer gegenwärtigen Krankenversorgung wird es, so hoffen wir, in Zukunft möglich sein, diesen beklagenswerten Zustand etwas zu mildern. Wir dürfen aber die Augen nicht davor verschließen, daß das Zeitproblem auch in einer optimal organisierten Krankenversorgung insbesondere für die ärztliche Praxis stets ein schwer zu bewältigendes Problem bleiben wird, denn für die Vielzahl der notleidenden Patienten und das nahezu unbegrenzte Informationsbedürfnis des Arztes, das dem Situationskreiskonzept zufolge nie endgültig ge-stillt sein kann, werden sich die Anforderungen mit den Möglichkeiten nie ganz zur Deckung bringen las-sen. Der niedergelassene Arzt, der den Patienten oft schon von früheren Kontakten her kennt und auch über die Familienverhältnisse desselben mehr oder weniger gut informiert ist, wird sowohl seine Kennt-nisse oft in fraktionierter Form erhalten, als auch sei-ne therapeutischen Interventionen auf mehrere Sprechstundenkontakte verteilen. Auf diese und an-dere Probleme werde ich in Kapitel 26 „Psychosoma-tische Medizin in der Praxis des niedergelassenen Arztes" ausführlicher eingehen.

17.2 Zwei exemplarische Krankheitsfälle

Die Problematik des ärztlichen Gesprächs in der Pra-xis des niedergelassenen Arztes soll nachfolgend am Beispiel zweier typischer Patienten aus der alltägli-chen Praxis dargestellt werden. Es sei dabei darauf hingewiesen, daß die Untersuchung und Behandlung beider Patienten unter großem Zeitdruck in der allge-meinen Sprechstunde erfolgen mußten, so daß Erst-interview und gründliche erste Untersuchung zusam-men nicht wesentlich länger als eine halbe Stunde, eher sogar etwas kürzer dauerten. Viele biographisch und psychodynamisch wichtige Informationen konn-ten daher erst im weiteren Verlauf der Behandlung gewonnen werden. Von Anfang an aber war es beson-ders wichtig, die jeweils bedeutendsten Informatio-nen herauszugreifen und zu bearbeiten. Für den Me-dizinstudenten ist es nicht einfach, die Bedeutsamkeit einzelner Informationen zu gewichten und in das Ge-samtbild, das wir uns vom Patienten machen, einzu-ordnen. Neben ständigem Sammeln von Erfahrungen soll nicht zuletzt dieses Lehrbuch eine Hilfe in dieser Richtung darstellen.

Das Sprechzimmer betritt erstmals ein 49jähriger etwas übergewichtiger Mann, der keinen schwerkranken Ein-druck macht. Er berichtet, daß er Handelsvertreter sei und ein ziemlich gehetztes Leben führe. Seit ungefähr einem halben Jahr bekomme er in zunehmendem Maße bei An-strengungen, insbesondere bei Treppensteigen, aber auch nach reichlicheren Mahlzeiten, drückende Schmer-zen hinter dem Brustbein, die ihn mehr belästigen als beunruhigen. Daß er bei längerem Gehen auch Schmer-zen in der linken Wade bekomme, so daß er oft stehen-bleiben müsse, bis die Schmerzen abgeklungen sind, be-richtet er erst auf direktes Fragen, nachdem bei der Unter-suchung abgeschwächte Fußpulse aufgefallen waren. Die EKG-Untersuchung ergibt dann das Bild eines nicht mehr ganz frischen, bis auf die Herzspitze übergreifenden Herzmuskelhinterwandinfarktes.
Dieser eher zur Dissimulation neigende Patient mußte also zunächst so versorgt werden, daß möglichst neuen Infarktschüben vorgebeugt wurde, um dann seine ganze Lebensweise von Grund auf umzustellen. Zunächst schien mir eine Krankenhauseinweisung die zweckmäßigste Form der Einleitung einer Behandlung zu sein. Auf lange Sicht aber mußte erreicht werden, daß der Patient seinen ganzen Lebensstil ändert. Er mußte abnehmen, das Rau-

chen einstellen, später genügend körperliche Bewegung haben, Ruhepausen einlegen und das Arbeitstempo auf ein vernünftiges Maß reduzieren. Diese für ihn sehr ein-schneidenden Änderungen des Lebensstils waren jedoch nur auf dem Boden eines Vertrauensverhältnisses zu sei-nem Arzt zu erreichen, der ihn neben der Überwachung verschiedener Kreislauf- und Blutbefunde im Gespräch ständig beraten konnte. Über Einzelheiten der Therapie dieses Patienten orientiere man sich in Kapitel 41.1.

1. Beratung. Das Sprechzimmer betritt erstmals eine et-was ängstlich und unsicher, aber körperlich gesund wir-kende 29jährige Patientin und berichtet, daß sie seit meh-reren Wochen schlaflos und unruhig sei und dauernd Herzklopfen habe. Die Beschwerden seien während des Umbaus des großväterlichen Hauses und den an-schließenden Putzarbeiten aufgetreten. Der Großvater sei nämlich vor drei Monaten verstorben, und jetzt werde sein Haus für den Bruder der Patientin umgebaut und herge-richtet. Da sie bei diesen Arbeiten viel Staub habe schluk-ken müssen, sei ihr Hausarzt der Meinung gewesen, sie habe sich eine Staubvergiftung zugezogen und habe ihr deshalb ein Sulfonamid verordnet. Davon sei es aber nicht besser, sondern schlechter geworden und sie habe noch zusätzlich Übelkeit, Brechreiz und Durchfälle bekommen.
Auf meine Frage, wie denn ihre Beziehung zum verstor-benen Großvater gewesen sei, berichtet sie, daß ihre Mut-ter sehr früh, als sie selbst erst vier Jahre alt war, gestor-ben sei und daß sie dann bei den Großeltern aufgewach-sen sei, zu denen sie ein herzliches Verhältnis gehabt ha-be. Nachdem die Großmutter schon vor mehreren Jahren gestorben sei, habe sie jetzt ihren Großvater bis zu seinem Tode gepflegt.
Sie berichtet dann spontan weiter, daß sie schon seit längerer Zeit in einem ihr selbst absonderlich erscheinen-den Drang alle Todesnachrichten mit besonderem Inter-esse verfolge und dann immer denken müsse: „So schnell kann es gehen!" Diese Mitteilung wird noch da-durch szenisch untermalt, daß sie nunmehr ängstlicher und hilfloser als zu Anfang wirkt und nur mühsam die Trä-nen unterdrücken kann. Auf mich macht sie dabei den Eindruck eines hilflosen verängstigten Kindes, dem ich gerne helfen möchte, ohne zunächst selbst so recht zu wissen wie.
Nach diesem einleitend-anamnestischen Teil des ärztli-chen Gesprächs wird am gleichen und darauffolgenden Tag eine gründliche internistische Untersuchung vorge-nommen, die außer einer erhöhten vegetativen Labilität und einer kleinen unverdächtigen Struma ein leises systo-lisches Geräusch links parasternal im 2. und 3. ICR ergibt, bei sonst völlig normalem Herzbefund. Da alle anderen somatischen Befunde völlig regelrecht sind, handelt es sich wohl nur um ein akzidentelles Herzgeräusch.
Abschließend wird der Patientin mitgeteilt, daß sie kör-perlich völlig gesund sei und die eingehende internistische Untersuchung nur zwei „Schönheitsfehler" ergeben ha-be, nämlich einen kleinen harmlosen Kropfknoten und ein ebenfalls harmloses Herzgeräusch, die beide mit ihren Beschwerden sicherlich in keinem Zusammenhang ste-hen. Diese seien vielmehr der Ausdruck eines Angstzu-standes, der ja schon normalerweise mit körperlichen Be-gleiterscheinungen wie allgemein nervöser Unruhe, Herz-klopfen, Zittern usw. einhergehe. Danach wird die Patien-tin mit einem Rezept für ein leichtes Sedativum (ein Bal-drian-Hopfen-Präparat) entlassen und nach 2 bis 3 Wo-chen zur Kontrolle wiederbestellt.
2. Beratung. Nach zweieinhalb Wochen erscheint die Patientin wieder in der Sprechstunde und berichtet, daß alles in Ordnung sei. Sie wird mit dem Hinweis, daß sie

mich jederzeit aufsuchen könne, wenn sie mich brauche, entlassen; diesmal ohne Rezept.

3. Beratung. Nach über fünf Wochen kommt die Patientin wieder in die Sprechstunde. Sie berichtet, daß sie jetzt keine Angst und auch keine Herzbeschwerden mehr habe, wohl aber immer etwas schwindelig sei. Da sie keine weiteren Informationen anbietet, wird sie wiederum mit einem Rezept für das oben erwähnte Sedativum entlassen.

4. Beratung. Nach knapp zwei Wochen erscheint sie wieder. Diesmal ist sie hochgradig verängstigt und macht einen geradezu verstörten Eindruck. Auf meine Frage, wie es ihr denn gehe, bricht sie sofort in Tränen aus und berichtet, daß sie immer weinen müsse, wenn sie jemand nach ihrem Befinden frage. Sie müsse auch immer dann weinen, wenn sie ihre arme kleine dreijährige Tochter ansehe. Vor einigen Tagen sei eine um zwei Jahre jüngere Arbeitskollegin an Brustkrebs verstorben und hinterlasse drei kleine Kinder. Seither könne sie selbst keinen klaren Gedanken mehr fassen, sitze nur noch da und grüble und werde selbst nachts von bösen Träumen verfolgt. So habe sie zum Beispiel in der vergangenen Nacht der verstorbene Großvater aus dem Sarg böse angesehen.

Während sie mir das alles berichtet, macht sie auf mich in noch viel stärkerem Maße als bei der ersten Beratung den Eindruck eines völlig verängstigten hilflosen Kindes. Ich sage ihr das und sage weiter, daß ich den Eindruck habe, daß die Todesfälle in ihrer Umgebung alte, nur schlecht vernarbte seelische Wunden wieder aufgerissen und alte Ängste in ihr wiedererweckt haben. All das Fürchterliche, das sie im Alter von vier Jahren beim frühen Tod ihrer Mutter habe erleiden müssen, werde jetzt wieder lebendig. Beim Anblick ihrer kleinen Tochter müsse sie unwillkürlich denken, jetzt werde die arme Kleine bald die gleiche Angst und Verzweiflung durchmachen müssen, die sie bei dem Tod ihrer Mutter erlebt und nur mühsam überwunden habe, denn sie selbst sei ja wohl davon überzeugt, in nächster Zukunft sterben zu müssen, d.h. zum Tode verurteilt zu sein.

Während ich das sage und diese Thematik noch mit ihr gemeinsam etwas vertiefe, beruhigt sie sich zusehends und fragt erstaunt: „Können Sie denn Gedanken lesen, Herr Doktor?!"

Nach diesem Gespräch hatten wir beide den Eindruck, einen zentralen Punkt ihres Krankheitsgeschehens getroffen und ein Stückchen positiver therapeutischer Arbeit geleistet zu haben. Eine weiterhin kontrollierte Beschwerdefreiheit („alles wie weggeblasen") bestätigte diesen Eindruck (Auszüge dieser Krankengeschichte wurden bereits auf Seite 252 als Beispiel eines Flashs gebracht).

17.3 Versuch einer informationstheoretischen Analyse

Auf einige grundsätzliche informationstheoretische Gesichtspunkte, vor allem die verbale, die nonverbale und die außersprachliche Kommunikation sind wir bereits in Kapitel 1 ausführlicher eingegangen. Hier wollen wir nur zum besseren Verständnis und zur besseren Strukturierung unserer ärztlichen Gespräche drei Informationsebenen aus der Sicht des Arztes und des Patienten voneinander unterscheiden, die im gesamten Interaktionsgeschehen natürlich miteinander verwoben sind:

– Die Ebene der objektiven Informationen.
– Die Ebene der subjektiven Informationen bzw. der Bedeutungen.
– Die Ebene der szenischen Informationen.

Auf der Ebene der objektiven Informationen berichten uns die Patienten über Tatbestände, die auch von anderen zumindest grundsätzlich nachprüfbar sind, wie z. B. die wichtigsten Lebensdaten. Zu den objektiven Informationen werden wir aber auch alle Befunde zählen können, die wir erhoben haben und die ebenso von anderen Ärzten erhoben und überprüft werden könnten. Hier zeigt sich bereits, wie unlösbar eng die Ebene der objektiven Information mit der subjektiven verbunden ist, denn nach den Überlegungen, die wir im Kapitel 1 nach der Diskussion des Situationskreismodells über den Unterschied zwischen der individuellen und der sozialen Wirklichkeit angestellt haben, ist das ja nicht weiter überraschend. Die objektive Ebene entspricht der sozialen, die subjektive der individuellen Wirklichkeit. Während die nur naturwissenschaftliche Medizin so gut wie ausschließlich an der objektiven Informationsebene bzw. an der „sozialen Wirklichkeit" im Sinne unseres Situationskreismodells interessiert ist und Bedeutungen nur im Sinne eines sozialen Konsensus zuläßt, versucht die Psychosomatische Medizin, die subjektive Informationsebene, d.h. die subjektive Bedeutung aller Informationen und Befunde bzw. die „individuelle Wirklichkeit" unserer Patienten, zu erreichen.

Für unseren ersten Patienten bedeuten seine stenokardischen Beschwerden demnach mehr eine Belästigung als eine Bedrohung. Die Dysbasiebeschwerden werden zunächst wohl als subjektiv bedeutungslos unterschlagen und erst auf ausdrückliches Befragen erwähnt. Er erwartet vom Arzt nur ein gründliches Check-up und eine möglichst rasche Beseitigung eventuell festgestellter Schäden. Der objektive EKG-Befund bedeutet für den Arzt, daß der Patient akut gefährdet ist. Der Patient aber muß erst vom Arzt dazu gebracht werden, dessen Interpretation sich zu eigen zu machen.

„Szenisch" vermittelt der Patient den Eindruck eines keineswegs besonders gefährdeten Kranken. Er bagatellisiert seine Beschwerden und sucht den Eindruck zu erwecken, es sei alles halb so schlimm. Diese szenische Information sagt über das Krankheitserleben und das Krankheitsverhalten des Patienten mehr aus als lange Gespräche.

Auch bei unserer zweiten Patientin können wir diese drei Informationsebenen recht gut voneinander unterscheiden. Die (grundsätzlich objektiv nachprüfbare) Mitteilung, daß sie mit 4 Jahren ihre Mutter, vor einigen Monaten ihren Großvater und vor wenigen Tagen eine Arbeitskollegin durch den Tod verloren habe, hat sicherlich für die Patientin und den Arzt nicht die gleiche Bedeutung. Der Arzt kann aber diese (objektive) Information nur richtig verstehen und werten, wenn er in seiner Interpretation dieser Ereignisse der Bedeutung nahekommt, die die Patientin diesen Ereignissen beimißt. Verwirft er die Interpretationsangebote der Patientin als „zeitraubendes, lästiges, subjektives Geschwätz", dann wird er sie nie verstehen und ihr auch nicht helfen können. Zum Verständnis der subjektiven Ebene der Patientin, d.h. zum Verstehen ihrer individuellen Wirklichkeit, trägt aber ganz besonders die szenische Information bei. Erst das Wahrnehmen und „Verstehen" der szenischen Information des völlig verängstigten und hilflosen Kindes ermöglicht dem Arzt unter Einbeziehung der objektiven Informationsdaten einen Flash zu erleben und ihn für die Patientin fruchtbar zu machen, wie im Kapitel 16 beschrieben.

Mit den Konstrukten objektive, subjektive und szenische Information bzw. Informationsebene gelingt es dem Arzt schon recht gut, die Vorgänge des ärztlichen Gespräches zu strukturieren und es damit besser zu handhaben. Diese Vorgänge werden aber noch durchsichtiger, wenn wir die von der Psychoanalyse herausgearbeiteten verschiedenen Übertragungs- bzw. Beziehungsebenen mit in unsere Strukturanalyse einbeziehen.

17.4 Die psychoanalytische Interpretation des ärztlichen Gesprächs

Im Gegensatz zur Informationstheorie, die im ärztlichen Gespräch einen Informationsaustausch sieht und diesen zu analysieren sucht, sieht die Psychoanalyse im ärztlichen Gespräch einen Teil des Interaktionsprozesses zwischen Arzt und Patient, der sich auf verschiedenen Übertragungs- und Gegenübertragungsebenen abspielt und der das Ziel verfolgt, „die für die Ich-Funktionen günstigsten Bedingungen" herzustellen (vgl. Freud, 1937). Unter Ich-Funktionen (Loch, 1967a) versteht die Psychoanalyse die Wahrnehmungsfähigkeit, die willkürliche Motorik, das Gedächtnis und die Intelligenz, Fähigkeiten also, die zur Lebensbewältigung erforderlich sind und bei Neurosen, Psychosen und psychosomatischen Erkrankungen immer mehr oder weniger gestört sind.

Hier kann natürlich kein Abriß der psychoanalytischen Theorie geboten werden (Freud, 1913, 1917; Kuiper, 1968; Loch, 1967a). Zum besseren Verständnis des hier Gesagten muß aber doch kurz auf die verschiedenen Übertragungs- und Gegenübertragungsebenen eingegangen werden, wie sie etwa von Loch (1967b) unter Berücksichtigung der wesentlichen Literatur herausgearbeitet wurden:

– Die Beziehung zwischen dem „fiktiven Normal-Ich" des Patienten und dem „fiktiven Normal-Ich" des Arztes, wobei es natürlich eine Frage der Definition bzw. der Wortwahl ist, ob man diese Beziehungsebene bereits als Übertragung bezeichnet. Es ist jene Ebene der therapeutischen Allianz und der „personalen Begegnung" von Arzt und Patient, von der in der ausgedehnten Literatur der „personal und anthropologisch" ausgerichteten Autoren sehr eingehend die Rede ist. Die Psychoanalyse hat sich mit dieser Ebene, ohne ihre Existenz zu leugnen oder ihre Bedeutung herabzusetzen, nicht eingehender

beschäftigt, weil sie nicht eigentlich zu ihrem Untersuchungsfeld gehört.

– Die Ebene der zielgehemmten Libido, der „milden" bzw. „unanstößigen" Komponente der Übertragung, der „anaklitisch-diatrophischen Gleichung" (Gitelson, 1962)[1]. Es ist jene „bewußtseinsfähige und unanstößige Komponente" der Übertragung, die nach Freud (1905) in der Psychoanalyse „ebenso die Trägerin des Erfolges wie bei anderen Behandlungsmethoden" ist, und die auch nach „Aufheben" bzw. „Vernichten" der neurotischen Übertragung bestehenbleibt. Es ist die Ebene jeder suggestiven Beeinflussung des Patienten durch den Arzt. Sie ist, wie Loch (1965), gestützt auf Gitelson (1962), schreibt, „eine primitive narzißtische Übertragung, mittels der über Besetzung einer pflegenden Person die Umwandlung narzißtischer Libido in Objektlibido in die Wege geleitet wird". Die frühe Mutter-Kind-Beziehung, die Spitz (1967) so eingehend studiert hat, ist der Prototyp dieser grundlegenden Übertragungsebene: „Alle späteren Beziehungen mit Objektqualität, die Liebesbeziehung, die hypnotische Beziehung, die Beziehung der Gruppe zu ihrem Führer und letzten Endes alle zwischenmenschlichen Beziehungen haben ihren ersten Ursprung in der Mutter-Kind-Beziehung".

– Erst die dritte Ebene ist die der neurotischen Übertragung und Gegenübertragung im eigentlichen Sinn. Es ist die Ebene der neurotischen Objektbeziehungen. Das heißt, der Patient „wendet dem Arzt ein Ausmaß von zärtlichen, oft genug mit Feindseligkeiten vermengten Regungen zu, welches in keiner realen Beziehung begründet ist und nach allen Einzelheiten seines Auftretens von den alten und unbewußt gewordenen Phantasiewünschen des Kranken abgeleitet werden muß" (Freud, 1913).

Was die Entstehungsgeschichte anbetrifft, müssen wir natürlich die drei Lochschen Beziehungs- bzw. Übertragungsebenen in anderer Reihenfolge sehen: Die Ebene der anaklitisch-diatrophischen Gleichung ist die früheste und entspricht unserem symbiotischen Funktionskreis. Im Umgang mit den ersten Beziehungspersonen – psychoanalytisch gesprochen den frühen Objektbeziehungen – bildet sich dann die neurotische Übertragungsebene. Die Ebene des „fiktiven Normal-Ichs" ist die jüngste und reifste Beziehungsebene, in der die anderen beiden als die genetisch älteren jeweils mitschwingen.

Die Analyse des ärztlichen Gesprächs unter dem Gesichtspunkt der hier kurz skizzierten Übertragungs- und Gegenübertragungsebenen ist deshalb so fruchtbar, weil das ärztliche Gespräch integrierender Bestandteil der Arzt-Patient-Beziehung ist und ohne Mitberücksichtigung dieser Beziehung weder theoretisch noch praktisch voll ausgeschöpft werden kann.

---

1 Unter der anaklitisch-diatrophischen Gleichung versteht man jene Beziehung auf Gegenseitigkeit, die auf der symbiotischen Mutter-Kind-Ebene zwischen den anlehnenden Bedürfnissen des Säuglings einerseits und den nährend-pflegenden Bedürfnissen der Mutter andererseits besteht.

Wird die Arzt-Patient-Beziehung auf der reifsten (der des „fiktiven Normal-Ichs" bzw. der personalen) Beziehungsebene verfehlt, oder, was in der Realität viel häufiger ist, gar nicht angestrebt (um der Fiktion einer falsch verstandenen Objektivität willen), dann wird der Patient zwangsläufig zum Objekt mehr oder weniger selbstsüchtiger wissenschaftlicher oder materieller Strebungen des Arztes mit allen daraus folgenden, erschreckenden Gefahren einer rein technischen Medizin. Bringt der Patient die ursprüngliche „milde" bzw. „unanstößige" Komponente der Übertragung nicht zustande, dann wird er für den Arzt psychotherapeutisch unerreichbar, unbehandelbar. Ein etwaiger physikalisch-chemischer Eingriff, der ja prinzipiell immer möglich ist, bleibt ohne jeden mutativen Effekt für die Gesamtpersönlichkeit des Kranken.

Auf der symbiotischen bzw. anaklitisch-diatrophischen Ebene der Übertragung entwickelt sich das für jede Behandlung so notwendige Vertrauen des Patienten zum Arzt.

Stellen also diese beiden eben genannten Übertragungsebenen die Voraussetzungen dafür dar, daß ein fruchtbares ärztliches Gespräch überhaupt zustande kommen kann, so ist es die Übertragungsebene der neurotischen Objektbeziehung, die uns durch die szenische Information Einblick in die tieferen psychodynamischen Vorgänge des Patienten gewährt.

Die von der Psychoanalyse erarbeitete Trennung in die drei Beziehungs- bzw. Übertragungsebenen ermöglicht uns auch eine Trennung der verschiedenen Dialogformen des ärztlichen Gesprächs, die allein durch informationstheoretische Analyse nicht möglich ist. Der wissenschaftliche Dialog und das Funktionsgespräch des Alltags beschränken sich auf die Ebene des „fiktiven Normal-Ichs". In das freundschaftliche oder seelsorgerische Gespräch ist die Ebene des Vertrauens, die anaklitisch-diatrophische Übertragung, mit einbezogen. Wenn das ärztliche Gespräch, wie es bisher meist der Fall war, nicht auf dieser Stufe stehenbleiben will, dann muß es die Übertragungsebene der neurotischen Objektbeziehungen mit einbeziehen und gewinnt damit eine neue fruchtbare diagnostische und therapeutische Dimension.

Erst wenn diese Übertragungsebene, die sich informationstheoretisch unter anderem als szenische Information beschreiben läßt, in das ärztliche Gespräch mit einbezogen wird, wird es zum psychoanalytisch orientierten ärztlichen Gespräch.

Ohne Einbeziehung dieser Übertragungsebene bleibt das ärztliche Gespräch eine sachliche Belehrung (Ebene des „fiktiven Normal-Ichs") oder ein philanthropisch-suggestiver Akt (Ebene der „infantil-narzißtischen Ichanteile", also der frühen Mutter-Kind-Beziehung). Diese beiden (Vor-)Stufen des ärztlichen Gesprächs sollen keineswegs gering geachtet werden, denn sie bilden nicht nur die notwendige Basis jedes darüber hinausgehenden therapeutischen Gesprächs, sondern genügen auch gewöhnlich zur Betreuung der vorwiegend akut somatisch Erkrankten. Zur Versorgung der funktionellen Syndrome, der Neurosen und der psychosomatisch Kranken, vor allem vieler chro-

nisch und lebensbedrohlich erkrankter Patienten aber reicht dieses verkürzte ärztliche Gespräch meist nicht mehr aus.

Strebt man eine konfliktlösende Therapie an, dann gilt für das (psychoanalytisch orientierte) ärztliche Gespräch bei allen sehr wesentlichen methodischen und technischen Unterschieden die gleiche Zielvorstellung wie für die Psychoanalyse selbst, die Freud (1937) folgendermaßen definiert hat: „Die Analyse soll die für die Ich-Funktionen günstigsten Bedingungen herstellen; damit wäre ihre Aufgabe erledigt."

Wie aber stellt die Psychoanalyse bzw. das psychoanalytisch orientierte ärztliche Gespräch „die für die Ich-Funktionen günstigsten Bedingungen" her? Durch die deutende Bearbeitung von Übertragung und Widerstand sowie in Grenzen auch durch die Flash-Technik. Um sich der ersteren Aufgabe zu unterziehen, ist es nötig, die technischen Grundregeln der Psychoanalyse zu beherrschen. Ohne auf diese Spezialfragen hier eingehen zu können (Greenson, 1973), möchte ich kurz darauf hinweisen, wann nach psychoanalytischer Theorie und Erfahrung eine Deutung erfolgreich ist und im Patienten einen „mutativen Effekt" hervorruft. Loch (1967b), dem ich hier folge, hat das folgendermaßen zusammengefaßt: „Die erfolgreiche, die ‚mutative Deutung' wird ermöglicht, wenn 1. drei Übertragungsdimensionen zur Konvergenz gebracht sind, die des ‚fiktiven Normal-Ichs', die der ‚infantil-narzißtischen Ichanteile' und die der ‚neurotischen' Objektbeziehung, und wenn 2. die Übertragungsdeutung den ‚dringlichsten Punkt' trifft."

Aufgrund von Einsichten, die uns die Theorie des Situationskreises (vgl. Kap. 1) bietet und die von Balint und Mitarbeitern bei der Erforschung der Flash-Phänomene gemacht wurden (vgl. Kap. 16), können wir die von Loch gemachte Feststellung noch durch einen weiteren Punkt erweitern und feststellen: Die erfolgreiche, die mutative Deutung wird ermöglicht, wenn

– die drei Übertragungsdimensionen zur Konvergenz gebracht sind,
– die Übertragungsdeutung den dringlichsten Punkt trifft und/oder
– in Form eines Aha-Erlebnisses (= Flash) der Patient innerhalb seiner „subjektiven Wirklichkeit" die Lösung „seines Problems" erfährt und entdeckt.

Kehren wir nun zu unseren eingangs geschilderten Fallbeispielen zurück. Bei dem ersten Patienten bewegen wir uns zunächst so gut wie ausschließlich auf der Ebene des „fiktiven Normal-Ichs". Wir müssen ihm die Diagnose „Herzinfarkt" mit allen Implikationen mitteilen. Vom Erreichen der zweiten, der infantil-narzißtischen Übertragungsebene wird es dann abhängen, wieweit der Patient die ärztlichen Ratschläge befolgen und seine Lebensweise ändern können wird. Die dritte Ebene der Objektbeziehungen wird zunächst nicht erreicht und auch nicht vom Arzt angestrebt. Sie wird möglicherweise zu einem späteren Zeitpunkt der Behandlung eine Rolle spielen, wenn man mit dem Patienten seine zwanghafte Fixierung an Leistung durcharbeiten wird. Viele Arzt-Patient-

Interaktionen brauchen diese Ebene nicht zu erreichen und führen trotzdem zu befriedigenden Resultaten.

Wenden wir uns nun dem zweiten Fallbeispiel zu. Auf der Übertragungsebene des „fiktiven Normal-Ichs" findet der Informationsaustausch zwischen Arzt und Patient statt, den ich weiter oben als Informationsaustausch auf der Ebene der objektiven Informationen beschrieben habe. Der Patient teilt Daten mit, der Arzt Ergebnisse der Befunderhebung und Diagnosen.

Unterhalb dieser Ebene des rationalen Gesprächs aber konstelliert sich die Ebene des Vertrauens, die zweite psychoanalytische Übertragungsebene, die „anaklitisch-diatrophische", die der „infantil-narzißtischen Ichanteile", die wir alle prototypisch in der frühen symbiotischen Kind-Mutter-Beziehung erleben und die in der Patient-Arzt-Beziehung wiederbelebt wird. Sie ist bei unserer Patientin so stark ausgeprägt, daß nach der ersten Beratung und gründlichen Untersuchung die Symptomatik zunächst verschwindet (siehe 2. Beratung). Dieses Phänomen – wir nennen es Suggestion – können wir ja immer wieder beobachten. Wir finden es nicht nur in der ärztlichen Sprechstunde, sondern genauso bei Kurpfuschern und Scharlatanen, „denn auf das Gemüt wirken könne jeder Prolet" (Ewald, zit. nach Schultz, 1952), vorausgesetzt, der Patient bringt das nötige Vertrauen auf.

Die 3. Beratung zeigt uns, daß sich eine Symptomverschiebung anbahnt. Die laute Symptomatik der 1. Beratung, die Herzbeschwerden und die Angst sind verschwunden. Das Aufsuchen des Arztes und die leichten Schwindelerscheinungen deuten jedoch darauf hin, daß die Patientin zwar oberflächlich beruhigt, ihr neurotischer Konflikt, ihre Angst aber nicht gelöst sind.

In der 4. Beratung ist dann, offenbar ausgelöst durch den Krebstod der Arbeitskollegin, der neurotische Grundkonflikt, nämlich die mit dem frühen Tod der Mutter zusammenhängende und nur unzureichend verarbeitete neurotische Problematik wieder voll aufgebrochen. Die Patientin bietet „in der Übertragung" erneut die szenische Information, das panisch verängstigte hilflose Kind, die dem Arzt jetzt nicht nur einen diagnostischen Zugang zum neurotischen Grundkonflikt, sondern auch seine deutende Bearbeitung ermöglicht. Der weitere Verlauf zeigt, daß hier offenbar der „dringlichste" Punkt getroffen wurde.

Was haben wir nun am Ende der 4. Beratung erreicht? Sind wir weiter als nach den vorhergehenden drei Beratungen, oder haben wir nach der akuten Verschlimmerung nur den status quo ante, wie er etwa zum Zeitpunkt der 2. Beratung bestand, wieder erreicht? Haben wir die Patientin gar von ihrer Neurose geheilt?

Um mit der zuletzt gestellten Frage zu beginnen: Nach unserem heutigen Wissen können wir eine neurotische Erkrankung, die auf primäre Traumen (Loch, 1970) in der Kindheit zurückzuführen ist und die zu entsprechenden psychischen Strukturveränderungen geführt hat, durch eine noch so treffsichere und gute Deutung nicht heilen. Dazu bedarf es einer langfristigen Durcharbeitung (Freud, 1914) der gesamten neurotischen Problematik, insbesondere der infantilen Neurose, die – wenn überhaupt – nur in einer psychoanalytischen Behandlung erfolgversprechend durchgeführt werden kann. Das heißt aber keineswegs, daß der Zustand der Patientin nach der 4. Beratung, in der eine konfliktbearbeitende Deutung vorgenommen wurde, mit ihrem Zustand nach den vorhergehenden Beratungen gleichzusetzen wäre, in denen lediglich (suggestiv) durch Vertrauen, durch die symbiotische bzw. die „anaklitisch-diatrophische" Ebene der Übertragung der Konflikt zugedeckt und damit die Symptomatik gebessert wurde. Wie labil dieses Gleichgewicht geblieben ist, zeigt nicht nur die Symptomverschiebung in der 3. Beratung, sondern auch die dramatische Exazerbation nach dem Tod der Arbeitskollegin in der 4. Beratung.

Der Unterschied nach der 4. konfliktbearbeitenden Beratung besteht gegenüber früher darin, daß die Ich-Funktionen der Patientin jetzt weniger eingeschränkt sind als vorher. Die Patientin ist jetzt imstande, ihre früheren ihr bisher unbewußten Ängste in ihre bewußte Wahrnehmung, Einsicht und Motorik einzubeziehen. Sie ist jetzt gesünder und gegen etwaige erneute belastende Auslösungssituationen widerstandsfähiger geworden.

17.5 Zusammenfassung

In den vorhergehenden Abschnitten haben wir das ärztliche Gespräch nach den Qualitäten der Information (objektive, subjektive und szenische) sowie nach den Dimensionen der Übertragung (Ebene des „fiktiven Normal-Ichs", der „infantil-narzißtischen Ichanteile" und der „neurotischen Objektbeziehungen") zu analysieren versucht. Abschließend wollen wir unser Augenmerk noch auf allgemeine Verhaltensweisen des Arztes richten, die weitgehend für das Glücken oder Mißglücken eines ärztlichen Gespräches verantwortlich sind.

Rogers und Mitarbeiter haben drei Verhaltensweisen von Psychotherapeuten ermittelt, die für das Zustandekommen eines psychotherapeutischen Erfolges von entscheidender Bedeutung sind. Es sind dies:

– Die Verbalisierung emotionaler Erlebnisinhalte der Patienten,
– die emotionale Wärme und positive Wertschätzung, die der Therapeut dem Patienten entgegenbringt, und
– die Echtheit bzw. Selbstkongruenz, die der Therapeut zwischen seinem Erleben, seinen Wertvorstellungen und seinen verbalen und nonverbalen Äußerungen herzustellen vermag.

Tausch drückt das folgendermaßen aus: „Je größer das Ausmaß angemessener Verbalisierung emotionaler Erlebnisinhalte von Klienten durch Psychotherapeuten, je größer das Ausmaß positiver Wertschätzung und emotionaler Wärme von Psychotherapeuten sowie innerhalb gewisser Grenzen das Ausmaß ihrer Echtheit-Selbstkongruenz, um so größer ist die Wahrscheinlichkeit gewisser konstruktiver Änderungen des Erlebens und Verhaltens von Klienten am Ende der Gesprächspsychotherapie und in der nachfolgenden Zeit" (1968). Rogers beschreibt die ideale Einstellung des Therapeuten wie folgt: „Am charakteristischsten: Der Therapeut ist imstande, vollkommen an der Kommunikation des Patienten teilzunehmen. Sehr charakteristisch: Der Therapeut befindet sich mit seinen Anmerkungen immer in Übereinstimmung mit dem, was der Patient mitzuteilen versucht. Der Therapeut betrachtet den Patienten als einen Mitarbeiter, mit dem er gemeinsam an einem Problem arbeitet. Der Therapeut behandelt den Patienten als seinesgleichen. Der Therapeut ist sehr gut imstande, die Gefühle des Patienten zu verstehen. Der Therapeut teilt durch die Modulation seiner Stimme seine vollständige Fähigkeit mit, die Gefühle des Patienten zu teilen" (1973). Dadurch hilft der Arzt dem Patienten, aus seiner individuellen Wirklichkeit in die soziale Wirklichkeit zurückzukehren (vgl. auch Kap. 1).

Hat der Arzt die richtige Grundeinstellung zum Patienten gefunden, dann wird ihm die Kenntnis der Qualitäten der Information und der Ebenen der Übertragung eine Richtschnur bieten, jedes ärztliche Gespräch zu strukturieren. Der kritische Leser wird hier wahrscheinlich einwenden, daß das ärztliche Gespräch ein so vielschichtiges und komplexes Gebilde ist, daß wir ihm Gewalt antun und es zu sehr vereinfachen, wenn wir es auf einige wenige Dimensionen reduzieren. Dieser Einwand ist prinzipiell richtig und gilt vor allem für den Erfahrenen und Geübten.

Wir wollen aber nicht vergessen, daß der Lernende doch vereinfachende Schemata braucht, um sich in der verwirrenden Vielheit der Interaktions-Phänomene zurechtzufinden. Wenn er bei entsprechender Grundeinstellung seine Aufmerksamkeit auf die Qualitäten der Information und die Ebenen der Übertragung richtet, dann weiß er, vergleichbar dem Chirurgen bei der Operation, in welcher Schicht er sich befindet, und er vermag, den Umständen entsprechend, zu entscheiden, bis zu welchem Ziel er das Gespräch fortzusetzen gedenkt.

18 Die Krankenvisite – Probleme der traditionellen Stationsarztvisite und Veränderungen im Rahmen eines psychosomatischen Behandlungskonzepts

Dirk Fehlenberg, Claudia Simons und *Karl Köhle*

18.1 Problemstellung: Die Visite als Gesprächssituation für Arzt und Patient

Diagnose wie Behandlung sind in einer psychomatisch-ganzheitlich praktizierten Medizin entscheidend auf die Verwirklichung eines intensiven Dialogs zwischen Arzt und Patient angewiesen.

Die anamnestischen Zielsetzungen sind hier weitreichend (vgl. Kap. 12; Adler und Hemmeler, 1988): Es geht darum, die Erkrankung als Teil der Lebensgeschichte, im Hinblick auf ihre Entstehung und auf die subjektiv geprägten Folgen zu verstehen und die „individuelle Wirklichkeit des Patienten" (v. Uexküll, 1982) zu erschließen. Darüber hinaus ist es wichtig, in welcher Art Patienten ihr Problem im Gespräch präsentieren, wie sie die Beziehung zum Arzt zu gestalten suchen. Die Arzt-Patient-Beziehung kann als Modell für andere Sozialbeziehungen des Patienten verstanden werden. In Verbindung mit biographischer Information ergibt sich für den Arzt ein differenziertes Verständnis des persönlichkeitsabhängigen Teils der Erkrankung, wenn er sich auf eine Reflexion der Beziehung einläßt (Balint, 1983).

Für jede medizinische Behandlung gilt, daß Gespräche, die die Erwartungen und Bedürfnisse des Patienten angemessen berücksichtigen, bereits selbst Entlastung schaffen können. Vor allem bilden sie eine notwendige Voraussetzung dafür, daß eine vertrauensvolle Arzt-Patient-Beziehung aufgebaut wird. Psychosomatische Medizin mißt dem ärztlichen Gespräch aber auch spezifischere therapeutische Aufgaben bei. Der Arzt kann unterstützend und korrigierend die Prozesse des Krankheitserlebens und der Krankheitsverarbeitung beeinflussen. Er kann dem Patienten die gewonnene Einsicht in die lebensgeschichtliche Bedeutung der Erkrankung rückvermitteln, um direkt oder indirekt eine Perspektive für eine psychosoziale Veränderung zu bahnen. Gespräche können auch unmittelbare Änderungen im psychischen wie im körperlichen Befinden des Patienten bewirken. Dieser Zusammenhang zwischen Arzt-Patient-Kommunikation und Befinden gilt wohl allgemein, in besonderer – manchmal dramatischer – Weise läßt er sich aber bei „klassischen" psychosomatischen Krankheitsbildern beobachten. Beispiele dafür sind die Verbesserung oder Verschlechterung der Asthmasymptomatik durch Gespräche, das Auftreten von Rezidiven bei Colitis-ulcerosa-Patienten sowie die Möglichkeit von Reinfarkten bei Herzinfarkterkrankungen durch Auslösung von Ängsten bzw. die Verhinderung von Rezidiven und die Stabilisierung des somatischen Heilungsprozesses durch therapeutisch geführte Gespräche.

Wir haben eine Zusammenfassung über die Bedeutung des Arzt-Patient-Dialogs deswegen an den Anfang dieses Kapitels über die Krankenvisite gerückt, weil sie genau die Perspektive bezeichnet, unter der wir die Visite[1] betrachten wollen: Welche Voraussetzungen für ein Arzt-Patient-Gespräch bietet die Visite? Wie lassen sich die Rahmenbedingungen für einen Dialog mit dem Patienten in der Visite verbessern? Gibt es Wege, die Visite für ein therapeutisches Gespräch zu nutzen?

Antworten auf diese Fragen zu finden, ist nicht zuletzt ein Erfordernis der klinischen Praxis: Die Visite ist die Hauptkontaktmöglichkeit für Krankenhausarzt und Patient. Mit einer Dauer von durchschnittlich ein bis zwei Stunden nimmt sie bis zu einem Viertel der täglichen Arbeitszeit des Arztes ein. Schließlich kommt ihr bei der Aus- und Weiterbildung von Medizinstudenten und Assistenzärzten eine herausragende Bedeutung im Hinblick auf den Erwerb sozialer Kompetenzen im Umgang mit Patienten zu.

18.2 Die Stationsarztvisite in ihrer traditionellen Form

18.2.1 Ergebnisse empirischer Forschung zur traditionell organisierten Krankenvisite

Die Krankenvisite ist eine vergleichsweise oft und methodisch vielfältig untersuchte Form medizinischer Kommunikation. Das Gespräch mit dem Patienten kommt in ihr zu kurz. In dieser Feststellung lassen sich alle Ergebnisse medizinsoziologischer,

---

[1] Wir beschäftigen uns hier nur mit der täglichen Stationsarztvisite. Oberarzt- und Chefarztvisiten, die im Stationsbetrieb andere Funktionen erfüllen, bleiben außer Betracht. Vgl. dazu v. Uexküll, 1977; Gück et al., 1981, 1983a, b.

medizinpsychologischer und kommunikationswissenschaftlicher Untersuchungen zusammenfassen (Übersicht bei Fehlenberg, 1983).

Quantitative Befunde der medizinsoziologischen und medizinpsychologischen Visitenforschung

Üblicherweise dauert eine Visite etwa 3,5 Minuten pro Patient. Während der Arzt einen Gesprächsanteil von etwa 60% hat, entfallen auf den Patienten nur 30%, der Rest verteilt sich auf die übrigen Teammitglieder (Pflegepersonal). Von den Patientenäußerungen ist der weitaus größte Teil wiederum reaktiv, d.h. sie bestehen zu ca. 80% aus Antworten auf Arztfragen. Umgekehrt stammt der überwiegende Teil der Fragen vom Arzt (82%); in einer durchschnittlichen Visite stehen 11 Arztfragen einer Patientenfrage gegenüber. 94% aller Unterbrechungen im Gespräch erfolgen durch den Arzt.

Krankheitsbezogene Informationen muß der Patient häufig (zu etwa 40%) in indirekter Weise dem Gespräch entnehmen, das das Team neben seinem Bett stehend über ihn führt. Die Information, die der Arzt dem Patienten direkt gibt, ist wiederum zum größten Teil durch die Arztinteressen bestimmt. Sie erfolgt zwei- bis dreimal so oft initiativ (Themenwahl nach den Prioritäten des Arztes) wie reaktiv (Themenwahl nach den Prioritäten des Patienten).

Besonders ungünstig sieht die Situation für schwerkranke Patienten aus. Bei ihnen steigt der Anteil indirekter Informationen noch einmal deutlich an. Während die Fragen leichter erkrankter Patienten in gut einem Drittel aller Fälle ausweichend beantwortet werden, erhalten Patienten mit infauster Prognose auf über 90% ihrer Fragen keine angemessene Antwort.

Es gibt auch viele Befunde zu Zusammenhängen zwischen dem Gesprächsverhalten von Arzt und Patient und zwischen Gesprächsmerkmalen und außerkommunikativen Variablen. Drei seien hier angeführt, die die Problematik der Situation besonders deutlich machen:

– Viele Arztfragen genauso wie ein häufiger Gebrauch unerläuterter Fachtermini gehen einher mit einer geringeren Redeaktivität des Patienten (quantitativ: Redeanteile, qualitativ: Fragen) und einer insgesamt kürzeren Visitendauer.
– Mit zunehmender Berufserfahrung des Arztes nimmt die Dauer seiner Visiten ab.
– Die einschätzungsmäßig erhobene „Patientenorientiertheit des Arztes", die mit einer Reihe von Sprachverhaltensmerkmalen positiv korreliert, wie längere Redezeit für den Patienten, weniger Fragen des Arztes, geht bei Patienten deutlich zurück, wenn sie mehrfach wegen einer Krankheit behandelt werden müssen.

Diese Zahlen zeigen den Arzt als die das Gespräch dominierende Person. Der Patient hat vergleichsweise geringe Einflußchancen. Je kränker er ist, desto schlechter stellt sich seine Situation dar.

Qualitative Ergebnisse gesprächsanalytisch-hermeneutischer Untersuchungen des Visitengesprächs

Neben dem rein quantitativ ausgerichteten Nachweis, daß die Patientenbelange in der Krankenvisite zu kurz kommen, haben sich „qualitative", gesprächsanalytisch-hermeneutische Untersuchungen[2] mit typischen Abwicklungsstrukturen beschäftigt, die zur Folge haben, daß Kommunikationsinteressen des Patienten übergangen werden. Dabei wurden drei Phänomengruppen beschrieben, die verhindern, daß der Patient – bei jeweils unterschiedlichen Graden aktiver Gesprächsbeteiligung – seine kommunikativen Interessen erfolgreich wahrnehmen kann.

Der Patient kommt nicht zu Wort: Was bewirkt den Ausschluß des Patienten aus dem laufenden Gespräch?

Mit dem fast vollständigen Ausschluß des Patienten aus dem Gespräch als Sprecher und als Adressat haben sich vor allem die Arbeiten von Nothdurft (1978, 1981, 1982) beschäftigt.

Nach den Kriterien einer „alltagsweltlichen Gesprächsmoral" müssen drei Bedingungen erfüllt sein, damit ein Zuhörer initiativ werden kann, so daß er zum aktiven Teilnehmer wird, ohne einen kommunikativen Konflikt zu riskieren: Die ablaufende Kommunikation muß beobachtbar, durchschaubar und absehbar sein. Nach den Ergebnissen von Nothdurft sind genau diese Kriterien für die zwischen dem Team stattfindende Kommunikation zumeist nicht erfüllt.

Unbeobachtbarkeit der Teambesprechungen wird beispielsweise dadurch erzeugt, daß die Lautstärke in den entsprechenden Passagen so weit abgesenkt ist, daß das Gespräch für den Patienten akustisch unverständlich wird. Weiter dadurch, daß am Bett eines Patienten im Wechsel über diesen oder auch über einen Nachbarpatienten gesprochen wird.

Undurchschaubarkeit, d.h. Unmöglichkeit, die aktuelle Gesprächsentwicklung mitzuverfolgen, stellt sich ein, wenn das Personal Jargon benutzt (vgl. auch Gück et al., 1983) oder in verkürzter, nur Insidern verständlicher Form kommuniziert. Die Kommunikation zwischen den Teammitgliedern ist darüber hinaus oftmals so verschachtelt und komplex in ihrer Themenentwicklung, sie wechselt so schnell über die verschiedenen Teammitglieder hinweg, daß der Patient einerseits dem thematischen Fokus nicht folgen kann und andererseits „einstiegsrelevante" Stellen nicht entdeckt.

Die folgende Passage bietet dafür ein typisches Beispiel (wiedergegeben bei Nothdurft, 1982, S. 28):

A 1:	Schreiben Sie mal auf, bitte: Äh, was wollte er noch mal haben?
MA:	Schädel
A 1:	Kreuzbein
MA:	Kreuzbein
A 1:	Kreuzbein, äh, Femur beiderseits und äh Tibia beiderseits und Schädel. Und machen wir noch mal die alkalische Phosphatase bei ihr (blättert in der Kurve). Bis jetzt nie gemacht,

> ja, machen wir mal die alkalische Phosphata-
> se und Bili dazu.
> **S:** Was war das jetzt noch „Kreuzbein"?
> **A 1:** Kreuzbein, Tibia, Femur und äh Schädel. Und
> dann die Atmung, a.P. und Bili. Ach, machen
> wir gleich die ganze Leber mit.
> **MA:** Die Prostata auch?
> **A 2:** (lacht)
> **A 1** = Arzt 1, **A 2** = Arzt 2, **MA** = Medizinalassistent,
> **S** = Schwester.

Der Patient beteiligt sich am Gespräch, kann aber nicht initiativ werden

Selbst wenn der Patient in das laufende Visitenge-spräch einbezogen wird, sei es als Adressat oder sei es, daß er selbst Beiträge liefert, geschieht dies viel-fach nur im Dienste medizinischer Aufgabensetzun-gen, nicht im Sinne persönlicher Kommunikation. Besonders deutlich wird das in einer Analyse von Gück et al. (1981, 1983) aus dem Bereich der Inten-sivmedizin. Die hier gemachten Beobachtungen las-sen sich vielleicht nicht im vollen Umfang auf die Regelvisite im üblichen Stations-Setting übertragen. Man kann aber davon ausgehen, daß die Bedingun-gen intensivmedizinischer Versorgung Auswirkungen auf den Patienten deutlicher sichtbar machen, die auch im alltäglichen Klinikbetrieb latent vorhanden sind. Die Autoren machen bei Oberarztvisiten auf ei-ner Berliner Intensivstation die Beobachtung, daß das am Bett stehende Team den Patienten in kurzen Frage-Antwort-Sequenzen in das ansonsten nur zwi-schen dem Medizinpersonal laufende Gespräch ein-bezieht. Dabei wird der Adressatenwechsel hin zum Patienten nicht, wie eigentlich erwartbar, deutlich markiert. Weiter werden die Patientenreaktionen nicht quittiert, der Patient erhält keine Rückmeldung vom Team. Thematisch gelten diese Frage-Antwort-Sequenzen hauptsächlich dem Komplex „Befinden des Patienten". Sie stehen typischerweise in Diensten der Behandlungskontrolle und sind sachlich durch die Abwicklung des Teamdiskurses motiviert. Das Team unterstellt also offensichtlich, daß der Patient dem für ihn größtenteils unverständlichen Gespräch zwischen den einzelnen Teammitgliedern mit ständi-ger Aufmerksamkeit folgt und sozusagen ständig be-reit ist, als Lieferant von Informationen zur Verfü-gung zu stehen, ohne selbst eigene Kommunikations-interessen einbringen zu können. Die Autoren kenn-zeichnen die beschriebene Situation als „permanente kommunikative Verfügbarkeit" des Patienten – eine besonders erschreckende Situation angesichts der auf intensivmedizinischen Stationen ohnehin unum-gänglich gegebenen körperlichen Verfügbarkeit des Patienten.

Quasthoff-Hartmann (1982) hat gezeigt, daß auch in einem Gespräch, in dem sich längere erklärende Diskurseinheiten direkt an den Patienten richten, der Arzt unter Ausnutzung konversationstechnischer Mittel (den Regeln der Abwicklung des Sprecher-wechsels; vgl. Sacks et al., 1978) weitergehende Akti-vitäten des Patienten zu verhindern weiß. An Stellen,

wo der Patient legitimerweise initiativ werden könnte (etwa wenn der Arzt eine längere Erklärung abge-schlossen hat), verhindert der Arzt eine Patientenak-tivität dadurch, daß er übergangslos ein weiteres komplexes Ablaufmuster initiiert; oder er bindet die Patientenaktivität, indem er eine Frage anschließt, deren Beantwortung Vorrang beansprucht.

Der Patient wird initiativ, hat damit aber keinen Er-folg beim Personal

Bereits in den quantitativen Untersuchungen zur Vi-sitenkommunikation wurden „transaktionsanalyti-sche" Merkmale (Siegrist, 1978, 1982) herangezogen, die verschiedene Typen des ausweichenden oder ab-wehrenden Gesprächsverhaltens von Ärzten gegen-über Patienteninitiativen klassifizieren, insbesondere im Gegenzug zu Fragen nach krankheitsbezogener Information: das „Nichtbeachten", der „Themen- oder Adressatenwechsel", der „Beziehungskommentar", die „Mitteilung funktionaler Unsicherheit". Wir geben hier zwei typische Beispiele für ein Ausweichen dadurch, daß der Arzt von der Sachebene auf die Beziehungs-ebene des Gesprächs wechselt, aus dem Material von Siegrist wieder:

> Eine schwerkranke Patientin versucht, den Zeitpunkt ihrer Entlassung zu erfahren.
> **Patientin:** Das geht bestimmt noch lange, bis ich ... noch ein paar Wochen, daß ich da oben bin. Nicht?
> **Arzt:** Wir wollen Sie ja nicht unnötig plagen? (Siegrist, 1982, S. 19).
>
> Im Anschluß an eine Röntgenuntersuchung fragt ein Patient den Arzt:
> **Patient:** Herr Doktor, haben Sie eine Vermutung, was es sein könnte?
> **Arzt:** Ich vermute nicht, ich sammle Fakten! (Siegrist, 1982, S. 19).

Eine vertiefte Behandlung hat der Komplex der Ent-wertung von Patienteninitiativen in den Arbeiten von Bliesener (1980a, b, 1982a) erfahren. Auf seiten der Patienten werden vor allem zwei Typen von Initiati-ven gebraucht:
– Das **Erzählen.** Der Konflikt zwischen Patient und Arzt resultiert bei Erzählungen – sofern sie überhaupt ansatzweise zugelassen werden – daher, daß der Arzt an den entsprechenden Stellen gemäß den für ihn im Vordergrund stehenden medizinischen Erfordernis-sen einen Bericht erwartet und sein Rückmeldungs-verhalten entsprechend ausrichtet. Das vom Patien-ten initiierte Muster – Erzählung[3] – wird in den Reak-

---

2 Wir benutzen den Terminus „gesprächsanalytisch" hier als Oberbegriff für verschiedene Arten der qualitativ verfahrenden mikrostrukturellen Kommunikationsanalyse, wie (ethnometho-dologisch) Konversationsanalyse oder Diskursanalyse usf.
3 Diese Spannung zwischen beabsichtigtem Erzählen durch den Patienten mit dem Zweck der Solidaritätsherstellung und erwartetem Bericht durch den Arzt mit dem Zweck von Infor-mationstransfer scheint auch in anderen Bereichen der Arzt-Patient-Kommunikation oder Berater-Klient-Interaktion ein grundlegendes Problem zu bilden (vgl. Quasthoff, 1979).

tionen des Arztes nicht aufgenommen (insbesondere fehlen expansionsstimulierende Momente), sondern durch Rückfragen, Minimalreaktionen und ähnliches zum Erliegen gebracht. Wie Bliesener (1980a) zeigen konnte, gelingt es Patienten dadurch, daß sie einerseits das Erzählmuster charakteristisch verkürzen und andererseits parallel das vom Arzt erwartete Muster bedienen, beide Kommunikationszwecke wenigstens rudimentär zu verwirklichen. Ein Phänomen, das, wie Bliesener betont, dafür spricht, daß der ursprüngliche Konflikt institutionelle Zielsetzungen versus individuelle Zielsetzungen vom Patienten zum innerpsychischen Zielkonflikt internalisiert wird.
– Die zweite Form von Patienteninitiativen sind **krankheits-** und **behandlungsbezogene Fragen.** Damit eine Frage Aussicht auf eine angemessene Reaktion hat, bedarf es in der Visite offensichtlich der besonderen Vorbereitung durch den Patienten. Bliesener (1980b) unterscheidet hier zwei Strategien: die Plazierung und die Lancierung. Plazieren einer Frage heißt, der Patient bringt seine Initiative an geeigneten Stellen des Gesprächs ein (etwa wenn er bereits in das Gespräch einbezogen ist, wenn er einer Antwortverpflichtung nachgekommen ist, wenn seine Frage thematisch „paßt", wenn gerade kein dringenderes Thema behandelt wird). Lancierung bedeutet dagegen, der Patient bereitet seine Initiative aktiv vor (etwa durch „direkte Kommentare", Vorfragen, Sondierungen, Ankündigungen usw.). Beide Strategien sind mit Risiken verbunden. Die erste erfordert vor allem ein hohes Maß an mentaler Aktivität und enthält die Gefahr, daß einfach keine passende Stelle auftritt; die zweite birgt durch ihre oft umständliche Vorbereitung die Gefahr, zu früh erkannt zu werden und damit auch zu früh angreifbar zu sein.

Am Siegristschen Material hat Bliesener 12 lokale Strategien der Abweisung von Patienteninitiativen herausgearbeitet („Abriegeln, Überfahren, Hinhalten, Leerlaufenlassen, Abwinken, Stillegen, Problematisieren, Abbiegen, Verlagern, Filibustern, Abgleiten, Sich-Rausreden"), die sich durch Elaboriertheit ihrer Planungsform und ihren zu unterstellenden Effekt auf den Patienten voneinander unterscheiden.

Als Beispiel für eine komplexere, in diesem Fall aus zwei Schritten bestehende Strategie, eine Patienteninitiative zu entwerten, sei hier der für die Strategie „Abbiegen" von Bliesener zitierte Text wiedergegeben:

A: Nur die eine Aufnahme. Wissen Sie, die Lunge kann man so schlecht …
P: Was ist es denn überhaupt?
A: … die Lunge kann man so schlecht beurteilen, wenn man liegt, gell?
P: Herr Doktor, gestern hätte ich bald einen Herzschlag gekriegt!
A: Warum denn? Warum haben Sie denn einen Herzschlag bekommen?
P: Sie haben mir gerade wollen eine Spritze geben.
A: ICH habe Ihnen eine Spritze gegeben?
P: Nein. Sie nicht, die Schwestern.
A: Was hat denn die Schwester gesagt?
P: (unverständlich)

A: Ne, Valium?
(Bliesener, 1982, S. 162)
A = Arzt, **P** = Patient.

Die Zeile 6 stellt einen typischen Auftakt für das Erzählen einer Geschichte dar. Der Arzt ermuntert die Patientin im Gegenzug auch zuerst mit seiner Reaktion. Nachdem die Patientin aber gerade mit der Erzählung begonnen hat (Zeile 10), unterbricht er sie mit einem Korrekturhinweis (11). Dieser Korrekturhinweis entwertet zwar nicht die Initiative, zwingt die Patientin aber zu einer „Selbst-Berichtigung", die das Thema in nicht beabsichtigter Weise präzisierend ausweitet. Der Arzt nimmt nun genau diesen auf seinen Korrekturhinweis erst eingeführten thematischen Aspekt in seiner nächsten Frage auf (13), womit er suggeriert, daß die Schwester das eigentlich aktuelle Thema sei. Das Thema der Patienteninitiative ist durch die Reaktion des Arztes soweit verändert („abgebogen") worden, daß es – wie Bliesener berichtet – der Patientin in der restlichen Visite nicht gelingt, bis zum Kern ihrer beabsichtigten Initiative, der Erzählung mit dem Thema der Beunruhigung, vorzudringen (für eine detailliertere und systematischere Beschreibung dieses Falles und weitere Textbeispiele und Analysen zu den anderen Strategien sei auf Bliesener (1982) verwiesen).

Allen 12 Abweisungsstrategien ist gemeinsam, daß sie auf eine unmittelbare lokale Unterbindung der Patienteninitiative abzielen. Sie lösen den grundsätzlichen Konflikt zwischen den Informations- und Mitteilungsbedürfnissen des Patienten einerseits und den Erfordernissen des medizinischen Arbeitsablaufs andererseits kurzfristig zugunsten der institutionellen Aufgabe. Damit bergen sie aber, obwohl oder gerade weil sie kurzfristig meist erfolgreich sind, die Gefahr der Eskalation in Folgekonflikten. Bliesener und Siegrist (1981) sprechen hier von Konflikten zweiter Ordnung, in denen die Motivation des Patienten von der des ursprünglichen Ausgangskonflikts weitgehend losgelöst ist. So finden sich als Reaktionsstrategien etwa „ausufernde Selbstdarstellung", „forciertes Rechthabenwollen" oder „Boykottieren durch minimale Reaktion" – ein Mittel, von dem auch eine Analyse von Gück et al. (1981) gezeigt hat, daß es offenbar die ultima ratio der Gesprächssteuerung für den Patienten darstellt. Diese Reaktionsstrategien sind dann weniger auf die nachträgliche Durchsetzung der abgewiesenen Initiativen gerichtet als vielmehr auf globalere Ziele, wie einen Ausgleich für eine persönliche Verletzung zu erlangen. Die Ärzte reagieren in Konflikten dieser Art, die die Beziehung umfassender in Frage stellen, anders als angesichts von Patienteninitiativen. Sie verwenden „Begründungen/Rechtfertigungen", machen „Versprechungen", „Zugeständnisse", schaffen „Entschädigungen" und stellen „Familiarität über Scherzen" her – Reaktionsstrategien, die als Konflikt-Befriedungsstrategien zusammengefaßt werden können (Bliesener und Siegrist, 1981). Wobei allerdings gilt, daß diese Strategien zwar auf einen Ausgleich zielen, oft auch erfolgreich sind, am zugrundeliegenden Konflikt zwischen Patientenbedürfnissen und der durch die Institution abverlangten passiven Rolle aber nichts zugunsten des Patienten verändern.

18.2.2 Ein zusammenfassender Problemaufriß: die traditionelle Krankenvisite – das gescheiterte Arzt-Patient-Gespräch

Durchgängig belegen die zitierten Untersuchungen die eingangs getroffene Feststellung: Das Gespräch mit dem Patienten kommt in der Krankenvisite zu kurz. Typische Kommunikationsmuster der Visite scheinen geradezu daraufhin angelegt zu sein, ihn aus dem Gespräch auszuschließen oder doch zumindest seine Kommunikationsinteressen einzugrenzen. Woran liegt das?

Zwei Perspektiven bieten sich zur Erklärung an. Die erste nimmt eine allgemeine Paradigmakritik medizinischer Theorie und Praxis auf. Die zweite wird durch eine organisationssoziologische Perspektive charakterisiert. Beide Erklärungsansätze ergänzen sich. Sie liefern eine allgemeine und eine institutionsspezifische Erklärung für die beobachtete Praxis.

Die Kritik an einer vorrangig und einseitig biomechanischen Ausrichtung der neuzeitlichen Medizin ist mittlerweile von Autoren mit sehr unterschiedlichen Standpunkten dargelegt worden. In der Öffentlichkeit wurden dabei vor allem negative und problematische Konsequenzen neuer medizinischer Technologien diskutiert (von den Kostenauswirkungen bis zu den inhumanen Behandlungskonsequenzen der die Grenzen des Machbaren immer weiter hinaus verlagernden Entwicklungen der „Apparatemedizin"). Weniger Resonanz hat dagegen die subtile und doch ungleich fundamentalere Kritik gefunden, die auf den Zusammenhang von medizinischer Alltagspraxis – wie sie sich beispielsweise in der sozialen Mikroorganisation („Mikropolitik") eines Arzt-Patient-Gesprächs spiegelt – und dem naturwissenschaftlichen Paradigma in den medizinischen Grundlagenwissenschaften aufmerksam gemacht hat. Die Arbeiten von Mishler (1981, 1984) zeigen anhand von genauen Analysen von Ambulanzgesprächen, daß sich der Arzt-Patient-Dialog als dialektische Auseinandersetzung zweier kontroverser Realitätsorientierungen verstehen läßt. Arztseitig wird die Gesprächsentwicklung durch Themen bestimmt, deren Relevanz sich aus dem biomedizinischen Krankheitsmodell ableitet, das einer möglichen Behandlung zugrunde liegt. Symptome müssen beispielsweise möglichst objektiv (d.h. auch nicht-individuell) beschrieben werden, um zu Syndromen zusammengefaßt und diagnostischen Klassifikationen zugrunde gelegt werden zu können. Patientenseitig bilden dagegen subjektive und damit individuell unterschiedlich erfahrene Erlebenszusammenhänge (Beschwerden, Einschränkungen usw.) den Anlaß zur Konsultation. Was heißt das für das Gesprächsverhalten? Der dem Ideal einer angewandten Naturwissenschaft verpflichtete Mediziner steht vor der Notwendigkeit, die vom Patienten angebotene psychosoziale Information (aus dem biomedizinischen Blickwinkel von „überschüssiger" und nicht relevanter Bedeutung) aus dem Gespräch zu eliminieren. Sagt ein Patient auf die Frage nach dem Beginn einer Beschwerde etwa, „seit Frühjahr, als meine Frau starb", so wird der Arzt allein die zeitliche Information verwenden, rückrechnen und feststellen, daß die Magenschmerzen vor ca. 6 Monaten zum ersten Mal auftraten. Noch weniger wird er psychosoziale Hintergründe aufgreifen, wo sie nur indirekt erwähnt sind, oder von sich aus thematisieren, wenn Patienten sich selbst einer biomedizinischen (Laien-)Sprache bedienen.

Was weiß man über die Abwicklungsprinzipien solcher Arzt-Patient-Gespräche? Konversation allgemein, d.h. nicht-zweckgebundene Alltagsgespräche, wird zum großen Teil in Beitragspaaren abgewickelt. Auf einen Gruß folgt ein Gegengruß, auf eine Frage eine Antwort usw. Die Gesprächspartner haben prinzipiell gleiche Rechte und Pflichten, Initiativen und zugehörige reaktive Paarteile zu benutzen. Für Arzt-Patient-Dialoge gelten demgegenüber, wie für andere zweck- und institutionsgebundene Kommunikationsformen, unterschiedliche Abwicklungsregeln (Fisher, 1984; Todd, 1984; Mehan, 1979). Die Kommunikation vollzieht sich in dreiteiligen Mustern: Einer Arztinitiative (etwa einer Frage) folgt eine Patientenreaktion (eine Antwort), die nachfolgend vom Arzt kommentiert bzw. quittiert wird. Die Beziehung zwischen den drei Schritten einer solchen Sequenz muß hierarchisch gedacht werden:

I. A.: Initiative   II. P.: Antwort   III. A.: Kommentar

Beispiel (übersetzt aus Todd, 1984): I. Arzt: „Also, Sie haben seitdem keine Blutung mehr gehabt?" II. Patientin: „Nein." III. Arzt: „In Ordnung..." Der dritte Schritt des Kommentars ist auf die vorausgegangene vom Arzt initiierte Sequenz bezogen. Sein Vorhandensein ist der interaktive Ausdruck der vorhandenen Asymmetrie. Er erlaubt dem Arzt, die Angemessenheit des Gesprächsablaufs zu kontrollieren und garantiert ihm am Ende des Musters das Rederecht, was er zur Initiierung einer weiteren 3-Schritt-Sequenz nutzen kann. Diese Form der Gesprächskontrolle ist (zumindest für viele Gesprächsaufgaben) struktureller Art, d.h. sie ist notwendig, um die Zwecke eines Experte-Klient-Dialogs zu erreichen. Der an einer biomedizinisch ausgerichteten Behandlung und Befunderhebung orientierte Arzt setzt dieses Abwicklungsmuster ein, um den psychosozialen „Bedeutungsüberschuß" der Patientenäußerungen auszublenden. Mishler verwendet dafür den anschaulichen Begriff des „context-stripping". Nur die für eine biomedizinische Konzeptualisierung relevanten Teile der Äußerung des Patienten werden als angemessen quittiert und in Folge-Initiativen vom Arzt weiter verfolgt. Im allgemeinen kooperieren Patienten mit ihrem Ge sprächsverhalten, um den Zweck der Kommunikation nicht grundsätzlich zu gefährden. Auf den ersten Blick wirken typische Arzt-Patient-Gespräche insofern geordnet und konfliktfrei. Eine subtilere Analyse zeigt aber, daß es auch eine thematische Kohärenz zwischen den vom Patienten als scheinbare Überschußinformation eingebrachten „Einsprengseln" gibt. Der Patient verfolgt offensichtlich ein eigenes Gesprächsskript. Die strukturell verankerte Kontrolle des Arztes garantiert zwar eine reibungsfreie Abwicklung des Gesprächs Zug um Zug, die Gesamtsituation ist aber richtiger beschrieben, wenn man sagt, daß das Gespräch eine Auseinandersetzung zwischen zwei Realitätsorientierungen darstellt: einer technokratischen, die inhaltlich gesehen am biomedizinischen Modell orientiert ist – sie bildet das Gesprächsskript des Arztes – und einer an alltagsweltlichen Sinnzusammenhängen (symbolisch) orientierten, die die eigeninitiativen Gesprächsanteile des Patienten kennzeichnet. Die beschriebene Form der Gesprächsabwicklung in 3 Schritten ist der Garant dafür, daß die technokratisch-biomedizinische Orientierung situationsbestimmend bleibt.

Für die Visite läßt sich feststellen, daß die im vorausgegangenen Abschnitt beschriebenen Abschottungs- und Abweisungsstrategien ebenfalls die Durchsetzung einer biomedizinischen Zweckausrichtung der Visitenkommunikation garantieren. Sie alle stellen Strategien zur Ausblendung individuell bestimmter Sinnbezüge des Patienten dar. Sie sind im Vergleich zur Zwei-Personen-Situation im Ambulanzgespräch zum Teil komplexer, weil die zweckrationale Orientierung hier von mehr als nur einer Person getragen wird. Diese letzte Feststellung führt zur zweiten, der organisationssoziologischen Erklärungsweise.

Die Behandlungspraxis in einer modernen Klinik ist in hohem Ausmaß durch Arbeitsteilung bestimmt. Die von unterschiedlichen Personengruppen für einen Patienten zu erbringenden Behandlungsleistungen müssen deswegen fortlaufend organisiert, koordiniert und kontrolliert werden. Die Visite übernimmt einen großen, zumeist den überwiegenden Teil dieser teaminternen Koordinierungsaufgaben. Dazu gehören vor allem:

- das ärztliche Konsil, d.h. die ärztliche Fachdiskussion über Diagnose, Diagnosemaßnahmen und über die Therapie,
- die Anordnung von diagnostischen und therapeutischen Maßnahmen gegenüber dem Pflegepersonal sowie
- an akademischen Lehrkrankenhäusern zusätzlich die Funktion der Ausbildung für Medizinstudenten.

Die Abwicklungsmuster der Visite müssen insofern eine doppelte Aufgabe erfüllen. Einerseits müssen sie einen – für die biomedizinische Zweckorientierung – störenden „Überschuß" an alltagsweltlichen Lebensbezügen aus dem Gespräch heraushalten. Andererseits müssen sie ein effizientes Funktionieren der teaminternen Kommunikation gewährleisten und diese gegen eine störende Beteiligung des Patienten

abschirmen. Im Ergebnis kommt es zu einer klaren Prioritätensetzung: Die arbeitsorganisatorischen Aufgaben und damit das Arzt-Team-Gespräch strukturieren den Gesamtablauf der Visite. Der Patient wird hauptsächlich dann beteiligt, wenn es um körperbezogene Informationen geht, die zur Abwicklung der Teamkommunikation unerläßlich sind. Die (alltagsweltlich bestimmten) Patienteninteressen erfahren dagegen eine Zurücksetzung durch Nicht- oder Minimalbeachtung. So gibt es für eine direkte Information des Patienten in der Visite keinen systematischen Ort. Zeitpunkt, Art und Ausmaß ihrer Durchführung liegen – sofern sie nicht von einer Initiative des Patienten eingeklagt werden – im Belieben des Arztes. Dieser Bereich hat die niedrigste Priorität. Die Behandlung von Patienteninitiativen zeigt, daß eine personenbezogene Arzt-Patient-Kommunikation tatsächlich nur in Minimalform vorliegt – und im wesentlichen nur erfolgt, um die Patientencompliance zu sichern.

Angesichts dieser für den Patienten negativen Ergebnislage haben sich viele der Autoren der dargestellten Untersuchungen bemüht, auf der Basis ihrer empirischen Analysen Alternativen zur herkömmlichen Praxis zu entwickeln. Die Diskussion der dabei beschrittenen Wege soll unseren Überblick zur Forschungslage für das traditionelle Setting abschließen.

Der vom medizinsoziologischen Ansatz beschrittene Weg der Veränderung geht dahin, im Konflikt die Position des Patienten zu stärken. Gemäß der wissenssoziologischen Analyse der Problemlage, „Patient im Krankenhaus zu sein", soll diese Stärkung vor allem über eine Versorgung des Patienten mit adäquater und ausreichender Information erfolgen. Diese soll dem Patienten dann eine Bewältigung (coping) der Anforderung der Kranken- und Krankenhaussituation aus eigenem Handlungsvermögen erlauben. Eine erfolgreiche kognitive Bewältigung schließt in dieser Sicht die Bewältigung emotionaler Probleme mit ein. Praktisch erprobt wurden mit dieser Intention ein Aufklärungsheft für Krankenhauspatienten, das an Patienten ausgegeben wurde, Informationen bereitstellte und zu aktivem Verhalten anregen sollte, sowie ein „Patientenfunk", der entsprechende Ziele über ein für den Zimmerrundfunk produziertes Programm verfolgte. Eine unter Evaluationszielsetzungen durchgeführte Befragung (von Troschke und Siegrist, 1977) zeigt einerseits, daß eine solche Informationshilfe von den Patienten grundsätzlich begrüßt wird, andererseits aber auch eine deutlich zurückhaltendere Beurteilung durch das Medizinpersonal. Damit zeichnet sich die Problematik dieses Weges ab: Er dürfte nur da erfolgreich sein, wo ein solches Informationsheft als vom Personal aktiv genutztes (zusätzliches) Mittel in einem patientenzentrierten Setting eingesetzt wird. Andernfalls besteht die Gefahr, daß hier eher Konfliktpotentiale geschaffen als abgebaut werden.

Die medizinpsychologischen und gesprächsanalytischen Autoren haben demgegenüber konkret am Sprachverhalten ansetzende Wege direkter und indirekter Veränderungsmöglichkeiten betont. Einerseits sollten die verwandten sprachlichen Defizit-Indikatoren direkt in Trainingsziele für kommunikatives Verhalten umgesetzt werden (medizinpsychologischer Ansatz). Dazu ist festzustellen, daß die erfaßten Merkmale kaum eine ausreichende Beschreibung aller sprachlichen Mittel liefern, die jedem kompetenten Sprecher zur Verfügung stehen, um beispielsweise die Themenentwicklung eines Gesprächs zu beeinflussen oder die Initiative eines Partners abzuwehren oder zu fördern. Trainingsziele wie „Unterbrich nicht!", „Frage weniger festlegend!", „Gib direkte und nicht ausweichende Antworten!", „Mache längere Visiten!", „Gebrauche weniger Fachtermini!" usw. sind sicher übbar, aber man kann unterstellen, daß ihre Verwirklichung allein noch keine die Patienteninteressen befriedigende Visitenführung sicherstellen wird. Der von den gesprächsanalytischen Autoren hergestellte Praxisbezug ist dagegen indirekter Art. Sie sehen ihn vor allem in der Sensibilisierung der beteiligten Ärzte, die durch Kenntnis der wissenschaftlichen Analyse ein tieferes Verständnis von den Auswirkungen der eigenen (unreflektierten) Gesprächspraxis erlangen sollen (vgl. Quasthoff-Hartmann, 1982; Bliesener, 1982a). Dazu bietet die ablauforientierte qualitative Textanalyse tatsächlich die besseren Voraussetzungen. Der hier verwandte Beschreibungsapparat ist auch aus der Perspektive potentiell beteiligter Interaktanten bedeutungshaltiger. Allerdings muß man sich darüber im klaren sein, daß individuelles Lernen einzelner Ärzte zwar wichtig ist, allein aber noch keine ausreichende Veränderung der Visitensituation erwarten läßt. Gerade in den Arbeiten von Gück et al. (1981, 1983a, b) sowie in den Erklärungsansätzen der älteren soziologischen Arbeiten wird deutlich, daß das vielgestaltige Übergehen von Patienteninteressen in der Visitenkommunikation durchaus funktional im Sinne der ungestörten Abwicklung der institutionsgebundenen Aufgaben ist. Diese wiederum besitzen ihre Fundierung in einer durchgehend technokratisch orientierten, am biomedizinischen Krankheits- und Behandlungsmodell ausgerichteten Praxis. Die Idee einer Veränderung durch (individuelle) Sensibilisierung kann höchstens für solche Kommunikationseigenschaften als erfolgversprechend gelten, die für den institutionellen Kontext nicht funktional oder zumindest nicht in zentraler Weise funktionsstabilisierend sind.

Eine Veränderung der Visiteninteraktion in Zielrichtung auf Patientenzentriertheit bedarf deswegen eines komplexen Ansatzes, der eine Erweiterung des allgemeinen medizinischen Behandlungskonzeptes auf psychosoziale Faktoren ebenso einschließt wie die organisatorische Umgestaltung der Visite und die Bereitschaft der beteiligten Ärzte und des Pflegepersonals, ihr eigenes Verhalten der kritischen Reflexion auszusetzen und im Sinne dieser Zielsetzung zu verändern. Im folgenden Abschnitt werden wir über die Erfahrung mit einem solchen Ansatz auf der Basis eines psychosomatisch-ganzheitlichen Krankheits- und Behandlungskonzeptes berichten.

18.3 Die Veränderung der Stationsarztvisite unter den Zielsetzungen einer psychosomatisch-ganzheitlichen Behandlung

Mit diesem Erfahrungsbericht beziehen wir uns auf die an der Universitätsklinik Ulm im Department für Innere Medizin in den Jahren 1972 bis 1979 geführte internistisch-psychosomatische Krankenstation und die hierzu durchgeführte Begleitforschung (Projekt B 5 SFB 129, Universität Ulm). Dabei handelte es sich um eine internistische Allgemeinstation mit 15 Betten, deren Belegung der einer typischen internistischen Station einer Universitätsklinik entsprach. Der konzeptuelle Hintergrund, die Stationsstruktur und Informationen zur Ausbildung der Ärzte und des Pflegepersonals sind in Kapitel 28 dargestellt (vgl. auch ausführlicher Köhle et al., 1980). Hier werden wir unsere Erfahrungen und Forschungsergebnisse zu der auf dieser Station erprobten Form der Visite darstellen. Wir sprechen im folgenden von der „psychosomatischen Visite", wobei man sich aber vor Augen halten muß, daß es sich um eine unter psychosomatischen Zielsetzungen geführte internistische Visite handelt. Weiter ist zu berücksichtigen, daß die Veränderung der Visite nur ein – allerdings entscheidendes – Teilelement im Rahmen einer umfassenden patientenzentrierten Umstrukturierung war.

18.3.1 Ein Konzept für eine patientzentrierte Visitenführung

Vermehrte Patientenzentriertheit der Visite sollte durch eine Reihe von Veränderungen erreicht werden, die sowohl organisatorisch-strukturelle Aspekte wie auch Änderung des individuellen Interaktionsverhaltens der beteiligten Stationsärzte und des Pflegepersonals umfaßten. In der Praxis stehen diese beiden Aspekte in einem gegenseitigen Bedingungsverhältnis; für den Versuch, eine Änderung reflektiert durchzuführen, und mehr noch für die Planung eines begleitenden und evaluierenden Forschungsansatzes, erweist es sich aber als notwendig, eine analytische Unterscheidung zu treffen. Dies bietet den Vorteil, konkrete Zielkriterien zu bestimmen, an denen sich die Beteiligten orientieren können und deren Erreichen später empirisch überprüfbar ist. Für die Änderung der Visite haben wir drei Zielbereiche unterschieden: die den allgemeinen Ablauf betreffende funktionale Entflechtung und zwei das Interaktionsverhalten der Ärzte betreffende Ziele, Symmetrie im Gesprächsverhalten und (psycho-)therapeutische Unterstützungsfunktionen.

Funktionale Entflechtung: Übergang vom Mehrgruppengespräch zum Arzt-Patient-Dialog

Bevor man an ein patientenzentriertes und therapeutisch unterstützendes Gespräch in der Visite denken kann, muß man das Arbeitsprogramm so verändern, daß die Visite von einem Mehrgruppen- zu einem Arzt-Patient-Gespräch wird. Die Visite am Bett des Patienten soll idealerweise ausschließlich für den Arzt-Patient-Dialog und die körperliche Untersuchung, also für die patientenbezogenen Funktionen der traditionellen Visite, genutzt werden. Andere Aufgaben (beispielsweise „Kurven-Visite", Supervisionsaufgaben und alle weiteren institutions- und arbeitsorganisatorischen Aufgaben) werden vor und nach der Kernvisite und außerhalb des Krankenzimmers wahrgenommen. Patientenzentriertheit im Sinne dieses ersten Zielbereichs bedeutet damit: Der behandelnde Stationsarzt wendet sich am Krankenbett idealerweise ausschließlich dem Patienten zu. Die Absicht, die Visite als ein Arzt-Patient-Gespräch zu führen, wird durch die räumliche Anordnung der Teilnehmer unterstrichen: Der visitenführende Arzt setzt sich üblicherweise auf einen Stuhl direkt an das Bett des Patienten. Er befindet sich damit in natürlicher Gesprächsposition (gleiche Augenhöhe wie ein im Bett sitzender Patient). Die übrigen Teammitglieder stehen dagegen mit einem größeren Abstand am Fußende des Bettes.

Weitere Faktoren aus der allgemeinen Arbeitsorganisation der Station wirken sich hierbei unterstützend aus:
– Die Ärzte der Station teilen sich die Zuständigkeit für die Patienten auf. Soweit es der Dienstplan erlaubt, wird die Visite – unter Begleitung der noch an der Versorgung des Patienten beteiligten Teammitglieder – täglich vom jeweils zuständigen Arzt geführt.
– Die Visite findet zu einer festgelegten Zeit statt. Dies unterstreicht einerseits ihren Stellenwert als eine zentrale Behandlungsmaßnahme und hilft die Teilnahme des gesamten Teams zu sichern. Andererseits erhält der Patient dadurch das Gefühl, eine sichere und verläßliche Kontaktmöglichkeit zu seinem Arzt zu besitzen; nicht zuletzt wird es ihm leichter, sich auf die Visite vorzubereiten, beispielsweise sich Fragen und Probleme zu überlegen.
– Der Zeitraum für die Visite ist mit ca. zwei Stunden (für 15 Patienten, einschließlich Vor- und Nachbesprechung) weiter gesteckt als normalerweise üblich.
– Außer der Vor- und Nachbesprechung gibt es eine Reihe von regelmäßigen Veranstaltungen (Morgenbesprechung, Stationskonferenzen, Entlassungs- und Organisationsbesprechungen – vgl. dazu Kap. 28, Abb. 28–5), die eine ausreichende Kommunikationsgelegenheit für das Personal bieten. Somit ist nicht nur gewährleistet, daß die Kernvisite für das Arzt-Patient-Gespräch frei bleibt, sondern darüber hinaus, daß die Vor- und Nachbesprechung ebenfalls „patientenbezogen" geführt werden kann. D.h. Schwestern können beispielsweise ihre Beobachtungen und Informationen einzelfallbezogen einbringen. Die Wahrnehmungen in der Visiteninteraktion können gemeinsam besprochen werden. Allgemeine Koordinationsaufgaben können zu anderen Zeitpunkten durchgeführt werden.

Symmetrie zwischen Arzt und Patient hinsichtlich der Einflußmöglichkeiten auf den thematischen Verlauf des Visitengesprächs

Mit Symmetrie in der Gesprächsführung ist der erste Zielbereich angesprochen, der eine Änderung des konkreten ärztlichen Gesprächsverhaltens impliziert. Der Arzt-Patient-Dialog am Krankenbett soll sich so gestalten, daß die Kommunikationsinteressen des Patienten, beispielsweise seine Bedürfnisse nach verständlicher, krankheitsbezogener Information, in angemessener Weise Berücksichtigung finden. Die Themen des Visitengesprächs sollten demnach gleicherweise durch die behandlungsbezogenen Aufgaben und Interessen des Arztes (Informationsgewinnung zum Befinden, zum Zweck von Diagnosestellung, Behandlungsüberwachung, Anordnung und Erklärung von Therapiemaßnahmen) wie durch die des Patienten (verständliche Information zu Prognose, Diagnose, zu Zielen und Zwecken therapeutischer Maßnahmen) bestimmt werden. Neben der Berücksichtigung von im engeren Sinne krankheitsbezogenen Informationswünschen gehört hierher auch das Eingehen auf Kommunikationsbedürfnisse des Patienten, die mit dem emotionalen Erleben der Krankheits- und Krankenhaussituation in Zusammenhang stehen (beispielsweise über seine Bedürfnisse, Entbehrungen und Ängste zu sprechen). Patientenzentriertheit impliziert im Sinne dieser zweiten Zielsetzung eine an den vorhandenen Bedürfnissen des Patienten orientierte Gesprächsleitung des Arztes, der als situationsspezifisch dominanter Gesprächspartner die Kommunikationsinteressen des Patienten nicht nur nicht übergehen, sondern aktiv und fördernd berücksichtigen sollte.

An einem Beispiel gesagt: Symmetrie im Sinne unserer Zielsetzung verlangt zweierlei. Einmal, daß ein Arzt einer „unbequemen" Patientenfrage nicht ausweicht; zum anderen aber darüber hinausgehend, daß der Patient ausdrücklich ermuntert wird, Fragen zu stellen, bzw. daß man Bemerkungen des Patienten daraufhin anhört, ob sie nicht eine – ungestellte – Frage enthalten.

Psychotherapeutische Unterstützung des Patienten

Nach unserer Auffassung bietet ein so bestimmtes symmetrisches Gesprächsverhalten noch nicht in jeder Hinsicht die Gewähr für eine umfassende Unterstützung des Patienten. In allen Fällen, in denen der Patient beispielsweise psychische Belastungen aus seinem Leben ausblendet (sie verleugnet oder verdrängt), kann es unter der Zielsetzung einer „psychotherapeutischen Unterstützung" notwendig sein, auch gegen die unmittelbaren Kommunikationsinteressen des Patienten gerade diese Aspekte des Erlebens zu verbalisieren. Die Rechtfertigung für eine solche aufdeckende therapeutische Strategie ist darin zu sehen, daß sie es dem Patienten möglich macht, über die bewußte Erfahrung und Integration seines abgewehrten Erlebens zu realitätsangemesseneren Formen der Bewältigung von Krankheit und Krankheitsfolgen zu gelangen. Weiter bietet sie die Gelegenheit, die Kommunikation gezielt für die Stabilisierung somatischer Heilungsprozesse einzusetzen.

Allgemeiner formuliert: Mit der Zielvorstellung psychotherapeutischer Unterstützung ist gemeint, daß die Visite in bestimmten Teilabschnitten für psychotherapeutische Interventionen im Sinne der von Balint vorgeschlagenen „Sprechstundentherapie" (Balint und Norel, 1975) genutzt wird. Ein Schwerpunkt liegt dabei auf den akuten krankheitsbedingten Problembereichen (Krankheitsverarbeitung und Krankheitsverhalten), die in der aktuellen Arzt-Patient-Beziehung besonders bedeutsam sind.

18.3.2   Resultate der Begleitforschung

Zielsetzungen der kommunikationsanalytischen Begleitforschung

Die von uns durchgeführte Begleitforschung stand unter drei Zielsetzungen. Zum ersten wurde eine Evaluation im engeren Sinne angestrebt, d. h. eine vergleichende Bewertung der Effekte, die die klinische Konzeption und Organisation hinsichtlich der Visiteninteraktion ergeben hat – soweit sich solche Veränderungen überhaupt mit Unterschieden zum herkömmlichen Setting nachweisen lassen. Das zweite Ziel lag darin, neu entstehende Interaktionsstrukturen ihrer Qualität nach zu beschreiben und ihre Relevanz für das veränderte Setting zu bestimmen (Objektivierung und Differenzierung der typischen Handlungs- und Kommunikationsstrukturen der psychosomatisch veränderten Visite). Die dritte Zielsetzung schließt sich unmittelbar an die zweite an und bezeichnet den Praxisaspekt: Kommunikationsforschung sollte bei der Weiterentwicklung einer praktikablen und klinisch angemessenen Gesprächsführung helfen. Bereits die differenzierte Beschreibung von Kommunikationsproblemen kann unter den günstigeren Bedingungen eines veränderten klinischen Settings tatsächlich zu einer Sensibilisierung der beteiligten Professionellen genutzt werden. Positive Erfahrungen wurden in Ulm mit der Diskussion von Videoaufzeichnungen, die gemeinsam von Klinikern und Kommunikationswissenschaftlern durchgeführt wurden, gemacht (vgl. z. B. Bliesener und Köhle, 1986; Bliesener, 1982b; Gaus und Köhle, 1982). Anhand von Video- und Audioaufzeichnungen und den daraus erstellten Transkripten kann darüber hinaus die vergleichende Beschreibung von im Feld vorgefundenen verschiedenen Möglichkeiten, mit einem bestimmten Problem umzugehen, die für die Arzt-Patient-Beziehung eher problematischen von den eher unproblematischen Interaktionsmustern abgrenzen. Solcherart gewonnene Analyseergebnisse lassen sich als Diskussionsgrundlage für Aus- und Fortbildungsprogramme nutzen. Im folgenden werden wir einige der Hauptbefunde aus diesen drei Stufen unseres Begleitforschungsansatzes darstellen. Wir sind hier gezwungen, eine Auswahl zu treffen und von Erörterungen der bei Kommunikationsanalysen oft schwierigen methodischen Fragen abzusehen. Wir verweisen deswegen im

Text jeweils auf die zugrundeliegenden Originalarbeiten.

Evaluation

An einer für die Station repräsentativen Stichprobe (7 Ärzte, 123 Patienten mit jeweils fünf Visitengesprächen vom Anfang ihres Aufenthalts) haben wir Gesprächsparameter untersucht, die in den medizinsoziologischen und medizinpsychologischen Untersuchungen als Indikatoren für eine die Patienteninteressen übergehende Interaktionsstruktur benutzt wurden.

Tabelle 18–1 stellt eine Auswahl aus zentralen Ergebnissen für Ulm den entsprechenden Werten für das traditionelle Setting gegenüber.[4] Bereits an den Redeanteilen sieht man, daß das Ziel der funktionalen Entflechtung weitestgehend erreicht wurde. Konsil, Therapie- und Pflegefestlegung finden in etwa drei Minuten vor und nach der Visite und nicht in Gegenwart des Patienten statt. Die Visite am Krankenbett wird trotzdem länger, sie beträgt durchschnittlich 6,7 Minuten.

Diese Zeit bleibt vollständig für Gesprächsbeiträge von Ärzten und Patienten. An einer kleineren, aber ebenfalls repräsentativen Stichprobe konnten wir mit Hilfe eines sequentiellen Auswertungsansatzes nachweisen, daß die psychosomatische Visite als ein – ty-

Tabelle 18–1. Gesprächsparameter für das traditionelle Visitengespräch und für die psychosomatische Visite. (Die Angaben zur traditionellen Visite stammen aus Raspe, 1979, 1983; Siegrist, 1978; Begemann-Deppe, 1978; Nordmeyer, 1982; Jährig und Koch, 1982.)

Variable	traditionelle Visite	psychoso-matische Visite
Zeit (min)	3,5	6,7 (am Bett) +3 (außerhalb)
Redeanteile		
in Sätzen (absolut)		
alle Teilnehmer (einschl. Pat.)	43	97
Patienten	13	44
in Sätzen (%)		
Arzt	59	51,5
Patient	30	45
andere	10	3,5
Anteil indirekter Informationsvermittlung in %	59	3
Anteil ausweichender Antworten (%) auf Fragen nach krankheitsbezogener Information		
Patienten mit günstiger	36	16
ungünstiger Prognose	92	15

[4] Ausführlichere Angaben finden sich in Westphale und Köhle, 1982a, 1982b; Safian et al., 1982; Fauler und Safian, 1983. Für eine Untersuchung zur Emotionalität in der psychosomatischen Visite (Gottschalk-Gleser-Analyse) vgl. Sodemann et al., 1982.

pischerweise ein- bis zweizügiges – Wechselgespräch zwischen Arzt und Patient abgewickelt wird (Fehlenberg, 1987). Die weiteren Teilnehmer schalten sich nur sporadisch, in ergänzender Funktion ein. Für Arzt und Patient ist die Visite tatsächlich zur Dialogsituation geworden. Diese Feststellung läßt sich für Visiten bei leichter erkrankten Patienten wie für solche bei schwerkranken Patienten treffen. Damit ist auch klar, daß die Koordinationsproblematik für die Ärzte bei dieser Form der klinischen Visite entfällt. Der Arzt kann, ohne Gefahr zu laufen, andere Aufgaben zu vernachlässigen, auf die Kommunikationsinteressen des Patienten eingehen. Erwartungsgemäß finden sich in der Ulmer Visite sehr viel weniger ausweichende, entwertende Reaktionen auf Patientenfragen. Safian et al. (1982, 1986) konnten auch im direkten Vergleich von Visiten aus einem Hamburger Akutkrankenhaus und einer nach Patientenmerkmalen parallelisierten Stichprobe aus dem Ulmer Material zeigen, daß Patienteninitiativen in der psychosomatischen Visite besser respektiert und die von Patienten geäußerten Affekte eher aufgegriffen werden.

Ergänzend zeigt auch hier eine sequentielle Analyse der Gesprächssteuerung, daß der direkte Informationsaustausch (Frage-Antwort-Sequenzen in beiden Richtungen zwischen Arzt und Patient) tatsächlich chancengleich verläuft (Fehlenberg, 1987). Fragen des Patienten zu einem vom Arzt zu beantwortenden Themenbereich werden in gleicher Weise angemessen beantwortet wie umgekehrt Fragen des Arztes zu einem vom Patienten zu beantwortenden Sachverhalt. Beide Sprecher besitzen gegenüber einer Informationsfrage des jeweils anderen Partners den gleichen Handlungsspielraum und zeigen keine unterschiedlichen Tendenzen in der Gesprächsfortsetzung. Ebenso sind ihre Fragemöglichkeiten vergleichbar. 85% der Abwicklungsstrukturen des psychosomatischen Visitengesprächs – soviel machen die um den Informationsaustausch zentrierten Zusammenhänge aus – werden damit im angestrebten Sinne symmetrisch abgewickelt. Darüber hinaus macht die sequentielle Analyse aber auch auf eine (weiter-)bestehende Asymmetrie aufmerksam. Dabei geht es um Entscheidungssequenzen. Die typische Form der Arztbeteiligung besteht in einer aktiven Umsetzung von Informationen (eigener oder vom Patienten gegebener) in (Be-)Handlungsentscheidungen, während die typische Form der Patientenbeteiligung einer Differenzierung bereits (durch den Arzt) eingeführter Entscheidungskontexte entspricht. Dazu vier Belegstellen aus unseren Texten (I = erster Zug; II = zweiter, fortsetzender Zug).

A: (I) Da ist immer noch 'ne ganze Menge (. . .) hier (. . .) (II) Ich glaub, wir setzen einfach mal das (Totocillin) ab.

P: (I) Des ist morgens nämlich ein scheußlicher, äh, (schmatzt) ja, Geruch, Geschmack, des ist doch wie ich aufwach ganz schrecklich.
A: (II) Ja (?) (.) Wir können höchstens versuchen, Ihnen jetzt mal 'ne Weile Nasentropfen zu geben

---

P: (I) Also das würd ich auf jeden Fall weglassen, die Nachmittagstablette (II) und die früh, soll ich noch nehmen oder?

P: Soll ich Schwimmen noch einen Tag lassen?
A: (I) Ja.
P: (II) Ich geh ganz gerne zum Duschen rüber (. . .)

....

aber ich weiß nicht, ob's gut wäre, wenn ich ins Wasser ginge, es ist zwar nicht sehr kalt, aber
A: (I) Wenn Sie so's Gfühl ham und ich find des ganz gut, / daß
P: Ja/
A: Sie sich dann / auf sich mal so
P: (II) Dann werde ich / morgen die Gymnastik machen und (drei Worte unverständlich)
A = Arzt, **P** = Patient

Zusammenfassend läßt sich feststellen, daß in den ersten beiden Beispielen (und in allen entsprechenden Belegstellen für die arzttypischen Sequenzen) Schlußfolgerungsrelationen vorherrschen. Der Arzt stellt eine behandlungsrelevante Information fest bzw. nimmt eine behandlungsrelevante Information des Patienten auf und legt (schlußfolgernd) eine Behandlungskonsequenz fest. Bei den patienttypischen Sequenzen finden sich dagegen – wie in den zweiten beiden Beispielen – anders zu beschreibende Relationen: Detaillierung, Konkretisierung bzw. Ergänzung eines (eingeführten) Handlungs- und Entscheidungskomplexes, Kombination von Aufforderung zu Direktiven, die an den Arzt gerichtet ist, und eigene Entscheidungen sowie Entscheidungen, die vom Arzt geforderte Direktiven bestätigen. Verallgemeinert drückt sich damit in der sprechertypischen Abfolgestruktur eine qualitativ unterschiedliche Beteiligung in den Entscheidungssequenzen aus: Der Arzt besitzt ein „Transformationsprivileg" für die Umsetzung von Informationen in Handlungen; die Rolle des Patienten ist dagegen eher in einer nachgeordneten Beteiligung, in der Konkretisierung bereits eingeführter Entscheidungskontexte, zu sehen. Die Interaktionsanalyse legt damit den klinischen Praktikern die Frage vor, inwieweit Chancengleichheit außer einer angemessenen Informationsmöglichkeit auch eine verstärkte eigeninitiative Beteiligung in Entscheidungskontexten bedeuten soll. Entsprechend zu den oben wiedergegebenen Beispielen geht es dabei nicht nur um die „großen" Entscheidungen der Behandlung, sondern auch (vielleicht sogar vor allem) um solche, die im häuslichen Lebensbereich des Patienten seiner Eigenverantwortung überlassen wären. Einer aktiveren Beteiligung des Patienten sind natürlich Grenzen gezogen, die im Flexibilitätsrahmen der Stationsorganisation, in ökonomischen Faktoren und in einem nichtreduzierbaren Anteil fachlichen Expertentums der Mediziner genauso begründet liegen wie in den „Versorgungs-Erwartungen" von Patienten. Nur im letzten Punkt kann kommunikatives Verhalten, kann eine gezielte Änderung des Gesprächsverhaltens der Ärzte mit zu einer Verbesserung beitragen (vgl. dazu das erste Fallbeispiel im übernächsten Abschnitt dieses Artikels). Ansonsten ist klar, daß hier ein Bereich angesprochen ist, in dem Kommunikationsanalyse zwar zur Evaluation faktisch bestehender Handlungsstrukturen beitragen kann, daß anzustrebende Änderungen aber der weitergehenden klinischen Diskussion und programmatisch umfassender Bemühungen bedürfen.

Mit zur Evaluation gehört die Frage, inwieweit die veränderte Situation auch von den Patienten als Verbesserung empfunden wird. Eine vergleichende Untersuchung, die diese Frage direkt beantworten würde, konnten wir aus methodischen Gründen nicht durchführen. Indirekt läßt sie sich aber aus den Ergebnissen einer Untersuchungsserie beantworten, die sich mit dem Zusammenhang von Gesprächsverhalten in verschiedenen psychosomatischen Visiten und der Beziehungsbeurteilung durch den Patienten (aus methodischen Gründen nachgebildet über ein 15 Items umfassendes intuitives Fremdrating durch medizinische Laien) beschäftigt hat. Die wichtigste (faktorenanalytisch gewonnene) Skala des Beziehungsurteils läßt sich mit „emotionaler Befindlichkeit des Patienten in der Beziehung zum Arzt" bezeichnen (Guth, 1985). Die von den Ratern auf dieser Skala vergebenen Einstufungen lassen sich zu immerhin 60% (dieser Wert entspricht einem multiplen R von .77 in einer schrittweisen Regressionsanalyse) aus Merkmalen des Gesprächs vorhersagen. Damit haben sich die von uns untersuchten Kommunikationsmerkmale als hochgradig bedeutsam für den Beziehungsaufbau erwiesen. Wie Ärzte und Patienten miteinander sprechen, bestimmt entscheidend mit darüber, wie gut sich ein Patient in emotionaler Hinsicht aufgehoben fühlt. Inhaltlich läßt sich das Ergebnis auf den Nenner bringen, daß die beste Beziehungsstruktur wohl am ehesten in denjenigen der psychosomatischen Visiten erreicht wurde, in denen sich die Handlungsrollen von Ärzten und Patienten am deutlichsten von denen unterscheiden, die in Arzt-Patient-Gesprächen mit dem Hintergrund eines naturwissenschaftlich-biologistischen Behandlungskonzeptes beobachtet werden. Die Beurteiler werten eine aktive Beteiligung des Patienten an den medizinischen Behandlungsaufgaben positiv und honorieren darüber hinaus die Bemühung des Arztes um solche Gesprächsthemen als beziehungsförderlich, die nicht im engeren Sinne traditionelle Behandlungsthemen sind, sondern auch das psychosoziale Lebensumfeld des Patienten mit einbeziehen (Fehlenberg, 1987).

Soweit hat die Evaluation der in Ulm erprobten Visitenführung also ein überwiegend positives Ergebnis. In der psychosomatischen Visite kann sich das Arzt-Patient-Gespräch entlastet von konkurrierenden Ansprüchen organisatorischer Routineaufgaben entwickeln, die Patienteninteressen finden eine insgesamt wesentlich bessere Beachtung als im herkömmlichen Setting, die intendierte Richtung der Veränderung wird von den Patienten emotional akzeptiert. Neben diesen allgemeinen Zielsetzungen zur Verbesserung der Visitenkommunikation war es aber auch ein erklärtes Ziel, die psychosomatische Visite als (psycho-)therapeutisches Gespräch zu nutzen. In den beiden folgenden Abschnitten wollen wir uns mit diesem spezifischen Anspruch der psychotherapeutischen Unterstützung beschäftigen. Die Besonderheit des Visitengesprächs liegt – beispielsweise gegenüber Therapiedialogen – in der Verbindung von einerseits praktisch orientiertem Handeln (Körperbehandlung bzw. dem Sprechen darüber) und andererseits einem reflektiv ausgerichteten Gesprächsverhalten, das ganz andere Formen der kommunikativen Kooperation und der Beziehungsdefinition erforderlich macht.

Objektivierung und quantitative Bedingungsanalyse psychotherapeutischer Interventionen in der psychosomatischen Visite

Unter psychotherapeutischen Interventionen verstehen wir solche Verbalhandlungen des Arztes, die auf der Grundlage klinischer Erwägungen eine Änderung im Erleben des Patienten bewirken sollen. Der erste Schritt unserer Auswertung zielte auf eine Beschreibung der von den Ärzten prinzipiell intendierten Interventionen. Dazu wurden auf der Station tätige Ärzte und die für die Supervision zuständigen Mitarbeiter befragt. Aus allen genannten therapeutischen Intentionen haben wir in gemeinsamer Diskussion mit den Klinikern 13 Typen psychotherapeutischer Interventionen spezifiziert (vgl. Tab. 18–2), die unter klinischen Gesichtspunkten jeweils einer von vier übergeordneten Interventionsstrategien zugeordnet werden können:

– Explorierende Interventionen,
– stützende Interventionen,
– konfrontierende und interpretierende Interventionen,
– direkte Führung bei psychischen Problemen.

Tabelle 18–2. Psychotherapeutische Interventionen im Visitengespräch

Explorierende Interventionen (Sondieren, Klären):
- Klärung emotionaler Erlebnisinhalte
- biographisch orientierte Klärung

Stützende Interventionen beim Umgang mit negativen Affekten durch Fördern von:
- Krankheitsverständnis
- differenzierter Leidenswahrnehmung
- Hinwendung zu positiven Aspekten der aktuellen Situation
- Hinwendung zu positiven Aspekten der Biographie

Konfrontierende und interpretierende Interventionen, Anregung zu einer vertieften Auseinandersetzung mit dem Bereich:
- Mitverantwortung des Patienten
- Zusammenhang von somatischer Erkrankung und psychischem Leiden
- Widersprüche innerhalb der Bereiche oder zwischen den Bereichen Erleben, Äußern, Verhalten des Patienten (konfrontierende Intervention)
- biographische Dimensionen aktueller Probleme
- Tod und Sterben des Patienten
- Beziehung Arzt-Patient (Übertragung)

Um einen Eindruck zu geben, wie eine Intervention konkret aussehen kann, seien hier zwei Belege aus unseren Texten angeführt.[5]

> Kategorie: **Klärung emotionaler Erlebnisinhalte**
> **P:** Und, und, ja, man möchte ja mal wissen, was (Ursache für Schmerzen) los ist
> **A:** Daß man immer wieder sich so ausgeliefert fühlt, oder was ist das, wenn des immer, plötzlich wieder aus heiterm so –, sozusagen heiterm Himmel der Schmerz kommt?
> Kategorie: **Konfrontieren mit Widersprüchen** innerhalb der Bereiche oder zwischen den Bereichen Erleben, Äußern, Verhalten des Patienten
> (Patient versucht seine traurige Stimmung hinter einem oberflächlich scherzhaften „Umgangston" zu verbergen)
> **A:** Nun es ist so, daß man bei Ihnen, Sie lachen ja immer sehr viel und man hat den Eindruck eigentlich, daß Sie sehr viel lachen, ja.
> **P:** Ja, ich bin doch ein lustiger Kerle, des is klar. Das, äh, lustig und johle und alles, was is, was, tati und tatüt, das hab ich alles bei mir, ja.
> **A:** Also dahinter können Sie doch auch gleichzeitig traurig sein.
> **A** = Arzt, **P** = Patient.

Der zweite Auswertungsschritt galt einer Bedingungsanalyse für den Gebrauch dieser Interventionen. Grundlage dafür war die inhaltsanalytische Auswertung von 296 auf Tonband aufgezeichneten und transkribierten Visitengesprächen, die von sieben Ärzten und 74 Patienten der Ulmer Station stammen. Jeder Patient ist mit vier Gesprächen vom Anfang seines Aufenthalts vertreten. 37 Patienten gelten hinsichtlich ihrer Prognose als schwer erkrankt,[6] 37 Patienten als vergleichsweise leichter erkrankt. Die beiden Gruppen wurden hinsichtlich Alters- und Geschlechtsverteilung der Patienten und der Beteiligung der sieben Stationsärzte parallelisiert.

In welchem Umfang wurde die Visite tatsächlich für psychotherapeutische Interventionen genutzt? In einer Visite, die für die untersuchte Stichprobe im Mittel etwa 7 Minuten dauert, werden durchschnittlich drei Interventionen realisiert. Anders gerechnet, auf 1000 Worte, die vom visiteführenden Arzt und anderen Teammitgliedern gesprochen werden, entfallen im Mittel 6,5 Interventionen. Dabei unterscheiden sich einzelne Visiten allerdings erheblich. In etwa einem Drittel der untersuchten Visiten werden keine Interventionen formuliert. Das Maximum findet sich in einer Visite mit 25 Interventionen. Zusammen mit der Gesamtverteilung (vgl. Abb. 18–1) machen diese Zahlen deutlich, daß die psychosomatische Visite nicht ausschließlich und nicht unter allen Umständen als ein psychotherapeutisches Gespräch gehandhabt wurde. In einer nicht geringen Zahl von Fällen war es den Stationsärzten überhaupt wichtiger, das Gespräch unter anderen als psychotherapeutischen Gesichtspunkten zu führen. Andererseits ist aber ebenso zu beobachten, daß die psychotherapeutische Zielsetzung in einzelnen Gesprächen in den Vordergrund rücken kann.

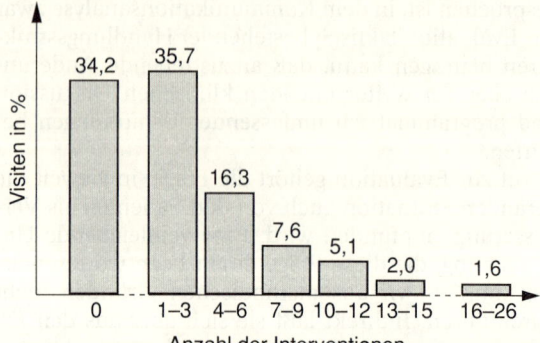

Abb. 18–1.   Verteilung der untersuchten Visiten (n = 296) auf Interventionshäufigkeiten.

Gibt es personengebundene Merkmale, die Ausmaß und Art der Interventionen beeinflussen? Untersucht haben wir auf der Patientenseite Alter, Geschlecht und Erkrankungsschwere, und außerdem wurde das Verhalten der 7 Ärzte untereinander verglichen.

Bei den Patientenmerkmalen zeigte sich, daß weder Alter noch Geschlecht Häufigkeiten und Art der von den Ärzten formulierten Interventionen beeinflussen.

Tabelle 18–3.   Interventionshäufigkeiten bei schwerkranken und nicht schwerkranken Patienten (Mann-Withney-U-Test, zweiseitig)

Variable	Interventionshäufigkeiten ($\bar{x}$)		Irrtumswahrscheinlichkeit
	nicht schwerkrank	schwerkrank	
Interventionen visiteführender Arzt pro Visite	3,29	2,24	0,045
Interventionen Gesamtteam (einschl. visitef. Arzt) pro Visite	3,39	2,54	0,141
Interventionen anderer Teammitglieder (ohne visitef. Arzt) pro Visite	0,10	0,30	0,035
Interventionen Gesamtteam pro 1000 Worte	7,35	5,63	0,043

---

5   Eine ausführliche Sammlung von Textbelegen zu allen aufgeführten Interventionstypen findet sich in Fehlenberg et al., 1982.
6   Als schwer erkrankt galten Patienten, auf die mindestens eine der zwei folgenden prognostischen Aussagen zutraf: 1. Die mittlere Überlebenszeit ist kürzer als zwei Jahre. 2. Die Wahrscheinlichkeit, innerhalb von drei Monaten nach Stationsaufnahme zu sterben, beträgt 30% oder mehr. Die Zuordnung wurde auf Basis der in der medizinischen Literatur berichteten mittleren Mortalitätsraten zu den entsprechenden Krankheitsbildern getroffen.

In Abhängigkeit von der Erkrankungsschwere der Patienten wurde dagegen ein quantitativ und qualitativ unterschiedliches psychotherapeutisches Gesprächsverhalten festgestellt. Hier ergab sich eine Anzahl von Unterschieden beim Vergleich von schwerer und leichter erkrankten Patienten (vgl. Tab. 18–3).

Betrachtet man global alle Teammitglieder, zeigt sich kein statistisch signifikanter Unterschied im Verhalten gegenüber den beiden Patientengruppen (Tab. 18–3: Gesamtteam). Anders sieht es aus, wenn man das Verhalten der verschiedenen Sprechergruppen des Teams getrennt analysiert. Der visiteführende Arzt interveniert häufiger bei den leichter erkrankten Patienten. Die übrigen Teammitglieder (d.h. die weiteren Ärzte und Schwestern) intervenieren dagegen häufiger bei den schwer erkrankten Patienten.

Vergleicht man nicht die absolute Zahl der Interventionen, sondern die auf die Anzahl der gesprochenen Worte relativierte Zahl von Interventionen, ergibt sich ebenfalls für das Gesamtteam eine signifikante Differenz: Im Gespräch mit den Schwerkranken finden sich, bezogen auf die Redemenge, weniger häufig Interventionen. Die nicht-psychotherapeutischen Funktionen des Gesprächs werden bei den Schwerkranken offensichtlich ausführlicher oder vielfältiger wahrgenommen.

Beim Vergleich der Unterdimensionen ergeben sich zwei Unterschiede: Bei den schwerkranken Patienten realisiert der Arzt weniger (absolut wie relativ berechnet) konfrontierende und interpretierende Interventionen und weniger Interventionen des Typs direkte Führung. Für die anderen Interventionstypen sind keine bedeutsamen Unterschiede festzustellen.

Zusammenfassend lassen sich diese Befunde so verstehen, daß in Abhängigkeit von der Erkrankungsschwere klinisch und interaktiv unterschiedliche psychotherapeutische Interventionsstrategien verwandt wurden. „Klinisch", insofern die Interventionen bei Schwerkranken weniger aktiv konfrontierenden und direkten Charakter haben und hinsichtlich ihres Gesamtumfangs in Relation zu den übrigen Gesprächsfunktionen vergleichsweise geringeren Raum einnehmen; „interaktiv" insofern, als bei der psychotherapeutischen Unterstützung schwerkranker Patienten das Team eine vergleichsweise aktivere Rolle spielt.

Der Vergleich zwischen dem Interventionsverhalten der einzelnen Ärzte erbrachte zum Teil erhebliche Unterschiede (vgl. dazu im einzelnen Fehlenberg et al., 1982). Die 7 untersuchten Ärzte unterscheiden sich hinsichtlich der Gesamthäufigkeit, mit der sie Interventionen formulieren, wie im Ausmaß, mit dem sie Unterschiede zwischen schwer- und leichtkranken Patienten machen. In der Tendenz gilt, daß die Ärzte, die überhaupt häufiger intervenieren, diese hohe Interventionshäufigkeit unabhängig von der Erkrankungsschwere sowohl bei schwerkranken wie bei nicht schwerkranken Patienten zeigen. Ärzte, die allgemein seltener psychotherapeutisch intervenieren, verhalten sich demgegenüber bei schwerkranken Patienten deutlich zurückhaltender.

Daß zwischen den Ärzten des untersuchten Feldes eine erhebliche Variation hinsichtlich ihres psycho-

therapeutischen Vorgehens bestand, spricht dafür, daß die Verwirklichung therapeutischer Gesprächsqualitäten nicht mehr allein durch allgemein organisatorische Veränderung zu gewährleisten ist, sondern daß für diese Zielsetzung verstärkt eine Schulung der kommunikativen und psychologischen Kompetenzen der Ärzte durchgeführt werden muß. Kommunikationsanalytische Untersuchungen können dabei die Funktion übernehmen, die in der Praxis von den beteiligten Ärzten analysierten Strategien psychotherapeutischer Interventionen zu beschreiben und der Reflexion zu erschließen.

Praktikable Gesprächsformen für die psychosomatische Visite

Anhand von zwei Texten aus unserem Material wollen wir zeigen, welche Aufgaben und Probleme sich für die Gesprächsführung einer veränderten Visitensituation ergeben und wie Kommunikationsanalysen dazu beitragen können, praktikable Gesprächsformen und Interventionsstrategien zu finden. Einerseits sind diese Aufgaben sozusagen natürlich in den Visitendiskurs eingebaut, d.h., es handelt sich um Schaltstellen des Gesprächs, die in jeder Visite auftreten und die bestimmte Steuerungsleistungen des Arztes und des Patienten notwendig machen. In einer typischen (internistischen) Visite gibt es zumindest drei solcher diskursstrukturellen, d.h. notwendig auftretenden Schaltstellen: die Gesprächseröffnung, Beginn und Ende der körperlichen Untersuchung und die Gesprächsbeendigung. Zum anderen handelt es sich um Aufgaben, die sich aus den beschriebenen Zielvorstellungen von „Symmetrie" und „Therapie" ergeben. Es sind also Aufgaben, die aus der aktuellen Gesprächs- und Beziehungsentwicklung, deren Wahrnehmung und (klinischer) Bewertung für den Arzt entstehen. In unseren Beispieltexten wird dabei das grundsätzliche Handlungsproblem einer psychosomatischen Gesprächsführung deutlich: Zielsetzungen symmetrischer Kommunikationen mit (aufdeckenden) therapeutischen Gesprächselementen zu verbinden.

Die in Abschnitt 18.3.2 beschriebenen Zielsetzungen von kommunikativer Symmetrie und psychotherapeutischer Unterstützung können unter bestimmten Umständen in Widerspruch geraten. Ihre Verwirklichung würde ein unterschiedliches Gesprächsverhalten des Arztes erforderlich machen. Dieser Zielkonflikt zeigt sich in unserem ersten Visitenausschnitt in deutlicher Form. Er wird hier zu einem Interaktionsproblem, zu einem Problem der Verständigung von Arzt und Patient.

Bevor wir das Gespräch unter diesem Aspekt betrachten, folgende Erklärungen zum Kontext:

Es ist eine Visite, die zu Beginn des Krankenhausaufenthaltes eines 73jährigen Patienten stattfindet. Es wurde die Diagnose „paroxysmale Tachykardie" gestellt, die zur Einweisung führte. Langfristig bestand eine Hypertonie beim Patienten. Nach der von den behandelnden Stationsärzten erhobenen Einschätzung wurde die Erkrankung des Patienten unter den

Gesichtspunkten Gefährdung und Prognose als vergleichsweise günstig eingestuft. Aus der Sicht des Patienten sind diese Feststellungen allerdings zu relativieren: Abgesehen von einer 24 Jahre zurückliegenden Blinddarmoperation ist der Patient zum ersten Mal so schwer erkrankt, daß er zur Behandlung in ein Krankenhaus eingewiesen werden muß.

Der Patient schläft sehr viel und verhält sich, auch wenn er wach ist, sehr inaktiv – das ergibt sich aus dem vorangehenden, nicht wiedergegebenen Teil der Visite. Diese aus medizinischer Sicht unbegründete und für das Behandlungsziel schädliche Inaktivität empfindet der Arzt offensichtlich als problematisch. Er versucht, dieses Verhalten einerseits mit der Belastung des Krankenhausaufenthaltes in Beziehung zu bringen. So etwa mit folgenden drei Äußerungen, die sich an verschiedenen Stellen der Visite befinden:

> Da wollte ich noch mal eins nachfragen, äh, schlafen Sie jetzt im Krankenhaus so viel oder war das zu Hause auch so?
> Hm, hat das (gemeint ist das häufige Schlafen) vielleicht mit dem Krankenhaus zu tun? Hm?
> Mhm, nun, ich meine, vielleicht ist Ihnen die Situation hier im Krankenhaus zu sein, nach diesen Herzanfällen, nicht so angenehm?

Zum anderen fordert der Arzt den Patienten an zwei Stellen nachdrücklich auf, sich aktiver zu verhalten, etwa mit:

> Ich würde Ihnen dann empfehlen, daß Sie ruhig weiter aufstehen, ja.
> Mhm, äh, ich glaube, eins sollte man noch einmal ganz deutlich sagen, Herr N., wenn die Anfälle nicht da sind (**P:** ja), dann können Sie Ihr Herz ganz normal belasten, so wie Sie das vorher getan haben. (**P:** ja) Da brauchen Sie, glaub ich, keine Angst zu haben.

Für den Arzt ist offensichtlich folgender Zusammenhang gegeben: Die Tatsache des Krankenhausaufenthaltes belastet den Patienten, macht ihn vielleicht sogar depressiv. Die Inaktivität des Patienten ist eine Reaktion auf diese Situation, sie stabilisiert ihrerseits das Schwächegefühl des Patienten usf. Die Antworten des Patienten zeigen, daß er eine andere Sicht des Zusammenhangs hat. Für ihn sind seine Schwäche und Müdigkeit das primär Bedrohliche. Weil er sich (körperlich) so geschwächt fühlt, verhält er sich – seinem Verständnis nach – angemessen vorsichtig. Diese beiden Positionen bleiben im ersten Teil des Gesprächs relativ unvermittelt nebeneinander stehen. Im zweiten Teil gibt der Arzt dem Patienten Erklärungen zur Medikation. Danach schließt sich die folgende Passage an.

> 1 (3 sec Pause)
> **A:** Ja, können wir noch was miteinander besprechen?
> **P:** Naja, ich wüßte an sich nichts.
> 5 (2,5 sec Pause)
> **A:** Wiedersehn

> **P:** Weil die, (.) was wollen Sie eben unternehmen, daß ich doch langsam wieder zu Kräften komme, ich meine, daß ich (.) beim Laufen mal so a
> 10 bißl mehr Sicherheit hab, und so?
> **A:** Nhä (. . .) Sie meinen das geht alleine nicht, daß Sie da noch ne Unterstützung brauchen?
> **P:** Na ich kann wohl laufen, aber
> **A:** Mhm
> 15 **P:** Vorsichtig natürlich und immer so, daß ich mich g'schwind eventuell wo anhalten kann.
> **A:** War denn das vor dem Krankenhausaufenthalt auch schon?
> **P:** Nein, nein das war nicht.
> 20 **A:** (nn)
> **P:** Das ist erst jetzt (..) durch die Behandlung.
> **A:** Durch die Behandlung?
> **P:** Naja, also nach, (.) (indem), nachdem ich eben liege im Krankenhaus.
> 25 **A:** Mhm
> (2 sec)
> **P:** Diese Schwäche
> **A:** Mhm (..) wir werden da unsere Krankengymnastin mal zu Ihnen schicken, daß die mit Ihnen
> 30 das mal trainiert.
> **P:** Ah-ja
> **A:** Aber es ist ja doch schon ein bemerkenswerter Punkt, daß Sie sich offensichtlich jetzt im Krankenhaus und nach der Behandlung so ge-
> 35 schwächt fühlen, daß Sie (äh), gar nicht mehr sich so alleine raustrauen aus dem Bett.
> **P:** Ja, also ich trau mich schon,
> **A:** Mhm
> **P:** aber ich bin halt vorsichtig.
> 40 **A:** Ja.
>
> **A** = Arzt, **P** = Patient; (. . .) = Pausen unter 2 sec.

Eine Äußerung wie in (2/3), „Ja, können wir noch was miteinander besprechen?", läßt sich als Pro-Initiative bezeichnen (Bliesener, 1982), d.h., Äußerungen dieser Art dienen dazu, dem Gesprächspartner das Recht auf eine eigene Initiative einzuräumen. Funktional äquivalente Formulierungen, „Haben Sie noch eine Frage?", „Haben Sie noch etwas auf dem Herzen?", „Kann ich/können wir noch etwas für Sie tun?", finden sich in etwa der Hälfte der Ulmer Visiten gegen Ende des Gesprächs. Sie wurden gebraucht, um dem Patienten noch einmal ausdrücklich Gelegenheit zu geben, noch nicht behandelte oder im Laufe des Gesprächs problematisch gewordene Fragen ansprechen zu können. Pro-Initiativen in dieser Form sind damit Gesprächsmittel, die bewußt im Sinne der Zielsetzung von Symmetrie eingesetzt wurden. Die hier vom Arzt gewählte Formulierung ist besonders wenig festlegend, was mögliche Gesprächsinteressen des Patienten angeht.

Daß diese Möglichkeit, initiativ zu werden – wie in unserem Beispiel –, auch bei bestehenden Problemen von den Patienten nicht so ohne weiteres spontan genutzt wird, ist in unseren Visiten ebenfalls häufiger zu beobachten. Ob sich hier die für die Krankenhauspatienten oft beschriebene rollenspezifische Schwierigkeit niederschlägt, gegenüber dem Arzt initiativ zu werden, Fragen zu stellen usf. (vgl. u.a. Måseide, 1981), oder ob es sich hier um eine allgemeine

Schwierigkeit im Umgang mit solchen vom Partner zugewiesenen Initiativerechten handelt, läßt sich schwer entscheiden. Auf jeden Fall kann man sagen, daß eine Pro-Initiative, damit sie tatsächlich im Sinne von Symmetrie wirken kann, durch ein entsprechendes Zuhörerverhalten begleitet werden muß. Der Arzt muß dem Patienten ausreichend Zeit zum Überlegen geben (Pause!), er muß durch nonverbales Verhalten (Blickkontakt, Körperhaltung) klarmachen, daß er tatsächlich auf eine Initiative des Patienten wartet und bereit ist, sie zu beantworten. Schließlich muß der Arzt damit rechnen, daß nach einer spontanen Ablehnung doch noch eine Initiative nachgeschoben wird, wie auch in unserem Beispiel. Es kann deswegen durchaus sinnvoll sein, eine Pro-Initiative noch einmal zu wiederholen.

Interessant für den Zielkonflikt Symmetrie versus Psychotherapie im ärztlichen Gesprächsverhalten sind nun die nachgeschobene Patienteninitiative (7–10) und die darauf bezogenen nachfolgenden Sequenzen (11–40). Die Patienteninitiative (7–10) ist ihrer Sprechaktform nach eine Frage an den Arzt, welche Maßnahmen von seiten der Mediziner zur Mobilisierung vorgesehen seien. Im Kontext des Gesprächs wirkt diese Frage, da bisher keineswegs besprochen wurde, daß überhaupt solche Maßnahmen erfolgen sollten, als Aufforderung an den Arzt, solche Maßnahmen zu überlegen und anzuordnen. Durch die mit der Pro-Initiative verbundene Selbstverpflichtung befindet sich der Arzt in einem verschärften Antwortzwang. Er muß eine entsprechende medizinische Maßnahme vorschlagen und ihre Durchführung in Aussicht stellen oder alternativ erläutern, warum solche Maßnahmen medizinisch nicht sinnvoll sind. Genau die erste Alternative wird in 28/30 realisiert, d.h. mit einer Verspätung von 6 Redebeiträgen. Die Berechtigung, von einer die Antwortverpflichtung verletzenden Verspätung zu sprechen, ergibt sich daraus, daß die Beiträge in 11–27 keine im Hinblick auf eine definitive Beantwortung notwendigen Zwischenschritte (Rückfragen, Verständigungssicherung usf.) sind, sondern andere Ziele verfolgen.

Was geschieht nun in 11–27? Mit 11/12 expliziert der Arzt implizite Voraussetzungen und den Aufforderungsgehalt der Patienteninitiative. Damit die Frage nach professionellen Maßnahmen berechtigt und sinnvoll ist, muß gelten, daß eine Besserung nicht durch die Eigeninitiative des Patienten erreichbar ist. Mit seiner Frage bittet der Patient den Arzt um professionelle Hilfe. Die Formulierung „Unterstützung" leistet dabei eine re-definierende Präzisierung des „was" des Patienten, der allgemein nach Maßnahmen des Arztes gefragt hatte. Eventuelle medizinische Maßnahmen können nur Beistand bei einer zuerst vom Patienten geforderten Aktivität sein.

Die in 11/12 vorgenommenen Explizierungen und Präzisierungen sind sicher nicht notwendig im Sinne der definitiven Beantwortung der Patienteninitiative. Der Arzt verfolgt hier ganz offensichtlich eine andere Zielsetzung als die Abarbeitung der durch die Patienteninitiative gesetzten Verpflichtung; er versucht, die Patienteninitiative nach den hinter ihr stehenden Vorstellungen zum Krankheitsgeschehen und zu den krankheitsbezogenen Erlebnisqualitäten zu hinterfragen. Form, d.h. Anbindungsfigur, Formulierung und Intonation entsprechen dabei ganz den in psychotherapeutischen Verbaltherapien wie Psychoanalyse oder klientzentrierter Gesprächstherapie üblichen Vorgehensweisen – allerdings die Patientenerwiderung nicht. Der Patient unterstützt seine Initiative mit Argumenten. Offensichtlich hat er die Arztäußerung eher als Kritik, als kritische Frage nach der Rechtfertigung seines Vorschlags verstanden.

Auch die in 17/18 gestellte Frage des Arztes kann nicht schlüssig als für die Anordnung von Therapiemaßnahmen notwendig angesehen werden. Bereits am Anfang der Visite hatte der Patient geantwortet, daß Müdigkeit und Schwäche erst seit der Krankheit und der Behandlung aufgetreten seien. Wahrscheinlich ist damit zumindest, daß der Arzt hier die negative Antwort des Patienten antizipiert und daß es ihm gerade darum geht, mit der Frage den Patienten auf diesen „bemerkenswerten Punkt", wie er es in 32/33 formuliert, aufmerksam zu machen.

Zusammengefaßt scheint in 11–22 der Versuch des Arztes vorzuliegen, den Patienten in ein reflektierendes Gespräch über seine Ängste, seine Passivität und die daraus resultierende inaktive Versorgungshaltung zu ziehen. Dieser Ansatz zu einem therapeutischen Diskurs wird dann in 28–30 mit der verpflichtungsgemäßen Beantwortung der Patienteninitiative unterbrochen und schließlich in 32–36 in einer expliziteren Form fortgesetzt.

Im hier vorliegenden schnellen Wechsel zwischen therapeutischen Interventionsversuchen und verpflichtungsgemäßer Beantwortung der Patienteninitiative finden wir die Struktur des allgemeinen Handlungsproblems der psychosomatischen Visite. „Psychotherapie" in der Visite richtet sich auf die Korrektur krankheitsbezogener realitätsangemessener Kognitionen (Emotionen, Werthaltungen, Wissensbestände). Immer dann, wenn diese – aus der Sicht des Arztes unbegründeten – Kognitionen zu einer realen Forderung des Patienten an den Arzt führen (Information, Prognosen oder wie im Fall oben Behandlungsmaßnahmen), befindet sich der Arzt in der Zwickmühle. Das Mittel psychotherapeutischer Intervention ist der nichtwertende, aber kritisch-reflektierende Dialog mit dem Patienten, der als notwendige Voraussetzung eine Modifikation gegenüber dem Alltagsdiskurs hat (vgl. Flader und Grodzicki, 1982; Koerfer und Neumann, 1982). Forderungen an den Therapeuten werden in einer psychotherapeutischen Situation prinzipiell nicht mit den nach den Formen des „Alltags-"Handelns erwartbaren Reaktionstypen beantwortet, sondern werden im Hinblick auf die mit dem Sprechakt assoziierten Einstellungen des Klienten verbalisiert. Verfährt der Arzt in der Visite in gleicher Weise, verstößt er notwendigerweise gegen die Zielsetzung der „Symmetrie". Vor allem läuft er Gefahr, da die Visite – jedenfalls bei primär körperlich erkrankten Patienten – aus der Patientenperspektive alles andere als eine psychotherapeutische Situation ist, die Behandlungs- und Kooperationserwartungen des Patienten zu frustrieren, ohne ihm den therapeutischen Zweck (und damit eine Kooperation „zweiter Stufe"; Ehlich, 1981) einsichtig zu machen.

Auch von Psychotherapie-Klienten sind solche Frustrationsempfindungen besonders zu Beginn der Therapie gut belegt. Psychotherapie-Klienten können mit der Situation aber leichter zurecht kommen. Durch bestimmte „Initiationsriten" (Mitteilung der Grundregeln in der Psychoanalyse, Offenlegung des Gesprächsverfahrens in der klientzentrierten Gesprächstherapie) oder durch die Konstanz der Verweigerung des Therapeuten lernen sie vergleichsweise schnell, ihre Erwartungshaltung zu modifizieren und die Bewertung des aktuellen Gesprächsnutzens zugunsten eines langfristigen therapeutischen Nutzens zu relativieren. Über eine psychotherapeutische Kooperation hinaus erfordert die Visite demgegenüber aber in jedem Fall auch eine alltagsweltliche Kooperation, die sich im Gegensatz zur Therapie nicht nur auf den Rahmen der Erhaltung von Kommunikation bezieht (Arbeitsbeziehung), sondern an realen Körperbehandlungsinteressen festmacht.

Das Hauptproblem einer psychosomatischen Gesprächsführung – in der der Arzt sowohl als Körpermediziner wie als Psychotherapeut professionell aktiv ist – liegt demnach darin, daß innerhalb einer Gesprächssituation ein Wechsel zwischen zwei Diskursformen mit tendenziell antagonistischen Kooperationsmustern notwendig ist. Dies ist eine Aufgabe, die nur interaktiv als wechselseitige Re-Definition der Kommunikationssituation lösbar ist. Der gelingende Fall besteht in einer Übereinstimmung des jeweiligen „kommunikativen Rahmens". In diesem Sinne ist auch die Forderung nach „Symmetrie" im Hinblick auf das ärztliche Gesprächsverhalten präziser zu fassen: Der Arzt muß seine eigenen Gesprächsintentionen für den Patienten transparent machen. Er muß verdeutlichen, wenn es ihm auf eine therapeutische Reflexion ankommt (beispielsweise durch explizite Metakommunikationen über seine Absichten und Zielsetzungen). Diese Offenheit stellt Symmetrie in einem umfassenderen Sinn her, weil sie es dem Patienten möglich macht, sich bewußt auf einen therapeutischen Dialog einzulassen (oder auch sich ihm zu verweigern). Kommt ein reflektierendes Gespräch zustande, fördert die Transparenz des Vorgehens ein „Reflexionswissen" des Patienten. Er gewinnt nicht nur Einsichten in einen bestimmten Krankheitszusammenhang, sondern lernt darüber hinaus den reflektierenden Umgang mit krankheitsbezogenen Problemen.

Warum mißlingt der Aufbau eines therapeutischen Dialogs in unserem Beispiel? Einerseits muß man sagen, daß die körperliche Abklärung der Müdigkeit zugunsten der Hypothese einer depressiven Verstimmung nicht ausreichend erfolgt, und daß die Wünsche des Patienten nach einer intensiveren körperlichen Versorgung einen berechtigten Kern haben. Andererseits ist es die nicht-transparente Gesprächsführung des Arztes, die diesen Ausgangskonflikt zu einem Beziehungsproblem werden läßt. In den Arztäußerungen unseres Textes sind die beiden Gesprächsmuster therapeutischer und alltagsweltlicher Koope-

ration extrem ineinander verschränkt. Damit entsteht die Gefahr, daß der jeweilige Wechsel für den Patienten uneinsehbar bleibt. Weil gleichlautende Äußerungen je nach dem kommunikativen Rahmen (dem Diskursmuster), in dem sie realisiert werden, unterschiedliche Funktionen erfüllen und unterschiedliche Antwortverpflichtungen, Fortsetzungsraster aufbauen, muß der Patient die Arztintention zwangsläufig mißverstehen, wenn seine Interpretation des kommunikativen Rahmens von der des Arztes abweicht. Die angemessene Antwort auf eine Rückfrage, wie sie der Arzt in unserem Beispiel oben gestellt hat (11/12), wäre im alltagsweltlich organisierten Gespräch eine Rechtfertigung und argumentative Stützung der Forderung, wie sie der Patient auch realisiert. In einem therapeutischen Bezugsrahmen würde eine solche Rückfrage aber dazu dienen, Sequenzen der Bedeutungsaushandlung zu initiieren (Kindt, 1984). Inwieweit sich der Patient darauf einlassen will, ist eine Frage, die sich unabhängig von den realisierten Kommunikationsstrukturen stellt. In unserem Beispiel spricht vieles dafür, daß es die Strukturen der ärztlichen Gesprächsführung sind, die eine Konfusion zwischen den zwei Kooperationsmustern herbeiführen und die es dem Patienten unmöglich machen, sich auf einen therapeutischen Diskurs einzulassen.

Nachdem wir gezeigt haben, wie der Zielkonflikt „Symmetrie" versus „Psychotherapie" für das Gesprächsverhalten des Arztes zu einem Verständigungsproblem in der Arzt-Patient-Interaktion werden kann, wollen wir anhand eines zweiten Textes zeigen, wie sich auch im beschränkten Zeitrahmen eines Visitengesprächs bei einer vergleichbaren Ausgangskonstellation, die sowohl körperbehandlungsbezogene Versorgung wie therapeutische Unterstützung erfordert, eine nach unserer Auffassung gewinnbringende Verbindung beider Elemente ergibt.

Der Patient ist hier ein 58jähriger Landwirt. Nach einem schweren Vorderwandinfarkt verlor er das Bewußtsein, mußte im Notarztwagen reanimiert und in die Ulmer Universitätsklinik eingeliefert werden. Auf der Intensivstation erlitt er einen Reinfarkt, und es trat eine Lungenentzündung auf. Nach 19 Tagen kann er dann auf die internistisch-psychosomatische Station verlegt werden. Die Visite findet dort am 2. Tag des Aufenthaltes statt. Der Patient wirkt auf der Aufzeichnung noch kurzatmig und insgesamt beeinträchtigt. Vor dem Gespräch hat die Ärztin von der Schwester erfahren, daß der Patient sich sehr gewünscht hat, daß ihn ein Pfarrer während des Krankenhausaufenthaltes besucht, und daß er sehr enttäuscht ist, weil dieser bisher nicht erschienen ist.

(Gesamtdauer 10 Minuten und 3 Sekunden)		
1	**A:**	Herr T., guten Morgen.
		(2 sec)
	S 1:	(Guten Morgen)
		(3 sec)
5	**S 2:**	hmhm
		(2 sec)
	P:	Mir geht's gut.
	A:	Ja?

P:	Hab gut geschlafen, (..) die Gymnastikleh-	
10	rerin hat heut den neuen Reiseplan aufge-	
	hängt, (Tag für Tag) draußen (Han i wieder	
	gesehe.)	
A:	Waren S' schon mal auf dem Flur?	
P:	Ja.	
15 A:	Und?	
P:	/(3 Worte unverständlich, setzt mit einer	
	Äußerung ein)	
A:	Wie gefällt's / Ihnen draußen?	
P:	Ach, bin halt noch a bißle osicher.	
20 A:	mhm	
P:	Aber es geht. Es steht noch a Weile an, des	
	merk i scho da.	
A:	Strengt Sie's an?	
P:	Ach, wenn i, des möcht i gar net bsonders	
25	sage. War auch heut s'erste Mal draußen auf	
	'em Klo.	
A:	mhm (..) Gibt's denn irgendwas, was Ihnen	
	nicht so, oder was Sie gern besser hätten?	
P:	I weiß net, warum daß ich manchmal, von	
30	hier bis hier, da (3 Worte unverständlich)	
	ich find da gar koin Ausdruck. (Anmerkung:	
	Patient präsentiert seinen Arm.)	
A:	Unverändert?	
	(Anmerkung: körperliche Untersuchung bis	
35	ca. Zelle 81)	
P:	Ja, / äh,	
A:	oder /	
P:	äh, / äh	
A:	Wenn / ich so entlangfahr', fühlt sich des	
40	überall gleich an?	
P:	Ja, des fühlt sich überall gleich an. Ha, des	
	isch manchmal a bißle schwer, i woiss net.	
A:	Aber nur an dem kleinen Stück da?	
P:	Ja.	
45 A:	Net der ganze Arm schwer?	
P:	Nein. Ich weiß net, wird's besser, wenn i	
	lauf, oder / wenn i	
A:	Tut's / weh, wenn man da hindrückt?	
	(4 sec)	
50 P:	/ Nein.	
A:	Da / haben S' doch den Katheter dringe-	
	habt, gell?	
P:	Ich weiß net.	
A:	doch, doch	
55 P:	Ja? (.) Daß' daher ist. Ja, do, wo Sie jetzt	
	grad nadrückt han.	
A:	Da hat's wehtan?	
P:	Ja, han i gspürt, das	
A:	Hier?	
60 P:	Ja.	
A 2:	Auch wenn man drückt, und auch wenn Sie	
	(ihn) bewegen?	
A:	(/ Wissen Sie)	
P:	Bitte?/	
65 A 2:	Wenn Sie den Arm bewegen auch?	
P:	Nein, des kann ganz in Ruhestellung /sein.	
A:	Mhm/ Also was mir am ehe, als erstes jetzt	
	dazu einfällt, ist, daß da wahrscheinlich 'n	
	kleiner Bluterguß drin ist. Wissen /Sie,	
70 P:	Ja./	
A:	die Vene ist ja ganz schön geärgert wor-	
	den, hm, durch den Katheter. Und daß da	
	noch a bißel so ein Bluterguß drin ist und	
	das drückt vielleicht auf den Nerven, der	
75	hier läuft, ja, und reibt den a bißel.	
P:	Des ischt aber net lang. Des kann nach a	
	paar Minuten kann alles wieder weg sein.	

A:	Ist Ihnen/ des aufgefallen, ob es in einer	
	bsonderen Lage so eher kommt?	
80 P	Ha, ich, i tu meinen Arm dann hoch, und	
	aber i kann's net sage.	
A:	Mhm. Und es ist nur an dem Arm?	
P:	Ja.	
A:	Am anderen net?	
85 P	Nein.	
	(5 sec)	
A:	Und das ist so alles, was es,	
P:	Ja.	
A:	Sie /beunruhigt?	
90 P:	Ja./	
A:	Ja?	
P:	Ja, ja.	
A:	Und sonst so, von der Betreuung her?	
P:	Gut. Gut.	
95 A:	Ja? Ich dachte, Sie hätten a bißl eigentlich	
	bedauert, daß noch kein Pfarrer bei Ihnen	
	war?	
P:	Ja, des, (.) weil, weil unser Gemeindepfar-	
	rer, der isch auch in ärztlicher Behandlung.	
100 A:	Ah ja.	
P:	Und die Frau Pfarrer hat gleich nach Ulm	
	agrufe, und (.) es hat sich bis jetzt nichts	
	getan. Und ich kenne a paar in Ulm, einige	
	Pfarrer, persönlich,	
105 A:	mhm	
P:	und meine Frau hat einen angesprochen,	
	der vor acht Tag bei uns a Predigt draußen	
	ghalte hat, soll doch er mal 'n Herrn T. be-	
	suche, aber er hat in der letzten Woch auch	
110	koi Zeit ghabt. (.) Aber heut soll ja	
A:	Der Pfarrer Sch., äh, Sch. heißt er glaub	
	ich? (zur **S**:)	
S:	ja.	
A:	wieder hier ist. Ond den möchten Sie dann	
115	gern sehen? (.) Ham'S viel mit ihm zu be-	
	sprechen?	
P:	Ah, ich, i war zwölf Jahre im Kirchenge-	
	meinderat und zwölf Jahre in der Bezirks-	
	synode.	
120 A:	Ah ja.	
P:	Und no interessiert man sich doch weiter-	
	hin auch. (drei Worte unverständlich)	
A:	mhm.	
P:	Ich fend, wenn ma so vierzehn Tag, drei	
125	Woche unten in der Intensivstation liegt	
	und dann so aufwacht, (..) würd's gar nix	
	schade, wenn da mal a Pfarrer a guts Wort	
	spreche tät.	
A:	Des bräuchten Sie doch, gell. Sie sagen	
130	zwar immer im ersten, wenn man rein-	
	kommt, daß es Ihnen so gut geht, aber wie	
	Sie mich da neulich nach, danach gfragt	
	ham, was jetzt mit dem Herz so los ist und	
	was da eigentlich jetzt so (.) passiert ist und	
135	passieren wird, da hab ich doch dacht, daß	
	Sie sich halt schon (.) Sorgen auch ma-	
	chen, oder wie's weiterläuft.	
P:	Ich meine, es ist halt, darüber bin ich mir	
	völlig im klaren, daß mein Gesundheitszu-	
140	stand nicht mehr ist wie er war und daß,	
A:	mhm	
P:	daß eben nicht viel dazukommen kann.	
	Daß ich da (.), damit rechnen muß, daß ich,	
	daß ich net alt werd.	
145	(3 sec)	
A:	Ja?	

	P:	Des isch mei Meinung.
	A:	mhm.
	P:	Je nachdem, des kommt drauf an.
150	A:	Mhm, ja, zumindestens, ich glaub auf der Intensivstation haben sie auch schon so ähnlich mit Ihnen gredet, (.) daß, ja /daß
	P:	Ja/
	A:	man damit rechnen muß. /Ja.
155	P:	Ja./
	A:	mhm
	P:	auch der, (..) ich weiß den Namen vom Arzt (ein Wort unverständlich) grad nicht.
	A:	Der B., der mit dem Bart?
160	P:	ja
	A:	Mhm. Doktor /B.
	P:	Der/ der hat zu meiner Frau gsagt, wenn ich mal rauskomm und Ihre Mann hat a Ga-bel in der Hand, no geht's ihm schlecht.
165		(Patient lacht)
		Weil ich soll scheinbar nicht mehr (.) viel arbeiten oder schwer überhaupt nicht.
	A:	Mhm.
	P:	Des muß mir klar sein, scheinbar.
170	A:	Und wie ist des für Sie, die Aus/sicht?
	P:	Ja,/ i war bislang sehr aktiv.
	A:	Mhm.
	P:	Gibt eine Umstellung, aber (.) /(ein Wort un-verständlich)
175	A:	Sagen Sie/, Sie ham vorher überhaupt kei-ne Beschwerden ghabt?
	P:	Nein.
	A:	Haben'S voll garbeitet?
	P:	Ja.
180	A:	Ja? (.) Oder /haben S'
	P:	War sechzig/
	A:	manchmal schon ein bißel sich schonen müssen?
	P:	Nein.
185	A:	Gar nicht?
	P:	Ich war sechzig Prozent kriegsbeschädigt, i war abends oft schon au scho arg müde.
	A:	Mhm, (.) ja.
	P:	Ich war allerdings, äh, vorher doch in Be-
190		handlung.
	A:	Ja, weshalb eigentlich, wenn Sie keine Be-schwerden hatten?
	P:	I hab immer rheumatische Schmerzen.
	A:	Ja, /aber
195	P:	(zwei Worte unverständlich)/
	A:	Sie haben doch schon auch dieses Marcu-mar bekommen ghabt.
	P:	Ja, und dann (..) isch der Arzt, der mir, der mir die Spritze gebe hat, isch dann in Ur-
200		laub, dann sagt mei Tochter, geh doch amal zu meim Arzt, die isch im Neu-Ulmer Krankenhaus. Und der hat dann festge-stellt, daß doch mei Blut zu dick sei. Und hat mir dann die
205	A:	mhm
	P:	Tabletten verschriebe. Und i glaub halt, daß es dann doch schon viel zu spät war.
	A:	Und Sie haben nie was gemerkt, (.) nie Be-schwerden gehabt, haben voll gearbeitet?
210	P:	Ja, ich, ich sag no ja zu meiner Frau, ich (ha i hab heut) wieder einen Arbeitsgeist. (la-chend gesprochen)
		(2 sec)
	A:	Wieder?
215	P:	Ja.

	A:	Also haben Sie eine Zeitlang doch schlech-ter arbeiten können?
	P:	Ja, manchmal war man scho müder, (gell des) auf die Witterungseinflüsse und so.
220	A:	Wie ist denn des? Haben Sie eigentlich schon übergeben? Das ist bei Bauern doch immer ein ganz wichtiger Zeitpunkt, wenn sie ein/en
	P:	Nein./
225	A:	Haben Sie einen Sohn?
	P:	Ja. Ja.
	A:	Und ham noch net übergeben?
	P:	Nein, aber (.) (des mach i gewieß), wenn ich wieder soweit hergestellt bin, daß des über die Bühne geht.
230	A:	Des werden'S jetzt machen?
	P:	Ja, ja. (.) Des hat ja keinen Wert. Wenn die andere Leut die Arbeit machen (er lacht kurz) und i bloß angeb (..). Des geht ja
235		nicht.
	A:	Fällt Ihnen das schwer?
	P:	Ja, das fällt schon schwer. (..) Ma hat fünf Kinder. (Aber) des geht schon.
	A:	Ja.
240	P:	Die sind alle so einsichtig.
	A:	Ja? (deutliche Frageintonation)
	P:	Ja, (..) hoffen wir's (Möglichste).
	A:	Ja? (deutliche Frageintonation)
	A 2:	Sind Sie nicht ganz sicher?
245	P:	Bitte?
	A 2:	Sie sind sich nicht ganz sicher?
	P:	Ach, aha (...), weil wir ham einen ganz neu-en Hof gebaut habet, nicht?
	A:	/Mhm.
250	P:	Der hat/ einiges gkostet und (.) damit, daß die andere vielleicht da e bißle ins Hinter-treffen kommen, dadurch.
	A:	Weil /man des halt
	P:	des isch ja/
255	A:	immer nur einem geben kann, gell.
	P:	jaja, jaja
		(2 sec)
	A 2:	Mhm.
	A:	Und des fällt Ihnen schwer, die Entschei-
260		dung?
	P:	Oh, noi. (..) Und es war doch so: i war jetzt doch die ganze Ernte weg und die hant alle gut zusammengschafft, (.) so gut (zwei Worte unverständlich)
265	A:	Und später gehört der Hof einem und die anderen werden weiter mitarbeiten?
	P:	Ja, die ham ja nachher ihren Beruf.
	A:	mhm
	P:	Aber sie werden in der Freizeit ihn do unter-
270		stütze.
	A:	mhm
	P:	Weil sie ja wisset von Jugend auf, wie's ebe isch, daß ebe doch Arbeitsspitzen gibt, die man allein nicht gut bewältige kann.
275	A:	Mhm (..) Aber Sie hoffen auf die Einsicht.
	P:	Ja.
	A:	Haben'S (.) Gründe dafür, daß Sie guter Hoffnung sein können, Anhaltspunkte?
	P:	Doch, /doch.
280	A:	Ja?/ Gibt's noch etwas, wo Sie a bißel im Zweifel sind, ob's klappt?
	P:	Ach, des glaub ich nicht.
	A:	Glauben S' nicht. (..) Gut, Herr T., dann las-sen Sie sich /Ihr Essen

285	**P:**	Vielen Dank/
	A:	schmecken, gell.
	A 2:	Mahlzeit
	A:	Wiedersehn.
	P:	Wiedersehn.

A = visiteführender Arzt; **A 2** = begleitender Arzt; **S 1**, **S 2** = Schwestern; **P** = Patient. (Text) = ungenau verständliche Passage; /Text/ = simultan gesprochene Passagen; (. . .) = Pausen unter 2 sec.

Wie man aus dem Gesprächsablauf sieht, hatte sich die Ärztin vorgenommen, über die Enttäuschung mit dem Patienten zu reden. Daß der Patient nicht spontan und offen negative Erlebnisse und Gefühle äußern kann, ist hierbei offensichtlich ein allgemeineres Problem. So ist er beispielsweise – bezogen auf sein aktuelles Befinden – zu Gesprächsbeginn deutlich bemüht, der Ärztin gegenüber eine Fassade von positiver Gelassenheit und Zuversicht aufrechtzuerhalten (7–12), obwohl sich später (29–31) zeigt, daß er über die Schmerzen in seinem Arm durchaus beunruhigt ist. Auch in bezug auf seine Krankheitsgeschichte (175 ff.) läßt sich feststellen, daß der Patient dazu neigt, Krankheits- und Schwächegefühle aus seinem Erleben zu verbannen. Diese grundlegenden Reaktionsmuster des Patienten scheinen der Ärztin klar zu sein, und damit stellt sich für sie unter klinischen Zielsetzungen die Aufgabe, der Tendenz des Patienten zur Krankheitsverleugnung entgegenzuwirken.

Um nachzuvollziehen, wie das in dieser Visite möglich ist, wollen wir den Ablauf des Gesprächs näher betrachten. Dazu ist es nötig, eine Ablaufstruktur, eine Gliederung des Gesprächs zu bestimmen. Unterschiedliche Einteilungsgesichtspunkte sind möglich. Zum einen kann man nach Einschnitten Ausschau halten, die durch das Verhalten der Gesprächsteilnehmer selbst als Zäsuren markiert werden (Pausen, Beendigungs-, Gliederungs-, Einleitungssignale oder Floskeln, Themenwechsel usf.). Zum anderen kann man ein Gespräch unter Gesichtspunkten einteilen, die aus theoretisch vorgegebenen, beispielsweise klinischen Analyseinteressen resultieren. Eine solche Einteilung muß nicht notwendigerweise der Perspektive aller Gesprächsteilnehmer entsprechen. Zur Analyse unseres Gesprächs verwenden wir beide Arten von Kriterien, wobei der theoretisch-klinische Gesichtspunkt durch unser Interesse gebildet wird, herauszufinden, wie die Ärztin und der Patient die beiden Aufgabenkomplexe einer patientzentrierten Visite abwickeln, und insbesondere mit welchen Mitteln und über welche Stufen es die Ärztin erreicht, den Patienten in einen therapeutisch-reflektierenden Dialog zu ziehen.

Legen wir diese Kriterien an, kommen wir zu folgendem Ablaufschema unserer Visite:

1–6	Begrüßung/Gesprächseröffnung
7	„Antwort" des Patienten auf die antizipierte Frage, „Wie geht es Ihnen?":
26		„Mir geht's gut" und Begründung dieser Feststellung
27 28	**A:**	mhm (..) Gibt's denn irgendwas, was Ihnen nicht so, oder was Sie gern besser hätten?
29 85		Patient schildert Körperbeschwerden und initiiert damit eine körperliche Diagnostik als Antwort der Ärztin.
86		(Pause: 5 sec)
87	**A:**	Und das ist so alles, was es,
88	**P:**	Ja.
89	**A:**	Sie /beunruhigt?
90	**P:**	Ja./
91	**A:**	Ja?
92	**P:**	Ja, ja.
93	**A:**	Und sonst so, von der Betreuung her?
94	**P:**	Gut. Gut.
95 96 97	**A:**	Ja? Ich dachte, Sie hätten a bißel eigentlich bedauert, daß noch kein Pfarrer bei Ihnen war?
98 128		Gespräch über das Ausbleiben des Pfarrers, in dem die Enttäuschung des Patienten anklingt.
129 137	**A:**	Des bräuchten Sie doch, gell. Sie sagen zwar immer im ersten, wenn man reinkommt, daß es Ihnen so gut geht, aber wie Sie mich da neulich nach, danach gfragt ham, was jetzt mit dem Herz so los ist und was da eigentlich jetzt so (.) passiert ist und passieren wird, da hab ich doch dacht, daß Sie sich halt schon (.) Sorgen auch machen, oder wie's weiterläuft.
138		Therapeutisch motivierter Diskurs: (a) Prognose (b) Krankheitsverleugnung (Krankheitsgeschichte) (c) Krankheitsfolgen, soziale Prognose
283– 289		Gesprächsbeendigung/Verabschiedung

An der Gesprächseröffnung (1–6) sind zwei Dinge bemerkenswert: Zum einen fehlt ein Gegengruß des Patienten, zum anderen gibt es eine längere, mit Hintergrundgeräuschen gefüllte Pause zwischen dem Gruß der Ärztin und der ersten Äußerung des Patienten.

Der erste Punkt läßt sich durch allgemeine Situationsbedingungen erklären. Die Begrüßung in der Visite ist eine komplexe Aktivität. Üblicherweise grüßen der visiteführende Arzt und auch die weiteren Teammitglieder beim Betreten des Zwei-Bett-Zimmers beide Patienten. Beide Patienten grüßen dann üblicherweise zurück. Danach wendet sich der visiteführende Arzt dem Patienten zu, mit dem er zuerst sprechen will und beginnt das Gespräch zumeist mit einem zweiten Gruß, der oft mit einer Anrede verbunden wird oder auch mit einer Partikel mit oder ohne Anrede („So, Herr . . .", „Ja, Frau . . ."). Daß der zweite Gruß durch solche Signale, die der Aufmerksamkeitsfokussierung dienen, ersetzt werden kann, zeigt, daß er nicht im konventionellen Sinn als Gruß gemeint ist. In Verbindung mit anderen Aktivitäten (Hinsetzen am Bett des Patienten, Blickkontakt) erfüllt er vielmehr tatsächlich die Funktion der Aufmerksamkeitslenkung. Aus dem gleichen Grund besteht keine Verpflichtung für einen Gegengruß des Patienten. Im allgemeinen schließt der Arzt eine erste thematische Initiative, eine Frage nach dem Befinden des Patienten an.

Dies passiert in unserer Visite nicht. Vielmehr schafft die Ärztin dadurch, daß sie nicht fragt „Wie geht es Ihnen (heute?)", eine interaktive Leerstelle. Die entstehende Pause bringt den Patienten in die Position, von sich aus die erste Initiative zu ergreifen. Man könnte diese Form der Gesprächsführung in der Eröffnungsphase deshalb als „offen" bezeichnen. Andererseits ist es nur eine relative Offenheit, das zeigt die erste Initiative des Patienten.

Diese Initiative (7) ist ihrer Form nach genau die Antwort auf die übliche Eröffnungsfrage. Für dieses Verhalten des Patienten gibt es zwei – sich ergänzende – Erklärungen: Zum einen ist es einfach Ausdruck davon, daß der Patient gelernt hat, wie eine Visite üblicherweise abläuft und wie seine Rolle in diesem Ablauf definiert ist. Nimmt man noch hinzu, daß der Patient hier eine beschönigende und übertrieben optimistische Darstellung seines Befindens gibt, läßt sich zum anderen sagen, daß die Bereitwilligkeit, mit der der Patient die Rolle eines „Muster-Patienten" erfüllt, als Ausdruck seiner Überangepaßtheit und seiner Konfliktvermeidungstendenzen verstanden werden kann. Auf jeden Fall kann die Ärztin nach diesem Gesprächsbeginn sicher sein, daß der Patient das für ihn belastende Thema „Enttäuschung über den ausbleibenden Pfarrer" nicht anschneiden wird. Sie muß, um ihr Gesprächsziel zu erreichen, selber aktiv werden. Wie macht sie das?

Die im Ablaufschema zwischen den nach Themen zusammengefaßten Blöcken wörtlich wiedergegebenen Passagen bezeichnen die verschiedenen Ansätze der Ärztin, das kritische Thema einzubringen. Es sind Versuche, die Visite von einem körperbezogenen Gespräch in einen psychotherapeutischen, ein psychosoziales Problem reflektierenden Dialog zu überführen.

Zwei Dinge fallen beim Vorgehen der Ärztin auf. Einmal steigert sie Deutlichkeit und Intensität der Versuche sukzessiv mit verschiedenen sprachlichen Mitteln, zum anderen macht sie diese Versuche nicht eher für den Patienten zwingend, als nicht dessen aktuelle Bedürfnisse (eine Erklärung für seine körperlichen Beschwerden zu bekommen) im Rahmen der gegebenen Möglichkeiten befriedigt worden sind.

Die anfänglich gebrauchte ambivalente Formulierung (27/28), „Gibt's denn irgendwas, was Ihnen nicht so oder was Sie gern besser hätten?" ist zwar vom späteren Gesprächsverlauf eindeutig als Versuch zu identifizieren, das kritische Thema anzuschneiden. Ihre Offenheit im Hinblick auf eine körperbezogene oder psychische Problematik erlaubt es dem Patienten aber durchaus noch, diese Frage auf sein körperliches Befinden zu beziehen, was dem momentan für ihn bedeutsamen Problem entgegenkommt (die durch die Venenkatheterisierung verursachten Schmerzen im Arm) und wohl eher seinen Erwartungen an die üblichen Themen eines Arzt-Patient-Ge-

spräches entspricht. Indem die Ärztin in der folgenden Sequenz ihren Plan zurückstellt und auf die vom Patienten geschilderten Körperbeschwerden eingeht, übernimmt sie dessen Situationsdefinition und seine Themenpräferenzen. Sie erfüllt also die Erwartungen des Patienten. Obwohl die Ärztin bei oberflächlicher Betrachtung die Gesprächssteuerung in dieser Phase des Gesprächs fast vollständig übernimmt (sie stellt viele „geschlossene" Fragen, denn sie will zu einer Diagnose kommen), verhält sie sich im Sinne einer übergeordneten Interaktionsstrategie vollständig – für den Patienten erkennbar – „patientzentriert", insofern dieses Verhalten genau die Erwartungen des Patienten nach einer gründlichen ärztlichen Untersuchung erfüllt. Die Plazierung der körperlichen Untersuchung erfolgt in dieser Visite also als Reaktion auf die Beschwerdeschilderung des Patienten. Neben der objektiven Funktion der Befunderhebung und somatischen Abklärung erhält sie damit einen positiven kommunikativen Stellenwert: Der Patient kann den Eindruck haben, daß die Ärztin seine Beschwerden ernst nimmt und adäquat beantwortet.

Der nächste Versuch der Ärztin, in den von ihr angestrebten psychotherapeutischen Dialog einzutreten, findet sich erst, nachdem die durch die Beschwerden des Patienten initiierte Episode als vollständig abgehandelt gelten kann. Dieser Versuch (87, 89) ist ähnlich wie der erste offen gehalten. Während er auf das von der Ärztin angestrebte Thema hinleitet, bietet er dem Patienten noch einmal die Möglichkeit, andere Dinge, die ihn belasten – auch körperbezogene –, zur Sprache zu bringen. Die Offenheit dieser „Intervention" der Ärztin erfüllt bei näherem Hinsehen, also genau wie dies bei der ersten der Fall war, eine dreifache Funktion. Eine psychotherapietechnische, etwa „Verbalisieren eines bestimmten emotionalen Erlebens beim Patienten", eine der allgemeinen Beziehungsförderung im Dialog, „explizites Angebot an den Gesprächspartner, eigene Bedürfnisse zu artikulieren", und eine Orientierungsfunktion für die Ärztin. – Eine Orientierungsfunktion in doppelter Hinsicht: Einmal kann die Ärztin anhand der Antwort des Patienten erkennen, ob sie in ihrem Versuch, auf das kritische Thema zu kommen, weitergehen kann oder ob beim Patienten momentan noch für ihn dringlichere Probleme vorliegen, die ihn in seiner Aufnahmebereitschaft vielleicht blockieren würden. Zum anderen erfüllt sie aber auch noch eine Orientierungsaufgabe im Rahmen der somatischen Funktionen der Visite. Erst wenn sichergestellt ist, daß die Körperbeschwerden keine unmittelbare somatische (Krisen-)Behandlung erfordern, kann die Ärztin sich den weitergehenden psychotherapeutischen Aufgaben der Visite zuwenden.

Trotz aller Offenheit enthält dieser zweite Ansatz der Ärztin auch eine Intensivierung gegenüber dem ersten. Beim ersten Mal versucht die Ärztin die vermutete Verärgerung in einer „Wunschformulierung" anzusprechen, also mit dem Komplement des negativen Affekts „Beunruhigung", „Verärgerung" zu arbeiten. Der zweite Versuch nennt ausdrücklich den unterstellten negativen Gefühlszustand.

In den folgenden Ansätzen gebraucht die Ärztin drei weitere Mittel der Intensivierung: Insistierende Nachfrage (91), thematische Eingrenzung, die gezielt im Hinblick auf das angestrebte Thema, „Enttäuschung des Patienten über den ausbleibenden Besuch des Pfarrers", erfolgt, das Thema aber noch nicht explizit benennt (93), Konfrontation (im Sinne der dritten Kategorie unserer konfrontierenden Interventionen: Ansprechen von „Widersprüchen innerhalb der Bereiche oder zwischen den Bereichen Erleben, Äußern, Verhalten des Patienten"), die einen Widerspruch zwischen verschiedenen Äußerungen des Patienten aufzeigt und gleichzeitig explizite Benennung des angestrebten Themas (95–97).

Der Patient geht in der folgenden Gesprächsepisode auf die Themenwahl der Ärztin ein und äußert in der Form eines ironisch abgemilderten Vorwurfs („würd's gar nix schade, wenn da mal a Pfarrer a guts Wort sprechen tät" – 126–128) sein Gefühl der Kränkung und Verärgerung. In dieser Passage ist das erste Kriterium für ein therapeutisches Gespräch erfüllt. Ärztin und Patient sprechen gemeinsam über ein „psychosoziales" Thema, ein begrenztes, den Patienten aber aktuell besonders bedrückendes Problem.

Wiederum über eine Konfrontation verwirklicht die Ärztin dann auch das von uns erwartete zweite Kriterium für ein psychotherapeutisches Gespräch: die Behandlung des allgemein zugrundeliegenden Themas der Verleugnung im Sinne einer kritischen Reflektion. Sie verallgemeinert die in der Visiteninteraktion im konkreten Fall aufgetretene Diskrepanz zwischen Empfindungen und beschönigender Darstellung des belastenden Sachverhalts durch den Patienten, indem sie an dieser Stelle (129–137) eine Diskrepanz im emotionalen Erleben des Patienten aufzeigt: sein Bedürfnis nach Unterstützung und Trost einerseits und seine Schwierigkeit, sich seine Hilflosigkeit und die daraus resultierenden Bedürfnisse zuzugestehen. An dieser Stelle ist der Übergang von einem an Alltagserwartungen orientierten Interaktionsverhalten zu einem an psychotherapeutischen Handlungsmustern orientierten vollzogen, ist der Prozeß der „Umdefinition der Gesprächssituation" zu einem vorläufigen Abschluß gekommen.

Das Ergebnis unserer Untersuchung im Hinblick auf die von der Ärztin benützten konversationellen und sprachlichen Mittel läßt sich im wesentlichen in drei Punkten zusammenfassen:

– Die Ärztin benutzt eine flexible, gegenüber den Interessen des Patienten sensible Strategie bei der Einführung der psychisch akzentuierten Themen, sowie eine Technik sukzessiver Intensivierung. Die verdeckten Erwartungen des Patienten werden zuerst geklärt und soweit wie möglich abgehandelt, sie können damit nicht zu Konflikten im späteren Gespräch führen. Die Intensivierung der Interventionsansätze geht mit dem Fortschritt dieses Klärungsprozesses einher.

– Die Ärztin arbeitet mit dem sprachlichen Mittel offener Formulierungen bei den verschiedenen Ansätzen, das Gespräch als psychotherapeutisches zu führen.

– Unter Gesichtspunkten „klinischer Strategie" benutzt die Ärztin ein begrenztes, aber aktuell besonders relevantes und situationsnahes Problem zum Einstieg für ein Gespräch über allgemeinere Probleme des Patienten.

18.4 Voraussetzungen der patientzentrierten Gestaltung einer psychosomatischen Visite

Abschließend wollen wir einige Voraussetzungen und Kriterien nennen, die nach unserer Auffassung für eine erfolgversprechende und dauerhafte Umgestaltung der Visite erfüllt sein müssen. Auf jeden Fall ist eine Änderung der Visitenablauforganisation wie in Abschnitt 18.3 dargestellt eine unbedingt notwendige Voraussetzung.

Die Textbeispiele des letzten Abschnitts haben ebenso deutlich gemacht, daß die Anforderungen an die Stationsärzte wachsen. Je mehr sie sich auf eine persönliche Beziehung einlassen, desto stärker werden die emotionale Belastung und der Arbeitsaufwand. Entsprechend steigen die Ansprüche an die therapeutischen und kommunikativen Kompetenzen der Ärzte. Sie brauchen zusätzliche Unterstützung, um ihre Aufgaben erfüllen zu können. Optimalerweise umfaßt eine Unterstützung dabei fünf Bereiche: Konzeptvermittlung (psychologischer, psychosomatischer und psychopathologischer Theorien), Hilfe beim Erwerb der Fähigkeit zur Reflexion von Sozialbeziehungen im Berufsfeld, Selbsterfahrung, Supervision für Interventions- und Therapieplanung und Techniken der Gesprächsführung.

Eine psychotherapeutische Weiterbildung für die Stationsärzte würde die breiteste Basis für den Erwerb psychotherapeutischer Kompetenzen bilden (vgl. Kap. 68). Läßt sich diese Möglichkeit nicht realisieren, so würden Balint-Gruppen (die unabhängig von einer psychotherapeutischen Weiterbildung angeboten werden) die Möglichkeit sichern, Beziehungen zu reflektieren und damit einen Teil der emotionalen Belastungen aufzuarbeiten. Eine Fortbildung in personzentrierter Gesprächsführung und/oder in personzentrierten Formen der Anamnese- bzw. Interviewtechnik (vgl. Kap. 12; Froelich und Bishop, 1973; Meerwein, 1984) könnte die Vermittlung basaler Fähigkeiten therapeutischer Kommunikation garantieren. Schließlich sollten auch ganz konkrete, auf die Behandlungsplanung und die Gesprächsführung in der Visite bezogene Hilfen bereitstehen. Klinisch ist hier die **Supervision** durch einen erfahrenen Psychosomatiker unerläßlich. Von der gesprächstechnischen Seite ist an eine fallbezogene Zusammenarbeit von Klinikern mit Kommunikationsspezialisten (Kommunikationspsychologen, Gesprächsanalytikern) in Video- oder Audioseminaren zu denken.

Es gibt sicher eine Reihe von Problemen in der Arzt-Patient-Interaktion, die tatsächlich visiten- oder zumindest krankenhausspezifisch sind oder in der Visite verschärft auftreten. Über das schon besprochene Grundproblem einer psychosomatischen Gesprächsführung hinaus denken wir dabei beispielsweise an die konkreten Problemsituationen:

Fokussieren des Gesprächs: Auch in der psychosomatischen Visite ist Zeit ein knappes Gut. Das Gespräch über psychische Probleme muß deswegen fokussiert, d.h. auf die aktuell besonders relevanten Aspekte eingegrenzt werden. Nicht viele Facetten eines Problems sollen angerissen werden, sondern die beim Patienten im Vordergrund des Erlebens stehenden Schwierigkeiten sollen punktuell vertieft werden. Die Möglichkeit für weitergehende Einzelgespräche außerhalb der Visite sollte bestehen; die Führung solcher Gespräche kann aber nicht die Regel sein.

Störung des Arzt-Patient-Verhältnisses und der Compliance: Die Compliance-Problematik hat im stationären Bereich zum Teil eine andere Prägung als in der ambulanten Versorgung. Die subjektive und objektive existentielle Bedrohung der Gesundheit, die die Krankenhauseinweisung nötig gemacht hat, die Situation der Hospitalisierung selber, die mit ihr einhergehenden Zwänge verschiedener Art, einschließlich des Zwanges, einem bestimmten zumeist nicht frei wählbaren Arzt zugewiesen zu werden, kann eine kritisch-aggressive, Behandlungsmaßnahmen abwehrende Haltung des Patienten auslösen. Wichtig ist es hier für den Arzt, nicht auf die Aggressivität und die mit ihr verbundene Kränkung durch den Patienten zu reagieren, sondern sie im Gespräch zu hinterfragen, zu versuchen, die dahinter liegende Angst und Verzweiflung zu verstehen, zu verbalisieren und gemeinsam mit dem Patienten zu reflektieren.

Die Problematisierung von Laientheorien: Eine im Krankenhaus gestellte Diagnose bedeutet zumeist, daß sich Patienten auch kognitiv mit in ihrem Körper ablaufenden Krankheitsprozessen auseinandersetzen müssen. Diese Auseinandersetzung ist immer mitgeprägt durch naive, vorwissenschaftliche Vorstellungen über bestimmte Krankheitsbilder, Körpervorgänge usf. Es ist wichtig, diese Laientheorien im Gespräch explizit zu machen, sie zu verstehen und – wenn nötig – zu korrigieren. Korrektur bedeutet hier mehr als Aufklärung im naturwissenschaftlichen Sinn. Laientheorien zur eigenen Erkrankung und emotionale Prozesse der Krankheitsverarbeitung bilden für den Patienten natürlich subjektiv eine untrennbare Einheit. Die Gesprächsführung muß diesem komplexen Zusammenhang Rechnung tragen. Aufklärung ohne Berücksichtigung der an vielleicht falsche Vorstellungen gebundenen Ängste wird ebenso erfolglos sein wie eine Exploration des emotionalen Erlebens, die nicht nach den beim Patienten bestehenden Vorstellungen zum Körpergeschehen fragt.

Aktivierung passiver Patienten und Abbau von Macht: Den Patienten zum aktiven Partner, zum Experten der eigenen Krankheit und Behandlung zu machen, ist im Krankenhaus besonders schwierig.

Diese Zielsetzung stößt an die durch die Struktur und Vorschriften der Institution gezogenen Grenzen. Eine wesentliche Aufgabensetzung der Gesprächsführung in der Visite besteht deswegen darin, zumindest unnötige Anteile von Machtausübung in Frage zu stellen und den Patienten – soweit es geht – in aktiver Weise in die Behandlung einzubeziehen, beispielsweise dadurch, daß man ihn bei der Medikation nach Vorerfahrungen fragt und die erhaltene Information berücksichtigt, daß man ihn bei der Entscheidung über Behandlungsalternativen beteiligt usf.

Ein besonderes Problem, das hier nicht besprochen wird, ist der Umgang mit unheilbar kranken und sterbenden Patienten (vgl. Kap. 66; Meerwein et al., 1976; Verwoerdt, 1966; Meerwein, 1981, 1984).

Bei allen genannten Problemen können die in der Literatur gesammelten Erfahrungen aus psychotherapeutischen Gesprächen, ambulanter Behandlung und aus Beratungsgesprächen wertvolle Anregungen und Orientierungshilfen geben. Sie ersetzen aber nicht die Forschung und die klinische Reflexion für den Handlungsraum des Visitengesprächs. Aus diesem Grund halten wir die fallbezogene Diskussion der Gesprächsführung in der Visite – zumindest im Kollegenkreis, besser noch unter Supervision eines Experten – für eine wertvolle Hilfe.

Aber nicht nur die Ärzte, sondern auch die Schwestern brauchen Hilfen bei einer Umgestaltung der Visite. Die Änderung der Visite in ein Arzt-Patient-Gespräch bedeutet natürlich gleichzeitig eine Verminderung der Beteiligung der Schwestern. Wichtig ist es, hier einen Ausgleich zu schaffen. Die Arbeitsorganisation für die Schwestern sollte so aussehen, daß über die pflegerische Versorgung hinaus genügend Möglichkeiten für Schwester-Patient-Gespräche bleiben (zimmerbezogenes Pflegesystem, Erstgespräch und Pflegevisiten durch die zuständigen Schwestern), andererseits muß es genügend Kommunikationsmöglichkeiten für Ärzte und Schwestern geben. Die Vor- und Nachbesprechung der Visite mit Beteiligung der Schwestern ist unter diesen Aspekten besonders wichtig. Darüber hinaus sollten feste Veranstaltungen für organisatorische und patientenbezogene Besprechungen bestehen (vgl. Kap. 28.2).

Insgesamt läßt sich natürlich eine Änderung der Visite um so leichter bewerkstelligen, je „patientzentrierter" die Gesamtorganisation einer Station angelegt ist. Daraus läßt sich die Empfehlung ableiten, eine Änderung des Visitenablaufs dann behutsam anzugehen, wenn die Möglichkeit zu einer grundlegenden Veränderung des Stationskonzepts nicht ohne weiteres besteht. Unter diesen Umständen ist die Strategie sinnvoll, schrittweise vorzugehen, d.h. die Visite zuerst nur bei einzelnen Patienten (zimmerweise) in der gewünschten Richtung zu ändern. Die Beteiligten haben so Gelegenheit, mit den auftretenden Effekten Erfahrungen zu sammeln und ihr eigenes Erleben zu reflektieren. Schwierigkeiten und Konflikte lassen sich begrenzen, und das Experimentieren mit der Visite gefährdet nicht den gesamten Arbeitsablauf auf der Station.

19 Verhaltenstherapie

Othmar W. Schonecke und *Christine Muck-Weich*

19.1 Einleitung

In den letzten Jahren hat sich im Bereich der psychosomatischen Medizin der Einfluß verhaltenstheoretisch orientierter Denkmodelle und Therapiemethoden erheblich verstärkt. Es gibt dafür eine Reihe von Ursachen. Unter einem historischen Blickwinkel sind zwei verschiedene Wege zu betrachten, die eine Verbindung von Lern- oder Verhaltenstheorie und Medizin betreffen. Diese Wege sind sehr eng mit wesentlichen Inhalten der Lernpsychologie verknüpft sowie dem biologischen Ausgangspunkt und Blickwinkel der Lerntheorie.

Der erste Weg besteht darin, daß die Verhaltens- oder Lerntheorie ihren Ausgangspunkt in den Arbeiten eines Physiologen hat, und autonome Reaktionen stets eine wesentliche Ebene von abhängigen Variablen im Bereich des klassischen Konditionierens bleiben. Dies hat sich bis heute nicht geändert, und Wolpe entwickelte (1958) eine verhaltenstherapeutische Behandlungsform der Angst, die zum einen wesentlich auf dem Modell des klassischen Konditionierens beruht, zum anderen „Angst" als ein psychophysisches Geschehen auffaßt, als „ein autonomes Reaktionsmuster, das charakteristischerweise Teil der Reaktion eines Organismus auf schädliche Reize ist". Diese Definition der Angst sieht in einer biologischen Denktradition die Beziehung eines Organismus zu seiner Umgebung als wesentlich für sein Verhalten, und dieses Verhalten betrifft den gesamten Organismus als psychophysische Gesamtheit.

Wolpe ging in seinen therapeutischen Überlegungen davon aus, es müsse ein der Angst entgegengesetzter psychophysischer Zustand hergestellt werden und der Patient, wenn er sich in diesem Zustand befindet, mit mäßigen angstauslösenden Reizen konfrontiert werden. Er ging von der Tatsache aus, daß Angst ein über das sympathische Nervensystem vermittelter Zustand ist, so daß ein damit inkompatibler Zustand parasympathisch vermittelt sein müsse. So kam er zur Anwendung der „progressiven Muskelrelaxation" nach Jacobson (1938).

Die Verhaltenstherapie der Angst war und ist zwar sehr wirksam, zumindest immer dann, wenn sie sich auf relativ gut überschaubare Störungen bezieht, die theoretische Diskussion der Angst ist jedoch auch heute noch nicht abgeschlossen (vgl. Kap. 8). Dabei geht es nicht so sehr um die Charakterisierung klinisch relevanter Angst, sondern eher um deren Zustandekommen und vor allem um deren langfristigen, d.h. chronischen Bestand. Es ist hier nicht möglich, die kontroversen Aspekte der Angstentstehung zu diskutieren; es ist unmöglich, klinisch relevante Angstzustände ausschließlich als konditionierte Furchtreaktionen zu betrachten, wie dies geschehen ist. Ein wesentlicher Grund besteht darin, daß einzelne physiologische Anteile konditionierter Furchtreaktionen phasischer Natur sind, nach kurzer Zeit, 10–20 Sekunden, wieder abklingen, zum zweiten recht rasch nach einigen Versuchsdurchgängen an Stärke abnehmen. Betrachtet man Tiere, die sich in einem Paradigma zur Konditionierung von Furchtreaktionen befinden, so fällt auf, daß sie recht ruhig sind, wenn kein konditionierter Stimulus dargeboten wird, sie haben gelernt, daß sie sicher sind, wenn der Stimulus nicht auftritt, der dem unkonditionierten Reiz vorausgeht. Im Gegensatz zur klinisch beobachtbaren Angst befinden sich die Versuchstiere in einer Situation relativer Sicherheit, was die Vorhersagbarkeit aversiver Ereignisse angeht, obwohl die Ereignisse nicht kontrollierbar sind. Dieses Problem wurde durch die Theorie der „Angstinkubation" (Eysenck, 1968) einer Lösung näher gebracht, nach der die physiologischen Anteile einer Angstreaktion nicht nur das Ergebnis von Angstreizen sind, sondern auch als angstinduzierende oder angstverstärkende Reize wirksam werden. Der Prozeß wird also nicht mehr linear, sondern dynamisch aufgefaßt. Dabei bilden die physiologischen Reaktionsanteile ihrerseits einen Reiz für einen Organismus, der als gefährlich und damit angstauslösend erlebt wird und so die Angst aufrechterhält. Angst wird damit als ein im eigentlichen Sinne psychosomatischer Prozeß aufgefaßt.

Der zweite historische Weg verlief zunächst gegensätzlich zu dem im Vorangegangenen geschilderten. In den Vereinigten Staaten wurde unter der Bezeichnung „Behaviorismus" eine empirisch sehr fruchtbare Denkrichtung entwickelt. In dieser wurden lediglich direkt der Beobachtung zugängliche Bedingungen und Verhaltensweisen als relevant für eine Verhaltensanalyse angesehen. Nach der Auffassung Skinners (1953) waren physiologische Reaktionen und physiologische Bedingungen von Verhalten für eine Analyse des Verhaltens nicht bedeutsam. Körperliche, nicht willkürliche, sondern autonome Reaktionen, die der direkten Beobachtung nicht zugänglich sind, wurden auf diese Weise von der funktionalen Analyse (vgl. Kap. 7) des operanten Denkmodells ausgeschlossen.

Entsprechend bestand zwischen diesen beiden Denkrichtungen eine zunächst unüberprüfte theoretische Übereinkunft, die darin bestand, daß ange-

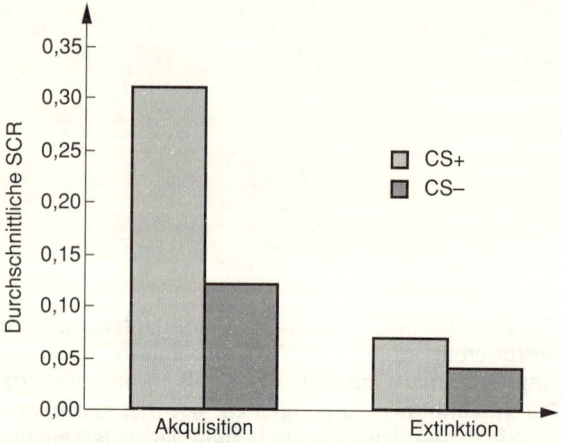

Abb. 19–1. Konditionierte Furchtreaktion, gemessen an der elektrodermalen Reaktionsamplitude (aus Frederikson, 1981).

nommen wurde, physiologische Reaktionen seien nur mit der Methode des klassischen Konditionierens zu beeinflussen und operantes Lernen betreffe nur die Ebene von willkürlichen Verhaltensweisen. Auf diese Weise wurde zwischen zwei Arten von Lernen unterschieden, die durch die jeweiligen Paradigmen der beiden Schulen definiert wurden, das operante Konditionieren als Lernen an der Beziehung des Verhaltens zu Ereignissen, die auf das Verhalten folgen, den positiven oder negativen Konsequenzen des Verhaltens, Belohnung oder Strafe, und das klassische Konditionieren als Lernen an den Beziehungen zwischen Elementen der Umgebung, konditioniertem und unkonditioniertem Reiz, auf die ein Organismus trifft und die einem Verhalten vorausgehen.

Die Meßebene beim klassischen Konditionieren blieb ihrem Ausgangspunkt, der Physiologie, bis heute verbunden. Dies sei am Beispiel der Operationalisierung von Furchtreaktionen in beiden Paradigmen verdeutlicht. Im Paradigma des klassischen Konditionierens wird die Stärke einer konditionierten Furcht-

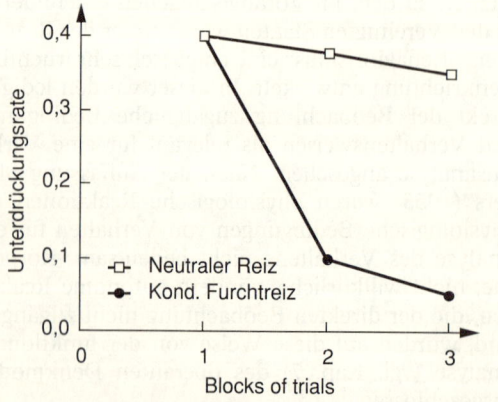

Abb. 19–2. Verhaltensunterdrückung durch einen konditionierten Furchtreiz oder einen neutralen Reiz (aus Rescorla, 1980).

reaktion durch Veränderungen etwa der Herzfrequenz, des Blutdrucks oder der elektrodermalen Aktivität überprüft, die die Präsentation eines konditionierten Furchtreizes, der vorher mit einem Schmerzreiz kontingent dargeboten worden war, hervorruft. Die abhängige Variable ist also ein physiologischer Reaktionsanteil und nicht eine Willkürreaktion.

Im Paradigma des operanten Lernens ist die Meßebene die der sog. Verhaltensunterdrückung. Die Darbietung eines mit einem aversiven Reiz konditionierten Reizes unterdrückt Verhalten, das gerade stattfindet, und die „emotionale" Stärke der Wirkung eines solchen Reizes wird nach diesem Paradigma von Estes und Skinner (1941) am Ausmaß der Verringerung des vorhandenen Verhaltens gemessen (vgl. dazu ausführlich Kap. 8).

Die Unterscheidung zwischen diesen beiden Arten des Lernens wurde jedoch auch in Zweifel gezogen, z.B. von Dollard und Miller (1950) oder Kimmel (1974). So wurde begonnen, einen möglichen direkten Einfluß operanter Bedingungen, also Belohnung oder Strafe, auf physiologische Prozesse nachzuweisen. Dieser Nachweis gelang zweifelsfrei. Dies hatte zwei Wirkungen. Zum einen war eine theoretische Frage vorangetrieben worden, zum anderen aber war die Ebene der Physiologie in die Denkrichtung des operanten Lernens eingeführt worden, der sie vorher fremd gewesen war. Mit diesen Ergebnissen wurde auch die Frage verbunden, inwieweit diese nachgewiesenen Effekte therapeutisch nutzbar gemacht werden könnten. Alles, was heute als „Biofeedback" bezeichnet wird, geht auf derartige Ergebnisse zurück. Weiterhin war der Bereich physiologischer Folgen operanter Versuchsbedingungen gleichsam entdeckt worden, und es folgte eine Vielzahl von Untersuchungen, die jetzt nicht mehr nur am Effekt der Verhaltensunterdrückung konditionierter emotionaler Reaktionen interessiert waren, sondern auch an ihren physiologischen Auswirkungen. So wurde auf mehrere Weisen klinisch geforscht, im Sinne möglicher therapeutischer Nutzen operanter Techniken auf physiologische Reaktionen, aber auch im Sinne einer Pathogeneseforschung, indem man versuchte, pathogenetisch relevante Bedingungen im Verhältnis eines Organismus zu seiner Umgebung zu ermitteln.

Wie wichtig die Vorhersagbarkeit auch von nicht aversiven Ereignissen ist, wurde schon im Labor von Pawlow in Erfahrung gebracht. Hunde wurden im Sinne des Diskriminationslernens trainiert, nach Darbietung eines Kreises Futter zu erwarten, sie reagierten entsprechend mit Speichelsekretion, dem Maß für die erfolgte Konditionierung, nach Darbietung einer Ellipse hingegen kein Futter zu erwarten. Auf diese Weise konnten die Tiere die Verabreichung des Futterpuders genau vorhersagen. Die Form der Ellipse wurde dann schrittweise der eines Kreises angenähert, bis zu einem Punkt der Ähnlichkeit, bei dem das Verhalten der Hunde schlagartig anders wurde, sie winselten, waren motorisch extrem unruhig und schienen alles vergessen zu haben, was sie gelernt hatten. Es war anschließend sehr viel schwieriger, das ursprüngliche Diskriminationsverhalten wiederherzustellen. Bei den geschilderten Umgebungsverhältnissen war ausschließlich die Vorhersagbarkeit des Futters für die Tiere unmöglich geworden, alles andere war gleichgeblieben. Dennoch reagierten sie viel heftiger als Tiere mit konditionierten Furchtreaktionen, und vor allem viel länger andauernd, und dies nur, weil die keineswegs aversiven Ereignisse nicht mehr vorhersagbar waren. An diesem Beispiel wird deutlich, daß Lernen ganz wesentlich die Funktion hat, dem Organismus eine Orientierung in seiner Umgebung zu ermöglichen, indem die Beziehung zwischen den Elementen der Umgebung erfaßt wird. Lernen impliziert also stets auch eine kognitiv zu nennende Leistung. Es wird weiterhin deutlich, daß die Orientierungsmöglichkeit an den Ereignissen der Umgebung extrem wichtig zu sein scheint, und die Unmöglichkeit dieser Orientierung im Hinblick auf bedeutsame Reize eine starke Belastung darstellt.

Lerntheoretische Modelle haben sich immer mit Vorgängen beschäftigt, die auch für die Medizin und vor allem die psychosomatische Medizin relevant sind. Diese Inhalte der Lernpsychologie sind auch die Grundlage für die wesentlichsten Konzepte der verhaltenstherapeutischen Verfahren. Im folgenden sollen die Grundlagen verhaltenstherapeutischer Verfahren sowie ihre Anwendungsmöglichkeiten im Bereich der Medizin dargestellt werden.

19.2 Prinzipien der Verhaltensmodifikation

19.2.1 Verhalten ist determiniert

Eine der wesentlichsten Grundannahmen der Verhaltensmodifikation besteht in der Annahme, daß menschliches Verhalten, wie das anderer Organismen, determiniert ist. Es gibt Ursachen dafür, wie ein Organismus sich zu einem gegebenen Zeitpunkt verhält, das Verhalten ist nicht zufällig. Diese Annahme wurde von Freud als „psychischer Determinismus" bezeichnet und bildet ebenfalls die Grundlage psychoanalytischen Denkens; eine Psychologie ohne diese Annahme ist nicht denkbar.

19.2.2 Die Methoden der Verhaltenstherapie stützen sich weitgehend auf empirisch gesicherte Befunde der Psychologie

Ein Unterschied zwischen Psychoanalyse und Verhaltenstheorie besteht jedoch in der Beantwortung der Frage, wie Verhalten erklärt werden kann. Die Verhaltenstheorie geht bei der Beantwortung dieser Frage davon aus, daß auch menschliches Verhalten hinreichend erklärt werden kann durch die Anwendung allgemeiner Prinzipien, die empirisch-wissenschaftlich ermittelt werden können, z.B. die Prinzipien des Lernens. Dies bedeutet nicht, daß sämtliche allgemeinen Prinzipien bereits bekannt sind, die notwendig sein können, um ein individuelles Verhalten ausreichend zu erklären. Es ist jedoch keine vom sonst üblichen wissenschaftlichen Vorgehen abweichende Denkmethodik notwendig, um zu erklären, warum jemand zu einem Zeitpunkt etwas Bestimmtes tut oder nicht. Aus intersubjektiv überprüfbaren Beobachtungen experimenteller oder anderer Art werden Regeln abgeleitet, die dann als allgemeine wissenschaftliche Prinzipien zur Erklärung eines spezifischen einzelnen Verhaltens angewendet werden. Die Aufgabe der Psychologie ist es, diese Prinzipien zu ermitteln, die wissenschaftlichen Ergebnisse der Psychologie bestehen u.a. in diesen Prinzipien.

Hieraus scheint sich jedoch eine sehr grundlegende Schwierigkeit zu ergeben, die der Individualität. Mit Recht kann gefragt werden, wie die Anwendung allgemeiner Prinzipien der Einmaligkeit menschlicher „Persönlichkeit", des Erlebens, das ja Verhalten beeinflußt, gerecht werden soll. Diese Frage soll lediglich im vorliegenden Zusammenhang behandelt werden, was sie nicht unter jedem Aspekt grundsätzlich klärt. Im Kapitel 7 wurde bereits darauf hingewiesen, daß die Prinzipien des Lernens einen formalen und weniger, oder oft gar nicht, inhaltlichen Charakter haben. Sie beziehen sich darauf, wie unter welchen Bedingungen gelernt wird, aber weniger darauf, was gelernt wird. Die Individualität eines der Hunde Pawlows bestand u.a. darin, daß ein akustischer Reiz einer bestimmten Frequenz Futter signalisierte, und das unterschied ihn von anderen Hunden. Lerntheoretische Prinzipien verallgemeinern diesen Reiz mit dieser individuellen Bedeutung als „konditionierten Reiz". Es ist Ergebnis des „formalen Vorgangs" des Konditionierens, daß dieser spezifische Reiz diese individuelle inhaltliche Bedeutung bekommt.

Selbstverständlich sind nicht alle Ergebnisse der Psychologie auf diese Weise formal. Im Kapitel 8 sind Ergebnisse empirischer Forschung dargestellt, die eher als inhaltlich zu bezeichnen sind, z.B. die Wirkung der formalen Bedingung der Unkontrollierbarkeit von aversiven Ereignissen auf die Magenschleimhaut. Der Gültigkeitsbereich dieses intersubjektiv überprüfbaren Sachverhalts muß ebenfalls empirisch ermittelt werden, beispielsweise, ob der Sachverhalt für alle Arten von Tieren gültig ist oder nur für einige. Der Umfang des Gültigkeitsbereichs definiert damit den Grad der Allgemeinheit von erklärenden Aussagen über einen individuellen Sachverhalt. Tritt bei einem Organismus ein Ulkus auf und kann ermittelt werden, daß der Organismus eine Zeitlang aversive Ereignisse nicht oder nicht mehr kontrollieren konnte, so kann das Auftreten des Ulkus damit erklärt werden. Da jedoch empirisch ermittelte Sachverhalte nie mit absoluter Sicherheit zutreffen, sondern aufgrund der Variabilität und Individualität von Organismen sowie der Variabilität von Meßverfahren lediglich mit einer bestimmten Wahrscheinlichkeit, enthält eine derartige Erklärung ein entsprechendes Maß von Unsicherheit. Man könnte sagen, diese Unsicherheit bei der Erklärung des Verhaltens eines Individuums zu vernachlässigen, bedeutet, der Individualität z.B. einer Person nicht gerecht zu werden.

Aus diesen Überlegungen ergeben sich weitere Grundsätze der Verhaltensmodifikation, das quasi-experimentelle Vorgehen sowie die Einzelfallorientiertheit.

19.2.3 „Experimentelle" Methode

Das experimentelle Vorgehen bedeutet, daß man aufgrund der oben genannten Unsicherheit, die empirischen Erklärungen eigen ist, solche Erklärungen fortlaufend überprüfen muß.

Ist man beispielsweise aufgrund der Analyse eines Symptoms (vgl. Abschn. „Verhaltensanalyse") der Auffassung, daß dieses vom Patienten aufgrund der Verstärkung durch ein Familienmitglied aufrechterhalten wird, so wendet man das Prinzip des operanten Lernens an, um den Sachverhalt zu erklären, warum das Symptom zu einem bestimmten Zeitpunkt besteht. Um das Symptom zu beseitigen, müßte anhand des Prinzips der operanten Löschung in der Behandlung diese Verstärkung beseitigt werden. Man könnte dem Familienmitglied mitteilen, daß seine gutgemeinte Zuwendung beim Auftreten der Symptomatik mit dazu führt, daß diese noch vorhanden ist, und daß es besser ist, dem Patienten die Zuwendung zu Zeitpunkten zukommen zu lassen, wenn es ihm gutgeht. So haben beispielsweise Flor et al. (1987) gefunden, daß die Intensität und Häufigkeit von Schmerzen bei Patienten mit chronischem Schmerz mit der geschilderten positiven Qualität der Zuwendung durch den Ehepartner korreliert. In diesem Beispiel wird das Auftreten selbstverständlich nicht ausschließlich durch die Verstärkung von seiten des Ehepartners erklärt, was deutlich macht, daß die Anwendung allgemeiner Prinzipien ein Phänomen teilweise, aber nicht vollständig erklären können muß, ebenso wie die meisten Phänomene nicht nur eine einzige Ursache haben.

Hat man mit dieser Mitteilung an das Familienmitglied Erfolg, und die Symptomatik verschwindet, so war die Hypothese über die Aufrechterhaltung der Symptomatik durch die Verstärkung durch den Ehepartner ausreichend zutreffend. Verändert sie sich nicht, oder nur geringfügig, so gibt es eine Reihe denkbarer Möglichkeiten, warum dies der Fall ist: Es ist beispielsweise denkbar, daß der Patient seinerseits die Zuwendung des Partners verstärkt und in diesem Zusammenhang ein großer Teil der ehelichen Kommunikation stattfindet. Die einfache Mitteilung an den Partner des Patienten ist also nicht wirkungsvoll, da das Verhalten aufgrund der Verstärkung nicht aufgegeben werden kann.

Dabei zeigt sich, daß eine Verhaltensweise nicht isoliert im Hinblick auf die unmittelbar vorausgehenden und nachfolgenden Reize betrachtet werden kann, sondern in einem „Handlungskontext" gesehen werden muß. Man würde also wie in einem Experiment davon ausgehen, daß die Hypothese zur Erklärung des beobachteten Phänomens falsch oder unzureichend war, da nicht alle Einflußfaktoren berücksichtigt worden sind.

19.2.4 Orientierung am Einzelfall

Das Beispiel kann hier nicht in alle denkbaren Einzelheiten weiter verfolgt werden, es wird daran jedoch auch die Einzelfallorientiertheit deutlich. Nicht jede Symptomatik wird durch die Verstärkung eines Partners aufrechterhalten, man muß klären, ob dies der Fall ist, und wenn es der Fall ist, so muß die Verstärkung der Zuwendung durch den Patienten keineswegs immer gegeben sein. Die Verhaltensanalyse als Grundlage des therapeutischen Vorgehens hat die Aufgabe, die individuellen Bedingungen der Symptomatik soweit zu erhellen, wie es zu ihrer dauerhaften Veränderung notwendig ist. Sie zielt ab auf die „Aufdeckung" der individuellen Lerngeschichte eines Patienten.

19.2.5 Orientierung am Symptom und an seinen Bedingungen

Das Kriterium für das Zutreffen einer den therapeutischen Maßnahmen zugrundeliegenden Hypothese ist das Eintreten der erwarteten Verhaltensänderung und weniger eine Einsicht von Patient und/oder Therapeut. Dieses Vorgehen beinhaltet auch, daß die Phase der Diagnostik die Therapie begleitet, d.h. im Grunde nie abgeschlossen werden kann, da es sich bei den meisten zu behandelnden Problemen nicht um isolierte, wirklich abgrenzbare Verhaltensweisen handelt. Aus diesem Grunde wird verhaltenstherapeutisches Vorgehen häufig als symptomorientiert betrachtet, was nur insoweit stimmt, als es die Kriterien betrifft. Verhaltenstherapeutisches Vorgehen zielt auf die Veränderung derjenigen Bedingungen, die das Symptom bedingen, seien es äußere, direkt beobachtbare, wie das Verhalten eines anderen Menschen, oder nicht direkt beobachtbare, wie die Gedanken eines Patienten. Dabei wird davon ausgegangen, daß die Gedanken eines Patienten ebenso wie sein beobachtbares Verhalten den Prinzipien des Lernens unterliegen, was jedoch wiederum nicht bedeutet, daß man Gedanken in allen Aspekten auf die Wirksamkeit dieser Prinzipien reduzieren kann. So können Gedanken eine wesentliche Bedingung für die Aufrechterhaltung eines Symptoms sein. Und daß das so ist, kann durch selektive Verstärkung verursacht worden sein, welchen Inhalt die Gedanken haben, ist jedoch durch die Verstärkung nicht beschreibbar.

19.3 Diagnostik in der Verhaltenstherapie

19.3.1 Aufgaben einer verhaltenstherapeutischen und Möglichkeiten der traditionellen Diagnostik

Nach Kanfer und Saslow (1976) sollte Diagnostik im Rahmen der Verhaltenstherapie die folgenden drei Fragen beantworten:

– Welche besonderen Verhaltensmuster verlangen eine Veränderung hinsichtlich ihrer Verhaltenshäufigkeit, ihrer Intensität, ihrer Dauer oder der Bedingungen, unter denen sie auftreten?

– Welches sind die Bedingungen, unter denen dieses Verhalten erworben wurde, und welche Faktoren halten es momentan aufrecht?

– Welches sind die praktikabelsten Mittel, um die erwünschten Veränderungen bei einem Individuum zu erzielen?

Danach wird deutlich, daß verhaltenstherapeutische Diagnostik in ihrer Zielsetzung auf das praktisch-therapeutische Handeln ausgerichtet ist (Schulte, 1980).

Der Ansatz der psychiatrischen Diagnostik

Traditionelle Ansätze zur Klassifikation psychischer Störungen finden sich zum einen innerhalb der psychiatrischen Diagnostik, zum anderen beschäftigt sich die herkömmliche psychologische Diagnostik in Weiterentwicklung der differentiellen Psychologie mit der Vorhersage menschlichen Verhaltens. Prinzipiell können Klassifikationen hinsichtlich der Ätiologie, der Symptome und im Hinblick auf die Prognose der Erkrankung vorgenommen werden.

Psychiatrische Diagnostik zielt auf die Zuordnung eines Patienten zu einer ätiologischen Kategorie (z.B. Angstneurose), wobei die Zuweisung im Bereich der Neurosen und Persönlichkeitsstörungen im wesentlichen aufgrund der vorliegenden Symptomatik erfolgt. Symptome werden als Hinweise auf die zugrundeliegende Erkrankung verstanden, d.h. aufgrund des Vorliegens bestimmter Symptome wird auf ätiologische Faktoren geschlossen, die für die jeweiligen Beschwerden oder Probleme als verursachend angesehen werden. Im Sinne des traditionellen medizinischen Krankheitsmodells wird angenommen, daß gleiche ursächliche Faktoren zu denselben Symptomen führen und auch dieselbe Behandlung erfordern. Ein solch umfassender Diagnosebegriff, der sowohl Kenntnisse über die Entstehung als auch den künftigen Verlauf bzw. die Behandlung der Krankheit einschließt, scheint gegenwärtig jedoch nur noch sehr eingeschränkt vertretbar zu sein. So wird heute in der Regel angenommen, daß Erkrankungen psychischer oder somatischer bzw. psychosomatischer Natur multifaktoriell bedingt sind, und auch die Entscheidung über die Therapie nach Diagnosestellung von zahlreichen weiteren Faktoren wie sozialen, ökonomischen oder ethischen Bedingungen abhängt. Eine eindeutige Zuordnung von Symptom, Ursache und Behandlung erscheint somit kaum noch möglich (vgl. auch Kap. 51).

Darüber hinaus erwiesen sich zumindest die früheren psychiatrischen Klassifikationssysteme als wenig objektiv und reliabel, so daß von daher bereits ihr prognostischer Wert angezweifelt werden mußte (vgl. Kanfer und Saslow, 1976). Schulte (1980) weist in diesem Zusammenhang auch auf die Probleme bei der Gruppierung oder Zusammenfassung der Symptome hin, die zum Teil aufgrund phänomenaler

Ähnlichkeit, dem Grad der Ausprägung oder dem häufigen koinzidenten Auftreten erfolgt.

In der **klassischen psychologischen Diagnostik** steht die Messung zeitlich stabiler und situationsinvarianter Persönlichkeitseigenschaften im Vordergrund, d.h. zur Vorhersage von Verhalten werden interindividuelle Persönlichkeitsunterschiede herangezogen. Es werden psychologische Tests konstruiert, um die Objektivität, Reliabilität und Validität (vgl. Kap. 14) der untersuchten Merkmale zu bestimmen bzw. zu erhöhen. Ähnlich wie bei der psychiatrischen Diagnostik liegt jedoch auch hier ein sog. „Zeichenansatz" vor: Aufgrund des Verhaltens, das ein Proband beim Ausfüllen eines Fragebogens zeigt, wird auf latente, diesem Verhalten zugrundeliegende Eigenschaften geschlossen. Damit werden die Antworten als Manifestationen zugrundeliegender Persönlichkeitscharakteristika angesehen, die dann ihrerseits wiederum zur Vorhersage des tatsächlichen Verhaltens herangezogen werden.

Der „Interaktionismus" der Persönlichkeitstheorie

Seit den 70er Jahren wird die Annahme der traditionellen, eigenschaftstheoretischen Persönlichkeitspsychologie, daß sich Individuen unabhängig von der Situation stabil, d.h. situationsinvariant verhalten, kritisiert (z.B. Mischel, 1968). Damals wiesen vor allem die „Situationisten", die die Ursachen von Verhalten nicht in Dispositionen der Person, sondern in Situationsfaktoren lokalisiert sehen, darauf hin, daß die Gleichartigkeit des Verhaltens um so geringer wird, je mehr sich die Situation wandelt. Heute erscheint allgemein ein interaktionistischer Standpunkt angemessen, dessen Vertreter (z.B. Magnusson und Endler, 1977; Bowers, 1973) davon ausgehen, daß Verhalten eine Funktion von Person und Situation ist, d.h. die Varianz im Verhalten am besten erklärt werden kann, wenn Person- und Situationsmerkmale diagnostisch berücksichtigt werden. Wie Heckhausen (1980, S. 22) es formuliert, nimmt die Annahme eines gegenseitigen Wechselwirkungsprozesses Abschied von der Vorstellung, die „Situation sei immer das zeitlich Vorauslaufende und damit Unbeeinflußte, worauf die Person dann reagiere".

Betrachtet man menschliches Verhalten unter einem interaktionistischen Blickwinkel, erscheinen auf der Personseite der Interaktion kognitive und motivationale Faktoren, auf der Situationsseite hingegen die psychologische Bedeutung, die eine bestimmte Situation für ein Individuum hat, als wesentliche Determinanten des Verhaltens (Magnusson und Endler, 1977). Diese Auffassung entspricht der des Situationskreises, in der davon ausgegangen wird, daß das Individuum die Situation aktiv z.B. durch vorhandene Bedürfnisse mitbestimmt, und die Situation dadurch ein System aus Umgebung und Individuum ist.

Diesen neueren Formulierungen bzw. Befunden wird die klassische Persönlichkeitsdiagnostik auch heute nur in Ausnahmefällen gerecht. Nach wie vor gilt das Hauptaugenmerk einer Interpretation der Testantworten und weniger einer repräsentativen

Auswahl der Testitems in dem Sinne, daß situative Determinanten des Verhaltens, wie beispielsweise im „S-R-Inventory of Anxiousness" von Endler et al. (1962), Berücksichtigung fänden. Insgesamt ist festzuhalten, daß die traditionellen diagnostischen Ansätze zur Gewinnung von Informationen, die für die Auswahl und Durchführung verhaltensmodifizierender Interventionen notwendig erscheinen, wenig geeignet sind.

19.3.2 Der funktionale verhaltenstherapeutische Ansatz

Diagnostik in der Verhaltenstherapie erhebt nicht den Anspruch, die gesamte Persönlichkeit eines Individuums zu erfassen, sondern beschränkt sich auf die Variablen, die für verhaltenstherapeutische Interventionen relevant sind (Schulte, 1980).

Darüber hinaus vermeidet der funktionale verhaltenstherapeutische Ansatz die Einführung von hypothetischen Konstrukten und beobachtet statt dessen das konkrete Verhalten, das eine Person in einer bestimmten Situation zeigt. Dabei schließt der Begriff „Verhalten" nicht nur beobachtbare Handlungen, sondern auch Kognitionen, Emotionen und körperliche Reaktionen ein. Verhaltenstheoretische Persönlichkeitsdiagnostik bemüht sich damit um eine direkte Messung der problematischen Reaktionsweisen eines Individuums in Situationen, in denen dieses Verhalten charakteristischerweise auftritt. Wird das symptomatische Verhalten in einer repräsentativen Auswahl kritischer Situationen untersucht (zu den verschiedenen Methoden der Informationsgewinnung vgl. Abschn. 19.5), wird angenommen, daß damit eine adäquate Stichprobe von möglichen Kriteriumsverhaltensweisen gewonnen wird, wobei die Reaktionen das Kriteriumsverhalten selbst darstellen. Im Sinne des sog. „Stichprobenansatzes" geht man davon aus, daß das Testverhalten eine Untergruppe des tatsächlichen problematischen Verhaltens ist. Bei dieser, auch als kriteriumsorientiert bezeichneten Art der Messung, der allgemein eine größere prognostische Validität zugeschrieben wird, reduziert sich somit die Anzahl der induktiven oder deduktiven Schlüsse samt der theoretischen Annahmen (zeitliche Stabilität von Verhalten, Situationsinvarianz etc.), die im Rahmen des Zeichenansatzes notwendig sind (Rückschluß vom Test auf ein zugrundeliegendes Konstrukt und weiter zu einer Verhaltensvorhersage unter spezifischen Bedingungen). Damit kann als eines der wesentlichen Merkmale verhaltenstheoretischer Diagnostik die erhöhte Ähnlichkeit zwischen der untersuchten Reaktion und dem tatsächlichen problematischen Verhalten angesehen werden (vgl. Goldfried und Kent, 1976).

Aus lerntheoretischer Sicht wird menschliches Verhalten nicht nur als Produkt der individuellen Lerngeschichte, sondern auch als durch aktuelle situative Bedingungen und/oder durch die Konsequenzen des betreffenden Verhaltens determiniert angesehen. Die Ursachen von Verhalten werden somit in den auf-rechterhaltenden Bedingungen lokalisiert, wobei zwischen historisch-genetischen Bedingungen, die zur Entstehung eines Problems geführt haben, und Bedingungen, die das Verhalten gegenwärtig aufrechterhalten, unterschieden wird, da beide nicht identisch sein müssen. Im Gegensatz zur klassischen psychiatrischen Diagnostik, bei der die Zuordnung zu einer Krankheitskategorie in der Regel bereits die Entscheidung für eine Therapie einschließt, geht die Verhaltenstherapie davon aus, daß für jeden Einzelfall ein eigenes Behandlungskonzept zu entwickeln ist, das „in seiner individuellen Komposition völlig individuell und in seiner Gesamtheit unvergleichbar ist mit jeder nächsten Behandlungsmethode" (Kaminski 1967, S. 128). D.h., in jedem einzelnen Fall wird nach den spezifischen aufrechterhaltenden Bedingungen gesucht, konstant bleibt nur das theoretische Modell, nach dem Verhalten erklärt wird, sowie das daraus abgeleitete (formale) Prinzip der Therapie. Welche Bedeutung bzw. welche funktionale Qualität spezifische situative Bedingungen für ein Individuum haben, muß im Sinne der obigen Ausführungen zum Interaktionismus (Determinanten des Verhaltens auf Situationsseite) jeweils ermittelt werden.

Wie Schulte (1976a, S. 69) ausführt, ist „das Kriterium für die Zielaussage und die Auswahl der Technik also nicht die Bezeichnung oder Klassifikation des Symptoms bzw. die Aussage über den Ausprägungsgrad eines bestimmten Verhaltens- oder Persönlichkeitsmerkmals, sondern die (zunächst hypothetische) Aussage über die funktionale Beziehung zwischen symptomatischem Verhalten und vorausgehenden und nachfolgenden Umweltbedingungen, d.h. das funktionale Modell des Symptoms". Ermittelt werden die funktionalen Reiz-Reaktions-Zusammenhänge im Rahmen der sog. Verhaltensanalyse (vgl. Abschn. 19.3.3) traditionell gemäß der Verhaltensgleichung:

$$S - O - R - C - K$$

von Kanfer und Saslow (1965, 1969). R (Reaktion) steht dabei für das symptomatische Verhalten, das es zunächst detailliert zu erfassen gilt, S (Stimulus = Reiz) für Ereignisse, die R vorausgehen, C für nachfolgende Konsequenzen, die sowohl situativer als auch organismischer Art sein können, und K für die Kontingenz, d.h. den jeweiligen Verstärkungsplan, nach dem die betreffenden Konsequenzen auf das Verhalten folgen. Kanfer und Saslow erweiterten die ursprüngliche Gleichung von Lindsley (1964), die nur diese Elemente einschloß, später um die Variable O, mit der biologische Bedingungen des Verhaltens, die insbesondere im Falle körperlicher Störungen in Betracht zu ziehen sind, gemeint sind.

Dieses allgemeine Modell bietet die Möglichkeit, sowohl Verhalten, das nach den Prinzipien des klassischen Konditionierens erworben wurde, wie auch Verhalten, das operant erlernt wurde, in seinen funktionalen Zusammenhängen darzustellen. Es ergeben sich als diagnostische Aufgaben, zunächst jedes Element der Gleichung einzeln empirisch zu bestimmen und auf dieser Grundlage ein Bedingungsmodell zu erstellen, das dem vermuteten funktionalen Reiz-Reaktions-Zusammenhang gerecht wird. Die Variablen, von denen das problematische Verhalten als abhängig angenommen wird, werden sodann zu therapeutischen Änderungspunkten, d.h. es wird überlegt, wie diese zu verändern sind, um eine Modifikation des Verhaltens in die gewünschte Richtung herbeizuführen.

19.3.3 Der diagnostisch-therapeutische Prozeß in der Verhaltenstherapie

In der Verhaltenstherapie werden psychische Störungen als unerwünschte Verhaltensweisen aufgefaßt, die infolgedessen verändert werden sollten. Beschwerden gelten nicht als Symptome, sondern werden als Probleme angesehen, die folglich auch einer Problemlösung zugänglich sind. So wird der diagnostisch-therapeutische Prozeß in der Verhaltenstherapie auch als ein Problemlösungsprozeß aufgefaßt, der unangemessenes Verhalten auf der Basis lerntheoretischer Gesetzmäßigkeiten verändern will.

Die diagnostischen Aufgaben, die sich im Rahmen der Planung und Durchführung einer Verhaltenstherapie stellen, sind Analyse des Verhaltens, Zielbestimmung und Therapieplanung, wobei die einzelnen Schritte eng aufeinander bezogen sind. So wird das funktionale Modell des symptomatischen Verhaltens, das am Ende der diagnostischen Phase erstellt wird, zunächst als noch nicht verifizierte Erklärung dieses Verhaltens betrachtet, deren Gültigkeit sich erst im weiteren Verlauf der Behandlung erweisen muß. Die Therapie bzw. der Therapieerfolg dient somit – ähnlich wie bei einem psychologischen Experiment, wenn auch weniger kontrolliert – einer Überprüfung der in der diagnostischen Phase aufgestellten Hypothesen. In diesem Zusammenhang erklärt sich die besondere Bedeutung, die der therapiebegleitenden Diagnostik in der Verhaltenstherapie zukommt. Nur wenn die Effekte der therapeutischen Interventionen kontinuierlich mit Hilfe geeigneter Parameter (vgl. Abschn. 19.5) kontrolliert werden, ist gewährleistet, daß die erarbeiteten Hypothesen in Frage gestellt werden, und das therapeutische Vorgehen gegebenenfalls entsprechend modifiziert wird.

Ein Schema zur Diagnose und Therapieplanung

Im deutschsprachigen Raum wurde von Schulte (vor allem 1976b, 1980, 1986) aufbauend auf den Arbeiten von Kanfer und Saslow (1965, 1969, 1976) ein Schema für Diagnose und Therapieplanung in der Verhaltenstherapie entwickelt, das einer Systematisierung und strukturierten Verarbeitung der zur Durchführung einer verhaltenstherapeutischen Intervention notwendigen diagnostischen Information dient. Da sich das Schema in der klinischen Praxis als sehr nützlich erwiesen hat, werden die wesentlichen diagnostischen Schritte, die danach zu erarbeiten sind, im folgenden kurz skizziert.

Verhaltensanalyse

Primäres Ziel der Verhaltensanalyse ist es, Hinweise auf das funktionale Bedingungsmodell zu gewinnen. Zu diesem Zweck werden insbesondere
– die Topographie des symptomatischen Verhaltens,
– die Reizbedingungen, die dem Verhalten mehr oder weniger regelmäßig unmittelbar vorausgehen und folgen,
– erfolgreiche oder erfolglose Selbstkontrollversuche des Patienten und
– die Genese des Symptoms und seine Veränderungen im Laufe der Symptomgeschichte differential diagnostisch herangezogen.

In Anbetracht der möglichen Komplexität von Patientenschilderungen empfiehlt Schulte (1976b), die aktuellen Beschwerden oder Probleme eines Patienten prinzipiell zunächst nach Problemkreisen zu ordnen und nach ihrer Relevanz in eine Rangreihe zu bringen. Die Verhaltensanalyse erfolgt dann für jeden Problembereich getrennt, bis am Ende der verhaltensanalytischen Phase der Zusammenhang zwischen den Einzelsymptomen untersucht wird. Die Analyse des symptomatischen Verhaltens beginnt zunächst mit einer Beschreibung der symptomatischen Verhaltensweisen. Dazu sollte aus den Beispielen eines Problemkreises ein möglichst charakteristisches, klar umschriebenes Verhalten ausgewählt werden, das hinsichtlich seiner Topographie und Intensität auf der motorischen, verbalen und physiologischen Ebene präzise (z.B. mittels Puls, Atemfrequenz etc.) beschrieben werden kann. Es sollte festgehalten werden, mit welcher Frequenz dieses Verhalten auftritt, und inwieweit Schwankungen der Frequenz in Abhängigkeit von äußeren Bedingungen auftreten. Danach sollte der Typ der vorliegenden Symptomatik:
– völlig unangemessenes Verhalten,
– an sich normales Verhalten, das zu häufig auftritt,
– normales Verhalten, das zu selten gezeigt wird,
– Verhalten fehlt völlig (im Sinne einer Verhaltenslücke) festgelegt werden.

Der Beschreibung des problematischen Verhaltens schließt sich die Analyse der vorausgehenden und nachfolgenden Bedingungen oder Verhaltensweisen an. Auch diese sind möglichst exakt zu beschreiben und wenn möglich mit ihrer Auftretenswahrscheinlichkeit zu belegen. Differentialdiagnostisch besonders zu berücksichtigen sind hier auch die Situationen, in denen das problematische Verhalten nicht auftritt. Werden verschiedene vorausgehende oder nachfolgende Reize eruiert, ist zu prüfen, inwieweit diese unter einem abstrakteren Merkmal zusammengefaßt werden können. Sollte das nicht möglich sein, ist zu untersuchen, ob die vorläufige Aufteilung der verschiedenen Problembereiche beim gegenwärtigen Informationsstand noch aufrechterhalten werden kann. Abschließend sollte die zeitliche Abfolge von Reizen und Reaktionen, die in Zusammenhang mit dem Symptom beobachtet werden, in der Form von „Zeitketten" dargestellt werden; auf eine funktionale Interpretation der Reiz-Reaktions-Zusammenhänge sollte jedoch noch verzichtet werden.

Bevor ein vorläufiges Bedingungsmodell erstellt werden kann, ist zuvor noch zu prüfen, inwieweit Organismus-Variablen (O der Verhaltensgleichung) als vorausgehende oder nachfolgende Bedingungen mit der Symptomatik in Verbindung stehen. Darüber hinaus ist differentialdiagnostisch zu verwerten, ob und unter welchen Bedingungen es dem Patienten gelingt, seine Beschwerden durch eigenes Verhalten, das selbst nicht wieder als Verhaltensstörung (z.B. Vermeidungsverhalten) angesehen werden muß, günstig zu beeinflussen, bzw. welche Selbstkontrollversuche nicht wirksam waren.

Sind diese Fragen beantwortet, kann mit der Aufstellung des vorläufigen funktionalen Bedingungsmodells begonnen werden. Damit stellt sich nun als zentrale Frage, welcher Art das symptomatische Verhalten ist, bzw. welches Lernprinzip zu seiner Erklärung herangezogen werden kann. Im einzelnen ist zu belegen, ob es sich um ein operantes Verhalten handelt, das von den nachfolgenden Reizbedingungen kontrolliert wird, oder um ein respondentes Verhalten, das durch vorausgehende Bedingungen ausgelöst wird. Liegt ein operantes Verhalten vor, ist zudem zu klären, ob es sich um ein Annäherungsverhalten, d.h. ein positiv verstärktes Verhalten, handelt oder ein Flucht- oder Vermeidungsverhalten, das negativ z.B. durch Angstreduktion verstärkt wird. Dabei sind diejenigen Reize, denen funktionale Qualität zugewiesen wird, im einzelnen genau auszuführen. Zu Zwecken einer Veranschaulichung hat es sich als sinnvoll erwiesen, das hypothetische Bedingungsmodell anhand der üblichen lerntheoretischen Symbolik (vgl. Verhaltensgleichung) mit den jeweils zugeordneten konkreten Reizen bzw. Reaktionen darzustellen.

Die Exploration der Symptomgenese dient einer nochmaligen Überprüfung des vorläufigen funktionalen Bedingungsmodells. Aus differentialdiagnostischen Gründen sollte hier besonders darauf geachtet werden, ob zum Zeitpunkt des Beginns der Symptomatik ein klassischer Konditionierungsvorgang stattgefunden hat. Des weiteren interessiert, inwieweit im Laufe der Symptomgeschichte Prozesse der Reiz- und/oder Reaktionsgeneralisierung wirksam wurden, was bei der Planung der konkreten therapeutischen Interventionen besonders zu berücksichtigen wäre. Auch die Bedeutung von Modellpersonen bei der Entwicklung der Störung sollte hier geklärt werden.

Die Verhaltensanalyse der einzelnen Symptome endet mit der vorläufigen prinzipiellen Therapieplanung, wobei zunächst nach dem Typ der vorliegenden Symptomatik (s.o.) die Richtung der gewünschten Veränderung festzulegen ist (Typ 1: Veränderung der Topographie oder völliger Abbau; Typ 2: Reduktion; Typ 3: Förderung; Typ 4: Aufbau). Aus dem vorläufigen funktionalen Bedingungsmodell wird abgeleitet, an welcher Stelle der funktionalen Reiz-Reaktions-Einheit eine Veränderung des Symptoms erreicht werden kann. Abschließend ist zu beantworten, mit welchem oder welchen lerntheoretischen Änderungsprinzipien eine solche Veränderung zu realisieren ist (z.B. mittels Löschung, einer Veränderung der operanten Verstärkungsbedingungen etc.).

Wurde jeder der im Einzelfall festzustellenden Problembereiche in Hinblick auf die angegebenen Kriterien analysiert, endet die Phase der Verhaltensanalyse damit, daß die Zusammenhänge zwischen den Symptomen analysiert werden. So ist es theoretisch möglich, daß zwei Symptome genetisch auf das gleiche Ereignis zurückgehen, oder daß ein Symptom als Folge des anderen entstanden ist, sich inzwischen aber verselbständigt hat. Verschiedene Symptome können auch durch gleiche Verstärkungsbedingungen aufrechterhalten werden, so daß therapeutisch an dieser Bedingung angesetzt werden müßte. Schließlich können sich Symptome gegenseitig bedingen, in dem Sinne, daß das eine Symptom das Auftreten des anderen Symptoms fördert. In jedem Fall ergeben sich Konsequenzen für die Auswahl der therapeutischen Ansatzpunkte.

Zielanalyse

Aufgabe der Zielanalyse ist es festzustellen, inwieweit, über die bislang bekannten Bedingungen des symptomatischen Verhaltens hinaus, gesellschaftliche, soziale oder ökonomische Bedingungen existieren, die zu einer Aufrechterhaltung der Symptomatik beitragen. Daneben soll die weiterreichende Bedeutung des symptomatischen Verhaltens für den Patienten analysiert werden. So sollte geklärt werden, inwieweit der Verhaltensspielraum des Patienten durch das symptomatische Verhalten eingeschränkt ist, welche Aktivitäten infolge der Symptomatik vermieden werden, welche positiven oder negativen Konsequenzen die Störung in Hinblick auf die sozialen Kontakte des Patienten zeigt etc. Zu berücksichtigen ist in diesem Zusammenhang auch, welche Bedeutung die Symptomatik für den Sozialpartner hat, um diesen gegebenenfalls direkt oder indirekt in die Therapie einbeziehen zu können. Schließlich sollte untersucht werden, welche Folgen eine Symptomveränderung, wie sie in der vorläufigen prinzipiellen Therapieplanung vorgeschlagen wurde, erwarten läßt. Inwieweit sind z.B. Partnerschaftsprobleme zu erwarten, wenn der Partner seine fürsorgende Rolle aufgeben muß; stehen dem Patienten alternative Verhaltensweisen zur Verfügung, wenn ihm die Möglichkeit fehlt, über das symptomatische Verhalten Zuwendung zu erzielen etc.?

Wurden diese Fragen geklärt, ist im Rahmen der Zielbestimmung zu entscheiden, welche Verhaltensweisen und/oder situativen Bedingungen therapeutisch verändert werden sollen, d.h. welche Punkte des funktionalen Bedingungsmodells oder aber auch andere, den Patienten nicht primär belastende Verhaltensweisen oder Verhaltensweisen von Bezugspersonen, die symptomverstärkend wirken, zu therapeutischen Ansatzpunkten werden sollen. Dabei ist in der Regel davon auszugehen, daß ein Ansatzpunkt allein nur in den seltensten Fällen ausreichend ist. Mehrere Ansatzpunkte sollten jedoch insbesondere dann gewählt werden, wenn ein respondentes Verhalten zusätzlich durch ein Flucht- oder Vermeidungsverhalten oder durch soziale Verstärkung aufrechterhalten wird. Gleiches gilt, wenn der Patient über kein Alternativverhalten verfügt und von daher mit einer Symptomverschiebung zu rechnen ist, wenn eine Verschlechterung der sozioökonomischen Situation oder eine Beeinträchtigung der sozialen Beziehungen zu erwarten ist.

Sind die therapeutischen Ansatzpunkte bestimmt, ist für das gewünschte Zielverhalten genau anzugeben, wie Topographie und Frequenz dieses Verhaltens aussehen sollen, unter welchen situativen Bedingungen es auftreten soll, und welche weiterreichenden, langfristigen Folgen von dieser Veränderung zu erwarten sind.

Therapieplanung

Im Rahmen der Therapieplanung ist zu klären, welche Änderungsprinzipien für die angestrebte Veränderung eines therapeutischen Ansatzpunktes herangezogen werden sollen (prinzipielle Planung), und wie die therapeutischen Maßnahmen im einzelnen gestaltet werden können (konkrete Planung). Sollten sich innerhalb der Zielbestimmung keine neuen therapeutischen Ansatzpunkte ergeben haben, kann die „vorläufige prinzipielle Therapieplanung", wie sie innerhalb der Symptomanalyse vorgenommen wurde, beibehalten werden. Anderenfalls sind aus dem Bedingungsmodell der neuen Ansatzpunkte entsprechende Änderungsprinzipien abzuleiten.

Zweck der konkreten Therapieplanung ist es, die gewonnenen Änderungsprinzipien in konkrete therapeutische Interventionen umzusetzen. Kann auf Standardverfahren der Verhaltenstherapie zurückgegriffen werden, sollte hier z.B. geplant werden, welche verschiedenen Angstreize verwendet werden, welche effektiven Verstärker eingesetzt werden können etc. Auch die Reihenfolge der einzelnen therapeutischen Maßnahmen ist an dieser Stelle festzulegen. Darüber hinaus sollte, um den Effekt der durchgeführten Maßnahmen kontinuierlich kontrollieren zu können, überlegt werden, welche therapiebegleitenden Kontrollmessungen durchgeführt werden. Ebenso sind zu diesem Zeitpunkt bereits die Erhebungsinstrumente auszuwählen, die zur Messung des Therapieerfolgs nach Beendigung der Behandlung herangezogen werden. In diesem Zusammenhang ist besonders hervorzuheben, daß nicht nur die Ausprägung des symptomatischen Verhaltens beurteilt werden sollte, sondern auch andere Lebensbereiche oder Persönlichkeitsmerkmale des Patienten, die nicht unmittelbar Gegenstand der therapeutischen Veränderung waren.

19.3.4 Erweiterungen der klassischen Verhaltensanalyse

Erweitert wurde die klassische Verhaltensanalyse, die sich vor allem auf die Analyse offen beobachtbaren Verhaltens konzentrierte, im wesentlichen in zweierlei Hinsicht. Zum einen wurden in den siebziger Jahren zunehmend die Selbstkontroll- bzw. Selbstregula-

tionsmöglichkeiten von Patienten registriert und nach dem Motto „Power to the Person" (Mahoney und Thoresen, 1974) auch therapeutisch genutzt. Damit wurde das Individuum nicht mehr nur als durch externe Kontingenzen gesteuert angenommen, sondern als ein „reflexives Subjekt" (Groeben und Scheele, 1977) angesehen, das aktiv gestaltend auf seine Umgebung einwirkt. Parallel zu dieser Entwicklung wurden die ersten kognitiven Verhaltenstheorien (Mahoney, 1974, dt. 1977; Meichenbaum, 1974, dt. 1979) formuliert. Diese Ansätze nehmen an, daß Kognitionen – als subjektiv-verbales Verhalten – in Form eines inneren Dialogs die Auseinandersetzung mit einer Situation begleiten und dementsprechend auch Verhalten beeinflussen. Da Kognitionen vom Individuum in verbaler Form aktualisiert werden (sog. „Selbstverbalisationen"), sind sie intersubjektiv erschließbar und einer therapeutischen Beeinflussung nach lerntheoretischen Prinzipien zugänglich.

In der jüngeren Zeit wurde im deutschsprachigen Raum vor allem von Fiedler (1979) und Bartling et al. (1979) darauf hingewiesen, daß nicht nur die Kognitionen, die in der spezifischen Situation aktualisiert werden, das Verhalten steuern, sondern Kognitionen auch situationsübergreifend, im Sinne von „Verhaltensdispositionen", Verhalten bestimmen. Dabei werden Verhaltensregeln und Verhaltenspläne, aber auch Stimuluserwartungen, Werthaltungen und Verhaltensbewertungen als zentrale Bedingungsvariablen unterschieden. Ziel einer solchen handlungstheoretisch orientierten kognitiven Verhaltensanalyse ist die Exploration des Regelsystems, das der jeweiligen Verhaltensstörung zugrunde liegt.

Lazarus (1971, 1976) geht in seinem Ansatz der „multimodalen" Verhaltenstherapie noch weiter und schlägt vor, auch Methoden anderer Therapierichtungen eklektisch zu berücksichtigen, sofern sie dazu beitragen, Verhalten auf den Ebenen **b**ehavior (Verhalten), **a**ffect (Affekt), **s**ensation (Empfindung), **i**magery (Vorstellung), **c**ognition (Kognition), **i**nterpersonal (soziale Beziehungen) und **d**rugs (Medikamente) – abgekürzt BASIC-ID – zu beschreiben und zu verändern.

Inwieweit die Berücksichtigung zusätzlicher potentieller Bedingungsvariablen (vgl. die Zusammenstellung bei Schulte, 1986) bei der Verhaltensanalyse jedoch tatsächlich eine Erweiterung der Analysemöglichkeiten bringt, ist bislang noch sehr unklar. So existieren bei verschiedenen Konstrukten Überlappungen, wie sie auch den gleichen Sachverhalt aus unterschiedlicher theoretischer Perspektive bezeichnen. Die Variable „Reaktion-Ergebnis/Folge-Erwartung" scheint z.B. den gleichen Sachverhalt wie Skinners Konzept des diskriminativen Stimulus zu erfassen. Darüber hinaus ist zu vermuten, daß verschiedene Variablen nicht nur mit dem symptomatischen Verhalten in funktionaler Beziehung stehen, sondern auch untereinander abhängig sind, so daß die Erhebung weiterer Variablen unter Umständen nur zu einem sehr geringen Beitrag an inkrementeller Validität führt. Bislang ist weder eine empirische Bestimmung des Beziehungsgefüges der verschiedenen Variablen untereinander durchführbar, noch erscheint eine Integration der verschiedenen Teiltheorien in Sicht (vgl. Schulte, 1986).

19.3.5 Methoden der Informationsgewinnung in der Verhaltenstherapie

Verhaltensdiagnostische Erhebungen sind zum einen notwendig, um die vorliegende Symptomatik samt der vorausgehenden und nachfolgenden Reizbedingungen möglichst exakt zu beschreiben – als Voraussetzung für die Erstellung des funktionalen Bedingungsmodells (explorative Funktion der Diagnostik). Zum anderen sind die Veränderungen der Symptomatik im Therapieverlauf systematisch zu dokumentieren, um eingetretene Veränderungen – als Abweichungen von der Baseline, die zu Beginn der Therapie erhoben wird – hinsichtlich ihrer Auftretenshäufigkeit, Intensität und Dauer, aber auch den auslösenden Bedingungen einschätzen zu können (therapiekontrollierende Funktion). Darüber hinaus wird angenommen, daß der Diagnostik in der Verhaltenstherapie, die sich um enge Einbeziehung des Patienten in den diagnostischen Prozeß bemüht, auch eine therapeutische Funktion zukommt. Nachdem die systematische Form der Informationserhebung zu einer Strukturierung des Problems auf seiten des Klienten führen sollte, wird er zu einer aktiven Selbstkontroll-Orientierung ermuntert und durch Rückmeldungen über den Behandlungserfolg für erfolgreiches Verhalten verstärkt.

Verhaltenstherapeutische Diagnostik strebt, wie oben beschrieben, eine kriteriumsorientierte Art der Messung an, so daß die direkte Beobachtung des symptomatischen Verhaltens zum wesentlichen diagnostischen Instrument wird. Dabei kann das Verhalten in der natürlichen Lebensumgebung des Patienten beobachtet werden, aber auch unter künstlichen (Labor-)Bedingungen im Sinne eines „Verhaltenstests" evoziert werden. Es kann das Problemverhalten selbst oder der Grad der Annäherung an das gewünschte Zielverhalten registriert werden, z.B. die Entfernung (in Metern), die der Patient zum angstauslösenden Objekt zu einem bestimmten Zeitpunkt tolerieren kann (sog. „Vermeidungstest"). Auch ist es möglich, die Reaktionsrate in Abhängigkeit von den auslösenden Bedingungen zu protokollieren. Wichtig erscheint bei allen Formen der Verhaltensbeobachtung (frei oder systematisch, kontinuierlich oder diskontinuierlich), daß die zu beurteilenden Verhaltenseinheiten klar definiert sind (z.B. Schmerzverhalten bei frischoperierten Patienten: Welche motorischen, mimischen oder verbalen Reaktionen werden als Indikatoren herangezogen?) und ein geeignetes Kodierungssystem vorhanden ist (Welche Verhaltensaspekte werden wann registriert? Beschränkt sich die Protokollierung auf Häufigkeiten, oder werden Intensitätsaspekte, und wenn ja, wie berücksichtigt, etc.?). Vielfach erweist sich ein Beobachtertraining als notwendig, dessen Erfolg durch eine Bestimmung der Beurteilungsübereinstimmung überprüft werden soll-

te. In den letzten Jahren wurden für verschiedenste Problembereiche – so z.B. auch für den Bereich zwischenmenschlicher Interaktionen – Beobachtungssysteme entwickelt, die genau vorgeben, nach welchen Regeln welches Verhalten auf welche Weise zu registrieren ist (vgl. z.B. Mash und Terdal, 1980).

Erweist sich im Einzelfall eine direkte Verhaltensbeobachtung als nicht durchführbar, kann ersatzweise auch mit Hilfe von Rollenspieltechniken versucht werden, die reale Situation zu simulieren. Dabei wird der Patient veranlaßt, sich so zu verhalten, als ob er sich tatsächlich in der betreffenden Situation befände. Wie nachgewiesen werden konnte, ähnelt das gespielte Verhalten in hohem Maße dem realen Verhalten, so daß bei dieser Methode der Reaktionserfassung nur sehr geringe Einbußen an Validität in Kauf genommen werden müssen.

Neben der Fremdbeobachtung des problematischen Verhaltens kommt der Selbstbeobachtung des Patienten eine besondere diagnostische Bedeutung zu. Selbstbeobachtungsmethoden werden – auch wenn sie weniger objektiv und zuverlässig sind – zur Datengewinnung vor allem dann eingesetzt, wenn eine Fremdbeobachtung, wie z.B. bei der Erfassung kognitiver Prozesse (Gedanken, Selbstverbalisationen, Selbstbewertungen etc.), nicht möglich ist. Sie werden auch angewandt, wenn dem Patienten das Problemverhalten nicht auf allen Verhaltensebenen deutlich ist, oder er über die Auftretenshäufigkeit des Symptoms in Abhängigkeit von verhaltenssteuernden Bedingungen keine Auskunft geben kann. Analog zur Fremdbeobachtung ist auch hier auf eine exakte Beschreibung der zu protokollierenden Verhaltensweisen und situativen Bedingungen, auf eindeutige Begriffe oder Symbole, die zur Registrierung verwendet werden, und eine genaue Definition der zeitlichen Bedingungen, unter denen die Aufzeichnung stattfinden soll, zu achten. Da sich gezeigt hat, daß der Selbstbeobachtung allein bereits therapeutische Funktion zukommt (Protokollierung vor dem Auftreten einer unerwünschten Verhaltensweise wird zum Hinweisreiz, dieses Verhalten zu unterlassen; positive Verstärkung durch Registrierung von Therapiefortschritten etc.), wurden in der letzten Zeit eine Vielzahl von standardisierten Selbstbeobachtungsinventaren zur Erfassung verschiedenster körperlicher Beschwerden entwickelt. Besonders bekannt sind die sog. Schmerztagebücher, die in Schmerzkliniken oder Schmerzambulanzen heute bereits routinemäßig eingesetzt werden.

Auch wenn der verhaltensdiagnostische Ansatz primär darum bemüht ist, das Verhalten unter den natürlichen Bedingungen direkt zu registrieren, werden auch Selbstberichte (self-reports) als Informationsquelle verwandt; d.h. der Bericht des Patienten, wie er sich in der betreffenden Situation seiner Erinnerung nach üblicherweise verhält, wird diagnostisch genutzt. Diese Art der Informationserhebung, die zwangsläufig weniger objektiv als eine direkte Messung sein muß, kommt beispielsweise innerhalb des verhaltensdiagnostischen Interviews zur Anwendung, wo bereits versucht wird, das symptomatische

Verhalten durch eine strukturierende Fragetechnik möglichst präzise auf den verschiedenen Verhaltensebenen zu erfassen. Zum anderen existieren inzwischen eine Vielzahl von standardisierten Fragebögen, die zur Identifikation von Problembereichen, aber auch zur Erfassung spezifischer Symptome entwickelt wurden (vgl. z.B. Schulte, 1980, 1986). Fragebögen erscheinen in Ergänzung zur Exploration insbesondere dann sinnvoll, wenn kognitive Prozesse wie Selbstverbalisationen oder Selbstbewertungen erfaßt werden sollen, die einer direkten Messung nicht zugänglich und dem Patienten eventuell auch nicht ausreichend bewußt sind, um über sie berichten zu können. Auch zur Feststellung wirksamer verstärkender Reize, die therapeutisch eingesetzt werden können, liegen inzwischen verschiedene sog. Verstärkerlisten vor (vgl. Schulte, 1980; Mash und Terdal, 1980). Fragebögen, die speziell für den Einsatz bei (psycho-)somatischen Störungen geeignet sind, wurden von Miltner (1986) zusammengestellt.

Verhalten äußert sich aus lerntheoretischer Sicht jedoch nicht nur auf der motorisch-handlungsmäßigen und der subjektiv-erlebnismäßigen Ebene, sondern auch auf der physiologischen Ebene, die besonders für Störungen mit hoher emotionaler Beteiligung bzw. bei körperlichen Symptomen von Relevanz ist. Um diese Ebene ebenso systematisch zu berücksichtigen, werden psychophysiologische Meßverfahren angewendet, die den Vorteil besitzen, durch entsprechende Ableitungen objektive und unmittelbar registrierbare Meßwerte zu liefern, die kontinuierlich erhoben und ausgewertet werden können. Physiologische Parameter, die im Rahmen verhaltenstherapeutischer Diagnostik verwendet werden, sind EKG, EEG, EMG, Hautwiderstand und Hautleitfähigkeit, Atemfrequenz, Blutdruck, Herzfrequenz, aber auch Augenbewegungen, biochemische Reaktionen etc. Ein genauer Überblick über die Anwendungsgebiete der einzelnen Parameter, einschließlich der Darstellung von Meß- und Interpretationsproblemen, findet sich z.B. bei Lang (1977).

19.4 Fallbeispiel

Eine 26jährige, verheiratete Patientin wurde wegen depressiver Stimmungslage und nur mäßigem Erfolg der somatischen Therapie eines Morbus Crohn zur psychosomatischen Untersuchung und Mitbehandlung überwiesen. Sie litt unter anfallsartigen Schmerzen im unteren Abdomen, die häufig durch „Aufregung" ausgelöst wurden. Sie hatte zudem an manchen Tagen bis zu zehn Stühle. Weder die Patientin noch ihr Ehemann waren über die chronische Natur der Erkrankung oder die Prognose informiert. Beide gingen von der Möglichkeit einer vollständigen, allerdings nur schwer erreichbaren Heilung aus. Die Patientin war seit zwei Jahren auf insgesamt fünf Jahre befristet arbeitsunfähig und war vorher, bis kurz nach der Eheschließung als kaufmännische Angestellte berufstätig. Erste Beschwerden waren während der Phase der Trennung von einem erheblich älteren, verheirateten Partner aufgetreten. Die Lösung von diesem Mann war ihr sehr schwer gefallen, und ist erst endgültig gelungen, als sie ihren Ehemann kennengelernt hatte. Im Rahmen dieser neuen Beziehung zog sie vom Wohnort der Mutter weg, um in der Nähe ihres Verlobten sein zu können, was zu einer Verschlechterung der Beschwerden führte und schließlich zur Diagnose des Morbus Crohn. Die befristete Berentung war von der Patientin nur widerwillig hingenommen worden, obwohl sie Schwierigkeiten am Arbeitsplatz mit einer Kollegin gehabt hatte; der Ehemann begrüßte die Berentung jedoch aus Gründen der Schonung für seine Frau. Tatsächlich trat nach der Berentung eine weitere Verschlechterung ein, die zu einer stationären Behandlung in einer internistisch-psychosomatischen Institution führte, bei der sie das autogene Training erlernte und psychotherapeutisch behandelt wurde. Es trat auch eine Besserung ein, die jedoch nach der Entlassung nur von kurzer Dauer war.

Neben der körperlichen Symptomatik bestand eine ausgeprägte Depression mit Antriebsverlust, niedergedrückter Stimmung und Selbstvorwürfen. Daneben war es im Laufe der Zeit zu einer fast totalen sozialen Isolation gekommen, sie hatte lediglich noch zu einer Mitbewohnerin aufgrund deren Initiative Kontakt. Die Hausarbeit wurde im wesentlichen vom Ehemann, der von Beruf Polizist war, geleistet. So war die Patientin nicht in der Lage, morgens zur selben Zeit wie ihr Mann aufzustehen, etwa um ihm das Frühstück zu machen. Sie stand gegen Mittag auf und verbrachte dann den Tag damit, kleinere Hausarbeiten zum Teil mehrmals zu wiederholen. Diese Tätigkeit wurde von ständigem Grübeln begleitet, dessen Inhalt fast ausschließlich aus Selbstvorwürfen bestand.

Die Patientin war jüngstes von vier Kindern, die Mutter hatte aus erster Ehe Zwillinge gehabt, die bei der Geburt der Patientin bereits 20 Jahre alt gewesen waren. Der Vater hatte aus erster Ehe eine Tochter, die kurz vor der Geburt der Patientin beim Baden ertrunken war. Da die Patientin dieser Tochter sehr ähnlich war, wurde sie vom Vater verwöhnt und bevorzugt worden, ebenso wie von der Mutter, die ihr alle Schwierigkeiten aus dem Weg geräumt hatte. Dies hatte in der Schule dann zu ersten Schwierigkeiten geführt, da sie nicht daran gewöhnt war, irgend etwas selbständig zu machen. Im Haushalt hatte sie nie helfen müssen, sie hatte es gut haben sollen. Der Vater hatte im Krieg eine Kopfverletzung davongetragen, die zunehmend zu Schmerzen geführt hatte, die er mit zunehmenden Mengen von Alkohol zu dämpfen versuchte. Dies hatte dazu geführt, daß er immer unberechenbarer und gewalttätig geworden war, und es zu häufigen tätlichen Auseinandersetzungen zwischen den Eltern gekommen war. Schließlich hatte auch die Patientin ihn nicht mehr beeinflussen können, was ihr anfangs noch gelungen war. Als die Patientin 20 Jahre alt war, verstarb der Vater in einer psychiatrischen Klinik. Die Patientin hatte sich wegen ihres Vaters stets geschämt, was in der Schule zu einer sozialen Isolierung geführt hatte. Sie war in der Schule mittelmäßig gewesen und hatte nach der Schule eine Lehre als kaufmännische Angestellte gemacht.

Art der Störung:
Im Gespräch mit der Patientin wurde sehr schnell deutlich, daß sie eine Beeinflussung bzw. Änderung ihrer Verhaltensprobleme wünschte. Sie litt anscheinend mehr unter ihrer

depressiven Symptomatik als unter den Symptomen des Morbus Crohn. Sie war zudem der Meinung, daß eine Veränderung dieser Probleme einen günstigen Einfluß auf die körperliche Erkrankung ausüben würde. So wurde das bereits oben geschilderte Verhalten als zu verändernde Störung definiert.

Auslösende Bedingungen:
Das geschilderte Verhalten trat hauptsächlich auf, wenn die Patientin sich allein in ihrer Wohnung aufhielt. Es gab keinen äußeren Reiz, der sie morgens zum Verlassen des Bettes veranlaßt hätte. Der Ehemann war der Auffassung, daß es für die Gesundheit seiner Frau günstig war, wenn sie sich schonte und lange im Bett blieb. Die zum Teil repetitiven Tätigkeiten wurden weniger durch Außenreize als durch Gedanken kontrolliert. Vorhandene Außenreize besaßen keine motivierende Qualität, dies galt auch für die seltenen Besuche der Nachbarin, deren Einladungen, diese zu besuchen, sie niemals nachgekommen war.

Biologische Bedingungen:
Es kann angenommen werden, daß der durch den Morbus Crohn insgesamt ungünstige körperliche Zustand der Patientin ihr antriebsarmes Verhalten begünstigt hat.

Reaktion bzw. Verhalten:
Untätigkeit, zielloses Verhalten, soziale Isolierung, quälende Gedanken der Insuffizienz und Selbstvorwürfe.

Konsequenz:
Wesentlich ist das Fehlen von Verstärkung für zielorientiertes Verhalten. Der Ehemann verstärkte das vorliegende Verhalten durch besorgte Zuwendung und „bestrafte" Ansätze von Initiative durch entmutigende Äußerungen, das sei zuviel für sie, das schaffe sie doch nicht usw. Aufgrund von im Verhältnis zum gegebenen Zustand überhöhten Zielen bestrafte sich die Patientin durch entwertende Gedanken von Mißerfolg für tatsächlich erbrachte Leistungen.

Kontingenz:
Die durch die Gedanken gegebene Selbstbestrafung erfolgt quasi verhaltensbegleitend. Der Einfluß des Ehemanns verstärkt vor allem die entwertenden Gedanken der Patientin, ihr Versagen und ihren Unwert betreffend. Auf diese Weise ist dieser Einfluß „mittelbar" ebenfalls während des Verhaltens präsent und wirksam.

Genese der Störung:
Das Selbstwertgefühl der Patientin wurde von seiten des Vaters dadurch gemindert, daß er ihr gegenüber seine besondere Zuwendung durch ihre „Ersatzfunktion" für die verstorbene Halbschwester begründete. Im weiteren Verlauf der Entwicklung verringerte sich ihr Selbstwertgefühl durch den schlechten Ruf ihres Vaters, was zu sozialer Isolierung führte.
Durch das extrem verwöhnende Verhalten und die massive Hilfe beider Eltern bei vielen Tätigkeiten, was Eigeninitiative verhinderte, wurde die Patientin in der Bewertung ihrer eigenen Kompetenz behindert. Statt dessen machte sie die Erfahrung, daß sie ohne fremde Hilfe nichts oder nur sehr wenig zu leisten imstande ist. Gleichzeitig dürften die mit der Hilfe der Eltern erbrachten Leistungen höher gewesen sein, als es ihrem damaligen tatsächlichen Leistungsvermögen entsprochen hätte. Der sich daraus ergebende überhöhte Leistungsanspruch an sich selbst mußte ebenfalls zum Gefühl eigener Unfähigkeit führen. Es kann zudem vermutet werden, daß beim geschilderten Erziehungsstil der Eltern eine Unabhängigkeit zwischen eigenem Verhalten und positiven Konsequenzen wie Lob bestanden hat, vermutlich

wurde sie für alles gelobt. Daraus ergibt sich die Tendenz, Erfolg als durch äußere Hilfe gegeben anzusehen, Mißerfolg jedoch eigener Unfähigkeit zuzuschreiben (depressive Attribution). So blieb sie stets auf andere Personen und sogar deren Präsenz angewiesen. Auf diese Weise konnte sie sich aus der problematischen ersten Beziehung erst lösen, als sie ihren späteren Ehemann kennengelernt hatte. Der berufliche Rahmen scheint ebenfalls, trotz der vorhandenen Belastung, eine stützende Funktion gehabt zu haben, zumindest erlebte sie bei dessen Fortfall durch die Berentung eine erhebliche Verschlechterung ihres Befindens. Das depressive Verhalten trat ebenfalls im vollen Umfang erst auf, als die Patientin berentet tagsüber in ihrer Wohnung allein war.

Therapieziele:
– Veränderung „depressiver" Einstellungen und Attributionen. Diese wurden als wesentlich vermittelnde Bedingung für die unangemessenen Verhaltensweisen angesehen. Sie hinderten die Patientin daran, alternative Verhaltensweisen auch nur probehaft durchzuführen.
– Erwerb von Selbstkontrolle für organisierte und zielgerichtete Tätigkeiten im Haushalt.
– Kontrolle des Auftretens störender Gedanken, die die Tätigkeiten der Patientin häufig unterbrochen hatten und die sie nicht kontrollieren konnte.
– Veränderung der Interaktion zwischen dem Ehemann und der Patientin, da dieser die oben genannten Einstellungen der Patientin verstärkte.
– Erwerb einer Technik der Entspannung. Die Patientin vermied eigene Verhaltensweisen und Situationen, die zu Aufregung geführt hatten, da sie realistischerweise befürchtete, daß es dann zu einer Verstärkung der körperlichen Symptomatik kommen könnte. Da sie bereits früher, während ihrer stationären psychosomatischen Behandlung das autogene Training erlernt, jedoch später nicht mehr angewendet hatte, sollte diese Technik erneut eingeübt werden.
– Abbau sozialer Ängste. Hierbei wurde jedoch davon ausgegangen, daß es unter Umständen nicht nötig sein würde, eine spezielle Angstbehandlung durchzuführen, da die Ängste weitgehend durch die oben genannten Einstellungen und Attributionen bedingt sein könnten, so daß deren Änderung zu einem Abbau der Angst führen müßte.
– Üben sozialer Verhaltensweisen wie aktiv Kontakt zu anderen Personen aufnehmen.

Therapieplan:
– Es wurden mehrere ausführliche Gespräche mit der Patientin und ihrem Ehemann geführt, in denen zu klären war, inwieweit der Ehemann zu beeinflussen war, sein Verhalten gegenüber der Patientin so zu ändern, daß er ihr „depressives" Verhalten nicht mehr verstärkte. Dazu war es notwendig, mit beiden die der Störung zugrundeliegenden Mechanismen zu erörtern. Die anzuwendenden Therapiestrategien wurden ebenfalls erörtert.
– Kognitive Umstrukturierung der inadäquaten vermittelnden Kognitionen. Diese sollten in ihrem Zusammenhang bei der entsprechenden Interpretation aktueller Situationen durch die Patientin erfaßt und alternative Sichtweisen argumentativ aufgezeigt werden.
– Verbalisierung dieser alternativen Sichtweisen in bezug auf konkrete Aufgabensituationen. Gemeinsam wurden mit der Patientin Gründe für erfolgreiches Handeln aufgezeigt, die zu einer die eigene Kompetenz beinhaltenden Aussage führten, die dann verbalisiert wurde.
– Es wurden detaillierte Tagespläne erstellt. Dabei wurde von einer Grundlinie, die ungefähr dem Status quo entsprach, ausgegangen, damit die Patientin beim Ausführen

dieser Pläne nicht überfordert wurde. Erlebnisse von Mißerfolg wurden auf diese Weise verhindert.

- Es wurden äußere Signale (Küchenwecker) zur Strukturierung und Limitierung von Zeitabläufen benutzt. Beim Erstellen der Tagesabläufe wurde für jeden Handlungsabschnitt der Zeitbedarf, ebenfalls am Status quo orientiert, geschätzt, der dann auf dem Küchenwecker eingestellt wurde. Beim Ertönen des Signals war die Tätigkeit abzubrechen. Die Zeiten wurden zunehmend verkürzt. Zur Vorverlegung des Zeitpunkts des Aufstehens wurde ein elektronischer Wecker beschafft, dessen Wecksignal so lange ertönte, bis er ausgeschaltet wurde. Es zeigte sich, daß dieser Wecker außerhalb der Reichweite der Patientin aufgestellt werden mußte, damit sie ihn nicht vom Bett aus ausschalten konnte.
- Die Patientin wurde dazu angehalten, durch Verbalisierung von Lob sich selbst zu verstärken.
- Das autogene Training wurde erneut eingeübt und dann vor und in Situationen angewendet, von denen die Patientin annahm, sie würden sie aufregen.
- In den Therapiestunden wurden soziale Verhaltensweisen geübt, die dann „in vivo" angewendet wurden, z. B. Anmeldung eines Besuchs bei der Nachbarin.

Kontrolle des Therapieverlaufs:
Zur Kontrolle des Therapieverlaufs dienten die Berichte der Patientin und gelegentlich die des Ehemanns. In den Therapiesitzungen konnte ebenfalls relevantes Verhalten beobachtet werden, z. B., ob die Patientin zur Stunde an die Tür klopfte, wie laut sie sprach und wie häufig sie „depressive" Gedanken äußerte. Die Planung der Tagesabläufe war täglich zu erstellen und schriftlich zu fixieren, wobei der Erfolg oder Mißerfolg für jede Tätigkeit festzuhalten war, genauso wie eine Skalierung, wieviel Spaß die Tätigkeit gemacht hatte und wie hoch der erlebte Erfolg eingeschätzt wurde.

Verlauf der Therapie:
Nach der Verhaltensanalyse wurde prototypisch ein Tagesplan aufgestellt, der für die einzelnen Wochentage variiert werden konnte. Die zeitlichen Einzelabschnitte mußten zunächst sehr klein gewählt werden. In den zunächst einwöchigen therapeutischen Sitzungen wurden Erfolge und Schwierigkeiten der Tagesabläufe erörtert. Dabei wurde die Patientin angehalten, im Hinblick auf Ereignisse laut positive Gedanken zu verbalisieren und dies auch in ihrer häuslichen Umgebung zu tun. Nach einigen Stunden konnten diese Verbalisierungen „verdeckt" durchgeführt werden. Zusätzlich wurde das „morgendliche" Aufstehen vorverlegt, zunächst in 15-Minuten-Intervallen. Nach einigen Sitzungen stellte sich heraus, daß es notwendig war, zusätzlich zur Selbstverbalisierung von Lob einen „äußeren" Verstärker in Form von Geld einzuführen. Für jeden erfolgreich durchgeführten Abschnitt konnte sie DM 0,50 aus einer vom Ehemann zur Verfügung gestellten Kasse entnehmen. Dabei zeigte sich die Notwendigkeit, die Patientin anzuhalten, eigene Wünsche zu formulieren und zu erfüllen; sie hatte das Geld zunächst ihrer Haushaltskasse hinzugefügt.
Durch den Fortschritt der Therapie entstand für die Patientin zunehmend mehr Freizeit, die sie zunächst nicht nutzen konnte. Zunächst begann sie mit Tätigkeiten, die sie als Hobbys empfand, die sie zu Hause durchführen konnte, wie Glasmalerei oder das Knüpfen eines Teppichs. In späteren Phasen der Therapie wurden die Hobbys mehr aus dem häuslichen Bereich herausverlegt, so besuchte sie Kurse an der Volkshochschule, zum Teil mit dem Ziel, ihre beruflichen Fähigkeiten wieder aufzufrischen. Dabei knüpfte sie zunehmend mehr soziale Kontakte, die schließlich auch zu gegenseitigen Besuchen führten, so daß ihre soziale Isolation auch hierdurch verringert wurde. Im Zuge dieser Veränderungen, vor allem nach der Verbesserung der körperlichen Symptomatik, entwickelte sich bei der Patientin ein Kinderwunsch, andererseits der Wunsch nach Berufstätigkeit. Beide Wünsche schlossen sich für die Patientin aus, da sie der Meinung war, mit der Erziehung eines Kleinkindes sei eine berufliche Tätigkeit nicht zu verbinden. Der Ehemann unterstützte ihren beruflichen Wunsch, da er durch eine Schwangerschaft die gesundheitliche Situation seiner Frau als gefährdet ansah. So setzte sich die Patientin mit dem Arbeitsamt in Verbindung, ließ ihre Arbeitsunfähigkeit vor dem durch die Befristung gegebenen Termin beenden und fand schließlich nach einigen Schwierigkeiten eine Halbtagsstelle als Schreibkraft.
Schwierigkeiten ergaben sich aus dem Umstand, daß der Ehemann sich nur schwer an die Veränderung seiner Frau anpassen konnte. Er empfand die zunehmende Selbständigkeit als Abkehr von seiner Person und als Gefährdung der Partnerschaft. Zu einer Eskalation dieses Problems kam es, als er nicht, wie geplant, zusammen mit der Patientin zu einer Karnevalseinladung gehen konnte und sie „rhetorisch" dazu aufgefordert hatte, doch alleine dorthin zu gehen, was sie zu seiner enttäuschten Überraschung dann auch getan hatte. Daraufhin nahm der Ehemann an zwei therapeutischen Stunden mit teil und schien danach diese Schwierigkeit überwunden zu haben, zumindest kam es zu keinen weiteren Schwierigkeiten mehr. Als die Patientin begann, sich eine Stelle als Schreibkraft zu suchen, erlebte sie zunächst Mißerfolg. So wurde sie wiederholt auch wegen ihrer Erkrankung abgelehnt, was sie wiederum an ihrem Wert als Person zweifeln ließ. Es ist nicht zu entscheiden, ob das therapeutische Vorgehen hinsichtlich dieser Schwierigkeit oder schließlich ihr Erfolg bei der Stellensuche ausschlaggebend war, zumindest hatte sie sich nicht „entmutigen" lassen, weiter nach einer Stelle zu suchen.

Erfolgskontrolle:
Es wurden insgesamt 34 therapeutische Sitzungen durchgeführt, von denen die ersten 18 in wöchentlichen Abständen, die folgenden 8 in 14tägigem Abstand und der Rest im Abstand von vier Wochen stattfanden. Die Patientin erschien, wie vereinbart, ein Jahr nach Beendigung der Therapie. Sie war berufstätig und hatte keine Schwierigkeiten, neben ihrer Berufstätigkeit ihren Haushalt zu erledigen. Die körperliche Symptomatik war gebessert geblieben, die Medikation mit Cortison war bereits während der Therapie abgesetzt worden und war nicht wieder erforderlich, die Therapie mit Azulfidine war reduziert geblieben. Für den Erfolg der Therapie war wohl nicht unwesentlich, daß der Ehemann die Therapie sehr unterstützt hatte und sich auch im Hinblick auf seine Einstellungen und Erwartungen an seine Frau sehr „einsichtig" gezeigt hatte.

19.5 Methoden der Verhaltensmodifikation in der Psychosomatik

Prinzipiell kommen in der Medizin sämtliche Techniken der Verhaltensmodifikation zur Anwendung, weil bei somatisch kranken Patienten potentiell jede psychische Störung auftreten kann. Im folgenden soll jedoch auf diejenigen Techniken und ihre Anwendungen eingegangen werden, denen in der Medizin besondere Bedeutung zukommt, da sie in Prozesse von Krankheit in irgendeiner Weise involviert sind.

Auch dies kann im Prinzip sämtliche Vorgehensweisen betreffen, es hat sich jedoch gezeigt, daß bestimmte Techniken besonders häufig angewendet werden. Es ist zudem zu bedenken, daß es in der Verhaltenstherapie zwar bestimmte „feststehende" Verfahren gibt, wie die systematische Desensibilisierung zur Behandlung der Angst oder das „Biofeedback" zur Beeinflussung physiologischer Reaktionsanteile, meistens jedoch ein individueller Behandlungsplan aus mehreren Komponenten zusammengesetzt wird.

Im folgenden Teil des Kapitels sollen kurz die einzelnen Techniken oder Komponenten verhaltenstherapeutischen Vorgehens dargestellt werden, die für die Psychosomatik oder für ein Verständnis des verhaltenstherapeutischen Ansatzes eine besondere Bedeutung besitzen. Es ist dabei im gegebenen Rahmen nicht möglich, sämtliche Anwendungsbereiche dieser Techniken im Bereich der Medizin zu erörtern, dies wird mit einem Schwerpunkt im Hinblick auf die Behandlung von Patienten mit Kreislauferkrankungen und Patienten mit akutem oder vor allem chronischem Schmerz geschehen. In den einzelnen, auf spezielle Erkrankungen bezogenen Kapiteln wird ebenfalls auf die Anwendung verhaltenstherapeutischer Techniken eingegangen, wenn sie in der Behandlung der jeweiligen Erkrankung eine besondere Rolle spielen, wie beispielsweise bei der Adipositas oder dem Kopfschmerz. Eine umfassende Übersicht über verhaltenstherapeutisches Vorgehen in der Medizin findet sich in Miltner, Birbaumer und Gerber (1986) „Verhaltensmedizin". Dort werden vor allem nicht nur therapeutische Aspekte, sondern auch diesen zugrundeliegende pathogenetische Modelle und Forschungsergebnisse ausführlich dargestellt.

19.5.1 Methoden des klassischen Konditionierens

Im Kapitel 7 wurde das Paradigma des klassischen Konditionierens ausführlich dargestellt. Es bezieht sich auf die Relation von Reizen, die einer Reaktion oder einer Verhaltensweise vorausgehen. So gibt es bei Angstreaktionen beispielsweise regelmäßig Reizkonstellationen, die die Reaktion auslösen. Diese Reaktion besteht aus mehreren Komponenten, von denen eine als physiologisch bezeichnet werden könnte. Geht man davon aus, daß eine Angstreaktion auf harmlose Reize eine klassisch konditionierte Reaktion darstellt, so bestünde die Technik zur Beseitigung dieser Reaktion in einem Verfahren der Löschung. Eine konditionierte Reaktion auf einen konditionierten Reiz wird dadurch gelöscht, daß der Reiz häufig ohne den unkonditionierten Reiz dargeboten wird. Da eine wesentliche Bedingung für die Aufrechterhaltung der Angst in einer angstverstärkenden physiologischen Reaktionskomponente besteht (vgl. Kap. 8), müssen die Reize in einer Intensität dargeboten werden, die diese Komponente nicht auslöst. Im so abgestuften Vorgehen bei der „systematischen Desensibilisierung" (SD) (Wolpe, 1958) wird diesem Sachverhalt entsprochen, indem eine Hierarchie angstauslösender Reize erstellt und dargeboten wird, wobei sichergestellt wird, daß es zu keiner ausgeprägten physiologischen und erlebnismäßigen Aktivierung kommt. Dabei ist es wichtig, daß die Abstände zwischen den Stufen der Hierarchie nicht zu groß sind.

> Bei der Behandlung einer Patientin mit einer sehr ausgeprägten Spinnenphobie konnte die Patientin zunächst ein zusammengefaltetes Blatt Papier, auf dem das Wort „Spinne" geschrieben stand, nicht in die Hand nehmen. Ebensowenig war sie in der Lage, das Wort angstfrei auszusprechen. Sie benutzte statt dessen den „Eigennamen" einer Spinne aus einem Kinderfilm. Da der Therapeut sich diesem Sprachgebrauch nicht anschloß, wurde die Patientin im Rahmen der Verhaltensanalyse sehr oft mit dem angstbesetzten Wort konfrontiert und konnte es selbst noch vor Beginn der eigentlichen Desensibilisierung angstfrei aussprechen. Ebenso war sie in der Lage, ca. 50 Dias, auf denen Spinnen in unterschiedlicher Größe und Deutlichkeit abgebildet waren, und die ein photographisch begabter und interessierter Bekannter angefertigt hatte, in ihrer Tasche mit in die Klinik zu bringen. Dieser Effekt war lediglich auf die milde Konfrontation mit dem ehemals angstauslösenden Reiz des Wortes „Spinne" zurückzuführen, und die Angstreaktion war für diese Klasse von Reizen dadurch gelöscht worden. Die eigentliche Desensibilisierung wurde dann von einem Prozeßrechner durchgeführt, der die Dias in abgestufter Reihenfolge darbot und dabei auf Anstiege der Herzfrequenz der Patientin von einem gleitenden Mittelwert und Verringerungen der Variabilität (Verringerung der respiratorischen Arrhythmie als Zeichen sympathischer Erregung) achtete. Zwischen den einzelnen „kritischen" Spinnenabbildungen war jeweils ein „Entspannungsdia" angeordnet, das angesteuert wurde, wenn die Spinnenabbildung für 30 Sekunden dargeboten worden war. War es bei einer dieser Darbietungen zu kritischen Veränderungen der Herzfrequenz gekommen, schaltete der Rechner auf das davor liegende Entspannungsdia und danach auf das vor diesem liegende „leichtere Spinnendia". Nach 11 derartigen Sitzungen, noch bevor das Ende der Hierarchie, die Makroaufnahme einer Tarantel, erreicht war, berichtete die Patientin, sie habe am Vorabend eine Spinne, die sie in ihrer Wohnung angetroffen habe, mit der Hand entfernt.

Die Wirkung der systematischen Desensibilisierung ist verschieden interpretiert worden, als Löschung im Sinne des klassischen Konditionierens (Wilson und Davison, 1971) bzw. als Habituation (Watts, 1979) sowie als Gegenkonditionierung einer neuen Reaktion (Entspannung) auf die angstauslösenden Reize, für die von Wolpe das neurophysiologische Prinzip der „reziproken Hemmung" als Grundlage angesehen wird. Die Methode der SD beinhaltet die graduelle Konfrontation mit angstauslösenden Reizen. Die Methode der Reizüberflutung (flooding) dagegen beinhaltet ebenfalls die Konfrontation mit einem angstauslösenden Reiz, jedoch in seiner stärksten Ausprägung. Der Patient wird dann daran gehindert, den Reiz zu vermeiden (response prevention; Baum, 1970), was dazu führt, daß die physiologischen Reaktionskomponenten habituieren. Es ist dabei darauf zu achten, daß es zu sehr hoher physiologischer Erre-

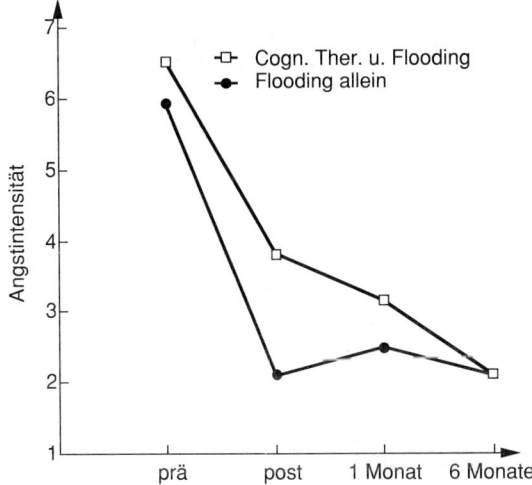

Abb. 19–3. Reduzierung der Angstintensität durch Konfrontation oder Konfrontation und kognitive Therapie bei Patienten mit Waschzwang (aus Emmelkamp, 1980).

gung kommen kann, so daß körperliche Erkrankungen, etwa eine koronare Herzerkrankung, zu berücksichtigen sind. Neben phobischen Störungen ist diese Methode sehr erfolgreich bei Patienten mit Waschzwang eingesetzt worden (Turner et al., 1979). Abbildung 19–3 zeigt die Wirkung von Flooding auf ein derartiges Zwangsverhalten, sie zeigt auch, daß das Hinzufügen einer Komponente, die in kognitiver Therapie bestand, den Effekt nicht erhöhte.

Im Vergleich zur ursprünglichen Technik der SD, bei der die angstauslösenden Reize meist in der Vorstellung „dargeboten" wurden, d.h. der Patient sollte sich eine bestimmte Situation vorstellen, wird bei der Reizüberflutung der Reiz „in vivo" dargeboten. Dieses Vorgehen hat sich in der SD zunehmend durchgesetzt, beispielsweise wird der Patient nach Durchführung der Methode bei einigen Reizen in der Vorstellung angehalten, sich diesen Reizen konkret in seinem Alltagsleben auszusetzen.

Im oben beschriebenen Fallbeispiel der Patientin mit Morbus Crohn war das Vorgehen bei dem Einüben sozialer Verhaltensweisen einer SD recht ähnlich. Kritische Situationen wurden dabei im Rollenspiel in den Therapiestunden geübt und dann das geübte Verhalten von der Patientin in ihrem Lebensbereich angewendet. So bestand eine Situation mit Variationen anhand von Schwierigkeit in Besuchen bei derjenigen Nachbarin, von der sie selbst gelegentlich Besuch bekam. Bevor die Patientin die Nachbarin aufsuchte, wurden verschiedene Varianten im Rollenspiel geübt, etwa daß die Nachbarin sie erfreut zum Tee einlädt, den sie sich gerade gemacht hatte, aber auch, daß diese ihr sagt, es passe ihr gerade nicht, da sie in die Stadt müsse, ob sie nicht in zwei Stunden oder an einem anderen Tag wiederkommen könne.

Komponenten der Angstbehandlung, wie sie hier dargestellt worden sind, lassen sich gut in der Behandlung von Patienten mit funktionellen Herz-Kreislauf-Störungen anwenden (vgl. Kap. 41). Eines der zentralen Symptome der Patienten ist bei dieser Störung eine auf die Funktionstüchtigkeit des eigenen Kör-

pers bezogene Angst. Die Patienten haben zudem meist die Erfahrung gemacht, daß die Beschwerden in bestimmten Situationen auftreten, die sie dann in der Regel wie einen Angstreiz vermeiden. Hier ist ein abgestuftes Vorgehen vorzuziehen, Methoden der Reizüberflutung sind weniger angezeigt. Es ist weiterhin wesentlich, die kognitiven Komponenten der Angst in die Behandlung mit einzubeziehen. Die Patienten haben meist genaue, wenn auch nicht zutreffende Vorstellungen über ihren körperlichen Zustand, die die Art der Beschwerden mehr bestimmen als physiologische Zusammenhänge. Diese Überzeugungen bestimmen das vermeidende Verhalten der Patienten mit, die beispielsweise häufig körperliche Belastung meiden, da die damit verbundenen und für sie spürbaren körperlichen Veränderungen von ihnen bereits als Krankheitszeichen und die Belastung entsprechend als schädigend oder gefährlich interpretiert wird. Werden diese Gedanken unbeeinflußt gelassen, so kann der Patient die Behandlung, in der er sich diesen dann immer noch als schädigend interpretierten Situationen aussetzen soll, nicht mitvollziehen. Die Patientin mit der Spinnenphobie war von der Harmlosigkeit der Spinnen überzeugt, schätzte ihre Angst also wie alle Phobiker als unrealistisch ein, Patienten mit funktionellen Herz-Kreislauf-Störungen hingegen sind zunächst davon überzeugt, daß ihre Angst realistisch ist. Dieser Unterschied darf beim therapeutischen Vorgehen nicht außer acht gelassen werden.

Inwiefern Prozesse des klassischen Konditionierens beim Phänomen Schmerz wirksam werden können, und somit Schmerz nach diesen Prinzipien auch behandelt werden kann, soll anhand zweier Beispiele demonstriert werden.

Ein Patient reagiert in einer spezifischen konflikthaften Situation mit Muskelanspannung (UCS), die langfristig zu Schmerz (UCR) führt. Nach dem klassischen Konditionierungsparadigma ist denkbar, daß spezifische Gedanken oder Vorstellungen, die im Laufe der kritischen Situation auftreten, zu konditionierten Reizen (CS) werden, die im weiteren die Schmerzreaktion auslösen, auch wenn zuvor keine Muskelanspannung (als ursprünglicher Auslöser des Schmerzes) vorlag.

Ein anderer Patient reagiert in einer subjektiv bedeutsamen Situation mit Angst, die zu muskulärer Verspannung im Sinne einer unkonditionierten Reaktion führt. Diese unkonditionierte Reaktion kann jedoch wiederum zum konditionierten Reiz (CS) für die Angst werden, die wiederum mit Verspannung (CR) beantwortet wird. Auf diese Weise kann es langfristig zu einem Aufschaukelungsprozeß mit der Konsequenz einer stetigen Schmerzzunahme als Folge der Verspannung kommen (sog. „Angst-Spannungs-Zyklus").

Eine wesentliche Komponente bei der SD stellt die Entspannung dar, als ein mit einer Angstreaktion inkompatibler Zustand. In zahlreichen Untersuchungen wurde die Wirkung von Entspannung für sich allein untersucht. Sie stellt vermutlich die am häufigsten angewandte therapeutische Komponente dar (vgl. Kap. 22). Die Anwendung reicht von Angststörungen (Levin und Gross, 1988) über Hypertonie (Benson et al., 1974; Agras et al., 1987; McCoy et al., 1988; Walter et al., 1988), Rückenschmerzen (low back pain) (Jamison et al., 1988), Diabetes (Lammers et al., 1984), Kopfschmerz (Budzynski et al., 1970), Tinnitus (Ireland, 1985), Asthma (Maß et al., 1988), Schlafstörungen (Espie et al., 1989) bis zur Epilepsie (Rousseau et al., 1985) und Chemotherapie bei Krebspatienten (Burish und Carey, 1986; Carey und Burish, 1988). Stets wurde eine positive Wirkung der verschiedenen Entspannungsverfahren festgestellt, wobei diese Wirkung größer war, wenn die Entspannung Teil eines übergreifenden Behandlungskonzepts war. Dabei wurde diese meist als alternative Reaktion in bestimmten aktivierenden Situationen im Sinne der Gegenkonditionierung eingesetzt. Dabei wird davon ausgegangen, daß bestimmte Reize eine physiologische Reaktion (Erhöhung des Blutdrucks, Anstieg der Muskelspannung usw.) auslösen, die durch die Gegenkonditionierung, ebenso wie bei der Behandlung der Angst die Angstreaktion durch die alternative Reaktion, gelöscht wird. Es zeigt sich dabei, daß das lerntheoretische Modell keine Unterschiede zwischen körperlichen und psychischen Verhaltenskomponenten einführt, was der ganzheitlichen Betrachtungsweise der Psychosomatik sehr entgegenkommt.

Zu den in der Schmerzbehandlung am häufigsten eingesetzten Entspannungsmethoden zählen die progressive Muskelrelaxation nach Jacobson (1938) und das autogene Training von Schultz (1932), deren Effektivität bei verschiedensten Schmerzzuständen (z.B. Spannungskopfschmerz, Geburtsschmerz, Rückenschmerz, Gesichtsschmerz etc.; vgl. Weisenberg, 1982; Turner und Chapman, 1982) untersucht wurde. Nach diesen Studien scheint Entspannung – ein entsprechendes Maß an Übung vorausgesetzt – ein wirksames Mittel zur Erhöhung der Schmerztoleranz zu sein, sie erweist sich allerdings nicht immer als die wirksamste Behandlungsart. Indiziert sind Entspannungsverfahren vor allem, wenn ein Angst-Spannungs-Zyklus vorliegt, wobei auch hier angenommen wird, daß Entspannung antagonistisch zur Angst wirkt. Durch den Einsatz von Entspannung kann somit ein Aufschaukelungsprozeß von Angst, Verspannung und Schmerz in der Regel verhindert werden. Andererseits wird im Falle von akutem Schmerz das Erleben von Angst allgemein als eine schmerzverstärkende Bedingung angesehen, die durch Entspannung erfolgreich reduziert werden kann.

19.5.2 Operante Methoden

Ein Teil der Methoden der Verhaltenstherapie nutzt Techniken des operanten Lernens. Es wird also versucht, Verhalten zu ändern, indem die Konsequenzen des Verhaltens geändert werden. Prinzipiell wird damit, wie im Kapitel 7 ausführlicher erörtert, die „Auftretenswahrscheinlichkeit" eines Verhaltens beeinflußt. Durch positive Konsequenzen (positive Verstärkung durch einen positiven Reiz oder negative Verstärkung durch Fortfall eines negativen Reizes) erhöht sich diese Wahrscheinlichkeit, durch negative Konsequenzen (Bestrafung: ein negativer Reiz folgt auf das Verhalten oder ein positiver Reiz unterbleibt) verringert sie sich. Tatsächlich bestimmt dieses Prinzip nicht nur therapeutisches Vorgehen oder das in einem erzieherischen Zusammenhang, sondern auch, meist unsystematisch, soziales Verhalten. Nicht nur Psychotherapeuten verstärken z.B. Klassen von Aussagen ihrer Patienten durch ein verstärkendes „Mmh", Kopfnicken oder zustimmenden Gesichtsausdruck, also selektive Zuwendung (Krasner, 1962). Im Rahmen der Verhaltenstherapie wird dieses Prinzip der Verstärkung systematisch angewendet, um erwünschtes Verhalten zu festigen oder unerwünschtes zu löschen, wobei Methoden der Bestrafung durch Strafreize kaum Anwendung finden.

Es gibt nun eine ganze Reihe von unterschiedlichen Möglichkeiten, operante Prinzipien in der Verhaltenstherapie anzuwenden.

Positive Verstärkung

Ein positiver Reiz folgt auf ein erwünschtes Verhalten. Im oben genannten Fallbeispiel wurde dies in den therapeutischen Sitzungen angewendet, zum einen im Hinblick auf Verhalten wie ausreichend lautes Sprechen, Verbalisieren sich selbst positiv bewertender Gedanken usw. Die Patientin verstärkte sich jedoch auch selbst durch „Eigenlob" nach erfolgreich ausgeführten Tätigkeiten im Haushalt, eine ihr bis dahin fremde Strategie.

Negative Verstärkung

Das im Fallbeispiel genannte Vorgehen impliziert ebenfalls eine negative Verstärkung, indem die negativen, sich selbst entwertenden Gedanken entfallen.

Operante Löschung

Dieses Prinzip besteht darin, daß Verhalten durch den Fortfall der Verstärkung gelöscht werden kann. Es wird immer dann sehr wesentlich, wenn unerwünschtes Verhalten durch Verstärkung, oft durch wohlgemeinte Zuwendung von Angehörigen oder sonstigen Bezugspersonen, aufrechterhalten wird. Es ist sehr wichtig, derartige, gelegentlich als „sekundärer Krankheitsgewinn" bezeichnete Bedingungen zu ermitteln und zu ändern. Im Fallbeispiel war dies durch das Verhalten des Ehemanns gegeben, der die Patientin nicht nur in ihrer entwertenden Einstellung bestätigte, sondern ihre Passivität durch Zuwendung ebenfalls direkt verstärkte. Für den Erfolg der Therapie war es ganz wesentlich, daß er sein Verhalten tatsächlich änderte. Nicht wenige Patienten mit funktionellen Syndromen erleben im Zusammenhang mit ihren Symptomen eine verstärkte Zuwendung durch ihre Angehörigen, aber auch bei Patienten mit chronischem Schmerz wurde ein derartiger Einfluß nachgewiesen (Flor et al., 1987).

Response Cost

Diese Verfahrenskomponente betrifft den Fortfall eines positiven Reizes beim Auftreten eines unerwünschten Verhaltens. Dies beinhaltet, daß der positive Reiz kontrolliert werden kann.

Bei einer Patientin mit einer psychogenen generalisierten Verkrampfung der Muskulatur im Kopf- und Rumpfbereich und der Arme und Hände sowie einem Blepharospasmus, der zu einer „Blindheit" geführt hatte, da sie schon seit Jahren die Augen nicht mehr hatte öffnen können, wurde dieses Verfahren angewandt. Eine Lieblingsbeschäftigung der Patientin bestand im Hören von Musik. Ihre Platten wurden auf einen Kanal eines Stereotonbands überspielt, auf die andere Spur ein mäßig lauter Sinuston von 1 kHz. Es war beobachtet worden, daß die spastische Symptomatik in der Verkrampfung der linken Hand gelegentlich nachließ und die Hand für kurze Zeit entspannt war. Die Patientin hörte zunächst ihre Musik zu den Zeiten, in denen die Hand entspannt war. Begann sie, die Hand zu verkrampfen, wurde der Kanal umgeschaltet und sie hörte statt der Musik den Sinuston. So wurde schrittweise vorgegangen, indem das Kriterium für das Hören der Musik erweitert wurde, also der Arm, der Hals, der rechte Arm usw. entspannt sein mußten. Auf diese Weise gelang es der Patientin, nach insgesamt 14 Jahren erstmals wieder, zunächst für Augenblicke, dann für Minuten, ihre Augen zu öffnen. Es soll nicht verschwiegen werden, daß die Behandlung damit nicht beendet werden konnte, da die Patientin sich zu diesem Zeitpunkt weigerte, dieses Verfahren weiter durchzuführen. Zudem handelte es sich bei dem spastischen Syndrom nur um einen Bereich einer übergreifenden Störung. Andererseits führte das angewandte Verfahren von „response cost" zu einer erheblichen Verringerung der motorischen Störung.

Verhaltensformung (Shaping)

Hierunter versteht man einen Vorgang, durch den ein zunächst nicht im Verhaltensrepertoire vorhandenes Verhalten in einzelnen Schritten angenähert und schließlich erreicht wird. In beiden genannten Fallbeispielen wurde diese Methode angewandt. Das Öffnen der Augen war zunächst nicht im Verhaltensrepertoire der Patientin mit dem spastischen Syndrom enthalten, konnte also auch nicht durch Verstärkung in der Auftretenswahrscheinlichkeit erhöht werden. Also wurde zunächst ein anderes, vorhandenes Verhalten verstärkt, das Entspannen der linken Hand usw. Im Falle der Patientin mit Morbus Crohn wurde auf diese Weise der Zeitpunkt des „morgendlichen" Aufstehens angenähert, ebenso das Durchführen von Tätigkeiten im Haushalt. Wichtig dabei ist, daß das Kriterium für die Verstärkung langsam erhöht wird, sonst erreicht man keine Annäherung an das gewünschte Verhalten. Ein weiterer Vorteil besteht darin, daß man das Kriterium für eine Verstärkung so wählen kann, daß der Patient mit großer Wahrscheinlichkeit eine Verstärkung erfährt, was seine Motivation erhöht. Im Falle der Patientin mit Morbus Crohn war dies besonders wichtig, wie bei allen Patienten, die unter depressiven Störungen mit Antriebsverlust leiden.

Münzverstärkersystem (token economy)

In verschiedenen Zusammenhängen kann es wichtig sein, daß die Verstärkung sofort erfolgt, jedoch eine primäre Verstärkung nicht zu jedem notwendigen Zeitpunkt möglich ist. In solchen Fällen wird ein Münzsystem benutzt, d.h. der Patient erhält eine Marke, die er zu einem späteren Zeitpunkt gegen einen primären Verstärker eintauschen kann. Man hat dieses Vorgehen vor allem bei psychotischen Patienten angewendet, um erwünschtes Verhalten zu erreichen (Ayllon und Azrin, 1965; Woods et al., 1984). Es liegt auf der Hand, daß der Wert des Geldes diesem System entspricht, und bei der Patientin mit Morbus Crohn war es zu einem Zeitpunkt notwendig, ihr Verhalten durch Geld zu verstärken, das ihr der Ehemann zur Verfügung gestellt hatte. Dabei zeigte sich eine Schwierigkeit: Die Motivation der Patientin schien dadurch nicht sonderlich beeinflußt zu werden, wobei sich herausstellte, daß die Patientin das Geld nicht für sich ausgab, sondern es sparte. Als sie es für angenehme Dinge ausgab, wirkte die Verstärkung.

Stimuluskontrolle

Betrachtet man die bisher referierten Komponenten verhaltenstherapeutischen Vorgehens, so könnte der Eindruck entstehen, als seien bei einer Maßnahme im Hinblick auf ein bestimmtes Verhalten stets alle lerntheoretischen Aspekte klar zu trennen. Tatsächlich sind Vorgänge des klassischen und operanten Konditionierens stets eng miteinander verknüpft. So kann es notwendig sein, in einem bestimmten Zusammenhang diejenigen Reize zu beeinflussen, die ein Verhalten auslösen, das man mit operanten Methoden ändern will. Ein wichtiges Anwendungsfeld hierfür ist das „Selbstkontrollernen", etwa bei der Behandlung der Adipositas (Stunkard, 1972; Kirschbaum et al., 1985). Aufgrund des Einflusses von Außenreizen auf das Eßverhalten, der bei Patienten mit Adipositas größer ist als bei Normalgewichtigen, ist ein wesentliches Element der Therapie, Nahrungsmittel aus den Wohnräumen zu entfernen, so daß sie als Reize nicht wirksam werden, wenn der Patient sich dort aufhält. Damit werden ein unerwünschtes Verhalten auslösende Reize im Sinne der Stimuluskontrolle beeinflußt (vgl. Kap. 38.1).

Auch Schmerz wird nicht nur durch die jeweilige Läsion, sondern auch durch die nachfolgenden Konsequenzen nach dem Prinzip des operanten Lernens geformt. Jede Verstärkung von Klagen und Passivität kann dazu führen, daß das Schmerzverhalten in Zukunft vermehrt auftritt. Demzufolge beschäftigen sich die operanten Methoden in der Schmerzbehandlung mit der Veränderung von Schmerzverhalten, indem die verstärkenden Konsequenzen dieses Verhaltens als therapeutische „Ansatzpunkte" gewählt werden. Da sich Schmerzverhalten bei chronischen Schmerzpatienten in der Regel in einer reduzierten Aktivität, in Medikamentenabhängigkeit und einer Tendenz, bei Bezugspersonen Zuwendung zu erzielen, äußert, wurde von Fordyce und Mitarbeitern (1976; vgl. auch Fordyce und Steger, 1982) ein operantes Therapieprogramm entwickelt, das dieses Verhalten zu verändern sucht. Es zielt im wesentlichen ab auf:

- eine Erhöhung des Aktivitätsniveaus, sowohl allgemein als auch bezogen auf die beeinträchtigten Verhaltensbereiche
- eine schrittweise Reduktion der Einnahme von schmerzreduzierender Medikation
- eine Beseitigung der Verstärkung für Schmerzverhalten in der unmittelbaren sozialen Umgebung des Patienten
- den Aufbau von „gesundem Verhalten", einschließlich einer Verbesserung der sozialen Fertigkeiten und der interpersonellen Kommunikation.

Die konkrete Behandlung besteht im wesentlichen aus einer Medikation nach Zeitplan und nicht nach Schmerzintensität, um die Beziehung zwischen Schmerzzunahme und Medikamenteneinnahme, die meist mit sozialer Zuwendung verbunden ist, zu entkoppeln, aus Bewegungstherapie und Buchführung über „up-time", d. h. die Zeit, die der Patient außerhalb des Bettes verbringt, und aus sozialer Verstärkung für nicht schmerzbezogenes Verhalten.

Voraussetzung für eine individuell angepaßte Therapie ist allerdings eine umfassende Verhaltensanalyse des Patienten. Besonders zu berücksichtigen ist auch hier die weiterreichende Bedeutung, die den Beschwerden im Leben des Patienten zukommt. So liegen häufig Verhaltensdefizite (z. B. mangelnde soziale Kompetenz) vor, die dazu beitragen, daß der Patient die Störung aufrechterhält. Da bei chronischen Schmerzpatienten in der Regel auch die unmittelbaren Bezugspersonen (Partner, Eltern) an der Aufrechterhaltung des Schmerzverhaltens beteiligt sind, sind diese sowohl im Rahmen der Verhaltensanalyse als auch in die Therapie einzubeziehen.

Beeinflussung physiologischer Funktionen durch operantes Lernen (Biofeedback)

Die Unterscheidung in klassisches und operantes Konditionieren betraf in der Lerntheorie für längere Zeit nicht nur das Paradigma des Lernens, sondern auch die dadurch beeinflußbare Ebene des Nervensystems. Man glaubte, operantes Lernen könne lediglich willkürliches Verhalten beeinflussen, klassisches Konditionieren dagegen autonome oder viszerale Funktionen. Diese Unterscheidung wurde in Zweifel gezogen (Dollard und Miller, 1950; Kimmel, 1974), d. h. es stellte sich die Frage, ob nicht auch autonome Funktionen direkt durch operantes Lernen beeinflußbar seien. Um diesen Nachweis zu erbringen, war es wesentlich, einen vermittelnden Einfluß anderer konditionierbarer Reaktionen zu kontrollieren und auszuschließen. Untersucht man beispielsweise die Modifizierbarkeit der Herzfrequenz durch operantes Lernen, so ist es notwendig, bekannte willkürliche Beeinflussungsmöglichkeiten wie die Atmung zu

kontrollieren. Aus diesem Grunde wurden die Versuchstiere in den anfänglichen Untersuchungen kurarisiert (z. B. DiCara und Miller, 1968).

Seit diesen Untersuchungen hat eine ganze Anzahl von weiteren Untersuchungen die direkte Beeinflussung autonomer Funktionen durch operantes Lernen für den tierexperimentellen Bereich belegt, auch wenn die experimentellen Effekte unterschiedlich ausgeprägt waren. Die Übertragung dieser Befunde auf den Menschen ist in dieser Stringenz ausgeschlossen. Experimentell läßt sich das tierexperimentelle Vorgehen aus ethischen Gründen nicht übertragen. Hinzu kommt die nicht kontrollierbare Möglichkeit der Beeinflussung autonomer Funktionen durch vermittelnde kognitive Prozesse, die sich in vielen Entspannungstechniken als wirksam erwiesen hat. Katkin und Murray (1968) schlagen deshalb wie viele andere sinnvollerweise vor, beim Menschen nicht vom operanten Lernen autonomer Funktionen zu sprechen, sondern vom Kontrollernen oder willkürlicher Kontrolle (vgl. auch Kap. 8).

Neben der Bedeutung dieser Befunde für pathogenetische Erklärungskonzepte, beispielsweise bei funktionellen Störungen, liegt die Relevanz für die Behandlung von Störungen nahe, bei denen physiologische Funktionsänderungen eine Rolle spielen. Leidet also ein Patient unter Tachykardien oder sonstigen Arrhythmien, so ist es hilfreich, wenn er eine Technik erlernt, mit der er deren Auftreten kontrollieren kann (Engel, 1973; Bleeker und Engel, 1973).

Es hat sich für das Erlernen einer willkürlichen Kontrolle über autonome Funktionen beim Menschen als sehr hilfreich erwiesen, wenn man der betreffenden Person eine möglichst kontinuierliche Information über die zu kontrollierende Funktion als Konsequenz ihres Verhaltens darbietet. Organismen orientieren sich an den Konsequenzen ihres Verhaltens, die sie über ihre Beziehung zu ihrer Umgebung informieren. Im Falle der Beeinflussung eigener Körperfunktionen besteht die „Umgebung" beim Menschen unter einem bestimmten Aspekt in einem Teilsystem des eigenen Organismus, d. h., im Hinblick auf die die Veränderung vermittelnden willkürlichen, meist kognitiven Prozesse ist die Umwelt der eigene Körper. Denkt eine Person an etwas Bestimmtes und erreicht damit eine Verringerung der Herzfrequenz als Konsequenz des kognitiven Verhaltens, so liegt die Konsequenz dieses willkürlichen kognitiven Verhaltens im eigenen Organismus. Im Tierexperiment ist dies anders, das Tier erlernt eine Reaktion, die Senkung der Herzfrequenz, und kann damit z. B. Schocks vermeiden, d. h. das Verhalten ist im Hinblick auf seine (experimentelle) Umgebung instrumentell. Die Rückmeldung der Körperfunktion beim Menschen ersetzt die Information, die im Tierexperiment durch das Auftreten oder Ausbleiben des Schocks gegeben ist. Insofern hat sie eine operante Bedeutung, d. h. sie hat einen verstärkenden Aspekt und wohl auch Effekt. Es ist jedoch fraglich, ob diese Information, wie im Tierexperiment durch den motivational wirksamen Reiz der Schocks, in sich motivationale Relevanz besitzt, oder ob diese nicht zusätzlich durch das Ziel, die Beeinflussung der Herzfrequenz, zustande kommt; das Tier „will" Schocks vermeiden, der Mensch seine Herzfrequenz senken. Man kann nun annehmen, daß die durch die Rückmeldung gegebene Information dadurch verstärkende, d. h. operante Wirkung bekommt.

In den 70er Jahren wurde die Wirkung von Biofeedback auf verschiedene Störungen eher isoliert untersucht. Budzynski et al. (1970) behandelten Patienten mit Spannungskopfschmerz mit Rückmeldung des Elektromyogramms und der Instruktion, dieses zu senken. Bei einer Kontrollgruppe wurden ebenfalls dieselben Signale wie bei der Rückmeldung verwendet, jedoch ohne daß diese eine Beziehung zur Muskelaktivität gehabt hätten. Die Ergebnisse zeigten nicht nur eine Verringerung der muskulären Aktivität, sondern auch eine Verringerung der Kopfschmerzen bei den Patienten, die Biofeedback erhalten hatten. Elder et al. (1973) untersuchten den Einfluß von Rückmeldung des Blutdrucks auf den diastolischen Blutdruck bei Patienten mit grenzwertiger Hypertonie. Dabei erhielt eine der Gruppen (III) zusätzlich zur Rückmeldung des Blutdrucks eine Verstärkung durch Lob, das je nach Erfolg abgestuft war. Die Behandlung dauerte vier Tage mit zwei Sitzungen pro Tag. Dabei wurde das Kriterium für die Rückmeldung von Erfolg vom Ausgangswert von 105 mmHg in Abhängigkeit vom Erfolg schrittweise gesenkt. Abbildung 19–4 zeigt den Verlauf des diastolischen Blutdrucks für die drei Gruppen. Der Blutdruck änderte sich am stärksten in derjenigen Gruppe, die zusätzlich zur Rückmeldung verstärkt worden war, bei der Kontrollgruppe veränderte er sich nicht.

Patel und North (1975) behandelten ambulante Patienten mit Hypertonie, die einer Behandlungs- oder Kontrollgruppe ohne Behandlung zugeteilt wurden. Die Behandlung bestand in Informationen über die Erkrankung, zwölf Sitzungen von 30 Minuten Dauer mit Relaxation, Meditation und Biofeedback des Hautwiderstands und der Muskelspannung. Nach

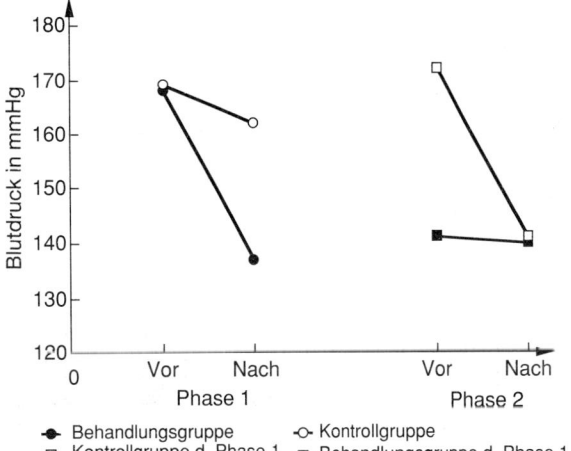

Abb. 19–5. Durchschnittliche Veränderung des Blutdrucks in zwei Phasen der Behandlung. In der zweiten Phase wird die Kontrollgruppe der ersten Behandlungsphase behandelt, die Behandlungsgruppe nicht (aus Patel und North, 1975).

zwei Monaten wurde die Behandlung beendet, nun erhielt die Kontrollgruppe die Behandlung. Die Ergebnisse belegen einen deutlichen Einfluß der Behandlung.

Abbildung 19–5 zeigt, daß der Effekt der Behandlung in der Behandlungsgruppe der Phase 1 auch während der 2. Phase, in der die Patienten nicht weiterbehandelt wurden, erhalten bleibt. In dieser Untersuchung wurde das Biofeedback als eine die Relaxation unterstützende Methode angewandt, es wurde ja nicht der Blutdruck, sondern Parameter der Entspannung rückgemeldet. Tatsächlich ist es sinnvoll, einen allgemeinen, entspannenden Einfluß des Biofeedback anzunehmen, der zum spezifischen, die rückgemeldete Funktion betreffenden, hinzukommt. Wie stark die unspezifisch entspannende Wirkung des Biofeedback ist, zeigt eine Studie von Steptoe und Ross (1982). Sie bildeten drei Gruppen, von denen in einer die Pulswellenlaufzeit rückgemeldet wurde, in der zweiten Relaxation geübt und in der dritten Gruppe keine Intervention durchgeführt wurde. Die Probanden wurden dann im Verlauf des Trainings mit Aufgabensituationen konfrontiert und das Ausmaß der Änderungen der Pulswellenlaufzeit durch die Aufgaben wurde gemessen.

Die Abbildung 19–6 zeigt deutlich, daß die aufgabeninduzierten Änderungen in den beiden Trainingsgruppen schnell kleiner wurden, in der Kontrollgruppe durch den Wiederholungseffekt zwar ebenfalls, aber viel langsamer und geringer. Es zeigt sich aber auch, daß in diesem Parameter kaum ein Unterschied zwischen der Gruppe, die eine Entspannungsmethode erlernt hatte, und der mit Biofeedback bestand. Anders ist dies jedoch bei der Wirkung beider Verfahren auf die Abstände der Herzschläge, also die Herzfrequenz, wie Abbildung 19–7 zeigt.

Im Hinblick auf die unterschiedliche Wirkung von Biofeedback und Relaxation auf das „interbeat interval" zeigt sich, daß Biofeedback diesen Parameter

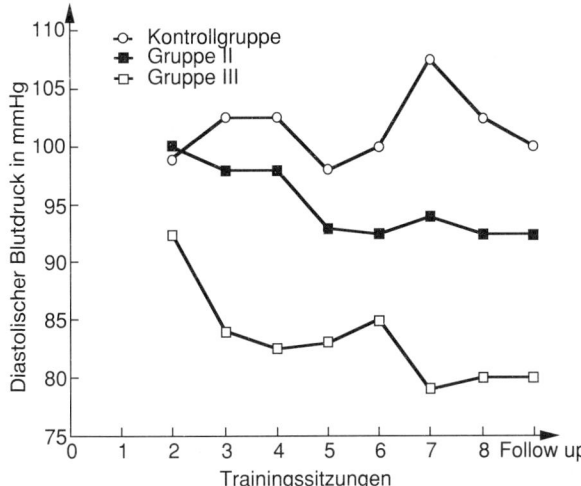

Abb. 19–4. Beeinflussung des Blutdrucks durch operantes Lernen (aus Elder et al., 1973).

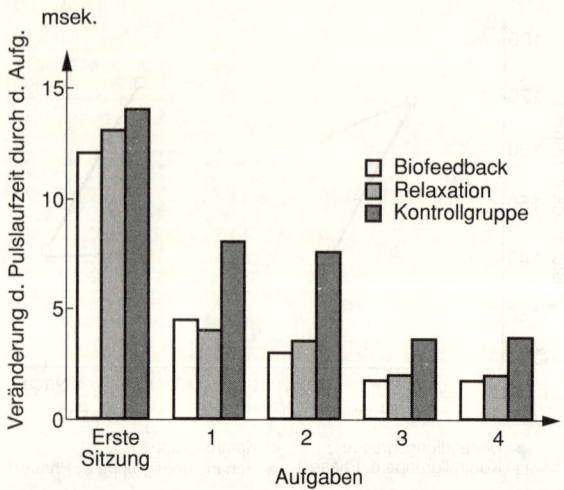

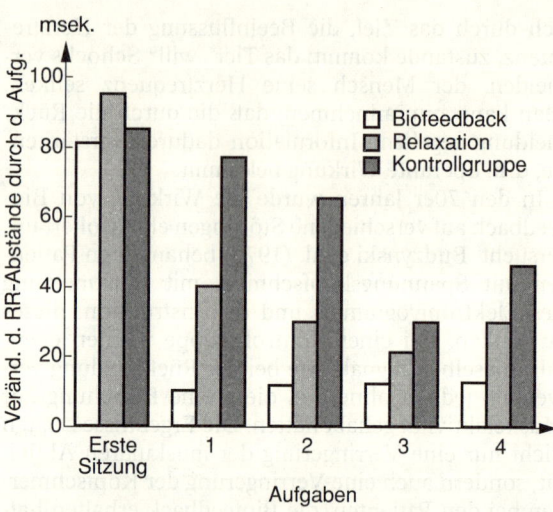

Abb. 19–6. Veränderungen der Pulslaufzeit durch Belastung. Beeinflussung dieser Veränderungen durch Biofeedback, Relaxation (aus Steptoe und Ross, 1982).

Abb. 19–7. Veränderungen des „interbeat interval" durch Belastung. Beeinflussung dieser Veränderungen durch Biofeedback, Relaxation (aus Steptoe und Ross, 1982).

sehr viel schneller und auch stärker beeinflußt als Relaxation. Diese Art der Studien legt den Schwerpunkt auf die unterschiedliche Wirksamkeit einzelner Komponenten, die in breiter angelegten Therapieplänen Anwendung finden. So hat es sich als wesentlich erwiesen, in den therapeutischen Sitzungen Eingeübtes in den Lebensbereich des Patienten zu übertragen, wie die Daten von Patel et al. (1985) im Hinblick auf das Üben von Entspannung eindrucksvoll zeigen (Abb. 19–8). Wichtig dabei ist die Tatsache, daß die unterschiedlichen Ausgangswerte dadurch zustande kommen, daß die Probanden, die die Entspannung nicht übten, lediglich ein erhöhtes Cholesterin aufwiesen und rauchten, während diejenigen Probanden, die über immerhin vier Jahre regelmäßig ihre Entspannungsübungen durchführten, mit der Diagnose einer Hypertonie konfrontiert waren, was in diesem Fall die Compliance gefördert hat. Es wurde keine zusätzliche medikamentöse Behandlung durchgeführt, so daß die Unterschiede auf die unterschiedlichen Übungsgewohnheiten zurückgeführt werden können.

Auch in der Behandlung von Schmerzzuständen wird Biofeedback eingesetzt, vor allem dann, wenn – wie z.B. beim Spannungskopfschmerz – autonome oder zentralnervöse physiologische Variablen in kausalem Zusammenhang mit dem Schmerzerleben stehen. Wie sich gezeigt hat, ist ein dauerhafter Effekt dieser Methode allerdings in hohem Maße davon abhängig, inwieweit es dem Patienten gelingt, die unter Laborbedingungen erlernte Reaktion auch ohne Rückmeldung in der natürlichen Lebensumgebung auszuführen. Die Anwendung von Biofeedback hat sich vor allem bei Spannungskopfschmerz, Migräne und Rückenschmerzen als effektiv erwiesen. Ein Biofeedbacktraining sollte in der Regel jedoch – wie andere Entspannungsverfahren – in ein umfassenderes verhaltenstherapeutisches Training eingebettet sein (vgl. Turk et al., 1982; vgl. Kap. 36).

19.5.3 Komplexe Therapiepläne

Es hat sich gegenüber den Anfängen der Verhaltenstherapie gezeigt, daß die Anwendung isolierter Techniken zwar durchaus einen Effekt hat (vgl. Fallbeispiel spastisches Syndrom), daß dieser Effekt jedoch erhöht werden kann, wenn Verhalten auf mehreren Ebenen betrachtet und beeinflußt wird. Im Beispiel der Patientin mit Morbus Crohn wurde bereits deutlich, daß bestimmte Gedanken ihr Verhalten in einzelnen Situationen beeinflußten. Eine Veränderung der Verhaltensweisen durch operante Kontingenzen von Verstärkung allein hätte sicherlich einen Einfluß gehabt, und der Verlauf der Therapie hat auch gezeigt, wie wichtig dieser Faktor gewesen ist, die langfristige Effektivität der Therapie hing jedoch auch von den Gedanken der Patientin ab, d.h. einer Verhaltensweise, die andere Verhaltensweisen beeinflußte bzw. kontrollierte. So wie Verhaltenstherapie nicht

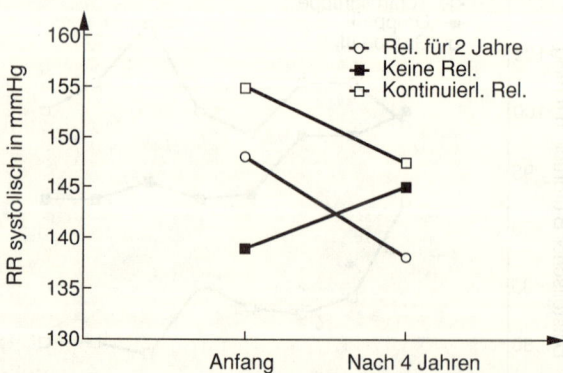

Abb. 19–8. Verlauf des Blutdrucks bei drei Behandlungsgruppen, von denen die Probanden in einer Gruppe über vier Jahre, die in der zweiten für zwei Jahre und die in der dritten keine Entspannungsübungen durchgeführt hatten (aus Patel et al., 1985).

auf die Änderung eines isolierten Symptoms als eine isolierte Verhaltensweise abzielt, sondern auf die Änderung eines Verhaltensgefüges, so werden verschiedene therapeutische Komponenten im Hinblick auf die Änderung verschiedener Komponenten dieses Gefüges angewandt.

Selbstkontrolle

Das Erlernen von eigener Kontrolle über Verhaltensweisen, über die eine Kontrolle zunächst nicht oder nicht mehr besteht, läßt sich als ein grundlegendes Ziel therapeutischen Vorgehens verstehen. Andererseits lassen sich verschiedene Gesichtspunkte von Selbstkontrolle zu einem Vorgehen zusammenfassen, dessen Anwendung sich bewährt hat. Um den Begriff „Selbst" zu erläutern, nennt Skinner (1953) eine funktionale Verhaltenseinheit ein System vereinheitlichter Reaktionen. Menschen würden sich also auch dann „verhalten", wenn sie sich oder ihr Verhalten kontrollieren würden. Sie würden sich ebenso kontrollieren, wie sie das Verhalten einer anderen Person zu kontrollieren versuchen, durch den Einfluß auf diejenigen Variablen oder Faktoren, die das entsprechende Verhalten beeinflussen, deren Funktion es ist.

Grundsätzlich wird eine Verhaltensweise, die dazu dient, eine andere eigene Verhaltensweise zu kontrollieren, als Ausübung von „Selbstkontrolle" angesehen. Steckt sich jemand keine Zigaretten in die Tasche, um unterwegs nicht zu rauchen, so ist dies ein Kontrollverhalten. Wichtig dabei ist, daß die Person diese Verhaltensweise selbst ausführt und damit die Reizbedingungen für das Auftreten der Reaktion, in diesem Falle das Rauchen, ändert. Der Erfolg der so ausgeübten Selbstkontrolle verstärkt das Kontrollverhalten. Der Erfolg ist jedoch von einer Reihe von anderen Faktoren abhängig und zunächst unter Umständen instabil, wenn z.B. aufgrund organismischer Bedingungen eine Bedürfnisspannung kontinuierlich anwächst und durch das zu kontrollierende Verhalten reduziert wird, was dann ebenfalls verstärkenden Einfluß hat. Personen unterscheiden sich jedoch in dem Ausmaß, in dem Bedürfnisspannungen für eine gewisse Zeit ertragen werden können, eine Fähigkeit, die in der Entwicklung (unterschiedlich ausgeprägt) gelernt wird (Mischel und Straub, 1965) und durch geeignete Methoden erweitert werden kann.

Kanfer et al. (1970) haben drei Komponenten des Selbstkontrolltrainings genannt:
- **Selbstbeobachtung (self monitoring):** Es erfolgt eine genaue Analyse des Verhaltens und seiner kontrollierenden Bedingungen sowie eine Aufzeichnung seines Auftretens.
- **Bewertung und Zielanalyse:** Es werden Zielvorstellungen erarbeitet und das tatsächliche Verhalten anhand der Aufzeichnungen des Verhaltens bewertet. Dabei muß darauf geachtet werden, daß beim schrittweisen Verändern des vorhandenen Verhaltens die Bewertung an „Zwischenzielen" erfolgt, die am ändernden Vorgehen orientiert sind und nicht am endgültig anzustrebenden Ziel, um

Mißerfolgserlebnisse zu vermeiden. Dieser Vorgang begleitet die gesamte Therapie und ist auch ein wesentlicher Bestandteil, um den Patienten an adäquate Bewertungen zu gewöhnen.
- **Selbstverstärkung:** Der Patient wird angehalten, sich nach einer erbrachten Leistung selbst zu loben, oder es werden äußere verstärkende Ereignisse eingeführt, beispielsweise Geld, wie im Beispiel der Patientin mit Morbus Crohn. Dabei ist die Selbstverstärkung ein Verhalten, das die Auftretenswahrscheinlichkeit des vorangehenden Verhaltens ändert und damit nach der oben angeführten Definition von Skinner Ausüben von Selbstkontrolle.

Im Kapitel 38.1 ist die Anwendung dieser Methode bei der Verhaltenstherapie der Adipositas sehr ausführlich dargestellt worden, so daß hier nicht weiter darauf eingegangen wird.

Streßmanagement

Im Kapitel 8 wurde der Streßbegriff als zu weitgefaßt kritisiert, jedoch auch angemerkt, daß eine, wenn überhaupt möglich, sinnvolle Definition Streß als eine Organismus-Umwelt-Interaktion betrachten müßte, die eine Auseinandersetzung des Organismus mit der Umgebung beinhaltet, die ihn aus unterschiedlichen Gründen überfordert, womit die Auseinandersetzung suboptimal wird, was negative Folgen für den Organismus hat. In der Verhaltenstherapie und dem Streßmanagement wird davon ausgegangen, daß es Bedingungen einer suboptimalen Auseinandersetzung gibt, die auf seiten des Organismus liegen und somit geändert werden können. Hat jemand vor einem Examen „zuviel" Angst, so kann dies in der Auseinandersetzung mit dieser Situation eine Reihe von Konsequenzen haben, die ein suboptimales Ergebnis zur Folge haben. Neigt er dazu, Angstreize zu vermeiden, so wird er sich unter Umständen zu selten als Vorbereitung mit dem Stoff beschäftigen, da ihn dies an das angstbesetzte Examen erinnert. Infolgedessen wird er in der eigentlichen Examenssituation intensiv Angst erleben, was dazu führt, daß er angstbezogene Information verarbeitet und nicht auf den Inhalt des Examens bezogene. Daraus wird sich dann ergeben, daß er Mißerfolg erlebt, was seine Angst weiter steigert usw. Fragt man sich, warum er zuviel Angst hat, so kann dies daran liegen, daß er seine eigene Leistungsfähigkeit unterschätzt oder die Wichtigkeit des Examens überschätzt.

Meichenbaum und Cameron (1983) nennen die folgenden Komponenten oder Phasen des Trainings:

Phase 1: Konzeptualisierung
- Datenerhebung und Integration
  Identifizierung der Problemdeterminanten durch Interview, gezielte Imagination und Verhaltensbeobachtung; Unterscheidung zwischen Leistungsversagen und Defizit an Fähigkeiten; Behandlungsplan und Aufgabenanalyse.
- Training der Fähigkeiten der Selbstbeobachtung
  Unabhängig vom Therapeuten soll der Patient

seine im Alltag auftretenden Probleme analysieren lernen.

Phase 2: Aneignen und Üben von Bewältigungsfähigkeiten
- Training zur Aneignung von Fähigkeiten
  beispielsweise der Kommunikation, Selbstsicherheit, Problemlösen, Arbeitsverhalten oder „palliativer" Fähigkeiten (z.B. bei Schmerz) wie Ablenkung, Suche nach sozialer Unterstützung, Relaxation, angemessener Gefühlsausdruck; Entwicklung eines flexibel einsetzbaren Repertoires dieser Fähigkeiten.
- Üben dieser Fähigkeiten
  im Rollenspiel, in der Imagination; Training von Selbstinstruktionen, um Bewältigungsreaktion kognitiv zu vermitteln.

Phase 3: Anwendung und „Durcharbeiten"
- Induzierung der Anwendung
  Bewältigung in der Vorstellung, wobei frühe Anzeichen von Streß als Hinweisreize zur Anwendung der Bewältigung dienen; Anwendung im Rollenspiel antizipierter Streßsituationen; abgestufte Anwendung in realen Situationen, denen sich der Patient aussetzt.
- Generalisierung und Pflege der Fähigkeiten
  In bezug auf belastende Situationen soll das Gefühl der erfolgreichen Bewältigung konsolidiert werden; Entwicklung von Strategien, um mit Mißerfolg umzugehen; Follow up.

Bei diesem Vorgehen ist es wichtig, den Patienten zu eigenem aktivem Teilnehmen zu motivieren. Die Abfolge muß daher so strukturiert werden, daß Selbstvertrauen durch Erfolgserlebnisse erhöht wird, was voraussetzt, daß Ziele realistisch aufgestellt werden. Wenn nötig sollten Angehörige einbezogen werden.

In der Medizin spielt eine suboptimale Auseinandersetzung mit der Umgebung (Streß) eine Rolle, da die negativen Folgen körperlicher Natur sein können, was bei einer Reihe von Erkrankungen vermutlich eine Rolle spielt (vgl. auch Kap. 41.1, 41.3 und 37). Vor allem für das sog. Typ-A-Verhalten als Risikofaktor (Booth-Kewley und Friedman, 1987; Friedman und Booth-Kewley, 1988; Matthews, 1988) für die koronare Herzkrankheit und Herzinfarkt konnten Verhaltensstile als wesentlich ermittelt werden, die beinhalten, daß Personen mit diesem Verhalten sich durch Aufgaben oder Beruf mehr belastet fühlen (Sims, 1988), mehr und intensiver arbeiten, kompetitiver und feindseliger sind, ein höheres Kontrollbedürfnis haben und sich häufiger Belastungen aussetzen. Es scheint eine Interaktion im Hinblick auf kardiovaskuläre Reaktivität zwischen genetischer Belastung durch hypertone Eltern und Typ-A-Verhalten zu bestehen (McCann und Matthews, 1988), die den Effekt des Typ-A-Verhaltens auf das Kreislaufsystem vermittelt.

Man hat nun versucht, mit der Methode des „Streßmanagement" das Typ-A-Verhalten zu beeinflussen, um dadurch eine Verringerung des damit verbundenen erhöhten Morbiditäts- und Mortalitätsrisikos durch Herzinfarkt zu erreichen. Dabei gibt es mehre-

re Arten von Studien. Zum einen wird versucht, durch eine derartige therapeutische Intervention die physiologische Streßreaktivität zu verringern. In einer Reihe von Studien war gezeigt worden, daß eine zwischen Typ-A-Verhalten und Krankheit vermittelnde Bedingung mit einiger Wahrscheinlichkeit in einer erhöhten kardiovaskulären Reaktivität besteht. So konnten beispielsweise Zurawski et al. (1987) durch ein Streßmanagement eine verringerte Reaktivität vor allem des diastolischen Blutdrucks nachweisen. Der systolische Blutdruck war ebenfalls in Ruhesituationen erniedrigt. Die Autoren verglichen diese Veränderungen mit den Effekten von Biofeedback der elektrodermalen Aktivität.

Wie aus Abbildung 19–9 ersichtlich wird, ist das Streßmanagement dem einfachen Biofeedback in der Veränderung der Blutdruckwerte überlegen. Andere Ergebnisse erzielte allerdings Roskies (1988). Sie behandelte 118 noch gesunde Personen mit ausgeprägtem Typ-A-Verhalten mit drei verschiedenen Behandlungstypen: körperliches Training, Streßmanagement und Gewichtskontrolle; innerhalb jeder Behandlungsart wurden drei Untergruppen gebildet. In der Fitnessgruppe wurden drei wöchentliche, in den beiden anderen Gruppen zwei wöchentliche Sitzungen über 10 Wochen durchgeführt. Die Ergebnisse der Reaktivität physiologischer Parameter zeigten keine Unterschiede zwischen den drei Behandlungsgruppen, das Streßmanagement war jedoch in der Reduktion des Typ-A-Verhaltens den anderen Verfahren deutlich überlegen. Friedman et al. (1984) berichten, daß vergleichbare Veränderungen des Typ-A-Verhaltens bei Postinfarktpatienten die Anzahl von Reinfarkten und die Mortalität senkten.

Walter et al. (1988) behandelten Patienten mit einer milden Hypertonie nach Absetzen der Medikation entweder durch Streßmanagement oder durch Fortsetzen der bis dahin erfolgreichen Medikation. Das

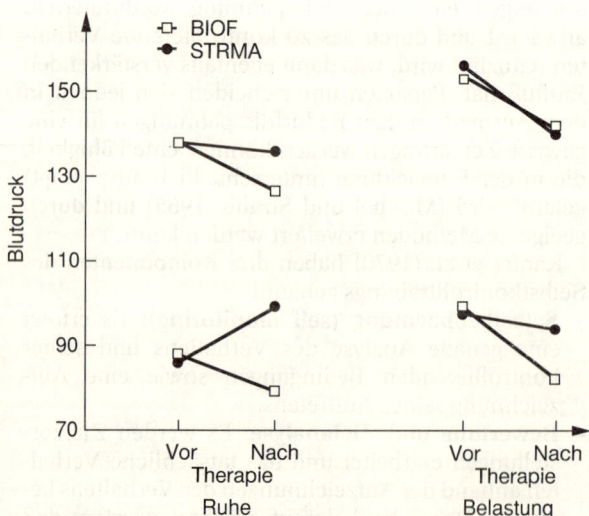

Abb. 19–9. Beeinflussung des Blutdrucks durch Streßmanagement (STRMA) und Biofeedback der elektrodermalen Aktivität (BIOF) unter Ruhebedingung und Belastung (nach Daten aus Zurawski et al., 1987).

Streßmanagement wurde in einer der beiden Gruppen durchgeführt. Die Ehepartner der Patienten nahmen an den Sitzungen teil, die insgesamt 2 Stunden dauerten und sechsmal in wöchentlichem Abstand, dann zweimal im Abstand von 14 Tagen und dann noch einmal nach 6 Wochen stattfanden. Das therapeutische Vorgehen beinhaltete das Üben des autogenen Trainings als Entspannungsmethode, kognitive Verhaltensmodifikation, Information über Blutdruck, Ernährung und körperliches Training. Die Ergebnisse zeigen eine deutliche Überlegenheit des Streßmanagements ohne zusätzliche medikamentöse Behandlung gegenüber der üblichen antihypertensiven Medikation. Dies galt vor allem für die mit einem tragbaren Blutdruckmeßgerät gemessenen systolischen und diastolischen Blutdruckwerte während eines normalen Arbeitstages, aber auch für übliche „klinische" Messungen. Dabei gab es eine deutliche Beziehung zwischen dem Ausmaß der Blutdrucksenkung und einem in der Therapie angestrebten Umgang mit Streß, der zusammengefaßt als „geringes Typ-A-Verhalten" bezeichnet werden kann. An dieser Studie ist besonders bemerkenswert, daß das verhaltenstherapeutische Vorgehen die Medikation ersetzt hat und dabei dieser überlegen war, bei sehr relevanten und genauen Messungen des Blutdrucks (im Alltag). Wichtig ist ebenfalls die Beziehung zwischen reduziertem Typ-A-Verhalten und anderen vergleichbaren, psychometrisch erfaßten Merkmalen und der erzielten Reduktion des Blutdrucks.

Nunes et al. (1987) führten eine Metaanalyse von 18 kontrollierten Studien zur therapeutischen Beeinflussung des Typ-A-Verhaltens durch. Die kombinierten Ergebnisse zeigen, daß sowohl das Typ-A-Verhalten reduziert werden konnte, als auch die Infarktmorbidität und Mortalität nach drei Jahren. Auch hier fand sich ein Zusammenhang zwischen der Reduktion des kritischen Verhaltens und der Reduktion der Morbidität und Mortalität. Weiterhin zeigte sich, daß der Erfolg einer Behandlung um so größer war, je komplexer sie angelegt war, d.h. je mehr Komponenten sie enthielt. Allerdings waren von den 18 Studien nur 5 methodisch soweit akzeptabel, daß sie im Hinblick auf die Mortalität über ein Jahr aussagekräftig waren und nur 2 für die von drei Jahren nach der Therapie. Zudem bestanden die Stichproben meist aus Männern im mittleren Alter.

Die therapeutischen Interventionen zielen also darauf ab, die Ursachen und die Art der suboptimalen Auseinandersetzung der Personen (hier mit Typ-A-Verhalten) mit ihrer Umgebung zu verändern.

Ähnliches gilt beispielsweise auch für Patienten mit Kopfschmerz (Spannungskopfschmerz und/oder migränoider Kopfschmerz). Die Patienten fühlen sich ebenfalls belasteter, sind leistungsorientiert, ängstlicher und feindseliger (Andrasik et al., 1982). Martin et al. (1988) konnten einen Zusammenhang zwischen Stimmung und Kopfschmerzaktivität feststellen, das Verhältnis von Schmerz und erhöhter Muskelspannung im Kopfbereich scheint auch für Patienten mit Migräne zu gelten, ebenso das Vorliegen eines erhöhten Ruhetonus für diese Muskelgruppen.

Da es im Falle des Stressors Schmerz nicht möglich ist, diesen direkt zu beeinflussen, kommt auch hier der kognitiven Bewältigung – als intrapsychischem Coping – eine besondere therapeutische Bedeutung zu. Von Selbstkontrolle spricht man, wenn Verhalten oder physiologische Reaktionen durch interne, informationsverarbeitende Prozesse modifiziert werden, und Individuen so in die Lage versetzt werden, ihr Verhalten unabhängig von äußeren Kontingenzen zu regulieren. Als wesentlichen Wirkfaktor nimmt man an, daß Methoden der Selbstkontrolle die wahrgenommene „Selbsteffizienz" (Bandura, 1977), d.h. Einschätzung der eigenen Bewältigungsmöglichkeiten, erhöhen. Verfügt eine Person über eine hohe Selbsteffizienzerwartung, ist es unwahrscheinlich, daß sie Angst und Hilflosigkeit, als schmerzverstärkende Bedingungen, erlebt.

Zu den wichtigsten kognitiven Schmerzbewältigungsstrategien zählen die Techniken der **Aufmerksamkeitsfokussierung, Ablenkung** und **Imagination.** Im einzelnen unterscheidet man die folgenden Strategien (zit. nach Birbaumer, 1984):

– Externale Aufmerksamkeitslenkung (Fokussierung auf Umgebungsreize)
– Internale Aufmerksamkeitslenkung (Konzentration auf Gedanken, Kopfrechnen etc.)
– Somatisierung (Lenkung der Aufmerksamkeit auf schmerzhafte Körperzonen bei gleichzeitiger Distanzierung durch Vorstellung, dieser Körperteil wäre unempfindlich; genaues Beschreiben der Körperreaktionen)
– Imaginative Unaufmerksamkeit (angenehme, schmerzinkompatible Phantasien)
– Imaginative Transformation des Schmerzes (Neuinterpretation der aversiven Reizung als willkommene Erfahrung, Autosuggestion schwacher Reizung)
– Imaginative Transformation des Kontextes der Schmerzerfahrung (Schmerzerfahrung in anderen Kontext einbauen, in dem der Schmerz eine andere Bedeutung erhält).

Obwohl bislang nicht geklärt ist, wie die sensorische und affektive Komponente der Schmerzerfahrung verarbeitet werden (parallel oder additiv), besteht doch Übereinstimmung darüber, daß Schmerz nicht nur Ausdruck des sensorischen Inputs ist, sondern durch Prozesse der Informationsverarbeitung kontrolliert wird. Die Schmerzerfahrung erscheint abhängig von der Aufmerksamkeit, die dem sensorischen Input zugewandt wird, und der Art und Weise, wie diese Information emotional verarbeitet wird. Da die Verarbeitungskapazität des Kurzzeitgedächtnisses begrenzt ist, kann Ablenkung zielgerichtet eingesetzt werden, um die Prozesse der Informationsverarbeitung dahingehend zu verändern, daß der sensorische Input – durch die Einführung zusätzlicher Reize – nur unzureichend verarbeitet und gespeichert wird. Der alleinige Einsatz von Ablenkungsstrategien scheint allerdings nur bei niedrigeren Schmerzintensitäten ausreichend (vgl. McCaul und Malott, 1984).

Im Zusammenhang mit Selbstkontrollmethoden in der Schmerzbehandlung ist auch auf die Modifikation der sog. Selbstverbalisationen hinzuweisen. Wie man weiß, ist Verhalten oft weniger von objektiven Merkmalen einer Situation abhängig, als vielmehr davon, wie die Situation vom Individuum wahrgenommen und interpretiert wird. Danach erscheint „Verhalten als Funktion der kognitiven Bewertung, die sich eine Person von einer streßhaften Situation macht" (McGrath, 1970). Wie die kognitive Verhaltenstherapie annimmt, ist Verhalten damit auch durch eine Veränderung der Kognitionen (Erwartungen, Handlungsziele, Motive, Vorstellungen, frühere Erfahrungen etc.), die in Form von internen Dialogen oder Selbstverbalisationen die Auseinandersetzung mit einer Situation begleiten, zu kontrollieren. Das Ziel einer kognitiv orientierten Schmerztherapie besteht somit darin, negative Selbstverbalisationen durch positive, d.h. solche, die die Schmerzbewältigung fördern, zu ersetzen. Dabei werden Selbstverbalisationen als positiv erachtet, wenn sie die Ernsthaftigkeit der Situation anerkennen, die physiologische Aktivierung als Schlüsselreiz für Bewältigungsverhalten verstehen, negative Gedanken kontrollieren, Mut machen in dem Sinne, daß die Situation bewältigbar ist, und den Erfolg bei gelungener Schmerzbewältigung betonen.

So behandelten Holroyd et al. (1977) Patienten mit chronischem Spannungskopfschmerz mit Streßmanagement, allerdings mit einem Schwerpunkt auf den kognitiven Komponenten der Streßreaktionen. Sie verglichen dieses Vorgehen mit Biofeedback der muskulären Aktivität (M. frontalis) und einer Wartegruppe. Abbildung 19–10 zeigt auch hier die Überlegenheit des Streßmanagements gegenüber dem Biofeedback und der Kontrollgruppe.

Das Schmerzbewältigungstraining von Bullinger und Turk (1982), das einer Modifikation des Streßbewältigungstrainings von Meichenbaum (1977) entspricht, integriert im Sinne eines komplexen Therapieplans verschiedene bereits angesprochene therapeutische Elemente. Das Training, das auch als Gruppenbehandlung durchgeführt werden kann, besteht aus drei Phasen, wobei dem Patienten sowohl antizipatorische – auf den Schmerz vorbereitende – als auch reaktive Schmerzbewältigungsstrategien vermittelt werden.

In der ersten „edukativen" Phase wird der Patient zunächst über die sensorische, affektive und evaluative Dimension der Schmerzerfahrung (vgl. Kap. 36), z.B. anhand des Modells von Melzack und Casey (1968), unterrichtet. Anhand dieses Modells wird diskutiert, was Schmerz auslösen kann, und wie durch

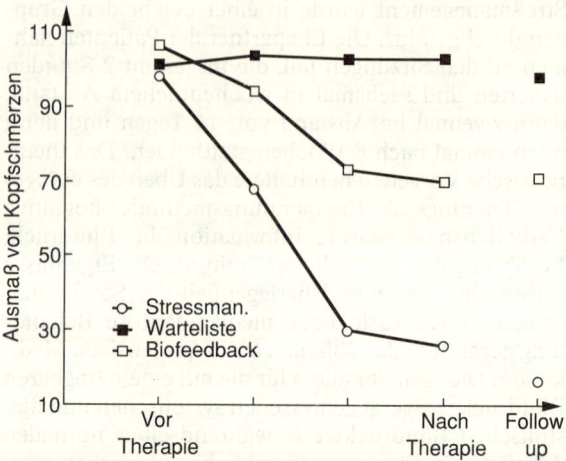

Abb. 19–10. Durchschnittliche Kopfschmerzaktivität in Blöcken von zwei Wochen und nach zwei Jahren Follow-up (aus Holroyd et al., 1977).

eine selektive Beeinflussung der Dimensionen Schmerz kontrolliert werden kann (Entspannungstechniken – Reduktion des sensorischen Inputs; Strategien der Aufmerksamkeitslenkung – motivational-affektive Komponente; Selbstverbalisationen – kognitiv-evaluative Komponente).

In der sich anschließenden Übungsphase wird durch eine Verhaltensanalyse geprüft, wie der Patient üblicherweise mit Schmerz umgeht, und welche Bewältigungsstrategien er dabei anwendet. Daraufhin werden ihm verschiedene Verhaltensstrategien (z.B. Entspannungs- oder Bewegungsübungen) und auch kognitive Bewältigungsstrategien (Techniken der Aufmerksamkeitslenkung oder imaginative Verfahren) vorgestellt, unter denen er sich die Methoden auswählen kann, die ihm zur Selbstregulation am geeignetsten erscheinen. In dieser Phase ist ein wesentlicher Aspekt des Trainings die vorbereitende Planung der Reaktion auf den Schmerzreiz und die Veränderung der Selbstverbalisierungen, die die Auseinandersetzung mit dem Schmerz begleiten. Es werden Selbstäußerungen für jede Phase der Auseinandersetzung mit dem Schmerz (Vorbereitung auf den schmerzhaften Reiz, Konfrontation mit dem Schmerz, Verhalten in kritischen Situationen, Verstärkung für erfolgreiche Bewältigung) erarbeitet.

Die abschließende Praxisphase dient einer Generalisierung, d.h. der Übertragung des am künstlichen Schmerzreiz gelernten Bewältigungsverhaltens auf die natürliche Schmerzsituation.

20 Gruppentherapiemethoden

Wolfgang Wesiack und *Wilfried Biebl*

20.1 Vorbemerkungen

In den letzten Jahren gewinnen gruppentherapeutische Methoden in der Psychotherapie und in der psychosomatischen Medizin zunehmend an Bedeutung. Es ist deshalb zweckmäßig, auch im Rahmen dieses Lehrbuches einen Überblick über die am häufigsten geübten Verfahren zu geben. Dies ist nicht einfach, da hier alles noch im Fluß und die Fülle der verschiedenen Ansätze und Publikationen kaum noch überschaubar ist. Die Autoren sind sich daher der Unvollständigkeit dieses Artikels voll bewußt. Wer eine eingehendere Information über dieses weite Feld sucht, sei deshalb auf die Spezialliteratur verwiesen und daran erinnert, daß man gruppentherapeutische, wie auch andere psychotherapeutische Methoden nur durch eigene langfristige Teilnahme an Gruppenprozessen erlernen kann.

Wenn man bedenkt, daß sich der Mensch in dem von Paläoanthropologen geschätzten Zeitraum von etwa 50 Jahrmillionen ohne Gruppenbildung und den Schutz der Kleingruppe gar nicht von seinen tierischen Vorfahren hätte lösen und zum Menschen entwickeln können, worauf z.B. Claessens (1980) überzeugend hinweist, dann kann man mit Recht feststellen, daß der Mensch ein soziales, also ein Gruppenwesen ist. Aristoteles bezeichnete deshalb den Menschen bereits als „Zoon politikon". Um so erstaunlicher ist es, daß die therapeutische Relevanz der Gruppe erst relativ spät entdeckt wurde.

Zunächst waren es wohl in erster Linie ökonomische Überlegungen, die zur Bildung von Therapiegruppen führten, weil insbesondere bei sehr lange dauernden und sehr zeitaufwendigen Therapien, wie z.B. einer psychoanalytischen Behandlung, die 1:1-Relation von Arzt und Patient ausgesprochen unökonomisch ist. Bei diesen Versuchen wurde dann bald der Eigenwert und die Reichweite der therapeutischen Gruppenprozesse entdeckt.

1906 führte J. Pratt bei Patienten einer Tuberkulosestation erstmals Gruppentherapien in Form von Schulklassen mit 80 bis 100 Personen ein. Es handelte sich um eine leiterzentrierte, ermutigende, stützende und pädagogische Gruppentherapie.

1910–1914 spielte J. Moreno mit Gruppen von Kindern in Wien Stegreiftheater und legte den Grundstein zur Rollenspielmethode und dem späteren Psychodrama. Von ihm stammt auch der Terminus „Gruppenpsychotherapie".

Nach dem 1. Weltkrieg arbeitete S. Bernfeld erstmals in psychoanalytisch orientierten Gruppen mit Jugendlichen, und in Wien begann A. Aichhorn seine Resozialisierungsarbeit mit jugendlichen Verwahrlosten in pädagogisch-psychotherapeutischen Gruppen.

1926 versuchte erstmals T. Burrow psychoanalytische Konzepte und Deutungsmöglichkeiten in Patientengruppen anzuwenden. Seither arbeiten viele Psychoanalytiker mit Gruppen und entwickelten verschiedene psychoanalytische Gruppenkonzepte (Slaoson, 1934; Bion, 1971; Foulkes, 1940; u.v.a.).

1931 behandelte L.C. Marsh, im Gegensatz zu Pratts leiterzentrierten Gruppen, Psychosekranke in Gruppen mit Peer-Orientierung (jedes Gruppenmitglied ist gleich verantwortlich und gleichberechtigt, aber auch gleich betroffen). Dabei tritt die Bedeutung des Gruppenleiters ganz in den Hintergrund. Dies kann als der Beginn der Selbsthilfegruppe gesehen werden, wie sie etwa bei den „Anonymen Alkoholikern" anzutreffen ist.

1946 prägte Main den Begriff „therapeutische Gemeinschaft", der dann von Maxwell Jones (1952) aufgenommen wurde und die gruppendynamische Interaktion im sozialpsychiatrischen Rahmen und in Krankenhäusern beschreibt (Kreeger, 1977).

1963 konzeptualisiert H. Argelander die Gruppe unter Zuhilfenahme des psychoanalytischen Strukturmodells. Er behandelt die Gruppen als Entität und verwandelt die multipersonalen Beziehungen in eine bipersonale zwischen dem Therapeuten und dem „Gruppen-Ich".

Einen guten Überblick über die psychoanalytisch orientierte Gruppenpsychotherapie findet der Leser bei Kutter (1985).

Neben den psychoanalytischen Theorienbildungen für die Gruppentherapie entwickelten sich unabhängige Theorien der Verhaltensmedizin, der Gruppendynamik und der Rehabilitation für die Gruppenarbeit.

In der psychosomatischen Medizin wird eine Vielfalt von gruppenpsychotherapeutischen Methoden angewandt, die zum Teil von recht unterschiedlichen theoretischen Konzepten ausgehen. Das Therapieziel einer qualifizierten Gruppentherapie in der psychosomatischen Medizin ist vor allem die Arbeit am Selbstwertgefühl und die Verbesserung der Kommunikationsfähigkeit. Dadurch kommt es vor allem zu einer verbesserten Integration in die natürlichen Gruppen der Patienten (Familien, Institutionen usw.).

20.2   Die in Gruppentherapien angewandten therapeutischen Prinzipien

Zunächst wollen wir einen Überblick über die wichtigsten therapeutischen Prinzipien geben, die in Gruppentherapien angewandt werden.

- Die pädagogische Gruppe: Hier wird nur Wissen und Information zu bestimmten Themen wie z.B. Gesundheits- und Krankheitsverhalten, Diät, Diagnose und Prognose bestimmter Krankheitsbilder usw. vermittelt.
- Die direktiv-suggestive Gruppenpsychotherapie: Hier geht es darum, die Patienten suggestiv zu beeinflussen, zu lenken und zu bestimmten Zielen hinzuführen.
- Die Aktivitätspsychotherapiegruppen: In diesen Gruppen steht das Ausagieren der Affekte im Mittelpunkt. Diese Methode wird vor allem in Kinderspielgruppen, aber auch in der Gestalttherapie nach Perls und in den Bioenergetikgruppen nach Lowen angewandt. Zu den Aktivitätsgruppen müssen wir auch jene „Sportgruppen" zählen, die, wie viele Gruppen mit Koronarkranken, ihr Hauptaugenmerk auf die wohldosierte körperliche Belastung legen (Brusis und Weber-Falkensammer, 1986).
- Die themenzentrierte Interaktion nach Ruth Cohn: Die Konzentration auf bestimmte relevante Themen ermöglicht die bevorzugte Bearbeitung besonderer Themenkreise.
- Die sozialkommunikative Gruppenpsychotherapie: Sie strebt eine Verbesserung der manifesten sozialen Wahrnehmung und Interaktion an. Hierzu sind vor allem die Encountergruppen nach Rogers zu zählen.
- Die Rollenspielmethode bzw. das Psychodrama nach Moreno: In diesen Gruppen werden persönlichkeitsspezifische Konflikte durch wechselseitige Übernahme bestimmter Rollen innerhalb der Gruppen dargestellt und bearbeitet.
- Verhaltenstraining in Gruppen: Diese Methode hat sich besonders bei der Raucherentwöhnung und bei der Arbeit mit Adipösen bewährt.
- Analytische Gruppenpsychotherapie: In ihr bemüht sich der Gruppenleiter, die latenten pathogenen Konflikte der Patienten mittels der freien Assoziation zu erfassen und durch deutende Bearbeitung von Übertragung und Widerstand einerseits, aber auch durch Bewußtmachen der symbiotischen Bedürfnisse und Phantasien andererseits einer besseren Lösung zuzuführen.
- Selbsthilfegruppen: Sie sind nach dem Vorbild der „Anonymen Alkoholiker" (AA)-Gruppen gebildet und arbeiten ohne Leiter, benötigen aber meistens sog. Berater. Um diese Gruppen hat sich in Deutschland vor allem Moeller große Verdienste erworben.

20.3   Über Auswahlkriterien von Mitgliedern in einer therapeutisch arbeitenden Gruppe

Die Art der Ausbildung, das Arbeitsfeld und das Forschungsinteresse des Gruppentherapeuten bestimmen im wesentlichen die Auswahlkriterien für die Aufnahme eines Patienten in eine bestimmte therapeutische Gruppe. In den Übersichtsartikeln von Silver und Mitarbeitern (1980) und Lubin und Mitarbeitern (1979) über Publikationen im Bereich der Gruppenpsychotherapie werden für die Jahre 1978 und 1979 jeweils mehr als 500 Arbeiten aufgeführt. Als Zusammenfassung davon kann ausgesagt werden, daß Gruppen in jeglicher Zusammensetzung arbeiten können, in unterschiedlichsten „Settings" mit unterschiedlichsten Zielsetzungen. Gesprächsgruppen, körperbezogene Therapiegruppen, Gruppen mit einem gemeinsamen Medium wie bei Musiktherapie, Maltherapie, Theatergruppen arbeiten effizient und weisen in katamnestischen Untersuchungen Erfolge auf.

Die Bewegung der Selbsthilfegruppen, die seit mehr als 10 Jahren mit zunehmendem Interesse von ärztlicher wie auch von politischer Seite unterstützt werden, weist darauf hin, daß im Gruppenprozeß selbst die Möglichkeiten der Entlastung, der Um- und Neuorientierung, jedoch auch der Symptomlinderung zu suchen sind.

Im Bestreben, dem breiten Spektrum der Literatur über Gruppen gerecht zu werden, ist es auch in einem Übersichtsartikel unvermeidlich, subjektive Wertungen, eigene Erfahrungen und Vorlieben einzubringen. Der Pluralismus der Methoden ist so vielfältig, daß daraus keine Auswahlkriterien verallgemeinernd dargestellt werden können. Zum Wesen eines jeden Gruppenprozesses gehört die Identifikation mit den Gruppeninteressen und die Projektion aggressiver Strebungen nach außen, also die Grenzziehung gegenüber Personen und Interessen, die als gruppenfremd erkannt werden. So darf es nicht verwundern, daß trotz der Vielfalt der möglichen Formen und Inhalte von Gruppen, trotz der erwiesenen Effektivität ganz unterschiedlicher Methoden eine Tendenz besteht, die eigene Gruppenerfahrung in schroffer Abgrenzung zu anderen darzustellen und zu verteidigen. So entsteht das interessante Paradoxon, daß zwar innerhalb einer Gruppe kreative Auseinandersetzung möglich ist und gefördert wird, bis hin zum demokratischen Ideal der Ebenbürtigkeit, Gruppen jedoch zu anderen Gruppen im Gegensatz stehen und im „Schulenstreit" ihre eigene Identität beziehen. Das vorhin Gesagte gilt nicht nur für gruppendynamische Bewegungen, sondern auch für Therapiegruppen innerhalb der praktisch angewandten Medizin, der Gruppenpsychotherapie.

Wir werden in der folgenden Darstellung verschiedener Gruppenpsychotherapiemethoden selbst in diesem Paradoxon der Abgrenzung verstrickt sein, besonders in der Würdigung der Selbsthilfebewegung.

Zur Einführung eine kurze Terminologie:

20.3.1 Das „Setting"

Dieser Begriff erfaßt als abstrakter Terminus die konkreten Fragen der jeweiligen Gruppenzusammensetzung, der Gruppengestaltung sowie der angewandten gruppentherapeutischen Technik. Beispiele dafür sind etwa eine psychoanalytisch orientierte Gruppenpsychotherapie stationärer Patienten, die zweimal in der Woche für 1½ Stunden zusammentreffen, oder eine ambulante Myokardinfarktgruppe, welche sich einmal pro Woche für 2 Stunden trifft, wobei das Programm aus ½ Stunde Gymnastik, 15 Minuten autogenem Training und anschließend einem Gruppengespräch mit internistischer Information und Erarbeitung psychosozialer Bewältigungsstrategien besteht.

20.3.2 Geschlossene – halboffene – offene Gruppen

Darunter sind die verschiedenen Modalitäten der Konstanz innerhalb einer Gruppe gemeint. In einer geschlossenen Gruppe einigen sich die Mitglieder darauf, den gesamten Gruppenprozeß gemeinsam zu durchlaufen. Ausgeschiedene Gruppenteilnehmer werden nicht ersetzt, scheiden zu viele Teilnehmer aus, endet die Gruppe (z.B. Gruppenpsychoanalyse). Im Gegensatz dazu strebt die offene Gruppe ein Arbeitsklima an, etwa in einer Ambulanz für Karzinomkranke oder auf einer Dialysestation, welches den Teilnehmern eine Atmosphäre anbietet, emotional und informativ zueinander zu finden, unter Verzicht auf den Gruppenprozeß einer bestimmten Teilnehmergruppe. Am häufigsten wird das Setting einer halboffenen Gruppe gewählt, der Wechsel von einzelnen Teilnehmern ist möglich, jedoch nach ausreichender Zeit, um der Gruppe als Ganzes die Verarbeitung des Verlustes sowie die Neuintegration von Teilnehmern zu ermöglichen, ohne daß der Gruppenprozeß darunter leidet.

20.3.3 Homogene – heterogene Gruppen

Die damit gemeinte Gleichartigkeit oder Verschiedenheit der Gruppenteilnehmer bezieht sich in der Praxis auf die Art der Organerkrankung (Asthma, Hypertonie, Ulcus duodeni) bzw. auf das neurotische Symptom (Phobiker, neurotische Depression) oder auf die Art des psychosozialen Hintergrunds (Schichtbegriff). Bei den Patienten wird dabei auch auf die Qualität der psychoneurotischen Abwehr, der Ich-Stärke und des Bewältigungsverhaltens geachtet. In der Literatur gibt es Beispiele für jegliche Form der Homogenität oder Heterogenität einer Gruppe. Unterschiede betreffen wiederum eher das Setting oder die Wahl der gruppentherapeutischen Methode.

Für die psychoanalytisch orientierte Gruppe bei Patienten mit organischen und funktionellen Störungen gilt, daß die meisten klinischen Erfahrungen mit halboffenen heterogenen Gruppen gemacht wurden.

20.3.4 Therapie der Gruppe – Therapie des einzelnen in der Gruppe

Diese beiden Pole des therapeutischen Ansatzes betreffen die Aufmerksamkeitsfokussierung des Therapeuten im Hinblick auf seine Interventionen. Gruppendeutungen beinhalten das Ergebnis der Assoziationen und Reflexionen des Therapeuten, welche aus den Einzelbeiträgen der Gruppenmitglieder herrühren. Auf die Einzelbeiträge wird nicht direkt eingegangen, vielmehr werden die Voten der Gruppenmitglieder als Auseinandersetzung vom Gruppen-Ideal-Ich, Gruppen-Über-Ich und Gruppen-Es der Reflexionsfähigkeit des Gruppen-Ichs anheim gestellt. In der Einzeldeutung, welche ein Teil des Eingehens auf die Problematik eines einzelnen Gruppenmitgliedes ist, wird der Gruppenprozeß durch die Bearbeitung und Erhellung der Konflikte und Bedürfnisse des einzelnen gefördert. In der pragmatischen Realität einer psychoanalytischen Psychotherapiegruppe werden die Pole des therapeutischen Ansatzes nicht starr eingehalten, so daß extreme Polarisierungen eher reinen Selbsterfahrungsgruppen oder Forschungsprojekten angewandter Therapietechniken vorbehalten bleiben.

20.3.5 Zum Begriff der „Gruppenfähigkeit"

Die Frage, ob ein Patient von einer Gruppenpsychotherapie profitieren könne, führt zu dem Problem der Indikationsstellung Einzeltherapie oder Gruppentherapie. In der Würdigung der publizierten gruppentherapeutischen Erfahrungen sollte diese Indikationsfrage nicht mehr in dieser Weise gestellt werden. Da die Indikation sich mehr auf das „Setting" einer Gruppe hin ausrichtet, kann die Indikation nicht mehr auf eine „Gruppenfähigkeit" an sich bezogen sein. Trotzdem ist es A. Heigl-Evers und F. Heigl (1970) in einem Übersichtsreferat gut gelungen, einige Kriterien für die Indikationsstellung zur Gruppentherapie herauszuarbeiten. Das spezifisch Neue im Setting der therapeutischen Gruppe, verglichen mit der Einzeltherapie, ist die Pluralität. Anhand von vielen Merkmalen der Pluralität: unter mehreren sein – Vielheit und Verschiedenheit – nicht souverän sein – Unabsehbarkeit der Folgen des eigenen Tuns – wird die Indikation für die analytische Gruppenpsychotherapie entwickelt.

In den Anfangsphasen einer Behandlung ist nach Heigl-Evers und Heigl meist eine Einzeltherapie angezeigt. Wenn sich jedoch der Patient auf die verstehende und tolerierende Haltung des Analytikers als Dauereinrichtung einzustellen beginnt, dann zieht er, wenn man seine Toleranzschwelle berücksichtigt, aus der Gruppentherapie vermehrten Gewinn. Die Gruppentherapie ist auch noch bei Patienten mit Organsymptomen, die über eine verminderte emotionale Ausdrucks- und Introspektionsfähigkeit verfügen, oft der Einzeltherapie vorzuziehen. Patienten mit Charakterneurosen sind nach den Erfahrungen des Ehepaares Heigl ebenfalls in der Gruppe erfolgreicher zu behandeln, weil sie es als weniger kränkend und demütigend erleben, zu sehen, daß auch andere Menschen ähnliche Schwierigkeiten haben.

Für depressiv strukturierte Patienten stellt die Gruppe einerseits eine tragende Matrix, andererseits aber eine besondere Frustration ihrer Verschmelzungstendenzen dar und bietet dadurch Anreize zur Identitätsfindung. Patienten mit Allmachtsphantasien lernen in der Gruppe besser als in der Einzeltherapie, daß ihre Möglichkeiten durch die anderen begrenzt sind. Andere wiederum lernen hier ein verbessertes Sozialverhalten, indem sie wahrnehmen, daß man in der Gruppe nicht nur konkurrieren, sondern auch koalieren kann. Auch bei Schwierigkeiten im Umgang mit Autoritäten und bei dissozialen Jugendlichen hat sich die Gruppentherapie, oft in Kombination mit Einzelsitzungen, sehr bewährt. Sowohl hysterische Patienten, die die Folgen ihres Tuns meist wenig berücksichtigen, als auch zwangsneurotische Patienten, die im Gegensatz dazu jede Veränderung fürchten, machen meist in einer gut geführten analytischen Gruppe gute Fortschritte.

Zusammenfassend kann man also feststellen, daß so gut wie alle Patienten „gruppenfähig" sind, mit Ausnahme jener so schwer narzißtisch gestörten, deren Toleranzschwelle die Pluralität noch nicht erträgt.

In Ergänzung dazu hat Kutter (1976) die Erfahrung gemacht, daß im Anschluß an eine Gruppentherapie oft erstaunliche Fortschritte in einer anschließenden relativ kurzen Einzeltherapie erzielt werden können.

Nach wie vor muß es also der Erfahrung des einzelnen Therapeuten überlassen bleiben, die Indikation für einen bestimmten Patienten und eine bestimmte Therapieform zu stellen.

Allgemeine Ausschlußkriterien nach Lonergan (1982) für eine Gruppe unabhängig vom jeweiligen Setting sind:
- extreme Isolierung,
- hochgradig destruktives Verhalten,
- Desorientiertheit oder manisches Verhalten.
- Eine weitere Eigenschaft schließt eine Gruppenfähigkeit aus: Wenn der Patient zu sehr im sekundären Krankheitsgewinn fixiert ist oder wenn er sich im kompletten resignativen Rückzug befindet. Wir denken dabei an jene Patienten, die sich mit ihrer Krankheit so sehr „resignativ abgefunden" haben, daß sie nicht mehr über die Möglichkeit verfügen, sich mit ihr auseinanderzusetzen.

Der Ansatz entspricht der Therapie des einzelnen in der Gruppe. So können Gruppenmitglieder mit unterschiedlicher Persönlichkeitsstruktur und Erkrankung in der gleichen Gruppe behandelt werden. Diese von Persönlichkeitsstruktur und Erkrankung der Mitglieder heterogenen Gruppen werden mit zunehmender „Gruppenfähigkeit" der Teilnehmer kohärent und in ihrer Beziehungsfähigkeit auch aufeinander hin orientiert. Durch die Zunahme des Selbstvertrauens und der Kommunikationsfähigkeit kann nun eine aufdeckende Psychotherapie einsetzen. Der Gruppentherapeut wird die pathogenen Konflikte der Patienten mittels der freien Assoziation und durch deutende Bearbeitung von Übertragung und Widerstand in der Gruppe bewußt werden lassen und mittels

Phantasieleistung und Durchdenken von Alternativen neues Bewältigungsverhalten ermöglichen. Stets wird der Gruppentherapeut auf die Stabilität des Selbstwertgefühls der einzelnen Gruppenmitglieder achten. Denn die „Arbeitsfähigkeit" der Gruppe ist stets abhängig von der Kohärenz der Gruppe, die wiederum in Abhängigkeit von der „Gruppenfähigkeit" der Teilnehmer steht.

In der Praxis mit somatisch Kranken und narzißtisch gestörten Patienten muß stets auf die reale somatische Situation, auf die momentanen psychosozialen Bedingungen geachtet werden. Nur unter Einbeziehung dieser Realität entsteht die nötige Geborgenheit und Sicherheit, sich selbst in seiner individuellen Wirklichkeit in Frage zu stellen und neue Sichtweisen zuzulassen. Die Realität darf auf keinen Fall als Widerstand oder Agieren mißinterpretiert werden. Selbstverständlich soll der Therapeut auch die Ebene des Widerstandes und der Übertragung reflektieren. In der Technik soll dies jedoch erst angesprochen werden, wenn die Bedingungen des biopsychosozialen Feldes verstanden werden.

Ein 36jähriger Patient leidet seit etwa 2 Jahren an einem funktionellen Leiden des distalen Kolons, einer sog. Colitis mucosa bzw. einem spastischen Kolon. Dabei wechseln Durchfälle mit Verstopfung, bestehen kolikartige Schmerzen vor allem im linken Unterbauch. Er hat außerdem Symptome einer Depression mit Antriebsschwäche, Gewichtsverlust, trauriger Gestimmtheit, nihilistischen Befürchtungen, Schlafstörungen sowie eine ausgeprägte Karzinophobie. Selbst nach ausgedehnten organischen Untersuchungen, wie Koloskopie mit Probeexzision, Sonographie etc., besteht weiterhin kaum Verständnis für die Möglichkeit einer funktionellen Erkrankung. Der Patient hat das Gefühl, ein funktionelles Krankheitsbild würde ihn in seiner Umgebung zum Simulanten, Hypochonder oder psychisch Kranken machen. Er hat einen technischen Beruf und ist der Überzeugung, ein tüchtiger Mensch habe einen gesunden Geist und einen belastbaren Körper. Die Seele melde sich nur bei Versagen oder Schwäche. Mit dieser Einstellung grenzt er sich ängstlich von seinen Familienangehörigen ab, die an Depressionen leiden oder litten. Seine Lebenssituation ist charakteristisch: Der Patient wollte als selbständiger Unternehmer arbeiten, wegen menschlicher Probleme – wie Streit, Unfreundlichkeit etc. – hatte er dabei keinen Erfolg. Er begann als Lehrer an einer technischen Schule. Er lebt abgekapselt, hat häufig Streit mit seiner Freundin. Vor allem seiner Mutter macht er massive Vorwürfe, daß sie schuld sei an seiner Erkrankung. Er hat bei sich selbst einen Dickdarmkrebs „diagnostiziert", der ihm – nach seiner Meinung – entweder von den untersuchenden Ärzten nicht mitgeteilt werde oder den die Ärzte wegen ihrer Vorurteile oder wegen ihrer oberflächlichen Arbeitsweise nicht diagnostizieren konnten. Er befürchtet, nurmehr wenige Monate zu leben.

Der Widerstand gegen die Annahme einer sozialen Unterstützung ist aus der Lebensgeschichte des Patienten zu verstehen. Er stammt aus einem Gebirgstal, hat enge Beziehungen zu seiner Mutter und dem heimatlichen Milieu. Er hat zwei ältere Brüder, die beide wesentlich kräftiger und durchsetzungsfähiger seien.

Er gleiche dagegen mehr seinem Vater, der zwar ein tüchtiger Kunsthandwerker, jedoch überhaupt kein Geschäftsmann sei. Er hatte immer den Eindruck, daß seine Mutter die beiden Brüder bevorzuge, da sie mit diesen ganz anders umgegangen sei. Diese beiden Brüder seien nicht abgeschoben worden wie er, diese durften zu Hause bleiben und seien heute tüchtige Geschäftsmänner. Er hingegen wird mit 10 Jahren von einem Priester zum Besuch eines Knabenkonviktes vorgeschlagen, da er begabt für ein Studium sei und eventuell auch Priester werden könne. Die Mutter greift diesen Vorschlag gerne auf, und so wird der Patient in das weit entfernte Heim gegen seinen Willen geschickt. Er darf nur zweimal im Jahr nach Hause kommen. Er erinnert die Schulzeit als Schreckenszeit in seinem Leben, durfte aber die Schule nicht abbrechen, da er von daheim keine Unterstützung erwarten konnte. Er begann, dank seiner guten intellektuellen Begabung, ein vorzüglicher Schüler zu sein, schaltete jedoch jede seelische Regung aus und wurde nurmehr „Körper" und „Geist". Er studiert dann auch als einziger der Geschwister. Dies gibt ihm die Hoffnung, die Anerkennung seiner Mutter zu erringen. Er erinnert sich, während seines Studiums immer gelacht zu haben, wenn Kollegen Heimweh oder Liebesschmerz gehabt hatten. Dies berührte ihn gar nicht. Er liebte niemanden und kümmerte sich um keine Gefühle außer der Schadenfreude, wenn andere Heimweh und Liebeskummer hatten. Er wollte ein anerkannter Fachmann sein, wenn er wieder nach Hause käme. Sein Körper ist ihm auch heute noch nur bis zum Zwerchfell vertraut, so sagt er, und lehnt jedes Vertrautsein mit dem unteren Körperabschnitt ab. Dies hat in der momentanen Situation Bedeutung, da seine Freundin, die er seit 9 Jahren kennt, nun gerne heiraten und Kinder haben möchte. Er kann sich dies nicht vorstellen. Auf die Frage, was er vorhabe nach seiner Genesung, sagt er: „Ich möchte weg, alles im Stich lassen und irgendwo anders neu beginnen".

Diagnostisch hat dieser Patient eine pseudounabhängige Charakterabwehr, ähnelt manchmal der Beschreibung der Alexithymie in seiner Unfähigkeit zur Phantasie und zum Herstellen von Zusammenhängen seiner Symptomatik mit der Lebenssituation und der seelischen Dynamik. Er befindet sich im narzißtischen Rückzug, ist voller Mißtrauen, versucht sich durch das Studium medizinischer Bücher abzusichern und selbst einen Heilungsweg zu eröffnen.

Dieses Fallbeispiel zeigt, wie sehr auf die narzißtische Abwehr eines Patienten geachtet werden muß, soll diesem geholfen und nicht wieder ein Arztwechsel herbeigeführt werden. Erst nachdem er sich in der Gruppe auf die anderen Gruppenmitglieder einlassen konnte, die er vorher ignorierte bzw. als psychisch krank bezeichnete, konnte er auch allmählich die Spaltung in „gesunden" Geist und „kranken" Körper überwinden und sich selbst auch ein Leben der Gefühle, Triebe und Wünsche erlauben. Dies äußerte sich darin zuerst, daß er nach etwa 6 Wochen begann, sich Gedanken zu machen, wie er auf die Mitpatienten wirke und sich erkundigte, ob sie sich vorstellen könnten, daß er ein Mann sei, der von seiner Freundin geliebt werde.

20.4 Die „pregroup technique" nach Lonergan

Diese psychotherapeutische Vorgangsweise ermöglicht es auch Patienten, die ansonsten nicht von einer Gruppenpsychotherapie profitieren können, in einer Gruppe zu verbleiben und zur „Gruppenfähigkeit" zu gelangen.

Solche Patienten reden direkt zum Gruppenleiter oder auch nur vor sich hin. Sie gehen auf andere Patienten nicht ein, wirken egoistisch, nur an der eigenen Leidensgeschichte interessiert. Sie zeigen sich fast unfähig, etwas Besonderes an ihrem Leben zu finden. Alles ist normal, andere würden gleich reagieren. Nemiah und Sifneos (1975) haben diese Sichtweise als Alexithymie beschrieben (vgl. Kapitel 5). Lonergan beschreibt ihre Vorgangsweise wie folgt:

- Akzeptieren der narzißtischen Abwehr. Alles, was erzählt wird in der Gruppe, ist richtig. Es gibt Platz für unterschiedliche Meinungen und Sichtweisen in der Gruppe.
- Betonen der „individuellen Stärken" der einzelnen Gruppenmitglieder. Menschen in akuten oder chronischen Belastungssituationen haben oft ein großartiges Durchhaltevermögen bewiesen. Dieses Betonen von Ich-Stärke führt zur Bereitschaft, auch über andere Verhaltensmöglichkeiten und Sichtweisen nachzudenken.
- Der Therapeut ist bereit, sich als reales Objekt besetzen zu lassen. Er hinterfragt Idealisierungen nicht. Youcha (1976) beschreibt den Wert, Patienten in dieser Therapiephase durch Briefe, Telefonate und auch Hausbesuche nachzugehen. Der Patient erlebt so, daß er seinem Therapeuten und der Gruppe wichtig ist. Es ist oft nötig, beim Wegbleiben aus der Gruppe die Patienten brieflich einzubestellen. Die Patienten empfinden dies als wichtig für ihren Stellenwert, den sie bei den behandelnden Ärzten haben. Viele sagen, sie würden sich spontan nicht mehr melden aus Angst, undankbar oder aufdringlich zu erscheinen oder von der Gruppe abgelehnt zu werden. Die reale Arzt-Patient-Beziehung ist ein wesentlicher Kontakt für die oft isoliert lebenden Patienten, da sie soziale Unterstützung entweder nicht haben oder davon keinen Gebrauch machen können.
- Das Einsetzen supportiver Methoden, etwa durch Ansprechen und Belohnen für „dem anderen zuhören". Der Gruppentherapeut beginnt mit den Erfahrungen eines Patienten und deren Bedeutung für die anderen, um die Möglichkeit zu eröffnen, aufeinander einzugehen.
- Förderung des Einander-Helfens. Wie bei Persönlichkeitsstörungen bereits Valliant (1977) den Altruismus als reifungsbegünstigenden Abwehrmechanismus beschrieben hat, fand auch Yalon (1980), daß Altruismus des Schwerkranken die bedeutsamste Einstellung war, das Leben sinnvoll zu gestalten. Es soll dabei erwähnt werden, daß die Patienten häufig auch dem Gruppenleiter helfen wollen und tatsächlich als „Hilfstherapeuten" eine wichtige Funktion erfüllen können. Dies kann

auch angesprochen werden, indem der Therapeut Wortmeldungen als besonders wichtig für den Gruppenverlauf und das Verstehen eines Gruppenmitgliedes hervorhebt.

– Zulassen des „Splittings": innerhalb der Gruppe sind alle „gut", draußen sind alle „böse". Daraus entsteht Gruppenkohäsion, welche genügend Angstfreiheit ermöglicht, um Beziehungen zueinander aufzunehmen und Probleme im eigenen Erleben bewußt werden zu lassen und zu berichten.

– Fördern des Vertrauens zu den behandelnden Ärzten, denn dadurch kann wieder Angst gebunden werden und der Sekundärprozeß des Durchdenkens von Alternativen, des Gewahrwerdens von Wünschen und Befürchtungen ermöglicht werden.

Ein junger Mann wird nach einer schweren Herzoperation plötzlich passiv-verweigernd, depressiv, er zieht sich zurück. Er wirkt ängstlich, abgekapselt. In der Gruppe berichtet er, wie ungerecht er sich behandelt fühle. Seine Ärzte erwarten sich von ihm mehr aktive Mitarbeit, er könne jedoch nicht mehr leisten. Die Gruppenteilnehmer sprechen von den Vor- und Nachteilen, wenn der Arzt mehr aktive Mitarbeit verlange. Der Patient beklagt sich, im Krankenhaus könne sich niemand vorstellen, wie anstrengend und schmerzensreich der postoperative Verlauf sei. Die Gruppenmitglieder gehen darauf ein und sprechen die Wut an, die so ein uneinsichtiges Verhalten auslöse. Darauf sagt der Patient ganz zornig: „Am liebsten wäre ich aufgestanden, hätte dem Arzt eine Ohrfeige gegeben und hätte das Zimmer verlassen". Darauf wird in der Gruppe lachend gesagt: „Am liebsten wärst Du gar nicht krank geworden". Der Patient kann mitlachen. Im Anschluß erzählt er zum erstenmal von seiner Betroffenheit, diese angstmachende Erkrankung zu haben. Bisher habe er immer gedacht, er würde sonst als Feigling bezeichnet werden.

Dies zeigt den Wert der Gruppe deutlich. Die „peer group" der Gruppenteilnehmer ermöglicht das Ansprechen von Aggressionen und von Ängsten, die sein „Ideal-Ich" kränken, da sich die Gruppenteilnehmer in der Gruppe geborgen und verstanden fühlt. Nach dieser katharischen Reaktion kann der Patient sein Basisvertrauen zu dem behandelnden Arzt wieder erneuern und ist zur aktiven Mitarbeit imstande. Er berichtet noch, wie sehr er das Leistungsvermögen seines Arztes bewundere. Mit der Wiedererrichtung der Idealisierung kann er Mut und Vertrauen schöpfen, seine Krankheit werde sich bessern, und er werde wieder nach Hause kommen.

20.5 Das psychosomatische Therapiekonzept in der Gruppentherapie

Das jeweilige „Setting" der Gruppe wird in erster Linie durch die Zielsetzung, die Ausbildung und Interessensrichtung des Gruppenleiters bestimmt. Gesprächsgruppen, körperbezogene Therapiegruppen, Gruppen mit einem gemeinsamen Medium wie bei der Musiktherapie, Maltherapie, Theatergruppen arbeiten effizient und weisen Erfolge auf.

Ziel des Gruppenprozesses sind die Möglichkeiten der Entlastung, der Um- und Neugestaltung des Erlebens und auch der Symptomlinderung. Zum Wesen eines jeden Gruppenprozesses gehört die Identifikation des Gruppenmitglieds mit den Gruppeninteressen, eine Erhöhung der Kommunikationsfähigkeit und die Projektion aggressiver Strebungen nach außen, also die Grenzziehung gegenüber Personen und Interessen, die als gruppenfremd erkannt werden.

Dies bezieht sich nicht nur auf das Bewahren von mitgeteilten Erlebnissen, dem Schweigegebot gegenüber Nichtmitgliedern, sondern auch auf das Erleben von Akzeptanz in der Gruppe. Fühlt sich nämlich ein Teilnehmer von der Gruppe abgelehnt, dann wird er nicht mehr aktiv am Gruppengeschehen teilnehmen oder sogar der Gruppe fernbleiben.

Einer der wichtigsten Prognosefaktoren ganz allgemein ist das Vorhandensein von sozialer Unterstützung und die Fähigkeit, davon Gebrauch machen zu können. 1959 zeigte dies Querido in seiner prospektiven Untersuchung von über 1500 Patienten mit den unterschiedlichsten Erkrankungen (vgl. Kapitel 28). Die soziale Unterstützung wird behindert durch Scham- und Schuldgefühle, Angst vor Kränkung und Zurückweisung und durch die Angst, vereinnahmt und gedemütigt zu werden. Diese Angst kann den Rückzug in das Größenselbst bis hin zu einem sensitiven Wahrnehmungsmodus und paranoiden Ideen verursachen. Das Vorliegen von Hilflosigkeit und Hoffnungslosigkeit, wobei sich ersteres auf Objekte und letzteres auf das Selbst bezieht, führt zur Verzweiflung. Diese wiederum erschwert oder verunmöglicht die Annahme sozialer Unterstützung. Daher ist es eine der wichtigsten Aufgaben der Gruppe, besonders des Gruppentherapeuten, soziale Unterstützung zu gewährleisten.

20.5.1 Das Sichern sozialer Unterstützung

Am einfachsten ist dies zu bewerkstelligen, wenn folgende Grundregeln in der Gruppe beachtet werden:
– Respekt vor dem anderen,
– ehrliche Information,
– Bestehenlassen unterschiedlicher Meinungen.
Diese Gruppenkultur muß der Gruppenleiter vorleben, er kann sich jederzeit als Hilfs-Ich, auch durch Identifikation mit einem Anliegen eines Gruppenmitgliedes und durch das Betonen der Gruppenregeln aktiv einbringen. Soziale Unterstützung heißt aber nicht Harmonisierung. Im Gegenteil, sie erlaubt Differenzierung in Geborgenheit. So erfahren die Gruppenmitglieder am besten, daß verschiedene Sichtweisen natürlich und fruchtbar für das „Miteinander" sind. Zusätzliche verbale Interventionen unterstützen diesen Prozeß.

20.5.2 Das Akzeptieren, sich in einer Belastungssituation zu befinden

In der psychotherapeutischen Arbeit bei Patienten mit funktionellen oder somatischen Erkrankungen wird häufig das Problem deutlich, welchen Stellenwert psychosoziale Probleme für das Selbstverständnis des Patienten und seine Angehörigen haben. Neben der Schuldfrage – anfänglich meinen viele Patienten die Verursachung der Erkrankung ist stets „verschuldet" – tritt häufig die Abwehr des „Besonders-sensibel-Seins" auf. Wichtig ist auch die Neigung zur Spaltung, d.h., auch bei schwerer körperlicher Erkrankung, etwa einem M. Crohn oder einem Karzinom, wird einseitig der seelisch-charakterlichen Schwäche die Alleinverantwortung zugeschrieben. In der Arbeit mit psychisch oder somatisch kranken Menschen ist es immer wieder entscheidend, sich dieser Abwehr, der Spaltung in Körper und Seele bewußt zu sein. Die Laien-Ätiologie und „ganzheitliche Konzepte", etwa das von Groddeck, neigen dazu, die Realität somatischer Funktionsebenen in einem falsch verstandenen Konversionskonzept aufgehen zu lassen. Die Folge davon ist eine schlechte Compliance auf der somatischen Ebene. Patienten mit Hochdruck, mit M. Crohn und Kolitis, aber auch Patienten, die an einem Karzinom leiden, vernachlässigen dann die somatische Therapie, da sie der Meinung sind, die Krankheit sei ja doch „seelisch verursacht". Auch bei den funktionellen Angstkrankheiten, etwa der Herzneurose, entstehen chronische Verläufe oft dadurch, daß der Arzt in die „dualistische Falle" gerät, indem er den Patienten nur als organisch Kranken oder nur als seelisch Kranken behandelt. Zur therapeutischen Verantwortung eines Gruppenleiters gehört daher auch die Akzeptanz körperlicher Störungen und deren Therapie. Wird in der Gruppenkultur die Einheit des biopsychosozialen Systems beachtet, dann kann auch der einzelne Patient bei sich selbst die häufig so schwierige, stets neu erforderliche Integrationsarbeit leisten. Denn stets müssen in der Realität eines Krankheitsverlaufs die belastenden Faktoren neu gewürdigt werden, zumal sie sich ja ständig verändern. Patienten in Gruppentherapie kommen gar nicht selten gerade dann nicht in die Gruppe, wenn sie vermehrt auf der körperlichen Ebene Beschwerden haben, also „krank" sind. Dies ist dann der Beweis für den Dualismus, der manchmal auch vom Psychotherapeuten induziert wird, als ob die Gruppe nur für den körperlich Gesunden geeignet wäre.

Verantwortlich für diese Fehlentwicklung ist häufig das Nicht-Realisieren einer akuten Belastungssituation, einer vermehrten Angst und einer narzißtischen Labilisierung durch die Erkrankung, die sich in Rückzug und Isolierung äußert. Gerade dabei kann jedoch die Gruppe nützlich sein, wenn in der Gruppenatmosphäre und im Setting dafür Platz ist.

20.5.3 Das Wiedererlangen der Stabilität des Selbstwertgefühls

Angst vor Trennung, vor Verletzung oder vor zu großer Nähe bedeutet eine Bedrohung des narzißtischen Gleichgewichts. Die Symptome der narzißtischen Störung sind bereits das Ergebnis eines unbewußten Selbstheilungsversuches (Kernberg, 1979; Valliant,

1977). Stolorow und Lachmann (1980) beschrieben den Narzißmus als jenen Bereich der seelischen Struktur, der das Selbstwertgefühl schützt, wiederherstellt, in Ordnung bringt und stabilisiert. Wird das Selbstwertgefühl bedroht, gerät der Patient in eine „narzißtische Labilisierung". In der Folge kommt es zum Auftreten narzißtischer Abwehr. Die beiden von Kohut (1972) beschriebenen Abwehrmechanismen der narzißtischen Störung sind die „Grandiosität", das ist der Rückzug in das Größenselbst, und die „Idealisierung". In der Entwicklung des einzelnen sind beide Abwehrmechanismen zunächst adaptiv. Nur bei einer Fixierung auf dieser Position entstehen psychische Strukturen, die bleibende Beziehungsstörungen verursachen. In Situationen akuter Belastung können narzißtische Regressionen auch adaptiv sein, da sie das Wiedererlangen von Selbstwert ermöglichen. Ihr Nachteil liegt in der Beeinträchtigung der Gemeinschaftsfähigkeit und dem Fehlen reifer Objektbeziehungen. Für Gruppenbildungen können diese regressiven Strebungen zunächst jedoch hilfreich sein. Der Gruppentherapeut soll daher im Anfang „Idealisierungen" zulassen und dem Patienten ermöglichen, sich in der Gruppe „grandios" zu fühlen. Dann können das erforderliche Sicherheits- und Selbstwertgefühl entstehen, die für das Einsetzen der Sekundärprozesse erforderlich sind. Körperliche Krankheit wird als Bedrohung erfahren und „erzwingt" beinahe regelhaft unreife Abwehrmechanismen. Es darf nicht vergessen werden, daß neben reifen Bewältigungsstrategien auch unreife Beziehungsmodi in schweren Belastungssituationen adaptiv sein können.

Patienten mit somatisch sich manifestierenden Krankheiten haben ein Anrecht auf einen supportiven Zugang, der auf die momentane Belastungssituation Rücksicht nimmt. Dieses Eingehen auf die jeweilige Realität des Gruppenmitgliedes ist auch in einer analytisch orientierten Gruppenpsychotherapie zu beachten, denn sonst entsteht eine narzißtische Kränkung mit Rückzug, oder es kommt sogar zum Therapieabbruch.

20.6 Therapie – Ziele und Ergebnisse

In einer Studie zeigten Altmann-Herz, Reindell et al. (1983) am Beispiel von Patienten nach Myokardinfarkt die Notwendigkeit, Therapieverfahren differenziert nach der Persönlichkeitsstruktur der Patienten anzubieten. Sie untersuchten 38 männliche Herzinfarktpatienten und erhoben bei ihnen folgende Daten:
– sozio-demographische Daten,
– somatische Befunde und labortechnische Daten,
– psychologische Daten.
Die psychologischen Daten wurden mittels eines konfliktorientierten Gesprächs von 40–50 Minuten Dauer und einer Testbatterie, bestehend aus den Persönlichkeitsskalen nach Hehl (PSS 25), dem Freiburger Persönlichkeitsinventar (FPI), dem Gießener Beschwerdebogen (GBB) sowie einem Fremdbeurtei-

lungsbogen mit 9 Skalen und einem Selbstbeurteilungsbogen mit 2 Skalen, erhoben. Auf statistischem Wege konnten die drei Autoren drei Gruppen von Herzinfarktpatienten unterscheiden: die impulsiv-aggressiven, die regressiv-ängstlichen und die scheinbar unauffällig „soziablen" Patienten. Die unterschiedlichen Verhaltensweisen der Herzinfarktpatienten waren vor allem durch die Art der Angstverarbeitung hervorgerufen. Das Verhalten der impulsiv-aggressiven Patienten konnte nach Erschöpfen ihres Verhaltens in das regressiv-ängstliche umschlagen, während die soziablen unverändert blieben. Die Autoren empfehlen in ihrer empirisch gut überprüften Studie, daß man für die impulsiv-aggressiven Patienten vor allem Sportgruppen, für die regressiv-ängstlichen Einzel- und Gruppentherapien und für die soziablen Patienten zunächst Gruppen mit autogenem Training anbieten müsse, die später in tiefenpsychologisch orientierte Therapiegruppen übergeführt werden können, um einen optimalen Therapieerfolg zu erzielen.

Die Ergebnisse der unterschiedlichen Therapieangebote wurden bisher nur aufgrund allgemeiner klinischer Beobachtungen beurteilt und noch nicht statistisch überprüft. Dies sowie die relativ kleine Patientenzahl schränken vorläufig die Bedeutung dieser Untersuchung noch etwas ein. Andererseits scheint dieser Ansatz ganz allgemein eine große Bedeutung für das Therapieangebot in der psychosomatischen Medizin zu haben, so daß ähnliche Differenzierungen sowohl bei anderen Erkrankungen als auch an größeren Patientenpopulationen vorgenommen werden sollten.

Neben diesen Untersuchungen, die sich bemühen, für Patienten mit den gleichen Erkrankungen je nach Persönlichkeitsstruktur und Angstabwehr unterschiedliche Therapieangebote zu machen, sind Untersuchungen wichtig, die auf klare therapeutische Ziele ausgerichtet sind. Diese Untersuchungen beschränken sich auf die fokussierte Bearbeitung von Konflikten und ermöglichen damit eine Verkürzung der Therapiedauer. Pohlen und Bautz (1974) zeigten in ihrer Psychiatrie-Studie, in der auch funktionelle Syndrome behandelt wurden, daß bezüglich der Erfolgsaussichten zwischen offenen und geschlossenen Gruppen kein nachweisbarer Unterschied besteht und daß Gruppentherapien, die zeitlich limitiert und in ihrem Arbeitsansatz fokussiert sind, auch bei chronifizierten und schwerkranken Patienten gute Therapieresultate erzielen.

Die leider nicht allzu zahlreichen methodisch sauberen, katamnestischen Studien sind überzeugende Hinweise, was von einer Gruppenpsychotherapie bei funktionell und körperlich Kranken zu erwarten ist. Am Beispiel des Asthma bronchiale zeigen dies drei Studien: Groen und Pelser (1960), Sclare und Crocket (1957) und Deter (1986) fanden übereinstimmend folgende Effekte einer Gruppenpsychotherapie: Der betroffene Patient kommt mit seinem Leben besser zurecht. Er ist seiner Erkrankung weniger ausgeliefert. Es verbessert sich seine Compliance, er nimmt die Medikamente verantwortungsbewußter und kann sich auf seine Erkrankung besser einstellen.

Die Folge davon ist eine flexiblere Lebensführung, weniger Krankenstände und kürzere Aufenthaltszeiten im Krankenhaus.

Wooley, Blackwell und Winget veröffentlichten 1978 eine 4-Jahres-Kontrolle von über 300 Patienten mit chronischen Krankheitsverläufen. Ihre Therapieziele waren: Verbesserung der Eigenständigkeit und Zunahme der sozialen Kontakte mit Annehmen von sozialer Unterstützung. Sie fanden nach vier Jahren folgendes: ein vermindertes Versorgungsverhalten, eine Erhöhung der Selbstkontrolle, einen verminderten Medikamentenverbrauch. Diese Ergebnisse waren in Abhängigkeit von intakten Familien und ausreichender sozialer Unterstützung deutlich besser als ohne dieses Umfeld.

20.7 Die Selbsthilfegruppen

Die Arbeit in und mit Selbsthilfegruppen ist seit den 70er Jahren sprunghaft angestiegen. Es zeigt sich, daß eine große Zahl leidender Menschen bei den professionellen Institutionen, die sich mit der Behandlung und Betreuung Kranker, insbesondere chronisch Kranker, befassen, nicht die Hilfe bekommen, die sie erwarten und erhoffen. Dersee (1982) meint auch, daß man zu diesem Zeitpunkt bereits „zurückhaltend geschätzt – über 5000 regionale Selbsthilfegruppen und Initiativen im deutschsprachigen Raum" zählen könne. Es ist vor allem das Verdienst von M. L. Moeller, daß er, ausgestattet mit psychoanalytischer und gruppenanalytischer Kompetenz, Selbsthilfegruppen beobachtet und untersucht und eine Arbeitsgemeinschaft für Selbsthilfegruppen ins Leben gerufen hat. Dadurch können entsprechende Initiativen wenigstens teilweise fachkundig unterstützt werden, was um so notwendiger erscheint, als hier, wie auch in manchen anderen Bereichen der Heilkunde, der Wunsch zu helfen nicht immer mit entsprechender Sach- und Fachkompetenz einhergeht. Es zeigt sich hier einmal mehr, daß Fachmann und Laie, Arzt und Patient aufeinander angewiesen und Partner sind und daher beide viel voneinander lernen können. Dies wurde nur allzulange von einer einseitig auf den Fachmann zentrierten Medizin übersehen. Eine Formel, auf die Selbsthilfegruppen in den USA gekommen seien, lautet: „Die Doktoren wissen besser als wir, wie die medizinische Behandlung für unsere Erkrankung aussieht. Wir aber wissen besser als sie, wie die beste Behandlung für uns als Menschen aussehen sollte" (zit. nach Moeller, 1983).

Ärzte können nach bisherigen Erfahrungen Vertrauen in die fördernden Kräfte von Selbsthilfegruppen haben (vgl. Moeller, 1978, 1981; Badura und Ferber, 1981; Kickbusch und Trojan, 1981). Sie sind jedoch verpflichtet, eine kooperierende Verantwortung mit zu übernehmen. Gruppendynamik ist wertfrei, Ideologie und Ausrichtung einer Gruppe jedoch nicht. Wahrscheinlich ist es im derzeitigen Stadium der Selbsthilfebewegung erforderlich, Berührungsängste abzubauen und zu ermutigen, Selbsthilfegruppen positiv anzusehen. Dennoch sollen aus eigener Erfahrung (Biebl) zwei Beispiele über Initiatoren von Selbsthilfegruppen gegeben werden, welche über die Not eben dieser Initiatoren berichten.

Eine 26jährige Frau beginnt mit großem Engagement eine Kolostomie-Selbsthilfegruppe. Sie ist initiativ, kontraphobisch und kontradepressiv. Sie bereitet ihren mitleidenden Gruppenmitgliedern Möglichkeiten zur Bewältigung, gibt Anleitungen und kooperiert mit Chirurgen, von denen sie sehr geschätzt wird. Eher durch Zufall wird bekannt, daß diese junge Dame während der Zeit ihrer verdienstvollen Tätigkeit für andere und Mitarbeit in einer Selbsthilfegruppe 3 Selbstmordversuche im Rahmen abnormer Trauerreaktionen gemacht hat.

Eine 48jährige Frau, die nach Brustamputation wegen eines Mammakarzinoms von ihrem Mann verlassen wurde, initiierte eine Selbsthilfeorganisation für geschiedene Frauen und eine für Brustkrebspatientinnen. Mit großem Geschick und organisatorischem Talent ist sie tätig und ermutigend. Sie selbst entwickelt eine paranoid-sensitive Reaktionsbereitschaft Männern gegenüber und verfehlt im privaten Bereich die Durcharbeitung des Trauergeschehens.

Diese beiden Beispiele sollen nicht entmutigen, Selbsthilfegruppen zu initiieren und ihnen als Arzt positiv zu begegnen. Sie scheinen jedoch auf etwas Wesentliches hinzuweisen. Initiatoren sollten nicht in ihrem Trauerprozeß über den Verlust von Unwiederbringlichem steckenbleiben und ihre Probleme auf die Ebene des Anderen-Helfens abschieben müssen. Auf die Bedeutung des Initiators einer Selbsthilfegruppe verweist auch Moeller (1983). Am Beispiel eines Selbsthilfegruppenmitgliedes namens Kelly zeigt er, wie notwendig es ist, daß der Initiator sein Leiden, seine Schwäche und den Tod angenommen hat und gerade deswegen lebensfähiger geworden ist.

Nach Schüffel (1983) verfolgen die medizinischen Selbsthilfegruppen das Ziel, durch wechselseitig angebotene Hilfe, also durch Laienhilfe zur Krankheitsbewältigung im psychischen und sozialen Sinne beizutragen, aber auch krankheitsbezogene Informationen zu vermitteln.

Die Selbsthilfegruppen kann man wiederum in Anlehnung an Schüffel einteilen in

- Rehabilitationsgruppen (Brustamputation, künstlicher Darmausgang)
- Gruppen mit dem Ziel der Verhaltensänderung und der Verhaltenskontrolle (AA, Drogenhilfe, Diätklubs)
- Gruppen, die der Primärversorgung dienen (Arthritiker, Diabetiker, MS-Patienten, Neurosekranke usw.)

In bezug auf das medizinische Versorgungssystem kann man Selbsthilfegruppen unterscheiden, die

- in medizinischen Systemen arbeiten,
- neben den medizinischen Systemen arbeiten (z. B. die AA),
- in Opposition zum System arbeiten (Gruppen mit para- und antimedizinischer Ideologie).

Das Ur- und Vorbild aller Selbsthilfegruppen sind die Anonymen Alkoholiker (AA). Sie wurden 1935 von dem Börsenmakler William Griffith Wilson und dem Chirurgen Dr. Robert Holbrook Smith, die beide schwere Alkoholiker waren, gegründet, haben seither eine große Verbreitung gefunden und sich auch sehr bewährt. Ihre Heilungsraten sind eindeutig besser als die der professionellen Suchttherapie.

Beim Lesen der Grundsätze der AA fallen vor allem drei Gesichtspunkte auf: eine restlose Akzeptierung des Leidens, der Entschluß zur Selbsthilfe und eine außerordentlich starke Ideologisierung.

Die ersten beiden Gesichtspunkte können wir bei allen erfolgreich arbeitenden Selbsthilfegruppen wiederfinden. Auch den dritten Gesichtspunkt können wir bei den meisten Selbsthilfegruppen beobachten, wenn auch nicht so stark ausgeprägt wie bei den AA. Je stärker die Selbsthilfegruppen in das medizinische Versorgungssystem integriert sind, um so weniger scheinen sie eine eigene Ideologie zu benötigen. Je mehr sie jedoch außerhalb oder gar im Gegensatz zu den medizinischen Versorgungssystemen stehen, um so stärker ausgeprägt ist ihre Ideologisierung.

Inzwischen haben sich Selbsthilfegruppen für die verschiedensten Erkrankungen gebildet, wie z. B. für psychosoziale Probleme, für Suchtkranke, für Krebskranke wie z. B. brustamputierte Frauen, für Patienten mit einem künstlichen Darmausgang (Ilco), Morbus-Crohn-Patienten (Hara), Multiple-Sklerose-Patienten (Amsel) usw., und für Angehörige Erkrankter. Aus dieser Aufzählung geht bereits hervor, welch große Bedeutung die Selbsthilfegruppen für die psychosomatische Medizin haben.

Die Frauenselbsthilfe nach Krebs hat sich z. B. folgendes Programm gegeben:

- Seelische Betreuung aller an Krebs Erkrankten
- Abbau der Angst vor weiteren Eingriffen
- Festigung der Widerstandskraft
- Ernährungsvorschläge zur allgemeinen Kräftigung
- Informationen über Versicherungsfragen und Behindertenrecht im Arbeitsprozeß.

In jeder Selbsthilfegruppe können wir charakteristische Bewegungen, einerseits des drohenden Zerfalls und andererseits des vermehrten Aneinanderrückens, beobachten. Je mehr in der Gruppe die Tendenz besteht, die Umwelt für die Probleme verantwortlich zu machen, um so mehr wird durch die projektiv-phobischen Mechanismen Gruppenkohärenz entstehen.

Besteht dagegen eher die Neigung, eigenes Fehlverhalten zu korrigieren, so wird die Gruppe stärker Gefühle der Solidarität und der teilnehmenden Unterstützung zulassen. Selbsthilfe im medizinisch-soziologischen Sinn wird mehr dem letzteren Gruppentyp zuzuordnen sein. Solche Gruppen können tragen und auch wieder freigeben.

So gesehen gelten für eine funktionierende Selbsthilfegruppe die gleichen Mechanismen wie für die Familie und die Arzt-Patient-Beziehung. In einer Atmosphäre der helfenden Anteilnahme wird Raum für persönliche Reifung und zum Aufarbeiten von Fehlverhalten sein. Weiter wird in einer solchen Beziehung auch die Verselbständigung und Reifung der

Persönlichkeit möglich. Die Gefahren des Scheiterns von Selbsthilfegruppen sind folgende:

– Wird für eine Gruppe die Abgrenzung nach außen wichtiger als die persönliche Entfaltung, die Reifung bedeutet, werden Angriffe nach außen oder Verstoßen eines Sündenbocks zur Regel.
– Kann sich eine Gruppe zu wenig auf ein Gruppenziel einigen, sind häufiger Wechsel der Teilnehmer oder krankheitsverstärkende Wirkung als Zeichen der Orientierungslosigkeit nachweisbar.
– Besteht zu wenig helfendes Zusammengehörigkeitsgefühl, so wird die Gruppe zerfallen oder sich radikalisieren.
– Bei Überbetonen eigenen Fehlverhaltens entsteht Angst und Aggression, was dann meist zu einer erhöhten Ideologisierung führt. Es kommt dann oft zu einer Stabilisierung einer Kerngruppe mit einem stark wechselnden Umfeld von Trabanten, die nur „emotionale Nahrung" für die Kerngruppe sind.

Was können Selbsthilfegruppen leisten?

Im Vordergrund steht der Wunsch, selbst aktiv-gestaltend in die Lebensbewältigung einzugreifen. Selbsthilfegruppen gibt es auf allen Ebenen menschlichen Interesses, z.B. Bürgerinitiativen, Emanzipationsbewegungen, Volksbegehren. Im medizinischen Bereich haben die Selbsthilfegruppen in erster Linie die Aufgabe, den Leerraum im Bereich der Rehabilitation und der präventiven Medizin auszufüllen und eine Aktivierung der Eigeninitiative zu bewirken.

Die Medizin ist gegenwärtig zu einseitig auf die Versorgung akuter Krankheiten ausgerichtet.

Verläßliche Untersuchungen über den Wert der Selbsthilfegruppen gibt es kaum. Der allgemeine Eindruck ist jedoch recht positiv. Stübinger (1977) hat mittels Fragebogen und Gießen-Test psychosoziale Selbsthilfegruppen (SH) mit expertengeleiteten Therapiegruppen (TH) verglichen und kommt zu dem Schluß, daß bezüglich der Therapieergebnisse zwischen Therapie- und Selbsthilfegruppen kein nennenswerter Unterschied besteht.

Ob diese zugunsten der Selbsthilfegruppen außerordentlich optimistischen Aussagen aufrechterhalten werden können, muß durch weitere Untersuchungen erhärtet oder modifiziert werden.

E. Hesse (1980), einem erfahrenen Allgemeinarzt in Stuhr bei Bremen, gelang es in Zusammenarbeit mit seiner Frau, einer Kinderärztin und einem Sozialarbeiter, um seine Allgemeinpraxis herum eine große Zahl von Selbsthilfegruppen zu initiieren, die sich zum Teil schwer Suchtkranker annehmen. Die Erfolge, über die er berichtet, sind sehr beeindruckend.

Unsere Erfahrungen mit Selbsthilfegruppen sind durchaus positiv. Wir nennen diese Gruppen im Unterschied zu den von Moeller beschriebenen echten Selbsthilfegruppen, die sich ja selbst organisieren und ganz ohne ärztliche Hilfe arbeiten, Semiselbsthilfegruppen. Sie treffen sich auf unsere Initiative hin in unseren Räumen einmal wöchentlich 2–3 Stunden und werden für kürzere Zeit jeweils von einem Thera-

peuten besucht, damit wir uns ein Bild über den Entwicklungsstand der Gruppe und ihrer Mitglieder machen und notfalls intervenieren können. Die Teilnehmer der Gruppen können uns auch jederzeit, insbesondere zum Zweck der Krisenintervention, in unserer Sprechstunde aufsuchen. In diese Semiselbsthilfegruppen werden vor allem jene Patienten aufgenommen, für die zur Zeit kein Therapieplatz bereitgestellt werden kann oder deren Motivation zur Therapie zunächst noch nicht abgeschätzt werden kann.

Viele Patienten, die früher entweder nicht sachgerecht behandelt werden konnten oder aber ununterbrochen, unerlöst von ihrer Not, die Sprechzimmer der verschiedensten Ärzte bevölkert haben, erhalten hier einerseits die für sie so notwendige emotionale Stütze und lernen darüber hinaus, besser mit ihren Schwierigkeiten umzugehen.

Söllner und Wesiack (1987) haben 107 in Semiselbsthilfegruppen behandelte Patienten mittels Fragebogen und Interview nach 3½ Jahren nachuntersucht und kamen zu folgendem Ergebnis:

In der Eigenbeurteilung bezeichnen die Teilnehmer ihre **Beschwerden** im Mittel als „gebessert" (Mittelwert 2,2 auf einer kontinuierlichen 5teiligen Skala von 1 = sehr gebessert bis 5 = sehr verschlechtert).

Besonders interessant ist, daß gerade Patienten mit Symptomen und Erkrankungen, die sich in der klinischen Praxis als besonders hartnäckig erweisen, wie Suchtprobleme, herzneurotische Beschwerden, körperliche Begleitsymptome depressiver Zustände und Kopfschmerzen, sich als überdurchschnittlich gebessert beschreiben. Neben diesen Beschwerden zeigten vor allem Selbstwertprobleme und Kontaktstörungen die deutlichste Besserung. Als am wenigsten verändert wurden sexuelle Störungen angegeben. Möglicherweise war dieser Bereich in den Semiselbsthilfegruppen zu sehr tabuisiert.

Gleichzeitig mit der Besserung der Symptome ist der **Verbrauch von Medikamenten** stark zurückgegangen. Dieser Effekt beruht vor allem auf dem starken Rückgang der Einnahme von Psychopharmaka. Es handelt sich dabei um einen stabilen Therapieeffekt im Sinne einer verminderten Abhängigkeit von Medikamenten, weil im gleichen Zeitraum nicht vermehrt andere Medikamente oder Drogen (Alkohol, Halluzinogene) eingenommen wurden.

Eine **Symptomverschiebung** sowohl auf andere körperliche oder psychische Beschwerden als auch auf Familienmitglieder (im Sinne einer „sozialen Symptomverschiebung") wurde nicht beobachtet.

Die Dauer der **stationären Aufenthalte** in Krankenhäusern oder Kurkliniken hat sich insgesamt gegenüber unbehandelten psychosomatisch Erkrankten (vgl. Ringel und Kropiunigg, 1983) nicht vermindert. Es ist aber eine Verschiebung dieser stationären Behandlungen weg von somatisch orientierten und vor allem im diagnostischen Bereich sehr kostenintensiven Krankenhäusern und Kliniken hin zu Behandlungen in psychosomatischen Kurkliniken eingetreten. Neben dem volkswirtschaftlichen Spareffekt ist das ein Hinführen zu einer patienten- und symptomgerechteren Behandlung, die hilft, einer Chronifizierung der Symptomatik vorzubeugen.

Im Vergleich der **Gießen-Test**profile nach Beckmann und Richter (1972) zu Beginn und am Ende der Semiselbsthilfegruppe zeigt sich eine signifikante Verminderung der Depressivität und der negativen sozialen Resonanz und eine Angleichung der Selbst- und Idealselbst-Bilder im Sinne einer Verminderung überstarker Über-Ich-Faktoren.

Aus den **Interviews** mit den Teilnehmern der Semiselbsthilfegruppen konnte eine deutliche Entwicklung zu mehr Autonomie von der Herkunftsfamilie und mehr Autonomie einerseits und Bindungsfähigkeit andererseits in der Partnerbeziehung beobachtet werden.

Aus dem Gesamtkollektiv der Teilnehmer an den Semiselbsthilfegruppen haben wir jene Gruppe, die **zusätzlich eine professionelle Psychotherapie** erhielt, der Gruppe, die nur in der Semiselbsthilfegruppe ihre Probleme bearbeitete, gegenübergestellt.

Signifikant andere Therapieergebnisse zeigte die Gruppe mit zusätzlicher Psychotherapie nur bei der Veränderung der Symptome. Alle anderen untersuchten Bereiche wie Medikamentenverbrauch, Krankenhausaufenthaltsdauer, Arbeitssituation, Beziehung zu Partnern und Familienangehörigen sowie die mit dem Gießen-Test beschriebenen Persönlichkeitsmerkmale zeigten keine signifikante Veränderung zwischen den beiden Gruppen. Innerhalb der Gruppe mit zusätzlicher Therapie wiesen jene Patienten, die eine analytische Therapie erhielten, erwartungsgemäß bessere

Ergebnisse auf als diejenigen, die ein autogenes Training durchführten.

Für manche Patienten reicht die Mitarbeit an einer Semiselbsthilfegruppe, eventuell noch kombiniert mit dem Erlernen des autogenen Trainings, völlig aus, um wieder allein zurechtzukommen. Manche Patienten benötigen die Selbsthilfegruppen als Dauerstütze und können dann mit einer viel geringeren Dosis der „Droge Arzt" im Sinne von Balint auskommen. Nicht wenige Patienten aber werden erst durch eine längere Teilnahme an einer Semiselbsthilfegruppe so weit im analytischen Sinne therapiefähig, daß sie dann in eine analytisch orientierte Therapiegruppe aufgenommen werden können.

20.8 Schlußbemerkungen

Die Chancen, Möglichkeiten und Grenzen der verschiedenen Gruppentherapien hinsichtlich der Behandlung psychosomatischer Erkrankungen sind derzeit noch nicht endgültig abzustecken. Hier ist die Forschung noch zu sehr im Fluß. Es gibt bisher zu wenig verläßliche, insbesondere statistisch abgesicherte Vergleichsuntersuchungen.

Die allgemeinen Erfahrungen und Eindrücke sind jedoch durchaus positiv. Schon heute kann ohne Übertreibung festgestellt werden, daß die Einführung der verschiedenen Gruppentherapiemethoden für die Behandlung und Betreuung unserer Kranken einen ganz wesentlichen Fortschritt darstellt.

21 Familiendynamik und Familientherapie

Michael Wirsching

21.1 Auf dem Wege zu einem ökologischen Krankheitsverständnis

Die Familiendynamik und die Familientherapie als Wege zum Verständnis und zur Veränderung menschlicher Beziehungen sind ein wesentlicher Teil der ökosystemischen Krankheitstheorie. In der Praxis leisten sie einen Beitrag zu einer ganzheitlichen (holistischen) Medizin, die im Rückgriff auf Familienmedizin, Patientenselbsthilfe und Naturheilverfahren Lösungen einer Krise unserer Gesellschaft, wie sie sich unter anderem als Krise des Gesundheitswesens zeigt, anstrebt. Das verbindende Element solch scheinbar entfernter Gebiete ist eine ökologische Sicht, die zugleich Grundlage weitreichender Veränderungen der naturwissenschaftlichen und psychosozialen Anteile unseres Weltbildes ist. Es geht in diesem Kapitel also ausdrücklich nicht nur um die Einführung veränderter diagnostischer Kategorien und neuer Behandlungstechniken, sondern vielmehr um die konsequente Anwendung der in den Anfangskapiteln ausführlich dargestellten erkenntnistheoretischen (epistemologischen) Überlegungen, wie sie sich aus der allgemeinen Systemtheorie herleiten lassen. Wir werden allerdings auch zeigen, daß manch altbekannte klinische Phänomene in neuem Licht erscheinen und sich manche Alternativen zu den vertrauten „gesicherten" Behandlungswegen ergeben, wenn eine veränderte Sicht zugrunde gelegt wird. Zwar hat Richardson (1948) bereits frühzeitig mit dem programmatischen Satz „Patienten haben Familien" auf die Bedeutung des Patientenumfeldes für die ärztliche Praxis hingewiesen, aber erst in jüngster Zeit sind unter den Titeln „Familienpsychosomatik" (Weakland, 1977), „Familienmedizin" (Huygen und Smits, 1983; Doherty und Baird, 1983), „Family Systems Medicine" (Bloch, 1983) oder „Systems Consultation" (Wynne et al., 1986) intensivere Anstrengungen unternommen worden, familiendynamische bzw. familientherapeutische Aspekte in die Psychosomatik einzuführen.

21.2 Zur Familiendynamik körperlicher Krankheiten

Beim nach wie vor schwachen Entwicklungsstand der familiendynamischen Theorie und Methode ist es sinnvoll, mit einer phänomenologischen Betrachtung zu beginnen (vgl. L'Abate, 1985; Ransom, 1986). Ausgehend von teilweise ganz unterschiedlichen Konzepten wird mit ganz verschiedenen Begriffen eine Reihe wiederkehrender Merkmale beschrieben, die in einer Vielzahl der Untersuchungen als Charakteristika von Familien mit chronisch körperlich kranken Mitgliedern beschrieben wurden (Übersicht bei Meissner, 1974; Grolnick, 1972; Milman und Todd, 1973; Waring, 1980; Smits, 1981; Campbell, 1986).

21.2.1 Fallbeispiel

Betrachten wir zunächst einen typischen Fall, Familie Kahl: Von der Medizinischen Klinik wurde ein psychosomatisches Konsil bei einer 20jährigen Patientin erbeten, die in den vergangenen 6 Monaten 12 kg an Gewicht verloren hatte (sie wiegt jetzt 43 kg bei einer Körpergröße von 1,69 m, die Periode blieb seit 5 Monaten aus).

Beim ersten Besuch auf der Station liegt die Patientin im Bett, anwesend ist auch die 52jährige Mutter, die den ganzen Tag bei ihrer Tochter verbringt. Beide verhalten sich sehr freundlich. Die Patientin Regina spricht wenig, ihre Mutter berichtet dagegen mit großer Besorgnis, „wie alles angefangen hat": Regina verließ vor einem halben Jahr das Elternhaus, um an einer ca. 90 km entfernten Universität zu studieren. Wenn sie an den Wochenenden heimkam, wirkte sie sehr unruhig und bedrückt. Sie sei mit dem Studium nicht klargekommen und fand keinen Kontakt zu ihren Kommilitonen. Ihr „Leistungsversagen" habe überrascht, da sie bislang eine sehr gute Schülerin war. Frage an die Tochter, wie es der Mutter ergangen sei nach ihrem Auszug? Sie habe sich wohl große Sorgen gemacht und beide hätten sie auch unter dieser ersten Trennung gelitten. Die Mutter sei tagsüber im großen Haus ganz allein. Ihr seit 20 Jahren bestehendes Colitis-ulcerosa-Leiden habe sich erneut verstärkt. Der Vater ist als Industriemanager 12 Stunden täglich außer Haus und verreist fast jeden Monat beruflich. Er leidet seit Jahren an „nervösen Herzbeschwerden" (offenbar einer Herzneurose), die jetzt unter dem Eindruck von Reginas Krankheit zunahmen. Dazu kommen noch schwere Berufsprobleme. Er habe das Gefühl, der täglichen Belastung kaum noch gewachsen zu sein.

Es gelingt einige Tage später, ein Familiengespräch mit allen drei Angehörigen zu vereinbaren. Hier bekommen wir zusätzliche Informationen zur Vorgeschichte: Herr Kahl verlor seinen Vater zu Beginn des 2. Weltkrieges. Bei Mutter und Großmutter sei er sehr behütet aufgewachsen, aber wohl auch in einer bedrückenden, von der Außenwelt abgeschlossenen Atmosphäre. Die beiden Frauen seien vor allem in der Trauer um den gefallenen Mann bzw. Sohn verbunden gewesen. Von dem Jungen wurde erhofft, daß er das Leben seines früh verstorbenen Vaters ersetzt. Als Herr Kahl 20 Jahre alt war, starb die Großmutter. Zwei

Jahre später wurde bei seiner Mutter ein Brustkrebs entdeckt, dem sie nach dreijährigem Leiden erlag. Als junger Mann sei er ganz allein gestanden, und war dann sehr froh, Anschluß an die weitläufig verwandte Familie seiner späteren Frau zu finden, die ihn fast wie ein weiteres Kind aufnahm.

Frau Kahl blieb als ältere von 2 Töchtern auch nach der Eheschließung (1959) stark ihrer Herkunftsfamilie verbunden. 1962 erkrankte Frau Kahl, während sie mit Regina schwanger war, an Colitis ulcerosa. Sie gab deshalb ihren Büroberuf auf, um sich ganz ihrem Kind zu widmen. Sie sorgte aber auch sehr für ihre eigenen Eltern, denn die Mutter litt an einer vorzeitigen Arteriosklerose, wurde immer hinfälliger und vorwirrter. Ganz schwierig wurde die Lage, als Herr Kahl 1966 in eine ca. 150 km entfernte Stadt versetzt wurde. Jedes Wochenende fuhr die Familie zu den Großeltern, damit die Tante, die während der Woche die Hauptlast übernommen hatte, bei ihrer eigenen Familie sein konnte. 1968 starb Frau Kahls Mutter, aber die Sorge um den hinterbliebenen Vater blieb, bis dieser 1980 an einem 2 Jahre dauernden Magenkrebsleiden verstarb.

Die Kahls schildern, wie ihr Familienleben über 20 Jahre durch die Versorgung der Großeltern geprägt war. Sie hätten kaum ein eigenes Leben entwickeln können, fanden zum Beispiel kaum Kontakte an ihrem neuen Wohnort. Sie fuhren niemals in Urlaub und hätten alle Kraft zur Erfüllung ihrer schwierigen Versorgungsaufgaben gebraucht. Auf der anderen Seite seien sie so auch einander sehr stark verbunden geblieben und hätten ein sehr harmonisches inniges Familienleben geführt. Zeit und Energie für andere Lebenserfahrungen fehlten allerdings.

Es fällt den Kahls in diesem ersten Gespräch schwer, einen Zusammenhang zwischen Reginas Erkrankung und der Situation der Familie zu sehen. Sie nehmen viel eher körperliche Ursachen an, z.B. die Umstellung der Ernährung durch den Auszug von zu Hause. Dennoch wird gegen Ende der ersten Sitzung allen Beteiligten deutlich, daß die Veränderungen der Familie, die Reginas Studienbeginn mit sich brachte, schwerer wiegen als ursprünglich angenommen wurde. Wir äußern die Vermutung, daß Regina es einerseits aufgrund ihrer bisherigen Lebenserfahrung schwer hatte, sich in der neuen Situation außerhalb der Familie zurechtzufinden, daß sie aber auch zu Hause, vor allem der Mutter, fehlte. Die Eltern machen sich Sorgen um Regina, die Tochter sorgt sich um ihre Eltern. Die krankheitsbedingte Studienunterbrechung hat diese Konflikte zunächst „neutralisiert", aber um den Preis eines anorektischen Symptoms und sicher nur für eine begrenzte Zeit. Weitergehende Konflikte, wie z.B. auch der aggressiv-ambivalente Anteil einer so starken Familienbindung oder Fragen von Reginas sexueller Entwicklung, können anfangs noch nicht besprochen werden. Aggression und Sexualität erscheinen zunächst als fehlende Dimensionen. Dennoch sind alle Beteiligten an einer Fortsetzung der Familiengespräche interessiert, zumal auch der Hausarzt und die Medizinische Klinik dringend zu einer Familienbehandlung raten.

21.2.2 Merkmale krankheitsanfälliger Familien

Schauen wir nun, wieweit in dieser Fallskizze bereits Elemente auftauchen, die auch sonst in der Literatur oder in unserer eigenen klinischen Arbeit als Charakteristika von Familien mit chronischen körperlichen Krankheiten angesehen werden.

Eingeengtheit

Ausgehend vom charakteristischen Kommunikationsstil psychosomatisch Kranker wurde auch für die Familie als Ganzes ein Muster qualitativ und quantitativ eingeschränkter Kommunikation beschrieben (Fleck, 1976). Es wird in diesen Familien weniger gesprochen und das wenige erscheint zwar logisch und klar, jedoch gefühlsärmer und mehr sach- als personenbezogen.

Titchener und Mitarbeiter sprechen von einem gespannten Zusammenhalt (tight cohesion), um eine Art von Familienleben zu beschreiben, „in dem die Individuen um den hohen Preis versteckter Angst nahezu verzweifelt ihren Zusammenhalt als Gruppe aufrechterhalten" (1974, S. 240).

Im gemeinsamen Rorschach- bzw. Zulliger-Test, bei dem sich die Familie auf eine Bezeichnung für das jeweilige Klecksbild einigen soll, fand sich bei Familien mit einem psychosomatisch kranken Kind das höchste Maß an spontaner Übereinstimmung, allerdings wieder nur auf dem Niveau sehr undifferenzierter vager Beschreibungen („irgendein Tier ...", eine Wolke ..., irgendeine Blüte" etc.), es handelt sich um Scheinlösungen durch Verallgemeinerungen.

Jackson und Yalom (1974) wählten in ihrer wegweisenden Arbeit über Familien mit kolitiskranken Kindern das Bild der eingeengten Familie (restricted family), um diese Situation zu beschreiben. Alle Mitglieder schienen darauf bedacht, sich gegen unvorhergesehene Ereignisse abzusichern und an einem bestimmten Verhaltensmuster festzuhalten, wie immer sich auch die äußere Situation ändern mochte. Die Kommunikation innerhalb der Familie schien leicht vorhersagbaren Regeln zu folgen, die von allen Mitgliedern strikt eingehalten wurden. Es läßt sich zum Beispiel gut vorhersagen, wer nach wem zu wem spricht (Haley, 1964).

Die sozialen Kontakte der Familie sind stark eingeengt, bei intensiven Bindungen an die erweiterte eigene Familie. Diese Eingeengtheit wird meist über Generationen weitergegeben (Prugh, 1963; Finch und Hess, 1962). Minuchin und Mitarbeiter (1982) hatten infolge solcher Familienstarrheit auf eingeschränkte Entwicklungsmöglichkeiten hingewiesen: Schwierigkeiten sind unvermeidbar, wenn Veränderungen von außen aufgezwungen werden (vgl. auch Olson et al., 1983). Die Adoleszenz, während der sich vielfältige Reifungs- und Ablösungsaufgaben stellen, wird nicht nur für die Jugendlichen selbst, sondern für die Familie als Ganzes zur Krisenperiode, in der es oft zu psychosomatischen Symptombildungen kommt (Zauner, 1978). Einmal erworbene Verhaltens- oder Beziehungsmuster werden beibehalten, auch wenn sich die Umstände verändert haben. Solch unzulängliche Versuche, das Familiengleichgewicht zu bewahren, haben dauerhafte Spannungen und Angst vor neuen Erfahrungen zur Folge. Gleiches zeigt auch die wiederholte Untersuchung von 45 Familien in den ersten beiden Jahren einer Bronchialkrebserkrankung (Wirsching, 1988). Die homöostatische Wirkung der einengenden Starrheit läßt sich nur um den Preis reduzierter Entwicklungs- und Konfliktverarbeitungsmöglichkeiten erzielen.

Konfliktvermeidende Harmonisierung

Von Minuchin und Mitarbeitern (1975) wurden frühzeitig die Schwierigkeiten psychosomatischer Familien bei der Konfliktlösung betont. Eine besonders unverbindliche Form der Kommunikation folgt dem Ziel, Stellungnahmen und Aussagen jederzeit abzuändern, zurückzunehmen oder in ihr Gegenteil zu verkehren, wann immer Konflikte drohen (Jackson und Yalom, 1974). Das affektive Klima ist gedämpft. Stärkere Gefühlsäußerungen, insbesondere von Ärger oder Wut, werden vermieden. Dabei ist die reale Situation der Familie keinesfalls harmonisch. Tiefgreifende existentielle Konflikte sind allen Beteiligten ständig bewußt. Jedoch scheint eine stillschweigende Übereinkunft darüber zu bestehen, diese Konflikte mit den anderen Familienmitgliedern oder gar mit Außenstehenden nicht zu besprechen (Jackson und Yalom, 1974). Ein weiteres Element der Konfliktvermeidung ist, daß offene Äußerungen über Beziehungen der Familienmitglieder vermieden werden. Offene Allianzen sind undenkbar, niemand erklärt sich offen zu einer Führungsrolle, alle streben das Ideal einer Generationen-, Geschlechter- oder Altersunterschiede negierenden Gleichheit an (Selvini-Palazzoli, 1976). Schließlich kommt dem jeweiligen „Patienten" der Familie noch eine wichtige, konfliktvermeidende Funktion zu, sei es, daß Kinder eine neutralisierende Vermittlerposition zwischen konfliktbelasteten Eltern einnehmen (Minuchin nennt dies „Triangulierung"), oder daß im Zuge neuer Erkrankungen bzw. Krankheitsrückfälle Ruhe eintritt und die Familie noch enger zusammenrückt (Minuchin et al., 1982).

In ihrem Verhältnis zum Umfeld strebt die Familie größtmögliche Anpassung an und erreicht so oft den Zustand einer überangepaßten Pseudonormalität. Meist vermittelt über Generationen ein strenger religiöser oder ethischer Kodex die Leitlinien der Konfliktvermeidung. Am Ende steht aber, daß nichts mehr offen besprochen werden kann. Vielfältige ungelöste Probleme wirken in einer Weise bedrohlich, die in keinem Verhältnis mehr zum ursprünglichen Anlaß steht. Die Familie hat sich in einem Teufelskreis gefangen, wo alltägliche, jederzeit zu erwartende unvermeidbare Aufgaben und Konflikte zur existenzbedrohenden zusätzlichen Belastung werden. Die unweigerlich in Krisenzeiten auftretenden körperlichen Krankheiten haben einen verzweifelten Rückgriff auf die immer gleichen untauglichen Bewältigungsversuche zur Folge und führen zur weiteren Einengung und Belastung. Dies belegt auch die oben erwähnte Studie der Bronchialkrebs-Familienanamnese (Wirsching, 1988).

Verschmelzende Bindung im Inneren, Isolation nach außen

Das starke Überwiegen zentripetaler, die Familie zusammenhaltender Kräfte und die Aufhebung bzw. Verwischung der Grenzen in der Familie, also ein Höchstmaß an Integration bei einem sehr niedrigen Grad an psychologischer Differenzierung (Wirsching und Stierlin, 1982), zählt zu den Hauptmerkmalen psychosomatischer Familien.

Minuchin und Mitarbeiter (1982) sprechen von einer Verfilzung: Jeder ist in hohem Maße empfänglich für die Probleme der anderen, jeder scheint in die Angelegenheiten der anderen verwickelt zu sein. Die Folge ist eine wechselseitige Abhängigkeit, individuelle Abgrenzungen werden ständig überschritten. Die Fähigkeit, zwischen eigenen und fremden Gefühlen, Gedanken und Wahrnehmungen zu unterscheiden, ist gering. Die Grenzen sind in diesen Familien so verwischt, daß kaum Eigenständigkeit herrscht. Jeder mischt sich in die Angelegenheiten des anderen ein und erlebt selbst, daß die anderen sich einmischen. Die Generationengrenzen sind weitgehend aufgehoben, die Kinder werden als Vermittler in eine Elternposition gedrängt, die sie überfordert (Parentifizierung). Solche Fusion erstreckt sich aber, wie unsere eigenen Untersuchungen zeigten (Wirsching und Stierlin, 1982), nur in den seltensten Fällen auf das gesamte Familiensystem. Viel häufiger ist die Verschmelzung zwischen Müttern und Kindern bei tiefer Entfremdung (Isolation) der Ehepartner. Immer aber fehlt ein ausgewogenes Verhältnis von Nähe und Distanz, die Fähigkeit zum Dialog. In der Mehrzahl der Fälle hatte bereits zwischen den Eltern und ihren eigenen Eltern eine Differenzierung und Ablösung nicht stattgefunden. Die Verschmelzung erstreckt sich dann über mindestens drei Generationen, läßt aber den jeweiligen Partner aus.

Noch ausgeprägter wird solche Mehrgenerationendynamik, wenn wir die Art der jeweiligen Bindungskräfte betrachten. Vor allem Bindungen auf der Über-ich-Ebene erweisen sich als Klammer, die die Familie über Generationen zusammenhält. Ein starres, rigides Familiengewissen wacht über die Einhaltung oft anachronistisch wirkender Ideale.

Charakteristisch sind weiterhin Bindungsprozesse auf der Es-Ebene, die von Minuchin und Mitarbeitern (1982) als „Overprotectiveness" beschrieben wurden: Die Familienmitglieder sind demnach auch über die Krankheit des jeweiligen Patienten hinaus stark um das Wohlergehen des anderen besorgt. Sie sehen einander häufig nur unter dem Aspekt der Sorge, die man sich aus verschiedenen Gründen umeinander machen muß. Selbst die Kinder machen sich Gedanken darüber, wie sich die Familie vor der feindlichen Umwelt schützen läßt. Verständlicher wird solche Überbesorgtheit wieder unter einer Mehrgenerationenperspektive (Sperling et al., 1982), denn die Eltern schildern ihre eigene Herkunftsfamilie meist als kalt, versagend und fordernd. Die Familie versucht heute, das, was die Eltern selbst einmal vermißten, einander zu geben.

Schließlich lassen sich auf der Ich-Ebene Bindungsprozesse beschreiben, die mit dem Austausch von Wahrnehmungen, Gedanken oder Gefühlen zusammenhängen. Im Extremfall gilt hier die von allen Familienmitgliedern geteilte Regel: Wir sehen die Welt durch die gleichen Augen. Wir denken, fühlen und empfinden alle das gleiche. Da solch ein Gedankenlesen nur selten gelingt, kommt es zu einer Vielzahl von Wahrnehmungsverzerrungen – projektiven und mystifizierenden Zuschreibungen. Keiner spricht für sich selbst. Alle versuchen, sich am Umfeld zu orientieren, intellektuelle Unabhängigkeit bleibt verwehrt.

21.2.3 Zusammenfassende familiendynamische Hypothese

Fassen wir die vorangegangenen Einzelbefunde zusammen, so läßt sich die folgende Hypothese zur Entstehung und Wirkung psychosomatischer Krankheiten in Familien formulieren:

In einem System, das seine Stabilität aus der rigiden Einhaltung vorgegebener Regeln, dem engen Zusammenschluß seiner Mitglieder und einer grenzenverwischenden Verschmelzung zu gewinnen sucht, kommt es schnell zu einer Häufung vielfältiger ungelöster Fragen und Konflikte, zu einer Ausbreitung familiärer Tabuzonen und einer Verkrustung der Entwicklungs-Anpassungsmöglichkeiten des einzelnen wie der Familie als Ganzes.

Werden Veränderungen aufgezwungen, etwa wenn Kinder im adoleszenten Reifungs-Ablösungsprozeß voranschreiten, oder wenn gar ein Mitglied der Familie stirbt (Perinelli und Günther, 1983), so liegt es nahe zu versuchen, durch eine Verstärkung der vertrauten Abwehrmechanismen das seelische oder körperliche Überleben der Familie zu retten.

Verstärkte Konfliktvermeidung, engeres Zusammenrücken, noch rigideres Festhalten am Status quo versprechen zwar eine kurzfristige Erleichterung, machen aber auf mittlere Sicht die Beteiligten noch verwundbarer. Die nächste Krankheitskrise scheint vorprogrammiert. Alltägliche unvermeidbare Ereignisse werden zu unerträglichen Krisenauslösern.

Die Problemlösungsversuche sind in diesem Teufelskreis selbst zum Hauptproblem der Familie geworden (Wirsching, 1988).

21.2.4 Gibt es „die" psychosomatische Familie – Fragen der Spezifität

Auch in der Familienpsychosomatik wird diskutiert, wie spezifisch die beschriebenen Störungsformen sind: Gibt es eine bestimmte psychosomatogene Familienstruktur, unterscheiden sich Familien mit verschiedenen Krankheiten voneinander, sind Untergruppen identifizierbar etc. (Loader et al., 1980).

Gegen eine Spezifitätsannahme sprechen vor allem zwei Gründe: Zum einen erscheint es falsch, Kausalbeziehungen herzustellen zwischen Prozessen, die sich in ganz verschiedenen Subsystemen abspielen, also z.B. zwischen einer bestimmten Familieninteraktionsform und einer neurodermitischen Läsion der Haut. Zum zweiten erscheint die Etikettierung einer Familie aufgrund klinischer Diagnosen einzelner Familienmitglieder immer etwas willkürlich, denn je weiter wir den Familienkreis ziehen, um so mehr werden wir fast immer eine ganze Reihe verschiedener körperlicher oder seelischer Störungen antreffen. Unser eingangs erwähnter Fall könnte z.B. je nach Standort des Untersuchers als anorektische, kolitische oder herzneurotische Familie angesehen werden.

Die vorliegenden Befunde wurden von einer größeren Zahl von Untersuchern mit ganz verschiedenen Methoden bei einer Vielzahl verschiedener Krankheitsbilder erhoben, zum Teil auch im Vergleich zu anderen neurotischen oder psychischen Störungen (Campbell, 1986; Liedtke, 1987; Overbeck, 1985). Das Spektrum umfaßt die Anorexia nervosa (Selvini-Palazzoli, 1982; Rosman et al., 1976; Minuchin et al., 1982; Buddeberg und Buddeberg, 1979; Petzold, 1979; Becker, 1980), die atopischen Leiden Asthma und Neurodermitis (Block, 1969; Pinkerton, 1970; Liebmann et al., 1974; Minuchin et al., 1982; Lask und Kirk, 1979; Overbeck und Overbeck, 1978; Bovensiepen et al., 1980; Haland-Wirth und Wirth, 1981; Wirsching, 1984; Liedtke, 1987), Colitis ulcerosa und Morbus Crohn (Finch und Hess, 1962; Jackson und Yalom, 1974; McMahon et al., 1973; Liedtke, 1987), juveniler Diabetes (Minuchin et al., 1982), Krebs (Sheldon et al., 1970; Thomas et al., 1974; Louhivuori et al., 1976; Bahnson, 1978; Wirsching et al., 1988; Wirsching und Stierlin, 1982; Stierlin et al., 1983; Buddeberg, 1985; Möhring, 1988), Tod eines Familienmitgliedes (Pattison, 1976; Baider, 1977; Hare-Mustin, 1979; Solomon und Hersch, 1979; Uhlenberg, 1980; Shanfield, 1983), Hämodialyse (Engel, 1978; Levenberg et al., 1978), Rheuma (Rekola, 1973), chronische Schmerzen (Roy, 1982) und Appendektomie (Hilpert, 1980). Danach läßt sich als gesichert annehmen, daß die beschriebenen Merkmale, wie sie unter den Oberbegriffen rigider Entwicklungsstillstand, harmonisierende Konfliktvermeidung und verschmelzende Bindung zusammengefaßt wurden, mit einer erhöhten Anfälligkeit für körperliche Krankheiten einhergehen (Wirsching et al., 1988), wie es z.B. die herabgesetzte Lymphozytenzahl nach dem Verlust eines Ehepartners vermuten läßt (Bartrop et al., 1977; D.D. Schmidt, 1983).

Darüber hinaus ist aber auch festzustellen, daß die Gesamtgruppe in bezug auf das vorherrschende Beziehungsmuster keineswegs homogen ist. So fanden wir bei einer Typenanalyse von 55 unausgelesenen Familien atopisch bzw. gastroenterologisch kranker Jugendlicher drei teilweise sehr unterschiedliche Grundmuster (Wirsching und Stierlin, 1982), die auch innerhalb der Krankheitsgruppen verschieden häufig vorkamen: Der in der Literatur besonders hervorgehobene Typus einer **gebundenen Familie** war zwar mit insgesamt 44% aller Fälle am häufigsten, kam aber vor allem in der Gruppe der gastroenterologisch kranken Jugendlichen vor, wo fast zwei Drittel der Fälle diesem Muster zugeordnet wurden. Es handelte sich dabei um fast geschlossene Systeme, die nichts und niemand hinein- oder herauslassen. Das Familienklima ist bestimmt von depressiven Affekten der Trauer um nicht bewältigte Verluste. Diese Konstellation war auch in den Familien Lungenkrebskranker am häufigsten (Wirsching, 1988).

In der atopischen Gruppe fanden wir hingegen in fast der Hälfte aller Fälle den Typus einer **gespaltenen Familie,** wo die verschmelzende Beziehung auf zwei voneinander isolierte oder gar verfeindete Lager beschränkt war. Das Familienklima war dann eher durch aggressive Auseinandersetzungen von beinahe mörderischem Ausmaß geprägt.

Der dritte Typus, der immerhin ein Viertel bis ein Drittel aller Fälle einschließt, wurde bislang in der Literatur kaum berücksichtigt, vielleicht weil die meisten Arbeiten von ausgelesenen, psychotherapeutisch zugänglichen Familien ausgingen. Zum Familientherapeuten kommt diese Gruppe, die wir **Familien in Auflösung** nannten, aber meist nicht. Die jugendlichen Patienten erschienen hier emotional vernachlässigt, in vorzeitige Selbständigkeit getrieben und erkrankten mit schwersten, oft lebensbedrohenden Verläufen. Eine charakteristische Situation ist etwa: Ein Elternteil steht in abhängiger Bindung an die Herkunftsfamilie und sucht die Befreiung durch einen pseudoselbständigen ausgestoßenen Partner, der seinerseits in eine „echte" Familie aufgenommen zu werden hofft. Solche widersprüchlichen Bestrebungen sind schnell zum Scheitern verurteilt und münden meist in einer langen Kette ähnlicher Wiederholungen. Hier fanden wir die größte Zahl zerbrochener unvollständiger Familien mit Kindern, die verschiedenen Verbindungsversuchen entstammten. Hier gab es auch eine besondere Häufung psychosomatischer Symptome neben vielfältigen anderen, darunter psychotischen oder delinquenten Störungen.

21.3 Familientherapie bei schweren und chronischen körperlichen Krankheiten

21.3.1 Indikationskriterien zur Einbeziehung des Patientenumfeldes

Gegen das diagnostische oder informatorische Familiengespräch gibt es im Grunde keine Einwände. Vor allem in der Anfangsphase einer Behandlung oder bei akuten Krankheitskrisen kann der behandelnde Arzt so die Situation des Patienten besser verstehen und auch sicherstellen, daß seine Mitteilungen alle Beteiligten in gleicher Weise erreichen. Es hat sich als günstig erwiesen, das Familiengespräch so früh wie möglich zu führen und eher später den Gesprächskreis einzugrenzen, als nach einer Reihe von Einzelgesprächen zu versuchen, das Umfeld hinzuzuziehen. Als besonders sinnvoll hat sich diese Reihenfolge auch im psychosomatischen Konsiliardienst erwiesen (Wirsching, 1983 a, b; McDaniel et al., 1980; Sluzki, 1986; Wellisch und Cohen, 1986; Munson, 1986). Mehr

spezifische Indikationen ergeben sich im Lichte der oben dargestellten verschiedenen Familientypen:

Bei gebundenen und verschmolzenen Familien hat der Versuch, ein Mitglied durch eine ablösungsbetonte Einzeltherapie herauszulösen, fast immer eine Krise des Gesamtsystems zur Folge, wenn nicht die Loyalität der Familie ohnehin überwiegt und der Behandlungsprozeß frühzeitig abgebrochen wird.

Eine etwas andere Indikationsstellung ergibt sich beim Spaltungstyp. Hier können sich in zwei Richtungen Probleme bei der Einzelbehandlung stellen: Begibt sich nur einer der Ehepartner zum Therapeuten, so steigt die Gefahr, daß die Familie ohne Chance der Bearbeitung ihres neurotischen Zusammenspiels (Kollusion) zerbricht. Ist eines der Kinder krank, so wird im Zuge der ablösungsfördernden Wirkung der Einzeltherapie auch dessen Einbindung in den Elternkonflikt aufgelöst, und eine Familienkrise droht.

Bei Familien in Auflösung sind gemeinsame Familiengespräche meist nicht realisierbar und auch nicht angezeigt. Hier muß als erster Schritt der Behandlung zunächst die Vertrauensbeziehung zu dem ausgestoßenen, vernachlässigten psychosomatisch Kranken gefestigt werden. Wird dieser Prozeß durch ein vorzeitiges Hereinzerren der Restfamilie gestört, so kommt meist überhaupt keine Behandlung zustande. Erst in späteren Phasen gelingt unter Umständen die Rekonstruktion eines Bezugssystems, nachdem die Beziehungs- und Vertrauensfähigkeit des Patienten sich im geschützten Einzelkontakt entwickelt hat.

Weitere Situationen, die gegen ein familientherapeutisches Vorgehen sprechen, sind solche, bei denen die Familie nicht das entscheidende Bezugssystem ist. Dies gilt z. B. für viele psychosomatische Konsiliarsituationen, wo im Mittelpunkt die Interaktionskrise der Station steht und somit das Teamgespräch Vorrang hat. Auch gibt es viele Situationen im Randgruppenbereich, wo außerfamiliäre Institutionen Aufgaben und Verantwortungen übernommen haben, die sonst der Familie vorbehalten sind. Hier wäre eine Ausweitung des Beziehungskontextes im Sinne einer Netzwerkintervention zu erwägen (Speck et al., 1976).

Schließlich sei noch eine Kontraindikation erwähnt, die im klinischen Alltag wohl am häufigsten zum Tragen kommt: Das Fehlen von erfahrenen Familientherapeuten und von Supervisionsmöglichkeiten. Die Behandlung von Familien mit schweren und chronischen psychosomatischen Symptomen gehört zu den schwierigsten und gefährlichsten Feldern der Familientherapie, vergleichbar der Arbeit mit psychotischen Familien, die nicht nur den erfahrenen einzelnen, sondern unter Umständen ein eingespieltes Behandlungsteam erfordert.

21.3.2 Verschiedene Behandlungskonzepte bedingen unterschiedliche Wege bei der Behandlung psychosomatischer Familien

Familientherapie steht oft nur als Sammelbegriff für ganz verschiedene Behandlungskonzepte und -methoden. Drei „Schulen" haben ausführliche Angaben zur Behandlung bei schweren chronischen körperlichen Krankheiten gemacht und haben teilweise sehr unterschiedliche Indikationsbereiche erschlossen. In der Reihenfolge ihrer historischen Entwicklung sollen hier die psychoanalytische, die strukturelle und die systemische Familientherapie vorgestellt werden (Übersicht bei Gurman und Kniskern, 1981; Simon und Stierlin, 1984).

Psychoanalytisch orientierte Familientherapie

Ausgehend vom psychoanalytischen Paradigma soll hier die Aufdeckung und Durcharbeitung langfristig angelegter Familienkonflikte zu einer schrittweisen und kontinuierlichen Veränderung der Familienbeziehungen führen (Richter, 1963, 1970; Bauriedl, 1980). Vor allem dem Bewußtsein der Betroffenen verborgene Komplexe bedingen, wenn sie auf andere Familienmitglieder übertragen werden, eine Vielzahl von Einstellungs- und Wahrnehmungsverzerrungen und Beziehungsdeformationen. Das analytische Konzept ist stark historisch orientiert. Die Aufarbeitung der Familiengeschichte und der Beziehungen über die Generationen steht meist im Mittelpunkt. Die Methodik wird von der familientherapeutischen Grundregel bestimmt: „Versuchen Sie soweit als möglich über die Themen zu sprechen, die Sie bisher vermieden haben, zum Beispiel Familiengeheimnisse, Enttäuschungen, Ungerechtigkeiten, Vertrauensbrüche" (Boszormenyi-Nagy und Spark, 1981). Der Therapeut wirkt als Mittler in einem konstruktiven Familiendialog. Die Deutung von Übertragungen und Widerständen ist das Hauptmittel, um einen Bewußtwerdungs- und Entwicklungsprozeß zur Entfaltung zu bringen. Oft werden auch mehrere Generationen in die Gespräche einbezogen (Sperling et al., 1982).

Aufgrund eigener langjähriger klinischer Erfahrungen hat sich gezeigt, daß die psychoanalytische Familienbehandlungsmethode vor allem in Anfangs- und Krisensituationen modifiziert werden muß (ähnlich wie die psychoanalytische Einzeltherapie), um in allen Fällen schwerer und chronischer körperlicher Störungen realisierbar und wirksam zu sein.

Die strukturelle Familientherapie

Sie ist vor allem in den USA am weitesten verbreitet. Ursprünglich aus der Beratungsarbeit im Randgruppenbereich entstanden, wurde sie durch die Aktivitäten von S. Minuchin und seiner Mitarbeiter an der Philadelphia Child Guidance Clinic schon bald zu einem der Hauptansätze in der Familienpsychosomatik (Combrinck-Graham, 1974).

Die konzeptuellen Wurzeln entspringen hier der Lerntheorie und der Kommunikationsforschung. In der Praxis wird versucht, das psychosomatische Problem mit irgendeiner Form falschen Verhaltens zu verbinden, das während der Sitzung durch aktiven Einsatz des Therapeuten so lange modifiziert wird, bis eine konstruktive Alternative in Szene gesetzt ist (Minuchin und Fishman, 1981). Beispielhaft demonstriert diese Gruppe ihr Vorgehen im gemeinsamen Familienlunch, bei dem die Familie eines anorektischen Mädchens unter heftigsten Auseinandersetzungen zu einer eindeutigen und klaren Kommunikation angehalten wird. Der Hungerstreik soll so seinen Sinn verlieren (Rosman et al., 1976).

Die Vorteile dieses Vorgehens im psychosomatischen Bereich liegen auf der Hand: Der Therapeut stellt eine starke positive Bindung zu allen Familienmitgliedern her (joining), er gibt klare und konkrete Verhaltensanweisungen und orientiert sich ganz an der Veränderung aktueller Nöte und Konflikte der Familie, ohne den teilweise sehr belastenden Konflikthintergrund zu berühren. Außerdem verschwinden auch chronifizierte Symptome frühzeitig und nachhaltig, womit in vielen Fällen bereits das Behandlungsziel erreicht ist. Als bisher einzige Gruppe haben die strukturellen Familientherapeuten ihre Behandlungsergebnisse katamnestisch überprüft. Bei der Anorexia nervosa, bei steroidabhängigen asthmakranken Jugendlichen und beim stoffwechsellabilen juvenilen Diabetiker wurden durchweg dauerhafte Symptombesserungen erreicht.

Als alleinig angewandte Methode erscheint die strukturelle Familientherapie jedoch zu eingeengt. Elemente der direkten Verhaltensänderung haben Eingang in alle anderen Schulen gefunden, so wie die strukturelle Familientherapie sich selbst auch der nun noch zu besprechenden systemischen Familientherapie angenähert hat (vgl. z.B. Andolfi, 1982).

Systemische Familientherapie[1]

Obwohl die Anfänge dieser Methode bis zu den Ursprüngen der Familientherapie zurückreichen (Watzlawick et al., 1969, 1974), hat sie sich erst in den letzten Jahren sprunghaft verbreitet. Das Schlagwort der paradoxen Intervention hat die Runde gemacht (Selvini-Palazzoli et al., 1977; Steinglass, 1978; Cade, 1980; Stanton, 1981; Madanes, 1981; Hoffmann, 1982).

Im Mittelpunkt steht hier ganz stark das systemtheoretische Paradigma. Eine Reihe behandlungstechnischer Besonderheiten, die diesem Vorgehen seine spezifische Form verliehen haben, ermöglichen einen besonders konstruktiven Umgang mit dem homöostatisch wirkenden Familienwiderstand (Selvini-Palazzoli et al., 1981): Anstatt einen Dialog innerhalb der Familie oder zwischen den Familienmitgliedern und den Therapeuten anzuregen, befragen die Therapeuten die einzelnen Mitglieder über die jeweils anwesenden anderen (zirkuläre Befragung). Anstelle von empathischer Allparteilichkeit oder „joining" streben sie eine strikte Neutralität an. Jegliche Stellungnahme, Wertung oder Interpretation wird vermieden. Seine Richtung gewinnt das Gespräch aufgrund be-

---

[1] Synonym wird der vor allem auch im angelsächsischen Sprachraum verbreitete Begriff der „strategischen Familientherapie" gebraucht.

stimmter familiendynamischer Hypothesen, die ständig, beginnend noch vor dem Erstgespräch, erweitert, modifiziert oder bestätigt werden, bis ein verläßliches Bild des zugrundeliegenden Familienkonfliktes gewonnen ist. Die jeweiligen Hypothesen sind auch Grundlage einer Schlußintervention, die meist in einer Pause im Gespräch zwischen Therapeuten und Co-Therapeuten, eventuell auch mit einem Beobachtungsteam, entwickelt wird. Von den oft paradox anmutenden „Verschreibungen" wird eine zusätzliche verändernde Wirkung erwartet, welche in den langen Intervallen (4 bis 6 Wochen und länger) zwischen den Sitzungen zum Tragen kommen soll. (Ein Beispiel solch paradoxer Verschreibung ist weiter unten wiedergegeben.)

Dieses Vorgehen ist verschiedentlich als zu technisch, manipulativ oder unaufrichtig kritisiert worden (Bauriedl, 1980). Dem ist entgegenzuhalten, daß auch eine paradoxe Intervention nur auf der Grundlage einer positiven, vertrauensvollen Beziehung zwischen der Familie und dem Therapeuten bei einem sehr weitreichenden Verständnis der Gesamtproblematik gelingen kann. Es handelt sich um keinen technischen Trick, sondern um eine einseitige, die regressiven Widerstandsanteile des jeweiligen Ambivalenzkonfliktes betonende Mitteilung des Therapeuten, die es den Familien nunmehr erlaubt, im Schutze des Therapeuten die eigene Situation in Frage zu stellen und somit selbst das Ausmaß und Tempo etwaiger Veränderungen zu bestimmen.

In der klinisch-psychosomatischen Arbeit hat die systemische Familientherapie in den vergangenen Jahren entscheidende Anregungen geliefert, die es möglich machten, auch bei schwersten chronischen Familienstörungen therapeutische Entwicklungen in Gang zu bringen. Die Behandlung psychosomatischer oder psychotischer Familienstörungen erscheint heute ohne die Anregungen der systemischen Familientherapie nicht mehr vorstellbar.

Abschließend sollen einige behandlungsmethodische Grundsätze dargestellt werden, die sich in der täglichen therapeutischen Arbeit herausgebildet haben. Sie sind unabhängig von den jeweiligen Schulbesonderheiten bei jeder Familienintervention im psychosomatischen Störungsbereich von Bedeutung, helfen aber ganz besonders wiederum in den Anfangs- und Krisenphasen der Therapie.

21.3.3 Einige Leitlinien zur Behandlung psychosomatischer Familien

Wir haben gezeigt, wie Familien, die sich mit wiederkehrenden Krankheitskrisen oder chronischen körperlichen Leiden einzelner oder mehrerer Familienmitglieder auseinandersetzen, oft versuchen, durch Konfliktunterdrückung, enges Zusammenrücken und Vermeiden von Veränderungen, ihr Gleichgewicht zu erhalten. Solche Bewältigungsversuche haben jedoch nicht nur eine ausgesprochen gesundheitsbelastende Wirkung, sondern sie stehen gerade auch im Gegensatz zu jeglichem Versuch einer unmittelbar verände-

rungsorientierten konfliktaufdeckenden Therapie. In der Regel begegnet ein Therapeut in seiner Sprechstunde also nur einer sehr kleinen (weniger als 10% umfassenden) ausgewählten Untergruppe psychologisch aufgeschlossener psychosomatischer Familien. Um in der Praxis dennoch in größerem Umfang sinnvolle familientherapeutische Hilfe geben zu können, haben sich die folgenden Gesichtspunkte bewährt (Stierlin et al., 1980; Wirsching, 1988).

Behandlungskontext

Entscheidend ist der Kontext, in dem die Behandlung stattfindet. Nur bei größtmöglicher Integration in den medizinischen Behandlungsablauf hat eine Familientherapie überhaupt Aussichten, zustande zu kommen. Das heißt, die Familiengespräche finden über weite Strecken vor Ort im ambulanten oder stationären medizinischen Rahmen statt (Kellner, 1963; Wellish et al., 1978; Eastman und Mesibov, 1981). Die Familien erleben, daß die Behandlung von allen Beteiligten mitgetragen wird, vor allem vom Stationspersonal und von den Hausärzten. Ein Austausch durch konsiliarische Kontakte ist oft unerläßlich (Wirsching, 1983).

Erstinformationen

Die ersten Informationen werden so gewonnen, daß eine Vertrauensbeziehung zu allen Beteiligten geschaffen wird. Eine etwas aktivere strukturierende Gesprächsform hat sich in der Anfangsphase bewährt. Wir versuchen, Verständnis für die Sicht der einzelnen Familienmitglieder zu entwickeln und zu zeigen. Etwaige Widersprüche werden zunächst nicht interpretiert. Damit folgen wir stärker dem von Boszormenyi-Nagy und Spark (1981) eingeführten Prinzip der Allparteilichkeit als etwa Minuchins bewußt kriseninduzierender Gesprächstechnik. Auch Selvinis (1981) Vorschlag, eine strikte Neutralität zu wahren, schien uns am Anfang oft nicht realisierbar.

Konstruktive Anteile der Familiendynamik (Familienressourcen)

Die Schwierigkeiten und tiefgreifenden Konflikte einer psychosomatischen Familie werden gerade im Lichte beschwichtigender Harmonisierungsanstrengungen sehr bald deutlich. Eine pathologie- oder störungszentrierte Gesprächsführung würde die brüchige Familienabwehr unzulässig gefährden. Je stärker das Familiengleichgewicht bedroht wird, um so mehr wird ein Gespräch als zusätzliche Belastung erlebt, gegen die sich die Familie zur Wehr zu setzen versteht und möglicherweise zu Recht von weiteren Kontakten absieht, wenn man bedenkt, daß etwa im Fall der schweren Kolitis oder des chronischen Bronchialasthmas die nächste Krise einen neuen, unter Umständen verhängnisvollen Krankheitsschub bedingen kann. Wir sind also gut beraten, wenn wir unseren Blick bewußt auch auf die intakten Familienbereiche richten, z. B. die Fähigkeit bei der Bewältigung vorangegangener Belastungen anerkennen. Auch die Umdeutung (positive Konnotation; Selvini-Palazzoli et al., 1977) scheinbar tief gestörter Einstellungen und Verhaltensweisen gehört hierher. Vielleicht zum ersten Mal erlebt eine überängstliche, besorgte Mutter, daß ihre Rolle nicht kritisiert, sondern tatsächlich verstanden wird. Die Kontrolle der Gegenübertragung des Therapeuten ist jedoch eine unerläßliche Voraussetzung, um nicht in vordergründiger Beschwichtigung steckenzubleiben.

Die Verknüpfung von Krankheit und Konflikt nicht zu früh deuten

Besonders vorsichtig sind wir bei der Deutung der krankheitsfördernden Konfliktdynamik. Da die Zusammenhänge den Beteiligten selbst ohnehin auf schmerzliche Weise bewußt sind, wirkt deren Hervorhebung durch den Therapeuten als zusätzliche, Schuldgefühle steigernde Belastung. Gerade dagegen versucht die Familie sich zu schützen, und gerade solche Schuldabrechnung wird in negativer Weise von einem Familiengespräch erwartet. Wir heben uns generell in der Anfangsphase jede Deutung des Familienkonfliktes für unsere abschließende Gesprächszusammenfassung auf.

Die Chancen des Gesprächsabschlusses nutzen

Wenn das Familiengespräch in der angedeuteten positiv konnotierenden, allparteilichen Weise geführt wurde, kommt der Punkt, wo sich die Beteiligten fragen, was wohl der Therapeut von ihrer Situation hält. Die Erwartungsspannung wird noch gesteigert, wenn das Familiengespräch vor dem endgültigen Abschluß unterbrochen wird. Eine Gesprächspause ist in stärker gestörten Familien sehr hilfreich, da der Therapeut hier regelmäßig besonders stark in die Familienabwehrprozesse (z. B. in eine harmonisierende Verschmelzung) hineingezogen wird. Im Gespräch mit einem bis dahin abwartenden Co-Therapeuten bzw. Beobachter kann bei räumlicher Trennung eine Metaperspektive gewonnen werden, eine Gesamtschau, die das Zusammenwirken aller Beteiligten erhellt. In der Anfangsphase konzentrieren wir uns vor allem auf die symptomfördernden Familienabwehrversuche. Wenn wir abschließend unsere Gesprächseindrücke zusammenfassen, so vermeiden wir jedes direkte Infragestellen der Familie. Statt dessen versuchen wir zu verstehen und zum Ausdruck zu bringen, auf welche Weise die einzelnen Familienmitglieder versucht haben, zur Stabilisierung der Situation beizutragen. Besondere Aufmerksamkeit verdient hier wieder der mit den jeweiligen Symptomen verknüpfte Teil der Familiendynamik.

Im Falle unserer oben dargestellten Familie Kahl, deren Tochter ein magersüchtiges Verhalten zeigte, betonten wir, wie alle Familienangehörigen seit langem versuchten, ein besonders harmonisches Zusammenleben zu verwirklichen und wie gerade die Jugendliche selbst versuchte, alle sonst in diesem Alter anfallenden Konflikte von ihren Eltern fernzuhalten, deren Belastung durch eigene Krankheiten und Sorgen um ihre eigenen Eltern sie seit langem gespürt hatte. Als Regina wegen ihres Studiums nicht mehr im Hause leben konnte, wurde ihre Sorge um die alleingebliebenen Eltern so stark, daß sie sich entschloß, nicht mehr zu essen. Die Familie war gern bereit, sie wieder aufzunehmen, und sie wird bei ihren Eltern bleiben, bis die Zeit für eine Trennung gekommen ist. Als Therapeuten wissen wir, daß solche Trennungen keinesfalls zu stark beschleunigt werden dürfen. Wir wollen deshalb abwarten und die Familie erst in einigen Wochen erneut sehen. Bis dahin empfehlen wir, nichts zu verändern.

Die Wirkung solcher Form von Abwehrdeutung ist, wie sich leicht vorstellen läßt, oft paradox. Es erweist sich als sehr schwer, ein Abwehrmuster, das bislang auf unbewußte Weise seine Wirkung entfaltete, bewußt fortzusetzen.

Die Möglichkeit eines langen Sitzungsintervalls nutzen

Statt tatsächlich „weiterzumachen wie bisher", begann die Familie Kahl, sich mit der einseitigen, überspitzten Mitteilung des Therapeuten auseinanderzusetzen. Damit setzte jedoch sofort auch ein Prozeß der Auseinandersetzung mit einem Teil ihrer eigenen problematischen Familiensituation (die ja vom Therapeuten zurückgespiegelt wurde) ein. Jedoch geschah dies auf eine Weise, bei der die Familie selbst Richtung, Ausmaß und Tempo der Entwicklung bestimmen konnte.

Wir hören in der folgenden Sitzung fünf Wochen später, daß Regina sich unmittelbar im Anschluß an unser Gespräch entschlossen hatte, wieder ein normales Eßverhalten anzunehmen. Die Eltern seien von unserer abschließenden Mitteilung sehr betroffen gewesen. Der Vater wußte nicht, wie sehr seine Frau unter dem Alleinsein gelitten hatte. Jetzt, wo Regina aus dem Haus geht, will er versuchen, eine Veränderung in seinem Arbeitsbereich herbeizuführen. Er hat bereits mit seinen Chefs gesprochen und stieß zu seiner eigenen Überraschung auf Verständnis. Man hatte sich in der Firma ohnehin seit längerem Gedanken gemacht, warum er in seinem Alter und bei seiner Position so viele Aufgaben selbst übernommen hatte, die durchaus an jüngere Kollegen delegiert werden konnten. Herr Kahl möchte versuchen, die nun gewonnene Zeit stärker seiner Familie zu widmen. Die Eheleute hatten darüber gesprochen, daß sie eigentlich in ihrer über zwanzigjährigen Ehe kaum eine Gelegenheit für gemeinsame Unternehmungen hatten.

Die Therapeuten äußern in diesem Gespräch ihre Überraschung über so weitreichende schnelle Veränderungen. Sie warnen sogar davor, eine Situation, die sich in so vielen Jahren entwickelt hat, nun in so kurzer Zeit so stark verändern zu wollen. Es wird vereinbart, daß die Eltern zum nächsten Termin allein kommen, um Gelegenheit zu haben, in Ruhe über ihre Situation weiterzusprechen. Auch Regina bieten wir an, daß sie sich an uns wenden kann, wenn sie über ihre eigenen Fragen mit uns allein sprechen will.

Damit die paradoxe Mitteilung nicht zum zynischen manipulativen Trick wird, was die Familie sofort spüren würde, ist es also wichtig, daß sie sich zum einen auf tatsächlich in der Sitzung bewußt gewordene Zusammenhänge bezieht, oft sogar nur noch etwas zusammenfaßt, was bereits in der Luft lag, und zum zweiten, daß der Therapeut tatsächlich hinter seiner Deutung steht, also das Dilemma der Familie auch emotional erfaßt hat und vor allem akzeptieren kann, wenn die Familie sich tatsächlich entschließt (dann aber bewußt), alles beim alten zu lassen, um gemeinsam auf einen günstigeren Entwicklungszeitpunkt zu warten. Um die volle Wirkung eines so geführten Familiengesprächs nicht zu mindern, eine Familie nicht zu stark zu bedrängen und vor allem auch oft unerwartete Selbstentwicklungsmöglichkeiten in der Familie zur Entfaltung zu bringen, haben sich vor allem in der Anfangsphase längere Sitzungsintervalle (4–6 Wochen) sehr bewährt. Vor allem, wenn die Gespräche in einem medizinischen Kontext stattfinden, ergeben sich so im allgemeinen keine Schwierigkeiten, auch zu sehr schwer gestörten, primär wenig motivierten Familien einen kontinuierlichen und vertrauensvollen Kontakt herzustellen.

Das Gesprächs-Setting flexibel halten

Familientherapie bedeutet keinesfalls, daß von Anfang bis Ende einer Behandlung alle Beteiligten zu gemeinsamen Gesprächen zusammenkommen. Vielmehr muß immer wieder neu entschieden werden, welcher Teil des Ganzen aus beziehungsdynamischer Sicht für die Bearbeitung der jeweiligen Konflikte entscheidend ist. Gerade in stark verschmolzenen Familien mit jugendlichen Patienten kann das gemeinsame Familiengespräch eine individuationshemmende Wirkung entfalten.

Im Fall der Familie Kahl hatten wir den Eindruck gewonnen, daß die Eltern nach der ersten Intervention eine Bereitschaft zeigten, die Tochter aus ihrer gebundenen Position zu entlassen und sich mit ihren

eigenen Konflikten auseinanderzusetzen. Auch Regina signalisiert mit der Aufgabe ihres Eßsymptoms ihren Entschluß, die zentrale, vermittelnde, alle Aufmerksamkeit absorbierende Stellung in der Familie zu verlassen. Sie wird sich später einer studentischen Selbsthilfegruppe anschließen. Wir sehen sie nur noch einmal vor dem endgültigen Abschluß der Therapie. Ihr ist es gelungen, von Symptomen frei zu bleiben und eine altersgerechte Selbständigkeitsentwicklung aufzunehmen. Sie hat jetzt einen festen Freund, den sie beim Studium kennenlernte.

Die Eltern durchlaufen einen intensiven paartherapeutischen Prozeß, in dem sie eine deutliche Wiederannäherung erfahren. Das ganze Ausmaß ihrer depressiven Existenz und einer inhaltlosen, freudlosen Beziehung, das von der Tochter offenbar schon immer wahrgenommen worden war, zeigte sich erst, nachdem die harmonisierende, konfliktvermeidende Abwehr schrittweise aufgegeben wurde. Als ganz zentral erwies sich hier die Bearbeitung der Verluste, die beide Eltern erlitten hatten (Perinelli und Günther, 1983). Im Zuge dieser Trauerarbeit wurde deutlich, wie beide Eltern versucht hatten, in Regina einen Ersatz für den nicht überwundenen Verlust der eigenen Eltern zu finden. Die krankmachende Wirkung der ungeleisteten Trauer war die seelische Grundlage von Frau Kahls Colitis-ulcerosa-Leiden. Sie fühlte sich bis heute schuldig, weil sie sich nicht genug um die alten und kranken Eltern gekümmert hätte. Erst jetzt wagt auch Herr Kahl seiner Frau mitzuteilen, wie enttäuscht er gleich zu Anfang der Ehe war, weil seine Frau den eigenen Eltern zuliebe ihn, den Ehemann vernachlässigte. Frau Kahl bestätigt, daß es ihr eigentlich immer klar gewesen sei, daß der scheinbar unvermeidbare berufsbedingte Umzug ein Versuch des Partners war, der ständigen Präsenz der Schwiegereltern zu entrinnen. Auf unausgesprochene Weise hatte zwischen den beiden Partnern über die vielen Jahre ein tiefgreifender Konflikt bestanden mit intensiven, gleichwohl verborgenen Gefühlen der Enttäuschung und des wechselseitigen Schuldvorwurfs.

In der Familientherapie gelingt es, Verständnis für die jeweiligen Positionen zu entwickeln. Die Partner erleben erstmals eine weitgehend von Symptomen befreite entspannte Partnerschaft. Sie nehmen auch die jahrelang unterbrochenen sexuellen Beziehungen wieder auf.

22 Suggestive und übende Verfahren

Reinhard Lohmann

„Die alten Komplexe sind wie große Steine im Flußbett, die bei tiefem Wasserstand störend über die Oberfläche kommen. Steigt der Pegelstand, so liegen dieselben Steine bedeutungslos auf dem Grund und die Schiffe fahren ruhig oben weg."
(E. Kretschmer, „Psychotherapeutische Studien", 1949)

22.1 Suggestive Verfahren

22.1.1 Vorbemerkungen zum Suggestionsbegriff[1]

Unter **Suggestion** verstehen wir „die Beeinflussung des Denkens, Fühlens, Wollens oder Handelns eines anderen Menschen unter Umgehung seiner rationalen Persönlichkeitsanteile auf der Grundlage eines zwischenmenschlichen Grundvollzugs, der zur affektiven Resonanz führt" (Stokvis und Pflanz, 1961). Die Suggestion spielt im Leben des einzelnen wie ganzer Völker eine große Rolle, die sich sowohl normalpsychologisch als auch psychopathologisch äußert. Man denke nur an ihre Bedeutung im Bereich der Religion, der Politik, der Pädagogik, der Reklame und der Mode, in der Geschichte wie im Zeitalter der modernen Massenmedien. Wir unterscheiden dabei eine „negative Suggestion" (= beunruhigende bzw. ängstigende Suggestion) von einer „positiven Suggestion" (= beruhigende bzw. ermutigende Suggestion). Für die Psychotherapie ist die „positive Suggestion" von Belang, welche die affektive Beziehung zwischen Arzt und Patient mitkonstelliert, u.a. in Form von Beruhigung, Ermutigung, Hoffnung, Tröstung, Vertrauen und Glauben. Weiter wird unterschieden zwischen einer „getarnten" oder „larvierten", meist indirekten und einer „gezielten", meist direkten Suggestion. Mit der larvierten Suggestion arbeiten viele naturheilkundliche und homöopathische Ärzte sowie nichtärztliche Heilpraktiker. Ihr Glaube an die Wirksamkeit der von ihnen angewendeten Heilmittel und/oder ihre Überzeugungskraft übertragen sich auf die Patienten und bewirken bei diesen mitunter verblüffende Anfangs- oder Teilerfolge. Aber auch im Bereich der Schulmedizin geschieht vieles „unter dem Schein körperlicher Therapie", was zur larvierten Suggestion zu rechnen ist. Das betrifft u.a. die Medikamentenwirkung, bei der ein „Placebo-Effekt" mit im Spiele sein kann, auch wenn es sich um eine pharmakodynamisch wirksame Substanz handelt. In diesem Zusammenhang muß auf das bahnbrechende Werk von M. Balint (1957) hingewiesen werden, das uns die große Bedeutung der „Droge Arzt" ganz allgemein und besonders bei der Verschreibung von Medikamenten gezeigt hat und das einen ersten großartigen Entwurf zu ihrer Pharmakologie wie auch Toxikologie darstellt. Als Beispiel für die Toxikologie soll die „iatrogene Neurose" hervorgehoben werden, die sich durch unkontrolliertes und unpsychologisches Verhalten des behandelnden Arztes nicht selten auf ein psychoneurotisches bzw. psychosomatisches Krankheitsbild aufpfropft und damit zu dessen Chronifizierung beiträgt.

Der russische Psychiater Bechterew (1904) gibt für den komplizierten Vorgang der Suggestion einen einfachen und anschaulichen Vergleich, wenn er davon spricht, daß sich die Suggestion wie ein Dieb des Nachts durch die Hintertür in ein fremdes Haus einschleiche, um es am kommenden Morgen als Hauseigentümer verkleidet durch die Vordertür wieder zu verlassen. Anhand dieses Vorganges können wir die drei Stadien des „erlebnismäßigen Gesamtgeschehens" bei der Suggestion (Stokvis, 1959) erkennen:

- Das **Stadium der Annahme** eines Ich-fremden Bewußtseinsinhaltes ohne nähere Motivation, genauere Kontrolle und eingehendere Realitätsprüfung = unauffälliges Eindringen in das fremde Haus.
- Das **Stadium der Verwirklichung,** in dem das für die Suggestion Typische stattfindet, nämlich das „subgerere" (= unterschieben, eingeben, einbilden). Ein Ich-fremder Bewußtseinsinhalt wird nun vom Suggerierten als dessen eigene Meinung, eigenes Gefühl, eigener Wille, eigene Handlung erlebt = Inbesitznahme des fremden Hauses.
- Das **Stadium der Handlung,** wobei es zur Ausführung des Ich-fremden Bewußtseinsinhaltes in Gestalt einer eigenen Vorstellung, eines eigenen Gefühls, eines eigenen Willensaktes kommt = Wiederverlassen des fremden Hauses als dessen Besitzer.

Ehe wir zur psychoanalytischen Theorie des Suggestionsvorganges kommen, müssen wir uns noch mit dem Begriff der **Suggestibilität** beschäftigen. Hierun-

---

1 Bezugnehmend auf das erste theoretische Kapitel sei hier angemerkt, daß sich die in der Literatur als Suggestion beschriebenen Vorgänge innerhalb des „symbiotischen Funktionskreises", wie er in Kapitel 1 beschrieben wurde, abspielen, wobei zwischen Arzt und Patient ähnliches geschieht wie in den ersten Lebenswochen und -monaten zwischen Mutter und Kind.

ter wird die affektive Empfänglichkeit für Bewußtseinsinhalte verstanden, die ohne Reaktionsprüfung oder Nachprüfung „untergeschoben" werden sollen. Die Suggestibilität kann bei ein und demselben Patienten, je nach seinem affektiven Zustand und seiner „Übertragung" schwanken. Sie ist von zahlreichen Einflüssen abhängig. Neben dem Lebensalter – Kinder sind besonders suggestibel – werden dem Geschlecht – Frauen werden aufgrund ihrer für gewöhnlich stärkeren affektiven Ansprechbarkeit als suggestibler beurteilt –, dem Triebleben, dem Temperament, der Intelligenz, der Rasse und auch der Volkszugehörigkeit Einflußnahmen zugeschrieben. Bekannt sind auch Verstärkungen der Suggestibilität in Zuständen von Ermüdung und Erschöpfung mit den ihnen zugehörigen Senkungen der Bewußtseinsschwelle. Auch hier wird zwischen einer negativen Suggestibilität, die bis zur Verwerfung der Suggestion (= „regressive Nein-Haltung") führen kann, und einer positiven unterschieden, die sich im Extrem als „masochistische Ja-Haltung" ausdrückt.

Psychoanalytisch gesehen, drückt die Suggestibilität in erster Linie die Bereitschaft zur Wiederbelebung infantiler Objektbeziehungen (Übertragung) aus (Freud und Breuer, 1895, 1970; Freud, 1921, 1970; Ferenczi 1910, 1911). „Die Allmacht des Suggestors wird als Wiederholung der Allmacht der Eltern über die Kinder erlebt. In beiden Fällen sind Liebe und Furcht die Beweggründe für das Zeigen einer Art von Gehorsam, die über das Normale hinausgeht. Frühere, lang vergessene Lebenssituationen wiederholen sich" (Stokvis, 1958, 1959). Die Fremd-(Hetero-)Suggestion wird von Jones (1936) auf die Vater-Imago zurückgeführt, die Selbst-(Auto-)Suggestion auf den Narzißmus. Fremdsuggestion besagt demnach, daß ein Stadium vorausgegangen ist, in dem das Ich-Ideal in das frühere Vater-Ideal aufgelöst wurde. Selbstsuggestion bedeutet in diesem Sinne, daß zwischen dem narzißtischen Ich-Ideal und dem realen Ich eine harmonische Vereinigung stattfindet. Diese wird durch Regression in autoerotische Richtung möglich, wenn der primäre Narzißmus aufgegeben und dann durch Konzentration auf die Vorstellung des Selbst wiedererlebt wird. In den meisten Suggestionstheorien sind die klassischen psychoanalytischen Gesichtspunkte wiederzufinden, nach denen immer eine Interaktion zwischen zwei Personen, zwischen zwei Ichs oder zwischen zwei Teilen einer Persönlichkeit stattfindet.

Von seiten der **Neurophysiologie** sind u.a. die Erklärungsversuche von Pawlow (1953–1954) und seiner Schule der „objektiven Psychologie" zu erwähnen. So erklärte B. Biermann (1929) die Suggestion als kortikal bedingten Reflex. Völgyesi (1938) bezog in diese Erklärung ebenso die Autosuggestion ein. Über erstaunliche körperliche Auswirkungen der Fremd- und Selbstsuggestion liegt ein großes Schrifttum vor, das sich seit O. Vogt (1897) insbesondere mit den Befunden der Hypnose als experimentell-psychologischer Methode befaßt. Hingewiesen sei in diesem Zusammenhang auf die verschiedenen motorischen, sensorischen und vegetativen Symptome, die dadurch hervorgerufen werden können. Wenn man z.B. einer

Versuchsperson suggeriert, daß sie Brot ißt, so ist die Zusammensetzung des Magensaftes anders als bei der Suggestion, Fleisch zu essen (Heyer, 1925). Eine hypnotische oder autosuggestive „Durchwärmung" der Leber ändert z.B. den Blutzuckergehalt und das Blutbild signifikant (Marchand, 1956).

Schließlich sind die „Stigmatisierten" (z.B. Therese von Konnersreuth) eindrucksvolle Beispiele für diese weitreichenden, hier insbesondere autosuggestiven Wirkungen.

Bei den Interaktionen zwischen dem **Suggestor** und dem **Suggerendus** spielt der zu Suggerierende die wichtigere Rolle. Das gilt vor allem hinsichtlich seiner Persönlichkeit. Begünstigend sind hier u.a. masochistische Charakterzüge, Ich-Schwäche, Intelligenzschwäche, aber auch Hingabefähigkeit, Duldsamkeit, Gläubigkeit bei guter Intelligenz. Von Bedeutung sind weiter die Einstellung des Patienten zum Arzt (Sympathie/Antipathie, positive „Übertragung"/negative „Übertragung") sowie zur Krankheit (Hoffnung auf Besserung bzw. Heilung/Selbstaufgabe). Demgegenüber tritt der **Suggestor** zurück. Dieser muß jedoch immer eine überzeugende Selbstsicherheit und Selbstvertrauen zum Ausdruck bringen, egal welcher Zweifel am eigenen Tun sich dahinter verbirgt, und muß so Vertrauen erwecken sowie Glauben an ihn als den „Arzt der Wahl" einflößen. Selbstunsicherheit und übertriebene Selbstkritik können andererseits hinderlich für den Suggestor sein und dessen suggestive Wirkungen beeinträchtigen. Schließlich sind noch der **Suggestionsinhalt** und die **Situation** zu erwähnen. Außer der Affektwirkung ist hier auch die moralische und ethische Seite des Suggestionsinhaltes zu beachten. So kann der Suggerendus die fremde Vorstellung nur mit Billigung seines Ideal-Ichs zur eigenen machen. Hinsichtlich der Formgebung ist die Wiederholung besonders hervorzuheben, weiter die Monotonie der Reizgebung. Mitunter wirkt aber auch ein einzeitiger aktiver Reiz (z.B. scharfes Kommando, energischer Befehl, eventuell in Verbindung mit faradischem Reiz) schon heilsam. Auch die Situation, in der die Suggestion gegeben wird, darf nicht unberücksichtigt bleiben. Hier können mitunter zahlreiche Imponderabilien eine Rolle spielen, so das Erscheinungsbild und das Auftreten des Suggestors, seine Kleidung, seine Überzeugungskraft, seine Fähigkeit zur Dramatisierung im gegebenen Falle, seine Geistesgegenwart, Intelligenz, nicht zu vergessen die Mitmenschlichkeit.

Das Problem **Heterosuggestion/Autosuggestion** ist in der Literatur viel diskutiert worden. Bei der Fremdsuggestion handelt es sich, wie gesagt, um zwei Personen, die in einem gegenseitigen affektiven Beeinflussungsverhältnis zueinander stehen. Bei der Autosuggestion handelt es sich demgegenüber um zwei Teile derselben einheitlichen Persönlichkeit, um das Ich und um das Selbst. Auch hier hat man es nach Stokvis (1959) mit einer erotischen (autoerotischen) Gefühlsbeziehung zu tun. Unter dem Einfluß triebhafter Bedürfnisse identifiziert sich das bewußte Ich mit dem eigenen Selbst (narzißtische Identifikation). Die Vorstellung, die zum Objekt der Autosuggestion werden soll, muß für den Betreffenden eine affektive Bedeutung haben. Jedes Erleben kann Inhalt einer Autosuggestion sein. Die **Autosuggestibilität** schwankt, ebenso wie die Suggestibilität, inter- und intraindividuell beträchtlich und ist von denselben Faktoren abhängig, wie sie dort erwähnt wurden. Auch hier unterscheiden wir eine negative von einer positiven Autosuggestibilität. Negative Autosuggestibilität kann z. B. durch Angst zustande kommen, daß gewisse vom Ideal-Ich nicht geduldete Triebneigungen bewußt werden. Dann fühlt oder tut der Betreffende das Entgegengesetzte von dem, was er sich selbst suggeriert hat („Gesetz der das Gegenteil bewirkenden Anstrengung" nach Baudouin, 1924).

Die **Wirkungsweise** der Suggestion kann nur ganz heitlich betrachtet werden. Zusammen mit den psychischen Einflüssen auf Denken, Fühlen, Wollen vollziehen sich, entsprechend dem psychosomatischen, „zweieinheitlichen Affektgeschehen" (Stokvis, 1958, 1959; Stokvis und Pflanz, 1961), auch solche auf die somatischen Funktionen. Von besonderer Bedeutung sind jene auf den Muskeltonus, die zum Ausgangspunkt zahlreicher Entspannungsverfahren geworden sind. Ihnen eng verbunden sind die Wirkungen auf das Gefäßsystem sowie in der Folge auf alle vom vegetativen Nervensystem innervierten inneren Organsysteme im Sinne eines „Plus-" bzw. „Minus-Effektes" bzw. einer Ruhigstellung in eutoner Mittellage. Dazu gehört auch die Haut. Schließlich muß in diesem Zusammenhang noch das System der endokrinen Drüsen erwähnt werden, das in vielfältiger und wechselseitiger Abhängigkeit vom vegetativen Nervensystem steht.

Leider ist der Begriff „Suggestion" zu einem Schlagwort geworden, das bei Medizinern wie bei Laien „in schlechten Ruf gekommen ist", wie Stokvis (1958, 1959) ausführt. Ebenso haben die suggestiven Verfahren bei vielen Ärzten die Prägung von etwas Minderwertigem erhalten, egal ob sie nun getarnt oder gezielt stattfinden. Doch sind sie aus der ärztlichen Heilkunde nicht wegzudenken, kommen doch viele Kranke aufgrund ihrer Persönlichkeit, ihrer Krankheit und ihrer Lebensumstände nicht für eine ursachengerechte, aufdeckende Behandlung in Frage. Zudem ist das „Heer der Nervösen" bzw. „seelischen Störungen" (Langen, 1969) viel zu groß, um den zeitintensiven tiefenpsychologischen Verfahren das Alleinvertretungsrecht in der Psychotherapie zuzugestehen. Auch kann vielfach schon eine Symptomheilung eine Krankheitsheilung bedeuten. Wir müssen uns hier ganz von der verführerischen Illusion freimachen, daß nur eine Änderung der Persönlichkeitsstruktur das alleinseligmachende Psychotherapieziel sei. „Vielmehr läßt sich günstigenfalls teils durch Sanierung der aktuellen Lebensverhältnisse, teils durch induktive Trainingsmethoden der affektive Pegelstand so einregulieren, daß die Störungen durch Affekt- und Spannungsentzug unterschwellig und bedeutungslos werden" (E. Kretschmer, 1949).

Mit Nachdruck verteidigt deshalb auch J. H. Schultz (1953), der unermüdliche Streiter für eine „Psychologisierung des Arztens" und Nestor der ärztlichen Psychotherapie in Deutschland aufgrund jahrzehntelanger Erfahrung als praktischer Nervenarzt in Berlin, in den verschiedenen Vorworten zu seiner „Hypnose-Technik" die suggestiven Verfahren, wenn er 1952 schreibt: „Allgemein verständnisvolle Menschenführung (Psychagogik), Unterstufe des autogenen Trainings und besonders auch methodisch einwandfreie Hypnotherapie richtiger Indikation können Wertvolles leisten (bei etwa 50–60% ‚funktioneller' Anomalien)"; und an anderer Stelle 1958: „Möge auch diese Auflage dazu dienen, der aktiv-klinischen ‚organismischen' Psychotherapie, die allein für den beschäftigten Allgemeinpraktiker und Facharzt anderer Gebiete in Frage kommt, neben der heute oft einseitig überheblich als allein wesentlich geschilderten ‚mentalen', insbesondere psychoanalytischen Therapie den gebührenden Platz zu sichern" (1952–1983).

22.1.2 Geschichtlicher Exkurs zu den Suggestivverfahren

Die Suggestion ist in der Geschichte der Heilkunst das älteste und am häufigsten verwendete Heilmittel zur Linderung menschlicher Not. Berichte über fremdsuggestive Behandlungsmaßnahmen reichen bis in die Frühzeit der Menschheit zurück. So heißt es im Papyrus Ebers (1552 v. Chr.), der ältesten Urkunde der Ägypter, die unter den Trümmern von Theben gefunden wurde:

„Lege die Hände auf ihn, um den Schmerz der Arme zu beruhigen, und sage, daß der Schmerz verschwinden wird."

Auch bei den Chaldäern blühte die weiße (gute) und schwarze (böse) Magie, wie zahlreiche Bibelstellen zeigen. So findet sich im Evangelium Marcus 16 (17, 18) folgende Stelle:

„In meinem Namen werden sie Teufel austreiben, mit neuen Zungen reden, Schlangen vertreiben, und so sie etwas Tödliches trinken, wird's ihnen nicht schaden; auf die Kranken werden sie die Hände legen, so wird's besser mit ihnen werden."

Erinnert sei auch an das Institut des Tempelschlafes, das noch bis ins 6. Jahrhundert n. Chr. existierte. Die Mönche traten später das Erbe der Tempelpriester an und vollzogen Wunderheilungen mittels Gebeten, Reliquien von Märtyrern und Weihwasser. Selbst die Päpste, Könige und Kaiser beteiligten sich bis ins Mittelalter an diesem Heilgeschäft. So behandelte z. B. Ludwig der XIV. mit Vorliebe Skrofulöse, denen er nach der Berührungsprozedur ein Geschenk von 2–5 Sous überreichen ließ mit den Worten: „Le Roi te touche, Dieu te guérit". Indessen, während die suggestiven Phänomene hier zu einem nützlichen therapeutischen Zwecke Verwendung fanden, machten sich daneben in immer steigendem Maße hypnotische und somnambule Erscheinungen spontaner und fremder Entstehungsart geltend, die die Grundlage der mittelalterlichen Massenepidemien sowie der Hexenverfolgungen bildeten. Visionen und Halluzinationen, lethargische und somnambule Zustände, Nymphomanien und Dämonomanien herrschten endemisch und epidemisch und wurden teils zur Grundlage abergläubischer Verehrung, teils zum Gegenstand grausamster Verfolgung gemacht.

In jüngerer Zeit ist vor allem Mesmer (1733–1815, 1952) mit seiner Lehre vom **„animalischen Magnetismus"** hervorgetreten. Er hat als erster geschulter Arzt der Neuzeit, wie J. H. Schultz (1952) ausführt, die Wirkung erlebt und immer wieder hervorgerufen, die von einer suggestiven Persönlichkeit, von ihrem Nahesein, Sprechen, Reden und Befehlen auf erschütterte Kranke heilsam ausgeht. Jedoch vermochte Mesmer diese Wirkung nicht zu erklären und sah in der ihm unverständlichen psychischen Mechanik noch mittelalterliche Magie. Ihm fehlte, wie allen seinen Zeitgenossen, der entscheidende Begriff der Suggestion. Dieser ist vor allem J. Braid (1843), einem englischen Augenarzt zu verdanken, der die Grundlagen der Hypnose gelegt und dieser heterosuggestiven Methode den Namen gegeben hat. In seiner Nachfolge ist vor allem die Schule von Paris (Charcot, 1889) zu erwähnen, die die Hypnose als Ausdruck eines „hysterischen Psychismus", als Erscheinung einer „kollektiven Pathologie" auffaßte. Die erste Schule von Nancy (Liébeault und Bernheim, 1888) stellte dieser pathophysiologischen Hypothese von Charcot eine psychologische, nämlich die der Suggestion, entgegen. Diese hat sich bis heute behauptet. In beiden Forschergruppen arbeitete Freud mit und schuf so die Voraussetzungen für die 1895 gemeinsam mit J. Breuer in den „Studien über Hysterie" (1970) beschriebene „Psychokatharsis", aus der er dann später die Psychoanalyse entwickelte. In Hypnose wurden für das betreffende Krankheitsbild entscheidende Erlebniszusammenhänge aufgedeckt und hinsichtlich ihrer „eingeklemmten Affekte" kathartisch abreagiert, mit dem Erfolg, daß die vorwiegend hysterischen Krankheitsbilder sich vorübergehend oder dauerhaft zurückbildeten.

Verfolgen wir den Entwicklungsgang suggestiver Verfahren am Beispiel der Hypnose bis in die Neuzeit weiter, so ist vor allem der später als Hirnforscher bekanntgewordene O. Vogt (1897) hervorzuheben, der die erste wissenschaftliche Bearbeitung des Hypnotismus durchgeführt hat. Von ihm stammen u. a. die Begriffe „hypnotische Hypermnesie" (= Bewußtmachen vergessener bzw. verdrängter Erlebnisse in Hypnose aufgrund des Fortfalls von Hemmung bzw. Verdrängung) und „seelisches Mikroskop" (= systematische Selbstbeobachtung in Hypnose zum Studium von Aufbau und Entwicklung der Neurosen sowie zu allgemeinpsychologischen Erkenntnissen). Von besonderer Bedeutung für die **moderne Hypnotherapie** war ihre Fortentwicklung in Gestalt der „fraktionierten Methode". Hierbei handelte es sich um eine systematische Erfassung der Selbsterfahrungen von Patienten in Hypnose, die zur Grundlage der nächstfolgenden Behandlung gemacht wurden, was die so notwendige Mitarbeit der Patienten stimulierte und insgesamt zu einer mehr gezielten und tiefer reichenden, der Individualität des jeweiligen Patienten besser angepaßten Form der Hypnotherapie führte. Nach Braid war es vor allem O. Vogt, der die praktische Bedeutung der **Selbst-(Auto-)Hypnose** erkannte und diese in Form von „prophylaktischen Ruhepausen" bei Labilen, Erschöpften und krisenhaft Asthenischen anwendete. Vogt faßte die Hypnose als ein „partielles Wachsein" auf, im Gegensatz zu Pawlow (1953/1954), der sie als „partiellen Schlaf" bezeichnete, hervorgerufen durch inadäquate, langanhaltende Reize, die bedingte Reflexe auslösen. Diese wirken dann „wie ein punktförmiger Reiz im Gehirn".

Um den weiteren Ausbau des **wissenschaftlichen Hypnotismus** in Deutschland haben sich vor allem J. H. Schultz (1952–1983, 1970), der Begründer der Selbsthypnose des autogenen Trainings und E. Kretschmer (1959) mit D. Langen (1969) durch die „gestufte Aktivhypnose" und die „zweigleisige Methode" verdient gemacht, weiter der Niederländer B. Stokvis (1955, 1958, 1959, 1965), der seine wichtigsten Hypnoseveröffentlichungen in deutscher Sprache geschrieben hat. Aus Raumgründen kann nicht auf die weitere Entwicklung der Hypnoseforschung in den übrigen Sprachräumen eingegangen werden. Hier liegen gerade in jüngerer Zeit aus anglo-amerikanischen, französischen, italienischen und spanischen, sowie sowjetischen und osteuropäischen Quellen so viele Forschungsergebnisse und Veröffentlichungen vor, daß tatsächlich zur Zeit von einer „Renaissance der Hypnose" gesprochen werden kann, wie J. H. Schultz dies vorausgesehen hat. Diese Entwicklung geht einher mit einer allgemeinen Wiederaufwertung der stützenden und insbesondere auch der suggestiven Methoden bei der Behandlung der Psycho- und vor allem Somato-Neurosen (= psychosomatische Störungen und Erkrankungen) gegenüber den lange Zeit beherrschenden psychoanalytischen. Dazu hat auch die neuere psychoanalytische Forschung selbst beigetragen, etwa mit dem Konzept der „Grundstörung" im Sinne von Balint (1957).

22.1.3 Hypnose

Bei der Hypnose handelt es sich nach Stokvis (1955) um einen mittels bestimmter Einleitungstechniken durch „affektive Faktoren hervorgerufenen Zustand einer (oftmals geringen) Senkung des zuvor eingeengten Bewußtseins, in dem ein Rückschritt der Grundfunktionen der Persönlichkeit (Denken, Fühlen, Wollen) sowie der animalischen Verrichtungen eintritt. Die Einsicht in die reale Situation geht höchst selten verloren. Seine Reaktionsweise in der Hypnose bleibt dem Hypnotisierten fast immer bewußt".

Das Wesentliche an der Hypnose sind der vor allem auf den Hypnotiseur eingeengte Bewußtseinszustand sowie die unterschiedlich starke Senkung des Bewußtseins, die mit einer erhöhten Suggestibilität einhergeht. Mit diesen Bewußtseinsveränderungen kommt es zu einer Regression, einem Zurückschreiten in frühere (infantile) Entwicklungsstufen mit Wiederherstellung von Verhaltensweisen, die diesen Perioden angehören. Vorrangig ist dabei das Wiederaufleben eines „Gefühlsprimitivismus". Abgesehen von der frühkindlichen Einstellung der Passivität und

Hingabe, kommt dabei das Magisch-Archaische dieser Entwicklungsstufe zur Wiederbelebung. Wie das Kind seinen Eltern, so traut der Patient seinem Hypnotiseur eine magische Potenz zu und gewinnt selbst mit Hilfe der Identifikation Anteil an dieser magischen Macht. Bedeutsam für das Gelingen der Hypnose ist eine positive affektive Beziehung zwischen dem Hypnotiseur und dem Hypnotisanden. Diese wird aus psychoanalytischer Sicht auch als „infantilerotische Bindung" bezeichnet. Unter ihrem Einfluß „nimmt die Neigung zur Identifikation mit dem Arzt stark zu, wodurch auf dem Wege der Introjektion die suggerierten Vorstellungen eher angenommen werden und der Psychismus der Suggestion im engeren Sinne sich leichter vollziehen kann. Diese Neigung zur Identifikation ist zuweilen derartig stark, daß der Hypnotisand in der Befolgung der Suggestionen einen narzißtischen Vorteil erblickt" (Stokvis, 1955), bzw. sich gelegentlich sogar masochistisch verhält. Dementsprechend wird der Hypnotiseur, je nach seinem mehr autoritären Auftreten als „Vater-Imago" oder mehr fürsorglichen als „Mutter-Imago" erlebt.

Bei der Hypnose kann man vereinfacht ein oberflächliches (leichtes) von einem tiefen Stadium unterscheiden. In den Zustand der **leichten Hypnose** kann jeder Mensch versetzt werden, der einsichtig genug und bereit dazu ist. Charakteristisch, wenn auch nicht spezifisch sind: Wach-Sein unterschiedlichen Grades bis zum „Hypnoid" (Zustand unmittelbar vor dem Einschlafen) mit Ruhe, Schwere, Wärme, Müdigkeit, Verlangsamung von Herzaktion und Atmung, eventuell Wachträumen. Therapeutische Suggestionen werden schon in leichter Hypnose mitunter besser akzeptiert und realisiert als in der **tiefen Hypnose.** Kennzeichnend sind hier Schlaferscheinungen unterschiedlichen Grades vom Pseudoschlaf bis zum „echten", hypnotischen Schlaf. Der Somnambulismus, in dem der Hypnotisierte schlafend mit geschlossenen oder offenen Augen umhergehen kann, in dem Katalepsie auftritt, posthypnotische Suggestionen meist verwirklicht werden und Amnesie besteht, kann sowohl im Pseudo- als auch im echten Schlaf auftreten. Er wird nur relativ selten und dann besonders bei dazu disponierten Hypnotisanden (z.B. besondere Fähigkeiten zur „Ideoplasie"; hysterische Charakterzüge bzw. Verhaltensweisen) oder aufgrund längerer Wiederholung und Vertiefung von Hypnosen erreicht. Der „Rapport" zwischen Hypnotiseur und Hypnotisand bleibt erhalten. Therapeutische Suggestionen und insbesondere posthypnotische werden realisiert, wenn die entsprechenden Voraussetzungen gegeben sind.

Voraussetzung der Hypnotherapie, wie jeder Psychotherapie, ist eine gründliche körperliche und psychische Untersuchung mit Diagnose und Indikationsstellung. Dabei braucht der Hypnotherapeut nicht auf eine körperliche Untersuchung zu verzichten, wenn diese erforderlich ist und soweit sie in seinen Kompetenzbereich fällt. Aus den oben geschilderten Gründen ist es aber verständlich, daß die körperliche Untersuchung mit dem erforderlichen Takt und mit mög-

lichst großer Zurückhaltung durchgeführt wird. Von besonderer Bedeutung ist eine ausreichende psychische Vorbereitung des Patienten auf die geplante Hypnotherapie. Dabei hat sich als günstig erwiesen, das ominöse Wort „Hypnose" zu vermeiden und statt dessen von „Ruhe- und Entspannungsbehandlung" zu sprechen. Ein eventueller Zweifel des Patienten, ob es sich bei der geplanten Behandlung nicht doch um Hypnose handele, sollte nicht negiert, aber dadurch entkräftet werden, daß kein automatenhafter Gehorsam, keine lächerliche Zurschaustellung, keine Entlockung von persönlichen Geheimnissen und keine Erinnerungslosigkeit stattfinden, sondern daß es bei dieser Form der Therapie ganz wesentlich auf die Mitarbeit des Patienten ankommt. „Die Leistung liegt bei dem Hypnotisierten, der Hypnotisierende kann ihm die Aufgabe nur erleichtern" (J. H. Schultz, 1952 bis 1983).

Hypnosen werden sowohl im Sitzen als auch im Liegen durchgeführt. Die Lagerung in der Horizontalen ist für therapeutische Zwecke vorzuziehen. Dabei ist der Patient weniger des Wärmeverlustes als seines Wohlgefühls wegen leicht zuzudecken. Das Äußere des Raumes, in dem Hypnosen stattfinden, ist weniger bedeutungsvoll als das Vertrauensverhältnis zwischen Arzt und Patient. Dieses wird z.B. durch eine anfängliche Pulskontrolle noch gestärkt. Die älteste Form der Einleitungstechnik ist die **Faszinations-Methode.** Der Patient fixiert dabei die Pupillen des Arztes, dieser jedoch eher die Glabella des Patienten, um nicht in die Gefahr einer eigenen Spontanhypnose zu kommen, wenn z.B. der Widerstand des Patienten zu groß und die eigene Ermüdung eventuell mit im Spiele ist. Diese Form der Hypnose-Einleitung wird vor allem von Laien-Hypnotiseuren geübt. Wegen ihres autoritären Unterwerfungscharakters wird sie heutzutage in der ärztlichen Praxis nur noch selten benutzt. Im Vordergrund stehen demgegenüber die Fixations- und die Farbkontrast-Methode mit begleitender Verbalsuggestion. Bei der **Fixations-Methode** wird ein kleiner, eventuell glänzender Gegenstand (z.B. Zeigefingerkuppe, Reflexhammerende, Kugelschreiberspitze) möglichst nahe fixiert, bis die üblichen Ermüdungserscheinungen der Augen und ein Undeutlichwerden des fixierten Gegenstandes auftreten, was durch begleitende, monotonisierende Verbalsuggestionen der Ruhe, des Wohlgefühls, der Geborgenheit, zunehmender Müdigkeit, Schläfrigkeit, Schwere usw. verstärkt wird, bis der Patient von selbst oder mit Hilfe des Arztes die Augenlider schließt und damit die für den weiteren Behandlungsgang notwendige „optische Subtraktion" und „Introversion" durchführt. Die **Farbkontrast-Methode** beruht auf dem physiologischen Simultankontrast der Farben und wird mit einer Farbtafel (Blau-Gelb- oder Rot-Grün-Kontrast) durchgeführt. Auftreten der Gegenfarbe im Umkreis der fixierten Farbe sowie damit verbundene weitere Farbsensationen werden verbal begleitet und als Suggestionshilfen benutzt. Die Allgemeinsuggestionen, die zum Augenschluß und zur Introversion führen sollen, sind im übrigen dieselben wie bei der Fixations-Methode.

Ist mit einer der geschilderten Techniken die **„neuroorganismische Umschaltung"** (J. H. Schultz, 1952–1983) zu Ruhe, Entspannung und Erholung erreicht worden, so bietet der Patient, äußerlich betrachtet, den Anblick eines ruhig Schlafenden, was zu der irreführenden Bezeichnung „Heilschlaf" geführt hat. In Wirklichkeit handelt es sich dabei nur um einen behaglichen Ruhezustand, in dem eine gesteigerte Suggestibilität gegenüber verbalen und haptischen Reizen besteht. Alle kontrollierbaren Körperfunktionen zeigen eine gelöste Mittelstellung. Die willkürliche Körpermuskulatur ist entspannt (= Schweregefühl). Die oberflächlichen Hautgefäße sind ebenfalls entspannt, erweitert und mehr durchblutet (= Wärmegefühl). Die Atmung ist nach anfänglicher Unruhe- und Angst-Beschleunigung ruhig und regelmäßig, ebenso der Herzschlag. Befragt man die Patienten nach einem solchen Zustand über ihr Selbsterleben, so geben sie in der Regel an, daß sie sich sehr ruhig und ausgesprochen wohl gefühlt hätten. Es ist verständlich, daß dieser Ruhe-, Entspannungs- und Erholungszustand in vielen Fällen von „konstitutioneller Nervosität" (Bumke, 1936; J. H. Schultz, 1928), „vegetativer Stigmatisierung" (v. Bergmann, 1926), „funktionellen Syndromen" (v. Uexküll, 1960, 1963) mit ihren vielfältigen psycho-vegetativen Symptomen als wiederholt durchgeführte Behandlungsmaßnahme mitunter allein ausreicht oder in schwierigen Fällen bis zu Schlafhypnosen unterschiedlicher Dauer vertieft bzw. verlängert werden muß. Bei speziellen Krankheitsindikationen und -symptomen wird die hypnotische Beeinflussung durch gezielte Verbalsuggestionen in die gewünschte Richtung weiter fortgesetzt und ausgebaut. Dabei ist die Redaktion der Suggestionen wichtig. Diese sollen ruhig, klar, sicher und insbesondere anschaulich gegeben werden, unter Einschluß der den Suggestionsvorgang allgemein begünstigenden Hilfsmittel der Wiederholung und der Monotonie.

Haptische Suggestionen, z. B. in Form von Handauflegen auf bestimmte Organbereiche, „passes" (= „Mesmerische Striche", die im Abstand einiger Zentimeter vom Körper des Patienten vorgenommen werden) sind aus den oben angeführten psychologischen Gründen in der Hypnose nur zurückhaltend zur Verstärkung der Verbalsuggestionen vorzunehmen (Vorsicht vor der „infantil-erotischen Bindung"). Die Reichweite dieser gezielten hypnotischen Suggestionen ist, entsprechend der ganzheitlichen Wirkungsweise von Suggestionen, wie sie weiter oben ausgeführt wurde, außerordentlich groß und erstreckt sich auf die gesamten vegetativen Regulationen innerer Organsysteme. Im Prinzip ergibt sich von daher eine ausgedehnte Wirkungs- und Verwendungsmöglichkeit der Hypnose. „Gleichgültig, ob es sich um einen organisch gesunden oder kranken Organismus handelt, kann grundsätzlich Hypnose (oder Psychotherapie anderer Form) Heilwirkungen entfalten, soweit funktionelles Geschehen reicht." Diese Bemerkung von J. H. Schultz (1952–1983) bezieht sich u. a. auch auf den Schmerz, von dem bekannt ist, daß er in besonderer Weise hypnosuggestiv beeinflußbar ist. Das

hat z. B. dazu geführt, daß Operationen und Entbindungen in hypnotischer Analgesie durchgeführt worden sind. Selbstverständlich ist dabei Rücksicht auf die Signalfunktion des Schmerzes zu nehmen.

Entsprechend der Reichweite hypnosuggestiver Wirkungsmöglichkeiten, ist es schwer, eine Indikationsliste aufzustellen. Ehe dies geschieht, sollen noch einige wichtige Punkte besprochen werden. Unter diesen steht die **Desuggestionierung** an erster Stelle, die in einer dem individuellen Verlauf der jeweiligen Hypnose angepaßten Form zu geschehen hat, bei der möglichst alle Suggestionen in einer schonenden Form wieder zurückgenommen werden, die eventuelle Störungen des Wohlbefindens nach dem Aufwecken verursachen können. Vor allem sind dies jene der Müdigkeit, Schläfrigkeit und Schwere, während Ruhe und Erholung posthypnotisch fixiert werden. Die Zurücknahme des hypnotischen Zustandes geschieht entweder drei- oder besser sechszeitig. Die Dauer der Hypnose wird unterschiedlich beurteilt. Zeiten von einer viertel bis zu einer halben Stunde reichen aber für gewöhnlich aus. Auch über die Anzahl der Wiederholungen bestehen Differenzen in der Literatur. J. H. Schultz rät, nicht zu früh abzubrechen. Als Anhalt mögen 30 bis 50 Hypnosen gelten, die je nach den Umständen in Form einer großen hypnotischen Kur (1. Woche 2 × tgl., 2. Woche 1 × tgl., 3. Woche 3 × wchtl., 4. Woche 2 × wchtl., 5. und 6. Woche 1 × wchtl., dann alle 14 Tage, alle Monate je eine Behandlung) oder 2- bis 3mal wöchentlich stattfinden. Auch in der Hypnotherapie gilt das alte Sprichwort, wie in der Psychotherapie insgesamt: „Langsam aber sicher!" Die besten Hypnosezeiten werden bestimmt durch den jeweiligen Tagesablauf des Patienten und die Zeiten ausreichender Konzentration und Hingabefähigkeit. Diese Zeiten sollten nach Möglichkeit bei den Wiederholungen beibehalten werden. Die Einzelbehandlung ist im allgemeinen der Gruppenbehandlung vorzuziehen, im Gegensatz zum autogenen Training. Das „Mehrkammersystem" ist unter seriösen Psychotherapeuten umstritten und birgt die Gefahr einer schablonenhaften statt individualisierten Hypnotherapie in sich. Dieser letzte Einwand wird auch gegen die „Ablations-Hypnose" vorgebracht. Dabei wird der Hypnosevorgang vom Hypnotiseur abgetrennt und durch verschiedene Hilfsmittel ersetzt (u. a. Signalhypnose mit Farbtafel und selbstgesprochener Formel, Schallplattenhypnose, Tonbandhypnose etc.).

Leuner und Schroeter (1975) haben einen aufschlußreichen Überblick über die **Indikationen** und **spezifischen Applikationen** der Hypnotherapie gegeben. In ihm beziehen sie sich hinsichtlich der psychosomatischen Krankheiten auf Spiegelberg (1968), der Indikationen 1. Grades (z. B. Asthma bronchiale, Ulcus pepticum, Colitis ulcerosa) von solchen 2. Grades (z. B. essentielle Hypertonie, Migräne, Adipositas) und denen 3. Grades (u. a. Anorexia nervosa, akute und chronische somatopsychische Krankheitszustände nach eingreifenden Operationen und bei sehr schmerzhaften oder malignen Erkrankungen) unterscheidet.

Bei der Hypnotherapie sollte man aber der symptomatischen Indikation den Vorzug gegenüber der kausalen geben, wobei Entspannung und Sammlung als wesentliche Ziele des Versenkungszustandes der Hypnose anzusehen sind, mit ihren Folgen der Ruhigstellung, der affektiven Resonanzdämpfung (insbesondere der Entängstigung), der Ermutigung, der Erholung, der Schmerzlinderung bzw. -abstellung, der Schlafförderung und mit den Möglichkeiten einer erweiterten und vertieften Innenschau zum Neu- bzw. Wiedergewinn seelischer Selbsthilfe. Dementsprechend gilt auch heute noch uneingeschränkt, was J. H. Schultz (1952–1983) immer wieder betont hat: „Die gesicherten Tatsachen hypnotischer Experimente und kritischer hypnotherapeutischer Erfahrung beweisen, daß überall, wo Funktionen abwegig sind, ein hypnotischer (= psychotherapeutischer) Zugriff möglich ist, auch wenn die Störung nicht im engeren Sinne ‚psychogen' ist."

Hier zeigt sich, daß die Hypnose eine weitreichende und wertvolle unterstützende Funktion im Rahmen einer mehrdimensionalen psychosomatischen Therapie ausüben kann. Wie J. H. Schultz (1952 bis 1983) jedoch ausführt, bleibt der Wach-Psychotherapie stets die führende Rolle. „Die Hypnotherapie ist nur eine spezielle Form, eine Unterstützung der gesamten Psychotherapie, in vielen Fällen darum besonders wirksam, weil die Hypnose einen Zustand ‚gesteigerter Suggestibilität' und, bei anderer Leitung, der Minderleistung darstellt, die der Erholung dienen kann." Bei einer Aufzählung und Besprechung der Indikationen der Hypnose darf eine Erwähnung der **Kontraindikationen** nicht fehlen. Hierzu gehören in erster Linie alle Formen von Psychosen sowie die paranoischen Entwicklungen. Eine Ausnahme macht jedoch die Involutionsdepression, bei der sich, entweder in Kombination mit einer antidepressiven Therapie oder in der Nachbehandlung, mitunter gute Heilerfolge zeigen. Auch bei larvierter Depression kann die Hypnose in Kombination mit anderen psychotherapeutischen Verfahren eventuell eine gute Hilfe sein. Weitere Kontraindikationen sind Vergewaltigungserlebnisse in der Vorgeschichte, soweit diese noch nicht aufgearbeitet worden sind und deshalb hypnotisch reaktiviert werden können. Es versteht sich von selbst, daß Patienten mit einer „negativen Suggestibilität" oder einer ablehnenden Einstellung zur Hypnose aus den verschiedensten Gründen sowie sehr Erregte und Ängstliche nicht dazu veranlaßt bzw. überredet werden sollten. Einen Hinderungsgrund stellen auch engere persönliche Beziehungen zwischen Hypnotiseur und Hypnotisand dar. Hysterisch strukturierte Patienten sind mit Vorsicht zu hypnotisieren, am besten in Gegenwart einer vertrauensvollen dritten Person.

22.1.4 Hypnokatharsis (Psychokatharsis) und Hypnoanalyse

Hier handelt es sich um „aufdeckende" Formen der Hypnose mit dem Ziel, unbewußte Konfliktsituationen freizulegen und je nach den Möglichkeiten zur Abreaktion zu bringen. Die **Hypnokatharsis** wurde zuerst von Freud und Breuer (1895) benutzt. Aus ihr entwickelte Freud dann später die Psychoanalyse. Durch den vom Arzt provozierten und auch geleiteten Affektausdruck des in der Hypnose erinnerten und wiederbelebten traumatischen Materials bzw. Komplexes kommt es unter günstigen Umständen zu einer „Reinigung der Seele" („Katharsis" im Sinne von Aristoteles) und damit nachträglich zu einer Konflikt- bzw. Komplexlösung unter therapeutischen Bedingungen. Stokvis (1955, 1965) legt dabei großen Wert auf eine nachfolgende Besprechung und Aufarbeitung des kathartischen Geschehens im Wachzustand („epikritische Nachschau"), weil das bloße Wiedererinnern und Abreagieren allein oft nicht für eine dauerhafte Besserung oder Heilung ausreichen, so daß es zweckmäßig ist, das zum Teil unbewußte Material dem Bewußtsein zur Bearbeitung zu übergeben. Wenn auch die Hypermnesie und gesteigerte Integrationsfähigkeit für frühere Konflikterlebnisse in der Hypnose günstige Voraussetzungen für einen Behandlungserfolg darstellen, so muß doch gerade bei psychosomatisch Kranken mit Organschäden auch auf die Gefahren hingewiesen werden, die mit einer affektiven Wiederbelebung und Abreaktion psychischer Traumen aufgrund der starken vegetativen Begleitwirkungen einhergehen. Das Risiko, welches darin enthalten ist, muß von dem eine Hypnokatharsis durchführenden Arzt stets sorgfältig beachtet werden.

Unter **Hypnoanalyse** verstehen wir ein 1940 von Hadfield (1940) eingeführtes Verfahren, das zwischen der Hypnokatharsis und dem klassischen analytischen Vorgehen liegt. Hier wird größerer Wert auf die hypnotische Hypermnesie als auf die affektive Abreaktion gelegt. Im Laufe der Behandlung werden auch Widerstand und Übertragung angesprochen und bearbeitet, in Verbindung mit darauf folgender suggestiver psychagogischer Führung. Sind trotzdem die affektiven Begleitwirkungen noch zu stark und gar gefährdend, so wird eine Indifferenzhaltung in der Hypnose angestrebt.

Die **Narkoanalyse,** welche in den letzten Kriegsjahren und in der ersten Nachkriegszeit als psychotherapeutische Kurztherapie eine vorübergehende Rolle gespielt hat, ist bald danach wegen ihres psychotherapeutisch nicht zu vertretenden Vorgehens wieder verlassen worden. Hierbei wurde in der Einschlaf- und besonders in der Aufwachphase einer medikamentös induzierten Kurznarkose eine Katharsis von pathogenetisch bedeutsamen psychischen Traumen durch eine gezielte Exploration angestrebt. In eine ähnliche Richtung zielten später auch Versuche mit Halluzinogenen (z. B. LSD[25], Psilocybin) zur sog. „psycholytischen Therapie" (Leuner, 1962).

22.1.5 Gestufte Aktivhypnose, zweigleisige Methode

Diese von E. Kretschmer (1959) entwickelte und von D. Langen (1969) weiter ausgebaute psychotherapeutische Methode stellt „die direkte Form einer auf Selbstübung aufgebauten Autohypnose dar. In der Praxis ist die gestufte Aktivhypnose immer gekoppelt an einen selbständigen, zeitlich befristeten analytischen Arbeitsgang, der außer einer detaillierten Aufarbeitung des aktuellen Konfliktes mit all seinen reaktiven und charakterogenen Verzahnungen auch mit einer Charakteranalyse sowie mit einer psychagogischen Endführung verbunden ist." Aus der Kombination beider therapeutischer Züge resultiert die „zweigleisige Methode" nach E. Kretschmer. Die Durchführung gliedert sich in vier Abschnitte:

1. Erlernen der psychotherapeutischen Grundübungen des autogenen Trainings (Ruhe und Schwere, Wärme), zum Teil mit Begleitsprechen des Arztes bei gemeinsamen Übungen.

2. Aktive Erlernung der hypnotischen Fixierübungen zur Erzielung hypnoider Zustände und anschließende Vertiefung der Hypnose durch den Arzt, bei zunehmender Verlängerung der einzelnen Hypnosen bis zu einer halben und einer Stunde.

3. Therapeutische Anwendung der Hypnose mit aus der Analyse gewonnenen „wandspruchartigen Leitsätzen". Die gleichzeitig durchgeführte „steuernde Analyse" konfliktzentrierten Charakters tritt jetzt zurück. Daraus gewonnene „wandspruchartige Leitsätze" (= formelhafte Vorsatzbildungen des autogenen Trainings) werden nun systematisch in die gestufte Aktivhypnose eingebaut. Zunächst sollen dabei störende Faktoren (z. B. Leitsymptome) zur „Indifferenz" gebracht werden. Dann wird versucht, die Charaktereigenschaften anzusprechen, die der Patient aus eigener Kraft nicht mobilisieren kann.

4. Selbsttätige Weiterführung der Entspannungsübungen und gelegentliche Hypnosen in größer werdenden Zeitabständen. Dieser Abschnitt stellt eine Kombination von den unter 1. aufgeführten beiden Grundübungen mit „wandspruchartigen Leitsprüchen" dar. Immer ist auf ein exaktes Zurücknehmen am Ende der Übungen zu achten, ähnlich wie bei Hypnosen bzw. beim autogenen Training.

Die Besonderheiten der gestuften Aktivhypnose gegenüber dem autogenen Training bestehen darin, daß der Lernvorgang einmal durch das dem jeweiligen Übungsstand des Patienten angepaßte begleitende Vorsprechen des Arztes beschleunigt wird, zum anderen, daß der hypnoide Zustand durch die Fixierübungen weiter vertieft wird. Im übrigen ist die gestufte Aktivhypnose dem autogenen Training sehr ähnlich, wie Langen mit Recht hervorhebt. Der Indikationsbereich der gestuften Aktivhypnose ist etwas enger als der des autogenen Trainings, weil er nicht so weit in den psychohygienischen Bereich hineinreicht. Von Ausnahmen abgesehen, wird sie vor allem bei „seelischen Störungen" (= Neurosen) mit Erfolg angewendet, die von den Fremdneurosen im Sinne von J. H. Schultz über die Rand- und Schichtneurosen, die abnormen Persönlichkeiten – einschließlich der Kernneurosen –, die Medikamentenabhängigkeit und den Alkoholabusus, die Störungen des Sexualtriebes bis zu den vegetativen Regulationsstörungen bzw. psy-

chosomatischen Krankheiten im engeren Sinne reichen. Die psychosomatischen Störungen und Krankheiten sind nach Stokvis und Langen (1965) „die Domäne kombinierter psychotherapeutischer Verfahren, da ausschließlich analytische Behandlungen aus mehreren Gründen nur bei etwa 10–20% der Kranken in Frage kommen". Für die Kontraindikationen gelten die gleichen Gesichtspunkte wie beim autogenen Training. Hinzu kommt: hysterisch strukturierten Persönlichkeiten wird nur selten vorgesprochen. Die Fixierübung unterbleibt oder wird, falls mehrere Ärzte zur Verfügung stehen, von einem anderen Arzt durchgeführt.

22.1.6 Wachsuggestive Verfahren

Allgemeines

Grundsätzlich ist bei jeder ärztlichen, insbesondere therapeutischen und erst recht psychotherapeutischen Tätigkeit die Suggestion in unterschiedlicher Weise und Stärke mit im Spiel, ob es sich dabei um die Verordnung von Medikamenten und Bettruhe, von Diät und Kuren, um die Anwendung von physikalischen Heilmitteln jeglicher Art oder um die in der Arzt-Patient-Beziehung immer wieder zu Worte kommende „Beeinflussung der inneren Haltung" des Patienten handelt, wie z. B. in Form von Beruhigung, Bagatellisierung, Ignorierung, Überredung (Persuasion), Tröstung und Ermutigung, oder um „Beeinflussung der äußeren Haltung", z. B. durch Ablenkung und Zerstreuung, Übung, Isolierung, Milieuwechsel usw. Das gilt auch für die Psychoanalyse, wo die Suggestion durch ihre ständige Bearbeitung bewußt klein gehalten wird, für die Verhaltenstherapie, wo dies durch eine möglichst rationale Versuchsanordnung geschieht, und erst recht für das Psychodrama, wo sie eine sehr große Bedeutung hat, um nur einige wenige Beispiele aus dem engeren Kreis psychotherapeutischer Methoden zu nennen. Es würde im Rahmen dieser Abhandlung zu weit gehen, auf die zahlreichen speziellen wachsuggestiven Verfahren einzugehen, die zum Teil auf uraltes ärztliches Allgemeingut gegründet sind, zum Teil nicht mehr in unsere Zeit hineinpassen. Das gilt auch für die in den 20er Jahren bekannt gewordene und vorübergehend in große Mode gekommene Methode von Coué (1966), eine passiv autosuggestive Methode, bei der zweifellos aber auch hypnosuggestive und massensuggestive Faktoren mit wirksam waren. Erwähnt werden hier nur zwei Methoden, die die Polarität wachsuggestiver Verfahren aufzeigen sollen.

Persuasion

Diese von P. Dubois (1910) um die Jahrhundertwende entwickelte Methode der „Überzeugung" des Patienten von der scheinbaren Widersinnigkeit seines Krankheitszustandes hat zum Ziel, mittels rationaler Psychotherapie „mit sanfter Geduld den Patienten aufzuklären, wobei man die Gesprächsform jeweils seinen geistigen Fähigkeiten anpaßt". In Wirklichkeit handelt es sich dabei aber auch, wie Stokvis mit Recht hervorhebt, in der Hauptsache um Suggestivwirkungen, die von der eindrucksvollen Persönlichkeit dieses Arztes, seiner Eloquenz sowie seiner eigenen, festen Überzeugung von der Echtheit und Richtigkeit seiner Methode ausgingen. Dubois hat auch versucht, Patienten mit sonst nicht beeinflußbaren Schlafstörungen zur inneren Annahme des Nicht-Schlafenkönnens zu bringen. Diese Idee findet später bei Baudouin (1924) ihre Bestätigung in dem Gesetz „der das Gegenteil bewirkenden Anstrengung" und zuletzt bei Frankl (1972) ihre Weiterentwicklung in Form der Technik der „paradoxen Intention" (= intensiver Willensentschluß, das leidvoll erlebte oder gefürchtete Symptom mit allen Kräften herbeizuführen, was unter günstigen Voraussetzungen dessen Verschwinden bewirkt). Auch die von Dubois empfohlene Nichtbeachtung von Krankheitssymptomen bzw. Indifferenzhaltung ihnen gegenüber findet sich bei Frankl wieder als Technik der „Dereflexion".

Protreptik

Als spezielle Methode der Behandlung funktioneller und insbesondere grobhysterischer Symptome hat E. Kretschmer (1959) am Modell der Dressur ein Verfahren entwickelt, das als „einzeitige aktive Reiztherapie" mittels kräftiger sensorischer, insbesondere faradischer Reize in Kombination mit verbalen Reizen und dadurch hervorgerufenen unangenehmen Affekten neurotische Fehlhaltungen in möglichst einer Sitzung in eine normale Verfassung überführen soll. Positive Indikationen haben nach W. Kretschmer (1959) folgende Leiden: Blindheit, Taubheit, Anästhesie, Aphonie, schlaffe und tonisierte Extremitätenlähmungen, Tremor, Steh- und Gehunfähigkeit bei Hysterie. Dabei spielt es keine Rolle, ob diese Störungen akut (etwa durch Schreck) oder langsam im Verlauf eines Lebenskonfliktes oder als hysterische Fixierung und Überlagerung früherer Organschäden entstanden sind. Gegenindikationen sind: hysterischer Dämmerzustand mit Bewegungssturm oder Stupor, Organneurosen, vegetativ-funktionelle und schmerzhafte Konversionssymptome, schwächende bzw. gefährliche organische Begleitkrankheiten, komplizierte Neurosen, insbesondere Psychoneurosen und Psychosen. Wichtig zur Erhaltung des Behandlungserfolges ist eine konsequente Nachbehandlung im unmittelbaren Anschluß an die Protreptik mittels Übungstherapie und Psychagogik.

22.2 Übende Verfahren

22.2.1 Vorbemerkungen zum Übungsbegriff

Übung ist nach Jaspers (1965) „die Steigerung der Leichtigkeit, Schnelligkeit und Gleichmäßigkeit einer Leistung durch deren Wiederholung. Diese geschieht zum Teil durch Mechanisierung ursprünglich mehr absichtlicher, willkürlicher seelischer Leistungen zu mehr reflektorischen, mechanisch ablaufenden". Ebenso wie die Suggestion ist die Übung ein allgemein menschliches Wirkungsprinzip, das sich von der „immanenten Übung" (= biologisch vorhandene, aber nicht durch Leistungssteigerung zutage tretende Übung mit dem Ziel, unter geringem Energieaufwand das leistungsübliche Maß zu erreichen bzw. zu erhalten) über die psychophysische Fähigkeits- und Fertigkeitsschulung in Schule, Universität und Beruf bis hin zur Rehabilitation (= Wiederherstellung verlorengegangener Fähigkeiten durch Übung und Umübung) erstreckt. Auch hier unterscheiden wir zwischen einer positiven (erfolgreichen) und negativen Übung („Paradoxübung" = Leistungsverschlechterung aus Unlust und Hemmung). In der **Psychotherapie** ist der Mechanismus der Übung und der mit ihm eng verbundene des Lernens in allen Methoden mit enthalten und wirksam. Besonders deutlich ist dies bei der Verhaltenstherapie. Die speziellen psychotherapeutischen Übungsmethoden haben zum Inhalt, daß der Kranke nach bestimmten Vorschriften an sich selber arbeitet, und zum Ziel, daß auf diese Weise therapeutisch „erwünschte Veränderungen der seelischen Haltung" erreicht und neue „Fähigkeiten erworben werden". Zwischen den übenden und den suggestiven, insbesondere autosuggestiven Verfahren sind die Grenzziehungen oft schwer durchführbar, und es bestehen fließende Übergänge, wobei im einen Falle der Akzent mehr auf dem Suggestiv-, im anderen mehr auf dem Übungsfaktor liegt.

22.2.2 Autogenes Training (Grundstufe)

Das autogene Training (A.T.) ist das klassische Beispiel einer psychotherapeutischen Methode, bei dem sich Autosuggestion und Übung zu einer Einheit verbinden. Dieses aus dem Selbst (griech. = autos) entstehende (griech. = genos) Üben (= Training) ist eine „aus alten und sicheren ärztlichen Erfahrungen der Hypnose" hervorgegangene Methode der Selbst-(Auto-)Hypnose.

Das A.T. wurde aus Anregungen von Vogt („prophylaktische Ruhepausen") und auf der Grundlage von Selbsterfahrungen hypnotisierter Versuchspersonen in den zwanziger Jahren von J.H. Schultz (1932, 1970, 1935, 1972) entwickelt und zu einer vorbildlich klaren, systematischen Selbstentspannungsmethode ausgebaut, deren weitreichendes Wirkungsspektrum von der Psychohygiene bis zur Psychotherapie reicht. So wird es z.B. in der Schule, im Betrieb, im Sport – einschließlich des Leistungssports – eingesetzt, um hier das Gleichmaß zwischen Spannung und Entspannung zu wahren und damit gesundheitlichen Störungen, die aus anhaltenden Überspannungen, aus Streß etc. herrühren, vorbeugend zu begegnen. Zahlreiche Volkshochschulen vermitteln das A.T. an Interessierte und Bedürftige. Der wichtigste Anwendungsbereich liegt aber zweifellos im Bereich der Heilkunde und hier besonders dem der Psychotherapie, wo es, wie Schultz ausführt, bei einem größeren

Prozentsatz funktionell Gestörter oft schon allein ausreicht, um wieder eine „eutone Mittellage" herzustellen, in anderen Fällen und insbesondere bei den psychosomatischen Krankheiten im engeren Sinne wie auch bei den Psychoneurosen eine wichtige und vielfach unentbehrliche psychotherapeutische Hilfe darstellt. „Das Prinzip der Methode ist darin gegeben, durch bestimmte physiologisch rationale Übungen eine allgemeine Umschaltung der Versuchsperson herbeizuführen, die in Analogie zu den älteren fremdhypnotischen Feststellungen alle Leistungen erlaubt, die den echten suggestiven Zuständen eigentümlich sind." Dabei kann das A.T. einmal als „übungsmäßig erworbene Umschaltung von Körpersystemen" betrachtet werden, „deren Funktionsänderung wieder den Gesamtzustand in erwünschter Weise beeinflußt", zum anderen als eine „Umkehrung des Ausdrucksgesetzes, indem Funktionen, die sonst unter dem Einfluß affektiver Erregungen sich verändern, nunmehr durch selbstgesetzte Veränderungen gewissermaßen einen rückwirkenden Einfluß ausüben". Entspannung im A.T. ist, wie Schultz weiter darstellt, „nicht nur Mittel und Weg zur Versenkungsruhe, sondern auch Wert an sich. Alles gesunde Lebendige wogt zwischen den Polen Spannung – Entspannung; das gesunde Tier ohne Tätigkeit schläft! Der Mensch von heute braucht ein Höchstmaß von spannender Leistung und spannender Selbstbeherrschung; daher ‚verkrampft' er sich leicht, so daß ganz grob mechanische Dinge wie die Atmung, Verdauung usw. notleiden, vom Lebensgefühl und höheren Seelenleben ganz zu schweigen. Das A.T. verlangt zwar unbedingten und ausdauernden Einsatz der inneren Sammlung (Konzentration), es benutzt aber nicht den bewußten Willen, sondern eine innere Hingabe an bestimmte Übungs-Einbildungen". – Die konzentrative Selbstentspannung des A.T. hat also den Sinn, mit genau vorgeschriebenen Übungen sich immer mehr innerlich zu lösen und zu versenken und so eine von innen kommende Umschaltung des gesamten Organismus zu erreichen, die es erlaubt, Gesundes zu stärken, Ungesundes zu mindern oder abzustellen. Wie der Mensch, der lesen gelernt hat, nun lebenslänglich lesen „muß", wenn er Schriftzeichen sieht, „muß" dem autogen Trainierten eine entsprechend gelassene Haltung zur zweiten Natur werden. Wir sprechen von einem „erworbenen Vollzugszwang im normalen Seelenleben". Dabei kann grundsätzlich alles aus eigener Leistung erreicht werden, was auch in der Hypnose hinsichtlich Entspannung und Versenkung erreicht wird. Allerdings ist zu berücksichtigen, daß es allgemein schwerer ist, sich selbst zu hypnotisieren als hypnotisiert zu werden, und daß die zur erhöhten Suggestibilität und Autosuggestibilität führende Bewußtseinssenkung und -einengung für gewöhnlich nicht so stark ist wie in der Hypnose. Das hat auch Kretschmer (1959) und Langen (1969) dazu gebracht, die Variante der „gestuften Aktivhypnose" zu entwickeln. Deshalb sind „sorgfältige und ausdauernde Mitarbeit und ausreichende Selbstverfügung der Versuchspersonen (V.P.)" wichtige Voraussetzungen erfolgreicher Selbstbehandlung, und es dauert immer

sehr viel länger, ehe sich die Heilerfolge einstellen, als bei der Hypnose. So schreibt Schultz aus eigener kritischer Grundhaltung heraus, die sich wohltuend von vielen unkritischen Propagierungen des A.T. in jüngster Zeit abhebt, einschränkend: „Je gesünder V.P. seelisch, um so besser gelingt die Einarbeitung und um so günstiger ist die Aussicht auf Erfolg."

Zu den erreichbaren **Zielen** des autogenen Trainings gehören:
– Selbstentspannung, insbesondere der willkürlichen Körpermuskulatur und der Blutgefäße, hier vor allem der oberflächlichen Hautgefäße.
– Selbstruhigstellung mit Entängstigung durch „Resonanzdämpfung des Affektes"; von daher auch Schlafförderung.
– Erholung mit Leistungssteigerung (z.B. Gedächtnis).
– Selbstregulierung sonst „unwillkürlicher" Körperfunktionen (z.B. Herz-Kreislauf-, Atmungs-, Verdauungssystem).
– Schmerzlinderung bzw. -abstellung.
– Selbstkritik und Selbstkontrolle durch Innenschau in der Versenkung.
– Selbstbestimmung durch in die Versenkung eingebaute formelhafte Vorsätze, die wie posthypnotische Suggestionen automatisch wirken.

Wie schon unter den Voraussetzungen zur Hypnotherapie erwähnt, ist u.a. eine ausreichende **psychische Vorbereitung** des Patienten unbedingt erforderlich, um ihm das nötige Verständnis und die erforderliche Motivation zur aktiven Mitarbeit an seiner Gesundung zu vermitteln. Wir tun dies in Gestalt eines 2stündigen Aufklärungsvortrages mit klaren Anleitungen zur praktischen Durchführung des Trainings für die jeweilige Patientengruppe. Wie schon im Abschnitt Hypnose ausgeführt, ziehen wir dort eine Einzelbehandlung, beim A.T. jedoch die Gruppentherapie vor, sowohl aus zeitlich-ökonomischen Gründen als auch im Hinblick auf didaktische Gesichtspunkte, insbesondere auf eine Verbesserung des Verständnisses durch die zahlreichen Zwischenfragen der Beteiligten, wie auf eine allgemeine Verstärkung der aktiven Mitarbeit durch die lebendigen und inhaltsreichen Gruppendiskussionen über die Selbsterfahrung der einzelnen Teilnehmer, über ihre Schwierigkeiten und Fehler im A.T. sowie die diesbezüglichen Erklärungen, Ratschläge und Hilfen zu deren Behebung.

Erfahrungen mit der Vermittlung des A.T. besonders in Gruppenkursen finden sich bei J.H. Schultz (1970) sowie in zahlreichen Veröffentlichungen u.a. von Binder (1964), Haring (1979), Hoffmann (1981), Iversen (1969), Kleinsorge (1974), Kraft (1982), Krapf (1976), Lohmann (1980), Mensen (1975), Rosa (1973, 1975), Thomas (1972), Wallnöfer (1979). Zur Theorie des A.T. hat Garcia (1983) einen wichtigen Beitrag geliefert. Das A.T. hat in den letzten Jahren auch zunehmend Eingang in die Kinderheilkunde gefunden (G. Biermann, 1975; Kruse, 1974, 1984).

Wie bei zahlreichen ostasiatischen Meditationspraktiken und insbesondere beim Yoga erleichtern bzw. ermöglichen bestimmte **Übungshaltungen** („Asanas") die Selbstentspannung und -versenkung. Diese sind für das A.T.:

- **Liegehaltung** in horizontaler Rückenlage mit den Armen über einer leichten Zudecke, mit im Ellbogen abgewinkelten Armen, pronierten Händen und mit Adduktoren-Entspannung der Beine.
- **Droschkenkutscherhaltung** auf einem Hocker oder Stuhl ohne Seitenlehnen (ca. 10 cm von der Rückenlehne entfernt) mit im Knochen-Gelenk-Bandapparat der Wirbelsäule möglichst entspannt ruhendem Rumpf, leicht nach vorn fallendem Kopf, zwanglos aufgestellten und gespreizten Beinen, auf deren Oberschenkeln die Unterarme ruhen, während die Hände locker in den Schoß fallen.
- **Passive Sitzhaltung** in einem bequemen Lehnsessel („Fernsehsesselhaltung").

Weitere Voraussetzungen für die A.T.-Übungen sind der Augenschluß, analog zum „reflektorischen Lidschluß" bei den Einleitungstechniken der Hypnose, ein abgedunkelter, möglichst geräuscharmer (notfalls ein sog. dunkler „Augentröster" vor den Augen, Oropax in den Ohren) und angenehm temperierter Raum. Notwendig ist mindestens 2–3maliges tägliches Üben in einer der angegebenen Haltungen, wobei ein Wechsel zwischen Liege- und Sitzhaltungen günstig ist. Die Übungen sollen möglichst zu den gleichen Zeiten stattfinden, die für die konzentrative Selbstversenkung am besten geeignet sind. Die Übungszeiten sollen im Anfang nicht zu lang sein, um die Übungen nicht durch ein „Zu-gut-machen-Wollen" in Gefahr zu bringen, brauchen aber doch nach unseren Erfahrungen zwischen 3 und 10 Minuten Dauer, um gelingen zu können. Eventuell empfiehlt sich bei starker Unruhe und Erwartungsspannung eine Zweiteilung in jeweils eine Vor- und eine Nachübung. Unentbehrlich, wie bei der Hypnose (vgl. Desuggestionierung!), ist das „Zurücknehmen" der einzelnen Übungen, das vor allem der Muskelentspannung (= Schwere) und allgemeinen Entspannung (= Ruhe), der optischen Subtraktion und Introversion (= Augenschluß und Selbstversenkung) gilt und dreizeitig durchgeführt wird mit dem kurzen Formelkommando:

1. **Arm(e) fest!** Dabei werden die Arme im Ellbogen ein paarmal kräftig gebeugt und gestreckt.
2. **Tief atmen!** Es wird tief ein- und ausgeatmet.
3. **Augen auf!** Die Augen werden wieder weit geöffnet.

Die Zurücknahme (Weckreiz) erfolgt jedoch nicht, wenn das A.T. als Einschlafhilfe benutzt wird. Dann dreht sich der Patient im Hypnoid in die gewohnte Schlafhaltung, um einzuschlafen, und das A.T. löst sich im Schlaf ohne Zurücknahme ganz von selbst wieder auf. Ausnahmen sind einzelne Perfektionisten, die nicht einschlafen können, ehe sie nicht eine ihnen gegebene Vorschrift von A–Z erfüllt haben.

Das A.T. hat einen sechsstufigen **Übungsaufbau**. Die einzelnen Übungen werden im Abstand von ge-

wöhnlich 14 Tagen erlernt, wozu die Patientengruppe jeweils wieder zusammenkommt, die Erfahrungen mit der vorausgegangenen Übung diskutiert und in die nächste Übung eingeführt wird.

1. Schwereübung. Sie dient der Muskelentspannung, die bekanntlich als schwer empfunden wird. Die Übung erstreckt sich zuerst auf den dominanten Arm, um die Konzentration nicht zu überfordern, und generalisiert sich für gewöhnlich im weiteren Übungsverlauf ganz von selbst auf die übrigen Körper, wobei die autosuggestive Formel dem jeweiligen Übungsstand angepaßt wird.
Zu Beginn heißt es:
1. „Der rechte (linke) Arm ist ganz schwer" (ca. 6 × wiederholt);
2. „Ich bin ganz ruhig" (ca. 2 × wiederholt).

2. Wärmeübung. Diese dient der Blutgefäßentspannung und verfährt analog zu 1.
Anfangsformel:
1. „Der rechte (linke) Arm ist ganz warm" (ca. 6 × wiederholt);
2. „Ich bin ganz ruhig" (ca. 2 × wiederholt).

Die Wärmeübung wird nicht gesondert zurückgenommen, da sich die Blutgefäße spontan wieder einregulieren.
Die Übungen 1 und 2 sind für sich allein schon vielfach ausreichend für eine Selbstentspannung, -ruhigstellung und -versenkung mit den oben angegebenen Leistungen der affektiven Resonanzdämpfung, der Erholung und der Leistungssteigerung. Auch kommt es dadurch schon zu einer induktiven Entspannungswirkung auf die in ihren Regulationen gestörten anderen Organsysteme. Deshalb haben Kretschmer und Langen auf die anderen Übungen des A.T. verzichtet. Für die psychosomatischen Funktionsstörungen und Krankheiten sind aber die weiteren autogenen Organübungen von Bedeutung und werden in der Folge ausgeführt. Außerdem führen sie zu einer fortschreitenden Verstärkung der trophotropen Umschaltung.

3. Herzübung. Sie dient zur selbständigen Beeinflussung der Herztätigkeit, insbesondere des Herzrhythmus. Die Formeln dazu lauten:
1. „Das Herz schlägt ruhig und regelmäßig" bzw. „ruhig und kräftig" (ca. 6 × wiederholt);
2. „Ich bin ganz ruhig" (ca. 2 × wiederholt).

4. Atemübung. Vgl. Herzübung. Die Formeln dazu lauten:
1. „Die Atmung ist ganz ruhig" oder „es atmet mich" (ca. 6 × wiederholt);
2. „Ich bin ganz ruhig" (ca. 2 × wiederholt).

Die beiden Übungen 3 und 4 sind Rhythmus-Übungen, die nicht solche Sensationen erleben lassen wie die Übungen 1 und 2, und stellen von daher in zahlreichen Fällen Übungen im Dunkelfeld des Körpers und seiner Lebensvollzüge dar. Darauf ist der Patient vorbereitend hinzuweisen, ebenso wie darauf, daß jedwede Art von Manipulation des Herzschlages wie des Atemrhythmus – etwa wie bei bestimmten Formen von Atemgymnastik – vermieden werden muß, um keine Störungen zu produzieren.

5. Sonnengeflechtsübung. Hier wird eine Beeinflussung der Abdominal- und Unterleibsorgane gesucht und erreicht, wozu sich die Konzentration auf den Plexus solaris als größtem sympathischem Nervengeflecht im Leib besonders gut eignet. Die Formeln dazu lauten:

1. „Das Sonnengeflecht ist ruhig, strömend warm" (6 ×);
2. „Ich bin ganz ruhig" (ca. 2 ×).

6. Kopfübung (Stirnkühleübung). Auf dem Umweg über die Stirnkühle (1. „Die Stirn ist angenehm kühl" – ca. 2–6 ×; 2. „Ich bin ganz ruhig" – ca. 2 ×) wird eine weitere Ruhigstellung der psychischen Funktionen angestrebt und erreicht, wobei das Motto wegweisend ist: Warmes Herz und kühler Kopf! Diese Übung bewirkt eine Minderdurchblutung der Stirn und ist von daher, wie alle Minderdurchblutungen, nicht so indifferent wie die Mehrdurchblutung bei der Wärmeübung und der Sonnengeflechtsübung. Sie muß deshalb mit Vorsicht eingeübt werden. Das gilt besonders für Patienten mit Herz-Kreislauf-Störungen, mit Neigungen zu Ohnmachts- und zu Migräneanfällen. Eventuell kann der Kopf deshalb oder überhaupt durch eine Ersatz- oder Zusatzformel angesprochen werden: „Der Kopf ist ganz ruhig, frisch, frei und klar." Aus den angeführten Gründen ist es verständlich, daß die einzelnen Übungen des A. T. nur unter fortlaufender ärztlicher Kontrolle durchgeführt werden sollten.

Hat der Patient nach ca. 3 Monaten auf diese Weise eine „selbstgesetzte Entspannungs-Spannungsumschaltung" mit den geschilderten Leistungen erworben („Vereinheitlichung"), so bietet sich als Ergänzung für den momentanen Gebrauch bzw. als Ersatz auch eine „Teilentspannung" an, zu der sich das Schulter-Nackenfeld besonders eignet (Formel: „Schulter-Nackenfeld ganz weich und warm; ich bin ganz ruhig"). Diese kann auch ohne die eingangs beschriebenen Übungshaltungen, im Sitzen, Stehen und Gehen ohne Augenschluß und Introversion an beliebigem Ort und zu verschiedenen Zwecken (z. B. zur Selbstruhigstellung, affektiven Resonanzdämpfung und kurzfristigen Erholung bei akuter Erregung bzw. Überlastung) durchgeführt werden und wird als **Schulter-Nackenfeld-Übung** bezeichnet.

Zur Technik des A. T. sind noch einige wenige Bemerkungen nötig. Für einen großen Teil der Patienten ist die Verbalisation der Übungsformeln gemäß, allerdings ohne diese laut oder leise vorzusprechen, was die Konzentration ablenken würde („innere Stimme"). Sog. Eidetiker müssen sich jedoch bildlicher Vorstellungen („Einbildungen") bedienen, um zu dem gewünschten Erfolg zu kommen, andere kommen wieder besser mit „Leuchtbuchstabenschrift im Dunkel der geschlossenen Augen", mit „Klangsprüchen" oder ähnlichem zurecht. Das muß ausprobiert und flexibel gehandhabt werden. Der stufenweise Aufbau des A. T. verlangt ferner, daß jede einzelne Übungsstufe

jeweils durchschritten wird, wobei sich mit zunehmender Einübung die bereits beschriebene „Mechanisierung" bzw. „Automatisierung" der einzelnen Teilübungen vollzieht, so daß es oft schon bei Einnahme der gewohnten Übungshaltung zur „Vereinheitlichung" der „organismischen Gesamtumschaltung" kommt. Demzufolge können im weiteren Übungsverlauf die erreichten Übungen weitgehend autosuggestiv verkürzt werden (z. B. „Schwere", „Wärme", „Herzschlag" usw.), während jede neue Übung zunächst einmal wiederholt und monotonisierend eingeübt werden muß. Zuletzt ein Wort zu den **„formelhaften Vorsatzbildungen"** (vgl. auch die „wandspruchartigen Leitsätze" in der gestuften Aktivhypnose). Diese wirken, in den Übungsaufbau der Unterstufe des A. T. günstig eingepaßt und persönlichkeitsgerecht formuliert, in der übenden Wiederholung automatisch wie posthypnotische Suggestionen und dienen zur besseren Selbstbeherrschung, Selbstbestimmung und Selbstverwirklichung. Beispiele dafür sind: „Angst geht vorüber", „Mut ist Sieg", „Ordnung ist Freiheit", „Arbeit macht Freude" usw. Die formelhaften Vorsatzbildungen sind auch auf bestimmte neurotische Zielsymptome ausgerichtet und wirken natürlich auf diese nur insoweit reduzierend, wie der betreffende Patient dazu bereit und einsichtig genug ist.

Die Oberstufe des A. T., die erst nach vollständiger und sicherer Beherrschung der zuvor geschilderten Grundstufe erlernt werden kann, ist ein vertiefter und systematisierter Einstieg in die Welt der inneren Bilder mit der Zielsetzung einer bewußten Einflußnahme auf die Innenerlebnisse und damit auf das Unbewußte des betreffenden Menschen. Von der Oberstufe des A. T. bestehen enge Beziehungen zum „Bildstreifendenken" (Happich, 1932; Kretschmer, 1971) sowie insbesondere zum „katathymen Bilderleben" (Leuner, 1970).

Hinsichtlich der **Indikation** ist hervorzuheben, daß das A. T., entsprechend seinem Wesen als selbstgesetzte „allgemein organismische Umschaltung", eine überaus weite Anwendung hat. War schon bei der Hypnose eine ausgedehnte Wirkungs- und Verwendungsmöglichkeit festzustellen, so trifft dies in noch weit größerem Maße für das A. T. zu. Es gibt praktisch keine psychosomatische Störung bzw. Erkrankung, bei der das A. T. nicht eingesetzt werden kann oder angewendet worden ist, meist im Rahmen eines mehrdimensionalen Behandlungskonzeptes zur Unterstützung der übrigen klinischen und insbesondere psychotherapeutischen Therapiemaßnahmen. Das 6bändige Handbuch des autogenen Trainings (Hrsg. W. Luthe, 1969–1973) referiert die einschlägigen Veröffentlichungen bis zum Jahre 1969 kritisch und umfassend. Sie reichen von den Krankheiten der Verdauungsorgane über die des Herz-Kreislauf-Systems, der Atmungsorgane, der inneren Drüsen und des Stoffwechsels, der Bewegungsorgane, der Hämophilie, des Urogenitalsystems bis hin zu Anwendungen im Bereich der Gynäkologie, der Geburtshilfe, der Dermatologie, der Ophthalmologie, der Chirurgie und der Zahnheilkunde. Hingewiesen sei in diesem Zusammenhang, quasi stellvertretend für die zahlreichen Anwendungsbereiche, nur auf die Methoden der schmerzarmen Geburt, bei denen die autogene Entbindungserleichterung ein wesentlicher Bestandteil ist.

Kontraindikationen sind nicht bekannt, wobei daran erinnert werden muß, daß psychisch Gesunde mit freier Selbstverfügung und ausdauernder Mitarbeit das A. T. am besten erlernen, Nervöse, psychosomatisch Gestörte und Kranke sowie chronisch körperlich Kranke dies schwerer lernen, während psychisch schwer Gestörte (z. B. Kernneurosen, psychopathische Persönlichkeiten), Geisteskranke und Geistesschwache das A. T. meist gar nicht lernen. Wegen der mitunter beträchtlichen Schwierigkeiten beim Erlernen des A. T. bei psychosomatischen und somatopsychischen Patienten, die mit der mangelnden freien Selbstverfügung und der vielfach nicht ausdauernden Mitarbeit zusammenhängen, sind hier Geduld, Geschick und Überzeugungskraft von seiten des Therapeuten besonders wichtig. Als wertvolle Hilfe hat sich für uns hier in der Einzelbehandlung oft der anfängliche Einstieg in die Hypnose erwiesen, mit sukzessiver Überleitung in das A. T. und mit wiederholtem „Begleitsprechen" im weiteren Verlauf der Übungen.

22.2.3 Katathymes Bilderleben

Jörg Michael Herrmann

Das von H. C. Leuner (1954, 1955) entwickelte, tiefenpsychologisch orientierte Verfahren des katathymen Bilderlebens (K. B., auch als Tagtraumtechnik oder Symboldrama bezeichnet) beruht darauf, daß es möglich ist, sich in einem hypnoiden, Ich-regressiven Zustand Bilder (Tagträume) vorzustellen.

Bereits 1932 publizierte Happich eine Arbeit, in der er über Imaginationen während einer meditativen Psychotherapie berichtete. Ähnliche Phänomene wurden als „Bildstreifendenken" von Kretschmer (1971), als „hypnagoge Imaginationen" von Holt (1964) und in der Oberstufe des autogenen Trainings von J. H. Schultz (1970) beschrieben.

Diese Fähigkeit des Menschen zur Imagination, d.h. innerseelische Zustände als Bilder, die Symbolcharakter besitzen, zu äußern, wurde von H. C. Leuner (1964) experimentell und systematisch untersucht und weiterentwickelt:

12 Standardmotive (Motivvorstellungen) dienen als Kristallisationskern, auf den der eigene innere Zustand projiziert werden kann, d.h., die imaginierten Bilder spiegeln gleichzeitig den individuellen – unbewußten – Konfliktbereich des Patienten wider.

Mit Hilfe einer Entspannungstechnik, z.B. der Grundstufe des autogenen Trainings, kommt der Patient in einen Zustand der Ich-Regression, eines leichten Hypnoids, in dem ihm vage formulierte Vorstellungsmotive als Kristallisationskern angeboten werden. Standardmotive der Grundstufe des katathymen Bilderlebens sind die Wiese (symbolischer Ausdruck der aktuellen Gestimmtheit), der Bachlauf (symbolischer Ausdruck der fließenden Dynamik des

seelischen Geschehens, der Entwicklung, des „Lebensflusses"), der Berg (symbolischer Ausdruck des eigenen Leistungsverhaltens und Anspruchsniveaus), das Haus (symbolischer Ausdruck der eigenen Persönlichkeit) und der Waldrand (symbolischer Ausdruck für das Unbewußte, verdrängte, verborgene Inhalte). Während die therapeutischen Techniken bei den Standardmotiven der Grundstufe die Entfaltung kreativer Imaginationen üben, stehen bei den Standardmotiven der Mittelstufe ein assoziatives Vorgehen, die Fokussierung akuter Konflikte, das Durcharbeiten und Übertragungsphänomene im Vordergrund. Zu den Standardmotiven der Mittelstufe gehören das Erscheinen naher Beziehungspersonen (Ausdruck der Familienbeziehungen und Familiendynamik), die Ermittlung des Ich-Ideals (Ausdruck von Identitätsproblemen), Einstellung gegenüber der Sexualität mit dem Motiv „Autostop" für Patientinnen und „Rosenbusch" für Patienten und schließlich die Prüfung aggressiver Impulse mit dem Motiv „Löwe".

Zu den Standardmotiven der Oberstufe gehören die „Höhle" (symbolischer Ausdruck unbewußter Fehlhaltungen und Probleme der Homoerotik und Rivalität), der „Vulkan" (Ausdruck stark andrängender aggressiver Impulse [Leuner, 1980] oder narzißtischer Wut [Roth, 1984]) und „Folianten" (symbolischer Ausdruck der kindlichen Sicht von „stereotypen Erwachsenen" [Kosbab, 1972] und von archaisch verschlüsselten Inhalten [Leuner, 1980]).

Ein 60jähriger Arzt mit ängstlich zwanghafter Struktur und essentieller Hypertonie, die bereits zu Organkomplikationen in Form von Augenhintergrundsveränderungen und Linksherzhypertrophie geführt hat, erlebt bereits in der Anfangsphase der Behandlung mit dem katathymen Bilderleben die Auseinandersetzung mit seinen eigenen aggressiven Anteilen:

Beim Motiv „Bach" wird er plötzlich, als er in einem Ruderboot flußabwärts fährt, von Schwänen angegriffen und möchte diese sofort mit einem Ruder abwehren und erschlagen. Da er damit eigene Ich-Anteile zerstören würde, muß dies bei der therapeutischen Intervention berücksichtigt werden. Der Patient wird aufgefordert, einen Schwan, der drohend hinter ihm aufs Boot fliegt, detailliert zu beschreiben und zu schauen, ob er nicht etwas zu essen bei sich hat. Er findet Weißbrotstücke, die er dem Schwan zuwirft, so daß sich dieser wieder entfernt.

Als therapeutische Technik wurde, da die Auseinandersetzung noch mit dem Standardmotiv „Bach" der Grundstufe erfolgte, die Entfaltung kreativer Imaginationen (Leuner, 1981) benutzt. Als „Regieprinzip", d.h. Operation am Symbol, stand „Versöhnen" und „Nähren" während dieser Sitzung im Vordergrund.

Die – für Hypertoniker charakteristischen – gehemmten aggressiven Impulse konnten anhand des Traumes anschließend mit dem Patienten bearbeitet werden, ohne daß Schuldgefühle auftraten, die entstanden wären, wenn der Patient die Schwäne erschlagen hätte.

Im weiteren Therapieverlauf konnten aggressive Inhalte (vor allem ärgerliche Gefühle gegen die überfürsorgliche Ehefrau und die ihn „im Stich lassenden", erwachsenen Söhne) geäußert und bearbeitet werden. Gleichzeitig sanken nach insgesamt 15 Sitzungen, in denen die Standardmotive der Grund- und Mittelstufe angewandt wurden, die Blutdruckwerte von durchschnittlich 180/100 mm Hg auf durchschnittlich 150/90 mm Hg. Eine medikamentöse antihypertensive Therapie war bei einer Nachbeobachtungszeit von 12 Monaten nicht mehr notwendig.

Die Interpretation des Symbolcharakters der Imaginationen erfolgt nach der allgemein gültigen Traumsymbolik (Freud, 1972; Jung, 1948), der individuellen Symbolik und einer phänomenologischen Betrachtungsweise.

Inzwischen liegen eine Reihe von Untersuchungen vor, die die therapeutische Wirksamkeit des katathymen Bilderlebens bei Patienten mit Colitis ulcerosa (Wilke, 1980), bei Patientinnen mit psychisch bedingten gynäkologischen Symptomen und Sexualstörungen (Roth, 1976) und Patienten mit Anorexia nervosa (Klessmann und Klessmann, 1978) belegen.

Gut dokumentierte Einzelfallstudien zeigen noch folgende Indikationsbereiche: Hyperhidrosis, Urtikaria (Pszywyj, 1980), Spannungskopfschmerzen (Roth, 1980), Betreuung von Malignompatienten (Landau, 1980; Szonn, 1980), Morbus Crohn (Simmet, 1980), Asthma bronchiale (Zepf, 1980; Wilke, 1984), funktionelle Herzbeschwerden (Steiner, 1982; Eibach, 1982) und Zwangsneurosen (Salvisberg, 1982).

Nach den bisherigen Untersuchungen und Erfahrungen ist – nach Leuner (1981) – das katathyme Bilderleben bei mangelnder Intelligenz, bei Psychosen und hirnorganischen Syndromen, schweren Depressionen, hysterischen Neurosen, Borderline- und narzißtischen Syndromen kontraindiziert.

Neuere Studien zeigen, daß der Anwendungsbereich des katathymen Bilderlebens kontinuierlich erweitert wird, z.B. durch Einführen des katathymen Bilderlebens in die Behandlung von Suchtpatienten (Stettler, 1984), als gruppentherapeutisches (Sachsse, 1984) und paartherapeutisches Verfahren (Kottje-Birnbacher, 1981).

Die therapeutischen Techniken des katathymen Bilderlebens entsprechen den tiefenpsychologischen Methoden (Wilke und Leuner, 1989):

Die Technik der freien Assoziation bezieht sich auf die imaginative Ebene, die Regression kann der Patient als konfliktbesetzte oder harmonische Kindheitsszene vor dem Konflikt erleben und dann zwischen konfliktzentrierten Imaginationen und der realen Situation eine Beziehung herstellen. Weiterhin kann ein akuter Konflikt fokussiert und imaginativ durch vorsichtig gelenktes Probehandeln ausgedrückt werden. Da Mißerfolge in der Therapie zumeist Folge einer nicht geklärten Gegenübertragung des Therapeuten sind, sind Selbsterfahrung und Ausbildung in Neurosenlehre für den K. B.-Therapeuten unabdingbare Voraussetzungen.

Die Kombination von katathymem Bilderleben und psychoanalytischer Therapie ist unter der Voraussetzung möglich, daß das K. B.-Material unter psychoanalytischen Aspekten bearbeitet wird, daß das katathyme Bilderleben als Hilfsmittel in der klassischen Psychoanalyse genutzt wird, um Widerstände zu bearbeiten, und daß die Imagination für die Psychoanalyse bei Patienten eingesetzt wird, die nicht träumen.

22.2.4 Progressive Relaxation

Von experimentell-psychologischen und psychophysiologischen Erfahrungen ausgehend, hat E. Jacobson (1938), ebenfalls in den zwanziger Jahren, seine Methode der progressiven Relaxation (P. R.) entwickelt. Hier liegt der Akzent eindeutig auf dem Faktor der Übung. Im Unterschied zum A. T. handelt es sich bei der P. R. um eine Selbstentspannungstechnik auf der Grundlage einer psychophysiologischen Muskelarbeit, während die systematische Erzielung eines Ruhe- und Versenkungszustandes durch konzentrative Vergegenwärtigung fehlt. Dementsprechend stellt die im A. T. zentrale Bewußtseinsveränderung des Hypnoids mit organismischer Gesamtumschaltung bei Jacobson (1938) nur einen Nebenbefund dar. Die P. R. besteht aus „willkürlich fortgesetzter Reduktion des Tonus oder der Aktivität von Muskelgruppen und von motorischen oder assoziierten Teilen des Nervensystems. Ist die Relaxation auf eine besondere Muskelgruppe oder einen Teil, wie ein Glied, begrenzt, wird sie lokal genannt; schließt sie so gut wie den ganzen lebenden Körper ein, so wird sie allgemein genannt". Auf die engen Beziehungen zwischen psychischer und motorischer Spannung sowie umgekehrt zwischen motorischer und psychischer Entspannung, die bis zur „Bewußtseinsleere" gehen kann, sei hier erklärend nur kurz hingewiesen. Von lokaler Muskelentspannung ausgehend, schreitet die Versuchsperson durch tägliches Üben zu den verschiedenen Hauptgruppen der Körpermuskulatur fort und kommt so mehr und mehr zu einer „Gewohnheit der Ruhe", die sich als Haltung automatisiert. Die Muskelarbeit von Jacobson (1938) ist darauf ausgerichtet, durch das systematische Erleben von muskulären Spannungs- und Entspannungszuständen quasi eine „Kultur des Muskelsinnes" herzustellen, mit dem Ziel, „einen glücklichen Durchschnitt" zwischen zu viel und zu wenig Aufmerksamkeit zu erreichen. Ist die Relaxation ausgebildet, so soll sie „am besten automatisch und mit weniger oder nicht deutlich bewußter Führung weitergehen".

Die psychische Vorbereitung ist knapp und allgemein gehalten. Rasch geht es zu den praktischen Übungen. Diese geschehen im Liegen unter Augenschluß und halbschwebender Aufmerksamkeit täglich 1–2 Stunden lang, unter Zuhilfenahme von ½–1stündigen gemeinsamen Übungssitzungen, die je nach der Aufgabe und den Schwierigkeiten 3mal wöchentlich für Wochen, Monate und eventuell Jahre stattfinden. Man beginnt mit einer Entspannung der Gliedmaßen in ihren einzelnen muskulären Anteilen und größeren Muskelgruppen, worauf in den folgenden Schritten die Muskeln von Brust, Stirn, Augen, Zunge, Lippen, Kehle und Kehlkopf zur lösenden Entspannung gelangen. Als besondere Geschicklichkeitsprobe dient die Augen- und Lidenspannung, einschließlich der Brauen. Hier sind die Vorbedingungen zu partiellen Einschlaferlebnissen besonders deutlich, was die Methode ebenfalls, wie das A. T., gut zur Linderung von Schlafstörungen geeignet sein läßt. Zur Überleitung auf „geistige Entspannung" führt Jacobson (1938) seine Versuchspersonen den Weg, sie die kleinen Kontraktionserlebnisse der Augen, des Sprachapparates und anderer hoch ausdruckswertiger Systeme kontrollieren zu lassen, die bei „Gedanken", „Gefühlen" usw. auftreten.

Sprachwerkzeugentspannung soll Gedankenruhe bringen. Vereinfacht und verkürzt kann das Verfahren werden, indem nur einige Muskelgruppen bearbeitet oder bilaterale Übungen angestellt werden; auch einheitliche Zusammenfassung von Muskelgruppen kann diesen Zweck erfüllen, ferner Auslassung der Ausbildung des Muskelsinnes. Im einfachsten Falle besteht die Arbeit lediglich in Anleitung zur Entspannung solcher Partien, die dem Arzt besonders verkrampft erscheinen, ohne daß irgendein allgemeines Training stattfindet. Es handelt sich dann um den „Relaxationsfall örtlicher Gymnastik". Eine differentielle, gewissermaßen nur graduelle Relaxation soll entspannte Aktivität vermitteln, wie sie im Unterricht der Körperbildung gymnastischer und künstlerischer Art als zentrale oder als Teilaufgabe angestrebt wird.

Die P. R. hat sich vor allem in den USA durchgesetzt, während sie bei uns nur wenig Zuspruch gefunden hat. Der Grund dafür ist vor allem in der einseitigen Ausrichtung auf die Muskelarbeit, dem komplizierten Übungsaufbau bis zum allgemeinen Training sowie dem erheblichen Zeitaufwand zu sehen. Erfolge werden, wie bei dem A. T., bei den psychosomatischen Störungen und Krankheiten beschrieben. In vereinfachter Form dient die P. R. als einleitende Entspannungstechnik bei der Verhaltenstherapie (s. Kap. 16 „Verhaltenstherapie").

22.2.5 Hypnotherapie nach Milton H. Erickson

Die Hypnotherapie von M. H. Erickson in einen knappen, klaren und das Wesentliche zusammenfassenden Lehrbuchbeitrag zu bringen, ist ein schwieriges Unterfangen. Entzieht sie sich doch dem Versuch, sie gleichzeitig im Extrakt und doch verständlich darzustellen weit mehr als die klassische Hypnose, welche zuvor abgehandelt wurde. Dies hängt wohl zusammen mit ihrer Komplexität und Unschärfe einerseits, mit der großen Variationsbreite ihrer Techniken und außerordentlichen Flexibilität des methodischen Vorgehens von Fall zu Fall andererseits sowie auch von Therapeut zu Therapeut. Halten wir uns an die von Erickson selbst initiierten Gedanken und Handlungsweisen, so ist zuvorderst eine fast radikal zu nennende Änderung der herkömmlichen Sichtweise im Hinblick auf das diagnostische wie therapeutische Vorgehen erforderlich, welche nicht ohne Widerspruch im einzelnen geschehen kann. Die Diagnostik wird so kurz wie möglich gehalten, die Therapie als „Prozeß des Erreichens und Nutzens von Ressourcen" nimmt einen viel breiteren Raum ein. Dies verstellt jedoch keineswegs den Blick auf die faszinierende

Persönlichkeit eines begnadeten Psychotherapeuten mit unzweifelhaft großen Heilerfolgen, der in der Hoch-Zeit seines beeindruckenden Wirkens in den mittleren Lebensjahren leider nicht die weltweite Aufmerksamkeit und Anerkennung gefunden hat, welche er zweifellos verdient gehabt hätte, und erst im nachhinein eine erstaunliche Aufwertung findet, welche ihrerseits wieder die Gefahr einer modischen Übertreibung enthält, die dem Erscheinungsbild und Wirken dieses im Grunde bescheidenen Mannes so gar nicht entspricht. Um die besondere Form dieser Hypnotherapie vorab zu kennzeichnen, wollen wir sie als eine systemische Kommunikationstherapie beschreiben, bei welcher die „formale Hypnose" eine relativ untergeordnete Rolle spielt, weil diese nur in ca. 20% der Fälle von Erickson eingesetzt wurde (Beahrs, 1971).

Für das Verständnis dieser Therapieform ist es hilfreich, die Persönlichkeit und die wirklich außerordentliche Lebens- und Krankheitsgeschichte von Erickson vorauszuschicken. Dies hebt auch Peter (1988) hervor, wenn er warnend ausspricht, „aus Ericksons Arbeit extrahierte Techniken zu lernen, ist eine Sache, sie in hypnotherapeutischer Arbeit sorgfältig und sinnvoll einzusetzen, wenn man nicht Erickson ist, ist etwas anderes".

Erickson wurde 1901 geboren und wuchs in einer kinderreichen Familie eines einfachen Bergarbeiters auf, der später nach Wisconsin übersiedelte und einen Bauernhof betrieb. Er war von Geburt an partiell farbenblind, tontaub und legasthenisch; auch litt er an einer allergischen Diathese, welche ihn z.B. nach harmlosen Insektenstichen und einer später einmal erforderlich gewordenen Tetanusprophylaxe komatös erkranken ließ. Mit 17 Jahren erlitt er eine erste schwere Kinderlähmung, welche ihn fast ein Jahr lang weitgehend bewegungsunfähig machte und die ihn nach einer Wiederholung im Alter von 48 Jahren mehr und mehr an den Rollstuhl fesselte, bis er nach einem reichen und erfüllten Leben mit 79 Jahren friedlich starb, das er mit nie erlahmendem Lebensmut, Lebensfreude und schöpferischer Kraft gelebt, ja man kann in Anbetracht der vielen Behinderungen durchaus sagen, bezwungen hat, wenn wir seinen Berufsweg vom abgeschlossenen Psychologiestudium zum Medizinstudium, zum Facharzt für Psychiatrie mit späterer außerordentlicher Professur, zu einer regen Lehrtätigkeit, Privatpraxis und zu vielbesuchten Lehrseminaren in seinem letzten Wohnort Phoenix (Arizona) mitsamt den zahlreichen Veröffentlichungen bedenken und dabei den privaten, in Form eines glücklichen Ehe- und Familienlebens sowie die vielen Freundschaften, welche er gepflegt hat, nicht vergessen.

Alle seine vielfältigen Behinderungen hat Erickson in einer erstaunlichen Weise genutzt und positiv umdeutend verwertet. Von daher ist es leicht verständlich, daß im Mittelpunkt seiner zukunftsorientierten Psychotherapie das **„Utilisationsprinzip"** steht. Danach ist jedes vom Patienten gezeigte Verhalten als Form der Kooperation mit dem Therapeuten anzusehen und als etwas, das letztlich zu einer konstruktiven Lösung führt (Schmidt, 1985). Den Legastheniker führte u. a. sein bewundernswerter Eifer und seine nie nachlassende Ausdauer beim mühevollen Erlernen der Schrift und des Lesens dazu, daß er als Schüler den Spitznamen „dictionary" bekam. Dabei waren ihm visuelle Halluzinationen autohypnoiden Charakters vielfach behilflich, welche auch richtungweisend für den späteren Gebrauch von Hypnosen sein könnten. Ein unlösbares Problem stellte es für Erickson als Kind z. B. dar, eine ‚3' von einem ‚m' zu unterscheiden. Plötzlich sah er nach seinem eigenen Bekunden „innerhalb eines blendenden Lichtblitzes die ‚3' und das ‚m' nebeneinander. Das ‚m' stand auf seinen Füßen und die ‚3' lag auf der Seite und streckte die Füße von sich" (Rossi und Erickson, 1977). So nutzte Erickson auch die fast einjährige Zeit der langsamen und mühevollen Rekonvaleszenz nach der ersten Poliomyelitis-Erkrankung, indem er die noch verbliebenen Fähigkeiten seiner natürlichen Beobachtungsgabe, seines Hör- und Sprachvermögens sowie seiner Sensorik systematisch steigerte und den Wiedergewinn der Motorik dank unermüdlichen Lernens in kleinen Schritten aufmerksam begleitete und in Erinnerung bewahrte. Von daher ist vielleicht sein besonderes Interesse für ideomotorische und ideosensorische Phänomene in der Hypnose, welche er reichhaltig einsetzte, nicht verwunderlich und seine in der Therapie bis zu 30 000mal angeblich benutzte Arm-Levitation (Zeig, 1988) gut verständlich, welche er meist freudig und auch staunend als erstes Zeichen einer positiven Veränderung seinen Patienten in der verschiedensten Weise mitteilte, als Nachklang eigenen langwierigen Bemühens zur Wiedererlangung einmal verlorengegangener und doch so kostbarer manueller Beweglichkeit. Mit dem Utilisationsansatz eng verbunden ist jener der **Veränderung.** Therapie ist danach hauptsächlich ein Prozeß des Erreichens und Nutzens von eigenen unausgeschöpften Ressourcen. Dazu ist Veränderung erforderlich, ohne daß Einsicht vorausgehen noch diese begleiten muß, wenn auch Einsicht und Verständnis nicht ausgeschlossen zu werden brauchen. Veränderung zu fördern, hat bei Erickson den absoluten Vorrang gegenüber dem Erhellen der Vergangenheit und der Bedeutung der Krankheitssymptome. Diese Veränderungen geschahen bei Erickson vielfach in einer Weise, welche seinen Patienten gar nicht bewußt wurde (Haley, 1988). Dazu diente auch die Fabulierfähigkeit und -freude von Erickson, im Verein mit einer treffsicheren Beobachtungsgabe und Kombinationsfähigkeit im Hinblick auf sein Gegenüber. Diese wurde oft mit Herstellung einer „Alltagstrance" verbunden, um den therapeutisch bedeutungsvollen Anteil der Anekdote dadurch tiefenwirksamer zu gestalten. Obwohl sich viele seiner Geschichten wiederholten, welche vorzugsweise aus dem ländlichen Milieu stammten, in dem Erickson aufgewachsen war, verstand er es in geschickter Weise, diese so individuell zu vermitteln, daß der Betreffende sie als für sich persönlich gegeben empfand, um dann jeweils seine eigenen Folgerungen daraus zu ziehen. Sehr zu Hilfe kam Erickson dabei seine ungewöhnliche sprachliche Ausdrucks-

kraft und auch Modulationsfähigkeit, welche zum Teil auch ihren kompensatorischen Ursprung in den sprachlichen Behinderungen nach der ersten Polioinfektion hatte. Durch den Einsatz solcher Geschichten in seinen therapeutischen Gesprächen gab Erickson Menschen mit ganz verschiedenartigen Ansichten und Verhaltensmustern Metaphern (= Redewendungen in einem übertragenen, meist bildlichen Sinne, welche mehrdeutig sind und zum Ziel haben, positive Veränderungen zu bewirken). In enger Verbindung damit steht die Methode des **schrittweisen Vorangehens** – man denke dabei an das eigene schrittweise Vorangehen von Erickson beim Wiedererlangen der postinfektiös verlorengegangenen Fähigkeiten, insbesondere denen motorischer Natur – und des **Aussähens** (Einstreuens) zukünftiger Suggestionen, was literarisch dem Erwecken von Vorahnungen ähnelt (Zeig, 1988). Bei dem schrittweisen Vorangehen werden Veränderungen durch kleinste strategische Schritte aufgebaut, beim Aussähen Schritte vorausgesagt, bevor diese ausführlicher beschrieben werden. Auch **paradoxe Interventionen** spielen eine bedeutende Rolle in dem Therapiekonzept von Erickson. Die Frage nach einer möglichst schnellen Veränderung bei seinen Patienten pflegte er gern zu beantworten, indem man langsam vorgehe; oder: „wenn Du eine große Veränderung willst, bitte um eine kleine"; bzw. „wenn Du teilst, eroberst Du". Eine der größten Fähigkeiten von Erickson war es, durch **indirekte Einflußnahme** (Suggestion) auf Menschen zu wirken. Zeig (1988) gibt ein sehr anschauliches Beispiel dazu im Hinblick auf eine angestrebte Levitation des rechten Armes in Hypnose: „Ich hätte gern, daß Sie wirklich wahrnehmen, auf eine erhebende Art wahrnehmen, daß Hypnose für Sie wirklich die rechte Erfahrung ist, in einer Weise, die Sie handlich finden können für sich." Die indirekte Art, Assoziationen zu lenken, schloß jedoch keineswegs aus, daß Erickson nicht auch direktive Techniken anwendete, insbesondere dann, wenn kein Widerstand zu umgehen oder zu unterlaufen war, welcher der anzustrebenden Veränderung im Wege stand. Wie Haley (1988) schreibt, vertrat Erickson hier schon frühzeitig eine gegenteilige Position zur herkömmlichen, nicht direktiven Therapie. „Er argumentierte, daß Veränderung dadurch bewirkt werde, daß der Therapeut direktiv sei. Er war der Ansicht, daß alles direktiv sei, was man in Gegenwart eines Klienten sage oder tue. Die Frage sei nur, wie geschickt man hierbei ist, man solle aber nicht annehmen, daß man nicht direktiv sei."

In dem fast unerschöpflichen und deshalb aber auch schwer abzuklärenden Reservoir innovativer Techniken, welche Erickson zum Teil intuitiv erfand, experimentell-spielerisch gebrauchte und strategisch handhabte, sind noch viele enthalten, welche hier nicht angeführt, geschweige denn dargestellt werden können, vom Gebrauch von Unlogik und Verwirrung (Konfusion), von Doppelbindung (= Beziehungsfalle; Stierlin, 1959) über das Lebendigmachen konstruktiver Emotionen mittels der Stilmittel der Überraschung, des Humors oder der Dramatisierung, die strukturierte Amnesie – oft in Verbindung mit fraktionierter Hypnose – bis hin zu therapeutischen Aufgaben mit Symbolcharakter in enger Verbindung mit der jeweiligen Lebenssituation der Patienten, zu Veränderungen des sozialen Kontextes (= Bezugsrahmen bzw. Zusammenhang, in dem verbale wie averbale Mitteilungen und Verhaltensweisen ihre Bedeutung erlangen; Simon und Stierlin, 1984) und zu zweckorientierten und mitunter weit in der Zukunft liegenden Zielen. Dazu gibt Zeig (1988) wieder ein schönes Beispiel aus der persönlichen Lebensgeschichte von Erickson. Auf dessen ausdrückliches Verlangen mußte Zeig ein Photo von Erickson zusammen mit seinem 26. Enkelkind Laurel anfertigen, bei dem Erickson darauf bestand, eine Eule aus Eisenholz in der Hand zu halten, welche er Laurel am Tage ihrer Geburt zum Geschenk gemacht hatte. Er erklärte Zeig dazu, „daß in 16 Jahren, wenn er schon lange tot sein würde, Laurel das Photo betrachten werde. Sie werde das Baby und die kleine Eisenholzeule sehen. Dies werde sich mit ihrem Empfinden vermischen, herangewachsen und in der high school zu sein. Er sagte mir, ich solle merken, wie Erinnerungen strukturiert sind. Erickson bemerkte, daß die Eisenholzeule dem Bild eine gewaltige Portion Menschlichkeit gebe." Und Zeig kommentiert dieses Erlebnis: „Eine Aussaat trägt nie unmittelbare Früchte, sie benötigt vielmehr Zeit zur Reife. Erickson als ‚Bauernsohn' verstand sich auf den Reifungsprozeß einer Aussaat."

Was war dies für ein Mensch, nach dem eine Psychotherapiemethode benannt worden ist, und von dem man sagt, daß sich viele Menschen in seiner Gegenwart „unwohl" gefühlt hätten, weil sie die Macht seines Einflusses fürchteten, der andererseits aber auch rücksichtsvoll und entgegenkommend sein konnte? Haley (1988), der viele Jahre intensiv mit Erickson zusammengearbeitet hat, sagt dazu: „Ich denke, es ist ein glücklicher Umstand, daß er bei aller Bereitschaft, Macht auszuüben und Einfluß zu nehmen, ein wohlwollender Mann war. Wenn diese Fähigkeit, Einfluß auf andere auszuüben, zu destruktiven Zwecken benutzt worden wäre, wäre das sehr nachteilig gewesen." Eine der herausragenden Eigenschaften von Erickson, die dies verhindert hat, ist zweifellos sein Sinn für Schalk und Humor gewesen. Er liebte einfache Witze, Wort- und Satzspiele sowie Rätsel, so wie er auch vieles in Rätseln ließ bzw. absichtlich vage formulierte. Haley (1988) faßt sein Urteil über die „mystische Persönlichkeit" von Erickson zusammen: „Weder hatte ich je einen Zweifel an seiner Ethik oder seinen guten Absichten." So kann auch das suggestive Motto für Mr. und Mrs. Martin J. Zeig (1980 und 1985) nicht mehr beunruhigend wirken, das dem letzten Lehrseminar von Erickson vorangesetzt war unter dem Titel: „Meine Stimme begleitet Sie überall hin, sie verwandelt sich in die Stimme Ihrer Eltern, Ihrer Lehrer, Ihrer Spielgefährten und in die Stimmen des Windes und des Regens ..."

Bei der Darstellung der Hypnotherapie von Erickson gerät man fast ungewollt in die Versuchung, das Zauberwort Hypnose gar nicht zu benutzen und näher darauf einzugehen. Hier wird meist von Trance und Trance-Induktion gesprochen. Es ist hier nicht der Platz, auf das breite Spektrum der Induktionstechniken einzugehen, die Erickson im Laufe seines Lebens entwickelt und verwendet hat, ebensowenig auf Definitionen dieses „Sonderzustandes", welche Erickson mit seinem ausgeprägten Sinn für praktisches Handeln statt theoretischem Überlegen weitgehend vermied. Hier stehen die **Symptomorientierung** und der **„autogene" Ansatz,** den Erickson ja selbst so vielfältig erfahren und genutzt hat, mit Recht im Mittelpunkt des Geschehens. Im Gegensatz zum analytischen Zeitgeist vertrat Erickson schon frühzeitig die Meinung, „daß man die Charakterstruktur dadurch ändert, daß man die Therapie auf das spezifische Problem konzentriert. Nach seinen Worten ist das Symptom wie der Griff eines Topfes; wenn man den Griff gut in der Hand hat, kann man eine Menge mit dem Topf machen. Er lehrte, daß man das Symptom nicht ignorieren, sondern sämtliche Details erfassen sollte" (Haley, 1988). Zu seiner Veränderung, welche ja auch vom Patienten gewünscht werde, diente ihm auch Hypnose (= Trance) in ihren vielfältigen Schattierungen von der Alltagstrance über die beiläufig eingestreute Trance bis hin zur tiefen Trance. Neben der autogenen Mitarbeit des Patienten an diesem Vorgang und dem durch ihn ausgelösten inneren Suchprozeß, ohne die keine Trance gelingen kann, legte Erickson außerdem großen Wert auf den Kontext, in dem Trance induziert wird bzw. geschieht. Trance war ein Mittel für Erickson, „die vorhandene zwischenmenschliche Ansprechbarkeit hervorzurufen und zu entwickeln" (Zeig, 1988), man könnte aber auch sagen, besser mit dem Unbewußten seines Gegenüber zu kommunizieren, um dieses in der therapeutischen Interaktion wirksamer zu beeinflussen. Dabei war Ericksons Sichtweise des Unbewußten das Gegenteil der psychodynamischen seiner Zeit. Er sah in ihm vor allem eine schöpferische Kraftquelle, welche mehr Wissen und Weisheit in sich birgt als das Bewußte („vertraue Deinem Unbewußten"). Wenn man bei einer Person einfach ihr Unbewußtes arbeiten lasse (= unbewußter Suchprozeß zu konstruktiven Lösungen), würde sich dieses um alles Erforderliche in einer positiven Weise kümmern (Haley, 1988). Schmidt (1985) geht sogar so weit, im Hinblick auf den hypnotherapeutischen Anteil in der Psychotherapie von Erickson festzustellen: „Auch Milton H. Erickson hat nicht eine einzige Person in Trance versetzt, sondern er hat in höchst gekonnter Weise Verhandlungsangebote gemacht, auf die viele Leute schließlich in autonomem Handeln eingingen. Das Resultat dieser Einigung wurde dann Trance genannt, und auch darüber einigten sich die Beteiligten."

Um Erickson in seinem therapeutischen Wirken besser verstehen zu können, muß man sich vergegenwärtigen, daß er in einem großen Land, den Vereinigten Staaten von Amerika, aufgewachsen ist und daß er dort sein Leben lang gelebt hat. Sein Weltbild ist dementsprechend amerikanisch geprägt von all dem Pragmatismus dieser relativ jungen dynamischen Nation, von ihrem Wirklichkeitssinn, ihrem Gemeinschaftsgeist und ihrer sozialen Hilfsbereitschaft, nicht zu vergessen aber auch von etwas therapeutisch durchaus positiv zu bewertender Naivität. Erickson vertrat eine andere Tradition als die, welche ihre Wurzeln in Europa hat. Dies macht es uns Europäern nicht immer leicht, ihm zu folgen. „Erickson beschäftigte sich nicht mit philosophischen Schulen, er richtete seine Aufmerksamkeit vor allem auf die reale Welt und reale Probleme" (Haley, 1978 und 1988). Seine Lebensphilosophie skizziert Rosen (1985) in zwei kurzen Sätzen: „1. Bleib in Bewegung; 2. Freu dich deines Lebens." Die überraschende Renaissance seiner Hypnotherapie, vor allem nach seinem Tode, hat viel zu tun mit unserem wachsenden Interesse an dem „geheimen Band zwischenmenschlicher Kommunikation" und an seiner großen Bedeutung für die individuelle wie auch kollektive Verhaltenssteuerung, deren Grundlagen die in Palo Alto tätige Gruppe um G. Bateson und Erickson gelegt hat, der eine durch seine theoretischen Überlegungen und Anstöße, der andere durch sein praktisches Handeln, wel-

ches seiner Zeit um „20 Jahre voraus war" und mit dem er wertvolle und ermutigende neue Anregungen gegeben hat, welche positive Perspektiven für die Zukunft enthalten, über die Individualtherapie hinaus für die Paar-, die Familien- und die Gruppentherapie. Wie Haley (1967) betont, ist es bei charismatischen Persönlichkeiten, welche eine Leitbildfunktion haben, für die Zukunft erforderlich, ihre Handlungsweisen (Techniken), welche innerhalb eines wohlbegründeten Systems ihren berechtigten Platz haben, von denen sorgsam zu scheiden, welche auf den Menschen zentriert sind, der sie ausgeführt hat. Erickson ist zweifellos ein Vorbild für alle, und seine Patienten haben dies wohl auch so empfunden, wie ein Mensch mit seinen Behinderungen, Hemmnissen und Hindernissen körperlicher und seelischer Art umgehen kann, sie zum Besten wendet und wie der „Geist" – mag er nun mehr unbewußt oder mehr bewußt sein – sich den Körper schafft, in dem er wohnt (Stefan Zweig, 1932).

23 Körperorientierte Psychotherapie

Hans Müller-Braunschweig

In den meisten unserer psychosomatisch-psychotherapeutischen Kliniken gehören körperorientierte Psychotherapieverfahren zum festen Bestandteil der stationären Behandlung. Auch in der ambulanten Behandlung werden sie zunehmend häufiger angewandt. Diese verhältnismäßig rasche Entwicklung einer Richtung der Psychotherapie, die noch in vieler Hinsicht Fragen aufwirft, aber auch neue Möglichkeiten eröffnet und eine Herausforderung für bereits etablierte Methoden darstellt, ist bemerkenswert.

Im folgenden soll der Versuch gemacht werden, Möglichkeiten und Grenzen einiger körperorientierter Verfahren darzustellen. Dazu ist einleitend auch ein kurzer Blick auf ihre Geschichte notwendig.

23.1 Zur Geschichte

Die heutigen Methoden haben ihre Wurzeln in Tendenzen, die sich in der Pädagogik (Gymnastik), der Kunst und der Psychotherapie besonders in den ideenreichen 20er Jahren unseres Jahrhunderts entwickelten. Die Gymnastik zeigte bereits seit der Antike den Doppelaspekt von Therapie und allgemeiner körperlich-seelischer Bildung. Die Entwicklung der heutigen körperorientierten Verfahren aus Heil- und rhythmischer Gymnastik, Ausdruckstanz („modern dance") und Psychotherapie bedeutet damit auch ein Ansprechen des Kranken auf verschiedenen Ebenen – von der physiologischen Funktion über den Ausdruck bis hin zu sprachlicher Symbolik. In diesem Sinne „verkörpert" die bereits 1869 geborene Bess Mensendieck in ihrer Person die Verbindung dieser verschiedenen Ansätze. Sie war Gymnastiklehrerin, hatte außerdem Bildhauerei, Atem- und Gesangstechnik studiert und sich ausgiebig mit Anatomie und Physiologie beschäftigt. Sie bildete bereits vor dem Ersten Weltkrieg Schülerinnen aus. Eine dieser Schülerinnen (Th. Malmberg) wirkte später an der „Schule für angewandte und freie Bewegung" in München, an der u.a. auch Marianne Fuchs (vgl. Abschn. 23.3) ihre Ausbildung erhielt. Zu nennen ist in diesem Zusammenhang auch der Tänzer, Tanzpädagoge und -theoretiker Rudolf v. Laban. Laban betonte freie Improvisation beim künstlerischen Tanz als Ausdruck seelischen Erlebens.

„Atmung, Stimme und Bewegung waren die somatischen Funktionsbereiche, die zu Beginn des 20. Jahrhunderts in ihrer Erlebnisbedeutung für den Gesunden und Kranken wieder entdeckt wurden" (Stolze, 1981).

Innerhalb der Psychotherapie zeigten sich zugehörige Tendenzen in der Atemtherapie, z.B. bei den Psychotherapeuten G.R. Heyer (1925) sowie Steger und Heyer-Grote. Aus der Psychoanalyse kam Wilhelm Reich und wies u.a. auf die Entsprechung von „Muskel- und Charakterpanzer" hin (Reich, 1933) (vgl. Abschn. 23.6). S. Ferenczi mit seiner „aktiven Technik" versuchte den Körper in gewissem Ausmaß in die psychoanalytische Arbeit mit einzubeziehen. Er setzte sich damit in Widerspruch zu Auffassungen Freuds (1914), der bestrebt war, alle verfügbaren Kräfte in das Wort zu lenken und jedes andere „Abreagieren" auszuschließen. So sah Freud Körperbewegungen während der Behandlung vor allem als „Widerstand" (Cremerius, 1984). Heute würden sie – wie das „Agieren" überhaupt – als „Sonderform der Kommunikations- oder Äußerungsweisen" betrachtet werden (Sandler et al., 1973; Becker, 1981; Thomae und Kächele, 1985). Beschreibend wies aber auch Fenichel (1928) in dieser Zeit auf die Bedeutung des Körperausdrucks hin (Grunert, 1977; Müller-Braunschweig, 1986). Während die Emigration von Psychoanalytikern aus Deutschland in die USA nach 1933 eher zu einem Verschwinden dieser frühen körperorientierten Ansätze führte, ging die Entwicklung durch Schülerinnen der bedeutenden Berliner Gymnastiklehrerin Elsa Gindler weiter, die erkannt hatte, „daß mit mechanischem Üben, und mag es noch so physiologisch aufgebaut sein, keine entscheidende Änderung im Gesamtverhalten zu erzielen" ist (Wilhelm, 1961). Ihre Schülerinnen G. Heller und C. Speads emigrierten nach England und in die USA. Dort kam die Methode Gindlers mit klinischer Arbeit und der Psychoanalyse in Verbindung. Die zuletzt genannten Linien führten zur Konzentrativen Bewegungstherapie (s.u.). Andere der frühen Ansätze kamen mit den Pionieren der Psychosomatischen Medizin in Deutschland in Kontakt, u.a. mit Viktor v. Weizsäcker (Funktionelle Entspannung).

In den 60er und besonders in den 70er Jahren förderte dann eine besondere soziokulturelle Situation in der Bundesrepublik Deutschland eine in Europa und den USA einmalige günstige Basis für das Entstehen stationärer Psychosomatik und Psychotherapie (Schepank, 1987). Damit verbunden war auch die Möglichkeit einer breiten klinischen Anwendung und Erprobung körperbezogener Psychotherapieverfahren im Rahmen stationärer Behandlung – also auch in der Zusammenarbeit mit verbaler Psychotherapie.

Im folgenden sollen zunächst ausführlicher zwei Methoden besprochen werden, die seit längerer Zeit im stationären und ambulanten Bereich in der Bundesrepublik angeboten werden und einen mehrjährigen geregelten Ausbildungsgang haben. Sie gehören damit auch zu einer integrierten stationären Psychosomatik (vgl. u.a. Krause, 1985). Es sind dies die Konzentrative Bewegungstherapie (KBT) und die Funktionelle Entspannung (FE).[1] Etwas kürzer wird auch

---

[1] Die Hervorhebung dieser beiden Methoden war der Wunsch der Herausgeber. Es erscheint auch sinnvoll, durch zwei ausführlicher dargestellte Verfahren typische Züge körperorientierter Psychotherapie darzustellen. (Es sind Verfahren, die der Autor außerdem durch Selbsterfahrung kennenlernen konnte, ebenso wie die körperorientierte Psychotherapie von G. Downing und die Methode von F. Besuden.)

die Bioenergetik behandelt, deren (anderer) Ansatz für eine praktische und theoretische Diskussion der körperbezogenen Methoden wichtig ist. Auf weitere, ebenfalls erprobte Verfahren mit geregeltem Ausbildungsgang kann aus Raumgründen nur hingewiesen werden. Überlegungen zu den theoretischen Grundlagen folgen am Schluß der Arbeit.

23.2 Falldarstellungen

Zwei stichwortartig vorgestellte Fallbeispiele sollen einen ersten Eindruck der Wirkung im ambulanten und stationären Setting vermitteln und typische Merkmale zur Diskussion stellen. Eine nähere Beschreibung der Methoden folgt.

Eine 41jährige Patientin sucht wegen ungeklärter ständiger Schmerzzustände eine Psychotherapeutin auf, die auch die Funktionelle Entspannung beherrscht. Die Patientin will keine verbale Psychotherapie, sondern „Entspannungsübungen". Sie nimmt ständig Medikamente und ruft wegen der Schmerzen oft den Notarzt. Organisch ist sie ohne Befund. Im Gespräch fällt auf, daß sie auf die Stuhlkante sitzt, die Schultern hochgezogen, die Beine aneinander gepreßt, die Arme unbewegt. Der Atem ist flach und kurz. Ebenso wie körperlich wirkt sie auch psychisch „verhalten".

In der Behandlung geht es u. a. um diese Einengung. Ihr Becken erscheint erstarrt (sie lokalisiert es anatomisch in der Bauchmitte), die Wirbelsäule unbeweglich. Sie kann sich nicht entspannt setzen oder hinlegen, ihr Gewicht nicht wirklich „abgeben". Im Verlauf der Behandlung wird klar, daß in der elterlichen Familie gehäuft Suizide auftraten, u. a. ein traumatisch wirkender Suizid der geliebten Großmutter im Nebenzimmer, als die Patientin ein kleines Kind war. Hinzu kam eine depressive Mutter. Die Patientin wagte nicht „ihren" Raum einzunehmen und eigene Emotionen auszudrücken, auch nicht körperlich, aus Furcht, damit weitere Katastrophen auszulösen. Nach 39 Behandlungsstunden mit Funktioneller Entspannung hatte sie selber ein Gefühl für diese körperliche Einengung entwickelt. Zu diesem Zeitpunkt war sie ein ¾ Jahr ohne Notarzt ausgekommen. Der Medikamentenverbrauch war stark zurückgegangen. Die Behandlung ging dann in eine verbale Psychotherapie über, in der sie nun zunehmend über seelische Schmerzen klagte, während die körperlichen Schmerzen weiter zurückgingen.[2]

Ein 39jähriger Patient mit 16jähriger Ulkusanamnese (⅔-Resektion des Magens vor 8 Jahren) beunruhigt das Behandlungsteam der Psychosomatischen Klinik durch übermäßige Aktivität. Er schlägt eine „völlige räumliche, architektonische und inhaltliche Umstrukturierung der Station" vor und fordert mehr politische Aktivität. Er wird mit dieser Einstellung auch in der verbalen Gruppe zum „anerkannten Führer und Co-Therapeuten". Durch Vorbehandlung und Literaturstudium verfügt er auch über eine gute Kenntnis der psychoanalytischen Theorie. Seine theoretische „Einsicht" in den Abwehrcharakter seiner Aktivität (auch aufgrund von früheren Deutungen) blieb ohne emotio-

nale Beteiligung und hatte keine Wirkung. Das Team ließ ihn aber zunächst seine Abwehrseite agieren. In der Konzentrativen Bewegungstherapie war dann eine erste Annäherung an die abgewehrten Impulse und passiven Wünsche möglich. Er rollte sich dort in der Stunde „auf dem Boden zusammen, klammerte sich geradezu an Gruppenmitglieder... In einer anderen Situation, wo er mit einem anderen Gruppenmitglied Rücken an Rücken saß, war es dem Mitpatienten kaum möglich, sein Anlehnungsbedürfnis zu ertragen. In diesem Erleben wurde für den Patienten erst deutende Arbeit möglich..." (Becker und Lüdecke, 1978).

23.2.1 Diskussion

Es gibt in diesen beiden Darstellungen einige Punkte, die für die weiteren Ausführungen besonders wichtig sind.
Zur 41jährigen Patientin:
– Eine Patientin, die sich nicht auf eine verbale Therapie einlassen will (und auch zu dieser Zeit dafür nicht geeignet erschien), kommt auf dem Wege über eine körperbezogene Methode auch zu einem langen verbalen psychotherapeutischen Prozeß.
– Mit der Körperarbeit verändert sich auch ihr Körperbild. Durch die Veränderung des Körperempfindens und der Körperhaltung verändern sich ebenfalls die seelische Haltung und der Atemrhythmus. Das Symptom bessert sich in der ersten Behandlungsphase vorwiegender Körperarbeit erheblich.
– Das „Durcharbeiten" findet dann vor allem in einer rein verbalen Therapie in der darauffolgenden Zeit statt.
Zum 39jährigen Patienten:
– Eine vom Patienten in der vorhergehenden ambulanten und zu Beginn der stationären Therapie durchgehaltene Abwehr wird erst in der partiell nonverbalen Körpertherapie erschüttert. Er erlebt „leibhaftig" seine bisher abgewehrte Seite.
– Es zeigt sich dabei eine emotional sehr intensive Annäherung an die bisher abgewehrten Inhalte. Das steht im Gegensatz zur bisherigen rein rationalen Einsicht.
– Eine Bearbeitung erfolgt nach Aussagen der Verfasser dann wieder längerfristig in der verbalen Gruppe. Wie noch zu zeigen sein wird, kann im Wechsel von Körperarbeit und verbaler Besprechung des Erlebten (oder durch ein neues körperbezogenes „Angebot") auch eine gewisse Bearbeitung in einem körperorientierten Verfahren erfolgen.
In diesem Feld gehen die Meinungen über die Reichweite der körperbezogenen Verfahren noch auseinander. Ein wichtiges Kennzeichen der hier besprochenen Methoden ist jedenfalls der „Wechsel der Ebenen" (vgl. Eberspächer, 1986): Auf der Körperebene

---

2 Frau Dr. med. Th. Woelk danke ich für die ausführliche Information über diesen Fall. Durch die Supervision der verbalen Therapie konnte ich den weiteren Verlauf in den letzten drei Jahren verfolgen.

wird unmittelbar erlebt. Häufig wird zwar dann erst beim Durchsprechen des Erlebten die durch die Körperarbeit ausgelöste emotionale Bewegung empfunden, aber hier auch schon mit dem Wort verbunden und damit berichtend und reflektierend auf eine andere Organisationsstufe gehoben. Der Körpertherapeut wird dabei über die Reaktion auf das vorhergehende Angebot auf der Körperebene informiert und kann sich in der Wahl seiner folgenden Körperangebote wiederum darauf einstellen. Dadurch kommt es zu einem fortwährenden Wechsel von unmittelbarem Erleben und ansatzweiser verbaler Verarbeitung – ein Wechsel, der natürlich auch zur verbalen analytischen Psychotherapie gehört, in diesen Verfahren durch das Einbeziehen des Körpers aber besonders intensiv erlebt wird.

Um einen konkreten Eindruck der Verfahren zu erhalten und die eben beschriebenen Abläufe besser zu verstehen, sollen die drei oben genannten Methoden zunächst eingehender beschrieben werden.

23.3 Funktionelle Entspannung

Nach ersten Erfahrungen, die Marianne Fuchs mit der Anwendung von Bewegungs- und Entspannungsübungen bei psychosomatisch erkrankten Patienten Ende der 20er Jahre in Marburg machte, wurde die Methode besonders nach dem Zweiten Weltkrieg in Heidelberg, vor dem Hintergrund der dortigen Anthropologischen Medizin (Siebeck, v. Krehl, v. Weizsäcker) entwickelt (vgl. Schüffel, 1988; Fuchs, 1985). Die Funktionelle Entspannung (FE) hatte damit von Beginn an unmittelbare Beziehung zur Praxis der sich in Deutschland in enger Verbindung mit der Inneren Medizin entwickelnden Psychosomatischen Medizin und ihren Pionieren. Der enge Praxisbezug und vor allem die besondere Begabung von Marianne Fuchs, einen intuitiven Zugang zu basalen körperlichen Abläufen und ihrem leiblichen Ausdruck zu finden, führten zu einer Methode, die vor allem in der Einzelarbeit mit dem Patienten (auch in der „Einzelarbeit in Gruppen") den Erkrankten leibhaftig Blockaden, Verspannungen oder entfremdete Körperregionen spüren ließ. Es gelang dann oft, Einschränkungen in schrittweiser Annäherung und Neuerfahrung aufzulösen und damit in das subjektive Körpererleben neu zu integrieren. Die Methode hat Eingang in die Arbeit psychosomatischer Kliniken gefunden, wird aber auch von niedergelassenen Psychotherapeuten und FE-Therapeuten angewandt.

Die 49jährige Abteilungsleiterin eines Großbetriebes wurde wegen einer medikamentös nicht beeinflußbaren Hypertonie des Kreislaufes auf die psychosomatische Station eines Akutkrankenhauses aufgenommen. Die systolischen Werte lagen bei 260 mm Hg und darüber. In der vorangegangenen sechsjährigen internistischen Therapie konnte der Blutdruck nie unter 200 mm Hg gesenkt werden.

Die Patientin war sehr leistungsbetont, mußte aber in der Konkurrenz mit Männern im Betrieb immer wieder Niederlagen hinnehmen. In privaten Beziehungen wurde sie meistens ausgenutzt. „Ich mußte immer geben." In einigen Gesprächen ließen sich diese aktuellen Haltungen auf frühkindliche Erlebnismuster zurückführen. Mit der weichen, depressiven Mutter war sie in ihren ersten zwei Lebensjahren allein. Sie erinnert sich aus späteren Zeiten, daß sie die Mutter eigentlich immer stützen mußte. Von dem aus dem Krieg zurückkehrenden Vater fühlte sie sich zunächst abgelehnt, später nur über Leistung anerkannt. Feindselige Impulse mußte sie verdrängen, um nicht die depressive Mutter zu verletzen und die Zuneigung der Eltern zu verlieren. Selbst eher subdepressiv suchte sie Halt und Anerkennung von seiten des Vaters. Dafür mußte sie aber Leistung zeigen und tüchtig sein. Daraus resultierte ein chronischer Spannungszustand. Der ärztliche Therapeut berichtet weiter, daß durch die Gespräche die ganz hohen Blutdruckwerte (bei gleichbleibender Medikation) zurückgingen. Da aber die Spannung im körperlichen Bereich persistierte und sich gleichzeitig eine starke Abwehr gegen Körpererleben zeigte (auch Ablehnung des Autogenen Trainings, AT), wurde sie in die FE eingeführt.

Aus dem Bericht der FE-Therapeutin:
In der FE war die außerordentliche Beherrschtheit der Patientin zu Beginn besonders auffällig. Ihrem Leben „mit zusammengebissenen Zähnen" entsprach die Spannung im Mundraum und Unterkiefer. Da diese Region viel mit ihrer Leistung und damit ihrem „Halt" zu tun hatte, wurde zunächst nur am „Aus"(atmen) gearbeitet, an dem Versuch, „den Atem strömen zu lassen". Ebenso wie Gähnen wurde das von der Patientin zunächst nicht gewagt. Bei der Arbeit am „unteren Kreuz" (s. u.) trat, während sie den Atem „herauspreßte", eine außerordentlich starke Hyperämie im Kopf auf, die sie nach einiger Zeit aber der Anregung folgend „nach unten wegpusten" konnte. Danach konnte auch die Verspannung im Unterkiefer bearbeitet werden. Ein „Loslassen" war erst allmählich möglich. Die bei einer Arbeit am Brustkorb angelegten Hände der Therapeutin empfand die Patientin als sehr wohltuend und hilfreich. „Wir arbeiteten anschließend vor allem im unteren Kreuz und im Becken-Bauch-Bereich. Dort zu lockern ist immer bei Hypertonikern außerordentlich wichtig, stauen sie doch ähnlich den Kopfschmerzpatienten viel zu viel nach oben, statt in der breiten, lockeren Mitte des Bauch-Becken-Bereiches zu ruhen." Nach der 7. Stunde FE konnte sie sich so gut entspannen, daß sie im Anschluß fast hypoton wurde. Sie erlebte plötzlich den Gegensatz zur bisherigen depressiven Gespanntheit. „Ach, das Leben ist schön!" Nach der 21. Stunde FE wurden die Übungen beendet, die sie instand setzten, den Blutdruck weitgehend selbst zu regulieren.

Aus dem Bericht des Arztes:
„Durch das weitere Üben der FE und in größeren Abständen durchgeführte Beratungsgespräche blieb der Blutdruck mit ganz geringer Medikation in normalen Grenzen um etwa 140–155 systolisch und 80–90 diastolisch..." (Bepperling und Klotz, 1978) (vgl. auch Kap. 41.3).

23.3.1 Zur Methode der FE

Die Möglichkeit, Blockaden und Verspannungen leibhaftig zu spüren, läßt sich beispielsweise so vorbereiten, daß man den Patienten bittet, einmal die Augenlider während des Ausatmens zufallen zu lassen und das gleiche dann nochmals während des Einatmens durchzuführen. Das Körpererleben in beiden Fällen kann dann vom Patienten verglichen werden. Ein Gleiches kann mit der Anspannung und Lösung der Kiefermuskulatur ausprobiert werden, ebenso parallel zum Nicken des Kopfes oder dem Hochziehen und Fallenlassen der Schultern – jeweils also parallel zum Ausatmen oder zum Einatmen. Dem nicht mit der FE vertrauten Leser wird empfohlen, diese Versuche zunächst selber durchzuführen, ehe er weiterliest, um mit diesem Minimum an Selbsterfahrung einen besseren Zugang zu den folgenden Beschreibungen zu gewinnen.

(Als „erlebniszentrierte Verfahren" (Petzold), die einen spezifischen Akzent auf nonverbale Abläufe legen, gilt für körperorientierte Psychotherapien in besonderem Maße, daß nur die Selbsterfahrung einen Eindruck der Methoden vermittelt und durch Lektüre nicht ersetzt werden kann.)

Übereinstimmend berichten Patienten bzw. die Teilnehmer eines Fortbildungskurses, daß das Nikken, Entspannen des Kiefers, Zufallen der Lider etc. beim Ausatmen wesentlich selbstverständlicher, „leichter", „organischer", „passender" usw. empfunden wird, als beim Einatmen, und dieses Gefühl nicht auf Lider, Kopf, Kiefer beschränkt bleibt, sondern Halsmuskulatur, Brustkorb und Becken miteinbezieht, ja, sich bis in die Beine fortsetzen kann. Das Ausatmen betone eher die Richtung des „Unten", des Abgebens von Gewicht, des Kontaktes mit der Sitz-/Bodenfläche und erzeuge ein Schweregefühl. Das Einatmen sei eher mit der Richtung des „nach oben" und mit Spannung verbunden. Die oben angeführten Bewegungsabläufe können nun auf andere Körperbereiche ausgedehnt werden, z.B. als kleine Bewegungen im Skelettsystem beim Ausatmen („Tun im Lassen"). So beispielsweise am „oberen Kreuz", das ist die „Querverbindung Schultergelenk zu Schultergelenk und Längsachse von der Schädelbasis bis zur Mitte der Brustwirbelsäule". Dieses gewichtige „inwendige Kreuz" wird so, wie es ist, „eingebildet" und zunächst wieder bodenwärts „abgegeben", in welcher Lage sich der Übende auch befindet (Fuchs, 1984).

So werden unter der Vorstellung des „oberen Kreuzes" z.B. Schultergelenke parallel zum Ausatmen bewegt. Das geschieht mit kleinen Bewegungen und soll höchstens zwei-, dreimal wiederholt werden, um die Bewegung nicht zu einem willensbetonten Üben werden zu lassen. Diese Anweisungen sind in den sog. „Spielregeln" enthalten:
- alle Reize (Tun/Spüren) werden an eine Phase des Atemrhythmus gebunden;
- der jeweilige Reiz wird nur 2–3mal wiederholt;
- Nichttun und Nachspüren;
- sich der autonomen Reaktion überlassen;
- verbalisieren.

Derartige Übungen, z.B. im Bereich des oberen Kreuzes, können dann zum Erleben einer Verspannung in den Schultergelenken führen, die vorher nicht bemerkt wurde. So berichtet Fuchs (1988) von einem Patienten mit Torticollis, der während der FE erstmalig bemerkte, daß er häufig die Schultern hochziehe. Dabei wurde ihm bewußt, daß die hochgezogenen Schultern Schutz geben sollen, d.h. ein psychisches Phänomen wurde bewußt und die mit dieser Körperbewegung verbundene Angst wurde spürbar. Weiter wurde ihm bei der Arbeit plötzlich klar, daß er bei offiziellen Anlässen den Bauch einzieht, „sich zusammennimmt". Aber dieses „Zusammennehmen" verhinderte gerade die psychische Flexibilität, die u.a. bei diesen Anlässen erwünscht war, und führte ihn von seinem Eigenrhythmus, d.h. von seinem autonomen Atemrhythmus, weg. Im weiteren Verlauf wurde auch das starre Becken bewußter. Statt – an der Basis – beweglich zu sein und sie als Kraftquelle zur Verfügung zu haben, hatte der Patient Schutz und Abwehr in den oberen Bereich (u.a. in die hochgezogenen Schultern) verlagert. Petenyi (1985) berichtet von einem Mädchen mit beginnender Magersucht und ausgeprägter rationaler Kontrolle, das nicht spüren konnte, wie der Stuhl ihre Oberarme und Unterschenkel berührt. Während der FE sagte es u.a.: „Mein Kopf kann dem Becken etwas sagen, aber nicht mein Becken meinem Intellekt."

Während der oben beschriebenen Übung am oberen Kreuz kann auch deutlich werden, wie starr die Wirbelsäule beim Auswärtsdrehen der Schultergelenke oder beim Zusammenschieben der Schulterblätter ist. Spürt der Patient das nicht, kann der FE-Therapeut durch leichtes Anlegen der Hand an dieser Partie beim Spüren helfen und damit eine (zugleich „haltende") **Rückmeldung** geben, die wiederum oft erstmalig ein zunächst diffuses Spüren und dann bewußteres Merken ermöglicht (vgl. v. Uexküll, 1986; Johnen und Müller-Braunschweig, 1988). Bei sehr gespannt lebenden Patienten findet sich oft diese Unbeweglichkeit der Wirbelsäule in Verbindung mit einer psychischen Haltung des „Durchhaltens", des „Aufrechtbleibens um jeden Preis", die schließlich beispielsweise zu Rückenschmerzen führen kann (psychoanalytisch könnte man hier auch vom „rigiden Über-Ich" sprechen). Auch bei Asthmatikern läßt sich diese Rigidität finden, die oft mit der erwähnten Haltung des „krampfhaften Durchhaltens und Aufrechthaltens" verbunden ist, die sich wesentlich im oberen Teil des Körpers abspielt und die den eigenen, gelassenen Rückhalt nicht kennt. In der Genese finden sich dann oft Defizite mütterlicher Zuwendung mit ambivalenter Haltung zum Kinde und seinem Körper, d.h., die kindlichen Impulse wurden nicht wohlwollend aufgenommen und rückgespiegelt. Das führt zu basaler Unsicherheit und Affektabspaltung, die sich körperlich als „Unbelebtheit" und Unsicherheit im Beckenbereich zeigen kann. Der fehlende mütterliche Rückhalt führt in der weiteren Entwicklung dann zu dem Versuch, dieses Defizit durch Verlagerung nach oben in den Brust- und Schulterbereich zu kompensieren. Der Rücken wird starr und wirkt in der FE oft wie „gepanzert". Der Brustkorb wird aufgebläht („stark sein wollen und müssen"). Die – auch im körperlichen Bereich – tieferliegende Labilität wird außerdem durch „Kopflastigkeit" kompensiert, die sich in der körperlichen Haltung oft durch einen hängenden oder auch starr erhobenen Kopf manifestiert; psychisch z.B. durch intellektualisierende Abwehr (vgl. auch Kap. 42).

Vor einer festen Zuordnung von körperlichen zu psychischen Haltungen ist allerdings zu warnen. Ähnlich wie die Bedeutung eines Traumes nur aus dem Kontext (aktuelle Situation, Lebensgeschichte etc.) verstanden werden kann, ist auch bei der Beurteilung körperlicher Haltungen die gesamte Situation mit einzubeziehen. Das sog. „body-reading" kann, übereilt angewandt, zu falschen Schlüssen führen. Um eine differenzierte Beurteilung zu ermöglichen, wird z.B. eine analytische Selbsterfahrung vor dem Beginn der KBT-Ausbildung verlangt. Diese wird dann während der Ausbildung durch mehrjährige körperorientierte Selbsterfahrung ergänzt.

Eine Auflockerung der oben genannten Haltung des „Nach-oben-Verlagerns", verbunden mit Lockerung des Thorax und der Verbindung zum Becken, mit Ausstrahlung auf das Zwerchfell, führt oft zu spontanem und freierem Durchatmen.

Neben dem erwähnten „oberen Kreuz" werden das „oberste Kreuz" (Querverbindung von Ohr zu Ohr), das „untere Kreuz und das Becken", Wirbelsäule, „Löcher und Innenräume", Grundrhythmus, Haut und Stimme als weitere Körperbereiche erwähnt, mit denen in der Funktionellen Entspannung gearbeitet wird.

Es ist nun entscheidend, daß derartige Fehlhaltungen oft erstmals während der FE gespürt werden. Diese Fehlhaltungen stellen die psychosomatische Einheit insofern überzeugend dar, als sie häufig auch erstmalig zu bewußtem Erleben vieler psychischer Phänomene führen, die bisher unbemerkt waren und erst mit der Konturierung der Körperrepräsentanzen auch eine Kontur, ein Profil erhalten, merkbar, erlebbar werden. So spürte auch der oben erwähnte Patient mit Torticollis erstmalig, daß er sich ständig „getrieben" und „unter Druck" fühle. Hier kann auch eine Anmerkung von De Boor (1965) angeführt werden, der auf die „ständig gespannte Haltung von Asthmatikern" hinweist, „deren extreme Inspirationsstellung des Brustkorbes wie das angemessene körperliche Korrelat (wirkt)... In dauernder Anspannung sind sie seelisch-körperlich auf rasches Reagieren gegenüber der (phantasierten) Gefahr vorbereitet."

Besonders bei der Behandlung von Kindern wird die Mutter häufig mit einbezogen werden müssen. Bei asthmatischen Kindern kann den Müttern dann eine andere Art des Umganges mit einem Anfall und eine andere Art des Anfassens der Kinder vermittelt werden. Diese Änderungen bewirken oft eine Angstreduktion der Mutter, die sich auch auf das Kind auswirkt (Fuchs, 1985).

Die Funktionelle Entspannung setzt im Vergleich mit anderen körperpsychotherapeutischen Richtungen am direktesten an den körperlichen Ausdrucks-

formen der gestörten vegetativen Regulation an, die als Dauerhaltungen über Störungen der Funktion oft zu schweren körperlichen Läsionen führen können (Ähnlichkeit besteht in einigen Punkten zur Eutonie, s. u.).[3]

Die FE ist eine eher „stille" Methode. Es kommt im Vergleich zu anderen körperbezogenen Psychotherapieformen seltener zu starken emotionalen Bewegungen, wie etwa in KBT-Gruppen oder – oft noch ausgeprägter – in der Bioenergetik und verwandten Verfahren. Dafür ist es aber möglich, auch ängstliche und abwehrende Patienten eher zu erreichen. In der Weiterentwicklung der FE dürfte sich eine stärkere Beachtung der Übertragungsvorgänge, aber auch die Verwendung von Elementen aus anderen Körpertherapieformen entwickeln, die bei FE und KBT im gewissen Ausmaß schon wechselseitig besteht.

23.3.2 Indikation

Detaillierte Falldarstellungen finden sich bei Fuchs (1984, 1985) sowie in Bepperling und Klotz (1978) u.a. über die Behandlung von Migräne, Hypertonie, Asthma bronchiale (besonders bei Kindern und Jugendlichen), Obstipation, Sprechstörungen, Beziehungsproblemen, Erythrophobie, beginnender Magersucht. Wiesenhütter (1984) nennt als Indikation u.a. Fehlspannungen im Bewegungsapparat, rheumatische und neurologische Beschwerden, Schluck- und Magenkrämpfe, Darmkrämpfe, psychogene Schlafstörungen, Sexualstörungen. Auch bei Magersucht könne die FE eine wichtige Rolle spielen – allgemein bei „...den sog. psychosomatischen Störungen..." Weiter werden u.a. Zwangsneurosen, Phobien, Depressionen nicht zu schweren Ausmaßes genannt.

Es wird weitere Behandlungserfahrung und Forschungsarbeit notwendig sein, um hier zu allgemeingültigeren Aussagen zu gelangen, dazu gehört auch die Frage, inwieweit die FE als alleinige oder als kombinierte Behandlungsform angewandt wird.

R. Johnen (1987) berichtet in einer Studie über neun „schwierige Patienten" im stationären Setting, die mit der laufenden Therapie unzufrieden waren und teilweise vor dem Abbruch standen. Sie hatten verschiedenartige psychosomatische Symptome und fanden keinen Zugang zur verbalen Behandlung. Die Patienten erhielten 12 FE-Sitzungen. Johnen zeigt in der Studie, daß die FE „zum Therapieeinstieg und zur

---

3 Noch ausschließlicher als reine Körperarbeit wird die Feldenkrais-Methode durchgeführt. Sie verzichtet weitgehend auf die Sprache und versucht – besonders in der Einzelarbeit („Funktionale Integration") – das Wahrnehmen eingefahrener Bewegungs- und Haltungsmuster zu ermöglichen. In einem schrittweisen Prozeß sollen die im Gehirn niedergelegten Bewegungsmuster aufgelöst und die Freiheit erreicht werden, neue, angemessenere Muster zu bilden, die dann auch das Denken und die Emotionen beeinflussen." Wenn die Körperhaltung nicht geändert wird, kommen die alten Gefühle zurück." (B. Walterspiel). Von dieser zentralen Stellung des Körpers und der Motorik aus gesehen, ist diese Methode ein Gegenpol zur Psychoanalyse. Sie wirkt offenbar auch bei der Arbeit an den Folgen stark somatisch beeinflußter Erkrankungen, z.B. im neurologischen Bereich (vgl. Feldenkrais, 1977, 1978).

Motivationsverstärkung" dienen kann. Bepperling (1978) betont, daß zuweilen erst in der FE ein Zugang zur Ebene der Dualunion gefunden wird.

Deter und Heintze-Hook (1986) untersuchten 90 Asthmapatienten, die eine analytische Gruppentherapie erhielten sowie jeweils entweder FE oder AT. Die Kontrollgruppe erhielt keine Behandlung. Nach einem Bericht von Heintze-Hook zeigte sich als Ergebnis der Behandlung durch analytische Gruppentherapie und FE die „Verringerung von Klinik- und Notfallbehandlung, Arztbesuchen und Medikamentengebrauch (Cortison!) sowie Gewinn von mehr Unabhängigkeits- und Selbstwertgefühl" (Intern, 1987).

Dr. I. Pachner-Knoll dokumentierte 60 Behandlungen mit FE. Diese Behandlungen wurden zum Teil als reine FE-Behandlung, zum Teil kombiniert mit verbaler Einzel- oder Gruppentherapie durchgeführt. Wie die dokumentierende Ärztin selber betont, handelt es sich nicht um eine empirische Untersuchung. Trotzdem sollen einige Daten auszugsweise wiedergegeben werden, die in der Dokumentation differenziert beschrieben werden, hier aber die klinische Symptomatik nur sehr verkürzt vermitteln können.

Es wurden insgesamt 32 Fälle von Asthma bronchiale, Migräne, psychogenen Magenbeschwerden, multiplen psychosomatischen Beschwerden, Zwangssymptomen, Angstneurose und neurotischer Depression behandelt. Davon hatten sich 12 gebessert, 1 leicht gebessert, 10 waren symptomfrei und 4 geheilt, in einem Fall kam es zu einem Symptomwandel und 4 Fälle blieben unverändert. Erfolgreich behandelt wurden auch einzelne Fällen von Hypertonie, Bulimie/Anorexie, Spannungskopfschmerz, Torticollis und Herzneurose.

Als Kontraindikation nennt Wiesenhütter u.a.: Psychosen, Hypochondrie und ausgesprochene Minderbegabung, mahnt zur Vorsicht bei „stark Schizoiden" und bei „narzißtisch rechthaberischen Patienten".

23.3.3 Zum Problem kontrollierter Untersuchungen in der körperorientierten Psychotherapie

Kontrollierte Untersuchungen über Behandlungserfolge der FE gibt es bisher kaum. Trotzdem ist zu hoffen, daß in der Zukunft nun auch im Bereich dieser Methoden, soweit es möglich und sinnvoll ist, häufiger Erfolgskontrollen durchgeführt werden (vgl. auch Kap. 25).

23.3.4 Exkurs: Übergreifend wirksame Faktoren in verschiedenen Entspannungsverfahren

Sehr viel häufiger gibt es diese Untersuchungen im Bereich der Verhaltensmedizin. Hier werden oftmals kognitive Methoden (z.B. Änderung angstbesetzter Vorstellungen) mit übenden Entspannungsverfahren (progressive Entspannung nach Jacobson) kombiniert, oder auch Biofeedback kombiniert mit progressiver Entspannung bzw. andere Kombinationen verwandt (so z.B. in der Untersuchung von van Dixhoorn et al., 1987, über Rehabilitation nach Herzinfarkt). Für unser Thema ist interessant, daß sowohl die **Kombination** von Verhaltenstherapie mit entspannenden Verfahren häufig gute Resultate zeigt, als auch die Klärung bisher unbewußter Anteile im analytisch orientierten Gespräch in Verbindung mit der Funktionellen Entspannung (d.h. also

auch die Kombination von Verfahren) (vgl. den eingangs referierten Fall einer hypertonen Patientin). Das verweist wieder auf das Ansprechen verschiedener Ebenen der Person (vgl. auch Kap. 19 und 22).

Bei aller Verschiedenheit der Methoden können folgende übergreifende Faktoren vermutet werden:
- die Sensibilisierung für körperliche Vorgänge zusammen mit dem Erfahren/Erlernen von Entspannen und damit dem Wissen um eine mögliche Beeinflussung des Symptoms beim Patienten;
- die physiologische Umstellung durch Entspannung;
- menschliche Nähe, Berührung, Rückmeldung, „holding function" (zumindest Beachtung des subjektiven Körpererlebens) durch den Therapeuten.

23.4 Konzentrative Bewegungstherapie

Die enge Verbindung des Körpererlebens mit psychischen Prozessen ist auch eine Voraussetzung der Konzentrativen Bewegungstherapie (KBT), die in vielen psychosomatisch-psychotherapeutischen und psychiatrischen Kliniken zum festen Bestandteil der stationären Therapie gehört. Außerdem wird sie von niedergelassenen Praktikern durchgeführt. An ihrer Entwicklung nach dem Krieg waren u.a. J.E. Meyer, H. Stolze und Myriam Goldberg beteiligt.

Ein älterer Pädagoge mit analytischer Vorerfahrung, narzißtischer Problematik, Beziehungsstörungen und psychosomatischen Symptomen liegt während einer Übung auf dem Boden. Auf Vorschlag der Leiterin liegt ein Teil der Gruppe, der andere geht umher. Der Teilnehmer sieht die anderen Teilnehmer also aus der liegenden Position und fühlt sich einen Moment unbeweglich und fremd, wie die in einen Käfer verwandelte Hauptperson in Kafkas „Verwandlung". Bei einem späteren Angebot, dem „Durchspüren des Körpers", wird dann angeregt, die Hand anzuspannen und wieder locker zu lassen. Der Teilnehmer sagt dazu später: „Als ich die Hand anspannte, hatte ich ein sicheres Gefühl. Als ich sie lockerte und eine Art zärtlicher Bewegung machen wollte, fühlte ich mich resignativ und wurde traurig." Die gleiche Folge von Aktivität/Aggression in Verbindung mit Sicherheit einerseits und weicheren Gefühlen, die mit Depression und Resignation verbunden sind, vollzieht sich am nächsten Tag, als ein Stab zur freien Verfügung der Teilnehmer steht. Der Teilnehmer rollt damit zunächst seinen Körper ab (spürt seine Körpergrenzen), schlägt dann aggressiv auf einen Ball und läßt den Stab pfeifend durch die Luft sausen, geht schließlich mit „drohendem" Rhythmus des Stabs (mit dem er rhythmisch auf den Boden klopft) durch den Raum. Mit geschlossenen Augen (wie die übrige Gruppe) trifft er auf eine sehr viel jüngere Teilnehmerin, die er bei einem kurzen Blinzeln erkennt. Es entwickelt sich ein zarter Kontakt über den Stab, dann über die Hände. Nach der Beendigung dieses Kontaktes durch den Teilnehmer selbst trifft ihn plötzlich Resignation mit „großer Wucht" und er ist durch die Trennung „wie gelähmt". Die Teilnehmerin hat die gleiche Trennung ganz anders erlebt: „Es war beinah wie mit einem Vater, der mich entläßt – ich durfte dann in die Welt (in den Übungsraum) hinausgehen." Das tut sie auch – zu neuen Kontakten.

Diskussion

Es wird deutlich, wie auch in diesem Fall durch Körperhaltung und Bewegung ganz verschiedene Seiten aus der persönlichen Entwicklungsgeschichte angesprochen werden: Die Lage des Pädagogen auf dem Boden mit dem Vorbeigehen der „Erwachsenen" ähnelt dem frühkindlichen Erlebnisraum und läßt in diesem Fall ein offenbar sehr früh erlebtes Fremdheitsgefühl anklingen, das der Patient später, für ihn evident, mit Erzählungen aus seiner Kindheit und seinen Beziehungsstörungen verbinden konnte.

(In der Erzählung Kafkas ist dieses Erlebnis im übrigen mit **Körperentfremdung** (Verwandlung in einen Käfer) verbunden. Die Beziehungsproblematik wird dann auch in der traurig-resignativen Regung deutlich, die die „zärtliche Handbewegung" begleitet, während der „harte Griff" eher das Gefühl der Sicherheit gibt. Diese harte männlich-phallische Seite zeigt sich zunächst auch im Umgang mit dem Stab. Sie geht dann aber in eine eher zärtliche Kontaktaufnahme über und endet mit starkem Trennungsschmerz. Diese Ergebnisse waren für den Teilnehmer sehr überraschend und hinterließen einen starken Eindruck. Spätere verbale Äußerungen und Reflexionen wiesen deutlich auf eine ungelöste Bindung zur Mutter hin, die auf das Verhältnis zur Tochter übertragen wurde. Deutlich wird hier auch eine Haltung, bei der mit „Leistung" Sicherheit verbunden ist, während z.B. Hingabe und Emotionalität verunsichern. Daraus resultiert wieder **Dauerspannung.**

23.4.1 Zur Methode der Konzentrativen Bewegungstherapie

Einer der Gründe für die häufige Anwendung der KBT im stationären Setting dürfte ihre bevorzugte Anwendung in der Gruppe sein (zur KBT-Einzeltherapie s.u.). Mit dieser Gruppenbezogenheit hängen auch noch andere Merkmale zusammen, die diese Methode von der FE unterscheiden: Die **Beziehung** der Teilnehmer untereinander spielt eine wichtige Rolle, so wie es auch im oben angeführten Beispiel des Ulcuspatienten der Fall war, häufig auch die Beziehung zum Gruppenleiter. Damit wird auch im psychoanalytischen Sinne das Moment der Übertragung wichtig und für die Therapie nutzbar. Die Kontaktaufnahme kann insbesondere bei Patienten mit besonderer Angst vor Nähe durch den Gebrauch von verschiedenen Materialien (Ball, Seile, Stäbe, kleine Sandsäcke etc.) erfolgen und ist ein weiteres Kennzeichen der KBT. Ihre Verwendung erschöpft sich natürlich nicht in der Unterstützung einer dosierten Kontaktaufnahme, sondern wirkt auch über den jeweiligen Symbolgehalt und regt die dazugehörigen psychischen Inhalte an, die sich dann im jeweils individuellen Umgang mit dem Material zeigen. Hinzu kommt weiterhin die Verwendung des Raumes und seiner Symbolik sowie grundlegender Entwicklungsphasen, wie es sich z.B. in der Verwendung von Haltungs- und Bewegungsformen wie Liegen, Aufstützen, Sitzen, Krabbeln, Stehen, Gehen zeigt. Wie schon oben im Fallbeispiel des Ulcuspatienten erwähnt, löst das Erleben dieser Körperhaltungen und -bewegungen sehr häufig starke Emotionen beim Patienten aus, die in ihrer Intensität und Farbigkeit sowohl für Patienten als auch für die Teilnehmer von Fortbildungskursen oft überraschend sind.

Die KBT bezieht also verschiedenste menschliche Dimensionen in ihre Arbeit mit ein, und ist deshalb nicht nur als Therapie für bestimmte Symptome zu sehen, sondern bietet sich auch als eine Möglichkeit an, die eigene Person mit ihren Beziehungen zur Umwelt und diese Umwelt selber differenzierter, sensibler zu erleben und neue Erfahrungen zu machen. Das gilt in etwas anderer Weise auch für die FE.

Mir scheint, daß der Schwerpunkt der FE als „Basismethode" bei mittelschweren und schweren Erkrankungen liegen kann, während die KBT in einem weiteren Anwendungsbereich auch gerade Beziehungsprobleme innerhalb ihres Settings gut erfaßt (vgl. auch Wiesenhütter, 1983).

Auch im Zusammenhang mit der KBT scheint hier eine Gelegenheit zu ansatzweiser körperlicher Selbsterfahrung für den Leser möglich zu sein. U. Kost hat in einem Vortrag den Zuhörern folgenden Vorschlag gemacht, der sich auch beim Lesen nachvollziehen läßt:

„Bitte ändern Sie jetzt nichts an Ihrer augenblicklichen Haltung. Schließen Sie die Augen für einen Moment, fragen Sie sich: Was spüre ich von mir? Wie sitze ich? Wo habe ich Kontakt zum Boden, wo ist mein Gewicht – wie ist meine Sitzfläche auf dem Stuhl, was fühle ich im (am) Rücken, wie halte ich meinen Kopf, die Arme und Hände? Was spüre ich von der Atmung – wieviel Raum habe ich in mir, wo enge ich mich ein..." Nach dieser ersten Beschreibung wird empfohlen, auf Verkrampftheit oder Entspannung zu achten (wo im Körper?), einem eventuellen Wunsch nach Veränderung nicht gleich nachzugeben, ihn zunächst in der Vorstellung zu vollziehen und dann – nach tatsächlicher Veränderung – sich zu fragen, was sich damit verändert hat. „Schließlich dann die Reflexion, die Erweiterung zu der Frage: Was bedeutet das für mich? Wie gehe ich mit mir um? Wie plaziere ich mich im Hier und Jetzt und sonst im Leben?"

In einer KBT-Stunde könnte in ähnlicher Weise ein Einstieg in die Arbeit vollzogen werden. Dabei können Körperhaltungen bewußt werden, die es vorher nicht waren. So z.B. die Anspannung des Kiefers, ein steifer Rücken, ein gepreßter Atem usw. (bis hierher ähnlich wie die FE). Es könnte die Aufforderung folgen, im Liegen sein Gesichtsfeld zu überprüfen, das gleiche mit leicht aufgestützten Armen zu tun (auf den Ellenbogen liegend), mit ausgestreckten Armen, im Liegen, im Sitzen usw. Es geht hier weniger um „Übungen" im gewohnten Sinne als um Angebote, die neue Erlebnismöglichkeiten eröffnen. Es geht dabei um den schon erwähnten Wechsel von Körperarbeit und Verbalisierung, die im Wechsel von Unmittelbarkeit und Reflexion durchaus Parallelen zum analytisch orientierten Gespräch hat. Die körperlichen „Angebote" können dabei mit den verbalen „Angeboten" des analytischen Psychotherapeuten verglichen werden. Das betrifft besonders die stärkere Aktivität des Therapeuten bci früh gestörten Patienten (vgl. auch Heigl und Streek, 1985).

Die Wirkung von derartigen Übungen ist oft erstaunlich. So hatte eine Teilnehmerin beim Umherblicken mit aufgestützten Armen das zwingende Gefühl, sie „dürfe" das eigentlich nicht und fand schließlich einen „Trick", doch umher zu schauen, ohne daß die Leiterin es merkte. Die Übertragung einer entsprechenden früheren Verbotssituation wurde im anschließenden Gespräch deutlich.

Ein wesentliches Kennzeichen körperbezogener Therapieformen ist auch hier vorhanden: das starke emotionale Erlebnis, dessen differenzierte Bedeutung dann in weiteren Schritten geklärt werden kann. Die starke emotionale Tönung des Erlebens hängt hier ganz offensichtlich mit der Beteiligung des Körpers zusammen (vgl. Kost, 1979). Weiterhin mit dem Erlebnis, daß man, wenn es notwendig wird, auch real „gehalten" werden kann, und mit Nähe zu frühen Vorgängen durch Motorik und die emotionsfördernde Anwesenheit der Gruppe. Eine intellektualisierende Abwehr hat es also in diesen Fällen schwerer (darf natürlich auch nicht „überrollt" werden). Auch lange verdrängte oder abgespaltene Erlebnisse werden in diesen Körpererlebnissen eher fühlbar und werden häufig emotional farbiger erlebt als in vorausgegangenen verbalen Psychotherapien. So bricht eine Teilnehmerin nach einer Übung, in der die Hände einbezogen waren, in verzweifeltes Weinen aus und schaut auf ihre eigenen Hände. Sie hat das überwältigende Gefühl, sie nie wieder bewegen zu können. Eine lange zurückliegende schwere Erkrankung mit der Gefahr dauernder Lähmung, die psychisch nicht bewältigt war, wurde hier als „handgreifliche" Erinnerung präsent. Das ist auch ein Hinweis auf die „zeitlose" Qualität dieser Erinnerungen und Emotionen (Freud, 1911; v. Uexküll, 1963) – hier in Verbindung mit dem Körpererleben. Mit der Körperarbeit können sie erlebt werden. Zur Bedeutung der unmittelbar emotional erlebten Körpererfahrung sagt Loewald: „Dies Aufsteigen zum bewußten emotionalen Erleben würde ... das psychoanalytische Element ausmachen, sofern es ‚Unbewußtes' bewußt macht, vom Gedanklich-Sprachlichen her gesehen eine nicht voll entwickelte Bewußtheit."[4] Es muß also noch die sprachliche Formulierung hinzukommen, die das Erlebte in eine sekundärprozeßhafte Organisation einbindet. Häufig werden in der KBT die Augen geschlossen. Das erlaubt eine Konzentration auf das Körpererleben mit seinen verschiedenen Komponenten, ohne durch optische Reize abgelenkt zu werden. Das Körpererleben wird damit eindringlicher und ontogenetisch frühe Phasen werden durch die Aktivierung der Nahsinne belebt. Auf dieser frühen Ebene zeigen sich dann auch die Ausgangspunkte für die spätere Wortbildung. Gräff (1983) bringt dazu Beispiele. Sie erwähnt das Liegen: „Wie ist die Lage?" („schlechte", „gute" Lage), „liege ich richtig?", man versetzt sich in die „Lage des anderen". Oder das Stehen: „Mit beiden Beinen im Leben stehen" bzw. „auf dem Boden der Tatsachen stehen". Hier ist ein Hinweis auf den Realitätsbezug und auf die notwendige „Basis" gegeben, für die man „Boden unter den Füßen haben" bzw. „standhaft" oder „standfest" sein muß.

Loewald (1986) hat aus psychoanalytischer Sicht überzeugend dargestellt, in welchem Umfang diese präverbalen Erlebnisse in die spätere Verbalisierung einfließen und wesentlich zur vollen Lebendigkeit, zur Intensität und Tiefe der Worte beitragen. Diese frühe Welt kann allerdings auch verlorengehen, wenn die „Sachvorstellungen" von den „Wortvorstellungen" (Freud, 1915) getrennt werden. Diese Wurzeln können u.a. in der KBT und FE noch einmal in ihrem ursprünglichen Sinn erlebt werden und können damit auch zu frühen Erlebnissen führen, in denen unsere Entwicklung ernsthaft und folgenreich behindert wurde. Die KBT ist dabei, trotz des erwähnten stärkeren emotionalen Erlebens, eine behutsame Methode, die nicht forciert und manipuliert, sondern eher „Anstöße" gibt und besonders in einer länger laufenden Gruppe auch zu Prozessen des Erinnerns, Wiederholens und Durcharbeitens führen kann (vgl. Stolze, 1984). Statt der verbalen Bearbeitung kann der KBT-Therapeut eventuell auch ein erneutes Übungsangebot geben, das der Konfrontation und Klärung dient und letztlich zur Deutung führt (Becker, 1981).

Abschließend ein Auszug aus einem Fallbericht, der auch die Doppelbedeutung von „Haltung" eindrucksvoll zeigt.

Eine Patientin, die wegen Knie- und Hüftgelenksbeschwerden in die KBT kam, sagte mehrmals, daß sie von den Knien abwärts nichts mehr von sich wahrnehmen könne.

Die stark angespannten Knie bezeichnet sie als ihre untersten spürbaren Punkte. Sie hätte das Gefühl, wie eine Marionette in der Luft zu hängen. Die sensitive junge Frau merkte bald, wie sich ihre Gefühle in den Knien widerspiegelten. „Lasse ich meine Knie los, habe ich Angst auseinanderzufallen. Es drängt sich immer eine Stimme in mir auf: ‚Gib ja nicht nach, werde nur nicht kniefällig'." Sie war ein „ungeliebtes widerborstiges Mädchen" gewesen, das die Bedürfnisse der Mutter, „eine Puppe aus ihr zu machen", mit Bubenstreichen und Ungezogenheiten quittierte ... Die Patientin erkannte, wie sie ihr Leben von der Mutter abhängig gemacht hatte, indem sie sich gezwungen fühlte, das Gegenteil von deren Wünschen zu tun. Das betraf sogar die Berufswahl und eine vermiedene Heirat, obwohl sie mit ihrem Partner ein gemeinsames Kind hatte. „Nur nicht nachgeben!" Sie stand nicht auf ihren eigenen Füßen; das nahm sie körperlich. Ihre Pseudo-Eigenständigkeit bestand in der Vermeidung der mütterlichen Erwartungen. Sie zahlte den Preis der Marionette. Die Patientin erlebte einen Wandel, als sie herausfand, daß „Nachgeben" nicht mit dem Gefühl „ausgeliefert zu sein" verbunden war. Sie spürte, wie die nachlassende Spannung die Knie durchlässiger machte und sie zunehmend mehr mit ihren Füßen und ihrem Boden verband. „Ich schlage Wurzeln", bemerkte sie eines Tages. Einerseits mußte sie zum Boden hinwachsen, ihn andererseits als festen Widerstand erleben, von dem man Kraft erlebt und sich abstößt (Gräff, 1983).

---

4 H. Loewald: persönliche Mitteilung

23.4.2 Zur Anwendung der KBT

Becker (1981) empfiehlt im stationären Bereich analytische Gruppe und KBT in getrennten Sitzungen bei einem oder zwei Therapeuten. In den Teamkonferenzen werden die Erfahrungen integriert.

Meistens werden in der Klinik KBT und verbale Therapie wohl von verschiedenen Personen durchgeführt werden. Bei Patienten, die sich in analytischer Psychotherapie oder einer Analyse befinden, habe ich in bestimmten Phasen gute Erfahrungen mit der Empfehlung zur Teilnahme an einer kürzeren (2–5tägigen) KBT oder KBT-ähnlichen Veranstaltung gemacht. Die dort gewonnenen Erfahrungen konnten in der verbalen Therapie durchgearbeitet werden. Es gibt allerdings auch längerdauernde parallele Teilnahme. Die auftretenden Übertragungsphänomene müssen noch genauer beschrieben werden.

23.4.3 Indikation

Becker (1982) nennt als Indikation: psychosomatische und funktionelle Beschwerden, Neurosen, geistige und korperliche Behinderung. Es wird die Bedeutung der Methode für Patienten mit frühen Störungsanteilen hervorgehoben.

Hinweise auf eine ausgesprochene Kontraindikation werden nicht gegeben. Bei psychotischen Patienten oder Borderline-Symptomatik sind gewisse Modifikationen der Technik notwendig (mehr Ich-Stützung, Abgrenzung, Realitätswahrnehmung). Von Fall zu Fall zu überlegen ist die Möglichkeit der Integration von Körperbehinderten in einer KBT-Gruppe.

H. Tammen (1988) fand in einer Studie über die „katamnestische Untersuchung von stationär oder ambulant behandelten Patienten mit Ulcus duodeni und/oder ventriculi in der Psychosomatischen Universitätsklinik Heidelberg", daß 64,3% der Patienten in der KBT wichtige Erfahrungen gemacht haben. Ebenfalls 64,3% beurteilten die KBT als sehr hilfreich bzw. hilfreich. 42,9% konnten über die KBT erstmals deutlich Konfliktbereiche wahrnehmen. Bei 21,4% wirkten sich die Übungen der KBT direkt und positiv auf die Beschwerden des Patienten aus. 28,6% bewerteten dieses therapeutische Angebot als für sie hilfreicher als die Gespräche.

In einer kontrollierten Studie wurden von Carl et al. (1982) die Verläufe in einer KBT-Gruppe und einer parallellaufenden analytischen Gruppe (AGT) untersucht. Es zeigte sich, daß „Prozeßverläufe bei AGT und KBT gleichsinnig" sind und bestimmte Gruppenphänomene in der KBT einen oder mehrere Tage früher auftreten als in der analytischen Gruppe.

Wichtig erscheinen hier auch die Erfahrungen, die aus Einzelbehandlungen mit KBT resultieren, über die L. Koch berichtet (1988).

23.5 Körperorientierte Psychotherapie nach F. Besuden und S. Damm

Mit Elementen der KBT, aber etwas stärker auf die Arbeit mit dem Körper konzentriert, arbeitet Frauke Besuden mit ihrer „körperorientierten Psychotherapie", die einen Schwerpunkt auf die Änderung des Körperbildes legt (vgl. Abschn. 23.10.2). Eigene Ansätze, Anteile der KBT, der Eutonie nach Alexander, Erfahrungen beim Unterricht mit Schauspielschülern (Stimme und Atem) sowie psychoanalytische Selbsterfahrung gehen in die Arbeit ein.

S. Damm verbindet in ihrer „Modellaufgabengruppe" Bewegung (Rhythmik), musikalische und bildnerische Elemente mit analytisch orientierten Interventionen.

23.6 Bioenergetik

Die Bioenergetik ist in den letzten Jahren gleichfalls vereinzelt als stationäres Verfahren eingesetzt worden. Sie wird in erster Linie von freipraktizierenden Therapeuten ausgeübt. Es gibt Übereinstimmungen, aber bislang auch grundsätzliche Unterschiede zu den bisher beschriebenen Methoden, die besonders den Grad des direktiven Arbeitens betreffen. Gleichzeitig versteht sich aber die „Bioenergetische Analyse" in der Tradition von Freud und Reich (Sebastian, 1985). In der BRD gibt es einen mehrjährigen Ausbildungsgang. Von Reich wurde das ganzheitliche Konzept des Menschen übernommen, seine Auffassungen über Charakteranalyse und vor allem „seine Begründung pathologischer Strukturen als Ergebnis von Hemmungen freifließender biologischer Energie, Bioenergie" (Sebastian, 1985). Schon früh in der Kindheit sieht die Bioenergetik den natürlichen Gefühlsausdruck häufig unterbrochen, seine Äußerung ist dann mit Angst (u.a. vor Liebesverlust) verbunden. „Die Folge ist ein Energiestau ... um sich zu schützen, reduziert das Kind ganz allgemein die Atmung als eine Möglichkeit, das sozial Unerwünschte einzudämmen, zu ‚blockieren'. Zusätzlich werden auch die mit den Impulsen verbundenen Muskelgruppen kontrahiert" (Sebastian, 1985). Wenn ein entsprechendes Verhalten der Umwelt in der Entwicklung über Jahre anhält, kommt es nach Auffassung der Bioenergetik zu einer „chronischen Panzerung" der Muskulatur in den betroffenen Bereichen. Dieser „Panzer" bindet auch die Energie, die bei ihrer Freisetzung Angst auslösen würde. In der bioenergetischen Praxis wird „das Aktivitätsprinzip auf der somatischen Ebene mit dem analytischen Verfahren auf der psychischen Ebene" kombiniert. Der Therapeut „wählt dem Patienten und der Situation angemessene Übungen und beobachtet die Reaktion des Patienten darauf" (Sebastian, 1985).

Es folgt ein Ausschnitt aus einer Behandlungsstunde, der aus Platzgründen kurz sein muß und nur einen sehr begrenzten Eindruck vermitteln kann. Interessierte Leser seien auf die Originalarbeit verwiesen.

„Es handelt sich um eine Patientin, die sich nach einer Pause wieder an den Therapeuten wandte, weil sie den Eindruck hatte, ‚in einer Sackgasse zu sein'. Die linke Seite gibt im folgenden das Gespräch wieder, die rechte Seite zeigt den Körperausdruck. Der durchgeschriebene Teil enthält Zusammenfassungen. Eigene Überlegungen und Interventionen sind in Klammern gesetzt.

Gesprächsverlauf	Körperausdruck
Ich: Wo bist du jetzt?	Der Körper wirkt voll und
Sie: Wenn ich einfach da-	kräftig, die Beine sind
von ausgehe, was ich	durchgedrückt … der
jetzt so fühle, merke ich,	Oberkörper aufrecht, die
daß ich kurz vor dem	Füße nach innen einge-
Weinen bin … ohne daß	knickt, die Augen beob-
ich jetzt so wüßte, womit	achtend. Der Gesichts-
dies zusammenhängt, ir-	ausdruck lächelnd, ins
gendwie bin ich so trau-	Weinen kippend beim
rig…	Gespräch. Energetisch
	gesehen, ein geladener
	Organismus, dessen
	Ausdruck – hier das Wei-
	nen – durch die flache At-
	mung und die soldatisch
	starre Haltung verhindert
	wird. Die Starre verhin-
	dert auch den Energiefluß
	zur Erde und damit auch
	das Gegründetsein.

(Ich schlage ihr eine einfache ‚grounding'-Übung vor: Gewichtsverlagerung auf ein Bein, um damit den Energiefluß in Bewegung zu setzen. Dabei verschieben sich Becken und Oberkörper seitlich zueinander, so daß ein Bruch in der Taille deutlich wird. Sie nimmt verschiedene Anspannungsbereiche in den Beinen wahr, besonders in den Waden, dann fühlen sich die Beine an ‚wie Beton'.)

Patientin: Ich habe jetzt	Sie beginnt zu weinen,
das Gefühl, irgend etwas	was sich gegen Ende zu
tragen zu müssen, oder –	starkem Schluchzen stei-
wenn ich mir meine Beine	gert. Der Körper gerät in
jetzt vorstelle, daß sie	Bewegung und zieht sich
mich vielleicht nicht tra-	im Zwerchfell krampfartig
gen können, daß, wenn	zusammen…"
sie nachgeben, daß ich	
dann zusammenbrechen	(Sebastian, 1985)
würde…	

In einer späteren Sitzung berichtet die Patientin, daß sie um das Ende des 1. Lebensjahres bis Anfang des 2. (etwa zur Zeit der Übungsphase nach M. Mahler) häufig im Kinderwagen lag, während ihre Mutter in der Nähe auf dem Felde arbeitete. Sie habe immer wieder versucht sich aufzurichten und zu ihr hinzulaufen und sei dabei öfters aus dem Wagen gefallen. Sie stand also schneller auf, als es von ihren Körperfunktionen her möglich war (Sebastian, 1985).

Wie schon angeführt, ist dieses Modell sichtlich direktiver als die bisher beschriebenen Methoden; prinzipiell kann aber das vorübergehende Einsetzen einer Körper-„Technik" bei manchen Patienten hilfreich sein, um in Kontakt mit dem Körper und zugehörigen Affekten zu kommen (s.u. Downing).

Die verbale Phase zu Beginn einer Bioenergetik-Stunde ist meist länger als oben beschrieben (J. M. Scharff, pers. Mitteilung).

23.6.1   Diskussion

In ihrer eher abwartenden Haltung und in vorsichtigen Angeboten steht die KBT der psychoanalytischen Technik näher, auch wenn sich die Bioenergetik auf theoretische psychoanalytische Positionen beruft, die dann allerdings eher eine Fortsetzung der frühen Triebpsychologie darstellen. In letzter Zeit gibt es Ansätze zur Verbindung mit moderneren psychoanalytischen Theorien (vgl. Sebastian, 1983).

23.7   Zum Problem der Katharsis

Es muß in diesem Zusammenhang kurz auf die Rolle starker affektiver Entladungen in den hier genannten körperbezogenen Verfahren hingewiesen werden. Seitdem die Psychoanalyse das Konzept der Katharsis aufgab, wissen wir, daß diese starke Entladungen keine dauerhaften psychischen Änderungen bewirken. Aber gerade die heftigen Ausbrüche galten einige Zeit bei Beobachtern und Teilnehmern verschiedener körperbezogener Verfahren als alleiniger Hinweis auf intensive Arbeit oder einfach auch als „chic". Es wurde Widerstand „gebrochen" und ein zuweilen verhängnisvolles Zusammenbrechen der Abwehr provoziert. Ein derartiges Vorgehen wird zu Recht kritisiert (vgl. Kind, 1985). Wenn hier auch wiederholt auf starke emotionale und affektive Verläufe hingewiesen wurde, so liegt der Vorteil der körperbezogenen Verfahren weniger in der Katharsis als im Evidenzerlebnis, mit dem bisher verdrängte, verleugnete etc. Gefühle, bzw. bisher farblose Erinnerungen während der Arbeit erlebt und neue Erfahrungen vermittelt werden, die der Patient vielleicht in einem „alexithymen" Elternhaus nicht erfuhr (vgl. Ahrens, 1988). „So schlimm war das also damals für mich? So intensiv war mein Gefühl?" „Wie gehe ich jetzt damit um?" Es ist nochmals zu betonen, daß nur „Anstöße" erfolgen dürfen, die die Persönlichkeitsstruktur (das Ich) noch verarbeiten, noch tolerieren kann. Anderenfalls verstärkt sich die Abwehr oder der Patient dekompensiert (vgl. Krystal, 1978). Auch hier bringen die kleinen Schritte mehr als spektakuläre Ausbrüche. Schließlich ändert der erlebte und auch in den Objektbeziehungen zur Verfügung stehende Affekt die Beziehungen zum anderen. Der Partner erhält eine eindeutige Mitteilung und kann eindeutige Antworten geben (vgl. Krause, 1988a).

23.8   Weitere körperorientierte Verfahren

Im Zusammenhang mit der Bioenergetik ist auch auf George Downing hinzuweisen, der in seiner „Körperorientierten Psychotherapie" zeitweilig direktive Körpertechniken (Reich, Lowen) mit verbaler, analytischer Arbeit und auch Elementen der Gestalttherapie (s.u.) verbindet. Übertragung und Gegenübertragung werden in die Arbeit einbezogen. Die verbale Arbeit hat stärkeres Gewicht als in der Bioenergetik, sie ist hier der „rote Faden", an dem sich der Therapeut orientiert. Körpertechniken werden bei Bedarf eingeschoben. Auch dabei zeigt sich die Wichtigkeit der Arbeit am Körperbild, besonders bei psychosomatischen und Borderline-Patienten, die die Individuationsphase nicht relativ störungsfrei durchliefen. Bei Borderlinefällen beobachtete Downing zuweilen eine Übertragungslinie, die sich eher auf den „Körpertherapeuten", und eine andere, die sich eher auf den „verbalen Therapeuten" richtet. (Bei Behandlung durch **einen** Therapeuten!)

In dieser zeitweilig eingeschobenen (und integrierten!) Einbeziehung des Körpers während verbaler Einzeltherapie scheint mir für die Zukunft eine wichtige Entwicklungslinie zu liegen. Es handelt sich dabei um die besondere Beachtung und evtl. zusätzliche Arbeit mit Haltung, Stimme, Atem, Raum (vgl. KBT).

So wurde z.B. einem Patienten mit Zwangssymptomatik vorgeschlagen, ein von ihm beklagtes Gefühl des „Gehetztseins" darzustellen, indem er sich im Raum entsprechend bewegte. Bei diesem Gehen erlebte er sich unmittelbar körperlich und emotional überzeugend als „fremdgesteuert", „wie ein Roboter". (Der Patient wies keine psychotischen Anteile auf!)

Voraussetzung ist dabei genügend Erfahrung sowohl in verbaler als auch in Körpertherapie – d.h. beide Elemente müssen im Therapeuten integriert sein. Moser (1988) beschreibt für die analytische Arbeit das Sitzen neben der Couch mit der Möglichkeit des Blickkontaktes und der (haltenden) Berührung.

Die **Eutonie** (G. Alexander) ist ein „übendes" Verfahren, das den Versuch macht, unbewußt verlaufende körperliche Vorgänge durch strukturierte Übungen bewußt zu machen (Brand, 1986).

23.9 Exkurs: Gestalttherapie

Da die Gestalttherapie in Theorie und Technik den Körper mit einbezieht und andererseits Techniken dieser Methode in körperorientierte Verfahren Eingang gefunden haben, sollen hier einige kurze Hinweise gegeben werden:

Prinzipien der Gestaltpsychologie (Wertheimer, Köhler, Koffka) und der Psychoanalyse wurden durch die von F. Perls begründete Gestalttherapie übernommen und in Richtung einer Persönlichkeitstheorie erweitert. So wird u.a. die Persönlichkeit als „Ganzheit", als „Gestalt" gesehen, in der die jeweils herrschenden Bedürfnisse als „Figur" vor einem „Grund" hervortreten. Der Neurotiker kann seine Bedürfnisse nicht mehr angemessen wahrnehmen und kann sie deshalb nicht regulieren. Aus „unvollendeten" Bedürfnissen entsteht ständiges Ungleichgewicht und Verarmung. Techniken der Gestalttherapie, die an das bewußte Wiedererleben des Nicht-Erlebten heranführen sollen, sind „Regeln und Spiele". So wird der Patient z.B. aufgefordert, mit imaginierten Elternfiguren oder auch mit getrennten Anteilen seiner Person einen Dialog zu führen bzw. auch wechselnde Rollen einzunehmen. So führt u.a. auch das Wahrnehmen von Dichotomien zwischen verbalem und averbalem Ausdruck zur Wahrnehmung bisher verborgener Tendenzen. Der Patient kann auch aufgefordert werden, über seinen Körper in der

Ich- statt der Esform zu sprechen (Thetford und Schumann, 1988). Im Versuch einer Integration auch des körperlichen Bereiches und der damit verbundenen psychischen Anteile liegt die Verwandtschaft zur körperbezogenen Psychotherapie. Jedoch spielt der verbale Anteil in der Gestalttherapie in der Regel eine größere Rolle (auch wenn es, wie ersichtlich, immer mehr Übergangs- und Zwischenformen gibt).

Nur hingewiesen werden kann hier auf die benachbarte „Integrative Bewegungstherapie" nach Petzold (1977) und die „Psychomotorische Therapie" von D. und A. Pesso (1986) sowie die „Körperarbeit" von Reich (1933). Zur Tanztherapie siehe Siegel (1986).

23.10 Theoretische Gesichtspunkte

23.10.1 Motorik und Emotion

Die enge Verbindung von Motorik und Emotion ist in der bisherigen Darstellung verschiedentlich betont worden. Bierbaumer (1983) schildert drei Experimente. Im ersten Experiment wurden Versuchspersonen verschiedene Gefühlszustände suggeriert (Trauer, Furcht, Freude etc.). Dabei lag der Mittelfinger der Versuchsperson auf einem empfindlichen Meßknopf auf, der die Ausschläge des Mittelfingers registrierte. Bei einer größeren Gruppe von Versuchspersonen ergaben sich typische Bewegungsformen für die einzelnen Emotionen, die man klar voneinander unterscheiden konnte. In anderen Experimenten wurden aus diesen Mikrobewegungen spezifische Kurvenverläufe für wesentliche Grundemotionen gewonnen. Clynes (zit. n. Bierbaumer, 1986) bat „die Versuchspersonen, die Kurvenverläufe für eine bestimmte Emotion auf einem Bildschirm mit ihrem Finger nachzufahren". Er berichtet, „daß bei einigen Versuchspersonen nach wiederholten Durchgängen ... rein muskulären ‚Nachzeichnens' ... das entsprechende Gefühl auch subjektiv" empfunden wurde (Bierbaumer, 1986). In einem anderen Experiment wurden z.B. kleinste, unsichtbare aber meßbare EMG-Änderungen der Gesichtsmuskeln in Richtung Depression verstärkt. Nach längerem „Training" trat schließlich das Gefühl der Depression auch bewußt auf. „Rückmeldung und operantes Training der EMG-Reaktionen oder Fingerausschläge führen also ohne Mitwirkung der Versuchsperson zu den jeweils in der physiologischen Rückmeldung repräsentierten Emotionen" (Bierbaumer, 1983). Es wird an den Satz von James/Lange erinnert: „Wir weinen nicht, weil wir traurig sind, sondern wir sind traurig, weil wir weinen." Im Anschluß an Clynes wird festgestellt, daß sich „die autonome und motorische Spezifität von Gefühlen ... nicht nur im Gesichtsausdruck nieder(schlägt), sondern in fast jedem Körpersystem ... Gesten und Körperhaltung sind ebenso den einzelnen Grundemotionen zuzuordnen wie Mikrobewegungen einzelner Muskeln..." (Bierbaumer, 1986).

Diese Befunde treffen sich mit den Feststellungen des Psychoanalytikers G.S. Klein. Klein betont, daß

Phantasien „zusammen mit Emotion und Handlung eine kognitiv-emotional-motorische Einheit" bilden. Diese Einheiten können der Verdrängung verfallen. „Trotz der Verdrängung bleibt die gesamte Einheit von Phantasie, Emotion und Handlung als zusammenhängendes ‚Pattern' aber aktiv…" (Klein, 1967; zit. nach Kutter, 1983).

23.10.2 Körperbild und Körperselbst

In diesem Zusammenhang ist nun auch auf den Begriff des Körperbildes und des Körperselbst einzugehen. Beginnend mit Schilder gibt es besonders in den letzten Jahren eine ausgedehnte Literatur mit experimentellen, psychotherapeutischen (psychoanalytischen) und psychosomatischen Fragestellungen (vgl. u.a. dazu Kiener, 1973; Joraschky, 1986 und die „Werkstattgespräche zum Thema Körperbild", 1983).

Abgegrenzt wird der Begriff vom „Körperschema", der sich auf neurologische Gesichtspunkte bezieht. Lichtenberg (1978) sieht im Begriff des Körperselbst den ganzen Umfang der Erlebnisse einbezogen, die sich um den Körper zentrieren, d.h. die Körperoberfläche und das Körperinnere mit den entsprechenden (unbewußten und bewußten) Vorstellungen.

Wenn wir an die oben erwähnten Experimente über den engen Zusammenhang von Motorik und Emotion denken, so wird auch deutlich, daß durch die persönliche Entwicklung hindurch eine ständige enge Koppelung von Motorik sowie anderen körperlichen Prozessen mit spezifischen Emotionen in jeder Entwicklungsphase erfolgen muß. Dieser Prozeß wird verstärkt durch die besondere Rolle, die die Sensomotorik zu Beginn des Lebens hat und damit in einer Phase besonderer Prägungsmöglichkeiten. Piaget (1976) hat auf diese Phase besonders hingewiesen. Der Körper ist zunächst „das Bezugsfeld für jedes Kind, das Zentrum seines Aktionsfeldes. Hand in Hand mit der Orientierung am eigenen Körper organisiert sich die Umweltwahrnehmung" (Joraschky, 1986; vgl. auch Bruner, 1971). Aber auch später im Leben können schwerwiegende körperliche Eingriffe stärkere psychische Krisen auslösen, z.B. eine Herzoperation (Möhlen und Davies-Osterkamp, 1979). Auch Besuden weist darauf hin, daß Defizite im Bereich der Ich-Struktur mit Defiziten im Bereich des Körperbildes verbunden sein können. Schütz, Besuden und Mitarbeiter gehen diesen vermuteten Verbindungen auch anhand von Körperzeichnungen ihrer Patienten nach (Schütz et al., 1988). In Analysen wird zuweilen die Entfremdung von Teilen des eigenen Körpers sehr deutlich erlebt. Bei weiblichen Patienten kann diese Entfremdung im Unterbauch lokalisiert sein. Diese Bereiche waren in eigenen Behandlungen in zwei Fällen assoziativ mit einem malignen Mutterbild verbunden. In diesem Zusammenhang traten Störungen von Konzeption und Schwangerschaft auf. Neben verbaler Therapie kann eine kürzere oder längere zusätzliche körperbezogene Methode in diesen Fällen sehr hilfreich sein. Maurer (1987) folgert aus Fragebogenuntersuchungen: „Wenn es gelingt, die Körperbesetzung zu … verbessern, werden sich auch Störungen im Selbst durch bessere Selbstbesetzung vermindern" (vgl. die oben erwähnte Behandlung von L. Koch).

Zurück zur Entwicklung. Auch für Bruner (1971) existiert für das Kind die Außenwelt im 1. Lebensjahr wesentlich durch den „handelnden Umgang". Beziehungen bestehen hier auch zum „Gestaltkreis" Viktor v. Weizsäckers (vgl. auch Wiesenhütter, 1983; Blankenburg, 1983). Nun erlebt das Kind nicht allein Greif- und Tasterlebnisse, sondern bei jeder Handlung auch Emotion. So ist schon das Stillen eine Sequenz, mit der Körperempfindungen, Berührung, Bewegung und – bei ungestörtem Ablauf – Lusterleben von Spannung zur Entspannung eingehen (zur „Sequenz" vgl. Müller-Braunschweig, 1975). Ebenso ist es später mit dem Gehen. Das Gehen setzt voraus, daß das Kind aufsteht, sich also vom tragenden Boden entfernt und im Vollzug des Gehens sein Gewicht von einem Bein auf das andere verlagert. Dieser labile Moment ist außerdem mit „Fort-Schreiten" von der Mutter verbunden und ist die Voraussetzung für Fortschritt. Je nach der Sicherheit, mit der die Mutter diese Versuche begleitet, werden sie eher von positiven oder auch von negativen Gefühlen begleitet sein (vgl. KBT; zur Theorie der KBT in diesem Zusammenhang wieder besonders Becker, 1981).

23.10.3 Frühe Konditionierungen

Aber schon früher, in den ersten Lebensmonaten, vor der Subjekt-Objekt-Differenzierung, können in der symbiotischen Beziehung offenbar vegetative Abläufe folgenreich beeinflußt, d.h. auch konditioniert werden. „Beeben, Stern und Jaffé fanden, daß im 3. Monat Mutter und Kind in einer Welt der ‚Mikroreaktivität' leben, in der jede Seite extrem sensitiv für die Körperbewegung des anderen ist und in weniger als einer Sekunde auf sie antwortet" (Krause, 1983). Krause weist auch darauf hin, daß in der Interaktion Mutter und Kind „offensichtlich … nur sehr feine Abweichungen in der zeitlichen Verlaufsstruktur ursächlich für das Zusammenbrechen der dialogischen Interaktion" (sind). So litt ein Säugling an einer zunächst unerklärlichen kindlichen Magenkolik. Als Ursache stellte sich heraus, daß die Mutter jeweils während des Stillens mit ihren Freundinnen telefonierte (Lempp, zit. nach Krause). Häufige Wiederholungen derartiger Abläufe könnten zu einer spezifischen Vulnerabilität des betreffenden Organs führen. Adler und Th. v. Uexküll (1986) sprechen von der „individuellen Physiologie (von) Organen … bei denen eine Konditionierung erfolgt" ist. „Die individuelle Physiologie kann nur biographisch … verstanden werden…" (v. Uexküll, 1986). Auf diese Weise kommt es zu einem erworbenen individuumspezifischen Erregungsmuster (Lacey et al., 1953; vgl. auch Müller-Braunschweig, 1980).

Sicher stellen diese mehr oder weniger stabilen Muster nur einen Teil der Faktoren dar, die zu einer psychosomatischen Erkrankung führen. So hat z.B.

H. Weiner (1986) Beispiele für die multifaktorielle Genese gegeben. Körperbezogene Psychotherapieformen haben aber in gewissen Fällen die Möglichkeit, diese sehr früh erworbenen und verbal schwer erreichbaren Muster und damit einen Faktor des pathologischen Systems zu beeinflussen, der Änderungen im Gesamtsystem nach sich ziehen kann (vgl. FE).

Eine Nichtbeachtung bestimmter Affekte (z. B. Wut oder Ärger) durch die Mutter kann auf Dauer auch zur Löschung des „Signalanteils" (also des Ausdrucksanteils) führen. Bei Fortfall der Ausdruckskomponente verstärkt sich aber nach Anderson der physiologische Anteil des Affekts – präziser: „Es besteht eine negative Korrelation zwischen motorisch-expressivem System und bestimmten physiologischen Abläufen" (Anderson, 1981; zit. nach R. Krause, 1988a). Ein Wiederbeleben des Ausdrucksanteils kann also auch aus diesem Grunde heilsam sein.

23.10.4 Motorisch-affektive Vorstellungsbilder

Mit der Differenzierung von Selbst- und Objektbildern in der weiteren Entwicklung können dann vorwiegend unbewußte Vorstellungsbilder entstehen, in die auch die affektiven und motorischen frühen Erlebnisse in der Interaktion eingehen (Kratzsch[5]; Müller-Braunschweig, 1975). Im negativen Fall können diese Vorstellungsbilder als „maligne Introjekte" relativ isoliert vom Gesamtorganismus existieren und als ständige Bedrohung der Integrität z. B. zu Dauerspan-

nung führen (Kernberg, 1981, spricht von „unverdauten internalisierten Objektbeziehungen").

23.11 Abschließende Bemerkungen

Alle körperbezogenen Psychotherapiemethoden versuchen u. a. diese verschiedenen Anteile des Individuums wieder zu reintegrieren, also auch die Verbindung verschiedener Stufen oder Ebenen zu verbessern. Die Übermittlung von Nachrichten zwischen somatischer, psychischer und sozialer Integrationsebene (v. Uexküll, 1988; v. Uexküll und Wesiack, 1986) wäre damit ein Ziel, das bei den Körpermethoden durch die Beteiligung des Körpers neben der Psyche besonders hervorgehoben wird. In diesem Sinne ist an die eingangs erwähnte Bess Mensendieck zu erinnern. Ihre Verbindung von künstlerischer Tätigkeit (die eine ungestörte Fluktuation zwischen unbewußten und bewußten Ebenen voraussetzt) mit naturwissenschaftlichen Kenntnissen in Anatomie und Physiologie sowie dem Umgang mit Ausdrucksformen wie Atem, Stimme und Bewegung, weist auf das gleichzeitige Arbeiten an verschiedenen Stufen einer Entwicklungshierarchie hin. Nimmt man die sprachliche Symbolisierung hinzu, zeigt sich in den körperorientierten Methoden eine Möglichkeit ganzheitlicher Psychotherapie, deren Weiterentwicklung lohnend erscheint.

---

5 persönliche Mitteilung

24 Psychopharmaka in der Psychosomatischen Medizin und in der Allgemeinmedizin

Gerhard H. Paar

24.1 Einleitung

Das folgende Kapitel gliedert sich in drei größere Teile. Im ersten wird nach einigen epidemiologischen Daten der Einfluß der Psychopharmakotherapie auf die Arzt-Patient-Beziehung erörtert. In dem folgenden Abschnitt „Spezielle Psychopharmakologie" werden kursorisch wichtige pharmakologische Daten zu den einzelnen Substanzklassen gegeben. Im dritten Teil werden in drei klinischen Anwendungsbereichen, der Pharmakotherapie der Angst, der Depression und des Schmerzes, spezifische Indikationen diskutiert. Es folgen noch eine kurze Fallgeschichte und eine Auseinandersetzung über die Problematik der Plazebogabe. Das Kapitel richtet sich ausdrücklich an Interessenten der Allgemeinmedizin und der Psychosomatik und vernachlässigt psychiatrische Fragestellungen.

24.2 Fallgeschichte

> In der psychotherapeutischen Ambulanz stellt sich eine 25jährige Patientin vor. Die mimisch starre Frau beginnt, sie wisse nicht, was sie erzählen solle. Nach einigem Zögern beschreibt sie ihre vielfältigen Beschwerden. Vor 6 Wochen sei sie auf der Straße ohnmächtig geworden und erst im Krankenwagen wieder zu sich gekommen. Sie demonstriert, daß ihre Hände in Pfötchenstellung gestanden hätten. Ferner klagt sie über Platzangst in Räumen und in der Straßenbahn. Die anfallsartigen Zustände erscheinen nicht eindeutig wie Hyperventilationsanfälle, zumal sie nicht mit einer Veränderung der Atmung einhergehen; zu denken ist auch an ernährungsbedingte Stoffwechselstörungen im Zusammenhang mit einer bestehenden Eßstörung. Nach Angaben der Patientin bestehen seit 1973 Freßanfälle; sie stopfe wahllos alles in sich hinein und müsse dann spontan auf der Toilette erbrechen. Ihrem bisher ausgeübten Beruf als Verkäuferin könne sie nicht mehr nachgehen, weil sie den Arbeitsplatz öfter wegen panischer Ängste verlassen mußte.
>
> Als Kind war sie normalgewichtig, in der Pubertät nahm sie massiv bis auf 80 kg zu. Hier sieht sie einen Zusammenhang mit der 1972 begonnenen Lehre. Im Geschäft sei es ihr schwergefallen, auf die Kunden zuzugehen und diese anzusprechen. Sie habe auch angefangen, Abführmittel zu nehmen. Schließlich entdeckte sie, daß sie willkürlich erbrechen konnte. Bei 1,67 m Körpergröße nahm sie bis auf 45 kg ab. Dann

entdeckte sie die Appetitzügler für sich, die sie „topfit" machten. Später nahm sie verschiedene Benzodiazepine, schließlich kam Alkohol dazu. Seit einer Entziehungskur 1980 blieb sie trocken. Zweimal war sie wegen Selbstmordversuchen in psychiatrischen Kliniken. Später begann sie wieder verstärkt zu essen und zu erbrechen. Seit 2 Jahren ist sie arbeitslos, lebt zurückgezogen in ihrer eigenen Wohnung, liegt auf der Couch, kann sich nicht konzentrieren. In solchen Augenblicken überfallen sie starke Angstzustände. Neben den geklagten agoraphobischen Zuständen sind es Dunkelangst und die Angst, auf der Straße von Männern angesprochen zu werden.

Der Vater ist Stahlarbeiter, die Mutter arbeitet als Näherin, die Patientin ist das mittlere von 3 Kindern. Zunächst berichtet sie wenig über ihre Familie und schildert ihre Kindheit als harmonisch. Mit 12 Jahren mußte sie für die berufstätige Mutter den Haushalt übernehmen (in dieser Phase begann sie, an Gewicht zuzunehmen). Ihren damaligen Tagesablauf schildert sie als eine für sie unproblematische Selbstverständlichkeit. Sie habe alles gerne gemacht. Dies von ihr entwickelte aktive Bild steht im Kontrast zur jetzigen Energielosigkeit. Freundschaften habe sie kaum gehabt. Während der Entziehung nahm sie eine intime Beziehung zu einem Mann auf. Sexualität habe ihr nichts bedeutet.

In der Diagnosekonferenz sehen wir, daß es im Leben der Patientin um Versorgen und Versorgtwerden geht. Als sie ihrer Mutter während der Pubertätsphase den Haushalt macht, ist sie hyperaktiv, kommt aber selber zu kurz. Es bestehen frühe Ängste und die Schwierigkeiten, orale Impulse zu steuern. In der Abwehr ihrer eigenen Sexualität hat sie eine Bulimie entwickelt. Tabletten haben im Leben dieser Frau eine vielfältige Funktion: beruhigende Aspekte, Ersatzaspekte für nicht vorhandene oder unbefriedigende Objektbeziehungen, selbstzerstörerische Aspekte.

Kurz nach Beginn der 6monatigen stationären psychotherapeutischen Behandlung geht die Einzeltherapeutin in Urlaub. Die Patientin fühlt sich alleingelassen, ohne ihre Wut wirklich äußern zu können. Sie entwickelt eine depressive Selbstentwertung mit Suizidphantasien. In der Klinik kann sie es kaum aushalten und möchte entlassen werden. In den ersten Wochen laufen viele wichtige Gespräche mit der Nachtschwester, der sie etwas von ihren Ängsten mitteilen kann. So habe sie seit früher Kindheit Angst, ihre Eltern könnten vor ihr sterben. Später kam die Angst hinzu, jemand könne nachts in ihr Zimmer eindringen. Die mütterlich beruhigende Hilfe der Nachtschwester kann sie zunächst annehmen. Bei dieser Annäherung steigert sich aber ihre Panik, sie kommt immer häufiger in ein Gerangel mit den Schwestern und möchte wieder die Klinik verlassen. Die Richtlinien der Klinik sieht sie als Beschränkungen, gegen die sie sich wehren muß. Die Schwestern erleben die Patientin immer wieder als suizidal, sie versuchen, mit der Patientin im Gespräch zu bleiben. Dann stellt sich heraus, daß die Patientin sich ständig außerhalb heimlich Medikamente besorgt und einmal einen ganzen Beutel voller Medikamente mitbringt. Sie kann aber dann diesen bei den Schwestern abgeben. Gegen die sie bedrohende Angst vor dem Alleinsein schützt sie sich mit Beruhigungs- und Schlafmitteln. Daneben wird auch ein anderes Einnahmemuster sichtbar. Als sie einmal das Gefühl hat, eine Mitpatientin werde ihr vorgezogen, verläßt sie die Station und besorgt sich Schlafmittel. Sie versucht damit das Personal in die Rolle bestrafender Eltern zu bringen. Ein weiteres Suchtmittel wird zusätzlich bekannt. Wir erfahren, daß die Patientin nach jeder Mahlzeit bis zu 13 Tabletten eines Abführmittels zu sich nimmt. Sie erinnert sich, wie sie früher gehänselt wurde wegen ihrer Übergewichtigkeit und kann dann sagen: bevor ich mich verletzen lasse, mache ich mich lieber selber kaputt. Bestrafungsphantasien tauchen in ihr auf, vielleicht würde sie entlassen oder in eine geschlossene Abteilung eingesperrt, wenn sie weiterhin in der Klinik Medikamente einnehme. Neben dem selbstzerstörerischen Aspekt sehen wir auch, wie sie versucht, über Tabletten ihre Autonomie zu bewahren. Dies zeigt sich in der Auseinandersetzung um die Kaliumsubstitution. Immer wieder ist sie infolge ihres Erbrechens und des häufigen Abführens hypokaliämisch. Sie weigert sich, ein kaliumhaltiges Medikament einzunehmen; sie phantasiert, sie falle um, und wir seien daran schuld. Schließlich gelingt es, den von der Patientin inszenierten Kampf mit der Therapeutin um die Abgabe der Abführtabletten und der Beruhigungsmedikamente in die Beziehung zu überführen. Aufschlußreich für den Versuch der Patientin, die Therapeutin zu einer kontrollierenden und bösen Mutter zu machen, ist der folgende Traum: Es liegt eine Frau auf ihr, die etwas in sie hineinstopft; die Patientin kann sich nicht rühren und fühlt sich erdrückt. Die Therapeutin versteht dies als Phantasie der Patientin, von einer mächtigen Mutter erdrückt zu werden. Es läßt sich dann aber herausarbeiten, daß es auch die eigenen selbstzerstörerischen Tendenzen der Patientin sind, die sie zu erdrücken drohen. Als Hinweis auf die Beendigung der Medikamenteneinnahme treten Entzugserscheinungen in Form merkwürdiger Schmerzbilder auf. Die Patientin klagt über massive neuralgische Beschwerden im Bereich ihres Kiefers und erreicht, daß ausführliche zahnärztliche und HNO-ärztliche Untersuchungen durchgeführt werden.

Erst zum Ende der stationären Behandlung kann sie auch ihre lockeren Seiten zeigen. Erstmalig erzählt sie von sich selber und von ihren eigenen Interessen. Wir glauben, daß eine Entwicklung erkennbar ist in Richtung auf ein Vertrautwerden und eine Auseinandersetzung mit einem Objekt bei allmählichem Überwinden des primären Mißtrauens. Die Patientin wirkt depressiv, kann weinen, sie zeigt ihre Sehnsucht nach körperlicher Nähe, Wärme und Geborgenheit. Die intensiven Einzelgespräche bereiten ihr aber auch Angst, sie muß auf Distanz gehen; auch in dieser Wegbewegung greift sie gelegentlich zu Medikamenten. Zum Ende der Behandlungsphase zeigt sie uns mehr von ihrer gewonnenen Autonomie, am Wochenende versucht sie auszuprobieren, inwieweit sie mit sich selber zurechtkommt. Auf die Entlassung reagiert sie dann psychisch, nicht mehr mit Medikamenten. Das erste

Nachgespräch kreist um Essen und Brechen, dann kommen mehr die Ängste vor dem Alleinsein ins Gespräch. Die Patientin nimmt ihre alte Arbeit wieder auf. Nach einigen Monaten kommt sie erneut auf die Therapeutin zu, als sie feststellt, daß sie sich für einen Arbeitskollegen interessiert. Sie möchte darüber mit ihr sprechen. Die Eßproblematik hat sich nicht noch völlig zurückgebildet, beherrscht sie aber auch nicht mehr. Sie nimmt keine Medikamente mehr ein.

24.3 Epidemiologische Aspekte

In allen Industrieländern steigt seit Jahren der Medikamentenverbrauch kontinuierlich an. Mit der Einführung des Chlorpromazins in die Psychiatrie Anfang der 50er Jahre begann eine neue Ära. Die Verschreibung zentralnervös wirksamer Medikamente steigerte sich sprunghaft und hat bis heute ihren Höhepunkt nicht erreicht.

Über den Verbrauch von Psychopharmaka in den Industrieländern gibt es mittlerweile verläßliche Zahlen, während wir über den Verbrauch in unterentwickelten Ländern kaum etwas wissen. Die unten beschriebenen Trends gelten allgemein für alle hochindustrialisierten Länder.

In der Bundesrepublik werden jährlich rund eine Milliarde Beruhigungspillen geschluckt. In der Zeit von 1970 bis 1982 wuchs das über öffentliche Apotheken verkaufte Volumen von 136 Millionen DM auf 559 Millionen DM (Der Spiegel, 1983). Die enorme Zunahme betrifft überwiegend die Substanzklasse der Benzodiazepine (Parry et al., 1973). In den letzten 20 Jahren nahm hingegen die Verschreibung von Hypnotika und Sedativa ab (Arznei-Telegramm, 1981). Alte Menschen nehmen besonders viele Medikamente (Murray et al., 1981; Abelson et al., 1977). Sie haben die höchste Inzidenz an organischen und psychischen Erkrankungen. Aber da sich im Alter physiologische Veränderungen in Aufnahme und Ausscheidung und im Ansprechen auf Psychopharmaka ergeben, treten bei ihnen Nebenwirkungen besonders häufig auf (Übersicht bei Wheatley, 1982).

Frauen nehmen psychotrope Substanzen doppelt so häufig ein wie Männer (Cooperstock, 1978). Gründe dafür seien die Doppelbelastung in Haushalt und Beruf, sowie der Versuch von Frauen, sich aus der Abhängigkeit von Männern zu lösen. Frauen im jüngeren und mittleren Alter nehmen vor allem Stimulantien ein. Das mag mit Gewichtsproblemen zusammenhängen (Mellinger et al., 1974). Männer neigen eher dazu als Frauen, ihre Beruhigungspillen auch ohne ärztliche Verschreibung einzunehmen. 70% aller Psychopharmaka werden von praktischen Ärzten, Internisten und Gynäkologen rezeptiert (Der Spiegel, 1983; vgl. Parry et al., 1973).

Eine Untersuchung von Pflanz, Basler und Schwoon (1977) befaßte sich mit Einnahmegewohnheiten 50jähriger Menschen hinsichtlich einiger sozialer Variablen. 15% der Männer und 21% der Frauen gaben an, in der letzten Woche ein Psychopharmakon eingenommen zu haben. Balter und Mitarbeiter (1974) fanden in einer internationalen Vergleichsstudie zu Einnahmegewohnheiten psychotroper Drogen, daß 14% der Bundesdeutschen (8,4% der Männer und 19,2% der Frauen) angaben, im letzten Jahr ein Medikament eingenommen zu haben. Auch andere Befunde der oben zitierten Studie decken sich mit der internationalen Literatur. Psychopharmaka werden in mittleren und höheren sozialen Schichten eher eingenommen als in unteren. Mittlerweile zeigt sich ein Trend zu einer verminderten Verschreibung von Anxiolytika, wahrscheinlich weil Ärzte kritischer die Verschreibungszeiten von Benzodiazepinen beachten (Griffiths und Sannerud, 1987). Prävalenzstudien zur Einnahme von Benzodiazepinen (BZP) in der Allgemeinbevölkerung ergaben 3,5% für BZP und Nicht-BZP-Hypnotika in Österreich (Lersch et al., 1986) und 8,2% für BZP in der BRD (Fichter et al., 1986). Schon lange ist bekannt, daß gerade in Krankenhäusern eine intensive Verschreibungspraxis von BZP ausgeübt wird. So wurden 1985 in einer Studie die Benzodiazepinbestellungen aller Abteilungen der Innsbrucker Universitätsklinik von der Krankenhausapotheke erfaßt. Die Ausgaben für BZP umfaßten 0,5% der gesamten Medikamentenkosten. Auf fünf großen Klinikabteilungen nahmen am Tag der Befragung 24,1% aller befragten Patienten und 20,5% der Patientinnen BZP ein! Etwa ein Drittel dieser Patienten nahmen BZP schon vor ihrer Einweisung ein (Fleischhacker et al., 1989).

Wenig ist bislang bekannt darüber, wie lange Patienten Psychopharmaka einnehmen. Nach einer amerikanischen Studie nahmen 20% der BZP-konsumierenden Patienten diese schon länger als drei Monate kontinuierlich ein (Mellinger et al., 1984).

Ebenfalls ist bislang wenig bekannt über die Sozialisation zum späteren Konsum von Psychopharmaka. Nach einer Untersuchung von Vogt (1977) scheint es nicht automatisch so zu sein, daß Mütter, die Psychopharmaka einnehmen, den späteren Konsum ihrer Kinder determinieren. Aufgrund unserer Kenntnis der wichtigen Rolle, die Mütter für die psychische Entwicklung der Kinder spielen, dürfen wir aber davon ausgehen, daß ihrem Verhalten Modellfunktion zukommt.

24.4 Interaktionsprobleme bei der Verordnung von Psychopharmaka

24.4.1 Historische Auffassung

Wichtige, bis heute gültige Erfahrungen über die psychodynamische Wirkung von Psychopharmaka wurden von einigen Pionieren in Psychotherapien und Psychoanalysen gemacht (Azima, 1959; Danckwardt, 1978; Gottschalk, 1968; Kubie, 1960; May, 1971; Ostow, 1962, 1979; Sarwer-Foner, 1960). Die dabei entwickelten metapsychologischen Wirkungsvorstellungen lassen sich in einem a) triebpsychologischen,

einem b) Ich-psychologischen und einem c) objektpsychologischen Konzept zusammenfassen.

a) Das triebpsychologische Konzept wurde besonders von Ostow entwickelt. Nach ihm wirken Psychopharmaka durch Ab- bzw. Zunahme libidinöser und/oder aggressiver Triebenergien. Azima hingegen nimmt an, daß sich die affektive Besetzung des Selbst und der Objekte durch die Psychopharmakawirkung umorganisiert.

b) Beim Ich-psychologischen Konzept wird die Wirkung von Psychopharmaka in der Stärkung der Abwehr und der Realitätsprüfung gesehen (Winkelman, 1960). Sarwer-Foner (1970) hat seine psychodynamischen Überlegungen auf die Veränderung der Abwehr fokussiert. Nach ihm werden Triebbedürfnisse nicht direkt durch Medikamente beeinflußt, sondern ihre Abfuhr verändere sich durch den Aktivitätszustand der Skelettmuskulatur und anderer auf den Austausch mit der Umwelt gerichteter Systeme. In einer der wenigen Doppelblindstudien versuchten Bellak und Mitarbeiter (1973), die Ich-stabilisierende Funktion eines Anxiolytikums im psychotherapeutischen Prozeß zu zeigen. Verschiedene Ich-Funktionen (Realitätsprüfung, Objektbeziehung, Triebkontrolle usw.) bei sog. normalen, neurotischen und psychotischen Patienten wurden durch unabhängige Beobachter im Verlauf einer Psychotherapie über sechs Monate ohne Katamnese überprüft. Die Anwendung von Diazepam verbesserte den Behandlungserfolg gegenüber den Plazebo-Kontrollen. Ich-psychologisch ließen sich die stabilisierenden Effekte beschreiben als Ich-Stärkung, Stärkung der Triebabwehr, Verminderung von Ängsten.

Aus der Sicht des Patienten beschreibt Sarwer-Foner (1975) die Ich-psychologische Wirkung folgendermaßen: „Wenn eine Medikamentenwirkung von dem Patienten erlebt wird als eine Verstärkung seiner Kontrollfunktionen über die ihm angstmachenden Impulse ..., dann ist die Möglichkeit gegeben für eine Verbesserung und eine Ich-Integration. Wenn die pharmakologische Wirkung nicht signifikant diese Kontrolle beeinflußt, kann eine signifikante Verbesserung nicht erwartet werden" (Übersetzung G. H. Paar).

Gottschalk und Mitarbeiter (1965, 1972) haben in mehreren neuropharmakologischen Studien die Wirkung von Psychopharmaka auf verbal geäußerte Affekte untersucht. Lorazepam, ein Benzodiazepin, verminderte dabei den Betrag an Angstäußerungen, während Imipramin im Vergleich zu Plazebos die nach außen gerichtete verbale Aggressivität verstärkte.

c) Das objektpsychologische Konzept zum Verständnis der Wirkung von Psychopharmaka wurde vor allem von Balint und Mitarbeitern (1975) entwickelt. Nach ihm ersetzt das Medikament symbolisch ein dringend benötigtes inneres Objekt und kann damit das Auftreten von Ängsten verschiedener Genese verhindern.

In diesem Zusammenhang gehören auch Überlegungen, in welcher Situation ein Medikament gegeben und eingenommen wird. Ist in einer Arzt-Patient-

Beziehung oder in einer stationären Behandlungssituation das Psychopharmakon die therapeutische Modalität, so werden im Erleben von Patienten, Ärzten, Schwestern und der Familie alle Vorstellungen über Veränderungen sich mit dem Medikament und seiner Wirkungsweise verbinden.

24.4.2 Zur Interaktion von Psychotherapie und Psychopharmaka

Wenn Psychopharmaka in einer Psychotherapie verordnet, oder umgekehrt, wenn eine Psychotherapie bei fortlaufender Psychopharmakotherapie begonnen wird, ergibt sich die Chance, Antagonismen und Synergismen genauer zu studieren. Interaktionsweisen müssen in vergleichenden wie kontrollierten Studien herausgearbeitet werden. Voraussetzung für ihre Erforschung ist die „Standardisierung" beider Behandlungsformen (Elkin et al., 1988). In der psychopharmakologischen Forschung kann man Blutspiegel eines Pharmakons bestimmen, um Aussagen über die Compliance des Patienten und über Medikamentenabsorption und Metabolismus zu gewinnen. Schwieriger ist die „Standardisierung" von Psychotherapien. Es wurden Behandlungsmanuale entwickelt, die spezifische Charakteristika, Behandlungsstrategien, Interventionsformen und Behandlungsziele zu operationalisieren suchten. Genannt seien das Manual von Klermann und Mitarbeitern (1984) zur Behandlung von Depressionen mit der interpersonellen Psychotherapie sowie die Manuale von Luborsky (1984) und von Strupp und Binder (1984) zur psychodynamischen Psychotherapie.

Aus Interaktionsstudien lassen sich Aussagen gewinnen zum zeitlichen Abtreten von Effekten wie zu den differenten Wirkungen der verschiedenen Behandlungsformen. Medikamenteneffekte sind oft innerhalb der ersten Wochen nachweisbar und betreffen insbesondere psychobiologische Funktionen wie Schlaf, psychomotorische Aktivität und Appetit. Die Wirkung von Psychotherapie tritt wesentlich später ein und betrifft eher Ich-Funktionen und die Fähigkeit, in Beziehungen befriedigender zu leben. Karasu (1982) hat die folgenden integrativen Thesen über die Wirkung von Psychopharmaka und Psychotherapie formuliert:
– Jede Therapieform hat differente Effekte und Wirkungsbereiche: Medikamente beeinflussen eher Symptome und affektive Spannung – Psychotherapie wirkt eher auf interpersonale Beziehungen und soziale Anpassungsfähigkeit.
– Jede Therapieform aktiviert und verläuft in einer differenten Zeitachse: Medikamente wirken schnell und kurzdauernd und können prophylaktisch eingesetzt werden – die Wirkungen einer Psychotherapie stellen sich später ein, halten aber dann länger an.
– Jede Therapieform bezieht sich auf verschiedene Erkrankungen und ihre Subtypen: Medikamente wirken auf zeitlich begrenzte und autonome

„State"-Erkrankungen – Psychotherapie bei langdauernden „Trait"-Erkrankungen.

Die mit den unterschiedlichen Behandlungssystemen arbeitenden Therapeuten sind unterschiedlich ausgebildet. Bei gleichzeitiger Anwendung hängt der Erfolg beider Behandlungsformen nicht zuletzt davon ab, wie die beiden unterschiedlichen Therapeuten miteinander kooperieren können.

Ergebnisstudien über alleinige und kombinierte Anwendung von Psychotherapie und Psychopharmaka zeigen drei Wirkungsspektren:
– Keine therapeutische Wirksamkeit: Die kombinierte Behandlung erbringt gleiche Resultate wie die jeweiligen individuellen Behandlungsformen.
– Positive Wirksamkeit: Die Wirksamkeit der kombinierten Behandlung übersteigt die der individuellen Behandlungsformen.
– Negative Wirksamkeit: Die kombinierte Behandlung zeigt schlechtere Resultate als die Wirkung jeder einzelnen Behandlungsform.

Bei der kombinierten Anwendung hofft der Therapeut auf additive oder synergistische Effekte. Häufiger scheint jedoch eine fördernde Interaktion („facilitative interaction"; Klermann, 1986) vorzuliegen. Demnach mildere die pharmakologische Wirkung einer Substanz die angenommenen zentralnervösen Dysfunktionen. Dies wirke sich wiederum auf Symptomatik, Psychopathologie und Affektstörung des Patienten aus. Die medikamenteninduzierte Beschwerdereduktion ermögliche nun dem Patienten, ungestörter zu kommunizieren und von der Psychotherapie zu profitieren. Dieser Überlegung folgend läßt sich ein Dosis-Wirkungs-Optimum zwischen beiden Behandlungsformen annehmen.

24.4.3 Das Eingreifen der Psychopharmaka in die Arzt-Patient-Beziehung

Die Verordnung eines Psychopharmakons geschieht immer in einem Interaktionsprozeß. Michael Balint und seine Ur-Balintgruppe haben sich ausführlich mit der „Pharmakologie der Droge Arzt" beschäftigt und uns damit auf die Beachtung der Beziehungen hingelenkt. Balint ging von folgender Erkenntnis aus, „daß das am häufigsten verwendete Heilmittel der Arzt selbst ist. Nicht die Flasche Medizin und die Tabletten sind ausschlaggebend, sondern die Art und Weise, wie der Arzt sie verschreibt – kurz die gesamte Atmosphäre, in welcher die Medizin verabreicht und genommen wird" (1965).

Die Wechselwirkungen zwischen Arzt und Patient sind nach Balint kein konstanter Faktor. Sie bestimmen sich jeweils durch den Einfluß von Übertragung und Gegenübertragung. In dieser Verschränkung gegenseitiger bewußter und unbewußter Einstellungen spielen sowohl die Psychodynamik des Patienten als auch die „Pharmakologie des Arztes" (Anwendungsbereich, Kontraindikation, Nebenwirkung und Dosierung) eine Rolle. Balint erfaßt die Beschwerden und Symptome des Patienten als Angebot an den Arzt, auf welches dieser reagiert.

Aus der Analyse der jeweiligen Beziehungskonstellation wird verständlich, warum der Arzt ein Psychopharmakon verschreibt oder nicht.

Grundsätzlich entwickeln sich in jeder Arzt-Patient-Beziehung folgende Beziehungsebenen: a) die reale der Arbeitsbeziehung, b) die neurotische der Übertragung und Gegenübertragung, sowie c) die symbiotische Beziehungsebene. Alle drei genannten Beziehungsebenen werden durch die Psychopharmakagabe beeinflußt.

a) Die Arbeitsbeziehung, „das Arbeitsbündnis kann als etwas aufgefaßt werden, das auf dem bewußten oder unbewußten Wunsch des Patienten nach Kooperation gründet und auf seiner Bereitschaft, die Hilfe des Therapeuten bei der Bewältigung innerer Schwierigkeiten anzunehmen" (Sandler et al., 1973). Psychopharmaka, insbesondere Tranquilizer, können hier störend eingreifen, indem sie zur Abhängigkeit führen. Noch gesunde Ich-Funktionen werden leicht durch Medikamentenwirkung beeinträchtigt (May, 1971). Insgesamt können alle Psychopharmaka dem Patienten seine persönliche Verantwortung für seine Lebenssituation nehmen: Sein Vertrauen, mit Krisen selbst fertig zu werden, vermindert sich. Stellt sich nun innerhalb der Behandlungssituation etwa eine Besserung seiner Beschwerden ein, so führt er dies eher auf das Medikament als etwa seine Beziehung zum Arzt zurück.

b) In einer Psychotherapie ist es Aufgabe des Therapeuten, die vielfältigen Aspekte von Beziehungen zu untersuchen, die sich in der Behandlung und gegenüber seiner Person einstellen. Eine besondere Rolle spielt dabei die Übertragung als eine „spezifische Illusion, die sich in bezug auf eine andere Person einstellt und die ohne Wissen des Subjekts in einigen ihrer Merkmale eine Wiederholung der Beziehung zu einer bedeutsamen Figur der eigenen Vergangenheit darstellt" (Sandler et al., 1973). Dies betrifft übrigens auch die unbewußten Vorstellungen des Patienten über die physiologischen Wirkungen des Medikamentes.

Sarwer-Foner (1970) hat ausführlich diese Übertragungsphänomene in Beziehung auf Pharmaka untersucht. In vielen Fällen kann die Medikamentenwirkung beim Patienten Konflikte reaktivieren, die er mit seiner Symptombildung und folglich mit seiner Abwehr gelöst hatte. Dies kann auch zu stärkeren Ängsten führen. So können Patienten mit einer Herzneurose, die bislang kontraphobisch ihre Angst abwehrten, durch den sedierenden Effekt von Sedativa und Tranquilizern verstärkt geängstigt werden; oder Patienten, die infolge einer Zwangsneurose oder einer Borderline-Persönlichkeit ängstlich um ihre Körperintegrität besorgt sind, können die physiologischen Wirkungen der Medikamente als eine Bedrohung ihres Körper-Selbst erleben etc.

Danckwardt (1980) vertritt eine eher skeptische Einstellung: „Bei der Verabreichung eines Medikamentes wird in Tat und Wahrheit gehandelt. Nichts bleibt in einem zwischen Patient und Therapeut zum Betrachten gleichsam ‚ausgespannten' Raum des Probehandelns. Die das Patienten-Verhalten steuernden, krankhaften, inneren Phantasien (und nur die kann man analytisch orientiert psychotherapeutisch behandeln) werden äußere Realität (die kann man nicht mit psychotherapeutischen Mitteln behandeln). Die Übertragung läßt sich also mit der Zeit nicht mehr virtualisieren (auflösen) und zum Schluß einer Behandlung ihrerseits als Widerstand gegen die Aufgabe früherer Objektbeziehungen darstellen."

Die Art, wie ein Therapeut seinen Patienten erlebt, wirkt sich ebenfalls auf die Wirksamkeit eines psychotropen Medikamentes aus. Unter der Gegenübertragung des Therapeuten verstehen wir „eine spezifische Gefühlsreaktion ... auf spezifische Qualitäten seines Patienten" (Sandler et al., 1973). Sarwer-Foner (1975) beschreibt Reaktionen von Ärzten, die aufgrund eigener Ängste durch eine Medikamentenverordnung auf Distanz gehen. Der Patient kann daraufhin die Medikamentenwirkung als sichtbaren Ausdruck der Zurückweisung seiner Person erleben. Das Medikament kann als unwirksam, schlecht, mit übler Nebenwirkung behaftet erlebt werden, wie (in der Übertragung) ein verfolgender, desinteressierter Elternteil, der sich entzieht oder das Kind im Stich läßt. Somit wird die psychologische Bedeutung der Arzt-Patient-Beziehung unbewußt auf das Medikament verschoben. Der Patient setzt unbewußt die objektive psychotrope Wirkung des Medikamentes mit seinem Gefühl dem Arzt gegenüber in Beziehung.

c) Unter der symbiotischen Beziehungsebene verstehen wir mit Loch (1965) eine tragende, positiv gefärbte narzißtische Übertragung. Sie beschreibt als Basis jeder menschlichen Beziehung den nährenden und pflegenden Umgang eines mütterlichen Objektes mit einem kindlichen Subjekt. Gerade bei körperlich kranken Patienten ist auf die Entwicklung einer positiven basalen Beziehung zu achten. Da diese Ebene nach Loch in einer Behandlung interpretatorisch nicht angesprochen werden darf, kann auch die Wirkung eines Psychopharmakons auf dieser Ebene nur schwer in den Beziehungs- und Verstehensprozeß mit einbezogen werden.

Die oben referierten Arbeiten zeigen, daß die beschriebenen Beziehungsebenen einer Arzt-Patient-Beziehung durch psychotrope Substanzen nicht irreversibel beeinflußt werden.

Jedes Psychopharmakon hat ein spezifisch klinisch-pharmakologisches Profil, wirkt aber therapeutisch unspezifisch. Dadurch entfaltet es seine Wirkungen therapeutisch auf unterschiedliche Weise (Sarwer-Foner, 1975). Gerade bei Patienten mit Ich-Strukturstörungen scheint die pharmakologische Wirkung zunächst nur physiologisch und vage koenästhetisch wahrgenommen zu werden und bleibt damit „nicht-psychisch". Erst durch die Arzt-Patient-Beziehung wird die pharmakologische Wirkung „psychisch" und erfährt ihre Wertung (Danckwardt, 1980). Somit kann das Medikament innerhalb der Arzt-Patient-Beziehung zum Teil eines Konfliktes werden, der sich nach folgenden Mustern strukturiert (Gauss et al., 1987):

– das gute Objekt (symbiotischer Modus);
– Übergangsobjekt (passagerer Ersatz einer lebenswichtigen Beziehung);
– das gespaltene Objekt (Modus der Spaltung);
– Angstmodus (Modus von Angst/Bedrohung);
– das unzuverlässige Objekt (Modus der mangelnden Objektkonstanz);
– das steuernde Objekt (Modus der Kontrolle und Fremdbestimmung durch das Objekt);
– das Objekt als Verfolger/Erlöser (paranoider Modus).

24.5 Spezielle Pharmakologie

Im folgenden sollen kursorisch einige Basisinformationen zu den Psychopharmaka vermittelt werden. Wer ausführlichere Darstellungen sucht, sei auf die entsprechenden Lehrbücher verwiesen (z.B. Goodman und Gilman, 1985; Langer und Heimann, 1983).

Im weiteren Sinne kann man als Psychopharmaka alle Medikamente bezeichnen, die auf das zentrale Nervensystem (ZNS) wirken. Im engeren Sinne sind es Substanzen, die nicht nur einzelne psychophysiologische Funktionen wie Schmerzempfinden, Schlaf, sondern auch komplexere seelische Vorgänge sowie verschiedene psychische Störungen beeinflussen. Über Einteilungskriterien der Psychopharmaka bestehen auch heute noch unklare und widersprüchliche Auffassungen. Das „Anti"-Klassifikationssystem (= antipsychotisch, antidepressiv) teilt die Medikamente nach den Krankheiten ein, die damit behandelt werden sollen. Eine andere Einteilung klassifiziert Psychopharmaka nach ihrer Wirkung auf Zielsymptome (antriebssteigernd, stimmungsaufhellend). Alle diese Klassifikationsversuche scheitern aber an dem Problem, daß die spezifischen integrativen und desintegrativen Prozesse, die den verschiedenen psychosomatischen, somatopsychischen und psychiatrischen Syndromen zugrunde liegen, bisher nur zum Teil verstanden sind. So ist weder die Einteilung nach „Krankheitsentitäten" noch die nach „Zielsymptomen" umfassend genug, um allen Tatsachen gerecht zu werden.

Wir klassifizieren psychotrope Psychopharmaka in folgenden Gruppen:
– Neuroleptika
– Antidepressiva
– Tranquilizer
– Beta-Blocker
– Sedativa
– Psychoanaleptika

Zur Wirkung der Psychopharmaka gibt es verschiedene biochemische und neuropharmakologische Wirkungshypothesen. Wir beziehen uns zunächst auf Hess (1958), der die Reaktionen eines Lebewesens auf Veränderungen in seiner Umwelt auf den Einfluß kortikaler und besonders subkortikaler Zentren bezog, welche somatische, autonome und psychische Funktionen integrieren. Er konzipierte zwei reziprok organisierte Systeme, die er „ergotroph" und „tropotroph" nannte.

Kognitive Prozesse, also auch Perzeptionen, seien sie exterozeptiv oder interozeptiv, unterliegen dem modulierenden Einfluß dieser subkortikalen Systeme. Neurophysiologische Bahnen und Neurotransmitter sind heute in groben Zügen bekannt.

Das ergotrophe, sympathische oder „Go-System" entwickelt alle Funktionen, die das Individuum für eine positive Aktion bereithält, charakterisiert durch Erregung, Wachheit, erhöhten Skelettmuskeltonus, Aktivität des sympathischen Nervensystems und Ausschüttung von Katecholaminen.

Das tropotrophe, parasympathische oder „Non-Go-System" integriert Systeme, die Energie zurückhalten und speichern: Erhöhung der Stimulusbarriere für perzeptuelle Einflüsse, Erniedrigung des Skelettmuskeltonus, Steigerung der parasympathischen Nervenaktivität und der Ausschüttung anaboler Hormone.

Acetylcholin wirkt als Neurotransmitter. Die Wirkung zahlreicher Psychopharmaka auf den Stoffwechsel dieser Neurotransmitter ist gesichert (vgl. Kap. 10).

Exemplarisch für das Bemühen, psychische Erkrankungen neurobiologisch zu verstehen, soll hier die Depression stehen. Die Katecholamin-Hypothese besagt, daß Depressionen aus einer funktionellen Verarmung von Katecholaminen im Gehirn resultieren. Trizyklische Antidepressiva werden an präsynaptischen Alpha-Rezeptoren gebunden. Sie bewirken durch Hemmung des Rücktransportes eine Konzentrationserhöhung von biogenen Aminen, welche die herabgesetzte Sensitivität bzw. die verminderte Zahl postsynaptischer Rezeptoren kompensieren. Dadurch wird das ergotrophe System reaktiviert. Umgekehrt kann Reserpin, das die zentralen Katecholaminspeicher entleert, eine Depression herbeiführen (Paioni et al., 1983).

24.5.1 Neuroleptika

Charakterisierung

Die Neuroleptika sind antipsychotisch wirksam durch Drosselung der halluzinatorischen, zwanghaften oder wahnhaften Erlebnisproduktion. In niedriger Dosierung wirken sie ähnlich wie Tranquilizer. Bei antipsychotischer Dosierung dämpfen die heute gebräuchlichen Mittel die emotionale Spannung (psychotische Angst), den Antrieb (psychotische Aggressivität) und sedieren durch Förderung der Schlafbereitschaft.

Im allgemeinen werden Neuroleptika gut durch den Gastrointestinaltrakt absorbiert. Ihre Halbwertszeit ist länger als 24 Stunden; Depotpräparate können über Monate wirken. Die Metaboliten werden über die Galle und die Nieren ausgeschieden. In ihrer Hauptwirkung blockieren Neuroleptika unspezifisch die Dopaminrezeptoren, ferner wirken sie anticholinergisch und blockieren die Histaminsekretion sowie die α-adrenergen Rezeptoren. Niedrig-potente Neuroleptika haben bei geringer antipsychotischer Potenz eine starke sedative, anticholinerge und histaminblockierende Aktivität. Hochpotente Neuroleptika haben geringere Nebenwirkungen dieser Art bei starken extrapyramidalen Nebenwirkungen.

Indikationen

Die Hauptindikation liegt in der Behandlung produktiver Psychosen, also paranoid-halluzinatorischer und katatoner Schizophrenien und manischer Phasen. Für uns sind sie auch interessant im Einsatz bei schwerkranken Patienten mit somatopsychischen Störungen.

Bei psychiatrischen Symptomen durch ein organisches Psychosyndrom können bei vorheriger Behandlung der zugrundeliegenden Störungen im Einzelfall auch Neuroleptika indiziert sein.

Stellen beispielsweise delirante Zustände eine Notfallsituation dar, so sind bei Abwägen der Nebenwirkungen die hochpotenten Neuroleptika die Therapie der Wahl (Boyer et al., 1984; Dubin et al., 1986). Dabei sollte die gewünschte Wirkung mit möglichst niedriger Dosis erreicht und diese schnell wieder abgebaut werden. Zur Einleitung wird eine intravenöse Gabe beispielsweise von Haloperidol empfohlen (Cassem und Hackett, 1987). Natürlich darf auf eine realitätsorientierende Unterstützung durch Personal und Angehörige nicht verzichtet werden.

Zur Behandlung der Demenz ist die Verordnung von Neuroleptika weit verbreitet, jedoch gibt es wenig valide Hinweise auf eine förderliche Wirkung (Raskind et al., 1987). Auch hier sind im Einzelfall hochpotente Neuroleptika vorzuziehen.

In der Intensivmedizin wird diese Substanzgruppe mit relativer Sicherheit bezüglich der kardialen Nebenwirkungen eingesetzt (Levinson und Simpson, 1986). Selbstverständlich sollten kardiale Notfallpatienten engmaschig überwacht werden.

Der Einsatz von Neuroleptika bei den klassischen psychosomatischen Erkrankungen ist durch oben genannte Nebenwirkungen begrenzt. Eher kommen Substanzen in Frage, die sich niedrig dosieren lassen und dabei ähnlich den Tranquilizern wirken. Die intramuskuläre Gabe von Fluspirilen bei „psychosomatischen Beschwerden" in regelmäßigem Rhythmus erfreut sich bei niedergelassenen Ärzten großer Beliebtheit. Hier wird sicher zu häufig bei unkritischer Indikationsstellung und geringer Beachtung von möglichen Spätdyskinesien rezeptiert.

Häufig werden Neuroleptika, insbesondere Prochlorperazin und Metoclopramid, als Antiemetika während einer Zytostatikatherapie verordnet (Goldberg und Cullen, 1986). Über den Einsatz von Neuroleptika in der kombinierten Schmerztherapie siehe Abschnitt 24.6.3.

Der Wirkungsverlauf von Neuroleptika ist mehrphasig: Nach Sedation, beginnender vegetativer Labilität und Blutdruckabfall folgt das mögliche Auftreten extrapyramidaler Nebenwirkungen, ehe die Hauptwirkung mit Distanzierung von psychotischen Erlebnissen eintritt.

Medikamenteninteraktionen

Veränderungen der Azidität und der Entleerungszeit des Magens können die intestinale Absorption von Neuroleptika beeinflussen. Aufgrund ihrer eigenen anticholinergen Wirkung verlängern sie die intestinalen Passagezeiten (Siris und Rifkin, 1981). Antipsychotika können andere ZNS-wirksame Substanzen wie Sedativa, Anticholinergika und Beta-Blocker potenzieren. Bei gleichzeitiger Beta-Blockereinnahme verstärkt sich die Orthostase. Durch Neuroleptika wird die blutdrucksenkende Wirkung von Clonidin, Guanethidin und Methyldopa blockiert (Bernstein, 1987). Gelegentlich wird ein Delir bei gleichzeitiger Einnahme eines Neuroleptikums mit Propranolol (Lima und Vannemann, 1983), ein dementes Zustandsbild durch die Kombination von Haloperidol und Methyldopa gefördert (Thornton, 1976).

Nebenwirkungen und Komplikationen

– Wirkungen auf das ZNS: Die Häufigkeit und Schwere von Nebenwirkungen variiert je nach Patient und Medikament. Wichtig ist, sie als Nebenwirkungen wahrzunehmen und nicht als Manifestationen der Krankheit fehlzudeuten. Die akuten extrapyramidalen Nebenwirkungen bestehen in parkinsonartigen Erscheinungen und Dyskinesien. Zu Beginn der Therapie kann es zur sog. neuroleptischen Wirkungsdissoziation kommen: Die charakteristischen Hauptwirkungen treten nach unterschiedlichen Zeitabständen ein und werden vom Patienten als quälend empfunden. Besonders bei älteren Patienten können bei Therapieeinleitung die psychotischen Symptome zunehmen. Bei Epileptikern kann die Krampfbereitschaft erhöht werden.

– Vegetative Wirkungen: Orthostatische Hypotension, selten kardiovaskuläre Nebenwirkungen wie Arrhythmien, kongestive Myokardiopathien und Herzinfarkte (Swett und Shader, 1977; Risch et al., 1981), Nasenverstopfung, Mundtrockenheit, Durchfall, Urinretention bei Prostatahypertrophie, Abnahme von Libido und Potenz, Ejakulationsstörungen, Glaukomanfall.

– Wirkungen auf Haut und Augen: Hautausschlag, Phototoxizität, Linsenveränderungen.

– Neuroendokrines System: Alle Neuroleptika erhöhen den Prolaktinspiegel durch ihre Dopaminblockade. Menstruationsstörungen, Gynäkomastie, Galaktorrhö werden häufig beobachtet. Da Brustkrebs teilweise in In-vitro-Studien prolaktinabhängig war (Salih et al., 1972), sollte die Verschreibung von Neuroleptika bei Patientinnen mit Brustkrebs in der Vorgeschichte mit Vorsicht geschehen (Schyve et al., 1978).

– Wirkung auf Schwangerschaft und Neugeborene: Hinweise auf mögliche teratogene Wirkungen von Neuroleptika sind inkonsistent (Levinson und Simpson, 1986). Allerdings kann deren Einnahme in der Schwangerschaft für das Neugeborene diskrete extrapyramidale Störungen und eine Atemdepression mit sich bringen (Ananth, 1976).

– Ferner wurden beschrieben: Medikamentenikterus, Agranulozytose und Thromboseneigung.

24.5.2 Antidepressiva

Charakteristika

Die Antidepressiva zeichnen sich durch unterschiedliche stimmungsaufhellende (thymoleptische), antriebssteigernde (thymeretische) und angstlösende (anxiolytische) Wirkung aus. Vor allem bei den trizyklischen Antidepressiva Imipramin, Amitriptylin und Doxepin überwiegt die dämpfende Komponente, während bei Desipramin und Nortriptylin die hemmungslösende, aktivierende Komponente überwiegt. Zusätzlich haben viele Antidepressiva vor allem am Therapiebeginn eine sedative Wirkung.

Mit den MAO-Hemmern und den sympathikomimetischen Stimulantien wollen wir uns hier nicht beschäftigen.

Wirkungsmechanismus

Die trizyklischen Antidepressiva (TCA) wirken am adrenergen Neuron und beeinflussen entweder direkt oder indirekt die Neurotransmitter Noradrenalin und Dopamin. Amitriptylin, Desipramin, Doxepin, Imipramin, Nortriptylin und Protriptylin hemmen in klinischen Dosen beim Menschen die Noradrenalin-Aufnahme. Eine Serotonin (5HT)-Aufnahmehemmung wurde für Amitriptylin, Clomipramin und Doxepin sowie in geringem Maße für Desipramin und Nortriptylin nachgewiesen.

Neben den beschriebenen trizyklischen Antidepressiva der ersten Generation wurden neuerdings nicht-trizyklische Antidepressiva mit teilweiser neurobiologischer Selektivität synthetisiert.

Einige von ihnen mußten mittlerweile wegen erheblicher Nebenwirkungen vom Markt genommen werden. Es wird hier wie auch bei anderen Psychopharmaka dazu geraten, auf die auch in ihren Nebenwirkungen vertrauten älteren Substanzen zurückzugreifen.

Indikationen

TCA sind indiziert bei verschiedenen Formen der Depression, wie neurotische, psychotische Depression, bei somatischen Äquivalenten der sog. larvierten Depression. Allgemein läßt sich sagen, daß der klinische Gebrauch von Antidepressiva bei körperlich Kranken sich nicht wesentlich unterscheidet von den psychiatrischen Indikationen – allerdings müssen die Nebenwirkungen noch sorgfältiger berücksichtigt werden. Eine besondere Indikation scheint für Panikattacken vorzuliegen (Gorman, 1987).

Bei den TCA läßt sich folgender Wirkungsverlauf beschreiben:
1. Woche: Sedation
2. Woche: Beginn der thymeretischen Wirkung
3. Woche: Beginn der thymoleptischen Wirkung.

Die TCA sollten zumindest für eine Periode von vier Wochen eingesetzt werden, ehe man entscheidet, daß dem Patienten mit diesem Medikament nicht geholfen werden kann. Danach ist es unwahrscheinlich, daß durch diese Substanz eine weitere Verbesserung eintreten kann. Da die Antriebssteigerung vor der Stimmungsaufhellung erfolgt, ist während der Therapieeinleitung mit einer Aktivierung suizidaler Tendenzen zu rechnen.

Dosierung

Das beste klinische Kriterium liegt in der Verminderung der depressiven Symptome und im Auftreten von pharmakologischen Wirkungen, wie oben beschrieben. Allgemein wird häufig mit niedrigen Dosen begonnen und schnelle Dosissteigerung entsprechend der Tolerierung des Patienten oder dem Auftreten therapeutischer Effekte angestrebt.

Nebenwirkungen

1. Zentrale Zusatz- und Nebenwirkungen: Die TCA können einen dauernden feinschlägigen Tremor hervorrufen, der besonders bei älteren Patienten in einen Parkinsonismus übergehen kann. Zuckungen, Dysarthrie, Parästhesien und Ataxien treten selten auf. Pharmaka mit depressiver Wirkung auf ZNS-Funktionen wie Neuroleptika werden potenziert. Appetit- und Gewichtszunahme sind häufig. Die Patienten klagen über Müdigkeit und innere Unruhe.
2. Periphere Zusatz- und Nebenwirkungen: Die autonomen Wirkungen sind primär anticholinergisch: es entstehen Mundtrockenheit, Miktionsstörungen, Obstipation, Mydriasis, verstärkte Schweißabsonderung und Darmatonien. Seltener treten toxische Schäden bis hin zum Medikamentenikterus auf. Über kardiotoxische Nebenwirkungen wurde häufiger berichtet (Veith et al., 1982; Glassmann et al., 1987). Therapeutische Dosen können reversible EKG-Veränderungen herbeiführen, Hypotension, hypertensive Reaktionen, Tachykardien oder verschiedene Herzrhythmusstörungen treten aber häufiger bei kardiovaskulären Erkrankungen auf. Als seltene Nebenwirkung wurden aber auch Herzinfarkte, kongestive Kardiomyopathien und plötzlicher Herztod beschrieben. Somit sollten bei Patienten mit kardiovaskulären Erkrankungen die TCA mit besonderer Vorsicht angewendet werden.

Vergiftungen: Hauptsymptome sind Krämpfe, hypertensive Krisen und Hyperthermie, die intensivmedizinisch behandelt werden müssen.

Medikamenteninteraktionen

Einige Reaktionen sind voraussagbar: so eine antagonistische Wirkung gegen das Antihypertensivum Guanethidin und eine Blockierung der vasopressorischen Wirkung von indirekt wirkenden Sympathikomimetika. Interaktionen mit Methyldopa, Clonidin, Reserpin und Digitalis wurden beschrieben (einen guten Überblick zu wichtigen Interaktionen zwischen somatischen Erkrankungen und Antidepressiva sowie mit anderen Medikamenten vermitteln Fava et al., 1988).

24.5.3 Tranquilizer

Hier soll nur die Gruppe der heute wichtigsten Tranquilizer, der Benzodiazepine (BZP), besprochen werden.

Charakterisierung

Tranquilizer erzeugen Gleichmütigkeit bei der Erlebnisperzeption, Entspannung bei der Erlebnisverarbeitung, Ausgeglichenheit bei der Reaktion auf eine Vorstellung oder ein Erlebnis. Diese Wirkungen begünstigen die Homöostase dysregulierter vegetativer Funktionen.

Wirkungsmechanismen

Vor kurzem wurden Benzodiazepin-Rezeptoren im ZNS nachgewiesen (Squires und Braestrup, 1977; Möhler und Okada, 1977). Auffallend hoch ist ihre Dichte in der Großhirnrinde, dem limbischen System sowie im Kleinhirn. BZP-Rezeptor und GABA-Rezeptor bilden einen „supramolekularen Rezeptorkomplex" und verstärken dessen Wirkung prä- und postsynaptisch (Enna, 1983). Der beschriebene Rezeptorkomplex mediiert auch die anxiolytische Wirksamkeit anderer Substanzen wie Barbiturate und Alkohol. Es werden mehrere Typen von Bindungsstellen angenommen, die die Trennung einer anxiolytischen von einer antikonvulsiven Wirkung ermöglichen (Squires et al., 1979). In Analogie zu den Endorphinen konnte ein endogener Peptidligand des BZP-Rezeptors isoliert werden. Dieser konnte wiederum durch einen selektiven BZP-Rezeptorantagonist blockiert werden (Hommer et al., 1987).

Darüber hinaus wird neuerdings die Beziehung zwischen Streß und Veränderungen des BZP-Rezeptorkomplexes untersucht (Biggo, 1983).

Alle heute erhältlichen Benzodiazepine wirken anxiolytisch, sedativ, hypnotisch, muskelrelaxierend und antikonvulsiv. Sie wirken auch auf dysphorische Stimmungen bei Ängsten.

Es gibt keine ausreichenden Daten, aus denen sich eine unterschiedliche anxiolytische Wirksamkeit der verschiedenen BZP begründen läßt. Wichtiger ist statt dessen die Beachtung unterschiedlicher Halbwertszeiten und aktiver Metaboliten.

Für den Arzt ergibt sich aus der unübersehbar gewordenen Substanzklasse das Problem, welches Medikament er für einmalige oder intermittierende Behandlung von Ängsten und Schlafstörungen, zur Prämedikation, im intensivmedizinischen Bereich, für den epileptischen Anfall oder einen agitierten Zustand wählen soll. Bei mehrfacher Anwendung gilt grundsätzlich: Benzodiazepine mit langer Halbwertszeit kumulieren langsam und extensiv. Besonders ist noch zu beachten, ob die Substanzen aktive Metaboliten haben. Bis zur vollständigen Eliminierung vergehen etwa vier bis fünf Halbwertszeiten (bei Stoffwechselgesunden). Umgekehrt gilt auch, daß zu Beginn einer Dauertherapie vier bis fünf Halbwertszeiten abgewartet werden müssen, bis ein Gleichgewicht entsteht. Die Medikamentenakkumulation hängt davon ab, wie oft das Medikament in Beziehung zur Halbwertszeit gegeben wurde. Außer Oxazepam und Lorazepam muß dies bei allen Benzodiazepinen beachtet werden. Bei schnell erwünschtem Wirkungseintritt wie Schlaf und bei akuter Angst sind Medikamente mit schneller Resorption und kurzer Halbwertszeit einzusetzen, während bei Patienten mit chronischen Angstzuständen eher eine einschleichende Therapie mit langsamer Anflutung und längerer Halbwertszeit wünschenswert sein kann (Übersicht bei Greenblatt et al., 1983).

Arzneimittelinteraktionen

Antazida behindern die Absorption von Benzodiazepinen (Greenblatt et al., 1977). Außer bei Oxazepam und Lorazepam verlängert Cimetidin die Halbwertszeit von Diazepam und Chlordiazepoxid und einigen aktiven Metaboliten (Klotz et al., 1979; Klotz und Reimann, 1980). Barbiturate rufen eine Enzyminduktion hervor und beschleunigen den Abbau der Benzodiazepine. Diazepam kann die Plasmabindung von Digoxin steigern und so bei verminderter Clearance den Digoxinspiegel erhöhen (Castillo-Ferrando et al., 1980). Benzodiazepine potenzieren natürlich die Wirkung anderer ZNS-dämpfender Substanzen wie Neuroleptika, Antidepressiva, Antihistaminika und Alkohol.

Indikationen

„Tranquilizer werden in erster Linie bei psychogenen Krankheitsbildern eingesetzt. Der Hauptindikationsbereich liegt gerade bei jenen Krankheitsbildern, bei denen psychotherapeutische Verfahren im weitesten Sinne den eigentlichen Behandlungskern darstellen sollten" (Bürke et al., 1983). Vor jeder Verordnung von Tranquilizern sollte an eine Psychotherapie gedacht werden. An klinischen Indikationen seien genannt: allgemeine Angstsymptome, Angst im Rahmen von Depressionen, Schlafstörungen, Erkrankungen des Muskelskelettsystems und Alkoholentzug (Übersicht bei Wheatley, 1988). Eine Therapieantwort zeigt sich nach einer Woche (Rickels, 1981). Prädiktoren einer positiven Wirksamkeit von BZP sind: keine vorherige anxiolytische Therapie oder ein früherer rascher Wirkungseintritt, eine State-Angst sowie ein supportiver und der Psychopharmakotherapie gegenüber positiv eingestellter Arzt (Elkin et al., 1988).

Eine maximale Symptomminderung mit BZP wird nach sechs Wochen erreicht (Rickels und Schweizer, 1987).

Nebenwirkungen

Die Benzodiazepine haben einen breiten therapeutischen Index. Bei einer über die empfohlenen sechs

Wochen hinausgehenden Einnahme können erhebliche Nachteile auftreten.

Die häufigste Nebenwirkung besteht in einer Depression zentralnervöser Funktionen, die sich in Müdigkeit, Dösigkeit, Schwindel und Koordinationsstörungen äußern kann. Herauszuheben sind Einflüsse auf die Kognition. Dies zeigt sich in einer in den ersten Stunden nach Einnahme von BZP markanten Beeinträchtigung verbaler und visueller Informationen. BZP in der vergleichbaren Dosis von 10 mg Diazepam beeinflussen die Verkehrstüchtigkeit. Insbesondere sind Autofahrer unfallgefährdet, die zum erstenmal ein Psychopharmakon einnehmen (Übersicht bei Taylor und Tinklenberg, 1987). Gefährliche Komplikationen treten bei gleichzeitiger Einnahme mit anderen psychotropen Substanzen auf. Die häufigste Nebenwirkung betrifft die zentrale Dämpfung. Sie reicht von affektiver Verflachung, Unfähigkeit zur Konfliktverarbeitung bis hin zu Alpträumen, Derealisationsphänomenen und Psychosen.

Vegetative und neurologische Nebenwirkungen: Hypotonie, Obstipation, Libidoverlust, Schwindel, Kopfschmerz, muskuläre Ermüdbarkeit.

Blutbildveränderungen, Medikamentenikterus, Zyklusstörungen und Galaktorrhö wurden selten gefunden.

Mißbrauch, Abhängigkeit und Entzugssymptome

Mittlerweile ist mehrfach beschrieben worden, daß auch die Benzodiazepine ein Suchtpotential enthalten, wenn es auch nicht das von Alkohol und Barbituraten erreicht. Dosissteigerungen sind selten und kommen vor, wenn Tranquilizer als Ersatzdrogen verwendet werden. Es liegen mehrere gut belegte Arbeiten vor, die nach längerem Gebrauch, auch in normalen Dosen, Entzugssymptome nachweisen (Owen und Tyrer, 1983; Schöpf, 1983). Rickels und Mitarbeiter (1983) fanden in einer Doppelblindstudie bei 43% ihrer 188 angstneurotischen Patienten Entzugssymptome nach über achtmonatiger Behandlung mit Diazepam. Bei kurzfristiger Behandlung gaben 5% Entzugssymptome an.

Die Entzugssymptomatik wird oft nicht als solche erkannt, da Arzt und Patient annehmen, das Auftreten der Symptome sei einzig Beweis für das Anhalten der früheren Störungen und indiziere eine weitere Medikation (vgl. Tab. 24–1).

Werden BZP abrupt abgesetzt, treten je nach Halbwertszeit die früheren Symptome wieder auf (de Figueiredo et al., 1981). Um Entzugssymptome oder Reboundphänomene zu vermeiden, ist neben der strikten zeitlichen Begrenzung der Medikation auf ein langsames Ausschleichen zu achten.

Bromazepam und Lorazepam sollen ein besonders hohes Abhängigkeitspotential aufweisen (Laux, 1982).

Es muß kaum noch erwähnt werden, daß nach Absetzen von BZP die alten Symptome wieder auftreten. Die Wiederauftrittsrate (relapse rate) bei chronisch angstneurotischen Patienten lag ein Jahr später zwischen 69 und 80% (Rickels und Schweizer, 1987).

Tabelle 24–1. Symptome bei Benzodiazepinentzug (modifizierte Darstellung nach Linder, 1983)

Angst
Psychovegetative Störungen
Schwitzen
Schwindel
Tremor
Appetitverlust — ähnlich den ursprünglichen Verschreibungsgründen
Abdominalbeschwerden
Übelkeit
dysphorische Stimmung
Rebound-Schlafstörungen
Untähigkeit, ruhig sitzenzubleiben (Akathisie)
Überempfindlichkeit auf Lärm, Licht und Geruch
Geschmacksmißempfindungen
Parästhesien
Kinästhetische, optische, akustische Mißempfindungen
Derealisation, Depersonalisation
Delirium
Generalisierte Krampfanfälle
Psychosen

24.5.4 Beta-Rezeptorenblocker

Charakterisierung

An den peripheren adrenergen Synapsen sind zwei Rezeptortypen beschrieben worden: Alpha- und Beta-Rezeptoren. Noradrenalin entfaltet vorwiegend eine Alpha-Aktivität, ruft Vasokonstriktion der Hautarteriolen, der Splanchnikusgefäße und eine Erhöhung des Blutdrucks hervor. Adrenalin, ein Beta-Sympathikomimetikum, bewirkt erhöhte Herzfrequenz, arterielle Dilatation der Skelettmuskelgefäße und Bronchodilatation. Zu den Alpha-Sympathikolytika gehören Phenoxybenzamin, Phentolamin – zu den Beta-Sympathikolytika Propranolol, Acebutolol, Atenolol, Oxprenolol, Pindolol und viele andere Präparate. Lange Zeit waren die Beta-Blocker nur gebräuchlich in der Inneren Medizin zur Behandlung von Angina pectoris, Hypertonie, Hyperthyreose etc., bis sich auch therapeutische Möglichkeiten in der Behandlung von Ängsten und Streßzuständen auftaten.

Wirkung

Die Beta-Sympathikolytika wirken als kompetitive Antagonisten. Hinsichtlich der Wirkung unterscheiden wir Beta$_1$-Adrenorezeptoren (kardial stimulierte Effekte, Reninfreisetzung) und Beta$_2$-Adrenorezeptoren (Erschlaffung der Bronchial- sowie der glatten Gefäßmuskulatur, Stoffwechselwirkungen, Tremor). Die bislang synthetisierten Substanzen wirken auf Beta$_1$-Rezeptoren 20- bis 25mal stärker als auf Beta$_2$-Rezeptoren (Wellhöner, 1982).

Hinsichtlich der Wirkung kann zwischen peripherer und zentraler unterschieden werden. Für die peripheren Effekte der Beta-Blocker dürfte die Antagonisierung des verstärkten Sympathikotonus verantwortlich sein. Zentrale Beta-Rezeptoren konnten erst in den letzten Jahren gefunden werden (Alexander et al.,

1975). Ihre pharmakologische Wirkung könnte sein: Verminderung der Lokomotion, Erhöhung der Krampfschwelle, Verminderung von Erregungszuständen verschiedener Genese (Emrich und v. Zerssen, 1983).

Indikation

Internistisch: Angina pectoris, Hypertonie, tachykarde Herzrhythmusstörungen, Prävention eines Reinfarktes, Hyperthyreose, Tremor und Migräneprophylaxe.

Anxiolytisch: Beta-Blocker sind wirksam in der Behandlung der somatischen Äquivalente von Angstzuständen (s. u.). Sie verbessern die Durchführung komplexer Aufgaben, die sowohl kontrolliertes Lampenfieber als auch eine koordinierte Aktivierung von Gedächtnis, Lernen und motorischen Funktionen erfordern (Dimsdale et al., 1989).

24.5.5 Sedativa

Charakterisierung

Sedativa sind Stoffe verschiedener Substanzgruppen, die relativ unspezifisch eine große Zahl von ZNS-Funktionen dämpfen. Hierzu gehören emotionelle Reaktionen, Aufmerksamkeit, Reaktionsschnelligkeit und andere sensomotorische Funktionen, sowie vegetative Reaktionen, die durch Sinneseindrücke oder Vorstellungen ausgelöst werden. Gegenüber den Tranquilizern unterscheiden sie sich, indem diese allein Reaktionen im emotionalen Bereich dämpfen, aber Aufmerksamkeit und kognitive Funktionen wenig beeinflussen.

Hypnotika wirken in niedriger Dosierung auch sedativ. Stoffe mit sedativem Nebeneffekt sind: Antihistaminika (Promethazin, Diphenhydramin), Parasympatholytika (Scopolamin), Beta-Sympatikolytika, Antisympathikotonika (Reserpin), Bromide.

24.5.6 Psychoanaleptika

Charakterisierung

Zu den Psychoanaleptika gehören Amphetamin, Methamphetamin, Methylphenidat und Fenfluramin. Sie fördern die Bewußtseinshelligkeit, rufen ein starkes Aktivitätsbedürfnis hervor und erleichtern körperliche Dauerleistungen.

Wirkung

Indirekte sympathikomimetische Aktion.

Indikation

Wegen der höheren Suchtgefahr und als unterstützende Droge bei Rauschgiftmißbrauch sind die Weckamine obsolet geworden. Gelegentlich werden sie heute als Appetitzügler in der Behandlung der Adipositas eingesetzt (vgl. Kap. 38.1).

24.6 Praktische Konsequenzen und allgemeine Richtlinien der Therapie

In dem hier vertretenen Ansatz wird zur Indikation von Psychopharmaka in der Psychosomatischen Medizin und auch für die Allgemeinmedizin ein eng begrenzter Katalog angegeben.

Nach May (1971) und Zauner (1972) ist die Indikation für Psychopharmaka nur in akuten Krisensituationen gerechtfertigt. Bei den Krisensituationen handelt es sich um akute regressive Zustände mit depressiven, suizidalen, körperlichen, mikropsychotischen akuten Symptomen.

Psychopharmaka sollen nur kurzfristig initial oder intermittierend gegeben werden.

Zu den weiteren Gesichtspunkten gehört, daß immer ein multitherapeutisches Vorgehen in der Behandlung gesucht werden sollte. Ferner sollte zunächst eine nicht-medikamentöse Therapie ins Auge gefaßt werden. Außerdem ist auf den psychosozialen Kontext der Symptome des Patienten zu achten. Selbstverständlich ist, daß wir den Patienten in den Behandlungsplan einbeziehen und ihn über die zeitliche Begrenzung der Pharmakotherapie informieren. Suchtgefährdete Patienten sollten nur in Ausnahmefällen Psychopharmaka erhalten, da sonst die Gefahr einer iatrogen erzeugten Polytoxikomanie besteht.

Bei den Überlegungen, ob die Verordnung eines Psychopharmakons indiziert ist, muß der Arzt gerade bei körperlichen Krankheiten eine Beziehung zwischen den psychischen Symptomen und der zugrundeliegenden Krankheit herstellen. Dabei sind auch Medikamenteninteraktionen mehrerer Pharmaka zu bedenken. Ferner ist zu fragen, wie Psychopharmaka den Verlauf der Krankheit beeinflussen und umgekehrt, ob die Krankheit und ihre Behandlung die Wirkung des Pharmakons beeinflussen (Übersicht bei Shader et al., 1978).

Psychopharmaka dürfen keinesfalls zur Behandlung von Problemen des allgemeinen menschlichen Lebens eingesetzt werden (Editorial, 1981).

Tabelle 24–2. Indikationen von Psychopharmaka in der Allgemeinmedizin und in der Psychotherapie

Ihre Ziele sind:
- eine Symptomreduktion, um eine Behandlung erst möglich zu machen,
- dem Patienten die Kontrolle über Aggression und Angst zu ermöglichen,
- Herstellung und Beibehaltung eines therapeutischen Kontaktes,
- Beseitigung neurotischer Aktivitäten, um Zuwachs an selbstbeobachtenden Ich-Funktionen zu ermöglichen,
- Überwindung psychotischer Episoden,
- Beseitigung vitaler Gefährdungen.

Ostow (1962) gibt darüber hinaus folgende Kontraindikationen an: Seiner Meinung nach ist es nicht berechtigt, Patienten, deren Selbstbeherrschung, Realitätsprüfung und Fähigkeit zur Selbstbeobachtung ausreichen und die den Mindestanforderungen des menschlichen Lebens gewachsen sind, mit Psychopharmaka zu behandeln. Absolut kontraindiziert ist der Versuch, durch Medikamente sog. Widerstände der Patienten zu überwinden.

24.6.1 Pharmakotherapie der Depressionen, bei Sterbenden und Trauernden

Depressionen sind ein geläufiges Problem in der medizinischen Praxis. Klagt der Patient über depressive Verstimmungen, Entschlußunfähigkeit, innere Unruhe, Leeregefühl, wirkt er in der Beziehung auf uns bewegungsarm oder andererseits gespannt, so läßt sich ein depressives Syndrom bei ihm leicht erkennen. Viele Patienten treten uns aber mit einer Flut verschiedenartiger Beschwerden entgegen oder klagen monoton immer wieder über die gleiche Symptomatik. Im Vordergrund dieser Funktionsbeschwerden und Organstörungen stehen klinisch Schlafstörungen, Kopfschmerzen, funktionelle Abdominalbeschwerden, Herzbeschwerden, seltener Schmerzen im Sinne eines Weichteilrheumatismus, Atembeschwerden, Parästhesien etc. (Kielholz, 1973). Die hintergründigen psychosozialen Probleme werden dabei oft weder vom Patienten geäußert, noch werden sie vom Arzt erfaßt. Diese Form des hypochondrischen Klagens oder der sog. larvierten Depression grenzt sich nosologisch ab von der psychogenen und der endogenen Form. Im Rahmen dieses Kapitels soll schwerpunktmäßig auf die Pharmakotherapie hypochondrischer Symptome eingegangen werden. Die diagnostische Situation erschwert sich dadurch, daß eine Reihe von vegetativen Symptomen auch als Begleiterscheinung bzw. Nebenwirkung bei der Pharmakotherapie auftreten können (Bushfield et al., 1962).

Viele Untersuchungen haben ergeben, daß unter den psychischen Störungen, sowohl bei ambulanten als auch bei stationären nicht-psychiatrisch Erkrankten depressive Symptome im Vordergrund stehen (Katon et al., 1982; Lipowski, 1967; Modestin, 1977; Weyerer und Dilling, 1984).

Eine Metaanalyse australischer und neuseeländischer Psychiater zeigte die Wirksamkeit von Psychotherapie bei neurotisch und die von Antidepressiva bei psychotisch Depressiven (Andrews et al., 1983). Die große Studie von Klerman und Mitarbeitern (1984) läßt eine Überlegenheit der interpersonellen Psychotherapie plus Antidepressiva im Vergleich zu jeder separaten Therapieform bei Patienten mit depressiven Episoden (nonpsychotic, nonbipolar) erkennen. Psychotherapie ist bei „milderen" Depressionsformen der Pharmakotherapie überlegen (Weissman et al., 1987).

Zwischen depressiver Verstimmung, funktionellen Beschwerden (somatoforme Störungen nach DSM III, 1984) und abnormem Krankheitsverhalten besteht eine innere Beziehung im Sinne der larvierten (somatisierten) Depression. Die Patienten klagen über vielfältige funktionelle Beschwerden und Schmerzen, geben aber spontan keine depressiven Symptome an. Es ist fraglich, ob ein Ansprechen auf Antidepressiva als „diagnostische Evidenz" gewertet werden darf (Lesse, 1983).

Es empfiehlt sich, vor jeder Behandlung einen Plan aufzustellen und ihn mit dem Patienten durchzusprechen, besonders ist bei der Verordnung von Psychopharmaka auf den Wirkungsverlauf und die Nebenwirkungen hinzuweisen.

Antidepressiva können hinsichtlich ihrer Zielsymptome in psychomotorisch aktivierende, psychomotorisch stabilisierende und psychomotorisch sedierende eingeteilt werden. Entsprechend der Phänomenologie des depressiven Zustandsbildes bei Patienten läßt sich die richtige Indikation zur Pharmakotherapie stellen. Ängstlich agitierte oder suizidale Patienten erhalten keine aktivierenden Antidepressiva, sondern sedierende. Bei suizidalen Tendenzen eignet sich auch die Kombination mit Neuroleptika, die ihrerseits antidepressive Eigenschaften aufweisen, wie Levomepromazin oder Thioridazin. Ist der Patient gleichzeitig ängstlich, können Antidepressiva mit Tranquilizern, Beta-Blockern oder Neuroleptika kombiniert werden.

Viele klinische, teilweise auch kontrollierte Studien zeigen, daß die Antidepressiva vom Typ des Amitriptylins, also die psychomotorisch sedierenden, das breiteste antidepressive Wirkungsspektrum haben (Hollister, 1978). Pharmakotherapeutisch ist somit diese Medikamentengruppe in der Behandlung hypochondrischer Symptome besonders geeignet.

Eine angstdämpfende, sedierende Wirkung ist bereits innerhalb von zwei bis vier Tagen zu beobachten, der „depressionslösende Effekt" tritt innerhalb von drei Wochen auf. Es empfiehlt sich, die psychomotorisch sedierenden Substanzen morgens niedrig, mittags höher und abends relativ hoch zu dosieren, um die schlafbahnende Wirkung auszunutzen. Hier bietet sich besonders auch die Retardform des Amitriptylins an.

Viele Befunde liegen vor über die Pharmakotherapie mit Antidepressiva bei Patienten mit gastrointestinalen Erkrankungen. Eine kontrollierte Studie über die Anwendung von Desipramin bei Patienten mit Colon irritabile ergab einen positiven Behandlungseffekt hinsichtlich der Unterbauchbeschwerden (Heefner et al., 1978). Kasuistische Berichte liegen vor über die Anwendung bei der Colitis ulcerosa (Kirsner, 1966; Spiegelberg, 1968).

Interessante Hinweise ergaben sich über das trizyklische Trimipramin hinsichtlich der Verminderung der Magensäuresekretion. Infolge der anticholinergen Wirkung blockiert es die H_2-Rezeptoren im ZNS (Green und Maayani, 1977). In einer kontrollierten multizentrischen Studie wurde ein günstiger Effekt von Trimipramin auf Beschwerden und Abheilung von Ulcera duodeni gefunden (Wetterhus et al., 1976).

Oft liegt bei Patienten mit Weichteilrheumatismus eine larvierte Depression zugrunde (Pöldinger, 1976). Konsequente supportive Psychotherapie und antidepressive Pharmakotherapie sind antirheumatischer Behandlung vorzuziehen. Bei bislang therapieresistenten Schmerzen im Bewegungsapparat zeigt das Ansprechen eines Antidepressivums auf die eigentliche Ursache (Storch und Steck, 1982; Weintraub, 1983). Antidepressiva sind auch phasenweise hilfreich in der Behandlung von Patienten mit rheumatoider Arthritis. Die Wirksamkeit antirheumatischer Therapien kann sich dabei dramatisch steigern, wenn eine Depression des Patienten angesprochen und gleichzeitig antidepressiv behandelt wird. Rimon (1974) konnte bei 60% von 37 reaktiv depressiven Rheumatikern zeigen, daß supportive Psychotherapie und Clomipramin eine wesentliche Erfolgssteigerung antirheumatischer Therapien ergaben. Ist der chronisch Kranke aber gerade in einem Prozeß der Auseinandersetzung mit seiner Krankheit (anhaltende Schmerzen, Verlust an Autonomie), darf diese Trauerarbeit psychopharmakologisch nicht unterbrochen werden (Zeitlin, 1977).

Psychopharmaka werden zunehmend auch bei Sterbenden und Trauernden gegeben. Bei den Hunderten von Interviews, die Frau Kübler-Ross (1973) mit Sterbenden machte, fiel ihr das Ausmaß der Verordnung von Psychopharmaka auf. Viele waren so benommen, daß sie kaum klar ihre Umgebung wahrnehmen und Kontakt aufnehmen konnten. Daraus folgerte sie, daß Psychopharmaka die Auseinandersetzung mit dem persönlichen Tod erschweren.

Wir stehen hier noch vor einem ziemlich unerforschten Gebiet. In einer retrospektiven Krankenblattanalyse über verordnete Analgetika und Psychopharmaka fanden Goldberg et al. (1972), daß nur 5% der Patienten innerhalb ihrer letzten 21 Lebenstage kein Psychopharmakon erhalten hatten!

Im Kontakt mit Sterbenden werden Ärzte, Pflegepersonal und Angehörige mit ihrem Selbstbild und ihren eigenen Todesvorstellungen konfrontiert. Dies kann bei den Betroffenen massive Affekte hervorrufen. So können geradezu Dosis und Anzahl verschriebener Psychopharmaka als Gradmesser der Gegenübertragungsreaktion des behandelnden Personals gewertet werden (Kübler-Ross, 1972). Aus dem Gefühl heraus, „nichts mehr tun zu können", dient das Medikament somit als Substitut eines schlechten Gewissens. Bewußtseinsklarheit ist ein von Ärzten und Pflegepersonal nicht unbedingt gewünschter Zustand des zu Tode kranken Patienten, wenn Affekte wie Depression, Angst und Ärger die Behandlungssituation „stören". Eher geschätzt sind ruhige oder ruhiggestellte Patienten, die unauffällig sterben.

Es wäre wichtig, daß Ärzte und Pflegepersonal sich im Umgang mit Sterbenden darum bemühen, die intellektuellen und emotionalen Fähigkeiten der Patienten so zu erhalten, daß sie sich selbst ihres kritischen Zustandes und ihres bevorstehenden Todes bewußt sein können. Diese Kapazitäten sind nämlich erforderlich in der persönlichen Auseinandersetzung, im Umgang mit den Angehörigen und im Verplanen der noch zur Verfügung stehenden Zeit (O'Connell, 1972).

„Psychopharmaka sollten immer nur eingesetzt werden als zusätzliche Therapie und nicht als Ersatz menschlicher Zuwendung" (Kübler-Ross, 1972). Auch bei der Pharmakotherapie Sterbender muß ein individueller Behandlungsplan aufgestellt werden. Die Indikation ist abhängig von Alter und Akutheit der Erkrankung. Besonders Jugendliche und junge Erwachsene, die entgegen des normal zu erwartenden Lebenszyklus sterben müssen und keine Zeit hatten, sich längerfristig damit auseinanderzusetzen, erleben Verzweiflung und Ängste angesichts ihres drohenden Todes. Wenn sie glauben, mit ihrem Leiden nicht fertig zu werden, sind Medikamente nach Absprache indiziert. So kann die Abwehr unterstützt werden. Der Sterbende sollte natürlich keinen Schmerz erleiden müssen (vgl. Kap. 24.6.3).

Im allgemeinen können dieselben Psychopharmaka verordnet werden, wie sie sonst gebräuchlich sind. Antidepressiva werden häufig verordnet, sind aber nach Kübler-Ross (1972) eher nicht für den sterbenden Patienten indiziert. Angesichts des eigenen Todes sei es normal, traurig zu sein. Könne diese Depression angesprochen werden, so verhelfe sie dem Patienten eher zu einer Akzeptation. Sollte er jedoch übermäßig leiden, mag ein Antidepressivum angebracht sein.

Bei der Verordnung von Anxiolytika muß das Kriterium zur Medikation sein, ob sich der Patient in einem für ihn unerträglichen Spannungszustand befindet. Oft leiden Schwerstkranke unter Schlafstörungen. Dann kann eine kurzfristige, adäquate Sedierung, besonders in einer Abenddosis, hilfreich sein.

Wegen ihrer anticholinergen, analgetischen und antiemetischen Komponente können Neuroleptika besonders bei Tumorpatienten indiziert sein (s. u.).

Zunehmend werden Psychopharmaka auch Trauernden verordnet. Normale Trauer ist keine Krankheit (Schmale, 1973). Aber es kann eine Phase sein, in der Angehörige sich hoffnungslos und hilflos fühlen und verstärkt vulnerabel reagieren können. Frau Kübler-Ross beschreibt in ihrem schon oben zitierten Buch, daß Ärzte und Pflegepersonal auch hier zu schnell Angehörigen Psychopharmaka anbieten. Sie nehmen ein Medikament ein, bevor sie sich nach ihrem ersten Schock äußern können und haben infolge des Drogeneinflusses später oft ungenaue Vorstellungen davon, was in den ersten Stunden ablief. Dies mag zur Befreiung von unkontrollierbaren Affekten beitragen, erschwert aber eher die Trauerarbeit. Anxiolytika zum Schlafanstoßen können kurzfristig angebracht sein. Viel wichtiger aber ist die menschliche Betreuung. Bei plötzlichen Todesfällen hilft eine dichte ärztliche Betreuung und Hilfe, um die Entwicklung pathologischer Trauer (Engel, 1961) zu verhindern. Gelegentlich können auch einmal Antidepressiva angebracht sein.

24.6.2 Psychopharmakotherapie bei Angstkrankheiten

Angst ist ein allgemeines Alarmzeichen bei Bedrohung des Organismus und hat als Signal des Ichs eine adaptive Funktion. Bei einigen Patienten gewinnt diese Angst jedoch eine eigenständige Bedeutung und wird zur Krankheit oder zum Begleitsymptom (Tab. 24–3).

Als deutliche Konsequenz neuerer psychophysiologischer Befunde ergibt sich die Tendenz, Angst in einer biologischen Dimension zu sehen (Hoehn-Saric, 1982; Shear, 1986). Das zeigt sich z.B. in dem neuen Klassifikationsschema der American Psychiatric Association, dem „Diagnostischen und Statistischen Manual Psychischer Störungen", dem DSM III (1984). Dort wird der Begriff der Neurose gar nicht mehr verwendet. Zur Begründung wird unter anderem angeführt, man wolle ätiopathogenetische Theorien nicht mehr präjudizieren.

Zu begrüßen ist sicher der Versuch einer Operationalisierung, der sich auf psychotherapeutische und psychopharmakologische Forschung begünstigend auswirken wird. Wir wollen jedoch in der folgenden Übersicht über die verschiedenen Angstformen den klassischen Neurosebegriff berücksichtigen.

Im Konversionssymptom ist der Angstaffekt in das Symptom eingebunden, und der Patient erlebt sich subjektiv angstfrei. Hingegen ist bei der Angstneurose die Bindung der Angst mißglückt.

Phobien zeichnen sich durch Angst und Vermeidung gegenüber bestimmten Objekten und Situationen aus. Der Agoraphobiker hat Angst vor dem Alleinsein oder Angst in öffentlichen Plätzen. Diese Störung kann mit oder ohne Angstanfälle auftreten. Bei den sozialen Phobien hat der Patient Angst, er könnte durch bestimmte, sozial anstößige Weise in der Öffentlichkeit auffallen. Die sog. einfachen Phobien beziehen sich auf bestimmte Objekte oder Tiere.

Die Angstneurosen im engeren Sinne schließen die Herzneurose mit ein. Die Diagnose „Angstattacke"

Tabelle 24–3. Einteilung der Angstneurosen (modifiziert nach Brown et al., 1984)

Phobien
  Agoraphobie ohne Angstattacken
  Agoraphobie mit Angstattacken
  einfache Phobie
  Sozialphobie
Angstneurosen
  Angstattacken (inkl. Herzneurose)
  generalisiertes Angstsyndrom
  Zwangsneurosen
Traumatische Neurose (post-traumatic stress disorder)
  akut
  chronisch

atypische Angstneurosen
somatisierte Ängste (Somatizer)
Angst im Rahmen von Depression
antizipierte Angst (z.B. Lampenfieber).

(Paniksyndrom) stellt sich, wenn mindestens vier Anfälle in vier Wochen mit mindestens vier der folgenden Symptome auftreten: Atemnot, Herzklopfen, Druckgefühl im Brustbereich, Globus, Schwindel, Taubheitsgefühl, Schwitzen, Ohnmachtsgefühl, Zittern, Gefühl der Unwirklichkeit, Hitze-/Kältewellen und Todesangst. Das generalisierte Angstsyndrom schließt neben frei flottierender Angst Beschwerden der Symptomgruppen motorische Spannung und autonome Hyperaktivität sowie Störungen der Vigilanz mit ein (DMS-III-R, 1989). Die Angstepisode dauert mindestens sechs Monate. Selbst bei der gleichen klinischen Symptomatik kann die Angst sich unterschiedlich psychisch oder psychosomatisch ausdrükken. Exemplarisch hat dies einmal Wesiack (1980) für die Herzneurose beschrieben. An drei Patienten beschreibt er herzneurotische Symptome als Funktionsstörung, als Angstneurose und als hypochondrisches Bild im Sinne einer Grundstörung (Balint). Je nach Grad des psychischen Erlebens der Angst lassen sich auch unterschiedliche therapeutische Konsequenzen angeben. Während bei der Funktionsstörung und der neurotischen Angst ärztliche Gespräche und spezifische psychotherapeutische Maßnahmen indiziert sind, kann bei der hypochondrischen Form oder auch bei psychotischen Ängsten eine Pharmakotherapie zusätzlich notwendig werden.

Zwangsneurosen zeichnen sich durch Denkstörungen, Zwangsimpulse und Zwangshandlungen aus. Bei dem Versuch, die Zwangssymptomatik zu durchbrechen, gerät der Patient in Angst und Panik.

Unerwartete Unfälle, Naturkatastrophen, Krieg, Mißhandlung, Haft und Konzentrationslager sind Erfahrungen, auf die fast alle Personen traumatisch reagieren. Traumatische Neurosen stellen Verarbeitungsmuster dar, die sich klinisch durch Schlafstörungen, Alpträume, Wiedererleben der traumatischen Situation (flashback), psychisches Ausgelöschtsein („psychic numbing" nach R. J. Lifton, 1968), vegetative Erregbarkeit etc. auszeichnen.

Bei den somatisierten oder hypochondrischen Ängsten bezieht sich die Furcht des Patienten auf einzelne Körperfunktionen oder Körperorgane, der psychische Aspekt ist dabei abgespalten (sog. Somatizer; vgl. auch Ford, 1983).

Zwischen antizipierter Angst, wie etwa der Prüfungsangst, und normalen Streßreaktionen bestehen fließende Übergänge.

Bei jedem Patienten hat die Angst einen affektiven, verhaltensmäßigen und physiologischen Aspekt. Die heutigen Behandlungsmethoden bestehen, zum Teil konkurrierend, aus Psychotherapie in ihren verschiedenen Modifikationen und Pharmakotherapie.

Die durch Klein (1964) mitgeteilte Beobachtung, daß Imipramin wirksamer als Chlorpromazin und Plazebo in der Behandlung von Patienten mit Angstattacken und Agoraphobie war, führte zu einer „pharmakologischen Dissektion" (Cassano et al., 1988). Seitdem bestehen Bemühungen, eine nosologische Einteilung der Angstkrankheiten nach dem unterschiedlichen Ansprechen auf Psychopharmaka vorzunehmen. Angstdämpfende Medikamente umfassen heute trizyklische Antidepressiva, Benzodiazepine, Beta-Blocker, MAO-Hemmer und andere Substanzgruppen wie Clonidin, nicht-trizyklische Antidepressiva und Carbamazepin (Ballenger, 1986; Hollister, 1986). Die Arbeitsgruppe um D. F. Klein hat sich der Untersuchung des Paniksyndroms wie der Agoraphobie gewidmet (Klein, 1987). Sie gehen von der Vorstellung einer biologisch fundierten Vulnerabilität aus und bestätigen die alten Freudschen Befunde, daß Angstpatienten unter zwei Typen von Angst leiden: der frei flottierenden Angst im Panikanfall und der Angst vor der Angst. Erstere soll eher auf TCA, letztere auf BZP reagieren (Klein, 1987). Die Bedeutung von TCA in der Behandlung von Panikattacken wurde in kontrollierten Studien bestätigt. In der ersten von Zitrin und Mitarbeitern (1978) wurden Imipramin plus Verhaltenstherapie, Imipramin plus supportive Psychotherapie sowie Plazebo plus Verhaltenstherapie miteinander verglichen. In einer zweiten Studie untersuchten sie Imipramin plus Expositionsbehandlung im Vergleich zu Plazebo plus Exposition (Zitrin et al., 1980). Die verschiedenen Psychotherapieformen ergaben keinen Behandlungsunterschied, ebenso zeigte sich keiner zwischen Imipramin und Psychotherapie. Panikattacken bessern sich dosisabhängig durch Imipramin in etwa 70–90% (Ballenger, 1986). Vermeidungssymptome bilden sich unter reiner Pharmakotherapie mit TCA nur verzögert und unter hohen Dosen zurück (Mavissakalian und Perel, 1989). Es gibt unterschiedliche Auffassungen darüber, ob TCA eine spezifische antipanische Wirkung, unabhängig von der antidepressiven, zukommt (Mavissakalian und Michelson, 1983; Sheehan et al., 1984).

Neuerdings wurde im Gegensatz zu früheren Untersuchungen mit BZP gezeigt, daß Alprazolam sowohl Panikattacken zu blockieren wie auch antizipatorische Angst zu reduzieren vermag (Sheehan et al., 1984; Ballenger et al., 1988). Der therapeutische Effekt von Imipramin entwickelt sich langsamer als jener von Alprazolam. Allerdings kann es bei Beendigung der Behandlung mit Alprazolam auch bei langsamer Dosisreduktion zu erheblichen Entzugssymptomen kommen (Pecknold et al., 1988).

Je mehr vegetative Symptome, je Ich-ferner die Angstsymptomatik, desto wirksamer sind Beta-Blocker (Hawkins, 1975; Freedman, 1980). Vorteilhaft ist, daß sie nicht sedieren, die Psychomotorik nicht negativ beeinflussen und auch keine Abhängigkeit hervorrufen. Beta-Blocker haben sich auch als günstig erwiesen in der Behandlung normaler Streßreaktionen wie Lampenfieber oder Examensangst. So können sie auch zur Therapie antizipierter Ängste als einmalige prophylaktische Gabe indiziert sein. Ein bis zwei Stunden vor dem befürchteten Ereignis können 40 mg Propranolol oder Oxprenolol gegeben werden. Eine vorherige Testung der Wirkung empfiehlt sich allerdings. In drei verschiedenen kontrollierten Studien konnte beispielsweise gezeigt werden, daß Beta-Blocker das Lampenfieber von Musikern senkten und damit zur Abnahme von Nervosität, Schwitzen, Tremor und Herzfrequenz führten. Darüber hinaus

verbesserte sich auch die Qualität der musikalischen Aufführung (James et al., 1977; Brantigan et al., 1982; Neftel et al., 1982).

So wie für Imipramin und Alprazolam antipanische und antiphobische Wirkungen postuliert werden, so beim Clomipramin ein Antizwangseffekt (Moeldner, 1980; Thoren et al., 1980; Insel und Zohar, 1987).

Es läßt sich beobachten, daß „reine" Pharmakotherapeuten zunehmend den „Randbedingungen" ihrer Behandlung Aufmerksamkeit schenken (Cassano et al., 1988; Elkin et al., 1988). Je mehr sie das Setting mitbeachten (was den psychoanalytischen Pionieren selbstverständlich war), desto eher werden sie mit Schlüssen vorsichtig sein, den Pharmaka Eigenschaften zuzuschreiben, die über die physiologischen Wirkungen hinausgehen. In Anbetracht vieler Unsicherheiten scheint es derzeit vernünftig, die Behandlung von Angstpatienten mit Psychotherapie zu beginnen und Medikamente nur dann einzusetzen, wenn sich innerhalb einiger Wochen keine Besserung einstellt und zusätzlich eine depressive Symptomatik besteht (Zitrin, 1983).

24.6.3 Psychopharmakotherapie bei Schmerzzuständen

Rolf Adler und *Gerhard H. Paar*

Bevor eine medikamentöse Schmerztherapie erwogen wird, müssen zuvor alle anderen Therapieformen durchdacht werden. Die Pharmakotherapie steht in einer Reihe therapeutischer Maßnahmen, die vom ärztlichen Gespräch, Notfallpsychotherapie, entspannenden Körperübungen, autosuggestiven Therapien zu verhaltenstherapeutischen und kognitiven Verfahren, von sorgfältiger Pflege bis zum Einbeziehen der Angehörigen reichen (Payk, 1983). Dazu sollten insbesondere auch lokale Maßnahmen wie Bestrahlung, Nervenblocks, chirurgische Interventionen in die therapeutischen Überlegungen einbezogen werden. Die folgende Übersicht zeigt die klinischen Schmerzsyndrome, bei deren Behandlung sich Psychopharmaka bewährt haben (Tab. 24–4):

Tabelle 24–4. Indikationen zur Psychopharmakotherapie bei Schmerzzuständen (Kocher, 1983)

Neurologische Schmerzsyndrome
Rheumatologisch-orthopädische Schmerzsyndrome
Chirurgisch-orthopädische Schmerzsyndrome
Karzinom-Schmerzen

Jeder Schmerzpatient verliert einen Teil seiner Aktivität und Unabhängigkeit, sein Selbstwertgefühl sinkt, seine Angst vor Isolation nimmt zu. Diese Angst trägt dazu bei, daß über das „motivierende affektive System" und das „zentrale Kontrollsystem" die afferenten nozizeptiven Impulse weniger gehemmt werden: Das Schmerzempfinden nimmt zu (Melzack und Wall, 1965).

Der akute Schmerz kann dank Analgetika und anästhetischer Methoden heute gut beherrscht werden. Hingegen bereitet die Behandlung chronischer Schmerzen noch große Probleme, zumal bei den Patienten inadäquate Schmerz- und Angstbewältigungsreaktionen zunehmend in den Vordergrund getreten sind.

Mäßig starke Schmerzen fand Bonica (1979) bei einem Drittel aller Tumorpatienten im mittleren und bei zwei Dritteln im Endstadium der Krankheit. Marks und Sachar (1973) beobachteten unter hospitalisierten Krebspatienten 40% mit mäßigen und 30% mit starken Schmerzen. Sie notierten auch eine Unterdosierung mit Analgetika. Bei Krebskranken fanden sich stärkere Schmerzen und weniger schmerzfreie Tage als bei anderen zu Tode Kranken (Oster et al., 1978). Nur ein Viertel der Tumorpatienten aus dieser Untersuchung waren in ihren letzten Lebenstagen schmerzfrei.

Aus eigener Erfahrung hat der amerikanische Pathologe Sanes, der an einem Sarkom erkrankte, den behandelnden Ärzten folgende Empfehlungen gegeben (1979) (Tab. 24–5):

Tabelle 24–5. Sanes' Vorschläge

Eine persönliche Beziehung zu Patient und Familie herstellen.

Verfügbar und pünktlich sein.

Sich Zeit nehmen.

Sich vorstellen.

Das Gespräch nach Ort und Zeit festlegen, damit unnötige Unterbrechungen ausbleiben.

Einfache Begriffe gebrauchen, medizinische Ausdrücke vermeiden.

Sich selbst beobachten und kontrollieren, um das Überspringen eigener Ängste und Sorgen auf den Patienten zu vermeiden.

Das Wort „Krebs" gebrauchen, die besondere Art des Krebses erklären, nötigenfalls mit Bleistift und Papier.

Sich den Fragen von Patient und Familie öffnen.

Die eigene Telefonnummer dem Patienten und seiner Familie angeben; mitteilen, wer im Fall von Unerreichbarkeit einspringt.

Sich über „unlogische" Fragen nicht ärgern oder über Fragen nach paramedizinischen Behandlungen.

Die Bedeutung der kommunikativen Situation der Schmerzerfahrung wird hier deutlich: Ein Arzt, der seinen Patienten warten läßt, bedeutet diesem, daß die Zeit des Patienten weniger wertvoll ist als die des Arztes. Der Patient registriert dies, vielleicht ohne es richtig zu merken, und er zieht daraus die Folgerung, daß er selbst weniger wert sei als der Arzt. Sein Selbstbewußtsein nimmt dadurch ab, und die Angst, verlassen zu werden, verstärkt sich, die Angst steigert den Schmerz usw.

Zur medikamentösen Behandlung steht heute eine Auswahl potenter Pharmaka zur Verfügung, die sich in vier Gruppen einstufen lassen (Tab. 24–6):

Tabelle 24–6. Auswahl von Pharmaka zur Schmerztherapie (Adler, 1978)

1. Antirheumatika (Prostaglandinhemmer): Acetylsalicylsäure, Indometacin, Flurbiprofen, Mefenaminsäure.
2. Nicht-alkaloide Schmerzmittel: Paracetamol, Metamizol, Glafenin.
3. Alkaloide (morphinartige): Morphinsulfat, Ketobemidon, Tilidin, Pentazocin, Dilaudid.
4. Psychopharmaka: Neuroleptika: Levomepromazin, Haloperidol, Chlorpromazin, Thioridazin; Antidepressiva: Imipramin, Amitriptylin, Clomipramin, Maprotilin.

Heute sind besonders zur Behandlung von Karzinompatienten Stufenpläne ausgearbeitet worden (Bruntsch und Gallmeier, 1980; Senn und Glaus, 1982; Abt. Innere Medizin, 1981).

Bei leichtem Schmerz werden Antirheumatika gegeben. Aspirin® in einer Einzeldosis von 500 mg bis zu 1 g wird auch heute von keinem anderen milden Analgetikum übertroffen (Adler, 1978). Bei Nebenwirkungen durch Acetylsalicylsäure (Rhinitis, Asthma bronchiale, Gerinnungsstörungen) kann auf Paracetamol in einer Einzeldosis von 500 mg und einer Tagesdosis von 3 bis 4 g ausgewichen werden. Bei stärkeren Schmerzen können Acetylsalicylsäure oder Paracetamol gut mit Codein (30–60 mg pro Dosis) kombiniert werden.

Alle hier aufgeführten Schmerzmedikamente sollten regelmäßig, d.h. in vier- bis sechsstündigen Intervallen dosiert werden! Viele Untersuchungen zeigten die Unzweckmäßigkeit der Analgetikatherapie auf Abruf (Marks und Sachar, 1973; Twycross, 1979). Wird gewartet, bis der Patient nach dem Schmerzmittel fragt, so haben die Schmerzen und mit ihnen Angst, Hoffnungslosigkeit und Verzweiflung zugenommen; die Spirale mit stärkeren Schmerzen ist in Gang gesetzt, höhere Dosen werden nötig (Adler, 1983). Allerdings bestehen gegen das Prinzip der Schmerzprophylaxe erhebliche Widerstände bei Ärzten und Pflegepersonal (Schreml et al., 1983). Die Bedenken richten sich gegen den höheren organisatorischen Aufwand, und es wird paradoxerweise eine schnellere Gewöhnung des Patienten befürchtet.

Die schmerzlindernde Wirkung des Neuroleptikums Chlorpromazin bei Krebspatienten wurde erstmalig 1959 beobachtet. Die Phenothiazine potenzieren Narkotika und scheinen deren Nebenwirkungen wie Brechreiz und Erbrechen zu dämpfen. Neben Chlorpromazin haben sich klinisch Levomepromazine und Haloperidol bewährt. Kontrollierte Studien, die auch die Neuroleptika untereinander vergleichen, liegen aber unseres Wissens bislang nicht vor. Chlorpromazin, Prochlorperazin und Haloperidol können günstig tagsüber, Levomepromazin und Propericiazin wegen ihrer dämpfenden Komponente nachts eingesetzt werden.

Die Phenothiazine wirken unterdrückend auf die Interneurone im Nucleus reticularis gigantocellularis und aktivieren somit absteigende Schmerzhemmbahnen. Das Haloperidol weist Strukturverwandtschaft

Tabelle 24–7. Neuroleptika

Name	Dosierung
Chlorpromazin (Largactil®)	100–300 mg/Tag per os 4 × 25 mg/Tag parenteral
Levomepromazin (Nozinan®)	15–250 mg/Tag per os und parenteral
Thioridazin (Melleril®)	30–250 mg/Tag per os
Propericiazin (Neuleptil®)	15–30 mg/Tag per os
Haloperidol (Haldol®)	2–3 × 0,5 mg/Tag per os
Prochlorperazin (Stemtil®)	10–30 mg/Tag per os oder rektal

Tabelle 24–9. Kombinationen von Neuroleptika und Antidepressiva

Namen	Dosierung
Levomepromazin und Imipramin	10 mg 3×/Tag 25 mg 3×/Tag parenteral
Levomepromazin und Clomipramin	10 mg 3×/Tag 50 mg i.v. und später 25 mg 3 ×/Tag
Propericiazin und Amitriptylin	15–30 mg/Tag per os 75 mg/Tag per os
Fluphenazin und Amitriptylin	max. 4 mg/Tag per os 75 mg/Tag per os

mit den Opiaten auf und bindet sich an Opiatrezeptoren (Creese et al., 1976). Daraus wurde auf einen opiatagonistischen Effekt geschlossen. Nach einer weiteren Hypothese beruht die analgetische Wirkung von Neuroleptika und Antidepressiva auf einer Interferenz mit den Endorphinen (Spencer et al., 1979). Tabelle 24–7 enthält die verwendbaren Neuroleptika und ihre Dosierung.

Unter den Antidepressiva haben sich diejenigen bewährt, die gleichzeitig eine stimmungsaufhellende, depressions- und angstlösende sowie eine sedierende Komponente haben. Nach Kocher (1984) wirken sie schmerzdämpfend in 67–90%, ihre Wirkung schwächt sich allerdings ab. Ein Wirkungsmechanismus der Antidepressiva in der Schmerzhemmung konnte belegt werden, der nicht über die Aufhellung einer Depression geht. Erkrankungen, die den zentralen Stoffwechsel des Serotonins beeinflussen, bewirken eine veränderte Schmerzwahrnehmung und verändern die Morphinwirkung. Allgemein führt eine Erniedrigung des Serotonins zu mehr Schmerz und umgekehrt (Koe und Weissman, 1966; Mayer und Price, 1976). Beaumont (1976) zeigte, daß Clomipramin, ein spezifischer Serotonin-re-uptake-Hemmer, die Enkephalinanalgesie verstärkt. Die Antidepressiva wirken aber nicht nur an sich analgetisch, sondern steigern auch die Wirkung der Opiate durch Verminderung des Morphinmetabolismus in der Leber (Goldstein et al., 1982; Beaumont, 1976; Liu und Wang, 1975; Maalseed et al., 1979).

Die folgende Tabelle 24–8 enthält die verwendbaren Antidepressiva und ihre Dosierung.

Tabelle 24–8. Antidepressiva

Name	Dosierung
Imipramin (Tofranil®)	50–200 mg/Tag per os und parenteral
Amitriptylin (Laroxyl®)	10–40 mg 4×/Tag per os
Clomipramin (Anafranil®)	50–150 mg i.v./Tag und später per os

Nach den ersten Berichten über analgetische Wirkungen von Neuroleptika und Antidepressiva lag es nahe, beide zu kombinieren. Sicher wird das Wirkungsspektrum erweitert. Durch niedrige Dosierung lassen sich auch stärkere Nebenwirkungen vermeiden (Kocher, 1983). Der Gebrauch stärkerer Analgetika kann somit vermieden oder hinausgezögert werden. Die Schmerzlinderung tritt allgemein schon nach einer Woche ein (Kocher und Schär, 1968). Die psychotropen Substanzen können auch im Stufenschema kombiniert mit Analgetika und Narkotika verabreicht werden. In der Tabelle 24–9 sind klinisch erprobte Kombinationen von Neuroleptika und Antidepressiva angegeben.

Narkotika werden insbesondere bei Krebspatienten von den Ärzten aus Angst vor Sucht und psychischen Veränderungen zuwenig gegeben (Houde, 1979), bei Wirkungsabnahme wird die Dosis nicht energisch genug gesteigert. Bei sehr starken Schmerzen empfiehlt sich, das Narkotikum in hohen Dosen zu geben und dann zu senken statt umgekehrt. Es gibt kein Morphinderivat, das eindeutig zu bevorzugen wäre (Catalano, 1975). Weder Heroin noch Methadon brauchen dem Morphinchlorid vorgezogen zu werden. Twycross (1977) fand beim Vergleich von Heroin mit Morphin bei Frauen keinen Unterschied in der Schmerzintensität, bei Männern waren die Schmerzen unter Heroin stärker. Bei Morphin besteht eine gute Korrelation zwischen Dosis und Plasmaspiegel, die interindividuellen Unterschiede im Ansprechen auf das Opiat bei gleichem Serumspiegel sind aber groß (Neumann et al., 1982). Das Methadon besitzt mit einer Halbwertszeit von 14 Stunden gegenüber dem Morphium mit 4 Stunden eine viel längere Halbwertszeit. Die erreichten Serumspiegel liegen bei ihm nach oraler Gabe auch höher als bei Morphium. Die Anweisung, beim Wechsel von Morphium auf Methadon per os das letztere Medikament in einem Drittel der Morphindosis zu geben, wird von Kepes und Thomas (1981) als zu hoch angegeben. Sie empfehlen, die Dosis noch zu vermindern.

Die seit kurzem bekannte epidurale Applikation des Morphins (z.B. Coombs et al., 1982) erlaubt die Verwendung viel kleinerer Dosen, erzeugt auch weniger allgemeine Nebenwirkungen. Die Anwendung gehört in die Hände eines Anästhesisten und der Einbau von Systemen zur Dauerapplikation in diejenigen des Neurochirurgen.

Die bei der medikamentösen Schmerzbehandlung verwendeten Substanzen werden am besten in einem Stufenschema gegeben (Tab. 24–10):

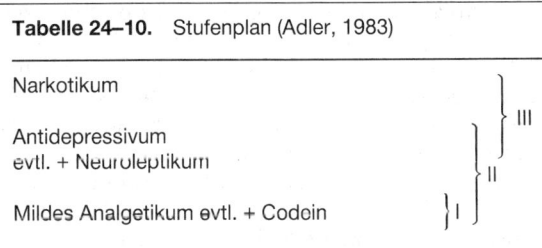

Tabelle 24–10. Stufenplan (Adler, 1983)

Narkotikum

Antidepressivum
evtl. + Neuroleptikum

Mildes Analgetikum evtl. + Codein

I, II, III

Fortwährend Berücksichtigung der psychischen Faktoren; Einsatz nicht-medikamentöser analgetischer Therapien, wenn möglich bzw. nötig.

I, II, III Übergang auf nächste Stufe, wenn die verabreichte Kombination in Maximaldosierung und vierstündlich gegeben nicht ausreicht.

Die Basis besteht in der dauernden Berücksichtigung der psychischen und sozialen Faktoren im Schmerzgeschehen und in der wiederholten Überprüfung, ob andere als medikamentöse Therapien angezeigt sind. Reichen milde Analgetika, eventuell kombiniert mit Codein, nicht aus, so kann ein Neuroleptikum und/ oder ein Antidepressivum dazugegeben werden. Wird damit keine genügende Schmerzlinderung erreicht, so empfiehlt sich die Kombination von Antidepressiva und Neuroleptika mit Morphin. Eine erste prospektive Studie (Schreml et al., 1983) über die Schmerzbehandlung von Tumorpatienten in einem onkologischen Zentrum zeigt, daß im Bereich supportiver Maßnahmen Behandlungspläne sinnvoll sind, wie sie schon längst für die Chemotherapie selbstverständlich wurden.

24.6.4 Plazebos

Der psychosomatische Konsiliarius wird zu einem 48jährigen Mann mit dilatierender Kardiomyopathie gerufen. Der Mann, seit 8 Wochen stationär und zweimal mit Lungenödem auf der Intensivstation, leide unter massiven Todesängsten. Diese verstärkten sich besonders gegen Abend und ließen nach, wenn seine Frau anwesend sei. Nach Auskunft des Stationsarztes sei die kardiale Situation derzeit stabil. Er werde mit verschiedenen Medikamenten kardial behandelt, erhalte tagsüber Bromazepam und abends Flunitrazepam, ferner wegen seit drei Tagen geklagter krampfartiger Abdominalbeschwerden „Spasmolex". Auf mein Nachfragen erfahre ich, daß sich hinter diesem Wundernamen ein Plazebo verbirgt.

Im Zimmer finde ich einen vom Tode gezeichneten Mann vor. Er zeigt auf seinen Oberbauch, da sei alles tot. Bitter beklagt er sich darüber, wie er unter Bauchschmerzen leiden müsse. Die Tabletten hätten ihm vor Tagen geholfen, jetzt würden sie nur noch für zwei Stunden die Schmerzen lindern. Dann versucht er mit

mir zu verhandeln, ob ihm etwa noch eine Herztransplantation helfen könne. Ich bin betroffen und versuche vorsichtig, mit ihm die Todesangst und seine reale Situation anzusprechen. Beim Nachgespräch mit Ärzten und Pflegepersonal wird mir die Verleugnung der realen Situation des Patienten überdeutlich. Das Personal fühlte sich durch den aufmerksamkeitfordernden Patienten irritiert: Die anfangs wirksame Plazebogabe wirkte als Entlastung. Nach der Besprechung erhält der Patient sofort Morphium. Stationsärzte und Konsiliarius sprechen mit der Ehefrau. Die Familie wird hinzugerufen, der Patient kann mit seinen Angehörigen sprechen, sich verabschieden und stirbt noch am gleichen Abend.

Diese kurze, bedrückende Erfahrung im Konsil zeigt eine unheimliche Dramatik. Zwischen der Angst des Patienten und der Angst der Mitglieder des Pflegeteams, der Angst des Arztes vor einem Verlust seiner Omnipotenz und seiner Autorität den Schwestern gegenüber, wird eine unausgesprochene Allianz sichtbar. In dieser sich zuspitzenden Krisensituation sorgt zunächst das verordnete Plazebo für Schmerzlinderung und Distanzierung. Erst als diese Abwehr bei Patient, Pflegeteam und Ärzten zusammenzubrechen droht, wird der psychosomatische Konsiliarius hinzugerufen. Von ihm wird eine Lösung der realitätsverleugnenden Situation vom Stationsteam erwartet.

Was bedeutet hier die Verordnung des Plazebos „Spasmolex"? Drückt es Bestrafung aus oder Rache für die Beunruhigung, die der Patient verursacht? Dieses Aneinander-vorbei-Handeln hätte vermieden werden können, wenn Pflegepersonal und Ärzte ihre vom Patienten ausgelösten Gefühle reflektiert hätten. Es ist ihnen nicht gelungen, die individuelle Wirklichkeit des Patienten zu erfassen.

In der Geschichte der Medizin haben schon immer – bewußt oder unbewußt – psychologische Faktoren in der Behandlung eine Rolle gespielt. Patienten wurden mit fast allem, was an organischen oder anorganischen Substanzen der Natur existiert, behandelt. Erstaunlich ist dabei, daß Medizinmänner und Ärzte trotz Verordnung von pharmakologisch unwirksamen oder gar gefährlichen Medikamenten Erfolge erzielten. Wir wissen seit einigen Jahrzehnten genau, daß die Wirksamkeit vieler Prozeduren auf dem Plazeboeffekt beruht. So gesehen, kann man die Geschichte der Medizin auch als eine Geschichte des Plazebos charakterisieren (Shapiro, 1964). Ein Plazebo ist jede Therapie oder jeder Therapiebestandteil, der wegen seiner unspezifischen, psychischen oder psychophysiologischen Wirkung eingesetzt wird, aber selbst keine spezifische Aktivität in bezug auf die Therapie entfaltet. Ein Plazeboeffekt wird definiert als Befindensänderung des Patienten, die dem symbolischen Gehalt des Heilungsvorganges und/oder der unspezifischen pharmakologischen oder physikalischen Wirkung zuzuordnen ist. Heute wird das Plazebo als Kontrolle in experimentellen Untersuchungen eingesetzt, um die Wirkung spezifischer Substanzen, spezifischer Therapieprozesse oder des Beobachters zu klären (Shapiro, 1978). Aus dieser Definition ergibt sich, daß jede aktive Behandlung Plazebokomponenten enthält!

Unsere Kenntnisse hinsichtlich der Basis der Plazebowirkung sind noch lückenhaft. Grundsätzlich kann sie bei der Verordnung eines schmerzhemmenden Plazebos über eine Aktivierung schmerzhemmender Mechanismen, z.B. über serotoninerge und/oder endorphine Systeme bewirkt werden (Fields und Levine, 1981). Levine und Mitarbeiter (1978) berichten, daß Schmerzpatienten, die auf Plazebos positiv ansprachen, nach dem Morphinantagonisten Naloxon vermehrt Schmerzen angaben.

Vermutlich wird der Effekt nicht durch ein einziges Phänomen hervorgerufen. Frühere Untersucher waren darauf aus, einen Plazebo-Reaktor herauszufinden, also Personen oder bestimmte Persönlichkeitsmerkmale, die voraussagbar auf Plazebos reagieren (Honigfeld, 1964). Beecher (1955) hatte in seiner oft zitierten Übersicht über die Plazebobehandlung von etwa 1000 Patienten, die unter verschiedensten Symptomen litten, einen durchschnittlichen Beschwerdenrückgang bei 35,2 ± 2,2% gefunden. Neuere Übersichten zeigen aber eine Varianz der Symptomrückbildung von 0 bis 100% unter verschiedenen Patientengruppen, Krankheiten und Behandlungsformen (Shapiro, 1978).

Die Annahme einer spezifischen Plazebo-Persönlichkeit läßt sich nicht mehr aufrechterhalten. Jedoch wissen wir um den Einfluß einiger Persönlichkeitsvariablen, die je nach Konzept unterschiedlich beschrieben werden: Suggestibilität, soziale Erwünschtheit, Extraversion, Hysterie.

Physikalische Charakteristika wie Farbe, Form, Größe, Geschmack und Geruch des Pharmakons erwiesen sich als Determinanten der Plazebowirkung. Lasagna (1955) schreibt: „Man glaubt, daß außerordentlich lange Pillen durch ihre Größe imponieren, eine besonders kleine durch ihren Wirkungsgehalt. Eine Injektion wird für wirksamer gehalten als etwas, was durch den Mund eingenommen wird. Vermutlich ist die Anwesenheit der Schwester oder des Arztes bei der Injektion eine wichtige Komponente des psychischen Effektes." Patienten verknüpfen die Wirksamkeit eines Plazebos mit der Person, die das Medikament verschreibt. Insoweit können Plazebos auch als Übergangsobjekte verstanden werden, die eine regulierende Selbstobjektfunktion übernehmen und damit symbolisch den abwesenden Arzt (das abwesende Objekt) kompensieren (Taylor, 1987).

Im folgenden Abschnitt sollen noch andere Charakteristika der „Pharmakologie des Plazebos" wie Dosis-Wirkungs-Beziehung, Nebenwirkungen und Kumulation aufgezeigt werden. Roberts und Hamilton (1958) behandelten 34 angstneurotische Patienten fünf Wochen lang mit Plazebos. Besserungen zeigten sich zunächst im Nachlassen von Angstsymptomen, später auch im Nachlassen vegetativer somatischer Symptome. Die Autoren schließen daraus, daß Plazebos zunächst eine Minderung psychischer und später auch somatischer Symptome bewirken.

Plazebos scheinen in ihren Wirkungen und Nebenwirkungen auch dosisabhängig zu sein. Als „Schlaf-mittel" zeigte eine doppelte Dosis annähernd den doppelten Effekt. In „Notfällen" konnten noch drei Einheiten gegeben werden. Wichtig war, die Patienten zur genauen Einhaltung der Dosis zu verpflichten (Kuschinski, 1975). Pogge (1963) wertete 67 Studien über Plazebowirkungen aus und unterschied 38 Nebenwirkungen. In fallender Häufigkeit fand er Apathie, Müdigkeit, Verwirrung (6%); Kopfschmerzen (3%); gesteigerte zentrale Erregung und Brechreiz (3%); Obstipation und Schwindel (2%); trockener Mund, Übelkeit, Brechreiz etc. Natürlich stehen die angegebenen Nebenwirkungen in Beziehung zu der Hauptwirkung des aktiven Pharmakons, das sonst gegen die betreffenden Beschwerden verordnet wurde. Die Höhe der Nebenwirkung kann auch indirekt Ausdruck der Unzufriedenheit des Patienten sein. Schätzungsweise reagieren 30% aller Patienten auf Plazebobehandlung (Gray und Flynn, 1981).

Wir haben ja schon angedeutet, daß das therapeutische Milieu, in dem eine Therapie abläuft, ein wichtiger Einflußfaktor auf die Medikamentenwirkung ist. Wir wissen um die positiven und negativen Einstellungen von Ärzten und Pflegepersonal auf die Wirksamkeit verordneter Maßnahmen. Ein auffälliger Mangel besteht in der Literatur allerdings an Untersuchungen, aus welcher klinischen Situation heraus Plazebos verordnet werden. Eine Befragung von Stationsärzten und Stationspflegepersonal ergab mehrere Verschreibungsmuster (Goodwin et al., 1979):

– Die Verordnung von Plazebos, um Patienten zu überführen, daß die geklagten Beschwerden „psychogener Natur" sind.
– Die Gabe an mißliebige und schwierige Patienten, wie Alkoholiker, Psychotiker, manipulierende Patienten etc.
– Beim Versagen von Standardbehandlungen.
– Die Verordnung als eine Gruppenaktivität des frustrierten und verärgerten Stationsteams.

Nach einer anderen Untersuchung wurden am häufigsten Schmerzpatienten mit Plazebos behandelt (Goldberg et al., 1979).

In der oben beschriebenen, sicherlich extremen Fallgeschichte kommen alle Verschreibungsmuster zur Geltung.

Leider wissen wir kaum Genaueres, in welchem Ausmaß Plazebos verordnet werden. Ein schon älteres Editorial des British Medical Journal (1952) schätzt, daß die englischen Praktiker bei 40% ihrer Klientel schon einmal Plazebos verordnet haben. Für uns zeigt die Plazebogabe zunächst ein gestörtes Arzt-Patient-Verhältnis auf mit einer negativen Einstellung des Arztes seinem Patienten gegenüber. Spricht der Patient mit seinen Symptomen an, so wird angenommen, diese seien „psychogener Natur", eingebildet oder gar vorgetäuscht. Somit entsteht ein Mythos, das Plazebo könne zur Differentialdiagnostik zwischen „organischen" und „psychischen" Symptomen eingesetzt werden (Brody, 1977). Dabei ist der Patient allemal Verlierer, denn ein Ansprechen sagt nichts über die Ätiopathogenese des Symptoms aus. Immer dann, wenn in der klinischen Routine ein Plazebo verordnet werden soll, muß sich der Arzt fragen: Warum gerade jetzt? Welche Beziehung habe ich gerade zum Patienten? Wie steht das Personal zu ihm?

Befürworter der Plazebotherapie gehen mehr von pragmatischen Gesichtspunkten aus (Fischer und Dlin, 1956). Nach ihrer Meinung ist alles gut, was dem Patienten hilft, zudem hätten Plazebos auch in Doppelblindversuchen ihre Wirksamkeit erwiesen. Sie seien allemal weniger schädigend, denn ein Schmerzpatient würde mit Analgetika weiter abhängig werden.

Aber wie steht es um die Arzt-Patient-Beziehung, wenn der Patient den Betrug entdeckt? Kann nicht gar der Mißbrauch von Plazeboverordnung zu einer verminderten diagnostischen Wachsamkeit führen (Editorial, 1954) und der mit Plazebo behandelte Patient mehr Widerstand gegen eine Psychotherapie entwickeln (Salfield, 1953)? In der modernen Medizin werden Plazebos zunehmend angewendet ohne Täuschung des Patienten, nämlich in jeder Doppelblindstudie mit seinem Einverständnis (informed consent). Recht betrachtet, führt die offene Ankündigung einer Plazebogabe hin zu den symbolischen Elementen der Arzt-Patient-Beziehung, zur Bedeutung der sozialen Unterstützung und dem Gefühl der Geborgenheit in der Behandlung (Adler und Hammett, 1979; Vogel et al., 1980).

25 Ergebnisforschung in der Psychosomatischen Medizin

Wolfgang Senf und *Michael von Rad*

25.1 Definition

Ergebnisforschung dient dem Ziel, die Wirksamkeit von Therapie anhand der Behandlungsergebnisse wissenschaftlich zu überprüfen. **Forschungsgegenstand** sind die durch die therapeutischen Maßnahmen angestrebten und erzielten Veränderungen, die nach Therapieende feststellbar und als „stabil" anzusehen sind. Demgegenüber zentriert sich die Verlaufsforschung, auf die hier nicht eingegangen wird, auf Bedingungen, Umstände und Entwicklungen therapeutischer Prozesse während der Behandlung (vgl. Orlinsky und Howard, 1986). Die **Veränderungsmessungen** in der Ergebnisforschung erfolgen in der Regel durch ein Prä-post-Design in einer Zwei- oder Mehr-Punkte-Erhebung (vgl. Tab. 25–1). Was als Erfolgskriterium definiert wurde, wird zu Beginn der Therapie und nach deren Beendigung gemessen und zueinander in Beziehung gesetzt (outcome). Darüber hinaus wird heute gefordert, daß der Behandlungserfolg in einem ausreichenden zeitlichen Abstand nach Behandlungsende überprüft wird (Katamneseforschung), um die Stabilität des Behandlungsergebnisses beurteilen zu können (follow-up). Veränderungen, die auf diese Weise erfaßbar sind, werden in der Regel auf die therapeutischen Einwirkungen zwischen Beginn und Ende der Therapie bezogen. Dieses Grundmuster wird je nach Forschungsfrage, Theorieansatz, Forschungsstrategie usw. modifiziert.

Die genannte klar und eindeutig erscheinende Vorgehensweise, die in dieser Form für die gesamte Medizin Gültigkeit beansprucht, entstammt den Befunden der Ergebnisforschung in der Psychotherapie. Sie wirft jedoch eine Reihe von schwierigen Problemen auf, die in besonderer Weise für die Ergebnisforschung in der Psychosomatischen Medizin relevant sind. Diese sollen nach einem grundsätzlichen Problemaufriß zur Evaluationsforschung in der Psychosomatischen Medizin anhand eines Fallbeispieles entfaltet werden.

25.2 Evaluationsforschung in der Psychosomatischen Medizin – ein Problemaufriß

Der Versuch, Ergebnisforschung in der Psychosomatischen Medizin zu umreißen, muß sich nach unserer Definition naturgemäß auf einen Bereich zentrieren, nämlich auf die psychosomatische Versorgungspraxis. Es geht also um das, was unter dem Etikett „psychosomatisch" im klinischen Alltag üblicherweise therapeutisch getan wird. Das betrifft in einem enger gefaßten Sinn insbesondere den „psycho"-therapeutischen Anteil der Psychosomatischen Medizin. In einem weiter gefaßten Sinn geht es aber um das, was heute mit dem Begriff „Evaluationsforschung" (Wittmann, 1985; Bengel und Koch, 1988) umschrieben wird, nämlich die generelle Prüfung der Effektivität, Effizienz und Adäquatheit von psychosozialen Versorgungsmodellen in von der öffentlichen Hand geforderten Gesundheitsprogrammen, Präventions- und Rehabilitationsmaßnahmen. Damit beginnt aber auch schon das Problem, das für die Evaluation psychosomatischer Versorgungspraxis mit folgenden Fragen lediglich angedeutet werden soll:

Geht es um psychosomatische Konzepte, Ansätze, Methoden in dem klinischen Handeln traditioneller Kliniken von Internisten, Gynäkologen, Dermatologen und anderen Fachgebieten?

Tabelle 25–1. Schematische Darstellung eines Katamneseprojektes

I. Erstkontakt	II. Therapiebeginn	III. Therapieende	IV. Katamnese
Wartezeit	Therapiezeit	Katamnesezeit	
Symptomatik	Symptomatik	Symptomatik	Symptomatik
Diagnostik	Psychodynamische Hypothesen	Psychodynamik	Psychologische Tests
Biographische Daten	Psychologische Tests	Psychologische Tests	Bewertung der individuellen Therapieziele durch Patienten
Psychologische Tests	Klinisches Rating	Bewertung der individuellen Therapieziele durch Therapeut	Katamnestisches Interview
	Individuelle Therapieziele	Abschlußbericht	

Oder handelt es sich um die Arbeit in eigenständigen psychosomatisch-psychotherapeutischen Kliniken mit einem speziellen therapeutischen Milieu für die Behandlung sog. „psychosomatischer Patienten"?

Sind die psychosomatischen Liaison- und Konsiliardienste gemeint? Steht die Effizienz etwa psychoonkologischer Versorgungsangebote, welcher Art auch immer, zur Diskussion?

Oder ist die psychosomatische Grundversorgung durch niedergelassene Fachärzte, durch die Hausärzte gemeint?

Oder soll ganz unabhängig von Ansatz, Struktur, Setting, institutionellem Rahmen usw. die prinzipielle Therapierbarkeit sog. „psychosomatischer Erkrankungen" ins Auge gefaßt werden?

Schon diese wenigen Fragen, die sich ohne Mühe fortsetzen ließen, zeigen ein zentrales Problem für die Evaluationsforschung: Psychosomatische Versorgungspraxis ist so vielfältig, unterschiedlich differenziert und facettenreich wie der Begriff „Psychosomatische Medizin" selbst. Zudem ist der Gedanke an eine Evaluationsforschung hierzulande noch recht jung; die psychosomatische Versorgung wird erst seit Mitte der 80er Jahre ernsthaft mit evaluativen Aufgaben konfrontiert, indem nach der Leistungsfähigkeit, dem Aufwand und der Angemessenheit psychosomatischer Versorgungskonzepte gefragt wird. Entsprechend sind für die Untersuchung von Effektivität und Effizienz psychosomatischer Versorgungsmodelle noch eine Reihe methodologischer und methodischer Probleme vorab zu klären. So ist es etwa eine erste zentrale Aufgabe, „die Faktoren, die den Prozeß der Implementierung eines psychosomatischen Dienstes beeinflussen, zu beschreiben und zu diskutieren", wobei hier ein „großer Mangel an gesichertem Wissen bezüglich der Implementierungsprozesse, aber auch bezüglich einer Forschungsmethodologie besteht", der sich zum Teil „aus der Komplexität der Bedingungen einer solchen Forschung erklärt" (Koch und Siegrist, 1988, S. 81). Alleine schon diese Frage, wieweit es überhaupt gelungen ist, einen psychosomatischen Dienst entsprechend der Konzepterklärung in einer klinischen Institution zur Anwendung zu bringen, zu implementieren, beinhaltet nach Koch und Siegrist (1989) eine ganze Reihe forschungsmethodischer Implikationen mit Forderungen an den Untersuchungsansatz (Mehrperspektivenansatz, Mehrzeitpunkteuntersuchungen zur Erfassung der Prozesse, multimodaler Zugang) und Problemen (Vielfalt und Komplexität, starke Variabilität, Integrierbarkeit der Einzelbefunde, Vergleichbarkeit der Institutionen, Fehlen valider Meßansätze, Akzeptanz der Forschung), die noch nicht zufriedenstellend gelöst sind.

Die Evaluationsforschung in der Psychosomatischen Medizin, entsprechend der eben gegebenen Definition, steht am Anfang, denn es mußten zuerst einmal geeignete psychosomatische Versorgungskonzepte entwickelt und an den Gegebenheiten der Praxis erprobt werden, und daß hier noch manches Problem zu lösen ist, darauf verweisen Tagungen, auf welchen alle an der psychosomatischen Kooperation Beteiligten zu Wort kommen (vgl. Bräutigam, 1988).

Grundlagen für die Forschung wurden in jüngerer Zeit durch die Ansätze zur Evaluationsforschung überhaupt in verschiedenen Bereichen des Gesundheitswesens geschaffen (Bengel und Koch, 1988). Die Entwicklung der Aufgaben und Konzepte der Evaluation hiesiger Versorgungspraxis orientiert sich an der Evaluationsforschung in den USA mit der ganzen Problematik der Transferierbarkeit. Brauchbare Forschungsansätze, wertvolle methodische Hinweise und konkrete Forschungserfahrungen zur psychosomatischen Versorgungspraxis in der BRD finden sich in der Evaluation des Modellprogrammes „Psychosoziale Betreuung krebskranker Kinder und Jugendlicher" (Koch und Schmied, 1989).

Für eine Ergebnisforschung in der Psychosomatischen Medizin, im enger gefaßten Sinne also für den „psycho"-therapeutischen Anteil psychosomatischer Therapeutik, ist die Ergebnisforschung in der Psychotherapie jedoch immer noch das am besten geeignete Vorbild. An ihr lassen sich gut einige grundsätzliche Gesichtspunkte und Problemstellungen verdeutlichen und diskutieren. Im folgenden konzentrieren wir uns auf diesen Bereich.

25.3 Fallbeispiel

Die Behandlung, über die hier berichtet wird, wurde in der Psychosomatischen Universitätsklinik Heidelberg durchgeführt und im Rahmen des „Heidelberger Katamneseprojektes" (Engel et al., 1979; Bräutigam et al., 1980; Kordy et al., 1983) wissenschaftlich untersucht. Das Forschungsprojekt, in dem eine größere Anzahl sowohl primär stationärer als auch rein ambulanter Einzel- und Gruppenpsychotherapien untersucht werden und das noch nicht abgeschlossen ist, kann gleichzeitig als ein mögliches Beispiel von Ergebnisforschung angesehen werden. Die Tabelle 25–1 gibt einen Überblick über den Ablauf des Projektes sowie die dabei zur Anwendung kommenden Instrumente.

Alle von 1978 bis 1980 in der Klinik begonnenen stationären und ambulanten Psychotherapien werden in einem sowohl prospektiven als auch retrospektiven Design wissenschaftlich untersucht. Der Forschungsansatz ist nicht experimentell, sondern praxisbegleitend und bezieht sich auf Therapieverfahren, die seit langem in der untersuchten Form an der Klinik ausgeübt werden. Der unterschiedliche Stand der psychoanalytischen Ausbildung sowie der klinischen Erfahrung bei den Therapeuten wurde in gewissem Umfang einbezogen. Die der Klinik angeschlossene Forschungsgruppe bemühte sich in engem Kontakt mit den klinischen Psychotherapeuten unter Berücksichtigung des Primates der Therapie vor der Forschung, den Ablauf der Diagnostik und Therapie so wenig wie möglich zu stören, obwohl natürlich auch die behutsamste Form der Begleitforschung Auswirkungen auf den therapeutischen Prozeß notwendig mit einschließt. Der Nachteil einer geringeren methodischen Klarheit wurde in der Hoffnung auf den Gewinn einer relativen Nähe zur alltäglichen psychotherapeutischen Praxis bewußt in Kauf genommen. Oder anders gesagt: Es wurde untersucht, was normalerweise getan wird und nicht etwas getan, um es dann zu untersuchen.

Der zeitliche Ablauf sowie die inhaltliche Gliederung der Gesamtuntersuchung (Tab. 25–1) sind daraufhin angelegt, mögliche Veränderungen des Patienten während der Wartezeit (erster und zweiter Untersuchungszeitpunkt), während der Therapiezeit (zweiter und dritter Untersuchungszeitpunkt) sowie während der Follow-up-Zeit (dritter und vierter Untersuchungszeitpunkt) zu erfassen.

- Direkt im Anschluß an den Erstkontakt in der Klinik wurden ein Symptombefund, verschiedene Fragebögen und Ratings, die Holtzmann-Inkblot-Technik sowie die Gottschalk-Gleser-Sprachprobe erhoben. Ergab sich in einem tiefenpsychologischen Interview die Indikation für eine ambulante oder stationäre Behandlung in der Klinik, dann folgte eine Wartezeit von mindestens drei Monaten. Durch Vergleich des Symptombefundes sowie der Testergebnisse bei Erstkontakt einerseits und Therapiebeginn andererseits sollte das ohnehin kaum lösbare Kontrollgruppenproblem umgangen werden (sog. „Eigenwartegruppe").
- Bei der zweiten Datenerhebung wurden der Symptombefund und die Testungen ergänzt durch eine psychodynamische Hypothese zur Störung des Patienten sowie durch individuelle Behandlungsziele in Form von „Katamnesefragen" (vgl. Abschn. 25.6) und ein klinisches Rating (Engel et al., 1979), das unter psychoanalytischen Gesichtspunkten vor allem ein Instrument zur Beurteilung des Standes der psychosexuellen Entwicklung, der Ich-Funktionen, Objektbeziehungen und der Über-Ich-Pathologie des Patienten ist.
- Die dritte Untersuchung (mit Hilfe der gleichen Instrumente) findet innerhalb eines angemessenen Zeitraumes kurz nach Abschluß der Therapie statt. Sie wird ergänzt durch einen strukturierten „Therapieabschlußbericht" des Therapeuten, der retrospektiv den Behandlungsverlauf sowie die Prognose enthält.
- Die vierte (katamnestische) Erhebung wird mindestens zwei Jahre nach Beendigung der Behandlung durchgeführt. Neben der bisherigen Diagnostik steht vor allem ein halboffenes tiefenpsychologisches Interview eines erfahrenen Psychoanalytikers im Vordergrund, das mit dem Ziel der Beurteilung der Entwicklung des Patienten, zur Beantwortung der Katamnesefragen und zur Überprüfung der psychodynamischen Hypothesen geführt wird. Darüber hinaus wird dem Patienten in diesem Gespräch Gelegenheit gegeben, seine persönlichen Erfahrungen mit der Therapie, seine Ansichten über die hilfreichen und ungünstigen Aspekte der Behandlung sowie seine Erfahrungen mit dem Therapeuten zum Ausdruck zu bringen.

Mit den Testungen 1 und 2 soll also die Stabilität der Störungen, mit den Testungen 2 und 3 die Veränderungen im Laufe der Behandlung und schließlich mit den Testungen 3 und 4 die Stabilität der dabei eingetretenen Veränderungen erfaßt werden. Bei der methodischen Analyse der gewonnenen Daten sind zwei Grundlinien wesentlich (Kordy et al., 1983):
- Da Diagnostik, Therapieverlauf und mögliche **Veränderungen des individuellen** Patienten im Zentrum des Behandlungsinteresses stehen, sollte die empirische Forschung auch die Möglichkeit von

Aussagen über den einzelnen Patienten bieten. Insofern war ein Kompromiß zu finden, bei dem die Notwendigkeit, Beobachtungen vieler Einzelfälle statistisch zu analysieren, trotzdem die Möglichkeit zu Aussagen über den Einzelfall offenhält.
- Da sich die Messung und Bewertung von Veränderungen im Verlauf einer psychotherapeutischen Behandlung auf Daten unterschiedlicher Qualität und Herkunft stützen muß (vgl. Abschn. 25.5), sollte die **Mehrdimensionalität der Veränderungen** nicht durch einen globalen, aber abstrakten Gesamterfolgswert verdeckt werden.

Auf eine vollständige Wiedergabe aller Untersuchungs- und Testergebnisse wird hier zugunsten einer übersichtlicheren und zusammenfassenden Darstellung eines Einzelfalles verzichtet.

25.3.1 Vorgeschichte

Die überaus zierliche, bei 1,60 m Größe nur 33 kg schwere Iris B. wirkt trotz ihrer 22 Jahre noch ausgesprochen kindlich. Bei ihrem ersten Besuch in der Psychosomatischen Universitätsklinik Heidelberg kommt sie – in jeder Hand eine große Tasche – beim Treppensteigen deutlich außer Atem, wirkt angestrengt und geschwächt. Sie hatte sich an die Klinik gewandt, da sie an einer „Anorexia nervosa" leide. Seit ihrem Abitur vor drei Jahren jedenfalls habe sie „Depressionen", Magenschmerzen, harten Stuhlgang, müsse sich ständig mit Diätplänen und kalorienarmer Nahrung beschäftigen, habe kontinuierlich an Gewicht verloren. Seit etwa fünf Jahren bestehe eine Amenorrhö; erbrochen habe sie nie. Spontan berichtet sie jedoch von ihrer „sehr verklemmten Beziehung zu ihrem Körper", von ihrer Hemmung im Sexualleben. Sie habe Angst und Ekel gegenüber dem männlichen Genitale. Ihre Beschwerden seien aufgetreten, als sie von den Eltern fort- und zu ihrem Freund in eine Wohngemeinschaft zog. Zuvor habe sie allerdings bereits mit diesem Freund in ihrem Elternhaus gelebt und in dieser Zeit schon langsam begonnen, sich mehr und mehr mit dem Essen zu beschäftigen und gleichzeitig an Gewicht zu verlieren. Bevor sie zu uns kam, war sie ein gutes Jahr in einer „Gesprächspsychotherapie" bei einer Psychologin, einer Freundin der Familie. Diese Therapie habe ihr erstmals überhaupt Zugang zu sich selbst verschafft, jedoch die bedrohliche somatische Entwicklung nicht aufhalten können, so daß die Psychologin von sich aus die Gespräche beendete und auf eine ärztliche Behandlung drängte.

Die Patientin ist die älteste von insgesamt vier Schwestern (−1½, −4, −6, −7) und einem besonders vom Vater sehnlich erwarteten Bruder (−11). Die zu ihrer Geburt führende Schwangerschaft war ungeplant: Die Eltern haben ihretwegen geheiratet und die Mutter, die von der Patientin als sehr ambitioniert und leistungsbezogen geschildert wird, mußte ihretwegen das Studium abbrechen. Sie hat es später dann wieder aufgenommen, so daß die Patientin als die Älteste immer wieder in die Rolle der „Ersatzmutter" kam. Aus beruflichen Gründen des Vaters mußte häufig umgezogen werden, und auf diese Weise seien Freundschaften mit anderen Kindern immer wieder auseinandergegangen; die Familie habe sich stark nach innen orientiert. Insbesondere zu der nächstjüngeren Schwester besteht eine enge, aber hochambivalente Beziehung.

Diese kam sehr viel früher in die Pubertät, begann früh sexuelle Beziehungen, blieb über Nacht von Zuhause weg und wurde immer wieder zum Mittelpunkt heftiger Auseinandersetzungen mit den Eltern, wobei die Patientin sich stets in der Vermittlerrolle wiederfand. Sie habe eben schon von früh an als Älteste die Sorgen der Eltern mitbekommen und „mitgelitten", ohne sie verkraften und verarbeiten zu können. Sie habe sich überhaupt damit alleingelassen gefühlt. Die Patientin vermutet, daß dies auch sicher mit der schlechten Beziehung zu ihrem Vater zu tun habe, der nie an ihr Interesse gezeigt, sich kaum mit ihr beschäftigt und statt ihrer sich wohl immer einen Sohn gewünscht habe. Im übrigen sei er eher „schwach", und die Patientin bezeichnet ihn als ein weiteres „Kind der Mutter" – man habe jedenfalls immer Rücksicht auf ihn nehmen müssen. So war die Mutter ganz die dominierende Figur in der Familie, deren Liebe die Patientin aber zu verlieren glaubte, seitdem ihre jüngere Schwester durch gute Schulnoten glänzte, während sie selbst einmal sitzenblieb.

25.3.2 Diagnose und psychodynamische Überlegungen

Wir stellten die Diagnose einer Anorexia nervosa und entschlossen uns zu einer (analytischen) Einzelpsychotherapie, auf deren stationärem Beginn wir jedoch entgegen dem ausdrücklichen Wunsch der Patientin bestanden. Innerhalb der ersten fünf Behandlungsstunden formuliert der Therapeut eine psychodynamische Hypothese und benennt die zentralen Störbereiche, an denen Erfolg oder Mißerfolg der Behandlung gemessen werden soll. Darüber hinaus füllt er ein klinisches Rating aus und schätzt Dauer und Intensität der verschiedenen Symptome ein.

Im einzelnen sieht der Therapeut psychodynamisch eine besonders enge symbiotische Beziehung zur Mutter, die aber ambivalent erlebt wird. Er beobachtet dieser gegenüber heftige Aggressionen wegen eines angeblichen Zuwenig an Zuwendung, andererseits starke Geborgenheitssehnsüchte. Darüber hinaus diagnostiziert er eine Fixierung im oralen Bereich, wobei die Einverleibung von Objekten als Bereicherung, jedoch auch als aggressiver Akt erlebt wird. Es bestünden rivalisierende Tendenzen mit der Mutter und dem Vater, der sich jedoch als schwach zeigt. In Enttäuschung darüber sei es zu einer negativen ödipalen Konstellation mit Werben um die Mutter gekommen. Aus sexuellen Ängsten heraus werde die weibliche Rolle abgelehnt mit heftiger Angst vor dem männlichen Glied, das einerseits begehrt, andererseits als bedrohlich erlebt wird. Im ganzen handele es sich um eine Regression auf die Latenzzeit. Dementsprechend sieht er im psychoanalytisch-klinischen Rating (Engel et al., 1979) eine Dominanz der oralen Stufe, eine ausgeprägt rigide Triebkontrolle sowie in den Objektbeziehungen eine symbiotische Abhängigkeit; in der effektiven Lebensgestaltung findet sich seiner Meinung nach eine völlig überschießende und unangebrachte Aktivität bei einem insgesamt strengen Über-Ich. Als zentrale Störungsbereiche werden die Problemkreise 1. des Gewichtes, 2. des Eßverhaltens, 3. der Selbständigkeit (Autonomie und Abgrenzung) sowie 4. der Objektbeziehungen bestimmt und festgelegt, was bei Therapieende als Behandlungserfolg anzusehen ist (s. S. 386).

25.3.3 Behandlungsergebnisse

Am Ende der insgesamt 7 Monate stationären und dann noch ein Jahr ambulant fortgesetzten Behandlung von etwa 200 Stunden kommentiert der **Therapeut** in seinem Abschlußbericht: Die anorektische Symptomatik habe sich bei einem Gewicht von 45 kg (bei einer Größe von 160 cm) nahezu vollständig gebessert. Während die Patientin früher gar nichts oder nur „abstruse" Dinge gegessen habe, nehme sie jetzt normale Nahrung bei regelmäßigen Eßgewohnheiten zu sich. Sie habe auch eine andere Einstellung zu ihrem Körper, fühle sich zunehmend weiblich und sei z. B. stolz auf ihren kleinen Busen. Die Patientin sei insgesamt selbständiger geworden, die früheren Hyperaktivitäten hätten nachgelassen, und sie zeige größere Konstanz in der Zielsetzung. Er beschreibt weiter eine Besserung der Beziehung zur Mutter, die sie jetzt als warmherziger, spendender, aber auch als begrenzender erleben könne. Er kommentiert: „Es war sicher nicht einfach für sie, ihre Magersucht aufzugeben, da deutlich wurde, daß sie mit dieser Erkrankung auch sehr viel Zuwendung, insbesondere von seiten der Mutter erhalten hatte". Den von der Patientin ausgehenden Wunsch, die Therapie zu beenden, sah der Therapeut davon getragen, daß sie „etwas Eigenes leisten, sich selbst auf die Probe stellen wollte". Abschließend ist er der Meinung, daß die Patientin eine Nachreifung mit zunehmender Autonomie und Fähigkeit zur Abgrenzung erlebt hat. Dagegen beurteilt er das Verhältnis der Patientin zu ihrer weiblichen Rolle eher skeptisch und äußert die Hoffnung, daß die Regelblutung mit weiterer Identifizierung der Patientin mit ihrer weiblichen Rolle wieder auftreten wird. Insgesamt meint er, daß die weitere Prognose günstig sei.

Bei der katamnestischen Untersuchung bestätigt sich im großen und ganzen – bei dieser Patientin – die Einschätzung des Therapeuten bei Therapieende: Das Gewicht hat sich bei etwas über 45 kg fast normalisiert („gute Besserung"); das Eßverhalten ist unauffällig („Heilung"); und im Gegensatz zum kritischeren Therapeuten erweist sich in der Einschätzung des Katamnestikers die Selbständigkeit der Patientin als ebenfalls regelrecht („Heilung"). Dagegen bleibt zwei Jahre nach Beendigung der Therapie die Entwicklung der Objektbeziehungen, insbesondere der sexuellen Beziehungen, nach wie vor schwierig und erreicht nicht den bei Therapiebeginn erwarteten Wert einer guten Besserung, sondern verbleibt in der Einschätzung des Katamnestikers lediglich bei einer „nur geringen Besserung". Auch die Erwartung des Therapeuten bei Therapieende, daß die Periode der Patientin in der Folgezeit spontan und regelmäßig einsetzen werde, hat sich nicht erfüllt: Die Patientin spricht zwar von einer mit Hilfe der Pille regelmäßig, aber schwach eintretenden Periode, von der sie aber selbst annimmt, daß diese spontan nicht auftreten würde – sie sieht dies als ein verbliebenes, in ihren Augen psychogen verursachtes Problem, das noch ungelöst und ihr unverständlich sei.

Im klinischen Rating (sowie in den hier nicht ausgeführten testpsychologischen Befunden) bestätigen sich diese Angaben: Der **Katamnestiker** sieht einen erheblichen Rückgang der oralen Fixierung bei der Patientin, eine gewachsene Realitätsprüfung und eine flexiblere Triebkontrolle, insbesondere auch eine verbesserte Ich-Integration und eine angemessene Lebensgestaltung. Allerdings scheinen dem Katamnestiker die Über-Ich-Funktionen kaum verändert und immer noch sehr intolerant und unflexibel – ein Befund, der gut mit den Angaben der Patientin über ihre sexuellen Schwierigkeiten korreliert.

Die **Patientin** berichtet zunächst im katamnestischen Gespräch, daß sie keine Gewichtsprobleme, keine Essensprobleme mehr habe und lediglich noch hinsichtlich ihrer Periode unsicher sei. Ihr Studium habe sie erfolgreich weitergeführt, sie denke an den Abschluß und freue sich schon auf eine Berufstätigkeit. Und auf Nachfrage: Seit Ende der Therapie sei sie nicht mehr ernstlich krank gewesen und habe keine ärztliche Hilfe mehr in Anspruch nehmen müssen. Allerdings habe sie noch immer „so eine sexuelle Angst vor Männern", und die Patientin deutet im Gespräch große Schwierigkeiten an. Der Katamnestiker respektiert ihre Scheu, darüber ausführlicher zu sprechen. Sie habe nur lockere Beziehungen, keinen festen Freund, es fehle ihr der Mut.

Die Beziehungen zu ihren Eltern haben sich positiv verändert: Sie fühle sich ihrer Mutter ähnlicher, spüre gleiche Interessen, könne aber auch eher ihre Schwächen sehen. Früher habe sie die Mutter immer nur als „den unerreichbaren Star" erlebt. Auch mit dem Vater habe sie inzwischen ein besseres Auskommen – sie könne ihn mehr achten. Allerdings habe sie bemerkt, daß sie in ihre alten Beziehungsprobleme zurückfalle, wenn sie längere Zeit bei den Eltern zu Besuch sei.

Hinsichtlich vieler ihrer Schwierigkeiten, derentwegen sie seinerzeit zur Behandlung kam, kann sie sich nicht mehr recht erinnern; sie meint auch (wohl zu Recht), daß sie infolge ihrer körperlichen Schwächung in der letzten Zeit vor ihrer Aufnahme hinsichtlich ihrer Konzentrationsfähigkeit und ihrer Merkfähigkeit eingeschränkt war. Besonders im Hinblick auf ihr Selbstwertgefühl habe sich sehr viel getan: Sie fühle sich sicherer, unabhängiger, könne zu ihren Entscheidungen stehen, traue sich mehr zu und wage mehr.

An der Veränderung ihres Selbstwertgefühls beschreibt die Patientin, was sie selbst als den entscheidenden Schritt in ihrer Therapie ansieht. Sie führt die positiven Veränderungen darauf zurück, daß sie gegen den Therapeuten und auch gegen ihre Familie durchgesetzt habe, die Behandlung zu beenden, „weil das zum ersten Mal eine Entscheidung war, gegen die eigentlich alle waren, ... daß ich das eben so durchgezogen habe, auch unter relativ schwierigen Umständen, das hat mir Auftrieb gegeben. ... Mein Selbstwertgefühl hat sich dadurch gebessert, daß ich einen eigenen Entschluß gefaßt und durchgeführt habe, ohne das alte Muster zu wiederholen, immer nur irgend etwas zu erleiden." Sie habe nicht mehr nur das gemacht, von dem sie glaube, daß es von ihr verlangt werde, sondern sie habe zu ihrer eigenen Entscheidung stehen können und diese durchgesetzt.

Auf Nachfragen, ob sie durch die Behandlung ein Verständnis ihrer Beschwerden erworben habe, meint sie zusammenfassend: „Die Beschwerden verstehe ich eigentlich erst mal so, daß es so war, ... ich war als Kind nie so der Mittelpunkt, um den sich alle gesorgt haben, ... ich war an sich meine ganze Kindheit über und auch die Pubertät über die, die unheimlich viele Sorgen mitgekriegt hat und eigentlich ertragen mußte, aber nie das Sorgenkind war. Ich war zwar öfters krank, hatte alle möglichen Kinderkrankheiten, aber ich hatte zwei Geschwister, die doch sehr dominiert haben. ... Ich bin, glaube ich, viel mehr auf meine Mutter fixiert, meine anderen Geschwister auch. Ich glaube, so dieses Gefühl da, immer was von der Mut-

ter zu wollen, aber aus Zeitgründen eben nicht bekommen zu können, das hat eine große Rolle gespielt." Und zu den Beschwerden: „... also da hatte ich das überhaupt nicht mehr in der Hand. Ich habe das nicht gemerkt. Ich habe das auch verlernt, glaube ich, normal zu essen und habe irgendwie so eine Angst davor aufgebaut, also ich hatte richtige Angst, was weiß ich, irgendwelche Dinge in mich rein zu –, was weiß ich, zu stopfen oder zu essen, weil ich dann eben mit Magenschmerzen oder so darauf reagiert habe. Das hat sich so entwickelt, daß ich eben, ja eigentlich wie so eine Angstvorstellung..., da habe ich überhaupt nur noch gewisse Dinge, die ich akzeptiert habe, essen können."

Auf die Frage, wie sie rückblickend überhaupt die Behandlung erlebt habe, schildert sie: Die Therapie sei für sie eine „Strapaze" gewesen. „Also mir ist es immer schwergefallen, mit jemand allein zu sein, dann noch mit jemand, der wenig redet, das ist eine richtige Qual, ... jetzt kann ich das viel besser, glaube ich, ... ich mußte immer reden, mich da quasi ausbreiten, das hat mich auch manchmal relativ aggressiv gemacht, glaube ich. Und natürlich, er hat dann öfters so Dinge angesprochen, die eben genau die Sache getroffen haben, und das war oft so. Also oft hat mich die Therapie sehr mitgenommen." Auf Krisen angesprochen, bemerkt sie: „Manchmal hat er mich in einem Zustand gehen lassen, das war ein paar Mal der Fall, ich weiß nicht, ob das immer bei jeder Person zu verantworten ist. Also bei mir schon, wahrscheinlich hat er das auch gewußt." Im Zusammenhang mit depressiven Verstimmungen, wenn der Therapeut sie mit seinen Deutungen getroffen hat, meint die Patientin: „Ich weiß nur noch, daß ich danach dermaßen deprimiert war und so mitgenommen, daß ich, was weiß ich, ich wollte überhaupt nicht mehr leben; oder ich habe, das war vielleicht zwei- bis dreimal, den ganzen Tag geheult ..." Im Zusammenhang mit dem (von ihr herbeigeführten) Therapieende und Abschied vom Therapeuten berichtet sie, wie wichtig es für sie gewesen sei, sich von dem Therapeuten zu trennen, aber das Gefühl zu haben, auf ihn zurückgreifen zu können (der Therapeut hatte der Patientin angeboten, sich wieder an ihn wenden zu können, um gemeinsam zu klären, ob eine Fortsetzung der Behandlung wünschenswert und sinnvoll sei). Die Patientin meint dazu, daß sie von diesem Angebot nicht Gebrauch machen würde; sie würde sich zunächst einmal an ihre Freunde wenden und versuchen, ihre Probleme allein zu lösen.

25.4 Die Aufgaben und Ziele empirischer Ergebnisforschung – eine Standortbestimmung

„Hat die Behandlung geholfen?" – Die Beantwortung dieser Frage ist Aufgabe und Ziel der Ergebnisforschung. Seit der provozierenden Skepsis von Eysenck (1952) allerdings war die globale Frage nach der generellen Wirksamkeit von Psychotherapie für lange Zeit die „Gretchenfrage" der empirischen Erfolgskontrolle psychotherapeutischer Maßnahmen. In dem Bemühen, psychotherapeutisches Handeln gegenüber kritischen Stimmen zu rechtfertigen, orientierte sich die Forschung als eine **Rechtfertigungsforschung** (Fürstenau, 1972) weitgehend nach außen mit dem Ziel, der wissenschaftlichen und gesundheitspolitischen Öffentlichkeit gegenüber die Notwendigkeit und den Nutzen von psychotherapeutischen Maßnahmen überhaupt aufzuweisen. Die Ernsthaftigkeit der wissenschaftlichen Bemühungen um dieses Kernproblem dokumentiert sich in einer sprunghaft wachsenden Zahl von Forschungsberichten. Es stellt sich die Frage, was nunmehr Jahrzehnte einer solchen Arbeit erbracht haben.

Die Eysencksche Provokation hat bedeutsame Effekte bewirkt, weniger für die therapeutische Praxis selbst als vielmehr für die empirische Forschung, die sich zu einem respektablen empirischen Wissenschaftszweig entwickelt hat (v. Zerssen et al., 1986; Hoffmann, 1987; Kächele, 1988; Strupp, 1986). Allerdings war auch ein Preis zu entrichten, nämlich der einer starken Einengung auf den globalen Nachweis der Effektivität von Psychotherapie überhaupt, was zu einer Vernachlässigung relevanter klinischer Fragestellungen geführt hat. Heute, d.h. seit gut zehn Jahren, wissen wir nun auch „empirisch", daß Psychotherapie hilft. Sichtet man über Einzelarbeiten hinaus zusammenfassende Darstellungen zu den Ergebnissen von Psychotherapie (vgl. Luborsky et al., 1975; Bergin und Lambert, 1978; Lambert et al., 1981; Rohrmeier, 1982), dann ist empirisch gesichert, daß Psychotherapie wirkt. Vor allem Smith et al. (1980) haben in ihrer vielbeachteten Metaanalyse überzeugend aufgezeigt, daß der durchschnittliche Patient mit verschiedensten Arten von Psychotherapie über verschiedenste Patienten-, Therapeuten- und Erfolgsmaße hinweg nach einer psychotherapeutischen Behandlung bessere Ergebnisse zeigt als 80% der Kontrollgruppenpatienten ohne eine Behandlung. Diese Ergebnisse sind nicht unwidersprochen geblieben, wie die Debatte im „Journal of Consulting and Clinical Psychology" (1983) zeigt. Landmann und Davis (1982) haben jedoch in einer sehr sorgfältig durchgeführten Re-Metaanalyse der Daten von Smith et al. mit etwas anderer Methodik das positive Ergebnis eindeutig bestätigt.

Mit diesen Veröffentlichungen kann die Rechtfertigungsfunktion der Psychotherapieforschung als erfüllt angesehen werden. Damit ist die Forschung frei für Fragen, die für die klinische Praxis relevant sind.

> **Tabelle 25–2.** Aufgaben empirischer Ergebnisforschung.
>
> **Legitimationsfunktion**
>
> **Psychotherapie-Rechtfertigungsforschung**
> Eysenck-Kontroverse; Frage nach der generellen Wirksamkeit von Psychotherapie überhaupt; globaler „objektiv" meßbarer Nachweis von Effektivität; Mangel an differentiellen klinischen Fragestellungen
>
> **Nachweis der Wirksamkeit**
> Wirksamkeit einer in der Versorgung angewendeten Psychotherapiemethode
>
> **Klinische Funktion**
>
> **Evaluation psychotherapeutischer Versorgungspraxis**
> bei Patienten der Regelversorgung durch professionelle Therapeuten
>
> **Differentielle Ergebnisforschung**
> mit der Frage nach der differentiellen Wirksamkeit des Therapieangebotes
>
> **Ansatzspezifische Angebotsexplikation**
> mit der Frage: „Was bietet eine konkrete Versorgungspraxis welchen Patienten im Hinblick auf was?"
>
> **Subjektive Erfahrung und Veränderungstheorie des Patienten**
> Ernstnehmen der subjektiven Erfahrungen und Bewertungen der Betroffenen, die die Therapie „am eigenen Leibe" erfahren haben
>
> **Konsequenzen für die Forschungspraxis**
>
> **Wahl der Therapieziele und Therapiebewertung**
> mehrdimensionale Therapiebewertung spezifischer Zielsetzungen des jeweiligen Behandlungsangebotes; verschiedene Beobachterstandpunkte; individuumorientierte Evaluation; a priori festgelegte Erfolgsbewertungsregeln usw.
>
> **Klassifikationen für Patienten, Therapeuten, Behandlungen**
> geeignete Beschreibungssysteme, welche die Variabilität der Behandlungsergebnisse in Relation zu den verwendeten Beschreibungsmerkmalen darstellbar werden lassen
>
> **Qualitative Untersuchungsansätze**
> Selbstauskünfte der Patienten als Ausgangspunkt wissenschaftlicher Erkenntnisgewinnung

25.4.1 Grundnotwendigkeiten für die Forschung

Für eine Standortbestimmung seien lediglich die folgenden Grundnotwendigkeiten für die empirische Ergebnisforschung (vgl. Tab. 25–2) festgehalten (Senf, 1989).

Evaluation psychotherapeutischer Versorgungspraxis

In einem kritischen Rückblick resümiert Frank (1979), daß trotz der bisherigen, zum Teil sehr aufwendigen Forschungsbemühungen klinisch relevante Ergebnisse „disappointingly meagre" seien. Ähnlich beklagt Parloff (1983), daß die Psychotherapieforschung oft allzuweit fernab der Praxis stattgefunden habe. Wegen der mangelnden Repräsentativität bisher vorliegender Forschungsdaten für die tatsächlich existierende psychotherapeutische Versorgung bedarf es aus unserer Sicht vor allem der Evaluation psychotherapeutischer Versorgungspraxis, wie sie üblicherweise bei Patienten der Regelversorgung durch professionelle Therapeuten stattfindet.

Differentielle Ergebnisforschung

Für die Evaluation psychotherapeutischer Versorgungspraxis lassen sich mit Strupp und Hadley (1977) sehr unterschiedliche Ebenen und Interessensphären präzisieren. Es sind dies

– die Perspektive der gesellschaftlichen Öffentlichkeit bzw. deren Repräsentanten, wie z.B. Gesundheitspolitiker, Kostenträger, Krankenversicherung etc.,
– die Perspektive der Patienten und
– die der Psychotherapeuten.

Strupp und Hadley haben auch die unterschiedlichen Interessen und Erwartungen dieser drei „Partner" herausgearbeitet. Die **gesellschaftliche Öffentlichkeit** interessiert vor allem die generelle Effektivität und die Wirtschaftlichkeit von Psychotherapie, was in neuerer Zeit vor allem unter dem Stichwort Kosten-Nutzen-Relation diskutiert wird. Dieser Erwartung kann durch den objektiv geführten Nachweis der generellen Effektivität eines angewendeten Behandlungsverfahrens weitgehend entsprochen werden. Für die Erwartungen und Interessen der Patienten und Psychotherapeuten ist eine solche Legitimationsfunktion der Forschung jedoch unzureichend. Damit ist die klinische Funktion der Forschung angesprochen. Die **Patienten** möchten wissen, was sie „kaufen" (Strupp, 1975), ob eine angebotene Therapieform für sie geeignet ist und bei welchem ihrer Symptome oder Probleme sie Erfolg verspricht, bzw. wo die Risiken liegen. Für diese Erwartung sprechen die in der letzten Zeit publizierten Erfahrungsberichte von Patienten und Wegweiser in dem Bereich der Psychotherapie. An die **Therapeuten** wird damit die Forderung herangetragen, ihr therapeutisches Angebot zu benennen und durchsichtig zu machen. Die Therapeuten wünschen von der Forschung Hilfen, z.B. für die Indikationsstellung und Therapieplanung, also zu der Frage, welche der verfügbaren Therapieformen bei welchen Patienten mit welchen Störungen, Voraussetzungen, Rahmenbedingungen usw. erfolgversprechend, nutzlos oder sogar schädlich sind. Werden die eben skizzierten Interessen und Erwartungen der Patienten und der Therapeuten ernstgenommen, dann liegt die vordringliche Aufgabe empirischer Ergebnisforschung über den unbestritten notwendigen Legitimationsaspekt hinaus in der Frage nach der differentiellen Wirksamkeit der therapeutischen Maßnahmen.

Ansatzspezifische Angebotsexplikation

Seit Kieslers (1966) grundlegender Kritik am „Uniformitätsmythos" in der Psychotherapieforschung gilt die Zielfrage: „Bei welchen Patienten mit welchen Störungen ist welche Behandlungsmaßnahme durch welchen Therapeuten zu welcher Zielsetzung wie effektiv?" als die Leitfrage der differentiellen Ergebnisforschung. Die daraus abgeleitete Strategie multifaktorieller, multivariater Untersuchungspläne stellt die „via regia" der Forschungsstrategie (vgl. Baumann et al., 1983) dar. Für eine solche differentielle Psychotherapieforschung sind zwei gegenläufige Strategien auszumachen, die sich beispielsweise in der neueren, von Baumann (1981) initiierten Indikationsdiskussion finden.

Zum einen wird die oben genannte Leitfrage als Aufforderung verstanden, durch akkumulativen Erkenntnisfortschritt durch eine Vielzahl experimenteller bzw. quasi-experimenteller Untersuchungen schulübergreifend nach allgemein gültigen Lösungen für klinische Fragen, wie z.B. der Differentialindikation, zu suchen. Zielsetzung ist die Erarbeitung verbindlicher Entscheidungsgrundlagen und Handlungsregeln für den Kliniker, die eben eindeutig vorgeben, welche Maßnahme für welchen Patienten unter gegebenen Bedingungen optimal ist. Die Forschungsstrategie ist, „sich allmählich voran zu arbeiten, indem man jeweils kleine Wissensaspekte zusammenträgt, die erst nach einem beträchtlichen Zeitraum möglicherweise größere Bedeutung gewinnen können" (Kiesler, 1966), mit dem Ziel, „maßgeschneiderte Psychotherapien" (Goldstein und Stein, 1980) zu erreichen.

Demgegenüber hat Westmeier (1979, 1981) aufgezeigt, daß solche nur auf den ersten Blick erstrebenswerten Zielsetzungen nicht realisierbar sind. Die Forschung sei schon auf der praktischen Ebene überfordert, wenn ihre Aufgabe darin gesehen werde, nach allgemeinen Lösungen klinischer Fragen und daraus abgeleitet nach optimalen Handlungsregeln für jedes therapeutische Tun zu suchen. Wollte man solchem Anspruch gerecht werden, dann müßten Tausende von Therapien über längere Zeit wissenschaftlich kontrolliert werden, und das bei einer fast unbegrenzten Zahl von Kombinationsmöglichkeiten einzelner Techniken, ganz abgesehen von der Tatsache, daß sich therapeutische Techniken im Laufe der Zeit ständig ändern. Westmeier (1981) bezeichnet die Hoffnung, therapeutisches Vorgehen empirisch lükkenlos planen zu können, als Relikt eines ansonsten überwundenen wissenschaftlichen Standpunktes des frühen logischen Empirismus, der die Entwicklung einer praxisrelevanten Forschung mehr behindert denn gefördert hat. An die Stelle allgemeiner und schulübergreifender Fragestellungen unter dem Ideal allgemeiner normativer Handlungsregeln, die zu künstlichen und praxisfernen Vereinheitlichungen führen, muß die empirische Begründung und Kontrolle des eigenen praktischen und theoretischen Arbeitsfeldes treten (Westmeier, 1979; Senf et al., 1984).

Ein konkretes Therapieangebot für einen Patienten ist dann begründet, wenn für ihn wichtige Ziele mit möglichst hoher Wahrscheinlichkeit nachweisbar erreicht werden. Unter dieser Perspektive geht es in der Forschung um eine ansatzspezifische Angebotsexplikation. Dabei kann der Forschung jedoch keine Entscheidungsfunktion für klinische Fragen zukommen, sondern lediglich die Rolle, brauchbare Argumente für den Begründungsdialog zwischen allen an der Psychotherapie beteiligten „Partnern" zu liefern (Westmeier, 1979).

Subjektive Sicht und Veränderungstheorie des Patienten

Unter dem Verdikt der Eysenckschen Provokation hat die Psychotherapieforschung dem Subjekt des therapeutischen Geschehens, nämlich dem betroffenen Patienten und seinem individuellen Erleben, kaum Gehör geschenkt. Wenn Patientenvariablen interessierten, dann in nomothetischer Forschungstradition mit (quasi-)experimentellen Ansätzen. Demgegenüber liegt eine weitgehend vernachlässigte Chance für die Forschung in der subjektiven Sicht der Patienten, die das therapeutische Geschehen sozusagen „am eigenen Leibe und an der eigenen Seele" erfahren haben. Das verlangt allerdings, daß wissenschaftliche Empirie auch die individuellen subjektiven Grunderfahrungen von Patienten ernst nimmt, sie als eine Quelle des Erkenntnisgewinns nutzt und sich dabei qualitativer Forschungsmethoden mit offenen Vorgehensweisen bedient.

25.4.2 Schlußfolgerungen für die Forschungspraxis

Unter der hier hervorgehobenen Perspektive einer ansatzspezifischen differentiellen Therapieevaluation liegt eine Hauptaufgabe empirischer Ergebnisforschung in der Beantwortung der Frage: Mit welcher Wahrscheinlichkeit erreicht ein Patient mit Hilfe einer therapeutischen Maßnahme das angestrebte Therapieziel?

Diese allgemeine Formulierung schließt eine Reihe von Teilaufgaben ein, wie sie schon in Tabelle 25–2 sowie im folgenden skizziert sind.

- **Wahl der Therapieziele und Bewertungsregeln:** Therapieevaluation im oben genannten Sinne präzisiert, welche Therapieziele mit welcher Wahrscheinlichkeit erreicht werden können. Unter dieser Perspektive ist es von besonderer Bedeutung, wie die Ziele inhaltlich bestimmt sind und welche Untersuchungsinstrumente in welcher Weise zur Bewertung der Therapieergebnisse eingesetzt werden. Wesentliche Aspekte sind in Tabelle 25–2 und 25–3 im Überblick zusammengefaßt.
- **Klassifikation von Patienten, Therapeuten, Behandlungen:** Die These, daß das Ergebnis einer Psychotherapie davon abhängig ist, welcher Therapeut mit welchem Patienten welche Art von Therapie durchführt, ist in dieser allgemeinen Form als Ausgangsthese für empirische Ergebnisstudien unbestritten. Strittig ist jedoch, wie die Patienten, Therapeuten, Behandlungen klassifiziert werden sollen. Gesucht sind geeignete Beschreibungssysteme für Patient, Therapeut und Behandlungstechnik, welche die Variabilität der Behandlungsergebnisse in Relation zu den verwendeten Beschreibungsmerkmalen darstellbar werden lassen. Dieses konzeptuell-theoretische Problem ist eine vordringliche Aufgabe in der empirischen Ergebnisforschung.
- **Qualitative Forschungsansätze:** Die via regia zur subjektiven Sicht von Patienten ist das persönliche Gespräch. Eine weitgehend vernachlässigte Chance für die Forschung liegt unseres Erachtens darin, Patienten als die Hauptbetroffenen des therapeutischen Geschehens in einer qualitativen Verfahrensweise direkt zu befragen und ihre subjektiven Erfahrungen und Sichtweisen mit den subjektiven Sichtweisen ihrer Therapeuten zu vergleichen

Tabelle 25–3. Beispiele verschiedener Beobachtungsebenen, Outcome-Variablen und Meßverfahren in der Psychotherapie-Ergebnisforschung

I. Symptomatik	
medizinisch	körperl. Untersuchung, Symptomrating, Beschwerdebögen, Labordaten, Arbeitsfehltage, Hospitalisierung etc.
psychologisch	subjektive Tests, Fragebögen, Selbstbeurteilungsskalen, projektive Tests, Interview, inhaltsanalytische Verfahren, Ratingverfahren, Selbst- und Fremdbeobachtung
soziologisch	soziographische Daten, Kosten-Nutzen-Analyse etc.
II. Patient	Symptomatik; Struktur; Erleben und Verhalten
Therapeut	Struktur; Erfahrung, Gegenübertragung; Ausbildung; Technik etc.
Arzt-Patient-Konstellation	Sympathie; Arbeitsbündnis; Krisen; Strukturgemeinsamkeiten bzw. -unterschiede etc.
Umwelt/Gesellschaft	sozioökonomischer Status; Bildung; Arbeitsbedingungen; Familie; Kosten-Nutzen-Analyse etc.
III. Objektive Daten	Arbeitsfehl- und Krankentage; Therapiedauer und Stundenzahl; Alter; Gewicht; Labordaten etc.
Subjektive Daten	Aussagen des Patienten, des Therapeuten oder eines unabhängigen Beobachters über den Befund (Selbsteinschätzungen, Therapieretrospektiven, Fremdeinschätzungen durch unabhängige Beurteiler)
Standardisierte Daten	(mehr subjektiv oder objektiv): Tests, Fragebögen, Ratings
Interaktionelle Daten	Erfassung der Arzt-Patient-Beziehung durch Beobachtungen (z. B. Video, Audio, Sprachinhaltsanalyse) oder Verlaufsberichte

(Senf, 1988 a, b; Senf und Schneider-Gramann, 1989). Methodologisch heißt das, die Selbstauskünfte der Patienten in das Zentrum des Forschens zu stellen. Als methodischer Zugang bietet sich etwa das Ablaufmodell von Mayring (1985) an.

Damit sind bereits die grundlegenden Bereiche einer Ergebnisforschung auch in der Psychosomatischen Medizin angesprochen. Im folgenden konzentrieren wir uns auf die Frage nach der Therapiebewertung und damit nach den Therapiezielen einer Behandlung.

25.5 Therapieziele und Therapiebewertung

Die zentrale Frage für die Ergebnisforschung lautet, was denn überhaupt als Erfolg oder Mißerfolg anzusehen ist, eine Frage, die anhand unseres Fallbeispieles im folgenden näher erläutert werden soll.

25.5.1 Was heißt „Erfolg"?

Einmal angenommen, so unterschiedliche Beurteiler wie Patienten, Therapeuten, Vertreter der Öffentlichkeit würden aufgefordert, das Behandlungsergebnis bei unserer Patientin (vgl. Fallbeispiel, 25.3) zu bewerten, dann könnte sich die folgende Diskussion ergeben:

Ist als ein Kriterium von ärztlicher Seite der Erfolg in der symptomatischen Besserung zu sehen, daß die Patientin ihre Eßstörung verloren, ein akzeptables Gewicht erreicht hat und nicht mehr vital gefährdet ist? Oder soll im Hinblick auf die Öffentlichkeit als Erfolg gelten, daß die Patientin keine ärztliche Behandlung und damit auch keine Kassenleistung mehr in Anspruch nehmen mußte, daß sie erfolgreich ihr Studium bewältigt, arbeitsfähig ist? Oder ist der letztlich entscheidende Gesichtspunkt die von dem Therapeuten festgestellte Nachreifung, die gewonnene Autonomie und Fähigkeit zur Abgrenzung, die bei allen verbliebenen Schwierigkeiten der Patientin eine weiterhin günstige Prognose erwarten lassen? Oder liegt das Wesentliche darin, daß die Patientin zwar noch an manchem leidet, aber insgesamt besser mit ihrem Leben zurecht kommt, sich wohler fühlt? Zu argumentieren wäre auch mit den Testdaten, die in manchen Bereichen durchaus positive Veränderungen erkennen lassen. Gegen einen Behandlungserfolg könnte nun aber argumentiert werden, daß zum Zeitpunkt der Katamnese noch eine Amenorrhö besteht und die Patientin noch immer unverändert an sexuellen Ängsten leidet. Analytische Psychotherapeuten mögen darauf hinweisen, daß die Patientin entgegen der Absicht des Therapeuten die Behandlung „abgebrochen" hat und deshalb zu vermuten sei, daß es nicht zu einer weiterreichenden Strukturveränderung kommen konnte, was unter psychoanalytischen Gesichtspunkten allein eine dauerhafte und kausale Heilung garantieren soll. Unter diesem Gesichtspunkt könnte dann auch gefragt werden, ob die symptomatischen Besserungen überhaupt nur als „Übertragungsheilung" oder „brauchbare Scheinlösung" (Malan, 1962) zu werten sind.

Diese Diskussion ließe sich fortführen. Je nachdem, welche Erwartungen und Interessen zum Tragen kommen, werden sich unterschiedliche oder sogar kontroverse Therapieziele und damit Erfolgskriterien ergeben. Damit ist ein Kernproblem der Ergebnisforschung angesprochen: die Festlegung, was als Erfolg oder Mißerfolg einer psychotherapeutischen Behandlung anzusehen ist.

Erfolgsbeurteilungen und damit die Erstellung von Erfolgskriterien sind ohne die vorherige Bestimmung der therapeutischen Ziele undenkbar. Die Methoden der Forschung – das Sammeln von Daten und das Messen von Eigenschaften – erlauben es, mehr oder weniger „objektive Tatsachen" und Sachverhalte aufzuzeigen, die aber für sich betrachtet noch wenig über ein Behandlungsergebnis aussagen. Erst durch Interpretation und Bewertung der Daten läßt sich Erfolg oder Mißerfolg festlegen. Interpretationen und Bewertungen sind aber eine Frage der Maßstäbe, die angelegt werden, und der jeweiligen Standorte, von denen aus beurteilt wird, und hier kann sehr Unterschiedliches und sogar Kontroverses zum Tragen kommen.

H. Strupp (1978) stellt rückblickend fest: „One of the great stumbling blocks in psychotherapy research in practice has been a failure to realize the importance of values."

Unterschiedliche Maßstäbe und Kriterien entstehen schon aus der Divergenz verschiedener psychotherapeutischer Schulen, mit ihren zum Teil sehr verschiedenen therapeutischen Zielsetzungen und Zielannäherungen auch im Hinblick auf die Inhalte der Ziele. Verhaltenstherapeuten z. B. verfolgen andere Zielsetzungen und legen andere Maßstäbe an als Psychoanalytiker. Während erstere eher klar definierte, spezifische Symptom- oder Verhaltensänderungen anstreben, sehen letztere das Ziel einer psychotherapeutischen Behandlung in einer tiefgreifenden und umfassenden Veränderung der Persönlichkeit, die zu einer Verarbeitung der zugrundeliegenden innerpsychischen Konflikte und damit erst zu einer Auflösung der manifesten Störung führt. So kann eine Symptombesserung für den einen Beurteiler der Erfolgsnachweis sein, während für den anderen (psychoanalytisch) Beurteilenden gilt, „daß die manifesten Syndrom- und Symptomschicksale (gleichgeblieben, verschwunden, verändert) nichts über Heilung, Besserung, Verschlechterung usw. aussagen" (Cremerius, 1978), sondern intrapsychische strukturelle Veränderungen erwartet werden.

Die Frage nach dem Erfolg einer Psychotherapie ist nicht zu trennen von der Frage nach den Erwartungen und Interessen der Adressaten, an die sich eine Untersuchung richtet. Strupp und Hadley (1977) nennen in ihrem „Tripartite Model of Psychotherapy" drei Perspektiven, von denen aus psychotherapeutische Behandlungen beurteilt werden: die Gesellschaft (Öffentlichkeit), der Patient und der Therapeut.

Die **Gesellschaft** erwartet vom Patienten soziale Integration sowie Stabilität im beruflichen und sozialen Verhalten. Psychotherapie ist auch zu einer wirtschaftlich bedeutsamen Dienstleistung geworden; die von den Folgekosten betroffene Öffentlichkeit (z. B. Sozialversicherungen, Krankenkassen) gewinnt hinsichtlich der Zielsetzungen von Psychotherapie zunehmende Bedeutung (Heim, 1981).

Der **Patient** erwartet von einer psychotherapeutischen Behandlung vor allem Symptomfreiheit und Wohlbefinden.

Der **Therapeut** wird sich an seinen mehr oder weniger theoriegeleiteten Vorstellungen von Krankheit und Gesundheit orientieren, die sowohl über soziale Anpassung wie auch subjektives Wohlergehen hinausgehen können.

Nur wenn die Erwartungen und Interessen dieser „Partner" berücksichtigt sind, kann es nach Auffassung der Autoren zu einer umfassenden Evaluation psychotherapeutischer Behandlungen kommen. Sie legen nahe, erst dann von einer Besserung oder Verschlechterung zu sprechen, wenn von diesen verschiedenen Perspektiven aus beurteilt wurde, und schlagen vor, standardisierte und allgemein akzeptierte Erfolgskriterien in den Bereichen Verhalten (Gesellschaft), Erleben (Individuum) und psychische Struktur (Therapeut) zu entwickeln. Der Wert dieser Diskussion von Strupp und Hadley ist darin zu sehen, daß sie auf die Unabdingbarkeit aufmerksam macht, bei der Evaluation psychotherapeutischer Behandlungen verschiedene Perspektiven und unterschiedliche Sichtweisen mit zu berücksichtigen.

Da „Forschungsergebnisse immer auch unmittelbar determiniert sind durch spezifische Fragestellungen einer Untersuchung", und da „die Entscheidung für bestimmte Kriterienmaße damit letztlich auch das Ergebnis einer psychotherapeutischen Untersuchung mitbestimmt" (Hartig, 1975), ist zu wünschen, daß für jede Untersuchung die zugrundeliegenden Standards, die therapeutischen Zielsetzungen, die angelegten Maßstäbe, die Erwartungen und Interessen durchsichtig gemacht werden.

Empirische Untersuchungen zur Evaluation psychotherapeutischer Behandlungen sollten auch Antworten auf die folgende Frage erlauben: Wer legt anhand welcher Kriterien und zu welchem Zweck fest, was Erfolg ist, und wie wird der Erfolg bestimmt? (Kordy und Scheibler, 1983).

25.5.2 Veränderung durch Psychotherapie ist multidimensional

Das Fallbeispiel macht deutlich, daß die Erfolgsbeurteilung psychotherapeutischer Behandlungen eine differenzierende Betrachtungsweise verlangt. Psychotherapie ist nicht als ein eindimensionaler Vorgang, sondern als ein multidimensionales Geschehen zu betrachten, das Auswirkungen und Resultate auf verschiedenen Ebenen zeigt. So können Bereiche wie z.B. Symptomatik, Konflikthaftigkeit, Selbstbild oder Selbstgefühl, Objektbeziehungen, soziale Veränderungen, Arbeitsfähigkeit, Zufriedenheit und andere, jeweils für sich betrachtet, unterschiedliche Richtungen und Ausmaße von Veränderungen zeigen, die keineswegs miteinander korreliert sein müssen. So war die anorektische Symptomatik der Patientin unseres Fallbeispieles zwar gebessert, sie klagte aber noch über die bestehende Amenorrhö, die Störungen im sexuellen Erleben und in ihren Objektbeziehungen. Werden nun einzelne Bereiche, wie z.B. die Symptomatik, für sich betrachtet, dann können verschiedene Beobachtungsebenen bzw. Datenquellen wie Selbstbeurteilungen durch den Patienten, Beurteilungen durch den Therapeuten, unabhängige Beobachter, wichtige Bezugspersonen oder Testinstrumente wiederum durchaus Unterschiedliches abbilden.

Empirische Untersuchungen zum Ergebnis psychotherapeutischer Behandlungen bei psychosomati-

schen Erkrankungen münden jedoch oft in eine globale Erfolgsbewertung, indem allgemein berichtet wird, daß sich der Zustand eines Patienten durch psychotherapeutische Behandlung gebessert oder verschlechtert hat. Das erscheint sinnvoll und hat seine Berechtigung, wenn das oben genannte erste Ziel der Ergebnisforschung, der Legitimationsaspekt, zugrundegelegt wird. Ein solches Vorgehen hat aber auch zur Folge, daß die Vielfältigkeit der Veränderungen von Patienten sowie die unterschiedlichen Ergebnisse verschiedener Datenquellen durch einen globalen und abstrakten Erfolgswert verdeckt werden. Unter der zweiten Zielsetzung der Ergebnisforschung, dem klinischen Aspekt, ist für empirische Untersuchungen zu wünschen, daß sich die Messung und Bewertung von Veränderungen auf Daten unterschiedlicher Qualität und Herkunft stützt, und daß die Mehrdimensionalität der Veränderungen datenmäßig repräsentiert ist. Dann können auch klinisch oder theoretisch interessante Aspekte empirisch weiter untersucht werden, wie z.B., ob es Veränderungen der Symptomatik auch ohne Veränderungen der Konflikthaftigkeit, der Objektbeziehungen oder des Selbsterlebens gibt, bzw. umgekehrt, oder welche spezifischen therapeutischen Interventionen welche spezifischen Auswirkungen auf einen Patienten haben.

Entsprechend der Vielfältigkeit, was aus der Sicht verschiedener Standorte durch eine psychotherapeutische Behandlung erreicht werden soll, findet sich in der bisherigen Ergebnisforschung ein breites Spektrum an unterschiedlichsten Kriterienmaßen und Erfassungsmethoden. Der Psychoanalytiker Knight hat schon 1941 Kriterien zur Erfolgsbeurteilung vorgeschlagen, an denen sich auch heute noch Kliniker wie Forscher orientieren können:

- Symptombesserung,
- erhöhte berufliche Produktivität,
- verbesserte Anpassung und bessere interpersonale Beziehungen,
- sexuelle Befriedigung,
- größere Fähigkeit, psychische Konflikte zu bewältigen und ein normales Maß an Alltagsstreß zu ertragen. Zumindest aber sollte heute zwischen Veränderungen der innerpsychischen Dynamik einerseits und der Symptomatik oder des Verhaltens andererseits unterschieden werden (Malan, 1962).

Die Tabelle 25–3 gibt einen exemplarischen Überblick über die Datenebenen und Meßverfahren, die heute in der Ergebnisforschung üblich sind. An dieser Stelle müßte die Thematik ins Detail gehend weiter vertieft werden. Die Diskussion methodologischer Probleme füllt jedoch inzwischen Bände einer Spezialliteratur, die gleichzeitig auch die Uneinigkeit sogar im Hinblick auf grundsätzliche Methoden der Psychotherapieforschung spiegelt. Wer selbst forschen oder in der Lage sein möchte, Forschungsergebnisse kritisch beurteilen zu können, dem wird nicht erspart bleiben, sich umfassender mit der Problematik und Methodik solcher Forschung auseinanderzusetzen: Angefangen bei definitorischen Problemen, was Forschungsgegenstand und Forschungsziel sein soll, über methodologische Probleme, Fragen der Validität empirischer Untersuchungen oder möglicher Operationalisierungen sowie Konzepte der Datenerhebung und Datenauswertung bis hin zu ethischen Gesichtspunkten, um nur einige Problembereiche anzudeuten. Einführende und weiterführende Hilfen sind bei den folgenden Autoren zu finden: Bastine, 1975; Beckmann et al., 1978; Bergin und Lambert, 1978; Hartig, 1975; Kächele, 1975; Köhnken et al., 1979; Kordy, 1982; Petermann, 1978; Lambert et al., 1981. Wir greifen deshalb Fragen der Meß- und Kontrollstrategien im folgenden mehr exemplarisch auf.

25.5.3 Individuelle Erfolgskriterien als Meßstrategie

Auch bei der Frage, wie „Erfolg" aufgezeigt werden soll, bieten sich verschiedene Strategien an:

Mit Hilfe der **allgemeinen Meßstrategie** werden bei allen Patienten die gleichen Eigenschaften gemessen, auch wenn die Stichprobe heterogen zusammengesetzt ist. Um ein umfassendes und objektives Bild über die Veränderungen durch die Therapie zu gewinnen, wird in der Regel eine große Zahl von Variablen und Meßinstrumenten eingesetzt, unter Beachtung einer hohen Standardisierung, Objektivität und Reliabilität.

Nachteilig ist, daß häufig nur sehr abstrakte Variablen erfaßt werden, und daß möglicherweise eine Vielzahl der Variablen für die Beurteilung des Therapieerfolges bei einzelnen Patienten keine Bedeutung haben. Demgegenüber werden bei der **gruppenspezifischen Meßstrategie** umschriebene Patientengruppen (z.B. Colitis-ulcerosa-Kranke oder „depressive Patienten") mit solchen Meßverfahren untersucht, die geeignet sind, die für diese Gruppen spezifischen Stör- und Problembereiche zu erfassen (z.B. Depressivitätsskalen für depressive Patienten). Dazu muß die untersuchte Gruppe aber u.a. hinreichend homogen sein. Bei der **individuellen Meßstrategie** schließlich wird darauf abgezielt, beim Einzelfall nur die Eigenschaften zu messen, für die eine Veränderung durch die Psychotherapie intendiert ist. So werden für den einzelnen Patienten jeweils spezifische Meßinstrumente oder eine spezifische Auswahl interessanter Variablen für diesen Fall herangezogen.

Die individuelle Meßstrategie ist von besonderem Interesse, da sich diese Methode unmittelbar an dem individuellen Patienten, seinem Beschwerdeangebot und seiner konkreten Behandlungssituation ausrichtet.

Strupp und Bergin (1969) schlagen drei Wege der individuellen Beurteilung von Therapieergebnissen vor:

– Bei allen Patienten werden zwar die gleichen Kriterien erfaßt, doch das Ausmaß und die Richtung der Veränderungen einzelner Kriterien werden für jeden Patienten individuell beurteilt.

– Bei verschiedenen Patienten werden unterschiedliche Meßverfahren eingesetzt, die den jeweiligen Störungen und Problemen der Patienten angemessen sind. Die Bewertung der Ergebnisse erfolgt individuell.

– Für jeden Patienten werden individuelle Therapieziele und Erfolgskriterien zu Behandlungsbeginn entwickelt.

Der Weg, individuelle Erfolgskriterien zu erstellen, wird am konsequentesten dem einzelnen Patienten gerecht und trägt der jeweils spezifischen Störung und dem jeweils individuellen Ausprägungsgrad am klarsten Rechnung.

Für das methodische Konzept der individuellen Therapieziele sind vor allem die Arbeiten zum „goal attainment scaling" (Kiresuk und Sherman, 1968) und zum „personal questionnaire" (Shapiro et al., 1975) sowie die Arbeiten von Malan (1962, 1973) zu nennen.

Auch in katamnestischen Untersuchungen des deutschsprachigen Raums kam diese Methode („Katamnesefragen") zur Anwendung (Engel et al., 1979; Ermann, 1974; Göllner et al., 1978). Anhand der individuellen Therapieziele (Katamnesefragen), die für unser Fallbeispiel aufgestellt wurden, soll diese Methode kurz erläutert werden (Kordy und Scheibler, 1984).

Das Grundprinzip ist, daß der Therapeut aus Kenntnis des Erstinterviews und aus der Erfahrung der ersten fünf Behandlungsstunden drei bis fünf Störungsbereiche für seinen Patienten formuliert, deren Bearbeitung er gerade bei diesem Patienten als zentral ansieht. Für jeden Störungsbereich werden sog. Katamnesefragen, die dem Patienten zum Zeitpunkt des follow-up vorgelegt werden, sowie dessen möglicherweise zu erwartende Antworten ausformuliert. Diese Antworten sind so gehalten, daß sie in einer Abstufung von verschlechtert, unverändert, leicht, gut bis sehr gut gebessert eine Bewertung des Behandlungsergebnisses darstellen. Ebenfalls zu Behandlungsbeginn legt der Therapeut in einer Konferenz fest, welche Antwort von dem Patienten zu erwarten ist. Damit wird a priori ein Therapieziel festgelegt, dessen Erreichen als Indikator für Therapieerfolg gilt. Der Katamnestiker stellt dem Patienten in der Nachuntersuchung die Fragen und kennzeichnet, welche Antworten zu diesem Zeitpunkt zutreffend sind. Für unser Fallbeispiel hat der Therapeut die folgenden Katamnesefragen formuliert (Th = Angabe des Therapeuten; K = Befundung des Katamnestikers).

I. Wie steht es heute mit Ihrem Gewicht?
    1. Mein Gewicht liegt unter 30 kg
    2. Ich wiege über 33 kg
    3. Ich wiege über 40 kg
(Th, K) 4. Ich wiege über 45 kg – fast normalgewichtig
    5. Ich wiege über 50 kg

Erwarteter Therapieerfolg für diese Therapieform $\boxed{3\text{–}4}$

II. Wie steht es mit Ihrem Eßverhalten?
    1. Meine Gedanken kreisen nur noch um das Essen und um das Gewicht. Habe große Ängste und Abscheu davor. Esse fast nichts mehr.
    2. Ich beschäftige mich den ganzen Tag nur noch mit dem Essen, der Zubereitung von Mahlzeiten und den Kalorien. Das Essen selbst löst bei

mir Unwohlsein, Ekel und Schuldgefühle aus. Esse deshalb ganz wenig und möchte auch nicht zunehmen.

3. Essen spielt für mich eine große Rolle, muß mich jedoch nicht mehr so zwanghaft damit beschäftigen, spüre weniger Abscheu vor der Mahlzeit.

(Th) 4. Ich kann ganz gut und regelmäßig essen, beschäftige mich zwar hin und wieder mit dem Essen und dem Zunehmen; aber das ist alles nicht mehr so problematisch.

(K) 5. Essen und Gewicht stellen kein Problem mehr für mich dar, ich esse regelmäßig und mit Appetit.

Erwarteter Therapieerfolg für diese Therapieform $\boxed{3-4}$

III. Wie steht es mit Ihrer Selbständigkeit?

1. Ich lebe wieder zu Hause, möchte auch dort bleiben, fühle mich hier am wohlsten.

2. Ich merke, daß ich mich allein nicht zurechtfinde. Ich möchte am liebsten wieder zu Hause bei Mutter und den Geschwistern bleiben: Fühle mich dort geschützt und geborgen. Habe große Angst vor der Selbständigkeit.

(Th) 3. Ich wohne zwar alleine; es zieht mich aber sehr ins Elternhaus zurück.

4. Ich finde es ganz gut, selbständig zu sein, wenngleich das gelegentlich ganz schön schwer ist.

(K) 5. Ich kann gut alleine zurechtkommen, ich kann mich gegen meine Familie abgrenzen; freue mich aber auch auf zu Hause.

Erwarteter Therapieerfolg für diese Therapieform $\boxed{3-4}$

IV. Wie steht es mit Beziehungen?

1. Partnerschaft kommt nicht in Frage und Sexualität schon gar nicht.

2. Eine Beziehung kommt im Augenblick nicht in Frage. Sexualität bereitet mir Ängste und Ekel.

(K) 3. Ich wünsche mir eine Beziehung, könnte mir das auch eigentlich vorstellen, habe aber noch Angst vor der Sexualität.

4. Habe gelegentlich eine Beziehung, die ich zuweilen auch im Sexuellen positiv erleben kann, obgleich es manchmal Schwierigkeiten gibt.

(Th) 5. Habe eine dauerhafte Beziehung, die mich auch sexuell befriedigt.

Erwarteter Therapieerfolg für diese Therapieform $\boxed{4}$

Wie zu sehen ist, stimmen die Prognosen des Therapeuten (Th) über den Therapieerfolg und die von der Patientin zum Katamnesezeitpunkt gegebenen Antworten (K) nicht immer überein. So hatte der Therapeut erwartet, daß die Patientin weniger selbständig ist, als sie sich selbst zum katamnestischen Zeitpunkt erlebte. Auch in bezug auf das Eßverhalten war der Therapeut pessimistischer in seiner Prognose, während er für die Objektbeziehungen der Patientin eine positivere Veränderung erhofft hatte.

Mit solchen Katamnesefragen und den zu Behandlungsbeginn festgelegten Prognosen werden Bedingungen gesetzt, an denen der Erfolg einer psychotherapeutischen Behandlung kontrolliert werden kann. Allerdings muß auch das Risiko in Kauf genommen werden, daß sich einige Katamnesefragen als ungeeignet erweisen, sei es, daß Problembereiche falsch eingeschätzt oder unpräzise formuliert wurden, sei es,

daß der Patient andere als die erwarteten Antworten gibt. Doch kann in der Abweichung von der erwarteten Antwort möglicherweise auch eine wichtige Information zu dem Behandlungsergebnis enthalten sein.

Auch das Konzept des Goal Attainment Scaling (GAS) ist nicht ohne Kritiker geblieben (Kordy und Scheibler, 1984; Lambert, 1983). Doch spiegeln sich in dieser Methode die Bemühungen der Ergebnisforschung, von einer globalen Erfolgsbewertung mit eher vagen Schlußfolgerungen wegzukommen, um spezifische klinische Fragestellungen und die verschiedenen Ebenen möglicher Veränderungen eines Patienten immer spezifischer ins Auge zu fassen.

25.6 Indikation und Prognose: Ist das Therapieergebnis vorhersagbar?

Im therapeutischen Alltag ist die Indikation die erste, oft entscheidende Weichenstellung. Deshalb liegt ein wichtiges Ziel der Ergebnisforschung darin, dem therapeutisch arbeitenden Praktiker Hilfen für seine Indikationsentscheidungen bereitzustellen. Allgemein gesagt liegt dem Indikationsproblem die Annahme zugrunde, daß unterschiedliche Patienten in verschiedenen Therapieverfahren unterschiedliche Behandlungschancen haben. Ziel der Indikationsstellung ist es, einen gegebenen Patienten einer erfolgversprechenden Therapie zuweisen zu können. Letztlich geht es um die Frage: Sind die Behandlungschancen eines Patienten vorhersagbar? Die Forscher gehen dabei den Weg, empirisch nach **Prädiktoren** für das Therapieergebnis zu suchen, d.h. nach Merkmalen der Patienten, der Therapeuten oder ihrer gemeinsamen Interaktion, die sich als Einflußgrößen für das Therapieergebnis erweisen. Die Psychotherapieforscher Orlinsky und Howard (1986) haben einen beeindruckend umfassenden Überblick über den gegenwärtigen Stand der empirischen Forschung gegeben. Geradezu akribisch haben sie über 1100 Einzelbefunde in Hunderten von Studien gesichtet und die Ergebnisse von mehr als 35 Jahren empirischer Forschung zu den unterschiedlichsten Psychotherapiemethoden zusammengetragen und ausgewertet. Ihr Fazit ist: Es gibt nur wenige signifikante Zusammenhänge zwischen Prozeßmerkmalen und dem Therapieergebnis; im Hinblick auf die für das therapeutische Geschehen interessanten Aspekte sind die meisten Befunde widersprüchlich. Zu welchen Ergebnissen haben die Forschungsbemühungen im einzelnen geführt?

Die Suche nach validen prognostischen **Patientenvariablen** ist bislang zwar enttäuschend verlaufen (vgl. Kächele und Fiedler, 1985), dennoch haben sich einige Merkmale als zuverlässige Einflußgrößen für das Therapieergebnis erwiesen. Dem **Schweregrad der Erkrankung** kommt bei aller Unbestimmtheit und Vagheit solcher Bezeichnungen aus übereinstimmenden Beobachtungen eine wesentliche limitierende Rolle für das Ergebnis psychotherapeutischer Behandlungen zu (Luborsky et al., 1971, 1978). Insoweit sollte sich die Aufmerksamkeit auf solche Patientenmerkmale richten, die für eine schlechte Behandlungsprognose sprechen (vgl. Strupp, 1983). Dabei finden sich Hinweise auf die Art und Ausprägung der Krankheitsbilder (z.B. Störwert, Chronifizierung usw.). **Diagnosen** allein für sich genommen sind nach fast einhelliger Auffassung von nur geringem prognostischem Wert. **Motivation,** meist definiert als Wunsch nach Veränderung, Einsicht und Behandlung, wird vielfach als besonders wichtiges prognostisches Kriterium angesehen, z.B. in der Weise, daß die Stärke der Motivation positiv mit Behandlungserfolg korreliert, dagegen die Art der Motivation (Erwartungen über die Art der Veränderung und die Erwartungen an den Therapeuten) kaum etwas über den zu erwartenden Behandlungserfolg aussagt (Luborsky et al., 1971; Sifneos, 1968). Manifeste **Angst** zu Behandlungsbeginn gilt ebenfalls als prognostisch günstig für das Therapieergebnis (Denford et al., 1983; Kernberg, 1972, 1973), was allerdings von anderen Autoren nicht bestätigt wird (Lolas et al., 1984; Sashin et al., 1975). Auch **Depression** vor Behandlungsbeginn wurde als prognostisch günstig angesehen (Frank, 1958; Luborsky, 1971). Interessant sind die Befunde von Sashin et al. (1975), daß „the greatest proportion of these (predictive) factors deals with the patients' family history" (S. 357). Aus psychoanalytischer Sicht sind bestimmte Konstrukte wie z.B. das Konzept der **Ich-Schwäche** (Kernberg et al., 1972) wertvoll, d.h. solche Kriterien, die sich auf die Persönlichkeitspathologie der Patienten beziehen. Allerdings finden sich kaum valide Versuche, etwa von Psychoanalytikern, im Rahmen empirischer Untersuchungen selbst Operationalisierungen und Untersuchungsinstrumente für persönlichkeitsbezogene Aspekte zu entwickeln. Daß sich auch „einfache" Instrumente zur Messung von Persönlichkeitsfaktoren wie z.B. der Ich-Schwäche eignen, zeigten Kordy und Mitarbeiter (1983).

Zur Frage der **Therapeutenvariablen** ist der Forschungsstand besonders unbefriedigend. Ergebnisse aus den letzten Jahren verweisen jedoch auf die zentrale Bedeutung der Fähigkeit des Therapeuten, eine spezielle zwischenmenschliche Beziehung zu entwikkeln, aufrechtzuerhalten und unter therapeutischen Zielsetzungen zu steuern (vgl. Strupp, 1986). Nach Orlinsky und Howard (1986) stehen ebenfalls alle erfaßten Prozeßmerkmale, die auf eine gute therapeutische Beziehung hinweisen, in einem signifikanten Zusammenhang mit guten Therapieergebnissen. Hinsichtlich der **interaktionellen Variablen** ist ein wesentliches Ergebnis der Literaturauswertung von Luborsky und Mitarbeitern (1971), daß Persönlichkeitsmerkmale des Patienten und des Therapeuten, die unabhängig von der Behandlungssituation erhoben werden, wenig für die Prognose der Behandlungsergebnisse hergeben, „crucial predictive factors may not be sufficiently apparent until the patient and the therapist have had a chance to interact" (Luborsky et al., 1980). D.h.: Informationen, die aus der dyadischen Interaktion zwischen Patient und Therapeut entstammen, tragen möglicherweise am ehesten zu einer Klä-

rung bei, warum manche Patienten profitieren, andere nicht. Mit den noch vorläufigen Ergebnissen zur prognostischen Valenz der „helping alliance" aus dem Penn-Psychotherapy-Project wurde dazu ein wichtiger Schritt geleistet (Morgan et al., 1982).

Die Aufzählung solcher prädiktiver Faktoren könnte fortgesetzt werden, doch ist nicht darüber hinauszukommen, daß sich nur wenige, zum Teil recht komplexe Variablen – wie z.B. die Ich-Schwäche und das Angstniveau – als brauchbar erwiesen haben. An dieser Stelle kommen wir auf unsere oben (vgl. Abschn. 25.4.2) aufgestellte These zurück: Gesucht sind geeignete Beschreibungssysteme für Patienten, Therapeuten und Behandlungen oder die Interaktion zwischen Patienten und Therapeuten im Rahmen einer bestimmen Behandlungstechnik, welche die Variabilität der Behandlungsergebnisse in Relation zu den verwendeten Beschreibungsmerkmalen darstellen lassen. Dieses bislang noch unbefriedigend gelöste konzeptuell-theoretische Problem ist gegenwärtig eine vordringliche Aufgabe der empirischen Ergebnisforschung. Kernberg (1976) hat in seinen Schlußfolgerungen aus den Erfahrungen des aufwendigen und sorgfältigen Menninger-Projektes die folgenden, unseres Erachtens für diese Aufgabe wertvollen Hinweise gegeben. Er schlägt vor, eher wenige, dafür klinisch und theoretisch komplexe Merkmale zu untersuchen, wie eben z.B. die Ich-Schwäche, die sich aus komplexen klinischen Beobachtungen zusammensetzt. Dabei sollen die Variablen Wesentliches des Therapieangebotes erfassen und konzeptuell klar sein, soweit es theoretische wie klinisch-pragmatische Aspekte betrifft; sie müssen praxisnah beobachtbar, klinisch relevant und schlußendlich auch ausreichend bei dem untersuchten Personenkreis („sample") vorhanden sein. Werden solche Gesichtspunkte ernstgenommen, dann lassen sich durchaus brauchbare klinische Gesichtspunkte herausarbeiten, die erlauben, die Behandlungschancen eines Patienten abzuschätzen (Kordy und Senf, 1987; Senf, 1989).

25.7 Spontanremission und Kontrollgruppen

Die in Outcome-Studien beobachteten oder gemessenen Veränderungen eines Patienten werden meist ganz selbstverständlich als Beweis der kausalen Wirksamkeit der angewandten Behandlungsmethode reklamiert. Dem steht gegenüber, daß eine Zwei- oder Drei-Punkte-Erhebung aber nur eine begrenzte Aussage darüber erlaubt, worauf die Veränderungen des Patienten nun tatsächlich zurückzuführen sind.

Außerhalb der Therapie liegende Ereignisse („life events"), Hilfestellungen aus der „natürlichen" Umgebung des Patienten oder die von Ernst (1959) hervorgehobenen „natürlichen Verläufe" psychogener Syndrome müssen berücksichtigt werden, obwohl sie in ihrem Gewicht oft nur schwer eingeschätzt werden können.

In diesem Zusammenhang wird immer wieder das Problem der sog. **Spontanremission** (Spontanheilung) diskutiert. Der Begriff Spontanremission entstammt der Beobachtung, daß sich psychogene Störungen auch ohne gezielte psychotherapeutische Behandlungsmaßnahmen bessern können. Dabei ist nicht gemeint, daß diese Beschwerden ohne irgendwelche Ursachen als eine bloße Funktion der Zeit verschwinden, sondern der Akzent liegt auf „ohne gezielte therapeutische Intervention" (Hartig, 1975).

Die Kontroverse um die Spontanheilung begann mit den provokativen Einwänden Eysencks (1952), der die Wirksamkeit vor allem psychoanalytischer Psychotherapie stark in Frage stellte. Anhand von Arbeiten, die über sog. Spontanheilungen berichteten, errechnete Eysenck eine „Spontanremissionsrate" von 75% nach drei Jahren und sogar von 90% nach fünf Jahren. Eysenck argumentierte, daß die Besserungsquote psychotherapeutischer Behandlungen über der Spontanremissionsrate liegen muß, wenn ihre Wirksamkeit als erwiesen gelten soll. Diese Ergebnisse Eysencks sind inzwischen widerlegt worden (zusammenfassend vgl. Bergin und Lambert, 1978).

Unter den vielen methodischen Argumenten gegen die Untersuchungen von Eysenck soll nur ein Aspekt hervorgehoben werden, da er ein Licht auf das oben diskutierte Problem der Erfolgsbewertung wirft. Eysenck stützte sich u.a. auf Untersuchungen von Denker (1946) an 500 Patienten, die wenigstens drei Monate wegen neurotischer Störungen arbeitsunfähig geschrieben waren und von praktischen Ärzten vor allem mit Beruhigungsmitteln und unterstützendem Zuspruch behandelt wurden. Das Kriterium für Besserung war die Wiederaufnahme einer bezahlten Beschäftigung, was bei diesem Patientenkollektiv dazu führte, daß innerhalb von zwei Jahren 52% der Patienten ihre Arbeit wieder aufnahmen und deswegen als „gebessert" betrachtet wurden. Daß dieses isolierte Kriterium kaum etwas zum tatsächlichen Verlauf aussagt, liegt auf der Hand.

Daß eine psychogene Symptomatik sozusagen spontan, d.h. ohne gezielte therapeutische Einwirkung, verschwinden kann, muß jedoch ebenfalls als gesichert angesehen werden. Dabei ist aber davon auszugehen, daß die spontane Remissionsrate bei verschiedenen Krankheitsbildern beträchtlich variiert. Sie wird für psychosomatische Erkrankungen sehr unterschiedlich angegeben, z.B. von Linneweh (1960) für die essentielle Hypertonie zwischen 40 und 60%, für das Ulcus duodeni zwischen 10 und 25%, Colitis ulcerosa 15 bis 24%, Hyperthyreose 20% und Asthma bronchiale 6,5%.

Auch innerhalb einzelner Krankheitsgruppen sind kaum einheitlich stabile Spontanremissionsraten zu finden. Symptombesserungen z.B. haben nun einmal unterschiedliche Ursachen. Verschiedene Patienten in einer Symptomgruppe unterscheiden sich oft erheblich, was Grad, Dauer ihrer Erkrankung, ihre Persönlichkeitsstruktur und ihre äußere Lebenssituation betrifft, Faktoren, welche die Heilungschancen beeinflussen. Schon unter solchen Gesichtspunkten ist es problematisch, eine sog. spontane Remissionsrate als Basislinie zur Erfolgsbeurteilung zu nehmen.

Auf die Frage, was eigentlich unter diesem „mysteriösen Prozeß" der Spontanremission zu verstehen ist, geben Bergin und Lambert (1978) die lapidare Antwort, „daß wir nicht wissen, was los ist". Sie weisen auf die Ergebnisse sozialpsychologischer Untersuchungen hin, daß therapeutisch wirksame Prozesse nicht notwendigerweise an den professionellen Therapeuten bzw. seine Technik gebunden sind und auch außerhalb des Sprechzimmers ablaufen können. So kann sich das Krankheitsbild eines Patienten spontan bessern, wenn er in seiner alltäglichen Umgebung ein notwendiges Maß an Unterstützung, Entlastung oder Hilfe gefunden hat. Spontanheilungen „may therefore result from seeking and obtaining therapeutic help from nontherapists!" (Bergin und Lambert, 1978). Hier spielt auch ein Problembereich eine Rolle, der gegenwärtig in der Psychotherapie-Prozeßforschung heftig und kontrovers diskutiert wird (Heim, 1981; Strupp, 1975 und 1978): Inwieweit sind die beim Patienten im Verlaufe einer Therapie eingetretenen Veränderungen Folge „spezifischer" oder „unspezifischer" Wirkfaktoren und welche Rolle spielen dabei objektive Lebenseinschnitte („life events")? Ist es für den Therapieerfolg wichtiger, ob ein Therapeut Wärme, verstehende Anteilnahme und Empathie ausstrahlt („unspezifische Faktoren") oder ob er z.B. die Abwehrstruktur des Patienten zutreffend einschätzt und im richtigen Moment eine Übertragungsdeutung gibt – also eine spezifische therapeutische Technik anwendet? Dieses wichtige und im Einzelfall äußerst schwierig zu bestimmende Problemfeld der **Prozeßforschung** wird auch noch durch den Streit der verschiedenen therapeutischen Schulen belastet, die oft gerade anhand dieser Frage die Überlegenheit des jeweils eigenen Vorgehens zu „beweisen" versuchen.

So kontrovers die Diskussion um das Problem der sog. Spontanheilung oder der unspezifischen Wirkfaktoren heute noch geführt wird, so ungelöst ist letztlich auch das Problem der **Kontrollgruppen,** womit versucht wird, Spontanremissionseffekte oder andere validitätsmindernde Variablen in der Ergebnisforschung zu erfassen. Kontrollgruppen tragen dem Umstand Rechnung, daß neben der Therapie auch nicht-therapeutische Faktoren die Veränderungen der Patienten bewirken oder mitbewirken können. Ihr Zweck liegt darin, alternative Erklärungshypothesen für ein Behandlungsergebnis aufzuzeigen oder auszuschließen. Das Grundprinzip ist, parallel zur behandelten Patientengruppe eine weitere Gruppe von Patienten zu beobachten, die keine Therapie erhalten, aber in allen relevanten Variablen vergleichbar sind. Da nicht zu vertreten ist, therapiebedürftige Patienten aus forschungsmethodischen Gründen nicht zu behandeln oder einer als weniger wirksam erachteten Therapie zuzuweisen, setzen sich Kontrollgruppen heute in der Regel aus Patienten zusammen, die auf eine Therapie warten. Hier gibt es prinzipiell zwei Möglichkeiten. Die erste ist, eine Gruppe von Patienten, die auf die Therapie warten, mit einer Gruppe von behandelten Patienten zu vergleichen (Fremdwarte-Kontrollgruppe). Angesichts der unübersehbaren Zahl von therapierelevanten Variablen (wie z.B. sozio-ökonomische Variablen, Symptome, Krankheitsgeschichte, Motivation oder Persönlichkeitsstruktur) ist es aber fraglich, ob diese an einer klassischen experimentellen Technik orientierten Bedingungen überhaupt zu verwirklichen sind. Wenn die Stichproben vergleichbar sein sollen und aus den Untersuchungsergebnissen generalisierende Schlüsse gezogen werden sollen, müßten die Stichproben relativ groß sein, was in der Psychotherapieforschung jedoch so gut wie nicht realisierbar ist. Auch wird in der Literatur zur Psychotherapieforschung darauf hingewiesen, daß solche Kontrollgruppen schon deswegen nicht aufgestellt werden können, weil diese Patienten nicht deswegen, weil sie zur Kontrolle dienen, bereit sind, ihre Leiden zu ertragen und auf andere, z.B. nicht-professionelle Hilfe zu verzichten. Ebenso wird argumentiert, daß die mit einem Forschungsprojekt verbundenen Interventionen wie Erstinterview, psychodiagnostische Untersuchung, Abschlußgespräche usw. schon therapeutische Einwirkungen auf den Patienten darstellen, so daß kaum davon ausgegangen werden kann, daß in den Kontrollgruppen wirklich unbehandelte Patienten sind. Dieser Einwand kann auch gegen die zweite prinzipielle Möglichkeit, die Eigenwarte-Kontrollgruppe (vgl. Tab. 25–1) geltend gemacht werden. Die Patienten, die eine Therapie erhalten sollen, warten einige Zeit, und die Veränderungen während der Wartezeit werden mit denen der Therapiezeit verglichen; d.h. die behandelten Patienten stellen ihre eigene Kontrollgruppe dar. Der Vorteil dieser Vorgehensweise liegt zum einen in der Ökonomie des Verfahrens, zum anderen in der optimalen Parallelität zwischen der Kontrollstichprobe und der Versuchsstichprobe, da es sich um die gleichen Patienten handelt. Nachteilig ist dagegen die Notwendigkeit einer mehrfachen Testwiederholung.

Die Diskussion um das Kontrollgruppenproblem in der Literatur ist kontrovers. Einige Autoren schlagen vor, Kontrollgruppen ganz wegfallen zu lassen und statt dessen nur noch verschiedene Behandlungsgruppen zu vergleichen. Andere Autoren sehen einen Ausweg darin, statt einer Kontrollgruppe unterschiedlich viel Therapie als unabhängige Variable zu verwenden, um spezifische gegen weniger spezifische Effekte isolieren zu können (Frank, 1958). Dann wieder wird argumentiert, daß Kontrollgruppen unbedingt notwendig sind, oft mit entsprechenden Vorschlägen.

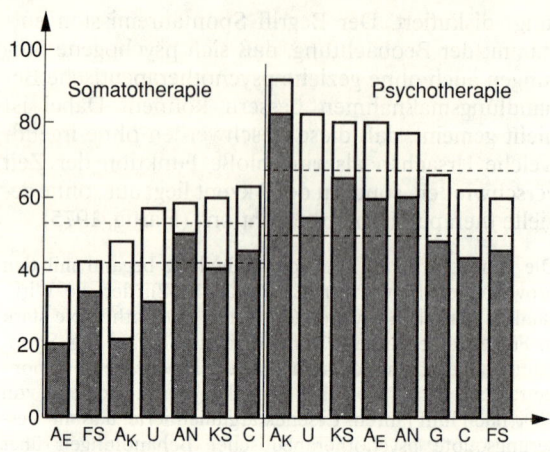

Weiße Säulen: Besserungsquoten der Zielsymptomatik (+)
Gerasterte Säulen: Quoten für „gute Besserungen" (++)

A_E = Asthma Erwachsene   FS = Funktionelles Syndrom
A_K = Asthma Kinder   G = gemischte Diagnosen
AN = Anorexia nervosa   KS = Kopfschmerz
C = Colitis ulcerosa   U = Ulkus

Abb. 25–1. Mittelwerte der Besserungsquoten nach Symptomgruppen (aus: Rohrmeier, 1982).

die Gefahr gegeben, daß eine Exaktheit durch Zahlen und Prozente vorgetäuscht wird, wo bestenfalls grobe Schätzungen und Tendenzen angenommen werden dürfen (dies gilt auch insbesondere für das hier benutzte Kriterium „Besserung" [+] und „gute Besserung" [++]). Der Leser sollte sich also der geringen Präzision und oft nur vagen Tendenz der Aussagen dieses Anhangs bewußt bleiben. Es schien uns aber dennoch sinnvoll, dem Interessierten Zahlenmaterial vorzulegen, das in seiner Vorläufigkeit die Kinderkrankheiten dieser noch jungen Forschungsrichtung nicht verleugnen kann. Wir stützen uns hinsichtlich der Zahlen u.a. auf Rohrmeier (1982; vgl. Abb. 25–1), bei dem Einzelheiten der verwandten Literatur, Bearbeitung und Verrechnung, insbesondere aber auch eine kritische Würdigung der methodischen Probleme der berücksichtigten Untersuchungen nachgesehen werden können.

25.8 Psychotherapeutische Behandlungsergebnisse bei verschiedenen Krankheitsbildern

Besinnt man sich auf die in Abschnitt 25.5 dargelegten Standards, an denen moderne Katamnesauntersuchungen gerade auch im Bereich der Psychosomatischen Medizin sich messen lassen müssen, dann ist es problematisch, hier noch eine notwendigerweise extrem verkürzende summarische Zusammenfassung über Behandlungsergebnisse bei einigen Erkrankungsformen anzufügen. Eine solche summarische Ergebnisdichte kann Gewicht und Wirkung der für jede einzelne Untersuchung ganz unterschiedlich bedeutsamen Variablen (wie z.B. Patientenselektion, angewandte Therapietechnik, Besserungskriterien und vieles andere) nicht mehr wiedergeben. Damit ist

25.8.1 Anorexia nervosa

Das Krankheitsbild (vgl. Kap. 38.2) ist hinsichtlich Ätiologie und Verlauf (Spontanheilung versus Chronifizierung) äußerst heterogen und wird von den meisten Autoren unter verschiedenen diagnostischen Kriterien untersucht. Es ist oft unklar, ob nur die somatischen Symptome (Gewichtsverlust, Amenorrhö, Obstipation, Erbrechen) oder auch die tiefgreifenden (aber schwerer operationalisierbaren) psychosozialen Störungen bei der Ergebnisforschung berücksichtigt wurden. Bei der Erstmanifestation haben ältere Patientinnen eine eindeutig ungünstigere Prognose als pubertätsnah erkrankte Jugendliche, bei denen z.B. Minuchin (1974) mit Hilfe seiner Familientherapie eine kaum glaubliche Quote von über 80% Heilungen berichtet hat, die in der Tendenz von anderen – wenn auch nicht so ausgeprägt – bestätigt wird (Morgan und Russel, 1975; Selvini-Palazzoli, 1974; Weber und Stierlin, 1989; Remschmidt et al., 1988;

Martin, 1985). Dazu paßt, daß in einem kontrollierten Vergleich von Familientherapie und „supportiver Einzeltherapie" bei Patientinnen ohne chronischen Verlauf und einer Erstmanifestation der Erkrankung vor dem 19. Lebensjahr tendenziell die Familientherapie günstigere Ergebnisse bewirkte als die supportive Einzeltherapie, die wiederum bei den älteren Patientinnen wirkungsvoller war (Russel et al., 1987). Insgesamt hat sich die Einbeziehung der Familie in den Gesamtbehandlungsplan bewährt (vgl. Petzold, 1977).

Die von Rohrmeier mitgeteilten Werte von 78,6% „Besserung" und 59% „gute Besserung" stützen sich auf 364 zumeist mit „eklektischer Psychotherapie" (vorwiegend tiefenpsychologisch fundiert) behandelte Patienten und scheinen uns zu optimistisch. Bei einer Nachuntersuchung aus der Heidelberger Psychosomatischen Klinik, in der sicher eine Selektion besonders schwer Erkrankter behandelt wurde (Becker et al., 1981), waren bei der Katamnese nach strengen Kriterien von 38 Patienten lediglich 6 als „geheilt", 8 als „gut gebessert" und 7 als „befriedigend gebessert" eingestuft worden, also insgesamt nur 56%. Dabei bestätigte sich die alte klinische Erfahrung, daß die Prognose um so schlechter wird, je länger der Zeitraum zwischen Krankheitsmanifestation und Beginn einer psychotherapeutischen Behandlung sich ausdehnt. (Dies ist aber erneut ein sehr heterogenes Kriterium.) Wichtige Aufschlüsse ergaben sich aus der 10-Jahres-Katamnese-Untersuchung von Engel et al. (1987), die 93% mit Hilfe eines speziellen Programms (Bettruhe, Phenothiazine, Sondennahrung und strikte „väterliche" Führung) stationär behandelter Anorexia-nervosa-Patienten einschließt. Das Ergebnis: 53% (114 Pat.) „praktische Heilungen"; bei 18% (39 Pat.) deutliche und bei 9% (9 Pat.) schwere Störungen. Die Zahl der Verstorbenen lag mit 14% (30 Pat.) erschreckend hoch. Interessant ist, daß eine Erweiterung der oben genannten Therapie um einige verstärkt psychotherapeutische Elemente bei einem anderen Patientenkollektiv (39 Pat.) nicht zu einer Erhöhung der Heilungschancen (ebenfalls 53% „praktische Heilungen") geführt hat.

Es wurden auch überzeugende Mitteilungen von erfolgreichen verhaltenstherapeutischen Ansätzen (Halmi et al., 1973) (insbesondere mit dem Ziel der Gewichtszunahme und des Essenstrainings) bekannt, die sich z.B. als initiale Phase im Rahmen einer stationären Therapie bewährt haben. Stationär eingeleitete Therapien scheinen erfolgreicher zu sein als primär ambulante; klassische Psychoanalysen kommen kaum noch zur Anwendung. Ein wichtiges Ergebnis: In der Gruppe der nicht psychotherapeutisch Behandelten findet sich – allerdings nach längerer Katamnesezeit – eine vergleichsweise deutlich höhere Zahl von Verstorbenen (vgl. Engel et al., 1987).

25.8.2 Bulimia nervosa

Für das erst seit kurzem neu in das Blickfeld geratene Krankheitsbild der Bulimie hat Garner (1987) in einer Übersicht herausgearbeitet, daß bislang über erste, vielversprechende Ansätze hinaus (z.B. Lacey, 1983; Brotman et al., 1988) kaum Katamnesestudien vorliegen, die einer kritischen Beurteilung standhalten. Insgesamt scheint hier eine Therapie günstig, die jeweils Elemente der Essensberatung, der Gruppenerfahrung und der konfliktzentrierten Einzelbehandlung vereinigt.

25.8.3 Asthma bronchiale

Nur bei etwa einem Drittel der erwachsenen Asthmatiker (vgl. Kap. 42) spielen seelische Einflüsse eine dominierende,

bei einem weiteren Drittel eine mitgestaltende Rolle bei der Entstehung und Aufrechterhaltung der Erkrankung (Rees, 1956). Das kindliche Asthma hat eine günstigere Prognose mit nicht seltenen Spontanremissionen. Je nach Schwere des Verlaufs wird nur in Ausnahmefällen auf eine die Psychotherapie begleitende medikamentöse Behandlung verzichtet werden können.

Die von Rohrmeier (1982) für das Asthma unvollständig zusammengestellten Daten stützen sich auf 444 Erwachsene und 180 Kinder und erreichen mit 80–90% Besserung und 60–80% guter Besserung die höchsten Werte von allen „klassischen" psychosomatischen Erkrankungen. Vermutlich spiegelt sich darin jedoch eher der Selektionseffekt, daß nämlich bei den schwer Erkrankten oft kein psychotherapeutisches Behandlungsangebot erfolgt. Hinsichtlich der angewandten therapeutischen Technik scheint sich besonders die psychoanalytische Einzel- (Baerwolf, 1958) und Gruppentherapie (Groen und Pelser, 1960; Deter, 1986) bewährt zu haben, aber auch entspannungsfördernde Maßnahmen, wie z.B. das autogene Training und/oder Atemschulung (Curtius, 1965).

Wertvoll ist die Studie von Groen und Pelser (1960), die in einem kontrollierten Gruppenvergleich die Überlegenheit einer zusätzlichen (psychoanalytischen Gruppen-) Psychotherapie gegenüber einer rein medikamentösen Therapie (jeweils mit und ohne Kortikoide) belegen konnten. Diesen Ansatz hat Deter (1986) aufgenommen und weiterentwickelt. Er konnte in einer umfangreichen Kosten-Nutzen-Analyse eindrucksvoll nachweisen, daß die Patienten, die (1 Jahr) mit einer „krankheitsorientierten Gruppentherapie" (zusätzlich) behandelt worden waren, im Vergleich zu der traditionell therapierten Kontrollgruppe wesentlich günstiger abschnitten. Bei der Katamnese ergab sich u.a. eine erhebliche Reduktion krankheitsbedingter Arbeitsfehltage und stationärer Behandlungsnotwendigkeit. Daraus errechnet sich nach Deter (1986) eine Kostenersparnis zugunsten der Psychotherapie-Gruppe von 1:5 – ein wichtiges Ergebnis, das in modifizierter Form auch für viele andere Krankheitsgruppen gelten dürfte.

25.8.4 Ulcus pepticum

Es ist erstaunlich, wie wenig Literatur – nur drei Arbeiten über insgesamt 51 Patienten bei Rohrmeier (1982) – zu einer der „klassischen" psychosomatischen Erkrankungen (vgl. Kap. 43) vorliegt, von der man geradezu sagen kann, sie sei aus der Psychotherapieforschung wieder ausgewandert (Nemiah, 1971). Dies liegt wohl an der relativ hohen Remissionsrate von etwa 80% nach Jores (1960) auf unspezifische Therapie hin (Trennung von Familie und Arbeitsplatz, Bettruhe, „Diät", Antazida), aber auch an den sehr optimistisch beurteilten Erfolgen chirurgischer Interventionen, die im Falle einer Chronifizierung meist vorgenommen werden. Bei rein internistisch-konservativer Therapie waren jedoch höchstens 20% der Erkrankten mehr als 5 Jahre beschwerdefrei (Nemiah, 1971; Römecke, 1954). Gegenüber der chirurgischen Ergebnisstatistik von Hackethal (1960) mit etwa 52% ausgeprägten und insgesamt 70% Besserungen bei einer (damaligen) Letalität von 10% (ungünstige Selektion!) schneiden die von Rohrmeier (1982) ermittelten Werte für (zumeist analytische) Psychotherapien mit etwa 55% guter und insgesamt 82% Besserung bei einer Katamnesezeit von fast 6 Jahren nicht schlecht ab. Es handelt sich aber dabei um ein nur kleines, prognostisch sicher günstigeres Kollektiv. Eine wichtige Untersuchung stammt von Orgel (1958), der 15 Patienten mit langjähriger Analyse im klassischen Setting behandelte: Drei von ihnen brachen innerhalb der ersten 21 Stunden, zwei weitere nach erfolgter Symptombesserung ab. Bei der Katamnese nach über 10 Jahren waren alle 10 Patienten, die die Analyse regulär abgeschlossen hatten, symptomatisch geheilt und „strukturell" gebessert, während alle anderen Rückfälle hatten und unter Beschwerden litten – davon drei trotz einer Operation. Interessant ist auch eine Untersuchung von Colgan et al. (1988), die in einem kontrollierten Vergleich von 30 Patienten mit häufig rezidivierendem Ulcus duodeni die rezidivprophylaktische Wirksamkeit einer zusätzlichen „Hypnotherapie" prüften. Bei der 1-Jahres-Katamnese hatten nur 53% der Psychotherapie-Gruppe, aber alle (100%) unbehandelten Patienten ein Rezidiv entwickelt. Immer mehr setzt sich die Erkenntnis durch, daß auch seelische Determinanten den Erfolg und die Prognose chirurgischer Maßnahmen entscheidend mitbeeinflussen. So erfaßten Möhlen und Mitarbeiter (1982) sowie Brähler und Möhlen (1988) für das Ulcus pepticum Persönlichkeitsmerkmale und typische Verarbeitungsmuster der Krankheitssituation, aus denen sich für Erfolg und Mißerfolg der Operation prädiktiv benutzbare Hinweise und Kriterien ableiten ließen.

25.8.5 Colitis ulcerosa

Der wechselhafte, sehr oft chronische Verlauf sowie die stets drohende, folgenreiche chirurgische Intervention in besonders schweren Fällen stehen auch bei der Colitis ulcerosa einer einfachen Ergebnisbeurteilung entgegen (vgl. Kap. 44.1). So spiegeln die von Rohrmeier (1982) zusammengestellten Übersichtszahlen mit den vergleichsweise niedrigen Erfolgswerten auch die Schwere des Krankheitsbildes. Bei insgesamt 518 Psychotherapiepatienten, die im Durchschnitt 8,2 Jahre nach der Behandlung untersucht wurden, waren nur 31,7% ohne Operation symptomfrei geblieben; 44% konnten als gut, 66% (ohne Operation nur 55,2%) als gebessert eingestuft werden. Die Vergleichswerte der nur somatisch Behandelten sind ähnlich, bis auf zwei wesentliche Unterschiede: nur 20,9% waren symptomfrei ohne Operation; trotz kürzerer Katamnesezeit (5,2 Jahre) waren mehr Patienten verstorben.

Eine der sorgfältigsten Katamneseuntersuchungen überhaupt verdanken wir der Arbeitsgruppe um O'Connor und Karush (Groen und Pelser, 1960; Karush et al., 1968; O'Connor et al., 1964; Daniels und O'Connor, 1962), die eine (nach Alter bei Krankheitsbeginn, Schwere der somatischen Symptomatik, Steroidtherapie und Geschlecht) parallelisierte Gruppe von jeweils 57 mit und ohne psychoanalytisch orientierter Therapie behandelte Patienten 25 Jahre kontinuierlich verfolgte. Die „Psychotherapie-Gruppe" mit insgesamt 19 Schizophrenie-Fällen bildete eine deutlich ungünstigere Selektion. Dabei ergab sich, daß in der „Psychotherapie-Gruppe" deutlich weniger Operationen und Todesfälle auftraten und die symptomatischen sowie rektoskopischen Befunde günstiger waren, wenn man die in allen Parametern ungünstiger verlaufende Untergruppe der Schizophrenen getrennt beurteilt. Eine andere wichtige Studie stammt von Weinstock (1962), der von 28 schweren chronifizierten Colitis-ulcerosa-Kranken, die je zur Hälfte mit analytischer Psychotherapie oder klassischer Psychoanalyse langfristig behandelt worden waren, nach durchschnittlich 9 Jahren immerhin 22 als völlig geheilt und nur 6 mit unbefriedigendem Ergebnis vorfand, wobei beide Verfahren vergleichbar gut abschnitten. Überhaupt scheint die Colitis ulcerosa eine Domäne der verschiedenen psychoanalytischen Verfahren zu sein. Bei weniger verbalisierungsfähigen Kranken können aber auch andere, vor allem entspannungsfördernde Techniken hilfreich sein (Feiereis, 1967).

25.8.6 Essentielle Hypertonie

Hinsichtlich der Therapieergebnisse bei der essentiellen Hypertonie (vgl. Kap. 41.3) stellt der Mangel an langfristigen Katamneseuntersuchungen das zentrale Problem dar, zumal diese heterogene Erkrankung situativ enormen Schwankungen unterliegt und offensichtlich auf nahezu jede Form von Psychotherapie, Suggestion oder Plazebobehandlung zunächst günstig reagiert, ohne daß der Erfolg stabil bleibt (Agras et al., 1987). Vor allem psychoanalytisch orientierte (Gaus et al., 1983; Quint, 1967; Wolff und Wolf, 1951; Wyss, 1955) oder entspannungsfördernde Techniken wie das autogene Training (Schultz, 1966), Biofeedback (Blanchard et al., 1979; Donald et al., 1957; Patel und Datey, 1974; Walsh et al., 1977) oder sogar transzendentale Meditation (Seer und Raeburn, 1980) bzw. Yoga (Patel und Datey, 1975) wurden (allerdings meist nach viel zu kurzer Nachuntersuchungszeit) als erfolgreich beschrieben. Bei kritischer Beurteilung der verschiedenen Untersuchungen läßt sich bestenfalls bei einem Viertel der Fälle eine gute und bei der Hälfte eine Besserung erzielen.

Vielleicht ist es auch wichtiger, die im engeren Sinne psychotherapeutische in die gesamte medizinische und soziale Beratung und Betreuung einzubetten. Dies haben neuerdings Kaluza et al. (1986) durch ein komplexes verhaltenstherapeutisches Kurztherapieprogramm (mit den Elementen medizinische Information, progressive Muskelrelaxation und autogenes Training als Entspannungstherapie sowie Verhaltenstherapie) eindrucksvoll nachgewiesen. Bei der 2-Jahres-Katamnese hatte die Psychotherapie-Gruppe gegenüber der Kontrollgruppe eine deutliche Blutdruckreduktion bei gleichzeitiger Reduktion antihypertensiver Pharmakotherapie erzielt, was auf die positive Wirkung der Verbindung von Pharmakotherapie mit Psychotherapie verweist (Lehnert et al., 1987).

Ein Beispiel für die Problematik und die Bewertungsschwierigkeiten katamnestischer Untersuchungen bei der essentiellen Hypertonie bildet die oft zitierte Arbeit von Wyss (1955), der 14 Patienten mit „juveniler Hypertonie" etwa ein Jahr lang mit Psychoanalyse behandelte. Bei sechs Patienten waren Hypertonus und neurotische Symptomatik verschwunden; bei fünf Patienten war der Hypertonus ebenfalls verschwunden, obwohl die neurotische Symptomatik weiterbestand und die Therapie vorzeitig abgebrochen worden war. Drei Patienten wurden hinsichtlich ihrer Neurose als erfolgreich behandelt eingestuft – hatten aber ihren Hypertonus nicht verloren. Dabei muß man noch zusätzlich die günstige Spontanremission der juvenilen Hypertonie mitbedenken.

25.8.7 Funktionelle Syndrome

Funktionelle Syndrome, mit deren Verlauf, Prognose und Behandlungsergebnissen sich vor allem Cremerius (1962, 1968) kritisch auseinandergesetzt hat, stellen eine negativ definierte, sehr heterogene Gruppe von organzentrierten Befindlichkeitsstörungen ohne morphologisch faßbaren Organbefund dar (vgl. Kap. 31–34 und Kap. 58). Sie sind in ihrer Spontanprognose (Symptomverschiebung!), aber auch hinsichtlich des psychotherapeutischen Behandlungserfolges keineswegs so günstig, wie dies weithin angenommen wird. Rohrmeier (1982) errechnet einen Wert von 45,7% guten und 61,1% einfachen Besserungen; nimmt man jedoch die wenig genau dokumentierten Zahlen von Laberke (1965) heraus, dann errechnet sich noch ein deutlich ungünstigerer Wert von nur 31,7% guten und 42,8% einfachen Besserungen (von 610 Patienten). Diese Zahlen liegen also in der Tendenz ungünstiger als bei den klassischen psychosomatischen Erkrankungen!

Für präzisere Aussagen fehlen bislang noch Untersuchungen der häufigeren Syndrome. So liegen z.B. über Therapieergebnissen bei der Herzphobie seit der klassischen Studie von Christian und Mitarbeitern (1966) keine katamnestischen Untersuchungen vor. Dagegen gibt es eine reiche Literatur zu Migräne und anderen Kopfschmerzformen, die eine gewisse Domäne des Biofeedbacks und anderer entspannungsfördernder Therapieformen zu sein scheinen. Günstige, zum Teil länger anhaltende Behandlungsergebnisse mit Hilfe des Biofeedbacks berichten Labbé und Williamson (1984), Diamond und Montrose (1984), Blanchard et al. (1987) sowie Ellertsen (1987), das nach der Vergleichsuntersuchung von Collet et al. (1986) dem autogenen Training überlegen sein soll. Aus der Vergleichsuntersuchung von Andrasik et al. (1984) läßt sich folgern, daß hier die Aufrechterhaltung eines regelmäßigen Kontaktes zum Patienten vielleicht der wichtigere therapeutische Wirkfaktor ist als die spezielle Therapieform (vgl. Abb. 25–1).

25.8.8 Myokardinfarkt

In letzter Zeit hat es tastende Versuche gegeben, die wachsende Kenntnis über die Bedeutung von Persönlichkeitsfaktoren für Entwicklung, Verlauf und Prognose des Myokardinfarktes auch therapeutisch zu nutzen (Fielding, 1980; Gruen, 1975; Hahn, 1971; vgl. Kap. 41.2). Die ersten katamnestischen Ergebnisse (Ibrahim et al., 1974; Rahe et al., 1975) sind ermutigend, bedürfen aber noch einer Überprüfung an größeren Kollektiven.

Ungeklärt ist auch noch die Frage, inwieweit die sog. „Koronargruppen" einen wirksamen Präventiveffekt herbeiführen können. Ist ein Infarkt eingetreten, so empfiehlt es sich nach Byrne (1975), möglichst zeitig mit einer psychotherapeutischen Hilfestellung einzusetzen und nicht erst eine weitgehende körperliche Wiederherstellung abzuwarten.

Van Dixhoorn et al. (1987) sowie Powell und Thoresen (1988) unterstreichen mit ihren Nachuntersuchungen die These, daß ein zusätzliches kombiniertes Therapieprogramm mit den Elementen Beratung, Entspannung und Psychotherapie, wenn es zu einer Reduktion des sog. „Typ-A-Verhaltens" führt, den vielleicht wirksamsten Schutz gegen das Auftreten eines Reinfarktes darstellt.

Differentielle, auf die persönlichkeitsspezifischen Merkmale bestimmter Untergruppen der heterogenen nosologischen Diagnose „Myokardinfarkt" zugeschnittene Therapieempfehlungen (Altmann-Herz et al., 1983) bedürfen noch katamnestischer Überprüfung.

26 Psychosomatische Medizin in der Praxis des niedergelassenen Arztes

Wolfgang Wesiack

26.1 Einleitung

Da der Erstkontakt zwischen Patient und Arzt in der Regel in der Praxis eines niedergelassenen Arztes stattfindet, ist es zweckmäßig, auch dieses Geschehen unter dem Gesichtswinkel der Psychosomatischen Medizin zu betrachten, denn wir werden bei diesem Arzt der ersten Linie, wie R. N. Braun ihn treffenderweise nennt, manche Phänomene beobachten können, die sich später nach längerer Patientenkarriere mehr oder weniger stark zu verändern pflegen. Außerdem wird die überwiegende Mehrzahl der Patienten nach wie vor von niedergelassenen Ärzten und nicht von Klinikärzten behandelt, die ein immer schon vorausgelesenes Krankengut zu Gesicht bekommen.

Für den Leser mag es vielleicht verwirrend sein, daß hier etwas verallgemeinernd von der Praxis des niedergelassenen Arztes und nicht von der des Arztes für Allgemeinmedizin gesprochen wird. Das hat seinen Grund darin, daß sich heute meist, vom flachen Land abgesehen, die Zahl der niedergelassenen Allgemeinärzte und die der Fachärzte die Waage hält, ja vielerorts bereits die Fachärzte zahlenmäßig überwiegen, und daß heute viele niedergelassene Fachärzte, z.B. Internisten, Gynäkologen und Pädiater, die gleiche Tätigkeit ausüben wie die Ärzte für Allgemeinmedizin, mit der einzigen Ausnahme, daß ihr Patientenkreis durch ihr Fachgebiet etwas eingeengt ist. Bei den nachfolgenden Betrachtungen werden wir also in erster Linie an das Tätigkeitsfeld des Arztes für Allgemeinmedizin denken, dabei aber nicht aus den Augen verlieren, daß auch viele niedergelassene Fachärzte faktisch Allgemeinmedizin betreiben und somit auch die gleichen Probleme wie die Ärzte für Allgemeinmedizin haben. Inzwischen ist in der BRD eine Gebietsbezeichnung „Arzt für Allgemeinmedizin" geschaffen worden, zu deren Erlangung, analog zu anderen Gebietsbezeichnungen, eine mehrjährige Weiterbildung erforderlich ist. Zur Zeit werden psychosomatische Lehrinhalte erarbeitet, die in Zukunft obligatorisch in die Weiterbildung zum Arzt für Allgemeinmedizin eingebaut werden sollen.

Nach R. N. Braun (vgl. Abb. 26–1) erlebt der „Durchschnittsmensch" täglich mehrmals flüchtige Symptome, die von ihm nicht als Krankheiten gedeutet werden (E). Im Laufe seines Lebens erlebt er etwa 600 „Gesundheitsstörungen" (D), von denen die meisten ohne ärztliche Hilfe ablaufen. Davon gelangen ungefähr 140 an die Ärzte (C). Etwa zwanzigmal im Leben kommt es zu spezialärztlichen Behandlungen innerhalb und außerhalb von Krankenhäusern (B). Schließlich führt dann eine Erkrankung den Tod herbei (A). Dieses Schema verdeutlicht meines Erachtens recht gut, welche große Bedeutung bei der Krankenversorgung dem Arzt der ersten Linie zufällt.

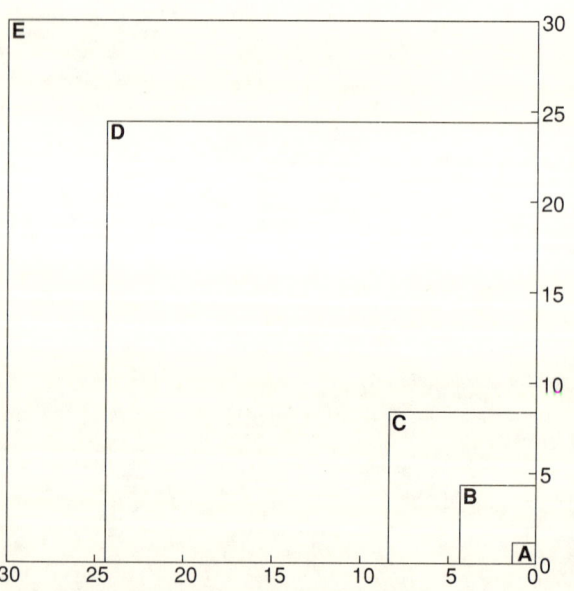

E Symptome, die von Patienten nicht als Erkrankungszeichen gewertet werden, und verschiedenartigste unentdeckte, aber an sich faßbare Gesundheitsstörungen

D Subjektiv als Krankheit empfundene Zustände, die ohne ärztliche Beratung ablaufen

C Diagnostische Versorgung durch den praktischen Arzt

B und Spezialisten bzw. das Krankenhaus

A Erkrankung führt zum Tod

Abb. 26–1. Schema über die Häufigkeit von Gesundheitsstörungen im Leben eines „Durchschnittsmenschen", der 70 Lebensjahre erreicht (aus Braun: Lehrbuch der ärztlichen Allgemeinpraxis, Urban & Schwarzenberg, München – Berlin – Wien 1970).

Obwohl natürlich zwischen den Praxen der niedergelassenen Ärzte erhebliche Unterschiede bestehen, die wir nicht nur zwischen den Ärzten für Allgemeinmedizin und den verschiedenen Fachärzten, sondern auch zwischen den Praxen der gleichen Fachrichtung feststellen können, abhängig vom Kenntnisstand und den Motivationen des jeweiligen Praxisinhabers, aber auch vom Standort der Praxis und vom Krankengut (z.B. Großstadt-, Kleinstadt- oder Landpraxis usw.), sei es wegen vieler Gemeinsamkeiten doch erlaubt, die sich in mancherlei Beziehungen ähnelnden Probleme der niedergelassenen Ärzte im Kontrast zu jenen der Klinik zu behandeln.

Zehn Punkte scheinen mir in dieser Hinsicht von besonderer Bedeutung und erwähnenswert zu sein:
1. Der Erstkontakt zwischen Arzt und Patient findet in der Regel in der Praxis des Arztes in der allen zugänglichen Sprechstunde statt, ausnahmsweise auch außerhalb der Sprechstunde und in der Wohnung des Patienten.
2. Die Anzahl der zu untersuchenden und zu behandelnden Patienten kann von vornherein nicht oder nur unzulänglich begrenzt werden, wodurch ein enormer Zeitdruck entsteht.
3. Die Patienten kommen noch weitgehend unausgelesen zum Arzt der ersten Linie, wodurch sich sein Tätigkeitsfeld enorm erweitert.
4. Er muß zunächst auf kompliziertere technische Hilfsmittel sowie auf kollegialen und fachärztlichen Rat verzichten und muß trotzdem weitreichende Entscheidungen fällen.
5. Dies zwingt ihn, das Wesentliche bzw. das Problem des Patienten zu erkennen und sich darauf zu konzentrieren.
6. Er steht dem Lebens- und Tätigkeitsbereich seiner Patienten in der Regel viel näher als der Kliniker und ist häufig von vornherein mit dem ganzen persönlichen Hintergrund der Patienten vertraut.
7. Bei einer großen Anzahl seiner Patienten kann er Langzeitbeobachtung und Langzeitbetreuung betreiben.
8. Dadurch entsteht eine sehr starke persönliche Beziehung und Bindung zwischen Arzt und Patient.
9. Diagnostik und Therapeutik stehen bei ihm in noch viel engerem Zusammenhang als beim Spezialisten und Kliniker. Der diagnostisch-therapeutische Zirkel (vgl. Kap. 1) bildet bei ihm noch eine Einheit.
10. Durch rechtzeitiges Eingreifen vermag er Chronifizierungen zu verhindern. Seine Bedeutung für die Prophylaxe, die Früherkennung und die Frühbehandlung von somatischen, psychosomatischen und neurotischen Erkrankungen ist enorm und noch keineswegs voll ausgeschöpft.

Diese kurze Aufzählung der Tätigkeitsbereiche der Ärzte der ersten Linie gibt uns schon einen Einblick in das Aufgabenfeld des niedergelassenen Arztes. Er ist dem Kliniker gegenüber einerseits im Nachteil, andererseits aber auch im Vorteil. Die große Zahl der unausgelesen zu behandelnden Patienten, der dadurch hervorgerufene Zeitdruck sowie das weitgehende Fehlen komplizierter technischer und spezieller personeller Hilfen ist sicher als Nachteil zu bezeichnen. Daß demgegenüber die Einheit von Diagnostik und Therapeutik, die Nähe der Lebensbereiche, die viel engere persönliche Beziehung, die Möglichkeit der Langzeitbeobachtung und -betreuung und die Möglichkeit, Prophylaxe, Früherkennung und Frühbehandlung zu betreiben, nicht als eindeutige Vorteile erkannt und genützt werden, liegt einerseits wohl an einer zu einseitig klinischen und vorwiegend somatisch orientierten Ausbildung auch der niedergelassenen Ärzte und andererseits an einem Krankenversicherungssystem, das diese Gesichtspunkte, zumindest soweit sie sich auf psychosoziale Aspekte beziehen, rigoros unterdrückt.

Die unter Punkt 5–10 angeführten Merkmale allgemeinärztlicher Tätigkeit lassen den niedergelassenen Arzt in besonderem Ausmaß als Psychosomatiker in Erscheinung treten, und zwar als einen Psychosomatiker, der nur über sehr wenig Zeit verfügt und sich deshalb nach Möglichkeit auf das Wesentliche bzw. das Problem des Patienten beschränken muß. Ich betone einschränkend „nach Möglichkeit", denn was das Wesentliche bzw. das Problem des Patienten ist, läßt sich meist nicht von vornherein und auf Anhieb feststellen.

Wir wollen nun diese zehn Punkte, in denen sich die Tätigkeit des Praktikers von der des Klinikers besonders unterscheidet, nachfolgend im einzelnen behandeln.

26.2 Der Erstkontakt zwischen Patient und Arzt (ad 1)

Wie bereits erwähnt, findet der Erstkontakt zwischen Patient und Arzt und damit die ärztliche Basisversorgung in der Regel bei einem niedergelassenen, meist bei einem Arzt für Allgemeinmedizin statt. Für die Psychosomatische Medizin kann die Bedeutung des Erstkontaktes gar nicht hoch genug veranschlagt werden, weil, unserem Situationskreismodell entsprechend (vgl. Kap. 1 und 17), erwartet werden muß, daß jede Änderung der Situation, also auch die erste Kontaktaufnahme zwischen Arzt und Patient, das Krankheitsgeschehen beeinflussen wird. Aus der Interaktion schlechthin, als integrierendem Bestandteil der „Situation", gewinnt aber der Erstkontakt, die erste Interaktion zwischen Arzt und Patient deshalb eine so herausragende Bedeutung, weil sie den Patienten und sein Krankheitsgeschehen oft noch in einem relativ plastischen Stadium der Krankheitsentwicklung, gewissermaßen in statu nascendi der Krankheit antrifft, und weil hier erstmals der „signifikante andere" (vgl. Kap. 1) in die Beziehung zum Patienten eintritt. Balint (1957) spricht deshalb vom noch „unorganisierten" Krankheitsprozeß, der sich dann allmählich um ein Symptom herum organisiert.

So wird z.B. der Krankheitsverlauf einer jungen Frau, die bei Abwesenheit des Ehemannes plötzlich an nächtlichen Herzbeschwerden und Angstzuständen erkrankt, ein anderer sein, je nachdem, ob sie ihre Symptomatik mit Hilfe des erstberatenden Arztes um ihre Beziehung zum Ehemann oder aber um ein ebenfalls bestehendes akzidentelles Herzgeräusch „organisiert". Im ersten Fall wird der Bearbeitung der Beziehungs- und Konfliktproblematik nichts im Wege stehen. Im zweiten Fall ist sie zunächst „herzkrank". Bei einer gründlichen körperlichen Untersuchung der Patientin erhalten diese Untersuchungsmaßnahmen dann im ersten Fall die Bedeutung von „um nichts zu übersehen", im zweiten Fall aber „um das Ausmaß des Herzfehlers" zu bestimmen. Das ist aber keineswegs dasselbe, sondern beinhaltet zwei für den Patienten und sein Schicksal oft entscheidend verschiedene Bedeutungen.

An diesem wie auch an ähnlichen Beispielen aus der ärztlichen Praxis, die an der „ersten ärztlichen Linie" sehr häufig sind, läßt sich die Verantwortung des erstbehandelnden Arztes gut erkennen, die dann entweder zu einer erfolgreichen Sofortbehandlung oder aber in eine den Patienten, die Gesellschaft und die Krankenkassen belastende Chronifizierung führen kann. Obwohl natürlich das Übersehen eines bedeutsamen Befundes und eines wichtigen diagnostischen Aspektes für den Patienten immer von Nachteil, manchmal sogar verhängnisvoll ist, besteht bei der gegenwärtigen Struktur der ärztlichen Versorgung meines Erachtens heute eine größere Gefahr, relevante psychosoziale als somatische Befunde zu übersehen: Während nämlich nichterkannte somatische Befunde, freilich für manche Patienten auch zu spät, an der zweiten oder dritten ärztlichen Linie, nämlich beim Facharzt oder in der Klinik doch noch aufgefunden und behandelt werden können, pflegen vom Allgemeinarzt zunächst übersehene wichtige psychosoziale Befunde von der vorwiegend technisch ausgerichteten Klinik – abgesehen natürlich von jenen wenigen, die psychosomatische Abteilungen haben – nicht mehr nachträglich entdeckt, sondern durch eine Unzahl weiterer, inzwischen mit Hilfe der reichlich eingesetzten technischen Hilfsmittel entdeckter Nebenbefunde eher zusätzlich verdeckt zu werden. So gesehen ist der Erstkontakt zwischen Arzt und Patient eine oft nie mehr wiederkehrende Chance, das Krankheitsgeschehen des Patienten psychosomatisch richtig zu verstehen (vgl. auch 26.4)!

26.3 Das Krankengut der ärztlichen Praxis
(ad 2, 3)

Der niedergelassene Arzt, insbesondere der Arzt für Allgemeinmedizin, ist sowohl bezüglich der großen Zahl seiner Patienten als auch bezüglich des sehr breit gefächerten Krankengutes einem besonderen Druck ausgesetzt.

Um Aussagen über die quantitative Belastung des Allgemeinarztes machen zu können, hat Häußler (1967) unter Mitarbeit von 71 praktischen Ärzten die Abrechnungsunterlagen der Kassenärztlichen Vereinigung Nord-Württemberg im 2. Quartal 1966 über-

prüft und dabei festgestellt, daß der praktische Arzt durchschnittlich pro Arbeitstag 80–100 Patienten, und davon 20 Neuzugänge, zu betreuen hat. Wenn andererseits R. N. Braun (1970) an seiner sehr kleinen Praxis mit einem Durchgang von nur ca. 30 Patienten pro Arbeitstag errechnet hat, daß ihm für die „nackte" Diagnostik und Therapie, also die unmittelbare Interaktion zwischen Arzt und Patient, nur durchschnittlich 3 ½ Minuten pro Patient zur Verfügung stehen, dann illustrieren diese Zahlen sehr wohl den enormen Zeitdruck, unter dem die Ärzte für Allgemeinmedizin arbeiten müssen. Zeit für längere Gespräche und Untersuchungen bei einzelnen Patienten kann der Allgemeinarzt nur dann gewinnen, wenn er einige der Patientenkontakte noch weiter einschränkt bzw. ärztlichen Hilfspersonen überläßt, oder aber sich zur teilweisen Anwendung gruppentherapeutischer Verfahren entschließt.

Bezüglich der qualitativen Belastung, also des recht breit gefächerten Krankengutes der Allgemeinmedizin, können wir der oben zitierten Untersuchung von Häußler entnehmen, daß über 70% der Beratungsursachen der Ärzte für Allgemeinmedizin in das Gebiet der Inneren Medizin entfallen, während sich die anderen, weniger als 30%, auf alle anderen Fachdisziplinen verteilen. Der gleichen Untersuchung kann ferner entnommen werden, daß die oben erwähnten 71 Ärzte nur 9,34% ihrer Patienten an Fachärzte und nur 1,61% zur stationären Diagnose oder Therapie in ein Krankenhaus oder in eine Universitätsklinik eingewiesen, also über 90% ihrer Patienten selbst versorgt haben.

Brandlmeier (1974) stellt fest, daß in der Allgemeinpraxis 200–250 Krankheitsbilder gesehen werden, davon 20–30 Krankheitsbilder mit überragender Häufigkeit. Er bezieht sich u. a. auf eine Umfrage unter amerikanischen Werkärzten, die als Grund für die Arbeitsunfähigkeit bei den 11 häufigsten Syndromen die in Tabelle 26–1 aufgeführten Diagnosen stellten.

Tabelle 26–1. Grund für Arbeitsunfähigkeit (nach Brandlmeier)

Infektionen der oberen Atemwege
Psychologische Probleme
Kleine Unfälle
„Rückenschmerzen"
Hieb-, Stich-, Platzwunden
Muskelzerrungen, Verstauchungen
Hautentzündungen
Gastrointestinale Beschwerden
Augenverletzungen
Persönliche oder Familienprobleme
Kopfschmerzen

Eine Zusammenstellung des eigenen Krankengutes bei den 352 Kassenpatienten im II. Quartal 1967 ist in Tabelle 26–2 gegeben (vgl. auch Wesiack, 1970).

Aus diesen Tabellen ist die Bedeutung zu ersehen, die psychische und psychosomatische Erkrankungen im Krankengut des niedergelassenen Arztes haben.

Tabelle 26–2. Eigene Untersuchung über die Häufigkeit der Erkrankungen in einer internistischen Praxis

Funktionelle Syndrome	119 (34%)
Chronische organische Erkrankungen	89 (25%)
Chronische organische Erkrankungen mit erheblicher psychischer Beteiligung	51 (15%)
Psychoneurosen	37 (10%)
Psychosomatische Erkrankungen im engeren Sinn	24 (7%)
Akute organische Erkrankungen	18 (5%)
Akute organische Erkrankungen mit erheblicher psychischer Beteiligung	8 (2%)
Psychoneurosen mit (unbedeutendem) vom Grundleiden unabhängigem Organbefund	3 (1%)
Psychosen	3 (1%)
	n 352 (= 100%)

1960 fand ich nach meiner Niederlassung unter den ersten 50 Patienten, die mich aufgesucht hatten und die ich deshalb ausgewählt habe, weil ihnen zu diesem Zeitpunkt mein besonderes Interesse für psychoneurotische und psychosomatische Erkrankungen noch nicht bekannt war, was dann später allerdings zu entsprechenden Selektionsprozessen geführt hat, folgendes:

1 „Bagatellfall", d.h. ein Patient, bei dem ich weder einen behandlungsbedürftigen somatischen noch einen behandlungsbedürftigen psychischen Befund fand.

12 „somatische Erkrankungen", d.h. Patienten mit behandlungsbedürftigen somatischen Befunden, aber ohne behandlungsbedürftige psychische Befunde.

7 „psychische Erkrankungen", d.h. Patienten mit behandlungsbedürftigen psychischen Befunden, aber ohne behandlungsbedürftige somatische Befunde.

30 „psychosomatische Erkrankungen", d.h. Patienten mit behandlungsbedürftigen somatischen und psychischen Befunden.

Teilt man dieses Krankengut weiterhin in leichte (die auch ohne ärztliche Behandlung abheilen würden), mittelschwere und schwere (die ohne ärztliche Hilfe unweigerlich zum Tode führen würden) Erkrankungen ein, dann findet man 5 somatische, 6 psychische und 20 psychosomatische Erkrankungen unter den mittelschweren und immerhin eine somatische und 6 psychosomatische Erkrankungen unter den schweren, unbehandelt in absehbarer Zeit zum Tode führenden Erkrankungen (vgl. Wesiack, 1975). Andere Untersucher kommen zu ähnlichen Ergebnissen. So konnte Keller an seinem allgemeinärztlichen Krankengut etwa 40% psychosomatisch Kranke finden (1975), und Vogt und Blohmke konnten bei der Analyse des Krankengutes einer Allgemeinpraxis feststellen, daß 40–50% der Patienten den Arzt aus psychosozialen Motivationen aufgesucht haben (1974).

Diesen Untersuchungen können wir entnehmen, wie groß die Belastung des niedergelassenen Arztes einerseits durch psychosomatische und psychoneuro-

tische Erkrankungen, andererseits aber auch durch schwerkranke Patienten ist.

26.4 Der Mangel an personellen und technischen Hilfsmitteln und die Notwendigkeit, sich auf das Wesentliche zu beschränken
(ad 4, 5)

Obwohl ärztliche Praxen heute im allgemeinen recht gut apparativ ausgestattet sind, kann auch die besteingerichtete Praxis weder mit den personellen, noch gar mit den technischen Möglichkeiten einer Klinik konkurrieren. Dies hat sich in der Vergangenheit immer wieder gezeigt und wird sich wohl auch in Zukunft nicht, auch nicht durch Errichtung von Gruppen- und Gemeinschaftspraxen, wesentlich ändern lassen. Das Übergewicht der Klinik bei den technischen Möglichkeiten bleibt bestehen und zwingt die Praxis apparativ und personell auf den zweiten Platz. Eine Werthierarchie, die sich ausschließlich am technischen Potential mißt, wird deshalb die Praxis des Allgemeinarztes immer auf den letzten Platz verweisen. Die Frage ist nur, ob für alle Zeiten die technische Effizienz der Maßstab für ärztliche Leistungen bleiben wird, wie das in den letzten Jahrzehnten gewesen ist.

In dem einleitenden Kapitel haben wir uns bemüht, ein neues Konzept der Medizin zu entwerfen, das technischen Fortschritt weder geringschätzt noch ablehnt, ihm aber im Rahmen des Situationskreismodells einen neuen und nicht mehr den bisher überragenden Stellenwert einräumt. Gerade unter dem Aspekt des Situationskreises gewinnt die Arbeit an der ersten ärztlichen Linie neues Gewicht. Anläßlich der Besprechung des Erstkontaktes wurde diese neue Bedeutung schon herausgestellt. Jetzt soll am Beispiel zweier Krankengeschichten (aus Wesiack, 1974) gezeigt werden, wie leicht der Einsatz technischer Mittel nicht nur aufklärend sein, sondern umgekehrt von einer richtigen Diagnose auch wegführen kann.

Ein Patient sucht vergeblich ärztliche Hilfe.

Ein 29jähriger, blasser, leptosomer Metzgermeister klagt über dauernde Magenschmerzen, die als Völlegefühl, drückend und bohrend geschildert werden und sich zu nächtlichen Koliken steigern. Dadurch sei sein Schlaf und in der Folge davon auch seine berufliche Leistungsfähigkeit schwer gestört. Durch Biertrinken könne er den Schmerz manchmal betäuben. Da er sich nichts mehr zu essen getraue, habe er stark abgenommen. Er habe bisher 18 bis 20 Ärzte erfolglos konsultiert, darunter drei Professoren, ferner mehrere Heilpraktiker, und habe bereits fünf Klinikaufenthalte und mehrere Heilkuren in Badeorten hinter sich. Die gestellten Diagnosen schwankten zwischen „Gastritis", „nervösen Magenbeschwerden", „Subazidität", „Ulkusverdacht", „Erkrankung der Bauchspeicheldrüse", „Leber- und Gallenleiden", „Porphyrieverdacht" und „vegetativer Dystonie". Jeder Arzt stellte, aus der Sicht des Patienten, eine andere Diagnose und ordnete die verschiedensten Kuren an, die manchmal zunächst eine gewisse Besserung brachten, im Endeffekt aber alle gleichermaßen wirkungslos waren. Auf Anraten eines Professors habe er sogar sein gutgehendes eigenes Metzgergeschäft aufgegeben und sei wieder zurück in das Angestelltenverhältnis gegangen. Eine Besserung sei dadurch aber nicht eingetreten. Jetzt sei er völlig verzweifelt, habe kein Vertrauen mehr und betrachte sich als „verlorenen Mann". Trotzdem aber sei er auf der Suche nach dem richtigen Medikament und nach Hilfe, falls sie überhaupt noch möglich sei.

Um die Psychodynamik dieses Patienten zu verstehen, will ich kurz die wichtigsten psychologischen Daten, gewonnen in zwölf psychotherapeutischen Sitzungen, berichten: Er stamme vom Lande, der Vater sei immer kränklich, aber sehr streng und jähzornig gewesen, so daß alle acht Kinder immer vor ihm Angst gehabt hätten. Die Mutter sei die Seele der Familie, unser Patient immer ihr Liebling gewesen. Die Lehrzeit wird als unmenschlich hart erlebt. Er habe sehr starkes Heimweh gehabt und wollte deshalb wiederholt heimlaufen, habe sich aber vor dem strengen Vater zu sehr gefürchtet. Als er nach vollendeter Lehrzeit an einem Wochenende nachts spät heimkam, sei es zu einer schweren Auseinandersetzung mit dem Vater gekommen. Daraufhin sei er zwei Jahre lang dem Elternhaus ganz ferngeblieben. Aber auch danach sei es bis zum Tode des Vaters zu keiner richtigen Aussöhnung mit diesem gekommen. Im Alter von 20 Jahren hatte er mit einer verwitweten Frau die ersten intimen Sexualbeziehungen, die er als große Schuld empfunden habe. Danach ging er als Geselle zu einem Metzger, der ihn in seine Familie aufnahm und ihn „wie einen Sohn" behandelte. In dieser Zeit starb sein Vater an einem Magenkarzinom. Mit seiner Chefin, die von ihrem Ehemann vernachlässigt wurde, kam es zu einem leidenschaftlichen Liebesverhältnis, und als es etwa zwei Monate nach dem Tode des Vaters zur ersten intimen Liebesbeziehung mit ihr kam, reagierte er nicht nur mit schweren Schuldgefühlen, sondern erstmals auch mit heftigen Magenschmerzen, die ihn von da an nicht mehr verließen. Dann begann der jahrelange Leidensweg mit vergeblicher Suche nach Hilfe. Inzwischen hatte der Patient geheiratet und mit seiner recht tüchtigen Frau mit gutem Erfolg ein eigenes Metzgergeschäft gegründet. Jetzt traten bei ihm Eifersuchtsängste auf. Er befürchtete, obwohl objektiv kein Anlaß dazu bestand, seine Frau werde ihn ebenso betrügen, wie seine frühere Chefin ihren Mann mit ihm betrogen hatte. Er erwartete das geradezu als eine Art gerechter Strafe für seine Verfehlungen. In dieser Situation verschlimmerten sich wiederum seine Magenbeschwerden. Da gründliche ambulante und klinische Untersuchungen keinen greifbaren somatischen Befund erbrachten, bekam er nun von ärztlicher Seite den völlig unbegründeten Rat, seine Existenz aufzugeben und wieder ins Angestelltenverhältnis zu gehen, ein Rat, den er auch befolgte. Der erhoffte Erfolg blieb, wie zu erwarten, aus.

Der Patient wurde wiederholt gründlich klinisch untersucht. Er wurde geröntgt und gastroskopiert, er wurde laparoskopiert und leberpunktiert und zuletzt noch mit dem Rat bedient, seine eigene wirtschaftliche Existenz zu vernichten. Es klingt fast wie ein Märchen, aber er hat es mir glaubhaft versichert: Keiner der vielen Ärzte, die er konsultierte, hat sich wirklich für seine Biographie und für die psychodynamischen Zusammenhänge interessiert oder ihm auch nur geraten, einen Psychotherapeuten aufzusuchen. Und dabei ist die Psychodynamik in ihren Grundzügen in diesem Fall keineswegs besonders kompliziert oder verborgen. Sie drängt sich dem einigermaßen Erfahrenen geradezu auf. Hier agiert ein Patient noch mitten im Ödipusdrama, er wird von Angst und Schuld erdrückt. Die Magenbeschwerden sind nicht allzu schwer als konversionsneurotische Identifikation mit dem magenkrebskranken Vater zu verstehen.

Ein Patient ist verbittert.

Vor zwei Jahren erlitt ein damals 33jähriger Mann in einer Drahtzieherei einen Arbeitsunfall mit linksseitigem Rippenbruch. Der Unfall ist passiert, weil er versucht hat, sich gegen einen stark schwingenden Drahtzug zu stemmen, der die ganze elektrische Schaltanlage dieses Arbeitsraumes zu zertrümmern drohte. Nach glatter Abheilung der Rippenfraktur wurde der Patient aber nicht beschwerdefrei, sondern bekam unbestimmte Oberbauchbeschwerden, die er selbst als „Wundsein" beschrieb und die er übrigens seit seinem 16. Lebensjahr immer wieder hatte. Da der Patient immer stärker über Beschwerden klagte und gleichzeitig an Gewicht abzunehmen begann, wurde er von seinem Hausarzt wegen Verdacht auf eine Magenerkrankung einem Internisten überwiesen. Dieser äußerte den Verdacht auf eine Pankreaserkrankung und veranlaßte die Einweisung des Patienten in eine namhafte gastroenterologische Klinik. Dort wurden bei einer Laparoskopie „breitflächige Verwachsungen zwischen Colon ascendens und Peritoneum parietale" festgestellt. Da man in diesem Befund die „Ursache" der Beschwerden des Patienten erblickte, wurde dieser sogar in eine Schweizer Universitätsklinik zur operativen Behandlung überwiesen. Der Versuch einer Adhäsiolyse war jedoch nicht erfolgreich, weil „die Verwachsungen offenbar zu weit dorsal lagen". Das Befinden des Patienten hatte sich – inzwischen waren eineinhalb Jahre seit dem auslösenden Unfall vergangen – durch die vielen diagnostischen und therapeutischen Eingriffe laufend verschlechtert. Durch Appetitmangel hatte er inzwischen fast 20 kg abgenommen und war nur bei Einnahme von Dolantinsuppositorien, von denen er oft zwei bis drei täglich nahm, einigermaßen beschwerdefrei. Um den sehr ordentlichen und fleißigen, inzwischen aber völlig verzweifelten Mann vor der drohenden Frühinvalidierung zu bewahren und um nicht doch „irgend etwas übersehen" zu haben, veranlaßte der sehr gewissenhafte und besorgte Hausarzt noch eine Durchuntersuchung des Patienten in der Klinik für Diagnostik in Wiesbaden. Dort wurde der Patient erstmals auch einem Psychosomatiker vorgestellt und neben einigen wohl mehr oder weniger bedeutungslosen Befunden der Verdacht auf eine „Konversionsneurose" geäußert.

Als ich den Patienten daraufhin erstmals sah, machte er zunächst einen sehr gedrückten und verschlossenen, ja fast versteinerten Eindruck, wobei an seiner Sprechweise eine eigenartig scharfe Artikulation und Betonung der Zischlaute auffiel. Wie gut gezielte Geschosse schleuderte er mir abgehackt einzelne Worte und Satzteile entgegen. Aus der Lebensgeschichte des Patienten war zu erfahren, daß er aus einer zerrütteten Ehe stammt, als Kind zwischen den geschiedenen Eltern, der Stiefmutter und den Großeltern hin und her geschoben wurde und sich ein Leben lang zurückgesetzt und zu kurz gekommen gefühlt hat. Nur seinen Lehrherrn hatte er in einigermaßen freundlicher Erinnerung behalten.

Zu seinem Betrieb hatte er ein recht ambivalentes Verhältnis. Da er sich dort durch Fleiß und Pünktlichkeit ein gewisses Ansehen erworben hatte, betrachtete er seinen Betrieb inmitten einer feindlichen und bedrohlichen Welt als relativ sicheren Festpunkt. Deshalb hat er auch versucht, ihn unter Einsatz aller seiner Kräfte und mit der Folge eines Rippenbruchs vor empfindlichem Schaden zu bewahren (was ihm übrigens auch gelungen war). Andererseits blieb er stets mißtrauisch und sah jetzt sein Mißtrauen durch den Unfall und alles, was er seither erfahren hatte, bestätigt. Jetzt fühlt er sich offenbar zwischen dem Betrieb und den verschiedenen Ärzten und Kliniken unverstanden und ungeliebt hin und her geschoben, wie seinerzeit in der Kindheit zwischen den Eltern und Großeltern.

Diese beiden Krankengeschichten, so verschieden sie im einzelnen auch sein mögen, zeigen eines sehr deutlich: Durch die Reduktion der diagnostischen und therapeutischen Bemühungen ausschließlich auf den organischen physikalisch-chemischen Aspekt des Menschen werden wichtige pathogenetische Faktoren übersehen und dadurch wird die nie wiederkehrende Chance verpaßt, eine sich anbahnende neurotische Entwicklung im Entstehen therapeutisch aufzufangen. Aufgrund von Erfahrungen mit sehr vielen ähnlichen Krankheitszuständen und der in der Literatur niedergelegten Berichte anderer (vgl. auch Kap. 31 und die dort zitierte Arbeit von Chester) möchte ich die Behauptung aufstellen, daß man bei sehr vielen Patienten im Entstehen begriffene neurotische Entwicklungen relativ leicht, d.h. durch wenige die Psychodynamik aufgreifende ärztliche Gespräche auflösen kann, die man später nach Chronifizierung überhaupt nicht mehr oder nur noch durch einen unverhältnismäßig großen therapeutischen Aufwand (z.B. Psychoanalyse) beeinflussen kann.

Hier zeigt sich bei beiden Patienten, daß die wesentlichen Momente der pathogenetischen Kette im psychosozialen Bereich zu suchen waren und daß immer wieder durchgeführte nur somatische Untersuchungen nicht zu einer diagnostischen Klärung, sondern von dieser immer weiter weggeführt haben. Die richtige diagnostische Erfassung des „Wesentlichen" ist natürlich unter dem enormen Zeitdruck der Praxis alles andere als einfach (vgl. auch die beiden therapeutischen Kapitel 16 und 17).

Hier muß darauf hingewiesen werden, daß sich das Zeitproblem für den Arzt für Allgemeinmedizin anders stellt als für den Kliniker oder Psychotherapeuten. Für die aktuelle Interaktion steht dem Arzt für Allgemeinmedizin meist nur sehr wenig Zeit zur Verfügung. Dieser Zeitmangel wird jedoch ausgeglichen durch die vielen Kontakte, die der Arzt in der Regel in der Vergangenheit bereits mit dem Patienten und oft auch mit seinen Angehörigen gehabt hat und die zu einer vertieften Kenntnis der gesamten psychosozia-

len Situation des Patienten geführt haben, wenn der Arzt darauf achtet und gelernt hat, diese Informationen im Sinne eines biopsychosozialen Modells neu zu ordnen und zu verwerten.

Im Allgemeinärztlichen Institut der Universität Nijmegen hat unter Leitung von F. J. A. Huggen eine interdisziplinäre Arbeitsgruppe von Verhaltenswissenschaftlern, Ärzten, Psychologen und Soziologen systematisch die Interaktionsprozesse zwischen Ärzten, Patienten und ihren Familien erforscht, um u.a. auch unnötigen Chronifizierungen, die heute noch an der Tagesordnung sind, vorzubeugen (vgl. Grol, 1985)

26.5   Die größere Nähe von Arzt und Patient in der Allgemeinpraxis und die Langzeitbetreuung   (ad 6–8)

Die größere Nähe von Arzt und Patient ist in der Allgemeinpraxis vor allem zwei Umständen zu verdanken. Die Wohn- und Arbeitsbereiche des Arztes und seiner Patienten berühren sich häufig, sind zumindest nicht weit voneinander entfernt, manchmal überschneiden sie sich geradezu. Ein anderer Grund für die Nähe ist darin zu sehen, daß immer wieder derselbe Arzt den Patienten über Jahre und Jahrzehnte hinweg auch mit unterschiedlichen Erkrankungen und in verschiedenen Notlagen behandelt. Dies schafft ein Vertrauensverhältnis besonderer Art, wie es zu den Klinikärzten nur selten bestehen wird. Wenn der Patient nämlich nach einiger Zeit wieder in die Klinik eingewiesen werden sollte, sieht er sich nicht selten anderen Ärzten und auch einem anderen Pflegepersonal gegenüber. Die Stabilität der Bezugspersonen ist in der Klinik nicht so gewährleistet wie in der Praxis.

Ist der niedergelassene Arzt, was nicht selten der Fall sein wird, auch der Hausarzt der Familie, oder hat er als Facharzt schon mehrere Mitglieder derselben Familie kennengelernt und behandelt, dann kann ihm diese Kenntnis des familiären Milieus und des sozialen Hintergrundes bei der diagnostischen Wertung der Symptome seines Patienten sehr helfen. So wird er z.B. die neurotischen Reaktionen eines Patienten anders und wahrscheinlich zutreffender beurteilen können, wenn er den Ehepartner und die Eltern des Patienten und ihre Verhaltensweisen ebenfalls gut kennt.

Der Patient wird daher im allgemeinen dem Klinikarzt mehr technisch-apparative und wissenschaftliche Kompetenz zugestehen, vom niedergelassenen Arzt aber, infolge seiner größeren Detailkenntnisse des psychosozialen Milieus, mehr psychosomatisches Wirken erwarten. Dies schließt natürlich nicht aus, daß er oft beiden Arztgruppen gegenüber stark magisch überhöhte und durch Vorurteile aus den verschiedensten publizistischen Quellen verzerrte Vorerwartungen hegt, die dann von vornherein die Interaktion zwischen Patient und Arzt beeinflussen.

Es wurde bereits erwähnt, daß der Arzt für Allgemeinmedizin bzw. der in freier Praxis tätige Arzt derjenige ist, der im allgemeinen auch die Langzeitbetreuung der Patienten übernimmt. Er ist demnach auch derjenige, der am ehesten dazu berufen wäre, über Langzeitverläufe, über die wir bisher noch so wenig wissenschaftlich Verbindliches wissen, zu berichten. Daß dies bisher nicht oder nur sehr vereinzelt geschehen ist, beruht darauf, daß sich in der Vergangenheit niedergelassene Ärzte kaum wissenschaftlich betätigt haben. Da der Kliniker nur die einzelnen Krankheitsepisoden, also gewissermaßen nur Querschnitte des gesamten Krankheitsgeschehens, sieht, entgehen so die Gesamtverläufe, also die Längsschnitte, weitgehend der wissenschaftlichen Bearbeitung. Dies trifft ganz besonders dann zu, wenn verschiedene Syndrome wechselweise auftreten, die aus der Querschnittsicht der Klinik einzelnen, voneinander unabhängigen „Krankheiten" anzugehören scheinen, im Grunde genommen aber wahrscheinlich doch die verschiedenen Auswirkungen eines tiefergelegenen Krankheitsprozesses sind. Eine typische Krankengeschichte aus der Praxis soll das verdeutlichen.

Solche und ähnliche Krankheitsgeschichten, die für das Krankengut der Praxis durchaus typisch sind, werfen eine Reihe wissenschaftlich bisher nur wenig bearbeiteter Fragen auf. Die wichtigste scheint mir im vorliegenden Fall folgende zu sein: In welchem engeren Zusammenhang stehen die einzelnen Erkrankungen des Patienten? Sind nicht hier die depressiven Verstimmungen, das Ulcus-duodeni-Leiden, der Alkoholabusus und die Familienprobleme Ausdruck ein und derselben Grunderkrankung? Im Kapitel 1 haben wir von allgemeinem Kranksein gesprochen, das auch allen spezifischen Erkrankungen zugrunde liegt, und vom symbiotischen Funktionskreis, dessen Störungen für spätere Erkrankungen von großer Bedeutung sind. Balint nannte diese Grunderkrankung treffenderweise „Grundstörung". Sie sollte behandelt werden, um wirklich kausale Therapie zu betreiben.

26.6 Die Besonderheit des diagnostisch-therapeutischen Interaktionsprozesses in der ärztlichen Praxis (ad 9)

Auf die grundsätzliche Verklammerung aller diagnostischen und therapeutischen Interaktionen im „diagnostisch-therapeutischen Zirkel" wurde schon im einleitenden Kapitel 1 hingewiesen. Diese Verklammerung wird in der Allgemeinmedizin besonders deutlich, wo diagnostische und therapeutische Maßnahmen zeitlich, personell und räumlich viel stärker miteinander verbunden sind als in der Klinik. Eine Trennung von Diagnostik und Therapeutik, wie sie in der Klinik meist vorgenommen wird, ist in der Praxis schon aus zeitlichen Gründen nur in Ausnahmefällen möglich. Ein Fallbeispiel soll das wiederum verdeutlichen.

Der bei Beobachtungsbeginn 44jährige Patient klagte damals über depressive Verstimmungen und Oberbauchbeschwerden, die in wechselnder Stärke seit über 10 Jahren bestünden. Ärztlicherseits seien mehrmals Zwölffingerdarmgeschwüre festgestellt worden. Die internistische Untersuchung ergibt einen narbig deformierten Bulbus duodeni mit einem frischen Ulkusschub. In psychischer Hinsicht ist eine depressive Grundstimmung nicht zu übersehen. Der Patient ist in kinderloser Ehe mit einer zwei Jahre älteren Frau verheiratet, mußte kürzlich wegen Unrentabilität seinen bisherigen Betrieb liquidieren und ist nun im Begriff, sich mit seiner Frau eine neue Existenz aufzubauen.

In den nächsten 1½ Jahren erfolgen noch mehrere Ulkusschübe, die mit der üblichen internistischen Behandlung zunächst jeweils relativ rasch abklingen, bis dann ein therapieresistenter, von pankreatischen Erscheinungen begleiteter Krankheitsschub zunächst zur konservativen stationären Behandlung, und da diese keine Besserung bringt, zur Resektion eines ins Pankreas penetrierenden chronischen Ulcus duodeni führt.

In den folgenden sechs Jahren erscheint der Patient nicht mehr zur Behandlung, sondern nur noch vereinzelt zu kurzen Nachuntersuchungen, insbesondere dann, wenn er irgendwelche Bescheinigungen, etwa fürs Finanzamt oder aber fürs Gericht wegen Verwicklungen in Verkehrsunfälle, benötigt. Bei diesen relativ kurzen und seltenen Kontakten mit dem Patienten fällt auf, daß er zu trinken begonnen hat und Anzeichen einer alkoholischen Leberschädigung aufweist. Zu einer geregelten Behandlung ist er jedoch nicht zu gewinnen. Eines Tages erscheint die verzweifelte Ehefrau und berichtet, daß der Patient Alkoholiker geworden sei, wegen eines erneuten Verkehrsunfalles zum dritten Mal den Führerschein entzogen bekommen habe, gelegentlich deliriumartige Zustände bekomme und das inzwischen wiederaufgebaute und florierende Geschäft gefährde.

Das Sprechzimmer betritt zum ersten Mal eine 52jährige Frau und berichtet, daß sie in den letzten 3 Wochen zweimal Anfälle von akuter Atemnot bekommen habe. Die Luft sei ihr weggeblieben und sie habe geglaubt, sterben zu müssen. Sie berichtet dann unter tiefem Seufzen weiter, daß sie mit einem Ausländer in schlechter Ehe verheiratet sei, der sie vernachlässige und oft nächtelang wegbleibe. Die so bedrohlich empfundenen Atemnotanfälle seien just in dem Augenblick aufgetreten, als ihr ältester 18jähriger Sohn ihr erklärt habe, er wolle sich nun von daheim trennen und wegziehen. Dabei bricht sie in Tränen aus.

Da es sich bei der sehr adipösen kurzatmigen Frau mit etwas zyanotischen Lippen, sie wog bei 161 cm Größe 108 kg, schon auf den ersten Blick hin um eine Fettsucht und Herzinsuffizienz handelte, konnte bereits aufgrund dieser Informationen noch vor jeder eingehenden körperlichen Untersuchung und ohne eine zeitraubende biographische Anamnese und ein umfangreiches Erstinterview die vorläufige diagnostische Hypothese „Fettsucht", „Herzinsuffizienz", „aktuelle Notsituation" durch drohenden Verlust des Sohnes und „chronische Notsituation" durch schlechte Ehe gestellt werden.

Diese diagnostischen Hypothesen führen uns zwangsläufig zu folgenden Handlungsanweisungen: Zunächst durch körperliche Untersuchung eine vielleicht lebensbedrohliche Kreislaufgefährdung rechtzeitig zu erkennen und das Ausmaß der medikamentösen Therapie festzulegen und andererseits zu der Notwendigkeit, auf die akute Notsituation der Patientin durch Bekundung mitfühlenden Verständnisses einzugehen. Zeigt sich, wie im vorliegenden Fall, daß eine akute Kreislaufgefährdung nicht besteht, und wirkt die Patientin schon nach kurzem Gespräch etwas erleichtert und entspannt, dann kann, je nach Praxissituation, der diagnostisch-therapeutische Zirkel hier schon zunächst unterbrochen werden, um ihn zu einem späteren Zeitpunkt fortzusetzen. Die Abklärung weiterer differentialdiagnostischer Überlegungen, die laborchemische Untersuchung des Blutes, aber auch das Erfahren weiterer biographischer Details, psychodynamischer Vorgänge und typischer Interaktionsformen kann auf später verschoben werden. Jede erneute Kontaktaufnahme mit der Patientin setzt dann den diagnostisch-therapeutischen Zirkel wieder in Gang, wobei es für den Arzt besonders wichtig ist, die wesentlichsten diagnostischen Hypothesen und therapeutischen Strategien auch über lange Zeiträume hinweg nicht aus den Augen zu verlieren.

Zusammenfassend läßt sich also zu diesem Abschnitt feststellen: In der Klinik müssen die diagnostische Abklärung und die therapeutischen Maßnahmen meistens voneinander getrennt und auf kurze Zeiträume zusammengedrängt werden. Sie haben deshalb zwangsläufig mehr begrenzten Charakter. In der Praxis begleiten diese Interventionen im allgemeinen den Patienten über die ganze Zeit seines Kontaktes hinweg, den der Patient mit seinem Arzt hat. Sie haben daher einen mehr betreuenden Charakter.

26.7 Die Bedeutung des niedergelassenen Arztes für die Prophylaxe, Früherkennung und Frühbehandlung psychosomatischer Erkrankungen[1] (ad 10)

Es besteht heute weitgehend Einigkeit darüber, daß zwar die Bereitschaft, mit psychoneurotischen oder psychosomatischen Erkrankungen zu reagieren, schon jeweils in den ersten Lebensmonaten und -jahren erworben wird, daß aber die Auslösung und die Aufrechterhaltung des Krankheitsgeschehens von bestimmten, das jeweilige Individuum besonders belastenden Situationen abhängen. Daraus können wir mehrere Ebenen der Prophylaxe ableiten:

- die Primärprophylaxe, die sich darum bemüht, daß Krankheitsbereitschaften gar nicht entstehen,
- die Sekundärprophylaxe, die das Ausbrechen der manifesten Erkrankung zu verhindern sucht, und
- die Tertiärprophylaxe, die sich darum bemüht, daß es nicht zu Chronifizierungen und Rezidiven kommt.

In der Primärprophylaxe kann der Arzt der ersten Linie vor allem indirekt wirksam werden, indem er die Eltern in bezug auf richtiges Verhalten bei der Kinderaufzucht und Erziehung berät und den jungen Müttern hilft, die Schwierigkeiten und Gefährdungen des symbiotischen Funktionskreises ohne nachhaltige Schädigung des Kindes zu meistern.

Viel weitgehender sind seine Möglichkeiten im Bereich der sekundären und tertiären Prophylaxe. Die Patienten, die den Arzt der ersten Linie aufsuchen, können nämlich in drei große Gruppen eingeteilt werden. Es sind dies erstens Patienten, die an einer akuten Störung ihres Befindens und ihrer Leistungsfähigkeit leiden, die sie gerne abgeklärt und nach Möglichkeit beseitigt haben möchten. Eine weitere große Gruppe von Patienten kommt zur Behandlung ihres oder ihrer chronischen Leiden wegen. Drittens wird der Arzt der ersten Linie auch noch von Menschen aufgesucht, die sich selbst nicht eigentlich krank fühlen, die aber aus Gründen der Vorbeugung und Gesundheitsberatung den Arzt aufsuchen. Der Arzt hat nun, oder genauer gesagt, hätte nun, gegenüber jeder dieser drei Gruppen besondere Aufgaben und Verantwortungen bezüglich der Früherkennung und Prophylaxe.

Am umfangreichsten und auch am folgenschwersten sind seine Aufgaben und Verantwortungen gegenüber der ersten Gruppe von Patienten, also gegenüber jenen, die ihn aus aktuellem Anlaß infolge irgendwelcher Befindensstörungen und Leistungsminderungen aufsuchen. Hier muß der Arzt mit dem Patienten gemeinsam zunächst sorgfältig die symptomauslösende oder die symptomauslösenden Situationen erarbeiten und sich im Erstinterview ein Bild von der Persönlichkeit des Patienten und seinen psychosozialen Beziehungen machen, ehe er ihn gründlich körperlich untersucht. Nur wenn dieser erste Interaktionsabschnitt zwischen Arzt und Patient – und das kann gar nicht stark genug betont werden – glückt, besteht die Chance, zu einer umfassenden Diagnose zu kommen, die uns dann Handlungsanweisungen für die Therapie gibt.

Mißglückt jedoch dieser erste Interaktionsschritt zwischen Arzt und Patient oder wird er gar – und dies dürfte heute die Regel sein – einfach übersprungen, weil der Arzt sich meist seiner Bedeutung nicht bewußt ist und in der Kassenpraxis auch keine Zeit dazu hat, dann sind eigentlich die Weichen für eine Fehlbehandlung und Chronifizierung des Krankheitsgeschehens schon gestellt.

Aber nicht nur bei den psychosomatischen Erkrankungen, sondern auch bei den reinen Neurosen, die ja gut ein Drittel, wenn nicht mehr unseres Gesamtkrankengutes ausmachen, wirkt sich diese eben beschriebene Informations- und Erlebnislücke, die wir bei den Ärzten und bei den Patienten gleichermaßen wiederfinden, verheerend aus. Die Patienten mit Neurosen und neurotischen Reaktionen haben ja sehr häufig organbezogene Beschwerden – man denke z. B. an die physiologischen Begleiterscheinungen der Angst wie Herzklopfen, Zittern, Schweißausbrüche usw. – und

---

[1] Vgl. Wesiack, 1977.

suchen deshalb zunächst einen Arzt für Allgemeinmedizin oder einen Internisten und nicht den Psychotherapeuten oder Nervenarzt auf. Hier werden sie nun – meist ohne klärendes Erstinterview – einer subtilen und aufwendigen klinischen Diagnostik unterworfen, wodurch das Krankheitsgeschehen nur zu oft nicht geklärt, sondern weiter verdunkelt, auf jeden Fall aber chronifiziert wird. Chronifiziert deshalb, weil das bei allen Menschen vorhandene neurotische Potential und die damit verbundenen Abwehrprozesse nun von ärztlich autoritativer Seite her verstärkt werden. So können sich immer wieder von neuem durchgeführte Organuntersuchungen und Behandlungen geradezu psychotoxisch und chronifizierend auswirken.

Wenn wir nun bedenken, daß psychische und psychosoziale Faktoren bei so gut wie jedem Kranken eine Rolle spielen, bei sehr vielen aber entscheidend für den ganzen Krankheitsverlauf sind, ist es dann noch verwunderlich, wenn unser einseitig auf die Organmedizin ausgerichtetes ärztliches Versorgungssystem trotz ungeheurer Aufwendungen immer unproduktiver wird?

Auch bei der Versorgung chronisch Kranker wird dieses Problem wieder deutlich. Oft lassen sich hier Rezidive und bedrohliche Verschlimmerungen vermeiden, wenn der Arzt die Psychodynamik des Krankheitsgeschehens begreift und seinem Patienten etwas von diesem Verständnis vermitteln kann. Die Behandlung von Koronarkranken und Herzinfarktgefährdeten ist ein schönes Beispiel dafür. Wenn es nämlich nicht gelingt, den zwanghaft an Leistung gebundenen Lebensstil dieser Patienten zu ändern, dann läßt der nächste Infarktschub, der schon das Ende bringen kann, meist nicht mehr lange auf sich warten. Ähnliches gilt für andere Erkrankungen. Die psychosomatische Forschung ist in zunehmendem Maße in der Lage, Risikopersönlichkeiten und Risikosituationen, wie z.B. die große pathologische Bedeutung des Verlustes wichtiger Beziehungspersonen, die dann in einem stark erhöhten Maße zu Erkrankungen führt, zu definieren. Durch rechtzeitige, gewissermaßen prophylaktische Behandlung dieser Risikogruppen müßte es möglich sein, die Erkrankungsraten wesentlich zu senken.

Wie sieht es aber nun bei der letzten großen Gruppe von Patienten aus, die den Arzt zum Zwecke der Vorsorge aufsuchen? Die bisher üblichen offiziellen Vorsorgeprogramme sind dürftig und im höchsten Maße unbefriedigend. Die meisten Ärzte sind deshalb, wie übrigens auch schon in der Vergangenheit, bereit, ihre Patienten unabhängig von und außerhalb der offiziellen Vorsorgeprogramme gründlich zu untersuchen. Infolge ihrer bisher ganz einseitig naturwissenschaftlich ausgerichteten Ausbildung beschränkt sich dies aber meist ausschließlich auf den organischen Bereich und berücksichtigt die psychodynamischen Determinanten und die daraus folgenden Risikofaktoren bisher viel zu wenig. Es genügt nicht, den Patienten nur über die Höhe seines Blutdrucks, die Beschaffenheit seines EKG's, seiner Röntgenbefunde, seiner Blutfettwerte, seines Rektal- oder Vaginalbefundes usw. zu unterrichten. Es ist darüber hinaus nötig, ihn auf die

Gefährdungen hinzuweisen, die sich aus seinem Lebensstil, seiner Psychodynamik und aus der Art seiner Objektbeziehungen ergeben. Nur so ist es möglich, auch vom psychosomatischen Standpunkt aus sinnvolle Prophylaxe zu betreiben.

Fassen wir unsere Überlegungen abschließend zusammen, dann können wir feststellen, daß es gegenwärtig nicht nur eine die psychodynamischen Determinanten berücksichtigende Vorsorgemedizin noch gar nicht gibt, sondern daß der Ausbildungsstand und die Arbeitsbedingungen der deutschen Ärzte im allgemeinen so beschaffen sind, daß psychoneurotische und psychosomatische Erkrankungen meist nicht erkannt und durch ärztliche Maßnahmen nur noch weiter chronifiziert werden. Diese Feststellung ist unerfreulich, ja deprimierend. Sie muß aber klar ausgesprochen werden, weil andererseits hier noch große Möglichkeiten einer sinnvollen Vorsorge bestehen – einer Vorsorge, die nicht nur den Patienten viel Leid und Not, sondern der Gesellschaft auch sehr viel Kosten ersparen würde.

Diese Form der Vorsorge setzt aber eine Neubesinnung und eine Neuorientierung der Medizin voraus. Sie ist nur dann realisierbar, wenn sich einerseits das Wissen, das Können und die Motivation zumindest eines großen Teils der Ärzteschaft im Sinne einer Psychologisierung der ärztlichen Tätigkeit ändern, und wenn andererseits die Gesellschaft und die Politiker bereit sind, dem diagnostisch-therapeutischen ärztlichen Gespräch die gleiche Bedeutung beizumessen wie subtilen und aufwendigen klinischen Untersuchungen. Diesbezüglich können wir sehr viel von den holländischen Ärzten lernen (vgl. Grol, 1985).

26.8 Schlußbetrachtungen

Schwierigkeiten und Widerstände, die einer Anwendung psychosomatischer Gesichtspunkte in der ärztlichen Praxis entgegenstehen

Nachdem in den vorhergehenden Abschnitten zu zeigen versucht wurde, daß der niedergelassene Arzt in ganz besonderer Weise dazu berufen ist, Psychosomatische Medizin zu betreiben, soll nun abschließend noch auf einige Schwierigkeiten und Widerstände eingegangen werden, die diesen Bestrebungen entgegenstehen.

Seinem ganzen Ausbildungsgang entsprechend ist der heute niedergelassene Arzt im allgemeinen im eigenen Selbstverständnis in erster Linie ein „Organiker" und „Mikrokliniker", der, wenn auch natürlich vergeblich, mit den Leistungen der Klinik zu wetteifern sucht. Dies ist eines der wesentlichsten Hindernisse einer Anwendung psychosomatischer Gesichtspunkte in der ärztlichen Praxis. Hinzu kommt, daß die wenigen bisher in der Praxis tätigen Psychosomatiker recht isoliert sind, weil sie weder bei den vorwiegend somatisch orientierten Allgemeinärzten, noch bei den Klinikern, aber auch nicht bei den Psychotherapeuten, zu denen sie ja wiederum nicht zählen, einen ausreichenden Rückhalt finden.

Bei den Patienten schwankt die Einstellung zur Psychosomatischen Medizin gegenwärtig zwischen der Ablehnung einer nicht rein organischen Therapie („Herr Doktor, ich hab's im Magen und nicht im Kopf!") und magisch überhöhten Riesenerwartungen. Die Widerstände gegen vor allem unbewußte persönliche Probleme sind ja aus der psychotherapeutischen Literatur so bekannt, daß sie hier nicht besonders aufgeführt werden müssen. Erwähnt sei nur, daß in der Psychosomatik diese Probleme durch die stets bestehenden körperlichen Symptome und das Alexithymieproblem noch potenziert werden.

Gesellschaftspolitische Probleme mit einem einseitig am organischen Krankheitsbegriff orientierten Versicherungssystem und einer Wertordnung, die den Kliniker und Organiker einseitig bevorzugt, sind weitere Schwierigkeiten, die erwähnt werden müssen. Sie müssen deshalb besonders erwähnt werden, weil sie die Tätigkeit des niedergelassenen Arztes in viel stärkerem Maß beeinflussen als die des Klinikers. Besonders nachteilig wirkt sich in diesem Zusammenhang unsere Gebührenordnung aus, die technische Leistungen prämiiert, den persönlichen Einsatz, den Zeitaufwand und das ärztliche Gespräch aber völlig unberücksichtigt läßt. Interessant sind in diesem Zusammenhang die Ausführungen eines amerikanischen Allgemeinarztes (vgl. Greco und Pittenger, 1966), der darauf hinweist, daß sein Einkommen nach seiner Ausbildung zum psychosomatisch arbeitenden Allgemeinarzt zunächst um über ein Drittel zurückging, dann zwar wieder anstieg, aber auch in späteren Jahren um über 20% unter dem Einkommen lag, das er als naturwissenschaftlicher Arzt erzielen konnte. Diese erschwerenden Rahmenbedingungen können natürlich nicht ganz ohne Einfluß auf den Stil der ärztlichen Tätigkeit bleiben. Hier muß angemerkt werden, daß sich allmählich auch in den Gebührenordnungen ein Wandel zu vollziehen beginnt, der den persönlichen und psychotherapeutischen Bemühungen des Arztes mehr, wenn auch immer noch viel zu wenig Spielraum einräumt.

Abschließend sei zusammenfassend festgestellt: Die Allgemeinmedizin litt in der Vergangenheit an dem Selbstmißverständnis, angewandte Klinik zu sein. Obwohl sie gegenwärtig im Begriff ist, sich davon zu befreien, hat sie bisher, zumindest in ihrem offiziellen berufspolitischen Teil, noch nicht den Weg zur Psychosomatik in vollem Umfang vollzogen, und damit auch noch nicht zu ihrer vollen Wirksamkeit gefunden.

27 Der Kliniker als Psychosomatiker

Rolf H. Adler und *Willi Hemmeler*

Dieses Kapitel ist eng mit dem Weg der beiden Autoren zu ihrer beruflichen Identität als internistischer Kliniker (R. H. Adler) und klinischer Psychologe (W. Hemmeler) verbunden. Die praktizierte Medizin, die internistische Psychosomatik in unserem Fall, kann und soll nicht von den Personen losgelöst sein, die sie ausüben. Diese Medizin findet im Situationskreis (Kap. 1) statt, in dem der Arzt beteiligtes Subjekt und nicht unabhängiger Beobachter ist, wie es das technische Modell der Medizin postuliert.

Das einleitende Fallbeispiel soll zeigen, welche praktischen Konsequenzen für Diagnostik und Therapie eine rein somatische Medizin (mechanische = M-Medizin), eine somatische, die psychosoziale Aspekte ergänzend addiert (additive = M'-Medizin), und eine Medizin, die somatische, psychische und soziale Aspekte integriert (integrative = I-Medizin), haben. Diese drei Stufen von Medizin(modellen) haben die beiden Autoren, der eine in der somatischen Medizin, der andere in der Psychologie beginnend, erlebt, bis sie in gemeinsamer Arbeit zu einem integrativen Konzept fanden. Der theoretische Hintergrund dieser Konzepte soll anhand des Fallbeispiels erläutert werden.

27.1 Fallbeispiel und Interpretation

Eine 1960 geborene Frau wird 1988 wegen „Chondropathia patellae beidseits mit hysterischer Aggravation der Kniebeschwerden, die zur 100%igen Invalidität mit Berentung geführt haben", vom Orthopäden zugewiesen.

Angaben aus den Unterlagen: Seit dem dritten Lebensjahr Temporallappenepilepsie bekannt. 1981 Computertomogramm und Elektroenzephalogramm wegen Kopfschmerzen. Kurz vorher psychiatrische Hospitalisation zum Medikamentenentzug. Nachher Betreuung durch psychiatrische Poliklinik, dabei psychogene Gangstörung, Schreianfälle, Episoden mit Atemnot. 1984 Hospitalisation in Lehrspital, medizinische Abteilung, wegen Kopfschmerzen.

Operative „lateral release" der Patellae beidseits. Postoperativ verschwindet Kopfweh, Kniebeschwerden treten auf. Führen 1986 zur Verlagerung der Tuberositas tibiae links, 1987 rechts, zu dieser Zeit atypischer Gesichtsschmerz.

Körperstatus: Unauffällig bis auf beidseitige habituelle Luxation der Patellae. Orthopäde empfiehlt Operation wegen Gefahr der Arthrose.

Lebenslauf: Scheidung der Eltern, als Patientin drei Monate alt war. Wegen Vernachlässigung durch Mutter zu Pflegeeltern. Mit drei Jahren Epilepsie, deswegen über ein Jahr lang in Anstalt für Epileptiker. Nachher zurück zu Pflegeeltern. Mit fünf Jahren ins Waisenhaus. Mit 14 Jahren zur zum zweiten Mal geschiedenen Mutter. Wegen Streit mit ihr ein Jahr später Rückkehr zur Pflegefamilie. Verkäuferinnen-Lehre, dann Verkäuferin, viele Stellenwechsel. Mit 23 Jahren Verlobung mit 14 Jahre älterem Mann.

Kommentar: Die Patientin wird wegen Kopfschmerzen, die nicht erklärt werden können, eingewiesen, aber wegen einer anatomischen Störung operiert, die nur zu geringen Beschwerden geführt hatte. Die Kopfschmerzen verschwinden und Kniebeschwerden treten auf, die seither anhalten. Psychische und soziale Angaben werden zwar in der Krankengeschichte festgehalten, aber es findet sich darin kein Versuch, diese mit den Kopfschmerzen in Zusammenhang zu bringen.

Medizintheoretische Aspekte: Die Patientin wird betrachtet, als wenn Medizin „Medizin an sich" wäre, unabhängig von einem medizinischen Weltbild, medizinischen Grundlagenwissenschaften, einer bestimmten Krankheitsdefinition und einer ebensolchen, was die Therapie betrifft. Die zur Anwendung gelangte Medizin besitzt aber Merkmale, die eine dahinter verborgene, bestimmte Theorie verraten. Sie wird von den sie ausübenden Ärzten nicht bewußt wahrgenommen. Die Theorie dieser Medizin ist die Biomedizin (Foss und Rothenberg, 1987), die besser Mechano-Medizin genannt werden sollte, weil sie alle nicht-mechanischen Eigenschaften des Lebens unberücksichtigt läßt. Sie ist **reduktionistisch** (Erklärung des Ganzen durch Erforschung der kleinsten Bausteine des Organismus), **deterministisch** (gleiche Ursachen bedingen gleiche Folgen), **unpersönlich** (der Körper funktioniert mit Seele gleich gut oder gleich schlecht wie ohne) und **ungeschichtlich** (Erfahrungen spielen keine Rolle). Der Arzt ist ein außenstehender, unabhängiger Beobachter. Diese Medizin hat wissenschaftliche Grundlagen, die nicht für die Medizin geschaffen wurden. Es sind die klassische Mechanik von Newton und die statistische Thermodynamik. Die Mechano-Medizin erwartet, daß durch Anwendung dieser Grundlagen die Forschung schließlich Bau- und Funktionsweise der kleinsten Einheiten des Körpers so weit zu analysieren vermag, daß daraus auf Bau und Funktion des gesamten Organismus geschlossen werden kann. Sie definiert Krankheit als Abweichung physikalischer und chemischer Parameter von der Norm, und Therapie entsprechend als Korrektur dieser Abweichungen durch physikalische und chemische Maßnahmen. Foss und Rothenberg verwenden die bekannte Untersuchung von Levine et al. (1981), um die Denkweise dieser Medizin zu beleuchten: Die Schmerzverminderung durch Placebo kann durch den Morphinantagonisten Naloxon aufgehoben werden. Sie interpretiert dieses Phänomen durch die Fähigkeit des Placebos, Endorphine freizusetzen. Die Möglichkeit, daß durch die Korrektur der Vorstellungen und Erwartungen „es ist nur ein Placebo" – die Endorphinausschüttung verhindert werden könnte, wird nicht berücksichtigt. Damit wird auch nicht erkannt, daß psychische Phänomene „materiell" ins Körpergeschehen einzugreifen vermögen.

Weitere Bemerkungen zum Fallbeispiel: Die Unterlagen enthalten neben Hinweisen auf organische Veränderungen auch solche auf psychogene Körpersymptome (Gangstörung, Schreianfälle, Atemnot, Medikamentenmißbrauch) und auf soziale Erfahrungen, wie den Verlust des Vaters mit drei Monaten durch Scheidung der Eltern, Hospitalisierung während mehr als einem Jahr, wechselnde Unterbringung bei Pflegeeltern und im Waisenhaus, später viele Stellenwechsel. Sie werden aber weder in die Diagnose noch in die Therapie integriert.

Die Mechano-Medizin hat hier zwar beachtet, daß es psychische und soziale Faktoren gibt, die bei dieser Patientin erwähnenswert sind, aber schließt sie nicht in ihre Medizin ein und leitet aus ihnen keine Handlungsanweisungen ab. Foss und Rothenberg (1987) nennen die Mechano-Medizin das „engineering model (E)". Es wird zum E'-Modell, wenn psychische und soziale Faktoren erwähnt, aber nicht integrativ verwendet werden, dies bezeichnen sie als additiv. Wir verwenden anstatt des englischen E ein M (Mechano)-Modell und für das additive ein M'. Seine Grundlagen sind die der Physik und Chemie des 17. bis 19. Jahrhunderts.

Das **konsiliarische Interview** (nach der in Kap. 11 dargestellten Technik durchgeführt) bringt folgende Informationen zutage: Die Patientin trug tiefschwarz gefärbtes Haar, war stark geschminkt, hatte sehr lange lackierte Fingernägel, jeder Nagel mit einem Herzchen versehen, dazu Krücken, mit Stofftierchen behangen. Sie wirkte burschikos, kindlich, offen und vertrauensselig. Ihre Ausführungen waren dramatisch, farbig, und sie gebrauchte häufig medizinischen Jargon. Ihre aufgeräumte Stimmung paßte oft nicht zum betrüblichen Inhalt ihrer Geschichte. Als Interviewer war ich mehrmals verwirrt über den zeitlichen Ablauf ihrer Angaben, die Dimensionen ihrer Symptome, und ich fühlte mich manchmal stark berührt, dann wieder weit entfernt. Sie gab an, mit 23 Jahren (im Jahr vor der Spitaleinweisung 1984) einen 14 Jahre älteren Alkoholiker kennengelernt zu haben. Er sei oft brutal zu ihr gewesen, habe z. B. ihren Kopf gegen die Wand geschlagen. Sie sei wegen seiner Verwahrlosung spät abends fast täglich von ihrem 35 km entfernten Arbeitsplatz – einer Bar – mit

einem Taxi zu ihm gefahren, um ihm zu helfen. Sie sei vom Gefühl her in einer verwirrenden Lage gewesen, einerseits geplagt und abgestoßen durch die Brutalität und andererseits innerlich gezwungen ihm beizustehen. Als sie sich zu wehren begonnen und gedroht hätte, ihn zu verlassen, habe er seine Brutalität aufgegeben und nicht mehr getrunken. Bei der Arbeit sei sie hin und wieder in den Knien eingeknickt, aber dadurch nicht eigentlich gestört gewesen. Jetzt aber könne sie nur noch an Stöcken gehen, und sie vermöge nicht mehr, die Knie zu strecken.

Mit sechs Jahren sei sie vom Stiefvater sexuell mißbraucht worden, mit 12 Jahren vom Freund der Mutter. Sie habe häufig Schläge erhalten. Im Waisenhaus sei die Atmosphäre kalt gewesen, bei Krankheit aber sei man in der Wohnung des Leiterehepaars untergebracht und liebevoll behandelt worden. Während der Zeit im Waisenhaus habe sie im Streit ein anderes Mädchen so stark gewürgt, daß ein Erwachsener sie habe wegreißen müssen, und sie habe ein anderes Kind mit einem Stein am Kopf verletzt. Beide Vorfälle hätten zu strengen Bestrafungen geführt. Mit 17 Jahren habe sie ihr Vater mit dem Kopf gegen die Wand des Wohnwagens geschleudert. Damals hätten die Kopfschmerzen begonnen.

In einer kontrollierten Studie verglichen Adler et al. (1988) Patienten mit psychogenen Schmerzen (Definition nach DSM-III) mit solchen mit organisch bedingten Schmerzen, weiter mit Patienten, die an körperlich erlebten, psychogenen Symptomen (aber nicht Schmerz) litten, und mit einer vierten Gruppe, die organische Krankheiten ohne Schmerzen hatte. Unter anderen Merkmalen fanden sich bei den Patienten mit psychogenem Schmerz signifikant häufiger Brutalität der Eltern dem Kind bzw. späteren Schmerzpatienten gegenüber, nur Zuneigung, wenn das Kind krank oder verletzt war, Blockierung des kindlichen Aggressionsausdrucks mit Entwicklung von Schuldgefühlen, sexueller Mißbrauch und später als Erwachsene häufiger schwere Operationen, Unfälle und Probleme am Arbeitsplatz sowie Mißbrauch von Medikamenten.

27.2 Wissenschaftlichkeit verschiedener Medizinmodelle

Korrelationen werden von der M-Medizin als valide erachtet, wenn der als ursächlich bezeichnete Faktor der Krankheit zeitlich vorangeht, die Beziehung nach Elimination von intervenierenden Variablen signifikant bleibt und eine Dosis-Wirk-Beziehung besteht (Morrison und Paffenbarger, 1981). Dies ist in der zitierten Studie der Fall, aber auch in manchen anderen, in denen psychische bzw. soziale Faktoren mit somatischen Parametern korrelieren. Eigenartigerweise würdigt die M-Medizin diese Faktoren aber höchstens als Risikofaktoren, nicht aber als ursächliche (Foss und Rothenberg, 1987). Dies zeigt, daß „Wissenschaft/lich" in der M-Medizin nicht vom Vorgehen abhängt, sondern vom Gebiet, dem die Forschung gilt. Dieser „Wissenschaftlichkeit" muß eine vom Fachgebiet unabhängige gegenübergestellt werden (Odegaard, 1986): „Science represents man's most persistent effort to extend and organize knowledge by reasoned efforts that ultimately depend on evidence that can be consensually (übereinstimmend, nach allgemeiner Übereinstimmung) validated." Engel merkt dazu an: „Note that this definition places no limits on what phenomena may be subject of scientific inquiry" (1988).

27.3 Ein integratives Medizinmodell

Die Patientin hätte viel eher vor invalidisierenden Operationen geschützt werden können, wenn die M'-Ärzte die Beziehung zwischen Kindheitserfahrungen und der Neigung als Erwachsene, aus psychischen Gründen Schmerz erleiden zu müssen („pain proneness"; Engel, 1959; s. Kap. 11) gewürdigt hätten. Das Beispiel verdeutlicht, daß ein Paradigmawechsel im Sinn Kuhns (1973) zu einem Modell der Medizin vollzogen werden muß, das bio-psycho-sozio-kulturelle Faktoren integrativ erfaßt und das M- und M'-Modell als eine Ebene unter vielen einschließt. Ein solches Modell ist von Engel (1977) entworfen und von Foss und Rothenberg (1987) weiterentwickelt worden. Es deckt sich weitgehend mit dem in Kapitel 1 dargestellten Konzept von v. Uexküll und Wesiack. Dieses Krankheitsmodell ist charakterisiert durch:
- sein **Offensein** (vgl. Situationskreis in Kap. 1),
- seine **kybernetischen Regelkreise** (vgl. Zeichen, Bedeutung, Merkmal und Wirkmal in Kap. 1),
- die **Indeterminiertheit** (eine Ursache zeigt nicht immer die gleiche Folge, sondern wird durch den Zustand des komplexen Organismus und dessen Programme mitbestimmt),
- die **hierarchische Schichtung** (mit ihren Auf- und Abwärtsbewegungen, s. Kap. 1),
- die **Emergenz** (Auftreten neuer Eigenschaften beim Zusammenschluß von Subsystemen zu Suprasystemen, s. Kap. 1),
- die **Selbstorganisationstendenz** lebender Systeme (Neigung, sich zu immer komplexeren Strukturen zu organisieren),
- den Einbau der **Lebensgeschichte** in die Programme des Organismus und
- die **Beteiligung des Arztes,** der in den Situationskreis eingeschlossen ist.
Wissenschaftliche Grundlagen bilden Quantenphysik und irreversible Thermodynamik. Die Krankheitsdefinition hängt nicht nur von äußeren Faktoren wie Bakterien oder inneren wie Enzymdefekten ab, sondern zusätzlich von den Regeln und Programmen, eingeschlossen psychische, soziale und kulturelle, des Organismus. Die Therapie greift multisystemisch an, kann also gleichzeitig chemisch, physikalisch, interaktionell und sozial sein.

Das I-Modell in der klinischen Forschung: Dieses Modell sei anhand eines Beispiels aus der Analgetikaprüfung mit experimentell erzeugtem ischämischem Muskelschmerz erläutert (Adler und Lomazzi, 1974):

Versuchspersonen wurden mit dem „submaximal effort tourniquet test" (randomisiert und doppelblind) vier Versuchen unterzogen, bei denen sie drei verschiedene Analgetika und einmal Placebo per os erhielten. Die Zeit-Schmerzintensitäts-Kurven unterschieden sich unter den vier Bedingungen nicht (Sicht vom Standpunkt des M-Modells aus). Nach jedem Versuch wurden die Versuchspersonen interviewt und mit offenen Fragen aufgefordert, über etwaige „Ängste" und „Bemühungen tapfer auszuhalten" zu berichten. Dann wurden die „ängstlichen" und „tapferen" Versuchspersonen ausgeschieden, und zwar aufgrund von Überlegungen, die auf dem Schmerzmodell von Melzack und Wall (1965; s. Kap. 35) beruhen. Es ist ein hierarchisch aufgebautes Interaktions-(I-)Modell und besteht aus dem „sensorisch-diskriminierenden" System, das afferente Impulse nach Qualität, Lokalisation und zeitlichem Muster analysiert, aus dem „motivierend-affektiven" System, das den Weh-Charakter zum Schmerzerleben und -verhalten beiträgt, und aus dem „zentralen Kontrollsystem", das die lebensgeschichtlich gesammelte Erfahrung mit Schmerzreizen in Schmerzempfinden und -verhalten integriert. Unsere Hypothese, die bestätigt wurde, nahm an, daß die vier Versuchsbedingungen nur von Versuchspersonen unterschieden werden können, bei denen das „sensorisch-diskriminierende" System relativ ungestört arbeiten kann, wenig beeinflußt vom „motivierend-affektiven" und vom „zentralen" Kontrollsystem, also klinisch nur von den ruhigen, nicht aber von den „angstvollen" und/oder „tapferen" Versuchspersonen. (Vom Standpunkt des I-Modells aus ließen sich also Analgetikawirkungen unter bestimmten psychosozialen Bedingungen nachweisen, die das M-Modell nicht erfassen konnte.) Diese Anwendung eines I-Modells illustriert, wie eine bestimmte Theorie entscheidet, welche Beobachtungen gemacht und zur Interpretation benutzt werden, und zeigt die Integration der Versuchsleiter-Versuchsperson (= Arzt-Patient)-Beziehung in die Untersuchung.

27.4 Wissenschaftliche Erfassung der Arzt-Patient-Beziehung

Es stellt sich die Frage, ob die im I-Modell berücksichtigte Arzt-Patient-Beziehung überhaupt wissenschaftlich erfaßt werden kann, oder ob sie als – zwar hochgeschätzte – ärztliche Kunst weiter aus der wissenschaftlichen Medizin ausgeklammert bleiben muß.

Diese Frage sei an der vom Kliniker häufig beobachtbaren Aufwärtsbewegung des Armes (mit Handflächen nach oben) besprochen: Engel und Schmale (1968) hatten diese Bewegung als Begleitzeichen des Affektes „Hilflosigkeit" interpretiert. Wie kann diese Interpretation wissenschaftlich untersucht werden? Verschiedene Möglichkeiten sind denkbar:
- Pantomimen könnten aufgefordert werden, den Affekt „Hilflosigkeit" in der Körpersprache auszudrücken.

– Versuchspersonen könnte in Hypnose die Suggestion gegeben werden, sich hilflos zu fühlen und dabei könnte die Gestik beobachtet werden.
– Versuchspersonen könnte ein Film gezeigt werden, der Hilflosigkeit erzeugen dürfte, und sie könnten beobachtet und nach ihren Gefühlen befragt werden.

Ein klinisches Beispiel soll die praktische Bedeutung des I-Modells zeigen, das die Arzt-Patient-Beziehung erfaßt und für wissenschaftlich analysierbar hält. Engel begab sich bei einem Besuch einer Universitätsklinik mit dem Gastroenterologen auf Visite. Dieser begrüßte eine im Bett wartende Frau und teilte ihr mit, die Leberbiopsie sei normal ausgefallen und sie dürfe am nächsten Tag nach Hause gehen. Als er sie fragte, ob sie sich freue, sagte sie „ja"; Engel entging die gleichzeitige Gestik der Hilflosigkeit (Heben des Unterarmes mit Handfläche nach oben und Fallenlassen des Armes) nicht, er trat an ihr Bett und wiederholte fragend „ja"? Daraufhin brach sie in Tränen aus, schilderte ihre Angst vor der Heimkehr in ihr leeres Haus, das der Ehemann vor kurzem verlassen hätte, weil er zur Freundin gezogen sei. Die soziale Situation dieser früher Alkohol mißbrauchenden Patientin, die Gefahr lief, ihre Sucht unter den trüben Lebensaussichten wieder aufzunehmen, konnte anschließend besprochen werden (1972).

27.5 Der Übergang vom M- und M'-Modell zum I-Modell

Der Kliniker kommt unserer Meinung nach um den Paradigmawechsel zum I-Modell nicht herum. Lehnt er ihn ab und bleibt er beim M- oder M'-Modell, so ist er an immer komplizierteren Patientenschicksalen beteiligt, wie dem zu Kapitelanfang geschilderten.

> Der Paradigmawechsel hat zur Folge, daß der psychosomatische Kliniker über eine Technik der Anamneseerhebung verfügen muß, die es ihm erlaubt, Daten der verschiedenen Ebenen des Patienten in **einem** Arbeitsgang zu erfassen und aufgrund seines bio-psycho-sozio-kulturellen Wissens ihre Bedeutung zu erkennen. Er bemüht sich, sie zu gewichten, ihre Wechselbeziehungen zu erarbeiten und sie in Diagnostik und Therapie einzubeziehen, ohne aus Bevorzugung der einen oder Abneigung gegenüber der anderen Art von Daten einer von ihnen mehr Aufmerksamkeit zu schenken als der anderen.

In unserem Fallbeispiel konzentriert er sich nicht einfach auf den einzigen von der Norm abweichenden Körperbefund – die Patellaluxation –, sondern er erfährt, daß die Patientin häufig Stellenwechsel vollzog, einen brutalen Alkoholiker als Freund wählte; er bemerkt, daß ihre Kindheitserfahrungen sie prädestinieren, als Erwachsene ein masochistisches Verhalten

und „pain proneness" zu zeigen. Vielleicht wird er ihr später raten, wegen der Gefahr der Arthrose einen Eingriff vornehmen zu lassen. Er wird sich aber vorher bemühen, ihr bei der Klärung der verworrenen sozialen Lage zu helfen, mit ihr die Beziehung zwischen ihren Kindheitserlebnissen und ihrer Partnerwahl zu verstehen und für sie eine zuverlässige, konstante Bezugsperson zu sein, bzw. eine solche zu finden, wie z.B. einen Hausarzt, eine Sozialarbeiterin, welche das I-Modell in ihrer Arbeit verwenden, das Symptom nicht isoliert betrachten und um jeden Preis zu beseitigen versuchen. Der somatische Kliniker, der keine Tätigkeit anstrebt, welche die Persönlichkeit des Patienten, seine soziale Situation, seine zwischenmenschlichen Beziehungen und diejenige zum Arzt versteht, läßt sich vom psychosomatischen Kliniker klar unterscheiden. Der somatische Kliniker, falls er sich zum M'-Modell bekennt, kann den Unterschied aber nicht erfassen, obwohl er sich ja um psychische und soziale Faktoren kümmert. Er geht mit ihnen nur additiv um, klärt zuerst somatisch ab, behandelt zuerst somatisch und – wenn dies nicht zum Ziel führt – beginnt er mit dem Patienten die psychosoziale Situation zu besprechen und überweist ihn zum Psychiater. Das Fallbeispiel zeigt, was dabei geschieht: Neben der somatischen Abnormität wird die psychische Problematik festgehalten und vom Psychiater zu behandeln versucht. Das M'-Modell integriert aber Sucht, Konversionssymptom, „pain proneness" und die Gefahr des sekundären Gewinns nach der allfälligen Knieoperation nicht.

Das Verhalten des Gastroenterologen in Engels Beispiel entspricht dem M-Modell. Er achtete nicht auf die averbale Mitteilung der Patientin, verpaßte ihre soziale Situation und setzt sie der Gefahr des Rückfalls ins Trinken aus.

Der (psychoanalytisch) geschulte Psychotherapeut als Konsiliarius ist vom Psychologischen her für die psychosomatische klinische Tätigkeit gut gerüstet. Er besitzt aber auf somatischen Gebieten wie der Differentialdiagnose, der Pathophysiologie usw. nicht die Erfahrung und das Wissen, um die verschiedenen Ebenen, die im I-Modell berücksichtigt werden müssen, sicher beurteilen zu können. Sein Denken im M'-Modell hinderte ihn, den Chirurgen vor dem sekundären Gewinn nach der Knieoperation zu warnen.

27.6 Das Bild des Patienten in Abhängigkeit vom verwendeten Modell

Einen weiteren Nachteil beim Praktizieren des M'-Modells bildet die Auswahl der Patienten, die der Somatiker dem Liaison- (Konsiliar-) Psychiater zuweist. Es werden ihm Patienten geschickt, die mit Suizid drohen oder einen Selbstmordversuch hinter sich haben, Patienten mit Depressionen, alterspsychiatrisch Kranke und solche, bei denen kein organischer Befund festgestellt wurde. Eine bedeutende Gruppe, bei der psychische und soziale Aspekte nicht ins Auge springen, deswegen aber nicht minder wichtig sind, entgeht ihm: z. B. Kranke, die ihren Körper im Bemühen, das psychische Gleichgewicht zu wahren, sekundär schädigen (Laxantienabusus, Schmerzmittelabusus), Patienten mit organischen Störungen, die durch psychischen Streß ausgelöst oder verschlimmert wurden (körperliche Begleit- und Folgeerscheinungen im Zusammenhang mit der Flucht-Kampf- und der Rückzug-Konservierungs-Reaktion), Patienten, bei denen die Symptome Folge einer Konversion sind (vgl. Kap. 30), Patienten mit chronischen und unheilbaren Leiden, denen das Leben mit ihrer Krankheit Schwierigkeiten bereitet.

27.7   Der Lehrer der psychosomatischen Medizin

Die Bedeutung des psychosomatischen Klinikers als Lehrer sei nicht vergessen. Der Somatiker, arbeite er nach dem M- oder M'-Modell, der annimmt, Intuition und „common sense" genügten, beschäftigt sich mit psychosozialen Aspekten nicht oder rein intuitiv. Da er sie nicht bewußt meistert, kann er sie schlecht erfassen, erklären und mitteilen. Wir stoßen auf das Bild des „guten" Arztes, gelangen aber nicht zu einem wissenschaftlichen Verständnis dieser grundlegenden Phänomene (Engel, 1975), wie sie am Beispiel der „Hilflosigkeit" oben erwähnt wurden. Zudem vermag weder er noch der psychiatrische Konsiliarius psychosomatische Medizin so vorzuleben, daß der Student ein Modell zur Identifikation vor sich hat, das die Spaltung in „eine Medizin für den Körper ohne Seele und eine für die Seele ohne Körper" erst gar nicht entstehen läßt. Die Erfahrung zeigt, daß sich psychosomatisches Arbeiten nicht von selbst einstellt, wenn man vom Somatiker in körperlichen und vom Psychiater in psychischen Aspekten der Medizin unterrichtet wird.

27.8   Die Arbeit des Klinikers im I-Modell

Die Tätigkeit des psychosomatischen Klinikers ist schwierig: Es werden ihm zum Teil Patienten zugewiesen, die hartnäckigste Probleme aufweisen, beispielsweise solche mit chronischen Schmerzen, bei denen unter Umständen „pain proneness" vorliegt, solche mit Medikamentenabhängigkeit usw. Da er heute noch Pionierarbeit leistet, und ihm Kollegen Patienten als letzte Instanz zuweisen, möchte er die Berechtigung seines Ansatzes unter Beweis stellen und Erfolge aufweisen. Dies kann bei den zuweisenden Kollegen und bei ihm selbst allzu große Erwartungen und entsprechende Enttäuschungen auslösen. Die Tatsache, daß sich die Auswirkungen seiner Arbeit erst nach längerer Bemühung zeigen und mit großem Einsatz erarbeitet werden müssen, macht die Aufgabe des psychosomatischen Klinikers auch nicht leichter. Zudem muß er es ertragen können, von den M- und M'-Ärzten als Psychiater und von den Psychiatern als Somatiker betrachtet zu werden. Er muß um eine eigene Identität ringen und aushalten, eine beruflich „marginale" Existenz zu führen (Friedman, 1988). Seine Konzepte bringen ihn in Gegensatz zu Kollegen, medizinischen Fachgesellschaften, Krankenkassen, Versicherungen, Gesetzgebung und Gesundheitspolitik. Nicht zuletzt hat dies ein geringeres Einkommen zur Folge, denn Labor- und Röntgenuntersuchungen werden meistens besser bezahlt als die Erhebung einer Anamnese, die die verschiedensten Ebenen des Patienten berücksichtigt.

Der psychosomatische Kliniker ist kein hochspezialisierter – in unserem Fall – Internist oder Psychotherapeut, und er muß die Grenzen seines Wissens annehmen können. Er wird aber durch die Integration der verschiedenen Ebenen und Arbeitsgebiete in sich selbst die Emergenz „neuer Eigenschaften" erleben und dadurch Befriedigung erfahren.

---

28 Die Institutionalisierung der Psychosomatischen Medizin im klinischen Bereich

Karl Köhle und *Peter Joraschky*

28.1 Ziele, Voraussetzungen, Bedarf und Konzepte

Karl Köhle

„Psychosomatische Medizin ist ein relativ neuer Name für eine Form der Medizin, die so alt ist wie die Heilkunde selbst. Es handelt sich um keine Spezialität, sondern um eine Betrachtungsweise, die alle Disziplinen der Medizin wie der Chirurgie betrifft; eine Betrachtungsweise, die nicht etwa dem Körperlichen weniger, sondern dem Seelischen mehr Beachtung schenkt" (Weiss und English, 1949).

„Der Naturwissenschaftler muß vor allem anderen darum bemüht sein, bei seinen Urteilen sich selbst auszuschalten" (Carl Pearson, 1892).

„Um Lebendes zu erforschen, muß man sich am Leben beteiligen. Man kann zwar den Versuch machen, Lebendes aus Nicht-Lebendem abzuleiten, aber dieses Unternehmen ist bisher mißlungen. Man kann auch anstreben, das eigene Leben in der Wissenschaft zu verleugnen, aber dabei läuft eine Selbsttäuschung unter" (V. v. Weizsäcker, 1946).

28.1.1 Die „psychosomatische Betrachtungsweise" in der Medizin – Grundfragen und Konsequenzen für die Institutionalisierung

Zielvorstellungen und Verständnisansätze

Ziel der klinischen Psychosomatik ist es, die „psychosomatische Betrachtungsweise" den übrigen klinischen Fächern zur Ergänzung ihrer Arbeitsansätze anzubieten. Bei jedem Kranken soll es dadurch möglich werden, die Wechselwirkungen zwischen Leiblichem, Seelischem und Sozialem zu berücksichtigen, unabhängig von der Zuordnung des Kranken zu einer medizinischen Spezialdisziplin. Psychosomatik wird damit zu einem Grundlagenfach der klinischen Medizin, nicht zu einer neuen Subdisziplin.

In „psychosomatischer Betrachtungsweise" versuchen wir, den Kranken in seiner Lebenssituation, in seinen Wechselbeziehungen mit seiner Umwelt zu verstehen. In einer solchen systemischen Betrachtungsweise benötigen wir sowohl Verständnisansätze, die speziell für einzelne Teilsysteme – etwa biologische, psychologische oder soziale Teilsysteme – entwickelt wurden, als auch Verständnisansätze für die Beziehungen zwischen diesen Teilsystemen.

Gefordert wird damit die Einführung neuer Grundlagenfächer in die wissenschaftliche Medizin: die Einführung psy-

chologischer und sozialwissenschaftlicher Verständnisansätze, vor allem aber auch die Einführung von Ansätzen der Kommunikationswissenschaften, die in einem systemtheoretischen Konzept die Regeln der Interaktion zwischen den einzelnen Systemen bzw. Subsystemen untersuchen (vgl. Th. v. Uexküll und Wesiack, Kap. 1).

Der Arzt gehört in dieser „psychosomatischen Betrachtungsweise" zur Umwelt des Kranken. Die Erforschung der wechselseitigen Beziehung zwischen Arzt und Patient und ihrer Folgen nicht nur für Diagnostik und Therapie, sondern auch für den wissenschaftlichen Erkenntnisprozeß wird zum zentralen Gegenstand klinischer Psychosomatik.

Gegenstand eines rein krankheitszentrierten biomedizinischen Verständnisansatzes sind pathologisch veränderte Organstrukturen und gestörte Organfunktionen. Die für Diagnostik und Therapie bedeutsamen Daten können dabei im Prinzip durch innerlich unbeteiligte, beliebig austauschbare Beobachter gewonnen werden. Einflüsse der Persönlichkeit des Beobachters müssen bei streng wissenschaftlichem Vorgehen sogar systematisch ausgeschaltet werden, beispielsweise im sog. Doppelblindversuch. Fragen nach Problemen im Interaktionsfeld zwischen Patient, Arzt und Pflegepersonal kommen in einer solchen Medizin nicht nur nicht vor, sie sind innerhalb des gewählten wissenschaftlichen Rahmens sinnlos. In einem solchen Verständnisansatz wird die Untersuchung des Kranken auf die Untersuchung einzelner Subsysteme reduziert; gleichzeitig werden diese Subsysteme aus ihrer Interaktion mit den übrigen Systemen herausgelöst. Dies stellt einen schwerwiegenden Eingriff in das untersuchte Objekt dar, der zu einer Verzerrung der Sicht führen kann und in jedem Falle bei der Interpretation der erhobenen Befunde zu berücksichtigen ist.

Exkurs: Veranschaulichung möglicher Folgen unreflektierter Reduktion in der Medizin

Wissenschaftliches Denken und Fortschritt der Wissenschaft haben auch in der Medizin häufig eine Isolation menschlicher Phänomene aus dem Gesamtzusammenhang und eine Reduktion der Betrachtung auf einzelne Teilaspekte zur Voraussetzung. Wir möchten versuchen, durch zwei Beispiele zu veranschaulichen, wie sehr es darauf ankommt, das wissenschaftliche Vorgehen im Gesamtzusammenhang zu reflektieren, um einseitige Gewichtung, Verfälschung und zerstörerische Auswirkungen einzuschränken.

Vogel (1961) machte auf ein anschauliches Beispiel der Isolierung eines Kranken aus dem geistigen Gesamtzusammenhang aufmerksam: „Raffael hat am Ende seines kurzen Le-

Abb. 28–1. Raffael: „Die Transfiguration"

bens ein herrliches Bild geschaffen, auf welchem er in einer Komposition zwei aufeinanderfolgende Erzählungen des Evangeliums vereinigte und aufeinander bezog. Das Bild zeigt in seiner unteren Hälfte die Vorstellung des epileptischen Knaben vor den Jüngern und darüber wie in einem transzendenten Raum – die Verklärung Jesu. Es hat den Namen ‚Die Transfiguration' erhalten" (Abb. 28–1).

Dieses Bild hat der verstorbene amerikanische Epilepsieforscher William Lennox seinem großen zusammenfassenden Werk über die Epilepsie als Leitbild vorangestellt – aber nicht in einer vollständigen Wiedergabe, sondern nur in einem Ausschnitt, der die Gesamtkomposition des Bildes nicht mehr erkennen läßt. Im Vordergrund steht nun ganz der epileptische Knabe und seine ratlose Umgebung; aber die obere Hälfte, die die Verklärung Jesu darstellt, ist völlig beseitigt. Bei Goethe können wir darüber lesen: „Wie will man nun das Obere und Untere trennen! Beides ist eins ... beides aufeinander sich beziehend, ineinander einwirkend. Läßt sich denn, um den Sinn auf eine andere Weise auszusprechen, ein idealer Bezug aufs Wirkliche von diesem trennen?"

Lennox beschränkt sich in der Wiedergabe des Raffaelschen Bildes auf das rechte untere Viertel des Bildes, „auf den Bereich dieser unteren Hälfte, dem alle Hinweise auf die obere fehlen. Das entspricht gleichsam einer Beschränkung der Forschung auf ein bloßes Viertel des an den Phänomenen Zugänglichen. Zudem fehlt meist ein Bewußtsein davon" (Kütemeyer, 1963).

Grenzt der Arzt seine Anschauung der Wirklichkeit des Kranken derart ein, so kann der Patient mit seiner Krankheit auch nur innerhalb dieses vom Arzt akzeptierten Bezugsrahmens in Erscheinung treten, nur diesem Bereich seiner Existenz Gewicht beimessen. Die machtvollen Auswirkungen interpersonaler Wahrnehmung hat die experimentelle Sozialpsychologie hinreichend aufgezeigt, für die Theorie der Heilkunde sind diese Befunde jedoch noch nicht ausgewertet.

Im Lehrbuch der Anatomie von Braus (1954) wird im Abschnitt „Allgemeine Gestalt des Menschen" das wissenschaftliche Vorgehen bei der Klärung des Verhältnisses von Gewicht und Größe bei der Gestaltung des „menschlichen Habitus" dargestellt: „Denkt man sich den Körper eines beliebigen Menschen zu Brei zerstampft und damit ein zylindrisches Gefäß von der Länge des betreffenden Individuums bis zum Rande angefüllt, wird man bei geringerer Gesamtmasse einen engeren, bei größerer einen weiteren Zylinder benützen müssen. Die mittleren und höheren Gewichtszunahmen des Menschen sind nach dieser Anschauung für ein gleichmäßiges Menschenmaterial berechnet worden." Es folgt eine entsprechende Graphik. Später wird auf die menschliche Entwicklung eingegangen: „Denkt man sich aus einem Zylinder des oben beschriebenen Schemas eine Scheibe von 1 cm Höhe herausgeschnitten, so wiegt sie beim Neugeborenen 60 Gramm, beim 17jährigen 330 Gramm und beim Erwachsenen 460 Gramm." In der dazugehörigen Abbildung wird noch einmal auf die besondere „Anschaulichkeit" dieses methodischen Vorgehens hingewiesen.

Ohne Schwierigkeit könnte der erwünschte Befund über ein eleganteres Vorgehen gewonnen werden. Das Beispiel macht jedoch Aufwand und Ziel einer solchen Reduktion deutlich. In der Graphik ist von dem erforderlichen affektiven Aufwand – das Zerstampfen beliebiger Menschen zu Brei fällt niemandem leicht – nichts mehr spürbar. Affektfrei kann nun mit den Daten des „Menschenmaterials" manipuliert werden. Der Kranke wird zu einem Objekt reduziert, dem alle Merkmale eines Subjektes, das „lebt und Seele hat",[1] fehlen. Die Handlungsmöglichkeiten des Forschers scheinen erweitert. Die gewonnenen Daten sind für fast beliebige Zwecke verfügbar.

Für die klinische Medizin hat die skizzierte „psychosomatische Betrachtungsweise" weitgehende Konsequenzen:

– Menschlicher Leib und körperliche Krankheit erhalten eine neue Position im Gesamtzusammenhang menschlicher Existenz.
– Der Beziehung zwischen Arzt und Patient, ihrem Umgang, kommt prinzipielle Bedeutung für den wissenschaftlichen Erkenntnisprozeß und die klinische Tätigkeit zu.

---

1 ... „In der Objektivität üben wir uns, das nur als Objekt zu nehmen, was doch ein Subjekt ist, lebt und Seele hat. So wird die Objektivität nicht nur ungenügend, sondern Fälschung. Die Gegenseitigkeit im Umgang geht zugrunde." V. v. Weizsäcker (1950d) verweist in diesem Zusammenhang auf das Dilemma und Verhängnis der in der wissenschaftlichen Forschung auch erforderlichen Objektivität und das durch sie bedingte Eingehen „unbewußter Schuld".

– Die Klinik wird zum entscheidenden Erfahrungsraum einer psychosomatischen Medizin.
– Struktur und Organisation medizinischer Institutionen sind auch unter dem Gesichtspunkt ihrer Auswirkungen auf die Kranken zu analysieren und zu modifizieren.

Wir möchten diese Gesichtspunkte im einzelnen näher ausführen.

Der menschliche Leib

„Ein junges Mädchen wird mit starker Angina, unfähig, auch nur zu sprechen, in die Klinik eingeliefert. Ein junger Arzt äußert nach der Untersuchung: ‚Ja, da haben Sie sich ja was Schönes geholt!‘, worauf sie spricht und sagt: ‚Das ist immer noch besser als ein Kind kriegen‘. Später stellt sich heraus, daß sie am Vortage dem Drängen eines Verehrers, welches solche Folgen hätte haben können, widerstanden hat" (V. v. Weizsäcker, 1946).

Die Rolle des Körpers zeigt sich in „psychosomatischer Betrachtungsweise" „in einem anderen Licht" und muß „nun anders dargestellt werden" (V. v. Weizsäcker, 1946). So kann der Körper eine wichtige Rolle in der Auseinandersetzung, in Konflikten zwischen dem Kranken und seiner Umwelt spielen. In solchen Konflikten können sich reales Verhalten, psychische Verarbeitung und körperliche Reaktionen zum Teil gegenseitig vertreten. Solche Möglichkeiten, solche Regeln der gegenseitigen Beziehung wären in der Medizin bis hinein in die Physiologie und Pathophysiologie zu berücksichtigen.[2]

Der Körper ist intensiv in das ganze Leben eingebunden: „Er ist nämlich jetzt einer, bei dem das Menschliche, welches die Psychoanalyse darstellt, mitredet, mitspricht, mitlügt und mitlistet, auch Wahres mitzeigt und Echtes mitfühlt; er handelt mit. Zu all dem müssen ihm Eigenschaften, Fähigkeiten erteilt werden, die er im Physikalismus und Chemismus nicht hatte" (V. v. Weizsäcker, 1950 b).

Gestalt und Verlauf körperlicher Krankheit werden somit abhängig vom Standpunkt und vom Verhalten des Untersuchers. Krankheit kann in einem Maschinenmodell „Betriebsdefekt sein" oder eine Funktion im Rahmen einer Lebenskrise haben. Krankheit begegnet uns nicht als etwas „objektiv Festgestelltes", sondern wir bestimmen das Wesen von Krankheit mit der Wahl des jeweiligen Bezugssystems mit.[3]

In diese Wahl gehen „Entscheidungen" mit ein: „Was als psychisches oder somatisches Phänomen erscheint, ist bereits Resultat der Parteinahmen und hat sich aus Verdrängung und Entscheidung abgeschieden" (V. v. Weizsäcker, 1927).

Dieses Konzept hat Folgen auch für die klinische Therapie. In einer entsprechend gestalteten Arzt-Patient-Beziehung oder unter den veränderten Lebensbedingungen einer Krankenstation können auch solche körperlichen Fehlfunktionen wieder in ihrem psychosozialen Zusammenhang sichtbar werden, die vorher in Eigengesetzlichkeit erstarrt waren und ihre Verbindung zum psychosozialen Kontext verloren zu haben schienen.

Ein 45jähriger, „hirngeschädigter", „debiler", seit Jahren in einer psychiatrischen Anstalt hospitalisierter Mann wurde wegen einer „fixierten", medikamentös nicht mehr behandelbaren Hypertonie bei beidseitiger Nierenarterienstenose operiert. Die erwartete Blutdrucksenkung trat zunächst nicht ein. Die Blutdruckwerte normalisierten sich erst, nachdem es dem Patienten gelang, freundliche Beziehungen zu Mitpatienten aufzunehmen; die Werte stiegen mehrfach wieder kritisch an, als der Patient sich von diesen Mitpatienten bei deren Entlassung trennen mußte.[4]

In „psychosomatischer Betrachtungsweise" können seelische und soziale **Konflikte** zu Erkrankungen des Leibes beitragen; Krankheiten können nur teilweise gelungene Lösungsversuche solcher Konflikte darstellen. **Widerstände** gegen die „psychosomatische Betrachtungsweise" haben zum Teil ihren Grund darin, daß bei einer psychosomatischen Behandlung diese Konflikte in ihrer ursprünglichen Vehemenz wieder in Erscheinung treten können.

Sowohl die Kranken selbst als ihre Umwelt leisten nicht selten gegen Behandlung und Gesundung lebhafte, zunächst paradox erscheinende Widerstände. Die Kranken sehen eine „Leistung" – den bisherigen Lösungsversuch – in Frage gestellt, die hierdurch entlastete Umwelt befürchtet, aufs neue durch den Konflikt irritiert zu werden. Beide, der Kranke und seine Umwelt, scheinen nicht selten die körperliche Erkrankung leichter ertragen zu können als die „äquivalenten" (V. v. Weizsäcker) seelischen und sozialen Spannungen.

Kranker und Arzt

Von der Art des Umganges zwischen Arzt und Patient hängt es ab, was von der Krankheit in Erscheinung

---

2 V. v. Weizsäcker (1949) weist darauf hin, daß es dabei nicht nur um die Addition psychologischer Aspekte, sondern um eine veränderte Betrachtungsweise geht. „Aber die Einführung von Psychologie ist dabei nur ein Symptom. Es handelt sich dabei um eine andere Auffassung des Menschen, des kranken Menschen, der Krankheit und der Therapie. Es kommt hier nicht darauf an, Beispiele und Argumente dafür zu bringen, daß in Pathologie und Therapie auch Seelisches mitspricht, sondern es handelt sich darum, daß eben, weil dies der Fall ist, jede rein anatomische Beschreibung, jede rein psychologische Analyse bereits einen Fehler enthält, wenn sie Tun und Leiden des menschlichen Subjekts nicht enthält."

3 „Hier kann nur angemerkt werden, daß mit der Wahl von Standpunkt und Bezugssystem auch eine Stellungnahme zur Frage der Bewertung menschlichen Lebens verbunden ist; es kommt darauf an, ob Medizin und die Ärzte sich der Bewertung des Betriebsstaates anschließen, oder ob sie den Wert des Menschenlebens ganz woanders, sagen wir einmal in seiner menschlichen Vervollkommnung sehen … Für den Betriebsstaat ist Gesundheit gleich beliebiger Verwertbarkeit, z. B. in der Wehrmacht, im Erwerbsleben, durch Arbeit, ja sogar im privaten Glücksgefühl …" (V. v. Weizsäcker, 1949).

4 Mitscherlich (1967) hat auf die Möglichkeit einer endgültigen „Zerreißung der psychophysiologischen Spontankorrelationen" aufmerksam gemacht. Diese Zerreißung kann aus einer fortschreitenden Eigengesetzlichkeit des somatischen Krankheitsprozesses herrühren. Sie kann aber auch durch einseitige medizinische Verständnisansätze unterstützt bzw. endgültig besiegelt werden, wenn diese die „somatischen" Anteile des Krankheitsprozesses aus dem Gesamtzusammenhang „herausreißen". (Daß es sich hierbei um eine aktive, ja gewaltsame Reduktion handeln kann, zeigt das angeführte Beispiel aus dem Lehrbuch von Braus.)

tritt. Mit dem Beginn ihrer Beziehung gehören Arzt und Patient jeweils der Umwelt des anderen Partners an. Stellt der Arzt sich dem Kranken flexibel als Person zur Verfügung, läßt er sich von ihm „benützen" (Winnicott), so kann der Kranke der Beziehung exemplarische Gestalt geben: In dieser Beziehung werden dann auch die Beziehungsformen des Kranken zu seiner übrigen Mitwelt deutlich. Für den Arzt wird die Reflexion seines Erlebens, seiner Reaktionen auf den Kranken zu einem wesentlichen Instrument von Erkenntnis und auch Therapie.

In einer solchen „psychosomatischen Betrachtungsweise" können Kranker und Krankheit nicht mehr unabhängig von den Interaktionsprozessen verstanden werden, die zwischen dem Kranken und seiner medizinischen Umwelt ablaufen. Die Einbeziehung der Wechselwirkungen zwischen Subjekt und Objekt führt zu einer Veränderung des Wissenschaftsbegriffes. „Wissenschaft gilt nämlich hier nicht als ‚objektive Erkenntnis' schlechthin, sondern Wissenschaft gilt als eine *redliche Art des Umganges von Subjekten mit Objekten*. Die Begegnung, der Umgang, ist also zum Kernbegriff der Wissenschaft erhoben" (V. v. Weizsäcker, 1950, S. 15).

Nicht die additive Einbeziehung psychologischer Gesichtspunkte in eine biologisch orientierte Medizin macht die „psychosomatische Betrachtungsweise" aus, sondern die konsequente Pflege und Benutzung dieses Umgangs zwischen Arzt und Patient; diese Begegnung nimmt zunächst vor allem im Gespräch Gestalt an.

„Im Gespräch steckt aber das Subjekt, die Seele der Sache. ... Die Anwendung der Psychologie bleibt auch seelenlos, wenn sie in kein Gespräch eingeschlossen ist, und es gibt auch eine Psychiatrie ohne Seele, eine Innere Medizin ohne Inneres" (V. v. Weizsäcker, 1947a).

Für die Heilkunde ergibt sich als Konsequenz die Aufgabe, heilsame und schädliche Auswirkungen in der Beziehung des Arztes und auch institutioneller Konstellationen auf Kranke in ihre wissenschaftliche Betrachtung einzubeziehen. Die Heilkunde steht vor der Aufgabe, all die Interaktionsprozesse, die schon immer zwischen dem Arzt bzw. den anderen Mitarbeitern von Institutionen und dem Kranken ablaufen, systematisch zu untersuchen und – falls erforderlich – ebenso systematisch ihre Entfaltung zu fördern.

Dies hat weitgehende Konsequenzen für den Arzt: „Am wichtigsten erscheint mir immer wieder, daß in einer umfassenden Therapie der Arzt sich selbst vom Patienten verändern läßt, daß er die Fülle aller Regungen, die von der Person des Kranken ausgehen, auf sich wirken läßt ..." (V. v. Weizsäcker, 1928).

Widerstände gegen eine „psychosomatische Betrachtungsweise" können sich daraus ergeben, daß es für den Arzt und andere Mitarbeiter in Institutionen des Gesundheitswesens ungewohnt ist, daß es dem bisherigen Rollenverständnis widerspricht, sich vom Kranken bewegen zu lassen, sich einer Situation auszuliefern, die vom Kranken und seinen wieder auftauchenden Konflikten mitkonstelliert werden kann.

Medizinsoziologische Untersuchungen haben gezeigt, daß Ärzte – von ihnen selbst häufig unbemerkt – das Eingehen einer Beziehung mit dem Patienten, den Eintritt in einen Umgang mit ihm verweigern können. So fand Siegrist (1978), daß Ärzte in 92% der von ihm untersuchten Situationen, in denen schwerkranke Patienten während der Visite Fragen zu ihrer Krankheit an sie richten, nicht die erbetene Auskunft geben, sondern dem Patienten ausweichen. Dieses Verhalten wird mitbestimmt von einem Patientenmerkmal, der Schwere des Krankheitsbildes bzw. der Prognose. Leichtkranken gegenüber fand Siegrist eine größere Bereitschaft der Ärzte, auf Fragen einzugehen. Weitere Beobachtungen und Untersuchungen sprechen dafür, daß das Arztverhalten gegenüber Schwerkranken von dem Versuch mitbestimmt wird, sich vor den Belastungen einer intensiveren Gesprächsbeziehung mit diesen Kranken zu schützen (vgl. Kap. 18 und 66).

Die Klinik

In der Klinik ist der Kranke stärker aus seinen sozialen Beziehungen herausgelöst, die Bedingungen der Institution bestimmen die Beziehung zwischen Arzt und Patient mit. Sie stehen der Verwirklichung einer „psychosomatischen Betrachtungsweise" vielfach entgegen, da die Strukturierung der klinischen Situation aufgrund anderer Konzepte erfolgte; eine systematische Berücksichtigung der Beziehungen zwischen Patienten und Mitarbeitern der Institution, die Möglichkeit zur Darstellung und gemeinsamen Reflexion der Konflikte des Patienten in der klinischen Situation wird nicht gefördert.

V. v. Weizsäcker hat bereits 1925 auf die schwierige Frage der Einbeziehung der Psychotherapie in die Praxis der Heilkunde aufmerksam gemacht und dabei auch auf strukturelle Aspekte der Klinik hingewiesen, die einer solchen Einbeziehung entgegenstehen: „Die Organisation der modernen Kliniken und Krankenhäuser mit großen Sälen und einer gewissen Hierarchie der Ärzteschaft erwächst aus einem völlig anderen Gedanken der Krankheit und des Heilprinzips als dem der Psychotherapie und dem einer Personalpathologie"; die Organisation der Klinik werde von den Methoden der rein somatischen Medizin bestimmt, „trotz allen Individualisierens in der Anwendung". Die Milieuwirkung dieser Klinikstrukturen gehe „als ein total unkontrollierbarer Faktor in die Behandlung" mit ein, während für eine Psychotherapie zu fordern sei, „die Tragweite aller Handlungen des Arztes zu kennen".

Es gibt auch heute noch Universitätskliniken ohne Räume auf den Stationen, in denen der Arzt mit dem Patienten sprechen könnte.

Bei „psychosomatischer Betrachtungsweise" nimmt der Umgang mit dem Kranken, insbesondere der therapeutische Umgang den Platz in der wissenschaftlichen Medizin ein, den sonst das Experiment innehatte.[5]

Die Krankenstation bildet für die klinische Psychosomatik den wichtigsten Erfahrungs- und Forschungsraum. Dieser Arbeitsbereich sollte entsprechend den

---

[5] „... Die Wucht ihrer Leidenschaft kann ja bei Versuchspersonen in einem physiologischen Institut ... auch im günstigen Fall nur sehr verkümmert zur Entfaltung kommen. Adel, Heroismus und Opfer, Neid, Faulheit und Feigheit sind nicht laboratoriumsfähig. Sie sind aber für die Entstehung, Verlauf und Behandlung von Krankheiten von entscheidender Bedeutung" (Kütemeyer, 1967).

Erfordernissen der „psychosomatischen Betrachtungsweise" gestaltet werden können. Insofern stellt die „psychosomatische Betrachtungsweise" auch herkömmliche Strukturen in Institutionen in Frage.

„Die recht verstandene Psychosomatische Medizin hat einen umstürzenden Charakter. In einer solchen Situation wird dann öfters gesagt, ehe man etwas einreiße, solle man etwas Besseres an die Stelle setzen. Dieser Rat ist nicht ganz anschaulich, denn es ist nicht zu sehen, wie an demselben Ort das Alte und das Neue stehen soll" (V. v. Weizsäcker, 1949).[6]

Werden neben theoretischen Konzepten auch Struktur und Organisation von Institutionen in Frage gestellt, so werden Widerstände sichtbar, die sich gegen eine solche Verunsicherung richten; dabei wird deutlich, daß wissenschaftliche Theorien nicht nur als Erklärungsmodelle, sondern auch als Hilfsmittel persönlicher Absicherung dienen, daß sie als Ideologien benutzt werden. In Institutionen wirken sich solche Ideologien gestaltbildend aus. Aus der Untersuchung eines bestimmten Realitätsausschnittes werden in der Überschreitung des Gültigkeitsbereiches Aussagen über das Ganze menschlicher Krankheit abgeleitet, Wissenschaft wird zur Ideologie, nicht selten zum Dogma, das nun nicht mehr dem Verständnis der Kranken, sondern nur der Absicherung ihrer Betreuer dient (Engel, 1977; Richter, 1978). Menschliche Krankheit kann so radikal, entsprechend den Verständnisansätzen des biomechanischen Konzeptes, reduziert werden, oder der Krankheitsbegriff selbst wird so gefaßt, daß etwa psychosoziale Phänomene nicht mehr in seinen Gültigkeitsbereich fallen. Entsprechende Strukturen der Institutionen können sowohl den reduktionistischen als auch den exklusionistischen Ansatz unterstützen. Ziel der „psychosomatischen Betrachtungsweise" wäre ein entgegengesetztes Vorgehen:

„Wenn also die Mahnung ergeht, die Psychologie neben der Somatologie, die Psychosomatik über Physiologie und Psychologie hinaus zu studieren, so ist es nur ein Kunstgriff, um das Unerforschliche in die Medizin *hineinzubringen* und *nicht* das bisher Unerkannte aus ihr *herauszubringen*" (V. v. Weizsäcker, 1947b).

Abschließend sei angemerkt, daß sich Widerstände gegen eine „psychosomatische Betrachtungsweise" auch daraus ergeben können, daß Wissenschaft nicht in einem den gesamtgesellschaftlichen und wirtschaftlichen Interessen und Ordnungen isolierten Freiraum betrieben wird. Dies ist bei allen Versuchen, eine solche Betrachtungsweise in der Medizin zu institutionalisieren, zu berücksichtigen, sollen sie nicht von vornherein zum Scheitern verurteilt sein.

„Es ist völlig aussichtslos, die Krise zu verstehen, wenn man sich der Erkenntnis verschließt, daß der Arztberuf ein Modus des Gelderwerbs oder, wie man in meiner Jugend schamhaft sagte, des Broterwerbs ist. Da Macht, Geld und Wissenschaft aber in einem Konnex stehen wie die drei Seiten eines Dreiecks, so kann niemand eine der drei Seiten zerschlagen, ohne die beiden anderen zu zerschlagen. Die naturwissenschaftliche Medizin ist also ganz präzise diejenige, welche mit der Machtordnung der bürgerlichen Gesellschaft und mit der Geldordnung, die Marx den Kapitalismus nannte, steht und fällt. Der Irrtum, man könne trotz der Vernichtung des Kapitalismus die Naturwissenschaft aufrechterhalten – dieser Irrtum ist inzwischen dadurch offenbar geworden, daß man im Osten Europas statt des Kapitalismus nicht die kommunistische Gesellschaft, sondern den Gewaltstaat bekam, der sich auf die Naturwissenschaft stützt." ... Die Krise „erscheint hier als eine politische, näm-

lich in der Unzertrennlichkeit von Macht, Geld und Naturwissenschaft. So ernst ist also die Lage, daß man das System von keinem dieser drei ändern kann, ohne auch die beiden andern zu ändern" (V. v. Weizsäcker, 1950c).

Klinische Erfahrungen

In der klinischen Situation befinden sich die Beteiligten – Arzt und Patient oder Schwester und Patient – aufgrund des zwischen ihnen abgeschlossenen therapeutischen Vertrages in einer „gemeinsamen Situation" (Th. v. Uexküll). Ihr Verhalten beeinflußt sich wechselseitig. Bei Zugrundelegung eines zirkulären Verständnisansatzes kann das Verhalten des Patienten in einem bestimmten Ausmaß als eine Funktion des Verhaltens etwa seines Arztes oder seiner Krankenschwester aufgefaßt werden und umgekehrt. Das Verhalten des einen beeinflußt die Erfahrung und damit auch das Verhalten des anderen (Laing et al., 1971) (Abb. 28–2).

Ein Experiment studentischer Nachtwachen (Geist et al., 1976) illustriert eine solche Wechselwirkung zwischen ihrem Verhalten und dem Verhalten von Patienten in der klinischen Praxis (Abb. 28–3).

Den Studenten war das häufige Läuten von Patienten während der Nacht aus „nichtigen Anlässen" aufgefallen; sie hatten den Eindruck, die Patienten würden vor allem aus innerer Beunruhigung so oft läuten; Beschwerden und Symptome würden von ihnen eher vorgeschoben, wohl, weil sie davon ausgingen, daß diese von den Nachtwachen leichter akzeptiert würden. Daraufhin variierten die Studenten ihr Verhalten während der abendlichen Stationsübernahme. Stellten sie sich beim Gang über die Station nur kurz vor, so lag die Häufigkeit der nächtlichen Rufe wesentlich höher, als wenn sie im Rahmen einer ausführlichen Pflegevisite länger auf die Sorgen und Befürchtungen der Patienten eingingen. Vor allem die vorher als „seelisch bedingt" eingestuften Anlässe zum Läuten gingen fast vollständig zurück.

Der Verhaltensänderung der studentischen Nachtwachen entsprach so eine Verhaltensänderung der Patienten, die zunächst die Häufigkeit des Rufs nach der Nachtwache betraf, aber auch – soweit durch zusätz-

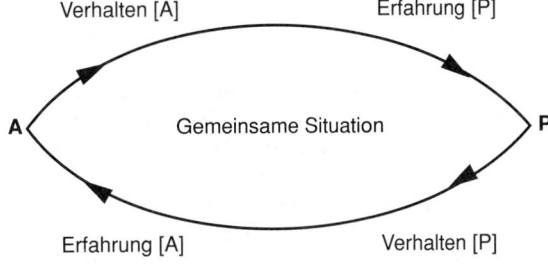

Abb. 28–2. Schema der gegenseitigen Beeinflussung von Verhalten und Erfahrung zweier Personen in einer gemeinsamen Situation (nach Laing et al., 1971).

---

6 V. v. Weizsäcker wendet sich dabei nicht gegen naturwissenschaftliche Forschungsmethoden, sondern er fordert eine Reflexion ihres Gültigkeitsbereiches und ihres Einflusses auf Institutionen: „Von einer Überwindung des Materialismus und Mechanismus im Sinne eines Aufgebens der naturwissenschaftlichen Forschungsmethoden und ihren logischen Grundlagen kann doch gar keine Rede sein. *Nicht* ihr *Prinzip*, sondern ihr *Gewicht* für die Medizin steht zur Debatte" (1925).

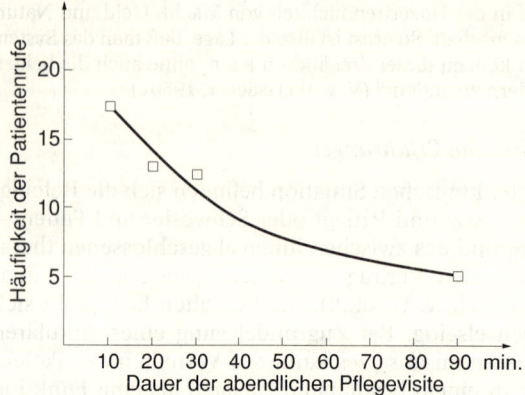

Abb. 28–3. Die Häufigkeit des Läutens von Patienten während der Nacht in Abhängigkeit von der Dauer der abendlichen Übergabevisite auf einer Station mit 15 Patienten (Geist et al., 1976).

liche Beobachtungen erkennbar – das Schlafverhalten der Patienten.

Andererseits ergab eine Untersuchung von Le Shan in New York (Bowers et al., 1971), daß die Reaktion von Krankenschwestern auf das Läuten von Patienten mit von der Prognose abhängt. Je aussichtsloser die Prognose der Kranken war, desto mehr Zeit verging, bis die Schwestern auf das Läuten reagierten. Es ist wahrscheinlich, daß das emotionale Betroffensein der Schwestern durch die Prognose der Kranken ihr Pflegeverhalten beeinflußte.

Im diagnostischen und therapeutischen Prozeß ist die klinische Psychosomatik entscheidend auf die Berücksichtigung der Wechselseitigkeit zwischen Arzt und Patient auch im emotionalen Bereich angewiesen, also auf Phänomene, die in der wissenschaftlichen Medizin bisher gerade systematisch von der Untersuchung ausgeschlossen worden sind. So hat der Soziologe Parsons (1958) „emotionale Neutralität" geradezu als ein wesentliches Merkmal der Arztrolle beschrieben.

„Reflektierte Emotionalität" wird nun im Gegensatz hierzu zu einem wesentlichen Instrument klinisch-psychosomatischer, d.h. wissenschaftlich-medizinischer Diagnostik. Die Konflikte des Patienten lassen sich nämlich nur zum Teil direkt erfragen oder aus objektiv erhebbaren Daten erschließen. Bei Patienten mit Krankheitsbildern, in deren Pathogenese psychosoziale Aspekte eine wesentliche Rolle spielen, hat häufig eine besonders tiefgehende Abwehr der Konflikte vom Bewußtsein stattgefunden. Der Untersucher ist deshalb darauf angewiesen, daß diese Konflikte in der Beziehung zwischen dem Arzt und dem Kranken bzw. in den Beziehungen im sozialen Feld einer Krankenstation wieder in Erscheinung treten und so direkt beobachtet oder über in den Beteiligten ausgelöste emotionale Bewegungen indirekt erkannt werden können. Dies ist möglich, weil Patienten entsprechend ihrer unbewußten Dynamik gegenwärtige Beziehungen analog zu früher internalisierten Beziehungsstrukturen zu konstellieren versuchen, ihre früheren konflikthaften Beziehungsmuster immer wieder neu inszenieren. Der Patient kann jedoch nur

dann seine Konflikte im gegebenen sozialen Feld darstellen und auf diesem Wege mitteilen, wenn Ärzte, Schwestern und die übrigen Mitarbeiter einer Krankenstation bereit sind – etwa wie ein eingestimmter Resonanzkörper – sich vom Patienten in Bewegung setzen zu lassen, mitzuschwingen, und über die Reflexion ihrer Mitbewegung versuchen, vom Kranken ausgehende Impulse zu erkennen und zu verstehen. Eigene Abwehrhaltungen der Mitarbeiter können das Wieder-in-Erscheinung-Treten der Konflikte beim Kranken behindern und sollten im Rahmen eines solchen Konzeptes deshalb der Reflexion zugänglich sein.

Ein solcher Ansatz erweist sich auch für die therapeutische Arbeit als fruchtbar. So kann der Patient im Umgang mit den, im Gegensatz zu seiner Alltagswelt, nicht unreflektiert mitagierenden Partnern seine Konflikte wahrnehmen und bearbeiten, neue emotionale Erfahrungen machen und oft erstmals neue Verhaltensweisen erproben.

Ein solch intensiver Umgang mit Kranken bringt für alle Beteiligten allerdings auch erhebliche emotionale Belastungen mit sich. Aus diesen Belastungen können Widerstände, Vermeidungs- und Rückzugsreaktionen beim einzelnen, aber auch beim Team und Schwierigkeiten in der interdisziplinären Kooperation resultieren.

Die Reflexion des eigenen Betroffenseins, der Mitbewegung des Arztes wird so zu einer Voraussetzung für den Erkenntnis- und Heilungsprozeß. Der Arzt wird während seiner Ausbildung auf diese Anforderungen bisher kaum vorbereitet. Anhand einiger Beispiele möchten wir sie im Rahmen dieses Abschnittes noch etwas verdeutlichen.

Eine 25jährige Patientin mit einer leichteren Form von Colitis ulcerosa malt im Anschluß an Psychotherapiestunden einige ihrer während der Stunden aufgetauchten Phantasien **(Farbtafel 1)**. Sie konnte nicht ertragen, daß andere Menschen, insbesondere Männer, in ihr Gefühle, Wünsche oder gar leidenschaftliche Bewegungen hervorrufen; deshalb wirft sie in der Phantasie alle Männer „in ein Höllenfeuer", „wie Don Giovanni" (Abb. 28–4). Anschließend spielt sie auf den verkohlten Gerippen Xylophon (Abb. 28–5). Diese Phantasien beziehen sich insbesondere auf mich, ihren Therapeuten, die Patientin betont: Ich hänge als größtes Gerippe ganz vorne. Dabei wird die Verbindung zwischen dem leidenschaftlichen Wunsch nach Beziehung und dem Abwehrvorgang durch die Darstellung einer „glühenden Zigarre" im oberen Bildteil, die assoziativ mit sexuellen Beziehungswünschen verbunden ist, deutlich; später im Therapieverlauf phantasiert dieselbe Kranke, wie sie – zunehmend im Mittelpunkt stehend – die anderen Menschen, einschließlich ihres Arztes, wie Marionetten „herumspringen" läßt, ihnen „Hörner und Masken" aufsetzt, u.a.m. (Abb. 28–6). Parallel zur Milderung der Aggressionen und zur Aufnahme der Beziehungen kann sie ihre eigenen aus früheren Entwicklungsphasen stammenden Abhängigkeits- und Versorgungswünsche hervortreten lassen: Sie möchte im Bett liegen wie ein krankes Kind und von den Eltern versorgt werden (Abb. 28–7).

Auf die Mitarbeiter der Krankenstation kommen gleichzeitig die infantilen Versorgungsansprüche der Patientin und die geschilderten abweisenden und aggressiven Reaktionen zu. Die sich hieraus ergebenden Konflikte und Reaktionsweisen der Teammitglieder lassen sich leicht vorstellen.

Ein 20jähriger Asthmatiker hielt während seiner häufigen stationären Aufenthalte Ärzte und Schwestern vor allem während des Nachtdienstes ständig in Atem. Er kompensierte extreme Angst- und Abhängigkeitsgefühle während der Asthmaanfälle bereits auf dem Weg ins Krankenhaus regelmäßig über folgende Phantasie: Jetzt könne er „wieder alle nach seiner Pfeife tanzen lassen".

Während der psychotherapeutischen Behandlung wurde im einzelnen deutlich, welche Probleme zu den zeitweise extremen emotionalen Belastungen für Ärzte und Schwestern im Krankenhaus beigetragen haben. Während der ersten zwei Jahre der Psychotherapie reagierte der Patient besonders empfindlich, wenn die Regulation von Nähe und Distanz seiner Kontrolle zu entgleiten drohte. Drohten seine eigenen Beziehungswünsche zu große Nähe herbeizuführen, so schlugen sie in heftige Aggressionen und Distanzierungsversuche um. Dann phantasierte der sonst sozial unauffällige, eher überangepaßte Kranke z.B., mich, seinen Therapeuten zu töten. Aber auch als Toten erlebte er mich dann noch als zu gefährlich; ausführlich phantasiert er, wie er mich in Stücke zerhacken, anschließend einem Hund zum Fraß vorwerfen und schließlich auch den Hund töten und beerdigen müsse, um mich wirklich unschädlich machen zu können.

Später im Behandlungsverlauf bemerkte der Patient bei sich Schwierigkeiten, meine Stimme zu hören. Über Wochen klang sie für ihn so, als säße er mehrere Räume von mir entfernt.

Das Verhalten des Kranken wäre für den Arzt unerträglich, ein solcher Umgang ohne therapeutischen Nutzen, gelänge es nicht, die Aggression und den Schmerz, den der Kranke jetzt in der Phantasie mir zufügt, zum Gegenstand der Behandlung zu machen und die Schwierigkeiten in der gegenwärtigen Beziehung zu mir als seinem Arzt durchzuarbeiten; im Rahmen einer solchen Durcharbeitung können die aktuellen Gefühle dann auch in Beziehung gesetzt werden zu den Empfindungen, die in der Pathogenese der Asthmaerkrankung wirksam waren, etwa zu den Schmerzen, die der Patient in der Beziehung zu seiner Mutter erlebte. Seine Mutter war ihm einerseits immer zu nahe: er lebte bis zum 20. Lebensjahr mit ihr in einem einzigen Zimmer; andererseits fühlte er sich von ihr als Kind immer wieder im Stich gelassen: die Mutter arbeitete abends als Kellnerin, wenn er nachts aufwachte, war er allein und fühlte sich hilflos seiner Angst ausgeliefert. Diesen äußeren Merkmalen in der Beziehung entsprachen erhebliche chronische emotionale Konflikte, die u.a. von der Art bestimmt wurden, in der die Mutter ihren Sohn als Objekt ihrer eigenen Bedürfnisse benutzte.

Für die Bearbeitung der emotionalen Probleme reicht die Kenntnis bzw. die Erinnerung der realen Situation und der realen Beziehung jedoch nicht aus; vielmehr kommt es darauf an, daß die Gefühle und Phantasien, die einst etwa der Mutter galten, in der gegenwärtigen Beziehung zum Arzt wiedererlebt und durchgearbeitet werden können. Dann besteht eine Chance, die Reaktionen auf der körperlichen Ebene (Allergie) mit den Reaktionen in den zwischenmenschlichen Beziehungen wieder verknüpfen und so günstigenfalls beide modifizieren zu können.

In welchem Grade die Störung der Beziehungen des Patienten zu seiner Umwelt, die Störung seiner zwischenmenschlichen Beziehungen, auch seine eigene Gefühls- und Phantasiewelt beeinträchtigen, ja zerstören, zeigt eine Phantasie, die nach etwa dreijähriger Behandlung zur Sprache kommt: Der Patient sitzt am Meer, die Wellen treiben eine Flaschenpost heran. Er öffnet die Flasche, entnimmt ihr ein altes, vergilbtes Pergament mit kaum lesbaren Schriftzeichen, die nur mühsam zu entziffern sind. Er weiß, daß der Inhalt der Botschaft für ihn außerordentlich wichtig ist. Er vergleicht den Inhalt dieser Botschaft mit dem Inhalt seiner Phantasien, die sich ihm, schon bevor er darüber nachdenkt, „immer in ein Nichts auflösen". Er spürt, daß er vom Inhalt dieser Botschaft gefühlsmäßig stark berührt wird; er fürchtet, überwältigt werden zu können und meint, dies könne oder möchte er nicht ertragen. Schließlich steckt er das Pergament in die Flasche zurück und sieht „mit tiefer Befriedigung" zu, wie die Flasche sich, auf den Wellen schaukelnd, langsam wieder entfernt, aufs weite Meer hinausgetragen wird. Mir gegenüber empfindet er ein triumphierendes Gefühl: auch mir bleibe dies nun alles vorenthalten.

An diesen Beispielen wird auch der heftige, oft verzweifelte Widerstand der Kranken gegenüber einer neuen Erfahrung und einer Modifikation ihres Verhaltens, ihr Widerstand insbesondere auch gegenüber dem Erwachen ihrer Affekte in der therapeutischen Beziehung deutlich.

In diesem Zusammenhang sei noch erwähnt, daß im Verlauf der Behandlung die Asthmasymptomatik von schwersten Angstzuständen und Depressionen im Sinne eines Symptomwandels abgelöst wurde. Diese Verlagerung der Symptombildung in den psychischen Bereich wurde vom Patienten jedoch als so unerträglich erlebt, daß er sich zunächst intensiv die für ihn in der Vergangenheit mehrfach lebensbedrohliche Asthmasymptomatik zurückwünschte.

Jeder Ansatz zur Institutionalisierung der klinischen Psychosomatik hat auch das Ausmaß dieser Widerstände zu berücksichtigen. Solche Widerstände können nur im Umgang mit Ärzten, Schwestern und den übrigen Bezugspersonen bearbeitet werden. Die Bearbeitung der Konflikte kann nur im Schutz einer tragfähigen Beziehung erfolgen; Agierenlassen bedroht die Beziehung und damit den notwendigen Schutz. Ein solcher Arbeitsansatz darf jedoch für die Mitarbeiter nicht unzumutbare Belastungen mit sich bringen, wie es vor allem dann der Fall ist, wenn nicht eine Bearbeitung der Bedürfnisse der Patienten erfolgt, sondern unreflektiertes Gewährenlassen und Möglichkeit zur Abreaktion gefordert wird.[7]

---

7 Die übliche Umwelt in medizinischen Institutionen, mitbestimmt u.a. durch die Übermacht des technischen Aufwandes, behindert im allgemeinen ein Wieder-in-Erscheinung-Treten der in der Krankheit abgewehrten Emotionen massiv. Es ist nicht genügend erforscht, inwieweit Phänomene wie „Alexithymie" oder „Psychosomatisches Phänomen" auch als Resultat der Interaktion zwischen medizinischer Umwelt und Patienten und nicht einfach nur als Patienteneigenschaften angesehen werden müssen.

> Eine 30jährige Patientin mit einer bedrohlich verlaufenden Form der Colitis ulcerosa rief durch ihr anspruchliches und auch trotziges Verhalten erhebliche Aggressionen bei den Schwestern der Station hervor. Zwar berichteten die Schwestern von der Mühe, die sie sich gaben, trotz ihrer Verärgerung freundliche Miene zum bösen Spiel der Kranken zu machen; beispielsweise versuchten sie beim Betreten des Zimmers trotz ihrer gegenteiligen Gestimmtheit eine heitere Miene aufzusetzen. Auf Dauer konnten sie dies jedoch nicht durchhalten und wurden gegenüber der schwerstkranken Patientin immer abweisender. Schließlich besprach der Stationsarzt mit ihnen die schwierige Situation. Die Schwestern gewannen nun auch ein weitergehendes Verständnis für die übergroßen Bedürfnisse der Kranken, verwöhnt zu werden; alle gaben sich nun wieder große Mühe und verhielten sich außerordentlich freundlich zu der Patientin. Bei diesem Vorgehen waren jedoch die eigenen Gefühle der Schwestern weitgehend unberücksichtigt geblieben. Die Folgen erlebten wir eine Woche später: Die Stationsschwester erschien mit verbundenen Handgelenken; es war ein Schub ihrer „chronischen Gelenkentzündung" aufgetreten. Die psychosomatische Hypothese, daß die Unterdrückung der Aggressionen und die hiermit verbundenen Spannungen im Berufsfeld mit zu diesem Beschwerderezidiv beigetragen haben, erschien uns sehr wahrscheinlich. Die bloße Mitteilung von Bedürfnissen der Patienten und die Verordnung einer Verhaltensänderung ohne die gleichzeitige Berücksichtigung der Emotionen der an der Behandlung Beteiligten ermöglichte in diesem Fall jedenfalls keine Lösung der bestehenden Interaktionsprobleme. Die Schwester wurde überfordert, der Konflikt zwischen ihren realen Verhaltensmöglichkeiten und dem von ihr mit der Rolle einer Krankenschwester untrennbar verbunden erlebten realen Forderungen wurde verschärft.

Als Ärzte und Schwestern fühlen wir uns von den in körperlichen Krankheiten oft verborgenen psychischen und sozialen Wirkkräften nicht selten intensiv bedroht und aktivieren verschiedenartige Schutzreaktionen. Hierzu gehört auch der Versuch, den Gültigkeitsbereich physikalischer und biologischer Methoden unzulässig zu erweitern und dogmatisch andere Verständnisansätze als das biomechanische Modell auszuschließen.

So meinte ein Kollege während der Diskussion über den Krankheitsverlauf bei dem oben zitierten Asthmatiker: Wenn unsere Schilderung zutreffe, fühle er sich bedroht, wie verloren auf einem wackelnden Stein inmitten eines reißenden Flusses stehend, und – inhaltlich unverständlich, jedoch den Bezug zum Dogma verdeutlichend – dann könne das ganze Christentum nicht stimmen.

Schließlich noch ein Beispiel für das diagnostische und therapeutische Vorgehen, wie es unserer Auffassung von klinischer Psychosomatik entspricht:

> Ein 33jähriger Patient ist vor einem halben Jahr an Diarrhö mit extremem Gewichtsverlust (von 110 kg auf 60 kg) erkrankt. Er wurde dem Psychosomatiker erst nach Abschluß sämtlicher, ergebnislos verlaufener organmedizinischer Untersuchungen zu Interviews überwiesen, blieb dann jedoch wegen der Schwere der Symptomatik noch mehrere Wochen in stationärer Be-

> handlung.[8, 9] In seinem sozialen Verhalten wirkte der Kranke auffallend angepaßt, psychologisch gesehen war er zunächst völlig „unauffällig". Die Mitarbeiter der Station kritisierten die Anwendung psychologischer Fragestellungen, man solle doch nicht in so unauffällige, „normale" Patienten auch noch Probleme hineintragen. Während der ersten konfliktorientierten Interviews kam es naturgemäß zu einer gewissen Beunruhigung des Patienten. Nach den Interviews nahm sich seiner auf der Station eine ältere, überfürsorgliche Schwester an. Herr B. hatte dieser Schwester immer bereitwilligst bei allen Arbeiten, vor allem beim Bettenmachen, geholfen. Von den Interviews zurück auf der Station, erhielt er jetzt mit folgender Bemerkung Wurstbrote: Solche psychologischen Gespräche seien sicher sehr anstrengend. Der konfliktvermeidenden Abwehr des Patienten entsprach diese selbstverständlich vom Patienten mitinduzierte bzw. mitaktualisierte Abwehrhaltung der Schwester und damit eines wesentlichen Teils der Station. Derartige Widerstände wurden noch deutlicher, als der Patient begann, sein Verhalten zu verändern; als er etwa nicht mehr alle ihm aufgetragenen Arbeiten widerspruchslos ausführte, vielmehr anfing, sich zu wehren oder an bestimmten Verhaltensweisen der Schwestern Kritik zu üben. Nun warnte auch der Stationsarzt den Psychosomatiker: „Machen Sie mir meine Patienten nicht unruhig!" Der Konflikt des Patienten drohte jetzt externalisiert als Konflikt zwischen den verschiedenen Ärzten und als Konflikt zwischen den Ärzten und Schwestern auf der Station ausgetragen zu werden.[10] Jetzt war es entscheidend, daß die Mitbewegung der an der Behandlung Beteiligten mit dem Ziel reflektiert werden konnte, den Kranken besser zu verstehen.[11]

8 Dieser Überweisungsmodus charakterisiert ebenfalls eine der Schwierigkeiten der interdisziplinären Zusammenarbeit zwischen klinischer Psychosomatik und den übrigen klinischen Fächern.

9 Die Behandlung dieses Kranken erfolgte auf der im Abschnitt 28.3.2. näher beschriebenen internistisch-psychosomatischen Krankenstation.

10 Nicht selten werden innere Konflikte der Patienten auf diese Weise externalisiert, z.B. im Rivalisieren zwischen Ärzten verschiedener Disziplinen oder Ärzten und Pflegepersonal ausgetragen. Ein Verständnis dieser Spannungen als vom Patienten induziert setzt eine enge Zusammenarbeit innerhalb eines Teams mit definierten Aufgaben und institutionalisierten Möglichkeiten zur Klärung von Spannungen voraus.

11 S. Freud hat bereits 1910 auf die Kooperationsschwierigkeiten zwischen Psychotherapeuten und somatisch orientierten Ärzten hingewiesen:

„Es war auch wirklich nicht bequem, psychische Operationen auszuführen, während der Kollege, der die Pflicht der Assistenz gehabt hätte, sich ein besonderes Vergnügen daraus machte, ins Operationsfeld zu spucken und die Angehörigen den Operateur bedrohten, sobald es Blut oder unruhige Bewegungen bei dem Kranken gab. Eine Operation darf doch Reaktionserscheinungen machen; in der Chirurgie sind wir längst daran gewöhnt. Man glaubte mir einfach nicht, wie man heute uns allen noch wenig glaubt; unter solchen Bedingungen mußte mancher Eingriff mißlingen. Um die Vermehrung unserer therapeutischen Chancen zu ermessen, wenn sich das allgemeine Vertrauen uns zuwendet, denken Sie an die Stellung des Frauenarztes in der Türkei und im Abendlande. Alles, was dort der Frauenarzt tun darf, ist, dem Arm, der ihm durch ein Loch in der Wand entgegengestreckt wird, den Puls zu fühlen. Einer solchen Unzugänglichkeit des Objektes entspricht auch die ärztliche Leistung. Unsere Gegner im Abendlande wollen uns eine ungefähr ähnliche Verfügung über das Seelische unserer Kranken gestatten. Seitdem aber die Suggestion der Gesellschaft die kranke Frau zum Gynäkologen drängt, ist dieser Helfer und Retter der Frau geworden."

Dies erwies sich als um so wichtiger, als bald auch die Ehefrau des Kranken erschien und im Wechsel entweder über die Unwirksamkeit der Behandlung oder über die durch die Behandlung verursachten und für sie selbst zum Teil unangenehmen Verhaltensänderungen des Patienten heftig zu klagen begann.

In den Besprechungen ermöglichten es uns diese Vorkommnisse jedoch, die Abwehrvorgänge des Kranken besser zu verstehen. Wir konnten sehen, wie er im Umgang mit der älteren, ihn überfürsorglich bemutternden Schwester unter Verzicht auf eigenständige Entfaltungsmöglichkeiten seine Abhängigkeitsbeziehung zur eigenen Mutter „szenisch reproduzierte", uns wurde allmählich deutlicher, warum dieser Patient bei seinem übergroßen Bedürfnis, akzeptiert zu werden, in seinem Beruf als Vertreter jede aggressive Regung sowohl gegenüber Kunden als auch gegenüber seiner Firmenleitung abwehren mußte. Nachdem der Patient sich auf der Station geborgen fühlte, eine tragfähige Beziehung zwischen ihm und den Teammitgliedern entstanden war, war es möglich, auch seine Widerstände langsam zu bearbeiten; dieser Absicht entsprach dann das Vorgehen der „psychosomatischen Schwester". Als davon die Rede war, daß der Patient sich durch das Ertragen der Belastung während der psychotherapeutischen Gespräche eine Medaille verdient habe – es war zur Zeit der Olympiade –, meinte diese Schwester: dieser Kampf sei wohl ein „olympischer Abwehrkampf". Eine solche Konfrontation mit seiner Abwehr war im Rahmen des Stationsmilieus möglich: Die Verbindung aus Behandlungsinteresse und Beunruhigung führte dazu, daß der Patient auch andere Schwestern, insbesondere die Nachtschwester, nach ihrer Einstellung zu einer psychotherapeutischen Behandlung zu fragen begann. Durch Hilfestellungen in diesen Gesprächen abgesichert, d.h. durch eine mütterliche Sicherheit bietende Funktion der Schwesterngruppe gestärkt, ließ sich der Kranke umgekehrt während der Einzeltherapiestunden ermutigen, auf der Station neue Verhaltensweisen zu erproben, neue Erfahrungen zu machen. So war er z.B. überrascht, ja tief betroffen, daß er auf der Station auch Forderungen stellen konnte und diese erfüllt wurden; daß er z.B. zu unüblichen Zeiten ohne besondere Gegenleistung etwas zu essen bekommen konnte.

Seine Symptomatik begann sich bald insofern zu ändern, als die Durchfälle nun zeitlich nicht mehr ungeordnet, regellos auftraten, sondern dieses Symptom bestimmten Situationen im Stationsleben, bestimmten Konflikten in Beziehungen und auch innerpsychischen Zuständen zugeordnet erschien oder, wohl zutreffender, daß aus der aus psychosozialer Sicht scheinbar regellosen Verselbständigung der pathophysiologischen Abläufe sich eine in der Pathogenese wohl vorausgegangene Verknüpfung mit Konflikten in zwischenmenschlichen Beziehungen und mit innerseelischen Zuständen wieder einstellte. Während der ambulant fortgeführten Einzelpsychotherapie trat das Symptom dann bald – und dieser Verlauf hatte für den Patienten selbst Evidenzcharakter – im Rahmen der Übertragungsbeziehung zum Therapeuten auf. Während der Patient ein Bewußtsein für seine Konflikte und die Notwendigkeit von Einstellungs- und Verhaltensänderungen entwickelte, wurden parallel zu den Widerständen auf der Station auch die Widerstände seiner Mitwelt, insbesondere die Widerstände seiner Ehefrau gegen solche Veränderungen deutlich; dies machte eine vorübergehende Einbeziehung der Ehefrau in die Behandlung erforderlich.

Während einer zweijährigen Behandlung bildeten sich die funktionellen Körperbeschwerden vollkommen zurück. Vor allem aber vermochte der Patient im weiteren Verlauf auch an seiner Einstellung gegenüber seiner Ehefrau, seinen Eltern und seinen Haltungen im beruflichen Bereich zu arbeiten. Es gelang ihm, in beiden Lebensbereichen neue Lösungsmöglichkeiten zu erarbeiten, die ihn innerlich unabhängiger und damit freier in der Wahl seiner Verhaltensmöglichkeiten werden ließen. Die Voraussetzung für eine erfolgreiche Einleitung dieser Behandlung war die dargestellte Abstimmung und Reflexion des Erlebens und Verhaltens aller Mitarbeiter der betreffenden Krankenstation. Aus unserer Sicht wäre dieser Kranke ohne die stationäre Behandlungseinleitung kaum für eine psychotherapeutische Behandlung zu motivieren gewesen.

Allgemeine Voraussetzungen für die Institutionalisierung

Die Institutionalisierung der klinischen Psychosomatik erfordert **Modifikationen der konkreten Arbeitssituation,** die der Erweiterung des theoretischen Bezugsrahmens entsprechen. Hierzu gehört nicht nur eine angemessene Verbesserung der zeitlichen und räumlichen Rahmenbedingungen, sondern auch die **Entwicklung von Organisationsformen,** die eine auf das psychosomatische Arbeitskonzept abgestimmte Zusammenarbeit aller Beteiligten erlauben. Eine weitere Voraussetzung ist entsprechende Weiterbildung für alle Teammitglieder. In Tabelle 28–1 sind die Voraussetzungen für eine Institutionalisierung der Psychosomatischen Medizin in die klinische Praxis zusammengestellt.

Fragen der **„Organisationsentwicklung"** wurden in der Vergangenheit bei Versuchen, den psychosomatischen Arbeitsansatz in die klinische Medizin zu integrieren, häufig unterbewertet oder überhaupt nicht berücksichtigt. Hierzu gehört insbesondere die Frage der Organisation einer interdisziplinären Zusammenarbeit in der Klinik. Die Folge war entweder ein Scheitern solcher Versuche oder eine Isolation der Psychosomatischen Medizin in ausgegliederten Spezialstationen oder in Spezialkliniken, in denen auf die Probleme der Kooperation mit den übrigen Klinikern weniger Rücksicht genommen werden mußte.

Forderungen nach Teamarbeit und interdisziplinärer Kooperation im Rahmen von Festreden und ähnlichen Veranstaltungen sind zwar üblich geworden, dabei wird jedoch häufig nicht deutlich genug gesehen, daß es sich bei Teamarbeit und interdisziplinärer Kooperation im Einzelfall meist um eine anspruchsvolle, schwierige und nur gegen zahlreiche Widerstände zu realisierende Innovation handelt. Die Widerstände resultieren vielfach aus ideologisch mißbrauchten theoretischen Konzepten und entsprechenden sozialen Interessen ihrer Vertreter.

Im Rahmen der **fachlichen Weiterbildung** sollte ein der Vielseitigkeit der Arbeitsanforderungen entsprechend differenziertes psychologisches Konzept angeboten werden. Unserer Auffassung nach kommt hierfür vor allem das psychoanalytische Konzept in Fra-

Tabelle 28–1. Voraussetzungen für die Institutionalisierung der Psychosomatischen Medizin in die klinische Praxis

1. Modifikation der Rahmenbedingungen
– Zeitliche Verhältnisse: Stellenplan
– Räumliche Verhältnisse
2. „Organisationsentwicklung"
– Teamarbeit:
  Kooperation zwischen verschiedenen Berufsgruppen (Ärzte und Schwestern)
– Interdisziplinäre Kooperation
– Abstimmung aller Maßnahmen und Veranstaltungen (Pflegesystem, Visiten u. a.) auf psychosomatisches Konzept
3. Weiterbildung
– Wissensvermittlung
– Verhaltensschulung (z. B. Interviewtraining)
– Berufsbezogene Selbsterfahrung

Tabelle 28–2. Arbeitsrichtungen der klinischen Psychosomatik

Orientierung an
1. Pathogenese und Krankheitsverlauf
2. Krankheitsverarbeitung (Coping)
3. Krankheitsverhalten (Compliance)
4. Psychischen Begleit- oder Folgeerkrankungen körperlicher Krankheitszustände (symptomatische Psychosen)

ge, da es ein dynamisches Verständnis sowohl der intrapsychischen Konflikte als auch der Interaktionsvorgänge zwischen Patienten und Mitarbeitern medizinischer Institutionen ermöglicht. Dieses Konzept sollte durch weitere psychologische Ansätze, vor allem solche, die sich aus der Lerntheorie ableiten, ergänzt werden.

Dagegen reicht es unserer Auffassung nach nicht aus, den naturwissenschaftlichen Ansatz lediglich durch einige sozialwissenschaftliche Gesichtspunkte zu ergänzen, wie es auch heute noch nicht ganz selten im Sinne einer „Psychologie des gesunden Menschenverstandes" versucht wird. Es ist nicht möglich, „in der Psychosomatik auf der somatischen Seite eine hochentwickelte Naturwissenschaft einzusetzen, auf der psychischen aber sich einer Trivialität zu bedienen, die denn doch kein so gutes Licht auf solche Forscher fallen läßt". Bei Berücksichtigung der nötigen Ergänzungen gilt unseres Erachtens auch heute noch: „Die Psychosomatische Medizin muß eine tiefenpsychologische sein, oder sie wird nicht sein" (V. v. Weizsäcker, 1949).

Die Weiterbildung aller im Bereich der klinischen Psychosomatik tätigen Teammitglieder kann sich entsprechend dem Grundverständnis der Psychosomatischen Medizin nicht auf eine reine Wissensvermittlung beschränken; sie muß vielmehr auch eine Schulung der erforderlichen Fähigkeiten, insbesondere auch eine Schulung konkreter Verhaltensweisen – z. B. Interview- und Gesprächsführung – sowie ein systematisches Training der Reflexion von emotionalen Erfahrungen im Berufsfeld („berufsbezogene Selbsterfahrung") einschließen, wie es z. B. in Balint-Gruppen möglich ist.

28.1.2 Aufgaben und Arbeitsgebiete der klinischen Psychosomatik

Im klinischen Feld hat die Psychosomatik entsprechend den hier gestellten Aufgaben sich bisher vor allem in vier Richtungen entwickelt (Tab. 28–2).

Orientierung an Pathogenese und Krankheitsverlauf
(Tab. 28–3)

Die Klärung einer Beteiligung psychischer und sozialer Faktoren an der meist als multifaktoriell aufzufassenden Pathogenese bzw. im Verlauf zahlreicher Erkrankungen ist Ziel dieser Arbeitsrichtung. Den einzelnen in Tabelle 28–3 dargestellten Verständnisansätzen kommt prinzipiell bei allen Erkrankungen eine, wenn auch im einzelnen oft erst noch zu gewichtende Bedeutung zu (vgl. Kap. 1 und den Teil Spezielle Krankheitslehre). Dies gilt auch für Erkrankungen, die epidemiologisch und in der Mortalitätsstatistik eine führende Rolle spielen, wie etwa für arteriosklerotisch verursachte Herz-Kreislauf-Erkrankungen (Schäfer, 1976). Gerade bei diesen Krankheiten steht der Arzt vor einer für ihn noch ganz neuartigen Aufgabe: Das Therapieziel beinhaltet die Veränderung menschlicher Verhaltensweisen, wie z. B. die Beeinflussung von Risikoverhaltensweisen.[12]

Tabelle 28–3. Orientierung an der **Pathogenese**

– Komplexe Verhaltensstörungen (z. B. Anorexia nervosa)
– Spezifische Risikoverhaltensweisen (Überernährung, Rauchen, Alkoholmißbrauch)
– Äußere Lebensbelastungen („psychosozialer Streß", Verlusterlebnisse)
– Innerseelische Fehlanpassung (z. B. Konversionshysterie)
– Psychophysiologische Fehlanpassung (z. B. Situationshypertonie)

---

12 Schäfer (1976) hat darauf hingewiesen, daß der Fortschritt der modernen Medizin vor allem durch solche Risikoverhaltensweisen begrenzt wird: Die Lebenserwartung für Männer zwischen 20 und 60 Jahren ist in der Bundesrepublik rückläufig; diese Entwicklung kommt durch die rasche Zunahme von nur sechs Krankheitsbildern zustande: koronare Herzerkrankungen, Lungen- und Bronchialkarzinom, Leberzirrhose, Bronchitis, Verkehrsunfälle und Diabetes. Der Anstieg der Sterblichkeit an diesen Erkrankungen sei vorwiegend durch Umweltfaktoren und falsche Verhaltensweisen bedingt, die in der Verantwortlichkeit des Menschen selbst liegen. Der Erfolg der modernen Medizin wird damit abhängig von der Berücksichtigung psychosozialer Faktoren auch bei organischen Krankheiten. Bisher sind jedoch weder die einzelnen Ärzte, noch die Institutionen darauf vorbereitet, die Modifikation von menschlichen Verhaltensweisen und Umweltfaktoren zu unterstützen.

Orientierung an der Krankheitsverarbeitung
(Tab. 28–4)

Das Verständnis **krankheitsreaktiver** psychologischer und sozialer Probleme in Diagnostik und Therapie, die Entwicklung von Hilfestellungen für Patienten im Prozeß der Adaptation an die Erkrankung („coping") ist Gegenstand dieses Arbeitsbereiches, dessen wichtigste Aspekte in Tabelle 28–4 zusammengestellt sind (vgl. Kap. 62 und 66).

Tabelle 28–4. Orientierung an der **Krankheitsverarbeitung**

- Innerpsychisch
  Verarbeitung der akuten Bedrohung
  (Angst und Depression des Infarktkranken)
  Adaptation an chronische Abhängigkeit
  (Dialyse)
  Trauerprozeß
  (unheilbar Kranke)
- Familie
  Rollenveränderungen
- Beruf
  Rehabilitation

Orientierung am Krankheitsverhalten
(Tab. 28–5)

In diesem Arbeitsbereich geht es um die Klärung und Optimierung des **therapeutischen Bündnisses** zwischen dem Arzt und den übrigen Mitarbeitern im klinischen Bereich einerseits und dem Patienten andererseits. Dem Patienten sollen Hilfestellungen angeboten werden, die ihm ein realitätsgerechtes Krankheitsverhalten und langfristige Kooperation bei seiner Behandlung ermöglichen. Hiervon hängt in hohem Ausmaß der heute prinzipiell oft mögliche Erfolg medikamentöser und apparativ aufwendiger sowie mit hohen Kosten verbundener Langzeitbehandlungsmaßnahmen ab. Beispiele hierfür sind die medikamentöse Dauertherapie von Hypertoniekranken oder die Dialysebehandlung von Patienten mit terminaler chronischer Niereninsuffizienz.

Die Beachtung der Qualität dieses „Arbeitsbündnisses" mit dem Patienten ist als ärztliche Basisaufgabe anzusehen, seitdem immer deutlicher wurde, in welchem Ausmaß dieses Arbeitsbündnis Voraussetzung jeder rationalen Behandlung ist.

Untersuchungen über die Einhaltung der verordneten Medikation durch den Patienten zeigten, daß die „compliance" der

---

Tabelle 28–5. Orientierung am **Krankheitsverhalten**

Arbeitsbündnis zwischen Arzt und Patient
u. a. **Compliance** (Einhaltung von Verordnungen)
mit abhängig von:
- **Kommunikationsverhalten** des Arztes
  Informationsverhalten
  affektiven Einstellungen: Interesse, Zuwendung
- **Persönlichkeitsvariablen** von Arzt und Patient

Patienten weitaus schlechter ist, als bisher angenommen, und daß sie in hohem Ausmaß mit abhängt von Persönlichkeitsvariablen und Verhaltensweisen sowohl der Ärzte als auch der Patienten.

Als Compliance wird die Befolgung ärztlicher Anordnungen durch den Patienten bezeichnet. Die Kooperation des Patienten betrifft die Medikamenteneinnahme, die Einhaltung von Diätvorschriften, die Änderung von Risikoverhaltensweisen bzw. des gesamten Lebensstils. Das Ausmaß von Non-Compliance ist überraschend hoch.

Eine Übersicht über 50 Studien ergab, daß 25–50% aller ambulanten Patienten die verschiedenen verordneten Medikamente überhaupt nicht einnehmen (Davis, 1971; Blackwell, 1973; Schmädel, 1976). Selbst gutverträgliche Medikamente wie Digitalispräparate werden nur von 36% der Kranken konsequent entsprechend der ärztlichen Anweisungen eingenommen; 40% nahmen die Digitalismedikation so unregelmäßig ein, daß sie nicht mehr therapeutisch wirksam war (Schettler, 1975). In Heidelberg nahmen nur 34% stationärer Patienten ein verordnetes Präparat regelmäßig ein, 66% der Patienten mußten als unzuverlässig eingestuft werden (Gundert-Remy, 1978). Diese Befunde stehen in deutlichem Gegensatz zur Einschätzung der Kooperation von Patienten durch Klinikärzte (Davis, 1966).

Die Compliance ist von zahlreichen Variablen abhängig. In diesem Rahmen sei lediglich darauf hingewiesen, daß auch hier eine Wechselseitigkeit der Beziehung zwischen Arztverhalten und Patientenverhalten von wesentlicher Bedeutung ist. Die Compliance von Patienten ist so u. a. abhängig von Merkmalen des **ärztlichen Kommunikationsverhaltens** sowie von Persönlichkeitsvariablen von Patienten und Ärzten.

Eine Voraussetzung für die Entwicklung eines guten Arbeitsbündnisses stellt die **Informiertheit** des Patienten dar. Aufgrund der vorliegenden Untersuchungen muß heute mit einem völlig unzureichenden Informationsstand gerechnet werden. Nur etwa 30% der Kranken einer internistischen Klinik sind über das Wesen ihrer Erkrankung, nur ca. 20% über die nach der Klinikentlassung erforderlichen Behandlungsmaßnahmen ausreichend informiert (Engelhardt, 1973; Raspe, 1976). Dieses Informationsdefizit hat mehrere Ursachen; neben dem mangelhaften Informationsverhalten der Ärzte spielen auch Abwehrvorgänge bei Patienten eine Rolle.

Auch das **sprachliche Verhalten** (Ley, 1977), vor allem aber die **affektive Einstellung** hat Einfluß auf die Compliance. So werden in den Untersuchungen von Korsch (1968, 1972; Schmädel, 1976) etwa Mangel an Wärme und Freundlichkeit des Arztes sowie die Nichtbeachtung der von Patienten ausgehenden Erwartungen als Grund für die Non-Compliance hervorgehoben. In einer pädiatrischen Ambulanz befolgten nur 42% der Mütter die ärztlichen Instruktionen in allen wesentlichen Einzelheiten. Die Befolgung der Instruktionen entsprach im wesentlichen der Einschätzung der Arzt-Patient-Beziehung durch die Mütter. Diejenigen Mütter, die den Besuch als „sehr unbefriedigend" bezeichnet hatten, befolgten nur in 17% die Anweisungen (Korsch, 1972).

Der Einfluß von **Persönlichkeitsvariablen** wird bei groben Störungen des Arbeitsbündnisses besonders deutlich; so gibt es Kranke, die sich nicht konsequent helfen lassen können und die Fähigkeiten ihrer Ärzte immer wieder in Frage stellen müssen. Charakteristisch sind Patienten, bei denen mit großer Regelmäßigkeit eine anfängliche Idealisierung des Arztes durch eine sich anschließende Entwertung abgelöst wird („Koryphäen-Killer-Syndrom"; Beck, 1977), wobei sich dieses Verlaufsmuster bei jedem der häufigen Arztwechsel wiederholt.

Orientierung an psychischen Begleit- oder Folge-
erkrankungen körperlicher Krankheitszustände:
Durchgangssyndrome und Funktionspsychosen

Dieser Arbeitsbereich der klinischen Psychosomatik
überschneidet sich mit Arbeitsbereichen der Psychiatrie.

Bei ca. 10% aller körperlich Kranken, die sich in stationärer
Behandlung außerhalb psychiatrischer Fachkliniken befinden, liegen bei genauer Diagnostik psychische Begleit- oder
Folgeerkrankungen im Sinne von Durchgangssyndromen
oder Funktionspsychosen vor. Diese Erkrankungen bedürfen einer sorgfältigen diagnostischen Bewertung und zum
Teil einer speziellen Behandlung (vgl. Kap. 63).

Die vielfältigen, in dieser Übersicht dargestellten Aufgaben der klinischen Psychosomatik erfordern in methodischer Hinsicht einen breiten Ansatz und die Bereitschaft zu interdisziplinärer Zusammenarbeit bei
der organisatorischen Eingliederung der Psychosomatik in die klinischen Institutionen.

Eine solche komplexe Auffassung von Psychosomatik hat
auch international die frühere, sehr enge Orientierung an
der Pathogenese einiger weniger Krankheitsbilder unter einseitiger Anwendung psychoanalytischer Konzepte abgelöst,
wie sie während der 40er und 50er Jahre vorherrschte (Engel, 1954, 1972, 1977; Lipowski, 1977; vgl. auch Richter,
1978).

28.1.3 Beteiligung der Psychosomatischen Medizin an der Krankenversorgung innerhalb der medizinischen Institutionen. Angaben zum Bedarf

Die vorliegenden Angaben schwanken in Abhängigkeit von der Intensität der diagnostischen Maßnahmen; das Fehlen entsprechender Institutionen und
die mangelhafte Ausbildung der Kliniker in diesem
Bereich tragen dazu bei, daß dieser Bedarf noch nicht
ausreichend wahrgenommen bzw. unterschätzt wird.

Poliklinischer Bereich

Im Bereich internistischer Polikliniken muß aufgrund
der bisher vorliegenden Untersuchungen bei ca. 50
bis 80% aller Patienten eine wesentliche Beteiligung
psychosozialer Faktoren am Krankheitsbild angenommen werden (vgl. Tab. 28–6).

Einen Minimalwert ergeben Untersuchungen über Patienten
mit „reinen funktionellen Syndromen", bei denen sich keinerlei pathologischer Organbefund und bei Laboruntersuchungen keine Abweichungen von Normwerten erheben
lassen. M. Pflanz (1962) fand bei 25,5% von 7825 Patienten
der Gießener Medizinischen Polikliniken solche „reinen
funktionellen Syndrome".

Stationärer Bereich

Bei 30–60% aller Patienten internistischer Kliniken
wird eine erhebliche oder entscheidende Beteiligung
psychosozialer Faktoren am Krankheitsbild angegeben (vgl. Tab. 28–6). Engelhardt (1973) fand, daß nur
bei 28% aller Patienten einer inneren Klinik kein Einfluß der persönlichen Vorgeschichte auf die Be-

Tabelle 28–6. Häufigkeit der Beteiligung psychosozialer Faktoren am Krankheitsbild

Stationäre Patienten Innere Medizin	Cupaln u. Davis	1960	USA	100 Pat.	51%
	Stoeckle	1964	Boston	101 Pat.	84%
	Kaufman u. Bernstein	1957	New York	1000 Pat.	81%
	Lipowski (Übersicht über 8 Studien)	1967	USA		50–80%
Ambulante Patienten Poliklinik	Kaufman	1959	New York	253 Pat.	66,8%
	Querido	1959	Amsterdam	1630 Pat.	46,6%
	Richman	1964	USA	184 Pat.	36,0%
	Rosen	1972	USA	1413 Pat.	22,0%
	Porot		Frankreich	604 Pat.	40,0%
	Porot (Übersicht über 7 Studien)	1972			30–60%
	Lipowski (Übersicht über 8 Studien)	1967			30–60%

schwerden nachweisbar ist. Wir selbst erhielten ähnliche Befunde auf einer internistischen Allgemeinstation (Tab. 28–7).

Stuhr und Haag (1989) untersuchten die Prävalenz
„psychosomatischer Erkrankungen" auf den internistischen Stationen von neun der elf staatlichen Krankenhäuser Hamburgs. Ziel war eine bedarfsgerechte
Planung stationärer psychosomatischer Versorgung.
In der methodisch aufwendig und sorgfältig durchgeführten Untersuchung wurden 38,4% der intensiv untersuchten Patienten als „psychosomatisch krank"
identifiziert: Bei diesen Kranken wurde „eine psychosomatische Genese der zur Einweisung führenden Erkrankung angenommen". Nicht enthalten sind hier also alle Patienten mit behandlungsbedürftigen psychischen Reaktionen oder sozialen Folgen nach körperlicher Erkrankung und Patienten mit Problemen des
Krankheitsverhaltens.

Wenn man berücksichtigt, daß es in Hamburg keine Möglichkeit zur stationären Versorgung psychosomatisch Kranker gibt und ein psychosomatischer
Konsultations-Liaisondienst nur an der Universitätsklinik zur Verfügung steht, wird ein krasser Versor-
gungsnotstand sichtbar.

Die genannten Befunde werden durch eine prospektive Studie von Querido (1959) in ihrer Bedeutung unterstrichen. Er

Tabelle 28–7. Einfluß der persönlichen Vorgeschichte auf Beschwerden

	Engelhardt 1973 (120 Pat.)	Internistische Allgemein-station Ulm (100 Pat.)
Kein Einfluß	28%	18%
Teilweiser Einfluß	26%	31%
Großer Einfluß	44%	51%
Unklar	2%	

untersuchte mit einem Team von Interviewern 1630 Patienten und stellte bei 46,6% einen erheblichen Einfluß psychosozialer Belastungsfaktoren fest; diesen psychosozialen Belastungsfaktoren kam ein signifikanter Vorhersagewert für die Prognose der Erkrankung zu: Nur 29,6% der Patienten, die gesund wurden, hatten unter ungünstigen psychosozialen Bedingungen zu leiden, während die Nichtgenesenden in 70,4% derartigen Belastungen ausgesetzt waren. Die günstige Prognose einer Genesung, die aufgrund des somatischen Befundes bei der Entlassung vorhergesagt worden war, trat entgegen den Erwartungen der Kliniker nur bei etwas über der Hälfte der Patienten ein; die anhand psychosozialer Belastungsfaktoren getroffene prognostische Aussage war im Vergleich wesentlich genauer. Weniger als ein Drittel der psychisch belasteten Patienten erholten sich nach dem Krankenhausaufenthalt. Nach der Auffassung von Querido unterstreicht dies die Notwendigkeit, aus den bereits vorliegenden Untersuchungsbefunden auch therapeutische Konsequenzen zu ziehen.

Eine Orientierung zur Verteilung der Problemstellungen auf die vier Arbeitsbereiche der klinischen Psychosomatik für die Patienten einer internistischen Allgemeinstation gibt Tabelle 28–8.

Tabelle 28–8. Art der wichtigsten psychosomatischen Probleme bei 123 Patienten in einer internistischen Allgemeinstation (Ulm 1974)

	Patienten in %
1. Psychosoziale Faktoren mit wesentlicher Bedeutung für die Pathogenese	30,1
2. Schwierigkeiten der Krankheitsverarbeitung	32,5
3. Gravierende Schwierigkeiten im Vergleich der Compliance	7,3
4. Durchgangssyndrome, Funktionspsychosen	7,3
– Keine psychosomatischen Probleme	14,6
– Nicht sicher einzuordnen	8,2

Inanspruchnahme vorhandener psychosomatischer Konsiliardiensteinheiten

Eine erste Orientierung über ein Bedarfsminimum an psychosomatischer Krankenversorgung im stationären Bereich erlauben Untersuchungen über die Inanspruchnahme vorhandener psychosomatischer Konsiliardiensteinheiten. In gut ausgestatteten Zentren werden solche Einrichtungen bei ca. 10% aller internistischen Patienten in Anspruch genommen (Tab. 28–9).

Die Beratungsfrequenz ist dabei abhängig vom Grad der Verfügbarkeit des Konsiliarius. Ist seine Tätigkeit in den Ablauf der Stationsarbeit weitgehend integriert ("Liaisondienst"), wurde er z.B. in Ulm bei 11% aller internistischen Patienten in Anspruch genommen; wird lediglich eine telefonische Rufbarkeit ("Konsultationsdienst") angeboten, so sinkt die Beratungsfrequenz auf 4% ab (Schüffel, 1973).

Zusammenfassend kann gesagt werden, daß heute ein großer Bedarf an einer intensiven Beteiligung der

Tabelle 28–9. Inanspruchnahme des psychosomatischen Konsiliardienstes (in % der Gesamtpatientenzahl)

Lipowski	1967	USA (Übersicht über 8 Studien)	9,0%
Crisp	1968	England	10,0%
Meyer u. Mendelson	1961	USA	8,8%
Goldenberg u. Sluzki	1971	Argentinien	17,9%
Papastamou	1970	USA	11,3%
Schüffel	1973	Ulm/BRD	11,0%

Psychosomatischen Medizin an der klinischen Krankenversorgung besteht. Soll dieser Bedarf auch nur einigermaßen befriedigt werden, so ist die Institutionalisierung eines angemessen großzügig ausgelegten und vielfältig abgestuften Versorgungssystems erforderlich.

28.1.4 Konzepte und Modelle für die Institutionalisierung der Psychosomatischen Medizin im klinischen Bereich

Für die Realisierung des psychosomatischen Arbeitsansatzes im klinischen Feld wurden mehrere Konzepte entwickelt. Sie unterscheiden sich vor allem hinsichtlich des Ausmaßes der fachlichen Spezialisierung und des Grades der Integration in das jeweilige klinische Fach. Beim heutigen Entwicklungsstand der klinischen Psychosomatik scheint die gleichzeitige Erprobung verschiedener Ansätze wünschenswert und jede einseitige Festlegung verfrüht; die verschiedenen Modelle vermögen sich gegenseitig zu dem benötigten vielfältig abgestuften Versorgungssystem zu ergänzen. Zur Zeit lassen sich vor allem folgende Modellansätze unterscheiden:

Der Kliniker als Psychosomatiker[13]

Der klinisch tätige Arzt integriert den psychosomatischen Arbeitsansatz in seine eigene klinische Tätigkeit. Er arbeitet als Internist, Pädiater, Gynäkologe usw. „patientzentriert" (Balint, 1965, 1969; Engelhardt, 1971).[14]

Das Konzept des Klinikers als Psychosomatiker entspricht der internistisch-psychosomatischen Tradition in Deutsch-

---

13 Dieser Arbeitsansatz ist ausführlicher dargestellt in Kap. 27.
14 Das Konzept einer „patientzentrierten" Orientierung wurde von Balint in England und Engelhardt in Deutschland der sonst in der Medizin üblichen „krankheitszentrierten" Orientierung gegenübergestellt. Mit diesen Begriffen sollten wissenschaftliche Konzepte und Interessen gekennzeichnet werden, nicht Art und Ausmaß des Engagements einzelner Ärzte für ihre Kranken. Darüber hinaus sollte mit diesem Begriff nicht wiederum die Einführung einer neuen Spezialdisziplin assoziiert werden, wie es bei „Psychosomatische Medizin" immer wieder der Fall gewesen war. Zumindest im deutschsprachigen Raum hat sich jedoch der Begriff „patientzentriert", vor allem im hochschulpolitischen Zusammmenhang, nicht bewährt; er erwies sich als mißverständlich und wurde von den Vertretern anderer klinischer Fächer häufig als persönlich kränkend aufgefaßt.

land und dem Ansatz der Arbeitsgruppe um G. L. Engel in Rochester/USA. In der BRD wurde bzw. wird es vor allem von v. Weizsäcker zusammen mit Christian und Kütemeyer (Heidelberg), von Jores (Hamburg), von Seitz (München) und von v. Uexküll (Gießen und Ulm) im Bereich der Inneren Medizin, von Römer (Tübingen) im Bereich der Gynäkologie, von Wallis (Hamburg) im Bereich der Pädiatrie, sowie einigen ihrer Mitarbeiter vertreten.

Diesem Konzept kommt dann besondere Bedeutung zu, wenn die Psychosomatische Medizin möglichst weitgehend in die klinische Routinetätigkeit des einzelnen Arztes integriert werden soll. Die Arbeitsgruppen um Balint und seine Schüler haben nachgewiesen, daß eine solche Integration im Rahmen der Tätigkeit niedergelassener Ärzte möglich ist.

Für den in der Klinik tätigen Arzt ist ein solches Arbeitskonzept aus verschiedenen Gründen heute noch schwieriger zu realisieren. Zwar wird dieser Ansatz in der Bundesrepublik durch die Einbeziehung der psychosozialen Fächer in den von der Approbationsordnung vorgesehenen Ausbildungsgang gefördert, es fehlen jedoch noch weitgehend jene klinischen Einrichtungen, in denen der Arzt entsprechend diesen in der Ausbildung erworbenen Kenntnissen tätig sein kann. Die bestehenden Institutionen behindern vielmehr eine solche integrierte Tätigkeit erheblich.

Es darf auch nicht übersehen werden, daß dieser Arbeitsansatz eine besonders breit angelegte Weiterbildung erfordert. Die Weiterbildung in der jeweiligen klinischen Spezialität ist durch eine psychotherapeutische Weiterbildung zu ergänzen. Hinzu kommt die hohe Forderung an die Integrationsfähigkeit des betreffenden Arztes; sie bildet die Voraussetzung für die Entwicklung einer tragfähigen beruflichen Identität.

Der psychosomatische Konsiliardienst

Der Psychosomatiker stellt sich als spezialisierter Berater den übrigen Klinikern zur Verfügung. Als Konsiliarius kann er dabei entweder selbst diagnostische und therapeutische Aufgaben beim Patienten übernehmen oder sich aber indirekt über die Beratung des behandelnden Arztes bzw. die Beratung des gesamten Behandlungsteams an Diagnostik und Therapie beteiligen (vgl. 28.2).

Klinisch-psychosomatische Krankenstationen

Im Rahmen dieses Konzeptes wird versucht, den psychosomatischen Arbeitsansatz möglichst weitgehend in die gesamte klinische Krankenversorgung zu integrieren. Von besonderer Bedeutung wird dabei die Einbeziehung des gesamten Behandlungsteams, die intensive Kooperation zwischen Ärzten, Schwestern, Sozialarbeitern, Psychologen und Krankenhausseelsorgern. Durch den Teamansatz lassen sich die Anforderungen an die Integrationsleistung des einzelnen auf ein leichter realisierbares Maß reduzieren. Im Rahmen eines solchen Konzeptes ist es im Unterschied zur konsiliarischen Beratung möglich, die psychosozialen Aspekte bei jedem Kranken von Anfang an ebenso systematisch zu untersuchen und zu gewichten wie die somatischen. In diesem Konzept entfallen außerdem die Probleme einer Patientenselektion im Zusammenhang mit der Überweisung (vgl. 28.3).

Psychosomatisch/psychotherapeutische Spezialstationen bzw. Spezialkliniken

Den verschiedenen Konzepten für solche Spezialstationen bzw. Fachkliniken, die eine intensive psychotherapeutische Behandlung bereits ausgewählter Patienten mit funktionellen Beschwerden und psychosomatischen Krankheitsbildern erlauben, sind einige Grundprinzipien gemeinsam. Die spezifische Behandlung erfolgt meist in psychoanalytisch orientierter Einzel- und/oder Gruppentherapie. Gleichzeitig kann in solchen Spezialeinheiten die ganze Vielfalt der heute entwickelten, die spezifische Psychotherapie unterstützenden paraverbalen und averbalen psychotherapeutischen Verfahren angeboten werden. Diese Behandlungsansätze sind eingebettet in ein therapeutisches Milieu, dem als einer Art alternativer Umwelt zur Lebenswelt der Patienten eine unspezifisch heilende oder doch die psychotherapeutischen Prozesse fördernde Wirkung zukommt (Janssen, 1987).

Für die Indikation zu einer solchen stationären psychotherapeutischen Behandlung ist mitentscheidend, daß ein Teil der psychosomatisch und auch der psychoneurotisch Kranken aufgrund ihrer schweren Zugänglichkeit für psychotherapeutische Methoden und aufgrund ihrer Ich-Schwäche oft nicht in der Lage sind, in einer ambulanten Behandlung eine therapeutische Beziehung aufzunehmen und aufrechtzuerhalten. Die Patienten benötigen das Milieu einer Krankenstation, um jenes Maß an Sicherheit zu gewinnen, das sie für die Einleitung eines psychotherapeutischen Prozesses benötigen. Ist diese Sicherheit gegeben, dann sind sie leichter in der Lage, von den vielfältigen Anregungen des psychosozialen Milieus einer Station oder einer Fachklinik sowie von den vielfältigen Psychotherapieformen zu profitieren, die für eine erfolgreiche Behandlung dieser Störungen meist erforderlich sind (Psychiatrie-Enquête, 1975).

Bei Patienten mit psychosomatisch mitbedingten Erkrankungen ist häufig eine längerfristige Nachbehandlung erforderlich. Wir möchten deshalb abschließend nachdrücklich darauf hinweisen, daß Einrichtungen der Psychosomatischen Medizin im klinischen Bereich nur dann voll wirksam werden können, wenn sie im Rahmen einer **Versorgungskette** institutionalisiert werden, die eine solche ambulante Nachbehandlung gewährleistet. Diese Nachbehandlung kann im Rahmen der Zusammenarbeit mit niedergelassenen Kollegen, durch Polikliniken und Ambulanzen erfolgen. Ein solches Versorgungssystem besteht heute nur an wenigen Orten in ersten Ansätzen. Die im Auftrag des Bundestages erstellte „Psychiatrie-Enquête"[15] enthält Vorschläge für ein solches Versorgungssystem und seine Realisierung. Anzustreben ist in jedem Fall die Nähe zu den Lebensbedingungen der Patienten („gemeindeorientiert") und – bei entsprechender Indikation – die Möglichkeit zur Einbeziehung der Familie in diagnostische und therapeutische Überlegungen. Von Bedeutung ist weiterhin die Möglichkeit einer kontinuierlichen Behandlung

---

15 Psychiatrie-Enquête: Bericht zur Lage der Psychiatrie in der BRD. Zur psychiatrischen und psychotherapeutisch psychosomatischen Versorgung der Bevölkerung. Dtsch. Bundestag, 7. Wahlperiode, Drucksache 7/4200 (1975).

bei gleichzeitiger Differenzierung des Behandlungssystems und Spezialisierung der in ihm tätigen Mitarbeiter. Die Kooperation psychosomatisch orientierter Einrichtungen mit anderen therapeutisch und beratend tätigen Institutionen des psychosozialen Bereiches innerhalb eines regionalen Versorgungssystems ist anzustreben (Dilling, 1978; Richter, 1978b).

28.2 Psychosomatische Konsultations- und Liaisondienste

Karl Köhle und *Peter Joraschky*

28.2.1 Historische Entwicklung und Definition der Liaison-Psychosomatik

Historische Entwicklung

Während sich in Deutschland die Psychosomatische Medizin in internistischen Universitätskliniken und an einigen wenigen psychoanalytischen Instituten entwickelt hat, erfolgte in den USA ihre Institutionalisierung in die Klinik auf breiter Basis durch die zunehmende Angliederung psychiatrischer Abteilungen an Allgemeinkrankenhäuser auch außerhalb der Universitäten (Lipowski, 1977). Dies trug einerseits entscheidend zu einer Öffnung der Psychiatrie, zu ihrem Heraustreten aus der Isolation innerhalb des medizinischen Versorgungssystems bei und ermöglichte andererseits eine Kooperation mit den übrigen medizinischen Fächern. Als Organisationsformen dieser Kooperation etablierten sich die Konsultations- und Liaison-Psychiatrie mit sowohl psychiatrischen wie psychosomatischen Aufgabenbereichen. Die fortschreitende Integration in die Klinik führte in den 60er und 70er Jahren zu einem Aufschwung klinisch-psychosomatischer Forschung und festigte den Stellenwert der Psychosomatik innerhalb der psychiatrischen Facharztausbildung in den USA. Im Gegensatz zu der raschen Entwicklung der Psychosomatik in der Forschung und Ausbildung sind jedoch die Organisationsformen und Konzepte der klinischen Kooperation mit anderen klinischen Fächern weniger klar umrissen.

In den USA wurde erstmals 1902 eine psychiatrische Abteilung einem Allgemeinkrankenhaus angegliedert. 1934 finanzierte die Rockefeller-Foundation die Einrichtung von 5 psychiatrischen „Liaison-Departments" an Universitätskliniken. Deren Unterstützung ermöglichte unter anderem die Pionierarbeit von F. Dunbar am Columbia University Medical Center. Von 1933–1942 entwickelten R. Kaufman und Mitarbeiter am psychiatrischen Department des Beth-Israel-Hospitals in Boston (Harvard) ein Arbeitskonzept, das gleichzeitig die psychotherapeutische Versorgung ambulanter Patienten, einen psychiatrisch-psychosomatischen Konsiliardienst für die stationär behandelten Kranken des übrigen Krankenhauses und ein entsprechendes Weiterbildungsangebot sowohl für die Spezialisten als auch für die Ärzte der übrigen klinischen Fächer enthielt (vgl. Zinberg, 1964). Nach dem Zweiten Weltkrieg wurden verschiedene Modelleinrichtungen aufgebaut, in denen besonderer Wert auf die enge Zusammenarbeit mit internistischen und chirurgischen Stationen gelegt wurde: G. L. Bibring am Beth-Israel-Hospital in Boston und R. Kaufman am Mount-Sinai-Hospital in New York (Zinberg, 1964; Kaufman, 1965). Der Liaison-Psychiatrie bzw. Liaison-Psychosomatik fiel aufgrund ihrer Aktivitäten in klinischer Versorgung und Aus- und Weiterbildung bald eine Schlüsselrolle als Bindeglied zwischen Psychiatrie und der somatischen Medizin zu. In diesem Zusammenhang wiesen Kaufman und Margolin (1948) nachdrücklich auf die Bedeutung des Weiterbildungsangebotes der Psychosomatiker für die Ärzte der übrigen klinischen Fächer hin.

In der Zwischenzeit hat die Konsultations- und Liaison-Psychiatrie und -Psychosomatik vor allem im Zusammenhang mit der Einrichtung zahlreicher psychiatrischer Abteilungen an Allgemeinkrankenhäusern auch außerhalb von Universitätskliniken erheblich an Bedeutung gewonnen. 1974 bestanden in den USA 800 solcher an Akutkrankenhäuser angegliederter psychiatrischer Abteilungen. In ihnen arbeiteten 22,4% aller Psychiater; 14,4% ihrer Arbeitszeit galt der Konsultation für andere klinische Fächer, 68,4% derjenigen Psychiater, die eine landesweite Umfrage in den USA beantworteten, waren auch im Rahmen eines Konsultationsdienstes tätig (Lipowski, 1975). 1976 führten 76% aller psychiatrischen Ausbildungszentren der USA Ausbildungsprogramme und Trainingsprogramme für Konsultationsarbeit durch, an den meisten psychiatrischen Kliniken gab es entsprechende Postgraduate-Trainingsprogramme für Residents, die einen Einsatz als Konsiliarius für internistische und chirurgische Abteilungen vorsehen.

Dieser Institutionalisierungsprozeß förderte das psychosomatische Denken in der Medizin insgesamt; im Zusammenhang mit der Übernahme von Versorgungsaufgaben fand eine Weiterentwicklung der Psychosomatik in Richtung auf eine ganzheitliche und komplexe Auffassung von Krankheit statt, die den ursprünglich hochspezialisierten, auf wenige Krankheitsbilder bezogenen psychoanalytischen Ansatz ablöste (Lipowski, 1977).

Leider hat sich inzwischen die Situation in den USA einschneidend negativ entwickelt. Eine einseitige biologische Ausrichtung der Psychiatrie verdrängt das psychodynamische Denken. Die Einbindung der Psychosomatik in psychiatrische Institutionen hat jetzt ausgesprochen negative Folgen, die Arbeitsbedingungen haben sich zum Teil drastisch verschlechtert.

In der Bundesrepublik hat die Einführung des Faches Psychosomatik/Psychotherapie in die Approbationsordnung von 1970 die Institutionalisierung des Faches zunächst sehr gefördert; es gelang die Entwicklung „vom Elend in die Armut" (A. E. Meyer). Die vorhandenen Institutionen entsprechen aber auch an den meisten Universitäten noch in keiner Weise dem klinischen Bedarf und sind in ihrer Mehrheit für die anstehenden Forschungsaufgaben noch unzureichend ausgerüstet. Diese „Armut" ist – neben inneren Widerständen – auch ein Hauptgrund dafür, daß die erforderliche Kooperation mit den übrigen klinischen Fächern bisher nur selten in vollem Umfang zufriedenstellend geleistet werden kann.

Definition: Konsultations- und Liaisondienste

Häufig wird nicht genügend scharf zwischen psychosomatischer Konsiliartätigkeit und sog. „Konsultations-Liaisondiensten" differenziert. Diese Ansätze unterscheiden sich in Konzeption und praktischer Durchführung erheblich.

Konsultationsdienste sehen sich vor allem mit psychiatrischen Dringlichkeitsanforderungen konfrontiert: Suizidversuche, Alkoholismus, organische Psychosyndrome, endogene Psychosen (Anstee, 1972). Der Konsiliarius übernimmt die psychiatrische Notfalldiagnostik, berät hinsichtlich der Therapie, meist einer akuten medikamentösen Therapie, oder führt auch eine Krisenintervention durch. Häufig vermittelt er die Überweisung von Patienten in psychiatrische Fachkliniken: In der Untersuchung von Anstee (1972) wurden 69% der Konsultationsfälle in psychiatrische Kliniken überwiesen. Die Konsultationsdienste arbeiten also im allgemeinen unabhängig von den Stationen, der Konsiliarius steht auf telefonische oder schriftliche Anforderungen zur Verfügung.

Liaisondienste haben weitergefaßte Arbeitsziele bzw. Aufgaben und stehen in engerer Verbindung mit den übrigen medizinischen Abteilungen. Der Liaison-Psychosomatiker ist dem klinischen Schauplatz näher; meist arbeitet er intensiv mit den Ärzten und Schwestern der Station zusammen, nimmt einmal oder mehrmals wöchentlich an der klinischen Visite teil; er ist für alle Mitarbeiter der Station möglichst jederzeit erreichbar. Der Liaison-Psychosomatiker versucht eine Brücke zwischen dem verhaltenswissenschaftlichen und dem naturwissenschaftlichen Verständnisansatz zu schlagen. Das Arbeitsfeld der Liaison-Psychosomatiker ist damit grundsätzlich verschieden von dem der Konsultations-Psychosomatiker bzw. -Psychiater, es umfaßt die klinisch-psychosomatische Tätigkeit im eigentlichen Sinn.

28.2.2   Tätigkeitsfelder und Inanspruchnahme von Konsultations- und Liaisondiensten

Tätigkeitsfelder

Der Liaison-Psychosomatiker hat Aufgaben in der Krankenversorgung, in der Aus- und Weiterbildung sowie in der Forschung. Er arbeitet dabei am häufigsten mit Internisten, aber auch mit Pädiatern, Gynäkologen, Neurologen, Dermatologen, Orthopäden und Chirurgen zusammen. Diese Zusammenarbeit hat sich während der letzten zwei Jahrzehnte im Bereich der medizinischen „Spitzentechnologie" und medizinischer Extremsituationen verstärkt: Intensivstationen der Inneren Medizin, Chirurgie und Anästhesie, Coronary Care Unit, Isolationsbettsysteme, Nierentransplantation, Operation am offenen Herzen. Die Belastungen von Ärzten und Schwestern im Umgang mit sterbenden Patienten führten zur Entwicklung der klinischen Thanatologie durch Mitarbeiter der Liaisondienste (vgl. Kap. 66).

Inanspruchnahme von Konsiliarien

Aus den **quantitativen** Unterschieden der Inanspruchnahme von Konsultations- und Liaisondiensten (vgl. Tab. 28–10 und 28–11) läßt sich die Intensität der Kooperation ablesen. Entsprechend den verschiedenen Organisationsformen manifestiert sich eine deutlich unterschiedliche Beratungsfrequenz.

Bei einer Untersuchung von Goldenberg und Sluzki (1971) in Argentinien stieg bei einer entscheidenden Vergrößerung der Mitarbeiterzahl und damit des Liaisondienst-Angebotes die Zahl der Anforderungen auf 17,9% aller stationären Patienten; in Ulm bezogen sich die Anforderungen im Rahmen eines stationsnahen Liaisondienstes auf 11% aller Patienten (Schüffel, 1973), ein Jahr später wurde ein stationsfernes Konsultationsangebot nur für 4% aller Patienten in Anspruch genommen.

Qualitativ stehen beim Konsultationsdienst diagnostische Dringlichkeitsanforderungen im Vordergrund.

Auch beim stationsfernen Konsultationsdienst stellt das Spektrum der Anforderungen an den Konsiliarius weder quantitativ (ca. 35% aller stationären Patienten werden als psychisch behandlungsbedürftig angesehen) noch qualitativ ein Spiegelbild der tatsächlichen psychosomatischen Morbidität im Krankenhaus dar; es ist vielmehr als das Ergebnis eines Interaktionsprozesses zwischen den behandelnden Ärzten und ihren Patienten zu betrachten. Zumeist handelt es sich um Kranke, die für den anfordernden Arzt oder die Schwestern oder das ganze Team „Problempatienten" sind, wie z.B. sozial oder psychologisch „auffällige" Patienten, die sich schlecht in den Stationsablauf einfügen, offen aggressiv sind oder in anderer Form die Kooperation stören. Dies hat u.a. zur Folge, daß der zurückgezogene, depressive, verhaltensunauffällige Patient gewöhnlich übergangen wird (Wilson und Meyer, 1962; Meyer und Mendelson,

Tabelle 28–10.   Inanspruchnahme von Konsultationsdiensten (Angaben in Prozent der gesamten Patientenpopulation des Krankenhauses)

Shepard et al.	1960	1,34
Eilenberg	1965	2,2
Bridges et al.	1966	2,8
McLeod u. Walton	1969	1,6
Anstee	1972	1,4

Tabelle 28–11.   Inanspruchnahme von Liaisondiensten (Angaben in Prozent der gesamten Patientenpopulation des Krankenhauses)

Meyer u. Mendelson	1961	8,8
Lipowski (Übersicht über 8 Studien)	1967	9,0
Crisp	1968	10,0
Papastamou	1970	11,3
Goldenberg u. Sluzki	1971	17,9
Schüffel	1973	11,0

1961). In dieser Gruppe finden sich andererseits viele im engeren Sinne psychosomatisch Kranke. Sie haben allenfalls dann eine Chance, an einen psychosomatischen Konsiliarius überwiesen zu werden, wenn sich kein organischer Befund erheben läßt; damit erfolgt jedoch eine einseitige Auswahl „funktioneller Syndrome".

Eine vergleichende Darstellung der von Konsultations- und Liaisondienst betreuten Patienten, etwa nach Diagnosen, ist schon wegen der uneinheitlichen Terminologie in der psychopathologischen Deskription schwierig; immerhin wird eine zunehmende Veränderung des Krankheitsspektrums im Verlauf der Jahre sichtbar.

Während Kaufman (1948) noch vorwiegend Patienten mit „klassischen" psychosomatischen Krankheiten behandelte, machen diese heute nach den durchschnittlichen Angaben in der Literatur nur noch etwa 7% des Patientengutes von Liaisondiensten aus; allerdings hängt dies auch damit zusammen, daß von den Internisten der Beitrag der Psychosomatik zu der Behandlung dieser Kranken noch immer unterschätzt wird (Stewart et al., 1962; Wilson und Meyer, 1962; Schwab, 1964; Spencer, 1964; Lipowski, 1967; Papastamou, 1970).

Bei Liaisondiensten verschiebt sich mit der zunehmenden Integration auch das diagnostische Spektrum beträchtlich: Der Schwerpunkt liegt nicht mehr auf einer psychiatrisch orientierten Selektion der Patienten, sondern bei Patienten, deren Diagnosen in etwa der durchschnittlichen Krankheitsverteilung im jeweiligen Fach entsprechen (Porot et al., 1972; Schüffel, 1973). Die Anforderung erfolgt verstärkt für Malignompatienten und andere chronisch Kranke mit emotionalen Problemen in der Krankheitsverarbeitung (30% nach Schüffel, 1973) und für Patienten mit funktionellen Beschwerden (20% nach Schüffel, 1973).

28.2.3 Organisationsformen

Organisatorisch wird der Konsiliardienst in den USA von einer psychiatrischen Abteilung, in der Bundesrepublik von einer Abteilung für Psychosomatische Medizin wahrgenommen. Größe und Zusammensetzung des Konsiliardienst-Teams sind abhängig von der Größe des Krankenhauses und der Verfügbarkeit von Psychosomatikern: vom „Ein-Mann-Dienst" bis zu großen Teams an manchen Universitätskliniken (Beigler et al., 1959; Richmond, 1965; Zinberg, 1964; Goldenberg und Sluzki, 1971).

In größeren Liaisoneinrichtungen hat sich eine Arbeit im Team durchgesetzt, wobei innerhalb des Teams eine klare Abgrenzung der Aufgabenbereiche zweckmäßig ist (Lipowski, 1975). Bewährt hat sich eine gemischte Besetzung mit Ganztags- und Teilzeitpsychosomatikern, die Beteiligung eines Sozialarbeiters, eines Psychologen und einer psychosomatischen Schwester („psychiatric liaison nurse").

Die psychosomatische Schwester vermag vor allem die häufigen Interaktionsprobleme zwischen Patienten und Pflegepersonal mitzubearbeiten, bzw. ihre

Kolleginnen und Kollegen bei der Lösung dieser Probleme zu beraten (Barton und Kelso, 1971).

Einsatzbereich und Tätigkeiten des Liaisondienstes werden entsprechend dem Bedarf und den Möglichkeiten gewählt. Der Liaison-Psychosomatiker nimmt gewöhnlich ein- bis zweimal pro Woche an den Stationsvisiten teil, er ist dem Stationspersonal gut vertraut, leitet regelmäßig Stationskonferenzen und kasuistische Seminare; darüber hinaus „sollte er den ganzen Tag jederzeit auf Abruf zur Verfügung stehen" (Lipowski, 1967a).

Die Effektivität eines Liaisondienstes ist abhängig von der Kooperation innerhalb des Teams, von der Koordination der einzelnen Tätigkeitssektoren (Krankenversorgung, Weiterbildung, Forschung) miteinander und von der Qualität der Kooperation mit den Vertretern der jeweiligen klinischen Abteilungen. Die Leitung des Liaisondienstes sollte bei der Vielfalt der Aufgaben und der Abhängigkeit der Kooperation auch vom Erfolgserleben der Kooperationspartner ein erfahrener Psychosomatiker übernehmen.

Entscheidende Bedeutung für eine stabile Etablierung eines Liaisondienstes wird dem Leiter der psychosomatischen Abteilung zugemessen. Er muß in der Lage sein, durch gute Kooperation mit den Leitern der anderen klinischen Abteilungen die Arbeit der am Liaisondienst beteiligten Mitarbeiter zu unterstützen und persönlich mitzutragen. Gelingt dies nicht, erweisen sich Liaisondienste als leicht verletzbar (Lipowski, 1975).

28.2.4 Funktion und Arbeitsweise von Liaisondiensten

Orientierung am Patienten versus Orientierung am Therapeuten

Wir unterscheiden drei sich ergänzende Ansätze:

Der patientzentrierte Arbeitsansatz

Gesprächspartner des Konsiliarius ist der Kranke selbst. Der Konsiliarius übernimmt die Untersuchung und eventuell auch die Behandlung des Patienten, allenfalls ist der anfordernde Kollege bei dieser Untersuchung anwesend. Ein solches Vorgehen wurde im einzelnen als „therapeutische Konsultation", etwa von Weisman und Hackett (1960), konzipiert. Es bewährt sich insbesondere bei speziellen differentialdiagnostischen Problemen, die eine entsprechend hochdifferenzierte Weiterbildung erfordern, und bei akuter Kriseninteryention.

Dieser Ansatz ist sehr zeit- bzw. personalaufwendig; er hat nur wenig Weiterbildungseffekt für die anfordernden Kollegen, die Effizienz im Sinne eines verstärkten Einflusses psychosomatischen Denkens in den klinischen Fächern ist gering. Deshalb wird von Liaisondiensten ein arztzentriertes oder – wenn möglich – ein teamzentriertes Vorgehen bevorzugt. Eine Zwischenform besteht darin, daß der konsultierende Arzt wenigstens bei der Untersuchung durch den Konsiliarius anwesend ist, die Untersuchungsergebnisse im Anschluß diskutiert und das weitere Vorgehen gemeinsam abgesprochen werden.

Der arztzentrierte Arbeitsansatz

Bei dieser Form der Beratung sieht der Liaison-Psychosomatiker den Patienten häufig nicht selbst; er versucht sich über die Darstellung des Kollegen, unter Einbeziehung von dessen emotionaler „Mitbewegung", ein Bild von dem Kranken zu machen. Insbesondere werden die Motive des anfordernden Arztes zur Konsultation, seine Erwartungen an die Konsultation und die Schwierigkeiten in der Behandlung bzw. im Umgang mit dem Patienten berücksichtigt. Der Psychosomatiker versucht, die zwischen dem behandelnden Arzt und seinem Patienten ablaufenden Interaktionsprozesse zu verstehen und eine Beziehung zu Daten aus der vom Arzt erhobenen biographischen Anamnese herzustellen.

Dieses Vorgehen orientiert sich möglichst konkret an der aktuell gegebenen Situation. Es ist die Aufgabe des Konsiliarius, seine Befunde und Hypothesen in einer Sprache zu formulieren, die dem Kollegen die Zusammenhänge deutlich werden läßt; es hat sich bewährt, Fachjargon und psychodynamische Spekulationen zu vermeiden. Der Kollege soll in die Lage versetzt werden, die Situation selbst zu bewältigen. Hierdurch bleibt eine ganzheitliche Betreuung des Patienten durch seinen Arzt gewährleistet, der Liaison-Psychosomatiker übernimmt lediglich Supervisionsfunktion.

Eine Verbindung zwischen arztzentriertem Vorgehen und systematischer Weiterbildung kann in regelmäßig über längere Zeit durchgeführten **berufsorientierten Selbsterfahrungsgruppen,** wie sie Balint beschrieben hat, erfolgen (vgl. Kap. 20 und 68). Hier diskutiert eine Gruppe von Ärzten über Probleme, die im Umgang von jeweils einem Arzt mit einem Kranken auftreten. Für niedergelassene Kollegen hat sich diese Form der Verbindung zwischen Beratung und Weiterbildung sehr bewährt; bei Durchführung solcher Gruppenarbeit mit Ärzten im Krankenhaus ist auf den Ausschluß hierarchischer Probleme sowie auf die Möglichkeit zu längerfristiger Patientenbetreuung durch die beteiligten Ärzte zu achten.

Der teamzentrierte Ansatz

Hier werden nach Möglichkeit alle an der Betreuung des „Problempatienten" Beteiligten – also auch das Pflegepersonal – in die Beratung einbezogen. Die Beratung erfolgt im allgemeinen im Rahmen einer Stationskonferenz. Dabei wird versucht, die Hauptkonflikte in den Interaktionsprozessen und die Art der affektiven Spannungen zu klären. Prinzipiell wird davon ausgegangen, daß diese Konflikte und Spannungen durch die spezifischen Probleme, die Persönlichkeitsstruktur des Patienten und die in ihm ablaufenden psychodynamischen Prozesse induziert wurden. In diesem Konzept berät der Psychosomatiker die „operationale Gruppe" (Meyer und Mendelson, 1961).

Er versucht, aus seiner distanzierten Position die zu beobachtenden Spannungen zu klären und zu interpretieren. Die unterschiedliche Wahrnehmung desselben Patienten durch die verschiedenen Stationsmitglieder, die Beobachtung der vielfältigen Interaktionsprozesse ermöglicht im allgemeinen

die Bearbeitung skotomisierender Wahrnehmungsvorgänge und erlaubt häufig ein Verständnis des aufgetretenen „Problems". Oft erweisen sich die Konflikte und Spannungen als eine für den Kranken charakteristische Neuinszenierung von Konflikten, die auch in seinen früheren und gegenwärtigen Beziehungen außerhalb des Krankenhauses auftreten. In vielen Fällen lösen sich die Spannungen schon dadurch, daß nach solchen Besprechungen der Patient von den Mitarbeitern der Station in veränderter Weise wahrgenommen und ihm dadurch die Möglichkeit zu einer Änderung seines Verhaltens eröffnet wird. Eine solche Veränderung des sozialen Feldes bewirkt oft unmittelbar überraschende und weitgehende Änderungen des Patientenverhaltens. Nicht selten ereignet es sich, daß Patienten nach einer solchen Konferenz „spontan" über die dort diskutierten Probleme zu sprechen beginnen. In anderen Fällen kann eine direkte Besprechung der Schwierigkeiten mit dem Patienten durch die Konferenz vorbereitet oder auch – was allerdings seltener vorkommt – die Indikation zu einer gezielten psychotherapeutischen Intervention gestellt werden.

Dieser Arbeitsansatz wird von vielen Autoren als besonders fruchtbar beschrieben; eine systematische Überprüfung seiner Effizienz erweist sich jedoch als schwierig.

Bei diesem teamzentrierten Arbeitsansatz wird jedem Beteiligten die Wirksamkeit eines zirkulären Verständnismodells, das die Gegenseitigkeit im Umgang mit dem Patienten berücksichtigt, unmittelbar erlebbar. Umgekehrt läßt sich aus dieser Sicht verdeutlichen, in welch ungeahntem Ausmaß in medizinischen Institutionen die Äußerungs- und Mitteilungsfähigkeit des Patienten nicht nur gefördert, sondern auch blockiert werden kann.

Ein Beispiel aus einer Stationsbesprechung auf der internistischen Intensivstation soll die Art des dargestellten Veränderungsprozesses illustrieren:

Zwei Schwestern der Intensivpflegestation klagten darüber, daß ein Patient, der erst am Vortag extubiert worden war und seit Wochen wegen einer aufsteigenden Lähmung unter teils kontrollierter, teils assistierter Beatmung auf der Station lag, immer noch nicht verlegt worden sei. Der Stationsarzt war von der Heftigkeit der Klage überrascht, objektiv erschien diese Verlegung verfrüht. Auf die fortbestehende Gefährdung des Patienten aufmerksam gemacht, begannen die Schwestern über sein ständiges Läuten und Quengeln zu klagen. Weiter habe der Patient paranoide Befürchtungen geäußert, u.a. behauptet, es werde ihm Gift in die Infusionsflaschen getan; und schließlich habe er geklagt, er sei der am schlechtesten versorgte Patient der Station. Die extreme Abhängigkeit des Patienten und seine Versuche, sie zu bewältigen, wurden mit den Schwestern durchgesprochen. Ohne daß die Schwestern nun ihr Verhalten für sie selbst erkennbar geändert hätten, erschien ihnen der Patient schon während der nächsten Schicht „wie verwandelt". Alle Klagen und paranoiden Befürchtungen waren verschwunden; der Patient bedankte sich dafür, daß er der am besten versorgte Patient auf der Station sei. Etwas am Verhalten der Schwestern hatte ihm offenbar gezeigt, daß er sich verstanden fühlen konnte.

Überraschende therapeutische Erfolge nach solchen Besprechungen – nicht selten als „Wunderheilungen" oder „Pfingstwunder" erlebt – verstärken die Bereitschaft der Konferenzteilnehmer, an diesem Verständnisansatz mitzuarbeiten. Ein Beispiel aus der Arbeit einer Schwesterngruppe soll dies etwas ausführlicher darstellen:

Eine 50jährige Geschäftsfrau war nach einem Schlaganfall auf die Privatstation eingeliefert worden. Die Symptome, Bewußtlosigkeit und Sprachstörungen, hatten sich rasch wieder zurückgebildet. Überraschenderweise entstanden bei der Rehabilitation Schwierigkeiten. Die Patientin blieb unbeweglich im Bett liegen und verhielt sich ablehnend allen Maßnahmen gegenüber, die eine Mobilisation zum Ziel hatten. Den Schwestern war die Patientin „furchtbar unsympathisch". Besonders irritierten sie das zugleich fordernde und kritische Verhalten der Kranken, die z.B. versuchte, beim Herrichten des Essens Vorschriften zu machen, eine Verhaltensweise, die nach Aussage der Schwestern zugleich Zorn und Schuldgefühle hervorrief. Die Schwestern berichteten, daß die Patientin, bei der sie einen hilflosen und zugleich trotzigen Gesichtsausdruck beobachteten, ihnen das Gefühl vermittle, sie wollte ihnen einerseits etwas mitteilen, vermeide aber andererseits das Gespräch mit ihnen. Eine Schwester schloß ihren Bericht: „Ich empfand gegenüber der Patientin eine nie gekannte tiefe Hilflosigkeit und Trostlosigkeit, ich hatte einfach das Gefühl, ich muß dies in der Gruppe erzählen".

Das Problem der Verärgerung durch die Patientin stieß in der Gruppe auf starke Resonanz. Zunächst wurden wir darauf aufmerksam, daß die Schwestern sich irgendwie gekränkt fühlten. Gekränkt in ihrem Selbstverständnis als Schwestern, zurückgewiesen in ihrem Hilfsangebot. Es stellte sich heraus, daß die Patientin eine selbständige Geschäftsfrau war, immer gewohnt, nicht nur ihre eigenen Angelegenheiten, sondern auch die ihrer ganzen Familie zu regeln. Sie war bisher nie krank gewesen. Die jetzige Krankheit brachte sie in eine Abhängigkeit, gegen die sie sich ihr ganzes Leben lang gewehrt hatte. Nun fiel einer Schwester auch noch ein, in welcher Weise sie versucht hatte, die Patientin zu loben: Sie hatte z.B. gesagt: „Ich bin heute ganz zufrieden mit Ihnen". Die Patientin, durch die vorübergehende Hilflosigkeit nach dem Schlaganfall in ihrem Selbstwerterleben bereits gekränkt, mußte diese Behandlung, die ja etwa dem Umgang mit einem Kind entspricht, als weitere Kränkung empfunden haben: sie reagierte hierauf mit Zurückweisung. Wir schlugen der Schwester vor, das Selbstvertrauen der Kranken zu stärken; sie sollte das Lob über den Fortschritt ihrer Rehabilitation nicht in Verbindung mit der Abhängigkeit vom Pflegepersonal erfahren. Wir empfahlen ihr daher folgende Formulierung: „Sie können aber heute mit sich zufrieden sein". Die Schwester hielt sich an unseren Vorschlag und berichtete in der nächsten Sitzung: „Ich hätte nie geglaubt, daß mit wenigen Worten so eine Wirkung erzielt werden kann. Ich habe das Gesicht der Patientin ganz bewußt beobachtet und ganz große Änderungen feststellen können".

Die Patientin machte in den nächsten Tagen fast unglaublich erscheinende Fortschritte. Während vorher die Hilfsangebote der Schwestern die Patientin in einem Circulus vitiosus weiter gekränkt hatten, die vermeintliche therapeutische Interaktion sie weiter in die Krankenrolle gezwungen hatte, nahm jetzt die Patientin ihre Rehabilitation selbst in die Hand. Wie sehr auch umgekehrt die Schwestern die Hilfsbedürftigkeit der Kranken für ihre eigene berufliche Befriedigung brauchten, kam in dem deprimierenden Gefühl zum Ausdruck, das die zuständige Schwester befiel, als sie eines Tages entdeckte, daß die Patientin wieder ohne fremde Hilfe gehen konnte. Einerseits freute sie sich über den Erfolg, andererseits verspürte sie schmerzlich, daß die Patientin sie nun nicht mehr benötigte.

Eine solche positive Bearbeitung von Problemen im Umgang zwischen Patienten und Mitarbeitern der Station ist nach unseren eigenen Erfahrungen bei vielen Patienten möglich, die unter der Abhängigkeit der Krankenrolle leiden oder die im Rahmen der Krankheitsverarbeitung verstärkt mit Angst und/oder Depressionen reagieren.

Ein zentrales Forschungsthema klinischer Psychosomatik wäre daher die Untersuchung, wie sich Einstellungsveränderungen der Mitarbeiter auf Verhaltensweisen und auch die Symptomatik des Patienten auswirken. Eine solche Untersuchung müßte die Erforschung der häufig averbalen Vermittlungsvorgänge in diesem Prozeß einbeziehen.

Der Beratungsverlauf

Die Anforderung von Beratung

Bereits die Formulierung der Anforderung gibt Hinweise auf die Wahrnehmung des Problems durch den konsultierenden Arzt. Verallgemeinernd läßt sich sagen, daß der Ungenauigkeit der Formulierung die Hilflosigkeit des Anfordernden entspricht. Der Liaisondienst hat zunächst die Problemstellung zu klären und eine neue Definition der Anforderung vorzunehmen.

Die Anforderung des Liaison-Psychosomatikers erfolgt etwa gleich häufig wegen differentialdiagnostischer Probleme wie wegen einer Beratung in therapeutischen Fragen. Die Angaben über Diagnosen als Anlässe für die Konsiliaranforderung sind nur schwer vergleichbar. Kaufman (1957, 1972) etwa gibt an, daß bei 61,4% aller Anforderungen differentialdiagnostische Probleme vorlagen, während andere Autoren diese Kategorie überhaupt nicht erwähnen.

Zumeist werden als Anforderungsgrund „ein psychisch auffälliger Patient", „sein psychiatrisches Problem", „z.B. Psyche", genannt, nicht selten auch das Vorliegen sozialer Probleme.

Als Auslösefaktoren für eine Anforderung spielen die „Klagsamkeit" der Kranken und Abweichungen ihres Verhaltens von den Erwartungen innerhalb der medizinischen Institution eine wesentliche Rolle. Im Rahmen der Krankheitsverarbeitung werden Angst, Depression und auch Panik als „normale Reaktionen" angesehen, dagegen werden Verhaltensweisen, die Ärzte und Schwestern persönlich tangieren, wie aggressive Äußerungen, zu enge persönliche Annäherung, Verliebtheit, als irritierend erlebt und als Anlaß zur Anforderung eines Liaison-Psychosomatikers genommen. Der stille, zurückgezogene Patient wird kaum überwiesen, dagegen eher Kranke, deren „Depressionen" von ausgeprägten körperlichen Klagen begleitet sind. Zusammengefaßt erfolgt die Konsultation weniger aufgrund des Krankheitsbildes oder der es begleitenden psychischen Auffälligkeiten, sondern eher entsprechend der Irritation der behandelnden Ärzte und Schwestern. Auffallend ist, daß die Anforderung häufig zu einem Zeitpunkt stattfindet, an dem die bisher stabile Arzt-Patient- bzw. Team-Patient-Beziehung aus einem nicht näher zu klärenden Grund gestört wurde. Bei

Patienten, deren Beschwerden auf keine somatische Ursache zurückgeführt werden können, erfolgt die Anforderung häufig spät, oft unmittelbar vor der Entlassung, mit dem Wunsch, der Psychosomatiker möge jetzt die Beschwerden rasch klären.

Das zur Anforderung führende Problem kann nur in seltenen Fällen tatsächlich allein aus den beim Patienten zu erhebenden psychischen Befunden verstanden werden. Die Anforderungssituation ist oft als eine „Interaktionskrise", der eine Entfremdung innerhalb des Arzt-Patient-Verhältnisses, ein Zusammenbruch der auf Verständnis und Toleranz gegründeten Beziehung vorausgeht.

Schwab (1964, 1965, 1966) fand bei 100 ihm als Liaison-Psychosomatiker überwiesenen Patienten in abnehmender Häufigkeit folgende Problemstellungen: unbefriedigendes Familienleben, trotz Therapie unveränderte Symptomatologie, „unbehandelbare Patienten", Häufung von Fragen an den Arzt, Verschlechterung einer chronischen Erkrankung. Nur jeder zweite Patient mit funktionellen Störungen wurde dem Liaison-Psychosomatiker überwiesen.

Der Anforderung eines Konsiliarius geht beim konsultierenden Arzt meist eine Verunsicherung durch die Probleme des Patienten voraus. Er wünscht nun die Verantwortung zu teilen oder zu delegieren. Bleiben die Interaktionsschwierigkeiten weiter ungeklärt bestehen, kann es zu einer zunehmenden Entfremdung kommen, und Abwehrmechanismen können das Verhalten des Arztes kennzeichnen: Interesselosigkeit, Stereotypie im Verhalten, Vermeidung, schließlich Verleugnung des Problems. Deshalb sollte im Beratungsprozeß möglichst rasch eine wenigstens vorläufige Klärung des Hauptproblems angestrebt und für den anfordernden Arzt Handlungsempfehlungen bzw. Handlungsanweisungen abgeleitet werden.

Die Beratung

Am Anfang steht die Klärung der Erwartungen des konsultierenden Arztes. Zumeist artikulieren sie sich in dem Wunsch, konkrete Ratschläge zu erhalten, wie „mit diesem Patienten", „mit diesem Charakter" umzugehen sei (Nadelson, 1971). Eine möglichst präzise Beschreibung des „Problems", der Arzt-Patient-Beziehung, der Team-Patient-Beziehung, der Psychodynamik des Patienten sowie seiner Lebenssituation in Beruf und Familie schließt sich an.

Wesentliche Arbeitsinstrumente des Liaison-Psychosomatikers sind neben der Interaktionsanalyse das psychosomatische Interview, klinische Beobachtung und verschiedene, flexibel zu handhabende psychotherapeutische Techniken, die sowohl direkt patientzentriert angewandt oder den anfordernden Ärzten im Rahmen eines Supervisionsangebotes vermittelt werden können.

Daneben ist häufig eine detaillierte Sammlung von Informationen von Familienmitgliedern, Hausärzten und weiteren Kontaktpersonen erforderlich, die meist ein Sozialarbeiter übernimmt.

Das psychosomatische Interview

Wenn im Rahmen der Beratung der Patient untersucht wird, kommt dem psychosomatischen Interview besondere Bedeutung zu. Der Arzt versucht in diesem Interview, die Beteiligung organischer, psychischer und sozialer Faktoren an der Krankheitsgenese ausgewogen und gleichzeitig zu berücksichtigen und verfolgt das Ziel, eine Gesamtdiagnose unter Einbeziehung der individuellen Krankheitsverarbeitung und des Krankheitsverhaltens zu stellen. Das Verhalten des Arztes während des psychosomatischen Interviews kann bereits die Angst und Depression des Patienten verringern und zu einer Verbesserung des Arbeitsbündnisses führen. Eine Klärung der gesamten Situation, das Ventilieren von gefühlsmäßigen Erlebnisinhalten und Hilfestellungen beim Verbalisieren von Konflikten und Emotionen können bereits während des ersten Interviews zu einer wesentlichen Entlastung des Patienten führen (F. Deutsch und Murphy, 1955; Kimball, 1970; Adler und Hemmeler, 1989).

Therapie

Meist nehmen die überwiesenen Patienten die Gelegenheit zu einem Gespräch gerne wahr, da sie sich sonst im Krankenhaus in ihren Kommunikationsmöglichkeiten sehr beschnitten fühlen. Therapeutisch können aktuelle Probleme, wie etwa präoperative Angst, Reaktionen auf die Mitteilung der Diagnose einer unheilbaren Krankheit, Trauerreaktionen und krankheitsbedingte Funktionseinbußen oder der Verlust von Angehörigen, Konflikte in der Familie, im Beruf und auf der Krankenstation herausgegriffen und häufig erfolgreich bearbeitet werden. Dabei ist es nur in seltenen Fällen sinnvoll, daß der Liaison-Psychosomatiker die Behandlung selbst übernimmt.

Besonders die **supportive Psychotherapie** bei chronisch Kranken kann erfolgreicher vom somatisch tätigen Arzt durchgeführt werden. Dieser kann für den Patienten überzeugend körperliches und seelisches Leiden gleichzeitig in die Behandlung einbeziehen. Der Patient wird so nicht „aufgeteilt", fühlt sich nicht an den Psychiater „abgeschoben", was nur unnötig Widerstände auslöst („Mir fehlt es doch im Bauch und nicht im Kopf").

Effizienz einer Beratung

Untersuchungen der Effizienz konsiliarischer psychosomatischer Beratungen liegen noch kaum vor. Aus den bisherigen Befunden können noch keine Hinweise für eine Indikationsstellung und die Auswahl geeigneter Patienten abgeleitet werden.

Lipowski (1975) untersuchte mittels Fragebogen und Interviews die Zufriedenheit der Patienten mit der Beratung. Insgesamt waren 30–60% der Patienten mit der Beratung zufrieden; das Ausmaß der Zufriedenheit hing dabei insbesondere von der Art und Weise ab, in der die Kranken von dem die Konsultation anfordernden Arzt auf die Beratung vorbereitet wurden. Daneben wirkten sich Persönlichkeitsvariablen und Aspekte der Interaktion auf die Zufriedenheit der

Patienten aus. Ein Drittel der Patienten empfand bereits die Konfrontation mit einem „Psychiater" als eine persönliche Diskriminierung.

Bisher existieren erst wenige systematische Untersuchungen zur Effektivität und Effizienz einer Beratung.

Bei Herzinfarktpatienten konnte die Wirksamkeit einer Einbeziehung des psychosomatischen Konsiliarius in die Behandlung im Sinne eines günstigeren Krankheits- und Rehabilitationsverlaufs mehrfach nachgewiesen werden (vgl. Kap. 41.2).

Die Mitarbeit eines psychosomatischen Konsiliarius auf einer operativen Station konnte die Verweildauer alter Patientinnen nach Schenkelhalsfraktur drastisch – um durchschnittlich 12 Tage – senken (Levitan und Kornfeld, 1981).

Eine additive psychotherapeutische Betreuung internistischer Patienten erwies sich als einflußreich auf die Einjahreskatamnese: so betreute Kranke blieben signifikant häufiger gebessert (Meyer III et al., 1982).

Solange derartige Effizienzuntersuchungen nicht in ausreichendem Umfang vorliegen, können die Zielvorstellungen für die Beratung gemäß der bisherigen klinischen Erfahrung in folgendem Aufgabenkatalog formuliert werden:

– Aufstellung von besonderen Richtlinien für die Betreuung unter Berücksichtigung der Bedürfnisse des Patienten: „Psychosoziales Behandlungs- bzw. Betreuungsprogramm".
– Lösung von „Managementproblemen": unkooperatives Krankheitsverhalten z.B. eines Infarktkranken, Zurückweisung einer Magenresektion bei lebensbedrohlicher Blutung, Verweigerung der Einnahme von Medikamenten usw.
– Beitrag zur Klärung der Differentialdiagnose und damit Beitrag zur Verkürzung der Behandlungszeit bzw. zur Verringerung der Wiederaufnahmerate.
– Beitrag zur Verminderung der Medikation mit Tranquilizern und Analgetika.

Aus den vorliegenden Erfahrungsberichten wird auch deutlich, daß eine kritische Analyse der Interaktionsschwierigkeiten zwischen Liaison-Psychosomatikern und Organmedizinern, d.h. eine kritische Analyse der interdisziplinären Kooperation, teilweise vermieden wird. Erfolgreiche und positiv erlebte Kooperation ist sicherlich stark persönlichkeitsabhängig; dies stellt wohl auch einen wesentlichen Grund für die Scheu von Liaison-Psychosomatikern dar, ihre Arbeitskonzepte offen zu diskutieren (Golden, 1975).

28.2.5 Der Prozeß der Integration des Liaison-Psychosomatikers in das klinische Setting – Interaktionsprobleme zwischen Liaison-Psychosomatikern und den zu beratenden Klinikern

Die Entwicklung der Berufsidentität des Liaison-Psychosomatikers

Die Arbeit an der Grenze zwischen zwei Berufen und zwei Verständniskonzepten von menschlicher Krankheit erschwert die Ausbildung einer Berufsidentität und erfordert große Kompromißbereitschaft,

die Fähigkeit zum Ertragen ständiger Ambiguität. Der Liaison-Psychosomatiker verläßt den sicheren Standort in seiner Spezialdisziplin zugunsten einer Tätigkeit, die sich auch auf ihn leicht verunsichernd auswirkt.

Lipowski (1967), Goldenberg und Sluzki (1971) und Pasnau (1975) beschreiben den Integrationsprozeß des Liaison-Psychosomatikers in das klinische Setting als einen langen Weg von Ignoranz und Isolation bis zu Anerkennung und Kooperation. Dieser Prozeß läßt sich in mehrere Phasen gliedern:

Goldenberg und Sluzki (1971) betonen für die Überwindung der „inneren Front", d.h. der aktiven und passiven Widerstände, die Notwendigkeit, erst einmal „eine Art Nützlichkeit" unter Beweis zu stellen. Die von den Autoren gegebenen Empfehlungen für „ein erfolgreiches Einschleichen" in die somatische Medizin spiegeln das Ausmaß der Widerstände, die überwunden werden müssen; Einsatz nur der erfahrensten Psychosomatiker, maximale Kooperationsbereitschaft, sofortige Beantwortung jeder Anforderung, schnelle therapeutische Intervention. Zusammengefaßt: Angebot eines „optimalen Service" (Lipowski, 1967).

Eine günstige Ausgangsbasis stellen gute Kenntnisse der Erfahrungswelt körperlich Kranker und Versiertheit auch im Denken der somatischen Medizin dar. Die zusätzliche internistische Kompetenz der Liaison-Psychiater bzw. Liaison-Psychosomatiker erwies sich vor allem in der Rochester-Gruppe (G. L. Engel) als förderlicher „carrier" für die Integration; reine Psychiater tendieren eher dazu, die Skepsis und Ignoranz ihrer Kollegen zu beklagen. Die Voreingenommenheit der Kliniker gegenüber der Psychiatrie veranlaßte Porot und Mitarbeiter (1972), ihren Liaisondienst „medizinpsychologischen Service" zu nennen. In Deutschland liegen ähnliche Erfahrungen vor.

Als Gradmesser der Integration wird vor allem auf die erfolgreiche Betreuung psychosomatischer und psychiatrischer Problempatienten hingewiesen. Dabei verlangt vor allem anfangs die Überweisung schwieriger bis aussichtsloser Kranker (Alkoholiker u. a.) vom Psychosomatiker hohe Einsatzbereitschaft und Frustrationstoleranz. Kommunikative Fähigkeiten und „persönliches Charisma" kommen dem Konsiliarius ebenso zugute wie das geduldige Bemühen um Annäherung an das Stationsteam. Außerordentlich hilfreich ist es, wenn es ihm gelingt, den zu Beratenden selbst Erfolgserlebnisse zu vermitteln – und zu ertragen, daß sie von diesen für sich verbucht und keineswegs auf die Beratung zurückgeführt werden.

Zu den äußeren Bedingungen gehört nach Pasnau (1975) auch, daß ein psychosomatischer Konsiliardienst einen Zeitraum von etwa 5 Jahren braucht, um akzeptiert zu werden.

Pasnau wählte während der ersten beiden Jahre vor allem ein patientzentriertes Vorgehen und erst nach zunehmender Vertrautheit mit den Ärzten und Mitarbeitern einen arzt- bzw. teamzentrierten Ansatz. Während des ersten Jahres nahm Pasnau an den Stationsvisiten teil und führte einmal wöchentlich Gruppenbesprechungen mit Stationsschwestern und Sozialarbeitern durch. Im Verlauf des zweiten Jahres begann er mit kasuistischen Seminaren und der Planung einzelner Konsultationsprojekte.

Wie auch bei der Entwicklung anderer Liaisondienste erwiesen sich gemeinsam mit den Ärzten anderer Abteilungen durchgeführte Forschungsprojekte als besonders förderlich für die Integration. Hier können sowohl Erfolge als auch Anerkennung als gemeinsame Erlebnisse gewonnen werden.

Im 4. und 5. Jahr akzeptierte Pasnau Aus- und Weiterbildungsanforderungen. Schließlich stellte sich seinem Bericht zufolge eine befriedigende Kooperation ein.

In ähnlicher Weise wird die Entwicklung anderer erfolgreich funktionierender psychosomatischer Liaisondienste beschrieben (Lipowski, 1967, 1968, 1975).

Frustration der Liaisondienst-Tätigkeit

Die Frustrationen der die Beratung anfordernden Kliniker haben mehrere Quellen:
- Die Notwendigkeit, einen psychosomatischen Fachkollegen zuzuziehen, kann eigene Insuffizienzgefühle auslösen und als kränkendes Scheitern erlebt werden.
- Die Konsultation führt zur Konfrontation mit emotionalen Problemen, die belastend sein können.
- Häufig werden die therapeutischen Erwartungen enttäuscht.

Das Ausmaß der Enttäuschung wird oft erst bei genauerer Kenntnis der Erwartungen verständlich: Viele Kliniker betrachten psychische Erkrankungen im Vergleich zu somatischen Krankheiten als leichter heilbar, u.a. deshalb, da sie nicht selten „psychisch krank" mit „willenskrank" oder „willensschwach" gleichsetzen. Zudem fallen diesen Klinikern häufig nur ausgeprägte, schwere psychische Störungen auf; die diagnostische Leistung wird mit Engagement für den Kranken gleichgesetzt, und vom Psychiater bzw. Psychosomatiker wird eine prompte und erfolgreiche Behandlung erwartet. Es fällt diesen Kollegen oft schwer zu akzeptieren, daß es im psychologischen Bereich ebenso wie im somatischen therapeutisch schwierige und prognostisch aussichtslose Erkrankungen gibt. Therapeutischer Mißerfolg löst besonders bei engagierten Kollegen oft scharfe Reaktionen gegenüber dem Psychosomatiker aus.

Auch für den Psychosomatiker gibt es zahlreiche Frustrationsquellen. Häufige Kränkungen und zusätzliche Identitätsprobleme können die Interaktion mit den Klinikern belasten und für die häufig nur geringe Attraktivität dieser Tätigkeit für Psychosomatiker und Psychiater, jedenfalls außerhalb bestimmter Forschungsprojekte, verantwortlich sein.

Zwar werden in der Literatur immer wieder die positiven Aspekte der Liaisontätigkeit herausgestellt: die Breite der zu gewinnenden diagnostischen und therapeutischen Erfahrung, die Fülle der wissenschaftlichen Aufgaben. Diesem „Angebot" steht auch in den USA eine offenbar nur geringe Motivation der in der Facharztweiterbildung begriffenen Psychiater gegenüber, diese Tätigkeit über die vorgeschriebene Zeit hinaus auszuüben. Relativ wenige Psychiater arbeiten längere Zeit in diesem Bereich. Wer länger als zehn Jahre als Liaison-Psychosomatiker tätig ist, wird als „grizzled veteran" nach Lipowski (1975) betrachtet.

Insgesamt sind die Interaktionsprobleme von der Seite des Konsiliarius kaum systematisch untersucht worden. Immerhin lassen sich für Frustration und Unbehagen einige Ursachen angeben:

- Die häufig verzweifelte Lage der Patienten, die dem Liaison-Psychosomatiker überwiesen werden, führt auch bei ihm zu starken emotionalen Belastungen. Dies gilt insbesondere für die zahlreichen Patienten mit unheilbaren Krankheiten, zu denen er wegen ungelöster Probleme der Krankheitsverarbeitung gerufen wird.
- Bei psychosomatisch Kranken im engeren Sinne trägt vor allem bei psychotherapeutisch orientierten Liaison-Psychosomatikern die distanzierende, affektisolierende Abwehr vieler dieser Kranken, ihr Widerstand gegen jeden psychotherapeutischen Behandlungsversuch, zu den Frustrationen bei. Der Widerstand der Patienten selbst gegenüber einer psychosomatischen Betrachtung ihrer Krankheit bildet eine weitere Frustrationsquelle. Vielfach stehen psychosomatisch Kranke nicht wie neurotisch Kranke unter psychischem Leidensdruck, vielmehr muß ihnen erst ein entsprechendes Problembewußtsein, oft gegen ihren Widerstand, vermittelt werden; der Psychosomatiker muß gewissermaßen um ihre Mitarbeit werben.
- Übereinstimmend wird die häufig negative Haltung der somatisch orientierten Kliniker bzw. die negative Haltung des ganzen Stationsteams als der am stärksten frustrierende Faktor beschrieben. Unter diesem Gesichtspunkt werden die Möglichkeiten einer Integration eher pessimistisch beurteilt. In der Literatur werden überwiegend Erfahrungen über das geringe Interesse von Ärzten an psychosozialen Faktoren geschildert, Einstellungen von Kollegen wiedergegeben, die Psychiater und Psychosomatiker als medizinisch ignorant, wissenschaftlich inkompetent, in der Praxis unbeholfen und umgeben von fragwürdigen Theorien und Praktiken ansehen (Eaton et al., 1965; Meyer und Mendelson, 1961).

  Lipowski (1967) weist nach damals bereits über zehnjähriger Tätigkeit in Liaisondiensten auf Indifferenz, Vorurteile, stereotypes Infragestellen der Kompetenz und Feindschaft der Kliniker gegenüber den Psychosomatikern hin. Selbst engagierte und erfahrene Liaison-Psychosomatiker schildern sich oft als „am Rande der Verzweiflung" (Lipowski, 1967; McKegney, 1972). In noch stärkerem Ausmaß sind hiervon die jüngeren Ärzte in der Facharztweiterbildung betroffen, die noch keine stabile Berufsidentität entwickeln konnten.
- In dieser „troubled marriage" (Pasnau, 1975) findet der Liaison-Psychosomatiker wenig Möglichkeiten, etwa durch befriedigende Kooperation seine eigene gefährdete Sicherheit zu stabilisieren. Er kann sich kaum auf berufliche Vorbilder berufen, es gibt noch keine Tradition des Rollenverständnisses, er muß sich von seiner erlernten Fachsprache lösen. Auch die äußeren Bedingungen seiner beruflichen Tätigkeit bringen eher Unsicherheit mit sich. Durch die Kommunikationsschwierigkeiten und die zum Teil negative Haltung der zu beratenden Kollegen werden auch in ihm affektive Prozesse in Gang gesetzt, die seine Fähigkeit zur Beratung einschränken können.
- Der Liaison-Psychosomatiker kann sich zwar um die Weiterbildung, um die Verbesserung der therapeutischen Fähigkeiten von Ärzten und Schwestern bemühen, er hat jedoch meist keinen Einfluß auf die äußeren Verhältnisse (Zeit und Raum für Gespräche, Stellenplan, Einfluß anderer Konsiliarien u.a.). Nur selten hat er die Möglichkeit, die gesamte Institution einschließlich der Führungskräfte zu beraten und mit ihnen gemeinsam einen „Prozeß der Organisationsentwicklung" einzuleiten. Seine Tätigkeit behält so fast immer das Merkmal des Bruchstückhaften, Ansatzweisen. Bis in die Architektur hinein entsprechen Krankenhäuser – im besten Falle – den Konzepten und Bedürfnissen lediglich der naturwissenschaftlich orientierten Medizin (Cleghorn, 1973).

– Der Faktor, der für die Unsicherheit, Verletzlichkeit und die Schwierigkeit einer Identitätsfindung des Liaison-Psychosomatikers am meisten verantwortlich ist, scheint jedoch das Fehlen einer sicheren Karriere und eines festen Berufsbildes zu sein. Hier fehlt eine entsprechende Weiterbildungsregelung (Facharzt).

Die Interaktionsprobleme im psychosomatischen Liaisondienst werden von der Haltung der ratsuchenden Ärzte einerseits und den Einstellungen und Reaktionen der Konsiliarien andererseits, ihren **Übertragungs- und Gegenübertragungsreaktionen** – im weitesten Sinne dieser Begriffe – mitbestimmt.

Die Reaktionen des Liaison-Psychosomatikers in der Interaktion mit unterschiedlichen Arzt-Typen haben Rotmann und Karstens (1974) eingehender untersucht. Sie weisen zunächst darauf hin, daß die Ausgangssituation – Notwendigkeit zu übergroßem Entgegenkommen, Anpassung bis zur Unterwürfigkeit, um akzeptiert zu werden, therapeutisches Überengagement beim Versuch, die Widerstände bei Patient und Institution zu überwinden – bei den Konsiliarien selbst häufig zu überschießenden Reaktionen führt. Die Verunsicherung wird nicht selten durch die Überzeugung kompensiert, „eine bessere Medizin" zu betreiben; der Konsiliarius hat also mit seinen eigenen Omnipotenzgefühlen zu kämpfen. Kränkungen werden nicht selten mit Rückzugsreaktionen, u.a. in die Rolle des psychotherapeutischen Fachmannes, der in seinem Spezialgebiet arbeitet, beantwortet.

Im einzelnen beschrieben Rotmann und Karstens (1974) vier charakteristische Interaktionsmuster zwischen Kliniker und psychosomatischem Konsiliarius, die sich bei bestimmten Einstellungen und Persönlichkeiten der Kliniker ergeben. Dieser Beschreibung liegen mehrjährige Erfahrungen mit 22 beratenen Fachinternisten bzw. Ärzten, die sich in der internistischen Facharzt-Ausbildung befanden, zugrunde.

Die Autoren beschreiben mit 4 dieser 22 Ärzte ein ausgezeichnetes, mit 3 ein gutes, mit 6 ein mäßiges und 9 ein schlechtes Konsultationsbündnis. Zusammengefaßt war mit einem Drittel der zu beratenden Ärzte eine sowohl erfolgreiche als auch befriedigende Zusammenarbeit im Rahmen des psychosomatischen Liaisondienstes möglich.

Der unsensible und offen ablehnende Arzt: Schon aus dem Mangel an Wissen dieses Arztes – er bringt selten die für die Beratung nötigen Basisdaten über den Patienten mit – wie auch aus seinen Zweifeln an der Nützlichkeit der psychosomatischen Konsultation ergeben sich erhebliche Schwierigkeiten. Bei diesen Ärzten bleiben alle Anforderungen an den Konsiliarius sofort aus, wenn der Konsiliarius nicht ständig auf der Station ist und seine Mithilfe aktiv anbietet.

Der ambivalente Arzt: Er schwankt zwischen Ablehnung und Interesse; der Liaison-Psychosomatiker kann mit einer Reihe komplexer Abwehrmanöver reagieren. Zunächst bemüht er sich um Freundlichkeit und verstärkte Mitarbeit, schon um seinen eigenen Ärger abzuwehren. Dies kann in Kritik am zu beratenden Arzt umschlagen, wenn dieser etwa Empfehlungen mißachtet und eine latente Aggression manifest wird. Die Abwehr der eigenen Enttäuschung und Wut kann beim Konsiliarius zu einer Einengung der Reaktionsbreite und Toleranz führen. Oberflächlich betrachtet herrscht Friede, die Wirksamkeit der Arbeit wird jedoch beeinträchtigt. Das Interesse des Konsiliarius kann sich nun auf die Schwestern oder direkt auf die Patienten verschieben, vor allem dann, wenn der zu beratende Kollege auch als

Kliniker Schwächen zeigt und der Konsiliarius selbst über eine klinische Ausbildung im betreffenden Fach verfügt. Die Tendenz des ambivalenten Arztes, besonders große Ansprüche an den Psychosomatiker zu stellen, kann zur Rivalität auf dem betreffenden klinischen Gebiet beitragen. Die Rollen- und Identitätsunsicherheit des Psychosomatikers läßt ihn auf die Identität des Somatikers ausweichen.

Der motivierte Arzt mit aktivem Interesse an seelischen Prozessen des Patienten: Mit diesem Arzt traten in der Konsultationsbeziehung nie größere Schwierigkeiten auf. Auch wenn der Konsiliarius von sich aus mit dem Arzt rivalisierte, indem er somatische Fragen diskutierte, handelte es sich letztlich doch mehr um ein Werben um diesen Kollegen, nicht um destruktive Rivalität; solche Reaktionen traten auch nur flüchtig und vorübergehend auf. Bei zeitweiligen Störungen der Kommunikation tendierte der Konsiliarius zu einer direkten, patientzentrierten Konsultation. Er konkurrierte mit dem zu beratenden Arzt um den Patienten.

Der freundliche, jedoch unerreichbare Arzt: Dieser meist als Internist sehr fähige Arzt stellt wenig Fragen, er verlangt selten nach einer Konsultation; er hört sich jedoch die Vorstellungen und Empfehlungen des Konsiliarius, etwa während der Visite, höflich an, beachtet sie jedoch nicht weiter. Seine Abwehr ist eher subtil, bei gleichbleibender Freundlichkeit, insgesamt jedoch erfolgreich. Er leistet vor allem passiven Widerstand bis hin zur Verleugnung der Tätigkeit des Konsiliarius. Er schätzt Konsiliarien lediglich als Fachkollegen, mißversteht aber ihr Konsultationsangebot, interpretiert es als Ausdruck ihres persönlichen „rein wissenschaftlichen" Interesses. Der Konsiliarius reagierte auf die Unerreichbarkeit dieser Kollegen nicht unwillig oder ärgerlich; offensichtlich bedeutet der narzißtische Rückzug dieses Arzt-Typs die wirksamste Form der Vermeidung.

Die Entwicklung der Arbeitsbeziehungen zwischen Konsiliarius und zu beratendem Kliniker ließ sich meist nach dem ersten Zusammentreffen voraussagen. Rotmann und Karstens empfehlen, bei Ambivalenz, Interesselosigkeit, scheinbarer Unempfindlichkeit des Klinikers für die Affekte der Patienten und Intoleranz gegenüber seinen eigenen Ängsten zunächst mit einer Verstärkung des Weiterbildungsangebotes in Form gemeinsamer Patienteninterviews zu reagieren, um die Nützlichkeit psychosomatischen Verständnisses zu demonstrieren. Bleibt dieser Versuch erfolglos, so sollte der Konsiliarius die Grenzen seiner Arbeitsmöglichkeiten anerkennen und weitere Versuche unterlassen.

Zusammengefaßt hat der Liaison-Psychosomatiker im Vergleich zu Konsiliarärzten anderer Fächer mit besonderen Schwierigkeiten von seiten der Kollegen, der Patienten und der Institution umzugehen. Hinzu kommt, daß er sich durch die methodische Einbeziehung emotionaler Reaktionen in den Beratungsprozeß persönlich exponiert und so verletzbarer wird. So ist die psychosomatische Konsultation im Vergleich zu anderen Konsultationen wesentlich störanfälliger.

28.2.6 Aus- und Weiterbildung

Beteiligung von Liaisondiensten an der Ausbildung von Medizinstudenten und an der Weiterbildung von Ärzten

Die Effektivität einer Beteiligung der Psychosomatischen Medizin in der ärztlichen Aus- und Weiterbildung hängt weitgehend von der Institutionalisierung der Psychosomatischen Medizin im klinischen Bereich und damit auch von der Integration eines psychosomatischen Liaisondienstes in die Arbeit der klinischen Abteilungen ab (Kimball, 1970; West, 1975).

Besonders bewährt hat sich die Einführung der Studenten während des praktischen Jahres durch Liaison-Psychosomatiker. Sie lernen selbst Anforderungen an die Beratung zu stellen und die Anforderungssituation zusammen mit dem Liaison-Psychosomatiker zu analysieren. Sie erhalten bei den von ihnen durchgeführten biographischen Interviews und Behandlungen die nötige Supervision. Nach unserer Erfahrung sind Internatsstudenten während des Übergangs von der Studenten- in die Arztrolle für die Annahme eines entsprechenden Weiterbildungsangebots besonders aufgeschlossen.

Besonders bewährt haben sich Interviewseminare mit Studenten über mehrere Semester, Balint-Gruppen mit Studenten („Junior-Balint-Gruppen") sowie die Beteiligung von Studenten an der supportiven Therapie Schwerkranker.

Die Ausbildung zum Liaison-Psychosomatiker

In den führenden US-amerikanischen Ausbildungszentren sind differenzierte Weiterbildungsprogramme zum Liaison-Psychiater bzw. Liaison-Psychosomatiker institutionalisiert. Aus unserer Sicht ist nach einer klinischen Weiterbildung in Psychiatrie und/oder Innerer Medizin eine intensive psychotherapeutische Weiterbildung Voraussetzung für die Qualifikation zur Liaison-Tätigkeit (vgl. Kap. 27).

28.2.7   Klinische Forschung

In der Literatur wird immer wieder auf die Befriedigung hingewiesen, die Liaison-Psychosomatiker aus der Vielfalt der klinischen Forschungsansätze ziehen können, als Ausgleich für die Frustration im klinischen Alltag. Neben der Untersuchung einzelner Krankheitsbilder steht die Untersuchung psychosomatischer Gesichtspunkte bei der Anwendung neuer medizinischer Technologien wie Hämodialyse, Operation am offenen Herzen, Implantation von Schrittmachern, Organtransplantationen, Chemotherapie in Isolierbettsystemen u. a. im Vordergrund. Viele Fortschritte der Psychopharmakologie und der Psychophysiologie psychosomatischer Störungen sind das Ergebnis enger Kooperation zwischen Psychosomatischer Medizin und den entsprechenden klinischen Fächern. Durch die wachsende Bedeutung von Präventivmedizin und Rehabilitation ergibt sich auch für den Liaison-Psychosomatiker ein neues Arbeitsfeld, das im Übergangsbereich zwischen Gesundheit und Krankheit liegt.

28.2.8   Zusammenfassung

Die Liaison-Psychosomatik stellt das Bindeglied zwischen den psychosozialen und den klinischen Fächern dar. Ihre Tätigkeit eröffnet die Möglichkeit, bei jedem Krankheitsgeschehen biologische, psychologi-

sche und soziale Daten gleichzeitig, und ihrer Bedeutung gemäß gewichtet, zu betrachten. Für diese Aufgabe benötigt die Liaison-Psychosomatik ein breitgefaßtes, komplexes und holistisches Verständnismodell für menschliche Krankheit und Gesundheit.

Bis heute hat sich jedoch die Prophezeihung F. Dunbars aus der Gründerzeit, daß von der Liaison-Psychosomatik eine entsprechende Veränderung der Medizin ausgehen würde, nicht – oder noch nicht? – erfüllt. Lipowski (1975) spricht von einem 40jährigen „Kampf". Die Probleme der Integration in die klinische Medizin blieben vielfältig; die Möglichkeiten, auf die klinische Praxis im Allgemeinkrankenhaus Einfluß zu nehmen, erwiesen sich meist als minimal. Dies liegt daran, daß noch kaum Modelle entworfen wurden, in denen der Liaison-Psychosomatiker voll integriert an der Krankenversorgung teilnimmt. Er ist abhängig von der Einstellung der Ärzte, auf deren Anforderung er reagiert. Er hat kaum Einfluß auf eine Weiterentwicklung des klinischen Settings im Sinne der theoretischen Konzepte Psychosomatischer Medizin. Es bleibt für ihn schwierig, dauerhaft eine stabile berufliche Identität zu entwickeln. Das Konsultationsbündnis mit den zu beratenden Ärzten und Teams ist meist labil und weitgehend auf gute persönliche Beziehungen angewiesen. Kleinere personelle Veränderungen in einem Liaisonstab, aber auch in der Gruppe der zu Beratenden, können das Bündnis gefährden und damit die Effizienz des konsiliarischen Dienstes in Frage stellen. Allerdings ist auch zu betonen, daß es, wenn auch persönlichkeitsabhängig, immer wieder gelingt, mit einigen Klinikern eine befriedigende Kooperation zu erreichen und auf diese Weise wenigstens den Freiraum für eine dann meist wesentlich befriedigendere Zusammenarbeit in der klinischen Forschung zu schaffen.

28.3   Klinisch-psychosomatische Krankenstationen

Karl Köhle

„Es gibt meines Erachtens Gegenstände der Erkenntnis, welche nur zu erkennen sind durch Akte des Handelns. Im Grunde ist jedes Experiment ein solcher Fall und vielleicht ist das ganze neuzeitliche naturwissenschaftliche Erkennen ein solcher Fall."

(V. v. Weizsäcker, 1955, S. 15)

Vorbemerkung

Vor der Darstellung eines eigenen Modellversuchs – „internistisch-psychosomatische Krankenstation" – gebe ich einen Überblick über die geschichtliche Entwicklung stationärer Behandlungsansätze in der Psychosomatik. Ich möchte hierdurch den derzeitigen Entwicklungsstand kennzeichnen und gleichzeitig auf Schwierigkeiten bei der Realisierung derartiger Konzepte hinweisen. Viele Kenntnisse der Grundlagen gehen bis in die 20er Jahre zurück und sind bis heute nicht zureichend rezipiert und in die Praxis umgesetzt worden.

In der Bundesrepublik wird die Institutionalisierung klinischer Psychosomatik häufig um den Kernbereich einer psychosomatischen Krankenstation herum versucht. Seit Kriegsende wurden verschiedene Modelle erprobt; während der letzten 20 Jahre hat sich bei den meisten Fachvertretern die Auffassung herausgebildet, daß eine solche Krankenstation für Krankenversorgung, Forschung sowie Aus- und Weiterbildung unentbehrlich sei.

28.3.1 Zur Geschichte stationärer Einrichtungen in der Psychosomatischen Medizin

Psychotherapie auf psychoanalytischer Grundlage wurde zunächst ambulant, meist in der Privatpraxis durchgeführt; Alternative war der Hausbesuch, nicht die Behandlung in der Klinik. Der Beginn Freuds psychotherapeutischer Tätigkeit fiel in die Zeit, als die ärztliche Tätigkeit allmählich vom Hausbesuch in die eigene Praxis verlagert wurde (de Swaan, 1978).

Im Rahmen der Privatpraxis wurden das „Setting" entwickelt und die äußeren Bedingungen formuliert, die den psychoanalytischen Prozeß erst ermöglichen. Zunächst blieben im Gegensatz zum Hausbesuch Störungen etwa durch Familienmitglieder der Patienten ausgeschlossen. Grundlegende Regeln für den Patienten (freie Assoziation) und den Arzt (Abstinenz, Diskretion) wurden entwickelt und konsequent angewandt. Setting und Regeln formten einen begrenzenden und schützenden Raum, in dem sich die Beziehung zwischen Arzt und Patient entwickeln konnte und in dem diese Beziehung für beide Partner erkennbar und verstehbar wurde. Setting und Regeln sollten es dem Patienten erlauben, aufgrund seiner Bedürfnislage und seiner Geschichte die aktuelle Beziehung zum Analytiker zu konstellieren. Diese Situation erlaubt eine „Wiederholung" früherer Beziehungsmuster unter gleichsam experimentellen Bedingungen; da der Arzt nicht mitagiert, werden Abweichungen von den sonst üblichen Beziehungsmustern erkennbar, wie sie entsprechend verinnerlichten Strukturen „geplant" werden, darin liegt die Chance für Erkenntnis und Therapie. Der Erkenntnisprozeß wird auch dadurch unterstützt, daß Schwierigkeiten bei der Einhaltung des Settings und der vereinbarten Regeln mit in die Analyse der Beziehung einbezogen werden können.

Es zeigte sich jedoch bald, daß nur ein Teil der psychisch Kranken dazu fähig war, unter den genannten Rahmenbedingungen eine Arbeitsbeziehung zum Psychoanalytiker aufzunehmen und längerfristig durchzuhalten. Neben solchen Patienten, die als Folge krankheitsbedingter Behinderungen den Arzt nicht in seiner Praxis aufsuchen konnten, waren viele Kranke mit Psychosen, schweren narzißtischen Neurosen, chronifizierten Charakterstörungen, Süchtige und auch „psychosomatisch" Kranke aufgrund krankheitsbedingter Einschränkungen und störender Umgebungseinflüsse nicht in der Lage, eine solche therapeutische Beziehung aufzunehmen. Die Vertiefung der psychodynamischen Verständnisansätze und die

Verbesserung der psychotherapeutischen Technik – vor allem die intensivere Nutzung der Beziehung zwischen Arzt und Patient für Verständnis und Therapie – erweiterten den Anwendungsbereich psychoanalytischer Psychotherapie. Vielfach fehlte jedoch ein Rahmen, der den verunsicherten Patienten jene Sicherheit hätte bieten können, die ihnen ein Sich-Einlassen in die therapeutische Beziehung erst ermöglicht hätte. Die Durchführung psychoanalytischer Behandlungen in herkömmlichen Kliniken war nicht möglich; der erforderliche, zugleich Schutz bietende und in seiner Wirksamkeit klar überschaubare Rahmen für psychotherapeutische Behandlungen war nicht herstellbar, u.a. weil mit einer konsequenten Kooperation der ärztlichen Kollegen nicht gerechnet werden konnte.

Mit der Weiterentwicklung der psychoanalytischen Methode war zunehmend deutlich geworden, daß sich Einstellung und Verhalten des psychotherapeutisch tätigen Arztes grundsätzlich von ärztlichen Einstellungen und Verhaltensweisen unterscheiden, wie sie in der naturwissenschaftlichen Medizin üblich und auch erforderlich sind.

Diese Unterscheidung bildete sich erst allmählich heraus; zum Zeitpunkt der ersten Entdeckung Freuds, bei der Anwendung der Hypnose und der Ableitung der ersten therapeutischen Techniken aus der Hypnose, war sie noch nicht deutlich erkennbar.

Die traditionellen Merkmale der Beziehung zwischen Arzt und Patient: „scheinbar allwissend, autoritär, hilfreich der eine; unwissend, völlig ergeben, hilfsbedürftig der andere" (Stone, 1973), entsprechen dem naturwissenschaftlichen Verständnisansatz, der auf ein „Beherrschen" der Natur zielt, und sind für die Anwendung der hieraus abgeleiteten therapeutischen Techniken auch erforderlich. Wissen und Tun liegen dabei ganz auf der Seite des Arztes. In der Psychotherapie dagegen versucht der Arzt, dem Patienten zu helfen, sich selbst entscheiden zu können, „Freiheit" zu gewinnen, „sich so oder anders zu entscheiden" (Freud, 1967). Jede dieser beiden Einstellungen und Verhaltensweisen hat im Rahmen des Gültigkeitsbereiches des jeweiligen Verständnisansatzes wissenschaftlich begründete Berechtigung. Schwierigkeiten entstehen dann, wenn wir versuchen, beide Verständniskonzepte menschlicher Krankheit nicht getrennt, sondern aufeinander bezogen, gleichzeitig in derselben Situation anzuwenden. Dann wird deutlich, daß sich hierbei Widersprüche sowohl auf der Ebene menschlicher Grundeinstellungen als auch auf der Ebene von Organisationsformen und Institutionen ergeben.

Für die Einbeziehung der Psychotherapie in die Klinik, für die Behandlung stationärer Patienten, entwickelten sich zwei Ansätze:

(1) Psychotherapeuten schaffen sich einen völlig neuen, erweiterten Rahmen für die Behandlung. Der einzelne Therapeut wird durch eine Gruppe von Therapeuten unterstützt, in neugegründeten Kliniken wird ein schutzbietendes „therapeutisches Milieu" entwickelt.

(2) In psychiatrischen, internistischen und neurologischen Kliniken entstehen Versuche, das herkömmliche klinische Setting kritisch zu überprüfen und so zu verändern, daß psychotherapeutische Methoden

in das klinische Handeln einbezogen werden können.

Mitarbeiter Freuds, vor allem Federn, begannen um 1920, auch Patienten mit schweren „psychiatrischen" Störungen psychoanalytisch zu behandeln. Die spezifische analytische Behandlung wurde dabei durch eine Gruppe von Therapeuten unterstützt, die den Raum für die Behandlung schützend absichern sollten. Federn (1978) bezog so u.a. Angehörige von Patienten als Hilfspersonen mit in die Behandlung ein; er betonte dabei die besondere Bedeutung mütterlicher Personen im Rahmen eines solchen Behandlungsansatzes.

Simmel richtete nach ersten Erfahrungen mit psychoanalytischer Psychotherapie bei Kriegsneurotikern in einem Militärkrankenhaus 1920 in Tegel bei Berlin eine psychoanalytische Klinik („Psychoanalytisches Sanatorium") ein (Bartemeier, 1978; Simmel, 1928, 1937). Simmel erarbeitete hier Regeln für die Zusammenarbeit u.a. von Psychoanalytikern und Krankenschwestern bzw. -pflegern im Rahmen eines strukturierten Teams mit dem Ziel, einen psychoanalytischen Prozeß auch für Patienten mit Psychosen, narzißtischen Neurosen, schweren Charakterstörungen und Süchten zu ermöglichen. Dabei erwies sich neben dem Wegfall von Störungen aus der Umwelt als wesentlich, daß das Agieren der Patienten, das bei ambulanter Behandlung außerhalb der therapeutischen Situation stattfindet und die analytische Therapie unwirksam macht oder gefährdet, in der Klinik innerhalb eines erweiterten therapeutischen Feldes, im Umgang mit Schwestern und Pflegern erfolgt und über die Zusammenarbeit im Team sofort wieder für die Behandlung des Patienten genutzt werden kann. Die Therapeutengruppe, das strukturierte Team, entspricht für den Patienten in vieler Hinsicht dem „Urtyp seiner Familie überhaupt" (Simmel, 1928); im Umgang mit diesem Team können sich die Konstellationen und Konflikte aus der Familiensituation der Kindheit wieder darstellen, die Zusammenarbeit des Teams macht ihre Bearbeitung möglich.

Freud hat das Tegeler Sanatorium mehrfach längere Zeit besucht, er hat diesen Arbeitsansatz unterstützt; Freud hat vor allem auch auf die Bedeutung einer solchen Klinik für die psychoanalytische Ausbildung hingewiesen; er war beeindruckt von den Möglichkeiten der direkten Zusammenarbeit zwischen erfahrenen Analytikern und in Ausbildung befindlichen Ärzten (Bartemeier, 1978).
    Das Tegeler Sanatorium mußte allerdings nach viereinhalbjähriger Arbeit wegen finanzieller Schwierigkeiten geschlossen werden.
    Ein ähnlicher Ansatz wurde unter Simmels Beratung in der Menninger-Klinik in Topeka, USA, entwickelt (Menninger, 1936; Simmel, 1937; Bartemeier, 1978). Hieraus entstand eines der führenden psychotherapeutischen Forschungszentren. Bekannt geworden ist auch die ähnlich strukturierte psychoanalytische Klinik in Chestnut Lodge bei Washington. Später entwickelten hier Frieda Fromm-Reichmann und Harry Stack Sullivan die Ansätze zur Psychotherapie von Psychosekranken weiter (Bartemeier, 1978; Foudraine, 1973). Die Arbeitsmöglichkeiten dieser Klinik lassen sich durch einen Hinweis auf den Stellenplan veranschaulichen: für 90 erwachsene Patienten und 32 Kinder bzw. Adoleszenten standen 29 Psychoanalytiker zur Verfügung (Bartemeier, 1978).

In diesen klinischen Arbeitssituationen gelang es auch, die Untersuchung der Arzt-Patient-Beziehung im therapeutischen Prozeß voranzutreiben. Die psychoanalytische Forschung hat zunehmend auf die Bedeutung der „Gegenübertragung" des Arztes für den Erkenntnis- und Heilungsprozeß aufmerksam gemacht (Heimann, 1950), nun konnte in dem geschützten Milieu der therapeutischen Gruppe auch die Gegenübertragung bei psychisch Schwergestörten näher untersucht werden (Searles, 1974).

Während in diesen psychoanalytischen Spezialkliniken Möglichkeiten einer weiteren Indikationsstellung und einer Intensivierung des psychoanalytischen Prozesses erprobt wurden, wurde die weitere Entwicklung vor allem in England und den USA dadurch gefördert, daß sich die Psychiatrie zunehmend neuen Verständnis- und Therapieansätzen öffnete. Neben dem psychodynamischen Verständnisansatz führte vor allem der kommunikationstheoretische Ansatz zu tiefgreifenden Veränderungen der klinischen Praxis. Etwa gleichzeitig wurde nach Ende des Zweiten Weltkrieges an psychiatrischen Kliniken versucht, die Behandlungssituation im Sinne einer „therapeutischen Gemeinschaft" (Jones, Main) umzustrukturieren.

Kommunikationstheoretische Ansätze verdeutlichten, ähnlich wie entsprechende Entwicklungen in der Psychoanalyse – vor allem die Untersuchung der frühen Mutter-Kind-Beziehung und die Entwicklung einer expliziten „Objektbeziehungstheorie"–, daß eine Betrachtung des Patienten als Einzelperson immer schon Ergebnis wissenschaftlicher Reduktion ist und daß bei einer solchen isolierten Betrachtung unverständlich wirkende Phänomene bei Berücksichtigung der Wechselwirkungen zwischen Patienten und ihrer Umwelt und der Kommunikationsvorgänge zwischen den Patienten und den für sie bedeutsamen Bezugspersonen verständlich und veränderbar werden können (Ruesch und Bateson, 1968; Ruesch, 1973; Foudraine, 1973). In einem solchen systemorientierten Ansatz wurden auch die Auswirkungen der „interpersonellen Wahrnehmung" auf das zwischenmenschliche Verhalten in ihrer Bedeutung für die Psychiatrie neu entdeckt (Laing et al., 1971).

Die Bewegung zur **„therapeutischen Gemeinschaft"** kam nicht zuletzt auch unter dem Druck der Frage zustande, wie die aus psychoanalytischer Psychotherapie gewonnenen Erkenntnisse größeren Patientengruppen zugute kommen könnten, als dies bei der Einzeltherapie der Fall ist. Nach dem Zweiten Weltkrieg versuchten vor allem Maxwell Jones (1976a,b) und Tom Main (1946), die klinische Situation als relevante Umwelt der Kranken so umzustrukturieren, daß sie in Form einer „therapeutischen Gemeinschaft" als Alternative zur bisherigen „pathogenen" Umwelt den therapeutischen Prozeß unterstützen kann. Main formulierte 1946 als Zielvorstellung:

„The socialization of neurotic drives, their modification by social demands within a real setting, the ego-strengthening, the increased capacity, sincere and easy social relationships, and the socialization of super-ego demands, provide the individual with a capacity and a technique for stable life in a real role in the real world."

Das Behandlungssetting im Krankenhaus wurde damit radikal geändert. So berichtet Main davon, daß die Patienten bei der Entlassung ihre Besserung auf den intensiven Kontakt mit Mitpatienten zurückführten und nur selten den Psychiater als Therapeuten erwähnten.

Die Arbeitsrichtung „therapeutische Gemeinschaft" hatte ihren Ausgang von der Erkenntnis therapiebehindernder, oft „pathogener" Umweltstrukturen und Interaktionsprozesse in der psychiatrischen Klinik genommen: der Tendenz, den Patienten vom realen Leben in der Gesellschaft zu isolieren; der Tendenz, ihn in Rollen zu halten, die mit Abhängigkeit und Unmündigkeit verbunden sind und in denen er von Ärzten und Pflegepersonal autoritär geführt wird. Daneben wurde auch darauf hingewiesen, daß sich im Krankenhaus Strukturen der primären Umwelt des Patienten reproduzieren können, wie sie auch an der Entstehung der Krankheit mitbeteiligt gewesen seien. In der „therapeutischen Gemeinschaft" wird Therapie auch als Lernprozeß und Lernen entschieden als sozialer Prozeß verstanden (Jones, 1976a). Der Patient soll nun möglichst selbstverantwortlich sein Leben in der Gruppe gestalten; die Schranke zwischen Ärzten und Pflegepersonal einerseits und Patienten andererseits soll abgebaut werden, so daß sich ein partnerschaftliches Verhältnis entwickeln kann. Im Rahmen dieses Settings werden dann die verschiedenen Formen der Einzel- und Gruppenpsychotherapie angewandt. Verschiedene Formen der therapeutischen Gemeinschaft für unterschiedliche Patientengruppen wurden inzwischen erprobt und beschrieben (Bettelheim, 1975; Cooper, 1971; Foudraine, 1973; Jones, 1976a,b; Kayser, 1973, 1974; Main, 1946; Ploeger, 1972; Talbot und Miller, 1968).

Maxwell Jones hat seine erste „therapeutische Gemeinschaft" 1947 in London aufgebaut. Für die Entwicklung seines Konzeptes waren jedoch Erfahrungen „maßgebend" (1976a), die er von 1940 bis 1945 mit der Leitung einer 100-Betten-Station für Soldaten gewann, die an Herzneurosen litten (im Rahmen des Maudsley-Hospitals in London); er konnte dabei nicht nur die psychogene Ätiologie dieses Krankheitsbildes („Effort-Syndrom") klären, sondern begann auch „das physiologische Forschungsmaterial" in Versammlungen sämtlichen 100 Patienten und dem Personal zugänglich zu machen. In diesen Versammlungen wurden auch „so wichtige Fragen wie Entlassung aus der Armee wegen Invalidität oder Rückbeorderung in den Aktivdienst" besprochen. Dabei begannen auch die Patienten Zusammenhänge der Symptombildung zu verstehen: „Nach und nach übernahmen die älteren Patienten, die inzwischen Zeit gehabt hatten, den Mechanismus der Symptombildung zu begreifen, die Verantwortung für den Lernprozeß und übertrugen ihn auf die neueren Patienten."

„Die Belastung des Personals nahm zusehends ab, und die meisten von uns erkannten – vielleicht zum ersten Mal – die potentielle Macht einer Gruppe von gleichgestellten Patienten." Hinzu kam der Therapieerfolg: „Nahezu 80% der auf unserer Station behandelten Patienten wurden erneut im Armeedienst eingesetzt, größtenteils jedoch als nichtkämpfendes Personal." Zu diesen Erfahrungen auf der „Psychoso-matischen Station" kamen für Jones dann Erfahrungen auf einer neuen Station hinzu, auf der die schwerstgestörten der aus den Lagern in Deutschland, Italien und im Fernen Osten heimkehrenden britischen Kriegsgefangenen behandelt wurden. Hier ging es vor allem um Probleme der Umstellung aus der oft mehrjährigen Anpassung an die soziale Ordnung in Kriegsgefangenenlagern an die Umwelt im eigenen Heimatland. Auch hier erwies sich eine strukturierte „Übergangsgemeinschaft" als außerordentlich hilfreich im Rehabilitationsprozeß.

Parallel zur Untersuchung pathologischer Strukturen in therapeutischen Institutionen wiesen einzelne Forscher auch auf pathologische Prozesse in gesellschaftlichen Strukturen außerhalb von Kliniken hin. Vor allem Sullivan hat darauf aufmerksam gemacht, daß solche Strukturen für das Individuum pathogen sein können. Die Psychiatrie hatte für die Behandlung ihrer Patienten vor allem im Zusammenhang mit der Entlassung aus „therapeutischen Gemeinschaften" hieraus Konsequenzen zu ziehen. Der Aufbau ambulanter Versorgungseinheiten hatte den Zusammenhang mit der „Empfangswelt" der Patienten, ihren Familien und ihrer Arbeitssituation zu berücksichtigen. Dies führte zur Entwicklung neuer Gesamtversorgungsmodelle und zu psychiatrischen Aktivitäten auch im politischen Raum. Mit einem solchen breiten Ansatz erarbeiteten seit 1960 Basaglia und später Pirella (1975) in Italien Konzepte, die zur Öffnung der psychiatrischen Anstalten, zur Rückkehr der Patienten in die Gesellschaft und zu einer Einbeziehung der therapeutischen Konzepte in die Gemeinschaft führten.

Psychosomatisch Kranke wurden in den bisher dargestellten stationären Einrichtungen nur gelegentlich behandelt. Simmel wies zwar bereits 1928 darauf hin, daß in der psychoanalytischen Klinik „auch eine nach psychoanalytischen Gesichtspunkten orientierte **systematische Psychotherapie organischer Krankheiten** ihre Stätte finden" müsse: „denn auch bei ihnen ist oft die Beziehung des Kranken zu seiner Umwelt ein ausschlaggebender Faktor". „Psychoanalytiker waren jedoch lange Zeit mit der Erforschung rein psychisch Kranker beschäftigt; von psychophysiologischen Zusammenhängen", schreibt Freud 1932 an V. v. Weizsäcker, habe er „die Analytiker aus erziehlichen Gründen fernhalten" müssen, „denn Innervationen, Gefäßerweiterungen, Nervenbahnen wären zu gefährliche Versuchungen für sie gewesen, sie hatten zu lernen, sich auf psychologische Denkweisen zu beschränken". Freud fügt allerdings hinzu, „dem Internisten können wir für die Erweiterung unserer Einsicht dankbar sein".

Erst der Versuch einiger Internisten, den psychoanalytischen Ansatz in die Innere Medizin einzubeziehen, führte zur Begründung einer **„Psychosomatischen Medizin".** Mit den theoretischen Fragen dieser Einbeziehung und den sich hieraus ergebenden praktischen Konsequenzen hat sich V. v. Weizsäcker besonders grundlegend befaßt. Auf einige seiner Überlegungen möchte ich hier etwas ausführlicher eingehen.

Für die Psychosomatische Medizin stellte sich von Anfang an die Frage nach dem Verhältnis zwischen Psychotherapie und Innerer Medizin. V. v. Weizsäcker stellte bereits 1926 fest, daß die additive Hinzufügung der Psychotherapie als Fach nicht genüge; wenn die Psychotherapie ein Gewinn für die ganze Medizin sei, dann müsse „sie auch auf den klinischen Brennpunkt wirken, indem aus beiden eine Einheit geschaffen werden muß". Er sah klar, daß die Einführung der Psychotherapie in die Innere Medizin eine Herausforderung an die Medizin zu einer Veränderung im Prinzipiellen darstellt. Er forderte in dieser Situation „die Neubildung eines geistigen Systems der Medizin als Grundform". „Die Abwehr und das Nichteinbringen der Persönlichkeitsmedizin" könne „zum Verlust des ärztlichen Gedankens überhaupt, zu einer Auflösung der Klinik in diagnostische und therapeutische Technisierung und Betriebshaftigkeit führen" (V. v. Weizsäcker, 1926).

Für die Beurteilung von Schwierigkeiten und Widerständen in der Entwicklung einer psychosomatischen Medizin erscheint mir folgende Überlegung V. v. Weizsäckers hilfreich:

Die Einführung des psychoanalytischen Verständnisansatzes in die Medizin sei nicht nur deshalb etwas prinzipiell Neues, weil nun auch der psychische Bereich mit einer ihm angemessenen Methode wissenschaftlich erforscht werde und so seinen Platz in der wissenschaftlichen Medizin erhalte, vielmehr enthalte der psychoanalytische Ansatz auch ein verändertes wissenschaftliches Grundverständnis mit prinzipiellen Folgen für das Verständnis der ärztlichen Rolle bzw. der Arzt-Patient-Beziehung im therapeutischen Prozeß.

Für v. Weizsäcker beinhaltet der psychoanalytische Verständnisansatz einen Versuch, das **Unerkennbare** in die Medizin hineinzunehmen; er beinhaltet das Eingeständnis, daß es nicht erkennbar Unbekanntes als im Menschen Wirkendes gibt; die Einführung des unbewußten Psychischen durch Freud, dessen „Umgang mit dem Menschen vom Hereinnehmen des Unerkennbaren durchtränkt ist", sieht v. Weizsäcker als entscheidenden Schritt an: das Unbewußte sei kein vorläufig Ungewußtes, ein prinzipiell rational Erkennbares, „sondern ein Block, der immer unerkennbar inmitten des Menschen liegenbleibt" (1947a). Der Rationalismus der naturwissenschaftlichen Medizin wird für v. Weizsäcker „im Begegnungssturm der Übertragung tatsächlich" gebrochen. So kann sich Krankheit nicht im Kranken unabhängig vom Umgang zwischen dem Kranken und seinem Arzt, von ihrem Verhältnis und ihrer Begegnung darstellen.

Die Bestimmung dieser Beziehung zwischen Arzt und Patient wird zur zentralen Aufgabe der Psychosomatischen Medizin.

„Ich habe immer gefunden, daß an dieser Stelle eigentlich eine Bruchlinie durch Ärzte wie Patienten geht, die heute geschichtlich vielleicht prägnant und unvermeidlich ist und die doch wohl allen Epochen der bekannten Medizin innewohnt. Quer durch psychotherapeutisch und nichtpsychotherapeutisch eingestellte Ärzte, quer durch die Erwartung der Patienten, ja quer durch jeden einzelnen von uns allen geht die Differenz von objektiver und von umfassender Therapie. Es ist ungeheuer schwer, diesen Unterschied ohne Erfahrung und Beispiele darzustellen. Am wichtigsten scheint mir immer wieder, daß in einer umfassenden Therapie der Arzt selbst sich vom Patienten verändern läßt; daß er die Fülle aller Regungen, die von der Person des Kranken ausgehen, auf sich wirken läßt; daß er sich nicht einengt in das System der Diagnostik und der systematischen Krankheitseinheit …" (V. v. Weizsäcker, 1928).

Diese Auffassung der Beziehung zwischen Arzt und Patient mache „einen Kanon der Zuwendung" erforderlich. Wesentlich sei zunächst, daß zwischen Arzt und Patient immer ein Austausch stattfinde. Hierbei weist v. Weizsäcker darauf hin, daß jede ärztliche Handlung, etwa auch ein Rat, „die Dynamik einer Bewirkung, einer Kraftübertragung vom Arzt auf den Patienten, tatsächlich" besitze. Ebensogut müsse die-

se Bewegung jedoch auch in umgekehrter Richtung, „als Kraftabgabe des Kranken zum Arzt, bewertet werden". „Jetzt wird manifest, was schon immer war, nämlich, daß beim Befehlen das Geben beim Gehorchenden ist. Befehlen ist Nehmen." Der also mit dem Wort wirkende Arzt nimmt „im Zureden eine Kraft vom Kranken weg, ebenso wie er, gäbe er ihm ein Ei zu essen, ihm Kraft hingibt".

Diese Beziehungsstruktur hat Folgen, gerade auch dann, wenn sie nicht berücksichtigt wird; dies wird heute z.B. am Problem der Non-Compliance deutlich, also an der Tatsache, daß bis zu 50% aller vom Arzt verordneten Medikamente von den Patienten nicht eingenommen werden. Die Folgen der Beziehung bleiben so auch bei einem rein somatischen Therapieansatz erhalten: „Jede Somatotherapie hat auch eine psychische Bilanz und umgekehrt" (V. v. Weizsäcker, 1927).

Damit werden Aufgabe und Rolle des Arztes neu bestimmt: „Nicht der Arzt heilt, sondern die organische Natur, nicht die Verordnung, sondern die Arznei. Nicht Reparation ist das letzte Ziel, sondern der Werdegang, der Stufengang des Kranken zu seinem metaphysischen Endziel, zu dem der Arzt aber als ein wahrer Sokratiker nicht hindeuten, nicht hinschieben, nicht hinzeigen darf. Denn er ist weder Führer noch Deuter, noch Weiser, sondern er ist ein Arzt, das heißt kein Bewirker, sondern ein Ermöglicher; er steht nicht über der Entscheidung, sondern mit dem Kranken in der Entscheidung" (1927).

Die **Beziehung** zwischen Arzt und Patient wurde damit zum zentralen Gegenstand von Erkenntnis und Therapie. Eine Analyse dieser Beziehung hat die möglichst genaue Klärung der Einstellungen und Handlungen des Arztes zur Voraussetzung. Die Reflexion der Wirkung des Arztes wird zur wichtigsten Aufgabe, mehr noch als die detaillierte Kenntnis des Patienten.

„Nicht darum handelt es sich in erster Linie, auch das außertherapeutische Milieu der Kranken zu verändern und zu beherrschen, sondern darum, die Tragweite aller Handlungen des Arztes zu kennen; nicht die Isolierung des Patienten, sondern die des Arztes gegenüber dem Patienten ist der wichtigere Zweck einer streng durchgeführten Einzelbehandlung" (V. v. Weizsäcker, 1925).

Dieses Ziel gilt sinngemäß auch für die stationäre Psychotherapie: Wichtig ist die genaue Kenntnis und die ständige Reflexion der Wirkungen des ganzen Settings, des Milieus und des Therapeutenteams auf den Patienten, nicht das Bemühen, den Patienten 24 Stunden lang mit verschiedenartigsten Methoden verändern zu wollen.

V. v. Weizsäcker dachte früh daran, Psychotherapie in die Klinik einzuführen, die er für den „Konfliktort von Kopf und Herz" als „Konfliktort von Theorie und Tätigkeit" und zugleich für die leitende Institution in der Entwicklung der Medizin hielt.

„Das Wesentliche der Klinik liegt ja nicht darin, daß sie betriebsmäßige Behandlungsanstalten sind, sondern, wie Krehl es einmal beschrieben hat, in derjenigen Verbindung von Theorie und Tätigkeit, welche sich ergab, als man im 17. Jahrhundert von der sinkenden Scholastik zur erfahrungs- und dann naturwissenschaftlich begründeten Medizin überging. Aber wesentlich ist weiter, daß diese Aufgabe hier in einer institutionellen Ordnung von einer Führerperson mit ihren Mitarbeitern gelöst wird und ferner, daß mit den staatlichen Approbationen die erste Prägung der Jugend in ihre Hände gelegt wurde. Darin liegen Kraft und Gefahr der Klinik" (1926).

Die konkrete Einbeziehung der Psychotherapie in die klinische Arbeit wird durch zwei Probleme erschwert, auf die v. Weizsäcker immer wieder zurückkam:

– Soll und kann Psychotherapie und somatische Therapie durch denselben Arzt durchgeführt werden, oder sollte diese Aufgabe von einer eng kooperierenden Gruppe von Ärzten – mit verteilten Rollen – übernommen werden? Das Problem ist die **Identität** des Arztes als Psychosomatiker.
– Die Eignung der **Organisationsstruktur** moderner Kliniken für die Psychotherapie bzw. die Frage nach den hierfür erforderlichen Veränderungen.

Zunächst spricht viel für die **Verbindung von Somatotherapie und Psychotherapie.** Seelische und körperliche Phänomene lassen sich beim Kranken nicht unabhängig von den Arzt-Patient-Beziehungen erkennen: „Was als psychisches oder somatisches Phänomen erscheint, ist bereits Resultat der Parteinahmen und hat sich aus Verdrängungen und Entscheidungen abgeschieden." Zwei Spezialisten werden zu unterschiedlichen Gewichtungen neigen, diese Gewichtungen werden in einer unterschiedlichen Beziehungsstruktur getroffen. Die Behandlung scheint nicht auftrennbar: „Wird er (der Neurosenarzt) vom Chirurgen oder Internisten ‚zugezogen', so ergibt sich regelmäßig, daß ein Behandeln zu mehreren nirgends aussichtsloser, ja schädlicher ist als hier" (1925). Hinzu kommt, daß ein „zweigleisiges" Vorgehen keineswegs immer vereinbar ist, sondern daß vor Aufnahme der Psychotherapie Entscheidungen fallen müssen.

Die psychotherapeutische Behandlung kann – dies leitet sich aus ihrem Grundverständnis her – ja auch im Gegensatz zur somatischen Behandlung stehen:

„Denn Übertragung als Heilmittel steht in scharfem Gegensatz zur Methode des Wirkens mit Mitteln, und es ist dabei gleichgültig, ob es sich um physikalische, diätetische, chemische, klimatische, hygienische oder selbst psychologische Bewirkungen handle. Bei einer Analyse soll der Kranke nur sich selbst darstellen, so wie er eigentlich ist, sie ist eine Identifikation. Bei einer kausalen und aktiven Therapie soll durch jene Mittel an ihm etwas bewirkt werden, sie ist eine Veränderung. Identifikation und Alteration, Gleichsetzung und Veränderung sind Gegensätze, welche im Therapeuten zu widerspruchsvollem Handeln führen können. Denken wir an die häufigste Mischung organischer und neurotischer Motive bei inneren Krankheiten ..."

V. v. Weizsäcker spricht davon, daß die Klinik hier in doppelter Beziehung vor einem Konflikt stehe. Sie müsse „wählen" zwischen der bisher gewohnten „internen" Therapie und andererseits einer „reinen Psychotherapie", die in vielen Fällen auch eine sog. große, „der Psychoanalyse nahestehende Psychotherapie" sein müsse". „Zweitens aber läßt sich der Internist hier neben dem Gehalt der Inneren Medizin dann leiten von den Grundsätzen der Psychotherapie und gelangt so zu einer kombinierten Verfahrensweise." Bei dieser kombinierten Verfahrensweise, so meint v. Weizsäcker, „aber liegen, wie ich glaube, die allerdringendsten

und zugleich erst im Status nascendi befindlichen Aufgaben". „Hier muß sich entscheiden, wie und was von der Psychotherapie Besitz und Methode auch des nicht-spezialistisch ausgebildeten Psychotherapeuten werden kann und soll" (1926). V. v. Weizsäcker plädiert mit Nachdruck für eine systematische Klärung des seinerzeitigen – und oft auch noch heute bestehenden Zustandes, bei dem „die Therapie tief unter der Theorie" stehe und „von Klinikern nur eine sehr oberflächliche Psychotherapie" praktiziert werden könne. Zur Entscheidung in der Therapie gehört der Umgang mit der Symptomatik. Beim Kranken hat z. B. der unbewußte Prozeß zu Herzbeschwerden geführt. Die herzbezogene Symptomatik kann Anlaß werden zu einem Versuch, Zugang zum innerpsychischen Prozeß, zum Konflikt zu finden, oder aber bei einer entsprechenden medikamentösen Therapie eine Fortführung der Abwehrvorgänge im Kranken unterstützen.

Allerdings komplizieren sich durch die gleichzeitige körperliche Behandlung oft auch die Übertragungsvorgänge des Patienten zum Arzt und erschweren so für den Patienten die Möglichkeit, seine Beziehungsprobleme in der Arzt-Patient-Beziehung zu reproduzieren. Bei intensiverer Psychotherapie wird deshalb eine Aufteilung der therapeutischen Funktion auf Mitglieder eines Teams, die in engem Kontakt stehen, oft unumgänglich. Allerdings ist es dann auch nötig, daß „hier diejenigen Bedingungen in Kraft" treten können, „unter welchen allein zwei oder mehrere Personen eine Anschauung, eine Erkenntnis als gemeinsame teilen, weil sie sich als Personen untereinander verstehen. Hier werden also die Lebensgesetze von Personengemeinschaften zu erkenntnistheoretischen Voraussetzungen. Drücken wir die Dinge praktisch aus, so läßt sich sagen, es komme hier darauf an, daß eine Gruppe von Ärzten, aber auch von Ärzten und Patienten sich untereinander verstehen und einigermaßen zusammen zufrieden sind". Hierfür ist jedoch ein umfassenderer theoretischer Ansatz Voraussetzung: „Eine solche Einigung läßt sich aber von der Theorie der naturwissenschaftlichen Medizin aus heute nicht mehr begründen, noch erzwingen" (1926).

V. v. Weizsäcker erkannte, daß aufgrund der institutionellen Gegebenheiten Versuche, die Psychotherapie in die klinische Behandlung einzubeziehen, auf größte, kaum zu überwindende Schwierigkeiten stoßen mußten:

„Die Organisation der modernen Kliniken und Krankenhäuser mit großen Sälen und einer gewissen Hierarchie der Ärzteschaft erwächst aus einem völlig anderen Gedanken der Krankheit und des Heilprinzips als dem der Psychotherapie und dem einer Personalpathologie ..." (1925).

In diesem Zusammenhang weist v. Weizsäcker auf das fehlende Bewußtsein der Medizin für „psychotherapeutische Kunstfehler" hin:

„Eine entschiedene Theorie und ein allgemein deutliches Bewußtsein der psychotherapeutischen Kunstfehler besitzt die heutige Medizin als legitimes Thema kaum." Gemeint sind die „Kunstfehler", die sich aus der Nichtberücksichtigung bzw. der Nichtreflexion der schon immer wirksamen Beziehungen zwischen Arzt und dem übrigen Personal einerseits und Patienten andererseits ergeben. „Pflegerin und Arzt erzeugen zugleich aber eine eigene Kategorie von seelischen Erlebnissen: Feindschaften, Freundschaften, Eifersüchten, Mißverständnissen, Intrigen und Kümmernissen des Krankenhausdaseins, eine Milieuwirkungsmasse, die jedenfalls als ein total unkontrollierbarer Faktor in die Behandlung um so mehr mit eingeht, weil er an die ärztlichen Handlungen unlösbar geknüpft ist. Dies letztere und eigentlich Entscheidende macht oft die Behandlung einer Neurose überhaupt zur Unmöglichkeit" (1925).

V. v. Weizsäcker hat von 1928 bis 1930 in der Heidelberger Neurologischen Klinik eine besondere Behandlungsstation für Rentenneurotiker geführt und eingehend beschrieben (1955). Dort wurde nach den Grundsätzen einer Psychotherapie behandelt, „die sich alles zunutze macht, was klinische, analytische und psychagogische Erfahrung bieten kann". Hierzu gehört vor allem auch „die umfassende Ergründung der Situation". Der Konflikt von Rentenneurotikern, die Wechselwirkung zwischen eigenen Ansprüchen und Umweltversagungen boten sich für einen solchen Verständnisansatz geradezu an. Ein Arbeitskonzept, wie es später von den „therapeutischen Gemeinschaften" beschrieben wurde, verbunden mit konsequenter Psychotherapie, erwies sich bei diesen Patienten als erfolgreich. V. v. Weizsäcker beabsichtigte, dieses Konzept auch auf die Behandlung internistischer Erkrankungen auszudehnen. Aufgrund der politischen Veränderungen konnte er diese Pläne zunächst nicht verwirklichen.

Nach dem Krieg war es v. Weizsäcker aufgrund seiner Freundschaft mit Siebeck möglich, in der Heidelberger Medizinischen Klinik Ansätze zu einer Integration der Psychosomatik in die Innere Medizin zu erproben. Er strebte eine enge Verflechtung zwischen Psychotherapie und Innerer Medizin an und nahm dafür methodische und institutionelle Schwierigkeiten in Kauf.

„Dem Fernstehenden kann es wie eine zufällige Äußerlichkeit oder Willkür aussehen, daß hier ein Laboratorium, eine Krankenvisite, eine schulmedizinische Therapie und zugleich biographische Anamnesen, Psychoanalysen und psychosomatische Forschungen, Arbeits- und Sozialtherapie angetroffen werden. Aber es ist kein persönlicher Zufall, keine äußerliche Willkür, daß dies alles unter demselben Dach und zu gleicher Stunde vor sich geht. Denn nur so entstehen die nötigen Reibungen, Kämpfe und produktiven Mißerfolge, die wir uns ersparen, wenn wir Feuer und Wasser getrennt halten, Psyche und Soma, Geist und Materie überall unterscheiden wollten."

Sein Ziel war eine **„allgemeine Medizin"**; die enge Verflechtung der verschiedenen Ansätze erschien ihm vor allem auch unter dem Gesichtspunkt wichtig, Arbeitsmöglichkeiten für Studenten und jüngere Ärzte zu schaffen:

„Die Beobachtung der jüngeren Ärzte und Studenten zeigt, daß unter ihnen immer einige sind, die gar nicht die freie Wahl mehr haben, sich entweder einer naturwissenschaftlichen Medizin *oder* einer psychotherapeutischen zuzuwenden. Ihre Lage ist unentrinnbar. Freilich nenne ich das weder Begabung noch Schicksal. Denn auch ihnen steht eine nicht endende Kette von Konflikten mit sich und ihrer Umwelt bevor, in denen sie noch ebensooft unterliegen oder

siegen können – unterliegen, indem sie einseitig werden, siegen, indem sie zweiseitig bleiben" (1947 b).

Mitarbeiter v. Weizsäckers wandten sich in der Folgezeit auch der psychosomatischen Behandlung von körperlich Schwerkranken zu. Sie erlebten nun auch in der täglichen Praxis, daß dabei der einzelne Arzt häufig den Kräften der psychischen Selbstzerstörung und der Abwehr in den Patienten alleine nicht gewachsen ist. Die Auseinandersetzung mit diesen Kräften erfordert vielmehr die Zusammenarbeit in einer „Gruppe von Therapeuten". Im Umgang mit einem strukturierten Team konnten sich dann die Konflikte des Kranken darstellen, und die Gemeinschaft der Therapeuten gab dem behandelnden Arzt den Halt, den er für den Umgang mit dem Patienten benötigte (Kütemeyer, 1963). Die Konflikte und Beziehungsstörungen der körperlich Kranken erwiesen sich dabei oft als besonders tief verborgen; die Kranken erschienen gleichzeitig sozial oft besonders gut angepaßt. Mitscherlich sprach von einer „zweiphasigen Verdrängung" des Konfliktes. Seiner Rückkehr aus der „Latenz der Latenz" (Kütemeyer) stehen besonders intensive Widerstände entgegen.

Diese Erfahrungen wurden auch vom Nachfolger Siebecks, Matthes, noch anerkannt.

„Für die Klinik bedeutet das allerdings, daß neben der somatisch-naturwissenschaftlichen auch der anthropologischen Forschung das ihr zukommende Gewicht gewährt werden muß. Wie die naturwissenschaftliche Forschung aufwendige Laboratorien und diagnostische Einrichtungen benötigt, so braucht die anthropologische Richtung eine – im Verhältnis zur Patientenzahl – recht große, auf Zusammenarbeit im gleichen Geiste eingestellte Ärztegruppe."

„Es bedarf der selbstverständlichen Unterstützung durch einen in gleicher Gesinnung verbundenen Kreis, der geeint ist durch die Überzeugung, daß das Menschliche im Kranksein ein ärztliches Problem, ja ein Problem der Forschung ist. Nur innerhalb eines solchen Kreises ist eine Erfahrungsbildung auf dem Gebiet der anthropologischen Medizin und eine entsprechende Therapie … möglich" (Matthes, 1963).

Aus der Entwicklung der letzten 20 bis 30 Jahre möchte ich drei Aspekte herausgreifen:
1. die Vertiefung und Intensivierung des psychoanalytischen Verständnisses psychosomatisch Kranker;
2. die Weiterentwicklungen der psychoanalytischen Theorie und Technik;
3. die verstärkte Institutionalisierung der Psychosomatik an den Universitäten in der Bundesrepublik.

1. Psychoanalytische Untersucher beobachteten bei Patienten mit körperlichen Erkrankungen nun in systematischerer Weise besondere gemeinsame Strukturmerkmale, auf die allerdings bereits Meng (1934) und Mitarbeiter v. Weizsäckers (Kütemeyer, 1963) aufmerksam gemacht hatten. Unter denjenigen Patienten aus der Gruppe der körperlich Kranken im allgemeinen und der „psychosomatisch" Kranken im speziellen, die in Institutionen untersucht wurden, fanden sich gehäuft Patienten, die stark symptomfixiert, phantasiearm und wenig introspektionsfähig erschienen; die Untersucher fanden sie kaum in der Lage, ihre Probleme zu verbalisieren, sie neigten vielmehr zum Agieren ihrer Konflikte und ließen wenig Voraussetzungen für ein psychotherapeutisches Arbeitsbündnis erkennen (zur Diskussion des sog. „Alexithymie"-Konzeptes vgl. Kap. 5).

Die einzelnen Untersucher hatten auf verschiedene, miteinander zusammenhängende Persönlichkeitsmerkmale dieser Patienten aufmerksam gemacht. Gemeinsamer Nenner aller Befunde ist die eingeschränkte Entwicklung und Differenzierung der Persönlichkeit. Ruesch (1948) nannte sie „infantile Persönlichkeit"; Analytiker wie Alexander (1950), Engel (1962), Margolin (1953) und Schur (1955) wiesen auf die intensive unaufgelöste Bindung dieser Patienten an frühe Bezugspersonen, auf die hierdurch verursachte Abhängigkeitsproblematik und mangelhafte Individuation hin.

In psychodynamischer Sicht hat sich bei diesen Patienten im Zusammenhang mit der Entwicklungsbehinderung ein **Mangel an psychischer Strukturbildung** ergeben, der sich deutlich von den spezifischen Konflikten bei gebildeter psychischer Struktur der Neurotiker unterscheidet. Aufgabe der Psychotherapie ist es in dieser Auffassung, Hilfestellung für die Entwicklung aus der Abhängigkeit von den primären Bezugsobjekten, für die **Individuation,** zu geben, die Bildung psychischer Strukturen zu unterstützen und so auch die durch den psychischen Mangel bedingte große Abhängigkeit von allen anderen Umwelteinflüssen zu vermindern und damit die bisherigen pathologischen Rückzugs- und Schutzhaltungen der Patienten überflüssig werden zu lassen (vgl. Kap. 5).

2. Zahlreiche Entwicklungen der **psychoanalytischen Theorie und Technik** verbesserten auch die Zugangsmöglichkeiten zu den psychischen Problemen körperlich Kranker. Sie können hier nur angedeutet werden.

Die Entwicklung der **Kinderpsychoanalyse** (vgl. Kap. 6 und 54) förderte das Verständnis der frühen Mutter-Kind-Beziehung und der in sie eingebetteten seelischen Entwicklung des Kindes zu seiner Individuation. Die differenzierte Darstellung der Entwicklung der Gegenseitigkeit in der Beziehung innerhalb der Mutter-Kind-Dyade zeigte in aller Deutlichkeit die Verfälschungen auf, die durch eine Reduktion auf eine Ein-Personen-Psychologie zustande kommen können. In engem Zusammenhang mit der Kinderpsychoanalyse steht die Entwicklung einer expliziten **Objektbeziehungstheorie** innerhalb der Psychoanalyse.

Für die allgemeine ärztliche Psychotherapie körperlich Kranker sind zunächst die bahnbrechenden Arbeiten M. Balints (1965) zu nennen, die ein systematisches Benutzen der Beziehung zwischen Arzt und Patient zum Ziel haben.

Fortschritte in der Benützung der Beziehung zwischen Arzt und Patient erweiterten auch die Möglichkeiten der psychoanalytischen Technik. Einerseits wurde durch die Untersuchungen der „psychoanalytischen Situation" (Stone, 1973) und der Arbeitsbeziehung zwischen Patient und Analytiker (Greenson, 1973) die Wirksamkeit des therapeutischen Raumes

geklärt und gefördert, andererseits auf dem Boden von Kinderanalyse, Objektbeziehungstheorie und Ich-Psychologie der Zugang zu genetisch früh entstandenen psychischen Störungen wesentlich erweitert (vgl. u.a. Balint, 1968; Winnicott, 1960, 1962; McDougall, 1974).

Eingehend stellt Loch diese Entwicklungen in den „nach-klassischen psychoanalytischen Untersuchungsfeldern", zu denen auch die beschriebenen psychischen Störungen bei körperlich Kranken gehören, dar (1974). Loch weist auf die Gemeinsamkeit im Umgang mit der „Grundstörung" (M. Balint), bei der Behandlung der Trennungs- und Individuationsphase (M. Mahler), der Ermöglichung eines kreativen Umgangs mit Objekten und der gleichzeitigen Entwicklung des „wahren Selbst" (Winnicott) sowie der Analyse des Settings bzw. des Rahmens der Therapie hin: Hier gehe es nicht um eine Bearbeitung von Konflikten über ein Symbolverständnis, sondern erst um die Ermöglichung einer Entwicklung von Symbolen und Sprache. „In allen diesen Fällen geht es nicht um das ‚Weg-Analysieren' (M. Balint) von Symptomen und Übertragungen, vielmehr kommt es primär auf die Ermöglichung einer Entwicklung, auf die Freilegung des Weges zur Entfaltung der Kreativität des Individuums an, was letztlich nur dem gelingt, der ein ‚unaufdringlicher Analytiker' (M. Balint) ist" (Loch, 1974, S. 449).

Bei der Diskussion von Konzepten von Einrichtungen zur stationären Psychotherapie wird immer wieder auf Winnicotts Begriff des **„facilitating environment"** hingewiesen. Die Überlegungen zu diesem Begriff werden dabei kaum je ausführlicher dargestellt. Winnicott selbst hat diesen Begriff auf eine Haltung des Psychotherapeuten in der Einzelpsychotherapie bezogen. Der Analytiker stellt für den Patienten unmittelbar oder in der Übertragung einen Ausschnitt des „facilitating environment" dar und übernimmt Funktionen dieser Umwelt, insbesondere das „holding", „handling" und „object presenting" (Winnicott, 1962). Der Analytiker schafft so einen Raum mit Strukturen, die denjenigen der kindlichen Entwicklungsbedingungen analog sind und die Entwicklung bzw. Nachentwicklung bestimmter Anteile der Persönlichkeit zu fördern vermögen.

In Winnicotts Auffassung geht die schwere psychische Erkrankung, u.a. die Psychose, auf ein Umweltversagen in einem frühen Stadium der emotionalen Entwicklung des Individuums zurück. Das Milieu der Psychoanalyse reproduziert Aspekte der frühesten Bemutterung. Es vermittelt Zuverlässigkeit und lädt zur Regression ein. In der Regression kehrt der Patient in organisierter Form in frühe Formen der Abhängigkeit zurück. Auch in dieser therapeutisch induzierten Abhängigkeitssituation tritt Umweltversagen auf; der Patient ist hier jedoch fähig, diese Situation zu bewältigen, ohne pathologische Abwehrmechanismen aufbauen zu müssen. Seine Reaktionen auf die frühen Umweltversagungen, seine Abwehrhaltungen können in der therapeutischen Situation nach und nach „aufgetaut" und abgebaut werden. Von Bedeutung ist, daß das therapeutische Milieu die Sicherheit vermittelt, daß es nicht durch die Wut, die sich auf das Umweltversagen in der frühen Kindheit bezieht und jetzt neu verspürt und geäußert wird, zerstört werden kann.

Winnicott (1960, 1962) hat das Halten (holding) als Eigenschaft einer die Entwicklung fördernden Umwelt, als Grundhaltung und Verhalten einer auf den Säugling optimal eingestellten Mutter ausführlich beschrieben.

Das Halten ist eine Form der Liebe, zu ihm gehört besonders auch das physische Halten und Tragen des Säuglings. Die Mutter unterstützt zunächst das Ich des Kindes durch ihre Fürsorge, vertritt es bzw. stabilisiert es durch ihre Kraft. So geschützt, wird das Ich mit zunehmender Entwicklung fähig, das Es zu beherrschen und allmählich auch im Umgang mit Einflüssen der Umwelt unabhängiger von der Mutter zu werden. In der Haltephase reduziert die Mutter die Übergriffe aus der Umwelt auf ein Minimum, so daß sich im geschützten Raum kontinuierlich die Existenz des Kindes, sein personales Sein („wahres Selbst") entwickeln und differenzieren kann; der Säugling ist so davor geschützt, ständig auf überfordernde Umweltreize reagieren und zu früh eine eigene Organisationsform für solche Schutzreaktionen („falsches Selbst") entwickeln zu müssen. Zunächst unterstützt die Mutter das Ich des Säuglings, übernimmt für ihn Ich-Funktionen, bietet Befriedigung der physiologischen Bedürfnisse, Zuverlässigkeit, Schutz vor Beschädigung, Pflege, und paßt diese Unterstützung fortlaufend an die Entwicklung des Säuglings an. Allmählich gelingt es dem Kind dann, von dieser Unterstützung der Mutter relativ unabhängiger zu werden, die störenden Einflüsse in den Bereich seiner eigenen Omnipotenz einzubeziehen und aus dem ihm adäquat von der Umwelt angebotenen Objekten sich seine eigenen subjektiven Objekte zu „erschaffen" (vgl. Kap. 61).

Verschiedene Konzepte für die Institutionalisierung stationärer Psychotherapie bei psychosomatisch Kranken versuchen, dieses von Winnicott für die psychoanalytische Behandlung einzelner Kranker differenziert entwickelte und ausführlich dargestellte Konzept auf Einstellung und Verhalten des therapeutischen Teams zu übertragen.[16] Zwei Formen der Anwendung eines u. a. aus den Winnicottschen Überlegungen abgeleiteten Konzepts zeichnen sich ab:

- Die Einleitung der Psychotherapie erfolgt im Schutze des stationären Settings, das die Aufnahme einer therapeutischen Beziehung im Zusammenhang mit dem Zulassen regressiver Prozesse und dem Abbau von Abwehrhaltungen ermöglicht. Die unter diesen Schutzbedingungen eingeleitete Therapie kann dann längerfristig ambulant fortgesetzt werden (Köhle et al., 1977), sowohl als Einzel- wie als Gruppenpsychotherapie.
- Das Stationsteam und seine Arbeit wird so strukturiert, daß der spezifische psychotherapeutische Prozeß während des längerfristigen stationären Aufenthaltes stattfindet. So können – vereinfachend skizziert – vor allem die Schwestern die Funktionen des „facilitating environment" übernehmen, während der Psychoanalytiker in seinen Deutungen stärker strukturierend dem Patienten hilft, sich aus seinen bisherigen Schutzhaltungen und Abhängigkeitsbeziehungen zu lösen (Janssen, 1987).

3. Die **Einführung der Fächer Psychosomatik und Psychotherapie** als Pflichtfächer in die medizinische Ausbildung hat die Einrichtung psychotherapeutisch-

psychosomatischer Krankenstationen an Universitätskliniken gefördert.

Der Schwerpunkt dieser psychotherapeutisch-psychosomatischen Stationen und Kliniken liegt in der Entwicklung und Anwendung psychotherapeutischer Behandlungsansätze bei bereits vorausgewählten psychosomatisch Kranken.

Nur wenige Berichte finden sich über Versuche, den psychosomatischen Arbeitsansatz in die allgemeine klinische Tätigkeit – z.B. in die internistische Krankenversorgung – zu integrieren und damit die medizinische Krankenversorgung etwa im Sinne v. Weizsäckers zu ergänzen bzw. zu erweitern (Filter et al., 1979; Hahn, 1975; Klagsbrun, 1970; Köhle et al., 1972, 1973, 1976, 1977). Die somatische Medizin hat ihre Ergänzungsbedürftigkeit bisher stärker in therapeutischen Extremsituationen als innerhalb der durchschnittlichen stationären Krankenversorgung artikuliert. Psychosomatiker wurden häufig gebeten, sich an der Arbeit in Dialyseeinheiten, Intensivstationen, Isolierbetteinheiten, bei der Vorbereitung und Nachbehandlung von Herzoperationen, also in der medizinischen „Spitzentechnologie" (A. E. Meyer) zu beteiligen, da hier die psychischen Belastungen für Patienten und Teammitglieder am deutlichsten spürbar und ihre Folgen am leichtesten sichtbar sind. Die Zusammenarbeit zwischen Psychosomatikern und übrigen Klinikern in diesen therapeutischen Extremsituationen hat sich vielerorts als sehr fruchtbar erwiesen; für die Realisierung einer psychosomatischen Medizin erscheint es mir jedoch mindestens ebenso wichtig, daß Modelle für die intensive Kooperation in der alltäglichen, „durchschnittlichen" stationären Krankenversorgung entwickelt werden:

Dort haben beide Partner, die Kliniker aus den verschiedenen Spezialfächern und die Psychosomatiker, mehr Freiraum für Lernprozesse als in der Notfallsituation und können so eher Konzepte zu einer partnerschaftlichen und integrierenden Zusammenarbeit in einem Team entwickeln.

In der im Auftrag des Deutschen Bundestages erstellten „Psychiatrie-Enquête" wird für Schwerpunktkrankenhäuser die Realisierung beider Stationsmodelle empfohlen (1975).

28.3.2 Ein Konzept für klinisch-psychosomatische Krankenstationen: Die internistisch-psychosomatische Krankenstation der Universität Ulm

Nach den theoretischen Vorüberlegungen und der Übersicht über mögliche Konzepte im Abschnitt 28.1 berichte ich hier über eigene Erfahrungen aus einem

---

16 Zu bedenken ist, daß die Anwendung von Winnicotts therapeutischem Konzept eine hochqualifizierte psychoanalytische Ausbildung und – seinen eigenen Aussagen zufolge – langjährige psychoanalytische Erfahrung voraussetzt.

Die Anwendung eines solchen Konzeptes durch eine Gruppe von Therapeuten unter klinischen Bedingungen, und oft nicht in Einzel-, sondern in Gruppenpsychotherapie, ist noch nicht ausreichend diskutiert und evaluiert. Deutlich wird dabei jedoch, welche Weiterbildungsansprüche an die Therapeuten auch im stationären Setting gestellt werden müssen.

konkreten Projekt, der Ulmer „internistisch-psychosomatischen Krankenstation", die wir in dieser Form von 1972 bis 1979 führten.[17]

Die Interessen bei der Begründung des Modellversuches

Es war uns in Ulm nicht gelungen, unser ursprüngliches Ziel zu verwirklichen: das psychosomatische Arbeitskonzept in die Krankenversorgung des gesamten Departments für Innere Medizin der Universität so weitgehend zu integrieren, wie es dem seinerzeitigen Wissensstand entsprach. Insoweit war der Konsultations-Liaisondienst unserer voll in dieses Department integrierten „Abteilung für Innere Medizin und Psychosomatik" gescheitert. 1972 zogen wir hieraus die Konsequenz; mit Hilfe von Drittmitteln entwickelten wir einen integrativen Arbeitsansatz in einem enger umschriebenen Arbeitsfeld, auf einer allgemein-internistischen Krankenstation mit 15 Betten.

Unser Ziel war es, den psychosomatischen Arbeitsansatz in der Versorgung aller internistisch Kranker zu erproben. Mit Ausnahme weniger Krankheitsbilder (u. a. Patienten mit Anorexia nervosa) wurden die Patienten dieser Krankenstation nicht speziell ausgewählt, die Aufnahme erfolgte wie bei allen anderen internistischen Allgemeinstationen des Departments für Innere Medizin über die Aufnahmestation bzw. die vorgeschaltete internistische Poliklinik.

Nur ein integriertes Konzept kann Leitbild und Herausforderung für Innovationen in anderen klinischen Fächern sein. Eine Vorauswahl von Patienten kann allenfalls im Sinne einer Indikationsstellung für das jeweilige psychotherapeutische Konzept erfolgen, nicht jedoch unter dem Gesichtspunkt der Bedeutung psychosozialer Faktoren für Entstehung und Verlauf eines Krankheitsbildes; ein Vorgehen, bei dem vor der Aufnahme auf der psychosomatischen Station „auf der internistischen Station oder ambulant die Basisdiagnostik abgeschlossen und die somatische Therapie weitgehend eingeleitet" wird (Freyberger, 1978), übersieht, daß hierbei Vorentscheidungen darüber gefällt werden, welche Anteile am Krankheitsgeschehen wie gewichtet werden, bzw. in welcher Form sie im institutionellen Rahmen in Erscheinung treten können. Im übrigen leitet sich eine solche krankheitsbezogene Selektion von der Annahme spezifisch „psychosomatischer Krankheiten" (Alexander, 1950) ab, die sich in dieser Form heute nicht mehr aufrechterhalten läßt.

In der Diskussion um die Reform der stationären Krankenversorgung spielt die psychosoziale Betreuung zum Tode kranker und sterbender Patienten eine wesentliche Rolle. Diese Probleme wurden lange Zeit aus der wissenschaftlichen Medizin, und zwar unter Einschluß der Fächer Psychotherapie und Psychosomatik, weitgehend ausgeklammert. Klinisch-psychosomatische Stationen mit unausgewähltem Patientengut bieten eine Chance, auch diese Probleme systematisch zu untersuchen und zu bearbeiten.

Auch zur Verwirklichung wesentlicher Zielvorstellungen der ärztlichen Approbationsordnung von 1970 erscheinen mir klinisch-psychosomatische Einrichtungen in der stationären Krankenversorgung wesentlich. Die Einbeziehung der Fächer des psychosozialen Bereiches in die ärztliche Grundausbildung ist nur dann zu verwirklichen, wenn das Angebot von Wissen und Erfahrungsmöglichkeiten während des Studiums ergänzt wird durch konkrete Arbeitsmöglichkeiten während des praktischen Jahres und der Facharztausbildung.

Zielvorstellungen des Modellversuches

Zielvorstellung bei der Veränderung des krankheitszentrierten Arbeitsansatzes einer Krankenstation zu einem patientzentrierten ist es, alle zwischen den Beteiligten ablaufenden Interaktionsprozesse auf ihre Bedeutung und Folgen für die Kranken hin zu untersuchen und umzugestalten. Daraus folgt die Aufgabe,
– Hilfestellungen für die Klärung und Korrektur von Einstellungen und Haltungen der beteiligten Mitarbeiter bereitzustellen;
– sämtliche „Veranstaltungen" – wie Anamneseerhebung, körperliche Untersuchung, Visiten, Pflegemaßnahmen, Konferenzen u. a. – entsprechend dem heutigen Kenntnisstand klinischer Psychosomatik weiterzuentwickeln und – soweit erforderlich – neue Veranstaltungen einzuführen;
– den Informationsaustausch und die Kooperation unter den Mitarbeitern zu fördern.
Gleichzeitig sollten Standard und Differenziertheit in der internistischen Krankenversorgung unvermindert aufrechterhalten werden.

Allgemeine Voraussetzungen

Äußere Arbeitsbedingungen

Stellenpläne und Raumverhältnisse sind den Erfordernissen eines psychosomatischen Arbeitsansatzes anzupassen. Die Notwendigkeit, diese Arbeitsbedingungen gegenüber der durchschnittlichen klinischen Arbeitssituation zu verbessern, läßt sich durch das Ergebnis zweier am Department für Innere Medizin in Ulm durchgeführter Untersuchungen verdeutlichen:

Auf einer 17-Betten-Station mit internistisch Schwerkranken beträgt bei einer Besetzung mit zwei Ärzten die durchschnittliche Zeit für Gesprächskontakte zwischen Patient und Arzt nur 4,4 Minuten pro Tag und Patient (Erdmann, 1974). Die Dauer der Visite liegt pro Patient bei 3,7 Minuten. Nur weniger als 30% dieser Zeit steht dem Patienten für eine seinen Bedürfnissen entsprechende Kommunikation mit dem Arzt zur Verfügung (Siegrist, 1975, 1976).

Fachliche Kompetenz der Mitarbeiter

Für das Pflegepersonal und die internistisch tätigen Ärzte ist eine intensive Weiterbildung in psychologi-

17 An der Entwicklung des hier dargestellten Konzeptes waren beteiligt: D. Böck, H. Bosch, A. Erath-Vogt, E. Gaus, M. Ginglmaier, A. Grauhan, H. Holl, M. Klingenburg, B. Kubanek, G. Paar, M. Rassek, Ch. Scheytt, K.-H. Schultheis, C. Simons, H. Urban, J. Zenz.
Der Zentraloberin der Ulmer Universitätskliniken, Frau I. Schulz, sowie dem seinerzeitigen Leiter der Abteilung Psychosomatik, Prof. Dr. Th. v. Uexküll, danke ich für ihre verständnisvolle Förderung und ihre eigene Beteiligung an dem Projekt.

scher Medizin zu fordern. Als theoretisches Grundkonzept für eine solche Weiterbildung bietet sich unserer Auffassung nach vor allem das psychoanalytische Modell für das Verständnis intra- und interpersoneller Konflikte an.

Eine wesentliche Vorbedingung für die Realisierung eines derartigen Stationskonzeptes ist die Möglichkeit zur Auswahl der Mitarbeiter, entsprechend ihrer Motivation und Eignung.

Einstellung der Mitarbeitergruppe gegenüber den Kranken

Wesentlich ist die Bereitschaft, sich im diagnostischen und therapeutischen Prozeß im Umgang mit dem Patienten in eine „Gegenseitigkeit" einzulassen, wie sie oben beschrieben wurde.

Aufgabe von Weiterbildung und Supervision ist es, immer wieder zu verdeutlichen, in welchem Ausmaß sich von Patienten ausgehende Kräfte auf die emotionalen Bewegungen der Mitarbeiter und des Teams auswirken können.

Entscheidend ist es, daß diese Bewegungen, einschließlich der Konflikte im Team, immer wieder vor allem auch auf ihre Verursachung durch den Patienten hin untersucht und so als Instrument von Erkenntnis und Therapie benützt werden können.

Kommunikation zwischen den Mitarbeitern

Die Organisationsstruktur einer internistisch-psychosomatischen Krankenstation hat Möglichkeiten zur intensiven Kommunikation zwischen allen Mitarbeitern der Station vorzusehen, in denen die Reflexion des eigenen Erlebens im Umgang mit den Patienten gefördert wird. Insbesondere gilt dies für die noch neu zu entwickelnden Formen der Zusammenarbeit zwischen Ärzten und Schwestern; hier stellt sich die Aufgabe, einem Verharren in traditionsbestimmten Rollenschemata mit hierarchisch bestimmten Funktions- und Machtzuteilungen im Sinne einer patientzentrierten Arbeit entschieden entgegenzuwirken.

Spezielle Zielvorstellungen für die Arbeit einer internistisch-psychosomatischen Krankenstation

Krankenversorgung

Unabhängig von der jeweils speziellen Problemstellung besteht das Hauptziel darin, mit den Kranken eine tragfähige Arbeitsbeziehung herzustellen, in der sie gegenüber Ärzten und Krankenschwestern zu möglichst selbständigen Partnern in der Behandlung ihrer Krankheit werden, eine therapeutische Beziehung, in der sie entsprechend ihren Bedürfnissen ausreichend Information, emotionale Stützung und Halt gewährende Führung bekommen.

In Tabelle 28–12 sind, entsprechend den Arbeitsrichtungen der Klinischen Psychosomatik (vgl. Tab. 28–2), Gesichtspunkte für das Vorgehen zur Förderung des erwünschten Krankheitsverhaltens, bei der Unterstützung der Krankheitsverarbeitung und bei der Einleitung der Bearbeitung pathogener psychischer und sozialer Konflikte zusammengefaßt.

Der psychosomatische Ansatz im Pflegebereich Kooperation zwischen Ärzten und Schwestern[18]

Krankenschwestern haben den zeitlich ausgedehntesten und – bestimmt durch den Charakter der Pflegetätigkeit – unmittelbarsten Kontakt mit dem Patienten. Die Aufgaben der Pflegepersonen umfassen die Sorge für das körperliche und seelische Wohl der Kranken oder, weniger anspruchsvoll, als Negativaussage formuliert: Es gibt keinen Bereich menschlicher Bedürfnisse, für den die Pflegepersonen ausdrücklich nicht zuständig sind. Für die Bestimmung ihrer Rolle ist es wesentlich, daß die Schwestern als einzige Mitarbeiter einer Station vom Patienten jederzeit mit der Klingel herbeigerufen werden können.

Den Krankenschwestern könnte somit eine zentrale Rolle in der Praxis einer psychosomatischen Medizin zukommen; in ihrer Tätigkeit ließe sich die pflege-

Tabelle 28–12. Spezielle Ziele und Vorgehen in der Krankenbehandlung auf einer internistisch-psychosomatischen Krankenstation

Generell: *Tragfähige Arbeitsbeziehung*	– Emotionale Unterstützung in halt- und schutzbietendem Milieu – Beziehungsangebot mit Merkmalen von „Gegenseitigkeit" – ausreichende Gesprächsmöglichkeit
Speziell: *Compliance und Krankheitsverarbeitung*	– ausreichende Information – Beachtung pathologischer Anpassungs- und Abwehrvorgänge – supportive Psychotherapie – Einbeziehung d. Angehörigen – Sozialarbeit
Arbeit an pathogenen psychischen und sozialen Konflikten	– Möglichkeit zur Darstellung von Beziehungsstörungen bzw. Konflikten im Rahmen der Station – Erarbeitung eines Bewußtseins von diesen Beziehungsstörungen bzw. Konflikten – Erarbeitung von Zusammenhängen zwischen diesen Störungen und der Symptomatik durch u. a. Konfrontation mit Zusammenhängen zwischen Situation und Symptom (Lebenssituation, Interaktion auf der Station, Übertragungsbeziehung) – Konfliktbearbeitung in längerfristiger vorwiegend poststationärer Psychotherapie – „Sozialtherapie": Einbeziehung der Konfliktpartner in Familie und Beruf

---

18 Ich danke Frau A. Grauhan für die Zusammenarbeit bei diesem Abschnitt.

rische Versorgung des Patienten mit dem Eingehen auf dessen Informations- und Kommunikationsbedürfnis und damit der Vermittlung emotionaler Sicherheit und Geborgenheit verbinden. Einem solchen patientzentrierten Ansatz in der Pflege stehen zunächst traditionelle Auffassungen des Arzt-Schwester-Verhältnisses entgegen.

Die Organisation der Pflege als „Funktionspflege" teilt die Pflegeaufgaben entsprechend der hierarchischen Gliederung auf verschiedene Pflegepersonen auf und behindert die Entwicklung einer intensiveren Arbeitsbeziehung zwischen Schwestern und Patienten entscheidend. Voraussetzung für eine „ganzheitliche" und patientenorientierte Pflege ist die Möglichkeit der Übernahme der Pflege einer kleinen Anzahl von Patienten durch jeweils eine Schwester oder eine Schwesterngruppe in voller Verantwortung. Gleichzeitig müssen neue Lösungen für die organisatorische Leitung der Station erarbeitet werden, die der neuen Funktionsaufteilung entsprechen.

Eine intensivere Einbeziehung von Krankenschwestern in die Krankenbehandlung wird jedoch auch durch das Verständnis der Ärzte und ihre Rolle und Funktion behindert. Ärzte gehen in der klinischen Arbeit – sowohl in der krankheitszentrierten Medizin als auch in herkömmlichen psychotherapeutisch-psychosomatischen Arbeitsansätzen – meist davon aus, daß die Dyade Arzt-Patient die Grundlage der Therapie bildet. Schwestern können im Rahmen dieser Auffassung lediglich Teilfunktionen übernehmen, vor allem als Mittler Anweisungen des Arztes ausführen und ihm Informationen über den Zustand des Patienten vermitteln. Im Rahmen der krankheitszentrierten Medizin impliziert diese Auffassung, daß die Interaktionen zwischen Schwester und Krankem im Rahmen der Mittlerfunktion therapeutisch bedeutungs- und wirkungslos sind; im klassischen psychotherapeutischen Modell wird die Wirksamkeit dieser Interaktionsprozesse gesehen, aber nur selten systematisch für die Behandlung genutzt; sie stellen vielmehr eine unerwünschte Störung der therapeutischen Beziehung zwischen Arzt und Patient dar. Die Ereignisse innerhalb der psychotherapeutischen Beziehung werden im allgemeinen vor den Schwestern „geheimgehalten", allenfalls werden die Informationen der Schwestern über den Patienten in der Behandlung berücksichtigt.

Wird einmal anerkannt, daß Beziehungen zwischen Schwestern und Patienten potentiell immer den Behandlungsverlauf beeinflussen und im klinischen Setting nicht neutralisiert werden können, stellt sich die Aufgabe, diese Beziehungen systematisch zu untersuchen und Krankenschwestern und -pfleger als Verbündete des Arztes in den therapeutischen Prozeß mit einzubeziehen.

Entsprechend den in der Tabelle 28–12 dargestellten Formen des Vorgehens im Rahmen einer internistisch-psychosomatischen Krankenstation könnten den Schwestern folgende Funktionen zukommen:

– Krankenschwestern orientieren sich über die Pflegebedürfnisse und die mit der Krankheit verbundenen emotionalen Probleme der Patienten. Sie stellen eine eigenständige „Pflegediagnose". Sie gehen auf die Gesprächswünsche der Patienten ein und informieren die Kranken im einzelnen über die geplanten diagnostischen und therapeutischen Maßnahmen; hierbei bietet sich häufig eine wesentliche Gelegenheit, Patienten emotional zu unterstützen. Die Schwestern tragen damit zur Ausbildung eines Stationsklimas bei, das speziellere Formen supportiver und konfliktbearbeitender Psychotherapie überhaupt erst ermöglicht, da es dem Patienten erst jene Sicherheit vermittelt, die er benötigt, um sich in einen therapeutischen Prozeß überhaupt einlassen zu können.

– Die Beobachtungen und Informationen der Schwestern tragen in der gemeinsamen Diskussion zum Verständnis der Patienten bei und vermehren die Interventionsmöglichkeiten während der Visite und therapeutischer Einzel- und Gruppengespräche.

– Die Schwestern sind daneben entsprechend den auf der Station gemeinsam entwickelten Zielvorstellungen auch selbständig therapeutisch tätig; sie versuchen mit den Kranken Probleme des Krankheitsverhaltens und der Krankheitsverarbeitung zu besprechen und bei pathogenetisch wirksamen psychologischen Konflikten durch entsprechende Konfrontation zur Bildung eines Konfliktbewußtseins beim Patienten beizutragen.

– Bei Patienten, die sich in psychotherapeutischer Einzelbehandlung befinden, können Schwestern die Therapie in vielfacher Hinsicht unterstützen, wenn sie um die ablaufenden Prozesse wissen; schon das Vermeiden kritischer Aussagen zur Therapie stellt eine wesentliche Hilfe dar; daneben können die Schwestern Fragen der Patienten zur Therapie klärend beantworten und die Kranken bei entsprechenden Verhaltensweisen auf den Zusammenhang mit dem Widerstand gegen den therapeutischen Prozeß aufmerksam machen.

– Nach einer entsprechend intensiven Weiterbildung können einzelne Schwestern auch unter Supervision eine selbständige Betreuung oder Nachbetreuung von Patienten übernehmen.

Die Realisierung einer derartig neu bestimmten Kooperation zwischen Schwestern und Ärzten ist davon abhängig, ob es den Beteiligten gelingt, eine gemeinsame Sprache zu finden, die ihnen erst eine gemeinsame Betrachtung der Probleme der Kranken ermöglicht. Oft scheitern organisatorisch gut vorbereitete gemeinsame Konferenzen von Ärzten und Schwestern auf der Station schon daran, daß eine Verständigung über bestimmte Phänomene deshalb extrem schwerfällt, weil eine gemeinsame Fachsprache fehlt.

Soll eine partnerschaftliche Zusammenarbeit zustande kommen, sind die Schwestern schließlich auf die Bereitschaft der Ärzte angewiesen, trotz ihres Bildungs-, Macht- und Statusvorsprungs zum Abbau der hierarchischen Verhältnisse zugunsten einer funktionsorientierten Zusammenarbeit beizutragen.

Die Durchführung des Modellversuches

Der institutionelle Rahmen

Die **internistisch-psychosomatische** Krankenstation war sieben Jahre lang eine von 12 internistischen Allgemeinstationen des Departments für Innere Medizin in Ulm. Die Mitarbeiter der Station waren der Abteilung Psychosomatik zugeordnet, die während des angegebenen Zeitraums als „Abteilung Innere Medizin und Psychosomatik" Teil des Departments für Innere Medizin war. Die Funktionsbereiche dieser Abteilung Psychosomatik können aus Abbildung 28–8 entnommen werden. Die Beratung der Stationsärzte der Allgemeinstationen hinsichtlich der Spezialprobleme einzelner internistischer Subdisziplinen erfolgte durch einen institutionalisierten Konsiliardienst der jeweiligen Fachabteilung oder Fach-„Sektionen" (unmittelbare Stationsversorgung = „Vertikaldienst"; Tätigkeit im Konsiliardienst = „Horizontaldienst").

Äußere Arbeitsbedingungen, Patienten, Mitarbeiter

Die Krankenstation

Die Station lag zentral im Klinikum, nahe bei Aufnahmestation, Dialyseeinheit und Intensivstation. Dies förderte eher eine Belegung mit internistisch Schwerkranken. Sie hatte insgesamt 15 Betten in sieben Zweibett- und einem Einbettzimmer. Je ein Arbeitsraum stand für Ärzte und Schwestern sowie eine Dachkammer als Aufenthaltsraum für die Schwestern und als Raum für sämtliche gemeinsamen Veranstaltungen und die Gruppengespräche mit den Patienten zur Verfügung. Der Flur der Station wurde als Aufenthalts- und Kontaktraum genützt.

Die Patienten

Die Häufigkeitsverteilung der Liegezeiten entsprach weitgehend derjenigen im gesamten Department für Innere Medizin; 50% der Patienten hatten eine Aufenthaltsdauer von nicht mehr als 14 Tagen. Die Mortalität auf der Station betrug 11,7%, im gesamten Department unter Einschluß der Intensivstation 12,8%. Diese Angaben zeigen, daß die auf der Station Tätigen überwiegend mit den psychosozialen Problemen internistisch Schwerkranker konfrontiert waren, wie es auch sonst der Situation in einer Universitätsklinik oder in einem Schwerpunktkrankenhaus entspricht.

Eine Typisierung der psychosomatischen Probleme bei diesen Patienten, entsprechend den Hauptarbeitsgebieten der Psychosomatischen Medizin, gibt Tabelle 28–13.

Die Mitarbeiter

Der Stellenplan der Station wurde im Vergleich zu den übrigen Stationen des Departments um eine Zweitarztstelle und die Stelle der „psychosomatischen Schwester" erweitert. Hinzu kamen Mitarbeiter der Abteilung Psychosomatik, die der Station mit einem Teil ihrer Arbeitszeit zusätzlich zur Verfügung standen, teils unmittelbar für die Krankenversorgung, teils für Supervisionsaufgaben (vgl. Tab. 28–14).

Rollen und Funktionen der Mitarbeiter veränderten sich im Lauf der Entwicklung des Stationskonzeptes. Nach einer anfänglichen Tendenz zur Rollendiffusion mit Angleichung der Funktionen wurde – auch im Zusammenhang mit der Verbesserung der Kooperation untereinander – eine stärkere Rollenspezifität und Aufteilung der Aufgaben erarbeitet.

Tabelle 28–13. Patientengut der internistisch-psychosomatischen Station (Diagnosen von 138 Patienten)

Diagnosegruppen	Patienten	Patienten (in % der Bettenbelegung)
Organerkrankungen insgesamt	76%	82%
Davon:		
Leukämien und andere Malignome	26%	39%
Zum Tode Kranke insgesamt	34%	54%
„Psychosomatische Krankheiten"	10%	13%
Funktionelle Störungen	6%	3%
Psychiatrische Erkrankungen	8%	2%

Tabelle 28–14. Stellenplan der internistisch-psychosomatischen Station

Mitarbeiter	
Der Station fest zugeordnet („Vertikaldienst")	Für Spezialaufgaben (Teilzeitmitarbeiter im „Horizontaldienst")
2 Ärzte	internistischer Oberarzt
1 Medizinalassistent bzw. Arzt einer anderen internistischen Abteilung (Rotation)	psychosomatischer Oberarzt
	Chefarzt
5 Schwestern, 1 Nachtschwester	
	Psychologin
2 Schwesternschülerinnen	Sozialarbeiterin
1 „psychosomatische Schwester"	Krankengymnastin
	Seelsorger

Abb. 28–8. Funktionsbereiche der Abteilung Innere Medizin und Psychosomatik des Departments für Innere Medizin der Universität Ulm.

Spezielle Rollen

Die „psychosomatische Schwester". In Anlehnung an entsprechende Versuche in psychiatrischen Kliniken (psychiatric consultation nurse; Barton und Kelso, 1971) übertrugen wir einer für diesen Arbeitsbereich bereits weitergebildeten Schwester spezielle Funktionen. Zu ihren Aufgaben gehörte es zunächst, die übrigen Schwestern, auf deren Auswahl wir zunächst keinen Einfluß hatten, für unseren Veränderungsversuch zu gewinnen. Gerade bei den in diesem Zusammenhang auftretenden Widerständen erwies sich die Zusammenarbeit mit der ständig auf Station tätigen „psychosomatischen Schwester" von entscheidender Bedeutung. Andererseits hatte diese Schwester die Interessen der übrigen Mitglieder des Pflegeteams gegenüber den Ärzten zu vertreten. Nicht selten hatte sie die Führung in Auseinandersetzungen zwischen beiden Gruppen zu übernehmen. Insgesamt fiel ihr so immer wieder die Aufgabe einer Vermittlerin zu.

Im Rahmen der Krankenversorgung beteiligte sich die „psychosomatische Schwester" an der Basispflege und versuchte, diese exemplarisch patientzentriert zu gestalten. Sie beriet die übrigen Schwestern bei Problemen im Umgang mit Patienten und versuchte in den gemeinsamen Besprechungen, z.B. bei der Schichtübergabe, die psychosozialen Probleme der Patienten herauszuarbeiten. Im Rahmen des „Weiterbildungskurses für Krankenschwestern in patientzentrierter Pflege/Psychosomatischer Medizin" übernahm sie die Supervision der Kursteilnehmerinnen während ihrer Tätigkeit auf der Station.

Die Sozialarbeiterin. Entsprechend dem gewandelten Verständnis der Sozialarbeit im Krankenhaus verstand sich die Sozialarbeiterin nicht mehr als bloße Fürsorgerin, die sich fast ausschließlich mit materiellen Hilfestellungen befaßt, sie bot vielmehr dem Kranken Hilfe zur eigenen Bewältigung seiner psychosozialen Konflikte an. Sie war ständige Mitarbeiterin im Stationsteam, nahm an den täglichen Aufnahmebesprechungen und an allen weiteren Konferenzen teil, um rechtzeitig ihre Auffassung über die Probleme der Patienten in die Diskussion einbringen und eventuell die Indikation für eigene Interventionen stellen zu können. Sie achtete vor allem auf die sozialen Beziehungen der Kranken und bemühte sich, sie bei der Verarbeitung der Krankheitsfolgen in materieller, psychischer und sozialer Hinsicht zu unterstützen. Ihr stellte sich auch die Aufgabe, den Patienten bei der Aufrechterhaltung ihrer sozialen Beziehungen während der jetzigen Krankheit und des Krankenhausaufenthaltes bzw. bei der Umgestaltung gestörter und belastender Beziehungen zu helfen. Bei entsprechender Indikation führte sie mit den Patienten und, falls erforderlich, mit deren Angehörigen Beratungsgespräche. Die Sozialarbeiterin leitete die wöchentlich stattfindende offene Patientengruppe mit dem Ziel, die Patienten darin zu unterstützen, intensivere Kontakte mit Mitpatienten aufzunehmen und auch so ihre soziale Isolation im Krankenhaus zu verringern.

Die Krankengymnastin. Eine weitgehende Einbeziehung einer Krankengymnastin in das Stationsteam hat sich insbesondere im Rahmen der Bemühungen um die Rehabilitation der Patienten als außerordentlich wertvoll erwiesen.

Der Krankenhausseelsorger. Ein Krankenhausseelsorger arbeitete im therapeutischen Team mit; er nahm regelmäßig an den Stationskonferenzen teil, bot den Patienten einmal wöchentlich während eines Besuches im Krankenzimmer die Möglichkeit zu einem Gespräch an und stand sonst jederzeit auf Wunsch für Gespräche zur Verfügung. Es ging dem Seelsorger dabei weniger um die aktive Vermittlung von Glaubensinhalten, sondern mehr darum, eine annehmende, mittragende Beziehung zum Gesprächspartner herzustellen.

Organisatorische Hilfen für den Informationsaustausch

Bei der großen Anzahl der Mitarbeiter kam einer optimalen Regelung der formalen Seite des Informationsaustausches eine wichtige Bedeutung zu. Wir hatten hierfür das „Kardex-System" gewählt. Durch dieses System werden Anordnungen überschaubarer und in ihrer Ausführung kontrollierbar; dieses System ersetzte die bisher üblichen Krankenblätter und -kurven. Im Kardex-System wurden auch psychologische Befunde, vor allem die Verhaltensbeobachtungen, dokumentiert und so allen Mitarbeitern zugänglich.

Das Therapiekonzept und die Supervision

Ärzte und Schwestern der Station bemühten sich in ihrer Arbeit mit Patienten um eine **Integration** des psychosomatischen Arbeitsansatzes in die internistische Krankenversorgung bzw. die Krankenpflege. Der Arzt verband das Gespräch mit dem Patienten mit den internistischen Maßnahmen. Im Falle spezifischer Psychotherapie wurde in jedem Einzelfall abgewogen, ob die Verbindung mit der allgemeinärztlichen Versorgung den Zugang zum Patienten erleichtert – wie das oft bei schweren Asthmatikern der Fall ist – oder durch die Komplizierung der Übertragungsbeziehung erschwert, wie z.B. im allgemeinen bei Anorexia-nervosa-Patientinnen; in letzterem Fall teilten sich aus methodischen Gründen zwei Ärzte die Behandlung.

Grundkonzept für die spezifische, konfliktbearbeitende Therapie war auf dieser Station die **analytisch orientierte Einzelpsychotherapie.** Für die Einzelpsychotherapie sprachen schon rein praktische Gründe, wie die Verpflichtung zu fortlaufender Patientenaufnahme; daneben erschien es uns gerade innerhalb eines solchen Stationskonzepts wichtig, daß therapeutische Prozesse auch in klar definierten und überschaubaren Zweierbeziehungen stattfinden, ganz abgesehen vom didaktischen Wert solcher Behandlungen für die Ausbildung der Mitarbeiter.

Bei entsprechender Indikation ergänzten wir die Einzeltherapie durch eine Beratung der Familie.

Auf der Supervisionsebene haben wir ein **Kooperationsmodell** entwickelt. Die internistische Oberarztfunktion übte ein erfahrener und für die psychosozialen Fragestellungen aufgeschlossener Internist aus. Die Supervision der psychosomatisch-psychotherapeutischen Arbeit übernahm jeweils ein entsprechend weitergebildeter Mitarbeiter der Abteilung Psychosomatik.

Die krankenhausübliche Pflegesupervision wurde durch die fachlich spezialisierte Supervision der „psychosomatischen Schwester" ergänzt.

Die Veranstaltungen

Abbildung 28–9 gibt eine Übersicht über die Veranstaltungen der Station. Ich berichte hierüber im Präsens. Der Ablauf der Stationsarbeit ist in einem Film („Wer will schon krank sein auf der Welt") dokumentiert.

Das Erstgespräch der Schwester

Am Tag der stationären Aufnahme versucht die Schwester im Erstgespräch Kontakt mit dem Patienten aufzunehmen und eine eigene „Pflegediagnose" zu erstellen (Köhle et al., 1983).

Dieses Erstgespräch, für das die Schwestern systematisch weitergebildet wurden, haben wir eingeführt, um dem Patienten zu vermitteln, daß er auf der Station „aufgenommen" wird und sich eine Schwester um seine Bedürfnisse als Kranker in einer für ihn zunächst fremden, oft beängstigenden Umwelt verantwortlich kümmert.

Das Interesse der Schwester gilt in diesem Erstgespräch vor allem dem subjektiven Krankheitsgefühl des Patienten, Art und Ausmaß seiner Hilfsbedürftigkeit, seinen subjektiven Vorstellungen über Wesen und Folgen der Erkrankung, seinen Erwartungen an

den Krankenhausaufenthalt, seiner sozialen Situation und den Umständen beim Beginn seiner Krankheit (vgl. Köhle et al., 1977).

Das Erstinterview des Arztes

Das Erstinterview des Arztes orientiert sich an der Technik der „assoziativen Anamneseerhebung" bzw. an der Technik des „klinischen Interviews" (vgl. Kap. 12).

Die Morgenbesprechung

In der Morgenbesprechung berichten Schwester und Arzt jeweils zusammenfassend über ihre Erstgespräche mit dem oder den am Vortag neu aufgenommenen Patienten; anschließend versuchen wir, vorläufige Hypothesen über die psychosozialen Probleme und psychosomatischen Zusammenhänge beim Patienten aufzustellen und einen ersten integrierten internistisch-psychosomatischen Behandlungsplan festzulegen. Daneben werden auch aktuelle Probleme anderer Patienten der Station diskutiert.

Dieses Vorgehen ermöglicht es uns, vom ersten Tag an auch die psychosozialen Probleme systematisch mit in den Behandlungsplan aufnehmen zu können.

Die Visite (vgl. Kap. 15)

Die Visite ist auf einer Krankenstation die zentrale gemeinsame Veranstaltung im Tagesablauf. Sie stellt gleichzeitig die hauptsächliche Gelegenheit zur Kommunikation zwischen Arzt und Patient dar. Die vorliegenden empirischen Untersuchungen (vgl. Kap. 18) zeigen, daß diese Möglichkeit zur Kommunikation meist nur unzureichend genutzt wird. Die Visite ist zu stark mit anderen Funktionen überfrachtet; z.B. findet am Krankenbett häufig der Informationsaus-

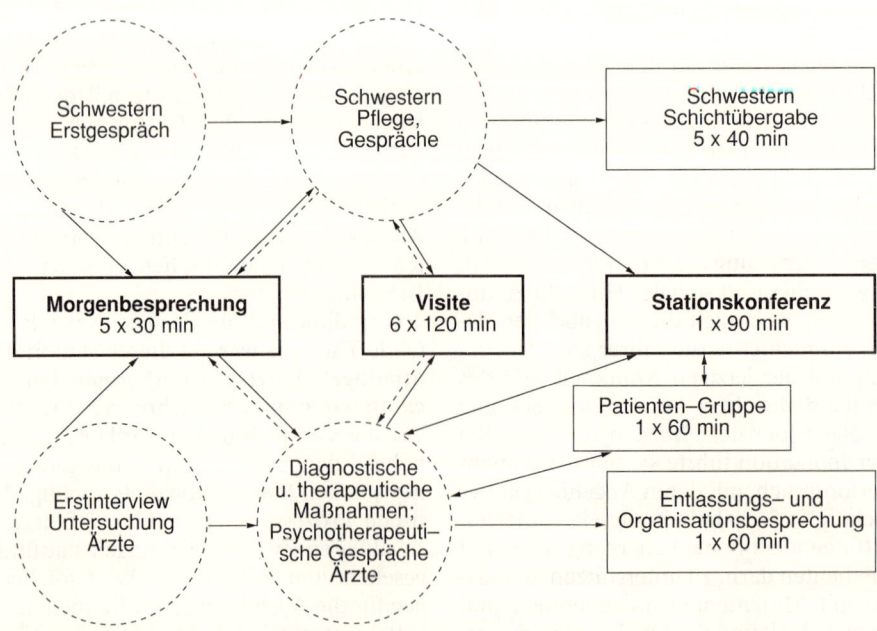

Abb. 28–9. Veranstaltungen auf der Station.

tausch zwischen verschiedenen Ärzten und Schwestern statt, so daß während dieser Zeit häufig mehr über den Patienten als mit ihm gesprochen wird. Wir hielten deshalb die Umgestaltung der Visite zu einer patientzentrierten Veranstaltung für besonders vordringlich. Eine solche Umgestaltung erschien uns auch deshalb wichtig, weil die Visite sich hervorragend für ein Training eines „Sprechstundengespräches" eignet, wie es der niedergelassene Arzt in seiner Praxis führt.

Wir versuchen, die Visite im Sinne unserer allgemeinen und speziellen Zielvorstellungen in den gesamten Behandlungsplan einzubeziehen und sie in bestimmten Abschnitten als **„therapeutische Visite"** zu führen. Während der Visite soll der Patient zunächst seine Fragen und Erwartungen möglichst ungehindert vorbringen können, er soll zur aktiven Beteiligung am diagnostischen und therapeutischen Prozeß ermutigt werden und hierzu in für ihn verständlicher Form ausreichend Information erhalten. Diese **Informationsvermittlung** kann bereits eine stützende und ermutigende Wirkung haben. Darüber hinaus kann die Visitensituation auch zu psychotherapeutischen Interventionen im Sinne der von Balint und Norell (1975) entwickelten „Sprechstundenpsychotherapie" genutzt werden. Je nach Indikation werden mehr emotional stützende oder mehr mit den anstehenden Konflikten konfrontierende Interventionsformen gewählt. Diese Interventionen sind dabei auf die jeweils sich zwischen Patient und Arzt einstellende Beziehungskonstellation bezogen.

Hinweise auf mögliche Zusammenhänge zwischen dem Auftreten von Beschwerden und auslösenden emotionalen oder situativen Faktoren auch während der Visite können dazu beitragen, zunächst psychotherapeutisch nur schwer erreichbare Kranke für eine **konfliktbearbeitende Psychotherapie** zu gewinnen.

Dies wird auch dadurch erleichtert, daß im Rahmen des therapeutischen Gespräches auf die Beschwerden und Symptome, unter denen der Kranke ja leidet, eingegangen wird. Beschwerden und Symptome werden gerade dadurch ernstgenommen, daß sie in einen Gesamtzusammenhang gestellt werden.

Während der täglichen Visite kann vor allem auch der Prozeß der **Krankheitsverarbeitung** gefördert werden; gelingt es dem Arzt, den Patienten in seiner Angst, Depression oder emotionalen Zurückgezogenheit, in seinem Zorn, seiner Enttäuschung über die Medizin oder einzelne Ärzte im Beisein des gesamten Teams während der Visite zu erreichen, entsteht hieraus nicht selten ein Prozeß, der bei weiteren Visiten und anderen Kontakten fortgeführt werden kann und der dem Patienten schließlich die Aufarbeitung seiner derzeitigen Situation ermöglicht. Besonders der „Trauerprozeß" schwerkranker Patienten kann so im unmittelbaren Kontext der medizinischen Versorgung unterstützt werden. Dies gelingt dem Arzt, der den Patienten auch sonst behandelt, im allgemeinen wesentlich leichter als einem spezialisierten Konsiliarius, der für „die seelischen Probleme" hinzugezogen wird.

Der Patient kann sich dem Arzt und dann auch der übrigen Welt leichter wieder zuwenden, seine „Objekte" wieder „libidinös besetzen", wenn er sich als kranke Person vom Arzt beachtet und wertgeschätzt fühlt.

Wir versuchen die Situation am Krankenbett während der Visite möglichst weitgehend für das Gespräch mit dem Patienten und die körperliche Untersuchung freizuhalten. Wir haben die Visite deshalb formal in drei Abschnitte gegliedert: Vorbesprechung und Austausch von Informationen vor der Tür des Krankenzimmers; die eigentliche Visite am Bett des Patienten; die Nachbesprechung wieder außerhalb des Krankenzimmers.

Die Krankenschwestern bringen ihre Informationen während der Vorbesprechung ein. Am Bett des Patienten spielt sich die Kommunikation vorwiegend zwischen dem Patienten und dem ihn behandelnden Arzt ab – die Ärzte der Station teilen sich die Zuständigkeit für die Patienten untereinander auf. Die Einbeziehung der für den Patienten zuständigen Krankenschwester und anderer Visitenteilnehmer in das Gespräch ist möglich; hierfür ist jedoch eine gute Abstimmung der Visitenteilnehmer untereinander Voraussetzung. Auf den Patienten kann eine aktive Beziehung von zu vielen Visitenteilnehmern auch verwirrend wirken; oft genügt für ihn schon die Gewißheit, daß seine Probleme auch dem übrigen Team vertraut sind.

Tabelle 28–15 gibt in schematisierter Form eine Übersicht über Ziele und Vorgehen während der Visite.

Für die Weiterentwicklung unseres Visitenkonzeptes und unser eigenes Verhaltenstraining erwies sich die Teilnahme eines erfahreneren Psychosomatikers einmal wöchentlich an der Visite und die Diskussion von Videoaufnahmen als sehr hilfreich. Wir waren immer wieder überrascht, wie schwierig es ist, im nachhinein erkennbare „Angebote" von Patienten in der Visitensituation zu verstehen und sinnvoll aufzugreifen.

Die Visite dauert durchschnittlich acht Minuten pro Patient, fünf Minuten am Krankenbett, drei Minuten außerhalb des Krankenzimmers.

Oberarzt- und Chefarztvisite

Wir versuchen, diese Visiten analog zu den Stationsarztvisiten so zu gestalten, daß die fachliche Supervision der Mitarbeiter die unmittelbare Beziehung des Oberarztes bzw. des Chefarztes zum Patienten, die diesem vor allem zusätzlich Sicherheit vermitteln soll, nicht zu stark beeinträchtigt.[19] Nach Möglichkeit erarbeiten wir vor dem Zimmer eine spezielle Zielvorstellung für die Chefarztvisite.

Die Pflegevisite

Gegen Ende der Nachmittagsschicht führen die Schwestern bei den ihnen zugeordneten Patienten ei-

---

[19] Eine ausführliche Darstellung und Diskussion der Chefarztvisite hat v. Uexküll (1977) vorgelegt.

Tabelle 28–15. Ziele und Vorgehen während der Visite

	Ziele	Vorgehen
Vorbesprechung außerhalb des Zimmers	Aufstellung diagnostischer und therapeutischer Ziele	– Verbindung von Vorwissen mit neuen Informationen: – Austausch zwischen Schwestern und Ärzten – Formulierung des engeren Visitenzieles
Visite am Bett des Patienten	Berücksichtigung von: Befinden; Bedürfnissen; Qualität des Arbeitsbündnisses; psychosomatischen Zusammenhängen; Interpretation und Gewichtung der Befunde	*Begrüßung* – „Wie geht es Ihnen heute?" – Evtl. eingehen auf Abwehr (kann der Patient die Situation für sich nutzen?) *Untersuchungsgang* – Open-ended-Interview (situationszentriert), evtl. mit Information, Interpretation und Stützung; – körperliche Untersuchung; – Diskussion der Kurvenwerte; – Einbeziehung der Umstehenden („wir"); – Angaben zur weiteren Diagnostik; – Zusammenfassung der Befunde und Bewertung für den Patienten; – Hinweise auf nächste Schritte; – Aufforderung an den Patienten, Fragen zu stellen
Nachbesprechung außerhalb des Zimmers	Ergebnisse der Visite; Aufgabenverteilung	– Kurze Diskussion der gemeinsamen Beobachtungen; Kritik am Vorgehen; – Absprache weiterer Maßnahmen

ne „Pflegevisite" mit dem Ziel durch, noch ausstehende pflegerische Maßnahmen und ärztliche Verordnungen durchzuführen und gleichzeitig mit den Patienten am Ende des Tages noch einmal ins Gespräch zu kommen. Die Schwester erkundigt sich nach dem Befinden des Kranken und etwaigen aktuellen Bedürfnissen, sie bespricht mit ihm die Ergebnisse bereits durchgeführter Untersuchungen und erläutert ihm bevorstehende diagnostische und therapeutische Maßnahmen sowie die Wirkungsweise verordneter Medikamente. Die Gespräche während der abendlichen Pflegevisite können unserer Erfahrung nach auch wesentlich dazu beitragen, dem Patienten die für eine entspannte Nachtruhe erforderliche Sicherheit zu vermitteln. Daneben haben die Gespräche während der Pflegevisite nicht selten auch die Funktion, den Patienten darin zu unterstützen, Informationen, Anregungen und psychotherapeutische Interventionen, die er während des Tages erhielt, zu verarbeiten.

Die Stationskonferenz

In der Stationskonferenz diskutieren alle Mitglieder der Station unter Leitung eines erfahrenen Psychosomatikers über Schwierigkeiten im Umgang mit einem einzelnen Patienten. Methodisch orientieren wir uns an der Balint-Gruppenarbeit. Ausgehend vom jeweils vorgeschlagenen „schwierigen" Patienten, bemühen wir uns darum, die Beziehung zwischen den einzelnen Stationsmitarbeitern und diesem Kranken für das Verständnis des Patienten zu benützen. Die unterschiedliche Wahrnehmung desselben Patienten durch verschiedene Stationsmitglieder und die Beobachtung der vielfältigen Interaktionsprozesse zwischen dem Kranken und den Mitarbeitern ermöglichen im allgemeinen die Bearbeitung skotomisierender Wahrnehmungsvorgänge und erlauben häufig ein Verständnis des aufgetretenen „Problems". Die Konflikte mit dem Kranken erweisen sich häufig als eine für den Patienten[20] charakteristische Neuinszenierung von Konflikten seiner früheren oder gegenwärtigen Beziehungen auch außerhalb des Krankenhauses. Ziel dieser Konferenzen ist es, die aufgetretenen Schwierigkeiten zu verstehen und Lösungsmöglichkeiten zu erarbeiten. Häufig lösen sich Spannungen schon dadurch, daß wir Patienten nach solchen Besprechungen in veränderter Weise wahrnehmen und ihnen damit die Möglichkeit zu einer Änderung ihres Verhaltens eröffnen; in anderen Fällen sind die Besprechung der Schwierigkeiten mit dem Patienten oder gezieltere psychotherapeutische Interventionen erforderlich.

Es überrascht uns immer wieder, in welchem Ausmaß bereits **Wahrnehmungsveränderungen** bei uns selbst zu Lösungen von „Schwierigkeiten" im Umgang mit Patienten beizutragen vermögen. Nicht selten er-

---

20 Entscheidend ist, daß die emotionalen Bewegungen der einzelnen Mitarbeiter und etwaige Spannungen zwischen ihnen im Team auf ihre Verursachung auch durch den Patienten hin untersucht werden; nur dann werden die Möglichkeiten einer solchen Arbeit für die Therapie ausgeschöpft.

eignet es sich, daß Patienten nach einer solchen Konferenz „spontan" über die in der Konferenz vermuteten, belastenden Probleme zu sprechen beginnen – wohl weil wir dies aufgrund unserer geänderten Wahrnehmung vom Patienten nun zulassen und u.a. über unser Ausdrucksverhalten mitteilen können.

In den Stationskonferenzen werden daneben auch Teamprobleme diskutiert.

Ein Beispiel soll die Arbeitsweise in der Stationskonferenz veranschaulichen.

Zu Beginn der Besprechung herrscht zwischen Schwestern und Ärzten der Station eine ungewöhnlich intensive aggressive Spannung; es wird vorgeschlagen, über „Teamprobleme" zu sprechen. Anlaß ist die entgegen ärztlichen Überlegungen vorgenommene Verlegung einer infektionsgefährdeten 45jährigen Patientin mit dem dritten Rezidiv einer akuten Leukämie aus dem Einzel- in ein Zweibettzimmer, u.a. mit dem Hinweis, daß sie dann nicht mehr so oft unnötig zu läuten brauche und mehr Kontakt habe. Nur gegen intensiven Widerstand gelingt es mir, als Leiter der Besprechung, Überlegungen einzubringen, ob die Spannung etwas mit der Patientin zu tun haben könnte. Langsam stellt sich heraus, daß die Patientin vor allem gegenüber den Schwestern ungewöhnlich aggressiv ist, ständig Forderungen an sie stellt, jedes Kontaktangebot von seiten der Schwestern jedoch zurückweist und versucht, die Schwestern gegeneinander auszuspielen. Die Versorgung der Patientin wird als Zumutung erlebt, die affektive Betroffenheit ist so heftig, daß die Schwestern zum Teil mit Empörung, Ablehnung und offenem Rückzug reagieren. Nur mit Mühe gelingt es in der Gruppe, Überlegungen zur Motivation der Patientin anzustellen. Allmählich werden dann ihre panische Angst und ihre Versuche, diese zu bewältigen, sichtbar. Sie bemüht sich, über ihre Aktivitäten ihr passives Ausgeliefertsein umzukehren: Wegen jeder Kleinigkeit ruft sie die Schwestern herbei und versucht, sie herumzukommandieren. Als ihr dies nicht ausreichend gelingt und die Schwestern sich eher zurückziehen, wird die Patientin inkontinent, dies zwingt die Schwestern zu vermehrtem Aufwand in der Pflege, verstärkt jedoch ihre gefühlsmäßige Ablehnung der Patientin. Angst und Hilfsbedürftigkeit, den ursprünglichen Anlaß des Läutens, können die Schwestern, von ihren Affekten behindert, gar nicht mehr wahrnehmen. Ähnlich verfährt die Patientin mit ihrer Familie: Ehemann und Tochter versucht sie übers Telefon zu dirigieren, was sie selbst als aufopfernde Fürsorge erlebt und darstellt.

Nun werden auch aggressive Gereiztheit und Zorn verständlicher: Die Patientin versucht, ihre tiefe Depression, ihre Verzweiflung angesichts von Tod und Verlassenheit zu kompensieren. Erstmals nach Wochen kann statt gereizter Ablehnung wieder Mitleid für die Patientin aufkommen. Nach der Konferenz ist ein völlig offenes Gespräch der Sozialarbeiterin mit der Kranken möglich: Die Kranke kann direkt ihre Befürchtungen, zu Hause schon abgeschrieben zu sein, äußern und erstmals ihre Situation mit allen Gefühlen ausführlich besprechen. Ihr „Trauerprozeß" ist jetzt nicht mehr blockiert, die aufkommende Depression ist einfühlbar, Ärzte und Schwestern können sie jetzt unterstützen, die Schwierigkeiten im Umgang treten nicht mehr auf; es entwickeln sich wirklich therapeutische Beziehungen. Der Konflikt zwischen Ärzten und Schwestern spiegelte einen Konflikt in der Patientin wider.

Die Schichtübergabe

Der mit dem Schichtdienst des Pflegepersonals verbundene Wechsel der Bezugspersonen der Patienten macht einen intensiven Erfahrungsaustausch und eine sorgfältige Absprache aller Betreuungsmaßnahmen während der Schichtübergabe erforderlich. Die tägliche Übergabebesprechung findet in der Gruppe statt, so daß alle Schwestern der Station bis zu einem gewissen Grad über alle Patienten informiert sind. Die „psychosomatische Schwester" kann während der Schichtübergabe die anderen Schwestern beraten; ihre Dienstzeit überbrückt den Wechsel zwischen beiden Tagschichten.

Die Entlassungs- und Organisationsbesprechung

Wir versuchen, einmal wöchentlich die mit einzelnen während der Vorwoche entlassenen Patienten gemachten Erfahrungen kritisch zu diskutieren. Daneben besprechen wir Probleme der organisatorischen Weiterentwicklung des Stationskonzeptes.

„Patientengruppe" und „Patientencafé"

Ziel der „Patientengruppe" ist es, den Kranken eine Möglichkeit zur Kontaktaufnahme miteinander und zu Mitarbeitern der Station auch außerhalb der „Routine"-Veranstaltungen der Station anzubieten. Daneben soll die Gruppe auch Möglichkeit dazu bieten, gemeinsam Kritik an Vorkommnissen auf der Station, am Stationskonzept oder am Verhalten einzelner Mitarbeiter äußern zu können. Während dieser Gruppenbesprechungen kann bei einzelnen auch das Bewußtsein für Konflikte im psychischen Bereich dadurch gefördert werden, daß auch andere Kranke von ihren Problemen berichten. Durch die Gruppenprozesse kann die Auseinandersetzung mit den Folgen der eigenen Krankheit, die nicht selten in stummer Anklage oder Depression (Rohde, 1974) erstarrt ist, dann wieder in Gang gebracht werden, wenn für die Patienten bei anderen Teilnehmern oder durch Intervention des Gruppenleiters Lösungsansätze sichtbar werden.

Während wir die Patientengruppe zunächst als reine Gesprächsgruppe führten, bezogen wir später auch nichtverbale Ansätze, vor allem gestalterische Methoden, mit ein. Die Gruppe wird von der auf der Station tätigen Sozialarbeiterin geleitet. Bei der kurzen durchschnittlichen Liegezeit der Patienten bemühen wir uns darum, jede Gruppensitzung als in sich geschlossene Einheit aufzufassen. Die Gruppenleiterin strukturiert die jeweilige Sitzung, indem sie der Gruppe Thema, Methode und Materialien vorschlägt. Über die inhaltliche Gestaltung des Themas und in den begleitenden Gesprächen gelingt es häufig, in lockerer Atmosphäre die gemeinsamen Probleme auf der Station und bei der Krankheitsverarbeitung zu besprechen.

Einmal wöchentlich findet auf dem Stationsflur ein „Patientencafé" statt. Wir versuchen auch hierdurch die informelle Kontaktaufnahme zu unterstützen. Die Patienten kommen zu dieser Veranstaltung besonders gerne.

Die Fort- und Weiterbildung der Mitarbeiter der Station

Die Ärzte der Station befinden sich zugleich in internistischer und psychoanalytischer Weiterbildung.

Die auf der Station tätigen Schwestern haben in der Mehrzahl an dem in Ulm institutionalisierten einjährigen „Vollzeitweiterbildungskurs für Krankenschwestern und -pfleger in patientzentrierter Pflege/Psychosomatischer Medizin" teilgenommen. Dieser Kurs besteht zur Hälfte aus einem Praktikum, das auf der geschilderten Station durchgeführt wird (Köhle et al., 1977).

Bisherige Ergebnisse des Modellversuches

Erfahrungen mit dem Stationskonzept

Die siebenjährigen Erfahrungen mit dem dargestellten Stationskonzept haben bei der Versorgung von insgesamt ca. 2000 Patienten gezeigt, daß es in einem solchen Rahmen möglich ist, den psychosomatischen Arbeitsansatz in die internistische Krankenversorgung zu integrieren. Alle Kranken können von Anfang an auch unter psychosomatischen Gesichtspunkten, d.h. gleichzeitig internistisch und psychotherapeutisch untersucht werden.

Psychotherapeutische Maßnahmen können frühzeitig, und eng mit der internistischen Behandlung verbunden, eingeleitet werden. Neben psychotherapeutischen Gesprächen fördert in dem dargestellten Stationskonzept der Umgang mit den Mitarbeitern im Rahmen aller dargestellten „Veranstaltungen" und bei informellen Gelegenheiten die Bildung eines Bewußtseins von Konflikten und Beziehungsproblemen im psychosozialen Bereich. Dieser Prozeß ist bei vielen Kranken wichtig, um ihnen eine psychotherapeutische Behandlung überhaupt zugänglich machen zu können. Das Stationsmilieu bildet dabei einen schützenden und Halt bietenden Raum, der umschriebene Regressionsprozesse als Voraussetzung einer inneren Umstrukturierung oder Wandlung fördert; er ermöglicht zugleich ein konfrontierendes Vorgehen. Den Patienten können Hinweise auf auch für sie beobachtbare Zusammenhänge zwischen bestimmten Situationen und dem Auftreten von Symptomen gegeben werden. Bei schwer gestörten, Ich-schwachen psychosomatisch Kranken können bedrohliche Dekompensationsvorgänge in diesem Milieu durch das Angebot zuverlässiger Objektbeziehungen und Ich-stützender Maßnahmen aufgefangen werden. Ein derartiges Stationsmilieu stellt meines Erachtens eine conditio sine qua non für eine längerfristige intensive „supportive" Therapie von Patienten mit schweren psychosomatischen Erkrankungen, wie etwa schweren Formen von Colitis ulcerosa, dar.

Das Problem der ambulanten Nachbehandlung konnten wir im Rahmen des Stationskonzepts nur zum Teil lösen; nur ein kleiner Teil der Patienten konnte durch Mitarbeiter der Station längerfristig nachbehandelt werden, zum Teil war eine Überweisung der Patienten an niedergelassene Psychotherapeuten oder an solche Ärzte möglich, die sich in Balint-Gruppen psychotherapeutisch weiterbildeten. Wir bemühten uns darum, die Zusammenarbeit mit solchen niedergelassenen Ärzten zu intensivieren, um die Isolation unseres Modellversuchs im medizinischen Versorgungssystem zu vermindern.

Für die **Aus- und Weiterbildung** hat sich die Station insofern bewährt, als hier die sonst nur theoretisch vermittelbaren Arbeitskonzepte der klinischen Psychosomatik in konkreter Tätigkeit eingeübt werden können. Dies galt sowohl für uns als Psychosomatiker als auch für solche Kollegen, die im Rahmen einer Rotation bzw. als Medizinalassistenten auf der Station mitarbeiteten, sowie für die beteiligten Psychologen, Krankenschwestern und Sozialarbeiter.

Für die klinisch psychosomatische **Forschung** bot die Station die Möglichkeit, Krankheitsbilder und Therapieansätze unter kontrollierbaren Bedingungen systematisch zu unterscheiden und darüber hinaus im Sinne einer „Aktionsforschung" neue Behandlungskonzepte zu erproben und gleichzeitig begleitend wissenschaftlich zu untersuchen.

Die **Grenzen unserer Arbeitsmöglichkeiten** wurden vor allem durch den Stand der psychotherapeutischen sowie der internistischen Weiterbildung der ärztlichen Mitarbeiter bestimmt. Die Schwierigkeit, somatisch und psychosomatisch Kranke für eine Anerkennung der Bedeutung psychosozialer Faktoren am Krankheitsgeschehen und eventuell für eine psychotherapeutische Behandlung zu gewinnen, die Notwendigkeit, psychotherapeutische Interventionen flexibel den situativen Gegebenheiten anzupassen, die Kürze der zur Verfügung stehenden Zeit sowie die oft schwierige Überschaubarkeit der Gruppenprozesse stellten besonders hohe Anforderungen an die Weiterbildung; hinzu kamen die ebenso berechtigten Anforderungen aus internistischer Sicht. Dies bedeutete für die meist am Beginn ihrer Weiterbildung stehenden Stationsärzte eine enorme zeitliche und emotionale Belastung, vor allem auch hinsichtlich der Integration der verschiedenen Gesichtspunkte zu einer eigenen beruflichen Identität.

Kasuistische Beispiele[21]

Um die therapeutischen Möglichkeiten der Station zu illustrieren, habe ich drei Beispiele ausgewählt:

Einer 50jährigen Patientin mit Hypertonie (Patientin A) fiel es schwer, ihre Erkrankung und die Folgen zu verarbeiten und ein angemessenes Krankheitsverhalten zu entwickeln; gleichzeitig ergaben sich auch

---

21 Hiltrud Bosch, M. Rassek und K.-H. Schultheis waren an der Behandlung und Darstellung dieser Patienten besonders beteiligt.

Hinweise dafür, daß diese Schwierigkeiten Einfluß auf den Verlauf der Erkrankung („essentielle Hypertonie") hatten.

Bei Patientin C, einer 15jährigen Kranken mit einem angeborenen, inoperablen Herzvitium, konnte eine bis dahin nicht mehr für möglich gehaltene Verbesserung des körperlichen Zustandsbildes und der Rehabilitation erreicht werden.

Bei Patient D, einem 58jährigen Leukämiekranken, erfuhren wir, welche Bedeutung auch bei Todkranken die Beteiligung des Patienten am Entscheidungsprozeß über die Fortführung einer als „aussichtslos" erlebten Behandlung für die noch verbliebenen Rehabilitationsmöglichkeiten und auch den Krankheitsverlauf selbst haben kann.

Ein vierter Patient, Herr B, mit funktionellen abdominellen Beschwerden, bei dem es um die schwierige Einleitung einer konfliktbearbeitenden Psychotherapie ging, wurde in Abschnitt 28.1 dargestellt.

Patientin A

Die 50jährige Kranke wurde wegen einer ambulant nicht mehr behandelbaren Hypertonie aufgenommen. Die Hypertonie war erstmals vor fünf Jahren im Zusammenhang mit der Operation eines Bandscheibenvorfalls aufgefallen, nach sorgfältiger Untersuchung war sie als „essentielle Hypertonie" diagnostisch eingeordnet worden. In der Zwischenzeit mußte Frau A. zweimal wegen apoplektischer Insulte und einer zusätzlichen Blutung in die Sehrinde stationär behandelt werden. Eine Behinderung beim Lesen (mnestische Störung) und ein Gesichtsfeldausfall (homonyme Hemianopsie) sind als Folge dieser Ereignisse zurückgeblieben. Vor sechs Monaten war nach schweren Blutungen eine Uterusexstirpation durchgeführt worden. Die jetzige Aufnahme erfolgte, nachdem die Patientin nach einer hypertonen Krise wieder kurzzeitig bewußtlos geworden war.

In den Erstgesprächen wirkte die große, massige Frau unruhig; sie ließ den Arzt kaum zum Sprechen kommen, stellte ihre Erklärungen aller Beschwerden in den Vordergrund. Der Arzt fühlte sich eher ängstlich, „wie vor einem Dampfkessel, der gleich explodieren könnte". Die Patientin klagte indirekt über die extreme Einschränkung in allen Lebensbereichen, die nur zum Teil auf die körperlichen Behinderungen zurückführbar ist. Seit der Uterusexstirpation hat sie sich weiter zurückgezogen; auch die sexuellen Beziehungen zum Ehemann hat sie abgebrochen. Jetzt verlasse sie kaum mehr das Haus, weil sie immer noch unter Narbenschmerzen leide.

In den Gesprächen wird ihre Enttäuschung über den Verlust von Selbständigkeit, Leistungsfähigkeit, Unabhängigkeit und die trotzig-depressive Form ihres Rückzugs deutlich. Ihre jetzige Situation charakterisiert sie im Kontrast zu ihrer Lebensgeschichte: In der CSSR aufgewachsen – sie betont ihre Zweisprachigkeit – ernährte sie nach dem Krieg als Lehrerin den verletzt zurückkehrenden Mann, den sie „aus Fairneß", „trotz seiner Verletzungen", geheiratet habe. Sie gebar vier Kinder; aus Zeitgründen konnte sie ihren Beruf nicht mehr ausüben. Später übernahm sie die Leitung einer Lebensmittelfiliale; sie betont, wie sie sich damals auf ihr phänomenales Gedächtnis verlassen konnte: Sämtliche Bilanzen habe sie sich nach einmaligem Lesen merken können. Invalidität nach der Bandscheibenoperation und Berentung führten zum Bruch in dieser Entwicklung.

Im Zusammenhang mit der jetzigen Erkrankung klagt sie über heftige Kopfschmerzen, „ein Druckgefühl"; sie befürchtet dabei, ihr „Röhrensystem" könne gleich platzen.

Die Schwierigkeiten im Krankheitsverhalten der Patientin stehen in enger Beziehung zu den Schwierigkeiten in der Krankheitsverarbeitung. Von den Ärzten ist sie enttäuscht, sie hat das Gefühl, diese könnten ihr auch nichts anbieten als „einen Tod auf Raten". Ein tragfähiges Arbeitsbündnis konnte sie nicht eingehen: Im Gespräch ergibt sich jetzt, daß sie die verordneten Medikamente nicht eingenommen hat; rationalisierend gibt sie als Grund die auf den Packungen angegebenen Nebenwirkungen an; sie hat es abgelehnt, sich den Blutdruck selbst zu messen – und fährt in dieser trotzigen Ablehnung auch zu Beginn des stationären Aufenthaltes fort.

Auf der Station werden allmählich Zusammenhänge zwischen bestimmten Situationen und dem krisenhaften Ansteigen des Blutdrucks sichtbar. Bei der Visite klagt die Patientin darüber, daß ihr Blutdruck gleich auf 220 mm Hg angestiegen sei, als auf dem Stationsflur ein älterer bewußtloser Patient an ihr vorbeigefahren worden sei. Sie habe angenommen – was zutraf –, daß der Mann einen Schlaganfall erlitten habe. Dabei erinnert sie sich an ihren Vater, der an einem Schlaganfall als Folge einer Hypertonie gelähmt habe liegen müssen und schließlich daran gestorben sei. „Dann ist es doch besser, gleich tot zu sein." Während der ersten Visiten fällt regelmäßig eine stark zunehmende Gesichtsrötung der Patientin auf, die sich angestrengt darum bemüht, Verständnisversuche für den situativen Kontext ihrer Symptome abzuwehren, „nichts auf sich sitzen zu lassen". In der sich anschließenden Visite klagt die Patientin über ihren „rebellierenden Magen". Der Arzt versucht eine Beziehung zwischen ihrem Gesamtverhalten und dieser Schilderung des Magens herzustellen, die Patientin fühlt sich grob mißverstanden. Sie schreibt dem Arzt einen Brief, in dem sie ihm vorwirft, ihre Äußerungen „zu wörtlich" zu nehmen. In dem sich anschließenden Gespräch kann sie erstmals über ein Gefühl der Ohnmacht während der Visite sprechen: Es falle ihr schwer, sich zu konzentrieren, sie stehe unter dem Druck, immer eine „intelligente Antwort" parat haben zu müssen, und fühle sich den Ärzten unterlegen.

Allmählich kommen auch ihre familiären Beziehungen ins Blickfeld: Ein weiterer krisenhafter Blutdruckanstieg tritt auf, als sich die Patientin mit einer jüngeren Mitpatientin, von Beruf Kindergärtnerin, unterhält. Wir erfahren dann von ihr, sie leide darunter, daß sie vor einem halben Jahr auch „ihre jüngste Tochter" habe „hergeben" müssen, die sich als Kindergärtnerin ausbildet. Die Patientin beschreibt sich selbst als „Löwenmutter" („Supermutter"), sie weist dabei darauf hin, daß sie all ihre Kinder eineinhalb Jahre lang gestillt habe. Heute kämen die Kinder noch mit allen Problemen zu ihr.

Eine Veränderung tritt nach einer Chefarztvisite ein. Frau A. hatte darüber geklagt, daß sie früher alle Telefonnummern im Kopf gehabt habe und sich jetzt kaum noch eine einzige merken könne. Die Antwort des Chefarztes, er wisse beispielsweise auch nicht, wann Napoleon geboren sei, so etwas könne man ja in Büchern nachlesen, imponiert ihr sehr. Sie geht auf den Vorschlag einer Schwester ein, sich vor dem Telefonat die Nummer aufzuschreiben. Allmählich scheint sie die hohen Idealvorstellungen, die sie erfüllen zu müssen meint, in einem Trauerprozeß reduzieren zu können. Wohl, weil sie erlebt, daß die Umwelt sie auch als Kranke achtet, beginnt sie, sich selbst wieder mehr zu akzeptieren. Jetzt kann sie auch eine Leistung ihrer Tochter annehmen und voll Stolz weiterberichten; dabei meint sie, sie könne diese Leistung anerkennen, ohne wie früher alle komplizierten Einzelheiten selbst verstehen und eventuell kritisieren zu müssen. Darauf müsse sie jetzt verzichten, ihr eigenes Wohlbefinden ginge ihr jetzt vor.

Der bisher auch in der Klinik nur schwer einstellbare Blutdruck sinkt. Die Abwehrhaltungen von Frau A. bilden sich zurück, während der Visite tritt auch die Gesichtsrötung nicht mehr auf. Sie meint schließlich, jetzt wisse sie selbst, wie es zu Hause wieder weitergehen könne. Sie müsse eben bestimmten belastenden Situationen ausweichen, insgesamt sich mehr um sich selbst kümmern. Sie hatte auch begonnen, selbst ihren Blutdruck zu messen. Wir wissen noch nicht, ob der Krankheitszustand hierdurch längerfristig ausreichend behandelbar geworden ist; das Beispiel läßt unserer Ansicht nach jedoch die Bedeutung der Beziehung zwischen Patient und medizinischer Umwelt – und die stellvertretende Funktion der medizinischen Umwelt für die alltägliche Mitwelt – für das Gelingen der Krankheitsverarbeitung und für die Entwicklung eines angemessenen Krankheitsverhaltens erkennen.

Patientin C

Die 15jährige Kranke leidet an den Folgen eines angeborenen, inoperablen Herzvitiums (Cor triloculare): pulmonale Hypertonie, AV-Block I. Grades mit anfallsweise auftretenden Kammerextrasystolen; häufig treten Synkopen auf. Die Patientin wurde seit der frühen Kindheit in einem Herzzentrum behandelt, nach der Ablehnung einer Operation waren mit Hilfe von Massenmedien Mittel für die Untersuchung in der Mayo-Klinik gesammelt worden; dort hatte sich eine Operation ebenfalls als undurchführbar erwiesen.

Die Patientin war jetzt nach einer besonders lang andauernden Synkope stationär aufgenommen worden; die Angehörigen hatten sie während dieses Ereignisses bereits als „tot" erlebt, der zufällig hinzukommende Hausarzt hatte sie „reanimiert".[22] In der Klinik wurde die Prognose als hoffnungslos eingeschätzt. Die Eltern, von der langen Krankheit und dem Miterleben der sich wiederholenden Anfälle von Bewußtlosigkeit zermürbt und durch ihre Hilflosigkeit verstört, weigerten sich, überhaupt noch über die Möglichkeit nachzudenken, daß sie ihre Tochter noch einmal nach Hause nehmen könnten. Die Patientin blieb „aufgegeben", „zum Sterben" in der Klinik, um ihr selbst, aber auch ihren Eltern weitere unzumutbare „Belastungen zu ersparen". Als sie von der Aufnahmestation auf unsere Station verlegt wurde, wirkte die Kranke außerordentlich zerbrechlich, ängstlich, anklammernd, hochgradig „infantil". Sie weckte Assoziationen an Klara im Roman Heidi, als Inkarnation von Schwäche, Zartheit und Hilflosigkeit: Die Anamnese ergab, daß sie in totaler Abschirmung aufgewachsen war, von Hauslehrern unterrichtet wurde und bis jetzt keinerlei Kontakt zu gleichaltrigen Jugendlichen hatte aufnehmen können.

Im Verlauf von vierzehn Tagen gelang bei gleichzeitiger entsprechender medikamentöser Therapie, vorsichtiger krankengymnastischer Arbeit und allmählicher Reduzierung der Sauerstoffzufuhr über die Na-

sensonde eine Teilmobilisierung der Kranken. Während dieser Zeit fiel allen Beteiligten immer stärker das unauffällige, zurückgezogene und schüchterne Verhalten der Kranken auf. Zunehmend gewann für sie die Beziehung zu ihrer Bettnachbarin, einer zehn Jahre älteren „ausbehandelten" Leukämiekranken mit hoffnungsloser Prognose, Bedeutung. Diese Kranke war trotz der Schwere ihres Zustandsbildes noch sehr aktiv, sie vermochte die Aufmerksamkeit von Schwestern und Ärzten stets auf sich zu lenken. Als sich mit der rapiden Verschlechterung ihres Befindens die Zuwendung aller Beteiligten ihr gegenüber noch einmal verstärkte, kam es gleichzeitig zu einer erheblichen Verschlechterung des Zustands von Frl. C. Es traten schwerste rezidivierende Rhythmusstörungen mit Zeichen zerebraler Hypoxie, Benommenheit, Amnesie und Krämpfen auf; eine Schiefhalsstellung – der Kopf war von der Mitpatientin abgewandt – blieb längere Zeit bestehen. Jetzt erst fielen panische Angstzustände auf, die die Verschlechterung begleiteten. Auf Drängen der „psychosomatischen Schwester" wurde in den Besprechungen erstmals ein möglicher Zusammenhang zwischen der beobachteten Angst und dem Auftreten von Rhythmusstörungen besprochen und bald auch allgemein über die Zusammenhänge von psychischer und sozialer Situation mit den verbliebenen physiologischen Kompensationsmöglichkeiten diskutiert. Die Patientin wurde in ein anderes Zimmer verlegt, die Mitarbeiter der Station begannen sich intensiv mit ihr zu beschäftigen. Es kam zu einer überraschend schnellen Besserung der Symptomatik. Die Leukämiepatientin starb, Frl. C. hatte über Mitpatienten davon erfahren, während das Stationsteam noch zögerte, mit ihr darüber zu sprechen. Zum Zeitpunkt der Beerdigung der Mitpatientin zündete sie Kerzen an; dabei konnte sie erstmals über ihre Gefühle für die Leukämiekranke und dann auch über ihre auf den eigenen Tod bezogenen Ängste sprechen; jetzt erst begann sie, zunächst die Schwestern und später auch die Ärzte über das Wesen ihrer eigenen Erkrankung und das Schicksal anderer Herzkranker auszufragen. Erstmals wurde dabei deutlich, daß die Patientin trotz ihrer langen Leidensgeschichte kaum Informationen über Art und Prognose ihrer Erkrankung erhalten hatte oder entsprechend den Spielregeln ihrer Umgebung das in Erfahrung Gebrachte weitgehend hatte verleugnen müssen: In ihrem Erleben waren Schweigen und Abriegelung vor Problemen und Ängsten bisher die Reaktionen der Erwachsenen gewesen.

Auf der Station bemühten wir uns nun darum, die nur ansatzweise vorhandenen Autonomiebestrebungen der Patientin zu unterstützen. Den Angehörigen fiel bald eine „Verwandlung" ihrer Tochter auf: Während sie früher von ihrer Mutter als ängstlich, kontaktarm und unfähig zur Formulierung eigener Gedanken erlebt worden sei, würde sie jetzt öfter wie eine Erwachsene wirken und an Selbständigkeit gewinnen. Überraschenderweise erwies es sich trotzdem als außerordentlich schwierig, den Widerstand der Familie gegen eine wieder möglich erscheinende Entlassung aus dem Krankenhaus zu überwinden. Die völlig verängstigte Mutter benützte dabei sogar eigene Beschwerden als Druckmittel, um die Patientin im Krankenhaus zu halten. Erst nachdem auch mit den Eltern ein Arbeitsbündnis aufgebaut werden konnte, ließ sich das therapeutische Ziel erweitern. Die Patientin sollte

---

22 Die Angaben „tot" und „reanimiert" stammen von den Angehörigen bzw. der Patientin.

unabhängiger vom Sauerstoffgerät werden, eine intermittierende Sauerstoffzufuhr wäre ja auch zu Hause möglich. Der Entwicklung autonomer Verhaltensweisen sollte die Entwicklung der Fähigkeit zur Aufnahme sozialer Kontakte entsprechen. Frl. C. wurde es allmählich möglich, an der wöchentlichen Patientengruppe und an Spielen mit anderen Patienten teilzunehmen. Es befriedigte sie dabei besonders, daß die anderen Patienten sie nicht – wie früher in der Familie üblich war – in den Spielen regelmäßig gewinnen ließen; sie registrierte es vielmehr mit Genugtuung, als „Todkranke" nicht nur verhätschelt zu werden, und lernte auch, solche und ähnliche Gefühle und Erlebnisse zu verbalisieren. Eine Gruppe von Studenten beteiligte sich schließlich an der Betreuung der Patientin und setzte diese Betreuung auch nach ihrer Entlassung fort: Die Studenten besuchten die Patientin daheim und nahmen sie auf kleinere Ausflüge mit. Der Patientin gelang es tatsächlich, zu Hause von zusätzlicher Sauerstoffzufuhr weitgehend unabhängig zu werden. Getragen durch den Schutz der Familie, vermag sie in begrenztem Umfang eigene Aktivitäten zu entfalten: Sie übernimmt schriftliche Arbeiten für eine religiöse Gruppe, der sie angehört, und konnte mehrfach an bis zu zweiwöchigen Ferienlagern teilnehmen. Sie beteiligte sich an Überlegungen zu einer etwaigen eigenen Berufstätigkeit. Fräulein C. lebte noch 4½ Jahre.

Patient D

Herr D., ein 58jähriger Maurer, weiß seit zwei Jahren, daß er Leukämie hat. Er kam bereits in kurzen Abständen ambulant zur Kontrolle und mußte dann wegen eines tiefergehenden entzündlichen Prozesses im rechten Sprunggelenk stationär aufgenommen werden. Nach zweimonatiger, erfolgloser Behandlung mußte Herr D. unterhalb des Kniegelenkes amputiert werden. Während der Operation kam es zu einem Herzstillstand mit anschließender erfolgreicher Reanimation. Er schien sich erstaunlich leicht mit seiner Amputation abzufinden und hoffte baldmöglichst so weit zu sein, sich mit Hilfe einer Prothese wieder fortbewegen zu können. Vielleicht half ihm der Gedanke an seinen Schwiegersohn, der, ebenfalls amputiert, sich wieder gut bewegen gelernt hatte.

Im Gegensatz zu solchen Hoffnungen verwirrte er allerdings durch Äußerungen, deren Bedeutung für die Schwester schwer verständlich war. Als sie dabei ist, ein Bild an die Wand zu hängen, schlägt er ihr z. B. vor, doch die Wasserwaage zu nehmen, sie liege ja neben ihr. Oder er sagt, er fühle sich längst tot, aber auch lebendig. Näher befragt, erzählt er dann von seinem Haus, das er als Maurer kurz vor seiner Einlieferung ins Krankenhaus fertiggestellt hatte. Diese Zustände der Verwirrung werden als Folge der Hirndurchblutungsstörungen während der Reanimation verstanden und angesichts seines wenig hoffnungsvollen Zustandes auch vom Arzt eher positiv erlebt und bewertet. Die Unfähigkeit, seine Phantasiewelt zu ordnen, schützt den Patienten gleichzeitig vor der Einsicht in die Ernsthaftigkeit seiner Situation. Dies akzeptiert der Arzt bereitwillig als gütige Laune des Schicksals.

Nachdem sich diese Verwirrungszustände allmählich verlieren und Herr D. sich für seine Prognose zu interessieren beginnt, äußert der Arzt besorgt gegenüber der Schwester, die sich über diese Entwicklung erfreut zeigt: „Der wird immer klarer; das ist schrecklich! – Ob der das bis zum Ende durchhält?" Die Schwester fragt sich ironisch, ob sich der Arzt nicht vielleicht selbst damit meint.

Trotz intensiver Bemühung verschlechtert sich der Zustand des Patienten, so daß eine Heilung der Wunde ausgeschlossen scheint und jede medizinische Therapie abgesetzt wird. Herr D. fragt nicht nach den Gründen, klagt selten über Schmerzen und hat allerlei kleine Wünsche, deren Erfüllung er zu genießen versteht. Er füllt den kleinen Lebensbereich aus, freut sich an seinem frisch gemachten Bett, am Obst, seinem Fläschchen Bier. Manchmal spricht er davon, daß es „wohl nichts mehr mit ihm werde". Er stellt sich vor, daß er so einfach einschläft, das Blut immer weniger wird, und vermittelt eine müde Traurigkeit. Er spricht davon, daß sein Haus fertig ist, die Familie eingezogen ist, während er bereits im Krankenhaus liegt, und erweckt damit den Eindruck, er sei bereit, mit dem Leben abzuschließen.

Bei der Visite verfolgt Herr D. aufmerksam das Gespräch der Ärzte und fragt auf die Bemerkung, daß sich sein Blutbild erstaunlich gebessert habe, ob ihm eine Bluttransfusion nicht guttun würde. Daraufhin läuft die Therapie wieder an: Bluttransfusionen, Antibiotika. Der Amputationsstumpf zeigt zwar keine Heilungstendenz, aber der Patient fühlt sich wieder wohler und nicht mehr so müde. Eines Tages äußert er, die Chirurgen hätten davon gesprochen, sein Bein noch einmal zu amputieren, aber sie wollten ihm die Entscheidung überlassen. Er wird nicht mehr darauf angesprochen, schiebt eine Entscheidung vor sich her, bis eine akute Blutung den plötzlichen Eingriff veranlaßt. In letzter Minute unterschreibt er die Einwilligung.

Nach der zweiten Amputation schöpft er neue Hoffnung. Er sieht sich mit zwei Krücken am Waldrand spazierengehen, später reduziert er die Vorstellung darauf, wenigstens allein, d. h. ohne Hilfe, in sein neugebautes Haus gehen zu können. Das Krankheitsgeschehen scheint stehenzubleiben. Schließlich schwankt der Patient zwischen der Hoffnung, noch einmal nach Hause zu kommen, und dem Wunsch, endlich sterben zu können. Einer Schwester gegenüber äußert er, es sei doch eigentlich nichts dabei, wenn er seinen Zustand mit Hilfe einer Überdosis Schlaftabletten entscheide.

In der Therapie läßt er willenlos alles mit sich geschehen. Inzwischen hat sich am Ohr eine Phlegmone gebildet. So bemühen sich abwechselnd die Chirurgen um den Stumpf, die Hämatologen um das Blutbild, der HNO-Arzt um das Ohr, und der Stationsarzt kommt täglich zur Visite. Einige Schwestern haben das Gefühl, daß keiner der Ärzte eigentlich die Initiative übernehmen will, eine gemeinsame therapeutische Konzeption zu entwickeln. Die Schwestern wissen nicht, auf welcher Seite sie stehen sollen: auf der Seite des Arztes, dessen Anweisungen zum Teil für fragwürdig gehalten werden, oder auf der Seite des Patienten, über dessen Ohnmacht sie betroffen sind.

In der Stationsgruppe diskutieren die Schwestern immer wieder den Sinn ihres Tuns: „Soll man nicht endlich die Therapie abbrechen?" – „Nun haben wir wieder angefangen, jetzt müssen wir auch weitermachen!" – „Welchen Sinn hat sein Leben, wenn er nur noch im Bett existieren kann?" – „Soll man ewig so weiterbehandeln?" – „Wir können aber doch nicht einfach gar nichts mehr tun!" – „Wir dürfen doch die Hoffnung nicht aufgeben. Wer kann schon entscheiden, wann es Zeit ist zu sterben?"

In dieser Diskussion, in der der Arzt dafür plädiert, die Therapie fortzusetzen, weil er meint, dem Willen des Patienten damit zu entsprechen, die Schwestern jedoch diese Meinung nicht teilen können, wird deutlich, daß der Patient in diesem Entscheidungsprozeß eine wesentliche Rolle spielen müßte. Es wird auch klar, daß man ihn nicht mehr, wie im Falle der Zweitamputation, mit der Entscheidung alleinlassen kann. Fast selbstverständlich kommt ihnen der Entschluß vor, dem Patienten in einem Gespräch mit Arzt und Schwester Zeit einzuräumen, in der er sich mit dem Für und Wider einer Therapie auseinandersetzen kann. In dieser Situation könnte er sein Informationsbedürfnis stillen und seine Ängste und Hoffnungen äußern.

Im anschließenden Gespräch zwischen Arzt, Herrn D. und der Schwester kann der Kranke das Angebot zur Mitentscheidung zunächst nicht annehmen. In seiner Frage an den Arzt: „Was würden Sie denn an meiner Stelle tun?" zeigt sich sein Wunsch, die Entscheidung von sich zu schieben. Der Arzt kann diese Frage nicht beantworten, worauf Herr D. sehr betont davon spricht, was er auf keinen Fall für seine Zukunft wünsche. So sieht er z. B. keinen Sinn darin, sich als pflegebedürftiger Krüppel zu Hause von seinen Angehörigen das Essen am Bett servieren zu lassen. Seine Hoffnung besteht eigentlich nur noch darin, sich mit Hilfe seiner Krücken in seinem eigenen Haus bewegen zu können. Am Ende des Gesprächs meint er dann: „Nun, dann versuchen wir's halt nochmal."

Der Patient, der bisher alle Rehabilitationsmaßnahmen abgelehnt hatte, begann nun mit der Krankengymnastin zusammenzuarbeiten und übte bald selbständig mit den zur Verfügung stehenden Geräten. Nach drei Wochen begann er mit Krücken zu gehen. Die Rehabilitation gelang weitgehend. Der Patient konnte noch in sein neues Haus einziehen, er lebte noch eineinhalb Jahre. Die in dreiwöchigen Abständen erforderlichen Bluttransfusionen wurden meist ambulant auf der Station durchgeführt; hierbei ergaben sich Möglichkeiten zu weiteren Kontakten und Gesprä-

chen. Für uns war erstaunlich, daß die Wunde am Amputationsstumpf und die eitrige Otitis, die bis dahin keine Heilungstendenz gezeigt hatten, nach der Entscheidung des Patienten, weiterbehandelt werden zu wollen, ohne weitere Veränderung der Therapie abzuheilen begannen. Inwieweit hier ein „psychosomatischer" Zusammenhang vorliegt, vermögen wir nicht abschließend zu beurteilen.

Ergebnisse der Begleitforschung: Ärztliches Interaktionsverhalten während der Visite

Wir stellen diese Ergebnisse in Kapitel 18 dar. Einzelfallanalysen finden Sie in Bliesener und Köhle, 1985.

Hier schließen wir mit der spontanen Äußerung einer 29jährigen Patientin mit Hirnmetastasen nach Mammakarzinom, die sich im terminalen Krankheitsstadium befand:

„Also auf der Station ist es ganz anders wie auf den anderen Stationen, wo ich eben war. Das Personal ist freundlicher und man kann Wünsche äußern und die Schwestern sind ganz anders, finde ich, die sind, die sprechen mit einem über die Krankheit und des, also ich fühle mich nachher immer erleichtert ein bißl, sonst könnte man das vielleicht manchmal fast gar nicht schaffen. Also mir geht es so. Bin da froh darüber, daß man sich aussprechen kann und daß einem nicht alles so verschwiegen wird, das finde ich auch, das find ich, ist auf anderen Stationen so nie gesagt worden, da konnte man auch keine Schwester fragen über seine Krankheit, da hat man nie Auskunft bekommen so und das ist eben hier, finde ich ganz schön, wenn man mit jemandem reden kann."

29 Die Bedeutung psychosozialer Faktoren für den Medizinischen Dienst der Krankenversicherung (ehemals Vertrauensärztlicher Dienst der Landesversicherungsanstalten)

Rüdiger Großpietzsch und *Doris Schmitt*

29.1 Gesundheitsreformgesetz, V. Sozialgesetzbuch, IX. Kapitel: Medizinischer Dienst der Krankenversicherung

Im Rahmen des Gesundheitsreformgesetzes (GRG), das am 01. 01. 1989 in Kraft trat, ist der bisherige Vertrauensärztliche Dienst von den Landesversicherungsanstalten abgelöst, gesetzlich neugeordnet und in die Trägerschaft der gesetzlichen Krankenversicherung überführt worden. Diese fünf gesetzlichen Krankenkassen, AOK (Allgemeine Ortskrankenkassen), BKK (Betriebskrankenkassen), IKK (Innungskrankenkassen), LKK (Landwirtschaftliche Krankenkassen) und VDAK (Verband der Angestellten-Krankenkassen), entsenden dazu je Kassenart 2 Mitglieder in den gemeinsamen Verwaltungsrat, der höchstens 16 Mitglieder umfassen soll. Später soll der Medizinische Dienst eine Arbeitsgemeinschaft werden.

Dem neu konzipierten Medizinischen Dienst (MDK) werden neben den bisherigen Begutachtungsaufgaben des Vertrauensärztlichen Dienstes (§ 275, 1–3) zusätzlich neue Beratungstätigkeiten der Krankenversicherung in grundsätzlichen Fragen im Gesundheitswesen aufgetragen (§ 275,4). Der Gesetzestext im GRG ist im IX. Kapitel: Medizinischer Dienst der Krankenversicherung, 1. Abschnitt, Aufgaben, § 275 Begutachtung und Beratung, nachzulesen.

Die vom Gesetzgeber vorgegebenen Aufgaben umfassen die Einzelfallbegutachtung mit sozialmedizinischer Exploration und körperlicher Untersuchung eines zu begutachtenden Patienten entweder in einer Untersuchungsstelle des Medizinischen Dienstes der Krankenversicherung, wie das die Regel ist, oder aber auch im Rahmen eines Hausbesuches zur Würdigung des häuslichen Umfeldes (z. B. zur Begutachtung von Schwerstpflegebedürftigkeit), am Arbeitsplatz (z. B. zur Erstellung eines arbeitsmedizinischen Leistungsbildes) und nicht zuletzt anläßlich einer Krankenhausbegehung mit interdisziplinärer Begutachtung eines stationären Behandlungsfalles (z. B. im Rahmen einer Stationskonferenz).

Für weitere, besonders übergreifende Beratungsaufgaben wird die Heranbildung einiger hochspezialisierter „Medizinmanager" erforderlich sein. Medizinisches Management wird die Zusammenführung von ärztlichem Sachverstand und gesundheitsökonomi-scher Kompetenz zur Grundlage haben müssen, wobei als Grundbedingung für dieses neu konzipierte Spezialgebiet – im Unterschied zum Management in der Wirtschaft – zu fordern ist, daß bei Anwendung gesundheitsökonomischer Steuerungsmittel ärztliche Ethik und medizinische Qualität ebenso gewahrt bleiben wie soziale Gerechtigkeit für die Solidargemeinschaft der Versicherten (Großpietzsch und Ihmann, 1988).

29.2 Das psychosoziale Defizit

Da das Problemlösungspotential der überwiegend naturwissenschaftlich ausgerichteten Medizin weitgehend ausgeschöpft ist, liegt die gewichtigste medizinische Problemstellung unserer Zeit auf psychosozialem Gebiet (Anschütz, 1988). Matern (1988) beklagt, daß auch fast alle ausbildenden Polikliniken zu hochspezialisierten Einheiten geworden seien, die dann junge Ärzte so spezialisiert in die Praxis entlassen würden. Diese Praxis habe in den letzten Jahrzehnten jedoch tiefgreifende Veränderungen erfahren, sowohl den somatischen als auch den psychischen Bereich betreffend. Schepank (1988) hat in einer Großstudie einer Großstadtpopulation nachgewiesen, daß 26% dieser „gesunden" Stichproben psychogen erkrankt und behandlungsbedürftig waren (vgl. Kap. 4). Untersuchungen aus Allgemeinpraxen haben ergeben, daß dort durchschnittlich 30% psychogene Erkrankungen im Patientengut vorliegen. Bei den 70% „organisch" Kranken können psychosoziale Faktoren ebenfalls eine Rolle spielen.

29.3 Aufgabenstellung für den Medizinischen Dienst der Krankenversicherung im psychosozialen Bereich

Der Medizinische Dienst der Krankenversicherung erhält von den Leistungsträgern überwiegend Begutachtungsaufträge von „Problempatienten", die eine längere „Patientenkarriere" hinter sich haben. Die festgestellten psychosozialen Defizite gewinnen somit für den sozialmedizinischen Gutachterdienst eine noch größere Dimension. Großpietzsch und Großpietzsch (1986) resümieren ihre Erfahrungen als sozialmedizinische Gutachter in diesem Zusammenhang so, daß, auch nach Würdigung zahlreicher vorliegender apparativer, biochemischer und sonstiger Befunde, das Zwiegespräch mit dem zu begutachtenden Patienten das grundlegende Informationsmittel aller gutachterlichen Erkenntnis bleibt. Dies gilt im besonderen Maße für den Bereich psychischer, psychosomatischer und Suchtkrankheiten, die einen erheblichen Anteil an den Begutachtungsaufträgen ausmachen.

Nur konsequente Hinwendung zum Patienten bei ausreichend ausführlicher Exploration kann für den zu begutachtenden Patienten weiterführende diagnostische, prognostische und nicht zuletzt rehabilitative Ergebnisse erbringen.

29.4 Sozialmedizinische Begutachtung im psychosozialen Bereich

Der Arbeitsbereich einer biopsychosozial ausgerichteten Gutachterstelle des Medizinischen Dienstes der Krankenversicherung muß entsprechend weit ausgelegt sein, da ein nosologisch unausgewähltes Patientengut von den ratsuchenden Leistungsträgern vorgestellt wird. Der vorgestellte Patient steht meist auch nicht am Anfang seiner Patientenkarriere, eine größere Anzahl verschiedenster organfachärztlicher oder auch psychiatrischer Berichte wird interdisziplinär zu würdigen und zu bewerten sein. Vor der interdisziplinären Würdigung sollte – idealerweise – die Integration im Interviewer stattfinden.

Gelingt es, einen auslösenden „Störfaktor" im beruflichen oder privaten Umfeld zu objektivieren, so kann durch gemeinsame Erörterung und Planung seiner zukünftigen Ausschaltung zuweilen ohne weitere medizinische Maßnahmen ein tragfähiges Behandlungskonzept oder besser Selbstbehandlungskonzept gefunden werden. Darüber hinaus hat der MDK eine Leitfunktion in der Rehabilitation vom Gesetzgeber übertragen bekommen, er kann diese zur Hilfestellung für den Patienten bei den Leistungsträgern nutzen.

29.5 Gutachterliches Vorgehen im psychosozialen Bereich

Die Aufgabe des ärztlichen Gutachters im Medizinischen Dienst der Krankenversicherung besteht darin, drei Wirklichkeiten (wenigstens teilweise) zu vereinen:
– die Wirklichkeit des Patienten,
– die Wirklichkeit des begutachtenden Arztes,
– die Wirklichkeit des Sozialversicherungsträgers.
Bekannt sind dem begutachtenden Arzt zwei Wirklichkeiten, seine eigene und die des Sozialversicherungsträgers. Die Wirklichkeit des Patienten muß er

erst kennenlernen. Läßt er sich in einem um echte Kommunikation bemühten Gespräch auf die subjektive Krankheitserklärung des Patienten ein, so erfährt er die subjektive Wirklichkeit des Patienten. Ein gemeinsames Erklärungsmodell muß dann zwischen Arzt und Patient „ausgehandelt" und dem Erklärungsmodell des Sozialversicherungsträgers gegenübergestellt werden. Die Grundlage für das Gutachten wird also von Arzt und Patient in gemeinsamer Strategie erarbeitet.

Exemplarische Krankheitsfälle, die im Medizinischen Dienst der Krankenversicherung begutachtet wurden, sollen nun nach den dargestellten Gesichtspunkten betrachtet werden.

> Ein 43jähriger Patient, der wegen einer schwer einstellbaren Hypertonie und eines stark schwankenden Diabetes mellitus über einen längeren Zeitraum arbeitsunfähig erkrankt war, wird von der Krankenkasse erstmalig zu einer sozialmedizinischen Begutachtung eingeladen.
>
> Der seit 5 Jahren bekannte Diabetes war bisher mit Diät und Tabletten problemlos zu behandeln gewesen. Jetzt schwankte er trotz einer erneuten klinischen Einstellung stark. Ebenso war es weder bei stationärer noch bei ambulanter internistischer Behandlung gelungen, den Hochdruck, der mit hypertonen Krisen einherging, befriedigend einzustellen.

Offensichtlich erlebte dieser Patient, der immer ein fleißiger Arbeiter gewesen war und während seiner Arbeitsunfähigkeit alle Anweisungen der Klinik und seines behandelnden Internisten ernsthaft befolgte, seine längere Arbeitsunfähigkeit schuldhaft und kam der Einladung der Krankenversicherung, sich zur Sicherung des Heilerfolges einer sozialmedizinischen Begutachtung zu stellen, nur widerwillig und aggressiv aufgeladen nach.

Bei solchen Patienten ist es vornehmste Aufgabe des begutachtenden Arztes, sich um eine gute Interaktion mit dem Kranken zu bemühen. Er muß berücksichtigen, daß diese Untersuchung keine vom erkrankten Versicherten gewünschte, sondern eine unfreiwillige Begutachtung ist, die häufig fälschlicherweise als Kontrollmaßnahme verstanden wird.

Nur in einem freundlichen Zwiegespräch kann der Sozialmediziner spüren, daß hinter der ärgerlichen, aggressiv gefärbten Haltung des Patienten auch ein Stück Verzweiflung über seinen therapieresistenten Zustand steckt.

> Bei Erhebung der biographischen Anamnese war zu erfahren, daß er seit 15 Jahren als Elektriker auf einem verantwortlichen Posten in einer kleineren Elektrofirma arbeitete. Seit der Juniorchef vor ca. einem Jahr die Betriebsleitung übernommen hatte, gab es laufend Terminschwierigkeiten. Da der Patient größten Wert auf Qualitätsarbeit legte und nicht ertragen konnte, wenn geschlampt wurde, entstanden Zeitprobleme, die er nicht lösen konnte und die fortwährend zur Konfrontation mit einem tobenden Juniorchef führten.
>
> Darüber hinaus bestanden auch familiäre Sorgen. Sein 17jähriger Sohn, dem er mit Mühe eine Lehrstelle

> besorgt hatte, erklärte kurzerhand, er ginge zurück in die Schule, um Abitur zu machen und später moderne Musik zu studieren. Dieser Sohn sei sein Unglück, er säße den ganzen Tag am Klavier und klimpere Jazz.

Bei dieser Schilderung wird deutlich, daß Stimmungen ansteckend sind. Sie können Wurzeln der Gemeinsamkeit zwischen Arzt und Patient sein, aus der dann die Fähigkeit entspringt, Worten und Gesten eine Bedeutung zu erteilen. Dies wiederum fließt in die Reaktion des Arztes ein und zeigt dem Patienten, daß er verstanden wird. Die Fähigkeit, diese Gemeinsamkeit zu erleben, bezeichnet man als Empathie (v. Uexküll).

Auf diesem Wege konnte der begutachtende Arzt die Stimmung unterdrückter Wut und Hilflosigkeit des pedantischen, etwas zwanghaften und sozial überangepaßten Handwerkers miterleben. Er konnte empfinden, wie die aggressive Bereitstellung, die sich weder am Arbeitsplatz noch zu Hause entladen durfte, zu einer permanenten, tiefgreifenden Verstimmung geworden war. Dies eröffnete den Zugang zum Patienten. Daraus ließ sich für den Gutachter ableiten, daß der Diagnose der Klinik und des Internisten „Hypertonie und Diabetes mellitus" noch die einer „depressiven Verstimmung in einer ausweglosen Situation" hinzuzufügen war.

Welche Bedeutung hatte nun die depressive Verstimmung für die körperlichen Erkrankungen Diabetes mellitus und Hypertonie? Welche körperlichen Symptome waren als Folge einer unlösbaren Problemsituation bzw. als Streß zu interpretieren?

Aggressive Bereitstellungen gehen mit einer vermehrten Ausschüttung von Katecholaminen einher, die sowohl zur Blutdruckerhöhung führen, als auch eine gegenregulatorische Wirkung gegen Insulin haben. Die somatopsychischen Aufwärtseffekte der Katecholamine bestehen in einer Verstärkung der aggressiven Stimmung, deren psychosomatische Abwärtseffekte dann wieder in einem Circulus vitiosus zu einer weiteren Stimulierung der Katecholaminproduktion führen können. Die Folge ist die weitere Verschlechterung der körperlichen Krankheit. Man kann sich also nicht mit einer medikamentösen Behandlung der Hypertonie und des Diabetes begnügen, sondern muß die intervenierenden Abwärtseffekte der aggressiven Verstimmung mit in Rechnung stellen. Es ist von einer Gesamtdiagnose und nicht von einer Teildiagnose auszugehen (v. Uexküll und Wesiack, vgl. auch Kap. 1).

Im Rahmen der Gesamtdiagnose bekommen Krankheit und Arbeitsunfähigkeit einen anderen Sinn. Hypertonie und plötzlich schwer einstellbarer Diabetes mellitus gaben dem Patienten die Möglichkeit, dem immer größer werdenden Streß an seinem Arbeitsplatz auszuweichen. Um auch der familiären Konfliktsituation mit dem ewig klavierspielenden Sohn zu entgehen, willigte er schnell in den Vorschlag des sozialmedizinischen Gutachters für ein psychosomatisches Heilverfahren ein. Die Krankheit wirkte hier wie ein Überdruckventil: nur durch sie war es dem Patienten möglich, dem Überdruck zu

Hause und am Arbeitsplatz zu entweichen. Erst die Krankheit gab dem übergewissenhaften Mann die Legitimation, die er brauchte, um sein Gewissen zu beruhigen, wenn er sich aus dem zermürbenden Kampf zurückzog. Laut Thure von Uexküll (1963) erhält die Krankheit einen anderen Verteilungsraum als den des festen und sichtbaren Körpers, den die Anatomie beschreibt (v. Uexküll und Wesiack, 1988). Der neue Verteilungsraum besteht aus einer individuellen Wirklichkeit, in welcher in diesem Falle soziale Ereignisse am Arbeitsplatz und in der Familie zwei unlösbare Problemsituationen geschaffen hatten, auf die der Körper des Patienten mit Kreislauf- und Stoffwechsel-Symptomen reagieren konnte.

Im Rahmen einer katamnestischen Nachuntersuchung sah der sozialmedizinische Gutachter diesen Mann, der im damaligen Vertrauensärztlichen Dienst (jetzt Medizinischer Dienst der Krankenversicherung) nie wieder als Arbeitsunfähiger aufgetaucht war, nach 5 Jahren wieder. Im psychosomatischen Heilverfahren hatten die etwas zwanghaften, überangepaßten Vorstellungen, nach denen der Handwerker seine individuelle Wirklichkeit aufgebaut hatte, psychotherapeutisch bearbeitet werden können. Somatisch erfolgte u. a. wegen der möglicherweise erhöhten Katecholaminproduktion als Folge der permanenten aggressiven Erregungszustände neben der Behandlung der Hypertonie und des Diabetes mellitus eine medikamentöse Senkung des Sympathikotonus.

Dem Elektriker ging es jetzt prächtig. Hypertone Blutdruckwerte konnte er nur selten feststellen und auch der Diabetes war gut eingestellt. Er hatte sich nach Rückkehr aus der psychosomatischen Klinik trotz seiner langen Betriebszugehörigkeit um einen Arbeitsplatz bei der Stadtverwaltung beworben und war aus einer großen Anzahl von Bewerbern ausgewählt worden. In seiner neuen Stelle konnte er ohne Termindruck arbeiten und fühlte sich wohl. Mit seinem Sohn verstand er sich bestens. Dieser war zur Zeit beim Musikkorps der Bundeswehr, verdiente nebenbei gutes Geld in einer Band und hatte bereits ein Stipendium für ein Studium moderner Musik am Konservatorium erhalten.

Diese Krankengeschichte zeigt, daß der sozialmedizinische Gutachter zu einer ganzheitsmedizinischen Betrachtungsweise finden muß, um zur Sicherung des Heilerfolges effektiv beizutragen. Unter dem Blickwinkel der Gesamtdiagnose können sowohl die Lebensqualität des Sozialversicherten verbessert als auch dem Sozialversicherungsträger enorme Kosten erspart werden.

Die Wirklichkeit des Patienten

Wegen eines stark schwankenden Diabetes mellitus und therapieresistenter hypertoner Krisen bestand erstmalig eine längerdauernde Arbeitsunfähigkeit. Sowohl stationäre als auch ambulante internistische Behandlung erbrachten keine Besserung. Der schon seit 15 Jahren in einer Firma arbeitende, fleißige, strebsame, auf Qualitätsarbeit bedachte Elektriker hatte nach Übernahme der Firma durch den Juniorchef, der mehr Wert auf Quantität legte, Arbeitsprobleme, die er nicht lösen konnte. Es gab aber auch unlösbare familiäre Konflikte mit dem heranwachsenden Sohn, für den er sich voll verantwortlich fühlte, der sich aber gegen den Vater auflehnte.

Die Wirklichkeit des Gutachters

Der Gutachter begegnet zunächst einem aggressiv aufgeladenen Mann, der sich unfreiwillig, auf Anordnung der Krankenkasse, von der er Krankengeld erhält, einer sozialmedizinischen Begutachtung unterziehen muß. Doch in einem freundlichen Zwiegespräch erkennt der Arzt bald, daß hinter der unterdrückten Wut des pedantischen, zwanghaften, sozial überangepaßten Mannes Hilflosigkeit und Verzweiflung über die ausweglosen Probleme am Arbeitsplatz und in der Familie stehen. Die ständige aggressive Bereitstellung, die sich weder am Arbeitsplatz noch zu Hause entladen durfte, führte zu einer anhaltenden psychischen Anspannung und somatisch vermutlich zu einer vermehrten Ausschüttung von Katecholaminen, was sowohl zu Blutdruckerhöhung führt als auch gegenregulatorische Wirkung auf Insulin hat. Der Gutachter erkannte, daß der Patient nur mit Hilfe seiner Hypertonie und seines entgleisenden Diabetes mellitus die Möglichkeit hatte, dem immer größer werdenden Streß am Arbeitsplatz zu entkommen. Diese Konfliktsituation, die noch durch den häuslichen Konflikt verstärkt wurde, war der Hintergrund für die Therapieresistenz.

Die Wirklichkeit des Sozialversicherungsträgers

Der 43jährige Elektriker war bereits ca. 9 Monate arbeitsunfähig erkrankt und bezog Krankengeld. Da sowohl internistische stationäre als auch ambulante Behandlung keine ausreichende Besserung erbracht hatten, bestand jetzt die Gefahr einer Frühinvalidität. Die sozialmedizinische Begutachtung hatte die Aufgabe, einen Rehabilitationsplan aufzustellen, der diese Entwicklung verhindern konnte.

Der ärztliche Gutachter hatte die vom Patienten ausgestrahlte Aggressivität hinterfragt und sich auf die Krankheitserklärung des Patienten eingelassen. In einem um echte Kommunikation bemühten Gespräch konnte ein gemeinsamer Code gefunden werden. Die Wirklichkeit des Patienten war dem Gutachter bewußt geworden und er konnte das Erklärungsmodell des Patienten annehmen.

Patient und Gutachter konnten in einem gemeinsamen Erklärungsmodell aushandeln, daß etwas geschehen müßte, um die Konflikte am Arbeitsplatz und in der Familie zu bereinigen und damit eine bessere Ausgangslage zur Behandlung des Diabetes mellitus und der Hypertonie zu haben.

Arbeitsfähigkeit und Vermeidung einer Frühinvalidität waren zu erreichen, wenn die therapeutischen Maßnahmen sowohl auf der sozialen, der psychischen und auf der somatischen Ebene erfolgten.

Hier bot sich als Strategie eine Rehabilitationsmaßnahme in einer Psychosomatischen Klinik zu Lasten der Rentenversicherungsträger mit vorübergehender Entfernung aus dem konfliktbeladenen häuslichen Milieu und nachfolgend ein Arbeitsplatzwechsel an.

Die katamnestische Untersuchung bestätigte die Richtigkeit dieses Vorgehens.

Bei der sozialmedizinischen Begutachtung der Arbeitsfähigkeit bzw. der Arbeitsunfähigkeit ist somit besonders zu bedenken, daß der Patient oft an einer Arbeitswirklichkeit gemessen wird, die primär nicht die von ihm erlebte individuelle Wirklichkeit ist. Häufig ergibt sich eine Diskrepanz zwischen dieser Arbeitswirklichkeit und seiner individuellen Wirklichkeit. Die Konflikte, die daraus entstehen können, sollte der begutachtende Arzt sehen und eine sicher häufig schwierige Vermittlerfunktion einnehmen. Für die Frage nach der pathogenetischen Rolle von Situationen ist ganz entscheidend, wie der Patient seine tägliche Arbeitswelt, seine Familie und sich selbst erlebt.

Eine weitere Krankengeschichte aus dem Alltag des Medizinischen Dienstes der Krankenkassen zeigt, daß bei Einbeziehung der individuellen Wirklichkeit des arbeitsunfähigen Versicherten auch manchmal jeder Leistungsanspruch an den Sozialversicherungsträger von vornherein ausgeschlossen werden muß. Hier handelt es sich um die Begutachtung eines sog. „mißglückten Arbeitsversuches", die klären sollte, ob der Patient überhaupt in der Lage war „Arbeit von wirtschaftlichem Wert" als Voraussetzung für das Zustandekommen eines Versicherungsverhältnisses zu leisten.

Ein 21jähriger Mann bricht am ersten Tage seiner Banklehre zusammen und macht am nächsten Tag einen Suizidversuch. Er stammt aus einer begüterten Familie und hat trotz guter Intelligenz eine schlechte Schulkarriere hinter sich gebracht. Der Lehrer empfiehlt ihm, statt des Abiturs eine Lehre zu machen. Der junge Mann, der schon während der Schulzeit zum Alkohol gegriffen hatte, verließ daraufhin die Schule, kam vom Alkohol los, entwickelte aber jetzt eine Magersucht. Trotz Untergewicht von 44 kg war er weiterhin körperlich aktiv und fühlte sich wohl. Der Vater konnte aufgrund seiner gehobenen Stellung dem Sohn ohne dessen Mithilfe einen Ausbildungsplatz bei einer Bank besorgen. Am ersten Arbeitstag mußten sich alle Lehrlinge gegenseitig vorstellen und ihren Werdegang schildern. Der junge Mann erlebte die schulische Überlegenheit der anderen Auszubildenden, die alle Abiturienten waren, als schwere narzißtische Kränkung, kollabierte und war von Stund an arbeitsunfähig.

Dieser junge Mann wurde durch die vom Lehrer empfohlene und vom Vater besorgte Banklehre in eine Situation hineinversetzt, die er als derartig kränkend und belastend erlebte, daß es ihm unmöglich war, diese Arbeit zu leisten. Es war die Aufgabe des Sozialmediziners der Krankenkasse mitzuteilen, ob dieser Patient aus ärztlicher Sicht überhaupt in der Lage war, an diesem Arbeitsplatz Arbeit von wirtschaftlichem Wert zu leisten.

Von der Beantwortung dieser Frage hängt die Verwaltungsentscheidung der Krankenkasse ab, ob ein Anspruch auf Leistung (Krankengeld, Krankenhilfe) aus dem Arbeitsverhältnis bestand. Dies mußte verneint werden. Der Arbeitsplatz als Teil der individuellen subjektiven Wirklichkeit des Patienten ist ein entscheidender Faktor in der Problematik der Arbeitsunfähigkeit und ihrer Begutachtung. Unter diesem Blickwinkel war der junge Mann nicht in der Lage, diese Ausbildung überhaupt zu beginnen.

In diesem Beispiel war die Diskrepanz zwischen Arbeitswirklichkeit und individueller Wirklichkeit, und der Konflikt, der sich daraus ergab, zu groß. Die akute Verletzung seiner individuellen Wirklichkeit mußte zur Dekompensation – hier Arbeitsunfähigkeit auf Dauer für diesen Ausbildungsplatz – führen.

Katamnestisch war zu erfahren, daß es dem jungen Mann erst nach einer längerdauernden stationären und ambulanten psychotherapeutischen Behandlung und Ablösung aus der Abhängigkeit des Elternhauses gelang, eine Ausbildung in einem von ihm selbst gewählten Beruf zu beginnen und ordnungsgemäß abzuschließen.

Da der Patient in seinem Zustand nicht in der Lage war, eine Ausbildung als Bankkaufmann zu beginnen, bestand auch für den Sozialversicherungsträger keine Leistungspflicht. Ein Pflichtversicherungsverhältnis war nicht zustande gekommen und es mußte vom Patienten bzw. dessen Vater ein anderer Kostenträger für die notwendige Behandlung gesucht werden.

Wie wichtig es besonders bei Gastarbeitern ist, sie in ihren Kulturkreis und ihre Sozialwelt eingebunden zu sehen, zeigt das folgende Beispiel:

> Eine 28jährige Sizilianerin hatte beim Vertrauensärztlichen Dienst (früher VÄD jetzt MDK) bereits eine dicke Krankenakte mit langen Arbeitsunfähigkeitszeiten. Sie erlitt immer wieder Fehlgeburten und war dann wegen einer Risikoschwangerschaft nach einer vorausgegangenen Totgeburt vom ersten Tag der neuen Gravidität an arbeitsunfähig. Sie sprach gut deutsch und bei Erhebung der biographischen Anamnese war zu erfahren, daß sie einen 10jährigen Sohn hatte, der in Sizilien bei den Schwiegereltern lebte und den sie im Urlaub nur knapp 4 Wochen im Jahr sah. Sie hatte ihr Kind im Alter von 2 Jahren verlassen müssen, um zusammen mit ihrem Mann in Deutschland Geld zu verdienen, damit das in Sizilien erbaute Haus, das jetzt die Schwiegereltern betreuten, abbezahlt werden konnte. Darum sollten auch alle Kinder, die sie noch gebären würde, von der Schwiegermutter aufgezogen werden. Die strenggläubige Katholikin, die jede Empfängnisverhütung ablehnte und für deren Familie Kinderreichtum ein Statussymbol war, konnte nicht nachvollziehen, daß in dieser Situation ihr Körper so reagierte, daß sie nur Fehl- oder nicht lebensfähige Frühgeburten haben konnte, um ein gesundes Kind nicht nach Sizilien abgeben zu müssen.

Gemeinsam mit dem behandelnden Gynäkologen empfahl der Sozialmediziner der Krankenkasse, die Versicherte für die gesamte Dauer der Schwangerschaft nach Sizilien zu beurlauben, wo sie dann ein lebensfähiges Kind gebar. Nach Ablauf der Mutterschutzfrist kam sie dann nicht mehr in die BRD zurück, sondern kündigte ihr Arbeitsverhältnis.

Die Einbeziehung der individuellen Wirklichkeit in die sozialmedizinische Begutachtung bewahrte diese Gastarbeiterin vor dem Schicksal einer Drehtürpatientin und ersparte dem Sozialversicherungsträger enorme Kosten.

Ein weiterer Begutachtungsfall veranschaulicht, daß unter dieser Betrachtungsweise der Gutachter nicht immer zugunsten des Versicherten entscheidet, wenn sich die Wirklichkeit des Patienten auch nicht teilweise mit der Wirklichkeit des begutachtenden Arztes und der Wirklichkeit des Sozialversicherungsträgers vereinen läßt.

> Zu Lasten der Krankenkasse wird für eine mitversicherte Ehefrau wegen schlechten Allgemeinzustandes ein Antrag auf Anschlußheilbehandlung nach Krankenhausbehandlung gestellt. Sie war zuletzt wegen einer Narbenhernie nach Cholezystektomie und einem Karpaltunnelsyndrom links operiert worden.
>
> Wie der Gutachter feststellte, wurde die Patientin in den letzten zwei Jahren insgesamt 17mal stationär behandelt. Den Berichten zweier Psychosomatischer Kliniken ist zu entnehmen, daß es sich hier um eine neurotische Dekompensation bei einer labilen, infantilen Persönlichkeit handelte, die durch die Offenbarung einer vorausgegangenen Erkrankung an einem Non-Hodgkin-Lymphom mit nachfolgender zytostatischer Behandlung überfordert war. Obwohl sich diese Erkrankung seit langem in stabiler, guter Remission befand, entwickelte sie immer neue somatische Erscheinungen, die zu aneinandergereihten stationären Aufnahmen führten.
>
> Der soziale Hintergrund war die bereits zweite scheiternde Ehe.

Sowohl im Interesse der Versicherten als auch der Versichertengemeinschaft, die bereits über zwei Jahre Krankenhausbehandlung finanziert hatte, wurde durch die Begutachtung klargestellt, daß eine weitere stationäre Rehabilitationsmaßnahme keine Besserung erwarten ließ, sondern daß hier nur eine Konfrontation mit der Realität des Alltages bei begleitender ganzheitsmedizinischer Betreuung eine Änderung bringen konnte. Denn neben den Interessen des einzelnen Kranken bzw. Sozialversicherten muß der Gutachter ja gleichzeitig die Ziele der Gesundheits- bzw. Sozialpolitik im Auge haben, also die Erkenntnisse der Sozialmedizin auch aus der Wirklichkeit des Sozialversicherungsträgers betrachten.

An diesen Beispielen sollte gezeigt werden, welche Chancen in der psychosozialen Versorgung der Medizinische Dienst der Krankenkassen bei einer ganzheitsmedizinischen Betrachtungsweise hat.

Neben den Interessen des einzelnen sozialversicherten Kranken muß der ärztliche Gutachter gleichzeitig die Ziele der Gesundheits- bzw. Sozialpolitik im Auge haben, also die Erkenntnisse der Sozialmedizin aus der Sicht der Versichertengemeinschaft beurteilen. Die Ärzte des Medizinischen Dienstes der Krankenversicherung müssen mit ihrer doppelten Aufgabe im heutigen Gesundheitsfürsorgesystem – psychosoziale Betreuung, aber auch Kontrolle – leben und an ihr wachsen (Silomon, 1983).

Teil III: Psychosomatik einzelner Erkrankungen und in verschiedenen Fachgebieten
Störungen von Funktions- und Verhaltensabläufen

30 Konversion

Rolf Adler

30.1 Definitionen

Konversion wird ein Symptom psychischen Ursprungs genannt, das einen Kompromiß zwischen einem bewußtseinsunfähigen Wunsch, einer Phantasie usw. und den ihn vom Bewußtwerden abhaltenden Strebungen in der Körpersprache ausdrückt. Breuer und Freud (1972) haben die Basis für das Verständnis des Konversions-Mechanismus gelegt. Das Konversions-Symptom stellt eine Möglichkeit dar, psychischen Streß zu bewältigen. Es ist der Preis, den das Individuum bezahlt, wenn es einen Konflikt und die mit ihm verbundenen Gefühle von Angst, Wut, Verzweiflung, Scham, Ekel usw. nicht erträgt, sondern ihn mit Hilfe eines körperlich erlebten Symptoms neutralisiert.

Freud und später Alexander (1971) hielten Konversions-Symptome nur im Gebiet der Willkürmotorik und des Sinneswahrnehmungssystems für möglich. Engel (1970) wies nach, daß sie im Bereich jedes Organsystems vorkommen können, falls der dem Symptom zugrundeliegende Funktionsablauf psychisch repräsentierbar ist. Von Uexküll (1963) bezeichnete das Konversions-Syndrom als ein Handlungsbruchstück mit Ausdruckscharakter. Der Patient ist darin Darsteller und Zuschauer in einem und bringt darin ein Stück individueller Wirklichkeit zum Ausdruck; das heißt, er erlebt sich als blind, heiser, von Schmerzen heimgesucht und berichtet seiner Umgebung und seinem Arzt über diese Wirklichkeit. Wie der Arzt als teilnehmender Beobachter versuchen muß, sich in die Wirklichkeit des Patienten einzufühlen und hineinzudenken, um die Programme zu entschlüsseln, die den Wahrnehmungs- (= Merk-) Bereitschaften und Verhaltens- (= Wirk-) Bereitschaften des Kranken zugrunde liegen, und wie sie historisch-biographisch entstanden sind, wird auf Seite 471 (Datenerhebung) besprochen (und im Kap. 12 über die Technik der Anamneseerhebung). Von den Konversions-Symptomen abzugrenzen sind psychogene Symptome, bei denen die körperlichen Erscheinungen Begleitzeichen von Affekten sind und keinen Ausdruckscharakter haben. Alexander (1971) hat sie Organeurosen genannt. Wir ziehen es vor, sie als psychophysiologische Symptome zu bezeichnen (S. 472).

Das Konversions-Syndrom kann mit Beschwerden organischen Ursprungs verwechselt werden, wenn der Arzt die Merkmale nicht kennt, die das Konversions-Symptom charakterisieren. Es wurde als Mimikry von körperlichen Symptomen bezeichnet. Das reduktionistische Denken in der Medizin, das alle Symptome letztlich einer organischen Veränderung zuschreibt, hat es den Ärzten bis heute schwergemacht, Konversions-Symptome von der ersten Begegnung mit einem Kranken an in die Differentialdiagnose einzubeziehen. Der Arzt geht von einer Wirklichkeit aus, die von anatomischen und physiologischen Gesetzen beherrscht ist. Er nimmt an, daß diejenige des Patienten auf den gleichen Gesetzmäßigkeiten beruht, also auch seine Symptome. Die Krankheits- und Körpervorstellungen des Patienten gehen aber von Erfahrungen aus, die er mit seinem Körper und demjenigen seiner Bezugspersonen im Laufe seiner Entwicklung von den ersten Tagen an als Säugling gemacht hat (Piaget, 1967). In bezug auf Lähmungen hat Freud diesen Sachverhalt schon früh ausgedrückt: „Die Konversion benimmt sich, wie wenn es keine Anatomie gäbe. Der Körper wird im populären Sinn verwendet." So hat das Kleinkind einst „gut" und „schlecht" danach unterschieden, wie sich etwas anfühlt, wenn es in den Mund genommen wird, und es hat die Linderung eines Schmerzes mit der Vereinigung mit der geliebten Mutter verbunden, und ein Erschöpfungsgefühl mit dem nutzlosen Schreien nach der Trennung von einer Bezugsperson. Beachtet der Arzt nicht, daß seine Wirklichkeit und die des Patienten nicht identisch sind, dann nimmt er ein Konversions-Symptom als Ausdruck einer verborgenen organischen Läsion an, und er veranlaßt Abklärungsuntersuchungen und führt Behandlungen, auch Operationen durch, die das Symptom weder erhellen noch lindern, sondern unliebsame Folgen nach sich ziehen können, eingeschlossen die Gefahr, zur Chronifizierung von Leiden beizutragen (Harding, 1962).

30.2 Analogien zum Mechanismus der Konversion

Das Studium von Geste, Pantomime, Scharadespiel, Umgangssprache und Traum fördert das Verständnis für den Mechanismus der Konversion. Die Geste läßt erkennen, daß ein Gedanke durch Konversion in Körpersprache ausgedrückt werden kann. Im Unterschied zum eigentlichen Konversions-Symptom ist der hinter der Geste stehende Gedanke dem sie Vollziehenden bewußt. Nemiah hebt deshalb hervor, daß nicht die „Konversion" das Wesentliche am Mechanismus sei, sondern die Abspaltung vom Bewußtsein, und möchte lieber von Dissoziation sprechen (Nemiah, 1974). Bei der Scharade und der Pantomime wird ein Gedanke in der Körpersprache ausgedrückt. Der Zuschauer muß wie der Arzt beim Konversions-Symptom das in der Körpersprache Ausgedrückte zurückübersetzen. Der Unterschied besteht darin, daß der Schauspieler den Gedanken kennt, während er dem Patienten nicht bewußt ist. Im Traum erlebte Körpersymptome, wie eine Lähmung, welche die Flucht verunmöglicht, das Gefühl von einer Last zusammengepreßt zu werden und Atemnot zu erleiden, zeigen auch dem Arzt, der nur anatomisch und physiologisch begründete Symptome anerkennt, daß sogar er selbst fähig ist, Körpersymptome zu erleben, ohne daß sein Körper organische Veränderungen erleidet. Die Umgangssprache schließlich belegt mit zahllosen Beispielen, daß der Körper Ausdrucksorgan für Gedanken sein kann. „Bei diesem Gedanken wird mir schwindlig", „es liegt mir auf dem Magen" usw. sind Hinweise dafür und umgekehrt, daß sprachliche Metaphern körperliche Vorgänge ausdrücken können (Sharpe, 1940).

30.3 Häufigkeit

Da Menschen das Konversions-Symptom in ihrer individuellen Wirklichkeit als körperlichen Ursprungs erleben, suchen sie als Arzt den Nicht-Psychiater auf. Die fehlende Vertrautheit mit dem Konversions-Mechanismus erschwert eine genaue Erfassung der Häufigkeit. Einzig der mit dem Konversions-Mechanismus vertraute Somatiker könnte Zahlen liefern. Einer dieser seltenen Ärzte (Engel, 1970) schätzt, daß bis 25% aller in einer allgemeinen internistischen Spitalabteilung hospitalisierten Patienten ein- oder mehrmals im Verlaufe ihres Lebens an einem Konversions-Symptom gelitten haben. Bei Patienten einer psychiatrischen Klinik waren es 24% (Woodruff et al., 1969), bei weiblichen Patienten mit organischen Krankheiten betrug die Zahl 30% (Woodruff, 1968). Bei aus einem Allgemeinspital dem Psychiater zugewiesenen Patienten lautete die Diagnose in 13% der Fälle Konversions-Symptom (McKegney, 1967). Das Konversions-Symptom kommt häufiger bei Frauen vor (McKegney, 1967; Axelrod et al., 1980; Stefansson et al., 1976; Raskin et al., 1966), findet sich aber bei Männern so oft, daß es differentialdiagnostisch nicht

vergessen werden darf. Die Konversions-Symptome kommen nicht nur bei psychisch auffälligen Persönlichkeiten vor, insbesondere solchen mit hysterischen Charakterzügen (eitel, egozentrisch, labil und erregbar, affektiv flach, dramatisch Aufmerksamkeit erregend, theatralisch, Sexuelles betonend, sexuell provozierend aber frigid, fordernd abhängig), sondern auch bei Menschen mit depressiven Zügen, schizophrenen Störungen und bei psychisch sonst unauffälligen Individuen (Freud, 1972; Stefansson et al., 1976; Raskin et al., 1966; Barnert, 1971; Ziegler et al., 1959; Lewis und Berman, 1965; Gatfield und Guze, 1963). Der Grundsatz, es seien möglichst alle Symptome eines Patienten durch eine einzige Krankheit zu erklären, kann beim Patienten mit Konversions-Symptom irreführen. Bei 10–50% aller Patienten mit Konversions-Symptomen liegt zusätzlich eine organische Störung vor (McKegney, 1967; Ziegler et al., 1959; Lewis und Berman, 1965; Captan und Nadelson, 1980; Merskey und Buknih, 1975; Marshall, 1949; Gatfield und Guze, 1963). Die Symptome beschränken sich auch nicht auf bestimmte soziale Schichten (Engel, 1970; Lewis und Berman, 1965). Sie sind von Kultur und Zeitepochen abhängig. Um die Jahrhundertwende fanden sich Symptome, die Konflikte auf dem Gebiet der Sexualität in oft sehr deutlicher Form ausdrückten, heute sind die Symptome mehr durch die Vorstellungen geprägt, die sich die Menschen von Krankheiten wie dem Herzinfarkt, einem Hirnschlag, einer Lungenembolie usw. machen. Konversions-Symptome sollen sich häufiger auf der linken Körperseite finden (Halliday, 1941; Edmonds, 1947; Merskey, 1967; Agnew und Merskey, 1976; Smokler und Shevrin, 1979). Die beträchtliche Zahl bilateraler und rechtsseitiger Symptome weist darauf hin, daß andere Faktoren als die Symbolisierung „links gleich schlecht/böse", und die Verantwortung der nicht-dominanten Hemisphäre für primärprozeßhafte Vorgänge (Smokler und Shevrin, 1979) eine Rolle spielen. Die an sich selbst und bei anderen erlebten organischen Symptome (siehe unten) spielen für die Wahl der Lokalisation eine bedeutsame Rolle (Freud, 1972; Engel, 1970; Axelrod et al., 1980; Fallik und Sigal, 1971).

30.4 Mechanismus der Symptombildung

Dem Konversions-Symptom liegt die Fähigkeit des psychischen Apparates zugrunde, Wünsche, Gedanken oder Phantasien symbolisch in der Körpersprache auszudrücken, die vom bewußten Teil des psychischen Apparates nicht akzeptiert werden können, und die nicht in einer entsprechenden Handlung Erfüllung finden. Der Träger des Wunsches stellt nur die Körperveränderung fest, weiß aber nicht, daß ein Wunsch dahintersteckt. Im Symptom wird zusätzlich die den Wunsch unterdrückende Strebung ausgedrückt. Gelingen sowohl symbolischer Ausdruck von Wunsch als auch von unterdrückender Tendenz, so verschwindet der psychische Streß, das seelische Gleichgewicht stellt sich wieder ein, der Preis dafür besteht im entstandenen Symptom. Gelingt die Konversion unvollständig, „so verbleibt ein Teil des Affektes als Komponente der Stimmung (z. B als Angst) im Bewußtsein" (Freud, 1972). Wie bei der Scharade der Zuschauer die Geste, muß der Arzt das Symptom in seine ursprüngliche Sprache zurückübersetzen. Der Unterschied zwischen Scharade und Konversion besteht darin, daß der betroffene Mensch bei der Konversion nicht weiß, daß und was für einen Wunsch, Gedanken usw. er ausdrückt, und der Arzt muß im Symptom nicht nur den Wunsch, Gedanken usw., sondern auch die ihn unterdrückende Strebung erkennen. Das Konversions-Symptom weist vier Ziele auf:

Ein unannehmbarer Wunsch, Gedanke usw. wird dennoch, wenn auch in modifizierter Form und verhüllt, in die Tat umgesetzt.
- Der betreffende Mensch wird dafür bestraft: das Symptom fügt ihm Leiden zu.
- Das Symptom enthebt den Menschen der psychischen Streßsituation (dies nennt man den primären Gewinn der Symptombildung).
- Das Symptom verhilft dem Träger zu neuen Beziehungsmöglichkeiten mit der Umwelt. Er wird jetzt beispielsweise von Bezugspersonen als körperlich Kranker umsorgt, nicht verlassen usw. (Dies entspricht dem sekundären Gewinn aus dem Symptom.)

Die Voraussetzung dafür, daß ein Körperteil oder eine Körperfunktion das Vehikel einer Konversion werden können, liegt in ihrer Fähigkeit, in zwischenmenschlichen Beziehungen eingesetzt zu werden. Diejenigen Körperteile und -funktionen, die früh im Leben in der Kommunikation zur Umwelt wichtig waren, kommen bevorzugt für Konversionen in Betracht. Deshalb finden sich diese Symptome häufig im muskuloskeletären Bereich, betreffen die Körpersensibilität, den oberen und unteren Darmtrakt, die Atmung, die Sprachwerkzeuge und die Sinnesorgane. Dabei werden Körperteile oder -funktionen nicht im anatomischen oder physiologischen Sinne verwendet, sondern so, wie sie vom Menschen im Verlauf seiner vor allem frühen Entwicklung erlebt wurden. Körperteile und -funktionen müssen zudem psychisch repräsentierbar sein. Eine schmerzhafte Magenkontraktion kommt beispielsweise für eine Konversion in Frage, die vom psychischen Apparat nicht registrierbare Sekretion der Magensäure hingegen nicht (Engel, 1970).

30.5 Wahl von Art und Lokalisation des Symptoms

Wenn ein Wunsch nicht zum Bewußtsein zugelassen werden kann, so ist es möglich, daß im Individuum ein körperlicher Vorgang aktiviert wird, der einst mit dem betreffenden Wunsch zusammen vorkam.

> Ein junger Mann, der seinem Vorgesetzten gegenüber heftige Wut empfinden müßte, erleidet intensive Stirnkopfschmerzen im Bereich einer Narbe an der Stirn –

alle Untersuchungen verlaufen negativ – die er sich als Junge beim Basteln mit Sprengstoff zugezogen hatte, zu einer Zeit, in der er mit seinem Vater aggressionsgetönte Auseinandersetzungen durchmachte. Dieses Beispiel läßt sich mit dem Situationskreismodell verstehen und beleuchtet die individuelle Erfahrung und Wirklichkeit des Patienten: Der junge Mann bemerkt Wut gegenüber einer Autorität (wenn auch unbewußt oder nur zum Teil bewußt). Erfahrungen haben ihn gelehrt, daß der unverhüllte Ausdruck seiner Wut verhängnisvoll war. Die Assoziation zwischen der von ihm als Junge erlebten Wut und der schmerzhaften Verletzung tritt jetzt als Programm auf, das sich zwischen „Merken" und unmittelbares „Wirken" (Ausdrücken seiner Wut) einschaltet. Das Symptom Kopfschmerz entsteht, es drückt stellvertretend seine Wut aus und läßt ihn für seinen Wunsch, das Gefühl auszuleben, leiden. Jetzt wirkt er für und auf seine Umgebung auf eine modifizierte Art und Weise, welche ihn z. B. durch Krankwerden in eine neue Beziehung zum Vorgesetzten treten läßt.

Es kommen nicht nur Körpervorgänge in Betracht, die das Individuum einst selbst, sondern auch solche, die es bei einer Bezugsperson erlebt hat.

> Eine junge Frau wird wegen Atemnot zur Abklärung ins Spital eingewiesen. Sie ist nach ihrer Rückkehr aus dem Ausland erkrankt, wo sie als Sekretärin gearbeitet hatte. Während ihres Auslandaufenthalts hatte sie mit ihrer herzkranken Mutter mittels Tonband korrespondiert. Sie hatte sich wiederholt Vorwürfe gemacht, daß sie nicht zur Pflege der inzwischen verstorbenen Mutter nach Hause zurückgekehrt war. Nach ihrer Rückkehr erkrankte sie an Atemnot. Hinter der Atemnot verbargen sich die Trauer um die Mutter, der Wunsch nach ihr und die Selbstbestrafung, sie im Stich gelassen zu haben. Das Symptom Atemnot der Tochter beruhte auf der beim Abhören der Tonbänder vernommenen Atemnot der Mutter.

Es kommen auch Körpervorgänge in Betracht, die einen Wunsch und die ihn unterdrückenden Strebungen symbolisch ausdrücken. Die Lähmung eines Armes kann den Wunsch zu schlagen und die Blockierung dieser Handlung bedeuten. Das Globusgefühl kann zum Ausdruck bringen, daß ein Gedanke, der widerlich ist, nicht geschluckt werden kann.

30.6 Diagnose

30.6.1 Beweisende Kriterien

Das Konversions-Symptom muß mit positiven Kriterien belegt werden. Als Ausschlußdiagnose genügt nicht, daß ein Symptom nicht durch anatomische oder pathophysiologische Veränderungen erklärt werden kann. Stützt man sich allein darauf, kommen schwerwiegende Fehldiagnosen vor (Adler, 1981). Die Symptomentstehung muß in eine Zeit fallen, in der ein Konflikt stattfand, der unbewußte Anteile hat, die im Symptom neutralisiert wurden. (Aber ausschließlich aus dem Vorliegen eines psychischen Konflikts vor dem Auftreten des Symptoms ein psychogenes Symptom abzuleiten, wäre falsch, denn psychische Konflikte können sowohl zum Entstehen psychogener als auch organischer Störungen beitragen.)

Bei einer Frau mit konversionsbedingten Schmerzen im Ellenbogenbereich stellte sich heraus, daß diese zu einer Zeit auftraten, während der der als Wirt tätige Ehemann die Patientin mit einer Angestellten betrog. Die Wahl gerade des vorliegenden Symptoms und seiner Lokalisation müssen belegbar sein. Die betrogene Frau hatte sechs Monate vor dem Auftreten des Symptoms einen Sturz mit Verletzungen des Ellenbogens erlitten, für die sie keine ärztliche Hilfe beanspruchte; zudem war ihre Mutter an Krebs gestorben und hatte an Knochenmetastasen mit Armschmerzen gelitten. Die Patientin betonte wiederholt, daß knotenförmige Wucherungen am Arm ihre Schmerzen verursachten. Die Armschmerzen hinderten die Frau im Betrieb mitzuarbeiten und zwangen das Ehepaar, die Gastwirtschaft aufzugeben, und zugleich bestrafte das Symptom die Patientin für ihre verhüllte Wut gegenüber dem untreuen Ehemann, den sie „mehr liebte als zuvor", weil sie ihn jetzt „realistischer sehe als vorher", während sie ihn in der Zeit des Sichkennenlernens als aufdringlich und ihre fehlende Liebe für ihn beschrieb.

Das Beispiel macht deutlich, daß ein einzelnes Symptom meistens mehrere Wurzeln hat – früher selbst erlebte Schmerzen, den Wunsch zu schlagen, das Sabotieren der gemeinsamen Arbeit im Betrieb und die Krankheit der Mutter – also überdeterminiert ist.

Wenn bei einem Symptom, das nicht durch anatomische und pathophysiologische Veränderungen erklärt werden kann, der Konflikt, seine Neutralisierung im Symptom (primärer Gewinn), die Wahl gerade dieses Symptoms und dieser Lokalisation und der sekundäre Gewinn nicht nachgewiesen werden können, so darf das Konversions-Symptom lediglich vermutet werden. Dies heißt, daß bei starkem Verdacht die somatische Abklärung nicht stur vorangetrieben werden darf. Der Arzt soll aber gegenüber der Möglichkeit einer organischen Ursache offenbleiben, und die weiteren Kontakte mit dem Patienten nach dem Erstinterview müssen der Suche nach den erwähnten positiven Kriterien zur Erhärtung der Diagnose gelten. Autoren, die sich lediglich darauf stützten, daß sie keine organische Erklärung für die Symptome fanden, stellten bei ihren Konversions-Patienten in ganz unterschiedlichem Maße Konflikte, primären und sekundären Gewinn und Modelle für die Symptomwahl und ihre Lokalisation fest (Raskin et al., 1966; Barnert, 1971; Lewis und Berman, 1965; Fallik und Sigal, 1971; Packard, 1980; Watson und Buranen, 1979). Es ist auch nicht verwunderlich, daß sie bei der Nachuntersuchung zum Teil eine erhebliche Zahl fälschlicherweise positiv diagnostizierter Konversions-Symptome entdeckten (Ziegler et al., 1959; Watson und Buranen, 1979; Carter, 1949; Liungberg, 1957; Slater und Glitherto, 1965; Guze, 1970).

30.6.2 Hinweisende Kriterien

Patienten, die zum Erleiden von Konversions-Symptomen neigen, weisen in ihrer Anamnese oft Beschwerden auf, die schwer einem bekannten organisch bedingten Krankheitsbild zugeordnet werden können. Oft sind Abklärungen und Eingriffe durchgeführt worden, die zu keiner eindeutigen Diagnose und zu keiner echten Besserung geführt haben. Der Arzt darf sich beim Erheben der Anamnese nie mit Beschreibungen wie „ich litt an Gallenkoliken" zufriedengeben. Bei näherer Erkundigung nach den Beschwerden erfährt er beispielsweise, daß die Patientin quer im Oberbauch einen Dauerschmerz verspürt hatte, der die Kriterien der Kolik überhaupt nicht erfüllt. Die komplexe und dramatische Krankengeschichte wird oft (fälschlicherweise) als diagnostisches Kriterium für die Konversion genommen, und nicht lediglich als Hinweis darauf (Woodruff et al., 1969; Raskin et al., 1966; Ziegler et al., 1959; Perley und Guze, 1962). Die Schilderung der Beschwerden erfolgt oft auf dramatisch-theatralische Weise (Engel, 1970; Stefansson et al., 1976; Raskin et al., 1966; Lewis und Berman, 1965), aber die Patienten wirken beim Beschreiben schwerster im Moment vorliegender Symptome eigenartig unbeteiligt. Dieses paradoxe Verhalten wird „la belle indifférence" genannt. Es wird von diagnostisch wertlos (Raskin et al., 1966), über in 8% vorhanden (Lewis und Berman, 1965) bis zu häufig (Barnert, 1971) beurteilt. Es ist eine Folge der Neutralisation des Konflikts durch das Symptom, das der Patient einerseits benötigt und unter dem er andererseits leidet, und es erklärt, warum Konversions-Patienten keine oder nur geringe Einsicht in den Zusammenhang zwischen Problem und ihrem Symptom zeigen (Engel, 1970; Barnert, 1971) und kaum je aus eigenem Bedürfnis heraus den Psychiater aufsuchen oder ihren Arzt bitten, sie dem psychiatrischen Konsiliarius zu überweisen. Der Konversions-Mechanismus bringt auch mit sich, daß die Patienten keine oder nur wenig Angst zeigen (Engel, 1970; Barnert, 1971; Lewis und Berman, 1965) und die nicht seltene Depression verhüllt ist. Immerhin wurde deutliche Angst (Lewis und Berman, 1965) klinisch und (Lader und Sartorius, 1968) experimentell bei Konversions-Patienten festgestellt und gegen die Theorie der Konfliktneutralisierung verwendet, ein nicht stichhaltiger Einwand, wie bereits erklärt wurde. Medikamentenabhängigkeit, Suizidversuche, Scheidungen, sexuelle Konflikte, viele Partnerwechsel, Frigidität usw. werden oft in der Anamnese gefunden (Engel, 1970).

30.7 Komplikationen des Konversions-Symptoms

Es kann organische und pathophysiologische Folgen haben. Das Erstickungsgefühl kann zur Hyperventilation und diese zur respiratorischen Alkalose mit Tetanie und Bewußtseinsverlust führen. Eine Muskellähmung kann Atrophie zur Folge haben. Diese Komplikationen sind nicht mit der Konversion zu verwechseln. Es sind Folgesymptome ohne symbolischen Gehalt. War die organische Veränderung, auf die durch den vom Bewußtsein ferngehaltenen Wunsch zurückgegriffen wird, damals von pathologischen Vorgängen begleitet, so können diese auch wieder auftreten. So können Erythem und Schwellung Konversions-Schmerzen begleiten, wenn die ursprüngliche schmerzhafte Läsion Rötungen und Ödem aufgewiesen hatte. Wiederum sind sie Begleitzeichen und haben keinen symbolischen Charakter (Engel, 1970; Barchilon und Engel, 1952). Forschungen auf dem Gebiet des instrumentellen Lernens an Mensch und Tier zeigen, daß autonome Funktionen lern- und kontrollierbar sind (Barr und Abernethy, 1977) und die frühere Einschränkung der Konversions-Symptome auf die Willkürmotorik und die Sensibilität unberechtigt ist.

30.8   Datenerhebung durch das Interview

Da der Patient mit Konversions-Symptomen den psychischen Konflikt vom Bewußtwerden fernhalten muß und auch nicht gewahr werden kann, auf welche eigenen früheren Körperveränderungen oder solche bei Bezugspersonen sich sein Symptom stützt, wird er auf direkte Fragen bezüglich Zusammenhängen zwischen Konflikten und seinem Symptom keine Auskunft wissen. Das Modell für das Symptom, die Motive für seine Wahl und Lokalisation findet der Arzt deshalb nur, wenn er den Assoziationen des Patienten folgt und beim Erwähnen früher vom Patienten selbst erlebter Symptome und solchen von Bekannten und Verwandten darauf achtet, ob sie identisch mit den derzeitigen Symptomen geschildert werden, beispielsweise ob eine Patientin bei der Beschreibung der Rückenschmerzen, die um die linke Flanke herum nach dem Oberbauch ausstrahlen, die gleiche Bewegung mit der Hand in bezug auf ihren Körper macht wie bei der Schilderung der Schmerzen, welche die Mutter bei ihrer Erkrankung an Leberkrebs aufwies. Der Arzt darf dabei dem Patienten nicht verraten, auf welche Zusammenhänge er achtet und nie die Frage stellen: „Hat jemand in Ihrer Familie die gleichen Schmerzen gehabt?" So kam die Atemnot bei der Mutter der jungen Frau (S. 469) zum Vorschein, als sich der Interviewer nach der Beziehung zwischen Patientin und Mutter und nach den Gefühlen der Patientin während des Getrenntseins erkundigte. Bezeichnenderweise lehnte die Patientin eine seelische Ursache ihrer Atemnot vehement ab. Das Betonen des körperlichen Ursprungs eines Symptoms durch den Kranken soll den Arzt geradezu als Faustregel einen psychischen Ursprung vermuten lassen, denn die Betonung der körperlichen Bedingtheit weist darauf hin, daß der Patient einen psychischen Hintergrund heftig abwehren muß. Betont ein Patient den psychischen Ursprung seines Symptoms, so denke der Arzt an eine organische Ursache, wie bei einer alten Frau, die ihre heftigen Schmerzen oberhalb des rechten Schlüsselbeins, die hinters Ohr und in die rechte Kopfseite ausstrahlten, auf ihren zu großen Einsatz bei der Gartenarbeit zurückführte, während die Abklärung schwerste Veränderungen der Halswirbelsäule ergab. Diese Faustregel wird von Engel (1970) erwähnt und ist indirekt aus Fallbeschreibungen (Raskin et al., 1966) ersichtlich.

30.9   Die häufigsten Konversions-Symptome

Motorik: Schwäche, Lähmung, Krämpfe, Tics, Tremor, Torticollis, Pseudokontraktur, Steifigkeit, Gangstörung, Aphonie, Heiserkeit, Blepharospasmus, Ptose, Schielen.
Sensorik: Schmerz aller Lokalisationen (besonders Kopf, Gesicht, Herzgegend, Bauch, Unterleib, Rükken), Anästhesie, Hypästhesie, Hyperästhesie, Pruritus, Brennen, Blindheit, tubuläres Sehen, Taubheit.
Oberer Magen-Darm-Trakt: Globusgefühl, Dysphagie, Anorexie, Polydipsie, Blähung, Brechreiz, Erbrechen.
Unterer Magen-Darm-Trakt: Inkontinenz, Verstopfung, Durchfall, Pruritus ani.
Atmung: Atemnot, Hyperventilation, Husten.
Harnwege: Harndrang, häufiges Wasserlassen, schmerzhaftes Wasserlassen, Inkontinenz, Harnverhalten.
Genitaltrakt: Pruritus vulvae, Dyspareunie, Impotenz, Frigidität, Ejaculatio praecox.
Haut: Erröten, Blaßwerden, hämorrhagische Stigmata.
Bewußtsein, geistige Funktionen: Schwindel, Synkope, Amnesie, Vergeßlichkeit, Konzentrationsschwäche.

Der Konversions-Patient mit Hyperventilation klagt über Enge beim Atmen, Erstickungsgefühl, Schwindel, Kribbeln um Mund, in Fingern und Beinen und Ohnmachtsanfälle. Vermutet der Arzt hinter bestimmten Beschwerden die Hyperventilation, kann sie im Interview aber nicht klären, so hilft ihm die Beobachtung des Patienten während der Körperuntersuchung. Er läßt ihn während der Lungenauskultation tief und lang atmen, ohne ihn zu informieren, daß er einen Hyperventilationsversuch durchführt. Nach Abbruch der Auskultation fragt er, was der Kranke eben verspürte. Nicht selten beschreibt dieser genau die Symptome, die er bei der Anamnese angegeben hat. Er darf aber nicht gefragt werden, ob er das gleiche gespürt habe wie bei seinen Störungen, die ihn zum Arzt gebracht hatten, sonst blockiert er und antwortet mit „nein".

Bei konversionsbedingten Bauchschmerzen zeigt der Patient hie und da eine betonte Abwehrspannung gegenüber der palpierenden Hand, während das die Darmgeräusche untersuchende Stethoskop gut unter das Niveau der Bauchdecke eingedrückt werden kann. Die Synkope auf Konversionsbasis führt selten zu Verletzungen, ihr Vorhandensein schließt eine Konversion aber nicht aus. Der Patient zeigt trotz mehrminütiger Bewußtlosigkeit keine Veränderungen von Blutdruck, Puls und Reflexen. Zungenbiß und Inkontinenz fehlen. Das Bewußtsein ist oft nicht ganz erloschen, die Lider können flattern. Das EEG ist immer normal (Engel, 1962).

30.10   Prognose und Therapie

Die Prognose des Konversions-Symptoms ist schwer vorauszusagen. Sie hängt von der psychischen Grundstörung und von der Lebenssituation ab, in der es aufgetreten ist. Konversions-Symptome, die bei akuten und großen psychischen Belastungen auftreten und einen psychisch sonst ausgeglichenen Menschen treffen, können sich in kurzer Zeit auflösen. Konversions-Symptome bei psychisch schwer gestörten Menschen können jahrelang andauern oder durch andere hartnäckige Konversions-Symptome abgelöst werden und psychische Krankheiten wie Alkoholismus, Zwangsneurose, Depression und Schizophrenie können in den Vordergrund treten (Ziegler et al., 1959; Carter, 1949; Gatfield und Guze, 1963; Liungberg, 1957; Slater und Glitherto, 1965; Perley und Guze, 1962).

Ein Konversions-Symptom stellt keine unbedingte Indikation für eine Psychotherapie, Psychoanalyse usw. dar. Die Indikation dafür richtet sich nach den für diese Therapien üblichen Kriterien. Da Patienten mit Konversions-Symptomen das Symptom als eines körperlichen Ursprungs erleben und die unterdrückten Wünsche usw. unbewußt sind, weigern sie sich meistens, einen Psychotherapeuten aufzusuchen (Barnert, 1971). Der behandelnde Arzt muß solche Patienten deshalb oft behalten und betreuen. Er darf nicht versuchen, das Symptom isoliert zu behandeln, denn er droht damit dem Patienten, den primären und sekundären Gewinn wegzunehmen. Der Patient würde als Folge von seinen Konflikten überwältigt. Auf Versuche des Arztes, das Symptom „wegzunehmen", kann er nur damit reagieren, daß er das Symptom verstärkt, ein anderes Konversions-Symptom entwickelt oder den Arzt wechselt. Am besten teilt der Arzt seinem Patienten mit, daß die Untersuchungen mit den heutigen technischen Mitteln das Symptom nicht erklärt hätten, und daß er bereit sei, den Patienten in regelmäßigen Abständen zu sehen, um mit ihm zusammen zu erfahren, wie es dem Patienten gehe und unter welchen Umständen er sich besser oder schlechter fühle. So könne mit der Zeit das Symptom verstanden werden. Fühlt der Patient, daß der Arzt ihm nicht mitteilen will, „ihm fehle nichts", oder ihm die psychische Entstehung des Symptoms aufzudrängen versucht, sondern ihn empathisch begleitet und sich als teilnehmender Beobachter in die Wirklichkeit des Patienten einzufühlen und hineinzudenken versucht, dann bleibt der Patient häufig in der Beziehung und drängt nur mehr gelegentlich auf neue Abklärungen und Eingriffe. Verfestigt sich das Arbeitsbündnis zwischen Arzt und Patient, dann

kommt dieser allmählich auf seine psychische und soziale Situation zu sprechen, und das Symptom tritt in den Hintergrund. Es kann dann geschehen, daß der Patient selber gewahr wird, wie sich sein Symptom unter gewissen Umständen verschlimmert oder mildert. Es verblaßt dann mit der Zeit, und der Patient kann beginnen, das reale Leid seiner Konflikte zu besprechen.

Bei Schmerz als Konversion (s. Kap. 36) kann sich der Schmerz aber hartnäckig halten, und der Arzt darf es sich schon als Erfolg anrechnen, wenn er seinen Patienten vor gefährlichen Abklärungen und vor Eingriffen zu schützen vermag. Verhaltenstherapie allein und gemischt mit psychoanalytischer Psychotherapie ist in besonderen Situationen (Militär) bei einigen Fällen zum Teil erfolgreich, bisher aber nur mit kurzer Nachkontrolle durchgeführt worden (Dickes, 1974).

30.11   Differenzierung von anderen psychogenen Symptomen

30.11.1   Psychophysiologische Symptome

Herzklopfen, Zittern, Schwitzen, Durchfall usw. können als physiologische Folgen der Aktivierung von biologischen Notfallsystemen durch starke Affekte wie Wut, Angst usw. auftreten. Sie beruhen auf autonom-nervösen und neuroendokrinen Mechanismen und besitzen keine symbolische Bedeutung (s. Kap. 46).

30.11.2   Hypochondrische Symptome

Sie sind durch intensive Beschäftigung mit dem eigenen Körper und seinen Funktionen gekennzeichnet. Der Patient achtet auf Hauteffloreszenzen, Herzschlag, Atmung, Stuhlentleerung, Windabgang, leichten Schwindel beim raschen Aufstehen aus kauernder Stellung usw., die normalerweise einfach zur Kenntnis genommen werden oder sich unbemerkt abspielen, und leitet von ihnen den Hinweis auf eine verborgene bedrohliche Krankheit ab. Beruhigende Erklärungen, negative Befunde bei der Körperuntersuchung und normale Laborresultate zerstreuen die Befürchtungen nicht. Die Kranken drücken ihre Angst vor gewissen Erkrankungen und die damit verbundenen Vorstellungen bereitwillig aus. Eine Frau mit rechtsseitigen Oberbauchschmerzen und Angst vor Leberkrebs war sich darüber klar, daß der Verlust ihres Sohnes an einem Sarkom und der Mutter an Bauchkrebs zu ihrer Angst beigetragen hatten. Im Gegensatz zur Konversion steht nicht das Symptom im Vordergrund, und der Kranke zeigt keine „belle indifférence", sondern die Befürchtung einer schweren Krankheit. Die Symptome reichen von noch einfühlbaren Beschwerden bis zu bizarren, irrealen Erscheinungen wie das Genitale schrumpft, der Darm verfault usw. Nase, Bauch, Genital- und Analgegend sind bevorzugte Lokalisationen. Hypochondrische Symptome finden sich bei verschiedensten Persönlichkeitsstrukturen. Ängstlich-zwanghafte, unreif-hysterische, sensitiv-schizoide, manisch-depressive und schizophrene Individuen sind beschrieben worden (Bishop, 1980; Kenyon, 1976). Psychodynamisch können hypochondrische Symptome als Mittel verstanden werden, um sich vor einer Depression oder dem Ausbruch einer Psychose zu schützen (Kenyon, 1976; Kohut, 1974). Feindseligkeit, Schuld, Abhängigkeitsbedürfnisse, Masochismus, Erotisierung von Organen sind als beteiligte Faktoren bei der Hypochondrie genannt worden (Kenyon, 1976). Die Behandlung kann äußerst schwierig sein. Nie lohnt es sich, dem Patienten die Symptome ausreden zu wollen. Viel besser versucht der Arzt zu verstehen, welcher Sinn sich im Symptom verbergen könnte, welche individuelle Wirklichkeit, ob der Patient beispielsweise einen wichtigen Mitmenschen verloren hat, den Anschluß an die Umwelt und seine Mitmenschen nicht mehr findet und all sein Interesse nur noch auf sich und seinen Körper konzentriert (Freud, 1972).

30.11.3 Simulation

Sie ist viel seltener als Ärzte gemeinhin annehmen. Das Fehlen von organischen Veränderungen beim Klagen über Beschwerden wird häufig mit ihr gleichgesetzt, vor allem beim Symptom Schmerz. Bei den meisten dieser „Simulationen" handelt es sich um Konversions-Symptome. Simulation findet sich selten und nur in Situationen, die rasch erfassen lassen, daß der „Kranke" mit dem Symptom einen ganz bestimmten Zweck verfolgt, beispielsweise will er im Spital bleiben, um nicht ins Zuchthaus zur Fortsetzung der Strafverbüßung zurückkehren zu müssen, oder er will sich mit der Simulation der Rekrutierung in die Armee entziehen, oder sich als Drogensüchtiger die Droge verschaffen. Die Simulanten präsentieren Symptome, die den in der Organmedizin beschriebenen Bildern viel näher kommen als die Konversions-Symptome, die viel eher ganz persönlichen Krankheitsvorstellungen entsprechen. Ihre Träger wissen, daß sie etwas vorspielen, und sie sind deshalb in der Beziehung zum Arzt befangen, vorsichtig, mißtrauisch, reizbar und abwehrend, während der Patient mit Konversion abhängiger, zutraulicher und bereitwilliger ist, sich befragen zu lassen (Engel, 1970).

Ein junger Mann behauptete an einer Horton-Neuralgie zu leiden und gab an, nur Dilaudid-Atropin unter allen Morphinderivaten würde seine Schmerzen lindern. Der Serotonin-Antagonist Deseryl habe geholfen, er dürfe ihn wegen Nebenwirkungen, für die sich in den Belegen früherer Hospitalisationen keine Hinweise fanden, nicht mehr einnehmen. Die Symptombeschreibung zeigte feine Abweichungen vom bekannten klinischen Bild. Der Patient gab an, bis zu vier Anfälle pro Tag zu erleiden. Er wurde zur Beobachtung hospitalisiert und dringend ersucht, beim Auftreten eines Anfalls sofort Schwester und Arzt zu rufen. Er betonte nach zwei

Tagen mehrere Anfälle durchgemacht zu haben, aber er habe Arzt oder Schwester nicht gerufen, um ihnen seinen Anblick und die Auseinandersetzung mit seinem Leiden zu ersparen.

Die mit der Interviewtechnik nach Engel (s. Kap. 12) erhobene Anamnese läßt die Abweichungen der simulierten Symptome vom organisch bedingten Symptom meist erfassen. Im Gegensatz zu Simulanten, die einen ganz bestimmten sekundären Gewinn erstreben, „simulieren" bestimmte Menschen Symptome und Krankheiten, ohne daß der sekundäre Gewinn klar zutage tritt. Sie sind sich über ihre Manipulationen im klaren, können aber nicht auf sie verzichten und nehmen schwere Selbstschädigungen in Kauf (Spiro, 1968). Sie werden unter dramatischen Umständen hospitalisiert, zwingen z. B. mit „Herzinfarkt-Schmerzen" wiederholt Verkehrsflugzeuge auf dem nächsten Flugplatz zu landen, werden immer wieder operiert und verlassen fluchtartig das Spital, wenn ihre Machenschaften entlarvt zu werden drohen. Für die Krankheiten dieser Menschen wurde der Begriff des „Münchhausen-Syndroms" geprägt. Es kommt bei Menschen vor, die viel mit Medizin zu tun haben oder hatten (Spiro, 1968), wie Patienten mit langwierigen Krankheiten, Krankenschwestern und -pfleger, Laborantinnen und Laboranten, Röntgenassistentinnen und -assistenten usw.

30.11.4 Künstlich erzeugte Symptome

Die Krankheiten umfassen das Beklopfen des Handrückens mit einem harten Gegenstand bis zur Ödembildung, das Abbinden einer Extremität bis zur Bildung einer grotesken Schwellung, das wiederholte Verletzen der Haut bis zum Entstehen „nichtheilender" Läsionen, das Einbringen von Watte oder anderen Fremdkörpern unter die Haut mit Erzeugung chronischer Abszesse, das Vermischen von ausgehustetem Schleim oder von Erbrochenem mit Blut zum Vortäuschen von Hämoptoe oder Hämatemesis, das Abzapfen von Blut bis zur schwersten Anämie, das Abzapfen von Blut und Trinken desselben zum Erzeugen des Bildes einer Blutung aus dem Magen-Darm-Trakt, das Spucken in den Urin zur Vortäuschung einer erhöhten Amylaseausscheidung und damit einer chronisch-rezidivierenden Pankreatitis als Ursache der Bauchschmerzen (Robinson et al., 1982), das Einnehmen von Diuretika, Digitalis, Thyreoideahormonen (die künstlich induzierte „Hyperthyreose" weist ein normales oder tiefes Thyreoglobulin auf im Gegensatz zur echten, bei der es erhöht ist) (Mariotti et al., 1982), das Selbstinjizieren von Insulin, das Verletzen von Rektum, Vagina und Harnröhre mit spitzen Gegenständen zum Erzeugen von Blutungen aus diesen Organen, oft unterstützt durch die Einnahme von Antikoagulantien, das Manipulieren der Senkung, des Thermometers usw. Auch wenn dieses Syndrom selten ist, so muß der Arzt daran denken. Sonst erwägt er bei hypoglykämischen Attacken einer Frau mittleren Alters nur die organische Differentialdiagnose der Hypoglykämie, und dies kann bis zur Resektion am Pankreas führen, anstatt daß er sich überlegt, ob die Patientin nicht etwa Krankenschwester ist, in psychosozialen Schwierigkeiten steckt und sich Insulin einspritzt. Patienten mit Münchhausen-Syndrom sind psychisch schwer gestört. Sie sind grandios, selbstzentriert, realitätsfremd, äußerst abhängig von der Aufmerksamkeit ihrer Umgebung und besitzen wenig Steuerung ihrer Impulse. Sie spalten

ihr krankhaftes Verhalten von ihrer übrigen Lebensweise ab und empfinden wenig Angst oder Schuld über ihre Verhaltensweisen. Sie gehören wohl meistens zur Gruppe der Borderline-Patienten mit entsprechend schlechter Prognose.

Die Betreuung dieser Patienten ist sehr mühsam. Der Arzt soll sich bemühen, die Hintergründe zu verstehen, die den Patienten zwingen, sich so zu verhalten, und sollte Kritik und Vorwürfe vermeiden.

31 Funktionelle Syndrome in der inneren Medizin

Thure von Uexküll und *Karl Köhle*

31.1 Exemplarische Falldarstellung[1]

Die Patientin wurde auf Überweisung des Hausarztes zweimal – mit einem Intervall von 2 Jahren – in der Ambulanz einer internistischen Universitätsklinik gesehen. Bei der ersten Überweisung klagte die damals 37jährige Frau seit mehreren Monaten über Druckgefühl im Hals, Herzklopfen, Kopfschmerzen, Schlafstörungen und nachts Taubheitsgefühl in den Armen. Sie hatte seit Jahren unter Appetitlosigkeit und Obstipation zu leiden. Frühere Krankheiten wurden nicht angegeben. Sie war Einzelkind, Flüchtling und verlor beide Eltern auf der Flucht nach dem II. Weltkrieg. Jetzt lebte sie auf dem Lande, war mit einem tauben Mann verheiratet und hatte eine fünfjährige Tochter.

Zur Orientierung hatte der Hausarzt Befundberichte aus den vergangenen Jahren mitgeschickt. Sie berichteten über eine Lungendurchleuchtung mit normalem Befund vor 8 Jahren, eine frauenärztliche Untersuchung vor 6 Jahren, bei der an die Möglichkeit eines Vaginismus gedacht und eine Schwangerschaft festgestellt wurde. Dann folgte der Bericht über die normale Geburt einer Tochter. Vor 3 Jahren wurde ein dermatologischer Befund, Ekzem beider Hände, angeführt.

Die ambulante Untersuchung in der Klinik stellte lediglich Zeichen einer vegetativen Labilität mit vermehrtem Dermographismus und leichtem Fingertremor sowie eine geringgradig vergrößerte Schilddrüse, jedoch keine Zeichen einer Überfunktion fest. Es wurde zu einer Behandlung mit leichten Sedativa geraten.

2 Jahre später wurde die Patientin noch einmal vorgestellt. Der Hausarzt schickte jetzt weitere Berichte zur Einsicht. Die Briefe des Hausarztes, die wir mit seiner Zustimmung wiedergeben, zeigen einen Aspekt der Krankheitsbilder, über die in diesem Kapitel berichtet wird, der in der Klinik gewöhnlich zu kurz kommt, nämlich den lebensgeschichtlichen Zusammenhang mit der Familie, der beruflichen Umwelt, dem Hausarzt und den anderen medizinischen Instanzen unseres Gesundheitsversorgungssystems. Die folgenden Auszüge aus den Berichten des Hausarztes zeigen, daß sich das zunächst relativ unerhebliche Beschwerdebild im Laufe der Zeit dramatisch gesteigert hatte.

Der erste Bericht stammt von dem Chefarzt eines städtischen Krankenhauses in X. Er datiert 7 Monate nach der ersten ambulanten Untersuchung in der Universitätsklinik. Er ist von lakonischer Kürze: „Patientin wurde wegen Herzbeschwerden bei uns aufgenommen. Blutdruck 125/90 mm Hg, BSG 5/20. Wir behandelten stationär mit 12mal Strophanthin-Cordalin® i.v., Megaphen®-Tabletten, Favistan®-Tabletten und Pandigal®-Tropfen. Nach 3 Wochen gebessert in hausärztliche Weiterbehandlung entlassen."

4 Wochen nach dieser Entlassung datiert ein Brief des Hausarztes an einen Chirurgen, Dr. Z. In ihm heißt es: „Frau L., die einen nicht ganz leichten Alltag hat (ehelich) und ihrer Persönlichkeit nach wahrscheinlich über dem Niveau des Lebens steht, zu dem sie sich nun einmal verpflichtet hat, klagt über starken Druck im Hals, den man nur mit der Schilddrüse in Zusammenhang bringen kann. Typenmäßig bestehen Kreislaufschwankungen, die ihr zu schaffen machen. Moderner ausgedrückt könnte man auch von „vegetativen Störungen" sprechen. Vor einiger Zeit erfolgte geradezu ein Kollaps des Kreislaufes, der es erforderlich machte, daß sie einige Zeit in das Krankenhaus in X einrückte. Auf meinen Vorschlag, sich Ihnen einmal vorzustellen, der an sich nur gering vergrößerten Schilddrüse wegen, willigte sie sofort ein. Die acht in meinen Händen befindlichen Unterlagen lege ich bei und bitte Sie, zu entscheiden, ob es für Sie etwas zu tun gibt oder welche Behandlung Sie vorschlagen."

Kurz darauf wurde Frau L. strumektomiert. Der Hausarzt schreibt bei der Überweisung an die Universitäts-Ambulanz 7 Monate nach der Operation: „Frau L. schicke ich Ihnen noch einmal. Heute war die Frau wieder einmal in der Sprechstunde, nachdem ihr tauber Ehemann ihren Besuch schon angekündigt und auch davon berichtet hatte, daß die Hebamme seiner Frau den Blutdruck gemessen habe. Bei den Ihnen beigelegten Befunden, die Sie zum Teil schon einmal gesehen haben, ließ ich einen Bericht an Dr. Z., weil ich darin von der Situation schrieb, in der die Patientin sich befindet. Bei solchen Patienten – schreibt der Arzt weiter – bei denen man zunächst schon nicht den Weg findet, auf dem unter Umständen ein Erfolg zu bekommen wäre, bei denen in der eigenen Unsicherheit einem stets eine gewisse Furcht bleibt, am Ende doch eine organische Veränderung übersehen zu haben, ziehe ich dann ab und zu an der Notleine und frage bei Ihnen an, weil mir die Gefahr des Übersehens einer organischen Veränderung bei Ihnen geringer erscheint."

Es folgt noch ein Hinweis auf die „innerbetriebliche Unordnung mancher Menschen, die sich zur Linderung ihrer Beschwerden in der Durchführung irgendwelcher Maßnahmen Heilkundiger erschöpfen".

Bei der ambulanten Untersuchung gibt die Patientin an, daß sich das Engegefühl im Hals nach der Schilddrüsenoperation gebessert habe. Sie leide jedoch jetzt unter zeitweilig auftretenden Angstgefühlen, Schwindel und Übelkeit. Sie nehme Nitropräparate. Vor 3 Wochen habe sie eine Halsentzündung durchgemacht, dabei sei der Hals „voll Eiter" gewesen. Sie baue oft ab, bleibe dann tagelang im Bett. Zur Zeit rausche es

---

1 Die folgenden Ausführungen sind z. T. eine Neufassung des Aufsatzes: „Die Bedeutung funktioneller Syndrome in der Allgemeinpraxis". Ärztl. Wochenschrift 14, Heft 30/31 (1957) 573.

im Scheitel, der Kopf schmerze, in der Brust verspüre sie Stechen. Der Befund der Ambulanz: Normaler Allgemeinzustand, reizlose Strumektomienarbe, deutlich verstärkte Hautschrift, Tonsillen, Herz, Lunge, EKG, Blutdruck, Blutbild, Laborstatus und Urin o. B.

Die Diagnose „funktionelles Syndrom" stützte sich in diesem Fall auf drei Argumente:
- Den trotz zunehmender somatischer Beschwerden – über zwei Jahre hinweg – negativen Organbefund.
  Wir werden noch darauf hinweisen, daß dieses Argument allein nicht ausreicht, sondern daß die positive Diagnose einer emotionalen Problematik dazu kommen muß. Dafür ergeben sich in diesem Fall zwei Hinweise:
- Die Ehe mit dem älteren, ihr geistig unterlegenen, tauben Mann. Der Vaginismus, der in der Vorgeschichte erwähnt wird, verstärkt den Verdacht, daß hier ein Problemfeld liegt.
- Die Interaktion mit dem zunehmend irritierten Hausarzt und dessen Überweisungsstrategien.

Diesen Hinweisen hätte in dem Gespräch mit der Patientin nachgegangen werden müssen. Das geschah auch in der Ambulanz der Klinik nicht. Auch das ist – wie wir noch sehen werden – für diese Patienten ein ziemlich charakteristisches Schicksal.

31.2 Symptomatologie

Der Bericht zeigt einige charakteristische Züge, die wir bei funktionellen Syndromen finden:

Das Beschwerdebild ist schwer abgrenzbar. Die Symptome reichen von relativ genau lokalisierbaren körperlichen Beschwerden wie Herz-, Hals- oder Kopfschmerzen bis zu vagen Gefühlen eines Bedrücktseins. Diese gehen oft ohne feste Grenzen in rein seelisch empfundene Spannungszustände wie Angst, Unruhe, Unlust usw. über.

Die Neigung zum Chronischwerden und die Wandlungsfähigkeit der Symptomatik: Bei der Patientin stand zuerst das Herz, dann der Hals und schließlich der Kopf im Vordergrund. Gleichzeitig traten Angstzustände auf.

Die Schwierigkeiten, die sie dem Arzt bereiten:
Diagnostisch: Zum Ausschluß organischer Krankheiten werden immer neue Untersuchungen durchgeführt.
Therapeutisch: Hier ist die Resignation des Hausarztes eindrucksvoll. Als Konsequenz sehen wir das, was Balint „Aufteilung der Verantwortung" genannt hat, die durch Überweisungen an Fachärzte oder Kliniken erreicht wird. Sie beschwört bestimmte Gefahren herauf:
Überbewertung von Teilbefunden durch den Spezialisten (z. B. der Struma durch den Chirurgen Dr. Z.).
Iatrogene Schäden: Durch die wiederholten Untersuchungen und durch nicht indizierte Behandlungen

wird bei den Patienten die Überzeugung fixiert, ein organisches Leiden zu haben (nach der ambulanten Untersuchung in der Universitätsklinik wurde die Patientin zu einer stationären Behandlung eingewiesen. Nach der stationären Herzbehandlung suchte sie ihren Arzt immer häufiger auf. Nach der Strumektomie wurden ihre Besuche immer dringlicher. Zwischen den Besuchen muß die Hebamme den Blutdruck messen. Die Patientin muß schließlich tagelang im Bett bleiben).
Hohe Kosten: Sie entstehen durch die vielen diagnostischen Untersuchungen, die nutzlosen oder schädlichen Behandlungen und – nicht selten – die Kurverschickungen. Zu diesen Kosten muß man den Arbeitsausfall durch die Krankheit und Frühinvalidisierung rechnen, zu der es nicht selten kommt.

Bei der Betrachtung dieser Krankengeschichte gewinnt man den Eindruck, daß alle diese Probleme zusammenhängen. Ja, daß sie sich vielleicht sogar nach Art eines Circulus vitiosus – in den Patient, Hausarzt, Spezialist und Krankenhaus eingeschlossen sind – gegenseitig hervorbringen und verstärken (vgl. Abb. 31–1).

Man wird einwenden, daß die Geschichte der Patientin einen Sonderfall darstellt. Das Gegenteil ist richtig: Nach epidemiologischen Untersuchungen gehören 30–50% aller Patienten, die einen Arzt oder ein Krankenhaus aufsuchen, in diese Gruppe. Auch zu der Frage, wie häufig bei solchen Kranken nicht indizierte Eingriffe vorgenommen werden, haben schon früh zwei Arbeiten gezeigt, daß diese Fälle keine Raritäten darstellen: Macy und Allen (1949) berichten über eine Nachuntersuchung von 235 Patienten, bei denen die Mayoklinik 6 Jahre zuvor die Diagnose „nervöser Erschöpfungszustand" („chronic nervous exhaustion") gestellt hatte. Die Nachuntersuchung sollte klären, ob das Beschwerdebild der Beginn eines damals noch nicht erkannten organischen Leidens war. Das Resultat bestätigte in 94% wiederum das Fehlen organischer Schäden, ergab aber, daß im Verlauf der 6 Jahre an 200 Patienten 289 verschiedene Operationen durchgeführt worden waren. Bennet

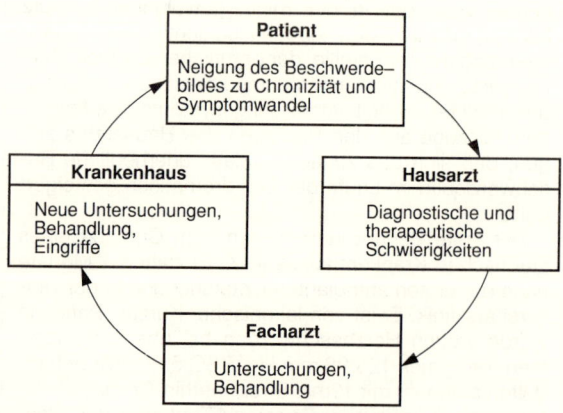

Abb. 31–1. Kreislauf der Überweisungen und seine Gefahren für den Patienten mit „funktionellen Syndromen".

(1936) stellt fest, daß 150 Patienten der psychiatrischen Abteilung eines Allgemeinkrankenhauses, die früher unter verschiedenen Diagnosen wegen organischer Krankheiten behandelt worden waren, 244 chirurgische Operationen durchgemacht hatten.

Hinter diesen Zahlen verbirgt sich ein Problem, vor dem der Arzt täglich steht: Bei der Diagnose „funktionelles Krankheitsbild" geht es nicht nur darum, einen organischen Befund auszuschließen – fast ebenso häufig geht es darum, einen Befund, der erhoben wurde, richtig einzuschätzen. Hier ergeben sich nicht selten große Schwierigkeiten.

31.2.1 Begriffsbestimmung

Die Schwierigkeiten beginnen bereits mit der Begriffsbestimmung. Wir glauben, wir präjudizieren am wenigsten, wenn wir diese Krankheitsbilder als „funktionelle Syndrome" bezeichnen. Dieser Terminus gibt uns die Möglichkeit, einen Oberbegriff für ein Krankheitsbild zu verwenden, hinter dem sich sowohl organische wie psychische Störungen als auslösende bzw. unterhaltende Faktoren verbergen können. Wir sprechen dann im ersten Fall von „symptomatischen", im zweiten von „essentiellen funktionellen Syndromen". Dieses Kapitel handelt von den weitaus häufigeren essentiellen Formen. Bei ihnen hat der Terminus „funktionell" eine doppelte Bedeutung: Er deutet an, daß die somatischen Symptome für den Patienten eine „Funktion" haben – d.h., daß sie für ihn nicht „sinnlos" sind, sondern im Sinne eines primären und/oder sekundären Krankheitsgewinns eine Aufgabe erfüllen, die dem Patienten zwar nicht bewußt ist, die der Arzt aber versuchen muß aufzudecken. Der Terminus macht darauf aufmerksam, daß es eine „funktionelle Anatomie" oder eine „Anatomie der Funktionen" gibt, und daß die verschiedenen Körperfunktionen bei psychosozialen Anforderungen z.B. im Rahmen von Bereitstellungsreaktionen unterschiedliche Aufgaben übernehmen.

Jores (1973) schlägt für das gleiche Krankheitsbild die Bezeichnung „psychovegetative Störungen" vor. Früher war der Begriff „vegetative Dystonie" am beliebtesten. H. Hoff bekam auf eine Umfrage, was „vegetative Dystonie" sei, von 10 verschiedenen Ärzten 10 verschiedene Antworten. Die gleiche Unklarheit zeigt sich, wenn wir die Fülle verschiedenartiger Bezeichnungen betrachten, die letzten Endes alle dasselbe meinen.

Die Zusammenstellung der Tabelle 31–1 zeigt eine Musterkollektion, die keinen Anspruch auf Vollständigkeit erhebt. In ihr finden wir die verschiedenen Lieblingsmythologien über den Sitz der Krankheit, die mit den Moden der Medizin wechseln. Einmal wird er in das vegetative Nervensystem, dann in das Endokrinium und schließlich in die Psyche verlegt.

Wesiack (1974) macht darauf aufmerksam, daß diese Krankheitsbilder die Medizin schon seit 300 Jahren beschäftigen und den Ärzten schon immer ähnliche Rätsel aufgaben wie heute. Er erwähnt einen Brief, in dem Thomas Sydenham 1681 diese Krankheitsbilder schildert und bereits darauf hinweist, daß sie infolge ihres „proteus- und chamäleonartigen" Charakters organische Krankheiten nachahmen würden. Sie seien außerordentlich häufig und machten über die Hälfte seines nicht fieberhaften Krankengutes aus.

Allen medizinischen Begriffen haften Mängel an. Sie sind um so größer, je größer ihr Gehalt an unbewiesenen Voraussetzungen ist. Auch der Begriff „funktionelle Syndrome" enthält unbewiesene Voraussetzungen, aber sie sind weniger apodiktisch und mehr auf die praktische Situation des Arztes zugeschnitten als die anderen Begriffe. Wir wollen sie der Reihe nach ansehen.

Der Begriff „funktionelles Syndrom" stellt drei Hypothesen auf:

1. Das Beschwerdebild soll das Resultat von Funktionsstörungen sein.
2. Diese Funktionsstörungen sollen nicht auf organischen Veränderungen beruhen.
3. Sie sollen im Falle der essentiellen funktionellen Syndrome durch seelische, vor allem emotionale Vorgänge ausgelöst und unterhalten werden.

Alle drei Hypothesen werfen Fragen auf, die wir im Einzelfall zwar oft nur schwer – manchmal überhaupt nicht – beantworten können. Sie umreißen aber die konkrete Problematik, vor der der Arzt bei diesen Krankheitsbildern immer wieder steht. Beginnen wir mit der ersten Hypothese:

1. Die Zuordnung körperlicher Beschwerden zu objektiv nachweisbaren Funktionsstörungen bleibt oft problematisch. Auch dort, wo wir Funktionsstörungen finden, ist die Diskrepanz zwischen dem objektiven Befund und der subjektiven Beschwerde oft erheblich. Diese Feststellung gilt aber auch für die organischen Krankheiten. Wir stoßen hier auf ein grundsätzliches Problem. Es betrifft die Frage, was ein „Symptom" eigentlich ist, und wie es sich bildet. Dabei zeigt sich, daß eine Betrachtungsweise nicht ausreicht, die das subjektive Beschwerdebild lediglich als kausale Folge körperlicher Störungen auffaßt, seien diese Störungen nun funktioneller oder struktureller Art. Wir kommen weiter, wenn wir Symptome als „Ausgleichsbestrebungen" auffassen, d.h. als aktive

Tabelle 31–1. Synonyme Bezeichnungen für funktionelle Syndrome.

1. a) Sympathikotonie
   b) Vagotonie
2. a) Sympathische Hypertonie
   b) Vegetative Areflexie
3. Vegetative Stigmatisation
4. Vegetative Dystonie
5. Vegetative Neurose
6. Vegetative Ataxie
7. Vegetativ-endokrines Syndrom
8. Funktionelle Erkrankung
9. Psychogene Syndrome
10. Organneurosen
11. Larvierte Depression

Leistungen der Gesamtpersönlichkeit des Kranken. Auf diese Weise können wir auch den Anteil ins Auge fassen, den die psychische und soziale Situation des Patienten an der Ausgestaltung der Symptomatik hat.

2. Die zweite Hypothese, daß die Funktionsstörungen nicht durch organische Schädigungen verursacht seien, wirft ein Problem von großer praktischer Bedeutung auf: Organische Schäden liegen nicht offen zutage, wir müssen nach ihnen suchen. Wann und aufgrund welcher Kriterien dürfen wir mit dem Suchen aufhören? Vor dieser Frage stehen wir vor allem deswegen immer wieder, weil funktionelle Syndrome eine große Ähnlichkeit mit Syndromen haben, die wir im Prodromalstadium einer Infektionskrankheit, bei chronischen Infekten, z.B. einer Tuberkulose, einer Endokarditis, bei chronischen Vergiftungen, aber auch im Anfangsstadium bösartiger Krankheiten beobachten. In solchen Fällen ist ein organisches Substrat oft nur schwer, manchmal überhaupt nicht zu finden. Die Diagnose „essentielles funktionelles Syndrom" kann also außerordentlich schwierig sein. Sie ist in jedem Fall sehr verantwortungsvoll. Der Gefahr, ein beginnendes organisches Leiden zu übersehen und damit den Zeitpunkt für eine erfolgversprechende Behandlung zu versäumen, steht die andere Gefahr gegenüber, durch zu lange fortgesetzte Untersuchungen und durch nicht indizierte Behandlungsverfahren einen Patienten auf die Diagnose eines organischen Leidens zu fixieren, und eine „Patientenkarriere" einzuleiten oder zu unterhalten. Schließlich kommt zu der Gefahr, einen Befund zu übersehen – wie bereits betont wurde – noch die Gefahr, einen Befund zu erheben und falsch zu deuten.

Unser Krankenbericht zeigt auch, welche Rolle eine bestimmte Einstellung der modernen Medizin und unserer Gesellschaft bei diesem Problem spielt: Ein organisches Geschehen zu übersehen, gilt als „Kunstfehler". Ein neurotisches Problem zu ignorieren, hat keine derartigen Konsequenzen. Der Arzt ist daher mit der Aufgabe, die „essentiellen" von den „symptomatischen" funktionellen Syndromen zu unterscheiden, oft überfordert.

3. Die dritte Hypothese, daß Funktionsstörungen seelisch ausgelöst sind, wirft ein ätiologisches und diagnostisches Problem auf. Hier gehen die Überlegungen des ärztlichen Alltags meist so, daß seelische Gründe angenommen werden, wenn man keine organischen Veränderungen findet. Auf diese Weise wird die Annahme eines funktionellen Syndroms zu einer „Exklusiv-Diagnose", aber nicht im Sinne der alten Wiener Schule, die festumrissene Krankheitsbilder gegeneinander abwog, sondern eher nach dem Motto: „Was ich nicht diagnostizieren kann, das sehe ich als ‚seelisch' an."

Hier ist die Forderung zu erheben: **„Seelische Störung"** oder **„emotionaler Konflikt"** darf keine Ausschlußdiagnose, sondern kann und muß eine positive Diagnose sein.

Dem stehen aber Hemmungen und Widerstände des Arztes entgegen, sich auf diese Probleme seiner Patienten einzulassen. Sie haben sehr reale Hintergründe: Die eingehende Exploration, die zur Stellung der Diagnose unerläßlich ist, kostet Zeit, die dem Arzt – vor allem im Vergleich zu physikalischen Untersuchungen – nicht entsprechend vergütet wird. Und dann: welche Konsequenzen soll er aus der Diagnose ziehen? Zur Behandlung dieser Patienten fehlt ihm nicht nur die Zeit, sondern – wie unser Bericht zeigt – auch die Erfahrung. Die Überweisung an einen Psychotherapeuten wird von dem Patienten in der Regel abgelehnt („Herr Doktor, es fehlt mir im Magen und nicht im Kopf!"). Selbst wenn die Zustimmung des Patienten erreicht werden kann, ist die Überweisung in den seltensten Fällen möglich.

31.2.2 Untergruppen und spezielle Erscheinungsformen

Es würde die Diagnose erleichtern, wenn es bestimmte Organe oder Organsysteme gäbe, die von funktionellen Syndromen verschont oder bevorzugt werden. Es gibt zahlreiche Statistiken, aus denen die Bevorzugung bestimmter Organsysteme hervorzugehen scheint. Übereinstimmung herrscht aber nur insoweit, als das kardiovaskuläre und das gastrointestinale System besonders häufig betroffen sind, daß es im übrigen aber kein Organ gibt, das nicht im Mittelpunkt funktioneller Beschwerden stehen könnte.

Dagegen gibt es unter den funktionellen Syndromen profilierte Beschwerdebilder, deren Kenntnis schon relativ frühzeitig eine Vermutungsdiagnose ermöglicht. In der Tabelle 31–2 sind Syndrome zusammengestellt, die sich zwar oft überschneiden, aber doch relativ gut abgrenzen lassen. Sie werden zum Teil in den folgenden Kapiteln besprochen.

Dieses Schema müßte – wie Cremerius (1968) hervorhebt – noch durch die funktionellen Syndrome ergänzt werden, die in den nichtinternistischen Fachbereichen, wie z.B. der Ophthalmologie, der Otologie, der Dermatologie usw., zur Beobachtung kommen.

31.2.3 Psychologische Symptomatik

Es gibt bei Patienten mit funktionellen Syndromen eine Reihe relativ charakteristischer Verhaltensmerkmale: Die Art und Weise, wie Befürchtungen geäußert

---

Tabelle 31–2. Einteilungsschema der „funktionellen Syndrome" nach Cremerius (1968).

1. Funktionelle Magensyndrome
2. Funktionelle Atmungssyndrome
3. Funktionelle kardiovaskuläre Syndrome
4. Funktionelle Kopfschmerzsyndrome
5. Funktionelle Hautsyndrome
6. Funktionelle Syndrome des Urogenitaltraktes
7. Funktionelle Syndrome des unteren Verdauungstraktes
8. Funktionelle diffus wechselnde, nicht dauernd an einem Organ lokalisierte Syndrome

oder unterdrückt werden, die emotionale Reaktion auf bestimmte Ereignisse in der Lebensgeschichte, die Einstellung dem Arzt gegenüber, können wertvolle Aufschlüsse über die Persönlichkeit des Kranken geben. Darauf wird unter 31.5 näher eingegangen. Hier sollen nur drei auffällige Verhaltensmerkmale aufgeführt werden: Es besteht häufig die Meinung, daß Patienten mit funktionellen Syndromen durch eine wortreiche, klagsame Theatralik zu erkennen seien. Das gilt jedoch nur für einen – nicht einmal häufigen – Typ, viel häufiger ist es, daß dem Patienten solche Ausdrucksmöglichkeiten fehlen. Auch die „Symptom-Pedanten" – wie man sie nennen könnte –, die ihre Beschwerden aus einer Liste vorlesen, um ja keine zu vergessen, findet man unter den Kranken mit funktionellen Syndromen immer wieder, aber auch sie sind nicht allzu häufig. Eine andere Gruppe sind die stillen, unauffälligen, depressiven, zu hypochondrischen Ideen neigenden Kranken, die mit großer Hartnäckigkeit immer wieder zur Schilderung ihrer Symptome zurückkehren.

In diesen Verhaltenstypen werden bestimmte neurotische Umgangsstile sichtbar: Hysterische Züge in der theatralischen Selbstdarstellung, zwanghafte Züge des Pedanten und die depressive Stimmungslage von Patienten, die Anklagen nur in Form von Klagen vorbringen können.

31.3 Epidemiologie

Wie häufig sind diese Krankheiten? Die epidemiologischen Untersuchungen lassen trotz großer Unterschiede ihrer Ergebnisse keinen Zweifel, daß die Gruppe der funktionellen Syndrome sehr groß ist. Sie ist sicher nicht viel kleiner, vielleicht sogar größer als alle anderen Krankheitsgruppen zusammen. Diese Tatsachen sind seit über 30 Jahren bekannt. Bekannt sind auch die großen menschlichen und gesellschaftlichen Kosten, die durch diese Krankheiten verursacht werden. Trotzdem hat die Medizin daraus keinerlei Konsequenzen gezogen. Im Gegenteil, man tut

weiter so, als würden sie nicht existieren. Sie passen nicht in das Bild, das die moderne Medizin von „richtigen Krankheiten" hat. Wir werden auf diesen Punkt noch zurückkommen. Er ist ein wichtiges („Krankheits"-)Symptom unserer Industriekultur und ihrer Medizin.

Um diesen Punkt zu unterstreichen, bringen wir die alten epidemiologischen Befunde. Sie sind in der Zwischenzeit immer wieder bestätigt worden. (Die Befunde neuerer Untersuchungen, insbesondere die Befunde der Mannheimer Studie, sind in Kapitel 4 ausführlich dargestellt.) Die folgende Abbildung 31–2 enthält eine Zusammenstellung aus 13 verschiedenen Arbeiten (Th. v. Uexküll, 1958).

Man sieht, daß die Angaben erheblich differieren. Dafür sind zweifellos die Auswahlkriterien verantwortlich, die so unterschiedlich sind, daß ein Vergleich kaum möglich ist. Eine genauere Besprechung erfordert die Arbeit von Kaufman und Bernstein (1957) wegen der überraschend hohen Zahl, die diese Autoren angeben, aber auch wegen der besonderen Sorgfalt der Untersuchungen, auf denen ihre Angaben beruhen: Es handelt sich um die Auswertung der Diagnosen von 1000 Patienten, die in einem bestimmten Zeitraum dem Ambulatorium des Mount Sinai Hospitals in New York wegen diagnostischer Schwierigkeiten überwiesen wurden. Alle Patienten wurden von einem Internisten und – soweit erforderlich – von anderen Spezialärzten untersucht. Durchschnittlich wurde jeder Patient 5mal bestellt und jedes Mal etwa zwei Stunden untersucht, so daß auf den einzelnen im Durchschnitt 10 Untersuchungsstunden kommen. Die Konsultation des Psychiaters, der die Diagnose „psychogenes Syndrom" stellte, erfolgte erst, wenn die Untersuchung der anderen Spezialisten keine organische Ursache aufdecken konnte, welche die Beschwerden hinreichend erklärt hätte. Die Autoren meinen, daß bei einer Konsultation des Psychiaters in allen Fällen der Prozentsatz noch höher liegen würde.

Eine Ursache für einen hohen Prozentsatz funktioneller Syndrome bei einer epidemiologischen Untersuchung kann also auch die Genauigkeit der Diagno-

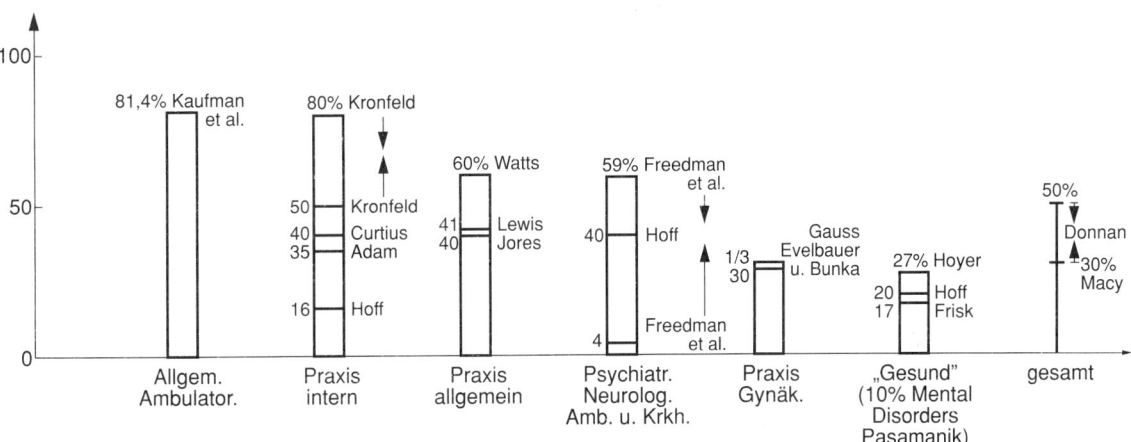

Abb. 31–2. Häufigkeit der „funktionellen Syndrome". (Die Autorennamen sind in Höhe der Häufigkeitsangaben eingezeichnet; unten stehen die ärztlichen Institutionen, aus denen die Angaben stammen; Th. v. Uexküll, 1958.)

stik sein. Wir stellen ja – wenn wir ehrlich sind – nicht ganz selten die Diagnose einer organischen Krankheit, ohne daß sich der Zusammenhang zwischen Beschwerden und einem mehr oder weniger eindrucksvollen Befund auf eine sehr strenge Beweisführung stützen kann. Bei genauer, eventuell wiederholter Kontrolle kann sich aber mancher – auch anfangs eindrucksvolle – Befund als belanglos erweisen.

Eine Zusammenstellung aus der Gießener Medizinischen Poliklinik (Th. v. Uexküll, 1958, 1960) hatte die Aufgabe, die Zahl der „reinen" funktionellen Syndrome zu ermitteln. Mit dieser Bezeichnung sind Patienten gemeint, bei denen sich überhaupt kein organischer Befund – auch belangloser Art, wie EKG-Veränderungen ohne pathologische Bedeutung, vorübergehende leichte Hypertonie usw. – erheben läßt. Es wurden also alle Fälle mit einer problematischen Bewertung der Bedeutung eines Befundes für das Beschwerdebild eliminiert. Die Patienten waren ausnahmslos von niedergelassenen Ärzten zur diagnostischen Klärung überwiesen. Die Zahl der Patienten, bei denen die Diagnose „funktionelles Syndrom" nach diesen sehr strengen Kriterien als Grund ihrer Überweisung gestellt wurde, stellt also einen Minimalwert dar. Sie betrug bei 7825 Kranken 25,5%, wobei die Frauen leicht überwogen. Jeder 4. Patient wurde also wegen Beschwerden überwiesen, die ohne jeden Zweifel nur funktioneller Natur waren (vgl. Abb. 31–3).

Weiss und English (1949) schlagen eine Gruppeneinteilung vor, die dieser Problemstellung Rechnung trägt. Sie unterscheiden:

– Eine Gruppe der „reinen" funktionellen Syndrome und schätzen, daß etwa ein Drittel aller Patienten, die einen Arzt aufsuchen, in diese Gruppe gehört.
– Ungefähr ebenso groß ist nach ihrer Schätzung die Gruppe der Patienten, die den Arzt wegen Beschwerden aufsuchen, die teilweise funktioneller Natur sind, bei denen aber organische Befunde erhoben werden. Sie betonen, daß diese Gruppe besonders wichtig ist, einmal weil sie dem Arzt die größten diagnostischen Schwierigkeiten bereitet, dann aber auch, weil Funktionsstörungen bei einem vorgeschädigten Organismus mehr Schaden anrichten können als bei einem gesunden.

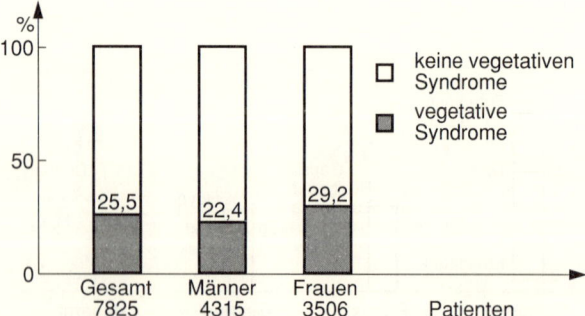

Abb. 31–3. Häufigkeit der „rein funktionellen Syndrome" unter den Patienten der Gießener Medizinischen Poliklinik. (Statt „funktioneller Syndrome" findet sich noch die Bezeichnung „vegetative Syndrome".)

– Die dritte Gruppe würde schließlich die Patienten umfassen, bei denen wir zum Schluß kommen, daß die organischen Veränderungen für ihre Beschwerden allein entscheidend sind.

Dölle (1984) zitiert drei Arbeiten, die in diesem Zusammenhang interessant sind:

– Eine Arbeit aus dem Jahre 1954 von Horder und Horder über sog. geringfügige Gesundheitsstörungen. 80% der Betroffenen werden nicht von einem Arzt gesehen, 15% suchen einen Arzt in seiner Praxis auf und nur 5% werden in einem Krankenhaus diagnostiziert und behandelt.
– Bezüglich der Verteilung der Beschwerden fand Anders (1979) bei einem internationalen Vergleich, daß in Deutschland Schmerzen mit 28,2% an erster Stelle standen. Danach kamen Kopfweh mit 22,7% und gastrointestinale Beschwerden mit 14,1%.
– Bei einer Untersuchung über Selbstbehandlung medizinischer Laien (Bundesminister für Arbeit und Sozialordnung, 1981) waren Erkältungskrankheiten mit 76,1% am häufigsten. Dann folgten Kopfschmerzen mit 47,8% und Magen-Darm-Symptome mit 34,6%.

Für die These, daß es sich bei den meisten der als „psychosomatische Symptome" aufgeführten Sensationen um Phänomene handelt, die bei vielen Menschen vorkommen, ohne einen Krankheitswert zu haben, spricht eine Arbeit von Wright und Wright (1981). Sie fanden in einer longitudinalen Untersuchung bei 11- bis 18jährigen gesunden Schülerinnen und Schülern, daß „psychosomatische Symptome" häufig (bei Mädchen häufiger als bei Knaben) angegeben wurden, und daß der Häufigkeitsgipfel bei 13 Jahren lag.

Es sieht also so aus, als ob die Beschwerden, unter denen Patienten mit funktionellen Syndromen leiden, einen fast ubiquitären Befund darstellen. Die Frage ist noch völlig offen, warum und wann solche Beschwerden mit einem Krankheitserleben gekoppelt werden. Dafür spielt nach H. Hoff die Intensität der Beschwerden keine entscheidende Rolle.

31.3.1 In welchen Altersklassen treten diese Krankheitsbilder auf?

Zu dieser Frage machten Pasamanik (1957) und Mitarbeiter eine interessante Feststellung: Sie fanden funktionelle Syndrome in der Altersklasse zwischen 14 und 34 Jahren doppelt so oft wie zwischen 35 und 64 Jahren. Jenseits des 65. Lebensjahres dagegen gar nicht mehr. Diese Feststellung deckt sich im wesentlichen mit den Ergebnissen der Untersuchung in der Gießener Medizinischen Poliklinik und anderen Erhebungen, allerdings mit dem Unterschied, daß auch jenseits des 65. Lebensjahres noch funktionelle Syndrome – wenn auch in geringerer Zahl – gefunden wurden (Abb. 31–4).

Franke und Mitarbeiter (1970) fanden in einer Studie über 148 Hundertjährige, daß die Zahl der 50jäh-

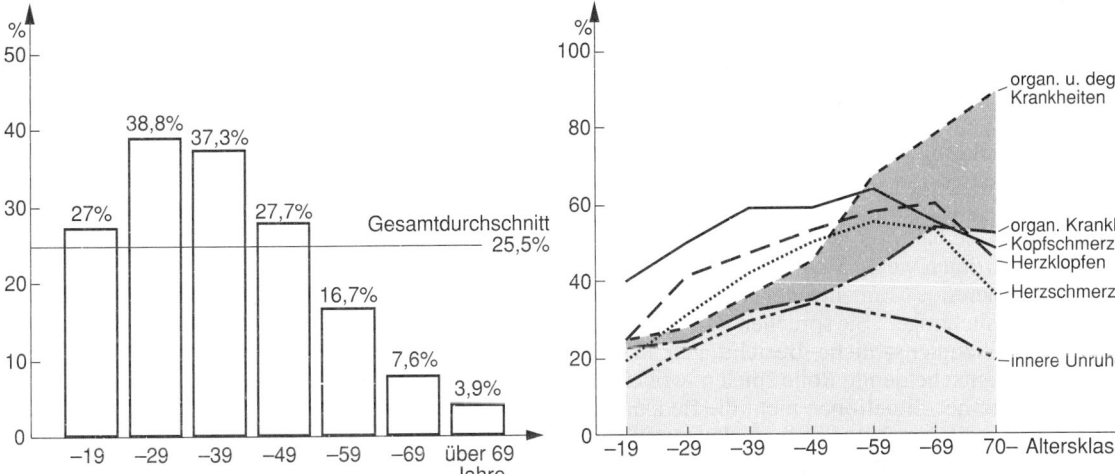

Abb. 31–4. Altersverteilung „funktioneller Syndrome".

Abb. 31–5. Altersabhängige Häufigkeit von organischen und degenerativen Krankheiten sowie von einzelnen funktionellen Symptomen (bei Männern).

rigen, die über Herzschmerzen klagten, höher war als die der 75jährigen, und daß bei den Hundertjährigen die Zahl am niedrigsten lag.

Für die Abnahme der funktionellen Syndrome im höheren Alter gibt es prinzipiell 5 verschiedene Erklärungsmöglichkeiten:

1. Die Störungen heilen mit zunehmendem Alter aus.
2. Patienten, die an solchen Störungen leiden, sterben früher.
3. Die früher funktionellen Störungen sind inzwischen in organische Krankheiten übergegangen.
4. Ein Patient, der ein organisches Leiden erwirbt, verliert sein funktionelles Syndrom.
5. Die Abnahme ist nur vorgetäuscht: Jenseits des 65. Lebensjahres werden die Symptome funktioneller Syndrome über den Beschwerdebildern der dann vorherrschenden organischen oder degenerativen Krankheiten übersehen.

Die unter 1. und 2. aufgeführten Möglichkeiten sind wenig wahrscheinlich. Funktionelle Syndrome sind ein chronisches Leiden, Spontanheilungen sind auch im Alter selten (Cremerius, 1968a). Die Lebenserwartung dieser Patienten ist nicht gemindert.

Die dritte Möglichkeit ist durch Untersuchungen der letzten Jahre weitgehend ausgeschlossen. Nur bei Patienten mit funktionellen Beschwerden des Oberbauches und bestimmten kardiovaskulären Syndromen wird überdurchschnittlich häufig das Auftreten von Magengeschwür und Hypertonie beobachtet.

Um die vierte und fünfte Möglichkeit zu prüfen, wurde die Häufigkeit funktioneller Beschwerden, organischer und degenerativer Leiden in den verschiedenen Altersklassen untersucht (Abb. 31–5).

Während die Zahl der Patienten, bei denen ein organisches oder degeneratives Leiden diagnostiziert wurde, zwischen dem 50. und 60. Lebensjahr stark zunahm und dann noch weiter stieg, fiel mit dem 60. Lebensjahr die Zahl der Patienten, die auf Befragen Kopfschmerzen, Herzklopfen, Herzschmerzen,

innere Unruhe angaben, in eindrucksvoller Weise ab. Der alte Spruch, daß das Alter die meisten Plagen habe, stimmt also für die funktionellen Plagen offenbar nicht.

Damit bleibt nur die vierte Möglichkeit als Erklärung für die Abnahme funktioneller Syndrome im Alter: Patienten, die an einem organischen oder degenerativen Leiden erkranken, verlieren ihre funktionellen Syndrome. Die wahrscheinlichste Deutung dafür scheint die Annahme, daß die Symptome organischer Krankheiten die funktionellen Beschwerden überflüssig machen. Wahrscheinlich verlieren sie den subjektiven Krankheitswert, den sie vor dem Auftreten der organischen Erkrankungen hatten.

Dieser Punkt ist auch differentialdiagnostisch von Bedeutung: Es wird immer wieder über Fälle berichtet, in denen der Arzt die Diagnose eines „funktionellen Syndroms" gestellt und die Beschwerden erfolgreich psychotherapeutisch behandelt hat – bis sich herausstellte, daß der Patient „doch" an einer organischen Krankheit litt. Diese Fälle gibt es natürlich und der Arzt muß dieser Möglichkeit durch exakte Diagnostik Rechnung tragen – aber sie sind offenbar außerordentlich selten. Sonst würde man in den Berichten nicht immer wieder den gleichen Paradefällen begegnen, z.B. der psychotherapeutisch erfolgreich behandelten Polydipsie, hinter der sich bei der Obduktion des durch einen Unfall ums Leben gekommenen Patienten ein Hirntumor fand, oder dem Patienten mit erfolgreich behandelten Durchfällen, bei dem dann „doch" ein Kolonkarzinom entdeckt wurde. In solchen Fällen muß man auch daran denken, daß eine organische Krankheit in ein vorher bestehendes, funktionelles Beschwerdebild „hineinwachsen" kann. Wir haben ja bereits (Seite 478) darauf hingewiesen, daß die Einstellung der modernen Medizin seltene Fehldiagnosen organischer Krankheiten ungleich schwerer bewertet als ungemein häufigere Fehldiagnosen, bei denen psychologische Probleme übersehen werden.

31.4 Ätiologie und Pathogenese

Bei der Frage nach Ätiologie und Pathogenese müssen wir – wie bei allen Krankheiten – zwischen konstitutionellen (genetischen), disponierenden (im Verlauf der Entwicklung erworbenen) und auslösenden Faktoren unterscheiden, die gemeinsam das multifaktorielle Wurzelgeflecht einer Krankheit bilden.

Über genetische Faktoren wissen wir bei den funktionellen Syndromen wenig Sicheres.

Über disponierende Momente gibt es eine Reihe gut belegter Untersuchungen. Sie sprechen dafür, daß pathogene zwischenmenschliche Beziehungen in der Kindheit eine entscheidende Rolle spielen, wobei die sog. „broken home"-Situationen nicht die Bedeutung haben, die man ihnen früher zumaß. Patienten mit funktionellen Syndromen stammen häufiger aus Familien, die durch eine „kohäsive" und „rigide" Struktur auffallen und sozial überangepaßt sind (Grollnick, 1972).[2]

Adler und Mitarbeiter (1989) fanden in einer Gruppe von Patienten mit psychogenen Schmerzen signifikant häufiger Störungen der zwischenmenschlichen Beziehungen in der Kindheit als in Vergleichsgruppen von Patienten mit organisch bedingten Schmerzen und organisch Kranker ohne Schmerzen.

Funktionelle Syndrome oder Verhaltensstörungen als Reaktion von Kindern auf Ereignisse, welche ihre Familie betreffen, sind Kinderärzten geläufig. Hodges und Mitarbeiter (1984) verglichen in einer retrospektiven Untersuchung die Zahl und die Schwere belastender Ereignisse in den Familien von 30 Kindern mit rezidivierenden Abdominalschmerzen, 67 verhaltensgestörten und 42 gesunden Kindern. Sie fanden in den Familien der 1. und 2. Gruppe signifikant mehr und belastendere Ereignisse als in der 3. Gruppe. In den Familien der verhaltensgestörten Kinder betrafen die belastenden Ereignisse in erster Linie Krankheit, Krankenhauseinweisung und Tod (genauere Einzelheiten vgl. Kap. 54).

Nach Jores (1973) haben bei der Entstehung einer Disposition für den Erwerb eines funktionellen Syndroms folgende Momente eine besondere Bedeutung:
– Spezielle Formen familiärer Einflüsse auf die Angstverarbeitung,
– unbewußte, auf das Kind gerichtete elterliche Erwartungen,
– bestimmte Erziehungsbilder und
– die sozio-dynamische Familienkonstellation.

Unter den Patienten mit funktionellen Syndromen lassen sich keine einheitlichen Persönlichkeitsstrukturen oder neurotische Krankheitsbilder finden. Bei entsprechender Belastung (s. u.) können offenbar alle Menschen ein funktionelles Syndrom entwickeln. Trotzdem findet man bei diesen Patienten häufig Unsicherheit und Kontaktschwierigkeiten als Ausdruck einer Störung des Selbstwerterlebens. Als Kompensation entwickeln manche ein extremes Bemühen, sich anzupassen und durch Leistung Zuneigung und Anerkennung zu erwerben. Solche Patienten können dann vor jeder Leistung Angst haben zu versagen.

Schließlich ist für überangepaßte Menschen jede Änderung der sozialen Umgebung beunruhigend.

Als auslösende Faktoren werden immer wieder „Belastungen" (Streß) oder „Verlusterlebnisse" beschrieben. Die Schwierigkeit dieser Begriffe liegt in der Unmöglichkeit, einem Ergebnis anzusehen, ob, für wen und unter welchen Bedingungen es eine „Belastung" oder einen „Verlust" bedeutet. Diese Definition ist meist erst post festum möglich. Aus diesem Grunde ist der neutrale Begriff der „psychosozialen Veränderungen" (psycho-social transition) hilfreich. Mit ihm lassen sich Ereignisse identifizieren, die Anpassungsleistungen an veränderte Situationen erfordern. Solche Anpassungsleistungen spielen als auslösende Faktoren funktioneller Syndrome (aber auch organischer Krankheiten) eine wichtige Rolle.

Ein psychosomatischer Verständnisansatz funktioneller Syndrome sollte zunächst die Diskrepanz zwischen objektivierbarem Befund und der Beeinträchtigung des Befindens, die Diskrepanz zwischen fehlender oder unbedeutender Funktionsänderung und den oft erheblichen subjektiven Beschwerden erklären.

Von Uexküll (1962) hat einen solchen Ansatz am Beispiel funktioneller Herz-Kreislauf-Störungen entwickelt (Abb. 31–7). In diesem Konzept kann eine Funktionsänderung oder Funktionsstörung als Folge einer Noxe, einer körperlichen Erkrankung, einer Anstrengung auftreten, aber auch die körperliche Begleitreaktion eines heftigen Affektes wie Angst oder Aggression sein. Die Ausprägung des Symptoms steht dabei – entgegen unserer gewohnten Denkweise – in keiner quantitativen Relation zu den auslösenden Bedingungen; entscheidend ist vielmehr die subjektive Wahrnehmung dieser körperlichen Funktionsänderung und deren individuelle emotionale Verarbeitung. In einem Circulus vitiosus können sich so z. B. Funktionsstörungen und Angst gegenseitig verstärken und hochschaukeln.

So fand Chester (1973) z. B. bei einer epidemiologischen Untersuchung geschiedener Frauen bei 85% nach der Scheidung Gesundheitsstörungen, die 75% von ihnen zum Arzt führten. Die meisten klagten über 4 bis 5 verschiedene Symptome, unter denen Kopfschmerzen, Schwindel, Hautausschläge, Abszesse, Haarausfall, Anorexie sowie Herz- oder Magenbeschwerden am häufigsten genannt wurden. Von diesen Patientinnen wurde nur ein Bruchteil psychotherapeutisch behandelt oder zu einer Sozialberatung überwiesen. Fast immer bestand die Therapie in der Verschreibung von Psychopharmaka oder Sedativa. Die Behandlungsdauer lag nur in 25% unter 6 Monaten, in 25% über 2 Jahren. In vielen Fällen entstand eine Drogenabhängigkeit.

Solche Beobachtungen weisen darauf hin, daß soziale Ereignisse eine bedeutende Rolle bei der Auslösung und der Unterhaltung funktioneller Krankheiten haben; sei es, daß Beschwerden, die bisher als geringfügige Gesundheitsstörungen kaum beachtet wurden, jetzt als Krankheitssymptome erlebt werden; sei es, daß als Disposition eine besondere Vulnerabi-

---

[2] S. auch Kap. 21 „Familientherapie und Familiendynamik".

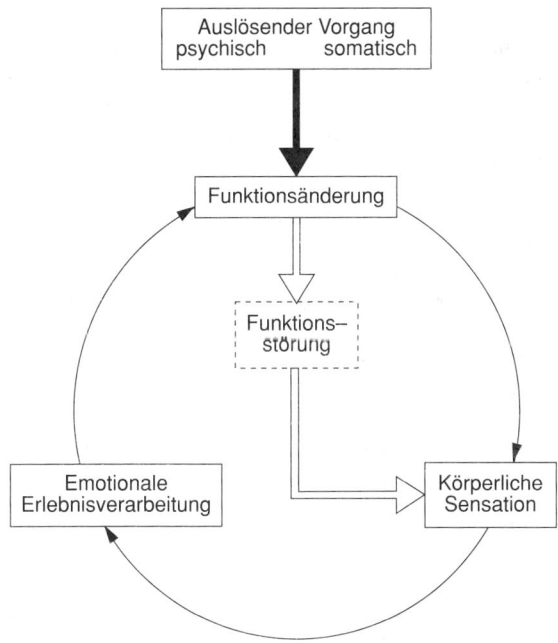

Abb. 31–6. Modell für die Symptombildung bei funktionellen Beschwerden (nach v. Uexküll, 1962).

lität für Schicksalsschläge bestand; sei es, daß eine soziale Belastung die Anpassungsfähigkeit auch robuster Naturen überfordert.

Die zahlreichen Untersuchungen über die Rolle sozialer Ereignisse als pathogene Faktoren handeln fast immer von organischen Leiden wie koronaren Herzkrankheiten, Tuberkulose, Asthma bronchiale, Hypertonie usw. Wir sprachen schon von dem Skotom der modernen Medizin für die funktionellen Krankheitsbilder und für die Probleme, die sie für die Betroffenen, deren Familien und für die Gesellschaft schaffen. Zu diesen Problemen gehört auch – und nicht zuletzt – die pathogene Rolle, welche die Medizin bei diesen Krankheiten spielt.

31.5 Psychologie, Psychodynamik

Die Frage, die wir zunächst besprechen müssen, ist die nach der Herkunft der Emotionen, welche die Wahrnehmung und deren Verarbeitung beeinflussen. Die in der Genese der Symptomatik wirksame Angst kann z.B. aus unmittelbar vorausgegangener Erfahrung, etwa dem Herztod eines nahestehenden Verwandten, herrühren. Häufig ist die Symptombildung jedoch komplexer. Der Patient erlebt einen intrapsychischen oder sozialen Konflikt als unerträglich und nicht lösbar; er muß die mit dem Konflikt verbundenen Affekte vom Bewußtsein abwehren. So kann es dazu kommen, daß nicht der Affekt, sondern nur die begleitende körperliche Funktionsänderung wahrgenommen und weiter verstärkt wird. Die im unlösbaren Konflikt aufkommende Angst fließt jetzt in die

Verarbeitung des körperlichen Symptoms ein; diese Angst verliert so ihren Bezug zum ursprünglichen Konflikt bzw. zur ursprünglichen Situation und wird an die Verarbeitung des körperlichen Symptoms gebunden. Die Angst bezieht sich jetzt z.B. ganz auf die Funktion des Herzens und nicht mehr auf die Unsicherheit, ob etwa der Partner seine Zuneigung zurückzieht, wenn man selbst aggressiv wird oder eigene Wünsche verfolgt. Der Konflikt wurde also zumindest entschärft, aus der sozialen Situation herausgelöst, in den eigenen Körper verlagert. Dies entlastet die Beziehungen und verspricht darüber hinaus weiteren Gewinn: Konfliktvermeidung und körperliche Beschwerden finden eher soziale Anerkennung, körperbezogene Klagen und passive Schonhaltung aktivieren eher die Hilfsbereitschaft der Umgebung. In diesem Sinn kann die **Symptombildung** auch als eine **Leistung,** als eine Art **Selbstheilungsversuch** des Betroffenen verstanden werden.

Die Gefahren dieses Selbstheilungsversuches liegen im Rückzug aus den sozialen Beziehungen in eine passive Schon- und Versorgungshaltung, die ihrerseits in der Regel zu weiteren Frustrationen führt und so die Chronifizierung des Prozesses begünstigt.

Was disponiert zu einer solchen Form von Konfliktlösung?

Es gibt Konflikte in Abhängigkeitsverhältnissen, aus denen eine Lösung nicht oder nur unter zu großen Opfern möglich ist. In der Regel gestalten die Patienten jedoch aufgrund ihrer Entwicklungsgeschichte die Situation nach dem Muster ihrer Abhängigkeitsproblematik in der Kindheit. Patienten mit schweren funktionellen Syndromen fühlen sich innerlich überstark von der Zuneigung und Versorgung ihrer Mütter oder entsprechender späterer Bezugspersonen abhängig. Sie kleben an dieser Beziehung, obwohl oder gerade weil diese während der Kindheit unbefriedigend war; in der Phantasie versuchen sie eine enge Beziehung aufrechtzuerhalten oder wiederherzustellen, um die in der Kindheit nicht erhaltene, aber ersehnte Zuwendung und Unterstützung doch noch zu bekommen.

Die Verhaltensforschung hat dafür eine eindrucksvolle Erklärung: Sie stellte fest, daß Strafreize in bestimmten Fällen das bestrafte Verhalten verstärken. Eine derartige Reaktion ist für kleine Tiere und Kinder adaptiv, „denn bei Schmerz sucht man am besten bei der Mutter Schutz. Daß diese selbst die Ursache des Schmerzes sein könnte, ist bei Tieren unwahrscheinlich und selbst in solchen Fällen wäre die Schutzsuche als Appell oder Beschwichtigung nützlich" (Eibl-Eibesfeld, 1984, S. 105). „Auch für Kinder ist die Mutter zweifellos eine sichere Basis, und auch hier gilt, daß von den Eltern mißhandelte Kinder im allgemeinen eine starke Bindung an die Eltern haben und zum Erstaunen der Fürsorge dagegen protestieren, wenn man sie zu ihrem Schutz aus dem Elternhaus in ein Heim bringen will" (Eibl-Eibesfeld, 1984, S. 107).

So ist es nicht erstaunlich, daß Menschen meist weitgehend verleugnen, wie enttäuschend die Beziehungen zu den Eltern in Realität verlaufen sind. Aber nicht nur physische Mißhandlungen, mehr noch psychische Mißhandlungen durch Gleichgültigkeit oder mangelnde Unterstützung der Entwicklung zur Selbständigkeit spielen bei den Müttern dieser Patienten eine Rolle. Mitunter waren die Mütter aufgrund körperlicher Krankheit oder depressiver Verstimmung nicht in der Lage, ihre Kinder ausreichend zu fördern, dominierten sie, schränkten expansive Entfaltung und aggressive Impulse ein, oft erwarteten sie ihrerseits von den Kindern zu früh Zuwendung, Verständnis, ja Hilfe. Verselbständigungstendenzen führten zur Androhung von Liebesentzug und Verlassenwerden zu einem Zeitpunkt, zu dem reale Unabhängigkeit mit dem Leben nicht vereinbar ist. Hierdurch wird die aufkommende Trennungsangst zur Todesangst.

Ermann (1980) konnte Patienten mit funktionellen Syndromen („psychovegetativen Störungen") testpsychologisch von Patienten abgrenzen, die an neurotischen Störungen litten. Unter ichpsychologischen Gesichtspunkten konnte er zwei Gruppen unterscheiden: Patienten, deren Störungen eine umgrenzte Ichpathologie zugrunde lag, und Patienten mit Borderline-Störungen, deren Ichpathologie viel umfassender war (1983).

Die **Manifestationsbedingungen** für die Symptomatik imponieren zum Teil als dramatisch: Die Patienten wurden unmittelbar vor Auftreten der Beschwerden mit fremder oder eigener Gefährdung konfrontiert. Zumindest bei Betrachtung von außen erscheint die auslösende Situation dann weniger dramatisch, wenn sich der Grundkonflikt zwischen Abhängigkeit und Versorgungswünschen einerseits und Verselbständigungstendenzen andererseits durch Entwicklungen wie beruflichen Aufstieg, Heirat, Geburt von Kindern oder die Enttäuschung von Versorgungserwartungen kritisch verschärft. Hier kommt es ja weniger auf die äußeren Ereignisse als solche, als auf die jeweils individuelle subjektive Bedeutung dieser Ereignisse an.

Für diese Zusammenhänge haben wir (in Kap. 1) als allgemeines Erklärungskonzept das Modell des Situationskreises entwickelt, das beschreibt, wie sich der Aufbau der individuellen Wirklichkeit, die jeden Menschen als eine für den außenstehenden Beobachter unsichtbare Hülle – eine zweite Haut – umgibt, unter normalen Verhältnissen vollzieht. Wir haben dargestellt, wie dieser Aufbau – wie im Zeitraffertempo – die Entwicklungsgeschichte wiederholt, in welcher der einzelne die Programme für den Aufbau seiner individuellen Wirklichkeit erworben hat; wie diese Entwicklungsgeschichte in den ersten Lebenstagen mit dem Aufbau einer primitiven Umwelt begann, und daß wir die kreative Fähigkeit, eine vorgefundene Umgebung in eine subjektiv bedeutungsvolle Umwelt zu übersetzen, als psychische Leistung verstehen müssen.

Systemtheoretisch handelt es sich um den „Sprung" von einer einfacheren (vegetativen) Organisationsstufe, wie sie vor der Geburt während der Embryonalzeit besteht, auf eine komplexere Integrationsebene, auf welcher der Organismus mit den für sein Überleben notwendigen Teilen der Umgebung (Luft, Nahrung, Wärme usw.) zu einem System zusammentritt. Die Organisationsform dieser komplexeren Integrationsstufe bezeichnen wir im Unterschied zu der einfacheren Organisationsform der vorgeburtlichen **vegetativen** Daseinsstufe, die als „bloßer Körper" noch keine Umwelt, sondern nur eine „Wohnhülle" kennt, mit dem lateinischen Wort für Seele = anima als **animalische** Organisationsform.

Der Übergang (Sprung) von der vegetativen Organisationsform als bloßer Körper zur komplexeren animalischen Integrationsebene als Körper, der von einer Umwelthülle umgeben ist, aber auch der Rückzug auf die frühere vegetative Stufe findet physiologischerweise während des ganzen Lebens immer von neuem statt: Wenn wir aus dem Schlaf erwachen oder in den Schlaf oder eine Ohnmacht versinken, erfahren wir den Aufbau einer Umwelt bzw. einer individuellen Wirklichkeit, durch Assimilation der Umgebung – oder den Rückzug aus unserer Wirklichkeit auf den Zustand als bloßer Körper, den die Umgebung nichts angeht. „Progression" und „Regression" wechseln aber nicht nur im 24-Stunden-Rhythmus wie Ebbe und Flut; sie können auch jederzeit bei der Lösung von Problemsituationen bzw. bei einem Versagen vor solchen Aufgaben eintreten.

Im Rahmen dieses Modells lassen sich die bunten Beschwerdebilder bei funktionellen Syndromen als unzeitgemäßer und unkoordinierter Wechsel von Progression zu Regression und umgekehrt deuten. Dabei kann bei einer pathologischen Frühgeschichte die Wirklichkeitsbildung mitunter bei infantil gebliebenen Programmen eines Wirklichkeitsaufbaus steckenbleiben, was mit Angst einhergehen kann, oder immer wieder von Phasen teilweisen Rückzugs unterbrochen werden, wenn die Situation emotionale Aufgaben stellt, für deren Lösung die erforderlichen Programme in früher Kindheit nicht gelernt werden konnten oder blockiert wurden.

Unter der Symptomatologie wurden als besondere Persönlichkeitsmerkmale die Patienten mit klagsamer Theatralik, die „Symptom-Pedanten" und die stillen, depressiven, zu hypochondrischen Ideen neigenden Kranken aufgezählt. Sehr viele Patienten sind psychisch bei oberflächlicher Betrachtung völlig unauffällig. Hier ist – auch darauf wurde bereits hingewiesen – der Interaktionsstil, den der Patient dem Arzt gegenüber an den Tag legt, oft aufschlußreich. Er kann zeigen, wie der Patient mit den Menschen seiner Umgebung – und mit sich selbst – umgeht. Dieser Interaktionsstil läßt sich auch in der Vorgeschichte verfolgen, wenn man darauf achtet, wie der Patient über Schlüsselfiguren in seiner Familie, in seinem Beruf, aber auch über Ärzte spricht, die ihn früher behandelt haben. Er sagt dem Arzt etwas über die Erwartungen, mit denen der Patient ihm gegenübertritt. Der Arzt muß diese Erwartungen kennen, um sich adäquat auf sie einstellen zu können.

31.6 Lebensgeschichte und soziale Interaktion

Es fällt immer wieder auf, daß diese Patienten Schwierigkeiten haben, ärztliche Hilfsangebote anzunehmen. Dies entspricht ihrem bisherigen Krankheitsverlauf: Sie haben ja zur Lösung ihres Problems gerade keine Hilfe finden können – vielleicht haben sie schon in der Kindheit erfahren, daß kein Dritter hilft – sie kommen meist erst, wenn ihr Selbstheilungsversuch zu scheitern droht. Dies erleben sie einerseits als beschämend, andererseits haben sie Angst davor, daß der Arzt die bisherige eigene Leistung in Frage stellt und damit den entschärften Konflikt wiederbelebt. So bekommt man als Arzt nicht selten das Gefühl, daß eigentlich keine Beziehung entstehen darf; neben der Wiederbelebung ursprünglicher Konflikte, der Infragestellung der Leistung bei der Symptombildung, trägt hierzu oft die Angst bei, daß der Arzt in die Beziehung zwischen Patient und Symptom bzw. in die Beziehung Patient – verlorene Bezugsperson der Kindheit als Dritter eindringen und damit diese Beziehung, die gerade bewahrt werden soll, gefährden könnte.

Kann der Patient dagegen das ärztliche Beziehungsangebot aufnehmen, so gestaltet er diese Beziehung in der Regel nach dem Muster seiner früheren Beziehungen; die Ambivalenz zwischen Anklammerungswünschen einerseits und Furcht vor Abhängigkeit und Einschränkung andererseits bestimmt jetzt auch die Probleme in der Beziehung zum Arzt. Hinter dem anklammernden Verhalten spüren wir dann häufig die frühen und lang hingezogenen Enttäuschungen der Kranken und ihre latenten Vorwürfe; oder aber sie verführen auch uns zu Einschränkungen, Vorschriften und Verboten.

Der oben skizzierte Circulus vitiosus zwischen Hausarzt, Patient, Spezialisten und Krankenhaus läßt sich als Ausdruck eines besonderen Interaktionsstils zwischen diesen Patienten und ihrer sozialen Umwelt verstehen: Psychodynamisch läßt sich das Mißtrauen, das diese Patienten gegen Diagnosen und Behandlungsformen ihrer Ärzte an den Tag legen, als „Übertragung" früherer Erfahrungen mit der Unzuverlässigkeit von Personen deuten, die Schutz und Hilfe leisten sollten. Das Verhalten der Medizin diesen Kranken gegenüber läßt sich als „Gegenübertragung" interpretieren. Dabei spielt beim Arzt die negative Einstellung allen Patienten gegenüber eine Rolle, deren Beschwerden sich nicht in die bekannten Kategorien organischer Krankheitsbilder einordnen lassen. Darüber hinaus muß man sich klarmachen, daß Symptome, die ein Patient äußert, für den Arzt „Auslöser" für bestimmte Verhaltensweisen sein können. Da die Beschwerden funktioneller Krankheitsbilder fast alle organischen Krankheiten nachahmen, veranlaßt der Arzt immer neue Untersuchungen. Weiss (1952) hat die sog. „big charts" – Krankengeschichten, die über 2 englische Pfund wogen – heraussuchen lassen und die Patienten nachuntersucht. Dabei stellte er fest, daß es sich in den meisten Fällen um Patienten mit funktionellen Krankheitsbildern handelte, die immer wieder von neuem in die Klinik kamen und dort immer von neuem von Kopf bis Fuß durchuntersucht wurden.

Soziologisch läßt sich der Gesamtzusammenhang zwischen Patient und unserem Gesundheitsversorgungssystem als „abweichendes Verhalten" des Patienten von sozial erwarteten Verhaltensschemata interpretieren, gegen das sich das soziale System mit „Sanktionen" verteidigt.

31.7 Funktionelle Syndrome als „politische Krankheiten"?

Wir werden Patienten mit funktionellen Syndromen nicht gerecht, wenn wir ihre Psychologie und Psychodynamik allein für ihr Versagen vor den Anforderungen der Gesellschaft und für ihr „Ausweichen" in die Krankenrolle verantwortlich machen. Der Hinweis auf den besonderen Interaktionsstil zwischen diesen Patienten und ihrer sozialen Umgebung macht bereits darauf aufmerksam, daß wir beide Seiten berücksichtigen müssen, wenn wir zu einem ausgewogenen Urteil kommen wollen. Die oben erwähnte Arbeit von Chester zeigt bereits exemplarisch, was alle Beobachtungen belegen: Die soziale Umgebung spielt für Gesundheit und Krankheit des einzelnen eine entscheidende Rolle (vgl. Kap. 2, 17, 18, 60, 61 und speziell: Lynch, 1977; Berkman und Syme, 1979; Weiner, 1985).

Wir haben in Kapitel 1 die These aufgestellt, daß es sich bei den funktionellen Syndromen um sog. „kulturgebundene Syndrome", d.h. um Krankheiten handeln könne, die von der Kultur erzeugt werden, in der Menschen leben. Wir haben dort dargestellt, daß kulturvergleichende Untersuchungen zu der Annahme geführt haben, jede Kultur würde Krankheiten erzeugen, die für sie spezifisch sind, und gleichzeitig eine Medizin entwickeln, die diese – für die Medizin einer anderen Kultur therapieresistenten – Krankheiten behandeln kann.

Die paradoxe Tatsache, daß die Bevölkerung der Industrienationen auf die spektakulären Erfolge der modernen Medizin bei lebensbedrohenden Krankheiten mit wachsender Unzufriedenheit und steigendem Unbehagen reagiert, wird verständlicher, wenn man einen Unterschied zwischen Krankheiten macht, die auf der einen Seite relativ unabhängig von kulturellen Einflüssen verlaufen – gleichgültig ob und wieweit diese Einflüsse zu deren Entstehen beigetragen haben – und Krankheiten auf der anderen, die von solchen Einflüssen unterhalten werden (etwa als Reaktionen auf ein „Unbehagen in der Kultur" (Freud) oder – nach der Metapher Levis – auf einen „schlecht passenden Schuh"). Die Feststellung, daß die moderne technologische Medizin dieser zweiten Gruppe von Krankheiten nicht nur hilflos gegenübersteht, sondern eine „iatrogene Gefahr" für sie ist, legt die Vermutung nahe, daß hier ein krankheitserzeugender Faktor der Industriekultur sichtbar werden könnte.

Wenn man von kulturellen Einflüssen als protektiven oder pathogenen Faktoren für die Gesundheit des einzelnen spricht, muß man eine Vorstellung von sozialer Wirklichkeit als einem Medium haben, in das sich der einzelne eingebunden bzw. als „Teil" erlebt, und das dessen Befinden und Reaktionen beeinflussen kann. Wenn diese Vorstellung nicht nur eine abstrakte Konstruktion bleiben soll, muß man davon ausgehen, daß soziale Systeme – wie alle Systeme – emergente Eigenschaften haben, d.h., daß ihre Eigenschaften nicht aus denen ihrer Elemente, z.B. als Summe der in sie integrierten individuellen Wirklichkeiten, ableitbar sind. Sie gehören einer komplexeren Systemebene an, auf der etwas Neues in Erscheinung tritt, von dem die individuellen Elemente höchstens profitieren können. Programmatisch haben Christian und Haas in ihrer Schrift „Wesen und Form der Bipersonalität" diese Feststellungen schon 1949 aufgrund ihrer subtilen Analyse der einfachen Zusammenarbeit beim Sägen getroffen (s.u.). Damals, noch vor Entwicklung der Systemtheorie, blieben ihre revolutionierenden Thesen weitgehend unverstanden und unbeachtet.

Sie schreiben, soziale Wirklichkeiten seien „Einheiten, die von Anfang an jenseits der anteiligen Subjekte ‚für beide' sind", und führen diesen Gedanken dann im einzelnen aus:

„Folglich ist z.B. die Partnerschaft nicht durch zwei (oder mehrere) Subjekte ausgefüllt, wie 2 = 1 + 1, sondern begreift die beteiligten Individuen wie Glieder – nicht wie Summanden – unter sich. ‚Bipersonal' meint also nicht Distributives, numerisch Zusammengefügtes. Deshalb ist das Zu-Dritt (Tripersonalität), Zu-Viert usw. keine quantitative und alsdann ableitbare Vermehrung des Zu-Zweit, sondern eine jeweils andersartige, qualitative Steigerung der Gemeinschaft."

Die Autoren betonen daher auch, daß dieses Phänomen einer sozialen Wirklichkeit „nicht etwa durch Akte oder Anstrengungen, ‚Beziehungsfunktionen' oder ‚Einfühlungsprozesse' einzelner erzeugt werden" könne. Man kann sie (die gemeinsame Wirklichkeit) „nur gewähren lassen, verändern, in die Krise treiben oder zerstören, d.h. von der Grenze her erfahren". Daher ist „ihre Erzeugung nicht Objekt der Erklärung, sondern sozialerkenntnistheoretische Voraussetzung".

Ihre Charakteristika fassen die Autoren in fünf Punkten zusammen:
– „Die Partnerschaft ist eine von vorneherein gegebene Tatsache. Ohne besondere Vereinbarung, ohne eigens hergestellte Anpassung beginnen die beiden Partner ... in einer zügigen Zusammenarbeit."
– Einigend ist das Gefühl für die Qualität des „Zusammen". „Jeder fühlt unmittelbar das harmonische Zusammenspiel oder die noch unausgeglichene Ungereimtheit."
– Fundierend ist die Gegenseitigkeit: „Was A für B tut, tut B für A: Das Spiel ist **für** beide und nicht **zwischen** beiden."
– Dementsprechend sind die beteiligten Subjekte

nicht „autonom": Der Anteil der Beteiligung kann objektiv erheblich schwanken, „aber der Unterschied wird unbemerkt vom Gegensubjekt vollkommen ausgeglichen".
– „Die Solidarität gründet in Selbstverborgenheit. Jeder ist Glied eines **Arbeitsganzen,** dessen Rolle er spielt."

Zeichentheoretisch läßt sich diese Zusammengehörigkeit als Abhängigkeit von einem gemeinsamen Kode bzw. als ständiges (unbewußtes) Bemühen beschreiben, individuelle Unterschiede der Kodierung empfangener und gesendeter Zeichen auszugleichen. H. G. Mead (1968) spricht von „intelligent gestures", die Gemeinschaftshandlungen begründen. Im Unterschied zu „unintelligent gestures", die z.B. nur Dressurakte ermöglichen, orientieren sie sich – genau wie Christian und Haas beschreiben – an den Reaktionen des Partners als dessen Interpretation der gesendeten (und von ihm empfangenen) Zeichen.

Störungen der gemeinsamen Wirklichkeit – und hier wird es speziell für unser Problem spannend – äußern sich für die Partner in einer Störung der eigenen Selbständigkeit; denn die Selbständigkeit, das souveräne Verfügen-Können über die eigenen Kräfte, setzt das vollkommene (aber unbemerkte) Gegenspiel des Partners oder der Partner voraus. Da jede Leistung der Ergänzung durch die passende Gegenleistung und jede Rolle der passenden Gegenrolle bedarf, kann das Ausbleiben oder das Ungenügen der Gegenleistung zu einer Lähmung des Handelnden führen. Das betonen Christian und Haas mit der Feststellung:

„Gerade dann, wenn beide Beteiligten sich auf dem Höhepunkt der Zusammenarbeit höchst selbständig erleben, zeigt die Analyse, daß beide objektiv in **strenger Gegenseitigkeit** der Abläufe verbunden sind. Das Erlebnis freier Selbständigkeit wird nur gewonnen, wenn die Gegenseitigkeit des Tuns objektiv erreicht ist."

Diese Beobachtungen können als Modell dienen, einmal, um ganz allgemein die Bedeutung kultureller Faktoren für das Wohl- oder Mißbefinden der Mitglieder einer Gesellschaft – und als dessen Folge auch für ihre Gesundheit und Krankheit – besser zu verstehen. Darüber hinaus kann es uns einen Hinweis auf „krankheitserzeugende Faktoren" der Industriegesellschaft liefern. Schließlich kann es uns noch die Fragen beantworten helfen, wie und warum die Effektivität ärztlichen Handelns so weitgehend von der vielberufenen – aber wenig verstandenen – Patient-Arzt-Beziehung abhängt.

Ein kürzlich im New England Journal of Medicine erschienener Artikel (Barsky, 1988) spricht von einem „Fehlschlag des Erfolgs" (failure of success) unserer modernen, technologischen Medizin. Er berichtet über Untersuchungen, nach denen in der Bevölkerung der USA das Gefühl „gesund zu sein" und sich „körperlich wohl zu fühlen" deutlich abgenommen habe, obwohl in dem gleichen Zeitraum bedeutende Fortschritte in dem objektiven Gesundheitsstatus festgestellt wurden.

Die wachsende Kluft zwischen objektivem Gesundheitsstatus und subjektivem Wohlbefinden könne nicht dadurch erklärt werden, daß eine höhere Lebenserwartung durch den Austausch akuter lebensbedrohender Krankheiten in den frühen Lebensjahren gegen chronische und zu Behinderung führende Leiden im späteren Leben erkauft wurde; daß die Menschen zwar länger leben, aber größere Zeiträume in schlechter Gesundheit verbringen würden.

Entscheidend sei eine fortschreitende Erniedrigung der Toleranzschwelle für Störungen auch leichter Natur und eine steigende Tendenz, bloße Unpäßlichkeiten (uncomfortable symptoms) als pathologisch – als Zeichen für Krankheit – zu erleben. Hand in Hand damit gehe eine steigende Bereitschaft, die Krankenrolle zu übernehmen und für isolierte Symptome ärztliche Behandlung in Anspruch zu nehmen. Im Gegensatz zu früher gebe man heute auch anderen gegenüber leichter zu, sich nicht gesund zu fühlen.

Das alles würde durch die Industrialisierung und Kommerzialisierung der Gesundheit gefördert, die den Gesetzen des freien Markts gehorchend zu einem ständig wachsenden Konsum an immer unentbehrlicher werdenden „Gesundheitsgütern" geführt habe, die mächtige Organisationen erzeugen und anbieten würden. Diese Analyse illustriert das, was Ivan Illich (1975) als „Kontraproduktivität" der modernen Medizin bezeichnet hat, die durch „Medizinalisierung" des ganzen Daseins die Tendenz fördern würde, jede Befindensstörung als „Krankheit" zu erleben.

Alle diese Feststellungen beschränken sich auf eine Beschreibung der Verhältnisse. Hinweise auf hinter dieser Entwicklung stehende Gründe oder Zusammenhänge erhalten wir durch sie noch nicht. Hier können uns die Beobachtungen von Christian und Haas weiterhelfen, weil sie Aufschluß darüber geben, wie „Abwärtseffekte" von einer gemeinsamen bzw. sozialen Wirklichkeit zustande kommen, die sich – „selbstverborgen" – in der individuellen Wirklichkeit eines Menschen auswirken.

Das hypothetische Modell, das sich daraus entwickeln läßt, sieht folgendermaßen aus: Beeinträchtigungen der Selbständigkeit, z.B. das Gefühl, nicht autonom über seine Kräfte verfügen zu können, werden als Befindensstörungen erlebt. Sie haben im sozialen Kontext eine wichtige Funktion: Sie sind Zeichen für eine „unausgeglichene Ungereimtheit im harmonischen Zusammenspiel" mit dem Partner oder den Partnern. Als solche werden sie unbewußt als Verhaltensdirektiven erlebt, die auf eine Harmonisierung der Interaktionen abzielen.

Die „Medizinalisierung" allen Erlebens beraubt diese Zeichen ihrer sozialen Funktion. Sie werden nicht als Verhaltensdirektiven, sondern als Gesundheitsstörungen erlebt. Die unbewußte Aufmerksamkeit wird von dem „Zusammen" mit anderen auf den eigenen Körper gelenkt, der jetzt zum Zentrum der individuellen Wirklichkeit geworden ist.

Natürlich können die Anlässe für Störungen des sozialen Integriertseins von Individuum zu Individuum – und sogar von Situation zu Situation – und natürlich erst recht von Kultur zu Kultur sehr ver-

schieden sein. Der gemeinsame „Mechanismus" ließe sich aber mit Hilfe dieses Modells und der allgemeinen Vorstellung, die wir (vgl. Kap. 1) entwickelt haben, als Fehlinterpretation von Zeichen beschreiben, deren Aufgabe es ist, das Verhalten lebender Systeme so zu lenken, daß die Nischenqualität der Umgebung erhalten bleibt oder wiederhergestellt wird.

Der Zusammenhang, den Christian und Haas in ihrem Experiment des Sägens zu zweit beobachtet und mit dem Satz beschrieben haben: Was A für B tut, tut B für A, ist ein exemplarisches Beispiel für einen Zeichenprozeß der Kommunikation (der „intelligent gestures" im Sinne G. H. Meads): Was A (mit seinem Verhalten bzw. mit seiner Leistung) als Zeichen sendet, ist gleichzeitig eine (unbewußte) Frage an B (nach der Adäquatheit des Zeichens). Was B (mit seinem Verhalten bzw. seiner Leistung) als Zeichen sendet, ist gleichzeitig eine Antwort (als „Gegen-Leistung") auf die Frage von A – und vice versa.

Die „Bedeutungserteilung" (Kodierung als Zeichen) und die „Bedeutungsverwertung" (als Bezeichnetes oder Signifikat) ergänzen sich ständig als gegenseitige Information über die Gemeinsamkeit des Kodes. Die semantische und die pragmatische Dimension der Zeichen sind rückgekoppelt. Diese Rückkoppelung als Basis für gemeinsame oder soziale Wirklichkeiten wird aufgebrochen, wenn die Zeichen nur als Symptome für den Zustand des eigenen Körpers erlebt werden.

Reaktionen auf ein „Unbehagen in der Kultur" oder einen „drückenden Schuh" werden nach diesem Modell als Folge und Ausdruck nicht mehr auszugleichender Disharmonien in sozialen Beziehungen gedeutet, wie sie in jeder Kultur das soziale Zusammenspiel bedrohen. Das „kommunikative Realitätsprinzip" (vgl. Kap. 1) gerät in Gefahr, wenn der einzelne die unbewußte Resonanz (das „Echo") des anderen nicht mehr oder falsch erlebt. Jede Kultur braucht ihre Medizin, um dem einzelnen helfen zu können, die kulturspezifischen Restriktionen zu ertragen und seine Resonanzfähigkeit zu erhalten oder wiederzufinden. „Bipersonalität" ist ein grundlegendes therapeutisches Prinzip – die „Droge Arzt" – in einer Patient-Arzt-Beziehung.

Die moderne Medizin, für die psychische und soziale Probleme keine relevanten Faktoren sind, versagt vor dieser Aufgabe. Statt dessen verbündet sie sich mit den pathogenen Tendenzen der Industriekultur, indem sie Gesundheit und Krankheit als individuelle Probleme eines Funktionierens oder Nicht-Funktionierens körperlicher Mechanismen deutet.

Was ist zu tun? Einseitige Schuldzuweisungen führen nicht weiter. Sie münden in unfruchtbare und jede Kreativität lähmende Polarisierungen. Sie helfen weder den Ärzten in ihrer Hilflosigkeit Patienten gegenüber, die als Folge sozialer Disharmonien Krankheitssymptome entwickeln, noch den Patienten, die sich unverstanden und alleingelassen erleben.

Was not tut, ist eine Erweiterung des Gesichtsfeldes unserer Medizin: Sie muß sich Rechenschaft geben, daß sie ein Erzeugnis der Kultur unserer Gesellschaft ist, und daß sich aus dieser Tatsache eine Verantwortung der Gesellschaft und ihrer Kultur gegenüber ergibt. Im einzelnen heißt das, daß Krankheiten auch Symptome für Störungen in dem gesellschaftlichen System sein können, in dem Ärzte und ihre Patienten leben, daß es, mit anderen Worten, in diesem Sinne auch „politische Krankheiten" gibt.

Ärzte sind zwar Werkzeuge der Kultur ihrer Gesellschaft, aber es handelt sich nicht um eine einseitige Beziehung, sondern um eine Beziehung auf Gegenseitigkeit. Ärzte dürfen daher nicht einfach die Modelle übernehmen, die sie vorfinden und die ihr Handeln und das Forschen der Medizin ihrer Zeit leiten. Sie haben auch Verantwortung für die Auswirkungen dieser Modelle auf die Gesundheitsvorstellungen und damit auf das menschliche Klima der Gesellschaft. Sie können diese Modelle ändern und auf diesem Wege die Gesundheitsvorstellungen und das menschliche Klima ihrer Gesellschaft beeinflussen.

Im Augenblick definiert die moderne Medizin den Körper als Maschine und orientiert ihren Gesundheitsbegriff an den Prinzipien eines technischen Überwachungsdienstes (TÜV). Dementsprechend versteht sie die ihr von der Industriekultur gestellte Aufgabe als Reparaturbetrieb. Dieses Modell läßt es nicht zu, daß psychosoziale Faktoren für Gesundheit und Krankheit gleiches Gewicht haben könnten wie physikalische, chemische und mikrobiologische Faktoren. Solange unsere Medizin an diesem Modell festhält, wird sie ihrer Verantwortung unserer Gesellschaft und Kultur gegenüber nicht gerecht. Patientenkarrieren sind nicht nur individuelle Tragödien. Sie sind auch Krankheitssymptome unserer Industriegesellschaft.

Die Probleme, mit denen wir konfrontiert sind, lassen sich nicht durch einen Rückfall in das vorindustrielle Zeitalter oder durch einen Rückgriff auf die Medizin einer früheren oder einer anderen Kultur lösen, sondern nur durch Überwindung der gefährlichen Einseitigkeit unserer reduktionistischen Modelle. Es geht um eine Revision der mechanistischen Theorie und ihres Maschinenmodells für den menschlichen Körper und die Entwicklung der Theorie einer Humanmedizin, in der die Bedeutung psychischer und sozialer Faktoren im Zusammenhang mit biologischen Einflüssen und Vorgängen gesehen und kritisch geprüft werden kann.

Vordringliche Aufgabe einer solchen Theorie ist die Entwicklung von Modellen, welche die Diagnose funktioneller Syndrome möglichst schon in statu nascendi erlauben, und aus denen sich Methoden für therapeutische Interventionen in allen Stadien, vor allem aber in der Frühzeit, ableiten und erproben lassen. Das Modell des Situationskreises für den Aufbau unserer individuellen Wirklichkeit versucht diese Aufgabe zu lösen. Nach ihm lassen sich funktionelle Syndrome als Folgen von Verletzungen unserer „zweiten Haut" interpretieren, die unsere individuelle Wirklichkeit durch ständige „Assimilation" der Vorgänge unserer Umgebung um unseren Körper legt.

Konzepte der Entwicklungspsychologie legen nahe, im Situationskreismodell eine zentrale Instanz zur Regulation der Bedeutungserteilung anzuneh-

men, die in der frühen Interaktion zwischen Kind und Mutter entstanden ist. Ihre Aufgabe ist es, Ereignisse der Umgebung unter dem Gesichtspunkt ihrer Bedeutung für die Problemlösungskapazitäten und das Selbstwertgefühl des Betroffenen zu beurteilen, d.h. ursprünglich für den Wert, den die Mutter den Reaktionen des Kindes auf ähnliche Ereignisse beimaß.

Bions Konzept des „container" und „contained" (1962) und Winnicotts Modell der mütterlichen Spiegelfunktion für das Erleben des Kindes (1973) gehen davon aus, daß die im Situationskreismodell enthaltene Rezeptor- (bzw. Bedeutungserteilungs-)Funktion aus einer Internalisierung dieser frühen Mutter-Kind-Interaktion entstanden ist. Die Stabilität und Zuverlässigkeit des Wohlwollens, mit dem die mütterliche „Rezeptor-Substanz" die Reaktionen des Ich beurteilt, sind die Quelle für das „Urvertrauen" (Erikson, 1965), mit dem der einzelne an seine Umgebung herantritt. Sie entscheiden, welchen Bedarf an Beistand durch eine Beziehungsperson und/oder an äußeren Erfolgen das Selbstwertgefühl des einzelnen hat.

Nach diesem Modell lassen sich Kränkungen und „Objektverluste", die häufig zu Beginn eines funktionellen Syndroms gefunden werden, als Ereignisse interpretieren, die zu einer Schädigung der „Rezeptor-Funktion" geführt haben, die sich u.U. durch eine einfache ärztliche Intervention im richtigen Augenblick oder durch eine supportive Psychotherapie als Angebot einer „Hilfs-Spiegel-Funktion" auffangen läßt.

Berkman und Syme (1979) haben gezeigt, daß die Einbindung des einzelnen in das primäre soziale Netz einer Familie, eines Freundeskreises usw. protektive Wirkung für die Gesundheit besitzt. In einem Zeitraum von 10 Jahren hatten von 6928 Erwachsenen die Personen ohne diese soziale Bindung eine dreifach erhöhte Mortalität. Das galt unabhängig von Risikofaktoren wie Alkohol, Rauchen, Bewegungsmangel usw.

Diese inzwischen vielfach bestätigten Zusammenhänge zeigen, wie wichtig es für den einzelnen ist, seine individuelle Wirklichkeit in soziale Wirklichkeiten integrieren zu können, die seinen Selbstwert bestätigen. Kulturgebundene Syndrome lassen sich unter diesem Gesichtspunkt als Krankheitszeichen einer Kultur interpretieren, in der die Fähigkeit der Gesellschaft zur Bildung sozialer Netze verlorengeht, in denen Menschen sich geborgen fühlen können.

31.8 Differentialdiagnose

Die Ähnlichkeit der Symptomatik funktioneller Syndrome mit der Symptomatik organischer Krankheitsbilder verursacht erhebliche differentialdiagnostische Schwierigkeiten. Deshalb sind gewisse allgemeine Kriterien hilfreich:

Die meisten funktionellen Beschwerdebilder weisen sog. „Randsymptome" auf, die bei organisch Kranken gewöhnlich nicht gefunden werden (Tab. 31–3).

Tabelle 31–3. Rand- oder Begleitsymptome funktioneller Syndrome.

Somatisch	Psychisch
Globus	Innere Unruhe
Parästhesien (an Mund,	Konzentrationsschwäche
Zunge und Extremitäten)	Erschöpfbarkeit
Atemhemmung	Depressive Stimmungs-
Herzsensationen	lage
Aufstoßen in Salven	Angstzustände
Anfallsweises Glieder-	Schlafstörungen
zittern	

Ein weiteres, diagnostisch wichtiges Kennzeichen ist die Länge der Anamnese: Ein funktionelles Beschwerdebild reicht gewöhnlich weit in die Vergangenheit zurück. Oft können die Patienten nicht angeben, wann es begonnen hat. Eine kurze Vorgeschichte ohne den Hinweis auf eine akute seelische Belastung in letzter Zeit spricht eher für eine organische Krankheit.

Die Zahl der Beschwerden ist ebenfalls ein wichtiger Hinweis: Je größer ihre Zahl, um so unwahrscheinlicher wird es, daß ein organisches Leiden vorliegt. Die Patienten, die über ihre Beschwerden Buch führen, um dem Arzt eine möglichst vollständige Liste vorlegen zu können, wurden schon erwähnt.

Viele dieser Krankheitsbilder zeichnen sich durch einen Wandel ihrer Beschwerden aus („Symptomwandel"). So kann es vorkommen, daß Patienten, die bei der ersten Konsultation über Herzschmerzen klagten, den Arzt später wegen Kopfschmerzen aufsuchen oder umgekehrt. Von der früheren Symptomatik ist dann häufig nicht mehr die Rede. Die differentialdiagnostischen Überlegungen und Bemühungen, die bei der ersten Untersuchung in eine bestimmte Richtung gegangen waren, erscheinen jetzt nutzlos. Man muß von neuem anfangen und fragt sich, ob man bei der ersten Erhebung der Vorgeschichte aufmerksam genug war oder ob man eine Krankheit übersehen hat, die in ihrem Fortgang neue Symptome macht. Wenn man weiß, daß dieser proteusartige Charakter für viele funktionelle Beschwerdebilder typisch ist, kann das ein wichtiger diagnostischer Hinweis sein. Auch bei der Patientin, über die am Anfang berichtet wurde, wandelte sich das Beschwerdebild im Laufe der Zeit erheblich.

Der zeitliche Zusammenhang mit einer einschneidenden Veränderung in der Lebensgeschichte zeigt manchmal den richtigen Weg. Auch die Mitteilung, daß Verwandte oder Bekannte im zeitlichen Zusammenhang mit dem Auftreten der Beschwerden erkrankt oder verstorben sind und daß sie an ähnlichen Symptomen gelitten hatten, kann ein wichtiger diagnostischer Hinweis sein. Die Identifikation mit solchen Personen ist ein relativ häufiges Ereignis.

Selbstverständlich können diese Informationen nur Hinweise, aber niemals Beweise sein, die ein „symptomatisches" funktionelles Syndrom ausschließen. Für das Vorliegen eines „essentiellen" funktionellen Syndroms gelten zwei Grundsätze:

1. Wir müssen mit der gleichen Gewissenhaftigkeit nach psychosozialen Störungen suchen und dürfen uns nicht mit Allgemeinplätzen wie „Streß", „Belastung", „Zivilisationsdruck" usw. zufriedengeben. Hier muß eine biographische Anamnese eine möglichst genaue Übersicht über disponierende und auslösende Faktoren sowie über die akuten und chronischen Probleme des Patienten geben. Gleichzeitig muß sich der Arzt einen Eindruck vom Umgangsstil des Kranken mit wichtigen Personen in seiner Umgebung einschließlich der Ärzte verschaffen.

2. Wir müssen durch eine genaue somatische Untersuchung das Vorliegen organischer Krankheiten ausschließen.

Dabei ist eine Differentialdiagnose zwischen folgenden Gruppen anzustreben:

– Reaktive Bilder als Folge akuter seelischer Belastungen. Diese Krankheitsbilder haben eine gute Prognose, wenn der Arzt das Problem mit dem Patienten durchspricht und erforderlichenfalls mit stützenden psychotherapeutischen Maßnahmen eingreift.

– Neurotische Störungen, die sich hinter einer somatischen Symptomatik verbergen. Die Prognose dieser Patienten ist je nach der neurotischen Erkrankung, den Möglichkeiten der Kranken, die Diagnose zu akzeptieren, und schließlich den Aussichten, eine psychotherapeutische Behandlung zu finden, verschieden.

– Eine große Kerngruppe von Patienten, die ständig auf ihre somatischen Beschwerden zurückkommen und die einen Zusammenhang mit auslösenden emotionalen Erlebnissen entweder nicht akzeptieren oder nicht erfahren können. Diese Patienten stellen an die Geduld und an den Optimismus der Ärzte die größten Anforderungen. Bei ihnen ist die Prognose „quoad sanationem" am ungünstigsten. Hier sind auch die Probleme der Therapie noch weitgehend ungelöst.

31.9 Therapie

Die Therapie beginnt – bei allen Patienten – bereits mit der Erhebung der Anamnese, die dem Patienten Gelegenheit geben muß, über sich und die Probleme zu sprechen, die möglicherweise hinter seinen Symptomen verborgen sind. Das geschieht am besten mit einer Technik, die, soweit möglich, „offene Fragen" verwendet und die dem Patienten erlaubt, seine Beschwerden mit seinen Worten und in der Reihenfolge, die ihm wichtig ist, zu schildern (vgl. Kap. 12). Mit „geschlossenen Fragen", die sich an den bekannten Diagnoseschemata der organischen Medizin orientieren, sind wir stets in Gefahr, die Symptomatik, die wir hören wollen, in den Patienten hineinzufragen. Darüber hinaus legen wir ihn auf diese Weise auf ein Kommunikationsschema mit Ärzten fest, in dem er dazu erzogen wird, seine Beschwerden in der Terminologie der Organmedizin auszudrücken. Schließlich versäumt der Arzt bei diesem Frage-Antwort-Spiel die

Gelegenheit, sich ein Bild darüber zu machen, wie der Patient seine Krankheit erlebt.

Meist wird der Patient mit seinen somatischen Beschwerden beginnen. Hier ist es von entscheidender Wichtigkeit, die einzelnen Symptome genau durchzusprechen. Damit erwirbt der Arzt das Vertrauen des Patienten und kann gleichzeitig die erforderlichen differentialdiagnostischen Überlegungen anstellen.

Es gilt also vom ersten Kontakt an und parallel zur sorgfältigen somatischen Abklärung gleichzeitig die psychischen Prozesse und die soziale Situation bei jedem Patienten individuell zu klären. An Stelle der sog. „Ausschlußdiagnostik" tritt ein Vorgehen, bei dem der Arzt die Stellung einer psychosomatischen Diagnose aufgrund positiver Befunde im psychischen und sozialen Bereich anstrebt. Nur so läßt sich sowohl eine einseitige diagnostische Fehleinschätzung als auch die Kränkung des Patienten vermeiden, die in der Regel mit der Mitteilung negativer körperlicher Befunde und der anschließenden Überweisung zu einem Spezialisten für psychische Probleme verbunden ist.

Neben der **Anamnese** stützt sich das diagnostische Vorgehen auf die **Analyse der Beziehung.** Die Rekonstruktion der Symptombildung erfolgt mit Hilfe der Erinnerung im Gespräch; der dem Bewußtsein nicht zugängliche Anteil des fortdauernden Beziehungskonfliktes und die selbstverborgenen Affekte werden dem Arzt nur über das Erleben der eigenen Beziehung zum Patienten zugänglich. Bekommt der Patient die Möglichkeit, diese Beziehung entsprechend seinen Bedürfnissen zu gestalten, so konstelliert er sie nämlich analog zu seinen alten Beziehungsmustern, und für uns als Ärzte werden damit die den Patienten bedrängenden Affekte spürbar. Bei günstigem Verlauf gelingt es, die im Rahmen der Symptombildung aus der sozialen Situation herausgenommenen Konflikte jetzt in der Beziehung zum Arzt wieder in eine soziale Beziehung einzubetten, die zugleich die Chance zu einer Bearbeitung und Veränderung des Konflikts eröffnet.

Anhand unseres Erlebens der Beziehung zum Patienten während der Sprechstunden können wir nun auch folgende Fragen klären: Haben wir für den Patienten lediglich die Funktion einer Klagemauer oder sucht er eine wirkliche Beziehung aufzunehmen? Wie benützt er uns in dieser Beziehung, sucht er überwiegend Versorgung und Befriedigung von Abhängigkeitsbedürfnissen? Erleben wir eine Möglichkeit, uns mit ihm dem alten Konflikt und den abgewehrten Affekten anzunähern und zusammen mit ihm den Weg der Wiederholung, der ihn gefangenhält, zu verlassen und zu neuen Verständnis- und Lösungsansätzen aufzubrechen? Die Analyse der Beziehung zwischen Arzt und Patient bildet so zugleich die Grundlage für die sich anschließenden therapeutischen Entscheidungen.

Die Differentialdiagnose zwischen Organerkrankungen und funktionellem Syndrom kann im Einzelfall immer wieder über längere Zeit schwierig bleiben. Für den Arzt ist es dabei wichtig, abwarten, sich offenhalten, Unsicherheit ertragen zu können. Nicht selten trägt der Wunsch nach eigener Sicherheit und rascher Festlegung zur Entstehung von Fehldiagnosen bei.

Ein überzeugender Nachweis für das Vorliegen funktioneller Beschwerden läßt sich von psychosomatischer Seite oft nur im Zusammenhang des Verlaufs der psychotherapeutischen Behandlung erbringen. Im Rahmen der therapeutischen Beziehung wird der Zusammenhang zwischen der Ausprägung der Symptomatik und dem Ausmaß der Konflikte oder der situativen Belastung oft deutlicher erkennbar. Klarheit für den Arzt und Einsicht beim Patienten entsteht dann, wenn die Symptomatik im Rahmen der therapeutischen Beziehung wieder ihren ursprünglichen Zusammenhang mit dem beziehungsbedingten Konflikt gewinnt, wenn etwa die Herzbeschwerden oder funktionell bedingte Durchfälle im Zusammenhang der Arzt-Patient-Beziehung auftreten, z. B. bei der Trennung am Ende der Sprechstunde oder vor einem Urlaub, oder bei aggressiven Phantasien gegenüber dem Arzt und vor allem dann, wenn diese Beschwerden bei der Bearbeitung der entsprechenden Ängste und Konflikte für beide Beteiligte einsehbar wieder abklingen.

Der **diagnostische Plan** muß alle Untersuchungsverfahren einschließen, die notwendig sind, um dem Arzt ein klares Bild über den körperlichen Zustand des Patienten zu geben und um organische Ursachen der Beschwerden auszuschließen. Dann muß man jedoch mit den Untersuchungen aufhören und darf – wenn der Patient früher oder später mit den gleichen oder anderen Klagen kommt und auf neue Untersuchungen drängt – nicht ohne zwingenden Grund wieder mit neuen Untersuchungen beginnen.

Ein entscheidender Moment ist die **Mitteilung der Diagnose** nach Abschluß der Untersuchung. Die Art, wie diese Mitteilung erfolgt, entscheidet zusammen mit anderen Faktoren darüber, ob der Patient sein Mißtrauen überwindet und ein therapeutisches Bündnis mit dem Arzt eingeht oder ob er den nächsten Arzt bzw. die nächste Klinik aufsucht.

In diesem Augenblick muß der Arzt sich klarmachen, daß seine Patienten unter ihren Beschwerden leiden, ganz gleich, ob sie eine organische Ursache haben oder ob sie psychisch ausgelöst sind. Weiter muß er sich vor Augen halten, daß es für einen Kranken beunruhigend ist, wenn für seine Beschwerden keine organischen Ursachen gefunden werden. Der Patient befürchtet, daß eine Krankheit, deren Ursachen so schwer zu finden sind, ein besonders unheimliches Leiden sein könnte. Da er meistens schon viele Ärzte aufgesucht, die widerspruchsvollsten Diagnosen und die verschiedensten Kuren mit nur vorübergehendem Erfolg durchgemacht hat, ist er besonders mißtrauisch. Auf der einen Seite beweist ihm jede neue körperliche Untersuchung, daß der Arzt ein organisches Leiden vermutet, auf der anderen Seite leidet er unter dem Gedanken, daß ein Kranker, bei dem die Ärzte keine organischen Ursachen für seine Beschwerden finden, für einen Simulanten oder Hypochonder gehalten wird, den man nicht ernst nimmt.

In dieser Situation ist es keine einfache Sache, dem Patienten mitteilen zu müssen, man habe für seine Beschwerden – wieder – keine organische Ursache finden können. Hier ist es entscheidend wichtig, dem Patienten Verständnis dafür zu zeigen, daß er über den negativen Befund nicht nur erleichtert, sondern auch enttäuscht sein wird, und daß man bereit ist, mit ihm über seine Enttäuschung zu sprechen. Dies ist – nach unserer Erfahrung – der beste Weg, um das Vertrauen des Patienten zu gewinnen, das auch die Voraussetzung für jede Art psychotherapeutischer Betreuung darstellt.

Die Psychotherapie wird bei jeder der drei Gruppen verschieden sein:

- Bei den Patienten mit reaktiven Beschwerden genügt häufig eine einmalige Aussprache mit dem Angebot des Arztes, dem Patienten für weitere Gespräche zur Verfügung zu stehen, wenn er dies wünscht. Wir haben darauf hingewiesen, daß der Arzt bei diesen Patienten eine besondere Verantwortung hat, die Entstehung chronischer Leiden zu verhindern.
- Bei den Patienten der zweiten Gruppe richtet sich die Indikation für die Therapie nach der neurotischen Grundkrankheit. Diese Patienten haben nicht selten Schwierigkeiten, zu akzeptieren, daß hinter ihren somatischen Beschwerden seelische Probleme verborgen sein können, sind aber früher oder später meist doch in der Lage, diesen Zusammenhang zu sehen.
- Die größten therapeutischen Probleme bieten die Patienten der Kerngruppe, die den Zusammenhang zwischen ihren Symptomen und emotionalen Problemen nicht wahrnehmen können, ja, denen häufig überhaupt der Zugang zu ihrem emotionalen Erleben verschlossen zu sein scheint. Hier gibt es noch viele offene Probleme, nicht nur der Therapie, sondern auch der Epidemiologie, denn hier scheint die Zugehörigkeit zur sozialen Schicht eine Rolle zu spielen – worauf auch amerikanische Untersuchungen hinweisen.

Bei diesen Patienten erlebt der Arzt aber immer wieder, daß seine Bereitschaft, die Klagen der Patienten anzuhören, eine therapeutische Wirkung erzielen kann. Manchen Patienten genügt es, den Arzt ein- oder zweimal im Jahr aufzusuchen und ihm ihre Beschwerden vorzutragen, um im beruflichen und familiären Bereich kompensiert zu bleiben. Die Tatsache, daß der Arzt für sie – im weitesten Sinne des Wortes – „erreichbar" ist, scheint für diese Patienten von großer Bedeutung zu sein. Hier muß der Arzt sich klarmachen, daß es sich bei diesen Patienten fast immer um chronisch Kranke handelt, bei denen schon viel gewonnen ist, wenn eine Verschlimmerung der Beschwerden verhindert wird. Spektakuläre Erfolge sind auch von einer psychoanalytischen Therapie – zu der die Patienten in den seltensten Fällen bereit sind – nicht zu erwarten.

Die besten Behandlungsergebnisse sind dann zu erwarten, wenn es im Rahmen der Arzt-Patient-Beziehung gelingt, im Laufe der Zeit im Zusammenhang mit aktuellen Frustrationen auch über die alten Enttäuschungen und Beziehungsprobleme zu sprechen.

In allen spezialisierten Formen der Psychotherapie geht es darum, mit dem Patienten systematisch die Situation zu klären und ihn darin zu unterstützen, für seine Probleme neue Lösungsansätze zu entwickeln. Dies gilt gleichermaßen für tiefenpsychologisch wie lerntheoretisch orientierte Therapieverfahren. Im tiefenpsychologischen Therapieansatz dient die Beziehung zwischen Arzt und Patient dem Wiedererleben und der Bearbeitung auch früherer Entbehrungen und Konflikte. Körperbezogene Therapieformen können den Zugang zu innerpsychischen Konflikten und den Beziehungsproblemen erleichtern und die genannten Therapieansätze ergänzen. Übende Verfahren wie das autogene Training kommen zunächst dem Bestreben des Patienten nach Unabhängigkeit entgegen.

Die Verschreibung von Medikamenten ist oft – nicht immer – sinnvoll. Wichtig ist dabei, daß man dem Patienten die Wirkungsweise der Medikamente erklärt und ihm deutlich macht, daß nicht kranke Organe (ein krankes Herz, ein kranker Magen usw.) behandelt werden sollen, sondern daß die Medizin dem Patienten hilft, mit sich und seinen Beschwerden besser fertig zu werden. Mit einem Wort: Die psychologische Bedeutung der Medikamente für den Patienten muß bei der Therapie berücksichtigt werden.

31.10 Prognose

Es gibt nur wenige katamnestische Untersuchungen über den Verlauf funktioneller Syndrome. Nach Cremerius (1968a) beträgt die Spontanheilung nach 10 bis 30 Jahren im Durchschnitt 8%. Christian (1979) fand nach 10 Jahren bei 12% eine Spontanheilung. Interessant ist, daß die Prognose bei den verschiedenen Syndromen verschieden ist, worauf bei den einzelnen Kapiteln eingegangen wird.

Sims (1984) bringt einen interessanten Gesichtspunkt in die Diskussion, der auch für die Prognose funktioneller Syndrome von Bedeutung sein dürfte. Er untersuchte in einer Follow-up-Studie die Mortalität neurotischer Patienten. Dabei stellte er eine deutliche Übersterblichkeit fest, die nicht mit der neurotischen Erkrankung selbst zusammenhing, aber ursächlich auf sie zurückging: Neben einer deutlich erhöhten Rate an Suizid und Unfällen fand sich auch eine deutlich erhöhte Rate „natürlicher Todesfälle". Als deren Grundlage diskutiert der Autor Arteriosklerose und erhöhten Zigarettenkonsum. Schwerere Fälle von Neurose zur Zeit der Behandlung hatten ein größeres Risiko, früher zu sterben.

Für den Zusammenhang zwischen psychosozialen Belastungen und Übersterblichkeit gibt es aber noch viele andere Erklärungen, die in den verschiedenen Kapiteln dieses Lehrbuchs besprochen werden. Wieweit diese auch für Patienten mit funktionellen Syndromen ein erhöhtes Risiko, früher zu sterben, bedeuten, ist noch nicht untersucht.

32 Das funktionelle kardiovaskuläre Syndrom

Othmar W. Schonecke und *Jörg Michael Herrmann*

Ein 36jähriger Patient leidet seit ca. 1 Jahr unter anfallsweise auftretenden Herz-Kreislauf-Beschwerden, wie Herzjagen, Schwächegefühle und Schweißausbrüche. Begleitet werden diese Beschwerden von einem intensiven Gefühl der Todesangst. Der Patient ist von Beruf Bankkaufmann und schildert vor allem seine berufliche Situation als außerordentlich belastend. Er sei häufig gezwungen, Überstunden zu machen und dann auch noch Arbeit mit nach Hause zu nehmen. Seine Frau, die zwar sehr verständnisvoll sei, habe ihm deswegen schon des öfteren Vorwürfe gemacht. Durch die Überlastung sei er insgesamt sehr reizbar. So würde er sich beispielsweise sehr aufregen, wenn seine Kinder nicht sofort gehorchten oder in der Schule Schwierigkeiten hätten.

Zu einem ersten „Herzanfall", der zu seiner Einweisung in das örtliche Kreiskrankenhaus geführt hatte, kam es, nachdem er nach einem intensiven Arbeitstag abends auf einer Veranstaltung eine Rede habe halten müssen. Als er anschließend in der Nacht nach Hause gekommen sei, habe seine Frau ihm Vorhaltungen gemacht. Etwas später sei er dann im Badezimmer zusammengebrochen, ohne jedoch das Bewußtsein zu verlieren. Er habe intensive Todesangst gespürt, Herzjagen, Schwäche, Schweißausbrüche usw. Der von der Ehefrau herbeigerufene Hausarzt veranlaßte die sofortige Einweisung in das örtliche Kreiskrankenhaus, wo die Diagnose eines Herzinfarkts gestellt wurde.

Bemerkenswert dabei ist, daß der Hausarzt, mit dem der Patient befreundet ist, ihm wenige Tage vorher in bezug auf sein berufliches Verhalten Vorwürfe gemacht und dabei geäußert hatte, wenn er so weitermache, würde es zwangsläufig zu einem Herzinfarkt kommen. Nach der Entlassung aus dem Krankenhaus konnte er sich in seinem beruflichen Verhalten nicht ändern, er habe wieder voll einsteigen müssen, vor allem, nachdem durch seine Krankheit viel Arbeit liegengeblieben sei. Nachdem es etwa ein halbes Jahr später zu einem erneuten Herzanfall gekommen war, wurden erneut ausführliche und genaue kardiologische Untersuchungen durchgeführt. Diese erbrachten keinen Anhaltspunkt dafür, daß der Patient je einen Herzinfarkt durchgemacht hatte. So sprachen auch die vom Patienten geschilderten Beschwerden nicht für das Vorliegen eines Herzinfarkts. Genau befragt gab der Patient dann später an, er habe ähnliche Beschwerden, wenn auch in geringerer Intensität, schon seit längerer Zeit vor seinem ersten „Herzinfarkt" gehabt, und zwar immer dann, wenn er öffentlich habe sprechen müssen. Er habe vor solchen Situationen immer recht starke Angst gehabt, hätte ihnen jedoch nicht ausweichen können, da seine berufliche Position solche Reden notwendig mache. Er habe damals jedoch das Gefühl von Mundtrockenheit und starkem Schwitzen sowie das Herzjagen auf diese Angst zurückgeführt und nicht das Gefühl gehabt, er sei in irgendeiner Form „herzkrank". Dieses Gefühl habe er erst nach seinem „Herzinfarkt" empfunden, als die geschilderten Beschwerden bei den entsprechenden Anlässen auch in größerer Intensität aufgetreten seien.

32.1 Begriffsbestimmung und Definition

Der Begriff „Herzneurose" wurde nach Richter und Beckmann (1969) erstmals vom Wiener Kliniker Oppenholzer (1867) verwendet. In der Folgezeit wurde den „nervösen Herz-Kreislauf-Beschwerden" zunehmend mehr Beachtung geschenkt. So beschrieb Da Costa 1871 Herzbeschwerden, die er bei Militärangehörigen gehäuft gefunden hatte, und die er als „irritable heart" bezeichnete. Aufgrund der untersuchten Population wurde auch die Bezeichnung „soldier's heart" verwendet.

In der darauffolgenden Zeit folgte eine Reihe von Veröffentlichungen, die sich mit dieser Erkrankung vom klinischen Gesichtspunkt aus, vornehmlich deskriptiv, auseinandersetzten. Es wurden Bezeichnungen eingeführt wie „effort syndrome", „neurozirkulatorische Asthenie", „Herzneurose", „Herzphobie", „Neurasthenie", „vasomotorische Neurose", „vegetative Dystonie", „funktionelle Herz-Kreislauf-Störungen" usw. In Freuds Arbeit: „Über die Berechtigung von der Neurasthenie einen bestimmten Symptomenkomplex als „Angstneurose" abzutrennen" (1971), findet man unter der Bezeichnung „Angstneurose" eine klinisch-phänomenologisch sehr treffende Beschreibung funktioneller Herz-Kreislauf-Beschwerden.

Es lassen sich insgesamt zwei verschiedene Strömungen bei der Auseinandersetzung mit diesem Krankheitsbild feststellen, die sich auch in den verwendeten Bezeichnungen widerspiegeln. Vor allem in der angelsächsischen Literatur wird vornehmlich der Begriff „neurocirculatory asthenia" verwendet. Diese Bezeichnung spiegelt auch eine bestimmte Auffassung zur Pathogenese wider. Es wird davon ausgegangen, daß wesentlich für die Entstehung der Erkrankung somatische Faktoren sind. So gibt es in diesem Bereich eine ganze Reihe von Arbeiten, die sich mit Besonderheiten des EKGs, mit Atemökonomie und körperlicher Leistungsfähigkeit beschäftigen.

Im deutschen Sprachbereich haben vor allem Delius und Fahrenberg (1963, 1964, 1966, 1972) einen Ansatz vertreten, der einerseits somatische, andererseits aber auch psychische Bedingungen mit einbezieht, und die Bezeichnung „psychovegetatives Syndrom" als Sammelbezeichnung für eine ganze Reihe von Störungen vorgeschlagen, u.a. auch für funktionelle kardiovaskuläre Störungen.

Delius und Fahrenberg (1966) teilten die „psycho- und neurovegetativen Herz- und Kreislauf-Störungen folgendermaßen ein:
– Dysrhythmische Störungen
– Dysdynamische Störungen
– Dysästhetische Störungen
– Vasodynamische Störungen

Christian und Mitarbeiter (1965, 1966) verstehen unter vegetativen Herz-Kreislauf-Störungen solche, die von Delius als dysdynam bezeichnet wurden. Sie unterteilen die Störungen in hyper- und hypoton labile Regulationsstörungen, wobei die hypertone Form als dynamisch, die hypotone Form als statisch bezeichnet wird. Wie Delius und Fahrenberg sind auch Christian sowie die amerikanischen Autoren Wheeler, Cohen, White und andere der Meinung, daß zur Erklärung dieses Krankheitsbildes die Annahme somatischer Faktoren notwendig sei. Schonecke (1987) konnte bei Patienten mit funktionellen Herz-Kreislauf-Störungen symptomspezifische kardiovaskuläre Reaktionsmuster ermitteln, anhand derer die Patienten in Untergruppen einzuteilen waren.

Auf der anderen Seite – vor allem auf der psychoanalytischen – wurde unter der Bezeichnung „Herzneurose" diese als grundsätzlich psychoneurotische Störung gesehen (Richter und Beckmann, Fürstenau, Studt, Wilke, Hahn u.a.m.). Es wird dabei gelegentlich angemerkt, daß möglicherweise somatische Bedingungen für die Pathogenese bedeutsam sein können, sie spielen bei Erklärungen jedoch keine oder nur eine untergeordnete Rolle. So räumen Zauner (1967) und Hahn (1976) ein „somatisches Entgegenkommen" (Freud, 1916) als eine Bedingung der Entstehung der Erkrankung ein, für vorgeschlagene therapeutische Interventionen spielt das jedoch, anders als etwa bei Delius, keine Rolle.

Der Begriff „Herzneurose" wird von psychoanalytischer Seite bevorzugt, „um die Relevanz des psychogenen Faktors zu unterstreichen (Richter, 1964). Zum Begriff der „Herzphobie" wiesen Fürstenau und Mitarbeiter (1964) darauf hin, daß der Begriff irrtümlich verwendet wird, da bestimmte Merkmale einer Phobie wie die Vermeidbarkeit des angstauslösenden Reizes nicht vorhanden sind. „Der Tierphobiker kann die betreffenden Tiere, der Agoraphobiker unter Umständen die freie Straße meiden, aber der Herzneurotiker kann um sein Herz keinen Bogen machen" (Fürstenau et al., 1964). Der Begriff „Herzphobie" war 1960 von Kulenkampff und Bauer vorgeschlagen worden und wird trotz dieser Kritik sehr häufig verwendet.

Von Uexküll (1962) verwendet den Begriff „funktionale Beschwerden", da „damit nichts präjudiziert wird, jedenfalls nichts Falsches über die Ätiologie und die Pathogenese dieser Zustandsbilder".

In den letzten Jahren werden funktionelle Herz-Kreislauf-Störungen zunehmend als Panikstörung nach DSM-III klassifiziert, zumal wenn eine phobische Komponente vorherrschend ist (z.B. Maier, 1985). Im vorliegenden Zusammenhang wird der Begriff „funktionelle Herz-Kreislauf-Störungen" beibehalten, da die Patienten, die über eine entsprechende Symptomatik klagen, der Auffassung sind, sie seien herzkrank und wegen ihrer Beschwerden internistisch-kardiologische Hilfe suchen, auch wenn es bei einem Teil von ihnen zu Panikattacken kommt.

So wird nach wie vor eine Vielzahl von Begriffen zur Bezeichnung dieses Krankheitsbildes verwendet, wobei den Begriffen durchaus, wie gezeigt wurde, gewisse Implikationen für die Auffassung zur Pathogenese anhaften. Andererseits macht diese Begriffsvielfalt auch die Unklarheiten der Vorstellungen zur Pathogenese deutlich, d.h., es haben sich keine hinreichend erklärenden Konzepte durchsetzen können.

32.2 Symptomatik

Patienten mit funktionellen Herz- und Kreislauf-Störungen zeichnen sich dadurch aus, daß sie über eine Vielzahl von Beschwerden klagen. Gibt man solchen Patienten Beschwerdelisten, so erreichen sie einen Perzentilrang von durchschnittlich 95, d.h. nur 5% der Eichstichprobe der jeweiligen Liste geben genausoviel und mehr Beschwerden an. Dies bedeutet aber auch, daß die Beschwerden der Patienten sich auf eine Vielzahl von Organfunktionen beziehen. Daneben spielen Beschwerden des psychischen Befindens eine sehr wichtige Rolle, die auch in Beschwerdelisten erfaßt werden. Der hohe Wert, der in diesen erreicht wird, wird oft als allgemeine Klagsamkeit bezeichnet, die z.B. bei Patienten mit organisch bedingten Herz-Kreislauf-Beschwerden viel weniger ausgeprägt ist. Weiterhin besteht ein wesentliches Merkmal der Beschwerden von Patienten mit funktionellen Herz-Kreislauf-Störungen darin, daß die Beschwerden durch die Vorstellungen der Patienten im Hinblick auf die entsprechende Körperfunktion geprägt sind, und mitunter nicht der tatsächlichen Funktion oder irgendwelchen Innervationsgebieten entsprechen.

Dies gilt auch für die Einschätzung der Belastungsabhängigkeit des Auftretens der Beschwerden durch die Patienten. Dabei treten die Beschwerden dem Eindruck der Belastung entsprechend auf, der der tatsächlichen Belastung keineswegs immer entspricht. Auch können sich die Patienten in Anwesenheit des Partners unter Umständen ohne Beschwerden körperlich belasten, und wenn der Patient alleine dasselbe unternimmt, treten die Beschwerden dann auf.

Das Beschwerdebild läßt sich nach v. Uexküll in 5 Hauptgruppen unterteilen:
– Auf das Herz bezogene Beschwerden: Herzklopfen, Extrasystolen, die als Herzstolpern empfunden werden, Herzjagen. Weiterhin Schmerzen, z.B. Drücken und Stechen in der Brust mit Ausstrahlung in den linken Arm, Beschwerden, die bisweilen an einen Infarkt denken lassen.

– Allgemeine Beschwerden: Klagen über Abgeschlagenheit, Schwarzwerden vor Augen, Müdigkeit, Erschöpfung, insgesamt Beschwerden, wie sie beim hypotonen Symptomenkomplex häufig gefunden werden.

– Auf die Atmung bezogene Beschwerden: Klagen über Beklemmungsgefühle, erschwertes Atmen, das bis zur ausgesprochenen Atemnot reicht und sowohl in Ruhe als auch bei körperlichen Belastungen auftreten kann. Diese Beschwerden werden zum Teil auch als eigenes Krankheitsbild definiert (nervöses Atmungssyndrom).

– Vegetative Beschwerden: z.B. Schlaflosigkeit, Parästhesien, Zittern, nervöses Kältegefühl, Schwindelgefühle, Schwitzen sowie Kopfschmerzen.

– Psychische Beschwerden: Häufig geben die Patienten an, unter Reizbarkeit, Angst, innerer Unruhe und niedergedrückter Stimmung zu leiden.

In verschiedenen Untersuchungen, in denen anhand von Beschwerdelisten und deren statistischer Bearbeitung durch Faktorenanalysen das Beschwerdebild in Grundkomponenten geordnet wurde, ergeben sich vergleichbare Beschwerdenkomplexe, die der klinischen Erfahrung entsprechen. Hier ist ebenfalls zu berücksichtigen, daß der Schwerpunkt der Beschwerden im Laufe der Zeit auch wechselt, so daß zu verschiedenen Zeiten unterschiedliche Beschwerden im Mittelpunkt der Befindensstörungen stehen. Dies ist auch wichtig für den Umgang mit den Patienten. Es wird gezeigt werden, daß die Patienten dazu neigen, die körperlichen Beschwerden zu betonen und durch eine schwere körperliche Erkrankung zu erklären und damit vermeiden, sich mit den zugrundeliegenden psychischen Problemen auseinanderzusetzen. Infolgedessen werden die psychischen Symptome als Folge der körperlichen Erkrankung interpretiert. Konzentriert man sich nun im Gespräch mit den Patienten auf die körperliche Symptomatik, so fixiert man damit auch die Befindensstörungen auf die körperlichen Beschwerden.

Die folgenden Beschwerden unterscheiden nach Richter und Beckmann Patienten mit funktionellen Beschwerden von Postinfarktpatienten:

– niedergedrückte Stimmung
– diffuse Ängstlichkeit
– Schonungstendenz
– innere Unruhe
– Herzklopfen
– Furcht, herzkrank zu sein
– Furcht vor Infarkt

Das bedeutet, daß Patienten mit funktionellen Beschwerden eine größere Angst vor Infarkt haben und auch mehr Herzbeschwerden angeben als Postinfarktpatienten. Es ist anzunehmen, daß diese Tatsache auch durch die Verleugnungstendenzen der Herzinfarktpatienten (vgl. Kap. 41.2) mitbedingt ist.

So fällt auch im Umgang mit diesen Patienten ihre allgemeine Ängstlichkeit auf. Diese kann sich auf sehr verschiedene Weise zeigen. Zunächst bezieht sie sich fast immer auf die Funktionstüchtigkeit des eigenen Körpers, vor allem des Kreislaufs. Auch bei geduldiger Information über die Ergebnisse von internistisch-kardiologischen Untersuchungen bleibt bei den Patienten ein Zweifel übrig; sie sind längerfristig nur sehr schwer von der Richtigkeit solcher Befunde zu überzeugen, oft auch gar nicht.

Immer wieder wird die Befürchtung geäußert, es könnte doch etwas übersehen worden sein, oder aber der jeweilige Arzt traue sich ganz einfach nicht, die in Wahrheit sehr ernste Diagnose mitzuteilen. Oft gibt dieser ängstliche Zweifel Anlaß zu immer erneuten, und dann eigentlich nicht mehr indizierten Untersuchungen. Oft auch werden aus diesem Grunde „sicherheitshalber", aber auch, um die Patienten vordergründig zu beruhigen, kardiologisch wirksame Medikamente verschrieben. Dabei wird häufig das Gegenteil erreicht. Durch die Diskordanz zwischen der Diagnose „Sie sind organisch absolut gesund" und der auf dem Beipackzettel des Medikamentes angegebenen Indikation treten eher Zweifel und weitere Beunruhigung ein.

Das bestimmende Moment der Angst zeigt sich aber auch in der Tendenz vieler Patienten, Situationen zu vermeiden, in denen die Beschwerden schon aufgetreten sind. Dies führt sehr häufig zu einer zunehmenden Einengung des Lebensraumes der Patienten, wie es auch bei der Agoraphobie der Fall ist. Das ist keineswegs bei allen Patienten, bei einer ganzen Anzahl jedoch sehr ausgeprägt. Es zeigen sich hier Ähnlichkeiten zu phobischen Verhaltensweisen, woraus sich therapeutische Konsequenzen ergeben (vgl. Abschn. 32.6), aber möglicherweise auch Konsequenzen für die Richtigkeit des Begriffs „Herzphobie".

Viele Patienten haben aufgrund ihrer ängstlichen Besorgtheit um ihre Gesundheit eine ausgesprochene Schonungstendenz. Sie muten sich immer weniger an körperlicher Belastung zu und geraten dadurch mitunter in einen erheblichen Trainingsmangel aus der Befürchtung heraus, sie könnten sich überlasten. Aus der Ängstlichkeit, aber auch aus der immer wieder bei diesen Patienten zu beobachtenden Zwanghaftigkeit leitet sich ein übermäßiges Kontrollbedürfnis her. So werden fast zwanghaft ärztliche Vorschriften beachtet, die Einnahme von Medikamenten peinlich genau eingehalten und vieles mehr. Auf der anderen Seite rührt von daher wohl auch der Wunsch der Patienten, nach einiger Zeit eine erneute körperliche Untersuchung durchführen zu lassen. Bisher Übersehenes könnte sich jetzt zeigen, oder Schädigungen könnten neu entstanden sein. Je aufwendiger die Untersuchung, desto besser. Dies erinnert mitunter durchaus an die Kontrollzwänge von Zwangskranken, die immer wieder überprüfen müssen, ob z.B. der Herd auch wirklich abgestellt oder die Tür verschlossen ist. Ist dann eine erneute Untersuchung durchgeführt worden, so beruhigt das Ergebnis für einige Zeit durchaus, der Zweifel erhebt sich allerdings nach einiger Zeit wieder. Diese Ängste erklären auch, daß es den Patienten meist in der Gegenwart eines Arztes schnell besser geht und es sehr selten unter ärztlicher Kontrolle zu starken Herzbeschwerden kommt. Für die Patienten ist dann eine Person vorhanden, von der angenommen wird, daß sie die Herzfunktion notfalls beeinflussen oder kontrollieren kann. Zum anderen repräsentiert der Arzt ganz allgemein eine schützende Bezugsperson, deren Anwesenheit ohnehin beruhigt.

Als Trennungsangst wurde die Angst der Patienten beschrieben, derart schützende Personen zu verlieren, was sich deutlich in der Arzt-Patient-Beziehung zeigt. Oft sind die Patienten enttäuscht, wenn der Arzt nicht ausgedehnt Zeit für sie hat, sie neigen dazu, sich anzuklammern. Dieses Verhalten zeigen sie aber auch anderen Personen, z. B. ihren Ehepartnern gegenüber. So kommt es immer wieder vor, daß Patienten fast nur noch in Begleitung ihrer Partner aus dem Haus gehen können, was nicht nur sie selbst, sondern auch ihre Partner erheblich einengt.

Ein weiteres, ganz wesentliches Merkmal besteht in der fast immer vorhandenen und oft sehr ausgeprägten Depressivität. Sie ist nicht grundsätzlich vorhanden, eine Reihe der Patienten, die durch ihre der Phobie ähnlichen Verhaltensweisen auffallen, haben sie weniger. Die Depressivität zeigt sich meist in einem verminderten Antrieb. Daneben grübeln diese Patienten häufig, wobei sich ihre Gedanken meist auf das körperliche Befinden richten und die sich daraus ergebenden Konsequenzen. Oft auch glauben sie, ihre Umgebung durch ihre verminderte Leistungsfähigkeit enttäuscht zu haben, wobei auffallenderweise das Verständnis der Partner fast immer gelobt wird.

Von Schwarz (1982) wurde auf eine ausgeprägte Leistungsorientiertheit hingewiesen, die sich bei 70% der von ihm untersuchten Patienten fand. Immer wieder hervorgehoben wurde eine versteckte Aggressivität der Patienten, die nur selten offen ausgedrückt werden kann. Sie ist entdeckbar in der mitunter spürbaren Genugtuung, mit der dem Arzt mitgeteilt wird, seine bisherigen Bemühungen hätten noch nicht geholfen. So äußerte ein Patient einmal: „Ich habe diese Krankheit jetzt etwa 10 Jahre. An mir haben sich viele berühmte Leute die Zähne ausgebissen, können Sie mich behandeln?"

32.3 Epidemiologie

Über die Häufigkeit funktioneller Herz- und Kreislauf-Beschwerden finden sich in der Literatur recht unterschiedliche Angaben. Cremerius (1963) fand 8% Patienten mit funktionellen Herz-Kreislauf-Störungen von 2330 Fällen einer medizinischen Poliklinik. Kannel und Mitarbeiter (1958) fanden bei über 1000 untersuchten Personen der Framingham-Studie 16% mit funktionellen Herz- und Kreislauf-Beschwerden.

Delius (1964) schätzt die Häufigkeit dieser Erkrankung in der Allgemeinpraxis auf 10–15%. Nach Cobb (1943) sind es im psychiatrischen Bereich etwa 27%. Jorswiek und Katwan (1967) ermittelten im Berliner Zentralinstitut für psychogene Störungen, daß die Zahl der Patienten mit Herzsymptomen sich in den Jahren 1945 bis 1965 verdoppelt hatte. Maas (1975) fand bei 162 332 Patienten der Deutschen Klinik für Diagnostik in Wiesbaden bei 20–25% Angaben von Beschwerden, die einen Verdacht auf das Vorliegen funktioneller Herz- und Kreislauf-Beschwerden rechtfertigten. Studt (1979) schätzt die Häufigkeit in der Gesamtbevölkerung auf 2–5%, in der Allgemeinpraxis auf 10–15%; bei 30–40% der Patienten mit Herz-Kreislauf-Beschwerden seien diese funktionell bedingt. Frauen haben häufiger als Männer funktionelle Herzbeschwerden.

Die Häufigkeit funktioneller Störungen des Herz-Kreislauf-Systems wird also sehr unterschiedlich angegeben. Neuere Angaben schwanken zwischen 2% und ca. 12% Häufigkeit (v. Weel, 1987; Schepank, 1987; Dilling et al., 1984) in der allgemeinen Bevölkerung. Die Schwankungen in den Zahlenangaben sind vermutlich durch Definitionsunschärfen sowie verschiedene Häufigkeitsmaße (Punktprävalenz, Inzidenz) bedingt. Der Anteil von Patienten mit psychischer bzw. psychiatrischer Symptomatik in Allgemeinpraxen kann auf ca. 30% geschätzt werden, wovon etwas mehr als die Hälfte neurotische und psychosomatische Symptome aufweisen (Zintl-Wiegand et al., 1988). So zeigt eine Untersuchung von Tress et al. (1989), daß bei 16% der Personen, bei denen in einer Prävalenzstudie psychosomatische Störungen diagnostiziert wurden, drei Jahre später das Vorliegen von neurotischen Störungen angenommen wurde. Umgekehrt wurden psychosomatische Störungen bei 38% der Patienten angenommen, bei denen drei Jahre vorher neurotische Störungen festgestellt worden waren. Damit wird deutlich, daß die Grenzen zwischen beiden Störungsformen nicht nur unscharf sind, sondern daß es unter Umständen sinnvoll sein kann, unter einem epidemiologischen Gesichtspunkt von einer Grundgesamtheit psychogener Störungen auszugehen, die bei wechselnder Symptomatik in unterschiedliche Klassen eingeordnet werden kann (vgl. Kap. 4).

Im Hinblick auf den sozioökonomischen Status fanden sich bisher keine nennenswerten Besonderheiten (Richter und Beckmann, 1973; Pflanz, 1962). Diese Erkrankung tritt eher bei jüngeren Menschen auf, wie auch v. Uexküll (1962) anmerkt. Jenseits des 40. Lebensjahres und mit fortschreitendem Alter nimmt die Häufigkeit der Diagnosestellung erheblich, fast schlagartig ab.

Die Gründe hierfür sind noch nicht untersucht worden. Es ließe sich jedoch annehmen, daß durch altersbedingte Veränderungen auch im Herz-Kreislauf-System möglicherweise „organische Diagnosen" auf der Grundlage geringer Befunde erhoben werden und damit auch eine Erklärung für das Vorliegen von Beschwerden liefern.

Hinze und Krüger (1981) sahen 502 Krankengeschichten einer gerontopsychiatrischen Poliklinik daraufhin durch, ob sich „Hinweise auf Herzangst nicht-organischer Genese fänden". Sie fanden bei 9%, das sind 46 Fälle, derartige Hinweise. Nicht berücksichtigt wurden dabei ausgesprochen depressive Patienten, bei denen es zu Herzbeschwerden kam, ohne daß das Moment der Angst im Vordergrund stand. Das Durchschnittsalter dieser Patienten betrug 66 Jahre, die Herzangstsymptomatik bestand im Mittel seit 15 Jahren. Da ebenfalls die Befunde internistischer Untersuchungen vorlagen, konnte auch die Frage geklärt werden, inwieweit organpathologische Veränderungen die Beschwerden hätten erklären können. Dies war von den 46 Patienten nur bei 5 der Fall, so daß bei 41 Patienten funktionelle kardiovaskuläre Störungen vorlagen. Dabei ist zu berücksichtigen, daß das Vorgehen der Autoren sehr strengen Maßstäben gerecht wurde, indem sie feststellten, daß auch bei weiteren Patienten trotz einer organischen Diagnose Beschwerden vorlagen, die nicht durch die Diagnose erklärt werden konnten. Interessant hierbei ist die Tatsache, daß die Häufigkeit des Vorkommens der „Herzangst", wie die Autoren es nennen, bei dieser Stichprobe etwa genauso hoch ist, wie sie von Cremerius (1963) in einer allgemeinen Poliklinik gefunden wurde. Allerdings ist hier die Stichprobe institutionell eingegrenzt, da es sich um Patienten einer gerontopsychiatrischen Poliklinik handelte. Dennoch verdient diese Untersuchung im vorliegenden Zusammenhang besondere Beachtung, da sie zeigt, daß das Vorliegen funktioneller kardiovaskulärer Störungen keineswegs ab einem bestimmten Alter schlagartig abnehmen muß.

Eine weitere Möglichkeit für die Abnahme der Häufigkeit funktioneller kardiovaskulärer Störungen besteht darin, daß sonstige Erkrankungen die Aufmerksamkeit der Patienten in Anspruch nehmen, so daß die funktionellen Beschwerden in den Hintergrund treten. In diesem Zusammenhang ist auch an die „Stellvertreterfunktion" der funktionellen Beschwerden zu denken, die überflüssig wird, wenn andere Beschwerden im Rahmen somatischer Erkrankungen auftreten.

Aus dem durchschnittlichen Alter der Patienten ergibt sich ein ökonomisch bedeutsamer Faktor, fast alle Patienten sind im erwerbsfähigen Alter, so daß sich die Frage nach krankheitsbedingten Ausfallzeiten stellt. Sturm und Zielke (1988) haben diese Frage an einer Stichprobe von 1155 Patienten einer psychosomatischen Fachklinik überprüft. Davon waren 35,9% bis zur Aufnahme in die Klinik ununterbrochen krank geschrieben und davon wiederum 38,29% über ein Jahr lang. 29,78% sind über 18 Monate arbeitsunfähig gewesen. Die Dauer der Krankheitsmanifestation betrug im Durchschnitt 7,04 Jahre, was auch bedeutet, daß die Patienten durchschnittlich 7 Jahre in irgendeiner Weise unzureichend behandelt worden sind. Aus einer Zusammenstellung der Autoren geht hervor, daß im Zeitraum eines Jahres (7/82–6/83) in der Bundesrepublik von praktischen Ärzten und Internisten 39,4 Millionen Verordnungen von Tranquilizern, Schlafmitteln und Antidepressiva durchgeführt wurden, davon 2,86 Millionen Antidepressiva. Kommt also ein Patient mit funktionellen Störungen nach Jahren in eine für seine Erkrankung fachspezifisch kompetente Behandlung, so stellt sich meist zusätzlich das Problem, eine Medikamentenabhängigkeit behandeln zu müssen.

Im allgemeinen wird angenommen, daß Frauen häufiger als Männer von funktionellen Störungen betroffen sind (z.B. Dilling et al., 1984; Schepank,

1987). In verschiedenen Studien, in denen Patienten mit funktionellen kardiovaskulären Störungen untersucht wurden, gab es jedoch mehr Männer in den Stichproben als Frauen (z.B. Richter und Beckmann, 1973; Schonecke, 1987; Nutzinger et al., 1987). Dies mag unter Umständen an der Institution liegen, in der die jeweiligen Untersuchungen durchgeführt wurden, bzw. in Verbindung damit am möglicherweise verschiedenen Inanspruchnahmeverhalten von Männern und Frauen.

32.4 Theorien zur Ätiologie und Pathogenese

Es wurde bereits bei der Behandlung der Frage der Begriffsbestimmung darauf hingewiesen, daß es verschiedene Ansätze der Betrachtung des vorliegenden Krankheitsbildes gibt.

32.4.1 Psychodynamische Erklärungsansätze

Bereits im Abschnitt 32.1 wurde darauf hingewiesen, daß von psychoanalytischer Seite das vorliegende Krankheitsbild als Form einer neurotischen Störung betrachtet wird. Dies gilt für „herzneurotische Störungen" im eigentlichen Sinne. Davon abgrenzbar sind nach Hahn (1965) „herzphobische Zustände". Für die Herzneurose gilt jedoch, daß sie eine Entstehungsgeschichte hat, die das Resultat einer inadäquaten Konfliktbewältigung darstellt.

Für die Entwicklung einer Disposition zur Herzneurose ist nach übereinstimmender psychoanalytischer Meinung (Fürstenau et al., 1964; Richter, 1964; Bräutigam, 1964; Baumeyer, 1966; Dieckmann, 1966; Zauner, 1967; Richter und Beckmann, 1973; Maas, 1975; Studt, 1979) eine bestimmte Form der Mutter-Kind-Beziehung wesentlich. Die Mütter der Patienten sind selbst sehr unsicher und haben auf dem Boden dieser Unsicherheit ihre Kinder in besonderer Weise an sich gebunden, so daß eine als „symbiotisch" bezeichnete Beziehung zwischen den späteren Patienten und ihren Müttern bestand.

Diese Bindung beinhaltet für das sich entwickelnde Kind einen übergroßen Schutz, der verhindert, daß das Kind eigene, unabhängige Strategien zur Bewältigung von Angst entwickeln kann. Diese überbeschützende Haltung der Mutter hat andererseits verwöhnenden Charakter, da sie das Kind vor allerlei Unangenehmem abschirmt. Richter (1964) weist entsprechend darauf hin, daß die Mutter für die Patienten immer noch die Bedeutung des „ersten Angstschutzes" besitze. Eine Konsequenz dieser Erfahrungen besteht für das Kind darin, daß es den Eindruck gewinnt, ohne die Mutter hilflos zu sein, auf sie angewiesen zu sein, z.B. um mit Angst umgehen zu können.

Dieser Aspekt der an sich selbst wahrgenommenen eigenen Hilflosigkeit steht in Beziehung zu der bei diesen Patienten immer wieder festgestellten Depressivität (vgl. auch Seligman, 1975). Mit fortschreiten der Entwicklung des Kindes werden Selbständigkeitswünsche zunehmend wichtig. In dem Maße, in dem das Kind versucht, diesen Bestrebungen nach Autonomie und eigener Kompetenz nachzugehen, stößt es auf die Grenzen, die die Haltung der Mutter setzt. Diese hat ihre Haltung ihrem Kind gegenüber nicht aufgrund irgendwelcher Überlegungen eingenommen, sondern als Resultat einer eigenen inadäquaten Konfliktbewältigung. Entsprechend unfrei ist sie, dem heranwachsenden Kind den notwendigen Spielraum für seine eigene Entwicklung, vor allem seine Loslösung von ihr einzuräumen. Hieraus ergibt sich ein Konflikt für das Kind und später für den Patienten: die Unvereinbarkeit eigener Selbständigkeitswünsche mit dem Angewiesensein auf die beschützende Mutter.

Dies hat einen weiteren Aspekt. Treffen die Autonomiewünsche auf Grenzen, wird Aggression der Mutter gegenüber ausgelöst. Die eigene Aggression bedroht so gerade die Person, die in besonderem Maße für das eigene Wohlergehen notwendig ist, ohne die das Kind hilflos ist. In diesem Sinne besteht damit eine Bedrohung der eigenen Person, indem die Ausübung der Aggression quasi das Fundament für das eigene „Leben-Können" vernichten würde. So bemerken Fürstenau und Mitarbeiter (1964): „Bei unterschiedlichen Graden von Anspruchsniveau, Aktivität und äußeren sozialen Erfolgen fanden wir bei diesen Patienten im Hintergrund die Phantasie von der frühen Mutter-Kind-Symbiose wirksam. Allerdings stets in der Weise, daß diese Symbiose zugleich als bedroht erlebt wird: Man darf nichts tun, was die Anklammerung an die Mutterfigur gefährdet, weil man ohne diesen Halt nicht existieren kann."

Die Angst vor dem Objektverlust gilt daher als ganz wesentliches Element der Bedingungen, die schließlich zur herzneurotischen Störung führen. Solange die Beziehung zur Mutter oder zu äquivalenten Personen gewährleistet bleibt und nicht durch eigene Wünsche nach Selbständigkeit gefährdet wird, bleibt das System in einem Gleichgewicht. Wird diese Beziehung jedoch gefährdet, so tritt die Angst vor der Trennung vom notwendigen und haltverleihenden Objekt in den Vordergrund. Die dadurch eintretende phantasierte Bedrohung der eigenen Existenz drückt sich möglicherweise in der mit der Herz-Kreislauf-Symptomatik verbundenen Todesangst aus.

Eine 30jährige Patientin litt seit folgender Begebenheit an einer intensiven Herz-Kreislauf-Symptomatik, die sie weitgehend in ihrer Bewegungsfreiheit einschränkte: Ein Mitbewohner ihres Hauses war von einer Wespe gestochen worden und hatte eine allergische Reaktion entwickelt, die bedingte, daß er vom Ersticken bedroht war.

Die Frau dieses Mannes kam zur Patientin und bat sie, den Mann zum nächsten Arzt zu fahren, da sie selbst kein Auto besaß. Die Patientin willigte sofort ein und fuhr den zumindest in ihrem Erleben fast sterbenden Mann zum Arzt. Während dieser Fahrt hatte sie ständig das Gefühl, vielleicht etwas falsch zu machen. Nach Hause zurückgekehrt – der Mann hatte alles überstanden – fühlte die Patientin sich sehr schlecht und legte sich ins Bett. Es kam zu einer Tachykardie, der Notarzt wurde gerufen usw. In den darauffolgenden Jahren zwangen die Symptome die Patientin mehr oder weniger zu Hause zu bleiben, aber auch im Haushalt konnte kaum noch bewältigt werden. Es gab eine ganze Reihe von Situationen, in denen es mit großer Wahrscheinlichkeit zu den tachykarden Beschwerden kam, so besonders deutlich während Besuchen bei der Mutter.

Nach einer verhaltenstherapeutischen Behandlung, mit der es gelungen war, die Symptomatik fast vollständig zu beseitigen, berichtete die Patientin, daß sie neuerdings beim Anblick des oben genannten Mitbewohners Angst spüre. Das störe sie nicht sehr, da sie ihn nicht sehr oft sehe, und wenn, auch nur für kurze Zeit, sie wolle es jedoch mitteilen. Im weiteren Verlauf des Gesprächs wurde die Patientin aufgefordert, den Vorgang, der zum Beginn ihrer Beschwerden geführt hatte, nochmals so detailliert wie möglich zu schildern. Während sie dies tat, brach sie in Tränen aus, was sie sich selbst nicht recht erklären konnte. Schließlich erinnerte sie sich auf die Frage, ob sie irgendwann jemanden gekannt habe, der Atembeschwerden gehabt hätte, daran, daß ihre Mutter als Folge einer später operativ behandelten Vergrößerung der Schilddrüse ganz erhebliche Atembeschwerden gehabt hätte, die bis zu Erstickungsanfällen gereicht hätten. Bei einer solchen Gelegenheit sei sie als kleines Kind einmal losgeschickt worden, um in der Apotheke ein schnell wirksames Mittel zu beschaffen. Unterwegs habe sie zu ihrem Entsetzen festgestellt, daß sie das Rezept vergessen hatte und nochmal nach Hause zurück mußte. Sie habe das entsetzliche Gefühl gehabt, daß ihre Mutter jetzt wegen dieses Fehlers möglicherweise sterben müsse, und sie dann schuld daran sei.

In früheren Gesprächen war die Mutter als rechthaberisch und bevormundend geschildert worden, die sich auch heute noch besserwisserisch in alles einmischen würde. Bei Besuchen bei der Mutter habe es häufig Streit gegeben, und in diesem Zusammenhang seien die Beschwerden häufig sehr heftig aufgetreten.

An diesem Beispiel wird deutlich, mehr als das gewöhnlich der Fall ist, wie die ambivalente Einstellung der Mutter gegenüber für die Patientin ein Problem darstellt, das in dem Augenblick zu einer Dekompensation führt, in dem durch ein „Resonanzereignis" die Thematik des Verlustes der Mutter ganz aktuell wird. Inwieweit das für die Patientin unerklärliche Gefühl, irgend etwas falsch zu machen, während sie den Mitbewohner zum Arzt fuhr, dem Gefühl entsprach, das sie damals gehabt hatte, als sie das Rezept vergaß, läßt sich nur vermuten. Ebenfalls nicht zwingend belegen läßt sich die Annahme, daß dadurch eigene aggressive Tendenzen der Mutter gegenüber aktualisiert wurden.

Hahn (1965, 1976) hat vor allem darauf hingewiesen, daß neben der typischen Herzneurose ein Syndrom bestehe, für das der Name „Herzphobie" verwendbar sei. Hierbei spiele ein „somatisches Entgegenkommen" (Freud, 1916) im Sinne einer Sympathikotonie eine wesentliche Rolle. Es gebe dabei somatische Reaktionsgegebenheiten, die entweder durch körperliche oder psychische Bedingungen ausgelöst werden können und bei ihrem Auftreten von intensiver Angst begleitet werden. Diese Angst fixiere sich und führe zu einer „Angst vor der Angst" im Sinne einer Phobie. Am Anfang einer solchen Herzphobie stehe immer genau datierbar ein sympathikovasaler Anfall, der aufgrund der erlebten Dramatik zu der phobischen Entwicklung führe. Von Bräutigam (1964) wurde ebenfalls der Begriff der Herzphobie verwendet, zur Kennzeichnung eines Krankheitsgeschehens, bei dem körperliche Bedingungen eine wesentliche Rolle spielen. Er geht jedoch davon aus, daß bei genauerer Analyse stets die oben skizzierte Konfliktproblematik eruierbar sei und für den Ausbruch der Erkrankung eine wesentliche und notwendige Bedingung darstelle. Ein ähnlicher Standpunkt wurde von Baumeyer (1966) eingenommen. Zauner (1967) unterscheidet dagegen „Herzfunktionsstörungen" mit und ohne spezifischen psychodynamischen Hintergrund, wobei es, wie von Hahn (1965) angemerkt wurde, bei Störungen ohne spezifischen psychodynamischen Hintergrund häufig sekundär zu einer phobischen Fehlverarbeitung komme. Er zieht die Folgerung: „Es liegt auf der Hand, daß bei allen Erkrankungen mit psychogener Teilursache, der keine neurotische Verarbeitung zugrunde liegt, die Anwendung psychotherapeutischer Behandlungsmethoden nicht sinnvoll erscheint" (Zauner, 1967). Die Auffassungen von Hahn und Zauner kommen in die Nähe psychophysiologischer Modelle im weiteren Sinne, indem sie annehmen, daß ein Zusammenspiel psychogener und somatischer Bedingungen als Prozeß eine Krankheit auslöst und stabilisiert (vgl. auch Mayer et al., 1973).

Aus den Überlegungen zur Psychodynamik der Patienten mit funktionellen Herz-Kreislauf-Störungen lassen sich bestimmte Hypothesen über das Vorhandensein von Persönlichkeitsmerkmalen ableiten. Ein wesentliches Merkmal scheint in einer erhöhten Angstbereitschaft zu bestehen.

Richter und Beckmann (1973) überprüften diese Hypothese und konnten bei Patienten mit funktionellen Herz-Kreislauf-Beschwerden zwei Typen der Angstabwehr feststellen, die sich mit Hilfe des MMPI (Minnesota Multiphasic Personality Inventory) unterscheiden lassen. Die Patienten des Typs A (nicht zu verwechseln mit dem Begriff Typ A aus der Herzinfarktforschung) zeichnen sich dadurch aus, daß sie ihre Angstproblematik nicht verleugnen und so auch im Test recht offen darstellen. Diese Patienten können ihre Ängste nicht abwehren und werden von ihnen „überflutet". Im Gegensatz dazu gelingt den Patienten vom Typ B eine Angstabwehr im Sinne der Verleugnung. Diese Tendenz zeigt sich auch bei der Beantwortung des Fragebogens vor allem in den Kontrollskalen, die das Ausmaß der Tendenz erfassen, in der Beantwortung der Fragen Probleme als gering darzustellen.

Die Angstabwehr gelingt jedoch nur oberflächlich und hat den Charakter des „krampfhaft Normalen". Im MMPI lassen sich beide Typen am deutlichsten anhand der sog. Validitätsskalen unterscheiden. Die L-Skala und die K-Skala sind bei den Patienten des Typs A erniedrigt, die F-Skala erhöht. Dadurch

kommt die Tendenz zum Ausdruck, Probleme offen darzustellen, auch wenn dies nicht als sozial erwünscht angesehen wird. Bei den Patienten des Typs B ist dies genau umgekehrt. Nach Richter und Beckmann lassen sich 84% der Herzneurotiker einem der beiden Typen zuordnen, wobei 48% zum Typ A zu zählen sind, 36% zum Typ B.

Mayer und Mitarbeiter (1973) konnten diese Befunde mit einer eigenen Stichprobe bestätigen, nicht jedoch Schüffel und Mitarbeiter (1972). Hierbei spielt möglicherweise eine Rolle, daß die Patientenstichprobe von Schüffel und Mitarbeitern auf andere Weise selektiert worden war.

Die Patienten beider Typen unterscheiden sich jedoch nicht im Hinblick auf das Beschwerdebild. Mayer und Mitarbeiter (1973) teilten in ihrer Untersuchung die Patienten anhand des Angsterlebens bei anfallsartigen Beschwerden ein. Sie fanden dabei zwei Gruppen, von denen die eine mit einem großen Ausmaß von Angst auf die Beschwerden reagierte, die andere weit weniger. Beide Gruppen unterschieden sich im Profil des FPI (Freiburger Persönlichkeitsinventar). Es zeigte sich weiterhin eine gewisse Übereinstimmung mit der Typ-A- und -B-Klassifizierung von Richter und Beckmann. Allerdings muß angemerkt werden, daß in der zitierten Arbeit keine quantitativen Aussagen über die Klassifikationsgüte oder das Ausmaß des Zusammenhangs mit den beiden Typen nach Richter und Beckmann enthalten sind.

Daß Patienten mit funktionellen Herz-Kreislauf-Beschwerden ein höheres Maß von Ängstlichkeit zeigen, kommt auch in einer Untersuchung von Oberhummer und Mitarbeitern (1979) zum Ausdruck, die die Patienten mit einer normalen Kontrollgruppe verglichen haben. Die Autoren benutzten dabei den MAS (Manifest Anxiety Scale).

Neben der Angst spielt die Depression eine wesentliche Rolle. Wie bereits aus der Untersuchung von Richter und Beckmann hervorgeht, läßt sich dieser Sachverhalt mit Hilfe des MMPI deutlich zeigen.

Schonecke und Mitarbeiter (1972) konnten mit den Aggressionsskalen nach Foulds und Caine zeigen, daß eine Gruppe unausgelesener Patienten mit funktionellen Herz-Kreislauf-Beschwerden sich signifikant von einer Kontrollgruppe im Hinblick auf die Tendenz, Feindseligkeit gegen sich selbst zu richten, unterschied. Die Patienten hatten dabei entsprechend höhere Werte. Eine nach außen gerichtete Aggressivität zeigte sich lediglich in der Möglichkeit, andere in Abwesenheit zu kritisieren. Mit Hilfe der auf Kelley (1955) zurückgehenden Technik des REP-Tests oder „Repertory Grid" konnte darüber hinaus gezeigt werden, daß die Wahrnehmung von Aggressivität bei anderen Personen in Abhängigkeit von der sozialen Distanz variiert. Nahe Bezugspersonen wie Eltern, Ehepartner oder ähnliche Personen wurden als nicht aggressiv erlebt, diese Eigenschaft wurde jedoch bei distanteren Personen wie Arbeitskollegen durchaus erlebt. Diese Ergebnisse können die Annahme stützen, daß das Thema der Aggressivität bei denjenigen Personen, auf die die Patienten angewiesen sind, ausgeblendet bzw. verringert ist.

32.4.2 Psychophysiologische Erklärungsansätze

Im folgenden werden Erklärungsansätze dargestellt, die zur Erklärung des vorliegenden Krankheitsbildes wesentlich das Zusammenwirken psychischer und somatischer bzw. physiologischer Faktoren betonen. Betrachtet man das Krankheitsbild als eine Form der Phobie oder allgemein der Neurose, so bleibt unerklärt, warum im Mittelpunkt der Symptomatik ein tatsächliches oder vermeintliches Körpergeschehen steht. Die folgenden Ansätze sind nur zum Teil im eigentlichen Sinne als psychophysiologisch zu bezeichnen, für alle gilt jedoch, daß sie somatische Bedingungen als wesentlich für die Entstehung des Krankheitsbildes annehmen.

Betrachtet man die Frage der Disposition aktivierungstheoretisch, so würde dies bedeuten, daß eine Gruppe von Personen ein individuell abgrenzbares Reaktionsmuster zeigt, etwa in dem Sinne, auf Belastungen vornehmlich mit kardiovaskulären Reaktionen zu antworten (vgl. Kap. 8). Diese Disposition würde dann bedingen, daß es bei diesen Personen unter einer Vielzahl von Bedingungen zu unter Umständen starken kardiovaskulären Veränderungen kommt, die von den Personen als unangenehm wahrgenommen und als krankhaft interpretiert werden. Das aktivierungstheoretische Konzept der individualspezifischen Reaktion (ISR) beinhaltet, daß es in verschiedenen Situationen zu gleichen Reaktionsmustern kommt.

Im angelsächsischen Bereich gibt es eine ganze Reihe von Arbeiten, die sich ausführlich mit körperlichen Bedingungen bei funktionellen Herz-Kreislauf-Störungen beschäftigt haben.

Cohen und Mitarbeiter (1947) fanden bei Patienten mit „neurozirkulatorischer Asthenie" (NCA) unter körperlicher Belastung höhere Blut-Laktatkonzentrationen als bei gesunden Kontrollpersonen. Dieser Unterschied war bei Patienten mit einem chronischen Verlauf der Erkrankung größer als bei akuten Erkrankungen. Die Autoren ziehen daraus den Schluß, daß der subjektive Eindruck der Patienten, sie seien weniger belastbar, zutreffend sei. Cohen und Mitarbeiter (1948) fanden darüber hinaus, daß Patienten mit NCA unter körperlicher Belastung deutlich höhere Anstiege der Herzfrequenz aufwiesen, eine geringere Fähigkeit zur Sauerstoffaufnahme besaßen, schneller dyspnoisch wurden und auch unter entsprechenden Beschwerden litten. Bereits unter Ruhebedingungen war die Atmung der Patienten schneller bei flacherer Amplitude. Die Herzgröße sowie EKG und Herzminutenvolumen waren jedoch nicht verändert gegenüber normalen Kontrollen. Cohen und Mitarbeiter (1951) untersuchten 139 Familien von Patienten mit NCA und stellten sie denen von gesunden Kontrollpersonen gegenüber. Sie fanden, daß die Häufung der Erkrankung innerhalb der Familien der Patienten größer war als bei den gesunden Kontrollen, besonders bei der chronischen Form. Die Autoren meinen, daß ein Imitationsmodell ihre Befunde nicht erklären könne, ein dominanter Erbgang, der ihren Daten am ehesten entsprechen würde, jedoch auch nicht „bewiesen" werden könne. Kannel und Mitarbeiter (1958) untersuchten im Rahmen der Framingham-Studie 203 Patienten mit NCA und 757 gesunde Kontrollpersonen. Die Autoren fanden keine mit der Gruppenzugehörigkeit verbundenen Abweichungen im EKG. Die Abweichungen, die bei den Patienten im EKG gefunden wurden, traten in der Kontrollgruppe mit derselben Häufigkeit ebenfalls auf. Dies waren vor allem ST-Streckenveränderungen, WPW-Syndrom und verlängerte PR-Zeit. Levander-Lindgren (1962) fand vergleichbare Veränderungen im EKG vor allem vor und nach körperlicher Belastung. Nach Belastung fanden sich bei 49% der weiblichen Patienten und in Ruhe bei 21% ST-Streckenveränderungen. Bei Männern lag dieser Prozentsatz erheblich niedriger mit 21 und 2%. In dieser Studie wird wiederum darauf hingewiesen, daß subjektive Beschwerden hauptsächlich bei den Patienten auftraten, bei denen auch die EKG-Veränderungen zu beobachten waren.

Insgesamt können diese Untersuchungen nicht stringent nachweisen, daß Veränderungen etwa des EKGs systematisch mit dem vorliegenden Krankheitsbild verknüpft sind. Zum Teil sind die genannten Untersuchungen methodisch sehr anfechtbar, so etwa die Arbeit von Levander-Lindgren, in der keine Kontrollgruppe untersucht wurde, so daß die Aussage, die gefundenen Besonderheiten seien für die Erkrankung spezifisch, nicht möglich ist. Die methodisch besseren Arbeiten legen eher den Schluß nahe, daß spezifische somatische Bedingungen nicht gefunden werden können (vgl. Kannel et al., 1958). Wichtig scheint vor allem die Unterscheidung zwischen chronischem und akutem Verlauf der Erkrankung. Zieht man diese Unterscheidung mit in Betracht, so zeigt sich in einem Teil der zitierten Untersuchungen, daß Abweichungen von einer Kontrollgruppe hauptsächlich Patienten mit chronischem Verlauf betreffen. Hierbei ist daran zu denken, daß die gefundenen Unterschiede möglicherweise als Folge eines allgemeinen Trainingsmangels zu interpretieren sind. Es wurde bereits darauf hingewiesen, daß die Patienten dazu neigen, sich zu schonen, um einer möglichen Gefährdung durch Überlastung vorzubeugen und dadurch in einen zunehmenden Trainingsmangel geraten. So könnten die Ergebnisse von Cohen und Mitarbeitern (1948) im Hinblick auf die verringerte Atemökonomie gedeutet werden.

In der letzten Zeit wurde wiederholt darauf hingewiesen, daß als körperliche Bedingung funktioneller Herz-Kreislauf-Störungen dem Mitralklappenprolapssyndrom (MVPS – mitral valve prolaps syndrome) eine wichtige Bedeutung zukomme (Strian et al., 1981; Pariser et al., 1978). „Beim MVPS liegt eine Verminderung der elastischen Elemente mit dünnen und verlängerten Chordae tendineae mit überschüssigen und verdickten Klappen vor" (Strian et al., 1981). Die angeborene Form dieses Syndroms ist gegenüber der erworbenen durch Herzinfarkt oder Myokarditis häufiger.

Von Pariser und Mitarbeitern (1978) wird ebenso wie von Strian und Mitarbeitern (1981) die Auffassung vertreten, daß das MVPS eine mögliche Ursache für auf das Herz bezogene Ängste darstellt. Strian und

Mitarbeiter sind der Meinung, daß die in früheren Untersuchungen gefundenen Abweichungen (welche oben zitiert wurden) unter körperlicher Belastung Folge eines MVPS sind. „Der Nachweis des MVPS und dabei häufiger Arrhythmien bei Patienten mit herzphobischer Symptomatik zeigt nicht nur die Notwendigkeit einer differenzierenden Diagnostik, sondern scheint auch die Bedeutung der vegetativen Wahrnehmung in der Entwicklung organphobischer Ängste zu bestätigen" (Strian et al., 1981). Stalmann et al. (1988) fanden jedoch keinen Zusammenhang zwischen dem Vorliegen eines Mitralklappenprolaps, funktionellen Herzbeschwerden, Güte der auf den Herzschlag bezogenen Körperwahrnehmung und Angst. Personen mit Mitralklappenprolaps nehmen Arrhythmien sogar signifikant seltener wahr als gesunde Kontrollpersonen. Devereux (1985) kam in einer Studie an 300 Patienten mit Mitralklappenprolaps zu dem Ergebnis, daß ein Zusammenhang zwischen Prolaps und Paniksyndrom oder „neurozirkulatorischer Asthenie" nicht besteht. Andere Autoren kamen jedoch zu anderen Ergebnissen, die allerdings meist nicht von einer Stichprobe von Personen mit Mitralklappenprolaps ausgingen, sondern von Patientenstichproben mit Angstsyndromen (Dager et al., 1987; Grunhaus et al., 1982; Liberthson et al., 1986) oder oft Paniksyndrom. Die unterschiedlichen Ergebnisse kommen möglicherweise durch die Art der Patientenstichprobe zustande und wohl auch durch die Unsicherheit der Diagnosestellung eines Mitralklappenprolaps. So legten Gorman et al. (1986) 15 Echokardiogramme von Patienten mit Paniksyndrom zwei Kardiologen zur Beurteilung vor. Einer der beiden Kardiologen fand bei neun einen Prolaps, der andere bei keinem Patienten.

Jenzer (1981) meint dazu: „All diese Patienten leiden, solange sie noch nichts davon wissen, nicht unter dem Problem oder dessen hämodynamischen Auswirkungen, sondern unter ihren auf das Herz zentrierten Ängsten und Auswirkungen ihrer konstitutionellen Schwäche. Die Behandlung hat sich demnach nicht nach einem Merkmal zu richten, das nur Kardiologen oder kardiologisch Interessierte fasziniert, weil nur die die Klicks oder Geräusche hören oder den Prolaps mit dem Echokardiographen registrieren können, die im Gesamtbild aber von untergeordneter Bedeutung sind."

Im folgenden werden Ansätze dargestellt, die für die Erklärung funktioneller Herz-Kreislauf-Störungen Dispositionen annehmen, die deshalb als psychophysiologisch anzusehen sind, weil sie wesentlich somatische und psychische Merkmale enthalten.

Delius und Fahrenberg gehen davon aus, daß eine Disposition vorliegt, die zu veränderten körperlichen Reaktionen führt. „Als psychovegetative Syndrome werden Ordnungsmängel im Befinden und Verhalten bezeichnet, die essentiell einen somatischen und einen psychischen Aspekt haben. Als Störmoment zeigen sie ein Mißverhältnis der psychosomatischen Regulation zum jeweiligen Erfordernis. Für diese Ordnungsstörungen ist charakteristisch, daß normalerweise wenig beachtete vegetative Funktionen in dis-

harmonischer Form im Erleben aktualisiert werden. Dem betroffenen Menschen eröffnet sich damit eine anomale Wahrnehmungswelt von Allgemeingefühlen und Organempfindungen" (Delius und Fahrenberg, 1966).

Sie gehen aus von einer „komplementären" Betrachtungsweise, als Stellungnahme zum Leib-Seele-Problem, die die psychosomatische Medizin fordere (vgl. auch Fahrenberg, 1979). „Organfunktion und Verhalten einerseits und Erleben und Befinden andererseits sind zwei in ihrer kategorialen Struktur verschiedene, inkommensurable Erscheinungsweisen derselben Lebensvorgänge. Was für sich als Inhalt der inneren unmittelbaren Erfahrung sinnbezogen erlebt wird, das erscheint der äußeren Erfahrung mittelbar als somatisch-nervöser Prozeß bzw. als Verhalten. Beide Ablaufreihen stehen untereinander nicht in einem Kausalzusammenhang, sie sind nicht sukzessiv korreliert (Wechselwirkung), sondern simultan korreliert (wie der Parallelismus annimmt) und außerdem komplementär" (Delius und Fahrenberg, 1966). Als eine Grundbedingung in der Pathogenese psychovegetativer Syndrome sehen sie eine grundlegende Regulationsschwäche, die „übergreifend-psychophysisch" zu verstehen ist. Die „psychovegetative Organisation" – nur sie ist für unser Thema wichtig – wird definiert als Leistungseinheit derjenigen biopsychischen Abläufe, welche an die (das Retikulär- und das limbische System einschließenden) vegetativen Strukturen des zentralen und peripheren Nervensystems gebunden sind. „Die psychovegetative Organisation verdichtet als dynamische Gliederung das somatische und das erlebnismäßige Geschehen in genau der gleichen Weise, wie das die psychomotorische Organisation in ihrem Bereich vollzieht" (Delius, 1964).

In Anlehnung an Eysenck (1947, 1958) gehen Delius und Fahrenberg von „in der Erbmasse vorgeprägten, von der somatischen und seelischen Lebensgeschichte unterschiedlich ausdifferenzierten Instabilitäten aus, die eine Anfälligkeit in der psychovegetativen Organisation schaffen", dem Konzept der „Dysthymie". Eysenck (1957) stellt die Dysthymie der inhibitorischen Natur der Hysterie in Anlehnung an Pawlow gegenüber. „Gegenteilig scheinen die Symptome der Dysthymie für ihn den Hinweis eines Übermaßes an exzitatorischen Potentialen zu zeigen, sowie das Fehlen der Entwicklung eines ausreichenden inhibitorischen Potentials. Dysthymische Symptome entwickeln sich bei Individuen, bei denen die Exzitation-Inhibition-Balance in Richtung auf eine übermäßige Exzitation verschoben ist" (Eysenck, 1957). Neben der vegetativen Übererregbarkeit zeichnen sich Dysthymiker durch eine höhere Streßanfälligkeit, leichtere Konditionierbarkeit, Introversion, Ängstlichkeit und Sensitivität aus. „Diese faktorenanalytisch verifizierbare Konstellation bildet eine wesentliche, wenn nicht prädisponierende Wurzelform der Syndromgenese bei Herz-Kreislauf-Störungen bzw. psychovegetativen Syndromen überhaupt" (Delius, 1964).

Für die Ausprägung eines funktionellen kardiovaskulären Syndroms gibt Delius drei Mechanismen an:

– Die Entstehung zusätzlicher vegetativer Reflexe.
– Die Entregelung, Umprogrammierung oder „Umstimmung" physiologischer Rückkoppelungs- und Steuervorgänge.
– Die Entwicklung besonderer, d. h. atypischer Regulationsmuster höherer Ordnung oder entsprechender Verhaltensweisen.

Die einzelnen Ansätze unterscheiden sich im Hinblick auf die Definition dieser Disposition. Wurde auf der einen Seite nach den Symptomen einer im Grunde rein somatisch verstandenen Disposition gesucht, wurde auf der anderen Seite von vorneherein von einer im eigentlichen Sinne „psychosomatischen Disposition" ausgegangen. Deren Annahme basiert einerseits auf einer Grundannahme zum Leib-Seele-Problem und ist andererseits orientiert an empirisch recht gut gesicherten Befunden aus der Psychologie.

In beiden Fällen jedoch beinhaltet diese Disposition eine Regulationslabilität vegetativer Funktionen. Diese wurde auch von Christian und Mitarbeitern (1965, 1966) als wesentliche Ursache funktioneller Herz- und Kreislauf-Störungen angesehen. Sie läßt sich hier zurückführen auf ein Konzept der Sympathikotonie. „Die Herzphobie hat pathophysiologisch fast stets eine permanente hypersympathikoton-ergotrope Grundverfassung zum Hintergrund" (Delius und Fahrenberg, 1966). Auch Hahn (1965) führt den Beginn der Herzphobie auf das Auftreten eines sympathikovasalen Anfalls zurück und geht dabei von einer gesteigerten ergotropen Reaktionslage aus.

Die Frage nach einer bestimmten Disposition im Sinne einer möglicherweise angeborenen psychophysischen Regulationsschwäche bleibt unklar. So kommt auch Myrtek (1980) in einer methodisch sehr aufwendigen Arbeit zu dem Ergebnis, daß „die niedrigen Korrelationen gegen das Konzept einer globalen vegetativen Labilität" sprächen. „Gäbe es eine globale Dimension vegetativer Labilität, so müßten die Reaktionswerte verschiedener Variablen über verschiedene Funktionsprüfungen hinweg hoch miteinander korrelieren, was aber nicht zutrifft. Vielmehr sind individual- und stimulusspezifische Reaktionsmuster anzunehmen. Das Ergebnis ist, wie schon aus der Diskussion über die körperliche Leistungsfähigkeit zu vermuten war, auch diesmal nicht im Sinne eines psychophysischen Zusammenhangs interpretierbar. Es lassen sich keine Beziehungen zwischen objektiven vegetativen Funktionsprüfungen und emotionaler Labilität nachweisen" (Myrtek, 1980).

Diese Befunde lassen immerhin die Möglichkeit offen, daß es Personen gibt, die im Sinne einer Individualspezifität etwa im kardiovaskulären Bereich besonders reagieren, was zum Ergebnis hätte, daß die Korrelationen mit anderen Funktionsbereichen niedrig wären. Auch wenn dies der Fall wäre, müßte anhand kontrollierter Studien geprüft werden, ob ein entsprechendes Merkmal in einer entsprechenden Patientengruppe häufiger anzutreffen ist als in einer gesunden Kontrollgruppe.

Einen möglichen Hinweis liefert ein Ergebnis des Orthostaseversuchs von Myrtek (Myrtek et al., 1974; Myrtek, 1980). Hier wurde gefunden, daß sich Personen mit einer bestimmten Blutdruckreaktion bzw. deren Verlauf, die sich in dem einen Fall im Sinne einer Regulationslabilität sympathikotoner, im anderen Fall vagotoner Art deuten lassen, von Personen mit normotoner Regulation auch im Hinblick auf Persönlichkeitsmaße unterscheiden. So waren sie im FPI psychosomatisch gestörter, depressiver, zurückhaltender, mehr reaktiv aggressiv, gehemmter, selbstkritischer, emotional labiler und weniger zuversichtlich.

Schonecke (1987) hat die Frage symptomspezifischer Reaktionsmuster bei Patienten mit funktionellen Herz-Kreislauf-Störungen untersucht. Er verglich die physiologischen Reaktionsmuster von Patienten mit denen gesunder Kontrollpersonen in einer Aktivierungsstudie, in der die Probanden verschiedenen Belastungssituationen ausgesetzt wurden, wobei kardiovaskuläre Parameter, solche der muskulären und der elektrodermalen Aktivität sowie solche der Atmung gemessen wurden. Die Ergebnisse zeigen, daß die Patienten ausschließlich in den kardiovaskulären Parametern stärker auf Belastung reagieren als die gesunden Kontrollpersonen, in denen der elektrodermalen und muskulären (M. frontalis) Aktivität sogar signifikant geringer. In den Parametern der Atmung gab es keinerlei Unterschiede zwischen beiden Gruppen. Diese Ergebnisse sprechen für das Vorliegen symptomspezifischer Reaktionsmuster, obwohl Unterschiede zwischen den Patienten und den Kontrollpersonen noch deutlicher im Niveau der kardiovaskulären Parameter vorhanden waren, also unabhängig von Belastung.

In einer Faktorenanalyse der physiologischen Meßwerte ergaben sich im wesentlichen 4 Faktoren, wobei der erste durch die Parameter der Atmung, der zweite durch Herzfrequenz und Pulswellengeschwindigkeit, der dritte durch die elektrodermale und muskuläre Aktivität und der vierte durch die Durchblutung am linken Unterschenkel bestimmt war. In den Faktorenwerten beider Gruppen wird die Symptomspezifität deutlich.

Bei der Inspektion der vor allem interessierenden Daten der Reaktivität der Herzfrequenz wurde deut-

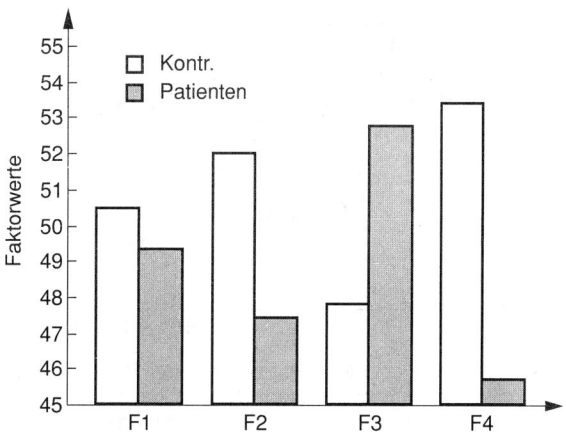

Abb. 32–1. Mittlere Faktorwerte von Patienten und Kontrollpersonen (aus Schonecke, 1987).

lich, daß es zwischen den einzelnen Probanden große interindividuelle Unterschiede gab. Es wurden infolgedessen über die Reaktivität der Herzfrequenz, also nicht die absolute Höhe der Herzfrequenz, sondern deren Veränderung durch Belastungen, Clusteranalysen gerechnet, um festzustellen, ob eine sinnvolle Gruppierung der Probanden anhand dieser Unterschiede möglich ist. Es ergab sich dabei eine Lösung, bei der die Probanden zwei Clustern zugeordnet werden konnten, die sowohl mit der Herzfrequenz reagierende von nicht-reagierenden Patienten und Kontrollpersonen unterschieden. In einem Cluster waren also diejenigen Patienten und Kontrollpersonen, die auf die Belastungen nicht mit Anstiegen der Herzfrequenz reagierten, im anderen diejenigen beider Gruppen, die mit Anstiegen der Herzfrequenz reagierten. Dabei zeigte sich, daß nur etwa ein gutes Drittel der Patienten auf die Belastungen reagierte, fast zwei Drittel jedoch kaum oder gar nicht. Letztere jedoch hatten die höchsten Niveauwerte, d.h. auch während der Ruhephasen lag ihre Herzfrequenz etwa im Niveau der Belastungen. Auch Ehlers et al. (1986) hatten zwischen Patienten mit Panikattacken und gesunden Kontrollpersonen ausschließlich Niveauunterschiede der Herzfrequenz, aber einen signifikant steileren Anstieg des systolischen Blutdrucks nach Laktatinfusion gefunden. Bei den Kontrollpersonen der vorliegenden Studie jedoch lagen die Niveauwerte der nicht-reagierenden Probanden sehr niedrig und veränderten sich während der Belastung nicht gegenüber den Ruhewerten. Dies wird in Abbildung 32–2 deutlich.

Interessant ist der Zusammenhang zwischen den verschiedenen Clustern und persönlichkeitsmetrischen Daten. So wurden verschiedene Verfahren zur Erfassung derartiger Merkmale angewendet und damit eine Faktorenanalyse gerechnet. Patienten und Kontrollpersonen unterscheiden sich in sehr hohem Maße in den Faktorwerten des ersten Faktors, der bestimmt ist durch Anzahl und Stärke von Beschwer-

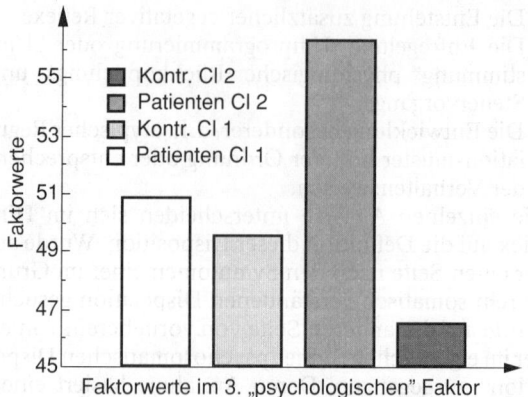

Abb. 32–3. Faktorwerte von Patienten und Kontrollpersonen im dritten „psychologischen Faktor" (aus Schonecke, 1987).

den, Depressivität und Angstbereitschaft. Dies ist unabhängig davon der Fall, in welchem der beiden Cluster sich die Patienten oder Kontrollpersonen befinden. Der dritte Faktor ist bestimmt durch hohe Ladungen der Skalen Ablenkung und Bagatellisierung, Projektion, mangelnde gedankliche Weiterbeschäftigung und Ersatzbefriedigung des Streßverarbeitungsbogens nach Janke (1978). Hier haben die Patienten des zweiten Clusters, die also mit starken Anstiegen der Herzfrequenz auf Belastung reagieren, die höchsten, die Kontrollpersonen dieses Clusters die niedrigsten Werte, wie Abbildung 32–3 zeigt.

Betrachtet man die Werte in der State-Trait-Angstskala nach Spielberger (1970), so ergibt sich, daß die Patienten im zweiten Cluster signifikant weniger Angst angeben als die Patienten im ersten Cluster. Bei den Kontrollpersonen besteht ein ähnlicher Unterschied, nur auf einem niedrigeren Gesamtniveau der Angstwerte. Dieses Ergebnis entspricht damit dem Konzept der repressiven vs. sensitiven Angstabwehr, das beinhaltet, daß Personen mit einer repressiven Angstabwehr in entsprechenden Situationen zwar weniger Angst erleben, aber physiologisch erregter sind als Personen mit einer sensitiven Angstabwehr.

Richter und Beckmann hatten ebenfalls ihre Patienten nach dem Stil der Angstverarbeitung unterschieden, allerdings anhand der Kontrollskalen im MMPI (Minnesota Multiphasic Personality Inventory). Dabei hatte u.a. die „Lügenskala" die Gruppen unterschieden, was in der vorliegenden Untersuchung nicht der Fall war, d.h. die Patienten in den beiden Clustern hatten annähernd dieselben Werte. Unterschiedlich ist ebenfalls das Beschwerdeniveau in den beiden Clustern, d.h. die Patienten im zweiten Cluster haben tendenziell (p=0,07) weniger Beschwerden als die im ersten Cluster. Richter und Beckmann hatten keine Unterschiede der Beschwerden zwischen ihren beiden Gruppen gefunden.

Mit den Faktorwerten der psychologischen und der physiologischen Faktoren wurde dann eine Faktorenanalyse zweiter Ordnung gerechnet, um Beziehungen zwischen physischen und psychischen Ordnungsdimensionen und mögliche Unterschiede zwischen Pa-

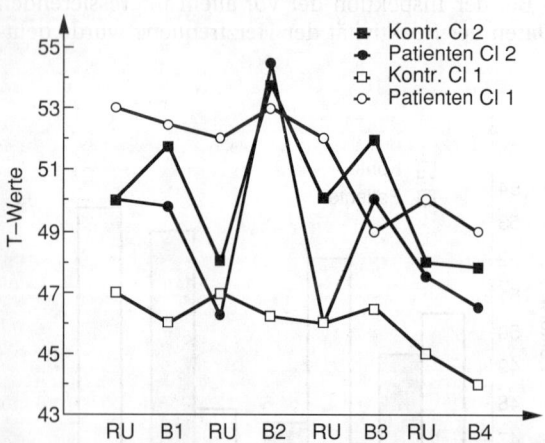

Abb. 32–2. Verlauf der Herzfrequenz von Patienten und Kontrollpersonen in den beiden Clustern. Cluster 1 enthält diejenigen Probanden, die nicht mit Veränderungen der Herzfrequenz auf die Belastungen reagierten, Cluster 2 diejenigen, die mit Anstiegen der Herzfrequenz reagierten (aus Schonecke, 1987).

tienten und Kontrollpersonen zu ermitteln. Es ergaben sich vier Faktoren zweiter Ordnung, wobei der erste durch die Atemparameter und repressive Angstvermeidung, der zweite durch den psychologischen Faktor hoher Depressivität, Angst und Beschwerdeintensität sowie den physiologischen Faktor hoher kardiovaskulärer Reaktivität, der dritte durch den psychologischen Faktor hoher Erregbarkeit, Aggressivität, mangelnder Gelassenheit, Dominanzstreben und geringer Beachtung sozialer Normen und den physiologischen Faktor hoher elektrodermaler und muskulärer Aktivität gekennzeichnet ist. Der vierte Faktor schließlich zeigt einen negativen Zusammenhang mit dem Faktor kardiovaskulärer Reaktivität und ist positiv bestimmt durch den psychologischen Faktor geringer sozialer Abkapselung, Neigung zu Intellektualisierung und Bedürfnis nach Aussprache. Wesentlich bei diesem Ergebnis ist die Tatsache, daß alle Sekundärfaktoren sowohl durch psychologische und physiologische Primärfaktoren bestimmt sind. Vor allem die Faktorwerte des zweiten und dritten Sekundärfaktors unterscheiden zwischen Patienten und Kontrollpersonen, wie Abbildung 32–4 zeigt.

Die Faktorwerte zweiter Ordnung trennen Patienten und Kontrollpersonen weitgehend unabhängig davon, in welchem Cluster sie sich befinden, d.h. weitgehend unabhängig von der Reaktivität der Herzfrequenz. Lediglich im ersten Sekundärfaktor unterscheiden sich die Patienten in den beiden Clustern voneinander. Die Patienten zeigen, wie schon oben erwähnt, eine repressive oder vermeidende Angstabwehr und eine stärkere Reaktivität der Atmung. Dies gilt für die Atemamplitude, die in allen Ruhebedingungen bei den Patienten des zweiten Clusters geringer war und vor allem während des Streßinterviews

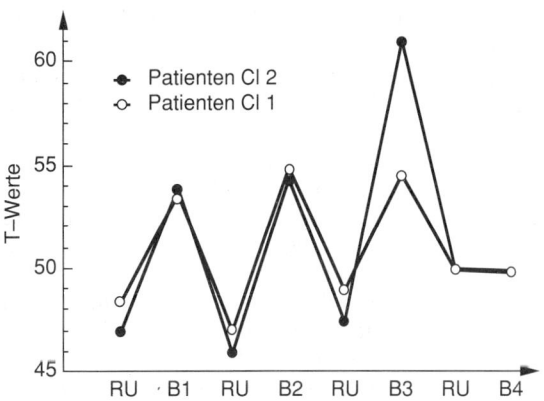

Abb. 32–5. Verlauf der Atemfrequenz in den Clustern (aus Schonecke, 1987).

mit einem emotional bedeutsamen Inhalt (detaillierte Sexualanamnese), das vor der Untersuchung mit den Patienten geführt worden war und dann während der Untersuchung vom Tonband abgespielt wurde. Nur während dieser emotional bedeutsamen Belastung unterscheidet sich die Atemfrequenz der Patienten in den beiden Clustern, wie Abbildung 32–5 zeigt.

Anhand dieser Ergebnisse erscheint es sinnvoll, die Patienten mit funktionellen Herz-Kreislauf-Störungen anhand ihrer Kreislaufreaktivität und Angstverarbeitung in zwei Untergruppen einzuteilen, da der vermeidende Umgang mit Angst und Belastung auch therapeutische Konsequenzen hat. Wenn die Bezeichnungen „Herzneurose" und „Herzphobie" unterschiedlich benutzt werden, so entsprechen die hier gefundenen Untergruppen der Patienten dieser Unterscheidung am ehesten.

Vergleicht man die in dieser Aktivierungsstudie gefundenen Werte mit denen, die in der Literatur bei Patienten mit Angststörungen gefunden werden, so ergeben sich Unterschiede vor allem in der Reaktivität und dem Niveau der elektrodermalen Aktivität und der muskulären Aktivität. In diesen beiden Parametern zeigen Patienten mit Angststörungen in allen Studien höhere Werte als gesunde Kontrollpersonen. Aufgrund dieser Befunde erscheint es sinnvoll, Patienten mit funktionellen Herz-Kreislauf-Störungen von Patienten mit neurotischen Angststörungen abzugrenzen.

Die bisher referierten, psychophysiologisch orientierten Ansätze versuchten im wesentlichen, eine psychophysische Disposition zu ermitteln, die die Entwicklung funktioneller Herz-Kreislauf-Störungen begünstigt. Stringent könnte eine solche Disposition jedoch nur in prospektiven Studien ermittelt werden, da beim Vorliegen des Krankheitsbildes die gefundenen Merkmale auch dessen Folge sein könnten. Dennoch stützen die vorliegenden Untersuchungen die Annahme einer derartigen psychophysischen Disposition.

Die folgenden Annahmen beziehen sich eher auf den Vorgang der Symptom- oder Krankheitsentstehung und beinhalten im wesentlichen keine Annah-

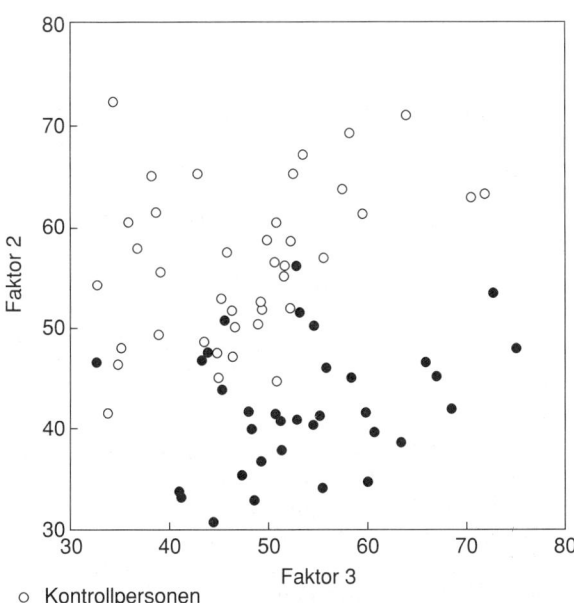

o Kontrollpersonen
• Patienten

Abb. 32–4. Scatterdiagramm der Faktorwerte zweiter Ordnung von Patienten und Kontrollpersonen im 2. und 3. Sekundärfaktor (aus Schonecke, 1987).

men über Dispositionen, die die Krankheitsentstehung begünstigen.

Bergold und Kallinke (1973) formulierten einen Erklärungsansatz, der lerntheoretisch orientiert ist, der jedoch in bestimmten Anteilen psychoanalytischen Überlegungen nahekommt. Die Autoren gehen vom Prinzip der negativen Verstärkung aus. Es beinhaltet, daß ein Verhalten wahrscheinlicher auftreten wird, wenn es von der Beendigung eines negativen Reizes gefolgt wird. Es wird nun angenommen, daß in einer Konfliktsituation, in der äußere, den Konflikt auslösende Reize nicht vermieden werden können, die physiologischen Reaktionsanteile einer konflikthaften Reaktion wahrgenommen werden, die nicht ausgeführt werden kann. Damit kann der eigentliche Konflikt vermieden werden. Die den Konflikt auslösende Situation erhält dadurch eine neue Bedeutung, in der der Konfliktinhalt nicht mehr oder nur noch in einer tolerablen Form vorhanden ist. Durch die Vermeidung des Konflikts wird die Wahrnehmung der physiologischen Reaktionsanteile verstärkt, was einer Sensibilisierung der Wahrnehmung für diese spezifischen Reize gleichkommt.

Als Beispiel führen die Autoren einen Arbeiter am Fließband an, der die Situation trotz ihrer Aversivität nicht vermeiden kann. Aber auch beispielsweise aggressive Reaktionen auf die Aversivität der Situation kann sich der Arbeiter nicht leisten, da er sonst Schwierigkeiten mit seinem Vorgesetzten bekommen oder sogar seinen Arbeitsplatz verlieren könnte usw. Unter bestimmten quantitativen Bedingungen (besonders großem Ärger über irgendein Ereignis, beispielsweise Tadel durch den Vorgesetzten) kann der in der Situation enthaltene Konflikt so verstärkt werden, daß die physiologischen Komponenten einer verhinderten aggressiven Reaktion, wie Herzfrequenz- oder Blutdruckanstieg, wahrgenommen werden. Durch den Tadel etwa ist das aggressive, ohnehin in der Situation enthaltene Potential dermaßen angestiegen, daß auch die Tendenz zur aus den oben genannten Gründen unmöglichen aggressiven Reaktion vergrößert wird, und damit auch die physiologischen Reaktionsanteile. Der in der Situation enthaltene Konflikt kann nun dadurch vermieden werden, daß die Körperwahrnehmungen nicht als Teil einer aggressiven „Bereitstellung", sondern als ungewöhnliche, möglicherweise als Krankheitssymptome interpretiert werden. Die Bedrohung geht jetzt nicht mehr von einem Verhalten (Aggression), verbunden mit negativen Verhaltenskonsequenzen aus (Verlust des Arbeitsplatzes), sondern von einer plötzlich befürchteten Erkrankung. Darüber hinaus können zusätzlich verstärkende Ereignisse eintreten. Die aversive Situation wird unterbrochen, der (spätere) Patient erhält Zuwendung usw.

Die relative Nähe zu psychoanalytischen Erklärungen ergibt sich aus der Annahme, daß der eigentliche Konflikt durch das Auftreten der Störungen vermieden wird, die Angst vor der Herzerkrankung eine sekundäre Angst darstellt, die eine Art von Stellvertreterfunktion besitzt, indem die primäre Angst, die sich aus dem Konflikt ergibt, vermieden und nicht mehr

erlebt wird. Es gibt zwei wesentliche Voraussetzungen für dieses Modell. Erstens ist es notwendig, daß autonome Reaktionen tatsächlich durch Vorgänge des operanten Lernens beeinflußbar sind. Die von Miller ausgehenden Untersuchungen (vgl. Miller, 1969, 1978) haben gezeigt, daß dies möglich ist. Für die vorliegende Fragestellung ist es dabei nicht so ausschlaggebend, ob die operante Kontrolle autonomer Funktionen vermittelt oder direkt stattfindet (vgl. Kap. 7). Zweitens setzt dieses Modell voraus, daß der Einfluß kognitiver Prozesse für das diskriminative Erleben eines Affekts bei relativ gleichen physiologischen Veränderungen von bestimmender Bedeutung ist. Die Arbeiten von Schachter und Singer (1962) haben zeigen können, daß es in Abhängigkeit von einer experimentellen Situation bei künstlich herbeigeführter Änderung physiologischer Parameter zu Affekten unterschiedlicher Valenz kommen kann (vgl. Kap. 8).

Unscharf bleibt in diesem Modell die Vorstellung der Quantität des Erregungsanstiegs, bzw. der Konflikthaftigkeit einer Situation, die zur Wahrnehmung der physiologischen Vorgänge führt. Präzisierungen hierzu wären wünschenswert. Deutlich wird auch, daß das Modell keine Annahmen über irgendwelche Dispositionen enthält, die als Voraussetzung dafür angesehen werden, daß der beschriebene Prozeß abläuft. Es läßt sich andererseits denken, daß ein derartiger Prozeß wahrscheinlicher ist bei Personen, bei denen stärkere Änderungen physiologischer Funktionen stattfinden, wie das etwa das Sympathikotoniekonzept nahelegt.

Von Uexküll (1962) entwickelte ein Modell, das ebenfalls einen dynamischen Prozeß der Symptomentstehung enthält. Er geht dabei davon aus, daß ein die Erregung steigernder psychischer oder somatischer Vorgang zu einer Funktionsänderung führt, die ihrerseits eine körperliche Sensation oder Empfindung hervorruft, die emotional verarbeitet werden muß. Bei einer bestimmten, nicht näher definierten Art der Erlebnisverarbeitung kann dies zu einer anhaltenden Funktionsänderung führen, die wiederum wahrgenommen wird, womit sich der Kreis schließt.

Den bisher dargestellten Ansätzen läßt sich folgendes entnehmen:

- Eine wesentliche Rolle für Entstehung und Aufrechterhaltung der funktionellen Herz-Kreislauf-Störungen spielen Angst und Angstverarbeitung. Fast alle Ansätze betonen diese Komponenten.
- Von der Mehrzahl der Autoren wird eine psychophysische bzw. im eigentlichen Sinne des Begriffs „psychosomatische" Disposition angenommen, die zumindest den Prozeß der Krankheitsentstehung und Perpetuierung begünstigt. Die Betonung dieser Komponente ist unterschiedlich.
- Von psychoanalytischer Seite wird betont, daß in der Entwicklung der Patienten spezifische Bedingungen zu einer psychischen Disposition führen, die beim späteren Auftreten von spezifischen Konflikten (Selbständigkeitsbestrebungen, Verlusterlebnissen) eine inadäquate Konfliktlösung bedingt.

– Der Körperwahrnehmung wird in allen Modellen eine wichtige Rolle zugemessen.

– Für die pathogene Wirkung der Körperwahrnehmung spielt offensichtlich ein Vorgang eine Rolle, der vielleicht am besten mit dem Konzept der Attribuierung, d.h. der Ursachenzuschreibung, benannt werden kann. Der Patient meint, daß eine Erkrankung die Ursache für die Körperwahrnehmung ist.

Von Uexküll (1962) hat darauf hingewiesen, daß mitunter zufällige, ängstigende Ereignisse am Beginn einer solchen Erkrankung stehen können. „Der auslösende Vorgang, sei er psychischer oder somatischer Natur, kann für das resultierende Symptom relativ belanglos sein." Wird dieses Ereignis auf eine Weise verarbeitet, die in bestimmten Anteilen einer Phobie entspricht, so kann die Symptomatik lange Zeit bestehenbleiben und sich weiter steigern. So lassen sich Patienten finden, die seit einer Infektionserkrankung oder nach übermäßigem Alkoholgenuß usw. an Herzbeschwerden leiden und zunehmend Situationen vermeiden, von denen sie die Erfahrung gemacht haben, daß die Beschwerden dabei aufgetreten sind. Richter und Beckmann (1973) nennen als unmittelbar auslösende Bedingungen für funktionelle Herz-Kreislauf-Beschwerden:

– Konfrontation mit Unfall, Krankheit oder Todesfällen (oft herzbedingt) in der Umgebung,
– beunruhigende Beobachtungen am eigenen Körper,
– induzierende ärztliche Diagnosen,
– psychische Konflikte.

Geht man nun davon aus, daß die Angst sich auf die Funktion des Herzens und des Kreislaufs bezieht, so stellen die vom Herz- und Kreislaufsystem ausgehenden Körpersensationen Gefahrensignale dar, die ihrerseits Angst auslösen. Bekanntlich führt Angst zur Veränderung verschiedener Körperfunktionen, u.a. auch zu Steigerungen der Herzfrequenz, wodurch die Gefahrensignale vermehrt werden. So fallen diese Patienten durch anfallsweise auftretende, sich steigernde Angstattacken auf, die von unter Umständen sehr heftigen Kreislaufreaktionen begleitet sind. Der Vorgang dabei entspricht dem der positiven Rückkoppelung, d.h. das Ergebnis des Vorgangs (Output: Körperreaktion) wird dem Eingang zugeführt (Input: Körperwahrnehmung, Gefahrensignal) und verstärkt die Prozesse (Angst, Aktivierung), die zu einer Vergrößerung des Outputs führen. So schaukelt sich das System zur Panik auf. Wichtig für diesen Vorgang ist, daß die Aufmerksamkeit, anders als bei anderen Phobien, nicht auf externe Signale gerichtet ist, sondern auf interne, solche, die von Körpervorgängen herrühren. Dies führt dazu, daß die „Beschwerden" mitunter ohne erkennbaren äußeren Anlaß auftreten, gleichsam aus „heiterem Himmel" beginnen.

Viele Patienten schildern, daß ihre Beschwerden hauptsächlich in Ruhe auftreten, einem Zustand, in dem durch das Wegfallen sonstiger Aktivitäten spontane Körperreaktionen deutlicher wahrnehmbar werden, da externe, aber auch interne Signale von Aktivität wegfallen. Zum zweiten ist beim Zustand der Ruhe zu bedenken, daß für die Patienten eine Erklärung für etwa auftretende Körperwahrnehmungen fehlt, die unter Umständen während irgendwelcher Aktivitäten gegeben ist, d.h. bei körperlicher Belastung kann die Körperwahrnehmung erklärt werden.

Bei genauerer Befragung der Patienten kann darüber hinaus eine wichtige Komponente des Vorgangs ermittelt werden: das Auftreten von immer wiederkehrenden Gedanken, die einerseits eine Hinwendung der Aufmerksamkeit auf Körpervorgänge vermitteln, andererseits unmittelbar ängstigend wirken, „hoffentlich fängt das jetzt nicht wieder an", oder Erinnerungen an eine gleiche Situation, in der die Beschwerden aufgetreten sind. Diese Gedanken erhalten ihrerseits gleichsam Signalcharakter und sind in der Lage, den oben genannten Vorgang auszulösen.

Damit wird auch verständlich, warum Patienten zunehmend mehr Situationen vermeiden. Gedanken der geschilderten Art treten in Situationen, in denen die Beschwerden bereits stattgefunden haben, mit größerer Wahrscheinlichkeit auf und führen dazu, daß durch die Gedanken die Wahrscheinlichkeit erhöht wird, daß die Beschwerden auftreten. Der Gedanke an die Angst führt zur Angst vor der Angst, diese wiederum zu Körpersignalen, auf die ohnehin verstärkt geachtet wird, und diese lösen schließlich den „Anfall" aus.

Der Vorgang der Beschwerdeentstehung läßt sich am besten anhand eines Modells der Verarbeitung von Informationen, die teilweise von Körpervorgängen, teilweise aus der Situation und ihrer Bedeutung gegeben sind, verständlich machen. Wesentliche Elemente dieses Modells, das sich an Ergebnissen der Angstforschung orientiert (vgl. Butollo, 1979), sind Kriterien für die Stärke von Erregungsanstiegen. Das erste Kriterium entscheidet in einer gegebenen Situation bei einem Erregungsanstieg darüber, ob dieser eine bestimmte Größe übersteigt und somit für die weitere Verarbeitung von Information berücksichtigt werden muß. Ist dies der Fall, so führt dies keineswegs notwendigerweise zur Unterbrechung der gegenwärtigen Aktivitäten. Es führt jedoch zu einer zusätzlichen Suche nach Information, die inhaltlich auf die Erregung bezogen ist, d.h. die Erregung erklären könnte. Da diese Informationssuche nicht zur Unterbrechung der gegenwärtigen Aktivitäten führt, ist sie „asynchron". Wird in der Situation Information gefunden, die die Erregung erklären kann, und bleibt die Erregung unter einem bestimmten Niveau (Kriterium 2), so wird die Situation im Hinblick auf ihre Aversivität, Bedrohung oder ähnliches beurteilt. Es finden dann die Prozesse der primären und sekundären Beurteilung statt, also auch im Hinblick auf die eigenen Möglichkeiten, mit der Situation umgehen (coping) zu können (vgl. Lazarus, 1975; Lazarus et al., 1980).

Ist die Information, die auf diese Weise gewonnen wird, nicht mit der Erregung konkordant (nicht-konkordant bedeutet, daß die aus der Umgebung gewonnene Information die Erregung nicht erklären kann), richtet sich die Informationssuche auf interne Bedingungen, die die Erregung erklären könnten. Auch hierbei wird zunächst nicht angenommen, daß diese

Vorgänge die Aufmerksamkeit auf sich ziehen, d.h., die sonstigen gegenwärtigen Aktivitäten laufen weiterhin ab. Daß eine derartige parallele Informationsverarbeitung möglich ist, haben Untersuchungen gezeigt, in denen verschiedene akustische Informationen über jedes Ohr dargeboten wurden (vgl. Broadbent, 1971).

Steigt die Erregung weiter, weil keine Möglichkeit der Erklärung für die Anfangserregung gefunden wurde, wird ab einem bestimmten Anstieg das Niveau des zweiten Kriteriums überschritten. Das beinhaltet, daß die bisherigen Vorgänge der Informationsverarbeitung auch fehlerhaft abgelaufen sein können, indem aus Gründen früherer Erfahrung (Lerngeschichte) bestimmte Inhalte von der Verarbeitung ausgeschlossen, gefiltert wurden.

Es ist weiterhin anzumerken, daß die Größe der Erregung ständig mit allen Kriterien verglichen wird. Mit der Dauer der Informationsverarbeitung, ohne ein erklärendes Resultat, steigt die Wahrscheinlichkeit eines weiteren Erregungsanstiegs, da die Ergebnislosigkeit Unsicherheit hervorruft und diese wiederum steigernd auf die Erregungslage wirkt (vgl. Berlyne, 1968). Wichtig ist die Tatsache, daß diese Vorgänge einerseits wirksam, andererseits ohne Beteiligung der Aufmerksamkeit ablaufen. Auch Lernen ist unter dieser Bedingung möglich. Nicht bewußt wahrgenommene Veränderungen der Muskelspannung in einer Hand z.B. können als diskriminative Reize dienen (Hefferline und Bruno, 1973). Weiterhin ist dabei wesentlich, daß dieser Vorgang ein Zusammenspiel zwischen der Bedeutung einer Situation und dem Verlauf der Erregung enthält. Von der Bedeutung ist die Art bzw. die Richtung der Informationssuche abhängig.

Das zweite Kriterium entscheidet über die Hinwendung der Aufmerksamkeit auf die Erregung. Wiederum wird versucht, durch Gewinnung zusätzlicher Information die nun wahrgenommene Erregung zu erklären und gegebenenfalls der Situation, falls diese als bedrohlich wahrgenommen wird, zu begegnen. Ist beides nicht möglich, wird Angst wahrgenommen. Diese führt zu einem weiteren Erregungsanstieg. Wenn er das Niveau des dritten Kriteriums überschreitet, wird die Aufmerksamkeit hauptsächlich auf die mit der Erregung verbundenen Körpervorgänge gerichtet, d.h. auch die Informationssuche konzentriert sich darauf. Diese Richtung der Aufmerksamkeit auf Körpervorgänge findet auch dann statt, wenn die Situation die Erregung nicht erklären kann, wie das oft unter Ruhebedingungen der Fall ist, unter denen es sehr häufig zu Beschwerden kommt. Ist die körperliche Erregung erklärlich, z.B. durch körperliche Belastung, so ist eine Bewältigungsmöglichkeit gegeben durch Beendigung der Belastung. Das bedeutet aber auch, daß die Erregung zumindest teilweise durch die Situation erklärt werden kann. Ist eine derartige Bewältigungsmöglichkeit nicht gegeben, ebenso auch keine Erklärungsmöglichkeit, so wird eine Störung der körperlichen Funktionen angenommen und damit die Erregung bzw. jetzt das wahrgenommene Körpergeschehen erklärt. Die Möglichkeit

einer Bewältigung ist durch die Hilfe eines Arztes zunächst gegeben.

> Das im vorangegangenen Geschilderte soll am Beispiel der oben genannten Patientin erläutert werden.
> Die Patientin wurde aufgefordert, einen sehr bedrohlich erkrankten Nachbarn zum Arzt zu fahren. Während sie dies tat, war sie sehr aufgeregt, d.h. die Erregung hatte die beiden, vermutlich alle drei Kriterien überschritten. Ihr war es jedoch möglich, die Erregung aus der Situation zu erklären, die Bedrohung des Nachbarn, die Notwendigkeit, ihn schnell zum Arzt zu bringen, also Zeitdruck usw. Während dies geschah, spürte sie Aufregung, vielleicht auch ein wenig Angst, im Hinblick auf das Gelingen des Unternehmens, aber es kam nicht zu einer Panik.
> Erst als alles vorüber und die Patientin in einer Ruhesituation war, spürte sie ihre Erregung, jetzt nicht mehr durch die Situation erklärlich. Durch die vorangegangene Erfahrung der Lebensbedrohung (vielleicht auch durch den erneut mobilisierten Konflikt mit der Mutter) wird die Entscheidung, ob die Erregung aus der „Rest"-Situation erklärlich ist, negativ beantwortet, was eine verstärkt nach innen gerichtete Informationssuche und Aufmerksamkeit bedingt. Die Wahrnehmung eigener Körpervorgänge, die jetzt nicht mehr erklärlich sind, tritt in den Vordergrund. Aus der Unsicherheit steigert sich die Erregung und es kommt zur Panik.

Geht man von einer „Konfliktdisposition" aus, d.h. der Annahme, daß eine Klasse von Konflikten gegeben sei, für deren Lösung einer Person keine Mittel zur Verfügung stehen, so ergibt sich zweierlei daraus: Zum einen kann eine derartige Person für längere Lebensabschnitte, in denen Konflikte der angenommenen Klasse nicht aktiviert werden, ohne weitreichende Störungen leben. Zum zweiten wird bei Auftreten von Konflikten durch den Mangel an Bewältigungsstrategien Angst ausgelöst, indem durch die Unmöglichkeit der Bewältigung des Konflikts eine Handlungs- bzw. Verhaltensblockade eintritt.

Die Angst wird in diesem Fall also nicht durch erlebte Körpersensationen ausgelöst, sondern durch die Unmöglichkeit, sich in einer konflikthaften Lebenssituation adäquat zu verhalten. Es läßt sich denken, daß ein gegebener Konflikt in einer konkreten Situation aktualisiert und das Verhalten blockiert wird. Die dadurch ausgelöste Angst führt zu Veränderungen physiologischer Funktionen, die wahrgenommen werden – vor allem des Kreislaufsystems. Dieser Vorgang läßt sich am ehesten anhand des von Butollo (1979) vorgestellten Modells der Angstverarbeitung konzeptualisieren.

Ausgehend von den kognitiven Angsttheorien (z.B. Lazarus et al., 1975, 1980) meint er, daß in Situationen, in denen der Organismus zu Aktivität mobilisiert wird, eine Verhaltensblockade eintritt. Es kommt zu einem Ansteigen der Erregung, die als Angst wahrgenommen wird und kognitive Prozesse, letztlich mit dem Ziel der Bewältigung, in Gang setzt. Lazarus ging von einer Stufe der primären Beurteilung der Gefahrensituation aus (primary appraisal), wobei von Butollo das Suchen zusätzlicher Information zur Beurteilung oder erneuten Beurteilung der Situation (reappraisal; Lazarus) betont wird. Dabei wird vom Organismus angestrebt, die mit der Situation verbundene Unsicherheit zu reduzieren. Wichtig dabei ist die Tatsache, daß während des Suchens zusätzlicher Information ein weiterer Erregungsanstieg toleriert werden muß.

Zum zweiten postuliert Butollo zwei verschiedene Richtungen der Informationssuche, interne und externe. Lazarus und Mitarbeiter hatten bereits darauf hingewiesen, daß zum Zwecke der Bewältigung einer Situation eine Uminterpretation der Situation stattfinden könne, als „interne" Strategie der Gefahrenbewältigung. Überlegungen dieser Art stehen in einer gewissen Nähe zu psychodynamischen Vorstellungen über Abwehrmechanismen. Für Butollo impliziert die interne Informationssuche auch die Hinwendung der Aufmerksamkeit auf Körpervorgänge. Damit könnte beschrieben werden, warum die Wahrnehmung von Körpervorgängen plötzlich die Aufmerksamkeit einer Person beherrschen kann. Hat ein derartiger Vorgang stattgefunden, gewinnt die Situation eine vollständig neue Bedeutung, dadurch wird die Bedrohung durch den ursprünglichen Konflikt reduziert. Es tritt eine neue Bedrohung auf, für die es jedoch Lösungsmöglichkeiten zu geben scheint, man kann einen Arzt aufsuchen oder ähnliches.

Geht man dabei von psychoanalytischen Vorstellungen einer Konfliktdisposition aus, die als Resultat Abhängigkeit und Angewiesensein auf andere beinhaltet, so zeigt sich, daß die Lösungsmöglichkeiten für diese neue Bedrohung diesen Merkmalen entsprechen. Der Patient kann sich wieder auf andere verlassen, die für ihn seine Ängste bewältigen, er hat einen neuen Angstschutz. Durch die Reduktion der Angst, die damit verbunden ist, werden Vorgänge dieser Art verstärkt: die Richtung der Informationssuche, das Wahrnehmen von Körpersignalen, die Interpretation der Situation. Damit wird die Wahrscheinlichkeit vergrößert, daß diese Prozesse beim erneuten Auftreten konflikthafter Situationen wiederum auf diese Art ablaufen.

Indem die Angst nicht mehr im Zusammenhang mit dem auslösenden Konflikt wahrgenommen wird, sondern sich auf die Körperfunktion richtet, kann im Hinblick auf die Körperfunktion eine „phobische" Entwicklung stattfinden, sich eine Angst vor der Angst entwickeln, wie es oben bereits geschildert wurde. Unter Umständen kann so die Angst bestehenbleiben, auch wenn durch äußere Umstände die konflikthafte Situation nicht mehr gegeben ist.

32.5 Differentialdiagnose

Das Vorliegen funktioneller Herz-Kreislauf-Störungen muß hauptsächlich von der relativen oder absoluten Koronarinsuffizienz differentialdiagnostisch abgegrenzt werden. Hierzu muß in der Anamnese in erster Linie die charakteristische Symptomatik herangezogen werden, sowie EKG und enzymatische La-

borwerte. Hegglin (1972) sowie v. Uexküll (1962) schildern folgende Charakteristika funktionell bedingter Schmerzen:

– Dumpfer Druck, Brennen, die Stunden bis Tage dauern. Es besteht kein eindeutiger Zusammenhang mit körperlicher Belastung. Oft besteht eine Hyperalgesie der linken Brustseite (Stethoskopschmerz).

– Kurze, Sekunden andauernde nadel- oder messerstichartige Schmerzen in der Herzgegend, meist unter der linken Brustwarze, von starker Intensität. Auch hierbei besteht keine Beziehung zu körperlicher Belastung.

Es muß dabei jedoch erwähnt werden, daß auch funktionell bedingte Beschwerden durchaus im Zusammenhang mit körperlicher Belastung auftreten können. Dies ist immer dann der Fall, wenn die oben genannten Attribuierungsvorgänge keine wesentliche Rolle spielen und Körpersensationen, auch wenn sie durch die körperliche Belastung erklärt werden könnten, Angst auslösen. Das immer wieder bei den Patienten zu beobachtende Schonverhalten rührt einmal von der Angst her, sich zu überlasten, zum zweiten gelegentlich aber auch daher, die mit der Belastung einhergehenden Körpersensationen zu vermeiden, weil sie Angst auslösen. Ist ein Patient darüber informiert, daß somatisch bedingte Beschwerden hauptsächlich im Zusammenhang mit körperlicher Belastung auftreten, so wird er entsprechend auftretende Körpersensationen ängstlich beobachten und überschätzen. Darüber hinaus ist zu beachten, daß das Auftreten und die Art der Beschwerden im allgemeinen den Vorstellungen der Patienten im Hinblick auf die Stärke der Belastung entsprechen und oft nicht der tatsächlichen, mit einer Tätigkeit verbundenen Belastung.

Häufige Fehldiagnosen sind Koronarinsuffizienz, Hyperthyreose, Myokarditis, Mitralinsuffizienz, Hypotonie und Tetanie. Von Uexküll (1962) weist jedoch darauf hin, daß ein negativer Herzbefund eine organische Krankheit keineswegs ausschließt, und betont in diesem Zusammenhang, daß auch an eine Fernwirkung von Organen, z. B. aus dem Abdominalbereich, gedacht werden muß (Gallenblase, Magen, Pankreas). Zu denken ist auch an Läsionen im Bereich der Halswirbelsäule.

Bei den genannten Störungen, z. B. der Koronarinsuffizienz, muß jeweils auch erwogen werden, inwieweit die Beschwerden tatsächlich durch die erhobenen Befunde erklärt werden können, oder ob nicht zusätzlich eine Fixierung auf die Beschwerden, durch psychische Faktoren bedingt, vorhanden ist.

Es wurde bereits darauf hingewiesen, daß in den letzten Jahren von einigen Autoren die „Herzphobie" als **Panikstörung** angesehen wurde. „Hauptmerkmale dieser Störungen sind wiederkehrende Panikattacken, d.h. abgrenzbare Episoden intensiver Angst oder Unbehagens, die zusammen mit mindestens vier charakteristischen Symptomen auftreten ... Panikattacken dauern üblicherweise nur Minuten, seltener auch Stunden. Die Attacken treten, zumindest anfänglich, unerwartet auf, d.h. nicht unmittelbar vor oder in einer Situation, die fast immer Angst auslöst (wie z.B. bei der einfachen Phobie)" (DSM-III-R, 1989). Es werden dann die folgenden Symptome genannt, von denen jeweils mindestens vier im Rahmen einer Attacke auftreten müssen, damit eine Attacke als Panikattacke klassifiziert werden kann:

- Atemnot (Dyspnoe) oder Beklemmungsgefühle;
- Benommenheit, Gefühl der Unsicherheit oder Ohnmachtsgefühl;
- Palpitationen oder beschleunigter Herzschlag (Tachykardie);
- Zittern oder Beben;
- Schwitzen;
- Erstickungsgefühl;
- Übelkeit oder abdominelle Beschwerden;
- Depersonalisation oder Derealisation;
- Taubheit oder Kribbelgefühle (Parästhesien);
- Hitzewallungen oder Kälteschauer;
- Schmerzen oder Unwohlsein in der Brust;
- Furcht zu sterben;
- Furcht, verrückt zu werden oder Angst vor Kontrollverlust.

Panikattacken als eigenständige und abgrenzbare Angststörung zu betrachten, geht auf eine Studie von Klein (1964) zurück, in der er 7 (!) Patienten mit Zuständen intensiver Angst, die als Patienten mit „affektiver Störung" klassifiziert worden waren, mit Imipramin behandelte und mit 6 Patienten, die mit Placebo behandelt worden waren, verglich. Er beobachtete, daß die Zustände intensiver Angst, die Panikattacken, durch die Behandlung mit Imipramin verschwanden, nicht jedoch die chronische antizipatorische Angst.

In der Folgezeit gab es eine Vielzahl von Studien, mit denen versucht wurde, zu zeigen, daß es sich bei der Panikstörung um ein eigenständiges Krankheitsbild handelt, das mit Imipramin behandelbar ist und sich von anderen Angststörungen unterscheidet. Inzwischen hat sich diese Auffassung fest etabliert. Barlow (1988) faßt sie zu einem ätiologischen Modell der Panikstörung zusammen. Es wird von einer biologischen Vulnerabilität im Sinne einer noradrenergen Überreaktivität auf Streß, vermittelt durch den Locus coeruleus, ausgegangen. Es kommt zu Alarmreaktionen, die jedoch unfunktional seien, im Sinne „falscher Alarme", da keine wirkliche intensive Bedrohung gegeben sei. Die angenommene biologische Vulnerabilität und Tendenz zur „Selbstbeobachtung" („self-awareness") sei durch Vererbung gegeben. Durch die Überreaktivität komme es zu starken vegetativen Reaktionsanteilen, die als interozeptive Angstreize dazu führen, daß die Attacken aufgrund ihrer Unfunktionalität nicht von selbst verschwinden. Dies führt dazu, daß auch die Beschwerden der Patienten mit Panikattacken durch körperliche Symptome geprägt sind und das Geschehen der Attacke als körperliche und unkontrollierbare Störung angesehen wird, ganz ähnlich, wie es bei Patienten mit funktionellen Herz-Kreislauf-Störungen der Fall ist.

So schreibt Barlow (1988) dann auch: „Von etwas verschiedenen Standpunkten bezeichneten Oppen-

heim (1918), Cohen und White (1950) im Rahmen epidemiologischer Studien in den 40er und 50er Jahren sowie Cohen, Badal, Kilpatrick, Reid und White (1951) etwa dasselbe Syndrom als ‚neurozirkulatorische Asthenie'. Dies ist der alte Begriff für Angstzustände mit ausgeprägten kardiovaskulären Merkmalen."

Im allgemeinen wird die Betonung körperlicher Symptomatik bei den Attacken, ihre Auslösbarkeit durch Laktatinfusion sowie die Behandelbarkeit mit Imipramin als Argument für die Annahme einer einheitlichen und abgrenzbaren Panikstörung angesehen. Dabei scheinen körperliche Sensationen am Beginn einer Attacke zu stehen, die im Sinne einer „Katastrophe" auch als unkontrollierbar interpretiert werden und dann aufgrund dieser Interpretation das System zur Panik aufschaukeln. Ein vergleichbarer Vorgang wird auch für das Auftreten anfallsartiger funktioneller Herz-Kreislauf-Beschwerden beschrieben, der Patient hat jedoch das Gefühl, dieses Geschehen sei die Folge einer ernsten körperlichen Erkrankung. Auch in Eysencks Theorie der „Furchtinkubation" wird der Wahrnehmung körperlicher Angstreaktionsanteile eine zentrale Bedeutung dafür zugemessen, daß die Angst vor eigentlich harmlosen Situationen nicht gelöscht wird, da das anfängliche Auftreten der vegetativen Funktionsänderungen als Gefahrensignal interpretiert wird.

Ehlers et al. (1986) untersuchten die unterschiedliche Wirkung von Laktatinfusion auf Patienten mit Panikstörung und gesunde Kontrollpersonen. Bei 10 Patienten mit Panikattacken und 10 Kontrollpersonen wurde eine Placeboinfusion oder eine Laktatinfusion durchgeführt. Laktat erhöhte subjektive Angst und die Herzfrequenz bei Patienten und Kontrollpersonen gleichermaßen. Nur im Blutdruck gab es leichte Unterschiede, wobei die Patienten geringfügig höhere Werte aufwiesen. Bei 44 Panikpatienten und 10 Kontrollpersonen wurde von Cowley et al. (1987) ebenfalls eine Laktatinfusion durchgeführt. Die Patienten, die eine typische Panikattacke durchmachten, hatten vor der Infusion höhere Angstwerte und Symptome angegeben. Die Herzfrequenz verlief bei beiden Gruppen gleich, wobei die Probanden, die eine Panikattacke erlebten, eine etwas höhere Herzfrequenz hatten. Es scheint also, daß die Auslösbarkeit von Panikattacken durch Laktatinfusion kein spezifisches Merkmal von Personen ist, bei denen eine Panikstörung diagnostiziert worden ist. Auch die unterschiedliche Behandelbarkeit von Panik mit Imipramin im Vergleich zu anderen Angststörungen scheint nicht gegeben zu sein. So konnten Kahn et al. (1986) zeigen, daß Imipramin einen anxiolytischen Effekt besitzt, auch bei Patienten mit anderen Angststörungen als Panikattacken. Telch et al. (1985) verglichen die Wirkung von Imipramin ohne sonstige Interventionen mit Imipramin mit Konfrontationstherapie und Konfrontationstherapie mit Placebo. Der deutlichste panikreduzierende Effekt kam durch Konfrontation mit Imipramin zustande, Imipramin allein hatte keine Wirkung auf Maße von Phobie und Panik, jedoch eine Wirkung auf dysphorische Stimmung. In dieser Gruppe waren die Patienten angehalten worden, sich neuen „phobischen" Situationen nicht auszusetzen, da in früheren Studien gezeigt worden war, daß Imipramin anscheinend „Erwartungsangst" reduziert, die Stimmung aufhellt und dadurch dazu führt, daß Patienten ein höheres Gefühl eigener Kompetenz (self efficacy) bekommen und sich mit angstinduzierenden Situationen konfrontieren. So scheint sich herauszustellen, daß die panikreduzierende Wirkung von Imipramin, die ursprünglich zur Abgrenzung der Panikstörungen von den anderen Angststörungen geführt hat, nicht spezifisch auf Panikattacken wirkt, sondern über eine von den Benzodiazepinen (Reduzierung des emotionalen Niveaus) verschiedene Wirkung (Erhöhung des Gefühls der eigenen Kompetenz) indirekt auf Vermeidungsverhalten einwirkt, was es auch plausibel macht, daß andere Angststörungen ebenfalls positiv beeinflußt werden.

Schließlich konnten Stalmann et al. (1988) zeigen, daß Patienten mit Panikstörung im Vergleich zu gesunden Kontrollpersonen ihre Herztätigkeit weniger gut interozeptiv wahrnehmen, so daß die Annahme, aufgrund der Interozeption einer vegetativen, vor allem kardiovaskulären Überreaktivität komme es zu Panikattacken, schwer haltbar ist. Dies bedeutet nicht, daß es keine Panikattacken gibt, sie können bei einer ganzen Reihe von Störungen auftreten, es ist jedoch fraglich, ob die Annahme, bei Panikattacken handele es sich um eine spezifische, von anderen Störungen abgrenzbare Störung, der eine hereditäre Disposition zugrunde liegt, notwendig ist. Auch bei Patienten mit funktionellen Herz-Kreislauf-Störungen kann es zu Panikattacken kommen, ohne daß es sinnvoll erscheint, diese Störungen nun als Paniksyndrom zu bezeichnen.

Will man funktionelle Herz-Kreislauf-Störungen nach DSM-III-R klassifizieren, so sind sie am ehesten als kardiopulmonales Somatisierungssyndrom einzuordnen. Nach DSM-III-R tritt das Somatisierungssyndrom selten bei Männern auf, der Beginn der Symptomatik fällt in die Zeit vor dem 20. Lebensjahr und nur selten oberhalb dieser Altersgrenze. Die diagnostischen Kriterien für das Somatisierungssyndrom treffen ansonsten zu. Andererseits werden Trennungsangst und plötzlicher Objektverlust als begünstigende Faktoren für die Auslösung der Panikstörung angesehen, Faktoren, die von psychoanalytischer Seite als typisch für die Auslösung einer Herzneurose gelten. Ein großer Teil, wenn nicht alle Symptome von Panikattacken treffen für einen Teil der Patienten mit funktionellen Herz-Kreislauf-Störungen ebenfalls zu, und bei vielen Patienten treten die Beschwerden ebenfalls anfallsartig, ohne zunächst ersichtliche auslösende Bedingung auf. Allerdings wird bei der Panikstörung angenommen, daß zwischen den Attacken Symptomfreiheit besteht, was kaum bei Patienten mit funktionellen Störungen der Fall ist.

32.6 Therapie und Prognose

Nach einer ausführlichen Anamnese ist zunächst auch eine gründliche körperliche Untersuchung notwendig, über deren Ergebnis der Patient zu informieren ist. Es ist erstaunlich, wie oft Patienten über ihre Krankheit so gut wie nicht informiert sind. Viele denken, daß das Auftreten funktioneller Beschwerden bei ihnen fast einmalig sei. Ebenso hilft es vielen Patienten, wenn ihnen anschaulich erklärt wird, daß es durchaus zu den normalen Reaktionsweisen eines Menschen gehört, im Zusammenhang mit emotionalem Erleben körperlich zu reagieren. Für Patienten ist es oft letztlich beunruhigend, auch wenn es ihrem momentanen Bedürfnis nach Beruhigung und Reduktion von Angst entgegenkommen mag, daß immer wieder körperliche Untersuchungen durchgeführt werden. Letztlich nützt dies den Patienten auch wenig, es kann im Gegenteil den Eindruck erwecken oder verstärken, daß doch eine ernsthafte organische Erkrankung vorliegt, die immer neue Untersuchungen notwendig macht. Bei den Mitteilungen an den Patienten ist es notwendig, auf Vermutungen über mögliche frühere kardiale Erkrankungen zu verzichten. Bemerkungen, die dahin gehen, es könnte früher einmal im Anschluß an eine Angina eine Myokarditis durchgemacht worden sein, hinterlassen bei den Patienten einen Eindruck, der kaum mehr rückgängig gemacht werden kann.

Die Diagnose eines funktionellen kardiovaskulären Syndroms sollte keine Ausschlußdiagnose sein, sondern eine positive Diagnose, indem die oben beschriebenen Merkmale nachgewiesen werden. Dazu ist es notwendig, in der Anamnese darauf zu achten, ob möglicherweise früherworbene Konfliktdispositionen vorliegen. Ist dies der Fall, so sollte ebenfalls deutlich werden, daß der Beginn der Beschwerden in einem Zusammenhang mit Konflikten der ermittelten Art steht. Allein aus der Tatsache, daß eine Person eine Reihe von Konflikten nicht adäquat bewältigen kann, kann noch nicht geschlossen werden, daß diese die Störung unterhalten. Dasselbe gilt auch für die Erhebung von Fragebogendaten. Das Vorliegen eines hohen Neurotizismuswertes erklärt das Auftreten der Beschwerden für sich noch nicht. In diesem Zusammenhang sei auf die Kapitel zur Anamneseerhebung (Kap. 12) sowie zur Verhaltensanalyse (Kap. 16) verwiesen.

32.6.1 Psychopharmaka

Tranquilizer sind dann sinnvoll einsetzbar, wenn die Angstintensität verringert werden soll. Bei entsprechender Dosierung besteht eine schlafanstoßende Wirkung, die die häufig begleitenden Schlafstörungen positiv beeinflußt. In diesem Zusammenhang sei ausdrücklich darauf hingewiesen, daß eine Medikation, gerade mit Tranquilizern, aber auch anderen psychotropen Substanzen, der geschickten Führung durch den Therapeuten bedarf. Als alleinige therapeutische

Maßnahme bewirken diese Medikamente zwar häufig eine zum Teil eindrucksvolle Besserung der Beschwerden, können jedoch langfristig zu Abhängigkeit führen. Medikamente dieser Art sollten therapiebegleitend eingesetzt werden, und es sollte von vornherein für den Patienten klar sein, daß das Medikament nur einen Teil der Therapie darstellt und die Einnahmezeit begrenzt ist.

Geht man davon aus, daß ein Teil der therapeutischen Maßnahmen zum Ziel hat, den Patienten dahin zu führen, daß er Vermeidungsverhalten aufgibt zugunsten von Verhaltensweisen, die ein aktives Umgehen auch mit angstauslösenden Reizen beinhalten, so wird einsichtig, daß die langfristige Einnahme angstlösender Substanzen eher eine passive Haltung fördert. Der Patient lernt, sich auf das Medikament zu verlassen und vermeidet, sich seinen eigenen Reaktionen zu stellen. Andererseits können derartige Medikamente, besonders bei sehr intensiver Angst, einen gewissen Schutz darstellen, der es dem Patienten gestattet, erste Schritte zur selbständigen Lösung seiner Probleme zu wagen.

Beta-Blocker können dann sinnvoll eingesetzt werden, wenn für den Patienten spürbare Rhythmusstörungen im Vordergrund stehen. Durch Beta-Blocker wird das angstauslösende Körpersignal verringert, was ebenfalls einen Schutz darstellt. Es gelten hier dieselben oben genannten Einschränkungen.

32.6.2 Psychotherapie

Tiefenpsychologisch orientierte Psychotherapie

Tiefenpsychologisch orientierte Psychotherapie ist vor allem in solchen Fällen angezeigt, bei denen die funktionellen Beschwerden Teil einer übergreifenden Störung sind, die Störungen der Objektbeziehung beinhalten, die die Anwendung sonstiger therapeutischer Verfahren unmöglich machen. Bei diesen Patienten sind mitunter auch länger dauernde psychotherapeutische Behandlungen notwendig. Andererseits können ebenfalls fokale therapeutische Behandlungen indiziert sein, und dies ist bei der überwiegenden Mehrzahl der Patienten der Fall, bei denen nicht auf eine umfassende Beeinflussung der Persönlichkeit abgezielt wird, sondern auf die Änderung eines spezifischen Konfliktes. Für die Dauer der Therapie spielt die Dauer der Erkrankung eine wichtige Rolle. Je kürzer die Erkrankungsdauer, desto weniger verfestigt sind die Strategien der Patienten, die auf der körperlichen Symptomatik aufbauen und der Vermeidung von Konflikten dienen. Zudem fällt es leichter, den Patienten den Zusammenhang zwischen psychischen Sachverhalten und ihren körperlichen Symptomen nahezubringen. Das zeitliche Umfeld des Beschwerdebeginns ist noch deutlicher in Erinnerung, die Situationen, in denen die Beschwerden auftreten, sind noch spezifischer.

Tiefenpsychologisch orientierte psychotherapeutische Verfahren sind also vor allem dann indiziert, wenn eine Störung vorliegt, in deren Zusammenhang die körperliche Symptomatik hauptsächlich der Vermeidung psychischer Konflikte dient. Der Umfang und die Dauer der Therapie richten sich nach der Schwere der zugrundeliegenden Störung.

Methoden der Verhaltenstherapie

Die Verhaltenstherapie hat ihren Ursprung und sicher auch ihre größten Erfolge bei der Behandlung von Angst. Infolgedessen sind verhaltenstherapeutische Methoden hauptsächlich dann angezeigt, wenn es sich um Störungen vom Typ der „Herzphobie" handelt. Hierbei steht die Behandlung der phobischen Komponente im Vordergrund, die darin besteht, daß die Patienten, wie oben dargestellt, dazu neigen, zunehmend mehr Situationen oder auch körperliche Belastung zu vermeiden. Die Behandlung der Angst sollte allerdings niemals als einzige Behandlungskomponente gewählt werden, sondern als ein Teil eines komplexen Behandlungskonzepts dienen. Als wesentlicher erster Schritt ist eine sehr genaue und detaillierte Verhaltensanalyse notwendig. Die Verhaltensanalyse dient der Gewinnung von Information über das „Umfeld" des zu ändernden problematischen Verhaltens. Dieses Umfeld beinhaltet unmittelbare Bedingungen des Verhaltens, auslösende Reize, Verhaltenskonsequenzen, dispositionelle Bedingungen usw. Darüber hinaus beinhaltet es aber auch soziale Bedingungen, in die das Verhalten eingebettet ist (für weitere Information zur Verhaltensanalyse vgl. Kap. 19). Die Gewinnung dieser Information durch die Verhaltensanalyse ist ein Vorgang, der sich entwickelt, auch dadurch, daß der Patient in zunehmendem Maße Fähigkeiten gewinnt, z.B. auf relevante Bedingungen zu achten, die seinen Beschwerden vorausgehen. Dabei wird er zunächst auf konkrete äußere Bedingungen abzielen und z.B. Gedanken, vielleicht als nicht relevant, übersehen. Es muß ihm dann gezeigt werden, daß ebenfalls Gedanken, sogar in einem hohen Maße, als auslösende Bedingung relevant sein können und meist auch sind. Der Patient muß von seiner Fixierung auf „körperrelevante" Bedingungen gelöst und zu einer im eigentlichen Sinne „psychosomatischen" Betrachtungsweise seines Erlebens hingeführt werden. Es wird hierbei deutlich, daß bereits im Zuge der Verhaltensanalyse im Grunde auf eine Einstellungsänderung abgezielt wird, die gleichzeitig ein wichtiges Ergebnis der Therapie überhaupt darstellt. Die Verhaltensanalyse begleitet also die Therapie und ist nicht als ein erster Abschnitt abgrenzbar.

Als weiterer wesentlicher Schritt ist eine geduldige und umfangreiche Information des Patienten notwendig. Hierbei sollte darauf abgezielt werden, dem Patienten ein Verständnis für psychophysiologische Zusammenhänge in verständlicher Form nahezubringen, vor allem die Tatsache, daß körperliche Reaktionen in jedem Fall einen Bestandteil normalen Erlebens darstellen. Die Patienten erleben häufig normale körperliche Reaktionen in den verschiedensten Zusammenhängen bereits als äußerst bedrohlich und haben die Tendenz, Situationen, in denen solche körperlichen Reaktionen auftreten, zu vermeiden.

Darüber hinaus sollten den Patienten Kenntnisse über eine „Psychologie der Angst" vermittelt werden, die sie die späteren Maßnahmen in einem größeren Zusammenhang verstehen lassen. Dies ist notwendig, da die genannten Maßnahmen keineswegs immer bequem und angenehm sind, so daß es wichtig ist, daß der Patient weiß, warum er die Mühe auf sich nimmt. Die Vermittlung der genannten Kenntnisse und Zusammenhänge erläutert man am besten an den konkreten Schilderungen des Patienten. Dabei wird der Patient angeleitet, bei sich selbst im Sinne der Verhaltensanalyse auf die vermittelten Zusammenhänge zu achten.

Die bisher genannten Vorgehensweisen enthalten bereits wesentliche Elemente der Therapie. Insgesamt könnten die Therapiestrategien aufgeteilt werden in solche, die direkt der Angstbewältigung dienen, und kognitive Verfahren, die Einstellungsänderungen bewirken sollen. Dabei sind Einstellungen dem eigenen Körper gegenüber gemeint, solche gegenüber nahen Bezugspersonen, der eigenen Leistung, eigenen Gefühlen usw., aber auch gegenüber der eigenen Kompetenz bei der Lösung von Problemen im Zusammenhang mit der Erkrankung.

Bereits die Durchführung einer sinnvollen Verhaltensanalyse zielt, wie eben erwähnt, auf Einstellungsänderungen ab bzw. bewirkt diese. Anfängliche Bemühungen, die klassische Methode der systematischen Desensibilisierung heranzuziehen (vgl. Kap. 19) (z.B. Rifkin, 1968), haben sich als wenig hilfreich erwiesen. Dabei wird dem Patienten aufgetragen, sich in der Vorstellung ängstigenden Gedanken und Phantasien im Zusammenhang mit der Angst vor einer Herzerkrankung auszusetzen. Methoden einer direkten Konfrontation mit Situationen, in denen nach der Erfahrung der Patienten die Beschwerden sehr wahrscheinlich auftreten, haben sich besser bewährt (vgl. Butollo, 1979; Schwarz, 1982).

Für die Methoden der Konfrontation ist es wesentlich, daß die Belastung so gewählt wird, daß die Wahrscheinlichkeit für die Vermeidung der Situationen bzw. die Flucht aus den Situationen nicht zu hoch ist, d.h., die Möglichkeit des Erfolgs sollte möglichst hoch sein. Der Patient sollte sich also den Situationen, wie bei der systematischen Desensibilisierung, graduell nach dem Maß der Schwierigkeit aussetzen. Damit unterscheidet sich dieses Vorgehen von Methoden der Konfrontation, wie der Reizüberflutung, bei der der Patient sehr ängstigenden Situationen ausgesetzt wird, ohne die Möglichkeit der Flucht zu besitzen. Dieses Vorgehen ist schon deshalb angezeigt, da sich der Patient meist ohne die Anwesenheit des Therapeuten den Situationen aussetzt, und damit jederzeit die Möglichkeit zur Flucht besitzt.

Ein weiterer wesentlicher Punkt bei diesem Vorgehen besteht darin, daß der Patient die Situationen, von denen er glaubt, sie möglicherweise bewältigen zu können, selbst aktiv, gleichsam zu Übungszwecken aufsucht. Er kann damit den Beginn, den Zeitpunkt und die Dauer der Situation von vornherein selbst kontrollieren, und wird nicht von äußeren Umständen gezwungen, sich der Situation auszusetzen. Damit wird Hilflosigkeit im Sinne der Unkontrollierbarkeit vermieden und die aktive Einstellung des Patienten zu seinem Problem gefördert. Es wird auch deutlich, daß ein solches Vorgehen eine aktivere Einstellung voraussetzt, indem der Patient bereit sein muß, die Situation freiwillig aufzusuchen. Diese Einstellung wird bei Erfolg aber auch verstärkt.

Bei der Verhaltensanalyse war bereits darauf hingewiesen worden, daß der Patient zur Beobachtung eigenen Verhaltens und wichtiger interner und externer Bedingungen angehalten wird. Auch jetzt soll der Patient sich gezielt beobachten, die Situation wie ein Experiment auffassen und protokollieren, was er empfindet, welche Bedingungen auftreten, welche Gedanken er hat, welche Tendenzen er spürt usw. Dadurch gewinnt er eine distanziertere Einstellung zu seinem Erleben in der Situation, wodurch diese in ihrer ängstigenden Eigenschaft entschärft wird. Es ist wichtig, darauf zu achten, daß tatsächlich schriftliche Protokolle vom Patienten angefertigt werden. Geschieht dies nicht, so sind die Beobachtungen meist sehr flüchtig und undifferenziert. Diese Aufzeichnungen müssen mit dem Therapeuten durchgesprochen werden. Mit ihrer Hilfe können Änderungsmöglichkeiten diskutiert werden, die sich jedoch keineswegs nur auf tatsächliches Verhalten beziehen, sondern vor allem auch auf Gedanken, die dabei auftreten und oft korrekturbedürftig sind (vgl. auch Schwarz, 1982).

Die im vorangegangenen kurz skizzierte Vorgehensweise hat sich vor allem bei Patienten mit einer Herzphobie bewährt. Je umfangreicher die gesamte Störung des Patienten ist, um so mehr rücken kognitiv orientierte Verfahren der Verhaltensmodifikation in den Vordergrund. Es ist auch wichtig, das soziale Umfeld der Patienten genau zu erfassen, zu wissen, inwieweit die Umgebung möglicherweise das Auftreten der Beschwerden ungewollt durch besondere Zuwendung verstärkt und ähnliches. Ebenso wichtig ist es, abzuschätzen, inwieweit Partner oder sonstige nahe Bezugspersonen in der Lage sind, die therapeutischen Bemühungen zu unterstützen.

33 Funktionelle Syndrome im gastrointestinalen Bereich

Wolfram Schüffel und *Thure von Uexküll*

33.1 Einleitung und historische Zusammenfassung

Funktionelle Syndrome im Gastrointestinalbereich stellen für den praktisch tätigen Arzt eine Herausforderung dar. Almy, einer der erfahrensten Ärzte dieses Jahrhunderts auf dem Gebiet funktioneller Abdominalbeschwerden sagt:

„In der Medizin gibt es nur wenige, gleichermaßen weitverbreitete und ökonomisch wichtige Bilder, bei denen Geduld und menschliches Interesse des Arztes derart viel bedeuten" (1973). „... Das Thema bedarf der Untersuchung auf jeder Stufe biologischer Organisation, beginnend mit intrazellulären Molekülen bis zur Gesellschaft und ihren kulturellen Werten. Noch haben diejenigen Forscher, die sich an einem entsprechenden interdisziplinären Unterfangen beteiligen, viel voneinander zu lernen, um es dem Allgemeinwohl zugute kommen zu lassen" (1983).

Funktionelle Abdominalbeschwerden wurden erstmals von Da Costa als ein eigenständiges Beschwerdebild beschrieben (1871). Dieser Autor stellte anhand von sieben Krankheitsverläufen Symptomatik, Behandlung und Entwicklung von Kranken dar, die unter dem Syndrom des **Reizkolons** oder des spastischen Kolons bzw. dem funktionellen Unterbauchsyndrom leiden, wie wir es heute bezeichnen würden. Da Costa selbst bezeichnete die Krankheit als „membranöse Enteritis". Bereits damals beschrieb er die Empfindlichkeit und Verletzbarkeit derartiger Patienten, obgleich er dieses Merkmal nicht zum Thema einer speziellen Untersuchung machte. Der Erstbeschreibung Da Costas folgte eine große Zahl von Arbeiten, in denen die Untersuchungsergebnisse von Teilaspekten des Beschwerdebildes bzw. Krankheitsbildes festgehalten wurden. Jedoch wurde erst 1939 durch die Autoren White, Cobb und Jones eine umfassende Darstellung des Beschwerdebildes in Form einer heute nahezu völlig vergessenen (!) Monographie vorgelegt, die auch für den Leser des ausgehenden 20. Jahrhunderts außerordentlich lesenswert ist. Die Autoren berichten über 60 konsekutiv in ihrer Klinik aufgenommene Patienten mit der damals üblichen Bezeichnung „muköse Kolitis". Für den Zeitabschnitt zwischen der Erstbeschreibung Da Costas und den Veröffentlichungen ihrer eigenen Arbeiten fanden sie nach ihrer eigenen Einschätzung lediglich vier klinisch weiterführende Studien. Die Autoren dieser Studien waren in der von White und Mitarbeitern aufgeführten Reihenfolge: Bockus und Mitarbeiter (1928), Friedenwald und Mitarbeiter (1929), Jor-

dan und Kiefer (1929), Bargen (1939). Für diese genannten Autoren war kennzeichnend, daß sie das Beschwerdebild sowohl in seinen somatischen wie psychosozialen Bezügen untersuchten und beschrieben. Sie beobachteten durchgehend psychosoziale Belastungen, die bei Auslösung und Unterhaltung **funktioneller abdomineller Beschwerden** auftraten. Unterdessen hatte also eine Ausweitung des Begriffes stattgefunden. Es ging nicht mehr ausschließlich um funktionelle Unterbauchbeschwerden bzw. -syndrome, sondern auch um Oberbauchbeschwerden bzw. -syndrome, allgemein funktionelle Abdominalbeschwerden genannt. – Nach dem Krieg wurde zunächst eine heute unverdienterweise vergessene Arbeit von Halsted und Mitarbeitern veröffentlicht (1946). Dort wurden gastroskopische und psychiatrische Befunde gegenübergestellt, die man bei Soldaten des zweiten Weltkrieges erhoben hatte, welche an einer nicht organisch bedingten Dyspepsie, d.h. generell unter funktionellen Abdominalbeschwerden litten. Bei diesen Soldaten verstärkte sich die Symptomatik in psychosozial belastenden Situationen. Für diese Situationen war kennzeichnend, daß sie als nicht erträglich erlebt wurden; die Soldaten konnten aber hiergegen nicht offen rebellieren, sondern der Körper reagierte mit Abdominalbeschwerden. Hiernach stellten Almy und Mitarbeiter in mehreren Arbeiten ihre wegweisenden Motilitätsuntersuchungen an Patienten mit spastischem Kolon vor. Sie beschrieben regelmäßig wiederkehrende Zusammenhänge zwischen Erlebnisweise und Motilität des Kolons (1947, 1949a, b, c, 1950). Chaudhary und Truelove zogen aufgrund ihrer Untersuchung bei 130 derartigen Patienten den Schluß, daß diese nur dann ausreichend behandelt werden, wenn der Arzt ihre Gesamtsituation berücksichtigt (1962).

Diese letztlich von der Biographie des Patienten geleitete Vorgehensweise ist in der Praxis, aber auch in der medizinischen Theorie kaum aufgegriffen worden. Den derzeitigen Zustand kann man etwa folgendermaßen zusammenfassen: 50 Jahre nach White, Cobb und Jones und über 100 Jahre nach der Erstbeschreibung durch Da Costa ist uns das Beschwerdebild wie die Problematik des Patienten mit funktionellen Abdominalbeschwerden einschließlich seiner biopsychosozialen Bezüge bekannt. Es ist bisher aber nicht gelungen, eine einigermaßen zufriedenstellende Therapie zu entwickeln. Vielmehr besteht der Eindruck, als seien derartige Patienten in den 20er und 30er Jahren dieses Jahrhunderts problemgerechter behandelt worden als in den Jahren zwischen 1950 und

1980. Erst in letzter Zeit scheint sich eine Wandlung abzuzeichnen, die maßgeblich aus einem psychotherapeutischen Verständnis heraus bestimmt wird. Sie manifestiert sich u. a. in einer Arbeit, die über günstige Ergebnisse bei einer speziell entwickelten Kurztherapie berichtet (Svedlund et al., 1983). Hierauf aufbauend könnte es gelingen, zu einem Fortschritt in der Behandlung dieser Patienten zu kommen (s. u.).

33.2 Exemplarische Fallgeschichte

Den Ausführungen soll die Fallgeschichte einer 31jährigen Patientin vorangestellt werden, die in kennzeichnender Weise Merkmale aufweist, die sich bei Patienten mit funktionellen Abdominalbeschwerden gehäuft finden. Der Kürze halber werden diese nachfolgend als FAB bezeichnet, funktionelle Oberbauchbeschwerden als FOB und funktionelle Unterbauchbeschwerden als FUB.

Die Patientin klagte über Beschwerden im Sinne der beiden Hauptsyndrome aus dem Bereich der FAB, nämlich über funktionelle Oberbauchbeschwerden (FOB) und funktionelle Unterbauchbeschwerden (FUB). – Die gesund aber ältlich wirkende Dame, die sehr dezent-korrekt gekleidet war, litt in unregelmäßigen Abständen an hochgradigen Unterbauchschmerzen, die nicht eindeutig lokalisiert wurden. Zeitweise waren sie mit Durchfall verbunden. Diese Zustände konnten länger anhalten und schließlich in einen drückenden Oberbauchschmerz, in Übelkeit und tagelang anhaltende Appetitlosigkeit übergehen. Die Durchfälle sistierten gewöhnlich, wenn es zur Appetitlosigkeit kam. – Die Patientin war ledig und von Beruf Heimerzieherin. Ihre Beschwerden schilderte sie in einer nachdrücklich-klagenden, fast anklagend wirkenden Form.

Die **Unterbauchschmerzen** traten gewöhnlich dann auf, wenn die Heimleiterin in Urlaub ging. Die Patientin hatte zu diesen Zeiten selbständig ein Heim für schwererziehbare junge Mädchen zu leiten. Nach außen gesehen kam sie den gleichen Verpflichtungen wie sonst nach; nur mußte sie nach den Statuten des Heimes die Heimleiterin in deren Abwesenheit vertreten. Das war den Heimbewohnern bekannt und die Patientin fühlte sich in diesen zeitlichen Abschnitten sehr genau beobachtet.

Über zum Teil schwerste **Oberbauchschmerzen** mit Brechreiz und Apathie, dagegen kaum über Unterbauchschmerzen, allerdings zeitweise verbunden mit Durchfall, klagte die Patientin dann, wenn außer der Heimleiterin auch zufällig die Wirtschaftsleiterin gleichzeitig abwesend war. Die Patientin hatte die Wirtschaftsleiterin als außerordentlich hilfreich, ja fast liebevoll-versorgend kennengelernt. Die Wirtschaftsleiterin hatte offensichtlich die Hilfsbedürftigkeit der Patientin erfaßt, wenn die Heimleiterin abwesend war: Sie kochte dann regelmäßg eine leichte Nahrung und sorgte sich darum, daß die Patientin das Essen zu sich nehmen konnte. Die Schilderungen der Patientin hinterließen beim Untersucher den Eindruck, daß die schlimmsten Beschwerden von dieser Wirtschaftsleiterin verhindert werden konnten.

33.3 Symptomatologie der funktionellen Abdominalbeschwerden

Um die vielfältigen Beschwerden zu ordnen, sind die verschiedensten Unterteilungsversuche durchgeführt worden. Im allgemeinen werden hierzu folgende Kriterien benutzt: Die Beschwerden werden aus differentialdiagnostisch-anatomischer Sicht auf einzelne Abschnitte des Verdauungstraktes einschließlich der Speiseröhre bezogen. Psychosoziale Verhaltensweisen werden nicht für eine Klassifizierung berücksichtigt, was bereits an dieser Stelle problematisiert werden und im Abschnitt Therapie wieder aufgegriffen werden soll. Da der Verdauungstrakt in einen oberen und einen unteren Trakt unterteilt wird, findet sich dementsprechend die erwähnte Unterscheidung in FOB und FUB. Funktionelle Abdominalbeschwerden (FAB), gleichgültig ob solche vom Oberbauch- oder Unterbauchtyp, weisen auf einen im Erleben des Patienten störanfälligen Verdauungstrakt hin. – Von den genannten Hauptbeschwerdegruppen lassen sich eine Reihe anderer, mehr oder weniger umschriebener Beschwerdegruppen trennen. Die chronische Obstipation wird häufig als gesondertes Beschwerdebild aufgeführt. Dies erscheint vom klinischen wie möglicherweise auch psychodynamischen Gesichtspunkt aus gerechtfertigt. Ebenfalls als gesondertes Beschwerdebild wird häufig die Aerophagie beschrieben. Daneben sind „Randsymptome" zu beachten, die sich fast regelmäßig mit den FAB vergesellschaftet vorfinden (von Uexküll, 1963; vgl. Kap. 31).

33.3.1 Funktionelle Oberbauchbeschwerden

Sie zeigen als Leitsymptome Schmerz- und/oder Völlegefühl im Oberbauch, rechts oder links und im Rippenwinkel (Hill und Blendis, 1967; Edward et al., 1968; Almy, 1983). Die Schmerzen werden als brennend oder dumpf geschildert, sie strahlen selten aus und sind meist von mittlerer Intensität (Way, 1973). Manchmal werden sie durch das Essen oder unmittelbar danach verstärkt. Sonst lassen sich keine gehäuft auftretenden Beziehungen zu bestimmten Ereignissen feststellen; auch eine jahres- oder tageszeitliche Periodik oder eine zeitliche Abgrenzbarkeit, wie sie etwa bei Gallenkoliken besteht, ist meist nicht feststellbar. Die Symptomatik nimmt in der Regel gegen Abend ab. Nachts treten die Beschwerden selten auf (Kirsner, 1966). Ist dies doch der Fall, geschieht es gewöhnlich in den frühen Morgenstunden. Hierdurch unterscheiden sie sich vom Nachtschmerz des Ulcus duodeni, der gewöhnlich zwischen 1.00 und 2.00 Uhr nachts auftritt.

Die Schilderung der Beschwerden wird von den Patienten häufig schwallartig vorgebracht, und ein vorwurfsvoller Ton schwingt mit. Der Umgebung wird offen oder versteckt angelastet, für die Beschwerden verantwortlich zu sein (Harris, 1946). Obwohl die Patienten vielfach die Nahrung als Ursache der Beschwerden bezeichnen, bringen diätetische

Verordnungen keine Besserung. Bei längerer Unterhaltung mit den Patienten fällt ihre depressive Grundstimmung auf. Der unbestimmte Charakter der Beschwerden macht die Exploration zeitraubend und zähflüssig.

33.3.2 Funktionelle Unterbauchbeschwerden

Hier finden sich als Leitsymptome häufig kombiniert: abdomineller Schmerz (gewöhnlich unterhalb des Nabels), Diarrhöe und Obstipation (im Wechsel) (Almy, 1973; Rinaldo et al., 1963; Illig, 1961). – Bei Obstipation kann der Stuhl bleistiftartige oder bohnenartige Formen annehmen (spastische Obstipation); eine kleine Gruppe der Patienten (20%) leidet an schmerzloser Diarrhöe, eine noch kleinere (unter 10%) weist stark schleimhaltige Stuhlabgänge auf. Auch hier finden sich außer der Angabe, daß die Beschwerden häufig gegen Abend hin besser werden, keine Hinweise auf eine tageszeitliche oder jahreszeitliche Periodik.

Als signifikante Symptome wurden beim Vergleich mit einer organischen Erkrankung beschrieben: rigidgespannt und übertrieben-gepflegt erscheinendes Äußeres, Beschwerden über Monate bis Jahre im Charakter wechselnd, durch Emotionen und durch Streßsituationen verstärkt, Besserung im Urlaub; kein nächtliches Erwachen, im Stehen stärker als im Liegen, 0–30 Minuten postprandial und morgens auftretend; verbunden mit Migräne, Palpitationen, Globus, Atembeklemmung (Conen und Frey, 1982).

Gewöhnlich beschreiben die Patienten ihre Beschwerden äußerst minutiös. Obwohl ihre Schilderung einen sehr exakten und gewissenhaften Eindruck macht, lassen sich selten in den ersten Gesprächen auslösende, verstärkende oder erleichternde Ereignisse genauer herausarbeiten. Spontane Hinweise auf psychische oder soziale Zusammenhänge sind selten.

33.3.3 Vergesellschaftung mit anderen Beschwerdebildern

Randsymptome

Die Differentialdiagnose der FAB wird dadurch erschwert, daß sie praktisch alle organischen Krankheitsbilder im gastrointestinalen Bereich nachahmen können. Daher ist es wichtig zu wissen, daß alle funktionellen Syndrome häufig sog. „Randsymptome" aufweisen (von Uexküll, 1963).

FAB und ihre Beziehungen zu anderen Beschwerdebildern

Hier sind an erster Stelle die Beschwerden bei Aerophagie zu nennen, die ein umschriebenes Beschwerdebild darstellt (von Uexküll, 1960). Die Leitsymptome sind Schmerzen und rasch entstehendes Aufgetriebensein des Leibes. Beweisend ist salvenartiges, oft geräuschvolles Aufstoßen (6–7mal hinter-

einander und mehr). Nach diesem Aufstoßen muß gefragt werden, da es die Patienten meist nicht von selbst angeben. Sie sehen keinen Zusammenhang zwischen diesem Symptom und ihren Beschwerden. Das Symptom findet sich aber nur, wenn Aufstoßen und Schlucken von Luft reflexartig wechseln. Wenn die Luft nicht aufgestoßen, sondern in den Dünndarm weitertransportiert wird, sind Meteorismus und häufige, nicht riechende Winde charakteristisch. – Als Roehmheldscher Symptomenkomplex werden herzbezogene Schmerzen bezeichnet, die durch ein hochgedrängtes Zwerchfell bei überblähter Magenblase hervorgerufen sind.

Enge Beziehungen der FAB bestehen zu folgenden Störungen: Appetitstörungen, Schluckstörungen, gastro-ösophagealer Reflux, Dyspepsie (Almy, 1983). – Zu den umschriebenen Krankheits- bzw. Beschwerdebildern, nach Almy (1983) ebenfalls mit FAB gehäuft verbunden, gehören: das zyklische Erbrechen von Kindern, Migränezustände, Schwangerschaftserbrechen, psychogenes Erbrechen bei Erwachsenen, Erbrechen bei Anorexia nervosa. Auch viele Zustände des sog. „Postcholezystektomie-Syndroms" gehören hierher; sie haben nichts mit gestörten Funktionen der Gallenwege zu tun.

33.3.4 Synonyma

Im ärztlichen Alltag gibt es eine Reihe von Bezeichnungen für die FAB, die jedoch mehr über vorherrschende Konzepte und Auffassungen in der Medizin als über zugrundeliegende Störungen Aufschluß geben. FAB sind, um mit von Uexküll (vgl. Kap. 31) zu sprechen, „kulturgebundene Syndrome".
Für die FOB finden sich folgende Synonyma:
Dyspepsie, Gastritis, Gallenwegsdyskinesie, Magenneurose, Reizmagen, nervöser Magen, Gastropathie, Postcholezystektomie-Syndrom, vegetative Dystonie, vegetative Neurose.
Für die FUB finden sich folgende Synonyma:
Spastisches Kolon, Colon irritabile, Reizkolon, instabiles Kolon, spastische Obstipation, nervöse Kolitis, Colitis mucosa, Kolonneurose, Dissynergie des Kolons, membranöse Kolitis.
Zunehmend bürgern sich als Begriffe ein: Reizmagen bzw. Reizdarm (letzter Begriff sowohl für FAB allgemein wie für FUB speziell).

33.4 Epidemiologie

Patienten mit funktionellen Abdominalbeschwerden machen 40–60% aller Patienten mit Beschwerden im gastrointestinalen Bereich aus (Almy, 1973; Fahrländer, 1972; Hill et al., 1967). Thompson nimmt an, daß 30% der Normalbevölkerung (!) zeitweise eine gestörte Darmtätigkeit aufweisen und daß 10–15% der gesunden Bevölkerung unter Beschwerden im Sinne von FOB leiden (1983, S. 299). Wir finden Patienten mit funktionellen Abdominalbeschwerden gleichermaßen in der allgemeinärztlichen wie in der fachärztlich-internistischen oder -gastroenterologischen Sprechstunde und in den Ambulanzen der Kliniken. Häufig gestaltet sich die Abklärung des Beschwerdebildes so schwierig, daß sich derartige Patienten auch regelmäßig in hochspezialisierten Universitätskliniken finden. Allerdings wird dort seltener die Diagnose eines funktionellen Syndromes gestellt, da häufig von der Norm abweichende Befunde erhoben und die geklagten Beschwerden dann diesen zugeschrieben werden.

Männer und Frauen sind gleich häufig befallen. Obwohl üblicherweise ein Häufigkeitsgipfel zwischen dem 20. und 40. Lebensjahr angegeben wird, finden sich Patienten mit FAB in allen Altersgruppen. Die vielfach geäußerte Meinung, daß FAB im Alter abnähmen oder gar verschwänden, trifft also nicht zu. Vielmehr sind sie in gleicher Häufigkeit auch bei den Angehörigen der höheren Altersstufen zu finden (Berman und Kirsner, 1972; Sklar, 1979). Daß derartige Beschwerden im höheren Alter weniger verzeichnet werden, hängt nicht mit ihrer tatsächlichen Häufigkeit zusammen, sondern mit der Angst von Arzt und Patient, eine organisch bedingte oder gar maligne Erkrankung zu übersehen. Außerordentlich häufig sind FAB in den Kinderjahren:

Bei kindlichen Magen-Darm-Patienten werden in 90% funktionelle Abdominalbeschwerden angegeben (Apley, 1967; Carey, 1968; Green, 1967).

Über die Verteilung von FAB in Abhängigkeit von Sozialstatus und beruflicher Tätigkeit ist zur Zeit wenig bekannt. Transkulturelle Untersuchungen zur Häufigkeit liegen nicht vor; doch ist anzunehmen, daß die Häufigkeit derartiger Störungen in den industrialisierten Ländern im Vergleich zu Schwarzafrika besonders hoch ist. Die oben genannten Zahlen stammen aus angelsächsischen Ländern. Sie entsprechen aber mit großer Wahrscheinlichkeit weitgehend denen der deutschsprachigen Länder. Ein besonders häufiges Phänomen scheinen FAB in Japan darzustellen. Hier finden sie sich besonders unter Jugendlichen (Nakagawa et al., 1978).

33.5 Psychologie, Lebensgeschichte, Psychodynamik und psychosoziale Interaktion

Zur **Psychologie:** Bei Patienten mit FAB läßt sich keine einheitliche Persönlichkeitsstruktur finden, wohl aber treten gehäuft bestimmte Persönlichkeitszüge auf.

Bei Patienten mit FOB findet sich klinisch wie testpsychologisch bei durchschnittlich zwei Dritteln der Patienten eine neurotische Symptomatik (Palmer et al., 1974; Klumbies, 1983; Schüffel et al., 1971, 1976; Seward et al., 1965). – Bei Patienten mit FUB sind weniger häufig umschriebene neurotische Symptome zu beobachten. Die Patienten erscheinen viel eher „übernormal". Häufiger sind Persönlichkeitszüge zu finden, die ihnen die Äußerung von Affekten, insbesondere von Ängsten und unmittelbaren Aggressionen verbieten und die mit ihrem Wunsch nach Unabhängigkeit von anderen und ihrem Streben nach überdurchschnittlichen Leistungen zu tun haben (Klumbies, 1983; Schüffel et al., 1971).

Zur **Lebensgeschichte:** Zwei Drittel aller FAB-Patienten berichten bedeutsame persönliche Verluste, die offensichtlich mit ungelösten pathologischen Trauerreaktionen verbunden sind (Drossman, 1983). Bei einer konsekutiven Untersuchung von 330 Patienten mit FAB wurde gehäuft der Verlust eines Elternteils durch Tod, Scheidung oder Trennung gefunden. In 61% berichteten die Patienten über frostige und spannungsgeladene Beziehungen im Elternhaus (Hislop, 1980). Insgesamt scheinen Beschwerden dann gehäuft aufzutreten, wenn lebensgeschichtlich wichtige Umstellungen stattfinden, seien es Pubertät, Heirat, die Geburt von Kindern, Klimakterium und Alter (White et al., 1939; Chaudhary und Truelove, 1962; Almy et al., 1949), und wenn diese Umstellungen wichtige Entscheidungen erfordern (Paulley, 1982). Gehäuft werden FOB bei Gastarbeitern mit Erstmanifestation 3–6 Monate nach der Umsiedlung beschrieben (Meyer, 1981).

Zur **Psychodynamik:** Die vor über 50 Jahren von Alexander und Mitarbeitern vorgelegten Untersuchungsergebnisse gelten im Prinzip weiterhin (1933, 1934, 1968). Alexander und seine Mitarbeiter fanden bei ihren Patienten mit gastrointestinalen Beschwerden als psychologisch bedeutsames Grundmuster den Wunsch, gefüttert zu werden und gleichzeitig den Drang, diesen Wunsch in nachhaltiger Weise abzuwehren.

Die Beschreibungen der Alexanderschen Arbeitsgruppe muten als eine Art Ergänzungsreihe an, die durch das intrapsychische Abwehrverhalten der Patienten gekennzeichnet ist und sich durch die funktionellen Abdominalbeschwerden geradezu „etagenweise" hindurchzieht: Alexander und Mitarbeiter unterschieden ein gastrisches Verhaltensmuster von einem Kolon-Verhaltensmuster und beschrieben zusätzlich Verhaltensformen des Obstipierten.

Nach Alexander und Mitarbeitern läßt sich sagen, daß sich diese drei Formen der gastrointestinalen Störung durch die Funktionen des Nehmens, des Gebens und des Zurückhaltens kennzeichnen lassen. Relativ oberflächlich liegt beim FOB-Patienten der Wunsch nach Nehmen und Empfangen vor. Er beansprucht Versorgung. Beim FUB-Patienten wird dieser Wunsch stärker abgewehrt und statt dessen tritt die Abwehr in Form des Gebens in den Vordergrund: Dieser Patient ist der Meinung, ausreichend zu geben. Der obstipierte Patient kann praktisch nichts mehr geben. Die durchgehende Problematik einer oralen Nicht-Befriedigung führt zu drei Formen des Versuches einer Konfliktlösung, die folgendermaßen bezeichnet wurden:

– Gastrischer oder FOB-Typ: Es besteht der Wunsch, zu erhalten oder zu nehmen. Der Konfliktfall ist dann gegeben, wenn der Wunsch zu empfangen bzw. zu nehmen behindert wird.

– Der Kolon- oder FUB-Typ: Es besteht der Wunsch, zu geben oder auszuscheiden. Der Konfliktfall liegt dann vor, wenn das Geben behindert wird und die Elimination zu einem aggressiven Akt werden muß.

– Der Obstipationstyp: Er kann nichts mehr geben, d.h. eine Abwehr der FUB-Patienten. Der Konfliktfall ist dann gegeben, wenn das Zurückhalten als schuldhaft erlebt wird.

Diese idealtypische Anordnung psychodynamischer Abwehrvorgänge läßt sich in dieser reinen Form in der Praxis natürlich kaum wiederfinden. Schwerpunktmäßig lassen sich die klinischen Beobachtungsergebnisse häufig aber bestimmten Verhaltensmustern zuordnen, wie auch im Fall unserer Patientin. Diese kann immer dann – wenn auch unter Schwierigkeiten – arbeiten, wenn ihre Versorgungswünsche von der Wirtschafterin erfüllt werden. Die ersten nachhaltigeren Beschwerden treten dann auf, wenn ihre Form des Gebens während der Abwesenheit der Heimleiterin vermeintlich/tatsächlich kritisch unter die Lupe genommen wird. Während der Therapie mit dieser Patientin konnte beobachtet werden, daß sie in Zeiten einer Auseinandersetzung mit dieser Heimleiterin obstipiert war. Das waren solche Zeiten, in denen sich die Patientin nach außen gefestigt zeigte und ihre Umgebung ihr keine Unsicherheiten anmerkte.

Über die Zeitpunkte der Dekompensation könnte man aus psychoanalytischer Sicht zusammenfassend sagen: Zur Dekompensation kommt es beim FOB-Patienten, wenn er sich gemessen an seinen Erwartungen unterversorgt fühlt und diesen Eindruck seiner Umwelt aus Angst vor geringerer Zuwendung nicht mitteilen kann. Zur Dekompensation des FUB-Patienten kommt es, wenn sich dieser im Vergleich zu seinen unbewußt gestellten Versorgungswünschen als „nicht zahlungsfähig" sieht. Nochmals muß betont werden, daß diese analytischen Beobachtungsergebnisse in der Vielfalt der klinischen Phänomene oft schwer wiederzuerkennen sind. Sie werden von späteren Entwicklungsabschnitten der Persönlichkeit überlagert. So kann es vorkommen, daß viel stärker rivalisierende oder hysterische Momente im Vordergrund zu stehen scheinen. So sind bei Frauen häufig Schwangerschafts- bzw. Schwängerungsphantasien beschreibbar. Erst bei genauerem Einlassen auf diese Phänomene wird deutlich, daß nahezu regelmäßig bei den schwerer behinderten Patienten derartige frühe Entwicklungsstörungen auf der oralen Ebene und frühen Analebene vorliegen.

Zur Interaktion: Bei der Betrachtung des Interaktionsstiles ist zu beachten, daß Klagen über Körperbeschwerden auch Auslöser für soziales Verhalten der Mitmenschen sind. Das wird in der Interaktion mit Ärzten besonders deutlich. Sie ist auf der einen Seite durch die Klagen des Patienten über seine somatischen Beschwerden, seine affektive Verschlossenheit bzw. Unerreichbarkeit für Fragen bestimmt, die das Gefühlsleben betreffen; auf der anderen Seite durch die diagnostische Unsicherheit des Arztes, der

zwar weiß, daß psychosoziale Probleme vorliegen und organpathologisch keine Veränderungen bestehen, der aber dennoch fürchtet, eine organische Krankheit übersehen zu können.

In diesem Zusammenhang ist das „Überweisungsritual" oft charakteristisch, aber auch pathogenetisch bedeutsam: Der Hausarzt schickt den Patienten zum Spezialisten, der eine Reihe von Untersuchungen durchführt und den Patienten dann ohne überzeugende Diagnose an den Hausarzt zurücküberweist. Nach neuerlichen Überweisungen zu Fachärzten wird der Patient in die Klinik geschickt, durchuntersucht und wiederum ohne handfeste Diagnose und Therapievorschläge an den Hausarzt zurücküberwiesen. Je häufiger dieser Zirkel durchlaufen wird, um so mehr festigt sich in dem Patienten die Überzeugung, daß hinter seinen rätselhaften Beschwerden, die kein Arzt befriedigend deuten kann, eine geheimnisvolle und gefährliche Krankheit steckt: Das Beschwerdebild wird chronifiziert (vgl. Kap. 31).

Dabei spielen **Protest** und **Bestrafung** des **Protests** eine wichtige Rolle: Der Patient protestiert mit seinen Beschwerden – unter Umständen auch mit Arbeitsunfähigkeit – gegen die Ablehnung seiner (ihm selbst oft unbewußten) Wünsche durch die Gesellschaft. Die Gesellschaft reagiert mit Bestrafung (Ablehnung als „nicht richtig krank", Einstufung als „Simulant"), sozialem Abstieg usw., während die Ärzte mit Überweisungen an andere Kollegen reagieren. Zwischen Patient und Gesellschaft entwickelt sich ein „kalter Krieg", der mitunter in Form eines Rentenkampfes ausgetragen wird, häufig aber in die Iatrogenie (Allgemeinmedizin mit Tranquilizern, Psychiatrie mit Thymoleptika, Chirurgie mit abdominellen Eingriffen, Gynäkologie mit Exstirpationen, Innere Medizin mit Endoskopien und Spasmolytika) einmündet. Ein Chirurg beschreibt plastisch die von ihm vorgefundene Untersuchungssituation: „So verzehnfachen sich an Montagen die stationären Aufnahmen von jungen Patientinnen zur unnötigen Appendektomie. Die innerfamiliären Konflikte, ausgelöst durch die Adoleszenz der Töchter, werden ... an dem fordernden Verhalten der Mütter und dem Schweigen der Töchter sichtbar" (Hontschick, 1989).

Auf die Krankengeschichte der hier wiederholt zitierten Patientin angewendet: Ihrem Wunsch nach Nehmen und „Gefüttertwerden" kann sie dadurch nachkommen, daß sie durch ihre Tätigkeit als Erzieherin ihr eigenes „Geben" ausreichend unter Beweis stellt, also ein Entgelt liefert, das vom Heim auch anerkannt wird. Sind Heimleiterin und Wirtschaftsleiterin anwesend, so wird kein Arzt erforderlich. Der Arzt kann unter Umständen auch dann noch aus dem Spiel bleiben, wenn die Wirtschafterin in einer für die Patientin befriedigenden Weise agiert. Sind beide Personen abwesend, ist in der Regel die Bedingung der Dekompensation erfüllt und der Arzt muß in Erscheinung treten. Dieser nimmt, das ist offensichtlich, in dieser Situation eine entscheidende Rolle ein.

Ein Arzt, der im jeweiligen Beschwerdezustand Medikamente, Bettruhe verordnet und Arbeitsunfähigkeit attestiert, gewährt auch gleichzeitig Gratifikationen und Stütze, die ein persönliches Sich-Wiederfinden ermöglichen. Ein Arzt dagegen, der derartige Beschwerden als „eingebildet" oder „o. B." bezeichnet, d. h. diese Beschwerden in der medizinischen Alltagssprache als „nicht relevant" betrachtet, fordert den Patienten zum Protest bzw. in späteren Abschnitten zur Resignation heraus. In der vorliegenden Situation hatte eine Art „Liaison-System" (vgl. Kap. 28 „Psychosomatischer Konsiliardienst") ambulanter gastroenterologischer Versorgung bestanden. Der untersuchende Gastroenterologe war sich der möglichen psychosozialen Problematik bewußt gewesen, hatte die Patientin hierauf angesprochen und sie zunächst gemeinsam mit dem dann psychotherapeutisch intervenierenden Arzt gesehen.

33.6 Psychophysiologie

Ätiologie und Pathogenese der FAB werden allgemein abgehandelt im Kapitel 31. Hier soll auf spezielle psychophysiologische Untersuchungsergebnisse eingegangen werden, die bei FAB-Patienten erhoben wurden und von pathogenetischer Bedeutung sind.

Als gesichert kann angenommen werden, daß wir im Bereich des Magen-Darm-Traktes unter psychophysiologischen Gesichtspunkten zwei grundlegende Verhaltensmuster unterscheiden können. Diese Muster wurden für den oberen GI-Trakt von S. Wolf und H. G. Wolff herausgearbeitet (1947). Deren Arbeiten basieren auf einer 16jährigen fortwährenden Beobachtung eines Magenfistelträgers, des Laborangestellten Tom, sowie auf den Beobachtungen bei vier anderen Fistelträgern und der Diskussion von Literaturmitteilungen, die überwiegend von Beaumont stammen, einem kanadischen Arzt des vorigen Jahrhunderts (vgl. auch Kap. 43).

Tom war 1895 wegen einer Ösophagusatresie operiert worden, die er sich im Anschluß an das versehentliche Verschlucken eines kochenden Fischgerichtes zugezogen hatte. Während eines halben Jahrhunderts hatte er sich durch diese Fistel ernährt. Wolf konnte das äußere Erscheinungsbild dieser Fistel mit verschiedenen physiologischen Funktionen korrelieren und diese schließlich zum Gesamtzustand Toms in Beziehung setzen. Wolf beschrieb die zwei Verhaltensmuster folgendermaßen:

1. Tom klagte über epigastrische Schmerzen. Er war gereizt, ohne den Ärger recht ausdrücken zu können. Eine Vorwurfshaltung war spürbar. Im Magen fanden sich Zeichen der Hypermotilität und Hyperazidität. Der Magen verhielt sich wie in Augenblicken einer Essensaufnahme. Insgesamt lag eine verstärkte Magenaktivität vor. Nach längeren Perioden dieser Aktivität und damit verbundenen Hyperämien der Mukosa konnte diese unter Umständen blaß werden, blieb aber geschwollen und ödematös. Zu dieser Zeit ließ sich verschiedentlich eine hohe Azidität nachweisen. Biologisch schien diese Reaktion ein Äquivalent darzustellen für Nahrungsaufnahme und für den Wunsch nach Einverleibung allgemein.

---

2. Tom klagte über Übelkeit und Erbrechen. Er wirkte recht zurückgezogen und sein Verhalten erinnerte an Menschen, die sich mit einer Nahrungsmittelvergiftung plagen. Die Magenmukosa war blaß, die Säureproduktion ging ebenso wie die Motilität zurück. Es handelte sich um ein plötzliches Stoppen des gastrischen Verdauung. Diese Erscheinungsform wurde von Wolf als riddance- oder Auswurfmuster beschrieben.

Nach dem Krieg untersuchten Almy und Mitarbeiter Auswirkungen emotional belastender Interviews sowohl bei Normalpersonen wie bei Personen mit Colon irritabile (1950). Die Kolonreaktionen wurden entweder mit Hilfe direkter visueller Betrachtung oder durch ein Sigmoidoskop oder durch Aufzeichnungen mit Hilfe intraluminaler Ballontechnik sigmoidal durchgeführt. Etwa zur gleichen Zeit untersuchten Grace und Mitarbeiter während Interviewabläufen bei vier teilkolektomierten Patienten mit Anus praeter die Reaktion des einsehbaren Kolonsegmentes (1950). Auch diese beiden Arbeitsgruppen kamen zur Unterscheidung zweier grundlegender Verhaltensmuster im Sinne von Wolf und Wolff. Diese Verhaltensmuster lassen sich für den unteren GI-Trakt so zusammenfassen:

– Diskussionen über Themen, die Ärger, Vorwurf, im Englischen als „resentment" bezeichnet, auslösten, führten zu schmerzhaft empfundenen Kontraktionen des Darmes. Insgesamt fand sich eine vermehrte motorische Aktivität.
– Diskussionen über Gefühle, die Hoffnungslosigkeit und Hilflosigkeit auslösten, im Patienten ein Gefühl des Ungenügens und des Selbstvorwurfs bedingten, waren von einem Aussetzen der Kolonaktivität begleitet.

Diese beiden Grundverhaltensformen wurden bei den verschiedensten Probanden, also nicht nur bei FAB-Patienten gefunden. Chaudhary und Truelove fanden darüber hinaus bei einer Gruppe von 38 Patienten mit irritablem Kolon regelmäßig Veränderungen des Motilitätsmusters, die durch spezifische Themen ausgelöst wurden (1962). Teilte man die Patienten danach ein, ob sie sich in einem akuten Beschwerdezustand oder in einem symptomfreien Intervall befanden, so wiesen Patienten der ersten Gruppe in 40%, die der zweiten Gruppe in 13% bei entsprechenden Themen die beobachteten Motilitätsveränderungen auf.

Hinsichtlich des biologisch faßbaren Entgegenkommens des Organismus ist eine in der Literatur durchgehende Beobachtung wichtig: Patienten mit spastischem Kolon weisen im Vergleich zu übrigen Kranken wie im Vergleich zu Colitis-ulcerosa-Patienten eine erhöhte Empfindlichkeit gegenüber Prostigmin auf. Sie zeigen also eine verstärkte parasympathikomimetische Ansprechbarkeit.

In einer zusammenfassenden Arbeit stellte S. Wolf dar, wie sich der Einfluß zweier entgegengesetzter psychobiologischer Grundverhaltensmuster nahezu für den gesamten Gastrointestinaltrakt fundiert belegen läßt, im einzelnen für die Abschnitte des Magens, des Duodenums, des Jejunums und des Kolons (1971).

Versuche, zumindest einen Teil der FAB als eine allergische Reaktion zu erklären (Jones et al., 1982), konnten nicht bestätigt werden (Bentley et al., 1983). Auch der in den 70er Jahren viel diskutierte Laktasemangel, den man bei ca. 12–13% dieser Patienten findet, kann die Symptomatik nicht erklären (Whitehead und Schuster, 1985).

Besonderes Interesse beanspruchte die Frage, ob sich bei FAB gehäuft veränderte Motilitätsverhältnisse finden (Lasser et al., 1975; Sullivan et al., 1978). Kirsner faßte die derzeit vorliegenden Untersuchungsergebnisse zur Motilität beim Colon irritabile folgendermaßen zusammen (1981):

– Es findet sich im Vergleich zum Normalen ein verändertes slow-wave-Muster, das sich durch eine abnorme Zunahme der Kontraktionen nach Nahrungsaufnahme auszeichnet.
– Es kann aufgrund dieses Musters zu einer funktionellen partiellen Obstruktion kommen, die zumindest teilweise die Abdominalschmerzen erklärt.
– Es findet sich eine vermehrte Ansprechbarkeit der Kolonmuskulatur auf verschiedene Hormone (z.B. Cholezystokinin und Pentagastrin), sowie vermehrte Empfindlichkeit gegenüber Prostigmin, Morphin und dessen Derivate. Die vermehrte Aktivität wurde im Sigmoid, nicht in proximalen Anteilen des Kolons beobachtet.

Noch sind wir jedoch weit davon entfernt, die zugrundeliegenden Mechanismen auf zellulärer Ebene (Szurszewski, 1987) oder motorischer Ebene (Christensen, 1987) zu erklären.

In der deutschsprachigen Literatur charakterisiert Wienbeck das Problem folgendermaßen (1984): „Reizmagen, Colon irritabile und chronische Obstipation gehören zu den sogenannten funktionellen Syndromen, die durch das Fehlen eines morphologisch faßbaren Substrats gekennzeichnet sind. Alle drei Krankheitsbilder neigen zu einem chronischen Verlauf. Sie stellen an den betreuenden Arzt hinsichtlich Diagnostik und Führung des chronisch Kranken besonders schwierige Aufgaben."

Zusammenfassend läßt sich zum Stand unseres Wissens über die Entstehungsweise funktioneller Abdominalbeschwerden sagen, daß hierunter leidende Patienten eine Störung der Verarbeitung oraler Konflikte aufweisen, die in einem Teil der Fälle mit einer biologisch definierbaren Empfindlichkeit des Gastrointestinaltraktes verbunden sind. Es kommt erst dann zu Beschwerden, wenn die psychischen Belastungen das alte Konfliktpotential in einer spezifischen Weise reaktivieren.

33.7 Differentialdiagnose

33.7.1 Krankheiten, die ausgeschlossen werden müssen

Die Feststellung, daß funktionelle Abdominalsyndrome fast alle organischen gastrointestinalen Erkrankungen imitieren können, zwingt zu umfassenden differentialdiagnostischen Überlegungen. Dies gilt ganz besonders für die FOB; für die FUB finden sich etwas verläßlichere differentialdiagnostische Kriterien (vgl. Kap. 12).

Grundsätzlich gilt, daß bei derartigen Beschwerden nicht nur der Patient an Krebs denkt, sondern daß es der Arzt ebenso tun sollte. Dies gilt für alle länger dauernden Beschwerden, die erstmalig abgeklärt werden müssen. Das gilt aber auch für Verläufe nach einer Erstabklärung, wenn sich die Beschwerden ohne sichtliche Beziehung zu vorgenommenen Maßnahmen ändern. Ganz besonders im Bereich der funktionellen Oberbauchbeschwerden kann das frühe Karzinom des Magens zunächst funktionelle Beschwerden vortäuschen (!)

Darüber hinaus sind gegenüber den FOB wichtige Differentialdiagnosen: Hiatushernie, Achalasie, Ulcera ventriculi et duodeni, Gallenwegs-, Pankreas- und Lebererkrankungen, Kardiospasmus, Neoplasien im Bereich von Pankreas und Leber, eine diabetische Enteropathie, postoperative Folgezustände, intestinale Pseudoobstruktion und schließlich Erkrankungen des ZNS einschließlich eines Hirntumors. Ferner ist die Nebenwirkung von Medikamenten zu beachten, wie Opiate und Digitalis. Schließlich können durch Sonden virusähnliche Infektionen übertragen werden, z.B. nach einer Endoskopie (Wienbeck, 1984).

Gegenüber den FUB sind wichtige Differentialdiagnosen: Nahrungsmittelunverträglichkeiten, Laktasemangel (wenngleich gerade hier sehr sorgfältig abgewogen werden muß, was auf den Laktasemangel und was auf psychosoziale Belastungen rückführbar ist), Malabsorptionssyndrome, Medikamentenunverträglichkeit (z.B. gegenüber Magnesium, Laxantien, den Gallensäuren vieler Pankreasenzympräparate) (Wienbeck, 1984). – Erkrankungen des Dünndarms (z.B. Morbus Crohn), Erkrankungen des Dickdarms (z.B. Kolitis, Divertikulitiden), Entzündungen im kleinen Becken und schließlich Erkrankungen der Thoraxorgane, die abdominelle Beschwerden verursachen können (z.B. Hinterwandinfarkt) und natürlich wiederum Karzinom.

Gegenüber der chronischen Obstipation sind wichtige Differentialdiagnosen (nach Wienbeck, 1984): Durch Medikamente hervorgerufene Obstipation wie z.B. durch Psychopharmaka, aluminiumhaltige Antazida, Opiate; Hypothyreose, diabetische Enteropathie, eine Analfissur; stenosierende Prozesse, ein rudimentär angelegter Morbus Hirschsprung.

33.7.2 Anamnese

Es gilt die somatischen, psychischen und sozialen Faktoren, die zum Krankheitsgeschehen beitragen, und ihr Ineinandergreifen bei der Symptombildung zu erfassen. Dies ist am ehesten möglich, wenn der Patient ermuntert wird, seine Beschwerden, die Lebensumstände, die Beziehung zur Umgebung und seine Entwicklung in seinen eigenen Worten zu berichten. Hierzu ist am besten die von Engel beschriebene Anamneseform geeignet (vgl. Kap. 12). Die Anamneseerhebung schließt selbstverständlich die sieben Dimensionen jedes Symptoms ein und auch ein Abfragen bestimmter Symptome mit dem Ziel der differentialdiagnostischen Abgrenzung gegenüber organisch bedingten Leiden (Conen und Frey, 1982). Deutlich für ein Reizkolon sprechen: Schmerzverstärkung durch Emotionen und Streß, Beschwerden im Charakter nicht wechselnd über Monate bis Jahre, Besserung im Urlaub und ein als rigide-gespannt empfundenes Äußeres. Mit der Diagnose des Reizkolons verbunden sind: kein nächtliches Erwachen durch Schmerz, Symptome morgens, unmittelbar postprandial, im Stehen stärker als im Liegen, normale Konsistenz des Stuhles. – Typisch für organische Beschwerden sind: Schmerz verstärkt durch leichten Druck auf das Abdomen, nächtliches Erwachen und Schlaflosigkeit wegen des Schmerzes, der krampfartig bohrend-nagend, Minuten bis Stunden anhält, keine Lageabhängigkeit, keine Besserung durch äußere Umstände, ein bis zwei Stunden postprandial und unter Umständen Nüchternschmerz, Besserung durch Defäkation oder Flatus, Blutbeimengung im Stuhl, imperativer Stuhldrang. Conen und Frey schließen aus ihren Untersuchungen, daß ein organisches Leiden mit sehr hoher Wahrscheinlichkeit diagnostiziert werden kann, wenn mindestens zwei der organischen Symptome und höchstens eines der „funktionellen" Symptome erhoben wird; ein Colon irritabile sei um so wahrscheinlicher, je breiter die Streuung der „funktionellen" Symptome sei.

33.7.3 Untersuchungsplan

Nach Anamneseerhebung, Körperuntersuchung, Einbeziehen der Befunde voruntersuchender Ärzte und Aufstellen der Diagnose FAB soll eine Therapie in die Wege geleitet werden, wie sie im Abschnitt 33.8 skizziert wird.

Anfänglich sollte eine häufigere Wiedervorstellung erfolgen und diese sollte monatlich, schließlich alle 2–3 Monate durchgeführt werden. Keinesfalls aber darf eine Rezeptverschreibung ohne Wiedersehen des Patienten erfolgen. Eine telefonische Erreichbarkeit des Arztes sollte gegeben sein.

Wird mit diesem Vorgehen kein Erfolg erzielt, dann sind mögliche Ursachen für die Erfolglosigkeit:
- eine nicht ausreichende Mitarbeit des Patienten;
- die – zumeist psychogenen – Ursachen wurden übersehen;
- die Diagnose „FAB" war nicht korrekt (erst an dritter Stelle sollte diese Überlegung angestellt werden!).

Die bisherigen Befunde sind zu überprüfen, ohne daß neue Endoskopien oder radiologische Untersuchungen veranlaßt werden. Aber: anamnestisch und körperlich ist nachzuuntersuchen. Bereit sein für eine Revision der bisherigen Diagnose, wenn sich neuartige Symptome ergeben, wenn Unregelmäßigkeiten längerfristig bestehen oder nächtliche Beschwerden auftreten (Kirsner, 1981).

Dölle (1984) weist zusätzlich auf die Möglichkeit eines Symptomwandels hin.

33.8 Therapie

Jeder Patient mit FAB muß als unter psychosozialer Belastung stehend angesehen werden – oder mit von Uexküll (vgl. Kap. 31): „Nach ihm (dem Situationskreis, Anm. W. Sch.) lassen sich funktionelle Syndrome als Folgen von Verletzungen unserer ,zweiten Haut' interpretieren, die unsere individuelle Wirklichkeit durch ständige ,Assimilation' der Vorgänge unserer Umgebung um unseren Körper legt." Er befindet sich in einer Konfliktsituation, die er derzeit oder bereits über lange Zeitabschnitte hinweg nicht allein lösen kann bzw. konnte. Die therapeutische Chance des Arztes liegt somit in erster Linie darin, bereits während der Anamneseerhebung, d. h. während der Diagnostik zu versuchen, ob es gelingt, den Patienten auf die Schwierigkeiten eines ungestörten zwischenmenschlichen Miteinanders anzusprechen. Die Schwierigkeiten, denen der Patient begegnet, spiegeln sich in der Arzt-Patient-Beziehung wider, in der paradigmatisch die so häufig erwähnte und in der Literatur so oft beschriebene Vorwurfshaltung auftaucht. Auch dem Arzt gegenüber wird offen oder implizit ausgedrückt, daß er zu wenig für einen tut, bzw. daß man zu wenig für seine Leistungen erhalte. Dementsprechend wird der Arzt den entscheidenden Schritt dann tun, wenn er es vermeidet, den Patienten weiterhin in seiner schon so oft bestehenden Erwartung eines mangelnden Aufgeschlossenseins für sein Leiden oder in seiner somatischen Fixierung zu bestärken.

33.8.1 Mitteilung der Diagnose

Die Mitteilung der Diagnose ist der kritische Punkt, bei dem sich entscheidet, ob ein therapeutisches Bündnis zustande kommt.

Viele Patienten äußern bei der Mitteilung der Diagnose mit halbem Herzen ihre Erleichterung, daß keine organische Krankheit vorliege. Hier sollte der Arzt sein Verständnis für die Enttäuschung des Patienten zeigen, daß trotz dessen Beschwerden wieder „nichts Organisches" gefunden wurde. Ein pathologischer Organbefund könnte aber für den Patienten unbewußt bedeuten, endlich von seiner Umgebung als „Kranker" akzeptiert zu sein. Vor der Besprechung der fehlenden organischen Ursachen sollte man versuchen, Beziehungen zu bestimmten Lebenssituationen oder psychischen Belastungen (sofern sich diese aus der Anamnese ergeben haben) aufzuzeigen. Dabei muß man sehr behutsam vorgehen, um den Patienten nicht zu kränken und das bei diesen Erkrankten ohnehin sehr prekäre Arbeitsbündnis zwischen Arzt und Patient nicht zu gefährden.

33.8.2 Unspezifische ärztliche Maßnahmen: Begleitung und Abgrenzung

Dem Patienten mit FAB ist bereits dann wesentlich geholfen, wenn er sich der Aufmerksamkeit seines Arztes sicher ist. So betonen Waller und Misiewicz den therapeutischen Wert regelmäßiger Konsultationen, die dem Patienten das Gefühl der Verfügbarkeit einer Hilfsquelle ermöglichen (1969). Mit dieser Ansicht stehen sie in Übereinstimmung mit Apley und Hale (1973), Hislop (1971), Möller (1965), Dölle und Wiedmann (1978). Deren Grundannahme ist, daß die Behandlungsergebnisse um so günstiger werden, je tragfähiger sich das Arzt-Patient-Verhältnis entwickelt.

Im Sinne unspezifischer Maßnahmen sind auch die Beobachtungen zu verstehen, die aus der Pädiatrie, Gynäkologie und Chirurgie stammen, wenngleich andere als stützende Maßnahmen primär geplant waren: Stone und Mitarbeiter berichten über den Verlauf bei 102 stationär wegen FAB aufgenommenen pädiatrischen Patienten. Bei 53% der Kinder sistierten die Beschwerden ohne jegliche spezifische Therapie total, bei insgesamt 81% war eine deutliche Besserung des Befindens zu verzeichnen. Ingram und Evans (1965) beobachteten bei 45% jüngerer Frauen mit FAB, die fälschlicherweise wegen Appendizitis operiert worden waren, ein Sistieren der Beschwerden. Bei Probelaparotomien wird wiederum in ca. einem Drittel der Fälle Beschwerdefreiheit postoperativ angegeben, obwohl kein organpathologischer Befund erhoben wurde (Devor und Knauf, 1968).

Alle diese Beobachtungen dürften damit zu erklären sein, daß die Patienten sich aufgehoben fühlten, der Verlauf ihrer Beschwerden beobachtet wurde und ihre Ärzte sie während einer Phase begleiteten, in der sie aus psychosozialer Sicht entsprechend den oben dargestellten lebensgeschichtlichen Überlegungen dekompensiert waren.

Immer wieder tritt während der Behandlung der Wunsch in den Vordergrund, erneut untersucht zu werden oder die Behandlungsweise zu wechseln. Hiermit kann der Patient ausdrücken, daß er einerseits die bisher erzielte Konfliktlösung noch immer für unbefriedigend hält; andererseits kann dies aber auch bedeuten, daß er den Arzt auf die Probe stellen will. Es ist wichtig, in diesem Stadium bei seiner bisherigen Verhaltensweise als Arzt zu bleiben, bzw. diese nur nach sehr eingehender Überlegung zu ändern. Vielmehr sollte man die Gründe für den Wunsch des Patienten nach Änderung aufgreifen und durchsprechen. Dabei ergibt sich häufig, daß die alte Angst des Patienten durch ein Erlebnis aktiviert worden ist. Der Arzt sollte nicht der Meinung sein, daß er alle Beschwerden beheben kann. Gerade derartige Patienten provozieren therapeutische Allmachtswünsche. Es ist Dölle zuzustimmen, daß dem Patienten mitunter sein Symptom gelassen werden sollte, statt ihn auf eine nicht lösbare psychosoziale Problemsituation hinzuführen (1984).

Im Grenzgebiet allgemeiner therapeutischer und speziell psychotherapeutischer Maßnahmen bewegt

sich J. Paulley. Er beschreibt die FAB-Patienten als in einer Situation des „fence-sitting" befindlich. Sie können sich nicht zu einer notwendigen Entscheidung durchringen, obwohl sie von der derzeitigen Lebenssituation verlangt wird. Er schlägt vor, einen solchen Konfliktfall als die Regel bei FAB-Patienten anzusehen und ihn schon bei der ersten Konsultation nach eingehender anamnestischer und körperlicher Untersuchung anzusprechen. Fragen können dann sein: „Sehen Sie sich als Mensch, der sich schwer entschließen kann, z. B. die Arbeitsstelle zu wechseln, umzuziehen usw.? – Hat sich derartiges in letzter Zeit ereignet?" (Paulley, 1984).

33.8.3 Spezielle psychotherapeutische Verfahren

Der Mangel an spezifisch psychotherapeutischen Arbeiten über Patienten mit FAB ist auffallend. Svedlund (1983) und seine Arbeitsgruppe verfolgten das Beschwerdebild bei 101 ambulanten Patienten. Sie wurden zwei verschiedenen Behandlungsgruppen nach dem Zufallsprinzip zugewiesen. Bei gleicher medizinischer Standardtherapie erhielt die Untersuchungsgruppe eine dynamisch orientierte Einzeltherapie mit maximal zehn Sitzungen über drei Monate. Es zeigte sich eine Überlegenheit der kombinierten Therapie im Vergleich zur medizinischen Standardtherapie. Die Autoren interpretieren den Erfolg ihrer Kombinationstherapie damit, daß die Patienten ein geändertes Selbstverständnis aufwiesen, das sich auf den Umgang mit ihren Beschwerden wie mit generellen Lebensproblemen bezog. Nach einem Jahr wiesen die Mitglieder der kombinierten Behandlungsgruppe signifikant reduzierte Beschwerden auf. Die Therapeuten bezogen sich in ihrem Therapieansatz auf Karasu (1979). Es handelt sich um die erste Arbeit, die in vergleichender Weise den Einfluß psychotherapeutischer Interventionen bei Patienten mit Reizkolon untersucht.

Von 60 nacheinander aufgenommenen Patienten mit funktionellen Abdominalbeschwerden behandelte Hislop 52 mit zum Teil nachhaltigem Erfolg (1980). Im Schnitt wendete er 2,2 Stunden einer unterstützenden/beratenden Psychotherapie an. Über positive Erfolge einer Gruppentherapie mit Verhaltenstherapie und beratenden Elementen, die zur Besserung emotionaler Probleme beim Fortbestehen körperlicher Symptomatik führte, berichteten Wise und Mitarbeiter (1982). A. Johannsen beschrieb ebenfalls positive Auswirkungen einer verhaltenstherapeutisch orientierten Gruppenarbeit mit zehn wöchentlichen Sitzungen (1984).

Über günstige Ergebnisse bei der Anwendung der Biofeedback-Therapie berichtet Kläger in einer experimentellen Arbeit (1984).

33.8.4 Medikamentöse Behandlung

Medikamente haben bestenfalls eine ergänzende Bedeutung bei der Behandlung der Beschwerden des

Patienten mit FAB. Handelt es sich überwiegend um Beschwerden vom FOB-Typ, bei denen Zeichen der Säureüberproduktion in Form von Nüchternschmerz und Brennen im Vordergrund stehen, so sollten Antazida gegeben werden. Keinesfalls sind hier Histamin-H$_2$-Rezeptorenblocker indiziert. Stehen Blähungen und Völlegefühl stärker im Vordergrund, so ist Metoclopramid angezeigt (Dölle, 1984). – Bei FUB steht ganz die Anregung zur Aufnahme faserreicher Kost einschließlich Kleie im Vordergrund. Hier wird eine tägliche Zufuhr von 20 g empfohlen. Dölle verweist darauf, daß erst in zweiter Linie spezielle Präparate indiziert sind, hierunter Spasmolytika (1984). Genannt werden Mebeverin (Duspatalin®) und Pinaverionbromid (Dicepel®) (Schmidt, 1983). Nur in Ausnahmefällen sind Psychopharmaka angezeigt. Keinesfalls dürfen diese (entgegen immer noch weitverbreiteter Praxis) dauernd gegeben werden. Bei Durchfällen sind Loperamid (Imodium®) und Diphenoxylat (Reasec®) die Mittel der Wahl (Dölle, 1984).

33.9 Zur Prognose

Sieht man von den akuten, einmalig auftretenden Abdominalbeschwerden ab, so läßt sich eine starke Neigung zu chronisch rezidivierenden Verläufen feststellen (Chaudhary und Truelove, 1962; Waller und Misiewicz, 1969; Apley und Hale, 1973). Diese sind stark abhängig von der sozialen Situation sowie vom Arzt-Patient-Verhältnis. Um Enttäuschungen vorzubeugen, sollte sich der Arzt vor Augen halten, daß schon viel erreicht ist, wenn es gelingt, den Patienten einigermaßen zu rehabilitieren und ständige Neuuntersuchungen zu verhindern.

Ein statistisch gehäufter Übergang in organische Erkrankungen findet sich nicht. Ganz besonders ist hervorzuheben, daß sich bösartige Veränderungen bei Patienten mit FAB nicht häufiger finden als im Durchschnitt der Bevölkerung. Ebenfalls wichtig ist, daß praktisch nie ein Übergang von funktionellen Darmbeschwerden in eine echte chronisch-entzündliche Darmerkrankung (M. Crohn, Kolitis) beschrieben wurde.

Herrn Prof. Dr. W. Dölle (Tübingen) danken wir für die kritische Durchsicht des Manuskriptes der 3. Auflage.

34 Das Hyperventilationssyndrom

Jörg Michael Herrmann, Othmar W. Schonecke, Andreas Radvila und *Thure von Uexküll*

Die 41jährige Patientin litt seit einer Operation, die vor etwa 4 Wochen in der Urologischen Klinik wegen Harnröhrenstriktur durchgeführt worden war, an wiederholt auftretenden Anfällen, die mit Verkrampfung der Hände, Atemnot und Herzbeschwerden einhergingen.

Die Familie stammte aus Ostpreußen. Der Vater war Anfang der dreißiger Jahre in die Sowjetunion übergesiedelt. Dort wurde die Patientin 1937 geboren. Nach Ausbruch des deutsch-sowjetischen Krieges wurde der Vater der Patientin in der Ukraine von den Russen verhaftet und später getötet. Die Patientin gelangte mit ihrer Familie bei dem Rückzug der deutschen Truppen aus Rußland nach Deutschland und wurde dort während eines Bombenangriffes verschüttet. Dabei verlor sie die Sehkraft des rechten Auges.

Mit 21 Jahren heiratete sie einen Landwirt, ein Ereignis, das sie als einen sozialen Abstieg erlebt. In den Jahren von 1956 bis 1960 bekam sie 4 Kinder; bei allen Geburten habe es erhebliche Komplikationen gegeben. Dabei sollen auch die Vernarbungen am rechten Auge aufgebrochen sein.

Etwa seit dem 36. Lebensjahr leidet sie zunehmend unter Atemnot nach körperlicher Belastung und unabhängig davon an anfallsweise auftretendem Herzjagen. In den letzten 1½ Jahren fühle sie sich in ihrer Leistungsfähigkeit durch Zunahme der Atemnot und des Herzklopfens erheblich beeinträchtigt. Außerdem leide sie unter starker Müdigkeit.

Die oben geschilderten Anfälle treten in seelisch belastenden Situationen auf. Die Belastungen können – von außen betrachtet – geringfügig sein, so kam es auch bei der Ankündigung der Untersuchung in der psychosomatischen Abteilung zu einem Anfall. Sie wird im Rollstuhl zum Interview gebracht und hat versteifte Hände (Pfötchenstellung), als sie in das Zimmer gefahren wird. Im Laufe des Gespräches löst sich die Verkrampfung der Hände und die Patientin wirkt entspannter. Bei der Schilderung von Ereignissen, die sie emotional bewegen – als sie von dem Tod des Vaters spricht oder die Verschüttung während des Bombenangriffes schildert – beginnt die Patientin auffallend tief und schnell zu atmen und es kommt erneut zu einer Versteifung der Hände.

Für ihre gegenwärtige Lebenssituation scheint es besonders bedeutsam zu sein, daß sie die Ehe mit dem ihr sozial und intellektuell unterlegenen Mann als außerordentlich bedrückend erlebt. Sie hat ihn offenbar nur unter dem Zwang äußerer Verhältnisse geheiratet. Dabei hat das Gefühl eine Rolle gespielt, daß sie sich wegen der Erblindung des rechten Auges als Krüppel fühlte, der froh sein mußte, überhaupt einen Mann zu finden. In der Ehe komme es ständig zu Reibereien, vor allem über die Erziehung der Kinder. Der Mann verstünde nichts von Erziehung; er habe nur einen „verdummenden" Einfluß. Im Gegensatz zu ihrem Mann fühlt sie sich dafür verantwortlich, daß die Familie ein gewisses soziales Niveau aufrechterhält. Sie scheint das Gefühl zu haben, ihr eigentliches Leben sei durch den Krieg zerstört worden, ihre körperliche Beeinträchtigung und als Folge davon die Heirat hätten sie zu einer Art von „Schattendasein" verdammt. Ihre Kinder aber sollten einmal das Leben führen können, das das Schicksal ihr vorenthalten habe. So lebt sie in dem Konflikt zwischen beherrschen, d.h. unbedingt die Kontrolle haben zu müssen, und der Ohnmacht ihren eigenen Lebensumständen, z.B. auch dem Mann gegenüber.

Sehr wahrscheinlich kam es durch das Erlebnis der Operation zu einer „resonanzhaften" Reaktivierung der offenbar mit heftiger Angst erlebten Ohnmachtsgefühle während des Verschüttetseins in den letzten Kriegstagen. Man könnte auch daran denken, daß – in Zusammenhang mit der urologischen Operation – Kastrationsphantasien und Ängste eine Rolle spielten. Die Frage, was der Tod des Vaters für die Phantasieentwicklung der Patientin und für ihr Verhältnis mit dem Ehemann bedeutete, läßt sich aus den Informationen, die während des Interviews gewonnen wurden, nicht eindeutig beantworten.

Die tetanischen Symptome im Zusammenhang mit verstärkter Atmung sind typisch für ein Hyperventilationssyndrom.

34.1 Klinik und Symptomatologie

Die Beschwerden dieses Krankheitsbildes, das definiert ist durch eine über das physiologische Bedürfnis hinausgehende Beschleunigung und Vertiefung der Atmung, die zur Verminderung des Kohlendioxids im Blut führt, lassen sich folgendermaßen einteilen:

34.1.1 Neuromuskuläre Symptome

Recht charakteristisch sind Beschwerden wie „Ameisenlaufen", Gefühllosigkeit und Zittern an Händen (besonders in den Fingerspitzen) und Füßen, wobei in manchen Fällen die Parästhesien als einziges Symptom geschildert werden und zum Arztbesuch führen. Häufig ist aber auch ein Kribbeln um die Mundregion, wobei vor allem die Lippen, manchmal auch die Zunge betroffen sind. Ausgeprägtere Tetaniesymptome, die auch einseitig auftreten können, wie Verkrampfungen der Akren, Lähmungen und eine motorische Unfähigkeit zu sprechen, sind eher selten. Zittern und Muskelschmerzen werden ebenfalls angegeben.

34.1.2 Zentrale Symptome

Häufig wird über Sehstörungen berichtet und das Gefühl wie auf „Wolken zu gehen", ein Zustand, den der Volksmund als „Mattscheibe" bezeichnet. Darüber hinaus klagen viele Patienten über Benommenheit, Kopfschmerzen und Schwindel, der weder ein Dreh- noch ein Schwankschwindel ist.

34.1.3 Respiratorische Beschwerden

Sehr häufig wird über Atemnot geklagt, wobei die Hyperventilation selbst selten bemerkt wird. Die Atemnot ist auch meistens der Grund für den Arztbesuch. Meist besteht eine Tachypnoe, die von wiederholtem Seufzen, Gähnen oder Schnupfen, gelegentlich auch einem eigentümlichen, trockenen abgehackten Hüsteln (Hoff et al., 1952) begleitet sein kann. Klagen über Lufthunger und den Zwang, tief atmen zu müssen, verbunden mit einem Engigkeitsgefühl über der Brust (Gürtel- oder Reifengefühl), bzw. das Gefühl „nicht richtig durchatmen zu können" werden häufig angegeben. Die Patienten zeigen mitunter auf einen Punkt unter dem Zwerchfell, den sie bei der Atmung erreichen wollen, aber nicht erreichen können.

34.1.4 Kardiale Beschwerden

Die Hyperventilation wird nicht selten von Herzklopfen und Herz- und Thoraxschmerzen begleitet, die oft schwer von einer Angina pectoris abgrenzbar sind. Die Kombination eines Hyperventilationssyndroms mit einem funktionellen kardiovaskulären Syndrom ist häufig (Hoff et al., 1952), oder das Hyperventilationssyndrom wird im Sinne eines Symptomwandels von einem funktionellen kardiovaskulären Syndrom abgelöst (Weimann, 1968). Wie bei diesem Syndrom wird der Thoraxschmerz von den Patienten mit Hyperventilation in zwei Qualitäten erlebt: entweder als stechender, kurzanhaltender Schmerz oder als dumpfes Druckgefühl über dem Herzen, eventuell mit Ausstrahlung in den Oberbauch und in den linken Arm (vgl. Kap. 32).

34.1.5 Neurovegetative Beschwerden

Sehr häufig sind Klagen über kalte Hände und Füße, die Patienten fühlen sich dadurch oft sehr belästigt, sowie Schwitzen und Harndrang.

34.1.6 Gastrointestinale Symptome

Viele Patienten leiden unter funktionellen Oberbauchbeschwerden (vgl. Kap. 33), die meist durch eine Aerophagie bedingt sind und zu Aufstoßen, Meteorismus, Flatulenz und Dysphagie bis hin zu Anorexie und Nausea führen (Brashear, 1983; Radvila, 1984).

34.1.7 Allgemeine und psychische Beschwerden

Fast immer wird über Müdigkeit, Schlappheit, Schläfrigkeit und Wetterfühligkeit geklagt. Viele Patienten haben Mühe, sich zu konzentrieren, sind vergeßlich, reizbar und gespannt. Affektiv sind sie oft ängstlich und depressiv, nicht selten entwickeln sie Phobien, Agora- und Klaustrophobien vor allem, oder leiden unter Panikzuständen (Sheehan, 1982).

34.1.8 Der akute Anfall

Der Arzt, der zu einem Patienten mit einem akuten Hyperventilationsanfall gerufen wird, findet einen ängstlichen und unruhigen Patienten mit schneller, unregelmäßiger Atmung, der über folgende Beschwerden klagen kann:

Seine Finger, Hände, Füße und Beine würden absterben. Das Herz würde klopfen, Atemnot und ein Druck auf der Brust würden ihn zwingen, dauernd und schnell zu atmen, um nicht zu ersticken (Hayn, 1974). Die Lippen seien taub, der Mund nicht richtig beweglich, das Gesicht würde sich steif anfühlen. Er hätte Schwindel, Druck im Kopf und im Oberbauch (viszerale Tetanie), Aufstoßen, Übelkeit, trockenen Mund und Kraftlosigkeit.

Sehr ausgeprägt ist meist die Ängstlichkeit der Patienten, wobei die Angst ansteckend auf die Umgebung wirkt. Auch bei der oben geschilderten Patientin war diese Dramatik vor und während des Interviews deutlich feststellbar.

34.1.9 Das chronische Hyperventilationssyndrom

Patienten, die chronisch hyperventilieren, haben oft keine eindeutig abgrenzbaren akuten Anfälle. Ihre Beschwerden sind unspezifischer und vage, beinhalten selten Atemstörungen oder Tetaniezeichen, so daß der Arzt Mühe hat, sie einem bestimmten Krankheitsbild zuzuordnen. Leitsymptome sind: Schwindel, Thoraxschmerzen, kalte Extremitäten und die oben beschriebenen, psychischen Störungen, welche oft zu neurologischen, kardiologischen und psychiatrischen Abklärungen führen (Radvila, 1984). Der Hyperventilations-Provokationstest führt häufig nicht zu den typischen Symptomen, da sich der Organismus an die chronische respiratorische Alkalose gewöhnt und sie teilweise metabolisch kompensiert.

34.2 Epidemiologie

Die meisten Untersuchungen geben an, daß das Krankheitsbild bei Frauen etwa dreimal so häufig (Weimann, 1968) vorkomme wie bei Männern. In einer Untersuchung, die mehr als 700 Fälle umfaßt, war die Häufigkeit gleichmäßig auf die beiden Geschlechter verteilt (Lum, 1976). Mit fortschreitendem Alter nimmt die Häufigkeit bei beiden Geschlechtern ab:

Wie bei den Patienten mit anderen funktionellen Syndromen findet sich ein Hyperventilationssyndrom vor allem im 2. und 3. Lebensjahrzehnt, bei über 60jährigen ist es eher selten (Weimann, 1968). Nach der oben erwähnten Untersuchung von mehr als 700 Fällen ist die Altersverteilung etwas anders. Nach ihr sind die Jahrgänge zwischen 30 und 60 Jahren etwa gleich häufig betroffen, während die Hyperventilation zwischen dem 60. und 69. Lebensjahr zwar seltener, aber doch noch ebenso häufig vorkommt wie zwischen dem 20. und 29. Lebensjahr.

Die Häufigkeit der Hyperventilation wird mit 6 bis 10% der Patienten eines internistischen Ambulatoriums angegeben (Lum, 1976). Obwohl uns diese Zahl etwas zu hoch erscheint, zeigt die Erfahrung, daß es sich bei der Hyperventilation um eine häufige, alltägliche Erscheinung handelt, die zu einer erheblichen Morbidität führt (Radvila, 1984).

34.3   Theorien zur Ätiologie und Pathogenese

34.3.1   Psychische Faktoren

S. Freud beschrieb 1894 unter den klinischen Symptomen der Angstneurose auch Störungen der Atmung, die er als „nervöse Dyspnoe" bezeichnet. Er hebt hervor, „daß selbst diese Anfälle nicht immer von kenntlicher Angst begleitet sind".

Als häufigste ätiologische Faktoren der Hyperventilation werden Emotionen, vor allem Angst und Aufregung beschrieben (Adlersberg und Porges, 1924; Cannon, 1928; Kerr et al., 1937; Stead und Warren, 1943; Christian et al., 1955; Dudley et al., 1964).

Alexander und Mitarbeiter haben Anfang der vierziger Jahre darauf hingewiesen, daß sexuelles Verlangen und Abhängigkeitsgefühle einen spezifischen Einfluß auf die Atmung ausüben. Nach ihnen ist Hyperventilation symbolischer Ausdruck eines gefühlsbesetzten, neurotischen Konfliktes.

Hoff und Mitarbeiter (1952) beschrieben, daß Patienten mit Hyperventilationssyndrom einen psychischen Konflikt „nicht lösen", sondern nur „abatmen - ausseufzen" können. Es fände bei ihnen eine neurotische Flucht vor Entscheidungen statt, die Flucht in eine beschleunigte Atmung oder Hyperventilation sei ein Ausweichen vor einer direkten Auseinandersetzung mit realen Gegebenheiten. So werde die funktionelle Störung fixiert und es komme später schon bei geringen psychischen Belastungen zu Hyperventilation. Zuerst sei die beschleunigte Atmung somit Antwort auf Schmerz, Wut oder Angst, in der auch Hoff den bedeutendsten Faktor für die Auslösung dieses Krankheitsbildes sieht. In der späteren Entwicklung würden die Atembeschwerden dann in jeder unangenehm erlebten Situation auftreten.

Lum (1976) betont, daß die Hyperventilation in einer Vielzahl klinischer Situationen und in Verbindung mit verschiedenartigen Persönlichkeitsfaktoren und emotionalen Störungen vorkomme. Er legt besonderes Gewicht auf die Feststellung, daß es sich um

eine Gewohnheit handle, die wie alle einer willkürlichen Beeinflussung zugänglichen Funktionen durch Konvention, Training oder auch Vorstellungen über Gesundheit, Tüchtigkeit usw. zustande kommen könne. Die gewohnheitsmäßige Hyperventilation würde dann zu einer ständigen Disposition für das Auftreten der typischen Beschwerden in Situationen führen, in denen emotionelle Faktoren der verschiedensten Art eine weitere Steigerung der Atmung hervorrufen.

Sheehan schlug 1982 vor, das Hyperventilationssyndrom wie die Herzneurose und das Colon irritabile einem endogenen Angstsyndrom mit Panikzuständen zuzuordnen. Die bei Hyperventilationspatienten häufigen Phobien erklärte er mit durch Angst und Panik konditionierten Reaktionen auf eine bestimmte Situation wie Menschenansammlungen, Lifte oder Autofahren (vgl. Kap. 15 und 32).

Die Symptome können Todesangst hervorrufen und als Strafe für bewußte oder unbewußte ambivalente Gefühle gegenüber verlorenen oder entfremdeten Bezugspersonen der Kindheit erlebt werden.

34.3.2   Pathophysiologie

Die auslösende Ursache ist die verstärkte und beschleunigte Atmung.

Zwei charakteristische Atemtypen werden beschrieben:

- Die Angstpolypnoe mit unruhiger Hyperventilation. Sie wird als spezifischer Ausdruck von Angst (Angstneurose, Angsthysterie) aufgefaßt.
- Die flachfrequente Polypnoe mit Seufzerzügen, die Christian (1957) beschrieben und als Ausdruck einer persönlichen Situation gedeutet hat, die durch Abgespanntheit und Resignation gekennzeichnet sei, in der „trotz Anstrengung gesteckte Ziele nicht mehr erreicht werden können".

Von besonderer Bedeutung ist jedoch die Beobachtung, daß der Atemtyp bei diesen Patienten verändert ist: Sie atmen hauptsächlich mit dem Thorax und kaum mit dem Zwerchfell. Bei Patienten mit chronischem Hyperventilationssyndrom wird eine Zwerchfellatmung in weniger als 1% gefunden (Lum, 1976). Diese Feststellung hat erhebliche diagnostische Bedeutung. Pathogenetisch ist die Feststellung wichtig, daß bei Menschen, die mit dem Thorax atmen, der P_{CO_2} unter 40 mm Hg zu liegen pflegt.

Beim Hyperventilationssyndrom kommt diese gesteigerte Ventilation durch eine hohe Atemfrequenz mit inspiratorischer Verschiebung der Atemruhephase – vor allem auch durch den veränderten Atemtypus zustande. Bei der Gasanalyse des arteriellen Blutes findet sich eine respiratorische Alkalose mit herabgesetzter CO_2-Spannung. Lum (1976) fand bei 200 Patienten mit Anfällen von Hyperventilation in der anfallsfreien Zeit einen P_{CO_2} von im Durchschnitt 33 mm Hg im Unterschied zu 152 Normalpersonen mit einem P_{CO_2} von 40,7 mm Hg, ein Befund, der von Radvila und Mitarbeitern (1983) bestätigt werden konnte.

Es lassen sich zwei Formen der Hyperventilation unterscheiden:
– Eine vergrößerte alveoläre Ventilation und
– eine bloße Zunahme der Ventilation des Residualvolumens. Dies kommt beispielsweise bei dem Hecheln der Hunde vor.

Nur die alveoläre Hyperventilation führt zu einer Senkung des arteriellen CO_2-Partialdruckes, die sekundär folgende Veränderungen nach sich zieht:
– Respiratorische Alkalose mit Abfall des ionisierten Phosphors und der organischen Phosphate, sowie inkonstant mit anderen Störungen des Mineralhaushaltes (Kalzium, Magnesium). Die chronische respiratorische Alkalose kann renal durch vermehrte Bikarbonatausscheidung teilweise kompensiert werden (Magarian, 1982).
– Neuromuskuläre Übererregbarkeit mit tetanischen Symptomen (Parästhesien, Chvostek- und Trousseau-Phänomen, Karpopedalspasmen, Pfötchenstellung), hervorgerufen durch die Alkalose per se oder eine andere Ionenverschiebung.
– Änderungen der regionalen Durchblutung: Bei akuter alveolärer Hyperventilation nimmt die Gehirndurchblutung ab, was klinisch zu einem Präkollaps oder sogar zu einer Ohnmacht führen kann. Die verminderte Durchblutung im Hyperventilationsversuch läßt sich im Elektroenzephalogramm nachweisen, aber auch im Arteriogramm unmittelbar anschaulich machen (Lum, 1976). Das wird verständlich, wenn man sich klarmacht, daß der stärkste Reiz für die Gehirndurchblutung Änderungen der CO_2-Konzentration im Blut sind. Engel und Mitarbeiter (1947) konnten nachweisen, daß die Schwere der EEG-Veränderungen mit dem Grad der Bewußtseinsstörung annähernd parallel geht. Die Klagen über Schwindel, „Mattscheibe" und andere zerebrale Symptome werden dadurch verständlich.
Aber auch die Hautdurchblutung wird durch Hyperventilation verändert. Ihre Abnahme kann zu einem deutlichen Abfall der Hauttemperatur und zu einer Akrozyanose führen (Weimann, 1968).
– Aktivation des Sympathikus: Hyperventilation aktiviert das sympathische System. Dadurch kommt es zu einem Pulsanstieg und unter Umständen zu EKG-Veränderungen mit Senkung der ST-Strecke, T-Inversion und mit Extrasystolen. Der Mechanismus dieser EKG-Veränderungen ist unklar geblieben; möglicherweise kommen sie durch eine verminderte Koronardurchblutung zustande (Lary und Goldschlager, 1974).
– Stoffwechselveränderungen, z.B. der Laktatkonzentration im Serum.

Lewis (1957) hebt hervor, daß nicht nur Angst zu Hyperventilation führt, sondern daß die dadurch ausgelösten Symptome wieder die Hyperventilation verstärken und verlängern. Dadurch entsteht ein Circulus vitiosus, wobei die Patienten das Hyperventilieren meist nicht bewußt wahrnehmen. Schematisch wird der Ablauf des Geschehens, das zu Symptomen führt, in Abbildung 34–1 dargestellt.

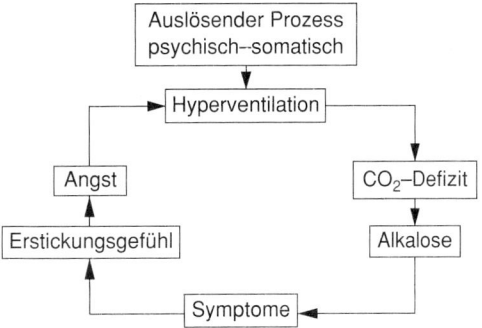

Abb. 34–1. Schema des Ablaufs des zu Symptomen führenden Geschehens (modifiziert nach Lewis und Siegenthaler).

34.3.3 Hyperventilation als Störung im Funktionskreis der Atmung

Bei jeder Störung der Atmung stellt sich zunächst die allgemeine Frage, welche Rolle die Atmungsfunktion in der Gesamtsituation des Patienten, d.h. in seiner Auseinandersetzung mit inneren (zum Teil auch erlebten Bedürfnissen, Sensationen usw.) und äußeren (Umgebungs-)Faktoren spielt. Das in Kapitel 1 entwickelte Modell des Funktions- und Situationskreises hilft uns den Zusammenhang zwischen all diesen Einzelfaktoren zu sehen, zu ordnen und entsprechend ihrer Bedeutung für den einzelnen Fall zu gewichten.

Wir haben in Kapitel 1 darauf hingewiesen, daß der Funktionskreis der Atmung im Unterschied zu den Funktionskreisen der Nahrungsaufnahme, der Ausscheidung und der Sexualität in der normalen menschlichen Entwicklung nicht sozialisiert wird und daher zeitlebens eine archaische, gewissermaßen primärprozeßhafte Dynamik beibehält. Das bedeutet jedoch nicht, daß dieser Funktionskreis von anderen gleichzeitig ablaufenden Vorgängen während der Auseinandersetzung des Individuums mit den inneren und äußeren Faktoren isoliert sei. Er ist über den O_2-Bedarf des Körpers, die CO_2-Spannung, den pH-Wert des Blutes direkt und indirekt von zahlreichen somatischen Abläufen abhängig. Er ist reflektorisch (durch angeborene Verbindungen) mit dem Kreislauf und dem Schmerzgeschehen verflochten. Er nimmt an Stimmungsschwankungen teil, wobei die Sexualität eine besondere Rolle spielt, und er ist aufs engste mit dem Ausdrucksgeschehen verknüpft. Schließlich wird die Atmung – wie Haltung und Gang – durch persönliche Gewohnheiten geprägt.

Es gibt also zahlreiche Faktoren, welche die Atmung direkt und indirekt beeinflussen. Um Störungen, die den Funktionskreis der Atmung selbst betreffen, von Störungen abzugrenzen, die auf seiner Verbindung mit anderen Funktionen beruhen, ist folgende Überlegung nützlich: Der Funktionskreis der Atmung wird durch ein Bedürfnis in Gang gesetzt, unangenehme Sensationen zu beseitigen, die sich bei zunehmender Intensität rasch zu dem Gefühl unmittelbarer Lebensbedrohung („Atem-Not") steigern. Unter diesem Gesichtspunkt lassen sich für differentialdiagnostische und therapeutische Überlegungen die verschiedenen Störungsmöglichkeiten nach folgendem Schema ordnen:

1. Störungen, die in dem Funktionskreis Atmung selbst angreifen (z.B. O_2-Bedarf; CO_2-Gehalt des Blutes; reflektorische Hemmungen, z.B. durch Reizung der Bronchial- oder Nasenschleimhaut; Entzündungen, Fremdkörper, Gerüche usw.).

2. Störungen durch (angeborene oder erworbene) Koppelungen an andere Funktionskreise, z.B. im Rahmen von Bereitstellungen zu Kampf oder Flucht.

3. Pathologische Entwicklung des Funktionskreises durch Einbau von Beziehungspersonen oder Objekten (Apparaten etc.), von denen der Ablauf der Atmung abhängig wird.

4. Indienstnahme der Befreiungsfunktion von Not durch andere Nöte, deren man (aus verschiedenen Gründen) nicht Herr werden kann und die man nun – etwa im Sinne einer Übersprunghandlung – „abzuatmen" versucht. Bei dem Hyperventilationssyndrom handelt es sich meistens um Störungen der Kategorie 2 oder 4, bzw. um eine Kombination aus beiden.

Von Störungen des Funktionskreises der Atmung, die unter den Ziffern 2–4 angegeben sind, macht die Feststellung der unter 3 genannten keine Schwierigkeiten. Aber auch hier müssen, wie bei den Störungen, die unter 3 und 4 aufgezählt sind, Persönlichkeitsfaktoren des Patienten und seine psychosoziale Situation erkundet werden.

34.4 Diagnose

Wie bereits oben erwähnt, ist der Atemtyp – nämlich die Thoraxatmung und die geringe oder fehlende Bauchatmung – ein besonders wichtiges Merkmal: Das hilft dem Arzt seinen Verdacht auf das Vorliegen eines Hyperventilationssyndroms auch in anfallsfreien Zeiten zu untermauern, in denen die Patienten nur über die oben erwähnten subjektiven Beschwerden klagen.

Der akute Anfall mit hoher Atemfrequenz ohne Zyanose, inspiratorischer Verschiebung der Atemlage, Tonuserhöhung der Muskulatur, die bis zur Tetanie führen kann, gesteigerten Reflexen und positivem Chvostek und Trousseau, Karpopedalspasmen, Karpfenmaul, Tremor und Hypothermie der Akren in wechselndem Ausmaß bietet kaum diagnostische Schwierigkeiten. In der arteriellen Blutgasanalyse findet sich eine respiratorische Alkalose mit stark erniedrigter CO_2-Spannung. Eine Erniedrigung des P_{CO_2} kann – wie oben erwähnt – auch im anfallsfreien Intervall gefunden werden. Zur Messung des P_{CO_2} über einen längeren Zeitabschnitt eignet sich besonders die nichtinvasive Methode mit kutaner Hautelektrode (Radvila et al., 1983).

Nach Weimann (1968) ist die Beobachtung, daß das Hyperventilationssyndrom gehäuft bei jungen Frauen vorkommt, ebenfalls ein diagnostischer Hinweis. Die Beobachtungen Lums (1976) sprechen jedoch dafür, daß man auch bei Männern häufiger an dieses Syndrom denken muß, als es bisher üblich war.

Der Verdacht läßt sich durch den Hyperventilationsversuch objektivieren: Durch bewußte Hyperventilation lassen sich bereits nach wenigen Minuten Pupillenerweiterung, kalte Extremitäten, Schwitzen an Handflächen und in den Achseln sowie Tachykardie bei allen Menschen hervorrufen. Für das Vorliegen eines Hyperventilationssyndroms ist beweisend, daß der Patient das Auftreten der ihm bekannten Beschwerden feststellt: Kribbeln an den Fingern, Armen und Füßen, Verkrampfung der Finger bzw. Hände, Verkrampfung des Mundes, Benommenheit, Schwindel, Schwarzwerden vor den Augen, Unvermögen durchzuatmen, Herzdruck, Herzklopfen, Globusgefühl und Angst. Wie erwähnt, können beim chronischen Hyperventilationssyndrom oft nur schwache oder keine Symptome durch willkürliche Überatmung ausgelöst werden.

Der manchmal nach Strumektomie oder auch spontan auftretende Hypoparathyreoidismus, bei dem im Hyperventilationsversuch die oben geschilderten Symptome ebenfalls rasch auftreten, kann leicht durch das – im Gegensatz zum funktionellen Atmungssyndrom – erniedrigte Serum-Kalzium ausgeschlossen werden.

Wie bei allen funktionellen Syndromen, so gilt auch hier, daß die Diagnose nicht nur durch den Ausschluß einer organischen Krankheit, sondern auch durch positive psychische, vor allem depressive und angstbezogene Symptome gestellt werden muß. Allerdings haben bei der Polyätiologie dieses Krankheitsbildes und bei den charakteristischen subjektiven und objektiven Beschwerden psychische Symptome für die Diagnose häufig nur einen ergänzenden Charakter. Sie sind jedoch unerläßlich für die Aufstellung eines vernünftigen Therapieplanes.

34.4.1 Differentialdiagnose

Grundsätzlich müssen bei jeder Hyperventilation somatische Krankheitsbilder (Störungen des Funktionskreises der ersten Form) in die differentialdiagnostischen Überlegungen eingeschlossen werden, obwohl bei über 95% der Patienten ein psychisch bedingtes Hyperventilationssyndrom vorliegt.

Die folgende Zusammenstellung gibt einen Überblick über die Krankheitsbilder, an die bei tetanischen Symptomen gedacht werden muß:

Tetanische Symptome mit alveolärer Hyperventilation
– psychisch bedingte Hyperventilation
– direkte Stimulierung des Atemzentrums durch lokale Prozesse (Enzephalitis, Tumor)

Alveoläre Hyperventilation ohne tetanische Symptome (kompensatorische Hyperventilation)
– Gewebshypoxie
– arterielle Hypoxämie (atmosphärisch, pulmonal, kardial)
– arteriovenöse O_2-Differenz vergrößert (z.B. bei Anämie)

– vermehrter peripherer O_2-Bedarf (z. B. bei Muskelarbeit, Fieber)
– metabolische Azidose

Tetanische Symptome ohne alveoläre Hyperventilation (Ziegler, 1976)
Bei Normokalzämie
– relativer Parathormon-Mangel: latenter Hypoparathyreoidismus
– Magnesiummangel
– Hyperkaliämie
– Infektionskrankheiten (Tetanus!)
– Intoxikationen
– Alkalose (z. B. HCl-Verlust)
Bei Hypokalzämie
– strumipriver Hypoparathyreoidismus
– idiopathischer und sekundärer Hypoparathyreoidismus
– verminderte Ca-Aufnahme (Mangelernährung, Malabsorption)
– Ca-Sog in die Knochen (Heilungsphase der Rachitis, insbesondere Osteomalazie, Zustand nach Operationen bei primärem Hypoparathyreoidismus)

34.5 Therapie

34.5.1 Symptomatisch

Während des Anfalls sollte zunächst versucht werden, den Patienten zu beruhigen. Das Auftauchen des Arztes genügt bereits häufig, um den Anfall zu beenden. Als einfachste Behandlungsmethode – auch in der Klinik – hat es sich bewährt, den Patienten in eine Tüte atmen zu lassen, um so die Kohlensäurespannung im Blut wieder zu erhöhen. Nach Hayn (1974) soll es jedoch vorkommen, daß diese Methode die Angst des Patienten vermehrt. Er schlägt daher vor, statt dessen die untere Thoraxapertur mit beiden, flach angelegten Händen von beiden Flanken nach der Wirbelsäule zu kräftig zu komprimieren und den Thorax in dieser Kompressionsstellung etwa 2–3 Minuten fest zusammengedrückt zu halten. Dadurch werden die Atemexkursionen des Brustkorbs und das Atemvolumen wesentlich vermindert und die forcierte Atmung des Patienten unterdrückt. Wenn der Patient etwa nach einer Minute bereits eine Besserung verspürt, kann die Kompression gelockert und nach etwa 3 Minuten die Atmung völlig freigegeben werden. Bei dieser Maßnahme ist allerdings die symbolische Bedeutung zu beachten, die ein derartiger in engem körperlichem Kontakt ausgeübter Zwang für den Patienten haben kann. Die Methode sollte daher nur zur Anwendung kommen, wenn andere Maßnahmen versagen. Die häufig geübte Praxis der intravenösen Applikation von 10 ml einer 10%igen Kalziumlösung ist nicht gerechtfertigt, da es sich hierbei im wesentlichen um einen Placeboeffekt handelt: die Kalziuminjektion führt zu einem subjektiven Wärmegefühl, das sich bei Patienten, die sich vom Absterben ihrer Hände und Füße bedroht fühlen, günstig auswirken kann.

34.5.2 Behandlung der gewohnheitsmäßigen Thoraxatmung

Da die Mehrzahl der Patienten zu wenig Kontrolle über ihr Zwerchfell haben und es bei der Atmung zu wenig benutzen, ist eine konsequente Atemtherapie von großer Bedeutung. Der Therapeut muß die Patienten lehren, mit dem Zwerchfell zu atmen und vor allem in Ruhe ausschließlich die Zwerchfellatmung zu betätigen. Lum (1976) empfiehlt, die Patienten täglich 2mal 20 Minuten die Atemübungen durchführen zu lassen und während der übrigen Zeit dauernd auf ihre Atmung zu achten. Nach seinen Erfahrungen stellt dann die Mehrzahl der Patienten bald fest, daß die anfangs für sie selbstverständliche Thoraxatmung für sie schwierig wird, was ein günstiges Omen bedeutet. 70% der von ihm nach dieser Methode behandelten Patienten wurden vollständig asymptomatisch und haben ihre Ängstlichkeit verloren. 25% behielten zwar einige Symptome, konnten sie aber durch Überwachung ihrer Atmung unter Kontrolle halten. Nur 5% zeigten keinerlei Besserung.

34.5.3 Psychotherapie

Eine effektive Behandlung verlangt, daß der Patient die Beziehungen zwischen auslösender Situation, Emotion und Hyperventilation durchschaut. Dazu ist es notwendig, die emotionalen Probleme mit dem Patienten zu besprechen, was wiederum eine gute Arzt-Patient-Beziehung voraussetzt, bei der der Arzt vor allem die Geduld nicht verlieren darf. Hoff und Mitarbeiter (1952) schlagen vor, die Patienten über das Wesen der Störung als „eine Gewohnheitsreaktion" aufzuklären und auch die Angehörigen und die Umgebung des Patienten mit in die Behandlung einzubeziehen. Dadurch, daß man den Patienten darauf aufmerksam macht, wenn er zu hyperventilieren beginnt, könne ihm geholfen werden, auf seine Atmung zu achten. Dabei ist allerdings wichtig, daß man dem Patienten und den Angehörigen klarmacht, daß es sich bei seinen Beschwerden nicht um bloße Einbildungen, sondern um echte Symptome handelt, die auch ihre somatischen Äquivalente haben.

Eine tiefenpsychologisch fundierte Therapie ist nur bei schwereren neurotischen Störungen indiziert.

Mit Hilfe spezieller psychotherapeutischer Techniken (z. B. Hypnose oder „Symptom-Verschreibung") ist es möglich, die Angst des Patienten zu vermindern und die Beschwerden und die Häufigkeit der Anfälle zu reduzieren (Compernolle et al., 1979; Wilkinson, 1981). Auch verhaltenstherapeutische Verfahren werden beim Hyperventilationssyndrom mit Erfolg eingesetzt (Walker, 1978; van Doorn et al., 1982).

34.5.4 Psychopharmaka

Drei Gruppen von Psychopharmaka können nach den oben geschilderten psychophysiologischen Mechanismen zur Behandlung des Hyperventilationssyndroms beitragen: Anxiolytika, Antidepressiva und Beta-Blocker. In der Literatur findet man nur wenige kontrollierte Studien über den Einsatz dieser Medikamente beim Hyperventilationssyndrom. Folgering und Cox (1981) behandelten 16 Patienten erfolgreich mit einem kardioselektiven Beta-Blocker in einem doppelblinden, gekreuzten Versuch. Bei ausgeprägten Angst- oder Panikzuständen drängt sich ein Behandlungsversuch mit Benzodiazepinen oder – wie kürzlich propagiert – mit Antidepressiva auf (Sheehan, 1982). Letztere sind sicher auch indiziert bei hyperventilierenden Patienten mit ausgeprägter Depression, ein Bild, das man recht häufig findet (Radvila, 1984).

Allerdings sollte sich der behandelnde Arzt bewußt sein, daß durch eine medikamentöse Therapie mit Psychopharmaka die Arzt-Patient-Beziehung verändert und der psychotherapeutische Zugang erschwert werden kann (vgl. Kap. 24).

34.6 Prognose

Die Prognose des akuten Hyperventilationsanfalles, der nach unterschiedlicher Dauer spontan abklingt, ist immer gut. Organschädigungen durch Hypokapnie sind bisher nicht beschrieben worden. Anders steht es mit der chronischen Hyperventilation. Weimann (1968), der insgesamt 121 Patienten mit einem Hyperventilationssyndrom nach einem Zeitraum von 1–7 Jahren katamnestisch nachuntersucht hat, berichtet über eine Besserung in 65% und ein Verschwinden der charakteristischen Symptomatik in 26% der Fälle. Das Ergebnis fiel bei Kranken, die über den Ventilationsmechanismus nicht aufgeklärt worden waren, wesentlich schlechter aus: Die hyperventilationsabhängigen Symptome waren in 78% unverändert oder sogar verschlechtert. Wie oben erwähnt, kann die Prognose durch konsequente Atemtherapie und Besprechung der emotionellen Probleme des Patienten wesentlich verbessert werden.

Wie bei allen funktionellen Syndromen sind iatrogene Verschlimmerungen dadurch möglich, daß organische Leiden, wie z.B. ein Hypoparathyreoidismus oder eine koronare Herzkrankheit, diagnostiziert und damit die Beschwerden fixiert werden. Bei Patienten mit einer neurotischen Störung ist ohne Psychotherapie eine Chronifizierung des Leidens zu erwarten (Cremerius, 1968; Delius und Fahrenberg, 1966).

35 Synkopen

Claudia Simons und *Karl Köhle*

35.1 Definition und Symptomatik

Flüchtiger Bewußtseinsschwund, der mit einem Verlust sämtlicher Reaktionen, des Tonus der Skelettmuskulatur und des Stehvermögens einhergeht und von dem sich der Patient in der Regel spontan und ohne Notwendigkeit von Wiederbelebungsmaßnahmen erholt (Gurtner, 1984). Eine erweiterte Definition schließt Symptome wie Schwindel, Ohnmachtsneigung, Kraftlosigkeit und Beschwerden im Sinne einer reduzierten Bewußtseinslage mit ein. Der Wechsel von Bewußtseinstrübung zu völliger Bewußtlosigkeit erfolgt nicht plötzlich. Es gibt alle Übergänge, die nur rascher oder langsamer durchlaufen werden, wobei auf jeder Stufe Rückbildung möglich ist. Die Diagnose einer „Synkope" sollte nicht nur bei Bewußtseins- und Tonusverlust gestellt werden, da die Vielzahl von Beschwerdebildern, bei denen es zu keinem vollständigen Bewußtseinsverlust kommt, sonst der Diagnostik und Therapie entgeht. Im einzelnen variiert die Symptomatik entsprechend den zugrundeliegenden Krankheitsbildern, die angegebenen Leitsymptome sind jedoch regelmäßig vorhanden.

Synonyma: „Ohnmacht", „Fainting".

35.2 Pathogenetische Mechanismen

Nach den pathogenetischen Mechanismen bietet sich folgende Einteilung an (nach Engel, 1962):
– Verminderung des Gehirnstoffwechsels als Folge unzureichender Durchblutung bzw. Sauerstoffversorgung des Gehirns;
– Verminderung des Gehirnstoffwechsels als Folge allgemeiner oder lokaler Stoffwechselstörungen;
– direkte oder reflektorische Einwirkung auf Teile des zentralen Nervensystems, die mit der Regulation des Bewußtseins und des körperlichen Gleichgewichts zu tun haben;
– psychische Mechanismen, die den Bewußtseinszustand und die Wahrnehmungsfunktionen beeinträchtigen.

35.3 Klassifikation

Bei den einzelnen Krankheitsbildern sind häufig mehrere dieser Mechanismen beteiligt. Deshalb wird im folgenden eine Klassifikation nach klinischen Gesichtspunkten gewählt (nach Engel, 1962).
1. Synkopen bei Insuffizienz der peripheren Kreislaufregulation verschiedenster Genese (z.B. akute Blutung, Exsikkose, postinfektiöse Kreislaufschwäche und andere Formen der Orthostase)
2. Synkopen kardialer Genese
   – bradykarde und tachykarde Rhythmusstörungen
   – Koronarinsuffizienz und Herzinfarkt
   – Aortenstenose, Mitralstenose
   – angeborene Herzvitien
3. Synkopen bei Störungen der Atmungsfunktion und bei Lungenerkrankungen
   – Hyperventilation
   – Verletzungen des Larynx
   – Synkopen nach Husten, Niesen oder Lachen (der „Lachschlag"); analog (postpressorisch): Synkopen nach Defäkation und Miktion
   – Lungenerkrankungen: Lungenembolie, pulmonale Hypertension
4. Synkopen bei Erkrankungen des Gehirns
   – Verschlüsse im Bereich des Systems der Arteriae carotis, vertebralis, basilaris
   – intrakranielle Gefäßerkrankungen
   – intrakranielle raumfordernde Prozesse
   – Überempfindlichkeit des Karotissinus
   – Stoffwechselstörungen (Hypoglykämie)
5. Synkopen primär psychischer Genese
   – vasovagale Synkope („Vasodepressor-Synkope")
   – konversionsneurotische Synkope
   – Synkope unklarer Genese bei akutem Streß

Diese Einteilung ist als idealtypisch anzusehen; klinisch sind die einzelnen Formen oft nicht scharf voneinander zu trennen.

In diesem Kapitel werden die unter psychosomatischen Gesichtspunkten besonders relevanten Synkopen primär psychischer Genese besprochen.

35.4 Inzidenz und Prognose im klinischen Bereich

Da es sich bei Synkopen oft um einmalige oder seltene Ereignisse handelt, fehlen genaue epidemiologische Angaben. Die Angaben für den klinischen Bereich beziehen sich auf Inzidenz und Prognose von Synkopen verschiedener Genese.

Eine relativ große Zahl von Synkopen bleibt trotz intensiver Diagnostik ungeklärt.

Kapoor und Mitarbeiter (1983) konnten in einer prospektiven Studie nur bei 107 von 204 Patienten erklärende Ursachen für die Synkope finden. Es handelte sich um ein gemischtes Krankengut (Ambulanz, Aufnahmestation, „emergency room"; Alter: 55,8 ± 16,6 Jahre). Von den 107 diagnostisch geklärten Fällen entfielen 53 auf kardiovaskuläre, 54 auf nicht-kardiovaskuläre Synkopen. In der letzten Gruppe befanden sich 9 Patienten mit vasovagaler Synkope. Diese Form wurde nur dann diagnostiziert, wenn ein auslösendes Ereignis (Angst, Schmerz, medizinische Maßnahmen) eruiert werden konnte. Die kardiovaskuläre Gruppe zeigte ein höheres Durchschnittsalter und eine signifikant höhere Todesrate nach 12 Monaten.

Day und Mitarbeiter (1982) fanden bei fast 40% von 198 Patienten, die im „emergency room" gesehen wurden, vasovagale oder psychogene Synkopen. Bei 85% der Patienten konnte die Diagnose aufgrund der Anamnese und der körperlichen Untersuchung gestellt werden.

Silverstein und Mitarbeiter (1983) berichteten über eine Gruppe von 108 Aufnahmen auf Intensivstationen wegen Synkopen (Durchschnittsalter 67 Jahre). Auf kardiovaskuläre Synkopen entfielen 36%, auf nicht-kardiovaskuläre 17%, 47% der Fälle blieben ungeklärt. Die Autoren gehen ebenfalls davon aus, daß aufwendige Untersuchungsmethoden nur in einer geringen Zahl der unklaren Fälle weiteren Aufschluß bringen. Mortalität nach Entlassung (Ein-Jahres-Katamnese): In der kardiovaskulären Gruppe betrug die Mortalität 18,5%, in der nicht-kardiovaskulären Gruppe 5,9% und 6,3% in der Gruppe der ungeklärten Fälle. Die altersstandardisierte Mortalitätsrate der ungeklärten Fälle betrug etwa das Anderthalbfache der weißen US-Bevölkerung. Entsprechend dieser Stichproben besteht kein signifikant erhöhtes Risiko eines plötzlichen Todes bei ungeklärten Fällen.

Eagle (1983) schließt alle Fälle von Synkopen mit klassischen Prodromalerscheinungen wie Nausea in die Gruppe der vasovagalen Synkopen ein. Bei einer Untersuchung von 178 Patienten, die im „emergency room" gesehen wurden, kommt er zu folgender Verteilung: 54% vasovagale, 22% unklare Synkopen, bei letzteren bestand eine geringfügig höhere Mortalität nach 10 Monaten.

Bertel et al. (1985) untersuchten an der Medizinischen Poliklinik der Universität Basel innerhalb von 32 Monaten 105 konsekutive Patienten (56 Männer, 49 Frauen im Alter zwischen 15 und 87 Jahren) mit Synkopen. Dies waren 0,75% aller Erstkonsultationen während dieses Zeitabschnittes. Bei 23% gelang keine ätiologische Zuordnung. Eine rein kardiale Ursache fand sich bei 15%, ein vasovagaler Mechanismus bei 17%, eine Orthostase bei 17%, eine Epilepsie bei 12%. Hyperventilation und Miktion wurden bei je 6% als ätiologisch entscheidend angesehen; der Rest verteilte sich auf seltenere Krankheitsbilder.

Gegen die generelle Anwendung aufwendiger Untersuchungsmethoden wenden sich auch Clark und Mitarbeiter (1980). Von 98 Patienten (25 bis 82 Jahre), die ambulant wegen Schwindelgefühl oder Synkopen gesehen wurden, zeigten zwar 42% Symptome

während der Langzeit-(24 Std.-)EKG-Messung, aber nur in 3 Fällen bestand ein zeitlicher Zusammenhang zwischen den angegebenen subjektiven Beschwerden und den Arrhythmien.

Die Untersuchungsergebnisse basieren auf unterschiedlichen Populationen, verschiedenen Untersuchungsmethoden sowie Definitionen der vasovagalen Synkope. Gemeinsam sind drei Ergebnisse:
– Eine relativ große Zahl der Fälle bleibt ungeklärt.
– Wichtigste diagnostische Instrumente zur Abklärung der weitaus meisten Fälle sind Anamnese und körperliche Untersuchung.
– Der Anteil der psychogenen Synkopen beträgt bei Autoren, die diese Möglichkeit gezielt in ihre diagnostischen Überlegungen einbeziehen, zwischen 30 und 50% der Fälle.

35.5 Vasovagale Synkopen

35.5.1 Symptomatik und Klinik

Die vasovagale Synkope tritt charakteristischerweise bei Personen auf, die sich in einem akuten Angstzustand befinden, dem sie sich nicht entziehen können. Erste Anzeichen sind Muskelschwäche, gefolgt von Nausea, Schweißausbruch, Unruhe, Blässe, Seufzeratmung und Gähnen. Das Ablaufen dieser Sequenz kann zu jedem Zeitpunkt durch Einnehmen der horizontalen Lage unterbrochen werden und ist voll reversibel. Wird der Ablauf nicht unterbrochen, kann eine Abnahme des Muskeltonus und plötzlicher Bewußtseinsverlust innerhalb weniger Minuten, sogar Sekunden, folgen. Dauert die Bewußtlosigkeit länger als 10 bis 20 Sekunden an, so können klonische Muskelkrämpfe auftreten. Die Symptomatik ist auch zu diesem Zeitpunkt noch rasch reversibel, wenn der Patient in die horizontale Lage gebracht wird.

Die vasovagale Synkope zeigt charakteristische Kreislaufveränderungen: Der arterielle Druck sinkt ab, bei systolischen Werten zwischen 60 und 55 mm Hg tritt Bewußtlosigkeit ein. Der Blutdruckabfall wird zunächst von einer Pulsbeschleunigung begleitet; bei Erreichen des kritischen systolischen Wertes sinkt die Pulsfrequenz dann plötzlich auf 30 bis 60 Schläge pro Minute ab. Kontinuierliche EEG-Messungen ergaben Veränderungen der elektrischen Erregung als Folge der Minderdurchblutung des Gehirns bei Beginn des Bewußtseinsverlustes: Verlangsamung der Frequenz auf 2 bis 4 pro Sekunde.

Gegenüber Synkopen, die plötzlich nach starken Schmerzreizen oder Schreckerlebnissen auftreten, zeichnet sich die vasovagale Synkope meist durch einen protrahierten Verlauf aus, bei dem die beschriebenen Stadien der Symptomatik relativ deutlich ausgeprägt sind. Retrospektiv wird für diese Zeit das Erleben ansteigender Angst beschrieben sowie der Wunsch, der beängstigenden Situation zu entfliehen. Entweder unmittelbar vor Auftreten der Bewußtlosigkeit oder auch schon in früheren Stadien bildet sich plötzlich zunehmende Gleichgültigkeit aus, die ein

Umschlagen der Fluchttendenzen in das „Sich-der-Situation-Ausliefern" („Rückzug-Konservierungsreaktion"; Engel, 1962) einleitet. Der Übergang zur Bewußtlosigkeit kann dabei als durchaus angenehm geschildert werden.

Von Uexküll (1952) hat gezeigt, daß die Nausea eine „Abkoppelung von den Affekten" bereits in einem Stadium bewirkt, in dem noch keine Übelkeit auftritt: Zunächst tritt eine ausgeprägte Gleichgültigkeit auf. Die in ihrem funktionellen Muster der Nausea entsprechende vasovagale Synkope könnte im Sinne eines „Abwärts-Effektes" als regressiver Problemlösungsversuch in einer angsterregenden, dem Individuum aussichtslos erscheinenden, jedenfalls seine „höheren" Problemlösungsprogramme überfordernden Situation aufgefaßt werden. Dies würde auch das Auftreten vasovagaler Synkopen in ausweglosen Kampfsituationen während des Krieges (Marshall, 1951) erklären.

Sämtliche, oft als bedrohlich imponierenden Symptome und Befunde sind in der Regel rasch reversibel, zu einem protrahierten Verlauf im Sinne eines Kreislaufschocks kommt es selten.

35.5.2 Epidemiologie

Die Erhebung zuverlässiger epidemiologischer Daten wird durch den Umstand erschwert, daß es sich bei vasovagalen Synkopen oft um seltene oder einmalige Ereignisse handelt, denen häufig kein Krankheitswert zugeschrieben wird. Vorwiegend betroffen sind junge Männer. In größeren, unausgelesenen Stichproben werden von 15 bis 20% der Befragten anamnestisch eine oder mehrere Episoden von Bewußtseinsverlust seit der Pubertät angegeben. Der weitaus größte Teil dieser Attacken tritt während medizinischer Maßnahmen (Blutabnahme, Injektion, zahnärztliche Behandlung etc.) auf. Daneben sind vasovagale Synkopen gehäuft in überfüllten Räumen und – besonders bei Jugendlichen – während des Gottesdienstes in der Kirche zu beobachten.

35.5.3 Pathogenetische Konzepte

Beteiligung emotionaler Faktoren an der Pathogenese

Vasovagale Synkopen treten bei sonst körperlich gesunden Personen zumeist in Situationen auf, in denen sich die Betroffenen extrem bedroht fühlen, die Situation aber weder verändern noch aus ihr entfliehen können. Das in der Situation auftretende Angstgefühl wird aus sozialen Gründen nicht geäußert. Eigene Aktivität sowie Vertrautheit mit der Situation vermindern das Risiko, ohnmächtig zu werden.

Kinder werden bei Blutentnahmen selten ohnmächtig, was damit zusammenhängen könnte, daß sie Angstgefühle und Ansätze zu Flucht- bzw. Kampfreaktionen in der Situation nicht unterdrücken. Auf die Bedeutung der Äußerung von Emotionen weist in diesem Zusammenhang auch eine Be-

obachtung von Engel (1962) hin: Bei einem Probanden wurden Synkopen durch Aufblähen eines ins Rektum eingeführten Ballons ausgelöst; als der Patient auf den zunehmenden Schmerzreiz hin wütend aufschrie, kehrten Puls- und Blutdruckwerte zur Norm zurück und die Synkopensymptomatik klang ab.

Exkurs: Es läßt sich ein Bezug herstellen zu Berichten über solche Häftlinge und Kriegsgefangene, die in der Situation der subjektiven Hoffnungslosigkeit Nahrung und Flüssigkeit verweigerten und innerhalb von Stunden oder Tagen verstarben. Gelang es, den Sterbenden zu einer erneuten Kontaktaufnahme mit seiner Umwelt zu aktivieren, ließ sich der Tod abwenden (Stumpfe, 1973).

Apathie und rascher körperlicher Verfall bei Verweigerung der Nahrungs- und Flüssigkeitszufuhr werden ebenfalls beim sog. Voodoo-Tod (nach Tabuverletzung) beschrieben (Kächele, 1970).

Weiss (1940) sieht einen engen Zusammenhang zwischen dem Mechanismus des plötzlichen (ungeklärten) Todes und dem des synkopalen Anfalles. Bilz (1966) vertritt die Auffassung, daß es sich beim psychogenen Tod, d.h. bei jenen Todesfällen, die ohne relevanten pathologisch-anatomischen Befund einhergehen, um einen Vagustod handelt.

In luftfahrtmedizinischen Versuchen mit Überdruck-Unterdruck-Kammern ließ sich nachweisen, daß die Tendenz von Versuchspersonen, in angsterregenden Situationen mit Synkopen zu reagieren, vom Grad der Vertrautheit mit der Situation abhängt: Die Häufigkeit von Synkopen nahm vom ersten bis zum siebten simulierten Flug von 18 auf 2% ab, obwohl Auftreten und Schweregrad der übrigen durch den Druckwechsel hervorgerufenen Symptome unverändert blieben (Romano, nach Engel, 1962). Entsprechende Beobachtungen an Tieren liegen vor (Richter, 1957, zitiert nach Kächele, 1970): Beim Schwimmen in einem Glaszylinder, aus dem es kein Entkommen gab, ertranken wilde Ratten innerhalb weniger Minuten, zahme, mit der Situation vertraute Ratten schwammen bis zu 80 Stunden. Das Beschneiden der Barthaare führte zu einem beschleunigten Eintritt des Todes.

Die größere Häufigkeit vasovagaler Synkopen bei Männern läßt sich hypothetisch darauf zurückführen, daß Männer tiefergehende Ängste vor Verletzungen haben (Kastrationsangst) und daß in unserer Kultur die Äußerung von Angstgefühlen (die für das Vermeiden der Synkopensymptomatik entscheidend ist, s.o.) bei Männern weniger toleriert wird.

Dem wiederholten Auftreten vasovagaler Synkopen bei organisch Gesunden liegt im allgemeinen eine psychische Störung zugrunde. In diesen Fällen entspringt die Bedrohung einem inneren Konflikt. Der innere Konflikt kann durch äußere Gefahrensignale mobilisiert werden, die dem unbeteiligten Beobachter trivial erscheinen mögen, jedoch eine individuelle, im einzelnen zu klärende Bedeutung haben. Die betroffenen Personen zeigen zumeist bei sorgfältiger Exploration auch andere Zeichen vermehrter Angst. Synkopen treten häufig dann auf, wenn andere Abwehrformen, wie z.B. eine Konversionssymptomatik oder eine phobische Abwehr, zusammenbrechen.

Engel (1962) berichtet von einem Farmer, der wegen einer konversionsneurotisch bedingten beidseitigen Lid-Ptose nicht in der Lage war, Tiere zu erschießen. In die konversionsneurotische Symptombildung war u.a. die Abwehr aggressiver Impulse zusammen mit der Kastrationsangst eingegangen. Der Versuch, diese Symptombildung bei einer Vorstellung vor Studenten dadurch aufzuheben, daß er dem Farmer beide Augenlider hochzog, führte zu einer vasovagalen Synkope: Das Durchbrechen der mit der Symptombildung verbundenen Abwehr mobilisierte massive Angst, weder Flucht noch Affektäußerungen waren möglich.

In Beschleunigungsversuchen in Zentrifugen ließ sich experimentell nachweisen, daß die Synkopensymptomatik bei solchen Versuchspersonen schon bei niedrigen Beschleunigungswerten auftritt, die testpsychologischen Befunden zufolge besonders ängstlich und selbstunsicher sind.

Psychophysiologie

Phänomenologisch läßt sich das Gesamtgeschehen bei der vasovagalen Synkope als Rückzug-Konservierungsreaktion (Engel) beschreiben; diese ist das Gegenstück zur fight-flight- oder defense-reaction (Folkow, 1955; Folkow und Uvnas, 1966), die mit gesteigerter Wachsamkeit und Erhöhung des Blutdruckes einhergeht (vgl. Kap. 41.3). Soweit Untersuchungen der Kreislaufparameter bei der Synkope vorliegen, scheint der Blutdruckabfall mit einer Verminderung des peripheren Widerstandes und des Herzzeitvolumens einherzugehen (Barcroft et al., 1944).

Mechanistische Interpretationen, die ein „Versakken des Blutes in der Peripherie" infolge akuter Vasodilatation als Ursache annehmen, bleiben die Antwort auf die Frage nach der Ursache der Vasodilatation schuldig. Sie übersehen auch, daß es sich nicht um ein bloßes Versagen von Regulationsvorgängen handelt, sondern um ein offensichtlich reguliertes und koordiniertes, in bestimmten Phasen ablaufendes Gesamtgeschehen.

Neurophysiologisch lassen sich zwei Reaktionsweisen unterscheiden:
- die „histiotrope" (Körpergewebe schützende, Energie sparende) und
- die „ergotrope" (auf Handlung und Energieverbrauch ausgerichtet) (Hess, 1948).

Die Synkope läßt sich – ähnlich wie die ebenfalls wenig erforschten hypotonen Kreislaufzustände – dem histiotropen Reaktionsmuster zuordnen, dessen psychologische und physiologische Komponenten sehr viel weniger gut untersucht sind als die des ergotropen Reaktionsmusters. Vor allem wissen wir noch so gut wie nichts über ihre Beziehungen zu Zuständen der Hilf- und Hoffnungslosigkeit, denen Schmale und Iker (1966, 1971) eine überragende Bedeutung für die Pathogenese vieler Krankheiten zumessen.

Als psychophysiologische Gesamtreaktion hat die Synkope große Ähnlichkeit mit der Nausea, die nach Th. von Uexküll (1952) der Prototyp einer histiotropen Reaktion ist, die psychologisch durch Rückzug, Minderung der emotionalen Spannung und Gleichgültigkeit gekennzeichnet ist und physiologisch den Prodromalstadien der Synkope entspricht.

Von allgemeiner Bedeutung wäre eine genauere Klärung des „Abwärts-Effektes": Ist eine Problemlösung auf der Verhaltensebene (Flucht oder Angriff) und auf der Ebene innerpsychischer Verarbeitung und damit eine Verminderung der Angst nicht mög-

lich, erfolgt über eine physiologische Umschaltung eine Art „Ausklinken" aus der Situation; es erfolgt ein Umschlag aus dem offenen System, das die Beziehung zur Umwelt als jeweiliger individueller Wirklichkeit enthält, „in das relativ geschlossene System eines umweltlosen Körpers" (v. Uexküll und Wesiack, 1988). Entscheidend für das Verständnis dieses „Umschlages" ist die Kenntnis der jeweiligen subjektiven Bedeutung der Situation bzw. des in ihr stimulierten intrapsychischen Konfliktes.

35.5.4 Therapie und Prognose

Bei Auftreten einer vasovagalen Synkope genügt im allgemeinen die horizontale Lagerung des Patienten, um die Symptomatik rasch zum Abklingen zu bringen. Aktive Bewegung der Beine verbessert den Rückstrom des Blutes.

Die therapeutische Aufgabe des Arztes besteht in einer beruhigenden Information des Patienten, der Angehörigen und der Umgebung über die Ungefährlichkeit der bedrohlich erscheinenden Symptomatik. Darüber hinaus ist es unerläßlich, mit dem Patienten über das der Symptomatik zugrundeliegende Reaktionsmuster zu sprechen. Dabei ist besonders sorgfältig darauf zu achten, daß der Patient durch eine solche Mitteilung nicht gekränkt wird (etwa: „Wir haben bei Ihnen nichts gefunden"), d.h., daß er sich nicht als Simulant eingestuft erlebt. Auch wenn in der Symptomatik ein Konflikt bewältigt wird (dies könnte auch als „Leistung" des Organismus aufgefaßt werden), fühlt sich der Patient den Anfällen ja „ohnmächtig" ausgeliefert.

Eine Klärung der auslösenden Situation und ein Durchsprechen damit verbundener Ängste kann von prophylaktischer Bedeutung sein. Bei rezidivierenden vasovagalen Synkopen ist eine psychotherapeutische Bearbeitung des zugrundeliegenden Konfliktes dann zu empfehlen, wenn die Synkopensymptomatik allein oder im Zusammenhang mit Symptomen im psychischen Bereich für den Patienten den Charakter eines Leidens anzunehmen droht oder bereits angenommen hat.

35.6 Konversionsneurotische Synkopen

35.6.1 Exemplarische Krankengeschichte

Eine 17jährige, leicht adipöse Patientin wird innerhalb von 10 Tagen sechsmal mit dem Notarztwagen in die Klinik gebracht. Bei jedem dieser Ereignisse war sie am Arbeitsplatz plötzlich bewußtlos geworden.

Symptomatik und Klinik: In bestimmten Situationen wird die Patientin unruhig, zum Teil beobachtet sie Zittern und Flimmern vor den Augen, innerhalb kürzester Zeit tritt dann Bewußtlosigkeit ein, die meist ca. 15 Minuten, gelegentlich aber auch bis zu 2 Stunden andauert. Beim Erwachen fällt ein heftiges und rasches Atmen auf, die Patientin klagt über Kopfschmerzen, die innerhalb von ein bis zwei Stunden wieder abklingen.

Die klinische Anamnese ist unauffällig. Im Anfall findet sich ein Blutdruck von 130/80 mm Hg und eine Pulsfrequenz von 100 Schlägen pro Minute.

Die fachneurologische Untersuchung einschließlich EEG, Schlafentzugs-EEG, Hirnszintigramm und Hyperventilationsversuch ergibt keinen pathologischen Befund.

Entwicklung der Symptomatik und auslösende Situation: Zustände von anfallsartig auftretender Übelkeit, Zittern am ganzen Körper, Weichwerden in den Knien und Flimmern vor den Augen beobachtete die Patientin erstmals vor zwei Jahren.

Damals trat die Symptomatik plötzlich bei einer Auseinandersetzung mit den Eltern auf, in deren Verlauf diese ihr verboten hatten, weiter mit ihrem Freund auszugehen. Die Beschwerden klangen nach kurzem Ausruhen wieder ab, es kam damals zu keinem Bewußtseinsverlust. Zum jetzigen Zeitpunkt besteht ein Zusammenhang zwischen dem Auftreten der Beschwerden und starken „Aufregungen", die mit dem Gefühl von Hilflosigkeit und ohnmächtiger Wut verbunden sind. Die Anfälle sind jetzt gekennzeichnet durch Unruhezustände mit Zittern und Flimmern vor den Augen, auf die rasch Bewußtlosigkeit folgt. Die zur stationären Aufnahme führenden Synkopen traten am Arbeitsplatz auf, wo die Patientin als kaufmännische Angestellte beschäftigt ist. Auslösend sind Auseinandersetzungen mit ihrem Vorgesetzten, gegen den sie sich nicht durchsetzen kann. Zeitlich fällt die Symptomhäufung mit der Kündigung einer älteren Arbeitskollegin zusammen, zu der die Patientin ein besonders gutes Verhältnis hatte und von der sie sich beschützt fühlte. Seit deren Weggang fühlt sie sich alleingelassen und den Angriffen der Mitarbeiterinnen und Vorgesetzten ausgeliefert. Sie schildert sich als besonders erfolgreich, aber isoliert am Arbeitsplatz, Neidreaktionen älterer Kolleginnen ausgesetzt.

Familiäre Situation: Die Patientin lebt in der elterlichen Familie, fühlt sich dort jedoch unterdrückt. Nach ihrer Schilderung wird sie besonders in der Kontaktaufnahme mit Gleichaltrigen durch die rigiden elterlichen Moralvorstellungen behindert, die zu verschiedenen Einschränkungen – so etwa des abendlichen Ausgehens – führen. Anstelle einer sexuellen Aufklärung wurden der Patientin immer wieder in Auseinandersetzungen anhand negativer Beispiele die Gefahren sexueller Beziehungen vor Augen geführt.

Besonders die Mutter übt Druck auf die Patientin aus, indem sie ihren Vorhaltungen durch Hinweis auf ihren Gesundheitszustand (sie leidet an nicht näher abgeklärten Herzbeschwerden) Nachdruck verleiht. Die herzbezogenen Symptome der Mutter verstärken sich in Auseinandersetzungen mit der Tochter. In einer Zeit besonders heftiger Auseinandersetzungen mit der damals 16jährigen Tochter zog sich die Mutter ohne Hinzuziehung eines Arztes mit der selbstgestellten Diagnose „Herzinfarkt" für drei Monate ins Bett zurück.

Zum Vater, der aufgrund einer chronischen Erkrankung Frührentner ist und sich viel um die Patientin kümmern konnte, bestand in der Kindheit eine enge Beziehung. Seit der Pubertät erlebt sie ihn als distanziert. In Auseinandersetzungen ist er letzte Instanz, wobei er meist Partei für die Mutter ergreift. Das Urteil des Vaters ist der Patientin sehr wichtig; ihre

Unsicherheit, was er von ihr halten könnte, wird im Gespräch deutlich.

Den einzigen Bruder (+4) schildert sie als jähzornig; sie fühlt sich ihm gegenüber von den Eltern benachteiligt. Zum Zeitpunkt der Symptomhäufung stand die Rückkehr des Bruders von der Bundeswehr unmittelbar bevor.

Da die Patientin unter der familiären Konfliktsituation deutlich litt und im ersten Gespräch gemeinsam mit ihr wesentliche Probleme erarbeitet werden konnten, wurde ihr eine psychotherapeutische Behandlung vorgeschlagen.

Verlauf der Kurzpsychotherapie: Während der ersten Wochen stehen Probleme am Arbeitsplatz und Auseinandersetzungen mit der Mutter ganz im Vordergrund. Eine Synkope tritt jetzt erstmals auch zu Hause in folgender Situation auf: Die Mutter hatte einen abendlichen Spaziergang mit einem Freund verboten. Danach entwickelt sich ein Streit über die von der Patientin als unnötig erlebten Einschränkungen, der sich so weit zuspitzt, daß die Patientin droht, auszuziehen. Sie erlebt eine unerwartete Enttäuschung, als der Vater ihr dies ganz gelassen konzediert. Als darauf der Bruder der Patientin ankündigt, dann ebenfalls auszuziehen, verliert sie ganz plötzlich das Bewußtsein. Nach diesem Ereignis grübelt sie lange über ihre Beziehung zum Vater nach.

Das Zustandekommen einer Übertragungssituation kündigt sich an, als die Symptomatik erstmals in der Beziehung zum Therapeuten auftritt. Wiederholt treten Synkopen kurz vor der Therapiestunde auf, so daß der behandelnde Arzt zum Zeitpunkt der Stunde auf die Aufnahmestation gerufen wird, wohin die Patientin gebracht worden war.

Im Verlauf der Behandlung wird deutlich, daß der neurotische Konflikt die Beziehung zum Vater betrifft. Er aktualisierte sich in einer beruflichen Parallelsituation, die durch das Ausscheiden einer mütterlichen Person und die Schutzlosigkeit gegenüber dem als bedrohlich erlebten Chef gekennzeichnet ist. Im Elternhaus schützt die Anwesenheit der Mutter die Patientin vor dem Ausleben ihrer ödipalen Wünsche gegenüber dem Vater. Die Ambivalenz gegenüber der Mutter wird in der Angst deutlich, ihren Tod zu verschulden. Nach dem Tod der Mutter könnte der ödipale Wunsch, den Vater ganz für sich zu gewinnen, in Erfüllung gehen. In der unbewußten Phantasie verbindet sich mit dem Ausscheiden der älteren, Schutz gewährenden Kollegin deren Beseitigung (Tod der Mutter), sie bleibt mit dem Chef (Vater) allein.

Die Patientin gewinnt schließlich zunehmend Einsicht in ihren Beitrag zum regelhaften Ablauf der häuslichen Auseinandersetzungen. Sie kann ihre Unabhängigkeitswünsche besser durchsetzen und gewinnt einen Bewegungsfreiraum in der Familie, auf den sie zunächst depressiv reagiert, da sie das Nachgeben besonders des Vaters als Desinteresse interpretiert. Ihre Bindung an den Vater wird ihr zunehmend bewußt. Sie unternimmt Schritte zur Verselbständigung im Arbeitsbereich, läßt sich nach einer Übergangsphase des Rückzugs freier in Kontakte mit Gleichaltrigen ein und bleibt über mehrere Monate nach Abschluß der insgesamt zehnstündigen Kurztherapie beschwerdefrei.

35.6.2 Symptomatik und Klinik

Die Symptomatik tritt unabhängig von der körperlichen Lage auf. Bei Eintritt der Synkope können die Patienten langsam zu Boden sinken oder auch abrupt fallen; sie verletzen sich selten. Gelegentlich fallen bizarre Haltungen oder Bewegungen auf. Die Anfallsdauer schwankt zwischen wenigen Sekunden bis zu (seltener) mehreren Stunden.

Bei konversionsneurotischen Synkopen finden sich weder Kreislaufveränderungen noch EEG-Veränderungen. Die Pupillen sind unverändert und reagieren normal auf Licht, auch der übrige neurologische Status ist unauffällig. Konversionsneurotische Synkopen treten fast nur in Gegenwart anderer auf; es kann ein theatralisch-dramatisches Gehabe auffallen, was oft zum unberechtigten Vorwurf der Simulation führt.

Sowohl das Erscheinungsbild psychischer Störungen als deren Einordnung in ärztliche Verständnissysteme werden von kulturellen Entwicklungen mitbeeinflußt. Die Diagnose „Hysterie" wird heute seltener gestellt; konversionsneurotische Symptome sind jedoch wahrscheinlich nicht seltener geworden. Mit zunehmendem Bekanntheitsgrad der klassischen „hysterischen" Symptome ist das Erscheinungsbild konversionsneurotischer Symptombildung z.T. differenzierter und „unauffälliger" geworden. Vom Arzt verlangt diese Entwicklung differenziertere Kenntnisse und sorgfältige Beachtung psychodynamischer Prozesse, sollen diese Krankheitsbilder nicht zu häufig unzureichend begründet vermutet werden (vgl. Kap. 30 „Konversion").

Das Auftreten konversionsneurotischer Synkopen im Zusammenhang mit ärztlichen Untersuchungen hängt mit der unbewußten erotischen Bedeutung zusammen, die diese Patienten der Untersuchungssituation beimessen.

Die Symptomatik beginnt zumeist in der Pubertät, oft in zeitlichem Zusammenhang mit der Menarche. Vielfach finden sich weitere Konversionssymptome, insbesondere abdominelle Schmerzen, die gehäuft zu Appendektomien ohne pathologischen Befund führen. Für differentialdiagnostische Überlegungen ist es wichtig, daß aus klinischen und anamnestischen Daten die positive Diagnose einer konversionsneurotischen Synkope gestellt werden kann. Hierfür ist neben dem Fehlen pathophysiologischer Befunde die Klärung des Konfliktes, der Bedeutung der Symptomwahl und des primären und sekundären Krankheitsgewinns erforderlich. Die Diagnose einer hysterischen Persönlichkeitsstruktur kann nur als Hinweis, nicht aber als Beweis gelten (vgl. Kap. 30 „Konversion").

35.6.3 Epidemiologie

Die konversionsneurotische Synkope ist neben der vasovagalen Synkope die häufigste Form der Ohnmacht bei Adoleszenten und jungen Erwachsenen. In der überwiegenden Mehrzahl sind Frauen betroffen.

Bei den an konversionsneurotischen Synkopen leidenden Männern finden sich Hinweise auf Geschlechtsidentitätsstörungen.

35.6.4 Pathogenetisches Konzept

Im psychoanalytischen Verständnis ist das Konversionssymptom Ergebnis eines „Selbstheilungsversuches" (Kahn, 1977). Das Symptom stellt einen Kompromiß zwischen den ursprünglichen (Trieb-)Wünsche bzw. Bedürfnissen und den ihnen entgegenstehenden Forderungen der äußeren Realität oder der intrapsychischen Zensurinstanzen dar. Über verschiedene Abwehrmechanismen gelingt es, die ursprünglichen Wünsche unter Kontrolle zu bekommen und vom Bewußtsein fernzuhalten. Zugleich drückt das Symptom in symbolisierter Form noch etwas vom ursprünglichen Wunsch aus. Insgesamt stellt das Symptom so das Endresultat einer mehr oder weniger komplexen Ich-Leistung dar. Solange dem Ich diese Form der Konfliktlösung gelingt, sind die Patienten relativ angstfrei, wenn auch in ihren kommunikativen Möglichkeiten eingeschränkt; nach außen können sie so ruhig und ausgeglichen, wenn auch ein wenig gleichgültig oder fremd („la belle indifférence") wirken. Diese stabilisierende Funktion der Symptombildung erklärt auch den Widerstand, der einer psychotherapeutischen Bearbeitung des Konfliktes entgegensteht: eine vom Patienten selbst geleistete „Lösung" des Konfliktes wird in Frage gestellt. Während die Ohnmacht für den Zuschauer häufig dramatisch und alarmierend wirkt, sind an konversionsneurotischen Synkopen leidende Patienten über ihre Symptomatik meist nur wenig beunruhigt. Im Gegensatz hierzu ist sich der Patient mit vasovagaler Synkope seiner Angst – etwa vor einem medizinischen Eingriff – durchaus bewußt.

An Konversionssymptomen leidende Patienten sind meist in ihrer sexuellen Erlebnisfähigkeit eingeschränkt. Es kommen extreme Hemmung und Vermeidung sexueller Kontakte ebenso vor wie Promiskuität. Frigidität ist sehr häufig. Die spezielle Bedeutung der Synkope variiert individuell. Eine sexuelle Bedeutung des Anfalls wird jedoch oft deutlich, wenn das Verhalten während der Ohnmacht genau beobachtet wird.

In der konversionsneurotischen Synkope kann sich der Wunsch nach sexueller Hingabe ausdrücken – allerdings nur um den „Preis" des Bewußtseinsverlustes. In dieser Symptomwahl wird auch die häufig große Hilflosigkeit der Betroffenen im Umgang mit ihren (Trieb-)Bedürfnissen und Wünschen deutlich, oft als Folge einer Störung der Ich-Entwicklung oder – vor allem bei Adoleszenten – einer noch ungenügenden Stabilität erwachsener Ich-Funktionen mit größerer Anfälligkeit für regressive Prozesse unter entsprechender Belastung.

Bei der eingangs beschriebenen Patientin ist im Zusammenhang mit dem Wunsch, vom Vater akzeptiert zu werden, auch der ödipale Konflikt deutlich; er wiederholt sich in der

Übertragungsbeziehung zum Arzt. Die Symptomwahl kann hier als Identifikation mit der Mutter, die in ihrer Jugend ebenfalls an Synkopen litt, verstanden werden.

35.6.5 Therapie

Die Synkope selbst bedarf keiner speziellen Behandlung, wohl aber in bestimmten Fällen die zugrundeliegende psychische Störung. Die Gabe blutdrucksteigernder Medikamente erzielt keinen Effekt und ist deshalb nicht indiziert.

Beim Gespräch über die Diagnose (vgl. Kap. 31 „Funktionelle internistische Beschwerdebilder" und 66 „Zum Umgang mit unheilbar Kranken") ist es wichtig, eine Kränkung der oft sehr selbstunsicheren Patienten zu vermeiden. Handelt es sich, vor allem im Rahmen der Pubertät und Adoleszenz, um leicht verständliche, entwicklungsbedingte, bewußtseinsnahe sexuelle Konflikte, so sind klärende Gespräche oft ausreichend, aber auch indiziert (Dührssen, 1974).

Bei rezidivierenden konversionsneurotischen Synkopen ist eine psychoanalytisch orientierte Bearbeitung des zugrundeliegenden Konfliktes sinnvoll. Aufgabe des primär versorgenden Arztes ist es, den Patienten hierzu zu motivieren.

Insbesondere bei dieser Patientengruppe hat die Wahl der psychotherapeutischen Methode eine sorgfältig differenzierende psychodiagnostische Abklärung zur Voraussetzung. Zu berücksichtigen ist dabei die Ausprägung der Störung sowie das Verhältnis von konflikthaften und konfliktfreien Persönlichkeitsanteilen. Nicht ganz selten finden sich rezidivierende Synkopen bei Patientinnen, die an schweren Identitätskrisen während der Adoleszenz leiden, sich zum Zeitpunkt der notwendigen Ablösung von den Eltern vom Scheitern bedroht fühlen und in dieser Situation sozusagen vor einer negativen Gesamtbilanz immer wieder „ohnmächtig" zusammenbrechen. Es können gravierende Störungen im Sinne des Borderline-Syndroms vorliegen, die Synkopensymptomatik kann von schweren depressiven Reaktionen mit Suizidversuchen und psychosenahen Bildern abgelöst werden. Das therapeutische Vorgehen muß entsprechend flexibel sein.

35.7 Psychogene Synkopen unklarer Genese

Schreck, Streß und akute Angst können sowohl zu vasovagalen Synkopen als auch zu rein psychogenen Synkopen führen. Die Betroffenen fühlen sich typischerweise plötzlich schwindlig und verlieren oft rasch das Bewußtsein. Da es sich im allgemeinen um einmalige Ereignisse handelt, liegen keine detaillierten Untersuchungen über diese Form der Synkope vor. Der in der Synkope enthaltene Rückzug (conservation-withdrawal) kann sich wahrscheinlich auch auf überwältigende oder sich zuspitzende innere Konflikte beziehen; so sind bei neurotischen und psychotischen Krankheitsbildern plötzlich auftretende psychogene Synkopen (ohne Kreislaufveränderung) beobachtet worden.

35.8 Differentialdiagnostische Überlegungen

Die Differentialdiagnose hat Anfallsleiden (Epilepsie, Adams-Stokes-Anfälle, Hypoglykämie etc.) auszuschließen.

Die Diagnose läßt sich häufig aufgrund der genauen Beschreibung der Synkope stellen. Dabei sind in der Anamnese besonders zu beachten.
– die Situation, in der die Symptome auftreten;
– angsterregende Momente in der Situation;
– vorausgehende Belastungen und Konflikte;
– Anwesenheit anderer Personen;
– Atemverhalten;
– Charakter und Dauer prämonitorischer Symptome;
– die Körperlage des Patienten bei Beginn des Anfalls;
– die Zeit bis zum Eintreten des Bewußtseinsverlustes;
– die Dauer des Bewußtseinsverlustes;
– das Auftreten von Krämpfen oder anderen neurologischen Symptomen;
– das Auftreten von Verletzungen oder Zeichen von Inkontinenz;
– der Verlauf der Rekonvaleszenz;
– persistierende Symptome.
Die Befragung von Zeugen des synkopalen Anfalles ist wichtig.

Die meisten Fälle der in diesem Kapitel beschriebenen wiederholten Anfälle von Bewußtseinsverlust werden durch wenige Synkopenformen verursacht. Vor allem bei jüngeren Patienten handelt es sich bei wiederholtem Auftreten von Synkopen in der Mehrzahl der Fälle um psychogene Synkopen. Differentialdiagnostisch kommen hier die vasovagale Synkope, die konversionsneurotische Synkope und die Synkope bei Hyperventilation in Frage. Letztere tritt als Folge der Hypokapnie bzw. der resultierenden respiratorischen Alkalose auf. Das psychogene Hyperventilationssyndrom (vgl. Kap. 34 „Hyperventilation") kann sowohl als Folge neurotischer Angst als auch als Konversionssymptom auftreten. Die Abgrenzung aufgrund psychologischer Daten von Patienten mit rezidivierenden vasovagalen Synkopen gegenüber Patienten mit konversionsneurotischen Synkopen ist oft nicht streng möglich. Mischformen zwischen konversionsneurotischen Synkopen und Synkopen bei Hyperventilationssyndrom sowie alternierendes Auftreten dieser beiden Formen sind häufig. Synkopen psychogenen Ursprungs beginnen fast immer während der Jugend, meist während der Pubertät. Männer leiden öfter an vasovagalen Synkopen, während bei Frauen die konversionsneurotischen Synkopen überwiegen und auch Synkopen infolge eines Hyperventilationssyndroms häufiger sind. Die Differentialdiagnose zwischen diesen drei Synkopenformen läßt sich aufgrund der klinischen Untersuchung einschließlich EEG-Untersuchung stellen. Provokationstests können nötigenfalls herangezogen werden: orthostatische Belastung durch Stehen bzw. auf dem Kipptisch, eventuell Simulation der anamnestisch berichteten Angstsituation bei der vasovagalen Synkope, Hyperventilationsversuch bei Verdacht auf Hyperventilationssyndrom.

Ein Hyperventilationsversuch kann bereits – ohne daß der Patient die Absicht des Arztes bemerkt – während der Auskultation vorgenommen werden. Wenn dabei entsprechende Symptome auftreten, kann der Patient sie schildern, ohne damit die Abwehr des unbewußten Konfliktes aufgeben zu müssen. Patienten, die zur Hyperventilation neigen, sind oft schon am Atemtyp zu erkennen (vgl. Kap. 34 „Hyperventilation").

Wiederholte Synkopenanfälle als Folge organischer Erkrankungen sind gegenüber psychogenen Synkopen selten. Es ist jedoch daran zu denken, daß eine organische Verursachung jenseits des 50. Lebensjahres häufiger wird und daß emotionale Belastungen auch vorwiegend organisch bedingte Synkopen auslösen können, ferner, daß beim gleichen Patienten psychogen bedingte Synkopen in der Jugend in späteren Jahren durch organisch bedingte Synkopen abgelöst werden können (vgl. Kap. 41.1 „Arterielle Verschlußkrankheit").

36 Schmerz

Rolf Adler

Schmerz könnte als unangenehme Empfindung definiert werden, die dem Leiden entspricht, das durch die Wahrnehmung einer Verletzung hervorgerufen wird. Diese Definition würde folgenden Beobachtungen nicht gerecht werden: Schwere Schmerzen können ohne Gewebsverletzungen bestehen; schwere Verletzungen brauchen nicht mit Schmerz verbunden zu sein (Beecher, 1956). Angst kann Schmerz verstärken, Ablenkung kann ihn lindern; Placebos vermögen analgetisch zu wirken. Für eine Schmerzempfindung ist also die Reizung einer peripheren schmerzempfindlichen Struktur weder notwendig noch hinreichend. Auch wenn bei Schmerzen, zu denen Gewebsverletzungen beitragen, an peripheren Rezeptoren und afferenten Nerven Ionenverschiebungen und elektrische Potentiale mit Instrumenten erfaßt werden können, handelt es sich beim Schmerz um ein psychisches Phänomen, das nur das erleidende Subjekt fühlt und nur von diesem erfahren werden kann, also um eine Empfindung, die zu der individuellen Wirklichkeit eines Menschen gehört. Ein anonymer Versschmied hat dies folgendermaßen ausgedrückt: „There was a young lady of Deal, who said: although pain isn't real, if I sit on a pin, and it punctures my skin, I dislike what fancy I feel."

Die individuelle Wirklichkeit, mit der Erfahrung eines Menschen eng verbunden, trägt also zu Schmerzempfinden und -verhalten bei. In sensorischer Isolation aufgezogene Hunde stecken später ihre Schnauze in eine Flamme, ohne Schmerzreaktion zu zeigen (Melzack und Scott, 1957). Erwachsene, die viele Schmerzzustände durchmachen, die zum Teil nicht erklärt werden können, und die häufig operiert werden, haben als Kinder oft Brutalität zwischen ihren Eltern und auf sie selbst gerichtete Aggressionen erfahren (Engel, 1959; Adler et al., 1989). Die Nähe eines Vertrauen ausstrahlenden Menschen kann beim Patienten Schmerz lindern. Diese Beobachtungen beleuchten, daß zwischen den neurophysiologischen Abläufen bei der Schmerzentstehung, also Zeichen, die zwischen Zellen, Geweben und Organen „verständigen", und intrapsychischen Vorgängen, die als Zeichen zwischen Bedürfnissen des Organismus und Anforderungen der Umgebung vermitteln, Zusammenhänge bestehen. Die Zeichen in den beiden Systemen sind völlig verschiedene und haben primär nichts miteinander zu tun. Es besteht also ein „Bedeutungssprung" zwischen den Systemen oder zwischen zwei Integrationsebenen. Die zweite ist die komplexere und schließt Organismus und Umwelt zusammen. Diesem Konzept liegt zugrunde, was Pawlow mit seinen Beobachtungen der Bildung bedingter Reflexe oder Konditionierungen, also der Koppelung von Zeichen verschiedener Integrationsebenen, beschrieben hat. Der unbedingte Reiz beim Schmerz entspricht dann der Reizung des peripheren Rezeptors, der bedingte Reiz beispielsweise der Drohung mit Strafe durch einen Elternteil. So entsteht dann eine persönliche, individuelle Schmerzphysiologie, deren Erfassung für das Verstehen und Behandeln klinischer Schmerzzustände beim Patienten unerläßlich ist.

36.1 Grundlagen des Schmerzes

36.1.1 Peripherer Schmerzapparat

Zu den peripheren kutanen Endorganen, die bei der Schmerzentstehung eine Rolle spielen, gehören mechanosensitive, hitzesensitive und polymodale Nozizeptoren. Letztere werden durch starke mechanische und Hitzereize erregt. Die Spezifität der Rezeptoren für bestimmte Reize ist eine relative. Sehr starke Reize erregen alle Rezeptoren (Handwerker und Zimmermann, 1976). Diese Rezeptoren sind durch markarme A-δ- und marklose C-Fasern mit dem Rückenmark verbunden. Die viszeralen Nozizeptoren werden durch Dehnung, chemische Reize und Ischämie erregt. Freie marklose Nervenendigungen dienen ebenfalls der Nozizeption.

Bei der Reizung eines Rezeptors entsteht ein Generatorpotential, dessen Amplitude eine Funktion der Reizstärke ist. Erreicht sie eine bestimmte Höhe, entstehen an der Membran des Nerven Aktionspotentiale nach dem „Alles oder Nichts"-Gesetz. Ihre Frequenz stellt eine Potentialfunktion der Generatorpotentialhöhe dar. Die Erregung der schneller leitenden Fasern löst Sensationen wie Druck aus, die der dünnen A-δ-Fasern hellen, schnell abklingenden, sog. Erstschmerz, die der C-Fasern dumpferen, brennendbohrenden Zweitschmerz, der von vegetativen Zeichen wie Übelkeit, Schwitzen usw. begleitet sein kann (Casey, 1973).

36.1.2 Schmerzapparat im Rückenmark

Im Hinterhorn konvergieren die peripheren Fasern auf Nervenzellen, hauptsächlich in den Schichten 2 und 3 der Substantia gelatinosa, welche die eintreffenden Impulse zentralwärts leiten. Zusätzlich konvergieren auch markreiche A-α- und -β-Fasern auf die Synapsen der A-δ- und C-Fasern mit den Hinterhornzellen. Die Entladungsschwelle an den Synapsen wird von ihnen und von zentrifugal in der weißen Substanz deszendierenden Fasern moduliert. Sie stammen aus dem periaquäduktalen Grau, dem Nucleus raphe, dem Nucleus coeruleus, lateralen Teilen der Formatio reticularis, dem Nucleus gigantocellularis und dem sensomotorischen Kortex (Mayer und Price, 1976).

36.1.3 Das sensorisch-diskriminierende System

Von den Hinterhornzellen (T-Zellen) leitet eine zum Tractus spinothalamicus anterolateralis zusammengefaßte Neuronengruppe, die auf Rückenmarksniveau kreuzt, die Impulse via Lemnicus medialis zu ventrokaudalen und rostralmedialen Thalamuskernen, letztere erhalten auch Fasern aus der Formatio reticularis. Vom ventrokaudalen Thalamus führt das dritte Neuron zum somatosensorischen Kortex, vom medialen Thalamus zum Assoziationskortex. Das spinothalame Faser- und Kernsystem projiziert die Körperregionen nach zentral und erlaubt die Analyse der auf den Reiz folgenden Impulse bezüglich Raum, Zeit und Intensitätsmerkmalen der durch den Stimulus betroffenen Körpergebiete. Dieses System wird deshalb das „sensorisch-diskriminierende" genannt (Melzack, 1970).

36.1.4 Das motivierend-affektive System

Von den T-Zellen steigen paramedian die Neuriten der phylogenetisch alten Schmerzfasern auf. Sie zweigen Fasern an die Formatio reticularis, an mediale intralaminäre Thalamuskerne und an das limbische System ab. Dieses Faser- und Kernsystem kennt keine präzise topographische Projektion der Peripherie nach zentral. Es ist für den Weh-Charakter des Schmerzes verantwortlich und bewirkt, daß das Individuum sich dem schmerzerzeugenden Stimulus zuwendet oder sich vor ihm zurückzieht. Dieses System wird deshalb das „motivierend-affektive" genannt (Melzack, 1970).

36.1.5 Das zentrale Kontrollsystem

Das „sensorisch-diskriminierende" und das „motivierend-affektive" System sind untereinander verbunden und stehen beide unter dem Einfluß neokortikaler Zentren. Diese bewirken, daß die Bewußtseinslage, die Aufmerksamkeit, die Erfahrung mit Schmerz, die jeweilige Situation und der symbolische Gehalt des Stimulus zum Schmerzempfinden und -verhalten beitragen. Das neokortikale System wird durch schnellleitende Hinterstrangbahnen benachrichtigt und wirkt hemmend und bahnend für die afferenten Impulse bis hinab auf Rückenmarksniveau. Es wird deshalb „zentrales Kontrollsystem" genannt.

36.1.6 Das Spinal Gate Control System

Die Hypothese von Melzack und Wall (1965), daß im Rückenmark in der Nähe der T-Zellen gelegene Substantia-gelatinosa-Zellen die über die C-Fasern hereinströmenden Impulse modulieren, hat sich auf die Schmerzforschung befruchtend ausgewirkt. Die Autoren postulierten, daß die Reizung myelinisierter dicker A-β-Fasern in der Peripherie via Substantia-gelatinosa-Zellen den Impulsstrom von den A-δ- und den C-Fasern zur T-Zelle präsynaptisch hemmt, während die über die C-Fasern hereinströmenden Impulse die Substantia-gelatinosa-Zellen hemmen und deren blockierenden Einfluß auf die Synapse zwischen C-Faser und T-Zelle aufheben. Dieses System nannten sie das „Spinal Gate Control System". Mit ihm lassen sich schmerzhemmende Wirkungen von Eis, Akupunktur, Vibrationsmassage, transkutaner Elektrostimulation usw. erklären. Es sind nicht alle Befunde von Mendell und Wall (1964) verifiziert worden, auf die sich die Theorie stützt. Nach ihnen führt die Reizung markreicher, dicker peripherer Fasern zu einem Hemmung anzeigenden negativen Hinterwurzelpotential und diejenige von C-Fasern zum positiven Hinterwurzelpotential. Zimmermann (1968) hingegen fand bei C-Faserstimulation auch negative Potentiale. Die Konvergenz noxischer und nicht-noxischer Afferenzen an zentralen Neuronen und die Hemmung noxischer Reize durch vorausgehende nicht-noxische ist aber belegt (Larbig, 1982).

36.1.7 Hormone und Überträgersubstanzen

Im Schmerzempfinden und -verhalten wirken lokale Hormone und Überträgersubstanzen im Zusammenspiel mit den peripheren und zentralen Anteilen des Nervensystems. Im peripheren Gewebe wird durch Schädigung (Entzündung, Verletzung) Kallikrein frei, das Bradykinin aktiviert, welches die Schmerzrezeptoren reizt. Es wirkt vasodilatierend und erhöht die Kapillarpermeabilität, so daß weitere, die Rezeptoren reizende Substanzen ins Gewebe ausfließen können, wie H^+- und K^+-Ionen etc. Bei Gewebsschädigung entstehen aus Arachidonsäure unter der Einwirkung von Bradykinin via Aktivierung der Phospholipase A_2 Prostaglandine, welche die Schmerzrezeptoren ebenfalls sensibilisieren. Der am längsten bekannte algetische Stoff, ein Neuropeptid, wird „Substanz P" genannt. Sie wirkt in spinalen Neuronen als Neurotransmitter und wird in der Synapse durch die den Morphinen ähnlichen endogenen Substanzen (Endorphine) blockiert. Serotonin, das aus Tryptophan via 5-Hydroxytryptophan entsteht, findet sich im Nucleus raphe, Mittelhirn und im Seitenhorn des Rückenmarks. Es ist vermutlich für deszendierende Hemmung von Hinterhornneuronen verantwortlich (Reubi, 1980). Wird die Serotoninsynthese blockiert, so nehmen bei Kopfschmerzpatienten die Symptome zu (Sicuteri et al., 1973). Die schmerzhemmende Wirkung des Antidepressivums Amitriptylin (Sternbach et al., 1975) beruht vermutlich auf seiner Fähigkeit, den Serotoninspiegel in deszendierenden Hemmbahnen zu erhöhen.

Endogene opiatähnliche Substanzen (Endorphine) wurden nach der Entdeckung spezifischer Opiatrezeptoren im Nervensystem vermutet (Pert und Snyder, 1973). Radioaktive Opiate besetzen Hirnregionen, wo gehäuft Rezeptoren für l-konfiguriertes Morphin vorkommen. Die elektrische Reizung in diesen Gebieten, z.B. des periaquäduktalen Höhlengraus, des Mandelkerns, kann zur Schmerzhemmung führen. Die vermuteten Substanzen wurden von Hughes und Mitarbeitern (1975) und Terenius (1975) in Form der Pentapeptide Methionin- und Leukinenkephalin entdeckt. Sie finden sich im Striatum, dem Nucl. amygdalae, dem Nucl. caudalis. Das größere Molekül (3500) mit 31 Aminosäuren, β-Endorphin genannt, wurde von Cox und Mitarbeitern 1975 aus der Hypophyse isoliert. Seine Aminosäuresequenzen entsprechen dem β-Lipotropin aus der Hypophyse, aus dem ACTH und Endorphine entstehen (Mains et al., 1977). β-Endorphine kommen auch im Hypothalamus und Mesenzephalon vor. Rezeptoren für die β-Endorphine finden sich vor allem entlang der paläospinothalamen Bahnen. Die Wirkung dieser Substanzen ist wie die des Morphins. Sie stimulieren die schmerzhemmenden deszendierenden Bahnen, sie werden durch Selbstreizung zur Schmerzlinderung im zentralen Höhlengrau freigesetzt (Reynolds, 1969; Liebeskind et al., 1974; Boethins et al., 1976) und finden sich bei Ratten, deren Füße durch Elektroschocks gereizt wurden, im Plasma mehrfach erhöht, wobei die Schmerzempfindlichkeit längere Zeit nach Applikation des Stresses noch erniedrigt ist. Nach verlängertem Streß nimmt die Schmerzempfindlichkeit zu, während die Endorphinaktivität im ZNS und peripher abnimmt (Madden et al., 1977; S. Amir und Z. Amir, 1978). Da Naloxon, der Morphinantagonist, die schmerzhemmende Wirkung der Akupunktur, der transkutanen Nervenstimulation und von Placebos zumindest partiell aufhebt, müssen die Endorphine bei diesen schmerzlindernden Maßnahmen eine Rolle spielen (Mayer et al., 1977; Levine et al., 1981).

36.1.8 Psychische Entwicklung und Schmerz

Die Beobachtung von Melzack und Scott (1957), daß in Isolation aufgezogene Hunde später ihre Schnauze in eine Flamme stecken, ohne Schmerzreaktionen zu zeigen, weist auf den engen Zusammenhang zwischen der Funktionsweise des „Schmerzapparates" und der psychophysiologischen individuellen Entwicklung hin. In dieser stehen die ersten Erfahrungen mit afferenten Impulsen, die zum Erlebnis Schmerz führen, mit dem Aufbau des intrapsychischen Körper-Selbst in Zusammenhang. Man muß annehmen, daß das Neugeborene unfähig ist zu erkennen, ob Reize, die auf es eindringen, aus seinem Körperinnern oder von außen stammen. Es empfindet von einer gewissen Reizstärke an diffuses Unbehagen. Es befindet sich im Stadium des primären Narzißmus (Freud, 1969). Die Begriffe des primären Autismus (Mahler, 1952) oder der objektlosen Phase (Spitz, 1969) meinen weitgehend dasselbe. Unter der Voraussetzung einer intakten Beziehung zu einer Pflegeperson (Mutter), die Mißempfindungen des Säuglings behebt, indem sie ihn z.B. stillt und trocknet, und seines reifenden Nervensystems, erwirbt er allmählich die Fähigkeit, innere Reize von äußeren zu unterscheiden und wahrzunehmen, daß in seiner Umgebung etwas geschieht, das mit dem Verschwinden seiner Mißempfindungen zu tun hat. Diese Wahrnehmung des eigenen Körpers und der Gestalt der Mutter geschieht anfänglich nur bruchstückhaft. Je vollständiger sie wird, desto mehr tritt das Kind in die prä- oder teilobjektale Phase ein (Klein, 1962). Die Begriffe symbiotische Phase (Mahler, 1952) und anaklitische Phase (Spitz, 1969) sind für die gleiche Periode verwendet worden. Man stellt sich vor, daß die in der Phase des primären Narzißmus diffus im Körperinnern verteilte Energie in der symbiotischen Phase mehr und mehr die Körperoberflächenstrukturen und die Sinnesorgane besetzt, und daß das Ausmaß, in dem diese Besetzung[1] erfolgt, für die Schärfe verantwortlich ist, mit der der eigene Körper differenziert wahrgenommen, als von der Umgebung getrennt erlebt und Reize intensiv empfunden werden. D.h., daß Reize (Schmerzreize) nötig sind, damit die Wahrnehmungsorgane besetzt werden, und daß diese Besetzung ihrerseits dafür verantwortlich ist, wie intensiv Reize später empfunden werden. Das Ausmaß der Besetzung der Wahrnehmungsorgane kann testpsychologisch erfaßt werden (Witkin et al., 1962; Petrie, 1967; Fisher, 1968). Eine hohe Besetzung der Wahrnehmungsorgane geht mit hoher Schmerzempfindlichkeit einher (Petrie, 1967; Adler und Lomazzi, 1973; Adler et al., 1973; Adler und Lomazzi, 1974). Da schmerzunempfindliche Individuen bei identischer organischer Läsion andere klinische Schmerzbilder zeigen als schmerzempfindliche Personen (Libman, 1934), scheint die Stärke der „Besetzung" für das entstehende klinische Bild und damit die Differentialdiagnose von Schmerz bedeutsam zu sein (Breuer und Freud, 1895).

Die oben erwähnte Beobachtung von Melzack und Scott (1957) und diejenige von Mahler (1952), daß autistische Kinder, die sich selbst schlecht als von ihren Bezugspersonen losgelöst und getrennt erfahren, schmerzunempfindlich sind, passen zur Vorstellung vom Zusammenhang zwischen Schmerzempfindlichkeit und Ausmaß der Besetzung der Wahrnehmungsorgane. Die Besetzung der Wahrnehmungsorgane ist also einerseits mit neurophysiologischen Vorgängen verbunden, und andererseits mit Erlebnissen zwischen Individuum und Umwelt. Neurophysiologische Abläufe und Interaktionen zwischen Subjekt und Umwelt gehören verschiedenen Ebenen an und werden mit Begriffen aus verschiedenen Wissenschaftssprachen bezeichnet. Sie werden

---

1 Der Begriff „Besetzung" wird im psychoanalytischen Sinn verwendet (eine bestimmte psychische Energie, die an eine Vorstellung oder Vorstellungsgruppe, einen Teil des Körpers, ein Objekt gebunden ist) (vgl. Laplanche, J. L., J. B. Pontalis, Suhrkamp, Frankfurt a. M. 1973).

aber durch ihr gleichzeitiges Vorkommen verbunden, eine Bedeutungskoppelung findet statt, und sie beeinflussen sich wechselseitig während der Entwicklung und Reifung des Individuums.

Die in früher Kindheit erlebten Schmerzen werden in die Beziehung des Kleinkindes zu seinen Pflegepersonen integriert. Leidet es unter Schmerz und drückt es ihn in seinem Verhalten aus, so führt dieses Signal zur Zuwendung der geliebten Person, des sog. Objektes. Zuwendung des Objektes und Abklingen der Schmerzen verbinden sich erlebnismäßig. Es entspricht einer geläufigen Beobachtung, daß ein Kind, das sich beim Sturz weh getan hat, zu weinen beginnt, wenn es die Mutter gewahrt und zu weinen aufhört, sobald sie es aufnimmt und tröstet. Darauf beruht ein wesentlicher Anteil der Wirkung von Placebos: Sie besitzen eine viel größere analgetische Wirkung beim Vorliegen von Erregung als beim ruhigen Patienten (Becher, 1962), weil zum Medikament das schutzbietende Objekt kommt. Die Placebowirkung kommt vermutlich durch Freisetzung von Endorphinen im Zentralnervensystem zustande (Levine et al., 1981).

Wiederum können wir die Bedeutungskoppelung zwischen Körpervorgängen einerseits und Interaktionen zwischen Subjekt und Umwelt erkennen. Die Verletzung, die zu neurophysiologischen Vorgängen führt, z.B. Ausschüttung von lokalen Gewebshormonen, Na^+-Ionenverschiebung usw., kann als unkonditionierter Reiz, die Reaktion im Körper als Antwort darauf bezeichnet werden. Die Zuwendung der schutzbietenden Person stellt den konditionierten Stimulus dar, der später allein, durch Bedeutungskoppelung, die Körperantwort auszulösen vermag.

In der frühen Entwicklung verbinden sich auch Aggression und Schmerz im Erleben. Das Kind lernt, daß es durch aggressive Handlungen anderen Schmerz zufügen kann, und daß die anderen auf sein Verhalten mit Zufügen von Schmerz zu antworten pflegen, den es dann als Strafe empfindet. Da es nicht wie der Erwachsene Gedanken und Handlung auseinanderzuhalten vermag, verbinden sich für es nicht nur aggressive Verhaltensweisen mit Bestrafung in Form von Schmerz, sondern es verquickt schon aggressive Gedanken mit Schmerz, den es als Buße für solche Gedanken zu erleben beginnt. (Dies wird für die Entstehung psychogener Schmerzen in Form sog. Konversionssymptome und die Lebenshaltung des „Schmerz-Erleiden-Müssens" vgl. bedeutsam; Engel, 1970.) Moderne Sprachen scheinen Zeugnis für diese Zusammenhänge abzulegen. Im Englischen besitzen z.B. die Wörter „pain" (Schmerz) und „punishment" (Bestrafung) oder „penalty" (Buße) eine gemeinsame Wurzel. Wiederum läßt sich eine Bedeutungskoppelung zwischen Interaktionen des Subjekts mit der Umwelt und Körpervorgängen erkennen.

Die Verbindung zwischen Schmerz und Vereinigung mit dem geliebten Objekt und zwischen Schmerz und Sühne läßt verstehen, daß Schmerz auch in der sexuellen Beziehung eine Rolle spielen kann. Dabei ist nicht das Zufügen von Schmerz im Sadismus lustvoll und sein Erleiden im Masochismus,

sondern er stellt die Vorbedingung für die sexuelle Befriedigung dar.

36.2 Klinische Schmerzbilder und der Einfluß psychischer Faktoren

36.2.1 Der konversionsneurotische Schmerz und die „Neigung, Schmerz erleiden zu müssen"

Anhand von Ausschnitten aus Anamnesen von zwei Patientinnen sollen die Kriterien, die den konversionsneurotisch bedingten Schmerz und die „Neigung, Schmerz erleiden zu müssen", diagnostizieren lassen, herausgearbeitet werden. Das Material entstammt gewöhnlichen beim Spitaleintritt durchgeführten Erhebungen der Anamnese. Die verwendete Anamnesetechnik ist im Kapitel 12 dargestellt.

Das Symptom, seine Beschreibung und das Verhalten des Patienten während der Erhebung der Anamnese

H. Q., eine 37jährige unverheiratete Frau leidet an Blähungen, Erbrechen und Schmerzen im Oberbauch, die zehn bis zwanzig Minuten nach jeder Mahlzeit auftreten, nach rechts in die Flanke und den Rücken, in den Unterbauch und in die Vorderseite der Oberschenkel ausstrahlen. Sie sind unabhängig von der Körperstellung, der Art der Speisen und der Stuhltätigkeit. Zusätzlich bestehen Schmerzen im Nacken, beiden Schultern und Ellbogen. Die Patientin klagt über Müdigkeit und Mühe, sich zu konzentrieren. Sie gibt spontan, bescheiden und zurückhaltend Auskunft. Sie betont, daß es sich um organische Störungen handeln müsse, obwohl sie durch Aufregung verstärkt würden. Die Angaben erfolgen detailliert, bereiten dem Arzt aber Mühe, sie in ihm vertraute Kategorien einzuordnen. Die Kranke erzeugt im Arzt das Gefühl, ein schweres Schicksal tapfer zu ertragen. Bei der Schilderung der Gelenkschmerzen ihres Vaters im Bereich des Nackens, der Schultern, der Ellbogen und der Oberschenkel sowie bei der Beschreibung der Oberbauchschmerzen ihrer leberleidenden Mutter macht die Patientin die gleichen Handbewegungen und legt die Hände an die gleichen Körperstellen wie bei der Darstellung ihrer eigenen Schmerzen.

D. S., eine 32jährige unverheiratete Mutter eines Kindes erleidet anderthalb Stunden nach einer Zahnbehandlung in Lokalanästhesie heftigste Schmerzen im Bereich des linken Unterkiefers. Sie strahlen in die linke Gesichtshälfte aus, hinter das linke Ohr, in den Nacken, die linke Schulter und den linken Arm. Nach zwei Tagen gesellen sich Erbrechen, Schwindel und Kraftlosigkeit hinzu. Die Schilderung erfolgt spontan, farbig und dramatisch. Die Patientin wirkt auf den Arzt schutzlos und schweren Schicksalsschlägen ausgesetzt. Bei der Beschreibung gerade vorliegender, heftigster Schmerzen lächelt sie.

Ausführliche klinische und Laboruntersuchungen ergeben bei beiden Patientinnen keine Anhaltspunkte für organische Veränderungen, die die Symptomatologie erklären würden. Bei beiden sind die Symptome keinem bekannten Krankheitsbild zuzuordnen. Die nähere Befragung läßt die Unfähigkeit der Patientinnen hervortreten, die Symptome in bezug auf Lokalisation, Ausstrahlung, verschlimmernde und lindernde Faktoren, Begleitsymptome usw. präzise zu umreißen. Die ausgesprochen schüchterne, zurückhaltende Art der Schilderung der ersten Patientin und die farbig-dramatische Darstellung der zweiten, die beim Beschreiben intensivster Schmerzen lächelt, fallen auf.

Die persönliche Anamnese

H. Q. erkrankt mit sechs Jahren an Diphtherie. Mit vierzehn Jahren setzen Schwächezustände ein, die sie zwingen, während fast allen Schulferien das Bett zu hüten. Die Menses sind von Beginn an sehr schmerzhaft. Zwischen dem zwanzigsten und vierundzwanzigsten Lebensjahr unterzieht sie sich vier gynäkologischen Operationen wegen Schmerzen und Blutungen. Diese Beschwerden verschwinden erst mit zweiunddreißig Jahren nach der Hysterektomie, der partiellen Ovarektomie links und der totalen rechts.

Schon kurz nach dem Eingriff setzt das jetzige Beschwerdebild ein. In den fünf Jahren seines Bestehens fallen verschiedene, vom Gastroenterologen durchgeführte Abklärungen normal aus.

D. S. erleidet mit zehn Jahren einen Nasenbeinbruch. Mit elf und zwölf Jahren erfolgen Eingriffe am Nasenseptum. Mit vierzehn Jahren wird die Appendektomie durchgeführt. Mit zwanzig verunglückt sie beim Skifahren und zieht sich einen Oberkieferbruch links zu. Wegen dauernder Schmerzen im Bereich der linken Wange wird der N. infraorbitalis in den folgenden zwei Jahren zweimal dekomprimiert und ein Neurom entfernt, ohne daß die Schmerzen nachlassen. Mit siebenundzwanzig wird der Nerv reseziert, und die Patientin bleibt zwei Jahre lang beschwerdefrei. Mit neunundzwanzig setzen die gleichen Schmerzen wieder ein. Während des letzten Jahres klingen sie überhaupt nicht mehr ab. Zu ihnen kommen in den letzten Monaten Oberbauchschmerzen rechts, die sich nicht näher charakterisieren lassen. Das Cholezystogramm fällt normal aus.

Die Anamnese beider Patientinnen ist gekennzeichnet durch jahrelang anhaltende Beschwerden, vorwiegend Schmerzen, die in der Pubertät begannen und zu den verschiedensten Interpretationen, Abklärungsuntersuchungen und Behandlungen führten, ohne daß eine Heilung erzielt werden konnte.

Die Beziehung zu den Eltern und
die persönliche Entwicklung

H. Q. hat einen jähzornigen und brutalen Vater, der die Patientin bis fast ins Erwachsenenalter körperlich züchtigte und auch seine Ehefrau schlug. Vom Arzt wegblickend schildert die Patientin, wie der Vater sie vom fünfzehnten Lebensjahr an sexuell mißbraucht ha-

---

be. Sie sei in fürchterlichen Zwiespalt geraten, denn einerseits habe sie ihren Vater überaus gern gehabt und liebe ihn heute noch, und andererseits habe sie ihn gehaßt und über sein Verhalten aus Angst vor seiner Rache niemandem etwas anvertraut. Mit zwanzig Jahren hätte sie sich in einen bescheidenen, zuverlässigen Burschen verliebt, seine Annäherungsversuche aber nicht ertragen und deshalb mit ihm gebrochen. Sie habe ihn noch nach seiner Verheiratung und bis zu seinem Tode an Lungenkrebs geliebt. Spätere Freunde hätten sie lediglich auszunutzen versucht.

Mit zweiundzwanzig unternimmt sie einen Selbstmordversuch. Im Einweisungszeugnis für den jetzigen Spitalintritt bemerkt der Hausarzt, daß die Patientin auffällig viele Analgetika und Spasmolytika braucht.

D. S. wuchs bei Eltern auf, die sie wiederholt mit Lederriemen und Stöcken züchtigten. Sie erleidet dabei Verletzungen im Bereich der Nase. Sie fragt sich in der Kindheit oft, ob sie wohl ein so böses Kind sei, daß sie so häufig geschlagen werden muß. Sie verläßt das Elternhaus kurz nach dem Schulabschluß und zieht in eine 200 km entfernte Stadt. Dort befreundet sie sich mit einem Mann, mit dem sie eine sexuelle Beziehung aufnimmt. Sie wird schwanger. Erst dann erfährt sie, daß ihr Freund bereits verheiratet ist. Er will für die Unterbrechung der Schwangerschaft aufkommen. Sie lehnt ab, nimmt die Schwangerschaft auf sich und zieht nach der Geburt des Kindes in eine andere größere Stadt. Hier wohnt sie isoliert von ihrer weit weg lebenden Familie, ohne Freunde und Bekannte. Sie arbeitet ganztägig, gibt das Kind tagsüber in eine Krippe und betreut es abends und besorgt den Haushalt, ganz erschöpft von der Tagesarbeit. Sie hat keine Beziehungen zu den anderen Hausbewohnern und kennt niemanden, der das Kind abends einmal hüten würde. Am Arbeitsplatz leidet sie stark. Sie fühlt sich von den Vorgesetzten zu Unrecht häufig geplagt.

Beide Patientinnen haben in der Kindheit unter der Brutalität ihrer Eltern gelitten, die nicht nur gegenüber dem Kind, sondern auch gegenseitig körperlich tätlich wurden. Die Mutter der ersten Patientin war dabei selbst dauernd krank. Die spätere Entwicklung war bei beiden gekennzeichnet durch Beziehungen zum anderen Geschlecht, die nie in eine harmonische Verbindung mündeten, sondern zusammenbrachen und die Patientinnen vereinsamt zurückließen. Beide leben eingeengt, gequält, leiden an ihrem Arbeitsplatz und tragen ihr Los, ohne aus ihrer mißlichen Lage herauszukommen. Oberflächlich betrachtet scheint sich das Schicksal gegen sie verschworen zu haben. Eine genauere Prüfung der Lebensgeschichte zeigt hingegen, daß Kräfte, die in diesen Kranken selbst liegen, sie immer wieder in schwere Lebenslagen treiben. Die Neigung zu Depressionen, zum Selbstmord und die Abhängigkeit von Medikamenten fallen auf.

Zeitpunkt des Auftretens der Schmerzen, Wahl
der Art des Symptoms und seine Lokalisation

Bei H. Q. setzen die Schmerzen im Unterleib zu einer Zeit ein, in der sie sich erstmals mit einem Burschen befreundet und die Beziehung scheitern lassen muß.

Die Oberbauchschmerzen beginnen, nachdem die gynäkologischen Eingriffe mit der Hysterektomie geendet haben. Sie werden zweimal während Konflikten mit Kollegen am Arbeitsplatz intensiver.

Bei D. S. intensivieren sich die Schmerzen in der Zeit einer Auseinandersetzung mit einer älteren Vorgesetzten am Arbeitsplatz.

Bei H. Q. lokalisieren sich die Schmerzen im Genitalbereich, wobei sexuelle Konflikte auf der Hand liegen. Für die Wahl der Art und Lokalisation der Gelenkbeschwerden stehen diejenigen des Vaters Modell, zu dem eine ambivalente Beziehung besteht. Die Oberbauchschmerzen stützen sich auf das Vorbild der leberleidenden Mutter.

Bei D. S. erfolgt die Wahl der Symptome und ihre Lokalisation entsprechend den Schmerzen, die sie in der Kindheit wiederholt erlitten hat und die ihr von bedeutsamen Bezugspersonen zugefügt worden sind.

Der konversionsneurotische Schmerz tritt demnach in Konfliktsituationen auf, in denen verpönte Wünsche, Affekte usw., z.B. Aggression, vom Bewußtsein abgehalten werden müssen (Freud, 1972), bei drohendem oder realem Verlust einer (meist ambivalent) geliebten Person, einer hochgeschätzten Tätigkeit, eines Besitzes und anderes mehr. Ihre Wahl und Lokalisation erfolgt aufgrund von früher selbst erlebten Schmerzen, von Schmerzen, die von bedeutsamen Bezugspersonen erlebt wurden, oder in einem Körperbereich, der dem Ausdruck von (verpönten) Wünschen dient (Engel, 1951, 1959, 1970). Das Symptom Schmerz drückt dann diese Strebungen sowie die sie unterdrückenden Tendenzen der ethischen und moralischen Teile der Persönlichkeit im Sinne eines Kompromisses aus. Schmerz eignet sich für solche symbolischen Ausdrucksweisen besonders gut, denn er bringt die verpönte Regung zum Ausdruck und bestraft das Individuum gleichzeitig dafür.

Die Schilderung des Symptoms erfolgt nach der intrapsychischen Vorstellung des Patienten und entspricht nicht anatomischen und pathophysiologischen Gegebenheiten. Eine Ausnahme bildet der konversionsneurotische Schmerz, der als Modell einen vom Individuum selbst früher erlebten organisch bedingten Schmerz verwendet. Als Beispiel sei der Patient mit Pseudoangina pectoris genannt, der einen Myokardinfarkt erlitten hat, in dessen Folge er seine berufliche Situation nicht mehr meistern kann und in einen Konflikt gerät.

Konversionsneurotische Schmerzen bedürfen zur Diagnose wie jede andere Krankheit positiver Kriterien. Wir benötigen zur Diagnose den Nachweis, warum das Symptom „Schmerz" gewählt wurde und nicht ein anderes, wie z.B. Atemnot, wir müssen belegen, warum es gerade jetzt auftritt, und wieso es diese Lokalisation wählt. Wir müssen zeigen, daß ein bestimmter Konflikt im Symptom neutralisiert wird (pri-

märer Krankheitsgewinn) und daß dem Patienten aus dem Kranksein heraus neue Möglichkeiten der Beziehung zu Objekten erwachsen (sekundärer Krankheitsgewinn). Als Hinweise, daß es sich beim vorliegenden Symptom um einen konversionsneurotischen Schmerz handeln könnte, dienen die auffallend gehemmte, bescheidene oder aber dramatisch-theatralisch wirkende Schilderung, die Vagheit, mit der das Symptom beschrieben wird, die Neigung, den Arzt gefühlsmäßig zu fesseln, ihm zu gefallen, seine Zuneigung zu gewinnen, die Suggestibilität, die Übernahme verschiedener Rollen innerhalb kurzer Zeiträume, die übermäßige Abhängigkeit von Bezugspersonen, die Neigung zu depressiven Reaktionen, Suizidversuchen, zur Abhängigkeit von Medikamenten, sowie eine Anamnese mit Symptomen, die sich keinem organischen Krankheitsbild zuordnen lassen. Häufig finden sich Störungen im sexuellen Verhalten wie häufiger Partnerwechsel, Scheidungen, Frigidität und Impotenz. (Diese Hinweise können unter dem Begriff des hysterischen Charakters zusammengefaßt werden.)

Wie die zwei Fallbeispiele zeigen, finden sich konversionsbedingte Schmerzen oft (aber nicht nur) bei Patienten, die „Schmerz erleiden müssen". Sie wirken oft traurig, schuldbeladen, schwernehmerisch, selbstquälerisch. Sie weisen Lebensgeschichten auf mit vielen Schmerzzuständen, die schwer einem organischen Krankheitsbild zugeordnet werden können, belastende Erfahrungen während der Kindheit wie z.B. Mißhandlungen (Merskey und Boyd, 1978; Adler et al., 1988), viele Operationen mit zum Teil unverständlicher Indikation, frustrierende zwischenmenschliche Beziehungen, die immer wieder auseinanderbrechen, und sie versagen häufig, wenn ihnen die Umwelt günstig gesinnt ist, um andererseits stoisch schwerste Lebenssituationen zu ertragen. Engel hat diese Züge unter dem Begriff der „pain proneness" („Neigung, Schmerz erleiden zu müssen") zusammengefaßt (Engel, 1959, 1970).

Behandlung konversionsneurotischer Schmerzen

Die Schmerzen können bei Änderungen in der Lebenssituation des Patienten ohne spezifische Therapie verschwinden. Sie können aber auch hartnäckig anhalten. Ihre Beeinflussung wird dadurch erschwert, daß das Symptom „Schmerz" den intrapsychischen Konflikt neutralisiert. Damit fehlt der Leidensdruck und die Motivation für eine Therapie. Dazu kommt, daß Schmerz bei Patienten mit konversionsneurotischem Schmerz schon früh in der Kindheit für die Erhaltung des psychischen Gleichgewichts bedeutsam geworden sein kann und diese Funktion später beibehält (Sorvant et al., 1988). Daraus ergibt sich, daß eine Behandlung, die auf die Befreiung vom Symptom hinzielt, den Patienten des für ihn möglichen Wegs der Erhaltung seines psychischen Gleichgewichts beraubt. Die Beziehung zum Arzt wird für einen solchen Patienten damit bedrohlich. Er muß sie entweder scheitern lassen oder sein Schmerz exazerbiert, oder es tritt ein neues Schmerzsyndrom mit neuer Lokalisation auf. Der Arzt muß deshalb geduldig warten können, bis die Arzt-Patient-Beziehung so stabil geworden ist, daß der Kranke es riskieren kann, sein Symptom aufzugeben. Es kann nützlich sein, als Überbrückung der dafür nötigen Zeitspanne eines der mehr somatisch bedingten Symptome des Patienten mit physikalischen Mitteln oder medikamentös zu behandeln. Dabei muß aufgepaßt werden, daß der Arzt nicht „mitagiert", d.h. dem Wunsch nach somatischer Behandlung des Patienten nicht immer weitgehender entgegenkommt. Er soll im Gegenteil keine Gelegenheit verstreichen lassen, den Patienten im Sinne von Konfrontationen (Greenson, 1967) auf Lebensschwierigkeiten und emotionale Probleme hinzuweisen, auf die der Patient im Laufe der Zeit oft spontan zu sprechen kommt, wenn er im Arzt einen wohlwollenden und verständnisvollen Zuhörer kennenlernt. Die Aufgabe, den Patienten die zeitlichen und inhaltlichen Zusammenhänge zwischen Schmerzentstehung oder -exazerbation und Wünschen, Phantasien und Affekten erleben zu lassen, stellt in bezug auf Wahl des Zeitpunktes, der Worte, der Stimmlage usw. an den Arzt Anforderungen, denen zu genügen er der Anleitung und Übung bedarf. Die „Neigung, Schmerz erleiden zu müssen" kann aber bei gewissen Patienten so ausgeprägt sein, daß der Arzt schon viel erreicht, wenn er dem Kranken helfen kann, seine Schmerzen zu akzeptieren und ihn vor neuen, unnötigen Abklärungsuntersuchungen und therapeutischen Eingriffen schützt.

36.2.2 Hypochondrie und hypochondrische Reaktion

Sie stellt einen Zustand dar, bei dem sich der Patient übermäßig mit Veränderungen in seinem Körper und Krankheiten beschäftigt. Sie reicht von der vorübergehenden, leichten Störung (hypochondrische Reaktion), z.B. des Studenten, der erstmals in der Klinik mit einem Schmerz beinhaltenden Krankheitsbild in Berührung kommt, oder des Menschen, der eine wichtige Bezugsperson an einem mit Schmerz verbundenen Leiden verloren hat, bis zur hartnäckigen, invalidisierenden Überzeugung krank zu sein (Hypochondrie). Solche Patienten beachten leichte Störungen, die mit Schmerz einhergehen, Flecken auf der Haut, mit der Darmtätigkeit in Beziehung stehenden leicht schmerzhaften abdominellen Druck und anderes mehr übermäßig und leiten aus ihnen schwere Krankheiten ab wie Krebs usw., vor denen sie sich fürchten. Von der Konversion unterscheidet sich das hypochondrische Symptom hauptsächlich dadurch, daß es noch vager ist, wechselnden und nagenden, quälenden Charakter hat, während sich der Patient mit Konversion durch seine Unbeteiligtheit – „belle indifférence" – seinem Symptom gegenüber auszeichnet. Er beantwortet beispielsweise die Frage, an was er leide, mit „ich weiß es nicht", während der Patient mit hypochondrischem Schmerz antwortet „ich habe Angst, daß der Schmerz Krebs usw. bedeutet".

Prognose und Therapie entsprechen weitgehend dem unter Konversion Gesagten. Es sei speziell erwähnt, daß der Arzt nur diejenigen Patienten gezielt beruhigen und ihnen die Harmlosigkeit ihrer Befürchtungen bestätigen soll, die psychisch weitgehend ausgeglichene Individuen sind, bei denen eine akute, belastende äußere Situation hypochondrische Befürchtungen ausgelöst hat. Bei neurotischen und psychotischen Patienten führt die aufmunternde Beruhigung vielleicht zu einer kurzdauernden Besserung, dann aber zum Wiederauftreten der Befürchtungen und zum Aufsuchen eines neuen Arztes. Der geduldige Aufbau einer soliden Beziehung zum Patienten ist lohnender.

36.2.3 Depressive Reaktionen

Sie sind gekennzeichnet durch Gefühle der Entmutigung, abnehmendes Interesse an der Arbeit und den Mitmenschen, Mangel an Energie und Lebenslust, Müdigkeit und Apathie. Weinanfälle „ohne Grund" sind häufig. Inaktivität und Schlafbedürfnis einerseits oder gespannte Unruhe und Schlaflosigkeit andererseits sind typisch. Gewisse Patienten verlieren den Appetit und nehmen ab, andere essen vermehrt, aber ohne Vergnügen und nehmen zu. Frigidität und Impotenz können sich einstellen. Somatische Symptome sind häufig, unter ihnen auch Schmerz, beispielsweise im Bereich des Bewegungsapparates. Die Hauptrolle bei der Schmerzentstehung spielt wahrscheinlich der veränderte Muskeltonus, denn der depressiv reagierende Patient, wie schon aus seiner vornübergebeugten Haltung, den hängenden Schultern usw. hervorgeht, zeigt eine andere Innervation bestimmter Muskelgruppen, wie elektromyographische Untersuchungen bei psychisch Ausgeglichenen und depressiven Patienten mit Ableitung aus Gesichtsmuskeln gezeigt haben (Schwartz et al., 1974). Depressive Reaktionen finden sich häufig nach Verlust einer bedeutsamen Bezugsperson, einer geschätzten beruflichen Tätigkeit, beispielsweise nach der Pensionierung usw. Schmerzzustände bei reaktiv deprimierten Menschen können selbstverständlich auch konversionsneurotischer Natur sein.

Die Behandlung des Schmerzes soll nicht symptomatisch erfolgen, auch wenn als Brücke zur kausalen Therapie der reaktiven Depression das Symptom „Schmerz" beispielsweise mit physikalischer Therapie angegangen wird. Bei einer reaktiven Depression, die vor allem durch äußere schwere Lebensumstände ausgelöst worden ist, bildet das patientenorientierte Interview mit seinen psychotherapeutischen Aspekten den Schlüssel zur Behandlung. Die Unterstützung der Trauerarbeit, die der Patient zu leisten hat, durch den Arzt stellt die Hauptaufgabe dar. Liegen neben äußeren Umständen, die nicht nur den psychisch verletzlicheren, sondern auch den ausgeglicheneren Menschen getroffen haben würden, Konflikte der Depression zugrunde (depressiv-neurotische Entwicklung), dann stellt sich die Frage der Indikation zur Psychotherapie. Ihre Einleitung kann in manchen

Fällen nicht sofort erfolgen, sondern der Patient bedarf der Vorbereitung, wie sie beim konversionsbedingten Schmerz beschrieben worden ist.

36.2.4 Endogene Depression

Sie kann ebenfalls mit Schmerzsymptomen einhergehen. Sie unterscheidet sich von der depressiven Reaktion und der depressiv-neurotischen Entwicklung hauptsächlich durch den Verlust der Realitätsprüfung und das unlogische Denken. Die Gefühle von Wertlosigkeit, Scham, Schuld usw. entsprechen nicht der realen Situation. Stimmung und die Intensität der körperlichen Symptome zeigen die bekannte Tagesschwankung mit morgendlicher Verschlimmerung und Besserung am Abend. Dem Schmerz kann eine wahnhafte Denkstörung zugrunde liegen im Sinne des Kompromisses, die dem Kranken ermöglicht, sich für seine vermeintlichen Verfehlungen zu bestrafen, vergleichbar der Vorstellung zu verhungern, zu verarmen usw. Schmerz kann auch über Veränderungen des Muskeltonus zustande kommen.

Die Behandlung richtet sich nicht nach dem Symptom, sondern nach der zugrundeliegenden Störung, nämlich der endogenen Depression. Der Leser findet sie in Lehrbüchern der Psychiatrie. Das Auftreten von Schmerzen bei endogener Depression ist umstritten, steht aber nicht im Vordergrund (Merskey, 1980). Sie lassen unter der Behandlung der Depression nach (Bradley, 1963; Ward et al., 1979; Webb und Lascelles, 1962).

Chronische organisch bedingte Schmerzen können zu psychischen Folgen führen, die depressive Züge aufweisen. Sie schwächen sich ab, wenn der Schmerz nachläßt (Kissen, 1964; Merskey und Boyd, 1978; Sternbach und Timmermans, 1975; Armentrout, 1979).

36.2.5 Schmerz als körperliches Begleitzeichen von Affekten

Unangenehme Affekte wie Angst, Furcht, Scham, Schuld, Ekel, Ärger und Wut usw. treten auf, wenn Veränderungen im Bereich des eigenen Körpers, z.B. bei einer Krankheit, im Bereich der Psyche, z.B. bei einer bedrohlichen Phantasie, oder in der Umwelt den Menschen vor eine Problemsituation stellen, zu deren Lösung er kein Programm gebrauchsfertig zur Verfügung hat. Die sog. Affekte stellen dann Signal- oder Prüfaffekte dar, die das Individuum zu einer Bereitstellungsreaktion veranlassen. Diesen Affekten sind als Bereitstellungsreaktionen die zwei biologischen Grundmuster zugeordnet, von denen das eine von Cannon (1920) als „Flucht-Kampf"-Muster, das andere von Engel (1976) als „Rückzug-Konservierungsmuster" bezeichnet worden ist. Zu den Affekten Ärger und Wut gehört das erstere, mit den körperlichen Zeichen wie Herzklopfen, Zittern, Schwitzen, Schwindel und für uns hier wichtig, Muskelanspannung, zum zweiten Nausea, Erbrechen. Auf Nausea

und Erbrechen im Zusammenhang mit dem Rückzugs-Konzept hat v. Uexküll unabhängig von Engel hingewiesen (1952).

Patienten mit Schmerz durch Muskelanspannungen als Begleitzeichen der Affekte Ärger und Wut begeben sich mit ihrem Symptom zum Hausarzt, Internisten usw., weil sie die auslösende äußere Situation und/oder intrapsychische Konflikte, die zu den genannten Affekten geführt haben, oder sogar die vorliegenden Affekte selbst verleugnen oder verdrängen. Solche Patienten klagen dann über Herzklopfen, Zittern und Schmerzen im Bereich der Schläfen, des Nackens, des Unterkiefers im Gebiet der Kaumuskulatur, der Schultern und des Rückens (vgl. Kap. 45.3). Dem Angebot von Muskelschmerzen als Leitsymptom an den Arzt liegt nicht immer Verleugnung und Verdrängung zugrunde, sondern es kann auch darauf beruhen, daß der Patient erfahren hat, daß sich Ärzte wohl mit körperlichen Symptomen abgeben, nicht aber mit quälenden Vorstellungen, Affekten und Erlebnissen. Patienten mit Muskelschmerzen, die Affekte von Ärger und Wut selbst nicht gewahren, verraten sich häufig durch zusammengepreßte Lippen, geballte Kiefermuskulatur, verkrampfte Fäuste, gezwungenes Lächeln, Verneinen jeglicher Ärgergefühle und die Schilderung ärgerlicher und gespannter mitmenschlicher Beziehungen mit Worten ausgewählter Freundlichkeit und Liebenswürdigkeit.

Wiederum liegt der therapeutische Schlüssel im richtig gehandhabten Interview (vgl. Kap. 12), das erlaubt, den Patienten auf seine Affekte aufmerksam zu machen und die mit ihnen verbundenen Erlebnisse usw. zu klären. Oft sind die Affekte so bewußtseinsnah, daß die therapeutische Arbeit im Sinne der konfliktgerichteten Kurztherapie geleistet werden kann. Schon die Entdeckung des Patienten, daß er beim Arzt auf einen Menschen trifft, dem Affekte von Ärger und Wut wichtig sind, und denen er nicht ausweicht, sondern sie beim Kranken sogar erwartet, kann zum raschen Abklingen der Symptome führen. Voraussetzung für ein solches ärztliches Verhalten ist, daß der Arzt seine eigenen Gefühle bemerken und verstehen kann. Bei zu Schmerz führenden Muskelspannungen kann das autogene Training zur Entspannung beitragen. Es muß aber die Therapeut-Patient-Beziehung zur Klärung der Affekte und den damit verbundenen Umwelt- und intrapsychischen Problemen gleichzeitig gehandhabt werden, denn das Training verlangt vor allem ein Aufgebenkönnen einer gewissen Kontrolle (über die Motorik), die gerade bei zu Dauerverspannungen der Muskeln neigenden Individuen einen der hervorstechenden Mechanismen der Streßbewältigung darstellt, also einem Schutz vor psychischer Dekompensation gleichkommt.

Bei Spannungskopfschmerzen, die über Jahre andauern und zum chronischen Analgetikagebrauch geführt hatten, wurde die Spannung des M. frontalis elektrisch abgeleitet und dem Patienten in Form akustischer Signale so zugeführt, daß einer erhöhten Spannung eine Zunahme und einer verminderten Spannung eine Abnahme in der Frequenz der Signale entsprach. Dabei lernte eine erste Gruppe von Patienten sich so einzustellen, daß sie bewußt eine Entspannung des Frontalmuskels herbeiführen konnten. Diese Gruppe zeigte eine erstaunliche Verminderung der Kopfschmerzen und des Analgetikaverbrauchs, die noch nach 18 Monaten nachweisbar waren, während eine durch falsche Signale benachrichtigte zweite Gruppe und eine dritte, die auf die Warteliste für die Behandlung kam, keinerlei Erfolge zeigten (Budzynski et al., 1973). Hier scheint die erlernte Koppelung intrapsychischer Vorstellungen und Haltungen mit einer Verminderung der Muskelspannung zur Schmerzabnahme beigetragen zu haben, möglicherweise über eine Dämpfung des Beitrages des „motivierenden affektiven Systems" zum Schmerzempfinden. Da Entspannungsübungen bei Spannungskopfschmerz erfolgreich sein können (Warner und Lance, 1975), spielen vermutlich noch weitere unspezifische Faktoren eine therapeutische Rolle (Jessup et al., 1979).

36.2.6 Simulation

Die willentliche Vortäuschung einer Krankheit, hier von Schmerz, trifft man in der zivilärztlichen Tätigkeit selten. Sie kommt im Militärdienst, bei Gefangenen und bei Kindern vor und verschwindet meist, wenn die List entdeckt wird. Der Laie als Simulant ist leicht zu erkennen, denn er besitzt nicht genügend Kenntnisse über das dem vorgegebenen Schmerz zugrundeliegende Krankheitsbild. Schwieriger bis unmöglich sind Simulanten zu erkennen, die beruflich mit Kranken zu tun haben. Zu ihnen gehört das gesamte paramedizinische Personal, wie Laboranten, Physiotherapeuten, Röntgenassistenten, Krankenschwestern usw. Heute gesellen sich noch die Drogensüchtigen dazu, die durch Vortäuschung einer schmerzhaften Erkrankung zu ihren Suchtmitteln zu gelangen versuchen. Hier können sich diagnostische Schwierigkeiten ergeben, wenn der Simulant einen Schmerz anbietet, der einem Symptom entspricht, das er in der Vergangenheit anläßlich einer organischen Störung einmal erlebt hatte. Sind rechtliche und finanzielle Aspekte im Spiel, so kann der Simulant dem Arzt die für die Diagnose nötigen Informationen vorenthalten.

Die Unterscheidung von Simulation und Konversion kann schwer sein. Als Hilfszeichen kann verwendet werden, daß der Simulant meist mürrisch, verstimmt, verschlossen, geheimnisvoll und abweisend auftritt, der Konversionskranke hingegen ist offener, freundlicher, zugewandter, anhänglicher und unbekümmerter.

Das Stellen der Diagnose beim Symptom „Schmerz" beruht auf der in Kapitel 12 dargestellten Technik der Anamneseerhebung, auf der Kenntnis organisch bedingter Schmerzsyndrome, auf derjenigen der entwicklungspsychologischen Bedeutung von Schmerz und auf dem Wissen um die Mechanismen der Symptombildung bei der Konversion. Es kann nicht genug betont werden, daß „Schmerz" nie nach dem Aus-

schlußverfahren als „psychogen" oder „funktionell" bezeichnet werden darf. Fehlende organische Veränderungen sind nur einer unter vielen Hinweisen auf das mögliche Vorliegen eines vorwiegend oder ganz psychogen bedingten Schmerzzustandes, aber noch lange kein Beweis dafür. Man denke nur an Schmerzsyndrome, bei denen die organischen Veränderungen diskret sind, oder gar nur auf biochemischen Störungen beruhen, wie bei der akuten, intermittierenden Porphyrie, dem Fabry-Syndrom usw.

36.3 Hilfsmittel für die Differentialdiagnose zwischen vorwiegend psychogenen und organisch bedingten Schmerzen

Aus den einleitenden Bemerkungen über Neurophysiologie/Anatomie und Schmerz in der psychischen Entwicklung sollte klar geworden sein, daß jede Trennung in „psychogenen Schmerz" und „organischen Schmerz" eine künstliche sein muß. Aus diagnostischen und therapeutischen Gründen ist eine Gewichtung der jeweiligen sozialen, psychischen und somatischen, zum Schmerz beitragenden Faktoren jedoch unerläßlich. Der Arzt muß aus der Diagnose (die mehrere Ebenen – vom biochemischen bis zum sozialen Bereich – betreffen kann) die Anweisung ableiten können, ob er beispielsweise auf weitere somatische Tests verzichten darf und ob er bei der Begleitung des Schmerzkranken sich um die Klärung sozialer und psychischer mitverursachender Faktoren bemühen soll, oder ob die Suche nach einer wahrscheinlich noch verborgenen organischen Komponente weitergetrieben werden soll.

36.3.1 Der Libman-Test

Patienten mit identischen organischen Läsionen zeigen häufig Unterschiede in Schmerzempfinden und -verhalten. Die zentralnervös bedingte, individuell unterschiedliche Lebhaftigkeit, mit der die beim Einwirken eines Reizes entstehenden afferenten sensorischen Impulse registriert und verarbeitet werden, scheint dafür bedeutsam zu sein, mit anderen Worten also der „perzeptuelle Stil" (Sweeney und Fine, 1965; Adler und Lomazzi, 1973; Adler et al., 1973). Ein einfacher, ohne besondere Hilfsmittel während der Körperuntersuchung durchführbarer Test der individuellen Schmerzempfindlichkeit, des Stils der Schmerzperzeption, ist der Styloiddrucktest (Libman, 1934): Der Daumen wird zuerst als Kontrolle auf das linke Mastoid gepreßt, dies löst beim gesunden Mastoid keinen Schmerz aus. Dann wird er unterhalb des linken Ohrläppchens zwischen Mastoid und aufsteigendem Unterkieferast in Richtung des Processus styloideus gepreßt, so daß auf einen Ast des Nervus auricularis magnus ein Druck ausgeübt wird. Das gleiche Vorgehen wird rechts wiederholt. Der Test wird in die körperliche Untersuchung einbezogen, ohne daß der Patient darauf aufmerksam gemacht wird, daß es um die Erfassung seiner Schmerzempfindlichkeit geht. Sonst fließen emotionale Faktoren wie Angst, Verleugnung usw. störend in die genuine Schmerzempfindlichkeit ein. Aufgrund der Reaktion auf den Styloiddruck werden die Patienten in drei Gruppen eingeteilt: O sensitiv, + sensitiv und +++ sensitiv. Zur Gruppe O zählen Personen, die beim Styloiddruck weder mimisch noch verbal eine Schmerzreaktion zeigen und auf die Frage „Was haben Sie gespürt" höchstens von Druck, nicht aber von Schmerz sprechen. In die Gruppe + werden Patienten eingereiht, die mimisch nicht reagieren, spontan verbal höchstens „leichten Schmerz" nennen und auch die erwähnte Frage höchstens mit „leichtem Schmerz" beantworten. Die Gruppen O und + werden der Gruppe +++ gegenübergestellt, die deutliche mimische und/oder verbale Schmerzreaktionen zeigt. Sie wird „hypersensitiv" genannt. Die „hyposensitive" (O-) Gruppe fällt dadurch auf, daß die Patienten bei vorhandener Läsion nur wenig Schmerz oder an seiner Stelle „Substitutionssymptome" äußern, die Schmerz repräsentieren, wie Brennen, Völlegefühl, Druck, Ameisenlaufen usw., oder „Ersatzsymptome" wie z.B. Dyspnoe, Husten, Aufstoßen und anderes anstelle von Schmerz. Auch findet sich der Schmerz, falls er überhaupt verspürt wird, nicht selten auf der kontralateralen Körperseite der Läsion und strahlt von der Peripherie zur Läsion. Bei den „hypersensitiven" Patienten stellt der Schmerz hingegen das Leitsymptom dar, findet sich am Ort der Läsion und strahlt eventuell in die gleichseitige Headsche Zone aus. Auch wenn dieser Test nur eine grobe klinische Prüfung darstellt, so korreliert die mit ihm festgestellte Schmerzempfindlichkeit befriedigend mit derjenigen, die mit feineren Methoden gemessen wird (Pelner, 1941; Sherman, 1943; Keele und Lond, 1954). Die mit dem Libman-Test erfaßte Schmerzempfindlichkeit korreliert auch mit dem perzeptuellen Stil z.B. des visuellen Systems. Er erlaubt in Fällen, wo Anamnese und Körperuntersuchung keinen Entscheid zulassen, ob eine organische Läsion zum Schmerzempfinden und -verhalten beiträgt, eine zusätzliche Orientierung: Bei einem „hypersensitiven" Patienten, der über Schmerz klagt, müssen die Beschwerden „verkleinert" werden, um Bedeutung und Ausdehnung der organischen Läsion beurteilen zu können. Klagt hingegen ein „hyposensitiver" Patient über Schmerz, so müssen seine Beschwerden „vergrößert" werden, damit eine Vorstellung von der Läsion gewonnen werden kann. Besondere Vorsicht sollte man walten lassen, wenn ein im Libman-Test hyposensitiver Patient über andere Beschwerden als typische Schmerzen klagt. Es sollte auch berücksichtigt werden, daß die Läsion auf der kontralateralen Körperseite lokalisiert sein kann und auch an der Stelle, die der Schmerz bei seinem Ausstrahlen erreicht.

36.3.2 Weitere Hilfsmittel

Für die Unterscheidung zwischen psychisch und organisch bedingtem Schmerz wurden weitere testpsy-chologische Methoden gesucht. Eine Reihe von Studien vermochte mit dem MMPI (Minnesota Multiphasic Personality Inventory) organische von psychogenen Schmerzen zu trennen (Calsyn et al., 1976; Cox et al., 1978; Freeman et al., 1976; Hanvik, 1951; McCreary et al., 1977). Die Überlappung beider Gruppen war aber beträchtlich und das Instrument dadurch für den Einzelfall schlecht brauchbar. Andere Autoren konnten diese Beobachtungen nicht bestätigen (Fordyce et al., 1978; Sternbach et al., 1973). Aus dem MPQ (Maudsley Personality Questionnaire) abgeleitete Skalen vermochten die organischen und möglicherweise organischen Schmerzpatienten von Patienten mit Schmerzen psychogenen Ursprungs zu trennen (Leavitt et al., 1979). Eine „Back Pain Classification Scale" ließ beobachten, daß Patienten ohne nachweisbare somatische Läsion, die auf der genannten Skala denen mit Läsion glichen, im Unterschied zu Patienten mit sicher psychogenen Rückenschmerzen sich erfolgreich einer Physiotherapie unterzogen. Unterschiede im Plasmacortisolspiegel zwischen organischen und psychischen Schmerzpatienten wurden festgestellt (Johansson, 1982; Lascelles et al., 1974; Shenkin, 1964). Wiederum war die Überlappung störend. Im Liquor organischer Schmerzpatienten wurden niedrigere Endorphinspiegel gefunden als bei psychogenen Schmerzpatienten (Almay et al., 1968). Das ausgefeilteste Instrument zur Schmerzerfassung wurde von Melzack und Torgerson (1975) entwickelt, der McGill Pain Questionnaire (MPQ). Sein Kern umfaßt schmerzbeschreibende Wörter, in Gruppen von drei bis fünf verwandten Adjektiven – z.B. „einschießend", „elektrisierend", „durchzuckend" – geordnet und in drei übergeordneten Dimensionen klassifiziert: sensorisch, affektiv und evaluativ. Diese drei Dimensionen konnten mit Faktoranalysen bestätigt werden, u.a. im „Berner Schmerzfragebogen" (BSF), der deutschen Übersetzung des MPQ (Radvila et al., 1987, 1989). Einigen Forschern gelang es mit dem MPQ bestimmte Schmerzsyndrome, teilweise auch psychogene, von organischen zu trennen. Nach eigener klinischer Erfahrung können im Interview, neben Konversionsschmerz beweisenden Kriterien, solche Kriterien beobachtet werden, die auf Psychogenie respektive organische Genese von Schmerz hinweisen (Radvila et al., 1989).

Tabelle 36–1 stellt die Merkmale vorwiegend organisch bedingter denjenigen vorwiegend nichtorganisch bedingter Schmerzen gegenüber. Die meisten Unterscheidungskriterien verstehen sich leicht. Unter dem Merkmal „Abhängigkeit von Willkürmotorik" wird die Beobachtung verstanden, daß Schmerzen mit organischer Läsion im Hintergrund in Abhängigkeit von der Willkürmotorik des Patienten eine Zunahme oder Linderung aufweisen. Ein Patient beispielsweise, der sich auffällig benimmt, schon ein halbes Jahr lang über Schmerzen in der rechten Hüfte klagt, ohne daß die differenziertesten technischen Untersuchungsmethoden eine Läsion nachweisen lassen, und der erzählt, daß seine Schmerzen abnehmen, wenn er im Sitzen das Knie nach außen rotiert, hat wahrscheinlich eine organische, nur noch nicht

Tabelle 36–1. Merkmale „vorwiegend organisch" und „vorwiegend nichtorganisch" bedingter Schmerzen

Merkmal	organisch	nichtorganisch
Schmerzlokalisation	eindeutig, umschrieben	vage, unklar, wechselnd
Affekte des Patienten	passen zu geschildertem Schmerz	inadäquat
Zeitdimension	eindeutige Phasen von Präsenz und Fehlen bzw. deutlicher Abnahme	dauernd da, etwa gleich intensiv
Abhängigkeit von Willkürmotorik	vorhanden	fehlt
Reaktion auf Medikamente	pharmakokinetisch plausibel	nicht verständlich
Schmerz u. mitmenschliche Beziehung	unabhängig davon	damit verbunden
Schmerzschilderung	Bild paßt	Bild inadäquat, z.B. dramatisch
Betonung der Ursache	psychische betont	organische betont
Sprache	einfach, klar, nüchtern	intelligenzlerisch, Ärztejargon
Affekte des Arztes beim Zuhören	ruhig, aufmerksam, einfühlend	Ärger, Wut, Langeweile, Ungeduld, Lächeln, Hilflosigkeit, Verwirrung

erfaßte Läsion (Adler, 1981). Beim Merkmal „Betonung der Ursache" läßt sich eine Faustregel aufstellen: Patienten mit organischem Hintergrund für die Schmerzen ziehen als Erklärung psychische Gründe heran, und solche mit psychogen bedingten Schmerzen, beispielsweise einem konversionsneurotischen Schmerz, betonen, daß hinter ihrem Schmerz eine nur noch nicht erfaßte organische Störung liegen müsse (Engel, 1970).

36.4 Schmerzbehandlung

Zahlreiche Untersuchungen mit experimentellem Schmerz und klinische Beobachtungen zeigen, daß Angst die Schmerzempfindlichkeit steigert. Deshalb soll der Arzt bei jedem Patienten, der Schmerzen hat, die Beziehung so gestalten, daß Angst vermindert wird. Dadurch lassen sich Analgetika einsparen und es kann gelingen, mit milden Analgetika auszukommen, wo bei gesteigerter Angst schon zu Narkotika gegriffen werden muß. Daß dabei beruhigende und ermutigende Worte nicht in jedem Fall genügen, geht aus Untersuchungen hervor (Lazarus und Alfest, 1964; Egbert et al., 1964), die zeigen, daß der Arzt die Persönlichkeit des jeweiligen Patienten und die individuelle Situation berücksichtigen sollte. Der Patient mit einer dramatisierenden, ausdrucksvollen, suggestiblen, anklammernden Persönlichkeit bedarf eines aufmerksamen, warmen und Mitgefühl zeigenden Arztes, der sich aber davor hüten muß, übermäßig nachzugeben und sich vom Patienten manipulieren zu lassen. Der zu Ordnung, Kontrolle und intellektuellem Verstehen neigende Patient bedarf der sachlichen, kurzen Information. Er sollte nur im Notfall Sedativa erhalten, da sie seine Bemühungen um Kontrollieren und Meistern der Situation beeinträchtigen können.

Antidepressiva und **Neuroleptika** sollen bei psychogenen Schmerzen nur im Rahmen einer Depression gegeben werden. Ihre Verwendung als Analgetika im weiteren Sinne spielt bei organisch (z.B. Karzinom) bedingten Schmerzen eine Rolle. Sie werden deshalb im Kapitel 24 besprochen.

Verhaltenstherapeutische Maßnahmen machen sich die Erkenntnis zunutze, daß Schmerz nicht allein ein vom Stimulus abhängiges neurophysiologisches Geschehen ist, sondern sich bei chronischem Vorliegen mit gewissen Reaktionen der Umwelt verknüpft. Im Laufe der Zeit lernt der Patient, daß sein Schmerzverhalten gewisse Reaktionen bei den Mitmenschen auslöst. Als „operant" wird die als Signal dienende, Schmerz einschließende Verhaltensweise bezeichnet, als „positiver Verstärker" eine das Verhalten fördernde Umweltreaktion. Die Einflußnahme setzt sich zum Ziel, die Beziehung zwischen Schmerzverhalten und positiven Verstärkern zu lösen. Ein Aktivierungsprogramm, z.B. auf dem Gebiet der Physiotherapie, wird in Schritte zerlegt und immer nur so weit geübt, daß noch keine Schmerzen auftreten. Auf Schmerzverhalten wird möglichst nicht eingegangen. Die Schmerzmittelart, -dosierung und das Zeitintervall der Verabreichung muß der Patient ganz dem Arzt überlassen, der die Reduzierung nach einem unabhängig vom Wissen des Patienten festgelegten Schema vornimmt. Dem Patienten wird geholfen, soziale Beziehungen aufzubauen, und die Angehörigen werden trainiert, Schmerzverhalten des Patienten nicht zu fördern und auf „gesundes" Verhalten vermehrt zu reagieren (Fordyce, 1978). Gegenüber Kontrollpatienten (Roberts und Reinhard, 1980) und gemäß subjektiven Angaben (Ignelzy et al., 1977; Seres und Newman, 1976) scheint sich diese Therapieform zu bewähren.

Von Turk und Meichenbaum (1984) wurde ein Behandlungskonzept entwickelt, das die kognitive, sensorische und affektive Erlebensweise von Schmerz zu beeinflussen versucht. Es ist abgeleitet aus der sog. „kognitiven Verhaltenstherapie". Die Anwendung dieser Verfahren auf den Umgang mit Schmerz beinhaltet sechs Behandlungsstufen, die jedoch nicht strikt zu trennen sind.

Nach einer Phase der Erfassung des gegenwärtigen Zustands im Hinblick auf Annahmen des Patienten über Schmerz, sein Schmerzverhalten (Einnahme von Medikamenten, Arztbesuche, unmittelbare Reaktionen auf Schmerz usw.) sowie seine sonstigen Lebensbedingungen (Ehe, Familie, Arbeit usw.) wird begonnen, diese Aspekte im Hinblick auf ihre Bedeutung für das Schmerzerleben mit dem Patienten zu erörtern. Dabei wird der Patient damit konfrontiert, daß bestimmte Aspekte seiner Gedanken und seines Verhaltens eher das Schmerzerleben verstärken. Mit ihm werden gegenwärtige wissenschaftliche Konzepte den Schmerz betreffend besprochen, so etwa auch der Gedanke, daß Bedingungen, die er als angenehm erlebt, das Auftreten der Schmerzen verstärken können. In diesem Zusammenhang werden die Therapieziele erarbeitet.

In der nächsten Stufe werden diese Konzepte auf seine konkrete Situation angewandt, es werden Entspannungs-, Atmungs- und Aufmerksamkeitsübungen durchgeführt. Die weiteren Stufen dienen der Konsolidierung dieser Fertigkeiten und der Generalisierung im Lebensbereich des Patienten.

Durch eine Fülle von Studien wurde die Effektivität dieses Vorgehens für sehr verschiedene Schmerzbereiche (rheumatische Arthritis, Verbrennungsschmerz usw.) belegt.

Der Beitrag der **Hypnose** zur Schmerzbehandlung ist alt und heute unbestritten. Übermäßige Hoffnung in sie und ihr Gebrauch bei dafür ungeeigneten Patienten und Läsionen haben zu ihrer wechselhaften Beurteilung geführt. Die experimentelle Schmerzforschung der letzten Jahre hat ihren Wert und ihre Grenzen besser erkennen lassen. Die Hypnotisierfähigkeit ist von Mensch zu Mensch ganz unterschiedlich, aber recht konstant über die Zeit (Shealy et al., 1967) und korreliert mit der hypnotischen Schmerzverminderung im experimentellen Schmerztest (Evans und Paul, 1970; Hilgard, 1967).

Die Macht des Placebos zur Linderung von Schmerz ist ein so eindrückliches Phänomen, daß eine Prüfung eines Analgetikums ohne Doppelblindbedingung heute wissenschaftlich nicht mehr akzeptiert wird. Die Wirkung von Hypnose wurde oft der Placebowirkung zugeschrieben. McGlashan und Mitarbeiter (1969) konnten aber nachweisen, daß sich die beiden unterscheiden. Gut hypnotisierbare Subjekte zeigten unter hypnotischer Analgesie eine starke Zunahme der Schmerztoleranz in einem ischämischen Muskelschmerztest, unter Placebo aber überhaupt keine analgetische Wirkung. Schlecht hypnotisierbare Individuen zeigten unter beiden Bedingungen eine identische, leichte Zunahme der Schmerztoleranz. Hypnose ist auch nicht lediglich der Angstverminderung zuzuschreiben (Hilgard, 1975).

Obwohl die Hypnose die Schmerzempfindung bei gewissen Individuen vermindert, zeigen die physiologischen, den Schmerz normalerweise begleitenden Funktionen eine Mitreaktion. Dies besagt, daß durch die Hypnose die Schmerzempfindung im Bereich des „zentralen Kontrollsystems" beeinflußt wird, nicht aber in Systemen, die mit den physiologischen Begleitreaktionen eng gekoppelt sind, wie z.B. im „motivierend-affektiven System". Die Beobachtung, daß die „offene" Angabe des Subjekts unter Hypnose über die Schmerzintensität nicht mit der „verdeckten" übereinstimmt, legt diese Interpretation nahe (Hilgard, 1975). So gibt die Versuchsperson in hypnotischer Analgesie im experimentellen Schmerztest verbal nur geringe Schmerzen an, während sie mit der Hand, die in der Hypnose zum automatischen Schreiben aufgefordert wird, eine Schmerzintensität angibt, die nur wenig unter derjenigen liegt, die im Wachzustand im gleichen Experiment verspürt wird. Nach Hilgard stört die Tatsache, daß es „verdeckten" Schmerz in der Hypnose gibt, ihren klinischen Gebrauch nicht, solange die hypnotisierbaren Individuen und die klinischen Zustände sorgfältig ausgewählt werden (Hilgard, 1978).

Die Schmerzbehandlung verlangt das diagnostische Erkennen organischer Veränderungen, der „Neigung, Schmerz zu erleiden" und des konversionsneurotischen Schmerzsyndroms, ferner die Erfassung psychischer Faktoren wie Angst usw., welche Schmerzempfinden und -verhalten beeinflussen. Die Handhabung der psychischen Faktoren ist für jede Form von Schmerz wichtig, sei er konversionsneurotischer oder vorwiegend organischer Natur, weil es bei der Schmerzentstehung und -behandlung nicht um ein „psychisch oder somatisch", sondern um ein „sowohl als auch" geht. Das multifaktoriell bedingte Entstehen von Schmerz verlangt in bestimmten Fällen ein ebenso **multifaktorielles therapeutisches Vorgehen.** Die Vorstellung, daß ein bestimmtes Medikament, ein spezifischer Eingriff, die **Ursache** eines Schmerzes ausschalten wird, ist Folge veralteter Schmerzkonzepte. Welche Therapieform oder **Kombination von Therapien** beim einzelnen Schmerzkranken angewandt werden soll, muß bis jetzt von Patient zu Patient entschieden werden. Bei gewissen „zum Erleiden von Schmerz neigenden" Patienten muß überlegt werden, welcher therapeutische Aufwand sich lohnt. Schmerzen, die mehr als vier Jahre dauern, Arbeitsunfähigkeit seit mindestens anderthalb Jahren, mit drei oder mehr Operationen wegen des geklagten Schmerzes, mit einer Intensität von sieben und mehr auf einer Visual-Analogskala von 0–10 und hohe Werte auf der Hypochondrie- und Hysterie-Skala im MMPI sind ungünstige prognostische Zeichen (Maruta et al., 1979). Mit kombinierten Programmen – operantes Konditionieren, Physiotherapie und Entspannungstraining (Kockott, 1983) – lagen die Behandlungserfolge ein bis drei Jahre nach Abschluß bei 60–70%, beurteilt anhand der subjektiven Schmerzintensität, Reduktion der Einnahme von Schmerzmitteln, gesteigerter Aktivität und beruflicher Reintegration; bei Swanson (1979) konnten jedoch nur 39% der behandelten Patienten den Erfolg halten.

Nach unserer Erfahrung führt keine Methode allein oder in Kombination mit anderen zum Erfolg, wenn nicht eine Arzt-Patient-Beziehung aufgebaut wird, in der **ein** Arzt die Therapieverantwortung für Monate bis Jahre übernimmt.

37 Kopfschmerz

Claus Bischoff, Helmuth Zenz und *Harald C. Traue*

In diesem Kapitel werden Diagnose, Psychophysiologie, Ätiologie und Therapie von Migräne, Spannungskopfschmerz, kombiniertem Kopfschmerz und von „psychogenen" (wahnhaften, konversionsneurotischen und hypochondrischen) Kopfschmerzen behandelt.

Herr M., 29 Jahre, verheiratet, zwei Kinder, als Sachbearbeiter tätig, leidet seit 8 Jahren an Kopfschmerzen. Seit einem Suizidversuch vor zwei Jahren im Zusammenhang mit der Kündigung seiner ersten Stelle treten die Kopfschmerzen verstärkt auf, gegenwärtig im Durchschnitt zweimal pro Woche mit einer Intensität von 50–80 auf einer 100-Punkte-Skala. Die Kopfschmerzen beginnen einseitig in der rechten Schläfe bzw. der rechten Stirnseite, werden dann meist beidseitig, ziehen über das Schädeldach in die Nacken- und Schulterregion und erreichen ihre maximale Intensität innerhalb von 3–5 Stunden. Ort der intensivsten Schmerzen bleiben die rechte Schläfe und Stirnseite. Die Schmerzqualität ist ziehend, drückend und spannend. Die Schmerzen halten bis zum Einschlafen in voller Stärke an, unabhängig von der Tageszeit ihres Beginns. Begleitsymptome sind Übelkeit und Erbrechen, vor allem bei starken Kopfschmerzen. Der Patient nimmt ausdrücklich keine Schmerzmittel – seit dem Suizidversuch mit Schlaftabletten fürchtet er sich vor der Einnahme von Medikamenten jedweder Art. Die somatische Abklärung erbringt keinen Befund.

Der Patient hat zwei Brüder. Der eine, um 10 Jahre ältere Bruder stammt aus der geschiedenen ersten Ehe des Vaters; der andere Bruder ist vier Jahre älter als der Patient und unehelicher Sohn der Mutter. Der Patient hat zunächst eine Elektrikerlehre begonnen, diese jedoch wegen einer starken Sehschwäche auf dem rechten Auge abgebrochen, danach eine Zimmermannslehre abgeschlossen und erfolgreich eine Technikerschule besucht. Mit 26 Jahren heiratete er – seine Frau hatte er mit 16 Jahren kennengelernt –, trat nach einer Zeit der Arbeitslosigkeit seine 1. Stelle an, die ihm jedoch bereits ein halbes Jahr später gekündigt wurde. Darauf folgte eine starke depressive Reaktion, die in dem Suizidversuch gipfelte. Seine jetzige Stelle vermittelte ihm die Ehefrau in derselben Firma, in der sie seit vielen Jahren arbeitet. Als wesentliche Kindheitserinnerung berichtet er, sein Vater habe ihn in der ersten Schulzeit wegen Schulschwierigkeiten oftmals erbarmungslos und für ihn unvorhersehbar verprügelt, vor allem durch Schläge auf den Kopf und ins Gesicht. Damals habe er auch öfters Kopfschmerzen gehabt. Seit dieser Zeit versuche er, möglichst alles mit sich selbst abzumachen.

Herr M. erlebt seine Kopfschmerzen als fast ausschließlich wetterbedingt. Erstes diagnostisches und therapeutisches Ziel ist deshalb eine Wahrnehmungs-

schärfung des Patienten für solche Ereignisse, die in zeitlicher Nähe zum Kopfschmerz auftreten. Der Patient wird angeleitet, ein Kopfschmerztagebuch zu führen, in das er u. a. eine Tageszeit-Intensitäts-Kurve und die Geschehnisse einträgt, die mit dem Einsetzen und dem weiteren Verlauf der Kopfschmerzen in Zusammenhang stehen.

Im Laufe der nächsten beiden verhaltensanalytischen Sitzungen schälen sich aufgrund dieser Aufzeichnungen typische kopfschmerzauslösende Situationen heraus, die durch Gefühle von Kontrollverlust gekennzeichnet sind und mit starker körperlicher Erregung einhergehen. Ein Beispiel am Arbeitsplatz: Eine Terminarbeit ist zu erledigen, der Vorgesetzte gibt einen zusätzlichen Auftrag. Der Patient nimmt an, daß ihn der Vorgesetzte aus der Fassung bringen will. Mit großem Kraftaufwand zwingt er sich, seine starke innerliche Aufregung und sein Zittern zu unterdrücken und langsam weiterzuarbeiten. Die ersten Kopfschmerzen deuten sich an. Entstehungsbedingungen dieser Art decken einen großen Teil der Auftretensfälle von Kopfschmerzen ab. Von herausragender Bedeutung sind außerdem Interaktionsprobleme mit seiner Frau, die sich wegen akuter Angstzustände und herzneurotischer Beschwerden ebenfalls in psychotherapeutischer Behandlung befindet. Die Beschwerden der Ehepartner sind eng aufeinander bezogen: Geht der Patient auf die Angstzustände seiner Frau sofort fürsorglich ein, so bewirkt dies eine sofortige Entlastung bei ihr, führt aber bei ihm zu Kopfschmerzen. Versucht der Patient die Angstzustände seiner Frau zu ignorieren, so bleibt er zwar zunächst von Kopfschmerzen verschont, seine Frau jedoch „dreht fast durch", weshalb er dann meist doch noch „ein Erbarmen" hat, sie beruhigt, was wiederum die Entstehung von Kopfschmerzen begünstigt.

Die Kopfschmerzen sind selbst bei maximaler Ausprägung für den Patienten kein Anlaß, sich zu schonen; er treibt eine anstrengende körperliche Arbeit im Freizeitbereich auch dann bis zum bitteren Ende weiter, wenn ihm vor lauter Schmerzen inzwischen schwindelig und übel geworden ist. Zustände der Entspannung sind für ihn unlustbesetzt.

37.1 Klassifikation und diagnostische Probleme

Für die Definition chronischer Schmerzsyndrome liegt seit 1986 das Klassifikationssystem vor, das die ISP (International Association for the Study of Pain), Subcommittee on Taxonomy, erarbeitet hat (Merskey, 1986). Nach diesem System sind im Kopf- und

Tabelle 37-1. Formen chronischer Kopfschmerz-syndrome

─────────────────────────

- Neuralgien:
  Trigeminusneuralgien (Tic douloureux, akute trigemina-le Herpes-zoster-Neuralgie, postherpetische trigemina-le Neuralgie), Glossopharyngeus-Neuralgie
- Kraniofaziale Schmerzen muskuloskelettalen Ursprungs:
  akuter und chronischer Spannungskopfschmerz, temporomandibuläres Schmerzsyndrom
- Verletzungen von Ohr, Nase und Mundhöhle:
  Sinusitis maxillaris, Zahnschmerzen verschiedener Genese, Glossodynie, „dry socket"-Syndrom
- primäre Kopfschmerzsyndrome:
  einfache und klassische Migräne, Migränevarianten, kombinierter Kopfschmerz, Cluster-Kopfschmerz, posttraumatischer Kopfschmerz
- Kopf- und Gesichtsschmerzen psychologischen Ursprungs:
  wahnhafte Kopfschmerzen, hypochondrische Kopfschmerzen

Gesichtsbereich fünf Gruppen von Schmerzsyndromen zu unterscheiden, die in Tabelle 37–1 mit ihren wichtigsten Unterformen genannt werden. Das vorliegende Kapitel befaßt sich fast ausschließlich mit solchen Kopfschmerzformen, bei denen **psychologische** oder **psychophysiologische** Ursachen für denkbar oder wahrscheinlich gehalten werden:
- die verschiedenen Formen der Migräne
- akuter und chronischer Spannungskopfschmerz
- kombinierter Kopfschmerz
- wahnhafter, konversionsneurotischer oder hypochondrischer Kopfschmerz.

Diese Kopfschmerzformen[1] werden im folgenden definiert. Für die Differentialdiagnose gegenüber anderen Kopfschmerzformen verweisen wir auf Mumenthaler und Regli (1981) und Soyka (1985).

Die wichtigste Form chronischer Kopfschmerzen sind die primären Kopfschmerzsyndrome: **Migräne** ist allgemein definiert durch sich wiederholende Attacken, die gewöhnlich einseitig beginnen und in der Regel mit Appetitlosigkeit, manchmal mit Übelkeit und Erbrechen verbunden sind. Sensorische, motorische und Störungen der Stimmungslage gehen den Attacken in einigen Fällen voraus oder begleiten sie. Die **klassische Migräne** (auch ophthalmische oder Augenmigräne) ist in der Prodromalphase zunächst durch verschiedene Sehstörungen charakterisiert (Flimmerskotom, Hemianopsie, Amaurose). Erst nach deren Abklingen beginnt die eigentliche Kopfschmerzphase. Die Schmerzen werden meist als hämmernd und pulssynchron pochend beschrieben und sind meist einseitig in der Schläfenregion oder über einem Auge lokalisiert. Begleiterscheinungen sind oftmals auch Gesichtsblässe bei gleichzeitig starker Blutfüllung der oberflächlichen Gefäße. Die typische Dauer einer klassischen Migräne beträgt 6–8 Stunden; sie kann sich aber auch über Tage hinziehen. Nach schweren Attacken können schwächere, diffuse Schmerzen „nachhallen". Im Unterschied zur klassischen Migräne kommt es bei der **einfachen (gewöhnlichen) Migräne** nicht zur visuellen Aura, Anfallsbeginn und -ende sind nicht so scharf markiert (Serratrice, 1976), der Anfall zieht sich länger hin und ist dadurch beeinträchtigender. Der Schmerz ist häufiger auch beidseitig. Die

hemiplegische Migräne (Migraine accompagnée, komplizierte Migräne) ist pathophysiologisch vor allem Ausdruck passagerer Ischämie in einem lokalisierten Gehirn- oder Retina-Areal (Mumenthaler und Regli, 1981), die zu Parästhesien, aphasischen oder paraphasischen Störungen oder motorischen Ausfällen führt. Als seltenere Formen werden beschrieben: die ophthalmoplegische Migräne mit den charakteristischen Augenmuskellähmungen als Begleitsymptom, die basiläre und die abdominelle Migräne (für Einzelheiten vgl. Heyck, 1975). Zu den primären Kopfschmerzen zählt auch der **Cluster-Kopfschmerz** (Horton-Neuralgie, Erythroprosopalgie), der sehr rasch einsetzt und fast ausnahmslos halbseitig mit größter Intensität in einem bestimmten Zeitraum oftmals täglich „fahrplanmäßig" zu einer bestimmten Zeit („cluster") in bis zu zweistündigen Attacken auftritt, dann aber wieder monateweise völlig verschwinden kann. Typisch im Anfall ist das Horner-Syndrom: Zurücksinken des Augapfels, schmale Lidspalte und kleine Pupille. Bei allen primären Kopfschmerzformen zieht man das Nervensystem als das betroffene Organsystem in Betracht, insbesondere seinen peripher-vaskulären Ast. Die Ätiologie wird bei den Migränen in der dysfunktionalen – dabei auch psychophysiologisch beschreibbaren – Arbeitsweise dieses Systems vermutet.

Unter **chronischem Spannungskopfschmerz** (auch: scalp muscle contraction headache) versteht man einen üblicherweise symmetrischen, beinahe ständig vorhandenen, auch hinsichtlich der Intensität variierenden, in der Qualität dumpfen Kopfschmerz, der mit Muskelspannung, Angst und Depression einhergeht. Lokalisiert ist chronischer Kopfschmerz orbital, frontal, fronto-okzipital, okzipital, subokzipital oder im gesamten Kopfbereich. Der Schmerz ist diffus oder legt sich wie ein Hut- oder Stahlband um den Kopf. Charakteristisch sind Exazerbationen mit verstärktem, hämmerndem Schmerz, Photo- und Phonophobie, Übelkeit und Erbrechen, in welchem Fall der Schmerz auch einseitig auftreten kann. **Akuter Spannungskopfschmerz** nimmt im Gegensatz zum chronischen die Form diskreter Ereignisse an. Ein Kopfschmerzereignis beginnt mit einer allmählichen Intensitätssteigerung, verbleibt dann auf einem bestimmten Intensitätsplateau und geht langsam zurück. Akuter Spannungskopfschmerz hat meist eine Dauer von wenigen Stunden und ist in seiner Qualität selten pochend. Bei akuten und chronischen Spannungskopfschmerzen wird als betroffenes Organsystem das muskuloskelettale System und das Bindegewebe angenommen. Kopfschmerzen entstünden aufgrund der dysfunktionalen psychophysiologischen Arbeitsweise dieser Systeme (zur Kritik dieser Annahme vgl. Abschn. 37.6.2). Beim **kombinierten Kopfschmerz** treten Symptome sowohl der Migräne als auch von Spannungskopfschmerzen auf.

Genuin psychischen Ursprungs sind wahnhafte, konversionsneurotische oder hypochondrische Kopfschmerzen. **Wahnhafter Kopfschmerz** ist dadurch charakterisiert, daß der Patient eine wahnhafte Ursache für ihn verantwortlich macht, der Patient mit einem messianischen Wahn z. B. Dornenkrone. Den Hintergrund bildet eine psychotische Erkrankung (Zyklothymie, Schizophrenie). **Konversionsneurotischer** oder **hypochondrischer Kopfschmerz** wird kognitiven Prozessen, emotionalen Zuständen oder der Persönlichkeit des Patienten zugeschrieben, wobei weder organische, wahnhafte noch muskuläre Mechanismen am Werk sind. Das Hauptgewicht des Schmerzes liegt auf seiner affektiv-motivationalen Komponente. Er wird vom Patienten oft weniger als Schmerz denn Kopfdruck erlebt, der das

─────────────────────────

1 Das Kapitel kann nur einen Überblick geben. An Monographien und Übersichtsartikeln seien empfohlen: Bakal (1982), Bischoff et al. (1989) und Gerber (1986).

Denken behindert und die Entschlußkraft lähmt. Meist tritt er tagsüber als ein in seiner Intensität variierender Dauerkopfschmerz auf, ohne den Schlaf zu beeinträchtigen.

Das Subkomitee gibt zu bedenken, daß der Kopfschmerz des **depressiven** Patienten ganz verschiedener Provenienz sein kann: er kann in die letztgenannte Gruppe fallen, wahnhafter Natur sein oder auf Muskelverspannungen zurückgehen.

Alle bislang vorgestellten Schemata zur Klassifikation von Kopfschmerzen, auch das Schema des Subkomitees, werfen Probleme auf: Die Problematik der Differentialdiagnose von migränoidem und Spannungskopfschmerz drückt sich schon darin aus, daß neurologische und psychologische Experten nur bei ca. 70% der Patienten zu übereinstimmenden Diagnosen gelangen (Blanchard et al., 1981). Dazu paßt, daß mehrere empirische Untersuchungen nahelegen, in Migräne und Spannungskopfschmerz keine distinkten Krankheitseinheiten, sondern verschiedene Ausprägungen ein und derselben Erkrankung auf einem Kontinuum des Schweregrades zu sehen (Bakal, 1982).

In Übereinstimmung mit dem Modell der Eindimensionalität (severity model) stehen Befunde, nach denen sowohl tonische als auch phasische EMG-Werte der Stirn-, Schläfen- und Nackenmuskulatur bei Patienten mit Migräne genauso hoch, wenn nicht höher liegen als die von Patienten mit Spannungskopfschmerz, wobei in beiden Gruppen die Werte höher ausfallen als in Kontrollgruppen (vgl. Traue et al., 1984). Befürworter der traditionellen Unterscheidung weisen indes vor allem auf Untersuchungen zum Vasotonus der Temporalarterie hin: Migräniker haben sowohl im anfallsfreien Intervall, besonders aber im Anfall eine gegenüber Normalpersonen verstärkte Dilatation der Temporalarterie, Patienten mit Spannungskopfschmerz dagegen nicht (Tunis und Wolff, 1953, 1954; Sakai und Meyer, 1978, 1979). Die einzige methodologisch überzeugende faktorenanalytische Studie zur Kopfschmerzsymptomatik spricht ebenfalls für die kategoriale Unterscheidung: Kröner (1983) identifizierte zwei orthogonale Faktoren, die mit großer Eindeutigkeit vom Subkomitee als typisch für Migräne bzw. für Spannungskopfschmerz bezeichnet würden.

Dicht mit dem differentialdiagnostischen Problem verwoben ist das Problem der Diagnosepraxis. Das Subkomitee versteht Spannungskopfschmerzen als Muskelkontraktionskopfschmerzen. Die Muskelaktivität wird jedoch vom Kliniker in der Regel nicht als diagnostisches Kriterium erhoben; der Kliniker diagnostiziert Spannungskopfschmerz vielmehr per exclusionem als nicht-organischen und nicht-migränoiden Kopfschmerz (vgl. Bischoff und Traue, 1983). Alle empirischen Arbeiten, bei denen die Versuchsgruppen aufgrund von klinischen Urteilen zusammengestellt wurden, kranken an diesem Umstand (vgl. Traue et al., 1984).

Ein weiterer Kritikpunkt betrifft die Kriterien für die Klassifikation von Kopfschmerzformen. Mit Ausnahme der Kopfschmerzen mit psychischem Ursprung werden organische Ursachen genannt, die jedoch selbst wiederum psychische Ursachen haben können. Peripher-physiologische Mechanismen stehen ganz unvermittelt neben möglichen psychologischen Annahmen. Hier wird der Mangel an Theorie in den Klassifikationen deutlich.

> Die Kopfschmerzen von Herrn M. haben phänomenologisch sowohl vaskuläre als auch muskuläre Komponenten: Der einseitige Beginn der Attacken, Übelkeit und Erbrechen – allerdings ohne Aura – weisen in Richtung einfache Migräne, Schmerzqualität und helmartige Ausbreitung der diskreten Schmerzereignisse in Richtung akuter Spannungskopfschmerz. Insofern wäre die Diagnose: kombinierter Kopfschmerz. Man könnte allerdings auch für die Diagnose: ausschließlich „akuter Spannungskopfschmerz" plädieren, d.h. die Dominanz der Schmerzen auf der rechten Seite auf die Sehschwäche zurückführen und Übelkeit und Erbrechen als vegetative Reaktionen auf starken Schmerz interpretieren.

37.2 Epidemiologie

Nach Untersuchungen mit dem Gießener Beschwerdebogen sind Kopfschmerzen für ein Drittel der deutschen Bevölkerung ein nicht zu vernachlässigendes und für 14% ein erhebliches Problem (Brähler und Scheer, 1983). Übereinstimmend kommen verschiedene Autoren zu dem Schluß, daß Frauen etwa im Verhältnis 4:3 häufiger und stärker betroffen sind als Männer (Newland et al., 1978).

Die Belästigung durch Kopfschmerzen nimmt bis zum mittleren Alter zu und danach bei älteren Menschen wieder ab (Serratrice et al., 1985). Nur 25% der Personen mit chronischem Kopfschmerz sind frei von für Migräne charakteristischen Begleitsymptomen, 15% haben mehr als zwei Begleitsymptome und ließen sich somit eindeutig als Migräniker bezeichnen (Newland et al., 1978). Übereinstimmend geben alle Untersuchungen die Prävalenz von Migräne als geringer an als die von Spannungskopfschmerz. Die Prävalenz der Migräne wird mit 2,5% der Bevölkerung für den österreichischen Raum (Barolin, 1969), über 8% für die USA (Friedman, 1979) und bis zu 20% für Großbritannien (Waters und O'Connor, 1975) beziffert. Von den Migränepatienten leiden etwa 10% an klassischer Migräne, über 70% an einfacher Migräne und ca. 12% an Cluster-Kopfschmerz und anderen komplizierten Migräneformen (vgl. Mumenthaler und Regli, 1981).

Nur ein Fünftel der Personen mit Kopfschmerz geht zum Arzt (Andrasik et al., 1979). Dabei sind die von stärkeren Schmerzen Betroffenen, besonders also die Migräniker, eher zum Arztbesuch bereit. Obwohl bezogen auf die Gesamtheit der an Kopfschmerz **Leidenden** eine Minderheit, spielt die Gruppe der Kopfschmerz**patienten** im medizinischen System zahlenmäßig eine bedeutsame Rolle. Nach Schätzungen der National Migraine Foundation von 1973 kam es in den USA zu 12 Millionen Konsultationen aufgrund von Kopfschmerzsymptomen (Adams et al., 1980).

37.3 Physiologische Grundlagen des Kopfschmerzanfalls

37.3.1 Migräne

Vasomotorik extra- und intrakranialer Blutgefäße

Seit der Pionierarbeit von Tunis und Wolff (1953) hat man mit den verschiedensten Meßverfahren (Volumen-Plethysmographie, Infrarotthermographie, Natrium-Clearance-Methode, ^{133}Xe-Inhalations- und -Injektionsmethode) bei der Untersuchung intra- und extrakranialer Blutgefäße immer wieder einen charakteristischen biphasischen Durchblutungsverlauf vor und während der Migräneattacke festgestellt: Sowohl bei der klassischen wie auch bei der einfachen Migräne kommt es vor Beginn der Schmerzphase zu einer Vasokonstriktion der Arteria carotis interna und externa. Der rCBF (regional cerebral blood flow) ist bei der klassischen Migräne intrakranial wahrscheinlich in jenen Gehirnarealen besonders reduziert, die für die Aurasymptome verantwortlich sind. Während der Schmerzphase dilatieren die beiden Carotis-Äste um 20–50% gegenüber der anfallsfreien Phase. Ausnahmen vom biphasischen Muster bilden Fälle von hemiplegischer Migräne, die während der Schmerzphase ein gemischtes Bild von Vasokonstriktion und -dilatation zeigen (Sakai und Meyer, 1979; Skinhøj, 1973).

Der biphasische Verlauf schließt nicht die peripheren extrakranialen Arteriolen mit ein, was sich schon an der typischen Gesichtsblässe des Migränepatienten vor und während der Attacke ablesen läßt: Besonders auf der Kopfschmerzseite ist die Durchblutung des oberflächlichen Hautgewebes auch in der Schmerzphase reduziert. Heyck (1975) hat durch den Nachweis arteriovenöser Anastomosen (av-Shunts) Licht in dieses Geschehen gebracht: Trotz des starken Blutflusses in den Arterien wird während des Migräneanfalls besonders auf der schmerzenden Seite weniger O_2 an die Kapillaren abgegeben als in der anfallsfreien Zeit. Das Blut wird unter Umgehung der Kapillaren über Kurzschlüsse (Shunts) direkt an die Venen weitergeleitet, dem peripheren Kapillarnetz somit „weggestohlen" („steal-effect"). Wahrscheinlich ist die Vasokonstriktion der Kapillaren die Ursache für die Eröffnung von av-Shunts.

Physiologisch-biochemische Vorgänge

Die **initiale Vasokonstriktion** der extra- und intrakranialen Gefäße geht hauptsächlich zu Lasten einer exzessiven Sympathikusaktivität zu Beginn der Prodromalphase, abzulesen unter anderem an der Herzfrequenzsteigerung in diesem Stadium (vgl. Abb. 37–1). Die Vasokonstriktion führt zerebral zu hypoxischen Verhältnissen. Wahrscheinlich sind diese eine Mitursache des Migräneschmerzes. Die hypoxischen Verhältnisse bewirken außerdem eine Freisetzung von Serotonin aus den Thrombozyten.

Serotonin spielt im Migräneanfall eine herausragende Rolle. Dafür, daß es aus den Thrombozyten freigesetzt wird, ist neben der zerebralen Hypoxie die am Beginn der Serotoninfreisetzung zu beobachtende gesteigerte Thrombozytenaggregation verantwortlich, die ihrerseits maßgeblich durch Streß, Östrogenveränderungen und diätetische Faktoren bedingt ist.

Serotonin hat seinerseits auf die Arterien vasokonstriktive Wirkung, die allerdings – im Gegensatz zu früheren Annahmen – die initiale Vasokonstriktion allein nicht erklärt. Wichtiger ist, daß Serotonin Mastzellen und basophile Leukozyten zur Degranulation (Zellkörperabstoßung) und damit zur Freisetzung der Schmerzsubstanzen Histamin, Prostaglandin, Bradykinin und Plasmakinin veranlaßt. Histamin und Serotonin zusammen steigern die Permeabilität der Gefäßwände, so daß es zur Transsudation von Plasmakinin in die Gefäßwände und in das perivaskuläre Gewebe sowie zu perivaskulärer Ödembildung kommen kann. Serotonin und Plasmakinin zusammen reduzieren die Schmerzschwelle der Rezeptoren in den betroffenen Arterienwänden.

Während der Attacke kommt es zu einem deutlichen Abfall des Plasmaserotonins – das Thrombozytenserotonin erschöpft sich; MAO, die das Serotonin zu 5-HIE abbaut, wird parallel zur und als Folge der Serotoninfreisetzung aktiv. Der Serotoninabfall ist die Hauptursache für die Dilatation der extra- und intrakranialen Arterien und die gleichzeitige Konstriktion der Arteriolen. Auch diese beiden Konsequenzen tragen zur Entstehung von Schmerz bei: Die Bedeutung der Dehnung extrakranialer Arterien als Schmerzursache hat schon Wolff (1963) betont – wahrscheinlich ist sie jedoch überschätzt worden. Andererseits hat Heyck (1975) darauf hingewiesen, daß durch die Konstriktion extra- und intrakranialer Kapillaren ischämische Bedingungen hergestellt und dadurch schmerzerzeugende Azidosen verursacht werden.

Alle bisherigen Theorien mit Ausnahme der Heyckschen Shunt-Theorie geben übrigens keine befriedigende Erklärung für die häufig beobachtete Einseitigkeit der „Hemikranie".

37.3.2 Spannungskopfschmerz

Voraussetzung für eine kausale Verknüpfung von Muskelspannung und Schmerz ist die Existenz nozizeptiver Strukturen im Muskel und in der Umgebung der den Muskel versorgenden Gefäße. Die mechanische Schwelle dieser Nozizeptoren liegt im Normalfall nahe bei der Maximalkontraktion des Muskels. Dies ändert sich bei relativer oder absoluter Ischämie, zu der es aufgrund von erhöhter Muskelkontraktion (vgl. Abb. 37–2) kommen kann (Rasmussen et al., 1977)[2].

---

2 Muskelkontraktion ist nicht die einzige mögliche Ursache von Schmerz im Muskel. Schon Wolff (Tunis und Wolff, 1954) hielt es für wahrscheinlich, daß durch ANS-Aktivität vermittelte vasomotorische Instabilität zu ischämischen Verhältnissen führen kann. Mit wenigen Ausnahmen (Onel et al., 1961; Martin

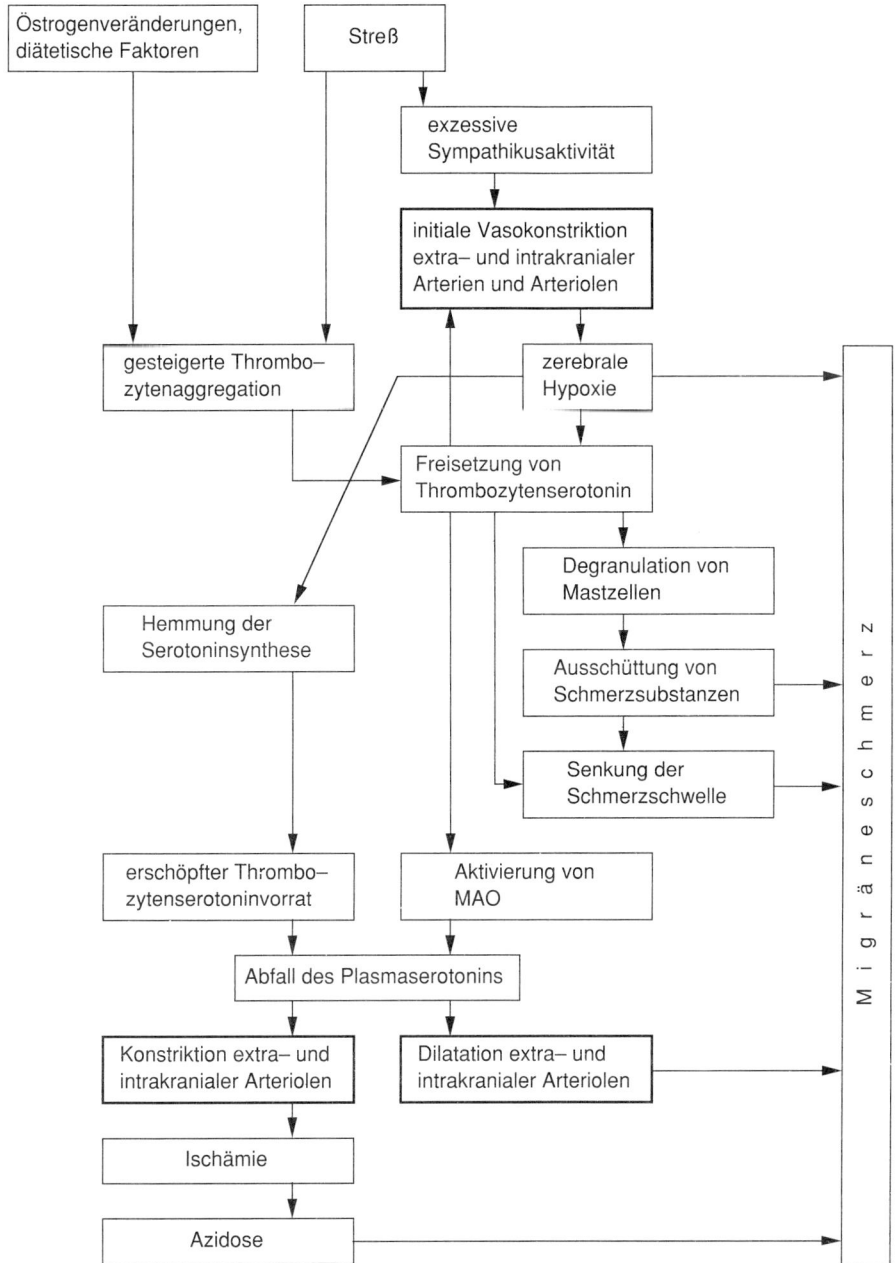

Abb. 37–1. Physiologisch-biochemische Vorgänge während der Migräneattacke.

Die Ischämie bewirkt über die Erniedrigung des pH-Levels die Freisetzung von chemischen Schmerzstoffen (Bradykinin, Serotonin, Prostaglandine), die einerseits die chemosensiblen Nozizeptoren reizen, andererseits die Schwelle der mechanosensiblen Nozizeptoren senken (Mense, 1977), so daß bereits Muskelspannungen geringerer Kontraktionsstärke Schmerzen auslösen können. Schmerz, auch Kopfschmerz, als Folge von Ischämie ist mehrfach experimentell demonstriert worden (z.B. Myers und McCall, 1983). Ist erst einmal Schmerz entstanden, dann besteht die Gefahr, daß durch positive Rückkopplung ein Teufelskreis von Muskelspannung und Schmerz entsteht. Was Dalessio (1974, 1978) postu-

lierte, konnten Schmidt, Kniffki und Kollegen (vgl. Schmidt, 1981) experimentell belegen: Schmerzreize führen reflektorisch zu Muskelkontraktionen. Erstens erzeugen sie ein Kontraktionsmuster in Form eines Flexor-Reflexes (eine Wegziehbewegung). Zweitens kommt es über die Gamma-Schleife zu einer generellen Erhöhung der Muskelspannung in den beteiligten Muskeln. Außerdem führt Schmerz als psychophysiologische Belastung – ob tatsächlich er-

und Mathews, 1978) belegen experimentelle Arbeiten, daß bei Spannungskopfschmerzpatienten die relevanten Arterien tatsächlich im Vergleich zu schmerzfreien Kontrollpersonen stärker konstringiert sind (z.B. Friedman und Merritt, 1959; Bakal und Kaganov, 1979).

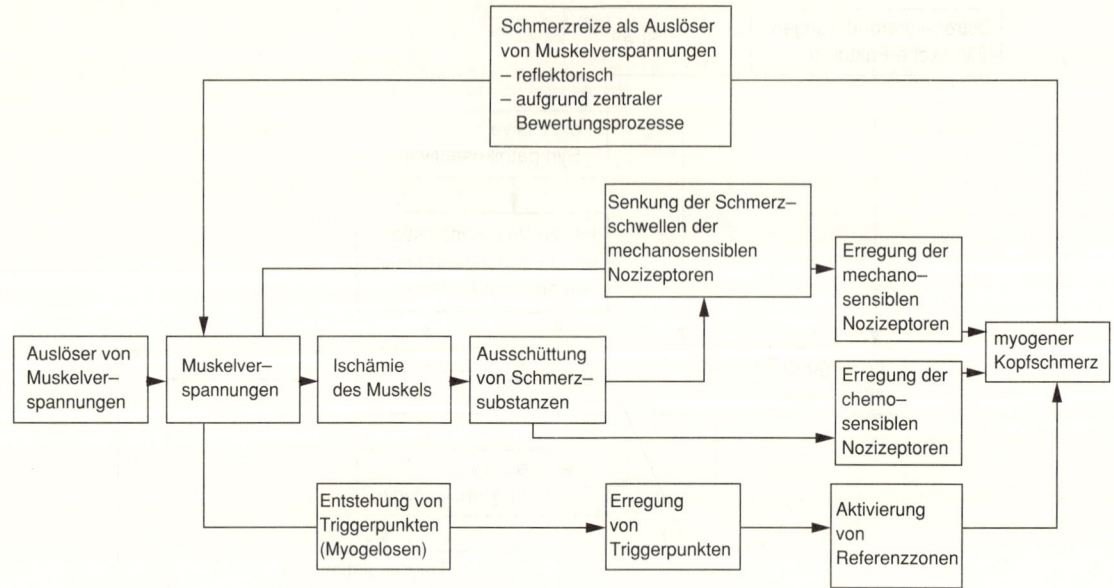

Abb. 37–2. Physiologisch-biochemische Vorgänge beim Muskelkontraktionskopfschmerz.

lebt oder auch nur erwartet – u.a. zu bedeutsamen Muskelspannungsanstiegen (Bischoff et al., 1982).

Orte der größten Muskelverspannung sind nicht notwendigerweise die Orte, an denen Schmerz erlebt wird (Philips, 1977). Myers und McCall (1983) fanden, daß nicht unbedingt der Muskel schmerzt, dessen Blutzufuhr experimentell abgestellt wird. Die Gründe dafür sind nicht vollständig klar. Es steht allerdings fest, daß im gespannten Muskel mit der Zeit durch Degeneration von Muskelgewebe sog. Triggerpunkte entstehen, die auf mechanische Belastung hin einen übertragenen Schmerz in „Referenzzonen" entstehen lassen (Travell, 1976).

37.3.3 Kombinierter Kopfschmerz

Aufgrund der physiologisch-biochemischen Gesetzmäßigkeiten ist es aus mehreren Gründen hochwahrscheinlich, daß zu migränoidem Kopfschmerz Spannungskopfschmerz hinzutritt: Wenn die initiale Vasokonstriktion auf einen Sympathikusexzeß zurückgeht, bewirkt dieser simultan eine Erhöhung des Muskeltonus und eine Erniedrigung der Schmerzschwelle in den den Muskel versorgenden Arterien infolge von Vasokonstriktion (vgl. 37.3.2). Außerdem sind die Migräneschmerzen Reize, von denen zu erwarten ist, daß sie im Sinne der oben beschriebenen Reflexmuster Muskelverspannungen provozieren (vgl. Schmidt, 1981).

37.3.4 Kopfschmerz psychischen Ursprungs

Die physiologischen Grundlagen von Kopfschmerzen dieser Kategorie sind weitgehend ungeklärt. Möglicherweise handelt es sich um Störungen des Hirnstoffwechsels. Sicuteri (1981) hat eine Theorie des „antinozizeptiven Systems" vorgeschlagen, die in diesem Zusammenhang von Bedeutung sein könnte. Er postuliert ein autoanalgesierendes System, das bei Stimulation durch die verschiedensten Stressoren Enkephaline und Endorphine freisetzt und damit die Schmerzempfindlichkeit verringert. Durchschnittliche Stimulation wirkt „abhärtend" und ist deshalb sinnvoll. Bei überstarker Stimulation, etwa bei chronischer emotionaler Belastung oder infolge genetischer Verankerung, kann das antinozizeptive System allerdings auch zusammenbrechen, wodurch eine allgemeine Hyperalgesie entsteht. Abgesehen von Belegen über den Zusammenhang von Depression und Endorphinhaushalt (Larbig, 1982a) ermangelt diese Theorie einer empirischen Fundierung.

37.4 Auslöser von Kopfschmerzen

Es gibt eine Vielzahl empirischer, obschon häufig widersprüchlicher Studien zur Auslösung von vaskulärmigränoiden Kopfschmerzen, im Gegensatz zu den spärlichen Untersuchungen zu anderen Formen nicht-organischer Kopfschmerzen. Letztere beruhen meist auf klinischen Urteilen.

37.4.1 Auslöser von Migräne[3]

Hormonelle Veränderungen bei Frauen werden oftmals mit Migräneanfällen in Verbindung gebracht. Migräneanfälle treten gehäuft kurz vor der Menstruation auf, bei manchen Patientinnen werden sie durch

---

[3] Ausführliche Darstellungen finden sich bei Adams et al. (1980); Dalessio (1980); Gerber (1986); Heyck (1975); Murray (1981); Knapp (1983a); Raskin und Appenzeller (1980) und Rose und Gawel (1979).

Kontrazeptiva ausgelöst. 60–80% der Patientinnen geben eine deutliche vorübergehende Besserung ihres Leidens während der Schwangerschaft an, während sich das Leiden im Klimakterium sowohl verbessern als auch verschlimmern kann. Bei den **alimentären Substanzen** sind es vor allem Tyramin (enthalten in Rotwein, Schokolade und bestimmten Käsesorten), Phenyläthylamin, Alkohol und Histamin, von denen behauptet wird, sie lösten Migräneattacken aus. In neuen und empirisch besseren Studien wurden solche Zusammenhänge allerdings nicht mehr identifiziert (vgl. Kohlenberg, 1982). Man weiß auch wenig über die immer wieder angeführten **physikalischen Auslöser**: körperliche Belastungen (z. B. Migräne aufgrund von Koitus bei Männern) und optische Reize (grelles Licht, Lichtblitze etc.). **Kurzfristige Veränderungen der bioklimatischen Verhältnisse** (insbesondere Föhn) und bestimmte **Jahreszeiten** (Frühjahr und Herbst) werden nicht nur von Patienten, sondern auch von vielen Kopfschmerzforschern als Auslöser von Migräne genannt (z. B. Cull, 1981; Dirnagl und Kugler, 1981). Die empirischen Studien sind jedoch meist methodisch anfechtbar.

Auch wenn die neueren Untersuchungen gegen die Wirksamkeit von alimentären und bioklimatischen Faktoren sprechen, halten wir es für verfrüht, über sie abschließend zu urteilen. Von praktischer Relevanz bleiben sie in jedem Fall, weil viele Patienten an ihren Einfluß glauben und der Therapeut sich mit diesen Laienkonzepten auseinandersetzen muß, wenn er einen Zugang zum Patienten finden will.

Im allgemeinen wird angenommen, daß **psychische Auslöser** mehr Gewicht haben als hormonelle, physikalische und alimentäre; sie sollen bei über 60% der Patienten die Hauptauslöser sein. Zur Erklärung wird meist die Streßtheorie herangezogen: Migränepatienten leiden heftiger unter Streß als beschwerdefreie Kontrollpersonen. Dies trifft vor allem für Patienten mit täglichem Kopfschmerz zu. Dabei ist Streß keine „objektive" Größe – Migränepatienten sind im Alltag nicht mehr Streßereignissen ausgesetzt als Kontrollpersonen (Henryk-Gutt und Rees, 1973) –, sondern die Folge einer individuell andersartigen Bewertung derselben Situationen als belastend – ein Befund, der auf die Bedeutung der „individuellen Wirklichkeit" (v. Uexküll) verweist.

37.4.2 Auslöser von Spannungskopfschmerz

Alle Versuche in den fünfziger Jahren, Spannungskopfschmerz experimentell zu provozieren, erschöpften sich in Versuchsanordnungen von geringer ökologischer Plausibilität (die zudem an Frankenstein [1897] erinnern): Wolff und seine Kollegen (vgl. Simons et al., 1943) injizierten wiederholt hypertone Kochsalzlösung in die Halsmuskulatur, reduzierten die zerebrospinale Flüssigkeit und erzeugten Schmerzen durch das Tragen einer 3-diopter-Prismenbrille. Die beobachtete Korrelation von Muskelspannungsanstieg und Schmerz ist jedoch kein Beweis für die Muskelspannung-Schmerz-Annahme, weil die Muskelspannung auch sekundär als Folge der Schmerzen angestiegen sein kann.

Der Vergleich mehrerer optischer Stressoren führte nur bei Lichtblitzen zu einem signifikanten Anstieg der Muskelspannung in der Stirnmuskulatur, während die Nackenmuskulatur keine erhöhte Aktivität zeigte. Nur die Gruppe der Spannungskopfschmerzpatienten reagierte auf diese Stressoren mit Schmerzen, die Kontrollpersonen blieben beschwerdefrei. Die Befunde dieser Auslöserstudie sprechen für differentielle Reaktionen der Muskulatur auf spezifische Belastungen hin (Traue und Lösch, 1985).

Obwohl Alltagsstreß immer wieder in den verschiedenen Definitionen von Spannungskopfschmerz als Auslöser genannt wird, ist ein empirischer Nachweis bisher nicht gelungen. In der Feldstudie von Schlote (1989) korrelierten die Streßeinschätzungen der Patienten in Kreuzkorrelationen weder mit den Kopfschmerzen noch mit den tatsächlichen Muskelspannungen. Allerdings schätzten die Kopfschmerzpatienten ihren Alltagsstreß auch signifikant geringer ein als die Kontrollpersonen, so daß ein direkter Zusammenhang zwischen subjektivem Streßerleben und Schmerz nicht wahrscheinlich ist. Bischoff und Traue (1983) schlagen vor, den schmerzauslösenden Streß, unter den sich diese Patienten setzen, als einen Streß zu spezifizieren, der sich auf die Muskulatur in ihrer **motorischen** Funktion bezieht. Algogene Muskelspannungen entstehen, wenn muskelarbeitaktivierende Handlungen, exzessiv oder unökonomisch praktiziert, ihrerseits erhöhte Muskelspannung zur Folge haben, wenn Handlungsimpulse unter Aufbietung von Muskelkraft unterdrückt werden oder wenn Personen nach muskulärer Arbeit kein Verhalten entwickeln, welches bestehende Muskelspannungen reduziert.

37.5 Somatische und psychosomatische Dispositionen für Kopfschmerz

37.5.1 Migräne

Auch im anfallsfreien Zeitraum weisen Patienten mit Migräne eine Reihe von physiologisch-biochemischen Auffälligkeiten auf (vgl. Knapp, 1983a): einen reduzierten Serotoninspiegel und reduzierte MAO-Aktivität im Gehirn; eine gesteigerte Impulsrate serotoninerger Neuronen, eine beschleunigten Abbau von Transmitter-Serotonin im synaptischen Spalt, eine gedrosselte Serotoninsynthese, erhöhte Thrombozytenaggregationstendenz, reduzierte Serotoninbinde- und -speicherfähigkeit der Thrombozyten, eine erhöhte Adrenalinfreisetzung und eine gesteigerte Prostaglandinsynthese – was die migränespezifische zerebrovaskuläre Hyperreaktivität auf CO_2-Zufuhr belegt – und schließlich gesteigerte Sensitivität adrenerger kranialer Vasorezeptoren. In verwandelter Form treffen wir bei der Vasomotorik auf dieselben Auffälligkeiten: Seit den Studien von Tunis und Wolff (1953, 1954) ist bekannt, daß Migränepatienten auch im anfallsfreien Intervall eine für sie spezifische, bis zu 50% größere Dilatation der Temporalarterien haben. Dem korrespondiert eine gegenüber beschwerdefreien Personen reduzierte av-O_2-Differenz, was auf eine dispositionell erschwerte Sauerstoffabgabe an die kleinen Blutgefäße hinweist (Heyck, 1975).

Migränepatienten haben einen erhöhten vasokonstriktiven Sympathikotonus, ablesbar an der niedrigeren Handtemperatur und der erhöhten Herzrate auch in der schmerzfreien Zeit. Die Klagen vieler Patienten über kalte Hände und Füße weisen in dieselbe Richtung. Die Annahme einer Dysfunktion des autonomen Nervensystems wird ferner durch das Auftreten von zyklischem Erbrechen, Reisekrankheit und Ohnmachtsanfällen in der Kindheit von Migränikern gestützt. Die erhöhte Sympathikusaktivität ist mit einiger Wahrscheinlichkeit nur für Migräne, nicht aber für die Spannungskopfschmerzen spezifisch.

37.5.2 Spannungskopfschmerz

Nach der Definition des Subkomitees geht Spannungskopfschmerz auf eine überdauernde Kontraktion der Skelettmuskeln im Kopf- und Nackenbereich zurück. Also würde man in diesem Bereich ein erhöhtes tonisches Niveau der Muskelspannung als Disposition des Patienten mit Spannungskopfschmerz erwarten. Es gibt jedoch fast ebenso viele experimentelle Untersuchungen, die für Ruhesituationen ein erhöhtes, wie solche, die ein unverändertes Niveau fanden, was wahrscheinlich darauf zurückzuführen ist, daß in den früheren Untersuchungen Ruheniveaus in eher unstrukturierten Situationen gemessen wurden, in den neueren Untersuchungen dagegen nach Entspannung (Traue et al., 1984). Wie wir gesehen haben, ist ein erhöhtes Basisniveau der Muskelspannung nicht spezifisch für Spannungskopfschmerz, sondern auch bei Migräne anzutreffen.

Auch die Befunde zur Muskelspannungsaktivität in Belastungssituationen sind uneinheitlich. Als generelle Linie zeichnet sich ab: Es gibt eher geringe Unterschiede zwischen Personen mit Spannungskopfschmerz und schmerzfreien Kontrollpersonen, wenn die Höhe der Muskelspannung als Reaktion auf physikalische oder neutral-mentale Stressoren zur Debatte steht (Bakal und Kaganov, 1979; Andrasik et al., 1982; Feuerstein et al., 1982). Personen mit Spannungskopfschmerz haben erhöhte EMG-Werte, je emotionaler und interaktionaler die Streßsituationen sind (Traue et al., 1981; Traue et al., 1984; Traue, 1989; Thompson und Adams, 1984), oder wenn motorische Reaktionen durch ganz spezifische physikalische Stressoren ausgelöst werden (Bischoff et al., 1982; Traue und Lösch, 1985). In der Regel unterscheidet sich auch die EMG-Reaktivität von Personen mit Spannungskopfschmerz nicht von der von Migränikern – auch diese reagieren mit überhöhter Muskelspannung; in der sorgfältig geplanten Studie von Thompson und Adams (1984) ergab sich allerdings eine deutlich größere Reaktivität von Personen mit Spannungskopfschmerz.

Die immer wieder erhobene Forderung nach Muskelspannungsmessungen über einen längeren Zeitraum im Alltag der Patienten wurde von Schlote (1989) erfüllt. Ihre psychophysiologische Feldstudie

wies Tagesmittelwerte im Elektromyogramm des M. trapezius zwischen 26 und 28 μV für Patienten mit Spannungskopfschmerzen nach, während die Kontrollpersonen nur Spannungen von 15–18 μV erreichten. Diese im Alltag erhobenen Daten bieten erste Möglichkeiten, schmerzrelevante Spannungswerte auch quantitativ zu schätzen.

Eine eingeschränkte Variabilität der Muskelspannung wurde bei Spannungskopfschmerzpatienten ebenfalls beobachtet (Bakal und Kaganov, 1979; Bischoff und Sauermann, 1985); außerdem eine verzögerte Rückbildung der Muskelspannung nach Streßereignissen (Wilker und Bischoff, 1979; Traue et al., 1981; Traue et al., 1985). Die Studien zeigen, daß es entgegen der Subkomitee-Definition nicht immer das tonische Niveau der Muskelspannung ist, das den Patienten mit Spannungskopfschmerz charakterisiert. Um die Idee eines Kopfschmerzes aufgrund von Muskelspannung weiter zu explizieren, haben Bischoff und Traue (1983) deshalb vorgeschlagen, mit dem Begriff „myogener Kopfschmerz" all die Kopfschmerzen zu bezeichnen, welche aufgrund einer Muskelmehrarbeit pro Zeiteinheit entstehen, die ein bestimmtes kritisches Niveau übersteigt.

37.5.3 Genetische Faktoren beim Kopfschmerz

Die Forschungsergebnisse zur Vererblichkeit sind widersprüchlich und methodisch häufig unsauber. Für eine ausführliche Erörterung der Vererbungsfrage verweisen wir auf Knapp (1983a) und Gerber (1986). Ein Beispiel sei dennoch angeführt: Selby und Lance (1960) geben an, daß bei 55% der Migräniker Migräne auch in der näheren Verwandtschaft auftritt. Mit dieser Zahl wird die Prävalenzrate aber wahrscheinlich weit überschätzt. Befragt man nämlich die Angehörigen selbst und nicht die Migräniker, so sinkt sie auf 10%, für Spannungskopfschmerz liegt sie bei 5%, bei beschwerdefreien Personen ebenfalls bei 5%, wobei die Differenz von 5% statistisch nicht bedeutsam ist (Waters, 1971).

37.6 Psychische Dispositionen für Kopfschmerz

Die Auffassung, es gäbe eine bestimmte Kopfschmerzpersönlichkeit, wird von Klinikern, insbesondere psychoanalytisch orientierten Therapeuten, mit größerer Überzeugung vertreten (z.B. Peters, 1983) als von der empirisch-psychologischen Kopfschmerzforschung (vgl. z.B. Andrasik et al., 1982).

Nach klinischen Erhebungen gelten Migränepatienten als ehrgeizig, erfolgsorientiert, überordentlich, perfektionistisch, ausdauernd, leicht irritier- und kränkbar; sie gestalten soziale Beziehungen eher unpersönlich, ihre Sexualität ist gehemmt (Wolff, 1963; Anderson, 1980). Nach anderen Untersuchungen finden sich auch häufig Passivität und geringe Frustrationstoleranz; Peters (1983) faßt diese Persönlichkeitszüge als „Typus migränicus" zusammen. Feindseligkeit, Abhängigkeit, Depressivität und eine Häufung psychosozialer Konflikte kennzeichnen dagegen den Patienten mit Spannungskopfschmerz (Philips, 1976; Diamond, 1983).

Kritische Einwände gegen die meisten klinischen Studien machen darauf aufmerksam, daß erstens die Beobachtungen aus der stark selektierten Gruppe von Patienten stammen, die den Arzt aufsuchen, und zweitens retrospektiv sind, so daß unklar bleibt, ob die Persönlichkeitszüge nicht als Folge des Schmerzleidens auftraten (vgl. z. B. Larbig, 1982 b).

Knapp (1983 a) bietet einen Überblick über die Befunde von 17 testpsychologischen Studien zu Persönlichkeitszügen bei Kopfschmerzpatienten. Empirische Erhebungen mit Hilfe von bekannten psychometrischen Persönlichkeitsfragebogen (vorwiegend dem MMPI, dem FPI und dem Eysenckschen Fragebogen) liefern ein widersprüchliches Bild der Kopfschmerzpersönlichkeit, in dem sich Migränepatienten und Patienten mit Spannungskopfschmerz nur undeutlich voneinander abheben. Die empirischen Erhebungen mit Hilfe von Befragungstechniken, die sich an psychodynamischen Erklärungsansätzen orientieren (z. B. Rorschach-Test oder klinische Tiefeninterviews), scheinen den klinischen Eindruck einer differentiellen Migränepersönlichkeit eher zu bestätigen. Jedoch sind die vorliegenden Studien, die sich dieser Befragungstechnik bedienen, methodisch fragwürdig. Es stellt sich heraus, daß die Gesamtgruppe der Kopfschmerzpatienten im Sinne des Neurotizismuskonzepts von Eysenck nervöser ist als schmerzfreie Kontrollpersonen und häufiger außer Kopfschmerz weitere Körperbeschwerden hat, besonders im Bereich des Haltungs- und Bewegungsapparates (Brähler, 1983). Gibt man tendenziellen Unterschieden eine Bedeutung, so scheinen die psychischen Störungen der Migräniker sozial besser kompensiert, Impulse von Feindseligkeit besser kontrolliert als die von Patienten mit Spannungskopfschmerz; am besten gesichert ist auch in Fragebogenstudien die größere autonome Hyperreaktivität der Migräniker. Dem gegenüber treten bei Patienten mit Spannungskopfschmerz die Persönlichkeitszüge der Feindseligkeit, der Ängstlichkeit und der psychischen Desorganisation manifester in Erscheinung (Andrasik et al., 1982; Passchier et al., 1984). Passchier und Kollegen kommen aufgrund ihrer Untersuchung zu dem Schluß, daß für beide Gruppen von Kopfschmerzpatienten handlungsbezogene Merkmale im Leistungsbereich aussagekräftig sind: Sowohl Migräniker als auch Personen mit Spannungskopfschmerz haben im Vergleich zu einer schmerzfreien Kontrollgruppe eine höhere Leistungsmotivation und mehr Furcht vor Mißerfolg.

37.7 Psychogenetische Modelle der Migräne und des Spannungskopfschmerzes

37.7.1 Verhaltensmedizinische Ansätze

Dieser Abschnitt stellt exemplarisch vier psychobiologische Modelle der Kopfschmerzgenese und -aufrechterhaltung vor. Zunächst werden zwei wichtige Theorien zur Migräne abgehandelt (Cinciripini et al., 1981; Knapp, 1983 a, b), dann unser Ansatz zum myogenen Kopfschmerz (Bischoff und Traue, 1983) und abschließend die Einheitstheorie des Kopfschmerzes von Bakal (1982).

Als wesentliche Entstehungsbedingung der Migräne sehen Cinciripini et al. (1981) erhöhten Streß (Arbeit, gefürchtete Ereignisse), der, besonders wenn es an Bewältigungsfähigkeiten zur Reduktion fehlt, Vasokonstriktion mit den bekannten physiologisch-biochemischen Folgen auslöst, die bei nachlassendem Streß (Ende des Arbeitstags, Wochenende) in schmerzerzeugende Vasodilatation umschlägt. Intensiviert werden die „Klagen", wenn aus dem inadäquaten Bewältigungsverhalten Mißerfolge, vermehrte Angst und weniger Selbstvertrauen resultieren, die zusammen als Stimmungslage die Kopfschmerzäußerung „tönen". Die Chronifizierung (und damit die Aufrechterhaltung) betrifft einerseits strukturelle physiologische Veränderungen (reduzierter vasomotorischer Tonus, gesteigerte Reaktivität der Gefäße). Diese sind ein Grund für eine erniedrigte Schmerzschwelle nach wiederholten Attacken. Entsprechend den Vorstellungen zur Konditionierbarkeit von Schmerzen (Fordyce, 1976) sehen die Autoren andererseits Umweltfaktoren in Form von Zuwendung, Massage, Medikation oder Entpflichtung von unangenehmen Aufgaben als verstärkend für die Kopfschmerzäußerung an. Durch diese Prozesse kann auch die Schmerzschwelle sinken, so daß der Patient gegebenenfalls ohne parallele pathophysiologische Vorgänge Schmerz äußert.

Ein sehr elaboriertes Bedingungsmodell der Migräne hat Knapp (1983 a) vorgelegt. Knapp geht von dem klassischen Diathese-Streß-Modell aus: Bei aufgrund von angeborenen oder erworbenen Defekten vorbelasteten Personen ist vor allem psychischer Streß zentraler Auslöser der Migräne. Psychischer Streß wird subjektiv definiert als Hilflosigkeits- und Kontrollverlusterleben. Psychischer Streß ist Resultat einer spezifischen Informationsverarbeitung, bei der der Betreffende eine Diskrepanz zwischen selbstgesetzten und/oder fremdbestimmten Anforderungen und den behavioralen Bewältigungsstrategien erlebt, die er zur Verfügung zu haben glaubt. Sowohl bei Über- als auch Unterforderung mündet der dadurch bewirkte Streß in gesteigerte Sympathikusaktivierung und setzt beim diathetisch Vorbelasteten den bekannten physiologisch-biochemischen Prozeß in Gang.

Ausgehend von den physiologisch-biochemischen Mechanismen des Spannungskopfschmerzes (vgl. auch Abschnitt 37.4.2) stellen Bischoff und Traue (1983) dysfunktionale Muskelmehrarbeit in das

Zentrum ihrer Theorie myogener Kopfschmerzen. Diese dysfunktionalen Muskelspannungen können durch übermäßige Anstiege, verlängerte Rückbildung, erhöhte Verspannungen in Ruhe und durch gehäufte oder übermäßig lange Belastungen akkumulieren.

Lerntheoretisch läßt sich die Entstehung dysfunktionaler Muskelmehrarbeit als Folge klassischer und operanter Konditionierung verstehen. Dysfunktionale Muskelspannungen sind besonders dann konditionierbar, wenn sie als motorische Aktivität die physiologische Basis von Handlungen und Bewegungen sind. So ist Muskelmehrarbeit als Korrelat von beruflicher Tätigkeit direkt durch positive Verstärkung oder indirekt durch die Bestrafung von Ruhepausen konditionierbar. Solche Vorgänge sind für berufsbedingte Kopfschmerzen verantwortlich. Durch die Bestrafung von emotionalem Ausdrucks- und Bewegungsverhalten kann die durch übermäßige Anspannung realisierte Ausdruckshemmung operant konditioniert werden. Das Bedingungsmodell myogener Kopfschmerzen (Abb. 37–3) berücksichtigt aber auch die Möglichkeit operanter Kontrolle von Kopfschmerz als Schmerzverhalten, das partiell oder vollständig ohne physiologische Beteiligung denkbar ist. Unseren klinischen Erfahrungen nach wird diese Möglichkeit jedoch zumeist überschätzt.

Eine direkte Bestätigung der Muskelmehrarbeitshypothese erbrachte die Feldstudie von Schlote (1989), in der eine sorgfältig ausgewählte Gruppe mit Spannungskopfschmerzpatienten über eine Woche hinweg bei ihrer normalen Arbeitstätigkeit elektromyographisch erfaßt wurde. Die Personen mit Spannungskopfschmerz akkumulierten – auch während der Arbeitspausen – nahezu doppelt soviel Trapeziusverspannungen wie die Kontrollpersonen. Die Herzaktivität war nicht unterschiedlich.

Klinisch werden als Auslöser für Kopfschmerzen häufig interpersonale Belastungen von Patienten genannt. Solche aversiven sozialen Stressoren waren auch in Laboruntersuchungen besonders gut geeignet, Patienten mit Spannungskopfschmerzen von Kontrollpersonen in ihren muskulären Reaktionen zu trennen. Die Patienten waren durch größere Anstiege, höhere Absolutspannungen und verzögerte Rückbildungszeiten gekennzeichnet (Traue et al., 1985; Traue et al., 1986). Ebenfalls bedeutsam sind Schmerzreizung und Schmerzerwartung als bedingte und unbedingte Stimuli, die bei Personen mit Spannungskopfschmerz stärker zu dysfunktionaler Muskelmehrarbeit führen als bei gesunden Kontrollpersonen (Bischoff et al., 1982). Diese Kausalbeziehung zwischen Schmerzerwartung und Spannungsanstieg kann ein wichtiges Bindeglied zum Verständnis der Aufrechterhaltung von Spannungskopfschmerzen sein.

Emotionale Reaktionen in sozialen Situationen, die unter Bestrafungsbedingungen geraten, bleiben mit ihren motorischen und autonomen Komponenten erhalten, wenn das offene Ausdrucksverhalten unterdrückt wird. In solchem Hemmungsverhalten sehen wir eine wichtige Quelle schmerzerzeugender Muskelaktivität. Tatsächlich zeigen Patienten mit Spannungskopfschmerzen unter sozialem Streß verminderte Expressivität und reduzierte kommunikative Bewegungen der Arme und des Kopfes. Gleichzeitig korreliert die Hemmung mit den erhobenen Muskelspannungswerten (Traue et al., 1985). Die Hemmung expressiven Verhaltens führt jedoch nicht nur zur Akkumulation von Muskelspannung, sondern stellt eine ineffiziente Strategie zur Bewältigung von sozialem Streß dar und behindert den Aufbau eines sozialen Unterstützungssystems (Holm et al., 1986; Traue, 1989).

Wird aufgrund klassischer und operanter Konditionierung ein Verhalten verstärkt, das mit dysfunktionaler Muskelaktivität einhergeht, so beeinflußt das die propriozeptive Wahrnehmung der muskulären Aktivität. Diese hat normalerweise eine handlungsregulierende Funktion, indem sie dem Individuum seine Belastung signalisiert und Hinweisreize zur Erholung gibt. Bei emotionaler Stimulierung werden über das muskuläre Feedback qualitative emotionale Informationen verarbeitet. Wenn motorische Aktivität, die mit Muskelmehrarbeit einhergeht, positiv oder negativ verstärkt wird, verliert die Wahrnehmung der Muskelspannung diese Funktionen und wird gelöscht. Bischoff (1989) konnte ein solches hypostasiertes Wahrnehmungsdefizit bei Patienten mit Spannungskopfschmerzen experimentell nachweisen. Die-

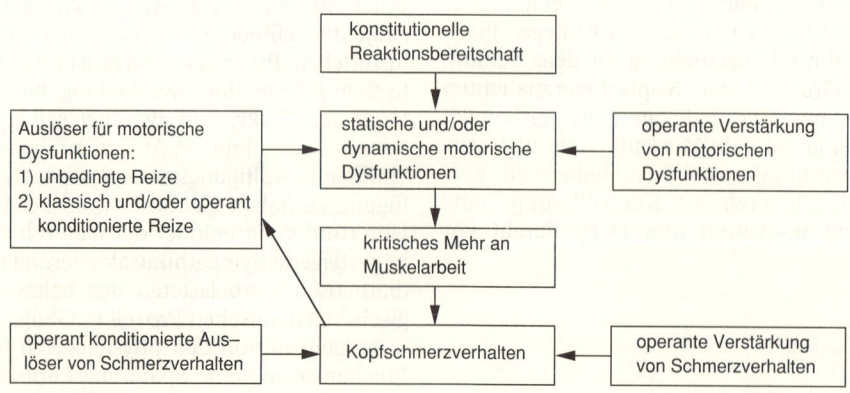

Abb. 37–3. Vereinfachtes Bedingungsmodell des myogenen Kopfschmerzes nach Bischoff und Traue (1983).

ser Befund wiegt um so schwerer, als Migränepatienten ihre Stirnmuskelspannung nicht schlechter wahrnehmen konnten als die Kontrollpersonen. Es handelt sich also um einen Mechanismus, der speziell Patienten mit Spannungskopfschmerzen betrifft.

Das Modell läßt sich an einem Beispiel aus der Krankengeschichte von Herrn M. illustrieren. Einerseits können die Kopfschmerzen betrachtet werden als physiologisch erklärliche Folge klassisch-konditionierter dysfunktionaler Muskelspannung. Die dysfunktionale Muskelspannung ist die konditionierte Reaktion, die durch den Vorgesetzten in einer Leistungssituation auslösbar ist. Es ist anzunehmen, daß also Reizgeneralisierung vom ursprünglichen konditionierten Reiz stattgefunden hat: Der Vater in einer Leistungssituation ist die ursprüngliche komplexe konditionierte Reizsituation, in der durch den unkonditionierten Reiz „Prügel" die unkonditionierte Reaktion „Muskelverspannung", wahrscheinlich in Form von Kopfeinziehen, hervorgerufen wurde.

An diesem Patienten läßt sich auch die Möglichkeit operanter Konditionierung von dysfunktionaler Muskelspannung veranschaulichen. Nicht genug, daß der Vater ihn schlug, er verbot unter Androhung weiterer Prügel dem Patienten ja auch zu weinen. Wir haben es beim Weinen und Sich-zur-Wehr-Setzen mit angeborenen Impulsen zu tun, die mit vermehrter Muskelspannung einhergehen, und die der Patient unter Aufbietung von Muskelspannung aktiv unterdrückt. Die Unterdrückung ist operant gelernt, insofern durch sie der aversive Reiz der Prügel vermieden werden konnte. Die dysfunktionale Muskelspannung setzt sich in solchen Fällen aus zwei Komponenten zusammen: Der ursprüngliche Handlungsimpuls bleibt bestehen, da er nicht in Handlung umgesetzt wird. Außerdem kostet es Muskelarbeit, den Handlungsimpuls zu hemmen.

Der Patient kann bei der Arbeit nicht aufhören, auch nicht in der Freizeit. Auch hier läßt sich operant konditionierte dysfunktionale Muskelspannung annehmen. Für Herrn M. sind Nichtstun und Entspannung aversiv. Depressive oder Angstgefühle steigen dann auf. Da muskelarbeitfordernde Aktivitäten diese Gefühle zu beenden in der Lage sind, hat sich der Patient diese aufgrund von negativer Verstärkung angewöhnt.

Im Rahmen des Situationskreiskonzepts (v. Uexküll, vgl. Kap. 1) lassen sich die beschriebenen Zusammenhänge als spezifische Bedeutungskopplungen zwischen Individuum und Umwelt interpretieren. Es ist auch nicht zu übersehen, daß die Interpretation des myogenen Kopfschmerzes Parallelen mit Reichschen Vorstellungen aufweist (1969, S. 260 ff.). Bestimmte frühkindliche Lernbedingungen – so Reich – hemmen die Affektabfuhr und begünstigen dadurch die Entwicklung von Muskelpanzerungen, d. h. von chronischen Muskelverspannungen, die im Extremfall und – in bestimmten Körperregionen bevorzugt – Schmerzen zur Folge haben. Reichs Beispiele und Überlegungen können in Begriffen der klassischen und operanten Konditionierung dargestellt werden.

Bakal (1982) vertritt, abgeleitet aus seinen empirischen Studien (vgl. 37.1), ein eindimensionales Konzept, das Kopfschmerzen nach ihrem Schweregrad ordnet. Seine Überlegungen basieren auch auf dem Diathese-Streß-Modell, das jedoch durch eine Komponente der Krankheitsentwicklung erweitert ist: Kopfschmerzen haben immer eine vaskuläre und eine muskuläre Seite. Am Anfang mag eine Prädisposition zu erhöhter Muskelaktivität im Kopf-/Nackenbereich das Übergewicht haben. Wenn der Kopfschmerzleidende nicht in der Lage ist, daraus erwachsende Kopfschmerzen zu bewältigen, entstehen schwerere Kopfschmerzen, die eine immer stärker ausufernde vaskuläre Komponente erhalten. Mit zunehmender Schwere ändert sich auch strukturell die Disposition zum Kopfschmerz dahingehend, daß sie immer mehr physiologische Systeme involviert. Je schwerer die Kopfschmerzen, desto eher treten sie unabhängig von psychosozialen Stressoren, meist schon morgens auf das Leiden verselbständigt sich.

37.7.2 Psychodynamische Erklärungsansätze

Aus psychodynamischer Sicht bildet in der Regel ein intrapsychischer Konflikt die Basis für Kopfschmerz. Dem Symptom kommt in diesem Konflikt die Funktion einer neurotischen Bewältigung zu.

Kopfschmerz als Konfliktbewältigung hat eine **konversionsneurotische Form,** wenn er symbolischer, körpersprachlicher Ausdruck für den zugrundeliegenden Konflikt oder ein einzelnes seiner Elemente ist (vgl. Kap. 30). In Anwendung der Konzeption Adlers (1981) kann der psychische Apparat im Kopfschmerz symbolisch Wünsche, Gedanken und Phantasien ausdrücken, die vom bewußten Teil des psychischen Apparates nicht akzeptiert werden und die nicht in einer entsprechenden Handlung Erfüllung finden.

Nach Alexander (1951, S. 117) beruht der konversionsneurotische Kopfschmerz auf einem „Vorgang in den höheren sensiblen Zentren des Gehirns", bei dem „keine lokalen Veränderungen" zu beobachten sind. Da die Migräne auf der Dilatation der kranialen Gefäße und somit auf peripheren Veränderungen beruht, ist sie theoretisch vom Konversionskopfschmerz abzugrenzen. Konsequent ordnet Alexander die Migräne den vegetativen Neurosen zu.

Alexander (1951) konzipiert Migräne als Folge eines zu Beginn der vegetativen Vorbereitung gehemmten aggressiven Aktes, der mit einer erhöhten Blutzufuhr zum Gehirn bei gleichzeitiger Blockierung der Muskelaktion einhergeht. Diesem Konzept liegt ebenfalls ein Konfliktmodell zugrunde; der Konflikt besteht in dem aggressiven Akt und den gegengerichteten Impulsen, die den Akt hemmen.

Nach psychoanalytischen Studien charakterisiert die biographische Entwicklung von Personen mit Migräne in der frühen Kindheit eine affektive Mangelkonstellation, die durch mütterliche Kühle, Zwanghaftigkeit und Härte bedingt ist (Larbig, 1982b). Bei den Müttern der Migräniker fanden sich vermehrt trieb- und sexualfeindliche Strebungen, bei den Vätern Weichheit und Nachgiebigkeit, insgesamt eine eher verwöhnende Haltung gegenüber den Patienten. Die Migräniker übernehmen sehr hohe Leistungsideale von ihren Eltern in Form einer positiven Identifikation. Dies trägt dazu bei, ihre starke Erfolgsorientiertheit zu erklären.

Die psychoanalytische Forschung hat dem **Spannungskopfschmerz** kaum Aufmerksamkeit geschenkt. Janus (1978) kommt aufgrund von EMG-Ableitungen während einer analytischen Behandlung von Zervikalsyndrom-Patienten zu der Deutung, daß dieser Schmerz als somatisches Äquivalent einer gehemmten Kampf-/Fluchtreaktion zu verstehen sei.

37.8 Therapieverfahren

37.8.1 Somatisch-medizinische Therapieangebote

Pharmakologische Behandlung und Abusus-Problem

Als symptomatischer Therapieansatz ist die medikamentöse Kopfschmerzbehandlung[4] eine Gratwanderung zwischen dem Anspruch des Patienten, von seinem Schmerz befreit zu werden, und der Gefahr des Schmerzmittelmißbrauchs. Da Mischpräparate mit psychotropen Substanzen einen Abusus begünstigen, empfiehlt sich die Verschreibung reiner Substanzen. Als solche kommen für **akute Anfälle** Acetylsalicylsäure (Aspirin) und Paracetamol in Frage. Intensive Schmerzen bei Migräne können zu Beginn eines Anfalles mit Ergotamin bekämpft werden. Zur **Prophylaxe** (Intervallbehandlung) häufiger Migräneanfälle werden Beta-Blocker, Pizotifen, Methysergid sowie dehydrogenierte Ergotalkaloide angewendet. Alle diese Medikamente haben jedoch Nebenwirkungen (Langbein et al., 1983). Die Gefahr von Schmerzmittelabusus besteht aus drei Gründen: 1. Werden Schmerzmittel „bei Bedarf" eingenommen, so wird ihre Einnahme dadurch, daß die Schmerzen zurückgehen und somit ein aversiver Reiz wegfällt, negativ verstärkt. 2. Tranquilizer in Mischpräparaten erzeugen Abhängigkeit und 3. wirken Analgetika, insbesondere Mutterkornalkaloide, bei längerem Gebrauch selbst schmerzerzeugend – der Patient bekämpft mit ihnen Schmerzen, die er durch sie erzeugt (rebound-headache; Wörz und Lendle, 1980; Dichgans et al., 1984; Wilkinson, 1984). Bei Abusus ist ein Schmerzmittelentzug notwendig, der in der Regel stationär durchgeführt werden muß. Der Erfolg der Entzugsbehandlung ist einerseits abhängig von der psychotherapeutischen Unterstützung des Patienten durch die Klinik, andererseits von der Auswahl einer geeigneten Entzugsmethode. Bewährt hat sich z.B. neben dem Totalentzug, bei dem mit Entzugserscheinungen gerechnet werden muß, eine schrittweise Reduktion, bei der die Medikamente in einem Schmerzcocktail nach festgelegtem Zeitplan verabreicht werden (Kontingenzmanagement nach Fordyce, 1976).

Alternative Behandlungsverfahren

Besonders bei übertragenem Schmerz (Travell und Simons, 1983) wird häufig intramuskuläre **Infiltra**-tion von **Lokalanästhetika** in die Triggerzone vorgenommen. Viele Patienten erleben eine deutliche Verminderung ihrer Schmerzen oder sogar Schmerzfreiheit, die auch nach mehreren Monaten noch anhält. Eine andere Methode ist die **transkutane Nervenstimulation** (TNS), bei der die Oberfläche der schmerzenden Muskulatur elektrisch gereizt wird. Diese Behandlung kann ambulant vorgenommen werden. Es besteht auch die Möglichkeit, den Patienten ein Gerät mit nach Hause zu geben. Die Erfolge sind ähnlich wie bei der Infiltration mit Lokalanästhetika.

Bei Spannungskopfschmerz und Migräne wird neuerdings zunehmend **Akupunktur** angewandt. Sie erweist sich auch in sorgfältig kontrollierten Studien als eine erfolgzielende Methode (Wittchen, 1983). Die physiologische Basis dieser Erfolge ist nicht hinreichend geklärt. Während Schmidt und Struppler (1982) sie noch als Suggestionseffekte abtun, wird es neuerdings für möglich gehalten, daß Akupunktur ihre Wirkung über die Stimulation der Endorphinproduktion entfaltet.

Physiotherapie und Sporttherapie

Peters (1983) gibt in bezug auf Massagen und Gymnastik bei Kopfschmerz diese Hinweise: Beim Migräneanfall lindert Rotlichtbestrahlung und sanfte Massage von Myogelosen in der Wirbelsäule nahegelegenen Muskelpartien des Rückens und der Schultern sowie im Ansatz des M. sternocleidomastoideus den Schmerz erheblich. Mehrfaches Massieren von Myogelosen im Ansatz des M. trapezius trägt zur Beseitigung von Spannungskopfschmerzen bei. Für bedeutsamer im Hinblick auf längerfristigen Erfolg hält Peters jedoch ein von Patienten selbständig auszuführendes krankengymnastisches Programm zur Lockerung der Schulter-Nacken-Muskulatur, für das er in seiner Monographie eine detaillierte Beschreibung gibt.

Gerber et al. (1987) schlagen als flankierende Maßnahme bei der Behandlung von chronischem Kopfschmerz Sport in Form eines Jogging-Programms vor.

Eine Maßnahme im Grenzbereich von Physiotherapie und Psychotherapie ist die „funktionelle Entspannung" nach Fuchs (1974), die sich bei der Behandlung von vegetativen Neurosen als wirksam erwiesen hat. Die Übungen bestehen in minimalen Körperbewegungen, vor allem während der Phasen des Ausatmens. Sie führen einerseits zu einer umfassenden vegetativen Entspannung, andererseits zu Körperempfindungen, die die interozeptive Sensibilität steigern. Förderlich für den Erwerb dieser Sensibilität ist die Akzeptierung einer neuen „Körperphilosophie", die dem Patienten mit den Übungen auch kognitiv vermittelt wird.

---

4 Eine ausführliche Erörterung pharmakologischer und alternativer Therapieverfahren findet sich bei Bowdler und Kossmann (1983), Kossmann und Bowdler (1983) und Gerber (1986). Siehe auch den Beitrag von Adler über Schmerz in diesem Lehrbuch in Kapitel 36.

37.8.2 Psychotherapeutische Möglichkeiten

Psychodynamisch orientierte und klientzentrierte Psychotherapie

Entsprechend der hohen Prävalenz von Kopfschmerzen in der Gesamtbevölkerung spielen bei Patienten, die sich wegen **psychischer Konflikte** in Psychotherapie begeben, Kopfschmerzen als zumindest gelegentlich auftretendes Ereignis während der Behandlung eine Rolle. Da diese Patienten psychische Störungen als Quelle ihres Leidensdrucks betrachten, wird für sie wie für den Psychotherapeuten ein passager auftretender Kopfschmerz die gleiche Bedeutung haben wie etwa Fehlleistungen, die für das Verständnis der psychischen Störung wichtig sind.

Diese Konstellation ist jedoch für Kopfschmerzpatienten untypisch. Selbst Kopfschmerzpatienten mit sprachlichen Fähigkeiten und ausreichender geistig-seelischer Differenziertheit lehnen nicht selten psychotherapeutische Behandlung ab. Damit ist sogar zu rechnen, da sich ihnen ihr Leiden vor allem von der körperlichen Seite bemerkbar macht. In ihrer „psychotherapiefeindlichen" Haltung werden sie dadurch bestärkt, daß seitens der Medizin meistens gefordert wird, Kopfschmerzen besonders gründlich auf Organbefunde abzuklären. Dies führt bei den Patienten zu einer iatrogenen Fixierung ihrer ohnehin überwiegend naturalistischen Laientheorie (Harris, 1973). Da die umfassenden Diagnoseprozeduren in aller Regel keinen körperlichen Befund erbringen, wird der von Kopfschmerz Gequälte zum „Problempatienten", der, obwohl ihm „nichts fehlt", weiterhin auf Ursachenklärung seiner Schmerzen besteht. Den Hinweis auf den psychogenen Hintergrund seines Leidens empfindet er oftmals als stigmatisierend. Hinzu kommt, daß ihn die Frustration durch chronische Schmerzen in einen Zustand aggressiver Gereiztheit oder depressiver Hilflosigkeit als einem algogenen Psychosyndrom (Wörz und Lendle, 1980) versetzt.

Die Indikation zu einer „rein" verbalen Psychotherapie ist aus diesen Gründen meist fraglich. Volger (1983) kommt in ihrer Indikationsstudie zu dem Ergebnis, daß mehr als 50% der chronisch erkrankten Kopfschmerzpatienten nicht die Voraussetzungen erfüllen, die für eine erfolgreiche klientzentrierte Gesprächspsychotherapie notwendig sind. Eine ähnlich ungünstige Indikationsstellung dürfte für analytisch orientierte Interventionsverfahren gegeben sein. Der Umgang mit Kopfschmerzpatienten gilt in der Psychotherapie dementsprechend als schwierig (Lamprecht, 1979; Sommer und Overbeck, 1977).

Angesichts dieser Ausgangslage ist es empfehlenswert, sich in die Psychotherapie von Kopfschmerzpatienten gewissermaßen einzuschleichen (vgl. auch Brenner et al., 1949) – eine Aufgabe, vor die sich im übrigen nicht nur der Psychotherapeut selbst, sondern noch viel mehr der niedergelassene Arzt als erste Anlaufstelle und „Verteiler" gestellt sieht. Dazu kann man z.B. den psychosozialen Kontext einzelner Kopfschmerzanfälle explorieren und diesen dem Patienten klarmachen. Als Datenbasis solcher Explorationen sind Eintragungen in Kopfschmerztagebüchern wertvoll, die der Patient nach einem festen Zeitmuster mehrmals am Tag vornehmen soll. Der Patient hält z.B. fest, wie stark die Schmerzen gerade sind, was er jetzt und unmittelbar zuvor getan, gedacht, gefühlt hat, wie er starke Schmerzen zu bewältigen versuchte, wie wichtige Bezugspersonen auf Schmerzäußerungen reagiert haben usw. Über die Realitätskontrolle erhält er die Gelegenheit, seine naturalistische Laientheorie zu falsifizieren. Eine andere Möglichkeit, die Mitarbeit des Patienten zu gewinnen, besteht darin, ein Entspannungstraining anzubieten und über seine Erfahrungen mit der Entspannung seine „Selbstexploration" in Gang zu bringen. Volger (1983) berichtet von guten Erfolgen bei der Integrierung von progressiver Muskelentspannung nach Jacobson (1938) in die klientzentrierte Gesprächspsychotherapie. Der analytisch orientierte Psychotherapeut wird möglicherweise den Patienten lieber Techniken wie die funktionelle Entspannung nach Fuchs (1974) außerhalb der Gespräche praktizieren lassen.

Verhaltenstherapeutische Verfahren

Zur **Behandlung migränoider Kopfschmerzen** sind verschiedene verhaltenstherapeutische Verfahren gebräuchlich:
- Entspannungstraining ohne apparative Unterstützung: progressive Muskelentspannung nach Jacobson (1938); großes autogenes Training nach Schultz (1976); seltener: transzendentale Meditation und Hypnose.
- Entspannungstraining mit apparativer Unterstützung (Biofeedback): Handerwärmungstraining, autogenes Feedbacktraining, Feedbacktraining der Temporalarterie (Vasokonstriktionstraining), seltener: Alpha-, EEG-, EDA- und EMG-Feedbacktraining.[5]
- „Behandlungspakete" und multimodale Therapie: z.B. Konkordanztherapie (Haag et al., 1982; Gerber, 1982); das situationsbezogene Muskel- und Gefäßempfindungsprogramm (SEP; Wittchen, 1983); das „kognitiv-behaviorale Streßbewältigungs-Training" (KBST; Knapp, 1981).

Zwei der Biofeedbackverfahren sind speziell für die Therapie von Migräne entwickelt worden:

Handerwärmungstraining – das Verfahren geht auf Green und seine Kollegen zurück (Sargent et al., 1972) – ist ein Verfahren, bei welchem dem Patienten durch Messung der peripheren Durchblutung mit Hilfe eines Plethysmographen das Ausmaß der digitalen Durchblutung zurückgemeldet wird. Aufgabe des Patienten ist es, die Durchblutung der Hand zu steigern. Beim **autogenen Feedbacktraining** wird die Handerwärmung durch formelhafte Vorsatzbildun-

---

5 Für eine Beschreibung von progressiver Muskelentspannung, autogenem Training und Hypnose vgl. Kapitel 22 von R. Lohmann über „Suggestive und übende Verfahren", für eine Beschreibung des Prinzips von Biofeedback vgl. Kapitel 19 von D. Schwarz über „Verhaltenstherapie".

gen aus dem autogenen Training unterstützt. Die therapeutische Wirkung des Handerwärmungstrainings bei Migränepatienten ist indirekt: es dämpft die Sympathikusaktivität (Sovak et al., 1980).

Beim **Feedbacktraining der Temporalarterie** (vgl. Friar und Beatty, 1976) übt der Patient eine mentale Strategie zur Konstriktion der A. temporalis superficialis ein. Gemessen und zurückgemeldet werden entweder das Gefäßkaliber, die Blutfließgeschwindigkeit oder der Blutvolumenpuls. Viele Patienten können sich den Erwerb der Kontrolle über das Gefäßkaliber durch die Vorstellung, in einen Tunnel zu fahren oder durch die Vorstellung der Enge erleichtern. Wenn sie in der Lage sind, die Vasokonstriktion zuverlässig zu kontrollieren, werden sie angeleitet, die Gefäße auch ohne apparative Hilfe bei den ersten Anzeichen eines Migräneanfalls engzustellen. Das Behandlungsverfahren zielt in Analogie zur medikamentösen Anfallsbehandlung also auf die Kupierung der Anfälle ab. Leider ist es für Migräniker schwierig, die gewünschte Vasokonstriktion zu erlernen (vgl. Gerber et al., 1983). Die Erfolge sind dementsprechend schwankend.

Blanchard und Andrasik (1982) analysierten 23 methodisch einwandfreie empirische Studien, die die Wirksamkeit des apparativen und nicht-apparativen Entspannungstrainings zur Migränebehandlung überprüften. Sie kommen zu dem Ergebnis, daß von diesen Verfahren das autogene Feedbacktraining gegenwärtig die effektivste nicht-pharmakologische Therapieform für Migräne ist. Es erbringt signifikant bessere Erfolge als andere Verfahren mit Ausnahme der progressiven Muskelentspannung. Handerwärmungstraining, Feedback der Temporalarterie und Muskelentspannung unterscheiden sich in ihrer Effektivität nicht bedeutsam voneinander, wohl aber sind sie alle den Nichtbehandlungsbedingungen klar überlegen.

Die **Tübinger Konkordanztherapie** (Gerber, 1982) geht von der auch im Alexithymiekonzept anklingenden Überlegung aus, daß für Patienten, die an Migräne leiden – wie für Patienten mit psychosomatischen Beschwerden überhaupt – Diskordanzen zwischen den drei Ebenen menschlichen Verhaltens, der subjektiv-verbalen, motorisch-verhaltensmäßigen und physiologischen Ebene, bestehen.

Ziel der Konkordanztherapie ist es, den Klienten zur Wahrnehmung solcher Inkonsistenzen zu führen und ihm Strategien an die Hand zu geben, diese aufzulösen. Die Therapie wird in kleinen Gruppen durchgeführt, in denen die Problematik der Erklärung der Migräne durch den Patienten, das Umgehen mit Lob, Körperkontakt, Forderungen-Stellen und -Ablehnen, Kritik-Äußern und -Anhören, Aggressionen, Ertragen von Ambivalenzen, Partnerschaft und Sexualität und Krankheitsbewältigung thematisiert, bearbeitet und trainiert werden. Der Patient übt zunächst in der Gruppe und wird dann angeleitet, das neu Gelernte in für ihn kritischen Alltagssituationen einzusetzen. Verglichen mit Vasokonstriktionstraining der Temporalarterie und einer Jacobson-Entspannungsgruppe ist die Konkordanztherapie ähnlich effektiv, bewirkt jedoch zusätzlich eine Veränderung von Einstellungen und Verhaltenstendenzen.

Da die Hemmung expressiven Verhaltens in unmittelbarer Beziehung zu dysfunktionaler Muskelaktivität steht und darüber hinaus die Bewältigung sozialer Belastungen sowie das soziale Unterstützungssystem beeinträchtigt, ist ein **Expressionstraining** erfolgversprechend, wenn ein dementsprechendes Defizit diagnostiziert wurde. Hudzinski (1984) trainierte Spannungskopfschmerzpatienten, verschiedenes Ausdrucksverhalten zu zeigen, um die muskuläre Diskriminationsfähigkeit zu üben. Obwohl eigentlich nicht als Ausdruckstraining konzipiert, wurde der therapeutische Effekt seiner Intervention möglicherweise zum Teil durch emotionale Ausdrucksübungen erzielt. Als verhaltenstherapeutische Basisübungen für ein Expressionstraining können Patienten angewiesen werden, auf Bildvorlagen dargestellte Gesichter mit hinsichtlich Qualität und Intensität unterschiedlichen Emotionen zu imitieren. In einer zweiten Stufe stellen Patienten sich Situationen mit entsprechenden Emotionen vor und üben dazu passendes Ausdrucksverhalten mit Hilfe des Therapeuten. Ein solches Therapieprogramm ist derzeit in Vorbereitung (Traue und Kraus, 1988; Traue und Bischoff, in Vorbereitung).

Das **situationsbezogene Muskel- und Gefäßempfindungsprogramm** (SEP) (Wittchen, 1983) basiert auf dem psychobiologischen Migränemodell von Cinciripini und Mitarbeitern (1981, vgl. 34.7.1).

Der erste Teil des 10 Sitzungen umfassenden Gruppenprogramms fokussiert vor allem auf die psychophysiologischen Aspekte der Migräne, der zweite Teil stärker auf die kognitiven und Verhaltensaspekte. Die Patienten werden über die Pathophysiologie der Migräne informiert, erhalten ein Wahrnehmungstraining zur besseren Diskriminierung physiologischer Veränderungen und Entspannung nach Jacobson zur willkürlichen Kontrolle dieser Vorgänge. Der zweite Schritt besteht in individuellen Verhaltensanalysen, in denen mit Hilfe der täglichen Aufzeichnungen der Patienten die physiologischen, psychologischen und Umweltreize identifiziert werden, die die Attacke triggern. Die Patienten werden darin unterrichtet, diese Verhaltensanalyse selbständig durchführen zu können. Im dritten Schritt werden kritische Trainingssituationen zusammengestellt, in denen die Patienten ihre neu erworbene Kontrollfähigkeit systematisch üben. Das Training wurde an einer Gruppe chronischer Migränepatienten mit sehr gutem Erfolg erprobt und erwies sich einer Akupunkturbehandlung gegenüber als deutlich überlegen, vor allem hinsichtlich der Stabilität der Behandlungserfolge im follow-up nach eineinhalb Jahren.

Neben diesen multimodalen Therapien wurden in jüngster Zeit auch Therapieprogramme erprobt, die die **kognitive** Seite des Streßerlebens und der Streßbewältigung ins Zentrum rücken (vgl. Gerhards et al., 1983; Knapp, 1983b; Bakal, 1982). Gemeinsam ist diesen Ansätzen die Annahme, daß die Patienten unter anderem Defizite der Streßbewältigung aufweisen, weil sie sich irrationale und damit „sympathikusaktivierende" Gedanken machen. Ziel der Therapie ist die Sensibilisierung für diese Gedanken, die Erprobung alternativer Gedanken und der aus ihnen folgenden Verhaltensweisen. Alle vier Arbeiten berichten über sehr ermutigende Behandlungserfolge.

Die bisher am besten untersuchten und am häufigsten angewandten **verhaltenstherapeutischen Verfahren zur Behandlung des Spannungskopfschmerzes** sind progressive Muskelrelaxation und EMG-Biofeedback.

Progressive Muskelentspannung wird in der Regel in 4–10wöchigen Kursen mit ein oder zwei Sitzungen pro Woche und täglichen Hausaufgaben gelehrt.

Manchmal wird ein verbaler Hinweisreiz an den Zustand der Entspannung gekoppelt, damit der Patient die Entspannungsreaktion später besser abrufen kann (cue-controlled relaxation; Russel und Sipich, 1973); manchmal werden auch Instruktionen zur differentiellen Entspannung (vgl. Bernstein und Borkovec, 1975) gegeben: Der Patient soll alle Bewegungen und Haltungen mit dem geringstmöglichen Kraftaufwand durchführen; vereinzelt wird die Entspannung auch in eine systematische Desensibilisierung eingebunden.

EMG-Biofeedback wird in der Regel als Feedback der Spannung des Stirnmuskels praktiziert, wobei die Patienten, in einem Entspannungsstuhl sitzend, lernen sollen, das Spannungsniveau zu senken.

Es gibt jedoch auch Varianten:
- Biofeedback wird gemeinsam mit progressiver Muskelentspannung gelehrt.
- Es wird durch Biofeedback die Entspannung eines anderen Muskels gelehrt, z.B. des M. trapezius oder des M. temporalis.
- Es wird vom Patienten verlangt, die Spannung auf einem bestimmten Niveau zu halten.
- Die Patienten erhalten Biofeedback auch in sozialen Situationen, z.B. während eines Gesprächs mit dem Therapeuten oder während sie sich für sie schwierige Situationen vorstellen oder im Alltag.
- Die Patienten erhalten Biofeedback in unterschiedlichen Körperhaltungen oder während dynamischer Körperbewegungen (vgl. Andrasik, 1989; Bischoff und Müller, 1989).

Die Metaanalyse der zahllosen empirischen Studien ergibt, daß sowohl progressive Muskelentspannung als auch EMG-Biofeedback gegenüber Kontrollgruppen jeder Art eine signifikante Reduktion der Kopfschmerzaktivität bei Patienten mit Spannungskopfschmerzen einleiten, daß die Symptomatik bei Therapieende um etwa 60% gelindert ist und dieser Therapieeffekt zumindest über ein Jahr bestehenbleibt (Blanchard et al., 1980; Holroyd und Penzien, 1985). Da Biofeedback apparativen Aufwand bedeutet, könnte man der Auffassung sein, daß der progressiven Muskelentspannung der Vorzug zu geben ist. Es ist allerdings zu bedenken, daß diese Auffassung auf einem Uniformitätsmythos (Andrasik, 1989) beruht. Nicht unbedingt dieselben Patienten sprechen auf Biofeedback und Entspannung positiv an. Das kann daran abgelesen werden, daß die Kombinationsbehandlung mit Entspannung und Biofeedback der jeweiligen Einzelbehandlung tendenziell überlegen ist.

EMG-Biofeedback ist auch aus einem anderen Grund kritisch hinterfragt worden. Grundannahme des Verfahrens ist, daß die Reduktion der Kopfschmerzen durch die Reduktion der Muskelspannung erzielt wird. Empirisch sind Korrelationen zwischen Veränderungen der Muskelaktivität und parallelen Veränderungen der Muskelspannung allerdings niedrig oder gar nicht nachweisbar (Kröner-Herwig und Weich, 1988). Darüber hinaus konnte gezeigt werden, daß der Glaube der Patienten, die Biofeedback-Aufgabe gut gemeistert zu haben, die Minderung der Kopfschmerzaktivität besser vorherzusagen erlaubt als die Veränderung der Muskelspannung selbst. Entfaltet Biofeedback seine Wirkung aufgrund einer Verbesserung der Selbsteffizienzerwartung der Patienten, also aufgrund von psychologischem, nicht von physiologischem Lernen (Holroyd et al., 1984)?

Physiologisches und psychologisches Lernen schließen sich nicht aus. Physiologisches Lernen wird dann bedeutsam, wenn der Patient genuin myogene Kopfschmerzen und ein Wahrnehmungsdefizit für Muskelverspannungen hat und wenn das therapeutische Setting relevantes physiologisches Lernen erlaubt. Dazu muß die Aktivität der für die Schmerzen verantwortlichen Muskulatur zurückgemeldet werden – und dies am besten genau dann, wenn sie in unphysiologischer Weise verspannt ist. Wir versuchen derzeit, das physiologische Lernen durch Biofeedback mit einem tragbaren Gerät zu optimieren. Dem Patienten werden in seinem Alltagsleben über einen taktilen Reizgeber am Oberarm statische Muskelverspannungen zurückgemeldet. Statische Verspannungen werden über zwei Biofeedbackschwellen operationalisiert: die Muskelaktivität überschreitet länger, als durch eine Zeitschwelle festgelegt, ununterbrochen eine bestimmte obere Amplitudenschwelle. Der Patient hat die Aufgabe, sich mit Hilfe des taktilen Reizes, der die Muskelaktivität analog abbildet, zu entspannen. Wenn er eine zuvor definierte untere Amplitudenschwelle unterschreitet, hört die Rückmeldung auf (Bischoff und Müller, 1989).

Auch bei der Therapie der Spannungskopfschmerzen haben in der letzten Zeit die **kognitiven Verfahren** Einzug gehalten. Holroyd und seine Kollegen (Holroyd und Andrasik, 1982) gehen wie Knapp (1981) bei der Migräne davon aus, daß Spannungskopfschmerz durch psychischen Streß verursacht wird und psychischer Streß von Kognitionen herrührt.

Die Patienten müssen zunächst lernen, daß es solche Kognitionen von Situationen sind, die Kopfschmerz verursachen und nicht die Situationen „an sich" oder persönliche Dispositionen (kausale Reattribution). Der zweite Schritt besteht in der Selbstüberwachung. Der Therapeut erarbeitet mit dem Patienten eine Liste streßerzeugender Ereignisse:
- Er versucht, die Hinweisreize in diesen Situationen zu identifizieren, die Spannung und Angst beim Patienten auslösen;
- er stellt fest, wie der Patient reagiert, wenn er ängstlich ist;
- er erfragt die Gedanken, die sich der Patient macht, ehe, während und nachdem er seiner Spannung gewahr wird;
- er erforscht wie diese Kognitionen vermutlich zu Spannung und Kopfschmerz beitragen.

Der dritte Teil des Trainings besteht im Erwerb von Bewältigungsstrategien: Der Patient soll lernen, die Gedankenkette, welche ein Ereignis zum Streßereignis werden läßt, so früh wie möglich zu unterbrechen und solche Kognitionen zu produzieren, die mit den ursprünglichen unvereinbar sind. Dieses kognitiv-verhaltenstherapeutische Programm hat sich sowohl in Form von Einzel- als auch von Gruppentherapie als äußerst effektiv erwiesen.

Berichte über **Therapieverfahren für kombinierten Kopfschmerz**[6] werden trotz seiner hohen Prävalenzrate selten gegeben, was u.a. darauf zurückgehen

---

6 Eine Literaturübersicht geben Blanchard und Andrasik (1982).

mag, daß die Diagnose mit größerer Unsicherheit seitens der Kliniker behaftet ist und deshalb selten gestellt wird.

Adams und seine Forschungsgruppe (Feuerstein et al., 1976) beschreiben die erfolgreiche Therapie von drei Patienten mit kombiniertem Kopfschmerz mit Frontalis-EMG-Feedback und Feedback der Temporalarterie, wobei die vaskulären Symptome eher auf Feedback der Temporalarterie, die muskulären Symptome eher auf EMG-Biofeedback ansprachen. Andere Autoren berichten nur mäßige Schmerzlinderung, auch wenn sie vergleichbare kombinierte Verfahren einsetzen.

Wenig ermutigend sind die Erfolge von **Verhaltenstherapie bei Cluster-Kopfschmerz**[7]: Die Erfolge sind nur temporär, die Abbruchquote hoch.

Zur Stabilität der Behandlungserfolge[8]: Erfolge von Entspannungstherapie scheinen am ehesten vorzuhalten, wenn der Patient die Übungen zu Hause regelmäßig weiterpraktiziert. Günstig scheint es zu sein, die Sensibilität der Patienten für kopfschmerzauslösende Situationen durch das Führen von Kopfschmerztagebüchern oder kurze Besuche beim Therapeuten in längeren Zeitabständen auch über den Zeitraum der eigentlichen Behandlung hinaus wachzuhalten.

Insgesamt zeichnet sich in der verhaltenstherapeutischen Kopfschmerztherapie die Tendenz ab, die Patienten entweder mit zuvor festgelegten Behandlungspaketen zu therapieren, in denen alles vereinigt wird, was sich beim „Durchschnittspatienten" als bedeutsam herausgeschält hat, oder aber eine aufwendige individuelle verhaltensanalytische und psychophysiologische Diagnostik zu betreiben, die erlaubt, die Therapie noch stärker an den Problemstellen des einzelnen Patienten zu orientieren – letzteres in der Hoffnung, die therapeutische Effektivität zusätzlich zu steigern. Ob dies gelingt, ist derzeit eine offene Frage.

Ziel bei der Therapie der Kopfschmerzen von Herrn M. war die Sensibilisierung für spannungserzeugende und spannungsreduzierende Ereignisse. Dieses Ziel erreichte der Patient einerseits durch regelmäßige Tagebuchführung – die ja auch schon für die Diagnose Bedeutung hatte –, andererseits durch Feedback der Muskelspannung auf der Stirn in der klassischen Feedback-Anordnung und während zweier verhaltensanalytischer Gespräche mit dem Therapeuten.

Der zweite Schritt bestand in der Analyse der beiden Situationsformen, die nach der Verhaltensanalyse für die meisten Auftretensfälle von Kopfschmerzen verantwortlich waren. Sowohl für die Interaktion mit dem Vorgesetzten als auch mit der Ehefrau wurden die Kognitionen exploriert, alternative Kognitionen erprobt und alternative Verhaltensweisen als Konsequenz aus den geänderten Kognitionen im Rollenspiel mit dem Therapeuten eingeübt.

Diese Übungen trugen wesentliche Züge eines Selbstsicherheitstrainings. Parallel dazu praktizierte der Patient progressive Muskelentspannung nach Jacobson, ein von ihm begeistert aufgenommenes Mittel zur Selbstkontrolle. Nachdem er die Technik beherrschte, wiesen wir ihn an, sie auch in solchen streßerzeugenden Situationen einzusetzen, zu deren Bewältigung kein anderes alternatives Verhalten möglich war. Der letzte Teil der Therapie bestand in drei gesprächstherapeutisch orientierten Sitzungen – dies gemäß seinem Wunsch, die Beziehung zu seiner Frau für sich besser zu verstehen.

In der Therapie von Herrn M. finden sich also zahlreiche Elemente der beschriebenen, empirisch validierten Therapieformen. Neben den klassischen verhaltenstherapeutischen Maßnahmen – EMG-Biofeedback und progressive Muskelentspannung – wurden Methoden eingesetzt, die vor allem an das SEP von Wittchen, aber auch an die kognitiven Verfahren (Holroyd und Andrasik, 1982a; Gerhards et al., 1983) und an Volgers (1983) Kombination von Verhaltens- und Gesprächstherapie erinnern. Spezifikum der Therapie ist ihre strenge Ausrichtung an der individuellen Verhaltensanalyse.

Nach 12 wöchentlichen Kontakten war der Patient beschwerdefrei und äußerte den Wunsch, die Therapie zu beenden, da er glaubte, die Beschwerden unter Kontrolle zu haben. Im Nachgespräch nach fünf Monaten schilderte er, daß seine 16 Monate alte Tochter nach 14tägigem Krankenhausaufenthalt an Enzephalitis gestorben sei. Der Patient fühlte sich in dieser Zeit, in der auch schwerwiegende Entscheidungen von ihm gefordert wurden (Fragen der Sterbehilfe für seine Tochter), vor allem von seiner Frau und von seinen Schwiegereltern völlig alleingelassen. Er hatte in den 14 Tagen bis zum Eintritt des Todes der Tochter nahezu immer sehr starke, unkontrollierbare Kopfschmerzen; in den Wochen nach dem Tod jedoch nur noch zweimal. Ein Bedürfnis nach weiterer Therapie hatte er zu diesem Zeitpunkt aufgrund der relativen Beschwerdefreiheit nicht.

Eineinhalb Jahre später begab sich Herr M., weil er über den Tod der Tochter nicht hinwegkommen konnte, erneut in psychotherapeutische Behandlung. Diagnose: pathologische Trauerreaktion. Herr M. konnte seine Problematik in sechs psychodynamisch orientierten Gesprächen für sich zufriedenstellend bearbeiten. Kopfschmerzen waren in der Zwischenzeit nur sporadisch aufgetreten.

Was die Kopfschmerztherapie offensichtlich erreicht hat: Der Patient kann jetzt körperliche Vorgänge, insbesondere Muskelreaktionen, mit mentalen Ereignissen, mit der Wahrnehmung von sozialen Situationen und mit Gefühlen in Zusammenhang bringen. Er hat seine rein naturalistische Kopfschmerztheorie durch eine psychosoziale Theorie ersetzt, und er hat feststellen können, daß er mit dieser neuen Theorie eher Ansatzpunkte für wirksame Gegenmaßnahmen zur Kontrolle der Kopfschmerzen hat.

7, 8 Eine Literaturübersicht geben Blanchard und Andrasik (1982).

38 Eßstörungen

38.1 Adipositas

Albert J. Stunkard und *Volker Pudel*

38.1.1 Definition

Adipositas ist durch übermäßige Anhäufung von Fett im Körper charakterisiert. Der ebenfalls gebräuchliche Terminus „Fettsucht" stellt die Adipositas unzureichend als polaren Gegensatz zur Magersucht heraus, kennzeichnet jeden Adipösen vorschnell als „Süchtigen" und wirkt überdies im Sprachgebrauch diskriminierend. Die Bezeichnung „Adipositas" wird daher vorgezogen (Pudel, 1982).

Gewöhnlich spricht man von einer Adipositas, wenn das Körpergewicht ein bestimmtes Standardgewicht um 20% übersteigt, wobei zumeist das Broca-Referenzgewicht im Sinne des Normalgewichts als Bezugsgröße verwendet wird (Broca-Referenzgewicht in kg = Körpergröße in cm − 100). Dieser Index gibt allerdings im Einzelfall nur eine grobe Orientierung. In Zukunft wird die Diagnose wahrscheinlich auf neueren und genaueren Methoden, das Körperfett zu messen, basieren. Einstweilen ist die simple Regel: „Menschen, die fett aussehen, sind fett", für die meisten klinischen Zwecke ausreichend.

38.1.2 Epidemiologie

Selten, wenn überhaupt, hat in der Geschichte der Menschheit ein Volk für längere Zeit mehr als gerade genug zu essen gehabt. Infolgedessen ist Adipositas im Verlauf der Jahrhunderte, wie in vielen unterentwickelten Ländern noch heute, auf die privilegierten Schichten beschränkt. In vielen Kulturen ist sie sogar ein Statussymbol. Unter diesen Umständen müßte man annehmen, daß Adipositas in privilegierten Gruppen häufiger ist. Wie wir sehen werden, trifft dies für eine Reihe von Kulturen zu. Aber bei westlichen Überflußgesellschaften ist meist das Gegenteil richtig. Die Beziehung zwischen sozialen Faktoren und Adipositas ist faszinierend und zeigt den außergewöhnlichen Einfluß dieser Faktoren in eindringlichen Farben.

Aber zunächst wollen fragen, was überhaupt über die Häufigkeit von Adipositas bekannt ist. Die Antwort lautet: erstaunlich wenig.

Da die meisten verläßlichen diagnostischen Methoden für die Untersuchung größerer Bevölkerungsgruppen zu aufwendig sind, basieren unsere Informationen überwiegend auf Größen und Gewichtsangaben zweifelhafter Qualität, die über die gesamte Bevölkerung gemittelt und lediglich den üblichen Kriterien von 20% über Referenzgewicht unterworfen wurden. Die verfügbaren Daten lassen vermuten, daß in den USA bei 35% der Männer und bei 40% der Frauen der Häufigkeitsgipfel für die Adipositas bei 40 Jahren liegt. Während der letzten 30 Jahre hat die Häufigkeit bei den Männern zugenommen, während sie bei den Frauen anscheinend unverändert blieb. Untersuchungen an kleineren Gruppen mit zuverlässigeren Daten zeigen, daß das Alter von erheblichem Einfluß auf die Häufigkeit der Adipositas ist, mit einem monotonen Anstieg der Häufigkeit zwischen der Kindheit und dem 40. und einem doppelt so großen Anstieg zwischen dem 20. und dem 50. Lebensjahr. Mit 50 fällt die Häufigkeitskurve steil ab, wahrscheinlich weil bei älteren adipösen Menschen eine sehr hohe Mortalität durch kardiovaskuläre Erkrankungen besteht. Hier muß betont werden, daß Untersuchungen, die nur das Größe-Gewichts-Kriterium anwenden, ziemlich sicher die Häufigkeit der Adipositas bei älteren Menschen unterschätzen, weil der Fettgehalt des Körpers mit dem Alter pro Gewichtseinheit zunimmt. Neuere und genauere Methoden, das Körperfett zu bestimmen, wie die Messung der Hautfaltendicke, werden vermutlich bald zuverlässigere Angaben liefern. Bei Vergleichen der beiden Geschlechter zeigt sich immer wieder eine größere Häufigkeit der Adipositas bei Frauen. Dieser Unterschied ist besonders nach dem 50. Lebensjahr deutlich, weil adipöse Männer dieser Altersgruppe eine höhere Mortalität haben. Ein großer Teil unserer Kenntnis und die ersten Daten über die Beziehung zwischen sozialen Faktoren und Adipositas stammen aus der Midtown-Manhattan-Studie, einer umfassenden epidemiologischen Untersuchung über Geisteskrankheiten (Srole et al., 1962). Diese Studie ist an anderer Stelle so detailliert beschrieben worden, daß hier eine kurze Übersicht über ihre Methodik genügt.

Die erfaßte Bevölkerungsgruppe bestand aus 110 000 Erwachsenen zwischen 20 und 59 Jahren eines Gebietes auf Manhattan, das unter dem Gesichtspunkt extremer Unterschiede des sozioökonomischen Status seiner Bewohner (von besonders hoch bis besonders niedrig) ausgewählt war. Aufgrund statistischer Kriterien wurden 1660 Personen als Repräsentativgruppe ausgewählt. Erfahrene Interviewer führten mit diesen Personen in deren Wohnungen zweistündige Gespräche und erhielten Informationen über ihren sozialen und ethnischen Hintergrund, eine Anzahl von Einzelangaben über psychologische und zwischenmenschliche Beziehungen, sowie über Körpergröße und Gewicht.

Die Midtown-Studie enthüllte auf dramatische Weise einen unerwarteten Einfluß sozialer Faktoren auf die Häufigkeit der Adipositas (Stunkard, 1975). Diese sozialen Faktoren waren von so starkem Einfluß, daß jede einzelne der untersuchten Variablen zu der Häufigkeit von Adipositas in Beziehung stand. Den größten Einfluß hatte die soziale Schicht bzw. der sozioökonomische Status. Diese Variable wurde durch eine einfache Bezugsgröße ermittelt, die auf der Beschäftigung, der Schulbildung, dem Einkommen und der monatlichen Miete basierte und in „niedrig", „mittel" und „hoch" unterteilt war.

Zwischen der Höhe des sozioökonomischen Status und der Häufigkeit von Adipositas wurde eine deutlich gegensätzliche Beziehung festgestellt. Abbildung 38.1–1 zeigt, daß 30% der Frauen mit niedrigem sozioökonomischem Status adipös waren, 16% der Frauen mit mittlerem Status und nur 5% der Gruppe mit dem höchsten Status. Einfach ausgedrückt, Adipositas ist bei Frauen mit niedrigem Status sechsmal häufiger als bei Frauen mit hohem Status! Als man den sozioökonomischen Status in 12 Klassen teilte, was die Fülle der Daten erlaubte, wurde der Unterschied zwischen der untersten und der obersten Klasse sogar noch größer – von etwa 2% in der obersten zu 37% in der untersten Klasse. Eine ähnliche, aber weniger eindrucksvolle Relation fand sich bei den Männern. Bei Männern mit niedrigem sozioökonomischem Status fand man z.B. bei 32% eine Adipositas im Vergleich zu 16% bei Männern der Oberschicht.

Zwei Feststellungen legen nahe, daß diesen Korrelationen eine kausale Beziehung zugrunde liegt. Erstens, wie Abbildung 38.1–1 zeigt, war die soziale Schicht der Eltern fast so eng mit Adipositas verbunden wie die soziale Schicht der Person selbst. Obwohl die Adipositas eines Menschen natürlich einen Einfluß auf seinen sozialen Status gehabt haben kann, ist es kaum denkbar, daß sie den sozialen Status der Eltern beeinflußt hat. Zweitens, Adipositas ist bereits bei Kindern der Unterschicht viel häufiger als bei Kindern der Oberschicht; signifikante Unterschiede

zeigen sich schon bei Sechsjährigen (Stunkard et al., 1972).

Die Untersuchung der sozialen Variablen erstreckte sich außer auf den sozioökonomischen Status auf die soziale Mobilität, die Zahl der Generationen einer Familie in den USA und die ethnische und religiöse Zugehörigkeit. Adipositas war bei sozialem Abstieg häufiger (22%) als bei denen, die in der sozialen Schicht ihrer Eltern blieben (18%), und viel häufiger als bei sozialem Aufstieg (12%). Noch auffallender war die enge Beziehung zwischen der Zahl der Generationen der Familie in den USA und Adipositas. Die Untersuchten wurden in vier Gruppen geteilt, je nach der Anzahl der Generationen, die ihre Familien in diesem Land gelebt hatten: Generation I bestand aus nicht in den USA geborenen Einwanderern; Generation II aus allen im Land Geborenen mit mindestens einem nicht in den USA geborenen Elternteil; Generation III aus allen im Land Geborenen, die von im Land geborenen Eltern abstammten, aber mindestens einen nicht in den USA geborenen Großelternteil hatten; und Generation IV aus allen, die in den USA geborene Großeltern hatten und im übrigen der Generation III entsprachen. Die Ergebnisse zeigten, daß Frauen um so seltener adipös waren, je länger ihre Familien in den USA gelebt hatten. In der Generation I hatten 24% Übergewicht im Gegensatz zu nur 5% in der Generation IV. Da die Zahl der Generationen einer Familie in den USA eng mit ihrem sozioökonomischen Status verbunden ist, wurden die Beziehungen zwischen Generation und Adipositas nochmals analysiert, wobei der sozioökonomische Status konstant gehalten wurde. Diese Analyse zeigt, daß die Beziehung zwischen Generation und Adipositas unabhängig vom sozioökonomischen Status war.

Da die Midtown-Studie neun verschiedene ethnische Gruppen erfaßt, läßt sich auch der Einfluß der ethnischen Zugehörigkeit auf die Adipositas beurteilen. Obwohl die Gruppengröße nicht zur Errechnung statistischer Signifikanz ausreichte, enthüllte die Analyse doch überraschende Unterschiede der Häufigkeit von Adipositas in den neun Gruppen, vor allem bei Frauen. Am deutlichsten war der Einfluß ethnischer Faktoren bei Personen mit niedrigem sozioökonomischem Status, mit einem Mittelwert von 30% bei großer Streuung. Ethnische Zugehörigkeit ließ jedoch keine ganz so zuverlässige Vorhersage zu wie der sozioökonomische Status mit seinem 6 : 1 Unterschied zwischen Unter- und Oberschicht. Trotzdem, wenn man nur die Unterschicht in Betracht zieht, zeigen sich einige bemerkenswerte Unterschiede. So war z.B. Adipositas bei Ungarn und Tschechen der Unterschicht dreimal so häufig wie bei Amerikanern der Generation IV in der gleichen Schicht. Diese Daten enthüllten auch eine interessante Wechselbeziehung zwischen ethnischer Zugehörigkeit und sozioökonomischem Status.

Unter Ungarn und Tschechen fand sich ein großer Unterschied der Häufigkeit in Unter- und Oberschicht: 40% der Unterschicht waren adipös, keiner der Oberschicht. Im Unterschied dazu betrug der entsprechende Unterschied bei Amerikanern der Generation IV nicht mehr als 9%.

Religiöse Zugehörigkeit ließ sich als ein weiterer sozialer Faktor, der mit Adipositas in Beziehung stand, identifizieren. Wiederum folgten die Ergebnisse dem zu erwartenden Muster: Adipositas war am häufigsten bei Juden, gefolgt von Katholiken und Pro-

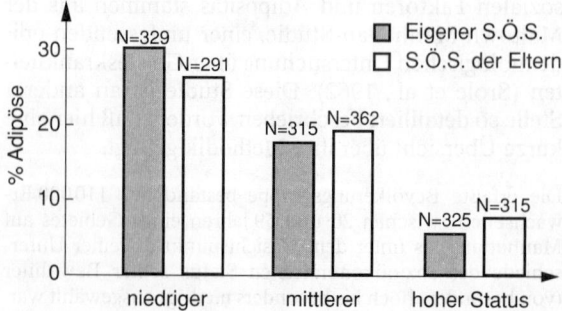

Abb. 38.1–1. Adipositas und sozioökonomischer Status (SÖS) bei Frauen.

Tabelle 38.1–1. Verteilung des relativen Übergewichts in der Bundesrepublik Deutschland in Abhängigkeit der Schulbildung als Indikator der sozialen Schicht. Die Daten basieren auf einer repräsentativen Erhebung einer Stichprobe von 1920 Personen, die gewogen und gemessen wurden. Zeitpunkt der Erhebung: 1979.

Schulabschluß	Geschlecht	unter 15% Broca-Ref.Gew.	−15% bis −5% Broca-Ref.Gew.	−5% bis +5% Broca-Ref.Gew.	+5% bis +15% Broca-Ref.Gew.	über 15% Broca-Ref.Gew.
Volksschule	männlich	7,4%	17,2%	29,0%	26,3%	20,1%
	weiblich	12,9%	21,6%	23,7%	17,9%	23,9%
	Gesamt	10,7%	19,6%	25,8%	21,4%	22,5%
weiterführende Schule	männlich	21,0%	38,0%	20,0%	8,3%	12,7%
	weiblich	25,0%	36,2%	20,5%	12,6%	5,7%
	Gesamt	23,1%	37,2%	20,4%	10,8%	8,5%
Abitur/Hochschule	männlich	18,8%	34,7%	23,2%	17,0%	6,3%
	weiblich	50,1%	33,3%	10,6%	4,5%	1,5%
	Gesamt	30,3%	34,3%	18,5%	12,4%	4,5%

testanten. Unter den Protestanten war Adipositas am häufigsten bei den Baptisten, dann folgten mit abnehmender Häufigkeit Methodisten, Lutheraner und Angehörige der Episcopanischen Kirche.

Mehrere Untersuchungen über die Verbreitung von Adipositas in anderen Ländern zeigten, daß der Einfluß der sozialen Faktoren kein ausschließlich amerikanisches Phänomen ist. Zwei Arbeiten aus England beschreiben die gleiche Beziehung zwischen sozioökonomischem Status und Adipositas bei Frauen wie die Manhattan-Studie. Silverstone et al. (1969) fanden, daß Adipositas fast doppelt so häufig bei Frauen mit niedrigem sozioökonomischem Status war wie bei Frauen mit hohem Status. Baird et al. (1974) bestätigen diese Beziehung in einer Studie an 1334 Personen in London.

Zu nahezu gleichen Ergebnissen kommt eine repräsentative Studie für die Bundesrepublik Deutschland aus dem Jahre 1979 (Ernährungsbericht 1980). Tabelle 38.1–1 belegt, daß insbesondere Frauen mit Volksschulabschluß in 40% aller Fälle zum Übergewicht neigen, während nur 6% der Frauen mit Abitur unter Übergewicht leiden. Ebenfalls, aber weniger deutlich ausgeprägt ist diese Beziehung bei der männlichen Bevölkerung.

Die auffallend feste Beziehung zwischen sozialen Faktoren und Adipositas hat drei Gruppen angeregt, die entscheidende Frage nach dem Alter zu stellen, in dem diese Beziehung sich geltend macht. Eine Untersuchung in Kalifornien fand die negative Korrelation zwischen sozioökonomischem Status und Adipositas schon in der Adoleszenz, und eine in London stellte das gleiche Verhältnis schon bei Jungen und Mädchen im Alter zwischen 7 und 11 Jahren fest. Es ist von Interesse, daß diese letztere Studie keine Beziehung zwischen dem sozioökonomischen Status und durchschnittlicher Hautfaltendicke fand. Das läßt vermuten, daß der Einfluß sich nicht unterschiedlos auf alle Kinder, sondern besonders deutlich auf die offensichtlich fetten Kinder beschränkte.

Die umfassendste Untersuchung sozialer Faktoren und Adipositas wurde an 3344 weißen Schulkindern im Osten der USA durchgeführt (Stunkard et al., 1972). Sie lieferte einen überzeugenden Beweis und außerdem alarmierende Anzeichen dafür, wie früh-

zeitig dieser Einfluß sich auswirkt. Abbildung 38.1–2 zeigt die Beziehung zwischen sozioökonomischem Status und Adipositas bei Mädchen, wobei die Unterschiede hoch signifikant sind. Bei den Sechsjährigen waren in der niedrigen sozioökonomischen Gruppe 8% adipös, während sich in der Oberschicht weder unter den sechs- noch den siebenjährigen Mädchen Adipöse fanden. Dieser Unterschied blieb bis zum 18. Lebensjahr bestehen, wobei mit Zunahme des Alters die Häufigkeit der Adipositas in beiden Gruppen stieg.

Abbildung 38.1–2 zeigt außerdem, daß die Zunahme in der Ober- und Unterschicht unterschiedlich stark ist, mit dem größeren jährlichen Zuwachs des Prozentsatzes von Adipösen bei den Mädchen der Unterschicht. Adipositas ist also nicht nur häufiger bei armen Menschen, sondern ihre größere Häufigkeit tritt auch früher auf und steigt rascher an als bei der Oberschicht.

In dieser Studie wurde Adipositas als die 10% jedes Geschlechts in der Gesamtpopulation mit den dicksten Haut-

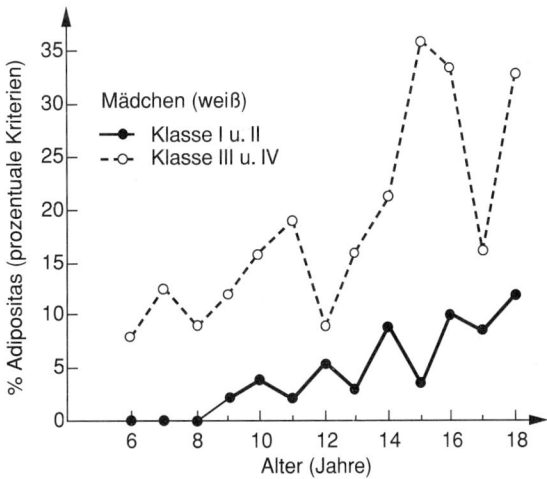

Abb. 38.1–2. Beziehung zwischen sozioökonomischem Status und Adipositas bei Mädchen.

falten definiert, und die geringste Hautfaltendicke wurde benutzt, um Adipositas innerhalb jeder Altersgruppe zu definieren. Diese empirisch abgeleiteten Werte für Adipositas betrugen 23 mm bei Mädchen und 18 mm bei Jungen. Jungen der Unterschicht zeigten eine größere Häufigkeit von Adipositas als diejenigen der Oberschicht, mit Unterschieden, die denen bei Mädchen im Alter von 10 Jahren vergleichbar waren. Im Unterschied zu dem kontinuierlichen Anstieg der Adipositas bei den Mädchen waren die Unterschiede bei den Jungen bis zum Alter von 18 Jahren nicht kontinuierlich. Jungen der Unterschicht zeigten größere Häufigkeit von Adipositas als diejenigen der Oberschicht, aber wie bei den Männern waren diese Unterschiede nicht so groß und nicht so übereinstimmend wie bei den Mädchen.

In deutlichem Kontrast zu dem negativen Verhältnis westlicher städtischer Gesellschaften zeigen die ersten Daten über die Beziehung zwischen sozialen Faktoren und Adipositas in einer weniger wohlhabenden Gesellschaft, daß unter diesen Umständen der Wohlstand in direkter Beziehung zur Häufigkeit der Adipositas steht. Eine der mächtigsten sozialen Kräfte in der Navaho-Gesellschaft ist die Akkulturation in die umgebende „englische" (weiße) Kultur, ein Faktor, der deutlich mit dem relativen Wohlstand korreliert. Bei einer Untersuchung über das Ausmaß der Akkulturation von 690 Navaho-Kindern im Alter von 7 bis 11 war Adipositas bei den akkulturierten Jungen wesentlich häufiger als bei denen mit alter Tradition; und noch etwas häufiger bei den akkulturierten Mädchen. Außerdem war im Gegensatz zu den Feststellungen in westlichen städtischen Gesellschaften die schlanke Körperform unter den traditionsgebundenen Kindern häufiger. Vereinzelte Informationen aus früheren Untersuchungen über den Zusammenhang zwischen sozialen Faktoren und dem durchschnittlichen Körpergewicht oder der durchschnittlichen Hautfaltendicke (nicht Adipositas!) in Entwicklungsländern haben ebenfalls eine Beziehung festgestellt, die der in westlichen städtischen Gesellschaften genau entgegengesetzt ist: Wachsender Lebensstandard ist verbunden mit steigendem Körpergewicht oder Zunahme der Hautfaltendicke.

Ein Vergleich des Lebensstandards der Navaho-Kinder mit dem von Stadtkindern im Osten der USA

zeigt, daß zumindest die wohlhabenden Stadtkinder sich eines beträchtlich höheren Lebensstandards erfreuen als die meisten akkulturierten und wohlhabenden Navahos. Diese Feststellungen erlauben uns eine allgemeine Hypothese aufzustellen, die Wohlstand und die mit ihm verbundenen sozialen Faktoren zu der Häufigkeit von Adipositas in Beziehung setzt. Abbildung 38.1–3 zeigt die größte Häufigkeit von Adipositas bei den ärmeren Mitgliedern der städtischen Gesellschaften in den westlichen Ländern. Diese Häufigkeit sinkt sowohl bei abnehmendem als auch bei zunehmendem Wohlstand, aber die Gründe dafür sind ungeheuer verschieden. Mit abnehmendem Wohlstand verhindert der Mangel an Nahrung die Entwicklung von Adipositas; nur die Privilegierten können sie sich noch leisten.

Mit wachsendem Wohlstand übernehmen Liebhabereien und Moden die Kontrolle. Im Augenblick haben wir noch nicht genug Informationen, um dieses qualitative Modell in ein quantitatives zu verwandeln, aber der Tag ist nicht mehr fern, an dem dies möglich sein wird. Es ist von Interesse, daß, obwohl weniger detailliert, Informationen über die Beziehung zwischen Wohlstand und schlanker Körperform ein spiegelbildliches Muster zur Adipositas zeigen.

Die Konsequenzen dieser Befunde für unser Verständnis der Adipositas und vor allem für einen Weg, sie unter Kontrolle zu bringen, müssen jedoch noch gezogen werden. Denn sie bedeuten, daß Adipositas, was auch immer ihre genetischen Determinanten und ihre biochemischen Besonderheiten sein mögen, in ungewöhnlichem Maß durch die soziale Umgebung bestimmt wird. Damit ist das nächste Forschungsziel klar: Wie wirkt die soziale Umgebung? Man darf vermuten, daß ein erfolgreicher Angriff auf die Adipositas nicht auf das Verständnis weiterer biochemischer Determinanten zu warten braucht. Das Verständnis der sozialen Determinanten könnte ausreichen.

38.1.3 Der traurige Saldo der traditionellen Adipositasbehandlung

Gewichtsabnahme bringt so vielen Adipösen so große Wohltaten und ist scheinbar so einfach, daß es eigentlich viele Menschen geben müßte, die früher adipös waren. Aber es gibt sie nicht. Hilfe in der Behandlung von Adipositas ist dringend nötig, von welcher Seite sie auch kommen mag. Wenn soziale Faktoren für die weite Verbreitung und ständige Zunahme der Adipositas in der westlichen Welt mitverantwortlich sind, dann müßte es möglich sein, diese Faktoren auch für eine Behandlung der Adipositas zu nutzen. Denn diese Behandlung war bisher – ohne Zweifel – wenig erfolgreich: In den letzten Jahren hat die systematische Untersuchung der Ergebnisse traditioneller Behandlungsformen von Adipositas bestätigt, was jeder praktische Arzt und die meisten Adipösen längst wissen: Die Resultate sind entmutigend. Der Stand der Dinge in jüngster Zeit läßt sich in fünf Sätzen zusammenfassen.

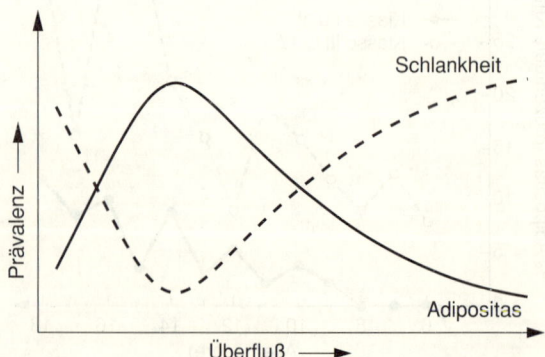

Abb. 38.1–3. Zusammenhang zwischen Lebensstandard und Adipositas.

– Die meisten Adipösen kommen nicht in die Behandlung.
– Von denen, die sich behandeln lassen, brechen die meisten die Behandlung ab.
Eine noch immer oft zitierte Übersicht über die medizinische Literatur aus dem Jahre 1959 zeigt bei der ambulanten Behandlung von Adipositas eine Schwundrate von 20–80% (Stunkard und McLaren-Hume, 1959).
– Von denen, die in Behandlung bleiben, verlieren die meisten kaum Gewicht. Die oben erwähnte Literaturübersicht zeigt, daß nicht mehr als 25% der Personen, die eine ambulante Adipositasbehandlung begannen, 20 Pfund[1] und nur 5% 40 Pfund verloren. Eine anspruchsvollere mathematische Behandlung der Daten einer noch größeren Serie aus dem folgenden Jahr zeigt ein detaillierteres Bild der Fehlschläge, und eine Studie über die Resultate medizinischer Routinebehandlungen enthüllte im Kontrast zu den Ergebnissen von Ärzten, die über ihre Forschungsarbeiten publizierten, ein noch traurigeres Bild: Nur 12% brachten es fertig, etwa 20 Pfund abzunehmen.
– Die meisten, die Gewicht verlieren, nehmen es wieder zu. Von den 12% Ausnahmen, die es nach der eben erwähnten Studie fertigbrachten, 20 Pfund zu verlieren, hatten ein Jahr später nur 2% ihr reduziertes Gewicht gehalten.
– Viele müssen einen hohen Preis für den Versuch bezahlen. Eine kürzliche Analyse der medizinischen Literatur über ungünstige Reaktionen auf Diätmaßnahmen zeigte, daß emotionale Symptome bei ambulant behandelten Adipösen in großer Häufigkeit auftreten, und daß sich solche Symptome auch bei längerer stationärer Behandlung mit Diätmaßnahmen oder Fastenkuren nicht vermeiden lassen (Stunkard und Rush, 1974).
Diese Ergebnisse waren nicht etwa wegen des Unterlassens einer einfachen Therapie von bekannter Effektivität so schlecht, sondern weil keine einfache oder effektive Therapie existiert. Adipositas ist ein chronisches Zustandsbild, das resistent gegen Behandlungsversuche ist und zu Rückfällen neigt.

38.1.4 Körperliche Aktivität als Determinante für Adipositas

In den meisten unterentwickelten Ländern ist Adipositas nicht nur wegen unzureichender Ernährung eine Seltenheit. Vor allem in ländlichen Gebieten ist ein hoher Grad an körperlicher Aktivität zumindest ebenso wichtig, um Adipositas zu verhindern. In der westlichen Gesellschaft gehört eine so hohe Aktivität bekanntermaßen zu den Ausnahmen; tatsächlich scheint der Mangel an körperlicher Bewegung am meisten dazu beizutragen, daß Adipositas in der Wohlstandsgesellschaft zu einem öffentlichen Gesundheitsproblem geworden ist. Wenn der Trend, illustriert durch automatische Büchsenöffner und elektrische Zahnbürsten, anhält, werden wir den Auf-

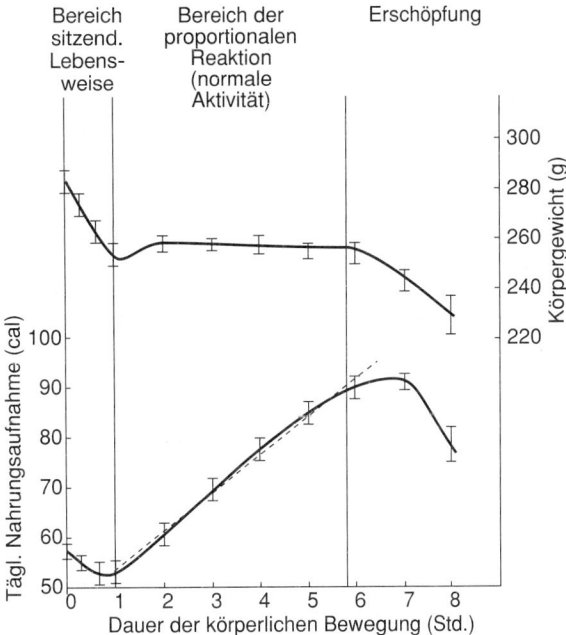

Abb. 38.1–4. Kalorienaufnahme und körperliche Bewegung.

wand an Bewegungsenergie bald nahezu auf das Ruhe-Basisniveau gesenkt haben. Adipöse Frauen haben es darin schon sehr weit gebracht. Das abendländische Volk ist in den letzten 70 Jahren bei einer Ernährung fett geworden, die um 1000 Kalorien reduziert wurde. Die Ursache dafür ist verminderte körperliche Aktivität.

Bis vor kurzem war man der Meinung, daß Mangel an körperlicher Bewegung nur über eine Verminderung des Energieverbrauchs zu Adipositas führe. Mayer und Thomas (1967) haben gezeigt, daß Inaktivität, zumindest bei Tieren, auch zu einer vermehrten Nahrungsaufnahme beiträgt. Obwohl über einen weiten Bereich mit wachsendem Energieaufwand die Nahrungsaufnahme ansteigt, vermindert sie sich, wie Abbildung 38.1–4 zeigt, nicht proportional, wenn die körperliche Aktivität unter ein gewisses Minimum fällt. Wenn dieses Niveau erreicht ist, kann eine weitere Einschränkung körperlicher Aktivität die Nahrungsaufnahme im Gegenteil sogar erhöhen. Umgekehrt kann bei einer Zunahme körperlicher Bewegung die Nahrungsaufnahme abnehmen.

Diese Beobachtungen legen die Vermutung nahe, daß in technisch fortgeschrittenen Gesellschaften die körperliche Bewegung auf das Niveau abgefallen sein könnte, das für den Menschen dem Punkt entspricht, den Mayer bei Ratten als Inaktivitätsschwelle bezeichnet hat, jenseits derer die Nahrungsaufnahme zunimmt. Wie bei den inaktiven Ratten könnte verminderte körperliche Bewegung also auch bei uns eine Zunahme der Nahrungsaufnahme begünstigen.

Wenn diese Annahme auch nur annähernd richtig ist, könnte ein bedeutender Fortschritt in der Be-

---

1 Die Gewichtsangaben in diesem Kapitel beziehen sich auf englische Maßeinheiten: 1 Pfund = 450 g.

handlung von Adipösen allein durch eine Erhöhung ihrer körperlichen Aktivität auf das Niveau erreicht werden, auf dem wieder normale Regulationsmechanismen für die Beziehungen zwischen Nahrungsaufnahme und körperlicher Bewegung zu wirken beginnen. Der Versuch, Steigerung der körperlichen Aktivität in eine Therapie einzubauen, muß natürlich Verhaltensfaktoren berücksichtigen, die Adipösen eine Änderung ihres Bewegungsverhaltens möglicherweise ebenso schwer machen wie eine Änderung ihrer Diät. Sicher ist auf diesem Gebiet noch eine Menge Forschungsarbeit notwendig; aber die vorläufigen Ergebnisse lassen doch schon verschiedene Forschungsansätze als vielversprechend erscheinen. Von den drei Möglichkeiten, körperliche Aktivität zu beeinflussen – durch emotionale, biologische und soziale Faktoren – scheinen nach unseren Beobachtungen die sozialen Faktoren bei weitem die effektivsten zu sein.

Vermehrte körperliche Bewegung wird zwar häufig als Teil von Maßnahmen zur Gewichtsreduktion empfohlen, aber die hier diskutierten Beobachtungen lassen vermuten, daß ihr Nutzen bisher sogar von ihren überzeugtesten Verfechtern unterschätzt worden ist. Berücksichtigung der Bedeutung möglicher regulatorischer Umschaltungen bei körperlich sehr indolenten Personen und Verständnis der Faktoren, die den einzelnen Patienten veranlassen können, mehr oder weniger aktiv zu sein, müßten die Effektivität einer Therapie wesentlich erhöhen können.

38.1.5 Emotionale Determinanten

Adipositas war eine der ersten Störungen, für die man „psychosomatische" Faktoren verantwortlich machte, und diese Meinung hat sich bei den meisten Ärzten und Laien bis heute gehalten. Die Gründe dafür sind nicht schwer zu finden. Viele Adipöse berichten, daß sie in Augenblicken, in denen sie seelisch beunruhigt sind, oder bald danach, zuviel essen. Unglücklicherweise berichten viele Nicht-Adipöse von ähnlichen Erfahrungen, und es ist schwierig, in kurzen Zeiträumen die Spezifität eines solchen Zusammentreffens von Ereignissen für Adipositas zu beurteilen. Langzeitberichte über Zusammenhänge zwischen emotionalen Faktoren und Adipositas scheinen spezifischer. Adipöse verlieren häufig viel an Gewicht, wenn sie sich verlieben, und nehmen an Gewicht zu, wenn sie einen geliebten Menschen verlieren. Solche Veränderungen ereignen sich so völlig unbeeinflußt durch den Willen, ja sogar so außerhalb jeder Möglichkeit einer Kontrolle, daß man annehmen könnte, sie seien das Ergebnis einer fundamentalen Änderung der Regulation des Körpergewichts, vielleicht durch Verstellung des Sollwerts.

Die meisten Eßgewohnheiten Adipöser ähneln den Mustern, die man bei verschiedenen Formen experimenteller Adipositas gefunden hat. Die Beeinträchtigung des Sättigungsgefühls ist eine besonders wichtige Beobachtung. Solche Menschen klagen charakte-

ristischerweise darüber, daß sie nicht aufhören können, zu essen. Dagegen kommt es selten vor, daß Adipöse über einen sehr starken Trieb oder gar eine Gier zu essen berichten. Adipöse scheinen ungewöhnlich verführbar zu sein; schon Hinweise auf Nahrung in ihrer Umgebung, aber auch deren Schmackhaftigkeit macht sie unfähig, zu essen aufzuhören, solange Nahrung erreichbar ist.

Bruch (1973) hat viele dieser Themen unter einem mehr klinischen Gesichtspunkt diskutiert und eindrucksvolle Beschreibungen der Fehlwahrnehmung wichtiger viszeraler Prozesse bei Adipösen gegeben. Sie betont, daß einige, vor allem die emotional Gestörten, Schwierigkeiten haben, Hunger und Sattsein zu erkennen und oft unfähig sind, zwischen Hunger und anderen Zuständen von Unbehagen zu unterscheiden. Sie hat diese „Begriffsverwirrung" mit ernsthaften Störungen der Identität und mit Gefühlen persönlichen Versagens in Beziehung gebracht und überzeugend beschrieben, in welchem Maße solche Menschen auf äußere Signale angewiesen sind, die ihnen sagen, wann sie essen und wann sie damit aufhören sollen. Die Autorin hat diese Theorie experimentell untermauert: Sie zeigt, daß bei neurotischen Adipösen im Unterschied zu nichtneurotischen die Wahrnehmung der Magenkontraktionen durch Reaktionsvorurteile beeinträchtigt ist, die auch ihre Interpretation von Hunger verändern.

In den Anfangsjahren der Psychosomatischen Medizin wurde Adipositas, wie die meisten anderen „psychosomatischen Störungen", als ein einheitliches Zustandsbild aufgefaßt und alle Fälle von Hyperphagie auf emotionale Determinanten zurückgeführt. Es schien nichts auszumachen, daß verschiedene Autoren verschiedene emotionale Determinanten beschuldigten; da Adipositas eine psychosomatische Störung sein sollte, mußte Zuvielessen neurotisch bedingt sein. In gewisser Hinsicht folgt auch die bisherige Darstellung dieser Tradition, und Schachter, Bruch und die Autoren haben oft über „Fettsüchtige" geschrieben, als ob es sich um ein einheitliches Krankheitsbild handeln würde, auch wenn sie sich bewußt waren, daß dies wahrscheinlich nicht stimmt. Das Problem bestand darin, daß es keine einfache und wissenschaftlich begründete Methode gab, verschiedene Arten von Adipositas zu unterscheiden.

Im Laufe der Zeit machte diese simplifizierende Darstellung des Adipositas-Problems einer weniger eindeutigen, bescheideneren, aber wahrscheinlich realistischeren Darstellungsweise Platz. Obwohl es noch keine fundierte Nosologie gibt, glaubt man heute, daß Adipositas eine heterogene Störung sei, und einige Autoren ziehen es vor, von „den Fettsüchtigen" ähnlich wie von „den Schizophrenen" zu sprechen. Wenn diese Auffassung auch nur auf einem Glauben beruht, so stimmt sie doch besser mit empirischen Beobachtungen überein. Solche Beobachtungen gibt es noch sehr wenige. So wissen wir z. B. noch nicht einmal, ob emotionale Störungen bei Adipösen häufiger sind als bei Nicht-Adipösen.

Eine Studie aus England (Crisp, 1975) berichtet, daß Adipöse nicht nur nicht häufiger emotionale Störungen haben als Nicht-Adipöse, sondern sogar etwas seltener. Diese Studie ist eine sorgfältig kontrollierte Untersuchung der Zufallsauswahl einer Bevölkerungsgruppe ohne evidente Verzerrung. Sie steht in Widerspruch zu einer Untersuchung der Daten der oben erwähnten Midtown-Studie (Stunkard, 1975). Diese Studie fand, daß Adipöse in sieben von acht psychopathologischen Indizes höher und in den Kategorien Unreife, Mißtrauen und Rigidität signifikant höher rangierten als Nicht-Adipöse. Diese Unterschiede waren zwar statistisch signifikant, aber für die Gesamtpopulation relativ klein. In bestimmten Untergruppen schien der Unterschied jedoch erheblich zu sein. In diesen fielen junge Frauen mit hohem und mittlerem sozioökonomischem Status auf. Die Gründe für die Auffälligkeit dieser Gruppen sind interessant.

Da sowohl Adipositas wie emotionale Störungen bei Personen mit niedrigem sozioökonomischem Status häufig sind, ist in dieser Schicht das Zusammentreffen beider Faktoren wahrscheinlich zufällig. Höher oben auf der sozialen Leiter ist Adipositas viel weniger häufig und die Sanktionen gegen sie sind viel strenger. Da hier auch emotionale Störungen viel seltener sind, ist die Wahrscheinlichkeit, daß das gleichzeitige Auftreten von Adipositas und emotionalen Störungen in dieser Gruppe einen Zusammenhang hat, viel größer. Bei jungen Frauen der Oberschicht ist Adipositas sehr häufig mit Neurose verknüpft. Was steckt hinter dieser Verknüpfung?

Von den verschiedenen emotionalen Störungen, unter denen Adipöse leiden, haben nur zwei eine spezifische Beziehung zur Adipositas. Die erste ist Hyperphagie (overeating); die zweite eine Störung des Körperschemas (body image) (Stunkard, 1976).

38.1.5.1 Die Hyperphagie-Syndrome

Die überzeugendsten Hinweise dafür, wie emotionale Faktoren Adipositas beeinflussen, erhielten wir von zwei kleinen Untergruppen Adipöser, die beide durch abnorme und stereotype Verhaltensmuster ihrer Nahrungsaufnahme charakterisiert sind. Wir fanden bei etwa 10% der Adipösen, gewöhnlich Frauen, das **Syndrom nächtlichen Essens** (night-eating syndrome), charakterisiert durch Anorexie am Morgen, Hyperphagie am Abend und Schlaflosigkeit. Dieses Syndrom scheint durch Streßsituationen ausgelöst zu sein und tendiert, wenn es einmal da ist, bis zur Lösung des Stresses zu täglichen Wiederholungen. Versuche, das Gewicht zu reduzieren, solange das Syndrom besteht, haben ungewöhnlich dürftige Ergebnisse und können sogar zu ernsteren psychischen Störungen führen.

Das „**Syndrom der Freßorgien**" (binge eating), das wir bei weniger als 5% der Adipösen fanden, ist eine der seltenen Ausnahmen von dem Muster gestörten Sättigungsgefühls. Es wird charakterisiert durch plötzliches, zwanghaftes Verschlingen sehr großer Nahrungsmengen in sehr kurzer Zeit, gewöhnlich gefolgt von großer Erregung und Selbstverdammung.

Auch dieses Syndrom scheint eine Reaktion auf Streß zu sein. Aber im Gegensatz zu dem Syndrom nächtlichen Essens sind diese Anfälle von Hyperphagie nicht periodisch und viel öfter mit auslösenden Ereignissen verbunden. Die Veranstalter einsamer Freßorgien (binge eaters) können manchmal durch Strenge und unrealistische Diät sehr viel Gewicht verlieren, aber der Erfolg solcher Anstrengungen wird fast immer durch einen Rückfall zunichte gemacht.

38.1.5.2 Störung des Körperschemas (body image)

Die zweite für Adipöse typische Form emotionaler Störungen ist eine Störung des Körperschemas. Der Adipöse, der daran leidet, erlebt charakteristischerweise seinen Körper als grotesk, ekelerregend und sich selbst von anderen mit Feindseligkeit und Verachtung betrachtet. Dieses Erleben ist eng mit extremer Unsicherheit und gestörtem sozialem Verhalten verbunden. Obwohl man annehmen könnte, daß alle Adipösen solche entwürdigenden Gefühle über ihren Körper haben, trifft dies nicht zu. Adipöse, die emotional gesund sind, haben keine Störungen des Körperschemas, wie sie hier beschrieben werden. Sie finden sich nur bei einer Minorität neurotischer Adipöser und zwar nur bei solchen, die schon seit der Kindheit adipös sind; sogar von diesen leiden weniger als die Hälfte darunter. In dieser Gruppe findet sich eine Majorität Adipöser mit spezifischen Eßstörungen.

38.1.5.3 Störung des Sättigungsgefühls

Klinisch-experimentelle Studien des spontanen Eßverhaltens adipöser Probanden im Eßlabor, bei denen gewöhnlich ein Food-Dispenser verwendet wurde, haben wiederholt Hinweise auf eine Störung in der Sättigungsregulation ergeben (Pudel, 1982).

Food-Dispenser sind apparative Anordnungen, die es gestatten, verschiedene Aspekte der Nahrungsaufnahme (Geschwindigkeit, Volumenaufnahme pro Zeit, Einfluß bestimmter Bedingungen) zu registrieren und zu messen. Häufig wurde – ohne daß der Proband die aufgenommene Nahrungsmenge visuell kontrollieren kann – Flüssigkost über ein Trinkröhrchen verabreicht.

Als charakteristisch für Adipöse hat sich herausgestellt, daß sie im Verlauf einer Mahlzeit ihre Nahrungsaufnahme nicht kontinuierlich reduzieren, wie es bei normalgewichtigen Erwachsenen und Kindern beobachtbar ist. Sie nehmen konstant gleichgroße Mengen bis zur Beendigung auf.

Bei der Nahrungsaufnahme aus einem Teller, der unbemerkbar immer wieder aufgefüllt wird, „überessen" sie. Bei der Flüssigkeitsaufnahme aus einem sichtbaren Vorratszylinder, in dem der absinkende Flüssigkeitsspiegel durch technische Vorrichtungen manipuliert werden kann, nehmen adipöse Probanden Mengen zu sich, die „pegelorientiert" und nicht – wie bei normalgewichtigen Personen – durch die Magenfüllung determiniert sind.

Diese und weitere Befunde lassen den Schluß zu, daß die physiologische Regulation des Sättigungsgefühls nicht adäquat arbeitet, daß es bei adipösen Probanden zu einer Fehlwahrnehmung des „Stop-Signals" für die Nahrungsaufnahme kommen kann, woraus unter den beschriebenen Laborbedingungen eine erhöhte Nahrungsaufnahme resultiert.

38.1.6 Therapie

38.1.6.1 Allgemeine Probleme

Das Prinzip jeder Gewichtsreduktion ist äußerst einfach – man muß nur die Zufuhr von Kalorien unter ihren Verbrauch senken. All die vielen Diätbehandlungen haben diese scheinbar einfache Aufgabe zum Ziel. Vielleicht war die Einfachheit dieser Formel für einen unglücklichen Behandlungsaspekt mitverantwortlich: Da der Arzt oft unfähig war zu verstehen, warum seine Patienten diese Vorschrift nicht einhalten konnten, verhielt er sich ihnen gegenüber strafend. Es ist uns erst seit kurzem klar, wie oft Adipöse Opfer ebenso intensiver Diskriminierung sind wie andere Minderheiten. Der Arzt sollte sich daher, bevor er die Behandlung eines Adipösen beginnt, klarmachen, ob er zu der vorhandenen eine neue Belastung hinzufügen darf.

Der einleuchtendste Weg, die Kalorienzufuhr zu reduzieren, ist eine energiereduzierte Diät; die besten Langzeiterfolge wurden mit einer ausgewogenen Diät aus überall erhältlichen Nahrungsmitteln erzielt. Vielen Menschen scheint es am leichtesten, eine Diät einzuhalten, die aus ihren gewohnten Nahrungsmitteln besteht und deren Menge mit Hilfe von Kalorientabellen bestimmt wird. Aber dies ist genau die am schwersten durchzuhaltende Diät: Für die meisten Adipösen ist es leichter, eine der neuartigen oder sogar bizarren Diäten durchzuhalten, von denen in den letzten Jahren genügend angeboten wurden. Aber welchen Erfolg diese Diätkuren auch haben mögen, er ist größtenteils auf ihre Monotonie zurückzuführen – fast jeder Mensch wird fast jeder Nahrung überdrüssig, wenn er nichts anderes als sie zu essen bekommt. Infolgedessen ist die Verführung, wieder zuviel zu essen größer als vorher, wenn diese Diät zugunsten der abwechslungsreicheren gewohnten Kost aufgegeben wird.

Hungern, das zu schnellem Gewichtsverlust führt, hat in den letzten Jahren ebenfalls beträchtlich an Popularität gewonnen. Viele Adipöse finden es relativ leicht zu hungern: Nach einigen Tagen ohne Nahrung läßt das Hungergefühl stark nach und der Patient kommt ganz gut zurecht, solange die Umgebung keine Forderungen an ihn stellt. Für einige massiv Adipöse oder für die seltenen Patienten, die rasch abnehmen müssen, hat Hungern eine begrenzte Indikation. Nachuntersuchungen von Patienten, die über längere Zeit gehungert haben, zeigen jedoch, daß fast alle mindestens das verlorene Gewicht wieder zugenommen haben.

38.1.6.2 Psychodynamische Therapie

Informationen über Abmagerungsdiäten sind so verbreitet, daß nur Menschen in die Sprechstunde des Arztes kommen, denen es nicht gelungen ist, selbst ihr Gewicht zu reduzieren; und nur die Patienten, bei denen die ärztliche Therapie erfolglos war, landen beim Psychiater. Auf diesem Hintergrund ist es leicht zu verstehen, warum es keinen Beweis dafür gibt, daß psychodynamische Therapie effektiver ist als andere, weniger kostspielige Kuren zur Gewichtsreduktion. Weniger verständlich ist der weitverbreitete Glaube an die Wirksamkeit dieser Psychotherapie. Viele Menschen sind überzeugt, daß sie die einzig dauerhaft erfolgreiche Behandlung für Adipositas sei. Es gibt keinen ersichtlichen Grund, der diese Vorstellung stützt.

Ebenso unbewiesen ist die Vorstellung, daß die Aufdeckung unbewußter Ursachen des Zuvielessens von Nutzen sei, weil der Patient danach nicht länger zu dieser Reaktion seine Zuflucht zu nehmen brauche (entsprechend dem psychoanalytischen Modell für die Auflösung neurotischer Symptome). Adipöse können interessante Phantasien und Erinnerungen als Antwort auf das Interesse des Therapeuten produzieren, aber mit einer Ausnahme, auf die wir später zurückkommen, führen solche Enthüllungen selten zu positiven Verhaltensänderungen. Viele Adipöse scheinen überdies besonders gefährdet, vom Therapeuten abhängig zu werden und während der Psychotherapie unkontrollierbar zu regredieren. Wir glauben, daß Psychotherapie das Basismuster des Zuvielessens als Reaktion auf Streß, das bei Adipösen so häufig ist, nicht ändern kann. Auch Jahre nach erfolgreicher Psychotherapie und erfolgreicher Gewichtsabnahme fallen Menschen, die unter Streß zuviel aßen, wieder in das alte Reaktionsmuster zurück.

Hat Psychotherapie also keinen Platz in der Behandlung von Adipositas? Entschieden doch! Obwohl aller Wahrscheinlichkeit nach das Muster des Zuvielessens als Reaktion auf Streß nicht zu ändern ist, kann Psychotherapie Adipösen helfen, weniger streßvoll und zufriedener zu leben. Wenn das erreicht ist, sind sie weniger gefährdet, zu viel zu essen. Sie können sich einschränken und sogar dabei bleiben. Diese Erfolge sind nicht weniger bedeutsam, weil sie keine spezifischen Behandlungseffekte sind.

Darüber hinaus kann Psychotherapie recht erfolgreich bei der Behandlung der beiden oben besprochenen Krankheitsbilder sein: Störungen des Körperschemas (body image) und Freßorgien. Beide wurden erfolgreich psychotherapeutisch behandelt, und die Patienten haben eine dauerhafte Gewichtsabnahme erreicht. Keines der beiden Krankheitsbilder konnte durch andere Behandlungsformen, auch nicht durch drastische Gewichtsreduktion, beeinflußt werden. Aber es muß betont werden, daß die Psychotherapie solcher Patienten häufig Jahre braucht, um dauernde Erfolge zu sichern. Der Prozeß kann durch Modifikationen der traditionellen psychoanalytischen Technik gefördert werden. Es sollten alle Anstrengungen gemacht werden, Intellektualisierungen und Regression zu minimalisieren, um den Patienten bei ihren Begriffsverwirrungen zu helfen und ihr oft schwer beeinträchtigtes Selbstwertgefühl zu stärken. Bruch bringt beispielhafte Beschreibungen solcher Maßnahmen in ihren zahlreichen Schriften.

Die erste Frage, die vor der Entscheidung zu einer Psychotherapie beantwortet werden muß – die aber nur allzu oft ignoriert wird – ist die, ob sie überhaupt notwendig ist. Die Tatsache, daß die Hälfte der von Hausärzten behandelten Adipösen Angst und Depression entwickeln und daß auch bei einer stationären Langzeitbehandlung (Stunkard und Rush, 1974) in einem hohen Prozentsatz emotionale Störungen auftreten, könnte von Befürwortern einer Psychotherapie ins Feld geführt werden. Aber Psychotherapie – die zwar ungünstige Effekte einer Behandlung eher wahrnimmt – ist nicht unbedingt harmlos. Da wahrscheinlich bei jeder Behandlung die Fähigkeit zu heilen der Möglichkeit zu schaden entspricht, muß auch bei der Indikation zu einer Psychotherapie beides sorgfältig abgewogen werden. Auch wenn kein direkter Schaden entsteht, kann die Behandlung doch durch lange Perioden quälenden Stillstandes führen, die dem Patienten nicht helfen und die seine Energien und seine Aufmerksamkeit oft von nützlicheren Tätigkeiten ablenken. Patienten und Therapeuten, die in einer solchen Sackgasse gefangen sind, haben es oft schwer, das Problem zu erkennen, geschweige denn, die Behandlung zu beenden.

Als ersten Schritt bei der Abschätzung der Zweckmäßigkeit einer Behandlung sollte der Psychotherapeut klären, was der Patient für seine Probleme hält, wie es mit seinen Kräften bestellt ist, sich mit diesen Problemen auseinanderzusetzen und was er von der Behandlung erwartet. Während er sich mit der Klärung dieser Fragen befaßt, erfährt er auch sehr viel über andere Probleme, über die sich der Patient nicht völlig im klaren ist. Die Entscheidung über die Indikation zur Behandlung sollte der Therapeut dann von der Einschätzung seiner eigenen Fähigkeiten abhängig machen, dem Patienten bei der Erreichung vernünftiger Ziele ohne unnötiges Risiko und ohne übermäßigen Aufwand an Zeit und Mühe zu helfen.

Die Wichtigkeit eines frühzeitigen Übereinkommens zwischen Patient und Therapeut über die Ziele der Behandlung sollte nicht unterschätzt werden. Wir glauben, daß die Psychotherapie durch das Versäumnis, rechtzeitig zu solchen Übereinkünften zu kommen, ebensooft in Schwierigkeiten geraten ist wie durch andere Probleme. Andererseits ist Psychotherapie auch angesichts bedrohlicher Schwierigkeiten oft erfolgreich gewesen, weil Patient und Therapeut zu Beginn eine klare Verständigung über die Ziele der Behandlung erreicht hatten.

Es gibt nur wenige Indikatoren, die es erlauben, den Erfolg einer Therapie vorauszusagen. Realistische Ziele, die der Patient klar artikulieren kann, sind ein gutes Zeichen. So wird der seit Kindheit adipöse Patient, der in die Behandlung kommt, um in einen „Adonis" verwandelt zu werden, fast sicher ebenso enttäuscht werden wie der Therapeut, der hofft, ihm bei dieser Verwandlung helfen zu können. Die Reaktion des Therapeuten auf den Patienten ist ebenfalls ein wichtiger Faktor: Sympathie für den Patienten ist für den Erfolg der Behandlung ebenso wichtig wie positive Erwartungen. Ein anderes prognostisches Kriterium ist die Fähigkeit des Patienten, seine Symptome und seine Verhaltensstörungen zu bestimmten Problemen seines Lebens in Beziehung zu setzen. Da Psychotherapie zu einem großen Teil gerade in der Aufdeckung solcher Beziehungen besteht, sind natürliche Fähigkeiten des Patienten, solche Zusammenhänge zu sehen, ein Indikator für den Erfolg der Behandlung. Wenn der Patient hingegen zu Beginn der Behandlung keinerlei Fähigkeit zeigt, solche Verknüpfungen zu sehen, wird er sie wahrscheinlich auch während der Behandlung nicht entwickeln. Bis zu einem gewissen Grad sind zwar alle neurotischen Patienten außerstande, solche Verbindungen zu sehen; aber Patienten mit guten Chancen für eine Therapie scheinen es doch wenigstens zu einem rudimentären Verständnis ihrer Schwierigkeiten zu bringen. Schließlich spricht für eine positive Erwartung die Beobachtung, daß Symptome oder ein gestörtes Verhalten des Patienten durch ein spezifisches Ereignis, das der Therapeut eventuell reproduzieren kann, unterbrochen werden können.

Eine Voraussetzung, die in gewisser Hinsicht die Basis für jede Psychotherapie bildet, ist schließlich supportiver, akzeptierender, nicht strafender menschlicher Kontakt.

Wenn nach Abwägung der Erfolgsaussichten und der Risiken für den Patienten eine Psychotherapie begonnen wird, was geht dabei vor?

Jerome Frank hat vorgeschlagen, bei dem psychotherapeutischen Prozeß zwei klar voneinander abgrenzbare Elemente zu unterscheiden (Frank, 1961). Eines ist „Überredung" mit dem Effekt eines Erlernens neuer Verhaltensformen, neuer Lebensaspekte, die besser an die Lebenssituation des Patienten angepaßt sind und ihm erlauben, besser mit Streßsituationen fertig zu werden. Das andere ist „Heilen", das vor allem zu einer Stärkung des Wohlbefindens und der Selbstachtung sowie zu einer Abnahme unangenehmer Gefühle, wie Angst und Depression, führt. „Überredung" erfordert in dem therapeutischen Prozeß meist eine gewisse Zeit; „Heilung" kann sich dagegen während einer Sitzung ereignen und ereignet sich wahrscheinlich überhaupt nicht, wenn sie nicht sofort einsetzt. „Überredung" übt ihre Wirkung weitgehend durch einen primären Impuls auf den Verstand aus; „Heilen" baut stärker auf irrationale Elemente. Beide hängen bis zu einem gewissen Grad von Erfahrung und Geschick des Therapeuten ab, aber „Heilung" ist ein viel unberechenbareres Ereignis, das enger mit der Persönlichkeit des Therapeuten zusammenhängt.

Es gibt offenbar so viele Konzepte, um Psychotherapie zu erklären, wie Therapeuten, und es ist nicht leicht, eine Basis für eine Auswahl zu finden. Nach einem Konzept, das einleuchtet, besteht Psychotherapie aus fünf Komponenten:
– einer besonderen, persönlichen Beziehung, die es ermöglicht,

– die Probleme und die Kräfte des Patienten zu erkennen, sowie
– die verschiedenen Elemente seiner Probleme zu isolieren, um
– Strategien für den Umgang mit diesen Elementen zu entwickeln und dem Patienten zu helfen, diese Strategien erfolgreich einzusetzen, mit dem Ziel,
– zu besserer Beherrschung vor allem im Bereich seiner persönlichen Beziehungen und zu erhöhter Selbstachtung und Verminderung seiner schmerzhaften Gefühle zu kommen. Dieses Konzept ist an anderer Stelle ausführlicher beschrieben (Stunkard, 1976).

38.1.6.3 Verhaltenstherapie

Allgemeine Überlegungen

1967 löste ein kleiner Artikel in einer unbekannten Zeitschrift über Verhaltensmodifikation eine nie dagewesene Aktivität auf dem Gebiet psychologischer Behandlung von Adipositas aus und führte zu bedeutenden Fortschritten ihrer wissenschaftlichen Erforschung. In den fünf Jahren nach der Veröffentlichung von „Verhaltenskontrolle des Zuvielessens" (Stuart, 1967) erreichte dieses Thema eine Popularität, die an modische Überschätzung grenzte. Es erschienen 50 Berichte über kontrollierte klinische Untersuchungen, die in einer Übersicht zusammengefaßt wurden (Stunkard, 1975).

Was versteht man unter Verhaltensmodifikation oder Verhaltenstherapie (Begriffe, die wir synonym verwenden werden)? Später sollen einige Charakteristika dieser relativ neuen Behandlungsform von Adipositas beschrieben werden. Im allgemeinen ist das besondere Kennzeichen der verschiedenen Methoden, die man Verhaltensmodifikation nennt, die Annahme, daß Verhaltensstörungen der verschiedensten Art teilweise erlernte Reaktionen sind, und daß die modernen Lerntheorien uns viel über das Entstehen und Vergehen dieser Reaktionen lehren können. Außerdem zeichnen sich die Verfechter der Verhaltensmodifikation durch die detaillierten Beschreibungen ihrer Methoden und deren Ziele aus, sowie durch ihre Bereitschaft, ihre Ergebnisse mit denen anderer Behandlungsmethoden zu vergleichen. So waren z. B. Verhaltenstherapeuten unter den ersten, welche die Bedeutung der Gewichtsveränderung in Pfund als eine abhängige Variable für die Psychotherapie-Forschung erkannten, und sie haben sich in wachsender Zahl der Behandlung von Adipositas zugewendet, um die hier liegende Möglichkeit nutzbar zu machen. Es ist eine Ironie, daß die Psychiatrie, die so dringend quantitative Maßstäbe nötig hat, um therapeutische Wirkungen beschreiben zu können, so lange brauchte, um die Empfindlichkeit, Verläßlichkeit und Gültigkeit von Gewichtsveränderungen als einen derartigen Maßstab zu erkennen.

An dieser Stelle scheint eine Richtigstellung nötig: Der Terminus Therapie bedeutet den Versuch, eine gestörte Funktion zu normalisieren: Einige der enthusiastischeren Verhaltenstherapeuten ziehen aus ihrer Beschäftigung mit Adipositas derartige Schlußfolgerungen. Sie behaupten, gestörte Eßgewohnheiten müßten die Ursache für Adipositas sein, weil Adipöse durch Veränderung ihrer Eßgewohnheiten Gewicht verlieren können. Es gibt keinen Beweis für eine solche Behauptung. Es könnte ebenso sein, daß Verhaltensmodifikation nur einem aus biologischen Gründen adipösen Menschen hilft, in einem halbverhungerten Zustand zu leben. Eine solche Sachlage würde nicht für eine ätiologische Wirksamkeit von Verhaltenstechniken sprechen. Es wäre ein größerer therapeutischer Triumph, eine solche biologische Tendenz zu überwinden, als lediglich schlechte Gewohnheiten zu ändern. Wie dem auch sei, die im folgenden beschriebenen Effekte einer speziellen therapeutischen Umgebung lassen keinen Schluß auf die Ätiologie von Adipositas zu. Sie lehren uns jedoch etwas über die soziale Kontrolle einer biologischen Funktion, und sie beschwören die verlockende Möglichkeit, die planlosen und oft unberechenbaren Einflüsse, welche die komplexen Faktoren der jeweiligen sozialen Umgebung auf unsere Nahrungsaufnahme ausüben, durch eine durchdachte Kontrolle über Essen und Adipositas zu ersetzen. In der Vergangenheit war es ziemlich einfach, ambulante Adipositasbehandlungen zu beurteilen. Die Ergebnisse waren so einheitlich dürftig und die Behandlungsweisen so offensichtlich inadäquat. (Stationäre Behandlungen mit der Möglichkeit einer schärferen Kontrolle der Patienten waren natürlich für eine Gewichtsabnahme effektiver. Aber ihr Nutzen war begrenzt, weil fast alle Patienten nach der Entlassung das verlorene Gewicht wieder zunahmen.) Wie oben erwähnt, verloren nur 25% der Patienten, die ihre Adipositas behandeln ließen, 20 Pfund und nur 5% 40 Pfund.

Vor diesem Hintergrund heben sich die Ergebnisse Stuarts von 1967 als etwas absolut Ungewöhnliches ab. Denn er beschreibt die besten Resultate ambulanter Adipositasbehandlung, die bis zu diesem Zeitpunkt berichtet worden waren, und liefert den Ansatz für ein besseres Verständnis dieser Störung. Selbst das Fehlen einer Kontrollgruppe beeinträchtigt die Bedeutung seiner Feststellungen nicht, denn von seinen 10 Patienten, die das 12-Monats-Behandlungsprogramm begannen, nahmen 3 mehr als 18 kg und 6 mehr als 15 kg ab.

Das von Stuart entwickelte Programm bildete die Basis für ein ständig wachsendes Angebot immer weiter verfeinerter Verhaltenstechniken. Ihr entscheidendes Merkmal ist die genaue Ausarbeitung einer standardisierten Auswahl von Verhaltensleitlinien für jeden einzelnen Kontakt mit dem Patienten. In diesem Sinne ist die Verhaltenstherapie der Adipositas mehr ein pädagogisches Konzept als eine traditionelle Therapie. Innerhalb der festgelegten Rahmenbedingungen bietet sich jedoch eine überraschende Möglichkeit für die Entfaltung von Kreativität sowohl des Patienten wie des Therapeuten. Stuart konstatierte, daß z. B. für Patienten, die unter „Verhaltensdepression" leiden, Essen der einzig verfügbare positive Verstärker sein kann. Eine wirksame Behandlung ist in diesen Fällen darauf angewiesen, ein Repertoire von alternativen, ebenfalls verstärkenden Verhaltensweisen zu entwickeln, um den Patienten zu helfen. Für zwei Patienten in Stuarts Programm waren solche hilfreichen Alternativverhaltensweisen: Interesse an Veilchenzucht und Pflege von Zimmervögeln.

Schon bald nach Stuarts erstem Bericht entstand eine Welle sorgfältig kontrollierter klinischer Studien zur Verhaltenstherapie der Adipositas. Diese Behandlungskonzepte spielten eine bedeutende Rolle in der Entwicklung wissenschaftlicher Konzepte zur Fundierung psychologischer Behandlung generell. Aber sie erwiesen sich als weitaus weniger effektiv hinsichtlich der klinisch bedeutsamen Gewichtsreduktion. So ergab sich der Verdacht, daß die Verhaltenstherapie der Adipositas mehr als Instrument zur Erforschung von psychotherapeutischen Faktoren schlechthin eingesetzt wurde, denn als Instrumentarium zur tatsächlichen Behandlung übergewichtiger Patienten. Drei neuere Studien haben allerdings diesen Verdacht ausgeräumt, indem sie aufzeigten, daß Verhaltenstherapie auch zu klinisch relevantem Gewichtsverlust beitragen kann; mehr noch, daß diese Erfolge auch über ein Jahr hinaus gesichert werden konnten.

Die erste dieser Studien bestand in einem weitgefaßten Ansatz, die relative Wirksamkeit von Verhaltenstherapie, von Pharmakotherapie (mit Fenfluramin) und Gruppensitzungen sowie der Kombination von beiden festzustellen (Stunkard et al., 1980). 145 Patienten mit einem durchschnittlichen Übergewicht von 60% über dem Idealgewicht wurden in wöchentlichen Gruppensitzungen über 6 Monate behandelt. Neben diesen drei „experimentellen" Gruppen wurden zwei Kontrollgruppen gebildet; eine traditionelle „Wartelistengruppe", eine zweite als „Arztpraxisgruppe". Diese „Arztpraxisgruppe" sollte es erlauben, die Wirksamkeit der üblichen Adipositasbehandlung in der Arztpraxis (mit Medikamenten und Diätempfehlungen) abzuschätzen.

Abbildung 38.1–5 veranschaulicht die Therapieergebnisse und die Resultate der Nachkontrollen. Die Patienten der Verhaltenstherapie verloren 10,9 kg. Die Patienten unter Fenfluramin und Gruppensitzungen reduzierten ihr Gewicht um 14,5 kg. Die Kombination von Verhaltenstherapie und Medikation erzielte mit 15,3 kg Gewichtsreduktion keinen zusätzlichen Effekt. Alle drei Behandlungsformen überstiegen den Effekt der üblichen Behandlung in der Arztpraxis, der bei durchschnittlich nur 6 kg lag.

Ein Jahr nach Abschluß der Behandlung allerdings ergibt sich ein grundsätzlich neues Bild. Patienten, die mit Verhaltenstherapie behandelt wurden, stabilisierten ihren Gewichtsverlust sehr gut, da sie nur 1,9 kg zunahmen. Jene Patienten, die zuvor Appetitzügler erhielten, nahmen um 8,2 kg zu, und die Patienten, die die „kombinierte" Behandlung durchliefen, zeigten das überraschendste und zugleich enttäuschendste Ergebnis: Sie nahmen um 10,7 kg zu und waren am Ende des Nachkontrolljahres nur um 4,6 kg leichter als zu Beginn des Behandlungsprogrammes.

In dieser Studie stellt sich die Verhaltenstherapie als die effektivste Behandlung der hier verglichenen Möglichkeiten heraus, was therapeutischen Optimismus für die Zukunft der Verhaltenstherapie stimulieren kann.

Eine zweite Studie begründet ebenfalls einen realistischen Optimismus für die Wirksamkeit der Verhaltenstherapie auch für die Behandlung der jugendlichen Adipositas (Brownell et al., 1983).

Die Ergebnisse nach der Behandlung jugendlicher Adipöser waren bislang ebenso enttäuschend wie die bei adipösen Erwachsenen, zudem ein noch stärkerer Pessimismus sich für die Therapieprognose dieser Altersgruppe breitgemacht hatte. Der Erfolg der Verhaltenstherapie gerade bei Jugendlichen stellt daher eine hoffnungsvolle Zukunftsaussicht dar.

42 Patienten im Alter zwischen 12 und 16 Jahren mit einem durchschnittlichen Übergewicht von 58% wurden in dieser Studie behandelt. Die verhaltenstherapeutischen Methoden wurden variiert hinsichtlich drei verschiedener Möglichkeiten, die Patientenmütter miteinzubeziehen. Ein Teil der Jugendlichen wurde allein, ohne ihre Mütter, in Gruppen behandelt, ein anderer Teil zusammen mit ihren Müttern in der Gruppe, beim letzten Teil der Jugendlichen wurden schließlich Mütter und Kinder in verschiedenen Gruppen behandelt.

Alle Patienten verloren signifikant an Gewicht: 3,3 kg unter der „Kind-Allein-Bedingung", 5,3 kg unter

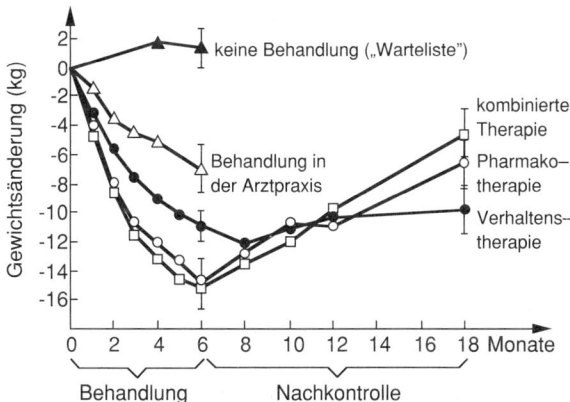

Abb. 38.1–5. Gewichtsveränderungen während der 6monatigen Behandlung und bei der 12monatigen Nachkontrolle. Die drei Therapiegruppen nahmen erfolgreich ab: Verhaltenstherapie (schwarze Kreise): 10,9 kg; Pharmakotherapie (weiße Kreise): 14,5 kg; kombinierte Therapie (Quadrate): 15,3 kg. Die Verhaltenstherapie-Gruppe verringerte für zwei Monate nach der Therapie noch ihr Gewicht, nahm dann wieder allmählich etwas zu, während die Pharmakotherapie und die kombinierte Behandlung zu einer raschen Wiederzunahme des Gewichts führten. Die Kontrollbedingung „Warteliste" (schwarze Dreiecke) führte zur Gewichtszunahme, die Behandlung in der Arztpraxis (weiße Dreiecke) erzielte 6,0 kg Gewichtsverlust. Die Patienten dieser Kontrollbedingungen wurden anschließend für 6 Monate behandelt und standen somit für eine Nachkontrolle nicht zur Verfügung. Die senkrechten Linien entsprechen dem Standardfehler des Mittelwerts (aus Stunkard et al., 1980).

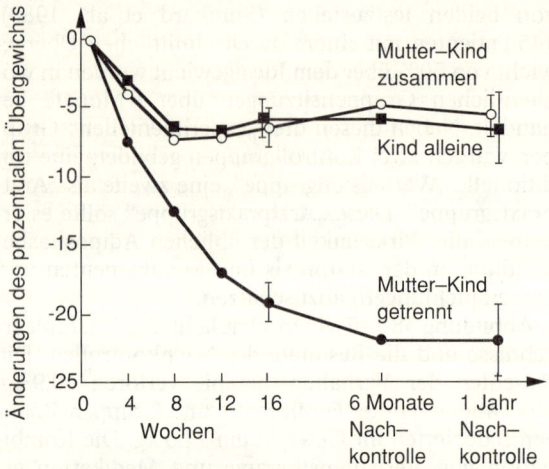

Abb. 38.1–6. Durchschnittliche Veränderungen (± SEM) des prozentualen Übergewichts während drei verschiedener Behandlungsformen (Kind alleine; Kind–Mutter zusammen; Kind–Mutter getrennt) und zum Zeitpunkt einer Ein-Jahres-Nachkontrolle (aus Brownell et al., 1983).

der „Kind-Mutter-Bedingung" und 8,4 kg in den Fällen, in denen die Jugendlichen und ihre Mütter in separaten Gruppen behandelt wurden.

Weiterhin ergab sich eine signifikante Senkung des Blutdrucks vor allem bei Kindern mit Grenzwerten zur Hypertonie. Die Besserung der Adipositas wurde, wie in Abbildung 38.1–6 dargestellt, auch während der einjährigen Nachkontrolle aufrechterhalten.

Eine gute Erfolgsbilanz legt auch eine neuere deutsche Studie vor. Im Rahmen des 7 Jahre dauernden Projektes wurden insgesamt 265 Patienten behandelt. Der ambulante Therapieansatz basiert auf einer interdisziplinären Konzeption, d.h., es wurde versucht, die Wirksamkeit des Behandlungsteams durch die Berücksichtigung dreier Berufsgruppen – nämlich eines klinischen Psychologen (in der Regel verhaltenstherapeutisch orientiert), eines Internisten und eines Diätassistenten – zu verbessern.

In den Gruppen wurde als Rahmenkonzeption ein Selbstkontrollprogramm verwirklicht. Es wurde versucht, den Partner des Patienten einzubeziehen, ohne daß dieser permanent präsent war, und eine selbsthilfeorientierte Nachsorge im Anschluß an die 6 Monate dauernde Therapie wurde angestrebt. Die Patienten wiesen ein durchschnittliches Übergewicht von 43% über Normalgewicht nach Broca auf, hatten vielfältige erfolglose Behandlungserfahrungen mit anderen Methoden der Gewichtsreduktion, das Alter lag zwischen 20 und 55 Jahren, etwa zwei Drittel der Patienten kamen aus den unteren Schichten. Die Abbrecherquote lag bei 23% (vor allem sog. Frühabbrecher, die im ersten Drittel der Therapie ausschieden). Die verbleibenden Patienten (n = 205) nahmen im Durchschnitt 11,4 kg während der Behandlung ab. Im 6-Monate-Follow-up lag die durchschnittliche Gewichtsreduktion gegenüber Therapiebeginn bei 10,5 kg, zum 12-Monate-Follow-up-Zeitpunkt bei 8,0 kg.

In der Untersuchung wurde auch der Einfluß der Therapie auf andere psychische und somatische Parameter untersucht. So zeigte sich, daß

– 82% am Ende der Therapie von einer Steigerung des Genusses von Mahlzeiten berichteten,
– 35% „unangemessene Eßverhaltensweisen" wie hastig essen oder Nebentätigkeit während des Essens verändern konnten,
– eine Veränderung der Nahrungszusammensetzung entsprechend den diätetischen Empfehlungen erreicht wurde,
– eine Abnahme von sozialer Angst und Depressivität sowie eine Zunahme von Arbeitszufriedenheit und Selbstakzeptanz zu verzeichnen war,
– eine klinisch relevante Verbesserung verschiedener wichtiger somatischer Parameter wie Blutdruck und Lipoproteine erreicht wurde, und
– der Behandlungsansatz einen Kostenvorteil gegenüber stationär durchgeführten Reduktionsdiäten im Verhältnis von 1:8 aufwies (Gromus et al., 1984).

Diese drei neueren Darstellungen müssen den Informationen hinzugefügt werden, die bereits von nahezu hundert klinischen Studien vorliegen, um eine begründete Beurteilung des gegenwärtigen Standes der Verhaltenstherapie vorzunehmen. Es besteht demnach Einvernehmen über 6 Punkte:

– Die Abbrecherquote bei ambulanter Therapie konnte weitgehend von Prozentsätzen um 75% auf nicht mehr als 15% verringert werden. Therapieverträge ebenso wie Kautionserhebung zu Beginn der Behandlung haben in dieser wünschenswerten Entwicklung eine bedeutende Rolle gespielt.
– Eine gefühlsmäßige Ablehnung gegenüber den verhaltenstherapeutischen Programmen ist selten und die meisten der Programme verbessern das psychische Befinden und reduzieren Symptome bei den Teilnehmern. Diese Feststellungen stehen im deutlichen Gegensatz zu den bis 50% häufigen psychischen Problemen während traditioneller Diätprogramme.
– Die erreichbaren Gewichtsverluste tendieren dazu, eher von mittlerer Größe zu sein, obschon neuere, eindringlichere Programme – wie oben beschrieben – höhere Gewichtsverluste erbrachten und – mindestens ebenso wichtig – eine gute Stabilität der Reduktionserfolge aufwiesen.
– Es besteht eine bemerkenswerte Variabilität der Gewichtsverluste bei verschiedenen Patienten.
– Eine Erfolgsprognose über die Verhaltenstherapie der Adipositas ist bislang noch unsicher. Es sollte indes gesehen werden, daß die Schwierigkeit solcher Prognose hier nicht größer ist als in anderen Behandlungsverfahren der Adipositas auch.
– Der wichtigste Aspekt der Verhaltenstherapie besteht wahrscheinlich in der Tatsache, daß ihre Techniken klar beschreibbar und schnell erlernbar sind.

Da die Verhaltenstherapie der Adipositas ohne Schwierigkeit erlernt werden kann, stellt sich die Frage, ob sie nicht auch einem größeren Kreis von Patienten angeboten werden kann als nur jenen, die in die Klinik kommen und für die sie primär entwickelt wurde. Die Antwort ist ein uneingeschränktes „Ja".

Eine der ersten Entwicklungen in verhaltenstherapeutischen Programmen war die Konzeption von Handanweisungen, die detaillierte Instruktionen für die Führung von Therapiegruppen anboten. In deren Folge wurden immer mehr Gruppen von Personen mit immer weniger professionalisierter Ausbildung geleitet und in dieser Weise angelernte Laien haben in weitem Umfang die Ausübung der verhaltenstherapeutisch orientierten Behandlung bei mittleren Graden der Adipositas übernommen.

Eine offensichtlich zweckmäßige Möglichkeit zum Angebot dieser Behandlungsmethoden stellt die ärztliche Praxis dar. Obschon wenig Ärzte selbst diese Verhaltensprogramme zur Gewichtskontrolle durchführen, gehen mehr und mehr Ärzte dazu über, Ernährungswissenschaftler, Diätassistenten und Schwestern mit Gewichtsreduktionsprogrammen innerhalb ihrer Sprechstunde zu beauftragen. Ebenfalls beginnen Krankenhäuser und Kliniken, solche Programme in ihren Ambulanzen einzurichten. Verhaltenspsychologen haben solche Programme vor Jahren bereits entwickelt. Bei weitem die größte Anzahl adipöser Personen wird schließlich in Gruppen mit Laientherapeuten behandelt.

Selbsthilfegruppen mit Laientherapeuten

Das umfangreichste wissenschaftlich kontrollierte „Gruppenprogramm für gesunde Ernährung" wird in der Bundesrepublik von der Agrarsozialen Gesellschaft in Zusammenarbeit mit gesetzlichen Krankenkassen seit 1977 angeboten, bei dem bis 1983 in etwa 800 Gruppen mehr als 10 000 Klienten bei einer Programmdauer von 9 Monaten betreut wurden. Jeweils etwa ein Drittel der Teilnehmer erreicht nach Abschluß des neunmonatigen Programms keinen nachhaltigen Erfolg (Abnahme weniger als 4,5 kg), durchschnittlichen Erfolg (4,5–9 kg Abnahme) oder guten Erfolg (über 9 kg). Nur 8% der Teilnehmer brechen vorzeitig ab. Nahezu alle Gruppen bleiben nach Ende des „offiziellen" Programms häufig mehr als drei Jahre zusammen. Grundlage des Trainings bildet ein verhaltenstherapeutisch orientiertes Programm (Kappus, 1982). Teilnahmeverträge und Kaution werden angewendet. Eine Ernährungsberaterin steht allen Gruppen für 10 Sitzungen zur Verfügung. Follow-up-Untersuchungen nach 3,5 Jahren lassen erkennen, daß 35% der Teilnehmer ihr reduziertes Gewicht halten oder weiter verringern können. Ein weiteres Drittel nimmt bis zu 5 kg wieder zu, bleibt jedoch unter dem Ausgangsgewicht. Im Durchschnitt ist 42 Monate nach Ende des Programms eine Gewichtsabnahme von 4,8 kg im Vergleich zum Ausgangsgewicht festzustellen.

Weitere Versuche, verhaltenstherapeutische Methoden für nicht ausgebildete Personen nutzbar zu machen, sind in den letzten Jahren häufig unternommen worden. Im Max-Planck-Institut für Psychiatrie wurde von Brengelmann die sog. „Bibliotherapie" oder „Brieftherapie" entwickelt und auf ihre Wirksamkeit hin überprüft. Das Zweite Deutsche Fernsehen produzierte sieben Sendefolgen zur Gewichtsabnahme, ebenfalls mit Unterstützung von verhaltenstherapeutischen Techniken (Pudel, 1979).

Ein dezidiertes Behandlungsprogramm nach Muster des programmierten Lernens, geschrieben speziell für den Adipösen, wurde von Hautzinger (1978) entwickelt. Weitere solche oder ähnliche Angebote werden inzwischen häufig auf Ton- und Bildkassetten von den Krankenkassen und auch den Volkshochschulen bereitgestellt.

Wie sieht nun das Basisverhaltensprogramm aus, das so viel Aktivität und Forschung auf sich gezogen hat?

Verhaltenstherapie, Beschreibung eines Programms

Eine kurze Beschreibung eines Behandlungsprogramms kann nur allgemeine Richtlinien geben, und Leser, die genaueres erfahren möchten, werden auf die Bücher von Mahoney und Mahoney (1976) und von Hautzinger (1978) verwiesen. Ein typisches Verhaltensprogramm besteht aus fünf Elementen:
1. einer Beschreibung des Verhaltens, das kontrolliert werden soll;
2. der Kontrolle der Stimuli, die dem Essensakt vorausgehen;
3. Verlangsamung des Eßvorgangs;
4. Verstärkung vorangehender Aktivität;
5. kognitive Therapie.

1. Beschreibung des zu kontrollierenden Verhaltens

Die Patienten werden aufgefordert, über ihre Nahrungsaufnahme sorgfältig Buch zu führen: Jedesmal, wenn sie essen, schreiben sie genau auf, was es war, wieviel, zu welcher Tageszeit, wo sie sich befanden, mit wem sie zusammen waren und wie sie sich fühlten. Die unmittelbare Reaktion der Patienten auf diese zeitaufwendige und lästige Prozedur war Murren und Klagen. Rückschauend stellten wir jedoch fest, daß solche Reaktionen häufiger waren, als wir mit dem Programm begannen, so daß sie möglicherweise unserer eigenen Unsicherheit bei der Handhabung seiner Technik zuzuschreiben waren. Seit wir von der Wirksamkeit des Programms überzeugt sind und seine Wichtigkeit vorbehaltlos betonen, reagieren die Patienten positiver. Viele kamen zu der Überzeugung, daß die Aufzeichnung der allerwichtigste Teil des Verhaltensprogramms sein könnte. Sie bringt die Patienten dazu, bewußter zu essen. Wenn sie erst einmal begonnen haben, Buch zu führen, sind sie, ungeachtet ihres jahrelangen Kampfes mit dem Problem, überrascht, wieviel und unter welchen Umständen sie essen.

Drei Beispiele sollen die Effektivität der Buchführung illustrieren: Ein Geschäftsreisender mittleren Alters begann zum ersten Mal zu realisieren, daß er nur im Auto zuviel aß, in dem er einen großen Vorrat von

Süßigkeiten, Erdnüssen und Kartoffelchips angelegt hatte. Er dachte über das Problem nach, entfernte die Vorräte aus seinem Auto und verlor prompt an Gewicht. – Eine 30 Jahre alte Hausfrau berichtete nach zwei Wochen Buchführung, es sei ihr zum ersten Mal in ihrem Leben klargeworden, daß Ärger sie zum Essen trieb. Sie reagierte prompt auf diese Entdeckung: Sobald sie sich ärgerte, versuchte sie, die Küche oder Orte, an denen Essen erreichbar war, zu verlassen und schrieb ihre Empfindungen nieder. Sie lernte es immer besser, essen als Reaktion auf Ärger zu vermeiden und begann ebenfalls Gewicht zu verlieren. – Ein eindrucksvolles Beispiel für den Erfolg der Buchführung war eine adipöse Frau mittleren Alters, die seit drei Jahren in einer intensiven psychotherapeutischen Behandlung war, die sie und ihr Therapeut für ziemlich erfolgreich hielten. Sie meinte, zwei Wochen Buchführung hätten sie mehr über therapeutisch wichtige Aspekte ihres Essens gelehrt als die ganze frühere Behandlung.

2. Kontrolle der Stimuli, die dem Essen vorausgehen

Eine Verhaltensanalyse beginnt traditionsgemäß mit der Untersuchung der Ereignisse, die dem Verhalten, das kontrolliert werden soll, vorausgehen. Die sog. Stimulus-Kontrolle umfaßt viele Maßnahmen, die in Programmen für Gewichtsreduktion seit langem üblich sind. Es wird keine Mühe gescheut, den Vorrat an hochkalorischer Nahrung im Haus zu begrenzen und den Zugang zu Nahrungsmitteln zu erschweren. Für Zeiten, in denen der Patient dem Trieb zu essen nicht widerstehen kann, werden genügend niedrigkalorische Nahrungsmittel wie Sellerie und rohe Karotten bereitgehalten. Zusätzlich dazu hat das Verhaltensprogramm neue und spezifische Maßnahmen eingeführt. Die meisten Patienten berichten z. B., daß sie an den verschiedensten Orten und zu den verschiedensten Tageszeiten essen. Einige, die während des Fernsehens aßen, hatten festgestellt, daß sie erst seit kurzer Zeit vom Fernsehen zum Essen stimuliert wurden. Es sah so aus, als ob die verschiedenen Zeiten und Orte zu sog. diskriminierenden Stimuli, die essen signalisieren, geworden waren.

Der Begriff „diskriminierender Stimulus" kommt von Tierversuchen, wo Stimuli wie das Aufleuchten einer Lampe oder ein Ton einem Tier signalisieren, daß es beim Drücken eines Hebels mit Futterbrocken oder etwas anderem belohnt wird. Da die Belohnung nie ohne den diskriminierenden Stimulus erfolgt, wird dieser in der Sprache der Lerntheorie zum „Kontrolleur" von Verhalten. Um die Anzahl und die Stärke der diskriminierenden Stimuli, die ihr Essen kontrollieren, zu vermindern, wird den Patienten geraten, alle Mahlzeiten, auch kleinere Naschereien, nur noch an einem Ort einzunehmen. Um die häuslichen Gewohnheiten nicht zu stören, ist dieser Ort meist die Küche.

Gleichzeitig bemüht man sich, neue diskriminierende Stimuli für essen zu entwickeln und ihren Einfluß zu verstärken. So wird den Patienten z. B. vorgeschlagen, aparte Tischgarnituren zu verwenden, wie Sets und Servietten in auffallenden Farben und besonderes Silber. Anstelle von Versuchen, die Menge der Nahrung zu verringern, wurden die Patienten angehalten, die aparten Tischgarnituren in jedem Fall auch bei kleinen Zwischenmahlzeiten zu benutzen. Eine Frau mittleren Alters war von der Wichtigkeit dieser Maßnahme so überzeugt, daß sie ihre Tischgarnituren mitnahm, wenn sie auswärts aß. Sie war einer unserer frühesten Erfolge.

3. Die Entwicklung von Techniken, um die Eßhandlung selbst zu kontrollieren

Wir haben spezifische Techniken verwendet, um den Patienten zu helfen, langsamer zu essen und sich der verschiedenen Komponenten bewußtzuwerden, aus denen sich die Eßhandlung zusammensetzt. Um Kontrolle über diese Komponenten zu gewinnen, üben die Patienten, während des Essens die einzelnen Bissen, jedes Kauen oder jeden Schluck zu zählen. Sie werden angehalten, ihr Eßbesteck nach jedem dritten Bissen so lange hinzulegen, bis sie den Bissen gekaut und hinuntergeschluckt haben. Schließlich werden längere Pausen eingeführt, zunächst von einer Minute Dauer und gegen Ende der Mahlzeit, wenn solche Pausen leichter erträglich sind. Allmählich werden die Pausen häufiger, länger und finden früher im Verlauf der Mahlzeit statt.

Weiter wird den Patienten klargemacht, wie wichtig es ist, alle Nebenbeschäftigungen, wie Zeitunglesen oder Fernsehen, während des Essens zu unterlassen und sich bewußt auf den Versuch zu konzentrieren, ihr Essen wirklich zu genießen. Sie werden aufgefordert, alles zu tun, um eine behagliche und entspannte Atmosphäre für die Mahlzeiten zu schaffen und vor allem Diskussionen über alte Argumente und neue Probleme bei Tisch zu vermeiden. Sie sollen bewußt versuchen, sich auf den Geschmack des Essens zu konzentrieren, auf ihr Kauen und Schlucken zu achten und die Wärme und das angenehme Völlegefühl ihres Magens zu genießen. Sobald sich der Erfolg einstellte, aßen sie weniger und hatten mehr Spaß.

Mr. Ives, ein Mammut von einem Mann, der sich solche Ermahnungen zunächst mit Herablassung anhörte, berichtete einen Monat später: „Wissen Sie – es ist erstaunlich, als ich all dies so machte, habe ich zum ersten Mal in meinem Leben etwas geschmeckt. Ich schmeckte wirklich, was ich aß. Ich kann es gar nicht verstehen. Wenn mich früher jemand fragte, ob ich gerne essen würde, antwortete ich immer: sicher, ich esse gern. Aber ich hatte es nie wirklich geschmeckt. Jetzt esse ich einen Löffel Eis und schmecke es wirklich und es macht mir soviel Spaß wie früher eine Riesenportion."

4. Die Modifikation der Folgen des Essens

Die im einzelnen nicht festgelegten, gelegentlichen Belohnungen, welche die Patienten durch das Verhaltensprogramm erhalten, werden durch ein System von formalen Belohnungen ergänzt. Das jetzt von uns verwendete Programm unterscheidet sich von bisherigen darin, daß es zwei verschiedene Belohnungslisten, eine für Veränderungen des Verhaltens und eine andere für Gewichtsabnahme, aufstellt. Von diesen beiden scheinen Belohnungen für Verhaltensänderungen effektiver.

Um die Zeitspanne zwischen dem spezifischen Verhalten und der erwarteten Belohnung zu verkürzen, entwickelten wir ein System, durch das der Patient für jedes Verhalten, das er einübt, eine bestimmte Anzahl von Punkten erhält: Buchführen, Zählen von Kauen und Schlucken, Pausen während der Mahlzeit, nur an einem Ort essen usw. Die Patienten können auch Extrapunkte verdienen, etwa ihre Anzahl verdoppeln, wenn sie sich angesichts großer Versuchung eine Alternative zum Essen ausdenken.

Diese Punkte, die eine unmittelbare Verstärkung des Verhaltens bewirken sollen, werden gesammelt und oft mit Hilfe des Ehepartners in greifbare Belohnungen verwandelt. Beliebte Belohnungen waren z. B. Kinobesuche oder Entlastung von Hausarbeit. Eine unpersönlichere Form war die Umwandlung von Punkten in Geld, das die Patienten zur nächsten Zusammenkunft mitbrachten und der Gruppe schenkten. In unserem ersten Programm schenkte die erste Gruppe das angesammelte Geld der Heilsarmee, die zweite schenkte es dem notleidenden Freund eines Mitglieds.

Prompte Verstärkung schien der Schlüssel zum Erfolg zu sein. Eine Hausfrau mittleren Alters sagte: „Mein Mann versprach immer, mir ein Auto zu schenken, wenn ich 50 Pfund abnehmen würde. Ich strengte mich bis zur Erschöpfung an und verlor 30 Pfund. Das war eine Menge! Aber was bekam ich dafür? Ich bekam keinen halben Wagen. Ich bekam gar nichts. In diesem Programm habe ich nur acht Pfund abgenommen, aber ich habe dafür schon eine Menge von ihm bekommen."

5. Kognitive Therapie

In den letzten Jahren ist die Verhaltenstherapie durch ein wachsendes Interesse an den kognitiven Inhalten und durch das allgemeine Gebiet der kognitiven Verhaltensmodifikation bereichert worden. Den Kognitionen hatte man in der Adipositastherapie weniger Beachtung geschenkt als den mehr traditionellen operanten Ansätzen, und bis heute liegen nur begrenzte experimentell gesicherte Erkenntnisse über die Wirksamkeit kognitiver Behandlungsstrategien für die Adipositas vor.

Dennoch erreichen diese Behandlungskonzepte zunehmende Beachtung und mehrere Kliniker sind überzeugt, daß hierdurch ein nützlicher Beitrag zu einem umfassenden Behandlungsprogramm der Adipositas geleistet wird.

Die Brücke, über die Kognitionen den Verhaltenstherapeuten so nahe gebracht wurden, war die Entdeckung, daß Patientenmonologe, die soviel unserer Zeit beanspruchen, leicht zugänglich sind. Sie lassen sich darüber hinaus sowohl quantifizieren als auch beeinflussen wie jedes operante Verhalten durch Verstärkung, Löschung etc.

Mahoney und Mahoney (1976) stellten die entscheidende Bedeutung von Kognitionen und Selbstgesprächen heraus, die diese für die Manifestation und die Stabilisierung einer Adipositas haben. Der erste Schritt zur Anwendung kognitiver Strategien in der Adipositasbehandlung besteht darin, dem Patienten zu helfen, seine häufigsten negativen Selbstgespräche oder Selbstreferenzen zu entdecken, sowie ihre Häufigkeit zu bestimmen.

Daraufhin werden Patienten, wie von Beck (1976) beschrieben, angeregt, mit Gegenargumenten gegen ihre Monologe vorzugehen. Da diese negativen Selbstgespräche mehr stereotypisiert und auch in ihrer Anzahl begrenzt sind, stellt es sich gewöhnlich nicht schwierig dar, Gegenargumente zu konzipieren. Danach lernt – und „überlernt" – der Patient mit Unterstützung, diese mehr adäquaten Selbstreferenzen als Antwort auf die negativen Denkinhalte mehr und mehr automatisch zu geben.

Das Einüben dieser Art der „Gegenargumentation" mit sich selbst scheint positiv zu wirken und das Durchhaltevermögen zu stärken. Dies wurde für depressive Patienten überzeugend belegt und wirkte sich wahrscheinlich auch günstig auf die Gewichtsabnahme aus.

Das einfache Memorieren der Gegenargumente über eine gewisse Zeit mag bereits helfen, selbst wenn der Patient zu Beginn dem Inhalt dieser neuen Kognitionen noch nicht vollständig zustimmt.

Fünf Beispiele für diese negativen Selbstreferenzen und ihre dazugehörenden Gegenargumente wären:

- Betrifft Gewichtsverlust: „Abnehmen dauert so lange"; Gegenargument: „Aber ich nehme doch ab. Und dieses Mal lerne ich auch, das Gewicht zu halten".
- Betrifft Fähigkeit zum Abnehmen: „Ich habe niemals Erfolg gehabt. Warum sollte ich jetzt Erfolg haben?"; Gegenargument: „Es gibt immer ein erstes Mal. Und jetzt habe ich ein erfolgreiches Programm, das mich unterstützt".
- Betrifft Ziele: „Ich kann nicht aufhören zu naschen"; Gegenargument: „Das ist auch unrealistisch. Ich versuche daher, weniger zu naschen".
- Betrifft Gedanken ans Essen: „Ich entdecke mich, wie meine Gedanken um den herrlichen Geschmack der Schokolade kreisen"; Gegenreaktion: „Stop! Dieser Gedanke frustriert doch nur. Ich denke lieber daran, am Strand in der Sonne zu liegen" (oder an eine Aktivität, die der Patient als besonders angenehm empfindet).
- Betrifft Ausflüchte: „In meiner Familie sind alle dick. Bei mir ist das vererbt"; Gegenargument: „Das macht's Abnehmen schwer, aber nicht unmöglich. Wenn ich durchhalte, muß ich Erfolg haben".

Verhaltensmodifikation

(abschließende, einigermaßen wissenschaftliche Nachschrift)

Zwei weitere Aspekte der Verhaltensmodifikation sind nicht genügend gewürdigt worden: Sie geben sowohl dem Patienten wie dem Therapeuten die Möglichkeit, große Kreativität zu entfalten und dem Patienten die Ermutigung zur Übernahme eines ungewöhnlich hohen Maßes von Eigenverantwortung für die Behandlung.

A. Kreativität in der Verhaltensmodifikation

Eine wenig schmeichelhafte Vorstellung von Verhaltensmodifikation sieht darin mechanische Manipulationen, die auf Kosten von innerer Freiheit und einzigartiger menschlicher Qualitäten triviale Veränderungen des äußeren Verhaltens erreichen sollen. Solche Polemiken zitieren gewöhnlich Beispiele systematischer Desensibilisierung von Phobien oder die Therapie konditionierter Aversion für Alkoholismus, Behandlungen, die auf der Methode der klassischen Konditionierung (Pawlow) beruhen. Wieweit die Kritik an diesen Behandlungen berechtigt ist, sei dahingestellt – für verhaltenstherapeutische Maßnahmen, wie sie hier dargestellt werden, ist sie kaum relevant: denn diese beruhen in erster Linie auf operanter Konditionierung (Skinner). Eine genauere Betrachtung dieser Maßnahmen wird das verdeutlichen.

Eine Verhaltensanalyse beginnt mit einer sorgfältigen Untersuchung der Frage, welche Variablen in der Umgebung des Patienten sein krankhaftes Verhalten kontrollieren. Diese Variablen sind einzuteilen in vorangehende Ereignisse, die das Verhalten auslösen, und in nachfolgende, die es belohnen und erhalten. Um die Episode einer Freßorgie zu verstehen, muß man z. B. genaue Informationen darüber haben, was dem Patienten kurz vor dem Zusammenbruch seiner Kontrolle über das Essen zustieß. Ebenso wichtig ist es, die unmittelbaren Auswirkungen auf die Gefühle des Patienten, auf das Verhalten des Ehepartners, ja auf jeden wichtigen Aspekt seines Lebens in Erfahrung zu bringen.

Der Wert dieser Methode wurde uns klar, als wir uns Rechenschaft gaben, in welchem Ausmaß sie die Behandlungstechniken für Adipositas, die wir seit etwa 20 Jahren empirisch entwickelt haben, differenziert hat. Diese Methode steht in deutlichem Kontrast zu den Empfehlungen für Behandlungen, die von psychodynamischen Theorien abgeleitet sind. Diese Theorien warnen z. B. davor, die einzelnen Eßsünden und die darauf folgende Gewichtszunahme mit den Patienten zu besprechen. Denn für sie sind diese Sünden nur Symptome eines zugrundeliegenden Konflikts. Sie lehren, daß jede kleine Wohltat, die aus der Beschäftigung mit diesen Symptomen entstehen könnte, wahrscheinlich bei weitem durch die Ablenkung des Patienten und des Therapeuten von ihrer primären Aufgabe, der Lösung der pathogenen Konflikte, aufgewogen werde; denn nur eine solche Lösung könne zu einer dauerhaften Heilung führen.

Erfahrungen mit Patienten überzeugten uns schon sehr frühzeitig, daß ein solches Vorgehen antitherapeutisch ist, und wir begannen, mit ihnen ihre Eßgewohnheiten zu besprechen und unsere Aufmerksamkeit immer detaillierter auf spezifische Episoden zu vielen Essens zu konzentrieren. Wir waren bereit

darzulegen, daß diese Abweichungen von der akzeptierten Lehre eine notwendige Anpassung an die speziellen Umstände von Eßstörungen darstellen. So leuchteten uns auch Empfehlungen, die unserem Vorgehen entsprachen, unmittelbar ein. Darüber hinaus schlagen diese Empfehlungen außerordentlich effektive neue Maßnahmen von großer Kreativität vor.

Was ist die Quelle dieser Kreativität? Im Mittelpunkt jeder Verhaltensanalyse steht die Suche des Patienten und des Therapeuten nach Lösungen für Probleme, die gleichzeitig relativ bescheiden und potentiell erreichbar sind. Dieses Vorgehen engt durch Begrenzung des therapeutischen Interesses auf einzelne, klar abgrenzbare Verhaltensweisen das prinzipiell grenzenlose Feld therapeutischer Begegnung ein und erlaubt dem Patienten und dem Therapeuten, ihre Anstrengungen auf eine überschaubarere Anzahl von Variablen zu konzentrieren, als es in traditionellen Therapien möglich ist. Durch Konzentration eines großen Teils der Aufmerksamkeit auf relativ kleine Probleme steigt die Wahrscheinlichkeit ihrer Lösung erheblich und der Erfolg, wenn auch in kleinen Dingen, ermutigt den Patienten sein Bemühen fortzusetzen, lösbare Probleme zu definieren und für sie nach Lösungen zu suchen. Man kann das Ausmaß an Kreativität, die sich im Laufe solcher Bemühungen entfalten kann, kaum überschätzen. Die wenigen in diesem Bericht erwähnten Beispiele können die Art, den Umfang und die Mannigfaltigkeit innovativer Maßnahmen nur andeuten, die in der Behandlung von Adipositas zur Anwendung gekommen sind.

B. Verantwortung in der Verhaltensmodifikation

Die zentrale Wichtigkeit bei der Behandlung von Adipositas, die Beobachtung auf Umgebungsvariablen zu konzentrieren, wird durch Schachters Beobachtungen unterstrichen. Er konnte zeigen, in welchem Ausmaß das Eßverhalten vieler Adipöser von der Umgebung kontrolliert wird. So stellte er z. B. fest, daß Adipöse durch „äußere" Faktoren wie Schmackhaftigkeit, Tageszeit und Erreichbarkeit von Nahrung viel mehr beeinflußt wurden als Nicht-Adipöse, und viel weniger durch „innere" Faktoren wie Hunger, den er durch Selbstbeobachtung und die Länge der Zeit seit dem letzten Essen gemessen hatte.

Das ungewöhnliche Ausmaß, in dem die Umgebung die Nahrungsaufnahme von Adipösen konstelliert, hilft uns zu verstehen, warum die üblichen medizinischen Maßnahmen ebenso versagen wie die traditionelle Therapie, die auf Introspektion des Patienten basiert. Adipöse passen sich zunächst leicht an das in der somatischen Medizin allgemein übliche autoritäre Patient-Arzt-Verhältnis an und nehmen ab, um dem Arzt zu gefallen. Durch die Vernachlässigung der Umgebungsvariablen, die einen so großen Einfluß auf ihr Essen ausüben, bleiben die Patienten jedoch diesem Einfluß hilflos überlassen und werden früher oder später rückfällig. Solche Übertretungen ärztlicher Vorschriften tangieren dann die speziellen Eigenschaften dieser Art von Arzt-Patient-Verhältnis:

Es verliert seine Macht, das Zuvielessen beginnt von neuem und es entsteht ein Circulus vitiosus.

Den Adipösen in psychotherapeutischer Behandlung geht es häufig nur wenig besser. Eine Therapie, die ihr Augenmerk auf Triebe und innere Konflikte konzentriert, ignoriert die Faktoren der Umgebung, welche die Nahrungsaufnahme des Patienten kontrollieren, allzuoft ebenso vollständig wie die übliche medizinische Behandlung. Außerdem kann die magische Erwartung einer Heilung der Adipositas durch Lösung von Konflikten den Patienten von näherliegenden Problemen mit größeren therapeutischen Möglichkeiten ablenken.

Im Gegensatz zu einer solchen Vernachlässigung der Umgebungseinflüsse, gegen die der Adipöse offensichtlich so wehrlos ist, hilft ihm die Verhaltensmodifikation, seine Aufmerksamkeit auf diese Einflüsse zu konzentrieren. Er wird nicht nur ermutigt, die Einflüsse der Umgebung auf seine Eßgewohnheiten zu beobachten und detailliert über sie Buch zu führen; es wird ihm auch gezeigt, wie er die so gewonnenen Informationen nutzen kann, um Aufgaben zu planen und auszuführen, die ihm helfen, Kontrolle über seine Gewohnheiten zu gewinnen. Der Unterschied zwischen Verhaltensmodifikation und den traditionellen Therapieformen besteht daher in erster Linie in dem Ausmaß an Anforderungen, die in den Zeiten zwischen den Therapiestunden an den Adipösen gestellt werden. Im Gegensatz zu den begrenzten Erwartungen traditioneller Therapien an die Eigeninitiative des Patienten gibt die Verhaltensmodifikation dem Patienten die Möglichkeit, in erheblichem Ausmaß eigene harte Arbeit in seine Behandlung zu investieren. Der offensichtliche Zusammenhang zwischen Erfolg und den eigenen Anstrengungen ermutigt die Patienten, Verantwortung in einem sonst ungewohnt hohen Ausmaß für ihre eigene Behandlung zu übernehmen. Dieser Zuwachs an Möglichkeiten, die Übernahme von Verantwortung für sich selbst einzuüben, könnte sich als der wichtigste Beitrag der Verhaltensmodifikation für die Psychotherapie erweisen.

38.2   Anorexia nervosa

Karl Köhle und *Claudia Simons*

38.2.1   Zur Bedeutung des Krankheitsbildes

Mit „Anorexia nervosa" wird eine schwere Störung des Eßverhaltens bezeichnet, die fast ausschließlich Jugendliche, überwiegend Mädchen betrifft. Diese **Verhaltenskrankheit** stellt ein schweres Leiden dar, das häufig zu chronischer körperlicher und psychosozialer Invalidität, nicht selten zum Tode führt. Ätiologie und Pathogenese des Krankheitsbildes sind bis heute ungeklärt: Die vorhandenen Modellvorstellungen zum Verständnis der Entstehung und Aufrechterhaltung des Krankheitsbildes sind noch bruchstückhaft und lassen sich noch nicht zu einem Gesamtkonzept verbinden. Es ist auch noch ungeklärt, ob es sich bei der Anorexia nervosa um eine Krankheitsentität handelt, oder um ein Syndrom, das sich aus Krankheitsuntergruppen zusammensetzt (vgl. Weiner, 1982; zur Forschungsmethodik A. E. Meyer, 1984) – wie dies heute auch bei anderen „klassischen" Krankheitsbildern erkannt wird.

Die Radikalität der Verneinung der Nahrungsaufnahme als einem menschlichen Grundbedürfnis und die ebenso radikale Verweigerungshaltung in den Beziehungen wirkt auf Ärzte wie auf Laien immer wieder gleichzeitig intellektuell faszinierend und emotional bewegend; besonders eindrucksvoll dargestellt ist diese Verweigerungshaltung in Franz Kafkas Erzählung „Ein Hungerkünstler".

Die heute zum Krankheitsbild der Anorexia nervosa vorliegenden Befunde sprechen für ein Zusammenwirken biologischer, psychischer und sozialer Faktoren; das Krankheitsbild stellt in besonderem Maße eine Herausforderung an interdisziplinäre Zusammenarbeit in Forschung und Klinik dar. Die starke Zunahme und der Inhalt der Veröffentlichungen zu diesem Krankheitsbild während der letzten Jahre läßt erkennen, daß diese Herausforderung auch vermehrt angenommen wird.

Die 18jährige Franziska kommt erst zur Aufnahme in die Klinik, nachdem sie zunehmend unter Schwächezuständen bis hin zu Schwindelanfällen litt; bei der Aufnahme wiegt sie 41 kg bei einer Größe von 172 cm.

Die stark abgemagerte Patientin wirkt im Gespräch sehr gehemmt: sie hält ihre Schultern hochgezogen, den Kopf schief. Der Untersucher wird bei der Betrachtung ihres hübschen und klaren Gesichts durch eine alte Narbe und durch ticartige Bewegungen des Kopfes irritiert. Die Patientin äußert sich kaum spontan; im Interview verhält sie sich passiv-reaktiv; ihre Antworten wirken hinhaltend-abwehrend, ihre Mitteilungen intellektualisierend.

Die Patientin ist das mittlere von fünf Kindern. Ihre Mutter arbeitet als Kindergärtnerin, der Vater ist mittlerer Angestellter. Aus ihrer Kindheit wird berichtet, daß sie als „extrem braves Musterkind" beispielhaft bei Verwandten herumgereicht wurde. Allerdings habe es bei bestimmten Speisen schon immer Essensschwierigkeiten gegeben, z.B. bei Reis, Sago, Joghurt, Milch. Die Beziehungen in der Familie werden als ständig gespannt geschildert: zum Vater war es schwierig, überhaupt eine engere Beziehung herzustellen; zur Mutter war die Beziehung zwar enger, doch konnte Franziska auch mit ihr kaum über gefühlsmäßige Probleme sprechen. Die Mutter übte Kontrolle aus, indem sie Schuldgefühle erzeugte; sie kommunizierte mit Franziska vorwiegend „analog", weniger „digital", d.h. es wurde in der Kommunikation weniger direkt verbalisiert als indirekt durch bestimmte Haltungen, Gesten, feste Gebräuche etwas ausgedrückt oder mitgeteilt. Auch sonst erscheint die Kommunikation innerhalb der Familie eingeschränkt; die Patientin spielt hierbei eine besondere Rolle: sie erlebt sich als „Informationsvermittlerin" zwischen den Eltern.

Aus der Zeit vor der Krankheitsmanifestation berichtet die Patientin von einigen äußeren und inneren Belastungen. Zunächst zog die Familie um, sie selbst mußte die Schule wechseln und verlor damit ihren Freundeskreis. Ein Jahr vor der Krankheitsmanifestation erkrankte die Großmutter, bei der Franziska als Kind zeitweise gelebt hatte, an einem Karzinom. Die dann im Haus lebende Großmutter starb schließlich unter dem Bild einer Tumorkachexie ein Jahr vor der stationären Aufnahme der Patientin. Franziska hat das Bild der Abmagerung und Auszehrung bei der Großmutter sowie deren durch die Krankheit neu gewonnene Fähigkeit, vieles in der Familie zu bestimmen, sehr beschäftigt. Im Zusammenhang mit dem Tod der Großmutter begann Franziska erstmals, ihr Gewicht genauer zu kontrollieren. Neun Monate vor der Klinikaufnahme nahm die Mutter ihre Berufstätigkeit wieder auf, wodurch sich die häuslichen Gewohnheiten veränderten: die Versorgung durch die Mutter nahm ab, Franziska wurde stärker im Haushalt beansprucht. Jetzt begann die erste stärkere Gewichtsabnahme (Ausgangsgewicht 60 kg). Sechs Monate vor der Klinikaufnahme, bei einem Gewicht von 53 kg, sistierten die Menses (Menarche mit 13 Jahren). Vier Monate vor Klinikeintritt wurde die Gewichtsabnahme während eines Landschulaufenthaltes erneut stärker; Franziska erlebte diesen Landschulaufenthalt als Versuchungssituation, der sie nicht gewachsen war: heftige Ängste vor Kontrollverlust (Alkohol, Sexualität) und als Reaktion hierauf Rückzug aus der Gemeinschaft kennzeichneten ihr Erleben und Verhalten. Ärztliche Behandlung wurde zunächst wegen der Amenorrhoe in Anspruch genommen; stationäre Aufnahme erfolgte jedoch erst, als Schlaflosigkeit und zunehmende Schwäche dazu geführt hatten, daß sie auch in der Schule kaum mehr mitarbeiten konnte.

Innerhalb der psychologischen Symptomatik stand vor allem die auffallende **Verleugnung des Krankheitszustandes** im Vordergrund. Die Patientin erlebte sich weder als zu mager noch irgendwie als krank; sie selbst wollte nur kurzdauernde Hilfe für ihre Schwächeanfälle in Anspruch nehmen. Trotz des reduzierten Allgemeinzustandes fiel ihre **motorische Aktivität** auf: bis zuletzt hatte sie ausgedehnte Spaziergänge unternommen, war schwimmen gegangen und hatte im Rahmen ihres Putzzwanges zu Hause unter großer Anstrengung die ganze Wohnung gesäubert; schon nach wenigen Tagen Klinikaufenthalt bestieg sie das Ulmer Münster. Bemerkenswert war auch ihr enormer **Leistungsanspruch**: Trotz Schulwechsel war sie nach einem halben Jahr bereits wieder Zweitbeste der Klasse und wehrte sich gegen eine längere stationäre Behandlung auch aus der Angst heraus, in der Schule ins Hintertreffen zu geraten.

Charakteristisch war ihr **Eßverhalten**: Sie bevorzugte kalorienarme Speisen; obwohl sie unter starken, ständig spürbaren Hungergefühlen litt und den ganzen Tag über fast ausschließlich mit Vorstellungen vom Essen beschäftigt war, entwickelte sie schon nach dem Essen nur geringster Nahrungsmengen Schuldgefühle, die mit monotonen Selbstanklagen einhergingen: „Mußte das denn wieder sein?" oder „Das wäre doch nicht nötig gewesen". Hungrig bewegte sie sich oft stundenlang um den Eisschrank herum, um schließlich eine halbe saure Gurke zu sich zu nehmen und danach wieder tiefe Schuldgefühle zu empfinden.

In der Absicht, das Gegessene wieder aus dem Körper zu entfernen, gebrauchte sie regelmäßig **Laxantien**.

Franziska war besessen von der **Vorstellung, zuviel Raum einzunehmen** und dadurch „angreifbar" zu werden. In diesem Zusammenhang entwickelte sie eine Lieblingsvorstellung: bei vollständigem Verzicht auf das Essen könne sie so leicht und frei werden, daß sie schweben könne. Das Essen rief auch deshalb Angst bei ihr hervor, weil es ihrer Vorstellung zufolge bei ihr „im Busen" verschwand und sie befürchtete, dann wieder den spöttischen Bemerkungen ihrer Mitschüler ausgeliefert zu sein. Ihre Körperoberfläche erlebte sie als so porös, daß beispielsweise grelle Farben oder laute Geräusche direkt in sie eindringen könnten. Während des Höhepunktes der Erkrankung konnte sie deshalb z. B. keine Musik mehr hören.

38.2.2 Definition des Krankheitsbildes unter klinischen Gesichtspunkten

„Anorexia nervosa" bezeichnet eine schwere psychische Erkrankung, in deren Symptomatik eine Störung des Eßverhaltens eine zentrale Rolle spielt, die ihrerseits zu einem oft bedrohlichen Zustand von Unterernährung führt.

Synonyme: „Pubertätsmagersucht", „endogene Magersucht", „psychogene Magersucht", „hysterische Anorexie", „weight-phobia".

Für die klinische Arbeit relevante Verständnismodelle berücksichtigen, daß die Erkrankung regelmäßig im Rahmen einer Entwicklungskrise während der Adoleszenz auftritt. Die Krankheit wird als ein Versuch aufgefaßt, die mit der Adoleszenz verbundenen und/oder subjektiv befürchteten körperlichen Veränderungen und sozialen Anforderungen zu vermeiden. Eine zentrale Rolle bei von der Erkrankung Betroffenen spielen die panische, oft wahnhafte Befürchtung, übermäßig „fett" zu werden und das verzweifelte Bemühen, eigene triebhafte Bedürfnisse unter Kontrolle zu halten (Crisp, 1980, 1983 b). Die Patienten – ganz überwiegend Mädchen – sind durch die mit der Entwicklung verbundenen Veränderungen überfordert und den mit der sozialen Rolle in der Adoleszenz gestellten Aufgaben nicht gewachsen. Sie leiden zwar meist bereits seit vielen Jahren unter einer tiefgehenden Selbstunsicherheit mit ausgeprägten Insuffizienzgefühlen – oft als Folge mangelnder Empathie ihrer früheren Bezugspersonen –, es war ihnen jedoch bis zur Krise vor oder während der Adoleszenz gelungen, sich an die Anforderungen ihrer Umwelt ausreichend anzupassen. Von außen oft nicht erkennbar, hatten die späteren Patienten ihr psychisches und soziales Gleichgewicht um den Preis eines Verzichts auf eine autonome Entwicklung scheinbar aufrechterhalten. Jetzt treffen sie die eigene psychosexuelle Entwicklung und die mit dem Erwachsenwerden verbundenen Aufgaben weitgehend unvorbereitet. Die Patienten geraten in intensive Ängste und befürchten, in dieser Situation die Kontrolle über sich selbst und die anderen total zu verlieren. Insuffizienzgefühle und entsprechende Ängste verzerren wiederum die Wahrnehmung der sich in der neuen Situation stellenden Anforderungen, in einem Circulus vitiosus können hierdurch Ängste und Insuffizienzgefühle verstärkt werden.

Beim Krankheitsbild der Anorexia nervosa läßt sich im einzelnen verfolgen, wie die bio-psycho-sozialen Entwicklungs- und Lernprozesse zum Aufbau einer individuellen Wirklichkeit führen, die ab einem bestimmten Zeitpunkt zwar auf der psychischen Ebene noch sinnvolle Funktionen erfüllen kann, das soziale Leben der Betroffenen jedoch entscheidend behindert: Diese individuelle Wirklichkeit hat vor allem die Funktion einer Abschirmung und ist mit der „Wirklichkeit der anderen" nicht mehr kompatibel, läßt den Aufbau einer für das soziale Leben ebenfalls unabdingbaren „gemeinsamen Wirklichkeit" nicht mehr zu.

Die Organisation der einzelnen Befunde entsprechend dem Situationskreismodell kann das theoretische Verständnis und die Konzeptualisierung der klinischen Arbeit unterstützen.

Der Leser steht damit vor der Aufgabe, die Symptomatik der Erkrankung hinsichtlich ihrer Funktion für die Betroffenen in ihrer aktuellen Lebenssituation und in ihrer geschichtlichen Entwicklung zu betrachten. Bei dieser Betrachtung stellt sich die Frage nach der Wechselwirkung zwischen individueller Disposition und den äußeren Bedingungen, die zur Entwicklung einer „prämorbiden Persönlichkeit" führten, die aufgrund ihrer individuellen Erlebnis- und Verhaltensmöglichkeiten („Programme") eine bestimmte Lebenssituation nicht mehr problemlos bewältigen konnte.

Im einzelnen wird jeweils zu klären sein, in welchem Ausmaß Anorexiepatienten unvorbereitet mit der Adoleszenzkrise konfrontiert werden und warum sie weder selbst in der Lage sind, noch ausreichend von ihrer Umwelt darin unterstützt werden können, der neuen Situation angemessene Verhaltensformen zu entwickeln. In dieser Verunsicherung läßt sich die Erkrankung als ein Versuch auffassen, über die Wiederherstellung einer früheren Form der Lebensbewältigung wieder Sicherheit zu gewinnen. Die Abmagerung soll die wahrgenommene körperliche Realität vor der Pubertät wiederherstellen. Hinzu kommen jetzt jedoch „Abwärtseffekte" (vgl. Kap. 1), vor allem im hormonalen System (Hypothalamus-Hypophyse-Ovarien; vgl. Weiner, 1982), denen in dieser Sicht der Charakter einer „Regression" auf ein präpubertäres Funktionsmuster im körperlichen Bereich zukommen könnte. Die wahnhafte Verleugnung des eigenen Krankheitszustandes versucht dieses Ergebnis abzusichern, die weitgehende soziale Isolation eine erneute Konfrontation mit den nicht bewältigbaren Entwicklungsproblemen zu vermeiden, das arrogant überhebliche Verhalten quälende Verunsicherung und Minderwertigkeitsgefühle zu überdecken.

„Aufwärtseffekten" (vgl. Kap. 1) kann umgekehrt krankheitserhaltende und/oder -verstärkende Funktion zukommen: Aus dem extremen Gewichtsverlust ergeben sich sowohl pathophysiologische Veränderungen als auch Konsequenzen auf der psychischen und sozialen Ebene; z.B. können die mit der extremen Unterernährung verbundenen intensiven Hungergefühle in einem Circulus vitiosus die Angst vor dem Kontrollverlust und damit den verzweifelten Kampf um die Aufrechterhaltung der Kontrolle verstärken. Dieser Verständnisansatz hat Konsequenzen für die Entwicklung therapeutischer Konzepte: Sie sollten die Funktion der Erkrankung und ihre Konsequenzen auf den verschiedenen Ebenen berücksichtigen. Die Ergänzungsbedürftigkeit einseitiger Ansätze – etwa „Auffütterung" ohne Psychotherapie oder Psychotherapie ohne Berücksichtigung der sekundären Folgen der Kachexie – wird sofort evident.

38.2.2.1 Zur Geschichte der Beschreibung des Krankheitsbildes

Die Anorexia nervosa wurde als Krankheit zwar erst in neuerer Zeit beschrieben, ihr Vorkommen zumindest bis zurück ins Mittelalter scheint jedoch ausreichend dokumentiert. So berichtet Halmi (1982) über die Erkrankung von Prinzessin Margaret von Ungarn im 13. Jahrhundert. In der Folge eines Gelübdes ihres Vaters wurde die 1245 geborene Margaret von Geburt an Nonnen zur Erziehung übergeben; als ihr Vater später seine Ziele änderte und ihre Hochzeit mit einem geeigneten Nachfolger wünschte, wurde sie zunächst erregt, bemühte sich dann darum, sich so unattraktiv wie möglich zu machen; sie begann zu fasten, arbeitete bis zur Erschöpfung, mied den Schlaf und verrichtete in besonderem Maße Bußübungen. Während sie im Refektorium die anderen bediente, fastete sie selbst unerbittlich; während die

Mitschwestern aßen, schlich sie in die Kirche, um vor dem Kruzifix zu beten. Ihr Körper wurde als armselig beschrieben, sie starb 1271 im Alter von 26 Jahren einige Tage nachdem sie heftiges Fieber entwickelt hatte. Aus den vorhandenen vatikanischen Unterlagen (der Heiligsprechungsprozeß wurde nach ihrem Tode eingeleitet) geht das kontinuierliche und nicht in Beziehung zu religiösen Vorstellungen stehende Fasten, ihre Weigerung, das Körpergewicht im Normalbereich zu halten und die Kombination von Überaktivität mit extremer Magerkeit hervor (Halmi, 1982).

Morton (1689) gab anhand einer Kasuistik die bekannteste Erstbeschreibung des Krankheitsbildes. Gull (1868) in England und Lasègue (1873) in Frankreich beschrieben im einzelnen die Symptomatik der Erkrankung, die sie als psychogene Störung auffaßten. Ihre Beschreibung der Symptomatik ist bis heute gültig. Gull beschrieb die seelische Störung bei Magersüchtigen bereits als eine wahnähnliche „perversion of the ego" (nach Thomä, 1961). Beide Autoren wiesen auch bereits auf den therapeutischen Wert einer Isolation der Patienten von ihrer bisherigen Umgebung hin.

Die intrapsychischen Prozesse, die der Symptomatik und dem Verhalten zugrunde liegen, konnten erst durch psychoanalytische Untersuchung und Behandlung von Anorexiepatienten seit etwa 1930 näher geklärt werden (u.a. F. Deutsch, H. Meng, F. Alexander, V. v. Weizsäcker, M. R. Kaufmann, H. Thomä, J. Nemiah, H. Bruch, M. Selvini Palazzoli).

Vor allem in Deutschland wurde die Anorexia nervosa zu Unrecht lange Zeit als „Endokrinopathie"(„Simmondssche Kachexie") aufgefaßt. Der Pathologe Simmonds hatte 1914 zunächst einen Fall von Kachexie als Endzustand nach einem organisch bedingten Hypophysenvorderlappenausfall beschrieben. Diese Auffassung der Erkrankung wurde durch den Erfolg hieraus abgeleiteter Therapieversuche, z.B. die Implantation von Tierhypophysen, scheinbar gestützt; übersehen wurde, daß es sich hierbei um eine Placebowirkung bzw. um eine Wirkung aus der Arzt-Patient-Beziehung heraus handelte. Ein weiterer Grund für das überlange Festhalten an dieser unzutreffenden Auffassung dürfte darin liegen, daß es Medizinern schwerfällt, einen so schweren „körperlichen" Krankheitszustand, wie er bei magersüchtigen Patienten vorliegt, als seelisch verursacht anzuerkennen.

Eine Anekdote möge die Wissenschaftsgeschichte veranschaulichen:

Auch Gustav von Bergmann war einige Zeit Anhänger der Therapie durch Hypophysenimplantation. Solche Behandlungen erforderten einen großen und für die Patientinnen sehr eindrucksvollen Aufwand; so mußten Hypophysen aus den Köpfen frischgeschlachteter Kälber in der Klinik gewonnen werden. Von Bergmann hatte eine seiner Patientinnen zur weiteren Beobachtung in sein Wochenendhaus mitgenommen. Er verwarf die Theorie einer „Endokrinopathie", nachdem er durch die dünnen Wände des Wochenendhauses mitangehört hatte, wie seine Patientin ihr Erbrechen selbst induzierte (mündliche Mitteilung durch Th. von Uexküll).

38.2.3 Symptomatologie

38.2.3.1 Die Störung des Eßverhaltens

Die Störung des Eßverhaltens steht im Mittelpunkt: Die Patientinnen nehmen zu wenig Kalorien zu sich (Weglassen von Mahlzeiten, Auswahl kalorienarmer Nahrungsmittel wie Obst und Salat), beseitigen die Nahrung häufig wieder durch selbstinduziertes Erbrechen und behindern die Resorption durch Laxantienabusus.

Gestört ist dabei die Motivation zu essen: die Patientinnen können nicht essen wollen; sie wollen vielmehr abmagern und dünn bleiben. Dabei sind sie jedoch mit Nahrungsfragen präokkupiert, haben intensives Interesse an allem, was mit dem Essen zusammenhängt; eine Patientin beschrieb diesen Zustand als „living in a food world" (Walton, 1975).

38.2.3.2 Gewichtsverlust

Ein starker Gewichtsverlust bis zur Kachexie ist die Folge dieses Eßverhaltens: Die Patientinnen unterschreiten oft ein Gewicht von 30 kg, gelegentlich auch von 25 kg.

Die körperlichen Folgen des Gewichtsverlusts stellen wir im Abschnitt „Differentialdiagnose" dar.

38.2.3.3 Amenorrhoe

Fast regelmäßig besteht eine sekundäre, selten eine primäre Amenorrhoe; diese ist nicht notwendig Folge der Gewichtsabnahme. Bei 50–70% der 79 Kranken Halmis (1974) und 25–50% der 83 Patientinnen Frahms (1973) setzte die Amenorrhoe bereits vor oder mit Beginn der Gewichtsabnahme ein.

38.2.3.4 Obstipation, Laxantien- und Diuretikaabusus

Oft findet sich ein ausgeprägter Laxantien- und/oder Diuretikaabusus. Die Patienten geben häufig eine Obstipation an. Sie klagen über abdominelle Beschwerden, ihre Gedanken kreisen um den „zu dicken Bauch", sie versuchen Leibesumfang und Gewicht mit allen Möglichkeiten zu vermindern.

Bei vorwiegend organisch orientierter Betrachtungsweise wird für Magersüchtige immer wieder die Bedeutung folgender **„Trias"** hervorgehoben: **„Abmagerung, Amenorrhoe, Obstipation".**

Frahm (1973) hat für 83 der von ihm untersuchten Patientinnen die Häufigkeit des gemeinsamen Auftretens von Eßstörung, Gewichtsabnahme, Obstipation, Menstruationsstörungen und Erbrechen zusammengestellt: alle fünf Symptome wurden bei 35, vier der fünf bei weiteren 33, nur drei Symptome bei weiteren 13 der 83 Patientinnen beobachtet.

Keine Menstruationsstörung fand sich nur bei 9% von Halmis (1974) Patientinnen. Eine primäre Amenorrhoe bestand bei 2 der 83 Patientinnen von Frahm.

38.2.3.5 Psychologische Symptome

Neben der Störung des Eßverhaltens finden sich regelmäßig weitere, für das Krankheitsbild pathognomonische psychologische Symptome.

Verleugnung des Krankheitswertes der Kachexie, panische Angst „fett" zu sein

Im Vergleich zu anderen Kranken mit extremem Gewichtsverlust fällt auf, daß die Patientinnen ihren kachektischen Zustand nicht als krankhaft erleben, ihn vielmehr hartnäckig und uneinsichtig als „normal" verteidigen. Anorektikerinnen sind mit ihrer skelettartigen Erscheinung identifiziert, sie versuchen alles, ja sie kämpfen darum, sich dieses Aussehen zu erhalten. Sie sehen in diesem Aussehen die einzige Möglichkeit, den von ihnen panisch gefürchteten Zustand des „Fettseins" auf Dauer von sich abzuwenden. Hierzu gehört auch das Bemühen, durch ständige Bewegung Kalorien zu verbrauchen.

> Entsprechende Ängste tauchen während der Behandlung auf und tragen zu Widerständen gegen die Behandlung bei. Franziska klagte während der Wiederauffütterungsphase über angstvolle Phantasien; zunächst befürchtete sie, „dick, aufgetrieben", „wie ein Luftballon", „wie eine häßliche Kröte" zu werden; später: „amorph" zu werden, „die Grenzen der eigenen Gestalt zu verlieren", „zu zerfließen". Panische Angst mit Zittern und Weinen begleitete diese Vorstellungen.

Körperschemastörung

Vor allem H. Bruch (1973) hat die verzerrte Wahrnehmung des eigenen Körpers bzw. das zugrundeliegende verzerrte Körperkonzept, das „Körperschema" („body image") bei Magersüchtigen als Kernstück ihrer Psychopathologie dargestellt. Die Verzerrungen in der Wahrnehmung bzw. im Erleben des eigenen Körpers stehen häufig im krassen Gegensatz zu den guten intellektuellen Kenntnissen der Betroffenen, denen diese Diskrepanz durchaus bewußt sein kann. So schreiben sie z. B. bestimmten Nahrungsqualitäten bestimmte körperliche Veränderungen zu. Sie befürchten u. a., daß von ihnen abgelehnte kalorienreiche Nahrung direkt dorthin im Körper wandern würde, wo sie Fettansammlungen besonders befürchten (Bruch, 1980). Bruch bringt die Störungen der Körperwahrnehmung mit dem „alles durchdringenden Gefühl der Ineffektivität" in Verbindung. Diese Patienten haben das Gefühl, die Kontrolle über ihren Körper, ihr Verhalten, ihre Bedürfnisse und ihre Impulse nicht selbst zu haben.

Vor dem Hintergrund der u. a. von Head und Schilder inaugurierten Forschungen zum „Körperschema" (Übersicht bei Joraschky, 1983) wurden durch diese klinischen Befunde Hilde Bruchs zahlreiche experimentelle Untersuchungen bei Anorexiepatienten angeregt.

Dabei werden vor allem drei Erfassungsmethoden angewandt: projektive Verfahren, Selbsteinschätzung über Fragebogen und der Vergleich von Schätzungen definierter Körperdimensionen durch den Probanden mit dem Realmaß (Übersicht: Hill, 1976; Fichter und Meermann, 1981; Freeman, 1983). Der Darstellung einiger Befunde aus diesen Untersuchungen sei vorausgeschickt, daß ihre Spezifität für magersüchtige Patienten u. a. als Folge unzureichender Zusammenstellung von Kontrollgruppen zum Teil noch nicht ausreichend gesichert erscheint.

Läßt man Magersüchtige mit Hilfe horizontal verschiebbarer Lampen ihre Körpermaße auf einer Skala einstellen, so überschätzen sie die Breite des eigenen Körpers, insbesondere im Gesichtsbereich, erheblich, während normgewichtige Vergleichspersonen eher dazu neigen, ihre Körpermaße zu unterschätzen. Im Vergleich zum eigenen Körper überschätzen Magersüchtige die Breitenmaße eines Modells weniger und die von Gegenständen ebensowenig wie die eigene Körperlänge (Slade, 1973; Slade und Russel, 1973; Askevold, 1975). Die Störung des „body image" scheint der Behandlung entgegenzustehen: Die Patientinnen mit der stärksten Fehleinschätzung ihrer Körpermaße tendieren stärker dazu, nach der Krankenhausentlassung wieder Gewicht zu verlieren (Slade und Russel, 1973).

Die Überschätzung der eigenen Körpermaße ist bei Patientinnen mit akuter Erkrankung im Vergleich zu Normalpersonen ausgeprägt, nicht dagegen bei chronisch an Anorexie Erkrankten (Fichter und Meermann, 1981). Die Überschätzung bildet sich mit der Gewichtsnormalisierung teilweise zurück (Crisp und Kalucy, 1974).

Die Störung des Körperschemas ist bei Anorektikerinnen am stärksten ausgeprägt, sie findet sich aber auch bei adipösen und bei gesunden jungen Frauen. Möglicherweise hängt dies mit der häufigen Präokkupation junger Frauen in unserer Kultur mit einem ausgeprägten Schlankheitsideal zusammen; hier zeigt sich auch ein Übergangsbereich zwischen kulturell bedingten Idealvorstellungen während der Pubertätsentwicklung („Twiggy") und einer pathologischen Entwicklung im engeren Sinne. Die Zusammenhänge sind im einzelnen jedoch ungeklärt. Bemerkenswert erscheint, daß die genannten Störungen des Körperbildes nach kohlenhydratreichen Mahlzeiten stärker ausgeprägt sind als nach kohlenhydratarmen Mahlzeiten; bei normalen Kontrollpersonen wurde dieser Unterschied nicht gefunden; bei Anorektikerinnen ist dieser Unterschied mit Normalisierung des Körpergewichts nicht mehr nachweisbar (Crisp und Kalucy, 1974).

Allerdings konnte dieser bei einer relativ kleinen Stichprobe erhobene Befund von Garfinkel und Mitarbeitern (1977), sowie von Fichter und Mitarbeitern nicht bestätigt werden (nach Fichter und Meermann, 1981).

Die Fehleinschätzung der eigenen Körpergrenzen bei Anorektikerinnen, aber auch bei Adipösen, korreliert mit anderen psychologischen Befunden, so mit dem Ausmaß von „Neurotizismus" (Eysenck: „Personality Inventory") und mit einer niedrigen Einschätzung der eigenen Fähigkeit zur Selbstkontrolle (Rotter: „Locus of Control Scale") (Garner et al., 1978).

Dabei haben Magersüchtige durchaus auch eine Idealvorstellung von einem „dickeren Selbst", dürfen diese jedoch wegen der hiermit verbundenen Ängste nicht erreichen (Fransella und Crisp, 1975).

Wahrnehmung, Interpretation und Kontrolle physiologischer Stimuli und Bedürfnisse

Bei phänomenologischer Betrachtung nehmen Magersüchtige vom eigenen Körper ausgehende Reize, insbesondere Hungerempfindungen, im Vergleich zu Normgewichtigen verändert wahr, bzw. sie interpretieren solche Stimuli und Bedürfnisse unterschiedlich. Hunger wird weitestgehend verleugnet. Eine Patientin von Bruch (1973) formulierte: „Ich brauche nichts zu essen".

Die Patientinnen versuchen, eine völlige Unabhängigkeit von körperlichen Bedürfnissen, insbesondere von Hunger, zur Schau zu tragen. Dieses Bemühen um Autarkie wird ständig von der inneren Auseinandersetzung mit den Themen Hunger und Nahrungszufuhr gestört. Aus diesem Konflikt resultieren die häufig bizarren Eßgewohnheiten; die Patientinnen geben gelegentlich dem sonst abgewehrten Bedürfnis heimlich nach, stopfen sich in der Speisekammer oder nachts voll, während sie bei Tisch jegliches Essen verweigern; anschließend induzieren sie oft wieder Erbrechen. Ständig werden sie dabei von der Möglichkeit des Kontrollverlustes über ihren Hunger und die mit ihm verknüpften Triebbedürfnisse beunruhigt: „Ich wage nicht zu essen; wenn ich nur klein bißchen in den Mund nehme, bekomme ich Angst, daß ich nicht mehr damit aufhören kann" (Bruch, 1973).

Die innere Getriebenheit und Not dieser Kranken zeigt das bizarre Verhalten einer 18jährigen Oberschülerin: im Kontakt arrogant distanziert, versucht sie in allen Lebensbereichen ihre Autonomie zu demonstrieren, von Mitmenschen gibt sie sich ebenso unabhängig wie von der Nahrungszufuhr. Die Mutter der Patientin wird in der Apotheke auf die umfangreichen Käufe von Babykost durch ihre Tochter angesprochen. Bald stellt sich heraus, daß die zum Skelett abgemagerte Kranke abends heimlich den Inhalt vieler Gläser Babykost in sich hineinschlingt und nachts mit Hilfe einer entsprechend verformten Kerze im Bad der Familie wieder erbricht. Das verschmutzte Bad bringt dann schon morgens die Affekte der Eltern in Bewegung.

Für das Verständnis der Patientinnen ist es wesentlich zu sehen, daß Hunger und die mit ihm verknüpften Triebimpulse um so stärker drängen, je mehr die Kranken hungern und an Gewicht verlieren; sie müssen diesen Bedürfnissen dann immer wieder stärkere Abwehrkräfte entgegensetzen. Mit zunehmender körperlicher Schwäche befürchten sie häufig einen

Triebdurchbruch bzw. einen Zusammenbruch ihrer Kontrollmöglichkeiten.

> Franziska führte deshalb immer einige Zuckerstückchen mit sich; sie meinte, bevor sie so schwach werde, daß sie ihren Hunger nicht mehr kontrollieren könne, weil auch ihre Hirnzellen unterernährt und damit funktionsuntüchtig seien, würde sie sich so den für ihre Funktion nötigen Zucker zuführen können.

Hyperaktivität

Die körperlich extrem hinfällig wirkenden Patienten verleugnen Schwäche und Müdigkeit. Oft kommen sie erst dann zur Behandlung, wenn sie vor Schwäche in Schule oder Beruf leistungsunfähig geworden sind. Häufig haben sie bis kurz vor der Aufnahme in die Klinik ein anstrengendes sportliches Training durchgeführt. In der Klinik sind sie kaum im Bett zu halten, unternehmen sportliche Übungen, weite Wanderungen, besteigen Kirchtürme usw.

In diesem Zusammenhang ist auf die oft ausgeprägte **Leistungsorientiertheit** der Kranken und das nicht seltene Vorkommen von **Verhaltenszwängen,** die im Zusammenhang mit der Triebabwehr verstanden werden können, hinzuweisen.

> Franziska unternahm trotz zunehmender Schwäche noch ausgedehnte Spaziergänge mit ihrem Hund, ging weiter schwimmen u.a.m. Gleichzeitig mit der Kachexie entwickelte sich bei ihr ein ausgesprochener Putzzwang. Häufig putzte sie schon frühmorgens vor dem Frühstück die gesamte Wohnung der Familie. Bei der Durchführung der Schularbeiten hielt sie einen ausgeklügelten Zeitplan starr ein: sie arbeitete trotz ihrer Schwäche bis zu 5 Stunden täglich, nach 2½ Stunden Schularbeiten gestand sie sich eine Unterbrechung von einer halben Stunde zu, um eine Tasse schwarzen Kaffee zu trinken.

Kontaktstörung

Die scheinbare Selbständigkeit der Kranken ist vor dem Hintergrund ihrer sozialen Isolation zu sehen. Sie halten sich in dieser „splendid isolation" künstlich von anderen unabhängig. Ihre Fähigkeit zu intensiverem Kontakt, zu emotionalem Austausch ist stark eingeschränkt. Ihre Selbstwertprobleme kompensieren sie in einer auf andere oft aggressiv-arrogant wirkenden Überheblichkeit.

Schon die Notwendigkeit einen Arzt in Anspruch nehmen zu müssen, verstärkt die Gefühle eigener Insuffizienz und Minderwertigkeit. So wird verständlich, daß Ärzte diese oft sehr irritierenden Kompensationsversuche der Patienten in besonderem Maße zu spüren bekommen.

> Bei Franziska fiel auf, daß sie sich während des Bettens auf der Station, während der Visiten und anderer Kontaktmöglichkeiten hinter ihren Schulbüchern (Buch mit englischen Vokabeln!) verschanzte und dabei so-

gar den Blickkontakt mied. Später konnte sie hierzu folgende Phantasie mitteilen und in der Stationsgruppe darstellen: sie sitze in einem Glashaus, das sie von den anderen isoliere; genüßlich warte sie darauf, daß sich die anderen bei dem Versuch, sich ihr zu nähern, die Hände zerschneiden würden.

Deutlich wird hier, wie Franziska ihre Angst reduziert, Vorsorge trifft, daß andere ihr nicht zu nahe kommen und ihre Abwehr gefährden.

38.2.4 Diagnose und Differentialdiagnose

38.2.4.1 Diagnostische Kriterien

Aufgrund der dargestellten Symptome läßt sich die Diagnose einer Anorexia nervosa im allgemeinen ohne Schwierigkeiten durch Anamnese und klinische Untersuchung stellen.

Zusammengefaßt wird die **Diagnose** aufgrund folgender Kriterien gestellt (DSM-III-R der American Psychiatric Association):

- **Weigerung,** das **Körpergewicht** über der unteren Normgrenze entsprechend dem Alter und der Größe zu halten: Abnahme auf ein Gewicht, das mindestens 15% unter dem zu erwartenden Wert liegt, oder – während der Wachstumsperiode – eine entsprechend geringe Gewichtszunahme.
- Intensive **Angst zuzunehmen** oder fett zu werden – sogar trotz Untergewichtes.
- **Störung der Wahrnehmung** von Gewicht, Maßen oder Gestalt des eigenen Körpers: die Betroffenen behaupten, sich „fett zu fühlen", selbst wenn sie abgemagert sind oder halten einen Körperbereich für „zu fett", selbst wenn auch dort das Untergewicht nicht zu übersehen ist.
- Bei Frauen: Ausfall von mindestens drei zu erwartenden Menstruationszyklen (primäre oder sekundäre **Amenorrhoe).** (Eine Amenorrhoe wird angenommen, wenn die Periodenblutung nur nach Hormongaben eintritt.)

Zur kritischen Diskussion dieses Klassifikationssystems vgl. Kapitel 15.

Die angegebenen Kriterien reichen für die klinische Arbeit aus. Für die Forschung, z.B. zur exakten Definition von in verschiedenen Kliniken behandelten Patientenpopulationen, sind sie nicht eng genug gefaßt. Hierdurch wird die klinische, aber auch die epidemiologische Forschung (vgl. dort) erschwert. Es gelang bisher nicht, ein befriedigendes diagnostisches Schema für Forschungszwecke zu entwickeln. Da sie vor allem zum Vergleich therapeutischer Ergebnisse immer wieder herangezogen werden, seien hier die Kriterien von Feighner und Mitarbeitern (1972) trotz der an ihnen geübten Kritik (z.B. Dally und Gomez, 1979; Hoppe, 1982; Halmi, 1983; Askevold, 1983) wiedergegeben:

A: Age of onset prior to 25.

B: Anorexia with accompanying weight loss of at least 25% of original body weight.

C: A distorted, implacable attitude towards eating, food, or weight that overrides hunger, admonitions, reassurance and threats;

e.g. – denial of illness with a failure to recognize nutritional needs,
– apparent enjoyment in losing weight with overt manifestation that food refusal is a pleasurable indulgence,
– a desired body image of extreme thinness with overt evidence that it is rewarding to the patient to achieve and maintain this state, and
– unusual hoarding or handling of food.

D: No known medical illness that could account for the anorexia and weight loss.

E: No other known psychiatric disorder with particular reference to primary affective disorders, schizophrenia, obsessive-compulsive and phobic neurosis. (The assumption is made that even though it may appear phobic or obsessional, food refusal alone is not sufficient to qualify for obsessive-compulsive or phobic disease.)

F: At least two of the following manifestations
– Amenorrhoe
– Lanugo
– Bradycardia (persistent resting pulse of 60 or less)
– Periods of overactivity
– Episodes of bulimia
– Vomiting (may be self-induced)

Für die Diagnose einer Anorexia nervosa sind die Punkte A–E zwingend, der Punkt F nur fakultativ notwendig.

Während der letzten Jahre wurden sowohl für die klinische Praxis als auch für Forschungszwecke vermehrt psychopathometrische Instrumente entwickelt: neben den Methoden zur Erfassung der Körperschemastörungen vor allem Selbst- und Fremdeinschätzungsskalen sowie erste Ansätze zur objektiven Messung einzelner Symptome (z.B. der körperlichen Hyperaktivität) (Übersicht bei Fichter und Meermann, 1981). Im deutschen Sprachraum findet vor allem ein von Fichter entwickeltes Anorexia-nervosa-Inventar zur Selbstbeurteilung (ANIS) sowie das für eine katamnestische Untersuchung vom selben Autor entwickelte teilstandardisierte Anorexie-Interview (mit in einem Manual festgelegten Definitionen und Einschätzungskonventionen (SIAN) Verwendung) (Fichter und Keeser, 1980; Fichter und Meermann, 1981).

Vor allem bei den Selbsteinschätzungsverfahren ist jedoch die Tendenz der Patienten zur Verleugnung ihrer Erkrankung kritisch zu berücksichtigen (Vandereycken und Vanderlinden, 1983).

38.2.4.2 Schwierigkeiten bei der Kontaktaufnahme und der Diagnosestellung

Die Aufnahme einer vertrauensvollen Beziehung beim ersten Kontakt zum Arzt ist für die Patienten meist besonders schwierig. Sie verleugnen ja den Krankheitswert ihres Zustandes und erleben ihr Verhalten als „normal". Gleichzeitig werden sie – zumindest bei zunehmender körperlicher Schwäche – mit dem Scheitern ihres bisherigen Lösungsversuches, ihres „Selbstheilungsversuches" konfrontiert. Hinzu kommen die durch die körperliche Auszehrung bedingten psychischen Folgen mit oft bizarren Verhaltensweisen, die sich aus der Zuspitzung des Konfliktes zwischen ihren elementaren Bedürfnissen und dem verzweifelten Wunsch nach Selbstkontrolle ergeben.

Aufgabe des ärztlichen Untersuchers ist es, die Patienten geduldig und ins einzelne gehend über das Wesen der Erkrankung zu informieren. Dabei erfahren die Patienten, daß der fachkompetente Arzt sie in ihren Grundproblemen, ihren Insuffizienzgefühlen, ihrem Mangel an Selbstwerterleben und ihren Entwicklungsproblemen versteht und sich von ihnen nicht im Rahmen ihrer Kompensations- und Abwehrbemühungen manipulieren läßt. Gegenüber solchen Manipulationsversuchen erweist es sich als günstig, die Patienten direkt mit ihren Problemen zu konfrontieren; die Konfrontation wirkt unmittelbar klärend, sie entlastet die Patienten, vermindert die mit solchen Manipulationsversuchen verbundenen Schuldgefühle und stärkt gleichzeitig das Vertrauen zum Arzt. Ein solches konfrontierendes Vorgehen empfiehlt sich auch gegenüber den häufigen Lügen der Patienten schon im Rahmen der Anamneseerhebung; in ihrer Not sind die Kranken „skrupellos unehrlich, sofern es um Essen, Gewicht etc. geht" (Fleck et al., 1965). Jedes verharmlosende Eingehen von seiten des Arztes auf solche Täuschungsversuche der Patienten kostet nicht nur unnötige Zeit, es bildet oft auch den Ausgangspunkt für eine spätere Eskalation der Auseinandersetzung im Verlauf der Therapie.

Ein entsprechendes, zugleich verständnisvolles und konsequentes Vorgehen empfiehlt sich auch für die möglichst frühzeitig anzusetzenden Gespräche mit den Familienangehörigen. Hier geht es vor allem um die Klärung derjenigen Kräfte, die einer Verselbständigung der Patienten entgegenstehen. Im Rahmen solcher Gespräche lassen sich auch ungenügende Angaben der Patienten zu ihren Symptomen klären.

Bei einer Gruppe der männlichen Patienten tritt die anorektische Symptomatik häufiger im Rahmen eines psychiatrischen Krankheitsbildes, zumeist aus dem schizophrenen Formenkreis, auf.

38.2.4.3 Differentialdiagnose

Somatische Erkrankungen

Die Abgrenzung der Anorexia nervosa gegenüber somatischen Erkrankungen mit Gewichtsverlust gelingt bei der typischen Anamnese im allgemeinen ohne Schwierigkeiten.

Es gibt keine endokrinologische Erkrankung – von Finalzuständen abgesehen –, die mit Kachexie einhergeht; insbesondere führen Erkrankungen der Hypophyse nicht zu einer Auszehrung in der beschriebenen Form, weder bei isoliertem noch bei komplettem Ausfall der Hypophysenhormone.

Die Symptomatik einer Thyreotoxikose oder einer primären Nebennierenrindeninsuffizienz läßt sich ebenfalls abgrenzen.

Rasche Gewichtsabnahme mit unstillbarem Erbrechen sollte an folgende Erkrankungen denken lassen: stenosierende Prozesse im Intestinaltrakt, Malabsorptionssyndrom, Nierenerkrankungen, zerebrale Prozesse.

Während der letzten Jahre wurde allerdings auch wiederholt über schwere körperliche Erkrankungen berichtet, bei denen die Fehldiagnose „Anorexia nervosa" gestellt worden war (Übersicht bei Hoppe, 1982). Bei verstorbenen Patientinnen wurden Hirntumoren, insbesondere Pinealome als Ursache des anorexieähnlichen Krankheitsbildes beschrieben (u.a. Hollatz und Ziolko, 1976; Heron und Johnston nach Fichter und Pirke, 1983, aber auch schon frühere Autoren wie Kalm und Magun, 1950 und Dally, 1969 nach Hoppe, 1982). Daneben fanden sich als Ursache für den Tod und wohl auch für das vorausgegangene Krankheitsbild u.a. eine chronisch rezidivierende Enteritis im proximalen Ileum mit starker narbiger Lumenverengung und prästenotischer Ileumerweiterung, eine nicht entdeckte Lungentuberkulose mit Dünndarmbefall, erheblicher Stenose im proximalen Ileum mit Ulzerationen und einer Perforation, die zu einer Peritonitis führte (Hollatz und Ziolko, 1976), ein faustgroßes Karzinom im Bereich des Magenausgangs mit Lebermetastasen. Eine Patientin mit einem Ulcus ventriculi verstarb nach Besserung dieses Befundes unerwartet an einer Mesenterialvenenthrombose unklarer Genese (Fichter und Pirke, 1983).

Bei einer von 37 Patientinnen ergab sich uns ein auch mit den Möglichkeiten einer Universitätsklinik nicht lösbares somatisches Problem: eine massive Blutsenkungsbeschleunigung; diese Patientin wurde nach Abschluß der stationären Behandlung mit ausreichender Gewichtszunahme entlassen, sie starb überraschend drei Monate später, ohne daß die Ursache hätte geklärt werden können.

Somatische Folgeerscheinungen der Anorexie mit Krankheitswert

Mangelernährung, selbstinduziertes Erbrechen, Gewichtsabnahme bis zur Kachexie, Abusus von Laxantien und Diuretika führen zu körperlichen Folgeerscheinungen, denen zum Teil Krankheitswert, gelegentlich lebensbedrohlicher Charakter zukommt (Übersichten: Fichter und Pirke, 1983; Silverman, 1983). Oft findet sich eine „vita minima" (Hypothermie, Bradykardie, Hypotonie). Bei der körperlichen Untersuchung fallen fast immer Hautveränderungen auf: Die Haut erscheint trotz der Reinlichkeit der Patientinnen dunkel und schmutzig, trocken und rauh, an feines Sandpapier erinnernd, schuppig mit flaumartiger Lanugobehaarung an Wangen, Vorderarmen und am Rücken. Bei extrem unterernährten Patienten finden sich Petechien und Ekchymosen (83 von 100 Patienten Silvermans zeigten solche Hautveränderungen).

Gleichzeitig fällt die ausgeprägte Kachexie (72% der Patienten Silvermans) auf; für die Beurteilung des Untergewichts erwies sich der klinische Eindruck bedeutsamer als die objektive Gewichtsangabe.

In neuroradiologischen Untersuchungen werden Veränderungen am Gehirn nachgewiesen: im Computertomogramm stellen sich erweiterte Windungsfurchen und vergrößerte Ventrikel dar. Auch diese Veränderungen scheinen sich mit Gewichtszunahme weitgehend zurückzubilden. Veränderungen der regionalen Hirndurchblutung ließen sich weder im Verlauf, noch gegenüber Kontrollpersonen nachweisen (Krieg et al., 1989).

Es kann also weder von einer „Hirnatrophie" im eigentlichen Sinn, noch von einer durchblutungsbedingten Funktionseinschränkung des Gehirns gesprochen werden; dennoch zeigen diese Befunde, wie weit die Folgen des Hungers reichen können. Schon aus Vorsicht – und im Bewußtsein unseres beschränkten Wissensstandes – sollten sie nicht zu gering bewertet werden.

Eine Hypothermie kann ausgeprägt sein (80%). Infektionen führen bei Magersuchtpatienten oft nicht wie sonst zu einer entsprechenden Temperaturerhöhung. Bradykardie (Frequenz unter 60 pro Minute bei 77%) wird oft von einer Bradypnoe (Frequenz 14 pro Minute oder weniger) begleitet

(61%). Entsprechend häufig findet sich eine Hypotension (systolischer Blutdruck 70 mm Hg oder niedriger: 63% der Patienten). Überraschend selten ist die in ihrer Pathogenese immer noch ungeklärte Ödemneigung (21%). Im Elektrokardiogramm bestehen bei der Mehrzahl der Patienten Abweichungen vom normalen Stromkurvenverlauf; in der Hauptsache handelt es sich um Sinusbradykardien, extreme Niedervoltage und niedrige oder negative T-Wellen. Entsprechend der Mangelernährung findet sich im Serum ein Vitamin-A-Mangel (62%), daneben eine Hyperkarotinämie (51%).

Diese Befunde sind sämtlich mit Gewichtszunahme reversibel und mit Ausnahme des letztgenannten auch in der sog. „Starvation-Literatur" (Hungerexperimente Freiwilliger während des 2. Weltkrieges, Fasten Übergewichtiger mit der sog. „Null-Diät"; Übersichten: Halmi, 1978; Ditschuneit et al., 1979) bekannt. Dies gilt ebenso für pathologische Befunde des Nüchternblutzuckers, des Blutzuckerverlaufs nach oraler Glukosebelastung, für erhöhte Serumwerte des Kreatinins, Harnstoffs, der Transaminasen und des Gesamtbilirubins sowie der Amylase.

Die Knochenmarkspunktion zeigt eine typische Form der Hypoplasie (Kubanek et al., 1977) bei 49 von 73 (67%) der von Silverman untersuchten Patienten. Sie führt nicht selten zu einer Leukopenie, gelegentlich zu einer ausgeprägten Thrombozytopenie mit generalisierter Blutungsneigung, dagegen nur selten zu einer Anämie.

Die Veränderungen im hormonalen System besprechen wir im Abschnitt „Pathogenetische Konzepte".

Trotz der ausgeprägten Mangelernährung findet sich in der Regel keine Erniedrigung des Serumalbumins und keine Erniedrigung oder Verschiebung der Zusammensetzung der Serumglobuline. Auch Elektrolytverschiebungen sind als Folgen der Mangelernährung selten; als Ergebnis eines chronischen selbstinduzierten Erbrechens finden sich erniedrigte Kalium- und Chlorwerte zusammen mit einer Alkalose; der Natriumverlust infolge von Erbrechen und Diarrhoe führt zu einem sekundären Hyperaldosteronismus mit Kaliumverlust durch die Niere. Mehrere der von Fichter und Pirke untersuchten Patienten entwickelten als Folge von Erbrechen und Laxantienabusus eine schwere Niereninsuffizienz. Kaufmann (1973) beschrieb eine dem Bartter-Syndrom ähnliche Symptomatologie nach Laxantien- und/ oder Diuretikaabusus bei Anorexiepatienten: schwere hypotone Kreislaufregulationsstörung mit gesteigerter Aktivität des Renin-Angiotensin-Aldosteron-Systems und kaliopenischer metabolischer Alkalose und gleichzeitiger Angiotensin-Resistenz der arteriellen Gefäße („Pseudo-Bartter-Syndrom").

Als **lebensbedrohlich** sind zu beachten (Silverman, 1983):
– akut rasch fortschreitender Gewichtsverlust (Gewichtskontrollen während psychotherapeutischer Behandlung!),
– Abnahme der vitalen Funktionen (insbesondere Blutdruck und Pulsfrequenz), sowie
– Zeichen eines Kreislaufschocks.

Diese Veränderungen treten dabei häufig in Kombination auf. Eine hypokaliämische hypochlorämische Alkalose wird nur in Verbindung mit dieser Entwicklung zum Problem. Infektionen verlaufen eher selten letal, können jedoch wegen der fehlenden Fieberentwicklung gelegentlich übersehen werden (Silverman, 1983).

Deter und Mitarbeiter (1983) haben die Todesursachen von 12 Patienten (aus einer Gruppe von 103 auf einer klinisch-psychosomatischen Station Behandelten) zusammengestellt. Bei allen Patienten spielte die Chronifizierung der Erkrankung (im Durchschnitt 8,4 Jahre) eine wesentliche Rolle. 3 Patienten verstarben während der ersten zwei Tage in der Folge aktiver Suizidhandlungen (Alkoholintoxikation, Tablettenintoxikation, Manipulation am Venenkatheter zur Verhinderung intravenöser Ernährung). Bei 6 Patienten lag das Gewicht unter 50%, bei 2 unter 60% des Sollgewichts. Bei 5 Patienten bestand eine schwerwiegende Nierenfunktionsstörung. Im einzelnen traten als Komplikationen auf: protrahierter Kreislaufschock (3 Patienten), Pneumonie (2 Patienten), paralytischer Ileus mit anschließendem Schockzustand (1 Patient), Darmperforation (2 Patienten), Sepsis (1 Patient), Thrombose und Lungenembolie (2 Patienten). Hieraus ergibt sich die Forderung der Autoren nach konsequenter Berücksichtigung der vitalen Parameter, nach der Untersuchung von Infektionen und zusätzlichen Erkrankungen (Magen-Darm-Bereich!) sowie eine kritische Indikationsstellung beim Legen von Venenkathetern und der Gabe von Infusionslösungen (hochprozentige Glukose, Lipide!).

Artefakte („factitious disease")

Bei ungewöhnlichen Symptomen, wie etwa einer ausgeprägten Anämie, muß auch an die Tendenz der Patientinnen zur Selbstschädigung gedacht und diese Möglichkeit mit aller Konsequenz abgeklärt werden.

Andere psychosomatische Krankheitsbilder mit Störung des Eßverhaltens

Neurotisch bedingtes Erbrechen: Es stellt häufig ein Konversionssyndrom dar.

„Anorektische Reaktion" als vorübergehende Reaktion während besonderer Belastungssituationen vom Vollbild der Anorexia nervosa zu unterscheiden.

Bulimie (Bulimia nervosa): (vgl. Kap. 38.3).

Andere psychopathologische Prozesse

Gelegentlich kann die Abgrenzung gegenüber schizophrenen Zustandsbildern und depressiven Erkrankungen bzw. die Gewichtung gleichzeitig bestehender psychopathologischer Krankheitsbilder Schwierigkeiten bereiten.

Untergruppen der Anorexia nervosa

Die klinische Erfahrung zeigt, daß sich Anorexiepatientinnen trotz einer relativ großen Homogenität der Symptomatik hinsichtlich der Schwere des Zustandsbildes, der pathologischen Abläufe in Psychodynamik und Familieninteraktion und der Zugänglichkeit für therapeutische Maßnahmen zum Teil erheblich unterscheiden. Aufgrund dieser klinischen Erfahrung wurden Vorschläge für die Differenzierung des Krankheitsbildes in Untergruppen gemacht. Besonders bewährt hat sich die Unterscheidung von „asketischen" gegenüber „impulsgetriebenen" Anorektikerinnen (A. E. Meyer, 1984b); diese Differenzierung erscheint aus klinischer Sicht sowohl für die Spontanprognose als auch für die Behandelbarkeit des Leidens von Bedeutung. Hilde Bruch (1973) unterschied

von der Kerngruppe eine Gruppe von Anorektikerinnen, deren Psychopathologie stärker derjenigen von „Borderline-Patienten" bzw. von Patienten mit psychotischen Zustandsbildern entspricht.

Andere Autoren gehen von mehreren Untergruppen, zum Teil von verschiedenen Krankheitsentitäten oder Syndromklassen aus (Übersicht bei Hoppe, 1982). So unterscheidet Dally (1969) mehrere „Anorexia-nervosa-like states" und folgende Anorexia-nervosa-Gruppierungen: „anxiety-depression, anxiety-hysteria, anorexia tardiva and phobic anxiety".

Die systematische Klärung der Bedeutung von Untergruppen dürfte von der Einführung methodisch komplexer Untersuchungskonzepte zu erwarten sein, die in der Lage sind, die verschiedenen Konstellationen der Befunde auf und zwischen den Betrachtungsebenen nach Vorkommen und Wertigkeit zu erfassen (z. B. nach dem von A. E. Meyer (1984) vorgestellten Taxonomie-Modell).

38.2.4.4 Früherkennung

Eine frühzeitige Diagnosestellung bei Anorexia-nervosa-Kranken könnte wahrscheinlich erheblich zur Verbesserung der Prognose bei diesen Patienten beitragen, da die verschiedenen sich gegenseitig verstärkenden Krankheitsprozesse (vgl. Abschn. 38.2.6.6) noch nicht angelaufen sind (Fries, 1974).

Garner und Mitarbeiter (1983) fanden in einer gesunden, jedoch mit Gewichtsproblemen präokkupierten Gruppe von College- und Ballett-Studentinnen in psychologischer Hinsicht zwei Untergruppen: Eine Untergruppe bestand aus psychisch unauffälligen, diäteinhaltenden Studentinnen, in der anderen Gruppe fanden sich ausgeprägte psychopathologische Züge, die denen von Anorexia-nervosa-Patientinnen sehr ähnlich waren.

Weeda-Mannak und Mitarbeiter (1983) fanden einige gemeinsame psychologische Charakteristika bei Patientinnen mit funktioneller sekundärer Amenorrhoe und Anorexiepatientinnen, insbesondere ein auffallendes Leistungsstreben, die sich in einer Kontrollgruppe Gesunder nicht in diesem Ausmaß nachweisen ließen; andererseits fanden diese Autoren einen „fundamentalen" Unterschied zwischen den beiden Gruppen hinsichtlich der Befürchtung, in der Leistungsdimension zu scheitern.

In der klinischen Praxis läßt sich erwarten, daß sich unter Kindern und Jugendlichen mit „Appetitstörungen" (mit Gewichtsverlust und Obstipation), bei Jugendlichen mit übertrieben auf Gewichtsprobleme gerichteter Aufmerksamkeit und jüngeren Patientinnen mit primärer und sekundärer Amenorrhoe immer wieder Kranke mit Anorexia nervosa finden werden.

38.2.5 Epidemiologie

Inzidenz und Prävalenz und deren Entwicklung über längere Zeiträume sind für das Krankheitsbild Anorexia nervosa noch unzureichend untersucht.

Die Untersuchung dieser Fragestellung wird durch methodische Probleme erschwert: Es gibt bisher keine international durchgängig akzeptierten diagnostischen Kriterien; die allgemein gebräuchlichen Diagnosekriterien eignen sich nicht ausreichend zur Erfassung leichterer Fälle; die Krankheits-verleugnung führt dazu, daß Einrichtungen des medizinischen Versorgungssystems erst in fortgeschrittenem Krankheitsstadium aufgesucht werden – ein großer Teil der Patienten, insbesondere leichtere Fälle werden so nicht erfaßt (Fichter, 1984). Vor allem fehlen weitgehend systematische epidemiologische Studien ganzer Bevölkerungsgruppen oder repräsentativer Stichproben (Übersichten: Hill, 1976, 1977; Crisp, 1977; Garner et al., 1983a; Schwartz et al., 1983; Fichter et al., 1983; Fichter, 1984; Lucas et al., 1983).

38.2.5.1 Inzidenz und Prävalenz

Untersuchungen zur Behandlungs- bzw. administrativen Inzidenz zeigen, daß die Anorexia nervosa in allen Ländern der westlichen Zivilisation vorkommt. Die jährliche Erkrankungsinzidenz beträgt nach diesen Untersuchungen zwischen 0,1 und 0,6 pro 100000 Einwohner bzw. zwischen 50 und 75 Patienten pro 100000 Personen der Risikopopulation (Frauen im Alter von 15 bis 25 Jahren).

Neuere epidemiologische Feldstudien (Crisp et al., 1976; Nylander, 1971; Szmukler, 1983) sprechen dafür, daß während der Adoleszenz, alters- und sozialschichtabhängig, ein Mädchen unter 100 bis 250 an Anorexia nervosa erkrankt. Legt man konservative Behandlungsergebnisse zugrunde – ein Drittel bis die Hälfte der Anorexia-nervosa-Erkrankungen heilen nicht aus – heißt dies, daß eine von 450 bis 750 erwachsenen Frauen lebenslang unter der Symptomatik einer Anorexia nervosa leiden dürfte (Schwartz et al., 1983).

Zusammengefaßt kann man nach Crisp (1977) grob geschätzt mit einer Häufigkeit der Anorexia nervosa von 1% der Frauen und 0,1% der Männer während der Adoleszenz rechnen. Tabelle 38.2–1 gibt eine Übersicht über die vorhandenen Untersuchungen.

38.2.5.2 Häufigkeitsentwicklung

Im allgemeinen wird über eine Zunahme der Behandlungsinzidenz während der letzten Jahrzehnte berichtet (Halmi, 1974). Hierfür sprechen u.a. die Befunde von Theander (1970): 1,1 Aufnahmen pro 100000 Bevölkerung zu Beginn der 30er und 5,8 Ende der 50er Jahre. Eine ähnliche Zunahme von Anorexiepatienten, die stationär behandelt wurden, ergibt sich aus den psychiatrischen Registeruntersuchungen von Kendall (1973) und Jones (1980) in Monroe County, New York, USA sowie von Willi und Grossmann (1983) im Kanton Zürich (vgl. Tab. 38.2–1).

Allerdings kann aufgrund der vorliegenden Untersuchungen noch nicht endgültig beurteilt werden, ob das Krankheitsbild der Anorexia nervosa – wenigstens in schwerer stationär behandlungsbedürftiger Form – tatsächlich zugenommen hat, oder ob lediglich die Diagnose häufiger gestellt wird und die medizinischen Versorgungseinrichtungen auch für ärmere Bevölkerungsschichten leichter zugänglich sind. Die Anorexia nervosa tritt fast ausschließlich während der Adoleszenz auf. Das Manifestationsalter verteilt sich bimodal mit zwei Häufigkeitsgipfeln für das Alter von 14 und 18 Jahren (Halmi, 1974); die meisten Patienten erkranken zwischen dem 13. und 25. Lebensjahr. Allerdings sind auch Erkrankungen vor dem 10. Lebensjahr (bis zu 8%) und – bei manchem Untersucher nicht ganz selten (u.a. Halmi, 1974) – auch nach dem 25. Lebensjahr beschrieben.

38.2.5.3 Geschlechtsverteilung

Die Anorexie ist überwiegend eine Erkrankung junger Mädchen. Die Häufigkeit der Erkrankung bei Männern wird unterschiedlich angegeben, im Verhältnis von 5:2 (Registeruntersuchung in Monroe County) bis 15:1 (Registeruntersuchung in Schottland; Crisp und Toms, 1972), bis 20 oder auch 30:1 in der übrigen Literatur (Crisp, 1977).

Bruch (1971), Beumont und Mitarbeiter (1972) und Crisp und Toms (1972) haben insgesamt 48 männliche Anorexiepatienten beschrieben. Auffallend ist, daß sich der Verteilungsgipfel des Manifestationsalters von dem bei den Patientinnen deutlich unterscheidet: zwölftes bei männlichen gegenüber siebzehntes bis achtzehntes Lebensjahr bei weiblichen Patienten.

In unserem eigenen Patientengut stehen 4 männlichen Patienten 45 Patientinnen gegenüber.

Während sich in Hsus (1980) Übersicht über die prognostischen Studien zur Anorexia nervosa nur 23 männliche Patienten unter insgesamt 778 Patienten (2,9%) finden, machten in der von Crisp geleiteten „Professorial Psychiatric Unit" im St. George's Hospital London die 36 männlichen Patienten 9% der 423 Patienten insgesamt aus. In diesem Patientengut fand sich für die Anorexia nervosa keine frühere Manifestation im Vergleich zu weiblichen Patienten (durchschnittliches Auftreten im Alter von 17 Jahren 2 Monaten) (Crisp und Burns, 1983).

38.2.5.4 Kulturelle und soziale Faktoren

Anorexia nervosa scheint in den Ländern der westlichen Zivilisation weit verbreitet zu sein. Untersuchungen liegen für Westeuropa, die USA, Kanada, Israel, Südafrika und Japan vor (Übersicht: Fichter, 1984). In Ländern der 3. Welt, wie in Nigeria und Malaysia, scheint sie selten zu sein (nach Fichter, 1984). In Ländern mit allgemeinem Nahrungsmangel finden sich zahlreiche Personen in der Gewichtsgruppe von „Magersüchtigen" (z.B. in Indien nach Pflanz, 1965), ohne daß diesem Zustand dort Krankheitswert zukommt.

Die Lebensbedingungen einer Konsumgesellschaft bzw. Überflußkultur könnten eine der notwendigen Bedingungen der Möglichkeit zur Erkrankung an Anorexia nervosa darstellen.

Von Bedeutung dürften **kulturell bestimmte Idealvorstellungen** vom Aussehen des weiblichen Körpers und – zum Teil damit verbunden – kulturelle Einflüsse auf die Pubertätsentwicklung der Frau sein.

So betrachten sich 50% der von Huenemann (1966) an High Schools in den USA untersuchten Mädchen als übergewichtig, obwohl nach objektiven Kriterien nur 25% übergewichtig waren. Ebenfalls 50% der von Nylander (1971) untersuchten 18jährigen Mädchen gaben an, sich zu irgendeinem Zeitpunkt als übergewichtig erlebt zu haben, gegen-

Tabelle 38.2–1. Untersuchungen zur Häufigkeit und zur Häufigkeitsentwicklung bei Anorexia nervosa

Autor	Inzidenz pro 100 000 Einwohner	Inzidenz in der Risikopopulation	Methodik
Theander, 1970 Schweden	0,24 (1931–1960) 0,45 (1951–1960)		Alle Patientinnen, die von 1931 bis 1960 in psychiatrische und allgemeine Krankenhäuser wegen Anorexia nervosa aufgenommen wurden. Gewichtsverlust mehr als 25% des Ausgangsgewichts.
		pro 100 000 Frauen, 15–34 J.	
Kendall, 1973 England	1,6 Nord-Ost-Schottland	10,0	Psychiatrische Patientenregister ausgewählter Gebiete (6 bis 9 Jahrgänge zwischen 1960 und 1971).
	0,37 Monroe County, NY/USA	0,8	
	0,6 Camberwell (District von London)	4,1	
		pro 100 000 Frauen, 15–24 J.	
Jones et al., 1980 USA	0,37 Monroe County, NY/USA (1960–1969)	0,55	Patientenregister von Monroe County (wie Kendall, 1973).
	0,64 (1970–1976)	3,26	
		pro 100 000 Frauen, 12–25 J.	
Willi und Grossmann, 1983 Schweiz	0,38 (1956–1958) 0,55 (1963–1965) 1,12 (1973–1975)	3,98 6,79 16,76	Behandlungsinzidenz für den Kanton Zürich.
Morgan und Silvester, 1976 England		2% aller 18jährigen Studentinnen	Fragebogenuntersuchung von 728 18jährigen Erstsemestern in Bristol. Fragestellung: vorausgegangene Episoden einer anorektischen Erkrankung.
Nylander, 1971 Skandinavien		14- bis 19jährige Minimum 0,65%, je nach Kriterien bis zu 2,7%	Repräsentative Stichprobe von 1241 Mädchen. Berücksichtigt wurden: Eßverhalten, Amenorrhoe, Gewichtsverlust, vorliegende psychiatrische Diagnose.
Crisp et al., 1976 England	Prävalenzrate 0,46% (ca. 1 von 200)	Mädchen an Privatschulen unter 16 J. 0,17%, über 16 J. 0,5% Staatliche Schulen 0,18% (1 von 550)	Feldstudien an neun öffentlichen bzw. privaten Schulen. Fallidentifikation über Schlüsselpersonen wie Lehrer und Mitarbeiter des schulärztlichen Dienstes.
Szmukler, 1983a England	0,83% (1 von 120)	Mädchen im Alter zwischen 14 und 18 J.	2stufige Feldstudie (screening) mit Fragebogen bei 1331 Schülerinnen von „day-schools" und 220 Schülerinnen an „boarding schools" in England mit Fragebogen zum Eßverhalten und zu anorektischer Symptomatik. Interview mit 143 so identifizierten Mädchen.
Fichter et al., 1983 BRD, Griechenland	0,35% bis 0,42% 1,10%	Griechische Schülerinnen in Griechenland Griechische Schülerinnen in München	Epidemiologische Feldstudie mit repräsentativen Stichproben (Messung von Körpergewicht und Größe, Menstruationsanamnese, Anorexianervosa-Inventar zur Selbstbeurteilung mit Zusatzfragen, Interview von so als vermutliche Patientinnen erfaßten Schülerinnen). Noch unklar, inwieweit Unterschied real vorhanden oder methodisch bedingt ist.

über nur 7% bei den 18jährigen Männern. 40% dieser Mädchen hatten zu irgendeinem Zeitpunkt eine Reduktionsdiät eingehalten; diese Reduktionsversuche fanden sich zwar häufiger bei übergewichtigen, jedoch auch bei normgewichtigen Mädchen. Unter den 18jährigen Mädchen war es besonders ungewöhnlich, sich als zu dünn einzuschätzen

Dabei erscheint die Motivation von Mädchen, die trotz der Warnzeichen von Amenorrhoe und Kachexie weiterfasten, deutlich „krankhafter" als die Motivation derjenigen, die lediglich aus „kosmetischen Gründen" fasten (Fries, 1974).

In den westlichen Kulturen nahm die Bedeutung der Schlankheit für das weibliche Schönheitsideal erheblich zu, wie neuerdings auch empirische Untersuchungen (Maße und Gewichtsangaben der Posterbeilagen im „Playboy"; Gewichtsangaben der Siegerinnen und Teilnehmerinnen bei den Miss-Amerika-Wahlen; Garner et al., 1983a) belegen; gleichzeitig findet sich eine zunehmend negative Einstellung zu Übergewichtigen bereits im Kindesalter (Wooley und Wooley, 1980). Parallel dazu stieg das allgemeine Interesse an Diätmaßnahmen (Untersuchungen von 20 Jahrgängen der sechs größten amerikanischen Frauenzeitschriften; nach Schwartz et al., 1983). Die durchschnittliche Gewichtsentwicklung steht allerdings in deutlicher Diskrepanz zur Entwicklung der Idealnorm. Das Durchschnittsgewicht erwachsener Frauen unter 30 Jahren hat während der untersuchten 20-Jahres-Periode, vermutlich in Zusammenhang mit dem verbesserten Nahrungsangebot, zugenommen (Garner et al., 1983a).

Es wird also ein Widerspruch zwischen einem Teil der gesellschaftlichen Anforderungen (Konsum) und möglicherweise biologischen Bedürfnissen einerseits und kulturspezifischen Erwartungen (Schlankheitsideal) andererseits deutlich. Möglicherweise verschärft sich dieser Konflikt bei selbstunsicheren, besonders akzeptationsbedürftigen Mädchen im Rahmen der gesamten Verunsicherung während der Pubertät krisenhaft. Auch Veränderungen des Wertsystems unserer Kultur mit Folgen für den Stellenwert von Berufstätigkeit, sozialem Status, Sexualität und Partnerschaft werden in diesem Zusammenhang diskutiert. Äußere Normen in ihrer auch haltgebenden Funktion haben an Bedeutung verloren; gleichzeitig wurden damit die Möglichkeiten für eine freie persönliche Entscheidung verbessert. Für innerlich unsichere, abhängige Personen können sich aus dieser Entwicklung jedoch zusätzliche Belastungen ergeben, wie sie Bruch (1978) für viele ihrer Patienten beschreibt. Diese verspüren einen starken Anforderungsdruck, die vielfältigen heutigen Möglichkeiten auch tatsächlich zu ergreifen und befürchten gleichzeitig, ihre jeweils getroffene Wahl könnte falsch sein. Die Abnahme von Außenleitung kann bei unsicheren Personen auch zu einem überstarken Bedürfnis nach eigener Kontrolle führen (nach Garner et al., 1983a).

Das Ergebnis (Gewichtsverlust, Schlankheit) und die damit verbundene soziale Anerkennung kann dieses Verhalten verstärken. Diskutiert wird auch, inwieweit das ausgeprägte weibliche Schlankheitsideal in Zusammenhang mit der generellen Emanzipation der Frau steht. Zentrale Ausdrucksform der neuen, emanzipierten Frau sei ihr schlanker Körper, der Athletentum, nichtreproduktive Sexualität und eine Art androgyner Unabhängigkeit symbolisiere (Bennett und Gurin, 1982).

Die Anorexia nervosa findet sich zwar in allen **Sozialschichten,** sie kommt jedoch häufiger in den sozioökonomisch höheren Schichten vor (Crisp, 1977; Sperling und Massing, 1972).

Crisp (1976) fand unter den 16- bis 18jährigen Mädchen auf englischen Privatschulen 1% Anorektikerinnen, dagegen nur eine einzige Patientin unter 550 Mädchen auf öffentlichen Schulen in der Grundausbildung.

38.2.5.5 Familienuntersuchungen und Zwillingsforschung

Das Auftreten der Erkrankung bei mehreren Kindern einer Familie wurde immer wieder beschrieben und findet sich auch in unserem Patientengut; nicht selten erkrankt ein Geschwister nach erfolgreicher Behandlung des ursprünglichen Patienten; leider liegen hierüber jedoch noch kaum systematische Untersuchungen vor. Theander (1970) berechnete ein Erkrankungsrisiko für Schwestern einer Patientin von 6,6%, was weit über dem oben angegebenen durchschnittlichen Krankheitsrisiko liegt. Besonders häufig scheint die Erkrankung von Geschwistern männlicher Anorexiepatienten zu sein (Crisp und Toms, 1972).

In den Familien von Anorektikerinnen werden bestimmte psychopathologische Befunde gehäuft beschrieben (Übersicht: Rakoff, 1983); deutlich erhöht ist für Väter von Anorektikerinnen die Erkrankungsrate an Alkoholismus (12–19%) und – besonders auffallend – die Erkrankungsrate der Mütter an Migräne (30% bei Kalucy, 1977, nach Rakoff, 1983). Schon diese Häufung psychischer bzw. psychosomatischer Erkrankung vermag einen Hinweis auf die hierdurch bedingten Belastungen der familiären Atmosphäre zu geben.

Inwieweit genetische Einflüsse bei der Anorexia nervosa eine Rolle spielen, ist noch nicht endgültig geklärt. Früher publizierte Sammelstatistiken über insgesamt 10 eineiige Zwillingspaare mit einem erkrankten Zwilling und 6 eineiige Zwillingspaare, bei denen beide an Anorexie erkrankt waren (Hill, 1976), sind aus methodischen Gründen anfechtbar (Schepank, 1982). Schepank (1982) hat 6 Zwillingspaare selbst untersucht und Informationen über weitere sieben Paare über Kontakte mit den Untersuchern überprüft. Für diese 13 Paare ergibt sich folgende Verteilung (Tab. 38.2–2):

Tabelle 38.2–2. n = 13 Zwillingspaare (nach Schepank, 1982)

	konkordant AN	diskordant AN
Eineiige Zwillinge	6	2
Zweieiige Zwillinge	0	5

Da keine Vorauslese nach Konkordanz oder Diskordanz bei diesen 13 Zwillingspaaren stattgefunden hat, interpretiert Schepank seine Befunde folgendermaßen: „Die Verteilung der Konkordanz/Diskordanz-Raten auf Eineiige und Zweieiige legt den dringenden Verdacht nahe, daß eine **erbliche Komponente** bei diesem Krankheitsbild angenommen werden muß".

Die weitere Absicherung dieser Ergebnisse ist schwierig, da zwei seltene Merkmale (eineiige bzw. zweieiige Zwillinge und Anorexia nervosa) in Kombination zusammentreffen müssen.

Auch Untersuchungen auf chromosomaler Ebene erscheinen bedeutsam: So fand Weiner bei 5 von 13 Anorektikerinnen chromosomale Besonderheiten (1976).

Auf den starken Einfluß **familiärer Interaktionsprozesse** weisen Crisp und Toms (1972) hin; sie beschreiben eine Familie mit einem männlichen Anorektiker, in der eine Adoptivtochter und ein Mädchen, das in der Familie zu Gast war, ebenfalls eine Anorexie entwickelten.

Einzelbeobachtungen von Zwillingspaaren können im besonderen Maße auch zum Verständnis der psychodynamischen Mechanismen einer Krankheitsmanifestation beitragen. Dührssen, Bruch, Crisp, J. E. Meyer haben solche Paare beschrieben (Becker, 1982).

38.2.6 Pathogenetische Konzepte

Es gibt heute noch kein geschlossenes pathogenetisches Verständniskonzept für die Anorexia nervosa. Die vorliegenden Konzepte beschäftigen sich mit der Bedeutung und zum Teil dem Zustandekommen von für die Krankheit prädisponierenden Faktoren. Wir stellen hier die Befunde zur Psychodynamik wegen ihrer Bedeutung für das Verständnis der in den Patienten ablaufenden Prozesse und wegen ihrer therapeutischen Konsequenzen besonders ausführlich dar. Bezüglich der Befunde zur Familiendynamik verweisen wir aus Raumgründen auf das Kapitel 21.

Die folgende Darstellung bezieht sich auf pathogenetische Überlegungen zur Anorexia nervosa bei Mädchen. Auf die Besonderheiten der Krankheitsentwicklung bei männlichen Patienten können wir nur anmerkungsweise hinweisen. Ausführlichere Angaben finden sich bei H. Bruch (1973) und Beumont und Mitarbeitern (1972) (Crisp und Burns, 1983).

38.2.6.1 Psychophysiologie

Bei Patienten mit Anorexia nervosa findet sich eine Vielzahl pathophysiologischer Veränderungen. Die Mehrzahl hiervon kann auf die Mangel- und Fehlernährung und ihre Folgen, vor allem den Gewichtsverlust zurückgeführt werden; andere pathophysiologische Befunde wie die Hypothermie und Veränderungen im Funktionskreis Hypothalamus-Hypophyse-Gonaden scheinen nur zum Teil mit diesen das

Krankheitsbild phänomenologisch bestimmenden Symptomen in Verbindung zu stehen (Weiner, 1982, 1983; Brown, 1983; Weiner und Katz, 1983; vgl. auch Kap. 10). Viele dieser Befunde, insbesondere die Anomalien im Bereich der hormonalen Steuerung, machen auf Ausmaß und Reichweite der somatischen Veränderungen bei Patienten mit Anorexia nervosa aufmerksam.

Für eine systematische Betrachtung wäre es natürlich wichtig herauszufinden, ob und inwieweit es sich bei diesen Anomalien um „Abwärtseffekte" (vgl. Kap. 1) einer im psychosozialen Bereich beginnenden Störung handelt (z.B. Amenorrhoe vor Gewichtsreduzierung) und ob umgekehrt auch „Aufwärtseffekte" eine Rolle spielen. Eine derart weitreichende Interpretation der heute vorliegenden Befunde erscheint jedoch verfrüht. Zunächst zeichnet sich erst einmal die Aufgabe ab, die wahrscheinlich existierenden Untergruppen des Krankheitsbildes näher zu charakterisieren und zu prüfen, ob sich hierdurch nicht ein Teil der heute noch existierenden Widersprüche und Unklarheiten im Bereich der Pathophysiologie bereinigen läßt (Weiner, 1982).

Neben Schlafstörungen und der Anomalie der Thermoregulation werden vor allem Besonderheiten in den Funktionskreisen Hypothalamus-Hypophyse-Gonaden, Hypothalamus-Hypophyse-Schilddrüse und Hypothalamus-Hypophyse-Nebennierenrinde beschrieben.

Im Bereich der **Schilddrüse** kommt es zu einem krankheitsunspezifischen Abfall der T_3-Spiegel, die hypothalamische Regulation der Schilddrüsenfunktion ist ungestört (Weiner, 1982; Fichter und Pirke, 1983). Möglicherweise läßt sich die Verminderung der T_3-Produktion als ein Selbstschutz im Zusammenhang mit der Aufgabe interpretieren, bei Mangelernährung den Energieverbrauch einzuschränken. Erniedrigter Grundumsatz, Bradykardie, trockene Haut und (zum Teil) die Obstipation spiegeln diese hypothyreote Stoffwechsellage wider, die sich auch bei fehlernährten und hungernden Patienten (Null-Diät, postoperativ) und einer Reihe chronischer Krankheiten findet (Fichter und Pirke, 1983; Weiner, 1982). Die niedrige T_3-Konzentration ist mit verantwortlich für Serumkonzentrationsveränderungen von Nebennierenrindenhormonen (Weiner, 1982).

Funktionskreis Hypothalamus-Hypophyse-Nebenniere

Bei Patienten mit Anorexia nervosa ist dieser Funktionskreis insgesamt aktiviert, die Nebennieren sind dabei durch ACTH in ungewöhnlichem Maß stimulierbar. Die hierfür ursächlichen Faktoren sind heute nicht bekannt. Die mittleren Plasmacortisolspiegel sind erhöht, während paradoxerweise die Urinausscheidung von 17-OHCS entweder normal oder erniedrigt ist. Für diesen paradox erscheinenden Befund gibt es folgende Erklärungsansätze: (1) langsamere Metabolisierung (Folge der niedrigen T_3-Werte) und Clearance des Cortisols; (2) exzessive Produktion von Cortisol vor allem als Folge der Aktivierung des Hypothalamus-Hypophysen-Nebennieren-Systems mit Störung der Tagesrhythmik und fehlender Supprimierbarkeit von Cortison durch Dexamethason (Weiner, 1983c). Für (2) bestehen wiederum folgende Erklärungsmöglichkeiten: eine ungewöhnliche Menge von Corticotropin-Releasing-Hormon oder eine ungewöhnliche Sensibilität der Hypophyse für normale Releasing-Hormonspiegel. Jede diese Veränderungen würde zu einer vermehrten ACTH-Produktion und -Sekretion führen. Andererseits reagieren die Nebennierenrindenzellen dieser Patienten exzessiv auf normale ACTH-Dosen. Wir wissen bis heute nicht, warum sich diese Veränderungen bereits bei leichter Gewichtszunahme von Anorexia-nervosa-Patienten zurückbilden (Weiner, 1983).

Funktionskreis Hypothalamus-Hypophyse-Gonaden

Wegen der zentralen Stellung der Amenorrhoe in der Symptomatik der Anorexia nervosa galt diesem Funktionssystem das intensivste Interesse endokrinologischer Forschung. Bei Anorexia-nervosa-Patienten findet sich die Kombination niedriger Werte von Östrogenen und Gonadotropinen im Plasma mit nicht altersentsprechenden Tagesrhythmen von LH (luteinisierendes Hormon) und FSH (follikelstimulierendes Hormon). Diese Kombination entspricht einer funktionellen Regression der Hormonmuster des Erwachsenen auf diejenigen der Pubertät oder Präpubertät. Die Gründe für diese Entwicklung sind unbekannt. Der Gewichtsverlust allein klärt nur 22% der Varianz für die niedrigen Östradiolspiegel und bis zu 50% der Varianz der veränderten LH- und FSH-Reaktion auf LH-Releasing-Hormon. Nicht altersentsprechende Tagesrhythmen von LH- und FSH-Plasmaspiegeln treten auch bei Frauen mit Bulimia nervosa auf, die normales Gewicht haben, von denen einige oligo- oder amenorrhoisch waren. Andererseits werden solche Hormonmuster bei anderen Formen der Amenorrhoe und auch bei chronisch untergewichtigen, aber menstruierenden Frauen nicht beobachtet.

Das Ausmaß der Reifestörungen des zirkadianen LH-Musters zeigt eine schwach positive Korrelation mit dem Ausmaß des Gewichtsverlusts.

Nach Gewichtszunahme normalisieren sich die Sekretionsmuster bei einigen, nicht aber bei allen Patienten. Altersunangemessene LH-Muster persistierten bei Anorexia-nervosa-Patientinnen, die zwar ihr Gewicht nahezu normalisiert haben, jedoch eine Bulimia nervosa entwickelten oder bei denen sonst die Anorexia-nervosa-Symptomatik fortbestand; dagegen bildeten sich die präpubertären LH-Sekretionsmuster zugunsten eines erwachsenen Musters bei solchen Patientinnen zurück, bei denen sich die Anorexiesymptomatik insgesamt entscheidend besserte, die Gewichtszunahme jedoch nur geringfügig war.

Das Körpergewicht trägt offenbar als ein Faktor zum Funktionieren der Hypothalamus-Hypophyse-Gonaden-Achse bei, die funktionale Integrität dieser Achse erfordert bei Anorexiepatienten jedoch Gewichtszunahme und Remission der übrigen Krankheitszeichen (Weiner, 1983).

Zusammengefaßt bleiben trotz der zahlreichen Befunde über dieses Funktionssystem viele Fragen unbeantwortet: Wie kommt die episodische LH-Releasing-Hormonsekretion zustande, wie wird sie aufrechterhalten? Warum führt die Veränderung der bei Erwachsenen üblichen Funktionsweise von LH-Releasing-Hormon bei Anorexiepatienten zu niedrigen Gonadotropinspiegeln, zu einer Umkehrung der altersangemessenen LH- und FSH-Muster und zu niedrigen Östrogenspiegeln? Unklar bleibt die Rolle des Körpergewichts und des Körperfetts für die Regulation von LH-Releasing-Hormon und die Rolle psychosozialer Faktoren bei der Störung dieses Funktionskreises.

Hieraus ergeben sich folgende Überlegungen (Weiner, 1983):
- Es ist möglich, daß Anorexia nervosa keine Krankheitseinheit ist, sondern Untergruppen mit verschiedener Ätiologie existieren.
- Die für die Entstehung mitverantwortlichen psychophysiologischen Faktoren können für die einzelnen Untergruppen des Krankheitsbildes unterschiedlich sein.
- Diejenigen pathogenetischen Mechanismen, die zur Entstehung der Anorexia nervosa führen, können sich von denjenigen unterscheiden, die das Krankheitsbild aufrechterhalten; z.B. könnten bei Patientinnen, bei denen die Amenorrhoe vor dem Gewichtsverlust auftritt, psychosoziale Faktoren (Trennung von zu Hause, sexuelle Versuchungssituationen, Furcht vor Schwangerschaft) zur Erniedrigung der LH-Releasing-Hormonproduktion führen, wie dies bei der „psychogenen" Amenorrhoe der Fall ist. Bei anderen Patientinnen können die hormonalen Veränderungen sekundär Folge des Hungerzustandes sein.
- Ist es einmal zu Gewichtsverlust und Unterernährung gekommen, verändert der niedrige T_3-Spiegel den Östradiolstoffwechsel. Die exzessive Produktion von Katecholöstrogen führt dazu, daß die präsynaptischen Neurone nicht auf Östradiol reagieren und die Gonadotropinsekretion auf LH-Releasing-Hormon ausbleibt und damit unabhängig von der Art der Auslösung das Krankheitsbild Anorexia nervosa unterhalten wird.

38.2.6.2 Angeborene Disposition, frühe Kindheitsentwicklung

Die psychoanalytische Untersuchung von Anorexiepatientinnen ergab, daß sich die Krankheit zwar im Rahmen der Pubertätsentwicklung manifestiert, eine Störung der innerseelischen Entwicklung jedoch regelmäßig bis in die frühe Kindheit zurückverfolgbar ist. In der Genese dieser Störung spielen neben den prägenden Einflüssen der Bezugspersonen, vor allem der Eltern, möglicherweise auch Besonderheiten der Konstitution des Kindes selbst eine Rolle, die bereits früh die Interaktion zwischen Mutter und Kind mitbestimmen.

Das Geburtsgewicht Magersüchtiger differiert signifikant vom Geburtsgewicht der Durchschnittsbevölkerung (Abweichung sowohl nach oben als auch nach unten; Halmi 1974). Es könnte so für die Eltern z.B. schwieriger sein, adäquat auf die Ernährungsbedürfnisse dieser Kinder einzugehen. Von übergewichtigen Säuglingen ist bekannt, daß ihre Ernährungsgewohnheiten stärker als bei normgewichtigen sich in Abhängigkeit vom Angebot einregulieren. Das höhere Durchschnittsalter der Eltern bei der Geburt der später an Magersucht Erkrankten könnte zur Entstehung solcher Interaktionsschwierigkeiten beitragen (Halmi, 1974).

Wir beobachteten bei Anorektikerinnen gelegentlich angeborene oder früh erworbene körperliche Anomalien – etwa eine Ichthyosis congenita, eine angeborene Kyphose, eine entstellende Gesichtsverletzung –, die ebenfalls früh die Interaktion zwischen Mutter und Kind und damit die psychische Entwicklung des Kindes mitbeeinflußt haben dürften.

Auf die Befunde zu chromosomalen Besonderheiten haben wir hingewiesen.

38.2.6.3 Die familiäre Situation

Im Umgang mit vielen Magersüchtigen fallen dem Untersucher die massiven Spannungen zwischen den Patientinnen und den übrigen Familienmitgliedern, aber auch die Spannungen der übrigen Familienmitglieder untereinander sofort auf. Das Eßverhalten der Patientinnen und die Folgen der Erkrankung tragen zu diesen Spannungen bei. Systematische Untersuchungen haben jedoch gezeigt, daß die oft ausgeprägten pathologischen Beziehungsstörungen nicht nur als Krankheitsfolge aufgefaßt werden können. Eine Systematisierung der beobachteten Beziehungsstörungen hat sich für die klinische Arbeit als wertvoll erwiesen; diese sind jedoch keineswegs so homogen, daß es berechtigt wäre, weiter von einer „Magersuchtsfamilie" (Sperling und Massing, 1972) zu sprechen. Zutreffender scheint die Beschreibung „psychosomatogener Familien" (Minuchin, 1974 a, b).

In der klinischen Arbeit fiel zunächst auf, daß die Eltern, insbesondere die Mütter, die Klinikaufnahme ihrer oft schwerstkranken Töchter behindern und eine begonnene psychotherapeutische Behandlung ihrer Kinder häufig von sich aus abbrechen. Für die Hypothese, daß die übrigen Familienmitglieder die Magersuchtskranken zur Aufrechterhaltung von familiären Gleichgewichtsprozessen benötigen und deshalb die Behandlung behindern, spricht auch, daß häufig ein anderes Familienmitglied erkrankt, wenn die ursprüngliche Patientin sich entweder von der Familie zu trennen vermag oder in der Psychotherapie Fortschritte macht. Nicht ganz selten erkrankt sogar ein Geschwister an Anorexie. „Psychosomatogene Familien" sind gekennzeichnet durch:
- eine enge „Verfilzung" der Beziehungen der Familienmitglieder untereinander,
- eine überprotektive Haltung der Familienmitglieder,
- eine ausgeprägte Rigidität der Familienorganisation,
- eine Unfähigkeit Konfliktlösungen zu erarbeiten.
Im einzelnen werden die familiendynamischen Verständniskonzepte in Kapitel 21 besprochen.

In Franziskas Familie wird viel „analog" kommuniziert: So kann der Vater etwa nicht verbal ausdrücken, daß ihm ein Teil des Frühstücks nicht geschmeckt hat, ebensowenig macht er Änderungsvorschläge; seine Kritik teilt er vielmehr dadurch mit, daß er vom Frühstückstisch aufspringt, die Wohnung verläßt und die Tür mit einem Knall hinter sich zuschlägt. Der Rest der Familie bleibt mit Schuldgefühlen sitzen, kann aber auch unter sich das Problem nicht besprechen.

In dieser Familie dominiert die Mutter, früher wurde sie darin noch von einer Tante und der Großmutter unterstützt. Entsprechend ihrer Leistungsideologie spielt für ihr Selbstwertgefühl die eigene Berufstätigkeit eine entscheidende Rolle. Im sexuellen Bereich vermittelte sie der Patientin kein Wissen, sondern nur die Angst vor einer bedrohlichen Umwelt: Die Patientin war nicht aufgeklärt worden, beim Eintritt der Menarche sagte die Mutter nur: „Von jetzt ab mußt du dich in acht nehmen". Die Beziehungen zur Außenwelt wurden auch sonst nicht gefördert: Traten in Freundschaften der Patientin Schwierigkeiten auf und wollte sie sich bei der Mutter Rat holen, so schlug ihr diese regelmäßig gleich den Abbruch der Freundschaft vor. Der Vater versucht sich zu Hause durch „Herumkommandieren und Schreien" Geltung zu verschaffen; man fürchtet mehr seine unangenehme Art, als Autorität im eigentlichen Sinn wird er dagegen nicht anerkannt.

Als sich Franziska gegen Ende der ambulanten psychotherapeutischen Behandlung anschickte, das Elternhaus zu verlassen, um in einer entfernten Stadt ein Studium zu beginnen, erkrankte die Mutter und drohte Franziskas weitere Entwicklung zu blockieren. Nur unter großen Mühen und unter Fortsetzung der psychotherapeutischen Behandlung (insgesamt 200 Std.) gelang schließlich die Trennung.

Die auf das Familiensystem bezogenen Verständnisansätze zeigen im einzelnen, wie und in welchem Ausmaß Familien am Aufbau und an der Fixierung der jeweiligen individuellen Wirklichkeit der betroffenen Patienten beteiligt sind und dazu beitragen, daß die Kranken zur Umformung dieser Wirklichkeit entsprechend den Anforderungen der Adoleszenz nicht in der Lage sind.

38.2.6.4 Prämorbide Persönlichkeitsstruktur

Die Patientinnen fallen häufig schon vor Krankheitsbeginn durch ein reserviertes, distanziertes Verhalten auf. Oft findet sich eine intellektualisierende Abwehrhaltung. Testpsychologisch erscheinen Magersüchtige häufig überdurchschnittlich intelligent, insbesondere Patientinnen aus der Kerngruppe der während der Pubertät erkrankten (Fleck et al., 1965). Gewöhnlich zeigen sie gute bis hervorragende Schulleistungen, was ebenfalls lange Zeit als Anzeichen hoher Intelligenz und Begabung interpretiert wurde. Diese guten schulischen Leistungen sind meist jedoch das Ergebnis einer sehr großen Anstrengung. Intensiver psychotherapeutischer Umgang mit Magersüchtigen hat nämlich gezeigt (Bruch, 1978/1980), daß Magersüchtige in ihren Denkfunktionen auf einem präadoleszenten Entwicklungsniveau verharren.

Hilde Bruch betont, daß Denkstil und Wertsystem einer Entwicklungsphase entsprechen, die Piaget die Phase der vorbegrifflichen oder konkreten Operationen, die Periode der Egozentrizität genannt hat. Eine große Rolle spielen in dieser Phase auch magische Vorstellungen. Die Fähigkeit zu formalen Operationen, zu abstraktem Denken und selbständiger Einschätzung und Beurteilung, die sich in der Adoleszenzphase normalerweise herausbildet, scheint bei Magersüchtigen zu fehlen oder doch nur mangelhaft entwickelt zu sein. Frau Bruch führt dies darauf zurück, daß die Patienten schon als Kinder nur passiv am Leben hatten teilnehmen können, die Dinge dieser Welt zwar aufnehmen, aber nicht aktiv integrieren konnten. Hierzu sei die Beziehung zu den Eltern viel zu eng gewesen, sie habe die notwendige Ablösung, Individuation und Differenzierung nicht zugelassen. Der äußeren Harmonie zuliebe habe sich das Kind übermäßig den an es gestellten Erwartungen gemäß verhalten müssen; dies habe die Entwicklung aus innerer Autonomie kommender eigener Aktivität entscheidend behindert.

Wie dies auch für Patienten mit anderen „psychosomatischen Erkrankungen" beschrieben wird (vgl. Kap. 5), waren Magersuchtspatienten also bereits als Kinder überstark an die Forderungen ihrer Umgebung angepaßt, erschienen oft als Musterkinder, die keinerlei Ärger verursachten. Dabei ist es wichtig zu verstehen, daß diese Anpassungsprozesse ein noch unreifes Ich zu leisten hatte, anders ausgedrückt, daß sie das Ergebnis eines „Dressates" sind. Das spätere unauffällige Verhalten wirkt dann auch fassadenhaft aufgesetzt, es entspricht der Leistung einer „Als-ob-Persönlichkeit", unter Umständen der eines „falschen Selbst" (Winnicott, 1965; vgl. Kap. 6).

> Franziska war ein ausgesprochenes Musterkind, das sogar bei Verwandten als Vorbild herumgereicht wurde. Besonders wurden ihre Sauberkeit und ihre Anständigkeit gelobt; positiv bewertet wurde, daß sie keine Freunde hatte, immer nur der Mutter half usw. Sie entwickelte früh selbst eine ausgesprochene Leistungsideologie, wozu die Rivalität mit dem nur 11 Monate jüngeren Bruder wohl das erste Motiv bildete: in ihrem Erleben hatte sie „der Bruder vom Schoß der Mutter verdrängt".

Die Mädchen werden entsprechend dieser Entwicklung auch vorwiegend als „ängstlich" und „nervös" beschrieben; sie sind schüchtern und gehemmt, zeigen starke Zeichen innerer Bindung und Abhängigkeit. Ausgeprägt ist auch ihre testpsychologisch objektivierte Tendenz zu sozialer Isolation (Stonehill und Crisp, 1977).

Versucht man, sie neurosenpsychologisch zu klassifizieren, so finden sich sowohl schizoide Persönlichkeitsstrukturen, bei der Mehrzahl der Patientinnen auch zwanghafte Züge, als auch – bei ca. 25% – eine hysterische Charakterproblematik (Halmi, 1974).

Zum Verständnis der Psychodynamik der Magersuchtspatientinnen reichen solche Befunde zu neurosenpsychologischen Klassifikationen nicht aus. Sie können jedoch zur Prognosestellung beitragen: Bei Überwiegen der schizoiden Anteile ist die Prognose ungünstiger zu beurteilen als bei Überwiegen hysterischer Anteile.

38.2.6.5 Lebenssituation zur Zeit der Krankheitsmanifestation

Entsprechend dem Erkrankungsalter finden sich zwei für die Adoleszenz typische Situationen: die tatsächliche oder phantasierte **Trennung** von den Eltern sowie **erotische** bzw. **sexuelle Versuchungssituationen.** Auf beide Situationen sind die Patientinnen aufgrund ihrer Entwicklung sowohl emotional als auch kognitiv nicht vorbereitet, für die mit beiden Situationen verbundenen Aufgaben finden sie in ihrer familiären Umgebung keine Unterstützung.

Der eigene Wunsch nach Verselbständigung und Trennung von der Familie oder eine Trennung aufgrund äußerer Gegebenheiten gefährdet zusätzlich das Gleichgewicht des Familiensystems. Zur Identitätskrise des Adoleszenten kommt die Krise des Familiensystems. Die Trennungsproblematik kann schon durch Ferienreisen, Auslandsaufenthalte, Besuch eines Internats o.ä. aktualisiert werden. Sie kann aber auch durch das Ausscheiden anderer Familienmitglieder – etwa von Geschwistern, durch

Krankheit oder Tod von Eltern oder Großeltern – oder andere Gleichgewichtsverschiebungen in der Familie – etwa Wiederaufnahme der Berufstätigkeit durch die Mutter – hervorgerufen werden.

Kontakte zu Männern, auch wenn diese von außen gesehen als nur oberflächlich erscheinen, können bereits als schwere Gefährdung erlebt werden. Die Entdeckung der Entwicklung der sekundären weiblichen Geschlechtsmerkmale am eigenen Körper oder Bemerkungen anderer hierüber – diese werden oft als kränkend empfunden – werden dann häufig zu dem von den Patienten selbst beschriebenen Anlaß, Gewicht zu verlieren. Sie verfolgen damit ursprünglich das Ziel, die geschlechtliche Entwicklung zu verbergen, ja rückgängig zu machen. Häufig fehlen den Patientinnen für die Bewältigung dieser Situation auch positive Identifizierungsmöglichkeiten mit ihrer Mutter oder anderen weiblichen Bezugspersonen.

> Eine 19jährige Patientin meinte: „Ich möchte wieder so werden wie mit 14 und dann immer so bleiben. Das Schlimmste wäre, so zu werden wie die Mutter. Die entwertet sich selbst als Frau, und vom Vater wird sie auch nicht akzeptiert".
>
> Franziska erlebte zur Zeit der Krankheitsmanifestation einen „Zwang", zu Hause stärker die weibliche Rolle zu übernehmen, da die Mutter wieder berufstätig geworden war. Diese Situation habe sie vor allem mit den „negativen Aspekten" dieser weiblichen Rolle konfrontiert: Sie sollte nun selbst als Erwachsene handeln, die anderen versorgen, und auf die Erfüllung ihrer eigenen starken, fortbestehenden Wünsche nach kleinkindhaftem Versorgtwerden verzichten. Das Krankheitsbild verschlechterte sich ein halbes Jahr später in einer typischen „Versuchungssituation": Beim Aufenthalt im Landschulheim kam es in Verbindung mit oralen Ausschweifungen („Besäufnissen"), die die Patientin bei anderen als übertrieben und ekelerregend erlebte und von denen sie sich distanzierte, auch zu ersten sexuellen Annäherungen. „Der Busen" spielte bei den Jungen in der Klasse als „erotisches Signal" eine besondere Rolle. Die Patientin beschloß, noch stärker abzunehmen, schon aus Angst, andernfalls zu große Brüste zu bekommen.

Für das Verständnis von Anorexiepatienten ist es wesentlich, die Berichte der äußeren Belastungen hinsichtlich ihrer subjektiven Bedeutung zu untersuchen; nur so kann die befürchtete Gefahr, das kränkende „Trauma" verstanden werden, denen sich der Patient mit seinen Verarbeitungsmöglichkeiten nicht gewachsen fühlt. Wird diese subjektive Bedeutung nicht sorgfältig geklärt, so werden die berichteten Ereignisse hinsichtlich ihres belastenden Charakters unterschätzt, zu harmlos eingestuft.

Die Patienten leiden am stärksten darunter, daß sie die äußere Entwicklung in der Adoleszenz und ihre eigenen Bedürfnisse nicht ausreichend kontrollieren können. Am stärksten nehmen sie diesen Kontrollverlust an den Veränderungen des eigenen Körpers wahr. Fettsein ist für sie gleichzeitig mit Kontrollverlust und Sexualität assoziiert. Mit dem Kontrollverlust ist wiederum eine extreme Minderung im Selbstwerterleben verknüpft, ein Gefühl weitgehender Insuffizienz. So führt die Entwicklung in der Adoleszenz zu einer extremen Diskrepanz zwischen dem idealen Wunschbild vom eigenen Körper und den eigenen Verhaltensmöglichkeiten einerseits und der Realität andererseits. In dieser Krise fühlen sich die Patienten oft vor Angst weitgehend gelähmt, häufig „unsicher, was sie für sich selbst wünschen oder erwarten" (Bruch, 1978/1980). Verunsicherung und Identitätsproblematik äußerte eine Patientin Bruchs, bevor sie das elterliche Haus verließ, um das College zu besuchen: „Mich beunruhigt, daß ich nicht weiß, welche Art von Mädchen ich sein soll. Soll ich Sportlerin werden, Partygängerin oder Intellektuelle?" (Bruch, 1978/1980).

38.2.6.6 Psychodynamik[1]

Die in dieser Adoleszenzkrise ablaufenden intrapsychischen Prozesse lassen sich unter zwei Perspektiven sehen: einerseits tragen sie zur Krisensituation bei, andererseits sind sie Bestandteile des oft verzweifelten Versuches, diese Krisensituation – vorwiegend über eine regressive Aufrechterhaltung präadoleszenter Funktionsweisen – zu bewältigen. Nach A. E. Meyer (1970) finden sich am häufigsten folgende vier Konstellationen, die auch miteinander kombiniert auftreten können.

Abwehr aller weiblichen sexuellen Bedürfnisse (Kampf gegen die Sexualität als Trieb)

Die Abwehr richtet sich gegen die Übernahme der weiblichen Rolle als solcher, besonders aber gegen die Inkorporationsaspekte weiblicher Sexualität, sowohl auf der genitalen als auch auf der oralen Ebene. Weibliche Sexualität zeigt Parallelen zum Essen (In-sich-Hineinnehmen von Glied und Samen; Dickwerden durch Schwangerschaft). So finden sich bei Kindern ja z. B. Phantasien über „orale Schwängerung". Über die Abwehrmechanismen **„Regression"** und **„Verschiebung"** werden die genital-sexuellen, jetzt in der Pubertät auftretenden Triebimpulse in den oralen Bereich zurückverlegt. Im Anschluß an die Regression erreicht die Abwehr jetzt neben dem innerpsychischen Erfolg (Angstreduktion) auch eine reale Wirkung: mit zunehmender Abmagerung schwinden die sekundären weiblichen Geschlechtsmerkmale und sistiert die Menstruation, und damit in der Phantasie und auch bald in der Realität die erotisch-sexuelle Anziehung. Über diese Wirkung gewinnen die Patientinnen auch wieder das für sie so wichtige Gefühl, sich selbst kontrollieren zu können.

Zum Zeitpunkt der Menarche (13 Jahre) hatte Franziska die Phantasie, durch Küsse geschwängert zu werden; daneben phantasierte sie auch, daß das Badewasser (im Hallenschwimmbad, beim Baden nach einem der Brüder zu Hause) sie schwängern könne. Zu diesem Zeitpunkt bezogen sich die Schwängerungsphantasien umschrieben auf den genitalen und auf den oralen Bereich. Zum Zeitpunkt der Krankheitsmanifestation erfolgte eine Generalisierung der Gefährdung

---

auf die gesamte Körperoberfläche (die gesamte Haut wurde „permeabel"), sowie eine Generalisierung der ursprünglich phantasierten Schwängerung über ein Eindringen eines männlichen Körperteils (Glied, Zunge) auf den Vorgang des „Eindringens" überhaupt. So wurden Blicke als stechend erlebt, ja starke Farben und Geräusche als gefährdend. Sie selbst bekam das Gefühl, „zuviel Raum einzunehmen" und damit eine zu große Angriffsfläche („wie ein aufgeblasener Luftballon, in den man nur hineinzustechen braucht") für die Umwelt zu bieten.

Das Beispiel zeigt, daß die Regression nicht nur die Triebabwehr betrifft, sondern auch das Ich im zunehmenden Konfliktdruck in seinen Funktionen ganz erheblich beeinträchtigt wird.

Die geringe Häufigkeit der Erkrankung bei Männern könnte dadurch mitbedingt sein, daß männliche Sexualität weniger mit Aufnehmen, sondern mit Eindringen und Ausstoßen zu tun hat. Weiter tritt mit der Gewichtsreduktion kein vergleichbarer Erfolg bezüglich der Attraktivität und der geschlechtlichen Entdifferenzierung ein.

Der Kampf um die Autonomie (Abwehr von Essen und Anstreben von Magerkeit als Kampf von Geist gegen Trieb)

Während der analen Entwicklungsphase lernen Kinder (oft in Form eines mit erheblichen Frustrationen verbundenen Sauberkeitstrainings), daß Körperbeherrschung höher bewertet wird und auch mehr Sicherheit verleiht als „Sich-treiben-Lassen" oder „Sich-gehen-Lassen". Entsprechende moralische Imperative („den inneren Schweinehund an die Leine nehmen") werden später auch gegen neue Triebgefahren verwendet, wenn sie die einmal, oft nur notdürftig gewonnene Autonomie zu gefährden drohen: gegen die Sexualität, aber auch (die Abwehrmechanismen Regression und Verschiebung vorausgesetzt) gegen die „Völlerei" als „eine der sieben Todsünden". Fasten gibt sich dann als „geistige", asketische Leistung aus und die erreichte Magerkeit dokumentiert dies gegenüber der Umwelt. Die Zuflucht zu asketischen Idealen findet sich während der Pubertät recht häufig („Pubertätsaskese", A. Freud). Auch im Rahmen christlicher Ideologie war solch ein asketisches Ideal oft gegen die Nahrungsaufnahme gerichtet: So schreibt etwa Tertullian (2. Jh. n. Chr.) in seiner Schrift über das Fasten: „Ein abgemagerter Leib wird hoffentlich leichter durch die Pforte des Heils eingehen, schneller wird ein leichter Körper einst auferstehen" (nach Schadewaldt, 1965). Das Abwerfen jedes irdischen Ballastes, das Anstreben eines engelhaften Zustandes findet sich immer wieder bei den Patientinnen. Schon dem Nervenarzt Hoffmann, dem Verfasser des Struwwelpeters, soll beim „Suppenkasper" eine magersüchtige Klavierlehrerin Modell gestanden haben, die glaubte, sich durch Fressen und Faulsein versündigt zu haben (nach Schadewaldt, 1965).

---

1 Zur Psychodynamik in der Perspektive der Selbstpsychologie vgl. Kapitel 54.

Mit dem Erleben der Fähigkeit eigener Triebkontrolle ist für die Patientinnen ein narzißtisches Hochgefühl verbunden. Sie erleben, daß es ihnen weitgehend gelungen ist, vom Essen und auch sonst von ihrer Umwelt unabhängig zu sein. Scheinbar haben sie keine Bedürfnisse mehr. Sie leben in einem wahnähnlichen Glauben an die eigene Autarkie, in dem sie die Abhängigkeit, „das passive Ausgeliefertsein des Ich an die Nähe und unerbittlich wirkende Gewalt des Hungers" (H. Kunz nach Thomä, 1961), „die Abhängigkeit des Ich von der Natur ...", und insbesondere auch von den sie (als Kind) versorgenden Personen verleugnen können (Thomä, 1961). Aus dieser Sicht wird jedes Hilfsangebot zu einer Gefahr, die die mittels Verleugnung erreichte Vollkommenheit und Sicherheit gefährden könnte (Thomä, 1961). Dieser sekundäre Krankheitsgewinn trägt wesentlich zu den starken Widerständen gegen die Behandlung bei.

Das mit dem Erleben der Fähigkeit eigener Triebkontrolle verbundene innere Hochgefühl wird in einem Essay von Kazantzakis über Spanien deutlich, in dem er einen jungen Spanier beschreibt: „Das ist Manola", sagte mein spanischer Freund und lachte dabei. „Den ganzen Tag liegt er hier ausgestreckt in der Sonne. Er will nicht arbeiten, sogar wenn das bedeutet, daß er wegen Hunger sterben muß." Ich ging auf ihn zu. „Ah, Manola", rief ich ihm zu, „die sagen mir, du bist hungrig. Warum stehst du nicht auf und arbeitest? Schämst du dich nicht selbst?" Manola starrte vor sich hin, dann hob er seine Hand in einer königlichen Gebärde: „Im Hunger bin ich der König". Als ob der Hunger ein grenzenloses Königreich wäre, und solange Manola hungrig blieb, konnte er das Szepter seines Königreiches in eigenen Händen halten (zit. nach H. Bruch, 1973).

Für den Umgang mit den Patienten ist es wichtig zu wissen, daß dieses narzißtische Hochgefühl für die Patienten eine Art regressive Ersatzlösung darstellt, der das Scheitern der Regulation ihres Selbstwertgefühls auf einer Erwachsenenebene vorausgegangen ist. Anstelle eines gesunden Selbstwertgefühls herrscht jetzt ein infantiles „Größen-Selbst". Der Patient hat sich, bildlich ausgedrückt, in diese Abwehrbastion zurückgezogen. Es hat keinen Sinn, ihn dort auch noch anzugreifen, vielmehr sollte ihm der Umgang ermöglichen, diese Position vorsichtig wieder zu verlassen.

So läßt sich auch arrogante Abweisung der Patientinnen gegenüber ihren Bezugspersonen, aber auch gegenüber Ärzten und Schwestern verstehen; aus der Position des infantilen „Größen-Selbst" „benützen" sie andere Personen allenfalls als ein unpersönliches Objekt, die anderen dürfen jedoch keine eigenständige Rolle spielen. Hier spiegelt sich ihre eigene Beziehungssituation in der Kindheit mit vertauschten Rollen wider.

Abwehr des Essens als Kampf gegen den Wunsch nach Annäherung (bis zur Verschmelzung mit der Mutterfigur oder anderen Personen)

Essen ist während der oralen Phase mit Nähe, mit Hautkontakt, mit Zusammensein überhaupt verbunden. Die Nahrungsaufnahme kann diese Bedeutung beibehalten (stärkere Fixierung) oder wiedergewinnen (Regression). Sprachlich drückt sich dies im

Wort „Kumpan" oder „Compagnon" (derjenige, mit dem man das Brot = panis teilt) aus (Thomä, 1961). Der Wunsch nach Annäherung und die zugehörigen Motive sind bei den Patientinnen in der Regel unbewußt und werden nur in der psychoanalytischen Behandlung wieder deutlich. Die Magersüchtigen erleben bewußt das Unbehagen, wenn ihnen andere Menschen „auf die Haut rücken".

Margaret Mahlers Konzept von der „psychologischen Geburt des Menschen", von der allmählichen Loslösung und Individuation des Kindes von der Mutter vermag zum Verständnis dieser Problematik Wesentliches beizutragen (Mahler et al., 1978).

> Bei Franziska stellte sich mit der Wiederauffütterung mit Sondenkost ein Gefühl ein, „die Form zu verlieren", „amorph" zu werden; dieses körperliche Erleben wiederholte sich im Gefühlsbereich später gegenüber der Therapeutin. Jetzt bedeutete Gefühle haben oder Gefühle zeigen gleichzeitig, „die Form verlieren", „sich aufzulösen", „nicht mehr vorhanden zu sein", das mit intensivem Angsterleben verbunden war.

Vorgänge, über die sich eine bereits angelaufene Anorexie selbst verstärkt bzw. perpetuiert

Circulus vitiosus zwischen Hungerbedürfnissen und Autonomiestreben

Wurde das Fasten einmal begonnen, so steigen Hungerbedürfnis und korrespondierende Triebimpulse ständig weiter an; dies wird zunehmend als eine immer bedrohlichere Macht erlebt, die die Aufrechterhaltung der Abwehr und damit die Selbstkontrolle zu überrennen droht: je hungriger, desto stärker die Abwehr, desto verhärteter die krankhaften Kontrollversuche.

Die Gewichtsabnahme fördert zwar ein auf die Möglichkeit zur Selbstkontrolle bezogenes Sicherheitsgefühl, führt jedoch gleichzeitig zu einer biologischen Instabilität, die zum Teil wiederum auch wahrgenommen wird (Crisp, 1983).

> Franziska befürchtete mit zunehmender Abmagerung, daß ihr Gehirn nicht mehr zur Kontrolle des Hungers in der Lage sein könnte und sie von unkontrollierbarer „Freßsucht" überwältigt werden könnte. In der Absicht, in einem solchen Notfall ihrem Gehirn die für den Zellstoffwechsel nötige Glukose zuführen zu können, trug Franziska immer einige Zuckerstückchen bei sich.

Soziale Isolation

Die Anorexiekranken geraten im weiteren Verlauf immer stärker in soziale und psychische **Isolation**. Ihre äußere Erscheinung macht den Verbleib in Gruppen Gleichaltriger schwierig; sie selbst verlieren die gemeinsamen Interessen mit ihrer Umgebung: Statt mit Freizeitaktivitäten, mit Freundschaften, mit der beruflichen Situation, mit Mode- und Kosmetikproblemen beschäftigen sich die Kranken lieber mit

Fasten und dem Thema der Selbstbeherrschung. Die hier erkämpften Erfolge werden im Erleben zum Beweis der eigenen Unabhängigkeit und der Überlegenheit gegenüber anderen umgemünzt: die anderen sind „primitiv", ihren Körpergefühlen ausgeliefert. Die eigene Abhängigkeit (ständige Beschäftigung mit dem Essen) wird dabei verleugnet. Das Gefühl der eigenen Effizienz beruht auf dem Autonomieerleben im Sinne der Triebkontrolle; im Sinne eines Circulus vitiosus muß es trotz oder gerade wegen der zunehmenden **Ineffizienz** (Bruch, 1973) in der Realität, die auch gespürt wird, nach Möglichkeit ständig noch gesteigert werden. Die Selbstwertprobleme wurden zum Teil über Größenphantasien kompensiert. Die entsprechende Arroganz, der Hochmut, die Anspruchlichkeit bekommen dann die Bezugspersonen, aber auch Ärzte und Schwestern im Umgang mit diesen Patientinnen zu spüren. Oft kann ihnen niemand gerecht werden, „in jeder Suppe" finden sie „ein Haar", die Mißerfolge der Behandlung bringen sie nicht mit ihrem Verhalten in Verbindung, sondern mit der Insuffizienz der Behandler bzw. deren Methoden. Ihre eigenen Ineffizienzgefühle werden zum Teil durch extreme ideologische Haltungen kompensiert.

> So phantasierte Franziska sich auf dem höchsten Hügel einer Kurstadt sitzend, die ganze Stadt zu ihren Füßen. Sie betrachtete die dort versammelten Kurgäste gleichzeitig: „Diese Leute, die nichts anderes im Kopf haben, als ihren Körper zu pflegen". „Die haben vielleicht Probleme". Der Ärger und die Verachtung gegenüber diesen „Abhängigen" führt bei Franziska zu einem ausgesprochenen Hochgefühl; danach kann sie, die sonst unter schweren Schlafstörungen leidet, leicht einschlafen.

Verhärtung der familiären Beziehungen

Auch die familiären Auseinandersetzungen, die früher Themen wie die gewährten Freiheiten und die erzwungenen Pflichten, das Taschengeld, die Bevorzugung von Geschwistern beinhalteten, engen sich immer mehr aufs Essen ein: „Esse ich, so wie die mich drängen, oder bleibe ich hart?" Essen würde bei dieser Konstellation Gesichtsverlust und Niederlage bedeuten.

Verstärkung des pathologischen Verhaltens in den sozialen Beziehungen

Die Umgebung (Familie, Ärzte, Krankenhauspersonal) kann durch ihr Verhalten die Lernprozesse der Patientinnen ständig in die falsche Richtung verstärken. Die Patientinnen haben einen unbewußten Wunsch nach Zuwendung und Versorgung, sie dürfen ihn jedoch nach außen nicht eingestehen. Sie bekommen nun aber immer dann Zuwendung, wenn die Krankheitssymptome zunehmen, wenn sie nicht essen oder abnehmen. Im doppelten Sinn wirkt dies als Belohnung: einmal erhalten sie die doch auch erwünschte Zuwendung, zum anderen können sie

dann gegen die Zuwendung der anderen ankämpfen, ohne ihre Bedürftigkeit zeigen zu müssen.

Hinzu kommt das heimliche Triumphgefühl, wenn es ihnen so gelingt, die eigene Ohnmacht wenigstens etwas dadurch zu kompensieren, daß sie die anderen unter Kontrolle halten.

> Bei Franziska wird deutlich, wie über das Eßverhalten die Kommunikation mit den Hauptbezugspersonen – wenngleich auch eingeschränkt – aufrechterhalten werden kann. In der Familie können Affekte nicht direkt kommuniziert werden. Dies wird entweder durch autoritäres Diktieren und/oder durch Erzeugen von Schuldgefühlen verhindert. So beantwortet die Mutter Angriffe auf ihre Person durch direkte oder indirekte Androhungen, die Familie zu verlassen. Zeichen der Trauer oder der Verzweiflung werden durch Bemerkungen wie „darüber brauchst du dich doch nicht aufzuregen" oder „das ist doch kein Grund zum Weinen" unterdrückt. Das Signal, das im affektiven Ausbruch enthalten ist, wird demnach durch einen Kommunikationsabbruch beantwortet. Lediglich auf das Eßverhalten wird dauernd mit intensivem Affekt reagiert, was zu einer ständigen Verstärkung (über diese Belohnung durch Zuwendung) dieses gestörten Eßverhaltens führt.

Wir erinnern in diesem Zusammenhang auch an das bereits früher referierte Beispiel einer 18jährigen Oberschülerin, bei der dieser Circulus vitiosus besonders deutlich wurde:

> Die nach außen sich von jeder intensiveren Beziehung distanzierende Kranke aß zu Hause überhaupt nichts, zu Besuch bei Bekannten leerte sie dort jedoch heimlich die Eisschränke. Schließlich begann sie, auf Rechnung der Mutter in der Apotheke Unmengen von Babykost zu kaufen, die sie heimlich verschlang; nachts erbrach sie alles wieder mit Hilfe einer zurechtgebogenen Kerze im Bad und verunreinigte dabei das Bad der Familie so sehr, daß die Familie weiter vollauf mit ihrer Krankheit beschäftigt blieb.

Zum psychoanalytischen Verständnis weiterer häufiger zu beobachtender Phänomene

Die „altruistische Abtretung" (A. Freud, 1959)

Dieser in der Adoleszenz häufige Abwehrmechanismus ermöglicht unakzeptable Triebansprüche wenigstens indirekt zu befriedigen. Die Patienten versorgen andere, nehmen an deren Befriedigung partizipierend – über Identifikation und Projektion – teil. So versorgen magersüchtige Patientinnen etwa Mitpatienten im Krankenhaus oder Familienmitglieder zu Hause mit Essen, beschäftigen sich (anscheinend „paradoxerweise") ausführlich mit Kochbüchern usw.

Die motorische Hyperaktivität

Sie ist mehrfach determiniert: zum Teil kann sie als Folge der Aufstauung des Hungerbedürfnisses verstanden werden. Die Kranken sind innerlich sozusagen ständig auf der **Suche** nach Objekten zur Befriedigung ihres Hungerbedürfnisses und der entsprechenden Triebimpulse, das „Appetenzverhalten" läuft jedoch leer, da es nicht zur triebverzehrenden Endhandlung (Essen) kommen darf (u.a. Thomä, 1961). Weiter gehen in die motorische Unruhe auch die aufgestauten aggressiven Triebimpulse mit ein.

Bei Franziska wird dieses Appetenzverhalten in dem oft Stunden in Anspruch nehmenden Kreisen um einen eßbaren Gegenstand deutlich. Bis zu zehnmal kann sie am Vormittag vom 1. Stock zur im Parterre liegenden Küche gehen, um dort neben einer halben Scheibe trockenen Brotes auf und ab zu gehen. Entsprechend intensiv ist ihre Beschäftigung mit Kochbüchern oder mit dem Kochen für andere. Auf der Station teilt sie – ein erschreckendes Bild: die zum Skelett abgemagerte Patientin mit der Nasensonde – das Essen für die Mitpatienten aus.

Amenorrhoe und Obstipation

Diese Symptome sind psychologisch teils im Sinne einer funktionellen Symptombildung als „körperliche Verhaltensstörung" im Sinne einer „Organneurose" aufzufassen, teils lassen sie sich als Folge der Ernährungspraktiken bzw. des starken Gewichtsverlustes erklären. Die Befunde zur Hormonregulation haben wir im Abschnitt 38.2.6.1 dargestellt.

38.2.7 Therapie

38.2.7.1 Allgemeine Zielvorstellungen

Jede Behandlung Magersüchtiger hat zugleich den lebensbedrohlichen körperlichen Zustand und dessen Ursache, die krankhafte Störung des Eßverhaltens, zu berücksichtigen. Jedes Behandlungsprogramm erfordert deshalb die Integration somatischer – internistischer bzw. pädiatrischer – Verfahren und psychotherapeutischer Methoden.

Die Beurteilung des Therapieerfolges ist aus methodischen Gründen schwierig; deshalb müssen die vorhandenen Therapieansätze noch nach zwei weiteren Kriterien beurteilt werden: ihrer Rationalität und ihrer Übertragbarkeit in andere als die im jeweiligen Projekt gegebene therapeutische Situation.

Rationalität

Therapieansätze sollten pathogenetischen Konzepten folgen; bei der Anorexia nervosa ist dies noch nicht ausreichend möglich. Beim heutigen Kenntnisstand sollte vor allem darauf geachtet werden, daß der gewählte Therapieansatz nicht den Zugang zu einem die Krankheit mitbeeinflussenden und heute schon weitergehend verstandenen Teilbereich blockiert; rein somatisch orientierte Behandlungsansätze sind daraufhin zu prüfen, ob sie den Zugang für eine psychotherapeutische Behandlung fördern oder blockieren; psychotherapeutische Ansätze darauf, inwieweit

sie berücksichtigen, daß die psychischen Veränderungen auch Folge der körperlichen Auszehrung sein können, d.h. inwieweit der Behandlungsansatz eine ausreichende Gewichtszunahme sicherstellt.

Die Durchführung der immer anspruchsvollen psychotherapeutischen Behandlung ist an eine entsprechende Ausbildung und – bei einem stationären Aufenthalt – an eine systematische und kontrollierte Zusammenarbeit aller Teammitglieder gebunden.

Übertragbarkeit

Es ist zu prüfen, inwieweit der Erfolg eines Behandlungsansatzes auf dem Engagement einzelner Forscher und einer speziellen Behandlungssituation beruht oder inwieweit dieser Ansatz lehr- und lernbar und in andere Arbeitssituationen übertragbar ist.

Allen erfolgreichen Therapieansätzen ist der Versuch gemeinsam, zunächst die akute Notsituation zu beherrschen und anschließend eine Korrektur des pathologischen Eßverhaltens herbeizuführen. Alle erfahrenen Therapeuten warnen davor, zugunsten einseitiger psychotherapeutischer Ansätze den kachektischen Zustand der Patienten und die aus ihm resultierenden psychischen Veränderungen zu vernachlässigen. Die Patienten selbst, die ihren Zustand ja verleugnen, versuchen auch ihre Umwelt zu dieser Verleugnung zu verführen! (Bruch, 1978/1980; Silverman, 1983).

38.2.7.2 Die Elemente eines Behandlungsplans

Konfrontation mit dem Ernst der Erkrankung; Herstellung einer therapeutischen Beziehung („Arbeitsbündnis")

Anorektische Patienten stehen einer Behandlung abwehrend gegenüber oder widersetzen sich ihr grundsätzlich (Bruch, 1978/1980). Sie verleugnen den Krankheitswert ihres Zustandes, haben oft das Gefühl, mit ihrer übermäßigen Schlankheit die perfekte Lösung all ihrer Probleme gefunden zu haben und neigen dazu, diesen Zustand zu verherrlichen (Bruch, 1978/1980). Oft werden sie von ihren Eltern in ihrer ablehnenden Haltung gegenüber den Behandlungsmöglichkeiten noch unterstützt.

Konsequenz dieser Abwehr ist, daß auch schwerkranke Patienten oft nur mit großer Verzögerung, gelegentlich gar nicht in Behandlung kommen. So beobachteten Sperling und Massing (1972) zwischen poliklinischer Vorstellung und stationärer Aufnahme einen „Schwund" von 40% der Patienten.

In den ersten Gesprächen geht es darum, Patienten und ihre Angehörigen einerseits nachdrücklich mit dem Ernst der Erkrankung, dem prognostischen Risiko sowie den Erfordernissen der Behandlung zu konfrontieren, andererseits sie für eine aktive Beteiligung an einer vertrauensvollen therapeutischen Beziehung zu gewinnen.

Zunächst ist es wichtig, sie ausführlichst über das Krankheitsbild, die bekannten Zusammenhänge und die therapeutischen Möglichkeiten zu informieren. Es geht darum, den Betroffenen das Gefühl zu vermitteln, daß sich die Krankheit verstehen läßt und es Hilfsmöglichkeiten für diese Krankheit gibt. Von Anfang an sollte kein Zweifel daran gelassen werden, daß im Zentrum der Problematik und dementsprechend der erforderlichen Psychotherapie nicht die Fragen des Gewichts und der Diät stehen, sondern die Probleme des inneren Selbstzweifels; ohne die Beteiligten zu kränken, sollte ihnen im ersten Gespräch doch deutlich werden, daß der fachkompetente Therapeut die Kernprobleme versteht und sich nicht durch die sekundär über Abwehrvorgänge entstandenen Symptome und Haltungen irritieren läßt.

In der ersten Besprechung sollte andererseits auch klargestellt werden, daß mit Übernahme der Behandlung auch die Verantwortung dafür übernommen wird, die Patienten „nicht an der Abmagerung sterben zu lassen" – bei allem Verständnis für die psychische Problematik. Gleichzeitig sollte allen Beteiligten erläutert werden, daß sich die psychischen Probleme erst nach einer Besserung der körperlichen Situation ausreichend beurteilen lassen.

Wenn irgend möglich, sollte im Anschluß an das Erstgespräch ein entsprechendes gemeinsames Gespräch mit dem Patienten und den Eltern stattfinden.

Wir halten es für erforderlich, daß Patienten und Eltern bis ins einzelne gehend über den therapeutischen Ansatz informiert werden und sich dann – soweit möglich nach einer gewissen Bedenkzeit, während der sie sich auch anderweitig orientieren können – für oder gegen den vorgeschlagenen Behandlungsplan entscheiden.

Stationäre Aufnahme zur Behandlungseinleitung

Ist das Krankheitsbild deutlich ausgeprägt, so ist in der Regel eine stationäre Behandlung indiziert. Nur unter stationären Bedingungen gelingt die Koordination der erforderlichen Maßnahmen; von einer ambulanten Behandlung kann – ausgenommen bei mit Magersüchtigen sehr erfahrenen Psychotherapeuten – im allgemeinen kein Erfolg erwartet werden.

Für die stationäre Behandlung sollte ein Zeitraum von zunächst mindestens 8 bis 12 Wochen eingeplant werden. Auch die durch den stationären Aufenthalt bedingte Trennung von der Familie kann sinnvoll sein. Allerdings ist es wichtig, daß das stationäre Setting so beschaffen ist, daß die Patientinnen in dem verständnisvollen Umgang mit allen Teammitgliedern auch eine wohltuende Hilfe erfahren können. Dies erfordert die Einbeziehung aller Beteiligten in ein systematisches Behandlungskonzept.

Notfallbehandlung

Anorexiepatienten mit ausgeprägtem Krankheitsbild sind nicht ganz selten zunächst als Notfallpatienten anzusehen, bei denen die Kontrolle und Aufrechterhaltung vitaler Funktionen im Vordergrund steht (Schockbekämpfung, Elektrolytsubstitution); bei entsprechender Gefährdung ist eine sorgfältige Überwa-chung der Patienten, notfalls analog zu dem Regime auf Intensivstationen, bzw. mit Hilfe von Sitzwachen nötig, auch um eine eventuelle weitere Selbstschädigung mit Sicherheit verhindern zu können.

„Wiederauffütterung"

Nächstes Ziel ist eine Gewichtszunahme auf ein Mindestgewicht: Sollgewicht minus 10%; hierdurch soll auch der im Abschnitt „Pathogenetische Konzepte" beschriebene Circulus vitiosus durchbrochen werden. Auf die Gewichtszunahme kann als Bestandteil des Behandlungsplans nicht verzichtet werden, die Gewichtszunahme ist auch kein Verhandlungsthema im Umgang mit den Patienten. Ohne Gewichtszunahme sind keine sinnvollen psychischen Veränderungen zu erwarten; „Psychotherapie ist für hungernde Patienten nutzlos" (Garfinkel und Garner, 1983). Gleichzeitig muß natürlich Sorge getragen werden, daß die mit der Gewichtszunahme verbundenen Ängste aufgefangen werden können.

Die Wiederauffütterung kann kurzfristig mit freiem Nahrungsangebot versucht werden. Unserer Erfahrung nach ist jedoch neben einem solchen Nahrungsangebot zumeist die Zufuhr von ca. 3000 Kalorien pro Tag über eine **Nasen-Magen-Dauersonde** notwendig. Während dieser Wiederauffütterungsphase wird den Patienten zunächst strenge Bettruhe verordnet, ihre Bewegungsfreiheit auf das Krankenzimmer beschränkt; Besuche von außerhalb und soziale Kontakte innerhalb der Klinik werden zumindest sehr stark eingeschränkt. Parallel zu diesen Maßnahmen sollte mit der Psychotherapie begonnen werden.

Mäßiggradige Amylase-Erhöhung im Serum ist Folge des Hungers und – ohne die typischen klinischen Symptome – nicht Zeichen einer Pankreatitis; sie stellt also keine Kontraindikation für enterale Nahrungsaufnahme dar.

Alternativ kann die Wiederauffütterung im Rahmen eines **verhaltenstherapeutischen Ansatzes** durchgeführt werden, wenn ein mehrstufiges Therapieprogramm unter stationären Bedingungen und ein entsprechend geschultes Team dies gewährleisten (Pierloot et al., 1975, 1982; Schaefer und Schwartz, 1974; vgl. Kap. 19).

Soziale Isolation wird vor allem auch in verhaltenstherapeutischen Ansätzen verwandt, auf ihre Bedeutung hatte bereits Gull (1888) hingewiesen. Gelegentlich erweist sich auch eine absolute Unterbrechung des Kontaktes zu den Eltern als notwendig. Diese sollten jedoch gleichzeitig in das therapeutische Konzept einbezogen werden; das Vorgehen sollte mit ihnen ausführlich besprochen und diskutiert werden.

Das festgelegte Therapiekonzept mit allen im Therapieplan vereinbarten Einzelheiten sollte für alle Teammitglieder verbindlich sein und schriftlich festgehalten werden. Dabei sollte für alle klar sein, daß das Ziel nicht darin besteht, die Kranken strafend einzuschränken, sondern eine klare Situation zu schaffen, die es ermöglicht, selbstschädigendes Verhalten der Kranken zu besprechen. Alle Versuche der Patienten, die Behandlung zu behindern (wie Erbrechen, Magenaushebung über die liegende Sonde, heimliche Einnahme von Abführmitteln u. a.) werden sofort angesprochen und mit aller Konsequenz geklärt.

In sehr schwierigen Fällen, vor allem bei stark ausgeprägter motorischer Unruhe, werden die Patienten konsequent durch Gabe von Psychopharmaka über die Nasen-Magen-Sonde sediert. Die Dosis wird entsprechend der Wirkung gesteigert.

In unserem eigenen Patientengut (internistische Klinik, internistisch-psychosomatische Station) war bei 90% der Patienten eine initiale Wiederauffütterung mit Hilfe der Sonde indiziert, dagegen nur bei ca. 10% der Patienten eine systematische Sedierung.

Modifikation des Eßverhaltens

Langfristig sollte immer eine Veränderung des Eßverhaltens angestrebt werden. Hierfür können zur Zeit folgende Methoden alternativ oder in Kombination eingesetzt werden:
– Psychoanalytisch orientierte Einzeltherapie
– Familientherapie
– Verhaltenstherapie
Vor der ausführlichen Darstellung dieser Behandlungsansätze möchten wir auf einige Voraussetzungen für die Durchführung der Akutbehandlung sowie auf häufige Schwierigkeiten und Fehler während der akuten Behandlungsphase näher eingehen.

38.2.7.3 Voraussetzungen für die Durchführung von Akutbehandlung und „Wiederauffütterung"

Therapeutisches Team

Die Absprache und Einhaltung sämtlicher Vereinbarungen durch alle Beteiligten – Ärzte, Schwestern, Pfleger, Hausmädchen, vorgesetzte Ärzte, Konsiliarien – ist unabdingbare Voraussetzung für das beschriebene Vorgehen. Gelingt dies nicht, kommt es regelmäßig dazu, daß die Patienten einzelne Teammitglieder gegeneinander ausspielen, bzw. die Teammitglieder im Rahmen eigener „Gegenübertragungs-Reaktionen" gegeneinander agieren; diese Aktionen werden häufig durch Kräfte in den Patienten, die sich intrapsychisch unvereinbar gegenüberstehen, stimuliert. Die Koordination im therapeutichen Team gelingt dauerhaft nur auf entsprechend organisierten psychosomatischen Spezialstationen (vgl. Köhle et al., 1977).

Freud beurteilte bereits vor etwa 70 Jahren die Indikation zur psychotherapeutischen Behandlung von Anorektikerinnen unter dem Gesichtspunkt der äußeren Rahmenbedingungen für eine solche Behandlung: „Natürlich wären die inneren Voraussetzungen, um einen solchen Fall von Anorexia nervosa psychoanalytisch zu behandeln, gegeben, aber es fehlten die äußeren Bedingungen, weil man diese Mädchen nicht ohne Zusammenarbeit mit einer Klinik und mit Internisten behandeln kann." Freud beklagte die fehlende Möglichkeit zur Zusammenarbeit und lehnte es seinerzeit ab, eine Patientin ohne den Rückhalt einer Klinik zu behandeln, weil „diese dem Tode nahen Patienten die Fähigkeit haben, den Analytiker so zu beherrschen, daß es ihm unmöglich gemacht wird, die Phase des Widerstands zu überwinden" (unveröffentlichte Bemerkung von Anna Freud, zit. nach H. Thomä; Fleck, Lange und Thomä, 1965).

Beachtung von „Gegenübertragungsreaktionen"

Unabhängig von der Art der durchgeführten Therapie ist für alle an der Behandlung Beteiligten wichtig, sich ihrer meist ambivalenten „Gegenübertragungsreaktionen" bewußt zu werden, die sich im Umgang mit Magersüchtigen sehr rasch, nahezu „automatisch" (Thomä, 1961) einstellen. Aufgrund ihrer dargestellten Persönlichkeitseigenschaften bringen Anorexiepatientinnen Therapeuten und Pflegepersonal regelmäßig in Konfliktsituationen. Ihr kachektischer Zustand mobilisiert zunächst Mitleid und Besorgnis; die Tatsache der „psychischen" Genese („wenn sie nur wollten, könnten sie auch essen"), der Kooperationsmangel und schließlich die Enttäuschungen über die Betrugsmanöver der Patientinnen sowie ihre Arroganz erwecken Zorn und Ablehnung. Auch gegenüber den Mitgliedern des therapeutischen Teams wiederholt sich oft der häusliche Kampf um Essen und Gewichtsentwicklung.

Eine gleichermaßen verständnisvolle und unbeirrbar das Therapieziel verfolgende Haltung rechnet von vornherein mit der Unfähigkeit der Patientinnen zu echter Kooperation und ihrer Tendenz, eine breite Palette von Mitteln einzusetzen, um die Behandlung zu hintertreiben. Schon um die hierdurch entstehenden – teils bewußten, teils unbewußten – Schuldgefühle der Patientinnen nicht noch weiter zu verstärken, ist es wichtig, sie deutlich mit ihren Verhaltensweisen zu konfrontieren. Dies vermag am ehesten die Beziehungen zu entlasten und die Möglichkeit zu einer weiteren Klärung der Situation zu eröffnen.

Psychologische Aspekte der Sondenbehandlung

Widerstände gegen die Therapie mittels Nasen-Magen-Sonde treten im allgemeinen bei Ärzten und beim Pflegepersonal in stärkerem Maße auf als bei den Patienten selbst; dieses Vorgehen wird häufig als aggressiv gegenüber den Patienten erlebt, die bei den Ärzten und Schwestern zunächst eher Mitleid erregen. Von den Patienten wird die Sonde nach gelegentlichem anfänglichem Widerstand jedoch meist bald ausreichend gut toleriert. Selbstverständlich ist darauf zu achten, daß das Einführen einer Magensonde eine jeweils individuelle, auch psychische Bedeutung hat bzw. haben kann: vom Überwältigtwerden (gewaltsames Durchdringen der Körpergrenzen) bis zur schuldfreien Befriedigung von abgewehrten symbiotischen Bedürfnissen. Gelegentlich trennen sich die Patienten von der Sonde nur unter Schwierigkeiten.

> Franziska entwickelte eine „Haßliebe" zur Sonde – wohl analog zur ambivalenten Beziehung zur Mutter bzw. der Ambivalenz in den oralen Triebbedürfnissen – und wollte schließlich noch länger, als vom Behandlungsplan her nötig, über die Sonde gefüttert werden.

Trotz der mit der Sondenfütterung verbundenen psychischen Probleme erscheint es uns nach wie vor in vielen Fällen günstiger, konsequent das notwendige Ziel der Gewichtszunahme anzustreben, als durch unzureichende Maßnahmen Zeit zu verlieren, die Patienten in ihrem qualvollen Zustand zu belassen und hinzunehmen, daß die Beziehungskonflikte zwischen Patienten und Behandlungsteam (und oft genug auch innerhalb des Teams) infolge der Enttäuschung über die Patienten eskalieren.

38.2.7.4 Häufige Fehler in der Behandlung von Anorexiepatientinnen

Durch insuffizientes, weil inkonsequentes Vorgehen wird vielfach Zeit verloren. Das Leid der Patienten, das unkalkulierbare Risiko von Komplikationen (bis zur „Pseudoatrophie" des Gehirns) und auch die erheblichen Kosten machen ein möglichst rationales Vorgehen erforderlich.

Inkonsequentes Vorgehen

Ärzte und Schwestern lassen sich zunächst oft auf das Handeln und Feilschen der Patienten ein; therapeutische Bemühungen eskalieren dann, sei es in Form besonderer Zuwendung oder in Form aggressiv geführter Auseinandersetzungen, bis endlich die Therapeuten resignieren und die Patienten triumphieren, wieder einmal „alle geschafft" zu haben. Ein erfolgreiches Vorgehen gerade bei schwierigen Patienten hat ein durchgängig klares Verhalten sämtlicher Teammitglieder zur Voraussetzung; Grundlage dieser Klarheit sollte die Einsicht in das Wesen des Krankheitsprozesses, und nicht ein aggressives Gegenagieren gegen die herausfordernden Kranken sein.

> Franziska konnte einen ganzen Tag lang nicht vor Lachen darüber an sich halten, daß es ihr beim ersten Aufenthalt gelungen war, mit 49 kg statt mit den vereinbarten 51 kg entlassen zu werden. Sie hatte das Gefühl, daß sie „alle geschafft" hatte.

Nicht selten unterstützen die Patienten ihr Feilschen durch eine Art „Pseudoflirt", den sie als scheinbar liebe kleine Mädchen mit den Ärzten zu führen versuchen. Das Eingehen auf diese äußere „Fassade" ist auch deshalb gefährlich, weil dann nur allzu leicht die schwere Pathologie der Ich- und Selbst-Entwicklung hinter dieser Fassade übersehen wird.

Unterlassen der Konfrontation bei Behinderung der Therapie durch die Patienten

Häufig werden die Patienten nicht mit ihren den Therapieerfolg behindernden Verhaltensweisen, wie Erbrechen, heimliche Einnahme von Laxantien, Verstecken von Essen usw. konfrontiert. Sie werden vielmehr zu lange als brave, freundliche Mädchen mit naiver Zuwendung behandelt. Auch in der neueren Literatur finden sich hierfür noch Beispiele. Der Erfolg eines solchen, psychologisch gesehen „naiven" Vorgehens besteht lediglich in einer weiteren Vermehrung der unbewußten Schuldgefühle der Patienten. Die klare Konfrontation – ohne Anklage – „wir wissen, daß Sie nicht anders können als . . .", entlastet die Kranken.

In schwierigen Fällen raten wir dringend dazu, Schubladen und Schränke usw. nach versteckten Medikamenten offen zu durchsuchen.

Zu früher Behandlungsabbruch

Soll die stationäre Behandlung eine Erfolgschance haben, darf sie nicht bereits nach einer Gewichtszunahme von einigen wenigen Kilogramm unterbrochen werden, auch wenn hierdurch die akute Lebensbedrohung beseitigt wurde. Die Patientinnen sollen vielmehr die Chance bekommen, soweit aufgefüttert zu werden, daß zumindest der mit der Abwehr des Hungergefühls verbundene Selbstverstärkungsprozeß unterbrochen wird. Ein besonderer Widerstand tritt oft mit der Annäherung an dasjenige Gewicht auf, bei dem die Menstruation sistierte. Hier spielen wahrscheinlich hormonale Prozesse eine Rolle. Auch dieser Widerstand sollte nicht zur Beendigung der Behandlung führen.

> Franziska konnte sich nach der Entlassung mit 49 kg gerade noch in ihre alten Kleider zwängen, was für sie konsequenterweise eine Aufforderung zu erneutem Abnehmen bedeutete. Nach der zweiten stationären Behandlung wurde sie mit 53 kg entlassen; diesmal besorgte sie sich neue Kleider, wozu sicher auch noch andere Motive, die sie jetzt bei fortgeschrittener Psychotherapie zulassen konnte, beigetragen haben dürften.

Die 50-Kilogramm-Grenze wird häufig als „magische Grenze" erlebt, die die Patientinnen nicht überschreiten möchten. Sie verbinden diese Gewichtsgrenze mit „endgültigem Erwachsensein", mit „die Kinderschuhe endgültig ausziehen" (Clauser, 1964), ähnliche Erfahrungen, zum Teil auch bei niedrigeren Gewichtsgrenzen, berichtet Crisp (1983).

Laisser-faire gegenüber exzessiver Flüssigkeits- und Nahrungszufuhr

Gelegentlich verändert sich das Eßverhalten der Anorektikerinnen während der Behandlung in sein Gegenteil. Exzessive Nahrungsaufnahme kann Erbrechen perpetuieren, exzessive Flüssigkeitszufuhr, vor allem bei Behandlungsbeginn, unter Umständen zu somatischen Komplikationen führen. In jedem Fall erleben die Patientinnen einen solchen unkontrollierten Durchbruch ihrer Bedürfnisse als beschämend und kränkend; ihr höchstes Ziel ist ja Selbstkontrolle und Selbstbestimmung. Aus dieser Kränkung können sich erneute Widerstände gegen die Behandlung, gelegentlich auch suizidale Tendenzen ableiten. Frühzeitige therapeutische Intervention und ein konsequentes Verhalten aller Beteiligten ist auch hier im Sinne eines schutzbietenden Halts für die Kranken erforderlich.

Fehler in der somatischen Therapie

Die Patienten werden in der Anfangsphase der Behandlung gelegentlich durch Infusionstherapie überwässert, was zu der sonst eher seltenen Ödembildung beitragen kann: in einem von uns beobachteten Fall eines 14jährigen männlichen Magersüchtigen führte dies zum Lungenödem.

Bei einer Zufuhr von täglich 3500 Kalorien können die Patienten durchschnittlich täglich ca. 220 g zunehmen (1500 Kalorien für Basisumsatz, bei einem Äquivalent von 9 Kal. für 1 g Fett). Die von uns zuletzt behandelten 28 Patientinnen nahmen im Durchschnitt 230 g/Tag zu. Im Behandlungsverlauf ist darauf zu achten, daß die tatsächliche Gewichtszunahme etwa der möglichen entspricht. Zu berücksichtigen ist allerdings, daß aus verschiedenen Gründen die Gewichtszunahme nicht in einem linearen Verhältnis zur Nahrungsaufnahme steht (Pertschuk et al., 1983). Nehmen die Patientinnen zuwenig zu, ist – die Korrektheit der Kalorienzufuhr durch Sondenkost vorausgesetzt – davon auszugehen, daß sie von sich aus die Behandlung behindern. Andererseits sollte nicht aus einem übertriebenen therapeutischen Ehrgeiz eine raschere Gewichtszunahme als die physiologisch mögliche erwartet werden; die Patientinnen sollten nicht für eine solche unphysiologische Gewichtszunahme, die auf vermehrte Wassereinlagerung zurückzuführen sein könnte, gelobt werden.

Bei der **Sondenbehandlung** ist auf folgende **Komplikationen** zu achten: Kontrolle der Lage der Sonde nach Einführen (Aspiration von Magensaft!); eine Lage der Sonde in den Bronchien kommt gelegentlich bei im Rachenraum anästhetischen Patientinnen (möglicherweise ein Konversionssymptom) vor, in der Folge als Komplikation bei der Einfuhr von Nahrung durch die Sonde schwere Aspirationspneumonien. Die Sonde sollte je nach Material nach ca. 14 Tagen gewechselt werden, da brüchiges Material zu stärkeren Reizungen der Schleimhaut führen kann. Es sollte auch nicht übersehen werden, daß die Patientinnen nicht selten den Mageninhalt durch Ansaugen wiederum über die Sonde entleeren.

Die **Wiege-Prozedur** nimmt in der Behandlung oft eine zu zentrale Bedeutung ein. Es genügt, die Patientinnen zweimal wöchentlich zu wiegen. Kleinere tägliche Gewichtsschwankungen führen nur zu Beunruhigung oder zu ständigem Feilschen.

38.2.7.5 Behandlungsansätze zur Beeinflussung des Eßverhaltens und der Persönlichkeitsstörungen

Psychoanalytische Therapieformen

Diese Therapieansätze entsprechen unseres Erachtens am weitgehendsten den von den Patienten erfahrbaren Konflikten.

Allerdings ist wegen der genannten Widerstände nur etwa ein Drittel der Erkrankten einer solchen Therapie zugänglich. Besonders schwierig ist die Initialphase der Therapie, die Entwicklung eines Arbeitsbündnisses. Sonst unterscheidet sich die Behandlung im wesentlichen nicht von der psychoanalytischen Behandlung anderer Patienten mit entsprechenden Neurosen (Thomä, 1961).

Wie auch sonst in der Psychotherapie von Adoleszenten ergibt sich häufig die Notwendigkeit einer gleichzeitigen Beratung oder auch psychotherapeutischen Behandlung der Eltern oder zumindest eines Elternteils.

Mit einer entscheidenden Besserung kann bei 30 bis 50% der so psychotherapeutisch behandelten Kranken gerechnet werden (Thomä, 1961; Sperling, 1965; Fleck, 1965). Diese Besserung bezieht sich sowohl auf das Eßverhalten als auch auf die Entwicklung im psychosozialen Bereich; auch die langfristige Prognose ist in diesem Fall als gut anzusehen (Thomä, 1972; Bruch, 1973). Allerdings gilt diese optimistische Einschätzung möglicherweise nur für die aufgrund ihrer Zugänglichkeit für konfliktorientierte psychoanalytische Verfahren bereits ausgewählte Patientengruppe.

Familientherapie

Die Untersuchungen zur Familiendynamik der Anorexiekranken forderten zur Entwicklung von Be-

handlungsansätzen heraus, die eine Veränderung des „Systems Familie" zum Ziel haben (vgl. Kap. 21).

Die familientherapeutischen Interventionen verändern den gesamten Lebenskontext der Kranken; sie können dadurch bei jüngeren, noch in der Elternfamilie lebenden Patientinnen überraschend schnell zu dauerhaften Krankheitsremissionen führen; der Krankenhausaufenthalt kann entscheidend verkürzt werden (8–25 Tage) (Minuchin et al., 1974a, b).

1978 berichteten Minuchin und Mitarbeiter über die Nachuntersuchung von 53 familientherapeutisch behandelten Anorexiepatienten. Das Durchschnittsalter dieser Gruppe (6 männliche Kranke) betrug 14½ Jahre. 60% der Patienten befanden sich im Adoleszentenalter. 3 Familien brachen die Therapie ab, die durchschnittliche Behandlungsdauer bei den übrigen Familien betrug 6 Monate (2–16 Monate). Die Behandlung wurde von insgesamt 16 verschiedenen Therapeuten („3 Therapeutengenerationen") durchgeführt. 1½–7 Jahre nach Abschluß der Behandlung fanden die Autoren 86% (!) geheilt. Dies galt sowohl für die Gewichtsentwicklung und das Eßverhalten als auch für die Situation im psychosozialen Bereich: das Verhalten und die Beziehungen in der Familie, mit Gleichaltrigen und in der Ausbildungs- bzw. Arbeitssituation. Ein Vergleich mit einer parallelisierten Gruppe von Kranken aus der Patientengruppe von H. Bruch (1973) spricht für eine Überlegenheit des familientherapeutischen Ansatzes bei dieser Patientengruppe.

Selvini (1974), die sich als Psychoanalytikerin viele Jahre lang mit der intrapsychischen Dynamik und den Objektbeziehungen von Anorektikerinnen befaßt hat, bei dieser Patientengruppe allerdings nur bescheidene therapeutische Erfolge erreichen konnte, berichtet über eine dramatische Verbesserung der Therapieerfolge nach Einführung des familientherapeutischen Ansatzes in die Behandlung. Bei 12 Patientinnen verschwand die Anorexie nach etwa 15 Sitzungen dauerhaft, gleichzeitig hatten sich ihre Familien tiefgreifend verändert.

Eine solche hochqualifizierte **familienpsychotherapeutische** Behandlung kann bisher nur in wenigen Zentren durchgeführt werden. Man kann sie nicht durch eine weniger spezifische Beratung der Familien ersetzen. Wirsching und Stierlin haben die Möglichkeiten des behandelnden Arztes in der Beratung von Familien im einzelnen ausgeführt (1982). Die Einbeziehung der Familien erweist sich schon unter der übergeordneten Zielvorstellung der Verminderung der Mortalität als wichtig:

Grolnick (1972) fand bei einer Nachuntersuchung von 115 Anorexiekranken als auffallendes Merkmal der 4 verstorbenen Patientinnen, daß nur bei ihnen die Familie jede Kontaktaufnahme mit den betreuenden Ärzten und Schwestern vermieden hatte. Über ähnliche Erfahrungen berichten Petzold und Mitarbeiter (1970): Scheitern der Behandlung sowie letale Verläufe wurden nur bei solchen Patienten gesehen, bei denen es nicht gelungen war, die Familie mit in die Behandlung, im Sinne einer „Familienkonfrontationstherapie", einzubeziehen.

Verhaltenstherapie

Ziel der Verhaltenstherapie ist die Beeinflussung des Eßverhaltens im Sinne eines systematischen Umlernens (vgl. Kap. 19).

Während die Patienten sonst zumeist Aufmerksamkeit und Zuwendung für ihre pathologischen Verhaltensmuster – Hungern und Gewichtsabnahme – erhalten, werden sie jetzt systematisch für vermehrtes Essen und Gewichtszunahme belohnt.

Zu Beginn der Behandlung werden die Patientinnen im allgemeinen weitgehend sozial isoliert. Als Behandlungsmethode wird das **operante Konditionieren** bevorzugt. Dabei wird entweder das Eßverhalten oder die Gewichtszunahme verstärkt. Als Verstärker werden vor allem die Möglichkeiten zu sozialem Kontakt (Gruppenaktivitäten, Beurlaubungen nach Hause u.a.) und die Möglichkeit zu körperlicher Aktivität gewählt. Belohnung und Verstärker werden dem jeweils erreichten Therapiefortschritt angepaßt, mit Erreichen des Sollgewichtes ist eine Umstellung von der Belohnung der Gewichtszunahme auf die Belohnung der Aufrechterhaltung des Sollgewichtes wesentlich (Schaefer und Schwartz, 1974), was anfangs häufig übersehen wurde.

In neueren Veröffentlichungen zum verhaltenstherapeutischen Vorgehen bei Anorexia-nervosa-Kranken wird neben der früher meist allein als Erfolgskriterium berücksichtigten Gewichtsentwicklung meist auch auf die Veränderung verschiedener, insbesondere sozialer Verhaltensweisen Wert gelegt (Fichter und Keeser, 1980; Fichter et al., 1983), dabei wird auch auf die Grenzen des verhaltenstherapeutischen Ansatzes aufmerksam gemacht.

Fichter und Mitarbeiter (1983) zeigen, daß während der stationären, breit angelegten verhaltenstherapeutischen Behandlung von 24 Anorexia-nervosa-Kranken sich neben dem Gewicht auch spezifische Haltungen und Verhaltensweisen von Anorektikerinnen signifikant verbesserten, während allgemeinere neurotische Probleme wie Insuffizienzgefühle, Ängste vor Sexualität und zwanghafte Züge sich nicht besserten.

Für eine abschließende Beurteilung des verhaltenstherapeutischen Ansatzes fehlen noch katamnestische Untersuchungen bei ausreichend vielen Patienten über eine ausreichend lange Zeit. Einzelne Befunde referieren wir in Abschnitt 38.2.9.

Der verhaltenstherapeutische Therapieansatz erfordert ausgebildete und hochmotivierte Mitarbeiter, insbesondere im Pflegebereich, sowie eine intensive und reflektierte Kooperation aller beteiligten Teammitglieder.

38.2.7.6 Gestufter Therapieplan in Kombination verschiedener Behandlungsverfahren

Zusammengefaßt erscheint es heute am aussichtsreichsten, Anorexiepatientinnen während der akuten Krankheitsphase stationär mittels Verhaltenstherapie oder Nahrungszufuhr über die Nasen-Magen-Sonde wiederaufzufüttern und gleichzeitig zu versuchen, eine psychotherapeutische Behandlung einzuleiten, die anschließend unter stationären oder ambulanten Bedingungen längerfristig fortgeführt wird. Dieses Vorgehen sollte durch Methoden der Familientherapie ergänzt werden. Entsprechende mehrstufige Ansätze

werden von verschiedenen Forschungszentren empfohlen (Liebman et al., 1974; Minuchin et al., 1974b; Wellens und Pierloot, 1975; Pierloot et al., 1982).

Zweiphasenprogramm von Pierloot (1975, 1982)

Die Arbeitsgruppe um Pierloot in Louvain berichtet seit über 10 Jahren über einen Behandlungsansatz, der immer wieder sorgfältig evaluiert und aufgrund der Befunde jeweils weiterentwickelt wurde.

Auch bei diesem Behandlungsansatz besteht das Ziel zunächst in einer Gewichtsnormalisierung und der Wiederherstellung eines normalen Eßverhaltens als Voraussetzung für weitergehende psycho- und soziotherapeutische Maßnahmen.

Das während der ersten Jahre angewandte Programm mit forcierter Nahrungszufuhr und Psychopharmakagabe wurde später vollständig durch verhaltenstherapeutische Techniken ersetzt. Operante Konditionierungsverfahren bewährten sich hinsichtlich der Gewichtsnormalisierung, betonten jedoch nach dem Eindruck der Autoren zu sehr die Bedeutung der äußeren Kontrolle bei der Symptomveränderung. Heute liegt die Betonung auf einer internen Kontrolle, mit den Patientinnen wird ein Behandlungsvertrag geschlossen, der eine wöchentliche Gewichtszunahme von 700 g (Maximum 2 kg) einschließt; ein operantes Konditionierungsprogramm wird nur noch als eine Art negativer Verstärker angewandt, wenn sich Patienten als unfähig erweisen, den Behandlungsvertrag einzuhalten. Die Patienten werden so in strukturierter Form mit einer stetigen, sie nicht überwältigenden Gewichtszunahme und den entsprechenden körperlichen Veränderungen konfrontiert. Sie beteiligen sich an einer speziellen Form psychomotorischer Behandlung, in deren Mittelpunkt das sich verändernde Körperbild steht und erlernen die progressive Muskelentspannung nach Jacobson (1944).

Erste, symptomorientierte Phase

Dieser Behandlungsansatz wird auf einer hierfür spezialisierten Krankenstation gleichzeitig bei 10–15 Anorexiepatientinnen angewandt. Neben explorativen und supportiven Einzelgesprächen finden wöchentliche Gruppengespräche unter der Leitung eines Therapeuten statt, der sonst nicht in das Therapieprogramm involviert ist. In diesen Sitzungen können die Patienten vor allem ihre Angstgefühle, ihr Erleben von Hilflosigkeit und ihre aggressiven Impulse, die das Behandlungsregime mit hervorruft, ausdrücken und untereinander austauschen. Als mitentscheidend für das Gelingen dieses Ansatzes hat sich vor allem auch die Kooperation mit den Familien der Patienten herausgestellt. Die Abbruchrate nahm erheblich ab, nachdem eine Gesprächsgruppe für die Eltern bzw. die Partner verheirateter Patienten (14tägig) eingeführt wurde. Dieser erste Behandlungsabschnitt ist beendet, wenn Gewicht und Eßverhalten normalisiert sind, er dauert im Durchschnitt 3 Monate.

Dieser Behandlungsansatz setzt die Zusammenarbeit mit entsprechend weitergebildeten Schwestern voraus und erfordert, daß auftretende Schwierigkeiten, insbesondere im Zusammenhang mit Gegenübertragungsproblemen, im Rahmen von Stationskonferenzen geklärt werden können.

Zweite, problemorientierte Phase

Ziel dieser zweiten Phase ist die soziale Reintegration. Jetzt sollen auch die psychischen Konflikte geklärt und bearbeitet werden.

Dieser Behandlungsabschnitt wird bei der Mehrzahl der Patienten auf einer Gruppenpsychotherapie-Station durchgeführt, bei einer kleineren Zahl ambulant mit individueller Psychotherapie und/oder Familientherapie.

Die Gruppenpsychotherapie-Station ist als therapeutische Gemeinschaft organisiert; hier haben die Anorexiepatienten Kontakt mit anderen Kranken mit unterschiedlichen neurotischen Problemen. Neben der Gruppentherapie finden psychoanalytisch orientierte Einzelsitzungen statt. Stationssetting und therapeutische Angebote bieten die Möglichkeit zu einer therapeutischen Regression und zur Entwicklung konstruktiver Alternativen.

Den Autoren erscheint es wichtig, die Behandlung nach Abschluß der ersten Phase selbst fortzusetzen, sie haben mit der Überweisung der Patienten nach Abschluß der ersten Phase schlechte Erfahrungen gemacht: Entweder nahmen die Patienten das weiterführende Therapieangebot nicht an, oder es ergaben sich Schwierigkeiten beim Versuch, zum neuen Therapeuten wieder eine gute Beziehung aufzubauen. Rückfälle waren häufig.

Die Behandlung auf der Gruppenpsychotherapie-Station dauerte in der Regel zwischen 9 und 12 Monaten. Eine individuell geplante Nachbehandlung schließt sich an.

Auf der Station ergaben sich mit Zunahme der Zahl der Anorexiepatienten (im Durchschnitt ein Drittel aller Kranken) neue Probleme. Überraschend traten gehäuft Rückfälle der anorektischen Symptomatik auf, die „epidemisches" Ausmaß anzunehmen drohten. Daraufhin wurde auch auf dieser Station ein Vertragssystem eingeführt, das für den Fall einer Gewichtsveränderung, die 10% des normalen Gewichts übersteigt, restriktive Maßnahmen enthält.

Pierloot und Mitarbeiter berichten über Katamnesen der von ihnen behandelten 145 Patienten (1975, 1981, 1982). Die Übersicht über die Patientenmerkmale bestätigte, daß es sich um schwer gestörte, zum Teil erheblich chronifizierte Anorexiepatienten handelte. Dies schlägt sich schon in der relativ hohen drop-out-Rate nieder: Von 145 Patienten hatten nur 67 die Behandlung bis zum Schluß durchgehalten, bereits während der ersten Behandlungsphase waren 29 Patienten (22%) aus dem Programm ausgeschieden. Die durchschnittliche Katamnesezeit betrug 4,2 Jahre. Von den 88 Patienten, für die katamnestische Daten gewonnen werden konnten, waren 10 gestorben (4 an Selbstmord, 4 an den Folgen von Unterernährung, 2 aus unbekannter Ursache). Für die übrigen 78 ergab die Nachuntersuchung hinsichtlich der anorektischen Symptomatik folgende Befunde: 20 waren leicht, 21 deutlich gebessert; 28 waren gesund (zusammen 78%); 9 waren weiter krank. In psychosozialer Hinsicht war die Entwicklung bei 8 Patienten schlecht, bei 25 ungünstig, bei 16 günstig, bei 29 gut (gut/günstig bei 51%).

Aufgrund ihrer großen Erfahrungen versuchen die Autoren heute während der ersten Phase Zielvorstellungen für das weitere therapeutische Vorgehen aufgrund der vorliegenden Problemkonstellation zu formulieren: Bei Patienten, deren Erkrankung als mißlungener Lösungsversuch von Interaktionskonflikten, insbesondere innerhalb des Familiensystems erscheint, empfehlen sie ambulante Familien- bzw. Ehetherapien. Für Patienten, bei denen die Nahrungsverweigerung die Unfähigkeit ausdrückt, mit der psychobiologischen Reifungskrise der Adoleszenz zurechtzukommen und Verantwortung für das Leben als autonomer Erwachsener zu übernehmen, empfehlen sie in der Regel die Behandlung auf der Gruppenpsychotherapie-Station; lediglich bei präadoleszenten oder adoleszenten Patienten mit kurzer Krankheitsdauer (meist ohne vorausgegangenen Behandlungsversuch), die in einer Familie ohne größere erkennbare psychische Probleme leben, wird eine ambulante individuelle Psychotherapie vorgeschlagen.

Pierloots Ergebnisse zeigen – ähnlich wie die Befunde Deters (1983), daß trotz eines sorgfältig geplanten Therapieprogramms ein Teil der schwer gestörten Anorexiepatientinnen auch heute noch eine ungünstige Prognose hat. Zur Zeit bemühen sich die Autoren darum, Kriterien für die Identifikation solcher Kranker zu erarbeiten.

Eigenes Vorgehen an einer internistischen Klinik

Parallel zur Notfallbehandlung und Wiederauffütterung – zumeist mit Hilfe der Sondenbehandlung – versuchen wir, die Patienten in ihren psychischen Konflikten zu erreichen und ihnen hierfür Verständnis anzubieten, sie nach Möglichkeit für eine analytisch orientierte Psychotherapie zu gewinnen. In jedem Fall führen wir mit den Eltern ein oder mehrere Gespräche. Nach der Entlassung bemühen wir uns darum, entsprechend der Motivation und Eignung der Patienten und der vorhandenen eigenen Kapazität, eine kurz- oder längerfristige analytisch orientierte Psychotherapie anzubieten. Wir versuchen – zusammengefaßt – eine differenzierende Haltung einzunehmen und bemühen uns mit Nachdruck darum, wenigstens diejenigen Patienten psychotherapeutisch zu erreichen, die einen Gewinn aus einer konfliktorientierten Behandlung ziehen können.

Zunächst behandelten wir Anorexiepatientinnen auf verschiedenen Stationen einer internistischen Klinik. Die Behandlung sollte vom jeweiligen Stationsarzt unter Hinzuziehung eines psychosomatischen Konsiliarius durchgeführt werden. Aufgrund der regelmäßig auftretenden Interaktionsprobleme zwischen Stationsteam und Patient und auch innerhalb des Stationsteams gelang es in vielen Fällen nicht einmal, die Wiederauffütterung mittels der Nasen-Magen-Sonde konsequent durchzuführen. Die Einleitung einer psychotherapeutischen Behandlung wurde durch diese Schwierigkeiten noch zusätzlich behindert.

Deshalb führten wir in der Folgezeit die Behandlung Anorexiekranker ausschließlich auf unserer internistisch-psychosomatischen Krankenstation (Köhle et al., 1977; vgl. Kap. 28) durch, auf der das Behandlungsregime von allen Mitgliedern des Stationsteams mitgetragen wurde und auftretende Probleme systematisch geklärt werden konnten. Selbst hier erwies sich die Koordination des Vorgehens noch häufig als schwierig genug.

In Übereinstimmung mit anderen Autoren wurde die somatische und psychotherapeutische Behandlung getrennt von zwei verschiedenen Therapeuten durchgeführt, da die konsequente somatische Behandlung Strenge und Konsequenz erfordert und nicht selten Gegenübertragungsreaktionen erzeugt, die den verständnisvollen psychotherapeutischen Ansatz behindern würden. Bei diesem Vorgehen ist die intensive Kommunikation zwischen den beiden Therapeuten und den übrigen Teammitgliedern von besonderer Bedeutung.

Bei Franziska trat während der ersten Monate der ambulanten Nachbehandlung ein Rezidiv auf, das eine zweite stationäre Aufnahme erforderlich machte. Im weiteren Verlauf der Behandlung – 200 Psychotherapiestunden im Verlauf von 2 Jahren – gelang es Franziska, ihre Probleme soweit zu bearbeiten, daß sie die anorektische Symptombildung nicht mehr zur Konfliktlösung heranziehen mußte. Trotz großer innerer und äußerer Schwierigkeiten gelang ihr die Trennung von der Familie und der Studienbeginn in einer entfernten Universitätsstadt; bis zuletzt drohte die Mutter, den Weggang der Tochter zu behindern: verschiedene funktionelle Beschwerden der Mutter verschlimmerten sich derart, daß sie bettlägerig wurde; die Familie war hierdurch so beeindruckt, daß sie bereits eine chronische Pflegebedürftigkeit der Mutter zu akzeptieren begann. Franziska begann am Studienort erstmals in ungezwungener Weise Kontakt mit Gleichaltrigen aufzunehmen und im Zusammenhang mit der Wahl ihres Studienfaches ihre eigenen Interessen und Lebensziele zu klären. Die bisher 6jährige Katamnese ergab eine weiterhin günstige Gesamtentwicklung.

Nach einer durchschnittlichen Katamnesedauer von fünf Jahren und drei Monaten konnten alle 36 Patienten, die zwischen 1972 und 1979 nach dem dargestellten Konzept behandelt worden waren, nachuntersucht werden. Das durchschnittliche Erkrankungsalter (33 Frauen, 3 Männer) betrug 16 Jahre 8 Monate, das durchschnittliche relative Aufnahmegewicht 70,3% des individuellen Idealgewichts, die mittlere prozentuale Gewichtsabnahme 31,9%. Eine Patientin war 3 Monate nach Behandlungsende an unbekannter Ursache verstorben (Mall, 1983; Köhle und Mall, 1983).

Die mit Interviews und Fragebogen erhobenen Befunde entsprechen den mit derselben Methodik gewonnenen Ergebnissen Pierloots:

Der Anteil der mit „geheilt" und „gebessert" Bezeichneten betrug 72% hinsichtlich der körperlichen Symptomatik und 80% hinsichtlich der Gesamtsymptomatik (17 Patienten ging es gut, 12 mittel, 6 schlecht). 6 Patienten hatten weniger als 75% ihres aktuellen Idealgewichts, 8 berichteten über rezidivierendes Erbrechen, 7 zeigten bulimisches Verhalten. 11 Patientinnen waren weiterhin amenorrhoisch, 3 hatten nur durch Ovulationshemmer stimulierte Blutungen.

Wichtig erscheint uns die retrospektive Bewertung der einzelnen Therapiemaßnahmen durch die Patienten: Zur Behandlung insgesamt äußerten sich 58% positiv, 6% negativ (die übrigen jeweils unentschieden). Die Sondentherapie wurde von 25% positiv, von 31% negativ bewertet. Ähnlich wie Körner (1978) waren wir über die hohe Zahl derjenigen Patienten überrascht, die im Nachhinein die durchgeführten bzw. versuchten psychotherapeutischen Maßnahmen positiv bewerteten (50% gegenüber 6% negativ). Viele Patienten berichteten davon, daß sie zwar seinerzeit dieses Angebot negativ beurteilt, später jedoch entweder diesen Hinweis aufgriffen und sich um eine entsprechende Behandlung bemüht hätten, oder doch ihre Ablehnung bedauerten.

38.2.8 Prognose

Die Angaben in der Literatur zur Prognose schwanken stark. Vor allem bei Therapiestudien sind vor einer endgültigen Bewertung des Therapieverfahrens zunächst immer die Zusammensetzung der untersuchten und behandelten Patientengruppen sowie Selektionsprozesse im Vorfeld und während der Therapie (drop out) sorgfältigst zu prüfen. Heftig geführte Kontroversen können hierdurch und durch Langzeitkatamnesen überflüssig werden.

38.2.8.1 „Spontanverlauf"

Die Prognose bei Spontanverlauf muß als schlecht bezeichnet werden: Die Mortalität beträgt bis 12%, die Anorexie chronifiziert bei ca. 40% der Patientinnen. Bei 20–30% findet sich eine „Spontanheilung" bezogen auf das Körpergewicht; schwere Störungen im psychischen und sozialen Bereich bleiben jedoch bestehen. Die Patientinnen leben in einer sozialen Randexistenz bzw. werden Dauerpatientinnen in psychiatrischen Kliniken. Bei ca. 10% der Kranken entwickeln sich chronische Psychosen.

Als Anhalt für eine Orientierung läßt sich zusammenfassen: „1/3 bleibt anorektisch und zeigt einen chronischen Verlauf, 1/3 wird psychisch schwer krank bzw. psychotisch nach Verlust der Anorexia-nervosa-Symptomatik, der Rest zeigt Syndromwandel und Besserung" (Cremerius, 1978).

Wir haben in Tabelle 38.2–3 (nach Mall, 1983) 19 katamnestische Untersuchungen zusammengestellt, die nicht als Therapiekatamnesen gekennzeichnet waren. Die Mortalität beträgt für die in diesen 19 Studien erfaßten 772 Patienten 12%.

Auch sorgfältige katamnestische Untersuchungen haben den Krankheitsverlauf bisher nur kurz- bis mittelfristig erfaßt. Langfristige Untersuchungen bei einem größeren Patientengut liegen nicht vor. Einzelbeobachtungen weisen jedoch darauf hin, daß die Prognose, langfristig gesehen, vor allem im psychosozialen Bereich noch schlechter ist, als bisher angenommen wurde.

Cremerius (1978) berichtet über eine Nachuntersuchung der von ihm zwischen 1947 und 1950 erstuntersuchten Anorexiepatientinnen. Aus den insgesamt 11 Katamnesen ergibt sich 26 bis 29 Jahre später folgendes Bild:

Von ursprünglich 13 Kranken (12 Frauen, 1 Mann) sind 2 im Zusammenhang mit der Erkrankung verstorben; bei 7 Patienten ist das ursprüngliche Syndrom nicht mehr nachweisbar. In dieser Gruppe finden sich 4 Kranke mit mehr oder weniger deutlich ausgeprägten Psychosen. 2 Patienten ohne Anorexia-nervosa-Symptomatik sind insgesamt wesentlich gebessert, 2 Patienten mit fortbestehender Anorexia-nervosa-Symptomatik sind im sozialen Bereich entscheidend gebessert.

Aus der Langzeitstudie von Theander (1970, 1983a, 1983b) ergibt sich ein ähnlich ungünstiges Bild bei insgesamt 94 Patienten. 22 bis 50 Jahre nach der ersten Krankenhausaufnahme der Patientinnen waren

Tabelle 38.2–3. Übersicht über 19 katamnestische Untersuchungen, die nicht als Therapiekatamnesen gekennzeichnet waren (nach Mall, 1983). (Jahr = Jahr der Veröffentlichung; n = Anzahl der einbezogenen Patienten; Quote = Prozentsatz der nachuntersuchten Patienten)

Autor	Jahr	n	davon ♂	Quote	Katamnese-dauer	geheilt	gebessert	krank	verstorben	(davon Suizid)
Meyer, J. E.	61	33	(–)	?	∅10J	9	–	20	4	(–)
Frazier	65	39	(4)	?	(5–20J)	–	15	21	3	(–)
Samuel-Lajeunesse	67	33	(?)	34%	(>4J)	10	8	13	2	(1)
Browning	68	36	(–)	?	(1–30J)	–	27	6	3	(–)
Seidensticker	68	53	(?)	88%	(>1J)	20	16	8	9	(2)
Warren	68	18	(–)	90%	(1–12J)	2	8	6	2	(–)
Ziegler	68	30	(?)	26%	?	16	8	–	6	(–)
Rowland	70	17	(?)	55%	?	–	8	6	3	(–)
Theander	70	97	(–)	?	(>6J)	50	28	7	12	(3)
Niskanen	74	46	(2)	?	∅5,2J	20	17	3	6	(1)
Davy	76	35	(–)	61%	∅6J	18	–	17	–	(–)
Halmi	73+75	79	(7)	83%	(1–35J)	41	7	11	20	(–)
Stonehill	77	39	(–)	87%	(4–7J)	26	26	12	1	(?)
Sturzenberger (+Cantwell 77)	76+77	26	(1)	29%	∅4,9J	12	10	4	–	(–)
Körner	78	32	(1)	78%	∅5,2J	6	8	13	5	(2)
Cremerius	65+78	13	(1)	100%	(26–29J)	3	–	7	3	(1)
Lauren	80	31	(3)	100%	?	30	1	–	–	(–)
Pavlovic (+Jörgens 80)	81	103	(3)	62%	?	30	40	22	11	(2)
Mester	81	12	(12)	71%	(1–36J)	1	1	9	1	(1)
insgesamt		772	(≧34)		>1J	64%	24%	12%		(2%)

13 an den Folgen der Anorexie, 4 an Suizid gestorben (zusammen 18%). Unklar bleibt, wieweit dieser ungünstige Verlauf Folge des mangelhaften Wissensstandes und der beschränkten therapeutischen Möglichkeiten zur Zeit der Ersterkrankung der Patientinnen ist.

Über einen ungewöhnlich günstigen Verlauf berichten Farquharson und Hyland (1966) bei 15 Patienten (3 Männer, 12 Frauen), die die Autoren 20 bis 30 Jahre nach der Erstbehandlung nachuntersuchten: 10 der Patienten waren symptomfrei, lebten ein „gut angepaßtes und nützliches Leben"; lediglich in einem Fall chronifizierte die Anorexie primär, 3 Kranke litten unter periodisch auftretenden schweren neurotischen Symptomen, in einem Fall kam es später zu einem Anorexie-Rezidiv. Die 4 zuletzt genannten Kranken waren jedoch während des größten Teils der Nachuntersuchungszeit in der Lage, ein einigermaßen befriedigendes Leben zu führen. Die ursprüngliche Behandlung bestand teils in einer erfolgreichen Ermutigung zu einer erhöhten Kalorienzufuhr unter stationären Bedingungen, teils in Sondenfütterung.

Exkurs: Suizid bei Anorexia-nervosa-Patienten

Über die Häufigkeit von **Suizidversuchen** gibt es in der Literatur nur wenige Angaben: 2 Versuche bei 12 Patienten (Nemiah, 1958), 5 Versuche bei 30 Patienten (Thomä, 1961). In unserem Patientengut beobachteten wir keinen Suizidversuch. Suizidgedanken beschäftigen Anorexiepatienten dagegen häufig, wie auch andere Adoleszenten in Krisensituationen. Über die Häufigkeit **vollzogener Suizide** bei Anorexiekranken finden sich folgende Hinweise:

In der von Theander (1970, 1983) dargestellten Gruppe von 94 Patientinnen begingen insgesamt 4 Selbstmord. In der Mitteilung von 1970 gehörten alle 4 Suizidpatientinnen zur Gruppe der atypischen Anorexia-nervosa-Kranken; zum Zeitpunkt des Suizids hatte sich bei allen die klinische Symptomatik erheblich zurückgebildet, eine Patientin war stark übergewichtig. Kay und Shapiro (1965) beobachteten bei 38 Patienten des Londoner Maudsley-Hospitals einen Suizid, bei 27 später von ihnen behandelten und nachuntersuchten Kranken in New Castle keinen Suizid. Crisp (1965) berichtet über einen Suizid bei 21 Kranken, Seidensticker und Tzagournis (1968) über 2 Suizide bei 60 Kranken (zitiert nach Theander). Von den 64 Patienten Hilde Bruchs (1971) beging keiner Selbstmord.

Damit liegt die Suizidrate der Magersüchtigen – soweit die kleinen Zahlen ein solche Einschätzung zulassen – in der Größenordnung der Suizidrate eines allgemeinen psychiatrischen Patientenguts: 0,2% pro Jahr (Rosman nach Theander, 1970); das Suizidrisiko wäre nach diesen Angaben aus Schweden 20mal höher als dasjenige der Durchschnittsbevölkerung.

Bei der Diskussion um das Suizidrisiko von Anorexiepatienten ist zu berücksichtigen, daß Anorexiekranke nicht, wie oft angenommen, mit ihrer Krankheit einen „Suizid in Raten" begehen. Ihr Ziel ist die Abmagerung und nicht die Selbstvernichtung; in der Abmagerung sehen sie nicht den Tod, sondern häufig eine besonders hohe Entwicklungsstufe, die Verwirklichung eines Lebensideals (Selvini, 1974, 1982). Die Kranken streben, oft in einem verzweifelten Kampf, die **Kontrolle** über ihre Bedürfnisse und ihre Körperfunktionen als Lebensziel an und wollen primär keineswegs sterben (Olstrup nach Theander,

Tabelle 38.2–4. Gewichtszunahme und Behandlungsdauer in Abhängigkeit vom Behandlungsansatz

Autor	n	Gesamt-Zunahme in kg	Dauer der stationären Therapie in Wochen
Williams, 1958	51	4,5	7,6
(Allg. Therapie)	44	3,5	9,6
(Sondenbehandlung)	7	10,8	9,8
Stafford-Clarc, 1958	13	9,9	16,9
(nur Psychotherapie)	6	8,9	23,17
(kombinierte Therapie)	7	11,6	11,6
Dally, 1969	19	10,1	6,1
(Sondentherapie)			
Thomä, 1961	19	4,6	30,1
(Psychotherapie)			
Clauser, 1964	19	12,8	23,3
(Kombination Sondentherapie und Psychotherapie)			
Frahm, 1973	88	13,9	7,6
(Sondentherapie, Phenothiazine, „strenges Regime")			(Sondentherapie, ohne Vor- und Nachbeobachtungszeit)
Minuchin, 1974b	25	13,2	2,3 stationär + 28 ambulant
(Familientherapie)			
Deter, 1983	103		
– internistisch stationär:	25		
– psychosomatisch ambulant:	14	2,8(!) (von 61,8% des Sollgewichts auf 67% des Sollgewichts)	12 (Gesamtdauer des Klinikaufenthaltes)
– internistisch-psychosomatisch stationär:	64		
(Sondenkost, Sedierung, Psychotherapie)			
Köhle und Mall, 1983	36	ca. 12 (von 70,3% auf 90,5% des Idealgewichts)	7,3 (Gesamtdauer des Klinikaufenthaltes)
(Sondenkost und Psychotherapie)			

1970; Bruch, 1971). Hieraus läßt sich zum Teil ableiten, warum in der Kerngruppe von Anorexiekranken Suizidversuche selten sind. Das Suizidrisiko ist dann erhöht, wenn die Kranken in ihrem Kampf um die Aufrechterhaltung der Kontrolle sich vom Scheitern, vom **Kontrollverlust** bedroht fühlen; der Suizid erscheint dann als letzte „autonome" Handlung. Zu diesem Erleben können einseitige, forciert nur die Wiederauffütterung anstrebende Therapieansätze beitragen.

38.2.8.2 Verlauf der Erkrankung nach systematischer Therapie

Intensive klinische Therapie

Die Prognose der **somatischen Symptomatik** läßt sich akut und – mit einiger Wahrscheinlichkeit – auch mittelfristig durch eine konsequente klinische Behandlung entscheidend verbessern.

Tabelle 38.2–4 gibt eine Übersicht über Gewichtszunahme und Therapiedauer bei verschiedenen Autoren. Die Ergebnisse beziehen sich auf die akute Behandlungsphase. Wir stellen diese Ergebnisse in der Absicht dar, wenigstens zu einer konsequenten somatischen Therapie der Anorexie

kranken zu ermutigen, da selbst dies bisher nur an wenigen Orten der Fall zu sein scheint (vgl. Ergebnisse von Deter, 1983).

Psychotherapie

Durch eine konfliktbearbeitende, d.h. psychoanalytisch orientierte Psychotherapie wird die Prognose bei denjenigen Patientinnen wesentlich verbessert, die für eine solche Therapie erreichbar sind (Thomä, 1972; Bruch, 1973). Entscheidend ist die möglichst frühe Einleitung einer solchen Behandlung. Vor allem bei jüngeren Patientinnen vermag ein konsequenter familientherapeutischer Ansatz (Minuchin, 1974b, 1978; vgl. Kap. 21), entweder allein oder in Ergänzung der individuellen Psychotherapie, die Prognose zu verbessern.

Im einzelnen ist noch nicht genügend geklärt, welche Form der psychotherapeutischen Behandlung für einzelne Patienten optimal ist, und welche Modifikationen der Psychotherapie eventuell im Vergleich zur Behandlung neurotisch Kranker erforderlich sind (Thomä, 1972; Bruch, 1978/1980).

In Tabelle 38.2–5 haben wir die Ergebnisse von 19 katamnestischen Untersuchungen von nach definierten Kriterien behandelten Patientengruppen zusam

Tabelle 38.2–5. Übersicht über 19 Therapiekatamnesen (n = nachuntersuchte Patienten; Quote = Prozent der nachuntersuchten Patienten im Verhältnis der zu behandelnden Gruppe)

Nr. Autor	Jahr	n	davon ♂	Quote	Alter bei Beginn der Krankheit	Alter bei Beginn der Behandlung	Katamnesedauer	geheilt	gebessert	krank	verstorben	davon Suizid	Rate geheilt gebessert
1 Beck	54	25	(–)	89%	17,7	21,3	11,3	20	–	4	1	–	80%
2 Frahm	66	49	(–)	100%	?	?	(0–4)	47	–	2	–	–	96%
3 Niederhoff	75	6	(–)	100%	12,7	13,5	2,3	4	1	1	–	–	83%
4 Lesser	60	15	(–)	100%	14,0	?	5,3	13	–	2	–	–	87%
5 Crisp	65	21	(2)	100%	19,0	24,0	1,4	9	6	3	3	1	71%
6 Morgan	75	41	(3)	100%	15,5	?	(4 10)	16	12	11	2	1	68%
7 Maloney	80	4	(–)	100%	?	15,0	0,8	4	–	–	–	–	100%
8 Deter	83	63	(?)	69%	17,7	20,4	3,6	24	27	11	12 (von 103)	3	70%
9 Pierloot	75,81	88	(–)	61%	17,4	20,5	4,2	28	41	9	10	4	78%
10 Thomä	61	19	(–)	79%	16,1	22,0	3,0	4	13	1	1	–	89%
11 Clauser	64	19	(–)	66%	18,1	20,6	3,6	14	2	3	–	–	84%
12 Willi	76	20	(–)	87%	19,5	22	11,0	7	7	5	1	–	70%
13 Goetz	77	30	(2)	91%	?	(9,5–16)	(5–20)	18	7	4	1	1	83%
14 Schäfer	74	8	(–)	100%	16,8	19,4	1,4	5	–	3	–	–	63%
15 Halmi	75	8	(1)	100%	?	23,3	0,6	4	3	1	–	–	87%
16 Pertschuk	77	27	(2)	93%	17,5	20,0	1,9	12	9	6	–	–	78%
17 Postpischel	81	23	(–)	100%	16,2	18,8	3,0	2	8	13	–	–	43%
18 Minuchin Rosman	75	50	(6)	94%	14,0	14,5	1,0	43	4	3	–	–	94%
19 Groen	66	6	(–)	86%	18,4	20,0	(1–6)	6	–	–	–	–	100%
20 Köhle Mall	83	36	(3)	100%	16,8		5,3	19	12	6	1	–	80%
insgesamt		558	(19)										

mengestellt. Dabei ist zu beachten, daß die Therapieverfahren auch innerhalb einzelner Studien von vornherein inhomogen und/oder unzureichend gewesen sein können (z. B. Deter, 1983) oder im Laufe der Zeit systematisch modifiziert wurden (z. B. Pierloot, 1975, 1981, 1982; Vandereycken und Pierloot, 1983).

Ein Teil dieser Untersuchungen ist in methodischer Hinsicht unzureichend, ein Vergleich ihrer Ergebnisse untereinander ist wegen der unterschiedlichen katamnestischen Kriterien, der unterschiedlichen Katamnesedauer und des jeweils unterschiedlichen Patientengutes nicht sehr sinnvoll.

Für die in diesen Studien insgesamt untersuchten 522 Patienten (davon mindestens 16 Männer) ergibt sich insgesamt ein Anteil von 80% geheilter bzw. gebesserter Kranker.

Wir möchten hier noch über einige wenige Untersuchungen berichten, deren Ergebnisse eher zur Vorsicht bei der Beurteilung mittel- bis langfristiger Verläufe mahnen. Nur wenige längerfristige Nachuntersuchungen liegen nach **konfliktbearbeitender Psychotherapie** vor.

Willi und Hagemann (1976) untersuchten 20 Patientinnen 8–16 Jahre nach einer stationären analytischen Psychotherapie. Fünf wiesen eine chronifizierte Anorexie auf, bei zwei kam es zum Übergang in eine endogene Psychose, eine Patientin ist im Beobachtungsintervall an Anorexiefolgen verstorben. Die Mehrzahl der übrigen Patientinnen wies nach wie vor typische Persönlichkeitszüge von Anorexiekranken auf; die Patientinnen litten unter überbewerteten Problemen der Nahrungsaufnahme und blieben untergewichtig. Willi

weist darauf hin, daß das Hauptgewicht der damaligen Behandlung auf einer analytischen Einzeltherapie lag, in der Regel ohne Einbeziehung der Familie und ohne intensivere Bearbeitung der auf der Station sich entwickelnden Gruppendynamik. Eine konsequente Wiederauffütterung wurde nur in Ausnahmefällen durchgeführt.

Von den vier vorliegenden katamnestischen Untersuchungen zur stationären **Verhaltenstherapie** (Schäfer und Schwartz, 1974; Halmi et al., 1975; Pertschuk, 1977; Postpischel, 1981) möchten wir die Untersuchung von Postpischel ausführlicher referieren.

Frau Postpischel untersuchte 23 Patientinnen 1975–1980, die im Max-Planck-Institut für Psychiatrie (München) behandelt wurden. Alle Patientinnen genügten den Feighner-Kriterien (vgl. Abschn. 38.2.4.1). Ihre Krankheit hatte im Mittel mit 16,2 Jahren begonnen (12–25), die Behandlung mit 18,8 Jahren. Bei fünf Patientinnen war diese Behandlung die erste Therapie. Im Mittel betrug ihr Gewicht 59% des Idealgewichts. Die Behandlung begann nach einer diagnostischen Woche mit einem Verhaltenskontrakt, dessen Übertretung zum „Privilegien“-Verlust führte. Später kamen individuelle therapeutische Maßnahmen, wie Rollenspiele zur Bewältigung der jeweiligen Problemsituation, hinzu. Gegen Ende der Behandlung lernten die Patienten „Selbstkontrollverfahren“. Teilweise wurden auch mit den Familien therapeutische Sitzungen durchgeführt. Pharmaka erhielten vier Patientinnen. 17 der 23 Patientinnen wurden nachbetreut, 14 vom Max-Planck-Institut. Zum Zeitpunkt der Entlassung wurden von den Therapeuten drei Patientinnen als geheilt, 15 als deutlich und 3 als wenig gebessert, 2 als unverändert beurteilt. Das mittlere Gewicht betrug jetzt 84,2% des

Idealgewichts. Während der Zeit bis zur katamnestischen Untersuchung unterzogen sich 16 der 23 Patientinnen 1–7 weiteren Therapieversuchen.

Die Nachuntersuchung erfolgte nach durchschnittlich drei Jahren (1,3–4,8); das mittlere Gewicht betrug jetzt 77,9% des Idealgewichts; das Gewicht von 4 Patientinnen wich weniger als 10%, von 13 Patientinnen weniger als 20% vom Idealgewicht ab. Für das Gewicht ergibt sich so zusammengenommen eine „geheilt/gebessert"-Rate von 57%. Bei Berücksichtigung von Kriterien im psychosozialen Bereich wurden 10 Patientinnen als geheilt oder nicht weiter behandlungsbedürftig (43%) beurteilt, bei 13 Patientinnen bestand eine ausgeprägte Anorexiesymptomatik fort. Nur 4 Patientinnen berichteten über eine regelmäßige Menstruation.

In den genannten vier verhaltenstherapeutisch orientierten Studien wird über Todesfälle nicht berichtet.

38.2.8.3 Einfluß von Behandlungs- und Patientenvariablen auf die Prognose

Therapievariablen

Die meisten Autoren stimmen darin überein, daß die Prognose heute durch die konsequente Anwendung eines abgestuften Behandlungsansatzes, der den körperlichen Zustand der Kranken ebenso berücksichtigt wie ihre seelischen Konflikte, entscheidend verbessert werden kann. Dies gilt auf jeden Fall für die Prognose quoad vitam.

Der Vergleich der jeweils 19 katamnestischen Untersuchungen von nicht systematisch und systematisch behandelten Patienten kann aufgrund der methodischen Probleme nur Hinweise ergeben (Mall, 1983):

In den „Therapiekatamnesen" waren 5%, in den übrigen 12% aller Patienten verstorben, ohne daß dies auf die mittlere Katamnesedauer oder den Zeitpunkt der Durchführung der Untersuchungen zurückzuführen wäre. Die Rate der „geheilt/gebesserten" Patienten beträgt in den „Therapiekatamnesen" 64% gegenüber 57% in den übrigen katamnestischen Untersuchungen.

Untersuchungen am St. George's Department in London haben früher gezeigt, daß sich die Ergebnisse einer zur Basistherapie (Wiederherstellung des Körpergewichts zusammen mit Einzel- und Familienpsychotherapie) zusätzlichen Behandlung mit trizyklischen Antidepressiva nicht von denen eines zusätzlichen Trainings sozialer Verhaltensweisen unterscheiden (Hall und Crisp, 1983).

Eine weitere Untersuchung an einer Serie von 100 Patienten (Hsu et al., 1979, nach Hall und Crisp, 1983) ergab einen günstigen Verlauf (71%) der ambulant psychotherapeutisch behandelten gegenüber nur 41% der stationär behandelten Patienten. Die stationär Behandelten erhielten einerseits zwar mehr intensive Einzel- und Familienpsychotherapie, andererseits handelt es sich bei ihnen sicherlich um eine Gruppe mit einem schwereren und chronischeren Verlauf der Anorexie.

Hall und Crisp (1983) führten erstmals eine kontrollierte prospektive Therapiestudie bei Anorexiepatientinnen durch.

30 Patientinnen wurden aus dem Gesamtkollektiv der überwiesenen Patientinnen nach folgenden Kriterien mit dem Ziel ausgewählt, eine relativ homogene Gruppe zu erhalten: weiblich, 13–27 Jahre, Sozialklassen I bis III, unverheiratet, Gewicht weniger als 85% des mittleren Gewichts einer parallelisierten Vergleichsgruppe Gesunder, Krankheitsdauer 6–72 Monate, amenorrhoisch. Alle Patientinnen wurden in zwei Gruppen randomisiert: psychotherapeutisch behandelte Gruppe und diätetisch beratene Gruppe.

Psychotherapie: 12 Sitzungen einmal wöchentlich oder vierzehntägig bei demselben Therapeuten; psychodynamische Einzelpsychotherapie in Verbindung mit einem familientherapeutischen Ansatz. Zusätzlich vier 15minütige Gespräche mit der Diätberaterin.

Diätberatung: 12 einstündige Sitzungen einmal wöchentlich oder 14tägig. Gelegentlich mit den Familienangehörigen. Die Beratungsgespräche betrafen die Diät, die Stimmung und alltägliche Verhaltensmuster. Ziel war, ein normales Eßverhalten mit normaler Ernährung wieder herzustellen, das Eßverhalten mit der Stimmung in Verbindung zu bringen und das Vertrauen der Patientinnen darin zu unterstützen, daß sie auch mit zunehmendem Gewicht die Kontrolle über ihr Eßverhalten behalten würden. Alle Patientinnen wurden vom Psychotherapeuten zu vier 15minütigen Interviews gesehen.

Ein unabhängiger Untersucher führte ein Jahr nach der Erstuntersuchung die katamnestische Untersuchung blind (d.h. nicht über das therapeutische Verfahren informiert) durch.

Neben ihrem Beitrag zu den methodischen Problemen erbrachte die Studie zu diesem Zeitpunkt folgende Ergebnisse: Die zwei Gruppen unterscheiden sich im Ergebnis signifikant in drei Punkten: Die Diätberatungsgruppe nimmt signifikant Gewicht zu, die Psychotherapiegruppe verbessert sich signifikant in den Bereichen Sexualität und soziale Beziehungen.

Patientenvariablen

Eine Übersicht über die in der Literatur zusammengestellten Angaben zum Zusammenhang von Patientenmerkmalen mit dem Krankheitsverlauf gibt Tabelle 38.2–6.

Auch für diese Zusammenstellung gelten die erwähnten methodischen Probleme. So zeigte kürzlich Swift (1982), daß sich das in der Tabelle übernommene Ergebnis einer günstigeren Prognose bei Patientinnen mit frühem Krankheitsbeginn, bei einer methodenkritischen Analyse der vorliegenden Veröffentlichungen, nicht halten läßt.

Der günstige prognostische Einfluß einer kurzen Krankheitsdauer ist hervorzuheben und verweist auf die Bedeutung einer frühzeitigen Diagnosestellung und einer konsequenten Einleitung angemessener Therapieverfahren.

Die übrigen psychischen Merkmale bei Patienten mit günstiger Prognose lassen sich dahingehend zusammenfassen, daß diese Kranken im Leben, im Interview und testpsychologisch bessere Ich-Leistungen erkennen lassen, daß sie eher unter einer abgrenzbaren traumatischen Situation gelitten haben, weniger tiefgehend neurotisch gestört sind und in einer günstigeren Umgebung leben. Insgesamt entsprechen die mitgeteilten prognostischen Kriterien den auch sonst in der Neurosenbeurteilung üblichen.

Zusammengefaßt kann gesagt werden, daß die Verbesserung der Behandlungsmöglichkeiten zu einer Abnahme der Mortalität und einer Verringerung der somatischen Komplikationen geführt hat. Viele Patienten mit Anorexia nervosa profitieren innerhalb gewisser Grenzen von intensiven Behandlungsansät-

Tabelle 38.2–6. Zusammenstellung prognostisch günstiger bzw. ungünstiger Merkmale von Anorexiepatientinnen aus der Literatur.

(1) Crisp et al. (1974), (2) Dally (1969), (3) Halmi et al. (1973), (4) Kay und Shapira (1965), (5) Morgan und Russel (1975), (6) Pierloot et al. (1975), (7) Samuel (1976), (8) Theander (1970), (9) Willi et al. (1976)

Prognostisch günstig	Prognostisch ungünstig
– Früher Krankheitsbeginn (zu Beginn der Adoleszenz) (3, 5, 6, 7, 8, 9)	– Später Krankheitsbeginn (Ende der Adoleszenz) (3, 5, 6, 7, 8, 9)
– Kurze Krankheitsdauer bei Behandlungsbeginn (5, 6)	– Längere Krankheitsdauer bei Behandlungsbeginn (5, 6)
	– Extremer Gewichtsverlust (5)
	– Vielzahl somatischer Beschwerden (3)
	– Unwillkürliches Erbrechen (1, 2, 3, 8)
	– Starker Laxantienabusus (3)
– Überwiegen hysterischer Anteile in der Persönlichkeitsstruktur (4, 7)	– Überwiegen schizoider, depressiver (3) oder zwangsneurotischer (3, 4, 6) Züge
	– Suizidversuche (6)
– Traumatisches Ereignis vor Krankheitsmanifestation (7)	– Kein traumatisches Ereignis vor Krankheitsmanifestation (7)
– Niedrige Werte im gesamten MMPI (6)	– Hohe Werte im MMPI (6)
– Niedriger Neurotizismus bei hoher Abwehr („Selbstschutz") (6)	– Hoher Neurotizismus (6)
	– Hohe Werte auf Schizophrenieskala im MMPI (6)
	– Vorausgegangene psychiatrische Hospitalisation (5)
	– Tendenz zu unkontrollierten Triebdurchbrüchen (Stehlen, Selbstverletzung) (6)
– Gute Anpassung in der Schule (5)	– Schlechte Anpassung in der Schule (5)
– Höheres akademisches Leistungsniveau (3)	– Höheres Alter der Mutter bei Geburt (8)
– Hohe Depressionswerte bei Eltern (1)	– Gestörte Beziehungen der Patientin zu anderen Familienmitgliedern (1, 6)
– Weniger Geschwister (7)	
– Heirat nach Erkrankung (9)	

zen – soweit heute belegbar – weitgehend unabhängig von den jeweiligen spezifischen Maßnahmen.

Die Anorexia nervosa ist jedoch nach wie vor eine Krankheit mit hohem Risiko zur Chronifizierung, mit hoher Mortalität und einem hohen Risiko zum Übergang in eine psychotische Erkrankung, meist aus dem schizophrenen Formenkreis.

Vordringlich erscheint es, die heute bereits an einigen Forschungszentren entwickelten Behandlungskonzepte in eine breitere klinische Praxis einzuführen, um wenigstens all denjenigen Patienten möglichst frühzeitige Hilfe anbieten zu können, die in der Lage sind, bei den heute vorhandenen Verfahren ausreichend gut mitzuarbeiten.

38.3 Bulimia nervosa

Hubert Feiereis

Die 26jährige Studentin E. S. kommt ambulant zu uns, weil sie unter Eßstörungen leidet. Nach dem Auszug von zu Hause vor 7 Jahren beginnt sie, durch eine Freundin angeregt, mit einer Diät, da sie sich zu dick und deshalb unattraktiv fühlt. Sie wiegt zu dieser Zeit 68 kg (161 cm). „Meine Mutter ist immer die Dorfschönheit gewesen."

Fortan wechseln etwa im Abstand von jeweils einem halben Jahr Zeiten strenger Diät mit erneuter Zunahme des „erhungerten Gewichtes"; entsprechend schwankt das Gewicht um jeweils etwa 10 kg. Nach den Diätkuren erwartet sie von der Umwelt Anerkennung, „die aber immer ausblieb". „Trotz aller Enttäuschungen waren Essen und Gewicht zu diesem Zeitpunkt noch kein zentrales Problem." Vor 3½ Jahren nimmt die Patientin erneut unter einer kontrollierten Diät bis auf 50 kg ab, „danach hat sich das Eß- und Gewichtsproblem selbständig gemacht".

Es setzen „Freßphasen" ein, „die ich nicht mehr kontrollieren kann". Das maßlose Essen von Süßigkeiten, vor allem Schokolade, steht in keinem Zusammenhang mit einem Hungergefühl.

Nach jedem Anfall ist die Patientin deprimiert, ratlos, geplagt von Völlegefühl und starken Gewissensbissen; ständig fürchtet sie sich vor erneut einsetzender Gewichtszunahme. „Ich kann mich nicht mehr ausstehen, ich mag nicht mehr unter Menschen sein." Sie zieht sich zurück.

Vor 2 Jahren beginnt sie, nach dem Essen zu erbrechen, bis zu dreimal täglich, durch Auslösen des Würgreflexes mit dem Finger. Sie verschafft sich damit Erleichterung, „ich fühle mich gereinigt".

Nur mit Hilfe des Erbrechens kann sie jetzt das Gewicht relativ konstant halten (57 kg). Sie nimmt keine Abführmittel, raucht nicht und trinkt keinen Alkohol. Die Periode setzt öfters aus, sie hat sie derzeit nur vier- bis fünfmal im Jahr.

Sehr bald legt sie sich gezielt Vorräte von Nahrungsmitteln an, vor allem Süßigkeiten, plant heimlich Freßattacken, „mein einziges Vergnügen". Sie sucht Tage und Situationen aus, in denen sie allein und ungestört ist, um das Risiko, entdeckt zu werden, auszuschließen.

Die Patientin unterscheidet betont zwischen dem Essen einer normalen Menge aus Hungergefühl, ohne zu erbrechen (seit 3 Jahren vegetarische Kost), und den Freßanfällen, die getrennt von den Mahlzeiten einsetzen oder sich aus einer Mahlzeit entwickeln. „Eine Banane oder ein Joghurt kann dann schon zu viel sein, und meine Lust auf Süßes ist nicht mehr zu bremsen." Bei den Freßanfällen verspürt sie kein Hungergefühl.

Auf die Frage nach konkreten Anlässen gibt sie an, daß „Alleinsein und hineingefressener Ärger" auslösend wirken und umgekehrt positive Erlebnisse, z.B. mit ihrem Freund, die Anzahl der Freßanfälle herabsetzen.

Die Grundstimmung sei „dumpf" und „deprimiert", ab und zu suche die Patientin Zuspruch bei einer Freundin, dann weine sie auch einmal, sonst mache sie alles mit sich allein aus.

Nach einer Erklärung für den Kontrollverlust befragt, meint sie, daß alles in der Pubertät begonnen habe; das ängstlich-beschützende Verhalten der Mutter habe dazu geführt, daß sie vieles verpaßte, sie sei „zu kurz gekommen" bei allem, was sich auf ihren Körper und ihr Aussehen beziehe. Sie trauere unerfüllten Wünschen nach, die sie sich wegen ihrer körperlichen Mängel nicht habe erfüllen können. Die seit der Pubertät ständig variierten Diäten hätten die negative Entwicklung nur verstärkt.

Die Patientin wuchs auf dem elterlichen Bauernhof mit den beiden 12 und 13 Jahre älteren Schwestern und ihrem 2 Jahre jüngeren Bruder auf. Zu ihren Schwestern hatte sie stets ein gutes, unkompliziertes Verhältnis, wozu wahrscheinlich der Altersunterschied beigetragen hat. Mit ihrem Bruder fühlte sie sich sehr eng verbunden. Ihm hätte sie als einzigem in der Familie ihre Konflikte schildern können, es aber nicht gewagt, da er ein „cooler Typ" sei und die Meinung vertrete, „ein ordentlicher Waldlauf löse alle Probleme".

Als die Patientin 19 Jahre alt war, starb ihr Vater im Alter von 61 Jahren an einem Herzinfarkt. Der Bruder mußte die Arbeit auf dem Hof übernehmen.

In ihrer Kindheit sei sie sehr behütet und unselbständig gewesen. Sie habe sich jede Freiheit viel schwerer erkämpfen müssen als z.B. ihr Bruder; mit 16 Jahren durfte sie erstmals allein ausgehen.

An Auffälligkeiten im Eßverhalten erinnert sie sich nicht. In der Kindheit sei das Essen für sie eine unerwünschte Unterbrechung des Spielens gewesen. Sie sei von den Eltern zum Essen ermuntert, aber nicht gezwungen worden. Allerdings sei zu unterschiedlichen Zeiten gegessen worden, weil sich die Mahlzeiten nach dem Arbeitsanfall auf dem Bauernhof richteten. Die Mutter habe sich dadurch gegängelt und überfordert gefühlt.

Die Patientin schildert sich als sehr schüchtern, in der Schule habe sie sich nicht recht behaupten können. Nach dem Auszug aus dem Elternhaus in einen süddeutschen Studienort lebte sie in einem großen Bekanntenkreis, der sich im Laufe der Erkrankung wieder verkleinerte. Einen Freund habe sie nie über einen längeren Zeitraum finden können, darüber möchte sie jedoch nicht sprechen; sie wirkt bei dieser Schilderung wortkarg und bedrückt, „da es nicht zum Thema gehört und es andere, wichtigere Fragen gibt".

Ihren Vater, stets ein Vorbild für sie, habe sie geliebt und verehrt. Er habe alles Wichtige in der Familie entschieden und dabei ebenso bestimmt wie ausgeglichen gewirkt. Auf seine Kinder sei er stolz gewesen.

Die Mutter wird als „typische Hausfrau" beschrieben; sie sei unselbständig, ihre Macht und ihren Willen setze sie auf indirekte Weise durch Erzeugung von Mitleid durch, sie übe psychischen Druck aus, indem sie „herumjammere". Nach dem frühen Tode des Mannes habe sie den Bauernhof leiten müssen, eine Rolle, für die sie nicht genug Selbstvertrauen aufgebracht habe und in der sie unzufrieden geworden sei. Sie brauche daher unaufhörlich Bestätigung und Zuwendung von der Familie. Andere Bezugspersonen würden für sie nicht existieren.

Im Gespräch wirkt die etwas übergewichtige Patientin zugewandt, freundlich und aufgeschlossen, betont lässig; sie antwortet rasch und ausführlich, solange biographische Daten erfragt werden, aber zurückhaltend und abwehrend über offenbar schmerzlich erlebte Versagungen und Kränkungen.

Wie beiläufig erwähnt sie ihre Einstellung, immer stark sein zu müssen, alle Lebenssituationen intellektuell kontrollieren zu können, damit niemand sie hintergehen, sie lächerlich machen könne. Den Preis hierfür zahle sie mit der Entbehrung von Liebe und Geborgenheit, von Anlehnung und intensiver Nähe. „Ich möchte stark, diszipliniert und erfolgreich sein und trotzdem geliebt werden."(!)

Die Patientin beschreibt präzise ihre Erwartungen an die Therapie:
- Sie möchte die Fixierung auf das Essen loswerden.
- Sie möchte, daß Erkenntnisprozesse über ihre Krankheit in Gang gebracht werden.
- Sie wünscht sich, die Vergangenheit aufzuarbeiten, vor allem die Mutterbeziehung, was sie unter dem Schlagwort „back to the roots" bereits selbst begonnen habe.

Obwohl bisher ohne jede therapeutische Erfahrung, äußert die Patientin auch qualitative Vorstellungen über den therapeutischen Weg: Am meisten erwartet sie von Einzelgesprächen, ihre – teils auch unbewußten – Konflikte und Probleme zu klären und zu verarbeiten. Der Gedanke an Gruppengespräche ist ihr suspekt, sie findet ihn „gruselig", wie eine „Auslieferung" an noch mehr Leute, die dann über sie Bescheid wüßten. „Ich fühle mich ohnehin ausgeliefert." Über andere Therapieformen, z.B. Mal- und Musiktherapie, hat sie keine Vorstellungen und sieht darin zunächst auch keinen Sinn, ist aber hierzu bereit.

38.3.1 Definition

Seit etwa zwanzig Jahren wird zunehmend über eine Krankheit berichtet, deren Hauptmerkmal eine Eßstörung ist und die inzwischen viele Namen hat (Feiereis, 1989; Potreck-Rose, 1987). Sie wurden meistens aus dem einen oder anderen Symptom der Krankheit abgeleitet oder phantasievoll gebildet, wie z.B.: Binge-eating-Syndrom; Binge-purge-Syndrom; Bulimarexia; Bulimia; Bulimia nervosa; Bulivomie; Dysorexia; Hyperorexie; paroxysmale Polyphagie; Stuffing-Syndrom; Thin-fat-people.

Am meisten durchgesetzt haben sich die Bezeichnungen Bulimia nervosa und Bulimie.

Auffälligstes Symptom der Krankheit ist die Sucht, große Nahrungsmengen unkontrolliert zu verschlingen („Fressen"). Die einsetzende Angst vor unaufhaltsamer Gewichtszunahme führt sehr bald zum Erbrechen, das zunächst selbst hervorgerufen wird („Kot-

zen") und schließlich reflektorisch eintritt. Betroffen sind vor allem Mädchen in der Pubertät und junge Frauen.

Die Krankheit setzt spontan oder allmählich ein, meistens nach dem Versuch, ein leichtes oder mäßiges Übergewicht mit Hilfe einer Diät zu reduzieren (Typ I), oder weist eine initiale bzw. intermittierende Phase einer Magersucht auf (Typ II), so daß wir in Anlehnung an Halmi (1985) diese beiden Gruppen ähnlich wie 2 Gruppen der Magersucht (passive-restriktive Form, Typ I; aktive Form, Typ II) unterscheiden.

38.3.2 Epidemiologie

Wegen unterschiedlicher Nomenklatur und Definition, schwer vergleichbarer Bezugsgrößen (Stichprobe, Form der Untersuchung, Altersgruppen, Region bzw. Land) und einer im Vergleich zur Magersucht erschwerten Diagnostik sind bisher exakte Zahlen nicht vorhanden oder schwer interpretierbar. Übereinstimmend wird von den meisten Autoren seit 20 Jahren eine Zunahme der Krankheit, vor allem während der letzten 10 Jahre, mitgeteilt (Cooper et al., 1987).

Befragungen in der Bundesrepublik Deutschland sprechen für eine Häufigkeit von 2–4% bei Frauen zwischen 18 und 35 Jahren (Fichter, 1985) und etwa 3% bei 12- bis 20jährigen Schülerinnen (Hänsel, 1987).

Soziodemographische Untersuchungen (Paul et al., 1984) ergaben, daß überwiegend Frauen zwischen 20 und 30 Jahren betroffen waren, nur 16% waren jünger und 22% älter; mehr als 60% hatten Abitur oder einen Hochschulabschluß.

In den USA fand man eine Häufigkeit von 13% der Collegestudentinnen (Halmi et al., 1981), in England bei Schülerinnen von 0,4% (Szmukler, 1985) und von 1,9% in einer Beratungsstelle für Familienfragen.

38.3.3 Das Krankheitsbild

38.3.3.1 Vorbemerkungen

Zwischen 1975 und 1989 kamen in unsere Klinik 837 Patienten mit den Eßstörungen Bulimie (n= 342) und Magersucht (n= 458) sowie mit Mischformen fraglicher Zuordnung (n= 37). 1989 waren es z.B. zehnmal mehr Bulimiekranke als 1980, d.h., die vielerorts berichtete Zunahme der Krankheit ist evident, auch unter Berücksichtigung einer gezielten Zuweisung.

38.3.3.2 Anamnese und Symptomatologie

Wie bei kaum einer anderen Krankheit sind Anamnese und Symptomatologie nahezu identisch, d.h., die anamnestisch erfahrbaren subjektiven Kennzeichen der Krankheit bilden gleichzeitig die Hauptmerkmale der Symptomatologie:

– Anfallsartig einsetzende heimliche Aufnahme großer Nahrungsmengen, die – häufig wahllos – innerhalb kurzer Zeit verschlungen werden („Freßanfall"). Der dem Anfall vorausgehende Heißhunger kann fehlen, die Kontrolle über die Nahrungsaufnahme geht im Anfall verloren, Hunger- und Sättigungsgefühl schwinden innerhalb des Anfalls.
– Panische Angst vor Gewichtszunahme; Übergewichtsphobie.
– Selbstinduziertes heimliches Erbrechen nach dem Anfall, teilweise auch außerhalb des Anfalls. Bei einem Teil der Patienten besteht ein Laxanzienmißbrauch.
– Initial oder intermittierend diätetische Nahrungsrestriktion oder Phase einer Magersucht.
– Rasch aufeinanderfolgende Gewichtsschwankungen bis zu 10 kg.
– Scham- und Schuldgefühle, Verzweiflung, depressive Verstimmungen. Die Krankheit wird lange Zeit gegenüber der Umwelt verheimlicht.
– Amenorrhoe.

Unter unseren 342 Patienten waren 47% dem Typ I zuzuordnen, 53% dem Typ II mit initialer oder intermittierender Phase einer Magersucht.

Die Häufigkeit der einzelnen Störungen im Eßverhalten zeigt Abbildung 38.3–1. Aus ihr geht hervor, daß 56% ständig, mehrfach täglich, und 41% häufig, d.h. mehrfach in der Woche bis täglich, Freßanfälle haben. 47% der Kranken erbrechen auch unabhängig vom Freßanfall und 36% haben keine Heißhungergefühle. Die aufgenommene Nahrungsmenge während eines Freßanfalles kann bis zu 12000 Kalorien enthalten, die einer bulimischen Episode bis zu 26000 Kalorien (Mitchell und Laine, 1985). Eine unserer Patientinnen z.B. „aß" fraktioniert 6 kg Süßigkeiten oder zwölf kg Nudeln pro Freßphase innerhalb einiger Stunden.

Fast zwei Drittel der Kranken sind darum bemüht, außerhalb der Freßanfälle die Nahrungsaufnahme zu reduzieren. Bei der ersten Untersuchung hatten 53% ein Idealgewicht, 19% ein Normalgewicht und 28% ein Übergewicht (Abb. 38.3–2).

Die Häufigkeit des Mißbrauchs von Laxanzien, Appetitzüglern, Tranquilizern, Alkohol und Nikotin zeigt Abbildung 38.3–3.

Die prozentuale Verteilung somatischer Befunde ergibt sich aus Abbildung 38.3–4; sichere Unterscheidungen gegenüber den Befunden bei Magersucht fanden wir nicht.

Bemerkenswert ist, daß 44% der Patientinnen eine sekundäre Amenorrhoe haben und 20% Zyklusunregelmäßigkeiten. Das bedeutet, daß bei 64% eine hormonelle Regulationsstörung besteht, die sich nicht wesentlich von der Magersucht unterscheidet. Die Häufigkeit der Zyklusstörungen dürfte noch größer sein, wenn man alle Patientinnen einbezöge, die infolge der Einnahme von Ovarialhormonen menstruieren. Die Amenorrhoe ist also nicht nur Folge des Untergewichtes, sondern eine Regulationsstörung zwischen Hypothalamus, Hypophyse und Gonaden

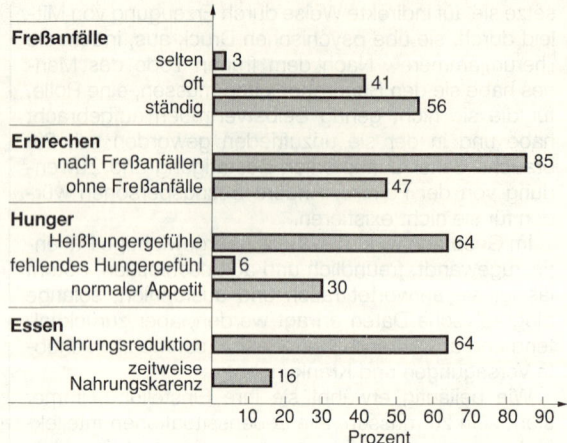

Abb. 38.3–1. Häufigkeit der Störungen im Eßverhalten.

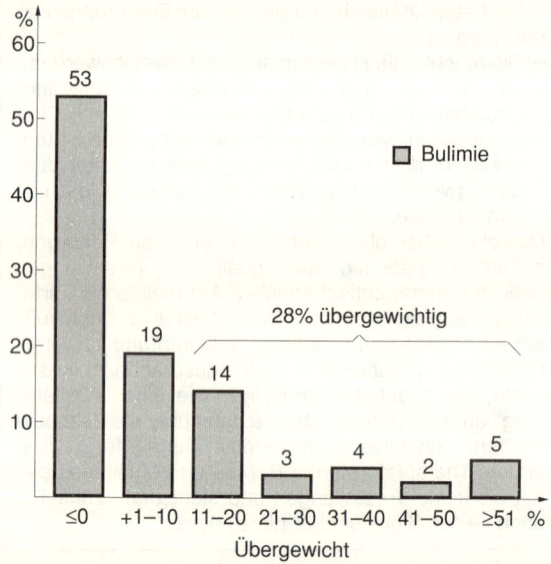

Abb. 38.3–2. Übergewicht der Patienten mit Bulimie, bezogen auf das Idealgewicht.

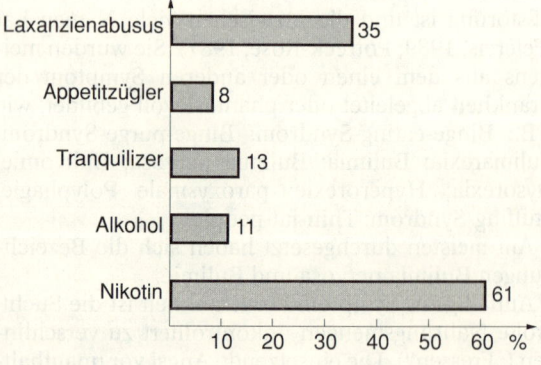

Abb. 38.3–3. Häufigkeit des Mißbrauchs von Medikamenten, Alkohol und Nikotin bei Patienten mit Bulimie.

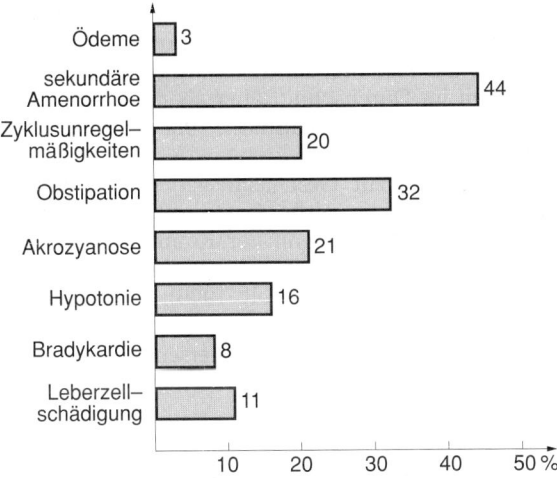

Abb. 38.3–4. Prozentuale Verteilung einzelner somatischer Befunde bei Bulimie.

mit verringertem LH und FSH (Fichter und Pirke, 1989; Pirke, 1989).

Die nahezu obligate Verbindung von Freßanfällen und Erbrechen kann vereinzelt fehlen, dennoch aber eine Bulimie vorliegen, wenn anfallsartig Freßanfälle und die weiteren Kriterien einer Bulimie bestehen, jedoch ohne Erbrechen: Bulimia sine vomitu. Die exzessive Gewichtszunahme bis zu extremer Adipositas ist dann die unausweichliche Folge (Feiereis, 1989).

38.3.3.3 Somatische Befunde

Im Gegensatz zur Magersucht ergibt die körperliche Untersuchung kaum Auffälligkeiten, da das Gewicht normal ist oder nur ein leichtes Übergewicht besteht. Hervorzuheben sind bei einzelnen Patienten als Folge des Erbrechens eine metabolische Alkalose (Mitchell und Bantle, 1983), Heiserkeit, Schmerzen in der Speiseröhre und funktionelle Magen-Darm-Störungen mit Obstipation (32%, Feiereis, 1989). Relativ häufig wird über Schäden an den Zähnen und am Zahnhalteapparat berichtet (Clark, 1985; Hurst et al., 1977; Trygstad, 1985), ebenso über eine Vergrößerung der Speicheldrüsen und über Dysphagie. In Verbindung mit dem Erbrechen sind Hautverletzungen am Handrücken (Russell, 1979) zu erwähnen.

Unter den Kreislaufveränderungen standen bei unseren Kranken Akrozyanose (21%), Hypotonie (16%), Bradykardie (8%) und Ödeme (3%) im Vordergrund (Abb. 38.3–4), Befunde, die überwiegend als Begleitsymptome der Krankheit aufzufassen sind.

Am Skelettsystem werden eine hypertrophische Osteoarthropathie und Trommelschlegelfinger beschrieben (Fichter und Chlond, 1988). Das kraniale Computertomogramm weist bei manchen Patienten hirnatrophische Veränderungen auf (Krieg et al., 1987). Die CT-Befunde unterscheiden sich nicht von den Ergebnissen bei Magersucht, d.h., sie sind offenbar unabhängig vom Körpergewicht. Der pathogenetische Mechanismus ist ungeklärt. Eine klinische Be-

deutung messen wir diesem Befund nicht bei; er ist sehr wahrscheinlich ebenso rückbildungsfähig wie bei der Magersucht.

Eine Reihe von Untersuchungen des Stoffwechsels und der neuroendokrinen Regulation erbrachte unterschiedliche Ergebnisse, so z.B. Störungen der Glukosetoleranz, einen möglichen Zusammenhang affektiver Störungen bei Bulimie mit Veränderungen der regulierenden Funktion zentraler Neurotransmitter (Noradrenalin, Serotonin) (Pirke, 1989) und eine herabgesetzte Suppression im Dexamethasonhemmtest oder eine verringerte Stimulierbarkeit von TSH durch TRH (Mitchell und Pomeroy, 1989; Fichter und Pirke, 1989).

Halmi (1989) weist auf pathophysiologische Befunde hin, die für eine Störung der Sättigungswahrnehmung sprechen (Chiodo und Latimer, 1986; Kissileff et al., 1986; Mitchell und Laine, 1985).

38.3.3.4 Psychische Befunde

Der dominierende psychische Befund bei Bulimiekranken ist die depressive Verstimmung, die unmittelbar vor Beginn einer Freßattacke auftreten kann und häufig nach dem Anfall besonders ausgeprägt ist. In Verbindung hiermit geben 30% der Patienten zeitweise Suizidgedanken an und 11% hatten in der Anamnese Suizidversuche (Feiereis, 1989).

Die Depressivität hat oft eine die Attacke auslösende Wirkung; sie kann aber auch eine Reaktion auf den Anfall sein. Unbeantwortet bleibt bisher, inwieweit genetische, prämorbide und – unabhängig vom Anfall – ähnliche oder andere affektive Störungen Teil oder Äquivalent der Krankheit Bulimie sind. In verschiedenen Untersuchungen fanden sich Zusammenhänge zwischen depressiver Symptomatik und Bulimie (Herzog, 1984; Hudson et al., 1983a; Laessle, 1989; Shaye, 1989; Walsh et al., 1985), allerdings mit unterschiedlicher zeitlicher Beziehung zum Beginn der Bulimie. In einer vergleichenden Untersuchung von Laessle (1987) zeigten sich symptomatologisch deutliche Unterschiede gegenüber primär depressiven Patienten, was mehr auf die enge zeitliche Zuordnung der depressiven Symptomatik zur bulimischen Attacke hinweist, wie auch von anderen Autoren hervorgehoben wird (Cooper und Fairburn, 1986). Schließlich wird von Laessle (1989) betont, daß auch die Diskontinuität der Remissionsmuster eher gegen eine gleiche ätiologische Wurzel der affektiven und bulimischen Störung spricht.

38.3.3.5 Schweregrade

Die Beurteilung des Schweregrades der Krankheit richtet sich in erster Linie nach der Häufigkeit der Freßanfälle und der Dauer der Erkrankung, der Ausprägung psychischer Befunde, dem Leidensdruck und der bisher erfolgten Therapie und ihren Ergebnissen. Zwischen der Häufigkeit des Erbrechens und den psychischen Veränderungen fanden wir eine

Wechselwirkung, die die Definition der verschiedenen Schweregrade mitkonstituiert. Mit diesen Kriterien lassen sich nach unseren Untersuchungsbefunden 3 Gruppen der Schweregrade bilden:

Schweregrad I: Freßanfälle 2- bis 3mal pro Woche, Krankheitsdauer mindestens 6 Monate, keine schweren psychischen Veränderungen, keine Suizidgedanken, Bereitschaft zur Therapie.

Schweregrad II: täglich Freßanfälle, Dauer der Krankheit 1 bis 2 Jahre, mittelschwere psychische Symptomatologie mit phasenhaft starker depressiver Verstimmung und Suizidgedanken. Falls Therapieversuche, so bisher ohne genügenden Erfolg.

Schweregrad III: täglich mehrfach Freßanfälle, Abusus von Medikamenten und/oder Alkohol, erhebliche Depressivität mit Suizidgefahr, großer Leidensdruck, absolute klinische Behandlungsbedürftigkeit.

In unserem Kollektiv konnten wir 32% dem Schweregrad I, 49% dem Schweregrad II und 19% dem Schweregrad III zuordnen.

38.3.3.6 Dauer der Krankheit

Im Vergleich zur Magersucht liegt bei Bulimiekranken oft ein jahrelanges Intervall zwischen Beginn der Krankheit und erster ambulanter Untersuchung bzw. der ersten stationären Therapie. Abbildung 38.3–5 zeigt, daß nahezu bei der Hälfte der Bulimiekranken 4 und mehr Jahre vergehen, ehe die Patienten ihre Krankheit offenbaren und eine Therapie einsetzen kann.

38.3.3.7 Alter und Geschlecht

Die Altersverteilung der Bulimiekranken unterscheidet sich nur unwesentlich von der Magersucht: 80% unserer Patienten waren bei der Manifestation der Krankheit 11 bis 20 Jahre alt, 16% zwischen 21 und 30 Jahren (Abb. 38.3–6). Andere Autoren hoben hervor, daß die Patienten mit Bulimie bei Beginn der Krankheit im Durchschnitt älter als die Gruppe der Magersüchtigen seien (Fairburn, 1982; Fairburn und Cooper, 1982).

Die Angaben zur Geschlechtsverteilung stimmen weitgehend überein, d.h., fast ausschließlich sind Mädchen und Frauen von der Krankheit betroffen (96,2%, Mitchell und Goff, 1984; 96%, Robinson und Holden, 1986; 93,8%, Russell, 1979); unter unseren Patienten waren 4% männlichen Geschlechts. Anamnese und Symptomatologie unterscheiden sich dabei nicht von den Krankheitserscheinungen bei Mädchen und Frauen.

Der 19jährige Student L.U. gibt beim Erstgespräch an, bis zum 15. Lebensjahr intensiv Sport betrieben zu haben. Nach einer Verletzung mußte er den Sport sehr einschränken. „Zum gleichen Zeitpunkt glaubte ich, etwas zu dick zu sein. Ich beschloß, einige Kilo abzunehmen, auch um meine Fußgelenke zu entlasten. Von montags bis freitags hielt ich ein Jahr lang streng Diät, wobei ich 2–4 kg im Durchschnitt abnahm. Am Wochenende bekam ich dann Heißhungeranfälle, so daß mein ursprüngliches Gewicht wieder auf der Waage war. So ging es Woche für Woche. Ich erbrach das Gegessene nicht."

Im 16. Lebensjahr wechselten Abmagerungskuren und Heißhungeranfälle in kürzeren Abständen. Zufällig oder automatisch „habe ich irgendwann nach einem Heißhungeranfall erbrochen. Von diesem Zeitpunkt an möchte ich von Freßanfällen bzw. Freß-Brech-Durchbrüchen statt von Heißhungeranfällen sprechen. Langsam aber sicher begann mich die ausweglos erscheinende Lage zu quälen. Schlimm fand ich, daß das Essen immer mehr mein Denken, Fühlen und Handeln bestimmte und daß zugleich das Leben für mich einen immer geringeren Stellenwert erhielt. Wie schon in den vorangegangenen Jahren und auch in den noch folgenden schlug sich jedes Defizit während einer Hungerkur im nächsten Freßanfall nieder. Die Freßanfälle

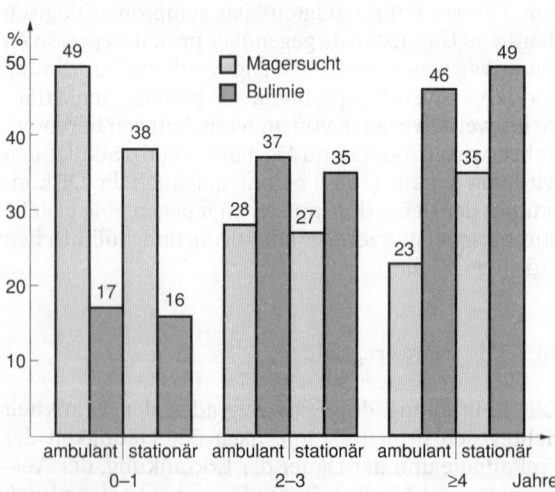

Abb. 38.3–5. Intervall zwischen Krankheitsbeginn und dem ersten ambulanten Termin bzw. der ersten stationären Aufnahme der Patienten mit Bulimie und Magersucht.

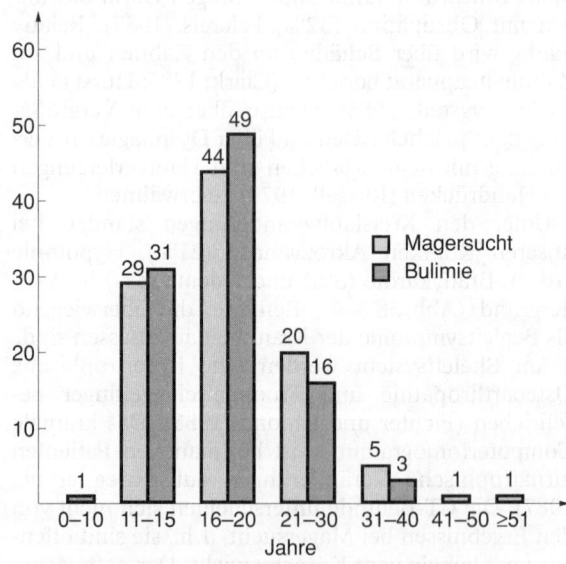

Abb. 38.3–6. Alter der Patienten bei Krankheitsbeginn.

haben aber im Laufe der Zeit bis heute immer weniger zur ‚seelischen Sättigung‘ geführt. Heute befriedigen sie überhaupt nicht mehr, im Gegenteil. Ich hielt mich in diesem Jahr nur noch zu ungefähr 20–40% an das herkömmliche Eßschema. Sonst hat mein Eßverhalten für mich fast unerträgliche Zustände angenommen: krampfhafte Versuche abzunehmen, Freßanfälle, Erbrechen. Obwohl der gesamte Teufelskreis größer und stärker wurde, glaubte ich immer noch, durch das ‚Herummanipulieren‘ an meinem Eßverhalten alles selbst in den Griff zu bekommen. Schließlich ist meine Gesamtsituation nunmehr unerträglich geworden, weil ich nicht mehr vernünftig und geregelt essen und leben kann. Ich akzeptiere mich so nicht mehr. Ich habe ungefähr 5 bis 6 Freßanfälle in der Woche, von denen jeder zweite mit einem Brechanfall abgeschlossen wird. Ich weiß nicht, was mich lähmt und hemmt, wovor ich Angst habe. Ich kenne nur ihre Stellvertreter: Ängste, die mich zu den Freß-Brech-Attacken treiben.“

„Großenteils sind meine Freßorgien bewußt geplant, d.h., daß ich oft einkaufen gehe, ganz bewußt mit dem Gedanken, zu Hause alles in mich hineinzustopfen und hinterher wieder auszubrechen. Ich bekomme dann häufig ein schlechtes Gewissen, oder ich verdränge meine Gewissensbisse, wobei ich mich genauso unwohl fühle. Ich habe in den letzten 2 bis 3 Jahren mehr für meine Freßorgien ausgegeben als beispielsweise für meine Hobbys, Kleidung usw. Ich habe viel Erspartes verpraßt.“

„Ich habe inzwischen verlernt zu leben, weil ich vieles durch die ‚Droge Essen‘ verdränge. Ich spüre das Leben daher nicht mehr in dem Maße wie vor Beginn meiner Schwierigkeit mit dem Essen. Meine Leistungsfähigkeit und meine Kondition haben sehr nachgelassen. Jeder erneute Anlauf, vernünftiger und geregelter zu leben, fällt mir sehr schwer. Ich habe mich in den letzten 10 Monaten immer weiter isoliert, so daß meine sozialen Kontakte sehr zurückgegangen sind. Ich möchte wieder selbstsicher sein oder ein starkes Selbstwertgefühl haben, so daß ich auch meine Schwächen den Mitmenschen offen mitteilen kann, statt unsicher oder aggressiv zu sein oder mich in der Maske einer starken Persönlichkeit darzustellen. Ich möchte mich wieder annehmen können und wissen, welche Signale hinter meinem ‚unersättlichen Hunger‘ stecken, damit meine Gedanken nicht mehr stets um das Essen kreisen.“

Auch Fichter und Hoffmann (1989) fanden keine wesentlichen Unterschiede des Eßverhaltens bei männlichen und weiblichen Patienten. Es gibt jedoch Hinweise auf eine geringere Gewichtsphobie und Körperschemastörung bei männlichen Patienten.

38.3.4 Bulimia nervosa und Diabetes mellitus Typ I

Während der letzten Jahre wird mehrfach über die Krankheitskombination Bulimie und Diabetes mellitus berichtet (Featherstone und Beitman, 1984; Feiereis, 1988; Giles, 1986; Hudson et al., 1983 c, d; Rodin et al., 1985; Szmukler und Russell, 1983). Von manchen Autoren wird die Bulimie wegen der potenzier-

ten Gefahr auftretender Schäden als lebensbedrohende (Hillard et al., 1983) oder tödliche Komplikation (Hillard und Hillard, 1984) des Diabetes bezeichnet.

In angloamerikanischen Ländern wird die Häufigkeit der Doppelkrankheit mit 20 bis 35% angegeben; bei unseren Patienten mit Bulimie beobachten wir eine Häufigkeit von 5,3% (Feiereis, 1988).

Das sonst bei der Bulimie nahezu obligate Erbrechen fehlt bei einem Teil der Patienten mit Diabetes mellitus, weil sie mit Hilfe verringerter Dosierung des Insulins den renalen Verlust von Glukose steigern und dadurch den gefürchteten Gewichtsanstieg verhindern („Erbrechen über die Niere“; Feiereis, 1988). Die bulimische Symptomatik tritt bei der Mehrzahl der Patienten erst längere Zeit nach der Manifestation des Diabetes auf, gelegentlich geht die Bulimie allerdings dem Diabetes voraus. Da die meisten Patienten die Zuckerkrankheit nicht akzeptieren können, begehen sie Diätfehler, um sich dadurch im Gefühl ihrer Freiheit zu bestätigen. Diese Art der Scheinhilfe, eine lebenslange Stoffwechselerkrankung nur so ertragen zu können, schlägt allmählich qualitativ und quantitativ um: Freßanfälle, falls sie nicht der Manifestation des Diabetes ohnehin vorausgingen, vermitteln das trügerische Gefühl einer Autonomie, sich nunmehr grenzenlos Nahrungsmittel einverleiben zu können. Die gestörte Compliance, die mißlungene Bewältigung (Coping) der Zuckerkrankheit, die psychischen Auswirkungen der Freßanfälle und deren Eigendynamik stehen dabei in enger Wechselwirkung.

Aus den Mitteilungen über diese bedenkliche Doppelkrankheit folgt, daß bei jedem schwer einstellbar wirkenden insulinpflichtigen Diabetes eines jugendlichen Patienten möglichst frühzeitig auch an die Möglichkeit dieser Kombination gedacht werden sollte. Vor allem ist der Verdacht auf eine Bulimie begründet, wenn der Zuckerstoffwechsel trotz sorgfältiger Kontrolle und Behandlung in Spezialkliniken immer wieder scheinbar unerklärlich entgleist.

38.3.5 Die Diagnose der Bulimia nervosa

Im Laufe der vergangenen 10 Jahre wurden verschiedene Kriterien zur Diagnose der Bulimie veröffentlicht. Die Mängel der Klassifikation der „International Classification of Diseases“ (ICD-9) und des „Diagnostic and Statistical Manual of Mental Disorders“ (DSM-III) der American Psychiatric Association führten zu neuen Vorschlägen, die im Entwurf der ICD-10 (Tab. 38.3–1) und in der Revision des DSM-III-R (Tab. 38.3–2) niedergelegt sind.

Zur ICD-10 ist anzumerken, daß eine Unterscheidung der Patienten mit „Bulimia nervosa“ und „Bulimia bei normalem Körpergewicht“ nicht begründet ist, will man an der Differenzierung zwischen der Gruppe der Bulimie mit initialer oder intermittierender Magersucht und der Gruppe der Magersucht mit phasenhaft auftretenden Heißhungeranfällen festhalten.

In der ICD-10 und ebenso im DSM-III-R erscheint uns nicht genügend berücksichtigt, daß die Gefühle von Schuld, Scham, Verzweiflung und Depression nach dem Freßanfall, die Verborgenheit seines Ablaufes und die oft lange Zeit bestehende Verheimlichung der Krankheit ebenso wichtige Kriterien darstellen wie die häufig vorliegenden Zyklusschwankungen und die Amenorrhoe.

Die Einnahme von Schilddrüsenpräparaten oder Diuretika ist nach unseren Erfahrungen sehr selten, ebenso das Bedürfnis nach übermäßiger körperlicher Belastung. Wir haben darum der Diagnose die im Abschnitt 38.3.3.2 angeführten Hauptkriterien zugrunde gelegt.

Tabelle 38.3–1. Richtlinien zur Diagnose der Bulimia nervosa nach dem Internationalen Diagnoseschlüssel (ICD) der WHO, Entwurf der 10. Revision 1987.

– Ständige übertriebene Beschäftigung mit dem Essen, unwiderstehlicher Heißhunger, Episoden übertriebenen Essens, Einnahme großer Nahrungsmengen innerhalb kurzer Zeit.
– Versuch, die Gewichtszunahme durch folgende Aktivitäten zu verhindern: selbstinduziertes Erbrechen. Laxanzienabusus. Phasen der Nahrungsrestriktion. Medikamentenmißbrauch, z.B. Appetitzügler, Schilddrüsenpräparate, Diuretika. Bei Diabetes wird eventuell die Insulintherapie vernachlässigt.
– Psychopathologisch krankhafte Furcht vor Gewichtszunahme. Es wird eine Gewichtsgrenze eingehalten, die weit unterhalb des prämorbiden Gewichts liegt, das als optimal anzusehen ist.
– Häufig, aber nicht immer, bestand in der Vorgeschichte eine Phase der Magersucht, die einige Monate bis mehrere Jahre gedauert haben kann. Diese Phase war nur latent mit geringem Gewichtsverlust und/oder vorübergehender Amenorrhoe vorhanden oder voll ausgeprägt.
– Davon zu unterscheiden ist die Bulimie bei normalem Körpergewicht oder leichtem Übergewicht und ohne frühere Magersucht. In dieser Gruppe werden wiederkehrende Heißhungeranfälle angegeben, der Gewichtszunahme wird durch selbstinduziertes Erbrechen, Laxanzienabusus, abwechselnde Phasen des Hungerns oder durch Mißbrauch von Medikamenten entgegengewirkt. Depressive Symptome können vorhanden sein.

Tabelle 38.3–2. Kriterien zur Diagnose der Bulimia nervosa nach dem Diagnostic and Statistical Manual of Mental Disorders III-Revision (DSM-III-R) der American Psychiatric Association (1987).

– Wiederholte Episoden von Freßanfällen (schnelle Aufnahme einer großen Nahrungsmenge innerhalb einer bestimmten Zeit).
– Gefühl, das Eßverhalten während der Freßanfälle nicht kontrollieren zu können.
– Um eine Gewichtszunahme zu verhindern, werden regelmäßig selbstinduziertes Erbrechen, Gebrauch von Laxanzien oder Diuretika, strenge Diät oder Fastenkuren oder übermäßige körperliche Aktivität angewandt.
– Durchschnittlich mindestens 2 Freßepisoden/Woche über mindestens 3 Monate hinweg.
– Andauernde übertriebene Beschäftigung mit Figur und Gewicht.

38.3.6 Differentialdiagnose

Die Differenzierung der Bulimie gegenüber der Magersucht, besonders der Form, die mit einer Neigung zu Erbrechen, Laxanzienabusus und Heißhungergefühl sowie Freßanfällen verbunden ist, erscheint am wichtigsten. Bei der Frage, ob Magersucht und Bulimie zwei voneinander abgrenzbare Eßstörungen sind oder lediglich Manifestationen einer einzigen Krankheit, bestehen noch immer unterschiedliche Auffassungen. Wenn sich auch die Grenzen im Einzelfall verwischen können und Mischformen vorkommen, die eine eindeutige Zuordnung nicht ermöglichen (unter den eigenen Patienten 4%), und epidemiologische (Häufigkeit, Alters- und Geschlechtsverteilung), pathophysiologische (Zyklusunregelmäßigkeiten, Amenorrhoe) und ätiopathogenetische Ähnlichkeiten bestehen, so sind jedoch heute Zweifel an der Krankheitsentität Bulimie (Bruch, 1982, 1985; Meermann und Vandereycken, 1987; Vincent und Kaczkowski, 1984) kaum mehr begründet. Unter unseren Magersuchtpatienten haben nur 2% ständig Freßanfälle, 15% erbrechen nach dem Anfall, 16% haben Heißhungergefühle. Andererseits haben 50% der Bulimiekranken (Feiereis, 1989) oder mehr (Fairburn und Cooper, 1984) keine Merkmale einer Magersucht. In Tabelle 38.3–3 sind wichtige Unterscheidungsmerkmale nochmals gegenübergestellt.

Die Bulimie ist differentialdiagnostisch auch abzugrenzen gegen das habituelle Erbrechen, gegen Krankheiten mit einer Passagestörung im Magen-Darm-Trakt und vor allem gegenüber der Polyphagie bei Adipositas. Der überschießende Appetit Adipöser beruht u.a. auf fehlendem Sättigungsgefühl, erhöhter Stimulierbarkeit durch Außenreize, reaktiver Steigerung des Appetits infolge von Konflikten, depressiven Verstimmungen und Unlustgefühlen. Berücksichtigt man die Unterschiede der Altersstruktur, der psychosozialen Prägungen und Befunde, so werden übermäßige Nahrungsaufnahme mit nachfolgendem Übergewicht und unmäßige Freßgier, ohne nachfolgendes Übergewicht, in der Regel eine Abgrenzung von Adipositas und Bulimie ermöglichen.

Tabelle 38.3–3. Gegenüberstellung wichtiger Symptome bei Bulimie und Magersucht.

Magersucht	Bulimie
„Anorexie"	Hyperorexie
Hypophagie	Hyperphagie
abnehmen wollen	nicht zunehmen wollen
Kontrollzwang	Kontrollverlust
Untergewicht	Normal-(Über-)Gewicht
Kachexie	Gewichtsschwankungen
Störung des Körperbildes	
kein Leidensdruck, Verleugnung	großer Leidensdruck, Schuldgefühle
geringes Therapiebedürfnis	starkes Therapiebedürfnis
geringe Compliance	gute Compliance
Amenorrhoe	

38.3.7 Ätiologie und Pathogenese

38.3.7.1 Vorbemerkungen

Die meisten Untersuchungen zur Klärung der Entstehung einer Bulimie sprechen ähnlich wie bei der Magersucht für eine plurikausale Pathogenese. Für bedeutungsvoll werden genetisch determinierte Faktoren, familiäre, soziale und soziokulturelle Einflüsse, die Psychodynamik einer gestörten Persönlichkeitsentwicklung und Merkmale der Persönlichkeitsstruktur gehalten. Potreck-Rose (1987) entwarf in Anlehnung an Johnson und Mitarbeiter (1984a) ein hypothetisches Modell, das disponierende Faktoren und aktuelle psychische Belastungen und Konflikte aufgliedert. Bei den disponierenden Einflüssen werden soziokulturelle, familiäre und genetische Faktoren unterschieden. Fichter (1989b) unterteilt die Ätiologie in biologische Faktoren, individuelle Defizite, soziokulturelle Einflüsse und chronische Belastungen; er sieht in der bulimischen Erkrankung einen (allerdings vergeblichen) „Lösungsversuch", mit Selbstunsicherheit, Ängsten und Stimmungsschwankungen durch einen einseitigen Kampf um körperliche Schlankheit fertig zu werden.

38.3.7.2 Prämorbide Disposition

Genetischer Faktor

Gegenüber den Untersuchungen bei Magersucht gibt es bisher bei der Bulimie nur wenige Studien zur Begründung einer genetischen Determiniertheit. Die konkordant festgestellte Krankheit bei eineiigen Zwillingen (Kaminer et al., 1988; Nögel, 1988) erhärtet die Annahme eines genetischen Anteils innerhalb der Pathogenese. Ob allerdings die Beobachtung eines vermehrten Auftretens psychischer Erkrankungen in den Familien der Bulimiepatienten (Gershon et al., 1984; Hudson et al., 1983b; Strober et al., 1982; Strober und Humphrey, 1987) auf eine genetische Wurzel der Bulimie schließen läßt, wie diese Autoren meinen, erscheint noch nicht genügend begründet. So hebt Laessle (1989) hervor, daß eine genaue Analyse der Befunde von Gershon und Mitarbeitern (1984) sowie Strober und Katz (1987) dafür spreche, daß Bulimie und affektive Störungen genetisch unabhängig voneinander sind.

Soziale Entwicklung

In der Literatur wurde wiederholt die Frage nach möglichen Zusammenhängen zwischen der Bulimie und dem Grad der Schulbildung, der beruflichen Entwicklung, dem sozialen Status, der Familienstruktur und partnerschaftlichen Bindungen untersucht.

Die meisten unserer Patienten besuchten das Gymnasium oder eine andere weiterführende Schule (56%); 11% erreichten den Hauptschulabschluß; in einer Kontrollgruppe (Ennulat, 1989) hatten 36% einen Fachhochschulabschluß bzw. Abitur und 24% den Hauptschulabschluß.

Das Berufsbild der Väter der Patienten zeigte fast ⅔ Akademiker, Beamte, selbständige Kaufleute oder Angestellte; nur 28% der Mütter unserer Bulimiekranken gegenüber 47% der Magersüchtigen waren nicht berufstätig und somit ausschließlich Hausfrauen. Möglicherweise findet sich hier eine Erklärung dafür, daß nur 41% der Bulimiekranken noch bei den Eltern leben, dagegen 79% der Magersüchtigen. In der Familienstruktur fällt auf, daß 19% der Eltern unserer Bulimiekranken geschieden sind oder getrennt leben, bei 12% der Vater und bei 5% die Mutter verstorben ist. Ein Suchtproblem des Vaters (meistens Alkohol) lag in 10%, der Mutter in 5% vor.

Soziokulturelle Einflüsse und Körperbild

Von den meisten Autoren wird die ätiopathogenetische Wirkung sozialer und kultureller Prägungen und Bindungen hervorgehoben (Klessmann und Klessmann, 1988; Rost et al., 1982; Schwartz et al., 1982), obgleich die Forschungsergebnisse noch keine bindende Aussage erlauben. Hohe Prävalenz und Inzidenz der Bulimie in den westlichen, hochindustrialisierten Ländern sprechen zweifellos für ein Zusammenspiel soziokultureller Einflüsse und Krankheitsmanifestation; dessen Analyse jedoch enthält Widersprüche, die wiederum Gegenstand weiterer Untersuchungen sind. Sicherlich läßt sich der paroxysmale „Fressen"-„Kotzen"-Vorgang nicht allein auf die symbolhafte Abwehr der vom materiellen Wohlstand bestimmten Forderung nach Leistung und Erfolg reduzieren. Eine die Symptomatologie der Krankheit programmierende und provozierende Wirkung besitzt ohne Zweifel die nahezu identifikatorische Kombination solcher Prägung mit der bis in den letzten Winkel reichenden informativen Macht der mit Text, Bildern und Ton für das Schlankheitsideal werbenden und die Schlankheitsleitbilder umwerbenden Medien. Erfolg, soziale Anerkennung und normierte Attraktivität gelten als synonym für das Idealbild der Frau, d.h., „Weiblichkeit" bzw. deren Ablehnung wird über den Körper, überspitzt ausgedrückt, nur noch über den Körper, entworfen. Das Ausmaß des Scheiterns an diesen Normvorstellungen korreliert mit der Stärke der Symptomatologie.

Laessle (1989) hebt hervor, daß – in Anlehnung an die Untersuchungen von Beck (1976) über die veränderte kognitive Struktur bei depressiven Patienten – sich auch bei Bulimiekranken inadäquate kognitive Konzepte finden ließen, die sich vor allem auf die Bewertung der Figur und des Gewichtes bezögen (Cooper und Taylor, 1989; Phelan, 1987; Ruderman, 1986). Das Selbstkonzept der Patienten definiert sich fast ausschließlich durch körperliche Attribute (Striegel-Moore et al., 1986). Fairburn und Mitarbeiter (1989) halten die genaue Beschreibung der Sorgen um die Körperform und die Kontrollmethoden des Gewichtes für die Diagnose der Bulimie für unerläßlich. Das wichtigste Instrument hierfür sei das klinische Interview, in standardisierter oder halbstandardisierter Form, der Erfassung mit Fragebögen überlegen.

Obwohl die meisten Patientinnen ein Idealgewicht haben (Fairburn und Cooper, 1984; Feiereis, 1989; Mitchell et al., 1985a; Pyle et al., 1981) und auch anamnestisch allenfalls Phasen eines nur leichten Übergewichtes vorliegen, bewirken die normativen äußeren Einflüsse den nachhaltigen Versuch, dem vorgegebenen Körper- und Selbstbild zu entsprechen. In Abhängigkeit davon wird mehr und mehr der Teufelskreis von Freßdurchbrüchen und Gewichtsphobie in Gang gesetzt. Die zwangsläufigen Versagenserlebnisse führen zum Dauerkonflikt, dessen Entwicklung und Manifestationsformen besonders innerhalb der Familie, Partnerschaft und sozialer Kontakte wiederholt beschrieben wurden (Ennulat, 1989; Johnson und Berndt, 1983; Johnson und Love, 1985; Norman und Herzog, 1984; Slade, 1982; Thompson und Schwartz, 1982).

Prämorbide Persönlichkeitsmerkmale

Nach bisher vorliegenden Befunden erscheint die prämorbide Struktur der Patienten mit Bulimie vor allem depressiv; wesentlich seltener kann sie als hysterisch bezeichnet werden. Bei den Mischformen sind die zwanghaften oder hysterischen Anteile weniger ausgeprägt als bei den Patienten mit Magersucht. Die depressive Struktur erklärt auch die häufige Auslösung eines Freßanfalles durch Verstimmung, Alleinsein oder erlebte Enttäuschung und umgekehrt die Abwehr der Depression mit Hilfe des „Fressen"-„Kotzen"-Anfalles. Als weitere Merkmale der prämorbiden Struktur werden autoaggressive Tendenzen, mangelnde Kontrollfähigkeit, Impulsivität und labiles Affektverhalten genannt.

Russell (1979, 1989) verneint das Vorliegen einer gestörten prämorbiden Persönlichkeitsstruktur, hebt aber „ausgeprägte Persönlichkeitseigenschaften" (1989) hervor, die das Verhalten formen und die Prognose beeinflussen könnten. Levin und Hyler (1986) beschreiben vorwiegend Borderline-Störungen mit zwanghaften Zügen, freilich ohne einen Zusammenhang mit der Schwere der Bulimiesymptomatik. Pathologische Auffälligkeiten im Verhalten, z. B. Ladendiebstähle, Anhäufung von Schulden, Selbstverletzungen oder auch Drogen- und Alkoholabhängigkeit stünden mit diesen prämorbiden Auffälligkeiten im Zusammenhang.

Die meisten Patienten mit Bulimie zeigen ebenso wie die Magersüchtigen gute bis hervorragende Schulleistungen, im Gegensatz zu den Magersüchtigen aber nicht als Folge des Anpassungsprozesses eines noch unreifen Ichs bei einem präadoleszenten Entwicklungsniveau (Bruch, 1982). Vielmehr wirken geistige Entwicklung, Introspektionsfähigkeit und entsprechend kognitive Leistungen eher akzeleriert, was den großen Leidensdruck ebenso erklärt wie die Therapiebereitschaft, sobald die eigenen Kompensationsmöglichkeiten ausgeschöpft sind oder die Krankheit offenbart werden konnte.

Psychodynamik

Stehen bei der Magersucht die intrapsychischen Triebkonflikte mit der Abwehr von Körperlichkeit und Sexualität, Verzerrung des Körperbildes und der Körperwahrnehmung, mit dem Bindungs- und Lösungskonflikt am Ende von Kindheit und Pubertät im Mittelpunkt der meisten Mitteilungen zur Psychodynamik, so sprechen die tiefenpsychologischen Befunde bei den Bulimiekranken für eine überwiegend frühe Störung der Selbstentwicklung. Bei den Magersüchtigen gehören die Verzerrung des Körperbildes und die Störung der Körperwahrnehmung zur primären Krankheitssymptomatik als Ausdruck des internalisierten Triebabwehrkonfliktes; „Auflösung des Körpers als Triumph über das Realitätsprinzip" (Export, 1987).

Bei den Bulimiekranken erscheint die zentrale Bedeutung von Körperform und Körpergewicht eher eine Folge der in früher Kindheit mißlungenen Ausgewogenheit in der narzißtischen Besetzung des Körper-Selbst und ebenso der mißlungenen Integration guter und böser Objektanteile, die auseinandergehalten werden müssen und zu dissoziierten Ich-Zuständen führen.

Die mangelhafte Subjekt-Objekt-Differenzierung hängt oft mit einer frühkindlichen Deprivation oder auch Überprotektion zusammen, die zu Störungen der Individuation und somit der Selbstentwicklung führen, also zu brüchigem Selbstwertgefühl und zu Selbstunsicherheit. In der Unsicherheit innerhalb des eigenen Selbstbildes sind auch Körperbild und Körperwahrnehmung einbezogen. Ähnlich wie bei Magersüchtigen liegt der Spannungsbogen der Auseinandersetzungen zwischen Abgrenzung, Autonomie und Abhängigkeit.

Im Gegensatz zu magersüchtigen Patienten, bei denen die Abwehr der Entwicklung zur Geschlechts- und Erwachsenenidentität Regression und Fixierung erklärt, scheinen bei den Bulimiekranken weitaus mehr erlebte Kränkungen, Versagungen und Enttäuschungen an der Schwelle zum Erwachsenwerden den Rückzug auf die Stufe des Größen-Selbst und die gleichzeitige Abkehr von der Außenwelt einzuleiten. Der Grund läge in der Angst vor weiteren Verletzungen und der Furcht vor Verlust an Autonomie und ebenso ersehnter wie schmerzlich erlebter Abhängigkeit und regressiven Verschmelzungswünschen. Alle reifen Objektbeziehungen werden deshalb abgewehrt, die Spannung zwischen Streben nach Autonomie und Symbiose kann so ausgehalten werden.

Die Psychodynamik der Bulimie kann demnach als eine frühe Störung der Selbstentwicklung verbunden mit der Verletzung des labilen Selbstwertgefühls angesehen werden. Die Verletzung steht meistens im engen zeitlichen Zusammenhang mit dem Beginn der Symptomatologie, wirkt also im Sinne der Auslösung. Der jedes Maß sprengende, als orale Aggression zu verstehende Freßanfall ist zugleich Korrelat und Ventil der Selbstwertkrise und entgleist sehr bald zur Sucht. Mit dem Anfall verbindet sich – psychodynamisch gesehen – eine der Grandiosität des Größen-

Selbst adäquate Triebbefriedigung, die gleichzeitig einen ausgeprägten autodestruktiven Akt darstellt, der mit dem Erbrechen sein entsetzliches und den Patienten entsetzendes, aber auch befreiendes Ende findet. Ein realer oder befürchteter Objektverlust, der Anspannung und Trauer verursacht hat, wird zunächst mit Hilfe des oral-destruktiven Impulses kompensiert und mit dem anschließenden Erbrechen wieder rückgängig zu machen versucht.

„Über die Einverleibung, mithin die Fusion, geraten die Kranken in einen Zustand der Abhängigkeit und Hilflosigkeit. Außerdem müssen sie durch die Fusion eine Identitätsdiffusion hinnehmen, welche Angst macht. Nach dem Essen bestimmt nun nicht mehr Trennungsangst das Bild, sondern Verschmelzungsangst. Abhängigkeit, Hilflosigkeit und Angst werden unerträglich" (Ettl, 1988).

In der von oraler Gier gesteuerten Sucht des „Fressen"-„Kotzen"-Vorganges drückt sich der verzweifelte Kampf um die Attribute eines von der Gesellschaft erwarteten vorgegebenen Idealbildes aus, gleichzeitig aber auch intensive Aggressivität gegen die Mutter, die mit ihrer Versagung oder Verwöhnung eine harmonische Selbstentwicklung verhindert. Dieser pathogenetische Anteil beruht nicht selten auf einer gestörten Eigenentwicklung der Mutter, die ihrerseits erhebliche Bindungs- und Lösungskonflikte hatte.

Die skizzierte Psychodynamik bei der Bulimie unterscheidet sich in manchen Zügen von der Entwicklung, die zur Magersucht führt und deshalb auch psychodynamisch eine sorgfältige Differenzierung beider Krankheiten begründet; sie hat ihr Korrelat in der unterschiedlichen Symptomatologie. Die Grenzen können allerdings bei der Form der Bulimie mit initialer oder intermittierender Magersucht fließend ineinander übergehen.

Die bisher vorliegenden Befunde deuten darauf hin, daß weniger die prämorbide Struktur als die Psychodynamik spezifische Merkmale besitzt, die allmählich mit dem Suchtpotential verschmelzen, wodurch Mühsal und Beschwernisse der therapeutischen Wege erklärbar werden.

Die Verknüpfung prämorbider, auch genetisch determinierter Persönlichkeitsanteile und der spezifischen Psychodynamik mit der Krankheitsentwicklung ist bei der Bulimia nervosa und Anorexia nervosa evident und wirkt unmittelbarer auf die körperliche und psychische Symptomatologie als bei psychosomatischen Krankheiten mit primär morphologischem Substrat. Viele Untersuchungen lassen deutlich werden, daß bei diesen Krankheiten keine Spezifität prämorbider psychischer Struktur und intrapsychischer Konflikte vorliegt.

Die eingangs geschilderte Patientin E. S. hat ihre Ängste und Aggressionen in der negativen Identifikation mit der Mutter rationalisiert. Diese wird als „typische Hausfrau mit typisch unselbständigen Zügen" charakterisiert. Schon seit frühester Kindheit habe die Mutter unter psychischen und physischen Störungen gelitten. Schlafprobleme, Tranquilizerabusus und ständige depressive pessimistische Grundhaltung werden hervorgehoben, schließlich sei die Mutter nahezu menschenscheu geworden.

Ihren Vater charakterisiert die Patientin als „dominant, unabhängig und entscheidungsfreudig"; lebensbejahende und starke Eigenschaften werden einer Vaterfigur zugeordnet, die ein Ich-Ideal entstehen läßt, das der Mutter feindlich gegenübersteht und sich genealogisch auf den Vater bezieht. Die geschilderten Ängste der Patientin, als erfolgreiche Frau keine Hoffnung auf Liebe hegen zu können, lassen sich auf die internalisierte Aufspaltung von Gefühl als „weibliche" sowie Verstand und Leistung als „männliche" Eigenschaften zurückführen. Bei dieser Patientin werden die Emotionen, vermittelt durch das Verhalten der Mutter, die durch ihr (über-)forderndes Leiden traditionell in der Familie ihr Kampfgebiet findet, weitgehend destruktiv erlebt. Die Vater-Tochter findet sich in einem dichotomisierenden Gegensatz zur Mutter wieder, deren autoaggressive Tendenz sich vor allem gegen das wendet, was die Tochter mit ihr verbindet: gegen den weiblichen Körper, den die Mutter durch Tranquilizer totstellt.

Die Patientin selbst empfindet ihre bulimischen Attacken als Ventil, um eine Leere zu füllen und Depressionen zu entgehen. Andererseits beschreibt sie ihr Eßverhalten als unproblematisch, wenn sie Gefühle ausleben und zeigen kann, wie beispielsweise beim Weinen. Allerdings muß sie solche Lebensäußerungen unterdrücken, da sie sich mit dem Vater identifiziert und versucht, sich in die symbolische paternale Ordnung einzugliedern.

In diesem Zusammenhang sind auch die zeitlich begrenzt aussetzenden Attacken der Patientin zu sehen, wenn sie „eine neue Beziehung" angefangen hat, die bald darauf wieder aufgegeben wird, also am Anfang einer Freundschaft, die noch keiner familiären Ordnung unterworfen ist und somit vorerst zugelassen werden kann.

Der Tod des Vaters kurz vor dem Beginn der Symptomatik kann als ein auslösender Faktor verstanden werden: Der reale Vater verschwindet zugunsten eines vermißten, verinnerlichten Ideals. Gleichzeitig führt die geänderte Familienkonstellation zwischen Mutter und Tochter zu einer anderen Beziehung, die durch den leeren Platz des Vaters an Nähe gewinnt, in der sich jedoch die Ambivalenzen zuspitzen. An dieser Stelle werden noch einmal die Bindungs- und Lösungskonflikte sowie die scheiternden Autonomiebestrebungen deutlich.

Böhme-Bloem und Schulte (1989) unterscheiden aufgrund ihrer Untersuchungsergebnisse (bei allerdings sehr kleiner Bezugsgröße) zwischen Bulimie mit neurotischem Entwicklungsniveau und Bulimie mit einer narzißtischen Persönlichkeitsstörung. Beiden Gruppen gemeinsam sei die orale Fixierung.

Bei den neurotischen Patientinnen entwickle sich ein narzißtisches Defizit mit libidinös unterbesetztem Selbst und dadurch starker psychischer Abhängigkeit von der Mutter. Über den Weg ödipaler Fixierung aktualisiere ein Partnerkonflikt die Probleme um Autonomie und Nähe. Die Symptomatologie der Bulimie sei ein Kompromiß unter dem Ansturm von Triebimpulsen.

In der Gruppe mit narzißtischer Persönlichkeitsstörung führe die Abweisung durch die Mutter zu einem Defekt im Selbst, die frühe Spaltung zwischen gutem und bösem Objekt bleibe erhalten. An den Vater werde die große Hoffnung geknüpft, das mütterliche „Defizit" wieder auszugleichen. In der Folgezeit trete eine pseudoödipale Stabilisierung ein. Schließlich werde das Selbstdefizit durch ein einschneidendes Erlebnis, meist einen konkreten Objektverlust, schmerzhaft aktualisiert. Die bulimische Symptomatologie sei Ausdruck einer Abwehr gegen die drohende Depression.

Bei der Hälfte der Patientinnen wurde in beiden Gruppen unabhängig von der Schwere der Störung eine durch Geschwisterrivalität gestörte ödipale Phase festgestellt.

Auslösung

Im engen zeitlichen Zusammenhang mit den ersten Freßanfällen, also dem Beginn der Erkrankung, stehen häufig erlebte, drohende oder imaginierte Kränkungen, die sich auf das Aussehen, vor allem auf das Gewicht der Patienten beziehen.

Andere Auslösungsmöglichkeiten sind phantasierte oder reale Verluste nahestehender Menschen oder enttäuschende Erfahrungen innerhalb erster Freundschaften mit erotischen oder sexuellen Beziehungen.

Im Gegensatz zu den Magersüchtigen identifizieren sich die Patientinnen mit Bulimie sehr mit der Weiblichkeit. Die Abwehr sexueller Wünsche ist oft Ausdruck der Furcht vor weiterer Kränkung und Entwertung.

Bei unserer Patientin stand der Tod des Vaters in enger zeitlicher Beziehung zum Beginn der Erkrankung. Schubweise trat auch stets eine Verschlechterung ein, wenn sich – kaum begonnene – Freundschaften wieder lösten, während die Symptomatik zu Beginn einer jeden neuen Beziehung, in der sie sich glücklich und akzeptiert fühlte, rückläufig war.

Aus den genannten Auslösungsvorgängen am Beginn der Krankheit formt sich allmählich ein Teufelskreis der Symptomatologie (Feiereis, 1989): Ein scheinbar bedeutungsloser Anlaß wie Ärger, Langeweile, Alleinsein, meistens am Nachmittag oder Abend, oder Unlustgefühl setzt ihn in Gang. In seinem Mittelpunkt steht gleichsam die Spaltung des Ichs, d. h., der Anfall wird bei vollem Bewußtsein und machtloser Selbstbeobachtung „wie von außen" erlebt; Freßanfall und Kontrollverlust sind Ausdruck der „Verselbständigung des Körpers". Das Ende der Spaltung signalisieren Erschöpfung, Verzweiflung, Scham, Schuldgefühle, Selbsthaß und Depression, die wiederum Impulse zu kontrolliertem Eßverhalten auslösen.

38.3.8 Kombinierte psychosomatische Therapie

38.3.8.1 Vorbemerkungen

Die Therapie der Bulimia nervosa gleicht oft einer sisyphusähnlichen Aufgabe – dem Begriff der Sucht gemäß.

Nicht anders als bei der Magersucht werden für die Bulimie unterschiedliche therapeutische Konzepte angeboten (Fairburn, 1985; Fichter, 1989a; Mitchell et al., 1985b; Meermann und Vandereycken, 1987). Die bisher gewonnenen und mitgeteilten Erfahrungen reichen nicht aus, um die Überlegenheit der einen oder anderen Behandlungsform oder der Kombination mehrerer Therapieverfahren zu begründen. Nach unseren Erfahrungen sollte die Behandlung bei Bulimie wie bei Magersucht offen sein für:
- Psychoanalytisch orientierte bzw. tiefenpsychologisch fundierte Psychotherapie ebenso wie für eine konfliktzentrierte stützende oder lerntheoretisch abgeleitete Psychotherapie.

- Einzeltherapie, Gruppentherapie und Familientherapie, die in keiner alternativen Beziehung zueinander stehen.
- Sorgfältig begründete körperorientierte Anteile der Therapie.

Die wichtigsten Grundsätze unserer Behandlung sind:
- Auf den Kranken und seine Krankheit individuell abgestimmte therapeutische Maßnahmen.
- Der **gleichzeitige** Beginn der somatischen und psychischen Therapie.
- Modifikation der Behandlung, je nach Indikation und Krankheitsverlauf.

Das therapeutische Konzept unserer kombinierten psychosomatischen Therapie bei Bulimie zeigt Abbildung 38.3–7.

Haben die Bulimiekranken nach oft jahrelangem verborgen gehaltenem und verborgen gebliebenem Krankheitsverlauf den Weg zur Therapie gefunden, so bedeutet das einleitende Gespräch bereits eine erste wirksame Hilfe gegen die von Depression und Hoffnungslosigkeit gekennzeichnete Stimmung. Das Ziel des diagnostisch-therapeutischen Gesprächs liegt ebenso auch in der Klärung offener Fragen, z. B. über Form (ambulant oder stationär), Inhalt und Dauer der Behandlung, Regelung sozialmedizinischer Schwierigkeiten (z. B. Schulausfall, Unterbrechung der Lehre oder des Studiums, gefährdeter Arbeitsplatz).

Primär ambulante Therapie

Indikationen:
- Schweregrad I.
- Dauer der Krankheit möglichst kürzer als 2 Jahre.
- Drohende schwerwiegende soziale Folgen einer längeren stationären Therapie, z. B. Verlust des Arbeitsplatzes, Gefährdung des Schulabschlusses, Unterbrechung der Vorbereitung auf wichtige Examina.

Für die ambulante Behandlung gilt gleichermaßen wie für die stationäre Therapie, daß sie innerhalb ihrer Methodenwahl individuell modifiziert sein sollte.

Schwerpunkte:
- Tiefenpsychologisch fundierte, konfliktzentrierte Einzeltherapie.
- Autogenes Training und progressive Relaxation.
- Lerntheoretisch abgeleitete Hilfen einschließlich der Ernährungsberatung zur Unterbrechung des bulimischen Teufelskreises.
- Familientherapie bei jüngeren Patienten.

Auch die ambulante Fortsetzung der primär stationären Therapie setzt sich im wesentlichen aus diesen vier Anteilen zusammen.

Der Leidensdruck der Patienten mit Bulimie ist groß (über 90%), die Bereitschaft zur Behandlung kaum weniger (79%).

In Übereinstimmung mit anderen Autoren (Fichter, 1989a) empfehlen wir nach langem Krankheitsverlauf die stationäre Einleitung der Therapie, vor allem dann, wenn vorausgegangene ambulante Behandlungen erfolglos geblieben sind (Schweregrad I) und/oder die Schweregrade II und III vorliegen.

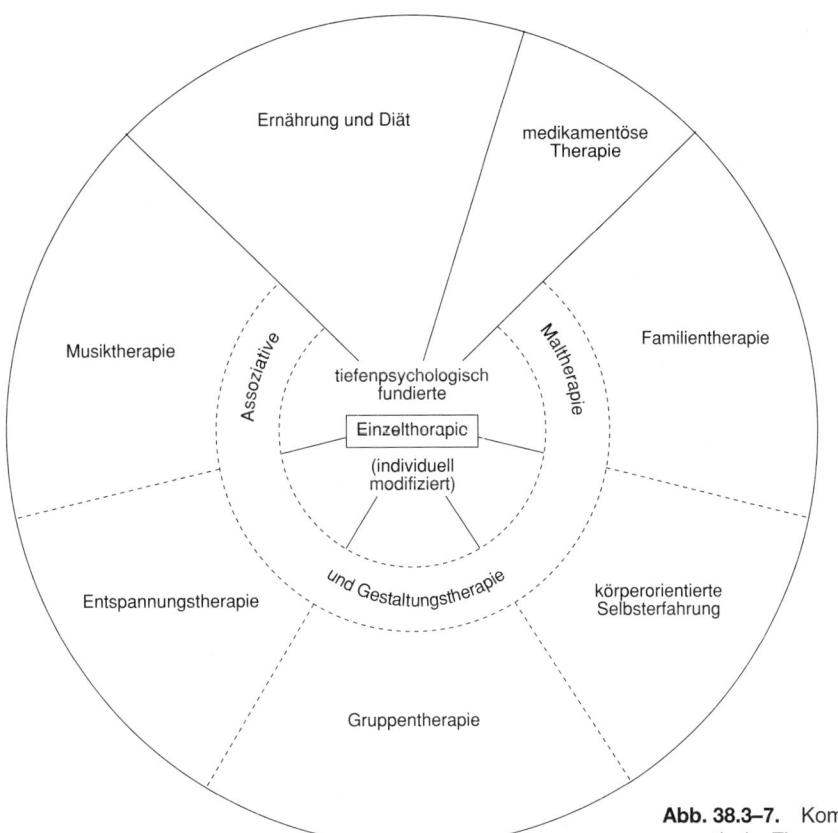

Abb. 38.3–7. Kombinierte psychosomatische Therapie bei Bulimie.

38.3.8.2 Psychoanalytisch orientierte konfliktzentrierte Einzeltherapie

Angesichts des hohen Leidensdruckes und der großen Therapiebereitschaft der Patienten mit Bulimie scheint die psychoanalytisch orientierte, konfliktzentrierte Therapie eher als bei der Magersucht indiziert zu sein, was sich aber spätestens dann als Trugschluß herausstellt, wenn deutlich wird, in welchem Ausmaß sich der Teufelskreis von Essen und Erbrechen als Sucht verselbständigt hat (Wilke, 1989). Im psychoanalytischen bzw. tiefenpsychologisch fundierten Ansatz ist ständig zu berücksichtigen, daß sich Suchtverhalten von selbst perpetuiert und die Suche nach Ursachen und ebenso die tiefenpsychologische Therapie scheitern müssen, solange dieses Suchtverhalten anhält. Am Beginn der Therapie muß daher die Stärkung der Motivation und der Bereitschaft zur konsequenten Therapie stehen, um zu tieferer Krankheitseinsicht zu gelangen, die wiederum eine Voraussetzung für die Behandlung der Suchterkrankung ist. Nach Wilke (1989) handelt es sich hierbei nicht um einen Prozeß linearer Weiterentwicklung, sondern um Kreisprozesse, in denen Motivation und Krankheitseinsicht durch die tiefenpsychologische Arbeit weiter verstärkt werden, besonders, wenn in ihr Selbstkonfrontationen, Selbstdarstellungen und Selbstbegegnungen mit den dazugehörigen Affekten möglich werden.

Innerhalb dieser so praktizierten Therapie sollten die Symptome Hungern, Fressen oder Kotzen wie auch die ständig um Gewicht, Waage und Körperbild kreisenden Gedanken sowie die minuziösen Aufzeichnungen über die einzelnen Mahlzeiten in den Gesprächen allmählich nur noch eine periphere Bedeutung erhalten, um den Blick für die Arbeit an der schwierigen Persönlichkeitsentwicklung und den damit verbundenen Konflikten zu öffnen. Mit der Aufgabe des pathologischen Eßverhaltens werden Affekte mobilisiert und Erinnerungen wach, die mit diesen Affekten verbunden sind.

Ein Hauptziel der psychoanalytisch orientierten Einzelbehandlung ist der Aufbau und die Konsolidierung des Selbstwertgefühls. Die im zeitlichen Zusammenhang mit dem Beginn der Krankheit aufgetretenen und noch unverarbeiteten Enttäuschungen, Kränkungen und Versagungen können integriert und überwunden werden. Die Therapie ermöglicht häufig Erinnerungen an frühkindliche Erlebnisse mit der defizitären Mutter, des Liebesentzugs wie der Überprotektion; die regressiven Verschmelzungswünsche können in Begleitung des Therapeuten weniger angstvoll, ja überhaupt wahrgenommen und gelebt werden und setzen die Autonomiestrebungen in ein anderes Licht. Durch die Übertragung kann das zerstückelte Selbst Hilfe zum Aufbau einer neuen Identität erhalten.

Erfahrungen mit der tiefenpsychologischen bzw. psychoanalytischen Behandlung Bulimiekranker wurden von verschiedenen Autoren mitgeteilt (Gonzalez, 1988; Ravenscroft, 1988; Schwartz, 1986, 1988).

Ähnlich wie bei der Magersucht hat sich für die

Behandlung von Bulimiekranken das imaginative Verfahren des katathymen Bilderlebens sehr bewährt. Mit ihm gelingt es, Selbstdarstellung und Selbstkonfrontation zu fördern, und in einer kontrollierten Regression, Affekte wieder zu beleben (Wilke, 1989). Einen wichtigen Schritt aus der Einsamkeit und Scheinautonomie dieser Patienten heraus bedeute, sich abbildende depressive, masochistische oder suizidale Tendenzen simultan bei ihrer Entstehung dem Therapeuten mitzuteilen.

38.3.8.3 Themenzentrierte Gruppentherapie

Innerhalb der themenzentrierten Gruppentherapie als interpersonelles und verbales Psychotherapieverfahren soll die Aufmerksamkeit gleichermaßen auf das Ich (die Persönlichkeit), das Wir (die Gruppe) und das Es (das Thema) gelenkt werden (Cohn, 1979, 1984). Durch den Therapeuten wird eine konstruktive thematische Arbeit bei gleichzeitiger persönlicher Autonomie der einzelnen Gruppenmitglieder angeregt.

Die themenzentrierte Gruppentherapie eignet sich besonders für Patienten, die ihre Gedanken und Meinungen, Gefühle und Ängste nur unter Schwierigkeiten äußern können oder wollen. Patienten, die sprachlich gewandt sind und zu Rationalisierungen neigen, werden weniger Nutzen aus der Teilnahme an der Gruppenarbeit ziehen. Im Unterschied zur Einzeltherapie stellt die Rückmeldung von Mitpatienten eine große Hilfe dar. Unterstützung und Zuwendung, aber auch Kritik können hilfreich sein, Positionen und Meinungen erneut zu bewegen und eventuell zu ändern.

Besonders zwei Verhaltensweisen fallen bei Patienten mit Bulimie noch stärker als bei Magersüchtigen innerhalb der gruppentherapeutischen Interaktionen auf (Drewes und v. Wietersheim, 1989): Entweder zeigen sie sich als äußerst zurückgezogen, bringen sich selten spontan ein und reagieren nur auf Nachfragen, oder sie wenden sich ganz den Problemen ihrer Mitpatienten zu, um die Aufmerksamkeit von eigenen Schwierigkeiten abzulenken. Sie stellen sich dadurch in den Mittelpunkt und versuchen, den Verlauf der Sitzung zu bestimmen.

Im Vergleich zu Magersüchtigen verhalten sich Bulimiekranke oft wesentlich aggressiver oder destruktiver. Die Schwere ihrer Krankheit, die eigenen Konflikte sowie ihre isolierte Situation bleiben lange Zeit hinter dem vordergründig extravertierten Verhalten verborgen.

38.3.8.4 Familientherapie

Über Ergebnisse der Kombination einer psychoanalytisch orientierten bzw. tiefenpsychologisch fundierten Einzeltherapie mit einer systemischen Familientherapie bei Bulimie liegen bereits eingehende Berichte vor (Andersen, 1985; Brownell und Foreyt, 1986; Emmett, 1985; Fichter, 1985; Jantschek et al., 1987; Powers und Fernandez, 1984; Schwartz, 1982).

Die Grundlage für die Indikation zur Kombinationstherapie bilden ebenso wie bei der Magersucht die gewachsenen Erkenntnisse über die Wechselwirkung intrapsychischer und systemischer Prozesse. Nach unseren Erfahrungen (Jantschek und Jantschek, 1989) sollte vor allem innerhalb der stationären Behandlung, nach Möglichkeit auch unter ambulanten Bedingungen, „der Einzeltherapeut einer von zwei Familientherapeuten sein, um die Einzelbehandlung nach den Informationen aus den Familiengesprächen zu modifizieren bzw. umgekehrt, diese aufgrund der Entwicklung und der Dynamik in der Individualtherapie beeinflussen zu können. Dies hat sich bis heute bewährt und läßt Rivalitäten und Spaltungen, die sowohl auf der Patienten- als auch auf der Therapeutenseite unweigerlich auftreten, besser aushalten".

Der Kontakt mit der Familie wird frühzeitig aufgenommen, also nicht der umgekehrte Weg bevorzugt, die Familie zunächst durch das Angebot des Rückzugs zu entlasten. Methodisch gilt als wichtigster Grundsatz, daß jede Familie ein spezifisches Problem aufweist und darum ein spezieller Ansatz nötig ist.

Praktisch hat sich bewährt, mit einer Hypothese über die familiäre Interaktion, die sich aus Vorinformationen und therapeutischen Überlegungen ergibt, die Familiensitzung zu beginnen. Mit den nunmehr gewonnenen Informationen werden die Hypothesen bestätigt oder von einer neuen Hypothese abgelöst. So ergibt sich ein kontinuierlicher und therapeutisch-diagnostischer Zirkel (Jantschek u. Jantschek, 1989).

Das Ziel der Therapie liegt in einer qualitativen Änderung des familiären Systems, um den familienatmosphärischen Anteil an der Entwicklung und Chronifizierung der bulimischen Erkrankung zu beeinflussen (Strober und Humphrey, 1987).

Die Voraussetzung für eine Familientherapie bildet die begründete Indikation, die nach Merl (1983) fünf Punkte umfaßt:

- Es muß ein Familienkonflikt, d. h. eine die Familienmitglieder involvierende Problematik bestehen.
- Der Bereinigung dieses Konfliktes kommt grundlegende oder mindestens wesentliche akzessorische therapeutische Bedeutung zu.
- Die Familie ist sich dieser Situation mehr oder weniger deutlich bewußt (Konflikt- bzw. Krisenbewußtsein) und an der Bereinigung interessiert.
- Alle in diesem Sinne betroffenen Familienmitglieder müssen erreichbar sein.
- Die aufgrund der Kapazität des Therapeuten und der Lage der Familie angestrebten Veränderungen müssen weitgehend verwirklicht werden können (Prognose).

38.3.8.5 Assoziative Mal- und Gestaltungstherapie

Im Unterschied zu verbalen und interpersonellen Psychotherapieverfahren lassen Mal- und Gestaltungstherapie andere Ausdrucksmöglichkeiten als Sprechen und Agieren zu.

In der **assoziativen Maltherapie** werden die Patienten angehalten, Vorstellungen, Phantasien, Träume (Abb. 38.3–10), Tagträume, Ängste, Konflikte,

Erinnerungen und Gedanken, ohne Vorgabe von Thema, Form und Material, im Bild darzustellen. Rationale Inhalte finden hier ebenso ihren Platz wie emotionale, vor- und unbewußte Prozesse. Bildhafte Speicherungen können direkt in nonverbaler Form mitgeteilt werden (Schuster, 1986).

Das Bild, verstanden als Modifikation und Realisation innerpsychischer Vorgänge, bietet dem Patienten einerseits die Möglichkeit, Zugang zu unbewußten Konflikten sowie zu sprachlichen Ausdrucksformen zu finden, öffnet andererseits für den Therapeuten schon in der Initialphase der psychoanalytisch orientierten Einzeltherapie einen Weg zum Verständnis der individuellen Konfliktlage des einzelnen Patienten (Franzke, 1977). Hilfreich sind dabei die schriftlich fixierten assoziativen Einfälle des Patienten zur Thematik seines Bildes.

In einer vergleichenden Studie (Feiereis et al., 1989) finden sich bei der Analyse des assoziativ gestalteten Bildmaterials von Bulimie- und Magersuchtpatienten, das nach Kategorien wie Aggression, Depression, Ambivalenz, Symptomatik u.a. geordnet wurde, Gemeinsamkeiten und Unterschiede in Thematik, inhaltlicher und formaler Gestaltung.

In beiden Gruppen fallen die zahlreichen Darstellungen auf zum Thema (Abb. 38.3–8 bis 38.3–11 s. Farbtafel 2):

- Ambivalente Haltungen gegenüber dem Selbst (Abb. 38.3–12 und 38.3–13), dem Essen, dem Gewicht, dem Körper (Abb. 38.3–18, 20, 21, 22), der Gesundheit und gegenüber der Krankheit (Abb. 38.3–15), die einerseits als Käfig, andererseits als Schutz empfunden wird;
- Widersprüchlichkeit zwischen Verstand und Gefühl (Abb. 38.3–17), Körper und Seele, Krankheit und Gesundheit – die quantitativ vielfältigste Problematisierung;
- Depression (Abb. 38.3–9, 12, 16) häufiger als Aggression und Autoaggression;

Abb. 38.3–12. 25j. Patientin C.T.: „Ich dachte daran, wie sehr ich mich durch meine Sucht vom Leben isoliert habe. Ich habe mir meine Seele eingesperrt vorgestellt, von mir selbst in die Flasche gestopft." (19. 11. 87)

- Abgrenzungs- und Identifikationskonflikte gegenüber den Eltern, Therapeuten und Mitpatienten (Abb. 38.3–11);

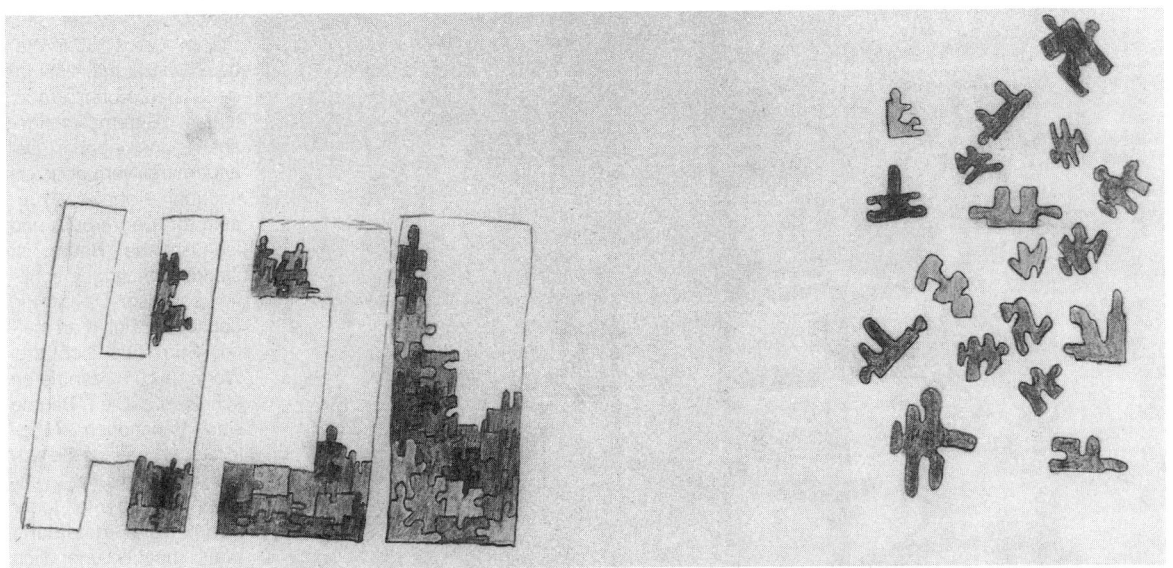

Abb. 38.3–13. 23j. Patientin T.Z.: „Ich weiß bis jetzt noch nicht, woher meine Probleme bzw. meine Krankheit kommen. Ich kann deshalb auch nichts Wirkliches dagegen tun. Ich glaube, daß mein ‚Ich' aus vielen Puzzleteilen zusammengesetzt ist, aus denen sich langsam ein Gesamtbild ergibt. Ich denke und hoffe, daß ich, wenn ich ‚das Bild' vervollständigt habe, auch wirksam gegen meine Krankheit ankämpfen kann." (22. 12. 87)

Abb. 38.3–14. 23j. Patientin J. L.: „‚Schwimmen'. Schwimme ich? – Tauche ich? – Wonach? Geh ich unter? – Worin? Fliege ich? – Wohin? Oder falle ich? – Tief?! –– Wonach kann ich greifen?" (9. 3. 89)

– als Isolation empfundene gegenwärtige Lage (Abb. 38.3–8, 9, 12) sowie die Abwehr von Bevormundung durch die Therapie und ihre Maßnahmen;
– Ängste vor „Normalität", vor Erwachsenwerden, vor Kontroll- (Abb. 38.3–14) und Symptomverlust.

Außer diesen Gemeinsamkeiten werden in den bildnerischen Darstellungen beider Krankheitsgruppen auch große Unterschiede deutlich (Tab. 38.3–4), die ebenfalls ein Indiz für die unterschiedliche Pathogenese sind.

Auch in der Darstellung von (Frauen-)Körpern (Abb. 38.3–20 bis 38.3–22, hier Gestaltungstherapie mit Ton) unterscheiden sich die Bulimie- und Magersuchtpatienten. Fragmentierungen, Verzerrungen, Halbierungen nehmen in beiden Gruppen eine zentrale Stellung ein. Bulimiekranke stellen vor allem feiste, unproportionierte Frauengestalten dar, Magersüchtige eher Strichmännchen. Vielleicht entspricht diese Beobachtung der These, daß bei Magersüchtigen der die Gewichtsphobie ablösende Wunsch, dünn zu werden, bei Bulimiekranken die konstante Angst, dick zu sein oder zu werden, die konfliktreiche Einstellung zum eigenen Körper bestimmt (Feiereis, 1989).

Auch auf der formalen Ebene sind bei beiden Krankheitsgruppen Unterschiede zu bemerken: Bei Magersuchtpatienten findet man Formen der Zentrierung, der Raumverengung und der zentripetalen Strebung, bei Bulimiekranken hingegen häufig die gegenteiligen Darstellungen: Die Formen streben nach außen, werden weiter, sind zentrifugal.

Der frei-assoziativen Gestaltung innerpsychischer Prozesse kann ein gleichzeitig diagnostischer und therapeutischer Wert zugesprochen werden, indem un- und vorbewußte Konflikte, Widerstände, Abwehr und Übertragungen dem Patienten und Therapeuten zugänglich werden und gleichzeitig auch bewußte,

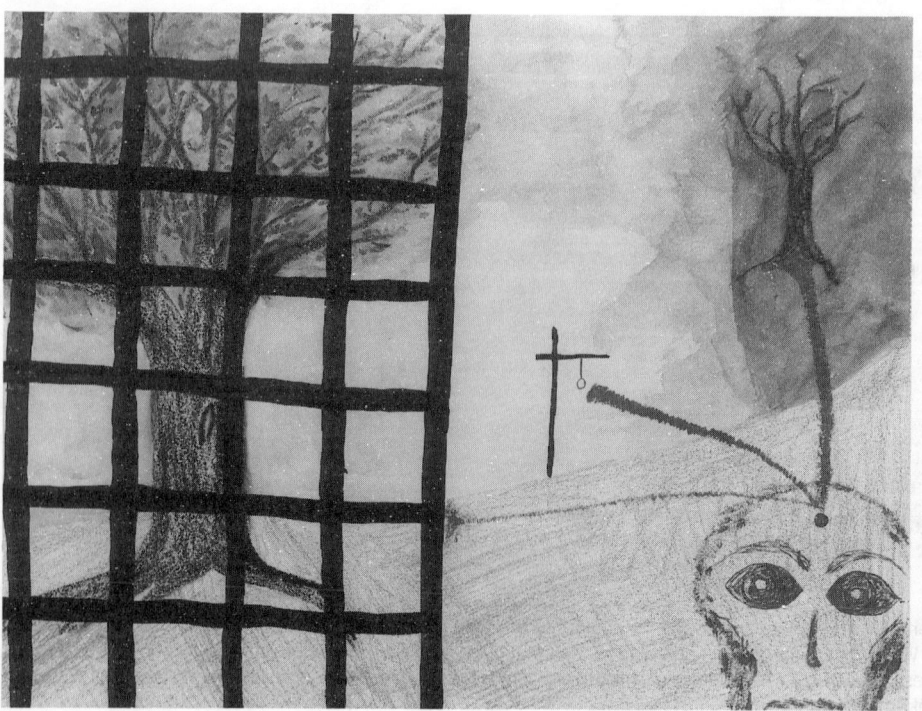

Abb. 38.3–15. 22j. Patientin D. Z.: „Zwei Bäume im Herbst oder Wege. Zwei Bäume, einer lebend, herbstbunt – aber hinter Gittern, eingesperrt, unerreichbar. Der andere frei, verfügbar, erreichbar – aber verkrüppelt, tot. Der farben- und lebensreiche Baum ist stärker, mächtiger, schöner sowieso. Davor: Ein denkender Punkt an meiner Stirn, wo sich drei Wege zu zentrieren scheinen. Die Bäume sind Variationen, Möglichkeiten meines Selbst. Die drei Wege, ihre eventuelle Begehbarkeit (oder nicht), die geforderte Wahl, machen unendlich hilflos: Welches ist das kleinere Übel? Aufgrund der Verfügbarkeit wohl eher die beiden rechten." (14. 10. 88)

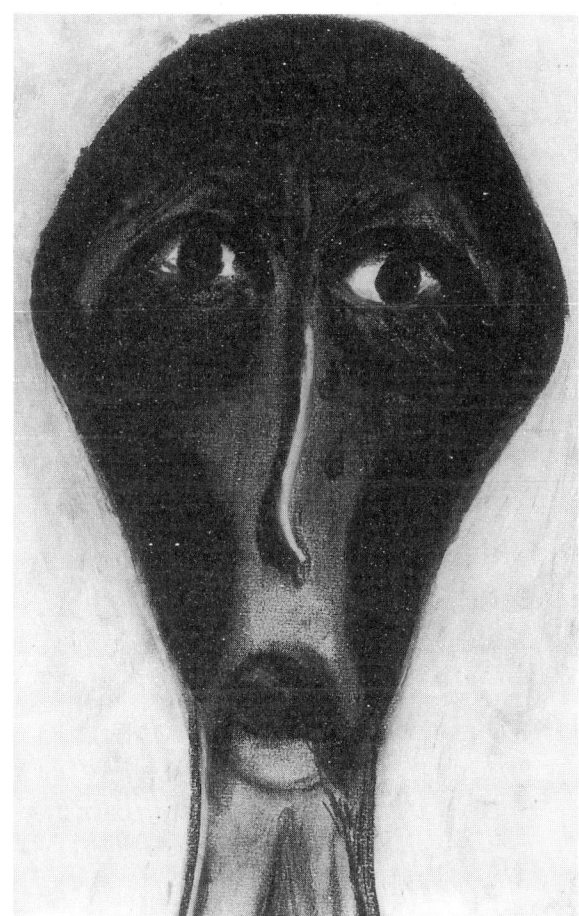

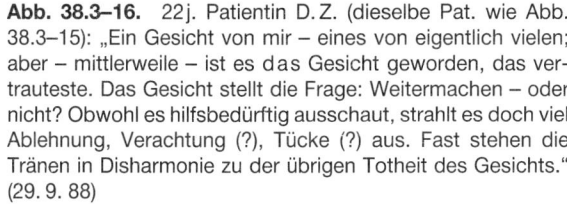

Abb. 38.3–16. 22 j. Patientin D. Z. (dieselbe Pat. wie Abb. 38.3–15): „Ein Gesicht von mir – eines von eigentlich vielen; aber – mittlerweile – ist es das Gesicht geworden, das vertrauteste. Das Gesicht stellt die Frage: Weitermachen – oder nicht? Obwohl es hilfsbedürftig ausschaut, strahlt es doch viel Ablehnung, Verachtung (?), Tücke (?) aus. Fast stehen die Tränen in Disharmonie zu der übrigen Totheit des Gesichts." (29. 9. 88)

angstbesetzte Inhalte vorsichtig und spielerisch angedeutet sein können (Biniek, 1982; Böhler, 1988; Feiereis et al., 1989; Franzke, 1977; Schuster, 1986).

Neben der Maltherapie können in der **assoziativen Gestaltungstherapie** mit Ton sowohl durch freies und spontanes Arbeiten als auch durch Anregungen eines Themas, wie z. B. „Körper-" oder „Kopfgestaltungen", über das geschaffene Werk Emotionen und Konflikte ausgedrückt werden. Für Bulimie- und Magersuchtpatienten erhält das thematische dreidimensionale Arbeiten einen besonderen Wert; Konflikte, z. B. bezogen auf den Körper oder die Sexualität, werden durch das figürliche Gestalten evident und thematisiert. Unbewußte Vorgänge werden externalisiert und damit dem Bewußtsein zugänglich.

38.3.8.6 Aktive und rezeptive Musiktherapie

Aktive und rezeptive klinische Musiktherapie übernehmen als präverbale Psychotherapieverfahren

Abb. 38.3–17. 24 j. Patientin R. G.: „Zwiespalt zwischen Gefühl und Logik." (Juli 1988)

wichtige Funktionen, besonders bei der Behandlung jener Krankheiten, an denen frühe seelische Störungen beteiligt sind, indem vor allem „emotional-affektive" Erfahrungen (Strobel und Huppmann, 1978) unter zeitweiligem Ausschluß von sprachlichen Ausdrucksformen erlebbar werden. Während die rezepti-

Tabelle 38.3–4. Unterschiede bildnerischer Darstellungen bei Bulimie und Magersucht.

Bulimie	Magersucht
Ängste vor Kontrollverlust, besonders beim Essen	Ängste vor dem Zusammenbruch der eigenen Kontrolle, vor Symptomverlust und Aufgabe der Autonomie
Selbstvorwürfe u. Versagergefühle bei Rückfällen	
Sehnsucht nach Symbiose und Geborgenheit	dumpfe Depressivität
Wunsch nach Verhaltenskontrolle	Todesvorstellungen Heimweh
eher positive Einstellung zur Therapie	Abwehr gegen die Therapie

38.3–18

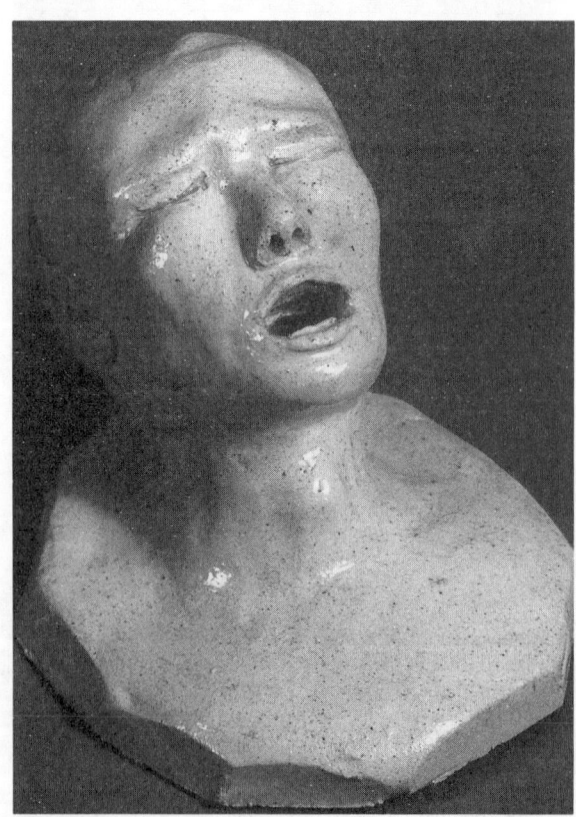

38.3–19

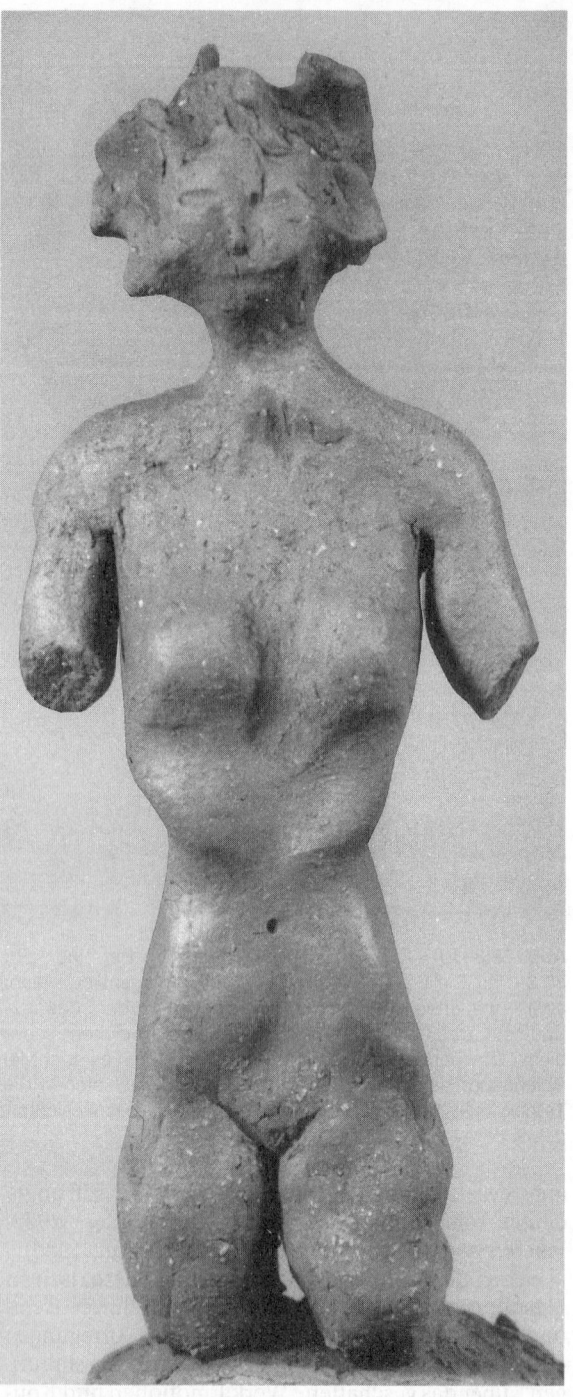

38.3–20

Abb. 38.3–18.   24 j. Patientin R. G. (dieselbe Pat. wie Abb. 17): „Figur mit extrem kleinen Händen. Hände bedeuten Handeln. Ich bin traurig, daß ich an meinem Zustand so wenig ändern kann." (Juli 1988)

Abb. 38.3–19.   24 j. Patientin D. Z.: „Ohne Text." (Juli 1989)

Abb. 38.3–20.   25 j. Patientin L. K.: „Beim Modellieren der Figur war es mir wichtig, daß eine ganz schmale, schlanke Gestalt entsteht. Ich denke, die Figur entspricht meinem Bild, das ich von meinem Körper habe. Ich möchte so aussehen wie die Tonplastik." (Februar 1989)

Abb. 38.3–21. 22j. Patientin T.S.: „Eine dicke, erotische, sitzende Frau, sich selbst, ihres Körpers, ihrer Ausstrahlung bewußt. Erotisch, gerade weil sie dick ist. Sinnlich. Schön und häßlich, ruhend und erregt, lustvoll und leidend, Frau und Geliebte, alt und jung, eine die gibt und die nimmt.
Diese Gedanken hatte ich vor, während und nach meiner Arbeit an dieser meiner ersten Tonfigur. Natürlich war ich bei einem so hoch gesteckten Ziel von dem Ergebnis enttäuscht. Einige Körperteile, vor allem der Kopf, mißfielen mir. Der ganze Ausdruck stimmte nicht mehr, und ich war froh, als ich die Frau endlich vollendet hatte. Noch immer bin ich traurig, daß sie mir nicht so perfekt gelungen ist, wie ich sie gern gehabt hätte. Über diese Traurigkeit habe ich mir Gedanken gemacht.
Diese Frau bin ich. Diese Frau habe ich am Anfang heiß und innig geliebt und später (bei der ‚Kopfarbeit') gewünscht, sie möge im Ofen zerspringen. Das Fazit wäre, wegen unterlaufener Fehler zu sterben!? Das ist nicht mein Ernst. Ich habe mir die Tonfrau noch einmal ange-

schaut und meine Gedanken noch einmal durchdacht. Diese Frau kann und muß neue Fehler machen, um daraus zu lernen und um damit zu leben. Sie strahlt das aus, was jetzt in mir vorhan-

den ist. Diese Frau aus Ton, die mich darstellt, wird sich nicht mehr verändern können – aber ich! Denn: Die Scham ist vorbei!" (15. 8. 89)

ve Musiktherapie das Hören von spannungsreicher sowie beruhigender Musik in den Mittelpunkt stellt, stehen bei der aktiven Musiktherapie musikalische Gruppenimprovisationen auf einfachen Instrumenten im Vordergrund, um im Patienten blockierte Spannungen zu lösen, sein Selbstwertgefühl innerhalb der Gruppe aufzubauen und zunächst averbale Gruppenkommunikation zu erfahren. Die 3 Grundelemente der Musik, nämlich Spannung, Schwingung und Struktur, erreichen spontan und nachhaltig affektive, emotionale leib-seelische Schichten des Menschen und regen die gestauten und chronisch blockierten Affekte, die defizitär verbliebenen Emotionen sowie die Kommunikationsbereitschaft an (Maler, 1989).

In dem der aktiven und rezeptiven Musiktherapie folgenden Nachgespräch können die Patienten ihre präverbalen Erlebnisse beim Musizieren und Musik-

Abb. 38.3–22. 26j. Patientin N.E.: „Liegende dicke Frau mit wallendem Haar. Ursprünglich wußte ich gar nicht, was ich beim Töpfern machen kann/will. Als ich dann an einem Körper modellierte, wurde fast von selbst eine dicke Frau daraus. Liegen sollte sie und entspannt wirken, nicht ausgeliefert, sondern hingebungsvoll. Sie gibt sich selbst hin und genießt ihren Körper. Form und Haltung des Körpers waren nicht konkret geplant; beides entstand beim Modellieren. Ich war überrascht, daß die Figur tatsächlich erkennbare menschliche Formen erhielt. So dick, zufrieden und schön – das habe ich nie erreicht. Dick konnte und wollte ich nie sein. Mit mir und meinem Körper zufrieden war ich nicht, seit ich denken kann; so breit und selbstverständlich entspannt dazuliegen – unvorstellbar!
In der letzten Töpferstunde hätte ich die Figur am liebsten zerstört. Mir sagte sie nichts mehr. Ich meinte immer noch, daß eine dicke Frau sehr attraktiv sein kann, konnte aber mit meiner Figur

nichts mehr anfangen. Es war, als wäre sie gar nicht von mir. Es war ein großes Problem, mir diesen Körper nackt vorzustellen. In der letzten Stunde konnte ich dann tatsächlich nicht mehr daran arbeiten. Ich habe vieles falsch gemacht.

Angefangen habe ich die Figur, als ich noch tief in meiner Krankheit steckte. Vielleicht stellt sie die Krankheit dar. Jetzt fühle ich mich besser, gesünder, normaler. Vielleicht habe ich wirklich nichts mehr mit der Figur zu tun?!" (Juli 1989)

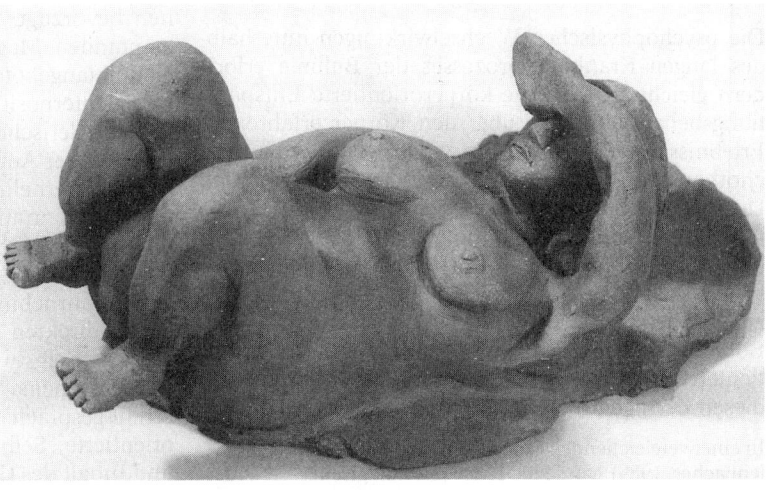

hören in Verbindung zur eigenen gegenwärtigen und biographischen Situation bringen und aussprechen.

In einer vergleichenden Studie (Maler, 1989) wurden innerhalb der aktiven Musiktherapie Unterschiede in der Ausdrucksdynamik von Bulimie- und Magersuchtpatienten festgestellt. Für den Vergleich wurden nicht nur Beobachtungen innerhalb der Improvisationen und des Nachgesprächs herangezogen; es wurden auch einige von Patienten zu den einzelnen Phasen der Musiktherapie gemalte Bilder ausgewertet. Auch hier fiel die Neigung der Bulimiekranken zur Symbolisierung von antagonistischen Gefühlsregungen, Zerrissenheit und Widersprüchlichkeit auf. Sie entwickeln einen heftigen Spannungsausdruck mit spontanem Körpereinsatz; im Laufe mehrerer Sitzungen pendeln sie sich dann auf kräftig schwingende, „warme" Holzklänge ein. Sie scheuen vorerst vor Kommunikation mit der Gruppe zurück und finden erst relativ spät einen Zugang zu ihr. Demgegenüber signalisieren Magersüchtige Situationen der Raumverengung, lassen sich schwer auf das Spielen affektbetonter Spannungsgefühle ein, bevorzugen vorerst „unterkühlte Töne" und neigen zu dominanter Gruppenführung.

Das Nachgespräch zeigt bei Bulimie- und Magersuchtpatienten unterschiedliche Introspektionsfähigkeit. Bulimiekranke ziehen spontan analoge Verbindungen von den musiktherapeutisch ausgelösten emotional-affektiven Erlebnissen zu individuellen Konflikten, vor allem mit den Eltern, besonders der Mutter. Magersüchtige scheinen ihre Introspektionsfähigkeit erheblich zu blockieren und finden schwerer Zugang zur Schilderung der präverbalen Erlebnisse.

Die auf phänomenologischer Ebene festgestellten Ergebnisse, gestützt auf die im Nachgespräch von den Patienten geäußerten Selbstreflexionen, werden als Hinweise auf eine unterschiedliche Konfliktlage innerhalb der Mutterbeziehung gedeutet: Bulimiepatienten scheinen häufig als völlig überforderte Vermittler im offen ausgetragenen Zerwürfnis beider Eltern mißbraucht zu werden, während Magersüchtige sich oft infolge einseitiger Leistungsüberforderung durch die Mutter in die Enge getrieben fühlen.

38.3.8.7 Entspannungstherapie und körperorientierte Selbsterfahrung

Autogenes Training und progressive Relaxation

Die psychophysischen Wechselwirkungen innerhalb des langen Krankheitsprozesses der Bulimie erfordern gleichermaßen eine körperorientierte Entspannungsbehandlung. Die über den Körper erfahrenen Erlebnisse der Ruhigstellung wirken gleichzeitig psychotherapeutisch, d.h., sie mindern Depressivität, Unruhe oder autoaggressive Gespanntheit.

Mit dem autogenen Training (Schultz, 1987) und der progressiven Relaxation (Jacobson, 1938; Bernstein, 1982; vgl. auch Kap. 23) stehen zwei Methoden zur Verfügung, die sich auch bei der Behandlung der Bulimie sehr bewährt haben (Gandras, 1989). Die Patienten nehmen bereits am Beginn der Therapie an diesen Übungen regelmäßig teil.

In einer vergleichenden experimentellen Untersuchung (Fallenbacher, 1989) wurde die Frage physiologischer Veränderungen während des autogenen Trainings und der progressiven Relaxation untersucht. Messungen der Hauttemperatur und des Muskeltonus erwiesen die Wirksamkeit beider Entspannungsmethoden als Bestätigung der von den Patienten mitgeteilten Wahrnehmungen.

Krankengymnastische Therapie

Wie autogenes Training und progressive Relaxation helfen auch krankengymnastische Übungen zur körperorientierten Selbsterfahrung bei der Korrektur des gestörten Körperbildes und veränderter Körperwahrnehmungen. Wir verbinden diese Behandlung mit einzelnen organbezogenen Übungen, besonders für die Bauchorgane, weil die Funktionen von Magen und Darm häufig angstbesetzt sind und zum Mittelpunkt der Sorge werden. In der körperbezogenen Psychotherapie „wird der Umgang mit dem Körper unmittelbarer als der Umgang mit der Sprache erlebt" (Schütz, 1983). Den Patienten mit Körperbildstörungen werde ein schnellerer Einstieg in die Therapie eher möglich als nur durch das Wort. Der Patient könne sich der direkten Auseinandersetzung mit sich selbst schwerer entziehen und dann körperlich direkt korrigieren, was er soeben verstanden habe.

Auch Vandereycken (1989) hebt die körperorientierte Therapie hervor, deren wesentliche Elemente Selbstkonfrontation und Selbstwahrnehmung sind; er bevorzugt hierbei homogene Gruppen.

Konzentrative Bewegungstherapie

Die konzentrative Bewegungstherapie (vgl. Kap. 23) ist ein überwiegend nonverbales psychotherapeutisches Verfahren; ohne vorgegebene Ziele oder Regeln kann sich der Patient seinen Assoziationen überlassen und sie in Bewegungen überführen. Die Interventionen des Therapeuten sind spärlich, nicht direktiv, jedoch Hilfestellung bei anfänglichen Hemmungen.

Nach Gandras (1989) fallen Bulimiepatienten von Anfang an durch eine ausgeprägte Aktivität auf; sie gestalten den Gruppenprozeß lebhaft mit und gehen schnell auf andere zu, wobei im Verhalten Unterschiede zur Musiktherapie deutlich werden. Aggressive und „chaotische" Verhaltensweisen werden von ihnen bevorzugt, Pausen und Ruhe als unangenehm empfunden. Magersüchtige wehren sich häufig gegen Kontaktangebote. Entweder ziehen sie sich auf eine Beobachterposition zurück oder stellen sich durch selbstquälerische, anstrengende Übungen in den Mittelpunkt der Aufmerksamkeit.

Die Wahrnehmung des eigenen Körpers in Ruhe und Konzentration, der Bezug zum Raum und den umgebenden Objekten und die Kommunikation in der Gruppe (Becker, 1987) bieten Möglichkeiten zu Selbstwahrnehmungen auf einer präverbalen Ebene, die im direkten Zusammenhang mit der oftmals verzerrten Selbstwahrnehmung bzw. dem gestörten „Körperschema" steht. Über das anschließende Abschlußgespräch in der KBT hinaus wird die körperorientierte Selbsterfahrung wiederum Gegenstand und Inhalt des Gesprächs in der Einzeltherapie.

Tanztherapie

Auch mit der Tanztherapie wird versucht, einen Heilungsprozeß über die Bewegung in Gang zu setzen (Klein, 1983). Grundlegende Bewegungselemente des Tanzes werden genutzt – frei von technischen Vorschriften und festgelegten tänzerischen Formen –, um Leib und Seele, Gefühl und Körperlichkeit zu integrieren. Ausgangs- und Ansatzpunkt ist das aktuelle Bewegungsmuster des Patienten. Seine Bewegungen werden vom Therapeuten aufgegriffen und übernommen. Das Ziel liegt in der Kommunikation über die authentische, selbstbestimmte Bewegung. Untersuchungen über das Bewegungsverhalten in der Tanztherapie (Lausberg et al., 1988) haben ergeben, daß die festgestellten Veränderungen des Bewegungsverhaltens den Erfolgskriterien anderer Therapien vergleichbar sind, z.B. Steigerung des Selbstbewußtseins, der Flexibilität und Beziehungsfähigkeit.

38.3.8.8 Verhaltenstherapie

Das Ziel der Verhaltenstherapie (Fairburn, 1985; Fairburn et al., 1986; Fichter, 1989a; Nutzinger und de Zwaan, 1989; Wilson et al., 1986) ist es, die Eßstörung durch Lernvorgänge in Verbindung mit Selbstkontrolle und Fremdkontrolle zu regulieren.

Als **direkte** therapeutische Hilfen werden angeführt:
- Zunächst Ausschaltung aller Einflüsse von außen (Besuchs-, Brief- und Telefonverbot);
- Vereinbarung qualitativ und quantitativ verbindlicher Mahlzeiten; „Stimuluseinengung", operantes Konditionieren mit positiver oder negativer Verstärkung;
- Selbstkontrolle (kognitiv und emotional) von Essen und Erbrechen durch Tagebuch und andere Techniken;
- Kontrolle auslösender Reize;
- Übungen zur differenzierten Wahrnehmung;
- Aufbau alternativen und sozial erwünschten Verhaltens, z.B. Einkaufen von Lebensmitteln, gemeinsame Restaurantbesuche.

Angegebene **indirekte** therapeutische Möglichkeiten:
- Selbstsicherheitstraining, z.B. Steigerung des Durchsetzungsvermögens und Abbau von Minderwertigkeitsgefühlen;
- Förderung der Kontakte;
- Beschäftigung mit der Identität als Frau;
- Ausschaltung irrationaler Vorstellungen über Nahrungsmittel, Gewicht und Aussehen.

Die Mitteilungen über eine Kombination von Verhaltenstherapie und tiefenpsychologisch fundierter Psychotherapie bei Bulimie (Dippel et al., 1988) lassen erkennen, daß lange Zeit bestandene Vorurteile über die Unvereinbarkeit mancher therapeutischer Strategien zugunsten sinnvoller, auf den einzelnen Patienten abgestimmter Behandlungsformen abgebaut werden können, vor allem dann, wenn der Erfolg nicht allein am Symptom gemessen wird, sondern auch am Einfluß auf innerseelische Prozesse und psychosoziale Befunde.

38.3.8.9 Ernährung und Diät

Da sich die meisten Bulimiekranken intensiv und mitunter selbstquälerisch mit der Zusammenstellung der einzelnen Mahlzeiten und deren Kaloriengehalt beschäftigen, häufig fehlerhafte Ansichten haben und daher schon bei Beginn der Behandlung Korrekturen notwendig sind, wird die Hilfe einer erfahrenen Diätassistentin benötigt (Beumont et al., 1989; Fairburn, 1985; Feiereis, 1989; Garner et al., 1985; Lacey, 1983; Laessle et al., 1988; Mitchell et al., 1985; O'Connor et al., 1988). Mit dem Abbau von Schuld- und Schamgefühlen durch verständnisvolle Gespräche über offene Fragen wird die Diätassistentin für die Kranken Übertragungsobjekt und Hilfs-Ich zugleich. Sie übernimmt auch nicht nur symbolisch, sondern real einen Teil der Verantwortung für das gesetzte Ziel, das Idealgewicht mit Hilfe der festgelegten Ernährungsform zu erreichen. In der Ernährung ist vor allem zu berücksichtigen, inwieweit die Freßanfälle reaktiver Art sind, d.h. Folge restriktiver Ernährung („enthemmtes Fasten"; Herman und Polivy, 1984). Sind die Freßanfälle trotz der kontinuierlichen Hilfe nicht zu beherrschen, so kann vorübergehend eine vollbilanzierte, hochmolekulare Formeldiät mit niedriger Osmolarität indiziert sein (z.B. Fresubin® flüssig, Nutrodrip®).

38.3.8.10 Medikamentöse Therapie

Die mögliche pathogenetische Beziehung der Bulimie zu affektiven Krankheiten und die verschieden ausgeprägte Depressivität mit Suizidgedanken, quälender Unruhe und hartnäckigen Schlafstörungen bilden oft die Indikation zur kurzfristigen medikamentösen Therapie (Hudson und Pope, 1989).

Psychopharmaka sollten sparsam angewendet werden und möglichst nur in Form von Antidepressiva oder niedrigpotenten Neuroleptika. Wir geben z.B. unter sorgfältiger Beachtung möglicher Nebenwirkungen zwei- bis dreimal täglich 25 mg Amitriptylin (Laroxyl®, Saroten®, Tryptizol®) und 75 mg retard zur Nacht, oder Thioridazin (Melleril®), zwei- bis dreimal 10–20 mg tags und 30–60 mg retard zur Nacht.

In verschiedenen Studien wurde über positive Ergebnisse auch mit anderen Antidepressiva berichtet (Agras et al., 1987; Hughes et al., 1986; Pope et al., 1983), ferner mit Antikonvulsiva (Kaplan, 1987). Für besonders wirksam wird von manchen Autoren (Ferguson, 1987) Fluoxetin infolge selektiven Einflusses auf Serotonin gehalten, eventuell in Kombination mit Lithiumkarbonat.

Liegen Zeichen einer metabolischen Alkalose vor, so ist die orale oder parenterale Substitution ebenso angezeigt wie eine symptomatische Behandlung von Motilitätsstörungen des Magens oder des Darmes, z.B. mit Metoclopramid (Paspertin®) oder Domperidon (Motilium®). Gegen hartnäckige Obstipation verordnen wir Weizenkleie, Leinsamen, Glyzerinzäpfchen oder für eine begrenzte Zeit Paraffinöl-Emulsion (Obstinol® mild).

38.3.9 Prognose

Im Vergleich zur Magersucht gibt es bisher nur relativ wenige katamnestische Studien, die eine Aussage über die Prognose erlauben (Abraham et al., 1983; Hsu und Holder, 1986; Lacey, 1983; Mitchell et al., 1986; Remschmidt und Herpertz-Dahlmann, 1989; Swift et al., 1987; Yager et al., 1987). Herzog und Mitarbeiter (1988, 1989) weisen auf methodologische Erschwernisse des Vergleiches hin, u.a. fehlende Zwischenbefunde, unterschiedliche Zeitdauer der Katamnesen und Mängel der Reliabilität und Validität der Meßinstrumente. Dennoch läßt sich bei summarischer Bewertung der mitgeteilten Ergebnisse feststellen, daß trotz unterschiedlichen Grades der Besserung die Mehrzahl der Patienten nicht geheilt war, sondern noch bei 29–87% gelegentlich Heißhungeranfälle auftraten. Meistens war die Besserung mit einem Rückgang der depressiven Symptomatik und zufriedenstellenderer sozialer Anpassung korreliert.

Eine prospektive Pilotstudie (Herzog et al., 1989) über 30 Patientinnen, die ambulant wegen ihrer Bulimie unterschiedlich behandelt wurden, ergab eine Heilung von 33% nach 6 Monaten; 23% hatten die Therapie vorzeitig abgebrochen.

Merkmale mit nach unseren Erfahrungen negativem Einfluß auf die Prognose sind in Tabelle 38.3–5 zusammengefaßt, die Kriterien der Behandlungsergebnisse in Tabelle 38.3–6.

38.3.10 Ausblick

Ein keltisches Märchen aus dem 14. Jahrhundert erzählt von Cathal, dem König von Munster, von seiner maßlosen Freßgier und deren wunderbarer Heilung. In der Brust des Königs Cathal hat ein wildes, gesetzloses Tier „Wohnung genommen" und verlangt nach unermeßlichen Mengen an Nahrung, was dazu führt, daß der Hunger selbst dann nicht gestillt ist, wenn der König „zum Frühstück ein Schwein, ein Rind, ein Kalb, 36 Kuchen aus reinem Weizen und ein Faß voller Aale" zu sich genommen hatte.

Ein junger Student, auf dem Weg, „sein eigenes Glück zu machen und zu schauen, ob er dabei auch dem König helfen könne", erlebt auf einem Fest einen für die dort anwesenden Gefolgsleute beängstigenden Freßanfall des Königs; da greift sich der Student einen Stein, steckt ihn in den Mund und beißt auf ihm herum. Der König ist beim Anblick dieses Verhaltens verwirrt und fragt den Studenten, was ihn wohl verrückt gemacht habe: „Ich werde traurig, wenn ich Dich allein essen sehe", war die Antwort. Das sinnlose wie schmerzliche Kauen des Steins wie die Antwort des Studenten werden dem kranken König zu einem Spiegel, in dem er seiner eigenen Verfressenheit, Traurigkeit und Einsamkeit angesichtig wird; in diesem Moment kann der König erstmals von dem abgeben, was das Tier in seiner Brust ausschließlich für sich verlangt, der König „schämte sich und warf dem Studenten einen Apfel zu".

Tabelle 38.3–5. Merkmale eines negativen Einflusses auf die Prognose der Bulimie.

Genetische Disposition
Hereditäre psychopathologische Belastung
Schwere prämorbide Entwicklungsstörung
Große Zeitspanne zwischen Krankheitsbeginn und Therapie
Mangelhafte Therapiebereitschaft
Geringe Introspektionsfähigkeit
Kombination mit anderen Krankheiten und Süchten
Ausgeprägte psychische Symptomatologie
Zwänge, Selbstbeschädigung, Suizidversuche
Abbrüche der Therapie
Somatische Folgen der Krankheit
Sozialmedizinische Folgen der Krankheit: sozialer Abstieg, keine berufliche Perspektive, Rentenverfahren, Hospitalisation
Fehlende Kontinuität der Langzeittherapie
„Therapieabriß" nach stationärer Behandlung

Tabelle 38.3–6. Kriterien des Behandlungsergebnisses.

Heilung	Besserung
Kein Freßanfall	Freßanfall seltener als 1–2 ×/Woche
Kein selbstinduziertes Erbrechen	Selbstinduziertes Erbrechen seltener als 1–2 ×/Woche
Ideales oder normales Gewicht	Ideales bis normales Gewicht
Gewichtsschwankungen unter 2 bis 3 kg	Gewichtsschwankungen unter 3 bis 4 kg
Kein Medikamentenmißbrauch	Kein Medikamentenmißbrauch
Keine weitere Sucht	Keine weitere Sucht
Keine Residuen psychischer oder psychosozialer Bulimiesymptomatik	Kontinuierliche Psychotherapie
Regelmäßige Menstruation	Keine Amenorrhoe

Das war der Anfang einer sonderbaren Heilung, in deren Verlauf der König auf phantastische Weise vom „gefräßigen Vieh" und damit von der Ursache seiner Krankheit befreit wurde; das Tier verschwand und „ward nie mehr gesehen" (Hetmann, 1975).

So lehrreich und tröstlich der Einfallsreichtum des therapeutisch wirkenden Studenten auch ist – die märchenhafte Heilung dieser offenbar uralten Krankheit gleicht einem Traum, nicht aber dem eingehend beschriebenen, mühevollen und langwierigen psychotherapeutischen Weg; in manchen Abschnitten ähnelt er einer Spiegelfunktion, die einen Erkenntnisprozeß in Gang setzen, den Heilungsverlauf einleiten und das Unvermögen aufheben kann, sich von dem „gefräßigen Tier" zu befreien.

Der Weg aber ist nicht frei von Beschwernissen und Rückschlägen, von Zweifeln und vielen Augenblicken der Verzweiflung. Trotzdem ermutigen uns die Erfahrungen, ihn immer wieder zu beginnen.

39 Sexuelle Störungen

Kurt Loewit

Sexuelle Störungen werden weitgehend in der ärztlichen Ausbildung und Praxis nicht in **der** Bedeutung gesehen und gewertet, die sie für die Betroffenen haben. Sie werden noch weniger in ihren größeren Zusammenhängen und Auswirkungen erkannt und verstanden. Dementsprechend unzureichend ist im allgemeinen das therapeutische Angebot. Fragt man nach den Ursachen dieser Situation, so ist unter anderem die in unserem Kulturkreis immer noch mächtige (und bis zu einem gewissen Grad unvermeidbare) Tabuisierung der Sexualität zu nennen, von deren Auswirkungen auch der Arzt nicht (genügend) frei ist, sowie der spezifisch „psychosomatische" Charakter menschlicher Geschlechtlichkeit. Er mag dem vorwiegend naturwissenschaftlich ausgebildeten Mediziner zusätzliches Unbehagen bereiten, zumal lange Zeit eine umfassende Theorie der Psychosomatik, wie sie etwa in den ersten Kapiteln dieses Buches entworfen wird (v. Uexküll, 1979), fehlte, und der Begriff „von vielen gerade dann verwendet wird, wenn klares und kritisches Denken aufhört" (Wesiack, 1984). Störungen der Sexualität zeigen besonders deutlich, daß nicht nur kranke Funktionen oder Organe zu behandeln sind, daß es noch nicht einmal genügt, den ganzen Menschen in seiner individualpsychischen und sozialen Dimension zu sehen, sondern daß ein interindividuelles Beziehungssystem in seinen Bezügen zur Umwelt und mit seinen eigenen Interaktions- und Kommunikationsmustern bzw. deren Störungen zu erfassen ist. Voraussetzung dieses interpersonalen Systems ist demnach der Aufbau einer neuen gemeinsamen sozialen Wirklichkeit zwischen bisher in ihrer individuellen Wirklichkeit eingeschlossenen Individuen.

Nach neueren Untersuchungen von Schnabl (1980), die sich größenordnungsmäßig mit früheren Angaben von Masters und Johnson (1970), Kaplan (1974), Wall und Kaltreider (1977), Springer-Kremser (1978) decken, kommen sexuelle Störungen, zumindest zeitweilig, bei fast jedem zweiten Paar bzw. bei 45–60% der untersuchten oder befragten Frauen und Männer vor. Dies fanden auch Frank und Mitarbeiter (1978) in einer als Kontrollgruppe rekrutierten Population von 100 „normalen" Paaren. Dabei bestehen für beide Geschlechter charakteristische Korrelationen zwischen sexuellen Funktionsstörungen und z.B. Lebensalter, monatlicher Koitusfrequenz, Koitusdauer oder Sexualfunktion des Partners bzw. mit lebensgeschichtlichen „Risikofaktoren" (Schnabl, 1980). In ähnlicher Größenordnung wird die Häufigkeit sog. psychosomatischer Erkrankungen angegeben: Strotz-

ka und Mitarbeiter (1979) haben dabei als vorrangige Ursachen psychosomatischer Störungen bzw. Krankheiten Probleme in den wichtigsten zwischenmenschlichen Beziehungen, d.h. in Partnerschaft und Familie, feststellen können.

Ebenso hat Husslein (1977) im Patientengut einer psychosomatisch-gynäkologischen Ambulanz bei 52% der Fälle Partner- und/oder Familienkonflikte zu Beginn der Erkrankung vorgefunden. In diesen engen zwischenmenschlichen Beziehungen in Partnerschaft und Familie spielt wiederum „Sexualität" direkt oder indirekt, primär oder sekundär eine wesentliche Rolle, im Bereich gynäkologisch- (vgl. Eicher, 1980; Frick-Bruder, 1981) und andrologisch- (vgl. Vogt, 1980; Pöldinger, 1981) psychosomatischer Erkrankungen kommt ihr häufig spezifisch pathogenetische Bedeutung zu. Die eingangs erwähnten sexuellen Funktionsstörungen stellen ja selbst eine Form psychosomatischer Erkrankungen dar.

Die aus den genannten Zahlen ersichtliche epidemische Verbreitung sexueller Funktionsstörungen und das noch größere Ausmaß sexueller Unzufriedenheit mit ihrem Gefolge von persönlichen, ehelichen, familiären und schließlich sozialen Problemen bzw. als Zeichen gestörter Beziehungen muß daher den Arzt alarmieren.

39.1 Ein exemplarischer Fall

Als Einführung zum Thema sei diesem Kapitel der verkürzte Fallbericht eines jungen Ehepaares mit sexuellen Problemen vorangestellt. Er spiegelt den derzeit für die meisten dieser Patienten typischen „Leidensweg" wider und illustriert zugleich, wie das im folgenden zu entwickelnde Konzept der Kommunikationsfunktion der Sexualität sowohl zur Erlangung eines umfassenderen Sinnverständnisses menschlicher Geschlechtlichkeit, als auch zur Erweiterung des therapeutischen Ansatzes innerhalb der Sexualtherapie herangezogen werden kann.

> Das seit zwei Jahren verheiratete, noch kinderlose berufstätige Ehepaar suchte gemeinsam die sexualmedizinische Sprechstunde auf. Die Frau war 26, der Mann 28 Jahre alt. Sie war mittelgroß und schlank, mit langem blondem Haar und eher elegant-damenhaft gekleidet, er ein leicht untersetzter sportlich kräftiger Mann in Lederbekleidung, der angibt, nur zur Begleitung mitgekommen zu sein und mit der Sache an sich

nichts zu tun zu haben. Das Problem bestünde in der „Frigidität" seiner Frau, die seit Jahren eine regelrechte Abneigung gegen alles Sexuelle entwickelt habe. Während der letzten sechs Wochen habe es überhaupt keinen Geschlechtsverkehr mehr gegeben. Auch die Frau sucht die Schuld allein bei sich und ist überzeugt, daß die ansonsten keineswegs gewünschte Scheidung ihrer Ehe nur noch eine Frage der Zeit sei. Sie wirkt eher niedergeschlagen und hat aus der bisherigen Erfahrung mit ärztlichen Behandlungsversuchen ihrer Störung die Hoffnung auf wirksame Hilfe aufgegeben. Eigentlich komme sie nur, damit man ihr nicht vorwerfen könne, sie habe nicht alles versucht. Offensichtlich leidet sie stark unter ihrem Unvermögen, wie eine „normale Frau" empfinden zu können. Sie kennt ihren Mann seit ca. 4 Jahren und hat bereits vor der Hochzeit wegen fehlenden sexuellen Verlangens und Empfindungslosigkeit beim Koitus einen Gynäkologen konsultiert. Die Untersuchung ergab keinerlei pathologischen Befund. Ein „Pillenwechsel" brachte keinen Erfolg, ebensowenig eine Kur mit Hormoninjektionen. Nach ca. einem halben Jahr suchte die Patientin einen anderen Gynäkologen auf, der sie angeblich mit der Erklärung, dies sei nicht sein Fall, an einen Psychiater verwies. Einige Gespräche und eine psychopharmakologische Behandlung über mehrere Monate führten ebenfalls zu keiner Besserung des Zustandes. Da dieser sich nach der Eheschließung verschlimmerte, suchte die Patientin gemeinsam mit ihrem Mann einen weiteren Gynäkologen auf. Laut Aussage des Paares habe dieser sie damit vertröstet, daß „die einen eben Musik lieben und die anderen nicht" und sich alles im Verlauf der Jahre ändern werde, sie solle ihren Mann nichts merken lassen. Von diesem Arzt fühlten sich beide nicht ernstgenommen, verletzt und abgestoßen und haben ihn nicht wieder aufgesucht. Monate später erlitt die Frau beim Jogging mit ihrem Mann einen Schwächeanfall. Obwohl ohne Schmerzen oder Fieber und bei gutem Appetit war sie in der Folge zwei Wochen lang bettlägerig und zu schwach, um aufzustehen. Der herbeigerufene Hausarzt behandelte mit Kreislauf- und Beruhigungsmitteln, verordnete Antibiotika und veranlaßte eine komplette Durchuntersuchung, die in allen Punkten optimale Werte ergab. Er motivierte die Patientin nochmals zu einem „Nervenarzt" zu gehen, der sie schließlich, knapp drei Jahre nach dem ersten Arztbesuch, wegen ihrer sexuellen Funktionsstörung an die sexualmedizinische Sprechstunde überwies. Hier mußte zunächst eine neue Vertrauensbasis geschaffen, die Frustration durch die bisherige Leidensgeschichte bearbeitet und soviel Hoffnung vermittelt werden, daß ein nochmaliger persönlicher Einsatz möglich wurde. Durch die erlebte Wirkungslosigkeit organisch-medikamentöser Behandlungsversuche war es in diesem Fall nicht schwierig, die psychische Seite ins Spiel zu bringen. Die Exploration der biographischen Anamnese erbrachte auf seiten der Frau große Verlust- und Trennungsängste (auf deren Hintergründe hier nicht näher eingegangen werden kann), deretwegen sie schon seit Beginn dieser Bekanntschaft viel Störendes an ihrem Partner tolerierte. Diese Ängste verboten ihr selber aktiv zu werden, eigene Wünsche und Forderungen anzumelden bzw. gewünschte Grenzen zu setzen. Aus der Lebensgeschichte des Mannes erschien eine zwei Jahre dauernde Beziehung, aus der auch ein Kind hervorgegangen war und die von seiner damaligen Partnerin abgebrochen wurde, bedeutungsvoll. Aus diesem Erlebnis resultierte ein ambivalentes Verhältnis aus Angst und Haßgefühlen bzw. Rachegelüsten gegenüber „der Frau", welche sich besonders am Anfang der Beziehung auch seiner jetzigen Partnerin gegenüber auswirkte. Ehe mit den sexualtherapeutischen Übungen (siehe später) begonnen werden konnte, war es notwendig, dem Mann seine Mitbeteiligung am Geschehen bzw. seinen Anteil am Problem einsichtig zu machen, sowie der Frau Hilfen zu einer neuen Sichtweise der Sexualität und Sinnfindung anzubieten: Auf die Frage nach der Bedeutung der Sexualität für sie, hatte sie in einer Mischung aus Verbitterung, Aggression und Resignation geantwortet, der eigentliche Zweck sei doch die Befriedigung des Mannes.

39.2 Übersicht über die direkten sexuellen Funktionsstörungen

39.2.1 Eingrenzung und Vorbemerkungen

Im folgenden soll ein kurzer Überblick über die häufigsten sexuellen Funktionsstörungen, mit denen der Arzt bzw. Sexualtherapeut konfrontiert wird, gegeben werden. Dabei handelt es sich um die eigentlichen sexuellen Funktionsstörungen bei Mann und Frau im engeren Sinn. Als weiterführende Literatur können Bancroft (1985), Buddeberg (1987), Eicher (1980), Kaplan (1979), Loewit (1988) und Money und Musaph (1978) empfohlen werden. Es geht hier nicht um beispielsweise organische und nicht-organische Störungen der Reproduktionsfunktion, Intersexualität, Mißbildungen, Homosexualität, Transsexualität oder Paraphilien. Insbesondere würden letztere (früher als Perversionen bezeichnet), also z.B. Exhibitionismus und Voyeurismus, Sado-Masochismus, Transvestitismus, Fetischismus, Pädophilie, Sodomie, Nekrophilie, Erotolalie und andere, zu weit vom Thema wegführen. Sie sind häufig ein Ausdruck früher Entwicklungsstörungen, oft ohne inneren Leidensdruck, daher der Therapie nur schwer zugänglich und von Ausnahmen abgesehen auch nicht durch Sexualtherapie, sondern durch langwierige analytische Psychotherapie beeinflußbar. Ebenso übersteigen die Themen der Homo- und Bisexualität den Rahmen dieses auf die sexuellen Funktionsstörungen im engeren Sinn zentrierten Beitrages. Während sich die Möglichkeit homo- oder bisexuellen Verhaltens generell aus der somatischen und psychosexuellen Anlage und Entwicklung menschlicher Geschlechtlichkeit ergibt, und sich nach 14jähriger klinischer Erfahrung von Masters und Johnson (1979) die sexuelle Funktionsfähigkeit Homosexueller nicht von der Heterosexueller unterscheidet, bzw. bei ihnen dieselben Funktionsstörungen auftreten können, erlaubt das immer noch unzureichende Wissen über Ätiologie und Genese von Homo- und Bisexualität im Einzelfall keine objektive Beurteilung und Einordnung. Die gesellschaftspolitische Betrachtung des Themas erschwert eine sachliche Diskussion noch zusätzlich. Sicher ist der Begriff der Homosexualität immer noch ein deskriptiver Sammelbegriff für Zustände unterschiedlicher Ausprägung und Genese. So erklären sich wohl die von Masters und Johnson angegebenen Therapieerfolge bei Homosexuellen, die unter ihrer Orientierung gelitten haben und sie verändern wollten, bzw. die als Homosexuelle unter sexuellen Funktionsstörungen zu leiden hatten. Wesentlich häufiger wird sich Beratung oder Psychotherapie Homosexueller mit nicht primär sexuellen Fragen befassen und die AIDS-Problematik miteinbeziehen (Haeberle, 1987). In diesem Kapitel geht es auch nicht um die rein organische Seite sexueller Funktionsstörungen oder um somatisierte Störungen aus der Sexualsphäre (Konversionssymptome), die sog. indirekten sexuellen Funktionsstörungen.

Gemäß der von Masters und Johnson (1966) erarbeiteten Erkenntnisse über den normalen Ablauf der sexuellen Reaktion lassen sich sexuelle Funktionsstörungen bei Mann und Frau pragmatisch anhand des physiologischen Reaktionsablaufes einteilen. Sie können jeweils:

– **primär** (vom ersten Versuch an)
– **sekundär** oder **reaktiv** (ab einem bestimmten Zeitpunkt oder in einer bestimmten Situation)
– **absolut** oder **obligatorisch** (in jedem Fall)
– **relativ** oder **fakultativ** (unter bestimmten Umständen, bei bestimmten Partnern) auftreten.

Allerdings würde eine Betrachtungsweise, die die jeweilige sexuelle Reaktion von Mann und Frau nur für sich und isoliert sieht, der Wirklichkeit nicht gerecht. Schon Matussek (1954) hat auf die gegenseitige Beeinflussung und Verzahnung der männlichen und weiblichen Sexualreaktion hingewiesen und ebenso auf die geschlechtsspezifisch unterschiedliche Störanfälligkeit im Koitusablauf:

Beim Mann sind Libido und Ejakulation wenig störbare, stabile Funktionen, während die Erektion eine sehr leicht störbare labile Phase des Geschehens darstellt. Bei der Frau hingegen sind Libido und Orgasmus leicht störbare Funktionen, während die vaginale Lubrikation, die der männlichen Erektion entspricht, eine stabile, wenig störbare Funktion darstellt.

Daraus ergeben sich sowohl Hinweise auf partnerschaftliche Interaktionsprobleme, wie auch auf vorwiegend zu erwartende Störanfälligkeiten bei Mann und Frau, nämlich Erektionsstörungen bzw. Libido- und Orgasmusstörungen. Im Sinne dieser gegenseitigen Beeinflussung finden sich gelegentlich Kombinationen sexueller Funktionsstörungen. Sie können bei einzelnen Patienten auftreten, wie z.B. Ejaculatio praecox und sekundäre Erektionsstörung beim Mann bzw. Dyspareunie und sekundäre Libidostörungen bei der Frau, oder innerhalb eines Paares, etwa als Kombination von vorzeitigem Samenerguß und weiblicher Anorgasmie oder von Vaginismus und männlicher Potenzschwäche.

39.2.2 Störungen beim Mann

Auf die eingangs erwähnte Einteilung der sexuellen Funktionsstörungen anhand des physiologischen Ablaufes zurückkommend, lassen sich beim Mann unterscheiden (Vogt, 1980):

– **Störungen der Libido** (Impotentia concupiscentiae): primär oder sekundär bzw. reaktiv oder auch nur als Schutzbehauptung vorgeschoben.
– **Störungen der Erektion** (Impotentia erectionis), wobei zu differenzieren ist, ob gleichzeitig eine ungestörte Libido vorhanden ist, ob Spontanerektionen auftreten bzw. Masturbation zur Erektion führt und diese nur in Gegenwart einer Partnerin fehlt, ob die Erektionsschwäche unmittelbar vor der Immissio penis (Impotentia coeundi im engeren Sinn) oder während des Koitus eintritt.
– **Störungen von Orgasmus und Ejakulation:** Obwohl im Regelfall Ejakulation und Orgasmus beim Mann als Simultangeschehen erlebt werden, weist Vogt (1980) darauf hin, daß die Ejakulation als Folge des vorausgehenden Orgasmus eintritt und somit die bisher gebräuchliche Terminologie der Ejakulationsstörungen im strengen Sinn und zum größten Teil Orgasmusstörungen bezeichnet. So müßte der vorzeitige Samenerguß als Orgasmus praecox statt Ejaculatio praecox bezeichnet werden, da die Ejakulation ja nicht gestört ist und regelrecht abläuft. Ähnliches gilt für die Ejaculatio retardata, den verzögerten Samenerguß bzw. Orgasmus. Eine in diesem Sinne echte Ejakulationsstörung wäre die Ejaculatio deficiens, das Fehlen eines Samenergusses bei vorhandenem Orgasmus, bzw. die Ejaculatio retrograda, das Ausschleudern des Samens in die Harnblase (Impotentia ejaculationis). Beim vorzeitigen Orgasmus bzw. Samenerguß, der häufigsten sexuellen Funktionsstörung vor allem junger Männer, kann nach Borelli (1971) wiederum differenziert werden, ob die Ejakulation bereits ante portas, bei Berührung des weiblichen Genitales durch den Penis, unmittelbar nach der Immissio oder nach wenigen Friktionsbewegungen in der Vagina eintritt. Definitorisch von Bedeutung ist, daß Orgasmus/Ejakulation unkontrolliert und – innerhalb der physiologischen Bandbreite – vor dem vom Mann gewünschten Zeitpunkt eintreten. Wenn bei der Ejakulation kein oder nur ein geringes Orgasmusgefühl entsteht, wird von orgastischer Impotenz (Impotentia satisfactionis) gesprochen, wenn der Orgasmus anstatt mit Zufriedenheit und Entspannung mit emotionalen Störungen verbunden ist, von einer Impotentia emotionis.

39.2.3 Störungen bei der Frau

In ähnlicher Weise lassen sich nach Eicher (1975) die sexuellen Funktionsstörungen der Frau einteilen in:

– **Libidostörungen,** wobei es sich um eine verminderte oder gänzlich fehlende Appetenz bzw. sogar um eine Aversion gegen alles Sexuelle handeln kann. Als Plusvariante einer Libidostörung, also

Hyperlibidimie, ist die sog. Nymphomanie zu nennen.

Entsprechend dem Ablauf der sexuellen Reaktion wären von Störungen der Libido solche der lustvollen körperlichen Erregbarkeit, der sog. **sexuellen Ästhesie** zu unterscheiden, wie sie sich etwa in gesteigerter Durchblutung der Genital- und Brustregion bzw. vaginaler Lubrikation äußern und wiederum als Hypo-, An- oder Hyperästhesien (subjektiv und/oder objektiv) in Erscheinung treten können (Bräutigam, 1977).

– In der Praxis wesentlich wichtiger sind, um wiederum Eicher zu folgen, die **Orgasmusstörungen,** Abnahme der Orgasmusfähigkeit bzw. -häufigkeit oder Fehlen orgastischen Erlebens (Anorgasmie), wobei diese wiederum primär oder sekundär bzw. situativ (Pseudoanorgasmie) sein kann. Der früher verwendete Begriff der Frigidität sollte wegen seiner Undifferenziertheit nicht mehr verwendet werden, auch ist er mittlerweile zu einem wertenden und diskriminierenden Etikett geworden.
– Von Libido- und Orgasmusstörungen abzugrenzen ist der **Vaginismus,** eine funktionelle Störung, bei der das Eindringen des männlichen Gliedes durch eine reflektorische Muskelverkrampfung und Abwehrspannung der Oberschenkel-, Beckenboden- und Vaginalmuskulatur und damit Verschluß des Introitus vaginae verunmöglicht wird.
– Unter Umständen nur durch eine fließende Grenze vom Vaginismus getrennt, ist die **Algopareunie,** das Auftreten schmerzhafter Mißempfindungen während und/oder nach dem Koitus, gelegentlich auch als **Dyspareunie** bezeichnet.

Es ist im Rahmen dieses Kapitels nicht möglich, mehr als einen allgemeinen Überblick über diese Störungen zu geben. Sie und ihre Behandlung sind in der sexualmedizinischen Literatur eingehend und ausführlich beschrieben (vgl. Masters und Johnson, 1970; Borelli, 1971; Schirren, 1971; Lo Piccolo, 1978; Kaplan, 1974/1979; Sigusch, 1975; Eicher, 1975; Bräutigam, 1977; Money und Musaph, 1978; Kolodny et al., 1979; Angermann, 1980; Arentewicz und Schmidt, 1980; Kockott, 1980; Vogt, 1980; Godow, 1982; Nadelson und Marcotte, 1983; etc.). Ähnliches gilt für die möglichen Wurzeln und Hintergründe sexueller Funktionsstörungen, die hier ebenfalls nur angedeutet werden können.

39.2.4 Hinweise auf Ursachen bzw. Genese sexueller Funktionsstörungen

Sexuelle Funktionsstörungen können grundsätzlich primär organisch als Folgen oder Begleiterscheinungen somatischer Genital- oder Allgemeinerkrankungen auftreten, wie z.B. Entwicklungsstörungen, Mißbildungen, infektiöse, vaskuläre, neurogene, endokrine Erkrankungen, als Folgen von Operationen, chronischen oder zehrenden Krankheiten etc. Häufiger sind sie primär psychisch bedingt, sei es nur als unrealistische Fehlvorstellungen oder -erwartungen oder als Störungen, die entweder hauptsächlich in der eigenen Lerngeschichte, der psychosexuellen Entwicklung bzw. Persönlichkeitsstruktur (z.B. Einstellung zur Sexualität an sich, Selbstwertgefühl, Körperbild etc.) liegen oder in der Beziehungsdynamik des Paares und im Zusammenspiel beider Faktoren wurzeln, z.B. neurotische Partnerwahl, Kollusionen nach Willi (1975), Kommunikationsstörungen, akute Partnerproblematik, Trennung oder Verlust des Partners etc. Dabei kann die sexuelle Funktionsstörung sowohl Anlaß bzw. Verstärker, als auch Folge einer Beziehungsstörung sein. Jedenfalls ist nach Richter (1979) die „Sexualität ein Barometer ersten Ranges zur Beurteilung der Kommunikationsfähigkeit von Paaren". Organische und psychische Genese schließen sich nicht gegenseitig aus, bzw. genügen je für sich allein nicht ohne weiteres zur Erklärung der sexuellen Störung, sie können vielmehr in komplexer Weise zusammenspielen (was etwa die Erektionsstörungen deutlich illustrieren; Wagner und Green, 1981). Eine sorgfältige organische Abklärung einschließlich der Frage nach Medikamenten (z.B. Antihypertensiva, Tranquilizer und Antidepressiva, Chemotherapeutika, Hormone) oder Suchtmittelgebrauch ist jedenfalls unerläßlich, auch wenn in der Mehrzahl der Fälle sexuelle Funktionsstörungen primär oder überwiegend auf psychische Ursachen zurückzuführen sind. Die Algopareunie ist bei Frau und Mann primär auf organische Ursachen verdächtig, der Vaginismus ist immer psychogen bedingt. Ebenfalls grundsätzlich kann gesagt werden, daß Leistungsdruck und Versagensangst vor allem nach tatsächlich erlebten, wenn auch objektiv unbedeutenden Versagenssituationen beim Mann, unbefriedigende Partnerbeziehung, mangelhafte Identifikation mit dem Partner bei der Frau geschlechtsspezifisch verschiedene Schwerpunkte der Psychogenese sexueller Funktionsstörungen charakterisieren.

39.3 Grundprinzipien der Sexualtherapie

39.3.1 Grundlegende Konzepte

Grundsätzlich ist von der psychosomatischen Einheit und psychosozialen Eingebundenheit des Menschen auszugehen und der dementsprechend psycho-soziosomatischen Natur seiner sexuellen Störungen. Daher müssen biologische und psychosoziale Momente gleicherweise beachtet und integriert werden. Paartherapie entspricht der Situation weit besser als Einzeltherapie. Masters und Johnson (1970) haben auf die therapeutisch bedeutsame Unterscheidung zwischen Sexualität als angeborener, natürlicher Funktion (z.B. Erektions- und Lubrikationsfähigkeit) und sozial erlerntem bzw. persönlichkeitsabhängigem Sexualverhalten hingewiesen. Vor diesem Hintergrund erklären sich die oft in kurzer Zeit erzielbaren dramatischen Veränderungen auch bei chronischen sexuellen Funktionsstörungen. Sexuelle Störungen müssen jedoch keineswegs immer Ausdruck tief verwurzelter Persönlichkeits- oder Beziehungsprobleme sein.

Dennoch ist Sexualtherapie Psychotherapie, die nach „gelernten" Psychotherapeuten verlangt und nicht nach „Kochbuch" durchgeführt werden kann (Kaplan und Langer, 1979). Letztlich braucht jedes Paar eine auf seine spezielle Situation zugeschnittene, „maßgeschneiderte" Therapie.

Zur Behandlung sexueller Funktionsstörungen respektive Beziehungsstörungen wurden verschiedenste Strategien entwickelt. Als wichtigste sind wohl die auf Freuds (1968) grundlegenden Arbeiten beruhenden psychoanalytischen Verfahren (vgl. Angermann, 1980; Willi, 1978) und die von Mandel und Mandel (1971, 1975) entwickelte Kommunikationstherapie sowie die Verhaltenstherapie (vgl. Kockott, 1980) zu nennen. Als therapeutische Grundhaltung könnte man die von Rogers (1951, 1977) in der klientzentrierten Gesprächspsychotherapie verlangten Elemente (empathisches Verstehen, emotionale Wärme und positive Wertschätzung, Echtheit und Selbstkongruenz sowie emotionales Durcharbeiten) anführen. Während diese Methoden nicht spezifisch für die Behandlung von Sexualstörungen sind, wurden sowohl von Masters und Johnson (1970) als auch von Helen Singer-Kaplan (1974) neue spezifische Therapieformen zur Behandlung sexueller Störungen entwickelt. Die neue Sexualtherapie von H. Singer-Kaplan versteht sich als eine Kombination verschriebener Sexualerfahrungen und Psychotherapie, welche verschiedene und in ihren theoretischen Ansätzen sehr unterschiedliche bis gegensätzliche Psychotherapieformen, im besonderen Psychoanalyse, Partner- bzw. Kommunikationstherapie und Verhaltenstherapie, in und zu einer neuen Therapieform „amalgamiert". Dabei spielen die verschriebenen sexuellen Erfahrungen, d.h. die nach Anleitung durch den Therapeuten in der Privatsphäre des eigenen Heimes durchzuführenden Übungen zwar eine wesentliche Rolle, stellen aber doch nur einen Aspekt des gesamten therapeutischen Prozesses dar (Kaplan, 1974). „Es ist die integrierte Verwendung von systematisch strukturierten sexuellen Erfahrungen zusammen mit psychotherapeutischer Exploration der unbewußten intrapsychischen Konflikte beider Partner sowie der subtilen Dynamik ihrer Interaktionen, die die Besonderheit und das Grundkonzept der Sexualtherapie ausmacht" (Kaplan und Langer, 1979).

39.3.2 Der sog. „Sensate Focus"

Masters und Johnson, die betonen, nicht gestörte sexuelle Funktionen, sondern gestörte Beziehungen zu behandeln, haben der Sexualtherapie eine besondere Form gegeben, indem sie ihre Patienten ausschließlich paarweise und in sozialer Isolierung einer zweiwöchigen Intensivbehandlung durch ein Therapeutenpaar unterziehen. Sie haben unter der Bezeichnung „sensate focus" (sensorische Fokussierung) erstmals jene strukturierten Übungen in die Behandlung der sexuellen Dysfunktionen eingeführt, die H. Singer-Kaplan als therapeutische Erfahrungen übernommen hat. Dieser sensate focus ist auch dort zu einem

wichtigen Bestandteil der Sexualtherapie geworden, wo nicht nach dem Konzept von Masters und Johnson vorgegangen wird (z.B. nur ein Therapeut, keine soziale Isolierung, wöchentliche Sitzungen etc.). Nach entsprechender Anamneseerhebung und psychotherapeutischer Vorbereitung wird das Paar aufgefordert, auf Koitus und genitale Stimulation zu verzichten und keinen Orgasmus anzustreben, sowohl um Leistungsdruck und (Versagens-) Angst zu reduzieren, als auch um Gelegenheit zu geben, neue Formen der nonverbalen Kommunikation und Sensualität zu erlernen. Jeder Partner wird sorgfältig instruiert, daß das Ziel dieser „Sensibilitäts-Übungen" nicht in sexueller Erregung, nicht im „Befriedigen" des Partners, sondern im bewußten Erleben der eigenen Empfindungen zu sehen ist (Kolodny, 1979). Sofern nicht zuerst eine Exploration des eigenen Körpers ohne Anwesenheit des Partners angezeigt erscheint, sollen sich die Partner in einem ersten Schritt unbekleidet abwechselnd unter Aussparung der Genital- bzw. Brustregion „von Kopf bis Fuß" erforschen, entdecken, streicheln und miteinander darüber sprechen. Die dabei gemachten positiven und negativen Erfahrungen bzw. die dabei aufgetretenen Widerstände werden in der nächsten Therapiestunde ausführlich und in die Tiefe gehend durchgearbeitet. Unter Umständen wird die Übung wiederholt. Wiederholung, Einsicht und Umgehung sind drei Strategien des Umgangs mit negativen Reaktionen (Kaplan und Langer, 1979). Als nächstes werden, eventuell in mehreren Schritten, die Genitalien und die weibliche Brust miteinbezogen. Wiederum geht es nicht um orgasmusorientierte sexuelle Erregung. Noch stärker als bisher kann es nun erwünscht sein, dem Partner Rückmeldung zu geben, Wünsche oder Unlustgefühle mitzuteilen, die Art der Berührung mitzubestimmen, ohne ihn dadurch zu verletzen. Dies kann nonverbal durch leichtes Führen seiner Hand geschehen („hand riding technique" nach Masters und Johnson). Weitere Stationen eines solchen Sensualitätstrainings können der nicht-fordernde Koitus, meist kniet dabei die Frau rittlings über dem Mann, und schließlich der Koitus mit Orgasmus sein.

Elemente aus dem sensate focus, erweitert um spezielle Vorgangsweisen, können für verschiedene Indikationen zu besonderen Behandlungsprogrammen zusammengestellt werden. Beim vorzeitigen Samenerguß (Ejaculatio/Orgasmus praecox) geht es um die sog. Stop-Start-Methode (Kaplan, 1974) oder die Squeeze-Technik (Masters und Johnson): Der Mann soll lernen, das präorgastische Unvermeidlichkeitsgefühl rechtzeitig zu erkennen und so die Ejakulation allmählich unter Kontrolle zu bringen. Während er von seiner Partnerin stimuliert wird, konzentriert er sich auf seine Gefühle und signalisiert ihr kurz vor dem Orgasmus mit der Stimulation aufzuhören bzw. unterhalb der Eichel oder an der Peniswurzel das Glied kräftig zusammenzudrücken. Beides resultiert in einem Nachlassen des Ejakulationsdranges und der Erektion. Diese Übung wird mehrmals (ca. 4×) wiederholt, bevor die Ejakulation zugelassen wird. Dabei kann die Reizintensität durch Gleitmittel, nicht-fordernden Koitus in der Frau-oben-Position bis zum fordernden Koitus stufenweise erhöht werden. Bei Erektionsstörungen soll zusätzlich zum sensate focus manuelles und später koitales „Teasing" (Reizspiele) Vertrauen in die Wiederkehr einer verlorengegangenen Erektion zurückgewinnen helfen und entängstigen. Wiederum erfolgt ein stufenweiser Aufbau einer Verhaltenskette von nicht-genitalem Streicheln bis zum intravaginalen Orgasmus.

Ähnlich können bei Empfindungsstörungen der Frau unter besonderer Berücksichtigung der Beziehungssituation zusätzliche Stimulation der Klitoris, Entwicklung vaginalen Empfindens durch nichtforderndem Koitus (wobei der Mann seinen Penis nur zur Verfügung stellt), Training der Beckenbodenmuskulatur (sog. Kegelübungen) angewendet werden. Bei Orgasmusstörungen können Masturbationserfahrungen hilfreich sein und die Kombination von Klitorisstimulation und Koitus eine „Brücke" zur koitalen Orgasmusfähigkeit bauen. Beim Vaginismus wird eine Aufhebung des bedingten Scheidenreflexes angestrebt, indem die Frau selbst und/oder ihr Partner Finger bzw. Gegenstände zunehmender Größe (z.B. Hegarstifte) in entspannter Atmosphäre in die Scheide einzuführen lernt.

39.3.3 Erfahrungen und Erfolge

Es besteht immer wieder die Gefahr, diese Übungen bzw. Programme rezeptartig und mechanistisch zu verstehen und für das Wesentliche an der Sexualtherapie zu halten. Tatsächlich sind sie nur eine Komponente eines viel umfassenderen psychotherapeutischen Prozesses. Deshalb verlangt die eigentliche Sexualtherapie auch nach ausgebildeten Psychotherapeuten. Es ist inzwischen erwiesen, daß die Kombination der genannten psychotherapeutischen Methoden in der Sexualtherapie wesentlich bessere Ergebnisse liefert als die Methoden für sich allein. Langer hat die neue Sexualtherapie als „dynamische Verhaltenstherapie" oder als „aktiv-verhaltensmodifizierende analytische Therapie" bezeichnet, womit sie nochmals als „Amalgamierung" verschiedener Methoden charakterisiert sei. Auch hier muß auf die umfangreiche Originalliteratur verwiesen werden, besonders was die zahlreichen Modifikationen im therapeutischen Setting, der Zeiteinteilung, der Methodenschwerpunkte etc. betrifft (Masters und Johnson, 1970; Kaplan, 1974; Kaplan und Langer, 1979; Lo Piccolo, 1978; Kolodny, 1979; Angermann, 1980; Arentewicz und Schmidt, 1980; Kockott, 1980; etc.).

Aus dem Masters-und-Johnson-Institut liegen die wohl umfassendsten Studien zur Erfolgskontrolle von Sexualtherapie vor: Bei insgesamt 1872 zwischen 1959 und 1977 behandelten heterosexuellen Männern und Frauen wird von 81,8% eindeutigen Therapieerfolgen berichtet.

Bei den sexuellen Funktionsstörungen des Mannes wies die primäre Impotenz mit 66,7% Behandlungserfolgen die geringste, die vorzeitige Ejakulation mit 96,1% die höchste Erfolgsrate auf. Bei den sexuellen Dysfunktionen der Frau standen an entsprechender Stelle die situative Anorgasmie mit 71,0% und der Vaginismus mit 98,8% erfolgreicher Behandlung (Kolodny, 1981). Wenn diese Zahlen auch als Ergebnis der spezifischen Behandlungsart am Masters-und-Johnson-Institut anzusehen sind und anderwärts kaum erreicht werden, so geben sie doch prognostische Hinweise, die auch von anderen Autoren bestätigt werden (vgl. Kaplan, 1974; Eicher, 1980). Im folgenden soll näher auf die bei uns praktizierte Variante der „kommunikationszentrierten Sexualtherapie" eingegangen werden.

39.4 Die Kommunikationsfunktion der Sexualität als spezifisch menschliche Dimension

39.4.1 Die Sinnfrage als pathogenetisches Element

Da auch bei der Sexualtherapie das letztlich Entscheidende die Frage nach dem Sinn ist, soll nun auf das eingangs erwähnte Konzept der Kommunikationsfunktion der Sexualität näher eingegangen werden, um auf diese Weise einen bewußteren Zugang zur Sinnfrage zu versuchen. Dies erscheint notwendig und hilfreich, da in vielen Fällen die nicht bzw. nur einseitig oder falsch beantwortete Sinnfrage ein wesentliches pathogenetisches Element darstellt, wie etwa die eingangs zitierte bittere Bemerkung der anorgastischen Ehefrau, die Sexualität diene doch nur zur Befriedigung des Mannes.

Sexualität wird weithin nur genital verstanden und diese genitale Sexualität wiederum isoliert und von Zuneigung und Zärtlichkeit abgespalten gelebt und erlebt. Dabei läßt sich fehlender Sinn auch durch keine „Techniken" ersetzen, vielmehr setzen diese Sinn voraus. So schreibt etwa Prill (1964): „Vollgültige Sexualität ist letztlich durch eine Sinnfindung charakterisiert, in der körperliches Miteinandersein, innerweltliche und umweltliche Situationen eingeschlossen sind. Wenn das, was Von Gebsattel als Geschlechtsleib definiert, zu einer echten, vollsinnigen Handlung kommt, ereignet sich das ganzheitliche Miteinandersein ..."

39.4.2 Der doppelte Aspekt der Geschlechtlichkeit

Wie an anderer Stelle ausführlicher dargestellt (Loewit, 1982, 1988), wird im Laufe der Evolution der Sexualität schon im vormenschlichen Bereich deutlich, daß diese zu ihrer reproduktiven Funktion zunehmend soziale Bedeutung bekommt und sich langsam und mit zunehmender Gehirnentwicklung immer deutlicher aus den Zwängen von Hormonspiegeln oder instinktgebundenen Reiz-Reflex-Schemata emanzipieren kann: „Die Geschlechtsbeziehung wurde zur psychischen Liebe" (Wendt, 1962). Mit dem Verlust von Brunstzeiten beim Menschen läßt sich mit Recht ein reproduktiver und ein kommunikativer

Aspekt der Geschlechtlichkeit unterscheiden (vgl. Ford und Beach, 1971; Wickler, 1969).

Diese Kommunikationsfunktion der Sexualität, also ihre Aussage- und Mitteilungsfunktion, hat eindeutig beim Menschen einen Entwicklungshöhepunkt erreicht und soll im folgenden als das spezifisch menschliche Sexualverhalten verstanden und auf ihre konkreten Mitteilungsinhalte hin untersucht werden (vgl. Loewit, 1980, 1981).

39.4.3 Menschliches Sexualverhalten als Körpersprache

Sexueller Ausdruck wird hier als Sprache, Sexualverhalten als Körpersprache verstanden, wobei der Mensch als das einzige Lebewesen mit Wort- und Zeichensprache, als digitalen und analogen Kommunikationsmöglichkeiten, in der Lage ist, die vieldeutigen Zeichen zu konkretisieren und begrifflich zu verstehen.

Vor diesem Hintergrund stellt sich menschliches Sexualverhalten als eine vorwiegend nonverbale, analoge Kommunikationsform dar. Ihre körpersprachlich gesendeten und empfangenen Botschaften werden im Gefolge des entwicklungsbedingten Übergangs von den ganzheitlichen „Gestaltsignalen der Körperhaltung und des Verhaltens" der frühen Mutter(Eltern)-Kind-Beziehung zu den „semantischen Signalen" der Sprache als den „Hauptinstrumenten des Ichs zur Vermittlung von Objektbeziehungen" (R. Spitz, 1985) beim Heranwachsenden und Erwachsenen, sowie im Gefolge der allgemeinen Tabuisierung des Sexuellen kaum bewußt und nicht mehr als Mitteilungen verstanden. Während die körpersprachliche Bedeutung von Mimik, Gestik, Parasprache etc. weithin geläufig und bereits zum Inhalt einer eigenen Wissenschaft von der Körpersprache, der Kinesik, geworden ist, scheint das menschliche Sexualverhalten weitgehend unbefragt auf biologisch-somatisch vorgezeichneten und zusätzlich psychosozial genormten Bahnen abzulaufen, ohne daß seine körpersprachlichen Aussagen überhaupt wahrgenommen und bewußt werden.

39.4.4 Die Inhalte der sexuellen Kommunikation

Beim Versuch, die durch die sexuelle Körpersprache vermittelten Inhalte in Begriffe zu übersetzen, muß zunächst auf die Vieldeutigkeit analoger Kommunikation hingewiesen werden, die jeweils der konkreten Deutung bedarf. Ebenso ist zu betonen, daß es hierbei keineswegs um ein symbolisches Hineininterpretieren, sondern um das konkrete Herauslesen und Bewußtmachen der tatsächlich und jedenfalls durch die Körpersprache übermittelten Botschaften geht. Schließlich ist darauf hinzuweisen, daß jede Übersetzung aus dem Analogen (Zeichensprache) ins Digitale (Wortsprache) einen Informationsverlust und eine Einschränkung der analogen Aussagekraft bedeutet (vgl. Watzlawick et al., 1969). In diesem Sinn könnte also das körpersprachliche Verhalten bzw. die Kör-

pererfahrung sexuell sich Begegnender konkret übersetzt werden mit: Aufeinander-Zugehen, Sich-Näherkommen, Sich-Nahestehen, Sich-zugeneigt-Sein, Sich-hüllenlos-zu-erkennen-Geben, Berührt-Sein, Sich-Einfühlen, Bergen, Annehmen, Sich-Öffnen, Aufeinander-Eingehen, Sich-in-den-anderen-Hineinversetzen, Vereint-Sein, Zusammenhalten, Sich-am-Herzen-Liegen usw., wenn positive Inhalte sexueller Gesten gemeint sind. Überwiegen ambivalente oder negative Inhalte, so könnte auch Angriff, Verfolgung, Überwältigung, Unterwerfung, Beraubung, Verletzung, auch Abwendung, Sich-Verschließen, Abweisung usw. zum Ausdruck gebracht werden. Diese „Begriffe" werden zur tatsächlich körperlich erfahrenen Wirklichkeit, stellen also konkretes leiblich-sinnenhaftes Erleben dar.

39.4.5 Sexuelle Kommunikation, menschliche Grundbedürfnisse und Partnerschaft

Die genannten positiven Inhalte stellen gleichzeitig notwendige allgemein menschliche Grundbedürfnisse wie die Angewiesenheit auf Annahme, Zuwendung, Wertschätzung, Zugehörigkeit, Sicherheit und Entfaltung bzw. Sehnsüchte oder Werte dar. Ihre ausreichende Erfüllung entscheidet in unserer Kultur wesentlich über Sinnfindung und damit psychische und somatische Gesundheit. Deshalb sind sie zugleich unverzichtbare Elemente und Voraussetzungen partnerschaftlicher Beziehungen, im besonderen von Liebesbeziehungen. Das besondere Gewicht und die Bedeutung der Sexualität ergeben sich daher nicht nur aus der Stärke des biologischen Geschlechtstriebes, sondern auf gesamtmenschlicher Ebene aus dem Umstand, daß die positiven Inhalte der sexuellen Körpersprache in der doppelten Bedeutung des Selbsterfahrens und des Dem-anderen-Vermittelns gleichzeitig unverzichtbare (prägenitale) Grundbedürfnisse darstellen und als solche durch die Sexualität verkörpert bzw. in der Sexualität erlebt werden können.

Das Bewußtmachen dieser Zusammenhänge kann zum Verständnis von Liebe als gegenseitiger Sorge um die ausreichende Erfüllung der Grundbedürfnisse des Partners und von Sexualität als potentester Mitteilungsmöglichkeit, weil Verkörperung dieser Inhalte, beitragen. Es kann die Integration einer zunächst isolierten und primär genitalen und egozentrischen Sexualität in die Gesamtpersönlichkeit und in eine ganzheitliche personale und sexuell-erotische Beziehung fördern. Die zunächst somatisch erfahrene genital-sexuelle Lust kann dadurch um die weitere Dimension der „Beziehungslust", d. h. der Lust am Nicht-Alleinsein, am Angenommensein und daraus Selbstwert und Geltung erfahren, erweitert werden. Mit anderen Worten: Aus der psychosomatischen Ganzheit und Bedürftigkeit des Menschen ergibt sich der Wunsch, daß die durch die sexuelle Körpersprache gemachten Erfahrungen sich nicht nur auf die Geschlechtsorgane oder die Momente der sexuellen Begegnung beschränken, sondern daß leiblich-sinnenhafte und geistig-sinnenhafte Ebene sich entsprechen. Deswegen ist das körperliche Geschehen und die körperlich-sinnenhafte Erfahrung gleichzeitig ein menschliches Versprechen und gibt Antwort auf Hoffnungen und Bedürfnisse des Partners. Letztlich entscheidend für die Erfahrung sinnvoller Sexualität ist damit die Frage nach der Echtheit und Wahrheit der körpersprachlich vermittelten Botschaften innerhalb der gemeinsamen Beziehungswirklichkeit.

Permanente Frustration, d.h. dauerndes Auseinanderklaffen von körperlichem Tun bzw. leiblich-sinnenhafter Erfahrung und partnerschaftlicher Wirklichkeit, wirkt pathogen und kann schließlich zu direkten Funktionsstörungen führen, die wiederum einen somatischen Ausdruck für die Beziehungswirklichkeit darstellen. In diesem Sinne können sexuelle Funktionsstörungen als psychosomatische Antwort auf die Diskrepanz zwischen Aussage der sexuellen Körpersprache und Beziehungswirklichkeit aufgefaßt werden, somit als „funktionale Dysfunktionen".

39.5 Kommunikationszentrierte Sexualtherapie

Bei der expliziten Einbeziehung dieses Kommunikationskonzeptes in die Behandlung sexueller Störungen handelt es sich nicht um eine neue Therapieform, sondern um die Bewußtmachung einer bisher meist unbeachteten, grundlegend neuen Denk- und Erlebensweise innerhalb bestehender Therapieformen.

Im konkreten Beispielsfall hat die Patientin als letzte von ihr uneingeschränkt positiv erlebte sexuelle Geste das Sich-an-ihren-Mann-Anschmiegen oder „Kuscheln" empfunden und konnte dieses Verhalten auch sofort als Ausdruck der Geborgenheit und Einheit für sich übersetzen. Die gemeinsame Erarbeitung der Inhalte der sexuellen Kommunikation ermöglichte es ihr dann, auch in der genitalen Sexualität ein noch intensiveres Zeichen solcher Einheit zu sehen. Damit wurde es ihr erstmals möglich, Zärtlichkeit und Sexualität nicht mehr als verschiedene Dinge zu begreifen: „So hab' ich das noch nie gesehen." Erst damit waren die Widerstände gegen die „Hausaufgaben" (keine Zeit gehabt, Besuch bekommen) zu überwinden, und auch der Ehemann lernte, seinen Anteil an der Störung zu verstehen. Der Mann, der anfänglich meinte, nichts damit zu tun zu haben, hat durch die erfolgreichen Sensate-focus-Übungen eine Menge über sich und sein Sexualverhalten gelernt und „kam aus dem Staunen nicht heraus". Die Frau gab an, jetzt „erstmalig selber dabei zu sein" und fühlte sich zum ersten Mal als vollwertige Frau, was für sie ein neues Lebensgefühl bedeutete und sich gleichzeitig positiv auf ihre Angst vor dem Verlassenwerden auswirkte. Sie hat ihren Partner von neuen Seiten kennen- und selber Wünsche äußern gelernt. Beide haben die Bedeutung des Sich-Zeit-Nehmens und des persönlichen Gespräches erkannt und begriffen, daß Sexualität als bewußter Ausdruck neue Erlebnismöglichkeiten bietet, aber auch Arbeit an der Beziehung voraussetzt. Nach 8 therapeutischen Sitzungen waren die Störungen soweit gebessert, daß die Therapie erfolgreich beendet werden konnte.

Ein anderes, seit 8 Jahren unter der Anorgasmie der Ehefrau leidendes Paar, Anfang Dreißig mit zwei Kindern, erlebte diesen durch die Sensate-focus-Übungen geförderten Lern- und Neuorientierungsprozeß folgendermaßen: Der Ehemann, der immer auf täglichem Geschlechtsverkehr bestanden hatte, sah Sexualität „jetzt in einem anderen Licht" und wurde „so zärtlich wie möglich". Dies verhalf seiner Frau, die immer streng zwischen Liebe und Sexualität unterschieden hatte, zu neuem Erleben, im Gegensatz zu früheren Versuchen mittels pornographischer Filme die Anorgasmie zu überwinden. Sie formulierte treffend: „Jetzt waren es mehr wir, die miteinander geschlafen haben, früher war es einfach eine normale sexuelle Handlung." Da nun „ein neuer Aspekt dazugekommen" ist und ich mich so richtig hineindenken kann, wie das auch noch einen tieferen Sinn hat", konnte sie sich auch gefühlsmäßig hingeben und erlebte gleichzeitig mehr Selbstbewußtsein, „weil ich spürte, daß sich mein Mann mit mir als Ganzem befaßte", und mehr Sicherheit und Freiheit, „weil man so weniger falsch machen kann und ich mir dabei nichts vergebe". Damit nahm auch die körperliche Erlebnisfähigkeit zu, und die Frau wurde mit ihrem Mann orgasmusfähig. Diese Wandlung erfolgte im Laufe von 10 Besprechungen, bei denen selbstverständlich auch andere Aspekte zur Sprache kamen und gegenseitiges Verstehen und Einsicht in lebensgeschichtliche Zusammenhänge auf vielen Ebenen angebahnt wurden.

Durch diese „neue" Sichtweise der sexuellen Kommunikation kann also der „sensate focus" in den meisten Fällen eine bewußte zusätzliche Dimension erhalten. Die Übungen können im positiven Fall als den eigentlichen Bedürfnissen und Wünschen entsprechend ganzheitliche persönliche Begegnungen vermitteln, im negativen als unausweichliche Konfrontation mit der Beziehungswirklichkeit erlebt werden. Die bewußte Miteinbeziehung des Kommunikationskonzeptes kann gleichzeitig mithelfen, Rollenfixierungen zu relativieren, Ängste zu vermindern und Leistungsdruck abzubauen, also insgesamt Verhalten umzustrukturieren. In unseren Händen hat die bewußte Berücksichtigung dieses Konzeptes zu einer Ausweitung der vor den Sensate-focus-Übungen gelegenen Therapiephase geführt. Eine etwas ausführlichere Bearbeitung der individuellen Lebens- und Lerngeschichte sowie partnerschaftlicher Probleme kann als Voraussetzung der gemeinsamen Erarbeitung dieser Kommunikationsfunktion der Sexualität notwendig sein.

Dadurch werden unter Umständen Widerstände, die sonst erst bei den ersten Übungen aktiviert werden, bereits vorher erkenntlich und bearbeitbar. Zur Erreichung des Zieles, gemeinsam eine umfassende Sichtweise der Sexualität zu erarbeiten, scheint uns weniger eine autoritativ durchgeführte Therapie mit viel Information geeignet, als vielmehr das Anknüpfen bei den letzten von den Patienten positiv erlebten und interpretierten Verhaltensweisen (vgl. Kockott, 1980), wobei das Tempo des Fortschritts von dem Patienten bestimmt wird. Nach unserer bisherigen Erfahrung können dadurch die Sensate-focus-Übungen an Erlebnistiefe und therapeutischer Wirksamkeit beträchtlich gewinnen.

Diese Betrachtung der Sexualität als Ausdruck und Verkörperung der in einer Beziehung wesentlichen menschlichen Werte erscheint zunächst meist ungewohnt und schwierig: „So haben wir das noch nie gesehen", „das war für uns ganz neu", „das haben wir bisher nicht gewußt" – kann aber dann als eigentlich selbstverständlich angenommen werden und: „als wesentlicher Bestandteil ins Bewußtsein übergehen, auch wenn man nicht immer davon redet". Eine ähnliche Brückenfunktion kann diese Betrachtungsweise auch für die Versöhnung von Schamgefühl und Sexualität, Ekel und Sexualität, Religion und Sexualität erlangen, womit einige unter Umständen pathogenetische Störfaktoren angesprochen sind (vgl. hierzu die Gedankengänge moderner christlicher Theolo-

gen, wie z. B. Greeley, 1977 und Nelson, 1978). Auch bei den Problemen der Sexualität im Alter, die häufig weniger aus altersbedingten Funktionsänderungen als aus der sozialen Ächtung der Sexualität in diesem Lebensabschnitt herrühren, könnte ein solches Verständnis mithelfen, irrationale Mythen und Widerstände zugunsten einer lebensgerechteren und humaneren Einstellung und Praxis aufzugeben (vgl. Scheingold und Wagner, 1976; v. Schumann, 1980).

Sexualität, sofern sie alle in ihr gelegenen Möglichkeiten ausschöpfen will, und Beziehung, als die dem sexuellen (Ausdrucks-)Geschehen zugrundeliegende interindividuelle Wirklichkeit, werden so zu gleichrangigen „Aufgaben" (vgl. Bodamer, 1970). Damit sind abschließend aber auch die Grenzen dieses Konzeptes bzw. erfolgreicher Paarberatung und Sexualtherapie in der Praxis angedeutet.

40 Schlaf und Schlafstörungen

Dietrich Schneider-Helmert

40.1 Der normale Schlaf

40.1.1 Neurophysiologische Grundlagen

Schlafen ist ein Ruhezustand, in dem das Bewußtsein und der Kontakt mit der Umgebung aufgehoben sind. Er ist jederzeit und rasch reversibel. Im Schlaf ist der Muskeltonus vermindert, und das Vegetativum befindet sich in einer trophotropen Einstellung. Daher sind Atemfrequenz, Herzfrequenz und Blutdruck tief. Der Metabolismus ist generell niedrig, und die Körpertemperatur sinkt ab. Aufbauprozesse, wie die Proteinsynthese, sind hingegen verstärkt. Die Hauptphase der Ausschüttung von Wachstumshormon ist mit dem Deltaschlaf verknüpft. Daher wurde vermutet, diese Schlafphase sei eine Voraussetzung für somatische Regenerationsprozesse. Die Reizaufnahme und -verarbeitung ist im Schlaf eingeschränkt, jedoch nicht ganz aufgehoben. Auf einer unbewußten Ebene findet eine Rezeption und Diskriminierung nach individueller Bedeutsamkeit statt. Die Verwebung äußerer (akustischer, olfaktorischer) Reize in einen Traum kann ebenso wie diejenige innerer Reize (z.B. Durst) der Abwehr dienen, indem eine halluzinatorische Bedürfnisbefriedigung im Traum das Aufwachen verhindert. Deshalb bezeichnete schon Freud den Traum als den „Hüter des Schlafs". Wird ein Reiz aber als individuell wichtig „erkannt", etwa für eine Mutter das Schreien ihres Neugeborenen, provoziert er ein sofortiges Erwachen.

Auf der zentralnervösen Stufe ist der Schlafverlauf durch charakteristische Bilder der elektrophysiologischen Aktivität gekennzeichnet. Die Einteilung in sog. Stadien wurde 1968 von einem Komitee unter Führung von Rechtschaffen und Kales festgelegt und wird seither international als verbindlich anerkannt. Der Wachzustand zeigt im kortikalen EEG ein Gemisch von hochfrequenten, niedrigamplitudigen Wellen, die bei psychischer Entspannung einem langsameren Rhythmus, den sog. Alphawellen, Platz machen. Beim Einschlafen entstehen zunächst etwas höhere sog. Thetawellen und langsame Pendelbewegungen der Augen. Dieses Stadium 1 ist als Dösen zu bezeichnen. In ihm treten oft kurze Träume, sog. hypnagoge Halluzinationen, auf. Nach wenigen Minuten folgt das erste eindeutige Schlafstadium, Stadium 2. Es zeigt im EEG spezifische Muster, nämlich K-Komplexe und Schlafspindeln. Hier ist der Schlaf leicht, d.h. die Weckschwelle ist von allen Schlafstadien am niedrigsten. Bald werden tiefere Stadien erreicht, die durch eine Dominanz sehr langsamer und hoher sog.

Deltawellen charakterisiert sind, und je nach deren Ausmaß als Stadium 3 oder 4, zusammen als Deltaschlaf, bezeichnet werden. Hier ist die Weckschwelle am höchsten, und beim Aufwachen aus diesem meist traumlosen Schlaf hat man oft das Gefühl, „von weither zu kommen". Diese Schlafstadiencharakterisierung geht auf Loomis (1937) zurück. Es vergingen danach 15 Jahre, bis von Aserinski und Kleitmann der REM-Schlaf (REM = Rapid Eye Movements) entdeckt wurde. Er zeigt elektrophysiologisch verschiedene Diskrepanzen, weshalb er auch als paradoxer Schlaf bezeichnet wurde (im Französischen bis heute üblich), nämlich im EEG das Bild eines Stadiums 1, aber eine Weckschwelle, die zwischen Stadium 2 und 4 liegt, und im Elektromyogramm den niedrigsten Muskeltonus aller Stadien, dazu aber schnelle, zuweilen in Salven auftretende Bewegungen der Augäpfel nach allen Richtungen. Das Faszinierendste war die Entdeckung, daß Leute, die man aus dem REM-Schlaf weckt, in einem hohen Prozentsatz berichten, sie seien gerade am Träumen gewesen, auch wenn sie „sonst nie träumen". Daher wurde er zuweilen auch „Traumschlaf" genannt. Inzwischen weiß man aber, daß Träume auch in den anderen Schlafstadien vorkommen, wenngleich man sie dort viel eher während des Aufwachens gleich wieder vergißt.

Die beschriebene Abfolge der Stadien umfaßt einen sog. Schlafzyklus, der eine Dauer von ca. eineinhalb Stunden hat. Ein normaler Nachtschlaf enthält 3–5 Schlafzyklen. Es ist aber kein wirklich zyklischer Ablauf im Sinne einer Wiederholung. Der Deltaschlaf konzentriert sich nämlich ausgesprochen auf den 1. und 2. Zyklus. Dementsprechend haben Messungen der Weckschwelle, die Kohlschütter schon 1862 in Leipzig durchführte, ergeben, daß der Schlaf zu Beginn am tiefsten ist und dann abflacht. Andererseits werden die REM-Phasen von Zyklus zu Zyklus länger und stabiler. Kurze Wachperioden, durchschnittlich in Abständen von 50–60 Minuten, sind normal. Sie stehen meist mit Lagewechseln in Zusammenhang und werden am Morgen nicht erinnert. An Wachepisoden erinnert man sich in der Regel nur, wenn sie sehr lang waren, wenn man aufgestanden, aus einem Angsttraum aufgewacht ist oder sich geistig intensiv mit etwas beschäftigt hat.

Alle beschriebenen Funktionen sind der Ausfluß der spezifischen Regulationsvorgänge im Bereich des Zwischenhirns und Hirnstamms. Sie sind auf der Stufe der Neurotransmitter, d.h. der Subsysteme der Wach- und der Schlaffunktionen, gut bekannt. Bei letzteren spielen serotoninerge Systeme mit ihren

Projektionen zum Kortex eine dominierende, aber keineswegs exklusive Rolle. Die Frage, wie diese antagonistischen Partialsysteme aufeinander abgestimmt und dirigiert werden, ist noch nicht annähernd beantwortet. Das Neuropeptid DSIP (Delta Sleep Inducing Peptide) scheint eine gewisse Führungsrolle zu spielen, denn für DSIP wurde beim Menschen nicht nur eine explizite Unterstützung der Schlaf-, sondern auch der Wachfunktionen nachgewiesen (Schneider-Helmert, 1988a). Für Einzelheiten der Schlafphysiologie sei auf die Monographie von Koella verwiesen (1988).

40.1.2   Schlafen und psychophysische Erholung im 24-Stunden-Zyklus

Das Schlafen ist in den Rhythmus des 24-Stunden-Zyklus eingebettet. Bei normal angepaßten Zirkadianrhythmen ist die Schlaftendenz zwischen 2 und 6 Uhr am stärksten, und ein Nebengipfel besteht am Frühnachmittag; niedrig ist sie am Vormittag, am geringsten am Spätnachmittag und Abend. Etwa zwischen 22 und 24 Uhr kommt es zu einem so steilen Anstieg, daß man hier von einem „Schlaftor" spricht (Lavie, 1986). Diese bimodale Kurve der Schlaftendenz ist ein sehr stabiler biologischer Rhythmus. Daher ist der Versuch mancher Schlafgestörter, ihr Problem durch besonders frühes Zubettgehen zu bekämpfen, zum Scheitern verurteilt. Die Grundlage dieser inneren Uhr ist noch nicht genau bekannt, aber einige Befunde sprechen dafür, daß das Neuropeptid DSIP bei der Steuerung dieses Rhythmus eine wichtige Rolle spielt (Ernst und Schönenberger, 1988). Die ontogenetische Entwicklung des Schlafmusters zeigt, daß sich dieser bimodale Verlauf der Schlaftendenz beim Kind noch deutlich manifestiert, indem es neben der nächtlichen Hauptschlafphase einen Mittagsschlaf braucht. Obwohl der Sozialisationsprozeß die mittägliche Schlaftendenz in der Adoleszenz unterdrückt, ist sie auch im Erwachsenenalter bei manchen Menschen sichtbar. Im Alter vermindert sich die nächtliche Schlafdauer im Mittel nur geringfügig, aber die interindividuelle Variabilität nimmt mit jeder Dekade zu (Miles und Dement, 1980; Williams et al., 1974). Der Grund dafür ist, daß es zwischen 50 und 70 Jahren zu einer Dichotomie im Schlafmuster kommt: Während etwa ein Drittel beim monophasischen Muster mit einer einzigen Schlafphase pro 24 Stunden bleibt, wie beim jüngeren Erwachsenen die Regel, bildet sich bei der Mehrzahl ein polyphasisches Schlafmuster heraus; der Nachtschlaf teilt sich in zwei oder drei Abschnitte auf mit u.U. recht langen, intermittierenden Wachperioden, in denen durchaus genügend Antrieb vorhanden sein kann, aufzustehen, zu lesen oder sich sonst einer ruhigen Tätigkeit hinzugeben (Webb und Schneider-Helmert, 1984a). Vor allem diese Menschen haben ein verstärktes Bedürfnis für einen Mittagsschlaf, womit sie die nächtliche Schlafverkürzung mehr oder weniger kompensieren. Die angebliche Regel, daß der ältere Mensch viel weniger Schlaf brauche, ist also unter

dem Blickwinkel des 24-Stunden-Zyklus nicht haltbar.

Schlafen dient der physischen und psychischen Erholung. Beim Erholungsschlaf nach körperlichem Streß setzt prioritär der Deltaschlaf ein, nach psychischem Streß treten Veränderungen im REM-Schlaf auf, die allerdings uneinheitlich und nicht auf einen Nenner zu bringen sind. Die psychische Erholung durch den Schlaf läßt sich an der kognitiven Leistungsfähigkeit am Tag nach experimentellen Eingriffen in den Nachtschlaf beurteilen. Der wichtigste Erholungsaspekt ist eine hohe Stabilität des Schlafs, denn Unterbrechungen in Intervallen unter 10 Minuten rufen bereits nach einer einzigen Nacht meßbare Defizite hervor (Bonnet, 1986a und b; Levine et al., 1987; Stepanski, 1984). Etwa 2 Stunden Delta- plus REM-Schlaf sind ein weiteres Erfordernis, unabhängig davon, wie lange der gesamte Schlaf dauert. Dies erklärt auch, warum es Kurzschläfer gibt, die mit 3 Schlafzyklen (4–5 Stunden Schlaf) sehr gut auskommen, sofern der Schlaf diese Effizienzkriterien erfüllt. In diesem Zusammenhang sprach Horne (1988) vom „obligatorischen" Schlaf, der mit dem „fakultativen" Schlaf ergänzt werden kann, aber nicht muß. Folgt man dieser Hypothese, wäre allerdings darauf hinzuweisen, daß Schlafen neben dem vitalen Anteil auch einen Lustaspekt beinhaltet, der für die psychische Äquilibrierung nicht ganz vernachlässigt werden kann (Schneider-Helmert, 1986). Dazu gehören auch die Träume, an die man sich spontan um so eher erinnert, je negativer ihr emotionaler Gehalt ist (Gnirss et al., 1978). Natürlich gibt es auch Menschen, die außerordentlich positive Träume haben und diese schätzen, besonders wenn sie kompensatorischen Charakter haben. Eine 40jährige Frau beispielsweise, die durch eine Poliomyelitis von Kindheit an körperlich behindert war, erlebte sich im Traum immer als gesund und konnte sich ebenso frei bewegen wie alle anderen. Sie beklagte ihre Benzodiazepinabhängigkeit in erster Linie deshalb, weil diese Medikamente ihre Träume unterdrückten und sie damit der kompensatorischen Erlebnisse beraubten.

40.1.3   Die problematische Selbstwahrnehmung des Schlafs

Bei der Exploration des Schlaf-Wach-Verhaltens sind wir auf die Schilderungen des Patienten angewiesen, die auf seiner Selbstwahrnehmung beruhen. Wie zuverlässig ist sie? Zunächst ist festzuhalten, daß man den Schlaf in seiner zeitlichen Dimension, auf die sich die Frage nach Schlafdauer, Ein- und Durchschlafen bezieht, nicht unmittelbar wahrnehmen kann, sondern man schließt aus den Lücken in der Wacherinnerung auf das Schlafen. Lediglich Träume sind beim Erwachen ein unmißverständlicher Hinweis auf Schlaf. Da Traumerinnerungen am ehesten vorhanden sind, wenn man aus REM-Schlaf erwacht, und die REM-Phasen gegen Morgen hin länger werden, steigt die Chance, beim Aufwachen erkennen zu können, ob man unmittelbar vorher geschlafen hat, im Verlauf einer Nacht an. Durchschnittlich liegt aber die korrekte Beurteilung bei öfterem Erwachen bei nur 57% (Sewitch, 1984). Es wundert daher nicht, daß bei Gesunden die subjektiv geschätzte Einschlafzeit und die nächtliche Wachzeit nur schlecht mit den objektiven Messungen korrelieren. Selbst die Wahrnehmung der Veränderungen von Nacht zu Nacht, Grundlage jeder Therapiebeurteilung in der Praxis, ist recht vage (Webb und Schneider-Helmert, 1984 b). Chronisch Schlafgestörte sind dagegen treffsicherer (Frankel et al., 1976; Schneider-Helmert und Kumar, 1986), was wohl auf ihre größere Ausrichtung auf diese Probleme und langjährige Selbstbeobachtung, eine fast unausweichliche psychologische Folge der chronischen Insomnie, zurückzuführen ist. Man kann daher als Faustregel sagen: Je schlechter der Schlaf geschildert wird, um so eher entsprechen die Angaben den Tatsachen. Natürlich sind im Einzelfall extreme Abweichungen möglich. Da es mehr darauf ankommt, wie erholsam als wie lange man schläft, ist es besser und klinisch relevanter, darauf zu schauen, wie man aufwacht, den Tag angeht und durchsteht.

40.1.4 Konzeptuelle Gesichtspunkte

Das manifeste Schlaf-Wach-Verhalten ist Ausdruck individueller Bedürfnisse und Strebungen. Sie formieren sich im Netzwerk der psychophysiologischen Funktionen, in das auch die innere Uhr eingebunden ist. In gewisser Weise ist der Schlaf mit einem Trieb vergleichbar. So wurde mit Bezug auf den Mittagsschlaf zwischen kompensatorischem Schlaf als notwendiger Ergänzung zu mangelhaftem Nachtschlaf (z.B. beim oben beschriebenen polyphasischen Schlafmuster) und appetitivem Schlaf als einem zusätzlichen Schlafen ohne biologische Notwendigkeit, aber mit einem gewissen Lustaspekt, unterschieden (Evans et al., 1977). Psychische Bedürfnisse manifestieren sich aber unmittelbarer in einer Stimulierung und Ausdehnung der Wachheit, die durch geistige Interessen, emotionale Erregung und willentliche Anstrengung in einem gewissen Rahmen erzwingbar ist. Hier besteht eine enge Verflechtung mit sozialen und kulturellen Faktoren, beispielsweise durch rhythmusbestimmende Arbeitszeiten, gesellschaftliche Aktivitäten, familiäre Einflüsse und Freizeitgestaltung. Der Schlaf hingegen läßt sich nicht willentlich herbeiführen. Im Gegenteil: Durch zuviel Wollen wird er verhindert. Bewußt kann man nur die Bedingungen zum Einschlafen passiv optimieren. Unbewußte Faktoren, wie vorausgegangene Konflikte und emotionale Stimulierung durch Antizipation – sie kann von positivem Gefühlsgehalt sein, wie vor einer Reise, oder belastend, wie vor einer Examenssituation – können ebenso störend sein wie Umgebungseinflüsse, z.B. Lärm, Wetter, schnarchender Partner. Für das Schlafen als Umsetzung der momentanen Bedürfnisse ist eine komplexe neurophysiologische Regelung erforderlich, die auch die vegetative Umstimmung beinhaltet, und der weitere Schlafablauf verlangt das richtige Zusammenspiel eines Bündels spezifischer Funktionen. Das manifeste Schlaf-Wach-Verhalten hat seinerseits Rückwirkungen auf den psychophysischen Bereich und damit auf die Interaktionen des Individuums mit seinem sozialen Umfeld, etwa dadurch, daß man sich nach einem guten Schlaf frisch und aktiv oder nach einem schlechten Schlaf müde und verstimmt fühlt.

40.2 Schlafstörungen

40.2.1 Übersicht und exemplarische Fälle

Bei der Klassifikation der Schlafpathologie folgen wir in zusammenfassender und vereinfachender Form den heute führenden Diagnosesystemen, die von den amerikanischen Fachgesellschaften für Schlafforschung (ASDC) und Psychiatrie (DSM-III-R) entwikkelt wurden. Wir gehen dabei von der in praxi wichtigsten Störung, der Insomnie, aus.

> Eine 45jährige Frau leidet seit 20 Jahren an einer persistierenden Durchschlafstörung. Diese hat nach der Geburt des 3. Kindes begonnen, das sie bis ins 2. Lebensjahr hinein, oft mehrmals pro Nacht, gestört hatte. Während ihr Mann selten darauf reagierte, wurde sie immer hellwach. Danach konnten weder Beruhigungsmittel noch autogenes Training die eingeschliffene Durchschlafschwäche beheben. Die Frau fühlte sich durch die Anforderungen des Alltags permanent überbeansprucht, war müde, energie- und freudlos.

> Ein 53jähriger Leiter eines Betriebszweigs einer größeren Firma bekam im Alter von 38 Jahren Schlafstörungen im Zusammenhang mit seinem Bemühen und Rivalisieren um den jetzigen Posten. Nach Erreichen des Ziels bestand die Insomnie unverändert fort, und er gewöhnte sich an den regelmäßigen Gebrauch von Hypnotika, oft auch mit Alkohol kombiniert. Bei langen Autofahrten, Konferenzen und Vorträgen mußte er gegen den Schlaf kämpfen. Ferner bestand eine Hypertonie. Die Abklärung ergab einmal eine chronische Durchschlafstörung, Benzodiazepinabhängigkeit und mäßige Schlafapnoe. Nach dem Hypnotikaentzug ging letztere auf marginale Werte zurück und die Hypertonie stabilisierte sich auf niedrigerem Niveau.

40.2.2 Psychosomatische Aspekte der Insomnie

Die Insomnie besteht in einer Diskrepanz zwischen dem Schlafbedürfnis und dem Schlafvermögen gemäß der je individuellen Situation. Sie äußert sich am deutlichsten durch die Symptome einer ungenügenden Erholung, also Erscheinungen am Tag wie Müdigkeit, Antriebsmangel oder Agitiertheit, Nervosität, verminderte Streßtoleranz, Depressivität, Abnahme der Spontanaktivitäten und abends vorzeitigem Energieverlust. Eine **akute Insomnie** kann als eine normale psychophysische Reaktion auf einen äußeren oder inneren Stimulus verstanden werden, nämlich als Verschiebung des Schlaf-Wach-Systems zugunsten einer höheren Vigilanz. Dies führt in erster Linie zu Einschlafstörungen. Kurzfristig erwächst daraus noch keine Konsequenz im Sinne der oben beschriebenen Symptome im Wachbereich, da in einer normalen Grundsituation genügend Reserven zur Kompensation eines Erholungsdefizits verfügbar sind. Doch kann eine akute Insomnie Unsicherheit, Ängstlichkeit und übertriebene Besorgnis des Patienten hervorrufen. Eine Beruhigung und damit Besserung der Insomnie kann mit psychischen Einflüssen erreicht werden, seien es spontane Umgebungseinflüsse, gezielte Maßnahmen, wie z.B. Entspannungstechniken, und sogar Placebos. Die Reaktion auf Placebos folgt jedoch nicht immer dem normalen Schema einer suggestiven Beeinflussung. In einem Experiment von Storms und Nisbett (1970) bekamen die Probanden die Placebotabletten das eine Mal mit dem Hinweis, es sei ein Schlafmittel, das andere Mal als angeblich anregende Substanz. Sie schliefen mit dem „Stimulans" schneller ein als mit dem „Hypnotikum". Diese paradoxe Suggestivwirkung wurde mit der Attributionstheorie erklärt. Sie besagt, daß die unwillkürliche Funktion des Einschlafens leichter vonstatten geht, wenn man die Verantwortung dafür nicht selbst zu tragen hat, sondern sie einer Tablette bzw. dem Therapeuten zuschieben kann. Allerdings blieben Replikationsversuche widersprüchlich, so daß vermutlich noch andere Faktoren, z.B. die Persönlichkeitsstruktur, mit eine Rolle spielen. Eine analoge Situation ist die paradoxe Intention: „Versuchen Sie wach zu bleiben", mit der ebenfalls ein schnelleres Einschlafen erzielt werden konnte (Ladouceur und Gros-Louis, 1986; Relinger et al., 1978).

Wenn eine die Insomnie auslösende Situation oder eine überängstliche Reaktion des Patienten lange anhält bzw. sich in kurzen Intervallen oftmals wiederholt, kann die **Insomnie chronifizieren.** Bislang gibt es keine Hypothesen, wie sich dieser Übergang vollziehen könnte. Doch ist in diesem Zusammenhang erwähnenswert, daß wiederholter Schlafentzug im Experiment nicht etwa zu einer Habituation, sondern im Gegenteil zu einer Sensibilisierung führte, in der Weise, daß die Folgen der Deprivation für die Wachfunktionen stärker wurden (Webb und Levy, 1984). Die Autoren dieser Studie vermuteten, eine motivationale Veränderung sei dafür maßgebend. Ähnliche Vorgänge könnten auch bei der Chronifizierung der Insomnie eine Rolle spielen (Zermürbung). Im weiteren zeigt dieser Prozeß typische Merkmale der Somatisierung eines psychophysischen Zusammenhangs: Die Insomnie nimmt die Form einer eigenständigen, persistierenden, funktionellen Störung an und wird vom Patienten als eine körperliche Dauerstörung erlebt, der er machtlos gegenübersteht. Sie ist der psychologischen Einflußnahme weitgehend enthoben. Hier sind auch Placebos unwirksam (Adam et al., 1976; Hartmann und Cravens, 1973; Mitler et al., 1984; Roehrs et al., 1985). Bei der chronischen Insomnie hat sich nämlich eine permanente, neurophysiologische Dysfunktion entwickelt und fest etabliert

(Schneider-Helmert, 1988b). Bezeichnend dafür ist, daß die Schlaftendenz nicht mehr den Bedürfnissen entsprechend umgesetzt wird, z.B. trotz permanent erhöhter Müdigkeit kein Mittagsschlaf möglich ist, oder der Patient zwar abends am Fernsehen einnickt, doch wieder hellwach wird, kaum daß er zu Bett gegangen ist. Fast immer liegt eine erhebliche Störung des Durchschlafens vor. Sie ist in erster Linie für die mangelnde Erholung und damit für die Beeinträchtigung der Wachfunktionen am Tag verantwortlich (Schneider-Helmert, 1987).

40.2.3 Klassifikation der Insomnien

Akute Insomnien

Nach ihrem Symptomverlauf werden akute auch als **transiente Insomnien** bezeichnet, was an den primär benignen Charakter dieser Formen erinnert. Um häufigen Wiederholungen und damit der Chronifizierung zuvorzukommen, sollten die Entstehungsbedingungen eruiert und im Sinne einer kausalen Therapie eliminiert werden. Nach der Entstehung kann man 4 Gruppen unterscheiden:
- **Psychophysiologische Insomnien** sind Reaktionen auf akute, stimulierende, psychische Faktoren, wie Streß, Frustration, Erwartungsspannung. Sie werden auch als primäre Insomnien bezeichnet.
- **Situative Insomnien** sind Reaktionen auf äußere Störfaktoren, wie Lärm, Wetter, Schichtarbeit.
- **Pharmakogene Insomnien** sind Akutreaktionen auf zentralnervös anregende Substanzen, wie Genußmittel, Stimulantien (inklusive Appetitzügler und gewisse Antidepressiva), oder aber auf das Absetzen von Suchtmitteln.
- **Symptomatische Insomnien** sind Schlafstörungen, die im Rahmen somatischer und psychiatrischer Erkrankungen als eine Begleit- oder Folgeerscheinung der Grunderkrankung entstehen (daher auch sekundäre Insomnien genannt).

Chronische Insomnien

Die chronische Insomnie ist durch eine hohe Symptomkonstanz von Nacht zu Nacht gekennzeichnet und wird daher auch **persistierende Insomnie** genannt. Die Schlafqualität unterscheidet sich im Vergleich zwischen Werktagen und Wochenende, Arbeitsperioden und Ferien kaum. Von den oben beschriebenen, akuten Formen neigen vor allem die beiden ersten zum Übergang in einen chronischen Verlauf, wenn die Ursachen nicht rechtzeitig behoben werden. Die pharmakogene und die symptomatische Insomnie jedoch entwickeln sich bei gegebenem Anlaß unmittelbar chronisch: Bei länger dauerndem, regelmäßigem Gebrauch von Hypnotika führt die Toleranzentwicklung zu einer speziellen Schlafpathologie (s.u.); bei der symptomatischen Insomnie sind chronische Grunderkrankungen verlaufsbestimmend. Hier kommt im psychiatrischen Bereich den Depressionen, im somatischen den rheumatischen Erkrankungen eine besonders große Bedeutung zu.

Schließlich gibt es auch eine **primär chronische Insomnie** ohne ersichtliche Ursache, bei der möglicherweise genetische Faktoren eine Rolle spielen.

Alle Formen chronischer Insomnie münden in ein psychophysisches Erschöpfungssyndrom, das durch folgende Charakteristika geprägt ist: Der Patient unternimmt alle Anstrengungen, die von der Umgebung erwarteten und von sich selbst geforderten Leistungen in der täglichen Arbeit zu erfüllen. Weil er dabei ständig gegen eine erhöhte Müdigkeit ankämpfen muß, erbringt er einen erhöhten Aufwand an psychischer Energie. Die Erschöpfung manifestiert sich daher besonders im „fakultativen" Lebensbereich, d. h. in einer verminderten Genußfähigkeit, geringem Antrieb für spontane Eigenaktivität, und oft auch in einem Gefühl, dies „nicht mehr lange durchstehen zu können", wobei auch Ängste vor einem körperlichen oder psychischen Zusammenbruch oder vage Suizidgedanken aufkommen können. Abbildung 40–1 zeigt einige Marksteine dieses Syndroms innerhalb des 24-Stunden-Zyklus und soll darauf hinweisen, daß diese Patienten in einen Circulus vitiosus verstrickt sind. Die auf den Nachtschlaf gerichtete Spannung und Leistungshaltung etabliert sich allerdings nicht in erster Linie wegen der Schlafstörung selbst, sondern entspricht der Hoffnung und dem Bemühen um einen besseren nächsten Tag. Manche vermögen ihre Wachdefizite aber noch einigermaßen zu kompensieren. Typische Schilderung dieser Situation: „Wenn man mich fragt, wie war's, muß ich immer sagen: schön aber anstrengend." Die Persönlichkeitsprofile von chronisch Schlafgestörten zeigten demgemäß in verschiedenen Untersuchungen ein Bild, das entweder normal oder durch typische Aspekte von psychosomatischen Störungen, Depressivität oder sozialer Introversion geprägt war (Bliwise et al., 1985; Edinger et al., 1988; Hermann et al., 1988; Levin et al., 1984; Schneider-Helmert, 1987; Zorick et al., 1984). In den letzteren Fällen ist nicht anzunehmen, daß es sich um disponierende Persönlichkeitszüge handelt, da beispielsweise das psychosomatische Profil auch bei chronischen körperlichen Krankheiten auftritt. Vielmehr zeigt dies, wie stark sich die psychische Struktur unter dem Druck einer permanenten Störung des Schlaf-Wach-Befindens im Laufe von Jahren verändern kann (Schneider-Helmert, 1987 und 1988 b).

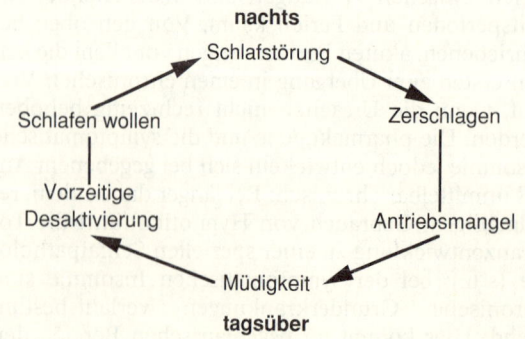

nachts

Schlafstörung

Schlafen wollen

Zerschlagen

Vorzeitige
Desaktivierung

Antriebsmangel

Müdigkeit

tagsüber

Abb. 40–1. Chronische Insomnie.

40.2.4 Hypnotikaabhängigkeit

Bei chronischer Insomnie ist heute in mehr als der Hälfte der Fälle mit einer Hypnotikaabhängigkeit zu rechnen, meist infolge zu langer, kontinuierlicher Verordnung. Die Abhängigkeit nimmt offensichtlich mit dem Alter der Patienten zu (Nolan und O'Malley, 1988). Nach einer Feldstudie an 1020 über 65jährigen Personen in England hatten 12% seit über einem Jahr Schlafmittel verordnet bekommen (Morgan et al., 1988), und in Allgemeinpraxen in Australien wurde bei 20% der über 70jährigen Patienten eine Benzodiazepinabhängigkeit diagnostiziert (Lyndon und Russell, 1988).

Der Grund für diesen Anstieg im Alter dürfte sein, daß dem weniger stabilen Altersprofil des Schlafs nicht genügend Rechnung getragen und damit zu häufig eine Insomnie angenommen wird, und daß man bei der mit steigender Morbidität zunehmenden sekundären Insomnie zu rasch ein Hypnotikum einsetzt, statt zunächst die Grunderkrankung zu behandeln. Die Niedrigdosisabhängigkeit von Benzodiazepinen, die heute im Vordergrund steht, ist also mehrheitlich eine iatrogene Komplikation. Sie betrifft nicht nur Hypnotika, sondern auch Tranquilizer und Anxiolytika sowie Kombinationspräparate, z.B. Benzodiazepine mit Kardiaka und Analgetika. Im Tierversuch kann eine körperliche Abhängigkeit schon in wenigen Tagen erzeugt werden (Cumin et al., 1982). Beim Menschen scheint diese Zeit variabler zu sein. Nach zuverlässigen Beobachtungen kann sie innerhalb von 3 Wochen liegen. Der Verlust der hypnotischen Wirkung infolge der Toleranzentwicklung ist ein Charakteristikum der Abhängigkeit. Wie für Hypnotika der älteren Generationen wurde er auch für die Benzodiazepine polygraphisch dokumentiert (Schneider-Helmert, 1988 c). Gleichzeitig zeigte sich, daß unter der Medikation die Vortäuschung einer hypnotischen Restwirkung besteht, wohl wegen der weiterhin vorhandenen Bewußtseins- und Gedächtnisdämpfung. Der Schlaf wird jedoch in diesem Stadium durch eine Suppression des Delta- und REM-Schlafs, also der für die Erholung wichtigsten Stadien, zusätzlich in seiner Qualität beeinträchtigt. Im Lichte dieser Erkenntnisse hat die Klasse der pharmakogenen chronischen Insomnien bei einer Niedrigdosisabhängigkeit eine wesentliche Präzisierung erfahren. Ihre gravierendsten Folgen liegen nicht im somatischen, sondern im psychischen Bereich: Die Patienten empfinden ein schwer zu definierendes Unbehagen, fühlen sich in ihrer Freiheit eingeengt, haben oft das Gefühl, nicht mehr richtig sie selbst zu sein (Persönlichkeitsveränderung), klagen über Gedächtnisschwund oder über eine emotionale Abstumpfung.

40.2.5 Differentialdiagnose

Im Bereich der Schlafpathologie sind die Störungen, die sich im typischen Fall mit einer **erhöhten Schläfrigkeit am Tag** manifestieren, eine wichtige Gruppe.

Die Tagesschläfrigkeit führt neben der Gefahr von Unfällen zu massiven Behinderungen der Sozialaktivitäten bis hin zu einer Isolierung. Die wichtigsten Syndrome sind die **Narkolepsie** und die **Schlafapnoe.** Pathognomonisch für die Narkolepsie sind die Kataplexie (akuter Tonusverlust als Reaktion auf starke Emotionen) sowie paroxysmal einschießender REM-Schlaf, für die Schlafapnoe sind es die Atemstillstände im Schlaf. Zusätzlich kommt es bei beiden Syndromen zu Störungen des Nachtschlafs im Sinne der Insomnie. Bei der Schlafapnoe stehen sie in einem direkten Zusammenhang mit der Grundstörung, da eine Aufwachtendenz die Voraussetzung für das Wiedereinsetzen der Respiration ist. Es ereignen sich also ebenso viele Schlafunterbrechungen wie Apnoen, in schweren Fällen bis zu Hunderten pro Nacht. Sie werden als die Ursache der abnormen Schläfrigkeit am Tag angesehen. Manche Patienten bemerken das häufige Aufwachen jedoch gar nicht. Daher bieten sie zuweilen ein paradoxes Bild dar, das neben scheinbar typischen Folgesymptomen einer Insomnie im Tagesbefinden die Klagen über Störungen des Nachtschlafs vermissen läßt. Ein weiterer Unterschied zur Insomnie ist, daß Narkolepsie- und Schlafapnoe-Patienten am Tag jederzeit rasch einschlafen. Die Schlafapnoe nimmt ab dem mittleren Lebensalter beim männlichen Geschlecht zu, erreicht bei spätem Auftreten aber meist keinen schweren Grad. Deutlich verstärkt wird sie durch Alkohol und Sedativa. In unserem Zentrum fanden wir bei männlichen Insomnie-Patienten über 45 Jahren in 15% eine begleitende Schlafapnoe, und davon standen 60% unter einer Dauerbehandlung mit Hypnotika, waren also einer erhöhten somatischen Gefährdung ausgesetzt. Hypnotika verstärken bei Schlafapnoe und Narkolepsie auch die Einschlafneigung am Tag.

Das **Syndrom der verspäteten Schlafphase** ist ein gegenüber der sozialen Umgebung verschobener Rhythmus. Seine Qualifizierung als krankhaft hängt stark von psychosozialen Faktoren ab. Die Interferenzen mit der Umgebung provozieren oft zusätzlich eine Insomnie. Umgekehrt kann das Syndrom bei stark Schlafgestörten nur scheinbar vorhanden sein, wenn sie nämlich am Morgen so zerschlagen sind, daß sie noch bis neun, zehn Uhr im Bett liegen bleiben.

Weil einerseits viele Patienten mit chronischer Insomnie eine gewisse Depressivität entwickeln, andererseits **Depressionen** oft eine Insomnie und Antriebsschwäche nach sich ziehen, werden primäre Insomnien oft als sekundäre Insomnien im Rahmen larvierter Depressionen eingestuft. In diesen Fällen haben Antidepressiva jedoch keine nennenswerte Wirkung, weder auf die Stimmungslage noch auf den Schlaf. Die Abgrenzung sollte aber idealerweise nicht ex juvantibus, sondern von Anfang an im diagnostischen Verfahren gesucht werden. Gleichförmigkeit der Symptome über lange Zeit spricht für Insomnie, phasenweises Auftreten für Depression. Ausgeprägte Stimmungsschwankung im Tagesverlauf mit dem Tief am Morgen spricht für Depression, Aktivitätsverlust am Abend für Insomnie. Hingegen ist das Früherwa-

chen kein zuverlässiges Unterscheidungsmerkmal (Hauri und Fisher, 1984). Weitere Faktoren, wie Persönlichkeit, Anamnese, auslösende Situation und Therapieerfahrungen, sollten in die differentialdiagnostischen Erwägungen einbezogen werden.

Mit relativ einfachen und daher kaum behindernden, **polygraphischen Registrierungen** kann der Verlauf des Nachtschlafs heute in einem gut eingerichteten Schlaflabor unter normalen Bedingungen registriert werden, so daß ein so gewonnenes Somnogramm den habituellen Verlauf im Sinne einer **Funktionsdiagnostik** repräsentativ wiedergibt. Sie wird heute vermehrt auf die Wachfunktionen ausgedehnt, so daß der ganze 24-Stunden-Zyklus im Zusammenhang beurteilt werden kann (Hermann-Maurer et al., 1989). Eine solche Spezialuntersuchung sollte nicht allzu zurückhaltend veranlaßt werden, auch wenn es derzeit in Europa noch zuwenig entsprechende Institutionen gibt. Sie ist sicherlich indiziert, wenn differentialdiagnostische Schwierigkeiten bestehen, bei langwierigem Krankheitsverlauf, bei Therapieresistenz, wenn die psychosomatische Situation undurchschaubar ist, Verdacht auf kombinierte Störungen aus dem Bereich der Schlafpathologie besteht, eine Diskrepanz zwischen den Klagen und den vermutlich objektiven Störungen besteht, oder wenn es um die Frage geht, welche anderen psychischen Störungen involviert sein könnten. Ein Verdacht auf Schlafapnoe oder Narkolepsie muß polygraphisch überprüft werden. Ferner erlauben die neuesten, oben referierten Befunde auch eine objektive Diagnose einer Benzodiazepinabhängigkeit und ihres Schweregrades auf der körperlichen Ebene und geben damit wichtige Richtlinien für eine Entzugsbehandlung (Schneider-Helmert, 1988c).

40.2.6 Psychosomatische Therapie

Aufgrund der dargestellten Interaktionen zwischen Schlafverhalten und Wachbefinden, zwischen willentlicher Einstellung und unbewußten Funktionen sowie zwischen evidenten Kausalzusammenhängen und schwer durchschaubaren, eigendynamischen Entwicklungen der Störungen des Schlafs ist es klar, daß man dem Patienten nur mit einer ganzheitlichen Sicht gerecht wird. Zunächst geht es darum, ihn aufgrund einer sorgfältig erstellten Diagnose über die Zusammenhänge aufzuklären. Dann ist ein Therapiekonzept zu erstellen. Liegen Komplikationen oder Begleitstörungen der Insomnie vor, müssen die therapeutischen Prioritäten erwogen und mit dem Patienten besprochen werden.

Die Psychotherapie, z.B. Gesprächstherapie, Fokaltherapie, hat absolute Präferenz, wenn abgrenzbare, aktuelle Störfaktoren und psychische Probleme die Symptomatik hervorgerufen haben oder weiter unterhalten, weil sie in dieser Situation die Kausaltherapie ist. Bei einer ausgesprochenen **Einschlafstörung** haben verschiedene Techniken der Verhaltenstherapie ihre Berechtigung, sei es das autogene Training (Nicassio und Bootzin, 1974), die progressive

Relaxation (Bootzin und Nicassio, 1978; Nicassio und Bootzin, 1974) oder die spezifischer auf Schlafstörungen ausgerichtete „stimulus control" (Bootzin und Nicassio, 1978), während sich das Biofeedback als wenig wirksam erwies (Hauri, 1981). Bei der Stimuluskontrolle wird der Patient angewiesen, im Bett nichts anderes zu tun als Sex und Schlafen; liegt er länger als 5 Minuten wach, soll er wieder aufstehen. Bei Patienten mit chronischer Insomnie und dem Hauptproblem einer **Durchschlafstörung** ist zu berücksichtigen, daß der oben beschriebene Circulus vitiosus weder mit psychologischen Mitteln noch mit sedativen Pharmaka allein zu sprengen ist, sondern nur mit Mitteln, die in der Lage sind, wieder physiologische Schlaffunktionen zu induzieren. Das ist möglich mit physiologischen Substanzen (vgl. Abschn. 40.1.1), nämlich L-Tryptophan, dem Baustein von Serotonin, und DSIP (Schneider-Helmert, 1986). Letzteres ist allerdings noch nicht als Medikament registriert, sondern zur Zeit in klinischer Prüfung. Beide Mittel haben keine zwingende, sondern mehr eine optionelle Wirkung. Daher sind flankierende psychotherapeutische Maßnahmen unabdingbar. Sie sollen Hindernisse, die der Option dieser Substanzen entgegenstehen, beseitigen und zugleich späteren Rezidiven vorbeugen. Auf der Ebene der inneren Einstellung des Patienten ist eine geduldige und entspannte Haltung anzustreben. Auch kleine Behandlungsfortschritte sollten ein Gegengewicht zur Hoffnungslosigkeit sein. Fehlverhalten bezüglich Schlaf ist zu korrigieren, besonders die verständliche, aber kontraproduktive Selbstbeobachtung (z.B. sollten Schlafgestörte die Uhr aus dem Schlafzimmer verbannen, mehr darauf achten, wie sie sich am Tag fühlen, als die Schlafdauer auszurechnen, und sich nicht vom subjektiven Eindruck des Nachtschlafs für den Tag bestimmen lassen). Es hat sich nämlich gezeigt, daß kein direkter Zusammenhang zwischen Schlafdauer in der Nacht und dem Befinden am folgenden Tag besteht (Berry und Webb, 1985; Taub und Berger, 1973). Der Patient soll daher lernen, seinen Tag unabhängig vom momentanen Schlaf möglichst gut zu strukturieren und am Entwurf auch dann festzuhalten, wenn er gerade wieder einmal besonders schlecht geschlafen hat (Hermann-Maurer, 1989). Dies hilft ihm überdies, der momentan zwar erleichternden, langfristig aber riskanten Zuflucht zu Sedativa zu widerstehen.

Die Behandlung der chronischen Insomnie ist also eine psychosomatische Therapie. Sie erfordert eine langfristige Perspektive, ein Konzept, das sich den Entwicklungen anpassen kann, ohne daß die Konsequenz und Geduld in der Führung des Patienten ins Wanken gerät.

Diese Richtlinien können an den exemplarischen Fällen folgendermaßen konkretisiert werden: Bei der Patientin ist die Ursache der Insomnie schon längst von selbst erloschen, ein relevantes, die Insomnie unterhaltendes Problem liegt nicht vor. Die einzige Chance, den Teufelskreis von Schlafstörung und beeinträchtigtem Tagesbefinden zu durchbrechen, bietet eine Therapie, welche die Schlafphysiologie unterstützt, also L-Tryptophan. Um dessen Wirkung zu ermöglichen, muß eine Entkrampfung der Patientin dem Schlaf gegenüber herbeigeführt werden, z.B. mit der Anweisung, sich bei Wachphasen in der Nacht abzulenken oder Entspannungsübungen zu machen. Ferner sollte sie einen regelmäßigen Schlaf-Wach-Rhythmus einhalten, unabhängig vom momentanen Befinden. – Beim Patienten (Betriebsleiter) ist nicht auszuschließen, daß berufliche Probleme immer noch eine Rolle für die Insomnie spielen, wenngleich sie gegenüber der auslösenden Situation wohl deutlich geringer sind. Doch diese Frage läßt sich erst angehen, wenn man in einem schichtweisen Abbau der aufgepfropften Komplikationen an die grundlegenden psychosomatischen Störbereiche gelangt ist. Der Patient sollte zunächst zur Einsicht kommen, daß sich ein Hypnotika- und Alkoholabusus entwickelt hat, der das prioritäre Behandlungsproblem darstellt. Daraus sollte er sowohl die Motivation als auch die Geduld für eine langfristige Behandlung gewinnen. Der Entzug erfolgt nämlich sehr langsam nach einem vom Arzt zusammen mit dem Patienten im voraus festgelegten Stufenplan. Beispielsweise wird die Dosis alle 3 Wochen um ein Viertel reduziert, jeweils eine Woche vor dem Konsultationstermin. So können allfällige Entzugssymptome in Grenzen gehalten werden (kleine Schritte, aber konsequent) und jeweils rasch nach ihrem Auftreten besprochen werden. Da in diesem Falle noch eine, ohne Hypnotika allerdings geringe, Schlafapnoe bestand, ist L-Tryptophan für die Behandlung des Grundproblems besonders indiziert, denn es kann die Schlafapnoe sogar günstig beeinflussen. – In der Praxis erfordert die Behandlung so langwieriger Störungen und komplexer Probleme, wie bei diesen beiden Patienten, meist viele Monate konsequenter Behandlung, so daß man sich in schweren oder gar therapieresistenten Fällen vielleicht eher für die 2wöchige, stationäre Intensivtherapie mit DSIP entschließt. Sie fügt sich gleichermaßen in den Rahmen einer ganzheitlichen psychosomatischen Therapie ein. DSIP ist aber zur Normalisierung der Schlaffunktionen im Vergleich zu L-Tryptophan zuverlässiger, stärker und schneller in der Wirkung. Außerdem hat es sich zur Entzugstherapie bei Hypnotikaabhängigkeit sehr gut bewährt.

41 Herz- und Kreislauferkrankungen

41.1 Arterielle Verschlußkrankheiten: koronare Herzkrankheit, Apoplexie und Claudicatio intermittens

Thomas H. Schmidt, Rolf Adler, Wolfgang Langosch und *Michael Rassek*

41.1.1 Exemplarischer Fall

Ein 57jähriger Schriftsetzermeister verspürt an einem Wochentag gegen 17.00 Uhr ein heftiges, schmerzhaftes Beklemmungsgefühl, das reifenförmig den Brustkorb umfaßt und mindestens 10 Minuten anhält. Nachdem es abgeklungen ist, fährt er mit dem eigenen Wagen zu seinem Sohn, um ihm beim Umzug in eine neue Wohnung zu helfen. Bei dieser Arbeit treten die gleichen Beschwerden wieder auf, diesmal begleitet von Schweißausbrüchen. Nach ihrem erneuten Abklingen begibt er sich nach Hause. Dort tritt heftiges Stechen und Druckgefühl über der linken Brustseite auf, das in den linken Arm ausstrahlt und nicht mehr verschwindet. Die Ehefrau benachrichtigt den Arzt, der den Patienten unter Verdacht eines Herzinfarktes in die Klinik einweist, wo die Diagnose bestätigt wird (EKG, Fermentablauf).

Der Patient ist das ältere von zwei Kindern. Die Mutter war nie berufstätig. Sie verstarb 72jährig an einem Schlaganfall. Der Patient erinnert sich vor allem daran, daß sie seine Versetzung als Soldat von der Ostfront nach Hause zur Fliegerabwehr durchgesetzt habe. Der Vater war Eisengießer. Er verstarb mit 44 Jahren nach 1½jähriger Bettlägerigkeit an „Venenentzündung" und Lungenkrankheit. Der Patient war zu diesem Zeitpunkt 15 Jahre alt. Der Vater sei ein fleißiger, strebsamer, sehr angesehener Mann gewesen, der ihn streng und gerecht erzogen hätte. Der um zwei Jahre jüngere Bruder sei immer der weichere gewesen. Er selber arbeite jetzt als Schriftsetzer seit 25 Jahren im gleichen Betrieb, in den letzten 15 Jahren als Abteilungsleiter. Seit 30 Jahren sei er verheiratet und habe einen jetzt 27jährigen Sohn. Sich selbst empfindet der Patient als korrekten Vorgesetzten, der darauf angewiesen sei, daß im Betrieb alles wie am Schnürchen laufe. Er könne auf nichts lange warten, sonst wäre er gleich aufgeregt, unruhig oder niedergeschlagen. Seine ganze Liebe gelte am Feierabend seiner handwerklichen Tätigkeit, die er sehr genau und sorgfältig ausführe.

Vor 1½ Jahren sei er als Leiter in eine andere Abteilung versetzt worden, wo häufig kurzfristig angesetzte Termine eingehalten werden müßten. Seit dieser Zeit wären bei Aufregungen Druck auf der Brust und Magenbeschwerden aufgetreten. Außerdem hätte die Potenz seither nachgelassen. Wegen dieser Beschwerden sei drei Monate vor dem jetzigen Ereignis eine 4wöchige Kur durchgeführt worden. Seitdem rauche er auch nicht mehr. 3 Tage vor dem Infarkt sei er einer außerordentlichen Terminhetze ausgesetzt gewesen. Ein großer Auftrag eines wichtigen Kunden hätte wegen der schlechten wirtschaftlichen Lage unbedingt in besonders kurzer Zeit von seiner Abteilung erledigt werden müssen. Das Gefühl habe ihn beherrscht, daß es nicht klappen werde, wenn er nicht hinter allem her sei.

Der Patient rauchte bis zur Kur vor 3 Monaten über 30 Zigaretten pro Tag, seine Blutdruckwerte betrugen während des Spitalaufenthalts bis zu 165 mm Hg systolisch und 100 mm Hg diastolisch, dazu bestand eine Hyperurikämie.

Bemerkungen zum exemplarischen Fall

Der Patient erlebt sich selbst als harten, fleißigen Arbeiter, der hohe Anforderungen an sich und seine Mitarbeiter stellt. Ein großes Verantwortungsgefühl kennzeichnet ihn und eine Neigung, seine Mitarbeiter zu kontrollieren. Er reagiert mit Verunsicherung, wenn ihm diese Kontrolle zu entgleiten droht. Dazu fällt sein rastloser Tätigkeitsdrang auf, der auf das Erreichen von Zeitlimits ausgerichtet ist. In den letzten 18 Monaten vor dem Infarkt wurde sein Verantwortungsgefühl besonders belastet. Das Bestreben, Zeitlimits zu erfüllen, erreichte in den letzten drei Tagen vor dem Infarktereignis einen Höhepunkt.

41.1.2 Historischer Rückblick

Berichte über den plötzlichen, unerwarteten Tod von Menschen in Situationen starker emotioneller Erregung wie Angst, Ärger oder Wut übten und üben immer wieder große Faszination aus. „Mein Leben liegt in der Hand eines jeden Rüpels, der es darauf anlegt, mich in Wut zu bringen!" klagte der berühmte englische Chirurg John Hunter, der an einer koronaren

Herzkrankheit litt. Er erkannte in seinen Gefühlen und in seinem Verhalten die Auslöser seiner Angina-pectoris-Schmerzen. Kurze Zeit später, 1793, starb er nach einer erhitzten Auseinandersetzung mit einem Kollegen im St. George's Hospital in London. DeBakey und Gotto (1977) zitieren Sir William Oslers Bericht über Hunters Tod: „In silent rage and in the next room he gave a deep groan and fell down dead". Eine Reihe anderer Ärzte im 18. und 19. Jahrhundert haben ebenfalls emotionale Ausbrüche als auslösende Faktoren für einen plötzlichen Herztod beschrieben, und sie sahen Ärger buchstäblich als Risikofaktor für dieses Ereignis an (Heberden, 1772; Fothergill, 1781; Wardrobe, 1851; Trousseau, 1882). Überhaupt ranken sich sehr viele und häufig zitierte Anekdoten um die Beziehung zwischen koronarer Herzkrankheit und insbesondere dem plötzlichen Herztod und Ärger, Feindseligkeiten und aggressivem Verhalten (Diamond, 1982).

Dies waren, historisch gesehen, aber nicht die einzigen Zusammenhänge, die für die Pathogenese der koronaren Herzkrankheit als wichtig erachtet wurden.

In seinem 1868 erschienenen Lehrbuch der Herzkrankheiten beschreibt der Heidelberger Kliniker Theodor von Dusch als Ursachen der zu seiner Zeit seltenen Angina pectoris neben erblichen Anlagen weitere Prädispositionen: „... doch übt hier die Lebensweise sicherlich einen großen Einfluß aus; denn namentlich leiden oft wohlhabende und reiche Leute an Angina pectoris, welche – den Genüssen einer reichen und luxuriösen Tafel ergeben, ohne zugleich die nötige körperliche Bewegung zu haben – zu einer bedeutenden Fettleibigkeit gelangen. Man hat ferner beobachtet, daß fortgesetzte leidenschaftliche Aufregungen, heftiges lautes Reden, Spiel, Nachtarbeiten und Nachtwachen zu dem in Frage stehenden Übel disponieren".

Osler beschreibt gefährdete Personen als temperamentvolle und außerordentlich ehrgeizige Menschen, die alles energisch anpacken: „whose engine is always at full speed ahead". 1910 hielt er seine Beobachtungen wie folgt fest:

„In a group of 20 men, every one of whom I knew personally, the outstanding feature was the incessant treadmill of practice; and yet if hard work – that ‚badge of all our tribe' – was alone responsible would there not be a great many more cases? Every one of these men had an added factor – worry; in not a single case under 50 years of age was this feature absent. Listen to some of the comments which I jotted down of the circumstances connected with the onset of attacks: ‚A man of great mental and bodily energy, working early and late in a practice, involved in speculations in land'; ‚troubles with the trustees of his institution'; ‚lawsuits'; ‚domestic worries'; and so through the list. At least six or seven men of the sixth decade were carrying loads light enough for the fifth but too much for a machine with an everlessening reserve".

Dunbar (1943, 1959) schilderte Patienten mit koronarer Herzkrankheit als hart arbeitend, ihrer Aufgabe ergeben, immer auf den Erfolg gerichtet, als Märtyrer der eigenen Ideale, die ihre Urteile unabhängig und selbständig fällen. Arlow (1949) betonte zusätzlich

die traumatische Wirkung des persönlichen Versagens in der Periode vor dem Infarktereignis. Schneider (1956) beschrieb seine Infarktpatienten als unter einem „eigenartigen" Druck stehend, immer in körperlicher Hast, die, wenn sie blockiert werden, zu Wutausbrüchen neigen. Sie waren von Ehrgeiz besessen, dabei aber außerordentlich verletzlich gegenüber Beschämung und Herabsetzung, was zu heftigen Gefühlen von Schuld und Wertlosigkeit führte. Wolf (1958) bezeichnete seine Myokardinfarktpatienten als Sisyphus-Typen, verurteilt, immer wieder die gleiche Arbeit leisten zu müssen, ohne Aussicht auf endgültigen Erfolg und Befriedigung, wenn sich Erfolge einstellen sollten. Van der Valk und Groen (1967) und van Heijningen und Treurniet (1966) gaben sehr ähnliche Beschreibungen ihrer Patienten.

Russek (1959) charakterisierte seine Patienten als Opfer eines nicht nachlassenden Antriebes, eines intensiven Sehnens nach Anerkennung und eines tiefen Gefühls von Verpflichtetsein. Am typischsten fand er eine Ruhelosigkeit in Mußestunden und ein Schuldgefühl in Zeiten der Entspannung.

Groen und Mitarbeiter begannen 1965 Studien mit ausführlichen biographischen Anamnesen von Herzinfarktpatienten. Sie stellten die Hypothese auf, daß das Infarktereignis gewissermaßen als das Ergebnis eines Zusammentreffens von drei Faktoren angesehen werden kann:

- Spezifische, durch Erbfaktoren und die individuelle Lerngeschichte bestimmte Persönlichkeitsmerkmale, die sich ähnlich den oben beschriebenen in Arbeitseifer, Ehrgeiz, Verantwortungsbewußtsein und Dominanzstreben ausdrücken.
- Diese führen bei einigen Individuen eher als bei anderen zu zwischenmenschlichen Konfliktsituationen in Familie oder Beruf, die als eigentlicher Stressor der Erkrankung unmittelbar vorausgehen.
- Hinzu kommt, wie die Betroffenen infolge ihrer Persönlichkeit auf diesen Konflikt reagieren. Sie zeigen ihren Ärger oder ihre Niedergeschlagenheit nicht offen, etwa durch kämpferisches Verhalten, Schimpfen oder Klagen; sie verbergen vielmehr ihre Gefühle, geben sich einen optimistischen oder gleichgültigen Anschein und verleugnen die innere Spannung.

Die psychosomatische Ätiologie der Erkrankung ist aufgrund dieser Hypothesen dreifach spezifisch: Personen mit den beschriebenen Persönlichkeitsmerkmalen werden keinen Herzinfarkt bekommen, wenn ihr übertriebenes Dominanzstreben in der Familie und im Beruf belohnt wird und von Erfolg gekrönt ist. Ein Konflikt am Arbeitsplatz oder in der Familie wird ebenfalls nicht zur Erkrankung führen, wenn die Gefühle frei ausgedrückt werden können. Erst das Zusammentreffen der drei Faktoren: **Persönlichkeit, zwischenmenschlicher Konflikt** und **unterdrücktes Verhalten** ruft – den Autoren zufolge – diese Krankheit hervor (Groen, 1976, 1985).

Wenngleich alle diese Berichte und Schilderungen auch aus heutiger Sicht wichtigen Hinweischarakter besitzen, blieben sie doch eher anekdotenhaft; immerhin konnten sie zur Hypothesenbildung herangezogen werden. Oft fehlten aber geeignete Meßinstrumente und verläßliche Untersuchungsstrategien. Erst die genaue und methodisch einwandfreie Erfassung der in Frage stehenden psychosozialen Faktoren und Verhaltensweisen kann in prospektiven Studien zu beweiskräftigeren, prädiktiven Aussagen führen. Die Ausarbeitung geeigneter Meßmethoden und Untersuchungsstrategien ist auch heute noch ein Zentralproblem dieser Forschungsrichtung. Dennoch sind Fortschritte erzielt worden. Insbesondere ermöglichte die Entwicklung quantitativer Techniken bei der Erfassung von Verhaltensweisen und psychosozialen Faktoren die Durchführung breit angelegter prospektiver epidemiologischer Studien. Einige der Hypothesen ließen sich klarer formulieren und konnten so auf ihre prädiktive Aussagekraft hin überprüft werden.

41.1.3 Multifaktorielles pathogenetisches Modell der koronaren Herzkrankheit und anderer arterieller Verschlußkrankheiten

Unser Wissen über das multifaktorielle Geschehen bei Ätiologie und Pathogenese der koronaren Herzkrankheit und anderer arterieller Verschlußkrankheiten ist lückenhaft. Die pathophysiologische Grundlage ist in der überwiegenden Mehrzahl der Fälle die Arteriosklerose. Zu thromboembolischen Verschlüssen kommt es vor allem an der arteriosklerotisch veränderten Gefäßwand. Aber auch Spasmen, beispielsweise im Bereich der Herzkranzgefäße (Prinzmetal-Angina) oder Schädigung der Hirngefäße, z.B. bei Blutdruckkrisen (sog. „break through"-Phänomen), spielen eine Rolle.

Eine der grundlegenden Theorien (Ross und Glomset, 1976) beschreibt die Entstehung der Arteriosklerose in folgenden Schritten:
– Endothelschädigung durch hämodynamische oder chemische Einflüsse (z.B. Katecholamine),
– Proliferation der glatten Muskelzellen in der Arterienwand als Reaktion auf die Verletzung,
– Akkumulation von Lipoproteinen und anderen Zellen (Plaques) am Ort der Verletzung.
Eine akute Erkrankung tritt gewöhnlich als Folge einer plötzlichen Unterbrechung des Blutflusses beispielsweise im Herzmuskel auf. Die wichtigsten, heute gesicherten Risikofaktoren für die Entwicklung einer Arteriosklerose sind Alter, männliches Geschlecht, Zigarettenrauchen, Erhöhung von Blutdruck und Serumcholesterin. Die meisten Untersuchungen mit Berücksichtigung auch psychosozialer Einflüsse liegen heute für die koronare Herzkrankheit vor, die deswegen bevorzugt dargestellt wird.

In den meisten Industrienationen kam es im zwanzigsten Jahrhundert zu einem erheblichen Anstieg der koronaren Herzkrankheit. Aufgrund vor allem amerikanischer epidemiologischer Studien wurde nach dem zweiten Weltkrieg das multifaktorielle ätiologische Konzept der sog. physiko-chemischen koronaren Risikofaktoren entwickelt (Schaefer und Blohm-

ke, 1977). Keys (1970) weist darauf hin, daß sich die unterschiedliche Inzidenz der koronaren Herzkrankheit (KHK) in verschiedenen Ländern aber mit diesem Konzept nicht zufriedenstellend erklären läßt. Epidemiologische Studien zeigen, daß die Beziehung zwischen diesen traditionellen Risikofaktoren und der KHK bei verschiedenen untersuchten Bevölkerungsgruppen nicht gleichartig ist. Obwohl beispielsweise Europäer und die Bewohner der Stadt Framingham (Massachusetts, USA), die in der bekannten Framingham-Studie untersucht wurden, die gleiche Verteilung traditioneller Risikofaktoren aufweisen, ist das KHK-Risiko in Framingham doppelt so hoch. Bei der Assoziation dieser Risikofaktoren mit der KHK bleibt deswegen ein großer Teil der Varianz ungeklärt (Burch, 1980; Werkö, 1976).

Die traditionellen koronaren Risikofaktoren klären nur etwa 50% der Varianz der KHK-Inzidenz auf (Keys, 1970; Marmot und Winkelstein, 1975). Deswegen liegt die Annahme nahe, daß es weitere Risikofaktoren gibt, die zu identifizieren Aufgabe der epidemiologischen Forschung ist (Jenkins, 1977). Die Hypothese einer multifaktoriellen Genese der KHK ist heute nicht mehr nur auf physiko-chemische Faktoren beschränkt, sondern schließt psychosoziale Variablen als wichtige Determinanten für die Entstehung einer KHK ein (Jenkins, 1971, 1976, 1982; Waltz, 1981). Die psychosoziale Herzinfarktforschung beschäftigt sich hierbei zum einen mit der Frage, inwieweit ein aufgrund bekannter traditioneller Risikofaktoren bestehendes Erkrankungsrisiko durch eine Akkumulation psychosozialer Belastungen erhöht wird (Langosch et al., 1982), und zum anderen, inwieweit psychosoziale Determinanten den Genesungsverlauf und die Rehabilitationsbemühungen beeinflussen (Waltz, 1981; Badura, 1981). Darüber hinaus ist eine wichtige Frage, inwieweit psychosoziale Faktoren traditionelle Risikofaktoren wie Bluthochdruck (vgl. auch Kap. 41.3), Höhe des Serumcholesterinspiegels, Rauchgewohnheiten oder Adipositas (vgl. auch Kap. 38.1) beeinflussen.

41.1.3.1 Ein psychosomatisches Modell

Kagan und Levi (1975) haben ein allgemeines Modell der Krankheitsentstehung unter dem Einfluß psychosozialer Faktoren vorgeschlagen, das insbesondere auch die Fragestellungen und Probleme der psychosomatischen Herzinfarktforschung veranschaulicht (Abb. 41.1–1). Die Erkrankung wird hier als das Ergebnis einer Reaktionskette dargestellt:

Den Ausgangspunkt bilden soziale Faktoren und Prozesse. Sie spielen sich in der Familie, am Arbeitsplatz, in der Schule oder Nachbarschaft ab, und sie betreffen die verschiedensten sozialen Aktivitäten in diesen Strukturen. Das Individuum erlebt und verarbeitet diese Vorgänge (1, psychosoziale Reize). Das Individuum ist durch psychobiologische Programme charakterisiert (2), die durch frühkindliche Umgebungseinflüsse bzw. die gesamte individuelle Lerngeschichte sowie genetische Faktoren bestimmt sind.

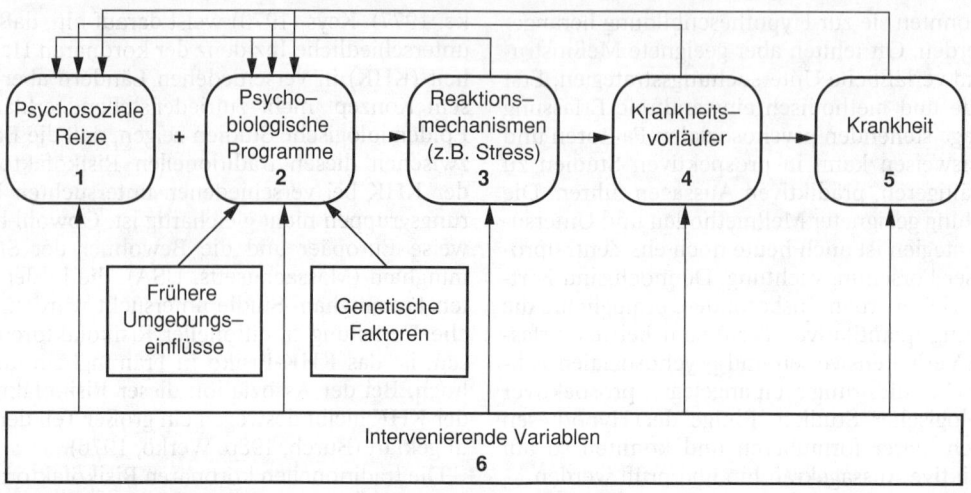

Abb. 41.1–1. Hypothetisches Modell der psychosozialen Krankheitsgenese (nach Kagan und Levi, 1975).

Die Wechselwirkung zwischen den entsprechenden sozialen Reizen und den psychobiologischen Programmen ruft Streßreaktionen hervor, die unspezifisch oder auch mehr oder weniger spezifisch sein können. Diese Streßreaktionen können zu Krankheitsvorläufern (4) und möglicherweise zur Krankheit selbst (5) führen. Eine Reihe von intervenierenden Variablen (6) können diesen Ablauf der Ereignisse modifizieren, sei es verstärkend oder mildernd. Es handelt sich um ein komplexes kybernetisches System mit vielfältigen Rückmeldesystemen und Einflußmöglichkeiten.

Bei der Erkrankung an Herzinfarkt beispielsweise wird heute angenommen, daß genetische und frühkindliche Umwelteinflüsse die Entwicklung der Persönlichkeit so beeinflussen, daß sie in bestimmten Situationen unter psychischen Streß gerät. Dieser zieht physiologische und biochemische Prozesse nach sich, die zu Krankheitsvorläufern und schließlich zur Krankheit führen. Bestimmte Organe erkranken möglicherweise, weil die Faktoren, die zur Ausprägung spezifischer Persönlichkeitsmerkmale beigetragen haben, auch die Verletzlichkeit des betreffenden Organs bedingen. So wird ein somatopsychisch-psychosomatisches Geschehen vermutet. Die Ursachen der koronaren Herzkrankheit liegen nach diesem Modell in einem schlechten „person-environment fit", wie Levi es nennt, der in Wechselwirkung mit genetischen Faktoren steht. Wie bei einem schlecht passenden Schuh entsprechen die Umgebungsbedingungen nicht den individuellen Bedürfnissen und Erwartungen; dies kann sowohl an den Umgebungsbedingungen als auch an den Eigenschaften des Individuums oder der sozialen Gruppe liegen. Der Nachweis, daß psychische Faktoren zur Entstehung arterieller Verschlußkrankheiten beizutragen vermögen, kann für einzelne Schritte erbracht werden. Da es methodisch kaum möglich ist, Menschen über ausreichend lange Zeit psychologisch und unabhängig davon physiologisch und biochemisch genügend genau zu begleiten,

werden die einzelnen Zusammenhänge wohl nie lückenlos analysierbar sein.

Zur wissenschaftlichen Erhärtung dieses hypothetischen Modells gilt es, folgende Fragen zu beantworten:
– Welche Lebensumstände sind es, die vom betreffenden Individuum als Stressoren erlebt werden, und die beispielsweise einen akuten Myokardinfarkt auslösen können?
– Von welchen Persönlichkeitsmerkmalen, von welchen Bewältigungsstrategien hängt es ab, daß entsprechende Belastungen pathogenetische Bedeutung bekommen? Welche genetischen und frühkindlichen Einflüsse bestimmen die Entstehung dieser Persönlichkeitsmerkmale?
– Welche physiologischen und biochemischen Folgen ziehen die Reaktionen des Individuums mit diesen speziellen Persönlichkeitsmerkmalen auf die besonderen Lebensumstände nach sich?
– Zu welchen Krankheitsvorläufern führen sie?
– Wie tragen diese Krankheitsvorläufer zur eigentlichen Krankheit bei?

Als mögliche Prädiktoren der arteriellen Verschlußkrankheiten werden im wesentlichen folgende psychosozialen Einflußgrößen diskutiert:
– ungünstige sozioökonomische Bedingungen und Mangel an sozialer Unterstützung,
– lebensverändernde Ereignisse,
– berufliche Überbeanspruchung,
– emotionale Probleme,
– sog. koronargefährdende Verhaltensweisen.
Klinisch treten vor allem drei Erscheinungsformen der KHK in den Vordergrund, die in vielen Studien mit einigen dieser Faktoren verknüpft wurden: Angina pectoris, Myokardinfarkt und plötzlicher Herztod. Erst wenige Untersuchungen liegen vor in bezug auf andere arterielle Verschlußkrankheiten wie Apoplexie und Claudicatio intermittens, die zunächst behandelt werden sollen.

41.1.4 Persönlichkeitsmerkmale, Lebensereignisse und periphere Arteriosklerose

Trotz der weitverbreiteten Meinung, daß zerebrovaskuläre Ereignisse durch heftige Emotionen, vor allem Wut, ausgelöst werden, und daß sie vornehmlich Individuen treffen, die gespannt sind und sich verhalten, wie wenn sie unter großem Druck stünden, liegen hierzu erst wenige und insbesondere keine kontrollierten Untersuchungen vor. Tuke (1872) beschrieb eine 56jährige Frau, die unmittelbar nach einer körperlichen Bedrohung eine Hirnblutung erlitt. Ecker (1954) erwähnte als erster derartige Zusammenhänge bei 20 Patienten mit Hirnschlägen: 13 hatten lang andauernde, durch ihre Persönlichkeit bedingte Schwierigkeiten hinter sich und große Schwierigkeiten mit aggressiven und feindseligen Gefühlen, und 15 hatten speziellen emotionellen Streß unmittelbar vor dem Hirnschlag durchgemacht. Sieben wiesen eine Thrombose auf oder eine Ischämie bei Spasmen und 13 eine intrazerebrale oder subarachnoidale Blutung.

Storey (1969) verfolgte 292 Patienten sechs Monate bis sechs Jahre nach Subarachnoidalblutung. Obwohl nicht speziell nach den Lebensumständen und dem psychischen Zustand unmittelbar vor dem Hirnschlag gefragt worden war, gaben 7 Patienten traumatische Umstände unmittelbar vor dem Ereignis an und 39 wiesen eine Anamnese mit emotionalen Störungen vor dem Hirnschlag auf. Ullman (1962), der Verhaltensänderungen nach dem Ereignis bei 300 Patienten untersuchte, beschreibt 6 Patienten, bei denen es unmittelbar vor dem ischämischen Hirnschlag zu tief aufwühlenden Lebenssituationen gekommen war. Daneben finden sich nur vereinzelte Fälle mit Beschreibung der psychischen Umstände vor dem Hirnschlag (Engel et al., 1953; Binger et al., 1955; Fisher, 1961; Hambling, 1961; Pool, 1961; Weiss, 1940).

Zusammenfassend entsteht der Eindruck, daß psychische Einflüsse gelegentlich zur Entstehung des Hirnschlages beitragen können. Zwei autobiographische Schilderungen von Schriftstellern (Hodgins, 1964; Wint, 1965), die Hirnschläge erlitten hatten, betonen tiefgehende emotionale Probleme in den Monaten vor dem Hirnschlag. In einer retrospektiven anamnestischen Studie von 32 Männern, die insgesamt 35 ischämische Hirnschläge erlitten hatten, fanden Adler und Mitarbeiter (1971), daß der Hirnschlag typischerweise in einer Periode anhaltender oder intermittierender und oft schwerer emotionaler Störungen auftrat, die wochen- bis monatelang gedauert hatten und manchmal kurz vor dem Ereignis intensiviert worden waren. Bestimmte Persönlichkeitsmerkmale fanden sich häufig. Sie umfaßten eine Verhaltensweise, die als „pressured" charakterisiert wurde: Ein Bedürfnis, selbstgesetzte Ziele zu erreichen mit einer Neigung, aktiv und ständig tätig zu sein. Eine Selbsteinschätzung als harter Arbeiter, hohe Ansprüche an sich selbst und hohes Verantwortungsgefühl, ebenso ein Gefühl der Zeitnot und ein Bedürfnis, Zeitlimits zu erfüllen und Ziele zu erreichen, Zielstrebigkeit

und starker Wille. Diese Persönlichkeitsmerkmale gleichen dem weiter unten beschriebenen sog. Typ-A-Verhalten, das als Risikofaktor der koronaren Herzkrankheit beschrieben worden ist. Bei den Apoplexiepatienten kamen Probleme in der Kontrolle von Ärgergefühlen hinzu, insbesondere in bezug auf Menschen, von denen sich der Patient abhängig fühlte. Die Beziehung zu diesen Menschen war bei ihnen durch starke Abhängigkeitswünsche charakterisiert. Entweder verleugneten sie diese durch selbstgenügsames und unabhängiges Verhalten oder entsprachen ihnen durch ein unterwürfiges Verhalten, mit dem der entsprechende Partner befriedigt oder besänftigt wurde. Der Hirnschlag trat in Zeiten auf, in denen der Patient mit Gefühlen von Ärger, Hoffnungslosigkeit und manchmal Scham reagierte, wenn er die an sich selbst gestellten Anforderungen nicht zu erfüllen vermochte, die anderen Personen nicht mehr kontrollieren konnte oder die Erwartungen anderer nicht mehr erfüllte. Eine Replikation dieser Studie durch Gianturco und Mitarbeiter (1974) konnte das „pressured pattern" im Verhalten von Schlaganfallpatienten nachweisen, die zuvor einen Myokardinfarkt erlitten hatten oder an Angina pectoris erkrankt waren.

Im Rahmen der prospektiven Framingham-Studie wurde eine Beziehung zwischen der 10-Jahres-Inzidenz des Schlaganfalls und verschiedenen psychosozialen Faktoren bei Männern und Frauen im Alter von 45–64 Jahren nachgewiesen (Eaker et al., 1981). Zwischen 1965 und 1967 erhielten 1317 gesunde Teilnehmer der Studie einen umfangreichen Fragebogen, der emotionale und psychische Einflußfaktoren erfaßte. Bei Frauen war die Schlaganfallinzidenz signifikant mit erhöhter emotionaler Labilität, Zurückhalten von Ärger sowie mit Symptomen von Spannung und Ärger verknüpft. Typ-A-Frauen erlitten häufiger einen Schlaganfall als Typ-B-Frauen; ihr relatives Risiko betrug 5,0 (p = 0,04). Dies galt auch für berufstätige Frauen in „blue collar"-Berufen im Vergleich zu „white collar"-Berufen oder Frauen, die als Büroangestellte arbeiteten. Berufstätige Frauen, die später einen Schlaganfall erlitten, berichteten auch vermehrt über Arbeitsüberlastung oder hatten Untergebene, von denen sie nicht unterstützt wurden.

Hausfrauen, die angaben, daß sie sich während des Tages nicht entspannen, hatten ein um mehr als 4mal höheres Risiko als Frauen, die sich entspannen konnten. Bei Männern bestand eine signifikante Beziehung zwischen Schlaganfallinzidenz und Angstsymptomen und grenzwertig auch zu Zeichen von Ärger. Typ-A-Männer wiesen im Vergleich zum Typ B ein mehr als 6mal höheres Risiko auf, wenn sie arbeitsmäßig überlastet waren. Dies galt auch für Männer, die glaubten, daß sie nur eine geringe Chance hätten, ein Einkommen in der angestrebten Höhe zu erhalten, im Vergleich zu denjenigen, die sich diesbezüglich gute Chancen ausrechneten. Mit Ausnahme der Symptome für Spannung bei Frauen und Ärger bei Männern und Frauen blieben alle diese Beziehungen zur Inzidenz des Schlaganfalls signifikant, auch wenn die Risikofaktoren Alter, systolischer Blutdruck, Serumcholesterin und Zigarettenrauchen rechnerisch berücksichtigt wurden. In einer kontrollierten, retrospektiven Untersuchung fanden Goetz et al. (1989) das „pressured pattern of behavior" bei 19 Frauen mit ischämischem Hirninfarkt (I) im Vergleich zu 19 Frauen mit nicht-vaskulären Leiden (II) und 19 weiblichen Spitalfreiwilligen (III) mit dem Instrument von Siegrist signifikant ausgeprägter, insbesondere die Dimensionen „Verausgabungsbereitschaft" (Einweg-Varianzanalyse, p = 0,01; I zu II: p = 0,001), „Perfektionsstreben" (p = 0,0001; I zu II: p = 0,01; I zu III: p = 0,0001) und „Distanzierungsunfähigkeit von der Arbeit" (p = 0,01; I zu II: p = 0,09; I zu III: p = 0,005).

In einer kontrollierten Studie (Cottier et al., 1983) wurden diese Persönlichkeitsmerkmale auch bei Patienten mit Claudicatio intermittens beschrieben, aber weniger ausgeprägt als bei Patienten, die neben der Claudicatio zusätzlich an einer koronaren, arteriellen Verschlußkrankheit litten. Dies galt sowohl für die mit Hilfe eines Interviews eingeschätzten Verhaltensweisen als auch für die mit Hilfe der Bortner-Skala, einem aus der Typ-A-Forschung bekannten Fragebogen, erhobenen Daten. Diskriminanzanalytisch unterschied der Bortner-Test die Gruppen unabhängig vom Rauchverhalten, das ebenfalls bei den Patienten am stärksten ausgeprägt war, die gleichzeitig an einer Claudicatio intermittens und einer koronaren Herzkrankheit litten. Stevens et al. (1984) fanden bei Männern und Frauen mit atheromatösen Läsionen der A. carotis signifikant häufiger das Typ-A-Verhalten als bei solchen mit intakter A. carotis. Joesoet et al. (1989) beobachteten eine signifikante positive Beziehung zwischen Arteriosklerose der Beinarterien und „hostility", einer aus dem MMPI abgeleiteten Komponente des Typ-A-Verhaltens.

41.1.5 Der plötzliche Herztod

Der plötzliche Herztod ist heute die häufigste Todesursache in den Industrienationen. Allein in den USA sterben täglich mehr als 1200 Personen daran; das sind jährlich rund 450 000 Menschen bzw. 25% aller Todesursachen (DeSilva und Lown, 1978). Bei der überwiegenden Mehrzahl aller Fälle besteht eine ausgeprägte koronare Herzkrankheit. Der plötzliche Herztod ist als unerwarteter natürlicher Tod kardialer Ursache definiert, der unmittelbar oder innerhalb von 6 Stunden (nach WHO-Definition 24 Stunden) nach dem Beginn der akuten Symptome auftritt (Eliot und Buell, 1983). Meist tritt der Tod aber in einem Zeitraum von 30 Sekunden bis zu einer Stunde auf und ist Folge von Rhythmusstörungen (Cobb et al., 1980; Friedman et al., 1973). Bei Erkrankten, die den Beginn der kardialen Symptomatik 24 Stunden überleben, liegt in der Regel ein Myokardinfarkt vor. In einer Studie von Rabkin und Mitarbeitern (1980) wird berichtet, daß der plötzliche Herztod montags besonders oft auftritt, nämlich, bezogen auf die Wochentage, in 26% aller Fälle. Dies macht den Einfluß deutlich, den hier wahrscheinlich Lebens- und Verhaltensweisen, emotionale Faktoren und akuter Streß

spielen. Opfer des plötzlichen Herztodes können nicht ohne weiteres von anderen KHK-Kranken unterschieden werden, da beide Gruppen eine hohe Prävalenz der gleichen Risikofaktoren miteinander teilen (Chiang et al., 1970; Doyle et al., 1976). Männer sind häufiger betroffen als Frauen und vor allem junge Männer weisen ein unverhältnismäßig hohes Risiko auf.

Cobb (1975, 1980) untersuchte Patienten, die im Gebiet von Seattle wiederbelebt und ins Krankenhaus gebracht wurden. Es ließen sich zwei Gruppen unterscheiden:

Bei der ersten Gruppe lag ein klassischer Myokardinfarkt vor; bei der zweiten Gruppe gab es keine Infarktzeichen in bezug auf EKG- und enzymatische Veränderungen. Bei der Infarktgruppe traten „wiederholte Episoden von Kammerflimmern" (recurrent sudden death) in etwa 2% auf. In der zweiten Gruppe ohne Infarktzeichen traten wiederholte Episoden von Kammerflimmern hingegen jährlich bei 20% auf. Dieser zehnfache Unterschied zwischen beiden Gruppen legt nahe, daß nicht nur das klinische Erscheinungsbild unterschiedlich ist, sondern auch die zugrundeliegenden pathophysiologischen Mechanismen.

41.1.5.1 Mechanismen des plötzlichen Herztodes

Gewöhnlich wurde angenommen, daß die akute und massive Schädigung des Herzmuskels die pathophysiologische Grundlage des plötzlichen Herztodes ist. Pathologisch-anatomische Studien aus den vergangenen 25 Jahren konnten zwar eine ausgeprägte Arteriosklerose der Koronargefäße dokumentieren, das regelmäßige Vorliegen einer frischen Koronarthrombose oder eines akuten Myokardinfarktes konnte jedoch nicht bestätigt werden (Adelson und Hoffman, 1961; Spain und Brades, 1977; Reichenbak et al., 1977). Baroldi (1978, 1979) wies bei 72% der infolge eines plötzlichen Herztodes Verstorbenen eine koagulative Myozytolyse und eine fleckförmige Fibrose des linken Ventrikels nach. Die koagulative Myozytolyse wird als Folge einer Katecholaminfreisetzung angesehen; sie ist bei Hunden bereits fünf Minuten nach einer Bolusinjektion von Katecholaminen nachweisbar. Werden die Tiere mindestens 72 Stunden am Leben erhalten, findet sich im linken Ventrikel eine fleckförmige Fibrose. Zentralnervöse Einflüsse können schnelle neurohormonale Veränderungen hervorrufen, die wahrscheinlich auch beim Menschen zu derartigen Veränderungen führen. Danach kann ein fortschreitender Verlust an Myokard buchstäblich als Folge einer Überdosierung der eigenen Katecholamine auftreten (Eliot und Buell, 1981). Klinisch kann sich hieraus auch ohne Hypertonie das Bild einer Dilatation des linken Ventrikels entwickeln. Eine Dilatation des linken Ventrikels steht in Beziehung zum gehäuften Auftreten des plötzlichen Herztodes (Hinckle, 1982). Wahrscheinlich bewirken derartige histologische und metabolische Veränderungen eine größere elektrische Instabilität des Herzens.

Durch die Entwicklung der Technik und durch klinische Anwendung der elektrischen Defibrillation des Herzens wurde es möglich, die elektrischen Mechanismen des plötzlichen Herztodes und vor allem den Stellenwert ventrikulärer Arrhythmien neu zu untersuchen.

Lown und DeSilva stellten vier Hypothesen auf, die sich im Laufe der Zeit erhärten ließen (Lown et al., 1976, 1977; DeSilva, 1978):

- Der plötzliche Herztod ist die Folge von Kammerflimmern.
- Die elektrische Instabilität des Myokards geht dem Beginn des Kammerflimmerns lange Zeit voraus.
- Bestimmte Typen ventrikulärer Extrasystolen kennzeichnen eine elektrische Instabilität des Myokards.
- Transitorische Risikofaktoren wirken auf das elektrisch instabile Myokard ein und lösen Kammerflimmern aus.

Beim Kammerflimmern besteht eine chaotische elektrische Depolarisation des Herzmuskels, die die mechanische Aktivität des Herzens, d.h. seine Pumpfunktion derart beeinträchtigt, daß es zum Stillstand des Blutflusses kommt. Der Tod tritt innerhalb von Minuten ein, wenn Wiederbelebungsmaßnahmen (z.B. Defibrillation) nicht unmittelbar die Funktionsfähigkeit des Herz-Kreislauf-Systems wiederherstellen. Eine elektrische Instabilität des Herzens ist Voraussetzung für das Auftreten von Kammerflimmern. Diese Instabilität ist nicht ohne weiteres im EKG erkennbar; sie kann sich allerdings in ventrikulären Extrasystolen ankündigen, die von Lown und Wolf (1971) in verschiedene Schweregrade klassifiziert worden sind (Tab. 41.1–1). Eine größere Inzidenz des plötzlichen Herztodes ist mit höheren Graden dieser ventrikulären Rhythmusstörungen verknüpft (DeSilva, 1978), wohingegen niedrige Grade (bis 2) keinen Hinweis für ein erhöhtes Risiko geben (Rubermann et al., 1977). Akuter psychischer Streß muß als transitorischer Risikofaktor angesehen werden, der bei entsprechender Vorschädigung über das Zentralnervensystem zu einer weiteren elektrischen Destabilisierung des Herzens führen kann.

In bezug auf Rhythmusstörung und klinisches Erscheinungsbild kann akuter psychischer Streß zu drei verschiedenen Reaktionsmustern führen (Tab. 41.1–2). Die gewöhnliche kardiovaskuläre Reaktion ist Erhöhung von Herzfrequenz und Blutdruck (Typ 1). Ein eher ungewöhnliches Muster (Typ 2) ist das Auftreten von Vorhofarrhythmien (supraventrikuläre Tachykardie, Vorhofflattern und -flimmern) sowie isolierte ventrikuläre Extrasystolen oder Couplets. Die häufigsten Symptome sind hierbei das Gefühl von Pochen in der Brust oder das Aussetzen von Herzschlägen. Bei anhaltenden Vorhoftachyarrhythmien kann Müdigkeit, Schwäche, Schwindel, das Gefühl des Ohnmächtigwerdens oder auch eine Synkope auftreten. Gelegentliche vasovagale Synkopen bei Streß sind mit Bradykardie, Hypotonie, kurzen Perioden von AV-Blockierung oder Asystolie verbunden.

Tabelle 41.1–1. Schweregrade ventrikulärer Extrasystolen (VES) (Lown-Klassifikation, Lown und Wolf, 1971)

Schweregrad	Art der ventrikulären Extrasystolen
0	keine VES
1a	gelegentliche isolierte VES (weniger als 30/Stunde)
1b	gelegentliche isolierte VES (weniger als 30/Stunde), aber mehr als 1 VES in einigen 1minütigen Perioden
2	häufige VES (mehr als 30/Stunde)
3	multiforme VES
4a	wiederholte VES (Couplets)
4b	wiederholte VES (Salven)
5	frühe, in den Bereich der vorhergehenden T-Welle einfallende VES (R-auf-T-Phänomen)

Tabelle 41.1–2. Kardiale Reaktionen auf akuten emotionalen Streß (nach DeSilva, 1984)

Reaktionstyp	EKG-Veränderung	Somatische Symptome	Klinische Folgen
Typ 1 gewöhnliche Reaktion	Sinustachykardie (Hypertonie)	Flush (Röte), Angst, Zittern, Schweiß, schneller Puls	spontane Erholung
Typ 2 ungewöhnliche Reaktion	A. Bradykardie (Hypertonie)	Blässe, Schwitzen, Ohnmacht, Synkope	spontane Erholung, ärztliche Hilfe kann erforderlich werden selten tödlich
	B. Vorhofarrhythmien	Palpitationen, Herzklopfen, Schwindel, Müdigkeit, Brustschmerzen, Synkope	spontane Erholung oder evtl. ärztliche Hilfe erforderlich
	C. Isolierte ventrikuläre Extrasystolen	Herzklopfen, Unwohlsein im Herzbereich, Müdigkeit	spontane Erholung oder evtl. ärztliche Hilfe erforderlich
Typ 3 extreme Reaktion	Arrhythmien, die zum plötzl. Herztod führen, (Kammerflimmern, Asystolie, elektromech. Dissoziation)	Herzstillstand und Tod	Tod, falls nicht behandelt

Arrhythmien, die die Pumpfunktion des Herzens unterbinden, führen zum Tod (Typ 3). Zu diesen Arrhythmien zählen die ventrikuläre Tachykardie, Kammerflimmern, Asystolie und elektromechanische Dissoziation. Führt akuter psychischer Streß zum plötzlichen Herztod, liegen gewöhnlich nicht Asystolie oder elektromechanische Dissoziation vor.

Vorhofarrhythmien wie supraventrikuläre Extrasystolen, Vorhofflimmern und -flattern sowie eine supraventrikuläre Tachykardie können durch akuten psychischen Streß, Angst und Furcht oder bei der Mitteilung erschütternder Nachrichten ausgelöst werden. Diese Rhythmusstörungen treten insbesondere dann auf, wenn gleichzeitig andere Faktoren das Herz für ihr Auftreten empfänglicher machen, wie beispielsweise Müdigkeit und Schlafmangel oder im Anschluß an mit Alkoholgenuß verbundene große Mahlzeiten oder auch Elektrolytverschiebungen im Blutserum, z.B. infolge von Laxantienabusus oder Diuretikabehandlung. Diese Arrhythmien sind meist nicht lebensbedrohlich, bedürfen jedoch gelegentlich ärztlicher Hilfe und medikamentöser Behandlung.

Ausgeprägte Bradykardie und Asystolie wurden von einigen Autoren als einer der Mechanismen angesehen, die zum plötzlichen Tod führen; beim Menschen wurde dies in Beziehung zu einer reflektorischen vagalen Hemmung des Herzens gebracht (Albutt, 1915). Tierexperimentelle Untersuchungen mit Ratten, die im Wasser schwimmen mußten und die nicht ertranken, sondern infolge von Bradykardie und Asystolie starben, sprachen ebenfalls für einen derartigen Mechanismus (Richter, 1957). Der „Voodoo"-Tod westindischer Eingeborener wurde demgemäß als Folge eines reflektorischen vagalen Herzstillstandes interpretiert. Wolff (1967) entwickelte diese Hypothese weiter und glaubte, daß das Auftreten der Asystolie und nachfolgender ventrikulärer Arrhythmien Teile eines angeborenen Reaktionsmusters, des sog. Tauchreflexes, sind, die durch emotionellen Streß ausgelöst werden. Der Tauchreflex kann beim Menschen durch Kältereize im Gesichtsbereich ausgelöst werden; er führt zu Bradykardie, gelegentlicher AV-Blockierung und Blutdruckanstieg als Reaktion auf den Kältereiz. Eine Literaturübersicht ergab, daß kein elektrokardiographisch dokumentierter Tod beschrieben worden ist, der als Folge dieses Reflexes angesehen werden kann (Lown und DeSilva, 1978); umfangreiche Untersuchungen bei Patienten mit ventrikulären Extrasystolen konnten diese Hypothese ebenfalls nicht stützen.

Ausgeprägte Angst und Furcht können Ohnmacht als Folge von Bradykardie und Hypotonie hervorrufen. Engel (1978) nimmt an, daß derartige Synkopen einem plötzlichen Tod zugrunde liegen können. Tritt der Tod durch derartige Mechanismen in seltenen Fällen ein, so ist die Ursache entweder Kammerflimmern, ausgelöst durch die lange Asystolie, oder es kommt zum vollständigen Stillstand der Schrittmacheraktivität des Herzens als Folge einer gleichzeitigen Schädigung des kardialen Reizleitungssystems (DeSilva, 1984).

Tod durch Bradykardie oder Asystolie ohne ventrikuläre Arrhythmien hat seine Ursachen gewöhnlich in den schweren hämodynamischen Veränderungen während eines akuten Myokardinfarktes, also bei ausgeprägter myokardialer Schädigung und kardialem Schock. Allerdings fanden Liberthson und Mitarbeiter (1974) bei 426 Patienten, die wegen eines plötzlichen Herztodes wiederbelebt wurden, nur in 10% eine vollständige AV-Blockierung oder Asystolie.

Bradyarrhythmien, Asystolie und Tod wurden in Untersuchungen an sog. Joch-Kontrollaffen (vgl. Kap. 8) beschrieben, die wiederholten elektrischen Schocks ohne eigene Einflußmöglichkeit ausgeliefert waren (Corley et al., 1975). Dies war wahrscheinlich die Folge einer Myokardschädigung. Bei einer Reihe von Tierarten kann psychischer Streß, hervorgerufen durch elektrischen Schock, Hitze/Kälte oder Lärm, zu einer Schädigung des Myokards und Tod durch Asystolie oder ventrikuläre Arrhythmien führen. Inwieweit dieser Mechanismus auch beim Menschen besteht, ist bisher ungeklärt. Cebelin und Hirsch (1980) untersuchten Todesopfer gewalttätiger körperlicher Auseinandersetzungen. In 15 von 497 Fällen konnten keine Verletzungen für die Todesursache verantwortlich gemacht werden. Alle 15 Todesopfer waren von Verwandten, Ehepartnern, Fremden oder anderen Personen geschlagen oder überfallen worden und in 10 Fällen sind verbale Auseinandersetzungen oder Kämpfe dem Tod vorausgegangen. In mikroskopischen Untersuchungen des Herzens fanden sich bei 11 der 15 Todesopfer myofibrilläre Degenerationen meist im subendokardialen Bereich. Als Kontrollgruppe dienten 15 Todesopfer von Verkehrsunfällen, bei denen derartige Veränderungen nicht nachweisbar waren. Die Autoren nehmen an, daß der psychische Streß eine akute Kardiomyopathie mit nachfolgenden Rhythmusstörungen und plötzlichem Tod hervorgerufen hat. Die Art der Rhythmusstörung, die hier zum plötzlichen Tod geführt hat, ist nicht bekannt, aber es erscheint wahrscheinlich, daß es sich um Kammerflimmern und nicht um eine Asystolie handelte.

In 85% der Fälle findet man zum Zeitpunkt des Kreislaufstillstandes Kammerflimmern. Die praktische klinische Bedeutung wird daraus ersichtlich, daß Kammerflimmern bei rechtzeitigem therapeutischem Eingreifen reversibel ist, und daß seine Prävention ein realistisches und erreichbares Ziel ist.

In der Mehrzahl der Fälle sind weder hämodynamische Faktoren noch eine akute myokardiale Schädigung für das Auftreten von Kammerflimmern verantwortlich. Mitunter können die hämodynamischen Veränderungen bei schwerer körperlicher Anstrengung Kammerflimmern hervorrufen, meist jedoch tritt der Tod in Situationen der Ruhe auf. In mehr als ⅔ der Fälle ist keine akute myokardiale Schädigung nachweisbar (Cobb et al., 1975). Eine akute strukturelle Schädigung des Herzmuskels ist deswegen keine Voraussetzung für das Auftreten von Kammerflimmern, wenngleich in den meisten Fällen eine koronare Herzkrankheit besteht. Auch beim Verschluß einer Koronararterie kann das Kammerflimmern unmittelbar unterbrochen werden. Dies spricht dafür, daß die irreversible Unterbrechung des Blutstroms der Koronararterie nicht als ausschließliche Ursache des plötzlichen

Todes anzusehen ist. Vielmehr weisen diese Untersuchungen darauf hin, daß das Auftreten von Kammerflimmern von situativen Faktoren abhängig ist, die zu einer Disorganisation der elektrischen Aktivität des Herzens führen. Es handelt sich gewissermaßen um einen „elektrophysiologischen Unfall" mit fast immer tödlichem Ausgang; durch eine unmittelbare Defibrillation des Herzens ist der Zustand jedoch sehr oft reversibel. Auf der Grundlage dieser Überlegungen haben Lown und Mitarbeiter (1976, 1977) transitorische Risikofaktoren postuliert, die beim elektrisch instabilen Herzen zum Kammerflimmern führen. Und als Quelle dieser Faktoren wird das Zentralnervensystem angesehen.

Neurophysiologische Studien haben den Verlauf der efferenten Bahnen vom Hypothalamus über das Rückenmark bis zum Herzen aufgeklärt (Beattie et al., 1930; Brow, 1930). Durch elektrische Stimulation über stereotaktisch eingeführte Elektroden konnten verschiedene Rhythmusstörungen einschließlich Kammerflimmern vom Hypothalamus, vom Dienzephalon, Mesenzephalon, von der Formatio reticularis und den Corpora quadrigemina ausgelöst werden. Eine Vagotomie hatte keinen Einfluß; vielmehr hob eine Reizung der peripheren Enden des Nervus vagus die Wirkungen der zentralen sympathischen Stimulation auf (DeSilva und Lown, 1978). Diese Untersuchungen weisen darauf hin, daß ventrikuläre Arrhythmien nicht über den Vagus, sondern über den Sympathikus ausgelöst werden. Im Tierversuch mit experimentellem Verschluß einer Koronararterie bewirkt die elektrische Reizung des posterioren Hypothalamus einen 6fachen Anstieg für die Auslösbarkeit von Kammerflimmern (Satinsky et al., 1971). Ventrikuläre Arrhythmien einschließlich Kammerflimmern können ebenfalls durch die Reizung peripherer sympathischer Strukturen hervorgerufen werden, wie beispielsweise des Ganglion stellatum oder kardialer sympathischer Nerven (Harris et al., 1971; Verrier et al., 1974; Klicks et al., 1975).

MacWilliam (1889, 1923) hat als erster bereits vor mehr als 60 Jahren den Verdacht ausgesprochen, daß der plötzliche Herztod eine Folge von Kammerflimmern ist, und daß der Einfluß des sympathischen Nervensystems bei der Auslösung der tödlichen Arrhythmie ein wichtiger Faktor ist. Cannon (1957) vermutete, im Gegensatz zu anderen Autoren, daß der „Voodoo"-Tod durch psychische Einflüsse über das sympathikoadrenale System ausgelöst wird.

41.1.5.2 Tierexperimentelle Untersuchungen

Eine Schlüsselfrage ist, inwieweit sich experimentell prüfen läßt, ob psychischer Streß im bewußten Zustand über eine Zunahme der sympathischen Aktivität bei einem Tier Kammerflimmern auslösen kann. Methodische Probleme erschweren eine direkte Beobachtung eines derartigen Zusammenhangs, da die traumatischen Folgen einer Wiederbelebung nach Kammerflimmern die fortlaufende Untersuchung psychischer Variablen unmöglich machen. Allerdings

konnten Indikatoren für die Empfänglichkeit von Kammerflimmern gefunden werden, die statt dessen in derartigen Studien untersucht wurden (Matta et al., 1976).

Die Schwelle für die Auslösbarkeit von Kammerflimmern wird gewöhnlich bestimmt, indem man das Herz mit geringem elektrischem Strom während der sog. vulnerablen Periode des Ventrikels stimuliert, die etwa mit dem Gipfel der T-Welle im EKG zusammenfällt. Mit der schrittweisen Erhöhung der Stromstärke kann die elektrische Schwelle bestimmt werden, bei der Kammerflimmern ausgelöst wird. Wiederholte ventrikuläre Extrasystolen (VES) gehen dem Kammerflimmern voraus; der elektrische Schwellenwert für die Auslösung wiederholter VES beträgt genau zwei Drittel der für die Auslösung von Kammerflimmern erforderlichen Stromstärke. Somit läßt sich der Schwellenwert für die Auslösung von wiederholten VES als Indikator für die elektrische Schwelle des Kammerflimmerns benutzen, ohne daß das Tier etwas bemerkt und ohne daß die traumatischen Wiederbelebungsmaßnahmen erforderlich werden.

Lown und Mitarbeiter untersuchten mit dieser Technik in einer Reihe von Studien den Einfluß von psychischem Streß bei Hunden. Als Stressoren wurden die Konditionierung mit unvermeidbaren Schocks in der Pawlow-Schlinge oder Sidmans bekannte Schock-Intervall-Schock-Methode benützt. Bei Tieren mit gesundem Herzen sank der Schwellenwert für die Auslösbarkeit von Kammerflimmern unmittelbar, nachdem sie erneut in die aversive Umgebung gesetzt wurden, um 40% selbst dann, wenn keine weiteren Schocks verabreicht wurden (DeSilva et al., 1978). Eine vermehrte sympathische Erregung war durch den Anstieg von Herzfrequenz, Blutdruck, Adrenalin und der Gesamtkatecholamine gekennzeichnet (Lown et al., 1973; DeSilva et al., 1978; Matta et al., 1974; Verrier und Lown, 1978). Wurden Hunde, die sich von einem akuten, experimentellen Myokardinfarkt erholten, in die gleiche Umgebung gesetzt, in der sie vorher elektrischen Schocks ausgesetzt waren, trat eine ventrikuläre Tachykardie auf (Corbalan et al., 1974).

Welche Mechanismen hier im einzelnen zur elektrischen Instabilität des Herzens und zur Auslösung von Kammerflimmern beitragen, bleibt weiter zu klären. Neben den biochemischen und hämodynamischen Veränderungen, die die Empfänglichkeit für Kammerflimmern erhöhen, spielt die Abnahme des vagalen Einflusses auf das Herz während psychischen Stresses eine wichtige Rolle (Verrier und Lown, 1978; Liang et al., 1979). Ein vermehrter vagaler kardialer Einfluß schützt das Myokard vor Kammerflimmern. Deswegen scheint die Abnahme des vagalen Einflusses bei aversiver Konditionierung als wichtiger Faktor zur größeren Anfälligkeit des Herzens für Kammerflimmern bei psychischem Streß beizutragen.

Mit einer Reihe von Methoden kann das Herz vor ventrikulären Arrhythmien, durch psychischen Streß ausgelöst, geschützt werden. Eine Reduktion des sympathischen Einflusses kann bei Hunden durch cholinerge Stimulation erreicht werden; Methacholingabe oder direkte Reizung des Nervus vagus erhöht den elektrischen Schwellenwert für die Auslösbarkeit von Kammerflimmern (Collmann et al., 1975; Rabinowitz et al., 1976). Auch Morphium erhöht den vagalen Einfluß und die elektrische Stabilität des Herzens (DeSilva et al., 1976). Eine Abnahme der sympathischen Aktivität wird auch durch eine Vermehrung der zentralnervösen Neurotransmitter-Substanzen erreicht, z. B. durch die Gabe ihrer Vorstufen wie Tryptophan (für Serotonin) und Tyrosin (für Adrenalin, Noradrenalin und Dopamin). Hierdurch wird ebenfalls die elektrische Stabilität des Herzens erhöht (Rabinowitz und Lown, 1978; Blatt et al., 1979; Scott et al., 1981). Beim Menschen senkt Tryptophan erhöhten Blutdruck; möglicherweise wird dies durch eine höhere zentrale Serotoninsynthese vermittelt (Feldkamp et al., 1984).

Eine Verringerung des sympathischen Einflusses durch Blockade der betaadrenergen Rezeptoren schützt das Herz ebenfalls vor ventrikulären Rhythmusstörungen. Bei Hunden wird die durch das Sidman-Schockvermeidungsparadigma hervorgerufene Empfindlichkeit für das Auftreten wiederholter ventrikulärer Extrasystolen durch Tolamolol aufgehoben (Matta et al., 1974); dies betrifft Hunde mit gesundem Herzen. Die Befunde beim ischämischen Herzen sind zum Teil widersprüchlich. Während Skinner und Mitarbeiter (1975) keinen Einfluß von Propranolol bezüglich der Auslösbarkeit von Kammerflimmern durch psychischen Streß bei Schweinen fanden, konnten Rosenfeld und Mitarbeiter (1978) eine Schutzwirkung durch Tolamolol bei Hunden nachweisen. Möglicherweise spielen hier Unterschiede zwischen den verschiedenen Tierarten, den verwendeten Medikamenten bzw. den experimentellen Bedingungen eine Rolle.

In den tierexperimentellen Untersuchungen konnte nachgewiesen werden, daß vertrauten Umgebungsbedingungen eine wichtige Schutzfunktion zukommt. Nach experimentellem Herzinfarkt kann allein der Aufenthalt in einer Umgebung, die für das Tier früher mit aversiven Reizen verbunden war, zum Auftreten von ventrikulären Arrhythmien führen. Wird das Tier in eine nicht bedrohliche Umgebung gebracht, verschwinden diese Arrhythmien (Corbalan et al., 1974). Entsprechend läßt sich ein Anstieg der elektrischen Schwelle nachweisen, bei der Kammerflimmern auslösbar ist (DeSilva et al., 1978). Nicht untersucht wurden bisher die Verhaltensänderungen und die sie begleitenden neurohumoralen Einflüsse, die bei diesen unterschiedlichen Umgebungsbedingungen auftreten. Die bisherigen Ergebnisse weisen jedoch darauf hin, daß die Umgebungsreize für die elektrische Stabilität des Herzens und für das Auftreten ventrikulärer Arrhythmien eine wichtige Bedeutung haben. Dies muß auch bei Patienten nach einem Myokardinfarkt berücksichtigt werden, da sie ein hohes Risiko für das Auftreten maligner ventrikulärer Rhythmusstörungen aufweisen.

41.1.5.3 Untersuchungen beim Menschen

Niedriger sozioökonomischer Status und Mangel an Schulbildung steht in Beziehung zu frühzeitigem Tod bzw. erhöhter kardiovaskulärer Mortalität (Kitagawa und Hauser, 1973). Weinblatt und Mitarbeiter untersuchten die Rhythmusstörungen von 1739 Patienten, die einen Myokardinfarkt überlebten. Patienten mit komplexen ventrikulären Arrhythmien wiesen das dreifache Risiko für einen plötzlichen Herztod auf, wenn sie nur acht oder weniger Jahre an Schulbildung hatten; längere Schulbildung war beim gleichen Schweregrad der Arrhythmien mit einer signifikant niedrigeren Mortalität verknüpft (33% versus 9%). Keiner der traditionellen Risikofaktoren oder andere klinische Unterschiede konnten für diesen Unterschied verantwortlich gemacht werden.

Rees und Lutkins (1967) konnten zeigen, daß der Tod der Ehefrau die Mortalität der hinterbliebenen Ehemänner im folgenden Jahr erhöht; sie betrug bei den Witwern 12,2% und bei der Kontrollgruppe 1,2%. Die „Broken Heart"-Studie (Parkes et al., 1969) bestätigte diesen Zusammenhang: 4486 Männer wurden nach dem Tod ihrer Frau neun Jahre lang bezüglich Mortalität beobachtet und mit einer Kontrollgruppe verglichen, die keinen Partner verloren hatte. In den ersten sechs Monaten starben 213 Witwer, das entspricht 40% mehr als der berechneten Erwartung für verheiratete Männer. Erhöht war vor allem die kardiovaskuläre Mortalität, die sich nach dem ersten halben Jahr wieder derjenigen der Kontrollgruppe anglich. Gemeinsam erfahrene Lebensbedingungen könnten zu einem Teil der erhöhten Mortalität beigetragen haben.

Depression wurde in einigen Studien in Beziehung zum plötzlichen Herztod gebracht. Greene und Mitarbeiter (1972) untersuchten während zwei Jahren bei einer Fabrikbelegschaft von 44000 Angestellten durch Befragung der Angehörigen die Lebenssituation, die dem akuten Herztod vorangegangen war. 54 Angestellte verstarben. 77% hatten vorher an koronarer Herzkrankheit gelitten. 76% waren eine Woche bis drei Monate lang vor ihrem Tod depressiv gewesen, meist wegen des Wegzuges eines Kindes oder der Enttäuschung durch ein Kind, das nicht den Erwartungen entsprochen hatte. Sie reagierten mit Überaktivität am Arbeitsplatz oder gerieten dort oder zu Hause in akute Aufregungen, kurz bevor der plötzliche Tod eintrat. Dieser Studie fehlt die Beobachtung einer Kontrollgruppe von nicht verstorbenen Angestellten bezüglich der genannten psychosozialen Faktoren. In einer prospektiven Studie fanden Bruhn und Mitarbeiter (1974) bei Patienten mit Myokardinfarkt eine Beziehung zwischen plötzlichem Herztod und erhöhten Werten auf einer Depressionsskala, Freud- und Lustlosigkeit bei der Arbeit und Typ-A-Verhaltensweisen. Orth-Gomer (1980) fand hingegen bei Koronarkranken keine Beziehung zwischen dem Auftreten ventrikulärer Arrhythmien und Depression oder Typ-A-Verhaltensweisen; allerdings fehlen hier nähere Angaben zur Häufigkeit des Typ-A-Musters; bei Koronarpatienten war die Höhe des systolischen

Blutdrucks das beste Unterscheidungsmerkmal für den Schweregrad ventrikulärer Rhythmusstörungen. Bei koronargesunden Männern erwies sich in dieser Studie ein depressiver emotionaler Zustand, nach dem Alter, als zweitwichtigster Faktor für das Auftreten von schweren ventrikulären Arrhythmien. In den Untersuchungen von Lown (Reich et al., 1981) wurden bei 25 von 117 Patienten psychophysiologische Auslöser für das Auftreten von ventrikulärer Tachykardie oder Kammerflimmern gefunden. Bei diesen Patienten fand sich in verschiedenen psychometrischen Tests ein außergewöhnlich hoher Depressionswert im Vergleich zur Kontrollgruppe.

Untersuchungen mit Fragebögen zur Erfassung der Lebensveränderungseinheiten (Rahe et al., 1974) bringen den plötzlichen Herztod in Beziehung zu einer Anhäufung verschiedener Arten psychischen Stresses, beispielsweise Tod eines nahen Angehörigen, Scheidung oder Arbeitsplatzverlust. Ähnliche Veränderungen fanden Theorell und Rahe (1975). Da dies retrospektive Untersuchungen sind, müssen die Resultate vorsichtig interpretiert werden, denn in der Rückschau mögen Patienten und Angehörige die Ereignisse kurz vor dem Infarkt oder Herztod anders beurteilt haben als Ereignisse, die weiter zurückliegen. Cottington und Mitarbeiter (1980) fanden in einer sorgfältigen Fallkontrollstudie, daß von verschiedenen lebensverändernden Ereignissen lediglich der Tod einer nahestehenden Person in Zusammenhang mit dem plötzlichen Herztod gebracht werden konnte. In den vergangenen sechs Monaten war bei 81 der infolge eines plötzlichen Herztodes Verstorbenen der Tod einer nahestehenden Person sechsmal häufiger aufgetreten als in der Kontrollgruppe. Andere lebensverändernde Ereignisse wie finanzielle, rechtliche, familiäre oder arbeitsplatzbezogene Probleme standen nicht in Beziehung zur Mortalität. Akute psychische Belastung war ebenfalls in einer Studie von Myers und Dewar (1975) bei 100 Patienten nachweisbar, die infolge einer koronaren Herzkrankheit plötzlich verstarben. Bei 23 dieser Personen trat die ausgeprägte Streßbelastung 30 Minuten vor dem Tod auf, bei 40 Personen innerhalb der letzten 24 Stunden. Die Art der Belastungen war breit gestreut: Aufnahme in eine chirurgische Klinik, eine Auseinandersetzung mit bissigen Hunden, Streit beim Spiel, Verwicklung in einen Verkehrsunfall ohne Verletzungsfolgen, Scheidungstermin usw. Ebenfalls in Beziehung zur Mortalität standen mäßige körperliche Aktivität, eine kurz vorher eingenommene Mahlzeit, insbesondere mit alkoholischen Getränken, sowie Tageszeit und Wochentag. Chronische psychische und anstrengende körperliche Belastungen, Jahreszeit und Umgebungstemperatur standen nicht mit dem plötzlichen Tod in Beziehung.

In einer Studie aus Helsinki wurden die Begleitumstände des plötzlichen Todes in 118 Fällen durch Interviews mit Angehörigen ermittelt (Rissanen et al., 1978). In 32% der Fälle trat eine ungewöhnliche Müdigkeit zuvor auf. Chronischer Streß bestand bei 25% und akuter Streß bei 19% der Patienten. Bei den 23 Patienten mit akuter Streßbelastung trat der Tod schnell innerhalb von 2 Stunden ein. 14 Patienten starben unmittelbar. Akuter Myokardinfarkt war in dieser Gruppe ungewöhnlich, trat aber häufiger bei Patienten auf, die chronischer Streßbelastung ausgesetzt waren. Die Ergebnisse dieser Untersuchung legen nahe, daß akute psychische Belastung bei Patienten mit koronarer Herzkrankheit eher zu einem plötzlichen Tod infolge von Kammerflimmern führt als zum akuten Infarkt. Bei Patienten mit chronischer Streßbelastung und Patienten ohne sichtbare Streßfaktoren waren die Prodromalsymptome von längerer Dauer, und es waren häufiger Zeichen eines akuten Myokardinfarktes nachweisbar.

Reich und Mitarbeiter (1981) untersuchten das Auftreten maligner Arrhythmien wie beispielsweise ventrikuläre Tachykardie und Kammerflimmern bei einer Patientengruppe, die ein hohes Risiko für das Auftreten des plötzlichen Herztodes besaß. Von den 117 Patienten waren 53% nach Kammerflimmern wiederbelebt worden. Die Gruppe setzte sich also aus Opfern bzw. potentiellen Opfern des plötzlichen Herztodes zusammen, die selbst Auskunft über das emotionale Geschehen unmittelbar vor dem Auftreten der Arrhythmien geben konnten. In zwei unabhängigen Interviews, von einem Kardiologen und einem Psychiater durchgeführt, wurden die Ereignisse 24 Stunden vor dem Auftreten der Arrhythmien untersucht. Bei 25 (21%) der 117 Patienten konnten psychische Einflüsse als Auslöser der malignen Arrhythmien identifiziert werden. Sie betrafen starke Emotionen wie Ärger, Furcht und Aufregung, die durch Konflikte in Ehe und am Arbeitsplatz oder durch den Tod eines Angehörigen ausgelöst wurden. Eine koronare Herzkrankheit bestand bei 66% der gesamten untersuchten Gruppe, aber nur bei 48% der Patienten, bei denen die ventrikuläre Tachykardie oder das Kammerflimmern auf psychophysiologischem Wege auslösbar war, lag eine ischämische Herzerkrankung vor. Bei 44% der Patienten ohne nachweisbare strukturelle Veränderungen des Herzens war die ventrikuläre Tachykardie oder das Kammerflimmern durch psychische Reize auslösbar.

Ventrikuläre Extrasystolen sind bei Koronarkranken mit einem erhöhten Risiko für den plötzlichen Herztod verbunden. Eine Anzahl von Studien hat das Auftreten ventrikulärer Extrasystolen in Zusammenhang mit psychischem Streß untersucht. Elektrokardiographische Veränderungen im ST-Streckenbereich und der T-Welle sowie das Auftreten ventrikulärer Extrasystolen können durch emotionale Einflüsse, durch traumatische Ereignisse oder durch Erinnerung an derartige Ereignisse, durch Angst, Furcht und Aufregung ausgelöst werden. Taggart und Mitarbeiter (1969) fanden bei 32 Patienten mit ischämischer Herzkrankheit unter dem psychischen Streß des Autofahrens 13 mit ST-Senkungen im EKG, von denen 6 massiv waren. Fünf weitere Patienten zeigten gehäufte polytope ventrikuläre Extrasystolen und zwei Angina pectoris. Von 32 Fahrern mit gesundem Herzen hatten nur drei eine ST-Veränderung. Es wurden auch Teilnahme an Autorennen, öffentliches Sprechen, der Einfluß lauter Geräusche und klinische Interviews in bezug auf vorangegangene emotionelle Traumata (Taggart et al., 1969; Wellens et al., 1972; Lown et al., 1976; Sigler, 1961) untersucht. Lown und DeSilva (1978) untersuchten 19 Patienten, von denen 8 nach Kammerflimmern wiederbelebt worden waren, in einem psychologischen Streßtest; die Anzahl ventrikulärer Extrasystolen stieg im Vergleich zur Ruheperiode um das Doppelte an. Bei einem dieser Patienten kam es zur ventrikulären Tachykardie, als der Patient weinend seine Furcht vor dem Tode beschrieb. Zu einem früheren Zeitpunkt war diese Tachyarrhythmie in Kammerflimmern übergegangen. Diese Arrhythmie konnte ohne medikamentöse Behandlung allein durch Beruhigung im Gespräch wieder zum Verschwinden gebracht werden. Diese Beobachtungen weisen darauf hin, daß die elektrische Stabilität des Herzens durch psychische Einflüsse verringert oder erhöht werden kann. Bei fünf der untersuchten Patienten trat während des Streßtestes keine Arrhythmie auf. Es ist wahrscheinlich, daß die elektrische Stabilität des Herzens nur durch sehr spezifische und für das Individuum psychisch bedeutungsvolle Einflüsse verändert wird. Beispielsweise kam es bei einem Patienten mit ausgeprägter KHK wiederholt zu Attacken ventrikulärer Tachykardie, wenn sich seine Frau bei ihren täglichen Besuchen im Krankenhaus wieder von ihm verabschiedete. Bei diesem Patienten waren die Rhythmusstörungen nicht durch einen psychischen Streßtest auslösbar; sie traten auch nicht auf, als er weinend über seine Sorge um die eigene Gesundheit und die seiner Frau sprach. Eine derartige Spezifität von Veränderungen der elektrischen Stabilität des Herzens bei psychischen Belastungen steht im Gegensatz zu den unspezifischen, durch viele Umgebungsreize auslösbaren Reaktionen des kardiovaskulären Systems z.B. mit Herzfrequenz- und Blutdruckveränderungen.

41.1.5.4 Die Bedeutung des sympathischen Nervensystems für das Auftreten ventrikulärer Arrhythmien und therapeutische Einflußmöglichkeiten

Die Aktivierung des sympathischen Nervensystems und die vermehrte Freisetzung von Katecholaminen erscheinen als ein wichtiger Faktor. In den Untersuchungen von Taggart waren die Katecholamine beim Auftreten ventrikulärer Extrasystolen beim Sprechen vor einem kritischen Auditorium, bei Autorennen oder beim Fahren im Londoner Verkehr erhöht. Nach Gabe des Betarezeptorenblockers Oxprenolol verschwanden die Rhythmusstörungen. Coumel und Mitarbeiter (1978) beschrieben vier Kinder mit organischen Herzerkrankungen, bei denen eine ventrikuläre Tachykardie und Synkopen durch emotionale Einflüsse oder Anstrengung ausgelöst wurden. Durch elektrische Stimulation mit Hilfe eines in das Herz eingeführten Katheters konnten diese Rhythmusstörungen nicht hervorgerufen werden; Stimulation der Betarezeptoren mit Isoproterenol hingegen führte regelmäßig zu diesen Störungen. Durch medikamentö-

se Behandlung, die auch die Gabe von Betarezeptorenblockern beinhaltete, gelang es, diese Arrhythmien günstig zu beeinflussen.

Eine vermehrte sympathische Aktivation löst maligne ventrikuläre Arrhythmien bei einer Reihe von Erkrankungen aus, die durch eine Störung der ventrikulären Repolarisation gekennzeichnet sind und sich in einer Verlängerung der elektrischen Systolendauer im EKG zeigen (Long Q-T Syndrome). Eine Reihe von Reizen wie emotionale Einflüsse, Lärm und Erschöpfung können hier ventrikuläre Tachykardie oder Kammerflimmern auslösen. Gewöhnlich sterben diese Patienten sehr frühzeitig, selbst wenn keine strukturellen Veränderungen am Herzen nachweisbar sind. Es wurde eine unterschiedliche sympathische Innervierung des Herzens über das rechte und linke Ganglion stellatum postuliert (Schwarz, 1976). Die chirurgische Entfernung dieser neuralen Strukturen schützt vor dem Auftreten der Rhythmusstörungen ebenso wie eine medikamentöse Behandlung mit Betarezeptorenblockern. Dies weist auf die Bedeutung sympathischer Einflüsse bei der Entstehung dieser Arrhythmien hin. Wahrscheinlich spielen aber auch andere Mechanismen eine Rolle. Die Wechselwirkung zwischen neuralen und humoralen Einflüssen wurde bislang nicht ausreichend untersucht. Lown und DeSilva (1978) untersuchten die Bedeutung autonomer neuraler Reflexe für das Auftreten ventrikulärer Arrhythmien. Die Aktivation sympathischer und parasympathischer Reflexe führte nur dann zu ventrikulären Arrhythmien, wenn gleichzeitig Streß vorlag. Auch die Auslösung des Tauchreflexes führte gewöhnlich nicht zum Auftreten von Arrhythmien. Lediglich bei einer Patientin kam es zu diesen Veränderungen, nachdem in diesem Test ihr Gesicht mit Eis in Berührung kam, worüber sie sich aufregte. Bei dieser Patientin war wiederholt Kammerflimmern durch emotionale Konflikte hervorgerufen worden. Aufgrund dieser Beobachtungen erscheint es wahrscheinlich, daß die passive Auslösung autonomer Reflexe nicht für das Auftreten ventrikulärer Arrhythmien ausreichend ist.

Das Auftreten von Arrhythmien kann durch eine Verringerung des sympathischen Einflusses vermindert werden; dies ist durch Untersuchungen mit Betarezeptorenblockern, durch chirurgische Intervention und Schlafstudien belegt worden. Während des Schlafes kommt es zusammen mit der Veränderung des Bewußtseinszustandes zu einer Verringerung von Herzfrequenz und Blutdruck, die einer Abnahme des sympathischen und einer Zunahme des vagalen Einflusses zugeschrieben werden. In einigen Untersuchungen wurde keine Beeinflussung ventrikulärer Arrhythmien während des Schlafes beobachtet, wohingegen andere Studien bei den meisten Patienten eine Verringerung feststellen konnten. Für diese unterschiedlichen Ergebnisse ist wahrscheinlich die geringe Anzahl der untersuchten Patienten verantwortlich zu machen, bzw. die Tatsache, daß Patienten oft unter medikamentöser Behandlung standen und in der Klinik auf der Intensivstation untersucht wurden. Untersuchungen mit ausreichend großen Fallzahlen bestätigen hingegen durchgehend eine Verringerung ventrikulärer Extrasystolen während des Schlafes (Pickering et al., 1977; Brodsky et al., 1977).

Lown untersuchte mit dem Holter-EKG 54 Patienten in ihrer häuslichen Umgebung (Lown et al., 1973). Bei 22 dieser Patienten waren die Rhythmusstörungen während des Schlafes um wenigstens 50% und bei weiteren 13 Patienten um 25–50% verringert. Auch der Schweregrad der Rhythmusstörungen, gemessen nach der Lown-Klassifikation, verringerte sich. Bei einer Anzahl von Patienten war der Schlaf wirkungsvoller als eine medikamentöse antiarrhythmische Behandlung in bezug auf Beeinflussung von Schweregrad und Häufigkeit ventrikulärer Rhythmusstörungen. Die ausgeprägteste Wirkung betraf die Schlafstadien 3 und 4, lediglich während der REM-Traumschlafphasen fand sich keine Verringerung im Vergleich zum Wachzustand (DeSilva et al., 1978). Pickering und Mitarbeiter (1977) berichten, daß bei 26% von 31 Patienten die Rhythmusstörungen während des Schlafes vollständig verschwanden und bei 71% teilweise verringert waren. Diese Untersuchungen können als Bestätigung von Studien angesehen werden, die belegt haben, daß der plötzliche Tod während des Schlafes sehr selten auftritt (Myers und Dewar, 1975; Friedman et al., 1973). Es erscheint wahrscheinlich, daß die Verringerung des nervalen und humoralen sympathischen Einflusses während des Schlafes vor einem plötzlichen Herztod schützt. Lown und Mitarbeiter (1976; Reich et al., 1981) beobachteten das Auftreten von ventrikulärer Tachykardie oder Kammerflimmern in zwei Fällen im Zusammenhang mit Angstträumen. Tritt während des Schlafes der plötzliche Tod ein, so geschieht dies wahrscheinlich während des REM-Schlafes; eine vermehrte sympathische Stimulierung ist hier vermutlich für die Zunahme der elektrischen Instabilität des Herzens verantwortlich.

Eine bewußte Veränderung des sympathischen Einflusses auf das Herz kann durch eine Reihe verschiedener Methoden wie Biofeedback und Meditation erreicht werden. Mit Biofeedback konnten Weiss und Engel (1971) und Pickering und Miller (1977) ventrikuläre Extrasystolen durch eine Beschleunigung der Herzfrequenz unterdrücken. Dies geschieht wahrscheinlich durch eine Unterdrückung ektopischer Reizzentren infolge der erhöhten Herzfrequenz. Meditation und Entspannungstechniken wurden ebenfalls zur Beeinflussung ventrikulärer Rhythmusstörungen verwendet. Die kardiovaskulären Wirkungen der Meditationstechniken liegen wahrscheinlich in einer Verringerung von Herzfrequenz und Blutdruck als Folge eines verminderten sympathischen Einflusses. Benson (1975, 1983) unterrichtete 11 Patienten mit koronarer Herzkrankheit mit einer von der transzendentalen Meditation abgeleiteten Technik über einen Zeitraum von 4 Wochen. Die Häufigkeit der ventrikulären Extrasystolen war im Wach- und Schlafzustand etwas verringert, wobei ein signifikanter Unterschied jedoch nur während des Schlafes bestand. Die Herzfrequenz war nicht signifikant verändert. Eine ähnliche Technik verwendeten Voukydis und Forwand (1977); sie fanden bei einigen Patienten eine Verringerung der Rhythmusstörungen, bei anderen traten keine Veränderungen oder sogar eine Zunahme der Rhythmusstörungen auf. Auch Lown und DeSilva (1982, 1985) weisen auf die individuell unterschiedliche Ansprechbarkeit der Patienten auf die von ihnen verwendete Meditationstechnik hin. Gelegentlich kommt es zu einer ausgeprägten Verringerung ventrikulärer Extrasystolen, und sogar die Unterbrechung einer ventrikulären Tachykardie ist möglich. Davidson und Mitarbeiter (1979) verwendeten Entspannungsmethoden; bei einzelnen Patienten korrelierte hier eine Abnahme der Herzfrequenz mit entsprechenden Veränderungen des Noradrenalinspiegels im Blut. Allerdings veränderten sich die Mittelwerte der Herzfrequenz zwischen Kontroll- und Entspannungsperiode nicht signifikant. Lynch und Mitarbeiter (1977) konnten zeigen, daß allein das Trösten und Beruhigen der bettlägerigen Patienten zu einer Abnahme ventrikulärer Extrasystolen führt. Auch hier fand sich keine gleichzeitige Veränderung der Herzfrequenz. Trotzdem kann vermutet werden, daß sich eine Verringerung des sympathischen kardialen Einflusses in einer höheren elektrischen Stabilität des Herzens auswirken kann, ohne daß die Herzfrequenz beeinflußt wird.

Allein die Tatsache, daß Biofeedback, verschiedene Meditations- und Entspannungstechniken sowie Beruhigung und psychologische Unterstützung der Patienten das Auftreten von ventrikulären Arrhythmien verhindern bzw. ihre Häufigkeit verringern kann, zeigt, in welchem Ausmaß verschiedenartige Umgebungssituationen über den Cortex und Hypothalamus das Herz beeinflussen können. Für weitere Forschungsstrategien erscheint es wichtig, diejenigen Gruppen zu identifizieren, die durch derartige Interventionsstrategien günstig zu beeinflussen sind. Eine andere Möglichkeit liegt darin, auf neuropharmakologischem Wege die spezifischen Einflüsse höherer Nervenzentren auf das Herz zu beeinflussen (Rabinowitz und Lown, 1978). Die wichtige therapeutische Bedeutung einer Verringerung des sympathischen kardialen Einflusses wird auch durch einige der großen Therapiestudien bei KHK-Patienten mit Betarezeptorenblockern deutlich. Eine Verringerung der Häufigkeit des plötzlichen Herztodes konnte in einigen Studien nachgewiesen werden. Dies gilt für die Betarezeptorenblocker Alprenolol (Wilhelmsson et al., 1974), Practolol (Multicenter International Study, 1975), Timolol (Norwegian Multicenter Study Group, 1981), Propranolol (Beta-Blocker Heart Trial Research Group, 1982) und Metoprolol (Herlitz et al., 1984). Eine Verminderung der kardiovaskulären Mortalität in der sekundären Prävention bzw. eine Verringerung der Häufigkeit von tödlichem Reinfarkt oder plötzlichem Tod liegt für unterschiedlich lange Zeiträume zwischen 3 Monaten und 3 Jahren zwischen 25 und 36%.

Zukünftige Präventionsstrategien zur Verringerung der großen Anzahl auch oft junger Opfer eines plötzlichen Herztodes erfordern eine Integration der verschiedenen bis heute nachgewiesenen Beeinflussungsmöglichkeiten; nur unter der gleichzeitigen Berücksichtigung der klinischen, pharmakologischen und psychophysiologischen Aspekte erscheint eine Verbesserung der gegenwärtigen Situation möglich.

41.1.6 Psychosoziale Faktoren bei Angina pectoris und Myokardinfarkt

41.1.6.1 Sozioökonomische Einflüsse

Der Myokardinfarkt hat sich in den letzten Jahrzehnten von einer Todesursache der oberen sozialen zu einer der unteren sozialen Schichten entwickelt (Marmot et al., 1978a, b; Walk, 1981). Neuere Studien aus England (Rose und Marmot, 1981), Finnland (Valkonen, 1982), Norwegen (Holme et al., 1982) und Japan (Kagaminori, 1981) zeigen eine erhöhte KHK-Mortalität bei Berufsgruppen mit niedrigerem sozioökonomischem Status. Auch ein niedriges Bildungsniveau erhöht das KHK-Mortalitätsrisiko (Hinckle et al., 1966; Kitagawa und Hauser, 1973; Rosenman et al., 1975) so wie Arbeitslosigkeit und niedriger Berufsstatus (Jenkins und Zyzanski, 1980). Personen mit höherem Bildungsniveau scheinen hingegen vermehrt unter Angina pectoris zu leiden (Shekelle, 1969; Blohmke et al., 1975). Die Gründe für die unterschiedlichen Mortalitätsraten verschiedener sozialer Schichten sind noch nicht hinreichend geklärt. Niedrige soziale Schichten weisen in den Industrieländern mehr traditionelle kardiovaskuläre Risikofaktoren auf (Blohmke et al., 1970; Schäfer und Blohmke, 1977; Marmot, 1981). Aber auch bei Kontrolle dieser Risikofaktoren findet sich bei den unteren Sozialschichten eine höhere KHK-Inzidenz bzw. -Mortalität (Koskenvuo, 1980; Rose und Marmot, 1981; Holme et al., 1982). Hierfür können mehrere Gründe verantwortlich sein (Waltz, 1981). Angehörige der unteren sozialen Schichten sind mehr lebensverändernden Ereignissen, physischen und psychosozialen Belastungen wie Unsicherheit des Arbeitsplatzes und Arbeitslosigkeit ausgesetzt (Myers et al., 1975). Brenner und Mitarbeiter (1980) wiesen auf den engen Zusammenhang zwischen Phasen ökonomischer Instabilität und dem Verlauf der KHK-Mortalität hin. Als weiterer Faktor könnte eine Rolle spielen, daß niedrige soziale Schichten über ein weniger effektives soziales Netzwerk verfügen. Sie haben weniger Kontakt mit Arbeitskollegen, Nachbarn und Verwandten (Marmot, 1982). Verminderte soziale Beziehungen und Aktivitäten sind mit einer erhöhten KHK-Mortalität verbunden, wie eine prospektive Längsschnittstudie nachweisen konnte (House et al., 1982); dies gilt auch nach Kontrolle der wichtigsten traditionellen Risikofaktoren. Dem entsprechen auch die Befunde von Siegrist und Mitarbeitern (1981), die bei Myokardinfarktpatienten im Vergleich zu einer Kontrollgruppe eine geringere soziale Unterstützung bei chronischen familiären Schwierigkeiten fanden. Ein funktionsfähiges soziales Netzwerk scheint hingegen eine wichtige Schutzfunktion auszuüben (Syme und Berkman,

1976). Angehörigen unterer sozialer Schichten scheinen weniger effektive persönliche Ressourcen zur Verfügung zu stehen. Dies drückt sich in geringerer Selbsthilfeaktivität (Pearlin und Schooler, 1978), fatalistischerer Einstellung (Wheaton, 1980) und niedrigerem Selbstwertgefühl (Rosenberg und Pearlin, 1978) aus.

Die Bedeutung von psychosozialem Streß und sozialer Isolierung für die unteren sozialen Schichten wurde in einer Studie an 2320 Überlebenden eines akuten Myokardinfarkts unterstrichen (Rubermann et al., 1984). Bei Teilnehmern der Beta-Blocker Heart-Attack Trial (BHAT) fand sich ein um viermal höheres Mortalitätsrisiko bei Patienten, die als sozial isoliert klassifiziert wurden und die einem hohen Grad an psychosozialem Streß ausgesetzt waren. Je höher der Bildungsgrad war, desto geringer war auch die Mortalität. Streßbelastung und soziale Isolierung waren am wenigsten bei höher Gebildeten und am stärksten bei Patienten mit niedrigem Bildungsgrad ausgeprägt. Die mit Streßbelastung und sozialer Isolation verbundene Erhöhung des Risikos galt sowohl für die gesamte Mortalitätsrate innerhalb von drei Jahren als auch für den plötzlichen Herztod; die Erhöhung des Risikos fand sich sowohl bei Patienten, die einen hohen als auch einen niedrigen Schweregrad ventrikulärer Extrasystolen während der Behandlungszeit im Krankenhaus nach dem akuten Infarkt aufwiesen. Andere für die Prognose nach einem Herzinfarkt wichtige Faktoren wurden kontrolliert; den Autoren zufolge muß das erhöhte Risiko den psychosozialen Faktoren zugeschrieben werden. Allerdings erschweren methodische Probleme bezüglich der gewählten Indikatoren für „Streß" und „soziale Isolierung" eine eindeutige Bewertung dieser Studie. Da sich die verglichenen Gruppen in einigen Merkmalen unterschieden, könnten z.B. auch höheres Alter und vermehrter Zigarettenkonsum für das erhöhte Risiko verantwortlich sein. Außerdem waren Patienten, die Items zu „Streß" und „sozialer Isolation" positiv beantworteten, häufiger unverheiratet und häufiger von schwarzer Hautfarbe (Myrtek, 1985).

41.1.6.2 Lebensverändernde Ereignisse

Auf die Bedeutung lebensverändernder Ereignisse insbesondere in bezug auf den Tod des Ehepartners wurde schon im Abschnitt über den plötzlichen Herztod hingewiesen. Die Grundannahme dieser Forschungsrichtung ist, daß unerwünschte, unerwartete, unbeeinflußbare und mit unangenehmen Konsequenzen einhergehende Ereignisse eine erhöhte Anpassungsleistung erfordern. Erweisen sich die individuellen Selbsthilfeaktivitäten als unzulänglich, kommt es zu exzessiven neurohumoralen und pathophysiologischen Reaktionen (Siegrist, 1980). In einer Untersuchung an Postinfarktpatienten konnten beispielsweise Theorell und Mitarbeiter (1972) aufgrund des Summenwertes für die pro Woche aufgetretenen lebensverändernden Ereignisse vorhersagen, ob die Adrenalinausscheidung innerhalb einer Woche angestiegen oder abgefallen war. Bei den meisten Arbeiten der Lebensereignis-Forschung handelt es sich jedoch um retrospektive Studien, in denen die Belastung durch Lebensereignisse, sog. LCU(Life Change Unit)-Werte, erst nach der Erkrankung erhoben wird. In derartigen Studien konnte beispielsweise ein kontinuierlicher Anstieg der LCU-Werte in den letzten 3 Jahren vor dem Herzinfarkt festgestellt werden, mit einem Maximum in den letzten 6 Monaten vor dem Krankheitsereignis (Theorell und Rahe, 1971). Das galt für Patienten ohne vorherige KHK-Symptome. Traten derartige Symptome vor dem Myokardinfarkt auf, wurde das Maximum bereits 2 Jahre vor dem Herzinfarkt erreicht, d.h. zu einer Zeit, als bereits deutliche Krankheitszeichen vorhanden waren. Connolly (1976) fand, daß Myokardinfarktpatienten im Vergleich zu Kontrollpersonen in dem Monat, der dem Krankheitsereignis voranging, vergleichsweise mehr Lebensereignissen unterworfen waren. Bengtsson (1973) zeigte, daß auch Frauen im Jahr vor ihrem Myokardinfarkt mehr Lebensereignissen ausgesetzt waren als eine Kontrollgruppe. Rahe und Mitarbeiter (1974) konnten zeigen, daß Patienten mit Vorerkrankungen in den letzten 2 Jahren vor dem tödlichen Myokardinfarkt höhere durchschnittliche LCU-Werte aufwiesen als die übrigen Patienten, und daß bei den bis zum Krankheitsereignis Gesunden die LCU-Werte im Vergleich zu ihren eigenen Vorjahresintervallwerten in den letzten 6 Monaten vor dem Krankheitsereignis wesentlich stärker angestiegen waren als bei den Patienten, die bereits an KHK-Symptomen litten. In einer Untersuchung mit ein- bzw. zweieiigen Zwillingspaaren fand DeFaire (1975), daß der Zwilling, der an einer KHK verstorben war, in den letzten 4 Jahren einen höheren LCU-Wert aufwies und dieser im letzten halben Jahr sein Maximum erreichte. Allerdings war nur für die 9 verstorbenen eineiigen Zwillinge der mittlere LCU-Wert vergleichsweise signifikant größer.

In einer Untersuchung von Siegrist und Mitarbeitern (1982) wird berichtet, daß Herzinfarktpatienten in den letzten 2 Jahren vor dem Krankheitsereignis signifikant mehr Lebensereignisse als Kontrollpersonen angaben, wobei in den letzten 3 Monaten bei den KHK-Kranken die Zahl der Lebensereignisse deutlich zunahm. Die Gruppe der Herzinfarktpatienten war darüber hinaus signifikant häufiger sehr belastenden Ereignissen ausgesetzt gewesen als die Kontrollgruppe, z.B. schwerer Erkrankung, Arbeitslosigkeit, Tod eines Angehörigen.

In gewissem Gegensatz hierzu stehen die Untersuchungen von Maschewsky (1982) und Lundberg und Mitarbeitern (1975), die keinen Unterschied zwischen Herzinfarktpatienten und Kontrollpersonen in der Häufigkeit der Lebensereignisse der „nächsten Vergangenheit" bzw. im letzten Jahr vor der Erkrankung fanden. Allerdings unterschieden sich beide Gruppen signifikant in der Bewertung bezüglich der Intensität der erlebten Anspannung durch diese Ereignisse. Auch in einer Untersuchung von Byrne und Whyte (1980) unterschieden sich Myokardinfarktpatienten und Patienten, die auf die kardiologische In-

tensivstation eingeliefert worden waren, nicht in der Anzahl der Lebensereignisse während des letzten Jahres. Die Infarktpatienten beurteilten jedoch das Ausmaß der emotionalen Belastung, das mit den Lebensereignissen einherging, signifikant höher. Auch im Vergleich mit altersparallelisierten Kontrollpersonen nannten die Myokardinfarktpatienten zwar nicht mehr Lebensereignisse, doch schätzten sie den daraus resultierenden Distreß ebenfalls signifikant stärker ein (Byrne, 1983). Dies entspricht dem Befund von Maschewsky (1982), der zeigte, daß sich Infarktpatienten im Bewältigungsverhalten von Kontrollpersonen dadurch unterscheiden, daß sie eher dazu neigen, psychisch-soziale Spannungen und Konflikte nicht auszutragen. Byrne (1983) zog die Schlußfolgerung, daß Myokardinfarktkandidaten Lebensereignisse als belastender erleben als andere Personen.

Beweiskräftige prospektive Studien gibt es bisher kaum. Die Untersuchungen von Bruce und Mitarbeitern (1976) und Siegrist und Mitarbeitern (1982) über den Zusammenhang von Lebensereignissen und kardialem Tod erlauben in Anbetracht der kleinen Fallzahlen von 8 bzw. 13 Personen kaum weitergehende Schlußfolgerungen. Theorell und Rahe (1975) untersuchten die Zusammenhänge zwischen Lebensereignissen und Herztod bei 36 Postinfarktpatienten, von denen 18 während einer Beobachtungszeit von 6 Jahren verstarben. Bei den Verstorbenen fanden sie höhere krankheitsabhängige LCU-Werte in den letzten 18 Monaten vor dem Tod, wobei die LCU-Werte zwischen dem 7. bis 12. Monat vor dem Tod ihr Maximum erreichten. In der Stockholmer Bauarbeiter-Studie (Theorell und Floderus-Myrhed, 1977; Theorell, 1979, 1981) konnte gezeigt werden, daß ein „Unzufriedenheits-Index", der u.a. ein Lebensereignis (Anzahl von Wohnortswechseln) enthält, auch in einer multivariaten Analyse mit Einschluß traditioneller Risikofaktoren einen signifikanten Beitrag zur Vorhersage leistete, welche Patienten innerhalb von 18 Monaten an einem Myokardinfarkt erkrankten. Ein weiterer Faktor „psychosoziale Belastungen im Beruf", der 2 Lebensereignisse im vorangegangenen Jahr (z.B. erhöhte berufliche Verantwortung, mehr als 30 Tage arbeitslos) einbezieht, trug nach 2jähriger Beobachtungszeit in einer multivariaten Analyse unter Einschluß der traditionellen Risikofaktoren am meisten zur Vorhersage von Myokardinfarkterkrankungen bei. Dies galt vor allem für Patienten, welche den Herzinfarkt überlebten. Allerdings war es nicht möglich, nach einem Jahr, mit Hilfe der LCU-Werte allein, die an einem Myokardinfarkt Erkrankenden vorherzusagen.

Insgesamt erscheinen die Ergebnisse der Lebensereignis-Forschung zwar wichtig in bezug auf eine Hypothesengenerierung, belegen aber einen Zusammenhang mit der koronaren Herzkrankheit bisher nicht hinreichend. Für künftige Untersuchungen erscheint es wichtig, die subjektive Belastung ausreichend differenziert zu erfassen. Prospektive Untersuchungen wären wünschenswert.

41.1.6.3 Berufliche Überbeanspruchung

Berufliche Überbeanspruchung ist wiederholt als koronarer Risikofaktor beschrieben worden. Arbeitsüberforderung infolge starken Zeit- und Termindrukkes und Verantwortung für die Arbeit anderer erhöhen das Erkrankungsrisiko (House, 1974). Jenkins (1982) wies darauf hin, daß die gleichzeitige Ausübung zweier beruflicher Tätigkeiten vor allem für Angina pectoris, aber auch Myokardinfarkt einen Risikofaktor darstelle. In der prospektiven schwedischen Bauarbeiter-Studie (Theorell und Floderus-Myrhed, 1977; Theorell, 1981) wurde festgestellt, daß 51 Arbeiter, die im Verlauf von 2 Jahren einen Herzinfarkt erlitten hatten, im vorangegangenen Jahr deutlich beruflich überbeansprucht waren im Vergleich zur Kontrollgruppe. Bei Arbeitern, die einen kardial oder zerebrovaskulär bedingten Tod erlitten hatten, war im Vergleich zu Kontrollpersonen nahezu sechsmal häufiger eine Kombination von „hohen beruflichen Anforderungen" und „geringem persönlichen Handlungsspielraum während der Arbeit" anzutreffen (Theorell, 1981).

Aus 4 verschiedenen Studien zogen Karasek und Mitarbeiter (1982) den Schluß, daß geringer beruflicher Handlungsspielraum mit erhöhter KHK-Morbidität oder -Mortalität assoziiert sei. Das relative Risiko entspricht mit 1,72 hierbei ungefähr dem des Serumcholesterins. Erhöhte berufliche Anforderungen seien ebenfalls, wenn auch weniger konsistent, mit der KHK assoziiert. Eine Kombination beider Faktoren ermögliche eine bessere Vorhersage der KHK als die isolierte Betrachtung beider Aspekte. In einer retrospektiven Kontrollgruppen-Studie fanden Siegrist und Mitarbeiter (1982), daß bei den KHK-Kranken am Arbeitsplatz sowohl psychosoziale Belastungen (Zeitdruck, Unterbrechungen) als auch physische Belastungen (Lärm, Hitze, Unfallgefahr) signifikant häufiger waren. Angehörige betrieblicher Zwischenpositionen (Betriebsmeister, Bauleiter, Vorarbeiter) und kaufmännische Angestellte, deren Status eng mit dem von ihnen erzielten Umsatz gekoppelt war (Handelsvertreter, Filialleiter), wiesen gehäufte psychosoziale Belastungen auf (Siegrist, 1980b). Langosch und Mitarbeiter (1983) stellten bei einer Gegenüberstellung von 30 jüngeren, männlichen Postinfarktpatienten und 30 männlichen Rheumatikern gleichen Alters und gleicher Berufstätigkeit fest, daß bei den KHK-Patienten die beruflichen Belastungen eher im Bereich fehlender Anerkennung und Beachtung lagen. Am häufigsten waren bei 46 jüngeren Myokardinfarktpatienten Unterbrechungen, Termindruck, Entscheidungsdruck, Aufgabenüberschneidungen und Anpassungsdruck zu beobachten gewesen. Siegrist und Mitarbeiter (1981) fanden, daß 69% von den 13 Patienten, die innerhalb von 18 Monaten an einem Reinfarkt verstarben, hohen beruflichen Belastungen ausgesetzt gewesen waren. Bei der Kontrollgruppe waren es hingegen nur 32%. Langosch und Mitarbeiter (1983) wiesen auf vergleichsweise erhöhte berufliche Anforderungen innerhalb der letzten 4 Jahre bei jüngeren Postinfarktpatienten hin, bei denen nach Ablauf dieses Beobachtungszeitraumes eine Progression der Koronarsklerose koronarangiographisch gesichert werden konnte. Kornitzer und Mitarbeiter (1982a) berichten in verschiedenen Studien über teilweise widersprüchliche Resultate in bezug auf einen Zusammenhang zwischen Aspekten beruflicher Überbeanspruchung und KHK. House (1972) fand, daß bei White-Collar-Beschäftigten über 45 Jahren Arbeitsunzufriedenheit signifikant mit einem Summenwert aus KHK-Risikofaktoren (Rauchen, Cholesterin, Blutdruck) korrelierte. Im Gegensatz dazu stellte er fest, daß bei jüngeren Personen Arbeitszufriedenheit mit den KHK-Risikofaktoren assoziiert war. Auch Michallik-Herbein und Mitarbeiter (1981) fanden bei Herzinfarktpatienten unter 40 Jahren vergleichsweise hohe Arbeitszufriedenheit.

Insgesamt muß festgestellt werden, daß die Beziehung zwischen beruflicher Überbeanspruchung bzw. Arbeitszufriedenheit und KHK-Inzidenz noch nicht hinreichend geklärt ist.

41.1.6.4 Emotionale Probleme und Schlafstörungen

In einer Reihe von prospektiven Untersuchungen wurden emotionale Probleme in Verbindung zu späterem Auftreten von Herzinfarkt und Angina pectoris gebracht. In einer Studie in Chicago erfaßten Shekelle (1969) und Ostfeld (1964) verschiedene Persönlichkeitsmerkmale mit dem MMPI (Minnesota Multiphasic Personality Inventory) und dem Catell-16-PF (Catell 16-Personality Factor Questionnaire). 1190 Männer aller sozialen Klassen wurden über 4½ Jahre untersucht. Die 50 in diesem Zeitraum an Angina pectoris Erkrankten wiesen auf der Hypochondrieskala des MMPI die höchsten und die 38 Herzinfarktpatienten die tiefsten Werte auf, während die Werte der Kontrollgruppe dazwischen lagen. Im Catell-16-PF wirkten die Infarktpatienten argwöhnischer, eifersüchtiger, selbstgenügsamer und zurückgezogener, die Angina-pectoris-Patienten emotional weniger stabil. In einer Studie aus Oklahoma beschrieben Bruhn und Mitarbeiter (1969) ebenfalls Unterschiede zwischen Angina-pectoris- und Myokardinfarktpatienten: Die erste Gruppe zeigte ausgeprägtere neurotische Züge, war unreifer und emotional instabil. Lebovits und Mitarbeiter (1967) fanden, daß Angst und Depressionen ausgeprägter bei Patienten waren, die später am Herzinfarkt verstarben, als bei Patienten, die den Infarkt überlebten. Nach dem Auftreten der koronaren Herzkrankheit fanden sich höhere Werte für die 3 Skalen Hypochondrie, Depression und Hysterie. In einer französisch-belgischen Inzidenzstudie konnte kein signifikanter Unterschied zwischen Myokardinfarktpatienten und Kontrollpersonen in bezug auf Neurotizismus nachgewiesen werden (French-Belgium Collaborative Group, 1982).

Hypochondrie und Neurotizismus beeinflussen die Bereitschaft von Personen, Krankheitssymptome mitzuteilen und ärztliche Hilfe zu suchen (Costa und

McCrae, 1985). Dem entspricht, daß in Untersuchungen mit Koronarpatienten verschiedene Neurotizismusmaße nicht mit dem Ausmaß der Koronarsklerose, wohl aber mit sog. „weichen" Symptomen wie Klagen über Schmerzen im Thoraxbereich assoziiert sind (Costa et al., 1985; Costa, 1986). Das trifft auch für die Befunde des Belgian-French Pooling-Projects (1984) zu. Hier stand Neurotizismus nicht mit der KHK-Morbidität und -Mortalität in Beziehung, allerdings fand sich ein Zusammenhang mit den „weichen" Krankheitsereignissen.

In einer prospektiven Studie mit 10000 männlichen Angestellten in Israel (Groen et al., 1968; Medalie et al., 1973) wurde nach einer Beobachtungszeit von 5 Jahren festgestellt, daß die Angina-pectoris-Inzidenz bei den Patienten, die bereits anfänglich die höchsten Angstwerte aufwiesen im Vergleich zu denjenigen mit den niedrigsten Angstwerten, insbesondere dann erhöht war, wenn sie keine Liebe und Unterstützung von ihren Ehefrauen erfuhren.

Für den Myokardinfarkt hatte der Angstindex keine prädiktive Bedeutung. Die Inzidenz der koronaren Herzkrankheit stand auch in Beziehung zu arbeitsbezogenen Problemen wie Konflikte mit dem Chef und Mitarbeitern, sowie zu familiären und finanziellen Schwierigkeiten. Männer, die derartigen Problemen ausgesetzt waren, wiesen eine höhere Inzidenz auf, wenn sie gleichzeitig ihre Ehefrauen als kalt und indifferent beschrieben. Bei denjenigen, die bei gleichen Problemen ihre Ehefrauen als liebevoll und unterstützend beschrieben, war dies nicht nachweisbar. Religiöse, orthodoxe Juden, die regelmäßig die Synagoge besuchten, hatten ebenfalls im Vergleich zu denjenigen, die selten oder nie zur Synagoge gingen, eine geringere KHK-Inzidenz.

1933 hatte Ludwig Aschoff in seinem Vorwort zu dem von E. Cowdry herausgegebenen Buch „Arteriosclerosis. A survey of the problem" seiner Überzeugung Ausdruck verliehen, daß neben Erbfaktoren und verschiedenen Umgebungseinflüssen

„... nicht nur die materielle Zivilisation, sondern auch die menschliche Kultur sogar in ihrer höchsten Form, der Religion, die Entwicklung des arteriosklerotischen Prozesses beeinflussen kann und sogar beeinflussen muß".

Jenkins und Mitarbeiter (1978) wiesen auf psychische Unterschiede zwischen Myokardinfarkt- und Angina-pectoris-Patienten hin. Zukünftige Angina-pectoris-Patienten sind sehr ungeduldig und leicht irritierbar, sehen sich selbst als ausgesprochen wettbewerbsorientiert, nehmen das Leben sehr ernst, sind sorgfältig und genau und üben oft mehrere Tätigkeiten gleichzeitig aus. Demgegenüber geben zukünftige Herzinfarktpatienten an, in der Konkurrenz mit anderen nicht primär eine Überlegenheit anzustreben, sie bezeichnen sich als beruflich sehr engagiert, und sie kontrollieren ihre Zeit sehr streng.

Vitale Erschöpfung und depressive Verstimmung wurden in mehreren Studien als Prodromi einer sich entwickelnden KHK beschrieben (Kuller et al., 1972; Wardwell und Bahnson, 1973; Alonzo et al., 1975; Rissanen et al., 1978; Nirkko et al., 1982). Auch auf

Schlafschwierigkeiten vor einem Myokardinfarkt wurde hingewiesen (Friedman et al., 1974; Thiel et al., 1973; Partinen et al., 1982). Zur Erfassung des Syndroms der vitalen Erschöpfung entwickelte Appels den „Maastricht-Fragebogen". Anlaß hierzu gab die Beobachtung, daß 60–70% aller Patienten, die später einen Myokardinfarkt oder einen plötzlichen Herztod erlitten, einige Monate vor diesem Ereignis ihren Hausarzt aufgesucht hatten (Falger und Appels, 1982). In einer prospektiven Längsschnittstudie mit Kontrollgruppenvergleich für verschiedene KHK-Manifestationen (Appels et al., 1979), in zwei Querschnittstudien mit Kontrollgruppenvergleich für Herzinfarktpatienten (Appels, 1980; Verhagen et al., 1980) und in einer Querschnittstudie mit Kontrollgruppenvergleich für Angina-pectoris-Patienten (Falger und Appels, 1982) konnten höhere Skalenwerte für zukünftige KHK-Kandidaten bzw. KHK-Kranke nachgewiesen werden, zum Teil ohne daß andere koronare Risikofaktoren den Unterschied zwischen Gesunden und Kranken erklären konnten. Mit Hilfe der prospektiven Längsschnittstudie entwickelten Appels et al. (1987) eine aus 21 Items bestehende Fragebogenskala. Das standardisierte Risiko, innerhalb von 4,2 Jahren einen Herzinfarkt zu entwickeln, betrug danach für den ersten, zweiten und dritten Tertil der Maastricht-Fragebogen-Scorewerte 1,00, 2,26 und 4,69. Diese Assoziation ist im wesentlichen unabhängig von traditionellen somatischen Risikofaktoren. Zu anderen Erkrankungen wie Krebs oder Ulcus duodeni besteht keine Beziehung. Aufgrund ihrer Ergebnisse trennen Appels et al. (1987) das Syndrom der vitalen Erschöpfung von Depression und definieren es folgendermaßen:

„A state which is present when an individual not only complains of unusual fatigue and decreasing energy but also by feeling dejected or defeated. Feeling exhausted when waking up is highly characteristic of the condition. Vital exhaustion is often associated with increased irritability and loss of libido. Symptoms of depression may be associated with vital exhaustion but not necessarily so. Usually self esteem is not lowered and guilt feelings are absent."

Möglicherweise handelt es sich bei dem Syndrom der vitalen Erschöpfung um die psychosomatischen Folgen einer sich länger hinziehenden Überforderung (Appels, 1981). Dies wird durch die Beobachtung gestützt, daß das Syndrom der vitalen Erschöpfung mit dem Typ-A-Verhaltensmuster korreliert ist (Appels, 1981; Schmidt, 1985) und darüber hinaus mit intensiven Problemen im familiären und sozialen Bereich zusammenhängt, die sich vor allem in den letzten 12 bis 18 Monaten ereignet haben (Falger und Appels, 1982). Dieses Syndrom könnte auch die Manifestation einer beginnenden KHK widerspiegeln. Die Ergebnisse zeigen, daß nur wenige Personen mit hohen Scorewerten im Maastricht-Fragebogen einen Myokardinfarkt in der nahen Zukunft entwickeln; nur 5% mit Werten im obersten Tertil erkrankten in der Follow-up-Periode an einem Myokardinfarkt, an Angina pectoris oder unterzogen sich einer Bypass-Operation. Die Autoren betonen, daß der Zustand einer vitalen Erschöpfung nicht eine ausreichende Ursache für die koronare Herzkrankheit ist. Solange das Herz nicht aufgrund einer Atherosklerose oder einer anderen Erkrankung besonders vulnerabel ist, würde ein „vitalerschöpftes" Individuum wohl lediglich durch eine Periode besonderer mentaler Belastung gehen, ohne aber krank zu werden. Der Zustand erinnert an die Rückzug-Konservierungs-Reaktion (Schmale und Engel).

Auch Schlafstörungen spielen in diesem Zusammenhang eine wichtige diagnostische Rolle (Siegrist et al., 1986; Siegrist, 1985). Durchschlafstörungen können zum einen bei bereits erkrankten Personen auf eine nächtliche instabile Angina pectoris hinweisen oder auch auf eine Schlafapnoe; zum anderen werden sie häufig durch emotionale Belastungen hervorgerufen (sog. distreßinduzierte Schlafstörungen). Alle drei Arten von Schlafstörungen sind von prognostischer Bedeutung.

Die nächtliche instabile Angina pectoris gilt als mögliches Prodromalzeichen eines akuten Myokardinfarktes (Cohn und Braunwald, 1984). Als pathophysiologischer Mechanismus kommt hier ein Mißverhältnis von Sauerstoffangebot und -bedarf infolge von Vasospasmus, Thrombozytenaggregation und Thrombusbildung z. B. während der Traum- bzw. REM-Phasen des Schlafes in Betracht. Sehr häufig treten schwere Schlafstörungen aufgrund einer nächtlichen Crescendo-Angina bei Patienten auf, die auf eine koronare Bypass-Operation warten (Jenkins et al., 1983).

Störungen der regulären Atemtätigkeit kommen relativ häufig während des Schlafes vor allem bei hypertonen und übergewichtigen Männern im mittleren bis höheren Alter vor; sie sind zentral oder durch obstruktive Veränderungen bedingt (Guilleminault et al., 1983; Guilleminault und Lugaresi, 1983; Peter et al., 1982): Pathologische Schlafapnoeaktivität (mindestens 30 Phasen pro Nacht mit über 10 Sekunden Dauer pro Phase) kann infolge einer Erstickungsreaktion zu abruptem nächtlichem Aufwachen führen. Mit den gelegentlich langanhaltenden (>1 Minute) Atemstillständen kann eine deutliche Verminderung der arteriellen Sauerstoffspannung verbunden sein sowie ein Anstieg des pulmonalarteriellen und des systemischen Blutdrucks, eine erhöhte Aktivierung des sympathiko-adrenergen Systems, Sinusarrhythmien, Bradykardie und ventrikuläre Rhythmusstörungen (Strohl et al., 1984). Schlafapnoe erhöht das Risiko für das Auftreten von malignen Rhythmusstörungen und plötzlichem Herztod sowie die Entwicklung einer essentiellen und pulmonalarteriellen Hypertonie und linksventrikulären Hypertrophie (Boundulas et al., 1983; Burack, 1984; Guilleminault et al., 1983; Schroeder et al., 1978).

Schlafstörungen ohne erkennbare pathophysiologische Grundlage können bei starker emotionaler Belastung auftreten. Diese distreßinduzierten Schlafstörungen können im Jahr vor der Erstmanifestation der koronaren Herzkrankheit gehäuft auftreten (Siegrist et al., 1984; Siegrist, 1984). Man vermutet, daß es sich hier um nicht mehr kompensierbare Zustände lang-

anhaltender sympathiko-adrenerger Aktivierung während des Tages handelt, die eine das normale Schlafverhalten kennzeichnende funktionelle Balance zwischen katecholaminergem und serotoninergem System stören. Dementsprechend fanden Siegrist und Mitarbeiter (1986) höhere Herzfrequenzwerte sowie eine signifikant erhöhte Variabilität der Herzfrequenz während des Schlafes bei Probanden mit entsprechenden Schlafstörungen (Siegrist, 1985). Ähnliche Befunde konnten tierexperimentell bei Tupaias erhoben werden (von Holst, 1986; Stöhr, 1986).

41.1.6.5 Koronargefährdende Verhaltensweisen und Typ-A-Muster

Definition

Als koronargefährdend werden alle Verhaltensweisen bezeichnet, die das Risiko für die Entwicklung einer koronaren Herzkrankheit erhöhen (coronary prone behavior). Sie beinhalten somit eine Vielzahl teils bekannter Verhaltensweisen, die mit dem Auftreten einer KHK assoziiert sind (Cooper et al., 1981). Das von Rosenman und Friedman beschriebene Typ-A-Verhaltensmuster ist nicht mit dem koronargefährdenden Verhalten gleichzusetzen, sondern stellt eine Untergruppe dieser Verhaltensweisen dar; einige Komponenten des Typ-A-Musters sind als koronargefährdend anzusehen.

Ausgangspunkt dieser Forschungsrichtung bilden die Untersuchungen der beiden amerikanischen Kardiologen Rosenman und Friedman, die seit Ende der 50er Jahre in vielen Studien ihr Konzept des Typ-A-Verhaltens ausgebaut haben. Sie waren überzeugt, daß die rapide Zunahme der koronaren Herzkrankheit, die sich in den letzten drei Jahrzehnten zur führenden Todesursache in den westlichen Industrienationen entwickelt hatte, nicht allein durch Veränderungen der Ernährungsweise, der Altersstruktur der Bevölkerung, durch Mangel an körperlicher Bewegung, durch das Rauchen, durch Änderung genetischer Faktoren oder Verbesserung der Diagnosestellung erklärt werden kann. Sie gingen vielmehr von der Annahme aus, die sich zunächst nur auf die Beobachtung ihrer Koronarpatienten stützte, daß ein Zusammenhang zwischen dem vermehrten Auftreten der KHK und der Entwicklung eines immer hastigeren und hektischeren Lebensstils bestand. Kennzeichnend dafür hielten sie bei ihren Patienten ein überdurchschnittliches Streben nach Anerkennung, Ungeduld, Hast und Eile, Reizbarkeit und Aggressivität. Ihnen fielen Merkmale im Sprachverhalten auf, eine laute, explosible Sprechweise, und übertriebenes psychomotorisches Verhalten als Reaktion auf Provokation durch andere Personen. Als diagnostisches Instrument entwickelten sie ein strukturiertes Interview, bei dessen Auswertung neben inhaltlichen Kriterien insbesondere die erwähnten Sprachcharakteristika berücksichtigt werden.

Rosenman und Friedman (1974) beschrieben das Typ-A-Verhaltensmuster folgendermaßen:

„An action-emotion complex that can be observed in any person who is aggressively involved in a chronic incessant struggle to achieve more and more in less and less time, and if required to do so, against the opposing efforts of other things or other persons."

Drohen Typ-A-Personen zu scheitern, so verstärken sie ihre Bemühungen und geben den Kampf nicht auf, wie es ängstliche Personen tun würden. Sie sind vielmehr aggressiv, wettbewerbs- und arbeitsorientiert, ungeduldig, stets in Eile und wachsam (Rosenman, 1981). Das Typ-A-Muster gilt als ein relativ stabiles Verhalten, mit dem Personen aufgrund entsprechender Persönlichkeitsmerkmale auf unterschiedliche situative Herausforderungen reagieren (Rosenman und Friedman, 1977).

Wenngleich häufig nur zwischen Typ-A- und Typ-B-Verhalten unterschieden wird, handelt es sich bei dieser Klassifikation nicht um eine echte Typologie, sondern vielmehr um eine Reihe von beobachtbaren Verhaltensmerkmalen, die bei empfänglichen Individuen durch geeignete, herausfordernde Umgebungsbedingungen ausgelöst werden – gewissermaßen ein Verhaltenskontinuum, das sich vom extremen Typ A bis zum extremen „nicht Typ A" oder Typ B erstreckt (Matthews, 1982; Sparacino, 1979). Neben dem von Rosenman und Friedman entwickelten strukturierten Interview (Rosenman, 1978) gelten als weitere Meßinstrumente Fragebogen wie der Jenkins Activity Survey (JAS) mit der Typ-A/B-Skala und drei weiteren faktorenanalytisch gewonnenen Skalen, sowie die Framingham-Typ-A-Verhaltensskala und die Bortner-Skala.

Epidemiologische Untersuchungen

Verschiedene Querschnittsstudien haben einen Zusammenhang zwischen dem Typ-A-Verhalten und/ oder einigen seiner Komponenten und der KHK belegt. Dies gilt auch für den deutschen Sprachraum (Cottier et al., 1983; Schmidt et al., 1983; Rüddel et al., 1985). Mehrere Studien zeigten auch einen Zusammenhang zwischen dem Typ-A-Muster und dem Schweregrad der koronarangiographisch nachweisbaren Koronarsklerose, andere konnten diesen Zusammenhang nicht nachweisen (Übersicht bei Schmidt 1982, 1988; Langosch, 1984). Eine Spezifität des Typ-A-Musters für die KHK scheint es nicht zu geben, da einige Studien keine Unterschiede vor allem im globalen Typ-A-Verhaltensmuster bei KHK-Kranken im Vergleich zu anderen chronisch Kranken nachweisen konnten.

Besondere Bedeutung hat das Typ-A-Muster erst durch prospektive Studien bekommen. Nachdem neben den Prävalenz- und koronarangiographischen Querschnittsuntersuchungen die ersten Ergebnisse von zwei prospektiven Studien, der Western Collaborative Group Study und der Framingham-Studie, vorlagen, kam ein unabhängiges Gutachtergremium des National Heart, Lung and Blood Institute in den USA zu dem Ergebnis, das Typ-A-Verhaltensmuster als kardiovaskulären Risikofaktor für berufstätige US-Bürger mittleren Alters anzusehen, der von anderen Risikofaktoren wie Alter, Erhöhungen von systolischem Blutdruck, Serumcholesterin sowie vom Rauchen unabhängig ist (Cooper et al., 1981).

Die Western Collaborative Group Study

Die Erhebungen der Western Collaborative Group Study (WCGS) wurden ab 1960 in verschiedenen kalifornischen Firmen begonnen. Etwa die Hälfte der untersuchten männlichen Teilnehmer wurde mit Hilfe des strukturierten Interviews als Typ A klassifiziert und die andere Hälfte als Typ B. Innerhalb von 8,5 Jahren entwickelten von ursprünglich 3154 gesunden Männern im Alter von 39 59 Jahren 257 eine koronare Herzkrankheit. Bei 11,2% der Typ-A-Männer, aber nur bei 5% der Typ-B-Männer fanden sich ein Myokardinfarkt oder Zeichen von Angina pectoris. Das Erkrankungsrisiko lag bei Typ-A-Personen 2,37mal höher als bei Typ-B-Personen. Darüber hinaus sagte das Typ-A-Muster den zweiten und dritten Infarkt vorher und stand unabhängig von der Todesursache in Beziehung zum Schweregrad der Arteriosklerose in Autopsiebefunden.

Die Bedeutung und Validität dieser Ergebnisse wurde zusätzlich dadurch erhöht, daß klassische Risikofaktoren wie Alter, erhöhter Serumcholesterinspiegel, Blutdruck und Zigarettenrauchen in gleicher Beziehung zur KHK standen wie in der Framingham-Studie (Rosenman et al., 1976). Das Typ-A-Muster trug zu einer Erhöhung des koronaren Risikos unabhängig von den traditionellen Risikofaktoren bei; nach statistischer Korrektur für diese vier Faktoren war das Erkrankungsrisiko bei Typ-A-Personen um 1,97mal höher als bei Typ-B-Personen. Das bedeutet beispielsweise, daß sich bei Typ-A-Personen mit erhöhtem Cholesterin im Vergleich zu Typ-B-Personen mit vergleichbaren Cholesterinwerten das Risiko ebenso verdoppelt, wie wenn normale Cholesterinwerte vorliegen (Brand, 1978). Aufgrund der Daten der WCGS läßt sich berechnen, daß das Risiko, zum Typ A und nicht zum Typ B zu gehören, einer Erhöhung des Blutdrucks um 31 mm Hg und einer Erhöhung des Serumcholesterinspiegels um 56 mg/dl entspricht; die entsprechenden Äquivalenzwerte beim Rauchen betragen mehr als eine Packung Zigaretten pro Tag und beim Lebensalter 7,6 Jahre (Rosenman et al., 1975). Hätte sich das Typ-A-Verhalten in der WCGS als Risikofaktor ausschalten lassen, so hätte dies eine Verminderung der KHK-Inzidenz um 31% zur Folge gehabt (Brand et al., 1978).

Im Zusammenhang mit der WCGS wurde ein Fragebogen zur Erfassung des Typ-A/B-Musters entwickelt, der sog. Jenkins Activity Survey (JAS). Er sollte die Nachteile des strukturierten Interviews wie höhere Kosten, größeren Zeitaufwand für das Training der Interviewer, Durchführung und Auswertung der Interviews sowie Probleme der Standardisierung umgehen. Der JAS enthält etwa 50 Fragen, ähnlich denen im strukturierten Interview, aus denen mit optimaler Skalierung und Regressionsanalyse vier Skalen entwickelt wurden:

Die Typ-A/B-Skala wurde aufgrund einer optimalen Übereinstimmung mit der Typ-A-Klassifikation in den Interviews der WCGS gewonnen. Die Übereinstimmung mit dem strukturierten Interview in der Typenklassifizierung betrug rund 70%, die Test-Retest-Korrelationen lagen nach 1–4 Jahren zwischen 0,6 und 0,7 (Jenkins, 1978).

Die drei weiteren Skalen wurden faktorenanalytisch ermittelt.

Die Skala S (speed-impatience) beschreibt als wichtigstes Verhaltensmerkmal Zeitdruck; Personen mit hohen Scorewerten neigen dazu, sehr schnell zu essen, sie werden ungeduldig, wenn die Gesprächspartner reden, bringen andere Leute dazu, sich zu beeilen, sind leicht aus der Fassung zu bringen sowie leicht irritierbar.

Die Skala J (job involvement) betrifft den Grad beruflichen Engagements. Personen mit hohen Werten berichten über starke berufliche Anspannung, sie machen häufig Überstunden und schaffen sich immer wieder Termine, deren Einhaltung ihnen äußerste Anstrengung abverlangt.

Die Skala H (hard driving-competitive) beinhaltet, daß sich Personen mit hohen Scorewerten als energisch und zielstrebig beschreiben, die alles besser machen als andere, die gewissenhaft, verantwortungsbewußt, ernsthaft sind, sich mit mehr Einsatz und Kraft engagieren und zu Rivalitätsverhalten neigen (Jenkins et al., 1979).

Bezogen auf die letzten vier Jahre der Studie sagte die JAS-A/B-Skala ebenfalls die KHK-Inzidenz voraus. Das oberste Drittel der Population mit hohen Skalenwerten in Typ-A-Richtung hatte ein um 1,7mal höheres Risiko, in diesem Zeitraum eine KHK zu entwickeln, als das untere Drittel mit niedrigen Werten in Typ-B-Richtung (Jenkins, 1978). Die drei faktorenanalytisch gewonnenen Skalen S, J und H, die positiv mit der Typ-A/B-Skala korrelierten, standen jedoch nicht prospektiv zur KHK-Inzidenz in Beziehung (Jenkins et al., 1974). Die JAS-A/B-Skala erwies sich auch als bester Prädiktor für den Reinfarkt, zu dem sie in einer „Dosis-Antwort"-Beziehung stand (Jenkins, 1974, 1981; Zyzanski et al., 1979), d.h., je stärker die Typ-A-Merkmale ausgeprägt waren, desto häufiger kam es zum erneuten Infarkt.

Bei der Verwendung von Fragebogen zur Erfassung des Typ-A-Musters können sich Probleme ergeben, da hier eine Selbstbeschreibung des eigenen Verhaltens vorgenommen wird. Typ-A-Personen sind sich ihrer eigenen Verhaltensweisen oft nicht bewußt. Das strukturierte Interview ist demgegenüber eher ein Verhaltenstest bzw. eine Verhaltensbeobachtung. Dabei kommt es weniger auf den Inhalt der Antworten als vielmehr auf die Art und Weise an, in der geantwortet wird. Die Hauptmerkmale für das Typ-A-Muster sind eine laute, explosible, schnelle und akzelerierte Sprechweise sowie eine kurze Antwortlatenz, Aggressionsbereitschaft und verbales Rivalitätsverhalten, die gewöhnlich aber deutlich niedrigere Korrelationen mit der globalen Typ-A-Klassifikation aufweisen als die anderen Komponenten (Schmidt, 1988; Dembroski, 1989).

Die Typ-A-Skala des JAS wurde konstruiert, um die Ergebnisse des strukturierten Interviews, wie sie in der ersten prospektiven Studie, der WCGS, erzielt wurden, möglichst genau vorherzusagen (Jenkins et al., 1979). Deswegen erhielt diese Skala ihre prädiktive Bedeutung vor allem über die Typenzuordnung im strukturierten Interview. In späteren Studien wurde jedoch deutlich, daß strukturiertes Interview und JAS im wesentlichen unterschiedliche Konstrukte messen (MacDougall et al., 1979; Matthews et al., 1981; Musante et al., 1983), was auch für den deutschen Sprachraum gilt (Myrtek et al., 1984; Schmidt, 1988). Nach Myrtek (1983) beträgt die gemeinsame Varianz für das im strukturierten Interview und im JAS bestimmte Typ-A-Verhalten 11%.

Den prospektiven Befunden zur KHK-Inzidenz in der WCGS stehen kürzlich publizierte Mortalitätsanalysen gegenüber. Die 257 Teilnehmer, die mit Hilfe des strukturierten Interviews in Typ-A- und Typ-B-Personen eingeteilt worden waren und die im Laufe der weiteren 8,5 Jahre an einer koronaren Herzkrankheit erkrankten, wurden bezüglich ihrer nachfolgenden Mortalität untersucht (Ragland und Brand, 1988a). Der Verhaltenstyp stand bei 26 Patienten, die im Laufe der nächsten ca. 12 Jahre innerhalb von 24 Stunden nach dem koronaren Ereignis verstarben, nicht zur Mortalität in Beziehung. Bei 231 Patienten, die die ersten 24 Stunden überlebten, betrug die KHK-Mortalität der 160 Typ-A-Patienten 19,1 pro 1000 Personenjahre und war damit signifikant niedriger (p = 0,04) als die der 71 Typ-B-Patienten, die 31,7 betrug. Auch nach Berücksichtigung unterschiedlicher Follow-up-Zeiten, der Art des ersten KHK-Ereignisses und anderer traditioneller Risikovariablen war das relative KHK-Mortalitätsrisiko bei Typ-A- im Vergleich zu Typ-B-Patienten mit 0,58 (p = 0,03) geringer. Die niedrigere Mortalität von Typ-A-Personen fand sich sowohl in einer jüngeren als auch in einer älteren Untergruppe der Patienten und war bei Patienten mit einem symptomatischen Herzinfarkt stärker ausgeprägt als bei Patienten mit einem stummen Infarkt oder mit Angina pectoris. Diese Studie berücksichtigt wichtige konfundierende Variablen, wie z.B. den Schweregrad des vorangegangenen Infarktes, den kardialen Zustand nach dem Infarkt, Bypass-Operationen etc. nicht bei den Analysen.

18,5% (584) der WCGS-Teilnehmer verstarben innerhalb des gesamten Untersuchungszeitraumes von 22 Jahren (1960/61–1982/83), davon 6,8% (214) infolge einer KHK (Ragland und Brand, 1988b). Die univariaten Mortalitätsanalysen über diesen Untersuchungszeitraum ergaben einen signifikanten Einfluß der traditionellen Risikofaktoren Alter, systolischer Blutdruck, Serumcholesterin und Rauchen (p < 0,001), aber nur einen grenzwertigen Befund für das Typ-A-Verhalten (3,91 vs. 3,09 Fälle/1000 Personenjahre; p = 0,08). Wurde mit Hilfe des Proportional-Hazard-Regression-Modells der jeweils von den anderen Variablen unabhängige Einfluß jedes einzelnen dieser Faktoren untersucht, fanden sich keine signifikanten Mortalitätsunterschiede zwischen Typ-A- und Typ-B-Personen, wohl aber für die traditionellen

Risikofaktoren. Bei Aufteilung des gesamten Zeitraums von 22 Jahren in vier aufeinanderfolgende Abschnitte mit je etwa gleicher Anzahl von infolge einer KHK verstorbenen Studienteilnehmern war die positive Assoziation zwischen systolischem Blutdruck, Serumcholesterin und Alter mit der KHK-Mortalität für alle Zeitabschnitte relativ konsistent. Zigarettenrauchen war ebenfalls in allen vier Intervallen positiv mit der KHK assoziiert, signifikant allerdings nur in den ersten beiden Zeitabschnitten. Das Typ-A-Verhalten war mit der KHK-Mortalität positiv, aber nicht signifikant im ersten und dritten Intervall assoziiert, jedoch signifikant negativ im zweiten Abschnitt und überhaupt nicht im letzten Intervall. Diese Ergebnisse weisen auf eine vorrangige Bedeutung traditioneller Risikofaktoren gegenüber dem Typ-A-Verhalten hin. Darüber hinaus legen sie nahe, daß das mit Hilfe des strukturierten Interviews erfaßte Typ-A-Verhalten in der WCGS stärker mit der nicht-tödlichen KHK-Inzidenz als mit der KHK-Mortalität verknüpft ist, und dies gilt auch nur für den kürzeren Zeitraum von 8,5 Jahren.

Die Framingham-Studie

Zwischen 1965 und 1967 wurden 1822 männliche und weibliche Teilnehmer der Framingham-Studie im Alter von 45 bis 77 Jahren mit einem umfangreichen Fragebogen untersucht, der verschiedene psychosoziale Faktoren und Verhaltensweisen erfaßte. Eine Expertengruppe hatte 10 Fragen ausgewählt, die als charakteristisch für das Typ-A-Verhalten angesehen und in einer Skala zusammengefaßt wurden. Diese Framingham-Typ-A-Verhaltensskala beschreibt Persönlichkeitsmerkmale und Verhaltensweisen wie Wettbewerbsverhalten, Dominanzstreben sowie Zeitdruck und fragt nach emotionalen Reaktionen auf einen durchschnittlichen Arbeitstag, wobei die arbeitsbezogenen Fragen für Hausfrauen etwas anders formuliert sind. Die Auswertungen wurden sowohl für die kontinuierlichen Scorewerte als auch für eine dichotome Typ-A/B-Aufteilung mit Hilfe des Medianwertes vorgenommen.

Hausfrauen und berufstätige Frauen mit koronarer Herzkrankheit wiesen unabhängig von den traditionellen Risikofaktoren signifikant höhere Werte auf der Typ-A-Skala und auf einer Skala für emotionale Labilität auf. Bei Männern bestand eine Beziehung zwischen Myokardinfarkt und Typ-A-Verhalten, Altersproblemen, täglichem Streß und Spannungen (Haynes et al., 1978 a, b). Die Ergebnisse dieser Studie veranlaßten Haynes, die Hypothese aufzustellen, daß für die Entwicklung und volle Ausprägung des Typ-A-Musters bei Männern und Frauen Faktoren und Bedingungen der amerikanischen Arbeitswelt verantwortlich sind. Berufstätige Frauen werden im Vergleich zu Hausfrauen häufiger als Typ A klassifiziert. Ausschlaggebend sind jedoch die Verhaltensweisen: Hausarbeit schützt Typ-A-Frauen nicht vor dem größeren KHK-Risiko.

1674 Personen waren zu Beginn dieser Untersuchung gesund. Eine Analyse nach 8 Jahren ergab, daß Frauen, die an einer KHK erkrankt waren, im Vergleich zu Gesunden höhere Werte auf der Typ-A-Skala aufwiesen, eher Ärger unterdrückten und häufiger Symptome von Angst und Spannung berichtet hatten. Typ-A-Frauen entwickelten zweimal häufiger eine koronare Herzkrankheit als Typ-B-Frauen und dreimal häufiger Angina pectoris. Nach Kontrolle der traditionellen Risikofaktoren in einer multivariaten Analyse blieben Typ-A-Verhalten und Unterdrücken von Ärger unabhängige Prädiktoren der KHK-Inzidenz bei Frauen. Männer besaßen ein höheres KHK-Risiko, wenn sie als Typ A eingestuft worden waren, Ärger unterdrückten, unter Arbeitsüberlastung litten und häufig befördert worden waren. Bei Berücksichtigung der traditionellen Risikofaktoren war Typ-A-Verhalten im Vergleich zum Typ-B-Verhalten bei 45- bis 64jährigen Männern mit einer Verdoppelung des Risikos für die Entwicklung von Angina pectoris, Myokardinfarkt und anderen Zeichen einer koronaren Herzkrankheit verknüpft. Dies galt allerdings nur für White-Collar-Berufe wie z.B. leitende Angestellte in Industrie und Wirtschaft, Rechtsanwälte, Ärzte, Zahnärzte etc.

Eine weitere Analyse wurde nach insgesamt 10 Jahren bei 750 Frauen und 580 Männern im Alter von 45 bis 64 Jahren durchgeführt. Hier war das Typ-A-Verhalten bei Männern nur mit solchen KHK-Ereignissen assoziiert, die auch Angina-pectoris-Symptome beinhalten. Typ-A-Männer erkrankten innerhalb von 10 Jahren zweimal so häufig wie Typ-B-Männer ausschließlich an Angina pectoris und an anderen mit Angina pectoris verbundenen KHK-Manifestationen. Auch bei den Frauen erhöhte das Typ-A-Verhalten lediglich die Inzidenz von KHK-Manifestationen mit Angina-pectoris-Symptomen.

Bei einer Unterteilung der männlichen Teilnehmer in Blue-Collar- und White-Collar-Berufe fand sich lediglich bei der letzteren Gruppe eine 2,5mal höhere KHK-Inzidenz bei Typ-A-Personen. Typ-A-Hausfrauen erkrankten ebenfalls 2,5mal häufiger an einer koronaren Herzkrankheit als Typ-B-Hausfrauen; berufstätige Typ-A-Frauen erkrankten 1,5mal häufiger als Typ-B-Frauen, wobei dieser Unterschied die Signifikanzgrenze verfehlte.

Eine multivariate Analyse der Beziehungen zwischen Typ-A-Verhalten, traditionellen Risikofaktoren und KHK-Inzidenz über einen 10jährigen Zeitraum ergab, daß das Typ-A-Verhalten das relative KHK-Risiko um so mehr steigerte, je stärker die traditionellen Risikofaktoren ausgeprägt waren; dies galt allerdings nicht für den Risikofaktor Rauchen. Bei Männern war diese Wechselwirkung am größten zwischen Typ-A-Verhalten und grenzwertigen Erhöhungen von systolischem Blutdruck (120–159 mmHg) und Serumcholesterin (220–259 mg/100 ml). Bei Frauen trat diese Interaktion im obersten Bereich jedes Risikofaktors auf. Ähnliche Interaktionen wurden auch in der WCGS in den obersten Bereichen der traditionellen koronaren Risikofaktoren gefunden (Rosenman et al., 1975; Brand et al., 1976). Diese Analysen weisen darauf hin, daß das Typ-A-Verhalten die Wirkung der traditionellen Risikofaktoren verstärkt und daß es vor allem bei denjenigen Personen als Risikofaktor anzusehen ist, deren Risiko schon anderweitig erhöht ist.

Die klinische Bedeutung dieser Befunde liegt in dem Prozentsatz neuer KHK-Fälle, die aufgrund des Typ-A-Verhaltens zusätzlich identifiziert werden konnten. Wurde das Typ-A-Verhalten in das logistische Modell für die Berechnung der erwarteten KHK-Inzidenz in der Framingham-Studie eingeschlossen, konnten 12% mehr White-Collar-Männer in den oberen beiden Risikodezilen entdeckt werden, die eine koronare Herzkrankheit entwickelten. Entsprechend wurden 7,4% mehr erkrankte Frauen in den obersten beiden Risikoquintilen (den oberen 40%) entdeckt, wenn das Typ-A-Verhalten in den Berechnungen berücksichtigt wurde. Dies bedeutet eine Verbesserung der Entdeckung von männlichen KHK-Kandidaten in White-Collar-Berufen, die bereits ein hohes Risiko (oberstes Quintil) aufweisen, um 37,5% (12 von 32); bei Frauen entspricht dies einer Verbesserung der Vorhersage, wer zukünftig erkranken wird, um 10,2% (7,4 von 72,3). Die Erfassung einiger psychosozialer Merkmale, die mit einem Fragebogen wie der Framingham-Typ-A-Verhaltensskala erhoben werden können, trägt also zur Aufklärung eines größeren Varianzanteils derjenigen bei, die zukünftig an einer koronaren Herzkrankheit erkranken (Haynes et al., 1983).

Die Fragen der Framingham-Typ-A-Verhaltensskala erfassen nicht direkt Gefühle wie beispielsweise Angst und innere Spannung; sie korrelieren aber positiv mit den Merkmalen Spannung, Angst, Neurotizismus, emotionale Labilität und täglichem Streß (Haynes et al., 1978a; Chesney et al., 1981). Das mittels dieser Skala definierte Typ-A-Verhalten erfaßt wahrscheinlich Individuen, die aufgrund der alltäglichen Belastungen dazu neigen, mit entsprechenden Symptomen zu reagieren. Die Autoren vermuten, daß die Arbeitswelt der in White-Collar-Berufen beschäftigten Typ-A-Männer für ihr erhöhtes Risiko verantwortlich ist. Es handelt sich um Berufe, die durch Zeitdruck und Wettbewerb gekennzeichnet sind. Männer, die in diesen Berufen arbeiten, entwickeln im Vergleich zu Büroangestellten und in Blue-Collar-Berufen Beschäftigten (z.B. Arbeiter, Handwerker, Polizisten und Angestellte in Dienstleistungsbetrieben) nach 8 Jahren am häufigsten Zeichen einer koronaren Herzkrankheit (Haynes und Feinleib, 1980). Frauen in White-Collar-Berufen – Lehrerinnen, Krankenschwestern, Bibliothekarinnen beispielsweise – sind gewöhnlicherweise nicht einem extremen Zeitdruck und Wettbewerb ausgesetzt; dementsprechend sind sie im Vergleich zu den anderen Gruppen keinem erhöhten Risiko ausgesetzt.

Koronares Risiko bei Typ-A-Ehemännern in Abhängigkeit von Verhaltensmerkmalen und sozialem Status ihrer Ehefrauen

Im Rahmen der Framingham-Studie wurden 269 Ehepaare über einen Zeitraum von 10 Jahren untersucht. Hierbei fanden sich Anhaltspunkte dafür, daß

eine unterschiedliche Inzidenz der koronaren Herzkrankheit bei Typ-A- und Typ-B-Männern möglicherweise nur dann auftritt, wenn bestimmte Eigenschaften ihrer Ehefrauen für sie belastend sind (Eaker et al., 1983). So entwickelten Typ-A-Ehemänner im Vergleich zum Typ B 2,5mal eher eine koronare Herzkrankheit, wenn sie mit Ehefrauen verheiratet waren, deren Schul- und Ausbildungszeit 13 oder mehr Jahre betrug. Sie wiesen ein 3,5mal höheres KHK-Risiko auf, wenn sie mit einer Frau verheiratet waren, die außerhalb ihres Hauses berufstätig war. Bei Berücksichtigung des Verhaltenstyps der Ehefrau fand sich bei Typ-A-Ehepaaren im Vergleich zu Typ-B-Ehepaaren eine Verdoppelung der KHK-Inzidenz.

Typ-B-Männer, die mit Typ-A-Frauen verheiratet waren, hatten ungefähr das gleiche Erkrankungsrisiko wie Typ-A-Männer, die mit Typ-A-Frauen verheiratet waren. Das höchste KHK-Risiko bestand jedoch bei Typ-A-Ehemännern, die mit einer Typ-B-Frau verheiratet waren; die Inzidenz der koronaren Herzkrankheit war bei ihnen mehr als dreimal so hoch wie bei Typ-B-Ehemännern, die mit einer Typ-B-Frau verheiratet waren (25% vs. 7,8%). Typ-A-Ehemänner entwickelten zweimal eher als Typ-B-Männer eine koronare Herzkrankheit, wenn sie mit einer nicht-ehrgeizigen Frau verheiratet waren (28,3% vs. 14,3%).

Wurden Wechselwirkungen zwischen Verhaltenstyp der Ehemänner und Merkmalen ihrer Ehefrauen berechnet, fanden sich bei Blue-Collar-Ehefrauen signifikante Effekte; das bedeutet, daß Typ-A-Männer in Blue-Collar-Berufen im Vergleich zum Typ B nur dann ein höheres Risiko aufwiesen, wenn ihre Ehefrauen außerhalb des Hauses berufstätig waren, dem Typ B zugeordnet oder nicht ehrgeizig waren. Typ-A-Männer in White-Collar-Berufen wiesen hingegen ein höheres KHK-Risiko unabhängig von diesen Merkmalen ihrer Ehefrauen auf. Auch bei Berücksichtigung der traditionellen Risikofaktoren des Ehemannes blieben die Wechselwirkungen zwischen Verhaltenstyp des in Blue-Collar-Berufen arbeitenden Ehemannes und dem Beschäftigungsstatus und Verhaltenstyp seiner Ehefrau signifikant. Die Wechselwirkung zwischen Verhaltenstyp des Ehemannes und Ehrgeiz der Ehefrau war nach Kontrolle der traditionellen Risikofaktoren hingegen nicht mehr signifikant; hierfür können die hohen negativen Korrelationen verantwortlich sein, die zwischen den Variablen „Ehrgeiz der Ehefrauen" und „Rauchverhalten" ihrer Ehemänner (r = −0,45) und zwischen „Ehrgeiz der Ehefrauen" und „Typ-A-Verhalten der Ehemänner" (r = −0,49) gefunden wurden.

Die Autoren folgern aufgrund dieser Ergebnisse, daß es für einen Typ-A-Ehemann in Blue-Collar-Berufen wahrscheinlich eine Bedrohung seines Selbstvertrauens bedeutet – und/oder seines Gefühls, seine Umgebung kontrollieren zu können – wenn er mit einer Frau verheiratet ist, die außerhalb des Hauses arbeitet oder die einen hohen Ausbildungsgrad besitzt. Nachdem die traditionellen kardiovaskulären Risikofaktoren nicht als Mechanismen in Betracht kommen, die als Mediatoren zwischen Verhaltensweisen und zukünftiger Erkrankung fungieren, spekulieren die Autoren, daß verstärkte neurohumorale und kardiovaskuläre Reaktionen bei Typ-A-Männern in Abhängigkeit von Eigenschaften ihrer Ehefrauen eine Rolle spielen könnten.

Eine Bestätigung einiger dieser Ergebnisse kommt von einer anderen Fallkontrollstudie (Carmelli et al., 1985). Im Rahmen der Western Collaborative Group Study wurden zwischen 1970 und 1971 insgesamt 130 Familien ausgewählt, die an einer Untersuchung familiärer Einflüsse auf die beobachtete Inzidenz der koronaren Herzkrankheit teilnahmen. 55 Familienväter hatten seit Beginn der WCGS eine koronare Herzkrankheit entwickelt. Wie in der Framingham-Studie fand sich eine Interaktion zwischen Ausbildungsgrad der Ehefrau und Typ-A-Verhalten des Ehemannes. Das relative Risiko für die Entwicklung einer koronaren Herzkrankheit betrug für Ehemänner, die mit einer Frau mit 13 oder mehr Jahren an Schulbildung verheiratet waren, 3,6, wenn sie als Typ A eingestuft worden waren; beim Typ B betrug das relative Risiko hingegen nur 0,4. Typ-A-Ehemänner waren ebenfalls häufiger erkrankt, wenn sie mit Frauen verheiratet waren, die sich in einem Fragebogen (Thurstone) als aktiver, dominanter und emotional weniger stabil beschrieben. Im Gegensatz zur Framingham-Studie fanden sich keine Unterschiede im KHK-Risiko der Ehemänner bezüglich einer Berufstätigkeit ihrer Ehefrauen. Weitere soziale Merkmale und Verhaltensweisen der Ehefrauen, die mit einem erhöhten KHK-Risiko des Ehemannes unabhängig von dessen Typ-A-Verhalten assoziiert waren, betrafen eine erhöhte soziale Mobilität (z.B. höhere Schulbildung der Ehefrau im Vergleich zu deren Mutter) und eine negative Selbsteinschätzung eigener Fähigkeiten.

Im Vergleich zur Framingham-Studie bestanden einige Unterschiede bei der hier ausgewählten Stichprobe, die für einige der unterschiedlichen Ergebnisse verantwortlich sein könnten: Die Ehemänner waren älter, rauchten häufiger, wiesen einen höheren Serumcholesterinspiegel auf und waren durch einen höheren sozialen Status gekennzeichnet; die Ehefrauen waren ebenfalls älter, häufiger in White-Collar-Berufen beschäftigt und besaßen eine höhere Schulbildung.

Multiple Risk Factor Intervention Trial

Im Multiple Risk Factor Intervention Trial (MRFIT), einer prospektiven multizentrischen Studie, in der der Einfluß von therapeutischen Maßnahmen bezüglich kardiovaskulärer Risikofaktoren auf das Erkrankungs- und Sterberisiko überprüft werden sollte, wurde auch das Typ-A-Verhalten untersucht. In dieser Studie konnte keine Beziehung zwischen Typ-A-Verhalten und der koronaren Morbidität und Mortalität nachgewiesen werden (Shekelle et al., 1983 b, 1986). Dies galt sowohl für das strukturierte Interview (n = 3110), das an acht der teilnehmenden Zentren durchgeführt wurde, als auch für den Jenkins Activity Survey, der an allen 22 Zentren verwendet wurde

(n = 12772), und betraf sowohl die Interventionsgruppe, die eine spezielle Behandlung bezüglich traditioneller Risikofaktoren erhielt, als auch die Vergleichsgruppe, die auf übliche Art und Weise von niedergelassenen Ärzten behandelt wurde. Die Teilnehmer dieser Studie waren im Gegensatz zur WCGS aufgrund eines erhöhten kardiovaskulären Risikos in bezug auf traditionelle Risikofaktoren ausgesucht worden. Das Typ-A-Verhalten sagte in der WCGS aber die KHK-Inzidenz für alle Risikofaktoren vorher, so daß ein entsprechendes Ergebnis auch im MRFIT zu erwarten gewesen wäre. Auch andere Studien, in denen ausschließlich „Hochrisikopatienten" untersucht wurden, ergaben alle negative Ergebnisse hinsichtlich des Typ-A-Verhaltens (Matthews und Haynes, 1986). Für diese negativen Ergebnisse werden zum Teil Selektionsmechanismen bei der Auswahl der Studienteilnehmer verantwortlich gemacht (Shekelle et al., 1986). Daß nur bestimmte, sehr motivierte Patienten bereit waren, am MRFIT teilzunehmen, wird auch dafür verantwortlich gemacht, daß die Mortalitätsraten insgesamt sehr viel geringer als erwartet ausfielen. Das hat insgesamt die Aussagekraft dieser Studie beeinträchtigt. Möglicherweise haben ungeduldige und immer in Zeitnot befindliche und kardiovaskulär gefährdete Typ-A-Personen sich seltener bereit gefunden, an der Studie teilzunehmen. Rosenman wies auch auf methodische Schwierigkeiten bei der Standardisierung und Durchführung der Interviews hin (Rosenman, 1988).

Weitere Studien zum Typ-A-Verhalten mit teilweise widersprüchlichen Ergebnissen

Zwei weitere prospektive amerikanische Studien, in denen der Jenkins Activity Survey bei Myokardinfarktpatienten verwendet wurde, erbrachten negative Ergebnisse hinsichtlich des Typ-A-Verhaltens. Bei 2314 Teilnehmern der Aspirin Myocardial Infarctions Study (AMIS), 2070 Männern und 244 Frauen, die einen Myokardinfarkt überlebt hatten, konnte keine signifikante Beziehung zwischen dem JAS-Typ-A/B-Score und dem Reinfarktrisiko nachgewiesen werden (Shekelle et al., 1985). Dies galt für Männer, Frauen sowie eine Untergruppe von Männern, die in höheren Berufsgruppen oder als Techniker oder Manager beschäftigt waren, und betraf sowohl die aufgrund der Rohdaten vorgenommenen Berechnungen, als auch diejenigen nach statistischer Berücksichtigung potentieller konfundierender Variablen. Allerdings wurde der JAS erst nach Randomisierung der AMIS ausgeteilt und von 14% der Teilnehmer waren keine Daten erhältlich; dies könnte zu einem Vorurteil in dieser Studie geführt haben. Auch könnte die Tatsache die Ergebnisse beeinflußt haben, daß es sich um Patienten handelte, die freiwillig an einer klinischen Interventionsstudie teilnahmen; dies könnte im Gegensatz zu nicht-ausgewählten Patienten mit systematischen Unterschieden in bezug auf Persönlichkeitsmerkmale verknüpft sein.

In einer Studie von Case und Mitarbeitern (1985) fand sich bei 516 Myokardinfarktpatienten des Multicenter Post-Infarction Project keine Beziehung zwischen dem JAS-Typ-A/B-Score und der Gesamt- oder kardialen Mortalität in einem ein- bis dreijährigen Zeitraum. Eine Funktionsbeeinträchtigung des linken Ventrikels erwies sich als wichtiger Risikofaktor, der wahrscheinlich alle Effekte, die das Typ-A-Verhalten auf die Langzeitprognose ausüben könnte, verdeckt. Bei dieser Studie wurden einige Einwände erhoben, die bei einer Bewertung berücksichtigt werden müssen: Der Follow-up-Zeitraum betrug ein bis drei Jahre, wobei nicht überprüft wurde, ob er im Mittel für Typ-A- und Typ-B-Patienten gleich lang war; deswegen könnten Typ-A-Patienten über einen kürzeren Zeitraum als Typ-B-Patienten nachverfolgt worden sein, was die Wahrscheinlichkeit für das Auftreten eines Reinfarktes bei Typ-A-Patienten eingeschränkt hätte; deswegen müßte die Dauer der Follow-up-Periode bei den Analysen als konfundierender Faktor berücksichtigt werden. Ein anderer Einwand ist, daß als Auswahlkriterium für eine Aufnahme der Patienten in die Studie ein Überleben des ersten Infarktes von nur zwei Wochen festgelegt wurde, d.h., daß Patienten mit einer relativ schweren myokardialen Schädigung und einer schlechten Prognose ausgewählt wurden, die den JAS-Fragebogen bereits zwei Wochen nach ihrem Herzinfarkt ausfüllten. Es ist nicht zu erwarten, daß bei diesen Patienten das Typ-A-Verhalten noch eine wichtige Determinante für das Fortschreiten der Erkrankung darstellt. Pickering (1985) folgert aus dieser Studie, daß die Hauptrisikofaktoren für das Überleben nach einem Herzinfarkt nicht die gleichen sind wie diejenigen, die den ersten Infarkt vorhersagen; nachdem in der Studie von Case offensichtlich auch ein Einfluß der drei wichtigsten Risikofaktoren – Erhöhung von Serumcholesterin und Blutdruck sowie Zigarettenrauchen – nicht nachweisbar war, sieht er es als nicht verwunderlich an, daß auch ein Einfluß des Typ-A-Verhaltens nicht demonstrierbar war. Ein von Friedman (1985) immer wieder erhobener Einwand betrifft die Verwendung des JAS als Meßinstrument des Typ-A-Verhaltens allgemein; der JAS erfaßt zwei wichtige Komponenten des Typ-A-Musters nicht, nämlich Feindseligkeit (hostility) und Ärger, denen wahrscheinlich aber die wichtigste prädiktive Bedeutung zukommt. Nach diesen Befunden erscheint der JAS nicht als Meßinstrument geeignet, die Prognose nach einem Herzinfarkt abzuschätzen.

Auch in Europa wurden mehrere prospektive Untersuchungen zum Typ-A-Verhalten durchgeführt. In der French-Belgium Collaborative Group Study (French-Belgium Collaborative Group, 1982) wurde das mit Hilfe der Bortner-Skala (Bortner, 1969) bestimmte Typ-A-Verhalten mit der KHK-Inzidenz in Verbindung gebracht. Drei Kohorten von gesunden Männern (insgesamt n = 2811) im Alter zwischen 40 und 60 Jahren (aus Brüssel, Genf, Marseille, Paris) wurden über einen Zeitraum von 35,5–74,1 Monaten nachverfolgt. Diejenigen Teilnehmer der Untersuchung, die während des Beobachtungszeitraumes an Angina pectoris erkrankten oder einen tödlichen oder nicht-tödlichen Myokardinfarkt erlitten oder am plötzlichen Herztod starben, wiesen im Vergleich zu den nicht KHK-Kranken ein ausgeprägteres Typ-A-Muster auf (allerdings nicht signifikant). Bezüglich der KHK-Untergruppe „harte kardiale Ereignisse" (tödlicher oder nicht-tödlicher Myokardinfarkt, plötzlicher Herztod) und der Gruppe aller KHK-Manifestationen (Angina pectoris oder hartes kardiales Ereignis) konnte ein signifikant höherer Skalenwert bei Erkrankten gesichert werden. In einer multivariaten Analyse, die die Bortner-Skala und die traditionellen Risikofaktoren Alter, Rauchen, Cholesterin und systolischer Blutdruck einschloß, wurde deutlich, daß das Typ-A-Verhalten sowohl für die harten kardialen Ereignisse als auch für alle KHK-Manifestationen ein eigenständiger Risikofaktor war. Auch nach Einschluß eines weiteren Persönlichkeitsmerkmals, Neurotizismus, in die multivariate Analyse, behauptete sich das Typ-A-Verhalten als eigenständiger Risikofaktor.

Im Rahmen des Belgian Heart Disease Prevention Project (BHDPP) fand sich nach 6 Jahren in der Interventionsgruppe keine Beziehung zwischen der Typ-A-Skala des JAS, der Bortner-Skala und der KHK-Inzidenz. Bei 10% der Kontrollgruppe wurden sowohl das strukturierte Interview durchgeführt, als auch der JAS eingesetzt. Hier standen die Skalen des JAS signifikant in Beziehung zur KHK-Inzidenz (hard events); lediglich die Job-Involvement-Skala des JAS wies einen negativen Gradienten auf, d.h. berufliches Engagement hatte hier möglicherweise eine Schutzfunktion. Mit dem strukturierten Interview bestimmte Typ-A-Personen wiesen ein 2,8mal höheres relatives Risiko im Vergleich zu Typ-B-Personen auf (p = 0,07). In der multivariaten Analyse, die die anderen Risikofaktoren einschloß, blieb die signifikante Beziehung zu den Skalen JAS A/B und Hard-Driving bestehen, wohingegen die Typ-A/B-Einteilung mit Hilfe des strukturierten Interviews die Signifikanzgrenze knapp verfehlte (Kittel, 1986; Kornitzer, 1985).

In einer holländischen prospektiven Studie über neuneinhalb Jahre wurden über 3000 Teilnehmer in Rotterdam mit dem JAS, sowie eine Untergruppe von 243 Teilnehmern mit dem strukturierten Interview untersucht. Für das im strukturierten Interview bestimmte Typ-A-Verhalten ergab sich kein signifikanter Zusammenhang, wenn tödliche Myokardinfarktereignisse zusammengefaßt wurden. Tödliche Infarkte traten nur in der Typ-A-Gruppe auf; hier war der Unterschied zwar signifikant, beruhte aber auf der Fallzahl von nur 4 Verstorbenen. Die JAS-Skalen konnten für ca. 2500 Teilnehmer ausgewertet werden. Auch hier fand sich auf keiner Skala eine signifikante Beziehung zu tödlichen und nicht-tödlichen Infarkten. Die holländische JAS-Adaptation zeigte aber zur Angina pectoris und zu koronaren Bypass-Operationen eine signifikante Beziehung, was auch für die JAS-Harddriving-Skala galt (Appels et al., 1986).

Williams et al. (1980) bestätigten den Zusammenhang zwischen Typ-A-Verhalten und Schweregrad der Koronarsklerose und wiesen eine Beziehung zu einem weiteren Faktor nach: das mit Hilfe des MMPI bestimmte Ausmaß von Feindseligkeit (hostility). Wurden beide Faktoren in einer multivariaten Analyse berücksichtigt, reduzierte sich die Signifikanz des Typ-A-Verhaltens von p = 0,01 auf p = 0,05, die Signifikanz der Feindseligkeit wurde von p = 0,02 auf p = 0,008 verstärkt. Bei über 2200 Patienten, die zwischen 1976 und 1980 im Duke University Medical Center, Durham, koronarangiographiert wurden und bei denen das strukturierte Interview durchgeführt wurde (Williams, 1988), zeigte sich eine signifikante Beziehung zwischen Typ A und Koronarsklerose, die altersabhängig war. Patienten unter 50 Jahren wiesen eine ausgeprägte Koronarsklerose auf, wenn sie dem Typ A zugeordnet waren; in der Patientengruppe ab 55 Jahre und älter fand sich hingegen bei den Typ-B-Patienten eine stärkere Koronarsklerose. Dies läßt sich so interpretieren, daß unter den Patienten, die zur Koronarangiographie überwiesen wurden, die jungen Typ-A-Patienten bereits eine stärkere Koronarsklerose entwickelt hatten, was ihre Überlebenschancen gegenüber den Typ-B-Personen verringerte; das Typ-B-Verhalten, so schließen die Autoren, schütze in jüngeren Jahren vor einer vorzeitigen Entwicklung einer Koronarsklerose, nicht jedoch in höherem Alter. Die jüngeren Typ-A-Patienten könnten danach eher sterben und es bleiben dann nurmehr die älteren resistenteren Patienten übrig. Entsprechende Erklärungen wurden auch in der Framingham-Studie herangezogen, um den sich stark abschwächenden und zum Teil sich umkehrenden Effekt der Risikofaktoren Rauchen und erhöhtes Serumcholesterin in fortgeschrittenem Alter zu erklären. In der Studie von Williams schwächte sich auch die Beziehung zwischen Serumcholesterin und Koronarsklerose mit zunehmendem Alter ab.

Die toxischen Komponenten des Typ-A-Verhaltens

Das Typ-A-Verhalten galt immer als multidimensionales Konstrukt.

Ein Individuum kann als Typ A beurteilt werden, ohne daß gleichzeitig alle Typ-A-Charakteristika vorliegen müssen. Nicht allen Komponenten kommt die gleiche Bedeutung für die Typenklassifikation zu. Im strukturierten Interview sind die Hauptkriterien laute, explosible, schnelle und akzelerierte Sprechweise und kurze Antwortlatenz. Die toxischen Komponenten sind aber eher Feindseligkeit/Aggressionsbereitschaft bzw. verbales Rivalitätsverhalten, die gewöhnlich niedrigere Korrelationskoeffizienten mit der Typ-A-Klassifikation aufweisen (Schmidt, 1988; Dembroski, 1989). Dies weist darauf hin, daß in der Typ-A-Beurteilung toxische und nicht-toxische Komponenten miteinander vermischt wurden.

In einer Teilstichprobe der WCGS (63 KHK-Fälle im Vergleich zu 124 Kontrollpersonen) wurden einzelne Komponenten des Typ-A-Musters untersucht (Matthews et al., 1977). Nur einige Komponenten konnten zukünftige KHK-Kranke von Gesunden signifikant unterscheiden. Der stärkste Prädiktor war Aggressionsbereitschaft (potential for hostility; p < 0,003), gefolgt von nach außen gerichtetem Ärger (anger directed outward; p < 0,01), Rivalitätsverhalten (competitiveness; p < 0,01), kraftvolle explosible Sprechweise (p < 0,05). Aktivitätsniveau, Anerkennungsbedürfnis und berufliches Engagement waren hingegen nicht mit der Herzinfarktinzidenz assoziiert. Diese Untersuchung wies erstmals auf die Bedeutung einer Komponentenanalyse des Typ-A-Musters hin. Die Vorgehensweise kann hierbei in Analogie zur Aufspaltung des Serumcholesterins in einzelne Lipoproteinfraktionen angesehen werden, die sich dann entweder als Schutzfaktor wie das HDL- oder als Risikofaktor wie das LDL-Cholesterin erwiesen.

In einer neueren Auswertung, die rund 250 KHK Fälle der WCGS 500 gesunden hinsichtlich Alter und Risikofaktoren vergleichbaren Kontrollpersonen gegenüberstellte, erwies sich „potential for hostility" als einzige Variable im strukturierten Interview, die in multivariaten Analysen signifikant mit der KHK-Inzidenz assoziiert war (Hecker et al., 1988). Dembroski bestätigte diese Befunde mit einer von ihm entwickelten Komponentenanalyse des strukturierten Interviews am gleichen Material.

Diese Komponentenanalyse des Typ-A-Verhaltens erlaubt eine differenzierte Erfassung von verschiedenen Komponenten, insbesondere auch der Merkmale Feindseligkeit/Aggressionsbereitschaft (Dembroski, 1978; Dembroski und MacDougall, 1983 a; Dembroski und Costa, 1987). Dembroski und Mitarbeiter (1985) untersuchten koronarangiographierte Patienten, bei denen das strukturierte Interview durchgeführt worden war, mit dieser Komponentenanalyse. Sie wählten aus dem aus über 2000 Patienten bestehenden Kollektiv von Williams in Duke University Durham jeweils 50 Patienten mit einer 3- und einer 2-Gefäßerkrankung aus sowie 50 Patienten, bei denen keine arteriosklerotischen Koronarveränderungen nachweisbar waren. 131 Patienteninterviews konnten mit der Komponentenanalyse ausgewertet werden. Im Gegensatz zu früheren Untersuchungen an Patienten, die aus der gleichen Population stammten (Blumenthal et al., 1978; Williams et al., 1980), konnte keine Beziehung zwischen globalem Typ-A-Muster und Schweregrad der Arteriosklerose gefunden werden. Nach multivariater Kontrolle der traditionellen Risikofaktoren waren allerdings Aggressionsbereitschaft und gleichzeitiges Zurückhalten von Ärger mit dem Schweregrad der Koronarsklerose, Angina pectoris und der Anzahl vorangegangener Herzinfarkte signifikant verknüpft. Wichtig erscheint, daß Aggressionsbereitschaft und globales Typ-A-Muster signifikant miteinander korrelierten (r = +0,57), obwohl letzteres nicht in Beziehung zur KHK stand. Jüngere Patienten (unter 40 Jahren) mit hohen Werten für Aggressionsbereitschaft hatten 3–4mal häufiger schwere Koronarveränderungen als solche mit niedrigen Werten (Williams, 1989).

Diese Befunde weisen darauf hin, daß „potential for hostility" wahrscheinlich **die** toxische Komponente des Typ-A-Verhaltens ist, und lassen vermuten, daß die Beziehung zwischen Typ-A-Muster und KHK in den verschiedenen Studien mit zum Teil sich widersprechenden Ergebnissen davon abhängig sein kann, welches Gewicht die Auswerter jeweils dieser Komponente bei der Typ-A-Klassifikation zugemessen haben. Diese Komponente ist in mehreren Studien sowohl mit der KHK als auch mit dem Ausmaß kardiovaskulärer Reaktionen in verschiedenen psychophysiologischen Testsituationen verknüpft worden (Arrowood et al., 1980; Barefoot et al., 1983; Dembroski et al., 1978, 1979 a, b, 1983b; Matthews et al., 1977; Shekelle et al., 1983 a; Williams et al., 1980). Es erscheint daher möglich, daß die Typ-A-Komponente Aggressionsbereitschaft/Feindseligkeit möglicherweise sogar auf genetischer Basis, wie eine Zwillingsstudie nahelegt (Matthews et al., 1983), gemeinsam mit erlernten Verhaltensmustern wie der Unterdrükkung von Ärger eine Basis koronargefährdender Verhaltensweisen darstellt.

Für eine Untergruppe von 80 Patienten der Studie von Dembroski et al. (1985) lagen gleichzeitig Werte der Cook-Medley-Skala des MMPI vor, das mit dieser Skala erfaßte Ausmaß an Feindseligkeit korrelierte 0,37 mit den Werten für Aggressionsbereitschaft im strukturierten Interview, was hinsichtlich der unterschiedlichen Methoden (Interview vs. Fragebogen) hinreichend gut erscheint. Mit anderen Fragebogen zur Messung von Feindseligkeit korrelierte die Cook-Medley-Skala beispielsweise auch nur 0,59 (Spielberger et al., 1983).

Die eben beschriebenen Resultate replizierten MacDougall et al. (1985). Diese Studie basiert auf den Interview-Ergebnissen von 126 Angiographiepatienten in einer Untersuchung von Dimsdale, in der erstmals keine Beziehung zwischen Schweregrad der Koronarsklerose und Typ-A-Muster sowohl im strukturierten Interview als auch im JAS gefunden wurde (Dimsdale et al., 1978). Die erneute Auswertung der Interviews mit der Komponentenanalyse durch ein anderes Team bestätigte erneut Dimsdales Befund, daß keine signifikante Beziehung zwischen Ausmaß der Koronarsklerose und globalem Typ-A-Muster im strukturierten Interview bestand. Nach Kontrolle des Alters in der multivariaten Analyse wurde jedoch wiederum ein signifikanter Zusammenhang zwischen Koronarsklerose und dem Zurückhalten von Ärger und Aggressionsbereitschaft gefunden.

Eine andere Forschergruppe verwendete die erste Version der Komponentenanalyse von Dembroski in einer Angiographiestudie (Arrowood et al., 1982). Auch sie fand keinen Zusammenhang zwischen Ausmaß der Koronarsklerose und globalem Typ-A-Muster im strukturierten Interview, wohl aber für die Komponenten explosible Sprechweise, verbales Rivalitätsverhalten und Aggressionsbereitschaft; hier wurden jedoch keine multivariaten Analysen vorgenommen, so daß die Unabhängigkeit dieser Faktoren nicht gesichert ist.

Insgesamt weisen diese Studien jedoch übereinstimmend auf die Bedeutung der Komponenten Aggressionsbereitschaft/Feindseligkeit und Unterdrükken von Ärger für den Schweregrad der koronarangiographisch nachweisbaren Arteriosklerose hin. Darüber hinaus stehen sie auch in Übereinstimmung mit prospektiven Untersuchungen, in denen diese Faktoren mit der KHK-Inzidenz assoziiert sind. Unterdrücken von Ärger war in der Framingham-Studie (Haynes et al., 1980) mit einer erhöhten KHK-Inzidenz verknüpft.

Feindseligkeit als Prädiktor der koronaren Herzkrankheit

Zwei prospektive Studien bringen die mit der Cook-Medley-Skala (Cook und Medley, 1954) des MMPI bestimmte Feindseligkeit in Verbindung mit der Entwicklung einer koronaren Herzkrankheit.

1877 männliche Angestellte der Hawthorne Werke der Western Electric Company in Chicago füllten im Zeitraum von 1957–1958 den MMPI-Fragebogen erstmals aus; 1653 Teilnehmer wiederholten dies 4 Jahre später zwischen 1961–1962 ein weiteres Mal. Bis 1969 wurden jährliche Untersuchungen zur Entdeckung einer möglicherweise neu aufgetretenen koronaren Herzkrankheit durchgeführt und zuletzt 1978 ein Follow-up bezüglich der Mortalität. Shekelle und Mitarbeiter (1983) untersuchten in der Western-Electric-Studie die Beziehung zwischen den Werten der Hostility-Skala und der 10-Jahres-Inzidenz der KHK sowie der 20-Jahres-Mortalität. Die Test-Retest-Korrelation für die Hostility-Werte über den vierjährigen Zeitraum betrug 0,84; das bedeutet, daß es sich zumindest bei der hier untersuchten Gruppe um ein sehr stabiles psychologisches Merkmal handelt. Die Inzidenz im zehnjährigen Zeitraum war nach Berücksichtigung anderer koronarer Risikofaktoren wie Alter, systolischer Blutdruck, Serumcholesterin, Zigarettenrauchen und Alkoholkonsum bei Personen mit Scorewerten über zehn 1,47mal größer als bei Männern mit niedrigeren Werten. Es fand sich ebenfalls eine positive signifikante Assoziation der Hostility-Skala mit dem Risiko, infolge einer koronaren Herzkrankheit oder einer malignen Krebserkrankung oder überhaupt innerhalb von 20 Jahren zu

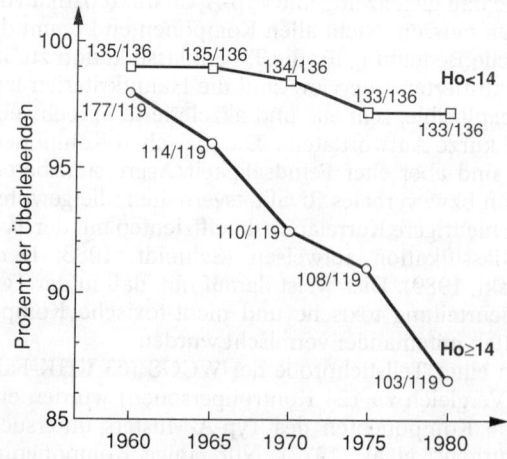

Abb. 41.1–2. Überlebenskurven von 136 Ärzten mit niedrigen (< Median) und 119 Ärzten mit hohen (≥ Median) Hostility-Werten (Ho) über einen Zeitraum von 25 Jahren (Barefoot et al., 1983).

sterben. Ein Unterschied von 23 Punkten auf dieser Skala entsprach der Differenz zwischen den Mittelwerten des untersten und des oberen Quintils der Scorewerte und war mit einem 42%igen Anstieg des Risikos zu sterben verknüpft.

In einer weiteren prospektiven Studie wurde die gleiche Skala bei 343 Medizinstudenten verwendet, die zwischen 1954 und 1959 den MMPI ausgefüllt hatten (Barefoot et al., 1983). 42 Studenten hatten diesen Fragebogen ein zweites Mal ein Jahr später beantwortet (Test-Retest-Reliabilität r = + 0,85). 1981 wurden alle Teilnehmer angeschrieben und nach ihrem Gesundheitszustand befragt. Die endgültige Stichprobe bestand aus 255 Ärzten, das sind 74% der Gesamtstichprobe. Die Todesursachen wurden mit Hilfe des Totenscheins, Aufzeichnungen der Alumnats der Medical School und Befragung von Familienangehörigen und Verwandten ermittelt.

Die KHK-Inzidenz war bei denjenigen, die Hostility-Scorewerte oberhalb des Medians von 14 aufwiesen, im Vergleich zu denjenigen mit niedrigeren Werten um das Fünffache erhöht. Da keine Autopsiebefunde vorlagen und die Bestimmung der Todesursachen somit ungenau gewesen sein konnte, wurden die Verstorbenen aus einer weiteren Analyse ausgeschlossen; es fand sich dann ebenfalls ein fast sechsfacher Unterschied bezüglich der Inzidenz klinisch relevanter KHK-Ereignisse zwischen Personen mit Hostility-Werten ober- und unterhalb des Medianwertes. Andere Risikofaktoren schieden als konfundierende Variablen aus. Innerhalb von 25 Jahren starben nur 2,2% aus der Gruppe mit niedrigen Hostility-Werten, aus der Gruppe mit hohen Werten waren es hingegen 13,4% (Abb. 41.1–2).

Eine Erklärung für die im Vergleich zu der jüngeren Ärztegruppe geringere Erhöhung des Erkrankungs- und Sterberisikos durch hohe Hostility-Werte bei den älteren Teilnehmern der Western-Electric-Studie liegt möglicherweise darin, daß diejenigen, die für die Folgen hoher Hostility-Werte besonders empfänglich waren, bereits gestorben waren.

McCranie und Mitarbeiter (1986) wiederholten die Untersuchung von Barefoot bei 478 Ärzten, die 25 Jahre zuvor den MMPI anläßlich ihrer Aufnahmeprüfung in ein Medical College beantwortet hatten. Es fand sich keine Assoziation zwischen den Hostility-Werten und der KHK-Inzidenz oder der Gesamtmortalität. Die Scorewerte der befragten Ärzte waren im Vergleich zu den anderen Studien deutlich niedriger, was die Autoren mit der Prüfungssituation in Zusammenhang bringen. Die Fragen waren hier anders als in der Barefootschen Untersuchung stärker im Sinne einer sozialen Erwünschtheit beantwortet worden, da von den Ergebnissen dieser Aufnahmeprüfung abhing, ob die Studenten in die Medical School aufgenommen wurden. Entsprechend waren auch die Werte auf der K-Skala des MMPI, die die Neigung anzeigen, sich selbst in einem möglichst guten Licht erscheinen zu lassen, deutlich erhöht. Diejenigen, die fälschlicherweise niedrige Werte angeben, können aber leider nicht von denjenigen unterschieden werden, die ehrlich sind.

Auch die Feindseligkeitsdimension umfaßt verschiedene Komponenten, von denen wahrscheinlich nur einige mit der koronaren Herzkrankheit assoziiert sind. In einer Angiographiestudie war ein Hostilitätsmaß, das nicht mit neurotischen Tendenzen korrelierte, bei Patienten unter 60 Jahren positiv mit dem Schweregrad der KHK assoziiert, und ein Feindseligkeitsmaß, das in Beziehung zu neurotischen Tendenzen stand, wies eine negative Assoziation auf (Siegman et al., 1987).

Von Williams (1989) wurden 118 Rechtsanwälte mit dem MMPI untersucht und 25 Jahre lang verfolgt. Es fand sich eine signifikante Beziehung zwischen den Cook-Medley-Hostilitätswerten und der Mortalitätsrate; sie lag bei der Gruppe mit niedrigen Werten unter 5% und bei der Gruppe mit hohen Werten bei 20%. Eine Unterteilung der Cook-Medley-Skala in verschiedene Unterskalen ergab, daß Zynismus, feindseliger Affekt oder aggressives Reagieren jeweils für sich genommen einen signifikanten Anstieg der Mortalitätsraten ebensogut vorhersagten wie die gesamte Skala; feindselige Attributionen, soziales Vermeiden und andere standen in keinem Zusammenhang mit der Mortalität. Wurden die Items der drei signifikanten Skalen zu einer Skala zusammengefaßt, so konnten die erhöhten Mortalitätsraten bei weitem reliabler und besser vorhergesagt werden als mit der ursprünglichen Skala. Eine ähnliche Verbesserung der Prädiktion wurde gefunden, wenn die Daten der Untersuchung von Barefoot et al. (1983) mit Medizinstudenten auf die gleiche Weise analysiert wurden. Diese Skala reflektiert eine zynische Sichtweise der Menschen, ohne Vertrauen, mit häufigem Erleben negativer Emotionen im Umgang mit anderen und dem häufigen Ausdruck von Ärger und Aggression bei Konfrontation mit Problemen.

Eine nachträgliche Analyse des MRFIT, die von Dembroski und MacDougall (1989) mit Hilfe ihrer Komponentenanalyse vorgenommen wurde, bestätigte die prädiktive Bedeutung der Hostilitätskomponente über einen durchschnittlichen Follow-up-Zeitraum von 7,1 Jahren. 192 KHK-Fälle (62 KHK-Todesfälle und 130 dokumentierte Myokardinfarkte) wurden mit 384 Kontrollpersonen verglichen. Von 8 verschiedenen Typ-A-Komponenten fanden sich in den logistischen Regressionsanalysen nur zwei signifikante Prädiktoren: das „potential for hostility" und die antagonistische interpersonelle Komponente „stylistic hostility". Nach Berücksichtigung der traditionellen Risikofaktoren und Stratifizierung in eine jüngere (< 47 Jahre) und ältere Gruppe (> 47 Jahre) betrug in der jüngeren Gruppe das relative Risiko für das dichotomisierte Merkmal Aggressionsbereitschaft 2,1 (p < 0,011) und für das ordinalskalierte Merkmal „stylistic hostility" 1,9 (p < 0,016). In der Gruppe der älteren Teilnehmer fanden sich hingegen keine signifikanten Zusammenhänge.

Welche Mediatoren sind zwischen den Komponenten einer ausgeprägten Feindseligkeit und einer erhöhten Sterblichkeit wirksam, über welche Mechanismen erhöhen sie das Krankheitsrisiko? Einer dieser Mediatoren könnte ein Mangel an sozialer Unterstützung sein. Berkman und Syme (1979) sowie andere Autoren haben gezeigt, daß Personen mit wenig sozialen Kontakten eine höhere Sterblichkeit bezüglich aller Todesursachen aufweisen. Diejenigen, die andere Menschen nicht mögen, ihnen mißtrauen und feindlich gesonnen sind, erfahren wohl nicht ausreichende Unterstützung. Diese Hypothese deckt sich mit der Beobachtung einer negativen Korrelation zwischen Hostility-Scorewerten und der Qualität aber nicht Quantität der sozialen Unterstützung bei etwa 1100 Koronarangiographiepatienten (Barefoot et al., 1983).

Weitere Ergebnisse des MRFIT weisen auf einen anderen prädiktiven Faktor hin (Scherwitz et al., 1986): Ichbezogenheit, gemessen mit Hilfe der Häufigkeit der Wörter „I, me, my" im strukturierten Interview (als Gesamtsumme und Dichte, d.h. Anzahl dieser Wörter pro Anzahl der Satzeinheiten bestehend aus Subjekt, Objekt und Prädikat). Frühere Untersuchungen hatten ergeben, daß Typ-A-Personen im Interview häufiger auf sich selbst Bezug nehmen (Scherwitz et al., 1978) und daß ein höheres Ausmaß der so bestimmten Ichbezogenheit mit stärkeren systolischen und diastolischen Blutdruckreaktionen verknüpft ist. Auch der Schweregrad der angiographisch nachweisbaren Koronarsklerose stand mit dieser Variablen in Beziehung (Scherwitz et al., 1983). 193 Teilnehmer der MRFIT-Studie, die im Zeitraum von sieben Jahren eine KHK entwickelten, wurden mit 384 parallelisierten gesunden Teilnehmern verglichen. Andere Risikofaktoren wie Alter, diastolischer Blutdruck, Cholesterin, Zigarettenrauchen und Typ-A-Verhalten wurden in einer multiplen logistischen Regressionsanalyse kontrolliert. Zukünftige KHK-Kranke verwendeten insgesamt die Wörter „I, me, my" häufiger im Interview, aber nicht in größerer Dichte. Diejenigen, die infolge einer KHK starben, benutzten diese Wörter häufiger (relatives Risiko [RR] = 1,62) und in größerer Dichte (RR = 1,54). Beide Variablen waren keine Prädiktoren für Angina pectoris und nicht-tödlichen Myokardinfarkt. Bei denjenigen, die an einem Infarkt erkrankten, erwies sich von allen erfaßten Risikofaktoren die Häufigkeit dieser Wörter im strukturierten Interview als stärkster Prädiktor der Mortalität (Scherwitz et al., 1986).

Metaanalysen

Die Ergebnisse vieler Studien zum Typ-A-Verhalten und seiner Komponenten machen deutlich, daß aufgrund von positiven oder negativen Ergebnissen einer einzigen Studie nicht weitreichende Schlußfolgerungen gezogen werden können. Eine zusammenfassende Bewertung der verschiedenen Studien in quantitativen Metaanalysen erlaubt eine bessere Übersicht und Einordnung der unterschiedlichen Befunde und eine gültigere allgemeine Beurteilung. Derartige quantitative Metaanalysen wurden kürzlich für die Bedeutung des Typ-A-Verhaltens, einzelner seiner Komponenten und verwandter psychologischer Konzepte als koronare Risikofaktoren durchgeführt

(Booth-Kewley und Friedman, 1987). Noch nicht berücksichtigt wurden hierbei die Ergebnisse des 22-Jahre-Follow-up der WCGS und des 4,5-Jahre-Follow-up des Recurrent Coronary Prevention Project (RCPP, Friedman et al., 1988), einer Interventionsstudie zum Typ-A-Verhalten, über die noch berichtet werden soll, sowie einige neuere Studien, die das Merkmal „potential of hostility" untersuchen.

Booth-Kewley und Friedman (1987) verwenden dabei die Metaanalysen zur Beantwortung einer Reihe von Fragen:

- Gibt es ausreichend Evidenz dafür, daß eine reliable Beziehung zwischen Typ-A-Verhalten und KHK besteht, und wenn ja, in welcher Größenordnung liegt sie?
- Bestehen reliable Unterschiede zwischen den Typ-A-Maßen im strukturierten Interview und im JAS in bezug zur KHK?
- Welche Komponenten des Typ-A-Musters weisen die stärkste Beziehung zur KHK auf?
- Stehen andere Persönlichkeitsmerkmale wie Ärger und Depression reliabel zur KHK in Beziehung, und wenn ja, in welcher Größenordnung?
- Stehen Persönlichkeitsvariablen in differentieller Beziehung zu verschiedenen KHK-Endpunkten wie Myokardinfarkt, Angina pectoris usw.?
- Sind die Ergebnisse bei prospektiven Studien und Querschnittsuntersuchungen unterschiedlich?
- Was ist bekannt über mögliche Mediatoren der beobachteten Beziehung zwischen Persönlichkeit und KHK? Stehen die prädiktiven Persönlichkeitsvariablen untereinander in Beziehung oder gibt es Zusammenhänge mit anderen Risikofaktoren?
- Hat sich die Assoziation zwischen Typ-A-Verhalten und KHK im Laufe der Zeit verändert?

Als Datenbasis für die quantitativen Analysen wurden insgesamt 87 Querschnitts- und Längsschnittstudien herangezogen, die die Berechnung eines Effektmaßes, dem Produkt-Moment-Korrelationskoeffizient zwischen Persönlichkeitsvariable und Krankheitsendpunkt, erlauben. Einzelne Metaanalysen wurden für 18 Kategorien von Persönlichkeitsvariablen bzw. deren Kombination (Tab. 41.1–3) und fünf Krankheitskategorien vorgenommen. Die Analysen wurden außerdem getrennt für prospektive und für Querschnittsstudien durchgeführt, sowie für Studien vor und nach 1977.

In Tabelle 41.1–3 sind die Ergebnisse der Metaanalysen, die alle Studien, d.h. alle Querschnittsuntersuchungen und alle prospektiven Studien, einbeziehen, wiedergegeben, wobei alle kardiovaskulären Krankheitsgruppen zusammengefaßt wurden. In der linken Spalte sind die Persönlichkeitsvariablen aufgeführt, in der nächsten Spalte folgt die Anzahl der Publikationen für jede dieser Variablen, sowie die Zahl der in der Analyse verwendeten unabhängigen Effektmaße, das durchschnittliche Effektmaß (r), das durchschnittliche normalstandardisierte Effektmaß (z), die dazugehörige Irrtumswahrscheinlichkeit (p) und das „fail-safe N", das die Anzahl nicht-publizierter Studien mit einem durchschnittlichen Nullergebnis

Tabelle 41.1–3. Metaanalyse und durchschnittliche Effektmaße von Querschnitts- und Längsschnittuntersuchungen zu Typ-A-Verhalten, verschiedenen Typ-A-Komponenten und anderen Persönlichkeitsvariablen für alle kardiovaskulären Krankheitskategorien zusammengefaßt (nach Booth-Kewley und Friedman, 1987)

alle Studien

Persönlichkeits-variable	Zahl der Publikatio-nen	Zahl der un-abhängigen Effektmaße	durch-schnittliches Effektmaß	durchschnitt-liches normal-standardisiertes Effektmaß		Fail-safe
			r	z	p	N
Typ A, alle Maße	55	51	0,136	12,86	<0,0000001	3066
JAS Typ A	25	23	0,067	4,59	0,000002	157
Speed/Impatience	8	8	0,058	2,99	0,0014	19
Job Involvement	7	7	−0,020	−1,22	0,1109	−
Hard Driving	7	7	0,153	4,08	0,00002	37
SI Typ A	22	20	0,221	11,19	<0,0000001	906
SI Time Urgency	3	5	0,095	2,42	0,0008	6
Speed/Impatience, Time Urgency, alle Maße	14	14	0,087	4,13	0,00002	75
Job Involvement, alle Maße	10	10	0,029	1,05	0,1459	−
Rivalitätsverhalten/ Hard Driving/ Aggressivität	15	14	0,198	8,01	<0,0000001	318
Ärger/Feindseligkeit/ Aggression	24	25	0,121	6,27	<0,0000001	339
Ärger/Feindseligkeit	19	20	0,138	6,83	<0,0000001	325
Ärger	8	7	0,077	1,81	0,0350	2
Feindseligkeit	12	14	0,160	7,31	<0,0000001	263
Aggression	5	6	0,071	1,68	0,0464	1
Depression	15	13	0,205	6,16	<0,0000001	170
Extraversion	12	15	0,071	2,80	0,0026	29
Angst	15	15	0,124	5,34	<0,0000001	144

Tabelle 41.1–4. Metaanalyse der wichtigsten Persönlichkeits- und Verhaltensvariablen in prospektiven Untersuchungen für alle kardiovaskulären Krankheitskategorien zusammengefaßt (nach Booth-Kewley und Friedman, 1987)

Persönlichkeits-variable	Zahl der Publikatio-nen	Zahl der un-abhängigen Effektmaße	durch-schnittliches Effektmaß	durchschnitt-liches normal-standardisiertes Effektmaß		Fail-safe
			r	z	p	N
Typ A, alle Maße	16	10	0,045	3,74	0,00009	42
SI Typ A	7	2	0,062	3,68	0,0001	8
Ärger/Feindseligkeit/ Aggression	6	6	0,074	2,23	0,0130	6
Feindseligkeit	3	3	0,135	3,94	0,00004	15
Depression	6	3	0,168	3,77	0,00008	13
Angst	3	2	0,136	5,45	<0,0000001	20

angibt, die vorliegen müßten, um zu einem nicht-signifikanten Ergebnis zu gelangen.

Das kombinierte Effektmaß für alle Typ-A-Maße aller berücksichtigten Studien beträgt r = 0,136 (p < 0,00000001). Nach den von Cohen (1977) angegebenen Kriterien ist dieser Effekt zwar klein, aber hoch reliabel und gewiß nicht trivial (Rosenthal und Rubin, 1982). Die Berechnung des fail-safe N ergibt, daß 3066 unpublizierte Studien mit einem durchschnittlichen Effektmaß von 0 vorliegen müßten, um den für alle Typ-A-Studien zusammengefaßten p-Wert auf ein nicht-signifikantes Niveau zu bringen. Unter der Annahme, daß je die Hälfte der Bevölkerung dem Typ A bzw. dem Typ B zuzuordnen ist und etwa 20 % der Bevölkerung diagnostizierbare Anzeichen einer KHK entwickeln, würden im Vergleich zum Typ B bei einer Korrelation zwischen Typ-A-Verhalten und KHK von 0,15 etwa zweimal soviele

Typ-A-Personen erkranken. Ein derartiges relatives Risiko ist unter epidemiologischen Aspekten für die öffentliche Gesundheit durchaus bedeutsam. Dies bedeutet, daß ein kleiner Effekt wichtige praktische Auswirkungen haben kann, wenn es sich um eine so ernsthafte und häufige Krankheit wie die KHK handelt.

Als wichtigstes Typ-A-Maß erscheint das strukturierte Interview (r = 0,221; p < 0,00001). Von den Typ-A-Komponenten finden sich die höchsten Effektmaße für Rivalitätsverhalten/hard driving/Aggressivität und Feindseligkeit. Von den weiteren Variablenkategorien ist vor allem Depression von Bedeutung (r = 0,205; p < 0,0000001). Dieser Befund ist bemerkenswert, zumal nur wenige Studien hierzu vorliegen, die Höhe des Effektmaßes aber dem des Typ-A-Verhaltens entspricht. Depression spielt also vermutlich eine Rolle bei der KHK-Entwicklung; darauf weisen auch die prospektiven Befunde hin. Aufgrund der nur wenigen Studien, die eine Analyse getrennt für beide Geschlechter erlauben, beträgt das Effektmaß für Typ-A-Verhalten bei Frauen r = 0,272 (p < 0,0000001); das bedeutet, daß die Beziehung zwischen Typ-A-Verhalten und Krankheit bei Frauen wahrscheinlich stärker ist als bei Männern. Eine Aufschlüsselung nach verschiedenen Krankheitsgruppen ergab mit wenigen Ausnahmen im wesentlichen die gleichen Ergebnisse wie die Zusammenfassung aller kardiovaskulären Erkrankungen. Im Vergleich zu den prospektiven Studien liegt eine systematische Überbewertung des Typ-A-Verhaltens in den Querschnittsuntersuchungen vor.

Als Kernvariablen verbleiben in den prospektiven Untersuchungen Komponenten des Typ-A-Verhaltens: die Kombination Ärger/Feindseligkeit/Aggression, Feindseligkeit allein sowie Depression und Angst. Von Bedeutung ist dabei auch, daß ein Maß der Feindseligkeit (hostility) mit Mortalitätsunterschieden assoziiert ist (Barefoot et al., 1983; Shekelle et al., 1983 a). Wahrscheinlich tragen diese Faktoren zur Krankheitsentwicklung bei. Für die letzten drei dieser Variablen liegen dabei die Effektmaße in etwa gleicher Größenordnung wie in Querschnittsstudien.

Einschränkend ist allerdings anzumerken, daß für diese Variablen bislang nur wenige prospektive Studien vorliegen, so daß die Zahl der unabhängigen Effektmaße auf zwei bis drei beschränkt bleibt. Allerdings sind in dieser Metaanalyse noch nicht die neueren Ergebnisse der von Dembroski et al. (1989) vorgenommenen Auswertungen des MRFIT und die Studien von Williams et al. (1989) berücksichtigt.

Die Beurteilung einzelner Komponenten des Typ-A-Verhaltens erscheint also auch aufgrund der Metaanalysen wichtiger als eine globale Typ-A-Klassifizierung. Einige Komponenten des Typ-A-Verhaltens wie „hard driving", Rivalitätsverhalten, Ärger und Feindseligkeit sind wahrscheinlich mit der KHK assoziiert. Für andere Komponenten wie berufliches Engagement, Hast und Eile trifft dies nicht zu. Einige Typ-A-Maße wie der JAS in seiner gegenwärtigen Form sollten den Autoren zufolge nicht mehr verwendet werden. Hingegen sollte Forschung gefördert werden, die sich darum bemüht, Verhaltensstile zu erfassen und

zu beurteilen. Depression korreliert mit der KHK in gleicher Größenordnung wie das Typ-A-Verhalten im strukturierten Interview. Für Angst findet sich eine schwache, für Extraversion keine Beziehung. Wenn es auch unterschiedliche Beziehungen zwischen verschiedenen Erkrankungsformen (z.B. Myokardinfarkt oder Angina) und Persönlichkeits- und Verhaltensvariablen geben mag, so sind sie aufgrund der bisher vorliegenden Forschungsergebnisse der Metaanalysen jedoch noch nicht gesichert. Wir wissen zu wenig darüber, inwieweit verschiedene Persönlichkeitsvariablen wie Typ-A-Verhalten, Depression und Feindseligkeit von einander unabhängig oder redundant in ihrer Beziehung zur KHK sind. Die durchschnittliche Assoziation zwischen Typ-A-Verhalten und Krankheit hat in letzter Zeit abgenommen; eine definitive Erklärung gibt es hierfür nicht. Das Bild eines durch seine Persönlichkeits- und Verhaltensmerkmale für arteriosklerotische Erkrankungen gefährdeten Menschen ist aufgrund der vorliegenden Analyse weniger durch Arbeitswut, Hast, Eile und Ungeduld, sondern eher durch negative Emotionen wie Depression, aggressives Rivalitätsverhalten, Feindseligkeit, Angst und Ärger oder verschiedene Kombinationen dieser Merkmale gekennzeichnet.

41.1.6.6 Pathophysiologische Mechanismen – Mediatoren zwischen Verhalten und koronarer Herzkrankheit

Neurohormonelle Einflüsse

Wenn es auch nach bald vier Jahrzehnten Streßforschung keine Definition von Streß gibt, die die Mehrzahl der Forscher auf diesem Gebiet anerkennen (Elliot und Eisdörfer, 1982), so wird doch übereinstimmend angenommen, daß der Beitrag psychischer Faktoren für die koronare Herzkrankheit wahrscheinlich über eine Aktivierung des sympathischen Nebennierenmarksystems und des Hypophysen-Nebennierenrinden-Systems erfolgt. Viele der pathophysiologischen Bedingungen, die Arterioskleroseentwicklung und die klinisch in Erscheinung tretende KHK in sich einschließen, wurden in Verbindung mit einer erhöhten Aktivität eines oder beider dieser Systeme in experimentellen Untersuchungen über den Einfluß von psychischem Streß beobachtet. Hämodynamische Veränderungen, beispielsweise Erhöhung von Herzfrequenz und Blutdruck, sowie biochemische Einflüsse wie Erhöhung der Blutspiegel von Katecholaminen, Cortisol und Cholesterin werden sowohl bei Tieren als auch bei Menschen als Folge langanhaltender oder starker Streßbelastung beobachtet (Frankenhaeuser, 1983; Henry, 1983; Herd, 1981; Schneiderman, 1983 a, b). Die pathophysiologischen Wirkungen der Katecholamine beinhalten:

– Hämodynamische Veränderungen, die für eine klinisch in Erscheinung tretende KHK bedeutungsvoll sind, wie Erhöhung von Blutdruck und Herzfrequenz und ein damit verbundener höherer Sauerstoffbedarf des Herzens;

- Auslösung von kardialen Arrhythmien;
- direkte Schädigung der Endothelauskleidung der Koronararterien;
- Lipidmobilisation;
- Erhöhung der Thrombozytenaggregation und
- direkte Schädigung des Myokards (Übersichten bei Henry, 1983; Herd, 1981; Raab, 1966; Schneiderman, 1983 a, b).

Dies gilt auch für den Einfluß des Hypophysen-Nebennierenrinden-Systems. Henry (1983) weist in einer Übersicht hierzu auf eine Reihe von wichtigen Befunden hin. So wurde beispielsweise über einen signifikanten Zusammenhang zwischen erhöhten morgendlichen Plasmacortisolwerten und einer mäßigen bis schweren Koronarsklerose berichtet (Troxler et al., 1977). Die Plasmacortisolwerte konnten fast ebenso gut wie das Serumcholesterin zur Unterscheidung verschiedener Schweregrade der angiographisch nachweisbaren Koronarsklerose herangezogen werden. Gut belegt ist auch die Beziehung zwischen einer Kortikoidbehandlung bei der rheumatoiden Arthritis und der beschleunigten Entwicklung einer Arteriosklerose (Kalbak, 1972). Aus tierexperimentellen Untersuchungen ist bekannt, daß ACTH und Cortison zu einer Erhöhung der Serumlipide (Skanse et al., 1959) und zu einer Beschleunigung der Blutgerinnung (Cosgriff et al., 1950; Rich et al., 1951) führen. Rosenfeld und Mitarbeiter (1960) konnten zeigen, daß ACTH zu erhöhtem Serumcholesterin und beschleunigter Arterioskleroseentwicklung führt. Adlersberg und Mitarbeiter (1950) berichteten über einen Anstieg des Serumcholesterins bei Patienten, die mit Cortison oder ACTH behandelt worden waren und von denen 36% Cholesterinwerte über 280 mg/100 ml erreichten. Stern und Mitarbeiter (1973) konnten diese Befunde bei Patienten bestätigen, die wegen rheumatischer Erkrankungen eine Steroidbehandlung erhielten. Sie berichteten über die Entwicklung von schweren Atheromen in einem transplantierten Herzen innerhalb von 18 Monaten bei einem Patienten, der mit Kortikoiden behandelt worden war, um die Abstoßungsreaktionen zu verhindern. Bulkley und Roberts (1975) beschrieben die schädigende Wirkung von Kortikosteroiden auf die Herzen von Patienten, die wegen eines systemischen Lupus erythematodes behandelt wurden. Hypertonie trat bei Patienten, die über einen Zeitraum von 12 Monaten Kortikosteroide erhielten, zweimal so häufig und fünfmal so häufig bei Lupus-erythematodes-Patienten im Vergleich zu nicht behandelten Patienten auf (Nixon und Bethell, 1974). Björkerud (1974) berichtete, daß Glukokortikoide die Zahl der toten und verletzten Zellen des Arterienendothels erhöhen. Er betrachtet dies als einen möglichen auslösenden Faktor für die Arterioskleroseentwicklung. Kortikosteroide sensibilisieren das Gefäßsystem für die Wirkung der Katecholamine (Schömig et al., 1976), was auf die synergistische Beziehung zwischen dem Nebennierenrinden- und dem Nebennierenmarksystem hinweist. In Kombination erhöhen beide Hormone zusammen wahrscheinlich auch das Risiko für nicht-obstruktive Myokardnekrosen (Selye, 1970).

Eine Überaktivität der Hypothalamus-Hypophysen-Nebennierenrinden-Achse ist für den depressiven Zustand des Menschen kennzeichnend (Schlesser et al., 1980), und es gibt Hinweise dafür, daß auch Depression mit der Arterioskleroseentwicklung und einem erhöhten kardiovaskulären Risiko verknüpft ist (Appels, 1983). So konnte Appels die Befunde von Dreyfuss und Mitarbeitern (1969) bestätigen, wonach Patienten, die wegen Depression stationär behandelt worden sind, eine höhere Prävalenz für Herzinfarkt aufweisen als psychiatrische Patienten ohne depressive Symptome (Appels, 1979). Studenten, die während ihres Medizinstudiums untersucht wurden und die 20 bis 30 Jahre später als praktische Ärzte eine KHK entwickelten, stammten aus einer für depressive Verstimmung anfälligen Gruppe (Thomas et al., 1975). Auf die Untersuchung von Parkes und Mitarbeitern (1969), die eine stark erhöhte kardiovaskuläre Mortalität bei Witwern innerhalb der ersten 6 Monate nach dem Verlust der Ehefrau nachgewiesen hatten, wurde bereits hingewiesen.

Henry (1983) beschreibt die psychische Bedeutung des sympathischen Nebennierenmarksystems und des Hypophysen-Nebennieren-Systems und weist darauf hin, daß beide in Beziehung zu unterschiedlichen emotionalen Reaktionsmustern von großer Bedeutung für das Verhalten von Tieren stehen. Das erste beinhaltet das Kampf-Flucht-Muster; es wird ausgelöst, wenn die Ausübung der Kontrolle über den Zugang zu Futter, Wasser, Zufluchtsorten, Geschlechtspartnern, abhängigen oder anderweitig bedeutungsvollen Partnern bedroht ist. Am einen Ende der Skala erstreckt sich das emotionale Verhalten von extremem Ärger bis zu ängstlich gefärbter Wachsamkeit und erhöhter Handlungsbereitschaft. Dieses Arousal ist von hohen Katecholaminspiegeln begleitet. Das entgegengesetzte Ende der Skala ist Entspannung, die durch niedrige Spiegel dieser Hormone gekennzeichnet ist und dann auftritt, wenn das Individuum alles unter Kontrolle erlebt und keine Bedrohung in Sicht ist.

Im Gegensatz dazu vermittelt das zweite System das Verhaltensmuster der Depression. Es findet seine volle Ausprägung dann, wenn das Individuum sich als unfähig erlebt, sich selbst und andere, an die es emotionell gebunden ist, zu verteidigen, wenn ihre Sicherheit oder ihr Status bedroht ist. Die begleitende Erregung des Hypophysen-Nebennierenrinden-Systems beinhaltet die Freisetzung des adrenokortikotropen Hormons (ACTH), vermindert aggressives Verhalten und erleichtert das schnelle Erlernen neuer Verhaltensmuster. Der Kortikosteronanstieg löst submissives Verhalten aus. Der entgegengesetzte emotionale Pol von Hilflosigkeit ist das Gefühl der Euphorie, das auftritt, wenn eine befriedigende Kontrolle ausgeübt werden kann. Das Meistern der Bedrohung stärkt die Rangposition, die Spiegel der Nebennierenrindenhormone nehmen ab, wohingegen die der Gonadotropine zunehmen (Henry und Stephens, 1977; Henry, 1983).

Experimentelle Befunde bei Mensch und Tier weisen darauf hin, daß unter entsprechenden Bedingungen Stimulation des Nebennierenmark- und des Nebennierenrindensystems unterschiedliche Verhaltensmuster repräsentiert (Lundberg und Frankenhaeuser, 1980; Ursin, 1980). Auch wurden beide Systeme herangezogen, um mögliche Mechanismen zu identifizieren, die bei Typ-A-Personen für ein erhöhtes KHK-Risiko verantwortlich sein könnten. Henry wies vom Standpunkt des Verhaltensforschers darauf hin, daß das Typ-A-Muster dem Verhalten von Tieren entspricht, wenn sie versuchen, die Kontrolle über ein Territorium im Angesicht einer Bedrohung zu erlangen oder zu behaupten. Das Wesentliche dieses Verhaltensmusters ist, daß die Ausübung der Kontrolle durch das Individuum bedroht ist und daß es die Kontrollfunktion, die es ausüben kann, als zu schwach erlebt (Henry und Stephens, 1977).

Glass hat die Situation in bezug auf den Menschen folgendermaßen analysiert: Wenn ein Individuum eine Bedrohung seiner Kontrolle über die Umgebung erlebt, kämpft es darum, eine bessere Kontrolle wiederherzustellen. Während dieser Periode kann man aktive Anstrengung und Bemühungen (active coping efforts) sowie begleitende Erhöhung des Noradrenalinplasmaspiegels erwarten. Solange keine Furcht auftritt, bleibt Adrenalin unverändert oder sinkt vielleicht ab. Bis hierhin betreffen alle pathophysiologischen Veränderungen lediglich eine Aktivierung des sympathischen Nebennierenmarksystems. Glass weist jedoch darauf hin, daß Typ-A-Personen, die in einem kompetitiven Milieu kämpfen, es als eine Bedrohung erleben, Kontrolle zu verlieren und daß sie, sobald sie scheitern, passiv werden. Jetzt nimmt wahrscheinlich der Noradrenalinspiegel ab und „zentrale cholinerge Einflüsse gewinnen die Oberhand". Er weist ferner darauf hin, daß der daraus resultierende Wechsel von Kontrollbemühungen und Aufgeben über die gesamte Lebensspanne eines Individuums hinweg ständig wiederholt wird und daß wahrscheinlich die Koronararterien um so mehr von arteriosklerotischen Veränderungen betroffen werden, je häufiger sich dieser Zyklus wiederholt. Er sieht demnach die Arterioskleroseentwicklung als Folge eines Einflusses wechselnder Stimulation beider Reaktionsmuster, sowohl des Nebennierenmark- als auch des Nebennierenrindensystems (Glass, 1977).

Neben Adrenalin, Noradrenalin und Cortisol wird ein weiteres Hormon als möglicher Mediator zwischen Verhalten und Arteriosklerose angesehen: Testosteron. Ähnlich wie Cortison fördert die zusätzliche Gabe von Testosteron im Tierexperiment die Arterioskleroseentwicklung (Uzunova et al., 1978). Frauen haben gewöhnlich höhere Spiegel der vor Arteriosklerose schützenden HDL-Cholesterinfraktion als Männer. Dieser Unterschied kommt erst während der Pubertät zustande: Wenn beim männlichen Geschlecht das Testosteron ansteigt, sinkt das HDL-Cholesterin. Testosteron ist nicht nur ein Hormon, das sexuelle Funktionen steuert. Es steht auch in engem Zusammenhang mit dem Dominanzverhalten (Henry, 1980). Bei Rangordnungskämpfen findet sich beim Sieger regelmäßig ein deutlicher Testosteronanstieg und beim Verlierer ein Abfall. Beim Menschen

spiegeln sich derartige Testosteronreaktionen auch in sportlichen Wettkämpfen wider (Kemper, 1989). Testosteron scheint das Fokussieren und Einengen der Aufmerksamkeit auf spezifische Elemente in der Umgebung zu erleichtern, z.B. in Situationen vermehrter Vigilanz, und es steht in Beziehung zu aggressivem Verhalten. Diese Wirkungen könnten die Ausübung sowohl von sexuellem als auch aggressivem Verhalten erleichtern.

In einer Untersuchung von Williams et al. (1982) veränderten sich die Testosteronwerte nicht bei jungen männlichen Typ-A-Probanden während Rechenaufgaben, die bei ihnen stärkere Muskeldurchblutung und höhere Adrenalin-, Noradrenalin- und Cortisolanstiege als bei den Typ-B-Probanden hervorriefen. Bei Vigilanzaufgaben, die ein starkes Beobachten der Umgebung erforderten, stieg Testosteron stärker bei den Typ-A-Personen an. Auch die Werte auf der Hostility-Skala differenzierten die Probanden hinsichtlich ihrer Testosteronanstiege: Bei Typ-A-Personen mit hohen Feindseligkeitswerten waren die höchsten, bei Typ-B-Probanden mit niedrigen Scorewerten die niedrigsten Testosteronanstiege zu verzeichnen. Nach Williams (1989) muß ein argwöhnisches, feindselig eingestelltes Individuum in vielen Situationen aufmerksamer sein und die Umgebung genauer beobachten, da es anderen nicht traut und sicherstellen muß, daß nicht etwas Schlimmes geschieht. In einer früheren Studie hatten bereits Friedman und Rosenman (1974) bei einigen Teilnehmern der WCGS gefunden, daß Typ-A-Personen tagsüber mehr Testosteron ausscheiden; während der Nacht fanden sich hingegen keine A/B-Unterschiede.

Psychophysiologische Reaktivität als potentieller koronarer Risikofaktor

Allgemeine Überlegungen und Modelle

Die Größe der physiologischen Reaktionen auf unterschiedliche Belastungssituationen variiert individuell sehr stark. Personengruppen, die ein erhöhtes Risiko für die Entwicklung einer koronaren Herzkrankheit aufweisen, zeigen sehr oft verstärkte kardiovaskuläre und neuroendokrine Reaktionen gegenüber herausfordernden Aufgabensituationen oder Stressoren im Laboratorium. Verschiedene Forscher vertreten die Hypothese, daß wiederholte Episoden akuter psychophysiologischer Reaktionen die Arterioskleroseentwicklung bei dafür empfänglichen Personen auslösen oder beschleunigen können. In Frage kommen hierfür hämodynamische Einflüsse, die mit plötzlichen Anstiegen von Herzfrequenz und Blutdruck verbunden sind (wie z.B. Turbulenzen und Scherungskräfte), oder die vermehrte Freisetzung neuroendokriner Hormone wie Katecholamine und Cortisol, die direkt toxisch auf die Gefäßwand einwirken können, oder auch hiermit in Beziehung stehende pathogene Prozesse, die die Thrombozytenaggregation beeinflussen.

Als Reaktivität wird die Abweichung von einem Vergleichs- oder Kontrollwert bezeichnet, die als Reaktion auf einen diskreten Umgebungsreiz resultiert.

Dieser Reiz kann primär physischer oder psychischer Natur sein, wie beispielsweise körperliche Arbeit oder das Lösen einer Rechenaufgabe; allerdings ist in vielen Testsituationen eine klare Trennung zwischen der physischen und der psychischen Komponente nicht möglich. Gemäß dieser Definition ist die Variabilität eines physiologischen Parameters unter Alltagsbedingungen (z.B. mehrfache Blutdruckmessungen während eines Tages) nicht als Reaktivität zu bezeichnen, es sei denn, der auslösende Reiz, auf den das Individuum jeweils reagiert, kann spezifiziert werden. Auch die Labilität beispielsweise von Ruheblutdruckwerten, die in einer bestimmten Zeitspanne gemessen werden, ist kein Maß der Reaktivität, es sei denn, ein auslösender Umgebungsreiz ist bekannt. Dies bedeutet nicht, daß Variabilität oder Labilität unwichtige oder uninteressante Größen für Physiologie oder Pathogenese kardiovaskulärer Krankheiten sind. Vielmehr ist der Begriff „Reaktivität" für Situationen reserviert, in denen ein auslösender Umgebungsreiz erkennbar ist (Matthews et al., 1984).

In den meisten epidemiologischen Studien beruht die Assoziation zwischen Risikofaktor und zukünftiger Erkrankung auf der Erfassung von Gelegenheits- oder Ruhewerten der physiologischen Risikovariablen wie beispielsweise beim Blutdruck. Die Erfassung der psychophysiologischen Reaktivität hingegen erfordert die Messung von Veränderungen der physiologischen Parameter, die in Abhängigkeit vom Verhalten oder in spezifischen psychologischen Reizsituationen auftreten. Das Ausmaß der individuellen Reaktionen kann gewöhnlich nicht aufgrund der Ruhe- oder Gelegenheitswerte vorhergesagt werden. Deswegen beinhaltet die Erfassung der Reaktivität Information über die physiologische Funktionsweise des Individuums.

Inwieweit die im Laboratorium erfaßten Reaktivitätsmaße Prädiktoren der unter Alltagsbelastungen gemessenen Veränderungen sind, ist eine kritische Frage, die bisher noch wenig untersucht ist (Fahrenberg et al., 1984, 1985). Theoretisch sind verschiedene Modelle möglich. So können z.B. Personen, die unter Laboratoriumsbedingungen starke physiologische Reaktionen zeigen, sog. Hyperreaktoren, auch unter Alltagsbedingungen ähnlich starke Reaktionen aufweisen. Und umgekehrt können Personen mit schwachen Reaktionen im Labor auch entsprechend geringere Reaktionen im Alltag zeigen. Dieses Modell der „wiederholten Aktivation" (recurrent activation) gilt als die gewöhnliche, meist vertretene Hypothese in bezug auf eine Generalisierung individueller Unterschiede der psychophysiologischen Reaktivität (Manuck und Krantz, 1984). Ein zweites Modell geht von der Vorstellung aus, daß „Ruhewerte" eher einem ungewöhnlichen Zustand des Organismus entsprechen, die nur zustande kommen, weil der Experimentator eine relativ reizarme Umgebung schafft. Dennoch haben die so ermittelten Werte Bedeutung, da sie den „physiologischen Zustand" des Individuums bei Abwesenheit spezifischer physischer oder psychischer Reize widerspiegeln. Allerdings tritt dieser Zustand während des Tages relativ selten auf, wenn die

Individuen ihrer alltäglichen Aktivität nachgehen. Die Reaktivität kann danach im Labor bestimmt werden, indem Reizsituationen des Alltags geboten werden. Die im Labor erfaßte psychophysiologische Reaktivität würde demnach eher den physiologischen Zustand unter Alltagsbedingungen widerspiegeln als wiederholte vorübergehende Episoden einer akuten Aktivation. Dieses Modell wurde von Manuck und Krantz (1984) als „prevailing state-model" (Modell des „vorherrschenden Zustands") bezeichnet.

Beide Modelle stellen wahrscheinlich nur grobe Vereinfachungen der Realität dar, die aber einen Teil der Komplexität des Problems veranschaulichen. Möglicherweise treffen sie jeweils nur für spezifische Untergruppen zu; auch kann für den einen Parameter (z.B. Blutdruck) das eine Modell gelten und für einen anderen (z.B. Muskelspannung) das andere. Vielleicht geben verschiedene Kombinationen dieser Modelle die Wirklichkeit besser wieder; vielleicht entspricht ihr auch keines dieser Modelle oder deren Kombinationen. Allerdings ist es zur Evaluation der Beziehung zwischen Reaktivität und Krankheit wichtig zu untersuchen, inwieweit sich individuelle Unterschiede einer psychophysiologischen Reaktivität im Alltagsleben ausdrücken, denn die kardiovaskulären Krankheiten entwickeln sich dort und nicht im Labor. Die im Labor induzierten Reaktionen sind typischerweise akute Veränderungen, während die unter Alltagsbedingungen erfaßten Maße eine Mischung von akuten und chronischen Einflüssen verschiedener Stressoren darstellen. Das Alltagsleben ist mit seinen relevanten Problemen im Labor nicht realistisch simulierbar. Dennoch stellt die Erfassung von Reaktivitätsmaßen im Laboratorium unter „künstlichen" Bedingungen eine wichtige, oft sogar die einzige Alternative dar, da Messungen unter Alltagsbedingungen einen ungleich größeren methodischen und zeitlichen Aufwand bedeuten und in Ermangelung der erforderlichen Technologien oft gar nicht durchführbar sind. Auch können der apparative Aufwand und die Belastung für den Probanden bei Alltagsmessungen so groß sein, daß hierbei ebenfalls kein „realistisches" Bild des Alltagslebens gewonnen werden kann.

Klinische Fragestellungen können prüfen, inwieweit die streßinduzierte oder vom Verhalten abhängige psychophysiologische Reaktivität Vorläufer, Korrelat oder Folge kardiovaskulärer Erkrankungen ist. Eine erhöhte kardiovaskuläre Reaktivität könnte der Krankheitsentwicklung auch ohne ursächlichen Zusammenhang gewissermaßen als Risikoindikator vorangehen. Aber auch ein ursächlicher Zusammenhang mit der Arterioskleroseentwicklung erscheint möglich, was gewissermaßen als eine der „Leithypothesen" dieses Forschungsgebietes gilt. Gemäß den Hypothesen der „response to injury"-Theorie (Ross und Glomset, 1976; Ross, 1986) vermutet Clarkson (1984), daß wiederholte situative Blutdruckanstiege je nach Höhe und Häufigkeit mechanisch immer wieder zu Verletzungen des Gefäßendothels führen und den normalen Heilungsprozeß dieser Läsionen verhindern, was ein Ausgangspunkt der Arterioskleroseentwicklung ist.

Hinweise dafür, daß individuelle Unterschiede der durch das Verhalten oder durch Umgebungsreize induzierten psychophysiologischen autonomen und neuroendokrinen Reaktionen mit der Entwicklung einer koronaren Herzkrankheit verknüpft sind, stammen im wesentlichen aus drei Quellen:

– Studien aufgrund entsprechender tierexperimenteller Modelle;
– prospektive und Fallkontrolluntersuchungen beim Menschen; sowie
– experimentelle Studien, die die physiologischen Korrelate koronargefährdender Verhaltensweisen untersuchen.

Tierexperimentelle Untersuchungen

Bei Ratten führt physische Immobilisation zu Blutdruck- und Herzfrequenzanstiegen. Hirsch und Mitarbeiter (1984) setzten Ratten einem Immobilisationsstreß aus. Diese Tiere erhielten ebenso wie die Kontrolltiere Infusionen mit radioaktiv markiertem Thymidin, das vorwiegend von den Zellkernen sich teilender Zellen aufgenommen wird. In der autoradiographischen Untersuchung der Aortenintima der gestreßten Ratten fanden sich im Vergleich zu den Kontrolltieren um 500 % erhöhte Teilungsraten der Endothelzellen. Zusätzlich wurde die Bedeutung des sympathischen Einflusses auf die streßinduzierte Erneuerung des Endothels untersucht, indem Tieren der gestreßten Gruppe Propranolol verabreicht wurde. Während des Immobilisationsstresses kam es unter der Betablockade nicht mehr zu signifikanten Anstiegen von Blutdruck und Herzfrequenz; die Replikationsrate der Endothelzellen war bei diesen Tieren gegenüber den unbehandelten Tieren der Kontrollgruppe nicht erhöht. Demnach schützte die Blockade der betaadrenergen Rezeptoren vor den hämodynamischen Auswirkungen des Immobilisationsstresses sowie vor einer Verletzung des Endothels der arteriellen Gefäße.

Beere und Mitarbeiter (1984) sehen in einer erhöhten Herzfrequenz einen wichtigen hämodynamischen Faktor, der zur Arterioskleroseentwicklung beiträgt und vor allem bestimmte arterielle Gefäßgebiete betrifft, an denen es zu abrupten Veränderungen in Richtung und Stärke des pulsierenden Blutflusses kommt, wie z.B. an der Karotisbifurkation oder den proximalen Koronararterien. In einer Gruppe von Cynomolgus-Makaken verringerten sie durch chirurgische Abtragung des Sinusknotens die Herzfrequenz. Diese Tiere erhielten ebenso wie die pseudooperierten Tiere der Kontrollgruppe über einen Zeitraum von sechs Monaten eine die Arterioskleroseentwicklung fördernde Ernährung. Die Koronarsklerose war bei Affen mit postoperativ niedriger Herzfrequenz nach dieser Zeit nur halb so stark ausgeprägt wie bei den Tieren mit höherer Herzfrequenz. Da sich die Gruppen mit niedriger und hoher Herzfrequenz nicht in der Höhe von systolischem und diastolischem Blutdruck, Serumcholesterin, Triglyzeriden und im Körpergewicht unterschieden, schließen die Autoren, daß die erhöhte Herzfrequenz allein für den Unterschied in der Ausprägung der Koronarsklerose verantwortlich ist; sie folgern, daß Herzfrequenzerhöhungen infolge von Umgebungseinflüssen und Verhaltensweisen für die Assoziation zwischen psychosozialen Variablen und der Entwicklung einer Koronarsklerose beim Menschen in Frage kommen.

Die Untersuchungen von Kaplan, Manuck und Clarkson an Cynomolgus-Makaken verdeutlichen das Zusammenspiel verschiedener Faktoren bei der Entwicklung der Koronarsklerose. Ernährungsweise, Verhalten in Abhängigkeit von der sozialen Rangposition und psychosozialer Streß wirken zusammen, wobei den kardiovaskulären Reaktionen wahrscheinlich ein wichtiger Stellenwert zukommt. Sie untersuchten die Wirkung des Stressors „Instabilität der sozialen Gruppe" in Abhängigkeit von der sozialen Rangordnung dieser Affenart auf die Arterioskleroseentwicklung. In dieser Untersuchung wurden 30 Tiere in 6 Gruppen zu je 5 Tieren eingeteilt. Sie erhielten eine Diät, ähnlich in der Zusammensetzung wie die amerikanische Durchschnittsernährung, die die Arterioskleroseentwicklung fördert (0,34 mg Cholesterin/cal oder 680 mg Cholesterin täglich). Bei 15 Affen wurden instabile soziale Bedingungen dadurch induziert, daß ihre Gruppenzugehörigkeit periodisch gewechselt wurde, indem die Tiere in den einzelnen Gruppen regelmäßig ausgetauscht wurden. Im Gegensatz hierzu blieb die Gruppenzugehörigkeit bei den übrigen 15 Affen unverändert. Sowohl unter der Bedingung der instabilen als auch der stabilen sozialen Gruppe wurden die einzelnen Tiere aufgrund ihres Aggressions- und Unterwerfungsverhaltens als dominant oder submissiv klassifiziert. Die dominanten Tiere der instabilen Gruppe sind durch Rangordnungskämpfe mit immer neuen Gruppenmitgliedern besonders belastet; durch diese Kämpfe versuchen sie ihre führende Stellung in der Gruppe zu behaupten; sie zeigen häufiger offene Kontaktaggression wie Beißen oder Schlagen. Nach 22 Monaten war bei den

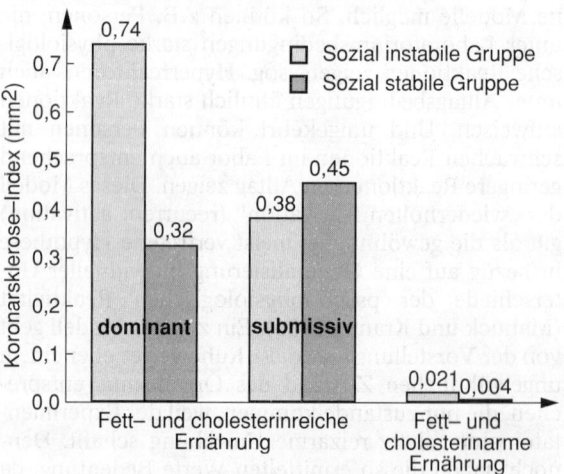

Abb. 41.1–3. Koronarskleroseentwicklung bei männlichen Cynomolgus-Makaken in Abhängigkeit von Ernährungsweise, sozialer Rangposition und psychosozialem Streß (nach Kaplan et al., 1984).

dominanten Tieren der instabilen Gruppen in den Koronargefäßen ein signifikant größeres Ausmaß an Arteriosklerose nachweisbar als bei den dominanten Tieren der stabilen Gruppen oder den submissiven Affen in beiden sozialen Gruppen (Abb. 41.1–3).

Die beobachteten Differenzen beruhten nicht auf Unterschieden im Serumcholesterinspiegel, dem HDL-Cholesterin, Blutdruck, Übergewicht oder dem Nüchternblutzucker. Die Autoren vermuteten, daß verstärkte kardiovaskuläre Reaktionen für die vermehrte Arterioskleroseentwicklung der dominanten Tiere verantwortlich sind. In einem nachfolgenden Experiment wurde die Herzfrequenz telemetrisch gemessen (Manuck et al., 1986). In den ersten Tagen nach der neuen Zusammensetzung der Gruppen fanden sich bei den dominanten im Vergleich zu den submissiven Cynomolgus-Makaken höhere Herzfrequenzwerte dann, wenn sie in scheinbarer Ruhe in Distanz zu den anderen Tieren saßen – eine Situation, die häufiger (in 29%) zu aggressiven Auseinandersetzungen führte, als wenn die Tiere in größerer körperlicher Nähe beieinander saßen (12%). Deswegen könnte auch hier bei den dominanten Tieren der instabilen sozialen Gruppen eine erhöhte Herzfrequenz zur Entwicklung der Koronarsklerose beitragen. Die Gabe eines Betarezeptorenblockers verhindert die stärkere Koronarskleroseentwicklung der dominanten Tiere (Kaplan et al., 1987). Dies weist darauf hin, daß eine vermehrte sympathische Aktivierung bei diesen Tieren für die stärkere Arterioskleroseentwicklung mitverantwortlich ist.

Wurde den Tieren eine Diät verabreicht, wie sie die American Heart Association für den Menschen empfiehlt mit niedrigem Gehalt an Cholesterin (0,05 mg/cal) und gesättigten Fetten, entwickelten die Tiere der instabilen Gruppe ebenfalls eine signifikant stärkere Koronarsklerose (Kaplan, 1983). Allerdings war das Ausmaß der Koronarsklerose um ein Vielfaches niedriger als bei der cholesterin- und fettreich ernährten Gruppe (Abb. 41.1–3). Diese Befunde weisen darauf hin, daß bei männlichen Primaten individuelle Unterschiede in der sozialen Rangstellung in Verbindung mit Ernährungsfaktoren vor allem dann zu der beschleunigten Entwicklung einer Koronarsklerose führen können, wenn die soziale Rangstellung der dominanten Tiere bedroht ist. Eine fett- und cholesterinreiche Ernährung verstärkt die psychosozialen Effekte auf die Atherogenese.

Im Gegensatz zu männlichen Affen entwickelten submissive weibliche Cynomolgus-Makaken bei einer cholesterinreichen Ernährung sowohl in der stabilen als auch in der instabilen sozialen Umgebung eine stärkere Koronarsklerose als die dominanten Weibchen (Kaplan et al., 1983; Hamm et al., 1983). Der submissive soziale Status ist durch eine Funktionseinschränkung der Ovarien gekennzeichnet mit häufigen anovulatorischen Zyklen und einem Progesteronmangel in der zweiten Zyklushälfte. Dies legt nahe, daß eine vom Rangordnungsverhalten induzierte Hemmung der ovariellen Funktion die weiblichen Tiere ihres gewöhnlichen Schutzes vor der Koronarsklerose beraubt. Weibliche Tiere mit häufigen hormonell abnormen Menstruationszyklen unterscheiden sich nicht im Ausmaß der durch die Ernährung induzierten Koronarsklerose von ovarektomierten Kontrolltieren (Adams et al., 1985). Auch diese Unterschiede sind wie bei den männlichen Tieren unabhängig von anderen Risikofaktoren wie Serumlipide und Blutdruck.

Manuck et al. (1986) untersuchten auch die Beziehung zwischen Ausmaß der Herzfrequenzreaktionen auf einen Stressor und Koronarsklerose bei ihren Cynomolgus-Affen. Der Experimentator zeigte diesen Affen demonstrativ und für sie in bedrohlicher, standardisierter Weise vor dem Käfig den „Fanghandschuh", mit dem die Tiere gewöhnlich gefangen werden. Die Herzfrequenz wurde telemetrisch in dieser Phase sowie während einer Ruheperiode gemessen. Die Herzfrequenz stieg dramatisch im Mittel von Ruhewerten von 128 bis auf 219 Schläge pro Minute an, was dem bedrohlichen Schauspiel zugeschrieben werden muß, da die alleinige Anwesenheit des Experimentators nur viel geringere Herzfrequenzänderungen hervorrief (144 Schl./min). Der mittlere Herzfrequenzanstieg betrug 91 Schläge pro Minute, variierte bei einzelnen Tieren aber zwischen +31 bis +123. Die Ergebnisse, die an weiteren Gruppen dieser Affen erhoben wurden, sprechen dafür, daß diese Reaktivitätsunterschiede stabile individuelle Merkmale widerspiegeln, die auch auf andere Situationen übertragbar sind. Zur Untersuchung der Arterioskleroseentwicklung wurden in einem neuen Experiment starke und schwache Herzfrequenz-Reaktoren ermittelt und miteinander verglichen – die oberen und unteren 30% der gesamten untersuchten Gruppe; beide Gruppen unterschieden sich nicht signifikant in der Herzfrequenz während der Ruheperiode (126 bzw. 123 Schl./min), wohl aber während der Streßperiode (236 bzw. 199 Schl./min). Die Untersuchung der Intima der Koronargefäße ergab einen zweimal so großen Schweregrad koronarsklerotischer Veränderungen bei den stark reagierenden Tieren im Vergleich zu den schwach reagierenden; der signifikante Unterschied bezog sich sowohl auf einen globalen „Koronarindex", als auch auf die Analyse der Intimaläsionen einzelner Koronargefäße. Keine Unterschiede bestanden zwischen den Gruppen bezüglich systolischem und diastolischem Blutdruck, Blutfetten und Körpergewicht; HDL-Cholesterin war bei den Hyperreaktoren geringgradig im Grenzbereich der Signifikanz erhöht. Das Verhalten der Tiere wurde über einen 22monatigen Zeitraum beobachtet und registriert; es fanden sich Unterschiede zwischen beiden Gruppen im Aggressionsverhalten, wobei die Gruppe der Hyperreaktoren häufiger zu direkter Kontaktaggression wie Beißen und Schlagen neigte. Die soziale Rangposition stand nicht in Beziehung zur so ermittelten Reaktivität der Herzfrequenz; auch waren sozialer Status und Aggressionsverhalten nur unvollkommen miteinander korreliert. Bei Ratten stehen individuelle Unterschiede im Aggressionsverhalten in Beziehung zur Blutdruckreaktivität (Koolhaas et al., 1986). Je aggressiver die Tiere sind, desto stärkere Blutdruckanstiege werden bei ihnen selbst in Ruhe einfach durch den Anblick eines Rivalen hervorgerufen, durch den sie zuvor besiegt worden sind.

Insgesamt weisen dieses Untersuchungen auf eine enge Koppelung zwischen Aggressionsverhalten und kardiovaskulärer Reaktivität hin und sie stehen in Übereinstimmung mit der Hypothese, daß eine verstärkte kardiovaskuläre Aktivierung mit einer beschleunigten Arterioskleroseentwicklung verbunden sein kann. Auch entsprechen diese Befunde den epidemiologischen und psychophysiologischen Studien beim Menschen, in denen Aggressionsbereitschaft und Feindseligkeit mit der KHK-Inzidenz und dem Schweregrad der angiographisch nachweisbaren Koronarsklerose bzw. mit verstärkten Herzfrequenz- und Blutdruckreaktionen verknüpft sind.

Untersuchungen beim Menschen

In einigen retrospektiven und Fallkontrollstudien wurden psychophysiologische Reaktionen bei Personen mit und ohne KHK untersucht (Übersicht bei Matthews et al., 1986). Die meisten dieser Untersuchungen zeigen erhöhte Reaktionen vor allem des systolischen Blutdrucks gegenüber Stressoren im Laboratorium bei Patienten mit Angina pectoris oder Myokardinfarkt im Vergleich zu anderen Patientengruppen oder gesunden Kontrollpersonen. Als Stressoren wurden u.a. Rechenaufgaben, psychomotorische Leistungstests oder das strukturierte Interview zur Erfassung des Typ-A-Verhaltens verwendet. Die Interpretation einiger dieser Studien ist aber aus methodischen Gründen erschwert; das resuliert z.B. aus der Vernachlässigung einer Kontrolle der Medikation bei Koronarpatienten oder daraus, daß andere chronische Erkrankungen, die die Reaktivität beeinflussen, wie beispielsweise das Vorliegen einer essentiellen Hypertonie, nicht berücksichtigt wurden. In einer Studie konnte der Schweregrad der Koronarsklerose nicht mit einer erhöhten kardiovaskulären Reaktivität in Verbindung gebracht werden (Krantz et al., 1981). Wenn auch die Mehrzahl der Untersuchungen auf einen möglichen Zusammenhang hinweist, könnten die erhöhten Reaktionen Folge und nicht Vorläufer oder gar Ursache der Erkrankung sein.

Bisher liegt eine prospektive Studie beim Menschen vor, die die kardiovaskuläre Reaktivität als Prädiktor für die KHK-Inzidenz untersucht (Keys et al., 1971). Die Stärke der diastolischen Blutdruckreaktionen im Cold-pressure-Test war hier der beste einzelne Prädiktor für die kardiovaskuläre Mortalität in einem Zeitraum von 23 Jahren; die prädiktive Aussage dieses einen Faktors schnitt besser ab als die der gleichzeitig mit erhobenen traditionellen Risikofaktoren. Nicht bestätigt hat sich hingegen die Vermutung, daß eine Hyperreagibilität im Cold-pressure-Test einen Risikofaktor für die Entwicklung einer Hypertonie darstellt (Übersicht bei Julius et al., 1986), d.h. es war bisher aus methodischen Gründen nicht möglich abzuschätzen, inwieweit die Hyperreagibilität und nicht lediglich ein unterschiedlicher Blutdruck bereits in Ruhe für die Hypertonieentwicklung verantwortlich

ist (bezüglich Blutdruckreaktivität und Hypertonieentwicklung vgl. Kap. 41.3).

Zwillingsuntersuchungen weisen auf eine genetische Komponente der kardiovaskulären Reaktionsbereitschaft hin (Rose et al., 1986). Normotone Kinder hypertoner Eltern, die wahrscheinlich genetisch für eine Hypertonie prädisponiert sind, reagieren auf Rechenaufgaben (Falkner et al., 1979) und Schockvermeidungstests (Light und Obrist, 1980) mit verstärkten kardiovaskulären Reaktionen. Nach Matthews und Rakaczky (1986) ist eine familiäre Hypertoniebelastung in der überwiegenden Mehrzahl der Untersuchungen mit signifikant stärkeren Blutdruck- und/oder Herzfrequenzreaktionen oder einer erhöhten Plasmareninaktivität assoziiert. Außerdem fand sich in vier Studien eine Wechselwirkung zwischen familiärer Belastung und Persönlichkeitsmerkmalen.

Diese Befunde legen einen Zusammenhang zwischen einer psychophysiologischen kardiovaskulären Hyperreaktivität und dem zukünftigen Krankheitsrisiko z.B. für eine Hypertonieentwicklung nahe. Es stellt sich die Frage, inwieweit sich beim Menschen auch spezifische Verhaltensweisen und Einflüsse der sozialen Umwelt identifizieren lassen, die zu verstärkten kardiovaskulären Reaktionen führen und mit einem erhöhten Erkrankungs- oder Sterberisiko verknüpft sind. Dieser Frage wurde in einer Anzahl von Studien in bezug auf das Typ-A-Verhalten und anderer psychologischer Merkmale nachgegangen. Eine Reihe von Studien untersuchte verschiedene psychologische Konstrukte und emotionale Merkmale in Beziehung zu kardiovaskulären Reaktivitätsmaßen mit zum Teil positiven und zum Teil negativen Ergebnissen (Houston, 1986); dies betrifft Ärger, Feindseligkeit und Aggression, Dominanz- und Rivalitätsverhalten, unterdrückte Machtmotivation (inhibited power motivation), Angst, Bewältigungsstrategien wie Verleugnung, externe versus interne Kontrolle (locus of control) und Intro-/Extraversion. Die meisten Studien betreffen das Typ-A-Verhalten und seine Komponenten. Sie nahmen ihren Ausgang von den ersten Untersuchungen von Rosenman und Friedman (1974), die unterschiedliche physiologische Veränderungen bei Typ-A- und Typ-B-Personen unter Alltagsbedingungen und spezifischen Testsituationen feststellten. So fanden sie bei Typ-A-Personen tagsüber eine höhere Noradrenalinausscheidung im Urin sowie höhere Plasmanoradrenalinspiegel, nicht jedoch nachts (Friedman und Rosenman, 1974). Je einer Typ-A- und Typ-B-Person wurde ein unlösbares Puzzle vorgelegt. Demjenigen, der das Puzzle als erster löse, wurde eine Flasche französischen Weins versprochen, die mitten auf den Tisch zwischen die beiden Kontrahenten gestellt wurde. Trotz gleicher Ausgangswerte reagierten Typ-A-Personen mit einem stärkeren Noradrenalinanstieg als Typ-B-Personen (Friedman et al., 1975). Diese Untersuchungen führten zu der Hypothese, daß bei Typ-A-Personen unter spezifischen Situationen ein stärkerer sympathischer Einfluß zumindest auf das kardiovaskuläre System nachweisbar sein müßte. Dembroski und MacDougall untersuchten die kardiovaskulären Reaktionen während des Typ-A-Interviews und während eines Quiz über amerikanische Geschichte bei Koronarpatienten und bei einer Vergleichsgruppe mit anderen chronischen Erkrankungen. Die Typ-A-Patienten beider Gruppen reagierten mit höheren systolischen Blutdruckanstiegen als die Typ-B-Patienten, wobei wiederum die Koronarkranken die stärksten Reaktionen zeigten (Dembroski et al., 1978, 1981). Nachfolgende Studien konnten einen Zusammenhang zwischen Typ-A-Verhalten und kardiovaskulärer Reaktivität nur zum Teil bestätigen. Signifikante Beziehungen fanden sich vor allem für die systolischen Blutdruckreaktionen (Myrtek, 1983; Krantz und Manuck, 1984; Wright et al., 1984). Ein positiver Zusammenhang zwischen Typ-A-Verhalten und kardiovaskulären Reaktionen wurde bei für dieses Verhaltensmuster relevanten Stressoren gefunden, wie Wettbewerbssituationen, die eigene Geschicklichkeit involvierende psychomotorische Tests und Aufgaben, die durch aktives Verhalten unter Zeitnot zu lösen waren. Passive Stressoren wie Kaltwassertest, platzende Luftballons und ähnliches mehr waren hingegen meist nicht geeignet, unterschiedliche kardiovaskuläre Reaktionen bei Typ-A- und Typ-B-Personen hervorzurufen. Wie in den epidemiologischen Untersuchungen zeigte sich auch bei den psychophysiologischen Studien die Notwendigkeit, einzelne Typ-A-Komponenten isoliert zu betrachten.

Im deutschen Sprachraum fanden sich bei männlichen Polizeibeamten mit stark ausgeprägtem Typ-A-Verhalten nur dann stärkere systolische Blutdruckreaktionen, wenn sie von einem männlichen Interviewer, nicht aber, wenn sie von einer Frau interviewt wurden (Schmidt et al., 1985, 1988), was die Hypothese stützt, daß Typ-A-Personen eher zu Rivalitätsverhalten gegenüber gleichgeschlechtlichen Artgenossen neigen. Verschiedene Sprachmerkmale, das Ausmaß von Aggressionsbereitschaft und verbales Rivalitätsverhalten korrelierten signifikant mit der Stärke der Blutdruckreaktionen.

Bereits bei 6–11jährigen Kindern finden sich geschlechtsspezifische Unterschiede im Verhalten und den sie begleitenden kardiovaskulären Reaktionen. Jungen werden von ihren Lehrern als ungeduldiger, zu mehr Rivalitätsverhalten neigend und dominanter eingestuft als Mädchen, und sie reagieren mit stärkeren Blutdruckanstiegen während eines kompetitiven Videospiels. Während dieses Spiels und während eines psychomotorischen Leistungstests erwies sich ein aus Verhaltensbeobachtungen faktorenanalytisch gewonnenes Verhaltensmuster als wichtigster Prädiktor für die systolischen und diastolischen Blutdruckreaktionen; es setzte sich aus den Komponenten Angespanntheit, Ehrgeiz, Engagement, Typ-A-Einschätzung, Ungeduld, Extraversion und emotionale Erregung zusammen. Wurden die einzelnen von zwei unabhängigen Beobachtern erhobenen Verhaltenseinschätzungen der Kinder gesondert in den Regressionsanalysen berücksichtigt, fand sich am häufigsten die Skala „freundlich-unfreundlich" als signifikanter Prädiktor der systolischen und diastolischen Blut-

druckreaktionen: freundliche Kinder waren vor starken Blutdruckanstiegen eher geschützt (Schmidt et al., 1986, 1988).

Blutdruckreaktionen im Cold-pressure-Test sind nicht nur vom physikalischen Kältereiz abhängig, sondern auch von den Instruktionen, die den Versuchspersonen gegeben werden. Wird der Test als starke Herausforderung und schwere Aufgabe dargestellt (Dembroski et al., 1979), so waren auch die Blutdruck- und Herzfrequenzreaktionen stärker und differenzierten deutlich zwischen Typ-A- und Typ-B-Personen; wurde die Aufgabe als einfach und leicht beschrieben, waren auch die Reaktionen geringer und unterschieden sich nicht zwischen beiden Gruppen. Probanden mit hohen Werten für die im Interview erfaßten Komponenten Aggressionsbereitschaft/Rivalitätsverhalten (potential for hostility) reagierten im Cold-pressure-Test mit maximalem systolischem und diastolischem Blutdruckanstieg unabhängig davon, ob die Belastung als leicht oder schwer dargestellt wurde. Typ-A-Personen mit niedrigen Werten für diese Komponenten zeigten derart extreme physiologische Reaktionen nur, wenn eine starke Belastung suggeriert wurde.

Eine psychische Variable ist keineswegs in allen Situationen mit einer erhöhten Reaktivität verknüpft; dies gilt nicht einmal für eine genetische Disposition. Die experimentelle Situation muß relevant für das jeweilige psychische Merkmal sein. So fanden Williams und andere Untersucher keine Effekte des Merkmals „potential of hostility" in bezug auf die kardiovaskulären Reaktionen während Rechenaufgaben. Wurden die Probanden bei diesen Aufgaben aber gestört, um Ärgerreaktionen auszulösen, wurde bei Probanden mit höheren Hostility-Werten stärkerer Ärger und eine stärkere Muskeldurchblutung ausgelöst (Williams, 1989). Aber erst als die Versuchspersonen während der Ausübung der Aufgaben das Gefühl bekamen, daß sie unfair kritisiert wurden, fanden sich dramatische Unterschiede, wobei Personen mit hohen Scorewerten höhere Anstiege von Blutdruck und Durchblutung aufwiesen sowie eine langsamere Rückbildung dieser physiologischen Größen. Alle Probanden gaben während des Experimentes starken Ärger an, aber nur bei den Personen mit hohen Hostility-Werten stiegen Blutdruck, Herzfrequenz und Durchblutung mit zunehmendem Ärger an (Suarez et al., 1988). Wird Ärger tagaus, tagein erlebt, kann dies bei feindselig eingestellten Menschen offensichtlich zu einem erhöhten kardiovaskulären Risiko beitragen.

41.1.7 Prävention und Therapie

Die Erkenntnis, daß psychosoziale Einflüsse und Verhaltensweisen wie das Typ-A-Muster bzw. einige seiner toxischen Komponenten die kardiovaskuläre Morbidität und Mortalität erhöhen können, hat zu einem wachsenden Interesse an Interventionsstudien geführt, die versuchen, das Risikoverhalten durch psychotherapeutische Maßnahmen zu beeinflussen.

Entsprechende Behandlungsstrategien beziehen sich sowohl auf die primäre Prävention, d.h. auf die Beeinflussung bekannter Risikofaktoren bei Gesunden zur Verhinderung der Erkrankung, als auch auf die sekundäre und tertiäre Prävention zur Verhinderung akuter Krankheitsereignisse wie Herzinfarkt oder plötzlichem Herztod bei KHK-Kranken bzw. Reinfarkt. Dabei können verschiedene theoretische Modelle zur Erklärung der Assoziation zwischen den entsprechenden Verhaltensweisen und den ischämischen Herzkrankheiten wie Myokardinfarkt, Angina pectoris und plötzlichem Herztod herangezogen werden (Cooper et al., 1981).

41.1.7.1 Theoretische Voraussetzungen

Das einfachste Modell geht wie das Modell von Kagan und Levi davon aus, daß Risikofaktoren direkt die koronare Herzkrankheit verursachen, also z.B. spezifische Verhaltensweisen durch streßbezogene Reaktionsmechanismen im neuroendokrinen und kardiovaskulären System. Hieraus läßt sich ableiten, daß eine erfolgreiche Modifikation der entsprechenden Risikofaktoren bzw. Verhaltensweisen die Auswirkung derjenigen pathophysiologischen Mechanismen verringern sollte, die in den Koronargefäßen die ischämische Herzkrankheit hervorrufen.

Ein zweites, alternatives Modell geht davon aus, daß Risiko-„Faktoren" lediglich Indikatoren sein können, die keine direkte Auswirkung auf die zukünftige Erkrankung haben, sondern lediglich von einem unbekannten Hintergrundsfaktor bedingt sind, der aber seinerseits eine ursächliche Beziehung zur Erkrankung hat. Verhaltensmuster und koronare Herzkrankheit könnten demzufolge zwei voneinander unabhängige, parallele Auswirkungen eines zentralen konstitutionellen Merkmals darstellen, das sich auf psychophysiologischer Ebene als Typ-A-Verhalten ausdrückt und im somatischen Bereich als fortschreitende Koronarsklerose.

Von diesen Modellen lassen sich wichtige Fragen in bezug auf Behandlung und Prävention ableiten. Sind beispielsweise die mit den relevanten Verhaltensweisen assoziierten psychophysiologischen Veränderungen für die Pathogenese der koronaren Herzkrankheit von zentraler Bedeutung, wie in Modell 1 angenommen, oder sind sie nur von untergeordneter Bedeutung und stellen vielleicht sogar nur ein Epiphänomen dar? Letzteres würde bedeuten, daß die Veränderung der relevanten Verhaltensweisen wenig oder keine Auswirkungen auf das koronare Risiko hätte. Die streßbezogenen psychophysiologischen Mechanismen könnten aber auch zusätzliche, additive pathogenetische Auswirkungen haben, d.h. ihre Wirkung würde sich zu der des konstitutionellen Merkmals, das in diesem Modell unabhängig vom Verhaltensmuster ist, addieren. Von Bedeutung sind diese Modelle vor allem für Therapie- und Interventionsstrategien. Interventionsstudien können auch zur Klärung beitragen, welches der Modelle eher für den jeweiligen zu modifizierenden Faktor zutrifft.

41.1.7.2 Therapiestudien im Rahmen der primären Prävention

Ausgehend von der Auffassung, daß pathophysiologische Streßmechanismen für den Zusammenhang von Verhalten und KHK-Entwicklung verantwortlich sind, bestehen derartige Behandlungsprogramme zum einen aus dem Training, Situationen zu erkennen, die entsprechenden Streß auslösen, und zum anderen aus der Modifikation der individuellen Reaktionen auf diese Situationen. Das „cardiac stress management training" von Suinn (1975, 1978) ist typisch für die Therapiestrategien im Rahmen der primären Prävention der KHK. Es beinhaltet das Training des Patienten, bei sich selbst physiologische Indikatoren der Streßreaktion wahrzunehmen und durch Entspannungstechniken zu vermindern, und die Einübung von Typ-A-Komponenten entgegengesetzten Verhaltensweisen in Situationen, die gewöhnlich das unerwünschte Verhalten auslösen. Dies kann beispielsweise in vorgestellten Situationen geschehen und in den Behandlungssitzungen diskutiert werden. In anderen Behandlungsprogrammen erhalten Patienten jeweils die „Hausaufgabe", eine aktuelle Verhaltensänderung vorzunehmen.

In einer kontrollierten Studie untersuchten Suinn und Bloom 14 gesunde Typ-A-Personen, die randomisiert einer Behandlungs- oder Kontrollgruppe zugewiesen wurden. In der in 6 Sitzungen behandelten Gruppe fand sich eine signifikante Verringerung von Angst und von Typ-A-Komponenten im JAS. Keine signifikanten Änderungen fanden sich hingegen bei Blutdruck und Serumcholesterin (Suinn und Bloom, 1978). Roskies und Mitarbeiter verglichen eine psychoanalytisch orientierte Gruppentherapie mit einem verhaltenstherapeutischen Trainingsprogramm. Neben der Beeinflussung von psychischen Symptomen, wie z.B. dem Gefühl von Zeitdruck, wurden in beiden Gruppen Blutdruck, Cholesterinspiegel und Anzahl der Überstunden gesenkt; der Effekt war in der Verhaltenstherapiegruppe ausgeprägter (Roskies et al., 1978, 1979).

Gewöhnlich berichten Studien zur Modifikation des Typ-A-Verhaltens über erfolgreiche Veränderungen einzelner Typ-A-Komponenten (Levenkron et al., 1983; Hart et al., 1984). Die Ergebnisse der Streßmanagement-Programme in bezug auf die Veränderung von physikochemischen Risikofaktoren sind nicht einheitlich (Suinn, 1982): In einigen Studien wird eine Verringerung dieser Faktoren gefunden, in anderen wiederum nicht. Dies drückt jedoch nicht unbedingt eine Schwäche dieser Programme aus, da das modifizierte Risikoverhalten unabhängig von anderen KHK-Risikofaktoren sein kann.

Auch der Einfluß von körperlichem Training auf das Typ-A-Verhalten ist untersucht worden. Dieser Ansatz wurde aufgenommen, da sich körperliches Training erwiesenermaßen neben seiner bisherigen anerkannten Bedeutung für die KHK-Prävention und -Rehabilitation (Hollmann et al., 1983) auch als wirksamer Bestandteil für das Streßmanagement erwiesen hat (Brammel und Niccoli, 1976; Eliot et al., 1976;

Folkins und Amsterdam, 1977; Froelicher, 1978). In zwei Studien wird über eine Reduktion des Typ-A-Verhaltens berichtet (Blumenthal et al., 1980; Lobitz und Brammel, 1981). Dies war beispielsweise nach einem 7wöchigen Trainingsprogramm im JAS in der Behandlungsgruppe nachweisbar, nicht jedoch in der Kontrollgruppe, die keine Behandlung erhielt, oder einer Vergleichsgruppe, die nur das „cardiac stress management training" durchführte.

Ein Zusammenhang zwischen einer Verhaltensmodifikation und der Verringerung des KHK-Risikos selbst läßt sich natürlich nur in prospektiven Untersuchungen sichern, die noch nicht vorliegen. Die bisher beschriebenen Untersuchungen basieren in der Regel auf Selbstbeurteilung des Typ-A-Verhaltens und müssen demgemäß mit Vorsicht interpretiert werden.

Gill und Mitarbeiter (1985) berichten über 118 gesunde Offiziere des U.S. Army War College im mittleren Alter, die ausgeprägtes Typ-A-Verhalten zeigten. Sie wurden auf freiwilliger Basis zufällig in zwei Gruppen eingeteilt: 62 Offiziere nahmen über einen Zeitraum von 9 Monaten an „Typ-A-Behandlungsgruppen" teil; 56 Offiziere, die keine derartige Behandlung erhielten, dienten als Kontrollgruppe. Zur Beurteilung des Typ-A-Verhaltens wurden neben einem auf Video aufgezeichneten strukturierten Interview zu Beginn und am Ende der Studie Fragebogen für die Teilnehmer und deren Ehefrauen verwendet. Diese Erfassungsmethoden des Typ-A-Verhaltens waren zusammen mit dem Behandlungskonzept bereits in einer Interventionsstudie mit Herzinfarktpatienten erprobt worden. Es galt zu überprüfen, inwieweit auch Gesunde motivierbar sind, ihr Typ-A-Verhalten zu verändern.

41,9 % der Typ-A-Behandlungsgruppe verringerten ihr Typ-A-Verhalten deutlich, in der Kontrollgruppe nur 8,9 %. Unterschiede einer Reihe von Verhaltensmerkmalen konnten als Folge der Behandlung zwischen beiden Gruppen festgestellt werden, wie beispielsweise ein besseres Verstehen von anderen, vermehrter Sinn für Humor, vermehrte Toleranz, eine Verringerung des Gefühls von Feindseligkeit und Zeitnot. Kollegen konnten keine Veränderung der Führungsqualitäten bei den „behandelten" Offizieren bemerken. Serumcholesterin und HDL-Cholesterin wurden monatlich gemessen. Ein signifikanter Anstieg des Serumcholesterinspiegels fand sich bei der gesamten Untersuchungsgruppe während eines Monats intensiver emotionaler Belastung und Spannung. Bei denjenigen Offizieren, deren Typ-A-Verhalten sich sehr deutlich verringerte, waren im Fortgang der Studie signifikant niedrigere Cholesterinwerte nachweisbar als bei denjenigen, deren Typ-A-Verhalten sich nicht veränderte; Ernährungsfaktoren und körperliches Training konnten für diese Veränderungen nicht verantwortlich gemacht werden.

Es kann bislang nicht beurteilt werden, ob die beobachteten Veränderungen über den Untersuchungszeitraum von 9 Monaten hinaus stabil sind; ebenso bleibt offen, inwieweit sie sich wirklich in einer Verminderung des koronaren Risikos niederschlagen.

41.1.7.3 Therapeutische Veränderungen des Typ-A-Musters bei Herzinfarktpatienten

Friedman meint, das Infarkterlebnis erleichtere dem Typ-A-Patienten jene Motivation aufzubauen, die Voraussetzung dafür ist, ein Verhalten zu verändern, das so viele Gratifikationen mit sich bringt. Unsere Kultur belohnt das Typ-A-Verhalten oft mit einem höheren sozialen Status (Shekelle et al., 1976). Auf der Verhaltensebene erscheint der Versuch nicht sinnvoll, alle ursprünglich zum Typ-A-Muster gehörenden Facetten zu beeinflussen. Die therapeutischen Bemühungen müssen sich vor allem auf die KHK-Prädiktoren konzentrieren – auf die Beeinflussung von Eigenschaften, die das Potential der Feindseligkeit und Aggression betreffen, Rivalitätsverhalten, Ungeduld und das Gefühl von Zeitdruck.

Friedman und Rosenman (1974) haben zahlreiche Vorschläge für ein Therapieprogramm gemacht, wie Verhaltensänderungen erreicht werden können. Sie führen verschiedene Methoden auf, die alle Berücksichtigung finden sollten:
– Aufklärende Unterrichtung des Patienten in bezug auf die koronare Herzkrankheit und das Typ-A-Verhaltensmuster.
– Methoden der kognitiven Verhaltenstherapie (z.B. bei einem Manager die systematische Umgestaltung des Tagesablaufs mit weniger Terminen; die zwanghafte Einhaltung vieler Termine führt gewöhnlich zum Gefühl von Zeitdruck, zu Hast und Eile und vermag Ungeduld und feindseliges Verhalten auszulösen; statt dessen soll die Arbeit durch mehr Ruhepausen und kontemplative Beschäftigung unterbrochen werden).
– Positive Verstärkung nicht pathogenen Verhaltens (z.B. Verkürzung der Zeit des Zusammenseins mit Personen, die Rivalitätsverhalten, Feindseligkeit und Ungeduld auslösen, und nachfolgende Belohnung durch angenehme Dinge oder Sozialkontakte).
– Streßmanagement-Programme mit systematischer Desensibilisierung, Gedanken-Stop- und Entspannungstechniken, dem Vorstellen kritischer, das Typ-A-Verhalten auslösender Situationen und dem Einüben von Typ-B-Verhaltensweisen in Rollenspiel und Alltag.
– Emotionale Unterstützung in Form einer Gruppenpsychotherapie.

Ein individueller Therapieplan fordert die Analyse der auslösenden und aufrechterhaltenden Bedingungen für den einzelnen. Da die soziale Umwelt hier eine wichtige Rolle spielt, kann auch die Einbeziehung von Ehepartner und Familie hilfreich sein.

In einer umfangreichen Therapiestudie, dem Recurrent Coronary Prevention Project (RCPP) untersuchten Friedman und Mitarbeiter (1986) bei 1013 Myokardinfarktpatienten die Wirksamkeit einer Verringerung des Typ-A-Verhaltens auf die kardiale Morbidität und Mortalität. Bei mehr als 95 % der Teilnehmer wurde Typ-A-Verhalten von mittlerer bis ausgeprägter Intensität diagnostiziert. Der vorgesehene Beobachtungszeitraum von 5 Jahren wurde auf 4,5 Jahre verkürzt, da die Wirksamkeit einer therapeutischen Veränderung des Typ-A-Verhaltens den Autoren zufolge bereits nach 3 Jahren statistisch erkennbar war (Friedman et al., 1984). 862 der Herzinfarktpatienten wurden zufallsmäßig 2 verschiedenen Behandlungsgruppen (Gruppe 1 und 2) zugeordnet. Die restlichen 151 Patienten, die keinerlei Gruppenbehandlung erhielten, dienten als Vergleichsgruppe (Gruppe 3). Die erste Behandlungsgruppe bestand aus 270 Patienten; sie wurden in 22 Einzelgruppen aufgeteilt und nahmen in 4,5 Jahren im Durchschnitt an insgesamt 33 Gruppensitzungen zu 90 Minuten teil, in denen sie „kardiologisch" beraten wurden (Friedman et al., 1982). Diese Behandlung war darauf angelegt, die Patienten bezüglich Diät, körperlichem Training, Medikamenteneinnahme und möglichen chirurgischen Maßnahmen zu beraten und sie über die ihrer Erkrankung zugrunde liegenden pathophysiologischen Mechanismen zu informieren. Die 592 Patienten der zweiten Behandlungsgruppe waren in 60 Einzelgruppen aufgeteilt und nahmen in einem Zeitraum von 4,5 Jahren an je insgesamt 62 Gruppensitzungen teil. Sie erhielten zusätzlich zur gleichen kardiologischen Beratung wie die erste Gruppe ein intensives Training zur Veränderung des Typ-A-Verhaltens. Beide Gruppen unterschieden sich zu Beginn dieser Behandlung nicht in traditionellen Risikofaktoren oder Schwere der Erkrankung. Als Maße des Typ-A-Verhaltens wurden das auf Videoband aufgezeichnete strukturierte Interview (VSI) sowie Fragebogen herangezogen, die von den Patienten selbst, ihren Ehefrauen und Arbeitskollegen beantwortet wurden.

Wurden die Analysen nach dem „intention-to-treat"-Prinzip vorgenommen, d. h. wurde die relativ große Gruppe der Patienten, die die Gruppenbehandlung vorzeitig verließen, mit in die Berechnungen aufgenommen, war nach 4,5 Jahren das Typ-A-Verhalten bei 35,1 % von den 592 Teilnehmern der Typ-A-Behandlungsgruppe deutlich verringert (um mindestens eine Standardabweichung im VSI und Patientenfragebogen) und unterschied sich signifikant (p<0,005) vom Typ-A-Verhalten der 270 Patienten der Vergleichsgruppe, das sich nur bei 9,8 % verringerte. Die Verringerung des Typ-A-Verhaltens war in der Behandlungsgruppe 2 bereits nach einem Jahr signifikant, nicht jedoch in der Behandlungsgruppe 1.

Die kumulative Reinfarktquote inklusive der kardialen Todesfälle betrug in der Typ-A-Behandlungsgruppe 2 nach 4,5 Jahren insgesamt 12,9 % und unterschied sich signifikant von der „kardiologischen" Behandlungsgruppe 1, deren Reinfarktquote bei 21,2 % lag. Die Reinfarktrate der Vergleichsgruppe, die keine Gruppenbehandlung erhielt, betrug 20,2 %. Die durchschnittlichen jährlichen Reinfarktraten der beiden Behandlungsgruppen betrugen 4,97 % bzw. 2,96 % (p<0,01). Wurden die vorzeitigen Therapieabbrecher ausgeschlossen, war der Unterschied der durchschnittlichen jährlichen Reinfarktraten inklusive kardiale Todesfälle zwischen den beiden Behandlungsgruppen noch deutlicher: 5,49 % versus 2,55 % (p < 0,001). Nicht-tödliche Infarkte traten in der Gruppe 2 (Typ-A-Behandlungsgruppe) innerhalb des 4,5jährigen Zeitraumes seltener auf als in Gruppe 1 (kardiologische Beratungsgruppe) (7,6 % vs. 14,9 %; p < 0,02). Wurden nur die letzten 3,5 Jahre berücksichtigt, also der Zeitraum, in dem sich eine im ersten Jahr erzielte Verhaltensänderung bereits positiv ausgewirkt haben sollte, unterschieden sich auch die Patienten der Gruppen 2 und 1 signifikant hinsichtlich der kardialen Todesfälle (3,4 % vs. 6,4 %; p<0,05).

Die Verringerungen von Mortalität und Morbidität werden der Verringerung im Typ-A-Verhalten zugeschrieben; Unterschiede bezüglich traditioneller Risikofaktoren oder medikamentöser Behandlung konnten nicht dafür verantwortlich gemacht werden. Die kumulative Reinfarktrate einschließlich kardialer Todesfälle derjenigen Patienten der Typ-A-Behandlungsgruppe (Gruppe 2), deren Typ-A-Verhalten sich innerhalb des ersten Jahres um mindestens eine Standardabweichung verringert hatte, war für die letzten 3,5 Studienjahre im Vergleich zu der Reinfarktrate von Patienten der kardiologischen Beratungsgruppe (Gruppe 1), deren Typ-A-Verhalten sich im gleichen Zeitraum nicht verringert hatte, weniger als halb so hoch (8,3 % vs. 21,5 %; p<0,002). Diese Studie liefert nach Ansicht der Autoren somit zusätzlich wichtige neue Befunde für die pathogenetische Bedeutung des Typ-A-Verhaltens. Es erscheint allerdings auch möglich, daß andere Faktoren eine Rolle gespielt haben. So hat sehr wahrscheinlich die Typ-A-Behandlungsgruppe ein größeres Ausmaß an sozialer Unterstützung erfahren, was sich zusätzlich positiv auf die Verringerung der Reinfarktquote ausgewirkt haben kann.

Es wurden auch weitere Unteranalysen nach dem „intention-to-treat"-Prinzip innerhalb des gesamten 4,5jährigen Studienzeitraumes vorgelegt; z.B. war die kumulative Reinfarktrate der mit Betarezeptorenblockern behandelten Patienten der Gruppe 2 (Typ-A-Behandlungsgruppe) deutlich geringer als die der entsprechend behandelten Patienten der Gruppe 1 (kardiologische Beratungsgruppe) (15,0 % vs. 27,4 %; p =0,05).

Einige klärungsbedürftige Unstimmigkeiten sind für den Zwischenbericht dieser Therapiestudie nach 3 Jahren (Friedman et al., 1984) bezüglich Anzahl und Reinfarktquote der vorzeitigen Therapie-Abbrecher deutlich geworden (Myrtek, 1986; Friedman, 1986). Danach scheint es, daß diejenigen Patienten, die beide Arten der Gruppenbehandlung vorzeitig verließen, eine besonders geringe Reinfarktquote hatten. Nach 4,5 Jahren allerdings fielen die Ergebnisse den Berichten der Autoren zufolge eindeutig zugunsten der Typ-A-Behandlungsgruppe aus. Die kumulative Reinfarktrate bei den 28 angegebenen Therapie-Abbrechern der Behandlungsgruppe 1 betrug 32,1 % und unterschied sich nicht signifikant von der der verbleibenden 192 Patienten dieser Gruppe 1 (21,4 %). Die Reinfarktrate dieser Therapie-Abbre-

Tabelle 41.1–5. Studien zur Modifikation des Typ-A-Verhaltens und Effektmaß seiner Veränderung (nach Nunes et al., 1987)[a]

Autor	Jahr	N	Alter	% Männer	% KHK	Typ A vB[c]	BDS	BBM	AK	AT	ET	CT	VK	VM	EU	PD	ABM[d]	Typ-A-Skala[e]	Typ-A-Effektmaß	methodische Probleme[f]
Ibrahim	74	99	50	84	100	0	75	1	−	−	−	−	−	+	+	−	2	F	0,09	−
Suinn	78	14	38	86	0	0	9	nz*	−	−	+	−	+	+	−	−	3	J	0,70	+
Jenni	79	29	43	64	17	2	9	nz	−	−	+	−	+	−	−	−	2	B	0,02	+
					17	2	9	nz	−	−	−	+	+	−	−	−	2	B	1,03	+
					17	1	9	nz	−	−	−	+	+	−	−	−	2	B	0,73	+
Rahe	79	44	52	90	100	0	9	1	+	+	−	−	−	+	+	−	4	F, A	1,22	−
Roskies	79	31	49	100	0	1	21	nz	−	+	−	−	−	−	+	+	2	F, A	0,28	+
					0	1	21	nz	−	+	−	−	−	−	+	+	3	F, A	0,58	+
					100	1	21	nz	−	+	+	−	−	+			0	F, A	0,70	+
Langosch	82	90	49	100	100	0	8	0	−	+	−	+	+	+	−	−	4	F	0,45	−
					100	0	8	0	−	−	+	−	−	−	−	−	1	F	0,23	−
Levenkron	83	38	35	100	0	1	12	nz	−	+	+	+	+	+	−	−	5	J, F	0,76	−
					0	1	12	nz	+	+	−	−	+	+	−	−	4	J, F	0,27	−
Friedman	84	1012	53	90	100	1	>100	6	+	−	−	−	−	−	−	−	1	SI, F	0,26	±
					100	1	>100	6	+	+	+	+	−	+	+	−	6	SI, F	0,79	−
Hart	84	9	20	70	0	1	17	nz	−	+	+	−	+	−	−	−	3	J	1,27	+
Gill	85	118	43	100	0	1	32	nz	+	+	+	+	−	+	+	−	6	SI, F	1,04	−
Korrelation mit Typ-A-Effektmaß:			−0,36	−0,17	−0,19	−0,05	−0,20	0,09	0,18	0,39	0,18	0,37	0,21	0,15	0,00	−	**0,48**[g]			

mittleres Effektmaß: **0,61** ± 0,20 [h,i]
korrigiertes mittleres Effektmaß: **0,57** ± 0,36 [h,j]

[a] Studien mit mehr als einer Zeile vergleichen zwei oder mehrere Behandlungsgruppen mit einer Kontrollgruppe. [b] Abkürzungen der Behandlungsmodalitäten: BDS = Behandlungsdauer in Stunden, BBM = Beginn der Behandlung nach Herzinfarkt in Monaten, AK = Aufklärung und Information bezüglich KHK, AT = Aufklärung und Information bezüglich Typ-A-Verhalten, ET = Entspannungstraining, CT = cognitive Therapieverfahren, VK = Vorstellen von Konflikten und Streßsituationen und Einüben von alternativen Bewältigungsmustern, VM = Verhaltensmodifikation, EU = emotionale Unterstützung, PD = psychodynamisch orientierte Behandlung. [c] Typ A vB (Typ-A-Verhalten vor Behandlung): 2 = extrem, 1 = mäßig stark, 0 = durchschnittlich oder ohne Angaben. [d] Anzahl der Behandlungsmodalitäten, ohne Berücksichtigung psychodynamischer Therapieformen (PD). [e] Typ-A-Skalen: F = Fragebogen nur in dieser Studie verwendet, J = JAS, B = Bortner, SI = strukturiertes Interview mit Videoaufzeichnung, A = Arbeitszeit verkürzt. [f] siehe Text. [g] p<0,05. [h] 95% Konfidenzintervall. [i] p<0,001. [j] p<0,01. * nz = nicht zutreffend.

cher der Gruppe 1 war jedoch signifikant (p<0,001) höher als die derjenigen Patienten der Gruppe 2, die 4,5 Jahre an der Typ-A-Behandlungsgruppe teilgenommen haben (13,0%). Weiterhin unterschied sich die kumulative Reinfarktrate der 63 Therapie-Abbrecher der Typ-A-Behandlungsgruppe (33,3%) zwar nicht signifikant von der Reinfarktrate der 192 Patienten, die 4,5 Jahre an der kardiologischen Beratungsgruppe teilgenommen haben, wohl aber von der der 4,5 Jahre aktiven Patienten der Typ-A-Behandlungsgruppe (13%; p<0,001).

41.1.7.4 Metaanalyse psychologischer Behandlungsmethoden der koronaren Herzkrankheit

Nachdem mehrere Therapiestudien zum Typ-A-Verhalten mit verschiedenen Behandlungsansätzen vorliegen, erscheint eine zusammenfassende und vergleichende Bewertung sinnvoll. Nunes und Mitarbeiter (1987) untersuchten in einem metaanalytischen Ansatz 18 kontrollierte Studien über die psychologischen Behandlungsmethoden der koronaren Herzkrankheit einschließlich des Typ-A-Verhaltensmusters. Die Ergebnisse jeder dieser Studien wurden in ein Effektmaß umgewandelt, das aus der Differenz zwischen Behandlungs- und Kontrollgruppe in Einheiten der Standardabweichung ausgedrückt gebildet wurde (Tab. 41.1–5). Bei der Beurteilung einer Veränderung des Typ-A-Verhaltens wurden 10 Studien mit insgesamt 17 Behandlungsgruppen berücksichtigt.

Das mittlere Effektmaß für eine Reduktion des Typ-A-Verhaltens betrug danach 0,61±0,20 (95% Konfidenzintervall) (p<0,001); das bedeutet, daß die Teilnehmer an diesen Behandlungsprogrammen über alle Studien gemittelt ihre Typ-A-Werte um ½ Standardabweichung verringert haben. Dies kann im Vergleich zur Literatur über therapeutische Interventionen als ein mäßig starker Effekt angesehen werden. Dieser Befund kann darüber hinaus als robust gelten, da für ihn das restriktive Bonferoni-Signifikanzkriterium (p<0,001) gilt, das Mehrfachtestung berücksichtigt. Außerdem wurde als Maß zur Abschätzung eines möglichen Publikationsvorurteils das sog. „fail-safe N" berechnet, das aussagt, wieviele (nicht-publizierte) Studien mit einem durchschnittlichen Ergebnis von Null durchgeführt worden sein müßten, um das Effektmaß auf ein nicht-signifikantes Niveau zu bringen. Danach müßten 35 weitere Studien mit negativem Ergebnis vorliegen.

Fünf Studien wiesen jedoch methodische Probleme auf. Im Recurrent Coronary Prevention Project (RCPP; Friedman, 1984) blieb es den Patienten selbst überlassen, ob sie an einer der beiden Behandlungsgruppen („kardiologische Beratung" oder „Typ-A-Behandlungsgruppe") oder einer Vergleichsgruppe teilnahmen, die keine zusätzliche Behandlung erhielt, sondern lediglich von Hausärzten in üblicher Art und Weise behandelt wurde. Nur die Zuordnung der Patienten zu den beiden Behandlungsgruppen war randomisiert. Ein korrigiertes Effektmaß wurde deswegen aus dem Vergleich dieser beiden Behandlungsgruppen berechnet. In die Metaanalyse gingen lediglich die Ergebnisse der Zwischenauswertung nach 3 Jahren, nicht jedoch die noch günstigeren Endergebnisse nach 4,5 Jahren ein. In drei Studien wurden Patienten zunächst zur Behandlung ausgewählt, dann wurde ihnen jedoch mitgeteilt, daß sie vorerst nur auf einer Warteliste berücksichtigt werden können. Bei diesen Patienten, die als Kontrollgruppe dienten, stiegen Angst und Typ-A-Maße während der Wartezeit deutlich an, weshalb diese Ergebnisse nur beschränkte Gültigkeit besitzen. In der Untersuchung von Roskies gab es keine Kontrollgruppe; hier wurden die Ausgangswerte vor der Behandlung zum Vergleich herangezogen. Werden die methodisch problematischen Studien nicht berücksichtigt und das korrigierte Effektmaß im RCPP verwendet, beträgt das mittlere Effektmaß 0,57 (p<0,01), das „fail-safe N" 5.

In den meisten Studien wird eine Selbsteinschätzung des Typ-A-Verhaltens verwendet. Deswegen muß als Einwand berücksichtigt werden, daß die Ergebnisse möglicherweise nicht so sehr „wahre" Veränderungen widerspiegeln als vielmehr einen Erwartungseffekt des Untersuchers. In zwei Studien wurde allerdings das auf Videoband aufgenommene strukturierte Interview verwendet, das von Auswertern beurteilt wurde, die die Behandlungsbedingungen nicht kannten; außerdem wurden Verhaltenseinschätzungen durch die Ehefrauen herangezogen (Powell et al., 1984; Gill et al., 1985). Hier sollte dieses Erwartungsvorurteil vermieden worden sein. Die Ergebnisse dieser beiden Studien stehen in Übereinstimmung mit den Gesamtergebnissen, was gegen einen starken Einfluß des Erwartungseffektes spricht.

Wenige Studien haben validierte KHK-Prädiktoren als Typ-A-Maße verwendet, wie das strukturierte Interview oder den JAS; die in diesen Studien erreichten Veränderungen entsprechen jedoch dem Effektmaß des Gesamtergebnisses. Für die Bewertung von Behandlungsprogrammen erscheint es wichtig, daß in künftigen Studien validierte KHK-Prädiktoren verwendet werden, da es letztlich um die Prävention oder Modifikation der koronaren Herzkrankheit geht.

Viele Untersuchungen messen Veränderungen des Typ-A-Verhaltensmusters nach einem kurzen Follow-up von Tagen oder Wochen (Suinn und Bloom, 1978; Jenni und Wollersheim, 1979; Roskies et al., 1979; Langosch et al., 1982; Levenkron et al., 1983; Hart, 1984; Gill et al., 1985). Im Recurrent Coronary Prevention Project (Powell et al., 1984) wurden deutliche Verbesserungen beim Typ-A-Verhalten nach 2 Jahren gefunden, und in einer Studie von Rahe et al. (1979) mit einer nur kurzen Behandlungsperiode waren substantielle Veränderungen noch nach 4 Jahren nachweisbar. Zukünftige Studien müßten Verhaltensveränderungen vor allem nach längeren Zeiträumen erfassen und prüfen, inwieweit dies von anhaltenden Interventionsmaßnahmen abhängig ist.

Ein weiterer begrenzender Faktor ist, daß die untersuchten Stichproben sich vorwiegend auf motivierte Freiwillige männlichen Geschlechts beschränkten. Deswegen ist eine Generalisierung auf Frauen, auf andere Altersgruppen oder weniger Motivierte bislang nicht möglich.

In der Metaanalyse wurde auch untersucht, ob Zahl und Art verschiedener Behandlungsmodalitäten einen Einfluß darauf haben, wie stark das Typ-A-Verhalten verändert wurde (Tab. 41.1–5). Die Zahl der Behandlungsmodalitäten korreliert mit dem Effektmaß für die Veränderung des Typ-A-Musters signifikant (r=0,48; p<0,05). Eine weitere Analyse ergibt, daß eine Behandlung besonders dann effektiv ist, wenn sie eine erzieherische Komponente (bezüglich KHK oder Typ-A-Verhalten) enthält, eine Coping-Strategie wie Entspannung oder kognitive Therapiemethoden (cognitive restructuring) und Verhaltensintervention (Vorstellen von Situationen, die Ärger, Ungeduld oder andere Komponenten des Typ-A-Verhaltens auslösen, und Erlernen von neuen Bewältigungsstrategien sowie Einüben von Typ-B-Verhaltensweisen). Bei Berücksichtigung nur dieser Therapiekomponenten erhöht sich die Korrelation zwischen Typ-A-Musterveränderung und Zahl der Behandlungsmodalitäten auf r=0,58; p<0,02. Dies bedeutet, daß ein umfassendes Behandlungsprogramm, das gleichzeitig viele verschiedene Facetten des Typ-A-Musters und unterschiedliche therapeutische Vorgehensweisen berücksichtigt, am besten geeignet ist. Die Ergebnisse zeigen allerdings auch, daß die stärksten Veränderungen dann zu erwarten sind, wenn erzieherische Maßnahmen bezüglich Typ-A-Verhalten und kognitive Therapieelemente enthalten sind (r=0,6; p<0,01). Diese Behandlungskomponenten beinhalten jedoch Information über das Typ-A-Verhalten auf einer intellektuellen Ebene. Das könnte bedeuten, daß die Teilnehmer ihr Typ-A-Verhalten lediglich bei der Selbsteinschätzung und Beantwortung von Fragebogen im erwünschten Sinne niedriger bewerten. Aus der kritischen Bewertung dieser Ergebnisse wird erneut deutlich, daß es sehr wichtig ist, bei künftigen Untersuchungen objektive Typ-A-Meßmethoden einzusetzen.

Beim Einsatz psychologischer Methoden zur Behandlung der koronaren Herzkrankheit ist eine Veränderung des Typ-A-Verhaltens nur ein Mittel für das eigentliche Ziel, die KHK-Morbidität und -Mortalität zu senken. Nachdem auch zu dieser Fragestellung eine Anzahl von Studien vorliegen (Tab. 41.1–6), kann mit Hilfe einer quantitativen Metaanalyse eine zusammenfassende Einschätzung der Wirksamkeit dieser Behandlungsmethoden vorgenommen werden; darüber hinaus läßt sich prüfen, welche Behandlungsmodalitäten am erfolgversprechendsten sind.

Es wurden 9 Studien mit insgesamt 11 Behandlungsgruppen analysiert. Die Daten ermöglichen, die Wirksamkeit an folgenden 4 Kriterien zu überprüfen: Mortalität nach einem Jahr, Reinfarkt nach einem Jahr, die Kombination Reinfarkt und Mortalität nach einem Jahr sowie die Kombination Reinfarkt und Mortalität nach 3 Jahren. Die mittleren Effektmaße sind für alle Wirksamkeitskriterien nach einem Jahr positiv (um 0,5), und die Konfidenzgrenzen sind für die 1-Jahres-Mortalität ziemlich eng, mit weitem Abstand zu Null, wie aus Tabelle 41.1–6 zu ersehen ist.

Tabelle 41.1–6. Studien zur psychologischen Behandlung der KHK und durchschnittliches Effektmaß (nach Nunes et al., 1987)

Autor	Jahr	N	Alter	% Männer	BDS	BBM	AK	AT	ET	CT	VK	VM	EU	Tod im 1. Jahr EM[b] (p)[c]	HI im 1. Jahr EM[b] (p)[c]	Tod + HI im 1. Jahr EM[b] (p)[c]	Tod + HI im 1. Jahr EM[b] (p)[c]	Angina in 3 Jahren EM[b] (p)[c]
Adsett	68	12	48	100	12,5	12	–	–	–	–	–	–	+	0,00	0,00	0,00		
Ibrahim	74	118	50	84	75	1	–	–	–	–	–	+	+	0,60(0,10)				
Gruen	75	75	54		10		–	–	–	–	–	–	+					0,10(050)
Wallace	77	81	57	78		0	+	–	–	–	–	–	–	0,29				
Rahe	79	61	52	90	9	1	+	+	–	–	–	+	+	0,69(0,24)	0,88(0,12)	1,22(0,02)	1,36(0,01)	0,60(0,12)
					9	1	+	+	–	–	–	+	+	0,62(0,31)	0,76(0,17)	1,08(0,05)	1,22(0,02)	0,00(0,50)
Fielding	80	10	50	100	15		+	–	+	–	–	–	+	0,00(0,50)	0,50(0,50)	0,50(0,50)		0,00(0,50)
Salonen	80			100		0	+	–	–	–	–	–	+	0,42	0,13			
Tulpule	80	205	55	100	1,5		–	–	+	–	–	–	–	0,74				
Ornish	83	46					–	–	+	–	–	–	–					1,48(0,0001)
Stern	83	64	54	86	13	9	+	+	+	–	–	–	+	0,37(0,45)	–0,34(0,50)	–0,13(0,50)		0,50
Friedman	84	1013	53	90	100	6	+	+	+	+	–	+	+	0,70	0,09	0,38		
					100	6	+	+	+	+	–	+	+	0,43(0,50)	0,54(0,05)	0,55(0,50)	0,34(0,005)	0,30(0,05)
mittleres Effektmaß: [d]														0,44±0,18	0,32±0,35	0,51±0,47		0,43±0,48
korrigiertes mittleres Effektmaß: [d]														0,34±0,41	**0,45±0,59**	**0,57±0,72**	**0,97±1,37**	
durchschnittliches Signifikanzniveau:														(0,15)	**(0,05)**	**(0,05)**	**(0,0001)**	**(0,0004)**

[a] Abkürzungen der Behandlungsmodalitäten: siehe Tab. 41.1–5.
[b] EM = Effektmaß
[c] p = Signifikanzwahrscheinlichkeit (nur für methodisch akzeptable Studien)
[d] ± 95% Konfidenzintervall

Allerdings gibt es bei den meisten Studien methodische Probleme, die eine derartige globale Betrachtungsweise als unzureichend erscheinen lassen. Werden ausschließlich methodisch akzeptable Studien berücksichtigt, verbleiben für eine bereinigte Analyse nur mehr fünf Studien. Danach ergibt sich, daß die Mortalitätsraten nach einem Jahr nicht signifikant verringert sind. Reinfarkt nach einem Jahr und die Kombination Reinfarkt und Mortalität nach einem Jahr erreichen gerade die schwächere statistische Standardsignifikanz von p<0,05. Nach drei Jahren erreicht das kombinierte Wirksamkeitskriterium Mortalität plus Reinfarkt eine Reduktion von rund 50%, für das das sicherere Bonferoni-Signifikanzkriterium von p<0,0001 gilt. Einschränkend muß gesagt werden, daß das letzte Ergebnis mit Vorsicht zu bewerten ist, da es auf nur zwei Studien beruht (Rahe et al., 1979; Friedman et al., 1984). Auch wurden hier nur motivierte männliche Teilnehmer in mittlerem Alter berücksichtigt; deshalb können keine Aussagen über andere Patientengruppen gemacht werden.

Diese Ergebnisse legen nahe, daß psychologische Behandlungsmethoden, die auf eine Modifikation des Typ-A-Verhaltens abzielen, die KHK-Morbidität und -Mortalität wirksam reduzieren. Das ist ein wichtiges Ergebnis, das nach Replikation in weiteren Studien verlangt. Derartige Studien sollten auf die Dauer von mindestens 3 Jahren angelegt sein, da eine Wirksamkeit nach einem Jahr weniger, nach drei Jahren wohl aber eindeutig nachweisbar ist.

Der Frage nach den wirksamsten Behandlungsmodalitäten kann durch eine Analyse der beiden Studien nachgegangen werden, die die überzeugendste Wirksamkeit nach drei Jahren demonstriert haben. In beiden Studien wurden erzieherische Aspekte in bezug auf die koronare Herzkrankheit und das Typ-A-Verhalten, d.h. Information und Aufklärung, betont, sowie Verhaltensänderung und soziale, emotionale Unterstützung in Form einer Gruppentherapie. Beide Studien verwendeten eine Kombination verschiedener Behandlungsmodalitäten, die sich für eine Verringerung von Typ-A-Verhaltensweisen als wirksam erwiesen haben, und in beiden Studien wurde eine deutliche Verringerung von Typ-A-Maßen erreicht. Unterschiede bestanden hinsichtlich der Verwendung von kognitiven Therapieelementen zur Veränderung von „Typ-A-Überzeugungen" und dem Einsatz eines Entspannungstrainings im Recurrent Coronary Prevention Project. In dieser Studie wurden viele Therapiesitzungen über mehrere Jahre hinweg durchgeführt, wobei eine mindestens sechsmonatige Wartezeit zwischen Infarkt und Behandlungsbeginn verging. Rahes (1979) Behandlungsprogramm war hingegen kurz und begann unmittelbar nach dem Herzinfarkt; hier wurden keine kognitiven Therapieelemente verwendet und kein Entspannungstraining. Es gibt aber auch Evidenz dafür, daß andere Behandlungsmethoden erfolgversprechend sein können; dies betrifft das gemeindebezogene Programm von Salonen et al. (1980) sowie den Therapieansatz von Tulpule (1980), der Yogamethoden verwendete; beide Studien weisen methodische Probleme auf. Es könnte allerdings nützlich sein, die hier verwendeten Techniken in künftigen Studien mitzuberücksichtigen und mit den anderen Therapieelementen zu integrieren.

Fünf methodisch einwandfreie Studien berichten über einen Rückgang der Angina-pectoris-Häufigkeit. Die kombinierte Wahrscheinlichkeit dieser Studien verfehlt das strenge Bonferoni-Kriterium knapp (p<0,004). In der Untersuchung von Rahe und im RCPP wird über eine geringe Verbesserung von Angina pectoris berichtet. In der Studie von Ornish et al. (1983) hingegen war Angina pectoris fast vollständig verschwunden, wobei die Patienten gleichzeitig weniger Medikamente einnahmen. Die Behandlungsmethoden bei Ornish umfaßten eine ausschließlich vegetarische Ernährungsweise, klassische Yogaübungen, Atem- und meditative Entspannungstechniken, die in einem streng strukturierten Programm über einen Zeitraum von drei Wochen in der Zurückgezogenheit ländlicher Umgebung gelehrt und ausgeübt wurden. Wenn auch bei der Beurteilung Vorsicht geboten ist, da Angina pectoris ein subjektives Maß ist, unterstreichen diese Untersuchungen die Möglichkeiten, die bei der Behandlung der koronaren Herzkrankheit durch Verhaltens- und Lebensstiländerungen erreicht werden können. In der Untersuchung von Ornish verbesserte sich gleichzeitig auch die körperliche Leistungsfähigkeit am Fahrradergometer erheblich: Bei unverändertem Anstieg des Druck-Frequenz-Produktes, einem Maß des myokardialen Sauerstoffverbrauchs, erhöhte sich die Dauer der körperlichen Belastbarkeit um 44 % und die geleistete Arbeit um 55 %.

Es stellt sich die Frage, ob psychologische Behandlungsmethoden vorwiegend über eine Modifikation traditioneller Risikofaktoren wirken wie Verringerung von Zigarettenkonsum, Serumcholesterin und Blutdruck und nicht über die Reduktion des Typ-A-Verhaltens. Im RCPP und bei Rahe et al. (1979) fanden sich nur geringe Unterschiede zwischen Kontroll- und Behandlungsgruppe in bezug auf das Serumcholesterin, die vernachlässigt werden können. Im RCPP wurde auch der Blutdruck gemessen, wobei keine wesentlichen Unterschiede festgestellt werden konnten, und es wurden nur Nichtraucher behandelt. Bei Rahe gab es in bezug auf das Rauchen keine Unterschiede. Ornish (1983) hingegen fand im Vergleich zur Kontrollgruppe bei der Behandlungsgruppe eine signifikante Verringerung des Serumcholesterins und des systolischen und diastolischen Blutdrucks in Ruhe und unter verschiedenen psychischen Belastungssituationen sowie eine geringe Reduktion beim Rauchen. Zukünftige Interventionsstudien sollten neben dem Typ-A-Verhalten das Spektrum der wichtigsten KHK-Risikofaktoren mit erfassen.

Neuere Untersuchungen von Ornish et al. (1988, 1989), bislang nur als Abstract publiziert, zeigen, daß mit ausschließlich verhaltensmodifikatorische Elemente enthaltenden Behandlungsmethoden sich im Vergleich zu einer auf übliche Weise behandelten Kontrollgruppe bereits nach einem Jahr mit Hilfe der quantitativen Koronarangiographie eine signifikante Reduktion der Koronarveränderungen nachweisen ließ, ein Ergebnis, das zu erzielen man bisher nicht für möglich hielt.

Studien, die eine Reduktion der KHK-Morbidität und -Mortalität zeigen, sind mit methodischen Problemen behaftet und es liegen nur wenige Studien vor; dennoch liefern diese genügend Evidenz dafür, daß verhaltensmedizinische und psychologische, auch Komponenten des Typ-A-Verhaltens einschließende Interventionsmaßnahmen bei der Behandlung der KHK wirksam sind. Eine Intervention erscheint vor allem dann effektiv, wenn sie gleichzeitig mehrere, d. h. die toxischen Typ-A-Komponenten berücksichtigt und verschiedene Behandlungsstrategien einschließt. Die Ergebnisse der Interventionsstudien

stützen somit die Bedeutung von Typ-A-Komponenten als unabhängigen Risikofaktor. Unter dem Gesichtspunkt von Kosten und Nutzen sind die verhaltensmodifikatorischen Therapien äußerst attraktiv. Die Behandlungen werden vornehmlich in Gruppen durchgeführt, wobei die Behandlungsdauer meist kurz ist (8–32 Stunden). Unklar bleibt noch, inwieweit die klinische Wirksamkeit von der Behandlungsdauer abhängig ist. Alle Studien, die eine klinische Bewertung mit Hilfe von Morbiditäts- und Mortalitätsdaten vornehmen, sind im Rahmen der sekundären und tertiären Prävention durchgeführt worden. Es sind aber auch Interventionsstudien erforderlich, die prüfen, inwieweit eine Verhaltensmodifikation eine eigenständige und wirksame Komponente bei der primären Prävention der koronaren Herzkrankheit darstellt. Psychologische Therapieansätze zur Modifikation der wichtigsten Komponenten des Typ-A-Verhaltens sollten bei zukünftigen, breit angelegten Interventionsprogrammen zur Verhütung der koronaren Herzkrankheit neben der Beeinflussung anderer Risikofaktoren mitberücksichtigt werden. Es erscheint an der Zeit, daß Ärzte die „Verschreibung" dieser relativ kostengünstigen Methoden bei der routinemäßigen Behandlung von Patienten nach einem Herzinfarkt oder von anderen KHK-Patienten ernsthaft in Erwägung ziehen.

41.1.7.5 Hinweise zur psychotherapeutischen Betreuung von Herzinfarktpatienten

Die Therapie nach erfolgtem Infarkt deckt sich bezüglich der bekannten Risikofaktoren mit der Prävention und Prophylaxe; es müssen aber zusätzlich das von den Persönlichkeitsmerkmalen des Infarktpatienten geprägte Krankheitsverhalten und die Probleme des Arztes im Umgang mit Patienten, die an möglicherweise tödlich ausgehenden Krankheiten leiden, in Rechnung gestellt werden. Dabei treten immer wieder ähnliche Probleme auf. Sie betreffen die akute Phase nach dem Infarktereignis; sie werden im folgenden Kapitel besprochen. In der an die Intensivpflege anschließenden Periode der Behandlung kann die Umgebung des Patienten, also Arzt, Krankenschwester und die Familie, Schwierigkeiten in die Behandlung bringen. Die den Patienten Pflegenden haben es mit einem Kranken zu tun, dessen Schicksal ungewiß ist. Nach Friedberg (1972) und Schettler und Nüssel (1974) sterben 30 % aller Herzinfarktpatienten innerhalb von 4–6 Wochen nach dem Infarktereignis. Von den Überlebenden sollen nach 10 Jahren noch 50–60 % derjenigen Patienten leben, die die ersten 4 Wochen überstanden haben, 20 Jahre nach dem Ereignis noch 26–31 % (Friedberg, 1972). Die für die Betreuung des Infarktpatienten Verantwortlichen können in zwei Richtungen ungünstig handeln: Durch die ungewisse und schlechte Prognose verängstigt, können sie überfürsorglich und restriktiv reagieren (Dominian und Dobson, 1969) und den Patienten dadurch mit ihrer Angst anstecken. Dieser gerät dann in ein hypochondrisches Verhalten hinein, wird übervorsichtig, passiv, abhängig und deprimiert, oder aber er verleugnet seine körperliche Schädigung zum Schutz vor seiner Angst und beginnt, sich unabhängig und expansiv zu verhalten, im Sinne seines früheren „pressured" Verhaltens. Wenn der Pflegende dem Infarktpatienten gegenüber die eigene Angst und Unsicherheit in Rechnung stellen darf, ohne ihr zu erliegen oder sie verleugnen zu müssen, erlaubt dies dem Patienten ein Verarbeiten der eigenen Angst und Trauer über den Verlust der Körperintegrität und deren Konsequenzen und bietet die beste Gewähr für eine vernünftige und an den körperlichen Zustand angepaßte Lebensgestaltung.

Von seiten des Patienten findet sich sehr oft ein verleugnendes, auflehnendes Verhalten gegenüber dem Verlust der körperlichen Integrität und dem überfürsorglichen Verhalten der Umgebung. Es ist einerseits sowohl für die Phase des Schocks und Unglaubens, wie für diejenige des Haderns in der Auseinandersetzung mit einer bedrohlichen Krankheit typisch (Kübler-Ross, 1969). Andererseits ist dieses Verhalten nicht nur Ausdruck einer vorübergehenden Periode im Trauerprozeß, sondern beruht auf den Persönlichkeitszügen des Patienten, der oberflächlich kooperativ erscheinen kann, unterschwellig aber zu Auflehnung gegen die Autorität und zu Steigerung seiner Aggressivität neigt. Dabei gilt die unbewußte Abwehr des Patienten weniger dem Inhalt der Anordnungen, die die Pflegenden ihm geben, sondern mehr dem emotionellen Ton und der Form, in der sie gegeben werden (Hahn, 1971). Arzt und Schwestern tun deshalb gut daran, den sich gegen seine tiefliegende Abhängigkeitsneigung mit einem pseudounabhängigen Verhalten wehrenden Infarktpatienten als Partner zu behandeln, ihm z. B. die ärztlichen Überlegungen so darzubieten, daß der Patient mitdiskutieren und zu ihrer Ausgestaltung in einer Form der praktischen Empfehlung beitragen kann, welche die Eigenleistung und Verantwortlichkeit des Patienten anregt (Belser, 1967). Die Schuldgefühle, die sich im Patienten beim Auftreten von aggressivem und dominierendem Verhalten sowie bei der Neigung, sich Muße und Entspannung hinzugeben, melden, bedürfen der besonderen Berücksichtigung (Bastians, 1968). Die Arbeit des Trauerns, und um sie geht es bei der Bewältigung des Verlusts der körperlichen Integrität, benötigt Monate bis Jahre. Deshalb empfiehlt es sich, Patienten nach Spitalentlassung regelmäßig zu bestellen, um sie bei ihrer Trauerarbeit, die Auflehnung, Hadern, Angst usw. einschließt, zu unterstützen. Patienten nach einem Infarkt ohne ambulante Kontrolluntersuchungen mußten 10mal häufiger hospitalisiert werden als solche, bei denen regelmäßige Kontrolluntersuchungen durchgeführt wurden (Dominian und Dobson, 1969).

Acker (1976) wies in einer nicht-randomisierten Studie nach, daß ein Rehabilitationsprogramm den Krankenhausaufenthalt und die Rekonvaleszenzzeit abkürzt, die Rückkehr an den Arbeitsplatz fördert und das Rauchen vermindert. Diese Einflüsse zeigten sich aber lediglich bei Patienten unter 50 Jahren, die niedrigen sozialen Schichten angehörten, wenig

Schulbildung besaßen und ungelernte oder gelernte Tätigkeiten ausübten.

Da die erfolgreiche Anpassung an das chronische Kranksein wesentlich von dem Verhalten und der Einstellung der Familie, vor allem der Ehefrau abhängt, ist es von Vorteil, ergänzende Maßnahmen durchzuführen, die geeignet sind, das Ausmaß an sozialer Unterstützung für den Patienten zu erhöhen (Frank et al., 1979; Razin, 1982). Adsett und Bruhn (1968) führten mit 6 Postinfarktpatienten im Verlauf von 6 Monaten 10 Gruppengespräche durch; Ehefrauen haben während dieser Zeit ebenfalls an 10, jedoch separaten Gruppentreffen teilgenommen. Die Behandlung wurde mit einer gemeinsamen Sitzung von Patienten und ihren Ehefrauen beendet. Das therapeutische Vorgehen war gesprächspsychotherapeutisch orientiert und die Patienten wurden ermutigt, ihre Gefühle, vor allem ihre Ängste vor einem Reinfarkt zu äußern, sowie sich wechselseitig bei der Bewältigung aktueller Probleme emotional zu unterstützen. Die psychosoziale Anpassung der Patienten und ihrer Ehefrauen wurde bei Abschluß der Behandlung als deutlich gebessert beurteilt. In keiner therapeutischen Sitzung waren Angina-pectoris-Beschwerden aufgetreten und auch klinisch bedeutsame EKG-Veränderungen waren nicht festzustellen gewesen. Sechs Monate nach Therapieende waren weder in der Behandlungs- noch in der Kontrollgruppe ernsthafte kardiale Komplikationen aufgetreten, doch hatte von den vier Patienten, die eine Therapiebeteiligung abgelehnt hatten, einer einen Myokardinfarkt, ein weiterer einen Schlaganfall erlitten.

In einer Familie schleifen sich im Verlauf der Jahre die Rollen ein, die die einzelnen Mitglieder zueinander einnehmen. So stellt sich beispielsweise die Ehefrau auf einen pseudounabhängigen Ehemann ein, indem sie ihre eigenen Unabhängigkeitswünsche und Selbstverwirklichungstendenzen zurückstellt und sich führen läßt. Mit der Erkrankung des Mannes kann dieses Gleichgewicht aus den Fugen geraten, wenn der Ehemann in seiner Verunsicherung die Führerrolle abgibt, seine Frau mehr Verantwortung übernehmen muß und damit ihre seit Jahren zurückgestellten Wünsche plötzlich aktualisiert werden. Dies kann dazu führen, daß sie den Mann zu dominieren und in die passive Rolle zu drängen beginnt, oder aber, daß sie sein Abhängigwerden nicht erträgt und ihn bewegt, sein früheres „pressured" Verhalten wieder aufzunehmen. Auch von Stichmann und Schönberg (1972) liegen günstige Berichte über die Betreuung beider Ehepartner vor. Das Gespräch über sexuelle Probleme des Infarktpatienten ist dabei besonders wichtig. Oberflächliche Bemerkungen und Empfehlungen erzeugen häufig mehr Angst, als daß sie Aufklärung bringen. Herzinfarktpatienten sind häufig ältere Menschen und es besteht oft die Auffassung, daß Sex nichts für Kranke sei und daß sich Herzerkrankungen und Sex nicht vertragen. Diese Auffassung stützte sich auf unbegründete Annahmen, auf unzulängliche oder in der künstlichen Laborsituation durchgeführte Untersuchungen (Wagner, 1976). Es wurde angenommen, daß der Geschlechtsverkehr zu Herz- und Kreislaufbelastungen führe, die einem Infarktpatienten unzuträglich sein müssen. Wir lernen heute, daß diese Annahmen wohl unbegründet sind. So zeigten Hellerstein und Friedman (1969) mit telemetrischen Beobachtungen an Infarktpatienten, die nicht wußten, daß bei ihnen auch physiologische Messungen während des Geschlechtsverkehrs vorgenommen wurden, bei sexueller Tätigkeit in der vertrauten Situation zu Hause und mit dem Ehepartner, daß die physiologischen Reaktionen erstaunlich gering sind. Beispielsweise stieg die Pulsfrequenz während des Orgasmus nur kurze Zeit auf 117 an, während sie bei den gleichen Menschen in beruflichen Situationen und bei leichter Arbeit auf 120 anstieg. Dorossiev (1976) wies nach, daß die sexuelle Aktivität bei Patienten in einem Rehabilitationsprogramm in einem Verhältnis von 6,5:1 erhalten blieb, während sie in einer unbehandelten Kontrollgruppe bei jedem zweiten Patienten abnahm. Ezra (1961) betont, daß die ganze Familie in die Anpassungsprozesse einbezogen werden sollte. Mulcahy und Hickey (1970) stellten fest, daß Entmutigung durch die Familie einen der Hauptgründe für das Versagen beim Versuch der Rückkehr zur Arbeit darstellt. Die Familie soll an Gesprächen über realistische Ziele für den Patienten teilnehmen (Jefferson, 1966).

41.2 Psychotherapie von Herzinfarktpatienten während der stationären und poststationären Behandlungsphase

Karl Köhle und *Ekkehard Gaus*

41.2.1 Zur Begründung eines psychosomatischen Arbeitsansatzes in der Behandlung von Infarktkranken während der akuten Behandlungsphase

Die Akutbehandlung von Infarktkranken erfolgt heute auf Intensivstationen mit hohem technischem und personellem Aufwand. Zumindest während der ersten Tage gilt die „Abschirmung von äußeren Reizen", die äußerste Schonung des Patienten, als therapeutische Maxime. Der Kranke soll „in Ruhe gelassen werden". Intensivere Gespräche mit ihm werden vermieden, vielfach aus der Befürchtung heraus, solche Gespräche könnten somatische Komplikationen, etwa Rhythmusstörungen, auslösen. Diese Auffassung erweist sich bei näherer Untersuchung jedoch zumindest als unvollständig, häufig als falsch und für den Patienten schädlich. Sie übersieht, daß für die Reaktion der Kranken nicht objektiv feststellbare „Reize", sondern deren subjektive Bedeutungen ausschlaggebend sind; die subjektive Bedeutung kann jedoch nur bei intensiver Kommunikation mit den Patienten, im Bemühen, Zugang zu ihrer individuellen Wirklichkeit zu bekommen, vom Arzt erfahren werden. Eine rationale ärztliche Beurteilung der Gesamtsituation des Patienten hat deshalb intensive Kommunikation zur Voraussetzung. Nur so kann der Arzt auch die Wirksamkeit innerer Reize, vor allem der Ängste und Phantasien der Patienten kennenlernen und beurteilen. Die genannten seelischen Vorgänge haben Konsequenzen für:
- das Befinden der Patienten und den Verlauf der Krankheitsverarbeitung,
- die Kooperation der Kranken mit Ärzten und Pflegepersonal,
- die Entstehung und den Verlauf der Erkrankung, sowie den Rehabilitationserfolg.

41.2.2 Übersicht über psychosomatische Befunde bei Infarktkranken während der stationären und der poststationären Behandlungsphase

41.2.2.1 Beeinträchtigung des Befindens, psychologische und psychophysiologische Reaktionen

Das Befinden

Das Befinden der Infarktkranken wird während der Intensivbehandlungsphase vor allem durch Angstzustände und Depressionen beeinträchtigt.

Bis zu 80% der Infarktkranken leiden unter mehr oder weniger starken Angstzuständen, bis zu 58% unter Depressionen (Gentry et al., 1972).

Cay (1972, 1976, 1982) beschrieb bei 65% der 203 von ihm untersuchten männlichen Infarktpatienten in Edinburgh während des stationären Aufenthaltes Angstzustände, bei einem Drittel der Patienten handelte es sich dabei um ernste Störungen. Hackett et al. (1968) beobachteten in Boston bei mindestens der Hälfte der Infarktpatienten Angstzustände und Depressionen während der Intensivbehandlungsphase.

Auf das Angebot eines psychosomatischen Konsiliardienstes hin wurde in Boston der Konsiliarius bei 32% von insgesamt 441 Infarktpatienten während der ersten Behandlungstage auf der Coronary Care Unit hinzugezogen (Cassem und Hackett, 1971). Die Konsiliardienstanforderung erfolgte aufgrund schwerer Angstzustände (32,4%), schwerer Depressionen (30,4%) und erheblicher Verhaltensprobleme der Patienten gegenüber Schwestern und Ärzten (20,6%). Im Rahmen dieses konsiliarischen Betreuungsprogrammes versuchten Cassem und Hackett (1971), den zeitlichen Ablauf der psychologischen Komplikationen bei 145 Herzinfarktpatienten näher zu bestimmen. Die Konsultation wegen Angstzuständen erfolgte durchschnittlich am zweiten Tag des stationären Aufenthaltes, die Konsultation wegen Depressionen am vierten und die Konsultation wegen Verhaltensstörungen, die vorwiegend als Folge von Verleugnungsprozessen zustande gekommen waren, am dritten Tag des Aufenthaltes. Aufgrund dieses Befundes entwickelten die Autoren eine hypothetische Darstellung des „natürlichen Verlaufes" der emotionalen Reaktionen nach einem Herzinfarkt (Abb. 41.2–1).

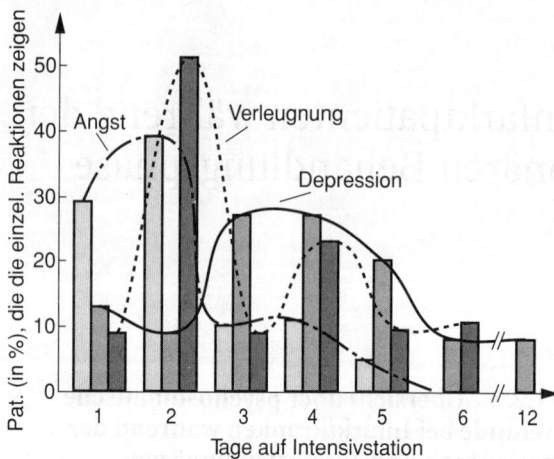

Abb. 41.2–1. Hypothetischer Verlauf emotionaler Reaktionen nach Eintritt eines Herzinfarktes (nach Cassem und Hakkett, 1971).

Bisher liegen kaum Untersuchungen vor, inwieweit diese emotionalen Reaktionen spezifisch für Infarktkranke sind und sich von den Reaktionen anderer lebensbedrohlich Erkrankter unterscheiden. Vetter und Mitarbeiter (1977) untersuchten in Edinburgh 308 Patienten mit ischämischer Herzerkrankung – darunter 272 Infarktkranke – bereits 30 Minuten nach Aufnahme in die Coronary Care Unit mit Hilfe eines kurzen Angstfragebogens. Die Angst der Infarktkranken entsprach zu diesem Zeitpunkt der Angst anderer, ebenfalls als Notfallpatienten, jedoch auf Allgemeinstationen aufgenommener Kranker. Die Angstreaktionen waren bei weiblichen Patienten und bei Kranken mit myokardialer Ischämie stärker als bei den übrigen Infarktkranken. Bemerkenswert schien eine Tendenz zu höheren Angstwerten zum Zeitpunkt der stationären Aufnahme bei denjenigen Kranken, die später während des Krankenhausaufenthaltes verstarben.

Psychophysiologische Reaktionen

Untersucht wurde vor allem der Zusammenhang zwischen psychologischen Variablen und der Katecholaminproduktion.

Klein und Mitarbeiter (1974) untersuchten bei 38 Infarktkranken während der ersten fünf Krankheitstage einerseits Angst, Depression und Feindseligkeit mittels Fragebogen und Einschätzungsskalen sowie die Katecholaminausscheidung im Urin. Patienten mit ausgeprägter und anhaltender emotionaler Unruhe hatten eine deutlich erhöhte Katecholaminausscheidung sowie ein erhöhtes Letalitätsrisiko. Letzteres ließ sich bei der geringen Patientenzahl allerdings noch nicht ausreichend statistisch sichern.

Miller und Rosenfeld (1975) fanden bei fortlaufender Registrierung somatischer und psychischer Parameter eine positive Korrelation zwischen psychometrisch ermitteltem Angstindex und Adrenalinausschüttung und dem Auftreten von Rhythmusstörungen.

Rhythmusstörungen können vor allem während der ersten 24 Stunden nach dem Infarkt, häufig im Zusammenhang mit verschiedenen emotionalen Belastungen, die auch schon durch einfache Interaktionen im Rahmen der medizinischen Versorgung oder während des Besuchs von Verwandten auftreten, aus-

gelöst werden (Theorell und Wester, 1973). Bemerkenswert ist der Zusammenhang zwischen Auftreten von Rhythmusstörungen und der Tendenz von Patienten, über unangenehme Erfahrungen auf der Station zu klagen (Theorell und Wester, 1973).

Die Wahrnehmung des Todes von Mitpatienten – entsprechend auch von Reanimationsmaßnahmen – in der kardiologischen Intensivstation führte regelmäßig zu Blutdruck- und Herzfrequenzanstieg bei den übrigen Kranken, auch wenn sie bei Befragung angaben, nichts bemerkt oder sogar geschlafen zu haben (Bruhn et al., 1970).

Differenzierte Untersuchungen über psychophysiologische Reaktionsmuster bei Infarktkranken, die auch die Abwehr- und Anpassungsmechanismen der Betroffenen mit in die Betrachtung einbeziehen, sind bis jetzt nicht durchgeführt worden. In diesem Zusammenhang sei jedoch auf die Bestimmung der Exkretion von Wachstumshormon und Cortisol während Herzkatheteruntersuchungen hingewiesen (Greene et al., 1970). Während der Katheteruntersuchung wiesen ängstlich-zurückgezogene Patienten stärkere Veränderungen der Hormonproduktion auf als Patienten, denen die Attribute „ruhig", „depressiv" oder „ängstlich-engagiert" zugesprochen waren.

41.2.2.2 Der Einfluß psychosozialer Faktoren auf Behandlungsmodalitäten und das Arbeitsbündnis

Eine rationale Behandlung von Infarktkranken erfordert die Berücksichtigung des Einflusses psychischer und sozialer Faktoren auf das Verhalten von Arzt und Patient und auf den Umgang der beiden Interaktionspartner miteinander. So führt die verleugnende Abwehr der Kranken beispielsweise häufig zu einer Bagatellisierung der Beschwerden durch den Kranken und einer Unterbewertung durch den Arzt; die Folge ist bei Infarktkranken dann eine zu niedrige Dosierung von Analgetika, Tranquilizern und Schlafmitteln (Hackett et al., 1969b).

Auch soziale Merkmale korrelieren mit den Behandlungsmodalitäten; so erhielten Infarktkranke auf der Privatstation signifikant mehr Schlafmittel als die übrigen Infarktpatienten desselben Krankenhauses (Hackett und Cassem, 1976).

Ziel jeder Infarktbehandlung ist die Entwicklung eines befriedigenden **Arbeitsbündnisses** zwischen Arzt und Patient, das eine aktive und langfristige Beteiligung des Kranken an der Behandlung seiner Erkrankung ermöglicht. Wesentliche Voraussetzung für die Entwicklung einer solchen Zusammenarbeit ist die ausreichende Information des Patienten über die Natur des Herzinfarktes, den Ablauf der Heilungsvorgänge und die Erfordernisse der Behandlung.

Empirische Untersuchungen haben gezeigt, daß der tatsächliche Wissensstand der Patienten selbst in führenden Kliniken mit langjährigen Bemühungen um eine effiziente Behandlung von Infarktkranken gering ist.

So verfügten im General Massachusetts Hospital in Boston nur etwa 25% der Infarktkranken über ein ausreichendes Wissen über das Wesen des Infarktes und nur 23% über ein entsprechendes Wissen über die Art der Heilungsvorgänge.

Im Krankenhaus der US-Marine in San Diego wußten die Infarktpatienten kaum mehr über das Wesen ihrer Erkrankung als eine vergleichbare Patientenpopulation, die keinen Herzinfarkt erlitten hatte (Rahe et al., 1975 a, b).

In den genannten Krankenhäusern lösten diese Befunde intensive Bemühungen aus, die psychosoziale Betreuung der Patienten unter besonderer Berücksichtigung der Vermittlung von krankheitsrelevanter Information zu intensivieren (Hackett, 1977; Rahe, 1975 c). Dabei konnte der Nachweis einer Effizienz psychotherapeutischen Vorgehens für die Vermittlung des relevanten Wissens erbracht werden (Rahe et al., 1975 a, b).

Soziale Merkmale haben Einfluß auf den Wissensstand von Infarktkranken. Die Zugehörigkeit zu einer niedrigeren sozialen Schicht („blue collar") korreliert mit weniger Wissen über die Ausrüstungsgegenstände der Intensivstation, insbesondere die Monitoren, mit mehr Angst vor diesen Geräten sowie mit einem niedrigeren Wissensstand über die Heilungsvorgänge am Herzen, ausgeprägteren regressiven Verhaltenszügen und schließlich einer geringeren Tendenz, Fragen über die eigene Zukunft zu stellen. Unterschichtpatienten waren auch seltener in der Lage, die Namen der sie behandelnden Ärzte anzugeben (Hackett und Cassem, 1976).

41.2.2.3 Die Wirksamkeit psychosozialer Faktoren im Krankheits- und Rehabilitationsverlauf

Die prinzipielle Bedeutung von **Risikoverhaltensweisen** und anderen Persönlichkeitsmerkmalen ist für die Pathogenese des Herzinfarktes in zahlreichen Studien nachgewiesen (vgl. Kap. 41.1).

In prospektiven Studien unterschieden sich die psychologischen Testergebnisse von Personen, die später an einem Herzinfarkt starben, bereits Jahre vorher von den Ergebnissen derjenigen, die später den Infarkt überlebten (Bruhn, 1974).

Die **psychotherapeutische Intervention** während der stationären Behandlungsphase scheint zu einer Verminderung der Mortalität beitragen zu können (Cassem und Hackett, 1971); während der poststationären Behandlungsphase kann sie zu einer signifikanten Verminderung von Reinfarkten und anderen Komplikationen der Koronarerkrankung (Ibrahim et al., 1974; Rahe et al., 1975 a, b) beitragen.

Psychosoziale Faktoren und **berufliche Rehabilitation:** Unter der Voraussetzung eines optimalen Rehabilitationsprogrammes können zwei von drei Infarktpatienten vier Monate nach dem akuten Ereignis die Arbeit wieder aufnehmen (Cay et al., 1976). Das Mißlingen der Arbeitsaufnahme ist nach Untersuchungen in England, den USA und Australien häufiger die Folge psychosozialer als körperlicher Behinderung (Cay et al., 1973).

Cay und Mitarbeiter (1973) untersuchten den Rehabilitationsverlauf bei 203 unausgewählten männlichen Patienten über ein Jahr; körperliche, psychologische und soziale Daten wurden während der Krankenhausbehandlung sowie vier Monate und ein Jahr nach der Entlassung erhoben:

Nach 4 Monaten hatten 66% der vor dem Infarkt arbeitenden Patienten die Arbeit wieder aufgenommen, davon arbeitete die Hälfte ebenso intensiv wie früher. Die Patienten dieser Gruppe hatten sich im Vergleich zur Zeit des stationären Aufenthaltes emotional stabilisiert, allerdings wurde die Hälfte von ihnen noch als ängstlich oder deprimiert, ein Viertel in einem deutlich pathologischen Sinn, eingestuft. Diese Befunde hatten sich auch ein Jahr nach dem Infarkt nicht wesentlich verändert.

Diejenigen Patienten, die die Arbeit nicht hatten wiederaufnehmen können, waren im Durchschnitt jünger und gehörten häufiger unteren sozialen Schichten an.

Die Art der Krankheitsverarbeitung bzw. der beobachtbaren emotionalen Reaktionen unmittelbar nach Infarkteintritt korrelierte mit dem späteren Rehabilitationsergebnis:

52% der Kranken, bei denen während des Krankenhausaufenthaltes keine ausgeprägte emotionale Erregung beobachtet worden war, befanden sich nach 4 Monaten wieder voll im Arbeitsprozeß, dagegen nur 36% derjenigen Patienten, die nach Krankheitseintritt unter Anzeichen von Angst und Depression gelitten hatten. Die Prognose für die Rehabilitation verschlechterte sich mit dem Schweregrad der emotionalen Störung: Nach einem Jahr waren lediglich 29% der Patienten mit schweren Angstzuständen und Depressionen wieder auf ihrem früheren Berufsniveau tätig, im Vergleich zu 45% der Patienten mit milden Störungen und 64% der Kranken ohne emotionale Probleme. Umgekehrt gehörten diejenigen Patienten, die die Arbeit nicht wieder aufnahmen, ganz überwiegend der Gruppe derjenigen Patienten an, die bei Krankheitsbeginn besonders ausgeprägt pathologisch reagiert hatten. Nach einem Jahr arbeiteten aus dieser Gruppe 31% noch nicht wieder, gegenüber 12% der übrigen Kranken.

Auch Merkmale der prämorbiden Persönlichkeit korrelieren mit dem Rehabilitationsverlauf.

58% der als „stabile Persönlichkeiten" eingeschätzten Kranken arbeiteten nach im Durchschnitt 68tägiger Rekonvaleszenz ebenso intensiv wie früher, im Vergleich zu nur 20% der „instabilen Persönlichkeiten" nach einer Rekonvaleszenzzeit von 92 Tagen (Cay et al., 1975, 1976).

41.2.3 Psychodynamik von Angst, Verleugnung und Depression bei Infarktkranken

41.2.3.1 Angst

Bei jedem Infarktkranken ist es erforderlich, individuell Ausmaß und Verlauf der Angst zu beurteilen und die Art ihrer seelischen Verarbeitung herauszufinden.

Klinische Beurteilung und quantifizierende Erfassung von Angstreaktionen

Die Angst des Infarktpatienten kann sich offen äußern – gelegentlich bis hin zu panischer Todesangst – und auch für den ungeschulten Beobachter erkennbar sein. Häufiger äußert der Infarktkranke seine Ängste jedoch nicht spontan und direkt. Sie sind dann nur aus seinem Ausdrucksverhalten oder aus testpsychologischen Reaktionen zu erschließen. Die klinische Beurteilung der Angst und ihrer Auswirkungen erfordert eine sorgfältige und längerdauernde

Beobachtung, wie sie auf Intensivstationen mit der hohen Zahl von Planstellen vor allem dem Pflegepersonal möglich sein sollte – eine entsprechende Schulung vorausgesetzt. Oft können die Patienten ihre Ängste leichter mitteilen, wenn sie gezielt darauf angesprochen werden. Die Kranken können ihre Ängste dann entweder direkt verbalisieren oder sie teilen sie indirekt mit über Äußerungen wie „Ich gäbe eine ganz schöne Leiche ab, meinen Sie nicht?" (Cassem und Hackett, 1971) oder „Ich habe keine Angst, daß mich während des Schlafs der schwarze Wagen holt". Innerpsychische Verarbeitungsprozesse bestimmen dabei die Formulierungen mit. So können Kranke versuchen, das eigene Betroffensein in ein Erschrecken des Arztes oder der Schwester zu verwandeln („Identifikation mit dem Aggressor"), wobei sie gleichzeitig an deren Reaktionen wiederum die Gefährlichkeit ihrer Situation testen.

Hierbei ist zu beachten, daß Ärzte in solchen „Krisensituationen" Fragen ihrer Patienten sehr häufig ausweichen, wie empirische Untersuchungen der ärztlichen Visite gezeigt haben (Siegrist, 1976). Gegenüber Infarktkranken spielt dabei die Bereitschaft von Ärzten, sich mit Patienten zu identifizieren („Ich an Ihrer Stelle"), insofern eine besondere Rolle, als Ärzte besonders häufig von koronaren Herzerkrankungen betroffen werden.

Eine quantitative Erfassung der Angst wird bei dem beeinträchtigten Allgemeinzustand der Patienten am günstigsten von Ärzten oder Schwestern nach einem Gespräch bzw. nach den Pflegemaßnahmen mit Hilfe von Einschätzungsskalen vorgenommen; leicht anwendbar ist beispielsweise die Anxiety-Depression Scale von Holland und Sgroi (1973), für die sich auch nach kurzem Training eine gute Interrater-Übereinstimmung erzielen läßt. Die Anwendung derartiger Skalen ermöglicht die Vergleichbarkeit der Ergebnisse bei verschiedenen Patienten sowie eine Verlaufsbeurteilung. Testpsychologische Untersuchung der Angst ist prinzipiell möglich, oft jedoch schwer durchführbar.

Froese und Mitarbeiter (1974) untersuchten mit Hilfe der Anxiety-Depression Scale von Holland und Sgroi den Angstverlauf bei Infarktkranken während der Zeit des gesamten stationären Aufenthaltes. Die höchsten Werte fanden sich bei der Aufnahme, während der ersten fünf Tage trat ein stärkerer Abfall, vor der Entlassung ein Wiederanstieg auf. Dabei wurde auch der Einfluß der Verleugnung als Abwehrmechanismus auf den Angstverlauf deutlich: In der Gruppe der „Verleugner" (Skala von Hackett und Cassem, vgl. Abschnitt 41.2.3.2 „Verleugnung") fielen die Angstwerte zu Beginn des Aufenthaltes signifikant schneller ab (Abb. 41.2–2).

Auch für den **Verlauf** der emotionalen Reaktionen nach Entlassung aus dem Krankenhaus ist das Ausmaß der Betroffenheit zu Beginn der Erkrankung mit entscheidend. In der Untersuchung von Cay und Mitarbeitern (1975) litten 72% derjenigen Kranken, die während des Krankenhausaufenthaltes verwirrt gewesen waren, 4 Monate nach Entlassung noch unter Angstzuständen und Depressionen, im Vergleich zu nur 28% derjenigen Patienten, die vom akuten Ereignis weniger stark betroffen waren.

Quellen der Angst

Schmerzen

Die Angst der Infarktkranken wird in den Lehrbüchern oft als direkte Folge der meist heftigen, zum Teil als „Vernichtungsschmerzen" beschriebenen Thoraxschmerzen dargestellt.

Die Zusammenhänge zwischen Thoraxschmerzen und Angstentstehung sind im einzelnen jedoch nicht ausreichend geklärt; in keinem Falle ist der Thoraxschmerz die einzige Angstquelle. Für den Zusammenhang zwischen Thoraxschmerz und Angst ist zu berücksichtigen, daß das Herz in Angstzuständen auch als Ausdrucksorgan benützt wird, sowie, daß auch als Begleiterscheinung von Angst bzw. als Angstäquivalent in der Herzgegend lokalisierte Schmerzen entstehen oder dort bereits bestehende Schmerzen dann verstärkt empfunden werden können.

Vorstellungen und Phantasien über Natur und Konsequenzen des Herzinfarktes

Spricht man eingehender mit Infarktkranken und befragt man sie nach den Vorstellungen und Phantasien über ihre Erkrankung, so findet man regelmäßig, daß Infarktpatienten auf der Intensivstation tagelang unter der eindeutigen Vorstellung leben, sie müßten zwangsläufig an den Folgen des Infarktes sterben. Diese Vorstellungen bestehen im allgemeinen auch dann noch, wenn die gefährlichste Zeitspanne der ersten Stunden nach Infarkteintritt überwunden ist. Die Kranken assoziieren zum Begriff Herzinfarkt bildhafte Vorstellungen vom Tod wie „Schnitter Tod", „Sensenmann" sowie Gedanken an den Infarkttod von Verwandten und Bekannten.

Solche Assoziationen werden nicht selten zum Teil auch durch die räumliche Gestaltung der Intensivstation – u.a. gekachelte Wände – mitgefärbt. („Leichenhalle", „Metzgerei"). In verstärktem Ausmaß gilt dies für Patienten, die als Folge einer zerebralen Ischämie nach kardiogenem Schock bzw.

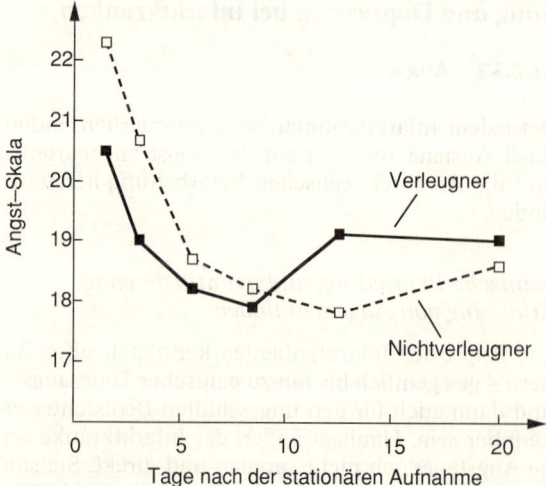

Abb. 41.2–2. Der Verlauf der Angst bei verleugnenden und nicht verleugnenden Infarktkranken während des stationären Aufenthaltes (nach Froese et al., 1974).

Reanimationsmaßnahmen in ihrer kognitiven Leistungsfähigkeit beeinträchtigt sind.

Die Vorstellungen und Phantasien, die sich Infarktkranke über das erkrankte Herz machen, entsprechen diesen Ängsten und können ihrerseits wiederum die Ängste verstärken.

Das Herz wird in diesen Phantasien vielfach aus dem selbstverständlichen Zusammenhang des Körperbildes herausgerissen und als verletztes, zerrissenes Organ vorgestellt (Dlin et al., 1966; Freyberger et al., 1969; Lipowski, 1967, 1968; Köhle et al., 1972). Die Vorstellungen über das Ausmaß des Organschadens sind häufig derart übertrieben, daß bei rationaler Betrachtung ein Weiterleben nicht mehr denkbar wäre. So phantasieren Patienten etwa: „Das Herz ist geplatzt", „zerrissen", „in zwei Hälften zerteilt" oder „alle Gefäße sind verstopft", „die großen Gefäße sind vom Herzen abgerissen".

Diese Vorstellungen und Phantasien teilen die Kranken oft erst nach längerem Bemühen des Arztes mit. Diese Mitteilung entlastet die Kranken, wenn ihnen die Beziehung zum Arzt und der Inhalt des Gesprächs mit ihm Sicherheit vermitteln. Dem Verschweigen dieser angstvollen Vorstellungen und Phantasien kommt nur sehr bedingt eine Schutzfunktion zu: Die Kranken befürchten, daß ihre Vorstellungen in der Realität oder durch die Aussagen des Arztes bestätigt werden können. Das „hartnäckige" Nachfragen durchbricht deshalb nicht gewaltsam einen sinnvollen Abwehrmechanismus, sondern macht eine in verzweifelter Situation aktivierte und nicht sehr effiziente Schutzmaßnahme überflüssig.

Bedrohung des Selbstwerterlebens

Weniger bekannt ist, daß sich die Angst der Infarktkranken oft nicht so sehr unmittelbar auf den körperlichen Tod bezieht, sondern ihre Quelle in einer Bedrohung des Selbstwertregulationssystems hat. Die Erkrankung und ihre Folgen werden als schwere narzißtische Kränkungen erlebt. Gleichzeitig wird der Verlust der sozialen Wertschätzung befürchtet. Patienten, die sich aufgrund ihrer Persönlichkeitsstruktur geradezu süchtig um soziale Anerkennung bemüht hatten, haben nun Angst, in der Folge der Infarkterkrankung jeglichen sozialen Wert und alle sozialen Kontakte zu verlieren, einen vorzeitigen „sozialen Tod" zu sterben (Huebschmann, 1966, 1967; Hackett und Cassem, 1968).

Die Kranken äußern oft in stereotyper Form die Furcht vor sozialem Abstieg, nicht selten finden sich diese Befürchtungen auch in ihren Träumen wieder. So befürchtete ein Arbeiter, der bisher als Kontrolleur in der Herstellung von Bremsen eine verantwortungsvoll erlebte Tätigkeit ausgeübt hatte, nun entweder ein „halbtotes Siechtum" als Frührentner führen oder – im günstigeren Fall – nur noch als „Laufbursche" im früheren Betrieb die ihm bisher unterstellten Kollegen bedienen zu müssen. Charakteristischerweise kompensierte er dabei die Angst vor Wertverlust gegenüber Ärzten und Schwestern durch nicht endenwollende Erzählungen über seine früheren Leistungen: seine Erfolge als Kriegsflieger und seine Fähigkeit, nach Abschüssen und Abstürzen zu überleben. In diese Erzählungen gingen zugleich sein Wunsch nach Anerkennung seines hohen Einsatzes, aber auch seine von Enttäuschung gefärbten Vorwürfe ein: Die jetzige koronare Herzerkrankung führte er mit auf die Kriegsbelastungen zurück.

Körperliche Beeinträchtigung als Krankheitsfolge

Neben dem Schmerz sehen Infarktkranke die krankheitsbedingte körperliche Schwäche oft als Beweis für den Ernst und die Irreversibilität der Herzerkrankung an. Aufgrund ihrer Persönlichkeitsstruktur benötigen sie Leistungsfähigkeit und Stärke in besonderem Maße für die Regulation ihres Selbstwerterlebens. Hinzu kommt die besondere Bedeutung motorisch-expansiven Verhaltens für die Abfuhr von Triebimpulsen und das Erleben intakter Ich-Funktionen bei Infarktkranken (Hahn, 1971). Körperliche Schwäche und der hierdurch mitbedingte Zwang zu Passivität wird von den Patienten besonders bedrohlich erlebt.

Eine Berücksichtigung dieser Zusammenhänge erweist sich als besonders wichtig im Falle der Verordnung von Tranquilizern, Sedativa und Hypnotika. Die Wirkung dieser Pharmaka erscheint oft unzureichend, nicht selten paradox; dies beruht gelegentlich darauf, daß diese Pharmaka zu einer weiteren körperlichen Schwächung und einer Einschränkung der Ich-Funktionen, wie Wahrnehmungs- und Denkvermögen, führen können. Die Patienten beginnen zu befürchten, nicht mehr „Herr im eigenen Haus" zu sein, die Kontrolle über geistige, seelische und körperliche Funktionen zu verlieren; dies steigert wiederum die Angst, führt nicht selten zu einer Erhöhung der Psychopharmakadosis durch den Arzt und damit zur Schließung eines Circulus vitiosus, der nur durch eine psychotherapeutische Intervention unterbrochen werden kann.

Eine solche paradoxe Wirkung sedierender Medikamente konnte bei Infarktkranken sowohl durch psychometrische als auch durch psychophysiologische Untersuchungen im einzelnen bestätigt werden (Williams et al., 1975).

Die Behandlungssituation im Krankenhaus

Während die Situation auf der Intensivstation von den Kranken im allgemeinen als unterstützend erlebt wird, kann vor allem die Beobachtung einer Verschlechterung der Erkrankung bei Mitpatienten oder das Miterleben von deren Tod zur Angstentstehung bzw. zur Steigerung bereits bestehender Ängste beitragen.

Der Tod von Mitpatienten wird regelmäßig direkt oder indirekt wahrgenommen, jedoch verleugnet. Werden die Patienten nach ihren Wahrnehmungen bezüglich des Todes von Mitpatienten befragt, so geben sie zumeist an, deren Tod nicht bemerkt zu haben, sie hätten zu dieser Zeit „geschlafen" o.ä. Psychophysiologische Untersuchungen, insbesondere der statistisch signifikante Anstieg von Herzfrequenz und Blutdruck zum Zeitpunkt des Todes des Mitpatienten stützen jedoch die Interpretation, daß es sich hierbei um ein Ergebnis von Verleugnung handelt (Bruhn et al., 1970). Nach Reanimation steigt in dem betreffenden Zimmer der Bedarf an Narkotika, die Überlebenden klagen über vermehrte Schmerzen und benötigen vermehrt Sedativa und Tranquilizer (Hackett, 1977). Bei einer Befragung von Patienten nach der Entlassung, ob sie lieber in einem Einzel-, Doppel- oder Vierbettzimmer untergebracht wären, wählten alle Patienten, die in einem Vierbettraum gelegen hatten, wieder diese Unterbringungsart mit Ausnahme derjenigen, die eine Reanimation miterlebt hatten; sie verlangten nach einem Einzelzimmer (Hackett, 1977) (vgl. Kap. 64).

Die Angst der Kranken wird vielfach durch die emotionalen Reaktionen und das Verhalten von Mitgliedern des Behandlungsteams verstärkt. Auch die hohe Mortalität auf Intensivstationen führt zu enormen emotionalen Belastungen und in der Folge zu Distanzierungstendenzen und anderen Abwehrhaltungen.

Medizinsoziologische Untersuchungen haben gezeigt, daß Ärzte besonders gegenüber Patienten mit unsicherer Prognose zur Distanzierung neigen, was u. a. eine mangelhafte Information gerade dieser Kranken zur Folge hat. Verkürzt formuliert, versucht der Arzt, durch die Aufrechterhaltung seines Wissensvorsprunges seine Überlegenheit und damit seine Sicherheit auf Kosten der Unsicherheit des Kranken zu festigen ("funktionale Unsicherheit", nach Davis, 1966; vgl. Siegrist, 1976; McIntosh, 1974, 1977 sowie Kap. 18).

Zu dieser Distanzierung tragen nicht selten auch Schuldgefühle bei, die im Zusammenhang mit aktiven Eingriffen, vor allem bei Reanimationsbemühungen, entstehen können.

So konnte eine sonst besonders mitfühlend auf Patienten eingehende Schwester das Zimmer eines schwerkranken Infarktpatienten nicht mehr betreten. Erst nach längerer Diskussion des Problems in einer Balint-Gruppe konnte erarbeitet werden, daß sie intensive Schuld empfand, weil während der von ihr durchgeführten externen Herzmassage Rippen gebrochen waren und im Anschluß daran eine Pneumonie auftrat.

Reaktionen der Angehörigen

Insbesondere die Ehefrauen von Infarktpatienten reagieren auf die Erkrankung häufig mit einer überprotektiven Haltung, die die Verängstigung der Kranken noch steigern kann. Diese Haltung kann sowohl aus der Verlustangst der Ehefrau als auch aus Schuldgefühlen unterschiedlicher Genese resultieren; ambivalente Einstellungen in der Partnerbeziehung, selbstkritische Vorwürfe im Zusammenhang mit der auslösenden Situation, aggressiv getönte Enttäuschung über den drohenden Tod des Partners im Sinne eines Verlassenwerdens können zu diesen Schuldgefühlen beitragen.

So beurteilte die Ehefrau eines 53jährigen Infarktkranken die objektiv hinsichtlich der Rehabilitation sehr günstige Prognose folgendermaßen: Beruflich habe er jetzt keine Aufstiegschancen mehr; seine Aktivitäten müsse er in jeder Beziehung einschränken; das Auto solle er verkaufen, denn sie habe Angst, er könne am Steuer einen weiteren Infarkt erleiden und einen Unfall verursachen. Um körperliche Belastung von ihrem Mann fernzuhalten, schlug sie einen Wechsel der Wohnung vor; sie benötigte eine Parterre-Wohnung, beim Treppensteigen könne ihr Mann bei einem Reinfarkt zusammenbrechen und neben ihr zu Boden sinken. In ihrer Vorstellung sollte der Ehemann in Zukunft zu Hause im Lehnstuhl sitzen, während sie ihn bediente. Für sie war es definitiv: Ihr Mann würde nie mehr seine volle Leistungsfähigkeit erreichen, "ein Topf, der einen Riß hat, wird nie wieder ein voll gebrauchsfähiger Topf", "eine Maschine, die kaputtging, wird nie wieder wie vorher funktionieren".

Dieses angstverstärkende Verhalten der Ehefrau wurde u. a. auch durch Schuldgefühle mitverursacht: Der Herzinfarkt des Mannes war aufgetreten, während sie im Ausland in Urlaub war.

Prämorbide Persönlichkeitsstruktur und die Lebenssituation zur Zeit der Krankheitsmanifestation

Das Verständnis der emotionalen Reaktionen von Infarktkranken, insbesondere ihrer Ängste, setzt die Kenntnis der wichtigsten Merkmale ihrer prämorbiden Persönlichkeitsstruktur voraus, wie sie weitgehend übereinstimmend in zahlreichen, zum Teil auch prospektiven Untersuchungen ermittelt wurden (vgl. Kap. 41.1). Von gleicher Bedeutung ist die Kenntnis der psychischen und sozialen Belastungen, die der Erkrankung unmittelbar vorausgingen (Ostfeld et al., 1964; Lipowski, 1967, 1968; Pelser, 1967; Rahe, 1973, 1975; vgl. Kap. 41.1).

Der oft geradezu süchtig nach Leistung, Erfolg, Anerkennung, nach absolut festem und jederzeit seiner Kontrolle unterworfenem Besitz seiner Objekte strebende Infarktkranke lebt in ständiger Hyperaktivität, unter Zeitnot und Termindruck, er definiert seine Ziele in der Arbeitssituation und auch im übrigen Leben nur unzureichend und er kann aus seiner Tätigkeit und sogar auch aus seinen Erfolgen nur ganz unzureichend Befriedigungserlebnisse gewinnen. Er muß seine eigenen Passivitäts-, Abhängigkeits- und Versorgungswünsche abwehren. Häufig hat er seine ständigen Verlust- und Trennungsängste jahrelang kompensiert gehalten, sein allgemeines Lebensgefühl entsprach jedoch der Angst des "Reiters über den Bodensee" oder wie es ein Patient formulierte: "So, als könnte ich jederzeit ins Moor einsinken". Psychodynamisch ist es von besonderer Bedeutung, daß diese Kranken unfähig sind, das Gute, das sie aufnehmen könnten, sich tatsächlich "einzuverleiben", innerlich zu "assimilieren", zu einem dauerhaften Bestandteil ihrer Person zu machen. Deshalb bleiben sie stark von der äußeren Zufuhr abhängig und müssen ständig Kontakt und Erfolg suchen.

In der "auslösenden Situation" kulminieren dann häufig die bereits chronisch bestehenden Belastungen und überfordern die Adaptationsmechanismen des Patienten. In der Regel erleben sie einen tiefgehenden Verlust, häufig in mehreren Lebensbereichen gleichzeitig, in Familie und Beruf, oder fühlen sich von solchen Verlusten bedroht. Nicht selten führt das in ihrem subjektiven Erleben zu einer tiefgreifenden Verunsicherung einem Schicksal gegenüber, dem sie sich weitgehend ohnmächtig – "hilflos und hoffnungslos" (Engel) – ausgeliefert fühlen. Sie befürchten ein endgültiges Scheitern (van der Valk, 1967; Bastiaans, 1968; Karstens et al., 1970).

Ein 58jähriger Büroleiter klagte am zweiten Morgen auf der Intensivstation während der Visite über stenokardische Beschwerden; die Nachfrage ergab, daß er die Schmerzen nach dem Erwachen aus einem Traum bemerkt hatte. Er habe geträumt, daß andere Angestellte in seiner Firma in seinem Büro einbrachen, ihn mit Gewalt verdrängten und seine Möbel vor die Tür stellten. Sein Bericht war von diffuser Angst begleitet. Das anschließende Gespräch ergab den realen Hintergrund: Der Patient hatte seinen Herzinfarkt unmittelbar nach dem Verlassen der chirurgischen Klinik erlitten, wohin er seine Frau zur Operation eines Kolonkarzinoms gebracht hatte. Der befürchtete Verlust seiner Ehefrau – seine erste Frau war 5 Jahre vorher am selben Leiden verstorben – wurde von ihm deshalb vollends als Katastrophe erlebt, da die Ehefrau zugleich seine Sekretärin war. Er fühlte sich während der letzten Monate seiner Arbeit in einem größeren Betrieb immer weniger gewachsen, er hatte die Arbeit jedoch mit Hilfe der jüngeren und tüchtigen Ehefrau bisher eben noch bewältigen können. Nachdem diese Problematik mit dem Patienten durchgesprochen worden war, nahmen seine Angst und seine Schmerzen rasch ab.

41.2.3.2 Verleugnung

Definition und Funktion des Abwehrmechanismus „Verleugnung"

Mit „Verleugnung" wird ein psychologischer Schutzmechanismus bezeichnet, der wie andere psychische Abwehr- und Anpassungsmechanismen weitgehend unbewußt abläuft. Verleugnung richtet sich gegen eine bewußte Wahrnehmung äußerer Gefahren, gegen die Wahrnehmung der möglichen oder tatsächlichen Folgen solcher Bedrohungen (etwa körperliche oder seelische Verletzungen), sowie gegen die bewußte Wahrnehmung der mit diesen Folgen einhergehenden Emotionen (etwa Kränkung oder Trauer nach Verletzungen oder Verlusten). Ziel des Abwehrmechanismus Verleugnung ist eine Verminderung von Angst; hierdurch soll die Funktionsfähigkeit des Ichs für die Aufgaben der Realitätsbewältigung aufrechterhalten, bzw. nach einer anfänglichen Schockphase als Folge der ersten vollen Konfrontation mit der Bedrohung wiederhergestellt werden.

Zusammengefaßt kann mit Weisman (1966) Verleugnung auch definiert werden als „bewußte oder unbewußte Zurückweisung der gesamten verfügbaren Bedeutung einer Erkrankung in der Absicht, Angst zu mindern und emotionalen Streß zu minimieren".

Verleugnende Abwehr beginnt entwicklungspsychologisch früh; eine Vorform findet sich beispielsweise im „Verschließen der Augen vor einer Gefahr" bei Kindern, zusammen mit der Phantasie, daß dann auch die angsterregende Person das Kind nicht mehr sieht.

Als Abwehrmechanismus ist Verleugnung vor allem von der Verdrängung abzugrenzen.

Im allgemeinen wird als Unterscheidungskriterium die jeweilige Arbeitsrichtung der beiden Abwehrformen genannt: Verdrängung richtet sich mehr gegen unbewältigte innere Gefahren, vor allem Triebimpulse aus dem „Es", während Verleugnung gegen die Wahrnehmung äußerer Gefahren eingesetzt wird (S. Freud, A. Freud).

Für die klinische Beurteilung scheint jedoch bedeutsamer, daß die beiden Abwehrmechanismen auf unterschiedlichen Funktionsniveaus und in zu differenzierender Weise im Zusammenwirken mit anderen Abwehrmechanismen in der Angstverarbeitung eingesetzt werden. Im Falle der Verdrängung wird durch Signalangst ein komplexer **Abwehrkampf** gegen die meist inneren Gefahrenquellen aktiviert, der Organismus setzt sich mit der Gefahr auseinander. Verleugnung wird eingesetzt, wenn eine solche Auseinandersetzung

nicht mehr oder nicht mehr in vollem Umfang möglich ist. Verleugnung aktiviert nicht eine Auseinandersetzung, die durch das Angstsignal ausgelöst wird, Verleugnung versucht vielmehr, das Angstsignal zu **negieren.** Das Individuum verhält sich – zumindest in Teilbereichen seiner Existenz – so, als bestünde gar keine Gefahr.

Aus dieser Analyse des Verleugnungsmechanismus läßt sich die für die klinische Tätigkeit wichtige **Gefahr totaler Verleugnung** ableiten, nämlich eine weitgehende oder völlige Lähmung situationsgerechten Anpassungsverhaltens (vgl. Kap. 62).

Einstufung und Verlauf verleugnender Abwehr bei Infarktkranken

Verleugnung konnte bis vor kurzem nur indirekt über ihren „Erfolg" eingestuft werden, d.h. komplementär zum Ausmaß der bei dem Patienten beobachteten Angstreaktionen. Erst die Entwicklung einer klinischen Einschätzungsskala durch eine Arbeitsgruppe in Boston („Hackett-Cassem Denial Scale"; Hackett und Cassem, 1974) ermöglichte eine verläßliche quantitative Einschätzung von Verleugnung und die systematische Erforschung dieses Abwehrmechanismus bei Infarktkranken.

Froese und Mitarbeiter (1974) untersuchten mit Hilfe dieser Einschätzungsskala 36 Patienten während des stationären Aufenthaltes. Die Kranken wurden entsprechend den Skalenwerten in zwei Gruppen, „Verleugner" und „Nicht-Verleugner", eingeteilt.

Es fanden sich keine signifikanten Korrelationen zwischen Verleugnung einerseits und Geschlecht, Alter der Patienten sowie der Dauer ihres Aufenthaltes auf der Intensivstation andererseits.

Abbildung 41.2–3 gibt den Verlauf der verleugnenden Abwehr der untersuchten Infarktkranken während des stationären Aufenthaltes getrennt für die Gruppen der „Verleugner" und „Nicht-Verleugner" wieder.

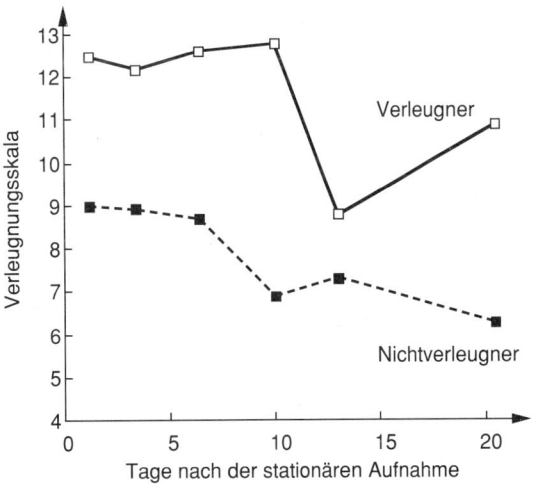

Verleugnungsskala

Verleugner

Nichtverleugner

Tage nach der stationären Aufnahme

Abb. 41.2–3. Der Verlauf verleugnender Abwehr von Infarktkranken während des stationären Aufenthaltes in den Gruppen „Verleugner" und „Nicht-Verleugner" (nach Froese et al., 1974).

Die als „Verleugner" eingestuften Kranken zeigten während des gesamten stationären Aufenthaltes – mit Ausnahme der Tage 11 bis 14 – signifikant höhere Verleugnungswerte als die „Nicht-Verleugner". Bei den „Nicht-Verleugnern" ging die Verleugnungsreaktion am 8. bis 10. Tag zurück, bei den „Verleugnern" erst gegen Ende der zweiten Woche; die „Nicht-Verleugner" konnten diesen Abwehrmechanismus flexibler handhaben und zurücknehmen („relaxation"). Bei den Verleugnern erfolgt diese Entspannung erst drei bis vier Tage später und auch nur vorübergehend; bei ihnen stieg die Verleugnungsreaktion vor der Krankenhausentlassung wieder an.

Die Wirkung der Verleugnung läßt sich am Verlauf der Angstreaktion verfolgen.

In der Studie von Froese und Mitarbeitern (1974) fielen die Angstwerte in der Gruppe der „Verleugner" während der ersten Woche signifikant rascher ab als in der Gruppe der „Nicht-Verleugner". Dabei fällt auf, daß bei den Verleugnern die Verleugnungsreaktion trotz abgefallener Angstwerte länger aufrechterhalten bleibt als bei den „Nicht-Verleugnern". Offenbar muß der Abwehrmechanismus der Verleugnung gegen die aus der Latenz noch wirkende Angst aktiviert bleiben. Dieser Hypothese entspricht der stärkere Anstieg von Verleugnung und Angst in der Gruppe der „Verleugner" vor der Krankenhausentlassung (Abb. 41.2–4).

Das Ausmaß der Inanspruchnahme des Abwehrmechanismus Verleugnung scheint für den einzelnen individuellen Infarktkranken charakteristisch und über den Zeitverlauf stabil mit der übrigen Persönlichkeitsstruktur verbunden zu sein. In der Studie von Froese und Mitarbeitern (1974) veränderte sich die Einstufung eines „Verleugners" nur sehr selten zu einem „Nicht-Verleugner" und umgekehrt. Die Anamnese von „Verleugnern" zeigte, daß sie auch früher in entsprechenden Lebensbelastungen meist ähnlich reagiert hatten. Auch die Reaktion zum Zeitpunkt der bevorstehenden Krankenhausentlassung bestätigte diesen Befund.

Die Mitglieder der Gruppe „Nicht-Verleugner" können offenbar diesen Abwehrmechanismus auch in einer Gefahrensituation flexibler handhaben. Bei den „Verleugnern" wird er schon bei geringgradigen Angstsignalen in Anspruch genommen. Bei den „Nicht-Verleugnern" wird er in der Gefahr, d. h. mit Eintritt des Infarktes, zwar ebenso schnell aktiviert, kann aber unterschiedlich rasch entsprechend der realen Situation wieder inaktiviert werden.

Verleugnung eignet sich nach den Ergebnissen dieser Untersuchung insbesondere zur Angstabwehr. Die Beziehungen zwischen Depression und Verleugnung sind weniger klar, in der Untersuchung von Froese konnte die Depression weniger erfolgreich durch Verleugnung gesteuert werden.

Korrelationen zwischen Verleugnung, Angst und Ausprägung psychogalvanischer Reflexe (Hautwiderstand) konnten ebenfalls nachgewiesen werden. Dabei fand sich, daß solche Korrelationen nur bei bestimmten thematischen Inhalten bestanden; dies weist darauf hin, daß in Zukunft die Konzepte über die Wirkungen des Abwehrmechanismus Verleugnung weiter differenziert werden müssen.

Der Verlauf der verleugnenden Abwehr kann beim einzelnen Patienten, ebenso wie der Verlauf der Emotionen Angst und Depression, von Tag zu Tag stark schwanken, da er u. a. mit von äußeren Ereignissen, der Ausprägung der körperlichen Beschwerden, dem Ausmaß der Information über die Erkrankung, der Zuwendung von Ärzten und Schwestern abhängig ist.

Folgen der Verleugnung für das Krankheitsverhalten und die Behandlungsmodalitäten

Verzögerungsverhalten („delay") in der prähospitalen Phase

Verleugnungsvorgänge verzögern insbesondere die Entscheidung der Patienten, fachkompetente Hilfe in Anspruch zu nehmen (Hackett et al., 1969b). Der Tod vieler Infarktkranker – etwa 50% der Todesfälle ereignen sich in den ersten vier Stunden – ist hierauf zurückzuführen (Hackett et al., 1969a, b; Goldstein et al., 1972; Greene et al., 1972). Die Hypothese, daß die Dauer dieser „Entscheidungszeit" im wesentlichen von Verleugnungsvorgängen und nicht durch Mangel an Informiertheit bestimmt wird, wird durch das Verhalten von Patienten mit einem Reinfarkt unterstützt: bei ihnen findet sich keine Verkürzung der Entscheidungszeit (Hackett et al., 1969b, 1974).

Die Patienten verzögern ihre Entscheidung, einen Arzt zu verständigen, obwohl sie aufgrund der massiven Beschwerden die Gefährlichkeit der Situation mehr oder weniger bewußt erkennen; so gaben 31 der 32 von Hackett und Cassem (1974) untersuchten Kranken Schmerzen als führendes Symptom an und bezeichneten diese Beschwerden fast immer als eine „der unangenehmsten Erfahrungen ihres Lebens": „als ob jemand auf meinem Brustkorb gestanden wäre", „als ob ein Loch in meine Brust gerissen wurde", usw. Trotzdem nahmen nur 21 dieser Kranken innerhalb von 5 Stunden medizinische Hilfe in Anspruch.

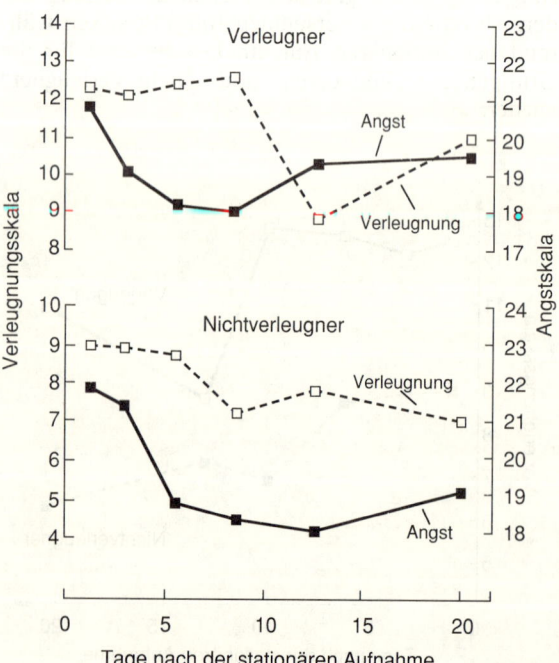

Abb. 41.2–4. Der Verlauf von Verleugnung und Angst bei Infarktkranken während des stationären Aufenthaltes bei „Verleugnern" und „Nicht-Verleugnern" (nach Froese et al., 1974).

Ein kasuistisches Beispiel soll den charakteristischen Verlauf dieses Verzögerungsverhaltens illustrieren.

> Ein 67jähriger Industrieller bekommt morgens um 5.30 Uhr, nach dem Aufstehen, heftige längerdauernde retrosternale Schmerzen. Seine Berufstätigkeit hat ihm genaueste medizinische Kenntnisse über Natur und Folgen des Herzinfarktes vermittelt. Der Patient entschließt sich jedoch, diese Schmerzen zunächst dadurch zu bekämpfen, daß er in seine Heimsauna geht und gymnastische Übungen macht. Als sich die Schmerzen nicht bessern, läßt er seine Schwester, die Ärztin ist, kommen. Sie stellt die Diagnose eines Hinterwandinfarktes und empfiehlt die sofortige Krankenhauseinweisung. Der Patient schickt jedoch den Krankenwagen wieder weg und begibt sich erst nach insgesamt 12 Stunden, als die Schmerzen für ihn unerträglich werden, am Steuer des eigenen Wagens in die Klinik. Es fand sich ein ausgedehnter Hinterwandinfarkt mit Rhythmusstörungen, die den Einsatz eines Schrittmachers erforderlich machten.

Erschwerung eines Arbeitsbündnisses zur rationalen Therapie der Erkrankung

Verleugnende Abwehr betrifft auch die Krankheitsfolgen, insbesondere die Schmerzen. Nicht selten verleugnen Patienten stenokardische Beschwerden sehr weitgehend, was ihre klinische Beurteilung sowie Indikation und Dosierung von Analgetika und Sedativa erschwert (Hackett et al., 1969).

> Während der Visite schildert sich ein Infarktkranker am zweiten Tag des stationären Aufenthaltes zunächst als völlig beschwerdefrei. Auf intensiveres Nachfragen hin klagt er über Beschwerden in der linken Großzehe. Erst nachdem wir mit ihm ausführlicher über seine Gesamtsituation und seine Ängste gesprochen haben, kann er über die noch fortbestehenden heftigsten stenokardischen Dauerschmerzen klagen. Zur verleugnenden Abwehr war eine **Verschiebung** der Beschwerden getreten.

Verleugnende Abwehr führt nicht selten zu Interaktionsproblemen zwischen dem Patienten einerseits und Arzt und Schwestern andererseits. Ein Hauptthema solcher Auseinandersetzungen ist die Einhaltung der verordneten Bettruhe. Infarktkranke können aufgrund ihrer Persönlichkeitsstruktur die körperliche Inaktivität und die damit verbundene Autonomieeinbuße oft nur schwer ertragen und übertreten nicht selten entsprechende therapeutische Vereinbarungen. Typisch ist der Infarktkranke, der bei der Visite im Trainingsanzug („stramm wie eine Eins") im Bett liegt und um die Erlaubnis zu körperlicher Aktivität bittet. In dieser Situation ist es wichtig, die Bedürfnisse der Kranken nach Autonomie, nach Wiederherstellung ihrer Leistungsfähigkeit verständnisvoll zu diskutieren.

F. Dunbar berichtete von einer zunächst erfolgreichen Behandlung eines Infarktpatienten, der die verordnete Bettruhe nicht einhalten konnte und dem deshalb die Benutzung eines Lehnstuhls und beschränkte Bewegung im Krankenzimmer erlaubt worden war. Derselbe Patient erlitt später einen Reinfarkt und starb, nachdem er in einer Auseinandersetzung mit einem anderen Stationsarzt seine Privilegien nicht mehr erhalten hatte.

Die Abwehr der tiefen Verletzung des Selbstwertgefühls mittels Verleugnung trägt dazu bei, daß Patienten die Rollenverteilung zwischen Arzt und Patient in Frage zu stellen oder umzukehren versuchen.

> Der erwähnte Industrielle schilderte bei der Chefarztvisite auf die Frage, wie es ihm gehe, nicht etwa seine Beschwerden, sondern stellte erst einmal fest, daß der Chefarzt auch Ostpreuße sei und berichtete über seine Leistungen als Soldat beim Kampf um Ostpreußen. Er verleugnete seine augenblickliche Situation und versuchte, seinen jetzigen Autonomieverlust durch den Rückgriff auf frühere Erfolge und den sich anschließenden langen Bericht über seine jetzigen beruflichen Leistungen zu kompensieren. Schließlich fragte er den Professor, ob dieser nicht etwa unter ihm – in einer Studentenkompanie in Ostpreußen – „gedient" hätte. Die Umkehrung der gegenwärtigen Rollen- und Machtverteilung sollte die krankheitsbedingte narzißtische Kränkung ausgleichen.

Vom Ausmaß der Verleugnung hängt es mit ab, wie weit Patienten Informationen auch über die Natur ihrer Erkrankung und die Erfordernisse der Behandlung während des stationären Aufenthaltes aufnehmen können, die sie für eine konsequente Durchführung der späteren Therapie und Rehabilitationsmaßnahmen benötigen.

Croog und Mitarbeiter (1971) untersuchten Infarktkranke zum Zeitpunkt der Entlassung aus einem Krankenhaus, in dem ihnen ausreichend Möglichkeit zur Information über das Wesen ihrer Erkrankung angeboten worden war. 20% von 293 Patienten antworteten auf die Frage, ob sie eine Herzattacke gehabt hätten, mit „nein" oder „weiß nicht". Croog stufte diese Patienten als „massive Verleugner" ein. Dieses Abwehrverhalten erwies sich bei Interviews im Abstand von einem Monat und einem Jahr als stabil. Die Mitglieder dieser Gruppe der „massiven Verleugner" hielten sich in der Folgezeit weniger an die ärztlichen Ratschläge; sie waren weniger als die anderen Kranken bereit zu versuchen, ihr Risikoverhalten zu ändern, insbesondere das Zigarettenrauchen einzuschränken. Die „Verleugner" befolgten seltener auf die Rehabilitation bezogene Ratschläge, etwa zum Termin der Wiederaufnahme der Arbeit und zur Wahl des Arbeitsplatzes. Bei ihnen fand sich gehäuft die Tendenz, körperliche Symptome zu bagatellisieren.

Levine und Mitarbeiter (1987) fanden mit Hilfe einer neuen Verleugnungsskala (Einschätzung nach Interview) unterschiedliche Funktionen von Verleugnung im Verlauf von Erkrankung und Rehabilitation. Sie untersuchten 30 von 45 Männern, die wegen eines Herzinfarkts oder einer koronaren Bypassoperation stationär behandelt wurden, ein Jahr nach Entlassung nach. Patienten mit hohen Verleugnungswerten lagen kürzer auf der Intensivstation, hatten weniger Zeichen partialer Dysfunktion während des Krankenhausaufenthaltes. Im folgenden Jahr jedoch war der Verlauf für Patienten mit hohen Verleugnungswerten ungünstiger: Ihre Compliance mit medizinischen Empfehlungen war schlechter, sie mußten länger als Patienten mit niedriger Verleugnung wieder ins Krankenhaus aufgenommen werden. Diese Ergebnisse weisen darauf hin, daß Krankheitsverleugnung während der Akutbehandlung im Krankenhaus (jedenfalls unter den dort normalerweise gegebenen Bedingungen!) adaptiv sein kann, nicht jedoch für den Langzeitverlauf nach Krankenhausentlassung.

Diese Befunde zeigen, wie wichtig die Beachtung von Verleugnungsvorgängen auch für die Durchführung von Rehabilitationsprogrammen ist.

Dabei soll noch einmal auf die Gefahr der Realitätsverkennung im Falle der Angstabwehr durch totale Verleugnung aufmerksam gemacht und darauf hingewiesen werden, wie subtil die einzelnen Verleugnungsstrategien wirksam sein können: Ein Infarktkranker, der sich zwar zunächst allen medizinischen Anordnungen exakt fügte und insgesamt kooperativ schien, lehnte die Diagnose „Herzinfarkt" mit folgender Begründung ab: Ärzte lernten alles aus Büchern, Patienten seien eben keine Bücher.

Klinische Beurteilung der Verleugnung; Indikation zu psychotherapeutischer Intervention

Verleugnung ist zunächst ein sinnvoller Abwehrmechanismus im Sinne einer physiologischen Schutzreaktion; bei entsprechender Disposition kann dieser Abwehrmechanismus in Belastungssituationen jedoch pathologisches Ausmaß annehmen, d.h. zu einer gefährlichen Einschränkung von Realitätsprüfung und Bereitschaft zu therapeutischer Mitarbeit führen. Eine Indikation zur psychotherapeutischen Intervention sehen wir dann als gegeben an, wenn hierdurch eine rationale Behandlung des Infarktkranken gefährdet wird.

Für die klinische Beurteilung von Verleugnungsprozessen erscheint uns vor allem auch die Mitberücksichtigung der sozialen Situation des Patienten in der Klinik wesentlich. Hat der Patient keine Möglichkeit zu entlastenden Gesprächen über das Wesen und die Prognose seiner Erkrankung, so muß er in verstärktem Ausmaß Abwehrmechanismen wie die Verleugnung benützen. In diesem Sinne kann Verleugnung auch als „sozialer Akt" angesehen werden.

Hackett und Mitarbeiter (1968) fanden für stabil verleugnende Infarktkranke eine Tendenz zu einer besseren somatischen Prognose. Die angesichts anderer Befunde aufzustellende Hypothese, daß dies nur für Patienten gilt, denen keine psychotherapeutischen Gesprächs- und Hilfsmöglichkeiten angeboten werden, müßte jedoch vor einer endgültigen Bewertung dieses Befundes überprüft werden. Bisher liegen keine Studien über psychotherapeutische Interventionen bei Infarktkranken unter Benutzung einer Verleugnungsskala vor.

41.2.3.3 Depression

Diagnostik (quantifizierende Einstufung) und Verlauf

Auch depressive Zustandsbilder werden bei Infarktkranken häufig nicht erkannt, da sie kein „lärmendes" Bild machen: Mimik und Gestik sind starr, der Patient wirkt insgesamt verlangsamt, interesselos, oft zurückgezogen. Hinter der stillen Unauffälligkeit findet sich nicht selten weitgehende Hoffnungslosigkeit bis hin zu Tendenzen der Selbstaufgabe.

Der Traum eines erfolgreichen Geschäftsmannes in der Nacht nach Infarkteintritt soll den krankheitsbedingten Selbstwertverlust und die hieraus resultierende Depression veranschaulichen: Der still und unauffällig wirkende Patient klagt bei der Morgenvisite über stenokardische Beschwer-

den. Im Gespräch ergibt sich, daß die Beschwerden nach dem Erwachen aus einem Traum aufgetreten seien: der Kranke sah sich im Traum, wie er „auf den Lumpenwagen" geworfen wurde. Von dieser bei ihm unbewußt vorhandenen Phantasie kann er jedoch erst nach einem längeren Gespräch ausführlicher berichten, nachdem er schon eine gewisse Stützung seines Selbstwerterlebens erfahren hatte.

Der Verlauf der Depressionswerte (Holland-Sgroi-Skala) für die Gruppe der „Verleugner" und „Nicht-Verleugner" während der Zeit des stationären Aufenthaltes ist aus Abbildung 41.2–5 ersichtlich.

Bei den „Verleugnern" sinken die Werte auf der Depressionsskala vom 3. bis 10. Tag im Vergleich zu den „Nicht-Verleugnern" signifikant ab, um allerdings später wieder anzusteigen. Die Autoren halten den Verleugnungsmechanismus gegenüber depressiven Emotionen für nicht so wirksam wie gegenüber Angst; als Grund hierfür diskutierten sie, daß es sich bei der Depression wahrscheinlich um ein psychologisch und biochemisch komplexeres Phänomen als bei der Angst handelt (Froese et al., 1974).

Nimmt man die Zeit des stationären Aufenthaltes zusammen, so lassen sich auch nach Untersuchungen anderer Autoren keine signifikanten Veränderungen der Depressionswerte nachweisen (Doehrman, 1977).

Ursachen der Depression

In der Psychodynamik von Infarktkranken können verschiedene Faktoren zur Entstehung einer depressiven Reaktion in jeweils individuell zu bestimmendem Ausmaß beitragen.

Unspezifische Reaktionen auf die Erkrankung

Zunächst handelt es sich bei den depressiven Verstimmungen um eine einfühlbare emotionale Reaktion auf die krankheitsbedingten Veränderungen der Lebenssituation. Die Erkrankung bringt für den Patienten körperliche Schwäche, Hilfsbedürftigkeit, Abhängigkeit und Ohnmachtsgefühle mit sich.

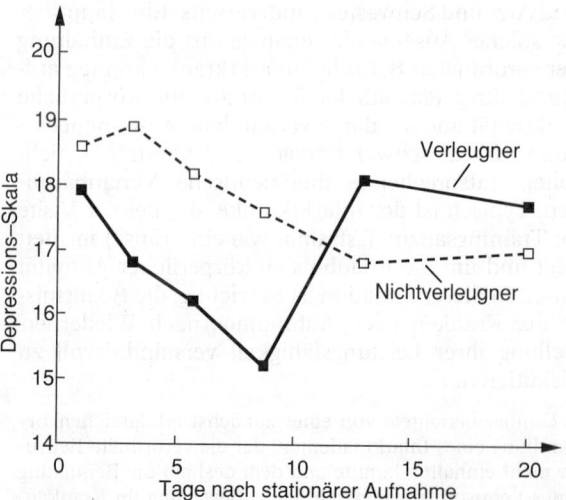

Abb. 41.2–5. Verlauf der depressiven Reaktion bei Infarktkranken, bei „Verleugnern" und „Nicht-Verleugnern" während der stationären Behandlungsphase (nach Froese et al., 1974).

Art der körperlichen Beeinträchtigung

Krankheitsbedingte Minderung des Selbstwerterlebens erfährt bei Infarktkranken häufig eine intensive Ausprägung und spezielle Akzentuierung. Aufgrund ihrer Persönlichkeitsstruktur trifft es sie besonders, daß „auf den eigenen Körper kein Verlaß mehr ist". Die erwähnten Phantasien über die Natur der Herzerkrankung und die Herauslösung des Herzens aus dem Körperbild setzen spezifische Akzente, die sich psychopathologisch auswirken können:

So berichtet Stein (1969) von einem Infarktpatienten, der unter der Vorstellung litt, daß das Infarktereignis ein zum Gehirn führendes Gefäß verstopft habe, was ihn völlig hoffnungslos machte. Der Patient klagte besonders über einen hochgradigen Gedächtnisschwund, ohne daß sich testpsychologisch Hinweise für ein organisches Psychosyndrom fanden. Die Störung besserte sich nach der Korrektur dieser Fehlvorstellung des Patienten über das Wesen seiner Herzerkrankung rasch.

Persönlichkeitsstruktur und Psychodynamik

Die tiefgehende Verunsicherung im Bereich der Selbstwertregulation, die bei infarktgefährdeten Personen beschrieben wird, findet durch den Eintritt der Erkrankung sozusagen eine reale Bestätigung; hinzu kommt die Befürchtung, die bisherigen Kompensationsmöglichkeiten, etwa in der beruflichen Tätigkeit, zu verlieren. Insofern stellt die Infarkterkrankung nicht selten eine sehr tiefgreifende „narzißtische Krise" (Henseler, 1974) dar, die nach außen in Form der depressiven Reaktion bzw. den zugehörigen Abwehr- und Kompensationsversuchen sichtbar wird.

Objektverluste zur Zeit der Krankheitsmanifestation

Die der Erkrankung häufig vorausgehenden belastenden Lebensereignisse beinhalten häufig Objektverlusterlebnisse: Verlust von Personen oder Sachen, endgültiges Nichterreichen von Lebenszielen, Frustration von Versorgungswünschen etc. Diese Verluste führen teils im Sinne von Trauerreaktionen unmittelbar zu depressiven Verstimmungen, teils aktivieren sie aggressive Impulse (Frustrationsaggressionen), die, gegen die eigene Person gewandt, zur Entstehung von Depressionen beitragen.

Aggressive Impulse und ihre Abwehr

Narzißtische Krise, Objektverluste sowie die Abhängigkeit von Ärzten und Schwestern und die krankheitsbedingte Hemmung der Motorik aktivieren aggressive Impulse. Aufgrund ihrer Angewiesenheit auf die Zuwendung von Ärzten und Schwestern und auch ihrer Angehörigen unterdrücken die Kranken diese aggressiven Regungen; dies führt nicht selten zu einer Situation der „feindseligen Abhängigkeit" und – zumindest zum Teil – zu einer Wendung der Aggressionen gegen die eigene Person; beides fördert das Entstehen depressiver Gefühle und ein entsprechendes Rückzugsverhalten. Spürbar wird die Abwehr der

Aggressionen gelegentlich indirekt über hartnäckige, zum Teil hypochondrische Klagen, vor allem aber in einem unterschwellig anklagenden Verhalten der Patienten (Freyberger, 1976a, 1977).

Klinische Beurteilung der Depression

Der Depression kommt bei Infarktkranken ein anderer Stellenwert zu als der Angst. Die depressive Reaktion ist ein notwendiger Bestandteil der Auseinandersetzung mit der durch die Erkrankung veränderten Lebenssituation. Die depressive Reaktion ist einfühlbar und verständlich. Die Patienten stehen vor einer längerfristigen Auseinandersetzung mit den Krankheitsfolgen: Sie befürchten Autonomieverlust, Verschlechterung der beruflichen Stellung und des Einkommens, reagieren auf die Infragestellung bisheriger Gewohnheiten und fühlen sich von vorzeitiger Alterung und Invalidität bedroht. Diese Auseinandersetzung ist vergleichbar der Arbeit, die während des Trauerprozesses nach dem Verlust eines Partners geleistet werden muß. Eine Verarbeitung der Erkrankung und ihrer Folgen ist Voraussetzung für eine erfolgreiche Rehabilitation. Sie ist ohne depressive Reaktion, ohne „Trauerprozeß", nicht möglich.

Wir weisen auf diesen Aspekt der depressiven Reaktion deshalb mit Nachdruck hin, weil häufig, insbesondere im Zusammenhang mit der vermehrten Einführung „antidepressiv" wirkender Psychopharmaka, versucht wird, das „Zielsymptom" Depression möglichst rasch und ohne Berücksichtigung seines Stellenwertes zu beseitigen.

Antidepressive Medikamente erweisen sich in dieser Situation im allgemeinen als nutzlos, ganz abgesehen von den eventuellen Nebenwirkungen bei kreislaufinstabilen Patienten.

Eine Indikation für psychotherapeutische Interventionen stellt sich dann, wenn die depressiven Reaktionen die rationale Behandlung der Erkrankung und ihre weitere Verarbeitung zu sehr behindern. So kann die Diskrepanz zwischen den idealen Forderungen an das eigene Selbst und der aktuellen Situation bei manchen Kranken zu einem Gefühl völliger Wertlosigkeit führen, sie können sich durch Gefühle des Versagens oder der Schuld erdrückt fühlen, es können Gefühle weitestgehender Hilf- und Hoffnungslosigkeit mit ihren bedrohlichen Folgen aufkommen. Solche Reaktionen behindern die Rekonvaleszenz und bedrohen einen erfolgreichen Rehabilitationsverlauf. Hier stellt sich die Indikation für supportive Psychotherapie, eventuell die Indikation zu einer sog. „Notfall-Psychotherapie" (Bellak und Small, 1972; Freyberger, 1976).

Wesentliche Voraussetzung für eine angemessene Verarbeitung der depressiven Reaktionen ist eine Einstellung bei Ärzten und Pflegepersonal, die depressive Gefühle und ihren Ausdruck auch im Krankenhaus akzeptiert. Schwierigkeiten treten hier insbesondere mit männlichen Patienten jüngeren und mittleren Alters auf: Die Äußerung von Traurigkeit ist oft weder mit ihrem Selbstbild noch mit den Erwartungen von Ärzten und Pflegepersonal vereinbar.

41.2.4 Besonderheiten emotionaler Reaktionen bei reanimierten Infarktkranken

Bei reanimierten Infarktpatienten wurden intensive und langanhaltende Angstzustände und eine Einschränkung der Ich-Funktionen durch einseitige Inanspruchnahme von Abwehrmechanismen beobachtet. Da es vor oder während der Reanimationsmaßnahmen häufig zu einer zerebralen Hypoxämie kommt, treten diese reaktiven Angstzustände in der Regel kombiniert mit Durchgangssyndromen auf. Zahlreiche psychische Störungen – vor allem Unruhe, Angst, Interessen- und Aktivitätsverlust und Alpträume – sind bei reanimierten Infarktpatienten oft noch nach Monaten nachweisbar und behindern die Rehabilitation.

Ärzte und Schwestern sollten bei reanimierten Patienten besonders sorgfältig auf Signale achten, durch die der häufig total verängstigte Patient seinen Wunsch nach einem Gespräch mitteilt, bzw. ihm immer wieder Gesprächsangebote machen. Die Kranken haben entweder in unterschiedlichem Ausmaß Erinnerungen an die Vorgänge der Reanimation oder sind durch die zeitliche Lücke in ihrem Erleben beunruhigt. Wenn sich Patienten scheinbar nicht an diese Situation erinnern oder nicht darüber sprechen, handelt es sich im allgemeinen um das Ergebnis psychischer Abwehrvorgänge, die nicht selten durch ärztliche Wunschvorstellungen („Amnesie") mitinduziert sind.

Systematische Befragung reanimierter Patienten sowie anderer Kranker, die sich in „Todesnähe" befunden hatten, ergab eine charakteristische Sequenz bestimmter Erlebnisinhalte (Moody, 1977). Diese „Todesnähe-Erlebnisse" scheinen weitgehend unabhängig von kulturellen Einflüssen zu sein (Kübler-Ross, 1976). Für das Verhalten der Mitarbeiter der Intensivstation ist es dabei wesentlich, zu wissen, daß nach klinischer Beurteilung „bewußtlose" Patienten sehr häufig Einzelheiten, insbesondere die Gespräche während der Reanimation, voll wahrnehmen (Moody, 1977) und später nur dann über diese „Todesnähe-Erlebnisse" sprechen, wenn sie sicher sind, damit ernstgenommen zu werden; auch hier spielt die erhöhte Sensibilität der Kranken für Abwehrvorgänge bei Ärzten und beim Pflegepersonal eine zentrale Rolle. Die Untersucher stimmen jedoch darin überein, daß die Patienten ein großes Bedürfnis haben, über diese Erlebnisse zu sprechen (Moody, 1977; Kübler-Ross, 1976).

Gelegentlich führen diese einfühlbaren Ängste zum Auftreten stenokardischer Beschwerden, worauf bei deren Beurteilung geachtet werden sollte.

> Ein 65jähriger Patient klagte jedesmal über heftigste stenokardische Beschwerden, wenn Schwestern und Ärzte nur in sein Zimmer traten, am ausgeprägtesten waren sie zu Beginn der Chefarztvisite. Der psychosomatische Konsiliarius wurde zugezogen, um die Ursachen dieser eher als „funktionell" aufgefaßten Beschwerden – im Gegensatz zu den eindeutigen stenokardischen Beschwerden des Patienten bei körperli-

cher Belastung – abzuklären. Der Patient war reanimiert worden und nach Aussage der Stationsärzte bestand für die ersten vier Tage des Krankenhausaufenthaltes eine „Amnesie".
> Im Gespräch zeigte sich, daß der Patient durchaus, wenn auch diffuse, Erinnerungen an diese Ereignisse hatte und für ihn das Erscheinen von Menschen in weißen Kitteln unmittelbar mit den Reanimationsmaßnahmen, unter anderem der Defibrillation, assoziiert war. Nach einigen Gesprächen und der Gabe von Betarezeptorenblockern ließen sich die Angstreaktionen und die stenokardischen Beschwerden abbauen.

Abschließend sei darauf hingewiesen, daß die Berücksichtigung der psychischen und sozialen Situation reanimierter Patienten auch während der Rehabilitationsphase dringend indiziert erscheint.

> Ein 60jähriger höherer Regierungsbeamter wurde nach infarktbedingtem Herzstillstand reanimiert und konnte trotz zahlreicher Komplikationen (Rippenserienfrakturen, Hämatothorax, organisches Psychosyndrom) und einer sich anschließenden vorübergehenden extremen Abhängigkeitshaltung schließlich erfolgreich rehabilitiert werden. Bei Wiedervorstellung nach einem halben Jahr klagte er lediglich über eine Impotenz, die jetzt die Ehe mit seiner wesentlich jüngeren Frau gefährde. Der Patient hatte Angst vor dem Verkehr aus der Vorstellung heraus, es könnte dabei ein Reinfarkt auftreten. Ein einmaliges Gespräch von ca. 30 Minuten Dauer genügte, um dieses Symptom dauerhaft zu beseitigen (vgl. Kap. 69).

41.2.5 Interaktionsprobleme mit Infarktkranken

41.2.5.1 Kooperation versus Auflehnung

Entsprechend ihrer Tendenz zu sozialer Anpassung wirken Infarktpatienten vordergründig zumeist kooperativ. Aus der Latenz werden jedoch vielfach Tendenzen zur Auflehnung spürbar und im Krankheitsverhalten sichtbar: Die Verleugnung der Krankheitsfolgen und die Auseinandersetzung mit der erzwungenen Passivität führen zur Übertretung von Verhaltensregeln im Behandlungsverlauf, insbesondere zu einer nicht ungefährlichen körperlichen Belastung. Dabei geht es den Infarktpatienten im allgemeinen weniger um den Inhalt der Verordnungen als um ihre Form (Hahn, 1971), in der klinischen Beurteilung solcher übertriebener Selbstbehauptungstendenzen sollte immer die besondere Sensibilität von Infarktkranken für Autoritätskonflikte berücksichtigt werden (Rosen und Bibring, 1966).

Auch die nicht seltenen sexuellen Anzüglichkeiten männlicher Infarktkranker gegenüber Krankenschwestern können im Zusammenhang mit diesen Selbstbehauptungstendenzen, d.h. als Versuch einer Stabilisierung des Selbstwertgefühls, verstanden werden. Die besondere Schwierigkeit liegt darin, daß sich in diesem Verhalten oft eine eher anklammernde Kontaktsuche mit sexuellem Imponiergehabe verbindet, was wiederum von den Schwestern als unmännlich erlebt wird; diese Fremdwahrnehmung kann im Umgang mit dem Patienten wiederum im Sinne einer Schwächung seines Selbstwerterlebens zurückwirken und dann entweder zu depressivem Rückzug oder zu noch unangemesseneren Versuchen führen, imponierend zu wirken.

41.2.5.2 Schwierigkeiten, Hilfe annehmen zu können

Hilfe annehmen ist bei Infarktkranken eng assoziiert mit Abhängigkeit. Hilfsangebote mobilisieren deshalb gleichzeitig Abwehrvorgänge und Selbstbehauptungsimpulse. Infarktkranke müssen die Kontrolle über sich und ihre Umwelt halten, können die Führung „nicht aus der Hand geben". Hinzu kommt, daß die Helfer all die aggressiven Impulse zu spüren bekommen, die durch die Objektverlusterlebnisse, die narzißtische Kränkung und die realen krankheitsbedingten Frustrationen mobilisiert werden; durch diese aggressiven Impulse werden die Angebote der Helfer häufig in Frage gestellt oder entwertet. Schließlich ist auch hier die bei Infarktkranken oft tiefgehende Störung der Fähigkeit zur Introjektion, zur wirklichen inneren Aneignung, zur Assimilation von Fremdem (Köhle und Simons, 1970) wirksam. Dies alles trägt dazu bei, daß die Kranken von dem Angebot der Helfer nichts oder nur wenig annehmen können, daß sie vielmehr versuchen, ihre eigene Überlegenheit aufrechtzuerhalten oder wiederzugewinnen. Die Helfer werden dann mit Berichten über frühere Leistungen überschüttet und fühlen sich mit ihrem Angebot mehr oder weniger abgelehnt. Schließlich erscheinen den Helfern dann oft nur noch solche Eigenschaften der Infarktkranken akzeptabel, denen eher pathogene Bedeutung zukommt, wie zwanghafte Perfektion und Bereitschaft zu sozialer Anpassung. Die Gefahr, daß im Interaktionsverlauf schließlich pathogene Verhaltensmerkmale durch die Helfer verstärkt werden, illustriert das folgende Beispiel:

Im Rahmen des psychosomatischen Praktikums soll ein Medizinstudent im 8. Semester einen 51jährigen Infarktpatienten interviewen. Bevor der sich noch unsicher fühlende Student überhaupt mit einer Frage beginnen kann, übernimmt der Patient die Führung: „Na, schießen Sie los!". Auch den übrigen Interviewverlauf bestimmt fast ausschließlich der Patient. Er stellt zunächst sich und seine Krankheit nach Art einer klinischen Demonstration für Anfänger dar, betont dann seine Überlegenheit recht aggressiv getönt, wenn auch in sehr wohlgeformte Sätze gekleidet: Er kenne die eigentlichen Interessen der Ärzte schon, sie seien häufig nur Geldschneider und hätten wenig Zeit für ihre Kranken. Er stamme selbst aus einer Arztfamilie. Immer wieder bezeichnet er die Studenten als „Anfänger".

Der Infarkt war während des Urlaubs der Ehefrau des Patienten – sie war allein ins Ausland gefahren – aufgetreten. Obwohl der Patient die Verdachtsdiagnose sofort stellte – sein Vater war ein Jahr vorher unter ähnlichen Beschwerden an einem Infarkt verstorben –, wartete er zunächst darauf, ob nicht ein Hausbewoh-

---

ner mehr oder weniger zufällig auf ihn aufmerksam würde – schon hier wirkte er ganz irrational enttäuscht, daß dieser Zufall nicht eintrat. Erst nachdem er auch die Ehefrau im Ausland telefonisch nicht hatte erreichen können, verständigte er schließlich Stunden später Arzt und Sanitätsdienst. Obwohl **er** den Beginn ärztlicher Hilfe hinausgezögert hatte, ist er jetzt im Gespräch noch voller Beschuldigungen gegenüber seinen Helfern: Die Sanitäter hätten ihn die Treppe hinunterlaufen lassen, anstatt ihn zu tragen; auf der Aufnahmestation habe er eineinhalb Stunden auf der Bank warten müssen und schließlich habe ihm der Stationsarzt auch falsche Tabletten gegeben – ausführlich schildert er die Kritik des Oberarztes am Stationsarzt.

Erst bei genauerem Zusehen wird deutlich, daß die verleugnende Abwehr des Patienten wohl auch auf der Aufnahmestation dazu geführt hatte, daß er nicht rasch genug beachtet worden war. Auch im Interview kann er Gefühle von Schwäche und Hilfsbedürftigkeit nicht mitteilen; er wehrt diese Gefühle ab, kann so nicht auf seine Hilfsbedürftigkeit aufmerksam machen und erhält auch im Interviewverlauf zuwenig Hilfestellungen.

Die anschließende Diskussion war zunächst durch die Verärgerung der Studenten über den Patienten bestimmt; sie hatten sich von ihm massiv entwertet gefühlt. Sie erlebten ihn als „Spielverderber", waren durch seine aggressiven Äußerungen stark irritiert. Die weitere Bearbeitung führte dazu, daß sie über ihn als „komische Figur" nur noch lachten. Vom Gruppenleiter nach vielleicht doch auch gleichzeitig vorhandenen Sympathiereaktionen befragt, erkannten die Studenten vor allem das Bemühen des Patienten um eigene Leistung – insbesondere im Beruf – und seine perfekte Darstellung des Krankheitsverlaufes im Interview an; darüber hinaus habe ihnen – im Vergleich zu anderen Patienten – seine Fähigkeit imponiert, die eigenen Affekte zu kontrollieren. Die weitere Diskussion ließ die Tragik dieser Entwicklung in der Interaktion erkennen: Leistungsbezogenheit und überstarke Affektkontrolle sind genau diejenigen Charakterzüge, um derentwillen der Patient auch sonst sozial akzeptiert wird, die ihn aber gleichzeitig darin behindern, von anderen etwas zu bekommen. Nach dieser Klärung änderte sich die Stimmung in der Studentengruppe: Aus dem Bild des lächerlichen Patienten, der „komischen Figur", war das Bild eines „armen Tropfes" geworden; erst diese emotionale Arbeit der Gruppe ermöglichte angemessenes Mitgefühl mit dem Kranken.

41.2.6 Ärztliche Psychotherapie bei Infarktkranken

41.2.6.1 Allgemeines zum methodischen Vorgehen

Das ärztlich-psychotherapeutische Gespräch (vgl. Kap. 17) bzw. supportive Psychotherapie (Freyberger, 1976; Kimball, 1975) soll Patienten bei der Verarbeitung ihrer Erkrankung Hilfe anbieten. Wesentlich ist das Angebot einer „hilfreichen Beziehung" (Luborsky, 1988): über die Entwicklung einer auch emotional tragfähigen Arzt-Patient-Beziehung dem Kranken zumindest vorübergehend einen Ausgleich für die erlittenen Objektverluste und narzißtischen Kränkungen zu vermitteln. Konstante Zuwendung, auch aktiv geäußertes Interesse, Anteilnahme und die Bereitschaft, den Patienten auch direkt emotional zu unterstützen, sollte der Arzt in die therapeutische Beziehung einbringen können.

Mit diesen Gesprächen wird zweckmäßigerweise möglichst bald nach der stationären Aufnahme zusammen mit oder parallel zur somatischen Behandlung begonnen. Hierdurch ist eine rechtzeitige Abklärung auch der emotionalen Probleme und eine Indikationsstellung für das weitere psychotherapeutische Vorgehen möglich. Schon im ersten Gespräch, in dem der Arzt zunächst versucht, die individuelle psychische und soziale Situation des Patienten kennenzulernen, kann auch mit therapeutischen Interventionen begonnen werden.

Die in jeder Phase patientzentrierten Gespräche nehmen ihren Ausgang von den vom Kranken direkt geäußerten, aus seinem Verhalten erschlossenen oder vom Untersucher über die Analyse seiner Gegenübertragung empfundenen Bedürfnisse des Patienten. Dabei werden die dem Kranken aktuell zur Verfügung stehenden Verarbeitungsmöglichkeiten und die bereits aktivierten Abwehrmechanismen berücksichtigt. Insgesamt geht es darum, die Gespräche an den aktuellen Problemen zu orientieren, den Patienten vor allem gegenüber seinen Ängsten zu unterstützen und ihm Hilfe im Trauerprozeß anzubieten; es ist nicht das Ziel supportiver Therapie in der akuten Krankheitssituation, die Persönlichkeitsstruktur zu verändern; angesichts der akuten Gefährdung sind, da alle psychischen Valenzen in Anspruch genommen werden, Entwicklungsprozesse kaum möglich. Im Verlauf dieser Gespräche wird der Patient auch über das Wesen der Erkrankung informiert.

Ein solches Gesprächsangebot ist für viele Kranke ungewohnt, manchen Patienten fällt es recht schwer, über sich und insbesondere über ihre Gefühle zu sprechen. Ein solches Vorgehen setzt deshalb eine Schulung voraus. Dabei macht die Persönlichkeitsstruktur des Patienten und die besondere Behandlungssituation gelegentlich eine Modifikation der sonst eher abwartenden Haltung des Arztes und seiner Gesprächsführung mit offenen Fragen erforderlich. So müssen etwa die Ängste der Patienten nicht selten ganz direkt und mit großem Nachdruck angesprochen werden, damit die Kranken davon berichten können. So bestehen wir etwa darauf, daß die Patienten ihre Vorstellungen, die sich auf die Verletzung des Herzens und auf das Wesen des Infarktes beziehen, äußern; wir formulieren etwa: „Sie sind krank und in dieser Situation macht sich jeder Mensch Vorstellungen über das Wesen und die Konsequenzen seiner Erkrankung". Ebenso nachdrücklich versuchen wir die Vorstellungen der Patienten über die Verursachung ihrer Erkrankung zu erfahren. Wir formulieren etwa: „Ein Infarkt kommt nicht aus heiterem Himmel, was haben Sie in letzter Zeit alles mitmachen müssen?"

Auch hier geht es nicht darum, eine Schutzfunktion des Patienten zu zerstören, sondern darum, ihm eine realitätsgerechte Verarbeitung seiner Erkrankung zu erleichtern.

Die Gespräche können, eine entsprechende Weiterbildung vorausgesetzt, vom Intensivmediziner selbst oder aber von einem Mitarbeiter des psychosomatischen Konsiliardienstes geführt werden.

In der Literatur wird eine ärztlich-psychotherapeutische Betreuung durch Intensivmediziner selbst nie beschrieben; möglicherweise hat dies prinzipielle Gründe: Die Tätigkeit des Intensivmediziners könnte Anforderungen an den Arzt stellen, die mit einer psychotherapeutischen Einstellung nicht oder nur sehr schwer in Einklang zu bringen sind. Falls dies zutrifft, sollte man sich davor hüten, den Intensivmediziner zu überfordern und versuchen, das Problem durch interdisziplinäre Kooperation zu lösen.

In jedem Falle erscheint es uns aufgrund der eigenen Erfahrung und den Empfehlungen zahlreicher Autoren wichtig, daß diese Gespräche in der Gesamtversorgung des Patienten nicht isoliert stehen, sondern daß die Kollegen der Intensivstation und die Schwestern-Pfleger-Gruppe auch in die psychologische Betreuung der Patienten einbezogen werden. Wie andere Autoren beobachteten auch wir, daß die Patienten oft gegenüber Schwestern und Pflegern leichter über ihre Ängste sprechen und Klagen äußern können als gegenüber Ärzten. Zumindest für das Erleben der Patienten kommt auch der gesamten therapeutischen Atmosphäre der Krankenstation wesentliche Bedeutung zu. Für die Gestaltung einer solchen therapeutischen Teamarbeit hat sich die Einführung regelmäßiger Stationskonferenzen bewährt, in denen unter Leitung eines psychosomatischen Konsiliarius über Schwierigkeiten, die im Umgang mit einem Kranken auftreten, sowie über Teamprobleme gesprochen werden kann.

Der psychosomatische Konsiliarius wird häufig erst in zugespitzten Krisensituationen zugezogen. Um die Situation für den Kranken nicht noch weiter zu belasten, sollte der Konsiliarius dem Patienten nicht als „Psychiater" oder „Psychotherapeut" vorgestellt werden, sondern als ein Arzt, „der über besondere Erfahrungen in der Behandlung von Infarktkranken verfügt" (Hahn, 1971), oder „der sich besonders mit Infarktkranken beschäftigt hat" und ähnliches.

Hierdurch wird vermieden, beim Patienten unnötige Widerstände gegen ein solches Vorgehen hervorzurufen.

Andernfalls könnten die Patienten die Zuziehung eines „Psychiaters" als Versuch zur Reglementierung erleben, wodurch der Konflikt eskalieren und der Zugang weiter erschwert werden könnte.

Die Gespräche haben eine Dauer von ca. 15 bis 20 Minuten. Bei einem Teil der Patienten genügt bereits ein einmaliges Gespräch, das in verkürzter Form dann bei den täglichen Visiten fortgesetzt werden kann, bei anderen Kranken ist eine Fortführung solcher stützend-psychotherapeutischer Gespräche während der ersten Behandlungstage erforderlich.

In der Untersuchung von Cassem und Hackett (1971) waren bei den 145 konsiliarisch betreuten Infarktkranken durchschnittlich 2,4 Gespräche pro Patient nötig, bei 29% der Kranken waren 5 oder mehr Gespräche erforderlich.

Im Gegensatz zu früheren Befürchtungen hat sich in den verschiedenen Untersuchungen keinerlei Hinweis dafür ergeben, daß derartige Gespräche bei Infarktkranken zu psychischen oder somatischen Komplikationen führen. Die Patienten begrüßen regelmäßig die Möglichkeit zu derartigen Gesprächen.

Bei den 80 von uns (Karstens et al., 1970; Köhle et al., 1972) betreuten Infarktkranken trat weder eine weitere psychische Beunruhigung auf, noch kam es zu irgendwelchen körperlichen Komplikationen wie Tachykardien oder einem vermehrten Auftreten von Rhythmusstörungen. Während für die Patienten die Gespräche deutlich entlastenden Charakter hatten, war es für uns anfangs recht belastend, uns den oft massiven Ängsten der Kranken auszusetzen. Vermutlich tragen solche Reaktionen auch dazu bei, daß solche Gespräche häufig vermieden werden.

41.2.6.2 Spezifische Hilfestellungen im ärztlich-psychotherapeutischen Gespräch

Das Angebot einer stabilen Arzt-Patient-Beziehung bildet die Grundlage jeder supportiven Psychotherapie; diese Beziehung soll dem Kranken jene Sicherheit vermitteln, die eine Entlastung seiner Ich-Funktionen von Abwehraufgaben erlaubt und ihm dadurch eine realitätsgerechte Anpassung an die gegebene Situation und die Kooperation bei der Behandlung ermöglicht.

Von dem Verlauf der emotionalen Reaktionen – Angst, Verleugnung und Depression – beim Infarktkranken lassen sich für ein spezifisches Vorgehen zusätzliche Gesichtspunkte ableiten.

Entängstigung

Angst wird durch das **Angebot einer tragfähigen Beziehung,** die auch die direkte Äußerung starker Affekte erlaubt, vermindert, sowie durch die eingehende Information über das Krankheitsbild und die Behandlungssituation.

Infarktpatienten benötigen Ärzte und Schwestern, die sich Zeit für Gespräche nehmen und die auch in der Lage sind, emotionalen Reaktionen Raum zu geben.

Eine kritische Situation entsteht nicht selten nach Anschluß des Patienten an den Monitor. Aus der Sicht des Personals ist er nun versorgt und kann alleingelassen werden. Die Patienten fühlen sich jedoch oft verlassen, ja vereinsamt (Hackett, 1977). Es ist wichtig, daß ein Teammitglied oder, noch besser, ein Familienmitglied beim Patienten bleibt oder wenigstens immer wieder nach ihm sieht. Die Erläuterung der Monitorfunktion („mechanischer Schutzengel"; Hackett, 1977) sowie der Möglichkeit eines „falschen Alarms" (locker sitzende Elektrode) sollte in jedem Fall ausführlich erfolgen.

Information vermindert Unsicherheit und damit Angst. Infarktkranke werden deshalb eingehend über die Diagnose und das Wesen ihrer Erkrankung sowie den Sinn aller therapeutischen Maßnahmen informiert. Wichtig ist auch die Mitteilung der Abnahme des Mortalitätsrisikos bei unkompliziertem Verlauf 24 Stunden nach Infarkteintritt. Bereits im ersten oder zweiten Gespräch werden den Kranken Zielvorstellungen und ein klarer Rahmenplan für die Behandlung und auch die spätere Rehabilitation mitgeteilt.

Bei der Erläuterung des Krankheitsprozesses empfiehlt es sich häufig, das Vorstellungsvermögen der Patienten durch kleine Skizzen zu unterstützen. Cassem und Hackett (1971)

fanden dies bei drei Viertel der von ihnen betreuten Kranken erforderlich.

Uns hat es sich bewährt, Herzinfarktkranken den Heilungsprozeß vom Infarkt zur Narbe am Beispiel einer Hautverletzung zu erläutern; etwa: nach einer Verbrennung werde funktionsfähiges Gewebe durch Bindegewebe ersetzt; die Narbe gewährleiste später wieder einen mit dem früheren Zustand vergleichbar festen Zusammenhalt des Gewebes. Dabei unterstützt auch die Verlagerung der Verletzung in der Vorstellung nach außen auf die Haut die Beruhigung des Patienten. Allerdings ist es wichtig, daß der Arzt hier nicht nur eine logische Gedankenfolge vermittelt, sondern gleichzeitig den Patienten nachhaltig emotional unterstützt. Unterbleibt die emotionale Unterstützung, so kann auch die Vorstellung einer Narbenbildung am Herzen wieder angstvolle Phantasien hervorrufen.

Auch bei reanimierten Patienten empfiehlt sich eine eingehende Information über Erkrankung, aufgetretene Komplikationen und durchgeführte Therapiemaßnahmen. Anderenfalls fühlen sich die Patienten, die die gefährliche Extremsituation doch mehr oder weniger wahrgenommen haben, mit ihren Ängsten alleingelassen. In einer Nachuntersuchung von Dobson und Mitarbeitern (1971) klagten Patienten und Ehepartner darüber, daß sie aufgrund des Informationsmangels oft monatelang – im Nachhinein gesehen, grundlos – unter ungewöhnlichen emotionalen Belastungen gelitten hatten.

Die direkte Äußerung von Gefühlsreaktionen – insbesondere Angst und Depression – wirkt sich oft sehr entlastend aus; nach einer solchen „kathartischen" Abreaktion bessern sich nicht selten auch stenokardische Beschwerden, die vorher gegenüber der pharmakologischen Therapie resistent geblieben waren.

Verminderung verleugnender Abwehr

Verleugnung kann die Anpassungsarbeit des Ichs blockieren und auf diese Weise den Patienten behindern, zu einem informierten, sich aktiv an seiner Behandlung beteiligenden Partner zu werden. Das Angebot von Information und die klare Orientierung über die Ziele der Therapie und Rehabilitation vermögen den Abwehrvorgang zu entschärfen; der Patient muß die Situation nicht mehr alleine bewältigen und findet einen Orientierungsrahmen. Zu berücksichtigen ist jedoch auch die Abwehr von Abhängigkeitsbedürfnissen durch Infarktkranke; deshalb weisen wir bei solchen Patienten ausdrücklich und anerkennend auch auf die besonderen Schwierigkeiten eines bisher aktiven Mannes hin, der jetzt die mit der Patientenrolle verbundene Passivität auszuhalten hat. Unter Umständen ist es günstig, das Ertragen dieser „Schwäche" als besondere „Leistung", als Ausdruck besonderer „Stärke" darzustellen (F. Dunbar). Der Patient kann auf die Notwendigkeit von „Entspannung", von „zur Ruhe kommen", von „Erholung" hingewiesen werden, dabei kann betont werden, daß er sich diese Entspannung nach all den Anstrengungen im Leben wohl auch verdient habe.

In jedem Falle ist das Ziel, die Mitarbeit des Patienten zu gewinnen, allen anderen therapeutischen Maßnahmen übergeordnet.

Bei der Verordnung von Sedativa und Tranquilizern wird die Wirkung dieser Medikamente den Patienten genau erklärt und ihre Verordnung mit der Notwendigkeit zu Entspannung und Erholung in Zusammenhang gebracht. Die Zustimmung des Patienten zur Medikation wird nach sorgfältiger Information eingeholt. So läßt sich weitgehend verhindern, daß die Sedierung als weitere unerwünschte Schwächung erlebt wird und zu einer erneuten Beunruhigung des Kranken führt (Cassem und Hackett, 1971).

Verleugnende Abwehr wird häufig auch durch die Vorstellungen der Patienten über zukünftige Krankheitsfolgen in Beruf und Familie aktiviert. Eine Klärung dieser Vorstellungen und erforderlichenfalls auch der realen Situation, etwa am Arbeitsplatz, kann hier weiterhelfen.

Bearbeitung der Depression

Das Gesprächsangebot beinhaltet Interesse und Zuwendung des Arztes und stützt schon dadurch das Selbstwertgefühl des Kranken. Im Gesprächsverlauf kann die krankheitsbedingte Minderung des Selbstwertgefühls sowie die Diskrepanz zwischen der jetzt gegebenen Realität und den Idealvorstellungen des Kranken thematisiert werden. Die depressive Reaktion kann auch durch eine Hervorhebung positiver Gesichtspunkte, etwa auch der bisherigen Leistungen des Kranken im Leben, gemindert werden. Die Äußerung depressiver Gefühle – derentwegen sich Infarktkranke immer wieder auch schämen – kann als „natürlich" und einfühlbar, d.h. für die Patienten auch als „normal", angesprochen werden. Gerade in Verbindung mit der verständlichen Trauerreaktion können auch die Möglichkeiten einer späteren Umstellung der Lebensweise ins Gespräch kommen. Auf die Chance einer vollständigen Rehabilitation sollte – bei bisher komplikationslosem Verlauf – ausführlich eingegangen werden. Dabei kann auf erfolgreich rehabilitierte Patienten, die dem Kranken bekannt sind, hingewiesen werden.

Insgesamt geht es darum, den Kranken einen angemessenen Ausdruck ihrer depressiven Gefühle zu ermöglichen und die krankheitsbedingten Einbußen im Laufe eines Trauerprozesses zu verarbeiten.

Es empfiehlt sich, vorschnelle Entschlüsse des Patienten zur Änderung seiner Lebensweise, insbesondere der Berufstätigkeit, als Ausdruck der Depression aufzufassen und den Patienten darauf aufmerksam zu machen, daß es zu solchen definitiven Entschlüssen in dieser Phase noch zu früh sei. Dies gilt insbesondere auch für den Umgang mit den Angehörigen, die mit ihren häufig übertriebenen Befürchtungen den Rehabilitationserfolg behindern können.

41.2.6.3 Analgetika und Psychopharmaka

Eine ausreichende Gabe von Analgetika und Sedativa – vor allem auch nachts – ist bei jedem Infarktkranken indiziert. Die Dosis sollte individuell ermittelt werden und dem Patienten ermöglichen, in einen relativ angenehmen Entspannungszustand zu kommen. Die Verordnung wird mit dem Patienten besprochen. Sie sollte jedoch immer nach einem festen Schema und nicht „nach Bedarf" erfolgen; andernfalls ergibt sich bei zahlreichen Patienten als Folge ihrer verleugnenden Abwehr eine zu niedrige Dosierung. Nach den Untersuchungen amerikanischer Autoren besteht der häufigste Fehler in einer Unterdosierung von Analgetika und Sedativa.

Empfohlen wird die Gabe von 5 bis 10 mg Diazepam (Valium®) drei- bis viermal täglich zu Beginn und eine Erhöhung oder eine Reduzierung (15 mg bzw. 2,5 mg) entsprechend der individuellen Wirkung. Alternativ kann Chlordiazepoxid (Librium®) in einer Dosierung von 10 bis 20 mg viermal täglich gegeben werden (Hackett, 1977).

Hingewiesen sei noch auf die Möglichkeit einer „paradoxen Wirkung" der Sedativa: Ein Teil der Patienten erlebt die Wirkung der Sedativa als zusätzliche Schwächung, vor allem wenn die Wirkungen nicht ausreichend mit ihnen besprochen wurden. Das Schwächegefühl führt zu gesteigerter Unruhe, diese in einem Circulus vitiosus zu einer Dosiserhöhung durch den Arzt, diese wieder zu einer Steigerung der Unruhe des Patienten. Gelegentlich werden Patienten dann schließlich so weitgehend „ruhiggestellt", daß sie künstlich beatmet werden müssen. Unterstützt werden kann ein solches Vorgehen durch das mitbeunruhigte Stationspersonal, so kann eine „Routine" aufkommen, unruhige Patienten auf Intensivstationen mit Sedativa „abzuschießen", wie es dann im Jargon heißt.

41.2.6.4 Die Verlegung von der Intensivstation

Die Verlegung von der Intensivstation auf die Allgemeinstation bedeutet für die Patienten den Verlust der ständigen Verfügbarkeit von Ärzten und Schwestern sowie der Monitorüberwachung. Dieser Verlust kann zu schmerzlichen Trennungsreaktionen und zu einer Reaktivierung von Ängsten führen (Hackett et al., 1969; Freyberger et al., 1969, 1976b). Ein „Verlegungsgespräch", in dem die eingetretene Besserung oder der Wegfall der unmittelbaren Bedrohung betont wird, kann den Patienten entsprechend beruhigen.

Hierbei ist zu beachten, daß initial stark verleugnende Patienten während der ersten Tage oft nur eine sehr geringe oder keine Besserung ihrer Symptome wahrnehmen (Gentry et al., 1972). Gelegentlich werden in ihrer Wahrnehmung bei Abnahme der Verleugnung die Beschwerden sogar stärker.

Durch das „Verlegungsgespräch" läßt sich die nicht ganz geringe Zahl von Rückverlegungen von der Allgemeinstation auf die Intensivstation vermindern, die oft dadurch zustande kommt, daß die entstandene Verunsicherung und Angst zu Schmerzen oder anderen Symptomen führen, die wiederum zur Verunsicherung des Stationspersonals beitragen. Klein und Mitarbeiter (1968) wiesen nach, daß praktisch nur solche Patienten mit erhöhter Katecholaminausschüttung auf die Verlegung von der Intensivstation reagierten, die auf diese Verlegung nicht vorbereitet worden waren.

41.2.6.5 Entlassung und Nachbetreuung

Es muß davon ausgegangen werden, daß bei einem größeren Teil der Patienten stärkere Probleme im emotionalen Bereich, insbesondere stärkere Angstzustände und depressive Reaktionen, noch Monate und zum Teil Jahre nach der Krankenhausentlassung fortbestehen. So berichten Cay und Mitarbeiter (1972), daß 42% der Infarktpatienten eines schottischen Krankenhauses mit sozialen Problemen nach der Entlassung rechneten; einem Drittel dieser Patienten erschienen diese Probleme bedrohlich. Dabei ist zu beachten, daß Patienten aus unteren sozialen Schichten während des Krankenhausaufenthaltes weniger Unterstützung und Ermutigung für die Zeit nach der Entlassung erfahren (Doehrman, 1977).

Für Patienten in der Rekonvaleszenz scheinen die stärksten emotionalen und sozialen Belastungen etwa vier Monate nach der Krankenhausentlassung aufzutreten, wenn sie zwar nicht mehr krank, aber doch noch nicht voll rehabilitiert sind und sich vor der Aufgabe sehen, die Arbeit wieder aufzunehmen (Doehrman, 1977).

Hieraus ergibt sich die Notwendigkeit, psychosoziale Gesichtspunkte im Rahmen von Anschlußheilbehandlungen und ambulanter ärztlicher Kontrolluntersuchungen entschieden mitzuberücksichtigen. Es sollte möglich werden, psychische Probleme, insbesondere langhingezogene Ängste in der Folge einer Herzinfarkterkrankung rechtzeitig zu erkennen und solchen Patienten frühzeitig psychotherapeutische Hilfsmöglichkeiten anzubieten.

41.2.7 Häufige Fehler im Umgang mit Infarktkranken in der akuten Behandlungsphase

41.2.7.1 Nichtbeachtung des „unauffälligen" Kranken

Der stille, zurückgezogene, „unauffällige", „pflegeleichte", für Ärzte und Schwestern oft angenehme Patient bleibt in der Regel unbeachtet. Gerade er leidet aber häufig unter großer Angst, scheut sich jedoch davor, diese zu äußern (Hackett et al., 1960, 1968). Oft ist dem Rückzug aus der Kommunikation bereits ein Konflikt zwischen Abhängigkeitserleben und aggressiven Impulsen vorausgegangen: Der Patient vermeidet die Äußerung feindseliger Affekte gegenüber Ärzten und Schwestern, um deren Zuwendung nicht zu verlieren. Solche aggressiven Impulse können als Reaktion auf die Erkrankung entstehen, aber auch als Reaktion auf die mit der Behandlungssituation verbundene Abhängigkeit.

41.2.7.2 Unnötige Frustration der Patienten

Den bereits aufs äußerste frustrierten Kranken sollten noch verbleibende Befriedigungsmöglichkeiten nicht ohne zwingende Notwendigkeit genommen werden

(Mandel, 1964; Pelser, 1967). So raten wir bei bestehendem Übergewicht dringend davon ab, bereits in den ersten Krankheitstagen eine Gewichtsreduktion einzuleiten. Da die Patienten als Folge von Erkrankung und Behandlungssituation häufig auf ein orales Organisationsniveau ihrer Bedürfnisse und Wünsche regrediert sind, werden solche Frustrationen um so stärker empfunden.

Die Kranken sollten vielmehr darauf aufmerksam gemacht werden, daß sie nach erfolgreicher Therapie und Rehabilitation weitgehend ohne Einschränkung ihrer Befriedigungsmöglichkeiten weiterleben können.

Selbstverständlich sollte dem Patienten später – falls erforderlich – ein entsprechendes Programm zur Gewichtsreduktion und zum Nikotinentzug vermittelt werden, am günstigsten zusammen mit einem „positiven" Programm für Bewegungstraining.

Ein Teil der restriktiven Empfehlungen von Ärzten für Infarktkranke hat seine Quelle nicht in kontrollierten wissenschaftlichen Untersuchungen, sondern in der Abwehr eigener Ängste.

Die Folge einer solchen eher restriktiven Einstellung ist wohl auch die Vernachlässigung der Bedeutung sexueller Beziehungen im Rahmen der Rehabilitation Infarktkranker. Neuere Untersuchungen zeigten jedenfalls, daß die Wiederaufnahme sexueller Beziehungen bei sonst gelungener körperlicher Rehabilitation unbedenklich empfohlen werden kann (Green, 1975; Halhuber, 1976). Bei den Patienten besteht in dieser Hinsicht ein hohes Bedürfnis nach Beratung, die nach Möglichkeit unter Miteinbeziehung der Ehefrau erfolgen sollte. Bei restriktiven Anweisungen ist auch zu bedenken, daß maximal verunsicherte Kranke häufig Anweisungen der Ärzte mißverstehen oder übertrieben stark zu befolgen suchen.

> Bei einem unserer Infarktkranken fiel auf, daß er tagelang völlig steif und bewegungslos im Bett lag. Auch seine Mimik wirkte wie versteinert. Als er im Rahmen des Rehabilitationsplanes mit kleinen gymnastischen Übungen beginnen sollte, zeigte sich der Patient dazu unfähig. Gleichzeitig fielen deutliche Zeichen von Angst auf. Auf sein Verhalten angesprochen, meinte der Patient, ein Arzt auf der Intensivstation habe zu ihm gesagt: „Wenn Sie sich bewegen, bekommen Sie noch eine aufs Dach". Danach war das merkwürdige motorische Verhalten des Patienten rasch abzubauen.

41.2.8 Beispiel eines Gesprächs mit einem Herzinfarktpatienten während der Intensivbehandlungsphase

Der 63jährige Herr B. hat einen Hinterwandinfarkt erlitten; bei der Aufnahme auf die Intensivstation wirkt er extrem mürrisch; er beklagt sich zunächst über den als beengend erlebten gekachelten Raum;

sein Verhalten wirkt gehemmt-aggressiv, unzufrieden, trotzig. Er scheint mit uns bzw. dem Schicksal zu hadern.

Bei der Visite nach Schmerzen gefragt, klagt er lediglich über Beschwerden im Bereich des Venenkatheters am rechten Arm. Im Anschluß an die Visite ergibt sich folgendes Gespräch:

A.: Wie es ihm denn nun ginge, ich hätte den Eindruck, er sei irgendwie unzufrieden.

P.: An sich nicht, doch er habe Beschwerden im rechten Arm, wo er so oft gestochen worden sei. Außerdem habe er im Schultergelenk alte arthrotische Veränderungen. Diese seien schon oft behandelt worden, hätten auch bereits Kurbehandlungen erforderlich gemacht.

A.: Ob ihn diese Schmerzen jetzt sehr belästigten?

P.: Nein, diesmal – er fährt mit der linken Hand über die rechte schmerzende Schulter zum Brustbein hin, dann etwas nach links – habe es hier geschmerzt, sich zusammengezogen, „gekrampft".

Der Patient hatte bisher überhaupt nicht über Schmerzen im Thoraxbereich geklagt, diese vollständig geleugnet bzw. in Arm- und Schultergelenke verschoben, wo er Schmerzen als ungefährlich kannte. Erst der Hinweis auf diese Schmerzen während des Gesprächs ermöglichte eine angemessene analgetische Therapie.

A.: Was er wohl für eine Krankheit zu haben meine?

P.: (zögernd) Er habe gedacht, es sei ein Herzinfarkt.

A.: Was er sich unter einem Infarkt vorstelle?

P.: (zögernd, überrascht, verlegen) „Ja, da kann man daran sterben". Er habe da jemand gekannt, der sei daran gestorben.

Unserer Erfahrung nach hat ein Großteil der Patienten ähnliche Vorstellungen über die Konsequenzen eines Herzinfarktes. Diese Ängste vor dem Tod werden jedoch meist nicht spontan geäußert; die Patienten liegen vielmehr ängstlich im Bett und warten auf den häufig als „Schnitter" oder ähnlich personifiziert vorgestellten Tod.

A.: Die Krankheit ist durchaus bedrohlich und ernstzunehmen; wenn man sie jedoch wie der Patient 36 Stunden ohne Komplikationen überstanden habe, sei trotz des Ernstes die Aussicht zu überleben sehr gut.

P.: Ja, er habe auch jemand gekannt, der sei zunächst nach dem Herzinfarkt nicht mehr recht arbeitsfähig geworden und dann später doch verstorben.

A.: Ich gehe zunächst auf die Arbeitssituation des Patienten ein und schildere dann Beispiele von rehabilitierten Herzinfarktpatienten, etwa den US-Präsidenten Johnson u.a.

Der soziale Wertverlust stellt auch hier das größte emotionale Problem dar. Die Vorstellung, nicht mehr arbeitsfähig und damit im sozialen Sinne nicht mehr wertvoll zu sein, belastet den Kranken außerordentlich stark.

A.: Wie er sich das nun vorstelle, so einen Herzinfarkt?

P.: (sehr zögernd, verlegen, sich abwendend) Das wisse er nicht so, da gäbe es etwas, wo die Gefäße zusammenkommen. Er spricht dann länger über seine Vorstellung von einem „gespaltenen Herzen", die er durch die Bewegung seiner beiden Hände zu illustrieren versucht. Der Hausarzt habe ihm einmal so etwas erklärt.

Die Vorstellung vom zerstörten Herzen ist mit Angst und auch mit einer Selbstwertminderung verbunden, die mit einem deutlichen Schamgefühl einhergeht.

A.: Ich erkläre ihm dann am Beispiel der Verbrennung der Haut den Untergang von Gewebe und den Prozeß der Narbenbildung. Die Narbenbildung mache zunächst eine körperliche Schonung erforderlich, andererseits würde eine feste Narbe später wieder eine volle Funktion des Herzmuskels gewährleisten.

P.: Er sei überrascht, daß das so zerstörte Herz hinterher wieder voll funktionieren könne. (Jetzt wirkt der Patient plötzlich wesentlich erleichtert.)

Wie häufig, äußert der Kranke jetzt auch die Angst, daß das Leben nach dem Infarkt nicht mehr lebenswert sein könnte.

P.: Er müsse eben jetzt ganz kurztreten, sich einschränken.

A.: Wie er das meine? (Die Stimmung läßt Vorstellung über weitere phantasierte Einschränkungen ahnen.) Zwar müsse er das Rauchen aufgeben, jedoch das Trinken keineswegs, vor allem das Spazierengehen sei jetzt sogar wichtig für ihn (der Patient hatte früher davon gesprochen), und auch alle anderen Genüsse seien ebenfalls erlaubt.

Es ergibt sich dann ein längeres Gespräch über die Zeitdauer des Klinikaufenthaltes, den späteren Rehabilitationsverlauf einschließlich der Behandlung in einer Rehabilitationsklinik. Schließlich:

P.: Ob er dann auch wieder Auto fahren dürfe? (Dabei kann bedacht werden, inwieweit Autofahren auch symbolisch für Autonomsein überhaupt steht und hierin auch die Frage nach der sexuellen Potenz enthalten ist.)

A.: Nach drei bis sechs Monaten, je nachdem. Ich betone wieder, wie wichtig in der Rehabilitationsphase die eigene Aktivität sei und daß mit Ausnahme des Rauchens sonst keine Einschränkungen der Lebensfreuden nötig seien.

Während des Gesprächs verlor sich zusehends der aggressiv-verspannte Gesichtsausdruck und die zurückgezogene, ablehnende Haltung des Patienten. Er lachte teilweise erleichtert, suchte später freundlich Kontakt, insgesamt wirkte er wesentlich entspannter.

Zusammengefaßt läßt sich sagen, daß Herr B. zwar um seine Diagnose wußte, die für ihn damit verbundene Todesbedrohung jedoch abzuwehren versuchte. Er verleugnete seine Beschwerden bzw. verschob sie in den ungefährlicheren Bereich des Armes; dies hatte eine zu geringe Dosierung von Analgetika zur Folge. Das aggressive Verhalten des Patienten stellte einen Versuch dar, seine Depression durch Wiederherstellung der eigenen Autonomie zu vermindern. Aggressive Affekte mußte er bei der gleichzeitigen Abhängigkeit von Pflegepersonal und Ärzten jedoch abwehren, was zum Rückzug aus der Kommunikation führte; dieser Rückzug stellte einen letzten, wenn auch verzweifelten Versuch dar, sein bedrohtes Selbstwertgefühl aufrechtzuerhalten.

Später konnte mit dem Patienten auch die Lebenssituation vor Manifestation der Erkrankung ausführlich besprochen werden.

41.2.9 Ergebnisse psychotherapeutischer Behandlungsansätze bei Herzinfarktkranken

41.2.9.1 Auswahl geeigneter Therapieverfahren

Aus den Ergebnissen der psychosomatischen Untersuchungen von Infarktkranken wurde immer wieder die Indikation zu psychotherapeutischer Behandlung abgeleitet. Therapieziele waren dabei:
- Modifikation der Persönlichkeitsstruktur (psychoanalytisch orientierte Einzeltherapie bzw. psychoanalytische Gruppenpsychotherapie, Verhaltenstherapie)
- Modifikation von Risikoverhaltensweisen (verschiedene gruppenpsychotherapeutische Ansätze)
- Unterstützung der Krankheitsverarbeitung (supportive Einzelpsychotherapie).

Versuche einer konfliktbearbeitenden psychoanalytisch orientierten Einzel- bzw. Gruppenpsychotherapie scheiterten – von wenigen Ausnahmen abgesehen – an den sich aus der Charakterstruktur von Infarktpatienten ergebenden Widerständen sowie an der Vorprägung der Patienten durch das somatische Krankheitserleben (Hahn, 1971; Ohlmeier et al., 1973). Eine längerfristige einzelpsychotherapeutische Behandlung ist nur dann sinnvoll, wenn die Patienten aus anderen Gründen als der Infarkterkrankung zu einer solchen Behandlung motiviert sind (Matussek, 1958). Gruppenpsychotherapeutische Ansätze haben sich dagegen dann als durchführbar erwiesen, wenn Probleme der körperlichen Erkrankung – Fragen der Krankheitsverarbeitung, des Risikoverhaltens – in den Gruppengesprächen Berücksichtigung finden. Eine Kombination solcher Formen von Gruppenpsychotherapie mit autogenem Training und auch systematischem Bewegungstraining scheint sich zu bewähren. In der Gruppe kann das Kontakt- und Informationsbedürfnis der Patienten eher befriedigt werden, die Widerstandsschwelle gegenüber einem introspektiven Vorgehen sinkt ab (Hahn, 1971). Auf Formen der Gruppentherapie, die auf die Bedürfnisse körperlich Kranker abgestimmt sind, hatte bereits Groen hingewiesen; für Infarktkranke hat vor allem Hahn (1971) einen solchen Ansatz erprobt, die Wirksamkeit dieses Verfahrens wurde neuerdings von Ibrahim et al. (1974) und Rahe et al. (1975) nachgewiesen.

Verhaltenstherapeutische Ansätze haben vor allem die Modifikation des Typ-A-Verhaltens zum Ziel. Sie werden im Kapitel 19 dargestellt.

41.2.9.2 Psychotherapie während der Krankenhausbehandlung

Gruen (1975) untersuchte die Auswirkungen einer kurzfristigen Psychotherapie während der Krankenhausbehandlung auf den Erholungsprozeß.

70 Infarktpatienten wurden in eine Behandlungs- und eine Kontrollgruppe randomisiert. Mit den Patienten der Behandlungsgruppe wurde während der Intensivbehandlungsphase sechsmal, anschließend fünfmal wöchentlich ein ärztlich-psychotherapeutisches Gespräch geführt. Die Auswirkungen der Behandlung sollten anhand objektiver Daten, der Aufzeichnungen von Schwestern und Ärzten, mit Hilfe psychologischer Tests sowie eines Interviews, das vier Monate nach Eintritt des Infarktes in der Wohnung des Patienten durchgeführt wurde, objektiviert werden.

In der Behandlungsgruppe war die durchschnittliche Aufenthaltsdauer sowohl auf der Intensivstation als auch insgesamt im Krankenhaus signifikant verkürzt; in der Behandlungsgruppe fanden sich signifikant weniger Patienten mit manifester Herzinsuffizienz; während der Aufenthaltstage 7 bis 11 wurden weniger supraventrikuläre Arrhythmien beobachtet. In der Kontrollgruppe beobachteten die Schwestern häufiger Zeichen allgemeiner körperlicher Schwäche, die Ärzte häufiger Depressionen. Die testpsychologische Untersuchung am 11. Tag ergab für die Gruppe der behandelten Patienten eine positivere Stimmungslage, größere soziale Aufgeschlossenheit und Kontaktfreudigkeit, geringere Deprimiertheit sowie niedrigere und homogenere Angstwerte. Die Auswertung der vier Monate nach Eintritt des Infarktes durchgeführten Interviews mit dem Patienten und seinem Hausarzt ergab für die Gruppe der behandelten Patienten geringere Angstwerte; die Patienten dieser Gruppe waren in weitaus stärkerem Ausmaß wieder zu ihren normalen Lebensaktivitäten zurückgekehrt.

Zusammengefaßt hatte die psychotherapeutische Behandlung während des Krankenhausaufenthaltes eine deutliche Wirkung im Sinne einer Verminderung des Auftretens von Herzinsuffizienz und Arrhythmien, einer kürzeren Behandlungszeit am Monitor, einer früheren Krankenhausentlassung; sie bewirkte die Entwicklung einer optimistischeren Gefühlseinstellung und verminderte exzessive Verleugnung und Angstreaktionen. Die psychotherapeutische Behandlung unterstützte die Entwicklung eines Gefühls von Hoffnung, das den Patienten erlaubte, ihre Angst zu bewältigen und ihre psychischen Energien wieder nach außen, auf Mitmenschen und die eigene Zukunft, zu richten.

Gleichzeitig war es den Kranken möglich, ihr Selbstkonzept zu erweitern und ihre Risikoverhaltensweisen durchzusprechen.

41.2.9.3 Psychotherapie nach Klinikentlassung und während der Rehabilitation

1. Hahn (1971) hat anhand der Behandlung von zwei Gruppen mit je 6–7 Teilnehmern über einenhalb Jahre die gruppentherapeutischen Arbeitsmöglichkeiten mit Infarktkranken erprobt und ausführlich dargestellt.

Bewährt hat sich die Einbeziehung einer ausführlichen Information der Patienten über die Infarkterkrankung und die Bearbeitung von Problemen des Gesundheitsverhaltens; die gruppenpsychotherapeutische Arbeit kann dann durchaus etwa auf persönlichkeitsbedingte Schwierigkeiten des Gesundheitsverhaltens fokussiert werden. Bewährt hat sich die Verbindung von Gruppentherapie mit autogenem Training.

Hahns Untersuchung läßt aufgrund des Therapieverlaufes der einzelnen Gruppenteilnehmer die Aussage zu, daß Infarktkranke ein Bewußtsein für die spezifischen, sie belastenden Konflikte entwickeln können;

darüber hinaus ist bemerkenswert, daß ein Teil der Patienten nach Abschluß der Behandlung in Krisensituationen den Psychotherapeuten aufsuchte und es mit Hilfe psychotherapeutischer Interventionen gelang, diese kritischen Situationen, in denen das Reinfarktrisiko stark erhöht erschien, zu bewältigen. Ergebnisse im Sinne einer Beeinflussung des Krankheitsverlaufes in der behandelten Gruppe liegen nicht vor.

Adsett und Bruhn (1968) behandelten 6 Infarktpatienten und deren Ehefrauen in zwei getrennten Gruppen mit je 10 Sitzungen. Die Autoren berichten über eine erfolgreichere Wiederanpassung der behandelten Patienten im Vergleich zu einer Kontrollgruppe.

Bilodeau und Hackett (1971) beschreiben die gruppentherapeutische Behandlung von 5 Infarktkranken unter Leitung einer psychiatrischen Konsiliarschwester. Die Patienten erlebten die Behandlung als sehr hilfreich im Wiederanpassungs- und Rehabilitationsprozeß; auf Wunsch der Kranken wurde die Behandlungsdauer von ursprünglich 12 auf 24 Sitzungen verlängert.

Ohlmeier und Karstens (1973) führten psychoanalytische Gruppeninterviews mit Infarktkranken durch, in denen vor allem Risikoverhaltensweisen und dem Infarkt vorausgegangene berufliche Belastungssituationen zur Darstellung kamen.

2. Ibrahim untersuchte systematisch die Wirksamkeit eines gruppenpsychotherapeutischen Ansatzes im Rahmen eines Nachbehandlungsprogrammes für Infarktkranke.

Insgesamt wurden 118 Infarktkranke in je 5 Therapie- und Kontrollgruppen mit je 12 Patienten eingeteilt; die Gruppenpsychotherapie wurde über ein Jahr mit einer wöchentlichen Sitzung von 90 Minuten Dauer durchgeführt.

Ziel der Gruppensitzungen war es, eine Atmosphäre zu bieten, in der die Infarktkranken ihre Lebensprobleme mitteilen und über ihre Gefühle und Einstellungen gegenüber ihren körperlichen und sozialen Lebensbedingungen sprechen konnten. Dabei wurden von den Gruppenteilnehmern auch Fragen der Medikation, der Diät und der körperlichen Aktivität diskutiert; der Therapeut nahm zu diesen Themen jedoch nicht Stellung, die Patienten wurden vielmehr gebeten, diese Fragen mit ihren jeweiligen Hausärzten definitiv zu klären.

Dieses Behandlungsangebot wurde von den meisten Kranken (84%) akzeptiert; die durchschnittliche Teilnahmequote an den wöchentlichen Sitzungen betrug 69%, die Drop-out-Rate 15,5%. Im Vergleich zur Gruppenarbeit mit Neurotikern erwies es sich als schwierig, persönliche Gefühle und Haltungen frei zu explorieren. Bald erfolgte eine Fokussierung der Gespräche auf den körperlichen Zustand der Patienten und die Realitäten des täglichen Lebens. Viele Stunden wurden damit verbracht, über Freizeitprobleme und Reaktionen auf Lebensbelastungen zu sprechen und konstruktive Ansätze in diesen Bereichen zu entwickeln. Dabei fiel die große Bereitschaft der Infarktkranken auf, sich gegenseitig zu unterstützen. Insgesamt wird die Einstellung derjenigen Patienten, die das ganze Jahr über am Programm mitarbeiteten, als extrem positiv beschrieben.

Zwischen behandelten Gruppen und Kontrollgruppen fanden sich zum Zeitpunkt der Nachuntersuchung signifikante Unterschiede sowohl hinsichtlich des sozialen Anpassungsprozesses als auch hinsichtlich des Gesundheitszustandes.

Bei den Patienten der Kontrollgruppen nahmen die Anzeichen für „soziale Entfremdung" während der eineinhalb Jahre dauernden Beobachtungszeit (Dauer der Therapie und ein halbes Jahr Nachbeobachtung) im Vergleich zu den behandelten Patienten signifikant zu. Die spezifische Wirkung der Gruppentherapie wird dadurch verdeutlicht, daß sich hinsichtlich dieses Merkmals die Ehefrauen aller Patienten ebenso verhielten wie die Kontrollpatienten.

Die Überlebenszeit nach einem Jahr lag in den Behandlungsgruppen 10% höher als in den Kontrollgruppen, bei den als schwerer eingestuften Infarkten betrug die Überlebensrate in den Behandlungsgruppen 93%, in den Kontrollgruppen 74%.

Der Prozentsatz der hospitalisierten Patienten war während der Beobachtungszeit in beiden Gruppen gleich groß, die durchschnittliche Dauer des Krankenhausaufenthaltes war jedoch bei den Mitgliedern der Behandlungsgruppen wesentlich kürzer: 26 gegenüber 36 Tagen bei den Mitgliedern der Kontrollgruppen. Hinsichtlich folgender Parameter fanden sich keine Unterschiede während der Beobachtungszeit: Blutdruck, Pulsfrequenz, Cholesterin, Triglyzeride, Glukose, Harnsäure im Serum; Körpergewicht.

Zusammengefaßt sieht Ibrahim in der geschilderten Form der Gruppenpsychotherapie eine wesentliche Ergänzung eines systematischen Behandlungsprogramms für Infarktkranke. Das Ich der Kranken kann unterstützt, Angst vermindert, soziale Anpassung und Gesundheitsverhalten können gefördert werden. Zur Förderung der Einbeziehung des emotionalen Bereiches und introspektiver Arbeit schlägt Ibrahim eine Intensivierung der Gruppentherapie mit zwei wöchentlichen Sitzungen oder durch einen zwei bis drei Wochen dauernden Behandlungsblock zu Beginn, sowie eventuell auch die Einbeziehung der Ehefrauen in die Behandlung vor.

Im einzelnen wäre dann zu klären, inwieweit sich tatsächlich dauerhafte Einstellungs- und Verhaltensänderungen, insbesondere eine Änderung des Wettbewerbsverhaltens, erreichen lassen, sowie welche spezifischen und/oder unspezifischen Wirkfaktoren der Gruppentherapie isoliert werden können.

3. Rahe und Mitarbeiter (1971, 1973, 1975 a, b, c, 1979) richteten am U.S. Naval-Hospital, San Diego, Kalifornien, eine Nachsorgeambulanz für Infarktkranke ein, in der den Kranken neben der kardiologischen Untersuchung eine Beratung hinsichtlich der Risikoverhaltensweisen und die Möglichkeit zu einer gruppenpsychotherapeutischen Behandlung angeboten wurde. Ziel der Untersuchung war die weitere Klärung des Krankheitserlebens und Krankheitsverhaltens von Infarktkranken sowie der Versuch, Rehabilitation, Risikoverhaltensweisen und nach Möglichkeit auch den weiteren Verlauf der koronaren Herzkrankheit durch eingehende Information der Kranken und gruppenpsychotherapeutische Maßnahmen zu beeinflussen. Rahe und Mitarbeiter berichten über insgesamt 60 Infarktkranke, die nach 18 Monaten nachuntersucht wurden.

Bei sonst identischer ambulanter Behandlung nahmen die 38 Mitglieder der Therapiegruppe an 4 bis 6 Gruppentherapiesitzungen teil. Die Kontrollgruppe umfaßte 22 Patienten; die Zuordnung Therapie- bzw. Kontrollgruppe war randomisiert erfolgt, die Kontrollgruppe mußte jedoch früher abgeschlossen werden, da es nicht mehr möglich war, einen bestimmten Teil des Behandlungsprogrammes (Informationsschrift) den Kontrollpatienten vorzuenthalten. Alle Patienten waren jünger als 60 Jahre, hatten ihren ersten Infarkt erlitten und schienen nach Beurteilung der Ärzte in der Lage, nach entsprechender Rehabilitation ihre Arbeit wieder aufzunehmen. Die gruppentherapeutischen Sitzungen wurden von einem Psychiater mit einer zusätzlichen zweijährigen internistischen Ausbildung geleitet, daneben nahmen weitere Mitarbeiter des Stabes, u. a. ein Kardiologe, teil. In der Behandlungsgruppe war auch eine Informationsschrift für Infarktkranke verteilt worden.

Die Ergebnisse des Nachbehandlungsansatzes von Rahe werden hier wegen ihrer grundsätzlichen Bedeutung und auch wegen der relativ leichten Übertragbarkeit dieses Modells ausführlicher dargestellt.

Themen der Gruppensitzungen: Die in der Literatur dargestellten psychologischen Merkmale und Charakteristika des Verhaltens von Infarktkranken kamen in den Gruppen deutlich zur Darstellung, insbesondere das Arbeitsverhalten mit den lebenslang durchgeführten zahlreichen Überstunden, der selbst auferlegten und als belastend erlebten Verantwortung, dem starken Ehrgeiz und der Tendenz zum Rivalisieren sowie dem Kampf gegen Termindruck; wesentlich erschien der von den Infarktkranken gleichzeitig erlebte Mangel an persönlicher Befriedigung. Die Infarktkranken gewannen Einsichten in die Belastungen, die der Krankheitsmanifestation vorausgingen, insbesondere auch in die unrealistischen Zielsetzungen in der Arbeit und in anderen Lebensbereichen. Im Zusammenhang damit trat bei einigen Kranken eine depressive Reaktion auf: sie fühlten sich am Eintritt des Infarktes mitschuldig. Bei der Erinnerung an den Krankenhausaufenthalt stand die verleugnende Abwehr während der ersten beunruhigenden Tage im Vordergrund. Deutlich wurde, daß während der zweiten und dritten Behandlungswoche in der Klinik die Patienten bereit sind, Information über die Erkrankung und den weiteren Umgang mit ihr anzunehmen, diese Bereitschaft von Ärzten und Schwestern jedoch offensichtlich stark unterschätzt wird: Zu Beginn der Gruppenbehandlung war der Wissensstand der Kranken zu den genannten Themen minimal. Nach der Entlassung traten zunächst Anpassungsprobleme in der Familie auf; hier fiel auch die unzureichende Information der Ehefrauen, insbesondere über Diätprobleme, auf. Im Zusammenhang mit der beruflichen Rehabilitation wurde das Fehlen von Information über später wieder mögliche körperliche Aktivität sowie das Fehlen eines Fitnessprogramms, das einerseits durchführbar ist, andererseits zu einer systematischen Leistungssteigerung führt, deutlich. Ein entsprechendes Trainingsprogramm (tägliche Spaziergänge mit Pulskontrolle) wurde eingeführt. Dieses Trainingsprogramm wurde von allen Kranken akzeptiert, im Gegensatz etwa zum Diätprogramm. Auch die Rückkehr zum Arbeitsplatz wurde in den Gruppensitzungen durchgesprochen und geplant, dabei konnten dann auch die Verhaltensmuster im Zusammenhang mit der Berufstätigkeit diskutiert werden, was viele Gruppenteilnehmer als den größten Gewinn der Sitzungen empfanden. Andere Risikoverhaltensweisen wie Überernährung und Rauchen traten oft mit Arbeitsbeginn wieder verstärkt auf und konnten in der Gruppe besprochen werden.

Die Gruppensitzungen erlauben so eine gründliche und korrekte **Information** über das Wesen der Erkrankung; günstig erscheint, daß diese Information gerade zu einem Zeitpunkt erfolgt, in dem die Patienten sich wieder auf das Leben in ihrer natürlichen Umwelt einstellen. Die Ergebnisse der Nachuntersuchung zeigen, daß die Infarktkranken tatsächlich das für ihre Dauerbehandlung erforderliche Wissen erwerben können. Das mit Hilfe eines Fragebogens quanti-

tativ untersuchte Wissen der Infarktkranken erwies sich in der Behandlungsgruppe als angemessen, während sich der Wissensstand der Kontrollgruppe nicht von demjenigen einer Vergleichsgruppe von Patienten ohne Infarkt unterschied. Gleichzeitig wurde deutlich, daß die Diskussion der Pathophysiologie von Infarkt und Heilungsvorgängen in den Gruppensitzungen die Motivation zu körperlichem Training, aber auch zur rechtzeitigen Einnahme von Nitroglyzerintabletten förderte. Der Herzinfarkt allein erwies sich dagegen nicht als ausreichender Stimulus, um sich Information über die Natur des Infarktes und die sich hieraus ergebenden Behandlungserfordernisse zu beschaffen. Der Wissensstand der Patienten, die an der Gruppentherapie teilgenommen hatten, erwies sich bei der Untersuchung auch als besser als der von Teilnehmern an einem intensiven Lehrprogramm zur koronaren Herzkrankheit ohne Gruppentherapie.

Die Autoren betonen, daß die gegenseitige, zum Teil auch humorvolle Unterstützung der Patienten eine realistische Einschätzung der Gefährdung erleichterte sowie das Selbstwertgefühl der Kranken weiter stabilisierte.

Krankheitsverlauf: Nach 6 Monaten war in der Kontrollgruppe die Hospitalisierungsrate wegen Koronarinsuffizienz bzw. koronaren Bypassoperationen signifikant größer. Nach 12 Monaten fand sich daneben eine signifikant höhere Reinfarkthäufigkeit in der Gruppe der Kontrollpatienten. Dieser Unterschied blieb auch nach 18 Monaten bestehen. Wurden alle Komplikationen, die als Folge der koronaren Herzkrankheit auftraten, zusammengerechnet (Koronarinsuffizienz, koronare Bypassoperationen, Reinfarkt und Tod), so bestanden zwischen der Kontrollgruppe und der Behandlungsgruppe nach 6, nach 12 und nach 18 Monaten jeweils signifikante Unterschiede. Diese schweren Komplikationen betrafen nach 18 Monaten 19% der behandelten Patienten gegenüber 58% der Kontrollpatienten.

Entsprechende Unterschiede fanden sich auch nach einer Nachuntersuchungsperiode von 36 Monaten (Rahe et al., 1979).

4. Cay und Mitarbeiter (1975, 1976) konnten zeigen, daß bei Einführung eines systematischen **Rehabilitationsprogrammes** 88% der Infarktkranken innerhalb von vier Monaten ihre Arbeit wieder aufnehmen können und viele der bisher nicht rehabilitierten Patienten durch ein entsprechendes Interventionsprogramm rehabilitierbar werden. Ein solches Programm muß somatische, psychische und soziale Aspekte ausreichend berücksichtigen. In einer kontrollierten Studie konnten Cay und Mitarbeiter in Edinburgh die Effizienz eines solchen Interventionsprogrammes bei ausgewählten Patienten mit gravierenden psychosozialen Problemen nachweisen.

Ein „Rehabilitationsteam", zusammengesetzt aus Kardiologe, Psychiater, Psychologe und Sozialarbeiter, diagnostizierte bei 58,5% (161 Patienten) der untersuchten Infarktkranken gravierende psychosoziale Probleme, von denen sie annahmen, daß sie die Rehabilitation behinderten. Diese Patienten wurden in eine Therapiegruppe und eine Kontrollgruppe geteilt; bei den Patienten der Therapiegruppe wurde während des Rehabilitationsverlaufes gezielt entsprechend den vordiagnostizierten Problemen interveniert. Innerhalb dieser Gruppe zeichnete sich ein günstigeres Rehabilitationsergebnis ab: frühere Wiederaufnahme der Arbeit, bessere Arbeitsfähigkeit, größere emotionale Stabilität (Cay et al., 1975).

Cay und Mitarbeiter (1975, 1976) weisen auf die Bedeutung einer frühzeitigen Diagnose körperlicher, psychischer und sozialer Behinderungen für das Rehabilitationsergebnis sowie auf die erforderliche Qualität der Kommunikation im Behandlungsteam hin.

5. Auch das Typ-A-Verhalten läßt sich therapeutisch beeinflussen; möglicherweise vermindert dies das Risiko eines Reinfarktes (vgl. Kap. 19).

Eine Metaanalyse von 18 kontrollierten Studien zur Behandlung des Typ-A-Verhaltens (meist Trainingsprogramme zur Verhaltensänderung kombiniert mit Entspannungsverfahren) ergab, daß die Teilnehmer aller Studien ihr Typ-A-Verhalten deutlich vermindern konnten. Ergebnisse von zwei Studien sprechen dafür, daß dieser Änderung des Verhaltens nach Psychotherapie eine erhebliche Reduzierung neuer koronarpathologischer Ereignisse (ca. 50%!) folgen könnte (Nunes et al., 1987).

41.2.10 Zusammenfassung

Die Erkrankung an einem Herzinfarkt hat während der akuten Krankheitsphase und auch während der Zeit der Rehabilitation schwerwiegende Folgen im psychischen und sozialen Bereich. Langandauernde emotionale Belastungen, familiäre Probleme und berufliche Anpassungsschwierigkeiten treten bei einer größeren Zahl der Patienten auf. Die Integration psychotherapeutischer Methoden in die Gesamtbehandlung scheint die Rehabilitation insgesamt zu erleichtern und auch zu beschleunigen. Dabei ist heute im einzelnen noch nicht geklärt, welche psychotherapeutischen Methoden am wirksamsten sind und als

fester Bestandteil der Routinebehandlung von Infarktkranken und ihrer Familien eingesetzt werden sollen (Doehrman, 1977).

Cay (1982) zeigte, daß es möglich ist, am Ende der ersten Woche des Krankenhausaufenthaltes aufgrund des psychologischen Befundes ein schlechtes Rehabilitationsergebnis, eine spätere eingeschränkte Fähigkeit zur psychosozialen Anpassung vorherzusagen. Da es sinnvoll ist, Ansätze zur psychotherapeutischen Intervention auf Patienten mit deutlichen psychischen Schwierigkeiten während der Akutphase zu konzentrieren (Cay, 1982; Horlick et al., 1984), wäre es für die klinische Praxis wichtig, vom Beginn des Krankenhausaufenthaltes an seelische Probleme der Patienten mitzuerfassen. Cay (1982) wies nach, daß bereits ein vom Patienten selbst in 5–10 Minuten ausfüllbarer Fragebogen (General Health Questionnaire) ein sehr nützliches Screening-Instrument sein kann, um Patienten mit psychischen Problemen, die später die Rehabilitation behindern können, zu identifizieren.

Die Orientierung auf die Rehabilitation macht eine Umstrukturierung der medizinischen Versorgung im Sinne einer Beteiligung von Vertretern neuer Fächer aus den Gesundheitsberufen im Team erforderlich (Cay, 1975; Aitken, 1975), wie sie heute erst an wenigen Zentren realisiert ist.

Ziel der dargestellten psychotherapeutischen Methoden ist es ganz überwiegend, die Patienten bei der Verarbeitung ihrer Erkrankung und deren seelischer und sozialer Folgen zu unterstützen. Daneben wird eine Veränderung von Risikoverhaltensweisen angestrebt, seltener weitergehende Veränderungen der Persönlichkeitsstruktur.

41.3 Essentielle Hypertonie

Jörg Michael Herrmann, Michael Rassek, Nikolaus Schäfer,
Thomas H. Schmidt und *Thure von Uexküll*

41.3.1 Exemplarischer Fall

Ein 25jähriger Patient wird mit einem Blutdruck von 220/150 mm Hg in der Klinik aufgenommen. Während des ersten Gespräches berichtet er, daß seine Mutter vor 5 Jahren gestorben sei. Ein halbes Jahr danach wurde bei ihm eine Hypertonie festgestellt. Sein eineiiger Zwillingsbruder hat ebenfalls einen Hochdruck. Bei dem 63jährigen Vater ist seit etwa 20 Jahren ein Hochdruck bekannt. Die 9 Jahre ältere Schwester und der 6 Jahre ältere Bruder haben keinen Hochdruck. Der Patient stottert sehr auffallend, der Zwillingsbruder weniger.

Zu dem sehr strengen und autoritären Vater haben beide Brüder kein gutes Verhältnis. Der Vater war so „eigen", daß er keine Haushälterin gebrauchen konnte. Der Patient mußte daher neben Schule und Lehre dem Vater den Haushalt führen. Die Zwillingsbrüder waren beim Vater allein zurückgeblieben, nachdem die beiden anderen Geschwister den elterlichen Haushalt verlassen hatten.

Der Vater lebt ganz für die von ihm aufgebaute Firma und nimmt an den Problemen der Familie keinen Anteil. Der Patient vermißt, daß der Vater ihn fragt, wie es ihm geht, wie er in der Schule und im Beruf zurechtkommt und ob er Hilfe braucht. Er bereitet dem Vater das Frühstück, bevor dieser morgens zur Arbeit geht, er richtet das Abendessen und bleibt fast immer abends zu Hause, damit der Vater nicht alleine ist.

Dieser Zustand wird für den Patienten unerträglich, nachdem sein Zwillingsbruder zur Ausbildung nach N. geht und er mit dem Vater allein bleibt. Jetzt sitzt er abends mit ihm zusammen und wagt nicht wegzugehen. Oft steht er auf, geht im Flur auf und ab und überlegt, wie er es anstellen kann, einmal wegzukommen. Dann aber setzt er sich doch zum Vater. Beide reden fast nie zusammen. Sie sitzen nur so da und der Vater trinkt. Seit die Mutter nicht mehr lebt, trinkt der Vater und ist jeden Abend mehr oder weniger betrunken.

Nachdem der Patient aus der Klinik entlassen war, verläßt er dann schließlich doch das väterliche Haus, um in einer Firma in Norddeutschland zu arbeiten. Bei gelegentlichen ambulanten Blutdruckkontrollen, die dort vorgenommen werden, liegen die Werte um 150/ 100 mm Hg. Nach zwei Jahren kommt der Patient zurück, um im elterlichen Betrieb zu arbeiten. Jetzt steigt der Blutdruck wieder auf die bei der ersten Untersuchung gemessenen Werte an. Trotz medikamentöser Behandlung wurden bei ihm immer wieder Werte bis zu 230/120 mm Hg gemessen. Der Zwillingsbruder kommt ebenfalls in den elterlichen Betrieb zurück, danach steigen auch seine Blutdruckwerte wieder an.

Der Alkoholkonsum des Vaters ist inzwischen noch weiter angestiegen. Der Patient und sein Zwillingsbruder müssen ihn häufig betrunken aus dem Betrieb entfernen. Zur Leitung der Geschäfte ist er nicht mehr fähig.

Nachdem der Patient wieder etwa ein Jahr im elterlichen Betrieb arbeitet, lernt der Vater eine zwanzig Jahre jüngere Frau kennen und heiratet sie. Um diese Zeit sinkt zwar sein Alkoholkonsum, die wirtschaftliche Situation des Betriebes ist aber inzwischen so desolat geworden, daß ein Konkurs droht. Der Patient versucht zusammen mit seinem Zwillingsbruder dies noch zu verhindern, es gelingt aber nicht mehr. Der Vater verliert den Betrieb und die Söhne müssen sich nach einer anderen Arbeit umsehen.

Der Patient findet durch die Vermittlung seines älteren Bruders eine Stelle als Betriebsingenieur. Die Arbeiten, die er verrichten muß, könnten auch durch einen Handwerker verrichtet werden. Er ist mit dieser Tätigkeit unzufrieden, sieht aber keinen Ausweg. Etwa fünf Jahre lang geht er nicht mehr zur Blutdruckkontrolle zum Arzt. Während einer Bergwanderung tritt dann ein akutes Lungenödem auf. Danach kommt er wieder zu einer Kontrolluntersuchung und der Blutdruck ist 220/140 mm Hg unter einer Monotherapie. Der Blutdruck des Patienten wird medikamentös wieder gesenkt auf Werte um 160/100 mm Hg. Trotzdem tritt einige Wochen später unter dieser Therapie erneut ein akutes Lungenödem auf.

Sein Zwillingsbruder hat eine Stelle als Handelsvertreter gefunden. Die Tätigkeit gefällt ihm gut, seine Blutdruckwerte, welche nur sehr selten gemessen werden, liegen um 160/100 mm Hg.

Ausschnitt aus dem letzten Teil des Erstinterviews mit dem Patienten:

A.: Haben Sie eine Frage an mich?
P.: Wie meinen Sie?
A.: Daß ich Ihnen vielleicht bei einer Frage helfen könnte, Ihnen etwas beantworten könnte.
P.: Sie können mir im Augenblick in meiner Lage zu Hause praktisch auch nicht helfen, Sie verstehen mich.
A.: Das wäre eigentlich Ihr Ziel?
P.: Das ist die Haupt...
A.: Haben Sie schon Vorstellungen, wie Sie diese Hilfe, wenn sie möglich wäre, sehen würden?
P.: Ja, wissen Sie, ich bin ein – ich bin ein – ich bin viel zu sehr – oder das Problem liegt bei mir so: Ich kann meinen Vater jetzt nicht – oder sagen wir, bisher konnte ich ihn nicht ganz allein zu Hause lassen. Er wäre doch – nach einem halben Jahr wäre er vollkommen erledigt.
A.: Erledigt?
P.: Vollkommen erledigt wäre er, oder wie sagt man in der Medizin dazu?
A.: Hätten Sie auch...
P.: Und zwar deswegen: Ich bin ein viel zu gutmütiger Mensch, das haben schon viele Leut' zu mir ge-

sagt, mein älterer Bruder usw. Viel zu gutmütig. Aber ich kann's aus menschlichen – aus menschlichen Beziehungen so, ich kann meinen Vater nicht – nicht so verkümmern lassen, wissen Sie.

A.: Haben Sie auch den Gedanken schon gehabt, daß er sterben könnte, wenn Sie weggehen?

P.: Also praktisch so nicht direkt, aber ich hab' also, das heißt jetzt also ganz ehrlich, ich habe schon mal gedacht, wo es jetzt so schlimm um ihn stand, wenn mein Vater die Augen zumachen würde, das wär – das wär besser für mich.

A.: Eine Erleichterung?

P.: Und für – und vielleicht noch für mehr Menschen. Vielleicht ist das kein christlicher Gedanke, aber man hat manchmal so – so komische Gedanken. (Der Patient lacht.)

41.3.2 Symptomatologie

41.3.2.1 Definition

Nach den derzeitigen Definitionen der WHO werden systolische Blutdruckwerte zwischen 140 und 160 mm Hg und diastolische Werte zwischen 90 und 95 mm Hg als „Grenzbereich", darunterliegende als „Normalwerte" und höhere als „Hypertonie" bezeichnet (WHO, 1959).
Es werden drei Stadien unterschieden:

Stadium 1: Hoher Blutdruck ohne Anzeichen von Organveränderungen im kardiovaskulären System.

Stadium 2: Hoher Blutdruck mit Linksherzhypertrophie ohne Anhalt für andere Organschäden.

Stadium 3: Hoher Blutdruck mit hypertoniebedingten Organschäden (WHO, 1962).

Das Risiko, infolge einer kardiovaskulären Erkrankung zu sterben, steigt kontinuierlich sowohl mit der Höhe des systolischen als auch des diastolischen Blutdrucks an (Kannel, 1975). Dabei wird das durchschnittliche Risiko bereits bei einem diastolischen Blutdruck von 85 mm Hg überschritten (Pflanz, 1977) (Abb. 41.3–1).

Nach diesen Ergebnissen muß man als Normalwert den niedrigsten Blutdruck (unter 120/80 mm Hg) bezeichnen, der mit körperlicher und geistiger Leistungsfähigkeit vereinbar ist.

Diagnostisch unterscheiden wir die „essentielle Hypertonie", bei der bis heute keine organische Ursache bekannt ist, die für den erhöhten Blutdruck verantwortlich gemacht werden kann, von den symptomatischen Hypertonieformen, die Komplikationen organischer Krankheiten sind (der Niere, der Nebenniere, wie beim Phäochromozytom oder dem Morbus Cushing usw.).

41.3.2.2 Allgemein-klinische Symptomatologie

Patienten mit essentieller Hypertonie klagen, im Gegensatz zu vielen Lehrbuchdarstellungen, nicht häufi-

ger über Beschwerden als die Durchschnittsbevölkerung. Meist wird die Hypertonie durch Zufall, z. B. bei einer Routineuntersuchung, entdeckt. Nach einer Untersuchung von Tibblin und Mitarbeitern (1972) werden von Hypertonikern sogar eher weniger Beschwerden angegeben als von Normalpersonen. Symptome wie Schwindelgefühle, Ohrensausen, Flimmern vor den Augen, Sehstörungen, Nachlassen des Gedächtnisses usw. sind – soweit organisch bedingt – bereits Symptome von Gefäßkomplikationen, die als Folge der Hypertonie auftreten.

41.3.2.3 Psychische Merkmale

Hypertoniker erwecken den Eindruck, „normaler" zu sein als Normalpersonen. Bastiaans (1963) beschreibt in Auswertung psychoanalytischer Interviews folgende „Fassadenstruktur":

„Nach außen erscheinen sie beherrscht, aktiv, ambitiös, perfektionistisch, gewissenhaft, zuverlässig, pflichtbewußt, genau, ehrlich, charmant, loyal und freundlich. Hinter dieser Fassade verbirgt sich jedoch in ausgeprägtem Maße Unsicherheit, Sensibilität,

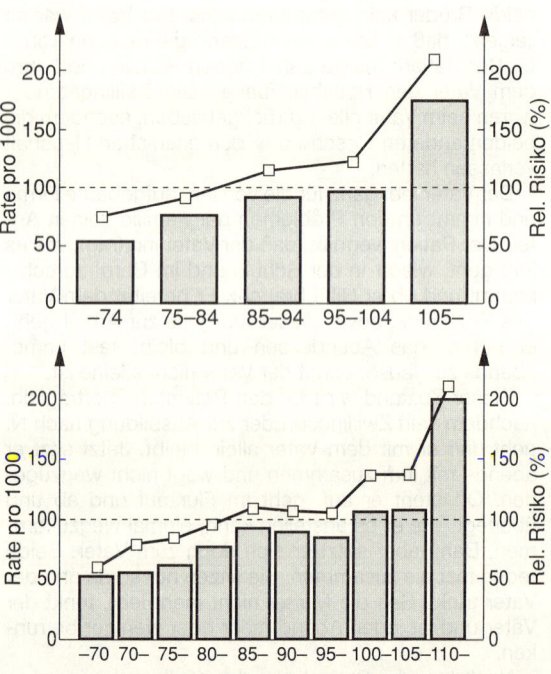

Abb. 41.3–1. Zwölfjahresraten an Herzinfarkt und kardiovaskulären Todesfällen pro 1000 nach diastolischem Blutdruck, berechnet nach dem Pooling-Projekt, welches die Daten aus Framingham, Albany, Chicago und Tecumseh zusammenfaßt. Die Erhöhung des Risikos für Herzinfarkt und kardialen Tod mit zunehmendem diastolischem Druck bei der Erstuntersuchung wird deutlich. Der systolische Druck steht in seiner prädiktiven Aussagekraft dem diastolischen kaum nach. Werden die diastolischen Werte von 10 zu 10 mm Hg zusammengefaßt, zeigt sich ein kontinuierlicher Anstieg des Risikos mit zunehmender Blutdruckhöhe; werden die Daten in Gruppen von 5 zu 5 mm Hg angeordnet, steigt das Risiko erst bei diastolischen Werten von 100 mm Hg stärker an (nach Pflanz, 1977).

Verletzlichkeit, Abhängigkeit und Unausgeglichenheit. So gefügig und friedliebend, wie sie nach dem äußeren Eindruck wirken, sind sie in Wahrheit nicht. Wollen sie auch bewußt Frieden stiften (wie sie das oft zwischen Vater und Mutter tun wollten), so sind sie auf einem weniger bewußten Niveau zu ,Streit und Krieg' bereit."

Linden (1983) sieht in einem lerntheoretischen Modell Hypertoniker wie folgt: Um die Symptome einer kardiovaskulären Hyperreaktivität zu vermeiden, die im Rahmen von Konflikten auftreten, schirmen sie sich gegen diese ab, indem sie unangenehme Wahrnehmungen sozusagen ausfiltern, diese Gefühle unterdrücken und sich auf der Verhaltensebene als angepaßt, konfliktvermeidend und „nett" präsentieren.

Auch unser Patient zeigt als äußere Fassade ein zuverlässiges, pflichtbewußtes Verhalten und großes Verantwortungsgefühl für den Vater, hinter dem sich Sensibilität, Verletzlichkeit und Kränkbarkeit verbergen. Auch er bemüht sich nach außen, Frieden zu stiften, während im Hintergrund Haß und Todeswünsche, aber auch große Schuldgefühle sichtbar werden.

41.3.3 Epidemiologie

Es wird geschätzt, daß an Hypertonie und ihren Folgen dreimal mehr Menschen sterben als an Krebs: Von insgesamt 722 200 verstorbenen Bundesbürgern hatten 1981 367 000 oder 50,8% eine Erkrankung des Kreislaufsystems; bei 84 100 war Herzinfarkt die Todesursache, 158 500 Menschen starben an einer Krebserkrankung (Süddeutsche Zeitung, 1982).

Nach einer Zusammenstellung von Laragh (1975) steht auch in den USA die essentielle Hypertonie mit den von ihr verursachten Komplikationen an erster Stelle der Todesursachen noch vor Malignomen und Unfällen zusammengenommen. Nach Ermittlungen der American Heart Association leiden etwa 20% aller 20- bis 80jährigen Menschen an einem Hochdruck. Davon sind aufgrund der Einschätzung der meisten Autoren 80% essentielle Hypertoniker, wobei von manchen bis zu 95%, von anderen nur 40% angegeben werden. Diese Unterschiede können damit zusammenhängen, daß bei fortgeschrittener Hypertonie kaum mehr zu unterscheiden ist, ob eine Nierenerkrankung Ursache oder Folge des Leidens ist. In Westdeutschland rechnet man, daß 6,3 Millionen Menschen an einer essentiellen Hypertonie und etwa 3,8 Millionen an einer hypertensiven Herzerkrankung als deren Folge leiden (Pflanz, 1977). Frauen werden etwas häufiger betroffen als Männer. Die essentielle Hypertonie kann schon in der Kindheit beginnen, sie ist jedoch im 3. bis 6. Lebensjahrzehnt am häufigsten.

Es gibt viele epidemiologische Untersuchungen, die Unterschiede der Hypertoniehäufigkeit zwischen verschiedenen Kulturen finden. Sie sind häufig nicht leicht zu interpretieren. So ist z.B. bekannt, daß Neger in den Nordstaaten der USA gegenüber den weißen Mitbürgern, vor allem aber auch gegenüber den Negern in den Südstaaten und in Afrika, häufiger an einem Hochdruck erkanken. Andererseits weiß man, daß auch in einigen sog. „unterentwickelten Gebieten" (z.B. in Südwestafrika) die Hypertonie häufig gefunden wird. Bei den Bantunegern ist die Hypertonie sogar häufiger und mit schwereren Komplikationen belastet als in der BRD. Auch auf den Bahamas und den Jungferninseln (Westindien) gibt es mehr Hypertoniker als in den westlichen Ländern. Auf das Problem der Interpretation dieser Unterschiede kommen wir später zurück.

41.3.4 Theorien zur Ätiologie und Pathogenese

Die Ursache der essentiellen Hypertonie ist bis heute nicht eindeutig bekannt. Es besteht nicht einmal Klarheit darüber, ob es sich

– um eine Krankheitseinheit handelt oder
– um ein Sammelbecken verschiedener Hypertonieformen, deren Ätiologie bisher nicht geklärt werden konnte (Sarre, 1971), oder
– um gar keine Krankheit, sondern das „Schwanzende der Normalverteilungskurve einer Bevölkerung" (Platt, 1960).

Die weite Skala der Theorien über die Entstehungsbedingungen reicht von der chromosomalen Heredität (Pickering, 1960; Platt, 1960) bis zur Plurikonditionalität (Stokvis, 1959) im Sinne einer Überschneidung von hereditären, konstitutionellen, sozioökonomischen und tiefenpsychologischen Gegebenheiten (Christian, 1960; Schunk, 1954; Michaelis, 1966). Theorien, die darauf abzielten, eine einzige isolierte Ursache für die essentielle Hypertonie verantwortlich zu machen, waren nicht geeignet, die Vielzahl der Befunde zu erklären. Der systemtheoretische Ansatz Guytons erlaubte eine genaue Analyse des komplexen physiologischen Geschehens bei der Blutdruckregulation. Guyton und Mitarbeiter (1975) betonen die zentrale Rolle der Niere als Langzeitbarostat, was für einige Hypertonieformen auch pathogenetische Bedeutung haben kann.

Sie unterscheiden 8 verschiedene miteinander in Wechselwirkung stehende Rückmeldesysteme:

– das Barorezeptorensystem;
– das Renin-Angiotensin-Vasokonstriktionssystem;
– das Nieren-Körperflüssigkeits-System;
– das Chemorezeptorensystem;
– die zentralnervöse ischämische Reaktion;
– den vaskulären Streß-Entspannungs-Mechanismus;
– den kapillären Flüssigkeitsveränderungsmechanismus;
– das Aldosteronrückmeldungssystem.

Störungen, die schließlich zu einer anhaltenden Blutdruckerhöhung führen, sind an vielen Stellen des Regelsystems möglich, wie auch die verschiedenen sekundären Hypertonieformen belegen. Weiner (1977) betont, daß

– die essentielle Hypertonie als heterogene Erkrankung angesehen werden muß; die Heterogenität manifestiert sich darin,

– daß verschiedene pathogenetische Mechanismen Blutdruckerhöhungen hervorrufen können;

– daß verschiedene physiologische Mechanismen in den verschiedenen Stadien während des Verlaufes der einzelnen Unterformen der essentiellen Hypertonie vorherrschen können;

– daß infolgedessen die essentielle Hypertonie als multifaktorielle Erkrankung anzusehen ist, bei der eben nicht nur ein einziger Faktor für Ätiologie, Pathogenese und Aufrechterhaltung zuständig ist.

Betrachten wir die Argumente, die für die verschiedenen Faktoren sprechen, im einzelnen.

41.3.4.1 Genetische Faktoren

Die Bedeutung der genetischen Faktoren für die Hypertonieentwicklung ist durch verschiedene Forschungsansätze deutlich geworden. Sie stammen aus tierexperimentellen Untersuchungen, aus der Zwillingsforschung und aus Familienuntersuchungen. Das spontane Auftreten der Hypertonie ist bei einer Reihe von Tierarten beobachtet worden (N. Alexander et al., 1954). Die umfangreichsten Untersuchungen bezüglich eines genetischen Einflusses auf die Hypertonieentwicklung liegen für verschiedene Rattenstämme vor (Folkow et al., 1977). Bei salzempfindlichen und salzunempfindlichen Rattenstämmen konnte gezeigt werden, daß für die genetische Blutdruckkontrolle wahrscheinlich nur zwei bis vier Gene verantwortlich sind. Dabei konnte ein Gen identifiziert werden, welches die Abgabe von 18-Hydroxydeoxycorticosteron aus den Nebennieren kontrolliert; es ist wahrscheinlich für etwa 16% der genetischen Blutdruckunterschiede zwischen salzempfindlichen und salzunempfindlichen Rattenstämmen verantwortlich (Rapp et al., 1973). Hierbei sind also endokrine Unterschiede das Ergebnis von genetischen Veränderungen. Spontan hypertensive Ratten reagieren gegenüber unterschiedlich starken Umgebungsreizen mit verstärkten Herzfrequenz- und Blutdruckanstiegen, die wahrscheinlich für die Hypertonieentwicklung verantwortlich sind. Situative Einflüsse lösen hierbei eine allmähliche strukturelle Veränderung der Widerstandsgefäße und des linken Herzens aus, die durch frühzeitige antihypertensive medikamentöse Therapie verhindert werden kann (Folkow et al., 1973).

Für den Menschen sind vor allem Zwillingsstudien aufschlußreich. Danach ist bei einem eineiigen Zwilling das relative Risiko für gleichhohe oder höhere systolische Blutdruckwerte im Vergleich zu seinem Bruder zwei- bis dreimal so groß wie bei einem zweieiigen Zwilling (Feinleib et al., 1975). Es wurde auch versucht, den Einfluß einer gemeinsamen und unterschiedlichen Umgebung bei Zwillingen gegenüber den genetischen Faktoren auf die Blutdruckvariabilität abzuschätzen. Danach soll den genetischen Faktoren die wichtigere Bedeutung zukommen.

In Familienuntersuchungen konnte eine quantitative Beziehung zwischen der Blutdruckhöhe bei Verwandten ersten Grades nachgewiesen werden. Die Ergebnisse legen nahe, daß die Blutdruckhöhe multifaktoriell vererbt wird. In einer Untersuchung von Mialle und Oldham (1963) wird der Einfluß von genetischen Faktoren mit Umgebungseinflüssen auf den Blutdruck verglichen: Nach diesen Schätzungen sind 36% der Varianz des systolischen Blutdrucks und 63% der Varianz des diastolischen Blutdrucks Umgebungsfaktoren zuzuschreiben.

Epidemiologische Untersuchungen aus Detroit (Chakraborty et al., 1977) belegen, daß nichtgenetische Variablen mehr zur Variation des Blutdrucks beitragen als genetische Unterschiede. Der Unterschied dieser Ergebnisse gegenüber früheren Untersuchungen wird damit erklärt, daß dort Effekte genetischen Ursachen zugeschrieben wurden, die in Wirklichkeit von einer gemeinsam geteilten Umgebung oder zeitlichen Trends abhingen.

Eine Reihe von Studien beschäftigt sich mit der Untersuchung kardiovaskulärer Reaktionsmuster in Belastungssituationen bei Kindern hypertoner und normotoner Eltern. Hierbei konnten bei normotonen Kindern hypertoner Eltern verstärkte Reaktionen von Blutdruck und Herzfrequenz nachgewiesen werden (Falkner et al., 1979; Collins et al., 1980; Manuck und Giordani, 1980). Nachdem Personen, bei denen ein Elternteil Hypertoniker ist, ein größeres Risiko haben, später an einer Hypertonie zu erkranken als Personen, deren Eltern Normotoniker sind, erhebt sich die Frage, ob eine verstärkte kardiovaskuläre Hyperreaktivität gegenüber Umgebungsreizen möglicherweise einer der Mechanismen für die zukünftige Hypertonieentwicklung ist.

Der genetische Anteil könnte in einer kardiovaskulären Hyperreagibilität bestehen, die erst unter Mitwirkung von Umgebungsfaktoren wirksam wird.

Zu einem zusammenfassenden Urteil kommen Havlik und Mitarbeiter (1980) und Stammler und Mitarbeiter (1975): In großangelegten Studien mit multipler Regressionsanalyse und der Berechnung von partiellen Korrelationskoeffizienten konnten eine genetische Prädisposition und biophysische Einflußvariablen wie relatives Körpergewicht, Herzfrequenz, Alkoholkonsum, Blutzuckerwerte und Hämatokrit nur 28 bis 34% der Blutdruckvariabilität in Ruhe erklären.

41.3.4.2 Umweltfaktoren

Die Tatsache, daß die essentielle Hypertonie familiär gehäuft auftritt, kann auch im Sinne einer „psychologischen Vererbung" (Freud, zit. n. v. Uexküll, 1963) interpretiert werden. Darunter ist zu verstehen, daß in der Familienatmosphäre von Hypertonikern bestimmte psychische Haltungen geprägt werden, die zum Erwerb einer Hypertonie disponieren. Ein Grund dafür könnte sein, daß die Persönlichkeitsstruktur eines hypertonen Familienmitgliedes das Interaktionsverhalten der ganzen Familie beeinflußt:

Bär et al. (1983) fanden in 16 Familien mit einem hypertonen Vater im verbalen und nonverbalen Verhalten signifikant mehr Ablehnung und negative Interaktion als in einer Kontrollgruppe (15 Familien mit einem normotonen Vater).

Das negative verbale Verhalten war charakterisiert durch Verneinung, Kritik, Unterbrechungen und Entschuldigungen. Die negative nonverbale Interaktion, die durch Videoaufnahmen dokumentiert wurde, zeigte sich in „Kopf-zur-Seite-Drehen", „keine Antwort", „Grimassieren" und „kein Blickkontakt", insbesondere bei konflikthaften Auseinandersetzungen. Dieser negative Interaktionsstil fand sich in den Hochdruckfamilien nicht nur bei den hypertonen Vätern, sondern auch bei den normotonen Kindern und Ehefrauen. So konnten Chazan und Winkelstein (1964) auch bei nichtverwandten Personen, die in einer Hypertonikerfamilie lebten, häufiger als bei der Durchschnittsbevölkerung eine Hypertonie finden.

Bekannt ist auch das Phänomen der positiven Korrelation zwischen Blutdruckwerten von Ehepartnern (Kannel, 1975). Das würde erklären, daß Umwelteinflüsse auch bei genetisch nicht belasteten Personen zu einer Hypertonie führen können. Diese Hypothese wird auch durch die Beobachtungen gestützt (Flynn et al., 1949; Friedmann und Kadanin, 1943), daß bei eineiigen Zwillingen der eine Zwilling aufgrund bestimmter emotioneller Belastungen eine Hypertonie bekam, der andere, bei dem diese Belastungen nicht gegeben waren, jedoch verschont blieb. Auch bei den beiden Zwillingsbrüdern, von denen eingangs berichtet wurde, waren Umwelteinflüsse – zumindest für das Ausmaß ihrer Hypertonie – von maßgebender Bedeutung.

41.3.4.3 Ernährungsbedingungen

Zwischen der essentiellen Hypertonie und Übergewicht bestehen aufgrund von statistischen Untersuchungen sehr enge positive Korrelationen, ein ursächlicher Zusammenhang konnte bisher allerdings noch nicht mit Sicherheit nachgewiesen werden.

Bei einer Reduktion des Körpergewichtes um 1 kg kommt es zu einer gleichzeitigen Blutdrucksenkung von 2 mm Hg systolisch und 1 mm Hg diastolisch (Holzgreve, 1980).

In der Hungerperiode nach dem Zweiten Weltkrieg ging die Zahl der Hypertoniker in der BRD zurück und stieg mit der Besserung der Ernährungsbedingungen und dem Auftreten von Übergewicht an. Die klinische Erfahrung zeigt, daß übergewichtige Patienten mit Hypertonie nach Reduktion ihres Gewichtes häufig normale Blutdruckwerte aufweisen. Die enge Beziehung zwischen Adipositas und Hypertonie ist jedoch vermutlich nicht allein auf Ernährungseinflüsse zurückzuführen, sondern hat auch eine genetisch-konstitutionelle Komponente. Nach der Framingham-Studie können Gewichtszunahme und Blutdrucksteigerung zwar gleichzeitig auftreten; Adipöse entwickeln aber häufiger als Normalgewichtige später einen Hochdruck und normalgewichtige Hypertoniker besitzen ebenfalls ein größeres Risiko als Normotoniker, später adipös zu werden (Kannel und Dawber, 1973).

Die Beziehung zwischen Körpergewicht und Blutdruckhöhe (r = 0,3) ist immerhin so ausgeprägt, daß die Hypertonie in einer durchschnittlich übergewichtigen Bevölkerung wie der unseren dreimal häufiger ist als in einer normalgewichtigen (Pflanz, 1977).

Bei Patienten mit Hochdruck und Übergewicht sollen sich vermehrt Mißtrauen, Fatalismus, geringe Schulbildung und niedriger Sozialstatus finden lassen (Pflanz, 1974).

Mögliche Ursachen, die Übergewichtige zu Hochdruckpatienten werden lassen, sind:
eine erhöhte Bildung von Kortikoiden in der Nebennierenrinde bei Übergewichtigen;
- eine vermehrte beta-adrenerge Ansprechbarkeit des kardiovaskulären Systems, möglicherweise durch Modulation der Rezeptorzahl unter dem Einfluß erhöhter Schilddrüsenhormonspiegel, ebenfalls bei Übergewichtigen (Dustan, 1980);
- eine vermehrte Kochsalzzufuhr als Folge größerer Mahlzeiten (ein Übergewichtiger mit einer durchschnittlichen Energiezufuhr von 4000 Kcal/Tag nimmt doppelt so viel Kochsalz – ca. 20 g/Tag – zu sich wie ein Normalgewichtiger – mit einer seinem Energiebedarf entsprechenden Aufnahme von nur 2000 Kcal/Tag);
- Hyperinsulinämie und verminderte Glukosetoleranz können nicht nur bei adipösen Patienten nachgewiesen werden, sondern auch bei einem Teil der normalgewichtigen Hypertoniker. Möglicherweise führt die erhöhte Aktivität des sympathischen Nervensystems – zumindest in den frühen Stadien einer essentiellen Hypertonie – über die Stimulation der Betarezeptoren im Pankreas zu einer gesteigerten Insulinsekretion mit der dadurch bedingten Zunahme des Körpergewichtes (Berglund et al., 1976).

Nach Dahl (1960) soll eine Hypertonie bei Bevölkerungsgruppen, die weniger als 5 Gramm Kochsalz am Tag zu sich nehmen, selten, dagegen bei Bevölkerungsgruppen mit einem durchschnittlichen Kochsalzverbrauch von 10 bis 15 Gramm pro Tag häufig sein. Bei einer epidemiologischen Untersuchung eines Betriebes teilte er die untersuchten Personen in drei Gruppen ein:
- Personen, die niemals einen Salzstreuer benutzen,
- Personen, die einen Salzstreuer nach dem Kosten der Speisen verwenden,
- Personen, die, ohne die Speisen zu kosten, gewohnheitsmäßig den Salzstreuer betätigen.

Er fand einen signifikanten Zusammenhang zwischen normalem Blutdruck in der ersten und hohem Blutdruck in der dritten Gruppe.

Diese Beobachtungen haben zu der Frage geführt, warum manche Personen mehr Salz essen als andere. Ein Unterschied der Geschmacksschwelle für Kochsalz zwischen Hypertonikern und Normotonikern konnte nicht nachgewiesen werden.

Die erhöhte Zufuhr von Kochsalz kann bereits im Säuglingsalter beginnen: Je nach Herstellerfirma variiert der Kochsalzgehalt von Fertigkost für Säuglinge und Kleinkinder erheblich und liegt zum Teil weit über deren täglichem Kochsalzbedarf (normal: 4 meq Natrium/kg Körpergewicht). Ist der Kochsalzgehalt der Fertigkost dem Kind angepaßt, kann möglicherweise die Mutter den „faden" Geschmack nicht akzeptieren und würzt mit Kochsalz (bis zu 1 g pro Portion) nach. Darüber hinaus enthält Kuhmilch viermal so viel Kochsalz wie Muttermilch. Es gibt also genug Gründe, warum bereits im Säuglingsalter in unseren sog. zivilisierten Ländern ein erhöhter Salzappetit entwickelt bzw. anerzogen werden kann.

Einen vielversprechenden Ansatzpunkt, um das Salzproblem bzw. das „Nachsalzen" zu lösen, könnten neuere Untersuchungen (Skrabal et al., 1981) ergeben, wonach Natriumsalze durch Kaliumsalze ersetzt werden.

Wird bei familiär belasteten Hypertonikern der Kochsalzgehalt der Nahrung von 200 auf 50 mmol pro Tag herab- und der Kaliumgehalt von 50 auf 200 mmol heraufgesetzt, so sinken die basalen Blutdruckwerte sowohl systolisch als auch diastolisch um 5 mm Hg. Noch deutlichere Ergebnisse fanden sich bei körperlicher und mentaler Belastung.

Der feste Zusammenhang zwischen Kochsalzmenge pro Tag und der Hypertoniehäufigkeit in einer Bevölkerung konnte allerdings nicht immer bestätigt werden. So wurden bei reisanbauenden buddhistischen Bauern in Thailand bei einem durchschnittlichen Kochsalzverbrauch von 18 bis 20 Gramm pro Tag und Kopf niedrige Blutdruckwerte – bei einer anderen Bevölkerungsgruppe mit niedrigem Kochsalzverbrauch dagegen häufig Hypertonien gefunden. Diese letzte Bevölkerungsgruppe zeichnete sich durch eine besonders schwierige Sozialstruktur aus.

Die früher oft geäußerte Ansicht, daß Alkohol nur als Kalorienträger, also vor allem bei übergewichtigen Hypertonikern, eine Rolle spielt, muß inzwischen differenziert beurteilt werden:

Ein täglicher Alkoholkonsum von 40 g (ca. 1 l Bier) läßt die Blutdruckwerte noch nicht meßbar ansteigen, sondern soll im Gegenteil den Blutdruck eher senken und durch eine Erhöhung des HDL-Anteils (High Density Lipoproteins, Lipoproteine hoher Dichte, d. h. mit hohem Protein- und niedrigem Lipidanteil) der Lipoproteine die Entwicklung einer Arteriosklerose eher behindern. Dagegen besteht eine – bisher noch nicht geklärte – positive lineare Korrelation zwischen Blutdruckwerten und Alkoholkonsum von mehr als 80 g/Tag (Saunders, 1987).

41.3.4.4 Soziale Faktoren

Statistiken großer Bevölkerungsgruppen zeigten, daß der Blutdruck mit zunehmendem Alter ansteigt. Diese Beobachtung wurde zunächst falsch interpretiert: Man hielt den Altersanstieg des Blutdrucks für physiologisch und normotone Blutdruckwerte im höheren Alter für pathologisch. Eine Aufschlüsselung der epidemiologischen Untersuchungsergebnisse ergab jedoch, daß im Alter lediglich die Zahl der Hypertoni-

ker zunimmt, daß aber normalerweise der Blutdruck während des ganzen Lebens gleichbleibt.

Das Auftreten einer Altershypertonie kann nach Pickering (1960) nicht allein auf genetische Faktoren zurückgeführt werden, da vergleichende epidemiologische Untersuchungen bei verschiedenen Kulturen bzw. Subkulturen feststellten, daß bei den einen der Blutdruck mit dem Alter ansteigt, bei den anderen jedoch nicht. Bei der Durchsicht der vorliegenden Daten gewann die bereits 1929 von Donnison aufgestellte soziokulturelle Hypothese, daß Gruppen mit hohem Blutdruck vermehrten sozialen Spannungen und Konflikten ausgesetzt sind, an Wahrscheinlichkeit. Scotch und Geiger (1963) fanden in Bevölkerungsgruppen mit häufigen Altershypertonien mehr soziale Spannungen als in Kontrollgruppen. Cruz-Coke (1960) konnte zeigen, daß ein niedriger oder normaler Blutdruck im Alter in Bevölkerungsgruppen zu finden ist, die in einer „ökologischen Nische" mit stabilen, nicht wechselnden sozialen Bedingungen leben. Daraus zog er den Schluß, daß ein Verlust dieser „Nische" durch den Einbruch soziokultureller Veränderungen vermehrt zu Krankheiten – und unter diesen vor allem zu Hypertonie – führt.

Henry und Cassel (1969) stellten Untersuchungen über den Blutdruck bei verschiedenen Altersgruppen in sog. „stabilen und instabilen" Kulturen zusammen. Abbildung 41.3–2 gibt einen Überblick über ihre Ergebnisse: Die Unterschiede zwischen den Blutdruckanstiegen mit zunehmendem Alter in den oberen und unteren Reihen entsprechen Unterschieden in der sozialen Struktur. Dabei scheint ausschlaggebend zu sein, ob Bevölkerungsgruppen eine feste Tradition haben, die während einer Generation stabil bleibt, oder ob die sozialen Strukturen sich wandeln.

Diese Untersuchungen lassen sich so interpretieren, daß die Altersanstiege des Blutdrucks mit der Unfähigkeit älterer Menschen zusammenhängen, sich entscheidenden Änderungen der Lebensweise anzupassen und in einer veränderten Welt der jungen Generation geeignete soziale Verhaltensmuster zu vermitteln (Donnison, 1929; Scotch und Geiger, 1963; Henry und Cassel, 1969).

Die Ergebnisse dieser Studie sind zweifellos kein Beweis für die soziokulturelle These. Cassel (1975) weist auf die Notwendigkeit hin, sorgfältig kontrollierte prospektive Studien durchzuführen und zu prüfen, inwieweit ein Blutdruckanstieg in Beziehung zur sozialen Geschichte einer Gruppe gebracht werden kann.

Ostfeld und d'Atri (1977) warnen davor, soziale Faktoren zu diskutieren, ohne die unterschiedliche Häufigkeit der Adipositas in einzelnen Sozialgruppen zu berücksichtigen. Andere Autoren nehmen an, daß Veränderungen und Anpassung der Lebensweise an eine andere Kultur (Akkulturation) erst zusammen mit zunehmendem Salzverbrauch einen Altersanstieg des Blutdrucks hervorrufen können (Friedman und Dahl, 1977; Freis, 1976).

In einer Studie in Polynesien und in Neuseeland wurden Eßgewohnheiten, Salzverbrauch und Körpergewicht bei Bewohnern der Tokelau-Atolle kon-

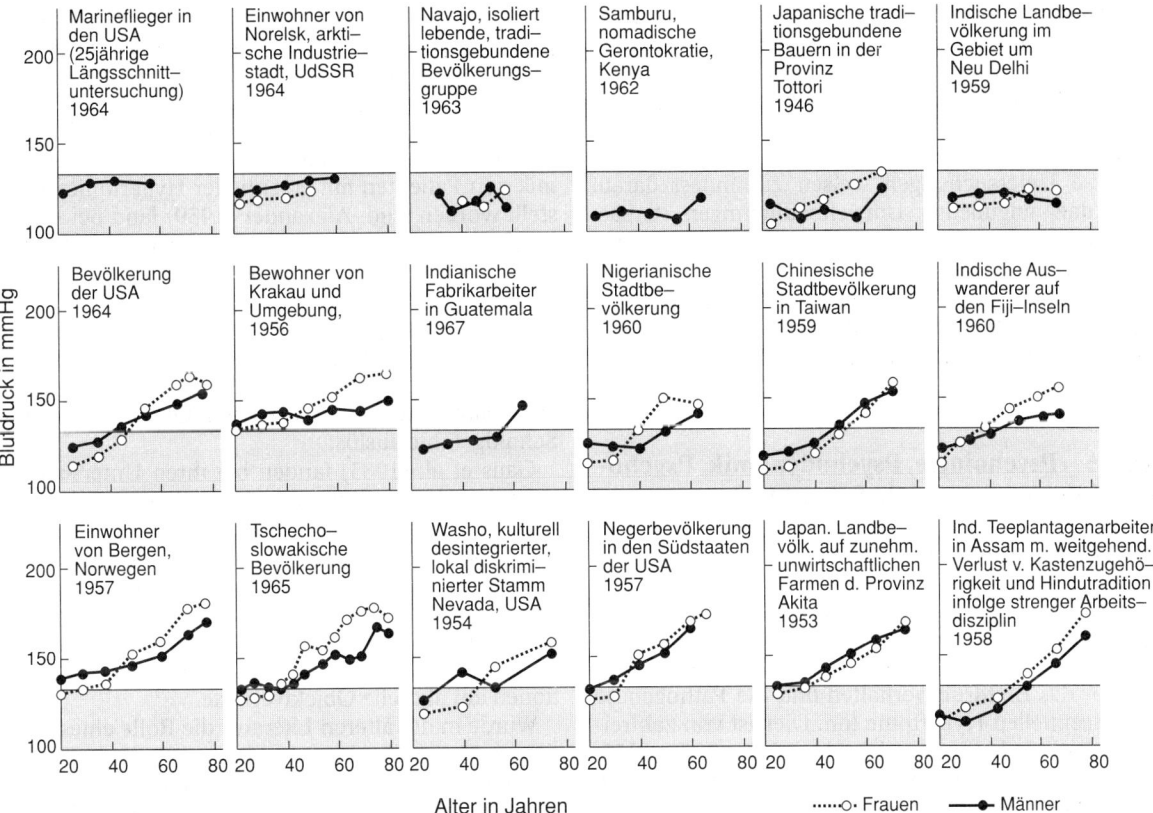

Abb. 41.3-2. Abhängigkeit des Blutdrucks vom Alter in verschiedenen Kulturen. Offene Kreise: Frauen; geschlossene Kreise: Männer (nach Henry und Cassel, 1969).

trolliert und der Einfluß der Emigration und Änderung der Lebensweise auf den Blutdruck untersucht. Wegen Überbevölkerung veranlaßte die neuseeländische Regierung 1966 ein Umsiedlungsprogramm für einen Teil der in ausgesprochener sozialer Balance lebenden Bevölkerung dieser Atolle. Rund 900 Erwachsene und ebenso viele Kinder siedelten im Laufe der Zeit nach Neuseeland um. Einige der Emigranten lebten in gut organisierten Gemeinden zusammen; sie konnten hier enge persönliche Kontakte mit anderen Auswanderern aus Tokelau sowohl zu Hause als auch bei der Arbeit pflegen – sie behielten ihre ethnische Identität. Andere Einwanderer hingegen – sei es durch Zufall, eigene Wahl oder Notwendigkeit – schlossen sich freizügiger der außerhalb ihrer eigenen Kultur stehenden neuseeländischen Bevölkerung an – ihnen fehlte der enge Kontakt zu ihrer eigenen Kultur und Tradition. Die Immigranten übernahmen alle – unabhängig vom Ausmaß der Akkulturation – neuseeländische (westliche) Eßgewohnheiten und hatten einen höheren Salzverbrauch als die Bewohner der Tokelau-Atolle. Es zeigte sich nun auch nach statistischer Kontrolle des Körpergewichtes und der Aufenthaltsdauer in Neuseeland ein signifikanter positiver Zusammenhang zwischen Blutdruckhöhe und Ausmaß der sozialen Interaktion der Immigranten mit anderen Neuseeländern; dies galt sowohl für Männer als auch für Frauen (Beaglehole et al., 1977; Prior, 1976).

Einige Studien weisen darauf hin, daß niedrige Sozialschichten höhere Blutdruckwerte zeigen (Reid et al., 1966; US National Center of Health Statistics, 1964). Bei Schulkindern der Innenstadt wurden höhere Werte im Vergleich zu Schülern der Vorstadt gefunden (Decastro et al., 1976).

Eine amerikanische Untersuchung bestimmte den Einfluß von geringer bzw. ausgeprägter Streßbelastung bei schwarzen und weißen Bewohnern in vier verschiedenen Wohngebieten von Detroit. Eine hohe Verbrechensrate, beengte Lebens- und Wohnverhältnisse, hohe Umzugshäufigkeit und eine große Anzahl gescheiterter Ehen kennzeichneten die meist von der sozialen Unterschicht bewohnten Stadtgebiete mit hoher psychosozialer Belastung – die Wohngebiete mit geringem Streßniveau wiesen entsprechend günstigere soziale Umgebungsbedingungen auf. Die höchsten Blutdruckwerte fanden sich bei der schwarzen Bevölkerung, die hoher Streßbelastung ausgesetzt war, wohingegen sich schwarze Detroiter in Wohngebieten mit geringer Belastung nicht im Blutdruckniveau von den weißen Stadtbewohnern unterschieden. Dies weist auf die Bedeutung einer Wechselwirkung von genetischen bzw. ethnischen Faktoren mit der sozialen Umwelt hin (Harburg et al., 1973).

Eine Vielzahl weiterer Faktoren wurde in Beziehung zur Entwicklung einer Hypertonie gebracht: Dazu gehören der Langzeiteffekt von Arbeitslosigkeit

(Kasl und Cobb, 1970), Überfüllung von Gefängnissen (d'Atri und Ostfeld, 1975), Fluglärm in unmittelbarer Nachbarschaft (Knipschild, 1977; Cohen et al., 1980; v. Eiff und Neus, 1980) sowie erhöhte Arbeitsbelastung bei Fluglotsen (Rose et al., 1978; Cobb und Rose, 1973).

Diese Untersuchungen weisen zumindest darauf hin, daß ungünstige Umweltbedingungen Anpassungsleistungen verlangen, die bei Überforderung Streßreaktionen hervorrufen und wahrscheinlich eine Krankheitsentwicklung einleiten können. Die Hypertonieentwicklung läßt sich somit auch als Adaptationskrankheit betrachten (vgl. Kap. 1).

41.3.5 Psychologie, Psychodynamik, Psychophysiologie und soziale Interaktion

41.3.5.1 Emotionale Faktoren

Der Zusammenhang zwischen sozialen Strukturen und der Häufigkeit des Auftretens einer Hypertonie weist bereits auf die Wichtigkeit emotionaler Faktoren für das Blutdruckverhalten und die Pathogenese der essentiellen Hypertonie hin. Dies ist von zahlreichen Autoren betont worden (Alexander, 1939, 1951; Bastiaans, 1963; Cannon, 1953; Cochrane, 1969, 1973; Delius und Fahrenberg, 1963; Jores, 1960; Quint, 1967; Reindell, Klepzig und Roskamp, 1971; Schunk, 1954; Thomas, 1967; v. Uexküll und Wick, 1962; Wyss, 1955).

Exkurs: Um die Wirkungsweise emotionaler Faktoren auf das Erleben besser zu verstehen, sollte man sich folgendes klarmachen:
– Emotionale Faktoren spiegeln Einflüsse unserer Vergangenheit wider, die unser gegenwärtiges Erleben prägen. Gefühle, die mit wichtigen Beziehungspersonen der Kindheit verbunden sind, werden jederzeit auf gegenwärtige Beziehungspersonen übertragen, also auch auf den Arzt.
– Emotionalität kennt keine Vergangenheit. Unabhängig davon, wie weit eine emotionale Erfahrung nach objektiver Zeitrechnung zurückliegt, ist sie in vergleichbaren Situationen jederzeit reaktivierbar und bleibt drängende Gegenwart.

So hat jede Situation eine besondere emotionale Bedeutung für das Subjekt. Neutrale Situationen gibt es nicht. Die Reaktion des Individuums auf eine Situation hängt von der bestehenden Konflikthaftigkeit und Verwundbarkeit für die Themen der verschiedenen Entwicklungsstufen wie Geborgenheit und Akzeptiertsein, Macht und Ansehen und Identität ab (vgl. Kap. 1).

Zur Annäherung an die Frage, welche Rolle emotionale Faktoren bei der essentiellen Hypertonie spielen, ist eine Beobachtung von Groen und Mitarbeitern (1971) besonders interessant:

Sie konnten in Israel an einem Patienten mit einer therapieresistenten schweren Hypertonie beobach-

ten, daß sich der Blutdruck während des 6-Tage-Krieges normalisierte, die Hypertonie aber nach der Rückkehr ins Zivilleben wieder in dem vorherigen Schweregrad auftrat.

Diese Beobachtung stimmt mit Thesen überein, welche von Psychoanalytikern über die Psychodynamik von Patienten mit essentieller Hypertonie aufgestellt worden sind. Alexander (1939) fand bei seinen psychoanalytisch untersuchten Hypertoniepatienten einen „unspezifischen Konflikt" zwischen aggressiven Tendenzen und innerer Abhängigkeit von den Objekten, denen die Aggressionen galten. In einem solchen Konflikt werden Gefühle von Wut, Neid und Haß gegen die Person, von der man innerlich abhängig ist, als Gefahr erlebt, die Angst vor Objektverlust und Schuldgefühle auslöst.

Gaus et al. (1983) fanden bei ihren Untersuchungen folgende Konfliktmuster besonders typisch: Konflikte mit dem Thema Aggression versus Unterwerfung; Gewährung versus Versagung, insbesondere oraler Bedürfnisse; Konflikte um das Selbstwerterleben, die durch krankheitsbedingte Entwertung und Einschränkung verstärkt sind, sowie häufig aus Ambivalenzkonflikten resultierende pathologische Reaktionen auf aktuelle Objektverluste.

Wurde in der älteren Literatur die Rolle eines autoritären Vaters für diese Konflikte betont, so fanden Perini et al. (1982), daß die Hypertoniepatienten, die Hinweise auf die oben beschriebenen Konflikte boten, auch in einengenden (overprotective) Familien aufgewachsen waren, also Hinweise auf eine gestörte Mutter-Kind-Beziehung.

Beide Situationen scheinen zur Aufrichtung eines strengen und starren Über-Ichs zu führen, das ein offenes Austragen von aggressiven Konflikten untersagt. Damit sind jedoch nicht nur die destruktiven Anteile des Aggressionstriebes gehemmt, die das lebensnotwendige Objekt bedrohen, sondern auch die positiven, die Unabhängigkeit vom Objekt, Entwicklung von Eigenständigkeit, Selbstbehauptung und das zu einer Leistung nötige Aktivitätspotential garantieren. Bastiaans (1963) spricht von einem „Law-and-order-Super-Ego". Dieses Über-Ich läßt sich sowohl im Leistungsverhalten als auch in einer veränderten Wahrnehmung wiederfinden.

So haben viele Autoren (Aresin, 1960; Enke und Gercken, 1955; Michaelis, 1966; Pflanz und v. Uexküll, 1962; Stern, 1958; Wyss, 1955) eine unrealistische, zwanghaft perfektionistische Einstellung der Hypertoniker zur eigenen Leistung beschrieben. Sie sind oft unfähig, die Ergebnisse ihrer Bemühungen objektiv zu beurteilen und empfinden ihre Tätigkeit mehr als eine von einer höheren Autorität auferlegte Pflicht als den Versuch, eigene Wünsche zu befriedigen. Diese Befunde decken sich weitgehend mit den Beobachtungen über das sog. Typ-A-Verhalten nach Rosenman und Friedman (1970, 1975) und Beobachtungen, wie sie Köhle (1969) bei Patienten mit peripherer arterieller Verschlußkrankheit gemacht hat.

Die veränderte Wahrnehmung ließ sich in einem Experiment von Sapira und Mitarbeitern (1973) zeigen:

Sie projizierten je einer Gruppe von Hypertonikern und Normotonikern einen Film, in dem zunächst ein Patient bei einem „bösen" Arzt gezeigt wurde, der kurz angebunden, in Eile und desinteressiert war und der sich sowohl über den Patienten als auch über dessen Blutdruck zu ärgern schien. Im zweiten Teil des Films benahm sich der gleiche Arzt dem gleichen Patienten gegenüber entspannt, gefällig, höflich, schien erfreut über dessen Blutdruck und an dem Patienten als Person interessiert. In dem anschließenden Interview konnten die Hypertoniker im Gegensatz zur Kontrollgruppe der Normotoniker keinen Unterschied zwischen den beiden Szenen beschreiben.

Die Allgemeingültigkeit der Beobachtungen zur Psychodynamik und Persönlichkeit von Hypertonikern wurde vor allem von Cochrane (1973) und Ostfeld (1973) bezweifelt. Cochrane stellte aufgrund epidemiologischer Untersuchungen die These auf, Hypertoniker, die psychotherapeutisch untersucht und behandelt werden, seien eine selektive Gruppe mit besonderen neurotischen Symptomen.

Neuere Untersuchungen von Esler und Mitarbeitern (1977) und Perini und Mitarbeitern (1982) unterschieden zwischen Hypertonikern mit hohem und mit normalem Plasmareninspiegel. Mit dem „Rosenzweig Picture Frustration Test" ließen sich zwischen Hypertonikern der zweiten Gruppe und Normotonikern keine Unterschiede feststellen. Die Gruppe mit einem hohen Plasmareninspiegel verhielt sich signifikant anders: Diese Hypertoniker konnten Frustrationen nicht wahrnehmen und sich schlechter behaupten, sie waren weniger aggressiv und eher geneigt, sich zu unterwerfen.

Ein Trend fand sich für ein dringendes Bedürfnis, Konflikte sofort zu lösen, gleichzeitig eine Neigung, dieselben zu verleugnen, eine Unfähigkeit, Gefühle auszudrücken, sich einer gegebenen Situation passiv zu unterwerfen und Aggressionen zu verinnerlichen. Auch hierin unterschieden sich die Hypertoniker mit normalem Reninspiegel nicht von den Normotonikern.

Diese Unterschiede ließen sich sowohl bei Grenzwerthypertonikern als auch bei Hypertonikern mit ausgeprägter essentieller Hypertonie sichern und scheinen damit eher ursächlich zur Entwicklung der essentiellen Hypertonie mit hohem Reninspiegel beizutragen als Folge davon zu sein (Perini et al., 1982).

Diese Befunde zeigen ebenfalls, daß essentielle Hypertoniker sowohl psychologisch wie physiologisch keine homogene Gruppe bilden.

41.3.5.2 Situative Faktoren

Von Uexküll und Wick (1962) konnten mit Hilfe eines – den Blutdruck (unblutig) automatisch registrierenden – Gerätes den Einfluß emotional belastender Situationen (z.B. des medizinischen Staatsexamens [vgl. Abb. 41.3–3] oder der Besprechung emotional belastender Themen) auf das Blutdruckverhalten sowohl bei Hypertonikern als auch bei Normotonikern demonstrieren. Für solche emotional ausgelösten

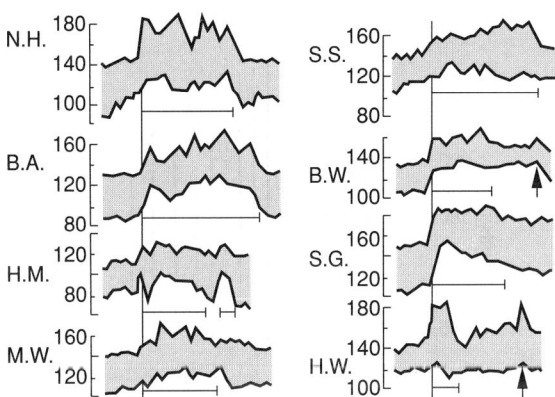

Abb. 41.3–3. Blutdruck- und Pulskurven von acht Medizinstudenten während des mündlichen Staatsexamens. Automatische Registrierung in Abständen von einer Minute. Der senkrechte Strich bezeichnet den Beginn der Prüfung, der waagrechte deren Dauer. Bei einem Pfeil wurde nach Abschluß der Fragen noch einmal eine Frage an den Kandidaten gestellt. Bei fünf Studenten war das Niveau des Blutdrucks schon erhöht, ehe der Examinator begann, Fragen an sie zu richten (nach v. Uexküll und Wick, 1962).

Blutdrucksteigerungen prägten sie den Ausdruck „Situationshypertonie". Ereignisse, die emotional als Bedrohung, Kränkung oder Beeinträchtigung erlebt werden, gegen die sich der Betreffende aus äußeren oder inneren Gründen nicht zur Wehr setzen kann, schienen der gemeinsame Nenner für Situationen zu sein, die zur Hypertonie führten.

Außer bei aktuellen Ereignissen kam es aber auch bei Personen zu Blutdruckanstiegen, die über emotional belastende Situationen ihrer Vergangenheit berichteten (vgl. Abb. 41.3–4). Dabei war es gleichgültig, ob sie während des Berichtes die emotionale Reaktion wiedererlebten oder verleugneten (vgl. Abb. 41.3–5).

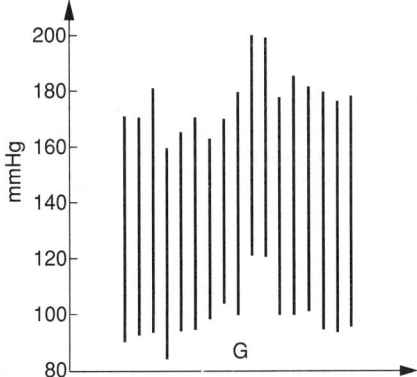

Abb. 41.3–4. Blutdruck- und Pulskurve eines 62jährigen Patienten mit essentieller Hypertonie. Automatische Registrierung in Abständen von 1 Minute. Bei der Marke G berichtet der Patient über einen Konflikt mit seinem Sohn, der statt das väterliche Geschäft zu übernehmen, ein Studium begonnen habe. Er erwarte vom Vater, der vorhatte, sich zur Ruhe zu setzen, ganz selbstverständlich weitere Opfer für seine Ausbildung (nach v. Uexküll und Wick, 1962).

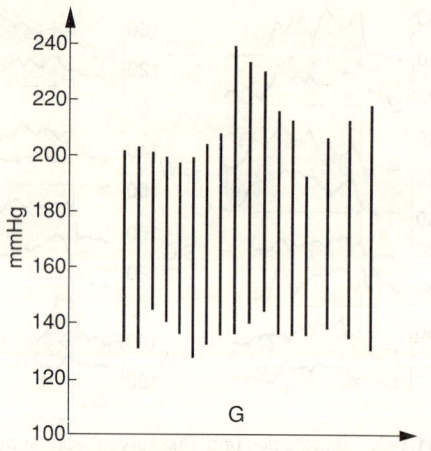

Abb. 41.3–5. Blutdruck- und Pulskurve einer 42jährigen Krankenschwester mit essentieller Hypertonie. Automatische Registrierung in Abständen von zunächst 1, dann 2 Minuten. Bei der Marke G berichtet sie über ihre Tätigkeit in der Psychiatrischen Klinik: Während der Nachtwachen wurde sie öfter von Patienten angefallen, bedroht und auch schon einige Male körperlich verletzt. Das mache ihr aber nichts aus, sie bleibe in solchen Situationen ganz ruhig und würde sich weder aufregen noch erschrecken (nach v. Uexküll und Wick, 1962).

Diese Beispiele zeigen nur relativ kurzzeitige Blutdruckreaktionen. Es gibt zwar noch keine systematischen Untersuchungen über die Frage, wie lange situative Blutdruckanstiege bestehenbleiben, es ist aber sicher, daß sie nicht nur Minuten und Stunden, sondern auch Tage, ja sogar Monate und Jahre dauern können. Ein ausführlicher Fallbericht dazu findet sich im Abschnitt 41.3.10.

Menzel (1961) hat auf die oft Tage anhaltenden Hypertonien aufmerksam gemacht, die als Reaktion auf eine Krankenhauseinweisung zu beobachten sind. Langanhaltende Blutdruckanstiege bei Soldaten stellte Graham (1945) unter den Belastungen des Fronterlebens im Kriege fest, Ehrström (1945) bei Rekruten als Reaktion auf die Belastungen des Kasernenlebens.

Die Bedeutung situativer Faktoren wird auch an den Beobachtungen deutlich, die Gaus und Mitarbeiter (1983) gemacht haben. Bei fast allen 66 Patienten mit starken Blutdruckschwankungen, die als „nicht einstellbar" überwiesen worden waren, resultierte das Scheitern der bisherigen antihypertensiven Behandlung aus der Nichtberücksichtigung psychosozialer Faktoren (vgl. Tab. 41.3–1).

Tabelle 41.3–1. Belastungen und Konflikte bei 66 Hypertoniepatienten mit besonders labiler Verlaufsform (Mehrfachangaben, modifiziert nach Gaus et al., 1983).

	(n = 66)
ausgeprägte berufliche Belastungssituation	33
Störungen der Arzt-Patient-Beziehung	32
chronische Konfliktkonstellation	30
ausgeprägte familiäre Belastungssituation	26
massives Selbstwertproblem	20
aktueller Objektverlust/Trauerreaktion	16

Wichtig in diesem Zusammenhang ist, daß eine Behandlung mit antihypertensiven Medikamenten situative Blutdruckreaktionen nicht unterbinden kann.

Für den Arzt und seinen Umgang mit Hypertonikern ist von Bedeutung, daß eine typische Auslösersituation für eine Verschlimmerung einer Hypertonie der plötzliche Verlust von Selbstsicherheit gegenüber Autoritätspersonen (z.B. auch Ärzten, von denen Patienten abhängig sind) darstellt (Binger, 1945). Nach Quint (1967) ist diese Situation gegeben, wenn der Patient fühlt, daß seine Kooperationsbereitschaft nicht anerkannt oder akzeptiert wird.

Dies soll folgendes Beispiel von Gaus und Mitarbeitern (1983) illustrieren:

> Ein 56jähriger Patient mit einer gestörten Vaterbeziehung hatte archaische Ohnmachtsgefühle wiedererlebt, als er während des Krieges, kaum von einer schweren Erkrankung genesen, vom Militärarzt sofort an die Front zurückgeschickt und wegen seines Protestes als „Simulant" beschimpft worden war. Bei der Vorstellung beim Vertrauensarzt 35 Jahre später nach einer Operation erlebte er erneut das Gefühl völligen Ausgeliefertseins und wurde wegen einer schweren Blutdruckkrise während der Untersuchung überwiesen. In der Sprechstunde konnte bei der Thematisierung dieses Erlebnisses und in Passagen, in denen es um Gewährung und Versagung ging, ein deutlicher Blutdruckanstieg beobachtet werden.

Die Annahme, daß das Verhalten von Ärzten den Blutdruck der Patienten beeinflußt, wird auch durch den Vergleich der Durchschnittswerte aller Blutdruckmessungen nahegelegt (vgl. Abb. 41.3–6), die 10 Ärzte einer medizinischen Poliklinik während eines Jahres durchgeführt hatten.

Für die Beurteilung dieser Umstände als Auslösersituation einer Hypertonie oder deren Verschlimmerung muß man sich klarmachen, daß Hypertoniepatienten ihre soziale Anpassung und ihre weitgehende Unauffälligkeit („Übernormalität") nur mühsam und oft nur unter Anspannung aller Kräfte aufrechterhalten können (Schuster-Erfmann, 1967).

Sobald das Gleichgewicht zwischen narzißtischem Gewinn aus Helferhaltung, Leistung als „Lastesel" und der Energie, die zum Neutralisieren feindseliger aggressiver Gefühle (Schur, 1974) benötigt wird, durch

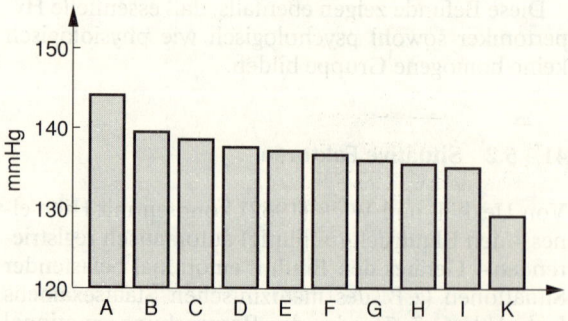

Abb. 41.3–6. Durchschnittswerte aller Blutdruckmessungen, die 10 Ärzte einer medizinischen Poliklinik (Gießen) während eines Jahres durchgeführt haben.

Veränderungen der Umgebung gestört wird, besteht die Möglichkeit für das Auftreten oder die Verschlimmerung des Symptoms.

Die Zusammenhänge zwischen Bluthochdruckverhalten und emotionalen Faktoren aufgrund bestimmter psychodynamischer Vorgänge wurden zu einem großen Teil mit psychoanalytischen Methoden beobachtet und beschrieben. Sie werden von schulpsychologischer, aber auch von statistischer Seite wegen ihrer Nichtquantifizierbarkeit, häufig zu geringer Fallzahlen, dem Überwiegen von Kasuistik und der Selektion der Patienten kritisiert. Deshalb ist es wichtig, daß die Beobachtungen, die an Menschen gemacht wurden, im Tierversuch eine zum Teil überraschende Bestätigung erfahren.

41.3.5.3 Tierexperimentelle Untersuchungen

Die spontan hypertensiven Ratten (SHR) von Okamoto und Oaki (1963) stellen mit einer multigenetischen Veranlagung zu hohem Blutdruck ein geeignetes Tiermodell für die essentielle Hypertonie des Menschen dar. Hier scheint die Tendenz, auf Umwelteinflüsse mit verstärkten, im Hypothalamus koordinierten neurohormonalen Reaktionsmustern zu antworten, genetisch festgelegt zu sein. Folkow und Mitarbeiter (1973) zeigten, daß die hypertensiven Ratten im Vergleich zu normotensiven eine Hyperreaktivität (auf Weckreize) der zentralen autonomen Strukturen aufweisen, die unter psychischem Streß das emotionale Verhalten in Alarmbereitschaft versetzen. Die spontan hypertensiven Ratten haben somit möglicherweise – genetisch bedingt – eine niedrigere Schwelle für die Auslösung spezifischer hypothalamischer Reaktionsmuster auf Umweltreize. Diese Reaktionen umfassen differenzierte neurohormonale Entladungsmuster, die den Organismus z.B. auf Kampf oder Flucht vorbereiten und die der motorischen Antwort vorausgehen können (Eliasson et al., 1951; Abrahams et al., 1960). Sie können in einer zentral ausgelösten vagalen kardialen Hemmung bestehen, verbunden mit einem vermehrten sympathischen Einfluß auf Herz, Venensystem und die meisten Gefäßgebiete, was mit Ausnahme der Skelettmuskulatur zu einer Vasokonstriktion führt: In den Muskelgefäßen tritt dann vielmehr durch den Einfluß von Adrenalin und über sympathische cholinerge Nervenfasern eine starke Vasodilatation auf (Folkow und Neil, 1971).

Hormonale Reaktionen bestehen in einer Freisetzung von Adrenalin aus dem Nebennierenmark (Grant et al., 1958), einer Aktivierung des ACTH-Kortikosteroid-Systems in Hypophyse und Nebennierenrinde (Folkow et al., 1967), sowie in einer neurogenen Reninfreisetzung und damit Beteiligung des Angiotensin-Aldosteron-Mechanismus (Davies, 1973; Zanchetti und Stella, 1975). Die Wirkungen psychosozialer Umweltreize auf den Organismus lassen sich zwei spezifischen Reaktionsmustern, der Ebene des emotionalen Verhaltens und der Ebene des neuroendokrinen Systems, zuordnen, die beide

für die Entwicklung einer Hypertonie bedeutsam zu sein scheinen (Henry, 1976; Henry und Stephens, 1977):

– Der Abwehrreaktion (Cannon und La Paz, 1911; Folkow und Neil, 1971) mit Beteiligung des sympathischen Nebennierenmarksystems als Bereitstellungsreaktion zu Kampf oder Flucht bzw. aggressivem Verhalten.

– Der Alarmreaktion (Selye, 1936) mit Beteiligung des Hypophysen-Nebennierenrinden-Systems mit dem Verhaltensmuster von Rückzug-Konservierung bzw. Depression.

Beide Reaktionsmuster sowie die beteiligten zentralnervösen Strukturen, Cortex, Hypothalamus und limbisches System, sind allen Säugetieren gemeinsam. Im Hypothalamus sind die Zentren für spezifische Verhaltensmuster (consummatory acts), z.B. Essen, Trinken und Sexualverhalten, sowie die für Angst und Freude typischen Reaktionsmuster repräsentiert, darüber hinaus werden hier die Aktivitäten des endokrinen und des autonomen Nervensystems, das u.a. auch das kardiovaskuläre System steuert, integriert (Folkow und Neil, 1971; Ganong, 1971). Das limbische System scheint vor allem für die emotionale Tönung des Verhaltens verantwortlich zu sein. Beide, Hypothalamus und limbisches System, sind an der Entstehung von Emotionen und dem damit verbundenen Verhalten beteiligt.

Auch die pathologischen Auswirkungen langanhaltender Abwehr- oder Alarmreaktionen scheinen bei verschiedenen Säugetieren mit denen des Menschen übereinzustimmen. Gleiche anatomische Verhältnisse der beteiligten zentralnervösen Strukturen, übereinstimmende physiologische und pathologische Reaktionsmuster bei vergleichbaren auslösenden Bedingungen der sozialen Umwelt lassen eine gewisse Übereinstimmung der an diesen Reaktionen beteiligten grundlegenden Emotionen bei Menschen und höheren Säugetieren vermuten (Henry et al., 1972).

Die Abwehrreaktion versetzt ein Tier in einen Zustand, der durch die Tendenz, entweder anzugreifen oder zu fliehen, charakterisiert ist. Diese Reaktion kann durch Umweltreize, und damit über den Cortex, ausgelöst werden, wobei dem Nucleus amygdalae die Rolle zukommt, die hypothalamischen Strukturen, die das entsprechende emotionale Verhalten kontrollieren, zu aktivieren, sobald die äußeren Bedingungen geeignet erscheinen. Situationen, die in der Erwartung bedrohlicher Ereignisse vermehrte Wachsamkeit verlangen, können diese Reaktionsmuster auslösen.

Dabei werden die sympathisch-cholinergen, vasodilatatorischen Nervenfasern zu den Arteriolen der Skelettmuskulatur und zum Herzen sowie die adrenergen Fasern zu den übrigen Gefäßen und ebenfalls zum Herzen unter gleichzeitiger vermehrter Katecholaminausschüttung aus dem Nebennierenmark aktiviert. Als Ausdruck eines vermehrten sympathischen Einflusses bei diesem Reaktionsmuster gilt ein erhöhter Plasmareninspiegel. Folkow und Rubinstein (1966) konnten durch intermittierende direkte Stimulation des hypothalamischen Abwehrzentrums bei Ratten nach mehreren Wochen eine ausgeprägte Hypertonie erzeugen. Schunk (1954) konnte zeigen, daß Katzen, die in einem Käfig täglich bellenden Hunden ausgesetzt waren, einen erhöhten Blutdruck entwikkelten, der alle Komplikationen einer malignen Hypertonie aufwies.

Früher wurde die sympathische Reaktion über das Nebennierenmark bei der Untersuchung psychophysiologischer Zusammenhänge in den Vordergrund gestellt. Psychische Einflüsse sind aber auch für das Hypophysen-Nebennierenrinden-System die stärksten der bekannten aktivierenden Reize (Mason, 1968). Die Erwartung neuer Situationen, insbesondere die Bedrohung des Territoriums oder des Platzes in der sozialen Rangordnung, führt zu einer vermehrten Freisetzung von Kortikosteroiden, vor allem wenn keine aggressive Auseinandersetzung möglich erscheint. An diesen Reaktionen ist der Hippocampus beteiligt; das Hypophysen-Nebennierenrinden-System steht unter seiner Kontrolle, vor allem in Streßsituationen. Angsterregende Bedingungen wie der Verlust des Status und der Kontrolle sind mit depressivem Verhalten, Immobilität, Unterwerfung und dem Gefühl der Hilflosigkeit verbunden – eine Strategie, die über den Hippocampus unter der Beteiligung des ACTH-Kortikosteron-Mechanismus initiiert wird. Dabei besitzt ACTH gleichzeitig eine aggressionshemmende Wirkung, erleichtert und festigt Vermeidungslernen. Die vermehrte Stimulierung der Nebennierenrinde kann zu ihrer Hyperplasie und zu Störungen der Mineralokortikoidproduktion führen, in deren Gefolge eine größere Empfindlichkeit der Gefäße gegenüber pressorischen Reizen entsteht. Mit diesem Reaktionsmuster sind niedrige Plasmareninspiegel verbunden. Die Hyperplasie der Nebennierenrinde ist charakteristisch für ältere Hypertoniker, wie in autoptischen Untersuchungen nachgewiesen werden konnte (Russel und Masi, 1973).

Die Bedeutung beider Reaktionsmuster (Alarm- und Abwehrreaktion) im Zusammenhang mit psychosozialen Reizen konnten Henry und Mitarbeiter (1967, 1971, 1976) demonstrieren, die den Einfluß der Bildung sozialer Verhaltensmuster in den ersten Lebenswochen auf die spätere Entwicklung einer Hypertonie und ihrer Folgeerkrankungen bei Mäusekolonien untersuchten: Mäuse wurden von Geburt an in einem durch Röhren verbundenen Boxensystem aufgezogen, das durch seine Konstruktion zwangsläufig zu häufigen sozialen Konfrontationen und Kämpfen führte. Es bildete sich eine stabile soziale Hierarchie mit dominanten und rangniederen Tieren. Die Blutdruckwerte der gesamten Population dieses Boxensystems waren leicht erhöht gegenüber einer Gruppe von normal aufgezogenen Mäusen in isolierten Boxen, in denen keine Kämpfe um das Territorium auftraten.

Wurden die Mäuse in den ersten 14 bis 16 Tagen der Säugeperiode – noch ehe sie die Augen öffnen konnten – von den Elterntieren getrennt und isoliert aufgezogen, so fanden sich bei ihnen besonders niedrige Blutdruckwerte. Wurden diese – in sozialer Deprivation aufgewachsenen – Tiere 4 Monate später in das Boxensystem gesetzt, so entwickelten sich heftige Kämpfe, die aber nicht – wie bei anderen Gruppen – zur Bildung einer stabilen Hierarchie und einem Respektieren des erkämpften Territoriums führten. In dieser Gruppe traten chronische Blutdruckerhöhungen, Gewichtsverlust, Hypertrophie der Nebennieren, Atrophie des Thymus sowie Kannibalismus an Neugeborenen auf. Die weiblichen Tiere waren unfähig, Junge großzuziehen. Nach 5 bis 6 Monaten konnten ausgeprägte pathologisch-anatomische Veränderungen wie Arteriosklerose, Myokardfibrose und interstitielle Nephritis nachgewiesen werden, die schließlich zu chronischem Nierenversagen mit Anstieg des Harnstoffes im Blut führte. Die Hypertonieentwicklung könnte hier möglicherweise auch durch die beeinträchtigte Nierenfunktion eingeleitet werden, die eine Folge des gestörten Territoriumverhaltens der Mäuse zu sein scheint: Die gewöhnliche Duftmarkierung des Territoriums unterbleibt, die Miktionshäufigkeit wird reduziert, es kommt zum Harnrückstau und zur interstitiellen Nephritis. Auf der Basis dieser Veränderungen kann das anhaltende Kampf- und Fluchtverhalten im Laufe der Zeit zum dauerhaften Blutdruckanstieg führen. Innerhalb der ersten Woche ist die Hypertonie noch durch Isolierung der Tiere reversibel, nach einem Monat ist sie jedoch stabil und nicht durch Isolierung der Tiere wieder zurückzubilden.

Den Einfluß der Stellung in der sozialen Hierarchie, die sich durch Beobachtung des unterschiedlichen Territorialverhaltens der dominanten Tiere, deren Rivalen und der rangniederen Tiere bestimmen läßt, auf Blutdruck, Plasmakortikosteron und die adrenalin- bzw. noradrenalinsynthetisierenden Enzyme des Nebennierenmarks untersuchten Henry und Mitarbeiter (1971) in einem für die stabile Hierarchiebildung geeigneten Käfigsystem zu verschiedenen Zeiten der Hierarchiebildung. In den ersten Wochen der intensiven sozialen Auseinandersetzungen wiesen die rangniederen Tiere höhere Kortikosteronwerte auf als die dominanten Tiere, bei denen jedoch das noradrenalinbildende Enzym im Nebennierenmark erhöht war. Die Adrenalinbildung stieg bei allen Tieren zunächst an, fiel jedoch nach Ausbildung der sozialen Rangordnung bei rangniederen Tieren im Gegensatz zu den dominanten wieder ab. Der Blutdruck der – in dieser Sozietät gut angepaßten – rangniederen Tiere stieg jedoch bis auf 150 mm Hg.

Der Verlust der Position in der sozialen Rangordnung führt zu einem Verhalten von „Rückzug" und „Depression". Dabei findet sich eine Erhöhung des Kortikosteronspiegels, Kämpfe um die Position in der sozialen Hierarchie führen dagegen zur Abwehrreaktion mit erhöhtem Katecholaminspiegel.

Wurden dominante Tiere in eine fremde Kolonie gesetzt und mußten sie sich jetzt wie rangniedere Tiere verhalten, die Kämpfe vermieden, so stiegen die Kortikosteronwerte an und das Enzym der Adrenalinsynthese sank bei weiter stark erhöhten Werten des noradrenalinsynthetisierenden Enzyms. Gleichzeitig zeigten die Blutdruckwerte jetzt einen besonders starken Anstieg bis auf 200 mm Hg (vgl. Abb. 41.3–7).

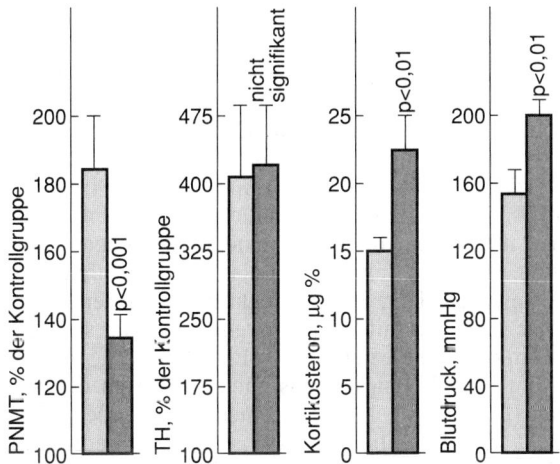

Abb. 41.3–7. Physiologische Auswirkungen des Verlustes der dominanten Stellung in der sozialen Rangordnung auf den Gehalt der adrenalin- und noradrenalinsynthetisierenden Enzyme PNMT (Phenylethanolamin-N-methyltransferase) und TH (Tyrosinhydroxylase) in den Nebennieren, auf den Plasmakortikosteroidspiegel und den Blutdruck bei Mäusen. Die offenen Säulen stellen diese Parameter bei normalen dominanten Kontrolltieren dar; die geschlossenen Säulen geben diese Größe bei ehemals dominanten Tieren wieder, die in eine fremde Mäusekolonie gesetzt wurden (nach Henry et al., 1971).

Das Bedeutsame dieser Ergebnisse liegt u. a. in der Möglichkeit, verschiedenartige emotionale Reaktionen zu differenzieren, die mit spezifischen neurohumoralen Reaktionen und Blutdruckanstiegen einhergehen, und die alle unter dem Generalnenner „Aggression" zusammengefaßt werden können: Zu Beginn des sozialen Wettbewerbs in der Mäusekolonie herrschen aggressive, mit Kampf und Flucht verbundene Verhaltensmuster vor, die mit einem Überwiegen der sympathischen Abwehrreaktion und Stimulation des Nebennierenmarks einhergehen. Mit zunehmendem Alter kämpfen die Mäuse weniger. Jetzt gewinnt das Vermeiden von Kämpfen mit einer Zunahme der Nebennierenrindenaktivität an Bedeutung.

Tierexperimentelle Untersuchungen von Dworkin und Mitarbeitern (1979) haben einen wichtigen Hinweis dafür geliefert, daß ein akuter Blutdruckanstieg einen biologisch selbstverstärkenden Charakter besitzen kann, daß er gewissermaßen eine Schutzfunktion bei der Wahrnehmung aversiver Umweltreize darstellt. Dieser Einfluß wird über die Barorezeptoren vermittelt. Die untersuchten Ratten versuchten seltener Schockeffekte zu beenden oder zu vermeiden, wenn ihr Blutdruck pharmakologisch durch Infusionen erhöht war. Operative Unterbrechungen des Barorezeptorenreflexes verhinderten diesen Effekt. Dieser Mechanismus bedeutet, daß unangenehme Umwelteinflüsse vermutlich weniger störend wirken, wenn der Blutdruck erhöht ist. Viele Untersuchungen belegen, daß Tiere – ebenso wie Menschen – lernen können, verschiedene Kreislauffunktionen (einschließlich Blutdruck) zu beeinflussen (Engel et al., 1981; Glasgow et al., 1982; Engel et al., 1983).

41.3.5.4 Pathophysiologische Grundlagen und psychophysiologische Untersuchungen am Menschen

Obwohl am Menschen so eingreifende Untersuchungen, wie sie an Tieren durchgeführt werden, nicht möglich sind, ist unser Wissen, insbesondere über die kardiovaskulären Reaktionen in emotional belastenden Situationen auch beim Menschen in letzter Zeit bereichert worden. Brod und Mitarbeiter (1959) untersuchten mit invasiven Methoden ein für die Abwehrreaktionen typisches hämodynamisches Muster:

Unter Zeitdruck und Belästigung durch ein tickendes Metronom traten Erhöhungen des arteriellen Drucks, Beschleunigung der Herzfrequenz, vermehrtes Herzminutenvolumen und stärkere Durchblutung der Muskeln auf, während die Durchblutung des Gastrointestinaltraktes, der Nieren und der Haut abnahm. Andere Untersucher bestätigten die Bedeutung dieses grundlegenden hämodynamischen Reaktionsmusters und wiesen zum Teil ebenfalls eine erhöhte Streßreagibilität bei Hypertonikern in verschiedenen Testsituationen nach (Richter-Heinrich et al., 1975; Schmidt, 1980, 1982, 1983; Schulte et al., 1981). Andere Autoren konnten zeigen, daß Hypertoniker bereits in Situationen zu Blutdruckanstiegen neigen, die bei Normotonikern noch keine Blutdruckänderungen hervorrufen (Hodapp und Weyer, 1982). Engel und Bickford (1961) untersuchten die Spezifität der Stimulusreaktionsmuster bzw. die intraindividuellen Reaktionsmuster bei Normotonikern und Hypertonikern (vgl. Kap. 8). Sie fanden, daß eine Hyperreaktivität des systolischen Blutdrucks für Hypertoniker weitgehend spezifisch und unabhängig vom Ausgangsstimulus ist.

Telemetrische Messungen unter Alltagsbedingungen sowie kontinuierliche Langzeitmessungen bei Bettruhe im Krankenhaus weisen auf den Zusammenhang vor allem einer höheren systolischen Blutdruckvariabilität bei höheren Durchschnittswerten hin (Krönig, 1976; Schmidt et al., 1984). Diese Untersuchungen stehen in Übereinstimmung mit der 1967 von v. Eiff formulierten Hypothese, daß die essentielle Hypertonie eine Krankheit sei, die durch eine angeborene Hyperreaktivität des hypothalamischen sympathischen Zentrums charakterisiert ist, wobei Persönlichkeit und Umwelt für die Manifestation des Hochdrucks und periphere Mechanismen für die Fixierung des erhöhten Blutdrucks eine Rolle spielen (v. Eiff et al., 1967 und 1977). Sie legen ebenso den Vergleich mit den tierexperimentellen Befunden nahe, bei denen größere Reaktionsbereitschaft des Blutdrucks gegenüber Umweltreizen bei hypertensiven Ratten von genetischen Faktoren abhängt. Allerdings kann die Blutdruckhyperreaktivität im Laufe der Hypertonieentwicklung auch allein Ausdruck geänderter morphologischer Verhältnisse in den Widerstandsgefäßen sein. Folkows Theorie der strukturellen Autoregulation (1975) macht die enge Verknüpfung von funktionellen und strukturellen Veränderungen am Gefäßsystem deutlich: Langanhaltende und auch häufig situativ ausgelöste Blutdruckerhöhungen werden in den Arteriolen, den Widerstandsgefäßen aufgefangen und führen hier zu einer allmählich zunehmenden Hypertrophie der glatten Muskulatur in der Media. Die Muskelmasse der Widerstandsgefäße nimmt zu; das Lumen der Arteriolen wird im Verhältnis zur jetzt dickeren Wandstärke kleiner. Das vergrößerte Wand/Lumen-Verhältnis ist dann die strukturelle Grundlage für den erhöhten Widerstand bei einer voll ausgebildeten essentiellen Hypertonie.

Die strukturelle Autoregulation wird als normale physiologische Adaptation des Gefäßsystems auf Druckimpulse angesehen; sie spielt eine wichtige Rolle bei jeder Hypertonieform, unabhängig von ihrer Genese. Schon in einem sehr frühen Hypertoniestadium, wie z. B. bei jungen Wehrpflichtigen mit grenzwertig erhöhtem Blutdruck, sind entsprechende Veränderungen im Gefäßsystem nachweisbar (Sannerstedt et al., 1976).

Die Vergrößerung des Wand/Lumen-Verhältnisses bewirkt eine Veränderung in der Reagibilität des Gefäßsystems, indem sie die reaktive Widerstandszunahme bei Vasokonstriktion enorm verstärkt. Danach wird ein gleicher nervaler oder humoraler pressorischer Reiz einen um so stärkeren Blutdruckanstieg hervorrufen, je größer das Wand/Lumen-Verhältnis der Widerstandsgefäße ist. Damit leistet ein peripherer Mechanismus, nämlich die morphologischen Veränderungen an den Widerstandsgefäßen, bereits im Frühstadium der Hypertonieentwicklung einen wichtigen Beitrag zur kardiovaskulären Reagibilität; diese wird dann bestimmt durch das Zusammenspiel von zentralen Mechanismen, wie dem Ausmaß der Veränderungen der sympathischen und vagalen Aktivität, und von peripheren Mechanismen, wie dem Wand/Lumen-Verhältnis der Widerstandsgefäße.

Der systemtheoretische Ansatz Guytons (1975) macht deutlich, daß bei dem komplexen physiologischen Geschehen der Blutdruckregulation Störungsmöglichkeiten an vielen Stellen auftreten können. Dabei betont Guyton die zentrale Stellung der Niere als Langzeitbarostat; die Niere gleicht kurzfristige Druckerhöhungen durch die Druckdiurese wieder aus. Nur wenn die Ausscheidungsfunktion der Niere auf ein höheres Blutdruckniveau eingestellt ist, kann der höhere Blutdruck auch aufrechterhalten werden. Das entspricht dem Phänomen des „Resettings" der als Kurzzeitbarostat wirkenden Barorezeptoren bei der Hypertonie. Das „Resetting" des Langzeitbarostaten Niere ist eine notwendige Voraussetzung, wenn es zur Hypertonieentwicklung kommen soll.

Tierexperimentelle Untersuchungen von Folkow und Mitarbeitern (1977) legen nahe, daß dem Phänomen des „Resettings" der Nieren strukturelle Gefäßveränderungen zunächst der präglomerulären Arteriolen zugrunde liegen, im fortgeschrittenen Stadium der Hypertonie lassen sich Strukturveränderungen jedoch auch im postglomerulären Bereich nachweisen. Eine wichtige Frage in diesem Zusammenhang ist, inwieweit situative Belastungen zu einem „Resetting" der Druckdiurese der Nieren führen können; nur

dann kann es zu einer anhaltenden Blutdruckerhöhung kommen, andernfalls würde der Volumenverlust den Blutdruck wieder senken. Neuere Untersuchungen bei Mensch und Tier haben inzwischen bestätigt, daß unter psychischem Streß die Niere weniger Natrium ausscheidet und damit Flüssigkeit im Körper zurückhalten kann (Anderson, 1984; Obrist, 1984). Beim Menschen sind hiervon vor allem Personen mit hohem Risiko für eine Hypertonieentwicklung, d. h. mit familiärer Belastung betroffen. Das Ausmaß der Herzfrequenzreaktionen während mentaler Aufgaben mit kompetitivem Charakter steht dabei in direkter Beziehung zur Einschränkung der Natrium- und Flüssigkeitsausscheidung durch die Nieren (Light et al., 1983). In einer umfassenden Übersicht der Literatur seit 1932 kommt Fredrickson (1986) zu dem Ergebnis, daß sich bei Hypertoniepatienten im Vergleich zu Normotonikern meist eine Blutdruckhyperreaktivität findet, die bestimmte Aufgaben betrifft und bei anderen nicht nachweisbar ist. Die Interkorrelationen der Blutdruckreaktionen zwischen verschiedenen Stimuli sind gering. Die Untersuchungen von Fredrickson zeigen weiter, daß diese Hyperreaktivität der Hypertoniker spezifisch das kardiovaskuläre System betrifft und nicht Ausdruck einer Aktivierung des gesamten sympathischen Nervensystems ist.

In einer Reihe von Untersuchungen wurde geprüft, inwieweit eine Hyperreagibilität im Cold-Pressure-Test einen Risikofaktor für die zukünftige Entwicklung einer Hypertonie darstellt – ein Verdacht der bislang nicht ausreichend belegt werden konnte (Thomas und Duszynski, 1982). Eine Studie mit gegenteiligen Ergebnissen (Wood et al., 1984) wies methodische Probleme auf (Julius et al., 1986), so daß es nicht möglich war abzuschätzen, inwieweit ein unterschiedlicher Ausgangsblutdruck in Ruhe und nicht die Hyperreagibilität für die Hypertonieentwicklung verantwortlich war. Verstärkte kardiovaskuläre Reaktionen auf einen Rechentest sind bei jugendlichen Grenzwerthypertonikern mit der schnelleren Entwicklung einer Hypertonie verknüpft (Falkner et al., 1981). Überhöhte Blutdruckreaktionen unter Ergometerbelastung sind auch bei Normotonikern (Franz et al., 1981) und Herzinfarktpatienten (Tammen et al., 1984) von prognostischer Bedeutung für eine zukünftige Hypertonieentwicklung.

Die Blutdruckhyperreagibilität auf mentale Belastung bei Hypertonikern wurde zuerst von Brod und Mitarbeitern (1959) mit Hilfe eines Rechentestes unter Zeitdruck demonstriert. Wenn auch die von Folkow (1977) beschriebenen sekundären strukturellen Gefäßveränderungen, die sog. strukturelle Autoregulation, für eine kardiovaskuläre Hyperreaktivität bei der Hypertonie verantwortlich sein können, müßten alle standardisierten Reize, die zu Blutdruckanstiegen führen, entsprechend starke Reaktionen auslösen. Bei der milden Hypertonie und bei der Grenzwerthypertonie finden sich verstärkte kardiovaskuläre Reaktionen bei mentalem Streß (Weder und Julius, 1985; Fredrickson, 1986), nicht jedoch beim Cold-Pressure-Test (Eliasson et al., 1983), bei statischer (Sannerstedt und Julius, 1972) und dynamischer körperlicher Belastung (Julius und Conway, 1968), bei orthostatischer Belastung am Kipptisch (Eliasson et al., 1983) und akuter Expansion des Blutvolumens (Julius et al., 1971). Deswegen erscheint bei diesen Patientengruppen hierfür nicht eine allgemeine Veränderung der Kreislaufregulation durch die strukturelle Autoregulation, sondern vielmehr eine Hyperreaktivität nur für spezifische Situationen verantwortlich. Auch tierexperimentelle Untersuchungen mit spontan hypertensiven Ratten haben verstärkte Blutdruckanstiege nur bei spezifischen Verhaltensweisen wie beim Trinken nachgewiesen, nicht jedoch bei Nahrungsaufnahme und Explorationsverhalten (Reis und LeDoux, 1987).

Groen und Mitarbeiter (1982) und Schäfer (1976) untersuchten mit invasiven Methoden das Verhalten von Blutdruck und Herzminutenvolumen während physischer Arbeit am Fahrradergometer und während der Besprechung emotional erregender Probleme. In beiden Situationen steigen sowohl der systolische wie der diastolische Blutdruck etwa gleich an. Während aber bei physischer Belastung diese Blutdruckerhöhung bei allen untersuchten Personen einheitlich durch eine Erhöhung des Herzminutenvolumens bei Absinken des totalen peripheren Widerstandes bedingt ist, sieht man bei psychischer Belastung zwei Gruppen. Bei der ersten steigt das Herzminutenvolumen ähnlich wie bei physischer Belastung, jedoch in geringerem Ausmaß, bei Absinken oder Gleichbleiben des Widerstandes. Bei der zweiten Gruppe steigt der Widerstand bei uneinheitlichen Reaktionen des Herzminutenvolumens, das ansteigen, gleichbleiben oder sogar absinken kann. Bedeutsam sind die Resultate dieser letzteren Gruppe unter dem Gesichtspunkt, daß ihr Kreislaufreaktionsmuster unter psychischer Belastung weitgehend dem der essentiellen Hypertonie entspricht. Durchschnittlich kommt es hier bei emotionaler Belastung zu einem Blutdruckanstieg, der im Gegensatz zu dem bei körperlicher Belastung durch veränderte Anteile von Herzminutenvolumen und totalem peripherem Widerstand bedingt ist: Bei vergleichbarem Blutdruckanstieg ist der Anstieg des Herzminutenvolumens im Mittel geringer bei gleichzeitig leicht erhöhtem totalem peripherem Widerstand. Das Fehlen der Möglichkeit zu muskulärer Aktivität macht emotionale Belastungen vermutlich gerade deswegen zu einem wichtigen Faktor in der Hypertonieentwicklung (Charvat et al., 1964).

Die Grenzwerthypertonie ist ein Prädiktor für das Auftreten einer stabilen Hypertonie (Julius und Schork, 1971), und Grenzwerthypertoniker stellen eine besonders interessante Gruppe zur Untersuchung psychophysiologischer Zusammenhänge bei der Hypertonieentstehung dar. Grenzwerthypertoniker sind durch eine veränderte Integration der autonomen kardialen Kontrolle mit einer vermehrten sympathischen kardialen Stimulation und einer verminderten vagalen Hemmung gekennzeichnet (Julius und Esler, 1975). Diese Veränderungen sind auch charakteristisch für die hypothalamische Abwehrreaktion, die unter psychischem Streß das emotionale Verhalten in Alarmbereitschaft versetzt. Diese Patienten schreiben

sich spezifische Persönlichkeitsmerkmale wie Submissivität, unterdrückten Ärger und Feindseligkeit zu (Julius, 1981; Esler et al., 1976), die in Zusammenhang mit der erhöhten kardiovaskulären Reaktivität in spezifischen Situationen stehen könnten (Eliasson et al., 1983). Julius et al. (1983) vermuten, daß derartige Persönlichkeits- und Verhaltensmerkmale einen beinahe permanenten Zustand einer „vermehrten Alarmbereitschaft" hervorrufen, der für die Veränderung der zentral integrierten autonomen Kontrolle des kardiovaskulären Systems verantwortlich ist. Folge ist ein vermehrter sympathischer Einfluß auf Herz und Gefäßsystem mit erhöhtem Blutdruck und Herzminutenvolumen. Die weitere Hypertonieentwicklung vollzieht sich über die strukturelle Autoregulation mit einer Hypertrophie der Mediamuskulatur der Widerstandsgefäße, einem Anstieg des peripheren Gefäßwiderstands und einer Abnahme des Herzminutenvolumens (Folkow, 1982). Diese Theorie würde demnach eher der sog. Reaktivitätshypothese nach dem sog. „prevailing state"-Modell („Modell des anhaltenden Zustands" nach Manuck und Krantz, 1984; vgl. auch Kap. 41.1) folgen, das besagt, daß eine erhöhte kardiovaskuläre Reaktivität gegenüber Laborstressoren den physiologischen Zustand kennzeichnet, der unter Alltagsbedingungen im Wachzustand vorherrscht.

Zwillingsuntersuchungen weisen auf eine genetische Komponente der kardiovaskulären Reaktionsbereitschaft hin (Rose et al., 1986). Normotone Kinder hypertoner Eltern, die wahrscheinlich genetisch für eine Hypertonie prädisponiert sind, reagieren auf Rechenaufgaben (Falkner et al., 1979) und Schockvermeidungstests (Light und Obrist, 1980) mit verstärkten kardiovaskulären Reaktionen. Matthews und Rakaczky (1986) haben kürzlich eine Zusammenstellung der Literatur vorgelegt, die die kardiovaskulären und neuroendokrinen Reaktionen auf psychische und physische Stressoren von Personen mit und ohne familiärer Hypertonie- oder KHK-Belastung untersucht. In 19 von 26 Studien war eine familiäre Hypertoniebelastung mit signifikant stärkeren Blutdruck- und/oder Herzfrequenzreaktionen oder einer erhöhten Plasmareninaktivität assoziiert. Studien, die einen derartigen Zusammenhang nicht nachweisen konnten, lagen entweder sehr lange zurück (vor 1943) und machten keine genauen Angaben über die angewandte Methodik, verwendeten ausschließlich den Cold-Pressure-Test oder konnten keine sicheren Angaben darüber machen, ob die Eltern normoton oder hyperton waren. Vier Studien zeigten eine Wechselwirkung zwischen familiärer Belastung und Persönlichkeitsmerkmalen.

Diese Befunde legen einen Zusammenhang zwischen einer psychophysiologischen kardiovaskulären Hyperreaktivität und einer Hypertonieentwicklung nahe, der allerdings nicht ursächlich sein muß. Einige Autoren vertreten die Hypothese, daß von der Streßbelastung und vom Verhalten abhängige wiederholte Episoden situativer Blutdruckanstiege unter Alltagsbedingungen in Abhängigkeit von Ausmaß und Häufigkeit derartiger Reaktionen zu kardiovaskulären und renalen Adaptationen führen, in deren Gefolge sich eine stabile Hypertonie entwickelt (Folkow, 1982).

Die bisher besprochenen Befunde lassen sich unter dem Gesichtspunkt der Cannonschen These (1953) zusammenfassen, nach der eine Erhöhung des Blutdrucks als „Bereitstellung" (emergency state) eines zusätzlichen Energievorrates für erwartete motorische Auseinandersetzung mit der Umgebung interpretiert wird. Die Blutdruckerhöhung hat psychophysiologisch aber wahrscheinlich nicht nur die Aufgabe, Energie bereitzustellen, sondern darüber hinaus die Funktion einer physiologischen Rückmeldung und Beeinflussung psychischer Funktionen über die Barorezeptoren. Hierfür sprechen die oben zitierten tierexperimentellen Befunde von Dworkin und Mitarbeitern (1979) sowie psychophysiologische Untersuchungen beim Menschen von Lacey (1967 und 1970). Lacey interpretiert die Erhöhung von Blutdruck und Pulsfrequenz in Situationen, die geistige Aktivität zur Lösung von Problemen in Anspruch nehmen, als Errichtung einer „Stimulusbarriere", die das Individuum vor einer Überflutung mit unbewältigten Informationen aus der Umgebung schützt. Dieser Mechanismus könnte die verschiedentlich berichtete Veränderung der Wahrnehmung aversiver Reize bei Hypertonikern erklären und gleichzeitig zur Aufrechterhaltung der Hypertonie bei ungelösten Problemsituationen beitragen.

Die unterschiedlichen funktionellen Änderungen im Frühstadium der Hypertonieentwicklung münden schließlich in die gemeinsame pathophysiologische Endstrecke der „strukturellen Autoregulation".

Durch sie können situative Blutdruckanstiege eine pathogene Bedeutung für eine Hypertonieentwicklung bekommen, unabhängig von den Entstehungsbedingungen der jeweiligen Hypertonieform. Hierbei kann sich ein Circulus vitiosus entwickeln, in dem eine stärkere Mediahypertrophie zu stärkeren Druckanstiegen und zu einer weiteren Zunahme des Widerstands führt. Daß dieser Circulus vitiosus für die Entwicklung der menschlichen essentiellen Hypertonie bedeutsam ist, zeigt eine Untersuchung aus Wales: Diejenigen Individuen einer Population, bei denen die höchsten Ausgangswerte des Blutdrucks zu Beginn der Studie gemessen wurden, entwickelten im Laufe der Jahre den höchsten Blutdruckanstieg (Miall und Lovell, 1967). Querschnittsuntersuchungen (Lund-Johansen, 1967) und die wenigen bis heute vorliegenden Längsschnittuntersuchungen (Lund-Johansen, 1976) bestätigen, daß das Folkowsche Modell der strukturellen Autoregulation auch für den Menschen gilt.

Danach kommt es mit zunehmendem Alter bei (Grenzwert-)Hypertonikern zu einem Abfall des Herzminutenvolumens und zu einem Anstieg des Widerstands. Normotoniker zeigen dagegen nur unwesentliche hämodynamische Veränderungen im Laufe ihres Lebens. Die zunehmenden strukturellen adaptiven Veränderungen in den Widerstandsgefäßen sind im Frühstadium noch reversibel, wenn die Druckerhöhungen ausbleiben; eine frühzeitige antihyperten-

sive Behandlung kann offensichtlich noch einen Rückbildungsprozeß ermöglichen (Hansson und Sivertson, 1977). Bleibt der Druck jedoch weiter erhöht, werden Ablagerungen von Mukopolysacchariden und Eiweißstoffen in der glatten Muskulatur der Media schließlich von Bindegewebe ersetzt: Diese Veränderungen sind irreversibel (Wolinsky, 1972), selbst wenn die pressorischen Reize nicht mehr auftreten. Die genaue Erforschung und Kenntnis der pathophysiologischen Zusammenhänge und Mechanismen legt also die Bedeutung psychophysiologischer Faktoren bei der Entwicklung der essentiellen Hypertonie nahe.

Zunehmende Bedeutung gewinnt das Blutdruck-Monitoring unter Alltagsbedingungen für Diagnostik und Behandlung der milden und der Grenzwerthypertonie. Nachdem die Höhe des einmal gemessenen Gelegenheitsblutdrucks im Sinne einer Dosis-Antwort-Beziehung mit dem zukünftigen kardiovaskulären Risiko verknüpft ist, der Blutdruck aber nachweislich eine außerordentlich variable Größe darstellt (Littler et al., 1972; Schmidt, 1985), erscheint es plausibel, daß wiederholte Messungen das Risiko genauer wiedergeben. Dies wurde entsprechend auch in epidemiologischen Studien bestätigt. So sagten in der Charlottesville-Survey acht Blutdruckmessungen, die sich auf vier Untersuchungstermine verteilten, die zukünftige kardiovaskuläre Morbidität besser voraus als eine einzige Messung zum ersten Untersuchungszeitpunkt (Carey et al., 1976). Untersuchungen mit dem ambulanten Blutdruck-Monitoring haben die Korrelation zwischen einzelnen klinischen Blutdruckmessungen und dem durchschnittlichen Blutdruckniveau innerhalb von 24 Stunden bestimmt; sie beträgt etwa 0,6, was bedeutet, daß eine einzelne klinische Blutdruckmessung rund 36% der Varianz des 24-Stunden-Blutdrucks erklärt (Pickering et al., 1985). Hohe Korrelationen zwischen Variabilitätsmaßen an zwei aufeinander folgenden Tagen weisen auf die Stabilität individueller Unterschiede vor allem der systolischen Blutdruckvariabilität ($r = 0,75$ bis $0,97$) hin. Die kardiovaskuläre Reaktionsbereitschaft ist ein relativ stabiles individuelles Merkmal (Giordani et al., 1981). Erhöhte kardiovaskuläre Reaktionen unter Alltagsbedingungen wirken sich auf eine Erhöhung des kardiovaskulären Risikos aus.

In Übereinstimmung mit diesen Überlegungen und Befunden stehen Untersuchungen, die zeigen, daß Blutdruckmessungen unter Alltagsbedingungen im Vergleich zu einzelnen klinischen Messungen bessere Prädiktoren für die kardiovaskuläre Morbidität sind. So korrelierten in einer Studie von Sokolow und Mitarbeitern (1966) die unter Alltagsbedingungen gemessenen systolischen und diastolischen Blutdruckwerte besser mit dem Schweregrad hypertensiver Komplikationen (linksventrikulärer Hypertrophie im EKG, Herzgröße im Röntgenbild und Augenhintergrundsveränderungen) als klinische Gelegenheitsblutdruckmessungen. Drei neuere Studien verwandten die methodisch überlegenere Echokardiographie zur Bestimmung einer linksventrikulären Hypertrophie: In allen drei Studien fanden sich im Vergleich zum klini-

Tabelle 41.3–2. Korrelationen zwischen linksventrikulärer Hypertrophie und Gelegenheitsblutdruckmessung bzw. 24-Stunden-Blutdruck beim ambulanten Blutdruck-Monitoring (nach Pickering et al., 1985)

Autor	Zahl der Patienten	24-Std.-SBD*	Gelegenheits-messung (SBD)*	24-Std.-DBD*	Gelegenheits-messung (DBD)*
Rowlands et al., 1981	50	0,60	0,51	0,43	0,30
Drayer et al., 1984	12	0,81	0,55	0,56	0,10
Devereux et al., 1983	100	0,50	0,24	0,39	0,20

* SBD = systolischer Blutdruck; DBD = diastolischer Blutdruck

schen Gelegenheitsblutdruck höhere Korrelationen beim 24-Stunden-Blutdruck (Tab. 41.3–2).

Aus diesen Untersuchungen ergeben sich viele Hinweise dafür, daß verstärkte kardiovaskuläre Reaktionen, die sowohl unter Laboratoriums- als auch Alltagsbedingungen erfaßt werden können, mit einem erhöhten kardiovaskulären Risiko verknüpft sind. Hieraus läßt sich die Frage ableiten, ob spezifische Verhaltensweisen und Einflüsse der sozialen Umwelt identifiziert werden können, die zu derartigen Hyperreaktionen führen. Dieser Frage wurde in einer Anzahl von Studien, insbesondere auch in bezug auf das Typ-A-Verhalten und seiner Komponenten sowie anderer psychologischer Merkmale nachgegangen (vgl. auch Kap. 41.1).

41.3.6 Ein psychosomatisches Modell: Der Situationskreis

41.3.6.1 Was ist eine Situation?

In den vorhergehenden Abschnitten wurden die Faktoren aufgezählt, die – nach unseren heutigen Kenntnissen – für Entstehung und Verlauf einer essentiellen Hypertonie verantwortlich sind: Auf der somatischen Ebene vor allem genetische Faktoren, auf der psychischen Ebene konflikthafte Einstellungen zu Aggression, Leistung und Autorität, auf der sozialen Ebene Anpassungsprobleme. Ein psychosomatisches Modell, das dem Arzt die Möglichkeit gibt, diese verschiedenen Aspekte bei seinen Patienten in Rechnung zu stellen und die für Diagnostik und Therapie erforderlichen Konsequenzen zu ziehen, muß daher das Ineinandergreifen somatischer, psychischer und sozialer Probleme beschreiben können. Dafür bietet sich an, von dem Phänomen der Situationshypertonie auszugehen.

Von Uexküll und Wick (1962) haben diesen Ausdruck geprägt, als sie feststellten, daß gleiche sensorische Reize, die in bestimmten Situationen (z.B. während eines Examens) zu Blutdrucksteigerungen führten, in anderen Situationen keine derartigen Folgen hatten. Sie analysierten daraufhin die Komponenten, die an der Konstellation des Beziehungsgeflechtes beteiligt waren, das wir „Situation" nennen, und fanden fünf verschiedene Anteile:

– **Physikalische** Anteile, die als Außenweltreize auf die Sinnesorgane treffen (z.B. Luftwellen, die bei einem Gespräch von dem Ohr aufgenommen werden, oder Lichtwellen, die unsere Retina treffen, etc.);
– **physiologische** Anteile, welche Außenweltreize in (subjektiv erlebte) Sinnesreize (z.B. Töne oder Licht- und Farbempfindungen) transponieren;
– **soziale** Anteile, die nach Art eines Codes Sinneszeichen in allgemein verständliche Signale verschlüsseln (z.B. Töne in Worte und Sätze einer Sprache, die vom Empfänger verstanden wird, oder Farb-, Form- und Tasteindrücke in uns geläufige Wahrnehmungen);
– **sozialpsychologische** Anteile, die nach Art eines Sub-Codes Worten und Sätzen oder anderen Wahrnehmungen außer ihrer für alle Menschen der gleichen Sprachfamilie oder Kultur verständlichen Bedeutung eine spezielle Bedeutung als Stichwort einer bestimmten Rolle (z.B. eines Berufes) erteilen;
– außer diesen vier Bestimmungsstücken einer Situation fanden sie noch einen **psychischen** Anteil: Die emotionale Verfassung des Empfängers verschlüsselt die empfangenen Signale für dessen jeweilige Erwartungen, Wünsche oder Befürchtungen. Sie prägt ihnen nach Art eines individuellen Codes zu ihrer sozialen und sozialpsychologischen Bedeutung noch eine individuelle Bedeutung auf.

Von diesen fünf Anteilen des Beziehungsgeflechtes zwischen uns und unserer Umgebung, das wir als „Situation" bezeichnen, hatte die emotionale Komponente für die Blutdruckreaktionen das größte Gewicht. Wir haben ausgeführt, daß sie jeder Situation eine Qualität verleiht, in der die biographische Vergangenheit des Empfängers als drängende Gegenwart erlebt wird.

41.3.6.2 Der Situationskreis

Die Feststellung, daß situative Faktoren zu Blutdruckerhöhung führen können, zeigt, daß das Regelgeschehen innerhalb des Organismus als Teilglied eines Regelgeschehens aufgefaßt werden muß, das den Organismus und seine Umgebung umgreift und die Umgebungsereignisse nach ihrer Bedeutung für das Individuum „mißt". Man kann sich dann vorstellen, daß es im Organismus in Abhängigkeit von der jeweiligen Umgebungssituation zu Sollwertverstellungen in dem Regelsystem für den Blutdruck kommt. Die resultierenden Blutdruckerhöhungen können kurzfristig oder langanhaltend sein.

Um diesen Zusammenhang zu verstehen, muß man sich vor Augen halten, daß ein solches „Messen" die Verschlüsselung der Außenweltreize unter physiologischen, sozialen, sozialpsychologischen sowie den emotionalen Gesichtspunkten des Empfängers voraussetzt und der Aufgabe dient, die jeweils vorgefundene Umgebung als Problem zu interpretieren, für dessen Lösung wir auf Deutungs- und Verhaltensprogramme zurückgreifen können, die wir in der Vergangenheit erlernt haben. Auf diese Weise baut jeder seine individuelle Wirklichkeit aus Situationen auf, welche die von ihm wahrgenommene Umgebung immer wieder als Problem deuten, das er auf einer sozialen, sozialpsychologischen und emotionalen Ebene lösen muß.

Für den Patienten mit essentieller Hypertonie haben aggressive Impulse und deren Blockierung durch ein Ideal größter Selbstbeherrschung eine besondere Bedeutung (Boss, 1949). Die Frage, wie die individuelle Wirklichkeit eines Menschen beschaffen ist, der von Wut und Haß und gleichzeitig von abgrundtiefer Angst erfüllt ist, diese Impulse könnten ihn überschwemmen, wurde ausgeklammert, weil es bisher noch keine systematischen Untersuchungen dieser Art gibt.

Es gibt jedoch die Beschreibung eines Selbstversuches, die uns bei dieser Aufgabe weiterhelfen kann: Goethe schildert die Veränderung seiner individuellen Wirklichkeit und seines Körpererlebens in einer Situation, der er sich anläßlich der Kanonade von Valmy am 19. September 1792 freiwillig aussetzte (Goethe, 1792):

„Ich hatte so viel vom Kanonenfieber gehört und wünschte zu wissen, wie es eigentlich damit beschaffen sei. Langeweile und ein Geist, den jede Gefahr zur Kühnheit, ja zur Verwegenheit aufruft, verleiteten mich, ganz gelassen nach dem Vorwerk La Lune hinaufzureiten ... Ich war nun vollkommen in die Region gelangt, wo die Kugeln hinüberspielten ... Unter diesen Umständen konnte ich jedoch bald bemerken, daß etwas Ungewöhnliches mir vorgehe; ich achtete genau darauf, und doch würde sich die Empfindung nur vergleichsweise mitteilen lassen. Es schien, als wäre man an einem sehr heißen Orte und zugleich von derselben Hitze völlig durchdrungen, so daß man sich mit demselben Element, in welchem man sich befindet, vollkommen gleich fühlt. Die Augen verlieren nichts an ihrer Stärke noch Deutlichkeit; aber es ist doch, als wenn die Welt einen gewissen braun-rötlichen Ton hätte, der den Zustand sowie die Gegenstände noch apprehensiver macht. Von Bewegungen des Blutes habe ich nichts bemerken können, sondern mir schien vielmehr alles in jener Glut verschlungen zu sein. Hieraus erhellet nun, in welchem Sinne man diesen Zustand ein Fieber nennen könnte. Bemerkenswert bleibt es indessen, daß jenes gräßlich Bängliche nur durch die Ohren zu uns gebracht wird; denn der Kanonendonner, das Heulen, Pfeifen und Schmettern der Kugeln durch die Luft ist doch eigentlich Ursache an diesen Empfindungen". Einige Zeilen weiter heißt es: „Es gehört übrigens dieser Zustand unter die am wenigsten wünschenswerten".

An dieser Schilderung sind folgende Punkte bemerkenswert:
– Die Mischung aus Aggressivität und kühler Berechnung.

Die Aggressivität war durch die Gefahr geweckt worden („ein Geist, den jede Gefahr zur Kühnheit, ja zur Verwegenheit aufruft"). Gleichzeitig war aggressives Verhalten durch ein Ideal äußerster Selbstbeherrschung blockiert, dessen Verwirklichung Goethe durch den Status des Nichtkombattanten möglich gemacht wurde. Er konnte sich „ganz gelassen" in die Region begeben, „wo die Kugeln hinüberspielten".
– Von besonderem Interesse ist die Schilderung der Veränderung seiner individuellen Wirklichkeit durch die Aufhebung der Grenzen zwischen Außenwelt und Körper: Goethe fühlte sich wie „an einem sehr heißen Orte und zugleich von derselben Hitze völlig durchdrungen, so daß man sich mit demselben Element, in welchem man sich befindet, vollkommen gleich fühlt" oder ihm „schien alles in jener Glut verschlungen zu sein". Dieser Zustand kann als Regression auf die Stufe der Stimmungen bezeichnet werden, auf welcher Körper und Umgebung noch ohne feste Grenzen ineinanderfließen und Motive (Programme) fehlen, um die Stimmung zu einer individuellen Wirklichkeit zu strukturieren, in der Gegenstände und Vorgänge dem aggressiven Drang konkrete Ziele bieten.

Hätte Goethe in der von ihm geschilderten Situation seinen Puls und seinen Blutdruck messen können, so hätte er zweifellos stark erhöhte Werte festgestellt. Wir haben bei essentiellen Hypertonikern beschrieben, daß sich ihre aggressive Stimmung nicht zu einer strukturierten Wirklichkeit differenzieren kann, weil die Handlungsmotive, die dafür erforderlich wären, so angstbesetzt sind, daß sie verdrängt bleiben (v. Uexküll, 1963). Der Unterschied zwischen der individuellen Wirklichkeit dieser Patienten und der, die Goethe erlebt, scheint lediglich der zu sein, daß Goethe als „Nichtkombattant" die Situation jederzeit beenden konnte, diese Möglichkeit den Hypertoniekranken jedoch verschlossen ist.

Diese in einer diffusen Stimmung aufgelöste Welt ohne feste Konturen und ohne Grenzen zwischen einem Ich und äußeren Gegenständen entspricht einer präverbalen Phase unserer Entwicklung. Sie ist daher mit den Begriffen unserer Wortsprache nur vergleichsweise zu beschreiben: „Das Ungewöhnliche", das mit ihm vorgeht, wird von Goethe als „Empfindung" bezeichnet, die sich nur „vergleichsweise mitteilen läßt".
– Bemerkenswert ist ferner die Tatsache, daß die raum-zeitliche Struktur der umgebenden Welt nicht in Mitleidenschaft gezogen ist. Goethe konnte sich trotz der Veränderung seiner Wirklichkeit weiter orientieren und blieb weiterhin Herr seiner Handlungen und Entschlüsse. Sein Ich war gewissermaßen gespalten: Ein Teil verschmolz mit der Stimmung des höllischen Feuers mit seiner Umgebung zu einem undifferenzierten Kontinuum; der andere Teil blieb zu kühler Beobachtung und Selbstbeherrschung befähigt. „Die Augen verlieren nichts von ihrer Stärke und Deutlichkeit". Entsprechend war auch die erlebte Wirklichkeit gespalten: Wie in einem brennenden Haus war der eine Teil in Feuer und Hitze aufgelöst, während der andere Teil, wie aus dem Feuer herausragende Mauern und Balken, seine feste Struktur behielt.
– Die Tatsache, daß die Wirklichkeit, die wir erleben, von der Phantasie aus Deutungen aufgebaut wird, welche an sich neutrale Sinneszeichen interpretieren, wird durch die Feststellung Goethes illustriert, es bleibe bemerkenswert, „daß jenes gräßlich Bängliche nur durch die Ohren zu uns gebracht wird". So sehr Goethes Deutungen auch der Realität seiner Umgebung entsprachen, so war es letztlich doch seine Phantasie, welche die akustischen Wahrnehmungen als „Heulen, Pfeifen und Schmettern der Kugeln durch die Luft" interpretierte, ehe sie zur „eigentlichen Ursache an diesen Empfindungen" werden konnten.

– Schließlich ist noch die Feststellung bemerkenswert, daß „dieser Zustand unter die am wenigsten wünschenswerten" gehört. Das macht es verständlich, daß Hypertoniker dazu neigen, Beobachtungen zu verleugnen, deren Inhalte aggressiver Art sind und die ihre prekäre Stimmung intensivieren würden. Da sie genug mit dem Feuer in sich selbst zu tun haben, vermeiden sie jeden Anlaß, der das Feuer noch schüren könnte.

Diese Vorsicht und die Tatsache, daß die raum-zeitliche Struktur der individuellen Wirklichkeit erhalten bleibt, erlaubt es den Hypertonikern, unauffällig für die Umgebung ihre Ziele zu verfolgen. Das bestärkt sie in dem allgemeinen Glauben, ihre Wirklichkeit würde sich in nichts von der ihrer Mitmenschen unterscheiden. So bemerken auch ihre Mitmenschen den Unterschied nicht, der zwischen ihrer individuellen Wirklichkeit und der der Hypertoniker besteht.

In jedem Fall ist entscheidend, ob die Programme, die uns in Form von Deutungs- und Verhaltensanweisungen zur Verfügung stehen, das Problem lösen können, mit dem die Situation uns konfrontiert. Ist das nicht der Fall, müssen wir ein vorhandenes Programm modifizieren oder ein neues entwickeln. Wir müssen, wie Piaget (1973) es genannt hat, Akkommodationsleistungen vollbringen, um unsere Umgebung als Situation assimilieren zu können, deren Problem lösbar ist.

Statt von Akkommodation sprechen wir gewöhnlich von Adaptation. Der Terminus „Adaptation" umschreibt also nur die Tatsache, daß wir die äußeren Faktoren unserer Umgebung ständig aufgrund innerer (physiologischer, sozialer, sozialpsychischer und emotionaler) Faktoren, die spezifische Bedürfnisse zum Ausdruck bringen, interpretieren, um jede Diskrepanz zwischen unseren Bedürfnissen und der Umgebung mit Hilfe unseres Verhaltens beseitigen zu können. Bei diesem Vorgehen testen wir gleichzeitig die Angemessenheit unserer Interpretation und unseres Verhaltens an dessen Resultaten. In Kapitel 1 „Wissenschaftstheorie und Psychosomatische Medizin" wurde das Modell des Situationskreises entwickelt, das anschaulich macht, wie der einzelne mit den für ihn (aufgrund seiner Bedürfnisse) relevanten Umgebungsfaktoren in einen großen Regelkreis zusammengeschlossen ist (vgl. Abb. 41.3–8).

Das große Regelsystem, das Individuum und Umgebung umfaßt, muß von dem kleinen Regelsystem innerhalb des Organismus unterschieden werden. Großes und kleines Regelsystem lassen sich nach den Vorstellungen der Systemtheorie als System und Subsystem einander zuordnen. Damit wird verständlich, daß sie zwei verschiedenen Integrationsebenen angehören, die zur Beschreibung ihrer Phänomene verschiedene Terminologien erfordern: Für Phänomene des kleinen Regelsystems ist die Sprache der Physiologie adäquat; die viel komplexeren Phänomene des großen Regelkreises erfordern dagegen eine verhaltensphysiologische und/oder psychologische Terminologie. Zwischen beiden Ebenen besteht ein Bedeutungssprung, der nur durch Bedeutungskoppelung überbrückt werden kann (vgl. Kap. 1).

Abb. 41.3–8. „Der Situationskreis" stellt dar, wie die Umgebung durch das Individuum (bzw. dessen innere Bedürfnisse) als Problemsituation interpretiert wird (rezeptorische Sphäre „Merken"). Dem entspricht eine Bedeutungserteilung, die auf der Stufe biologischer Bedürfnisbefriedigung automatisch ein Verhalten („Wirken") auslöst, das in der effektorischen Sphäre die Problemlösung herbeiführen soll. Dieses primärprozeßhaft ablaufende, zwanghafte Verhalten wird jedoch beim Menschen durch Zwischenschaltung der Phantasie modifiziert, in der Programme für Bedeutungsunterstellung vor der endgültigen Bedeutungserteilung (die dann das bedeutungsverwertende Verhalten in Gang setzt) durchgespielt und erprobt werden. Dadurch wird die Situation in der Phantasie experimentell (durch Probehandeln) vorstrukturiert.

Das Analoge gilt für die Ebene des Sozialen. Zwischen ihr und der Ebene des Psychischen besteht wieder ein Bedeutungssprung, dessen Überbrückung wiederum eine Bedeutungskoppelung (bzw. eine Übersetzung in eine andere Sprache) erfordert.

Übertragen wir dieses Modell auf das Problem der essentiellen Hypertonie, so läßt sich das kleine Regelsystem nach dem bekannten physiologischen Modell (Blutdruck als Regelgröße, Blutbahn als Regelstrecke, Pressorezeptoren als Fühler, Vasomotorenzentrum als Regler, Herz und Arteriolen als Stellglieder) beschreiben, in dem Abweichungen des Blutdrucks zum Sollwert zurückgeregelt werden. In diesem Modell wird jede Blutdruckgröße im Hinblick auf einen im Regler anzunehmenden Sollwert als „zu hoch" oder „zu niedrig" interpretiert.

Im großen Regelsystem, das wir als „Situationskreis" bezeichnen, würde das soziale, sozialpsychische und emotionale Erleben des Menschen die äußeren Faktoren der Umgebung im Hinblick auf seine Bedürfnisse (seine „Sollwerte") interpretieren und damit seine „Situation" bestimmen. Abweichungen von diesen Sollwerten würden durch das Verhalten des Individuums (das Stellwerk des großen Regelsystems) den Sollwerten angepaßt. In diesem Regelkreis bilden die Sinnesorgane die Fühler, die Situation die Regelstrecke, das Zentralnervensystem den Regler und die Gliedmaßen einschließlich Mimik und Sprache das Stellwerk. In dem Situationskreis kann z.B. ein Mensch mit aggressiven Triebbedürfnissen eine für den unbeteiligten Beobachter neutrale Umgebung als Herausforderung erleben (interpretieren). Damit entsteht für ihn eine Problemsituation, die durch sein aktives Verhalten – in diesem Fall Angriff oder Verteidigung – in eine Situation des gelösten Problems überführt werden muß. Die „inneren" oder psychischen Faktoren, die im Situationskreis Interpretation der Umgebungsfaktoren (= Erleben) und aktive Auseinandersetzung mit ihnen (= Verhalten) steuern, lassen sich, wie schon erwähnt, als „Programme" beschreiben, die teils genetisch ererbt, teils im Laufe des Lebens erworben (erlernt) wurden, d. h. sozial bestimmt sind. Die Adaptation an eine veränderte Umgebung gelingt, wenn Programme abgerufen werden können, die imstande sind, die Umgebungsfaktoren als eine Problemsituation zu interpretieren, für die Lösungsmöglichkeiten bereitstehen. Wenn derartige Programme jedoch nicht verfügbar sind und wenn es auch nicht gelingt, neue Programme aufzubauen oder verfügbare entsprechend zu modifizieren, entsteht eine Problemsituation, die nicht gelöst werden kann. Der spezifische Konflikt unterdrückter Aggression (Alexander, 1951), aber auch die unrealistische Einstellung zu Leistungszielen, lassen sich in diesem Modell als einander störende oder blockierende Programme beschreiben, die durch ihre widerspruchsvolle Interpretation der Umgebung und durch einander widersprechende Verhaltensweisen immer wieder unlösbare Problemsituationen konstellieren. Solche Konstellationen bezeichnet man mit dem Terminus „Streß". Sie entsprechen im Extremfall dem, was Engel und Schmale (1972) als „Zustand der Hilflosigkeit und Hoffnungslosigkeit" beschrieben haben, in dem sie ein typisches Merkmal für Situationen sehen, in denen die verschiedensten Krankheiten, u. a. auch Apoplexien auftreten können (Adler et al., 1971; Engel und Schmale, 1972).

Solche Extremsituationen können aber für gewöhnlich vermieden werden, weil Hypertoniker, wie wir gesehen haben, Kompromisse und Coping-Strategien gefunden haben, die ihnen erlauben, für sie bedrohliche oder aggressive Aspekte aus ihren individuellen Wirklichkeiten auszublenden. Auch der Mechanismus der „Stimulusbarriere" (Lacey und Lacey, 1970) spielt dafür eine Rolle.

Erst wenn diese Kompromiß- und Kompensationsmöglichkeiten überfordert werden, sei es, daß aggressive Tendenzen steigen oder daß Umgebungsereignisse außergewöhnliche Anforderungen stellen oder daß beides sich kombiniert, kommt es zu Situationen, deren Probleme durch keine Kompromisse mehr zu lösen sind. Dann können die Deutungs- und Handlungsentwürfe, mit denen die Phantasie die Situation vorstrukturiert, dem pragmatischen Realitätsprinzip nicht mehr genügen, d. h. ihre Vorhersagen und Vorerwartungen werden nicht mehr bestätigt, das von ihnen geleitete Verhalten wird nicht belohnt.

In solchen Fällen kommt es zu einer Verletzung und in besonders gravierenden Problemsituationen zu einem Zusammenbruch der individuellen Wirklichkeit. Die Reaktion darauf ist ein Rückzug, oder, wie wir es gedeutet haben, ein Umschlag aus der Organisationsform des offenen Systems (aus Organismus und Umwelt bzw. individueller Wirklichkeit) in die Organisationsform des semiotisch geschlossenen Systems eines bloßen Körper-Seins, wie sie bei Menschen in ausweglosen Situationen beschrieben wurde. Untersuchungen über die „psychosomatische" Stimmung bei der Nausea (v. Uexküll, 1951) haben

gezeigt, daß es dabei oft zu einem Blutdruckabfall auf hypotone Werte kommt.

Eine derartige situative Hypotonie kann, besonders bei arteriosklerotisch geschädigtem Gefäßsystem, zu Durchblutungsstörungen im Gehirn und/oder im Myokard führen. Das könnte ein Mechanismus sein, der das gehäufte Zusammentreffen von Situationen der Hilflosigkeit und Hoffnungslosigkeit mit dem Auftreten eines Schlaganfalles und/oder eines Herzinfarktes erklärt.

Das Modell des Situationskreises beschreibt also ein System aus Regelkreisen verschiedener Integrationsebenen. Dadurch ist es in der Lage, das Phänomen der Situationshypertonie und situativ ausgelöste Hypotonien, aber auch die so häufigen Blutdruckschwankungen zu deuten. Es interpretiert diese Ereignisse als „Abwärts-Effekte" in hierarchisch gegliederten Zusammenhängen, d. h. als Wirkungen, die von der komplexeren Integrationsebene auf eine weniger komplexe ausgeübt werden. Abwärts-Effekte würden psychosomatischen Wirkungen entsprechen, während „Aufwärts-Effekte", d. h. Auswirkungen von Ereignissen im Rahmen tieferer Integrationsebenen (z. B. eines Blutdruckabfalls bei einem Herzinfarkt) auf eine höhere Integrationsebene (das Erleben der Schwäche und das Schwinden der Bewußtheit) somatopsychischen Wirkungen entsprechen. Auch zwischen verschiedenen Integrationsebenen besteht wieder eine Rückkoppelung, weil Abwärts-Effekte wieder Aufwärts-Effekte auslösen und umgekehrt.

Nach diesem Modell lassen sich Verbindungen zwischen dem kleinen Regelkreis (auf der Integrationsebene von Organsystemen innerhalb des Körpers) und dem großen Regelkreis (auf der Integrationsebene von Organismus und Umgebung) an zwei Stellen vermuten:

– In der **Phase des Erlebens,** in der Umgebungsfaktoren als Problemsituation interpretiert werden, die durch aktives Verhalten gelöst werden muß. Hier würde der Sollwert des kleinen Regelkreises im Sinne einer Bereitstellung für Kampf, Flucht oder andere aktive Leistungen dem Sollwert des großen Regelkreises angepaßt werden (emergency state).

– In der **Phase des Verhaltens,** in der die aktive Auseinandersetzung mit der Umgebung stattfindet. Hier würde einmal die Muskelarbeit durch Öffnen der Gefäße in der Muskulatur in das kleine Regelsystem eingreifen und zum anderen eine Lösung der Problemsituation (durch aktive Veränderung der Umgebung) die Bereitstellung beenden und damit auch die Erhöhung des Sollwertes im kleinen Regelkreis aufheben (Bedeutungsentkoppelung).

Mit Hilfe dieses Modells ließe sich vorhersagen, daß ein Fehlen adäquater Programme zur Interpretation der Umgebung zu unspezifischen Störungen im Ablauf des großen Regelkreises (zu unspezifischem Streß) führt. Konstellationen, in denen Programme für Bereitstellung zu Kampf, Flucht oder allgemein Leistungen, die mit erhöhter Muskelaktivität und erhöhtem Blutdruck einhergehen, nicht in adäquate Programme integriert werden können, würden zu dem unspezifischen Streß noch eine spezifische hypertone Komponente hinzufügen. In beiden Fällen liegt die Störung in dem Bereich des Situationskreises, in dem in der Phantasie Bedeutungsunterstellungen durch Probehandlungen geprüft werden. Mit Hilfe dieses Modells lassen sich also die Zusammenhänge zwischen somatischen, psychischen und sozialen Faktoren beschreiben und Störungen „lokalisieren".

41.3.7 Differentialdiagnose

Die Differentialdiagnose hat zunächst die Aufgabe, die essentielle Hypertonie als Störung der Programme des großen Regelsystems von den sekundären Hypertonieformen als Störung in dem kleinen Regelsystem abzugrenzen. Wenn bereits hypertoniebedingte Organkomplikationen eingetreten sind, ist diese Abgrenzung häufig nicht mehr mit Sicherheit möglich.

Eine Differentialdiagnose unter psychodynamischen Gesichtspunkten ist für die Einstellung des Arztes dem Patienten gegenüber – aber auch für die Planung einer Therapie wichtig, die sowohl medikamentöse wie psychotherapeutische Erfordernisse berücksichtigt.

41.3.8 Prognose

Die Prognose der Hypertonie ist sowohl vom systolischen als auch vom diastolischen Blutdruck abhängig (Pflanz, 1977). Epidemiologische Studien zeigen, daß auch schon leichte Blutdruckerhöhungen die Lebenserwartung verkürzen. Die Kombination mit weiteren Risikofaktoren (Hyperlipidämie, Nikotinabusus, Diabetes mellitus, Übergewicht etc.) führt zu einer zusätzlichen Verschlechterung der Prognose. Die genetische Belastung spielt – wie erwähnt – ebenfalls eine Rolle. Ein weiterer ungünstiger Faktor für die Prognose scheint ein hoher Plasmareninspiegel zu sein, während essentielle Hypertonien mit normalem oder niedrigem Reningehalt eine bessere Prognose haben sollen. Über die prognostische Bedeutung der Persönlichkeitsfaktoren gibt es noch keine statistischen Untersuchungen. Einzelbeobachtungen (Reiser et al., 1950) und der klinische Eindruck sprechen jedoch dafür, daß sie erheblich sein muß.

Die Lebenserwartung eines Hypertonikers wird letztlich durch die hypertoniebedingten Organkomplikationen von seiten des Herzens, des Zentralnervensystems, der peripheren Gefäße und der Nieren bestimmt.

Die vaskulären Komplikationen der Hypertonie können durch eine rechtzeitige Behandlung weitgehend verhindert werden (Epstein, 1974). Die bekannteste kontrollierte Untersuchung ist die Veterans Administration Cooperative Study in den USA. Diese Studie mußte nach 18 Monaten abgebrochen werden, da bei Männern mit einem anfänglichen diastolischen Blutdruck von 115–129 mm Hg in der Placebogruppe etwa zehnmal so viele kardiovaskuläre Komplikatio-

nen und Todesfälle auftraten wie in der mit blutdrucksenkenden Mitteln behandelten Gruppe. Bei diastolischen Ausgangswerten von 90–114 mm Hg war zwischen den beiden Gruppen nach 40 Monaten ein etwa dreifacher Unterschied nachweisbar (Freis, 1967).

Aufgrund der bisher vorliegenden randomisierten Langzeitstudien erscheint es allerdings zweifelhaft, ob es dem Hypertoniker mit diastolischen Werten zwischen 95 und 104 mm Hg nützt, wenn er lebenslang behandelt wird, während bei höheren Werten eine medikamentöse Behandlung unerläßlich ist (Pflanz, 1977). Die modernen pathophysiologischen Vorstellungen der „strukturellen Autoregulation" legen aber doch eine möglichst frühzeitige Behandlung nahe.

Die Therapie mit antihypertensiven Medikamenten hat zu einer Verbesserung der Prognose der malignen Hypertonie, aber auch ganz allgemein zu einer Zunahme der Lebenserwartung aller lege artis behandelten Hypertoniker, auch bei sekundären Formen, geführt. Die Prognose ist daher entscheidend abhängig von der Qualität und Konsequenz der medikamentösen Therapie. Diese hängt wiederum von der Stabilität des therapeutischen Bündnisses zwischen Arzt und Patient ab.

41.3.9 Konsequenzen für die Therapie der essentiellen Hypertonie

Im Durchschnitt werden in allen Ländern mit „hohem medizinischem Standard" zur Zeit nur etwa 50% der Hypertoniker diagnostiziert und von diesen wiederum nur 25% ausreichend behandelt (Bühler et al., 1976). Von diesen 25% haben jedoch aus noch unerklärten Gründen nach 6 Monaten bereits fast die Hälfte die Behandlung abgebrochen (Sackett et al., 1975). Die Erklärung für diese enttäuschenden Ergebnisse scheint nicht nur die Tatsache zu sein, daß Hypertoniker zunächst beschwerdefrei sind, sondern auch in ihren Persönlichkeitsmerkmalen und dem Versäumnis der Ärzte zu liegen, ihnen Rechnung zu tragen. Es sieht so aus, als ob viele Hypertoniker in der Therapie nicht aus eigenem Interesse kooperativ sind, sondern um dem Arzt einen Gefallen zu erweisen. Wird der Patient dann durch den Arzt enttäuscht, so bricht er die Behandlung abrupt ab, ohne daß sich der Arzt erklären kann, weshalb dies geschieht. Psychodynamisch spielt sich dabei vermutlich folgendes ab: Den zur Abwehr aggressiver Tendenzen überangepaßten und überkooperativen Patienten macht Abhängigkeit – besonders von autoritären Ärzten – soviel Angst, daß sie früher oder später aus der Praxis verschwinden (Pflanz, 1969).

Der plötzliche Verlust von Selbstsicherheit gegenüber Autoritätspersonen stellt – wie bereits erwähnt – eine hypertonieauslösende Situation dar. Der Arzt sollte darauf gefaßt sein und versuchen, die schwierige Situation durch Verständnis zu überbrücken. Verärgerung auf seiten des Arztes kann zu einem endgültigen Abbruch der notwendigen Arzneimitteltherapie führen. Da die medikamentöse Behandlung die Prognose der Hypertoniker entscheidend verbessert, ist der Aufbau einer tragfähigen Arzt-Patient-Beziehung von entscheidender Bedeutung (Finnerty et al., 1973; Svarstad, 1976).

Dabei muß der Arzt wissen, daß die Behandlung sich über Jahre zu erstrecken hat und daß in ihrem Verlauf eine Reihe kritischer Situationen auftreten oder auftreten können, in denen die Kontinuität der Behandlung und der Erfolg der Therapie immer wieder in Frage gestellt werden: Zu Beginn der medikamentösen Therapie führt der Blutdruckabfall bei vielen Patienten vorübergehend zu orthostatisch bedingten Schwindel- und Schwächezuständen und dem Gefühl des Verlustes der Kontrolle über die Umgebung und über sich selbst. Daher ist es notwendig, den Patienten die Wirkung der Medikamente zu erklären und sie davon zu überzeugen, daß diese Phase nicht durch Abbruch der Therapie, sondern durch Anpassung des Organismus an das neue Blutdruckniveau überwunden werden muß.

Weiterhin ist wichtig, den Patienten die möglichen Ursachen ihrer Hypertonie zu erklären. Die Einsicht, daß Blutdrucksteigerungen unter bestimmten Bedingungen normal sind, kann die Patienten beruhigen. Wenn möglich, sollte auch die Einsicht vermittelt werden, daß es in bestimmten Situationen besser sein kann, Spannungen abzureagieren, als sie chronisch zu unterdrücken (Bastiaans, 1963). Nach Alexander (1951) reagiert der Hypertoniker während der psychotherapeutischen Sitzungen oft mit einem Absinken des mittleren Blutdrucks, wenn er merkt, daß es erlaubt ist, seine zurückgestauten feindseligen Regungen zum Ausdruck zu bringen, oder wenn er zur Selbstbestätigung in beruflichen oder familiären Situationen ermutigt wird. In manchen Fällen kann erst die Analyse von Schuldgefühlen und Abhängigkeitsbedürfnissen die Patienten in die Lage versetzen, ihre Strebungen zum Ausdruck zu bringen und für ihre Spannungen geeignete Abfuhrmöglichkeiten zu finden. Diese Hinweise, die Alexander für eine psychotherapeutische Behandlung gibt, lassen sich – wenn auch in abgeänderter Form – für den täglichen Umgang mit diesen Patienten verwerten. Sie können – wie weiter unten an einem Beispiel dargelegt wird – für die Behandlung von Patienten mit medikamentös schwer einstellbarer, labiler Hypertonie von zentraler Wichtigkeit sein.

Unter diesem Gesichtspunkt ist auch die Verordnung von Blutdruckmeßgeräten zu betrachten, mit denen die Patienten in der Lage sind, ihren Blutdruck selbst zu kontrollieren. Diese Geräte haben zwei Vorteile:

– Auf der einen Seite erlauben sie den Patienten, den Erfolg einer Arzneimitteltherapie selbst zu überwachen, und auf diese Weise mehr Autonomie und Selbstbestätigung zu gewinnen, sowie ihre Bereitschaft zur Kooperation mit dem Arzt zu bekunden.
– Auf der anderen Seite geben sie dem Patienten die Möglichkeit, den Zusammenhang zwischen Blutdruckreaktionen und Lebenssituationen zu beob-

achten – ein Zusammenhang, der von vielen zunächst geleugnet wird.

Sieht man sich solche Selbstmeß-Blutdruckprotokolle an, dann stellt sich heraus, daß stabile Blutdruckwerte bei einer essentiellen Hypertonie eine Rarität sind. Die Unterscheidung zwischen labiler und stabiler Hypertonie scheint in den meisten Fällen eine Illusion zu sein und die Folge ungenügend kontrollierter Blutdruckwerte.

Der psychosomatisch interessierte Arzt hat bei einem Gespräch mit dem Hypertoniker über seine Blutdruckwerte und die damit zusammenhängenden Lebenssituationen die Möglichkeit zu einer adäquaten „kleinen Psychotherapie" entsprechend den oben zitierten Prinzipien von Alexander (1951).

Psychotherapie beim Hypertoniker durchführen, heißt in Interaktion und Biographie die Situation aufsuchen, in der die pathogene Bedeutungskoppelung entstand, die von einer Bedrohung zum Auftreten des Symptoms Hypertonie als Alarm- und Abwehrreaktion (vgl. Abschnitt 41.3.5.3 und 41.3.5.4) führte, um auf diese Weise zu einer Bedeutungsentkoppelung zu kommen. Es geht darum, die Verarbeitungsmechanismen des Subsystems Psyche zu stärken (z.B. durch Entlastung am Über-Ich, am Ich-Ideal oder durch „holding function"), es kompetenter zu machen, psychosoziale Konflikte ohne krankmachende Reaktion des Subsystems „Körper" zu lösen. Körperzentrierte Therapie, wie die funktionelle Entspannung nach Fuchs (1974) oder die progressive Muskelentspannung nach Jacobson (1938), kann die Möglichkeit zu neuer, den psychischen Apparat entlastender Bedeutungskoppelung geben.

Der Wert einer lediglich psychotherapeutischen Behandlung wird unterschiedlich beurteilt. Weiss und Mitarbeiter (1952) äußern sich skeptisch über die Erfolgsaussichten. Reiser und Mitarbeiter (1950) fanden bei 98 Hypertonikern in 60 bis 80% der Fälle eine Besserung der Symptome, während der Blutdruck nur in 30 bis 60% und anatomische Befunde nur in 20 bis 45% beeinflußt wurden. Diese Befunde stammen aus einer Zeit, in der es noch keine effektiven Medikamente zur Senkung des Blutdrucks gab. Sie unterstreichen die Bedeutung einer konsequenten medikamentösen Behandlung.

Bei Frühfällen oder bei hypertonen Regulationsstörungen und den transitorisch-jugendlichen Hochdruckformen sind nach Bräutigam und Christian (1973) Heilungen durch eine Psychotherapie durchaus möglich. Hier besteht allerdings auch eine spontane Remissionsrate von 40 bis 60% (Linneweh, 1960).

Allerdings haben Gaus und Mitarbeiter (1983) über gute Erfolge mit psychotherapeutischen Interventionen (Krisenintervention, Kurztherapie, Sprechstundeninterview, Gruppengespräche) berichtet, gerade bei Hypertonikern, die an schweren Formen der Krankheit litten und vorher mit medikamentösen und diätetischen Maßnahmen nicht ausreichend gut einstellbar waren.

Darüber, daß übende und entspannende Verfahren wie autogenes Training, Yoga, Meditation, Biofeedback etc. hypertone Blutdruckwerte günstig beeinflussen können, besteht Einigkeit. Übersichten dazu finden sich bei Vaitl (1982), Schmidt (1982), Julius und Cottier (1983), Linden (1983) und Herrmann (1986).

Die physiologischen Reaktionen auf einen Entspannungszustand scheinen durch die Beeinflussung und Reaktion der hypothalamischen Zentren zustande zu kommen. Dabei wird eine generalisierte Abnahme des Sympathikotonus und möglicherweise eine Verstärkung der Parasympathikusaktivität erreicht. Diese physiologische Wirkung zeigt sich in einer Senkung des Tonus der Skelettmuskulatur, einer Verminderung des peripheren Gefäßwiderstands (Vasodilatation), einer Abnahme der Herzfrequenz, einer Hypoventilation, einer Reduktion des O_2-Verbrauchs und der CO_2-Abgabe, einer Zunahme der Alphawellen im EEG und einer Zunahme des Hautwiderstandes. Da die Blutdrucksenkung durch diese Verfahren im allgemeinen jedoch relativ gering ist (Abnahme des systolischen Blutdrucks um 10 bis 15 mm Hg, diastolisch um 5 bis 10 mm Hg; Seer, 1979), sind sie vor allem bei Patienten mit sog. labilen Hochdruckformen oder mit Grenzwerthypertonie indiziert.

Besonders eindrucksvolle Beobachtungen stammen von C. Patel (1983). Mit ihren Methoden einer Kombination moderner Verfahren wie Biofeedback mit alten Techniken der meditativen Yogapraxis und deren systematischem Einsatz unter Alltagsbedingungen und lernpsychologischen Gesichtspunkten konnten nach 8 Behandlungswochen (1 Treffen wöchentlich) nicht nur der Blutdruck, sondern auch das Serumcholesterin und der Zigarettenkonsum verringert werden. Renin- und Aldosteronspiegel wurden gesenkt; die Senkung von Blutdruck und Aldosteron korrelierten signifikant miteinander. Diese positiven Effekte ließen sich auch nach weiteren 8 Monaten beobachten. In dieser Zeit übten die Patienten ohne weitere Betreuung allein weiter. Die neuesten Befunde zeigen, daß der blutdrucksenkende Effekt selbst nach einem Zeitraum von 4 Jahren noch gegenüber der Kontrollgruppe nachweisbar ist (Patel et al., 1987). Eine Reihe anderer Studien haben Patels Ergebnisse bestätigt (Little et al., 1984; Johnston, 1984). Engel und Mitarbeiter (1981 und 1983) haben erfolgreich ein verhaltensmedizinisches Programm zur Blutdrucksenkung mit Hilfe von Biofeedback (Blutdruckselbstmeßgerät) und Entspannung entwickelt, das sich durch geringen ärztlichen Aufwand und gute Praktikabilität für Patienten auszeichnet. Bei einem Teil der mit einem Diuretikum behandelten Grenzwerthypertoniker konnte auf die Gabe dieses Medikamentes verzichtet werden. Diese Verfahren sind billiger als lebenslange Medikation, sie erfordern aber mehr persönlichen Einsatz des Patienten, d.h. Motivation bzw. Compliance: Hoelscher et al. (1986) untersuchten die Patienten-Compliance bei Entspannungstherapie: Den Hochdruckpatienten wurden Tonband-Kassettenrecorder mit Entspannungsbändern gegeben und sie wurden instruiert, jeden Tag über einen Zeitraum von 10 Wochen damit die Entspannung zu praktizieren. Die Patienten wußten

nicht, daß in den Tonband-Kassettenrecordern ein elektronischer Monitor eingebaut war, der die tägliche Anwendung überprüfte. Es zeigte sich, daß zwar 91% der Patienten berichteten, daß sie täglich mit der Tonband-Kassette Entspannung geübt hätten, aber nur 32% der Patienten hatten sich tatsächlich 1mal pro Tag entspannt.

Es erscheint sinnvoll, übende und entspannende Verfahren nicht isoliert, sondern zusammen mit medikamentöser Therapie und gegebenenfalls Einzeloder Gruppengesprächen anzuwenden. Die Wirksamkeit dieser Methoden weist gleichzeitig auch wieder auf den psychophysiologischen Zusammenhang bei der Hypertonieentstehung hin.

Zur Lebensführung bei essentieller Hypertonie können folgende Empfehlungen gegeben werden: Berufliche und sportliche Aktivitäten, die zu unphysiologischen Erhöhungen des Sympathikotonus führen, sollten vermieden werden, dazu gehören insbesondere Termin-, Akkord- oder Schichtarbeit. Arbeiten auf Gerüsten, an Maschinen und Hochöfen, insbesondere das Führen von öffentlichen Verkehrsmitteln sind nur mit Vorbehalt möglich, d.h. wenn keine Beschwerden durch hypertoniebedingte Organkomplikationen (z.B. Schwindel) eingetreten sind und sich die Reaktionen auf eine neu eingeleitete medikamentöse Therapie bereits übersehen lassen. Als sinnvolle Sportarten können Radfahren, Wandern, Dauerlaufen, Schwimmen (unter Vorbehalt), Gymnastik, Ballspiele ohne Wettbewerbscharakter und Skilanglauf angesehen werden. Isometrische Übungen der Muskulatur, wie z.B. bei Gewichtheben, Windsurfing etc., sollten vermieden werden. Jede Form von Leistungssport ist bei einer ausgeprägten Hypertonie ungünstig.

Zusammenfassend sei noch einmal darauf hingewiesen, daß eine stabile und vertrauensvoll-dauerhafte Beziehung, die der speziellen Psychodynamik des einzelnen Hypertonikers Rechnung trägt, Voraussetzung für eine konsequente Therapie ist, die ihrerseits wieder für das Leben eines Patienten entscheidend sein kann. Auch bei Begutachtungen ist es unserer Meinung nach notwendig, psychosoziale Risikofaktoren zu berücksichtigen.

Ein Fallbericht soll als Abschluß des Kapitels die Bedeutung der Interaktion zwischen Arzt und Patient für die Behandlung des Hypertonikers illustrieren.

41.3.10 Bericht über die psychosomatische Behandlung einer Patientin mit labiler Hypertonie und Blutdruckkrisen

> Die Patientin – Frau L. – war über die allgemeinen Risiken einer schweren Hypertonie hinaus durch zwei Komplikationen hochgradig gefährdet:
>
> Sie war im Gefolge einer hypertensiven Retinopathie bereits auf einem Auge erblindet. Auf dem anderen Auge war es zweimal während krisenhafter Blutdruckanstiege zu einer vorübergehenden Amaurose gekommen.

Bei der einseitig nephrektomierten Patientin bedeutete der Fortbestand der Hypertonie aufgrund der durch pyelonephritische Schübe geschädigten Restniere die Gefahr einer zusätzlichen vaskulären Komplikation.

Auch dieser Fall ist ein Beispiel dafür, wie problematisch die Einteilung in labile und stabile Hypertonieformen ist.

1972 wurde die damals 38jährige Patientin wegen einer therapieresistenten Hypertonie mit Blutdruckwerten von 280/160 mm Hg stationär aufgenommen. Die internistische Vorgeschichte ergab, daß 1956 eine Nephrektomie wegen multipler Nierenabszesse bei einer Sepsis durchgeführt werden mußte. Damals war ein erhöhter Blutdruck festgestellt worden, der seitdem in wechselnder Höhe weiterbestand. 1968 kam es nach einer Kur, die der Hausarzt veranlaßt hatte, zu einer ersten Blutdruckkrise mit Werten von 280/150 mm Hg und zur Erblindung des rechten Auges aufgrund eines Verschlusses der Zentralarterie.

Während der ersten Tage der klinischen Behandlung blieben die Blutdruckwerte trotz antihypertensiver Behandlung bei Werten, die zwischen 290 und 240 mm Hg systolisch und 160 und 130 mm Hg diastolisch schwankten. Am Augenhintergrund fanden sich auch auf dem gesunden Auge schwere Veränderungen im Sinne eines Fundus hypertonicus III bis IV. Die übrigen Untersuchungen ergaben pyelonephritische Veränderungen an der normal funktionierenden rechten Niere. Auffällig war, daß die Linksherzhypertrophie nicht ausgeprägt war.

Diagnostisch standen wir – wie so oft – vor folgender, nicht sicher zu entscheidender Alternative: Die Vorgeschichte der Nephrektomie und die pyelonephritischen Veränderungen sprachen für eine renale Hypertonie mit situativen Blutdruckkrisen und den entsprechenden Komplikationen. Die normale Nierenfunktion ließ an die Möglichkeit denken, daß es sich um eine essentielle Hypertonie handelte, die schon vor der Nierenoperation bestand und bei der die renale Komponente für die Blutdruckerhöhung nur untergeordnete Bedeutung hatte. Ein Phäochromozytom ließ sich ausschließen.

Die Patientin wurde von 1972 bis 1975 insgesamt viermal stationär aufgenommen. Bei jedem dieser stationären Aufenthalte wurden, außer bei den täglichen Visiten, jeweils 2 Stunden (insgesamt also 8 Stunden) psychotherapeutische Gespräche geführt.

Äußere Erscheinung und Verhalten

Die betont jugendlich zurechtgemachte Patientin, die einen fast jungmädchenhaften Eindruck machte, legte in auffallendem Gegensatz zu dieser äußeren Erscheinung ein recht unfreundliches Verhalten an den Tag. Sie war trotzig, schien ständig beleidigt und unzufrieden zu sein – ein personifizierter Vorwurf, der in dem Arzt Schuldgefühle hervorrief. Sein erster Eindruck läßt sich etwa folgendermaßen beschreiben: Ein sehr wohlerzogenes kleines Mädchen in Sonntagskleidern, das nicht spielen und herumtollen darf und das von dem Arzt erwartet, er solle ihr die verbotenen Spiele erlauben, das aber zugleich enttäuscht und trotzig überzeugt ist, daß seine Bitten abgeschlagen werden.

Auffällig war auch die Einstellung der Patientin zu ihrer Krankheit: Sie war tief beleidigt, als in ihrer Gegenwart die Frage diskutiert wurde, ob unter Umständen eine essentielle Hypertonie vorliegen könnte, weil sie der Meinung war, daß man ohne eine organische Ursache ihre Krankheit nicht ernst nehmen könne. Jede Möglichkeit eines Zusammenhangs zwischen psychischen Faktoren und Blutdruckanstieg lehnte sie strikt ab. Sie war überzeugt, daß ihr Blutdruck von der kranken Niere komme und daß die immer wieder auftretenden Blutdruckspitzen mit dem Wetter zusammenhängen würden.

Verlauf

Im ersten Interview berichtete die Patientin folgende Einzelheiten zur Vorgeschichte: Der Vater, den sie als „besten Vater der Welt" schilderte, mit dem sie sich „phantastisch verstand", starb in dem Jahr, in dem die Patientin an der Sepsis erkrankte und nephrektomiert werden mußte, an einem Schlaganfall. Die Mutter habe „immer ihren Willen durchgesetzt", sie war streng und brauchte die Kinder nur anzusehen, dann waren sie artig. Sie hat die Tochter immer unterdrückt und ganz selbstverständlich ihre Hilfe im Haushalt beansprucht. Zwei ältere Geschwister hatten das Elternhaus verlassen und kümmerten sich wenig um die Mutter. Als die Tochter schließlich trotz des Widerstandes der Mutter eine Berufsausbildung durchsetzte, hat sie hinter dem Rücken der Tochter so lange intrigiert, bis die Tochter entlassen wurde und wieder in den mütterlichen Haushalt zurückkehren mußte. Mehrere Versuche, sich zu verloben, wurden von der Mutter hintertrieben. Die Tochter hat nie aufbegehrt, sie war höchstens „verletzt". Die Mutter starb 1968 nach einem Schlaganfall, der sie völlig gelähmt ans Bett fesselte, unter qualvollen Umständen, von der Tochter gepflegt. Die anderen beiden Geschwister hatten sich auch dann nicht um die Mutter gekümmert.

Bei dem mißglückten Versuch einer Berufsausbildung geriet sie an einen sadistischen Vorgesetzten, der ihr nachstellte und sie ständig quälte. Nach einem Einbruchsdiebstahl in dem Geschäft, in dem sie arbeitete, wurde sie verdächtigt. Sie wagte aber weder sich zu beschweren, noch sich bei den Eltern auszusprechen. „Es war die Hölle."

Bei diesem ersten Interview wurde deutlich, daß die Patientin den offenbar sehr schwachen Vater in schwärmerischer Weise verehrt hat, ein Gefühl, das sie nun auf den Arzt übertrug. Im Verhältnis zur Mutter herrschten „Verletztsein", ohne daß sie aufzubegehren wagte, und bockiger Gehorsam vor. Das gleiche Verhalten legte sie später ihren Vorgesetzten gegenüber und jetzt wieder in der Klinik an den Tag. Dabei führte ihr unrealistischer Anspruch, verwöhnt und den anderen Patienten gegenüber bevorzugt zu werden, zu Schwierigkeiten im Umgang mit dem Pflegepersonal und den Ärzten. Bei dem Bericht ihrer häuslichen Situation – sie hatte nach dem Tod der Mutter einen sehr viel älteren Mann geheiratet – zeigte sich, daß die Einstellung der Patientin im täglichen Umgang mit ihrer Umgebung und ihrem Ehemann zu ähnlichen Spannungen führte.

Nach diesem ersten Gespräch, in dem die Patientin über die Vorwürfe sprechen konnte, die sie gegen die Mutter empfand, und bei dem ihre feindseligen Gefühle gegen die Umgebung deutlich wurden, über die sie sich immer wieder zurückgesetzt fühlte, sank der Blutdruck auf normale Werte. Gleichzeitig kam es zu erheblichen orthostatischen Beschwerden, über die die Patientin sehr vorwurfsvoll und anklagend berichtete. Trotzdem gelang es, mit der Patientin ein Arbeitsbündnis zu schließen, bei dem sie die orthostatischen Beschwerden als notwendige Umstellung ihres Organismus an den normalen Blutdruck – wenigstens rational – akzeptierte.

Dies überraschend befriedigende Behandlungsergebnis wurde dadurch getrübt, daß die Patientin vorzeitig die Klinik verließ. Bei ihrem nächsten Krankenhausaufenthalt – ein Jahr später, weil der Blutdruck wieder angestiegen war und wieder auf Medikamente nicht ansprach – gab sie an, der Grund für „ihre Flucht" aus der Klinik seien ihre Gefühle für den behandelnden Arzt gewesen, in dem sie den Vater wiederzuerkennen glaubte und der sie – ähnlich wie der Vater der Mutter gegenüber – dem Pflegepersonal und den anderen Ärzten der Klinik gegenüber nicht genügend in Schutz nahm. Sie sei auch zur verabredeten Zeit nicht zur Nachuntersuchung gekommen, weil der Arzt sie nicht persönlich bestellt habe.

Bei diesem Aufenthalt berichtete die Patientin Näheres über die Beziehung zu ihrem Ehemann, von dem sie sich nicht genügend verstanden fühlte. Besonders schlimm sei es, wenn er ihr, die doch alles tat, um ihm das Leben schön zu machen, ungerechtfertigte Vorwürfe machen würde.

Der Versuch, die Patientin darauf anzusprechen, daß sie außerstande sei, sich gegen Kränkungen und ungerechte Anforderungen zu verteidigen, beunruhigte sie offensichtlich sehr. Trotzdem war der Allgemeinzustand sehr viel besser als bei der ersten Untersuchung, der Blutdruck schwankte in erträglichen Grenzen, zeigte allerdings noch gelegentliche Steigerungen, die – wie sie weiterhin fest überzeugt war – mit Wetterumschlägen zusammenhingen.

1974 kam die Patientin erneut in die Klinik, diesmal wegen schwerer Angstzustände, die nach der Klinikentlassung vor einem Jahr aufgetreten waren und die sich etwa alle 6 bis 8 Wochen wiederholten. Sie wisse dann vor Angst nicht ein noch aus, müsse ihren Mann nachts wecken, der auch mit ihr sprechen würde, aber ohne ihr helfen zu können. In dieser Stunde fällt ihr ein, daß ihre Mutter sie mit 6 Jahren mit einem Stock so schlug, daß sie voller blutiger Striemen war und nachts nicht auf dem Rücken liegen konnte. Damals habe sie gedacht: „Das vergesse ich Dir nie!" Es fielen ihr dann nacheinander Episoden ein, in denen sie von der Mutter brutal behandelt und drangsaliert worden war. So wurde sie z. B. noch mit 22 Jahren eingesperrt, als sie zu ihrem Verlobten fahren wollte. Nach der Nierenoperation, bei der sie beinahe gestorben sei, habe die Mutter sie im Krankenzimmer beschimpft und ihr vorgeworfen, sie würde sich nur aus Faulheit ins Krankenhaus legen. Da habe die Patientin – zu ihrem eigenen Entsetzen – die Mutter geschlagen. „Die Hand rutschte aus." Diese Episode hatte sie völlig vergessen. Sie war lediglich von Angstträumen heimgesucht, in denen sie vor etwas Unheimlichem davonlief.

Offensichtlich lief sie im Traum vor dem Haß gegen die Mutter und den Schuldgefühlen, die diesen Haß unterdrückten, davon. Wahrscheinlich hat auch der qualvolle Tod der Mutter diese Schuldgefühle verstärkt. Jetzt wurde ihr Verhalten, d. h. die ständige Fürsorge für die Mutter trotz permanenter ungerechter und sadistischer Behandlung, ihr Verhalten den Ärzten und dem Ehemann gegenüber, sehr viel verständlicher. Nach dieser Aussprache sind die Angstzustände nicht wieder aufgetreten.

Vor dem letzten Krankenhausaufenthalt im Frühjahr 1975 waren die Blutdruckwerte wieder angestiegen. Die auslösende Situation ließ sich jetzt relativ leicht rekonstruieren: Der Ehemann wurde pensioniert und war jetzt den ganzen Tag zu Hause. Dabei kam es ständig zu Reibereien. Der Mann würde darunter leiden, nicht mehr die Anerkennung und Befriedigung seines Berufes zu haben. Es falle ihm schwer, sich mit der neuen Situation abzufinden. Das aber könne er nicht zugeben; er sei außerstande, Schwäche zu zeigen. Er werde dann ausfallend und kränkend. Das mache sie „krank". Sie fühle sich dann zu Unrecht beschuldigt, wie in ihrer Jugend von der Mutter oder während der Berufsausbildung, als man ihr einen Diebstahl zur Last legte. Die Patientin war über die Spannungen mit dem Ehemann so beunruhigt, daß sie über ein Wochenende kaum nach Hause zu fahren wagte.

Bei einer Aussprache konnte sie über ihre Aggressionshemmung freier berichten. Es fielen ihr folgende, besonders kränkende Situationen ein: Mit 4 Jahren, als sie untröstlich war, weil die Nachbarskinder, mit denen sie eng befreundet war, wegzogen, wurde sie von ihrer Mutter hart zurechtgewiesen. Mit 5 Jahren wurde sie von der Mutter zur Strafe in ein dunkles Zimmer gesperrt. Sie sei nicht nur mit 6 Jahren, sondern häufig von der Mutter mit dem Stock geschlagen worden.

Dann erinnerte sie sich an die Episode, in der man ihr den Diebstahl zur Last legte.

Die Patientin konnte jetzt den Rat annehmen, sich mit ihrem Mann auszusprechen und ihm zu erklären, was sie empfinden würde, wenn er ihr Vorwürfe macht. Sie würde ihn dann wie ihre Mutter oder den damaligen Vorgesetzten erleben. Trotzdem fuhr die Patientin nach diesem Rat besorgt ins Wochenende, kam aber sehr getröstet wieder. Es sei besser gegangen, als sie erwartet habe. Der Mann habe es verstanden und sie selbst verstünde jetzt, daß sie beide – der Mann und sie selbst – „das Wetter machen würden", das sie bisher für ihre Blutdruckschwankungen verantwortlich gemacht hatte.

Die Patientin war jetzt unter der Therapie mit einem Saluretikum und einem Betarezeptorenblocker ständig normoton. Die orthostatischen Beschwerden hatten sich völlig verloren.

Interpretation

Die „Situation", welche die Patientin immer wieder aufbauen mußte, folgte früh erworbenen – und im Laufe ihres Lebens immer neu „bestätigten" Programmen. Diese Programme legten das Szenarium für den Ablauf der Handlungen und die Rollen der Akteure in einem Drama fest, das sich – mit nur unwesentlichen Variationen – ständig wiederholte. In dem Drama ist die Patientin als kleines hilfloses Mädchen einer bösen Mutter ausgeliefert. Ihre brutale, ungerechte Behandlung und Quälereien muß sie mit zusammengebissenen Zähnen erdulden. Sie darf sich nicht offen auflehnen, ohne Gefahr zu laufen, die Gunst der Mutter, von der sie völlig abhängig ist, endgültig zu verlieren.

Der Vater ist ihre einzige Hoffnung. Von ihm träumt sie, er werde sie aus ihrer Abhängigkeit befreien und sich ihr zuwenden, wie der Prinz im Märchen der verkannten Prinzessin. Sie liebt ihn entsprechend schwärmerisch und umwirbt ihn. Aber er ist schwach und unzuverlässig. Er enttäuscht sie immer wieder, wenn er ihr die gehaßte und gefürchtete Mutter vorzieht. Gleichzeitig erwartet er von ihr, daß sie ein artiges und gutes Kind ist. Böse Gefühle und schlimme Gedanken würden ihn so erschrecken, daß er sich von ihr abkehren könnte.

Aggressive Gefühle und Gedanken waren also dreimal gefährlich: Man durfte sie nicht äußern, ohne von der Mutter verstoßen zu werden. Man durfte sie nicht einmal haben, ohne ein böses Kind zu sein, das die Zuneigung des Vaters nicht mehr verdient. Schließlich mußten in Gefühlen und Gedanken, vor denen die mächtigen Eltern so große Furcht hatten, nicht wirklich unausdenkbar gefährliche und zerstörerische Kräfte schlummern?

Erst als die Patientin erlebt, daß man Vorwürfe äußern und Haßgefühle nicht nur haben, sondern ihnen sogar nachgeben darf, ohne verstoßen zu werden, beginnt sich das starre Programm zu lockern. Es wird möglich, die Rolle für mütterliche und väterliche Repräsentanten (Objekte) zu differenzieren und damit auch die eigene Rolle freier zu gestalten. Mit der Kommunikation über die gefürchteten Themen wird deren Korrektur möglich, und in dem Maße, wie der Situationskreis seine Feindseligkeit, Gefahr und Ausweglosigkeit verliert, normalisiert sich der Blutdruck.

Die Biographie der Patientin ist ein Beispiel für eine Lebensgeschichte, in der es schon früh zu pathologischen Bedeutungskoppelungen zwischen dem großen und dem kleinen Regelkreis, d.h. zwischen situativem Erleben und Bereitstellungen zu Kampf oder Flucht kam, die mit einem erhöhten Blutdruck einhergehen. Die Hemmung der Exekutive, d.h. die Blockierung der aktiven Auseinandersetzung mit der Situation in offenem Kampf oder Flucht, kann dann dazu führen, daß die Bereitstellung schließlich zu einer Dauereinrichtung wird. Erlebnisse, die jetzt mit einer vermeintlichen oder realen Bedrohung oder Kränkung einhergehen, können dann zu krisenhaften Steigerungen des bereits erhöhten Blutdrucks führen.

Der Verlauf zeigt, daß es der psychotherapeutischen Bemühung des Arztes gelingen kann, selbst eine schon Jahre bestehende Bedeutungskoppelung zu lösen und damit eine Hypertonie zu normalisieren, die ebenfalls schon seit vielen Jahren besteht. Der Widerstand, den es dabei zu überwinden gilt, hängt offenbar auch damit zusammen, daß ein erhöhter Blutdruck im Sinne von Lacey (1970) eine Stimulusbarriere unterhalten und somit eine somatopsychische Schutzfunktion gegen das Überwältigtwerden von aggressiven Eindrücken haben kann.

42 Asthma bronchiale

Wolfram Schüffel, Jörg Michael Herrmann, Bernhard Dahme und *Rainer Richter*

Die Bedeutung psychosozialer Faktoren beim Asthma bronchiale wird kontrovers diskutiert. Die Kontroversen hängen damit zusammen, daß es schwerfällt, die Fülle der sich teilweise widersprechenden Untersuchungsergebnisse zu ordnen. Für den Leser ist dann hilfreich zu wissen, auf welche Autoren bzw. Autorengruppen sich die Verfasser von Übersichtsartikeln berufen. Hier berufen sich die Verfasser vor allem auf F. Alexander und seine Mitarbeiter, J. J. Groen und Mitarbeiter, A. Jores und die in seiner Tradition arbeitenden psychosomatischen Forschungsgruppen am Universitätskrankenhaus Hamburg-Eppendorf, P. H. Knapp und A. Mathe[1] sowie andere, von denen die ersten überwiegend Internisten, die anderen überwiegend klinische Psychologen und Psychotherapeuten sind.

42.1 Exemplarischer Fall

Die 35jährige Patientin, Ehefrau und Mutter zweier Töchter im Alter von neun und drei Jahren arbeitete aushilfsweise in einer Bäckerei. Dort fühlte sie sich unter ihrem Können eingesetzt. Sie hatte zwar eine kaufmännische Lehre hinter sich, konnte aber keine angemessene Stelle finden und mußte aus finanziellem Zwang die Stelle in der Bäckerei annehmen. Die Bäckersfrau hatte sie aus Gefälligkeit gegenüber der mit ihr befreundeten Mutter der Patientin eingestellt. – Mit Widerwillen, den sie nicht der Chefin mitzuteilen wagte, mußte sie erstmals an der halbjährlich stattfindenden Reinigung der Mehlstube teilnehmen. Mit Abscheu entfernte sie das Mehl, das mit Spinnweben verbacken in den Ecken hing. Im Raum hing ein muffiger Geruch. Noch nie hatte sich die Patientin in der Backstube, die sie zuvor häufig aufgesucht hatte, so unangenehm gefühlt. Sie empfand eine bedrückende Enge. Es kam zu Unruhe, zu Herzklopfen, zu Enge über der Brust und schließlich zu zunehmender Kurzatmigkeit. Die Patientin ging nach Dienstschluß mit leichten Atembeschwerden nach Hause. Während der Nacht geriet sie in einen ersten Status asthmaticus. Der Hausarzt mußte kommen und sie mit Cortison intravenös behandeln. – In den folgenden acht Jahren verspürte die Patientin immer wieder Phasen von Kurzatmigkeit. Sie mußte dann kurzfristig gegen asthmatische Beschwerden behandelt werden. Ein ausgesprochener Anfall trat jedoch nicht wieder auf. Sie litt zwar häufig unter Heuschnupfen, die Atemnot trat aber praktisch nie zusammen mit dem Heuschnupfen auf. Vielmehr ereignete sie sich in der Regel dann, wenn überhaupt kein Schnupfen da war. Sie arbeitete weiter in der Bäckerei und wechselte sechs Jahre später in einen anderen Betrieb über, in dem sie die Buchhaltung übernahm. Zwei Jahre nach diesem Wechsel und acht Jahre nach dem ersten Anfall wurde ihr angeboten, die bis dahin besetzte Halbtagsstellung in eine Ganztagsstellung umzuwandeln. Solle sie das Angebot nicht annehmen können, so wurde ihr gesagt, müsse man eine Ganztagskraft einstellen, der sie sich dann unterzuordnen habe. Die Patientin hätte das Angebot gern angenommen. Dies bedeutete aber, daß die eigene Mutter die Versorgung ihres Haushaltes hätte übernehmen müssen. Hierum wollte sie ihre Mutter nicht bitten. Andererseits konnte sie sich nicht entschließen, dem Firmeninhaber abzusagen. Sie bekam ihren zweiten schweren Atemnotanfall am Morgen des Tages, an dem sie dem Firmeninhaber ihre Entscheidung mitteilen wollte.

Mit schwerster Atemnot wurde sie ins Krankenhaus eingeliefert. Sie fühlte sich an ihre Atemnot aus der Zeit vor acht Jahren erinnert, die durch eine einmalige Cortisongabe beendet wurde. Sie wollte eine solche Cortisongabe auch jetzt. Dies wurde ihr verweigert. Sie wollte den Oberarzt sprechen; man erklärte ihr, daß man nach bewährten therapeutischen Schemata vorginge und eine Rücksprache mit dem Oberarzt nicht nötig sei. Die Patientin wollte aufbegehren und sich laut beschweren. Sie gab aber dann nach. Später sagte sie, sie hätte es aufgegeben zu kämpfen, obwohl in ihr alles voller Unruhe gewesen sei.

Ihr Zustand wurde bedrohlich. Aus den Unterlagen des Krankenhauses ging hervor, daß der O_2-Partialdruck zu dieser Zeit um 60 mm Hg lag, d.h. eine partielle Insuffizienz bestand. Von ferne hörte sie, daß ihr Zustand als sehr ernst anzusehen sei und sie auf die Intensivstation der benachbarten Großklinik verlegt werden solle. Die Patientin erinnerte sich, wie wenige Minuten hierauf die Atemnot nahezu verschwand. Offensichtlich wurde diese Entwicklung aber nicht bemerkt – bei der Aufnahme im neuen Krankenhaus war sie fast beschwerdefrei, das Exspirium war nur noch mäßig verlängert. Deutlich fielen jedoch ihre weit geöffneten Augen, ihre Unruhe, ihre ängstliche Gespanntheit auf. Es war, als könne sie den Arzt oder andere Besucher nicht loslassen. Näherte man sich ihr, kam dennoch kein Gespräch zustande, vielmehr war eine ärgerlich-gereizte Stimmung spürbar. – Zu einer deutlichen Besserung kam es im Verlauf von Atemübungen, während derer sie über die verweigerte Cortisongabe, dann über die Entwicklung daheim berichten konnte. Schon wenige Tage nach ihrer Aufnahme im Krankenhaus habe es die Mutter nicht mehr bei der stundenweisen Aushilfe daheim bewenden lassen wollen, sondern wolle dort selber einziehen, um den Haushalt zu versorgen. Der Ehemann habe in dieses Arrangement bereits eingewilligt.

Auf die hier wiedergegebenen Abläufe wird in späteren Abschnitten eingegangen. Insbesondere im Abschnitt „Interaktion" und „Therapie" wird auf diesen Fall verwiesen und die Kasuistik interpretiert.

---

1 in den USA

745

42.2 Definition

Ohne auf die vielfältigen Definitionen und Klassifikationsvorschläge einzugehen (vgl. Scadding, 1979), wird Asthma bronchiale nach internationaler Übereinkunft (vgl. Deutsche Liga zur Bekämpfung der Atemwegserkrankungen, 1980) folgendermaßen definiert:

„Die Krankheit ist durch Anfälle von Atemnot charakterisiert, begleitet von den Zeichen einer Bronchialobstruktion, die zwischen den Anfällen ganz oder teilweise reversibel ist. Den Anfällen entspricht ein akuter Anstieg des Atemwegswiderstandes."

Die Atemwegsobstruktion kann dabei durch Kontraktion der Bronchialmuskulatur, ödematöse Verdickung der Schleimhaut und/oder durch Schleimansammlungen (Hypersekretion) in den Atemwegen verursacht sein.

42.2.1 Beschwerden und Entwicklung des Krankheitsbildes

Das Leitsymptom des asthmatischen Anfalles ist die ausgeprägte bis lebensbedrohliche Atemnot des Patienten. Oft ist diese Atemnot begleitet von giemenden und pfeifenden Atemgeräuschen, von Husten und Auswurf. Obwohl es sich beim Asthma bronchiale um eine vorwiegend exspiratorische Obstruktion handelt, haben viele Patienten das Gefühl, zu wenig Luft zu bekommen. Dieser Lufthunger kann dann zu einer sekundären Hyperventilation und den damit verbundenen Symptomen der Hypokapnie (Parästhesien, Schwindel) führen. Untrennbar verbunden mit diesen körperlichen Symptomen sind jedoch Empfindungen und Stimmungen wie Angst, Unruhe, Gereiztheit oder Ärger. Auf diesen psychosomatischen Aspekt der Atemnot weisen auch Atemphysiologen hin, wenn sie Dyspnoe folgendermaßen definieren:

„Dyspnea is difficult, laboured, uncomfortable breathing, though it is not painful in the usual sense of the word. It is subjective and like pain, it involves both perception of the sensation by the patient and his reaction to the sensation" (Comroe, 1966).

Es lag nahe, die Struktur dieser unterschiedlichen körperlichen Beschwerden, Symptome und Empfindungen, die für die asthmatische Atemnot typisch sind, systematisch zu untersuchen. So berichtet Richter (1985) in Übereinstimmung mit Untersuchungen aus dem anglo-amerikanischen Sprachraum (Kinsman et al., 1974) über 5 Dimensionen der asthmatischen Atemnot, die sich aufgrund einer Faktorenanalyse von 79 typischen Symptomen, Beschwerden und Befindensstörungen an 338 Patienten mit Asthma bronchiale ergaben. Hieraus resultierte eine Asthma-Symptom-Liste (ASL) mit fünf Skalen:

I Nervöse Ängstlichkeit:
Patienten mit hohen Skalenwerten fühlen sich während asthmatischer Anfälle ängstlich, beunruhigt, bedrückt, hilflos und haben Angst, allein gelassen zu werden.

II Obstruktive Atembeschwerden:
Diese Skala beschreibt körperliche Beschwerden der obstruktiven Atemnot, wie erschwerte Atmung, Atemgeräusche, Engegefühl in der Brust, Erstickungsgefühle.

III Ärgerliche Gereiztheit:
Patienten mit hohen Testwerten fühlen sich während der asthmatischen Anfälle gereizt, ärgerlich, schlecht gelaunt, aufbrausend, zornig.

IV Hyperventilationssymptome:
Diese Skala beschreibt die typischen körperlichen Beschwerden der respiratorischen Alkalose, wie sie im Verlauf einer alveolären Hyperventilation auftreten, nämlich Schwindel, Kribbeln und Prickeln, Kopfschmerz, Gefühl von tausend Stecknadeln.

V Müdigkeit:
Die Skala besteht aus Beschwerden wie Müdigkeit, Trägheit, Schläfrigkeit.

Bei dem Syndrom „asthmatische Atemnot" handelt es sich um ein mehrdimensionales Beschwerdebild, das sich aus drei Befindlichkeitsdimensionen und zwei Dimensionen vorwiegend körperlicher Beschwerden zusammensetzt, deren individuelle Ausprägung mit der Asthma-Symptom-Liste (ASL) gemessen werden kann.

Die relative Bedeutung jeder dieser fünf Dimensionen geht aus Abbildung 42–1 hervor, aus der abzulesen ist, wieviele Patienten unter den jeweiligen Symptomen/Beschwerden während ihrer asthmatischen Anfälle oft bzw. immer leiden.

42.3 Epidemiologie

Es ist schwer, exakte Angaben über die Häufigkeit des Asthma bronchiale in der Bevölkerung zu erhalten. Dies hängt damit zusammen, daß die Diagnose „Asthma bronchiale" von den Kriterien abhängt, die der betreffende Untersucher anwendet. Diese Kriterien sind unterschiedlich beim praktischen Arzt, Kran-

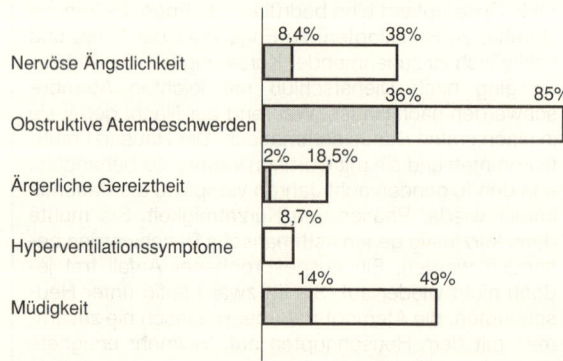

Abb. 42–1. Dimensionen der asthmatischen Atemnot. Prozent der 338 Patienten die immer (niedrigere Zahl) oder oft an den jeweiligen Beschwerden/Symptomen leiden (nach Richter, 1985).

kenhausarzt und beim Epidemiologen. Die britischen Angaben werden als die verläßlichsten betrachtet (Speizer, 1976): Im Vereinigten Königreich wird eine Prävalenz zwischen 0,1–1,0% in der gesamten Bevölkerung berichtet, während in den USA 3% der Bevölkerung an Bronchialasthma leiden sollen. Ähnliche Zahlen finden sich in den Niederlanden (Übersicht bei Schultze-Werninghaus, 1988).

Zum Geschlechterverhältnis: Bis zum Alter von fünf Jahren sind Jungen zweimal häufiger als Mädchen betroffen. Hiernach verändert sich das Verhältnis, so daß beide Geschlechter etwa gleich häufig betroffen sind. Bei Kindern unterhalb des zweiten Lebensjahres findet sich Asthma bronchiale so gut wie nicht; jenseits des 60. Lebensjahres ist die Häufigkeit der Erstmanifestation bei Männern im Vergleich zu Frauen wiederum leicht erhöht (Speizer, 1976).

Zur Prognose: Frühzeitige Erstmanifestation ist mit einer 60%igen, späte Erstmanifestation mit einer 30%igen Chance totaler Remission verbunden (Speizer, 1976). Insgesamt kann man sagen, daß 50–80% der Asthmatiker eine relativ gute Prognose haben, während 30% aller Patienten, insbesondere die älteren, mit schwerem Asthma weiterleben müssen. Der Tod durch Asthma scheint weder vom Alter der Erstmanifestation noch von der Schwere der Erkrankung abzuhängen. In den Vereinigten Staaten wird mit jährlich 2000, in der Altersgruppe von 5–34 Jahren mit 250 Toten gerechnet.

Zur Belastung des Gesundheitssystems: In den USA wird jährlich mit 28 Millionen Arztbesuchen und 183 000 Einweisungen ins Krankenhaus bei über 1 Million Krankenhaustagen wegen Asthma bronchiale gerechnet. Die direkten Kosten dieser medizinischen Versorgungsmaßnahmen belaufen sich auf 629 Millionen Dollar, die indirekten Kosten auf 435 Millionen Dollar pro Jahr (McCombs et al., 1979).

42.4 Psychosoziale Untersuchungsergebnisse

42.4.1 Soziale Interaktion

Über die psychologischen Befunde bei Asthmapatienten ist eine nicht mehr überschaubare Anzahl von Arbeiten geschrieben worden, noch mehr wurde hierüber diskutiert.

Es lohnt sich, wieder zum ursprünglichen Beobachtungsfeld zurückzugehen, d.h., die unmittelbare Arzt-Patient-Interaktion wahrzunehmen. Diese wird in einer erfrischenden Unmittelbarkeit von einem Internisten der Medizinischen Poliklinik des General Massachusetts Hospital in Boston wie folgt beschrieben (Rubenstein, 1976): Während der Arzt einen anderen Patienten untersucht, stecken sie plötzlich ihren Kopf in die Untersuchungskabine und beschweren sich, daß sie warten müssen. Sie fragen unvermittelt nach dem Sinn einer Behandlung. Am auffälligsten ist, daß sie im Gegensatz zu allen anderen Patienten der Poliklinik Termine nicht einhalten. – Der beobachtende Polikliniker kommentiert die Verspätung mit der Feststellung, daß man wohl mit seinem Arzt unzufrieden sei; denn sonst komme man nicht zu spät. Zur Behandlung insgesamt stellt er fest: „Asthmapatienten spicken ihre Versorgung mit Hindernissen, die ein Arzt erst lernen muß zu meistern, wenn er an guter ärztlicher Versorgung interessiert ist."

Die klinische Erfahrung zeigt, daß Asthmapatienten in der Regel nicht zu den beliebten Patienten gehören. Vor dem Hintergrund seiner tiefen Kenntnis des Asthmapatienten gibt Jores (1980) den Ausspruch eines seiner Patienten wieder: „Asthma, das ist Unzufriedenheit mit sich selbst und Protest dagegen."

42.4.2 Psychoanalytische Befunde

Im vorhergehenden Abschnitt beschreibt der dort erwähnte Bostoner Polikliniker die typische Unruhe des Asthmapatienten. Hinter ihr verbirgt sich Angst, die auch aus der Fallgeschichte unserer Patientin ersichtlich wird. Im Umgang mit dieser Angst sind zwei scheinbar widersprüchliche Aspekte erkennbar: Einerseits wird der Arzt dringend um Hilfe ersucht, andererseits fühlen sich die Patienten einfach von ihren Ärzten überrollt und meiden sie.

Psychoanalytische Untersuchungsergebnisse besagen, daß es sich hier nicht um ein krankheitsreaktives Verhalten handelt. Vielmehr ist dieses Verhalten in frühen Kindheitsjahren im Umgang mit der Mutter bzw. deren Vertretern erlernt:

„Im Zentrum steht die konflikthafte Beziehung zur Mutter, der Konflikt zwischen Anklammerungs- und Unabhängigkeitsbestrebungen zu ihr. Der drohende Verlust der Bindung an die Mutter veranlaßt das Ich (gemeint im psychoanalytischen Sinn als strukturierendes Funktionszentrum; Anmerkung d. A.) der Kranken zu verschiedenen Abwehrleistungen. … Bricht diese psychische Abwehrfunktion zusammen, kommt es unter Einschaltung regressiver Mechanismen zum Asthmaanfall" (de Boor, 1965, S. 218).

Zum Verständnis dieser psychischen Abwehrfunktionen ist es wichtig zu sehen, daß bei den Asthmapatienten mit schwerwiegender psychogener Komponente eine sehr frühe Störung der psychischen Entwicklung vorliegt, die ihrerseits die Entwicklung der Ich-Funktionen stark behindert. Um diese Behinderungen zu verstehen, sollen der analytischen Entwicklungstheorie folgend die Störungen der einzelnen Stufen aufgezeichnet werden. Hierbei ist es wichtig, sich auf drei Aspekte zu konzentrieren:

– Es besteht ein Ambivalenzkonflikt. Einerseits wird größtmögliche Nähe zur Mutter bzw. deren Repräsentanzen gesucht, andererseits werden diese Mutter bzw. ihre Repräsentanzen gemieden. Dieser Ambivalenzkonflikt wird symbolisch durch den Schrei und gleichzeitig dessen Unterdrückung ausgedrückt. Dieser Schrei bedeutet gleichermaßen den Appell an die Mutter, herbeizueilen, wie den Versuch, sich gegen die Mutter zu wehren.

– Die gestörte psychosexuelle Entwicklung (im psychoanalytischen Sinne) ist auf allen Reifungsstufen nachweisbar.

– Es besteht ein gestörtes Verhältnis zwischen Ideal-Ich und Über-Ich.

Traditionell-pneumologische Kollegen pflegen sich zuweilen distanzierend-abwehrend über den „unterdrückten Schrei nach der Mutter" (Alexander, 1968) zu amüsieren. Häufig erregen sie dann zustimmendes Gelächter in entsprechenden Vorlesungen und Fortbildungsveranstaltungen. Hier liegt Unkenntnis oder Unverständnis vor. Was mit dieser Kurzformel ausgedrückt werden soll, ist die Formulierung der symbiotischen Verdichtung eines spezifischen ontogenetischen Interaktionsmusters, das für den Asthmapatienten wie dessen Arzt von großer Tragweite ist.

Zur Entwicklung des Ambivalenzkonfliktes und der Störung der psychosexuellen Reifung

In Übereinstimmung mit heute vorherrschenden Auffassungen wird die grundlegende psychische Störung des Asthmatikers in die frühe orale Phase datiert (de Boor, 1965):

„Als das Fundament in der Pathogenese des Asthmas möchten wir ... die Störung in der oralen Phase der Triebentwicklung bezeichnen und die ihr zugehörende schwere Störung der Mutter-Kind-Beziehung. ... Die offenen oder abgewehrten oralen Impulse waren maßlos, unersättlich durch das Gefühl, niemals genug bekommen zu haben. Überdies zeigte die bei allen zu beobachtende Toleranzbreite, das Bestehen auf sofortige Befriedigung, das Alles oder Nichts der Forderung, die Fixierung der libidinösen Bedürfnisse auf ganz früher Ebene, in der das Ich als Regulativ noch keine Wirk-Funktion hatte. Pathognomonisch scheint die unauflösliche Ambivalenz gegenüber dem (Teil-)Objekt: Befriedigung bei gelungener Inkorporation eines ‚guten Objektes'; oder aber der Wunsch nach völliger Verschmelzung mit dem Objekt kann nach Frustration abrupt abgelöst werden von aggressiven Zerstörungsimpulsen, die sich gegen das gleiche Objekt richten. Wenn sie in der Phase noch fluktuierender Subjekt-Objekt-Grenzen auf das Objekt verschoben projiziert werden, wird dieses unmittelbar zum Verfolger, der nach erfolgter Introjektion des Objektes als verinnerlichte Gefahr zu vehementen Abwehranstrengungen Anlaß gibt" (S. 262). – Ferner führt de Boor aus, daß sich der Asthmatiker von Patienten mit andersartigen Störungen mit psychogener Komponente (z.B. Ulcus duodeni) dadurch unterscheidet, daß er mit einer besonderen Vehemenz das Befriedigung versprechende Objekt abstoßen muß. Wahrscheinlich geschieht es aus der Angst heraus, das Objekt vor seiner zerstörenden Wut nicht bewahren zu können.

Diese Überlegungen sind demjenigen nachvollziehbar, der entweder selbstreflektierend seinen Asthmazustand erlebt oder als Arzt mit dem wütend-verzweifelt-abwehrenden Asthmatiker durch geduldiges Abwarten und allmähliches Sprechen vom Status in einen erträglichen klinischen Zustand hinüberwechselt

oder der als Therapeut tiefgreifende Regressionsstufen erlebt hat. Aber auch die oben erwähnten Szenen aus der Bostoner Medizinischen Poliklinik sind in ihrer ganzen Zwiespältigkeit ebenso aus dieser Grundkonstellation ableitbar wie das Verhalten unserer Patientin. Im neuen Krankenhaus kommt sie erst dann zur Ruhe, als sie ihre Befürchtungen ausgesprochen hat, von „bewährten Therapieschemata" überrollt zu werden.

Aus psychoanalytischer Sicht ergeben sich aus diesen frühen Störungen Konsequenzen für die Entwicklung im analen Abschnitt. In diesem Abschnitt lernt der Mensch, libidinöse und aggressive Triebregungen in Verbindung zu bringen, d.h. in einem nahestehenden Menschen gleichermaßen positive und negative Anteile **akzeptierend** wahrzunehmen. Haben Asthmatiker die erwähnten oralen Schwierigkeiten hinter sich gebracht, so können sie diese positiven und negativen Anteile zwar sehen, aber nicht integrieren. Borderline-Menschen zeichnen sich im Vergleich zu Asthmatikern dadurch aus, daß sie das Problem der Aufteilung in Positiv und Negativ nicht wahrnehmen und infolgedessen sich auch nicht immer wieder neu in elementarer Weise damit auseinanderzusetzen haben. – Zur Entwicklung dieser Zeit sagt de Boor in bezug auf Asthmatiker:

„Allen Kranken gemeinsam war die mit dem Rückzug in die Phantasie und die Regression auf die anale Phase eingetretene Entmischung der libidinösen und aggressiven Triebregungen, die gegen das Objekt gerichtete Aggression und ihre Repräsentanz in den Phantasien. Die Abfuhr sadistischer Impulse in den Phantasien wurde vom Ich der Kranken als gefährlich, zerstörerisch erlebt und war Veranlassung zu umfassenden Schutz- und Abwehrmaßnahmen." – Gleichzeitig macht de Boor eine wichtige Beobachtung: „Wir konnten bezüglich dieser Phase keine Unterschiede zwischen Asthmatikern und anderen allergisch Kranken konstatieren. ... Wichtig scheint jedoch zu sein, daß die Allergiker ihre zwischenmenschlichen Konflikte mit den Objekten der analen Phase verschoben haben auf nicht-menschlich belebte (Tiere) oder unbelebte (Federn, Staub, chemische Substanzen) Objekte, die sie für ihre Allergie anschuldigen" (S. 260 f.).

Es ist nicht verwunderlich, daß auch die Entwicklung im nachfolgenden phallischen Abschnitt ausnahmslos gestört ist. Von verschiedenen Autoren wird eine Behinderung der reifen Sexualität beobachtet. Jores sagt: Der Asthmatiker sei **hingabegestört.** Aber diese Hingabestörung des Asthmatikers sei gleichzeitig verbunden mit dem starken Wunsch nach Hingabe und nach Liebe: „Ein Nein zu der Welt, ein Nein zu dem Schmutz, ein Nein zu den vielen sich inkorrekt verhaltenden Menschen, Wut darüber, daß diese sich so schlecht benehmen und gleichzeitig die Sehnsucht nach der Geborgenheit und nach der Liebe. De Boor hat diesen Zustand treffend geschildert als eine Karikatur von Zuwendung und Abwendung" (Jores, 1980).

Mit de Boor ist anzunehmen, daß die sexuelle Funktionsfähigkeit nur dann nicht gestört ist, wenn der Asthmatiker den sexuell erstrebten Partner dominieren kann. Andererseits lassen sich nahezu regelhaft Impotenz und Frigidität nachweisen. Beim Manne sind hierbei Entwicklungsstörungen im Sinne des „negativen Ödipus" festzustellen. Hiermit soll gesagt werden, daß sich der Sohn als Liebesobjekt des Vaters erlebt und sich hierbei an den starken Anteilen der Mutter in seiner Phantasie ausrichtet (partielle Identifikation mit einer als dominierend-phallisch erlebten Mutter). – Für die Frau besteht die analoge Entwicklung in einer partiellen Identifikation mit dem Vater der ödipalen Phase. Dieser wird als Repräsentant aggressiv und gefährlich phantasierter Sexualität gesehen.

Zum Verhältnis von Ideal-Ich und Über-Ich

Für die Beziehungsmedizin wichtige Aussagen macht de Boor, indem er Lampl-de Groot folgend zwischen Ideal-Ich und Über-Ich unterscheidet und deren Entwicklungsstörungen beim Asthmatiker verfolgt. Ideal-Ich und Über-Ich werden als „Substrukturen der Ich-Organisation" verstanden:

„Dem Ideal-Ich als Funktionszentrum wird eine bedürfnisbefriedigende, wunscherfüllende Aufgabe zugeschrieben, dem Über-Ich eine einschränkende und verbietende. Erstes beinhaltet: „Ich bin wie die Eltern", letztes dagegen: „Ich muß tun, was die Eltern verlangen". Ferner wird ausgeführt, daß beim Asthmatiker regelhaft die Entwicklung des Ideal-Ichs gestört ist. Einerseits fühlt sich der Patient äußerst klein; andererseits kann er sich aber auch dadurch wieder sehr groß fühlen, daß er die Eltern mit Omnipotenz ausstattet und sich dann mit ihnen identifiziert (de Boor, 1965, S. 269f.). Im stärker gestörten Fall findet sich die Verschmelzung mit dem idealisierten Objekt, an dessen Allmacht man blindlings teilhat. Bei der ausgeprägtesten Störung einer noch früheren Stufe (früheste „oral-passive Erlebnisebene") sind die „passiven Bedürfnisse des Kindes und die aktiven des Objektes Mutter noch in symbiotischer Entsprechung"!

Menschen mit derart gestörtem Ideal-Ich stehen nun einer Umwelt gegenüber, die in ihrem Empfinden ein außerordentlich hohes Anspruchsniveau hat.

„An diese frühen Quellen der Über-Ich-Bildung (die also im Gegensatz zur normalen Entwicklung prä-ödipal abgelaufen ist; Anmerkung d. A.) bleiben die Kranken fixiert, und sie sind weitgehend unfähig, durch spätere korrigierte Neuerfahrungen in ihrem Über-Ich andere Regulative einzubauen, sie bleiben die Sklaven der ungemilderten, prä-ödipalen Dressate (Trieb- und Ich-Einschränkung)". – An anderer Stelle: „Wir haben in unseren Krankengeschichten wiederholt die den Asthmaanfällen zugeordneten Konfliktsituationen interpretiert. Drohende (phantasierte) Frustrationen oder als bedrohend erlebte (libidinöse, aggressive) Triebregungen waren stets die Kristallisationskerne für die Entstehung des Asthmaanfalles" (de Boor, 1965, S. 273).

Dem Asthmaanfall ausgeliefert, fühlt sich der Patient klein oder auch als ein Teil des ihn behandelnden Arztes, der übermächtig ist. In diesem Arzt sieht er mütterliche Eigenschaften, die er benötigt, um seinen Omnipotenzansprüchen nachzukommen; die er gleichzeitig abwehren muß, um den Ängsten vor einer Überwältigung durch das drohende, böse Objekt zu entkommen. Über die Patienten wird gesagt: „Sie kämpfen mit beunruhigenden Phantasien und Impulsen, sie sind nicht verarmt, aber sie sind zutiefst in Konflikte verstrickt" (Knapp, 1980).

Hieraus sind wichtige Folgerungen für das therapeutische Vorgehen ableitbar: Der Arzt vertritt nicht nur die Gebots-, d.h. Über-Ich-Anteile, sondern er hat auch die Chance, Ideal-Ich-Anteile an den Patienten heranzuführen. Indem der Patient beide Anteile in dem ihn behandelnden Arzt wahrnimmt, kann eine Annäherung der bis dahin getrennten Inhalte von Ideal-Ich und Über-Ich erfolgen. Insbesondere im Umgang mit aggressiv-bestrafenden Impulsen kann der Patient erleben, daß er nach deren Äußerung keineswegs fallengelassen wird, weil er das stützende und tragende Mutterobjekt im Arzte verloren hätte; vielmehr bleibt ihm der Arzt erhalten und der Patient kann dann wahrnehmen, daß unter seiner Wut auch liebende und fürsorgliche Elemente lebendig sind, die dem Arzt gelten (vgl. Richter, 1988a).

42.4.3 Eine literarische Darstellung

Es gibt ein eindrucksvolles Zeugnis, wie das Verhalten der Mutter in der Kindheit von einem Menschen erlebt wurde, der später an Asthma erkrankte und dessen äußerst eingeschränkte Lebensführung das eigentliche soziale Verhalten des Asthmatikers illustriert: Marcel Proust, der in späteren Jahren fast nie sein Zimmer verließ – und wenn, dann nur bei Dunkelheit –, schildert die allabendliche Szene, in der er als Kind auf den Gutenachtkuß seiner Mutter wartete:

„Mein einziger Trost, wenn ich schlafen ging, war, daß Mama heraufkommen und mir einen Kuß geben würde, wenn ich bereits lag. Aber dies Gutenachtsagen dauerte nur so kurze Zeit, sie ging so bald schon wieder, daß der Augenblick, da ich sie heraufkommen und dann in dem Gang mit der Doppeltür das leise Rascheln ihres Gartenkleides aus blauem Musselin mit kleinen strohgeflochtenen Quasten hörte, für mich ein schmerzlicher Augenblick war. Er kündigte schon den nächsten an, der auf ihn folgen sollte, wo sie mich verlassen haben und wieder unten sein würde. Das ging so weit, daß ich mir beinahe wünschte, dies von mir so heiß ersehnte Gutenachtsagen möge erst so spät wie möglich stattfinden und die Gnadenfrist, in der Mama noch nicht gekommen wäre, zöge sich recht lange hin. Manchmal, wenn sie, nachdem sie mich geküßt hatte, die Tür öffnete, um zu gehen, wollte ich sie zurückrufen und ihr sagen: Gib mir noch einen Kuß, aber ich wußte, daß sie dann auf der Stelle ihr strenges Gesicht zeigen würde, denn das Zugeständnis, das sie meiner Trauer und Aufregung machte, indem sie heraufkam und mit diesem Friedenskuß Gutenacht sagte, verdroß jedesmal meinen Vater, der das Zeremoniell übertrieben fand; viel lieber hätte sie mich diesen Wunsch, diese Gewohnheit aufgeben sehen, als mich auch noch darin zu unterstützen, daß ich einen zweiten Kuß von ihr wollte, wenn sie schon an der Tür war. Hatte ich sie nun aber erzürnt, so machte das die ganze Beschwichtigung meines

Herzens, die sie mir einen Augenblick zuvor geschenkt hatte, als sie ihr liebevolles Antlitz über mein Bett neigte und es mir darbot, wie die Hostie einer Friedenskommunion, bei der meine Lippen ihre leibhafte Gegenwart und die Kraft, einzuschlafen, von ihr empfingen, zunichte."

42.4.4 Psychologische Befunde

Die Suche nach Zusammenhängen von bestimmten psychosomatischen Erkrankungen und Persönlichkeitstypus, also die Suche nach typischen Persönlichkeitsprofilen des Kolitikers, Rheumatikers und eben auch des Asthmatikers ist fast so alt wie die Psychosomatische Medizin selbst. Vermutlich geht dieser Forschungsansatz zurück auf Flanders Dunbar (1938, 1947), die derartige, letztlich statische Charakterstereotypien als Endergebnis frühkindlicher Erfahrungen und späterer lebensgeschichtlicher Einflüsse beschrieb. Es gibt eine nahezu unübersehbare Zahl von Publikationen, die – in der Regel mit Fragebogen gemessene – Persönlichkeitsauffälligkeiten des Asthmatikers beschreiben (Übersicht bei Kerekjarto et al., 1981). Es mag auch in der Handlichkeit und Ökonomie des Forschungsinstruments Fragebogen begründet sein, daß einerseits wenig theoriegeleitete Gruppenvergleiche durchgeführt wurden und man sich andererseits durch die Folgenlosigkeit derart erzielter Ergebnisse nicht beirren ließ: Aus geringfügigen (statistisch signifikanten) Mittelwertunterschieden zu einer gesunden „Norm"-Population wurde unbesehen die spezifische Persönlichkeit des Asthmatikers abgeleitet. Nur sporadisch wurde dem naheliegenden (und von Pneumologen zu Recht geäußerten) Gedanken Rechnung getragen, daß eventuell nachweisbare spezifische Persönlichkeiten letztlich auch die Folge der immerwährenden existentiellen Bedrohung durch die Krankheit sein könnten, d. h. das Ergebnis einer mehr oder weniger gelungenen seelischen Bewältigung der Angst vor dem Erstickungstod. So berichten etwa Meyer und Weitemeyer (1967) über positive Korrelation von Krankheitsdauer und Introversion bzw. Neurotizismus, was die These der Krankheitsdependenz von Persönlichkeitsauffälligkeiten des Asthmatikers stützt.

Die wenigen Untersuchungen (M. v. Kerekjarto et al., 1981), die diese methodischen Erfordernisse berücksichtigen, d. h. Kontrollgruppen einbeziehen und die Replizierbarkeit ihrer Untersuchungsergebnisse überprüfen, finden keine einheitlichen stabilen Persönlichkeitsauffälligkeiten, die für den asthmatischen Patienten spezifisch wären. Selbst die oft beschriebene „nach innen gerichtete Aggressivität" oder die seelische Überempfindlichkeit des Asthmatikers lassen sich auch bei Patienten mit anderen Erkrankungen in Abgrenzung zu gesunden Kontrollpersonen finden.

Vielleicht sind manche Untersuchungsergebnisse über die „typische Persönlichkeit" des Asthmatikers so resistent gegen empirische Überlegungen, weil sie unserem Stereotyp des Asthmatikers entgegenkommen. Die Ergebnisse von Kerekjarto und Mitarbeitern (1981) mit dem Gießen-Test (Selbst- und Fremdbild)

können dazu herangezogen werden, diese Hypothese teilweise zu bestätigen: Asthmatische Patienten werden von ihren psychotherapeutischen Interviewern depressiver, zwanghafter, phantasieärmer, kontaktscheuer, mißtrauischer, Konflikte stärker abwehrend, gehemmter und weniger hingabefähig beschrieben als sie sich selber im Vergleich zu anderen Menschen einschätzen.

Die mehr als 30jährige Suche nach dem typischen Persönlichkeitsprofil – nicht zu verwechseln mit der Suche nach spezifischen Konflikten und ihren Abwehrformen – des Asthmatikers muß heute als weitgehend erfolglos gelten. Sie sollte daher – auch aus den oben erwähnten forschungsmethodischen Schwierigkeiten (Krankheitsdependenz, Spezifität) – abgebrochen werden: „Dementsprechend finden wir unter Asthmatikern viele Persönlichkeitszüge: aggressive, ehrgeizige, streitsüchtige Menschen, waghalsige und auch überempfindsame, ästhetische Typen; manche Asthmatiker sind Zwangscharaktere, während andere eine mehr hysterische Natur zeigen. Der Versuch, ein charakteristisches Persönlichkeitsprofil zu definieren, wäre aus diesem Grunde vergeblich; ein solches Profil existiert nicht" (Alexander, 1971, S. 97). Etwas anderes ist es, Untergruppen von asthmatischen Patienten im Sinne einer Taxonomie etwa aufgrund von Daten zur Sozial- und Krankheitsanamnese sowie der subjektiven Symptomatik zu beschreiben (vgl. Richter et al., 1985).

42.4.5 Die Aufrechterhaltung des Asthma bronchiale durch psychosoziale Faktoren (sog. psychomaintenance)

Die Identifikation von psychologischen Faktoren, die eine bereits vorhandene chronische Erkrankung aufrechterhalten oder verschlimmern oder auch die subjektive Beeinträchtigung durch diese Erkrankung beeinflussen, führte in den letzten Jahren zu dem Konstrukt der „psychologischen Aufrechterhaltung" („psychomaintenance") auch des Asthma bronchiale.

So wurde gefunden, daß die mit den asthmatischen Beschwerden verbundene Angst (vgl. 42.2 mit dem Hinweis auf die Asthma-Symptom-Liste) außerordentlich bedeutsam ist für die Häufigkeit, mit der Asthmapatienten ihre Medikamente und hier insbesondere die Dosieraerosole benutzen. Bei den Untersuchungen diente das Ausmaß der Selbstmedikation als Indikator der subjektiven Beeinträchtigung durch das Asthma. Sollte die Beeinträchtigung also allein von dem Maß der obstruktiven Atemwegsbehinderung abhängen, könnten psychologische Faktoren keine wesentliche Rolle spielen (vgl. Abb. 42–2).

Dies gilt nun offenbar, wie Abbildung 42–2 zeigt, nur für diejenigen Patienten, deren asthmaspezifische Angst in einem mittleren Bereich liegt. Solche Patienten hingegen, deren asthmatische Atemnot von ausgeprägter nervöser Ängstlichkeit, ja zuweilen von panischer Angst begleitet ist, dosieren ihre Medikamente unabhängig vom Schweregrad der somatischen Symptomatik und bringen sich hierdurch unter Um-

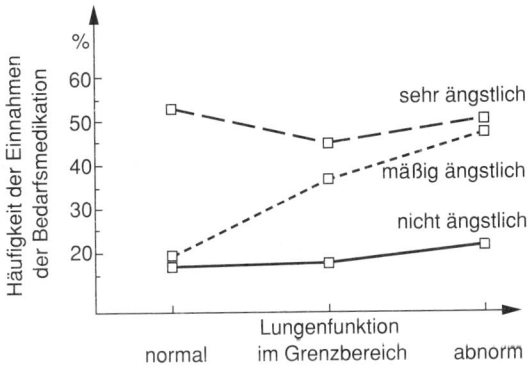

Abb. 42–2. Zusammenhang von Lungenfunktion, Häufigkeit der Einnahme der Bedarfsmedikation (in % der berücksichtigten Tage) und asthmaspezifischer Angst (Dahlem et al., 1977).

ständen in Gefahr. Eine noch größere Risikogruppe sind diejenigen Patienten, die ihre Atembeschwerden kaum beachten und auf die Atemnot eher gleichgültig reagieren, die „Unterdosierer". Sie setzen selbst dann die Medikation nicht ausreichend ein, wenn sie aufgrund der Lungenfunktionsbefunde notwendig wäre. Die Folge ist, daß die stärkeren Medikamente, in letzter Konsequenz Cortison, undifferenzierter verordnet werden müssen.

Die unterschiedlichen Formen der Angstwahrnehmung und -bewältigung wirken sich somit sowohl auf die subjektive Beeinträchtigung durch die Krankheit, als auch auf die medikamentöse Behandlung durch den Arzt aus. Ängstliche Patienten erhalten bei ihrer Entlassung aus der stationären Therapie eher stärkere Dosierungen (z.B. Cortisongaben) verordnet als nicht ängstlich wirkende Patienten. Diese Unterschiede können nicht durch die Schwere der Erkrankung erklärt werden (Dirks et al., 1977). Sowohl unsensible, an somatischen Befunden orientierte Ärzte als auch diejenigen, die empathisch auf ihre Patienten eingehen, unterliegen diesem „Fehler" gleichermaßen (Dirks et al., 1978).

Die besondere Bedeutung der Angst wurde von der Forschergruppe am National Asthma Center in Denver/Colorado wiederholt belegt. Es handelt sich hierbei sowohl um eine situative, asthmaspezifische wie um eine überdauernde persönlichkeitstypische Angst, die für die psychologische Aufrechterhaltung des Asthma bronchiale wichtig ist (Kinsman et al., 1981). Die unmittelbar auf die lebensbedrohlichen asthmatischen Symptome bezogene situative Angst könnte dabei durchaus als Angstsignal verstanden werden (vgl. Freud, GW XIV, S. 117f.), das vor dem Affekt der ursprünglichen traumatischen Situation warnt und hierdurch die Abwehrleistungen ermöglicht. Ihr so verstandener ökonomischer Nutzen würde sich dann in der besseren Prognose derjenigen Patienten widerspiegeln, die über diese Möglichkeit (im Sinne einer Ich-Funktion) verfügen, unlustvolle Affekte abzuwehren. Der möglicherweise darauf folgende asthmatische Anfall wäre dann als der Versuch zu verstehen, diesen Affekt auf der körperlichen Ebene abzuwehren, nachdem die Abwehr auf der psychischen Ebene versagte.

Die andere Angstform, die auch in diesen Untersuchungen ein überdauerndes Persönlichkeitsmerkmal darstellt, müßte dann eher mit den traumatischen Ereignissen in den frühen Phasen der psychosexuellen Entwicklung und den jeweiligen phasentypischen Angstformen in Zusammenhang stehen, wobei Trennungsängste von zentraler Bedeutung sind. In der Tat konnten in beschreibenden Untersuchungen bei Patienten mit hohen Angstwerten häufiger frühe Trennungen von nahen Bezugspersonen oder überfürsorgliche Versorgungsmuster exploriert werden (Dirks et al., 1979). Aufgrund dieser beiden Angstscores und weiterer psychologischer Tests (Einstellungen des Patienten zu seiner Krankheit) konnte die amerikanische Forschergruppe voraussagen, wie häufig innerhalb von sechs Monaten nach der Entlassung eine Rehospitalisierung der Patienten erforderlich wurde (vgl. Tab. 42–1).

Tabelle 42–1. Rehospitalisierung innerhalb von 6 Monaten nach Entlassung (in %) (Dirks et al., 1981).

		Asthmaspezifische Angst (ASL)		
		niedrig	mittel	hoch
allgemeine Ängstlichkeit (MMPI)	niedrig	43,8	33,3	*
	mittel	30,0	22,3	7,8
	hoch	*	37,9	33,3

Dieses Ergebnis zeigt eindrucksvoll die prädiktive Validität des Konstrukts der psychologischen Aufrechterhaltung des Asthma bronchiale. Insbesondere die Risikopatienten lassen sich mit hoher Genauigkeit erfassen: Patienten, die entweder hohe oder niedrige Werte der überdauernden, persönlichkeitstypischen Angst aufweisen, haben unabhängig von ihrer asthmaspezifischen Angst eine schlechtere Prognose und eine größere Rehospitalisierungsrate. Die beste Prognose haben diejenigen Patienten, die – verbunden mit mittleren Werten der persönlichkeitstypischen Angst – hohe Werte der asthmaspezifischen Angst aufweisen. Man könnte sagen, daß sie auf ihre asthmatischen Symptome ängstlich-nervös achten und bereits geringfügige Obstruktionen als Hinweis auf eine asthmatische Krise bewerten. Hinsichtlich ihres sonstigen Lebens nehmen sie jedoch eine ausgeglichene Haltung ein und bewerten weder überschießend noch zu träge.

Gerade aufgrund der unter Umständen erheblichen therapeutischen Tragweite dieser Befunde sollten diese erst dann in der therapeutischen Routine berücksichtigt werden, wenn sie von anderen Arbeitsgruppen unter vergleichbaren Bedingungen repliziert werden konnten. Es ist anzustreben, diesen Versuch der Identifikation von Risikopatienten, also einen taxonomischen Forschungsansatz, weiter zu verfolgen, um dann gegebenenfalls entsprechende therapeutische Interventionen zu entwickeln. Ein derartiges Vorgehen wäre sicherlich nicht nur für den betroffenen Patienten hilfreich, sondern auch kostensenkend (Dirks et al., 1980).

42.5 Psychophysiologie

Ziel klinisch orientierter psychophysiologischer Untersuchungen bei psychosomatischen Krankheiten ist es herauszufinden, ob sich bestimmte psychologische Bedingungen im betroffenen Organsystem besonders auswirken (nämlich im „locus minoris resistentiae") und dadurch möglicherweise zur Verstärkung des „symptomspezifischen" pathogenen Mechanismus beitragen.

Hinsichtlich des Bronchialasthmas führt dieses Konzept der Symptomspezifität zu folgender Fragestellung: Wirken sich psychische Belastungen in respiratorischen Funktionen stärker aus als in anderen Funktionssystemen (z.B. Herz-Kreislauf-System, ZNS oder Skelettmotorik) und bewirken sie damit eine zusätzliche Beeinträchtigung der Lungenfunktion des Asthmatikers?

Bei Asthmatikern müßte sich also eine Symptomspezifität darin äußern, daß sich physiologische Kennwerte der Atemwegsobstruktion (Atemwegswiderstand, Sekundenkapazität, Atemstoß oder auch die relative Exspirationsdauer) unter psychischer Belastung vergleichsweise stärker oder schwächer verändern als Kennwerte anderer Organsysteme.

Dekker und Groen (1957) stellten bei ihren Untersuchungen einen ihnen wichtig erscheinenden pathophysiologischen Faktor in der Genese des Asthma bronchiale fest: Sie fanden, daß die Erhöhung des Atemwiderstandes, der sich dann als exspiratorischer Stridor äußert, auch willkürlich – sogar von gesunden Versuchspersonen – durch Erhöhung des intrathorakalen Drucks ausgelöst werden kann. Jeder kann bei geöffneter Glottis und forcierter Ausatmung unter Anspannung der Bauchmuskulatur exspiratorischen Stridor erzeugen. Dabei kommt es zu einer Kompression der Hinterwand der intrathorakalen Trachea und der großen Bronchien. Dazu genügt bereits ein Überdruck von 1 mm Hg. Asthmatiker produzieren während des Anfalls viel höhere Drücke, nämlich bis zu 6 mm Hg. Dekker und Groen nehmen nun an, daß vielfach am Anfang der Atemschwierigkeiten des Asthmatikers ein solcher Prozeß zu beobachten ist. Erst hiernach kommt es zu einer Schleimproduktion und durch Lernprozesse bedingt auch zu einer Konstriktion der Bronchiolen. Diese Theorie ist vielfach diskutiert worden und bietet möglicherweise die Erklärung für häufig zu beobachtende schnelle Änderungen im Atemwegswiderstand der Asthmatiker.

42.5.1 Untersuchungen zur Symptomspezifität

In bisherigen psychophysiologischen Untersuchungen an Asthmatikern wurden Atemkennwerte oft nur aus dem Pneumogramm erhoben. Zur Identifikation obstruktiver Veränderungen bei Asthmatikern ist dann jedoch unbedingt eine gesonderte Betrachtung von In- und Exspirationszeiten vonnöten (vgl. Cohen et al., 1975). So läßt ein verlängertes Exspirium, wie dies Stevenson und Ripley (1952) sowie Cohen und Mitarbeiter (1975) bei Asthmatikern während belastender Gedanken bzw. negativer Emotionen berichten, auf eine obstruktive Veränderung der Atemwege schließen. Die Hamburger Forschergruppe fand eine signifikant verlängerte Exspirationszeit der Asthmatiker unter Ruhebedingungen im Vergleich zur nichtallergischen, nicht-psychosomatischen Kontrollgruppe und zu Neurodermitikern. Während unter emotionaler Belastung die In- und Exspirationszeit in der Kontrollgruppe und bei den Neurodermitikern abnahmen (die Atemfrequenz sich also erhöhte), veränderte sich bei den Asthmatikern die ohnehin schon verlängerte Exspirationszeit nicht gegenüber dem Ruhewert (Dahme und Richter, 1981). Etwas vereinfacht lassen sich diese Untersuchungen in dem Sinne interpretieren, daß sich das respiratorische System der Asthmatiker in seinem Zeitverhalten unter psychischer Belastung als „starrer" erwies.

Das Leitsystem des Asthma bronchiale ist jedoch der erhöhte Atemwegswiderstand. Seit jüngerer Zeit ist eine hinreichend valide Schätzung des Atemwegswiderstandes (im unteren und mittleren Bereich) mit der sog. forcierten Oszillationstechnik möglich (Nolte und Korn, 1979), einem offenen System, das auch unter ambulanten Bedingungen leicht eingesetzt werden kann und für die Probanden wenig belastend ist. Mit Hilfe dieser Methode konnte Levenson (1979) nachweisen, daß der Atemwegswiderstand bei Asthmatikern während der Darbietung von drei emotional belastenden Filmen, insbesondere während emotional bedeutender Szenen, anstieg, während er bei der gesamten Kontrollgruppe nahezu unverändert blieb. Im ersten Film wurde ein asthmatisches Kind im Krankenhaus gezeigt, im zweiten ein Industrieunfall und im dritten eine Mutter, die ihr kleines Kind zur Adoption freigibt.

Die größten und andauernden Veränderungen im Atemwiderstand – bezogen auf das Ausgangsniveau vor der Filmdarbietung – wurden bei den Asthmatikern während der Ankündigung und Darbietung des asthmatischen Kindes, also im ersten Film gefunden. Dieser Effekt ist zumindest hinsichtlich der untersuchten physiologischen Kennwerte (keine Veränderungen in der Herzfrequenz) und hinsichtlich der untersuchten Personengruppe (kein Effekt bei der Kontrollgruppe) spezifisch.

Im Film über einen Industrieunfall ergaben sich sehr unterschiedliche Effekte:
- Anstieg des Atemwegswiderstandes bei den Asthmatikern, aber keine Veränderung bei der Kontrollgruppe;
- Verringerung der Herzfrequenz nach Ankündigung und während der Darbietung des Films nur bei der Kontrollgruppe, nicht bei den Asthmatikern.

Keine signifikanten Effekte ergaben sich im dritten Film – der Adoptionsszene – mit einer spezifischen Ausnahme: In dem Moment, in dem im Film die Mutter ihr neugeborenes Kind an die Adoptionsstelle übergibt, steigt nur bei den Asthmatikern der Atemwegswiderstand deutlich an. Der interpretative Brückenschlag zu den analytischen Befunden, bei denen es sich im Kern um eine Trennung von oder eine Verschmelzung mit der Mutter dreht, liegt nahe.

Zu ähnlichen Ergebnissen gelangt die Untersuchung von Kuhn und Mitarbeitern (1981): Es wurde eine Gruppe von erwachsenen Asthmatikern und eine Gruppe von neurotischen Patienten in mehreren Belastungsphasen verglichen: S_1 = Interviewsituation, die den individuellen „relevanten unbewußten Konflikt beinhaltet" (der vorher exploriert worden war), unter S_2 wurden alle „psychodynamisch eher neutralen" Belastungssituationen zusammengefaßt (Lärmbelastung etc.), S_3 = Interviewsituation, in der „die eigene Hilflosigkeit angesprochen" wird. Im oszillatorisch gemessenen Atemwegswiderstand ergaben sich folgende signifikanten Effekte:

- Bei den Asthmatikern steigt der Atemwegswiderstand in der „Konfliktsituation" stärker an als in den „psychodynamisch neutralen" Situationen;
- in der Situation, in der der persönliche Konflikt angesprochen wurde, ist der Atemwegswiderstand der Asthmatiker deutlich größer (0,50 kPa/l/sec) als bei den Neurotikern (0,35 kPa/l/sec).

Die Ergebnisse dieser beiden Untersuchungen lassen vermuten, daß Asthmatiker also nicht generell auf psychische Belastungen mit der symptomatischen Erhöhung des Atemwegswiderstandes reagieren, sondern daß dies nur in bestimmten Situationen zu beobachten ist, die sicherlich durch weitere Untersuchungen noch überprüft und spezifiziert werden müssen.[2]

42.5.2 Operante Kontrolle und Modifikation des Atemwegswiderstandes (Biofeedback)

Wenn also – wie die obigen Untersuchungen vermuten lassen – der Atemwegswiderstand auch durch psychologische Faktoren beeinflußt wird, ist es dann nicht auch möglich, ihn mit psychologischen Mitteln – z.B. operantem Lernen – so zu modifizieren, daß es zu einer symptomatischen Besserung der asthmatischen Atemnot kommt?

Nachdem dies in einer ersten Untersuchung von Vachon und Rich (1976) versucht worden ist, konnten wir in einer ersten experimentellen Serie bei 3 von 8 Asthmatikern eine signifikante Erniedrigung des oszillatorisch gemessenen Atemwiderstandes durch Biofeedback erzielen (Leplow et al., 1985).[3]

42.5.3 Zur Wahrnehmung von Atemwegswiderständen (interozeptive Wahrnehmung)

Die Dyspnoe umfaßt nicht nur das subjektive Gefühl der Atemnot in ihren verschiedenen Erlebnisdimensionen (vgl. 42.2.1) oder die subjektive Bewertung der Obstruktion, sondern auch die sensorische Wahrnehmung der obstruktiven Atemwegsveränderung, d.h. die interozeptive Wahrnehmung des Atemwegswiderstandes (vgl. auch die Definition von Comroe, 42.2.1). Wie groß muß die Obstruktion sein, so lautet die Frage, damit ein Patient über Atemnot berichtet? Wie groß muß ein Strömungswiderstand sein, damit ein asthmatischer Patient ihn gerade schon als solchen wahrnimmt?

Zu dieser Frage nach der sensorischen Wahrnehmung von Widerständen im Atemstrom, d.h. die Frage nach der spezifischen Wahrnehmungsfähigkeit von asthmatischen Patienten, gibt es in der Literatur eine größere Zahl ausgezeichneter experimenteller Untersuchungen, deren Ergebnisse sich jedoch zum Teil widersprechen (Übersicht bei Richter, 1985). So kann es bislang nicht als gesichert gelten, daß Asthmatiker Atemwegswiderstände grundsätzlich besser oder schlechter wahrnehmen als Lungengesunde. Die größere Variabilität der Sinnesschwellen läßt jedoch darauf schließen, daß es innerhalb der Patientengruppe ausgeprägt hypersensitive bzw. hyposensitive Patienten gibt, die bereits geringfügige Erhöhungen ihres endobronchialen Atemwegswiderstandes bemerken oder selbst ausgeprägte Veränderungen nicht wahrnehmen.

Es gibt zwei weitere plausible Erklärungen für die Widersprüchlichkeit der Resultate, die auf einer Kritik an der experimentellen Methodik basieren.

- Asthmatiker unterscheiden sich möglicherweise in ihrer Antworttendenz, Änderungen des Atemwegswiderstandes mitzuteilen. Hierdurch kann die Beurteilung von physiologischen Sinnesschwellen beeinflußt werden.
- Bisherige psychophysiologische Untersuchungen zum Atemwegswiderstand beruhten auf extern in den Atemstrom eingeschalteten, also nicht endobronchialen Widerständen. Ein solches Vorgehen ist aus klinischer Sicht zumindest fragwürdig. – Erst mit der Ganzkörperplethysmographie durchgeführte Untersuchungen konnten zweifelsfrei nachweisen, daß nur geringe Zusammenhänge zwischen der subjektiven Einschätzung der Atemwegsbehinderung (Dyspnoe) und den unterschiedlichen Kennwerten der Lungenfunktion bestehen (Heim et al., 1972).

Zur Untersuchung dieser zweiten Frage ließ die Hamburger Forschergruppe (Richter, 1985) 33 Patienten mit allergischem Asthma bronchiale ihre subjektiv er-

---

2 Allerdings finden sich auch bei gesunden Probanden unter Experimentalbedingungen systematische Veränderungen des Atemwegswiderstandes (Richter et al., 1980); dagegen finden sich bei Asthmatikern wiederum durchgehend psychophysiologische Reaktionstendenzen im Sinne erhöhter Potentiale des M. occipito-frontalis unter emotionaler Belastung. Bei Kindern wurde diese Beobachtung auch erfolgreich therapeutisch verwertet (Kotses et al., 1978).

3 In einer Nachuntersuchung, diesmal unter Berücksichtigung des funktionellen Residualvolumens, konnte das Ergebnis wiederholt werden, allerdings mit einem wesentlich geringeren Effekt (Vachon und Rich, 1976a). Der Vergleich beider Untersuchungsergebnisse belegt u.a. die Schwierigkeit und Komplexität psychophysiologischer Untersuchungen im Bereich der Atmung.

lebte Atemnot fortlaufend während einer inhalativen Provokation im Ganzkörperplethysmographen einschätzen. Die intraindividuellen Korrelationen zwischen dem Atemwegswiderstand und der subjektiven Dyspnoe sind gering. Andererseits konnten in dieser Untersuchung einzelne Patienten als extrem gute Schätzer, andere als extrem hyposensibel beurteilt werden.

Erste Ergebnisse zu beiden oben genannten Problembereichen liegen vor. Hiernach stützen korrelationsstatistische Zusammenhänge zwischen Schätzgenauigkeit und testpsychologischen bzw. klinischen Variablen das Konstrukt der psychologischen Aufrechterhaltung des Bronchialasthmas (vgl. 42.4): Patienten, die ihren endobronchialen Atemwegswiderstand eher schlecht einschätzen können, haben höhere Werte in der Skala „nervöse Ängstlichkeit" der Asthma-Symptom-Liste (vgl. 42.2.1) und rufen häufiger den Notarzt als diejenigen Patienten, die den Grad ihrer Obstruktion in Übereinstimmung mit den Lungenfunktionsbefunden einstufen (Richter, 1985).

42.6 Ätiologie und Pathogenese; wissenschaftstheoretische Überlegungen

42.6.1 Ätiologie und Pathogenese

Als charakteristische Zeichen der Krankheit waren Atemnot und Symptome der Bronchialobstruktion genannt worden. Atemnot und Bronchialobstruktion treten anfallsweise auf und entsprechen einem akuten Anstieg des Atemwegswiderstandes (vgl. 42.2). Durchgehend wird beobachtet, daß die Atemwege des Asthmatikers **starrer** sind, auf einen **spezifischen** Konflikt bevorzugt reagieren und daß die resultierenden Atemwegswiderstände in einer für den einzelnen Menschen spezifischen Weise (hypo-, hypersensitiv) eingeschätzt werden. – Andererseits werden drei Hauptgruppen auslösender Faktoren beim Asthma unterschieden (Weiner, 1977): immunologische, infektiöse und psychogene Hauptfaktoren.

An anderer Stelle (Richter, 1986) wurde erhoben, daß bei der Auslösung eines Anfalls in 47% Allergene, in 80% ein Infekt und in 67% psychische Faktoren anamnestisch gesichert werden konnten. Auf die Vielzahl der einzelnen auslösenden Faktoren geht Weiner (1987) detailliert ein.

Bei den immunologisch bedingten Auslösern handelt es sich vorwiegend um eine Allergie vom Typ I. Der Hauptantikörper ist das Immunglobulin E. Dieses verbindet sich auf der Oberfläche der Mastzellen mit entsprechenden Antigenen. Der hierauf einsetzende biochemische Prozeß hat das Freisetzen von Histamin und anderen Mediatoren zur Folge. Es kommt zu Sofort- wie verzögerten Reaktionen. In der Sofortreaktion spielen hauptsächlich vasoaktive, in der verzögerten Reaktion chemotaktische Mediatoren eine Rolle. Die Einwirkung des Histamins bedingt eine reflexartige Kontraktion der glatten Muskulatur der großen und kleinen Luftwege; die Spätreaktion,

die durch chemotaktische Mediatoren vermittelt wird, hat vaskuläre Schädigungen zur Folge (Davies, 1981). Beim infektiösen Auslöser handelt es sich um virale oder bakterielle Infektionen. Es wird angenommen, daß diese zu einer erhöhten Empfindlichkeit der Atemwege gegenüber vasokonstriktiven Substanzen führen. Dies gilt insbesondere für Kinder. Am überzeugendsten scheint die Einwirkung von Viren belegt (Davies, 1981).

Zwischen 20 und 50% aller Asthmapatienten entwickeln eine Bronchokonstriktion, wenn sie Plazeboaerosol inhalieren, von dem sie annehmen, daß es ein bronchokonstriktives Reagens enthält. Diese Reaktion ist unabhängig von Reaktionen auf Histamin oder Methacholine. Es handelt sich um eine postganglionäre cholinergische Aktivität, die mit Atropinsulfat blockiert werden kann. Mit anderen Worten: Emotionale Reaktionen können in den efferenten vagalen Anteilen des parasympathischen Nervensystems vermehrte Aktivität auslösen, die eine Bronchokonstriktion mit entsprechenden Beschwerden erzeugt (Bonshey, 1981).

Wechselbeziehung der Hauptfaktoren

So klar sich diese Einteilung in die drei ätiologischen Hauptfaktoren anhört, so kompliziert wird die Angelegenheit bei der längerbestehenden Krankheit. Im oben beschriebenen Fall der Patientin in der Bäckerei liegt zwar eine allergische Diathese vor, doch sind die asthmatischen Anfälle unabhängig von allergischen Reaktionen. Auch findet sich keine Beziehung zwischen Infektionen und Atemnot. Ähnliches wird aus der Literatur berichtet: Ein 45jähriger Mann reagiert zunächst gegen eine Reihe von Antigenen mit schwachen Allergien. Erst als Schwierigkeiten am Arbeitsplatz auftreten, werden die allergischen Reaktionen heftiger, bis sich schließlich ein Asthmaanfall ereignet (Araujo et al., 1973). Auch über Wechselwirkungen wird berichtet: Eine Beurteilung von asthmatischen Kindern auf ihre allergische Diathese hin zeigte, daß psychopathologische Reaktionen in den Gruppen mit geringer Diathese häufiger, solche Reaktionen in der Gruppe mit starker Diathese aber seltener waren (Block et al., 1964). Interessanterweise zeigten die ebenfalls untersuchten Mütter eine ähnliche Verteilung ihrer psychosomatischen Eigenschaften: Eine schwächere Allergieausprägung war mit einer stärkeren Neigung zu überprotektiven Verhaltensweisen verbunden (zur Bedeutung psychosomatischer Aspekte bei allergischen Erkrankungen vgl. Richter und Ahrens, 1988).

Möglicherweise sind die einzelnen ätiologischen Faktoren in ihrer pathogenetischen Wirksamkeit genauer zu verstehen, wenn der Theorie gefolgt wird, daß beim Asthmatiker die Stimulierung intakter Betarezeptoren nicht in ausreichender Weise erfolgt und hierdurch etwa einer allergiebedingten Bronchokonstriktion nicht entgegengewirkt werden kann (Szentivanyi, 1968). Hierbei wird diskutiert, ob es sich bei dem Mangel an reaktiven Betarezeptoren um eine angeborene bzw. früherworbene Eigenschaft oder durch

Behandlung (adrenergische Substanzen) bedingte Eigenschaft des Organismus handelt. Unter Cortisongabe wird die Zahl verfügbarer Betarezeptoren erhöht, so daß die Wirkung von Cortison hierdurch erklärt werden könnte (Davies, 1981).

Es ist nun vorstellbar, und sehr viele klinische Beobachtungen sprechen hierfür, daß bereits unter der **Vorstellung** einer Inhalation bronchokonstriktiver Substanzen in der **Realität** eine Bronchokonstriktion ausgelöst wird. Das Atemwegssystem des Asthmatikers wird nachfolgend aus seinem labilen Gleichgewicht gebracht. Für derartige psychosoziale Auslöser kommen gleichermaßen vorweggenommene Zusammentreffen mit Allergenen wie mit Konflikten in Frage. Handelt es sich gar um eine der beschriebenen überängstlichen Persönlichkeiten, so wird es um so schneller zur Katastrophenreaktion kommen. Eine wesentliche Rolle spielt hierbei die (Auto-)Suggestion, auf die nachfolgend eingegangen werden soll.

Auf die Bedeutung dieses Faktors hat insbesondere die Arbeitsgruppe von Groen aufmerksam gemacht. Groen schildert die zugrundeliegenden Beobachtungen in einer plastischen Weise:[4]

Eines Tages kam Dekker – damals junger Assistenzarzt – zu Groen, um dessen psychosomatische Theorien zu widerlegen und zu beweisen, daß Asthma allein auf allergische Ursachen zurückgeführt werden kann. Dekker untersuchte Asthmapatienten mit einer Überempfindlichkeit gegen Pollen, Hausstaub und andere Allergene, indem er ein Aerosol inhalieren ließ, in dem das betreffende Allergen in verschiedenen Verdünnungsgraden enthalten war. Er konnte damit bei einem Teil der Patienten regelmäßig einen Asthmaanfall provozieren. Zu seiner Überraschung stellte er aber fest, daß zwei der Patienten bei der Wiederholung der Inhalation mit einem Aerosol, in dem sich kein Allergen mehr befand, auf die gleiche Weise reagierten. Bei ihnen ließ sich die Reaktion schließlich bereits bei der Einatmung von Sauerstoff, ja sogar durch die Einführung des sterilisierten Glasmundstücks des Inhalators hervorrufen, das nicht einmal mehr mit dem Inhalationsapparat verbunden war.

Lerntheoretisch könnten diese Abläufe folgendermaßen interpretiert werden: In diesen Experimenten wurden die spezifischen Allergene als unbedingte Reize (UCS) verwendet, während die neutrale wäßrige Lösung und der Sauerstoff, die beide inhaliert wurden, sowie das Mundstück des Inhalators als drei verschiedene zu bedingende Reize (CS) dienten.

Experimentell zeigten Dekker und Mitarbeiter (1957) dann, daß nach sukzessiver Darbietung der Reize – jeder für sich im Abstand von mehreren Tagen – die zu bedingenden Reize allein zu einer Reduktion der Vitalkapazität führten, und interpretierten diese Befunde im Sinne der klassischen Konditionierung. Schon 1965 bezweifelte Purcell jedoch die Validität dieser Experimente. Eine klassische Konditionierung von Asthmaanfällen würde dadurch nicht belegt werden. Folgende Bedenken müssen geltend gemacht werden (vgl. auch Weiner, 1977):

Außer dem Reiz „Mundstück" wurde kein zu bedingender Reiz simultan mit dem unbedingten Reiz „Allergen" entsprechend dem üblichen Paradigma der klassischen Konditionierung angeboten. Dieser Reiz weist aber keinen Lernerfolg

auf, wie er üblicherweise bei klassischer Konditionierung zu erwarten wäre, nämlich eine mit der Anzahl der Koppelungen zunehmende Reduktion der Vitalkapazität. – Beim Reiz „Sauerstoff" handelt es sich vermutlich um eine Pseudokonditionierung, d.h., dieser Reiz ist wahrscheinlich auch ohne Kombination mit dem UCS „Allergen" über die „irritant receptors" imstande, eine Bronchokonstriktion herbeizuführen. – Die „Konditionierung" ließ sich überhaupt nur bei zwei Personen durchführen, trotz vieler vergeblicher Bemühungen bei anderen. Sie ließ sich auch bei diesen, nach einer Suggestions- und Gruppentherapie zur „Entwöhnung" von den „konditionierten" Asthmaanfällen, nicht mehr wiederholen.

Viel wahrscheinlicher ist es, daß diese experimentellen Effekte durch **Suggestion** – also auch einem psychologischen Mechanismus –, nicht aber durch klassische Konditionierung hervorgerufen wurden. Bestärkt wird diese Interpretation durch experimentelle Untersuchungen über die Bedeutung suggestiver Effekte bei der Bronchokonstriktion.

42.6.2 Wissenschaftstheoretische Überlegungen[5]

Es besteht die Aufgabe, die vorgelegten klinischen Beobachtungs- und wissenschaftlichen Forschungsergebnisse in einen Gesamtzusammenhang zu bringen, d.h., sie unter konzeptionellen Gesichtspunkten zu verstehen.

Als wesentliche bio-psycho-soziale Merkmale des Asthmatikers waren herausgearbeitet worden: die „Starre" des Bronchialsystems (vgl. 42.5.1), der Ambivalenzkonflikt und die Angst; als krankheitsverursachende bzw. -reaktive Faktoren waren immunologische, infektiöse und psychogene benannt worden. Weiterhin war herausgearbeitet worden, daß diese Merkmale zeitüberdauernd, möglicherweise lebensbegleitend bestehen; doch die Perioden von Krankheit und Gesundheit schwanken in Abhängigkeit von der Lebenssituation des Patienten. Dies trifft auch für unsere Patientin zu, deren Krankengeschichte die Qualität ihrer Beziehungen zur Familie, zu den Menschen in der Bäckerei, in der Buchhaltung und im Krankenhaus widerspiegelt.

Damit werden die Beziehungen zur Orientierungshilfe bei der Sichtung vorliegender Beobachtungsergebnisse. Der Betrachter geht nicht mehr vom Organismus oder der Umgebung eines Patienten aus, sondern von den Beziehungen beider, durch die sie sich gegenseitig definieren. Damit entstehen zwei Hauptprobleme: Es sind die für den Asthmatiker wichtigsten Funktions- und Situationskreise zu benennen und es sind Integrationsebenen in einer Hierarchie von Subsystemen, Systemen und Suprasystemen darzustellen.

---

4 Herrn Prof. Dr. J.J. Groen danken wir herzlich für seine Diskussionsbeiträge bei der Erstfassung dieses Buchabschnittes (1. und 2. Aufl.).
5 Herrn Prof. Dr. Th. von Uexküll möchten wir sehr herzlich für seine Diskussionsbeiträge beim Abfassen dieses Abschnittes danken.

Die individuelle Wirklichkeit des Asthmatikers:
Funktions- und Situationskreise

Gehen wir von der Situation der hier vorgestellten
Patientin aus, so können wir drei Situationskreise
und einen Funktionskreis unterscheiden:

Situationskreis „Ich als Gesunder und meine individuelle Wirklichkeit"

35 Jahre lang war die Patientin gesund. Nach der vorübergehenden ersten Krankheitsepisode, die sich zur
Reinigungszeit in der Backstube abspielte, war sie
wiederum acht Jahre weitgehend gesund. Wir können
vermuten, daß die Patientin in dieser Zeit ihre Umgebung in einem Sinne wahrnimmt, daß hieraus eine zu
bewältigende individuelle Wirklichkeit entsteht, d.h.,
in der rezeptorischen oder Merk-Sphäre des Situationskreises werden Problemsituationen aufgebaut,
die gelöst werden können. Sie verfügt über zufriedenstellende Programme, um diese in der effektorischen
oder Wirk-Sphäre ihres Situationskreises zu lösen.

In diesem Situationskreis sind wesentliche Stellgrö-
ßen: Das Selbstwertgefühl der Patientin, das durch
eine zufriedenstellende Abstimmung zwischen den
Über-Ich-Forderungen und dem Ideal-Ich befriedigt
werden kann; ein Umgang mit der Mutter und Mutter-Substituten, der eine angstvermeidende Beherrschung der tatsächlichen Mutter bzw. deren Imagines
in verschiedenen Erscheinungsformen ermöglicht;
ein Umgangsstil, der Körpersensationen (im Merk-
Sektor) eine spezifische Bedeutung erteilt, die (im
Wirk-Sektor) mit den entsprechenden motorischen
Reaktionen verbunden sind (sensomotorische Zirku-
larreaktionen nach Piaget); der Umgangsstil ist durch
eine mittelgradig ausgeprägte ängstliche Persönlich-
keitshaltung und durch besorgte Reaktionen auf
durchbrechende Körperempfindungen verbunden.

Funktionskreis auf der pneumologischen Ebene

Die Umgebung wird entsprechend dem grundlegenden Ambivalenzkonflikt als Stimulus wahrgenommen, vermehrt atmen zu müssen und gleichzeitig in
der Abgabe der Atemluft behindert zu werden. Möglicherweise setzt der von Groen und Mitarbeitern
(1960) beschriebene Zirkel ein, in dem es zur Erhöhung des intrathorakalen Druckes mit der Einengung
der großen Atemwege und nachfolgender Schleim-
hautanschwellung und noch späterer Mukussekretion
kommt. Möglicherweise kommt es aber auch über vermehrte vagale Aktivität auf dem Boden einer Konditionierung bzw. **Bedeutungskoppelung** bereits primär
zu einer Bronchokonstriktion. Diese Entwicklung
spielt sich im effektorischen Bereich des Funktionskreises ab. Sie hat einen Zirkel zur Folge, in dem die
Umgebung als noch stärker bedrohlich erlebt wird.

Situationskreis „Ich als Kranker und meine Umgebung"

Die Umgebung wird als eine fordernde und gleichzeitig die Anerkennung (soziale Unterstützung) entzie-
hende Wirklichkeit wahrgenommen. Die Patientin ist

verunsichert. Bei unserem Falle fordern die Bäckersfrau und acht Jahre später der Geschäftsinhaber zu
viel von der Patientin; gleichzeitig befürchtet die Patientin, daß sich die Bäckersfrau wie der Geschäftsinhaber ihr entziehen könnten. Recht deutlich wird bei
der Bäckersfrau, daß diese ein Mutter-Substitut darstellt; im zweiten Fall wird befürchtet, daß die Mutter
sogar leiblich zu nahe kommt, indem sie daheim einzieht. Die Patientin fühlt sich vom Ehemann im Stich
gelassen, der sie bisher gestützt hat und sie gegenüber
der Mutter eine distanzierende Stellung einnehmen
ließ.

Im effektorischen Zirkel des Situationskreises fehlen Programme, die Problemsituation zu lösen. Die
Körpersensationen haben eine angsterregende Bedeutung (sie werden als Warnsignale vor Erstickung, Vernichtung erlebt). Sie führen zu einer Einschränkung
der Möglichkeiten, sich mit der Umgebung auseinanderzusetzen. Gleichzeitig werden von dieser Umgebung der Stridor und die Atemnot als Appell um Hilfe
wahrgenommen. Das hat einen sich selbst verstärkenden Zirkel zur Folge. In der rezeptorischen Sphäre
nimmt die Patientin nun wahr, daß sich ihre Umwelt
unsicher-helfend verhält und sie muß nun ihrerseits
mit noch größerer Unsicherheit reagieren usw.

Situationskreis „Rolle des Arztes in der Auslösesituation"

Hierunter ist die Situation der Kranken zu verstehen,
die diese mit den Ärzten aufbaut. Als 35jährige fand
die Patientin einen Hausarzt, der ihren Anfall mit
Cortisongaben erfolgreich bewältigte. Anders acht
Jahre später, als offensichtlich eine mangelhafte Interaktion zwischen Arzt und Patientin besteht. Sie erlebt
sich im Krankenhaus als „Routinefall", d.h. als nicht
ernstgenommen.

Sie gerät in einen panikartigen Zustand, der auch
in der Umgebung Angst auslöst. Allein die Gewißheit,
ja der fast magisch anmutende Glaube, in die große
Klinik verlegt zu werden, läßt den Status abklingen.
Möglicherweise spielen hier (auto-)suggestive Momente wie bei der Dekkerschen Patientin (s.o.) eine
Rolle.

Hierarchie von Subsystemen, Systemen und
Suprasystemen

Es geht um das Verhältnis der Situationskreise und
des Funktionskreises zueinander. Von zentraler Bedeutung sind hier: die Zeitgestalt, der Begriff der
Stimmung, das pragmatische Realitätskriterium, das
kommunikative Realitätskriterium.

Zur Zeitgestalt: Der Anschaulichkeit halber sollte
man sich die oben genannten vier Kreise so überein-
andergelegt im Raume vorstellen, daß sie einen Zylinder ergeben. Die durch ihren Mittelpunkt gezogene
Achse stellt den zeitlichen Ablauf eines Menschenlebens dar, hier den unserer Patientin.

Normalerweise fügt sich der Funktionskreis kon-
zentrisch in den Situationskreis „ich als Gesunder
und meine individuelle Wirklichkeit". Der Situations-

kreis umhüllt den Funktionskreis und stellt die „unsichtbare, aber reale Hülle" (v. Uexküll und Wesiack, vgl. Kap. 1) dar, mit der wir alle leben. Weitgehend unberührt von dieser äußeren Hülle läuft im Inneren des Atmungsgesunden der autonome Atem-Funktionskreis ab. Dies ist anders beim Asthmakranken. Hier überschneiden sich rezeptorische Schenkel des Atmungs-Funktionskreises und des Situationskreises „ich als Gesunder und meine individuelle Wirklichkeit". Es liegt hier eine Bedeutungskoppelung vor, die für den Asthmatiker kennzeichnend ist. Durch eine Bedeutungskoppelung wird die Umgebung zu einer Umwelt, die unmittelbar in den Funktionskreis eingreift, so daß dieser nicht mehr ausschließlich durch die Blutgaswerte bestimmt wird. Die Bedeutungskoppelung kann unterschiedlich intensiv erfolgen: Es können nur wenige Phantasien sein, in denen lebende Beziehungspersonen auftauchen und das asthmatische Atemmuster auslösen. Es können dann aber auch durch längere Lernvorgänge unbelebte Gegenstände (de Boor, 1965) und schließlich organische Antigene werden.

Im Status asthmaticus liegt möglicherweise eine extreme exzentrische Verlagerung des Funktionskreises aus dem erwähnten Situationskreis heraus vor: Es überschneiden sich jetzt rezeptorische Schenkel des Situationskreises und effektorische Schenkel des Funktionskreises in einem gegenläufigen Sinne. Es kommt zur totalen Diskontinuität aller bisherigen Lebensvorgänge.

Während sich beim Gesunden eine individuelle Zeitgestalt entwickelt, die praktisch eine geradlinige Achse verfolgen läßt, um die sich ohne große Reibungsverluste Funktions- und Situationskreise bewegen, ist dies beim Asthmatiker anders. Hier findet sich keine geradlinige Achse; bewegen sich die Situationskreise um sie, kommt es je nach Art des Situationskreises zu erheblichen Störungen in den Bewegungsabläufen. Es wird dann die Aufgabe von Arzt und Patient, die Stellen größter Abweichungen von einer gedachten geradlinigen Achse zu identifizieren.

Zur Stimmung: Nachfolgend ist zu sehen, durch welche Faktoren die Entwicklung der Zeitgestalt bestimmt wird. – Es war gesagt worden, daß sich für unser psychosomatisches Modell von Gesundheit und Krankheit der Begriff der „Stimmung" am ehesten eignet, um auf den Ursprung der individuellen Entwicklungsgeschichte zurückzugehen. Aus der Stimmung entwickeln sich Motive, aus den Motiven Programme, die zu Handlungen führen. Unsere Wahrnehmungen und entsprechend unsere Handlungen seien nur, so die Feststellung in den einführenden Kapiteln dieses Lehrbuches, die Abbilder der „eigenen sensomotorischen Akte".

Stimmungen gehen auf ganz frühe, aus dem symbiotischen Funktionskreis stammende Empfindungen zurück. Dieser symbiotische Funktionskreis hat die Aufgabe, eine neue Integrationsebene aufzubauen, die den Körper und die Umwelt umfaßt. Es ist eine „animalische" (vgl. Kap. 1) Integrationsebene.

Im symbiotischen Funktionskreis wird Körper (d. h. geschlossenes System ohne Umwelt, in etwa vergleichbar der Pflanze mit ihrer Wohnhülle) mit der Mutter als früheste Umwelt als Einheit erlebt. Körper und Mutter bzw. früheste Umwelt (es treten jetzt mit Auge und Ohr Fernsinne in Aktion) sind noch eins („Subjekt-Objekt-Identität" von Winnicott). Diese Einheit wird als Stimmung erlebt, die durch Stimmungssignale aufrechterhalten werden muß. Solche Stimmungssignale kommen aus der einfacheren Integrationsebene (Körper) oder der sich bildenden komplexeren. So ist z. B. Hunger ein Stimmungssignal, das die Gefährdung der Einheit signalisiert. Füttern und Wiegen sind Stimmungssignale, welche die Intaktheit der Einheit signalisieren usw.

Bei der Atmung läßt sich denken, daß Stimmungssignale, welche bei gesunder Interaktion von der Mutter adäquat beantwortet werden, keine adäquate Antwort finden – und daß dann Atmung und Vokalisation zu Signalen gekoppelt werden, die ihrerseits wieder eine Bedeutungskoppelung zur Angst haben.

Man könnte geradezu sagen: Exspiration bedeutet Ausatmen von Umwelt und Eintauchen in reines Körper-sein; vorausgesetzt Körper und Umwelt konnten als Identität erlebt werden. Es scheint so, daß beim Asthmatiker diese Identität nicht in ausreichender Weise erlebt wurde. Es ist auffallend, daß Asthma bronchiale erst jenseits des zweiten Lebensjahres beobachtet wird, wenn also das Kind die Trennungs- und Wiederannäherungsphase (Mahler et al., 1980) durchläuft. Exspirationshemmung läßt sich eventuell als ein Festhalten an der Atemluft in einer Situation verstehen, in der Ausatmen nicht zeitweise Rückkehr zu der Homöostase des „bloßen Körperseins" vor der Geburt, sondern Vernichtung der keimenden Individualität bedeutet. Von dort aus ist die Feststellung besonders interessant, daß Kinder unter zwei Jahren kein Asthma bekommen. Mit zwei Jahren beginnen die ersten symbolischen Besetzungen der Präobjekte, wenn die sensomotorische Umwelt, die im Sinne Piagets noch keine Trennung von Subjekt und Objekt kennt, durch die beginnende Vorstellung eine andere Struktur erhält, in der Subjekt und Objekt sich gegenübertreten. Jetzt kann die Umwelt symbolische Bedeutung für die Mutter und Umweltverlust Vernichtung der keimenden Subjektivität bedeuten. Von hier aus läßt sich auch ein Zugang zu der Bedeutung der Gerüche für den Asthmatiker gewinnen („Die erste Umwelt ist eine Geruchsumwelt", Th. von Uexküll).

Das Kind kann also nicht mit der notwendigen Intensität und Dauer die „proprio-interozeptive" Entwicklung durchmachen, sondern wird auf eine „sensorisch-perzeptive" nächste Entwicklungsphase verwiesen.

Zum pragmatischen Realitätskriterium: Dieses vermutlich aus der Alarmreaktion abgeleitete Realitätskriterium (vgl. Kap. 1) gibt darüber Auskunft, mit welcher Sicherheit Programme zur Bewältigung von Problemlösungen eingesetzt werden können. – Beim Asthmatiker geht es um die Erfahrung, inwieweit genügend Sauerstoff zur Verfügung steht. Im relativ geschlossenen System führt O_2-Mangel zu Sensationen (Merk-Zeichen), diese bewirken Atemtätigkeit (Wirk-

Zeichen), womit a) das relativ geschlossene System in ein offenes übergeht (Körper holt Luft = Umwelt) und b) die prognostische Erwartung eingelöst (oder nicht eingelöst) wird, daß das Merk-Zeichen (Atemnot) durch das Wirk-Zeichen ausgelöscht wird, womit das System wieder in den relativ geschlossenen Zustand übergeht. Atmung wird damit zu einem Pendel zwischen zwei Systemebenen. Beim Asthmatiker könnte das auf dem pragmatischen Realitätskriterium basierende Urvertrauen, d.h. eine Selbstverständlichkeit, in die Verfügbarkeit von Luft und Umwelt gestört sein. Die vielfach von Asthmatikern berichteten Erfahrungen, beim Tauchen nicht an die Wasseroberfläche zu kommen, überempfindlich gegen Gerüche zu sein (Jores, 1980), sind höchstwahrscheinlich Deckerinnerungen an diese ganz frühe Zeit.

Zum kommunikativen Realitätskriterium (vgl. Kap. 1): Im Unterschied zum pragmatischen Realitätskriterium, das ein Kriterium für die Richtigkeit einer Handlung darstellt und daher ein Nacheinander impliziert, setzt dieses Prinzip keine Handlung voraus. Es ist das Kriterium der Zuverlässigkeit, Sicherheit, in der man von dem anderen akzeptiert und verstanden wird, ebenso wie man ihn versteht und akzeptiert.

Winnicott (1974) macht die Unterscheidung zwischen Objekt-Beziehung und Objekt-Verwendung. Unter dem ersten versteht er die allerfrüheste Subjekt-Objekt-Identität, die kein Handeln, keine Triebdynamik enthält, sondern in reiner Form das Gefühl zu sein ist. Winnicott spricht von dem „weiblichen Anteil" in jedem Menschen. – Unter Objekt-Verwendung versteht er dagegen triebinduziertes Handeln, in dem der „männliche Anteil" wirksam ist. Das kommunikative Realitätskriterium würde also aus der Objekt-Beziehung erwachsen, die in der Mutter-Kind-Identität ihre Wurzel hat. Das pragmatische Kriterium würde der Objekt-Verwendung entsprechen.

Beim Asthmatiker scheinen ganz frühe Störungen im Aufbau der symbiotischen Einheit, d.h. der Subjekt-Objekt-Identität, vorzuliegen. Es ist zu fragen, wie das kommunikative Realitätskriterium die Zeitgestalt des Asthmatikers in deren Entwicklung bestimmt.

42.7 Diagnostik und Therapie

42.7.1 Grundmerkmale

Beschwerdebild: Vor dem Hintergrund der Überlegungen zur Ätiologie und Pathogenese werden die Dimensionen der Asthma-Symptom-Liste interpretierbar im Sinne einer gestörten Beziehung zwischen dem Patienten und seiner Umwelt: Der Patient fühlt sich nervös-ängstlich (Skala I), ärgerlich-gereizt (III), spürt Atemnot (II) und Hyperventilationszeichen (IV) und fühlt sich müde-träg (V).

Ambivalenzkonflikt: Der Arzt spürt seinerseits spiegelbildartig Signale der Ängstlichkeit, die zu unmittelbarer Zuwendung auffordern; andererseits Signale der ärgerlichen Gereiztheit, die zur Distanzierung vom Patienten drängen oder auch zur unterschwelligen bis offenen Aggression verführen. Damit ist stimmungsmäßig die Grundkonstellation des Ambivalenzkonfliktes gegeben. Er findet sich nahezu bei jedem Asthmatiker (vgl. Richter, 1988a).

Reaktion des Arztes: Jeder Arzt läuft Gefahr, für eine der beiden Seiten dieser ambivalenten Konfliktkonstellation Partei zu ergreifen. Im Falle unserer Patientin geschieht dies während des ersten Krankenhausaufenthaltes zugunsten der distanzierten Seite. Die Folge ist, daß sich der Patient nicht getragen und verstanden fühlt, d.h., es treffen Erwartungen entsprechend des kommunikativen Realitätskriteriums ein (s.o.). Eine Asthmatikerin formuliert ihre Bedürfnisse: „Ich brauche eine helfende Hilfe, keine einengende".

Der Ambivalenzkonflikt beginnt bereits mit den diagnostischen Anteilen des Erstgespräches. Es ist unerläßlich, den Patienten frei assoziierend seine Krankengeschichte berichten zu lassen und hierbei den von Engel entwickelten Schritten zu folgen (vgl. Kap. 12).

Während der Visite erzählt der Patient, wie er erstmals sechs Monate zuvor Atemnot verspürte, als er morgens den Pferdestall betrat. Der Arzt fragt nach Zeichen auffälliger Empfindlichkeit gegen Pferde. Diese liegen in der Tat vor. – Einem ebenfalls an der Visite teilnehmenden Arzt fiel die Mimik des Bauern während seines Berichtes über den erstmaligen Anfall auf. Er forderte den Mann auf zu berichten, was er gegen die Atemnot getan habe. Eigentlich habe er seinen Arzt holen wollen, so die Antwort. Aber … Nochmaliges Nachfragen: Eine Woche zuvor war sein einziger Sohn und Hoferbe kurz vor Übernahme des Hofes tödlich verunglückt, der fragliche Arzt hatte nur den Tod feststellen können. Trotz vielen Zuredens war der Sohn wieder ohne Sturzhelm gefahren.

Hier kommen zwei der drei erwähnten Hauptauslöser, d.h. der immunologische und der psychogene, in Frage. Der eine Arzt war der Gefahr erlegen, zu schnell „Partei" für eine der beiden Hauptgruppen zu ergreifen.

Angst und Angstbewältigungsformen: Anamnese und alle darauffolgenden Gespräche sollten das Ziel verfolgen, die Art und das Ausmaß der Angst zu erfahren. Dem Patienten sollte das Interesse des Untersuchers gerade an dieser Frage deutlich gemacht werden. Mit der Frage nach Bewältigungsformen, d.h. nach der spezifischen Ausprägung des pragmatischen Realitätskriteriums im Umgang mit der Angst, lenkt der Untersucher das Gespräch auf die wichtigsten Bezugspersonen. Man sollte sich zur Faustregel machen, daß nur derjenige Asthmapatient zufriedenstellend untersucht ist, dessen Beziehungen zur Mutter bzw. zu ihren Substituten erfaßt sind. Dabei wird die Mutter dem noch fremden Erstuntersucher in der Regel als positiv geschildert. Erst später kommt eine vorsichtige Kritik zum Vorschein, die eine „liebevolle Tyrannei" (J. J. Groen) durch die Mutter ahnen läßt. So sagte dann ein 30jähriger Mann mit chronischem Asthma: „Meine Mutter achtet auch bei 30 Grad im Schatten darauf, daß ich mich warm anziehe und nicht verkühle." – Anfangs hatte er gemeint, daß er die treusorgendste Mutter der Welt habe.

Zusammenfassend: Die Behandlung des Asthmapatienten erfordert, daß dessen Ambivalenz, Art und Ausmaß der Angst, Art der Beziehung zur Mutter/zu Mutter-Substituten und die Reaktionen des Arztes auf den Ambivalenzkonflikt erfaßt werden. – Um keine Mißverständnisse hinsichtlich psychosozialer Einseitigkeit aufkommen zu lassen: Selbstverständlich müssen immunologische und infektiöse Auslöser und die Reagibilität der Atemwege auf medikamentöse Substanzen abgeschätzt werden. Hierzu wird auf die entsprechende pneumologische Literatur verwiesen. Es gilt, daß wir drei Pathologien unterscheiden (W. Wesiack, 1984): die Beziehungspathologie, die Funktionspathologie und die Organpathologie. Die erste ist die übergeordnete; die beiden anderen Pathologien ordnen sich in den Rahmen ein, der durch die Beziehungspathologie vorgegeben wird.

42.7.2 Kombinierte Therapieformen

Erstmals wiesen Groen und Pelser (1960) an einem Kollektiv von 102 Asthmatikern der Amsterdamer Universitätsklinik nach, daß die kombinierte medikamentöse und Gruppentherapie der einfachen medikamentösen Therapie überlegen war. Die gruppentherapeutischen Sitzungen fanden wöchentlich zweimal über ein bis drei Jahre statt; die Katamnesen erstreckten sich über ein bis fünf Jahre. Diese Ergebnisse wurden von Ago und seinen Mitarbeitern an der Kyushu-Universität bestätigt (1976, 1980). Bei einer mindestens dreijährigen Katamnese kam es zu einer kompletten Remission oder zu einer deutlichen Verbesserung in 37 von 93 kombiniert behandelten, aber nur bei 6 von 73 traditionell behandelten Patienten (Ago et al., 1980). Es handelte sich um eine fünfstufige Therapie; sie umfaßte: somatische Behandlung und Aufbau einer psychosomatisch orientierten Arzt-Patient-Beziehung (I), eine psychologische Stabilisierung durch Abbau innerer Spannungen und durch das Erlebnis der Symptombehebung (II), das Erfahren psychosomatischer Korrelationen (III), eine Modifizierung verzerrter Wahrnehmungen und unangemessener oberflächlicher Anpassungsmechanismen (IV), die Beendigung der psychosomatischen Behandlung (V). – Die Therapieergebnisse hingen davon ab, welches Stadium erreicht wurde. Wurde das vierte oder fünfte Stadium erreicht, so waren 23 von 30 Patienten deutlich gebessert, während beim Abbruch im zweiten oder dritten Stadium lediglich 14 von 63 Patienten eine Besserung zeigten.

Im Prinzip ähnliche Ergebnisse, wenngleich mit Hilfe weitaus kleinerer Patientenzahlen erreicht, wurden in letzter Zeit auch in Deutschland vorgelegt (Deter, 1986). Deter bezeichnet als die wirksamen Elemente seiner von ihm so benannten „Coping-Gruppentherapie" für Asthmapatienten:

– Individuelle Information und Diskussion zur Krankheit.
– Autogenes Training nach J. H. Schultz oder funktionelle Entspannung nach M. Fuchs.
– „Die freie Interaktion der Mitglieder, die über die gemeinsame Erfahrung einer Krankheit zu einem lebhaften Erfahrungsaustausch und zu einer Gruppenkohärenz ... führt."

Die Ähnlichkeiten mit dem japanischen Vorgehen sind auffallend. Die psychodynamischen Elemente einer solchen Gruppentherapie werden vor allem unter Übertragungsaspekten ausführlich an anderer Stelle (Deter, 1988) diskutiert.

Eine wesentliche Verbesserung der Behandlungsergebnisse wurde durch die Änderung der institutionellen Abläufe erzielt (Davis et al., 1980). In einem britischen Kreiskrankenhaus wurde 1970 ein spezieller Versorgungsdienst für Asthmapatienten aufgebaut und die Behandlungsergebnisse im folgenden zehnjährigen Zeitraum bei 300 Asthmatikern erfaßt. Die einzelnen Versorgungsschritte wurden aufeinander abgestimmt: wöchentliche Ambulanz; eingehende Dokumentation; ständige Verfügbarkeit eines Krankenhausbettes; Zugriffsmöglichkeit auf die Intensivstation. – Wichtige Anregungen kommen aus der Pädiatrie. Unter der Mitwirkung von Psychiatern werden Versuche beschrieben, schwerkranke Asthmakinder zu versorgen. Es wirken Angehörige verschiedener Berufsgruppen zusammen, ohne die Beziehung zum niedergelassenen Arzt oder diejenige zur Familie zu unterbrechen (Maschia et al., 1976; Peshkin und Friedman, 1975; Piazza, 1981).

42.7.3 Spezielle psychotherapeutische Verfahren

Entspannende Verfahren: Muskelbezogene Entspannung allein bringt keine therapeutischen Effekte; kombiniert mit allgemein-entspannenden Methoden sind diese Verfahren mit Aussicht auf Erfolg einzusetzen (Erskine-Milliss und Schonell, 1981). In einer kontrollierten Studie an 54 Patienten war unter dem Einsatz von autogenem Training eine Verbesserung der Atemwerte (PEFR) nachweisbar, vergleichbar der Anwendung eines Bronchodilatators. – Nach 12 Monaten fühlten sich 70% der Patienten allgemein gebessert, 62% hinsichtlich ihres Asthmas. Die krankheitsbedingte Fehlzeit am Arbeitsplatz betrug vor der Behandlung 663 Tage, nach der Behandlung 77 Tage. – Wenngleich die methodischen Einschränkungen erheblich sind (Richter und Dahme, 1982), ist doch festzuhalten, daß sich hier weiterführende Schritte in der Therapieforschung abzeichnen. Es erscheint berechtigt, heute festzustellen, daß entspannende Maßnahmen zumindest kurzfristig, möglicherweise auch mittel- und langfristig Erfolge versprechen. Diese Form der Entspannung ist von tiefer Entspannung zu unterscheiden. Tiefe Entspannung (Meditation, Hypnose) wirkt günstig bei „stress-related diseases" wie Hypertonie oder Spannungskopfschmerz, aber beim Asthmaanfall ist diese Entspannungstherapie nicht von Vorteil: Die Stimulation des Parasympathikus führt zur Freisetzung von Acetylcholin und damit zur Bronchokonstriktion (Weiner, 1987). Nicht für jeden Asthmatiker sind entspannungsfördernde Maßnahmen geeignet (zur Bedeutung von Meditationsverfahren vgl. Goyeche et al., 1978, 1980).

Von zentraler Bedeutung dürfte bei den entspannenden Verfahren die Frage sein, was „Entspannung" in der individuellen Wirklichkeit eines Patienten bedeutet. Es könnte gut sein, daß das Gefühl des Versinkens für den Asthmatiker eine ähnlich panikerregende Bedeutung hat, wie unter Umständen das Ausatmen. Es sei auf die Überlegung im wissenschaftstheoretischen Abschnitt verwiesen, daß Exspiration möglicherweise gleichgesetzt werden könnte mit der Vernichtung keimender Individualität. Es ist wichtig, daß der atemgymnastisch vorgehende Therapeut eine Beziehung zum Patienten aufnimmt, in der die Bedeutung des Ausatmens geklärt und bearbeitet wird. Hierzu eignet sich in vorzüglicher Weise die funktionelle Entspannung, deren Urspünge aus der Behandlung eines asthmakranken Kleinkindes resultieren (Fuchs, 1989). Hier wird deutlich, wie psychotherapeutisch und körperlich fundierte Verfahren nicht mehr voneinander trennbar sind.

Verhaltenstherapie und Entspannungstherapie: Die optimistischen Folgerungen anderer Autoren (vgl. Knapp und Wells, 1978; Spevack, 1978) können wir nach einer kritischen Durchsicht der Literatur zu diesem Thema nicht bestätigen: Es gibt wenig empirisch begründbare Hinweise für die Effektivität verhaltenstherapeutischer Interventionen bei erwachsenen Asthmatikern (Richter und Dahme, 1982). Bei Kindern scheint es bessere Hinweise für den Nutzen derartiger Therapien zu geben, die jedoch hier nicht diskutiert werden sollen.

Systematische Desensibilisierung: Die in diesem Zusammenhang häufig zitierten Arbeiten von Moore (1965) und Yorkston und Mitarbeitern (1979) erweisen sich bei genauerer Betrachtung als methodisch derart angreifbar, daß die jeweils positiven Ergebnisse angezweifelt werden müssen (Richter und Dahme, 1982).

Der Schluß scheint erlaubt, daß es Hinweise für eine Verbesserung asthmatischer Beschwerden nach Meditation (TM) gibt, jedoch weder Entspannungsverfahren noch systematische Desensibilisierung generell zu positiven Therapieeffekten führen. Dennoch können sie, wie die klinische Erfahrung lehrt, im Einzelfall durchaus berechtigt sein. Hier soll lediglich vor der generellen, unreflektierten Verordnung angstreduzierender Verfahren (etwa autogenes Training) bei asthmatischen Patienten gewarnt werden. Wie die Ergebnisse der Denver-Gruppe vermuten lassen, kann die auf die asthmatischen Beschwerden gerichtete (Signal-)Angst durchaus schweren Krisen vorbeugen. Danach wäre es verhängnisvoll, diese asthmaspezifische Angst (medikamentös oder psychotherapeutisch) zu reduzieren. Wissenschaftlich gesicherte Erkenntnisse zur Indikation fehlen.

Analytische Einzeltherapie; Familientherapie: Die analytische Einzeltherapie erscheint in jenen Fällen indiziert, in denen auch ohne organische Krankheit ein solches Verfahren angebracht wäre. Zunehmend scheint sich herauszukristallisieren, daß Patienten mit akuten Asthmaschüben keine analytische Einzeltherapie empfohlen wird.

Mitte der 70er Jahre hat das Interesse familientherapeutisch orientierter Untersucher an der Behandlung von Asthmapatienten sprunghaft zugenommen. Familiendynamische Abläufe wurden kasuistisch beschrieben und hierbei auf die mangelnde Abgrenzung der einzelnen Generationen untereinander verwiesen (Neraal, 1980). Asthma bronchiale wird unter konzeptionellen Gesichtspunkten als pathologischer Lösungsversuch familialer Konflikte dargestellt (Overbeck et al., 1978). In Einzelfallbeispielen wird gezeigt, daß familientherapeutische Interventionen außerordentlich hilfreich sind und möglicherweise vor chronischer Krankheit bewahren können (Wirsching und Stierlin, 1982; Schönhals, 1984). Naturgemäß lassen sich noch keine endgültigen Beurteilungen abgeben; vielmehr sind diese widersprüchlich (Clark, 1981).

Der Asthmakranke und der Tod

Mit Absicht ist dieses Problem nicht bearbeitet worden. Sollte eine adäquate Darstellung erfolgen, würde der zur Verfügung stehende Platz nicht ausreichen. Zur allgemeinen Problematik sei auf die Kapitel „Schwerkranke" und „Sterbende" verwiesen; die speziellen Probleme sind in vorzüglicher Weise abgehandelt bei Knapp und Mitarbeitern (1966 a, b).

Herrn Prof. Dr. K. Lanser, Abt. Poliklinik des Zentrums für Innere Medizin (Leiter: Prof. Dr. P. von Wichert) der Philipps-Universität Marburg gilt unser besonderer Dank für eingehende Beratung beim Abfassen dieses Kapitels.

43 Ulcus duodeni

Wolfram Schüffel und *Thure von Uexküll*

Die Darstellung unserer Kenntnisse über Patienten mit ihrer Ulcus-duodeni-Krankheit gibt einen besonders guten Einblick in die Art, wie in der Psychosomatischen Medizin Konzepte entstehen. Sie gibt aber auch einen Einblick in die Schwierigkeiten der Hypothesenbildung und deren Überprüfung.

Zudem scheint es wenige Krankheiten zu geben, bei denen sich derartig schnell ein Panoramawechsel bemerkbar macht, wie dies beim Ulcus duodeni der Fall ist. Seit der Fertigstellung des Manuskriptes für die erste Auflage dieses Buches im Jahre 1978 hat sich folgende Entwicklung abgespielt:

- Das Cimetidin und andere Histamin-H$_2$-Rezeptorantagonisten wurden eingeführt und erzeugten einen ungeahnten therapeutischen Fortschritt.
- Der bereits vor der Einführung dieser Präparate beobachtete Rückgang der Häufigkeit des Ulcus duodeni setzte sich verstärkt fort.
- Unsere Kenntnisse über genetische und physiologische Besonderheiten dieser Krankheit haben einen wichtigen Zuwachs erfahren: Wir wissen heute, daß es sich um kein einheitliches Krankheitsbild handelt, sondern daß wir mindestens zwei Subgruppen unterscheiden müssen. Von diesen hat die eine einen normalen, die andere einen erhöhten Pepsinogenspiegel (Weiner, 1981). – Die wegweisenden Untersuchungen über bio-psycho-soziale Zusammenhänge von Weiner, Thaler, Reiser und Mirsky, sowie die von Mirsky auf der Basis der Ergebnisse dieser Untersuchungen entwickelten Hypothesen betreffen Patienten aus der Subgruppe mit erhöhtem Pepsinogenspiegel.
- Es ist zu keiner Nachuntersuchung dieser für die Theoriebildung so wichtigen Befunde gekommen. Zu der ebenso wichtigen Frage der bio-psycho-sozialen Zusammenhänge bei Patienten aus der Subgruppe mit normalen Pepsinogenwerten liegen bisher ebenfalls keine Untersuchungen vor.

Diese vier Feststellungen werfen folgende Fragen auf:

Lohnt es sich noch, auf psychosomatische Probleme des Ulcus duodeni einzugehen, wenn potente Pharmaka zur Verfügung stehen und dieses Krankheitsbild auch zahlenmäßig an Bedeutung verliert?

Spiegelt sich in den unterlassenen wissenschaftlichen Nachuntersuchungen ein unüberwindbares Desinteresse an psychosomatischen Fragestellungen, zumindest für das Ulcus duodeni, wider?

Wir werden auf diese Fragen eingehen und hierbei bemüht sein, die Ulkuskrankheit zumindest bei den Patienten mit erhöhtem Pepsinogenspiegel als den Ausdruck einer gestörten Lebenssituation auf dem Boden spezifisch geprägter Vorentwicklungen zu verstehen. Gleichzeitig soll versucht werden, Fragen zu formulieren, die wissenschaftlich-psychosomatischer Forschung zugänglich sind.

Eine 29jährige Patientin wird mit Oberbauchbeschwerden ins Krankenhaus eingeliefert. Sie treten gehäuft bei nüchternem Magen auf und bessern sich nach Nahrungsaufnahme. Zusätzlich berichtet die Patientin, nachts um 1.00–2.00 Uhr durch diese Schmerzen aufgeweckt zu werden. Die Beschwerden seien vor etwa vier Wochen aufgetreten; es seien die gleichen, die sie bereits früher während eines floriden Ulcus duodeni gespürt hatte. Nach mehrfachen Kuren seien damals keine Besserungen aufgetreten, und sie sei dann operiert worden.

Die Patientin ist alleinstehend. Sie ist Chefsekretärin und gilt als die rechte Hand ihres Vorgesetzten. Sie ist als älteste von vier Geschwistern von der alleinstehenden Mutter aufgezogen worden, nachdem sich der Vater von der Mutter getrennt hatte. Sie erinnert sich, die Kleider ihrer Geschwister gebügelt und für alle das Essen gekocht zu haben, um auf diese Weise die schwer arbeitende Mutter zu unterstützen. Sie hatten daheim in bescheidenen Verhältnissen gelebt. Weder die schwere Arbeit noch der häufige Verzicht auf das Spielen mit ihren Freundinnen, noch später der Verzicht auf Oberschule und Studium hätten ihr viel ausgemacht. Wichtiger sei ihr gewesen, der Mutter zur Seite zu stehen. Gleichzeitig habe sie ihre Freude daran gehabt, erfolgreich zu sein und später Anerkennung am Arbeitsplatz zu finden.

Die Familie war auseinandergegangen, und die Mutter war ins Ausland gezogen. Im letzten Jahr hatte die Patientin vieles getan, um die Mutter im langsam aufkeimenden Wunsch zu unterstützen, wieder aus dem Ausland zurückzukehren, um eine gemeinsame Wohnung mit der Tochter zu beziehen. Da entschloß sich die Mutter jedoch kurzfristig, weiterhin auf unbestimmte Zeit im Ausland zu bleiben.

Die Tochter akzeptiert diese Entscheidung mit Selbstverständlichkeit und ist, statt gekränkt zu reagieren, voller Verständnis für die Lage der Mutter. Sie erwägt sogar kurzfristig, der Mutter ins Ausland nachzuziehen und ihre jetzige Stellung gegen eine ungewisse Position einzutauschen. Einen Monat später wird sie mit den oben genannten Beschwerden ins Krankenhaus eingeliefert.

43.1 Klinik und Symptomatologie

Somatische Beschwerden und Befunde

Typischerweise schildert diese Patientin ihre Beschwerden als Nüchternschmerz. Anders ist das häufig beim Ulcus ventriculi, das nicht nur aus anatomischer und physiologischer Sicht, sondern auch aus psychosozialer Sicht vom Ulcus duodeni unterschieden werden muß. Beim Ulcus ventriculi treten die Schmerzen gehäuft in der Regel nach der Nahrungsaufnahme auf und werden besser, wenn der Magen leer ist. Gelegentlich unterscheiden sich die Schmerzen nicht von der Symptomatik des Duodenalulkus. Häufig finden sich beim Duodenalulkus jahreszeitliche Schwankungen zum Frühjahr und zum Herbst. Die Patienten geben oft Getränke und Speisen an, die ihre Beschwerden vermehren.

Röntgenologisch bzw. endoskopisch finden sich beim Ulcus duodeni in den meisten Fällen kurz hinter dem Pylorus ein oder mehrere Geschwüre im Bulbus duodeni. Bei narbiger Abheilung führen sie im Röntgenbild zur bekannten schmetterlingsförmigen Deformierung. Relativ selten sind pylorusferne Ulzera. – Laborchemisch fehlen bei unkomplizierten Fällen Entzündungserscheinungen. Eine beschleunigte Blutsenkung weist auf Komplikationen oder Begleiterkrankungen hin. Typisch für das Ulcus duodeni ist in den meisten Fällen der hyperazide Magensaft, der auch im Intervall gefunden wird. Allerdings lassen sich bei einer Gegenüberstellung von Basalsekretion und Sekretionsrate nach Stimulation verschiedene Sekretionsmuster unterscheiden (Ackerman und Weiner, 1977).

Verlauf

Meist heilen die Geschwüre in Tagen bis Wochen ab. Die Spontanheilungsrate ist sehr unterschiedlich und liegt in einzelnen Ländern zwischen 21 und 79% (Halter, 1978), wobei in Deutschland und in der Schweiz die 4-Wochen-Spontanheilungsrate ca. 50% beträgt (Sonnenberg et al., 1982). Bei 30% der Fälle kommt es zum progredienten Verlauf mit rezidivierenden Schüben, d.h. zur chronischen Ulkuskrankheit (Holtemüller, 1982). Es muß festgehalten werden, daß ein an Ulcus duodeni erkrankter Mensch in der Regel eine größere Chance hat, erneut zu erkranken als der Gesunde.

Vom chronischen Ulkus muß das akute Streßulkus unterschieden werden. Es wurde gehäuft z.B. bei jungen Männern zu Beginn des Krieges in Deutschland und während der Bombenangriffe in London beobachtet. Dabei kam es nicht selten nach einer sehr kurzen oder fehlenden Vorgeschichte von Beschwerden zu einer Perforation. Heute finden sich bei weitem die meisten Streßulzera nach schweren Verletzungen oder chirurgischen Eingriffen. Auch die meisten im Tierversuch erzeugten Ulzera (z.B. unter Bedingungen der Immobilisierung) sind akute Streßulzera. Die Frage, wieweit die Entstehung dieser Geschwüre, die ja bei Tierexperimenten bedeutsam sind, mit der Pathogenese der chronischen Ulcera duodeni verglichen werden kann, ist noch offen. Eingehend beschäftigen sich hiermit Ackerman und Weiner (1977) und im deutschsprachigen Bereich Wächter (1981).

Komplikationen

Dazu zählen die Perforation in die freie Bauchhöhle mit anschließender Peritonitis oder die Penetration in die benachbarten Organe, vor allem in das Pankreas; die Blutung, die vor allem bei Arrosion einer Arterie lebensgefährlich sein kann; die Stenose aufgrund narbiger Strikturen. – Im Unterschied zum Ulcus ventriculi zeigt das Ulcus duodeni keine Neigung zur malignen Entartung. In sehr seltenen Fällen verbirgt sich hinter einem Ulcus ventriculi, niemals aber hinter einem Ulcus duodeni, ein Malignom.

43.2 Epidemiologie

Die Prävalenz des Ulcus pepticum beträgt in der Bundesrepublik ca. 2%, wobei das Ulcus duodeni etwa doppelt so häufig auftritt wie das Ulcus ventriculi. Das Verhältnis Männer zu Frauen beträgt ca. 2:1. Diese Angaben entsprechen etwa der Häufigkeit in den anderen europäischen Staaten und den USA. So wird angenommen, daß mindestens 10% aller Männer an einem peptischen Geschwür erkranken. Die Häufigkeit erreicht um das 45. Lebensjahr einen Gipfel, um dann abzusinken. Gleichzeitig verschiebt sich das Verhältnis der Häufigkeit zwischen Ulcus duodeni und Ulcus ventriculi. Letztes nimmt im Alter zu.

In transkulturellen Untersuchungen zeigt sich, daß diese Zahlen nicht generell gelten. Vielmehr bestehen zwischen den verschiedenen Kulturen Unterschiede. Nicht eindeutig ist die Frage zu beantworten, wieweit dies real ist oder nur Ausdruck der in den einzelnen Ländern sehr unterschiedlichen ärztlichen Versorgung und diagnostischen Möglichkeiten. Sicher ist aber, daß die Ulkuskrankheit in allen Ländern vorkommt: in Grönland, in Belgisch-Kongo, in Äthiopien; in Südindien ist sie sehr häufig. In China wurde sie früher ebenso häufig angetroffen wie in den USA (Pflanz, 1962). – Es ist daher nicht sehr wahrscheinlich, daß Ernährungsbedingungen für die Häufigkeitsunterschiede verantwortlich sind. Unterschiede der politischen und sozialen Strukturen und damit soziale Faktoren scheinen eine größere Rolle zu spielen. Diese sozialen Faktoren scheinen auch für sich ändernden Relationen zwischen Ulcus ventriculi und Ulcus duodeni bedeutsam zu sein: In den meisten westlichen Ländern ist das Ulcus duodeni bei Männern häufiger als das Ulcus ventriculi. In einzelnen Gegenden, z.B. in Nordnorwegen, ist jedoch das Ulcus ventriculi häufiger und bei krassen Änderungen der sozialen Situation kann sich die Rate verschieben. So wurde während des 2. Weltkrieges in Deutschland das Ulcus ventriculi wieder häufiger (Pflanz, 1962).

Soweit man bei den verschiedenen Möglichkeiten der Diagnostik die vorliegenden Angaben verwerten kann, scheint bis zur Mitte des 19. Jahrhunderts das Ulcus ventriculi das Ulcus duodeni in der Häufigkeit überwogen zu haben, wobei sehr viel stärker die Frauen als die Männer befallen waren. Erst Ende des 19. Jahrhunderts kam es zur erwähnten Häufung an Ulcus-duodeni-Erkrankungen, wobei jetzt die Männer überwogen. Diese Entwicklung setzte sich in der ersten Hälfte dieses Jahrhunderts fort. Sie spiegelt sich wider in der Feststellung Hallidays, daß im Ersten Weltkrieg nur 709 Personen wegen eines Ulkus aus der britischen Armee entlassen wurden, während diese Zahl im Zweiten Weltkrieg 22 754 betrug (Pflanz, 1962). Bei den deutschen Truppen war die Ulkuskrankheit während des Zweiten Weltkrieges so häufig, daß man sog. „Magenkompanien" mit besonderer Diätversorgung aufstellte. Ab 1960 kann zumindest in den USA, später auch in Europa ein Rückgang des Duodenalulkus beobachtet werden; in den USA handelt es sich zwischen 1960 und 1972 um einen 50%igen Rückgang. Erklärt wird dies mit einem sich abschließenden „Urbanisationsprozeß" (Susser, 1976). Er besagt, daß in den westlichen Ländern weitgehend eine Umstellung auf die Bedingungen der postindustriellen Gesellschaft stattgefunden hat. Unterstützt wird diese These durch die Beobachtung, daß in diesen Ländern überdurchschnittlich gehäuft bei Gastarbeitern Duodenalgeschwüre gefunden werden (Horn und Herfarth, 1978; Jenny und Deyhle, 1976). Epidemiologische Untersuchungen unterstützen die These, daß soziale Isolierung für die Entstehung oder das Rezidiv eines peptischen Geschwürs von Bedeutung ist. Pflanz stellte aufgrund einer Untersuchung an fast 10 000 Patienten der Gießener Medizinischen Poliklinik und einer ebenso großen Vergleichsgruppe folgendes fest: Das Ulkus kommt häufiger bei Menschen vor, die aus einer Gemeinschaft ausgeschieden sind; bei verheirateten Männern, die eine Männerkameradschaft oder ihr Elternhaus aufgeben müssen; bei Heimatvertriebenen, die aus ihrem alten Gemeindeverband ausscheiden mußten, und bei Geschiedenen, welche die Gemeinschaft der Ehe verloren haben (1962).

43.3 Psychosoziale Untersuchungsergebnisse

43.3.1 Zur Biographie des Ulkuskranken; Anmerkungen zur sog. „Ulkuspersönlichkeit"

Zu Beginn dieses Abschnittes soll auf ein tiefgehendes Vorurteil eingegangen werden. Dieses besagt, es gäbe eine „Ulkuspersönlichkeit". Das trifft nicht zu. Diese Meinung wurde besonders von F. Dunbar vertreten und wurde während der 40er Jahre weitverbreitetes Überzeugungsgut (Dunbar, 1947). Neuen Auftrieb erhielt diese Ansicht durch die umfangreiche Forschung zum Typ-A-Verhalten des Herzinfarktpatienten. Die Hartnäckigkeit dieses Vorurteils ist um so beachtlicher, als ihr schon frühzeitig entgegengetreten wurde (Roth, 1955).

Als weitgehend gesichert können vielmehr gelten: eine Häufung von meist gruppenbezogenen Trennungsereignissen, die zeitlich mit Auslösung bzw. Schub eines Ulkus zusammenhängen (vgl. nachfolgende Ausführungen zum sozialen Umfeld einschließlich Familie); eine Häufung bestimmter sozialer Eigenschaften, die sich der analytischen Terminologie folgend auf Ich-Funktionen beziehen; psychoanalytisch definierte Konfliktsituationen.

In der Biographie von Ulcus-duodeni-Kranken lassen sich häufig Trennungserlebnisse finden. Die Trennung vom Elternhaus, der Eintritt in den Beruf, ein Berufswechsel, Heirat, Kinder, aber auch Scheidung, scheinen Krisenpunkte im Leben dieser Patienten zu sein. In exemplarischer Weise wird dies durch das oben geschilderte Patientenbeispiel dargestellt.

Wir können davon ausgehen, daß für die Ulkuspatienten die Beziehung zu der sozialen Gruppe, in der sie leben, eine besondere Bedeutung hat. Ruesch und Mitarbeiter fanden bei Marineangehörigen, die aus dem Mannschaftsstand zum Offizier emporgestiegen waren, eine besondere Häufung der Erkrankungen an Ulcus duodeni (1948). Die Autoren interpretierten diesen Befund damit, daß die Patienten die Gruppe, zu der sie gehört hatten, verlassen haben. Gleichzeitig werden sie in der neuen Gruppe aber nicht wirklich akzeptiert. Pflanz weist darauf hin, daß Werkmeister häufiger als andere Berufsgruppen an einem Ulkus erkranken (1962). Auch hier besteht eine soziale Isolierung, denn Werkmeister werden von den Arbeitern nicht mehr und von den Höherstehenden noch nicht als ihresgleichen anerkannt. Sie stehen zwischen zwei Gruppen.

Der Verlust der Zugehörigkeit zu einer Gruppe, die Anerkennung, Schutz und Verwöhnung verleiht, wurde von verschiedenen Autoren sowohl durch biographische wie durch epidemiologische Untersuchungen erfaßt und als auslösender Faktor für die Erkrankungen an einem Ulcus duodeni oder an einem Rezidiv dieser Erkrankung festgehalten. Am Beispiel der „life event"-Forschung sind diese Zusammenhänge herausgearbeitet worden (z. B. Giligan, 1987; Walker, 1988). Wir selbst fanden in einer Analyse von Biographien Ulkuskranker nach dem letzten Weltkrieg einen engen Zusammenhang zwischen dem Ausbruch eines Ulkus oder eines Ulkusrezidivs und dem Verlust der Zugehörigkeit zu einer relevanten Gruppe (1956). Dabei war die enge Beziehung zwischen Gruppensituation und der Verträglichkeit von Speisen bemerkenswert; es zeigte sich immer wieder, daß diese Patienten Hunger und schlechte Speisen ohne Beschwerden ertrugen, solange sie in Harmonie mit ihrer sozialen Gruppe waren. In sozialer Isolierung klagten sie jedoch auch bei guter Ernährung, selbst bei sorgfältiger Diät über Beschwerden. Die soziale Isolierung war in der Lebensgeschichte dieser Ulkuskranken ein spezifischer Faktor für die auslösende Situation. Dagegen hatten Todesgefahr und Katastrophen, die zu einer Zerstörung der wirtschaftlichen Existenz führten, also Situationen, die während des 2. Weltkrieges und kurz danach von vielen erlebt wurden, keinen Einfluß auf die Entstehung eines Geschwüres – sofern sie nicht zu einer sozialen Isolierung führten (Pflanz et al., 1956)!

Zur Illustration folgende Krankengeschichte:

P. war ein ehrgeiziger Mann, der in seinem Leben nach Anerkennung und Verantwortung strebte. Sein Beruf als Lehrer genügte ihm nicht. 1928 wurde er Mitglied der nationalsozialistischen Partei und aufgrund seines Eifers bald als Parteiredner herausgestellt. Das ging gut, bis er nach 1933 merkte, daß die Praxis der Partei im krassen Gegensatz zum Idealen stand, die er in seinen Ansprachen verkünden mußte. Es kam zu einer wachsenden Entfremdung von seinen Parteifreunden. Er hatte jedoch nicht den Mut, die Konsequenzen aus seiner Erkenntnis zu ziehen. In dieser Konfliktsituation bekam er Magenschmerzen, die er vordem nie gekannt hatte. Sie traten zunächst nur während seiner Ansprachen auf, bald aber auch zu anderen Zeiten, und er mußte schließlich nach jeder Rede erbrechen.

Der Arzt stellte ein Zwölffingerdarmgeschwür fest, das trotz intensiver Behandlung und strenger Diät nicht ausheilen wollte. Bei jedem Versuch, seine Tätigkeit wieder aufzunehmen, kam es zu einem Rückfall. Das zog sich mehrere Jahre hin, bis der Krieg ausbrach. Jetzt meldete sich P., um dem verhaßten Zwang zu entgehen, freiwillig zum Militär. Von diesem Augenblick an war er gesund und blieb während des ganzen Krieges, den er an der Front, oft unter großen Entbehrungen und Strapazen mitmachte, völlig beschwerdefrei. Er konnte gefrorenes Brot, rohe Kartoffeln und verdorbene Speisen essen, ohne daß sein Magen sich rührte. Das blieb auch so während seiner Gefangenschaft, aus der er 1947 zurückkehrte. Zu Hause angekommen, wurde er vor die Spruchkammer gestellt, die ihm wegen seiner politischen Vergangenheit die Tätigkeit im Lehrerberuf untersagte. Von da an begann das Magenleiden von neuem. Er stand vor einem Rätsel. Immer wieder fragte er den Arzt, warum er im Krieg und in der Gefangenschaft trotz größter Entbehrungen gesund war, während er jetzt, bei intensiver Pflege und strenger Diät von Schmerzen gepeinigt wurde. Die Antwort: Weil er jetzt von seiner Bezugsgruppe, den arbeitenden Lehrerkollegen, ausgeschlossen war.

43.3.2 Häufige Verhaltensweisen und Familienbindungen

Auf die Frage, warum die Zugehörigkeit zu einer Schutz und Verwöhnung gebenden Gruppe für diese Patienten so wichtig ist, gibt die Feststellung eine Antwort, daß Ulkuspatienten aus Familien stammen, die sich durch ihr soziales Muster von anderen Familien unterscheiden. Ruesch und Bateson stellten aufgrund von Kindheitserinnerungen ein spezifisches Familienmuster fest, das durch eine dominierende Mutter und einen wenig einflußreichen Vater charakterisiert war (1951). Goldberg legte 1958 eine der ersten Familienuntersuchungen vor, bei der er die Familien von 32 Patienten, die im Alter von 16 bis 25 Jahren an einem Ulcus duodeni erkrankt waren, mit 32 Familien einer Kontrollgruppe verglich. Dabei konnte eine Reihe von Hypothesen widerlegt werden, die über die Beziehungen, die Familienstruktur und

den Ausbruch der Erkrankung an einem Ulcus duodeni aufgestellt worden waren. So zeigte es sich, daß die Häufigkeit neurotischer Züge, die Größe der Familie, die Position des Patienten in der Geschwisterreihe, Desorganisation („broken home") – Faktoren, die in anderen Zusammenhängen bedeutungsvoll sein mögen – für Familien von Ulcus-duodeni-Patienten nicht charakteristisch waren. Statt dessen konnten folgende Hypothesen wahrscheinlich gemacht werden:

– Junge Männer, die an einem Ulcus duodeni erkranken, stammen aus Familien, in denen die Beziehungen zwischen den Mitgliedern besonders eng, aber auch besonders rigide sind.
– Die Mutter spielt eine dominierende Rolle, nicht nur als Hausfrau, sondern auch als Autoritätsfigur, mit der Gefahr, ihren Sohn zu eng an sich zu binden.
– Der Vater spielt mehr die Rolle eines älteren Bruders, selten die eines bewunderten Vorbildes.
– Beide Elternteile bemühen sich, ein „braves Kind" zu erziehen, das seine Aggressionen unterdrückt.

In einer sehr viel späteren Untersuchung zur Art der Partnerbeziehung von Ulkuspatienten werden gleichsam die Ergebnisse dieses Erziehungsprozesses beschrieben (Böttcher et al., 1980). Männer wurden von ihren Frauen als stabiler eingeschätzt, als diese es selbst taten. Hierbei scheint der Kranke seinen Partner durch „Sich-nicht-ganz-Zeigen" in die Irre zu führen. Häufiger als Gesunde geben sowohl die Kranken wie ihre Partner an, daß ihre intimen Beziehungen „weniger gelungen" seien. 20% der Ulkuspatienten stellten im Laufe der Jahre eine Verschlechterung, 0%(!) eine Verbesserung ihrer Ehebeziehungen fest. In einer größeren Untersuchungsreihe (127 Ulkuspatienten, 145 Gesunde, 79 Psychotherapiepatienten) wurden die Ulkuspatienten als „rigide" befunden, d.h. starrer auf Ziele und Gruppen eingestellt und störbarer als die Gesunden sowie die Psychotherapiepatienten (Bauer und Bergmann, 1981). Im Abschnitt „Psychoanalytische Untersuchungsergebnisse" wird dargestellt, daß sich gehäuft Eigenschaften der „Abhängigkeit" und „Pseudounabhängigkeit" finden. Diese Beobachtungsergebnisse würden sich gut in Böttchers Beobachtung einfügen, daß sich Ulkuspatienten weniger schwach zeigen dürfen, als sie sind.

Goldbergs Befunde, wie die von Böttcher und Mitarbeitern, sind nicht ausschließlich bei Ulcus-duodeni-Patienten, sondern auch bei anderen psychosomatischen und psychiatrischen Erkrankungen zu finden. Diese Beziehungsstrukturen gehören aber zu Faktoren (wie die Hyperazidität), deren Zusammenwirken eine spezifische Vorbedingung für das Ulcus duodeni zu sein scheint. Goldberg diskutierte die Bedeutung seiner Befunde auf dem Hintergrund der allgemeinen Entwicklung in den westlichen Industriekulturen und unter dem Gesichtspunkt der Zunahme der Krankheitshäufigkeit an Ulcus duodeni in den ersten 50 Jahren des 20. Jahrhunderts. Er hält es für möglich, daß das typische Beziehungsmuster in den Familien von Ulcus-duodeni-Patienten eine besonders prononcierte Antwort auf Forderungen der modernen

Gesellschaft sein könnte: In ihr würde gefordert, daß Mütter sich besonders intensiv um ihre Kinder kümmern und daß Väter besonders verständnisvoll und kameradschaftlich, aber nicht autoritär sein dürfen. Ferner würde gefordert, daß beide Eltern das Ziel verfolgen, ihre Söhne zu „kultivierten Gentlemen" zu erziehen (alles in der 1. Hälfte dieses Jahrhunderts!). Das würde dazu führen, daß die Kinder früh entmutigt werden, unerwünschte Ausdrücke triebhafter Regungen zu zeigen. Goldberg meint weiter, daß die Änderung der männlichen und weiblichen Rolle in unserer Zivilisation und die dadurch ausgelösten Unsicherheiten und Konflikte in den Familien eine besondere Bedeutung haben könnten, eine Annahme, die bereits vor ihm Halliday formuliert hatte (1943). Im Sinne vergrößerter Unsicherheit und vermehrter Konflikthaftigkeit der Lebenssituation wären auch die Ergebnisse von Ackerman zu interpretieren, daß sich bei Adoleszenten und Präadoleszenten gehäuft Trennungsereignisse vor Beginn der Ulkuskrankheit finden (1981).

Aufgrund dieser Hypothesen einer Rollenänderung in unserer Zivilisation muß man die sonst recht problematische Annahme von „Zivilisationskrankheiten" neu diskutieren. Goldberg geht von der Tatsache aus, daß in der industriellen Gesellschaft die Familie viele ihrer früheren Funktionen einbüßt und meint, daß dies zu einer pathologisch engen Verfestigung der emotionalen Beziehungen zwischen Kindern und Eltern – besonders im städtischen Milieu – führen könne. Er diskutiert dann folgende Möglichkeiten: Von „außen" gesehen scheinen moderne Eltern demokratischer und weniger autoritär zu sein als die Eltern früherer Generationen. Die Kinder scheinen mehr Freiheit zu haben und mehr als Persönlichkeiten behandelt zu werden. Von „innen" gesehen könnte es sich aber ganz anders verhalten: In der kleinen Familie könnte das übergroße Interesse an der emotionellen Entwicklung der Kinder angesichts der hohen Ansprüche der Gesellschaft an die Sorge für das kindliche Wohl zu einer wachsenden Ängstlichkeit der Eltern und damit zu einer Verfestigung der emotionellen Bindungen zwischen Eltern und Kindern führen, die schließlich eine größere Unfreiheit erzeugt als diejenigen Bindungen, die frühere Generationen durch Zwang und äußere Kontrolle erreicht haben.

Wir geben dieses Denkmodell deswegen so ausführlich wieder, weil es zusammen mit dem später zu besprechenden Konzept einer „somatopsychischen-psychosomatischen Genese" des Ulcus duodeni ein allgemeines Ordnungsschema entwirft, in dem die vielen Einzelbefunde über physiologische, soziale und psychologische einschließlich der nachfolgend zu besprechenden psychoanalytischen Faktoren in einen Zusammenhang gebracht werden können.

43.3.3 Psychoanalytische Untersuchungs-ergebnisse

Alexander hat aufgrund detaillierter psychoanalytischer Untersuchungen bei einer Gruppe von Patienten die These aufgestellt, daß Kranke mit Ulcus duodeni, im Gegensatz zu der Hypothese von Dunbar (1947), keinem bestimmten Persönlichkeitstyp zuzuordnen sind (1950). Charakteristisch für sie sei vielmehr ein spezifischer Konflikt, der sich bei sehr verschiedenartigen Persönlichkeiten entwickeln könne. Die Grundlage des Konfliktes ist nach Alexander eine übergroße Abhängigkeit von Belohnung, Zuwendung und schutzgebenden Instanzen (mütterlichen Objekten), d.h. der unbewußte Wunsch, in der kindlichen Situation, in der man geliebt und verwöhnt wird, zu bleiben. Dieser Wunsch gerät mit dem Streben des erwachsenen Ichs nach Unabhängigkeit und Erfolg in Widerspruch. Je nachdem, wieweit Patienten ihren unbewußten Wünschen nach Abhängigkeit nachgeben oder sie ablehnen oder durch überbetontes Streben nach Unabhängigkeit überspielen, erscheinen sie als offen abhängig, fordernd und unzufrieden oder erfolgreich, produktiv, ehrgeizig und bestrebt, andere von sich abhängig zu machen. Da das letzte Verhalten den Wunsch nach Abhängigkeit überkompensiert, hat man es als „Pseudounabhängigkeit" bezeichnet und der „offenen Abhängigkeit" gegenübergestellt.

Die Beobachtungen, auf die Alexander diese These gestützt hat, sind seitdem von vielen Seiten und mit verschiedenen Methoden bestätigt worden. Immer wieder fand man bei Patienten mit Ulcus duodeni diese Konstellation: Ihr Streben nach Unabhängigkeit und nach Anerkennung in Familie und Beruf verfolgt das Ziel, eine unbewußt unentbehrliche Verwöhnung in einer sozial akzeptierten Form zu sichern. Krisen, die zum Ausbruch der Krankheit und zum Auftreten eines Rezidivs führen, treten in Situationen auf, in denen das mühsam erreichte labile Gleichgewicht zusammenbricht. Dabei spielt die Bindung an die Mutter und an das Elternhaus bzw. an Personen oder Institutionen, die als Schutz und Anerkennung gewährende Mächte erlebt werden, eine zentrale Rolle. Schließlich kann eine Krise dadurch gekennnzeichnet sein, daß man in der eigenen Person nicht mehr wie bisher die Mutter, das Elternhaus oder jene Personen und Institutionen präsentieren kann, die den Schutz gewährten – mit anderen Worten, es kommt zu einer Krise des Selbstbewußtseins. Im Falle unserer Patientin stellte sich heraus, daß die ausschlaggebende Belastung der aufkeimende Entschluß war, der Mutter ins Ausland nachzuziehen und damit die schutzgewährende Institution „Chefsekretariat" aufzugeben. In dieser Situation erkrankte sie.

Kapp und Rosenbaum stellten fest, daß man außer den beiden größeren Gruppen der Pseudounabhängigen und offen Abhängigen eine dritte Gruppe von Ulkuspatienten unterscheiden kann, nämlich die „offen Parasitären". Bei ihnen sind schwere Charakterstörungen und psychische Defekte zu beobachten. Sie haben offenbar kaum eine Abwehr gegen ihre egoistisch-fordernden Tendenzen entwickeln können (1947).

Alle diese Beobachtungen fügen sich recht gut in die Erfahrungen ein, die man mit Ulkuspatienten bezüglich ihrer Einstellung zu einer relevanten Gemeinschaft gemacht hat. Wir konnten in der oben erwähnten Studie (von Uexküll et al., 1963) zwei Gruppen von Patienten unterscheiden:

– Personen, die keine Schwierigkeiten haben, Kontakt mit gleichgesinnten Menschen aufzunehmen. Sie fühlten sich isoliert, wenn sie keine derartige Gruppe fanden.
– Personen, die aufgrund ihrer Kontaktschwierigkeit nur schwer Beziehungen zu anderen Menschen knüpfen. Diese „konstitutionellen Individualisten" lehnten sich an anonyme Organisationen an. Sie versuchten, Schutz durch ihre berufliche Tätigkeit zu erwerben. Sie fühlten sich isoliert, wenn sie selbst versagten oder wenn die Organisation, auf die sie sich verließen, versagte.

Overbeck und Mitarbeiter haben diese auf konfliktpsychologischen Kriterien beruhende Einteilung durch Heranziehung ichpsychologischer und sozialpsychologischer Gesichtspunkte weiter differenziert. Sie kommen dadurch zu fünf Gruppierungen, die sie auch durch klinische Beobachtungen und testpsychologische Untersuchungen weiter absichern konnten (1975, 1977). Sie betonen, daß u.a. das kennzeichnende Problem der Aggressionsbewältigung einerseits zwar fortbesteht, aber unterschiedlich bewältigt wird (1989). Auch die Einteilung dieser Autorengruppe basiert letztlich auf der Alexanderschen These einer pathogenen Bedeutung der Abhängigkeitsproblematik für die Entstehung eines Ulcus duodeni. – Kritiker haben gegen diese These zwei Einwände geltend gemacht:

– Abhängigkeitswünsche können bei allen Menschen gefunden werden. Die Einschätzung ihrer Bedeutung für die Entstehung einer Krankheit würde von dem Urteil des Untersuchers abhängen.
– Alexander habe zunächst keine Untersuchungen an Kontrollgruppen durchgeführt. Vor allem hätten Untersuchungen gefehlt, bei denen der Untersucher nicht im voraus weiß, welche Patienten ein Ulkus haben und welche nicht.

Alexander und Mitarbeiter haben daher 1951 eine groß angelegte Kontrolluntersuchung begonnen, in der Interviews von Patienten aus sieben verschiedenen Krankheitsgruppen Internisten und Psychoanalytikern vorgelegt wurden, welche die Patienten nicht kannten. Beide stellten die Diagnose nur aufgrund der Interviews, aus denen zuvor alle Hinweise auf die Krankheit eliminiert waren. Dabei konnten Internisten und Psychoanalytiker bei Ulcus-duodeni-Patienten etwa gleich häufig die richtige Diagnose stellen. Bei anderen Krankheiten, z.B. Colitis ulcerosa, rheumatoider Arthritis oder Asthma, konnten die Analytiker sehr viel häufiger die richtige Diagnose stellen als die Internisten (1968).

Die Ergebnisse dieser Untersuchung konnten die Zweifel an der Richtigkeit der Hypothese der pathogenen Bedeutung eines spezifischen Konfliktes für die Entstehung des Ulcus duodeni also nicht beseitigen. Eine entscheidende Unterstützung dieser Hypothese fand sich aufgrund der Ergebnisse eines Untersuchungsprojektes, die zu der Aufstellung des sog. somatopsychischen-psychosomatischen Modelles führten. Es wird im folgenden Abschnitt wiedergegeben.

43.4 Theorien zur Ätiologie und/oder Pathogenese

Als Eingangsbemerkung zu diesem Abschnitt sollen drei Punkte festgehalten werden:

– Trotz des Vorhandenseins schlüssiger Vorstellungen über pathologische Mechanismen übersehen wir noch keineswegs die „pathogenetische Kette", welche die Entstehung der Ulkuskrankheit auch nur einigermaßen lückenlos erklären könnte.
– Wie eingangs betont, handelt es sich bei dem Ulcus duodeni nicht um eine einheitliche Krankheit, sondern um mindestens zwei verschiedene Subgruppen. Unsere Kenntnisse über bio-psycho-soziale Zusammenhänge beschränken sich auf Patienten der einen Gruppe.
– Wir wissen nicht, ob Ulcus ventriculi und Ulcus duodeni eine vergleichbare Ätiologie und Pathogenese haben oder ob es sich um zwei verschiedene Krankheiten handelt.

Hinsichtlich der dritten Feststellung werden fünf Punkte aufgeführt:

– Die Verteilung von Altersklassen, Sozialklassen und Geschlecht ist unterschiedlich.
– Beide Geschwürsgruppen weisen eine unterschiedliche Krankheitshäufigkeit auf.
– Nur beim Duodenalulkus der Gruppe 1 findet sich eine erhöhte sekretorische Aktivität.
– Die genetischen Faktoren beider Krankheitsbilder unterscheiden sich (Angaben zu AB0 und HLA in Abschnitt „Genetische Faktoren").
– Psychologische Merkmale scheinen bei den Vertretern der beiden Krankheitsgruppen unterschiedlich zu sein.

43.4.1 Pathophysiologie

Die älteste Theorie nimmt an, daß Durchblutungsstörungen die Vitalität der Schleimhaut beeinträchtigen. Virchow glaubte an Mikroembolien, die zu kleinen Infarkten führen würden. G. von Bergmann stellte die These auf, daß Motilitätsstörungen über Spasmen der Magenwandmuskulatur Durchblutungsstörungen verursachten. Es gibt Beobachtungen, die diese letzte These stützen. Da aber Methoden, welche die Durchblutung des Magens und Dünndarms messen können, noch unzuverlässig sind, ist ein exakter Beweis für das Ausmaß der Bedeutung von Durchblutungsstörungen bei der Entstehung von Magen- und Zwölffingerdarmgeschwüren vorläufig nicht zu führen. Änderungen der Magendurchblutung, über die berichtet wird, konnten bisher nur durch direkte Beobachtung der Magenschleimhaut bei Fistelträgern festgestellt werden. Interessant ist jedoch, daß eine Durchblutungsstörung über die Generationen hin immer wieder diskutiert wurde.

Heute wird von gastroenterologischer Seite das Vorliegen dreier pathogenetischer Mechanismen angenommen (Arnold, 1981, 1982).

Pathogenetisches Prinzip I

Es handelt sich um das „aggressive" Prinzip, das auf einer zu hohen Säure- und Pepsinogenkonzentration im Bulbus duodeni beruht. Es geht auf die alte Feststellung von Schwartz zurück, „ohne sauren Magensaft kein peptisches Geschwür" (1910). Heute wird allgemein anerkannt, daß die Träger eines Ulcus duodeni zumeist (jedoch wie betont nicht immer) eine erhöhte Sekretion von Salzsäure und Pepsin haben; daß Ulzera nur an Stellen des Magen-Darm-Traktes entstehen, die mit Salzsäure und Pepsin in Berührung kommen (Ulcus jejuni pepticum nach Gastroenterostomie-Operationen), und daß beim Zollinger-Ellison-Syndrom, bei dem eine abnorme Stimulierung der Magensäuresekretion durch gastrinproduzierende Tumoren des Pankreas vorliegt, gehäuft Ulzera im Duodenum und Jejunum auftreten. Dagegen wird bei Achlorhydrie, z.B. bei perniziöser Anämie, ein Ulcus duodeni praktisch niemals beobachtet. Die Hypersekretion wird durch eine größere Parietalzellmasse und eine höhere Parietalzelldichte erklärt. Die Steuerung der Sekretion erfolgt durch den Vagus sowie humoral durch das Gastrin und das Histamin. Die gastrointestinalen Hormone wie Sekretin, GIP, VIP, Somatostatin modulieren möglicherweise die steuernden Funktionen von Vagus, Gastrin und Histamin (Arnold, 1982).

Eine andere Überlegung zur Pathogenese des Ulcus duodeni macht eine beschleunigte Entleerung des Magens für eine länger anhaltende Säuerung im Duodenum verantwortlich. Dieser Mechanismus könnte aus psychophysiologischer Sicht vor allem für die Patienten aus der Subgruppe ohne erhöhte Pepsinogenwerte eine erhebliche Rolle spielen (vgl. auch die Darstellungen der Ergebnisse von Zander in Kap. 8).

Prinzip II

Es handelt sich um die Gesamtheit der zytoprotektiven, also defensiven Mechanismen. Hier werden u.a. eine ungenügende Schleimhautdurchblutung, eine mangelhafte Regenerationsfähigkeit des Schleimhautepithels und eine ungenügende Schleimsekretion als pathogene Faktoren diskutiert. Hinsichtlich der Regenerationsfähigkeit des Schleimhautepithels spielen wahrscheinlich die Prostaglandine eine hervorragende, bisher aber noch wenig untersuchte Rolle. – Allerdings sind unsere Kenntnisse über die Rolle der protektiven Faktoren in der Pathogenese des Ulcus duodeni noch äußerst mangelhaft (Arnold, 1982). Man gewinnt den Eindruck, daß die alten Überlegungen zu Durchblutungsstörungen durchaus wieder zu Ehren kommen könnten.

Zu den Faktoren, die das zytoprotektive Element stören, könnte auch Campylobacter pylori (C.p.) gezählt werden. Dieser säureempfindliche Keim findet sich vor allem in der Antrumschleimhaut und ist mit der chronisch aktiven Oberflächengastritis vergesellschaftet. Bei Ulcus duodeni ist er in 90–100% nachweisbar. Jedoch wird seine klinische Bedeutung überaus kontrovers diskutiert, da C.p. ubiquitär ist und

etwa ab dem siebzigsten Lebensjahr in 70% der Fälle vorkommt. – Nach Diskussion der Tatsache, daß C.p. unter antibiotischer Therapie verschwindet, nach Absetzen der Therapie aber nahezu regelmäßig wieder auftaucht, wird gastroenterologischerseits (Koop, 1990) festgestellt: „Die Entdeckung des Campylobacter pylori beeinflußt prinzipiell das Therapiekonzept des peptischen Ulkus nicht; der Einsatz insbesondere von H_2-Rezeptorantagonisten stellt weiterhin eine adäquate Behandlungsform dar, da es auch mit Wismutsalzen nur in einem kleinen Teil der Fälle gelingt, den Campylobacter pylori dauerhaft zu eliminieren."

Prinzip III

Es handelt sich um Einflüsse des zentralen Nervensystems. Im Kapitel 8 „Psychophysiologie" wird ausgeführt, wie Emotionen die Sekretion und die Motilität des Magens beeinflussen (vgl. Kap. 33). Versuche mit vorzeitig entwöhnten jungen Ratten, die zusätzlich immobilisiert wurden, führten zu gastrischen Erosionen (Ackerman, 1980), deren Häufigkeit von dem Zeitpunkt der Entwöhnung abhing. Ihr Auftreten scheint direkt mit einer Hypothermie zusammenzuhängen. Bemerkenswerterweise konnten die Erosionen weitgehend verhindert werden, wenn die Hypothermie extern vermieden wurde; bzw. es wurden vermehrt Erosionen gefunden, wenn unter den Bedingungen der Immobilisierung eine Hypothermie erzeugt wurde. Insgesamt wurde im Tierversuch mehrfach nachgewiesen, daß durch psychische Reize gastroduodenale Läsionen gesetzt werden können (Mikhail, 1973; Brodie und Hanson, 1960; Selye und Szabo, 1973; Gaskin et al., 1975; Lindholm et al., 1975; Kim et al., 1976; Sawrey, 1961; Weiss, 1968; Weitz, 1957). Das Entstehen von derartigen Läsionen scheint nicht von vermehrter Säuresekretion abhängig zu sein (Birnbaum, 1973; Ackerman, 1981).

Auch für die erwähnte Subgruppe, der etwa 50% aller Ulcus-duodeni-Patienten angehören (Rotter et al., 1979), die keinen erhöhten Pepsinogenspiegel haben, trifft das gleiche zu. Möglicherweise gewinnt das Prinzip II, also die Gesamtheit der protektiven Faktoren, in diesen Fällen an Bedeutung. Arnold diskutiert die Möglichkeit eines adrenergen Mechanismus, der im Erkrankungsfall zu einer verminderten Schleimhautdurchblutung, einer verzögerten Epithelregeneration oder fehlerhafter Schleimproduktion im Duodenalbereich führt (1982).

Es stellt sich die Frage, wie die drei pathogenetischen Mechanismen ineinander greifen. Bevor hierauf eingegangen wird, sollen einige Ergebnisse der Psychophysiologie angeführt werden.

43.4.2 Psychophysiologie

Die Beobachtung über Zusammenhänge zwischen emotionalen Vorgängen und der Entstehung von Magen-Duodenalkrankheiten ist schon sehr alt. Bereits die Umgangssprache drückt diese Beobachtung in Formen wie „es liegt mir im Magen", „das ist nicht zu verdauen" usw. aus. Die ersten ärztlichen Beobachtungen über einen Zusammenhang zwischen emotionalen Vorgängen und dem Verhalten der Magenschleimhaut wurden von dem nordamerikanischen Militärarzt William Beaumont (1785–1853) mitgeteilt (1833), der 1822 dem schwerverwundeten Frankokanadier Alexis St. Martin, der eine Schußverletzung in den linken Oberbauch erhalten hatte, das Leben rettete. Nach dieser Verletzung blieb eine Magenfistel zurück. Dadurch war die Möglichkeit gegeben, Magenschleimhaut und Magentätigkeit direkt zu beobachten. Beaumont war davon so fasziniert, daß er den Invaliden als Diener einstellte und während vieler Jahre Veränderungen der Durchblutung und der Sekretion des Magens im Zusammenhang mit den täglichen Erlebnissen seines Dieners beobachtete. In seiner 1833 erschienenen Schrift „Experimente und Beobachtungen über den Magensaft und die Physiologie der Verdauung" konnte er nachweisen, daß die Aktivität des Magens, also die Durchblutung, die Sekretion und die Motilität, sowohl von Nahrungsstoffen als auch von psychischen Einflüssen abhängig ist. Wesiack zitiert Diepgen, der feststellte, daß Beaumont dadurch zum Führer und Pionier der experimentellen Psychologie in Amerika wurde, und daß sein Buch bis Pawlow das wichtigste Werk über die Magenverdauung wurde (1974). Mitte dieses Jahrhunderts haben Wolf und Wolff die Versuche an dem Labordiener Tom unter Laboratoriumsbedingungen wiederholt und ebenfalls Durchblutungsstörungen, sogar das Auftreten von Erosionen, in emotional belastenden Situationen beschrieben (1943).

Über den Zusammenhang zwischen Emotionen und Magensekretion weiß man seit den klassischen Versuchen Pawlows über bedingte Reflexe um die Jahrhundertwende sehr genau Bescheid. Die Kenntnis dieses Zusammenhanges hat jedoch zunächst nicht viel zu unserem Verständnis der Ätiologie und Pathogenese der Ulkuskrankheiten beigetragen. Es gibt viele einander widersprechende Beobachtungen, die bei emotioneller Belastung einmal Steigerung, dann ein Versiegen der Sekretion festgestellt haben. Mahl stellte die Hypothese auf, daß Angst, gleichgültig ob durch unbewußte Konflikte oder durch bewußt erlebten Streß ausgelöst, zur Ulkusentstehung führen würde (1950). Demgegenüber betonen Engel und Mitarbeiter, daß bewußte emotionale Erlebnisse für die Entstehung und den Verlauf der Ulkuskrankheit weniger bedeutsam sind als psychische Vorgänge, die unbewußt ablaufen. Sie fanden bei dem Kind Monika mit einer Magenfistel, das sie vom ersten Lebensjahr bis heute (1989) als erwachsene und verheiratete Frau beobachteten, daß Gefühle, die mit Schuld, Ärger und Furcht einhergehen, sowohl zu einer Steigerung wie zu einer Verringerung der Peristaltik führten (1956). Der Grund für diese Unterschiede war, daß Gefühle, die offen ausgedrückt werden konnten, einen anderen Effekt hatten, als Gefühle, die unterdrückt werden mußten. Margolin beobachtete, daß es in derartigen Konfliktsituationen zu einer Dissoziation zwischen Säuresekretion und Durchblutung kommen kann, d.h. zur vermehrten Sekretion bei verminderter Durchblutung und damit erhöhter Verletzbarkeit der Mukosa (1951).

Zander und Mitarbeiter legten kontrollierte psychophysiologische Untersuchungen bei Ulcus-duodeni-Patienten vor (1977, 1978, 1981, 1982). In psychoanalytischen Interviews besprach Zander mit den Ulcus-duodeni-Patienten Lebensereignisse, die bei diesen Menschen in einer sehr persönlichen Weise Neid und Ärger auslösten. Gleichzeitig wurde röntgenologisch die Magenmotilität beobachtet (1978). Neid-Ärger-Themen lösten fast regelmäßig trichterförmige, spasmenartige Bewegungen im Antrumbereich aus; bei den magengesunden Kontrollpersonen war das praktisch kaum zu beobachten. Besprach man mit den Ulkuspatienten Themen, die für sie psychodynamisch nicht relevant waren, traten keine Spasmen auf. Die Frage, ob es bei dem Spasmus des Antrums zu einer vermehrten Säureentleerung in das Duodenum kommt, wurde nicht untersucht. Man kann aber annehmen, daß dies der Fall war. Wir werden auf die Zanderschen Befunde noch zurückkommen.

Wolcott und Mitarbeiter fanden bei 39 männlichen Ulcus-duodeni-Patienten eine signifikante Korrelation zwischen Serumgastrinspiegel und Meßwerten dreier Subskalen einer so bezeichneten „Family Environment Scale", FES (1981). Bei diesen drei Skalen handelt es sich um Unabhängigkeit (inwieweit die Familie selbständige Entscheidungen ermöglicht); Leistungsorientiertheit (inwieweit die Familie verschiedene Aktivitäten unter kompetitiven Gesichtspunkten sieht); Ausdrucksbereitschaft (inwieweit die Familie direkte Handlung und Ausdrucksmöglichkeiten fördert). – Die Autoren weisen darauf hin, daß ihre Ergebnisse der Überprüfung bedürfen, daß sie aber wichtige Hypothesen über Zusammenhänge zwischen psychologisch und physiologisch beobachteten Abläufen ermöglichen.

43.4.3 Das somatopsychische-psychosomatische Modell

Weiner und Mirsky haben eine Hypothese aufgestellt, welche die verwirrende Vielfalt der Beobachtungen bei Patienten mit Ulcus duodeni wenigstens für die etwa 50% mit erhöhtem Pepsinogenspiegel in einen ätiologischen und pathogenetischen Zusammenhang bringt (1957, 1958).

Die Hypothese einer psychophysiologischen Disposition

Die Hypothese besagt, daß eine biologisch definierte Anlage sehr früh in der Entwicklung des Individuums zu vermehrter sekretorischer Aktivität des Magens führt, die mit einem vermehrten Nahrungsverlangen des Säuglings einhergeht und so die frühesten Beziehungen zur Mutter belastet. Man kann sich vorstellen, daß schließlich auch die großzügigste Mutter auf ein unersättliches Baby mit Zurückweisung reagieren wird. Damit würden frühe Kränkungen des Nahrungsverlangens, d.h. orale Frustrationen, fast unvermeidbar. Solche Erfahrungen hätten wiederum bestimmte Erwartungen an die Umwelt zur Folge, die sich entlang der Schiene „Hunger bzw. nicht ausreichende Sättigung" entwickeln.

Für das Ineinandergreifen somatischer (genetisch bedingter) und psychischer (früh erworbener) Faktoren als pathogenetisch bedeutsame Konstellation (somatopsychisch) spricht auch eine Untersuchung von Eberhard an 30 eineiigen Zwillingspaaren, von denen wenigstens ein Zwilling an einem Ulkus erkrankt war (1968). Die Interviews aller Probanden wurden von einem unabhängigen Untersucher ausgewertet. Die Ergebnisse zeigten, daß die gegenwärtig oder in der Vergangenheit an einem Ulkus erkrankten Zwillinge statistisch signifikant durch höhere Empfindlichkeit für Streß und gestörte Abwehrmechanismen von den gesunden Zwillingen unterschieden waren. Daß die Neigung zur Übersekretion bereits sehr frühzeitig festgelegt ist, konnten Mirsky und Mitarbeiter bei einer Gruppe Neugeborener anhand eines erhöhten Pepsinogengehalts in dem Nabelschnurblut feststellen (1952). Diese Neugeborenen kommen also bereits als „Übersekretoren" auf die Welt.

Genetische Faktoren

Untersuchungen von Rotter und Mitarbeitern (1977 a, b) haben ergeben, hierauf wurde bereits verwiesen, daß es zwei Gruppen von Patienten gibt, die an Ulcus duodeni erkranken. Nur bei einer dieser Gruppen findet sich ein erhöhter Pepsinogenspiegel. Bei beiden Gruppen findet sich jedoch eine familiäre Häufung, was für eine genetische Determinante spricht.

Die 40–50% der Patienten mit erhöhten Pepsinogenwerten haben noch andere physiologische Besonderheiten, wie erhöhte Ansprechbarkeit auf die Stimulation mit Gastrin, postprandiale Hypergastrinämie, gestörte Inhibition der Säuresekretion durch Ansäuern des Mageninhaltes und eine beschleunigte Magenentleerung (Weiner, 1981). Erhöhte Pepsinogenwerte verhielten sich in den Untersuchungen von Rotter und Mitarbeitern (1977a, b) wie ein autosomal vererbtes dominantes Merkmal, das bei ungefähr der Hälfte der Kinder von Eltern mit dem gleichen Merkmal zu finden ist.

In der Gruppe der Patienten mit normalem Pepsinogenspiegel ist bisher kein genetischer Faktor gefunden worden. Man muß aber annehmen, daß er existiert (Weiner, 1981).

Bereits seit längerer Zeit ist bekannt, daß beim Ulcus duodeni überzufällig häufig die Blutgruppe 0 gefunden wird. Die Konkordanz für das peptische (!) Ulkus beträgt bei monozygoten Zwillingen 53%, diejenige bei dizygoten Zwillingen 36% (Gotlieb-Jensen, 1972); beim Ulcus duodeni findet sich das HLA-Antigen B5 um das 2,9fache gegenüber der Normalbevölkerung erhöht, was für eine genetische Heterogenität spricht (zitiert nach Sonnenberg et al., 1982). Völlig ungeklärt ist zur Zeit das genetisch vermittelte Muster der Kombination aggressiver und protektiver Faktoren beim Ulcus duodeni.

Zur Bedeutung der sozialen Situation

Die soziale Situation, die bei der mit erhöhten Pepsinogenwerten einhergehenden psychophysiologischen Disposition zur Ulkuskrankheit führen soll, ist aufgrund der vorausgegangenen Ausführungen zusammenfassend dadurch gekennzeichnet, daß Wünsche nach Versorgt-/Umhegtsein nicht mehr erfüllt werden.

Weiner, Mirsky und Mitarbeiter haben dieses hypothetische Modell in einer prospektiven Studie geprüft (1957/1958). Sie untersuchten 2073 zur Armee einberufene Rekruten und sonderten aus dieser Gesamtgruppe 63 Probanden aus, die einen hohen Serumpepsinogenspiegel des oberen 15%-Bereiches hatten. Sie wurden „Hypersekretoren" genannt. Die Kontrollgruppe bestand aus 57 Probanden mit niedrigem Pepsinogenspiegel aus dem unteren 9%-Bereich. Sie wurden als „Hyposekretoren" bezeichnet. Diese insgesamt 120 Probanden wurden vor und 8–16 Wochen nach Beginn der militärischen Grundausbildung testpsychologisch und röntgenologisch untersucht. Der Serumpepsinogenspiegel sowie die Gruppenzugehörigkeit blieben den testpsychologischen Untersuchern unbekannt. Die Autoren stellten zwei Hypothesen auf:

Hypothese I: Die „Hypersekretoren" können allein aufgrund psychologischer Tests von den „Hyposekretoren" unterschieden werden.

Hypothese II: Von 10 der insgesamt 120 Rekruten wurde aufgrund psychologischer Daten vorhergesagt, daß sie im Verlauf der Grundausbildung an einem Ulcus duodeni erkranken würden. Als psychologische Kriterien galten Bedürfnisse nach Abhängigkeit und Umsorgtsein, die mit großer Wahrscheinlichkeit während der Periode der Grundausbildung in der Armee (einer soziologisch definierten Situation) frustriert werden.

Die Ergebnisse

Zu Hypothese I: Aus der Gesamtgruppe der 120 Rekruten konnten aufgrund der psychologischen Tests 85% richtig den beiden Untergruppen der Hyper- und Hyposekretoren zugeordnet werden. 9 der psychologisch als besonders gefährdet identifizierten 10 Rekruten gehörten zu der Untergruppe der Hypersekretoren.

Zu Hypothese II: 7 der als gefährdet identifizierten 10 Rekruten erkrankten an einem Ulcus duodeni.

Mit anderen Worten: Sowohl die erste wie die zweite Hypothese wurden bestätigt. Zusätzlich erkrankten aus der Gruppe der Hypersekretoren zwei weitere Rekruten, während aus der Gruppe der 57 Hyposekretoren kein Rekrut ein Ulcus duodeni entwickelte. Es waren aber 9 der 63 Hypersekretoren unter den Bedingungen der Rekrutenausbildung an einem Ulcus duodeni erkrankt.

Damit wurde ein hypothetisches Modell untermauert, das die meisten der damals bekannten ätiologischen und pathogenetischen Faktoren in einen Zusammenhang brachte. In einer kritischen Übersicht über die „psychosomatischen Konzepte" für das Ulcus duodeni schrieb Fordtran (1973): „Die Genauigkeit, mit der Weiner und Mitarbeiter allein auf der Basis psychologischer Tests mit hoher Treffsicherheit die Erkrankung an einem Ulcus duodeni vorhersagen konnten, ist außerordentlich eindrucksvoll. Dieses Urteil wird nicht wesentlich dadurch beeinträchtigt, daß keine Gruppe von ‚Normalsekretoren' in die Untersuchung einbezogen wurde; denn sie konnten immerhin die besonders Gefährdeten aus einer großen Gruppe von Hypersekretoren herausfinden ... Die Studie stützt daher bestimmte Aspekte der psychosomatischen Theorie über die Pathogenese eines Ulcus duodeni bei jungen Männern. Sie legt außerdem nahe, daß Hypersekretion wenigstens in einem bestimmten Prozentsatz unabhängig von einem psychischen Konflikt auftreten kann und daß der psychische Konflikt, selbst wenn er schwer ist, bei Hyposekretoren kein Ulkus hervorruft."

Fordtran betont auch, daß über Frauen, die an einem Ulcus duodeni erkranken, erstaunlicherweise fast keine Untersuchungen vorliegen, und daß wir – worauf bereits hingewiesen wurde – über Persönlichkeitsfaktoren bei Patienten mit Ulcus ventriculi noch gar nichts wissen.

Wie bereits mehrfach erwähnt, müssen die Schlußfolgerungen, die Weiner und Mitarbeiter aus ihrer damaligen Untersuchung gezogen haben, auf die Gruppe der Ulcus-duodeni-Patienten beschränkt werden, die einen erhöhten Serumpepsinogenspiegel aufweisen. Damals nahm man an, daß fast alle Patienten mit Zwölffingerdarmgeschwür zu dieser erblich belasteten Gruppe gehören würden. Inzwischen haben Rotters Untersuchungen gezeigt, daß etwa 50% der Patienten einen normalen Pepsinogenspiegel aufweisen. Weiner kommentiert die dadurch entstandene Situation folgendermaßen (1981):

„Das führt zu der Frage, ob Mirskys Feststellung einer Korrelation von psychologischen Zügen und Erhöhung der Pepsinogenwerte auch für die 50% der Patienten mit normalen Pepsinogenwerten zutrifft oder nicht. Jeder künftige Versuch, Mirskys Befunde nachzuprüfen, muß dieses Problem in Rechnung stellen. Könnte es sein, daß Ulcus-duodeni-Patienten mit normalen Pepsinogenwerten eine andere Psychologie aufweisen? Oder sollten bei dieser Form der Krankheit psychologische Faktoren weniger wichtig sein?"

Er stellt dann fest, daß die Liste genetischer Faktoren und Schädlichkeiten der Umgebung, die für die Pathogenese des Ulkus von Bedeutung sind, sehr kurz wird, wenn man die psychischen Faktoren ausklammert. Als einziger Umgebungsfaktor, dessen Rolle in der Pathogenese dieser Krankheit gesichert sei, bleibe das Zigarettenrauchen übrig, da die Rolle der anderen Umgebungsfaktoren, die man für bedeutsam hielt, wie Koffein oder Alkohol, ebensowenig bewiesen wie ausgeschlossen sei. Unter genetischen Faktoren würden außer der Hyperpepsinogenämie nur noch die Blutgruppe 0, ein Fehlen der Blutgruppensubstanzen ABH im Speichel und das histokompatible Antigen HLA-B5 überdurchschnittlich oft gefun-

den werden. Personen mit diesem Merkmal hätten jedoch – im Unterschied zu Personen mit erhöhtem Pepsinogenspiegel, die eine 5–8mal so große Chance haben, an einem Ulcus duodeni zu erkranken als der Durchschnitt der Bevölkerung – nur weniger als ein doppelt so großes Risiko.

Als Resultat dieser Übersicht bleibt die Feststellung, daß eine Vermehrung des Pepsinogens nach wie vor die wichtigste der uns bekannten Determinanten dieser Krankheit darstellt – aber auch, daß die psychischen Faktoren, die im Zusammenhang mit diesen Determinanten erhoben wurden, für die betroffene Personengruppe eine Bedeutung haben, welche die aller anderen pathogenen Faktoren übertrifft.

Diese kritische Feststellung ist angesichts der weitverbreiteten Neigung, psychosomatische Modelle zu schnell zu verallgemeinern, notwendig. Ackerman und Weiner betonen nach einer Diskussion der neueren Befunde über die verschiedenen Kombinationsmöglichkeiten von Störungen einzelner Teilfunktionen (wie Sekretion, Motilität, Durchblutung und Wirkungsweise gastrointestinaler Hormone), daß die Psychosomatische Medizin die Verbindung zwischen den psychischen Faktoren, die sie als bedeutsame Variablen in der Pathogenese des Ulcus duodeni betrachtet, und den verschiedenen Störungen der Autoregulation im Verdauungstrakt aufzeigen müsse. Soziale und emotionale Momente könnten vor allem während der frühen Kindheit die neurale und endokrine Organisation des heranwachsenden Organismus beeinflussen. Die Natur der dabei stattfindenden Interaktion sei jedoch bisher nur rudimentär verstanden. Das gelte besonders für die gastrointestinalen Hormone. Die bisher vorliegenden Hinweise auf Zusammenhänge zwischen verschiedenen psychologischen Typen und bestimmtem Sekretionsverhalten sprächen dafür, daß der Zusammenhang zwischen psychologischen und physiologischen Vorgängen bei der Entstehung des Ulcus duodeni komplexer sei, als die verallgemeinernde These Alexanders von dem Zusammenhang zwischen einem spezifischen Konflikt und einer Funktionsstörung des Verdauungsgeschehens es darstellte.

Weiner meint, daß aufgrund der vorliegenden Untersuchungen drei Parameter für die Entwicklung eines Ulcus duodeni von Bedeutung sind:
- „ein physiologischer Parameter, der durch den hohen Pepsinogenspiegel angezeigt wird;
- ein psychologischer Parameter, der durch den Konflikt zwischen persistierenden intensiven infantilen Abhängigkeitswünschen und der Scham und dem Stolz des Erwachsenen, diese Wünsche zu zeigen, charakterisiert ist;
- ein sozialer Parameter, bei dem Umgebungseinflüsse durch psychischen Druck einen psychischen Konflikt mobilisieren".

„Dabei ist festzustellen, daß die beiden ersten Parameter spezifisch und miteinander verbunden sind, während der dritte nichtspezifischer Natur ist" (Weiner, 1981).

Er stellt fest, daß die Krankheit des Zwölffingerdarmgeschwürs kein einheitliches Leiden, sondern eine heterogene Gruppe verschiedener Krankheiten mit lediglich einer gemeinsamen Manifestation sei, nämlich einem Loch im Zwölffingerdarm. Er glaubt, dadurch einen besseren Zugang zu der Frage nach der Rolle, die die einzelnen pathogenen Komponenten dabei spielen, eröffnen zu können.

Inzwischen wurden einzelne Zusammenhänge zwischen psychologischen Variablen und pathophysiologischen Veränderungen, die das Ulcus-duodeni-Risiko bei Männern erhöhen, weiter untersucht. So fanden Walker et al. (1988) signifikante Beziehungen zwischen Scrumpepsinogenen und zwei psychologischen Variablengruppen (Mediatorvariablen): „negative Persönlichkeitszüge" wie „Feindseligkeit" und verminderte „Coping-Fähigkeit" („Ich-Schwäche").

43.5 Das somatopsycho-psychosomatische Modell unter wissenschaftstheoretischen Aspekten

Unabhängig von all diesen Einschränkungen ist das Modell unter wissenschaftstheoretischen Aspekten von allgemeiner Bedeutung: Es spricht vieles dafür, daß somatopsychisch-psychosomatische Zusammenhänge nicht nur beim Ulcus duodeni, sondern bei vielen anderen – ja vielleicht sogar bei allen – Krankheiten eine Rolle spielen. Mit anderen Worten: Das Weiner-Mirsky-Modell scheint die spezifische Variante eines allgemeinen psychosomatischen Konzeptes zu sein.

Wir haben im ersten Kapitel dieses Buches besprochen, daß sich die wichtigsten ungelösten wissenschaftstheoretischen Fragen der Heilkunde zwei großen Problemen zuordnen lassen:
– Dem Problemkreis der Beziehungen zwischen Organismus und Umgebung, d.h. der Beschreibung der individuellen Wirklichkeit;
– dem Problemkreis des Zusammenhanges zwischen physiologischen, psychologischen und sozialen Faktoren, d.h. der Beziehung zwischen Subsystemen, Systemen und Suprasystemen.
Wir haben ausgeführt, daß jede psychosomatische Hypothese auf diese beiden Problemkreise Antworten gibt, die mehr oder weniger vollständig und mehr oder weniger befriedigend ausfallen. Unter dem Gesichtspunkt der beiden Problemkreise können die hier diskutierten Hypothesen in ein gemeinsames Bezugssystem eingeordnet werden.

43.5.1 Die individuelle Wirklichkeit als Ausdruck der Beziehung zwischen Organismus und Umgebung

Im theoretischen Teil dieses Lehrbuches wurde ein Modell entwickelt, das auf beide Problemkreise eine Antwort zu geben sucht. Ausgangspunkt dafür ist die Vorstellung, daß jeder Mensch die (objektive) Umgebung aufgrund seiner subjektiven Bedürfnisse und

Erfahrungen interpretiert und daher in einer „individuellen Wirklichkeit" lebt, die sich von der Wirklichkeit seiner Mitmenschen mehr oder weniger unterscheidet. Da sie (und nicht die objektive Umgebung) sein Erleben und Verhalten bestimmt, muß sie auch einen Schlüssel zum Verständnis körperlicher Reaktionen in Gesundheit und Krankheit enthalten.

Die Schwierigkeit liegt darin, daß die individuelle Wirklichkeit eines Menschen dem außenstehenden Beobachter nicht ohne weiteres zugänglich ist und auch von dem Betroffenen selbst nur unzureichend geschildert werden kann. Das gilt vor allem für ihre vorbewußten und unbewußten Anteile.

Der außenstehende Beobachter kann jedoch – so lautete unsere These – die individuelle Wirklichkeit eines anderen Menschen rekonstruieren, wenn er davon ausgeht, daß sie aus aktuellen Situationen besteht, die aus Informationen aufgebaut werden, welche teils aus dem eigenen Körper, teils aus der Umgebung stammen und aufgrund von Programmen, die er in seinem Leben erworben hat, gedeutet werden.

Als Schema für den Aufbau der aktuellen Situation hatten wir das Modell des „Situationskreises" entworfen. In dessen Verlauf interpretiert das Individuum immer wieder die für seine Probleme relevanten Informationen aufgrund angeborener und erworbener Programme. Danach definieren wir „Erleben" als Aufbau einer subjektiven Situation durch Deutung der Umgebung als Potential für die Befriedigung von Bedürfnissen (= Problemsituation) und „Verhalten" als Nutzung dieses Potentials, entsprechend den gegebenen Deutungen (= Lösung der Problemsituation). Nach dem Modell des Situationskreises schreitet also die Entwicklung aus einer Problemkonstellation (die durch Bedeutungserteilung aufgebaut wird) in eine Konstellation fort, in der das Problem gelöst wird, bzw. in der die Richtigkeit der Bedeutungserteilungen geprüft wird.

Auf diese Weise gibt das Modell Antwort auf die Fragen des ersten Problemkreises, wie wir uns die Beziehungen zwischen Organismus und Umgebung vorstellen sollen. Auf die Frage des zweiten Problemkreises, wie der Zusammenhang zwischen physiologischen, psychologischen und sozialen Faktoren vorstellbar sei, gibt das Modell eine systemtheoretische und eine entwicklungspsychologische Antwort. Danach differenzieren sich die Beziehungen zwischen Organismus und Umgebung im Verlaufe der embryonalen und kindlichen Entwicklung. In ihr werden angeborene biologische Programme „sozialisiert" und nehmen dabei durch Bedeutungskoppelungen psychologische und soziale Dimensionen an.

43.5.2 Die Bedeutungskoppelung als Beziehung zwischen biologischen, psychologischen und sozialen Faktoren

Da dieser Teil des Modells für das Verständnis unseres Ansatzes von besonderer Wichtigkeit ist, erfahrungsgemäß aber gewisse Schwierigkeiten bereitet, sei er etwas breiter dargestellt.

Nach den Vorstellungen der Systemtheorie müssen wir zwischen verschiedenen Integrationsebenen unterscheiden. Sie entstehen dadurch, daß einfachere Systeme als Elemente (Subsysteme) in komplexere Systeme integriert werden (Zellen in Organe, Organe in einen Organismus, Organismen in soziale Systeme wie Familie etc.). Die auf diese Weise sich abzeichnenden Integrationsebenen entsprechen phänomenologisch verschiedenen Stufen; denn auf ihnen treten Phänomene auf, die es auf den einfacheren Stufen noch nicht gibt und die daher immer wieder eine neue wissenschaftliche Disziplin zu ihrer Beschreibung und Beobachtung erfordern. Auf diese Weise entsprechen Physiologie, Psychologie und Soziologie verschiedenen Integrationsebenen.

Dieses allgemeine Modell einer hierarchischen Organisation bekommt im Rahmen der Entwicklungsgeschichte vom befruchteten Ei über den wachsenden Embryo bis zum erwachsenen Individuum eine zeitliche Dimension, in der die Bildung immer komplexerer Systeme bestimmten Lebensphasen zugeordnet werden kann. So ist der Organismus am Ende der Embryonalzeit bereits ein hochkomplexes System, das aber verglichen mit dem Säugling nach der Geburt immer noch einer einfacheren Integrationsstufe zuzuordnen ist als dieser. Der Unterschied entspricht dem zwischen einer pflanzlich-vegetativen Organisationsform und einer Organisationsform für animalisches Leben, wobei der Begriff „animalisch" von dem lateinischen Wort „anima", also Seele, kommt (Bateson, 1982). Während Pflanzen mit ihrer Umgebung nur vermittels ihrer Oberfläche und der Wurzeln kommunizieren, bauen Tiere aus Sinneseindrücken eine subjektive Umwelt auf, in der ihre Kommunikation mit der Umgebung erfolgt. Pflanzen haben daher noch keine Umwelt, sondern nur eine „Wohnhülle" (J. v. Uexküll, 1940).

Solange die Bedürfnisse des wachsenden Organismus durch den Plazentarkreislauf befriedigt werden, haben für den Embryo Umgebungsfaktoren keine Bedeutung. Der Embryo hat noch keine Umwelt. Man kann daher von einer vegetativen Organisationsform als bloßer Körper sprechen, der nur von einer „Wohnhülle" umgeben ist. Die Vorgänge, die sich in ihm und zwischen ihm und dem mütterlichen Organismus abspielen, lassen sich mit physiologischen Begriffen beschreiben. – Diese Situation ändert sich nach der Geburt radikal: Jetzt wird der Aufbau einer subjektiven Umwelt durch die Sinnesorgane des Säuglings zu einer unabdingbaren Voraussetzung für sein Überleben; denn jetzt muß Umgebung in subjektive Umwelt verwandelt werden, um die Bedürfnisse des Organismus zu befriedigen. Damit findet ein Sprung von einer einfacheren zu einer komplexeren Integrationsebene statt: Das System eines bloßen Körpers muß sich jetzt in das komplexere System eines Körpers integrieren, der von einer subjektiven Umwelthülle umschlossen ist. Vorgänge, die sich zwischen dem Organismus und seiner Umgebung abspielen, lassen sich nicht mehr mit Begriffen der Physiologie beschreiben. Sie erfordern Begriffe, die psychologische („animalische") Phänomene beschreiben können.

Dem Wechsel von der vorgeburtlichen vegetativen zur animalischen Stufe nach der Geburt entspricht eine ebenso radikale Umstellung der Beziehung zwischen dem Kind und der nahrungsspendenden Umgebung, die auch nach der Geburt zunächst die Mutter bleibt. Aber die Mutter, die für den Embryo aus der Innenwand eines ernährenden und schützenden Organs bestand, wird jetzt Teil einer Umgebung, die für den Säugling nur als das existiert, was seine Sinnes- und Bewegungsorgane in seine subjektive Umwelt verwandeln können. Das Organsystem Plazenta und Nabelschnur, das vor der Geburt den Ausgleich der Homöostasestörungen im Körper des Kindes gewährleistet hat, wird nach der Geburt von Mund und Magen-Darm-Trakt abgelöst. Es muß allerdings bedacht werden, daß noch unmittelbarer als der Mund und der Magen-Darm-Trakt die Atemwege eingesetzt werden, um mit der Umgebung einen Kontakt herzustellen (vgl. Kap. 42). Der Kontakt zwischen dem kindlichen Organismus und der nahrungsspendenden Umgebung, der vor der Geburt innerhalb eines mütterlichen Organs (zwischen dem Endometrium des Uterus und der Plazenta) stattfand, muß jetzt zwischen der Brust der Mutter und dem Mund des Säuglings hergestellt werden. Dabei ist zu beachten, daß der Mund als sensorisches Organ des Magen-Darm-Trakts aus taktilen, olfaktorischen und Geschmackseindrücken sehr frühe Fundamente einer symbiotischen Umwelt aufbauen muß und daß für diesen Aufbau die Mutter als Prototyp für die spätere individuelle Umwelt fungiert.

Halten wir fest: Mit dem Übergang zu einer komplexeren Integrationsstufe sind neue Phänomene aufgetreten, zu deren Beschreibung die Sprache einer neuen Wissenschaft erforderlich wird. Die Vorgänge, die sich zwischen dem Säugling und der Mutter während des Stillens abspielen, lassen sich nicht mehr mit physiologischen Begriffen beschreiben. Dafür benötigen wir eine Sprache, die Begriffe für Sinneseindrücke, Gefühle und Triebregungen besitzt, d.h. wir brauchen die Sprache der Psychologie. Damit wird das Problem konkret, wie wir uns die Verbindung zwischen physiologischen und psychologischen Phänomenen vorstellen sollen.

Wir haben diese Verbindung als „Bedeutungskoppelungen" zwischen Zeichen verschiedener Zeichensysteme beschrieben: Zeichen, die Nachrichten zwischen Zellen und Organen im Körper vermitteln, werden an Zeichen gekoppelt, die den Organismus und dessen Sinnesorgane über dessen Umgebung informieren. Wir haben darauf hingewiesen, daß Pawlow mit seinem Konzept der unbedingten (angeborenen) und der bedingten (erworbenen) Reflexe solche Bedeutungskoppelungen nachgewiesen und gezeigt hat, wie in bestimmten Situationen neue Bedeutungskoppelungen (bedingte Reflexe bzw. Konditionierungen) entstehen.

Freud hat mit seinem Triebbegriff den gleichen Vorgang beschrieben; denn dieser Begriff bezeichnet ein Modell, das aus vier Komponenten, nämlich der Quelle, dem Drang, dem Ziel und dem Objekt besteht. Von diesen ist die Triebquelle ein physiologisches Geschehen im Körper. Freud spricht von einem interzellulären Chemismus, der bei entsprechender Konzentration in ein psychisches Drängen übersetzt wird. Es veranlaßt den Organismus, in der Umgebung nach einem Objekt zu suchen, mit dessen Hilfe die Befriedigung des Drängens (das Triebziel) und auf diesem Wege ein Abstellen der Quelle erreicht wird. Das Triebmodell beschreibt einen Regelkreis mit negativer Rückkoppelung, der zwei verschiedene Integrationsebenen (eine physiologische und eine psychische) durch eine Bedeutungskoppelung (zwischen physiologischen und psychischen Zeichen) verbindet.

Das Pawlowsche und das Freudsche Modell ergänzen sich gegenseitig. Freud interessierte sich nicht für die Frage, ob die Bedeutungskoppelung zwischen der physiologischen Quelle und dem psychischen Drängen angeboren oder erworben ist. Er konzentrierte sich auf die Entwicklung der psychischen Seite des Triebgeschehens im Laufe der Lebensgeschichte. Pawlow interessierte sich dagegen für die Frage, wann und unter welchen Bedingungen welche Bedeutungskoppelungen zustande kommen.

Dieser Punkt ist für unser Problem deswegen wichtig, weil die Verbindungen zwischen den sensorischen Eindrücken des Mundes, und später auch anderer Sinnesorgane, aus denen die symbiotische Umwelt des Säuglings aufgebaut wird (d.h. der psychische Bereich), und den nervalen, endokrinen und sekretorischen Zeichenprozessen im Gastrointestinaltrakt (d.h. der physiologische Bereich) beim Menschen und vielen Säugetieren nur zu einem geringen Teil durch angeborene Reflexe geregelt sind. D.h. mit anderen Worten: Die Bedeutungskoppelung zwischen Ereignissen in der Umgebung des Lebewesens und den Vorgängen im Magen-Darm-Trakt müssen zum weitaus größten Teil in Form von bedingten Reflexen und zwar in Abhängigkeit von bestimmten Situationen „gelernt" werden.

Pawlow hat gezeigt, daß ein Zustandekommen solcher Bedeutungskoppelungen einmal von einer Bereitschaft des Organismus (Appetenz), zum anderen von den Signalen seiner Umgebung abhängt; d.h.: Komplexe Programme zum Aufbau einer Umwelt und später einer individuellen Wirklichkeit entstehen nur bei dem Zusammentreffen kindlicher Bereitschaft und einem entsprechenden Entgegenkommen der mütterlichen Umgebung.

Wenn wir nach dieser Rekapitulation unseres allgemeinen Modells wieder auf das spezielle Konzept eines somatopsychisch-psychosomatischen Geschehens zurückkommen, wie es Mirsky für das Ulcus duodeni entwickelt hat, so läßt es sich jetzt im Sinne des Freudschen Triebmodelles folgendermaßen beschreiben:

Eine angeborene Hyperpepsinogenämie wirkt als somatische Triebquelle. Diese wird in den psychischen Drang übersetzt, Umgebungsreize für die Mund- und Zungenschleimhaut als nahrungsspendende (oder nahrungsverweigernde) Umwelt zu interpretieren, um mit ihrer Hilfe das Triebziel (die Befriedigung des Dranges) zu erreichen. Über Einverleibung von Umwelt (des Triebobjektes Milch) soll es zum Abstellen der somatischen Triebquelle kommen. Nach dem Pawlowschen Modell würden wir sagen: Einer angeborenen Hypersekretion des Magens entspricht eine erhöhte Appetenz des Säuglings, Umgebungsreize an gastrointestinale Zeichen zu koppeln. Beide Modelle beschreiben das Entstehen einer Umwelt, in der alles und jedes eine Fütterungsbedeutung, gewissermaßen einen „basalen Freßton" erhält.

In einer solchen Umwelt kann dann das Verhalten anderer Personen, in erster Linie natürlich der Mutter, unabhängig von der Bedeutung, die es sonst haben mag, eine unmittelbare Bedeutung für die Funktion des Magen-Darm-Traktes haben. Dabei ist es entscheidend, daß die Zuwendung der Mutter bei dem Vorgang des Stillens über die Bedeutung der Ernährung, des Wärmens, Wiegens usw. hinaus eine zusätzliche soziale Bedeutung hat: Es ist die Bedeutung des Stillens, das dem Säugling die Unversehrtheit der Dualsituation signalisiert, des frühesten sozialen Systems, von dessen Beständigkeit das Leben des Kindes abhängt. In der Terminologie Freuds handelt es sich um eine frühe sexuelle bzw. libidinöse Bedeutung.

Der „Freßton" der primitiven Umwelt, von der die Mutter ein integrierender Teil ist, bildet also zugleich einen „Überlebenston". Von dem Entgegenkommen der Mutter hängt es ab, ob die „Freß-Überlebens-Ton"-Umwelt Bestand hat oder ob sie trotz steigender Appetenz des Säuglings zerbricht. Im letzten Fall kommt es über Panik zum Rückzug auf die Organisationsform eines Körpers ohne Umwelt. Wir haben dann einen apathischen Säugling vor uns, den die Umgebung nichts mehr angeht.

In diesem Modell eines hierarchischen Systems immer komplexerer Integrationsstufen können wir uns vorstellen, wie einerseits „Aufwärts-Effekte" (Popper, 1977; Medawar und Medawar, 1977) von der Magenschleimhaut zur psychisch erlebten Umwelt und von dieser zu der sozialen Einheit mit der Mutter zustande kommen; wie andererseits „Abwärts-Effekte" von dem Verhalten der Mutter über das Umwelterleben des Säuglings zu dessen Magenschleimhaut entstehen. Vor allem können wir verstehen, daß solche „Aufwärts- und Abwärts-Effekte" aufgrund der in der individuellen Lebensgeschichte eines Menschen erfolgten Bedeutungskoppelungen gebahnt (oder nicht gebahnt) werden.

So würde das Mirsky-Weiner-Modell im Rahmen unseres allgemeinen bio-psycho-sozialen Modelles zu interpretieren sein. Dabei wissen wir natürlich noch sehr wenig darüber, wie in der Frühphase der menschlichen Entwicklung die einzelnen Bedeutungskoppelungen aussehen, die für eine normale Entwicklung erforderlich sind, und welche pathologischen Verbindungen entstehen können. Versuche mit frühzeitig von den Müttern getrennten Tieren sprechen jedoch eine eindrucksvolle Sprache für die Wichtigkeit solcher Bedeutungskoppelungen. Sie zeigen aber auch, wie subtil und komplex die Verbindungen sind, die dabei zustande kommen oder nicht zustande kommen. Vor allem machen sie klar, daß wir in solchen Bedeutungskoppelungen ein allgemeines Naturprinzip für den Übergang von einer pflanzlich-vegetativen zu einer animalisch-psychischen Organisationsform vor uns haben, wenn wir unter animalischem (psychischem) Leben den Aufbau subjektiver Umwelten verstehen.

43.5.3 Die Rolle des Pylorus

Fassen wir die Ergebnisse dieser Überlegungen zusammen, so zeichnet sich eine interessante Hypothese ab.
- Der Mund ist für den Säugling das wichtigste Sinnesorgan. Nach dem Erliegen des Plazentarkreislaufes kann der Säugling die abgerissene Verbindung zu der ernährenden und beschützenden Mutter nur mit seiner Hilfe in neuer Form wiederherstellen (von der Bedeutung der Atemfunktion sehen wir an dieser Stelle ab). Das geschieht durch den Aufbau einer symbiotischen Umwelt, deren Geschmacks-, Tast- und Geruchsempfindungen die Bedeutung eines Mediums für Nahrung und Überleben vermitteln.
- Der Mund ist darüber hinaus die Aufnahme-Öffnung des Magen-Darm-Trakts, dessen Funktionen zum Teil autonom, zum Teil über Bedeutungskoppelungen mit Umweltvorgängen (den „Triebobjekten") gesteuert werden.

Wir gehen von der Feststellung aus, daß der Magen-Darm-Trakt ein anatomisch und funktionell außerordentlich differenziertes Organ ist, dessen verschiedene Abschnitte in verschiedenem Maße und auf verschiedene Weise an den Bedeutungskoppelungen der Umwelt teilnehmen. Dabei kommen den drei Schließmuskeln (Sphinkteren) verschiedene, einander ergänzende Aufgaben zu: Der Mund (als erster Schließmuskel) reguliert und bewacht die „Einverleibung" von Umwelt, die dann als Nahrung den Magen-Darm-Trakt passiert. Der Pylorus (als zweiter Schließmuskel; möglicherweise wäre auch die Kardia zu berücksichtigen) reguliert und bewacht deren Weiterleitung in den Dünndarm, in dem der Austausch mit Bestandteilen des Organismus erfolgt. Der After (als dritter Schließmuskel) gibt schließlich die von dem Organismus nicht verwertete Umwelt einschließlich der Ausscheidungsprodukte wieder an die Umgebung zurück.

Von den drei Schließmuskeln und ihren verschiedenen Funktionen hat die Psychoanalyse nur den Mund und den After in Betracht gezogen und ihren Zusammenhang mit psychischen Vorgängen untersucht. Beide spielen in der psychoanalytischen Entwicklungstheorie eine wichtige Rolle. Der Pylorus und seine Funktion blieben unbeachtet, vor allem weil er keine dem Bewußtsein zugängliche Innervation besitzt.

Es könnte sein, daß wir hier auf ein fehlendes Glied in unseren Kenntnissen über die pathogenetische Kette des Ulcus duodeni stoßen; denn wir wissen, daß der Pylorus nicht nur die Weitergabe des Mageninhaltes, sondern auch die Säureverhältnisse im Bulbus duodeni überwacht, und daß er durch seine sensible Versorgung über das vegetative Nervensystem eng mit den Stimmungen und Verstimmungen des Organismus und dessen Umweltbeziehungen verbunden ist.

Der Pylorus schließt sich bei Ekel und Übelkeit und blockiert so den Weitertransport von Umwelt aus dem Magen in den Dünndarm. Er kann damit den Brechakt einleiten, der den Magen wieder von seinem Inhalt befreit, und der mit einem Rückzug des Organismus in den umweltlosen Zustand eines bloßen Körpers einhergehen kann, für den die Umgebung nichts mehr bedeutet. Darüber hinaus kann der Pylorus die Entleerungszeit des Magens in Abhängigkeit von den verschiedenen Stimmungen in weiten Grenzen variieren.

Wir können eine Reihe bilden, an deren einem Ende der atonische Magen bei geschlossenem Pylorus stehen würde, bei dem keine Entleerung zustande kommt. Das andere Ende dieser Reihe wäre ein gut tonisierter, peristaltisch aktiver Magen mit offenem Pylorus und sehr kurzer Entleerungszeit. Wenn wir versuchen, diese verschiedenen Funktionszustände bestimmten Stimmungen zuzuordnen, bekommen die oben erwähnten Untersuchungsergebnisse von Zander eine prinzipielle Bedeutung. Der von ihm beschriebene Funktionszustand des Magens mit spastisch-verengtem Antrum und offenem Pylorus würde ein Gegenstück zu dem schlaffen Magen mit geschlossenem Pylorus bilden, den wir in der Stimmung der Nausea beobachten.

Zander beschreibt die (unbewußte) Stimmung der von ihm untersuchten Patienten als Neid und als Gefühl, zu wenig zu bekommen. Die Stimmung der Nausea ist genau das Gegenteil. Sie entspricht dem Gefühl, etwas bekommen zu haben, das man nicht assimilieren kann und von dem man sich zurückziehen will. Es handelt sich neben der von Cannon beschriebenen fight-flight-Reaktion um die zweite der uns bekannten grundlegenden bio-psycho-sozialen Grundstimmungen des Menschen (v. Uexküll, 1952; Engel et al., 1956).

Im Fall der Nausea haben wir eine Situation vor uns, in welcher die somatische Triebquelle versiegt. Ihr entspricht eine Herabsetzung oder ein Sistieren der Säureproduktion im Magen. In dem anderen Fall sind wir mit einer Situation konfrontiert, in der es nicht gelingt, die somatische Triebquelle abzustellen. Daher ist in diesem Falle eine Hypersekretion zu erwarten.

Ein Säugling, der sich in einer Problemsituation befindet, für deren Lösung er noch über keine Programme verfügt – und das ist bei den Neugeborenen ständig der Fall – hat nur zwei Alternativen: Er kann durch Bedeutungskoppelung seine Programme zu erweitern und eine Umwelt aufzubauen versuchen, in der er sein Triebziel erreicht. Dann strömt die ernährende Umwelt durch den geöffneten Pylorus und neutralisiert die Triebquelle. – Oder er kann sich, wenn das mißlingt, in die Stimmung der Nausea und der Apathie zurückziehen. Dazwischen würden Situationen liegen, in denen mit steigender Appetenz versucht wird, die rettende Bedeutungskoppelung zustande zu bringen, in denen aber innere oder äußere Widerstände die Lösung des Problems verhindern. Diese Situation würde den von Zander beschriebenen Verhältnissen entsprechen. Man könnte sie eine „gastrointestinale Panikreaktion" nennen; oder eine spezielle Variante von „Streß".

Ein Erwachsener, der als Säugling nicht gelernt hat, seine primären Programme so zu modifizieren, daß er seine Bedürfnisse selber befriedigen kann, wird nur in einer sehr entgegenkommenden Umgebung vor solchen Panikreaktionen geschützt sein. Die ihm verfügbaren Verhaltensprogramme erlauben ihm keine Problemsituationen zu lösen, in denen die für ihn lebensnotwendige Verwöhnung ausbleibt. Unter diesem Gesichtspunkt läßt sich das Modell, das Weiner und Mirsky entwickelt haben, als spezifische Variante eines allgemeinen psychosomatischen Modelles verstehen.

43.5.4 Das Modell unter klinischem Aspekt

Die innere Medizin hatte zunächst ein Schema entworfen, nach dem die zahlreichen Vorgänge, die innerhalb des Organismus für die Pathogenese des Ulcus duodeni eine Rolle spielen, unter zwei Gesichtspunkten, als **aggressive** bzw. als **defensive** Mechanismen, geordnet sind. Da beide Mechanismen gemeinsam auf die Duodenalschleimhaut einwirken, soll eine Verstärkung der aggressiven und/oder eine Schwächung der defensiven Mechanismen zum Ulkus führen.

Die Verdauungstätigkeit des Magens ist aber nur eine Teilfunktion in dem Regelkreis „Nahrungsaufnahme". Daher muß dieses Schema erweitert werden. Der Regelkreis „Nahrungsaufnahme", der den Menschen und seine Umgebung umfaßt, besitzt einen endogenen 24-Stunden-Rhythmus. Ihm entspricht eine „periodische Sollwertvorstellung", die mit dem Auftreten und Abklingen des Nahrungsbedürfnisses einhergeht und die Einzelfunktion des Magens (HCl- und Pepsinsekretion, Schleimhautdurchblutung, Produktion von Schleim usw.) mit den Nahrungsangeboten der Umgebung koordiniert. Dementsprechend wurde von gastroenterologischer Seite ein drittes pathogenetisches Prinzip formuliert (vgl. Arnold, 1981).

In einem derart erweiterten Konzept bestehen Möglichkeiten, psychodynamische und biologische Aspekte in einen Zusammenhang zu bringen: Dem endogenen 24-Stunden-Rhythmus, in dem bei Neugeborenen Hunger, motorische Unruhe und Schreien, Gefüttertwerden, Sättigung und Schlaf periodisch wechseln, entspricht ein Rhythmus ansteigender und sich lösender psychischer Spannungen, deren Triebaspekt als „oral" bezeichnet wird. Dieser Rhythmus schließt – wenn er störungsfrei ablaufen soll – die Umgebung „kontrapunktisch" mit ein. Beim Säugling ist die Umgebung die Mutter, und deren Erleben und Verhalten ist gewissermaßen mit dem Erleben und Verhalten des Säuglings „verzahnt"; denn die motorische Unruhe und das Schreien des Säuglings werden

von ihr als Auslöser für ihr Stillverhalten erlebt, dem psychisch wiederum Spannungsanstieg und Spannungslösung entspricht. Das Stillverhalten der Mutter führt zur Sättigung (zum Stillen des Bedürfnisses) des Säuglings. Auf diese Weise haben wir einen umfassenden Regelkreis vor uns, der Erleben und Verhalten von zwei Individuen umschließt, von denen jedes die „Umwelt" für das andere darstellt (Mutter-Kind-Dyade).

In unserer Terminologie könnten wir sagen, daß ein Funktionskreis entsteht, in dem zwei Lebewesen zu einem System verbunden sind. Jedes der beiden wird von den angeborenen Programmen geleitet, die Deutungs- und Verhaltensanweisungen für Erleben und Beantworten der Umgebung (bzw. für Bedeutungserteilung und Bedeutungsverwertung) enthalten, und damit zwei spezifische, aber einander ergänzende Umweltsituationen aufbauen. Immer wieder löscht das „Wirkmal" des einen Subjektes das „Merkmal" des anderen aus, bis das ganze System zur Ruhe kommt.

Im Verlauf der kindlichen Entwicklung erfolgt im Rahmen der Auseinandersetzung mit der Umgebung eine Überwindung dieser Dyadensituation durch zwei Vorgänge:

- Es bildet sich ein Ich, in dem die angeborenen (primärprozeßhaften) sensomotorischen Programme in Auseinandersetzung mit der Umgebung umgeformt, differenziert und mit anderen erlernten Programmen verbunden werden.
- Es entsteht eine Außenwelt, in der andere Menschen (Objekte oder soziale Institutionen) die Erlebnisbedeutung einer nahrungsspendenden und schützenden oder enttäuschenden und zurückweisenden Instanz erhalten.

Eine Konfliktsituation ist dann gegeben, wenn die Umgebung aufgrund primitiver Abhängigkeitsbedürfnisse als nahrungsspendende Instanz erlebt wird, die Bedürfnisbefriedigung aus endogenen oder exogenen Gründen aber nicht zustande kommt; wenn es ferner nicht gelingt, Programme zu entwerfen, um die in dieser Situation gestellte Aufgabe zu lösen.

In diesem Schema ist dargestellt, wie der Säugling und später das Kind lernen, sich den gesellschaftlichen Bedürfnissen anzupassen; diesen Bedürfnissen haben sich nicht nur die oralen Bedürfnisse, also das triebhafte Verlangen nach Nahrung, Schutz und Verwöhnung, sondern auch die zugeordneten Magenfunktionen unterzuordnen. Psychodynamisch wird das orale Triebverhalten gesellschaftlich induzierten Bedürfnissen angepaßt, so daß das Nahrungsverhalten des Erwachsenen weitgehend zu einem Erziehungsprodukt wird. Diese „Sozialisierung" des Nahrungsverhaltens beobachtet man in Anfängen bereits bei Tieren, die z. B. bei der Fütterung die Rangordnung beachten.

Nach dem somatopsycho-psychosomatischen Modell – das in Wahrheit ein „somatopsycho-soziopsychosomatisches Modell" darstellt – ist die Situation, in der die Umgebung aufgrund überwältigender Bedürfnisse nach Nahrung, Schutz und Verwöhnung gedeutet wird, eine spezifische (orale) Variation einer allgemeinen Problemsituation, für deren Lösung keine Programme verfügbar sind. – Betrachten wir auf dem Hintergrund dieses Modelles die Krankengeschichte unserer Patientin, so lassen sich die Zusammenhänge besser verstehen. Die Kranke stammte aus einer Familie, in der die Mutter eine überragende Rolle spielte und der Vater fehlte. Eine genetische Disposition ist anzunehmen, wenngleich aufgrund der vorliegenden Angaben nicht zu rekonstruieren. Psychologisch ist eine enge Bindung an die Mutter eindrucksvoll, und psychodynamisch sehen wir den Versuch, den Wunsch nach Versorgt- und Umhegtsein durch die Mutter durch das gleichzeitige Streben nach Unabhängigkeit zu kompensieren. Hierdurch erhält die Patientin wiederum die Möglichkeit, für die Mutter zu sorgen und auf diese Weise die Mutter teilweise durch die Institution „Chefsekretariat" zu ersetzen. Das Beschwerderezidiv (nach chirurgischer Entfernung von ⅔ des Magens!) trat in einer Situation auf, in der die Mutter die Patientin wiederum enttäuschte, gleichzeitig der Verlust der Ersatzmutter „Chefsekretariat" und darauf auch ein Verlust der Gruppenzugehörigkeit drohte.

43.6 Diagnostisches Vorgehen

43.6.1 Anamnese

Im Vordergrund stehen für den Ulkuspatienten die Schmerzen. In der Regel ist er nicht darauf eingestellt, über psychosoziale Probleme zu sprechen (vgl. Abschn. 43.7). Das Erstgespräch sollte daher von den Symptomen ausgehen, um in einer assoziativen Weise zu den bio-psycho-sozialen Bezügen zu kommen. Die von Engel angegebene Anamneseform bietet sich an (vgl. Kap. 12). Es darf nicht als Regel erwartet werden, daß bereits im Erstgespräch die Konfliktsituation angesprochen wird. Anzustreben ist, daß
– der Beschwerdeverlauf erfaßt,
– die (psycho-)sozialen Bedingungen beschrieben und
– zeitliche Zusammenhänge zwischen Beschwerden und psychosozialer Situation herausgearbeitet werden.

Zum **Beschwerdeverlauf:** Die jahres- und tageszeitliche Periodizität der Schmerzen ist keineswegs in allen Fällen so charakteristisch, daß die Diagnose eines Ulcus duodeni schon bei der Erhebung der Vorgeschichte ausgesprochen werden kann. Charakteristisch ist die Neigung zur Chronizität. Auf sie sollte ebenso geachtet werden wie auf die oben beschriebene Symptomatik. Selten findet sich eine Gewichtsabnahme; es sei denn, die Patienten halten eine „Angstdiät" ein. Eine kurze Anamnese mit Gewichtsabnahme bei Patienten in höherem Alter ist immer verdächtig auf ein Karzinom anderenorts.

Zu den **psychosozialen Bedingungen:** Es empfiehlt sich, über eine „Ernährungsanamnese" auf dieses Ge-

biet überzugehen. Der Patient äußert sich über diejenigen Nahrungsmittel, die er gut bzw. schlecht verträgt. Er spricht darüber, zu welchen Tageszeiten er ißt, wie er das Essen auf das tägliche Arbeiten und auf die Abläufe in Familie und Umwelt abstimmt. Auf diese Weise wird es am ehesten möglich, die Beschaffenheit seiner Bezugsgruppe zu erfassen, sei es die Familie, sei es der Arbeitsplatz. Hier muß dann im Auge behalten werden, ob sich derzeit die oben beschriebene Loslösung aus seiner Gruppe vollzieht oder ob diese droht. Rechtzeitig ist auf Bewältigungsmechanismen zu achten, im Sinne der beschriebenen „offen-abhängigen" oder der „pseudounabhängigen" Form. Erst in einem fortgeschrittenen Gesprächsstadium, und dieses kann sich entweder sehr spät oder erst in anschließenden Gesprächen herausentwickeln, wird es möglich sein, auf gefühlsmäßig tiefergehende Themen einzugehen, die von „Scham und Stolz" (Weiner, 1981) handeln. Die von Zander beschriebene Angst-Neid-Problematik sollte hierbei im Auge behalten werden.

Zeitliche Zusammenhänge zwischen Beschwerden und psychosozialen Situationen: Immer wieder wird versucht, die assoziativ dargestellten Lebensbezüge in ihrem zeitlichen Ablauf möglichst genau zu erfassen und sie gleichsam vor dem inneren Auge mit den Beschwerden zeitlich zu verbinden. Die Versuchung ist groß, ursächliche Zusammenhänge zwischen Beschwerden und Lebensabläufen herzustellen. Es ist hier hilfreich, das Konzept der „Bedeutungskoppelung" als Hilfsmittel heranzuziehen. Hier ist das Grundproblem, daß eine solche Bedeutungskoppelung dem Patienten selbst nicht bewußt ist und der untersuchende Arzt sie zunächst nicht kennt. Dem Arzt ist jedoch mit dem Wissen um diesen Begriff grundsätzlich die Möglichkeit gegeben, den Assoziationen des Patienten und seinen eigenen folgend die Verbindungen zwischen Körpereindrücken und psychosozialen Abläufen herzustellen. In anderen Worten: Die günstigste diagnostisch-therapeutische Situation ist dann gegeben, wenn der Arzt bei Schilderungen von Schmerzzuständen an bedrohte Gruppensituationen, an Scham und Stolz, Neid und Ärger denkt; daß er umgekehrt an Schmerzen, gestörte Entleerungsvorgänge, vermehrte Sekretion und Entwicklung eines Ulkus denkt, wenn er entsprechende Themen aus dem zwischenmenschlichen Bereich hört. Gleichzeitig gehört zu diesem Vorgehen, daß der Arzt über diese Dinge nur allmählich und allgemein zu sprechen anfängt, d.h. berücksichtigt, daß der Patient diese Zusammenhänge in der Regel nicht kennt oder sie auch nachhaltig abwehren muß.

43.6.2 Differentialdiagnose

Das Spektrum auszuschließender Krankheiten ist relativ klein. Es beschränkt sich im wesentlichen auf das Ulcus ventriculi, maligne Prozesse und funktionelle Syndrome.

43.6.3 Der Untersuchungsplan

Entscheidend ist der endoskopische Befund. Die Untersuchung muß den Ausschluß von Komplikationen (Blutung, Stenosierung, Penetration) berücksichtigen.

Bei Verlaufsuntersuchungen ist es hilfreich, das Phänomen des „Syndrom-Wechsels" vor Augen zu haben: Ein erheblicher Prozentsatz der Patienten mit rezidivierenden Ulzera bzw. mit Zustand nach operativer oder wiederholter medikamentöser Therapie leidet unter Suizidimpulsen oder unter suchtähnlichen Zuständen, d.h. häufig unter Alkoholabhängigkeit. Es finden sich bei eingehendem Gespräch dann Angaben über psychiatrische Behandlungen (vgl. Abschn. 43.7).

43.7 Therapie

43.7.1 Therapieziele

Was ist das Ziel einer Therapie beim Ulcus-duodeni-Patienten? Diese Frage könnte je nach Befragtem unterschiedlich beantwortet werden.

Üblicherweise wird ein Sistieren von Schmerzen und nachfolgend ein Abheilen des Ulkus angestrebt. Das Therapieziel könnte auch darin bestehen, eine aktuelle Konfliktsituation zu beheben, die für das Ulkus mitursächlich oder gar primär verursachend ist. Schließlich könnte das Ziel sein, mit dem Patienten die Lebenssituation im Hinblick auf seine langfristige Gruppenzugehörigkeit und seine biographischen Vorstellungen zu verstehen, um dann eine Therapie zu beginnen, die kurativen wie sekundär-präventiven Charakter hat.

Andererseits sind die Standards der bisherigen Ulkustherapie zu berücksichtigen, um die Diskussion nicht im Luftleeren zu führen. Diese Standards richten sich daran aus, ob es um die Behandlung eines Patienten mit floridem Ulkus oder ob es um die Behandlung eines Patienten im Sinne einer Rezidivprophylaxe geht. Hieraus resultieren therapeutische Stufenpläne.

43.7.2 Therapeutischer Stufenplan – Das Cimetidin als „Goldstandard"; die substituierten Benzimidazole

Ca. 10–15% der männlichen Bevölkerung haben im Leben ein Ulkus. Ein erheblicher Prozentsatz geht nicht zum Arzt. Die Ulzera derjenigen, die zum Arzt kommen, heilen in ca. 50% der Fälle innerhalb von 4–6 Wochen auch unter Placebo ab; unter Cimetidin heilen im selben Zeitraum aber 75–80% ab (Holtermüller, 1982).

Hier hat sich eine dramatische Entwicklung der medikamentösen Möglichkeiten abgespielt: 1974 wurden 242 Ulkusmedikamente gezählt, von denen keines zu dem damaligen Zeitpunkt eine eindeutige Überlegenheit gegenüber Placebo bewiesen hatte (Blum, 1982). Heute liegt ein Medikament vor, das den Standard für alle anderen Medikamente abgibt. Cimetidin wird geradezu als eine Art „Goldstandard" der Ulkustherapie bezeichnet (Blum, 1982, S. 7). Der „Goldstandard" spiegelt sich in den Umsatzzahlen für dieses Präparat wider. Bis zum 23. 7. 82 betrugen diese weltweit 750 Millionen Dollar. – Stellvertretend für viele Patienten sagte ein Patient, der selber Arzt und langjähriger Ordinarius für Innere Medizin an einer Schweizer Universität ist, daß er sich erstmals in seinen 50 Jahren mit Ulcus duodeni nach Einnahme von Cimetidin gesund und genußfähig erlebt habe (Schlegel, 1982).

Dementsprechend wird heute eine Dreistufenleiter therapeutischer Maßnahmen befolgt: Im ersten und zweiten Schub wird mit der ersten Stufe in Form von Antazida, d.h. klassischen Präparaten begonnen. In einer zweiten Stufe kommen Cimetidin bzw. vergleichbare Präparate zum Einsatz. In der dritten Stufe ist das operative Vorgehen indiziert; der langfristige Einsatz von Histamin-H_2-Rezeptorantagonisten ist hiergegen abzuwägen.

Ein Vergleich des Cimetidins mit anderen modernen Medikamenten, so Ranitidin, Pirenzipin, Sucralfat, brachte keine eindeutige Überlegenheit über das Cimetidin (Henschel, 1984); Sucralfat kann jedoch wegen seines völlig andersartig lokal wirksamen Mechanismus als eine echte „Alternative" zum Cimetidin gesehen werden (Holtermüller, 1982). – Carbenoxolon, noch vor kurzem viel rezeptiert, ist aufgrund seiner Nebenwirkungen praktisch außerhalb der Diskussion.

Im Hinblick auf die dritte Therapiestufe sind folgende Erfahrungen wichtig: In einer langfristig angelegten Studie, an der sich 55 Patienten beteiligten, die sämtlich Kandidaten für eine elektive operative Therapie waren, blieben nach fünfjähriger Cimetidinbehandlung mehr als die Hälfte, nämlich 30, unoperiert. – Weitere Therapieperspektiven scheinen sich abzuzeichnen, indem von unseren derzeitig „säurelastigen" Therapiekonzepten (Arnold, 1982) abgegangen wird und stärker die protektiven Mechanismen berücksichtigt werden (Arnold, 1981; Fimmel et al., 1982).

Diesen außerordentlich optimistischen Einschätzungen der medikamentösen Möglichkeiten stehen folgende ungünstige Erfahrungen gegenüber: Cimetidin ändert nicht den Verlauf der Krankheit, deren Merkmale nach Absetzen des Medikamentes die gleichen bleiben (Halter, 1982); ca. 20% der Patienten sind gegenüber Cimetidin therapierefraktär (Hentschel, 1984; Massarrat et al., 1982).

Auch die Nebenwirkungen des Cimetidins müssen berücksichtigt werden. Diese wurden bisher als niedrig eingeschätzt, so die antiandrogene Wirkung bei über 1g/Tag, Verwirrtheit bei älteren Patienten, Hemmung des hepatischen Abbaues von Diazepam, Propranolol, Dicumarin. Mit größter Wachsamkeit sind bisherige Tierexperimente zu verfolgen, in denen aufgrund veränderter pH-Werte im Magen Nitrosaminentstehung beobachtet wurde. Nitrosamin ist ein Kanzerogen. Auch ca. 20% aller Vagotomie-Operationen sind therapeutische Versager (Alexander-Williams et al., 1977).

Mit den substituierten Benzimidazolen (Omeprazol) stehen noch wirksamere Säureblocker zur Verfügung, die eine komplette Achlorhydrie bewirken können (Classen et al., 1985). Sie haben ihre klinische Wirksamkeit erwiesen, wenngleich in multizentrischen Studien die Überlegenheit im Vergleich zum Ranitidin im Falle des Ulcus duodeni nicht so beeindruckend ist. Besondere Bedeutung scheinen sie im Falle der Refluxösophagitis zu haben (Koop, 1990).

Schließlich eine Stellungnahme zur Wismuttherapie: „Auch wenn Wismutsalze hinsichtlich der Abheilungsgeschwindigkeit H_2-Blockern annähernd ebenbürtig sind, bevorzugen viele Patienten H_2-Blocker wegen der rascheren Beschwerdefreiheit und des erheblich besseren Einnahmekomforts. Dagegen spricht für Wismut die geringere Rezidivrate innerhalb des ersten Jahres nach Schubtherapie, während die regelmäßige Schwarzfärbung des Stuhls und in Einzelfällen auch der Schleimhäute die Akzeptanz von Wismut weniger beeindruckt" (Koop, 1990).

43.7.3 Die Rolle der Psychosomatik bei der Behandlung des Ulcus duodeni – ein konzeptioneller Ansatz

Geht man von unserem im theoretischen Teil entwickelten Erklärungsmodell, dem Situationskreis aus, so lassen sich die Auswirkungen der verschiedenen therapeutischen Maßnahmen medikamentöser, chirurgischer und psychotherapeutischer Art auf den Kranken und das Erleben seiner individuellen Wirklichkeit besser verstehen. Das Modell beschreibt ja einen dynamischen Prozeß, mit dem in jedem Augenblick – wie Viktor v. Weizsäcker (1955) es ausdrückte – „Gesundheit erzeugt wird", und in dem die Krankheit bereits begonnen hat, „wenn sie nicht mehr erzeugt wird". Erzeugen von Gesundheit wird in unserem Modell als Aufbau einer individuellen Wirklichkeit beschrieben, in welcher der einzelne seine Bedürfnisse befriedigen kann.

In Situationen, in denen das aus inneren oder äußeren Gründen nicht gelingt, kommt es zu Störungen, die je nach der herrschenden Konstellation auf verschiedenen Ebenen eintreten und verschiedene Auswirkungen haben können. Unser Modell beschreibt Schutzmaßnahmen, die Schäden abwenden oder, wenn das nicht gelingt, begrenzen sollen. Sie bestehen darin, daß der Aufbau der individuellen Wirklichkeit auf einer Stufe innehält, auf der die dazu erforderlichen Programme in reduzierter Form noch ungestört abgewickelt werden können. Bildlich gesprochen schrumpft die Wirklichkeit sowohl räumlich wie zeitlich, so daß Gefahren, die aus der Entfernung oder der weiteren Zukunft drohen, ihre Bedeutung verlieren, ja sie können aus dem bewußten Erleben des Kranken verschwinden. Es kommt gewissermaßen zu einer Autonomie von mehr oder weniger großen Bereichen der individuellen Wirklichkeit des Kranken.

Wenn das nicht ausreicht, kann ein weitestgehender Rückzug auf die vorgeburtliche Organisationsform als „bloßer Körper" ohne Umwelt stattfinden. Wir haben beschrieben, wie es in der Nausea, in der Ohnmacht, im Koma und zeitweise auch im Schlaf zu einer Umstellung von der ergotropen, aktivierten Verfassung des normalen Wachzustandes zu einem histiotropen Schongang der Körperfunktionen kommt, in dem der Organismus von seinen Reserven lebt und die Umgebung ihn nichts mehr angeht (v. Uexküll, 1952).

Sowohl bei dem Rückzug von der ergotropen, umweltoffenen Einstellung in den histiotropen, umweltlosen Zustand, wie auch umgekehrt bei dem Übergang von dieser in sich abgeschlossenen Organisationsform in die offene, findet ein „Sprung" von einer Integrationsebene eines hierarchischen Systems in eine andere statt. Der Sprung in die Organisationsform des offenen Systems, das Außenwelt in Umwelt und später individuelle Wirklichkeit transponiert, verläuft zunächst immer wieder nach Programmen, die in der frühesten Kindheit erworben wurden. Damals entstand unmittelbar nach der Geburt mit den ersten Atemzügen eine „Luft-Umwelt", die sich bis in die Lunge des Säuglings ausdehnte. Damals schufen die ersten Lippenkontakte des Säuglings mit der mütterlichen Brust und das erste Trinken die Basisprogramme für eine orale Umwelt, die wahrscheinlich bis hinab zum Pylorus reicht.

Diese frühen Phasen der Wirklichkeitsbildung hat Freud mit seinem Triebmodell beschrieben, das zum ersten Mal in der Geschichte der Medizin ein System darstellt, in dem Organismus und psychische Umwelt eine Einheit bilden. Dieses Modell, das Freud später genauer als Zusammenspiel von vier Elementen, nämlich der Quelle, dem Drang, dem Objekt und dem Ziel des Triebgeschehens, definiert hat (vgl. oben), entspricht in unserem Situationskreis-Schema den ersten Phasen der sich immer wiederholenden Umwelt- bzw. Wirklichkeitsbildung (mit der „Gesundheit erzeugt wird"). Zeichnen wir uns dieses Modell schematisch auf, so können wir die verschiedenen therapeutischen Maßnahmen „lokalisieren" und in ihren Auswirkungen auf das Gesamtgeschehen verständlich machen (Abb. 43–1).

Prognostische Erfahrungswerte

Psychosomatisch orientierte therapeutische Langzeitstudien würden am ehesten die Frage beantworten helfen, auf welcher Seite des Situationskreises die wirkungsvollsten Faktoren zu finden sind. In einer der wenigen – zumindest den Autoren verfügbaren – prognostischen Langzeitstudien wird ein hohes Maß an Psychopathologie beschrieben (Knop und Fischer, 1981). Es wurden die Krankengeschichten von 1000 Billroth-II-Patienten 20–29 Jahre nach der Operation untersucht. Von ihnen waren 423 verstorben. Von diesen hatten 67, also 14%, Suizid begangen. Unter den Überlebenden waren 92 in eine psychiatrische Klinik eingeliefert worden, d.h. jeder 6. von ihnen. Der häufigste Einweisungsgrund war Alkoholismus.

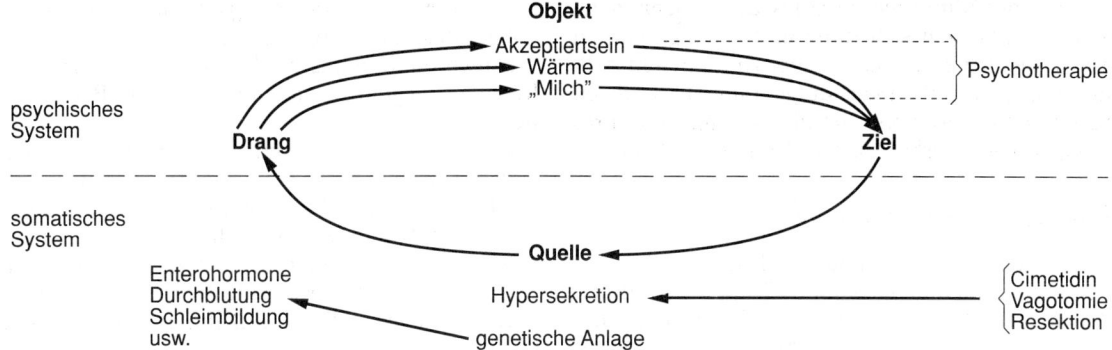

Abb. 43–1. Der somatische Teil des Modells besteht aus der Triebquelle, die Freud in einem „interzellulären Chemismus" vermutet. Hier wären die genetische Anlage zu einer Hypersekretion und eventuell andere Besonderheiten der aggressiven und protektiven Einrichtungen der Magenschleimhaut sowie die therapeutischen Maßnahmen anzuordnen, die diese Faktoren verändern.

Der psychische Anteil des Modells besteht aus dem Drang, der sich in der Außenwelt (durch Bedeutungserteilung) sein Objekt sucht, das ihm über eine Befriedigung seiner Triebbedürfnisse dazu verhilft, sein Ziel, die „Triebabfuhr" zu erreichen und im somatischen Bereich die Triebquelle zum Versiegen zu bringen.

Unter den vagotomierten Patienten finden sich – wie ausgeführt – 26% Versager. Etwa ⅓ von ihnen könnte nach chirurgischer Auffassung durch vorherige „sorgfältige präoperative psychiatrische Beurteilung" erfaßt werden (Alexander-Williams et al., 1977). Patienten, die präoperativ wenig Optimismus, depressive Züge und Ängstlichkeit an den Tag legten, zeigten schlechtere operative Ergebnisse als solche ohne derartige psychologische Befunde (Conron und Hardy, 1976).

Andererseits berichten deutschsprachige Autoren über günstige mittelfristige Ergebnisse nach selektiver Vagotomie. Möhlen und Mitarbeiter (1982) fanden bei ca. 80% ihrer Patienten eine gute bis sehr gute Beurteilung der Operationserfolge durch die Patienten selbst. Ähnliche Ergebnisse erhob Hess (1983) bei 50 männlichen Patienten 5 Jahre nach selektiver Vagotomie; Hier lag eine 20%ige Rezidivquote vor. Interessanterweise wird in beiden Arbeiten beschrieben, daß höchstens die Hälfte der Patienten eine ausreichende Reduktion der Säureproduktion zeigten; Hess (1983) fand bei ihren Patienten sogar in 69% (!) eine überhöhte Säureproduktion.

Insgesamt liegen also nur wenige langfristige Verlaufsuntersuchungen vor, welche die psychosoziale Dimension einbeziehen. Es dürfte aber berechtigt sein, die wenigen vorliegenden Angaben als ernsthafte Hinweise auf eine starke Häufung psychosozialer Probleme bei derartigen Patienten zu werten. Welches Vorgehen empfiehlt sich also?

Häufig zu beobachtende Merkmale

Häufig haben die Patienten von Zusammenhängen zwischen Lebenssituation und Geschwür gehört. Sie möchten dies zwar für andere, selten aber für sich selbst gelten lassen. Soziologische Untersuchungen scheinen diese Erfahrungen zu bestätigen (Ahrens, 1982, a, b): Bei einem Vergleich von 42 Ulcus-duodeni-Patienten mit 31 somatisch und 22 neurotisch

Kranken fällt auf, daß die Ulkuspatienten stärker als die organischen (!) oder gar die psychoneurotischen Patienten organische Erklärungsmuster für ihre Krankheit wählen, selten ihren Arzt wechseln und ein anklammerndes „Konsultationsverhalten" zeigen. Dies hat seitens des Arztes ein verstärktes Nicht-Einlassen auf die psychosoziale Konfliktebene zur Folge. Ahrens sagt sinngemäß, daß derartige Patienten zunächst darauf ausgerichtet sind, ihre Beschwerden und ihre Krankheit ausschließlich in einer somatisch-instrumentellen Weise zu verstehen und dementsprechend ihren Arzt beeinflussen, ausschließlich auf einer instrumentellen Ebene zu agieren. Dieser Prozeß wird durch ein Verhalten bekräftigt, das Ahrens als eine „Trennung von affektiven und funktionalen Zuschreibungen bei Ulkuspatienten" beschreibt (1981). Viele Ulkuspatienten können zwar wie andere Kranke zwischen einem Wunscharzt und einem abweisenden Arzt unterscheiden, sie koppeln aber häufig anders als andere Kranke den Idealarzt von Gefühlen ab. Hierbei handelt es sich gleichermaßen um Gefühle der Zuwendung wie Gefühle der Abwendung. Der Arzt steht praktisch als ein menschliches Neutrum da. Ahrens folgert, daß viele dieser Patienten ihre Beziehung zu ihren Ärzten überwiegend unter Leistungsaspekten beurteilen.

Es bedarf seitens des Arztes großer Beharrlichkeit, derartige Patienten mit dem Gedanken vertraut zu machen, daß psychosoziale Faktoren von pathogenetischer Bedeutung sein könnten. Am ehesten bietet sich eine Brücke in jenen Bereichen an, in denen über Wechselwirkungen zwischen Beschwerden und Tagesablauf gesprochen werden kann. Dabei scheint der Zugang zum Patienten phasenspezifisch unterschiedlich zu sein. Es finden sich Hinweise, daß in der Zeit des floriden Ulkus eine größere Abhängigkeit als in der Zeit des abgeheilten Ulkus zu bestehen scheint, die sich entsprechend älterer klinischer Beobachtungen in unterschiedlichen Verhaltensweisen gegenüber Ärzten und Pflegepersonal bemerkbar macht. In der Klinik wegen eines Ulkus behandelte Patienten verhielten sich ruhig und gefügig; wurden dieselben Patienten wegen anderer Krankheiten, z. B. Frakturen ins Krankenhaus aufgenommen, so fielen sie zum Teil wegen ihres Querulantentums und ihres unkooperativen Verhaltens auf (Cremerius, 1971).

Der Patient in der (Poli-)Klinik

Patienten der Poliklinik und der Klinik stellen mit großer Wahrscheinlichkeit eine Gruppe dar, deren Mitglieder zu Rezidiven neigen und bei denen die Therapie des niedergelassenen Arztes häufig versagt hat. Es sollte geradezu zur Regel gemacht werden, bei diesen Patienten dem psychosozialen Bereich besondere Aufmerksamkeit zuzuwenden und die im Abschnitt 43.6.1 aufgezählten Punkte zu berücksichtigen.

Unerläßlich ist, daß der Patient zunächst vom behandelnden Arzt befragt wird. Dies sollte nicht dem hinzugezogenen Konsiliarius und erst recht nicht einer Testbatterie überlassen werden. In dieser Initialphase beginnt der diagnostisch-therapeutische Zirkel (Wesiack, 1974), in dem nur der behandelnde Arzt die Einsicht vermitteln kann, daß psychosoziale Faktoren für die Entstehung und den Verlauf der Krankheit eine Rolle spielen. Hierauf verweisen auch Böttcher und Mitarbeiter bei einer ausgedehnten Untersuchung von Diagnostik und Therapie derartiger Patienten (1980). Anders ist die Lage, wenn diese Initialuntersuchung abgelaufen ist. Dann wird es häufig erforderlich, die spezialistische Hilfe des psychosomatischen Konsiliarius in Anspruch zu nehmen. Dieser wird auch zu erwägen haben, ob die Behandlung in einer psychosomatischen Fachklinik indiziert ist oder vor Ort geschehen sollte, bzw. ob lokale und überregionale Behandlungsmöglichkeiten aufeinander abzustimmen sind.

Unter den Therapiemöglichkeiten dürften von Bedeutung sein: Partnertherapie, Familientherapie, Gruppentherapie. In schwierigeren Fällen wären derartige Therapieformen innerhalb eines fachklinischen Behandlungsrahmens einzusetzen. Es gibt kasuistische Darstellungen hierzu; jedoch liegen, wie oben angedeutet, keine systematischen Untersuchungen vor, bzw. sind zumindest den Autoren nicht bekannt.

Für positive Auswirkungen des autogenen Trainings finden sich keine Hinweise.

Die Verhaltenstherapie bemüht sich, geeignete Methoden zu entwickeln, die auf Biofeedback-Verfahren beruhen. Klinisch ist sie derzeit noch nicht einsetzbar (Hölzl und Whitehead, 1983).

Therapeutische Folgerungen hinsichtlich interdisziplinärer Kooperation

Es ist der Aufforderung von gastroenterologisch-internistischer Seite zuzustimmen, zu einer engeren interdisziplinären Zusammenarbeit zu kommen (Rösch, 1984). Allerdings muß eine solche Zusammenarbeit bedeuten, daß im Gegensatz zur bisherigen Situation ein auf dem psychosozialen Bereich erfahrener Mitarbeiter hinzugezogen wird und daß er in voller Gleichberechtigung mitspricht.

Schwerpunktmäßig wird es bei einer derartig langfristig angelegten Kooperation um zwei Bereiche gehen:

– Primärversorgende Ärzte müssen in ihrer Arbeit den Gesichtspunkt einbeziehen, daß bei jedem Ulkuspatienten eine gestörte Lebenssituation besteht, und er das Recht hat, hierauf nicht nur hingewiesen, sondern auch beraten zu werden. Der primärversorgende Arzt hätte hier die Aufgabe, den Rat beim Spezialisten einzuholen, sofern notwendig.

– Es würde darum gehen, daß in einer kleineren Zahl von 20% der Patienten, zumeist Versager der konservativen Therapie, die Psychosomatik in einem spezialistischen Sinne tätig wird. Hier dürften sich derzeit auch die interessantesten Forschungsansätze in diesem Bereich finden.

43.8 Antworten auf eingangs gestellte Fragen und Ausblick

Die Fragen waren, ob es sich lohne, auf psychosomatische Probleme des Ulcus-duodeni-Patienten einzugehen, wenn potente Pharmaka zur Verfügung stehen und dieses Krankheitsbild auch zahlenmäßig an Bedeutung verliert; ob sich in den unterlassenen wissenschaftlichen Nachuntersuchungen ein unüberwindbares Desinteresse an der Psychosomatik, speziell des Ulkuspatienten widerspiegele.

Zur ersten Frage: Eine Beschäftigung mit psychosomatischen Problemen des Ulcus-duodeni-Patienten erscheint aus mehreren Gründen unerläßlich. 20% der heutigen Patienten kann auch mit sonst wirksamen Pharmaka und sonst wirksamer Chirurgie nicht geholfen werden. Aber auch die scheinbar geheilten Patienten haben offensichtlich in einem beachtlichen Ausmaß psychosoziale Probleme. Über das Ausmaß dieser Behinderungen liegen kaum Untersuchungen vor. Sie sind dringend erforderlich. Weiterhin muß davon ausgegangen werden, daß 10% der männlichen Bevölkerung ein Ulcus duodeni haben bzw. haben werden. Es handelt sich also um eine zahlenmäßig äußerst umfangreiche Gruppe. Unsere neuen Erkenntnisse haben gezeigt, daß wir zwei Subgruppen unterscheiden können, deren Unterscheidungsmerkmale zur Zeit nicht bekannt sind. Wir wissen lediglich, daß die eine Gruppe erhöhtes Serumpepsinogen hat, die andere nicht. Damit entstehen wissenschaftliche Fragen, die von unmittelbarer therapeutischer Auswirkung sein könnten.

Zur zweiten Frage: Das Desinteresse an wissenschaftlichen Fragestellungen in einem ganzheitlich psychosomatischen Sinne ist offensichtlich. Weiner schreibt hierzu (1981): „Eigentlich hätte man annehmen sollen, daß diese Berichte über den Zusammenhang zwischen erhöhtem Serumpepsinogenspiegel, meßbaren psychologischen Stigmata und der Disposition zur Erkrankung an einem Ulcus duodeni zu einer Flut von Nachprüfungen geführt haben würden ... aber das ist nicht geschehen. In den 23 Jahren seit der Publikation unserer Berichte ist mir keine Untersuchung bekanntgeworden, die versucht hätte, Pepsinogenspiegel, psychologische Züge und Zwölffingerdarmgeschwür zu korrelieren". Erst 30 Jahre darauf ist eine solche Arbeit erschienen, nämlich die oben referierte (Walker et al., 1988).

Umgekehrt weisen Wolff und eine Reihe namhafter nordamerikanischer Gastroenterologen anläßlich eines Symposions darauf hin, daß nach einer langen Phase therapeutischer Unsicherheit Konturen klarer werden, die prädiktive Studien beim Ulkuspatienten erlauben (Wolff et al., 1979a, b). Es lassen sich Risikopersonen beschreiben. Anhand derartig definierter Persönlichkeiten sind prädisponierende, auslösende und krankheitsunterhaltende Faktoren zu erheben: Einfluß der Pepsinogenkonzentration, der Säuresekretion, der Empfindlichkeit gegenüber biologischen wie psychosozialen Stimuli und die Rolle der Magenentleerung wären Fragen von besonderem Interesse. Ebenfalls wäre von größerer Bedeutung, welche Rolle das Phänomen der Stimmung bei Cimetidinversagern hat. Wesentlich erscheinen drei Punkte, denen Beachtung geschenkt werden müßte, um das Interesse an wissenschaftlichen Fragestellungen zu fördern:
– Einem allgemeinen Verdrängungsprozeß müßte entgegengewirkt werden, der sich in der trügerischen Annahme ausdrückt, daß mit der Gabe eines Pharmakons biologische, psychische und soziale Probleme gleichermaßen aus der Welt geschafft wären.
– Viele Patienten mit Ulcus duodeni sind per psychosozialer Definition im Zustand ihrer Erkrankung Mitglieder von Randgruppen. Weder der Patient selbst noch der Arzt möchten sich mit dieser Situation identifizieren, sondern suchen diese häufig zu übergehen.
– Eine klinik- und patientorientierte Forschung hat bisher zumeist an getrennten Orten stattgefunden: hier psychosozial orientierte Untersucher, dort organisch orientierte Untersucher. Nur durch einen über Jahre praktizierten Teamansatz wäre der sich ausdrückende Dualismus unserer Medizin wie unserer Gesellschaft allmählich angehbar.

Mit diesen abschließenden Bemerkungen soll verdeutlicht werden, daß nicht nur viele Patienten mit Ulcus duodeni zum Ausklammern psychosozialer Probleme neigen, sondern dies auch ihre Gesellschaft tut und deren Repräsentanten. Mit anderen Worten: Ulcus-duodeni-Träger haben sich in besonders eindrucksvoller Weise Lebensformen zu eigen gemacht, die wir in unserem Alltag allenthalben vorfinden. Erst insofern wird der Ulkusträger auch gleichzeitig zu einem Träger vieler Probleme der modernen Zivilisationsgesellschaft. In seiner Behandlung wie im Kenntnisstand über seine Krankheit drückt sich geradezu paradigmatisch die Unfähigkeit unserer Gesellschaft aus, mit ihren eigenen Problemen umzugehen.

Dank: Ein besonderer Dank gilt Herrn Prof. Dr. K. Arnold, Leiter der Abteilung Innere Medizin, Schwerpunkt Gastroenterologie des Zentrums für Innere Medizin der Philipps-Universität Marburg für die kritische Durchsicht des Manuskriptes der 3. Auflage.

44 Entzündliche Darmerkrankungen

44.1 Colitis ulcerosa

Hubert Feiereis

Der 44jährige Patient K. E. kommt zu uns, weil seit 3 Jahren wechselnd Durchfälle mit Blut und Schleim, 12–15mal täglich, Schmerzen im linken Unterbauch und in der Afterregion, besonders beim Stuhlgang, bestehen. Seit dieser Zeit ist er auch erheblich deprimiert und hat häufig Suizidgedanken. Die Schübe treten im Abstand von etwa einem viertel bis einem halben Jahr auf und dauern meist 4–6 Wochen. Bisher wurde er ausschließlich medikamentös mit Salazosulfapyridin, Cortison und Diät behandelt. Trotz der meistens bald einsetzenden Besserung unter dieser Therapie verbleiben Blutbeimengungen im Stuhl. Ebensowenig bilden sich hypochrome Anämie, erhöhte Blutsenkungsgeschwindigkeit und entzündliche Zeichen im Elektrophoresediagramm vollständig zurück. Bei den endoskopischen Untersuchungen wird während eines Schubes eine stärkergradige Colitis ulcerosa (makroskopisch und histologisch) im Rektum, Sigma und Descendens festgestellt, die auch im Intervall histologisch besteht und nur makroskopisch einen geringeren entzündlichen Befund aufweist.

Der Patient wirkt entmutigt wegen der langanhaltenden Krankheit mit wiederholt wochenlanger Dienstunfähigkeit und deren negativen Auswirkungen, ist klagsam, apathisch, antriebsarm, deprimiert. Er empfindet den Beruf nur noch als Zwang. Zu Hause fühlt er sich schlapp, schwach, „wie ausgelaugt". „Der Freitagabend ist für mich der schönste Moment; ich habe keine Energie und keinen Trieb, etwas zu unternehmen, seit 3 Jahren eigentlich nur rumgesessen." Seit Monaten schläft er schlecht, hat Konzentrationsschwierigkeiten, spürt einen Leistungsabfall und ist impotent. Sein Interesse hat ringsum nachgelassen.

Seit einer Ruhrerkrankung mit 18 Jahren hat er keinen festen Stuhl mehr, seit 10 Jahren häufiger wäßrige Durchfälle, zunächst jahrelang ohne Blut.

Der Patient wurde als 9. Kind von 11 Geschwistern in Ostpreußen geboren. Bis zum 14. Lebensjahr wohnte er im Elternhaus auf dem Lande. Er lebte gern dort. „Die Menschen waren einfach, ehrlich und unkompliziert." In der einklassigen Volksschule war er der beste Schüler. Während der Kriegsjahre konnte der einzige Lehrer zweier Klassen nur an 3 Tagen Unterricht geben, „deshalb kamen wir z. B. im Rechnen nicht viel weiter als bis zur einfachen Bruchrechnung und in Deutsch zu Satzgegenstand und Satzaussage. Der Lehrer lobte mich oft, und ich spürte, daß mir Lob guttat. Noch heute bin ich unsicher in Ausdruck und Rechtschreibung."

Die Eltern waren Auslandsdeutsche und sehr streng religiös. Sie hatten ihren Bauernhof 1918 in Rußland aufgegeben. Der Vater wurde Landarbeiter auf ostpreußischen Gütern. Wegen der Armut der großen Familie konnte sich der Patient selten sattessen. Seine Kleidung war abgetragen; deshalb wurde er oft gehänselt. An seinem 12. Geburtstag hatte er zum erstenmal Schuhe an den Füßen; bis dahin trug er nur Holzpantinen. Die Stimmung innerhalb des Elternhauses war oft äußerst gedrückt. „Ich hörte meine Eltern wegen ihres verlorenen Bauernhofes und des ganzen Daseins immer nur klagen. Sie haben nie eine Schule besucht. Aus der Bibel, die ihr einziges Buch war, lernten sie Schreiben und Lesen. Daher waren sie auch so streng religiös. Uns Kindern wurde mit erhobenem Zeigefinger immer gesagt, daß z. B. Ballspielen und Tanzen im Himmel auch nicht erlaubt sind. Oft wurde von der Hölle gesprochen." Von den Eltern wurde der Patient häufig getadelt. „Ich spürte schon damals, daß mir Tadel weh tat. Im Streit mit anderen Kindern durfte ich mich nicht wehren. Bekommst du eine Backpfeife, so halte auch die andere Backe hin." Selbst freudige Ausgelassenheit sei als Sünde ausgelegt worden. Wegen der Verbitterung der Eltern erinnerte er sich ohnehin an keine einzige fröhliche Stimmung. „Wir wurden immer zu Ernsthaftigkeit und zum Stillsitzen ermahnt. Als Kind mußten wir zur Strafe oft stundenlang in der Bibel lesen." Ihrer religiösen Einstellung wegen war die Familie auch politisch verfolgt worden und im Dorf isoliert.

Den Eltern wie auch Fremden gegenüber ist er stets sehr schüchtern gewesen. „Ich kann heute noch nicht im kleinen oder großen Kreis diskutieren. Ich habe schon vorher Angst, ob ich alles richtig tun oder sagen werde. Ich fürchte, der andere könnte keinen guten Eindruck von mir haben. Ich lasse mir die Meinung des anderen aufzwingen, umgekehrt gelingt es mir nicht."

Der Patient erlernte das Schmiedehandwerk und erwartete kurz vor Kriegsende seine Einberufung, die ihn wegen der Flucht aber nicht mehr erreichte. An der Weichsel wurde er festgenommen. Unter dem Vorwurf, sich nicht pflichtgemäß zum Fronteinsatz gemeldet zu haben, wollte man ihn erschießen. „Ich hatte keine Zeit, mich auf der Straße von meinen Eltern und Geschwistern zu verabschieden. Meinen Vater habe ich nie wiedergesehen; er ist in Rußland verstorben." Der Patient kam unmittelbar nach der Festnahme in russische Gefangenschaft und wurde ein halbes Jahr danach wegen Unterernährung und anhaltender Durchfälle nach einer Ruhrerkrankung entlassen.

Zwei Brüder waren gefallen, vier Geschwister wanderten in die USA aus.

Der Patient arbeitete dann 12 Jahre als Schlosser, besuchte Abendschulen, erwarb sich die Qualifikation eines Konstrukteurs und Erprobungsingenieurs. Eine feste Anstellung bei einer staatlichen Behörde gab ihm die äußere Sicherheit. Wegen seiner fehlenden akademischen Ausbildung habe er sich die Anerkennung seiner Vorgesetzten und Kollegen mühsam erkämpfen müssen. „Ich spüre im Leib immer ein Unbehagen, wenn es darauf ankommt, mich zu behaupten."

Mit 24 Jahren heiratet er. Seine Ehefrau schildert er zunächst als gütig und stets hilfsbereit, „obwohl ich ihr seit Jahren, seit Beginn der Abendschule vor 15 Jahren, nichts bieten kann". In seiner Familie fühlt er sich wohl; zu seinen beiden Söhnen hat er eine gute Beziehung. „Leider erkannte ich zu spät, daß ich sie zu streng erzogen, mich zu wenig um ihre Entwicklung gekümmert habe. Nun suche ich oft das Gespräch mit ihnen; Fremden gegenüber sind sie ebenso befangen, wie ich es gewesen bin."

Die Krankheit verlief parallel mit den steigenden Anforderungen im Beruf, die er mit Fleiß und Ehrgeiz — gerade wegen seines fehlenden Hochschulstudiums — besonders exakt zu erfüllen versuchte. Der letzte Schub fiel zeitlich eng zusammen mit seiner beruflichen Veränderung zum technischen Sachbearbeiter. Diese hatte ihm zwar gegenüber der früheren Konstruktionsarbeit Entlastung gebracht, jedoch assoziierte er hierzu ständig Begriffe wie „Verlust des mühsam erworbenen Status eines Konstrukteurs" und „Versager, Niederlage, du bist doch nichts wert".

44.1.1 Klinik der Colitis ulcerosa

44.1.1.1 Symptomatologie, Schweregrade und Verlaufsformen

Die Krankheit beginnt in der Regel schleichend, mitunter, besonders bei schwerer Verlaufsform, akut bis fulminant. Wir unterscheiden die drei Verlaufsformen akut, chronisch rezidivierend und chronisch kontinuierlich und die drei Schweregrade I bis III.

Schweregrad I: Geringe Beschwerden. Leichte Allgemein- und Lokalsymptome mit geringer Blut-Schleim-Beimengung im Stuhl, der selten dünn, oft breiig bis geformt ist und weniger als fünfmal täglich entleert wird. Leichter makroskopischer und histologischer Befund. Temperatur höchstens subfebril. Nur leichte Anämie. Serologisch nur leichte Entzündungszeichen. Keine oder nur geringe Minderung der Leistungsfähigkeit.

Schweregrad II: Zwischenstufe zwischen I und III. Die Stuhlentleerungen schwanken zwischen fünf- bis zehnmal täglich, die Konsistenz ist meist dünn bis dünnbreiig.

Schweregrad III: Schweres Krankheitsbild mit Fieber, starken Tenesmen, erheblichen Allgemeinsymptomen und häufigen (zehn- bis dreißigmal täglich) blutig-schleimig-eitrig-wäßrigen Durchfällen. Erheblicher makroskopischer und histologischer Schleimhautbefund. Serologisch ausgeprägte Entzündungszeichen. Anämie. Klinische Behandlung unumgänglich.

Die Verteilung der Verlaufsformen und Schweregrade im eigenen Krankengut von 2204 Patienten zwischen 1948 und 1989 zeigt Tabelle 44.1–1.

Tabelle 44.1–1. Schweregrad und Verlaufsform der Colitis ulcerosa bei 2004 Patienten des eigenen Krankengutes der Jahre 1948–1989

| Verlaufsform | Schweregrad | | | |
	I	II	III	zus.
akut	122	79	86	287 = 14%
chronisch-rezidivierend	360	547	430	1337 = 67%
chronisch-kontinuierlich	235	97	48	380 = 19%
insgesamt	717 = 36%	723 = 36%	564 = 28%	2004

Die Colitis ulcerosa befällt so gut wie immer das Rektum und breitet sich proximalwärts aus. In unserem Krankengut waren bei 38% unserer Patienten nur das Rektum und bei 24% das gesamte Kolon befallen.

44.1.1.2 Endoskopische und histologische Befunde

Die wichtigsten **makroskopischen akuten** bis **subakuten** Merkmale der Entzündung der Mukosa sind:
Rötung mit verwaschener und aufgehobener Gefäßzeichnung; Ödem; Granulierung; petechiale, kleinfleckige bis flächenhafte submuköse Blutungen; Spontanblutungen; Kontaktblutungen infolge stark erhöhter Vulnerabilität; streifenförmige, netzförmige oder flächenhafte fibrinöse, schleimig-eitrige Beläge und Nekrosen; Erosionen und Ulzera.

Demgegenüber sind Ausdruck der **chronischen** Schleimhautentzündung: verwaschene oder aufgehobene Gefäßzeichnung; Granulierung; erhöhte Vulnerabilität; narbige Einziehungen; entzündliche Polypen („Pseudopolypen"); Fibrose; Induration und Atrophie.

Die **histologischen** Befunde der **akuten bis subakuten** Entzündung sind gekennzeichnet durch eine der Schwere der Entzündung weitgehend parallele Infiltration des Stratum proprium mit neutrophilen Granulozyten, Plasmazellen und Lymphozyten. Die Infiltration kann das Oberflächenepithel durchbrechen (sog. Mikroabszesse) und auch bis in die Submukosa reichen; Hyperämie und Ödem; Defekte bzw. Ulzera des Oberflächenepithels mit fibrinösem und leukozytenhaltigem Schleim, dem auch Erythrozyten beigemengt sein können; Einbruch der Leukozyteninfiltration in die Krypten („Kryptenabszesse"); Irregularität des Oberflächen- und Kryptenepithels, irregulär gelagerte Drüsen; Reduktion der Becherzellen; kleinzystische Kryptenabschnürungen.

Die akute Entzündung kann bis zur restitutio ad integrum abheilen, rezidivieren oder in eine chronische Entzündung übergehen.

Die chronisch entzündete Schleimhaut kann „abheilen", in eine Atrophie übergehen, schließlich in unterschiedlicher Gestalt persistieren.

44.1.1.3 Röntgenologische Befunde

Die Röntgendiagnostik ist gegenüber der in der Regel zunächst bevorzugten und notwendigen koloskopischen Untersuchung während der letzten Jahre in den Hintergrund getreten. Sie ist aber ebenso ein wichtiger ergänzender Bestandteil einer vollständigen Befunderhebung bei Darmerkrankungen, wie sich die endoskopische und die röntgenologische Untersuchung des Magens ergänzen. Die röntgenologischen Charakteristika bei der Colitis ulcerosa setzen sich aus pathomorphologischen und pathofunktionellen Veränderungen zusammen. Im Vordergrund stehen Schleimhautschwellungen und Ulzerationen, die zunächst auf die Mukosa begrenzt sind und sich kontinuierlich vom Rektum proximalwärts ausdehnen. Sie können als feingranuliertes Relief mit feinster Zähnelung (Spicae), als dornartige Ulzeration, schließlich als tiefes Geschwür mit Unterminierung der Schleimhaut („Kragenknopfulzera") auftreten. Unter den pathofunktionellen Veränderungen dominieren Tonusstörungen mit Haustrenverlust und Motilitätsstörungen als Atonie oder Spasmus. Im chronischen Stadium treten Pseudopolypen und Atrophien, schließlich Fibrosierungen mit schlauchförmiger Verkürzung des Kolons als Merkmal einer „ausgebrannten Kolitis" auf: Mikrokolon. Bei 5–10% beobachtet man eine „Rückspül-Ileitis". Selten sind Fisteln und Stenosierungen.

44.1.1.4 Komplikationen

Zu unterscheiden sind lokale Komplikationen wie toxisches Megakolon, foudroyante Blutung, Perforationsperitonitis und (selten) Perianalabszeß und Fistel. Das gefürchtete postkolitische Karzinom ist seltener, als vielfach, vor allem in angloamerikanischen Ländern, angegeben wird. Die Gefährdung besteht in erster Linie bei totalem Befall des Kolons und länger als 10 Jahre dauernder chronischer Entzündung ohne Remissionen. Im eigenen Krankengut betrug die Karzinomfrequenz bei Patienten mit 8–20 Jahre langem Verlauf der Krankheit 3%.

Die Komplikationen an anderen Organen manifestieren sich vorwiegend als kutane Begleiterscheinungen, etwa Erythema nodosum oder Pyoderma gangraenosum, als rheumatoide Arthritiden, ankylosierende Spondylitis, Episkleritis und Iridozyklitis, ferner in allen Formen entzündlicher und nichtentzündlicher Lebererkrankungen, vor allem als primär sklerosierende Cholangitis oder Pericholangitis, reaktive Hepatitis und Cholestasesyndrom.

44.1.2 Differentialdiagnose

Breitgefächert ist das Spektrum der notwendigen differentialdiagnostischen Überlegungen. Sie umfassen vor allem die Enteritis regionalis Crohn, die ischämische vaskuläre Kolitis, die Enteritis pseudomembranacea („postantibiotische Kolitis", „postoperative Kolitis"), die Yersinienerkrankung, Salmonellosen, Campylobacter-Kolitis, Ruhrerkrankungen, schließlich die Purpura abdominalis, das solitäre Ulkus, die Amyloidose, die kollagene Kolitis, die Colitis cystica profunda, die Divertikulitis und die gut- und bösartigen Tumoren. Regelmäßig ist auch an funktionelle oder artifizielle Durchfälle mit den Kennzeichen des Abführmittel-Kolons zu denken.

44.1.3 Epidemiologie

Die Angaben zur Häufigkeit der Colitis ulcerosa schwanken zwischen 5 bis 10 Patienten auf 10000 Krankenhausaufnahmen in Mittel- und Westeuropa, während in den USA 50 bis 100 Kolitis-Kranke auf 10000 Einweisungen in internistische Krankenhausabteilungen angegeben wurden (Feiereis, 1970). Die jährliche Erkrankungsziffer beträgt etwa 5 bis 8 auf 100000 Einwohner.

Das männliche und weibliche Geschlecht ist etwa gleich häufig, überwiegend im 2. bis 4. Lebensjahrzehnt betroffen. Im eigenen Krankengut fanden sich unter 2004 Patienten 1026 männliche und 978 weibliche Kranke. 1398 Patienten (70%) waren zwischen 11 und 40 Jahre alt.

44.1.4 Ätiologie und Pathogenese

Ätiologie und Pathogenese der Colitis ulcerosa sind noch immer nicht geklärt, jedoch sprechen viele Befunde für eine plurikausale Entwicklung. Die meisten Untersuchungen beziehen sich auf Fragen nach der Bedeutung von immunologischen Faktoren, genetischer Disposition, Einflüssen der Ernährung (Milcheiweiß, Laktasemangel, Kohlenhydratüberschuß, Form des Zuckers in der Ernährung), infektiösen Agenzien (Virus? Ruhranamnese?) und vegetativpsychischen Anteilen. Für die genetische Disposition sprechen vor allem Beobachtungen der Colitis ulcerosa bei mehreren Mitgliedern der Familien dieser Patienten (Hoyer, 1983; Küster und Lenz, 1984; Nedbal und Mařatka, 1968; Purrmann et al., 1986). In angloamerikanischen Arbeiten wird die familiäre Disposition mit 5–10% angegeben (Evans und Acheson, 1965). Bei unseren Patienten betrug sie, bezogen auf Blutsverwandte 1. Grades, 2% (Hammer, 1968).

Die Annahme einer immunologischen Pathogenese liegt angesichts der nahen Beziehung zu anderen Autoimmunerkrankungen auf der Hand. Zahlreiche Untersuchungen und Beobachtungen erhärten die Bedeutung dieses ätiologischen und pathogenetischen Faktors (Bárta et al., 1964; Harrison, 1965a und b; Kirsner, 1960, 1961 und 1965; Raedler et al., 1982; Thiele et al., 1982). Bemerkenswert sind auch die Befunde über die verschiedenen Autoantikörper, z. B. gegen Kolonantigene, Zellkernsubstanzen und in 25% gegen Becherzellen (Otte et al., 1983 und 1989).

44.1.4.1 Neurovegetative Faktoren

Viele klinische und experimentelle Befunde belegen die enge Beziehung neurovegetativer Fehlsteuerungen und emotioneller Faktoren zu vaskulärem System, Motilität und Sekretion des Dickdarms (Almy et al., 1950; Chaudhary und Truelove, 1961; Grace et al., 1950b und 1951; Groen und van der Valk, 1956; Karush et al., 1955; Kirsner, 1960 und 1961; Paskuda et al., 1979; Wolf, 1966). Als pathogenetisch wirksam werden vor allem die nerval bedingte Ausschüttung von Acetylcholin und die Histaminfreisetzung mit Gefäßkontraktion, Ödembildung und muskulärem Spasmus angesehen.

In tierexperimentellen Untersuchungen wurden Änderungen der motorischen Darmtätigkeit (Hyperperistaltik und Spasmus) auch im Anschluß an eine Stimulierung verschiedener Hirnareale festgestellt, so daß bei der nicht selten zu beobachtenden gesteigerten Darmmotilität unter emotionaler Einwirkung, etwa von Schmerz, Schreck, Angst, aktuellem Konflikt, die Annahme einer zentral gesteuerten Pathogenese gestützt wird (Portis, 1949; Wener und Polonsky, 1950). In gleicher Weise wie für die Motilität gelten diese experimentellen Befunde auch für Sekretionsanomalien und Störungen der Zirkulation (Grace et al., 1950; Kirsner, 1960; Palmer, 1948).

Die häufige Angabe, daß dem Beginn der Kolitis manchmal jahrelang eine Neigung zu Durchfällen im Sinne nervöser Diarrhöen (Gromotka und Henning, 1966; Kühn und Nägele, 1967; Rosenblum, 1958; Sattler, 1960) (bei unseren Patienten 13%) oder umgekehrt eine Obstipation vorausging und somit eine funktionelle Darmstörung das Vorstadium der Kolitis gewesen sein kann, spricht ebenfalls für eine neurovegetative Komponente innerhalb des pathogenetischen Bündels.

44.1.4.2 Prämorbide Persönlichkeitsstruktur und Psychodynamik

Die psychopathologischen Befunde der Patienten mit Colitis ulcerosa werden einheitlicher beschrieben als die der Patienten mit Morbus Crohn.

Seit den Untersuchungen von Murray (1930), Sullivan und Chandler (1932) wurden von vielen Autoren psychische Auffälligkeiten der Kolitis-Kranken mitgeteilt (Alexander, 1965; Daniels, 1944 und 1948; Engel, 1958, 1961, 1962 und 1969; Groen, 1947, 1950; Groen und van der Valk, 1956; Karush et al., 1968 und 1969; Klußmann, 1979; Kollar et al., 1964; Krejci, 1962; Mohr et al., 1958; Studt und Mast,

1986). Engel (1955 und 1979) hat die Literatur über die psychischen Merkmale bei Kolitis-Kranken zusammengestellt und bereits das allzusehr vereinfachte Konzept kritisiert, daß die Colitis ulcerosa eine psychogene Erkrankung sei, verursacht durch psychische Störungen. Während die meisten Autoren charakterologische oder neurotische Merkmale als ätiologisch und pathogenetisch wirksam werten, äußern sich andere weitaus zurückhaltender oder deuten die psychischen Befunde lediglich als Reaktion auf Schwere und Dauer des Leidens (Hightower et al., 1958; Sloan et al., 1950; Spiegelberg, 1965; Svartz, 1956; Wildegans, 1959).

Viele, vornehmlich tiefenpsychologisch orientierte Untersuchungen galten daher der Frage nach der prämorbiden Struktur der Kolitis-Kranken. Überwiegend werden eine retardierte Entwicklung, Ich-Schwäche, Passivität, Konfliktvermeidung, Scheu vor Verantwortung, Abhängigkeit von einer dominierenden Bezugsperson und Schwierigkeiten, eine reife, flexible Beziehung zur Außenwelt aufzubauen, beschrieben (Barendregt und Groen, 1953; Drees, 1977; Mewes, 1973; Sandweg, 1986; Schucman und Thetford, 1970; Wijsenbeek et al., 1968).

Verschiedene testpsychologische Untersuchungsergebnisse (z. B. Gathmann et al., 1981; Kipnowski und Kipnowski, 1978, 1981 und 1982; Liedtke et al., 1972 und 1977; Mewes, 1973; Rabavilas et al., 1980; Reindell und Ferner, 1979; Reindell et al., 1981) bestätigen die klinischen Beobachtungen vieler Autoren (Bodman, 1935; Engel, 1955; Karush und Daniels, 1953; Karush et al., 1955; Paulley, 1956; Prugh, 1951; Rohrmoser, 1956).

Diese Anteile der prämorbiden Struktur werden meistens durch die jahrelange Dominanz einer versagenden oder überprotektiven Mutter erklärt, die herrschend, fürsorgend, streng, perfektionistisch kontrollierend, dabei wenig emotionale Wärme ausstrahlend, die Strebungen des Kindes nach eigener Entwicklung blockiert und somit die ursprüngliche symbiotische Bindung verfestigt.

Die Aussagekraft einiger dieser Studien ist jedoch begrenzt, da sie methodische Mängel aufweisen (relativ geringe Patientenzahl, Konzeption der Kontrollgruppe; Probst, 1989). Außerdem wird häufig vorausgesetzt, daß eine homogene Psychopathologie bei Kolitis-Kranken vorliegt. Die klinische Bedeutung solcher Studien ist für den einzelnen Patienten, besonders für die notwendige individuell modifizierte Therapie, begrenzt. Allerdings ergänzen und erhärten die testpsychologischen Befunde viele klinische Beobachtungen, aus denen hervorgeht, daß Anklammerungstendenzen an eine Schlüsselperson und ein ständig signalisiertes Bedürfnis nach Zuwendung, Geborgenheit und Verwöhnung vorliegen. Diese Verhaltensmodalitäten können in Verbindung mit der Infantilität auf eine prägenitale Organisationsstörung und daraus resultierende Ich-Schwäche zurückgeführt werden.

Unausgereifte infantile Züge der Primärstruktur lassen sich auch im zwiespältigen Verhältnis zu sich selbst erkennen. Daraus resultiert eine intensive Beschäftigung der Patienten mit sich selbst und einer libidinösen Besetzung der eigenen Person. Viele klinische Beobachtungen enthalten immer wieder Hinweise auf diese egozentrischen und narzißtischen Merkmale, das unsichere und widersprüchliche Selbstwertgefühl, mit dem Minderwertigkeits- und Schuldkomplexe verbunden sind (Groen, 1947; Groen und van der Valk, 1956; Paulley, 1950 und 1971; Sperling, 1946 und 1958).

Wir fanden aber bei unseren Untersuchungen (Feiereis, 1977; Wilke, 1978) narzißtische Züge nicht als das hervorstechende Merkmal der Kolitis-Patienten. Dennoch ergab sich in Verbindung mit therapeutischen Befunden (Wilke, 1980, 1983a und b), daß, entsprechend dem Konzept der Narzißmusstruktur als Ausdruck einer frühen Störung der Selbstentwicklung, Verschmelzungswünsche mit einem allmächtigen Objekt, Omnipotenzphantasien und Neigungen zur Idealisierung und Identifizierung als kontradepressiv wirkende Hilfe bei einer Reihe der Kolitis-Patienten festzustellen waren.

Untersuchungen von Sittaro (1980), Zepf und Mitarbeitern (1981a und b) bezogen sich auf die Hypothesen, daß

– bei Patienten mit Colitis ulcerosa zwischen der Repräsentanz ihres aktuellen und der Repräsentanz ihres idealen Selbst eine große Distanz besteht;
– Patienten mit Colitis ulcerosa ihre aktuelle Selbstrepräsentanz als unterschiedlich zur aktuellen Repräsentanz ihrer narzißtischen Schlüsselfigur erleben und
– die aktuelle Repräsentanz der Schlüsselfigur entweder in der Nähe der aktuellen Repräsentanz der Mutter oder der Idealform der mütterlichen Repräsentanz wahrgenommen wird.

Ihre mit einer vom Untersucher weitgehend unabhängigen Methode festgestellten Befunde bestätigen die Störung des Selbstwertgefühls der Kolitis-Patienten und eine narzißtische Beziehung zu einer Schlüsselperson; es fand sich eine Korrelation zwischen deren Verlust und dem Ausmaß der Darmsymptomatik. Die Autoren diskutieren die Möglichkeit eines allmählich eigengesetzlichen, somatischen Verlaufes der Krankheit.

Kolitis-Kranken fällt die Entwicklung affektiver Beziehungen oft schwer. Die Fähigkeiten zum Kontakt mit anderen sind reduziert, woraus Isolation und Absonderung resultieren. Konflikte werden strikt durch Anpassung, Versöhnlichkeit, überbetonte Freundlichkeit oder umgekehrt durch das Ausweichen vor Kontakten und durch Rückzug vermieden. Eng mit dieser Einstellung verbunden sind die starke Verletzlichkeit, Frustrationsintoleranz, Stimmungslabilität und Neigung zu überschießenden emotionalen Reaktionen (Barendregt und Groen, 1953; Krasner, 1953; Wijsenbeek et al., 1968).

Mewes (1973) hebt hervor, daß diese Reaktionsweisen in einer Abhängigkeit zu den Formen der Sozialkontakte stehen. Lassen sich Vorstellungen über soziale Beziehungen nicht verwirklichen, so fühlen sich die Patienten relativ rasch verletzt. Sobald ein Kolitis-Kranker Frustrationen zu verarbeiten hat,

wird ein aggressiver Impuls nicht uneingeschränkt ausgedrückt, sondern in den Mantel „verletzter Gefühle" gekleidet.

Die nach außen gerichteten Aggressionen sind bei den meisten Kolitis-Kranken gehemmt. Die Patienten geben sich freundlich, affektarm, sich unterwerfend (Fullerton et al., 1962; Krasner, 1953). Eng verbunden hiermit sind der ausgesprochene Mangel an Spontaneität, Ausdrucksfähigkeit von Gefühlen, besonders des Zorns, sowie die Introversion und auffällig demütig-gefügige Grundhaltung, mit der nicht selten Gereiztheit, Eigensinn und Überheblichkeit assoziiert sind (Feiereis et al., 1962; Leibig et al., 1985; Mewes, 1973; Schellack, 1954; Wittkower, 1938).

Weitgehend übereinstimmend wird die Depressivität bei Kolitis-Kranken beschrieben und vorwiegend der prämorbiden Struktur zugerechnet (Daniels, 1944; Engel, 1955 und 1961; Feiereis, 1970; Wijsenbeek et al., 1968).

Einige Autoren (Sperling, 1946) sprechen sogar von einer depressiven Grundstörung (dem inneren Kampf mit der introjizierten Mutter) als dem zentralen Konflikt. Auch verschiedene testpsychologische Befunde deckten depressive Züge auf, teilweise in einer größeren Streuung gegenüber einer Kontrollgruppe (Fullerton et al., 1962; Krasner, 1953; Mahoney et al., 1949). Andere Befunde (Mewes, 1973; Poddig, 1987) haben gegenüber einer Kontrollgruppe keine sichere Differenz ergeben. Die Depressivität wird, bewußt oder unbewußt, durch ein kontradepressives Abwehrverhalten (Freyberger, 1982) unterdrückt, oder sie ist mindestens überwiegend als reaktiv bzw. krankheitsdependent anzusehen.

Die klinischerseits beobachteten verschiedenartigen Zwangssymptome, z.B. peinliche Genauigkeit, Ordentlichkeit, Pünktlichkeit, Gewissenhaftigkeit, betontes Sauberkeitsstreben, pedantisches Verhalten, große Sparsamkeit und Bedürfnis nach Perfektionismus, entsprechen den von Schellack (1954) festgestellten Lücken im anal-retentiven Antriebserleben.

Vergleichende Untersuchungen der Persönlichkeitsstruktur von Patienten mit Colitis ulcerosa und Morbus Crohn

Lourens (1973) hält die Persönlichkeitsstruktur für sehr ähnlich; so seien bei beiden Gruppen Regression, Narzißmus, Abhängigkeit, Aggressionsgehemmtheit und Zwanghaftigkeit festzustellen. Bei 47% der Untersuchten findet er psychopathologische Auffälligkeiten (erhöhter „Cornell Medical Index Score"), bei 41% eine Neigung zur Regression, bei allen Patienten starke Abhängigkeitsbedürfnisse. Bei mehr als der Hälfte der Patienten wird eine Aggressionsgehemmtheit beschrieben, bei 80% eine zwanghafte Struktur.

Auch Paulley (1971) teilt große Ähnlichkeiten zwischen den beiden Patientengruppen in Abhängigkeit und Unreife, zwanghaften Zügen und Aggressionshemmung mit. Häufig seien bedeutsame Lebensereignisse („life events") zu finden; die Patienten mit Morbus Crohn seien aber insgesamt weniger abhängig als die Patienten mit Colitis ulcerosa.

Eine Gegenüberstellung beider Krankheitsgruppen findet sich ebenfalls bei Petzold und Reindell (1977). Die Autoren beschreiben die Persönlichkeitsstruktur der Patienten mit Colitis ulcerosa eher depressiv-zwanghaft, mit Morbus Crohn eher schizoid-hysterisch; das Verhalten bei Colitis ulcerosa sei abhängig von Bezugspersonen, bei Morbus Crohn eher pseudoautonom. Die Bereitschaft zur Psychotherapie sei deshalb bei den Kolitis-Patienten stärker ausgeprägt als bei Patienten mit Morbus Crohn. Die Autoren fanden in den Familien der Patienten mit Colitis ulcerosa einen „Bindungsmodus", d.h. den Typus einer gebundenen Familie mit einem fast geschlossenen System, bei Patienten mit Morbus Crohn einen „Ausstoßungsmodus", d.h. eine Familie, die sich in Auflösung befindet (vgl. Kap. 18).

Reindell und Ferner (1979) heben bei Patienten mit Colitis ulcerosa die Trias Abhängigkeit in Objektbeziehungen, Aggressionshemmung und Depressivität hervor, hingegen wirken die Patienten mit Morbus Crohn flexibler, weniger symbiotisch fixiert und auch früher vom Elternhaus gelöst, während Poddig (1987) keine wesentlichen Unterschiede beschreibt.

Faßt man alle diese Befunde zusammen, so ergeben sich gut begründete Hinweise auf die prämorbide Struktur einer narzißtischen, retardierten Persönlichkeit mit labilem Selbstwertgefühl, depressiv-zwanghaften Zügen, der Neigung zur Regression auf die frühkindlich-präödipale Entwicklungsstufe mit gehemmter Aggressivität, Abhängigkeit von einer nahen Bezugsperson und ausgeprägter Kränkbarkeit des Kolitis-Kranken. Die Depressivität kann ein prämorbides Merkmal der Struktur des Patienten sein oder vorwiegend reaktive Kennzeichen haben, vor allem als Folge langer und schwerer Krankheit oder häufiger Rezidive.

Die Wertung der empirischen Untersuchungsbefunde und der vergleichenden Studien ist nach wie vor divergent. Es wird besonders darüber diskutiert, ob die Colitis ulcerosa

– eine psychosomatische Krankheit ist,
– ob sich das psychosomatische Erscheinungsbild von anderen schweren chronischen Krankheiten, besonders auch vom Morbus Crohn, unterscheidet, und
– welche psychopathologischen Merkmale Krankheitsursache und Krankheitsfolge sind.

Unter Berücksichtigung aller kritischen Einwände (Feurle et al., 1988) läßt sich nach dem gegenwärtigen Forschungsstand für die klinisch-praktische Arbeit aussagen, daß weder eine **spezifische Persönlichkeitsstruktur** noch ein **spezifischer Konflikt** beim Kolitis-Kranken vorliegt.

Hingegen bestehen gut dokumentierte Befunde über die prämorbide Persönlichkeitsentwicklung, psychodynamisch wirksame Prägungen durch die Beziehung zu einer Schlüsselfigur und Merkmale depressiven, zwanghaften und aggressionsgehemmten Verhaltens. Diese Befunde erlauben aber **nicht** den Schluß auf deren ätiologische oder pathogenetische

Bedeutung, was leider oft nicht scharf genug getrennt wird. Jedoch sprechen sie für die Notwendigkeit einer **nichtselektionierten** psychodynamischen Diagnostik, die zeitgleich mit der körperlichen Diagnostik erfolgen sollte.

Ebenso wie die somatische Therapie aus den somatischen Befunden resultiert, so selbstverständlich beruhen psychotherapeutische Maßnahmen auf dem Ergebnis des psychodynamischen, psychoreaktiven und psychosozialen Befundes.

44.1.4.3 Krankheitsauslösung

Nach Curtius (1959) gelten als Kennzeichen eines Auslösungsvorganges:

– Die Auslösung, der eigentliche Anstoß zur krankhaften Reaktion, setzt eine Summe ätiologisch-pathogenetisch wirksamer Teilursachen als conditio sine qua non voraus.
– Im Gegensatz zu obligaten Teilursachen sind Auslösungsfaktoren individuell variabler Natur.
– Im allgemeinen folgt die krankhafte Reaktion der Einwirkung des Auslösungsfaktors unmittelbar.
– Der Auslösungsfaktor ist quantitativ unbedeutend; er wirkt oft analog einem Katalysator.

Weitaus am häufigsten gehen emotionale Einwirkungen dem Krankheitsbeginn bzw. einem Schub unmittelbar voraus, was der möglichen ätiologischen und pathogenetischen Bedeutung vegetativer und psychischer Faktoren für die Entwicklung der Colitis ulcerosa zu entsprechen scheint (Crismer und Drèze, 1961; Freyberger et al., 1980; Hönmann, 1982; Jörgens und Dieckhöfer, 1972; Kipnowski und Kipnowski, 1981; McKegney et al., 1970; Mörl und Matis, 1967).

Die psychischen Auslösungsvorgänge hängen oft mit Verlusterlebnissen zusammen, die real erfahren worden sind, drohend bevorstehen oder imaginiert werden. Sie können sich ebenso auf nahe Bezugspersonen oder eine soziale Bindung (Beruf, Wohnort) wie auf materielle oder ideelle Einbußen beziehen und zeitlich unmittelbar anschließend zum Ausbruch der Kolitis oder einer akuten Verschlechterung führen. In einer Studie (Scherl, 1978) wurden Verschlimmerungen infolge seelischer Belastungen von 59% der Patienten mitgeteilt. Besonders der sich in kindlicher Abhängigkeit innerhalb einer narzißtischen Objektbeziehung befindende oder kompensatorisch pseudounabhängig lebende Patient erfährt solche Trennungsängste oder Verlusterlebnisse als Einbruch in sein labiles Selbstwertgefühl und als eine tiefe Kränkung, Demütigung und unmittelbare Gefahr.

Wir behandelten z.B. einen 32jährigen Mann, der 5 Jahre lang keinen Schritt aus der Einzimmerwohnung tun und die Abwesenheit seiner Ehefrau nur für die 6 Stunden ihres Dienstes als Lehrerin ertragen konnte. Er erwartete sie stets voller Angst, sie könne nicht rechtzeitig wiederkommen. Bereits ein Telefonat seiner Frau mit einer Freundin führte bei ihm sofort zu Leibschmerzen und Durchfall.

Psychophysische Korrelationsbefunde (Zander et al., 1982a und b) erhärten die Bedeutung solcher

Auslösungskonflikte, die pathophysiologisch über antagonistische Funktionsstörungen („strain"), die schließlich zu morphologischen Veränderungen führen, erklärt werden können.

Mit der initialen Symptomatik sowohl der Ersterkrankung als auch eines Rezidivs hängt mitunter zeitlich auch ein Erkältungs- oder Darminfekt zusammen, wahrscheinlich ebenfalls im Sinne der Auslösung. Wir beobachteten dies bei 14%. Lokale Kälteeinflüsse auf den Rumpf oder die unteren Extremitäten können reflektorisch einen durchfallauslösenden Effekt haben. Selten hingegen standen schwerere Erkrankungen anderer Organe am Beginn der Kolitis oder eines Schubes. Auch die jahreszeitlich zu beobachtenden Schwankungen mit einem Frühjahrs- und Herbstgipfel können manchmal indirekt auf Erkältungsinfekte zurückzuführen sein.

44.1.4.4 Reaktive psychische Befunde

Wenn auch nach einer weiteren Untersuchung (Probst, 1989) die gegenüber einer Kontrollgruppe festgestellten Persönlichkeitsmerkmale, z. B. auch die Aggressionsgehemmtheit, von nahezu allen somatischen Symptomen unabhängig waren und damit die Hypothese einer besonderen prämorbiden Persönlichkeitsstruktur gestützt wird, so ist es dennoch schwierig, solche Veränderungen gegenüber reaktiven, d. h. krankheitsabhängigen Befunden abzugrenzen. Eine Aussage hierüber ist auch deshalb erschwert, weil in aller Regel ein umfassendes Bild psychischer Befunde erst nach Ausbruch der Krankheit gewonnen werden kann.

Aus diesem Grunde sind alle Untersuchungen wichtig, die in der Remission vorgenommen werden. Hierbei ergab sich z. B. in einer Studie von Leibig und Mitarbeitern (1985), daß Patienten mit Colitis ulcerosa im Intervallstadium keine erhöhten Depressivitätswerte aufwiesen. Die Ergebnisse der verschiedenen Mitteilungen der Literatur und auch die eigenen Erfahrungen sprechen dafür, beim einzelnen Kranken stets zu versuchen, prämorbide Strukturmerkmale von krankheitsabhängigen abzugrenzen und zu prüfen, ob sie, wie z. B. die Depressivität, nicht ebenso prämorbides Persönlichkeitsmerkmal wie Folge der eingetretenen Krankheit oder eines Schubes sind.

44.1.4.5 Psychosoziale Folgen der Krankheit

Der Beginn der Colitis ulcerosa mit der bevorzugten Inzidenz zwischen dem 15. und 35. Lebensjahr fällt meistens in die aktivste Lebensphase des Patienten. Während umfangreiche Studien über den Langzeitverlauf und die Prognose der Colitis ulcerosa vorliegen (Broström, 1983; Devroede et al., 1971; Hendriksen et al., 1985; Jalan et al., 1970; Lennard-Jones, 1983; Ritchie et al., 1978; Schröter, 1977), in denen vor allem die Letalität, das Karzinomrisiko, die Operationsrate und die Rezidivquote untersucht wurden, gibt es bisher kaum Informationen über die Auswir-

kung der Erkrankung auf Partnerschaft, Familie, Sexualleben, Beruf (Wyke et al., 1988) und Freizeitaktivitäten.

Nach einer Untersuchung von Mallet und Mitarbeitern (1978) gab ein Fünftel eine reduzierte Arbeitsfähigkeit an, ein Sechstel hatte nach Krankheitsbeginn die Arbeitsstelle wechseln müssen. Auch nach Besserung fühlte sich ein Viertel der Patienten in den Freizeitaktivitäten eingeschränkt. Die Patienten berichteten häufig über Eheschwierigkeiten und Störungen im Sexualleben. Demgegenüber teilten Hendriksen und Binder (1980) als Ergebnis eines strukturierten Interviews bei 122 Colitis-ulcerosa-Patienten und 83 akut erkrankten Kontrollpatienten mit, daß sich die Colitis-ulcerosa-Patienten gut an ihre chronische Krankheit adaptiert hatten. In beiden Gruppen wurden gleiche Ergebnisse für Familienstand, Frequenz familiärer und sexueller Probleme, Freizeitaktivitäten und Arbeitsfähigkeit gefunden.

Im Hinblick auf häufige Wechselwirkungen somatischer, psychischer und sozialer Anteile bei Entstehung und Verlauf einer Krankheit wurde der Zusammenhang zwischen körperlicher Krankheitssymptomatik, Persönlichkeitsstruktur und psychosozialer Beeinträchtigung bei 58 Patienten mit Colitis ulcerosa untersucht (Probst et al., 1989). Gegenüber der Kontrollgruppe konnten signifikante soziale Beeinträchtigungen nachgewiesen werden. Früher berichtete psychosoziale Einschränkungen (Mallet et al., 1978; Feurle et al., 1983) ließen sich bestätigen. Die Patienten hatten innerhalb ihres Arbeitslebens häufigere Fehlzeiten, geringeres Interesse, in der Freizeit weniger Freunde und seltenere gemeinsame Unternehmungen; sie fühlten sich unwohl in Gesellschaft und litten andererseits häufiger unter Einsamkeit und Langeweile. Es fiel ihnen schwer, über Gefühle und Probleme zu sprechen. Partnerschaftskonflikte bezogen sich vor allem auf die Sexualität. Mit zunehmender Krankheitsdauer ließ sich eine Adaptation beider Ehepartner an die chronische Erkrankung feststellen, die auch als sekundärer Krankheitsgewinn gedeutet werden kann, indem die chronische Krankheit zum Bindeglied in der Partnerschaft wurde. Auffallend war weiterhin, daß sich vor allem ein reduziertes Allgemeinbefinden mit dem Gefühl der Schlappheit und Müdigkeit auf soziale Interaktionen auswirkte. Es korrelierte nicht mit anderen somatischen Krankheitssymptomen. Schwäche, Müdigkeit und Schlappheit können ebenso als Folge der chronischen Entzündung mit einer Eisenmangelanämie gedeutet werden, wie Ausdruck einer depressiven Stimmung sein, die sich in der Tendenz zur Passivität, Hoffnungslosigkeit und Selbstaufgabe manifestiert und dadurch soziale Kontakte begrenzt. Die Besserung der körperlichen Symptomatik wiederum verlief in dieser kontrollierten Studie nicht parallel mit dem Rückgang der psychischen Symptomatologie. Darin zeigt sich erneut die Notwendigkeit, mögliche innere Konflikte und psychosoziale Auswirkungen der Krankheit in die Therapie einzubeziehen, d. h., daß bei der Colitis ulcerosa die Indikation zu einer kombinierten somatischen und psychosozialen Behandlung besteht.

44.1.5 Therapie

44.1.5.1 Einleitung

Auch aus psychodynamischer Sicht ist die Ätiologie der Colitis ulcerosa ungeklärt, d. h., die psychopathologischen und psychosozialen Befunde können nicht mit dem Grad der Wahrscheinlichkeit als Ausdruck einer Teilursache der Krankheit oder des einzelnen Schubes angesehen werden. Trotz der ungeklärten Ätiologie besteht aber bei vielen Patienten eine Indikation zur Entspannungs- und Psychotherapie, die selbstverständlich nicht als Alternative, sondern als Ergänzung und Erweiterung der gastroenterologischen Behandlung aufzufassen ist. Leider werden psychische Anteile am Krankheitsprozeß und deren psychosoziale Folgen innerhalb einer rein medikamentös-somatischen Therapie nach wie vor sehr oft bezweifelt oder negiert, mögliche Schäden, die hieraus resultieren, allzu leicht verleugnet, z. B. chronifizierte Konflikte bei der Krankheitsverarbeitung, drohender Verlust des Arbeitsplatzes, versäumte Rehabilitation. Wehe aber, wenn umgekehrt dem psychotherapeutisch engagierten Arzt, dem die psychosomatische Medizin zu einem wesentlichen Bestandteil seiner täglichen Arbeit geworden ist, mit seiner Therapie körperliche Prozesse entgleiten und dadurch Schäden eintreten! Weil also mit zwei Ellen gemessen wird, sollte sich der psychotherapeutisch tätige Arzt gemäß der Definition der psychosomatischen Medizin stets besonders auch um die Erweiterung seiner Kenntnisse über die körperliche Diagnostik und Therapie der Krankheiten bemühen, die er psychotherapeutisch behandelt.

Die interdisziplinäre Integration – durch Liaison – psychosomatischer und erst recht somatopsychischer Diagnostik und Therapie in die Stationsarbeit eines Krankenhauses oder in die tägliche Arbeit der Praxis ist weithin noch nicht erkennbar; auch in Universitätskliniken ist die Distanz manchmal noch so groß wie die räumliche Entfernung zwischen dem Krankenbett auf der somatischen Station und dem Sprechzimmer des psychosomatischen Arztes. Daher liegt auf vorerst noch nicht absehbare Zeit das Ziel einer für die körperliche, die psychische und die psychosoziale Dimension gleichermaßen kompetenten Liaison in der Personalunion ein und desselben Arztes innerhalb seines Alltages im Krankenhaus und in der Praxis. Der wissenschaftliche Fortschritt ist ohne Spezialisierung nicht denkbar, der Nutzen des Fortschritts für den Patienten aber ebenso nicht ohne ein hohes Maß an kontinuierlicher, verläßlicher Integration erkennbar.

Die Leitlinien für Integration, Form und Umfang der Entspannungs- und Psychotherapieverfahren sind:
- Prämorbide psychopathologische Befunde der Persönlichkeitsstruktur.
- Psychodynamik eines prämorbide vorhandenen chronifizierten Konfliktes, z. B. eines Abhängigkeits- oder Familienkonfliktes.
- Akute seelische Belastungen und Konflikte vor Beginn der Krankheit oder des einzelnen Schubes.
- Reaktive, sekundäre psychische Befunde als Folge der Krankheit.
- Psychosoziale Belastungen infolge der Krankheit oder des einzelnen Schubes.
- Eine Verschlimmerung der Krankheit durch psychische und psychosoziale Einflüsse im Sinne psychophysischer Wechselwirkung.

Die von uns entwickelte kombinierte konservative Therapie (Curtius, 1962, 1968; Curtius und Rohrmoser, 1955; Feiereis, 1970, 1975 und 1977) der Colitis ulcerosa setzt sich aus einer Reihe somatisch und psychisch wirksamer Einzelmaßnahmen zusammen. Aufgrund unserer Erfahrungen innerhalb von 40 Jahren mit inzwischen über 2000 Kranken lassen sich anhaltende Erfolge nur erzielen, wenn die Voraussetzungen zu dieser Therapie erfüllt sind. Eine der wichtigsten ist, daß die kombinierte Therapie möglichst konstant in der Hand eines Arztes oder einer therapeutischen Gruppe verbleibt, die über die notwendigen internistischen sowie über psychotherapeutische Kenntnisse und Erfahrungen verfügt. Eine Aufteilung der Kompetenzen im Sinne des Mottos „Ein Arzt für die Seele, ein Arzt für den Darm" läßt sich nach unseren Erfahrungen relativ selten und nur bei sehr guter und kontinuierlicher Kooperation verwirklichen (Daniels et al., 1962).

Die Therapie folgt wie kaum eine zweite keinem starren Schema. Schwere, Form und Dauer der Krankheit verlangen ebenso wie die individuellen Besonderheiten des Kranken und seiner Krankheit eine ständige Kontrolle der medikamentösen Ansätze und der einzelnen psychotherapeutischen Verfahren.

44.1.5.2 Medikamentöse Therapie

Zur Standardverordnung gehört die 5-Aminosalizylsäure, die in Verbindung mit einem Sulfonamid als Kombinationspräparat Salazosulfapyridin (Colo-Pleon®, Azulfidine®) angewendet wird, 3–4 g p.o. nach dem Essen. Mögliche Nebenwirkungen sind Übelkeit, Appetitlosigkeit, Erbrechen, Exantheme, Agranulozytose, Nephrolithiasis, gesteigerte Blutungstendenz, hämolytische Anämie, Oligo- oder Azoospermie, remittierendes Fieber. Bei Unverträglichkeit oder Nebenwirkungen hat sich alternativ die Monosubstanz Mesalazin (Claversal®, Salofalk®), 1,5 g/d, bewährt, eventuell lokal als Suppositorium oder Klysma.

Eine Indikation zur antibiotischen Behandlung besteht bei sekundären Infektionen und foudroyantseptischem Verlauf mit eitrigen Durchfällen.

Die rasche Wirkung der Glukokortikoide (Prednison, Fluocortolon) ist leider mit dem nicht selten verzögerten Heilungsverlauf und einer verringerten Resistenz verknüpft. Bei manchen Patienten verstärken sich sogar die Blutungsneigung und die Durchfälle. Die Dosierung richtet sich nach Schwere, Verlauf und Dauer der Krankheit. Trotz pathophysiologischer Gegenargumente kann nach unseren Erfahrungen bei chronisch rezidivierendem Verlauf das adrenokortikotrope Hormon des Hypophysenvorderlappens als Depot-ACTH (40–80 I. E.) oder Tetracosactid (Synacthen®-Depot), 0,5–1,0 mg i. m., 2–3mal in der Woche, mitunter noch hilfreich sein, wenn die übrigen medikamentösen Maßnahmen unbefriedigend bleiben. Bei der zusätzlichen Indikation zur immunsuppressiven Therapie bevorzugen wir Azathioprin (Imurek®), anfangs 3 mg/kg KG/d, nach 3–4 Wochen 0,5–1,0 mg/kg KG/d, eventuell 4–6 Monate und länger. Auf Nebenwirkungen und Gefahren dieses Zytostatikums ist stets sorgfältig zu achten.

Die symptomatische Zusatztherapie beruht auf Diphenoxylathydrochlorid und Atropinsulfat (Reasec®), Loperamidhydrochlorid (Imodium®), Aluminiumphosphat (Phosphalugel®) oder Opiumtinktur. Hier ist ebenso wie bei der Verwendung von Spasmolytika an die Gefahr einer toxischen Kolondilatation zu denken. Die Glukokortikoidtherapie wird meist mit einem Antazidum kombiniert.

44.1.5.3 Diät

Der oft reduzierte Allgemeinzustand und der schlechte Appetit der Kolitis-Patienten machen es erforderlich, für eine besonders abwechslungsreiche und kalorisch ausreichende Kost zu sorgen. Dabei sollte auf alle individuellen Wünsche eingegangen werden, besonders dann, wenn sich eine der Anorexia nervosa ähnliche Konstellation entwickelt, d.h. der Patient heimlich erbricht oder Nahrung beiseite schafft.

Diätassistentin und Stationsschwester haben mit der Abstimmung der Ernährung, der Absprache einer genügenden Eiweiß- und Kalorienzufuhr unter Einschluß aller notwendigen und schmackhaften Kohlenhydrate oft eine Schlüsselfunktion auf dem Weg zum Erfolg. Wir beschränken anfangs lediglich ungekochtes Obst, rohe und zellulosereiche Gemüse und verbieten kalte Speisen und Getränke sowie Alkohol. Eine Milchkarenz – bei Laktoseintoleranz – ist nur selten erforderlich.

44.1.5.4 Entspannungstherapie

Der beschriebene Einfluß vegetativer und psychischer Anteile auf Pathogenese und Verlauf der Krankheit führte erstmals durch Curtius 1943 (1962) zur Anwendung der Tiefenentspannung im Sinne der Hypnose. Sie vermittelt vor allem im akuten und subakuten Zustand dem schwerkranken, oft depressiven, durch Chronizität des Leidens und bisher fehlgeschlagene Therapie häufig entmutigten Patienten über den Weg der Regression ein Gefühl tiefgreifender Gelöstheit und Geborgenheit mit der Möglichkeit, „sich endlich fallenzulassen", ebenso die Besserung von Tenesmen und Schmerzen; das bedeutet, wieder primäre Sicherheit und Zuversicht erleben zu können (Widok, 1976). Darüber hinaus fördert diese enge, nahezu symbiotische therapeutische Beziehung die Übertragung und Gegenübertragung, hilft Ängste

abzubauen und den Weg zur weiteren Psychotherapie zu ebnen. Die notwendigen Voraussetzungen für diese Behandlung, nämlich Bereitschaft, Vertrauen und affektive Ansprechbarkeit, sind bei den Kolitis-Kranken so gut wie ausnahmslos vorhanden und ebenso unabhängig von der Schwere der Krankheit wie auch von Alter oder Intelligenz des Patienten.

Bereits Mohr (1930) hob hervor, daß gerade die Suggestionen am wirksamsten sind, die dem tatsächlichen physiologischen Vorgang am nächsten kommen und gleichzeitig anschaulich geschildert und dem Verständnis des Kranken angepaßt sind.

Bei chronisch rezidivierendem oder chronisch kontinuierlichem Verlauf bevorzugen wir unter den Methoden der Selbststeuerung zur Stabilisierung des psychischen und vegetativen Gleichgewichtes das von J. H. Schultz (1978) entwickelte autogene Training. Die Patienten erlernen diese Methode, solange sie bettlägerig sind, durch Einzelanleitung und üben später in der Gruppe. Die drei wichtigsten Faktoren dieses Hypnoids liegen in der aktiven Einflußnahme des Patienten über den Kortex, in der Entspannung des Organismus mit der Wirkung über die subkortikalen Zentren und in der Therapie und Prophylaxe mit der Wirkung auf Funktion und Struktur. Der Patient erlernt, Tonus- und Motilitätsschwankungen des Darmes mit reaktiven körperlichen und seelischen Folgen und die sich wiederholenden reflektorischen Spannungen selbst auszugleichen.

In die psychosomatische Behandlung des Kolitis-Kranken wird sehr bald eine Krankengymnastin eingeschaltet. Sie hat die Aufgabe, je nach Schwere der Krankheit, durch Atemtherapie, leichte Rücken-, Arm- und Beinmassage zur körperlichen Entspannung des Patienten beizutragen. Bei spastischen Beschwerden, Blähungen und Anspannungen der Bauchmuskulatur sowie Tenesmen werden zusätzlich leichte Leibmassagen angewendet. Abgestimmt auf die Struktur des Kranken, seine Belastbarkeit und Reaktion, erlernt er zunächst allein, bald innerhalb einer Gruppe, mit Hilfe krankengymnastischer Übungen nicht nur die Skelettmuskulatur wieder zu kräftigen, sondern auftretende Verkrampfungen und Tenesmen zu lösen.

Diese Behandlung wird, sobald der Patient umhergehen kann, mit den Übungen der progressiven Relaxation erweitert, die sich auch bei anderen Autoren bewährt hat (Kipnowski et al., 1988).

44.1.5.5 Stützende Gesprächspsychotherapie

In diese vorwiegend regressionsfördernden entspannungstherapeutischen Maßnahmen werden supportive bzw. anaklitisch wirksame, d.h. stützende verbale Hilfen eingefügt und allmählich, sofern es der körperliche Befund zuläßt, Fragen der psychischen Auslösung der Krankheit oder des einzelnen Schubes in die kombinierte Therapie einbezogen. Dem Patienten sind häufig solche Auslösungsfaktoren bewußt oder vorbewußt, so daß der Zugang zu diesen ihn unmittelbar belastenden Inhalten der Pathogenese relativ leicht ist und schließlich mit der Erweiterung zur biographischen Anamnese verbunden werden kann. Mit ihr werden auch die Übertragung stabilisiert, inadäquate Forderungen und Versagungen vermieden und gleichzeitig ein Einblick in die Familien- und Sozialanamnese, die frühkindliche Entwicklung und ihre möglichen Störungen, in die Latenzphase und Adoleszenz, in die Schul- und Berufsausbildung gewonnen. Auch das Bild über die gegenwärtigen Bezugspersonen des Patienten und über mögliche äußere Konflikte wird auf diese Weise erweitert. Wir verbinden die gesprächstherapeutischen Anteile jeweils innerhalb einer Sitzung gern mit den Einzelübungen der Entspannungstherapie, um möglichst jede negative Rückwirkung auf den somatischen Prozeß zu vermeiden.

Wir halten es im Gegensatz zu anderen Autoren (Freyberger und Müller-Wieland, 1966), die den psychosomatisch wirkenden Arzt anfangs in der akuten Phase „nur am Rande in das Blickfeld des Kranken rücken", da für sie die internistischen Maßnahmen ganz im Vordergrund stehen, für äußerst wichtig, daß die Kombinationstherapie unmittelbar einsetzt, vor allem auch gerade im akuten Stadium. Mit Hilfe dieser Form kombinierter Entspannungs- und Psychotherapie gelingt es meistens bereits von Beginn der Behandlung an, dem Patienten Geborgenheit und Zuversicht zu vermitteln und eine starke positive Übertragung zu entwickeln, die bei der somatischen Behandlung allein oft nicht möglich ist.

Ob und wie weit hierbei bereits eine Familienkonfrontationstherapie einzubeziehen ist, sollte äußerst behutsam geprüft werden. Es wird verschiedentlich auf die Bedrohung der symbiotischen Beziehung vor allem in der akuten und subakuten Krankheitsphase, mit erheblicher Gefährdung des Patienten hingewiesen (Petzold und Reindell, 1977). Allerdings wird auch darüber berichtet, daß Familiengespräche hilfreich sein können mit dem Ziel, die Mechanismen, Regeln und Beziehungskräfte innerhalb der Familie, die häufige Starrheit des Familiensystems, eine Art von Pseudogemeinschaft, und die eingeschränkten Beziehungen über die Familie hinaus (Jackson und Yalom, 1966) aufzudecken, durch die sich die Gefühle von Hilf- und Hoffnungslosigkeit bei Kolitis-Kranken ebenfalls erklären lassen (Ferner und Reindell, 1979; Reindell und Ferner, 1979).

Besonders auch die Untersuchungen von Minuchin und Mitarbeitern (1978) über die Familien psychosomatisch Kranker führten zu der Erkenntnis, die lineare mit der systemischen Therapie zu verbinden. Gleiches gilt auch für die Prüfung der Indikation zu einer Partner-Therapie, soweit erkennbar ist, daß interaktionelle Anteile auf Auslösung und Verlauf der Krankheit wirksam sind.

44.1.5.6 Assoziative Maltherapie

Als eine hilfreiche Psychotherapieform, vor allem für Patienten mit gehemmter oder mangelhafter sprachlicher Ausdrucksfähigkeit, hat sich bei uns die assozia-

tive Maltherapie bewährt. Die Patienten stellen ihre spontanen Einfälle, Phantasien, Gefühle oder Gedanken ebenso wie angstbesetzte Inhalte ihres eigenen Bildes von der Krankheit dar und formulieren anschließend Assoziationen zu dem so entstandenen Bild.

Eine vergleichende Studie (Feiereis et al., 1989) galt den Kategorien Aggression, Autoaggression, Angst, Depression, Regression, Symbiose und Abgrenzung. Für Patienten mit Colitis ulcerosa und ebenso für Patienten mit Morbus Crohn sind zentrale Themen der bildlichen Darstellung (Abb. 44.1–1 bis 44.1–4, siehe Farbtafeln 3–4; 44.1–5) und dazugehörende Assoziationen:
- Ängste vor der Krankheit, vor mit ihr verbundenen Untersuchungen und Operationen, vor Schmerzen (Abb. 44.1–1), aber auch vor Fremdbestimmung, vor anstehender Entscheidung, vor der Zukunft (Abb. 44.1–2), vor dem Leben nach der Entlassung aus der stationären Therapie. Viele Patienten fühlen sich der empfundenen Bedrohung durch die Außenwelt nicht gewachsen. Sie drücken ihre Hilflosigkeit als Versagens- und Isolationsängste (Abb. 44.1–3) in der Familie, in der Partnerschaft und auch im weiteren sozialen Leben aus, verbunden mit der Suche nach dem „richtigen Weg zur Gesundheit".
- Wünsche nach Ruhe, Entspannung, Wärme und Geborgenheit, nach einer stabilen Einbindung in eine Partnerschaft oder Familie.
- Von dem Darm ausgehende Einbuße der Lebensmöglichkeiten infolge der Symptomatologie (Abb. 44.1–4), vor allem des Schmerzes. Der Darm wird bei den Kolitis-Kranken zum bestimmenden Element aller Gedanken und Gefühle, ihm gilt ein hohes Maß an Aufmerksamkeit (Abb. 44.1–5), die jede Veränderung sorgfältig registriert, während die Patienten mit Morbus Crohn weitaus häufiger mit den Abwehrformen Verleugnung, Reaktionsbildung und Dissimulation reagieren.

44.1.5.7 Tiefenpsychologische Psychotherapie

Mitunter erst gegen Ende der klinischen Behandlung gelingt es, reaktive psychische Befunde, etwa eine erhebliche Depressivität, von der prämorbiden neurotischen Struktur des Patienten und der pathogenetisch mitwirkenden Psychodynamik eines inneren Konfliktes abzugrenzen. Hier erst ergibt sich dann die Frage der Indikation einer psychoanalytisch orientierten Psychotherapie, die im Remissionsstadium ambulant fortgesetzt wird. Für sie gelten die gleichen Voraussetzungen wie bei anderen Patienten mit neurotischer Entwicklung und psychosomatischen Krankheiten. Unsere Untersuchungen und Erfahrungen lassen keineswegs den Schluß zu, daß Kolitis-Kranke ein ungenügendes Introspektions- und Verbalisationsvermögen besäßen und die neurotische Fehlentwicklung eine irreversible Verfestigung besitze, die neue emotionale Haltungen nicht zulasse (Freyberger, 1976 und 1977), so daß eine interpretative Therapie, z.B. die

Man muß den Herrn bei Laune halten!

Abb. 44.1–5. 46jährige Patientin F. L. mit Colitis ulcerosa: „Man muß den Herrn bei Laune halten" (4. 11. 86).

Psychoanalyse, nicht in Frage komme (Freyberger, 1972 und 1976).

Die psychoanalytische Therapie kann vielmehr unter der Voraussetzung einer sorgfältig gestellten Indikation eine wichtige und erfolgreiche Hilfe sein und dazu beitragen, weiteren Schüben vorzubeugen. Eine der notwendigen Modifikationen kann darin bestehen, stets auf die besonderen Schwierigkeiten des Kolitis-Kranken, Trennungen zu ertragen, Rücksicht zu nehmen. Oft werden bereits sorgfältig abgewogene und behutsame Deutungen als kränkende Ablehnung empfunden. Die Behandlungstechnik besitzt Ähnlichkeiten mit dem zweistufigen therapeutischen Verfahren bei einer Grundstörung (Schöttler, 1981), bei dem eine stützende Therapie mit dem Ziel der Ich-Stärkung der psychoanalytischen Behandlung vorausgeht.

Das Ziel dieses tiefenpsychologisch-therapeutischen Anteils ist die adäquate Verarbeitung innerer Konflikte mit Abbau der Immobilität, Passivität, Depressivität und der aggressiven angstbesetzten Gehemmtheit sowie Korrektur narzißtischer Verhaltensmerkmale und struktureller Ich-Störungen (de Boor, 1964; Chambers und Rosenbaum, 1953; O'Connor et al., 1964, Daniels, 1948; Karush und Daniels, 1953; Karush et al., 1977; Knölker, 1986; Sperling, 1946; Weinstock, 1962).

Nach unseren Erfahrungen (Wilke, 1983 a und b) ist die Tagtraumtechnik des Katathymen Bilderlebens (Freiwald et al., 1975; Leuner, 1970 und 1978) mit ihrer regressionsfördernden Entspannung und der tiefenpsychologischen Aufarbeitung bisher unbewußt gebliebener oder abgewehrter Konflikte auch bei Patienten mit Colitis ulcerosa gut anwendbar. Der Tagtraum bietet dem Patienten ein weites Feld der Phantasie, auf dem er im Sinne einer Probehandlung regressiv-abwehrend und vorsichtig konfliktmobilisierend agieren kann. Er „phantasiert" dem Bewußtsein

bisher nicht zugängliche psychische Inhalte und erlebt sie so – bei behutsamer und einfühlsamer Einleitung durch den Therapeuten – nicht als angstbesetzt. Voreilige Deutungen und Konfrontationen durch den Therapeuten sollten in diesem Prozeß vermieden werden, um die Grenze der augenblicklichen Belastbarkeit des Patienten nicht zu überschreiten. Mehr als bei der psychoanalytischen Psychotherapie gilt es, die anerkennenden, ermunternden, schützenden Impulse flexibel einzusetzen, ohne daß die Entwicklung einer Übertragungsneurose blockiert wird.

Auch schwerkranke Kolitis-Patienten sind entgegen der im Alexithymiekonzept beschriebenen Phantasiearmut zum Erleben aus der Phantasie entwickelter Bilder und damit verbundener Emotionen fähig.

Wilke (1980) beobachtete in einer kontrollierten Studie bemerkenswerte Wechselwirkungen: Eine regressive Dynamik wirkte sich im Augenblick der Bewegung spannungsmindernd und günstig auf den Heilungsprozeß aus. Das Verharren in der Regression erzeugte dann nach einer Weile erneut ein Spannungsgefühl mit negativer Auswirkung auf den Krankheitsprozeß. Ebenso wie bei anderen Therapieformen gefährdete eine allzu forcierte Konfliktmobilisierung den Heilungsablauf, während eine der individuellen Reaktionsweise angepaßte psychische Progression keine körperliche Belastung darstellte. Die Progression wiederum schien eine Voraussetzung für eine dauerhafte somatische Remission zu sein. Die Dynamik im Katathymen Bilderleben paßte sich der körperlichen Belastbarkeit an, sofern eine schützende Übertragung bestand, die wiederum eine Voraussetzung für eine anhaltende Besserung war.

Die initiale Regression mit stets körperlicher Entlastung, Besserung körperlicher Symptome und klinischer Befunde verband sich bei vielen Patienten mit einer allmählich wachsenden Stärkung des Durchsetzungsvermögens, das durch Selbstvertrauen und realistischen, auch aggressiven Umgang mit der Umwelt gekennzeichnet war. Die tiefgreifende Wirkung dieser Behandlungsform ließ sich auch in Katamnesen nach 5 Jahren bestätigen (Wilke, 1990). Ergebnisse und Nachuntersuchungen zeigen, daß diese Kombinationsform analytischer Psychotherapie in der Hand des erfahrenen Therapeuten, der die möglichen Risiken abzuwägen versteht, ein wichtiger Baustein in unserem Therapiekonzept ist.

Auch die tiefenpsychologisch fundierte Musiktherapie (Maler, 1989) erbrachte nach ersten Erfahrungen Hinweise auf psychodynamisch mitwirkende Konfliktanteile, vor allem in der Beziehung zur Mutter. Im Vergleich zu den Ausdrucksformen in der Musiktherapie gegenüber Patienten mit Magersucht und Bulimie boten die Patienten mit Colitis ulcerosa Zeichen kindlicher Hilflosigkeit und Ängste vor Trennung und Verlust, woraus sich ein begründeter Anhalt für die Wirksamkeit solcher Erlebnisse und Erfahrungen in der frühen Kindheit ableiten läßt.

44.1.5.8 Therapie im Intervall

Eine der wichtigsten Aufgaben innerhalb der Therapie des Kolitis-Kranken liegt in der Nachbehandlung nach akutem Schub. Unsere Erfahrung spricht dafür, daß die Rückkehr zu den Anforderungen des Alltages, zu Familie und Beruf, eine kritische Phase im Krankheitsverlauf ist. Während dieses Überganges und auch nach gelungener Integration sind innerhalb der gesamten Zeit des Intervalls, d.h. selbst über 1–2 Jahre hinweg, unverändert einfühlendes Verständnis und ständige therapeutische Bereitschaft nötig, die als eine Gemeinschaftsaufgabe des bisher behandelnden Arztes und Psychotherapeuten, des Allgemeinarztes und der nächsten Bezugspersonen verstanden werden sollte (siehe auch Liedtke et al., 1972 und 1977; Feiereis, 1985b). Ein mühsam erreichter Behandlungserfolg kann schnell zunichte werden, wenn der Patient unvorbereitet oder forciert mit Belastungen konfrontiert wird, denen er noch nicht gewachsen ist. Die einzelnen Komponenten der Nachbehandlung sind:
- Regelmäßige Absprache über Art und Dosis der medikamentösen Langzeittherapie.
- Fortsetzung der erlernten Entspannungsübungen allein und von Zeit zu Zeit in Gruppen (z.B. einmal wöchentlich oder vierzehntägig).
- Fortsetzung der stützenden Psychotherapie mit ständiger Verfügbarkeit, auch telefonisch und brieflich, bei auswärtigen und berufstätigen Patienten auch am Wochenende. Das Ziel liegt in der Ich-Stärkung des Patienten, der Minderung seiner Ängste und Förderung seiner Autonomie mit Hilfe einer stabilen Arzt-Patient-Beziehung (Feiereis, 1985a).
- Fortsetzung und Abschluß der tiefenpsychologischen Psychotherapie.
- Regelmäßige indirekte Nachuntersuchungen (somatischer Status, Laborbefunde) und strenge Indikation für die oft als sehr belastend empfundenen

endoskopischen und röntgenologischen Kontrollen. Sie müssen auf die Struktur des Patienten und auf Stadium und Schwere seiner Krankheit abgestimmt sein.

44.1.6 Prognose und Ergebnisse

Folgende Faktoren bestimmen die Prognose der Colitis ulcerosa:
- Alter bei Beginn der Erkrankung,
- Schweregrad,
- Ausdehnung der Entzündung,
- Krankheitsdauer,
- Verlaufsform,
- prämorbide psychische Struktur und Psychodynamik,
- reaktive psychische Veränderungen und psychosoziale Folgen der Krankheit,
- Art, Lokalisation, Schwere von Komplikationen,
- Anzahl der Rezidive,
- Therapieform,
- therapeutische Ansprechbarkeit des Kranken und der Krankheit.

Manche Autoren (Banks et al., 1957; Bockus et al., 1956; Deucher, 1955) halten die Prognose trotz der gegenüber früheren Jahrzehnten differenten medikamentösen Therapie unverändert für fragwürdig. Die meisten stimmen darin überein, daß die höchste Letalität innerhalb des ersten Jahres nach erster stationärer Krankenhausaufnahme eintritt (Carleson et al., 1962; Demling et al., 1969; Edwards und Truelove, 1963/1964). Bargen und Mitarbeiter (1954) stellten eine Überlebenschance des Kolitis-Kranken 25 Jahre nach in der Klinik gestellter Diagnose mit 58% gegenüber 79% der Durchschnittsbevölkerung fest. Andere Autoren (Edwards und Truelove, 1963/1964) beobachteten eine Letalität von 22% innerhalb von 10 Jahren und 40% nach 20 Jahren, d.h., sie liegt wesentlich höher als in einer vergleichbaren Durchschnittsbevölkerung.

Die Berechnung der kumulativen Mortalität (Demling et al., 1969; Edwards und Truelove, 1963/1964) ergab ebenfalls eine ungünstige Prognose bei zwei verschiedenen Kollektiven. So waren bei einer Patientengruppe 9,7% innerhalb des ersten Jahres und nahezu 25% nach 14 Jahren verstorben.

Nach einigen Autoren sinkt die Anzahl der rezidivfreien Patienten zunehmend mit der Anzahl der Beobachtungsjahre (Weinstock, 1962). Wertvoll für die Beurteilung der Prognose und besonders der Rezidivhäufigkeit ist die Einteilung des Krankengutes nach Patientenjahren, d.h. nach der Anzahl der kontrollierten Jahre im Anschluß an den ersten Krankheitsschub (Goligher et al., 1968; Watts et al., 1966). Hiernach zeigte sich in einem englischen Krankenkollektiv, daß ein Rezidiv bei zwei Drittel der Patienten bereits im folgenden Jahre auftrat, eine Häufigkeit, die wesentlich größer als in unserem Krankengut ist.

Als besonders ungünstig wurde bisher die Prognose bei Kindern angesehen. Dies betrifft den Verlauf der

Krankheit, die Gefahren durch Komplikationen und vor allem die Karzinomgefährdung (Lagercrantz, 1960). In einer anderen Verlaufsstudie (Korelitz et al., 1962) wurde festgestellt, daß von 84 Kindern ein Drittel verstorben war, ein Drittel eine Ileostomie hatte und nur 15% ein günstiges Behandlungsergebnis aufwiesen.

Trotz mancher gegenteiliger Behauptungen konnten wir keine strenge Abhängigkeit des Behandlungsergebnisses und der Prognose von der Krankheitsdauer, der Lokalisation und der Verlaufsform feststellen. Auch ließ sich bei Patienten gleichen Schweregrades selbst nach mehrjähriger, teilweise bis 15 Jahre und länger zurückreichender Krankheitsdauer noch ein günstiges Behandlungsergebnis erreichen (Feiereis, 1980).

Neben den akuten Komplikationen wird von vielen Autoren die Karzinomgefährdung als besonders wichtiger Faktor für die Beurteilung der Prognose angesehen. Aus unseren Feststellungen innerhalb der vergangenen 40 Jahre ist abzuleiten, daß zwar die Karzinomgefahr bei einer Dauer von länger als 10 bis 15 Jahren zunimmt, jedoch keineswegs in solchem Maße, daß hieraus eine Operationsindikation als Karzinomprophylaxe abzuleiten wäre.

Hingegen stellt sich die Frage der Operationsindikation bei schweren Dysplasien, besonders wenn die Colitis ulcerosa seit dem Kindesalter besteht und das gesamte Kolon befallen ist. In dieser Krankheitsgruppe steigt das Entartungsrisiko nach einem Krankheitsverlauf von länger als 15 Jahren zunehmend an. Binder (1988) beschreibt hingegen das Risiko karzinomatöser Entartung nach 18 Jahren mit 1,4%, d.h. wesentlich geringer als in den bisher vorliegenden Berichten.

Faßt man die Kurzzeitprognose unseres Kollektivs zusammen, d.h. des einzelnen Schubes, so ergibt sich bei den konservativ behandelten Patienten eine Symptomfreiheit von durchschnittlich 11 Monaten pro Jahr, sofern die Krankheit zu Rezidiven neigte und nicht mit der ersten Behandlung abheilte.

Maßgebend für die Langzeitprognose und damit gleichzeitig für das Behandlungsergebnis sind Langzeitkatamnesen und Verlaufsbeobachtungen (Bötticher, 1977; Filler und Schwemmle, 1976; Rohrmeier, 1982). Die Bewertung der Ergebnisse gründet sich auf folgende Kriterien der Nachuntersuchung: Endoskopie, Biopsie, röntgenologischen Befund, Stuhlhäufigkeit und -beschaffenheit, Laborwerte, Gewicht, Nachbehandlung, Leistungsfähigkeit.

Bei den Behandlungsergebnissen ist zwischen 4 Heilungsgraden zu unterscheiden:
Heilungsgrad 1: Normalisierung aller klinischen Befunde, keine Nachbehandlung, volle Leistungsfähigkeit, insbesondere normaler histologischer Befund; von einer anhaltenden Abheilung der Kolitis sollte man erst sprechen, wenn der Heilungsgrad 1 rezidivfrei über mindestens 5 Jahre besteht.
Heilungsgrad 2: befriedigendes Behandlungsergebnis mit bioptisch rückläufiger Entzündung oder Restbefund bei sonst normalen Nachuntersuchungsergebnissen.

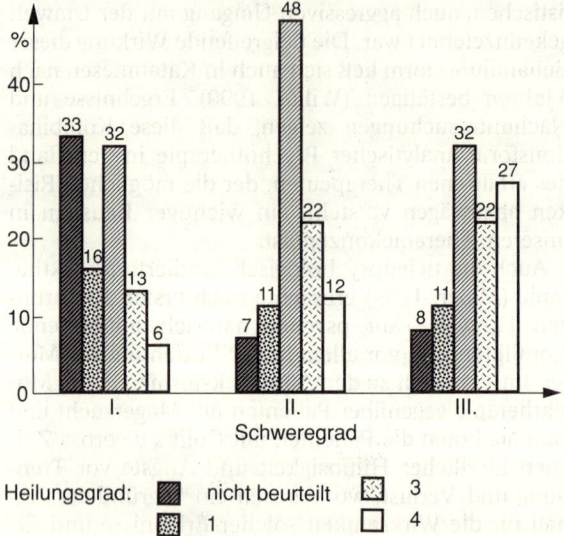

Abb. 44.1–6. Prozentuale Aufteilung der 4 Heilungsgrade bei 443 nachuntersuchten Kolitis-Kranken mit Schweregrad I–III.

Heilungsgrad 3: unbefriedigendes Behandlungsergebnis.
Heilungsgrad 4: bisher unbeeinflußter Verlauf.

Unsere Nachuntersuchungen bei 443 Kolitis-Kranken, die zwischen 1948–1972 bei uns waren, ergaben bei 49% (Heilungsgrad 1 und 2) einen günstigen Befund; bei 15%, überwiegend des Schweregrades III, lag ein bisher unbeeinflußter Verlauf vor (Abb. 44.1–6). Die Mehrzahl der Patienten mit Schweregrad I und II erreichte einen Heilungsgrad 1 und 2, hingegen beim Schweregrad III nur etwa die Hälfte der nachuntersuchten Kranken. Bei 17%, ganz überwiegend Patienten mit leichtem Verlauf, konnten Nachuntersuchungsergebnisse nicht gewonnen werden (Schröter, 1977). Die Letalität infolge der Kolitis oder ihrer Komplikationen belief sich auf 2,9%.

In einer weiteren Studie (Feiereis und Wetzel, 1989) wurde der Langzeitverlauf bei 268 Patienten mit dem Schweregrad III untersucht. Im Mittelpunkt stand die Frage nach Karzinomhäufigkeit, Operationsfrequenz, Letalität und sozialmedizinischen Auswirkungen. Die mittlere Beobachtungsdauer lag bei 12,2 Jahren und variierte von 0–33 Beobachtungsjahren. Bei 10% der Patienten war eine totale Kolektomie, bei 2% eine Hemikolektomie und bei 4% die Anlage eines Anus praeter bei belassenem Kolon notwendig geworden. Nach vorher schwerstem Krankheitsverlauf war die Operation von den Patienten als Heilung empfunden worden; sie konnten anschließend ein normales Leben führen. Den Heilungsgrad (Abb. 44.1–7) beurteilten wir für alle Kolitis-Kranken, ausgenommen diejenigen, die am Kolon operiert worden waren. Heilungsgrad 1 erreichten 7%, Heilungsgrad 2 43%. Somit heilte bei der Hälfte der Patienten die Kolitis ab, d.h. nach der stationären Therapie be-

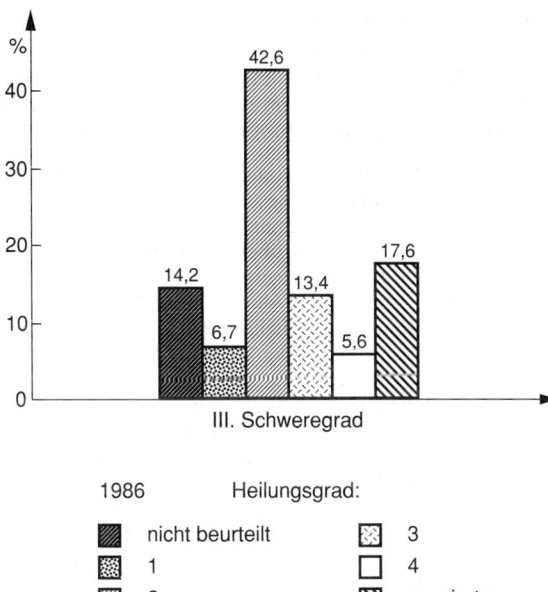

1986   Heilungsgrad:

▨ nicht beurteilt   ▧ 3
▨ 1   ☐ 4
▨ 2   ▧ operierte
                        Patienten

Abb. 44.1–7. Langzeitkatamnese von 268 Kolitis-Patienten des Schweregrades III nach mittlerer Beobachtungsdauer von 12,2 Jahren: Prozentuale Aufteilung der vier Heilungsgrade.

stand z. T. über 10 bis 15 Jahre hinweg keine Behandlungsbedürftigkeit mehr. Heilungsgrad 3 wiesen 13%, Heilungsgrad 4 6% auf.

Die sozialmedizinischen Auswirkungen untersuchten wir am Beispiel der Berufstätigkeit aller 180 Patienten, die unter 60 Jahre alt waren. Unverändert in ihrem Beruf waren 64% tätig, verkürzte Arbeitszeiten gaben 6% an, umgeschult wurden 3%. Zeitrenten erhielten 8%, Dauerrenten 3%. 22% dieser Patienten waren Hausfrauen, 12% Akademiker, 44% in einer Lehrlingsausbildung, 8% ohne erlernten Beruf und 3% Schüler bzw. Studenten. Zusammenfassend läßt sich feststellen, daß zwei Drittel der Patienten unverändert berufstätig sind und zusätzlich ein weiterer Teil mit verkürzter Arbeitszeit oder nach Umschulung tätig ist.

Nur ein Viertel der Patienten wurde noch fortlaufend mit Salazosulfapyridin bzw. Mesalazin behandelt, 6% mit Kortikosteroiden.

Eine kolitisbedingte Letalität bestand in knapp 15% bei dieser Gruppe mit Schweregrad III (postoperativ 4%; wegen eines Kolonkarzinoms 3%; wegen anderer kolitisbedingter Komplikationen 8%).

Anhand der kumulativen Mortalität konnte festgestellt werden, daß sich die Überlebenswahrscheinlichkeit für Kolitis-Patienten unter der kombinierten konservativen Therapie verbessert hat, was wir zu einem Teil auf die kontinuierliche Langzeitbehandlung der meisten unserer Patienten über viele Jahre hinweg unter Einschluß psychotherapeutischer und sozialtherapeutischer Maßnahmen zurückführen möchten.

Nach den von uns festgestellten Ergebnissen halten wir den Schluß für berechtigt, daß die kombinierte konservative Therapie eine günstigere Prognose erlaubt, als bisher häufig angenommen wird. Selten

wird sie bisher so extrem negativ beurteilt wie von einem Autor, der seine Erfahrungen in einer großen chirurgischen Klinik ebenso schlicht wie lapidar in drei Sätzen zusammenfaßte, ohne sie mit einem einzigen weiteren Satz zu begründen (Med. Klinik 68 [1973] 170): „Ein Drittel dieser Patienten ist für die konservative Therapie primär resistent. Ein weiteres Drittel kommt zum Chirurgen, weil immer wieder Rezidive entstehen. Das letzte Drittel schließlich kommt zu uns nach jahrelanger konservativer Behandlung, meist in kachektischem Zustand, hochgradig anämisch, eiweißverarmt und mit lokalen Komplikationen".

44.1.7 Epikrise und Katamnese zur einleitenden Krankengeschichte

Der Patient kam vor 11 Jahren körperlich und psychisch desolat zu uns. Leibschmerzen, blutige Stühle, Appetitlosigkeit, Schwäche, Hinfälligkeit und Schlaflosigkeit waren neben den Entzündungszeichen die hervorstechenden körperlichen Symptome. Depressivität, Resignation, Antriebsarmut, Apathie, Mattigkeit, Verzweiflung, Desinteresse an der Umwelt, Bedürfnis nach Alleinsein prägten das psychische Bild.

Innerhalb der siebenwöchigen stationären Therapie bildete sich die Entzündung zurück, der Stuhl war zuletzt nur noch blutig tingiert, Blutsenkungsgeschwindigkeit und Elektrophoresediagramm hatten sich weitgehend gebessert, die Anämie war beseitigt.

Unter der Tiefenentspannung, ferner mit Hilfe der stützenden Gesprächspsychotherapie sowie dosierter Bewegungstherapie, Atembehandlung und Entspannungsübungen in der Gruppe hellte auch die Stimmung auf, je mehr sich die körperliche Besserung stabilisierte. Der Patient konnte erstmals Kontakte zu Mitpatienten aufnehmen und ertragen, sich wieder den Ereignissen der Umwelt zuwenden, schließlich andere ermutigen, trotz jahrelanger Krankheit nicht aufzugeben.

Nachdem die Remission gefestigt erschien, bildete die tiefenpsychologische Therapie den Schwerpunkt der weiteren ambulanten Behandlung. Hierbei geschah innerhalb der Übertragungsneurose eine nahezu katharthische Befreiung. Dem Patienten wurde mehr und mehr bewußt, daß die bisher unbewußt gebliebene Scheinlösung seines intrapsychischen Konfliktes in ständiger Anpassung und ihn überfordernder Leistung bestanden hatte, um seine Ängste vor dem Verlust der ohnehin nur geringen Zuwendung, die er erfahren hatte und gegenwärtig erfuhr, abzuwehren. So wie seine Familie in seiner Kindheit in einer Enklave gelebt hatte und er um so mehr voller Ängste aufwuchs, als er auch innerhalb dieser Enklave weder konstante Zuwendung noch Geborgenheit erlebte, d.h. sich doppelt abhängig fühlte und die Angst vor dem Verlust selbst des geringsten Schutzes Bestandteil seines Selbst wurde, so war für ihn auch das spätere Dasein eine Wiederholung des in der Kindheit Erlebten: Die Objekte der Umwelt repräsentierten das negative Elternbild, seine Abhängigkeitszwänge und Verlustängste verstärkten sein Leistungsstreben; dessen mangelhafte Anerkennung wiederum erhöhte den Schmerz durch Abhängigkeit und Furcht vor Verlust. Gesellschaftliche Kontakte bestanden so gut wie gar nicht, die Familie lebte ebenfalls wie in einer Enklave.

Über den Weg der therapeutischen Regression und der symbiotischen Verflechtung mit der „therapeutischen Gruppe" erlebte der Patient den Abbau dieser Ängste, die das Ergebnis seiner Prägung und Erziehung gewesen sind. Aus Mißtrauen entwickelte sich Vertrauen, aus Vertrauen zunehmend Ich-Stärke, Streben nach Autonomie, Abbau der gehemmten Aggressivität und Gefühl für den Wert seines Selbst. Der Patient vermochte z.B. erstmals frei in einer Gruppe zu sprechen („ich hatte das Gefühl, eine kleine Heldentat vollbracht zu haben"), oppositionelle Einstellungen zu verteidigen, ohne Angst vor Sympathieverlust.

Auch die sozialmedizinischen Auswirkungen dieser Entwicklung waren nicht gering: Den ihm von der Behörde nahegelegten Rat, einen Rentenantrag zu stellen, um einem Jüngeren Platz zu machen, lehnte er ab, während er vorher meinte, diesem Rat natürlich ebenso folgen zu müssen wie sonst einer behördlichen Anordnung.

Seine Ehefrau, die er lange Zeit als „sachlich, frostig, nüchtern" erfahren hatte, mehr neben ihm, als mit ihm lebend, seit seiner Krankheit und Impotenz noch distanzierter, bestärkte ihn endlich in seiner Entwicklung, als sie nicht mehr die kühle, beschützende Mutter sein mußte. Eine wesentliche Ich-stärkende Bedeutung besaß schließlich auch die Wiederkehr der Potenz.

Die Kolitis blieb bis auf einen geringen Rückfall vor 7 Jahren rezidivfrei. Regelmäßige Nachuntersuchungen bestätigten die anhaltende körperliche und psychische Stabilität des nunmehr 59jährigen Mannes, der, wie er sagte, innerhalb der letzten 14 Jahre „ein anderer Mensch" geworden war.[1]

44.1.8 Zusammenfassung des diagnostischen und therapeutischen Weges

– Der psychosomatisch orientierte, gastroenterologisch und psychotherapeutisch tätige Arzt – innerhalb einer Arbeitsgruppe oder in Personalunion – bietet dem Kolitis-Kranken die Möglichkeit, ihn über den gesamten Krankheitsverlauf in einer verläßlichen und stabilen Beziehung zu begleiten. Nach Engel (1979) liegt in dieser Beziehung die Übernahme einer Rolle der bisherigen Schlüsselperson, deren phantasierter, drohender oder realer Verlust sich auf die Entwicklung und den Verlauf der Krankheit schädigend ausgewirkt hat. In der Übernahme dieser für den Patienten zentralen Funktion durch den Therapeuten liegen Chance und Gefahr zugleich: Die Chance, somatisch und psychotherapeutisch helfen zu können, aber auch

die Gefahr eines Rückschlages, wenn aus der Konstanz Inkonstanz wird und der Patient sich in seinen Erwartungen und Hoffnungen erneut enttäuscht erlebt. Auf diesem diagnostischen und therapeutischen Wege sind erste wichtige Schritte:

– Eingehende Anamnese der Krankheitsentwicklung einschließlich der Analyse auslösender Ereignisse.
– Objektivierung der Anamnese durch Beiziehung **aller** verfügbaren Unterlagen über bisher erfolgte Untersuchungen und Behandlungen.
– Ergänzende Diagnostik durch körperliche Untersuchungen, Kontrolle der Laborwerte, Sonographie, tägliche Stuhlkontrolle bei der Visite mit Registrierung der Anzahl der Stuhlentleerungen, der Form der Stühle und der Beimengungen (Blut? Eiter? Schleim?).
– Apparative Diagnostik: Sorgfältige Indikation! Eine endoskopische und histologische Untersuchung z.B. ist in der Regel nur dann notwendig, wenn die vorausgegangene Untersuchung mehr als ein halbes Jahr zurückliegt und dieser vorausgegangene Befund nicht durch eine gute Dokumentation überzeugend wirkt. Auch die Indikation zu einer radiologischen Untersuchung ist gründlich, vor allem unter Berücksichtigung der körperlichen und psychischen Belastbarkeit des Patienten, abzuwägen.

– Es folgt die eingehende biographische, tiefenpsychologisch fundierte und psychosoziale Anamnese. Sie bezieht sich besonders auf die Fragen nach genetisch determinierten psychopathologischen Dispositionsfaktoren; prämorbider Persönlichkeitsstruktur; Hinweisen auf eine neurotische Entwicklung; belastenden Auslösungsfaktoren; psychosozialen Konflikten als krankheitsverschlimmernden Faktoren; reaktiven psychischen Befunden; psychosozialen Folgen der Krankheit.
– Im Gespräch zu Diagnose und Therapie (Feiereis, 1985a) sollen belastende Informationen vermieden werden, vor allem, wenn sie den Patienten ängstigen und für die Therapie unerheblich sind (z.B. detaillierte Beschreibung möglicher Komplikationen, die ihm nicht drohen). Das Gespräch über die Therapie erstreckt sich auf die Einzelheiten der medikamentösen und diätetischen Maßnahmen ebenso wie der psychotherapeutischen Verfahren. Für diese ist nicht der „Wunsch" maßgebendes Kriterium, sondern die den Patienten überzeugende Darlegung der Indikation. Mit dem Gespräch wird zugleich das Ziel einer positiven Übertragung erreicht.

---

1 Wie wenig entgegen unseren Erfahrungen der psychosomatische Anteil in die Diagnostik und Therapie der Colitis ulcerosa einbezogen wird, mag folgendes Zitat aus dem Brief eines namhaften Gastroenterologen belegen (1981):
„Lange Zeit wurde die Colitis ulcerosa als psychosomatische Erkrankung aufgefaßt. Die diesbezüglichen Therapieergebnisse waren mehr als enttäuschend. Ich würde dringend abraten, Patienten mit einer Colitis ulcerosa in eine psychosomatische Klinik zu schicken, da auf diese Weise der organische Prozeß in keiner Weise beeinflußt werden kann, man unter Umständen notwendige Behandlungen unterläßt und unter Umständen sogar ein akutes, lebensbedrohliches Zustandsbild falsch interpretiert."
Diese Stellungnahme apostrophierte er als den „heute allgemein gültigen Gesichtspunkt in der Gastroenterologie"!

– Zur psychosomatischen Initialbehandlung gehören Entspannungsübungen (autogenes Training, vertiefte Entspannung, progressive Relaxation) sowie die stützende Gesprächspsychotherapie mit Förderung der Bedürfnisse nach Regression und mit Abbau der Gefühle von Hilf- und Hoffnungslosigkeit. Inhalte dieser stützenden Therapie sind besonders auslösende oder die Krankheit verschlimmernde emotionale Belastungen, die reaktive, sekundäre psychische Symptomatologie und soziale Folgen.

– Den therapeutischen Zugang zu psychodynamisch wirksamen Anteilen der Krankheit ermöglicht die tiefenpsychologisch fundierte Psychotherapie mit dem Ziel, neurotische Entwicklungen und deren Wechselwirkung im Krankheitsprozeß zu beeinflussen.

– Nach der Entlassung aus der stationären Therapie liegt ein weiterer entscheidender Schritt in der kontinuierlichen ambulanten synchronen Fortsetzung der somatischen und psychotherapeutischen Behandlung. Hierzu gehört auch die begleitende Beratung bei sozialmedizinischen Fragen (Arbeitsunfähigkeit; Schulschwierigkeiten; Umschulung; Heilverfahren; Zeitrente; Krankheitsbewertung nach dem Schwerbehindertengesetz).

– Die für die gesamte Therapie Verantwortlichen („therapeutische Gruppe") erfüllen ihre Schlüsselfunktion auch in der Remission der Krankheit und wiederum verstärkt mit Beginn eines Rezidivs. Erst nach Abheilung der Krankheit findet diese besonders intensive Beziehung ihr Ende.

44.1.9 Schlußfolgerung

– Die Colitis ulcerosa des Schweregrades I heilt unter der konservativen Therapie bei den weitaus meisten Patienten ab; die Prognose ist ebenso günstig, wenn Rezidive auftreten, die in der Regel wiederum therapeutisch gut beeinflußbar sind. Oft genügt eine ambulante Behandlung. Der Patient ist in seiner Leistungsfähigkeit meistens nicht eingeschränkt.

– Die Colitis ulcerosa der Schweregrade II und besonders III bedarf in der Regel einer mehrwöchigen stationären Therapie. Treten keine ernsten Komplikationen auf und kann eine ambulante Behandlung fortgesetzt werden, die auf die individuellen Gegebenheiten abgestimmt ist und sich mitunter über 1–2 Jahre erstrecken muß, so ist die Prognose, bezogen auf die Leistungsfähigkeit der Patienten und auch auf die Lebenserwartung, ermutigender, als andere Statistiken aussagen. Allerdings ist für eine gut begründete Beurteilung die Längsschnittbeobachtung über einen jahrelangen Zeitraum unerläßlich, da die Art der Krankheit mit ihrer vielgestaltigen Pathogenese und dem oft zunächst nicht vorhersehbaren Verlauf zur fortlaufenden Überprüfung der Aussagen zwingt. Dies ist allein durch Langzeitkatamnesen gewährleistet, für die exakte Nachuntersuchungen einschließlich endoskopischer und histologischer Dokumentation der adäquate Maßstab sind.

44.2 Morbus Crohn

Hubert Feiereis

Die 17jährige Patientin D. J. leidet seit 2 Jahren unter wechselnd starken Leibschmerzen. Der Schmerz wird als stechend bezeichnet und besonders in der rechten Seite empfunden. Die Schmerzattacken kommen ca. 20- bis 30mal täglich, mitunter auch als Dauerschmerz; oft Übelkeit, schließlich auch Magenschmerzen und häufig Erbrechen, „überwiegend bei Ärger". Sie ißt daher wenig und hat ca. 8 kg abgenommen. Stuhl breiig, in der letzten Zeit wäßrig, selten etwas blutig.

Befund (auswärts erhoben): Reduzierter Ernährungs- und Allgemeinzustand, 163 cm, 43 kg. Blässe. Leib gespannt. Im rechten Unterbauch war eine druckempfindliche Resistenz zu tasten.

Rektoskopisch kein krankhafter Befund. Die Magen-Darm-Passage zeigt keine sicheren Schleimhautveränderungen im Magen und im Duodenum, auch im Jejunum wird ein regelrechter Tonus und ein normales Relief festgestellt. Das terminale Ileum ist strangförmig gestreckt mit „flauer Anzeichnung", hier keine Peristaltik erkennbar, das Zökum wirkt erheblich geschrumpft.

BSG 27/44. Hb 108 g/l, Ery 3,9 T/l, Leuko 12,0 G/l. Im Differentialausstrich 2% Basophile, 3% Eosinophile, 12% Stabkernige, 43% Segmentkernige, 34% Lymphozyten, 6% Monozyten. Thrombozyten 520 G/l, Gesamteiweiß 62,0 g%, in der Elektrophorese Hypo- und Dysproteinämie, Transaminasen normal.

Von dem untersuchenden Gastroenterologen wird daraufhin (ohne histologische Sicherung) die Diagnose eines Morbus Crohn gestellt und dem Hausarzt empfohlen, Prednison sowie Salazosulfapyridin zu verordnen, zusätzlich Eisen und Vitamin B_{12}. Die Patientin solle Zucker meiden oder eine Elementardiät zu sich nehmen. „Zu überlegen wäre eine psychotherapeutische Mitbehandlung."

Die Patientin kommt 2 Jahre nach Auftreten der ersten Symptome erstmals ambulant zu uns. Sie sagt, daß sie sich schlecht fühle, antriebslos sei, sich abkapsele, obwohl sie seit Einnahme der Medikamente weniger Schmerzen habe. Sie leide auch unter Konzentrationsschwäche, ihre Gedanken seien „wie abgerissen", sie erlebe die Umwelt „wie einen Traum". Sie beobachte sich und ihr Handeln „wie von außen". Spontan äußert sie, daß sie in der letzten Zeit oft Selbstmordgedanken habe, „ich möchte die Eltern bestrafen", „ich glaube, meine Krankheit ist seelisch bedingt".

Bei unserer Koloskopie deutlich verengte Bauhinsche Klappe, eine Passage in das terminale Ileum gelingt nicht, erkennbar aber ist das narbig geschrumpfte terminale Ileum mit flachen Ulzerationen, die weißliche Fibrinbeläge zeigen. Im Zökum zahlreiche landkartenartig begrenzte Nekrosen, dazwischen einzelne kleinere rötliche Schleimhautinseln mit verdünnter atrophischer Schleimhaut. Übriges Kolon unauffällig. Histolo-

gisch im Zökum diskontinuierliche nonulzerative Entzündung mit Nachweis eines Mikrogranuloms, so daß in Verbindung mit allen anderen Befunden ein Morbus Crohn des terminalen Ileums und Zökums diagnostiziert wird.

Zur Vorgeschichte und biographischen Anamnese: Die Patientin hatte vor 3 Jahren eine starke Migräne mit stechenden halbseitigen Schläfenschmerzen und Flimmern vor den Augen. In diese Zeit fällt ein Suizidversuch mit Schlaftabletten. „Ich wollte mich von dem Schmerz befreien, und außerdem hat mein Vater die starken Migräneanfälle überhaupt nicht beachtet."

Die 47jährige Mutter litt seit einigen Jahren ebenfalls an einem Morbus Crohn, deshalb vor 2 Jahren Teilresektion der Ileozökalregion. Zur gleichen Zeit sei sie „nervenleidend", „depressiv" gewesen, habe „oft ein Kloßgefühl im Hals" empfunden und der Tochter gegenüber Selbstmordgedanken geäußert.

Der 64jährige Vater ist gesund, betreibt ein eigenes Geschäft. Nach zwei Scheidungen lebt er nunmehr in dritter Ehe, aus der die Patientin und ein 22jähriger Sohn stammen, der 1½ Jahre vorher nach einem Streit mit den Eltern in eine eigene Wohnung zog. „Er konnte sich immer gut durchsetzen."

Die Patientin wuchs in einem großen Dorf auf. „Ich denke gern an die Kindheit zurück." Sie habe sich sehr an die Mutter gebunden gefühlt, obwohl diese wegen eines eigenen Geschäftes wenig Zeit hatte. „Der Vater aber kümmerte sich gar nicht um mich. Er war immer autoritär, und wir alle mußten uns ihm unterordnen."

Sie besuchte die Grund- und Realschule und begann danach vor 1½ Jahren die Lehre als Bürogehilfin.

Ihr größtes Problem sei von jeher die schwierige Beziehung zum Vater, besonders auch während der letzten 2 Jahre. „Er kritisiert mich laufend, obwohl dafür keine hinreichenden Gründe vorhanden sind. Die Freunde kann ich mir nicht aussuchen, der Vater kontrolliert alle Kontakte. Wenn ich ausgehen will, macht er ironische Bemerkungen." Er meint häufig, sie habe keine Lust zur Arbeit; dies verursache ihm Magenbeschwerden. Die Patientin bezeichnet ihren Vater als „zerstörerisch".

Sie fühle sich nicht ernstgenommen, er ignoriere ihre Beschwerden, sie wiederum traue sich nicht zu, sich gegen ihren Vater zu wehren, aggressiv zu sein. Sie möchte von ihm geliebt werden und erlebe das Gegenteil.

Auch die Mutter verhalte sich anders als früher, auch sie äußere sich ironisch, wenn die Patientin Kontakt zu Gleichaltrigen habe oder aufnehme.

Zur wichtigsten Bezugsperson der Patientin wurde daher eine 8 Jahre ältere Frau, die sie vor 3 Jahren kennenlernte; damals war sie die Freundin ihres Bruders. Die Patientin beschreibt diese Freundin als Vorbild, „durch ihre Selbstsicherheit und vielseitigen Interessen". Sie stehe ihr bei, wenn sie Auseinandersetzungen mit den Eltern habe. Die Freundschaft zu ihr wurde noch intensiver, nachdem sich ihr Bruder und seine Freundin getrennt hatten.

Die Krankheitssymptome exazerbierten zu dem Zeitpunkt, als die Freundin aus beruflichen Gründen in eine 200 km entfernte Stadt umgezogen war.

Freundschaften zu Jungen waren demgegenüber für die Patientin bisher unverbindlich; nach kurzer Zeit brach sie die Beziehung ab, so auch zu ihrem ersten Freund, der ein Adoptivsohn der Eltern ihrer Freundin war. Verliebt habe sie sich noch nie; sie wünsche sich, wie sie sagt, möglichst einen Freund, der älter ist als sie.

Wir erleben die Patientin als freundlich-höflich, aber auch ängstlich-gespannt. Besonders auffallend ist das affektarme, formale, fast zwanghaft unbeteiligt wirkende Verhalten. Die Patientin benutzt oft stereotype Formulierungen. Selbst bei Schilderung sehr belastender Familienkonflikte zeigt sie wenig oder keine Emotionen, vor allem keine aggressiven Regungen. Allenfalls vermag sie rationalisierend Aggressions- und Haßgefühle anzudeuten, ohne sie je bewußt erlebt zu haben. Auffallend ist ihr psychologisierendes Sprechen über sich wie über eine dritte Person, z.B. „ich weiß, ich rede über mich, als wäre ich es überhaupt nicht..., ... einen Teil von mir habe ich schon lange begraben", nach einer Pause fortfahrend, „... der ist jetzt im Bauch und macht Schwierigkeiten".

44.2.1 Epidemiologie

Die Angaben über die Häufigkeit des Morbus Crohn differieren sehr, weil ihnen noch immer verschiedene

diagnostische Kriterien zugrunde liegen und die Zuordnung der Entzündung zur Enteritis granulomatosa regionalis Crohn oder zur Colitis ulcerosa nicht frei von unterschiedlicher Wertung der einzelnen Befunde ist. Dennoch zeigen die Untersuchungen vieler Autoren, daß die Inzidenz, also die jährliche Erkrankungsrate pro 100 000 Einwohner, während der letzten 20 Jahre angestiegen ist und derzeit etwa 4–9 beträgt, d.h. Inzidenz und Prävalenz etwa gleich häufig wie bei Colitis ulcerosa sind (Mendeloff, 1979, 1980; Küster und Lenz, 1984; Martini, 1988). Die Zunahme bei uns zeigt Abbildung 44.2–1. Im eigenen Krankengut überwog das weibliche Geschlecht (Abb. 44.2–2), manche Autoren beschreiben eine umgekehrte Verteilung. Mehr als die Hälfte unserer Patienten befand sich bei der Diagnose der Erkrankung zwischen dem 11. und 30. Lebensjahr (Abb. 44.2–2).

44.2.2 Klinik der Enteritis granulomatosa regionalis Crohn

44.2.2.1 Symptomatologie, Schweregrade und Verlaufsformen

Im Mittelpunkt der körperlichen Symptome stehen diffuse, mitunter auch lokalisierte abdominale Beschwerden. Die Krankheit kann dabei längere Zeit, mitunter jahrelang, infolge der uncharakteristisch anmutenden Schmerzen, die häufig mit Übelkeit und Brechreiz verbunden sind, maskiert sein. Einer

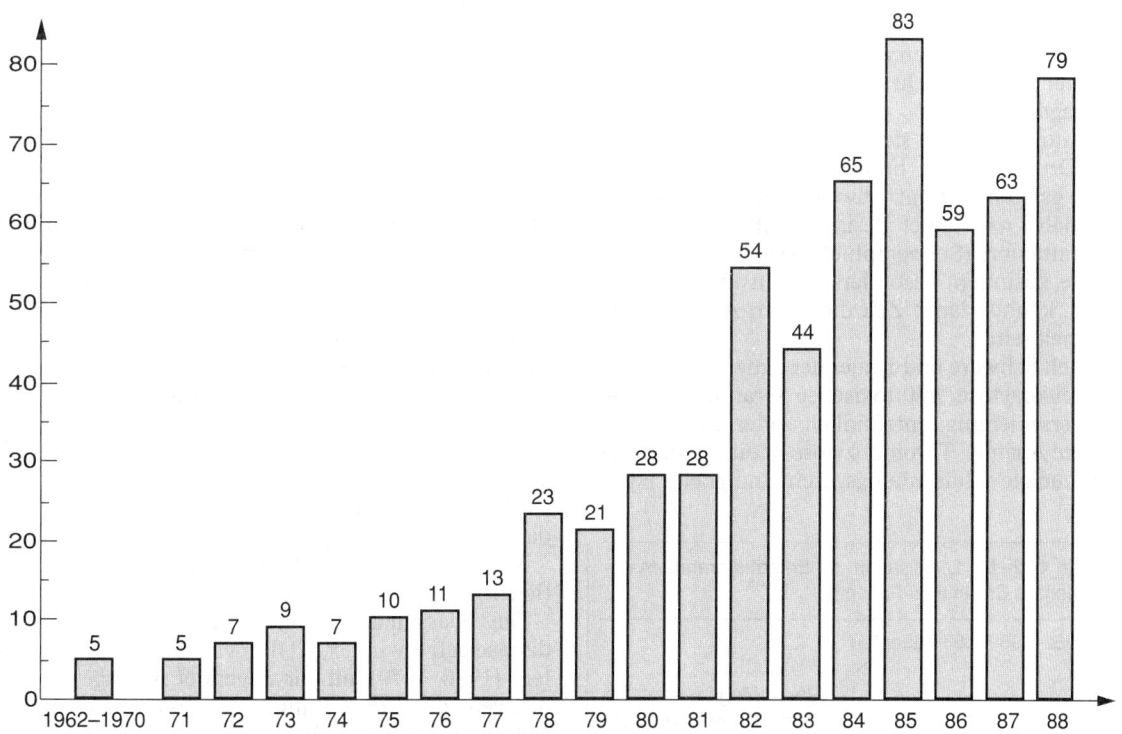

Abb. 44.2–1. Zunahme der stationären Einweisungen wegen Morbus Crohn innerhalb von 26 Jahren (1962–1988). n = 614 Patienten, 127 aus Lübeck = 21%, 487 von auswärts zugewiesen = 79%.

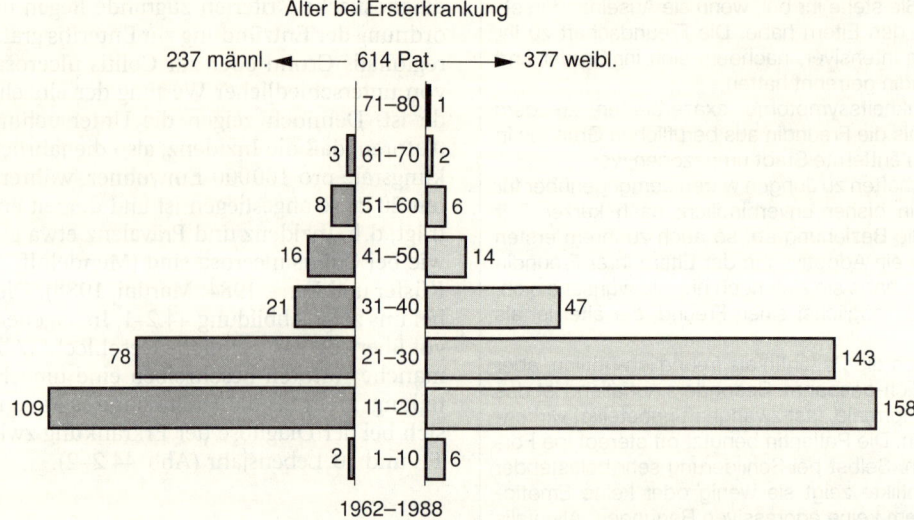

Abb. 44.2–2. Alters- und Geschlechtsverteilung bei 614 Patienten (237♂, 377♀) mit Enteritis granulomatosa regionalis Crohn.

manchmal kolikartigen Steigerung folgen Phasen leichterer oder auch fehlender ziehender oder druckartiger Beschwerden, die vorwiegend im rechten oder linken Unterbauch, aber auch im mittleren Oberbauch angegeben werden. Ebenso uncharakteristisch und abhängig von der Lokalisation der Krankheit sind die Stuhlveränderungen: Durchfälle mit Beimengung von Schleim und mehr oder weniger auch (wenngleich nicht obligat!) von Blut wechseln mit Phasen von Verstopfung oder normaler Stuhlentleerung, aber auch Blut ohne Stuhl.

Die Patienten haben oft subfebrile bis remittierende Temperaturen, nehmen ab, mitunter bis zu einem erheblichen Untergewicht zum Zeitpunkt der erstmaligen Diagnose.

Palpatorisch findet sich ein wechselnder abdominaler Druckschmerz, besonders im rechten Unterbauch, gelegentlich mit Abwehrspannung. Sorgfältig sollte dabei nach einer Resistenz infolge eines Konglomerattumors (Sonographie!) gesucht werden.

Anale Läsionen, besonders Fisteln und Abszedierungen, können lange Zeit die ersten Anzeichen der Krankheit sein.

Je nach Schwere und Dauer der Entzündung bestehen Leukozytose mit toxischer Granulierung und Linksverschiebung, entzündlich- oder eisenmangelbedingte Anämie, Thrombozytose (Zeitz et al., 1984), beschleunigte Blutsenkungsgeschwindigkeit, Hypal-

buminämie und Dysproteinämie, eventuell eine Hypokalzämie (Otte et al., 1983). Die Orosomukoidkonzentration und das C-reaktive Protein sind erhöht (André et al., 1980), das HL-Antigen-B27 häufig positiv. Die 25(=H)D$_3$-Werte liegen oft niedrig (Driscoll et al., 1982; Otte et al., 1983), dennoch war ein sekundärer Hyperparathyreoidismus nur ausnahmsweise nachweisbar. In 39% fanden sich in hohen Konzentrationen Autoantikörper gegen exokrines Pankreas im Serum (Stöcker et al., 1984b; Otte et al., 1989).

Die Krankheit kann entgegen der früher gebräuchlichsten Bezeichnung „Ileitis terminalis" an allen Stellen des Magen-Darm-Traktes, umschrieben oder ausgeprägt, lokalisiert sein, d.h. Mund und Ösophagus ebenso betreffen wie die Analregion, allerdings überwiegen weitaus die Befunde im unteren Dünndarm und im Dickdarm. Bei unseren Patienten fand sich die in Tabelle 44.2–1 aufgeschlüsselte Lokalisation. Das Rektum ist im Gegensatz zur Colitis ulcerosa oft frei von entzündlichen Veränderungen: Mekhjian und Mitarbeiter (1979) fanden eine Entzündung bei 14%, Malchow und Mitarbeiter (1978) sowie Malchow und Daiss (1984b) bei 21%; bei unseren Kranken sahen wir das Rektum bei 25% befallen.

Die Krankheit verläuft meistens chronisch rezidivierend, seltener chronisch kontinuierlich, gelegentlich akut oder subakut exazerbierend. Pathologischanatomisch unterscheidet man (Otto und Gebbers, 1982) das Frühstadium (ödematöse Schwellung) und das Spätstadium (Fibrose).

Man kann die Krankheit in die Schweregrade I (leicht), II (mittelschwer) und III (schwer) einteilen.

Die Klassifikation, nach den Indizes (Crohn's disease activity index, CDIA) von Best und Mitarbeitern (1976, 1979) mit vorwiegender Aussage über die Schwere der Erkrankung, oder von van Hees und Mitarbeitern (1980, 1984) vornehmlich über die Entzündungsaktivität, ermöglicht eine – wenngleich relativ grobe – Definition. Ein Index nach Best über 150

Tabelle 44.2–1. Lokalisation der Enteritis granulomatosa regionalis Crohn (665 Patienten)

1962–1989: 665 Patienten

20% Ileum	19% Kolon	Magen	1%
		Duodenum	1%
61% Ileum + Kolon		Jejunum	5%
		Rektum	25%

spricht für die Notwendigkeit einer Therapie, über 450 für einen schweren und sehr aktiven Verlauf.

44.2.2.2 Endoskopische und histologische Befunde

Die wichtigsten **makroskopischen** Befunde bei der Enteritis granulomatosa regionalis Crohn sind (Krauspe et al., 1972; Williams, 1979; Ottenjann, 1980; Müller-Wieland, 1982; Feiereis und Otte, 1983 a; Otte et al., 1989):
Diskontinuierliche, fleckige asymmetrische segmentale Entzündung („scip areas", „scip lesions"); umschriebene Nekrosen mit isoliert oder gruppenförmig angeordneten, aphthengleichen oder auch longitudinalen spaltförmigen, fissurähnlichen konfluierenden Ulzera; kopfsteinpflasterähnliches Relief („cobblestones") der Schleimhaut infolge scharfer Eingrenzung gesunder Schleimhautbezirke durch Hyperämie und Ödem; geringe Vulnerabilität, selten Kontaktblutung; geringe Sekretion von Schleim und Eiter; schließlich bei langem Verlauf Wandstarre, Wandverdickung, fibrotische Verkürzung, Einengung des Lumens bis zur Stenose und Striktur; Fistelbildung; Pseudopolyposis seltener als bei Colitis ulcerosa. Vereinzelt wird die „miliare" Form des M. Crohn beschrieben (Otto et al., 1975).

Die **histologischen** Befunde der Entzündung sind gekennzeichnet durch (Morson und Dawson, 1972; Morson, 1977; Oehlert, 1978; Schmitz-Moormann et al., 1979; Otto und Gebbers, 1982; Otte et al., 1989):
Erosionen, Ulzerationen, Ödem; fokale lymphoide Hyperplasie; transmurale diskontinuierliche und disproportionierte entzündliche, vorwiegend lymphoplasmazelluläre Infiltration; Persistenz der Becherzellen; gering veränderte Drüsenarchitektur, selten Kryptenabszesse; am wichtigsten und aussagekräftigsten, dennoch keineswegs pathognomonisch sind die epitheloidzelligen und eventuell Langhanssche Riesenzellen enthaltenden Granulome (50–70%, Otto et al., 1980; 25%, Otte et al., 1989), besonders in der Submukosa und in den vergrößerten regionalen Lymphknoten des Mesenteriums und Mesokolons; schließlich Verdickung der Mukosa und Submukosa; Fibrosierung aller Wandabschnitte bis zur Subserosa und Serosa.

44.2.2.3 Röntgenologische Befunde

Die beiden wichtigsten Untersuchungen sind die orale Doppelkontrastuntersuchung nach Sellink (1976) und die retrograde Darstellung des Kolons, ebenfalls im Doppelkontrast, nach Welin und Welin (1980). Die pathomorphologischen Befunde bestehen aus granuliertem marmoriertem Relief, Lymphfollikelhyperplasie, scharf begrenzter Vorwölbung der Schleimhaut durch entzündliche Infiltration und

Ödem („Pflastersteinrelief"), flachen, rundlichen bis longitudinalen, rinnenförmigen aphthoiden Ulzera.

Im Reparationsstadium imponieren Haustrenverlust und pseudopolypöse Umwandlung unter Bildung zum Teil einzelner großer Pseudopolypen. Mitunter sieht man eine pseudodivertikelartige Veränderung. Die Fibrose kann zur asymmetrischen Stenose bis auf Stricknadeldicke mit prästenotischer Dilatation führen. Bei der Darstellung von Fisteln mit tastbarem Konglomerattumor oder Abszeßverdacht ist ein Computertomogramm indiziert (Kerber et al., 1984).

Die Kombination endoskopischer und radiologischer Untersuchungen (Czembirek et al., 1983) erbringt auch nach unseren Erfahrungen bei den meisten Patienten die besten diagnostischen Resultate.

44.2.2.4 Komplikationen

Als **lokale** Komplikation kann sich die Entzündung in das umgebende Gewebe und Retroperitoneum ausbreiten und mit gedeckter Perforation zur Abszedierung und zu einem Konglomerattumor führen. Ebenso können auf dem Wege der expandierenden Entzündung Fisteln entstehen: entero-enteral, entero-vesikal, entero-vaginal, entero-kutan.

Eine zunehmende Stenosierung mit immer häufiger und immer heftiger werdenden kolikartigen Leibschmerzen verursacht schließlich einen Subileus und rezidivierenden Ileus (Krieg und Brünner, 1981). Seltener sind freie Perforation, foudroyante Blutung, toxisches Megakolon, Karzinom. Die Gefahr Morbus-Crohn-assoziierter Karzinome erscheint nach den bisher vorliegenden Studien gering (0,3% Dünndarm, 1,8% Dickdarm; Darke et al., 1973), obgleich ein erhöhtes Risiko vorzuliegen scheint (Zilkens und Peters, 1979; Otto et al., 1980; Kvist et al., 1986), das aber von anderen Autoren verneint wird (Binder, 1988).

Bei etwa einem Drittel der Patienten bestehen anale und perianale Komplikationen: Mazeration, Ekzem, Erosion, Ulzeration, periproktitischer Abszeß, Fistel, Fissur, Striktur, Stenose, Induration (Buchmann und Alexander-Williams, 1980).

Die Komplikationen im eigenen Krankengut zeigt Tabelle 44.2–2. Als wichtige extraintestinale bzw. metabolische Komplikationen sind die Folgen der Malabsorption von fettlöslichen Vitaminen (A, D, E, K), Elektrolyten, Vitamin B_{12} und Zink zu nennen.

Tabelle 44.2–2. Anzahl und prozentuale Verteilung der lokalen Komplikationen bei 665 Patienten mit M. Crohn

Erkrankung	n	%
mit Fistel/Abszeß	223	34
Operation/Resektion	251	38
Stenose	167	25
Ileus, Subileus	43	6
Anus praeter	24	4
toxisches Megakolon	1	0,15

Nicht selten beobachtet man auch einen Antithrombin-III-Mangel.

Als **systemische** Auswirkungen der granulomatösen Entzündungen können Erythema nodosum, Pyoderma gangraenosum, Stomatitis aphthosa, Episkleritis, Iridozyklitis, Uveitis, Entzündungen an den großen Gelenken, eine Sakroiliitis und Spondylitis ankylosans auftreten. Etwa zwei Drittel der Patienten mit positivem Histokompatibilitätsantigen HLA-B27 erkranken an einem Morbus Bechterew (McConnell, 1979; Malchow, 1983). Das Spektrum der Leberbeteiligung reicht ähnlich wie bei der Colitis ulcerosa von der Fettleber über entzündliche Leberveränderungen bis zur Cholestase, primär sklerosierenden Cholangitis und Pericholangitis.

Die Ureterobstruktion infolge eines entzündlichen Konglomerattumors bahnt die Entstehung einer Hydronephrose. Seltener wird eine sekundäre Amyloidose vom Typ AA beschrieben. Schließlich sind die allgemeinen Folgen der Mangelernährung (Oehler, 1984) ebenso hervorzuheben wie die retardierte Entwicklung der im Kindesalter Erkrankten (Shmerling, 1978; Bender, 1979; Booth und Harries, 1984).

44.2.3 Differentialdiagnose

Nach unseren Erfahrungen bei 665 Patienten mit Morbus Crohn beruht die oft jahrelange Verkennung der Krankheit auf der Fehlinterpretation der abdominalen Beschwerden, der extraintestinalen Komplikationen und einzelner Krankheitssymptome oder Krankheitsfolgen (Brandes und Eulenburg, 1976; Schmidt und Wölke, 1981; Rosahl, 1982; Feiereis, 1988).

Die abdominalen Beschwerden werden am häufigsten verkannt als
- Appendizitis („akut", „chronisch"),
- Adnexitis („rezidivierende Eierstockentzündung"),
- nervöse Durchfälle,
- Colon irritabile,
- Gastritis oder Ulkus,
- Cholezystopathie.
Zu den wichtigsten Fehlbeurteilungen extraintestinaler Komplikationen gehört die Annahme eines rheumatischen Fiebers oder einer chronischen Polyarthritis bzw. einer Spondylitis ankylosans als Krankheit sui generis, wenn deren Symptomatologie im Vordergrund steht, was mitunter bei zunächst gering ausgedehntem oder gering ausgeprägtem Darmprozeß 3–4 Jahre der Fall sein kann.

Schließlich werden manche Krankheitssymptome oder Krankheitsfolgen wie Inappetenz, Nahrungsverweigerung, Erbrechen, Durchfälle (letztere beide als artifiziell gedeutet) und Gewichtsabnahme bis zur Kachexie in Verkennung der entzündlichen Darmerkrankung längere Zeit als Anorexia nervosa angesehen und fehlbehandelt.

Als besonders schwer erweist sich manchmal die differentialdiagnostische Abgrenzung des Morbus Crohn gegenüber der Colitis ulcerosa, vor allem dann, wenn die Entzündung im Kolon lokalisiert, das Rektum mitbetroffen ist und der histologische Befund keine eindeutige Aussage erlaubt. Nach unseren Befunden ebenso wie nach Erfahrungen von Malchow und Daiss (1984a) verbleiben weniger als 5%, bei denen eine differentialdiagnostische Differenzierung nicht möglich ist. Zur Differentialdiagnose am Operationspräparat und zu bioptischen Untersuchungen wurden Diagnoseschlüssel zusammengestellt, deren Kriterien die Unterscheidung erleichtern können (Otto und Gebbers, 1982).

Ein positiver Befund der Antikörper gegen exokrines Pankreas spricht ohne Zweifel für einen Morbus Crohn, Antikörper gegen Becherzellen sprechen dagegen für eine Colitis ulcerosa.

Eine weitere wichtige differentialdiagnostische Abgrenzung ist notwendig gegenüber infektiösen Darmerkrankungen, besonders durch Yersinia enterolitica oder pseudotuberculosis, vor allem bei akutem Beginn, da hier mitunter, selbst wenn wegen der Vermutung einer akuten Appendizitis operiert wurde, ein Morbus Crohn angenommen wird, in Wirklichkeit aber eine Yersiniose vorliegt. Ebenso sind differentialdiagnostisch andere Infektionen, vor allem Salmonellosen (Giger et al., 1979), Amöbenerkrankung (Höfler, 1982; Kappeler und Halter, 1985) und, auch heute noch, Tuberkulose zu erwägen.

An die postantibiotische bzw. pseudomembranöse Enterokolitis (Werner, 1979; Jaeger et al., 1981), die ischämische Enterokolitis (Brown, 1972; Hermanek und Tonak, 1976; Parks, 1980; Pimpl und Umlauft, 1983), die kollagene oder auch die radiogene Kolitis (Kutzner et al., 1979; Baer et al., 1981; Stein, 1983) ist ebenso differentialdiagnostisch zu denken wie an Maldigestionssyndrome (exokrine Pankreasinsuffizienz, Laktasemangel), an ein globales Malabsorptionssyndrom (Glutenenteropathie, Morbus Whipple) oder hormonproduzierende gastrointestinale und pankreatische Tumoren (Vipom, Gastrinom, Karzinoid, Somatostatinom), auch segmentale Metastasierung in der Darmwand (Madeya und Börsch, 1989). Schließlich kann differentialdiagnostisch auch eine Bilharzioseerkrankung des Darmes wie das solitäre Ulkus des Rektums Schwierigkeiten bereiten (Feiereis und Otte, 1983b).

44.2.4 Ätiologie und Pathogenese

44.2.4.1 Genetik – Immunologie – Umwelt

Ätiologie und Pathogenese des Morbus Crohn (Crohn et al., 1932; Crohn, 1949, 1967; Crohn und Yarnis, 1966) sind bisher – wie bei der Colitis ulcerosa – ungeklärt (Fahrländer, 1979). Fragen nach einer genetischen Disposition, immunologischen Prozessen, dem Einfluß der Ernährung, infektiösen Faktoren und vegetativ-psychischen Anteilen (primär? sekundär?) bilden die Schwerpunkte der Forschungen seit vielen Jahren (Ward, 1977; Straub, 1980; Koflach, 1986).

Für eine **genetische** Disposition sprechen die erhöhte Konkordanz (Almy und Sherlock, 1966; Kirsner, 1973) bei monozygoten Zwillingen (unter unseren Patienten ein eineiiges Zwillingspaar mit sogar identischer Symptomatik einschließlich der Komplikationen) und die größere Häufigkeit bei Verwandten ersten Grades gegenüber der in großen Statistiken festgestellten Inzidenz- und Prävalenzrate (Mörl et al., 1976; Kirsner und Shorter, 1980; McConnell, 1979; Küster et al., 1987). Singer und Mitarbeiter (1971) stellten fest, daß eine positive Familienanamnese in 17,5% bei Patienten mit Colitis ulcerosa oder Morbus Crohn des Dickdarms vorlag, dagegen in einer Kontrollgruppe bei 4%. Nach Müller-Wieland (1982) ist das Risiko eines Verwandten ersten Grades 5- bis 6mal so groß wie in der übrigen Bevölkerung. Für eine polygenetische Komponente spricht das Vorliegen von Morbus Crohn und Colitis ulcerosa in der gleichen Familie (Kirsner, 1978). Dennoch kann beim Morbus Crohn ebensowenig wie bei der Colitis ulcerosa von einer Erbkrankheit gesprochen werden.

Zahlreiche Befunde erhärten die pathogenetische Bedeutung von **Immunreaktionen** bei Entstehung und Verlauf des Morbus Crohn. Nach eingehender Darstellung vieler einzelner Befunde stellen Thiele und Mitarbeiter (1982) in einer kritischen Schlußbemerkung fest, daß diese Frage trotz verschiedener humoraler (Auer, 1979; Brandtzaeg und Baklien, 1979) oder zellvermittelter (Shorter et al., 1979; Otto und Gebbers, 1982) Befunde noch völlig offen sei.

Stöcker und Mitarbeiter (1984a) begründen ihre Hypothese, daß die Entzündung des Darmes durch Immunreaktionen gegen Bestandteile des Pankreassekretes hervorgerufen werde, wie folgt: Sobald eine Sensibilisierung gegen Pankreassekret eingetreten ist, entstehe eine Reaktion der in der Darmwand diffundierten Antigene des Pankreassekretes mit Autoantikörpern. Die gebildeten Immunkomplexe lagern sich in Mukosa und Submukosa ab und aktivieren das Komplementsystem. Die Enzyme der durch Mediatoren angezogenen Granulozyten zerstören das Gewebe. Ähnlich könnten die Ablagerungen dieser Immunkomplexe die extraintestinalen Manifestationen des Morbus Crohn erklären. Die Autoren fanden bei 39% der Patienten mit Morbus Crohn Autoantikörper gegen exokrines Pankreas im Serum, bei der Mehrzahl mit einem Titer von 1:100 bis 1:1000. Die Antikörper gehörten vorwiegend den Immunglobulinklassen IgA und IgG an.

Aus diesen Befunden ergibt sich die Frage nach der pathogenen Bedeutung der **Ernährung.** Die Zunahme des Morbus Crohn während der letzten 30 Jahre und die ethnischen Befunde lenkten die Aufmerksamkeit auf die Ernährungsgewohnheiten der Erkrankten. Eine Reihe von Autoren (Miller et al., 1976; Martini und Brandes, 1976; Kasper und Sommer, 1979; Järnerot et al., 1983) wiesen auf den erhöhten Verbrauch von Zucker und Stärke bei Patienten mit Morbus Crohn hin, wodurch weniger Pankreassekret sezerniert werde und hierdurch die Unverträglichkeit gegen Pankreassekret vom Patienten unbewußt zu

korrigieren versucht wird. Andererseits stellten Heaton und Mitarbeiter (1979) fest, daß bei einer über 4 Jahre eingehaltenen Diät mit wenig raffiniertem Zucker und wenig weißem Mehl, dagegen erhöhter Zufuhr von ballastreichen Nahrungsmitteln (Früchten, Gemüse, Vollkornmehl), weniger häufig stationäre Behandlungen nötig waren und bei keinem ihrer 31 Patienten eine Resektion erforderlich wurde.

Guthy (1983) beschrieb eine Häufung des Morbus Crohn in Ländern mit hohem Margarineverbrauch und schloß hieraus ebenfalls auf einen ernährungsbedingten Faktor innerhalb der Pathogenese. Heckers und Mitarbeiter (1984) stellten einen erhöhten Gehalt an trans-Fettsäuren im Unterhautfettgewebe Crohn-Kranker fest.

Das Gewicht **infektiöser** Agenzien bei Morbus Crohn stand längere Zeit im Mittelpunkt des Interesses (Kasper, 1977; Mitchell und Rees, 1979; Beeken, 1979; Lennard-Jones, 1979; Riemann und Demling, 1979). Die Isolierung von Bakterien und Viren (Schneierson et al., 1962; Järnerot und Lantorp, 1972; Whorwell et al., 1977) oder deren Partikeln (Gitnick et al., 1976; Riemann, 1977a, b) führte zur Annahme, daß der Morbus Crohn eine Infektionskrankheit sei (Whorwell, 1981), was aber bisher nicht bewiesen ist.

44.2.4.2 Prämorbide Persönlichkeitsstruktur und Psychodynamik

Da immer wieder angenommen wird, daß prämorbiden psychopathologischen Strukturmerkmalen oder chronifizierten Konflikten die Bedeutung einer Teilursache für die Krankheit zukomme, sei nochmals ausdrücklich darauf hingewiesen, daß die Ätiologie auch beim Morbus Crohn bisher ungeklärt ist. So, wie es für die meisten Krankheiten keine spezifische Ursache oder monokausale Pathogenese gibt, so besteht auch meistens keine Beziehung einer Krankheit zu einer spezifischen Persönlichkeitsstruktur oder einem spezifischen Konflikt. Diese Feststellung entbindet jedoch nicht von einer auch psychische Befunde einbeziehenden individualisierenden Diagnostik. Von ihrem Ergebnis hängt es ab, ob die Indikation zur Psychotherapie besteht, die die körperliche Therapie erweitert und ergänzt.

Die Bedeutung psychischer Anteile für die Ätiologie und Pathogenese wie auch für die Auslösung eines Schubes und schließlich für den Verlauf der Krankheit ist bisher umstritten. Nach allen vorliegenden, meistens retrospektiven Untersuchungsergebnissen ist es gerechtfertigt, ebenso wie bei der Colitis ulcerosa, die bei Patienten mit Morbus Crohn beobachteten psychischen Befunde in Anteile der prämorbiden Struktur und Psychodynamik, der Auslösungsfaktoren und der krankheitsabhängigen Merkmale zu differenzieren.

Während man bei der Colitis ulcerosa weitaus umfangreicher und genauer die prämorbide Struktur des Patienten untersucht hat, läßt sich vorerst für die Gruppe der Patienten mit Morbus Crohn nur begrenzt etwas aussagen. Manche Autoren (Feldman et al., 1967; Monk et al., 1970; Goldberg, 1970; Helzer et al., 1987) fanden überhaupt keine pathogenetisch bedeutsamen psychischen Strukturmerkmale oder Auslösungsfaktoren, was freilich mit fragwürdigen Erhebungs- und Kodierungsverfahren zusammenhängen kann. Dagegen beobachtete Bockus (1945) als einer der ersten psychische Veränderungen mit emotionaler Unreife und Ängstlichkeit, Erregbarkeit und Sensibilität. Stewart teilte bereits 1949 bei 36 von 47 untersuchten Patienten psychische Auffälligkeiten, z.B. gehemmte Aggressivität, mit.

Untersuchungen über die Selbsteinschätzung anhand des Körperbildes ergaben, daß sich die Patienten selbstzerstörerischer darstellten und stärker ängstigende Phantasien entwickelten als Patienten mit Colitis ulcerosa (Schütz et al., 1988).

Verschiedene Autoren (Stewart, 1949; Reinhart und Succop, 1968; Ford et al., 1969; Glasmacher et al., 1988) beschrieben die Mütter ihrer Patienten mit Morbus Crohn als sehr ängstlich-depressiv, zwanghaft, kontrollierend, unfähig, die Kinder mit Wärme und Geborgenheit zu umgeben. Deren Angepaßtheit, Wohlverhalten und Aggressionsabwehr wiederum führten zur Verzögerung der eigenen Entwicklung, wodurch die Abhängigkeit bestehenblieb, im Gegensatz zu den gesunden Geschwistern, die nicht in dieser Entwicklungsphase zurückblieben (McMahon, 1973; Teufel et al., 1988). Die Analyse der Lebenssituation von 20 Familien ergab erhebliche Konflikte bei den erkrankten Kindern, woraus Schmitt (1985) die Indikation einer begleitenden Psychotherapie ableitet, ähnlich wie Lask (1986).

Wirsching und Stierlin (1982) und Wirsching (1984) fanden bei zwei Drittel der untersuchten Familien, daß der familiäre Zusammenhalt besonders ausgeprägt war (Bindung), die psychologischen Grenzen zwischen den einzelnen Familienmitgliedern und zwischen den Generationen weitgehend aufgehoben erschienen (Fusion), die Familie sich gegenüber der Umwelt stark abgrenzte (Isolation) und die Entwicklungsfähigkeiten eingeschränkt blieben (Rigidität). Ferner und Reindell (1979) beschrieben – wie auch bei anderen psychosomatisch Kranken – ein starres Familiensystem, das – als Ausdruck tiefer Angst – Unerwartetes nicht zulasse und auf Einflüsse von außen heftig reagiere. Andererseits würden Gefühlsäußerungen tabuisiert; Gefühle der Hilf- und Hoffnungslosigkeit seien vorherrschend. Hierin sehen die Autoren den Ansatz zur Familientherapie, um den Zugang zur offenen Aussprache zu vermitteln.

Die Untersuchungen von Jantschek und Mitarbeitern (1989) an Familien der Patienten, die an einer chronisch entzündlichen Darmkrankheit leiden, zeigen im Gruppenvergleich mit anderen Familien unterschiedlicher Diagnosen mehr Ähnlichkeiten als Unterschiede, was als Hinweis darauf zu werten ist, daß es „keine spezifischen krankheits- oder diagnoseassoziierten Familien gibt". Gleichzeitig wurde sichtbar, daß die Familien eine eigene und differenzierte, jedoch andere Auffassung von Krankheit und Beziehung haben als die außenstehenden Beobachter.

Auch die Untersuchungsergebnisse von Glasmacher und Mitarbeitern (1988) sprechen für eine gestörte Struktur der Primärfamilie bei Patienten mit Morbus Crohn. Abzugrenzen hiervon sind die Interaktionsformen der Familie, die sich als eine Reaktion auf die chronische Krankheit eines Familienmitgliedes zurückführen lassen (Wilke, 1989).

Faßt man die bisher aufgrund psycho-biographischer und testpsychologischer Untersuchungen beschriebenen Strukturmerkmale (Engel, 1969; McKegney et al., 1970; Sheffield und Carney, 1976; Gazzard et al., 1978; Fürmaier, 1979; Steinhausen und Kies, 1982; Zacher und Weiß, 1985; Weiß und Zacher, 1986) zusammen, so stehen Kennzeichen betonter „Selbstsicherheit", hinter der sich oft ein pseudounabhängiges Verhalten ähnlich dem des Ulkuskranken verbirgt, im Vordergrund (Ford et al., 1969; Reindell und Ferner, 1979, 1981; Biebl et al., 1984). Weiterhin werden rigide-zwanghafte wie auch zwanghaft-hysterische Anteile beobachtet (Studt und Mast, 1986). Übereinstimmend stellte man eine Aggressionsgehemmtheit sowie emotionale Unreife und Labilität fest (Lourens, 1973; McMahon et al., 1973; Schultheis und v. Uexküll, 1979; Zander et al., 1982a und b).

Während frühe Studien (Whybrow et al., 1968; Ford et al., 1969; Cohn et al., 1970) wegen fehlender Kontrollgruppen nur begrenzt aussagefähig sind, sprechen die Befunde von Helzer und Mitarbeitern (1984), die bei 50 Patienten mit Morbus Crohn im Vergleich zu 50 anderen chronisch internistisch Kranken erhoben wurden, für die Notwendigkeit einer nicht nur sorgfältigen körperlichen, sondern auch psychischen Diagnostik. Die Autoren teilten eine Reihe auffälliger psychopathologischer Befunde mit und fanden ebenso auch gegenüber der Kontrollgruppe häufiger Zeichen einer depressiven, zwanghaften oder phobischen Symptomatologie. Die Depressivität war mit höherer Krankheitsaktivität korreliert, was für eine sekundäre Pathogenese spricht.

Da sich häufig bei der Interpretation abweichender Persönlichkeitsmerkmale die reaktiven, sekundären psychischen Veränderungen von der prämorbiden Struktur nur schwer abgrenzen lassen, haben wir 30 Patienten mit Morbus Crohn in der Remission untersucht (Leibig et al., 1985). Die hierüber mit dem FPI (Freiburger Persönlichkeitsinventar) festgestellten Ergebnisse einer deutlicheren Aggressionsgehemmtheit, Selbstsicherheit, Nachgiebigkeit und emotionalen (Pseudo-?) Stabilität der Patienten mit Morbus Crohn gegenüber der Normalbevölkerung stimmen teilweise mit den Befunden anderer Autoren überein (Cohn et al., 1970). Die Patienten erweisen sich als nachgiebiger, gemäßigter, unfähiger im Durchsetzen ihrer Interessen.

Die von anderen Autoren beobachtete Depressivität, Ungeduld und emotionale Labilität (Ford et al., 1969; Paulley, 1971; Gerbert, 1980) fanden wir bei unseren Patienten nicht. Vielmehr schildern sie sich als zufriedener und emotional stabiler gegenüber der Normalbevölkerung. Als Grund hierfür kann die erreichte Remissionsphase der Krankheit und ebenso das Dissimulationsbedürfnis angesehen werden, wodurch die Selbstschilderung im Test von den klinischen Beobachtungen während des einzelnen Schubes abweicht.

Bestätigen ließ sich diese Annahme auch durch die Antworten in einem standardisierten Interview, bei dem 70% eine innere Unruhe mit Spannungsgefühlen äußerten, ein Drittel eine Neigung zu häufigeren depressiven Verstimmungen.

Die normalen Werte auf der Skala 1 des FPI (Nervosität im Sinne von psychosomatisch gestört), Skala 4 (Erregtheit), Skala 6 (Gelassenheit) und Skala 8 (Gehemmtheit) lassen sich mit dem Remissionsstadium erklären, da diese Faktoren gerade innerhalb der akuten Symptomatik am ehesten verändert werden.

Nach unseren Befunden, die bei je einer Gruppe von Patienten mit Morbus Crohn und Colitis ulcerosa während des Remissionsstadiums, d.h. nicht innerhalb eines aktiven Krankheitsprozesses erhoben wurden, fanden sich keine erheblichen Unterschiede der prämorbiden Struktur (Leibig et al., 1985). Die Patienten beider Krankheitsgruppen sind deutlich aggressionsgehemmter als die gesunde Kontrollgruppe, ein Befund, den bereits Paulley (1971, 1974) als auffallende Störung bei beiden Krankheiten hervorgehoben hat. Unsere Ergebnisse bestätigen die Mitteilungen von McKegney und Mitarbeitern (1970), die bei beiden Krankheiten aufgrund sehr sorgfältig ermittelter Daten einschließlich der Benutzung des Cornell Medical Index keine sicheren prämorbiden Unterscheidungsmerkmale feststellen konnten. Ahrens und Mitarbeiter (1986) verglichen bei je 22 Patienten mit Morbus Crohn und Colitis ulcerosa die eingehend erhobenen testpsychologischen und experimentalpsychologischen Befunde, die keine wesentlichen Unterschiede beider Gruppen ergaben. Die Colitis-ulcerosa-Patienten wiesen lediglich eine stärkere kognitive Ausblendung von Affekten und eine stärkere Affinität zur Psychotherapie auf. Reindell und Mitarbeiter (1981) beschrieben demgegenüber die Patienten mit Morbus Crohn als auskunftsbereiter, depressiver, ängstlicher und rigider als die Patienten mit Colitis ulcerosa. Wir fanden die Patienten mit Morbus Crohn ebenfalls auskunftsbereit, aber ohne sicheren Unterschied gegenüber den Kolitis-Kranken.

In einer weiteren kontrollierten Studie (Probst et al., 1989) fand sich bei unseren Patienten eine Abhängigkeit der im FPI (Fahrenberg et al., 1984) gemessenen Persönlichkeitsfaktoren von der Krankheitssymptomatik, was auf eine vorwiegend sekundäre, d.h. krankheitsabhängige Genese psychischer Befunde hinweist.

Somit läßt sich zusammenfassen, daß sich in der Remissionsphase einige Persönlichkeitsmerkmale verändern und Unterschiede, die während des aktiven Krankheitsstadiums aufgefallen waren, nunmehr verwischen.

In Tabelle 44.2–3 sind die psychischen Befunde bei Morbus Crohn und Colitis ulcerosa zusammengefaßt, vergleichende Befunde sind darüber hinaus im Kapitel 44.1 beschrieben.

Tabelle 44.2–3. Übersicht psychischer Befunde bei Morbus Crohn und Colitis ulcerosa

Morbus Crohn	Colitis ulcerosa
Struktur:	**Struktur:**
selbstsicher, „pseudo-unabhängig"?	Selbstentwicklung retardiert
zwanghaft-rigide	Selbstwertgefühl labil
aggressionsgehemmt	aggressionsgehemmt
nachgiebig	depressiv-zwanghaft
emotionale (Pseudo-?) Stabilität	
Auslösung:	**Auslösung:**
Abhängigkeits-/ Trennungskonflikt	Verlusterlebnisse (real, drohend, imaginiert)
Ambivalenzkonflikt	
Überforderung (in-between-Situation; Streß)	Versagungen
Krankheitsfolge:	**Krankheitsfolge:**
Depressivität	Depressivität, Kränkbarkeit
Stimmungslabilität	verstärktes Bedürfnis nach
Dissimulationstendenz	Regression und Abhängigkeit
anorektische Entwicklung	hypochondrisches
kontraphobisches Verhalten	Agieren
psychosoziale Desintegration	zeitlich begrenzte psychosoziale Einschränkungen

44.2.4.3 Krankheitsauslösung

Für die Auslösung der Krankheit oder eines Schubes scheinen bei Patienten mit Morbus Crohn ebenso wie mit Colitis ulcerosa emotional belastende Konflikte („life stress"; Grace, 1953) ein wichtiger Faktor zu sein (Whybrow et al., 1968; McKegney et al., 1970; Whybrow und Ferrell, 1973; Hislop, 1974; Lask, 1986). Ford und Mitarbeiter (1969) stellten fest, daß ein deutlicher Zusammenhang zwischen emotionalen Belastungen und Manifestation der Krankheit oder des Krankheitsschubes vorlag. Er beruhte oft auf ungelösten, nunmehr exazerbierten Konflikten, deren Bewältigung wesentlich zur Besserung der Symptomatik beitrug.

Stehen bei Patienten mit Colitis ulcerosa eher Verlusterlebnisse, die phantasiert werden, drohen oder eingetreten waren, im Vordergrund, so überwiegen nach unseren Erfahrungen bei Patienten mit Morbus Crohn eher Trennungsängste, nahende oder erlebte Trennungen und die damit verbundenen Konflikte. Ebenso häufig sind auch ambivalente Haltungen, die bewußt oder unbewußt konflikthaft erlebt werden, weiterhin Überforderungen, real oder innerhalb intrapsychischer Auseinandersetzung mit fremdbestimmten Entwicklungen, z.B. der vom Vater diktierten Berufswahl bei unserer eingangs geschilderten Patientin.

Auch bei anderen Untersuchungen wird der zeitliche Zusammenhang zwischen einem Konflikt und dem folgenden Krankheitsschub eingehend beschrieben (Paulley, 1958; Mersereau, 1963; Rechlin, 1977). Hervorgehoben werden besonders Abhängigkeitsprobleme (Ford et al., 1969; Cohn et al., 1970), überwiegend infolge einer gestörten Mutter-Kind-Beziehung, Rivalität mit den Geschwistern und erheblicher familiärer emotionaler Belastung während der Kindheit und Adoleszenz.

Im standardisierten Interview bestätigt die Mehrzahl der Patienten den engen Zusammenhang zwischen Konflikt und Krankheitsschub. Bei etwa der Hälfte bestanden Ambivalenzkonflikte, Überforderungs- und Trennungserlebnisse. In der Selbsteinschätzung der Patienten wird das Bestreben deutlich, sich im Sinne eines „Ideal-Selbst" darzustellen: 90% haben keine Kontaktschwierigkeiten, 80% fühlen sich als selbständig. Angesichts der demgegenüber häufigen Trennungskonflikte kann die von manchen Autoren (Reindell et al., 1981; Ullmann, 1982) hervorgehobene „Pseudounabhängigkeit" als Abwehr der eigentlich passiven Wünsche und als Enttäuschungsprophylaxe verstanden werden.

Paulley (1974) erwähnt die einem Krankheitsschub oft vorausgehende „in-between"-Situation, die von den Patienten vermieden wird, indem sie innerhalb einer krisenbedrohten Familie zu vermitteln versuchen, hier aber Aggressionen beider Parteien erleben. Für dieses Bemühen, als Friedensstifter Konflikten aus dem Wege zu gehen, spricht auch, daß 70% unserer Patienten sich in dieser Rolle erleben.

Einen signifikanten Zusammenhang zwischen psychosozialem Streß und nachfolgenden Krankheitsschüben stellten Paar und Mitarbeiter fest (1988).

44.2.4.4 Reaktive psychische Veränderungen

Als krankheitsdependente psychische Veränderungen dominieren Depressivität (Ford et al., 1969; Whybrow und Ferrell, 1973; Rechlin, 1977; Latimer, 1978) mit Stimmungslabilität, Dissimulationstendenz, andere unterschiedliche Abwehrformen und anorektische Entwicklung; diese sollte zur Differenzierung von der Anorexia nervosa sui generis besser als Pseudo-Anorexia nervosa bezeichnet werden.

Auch die von Probst (1989) festgestellte Abhängigkeit veränderter Persönlichkeitsstrukturen von der körperlichen Symptomatologie und deren Aktivität spricht für reaktive, sekundäre somatopsychische, d.h. infolge der Krankheit eingetretene psychische Befunde, wodurch – entgegen den Ergebnissen bei Colitis ulcerosa – die Annahme prämorbider pathologischer Strukturmerkmale in Frage gestellt wird. Hierbei geht es aber nicht darum, daß „psychische Veränderungen die Folge und nicht die Ursache der Erkrankung darstellen" (Fahrländer, 1979), sondern um die Trennung prämorbider von reaktiven Veränderungen. Selbstverständlich impliziert die Feststellung einer auffälligen prämorbiden Struktur nicht gleichzeitig die einer Kausalität mit der Krankheit. Am besten lassen sich krankheitsabhängige von prämorbiden psychischen Veränderungen und Strukturmerkmalen durch die Untersuchung in der Remission

abgrenzen. Gleichzeitig kann dann auch ein Bild über bestehende Wechselwirkungen körperlicher, psychischer und sozialer Einflüsse auf die Pathogenese, die Ausgestaltung eines Krankheitsschubes und die Verarbeitung der Krankheit (coping) gewonnen werden. Ein wichtiges Instrument weiterer Untersuchungen bildet hierbei sehr wahrscheinlich der intraindividuelle Vergleich der psychischen Befunde des Patienten innerhalb des akuten und chronischen Stadiums sowie in der Remission (Küchenhoff et al., 1989).

44.2.4.5 Psychosoziale Folgen der Krankheit

Bisher gibt es nur wenige Berichte über die psychosozialen Auswirkungen der Krankheit, vor allem auf Beruf, Familie, Partnerschaft, Sexualität und Freizeitaktivitäten. Während Balzer und Mitarbeiter (1985) gegenüber 2 Kontrollgruppen keine Unterschiede der Berufsausbildung und des Berufsstatus fanden, stellten Feurle und Mitarbeiter (1983) fest, daß 10% der Patienten ihrer Erkrankung wegen den Beruf wechseln mußten, 23,5% konnten nicht mehr ganztags arbeiten, 10% bezogen eine Zeit- oder Dauerrente und 3,8% waren arbeitslos. Bei 44% waren die Beziehungen zur Familie oder zum Partner beeinträchtigt, bei 36% die Freizeitaktivitäten behindert. Die sozialen Konsequenzen korrelierten mit der Dauer der Erkrankung, dem Zeitpunkt des Beginns und der Anzahl der Operationen. Bei dieser Studie mangelte es allerdings ebenso wie bei anderen (Gazzard et al., 1978; Olbrisch und Ziegler, 1982b) an der Kontrollgruppe. In einer anderen Studie wird eine gegenüber 2 Kontrollgruppen (3,3% bzw. 1,7%) gesteigerte Erwerbsunfähigkeit mit 17,5% angegeben (Balzer et al., 1985).

In einer kontrollierten Untersuchung des Zusammenhanges zwischen psychosozialer Beeinträchtigung und körperlichen Krankheitssymptomen, Krankheitsaktivität und Krankheitsverlauf bei unseren Patienten mit Morbus Crohn (Probst, 1989) fand sich eine gravierende Einschränkung in den Bereichen Familie, Partnerschaft, Sexualität, Beruf und Freizeit. Die Einbußen standen vor allem in einer engen Beziehung zur Krankheitsaktivität; mit zunehmender Krankheitsdauer wurden bei den Patienten ein stärkeres Selbstbewußtsein und eine bessere soziale Kompetenz beobachtet. Ein gutes Arzt-Patient-Verhältnis wirkte sich positiv auf die Adaptation an das Leben mit der Erkrankung aus. Auch die Partnerschaftsprobleme nahmen mit der Krankheitsdauer ab. Patienten, die wegen ihrer Krankheit eine Darmresektion bzw. Fisteloperation überstanden hatten, fühlten sich nach einem postoperativen Intervall von mehr als einem Jahr im Berufsleben weniger eingeschränkt als die übrigen. Diese Befunde stimmen mit den Ergebnissen von Meyers und Mitarbeitern (1980) überein; bei 71% der Patienten bestand präoperativ eine Beeinträchtigung im Berufsleben, postoperativ nur noch bei 29% der Patienten mit Morbus Crohn.

44.2.4.6 Zusammenfassung

Stärker als bei der Colitis ulcerosa sind die Ergebnisse der Einzelbefunde über Struktur und Psychodynamik, Auslösung, reaktive psychische Veränderungen, Krankheitsverarbeitung des Patienten und seiner Familie sowie psychosoziale Auswirkungen teilweise sehr unterschiedlich. Eine spezifische, für den Morbus Crohn typische Persönlichkeits- oder Familienstruktur läßt sich bisher nicht verifizieren. Ebensowenig bestehen kennzeichnende Auslösungserlebnisse gleichermaßen für die gesamte Gruppe der Patienten. Dagegen weisen voneinander unterscheidbare Untergruppen (v. Wietersheim, 1989) darauf hin, daß die verschiedenen Anteile prämorbider Psychodynamik und schubauslösender Ereignisse einer differenzierenden Analyse bedürfen. Hierzu gehören Einflußgrößen wie Akuität oder Chronizität der Krankheit, Remissionsphase oder postoperatives Stadium. Diese Analyse ermöglicht die Antwort auf Art und Umfang der allemal individuell zu konzipierenden und modifizierenden ergänzenden Psychotherapie.

44.2.5 Therapie

44.2.5.1 Voraussetzungen

Die Behandlung der Patienten mit Enteritis granulomatosa regionalis Crohn beruht auf einem mehrdimensionalen Ansatz, der aus medikamentösen, diätetischen, entspannungs- und psychotherapeutischen, operativen, rehabilitativen und nachsorgenden Verfahren besteht. Die Koordination dieser Therapie setzt eine eng konsiliarisch miteinander arbeitende Gruppe voraus (Hausarzt, gastroenterologisch kompetenter Internist, Psychotherapeut, Chirurg, Diätassistentin, Krankengymnastin). Die Therapie sollte ebensowenig wie bei der Colitis ulcerosa von einem Schema bestimmt sein, d.h., der individuell sehr unterschiedliche Krankheitsverlauf erfordert eine auf den einzelnen Kranken und alle Besonderheiten seiner Krankheit abgestimmte Behandlung, die wiederum einer ständigen gemeinsamen Kontrolle und Überprüfung unterliegt. Bei uns hat es sich auch sehr bewährt, daß der internistisch-psychotherapeutische Anteil innerhalb ein und derselben Klinik und Poliklinik liegt, so daß bei der stationären wie auch ambulanten Behandlung eine Aufteilung in verschiedene Kompetenzen vermieden wird.

Der Stellenwert psychotherapeutischer Hilfen zur Besserung der Krankheit und Überwindung somatopsychischer und psychosozialer Folgen der Krankheit wird allerdings nach wie vor sehr unterschiedlich beurteilt, trotz aller Bemühungen um eine Verständigung (Bartels et al., 1989; Feurle et al., 1988).

44.2.5.2 Medikamentöse Therapie

Bei isoliertem Befall des oberen Gastrointestinaltraktes und somit vor allem des Dünndarmes sind Kortikosteroide das Mittel der Wahl: Wir geben (Feiereis, 1985) Prednison oder Prednisolon, 40–60 mg/Tag per os, und reduzieren nach Rückgang der akuten Krankheitszeichen allmählich auf etwa 20 mg jeden 2. Tag. Antazida oder H_2-Rezeptor-Antagonisten senken die ulzerogene Nebenwirkungsgefahr.

Für den isolierten Befall des Dickdarmes ist bei geringem entzündlichem Befund Salazosulfapyridin (Azulfidine®, Colo-Pleon®), 4–5 g/Tag, in magen- oder dünndarmlöslicher Form angezeigt, bei mittelschwerer oder schwerer Entzündung Prednison oder Prednisolon, wie bei der isolierten Entzündung des Dünndarmes. Bei Unverträglichkeit oder Anzeichen von Nebenwirkungen hat sich auch die Monosubstanz Mesalazin (Claversal®, Salofalk®) bewährt.

Die gleichzeitige Erkrankung des Dünn- und Dickdarmes erfordert die Kombination aus Salazosulfapyridin und Prednison bzw. Prednisolon.

Bei enteralen oder perianalen Fisteln kann die Behandlung der oft durch Anaerobier verursachten Infektion mit Metronidazol (Clont®, Flagyl®) hilfreich sein: 800–1500 mg/Tag, 4–6 Wochen lang (Bernstein et al., 1980; Brandt et al., 1982). Bei mittelschwerer oder schwerer Entzündung kombinieren wir es mit Prednison oder Prednisolon. Sekundäre Infektionen, Superinfektion und Sepsis erfordern Breitspektrumantibiotika.

Die Wirkung von Azathioprin ist umstritten, die Gabe von zunächst 3 mg/kgKG, dann 2 mg/Tag erscheint dennoch begründet, wenn ein Rezidiv der Dünn- oder Dickdarmentzündung vorliegt, die Therapie bisher unbefriedigend verlief und somit der Versuch einer Kombination von Azathioprin und Kortikosteroiden angezeigt erscheint. Diese kombinierte Therapie sollte dann langfristig fortgesetzt werden.

Die medikamentöse Behandlung kann durch eine Reihe von Nebenwirkungen kompliziert sein:

Kortikosteroide: Cushing-Gesicht, Ulkus, Diabetes, Osteoporose, Katarakt, Nebennierenrindeninsuffizienz, Depression, Exazerbation von Infektionen (Tbc!).

Salazosulfapyridin: Fieber, LE-Syndrom, allergische Hautveränderungen, Alopezie, Knochenmarkdepression, Hämolyse, Methämoglobinämie, gastrointestinale Symptome, Lungeninfiltrate, Nephrolithiasis, Hemmung der Spermiogenese, Schwindel und Ohrensausen, reaktive Verstimmung.

Metronidazol: Allergien, Schwindel, Polyneuropathie, Alkoholunverträglichkeit.

Azathioprin: Knochenmarkdepression, Cholestase, Infektdisposition, Kanzerogenität.

Die medikamentöse Zusatztherapie gilt der Substitution von Flüssigkeit, Elektrolyten und Eiweiß. Die Behandlung einer Anämie mit Eisen, Vitamin B_{12}, Folsäure oder auch Erythrozytenkonzentraten richtet sich nach deren Form und Schweregrad.

Bei Diarrhöen – infolge des Krankheitsprozesses oder auch postoperativ, z.B. nach ausgedehnter Resektion – sind Substanzen wie Loperamid (Imodium®), Diphenoxylat, kombiniert mit Atropin (Reasec®) oder auch Opiumtinktur hilfreich. Sekretorische, chologene Durchfälle sprechen auf Colestyramin (Quantalan®), je 2g vor und nach dem Frühstück, an.

44.2.5.3 Diät

Die Notwendigkeit einer hochkalorischen Kost steht bei akutem Schub wie bei chronischem, kompliziertem Verlauf mit erhöhtem Nährstoffbedarf, Folgen der Malabsorption und intestinalen Verlusten (Oehler, 1984) oft in Kontrast zur geringen Bereitschaft der Patienten, regelmäßig und reichlich zu essen. Inappetenz, Übelkeit, Brechreiz, abdominale Schmerzen, dysphorische, resignierende Stimmung bis zur sekundären Depression können die Ursachen einer anorektischen Entwicklung sein, die mitunter zunächst nur mit einer niedermolekularen Peptiddiät (z.B. Nutricomp® Peptid F, Peptisorb®, salvipeptid®) über die naso-duodenale Verweilsonde oder eine parenterale Ernährung überwunden werden kann. Ob die hierbei beobachteten Besserungen (Lochs et al., 1984) ebenso wie unter der Ernährung mit Elementardiäten (Morin et al., 1980; Stober et al., 1983; O'Morain et al., 1983) als Folge einer geringeren Pankreassekretion (Stabile et al., 1984) zu erklären sind, bleibt vorerst noch offen.

Manche Patienten akzeptieren in diesem schweren Krankheitsstadium auch eine hochmolekulare Trink- oder Sondennahrung (z.B. Fresubin® flüssig, Nutricomp® F, Nutrodrip® Standard), die möglichst bald mit einer eiweiß- und kalorienreichen Kost kombiniert werden sollte. In ihr sind nur wenig aufgeschlossene Kohlenhydrate enthalten. Die Kalorien verteilen sich auf etwa 45% KH, 25% Eiweiß, 30% Fett.

Ähnlich wie bei Patienten mit Colitis ulcerosa wird die Ernährung sorgfältig mit einer Diätassistentin besprochen, die dem Patienten auch stets zur Verfügung steht, um Wünsche nach Erweiterung und Ergänzung seiner Ernährung rasch berücksichtigen zu können.

44.2.5.4 Entspannungs- und Psychotherapie

Einleitung

Während entspannungs- und psychotherapeutische Verfahren bei Patienten mit Colitis ulcerosa bereits seit etwa 40 Jahren erfolgreich angewendet werden und umfassende Ergebnisse katamnestischer Untersuchungen vorliegen (Curtius, 1962; Feiereis, 1970, 1977, 1979; Wilke, 1980), wurden für die Enteritis granulomatosa regionalis Crohn bisher nur vereinzelt Erfahrungen mitgeteilt (Feiereis, 1984; Lempa et al., 1984; Zacher und Becker, 1988). Der Grund hierfür liegt in der gegenüber der Colitis ulcerosa weit weniger gut dokumentierten Mitwirkung psychodynami-

Tabelle 44.2–4. Begleitende entspannungs- und psychotherapeutische Verfahren bei Morbus Crohn

Gespräch zur Diagnose
Autogenes Training
Tiefenentspannung
Krankengymnastische Einzel- und Gruppenbehandlung
Konzentrative Bewegungstherapie
Stützende Gesprächspsychotherapie
Postoperative Psychotherapie
Tiefenpsychologische Therapie (Einzel-, Paar-, Familien-, Gruppentherapie)
Assoziative Maltherapie
Intervall-Psychotherapie

scher und psychoreaktiver Anteile bei Pathogenese oder Verlauf dieser Krankheit. Dennoch sprechen die mitgeteilten funktionellen und psychischen Faktoren, seien sie primärer, disponierender Art oder als Folge der Erkrankung aufzufassen, für die Indikation der entspannungs- und psychotherapeutischen Maßnahmen; sie erweitern und ergänzen nach unserer 20jährigen Erfahrung die medikamentöse und diätetische Behandlung ebenso, wie sie nach mehr oder weniger eingreifenden Operationen hilfreich sein können (Tab. 44.2–4).

Die meisten Patienten bringen Bereitschaft und Vertrauen in diese Behandlung mit, besonders dann, wenn der lange Leidensweg mit Schmerzen, Durchfällen, Gewichtsabnahme und wiederholten Krankenhausaufenthalten psychische und psychosoziale Auswirkungen nach sich zieht, die das Bedürfnis nach Regression in der Entspannung und auch nach umfassender stützender Psychotherapie längst hervorgerufen haben. Wir können dabei die Erfahrungen von Freyberger und Mitarbeitern (1983) nicht bestätigen, daß 83% der Patienten mit Morbus Crohn Psychotherapie von vornherein ablehnten oder wieder abbrächen, nur eine sehr kleine Gruppe von Patienten ähnliche seelische Modalitäten aufweise wie die Colitis-ulcerosa-Patienten und daher für eine Psychotherapie in Betracht käme, die Mehrzahl aber einer psychotherapeutischen Zusammenarbeit ablehnend gegenüberstünde und „griffige Ansatzpunkte für eine Psychotherapie" noch nicht zu sehen seien (Freyberger et al., 1980; Malchow, 1983).

Gespräch zur Diagnose

Je nach Schwere der subjektiven und objektiven Symptomatologie steht am Anfang ein sorgfältig abgewogenes diagnostisches Gespräch, das Zuwendung und Geborgenheit vermitteln soll und mit einer konfliktzentrierten Initialbehandlung verbunden werden kann. Bei dieser bereits stehen oft die psychosozialen Folgen der Krankheit, etwa lange Unterbrechung der Ausbildung oder die der Aussteuerung sich nähernde Arbeitsunfähigkeit im Vordergrund. Sie können bei den Patienten destruktive Phantasien auslösen, ähnlich wie ein bereits irreparabler körperlicher Defekt.

Das Gespräch dient der Übertragung, dem Aufbau einer stabilen Arzt-Patient-Beziehung, in die auch die konstruktive Aussprache zur Diagnose einbezogen ist, worauf Engelhardt (1982) deutlich hingewiesen hat. Dabei sollte bei aller Offenheit über die Schwere der Erkrankung vermieden werden, den Patienten mittels apodiktischer Formulierungen zusätzlich zu ängstigen. Auch die Ausbreitung aller Informationen über mögliche Komplikationen oder die Prognose der Krankheit ist dem Patienten innerhalb dieses diagnostischen Gesprächs nicht hilfreich.

Entspannungstherapie

Sehr bald beginnen wir mit den täglichen Übungen des autogenen Trainings. Damit werden die Patienten, sofern sie bettlägerig sind, zunächst in Einzelübungen vertraut gemacht, um später an Gruppenübungen teilzunehmen. Wir kombinieren diese Form der Übungen mit der Tiefenentspannung im Sinne des Hypnoids bei ausgeprägter funktioneller Komponente der abdominalen Beschwerden. Tenesmen, Schmerzen und Durchfälle, Hypermotilität und Spastik des Darmes lassen sich gut beeinflussen. Zeichen der reaktiven Verstimmung und Affektlabilität als Ausdruck eines hyperästhetisch-emotionellen Syndroms bilden sich oft gleichzeitig zurück.

Wie bei Patienten mit Colitis ulcerosa halten wir die Mitwirkung einer Krankengymnastin für unentbehrlich. Atemtherapeutische Hilfen, besonders für die Zwerchfellatmung, leichte abdominale Massagen, Entspannungsübungen in der Gruppe zur Beruhigung von Hypermotilität und Spastik des Darmes erscheinen uns ebenso wichtig wie der psychologische Einfluß einer einfühlsamen krankengymnastischen Therapeutin.

Sobald die Patienten ausreichend mobilisiert sind, beziehen wir die konzentrative Bewegungstherapie in die kombinierte Behandlung ein. Sie ist ebenfalls eine körperorientierte Therapie, die hilft, Erlebnismöglichkeiten innerhalb der Wahrnehmung des eigenen Körperbildes zu erweitern und zu vertiefen. Gleichzeitig erleichtert sie den Zugang zur weiteren therapeutischen Einflußnahme auf die primären und sekundären psychischen Veränderungen.

Stützende Gesprächspsychotherapie

Die geschilderten Behandlungen, die beruhigend, ermutigend, entspannend und regressionsfördernd wirken, können je nach Indikation sowie Form und Schwere des somatischen Prozesses mit den weiteren psychotherapeutischen Möglichkeiten kombiniert werden. Vor allem lassen sich mit der stützenden Gesprächspsychotherapie die Übertragung stabilisieren, die Depressivität und oft als Abwehr zu deutende Leere des Gefühls positiv beeinflussen und Ansätze zu einer konfliktzentrierten Therapie finden (Künsebeck et al., 1987).

Im Mittelpunkt der Gespräche stehen Abhängigkeits- und Trennungskonflikte, ambivalente Haltungen, Überforderungen, die in engem zeitlichen Zusammenhang mit der Auslösung der Krankheit oder dem einzelnen Schub stehen. Das Ziel ist eine adäquate Verarbeitung der Konflikte, die aber nur erfolgen kann, wenn sich der Therapeut – in Kenntnis der biographischen Anamnese – ein Bild von der Entwicklung des Patienten, von früher Kindheit bis zum Ausbruch seiner Krankheit, machen kann.

So kann der Patient bereits frühzeitig, meist innerhalb der ersten klinischen oder ambulanten Therapie, Zugang finden zu den ihn belastenden bewußtseinsnahen Konflikten in den einzelnen Abschnitten seines Lebens und zu den psychischen und sozialen Auswirkungen der Krankheit. Für viele Patienten stellen die Einbußen der Ausbildung in Schule und Beruf, ständige oder wiederholte Arbeitsunfähigkeit, sozialer Abstieg, Fragen einer Umschulung oder gar eines Antrages auf Zeit- oder Dauerrente eine erhebliche seelische Belastung dar, wie auch die Versagungen innerhalb der Freundschaft, Ehe und Familie infolge der verschiedenen lokalen (Abszedierung, Fistel, ausgedehnte Narben) oder der generalisierten Manifestation der Krankheit.

Gleiches gilt auch für die postoperative Psychotherapie. Unter unseren Patienten waren 30% abdominal voroperiert oder mußten von uns der chirurgischen Klinik überwiesen werden. Die öfter nach einer Resektion anhaltende Diarrhö mit funktioneller Komponente bildete eine Indikation für die psychotherapeutische Mitbehandlung, ähnlich wie die Verarbeitung des passageren oder definitiven Anus praeter oder die seelische Anspannung angesichts des Bewußtseins der unvorhersehbaren Gefährdung durch ein Rezidiv (nach einzelnen Autoren bei Jugendlichen in 80–100%; Fahrländer, 1983).

Eine besonders schwere psychotherapeutische Aufgabe stellt sich bei den Patienten mit anorektischer Entwicklung (Pries et al., 1981). Gerade bei jugendlichen Patienten wird der Morbus Crohn mitunter lange Zeit als Anorexia nervosa verkannt und fehlbehandelt, da Inappetenz bis zur Nahrungsverweigerung neben den abdominalen Beschwerden die Symptomatologie des chronischen Verlaufes des Morbus Crohn beherrschen kann. Von einigen Autoren (Wellmann et al., 1981) wird die Koinzidenz von Morbus Crohn und Anorexia nervosa nicht im Zusammenhang mit der somatischen Krankheit gesehen, sondern das Untergewicht zu einem erheblichen Teil auf psychische Ursachen zurückgeführt. Wir haben die Kombination des Morbus Crohn mit einer Anorexia nervosa nur selten gesehen und halten deshalb die Bezeichnung Pseudo-Anorexia nervosa für treffender, weil hiermit die unmittelbare oder zumindest mittelbare Abhängigkeit der gestörten Nahrungsaufnahme von dem Grundleiden gekennzeichnet wird. Dem psychotherapeutischen Ansatz einer stützenden Gesprächspsychotherapie, verbunden mit verhaltenstherapeutischen Anteilen, kommt allerdings ähnlich wie bei der Anorexia nervosa eine entscheidende Bedeutung im gesamten Therapieplan zu.

Oft ist es nötig, in die Psychotherapie auch die Familie einzubeziehen (Reindell und Ferner, 1979). Die Indikation ergibt sich aus den Untersuchungsergebnissen. Wie in der Einzeltherapie ist das Ziel anfangs der Aufbau einer „positiven vertrauensvollen Übertragung" (Wirsching, 1984). Zunächst ist der Gesprächsinhalt krankheitszentriert und – geleitet von einer Hypothese über die zugrundeliegende Beziehungsdynamik – vom Therapeuten aktiv strukturiert, ohne Interpretation oder Wertung durch ihn. Auf dem Wege über abschließende positive Äußerungen zum Beziehungsmuster und nur verdeckt vorgebrachte Zweifel sowie Empfehlungen zum weiteren Verhalten lassen sich manchmal Veränderungen in Gang bringen, die dem Kranken eine wichtige Hilfe, besonders für seine Selbstentwicklung, bedeuten können.

Jantschek und Mitarbeiter (1989) fanden bei ihren Familienuntersuchungen eine große Bereitschaft zur therapeutischen Mitarbeit, nicht zuletzt auch wegen der psychosozialen Folgen bei den jungen Patienten. Auch bei anderen Autoren (Bruce, 1986) hat sich eine ähnlich umfassende Therapie sehr bewährt.

Assoziative Maltherapie

Dissimulationstendenz und Abwehrhaltungen der Patienten mit Morbus Crohn erschweren vor allem am Anfang häufig den Zugang zur verbalen psychotherapeutischen Hilfe. Dagegen können die Patienten meistens von Anfang an körperorientierte therapeutische Angebote annehmen. Relativ leicht finden sie auch den Zugang zu primär nonverbalen Psychotherapieverfahren als Möglichkeiten für die Verarbeitung innerer oder äußerer Konflikte. Hierbei hat sich bei uns die assoziative Maltherapie (Feiereis et al., 1989) sehr bewährt. Im Abschnitt über die Colitis ulcerosa (vgl. Kap. 44.1) wurden die zentralen Themen

der Patienten mit Colitis ulcerosa und Morbus Crohn und ihre damit verbundenen Assoziationen beschrieben.

Einige Beispiele: Angst im Bewußtsein des somatischen Prozesses (Abb. 44.2–3, siehe Farbtafel 4); Sehnsucht nach der körperlichen Integrität (Abb. 44.2–4); Resignation und Depressivität (Abb. 44.2–5 und 44.2–6, siehe Farbtafel 4); Krankheit als Aggression, Hoffnung auf Hilfe (Abb. 44.2–7, siehe Farbtafel 4); das Bild als Auslöser kontradepressiver, abwehrender Assoziationen (Abb. 44.2–8). Sie mögen andeuten, daß assoziativ gemalte Bilder und die Gedanken, die sie auslösen, eine wertvolle Möglichkeit darstellen, der verbalen, konfliktzentrierten, tiefenpsychologisch fundierten Therapie den Weg zu ebnen. Bei vielen dieser körperlich oft sehr kranken Patienten bilden diese psychotherapeutischen Möglichkeiten zusammen mit der körperorientierten Therapie eine untrennbare Einheit.

Tiefenpsychologische Psychotherapie

Nach unseren bisher vorliegenden Erfahrungen bei 665 Patienten mit Morbus Crohn bedarf nur die Minderzahl einer psychoanalytischen oder tiefenpsychologisch fundierten Psychotherapie, über die bisher auch nur vereinzelt berichtet wurde (Riemer, 1960; Sperling, 1960; Gerich, 1980). Die Indikation sollte besonders sorgfältig gestellt werden und ist ohnehin erst dann zu erwägen, wenn der Entzündungsprozeß beherrscht ist und eine ausreichende körperliche Stabilität vorliegt. Andererseits trifft es nicht zu, daß diese Patienten ein ungenügendes Verbalisations- und Introspektionsvermögen besitzen oder die neuroti-

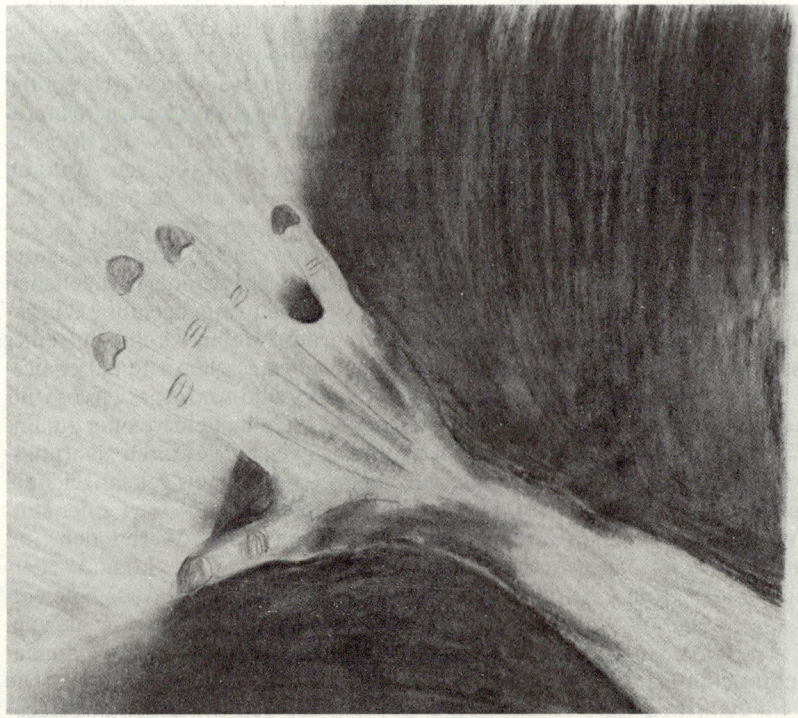

Abb. 44.2–4. 31j. Patientin C. T. mit Morbus Crohn: „Die Hand, die nach Freiheit und Unabhängigkeit greift" (27. 6. 85).

Abb. 44.2–8. 19j. Patient B.U. mit Morbus Crohn: „Als ich das Bild malte, habe ich an gar nichts gedacht – nichts Konkretes jedenfalls. Mein Feeling war wie immer – bestens! Mehr fällt mir nicht ein – ich hoffe, es genügt (ich hoffe es wirklich!)" (22. 2. 88).

sche Fehlentwicklung so verfestigt ist, daß somit eine tiefenpsychologische Therapie wenig Erfolg verspricht. So läßt sich auch aus der eingangs geschilderten Krankengeschichte und der zusammenfassenden Beurteilung der psychodynamischen Inhalte die Notwendigkeit einer tiefenpsychologischen Therapie im Intervall des Krankheitsprozesses ableiten. Ihr Ziel besteht im Abbau der vielfältigen ambivalenten Haltungen einschließlich der bisher unbewußten Abwehrformen und in der Entwicklung des Selbst, dem Aufbau der Identität, für die z. B. unsere Patientin weder im Bilde der Mutter noch des Vaters Identifizierungsmöglichkeiten sah und daher Ersatzobjekte suchte (real als ältere Freundin, phantasiert als älteren Freund).

Die Patienten, die unter der stützenden und regressionsfördernden Therapie innerhalb der aktiven Erkrankung oral befriedigt und emotional stabilisiert wurden, waren mit der gleichzeitigen Besserung des körperlichen Prozesses für die weitere Therapie auch belastbar geworden, sofern Indikation und Voraussetzungen für die tiefenpsychologische Psychotherapie als Einzel-, Paar-, Familien- oder Gruppentherapie gegeben waren.

Ähnlich wie bei der Colitis ulcerosa sprechen unsere Erfahrungen (Wilke, 1983, 1990) auch für die Anwendung der Tagtraumtechnik des Katathymen Bilderlebens. Sie enthält die Möglichkeit einer einleitenden Entspannung mit tiefenpsychologischer Aufhellung der bisher unbewußten und abgewehrten inneren Konflikte. Allerdings fanden mehr Patienten als bei der Gruppe mit Colitis ulcerosa keinen tragfähigen Zugang zur Handlungs- und Erlebnisebene des Tagtraumes. Die Inhalte der Tagträume blieben ohne adäquate Affekte, die Inkongruenz sprach für Verleugnung und Abspaltung. Dennoch war bei der Mehrzahl der Patienten die Fähigkeit zur Imagination gegenüber einer Kontrollgruppe nicht eingeschränkt.

44.2.5.5 Operative Therapie

Der unberechenbare Verlauf der Krankheit, die Vielfalt der möglichen Komplikationen und die nach einer Darmresektion große Gefahr eines Rezidivs (50–80%; Mappes, 1979) erfordern eine besonders enge Kooperation zwischen Internisten und Chirurgen (Koch und Kriener, 1981), um Vor- und Nachteile der Operation, Risiken und Gefahren bei weiter abwartender Haltung sorgfältig abzuwägen (Siewert und Isemer, 1983; Gall et al., 1983). Die Bedeutung des chirurgischen Anteils der Therapie des Morbus Crohn zeigt die große Anzahl operierter Patienten in Tabelle 44.2–2. Unsere Erfahrungen sprechen allerdings nicht dafür, daß „bis zu 80% der Patienten mit Morbus Crohn sich im Laufe ihrer Erkrankung eines operativen Eingriffes unterziehen müssen" (Singe und Ewe, 1985) bzw. 60–90% (Eßer, 1981).

Absolute Operationsindikationen sind: Perforationsperitonitis, Ileus, foudroyante, therapieresistente Blutung, Sepsis, Konglomerattumor mit Abszedierung und/oder Ureterstenose bzw. Blasen-Darm-Fistel.

Relative Operationsindikationen: Toxisches Megakolon, Subileus, gedeckte Perforation mit Abszedierung, Fistelung (abdominal blind endend oder enteroenteral, entero-vaginal, entero-kutan, perirektal) (Alexander-Williams und Buchmann, 1983), Konglomerattumor, rezidivierende infektiös-toxische Schübe. Eine relative Operationsindikation ist auch gegeben bei Progredienz mit häufigen Rezidiven und zunehmenden Auswirkungen auf Allgemeinbefinden und den metabolischen Stoffwechsel sowie auf sozialmedizinische Konsequenzen (lange Arbeitsunfähigkeit, Berufsunfähigkeit, Erwerbsunfähigkeit), ferner bei ausgedehnten analen Läsionen, zunehmenden Steroidnebenwirkungen und schließlich bei Kindern mit Wachstumsstörungen.

Gegenüber früher sprechen sich gegenwärtig die meisten Chirurgen für eine sparsame Resektion aus, da die Rezidivrate hierbei nicht höher liegt als bei der früher bevorzugten Operation weiter im Gesunden.

44.2.5.6 Langzeittherapie und Prognose

Auch bei den Patienten mit Morbus Crohn hängen Verlauf und Prognose von kontinuierlicher Weiterbehandlung und Kontrolle nach eingetretener Besserung ab. Manche Patienten entziehen sich dieser regelmäßigen Therapie und Kontrolle, sobald sie beschwerdefrei sind (Feiereis, 1985). Nicht selten aber weist auch der behandelnde Arzt nicht eindringlich genug auf die damit verbundenen Gefahren hin oder überläßt es dem Patienten, wieder in die Poliklinik oder Sprechstunde zu kommen, anstatt ihm feste Termine zu geben. Es erscheint mitunter unbegreiflich, daß ein Patient erst mit einem über faustgroßen Konglomerattumor als dem Produkt des Rezidivs, womöglich mehrere Jahre nach der ersten Diagnose und Behandlung, in die Klinik kommt mit dem Bemerken, er habe nichts von der regelmäßig notwendigen zwischenzeitlichen Kontrolle seiner Darmkrankheit gewußt.

Die einzelnen Teile der kontinuierlichen Behandlung sind:

– Regelmäßige termingebundene Besprechung von Art und Dosis der medikamentösen Langzeittherapie.
– Fortsetzung der Entspannungsübungen und der eingeleiteten Psychotherapie, eventuell Kontaktaufnahme zu einer Selbsthilfegruppe (Tecker, 1989).
– Regelmäßige Nachuntersuchung: Somatischer Status, gründliche Palpation, besonders auch mit Fahndung nach umschriebenen abdominalen Schmerzen oder Resistenzen. Kontrolle der Entzündungsbefunde im Blut; hierbei ist es hilfreich, den Aktivitätsindex einzubeziehen. In Übereinstimmung mit den Ergebnissen der europäischen (Malchow et al., 1984 c) und der amerikanischen (Singleton, 1979; Summers et al., 1979) Crohn-Studie ist auch nach unseren Erfahrungen bei geringem entzündlichem Prozeß eine medikamentöse Therapie mit Kortikosteroiden oder Salazosulfapyridin selten indiziert.
– Die Indikation zur endoskopischen und röntgenologischen Verlaufsdiagnostik eines entzündlich veränderten Darmsegmentes, anastomosierter Schleimhautbezirke oder der Schleimhaut im Gebiet einer Anastomose ist sorgfältig abzuwägen.

Die wichtigsten Faktoren zur Beurteilung der Prognose sind:

– Alter bei Beginn der Erkrankung,
– Schweregrad,
– Ausdehnung und zeitliche Entwicklung des lokalen Prozesses und dessen lokaler Komplikationen,
– Form und Schwere extraintestinaler Komplikationen,
– Anzahl und zeitlicher Abstand der Rezidive nach konservativer und operativer Therapie,
– primäre psychische Struktur und reaktive psychische Veränderungen sowie
– kontinuierliche therapeutische Mitarbeit des Patienten und Art und Inhalt seiner Krankheitsbewältigung (Compliance und Coping).

Insgesamt wird die Prognose bisher sehr unterschiedlich beurteilt (Hefti, 1981). Malchow und Mitarbeiter (1981) beschrieben eine um das 3,7fache erhöhte Sterberate. Die Überlebenswahrscheinlichkeit korrelierte negativ mit einem steigenden Aktivitätsindex. Aber selbst bei Krankheitsbeginn im Kindes- und Jugendalter kann die Prognose günstiger sein (Puntis et al., 1984), als angenommen oder errechnet wird.

Von manchen Autoren (Feurle, 1986) wird hervorgehoben, daß nach zehnjährigem Verlauf des Morbus Crohn in der Letalität gegenüber der Normalbevölkerung kein sicherer Unterschied mehr bestehe. Daraus wird der Schluß gezogen, daß die entzündliche Aktivität nach jahrelangem Verlauf doch zur Ruhe zu kommen scheint. Ein mitunter eigengesetzlicher Krankheitsverlauf läßt sich auch aus Berichten über Besserungen akuter Entzündungen bei über 40% der Patienten ohne spezifische medikamentöse Therapie ablesen (Meyers et al., 1984).

Die Rezidivgefahr erscheint höher bei primär ileozökalem Prozeß gegenüber ausschließlicher Entzündung des Dünn- oder Dickdarms.

Eine medikamentöse Prophylaxe zur Verhütung eines Rezidivs ist wahrscheinlich wirkungslos.

44.2.6 Epikrise zur einleitenden Krankengeschichte

Das 17jährige Mädchen leidet an einer Enteritis granulomatosa regionalis Crohn, die sich subjektiv bereits 2 Jahre lang abzeichnete, bevor die Diagnose gestellt wurde. Innerhalb der Ätiopathogenese ist ein genetischer Anteil anzunehmen, da auch die Mutter die gleiche Krankheit hat, weshalb bei ihr eine Ileozökalresektion notwendig geworden war.

Für einen prämorbiden psychodynamischen Faktor spricht die seit ihrer frühen Kindheit bestehende Labilität der emotionalen Beziehungen zu den Eltern. Der Vater wird stets als pedantisch bevormundend und unberechenbar erlebt, verbunden mit jederzeit drohendem Liebesentzug, wenn die Bedingungen seiner Zuneigung von der Patientin nicht erfüllt wurden. Jeder auch nur andeutungsweise gewagten Aggressivität standen als Antwort des Vaters Rückzug, Liebesentzug oder Verachtung gegenüber. „Wenn ich mich manchmal weigerte oder etwas nicht so lief, wie er es wollte, legte er sich mit Kopfschmerzen ins Bett, und ich bin dann immer wieder zu ihm hingegangen und habe mich entschuldigt, bis er wieder mit mir redete." Voll innerer Hemmungen war die Patientin chancenlos, gestört in ihrer Selbstentwicklung und bemüht um ein zwanghaftes Wohlverhalten, das geprägt war von der Angst vor den unberechenbaren Reaktionen des Vaters.

Auch die Mutter wird von der Patientin, besonders während der letzten Jahre, zwiespältig erlebt: als duldsame, wehr- und hilflos die Attacken die wesentlich älteren Ehemannes ertragende Frau – „sie schluckt alles" und der Patientin gegenüber als wankelmütige, emotional labile, alles andere als Ich-stärkende Mutter – „erst kommt sie zu mir und redet mit mir und tröstet mich, und eine Stunde später fällt sie mir in den Rücken und hackt mit meinem Vater auf mir herum". Ähnliches erlebt sie, wenn sie bei offenen Auseinandersetzungen der Eltern zu vermitteln versucht, z.B. der schweigend duldenden Mutter helfen möchte. Die Aggression des Vaters richtet sich dann plötzlich gegen sie. In ihrer „in-between-Situation" wird die „Friedensstiftcrin" so unvermittelt zum Opfer einer pervertierten Aggression, der sie hilflos ausgeliefert ist und die sie demütig erträgt.

Tief gestört in ihrer Emotionalität, rigide, selbstsicher wirkend hinter der Maske formalen Wohlverhaltens, freilich ohne Perspektive eines für sie akzeptablen Lebensentwurfes, erscheint ihr Suizidversuch, den sie früher wegen der Migräne unternommen hatte, als der Wunsch, nicht nur von der Krankheit, sondern zugleich von den Eltern und dem Leben befreit zu sein. Als sie erneut den Suizid erwägt, liegt der Grund wiederum weniger in der Krankheit, sondern darin, endlich die „Bestrafung der Eltern" zu verwirklichen: Autoaggression als Instrument einer Heteroaggression, die ihr real auf andere Weise nie möglich gewesen war, mit dem Morbus Crohn als Angst- und Schuldgefühl minderndes Alibi.

Der Beginn ihrer Krankheit steht wahrscheinlich in engem zeitlichem Zusammenhang mit dem Abschluß der Schulausbildung und dem Beginn einer Lehre, mit der die Patientin nicht einverstanden war. Art und Ort der Ausbildung waren vom Vater bestimmt. Zur gleichen Zeit wurde die Mutter wegen ihres Morbus Crohn operiert, war depressiv, „nervenleidend". Außer diesen beiden äußeren, bei der Auslösung der Krankheit möglicherweise mitwirkenden Ereignissen wurde der Patientin zu dieser Zeit bewußt, weder zum Vater noch zur Mutter eine befriedigende emotionale Beziehung zu haben, andererseits aber unfähig zu sein, sich konsequent von den Eltern zu trennen, wie es dem Bruder gelungen war. Statt dessen entwickelt sich eine enge, fast symbiotische Beziehung zu einer 8 Jahre älteren Freundin, einer Ersatzmutter, die Vorstellungen und Wünsche einer konstanten, verläßlichen Beziehung erfüllt. Die Exazerbation der Krankheit, wegen der die Patientin dann erstmals zu uns kam, trat ein, als die Freundin sich räumlich von ihr trennte.

Gegenüber den Merkmalen der psychodynamischen prämorbiden Entwicklung und Struktur der Patientin sowie den Auslösungsfaktoren sind Depressivität, anorektische Symptomatik und Dissimulationstendenzen und auch die angedeuteten Depersonalisations- und Derealisationsphänomene („beobachte mich wie von außen", „Umwelt wie ein Traum") als krankheitsabhängige Folgen anzusehen, also sekundärer Art.

44.2.7 Schlußfolgerung

– Die Ätiopathogenese der Enteritis granulomatosa regionalis Crohn steckt noch immer voller Rätsel. Nach bisher vorliegenden Erkenntnissen beruht sie wahrscheinlich auf einem Bündel verschiedener Faktoren, das aus genetischen, immunologischen, umweltbezogenen (Ernährung? Infektionen?) und psychischen Anteilen besteht.

– Die Diagnose kann mitunter jahrelang hinter der Fehlannahme anderer abdominaler und nichtabdominaler Krankheiten verborgen sein. Beispiele: chronische Appendizitis, Adnexitis, Arthritis, Anorexia nervosa.

– Die Differentialdiagnose kann auch dann erschwert sein, wenn an den Morbus Crohn gedacht wird, z.B. bei akutem Verlauf gegenüber der Yersiniose, bei chronisch rezidivierendem Verlauf gegenüber der Colitis ulcerosa. Gelegentlich ist selbst nach wiederholter Untersuchung die Diagnose Colitis ulcerosa oder Morbus Crohn nicht endgültig (Otte et al., 1989).

– Der unberechenbare, mit vielen möglichen Komplikationen und nicht selten erheblichen sozialmedizinischen Folgen verbundene Verlauf der Krankheit erfordert von allen Beteiligten ein hohes Maß an Kooperation in Diagnostik und Therapie, besonders auch vom Kranken selbst.

– Der psychosomatisch und psychotherapeutisch orientierte Arzt hat die Aufgabe, für Pathogenese und Verlauf mögliche und wichtige prämorbide Persönlichkeitsmerkmale und psychodynamische sowie psychosoziale Anteile zu untersuchen und auszuwerten (mit Hilfe der biographischen Anamnese, standardisierter Interviews, testpsychologischer Verfahren), Auslösungsfaktoren zu differenzieren und krankheitsabhängige, sekundäre psychische Veränderungen abzugrenzen.

– Da es bisher keine kausale Behandlung gibt, setzt sie sich aus verschiedenen und individuell sorgfältig abgestimmten Hilfen zusammen. Die Psychotherapie, die hierbei die medikamentöse, diätetische und operative Therapie ergänzt und erweitert, beginnt bereits bei Form und Inhalt des Gespräches zur Diagnose. Der Schwerpunkt der Psychotherapie beruht auf entspannungstherapeutischen Verfahren, der stützenden, konfliktzentrierten Therapie, der tiefenpsychologischen Behandlung und der Einflußnahme auf psychosoziale Folgen einschließlich Fragen nach den Möglichkeiten der Rehabilitation, z.B. nach Indikation, Zeitpunkt und Ort eines Heilverfahrens, einer Umschulung oder eines Antrages auf Zeitrente.

– Die Psychotherapie muß nach erreichter Remission oder auch postoperativ im Intervall fortgesetzt werden, um die körperliche und psychische Stabilität zu festigen; dem steht nicht selten eine neue Belastung des Patienten gegenüber, nämlich das Bewußtsein der Gefahr eines Rezidivs und die damit verbundenen Ängste.

– Nach unseren Erfahrungen sind Bereitschaft und Mitarbeit der Patienten weitgehend vorhanden. Die bisher erkennbaren Ergebnisse sprechen sehr für die Anwendung dieser kombinierten Therapie. Ob und wie weit sie vermag, die Krankheit oder ihre Rezidive zu verkürzen, Remissionen zu verlängern, eine klinische Heilung zu ermöglichen, somit die Prognose zu verbessern, wird derzeit in einer prospektiven kontrollierten Studie untersucht.

– Die Klärung der Ätiologie und die Suche nach der bestmöglichen Therapie des Morbus Crohn werden wahrscheinlich noch lange Zeit ein Feld intensiver Arbeit, fruchtbarer wissenschaftlicher Diskussion und regen, möglichst emotionsfreien Austausches der Erfahrungen sein – eine interdisziplinäre Aufgabe ersten Ranges. Sie wird nur erfüllt werden können, wenn allseits die Bereitschaft zur Kooperation jeder Art der Versuchung zur Konfrontation oder gar Mißachtung widersteht.

45 Erkrankungen des Bewegungsapparates

45.1 Chronische Polyarthritis

Hans-Heinrich Raspe

Die chronische Polyarthritis (cP) oder rheumatoide Arthritis (rA) ist eine entzündliche Allgemeinerkrankung der mesenchymalen Gewebe, die sich vor allem an den Gelenken (als Synovialitis) manifestiert.

Die klassische (psychoanalytische) Psychosomatik Alexanders (1977) verstand sie als eine der sieben psychosomatischen Erkrankungen im engeren Sinne. Unter allen rheumatologischen Krankheitsbildern fand sie in der Folge die größte psychosoziologische Aufmerksamkeit. Eine besondere Herausforderung für psychosomatische Theoriebildungen wurde in ihrer biomedizinisch ungeklärten Ätiologie, ihrem unvorhersehbaren (oft schubhaften) Verlauf und ihrer ungewissen Prognose quoad rehabilitationem gesehen. Auch der anscheinend auf zentralnervöse Mechanismen hindeutende symmetrische Befall kleiner Gelenke gab zu psychogenetischen Überlegungen Anlaß.

Die Literatur zur Psychosomatik der chronischen Polyarthritis ist kaum noch zu übersehen und kann hier nicht im einzelnen referiert werden. Seit 1955 sind wenigstens 12 Übersichtsreferate erschienen, die eine erste Orientierung ermöglichen. Manche diskutieren vor allem methodische Probleme (King, 1955; Scotch und Geiger, 1962; Rutter, 1979), während andere daneben auch die wesentlichsten Einzeluntersuchungen detaillierter darstellen (Moos, 1964; Wolff, 1971/1972; Hoffman, 1974). Besonders kompetent und instruktiv sind die Arbeiten von Meyerowitz (1971), Weiner (1977), Solomon (1981) und Anderson et al. (1985). 1983 erschien Weintraubs deutschsprachige „Psychorheumatologie" mit einem kürzeren Kapitel zur chronischen Polyarthritis.

45.1.1 Epidemiologische und sozialmedizinische Hinweise

Die cP ist eine relativ häufige Erkrankung. Ihre Punktprävalenz liegt – in unseren Breiten – zwischen 0,5 und 1,0%. Sie nimmt mit steigendem Lebensalter zu. Frauen sind wenigstens zweimal so häufig betroffen wie Männer. Pro 2000 Einwohner und Jahr erkrankt wahrscheinlich eine Person neu an cP. Diese Inzidenzrate steigt ebenfalls mit zunehmendem Lebensalter; sie ist – säkular betrachtet – möglicherweise rückläufig (Hochberg, 1981). Trotz ihrer Prävalenz wird ein Arzt der Primärversorgung jeweils nur wenige cP-Patienten zu betreuen haben. Kaum mehr als 1% seiner Diagnosen entfallen auf diese Erkrankung.

Auch heute noch ist mit einem Anteil von 5–30% gänzlich unbehandelter cP-Kranker zu rechnen. Von den primärärztlich betreuten Kranken wird ein erheblicher Teil nicht an Rheumatologen überwiesen, nur wenige werden in Krankenhäusern und Spezialkliniken behandelt (Tab. 45.1–1).

Die sozialmedizinische Bedeutung speziell der cP ist heute kaum abzuschätzen; es fehlt an zuverlässigen krankheitsspezifischen Daten der Krankenkassen und Rentenversicherungsträger (Wasmus und Raspe, 1988). Allein für das Jahr 1978 errechneten Blohmke und Neipp (1981) Gesamtausgaben für Arbeitsunfähigkeitstage, Krankenhaustage, stationäre Heilbehandlungen und neubewilligte Erwerbs- und Berufsunfähigkeits-Renten in Höhe von 200 Millionen DM; das waren 5% der entsprechenden Gesamtausgaben für alle Wirbelkörpersyndrome, Arthrosen, Osteochondrosen und entzündlich-rheumatischen Erkrankungen. Die cP verursachte überproportionale Kosten bei den Erwerbsunfähigkeits-Renten und den Krankenhausbehandlungstagen.

45.1.2 Die körperliche Krankheit

Die **Ätiologie** der cP ist noch unbekannt. Von Bedeutung sind vermutlich eine genetische Disposition, ein (Fremd-?)Antigen und eine auslösende Situation, um

Tabelle 45.1–1. Inzidenz, Prävalenz und medizinische Versorgung der chronischen Polyarthritis in der Gemeinde (Fälle pro 10 000 Einwohner)

Inzidenz/Jahr	7
Punktprävalenz	100
In primärärztlicher Behandlung*	70–90
In kontinuierlicher Betreuung*	60–80
Überweisungen an andere Ärzte*	20–40
Krankenhausbehandlung*	3–8

* in den letzten 12 Monaten
Zahlen geschätzt nach 10 Studien aus D, GB, S, SF, USA

die – dann besser bekannten – akuten und chronisch-entzündlichen Prozesse in Gang zu setzen. Meist entwickelt sich die Krankheit schleichend im Laufe von Monaten; möglich ist aber auch ein hochakuter Beginn in wenigen Stunden bis Tagen.

Bevorzugt befällt die cP zuerst symmetrisch kleine, stammferne Gelenke, oft zuerst die Hand-, die Fingergrund- und Fingermittelgelenke. Eine Einbeziehung der Halswirbelsäule ist nicht selten.

An den Gelenken manifestiert sie sich mit den klassischen Zeichen einer Entzündung: mit Schwellung, Überwärmung, lokalem Druck-, Bewegungs- oder Ruheschmerz und Funktionsbehinderung. Oft zieht sie an den befallenen Gelenken und Wirbelsäulenabschnitten Destruktionen des Gelenkknorpels, der subchondralen knöchernen und der periartikulären Strukturen nach sich. Die Folgen sind sichtbare Deformationen und irreversible Funktionsverluste. Zusätzlich sind extraartikuläre Manifestationen zu befürchten.

Der **Krankheitsverlauf** ist äußerst variabel und im Einzelfall kaum vorherzusehen. Wir unterscheiden neben der ungünstigen, eventuell „malignen" chronisch-progredienten Form mit oder ohne abgrenzbare Schübe eine chronisch-symptomatische und eine gutartige remittierende, eventuell nur episodische Form. Übergänge eines über Jahre (scheinbar) stabilen Verlaufstyps in einen anderen sind zu beobachten.

Die **Prognose** der cP ist quoad vitam scheinbar günstig. Sie taucht in Mortalitätsstatistiken in aller Regel nicht auf. Dennoch verkürzt sie das Leben der Kranken um etwa 5 bis 10 Jahre. Das relative Mortalitätsrisiko liegt ungefähr bei 2,0%. Es läßt sich eine kleinere Patientengruppe identifizieren, die ein deutlich höheres Letalitätsrisiko aufweist. Diese Patienten sterben einmal an Komplikationen der cP (z.B. Amyloidose, Myelonkompression), zum anderen an überzufällig häufigen Zweit- bzw. Begleiterkrankungen wie bakteriellen Infektionen.

Ernster ist die Prognose quoad rehabilitationem, vor allem wenn wir die Erfahrungen aus klinischen Stichproben berücksichtigen. Dagegen ist aus epidemiologisch-prospektiver Sicht damit zu rechnen, daß bis 50% der eingangs gestellten cP-(Verdachts-)Diagnosen nach vier Jahren nicht mehr aufrechtzuerhalten sind. Diese „Remissionsrate" reduziert sich in klinischen Studien auf 13–26%. Dazwischen liegen die Raten ambulant betreuter Patienten.

Einen ungünstigen progredienten Verlauf erleben zwischen ein und zwei Drittel der klinisch Betreuten. Als dessen Ergebnis stellt sich bei etwa 40% eine gravierende Bewegungsbehinderung ein; etwa 5–10% sind schließlich an den Rollstuhl oder das Bett gefesselt.

Die **Diagnose** der cP gelingt in typischen Fällen ohne Schwierigkeiten. Hilfreich sind dabei diagnostische Kriterienkataloge, z.B. die „1987 revised criteria for the classification of rheumatoid arthritis" der American Rheumatism Association (ARA) (Arnett et al., 1988).

Entscheidend für die Diagnose sind die Anamnese (mit etwa 70% später bestätigten Verdachtsdiagno-

sen) und die klinische Untersuchung. Immunologische, biochemische und röntgenologische Befunde spielen eine nachgeordnete, nuklearmedizinische, zytologische und histologische Befunde eine untergeordnete Rolle.

Angesichts der sehr unterschiedlichen Manifestationsformen und Krankheitsverläufe ist es unsicher, ob die cP als eine nosologische Einheit aufgefaßt werden darf. Es ist denkbar, daß unser diagnostisches Etikett heterogene Krankheiten oder verschiedene Syndrome deckt. Allein dieser Umstand könnte für inkonsistente Ergebnisse verschiedener (psycho-)somatischer Untersucher verantwortlich sein.

Eine kausale Therapie steht bisher ebensowenig zur Verfügung wie eine zuverlässig wirksame und gefahrlose symptomatische. Die Behandlung ist in aller Regel polypragmatisch und beinhaltet die oft sehr erfolgreichen rheumachirurgischen Eingriffe. Die internistisch-rheumatologischen Anstrengungen zielen zuerst auf die Remissionsinduktion. Dies gelingt bei etwa einem Drittel der Patienten mit Hilfe der sog. Basistherapeutika – wenigstens für Monate. Es handelt sich dabei um langfristig zu verordnende differente Substanzen wie Hydroxychloroquin, Gold, D-Penicillamin, Azathioprin, deren Einnahme häufig mit Nebenwirkungen, sehr selten mit tödlichen Komplikationen belastet ist. Sie erfordern daher eine sorgfältige Indikationsstellung und engmaschige Kontrolluntersuchungen im Rahmen einer verläßlichen Zusammenarbeit von Arzt und Patient.

Ehe die Wirkung der Basistherapeutika nach mehreren Wochen einsetzt und ebenso, wenn die erreichten Besserungen unvollständig bleiben, wird eine begleitende, individuell angepaßte symptomatische Therapie mit nonsteroidalen Antirheumatika und/oder Steroiden notwendig. Es gibt keinen Anhaltspunkt dafür, daß diese Medikamente den Krankheitsverlauf günstig beeinflussen.

Es ist für die spätere Diskussion wichtig, daran zu erinnern, daß seelische oder Verhaltensstörungen auch das Ergebnis der medikamentösen Behandlung (z.B. durch zentralnervöse oder periphere Wirkungen aller Antirheumatika) und auch Folge cP-assoziierter somatischer Prozesse (Vaskulitis, Hyperviskosität) sein können.

Ohne Zweifel ist die konservative Behandlung des chronischen Polyarthritikers ohne die genannten Medikamente nicht mehr denkbar – ihr Zentrum bildet jedoch das Ensemble jener Ansätze und Verfahren, die wir als die „eigentliche Basisbehandlung" bezeichnen wollen. Es handelt sich um:

– Die begleitende ärztliche Aufklärung des Patienten im dreifachen Sinne kognitiver Informierung, lebenspraktischer Beratung und emotionaler Stützung (Raspe, 1983). Hierbei kann der Mitwirkung einer Krankenschwester oder Arzthelferin eine besondere Bedeutung zukommen.
– Krankengymnastisch angeleitete aktive und passive Bewegungsübungen.
– Lokale physikalische Anwendungen, z.B. von Wärme und Kälte.
– Ergotherapeutische Beratung und Behandlung mit

den Hauptbestandteilen Gelenkschutz, funktionelle Übungsbehandlung, Haushalts- und Arbeitsplatzanpassung, Versorgung mit Hilfsmitteln und Orthesen.
– Sozialrechtliche Beratung und Rehabilitation.
– Laienhilfe in Selbsthilfegruppen z.B. der Deutschen Rheumaliga mit den drei Schwerpunkten Bewegung, Beratung, Begegnung.
Alle diese therapeutischen Zugänge sollten am Wohnort und im ambulanten Versorgungsbereich zur Verfügung stehen. Sie erfordern eine kontinuierliche und koordiniert-interdisziplinäre Arbeitsweise (Bundesminister Forschung und Technologie, 1988).

45.1.3 Die chronische Polyarthritis aus psychosomatischer Sicht

Im Horizont einer „integrierten Psychosomatischen Medizin" (v. Uexküll, 1981) sind drei Schwerpunkte der bisherigen und gegenwärtigen Diskussionen zu erkennen.
– Die älteste Frage lautet: Wieweit läßt sich die cP als psychosomatisch entstanden auffassen? Wie kann man sich eine spezifische psychosomatische Ätiologie der cP vorstellen? Diese Frage hat das breiteste wissenschaftliche Interesse gefunden. Die sich aus den vorliegenden Antworten ergebenden klinisch-praktischen Konsequenzen für die Betreuung bereits an cP Erkrankter sind dagegen gering.
– Wie weit werden Krankheitsausbruch, Krankheitsverlauf und Krankheitsausgang von seelischen und sozialen Einflüssen mitbestimmt? Ergeben sich Hinweise auf eine spezifische oder unspezifische Pathoplastik? Es ist sehr wahrscheinlich, daß Kenntnisse über psychosoziale Verlaufsrisiken praktische (tertiärpräventive) Bedeutung gewinnen könnten.
– In jüngster Zeit konzentrieren sich die Untersuchungen auf die Frage, welche psychosozialen Implikationen und Folgen eine cP mit sich bringt. Es liegt auf der Hand, daß ein genaueres Wissen um die Lasten und Leiden von cP-Kranken unsere therapeutischen Bemühungen vertiefen und erweitern müßte.

Frau U. war 45 Jahre alt, als sie im November 1983 subakut mit Schmerzen, Schwellungen, Kraftlosigkeit, Morgensteifigkeit und Bewegungsbehinderungen im Bereich der rechten Hand erkrankte. In der Anfangsphase wurden, rasch wechselnd, weitere große und kleine Gelenke in Mitleidenschaft gezogen.

Die Erkrankung manifestierte sich, kurz nachdem ihr einziger Sohn das Abitur abschloß und nicht sofort den gewünschten Ausbildungsplatz bei der Polizei erhielt.

Es dauerte 7 Monate, bis sie auf eigene Initiative den Weg in die rheumatologische Sprechstunde fand. Vorher hatte sie zwei Internisten und einen praktischen Arzt konsultiert, ohne daß eine spezifische Therapie eingeleitet worden wäre. Zur „Fokalsanierung" ließ sie eine – letztlich erfolglose – Tonsillektomie über sich ergehen.

Wir diagnostizierten im Juni 1981 eine hochaktive seropositive cP mit röntgenologisch nachweisbarer gelenknaher Osteoporose. Die Familienanamnese blieb leer.

Die Patientin vermutete als Mitursache ihrer cP „Arbeitsstreß" und sprach damit gleich anfangs einen Problembereich an, der sie bis heute (1984) nicht losließ und die Quelle einer schweren Kränkung wurde.

Der Vater der Patientin war Zimmermann und wurde 1943 ein Opfer des Krieges. An ihn hat Frau U. kaum noch Erinnerungen. Eindrucksvoll blieb für sie der letzte Abend, bevor er wieder ins Feld mußte. Sie habe die ganze Nacht geweint. Bis 1949 lebten sie und ihre beiden älteren Schwestern allein mit der bei ihrer Verwitwung 30jährigen Mutter. Diese, selbst elternlos aufgewachsen, zog ihre Kinder äußerst streng, nicht selten mit Wutausbrüchen und Schlägen groß.

Schon als Kind hatte die Patientin Probleme mit einer Schwester, die bis heute das „schwarze Schaf" der Familie darstellt. Sie wird als in jeder Hinsicht unkontrolliert geschildert. 1949 fand die Mutter einen neuen Lebenspartner, einen Flüchtling, der von der etwa 14jährigen Patientin heftig und „total" abgelehnt wird. Sein Benehmen sei, verschärft durch äußerst beengte Wohnverhältnisse, kaum zu ertragen gewesen. Es habe Situationen gegeben, in denen sie ihre sonst immer gewahrte Fassung verloren und mit Gegenständen geworfen habe.

Mit 15 beginnt sie nach Abschluß der Volksschule eine Lehre als Versicherungskaufmann, nachdem die Mutter sie – gegen den Rat des Lehrers und gegen ihren eigenen Willen – aus der Schule genommen hat.

Mit 18 lernt sie einen 9 Jahre älteren, verheirateten Mann kennen, den sie – gerade 21 und volljährig geworden – heiratet. Uneins mit ihrer Situation zu Hause hatte Frau U. es zudem nicht leicht, sich in der weniger ärmlichen und sehr lebendigen Familie ihres Mannes zu behaupten.

Vier Jahre später wird das einzige Kind, ein Sohn, geboren. Vorher mußte von beiden Ehepartnern erst für gesicherte Verhältnisse gesorgt werden. Heute bedauert die Patientin, daß es bei einem Kind geblieben ist. Sie fühlt sich „irgendwie armselig".

Ihr Sohn wächst, nicht ohne ihr Sorgen zu machen, heran. Seine schulischen Leistungen bleiben hinter den „hohen Maßstäben" der Mutter zurück. Trotz ihrer Hilfe und zum Teil strenger Kontrollen muß er das Humanistische Gymnasium aufgeben. Er macht schließlich das Abitur auf einer Integrierten Gesamtschule; dort sei es ihm „geschenkt" worden.

Die cP wird von Frau U., soweit das möglich ist, nicht wahrgenommen. Ihre anfänglich offenen Informationsbedürfnisse scheinen nach kürzerer Zeit befriedigt, obwohl ihr objektives Krankheitswissen gering bleibt. Abgesehen von Krankenhausaufenthalten hat sie keinen einzigen Arbeitstag versäumt, obwohl sie sich manchmal kaum anziehen und fortbewegen konnte. 5 Jahre lang hat sie statt der abzuleistenden 6 oft 8 und 9 Stunden für einen schwierigen und ungeliebten Chef gearbeitet – immer äußerst korrekt und mit hohem Einsatz. Auch in der Familie ist sie „der Motor" geblieben, der sie immer war. „Daher darf ich keine Probleme haben. Insofern störe es sie, daß ihr Mann ihre Krankheit – auch vor anderen – dramatisiere und sie ebenso wie ihre Mutter mit paramedizinischen Ratschlägen überschütte. Wenn man diesen dann nicht folge, sei man am Ende „selbst schuld, daß man noch krank ist".

Die Beziehung zu ihrem Mann bleibt für uns blaß. Über sexuelle Probleme wissen wir nichts. Beruflich ist er oft länger von zu Hause fort; die Patientin hat sich „daran gewöhnt".

Seit Mitte 1981 wird sie mit einem oralen Goldpräparat behandelt. Zusätzlich nimmt sie – nach Bedarf und ebenfalls ohne schwere Nebenwirkungen – nonsteroidale Antirheumatika. Die immer wieder auftretenden Kniegelenksentzündungen wurden mit lokalen Anwendungen, intraartikulären Injektionen und schließlich auf der linken Seite mit einer Synovektomie behandelt. Die cP ist trotz eines eher gutartigen Krankheitsverlaufes weiter kontroll- und behandlungsbedürftig. Fast nie ist Frau U. schmerzfrei, auch beim Gehen ist sie behindert. Wir mußten sie zweimal kurzfristig stationär aufnehmen (einmal wegen des Verdachtes auf eine iatrogene bakterielle Gonarthritis); wir haben sie mehr als zwanzigmal ambulant gesehen.

Im Lichte psychometrischer Untersuchungen und standardisierter Befragungen erscheint die Patientin kaum aggressiv und nur grenzwertig depressiv bei geringer allgemeiner Klagsamkeit. Deutlich hat sich im Krankheitsverlauf eine Ängstlichkeit ausgeprägt. Auch die emotionale Isolation hat zugenommen. Nach wie vor verfügt sie bei einem großen Bekanntenkreis über nur wenige nahestehende Personen (den Ehemann, den Sohn und mit Einschränkungen die Mutter); und nach wie vor fehlt es an einem Vertrauten. Am nächsten stehe ihr der Sohn. Inzwischen fühlt sie sich durch die cP, von der sie ihr Leben lang begleitet zu werden glaubt, deutlich im Aussehen beeinträchtigt und weniger anziehend.

Ihrem Arzt ist sie eine sehr sympathische und nicht schwierige Patientin. Sie wirkt offen und lebhaft, gepflegt und sportlich, jünger aussehend. In der Behandlung erscheint sie selbständig und zuverlässig. Bei Komplikationen oder in Schubsituationen hat sie sich immer von sich aus gemeldet. Der Arzt hatte nie das Gefühl, manipuliert zu werden, auch wenn ihm klargeworden ist, daß Frau U. das, was sie mitteilt, vergleichsweise gut kontrolliert und emotional neutralisiert. Über ihre augenblickliche psychosoziale Situation spricht sie selten spontan; ihre Biographie lag jahrelang im dunkeln. Diese Verschlossenheit blieb lange unbemerkt.

1983 wurde Frau U. von heute auf morgen gekündigt. Zur Begründung wurde ihr die ungewisse Prognose ihrer Erkrankung genannt.

Nach allem Gesagten ist verständlich, daß die Patientin hier an einem Lebensnerv getroffen wurde – um so mehr, als sie sich völlig sicher fühlte. „Etwas Schlimmeres habe ich noch nie erlebt." Dennoch blieb sie auch in dieser Situation beherrscht, sie wurde nicht offen zornig und weinte erst zu Hause.

Nach einer kurzen arbeitsgerichtlichen Auseinandersetzung wurde die Kündigung überraschend zurückgenommen. Frau U. hatte also ihren Arbeitsplatz wieder.

Der bisher letzte Kontakt erfolgte Anfang 1989, nach einer mehr als dreijährigen Pause und auf unsere Initiative. Sie sei nicht mehr gekommen, weil es immer irgendwie gegangen wäre und wir uns nicht gemeldet hätten.

Abgesehen von einer kurzen Unterbrechung ist die orale Goldtherapie bis heute fortgeführt worden, im wesentlichen ohne hausärztliche Mitwirkung.

Die jetzt grenzwertig seropositive cP ist im Bereich des rechten Handgelenks und vor allem des linken Kniegelenks weiter aktiv. Röntgenologisch ist es zu einer Gelenkspaltverschmälerung, zu multiplen Erosionen und zu Ankylosen (im Handwurzelbereich) gekommen.

Möglicherweise wird bald eine Reoperation des linken Kniegelenks (partieller oder totaler Gelenkersatz?) notwendig. Seit längerem erfolgt keine konsequente Physiotherapie. Die Muskulatur des linken Oberschenkels ist deutlich atrophisch.

Frau U. hat Wohnung und Arbeitsplatz nicht gewechselt; sie arbeitet in der alten Firma mit jetzt 30 Wochenstunden und will dies noch ein bis zwei Jahre fortführen.

Ihr Mann ist inzwischen im Vorruhestand, der Sohn hat geheiratet, sie sind Großeltern geworden.

Ihrem Arzt begegnet sie weiter ruhig und freundlich, aber auch verschlossen und zurückgezogen.

Frau U. hat, scheint es, ein bewegungsarmes Gleichgewicht gefunden. Die Tendenz, die Krankheit zu übersehen und therapeutisch zu vernachlässigen, besteht weiter.

Subjektiv geht es ihr „mittelmäßig", „ich kann nicht mehr normal gehen".

Ein deutlicheres Bild zeichnen die klinimetrischen Befunde: Die Schmerzen sind ausgeprägt, die „gesamte Verfassung" wird als schlecht und unbefriedigend eingestuft, das Befinden habe sich seit dem letzten Kontakt stark verschlechtert.

Ängstlichkeit und Depressivität haben deutlich zugenommen.

45.1.3.1   Die chronische Polyarthritis – eine psychosomatische Erkrankung?

Ausgangspunkt dieser Fragestellung ist die weit zurückzuverfolgende und immer wieder berichtete Beobachtung bestimmter Persönlichkeitszüge bzw. eines abgrenzbaren Charakters von cP-Patienten. So schreibt Lichtwitz über eine seiner Patientinnen:

„Das Gesicht als Schaufenster der Seele zeigt an, was das nähere Studium des Charakters der Arthritiker ergibt. Frauen im späten Stadium der deformierenden Arthritis gleichen in ihrem Wesen Anna Scheede. Es gibt nicht freundlichere und geduldigere Patienten als diese. Sie klagen nicht, sie machen keine Vorwürfe, wenn nichts hilft. Ich habe immer den Eindruck, als ob sie im Sinne hätten, den Doktor zu trösten und um Verzeihung zu bitten, daß alle seine Bemühungen erfolglos sind. Sie verlieren nie das Vertrauen, grüßen jeden Morgen mit dem selben stillen Lächeln und scheinen glückliche Menschen zu sein, wenn der Doktor die Handarbeiten bewundert, die sie mit ihren armen Händen vollbringen" (Lichtwitz, 1936, S. 120).

Dieses Verhalten ist – historisch gesehen – Gegenstand zweier unterschiedlicher Erklärungsversuche geworden, eines tiefenpsychologisch-psychogenetischen – verbunden mit dem Namen des ärztlichen Psychoanalytikers Alexander – und eines antipsychosomatisch-anthropologischen – verbunden mit dem Namen des Internisten Plügge.

Beide Autoren vertiefen das Verständnis dessen, was Lichtwitz an seiner Patientin als Güte, stille Freundlichkeit und Geduld auffiel.

Plügge legt diese Persönlichkeitszüge als (radikal verstandene) Selbstlosigkeit aus; Alexander begreift sie als Selbstbeherrschung.

Die Auffassung von F. Alexander

„Im Erwachsenendasein beweisen sie *(die cP-Kranken)* starke Beherrschung in bezug auf jeden emotionalen Ausdruck ... In Ergänzung zu dieser Neigung, ihre Gefühle zu beherrschen, neigen diese Kranken auch dazu, ihre menschliche Umgebung, ihre Ehemänner und Kinder zu beherrschen" (1977, S. 156).

Selbst- und Fremdbeherrschung scheinen auf allen Lebensgebieten, auch dem der Sexualität, miteinander verwoben. Was steht nach Alexander hinter diesem Verhalten?

„Der allen Fällen gemeinsame psychodynamische Hintergrund besteht in einem chronisch gehemmten, feindseligen, aggressiven Zustand, einer Aufständischkeit gegen jede Form von äußerlichem oder innerlichem Druck..." (S. 158).

Er läßt sich „bis auf eine höchst charakteristische frühe Familienkonstellation" zurückverfolgen: Einer starken, beherrschenden und fordernden Mutter stand ein mehr anlehnungsbedürftiger nachgiebiger Vater gegenüber. So ergibt sich als Kern der feindseligen Impulse der Patientinnen eine „gehemmte Aufsässigkeit gegen die Mutter", die später auf sämtliche Familienmitglieder und alle Männer „übertragen" werde. Diese finde um die Zeit der Pubertät Ausdruck und äußerliche Entlastung in körperlicher (männlicher) Aktivität - z.B. in Kampfsportarten. Im Inneren müsse das Gewissen beruhigt werden. So lasse sich die auffällige „dienende Haltung anderen gegenüber" verstehen.

Es wundere daher nicht, daß sich die Arthritis vor allem in solchen biographischen Situationen manifestiere, die dieses labile Gleichgewicht von Dienen und Herrschen bedrohen oder auslenken.

Als somatischen Pathomechanismus, der den emotionalen Konflikt in die manifeste Arthritis überführt, faßt Alexander einen gesteigerten Muskeltonus ins Auge:

„Wir nehmen an, daß gehemmte feindselige Antriebe zu gesteigertem Muskeltonus führen. Die feindseligen Antriebe suchen Abfuhr durch Muskelkontraktionen, aber ihre Hemmung führt zu gleichzeitiger Steigerung des Antagonistentonus. Diese gleichzeitige Erregung des Antagonisten kann für die Gelenke ein Trauma bedeuten und einen bereits in Gang befindlichen Krankheitsprozeß fördern, der vielleicht eine noch unbekannte somatische Grundlage besitzt" (S. 162).

So vollendet sich Alexanders relativ geschlossener Versuch der psychogenetischen Erklärung der cP: Frühe Familienkonstellation → prädisponierender Persönlichkeitsfaktor + auslösende emotionale Faktoren → gesteigerter Muskeltonus → Arthritis.

Alexander übersieht nicht die hypothetischen und spekulativen Elemente seines Versuches und äußert sich auch vorsichtig über die von ihm dennoch behauptete Spezifität der Zusammenhänge. Diese sieht er vor allem:

„In der Konfliktsituation ..., in der die verschiedenen Faktoren in Erscheinung treten ... Weiterhin findet sich eine Spezifität in der Art, in der sich eine motivierende psychologische Kraft ausdrückt" (S. 45).

So hätten sorgfältige psychodynamische Untersuchungen gezeigt,

„daß gewisse Störungen vegetativer Funktionen sich mit spezifischen emotionalen Zuständen direkt korrelieren lassen, viel stärker als mit oberflächlichen Persönlichkeitsbildern, wie sie in den Persönlichkeitsprofilen beschrieben werden" (S. 49).

So wie Alexander sich auf vorhergehende Untersuchungen anderer, nicht-psychoanalytisch arbeitender Autoren beruft (z.B. Booth, 1937; Halliday, 1942) und sie glaubt bestätigen zu können, haben sich später andere Autoren in seiner Nachfolge gesehen. Cobb (1959) formulierte - nach der sehr ausführlichen Untersuchung eines etwa 50jährigen männlichen promovierten Patienten mit einer cP und einem Ulcus duodeni - seine Hypothese der unterdrückten Feindseligkeit („contained hostility").

Unter den deutschsprachigen Autoren scheinen uns besonders Cremerius (1955), Jores (1960), Pieringer (1978) und Jordan (Jordan und Schmidt, 1988) von den Überlegungen Alexanders geprägt, auch wenn sie im einzelnen andere Akzente (wie z.B. Cremerius auf das „retentive Antriebserleben") setzen.

Die Auffassung von H. Plügge

In bewußter und scharfer Absetzung gegen diese tiefenpsychologische Auffassung von der Psychogenese der cP verdeutlicht Plügge (1953) seine „anthropologischen Beobachtungen". Er hebt die Gefahr hervor,

„daß man meinen könnte, daß die Analyse unbewußter triebhafter Prozesse das Substrat einer inneren Erkrankung zutage fördern könne". - „Das pathologisch-morphologische, also leibliche Substrat *(der Erkrankung)* geht ja im Prozeß einer solchen analytischen, rein psychologischen Untersuchung weitgehend verloren" (S. 232/233).

Plügges Ansatzpunkt ist die angeblich schon prämorbid ausgeprägte Selbstlosigkeit der späteren cP-Kranken:

„Es handelt sich ... da um Menschen, die vor ihrer Erkrankung in einer stillen und unscheinbaren Art in einem besonderen Maße aktiv, tüchtig, unermüdlich, unentwegt tätig und entschlossen zupackend waren. Immer wieder hört man von den Angehörigen, daß es so etwas wie ‚famose' Menschen waren" (S. 236).

Sie waren

„selbstlos. So selbstlos, wie man es eigentlich nur sein kann, wenn das Selbst nicht recht zu Worte kommt, oder wenn das eigene Selbst nicht recht bemerkt wird" (S. 239).

Genau diese Selbstlosigkeit trage auch ihre spätere

„Geduld, ihre Bescheidenheit und ihre Genügsamkeit. Sie sind meist ganz und gar nicht schwierig, nicht unzufrieden, nicht nörgelig. Das bedeutet aber in diesem Falle, daß sie auf eine auffällige Weise wenig an ihre Symptome gebunden zu sein scheinen" (S. 238).

In beiden Verhaltensweisen – vor und in der Erkrankung – finde sich „das gleiche auffallende Minus an Anerkennung der eigenen Leiblichkeit, die gleiche Verkümmerung" (S. 239). Die Frage ist nun,

„ob nicht diese Arthritis einfach auch ein Ausdruck dieser Kümmerform ist, eine somatische Teilerscheinung der gesamtpersonalen Dürftigkeit" (S. 242).

Plügge bejaht und verweist darauf, daß man „ohne die Wirklichkeit zu verlassen"

„die chronische Arthritis auch als eine Erkrankung des ausgestaltenden, formgebenden und gliedernden Gewebes ansehen (kann); und zwar in dem Sinne, daß bei dieser Erkrankung der Gliederungsprozeß entweder nicht bis zur vollkommenen Differenzierung gelangt oder geschädigt wird. Dieser Gliederungsschaden spielt sich nun auffälligerweise gerade dort an unseren Extremitäten ab, wo die Gliederung am differenziertesten durchgeführt ist, an den distalen Partien, an den kleinsten Gelenken" (S. 243).

Das heißt weitergeführt,

„an einem ausgezeichneten Ort der Repräsentanz des Menschlichen … einem ausgezeichneten Organ unseres Ausdruckvermögens" (S. 244). – „Dort also, wo Personales und Individuelles am ungebrochensten sich leiblich zu erkennen geben, ist der morphologisch erkennbare Schaden am deutlichsten" (S. 244).

Offen und nicht grundsätzlich beantwortet bleibt seine Frage, ob die cP einer „ab ovo angelegten Kümmerform" entstammt oder ob ein „postnataler Schaden", z.B. im Sinne eines „frühkindlichen Traumas", die Erkrankung bahnt, die sich dann manifestiert, wenn besondere Entwicklungs- und Differenzierungsanforderungen, z.B. an der Schwelle des Klimakteriums, an den Menschen herantreten.

Auch Plügge denkt sich also die Entstehung der (adulten) cP als Ineinanderwirken konstitutioneller und situativer Momente. Im scharfen Gegensatz zu Alexander ist er jedoch der auf Viktor v. Weizsäcker zurückgehenden Auffassung, daß Körperliches und Seelisches zwei Teilerscheinungen des einen Lebens seien, daß – anders gesagt – Körperliches Seelisches und Seelisches Körperliches „interpretieren" könne. Psychopathologisches bewirke nicht Pathomorphologisches, sondern erläutere oder vertrete es; in beiden Erscheinungsformen drücke sich das gleiche zentrale Defizit, „die personale Ärmlichkeit des chronischen Arthritikers" aus. Neben und nach Plügge haben sich in verwandter Weise vor allem W. Kütemeyer (1963) und A. Weintraub (1967) mit der cP beschäftigt.

Zusammenfassung

Bei allen Unterschieden zwischen den beiden eben referierten Theorien lassen sich auch eine Reihe von Gemeinsamkeiten formulieren:
- Beide Autoren gehen von einem ähnlich beschriebenen Verhaltensmuster der cP-Kranken aus; beide wollen es „mit großer Regelmäßigkeit" oder in einem „sehr hohen Prozentsatz" beobachtet haben.
- Dennoch kommen sie zu sehr unterschiedlichen Auffassungen über die spezifischen Entstehungsbedingungen und Hintergründe dieses Verhaltens. Da Plügge sich bewußt gegen die psychogenetischen Theorien Alexanders abgegrenzt hat, wollen wir nicht versuchen, beide in sich geschlossenen Ansätze auf einer höheren Ebene zu integrieren.
- Die genannte Unterschiedlichkeit gründet offensichtlich in unterschiedlichen theoretischen Vorentscheidungen: Alexander bewegt sich im Rahmen einer psychogenetischen, Plügge im Rahmen einer weitergefaßten und, wie er meint, vorrangigen anthropologischen Fragestellung. In der Ablehnung konversionsneurotischer Interpretationen verfolgt und prüft Alexander (vgl. Alexander et al., 1968) Spezifitätsannahmen über die Bedeutung bestimmter Persönlichkeitskonflikte, Emotionen und sozialer Situationen für die Bahnung und Auslösung bestimmter körperlicher Reaktionen bzw. Erkrankungen. Plügges Haltung bleibt hier undeutlich: Wieweit sich die von ihm herausgearbeitete „gesamtpersonale Dürftigkeit" auch bei Nicht-Arthritikern findet und warum sie dort zu keinem Glieder(ungs)schaden führt, wird nicht behandelt. Wichtig ist ihm vor allem das von V. v. Weizsäcker herausgearbeitete Prinzip der Äquivalenz zwischenmenschlichen Verhaltens, seelischer Befindlichkeiten und körperlicher Prozesse. In diesem Sinne scheint seine theoretische Voreingenommenheit ausgeprägter.
- Weder Alexander noch Plügge formulierten eine Theorie mit Ausschließlichkeitsanspruch. Beide lassen ausdrücklich Platz für zusätzliche biomedizinische oder psychosoziologische Hypothesen; beide sind sich der spekulativen Elemente ihrer Überlegungen bewußt.
- Beide Autoren gleichen sich darin, daß ihre Theorien kaum fortentwickelt wurden. Weitere anthropologische Studien sind uns nicht bekannt. Auch psychoanalytische Folgeuntersuchungen sind rar (vgl. Ludwig, 1955; Barchilon, 1963; Lefer, 1972; Levitan, 1981; Jordan et al., 1987); sie haben nicht wesentlich weitergeführt.

Wenn wir uns aber über das Bedenken (s.u.) hinwegsetzen, ob ein psychoanalytisch gewonnenes Verständnis in Test-, Interview- oder Fragebogenergebnissen empirische Belege finden kann, dann hat Alexanders Theorie zwar keine Weiterentwicklung, wohl aber für verschiedene ihrer Elemente eine Stütze gefunden – unter anderem durch die methodisch ganz unterschiedlichen Studien von Cormier et al., 1957; Cobb, 1959; Moos und Solomon, 1965; Meyerowitz et al., 1968; Shochet et al., 1969; Levitan, 1981.

Auch wir fanden mehrfach eine unterdurchschnittliche spontane Aggressivität, Erregbarkeit und Offenheit (Mattussek und Raspe, 1988). So ist es wahrscheinlich, daß ein Teil der untersuchten cP-Patienten eine Psychodynamik erkennen läßt, wie sie Alexander „mit großer Regelmäßigkeit" beobachtete. Der zentrale Konflikt dieser Kranken scheint in einer ausgeprägten, aber schuldhaft erlebten und ängstlich abgewehrten Aggressivität zu liegen.

Dennoch fällt es fast allen oben genannten Autoren der Übersichtsreferate (Ausnahme: Solomon, 1981) schwer, diesem die Persönlichkeit (wie?) vieler cP-Patienten prägenden Konflikt eine ätiologische Signifikanz zuzuerkennen (Krüskemper, 1985; Köhler, 1987).

Die Skepsis hat vor allem drei Gründe: Zuerst einen empirischen, nachdem es klarwurde, daß bei weitem nicht alle cP-Patienten die herausgearbeiteten Persönlichkeitszüge erkennen lassen. Wie ist die Krankheit bei den Arthritikern entstanden, die ganz andere seelische Konflikte zeigen; und was ist mit jenen, die „trotz" entsprechender Konflikte gesund bleiben bzw. an einer Hypertonie oder einer anderen Erkrankung leiden?

Der zweite Einwand ist methodischer Art: Fast alle Studien sind Querschnittsuntersuchungen an langjährig und schwer erkrankten cP-Patienten aus klinischen oder poliklinischen Einrichtungen. Nur selten konnten epidemiologisch identifizierte, gerade erkrankte, jüngere Patienten mit gutartigen Verläufen prospektiv beobachtet werden. Damit steht der Geltungsbereich vieler Ergebnisse zur Diskussion.

Der dritte Einwand ist theoretischer Natur: Kennzeichnet die Suche nach ätiologisch signifikanten emotionalen Konflikten nicht eine inzwischen überwundene Stufe psychosomatischen Denkens? Befestigt sie nicht einen psychosomatischen Dualismus? Geht in ihr nicht, um noch einmal Plügge zu zitieren, „das leibliche Substrat ... (der cP) weitgehend verloren"?

– Schließlich ist beiden Theorien gemeinsam, daß sich aus ihnen keine sicheren therapeutischen Indikationen für die Behandlung bereits Erkrankter ergeben. Sie beinhalten vielleicht sogar ein belastendes Moment: Beide Autoren verführen zu der (allerdings so nicht ausgesprochenen) Annahme, daß gelungene, ausgewogene, glückliche Menschen von Arthritis verschont bleiben müßten. Jedenfalls zeige das Auftreten dieser Krankheit einen verborgenen Mangel, eine (bei Kütemeyer auch moralische) Unzulänglichkeit an. Vielen Kommentatoren ist aufgefallen, daß vor allem nach negativen Persönlichkeitsmerkmalen gesucht worden ist oder daß die hervorgehobenen Charakterzüge negativ interpretiert worden sind.

Die psychometrische Forschung

Die Suche nach cP-typischen Persönlichkeitsmerkmalen bediente sich nach dem Zweiten Weltkrieg weniger der früher eingesetzten offenen Interviews oder projektiver Verfahren (z.B. des Rorschach-Tests), sondern vor allem hochstandardisierter psychometrischer Instrumente, wie z.B. des MMPI (s.u.) oder des 16-PF von Catell (Moldofsky und Rothman, 1971) oder des MPI von Eysenck (Gardiner, 1980). Damit ist eine Kontinuität der Forschung nur scheinbar gewahrt.

Die psychometrischen Persönlichkeitsfragebogen können ihrem Anspruch und ihren Ergebnissen nach weder die tiefenpsychologisch aufgedeckten emotionalen Konflikte noch die anthropologisch beobachteten personalen Unzulänglichkeiten abbilden, die Alexander oder Plügge beschrieben haben. Sie richten sich vielmehr auf die quantifizierende Erfassung von einigen wenigen, analytisch getrennten und zuverlässig erhebbaren Persönlichkeitseigenschaften.

Die Inventare sind damit nicht in der Lage, cP-**spezifische** Konflikte, Charakterzüge oder Verhaltensweisen zu eruieren. Ihre Anwendung und Auswertung führt zu Meßwerten, die die individuelle Ausprägung ganz allgemeiner und weitverbreiteter Persönlichkeitseigenschaften im Vergleich mit der statistischen Norm abschätzen lassen. Diagnostische Spezifität kann nur von solchen Verfahren beansprucht werden, die bei (den meisten) Gesunden oder den nicht an cP-Erkrankten negative Resultate ergeben. Es liegt auf der Hand, daß so etwas mit den bisher eingesetzten Inventaren nicht möglich sein kann.

Äußerstenfalls ließe sich eine Spezifität in einer cP-typischen, bei anderen Krankheiten aber vermißten **Kombination** solcher einzelnen Eigenschaften erblicken.

Die 1949 mit dem MMPI begonnenen Studien geben hierfür keine Anhaltspunkte. Fast regelmäßig zeigen die untersuchten Gruppen von cP-Kranken über die Norm erhöhte Mittelwerte auf den Skalen der sog. neurotischen Trias (Hypochondrie, Depression, Hysterie; vgl. Polley et al., 1970; Moos und Solomon, 1964a). Solche Mittelwerterhöhungen liegen selten weit oberhalb der Referenzbereiche; und sie liegen unterhalb derer von „neurotischen" Kontrollpersonen (Ward, 1971; Crown und Crown, 1973b). Damit dürfte nur ein Teil der Patienten signifikante Auffälligkeiten zeigen.

Kontrollierte Studien haben es wahrscheinlich gemacht, daß sich solche Veränderungen (manchmal sogar stärker) auch bei anders chronisch Kranken ausprägen (Spergel et al., 1978). Sie repräsentieren offenbar ein **Syndrom des chronisch Kranken** (Krüskemper und Schejbal, 1980).

Die zugrundeliegenden Einzelfragen der Skalen der neurotischen Trias des MMPI weisen darüber hinaus eine besondere Nähe zu körperlichen cP-Symptomen auf (Polley et al., 1970; Nalven und O'Brian, 1964; Pinkus et al., 1986).

Es erschien vielversprechend, die insgesamt heterogene Gruppe der cP-Kranken nach bestimmten Gesichtspunkten zu unterteilen und homogenere Untersuchungsgruppen zu bilden. Seropositive Patienten – in ihrem Serum läßt sich der Rheumafaktor, ein 19S-Antiglobulin der Klasse IgM nachweisen – zeichneten sich in mehreren Studien durch niedrigere Neurotizismuswerte aus als seronegative (Crown und Crown, 1973a; Rimón, 1973; Gardiner, 1980; Vollhardt et al., 1982). Dies konnten Kiviniemi und Lyytikäinen (1982) sowie Raspe et al. (1986) nicht bestätigen.

Vergleichbares wurde bei Patienten gefunden, die bisher (noch) nicht mit Steroiden behandelt worden waren (Moldofsky und Rothman, 1971).

Patienten mit einem akuten Beginn ihrer Erkrankung kontrollierten ihre aggressiven Gefühle nach der Studie von Rimón (1969) weniger stark als solche mit einem schleichenden. Ähnliches haben Moos und Solomon (1964b) in Zusammenhang mit einer raschen Krankheitsprogredienz beobachtet.

Zusammengefaßt zeigen die jüngeren (seit Mitte der siebziger Jahre seltener werdenden) psychometrischen Studien, daß vor allem länger erkrankte cP-Patienten häufiger „neurotische" Züge aufweisen als gesunde Kontrollpersonen. Verglichen mit körperlich gesunden Psychoneurotikern erscheinen sie allerdings weniger gestört, während sie sich von anders chronisch Kranken nicht wesentlich unterscheiden.

Die Hypothese einer spezifischen Rheumapersönlichkeit kann in den psychometrischen Untersuchungen, gleichgültig was ihre Ergebnisse sind, keine Stütze finden. Es ist noch einmal hervorzuheben, daß hochstandardisierte Persönlichkeitsinventare keinen Zugang zu den Persönlichkeitsschichten gewinnen können, die in einer biographischen Anamnese oder einem psychoanalytisch orientierten Interview zu erreichen sind. So würden ihre Ergebnisse auch weit überinterpretiert, wenn sie ernsthaft als Beleg für eine Alexithymie von cP-Kranken herangezogen würden (wie bei Vollhardt et al., 1982; vgl. Shands, 1975).

Die psychometrische Diagnostik erlebt in allerjüngster Zeit eine überraschende Renaissance:

Es wurden eine Reihe von Fragebögen entwickelt, um systematisch zu erfassen, wie der Kranke sich **präsentiert** und mit sich, seiner Krankheit und den Therapeuten **umgeht.**

Die Instrumente zielen unter der Annahme stabiler Persönlichkeits- und Verhaltenszüge auf die Untersuchung einer gelingenden oder mißlingenden Krankheitsbewältigung („coping"; vgl. Beutel, 1988).

Nicassio et al. (1985) haben einen „arthritis helplessness index" entwickelt und später zu einem „rheumatology attitudes index" erweitert (Callahan et al., 1988; vgl. Roberts et al., 1986).

Im Anschluß an Felton et al. (1984) erprobten Parker et al. (1988) einen Fragebogen zur Untersuchung eines breiten Spektrums von „coping strategies in rheumatoid arthritis" (vgl. Schüßler et al., 1988).

Flor und Turk (1988) konzentrierten sich speziell auf Krankheits- und Selbstkognitionen, wie sie als „self evaluation process" auch von Blalock et al. (1988) untersucht wurden.

Im Rahmen von Quer- und Längsschnittstudien ist gezeigt worden, daß die Coping-Variablen sowohl mit einer Reihe von Persönlichkeitszügen der Kranken (z.B. Depressivität, Ängstlichkeit) als auch mit der Selbstschilderung der rheumatischen Beschwerden (z.B. Schmerz) und Behinderungen korreliert sind (vgl. Spiegel et al., 1988; Mason et al., 1988).

Diese Befunde haben keine Relevanz für die bisher behandelte psychoätiologische Problematik. Möglicherweise können sie für die nächste Fragestellung (Abschn. 45.1.3.2) Bedeutung gewinnen, auch wenn bis heute keine überzeugenden Daten vorliegen.

45.1.3.2 Psychosoziale Einflüsse auf Ausbruch, Verlauf und Ausgang der chronischen Polyarthritis

Schon früh ist der Einfluß von Lebenskrisen auf den Ausbruch einer cP untersucht worden (Thomas, 1936). 1939 haben Cobb und Mitarbeiter in einer kontrollierten Studie zu zeigen versucht, daß Beginn und Verschlimmerungen einer cP sich zu Zeiten persönlicher Krisen ereignen: Einen signifikanten „synchronism of social factors and arthritic symptoms" stellten sie bei 62% ihrer Fälle und 12% ihrer Kontrollen fest. Interessanterweise bedienten sie sich der von dem Psychiater A. Meyer eingeführten „life chart", dem wesentlichen Vorläufer der später von Holmes und Rahe (s.u.) ausgearbeiteten „Social Readjustment Rating Scale" zur Belastungsmessung von lebensverändernden Ereignissen.

Die bisher letzte solcher Studien wurde 1981 von Baker und Brewerton veröffentlicht: 15 von 22 Patientinnen mit einer frühen cP berichteten von negativen Ereignissen im Jahr vor Ausbruch der Erkrankung (vs. 8/22 unter den Kontrollpatientinnen).

Beide Untersuchungen unterstellten keine spezifische Bedeutung der Ereignisse, etwa vor dem Hintergrund der oben genannten „prädisponierenden Persönlichkeitsfaktoren" im Sinne Alexanders.

Zwischen ihnen liegt eine Reihe methodisch wenig befriedigender Studien (Pancheri et al., 1978; Rimón, 1969; Meyerowitz et al., 1968). Die Untersuchung von Hendrie und Mitarbeitern (1971) erbrachte ein negatives Ergebnis: Die mit der „Social Readjustment Rating Scale" (vgl. Rahe, 1978) bestimmten mittleren Belastungswerte trennten nicht zwischen 21 frühen Arthritikern und 37 gesunden Kontrollpersonen.

Neben allen bekannten Problemen der Life-event-Forschung ist hier noch zu berücksichtigen, daß sich die cP nur in etwa einem Drittel der Fälle akut, d.h. innerhalb von Stunden bis Tagen manifestiert. Häufiger beginnt sie schleichend, im Verlaufe von Wochen und Monaten. In wenigstens einem Viertel der Fälle gehen den Gelenksymptomen uncharakteristische Prodromi (s.o.) voraus, die jede bestimmte Aussage über den Zeitpunkt des Erkrankungsbeginns unsicher machen.

Zusammenfassend scheint es uns bisher nicht gesichert, daß akute lebensverändernde Ereignisse dem Ausbruch einer cP als ein unspezifisch-präzipitierender Faktor überzufällig häufig vorausgehen (vgl. Köhler, 1987; Wallace, 1987). Zur Bedeutung bestimmter Ereignisse/Ereignisklassen und zur Wirkung bestimmter Ereignisse bei bestimmten Patienten (-gruppen) liegen keine verallgemeinerungsfähigen Ergebnisse vor – auch wenn im einzelnen ebenso beeindruckende wie stimulierende Fallgeschichten berichtet worden sind (Kütemeyer, 1963).

Wir finden nur wenige prospektive Beobachtungsstudien zum Einfluß psychosozialer Faktoren auf den Verlauf und Ausgang einer cP: Crown und Mitarbeiter (1975) konnten in der Erstuntersuchung von 102 Patienten mit einer frühen cP psychometrisch (mit dem Middlesex Hospital Questionnaire, MHQ) keine Merkmale identifizieren, die nach 2 und 4 Jahren als

Prädiktoren eines ungünstigen klinischen Verlaufes hätten gelten können. Gardiner (1980) fand, „daß keine einzige psychologische Variable eine signifikante Beziehung zur Krankheitsaktivität" am Ende eines dreijährigen Beobachtungsintervalls aufwies. Auch in der dreijährigen Kohortenstudie von McFarlane und Brooks (1988a) ließ sich kein konsistenter Einfluß von Persönlichkeits- und Befindensvariablen auf die spätere somatische Situation von 30 ambulant betreuten cP-Kranken herausarbeiten. Die psychologischen Merkmale scheinen jedoch die spätere Selbsteinschätzung der funktionellen Kapazität mitzubestimmen (McFarlane und Brooks, 1988b).

Risiken für den Krankheitsverlauf können auch im Verhalten des Kranken liegen. An dieser Stelle wollen wir beispielhaft zwei (künstlich isolierte) Facetten des Krankheitsverhaltens behandeln.

Wenn wir der ärztlich vorgeschlagenen Therapie (z.B. mit Basistherapeutika oder operativen Eingriffen) eine den natürlichen Verlauf beeinflussende Wirkung zuschreiben – und nach unserer Auffassung sind wir dazu berechtigt –, dann müssen wir in diesem Abschnitt auch die Probleme der therapeutischen Kooperation („non-compliance") und der Inanspruchnahme außerschulischer (paramedizinischer) Heilmethoden streifen. Mit beidem ist – bezogen auf den gesamten Krankheitsverlauf – bei wenigstens 50% der cP-Kranken zu rechnen (Deyo et al., 1981; Jette, 1982; Ulreich et al., 1982; Pullar et al., 1982; Kronenfeld und Wasner, 1982; Raspe und Ritter, 1982; Struthers et al., 1983).

Bisher liegen keine kontrollierten Studien über die tatsächliche Bedeutung dieser Verhaltensweisen für den Krankheitsverlauf vor. Aufgrund theoretischer Überlegungen und kasuistischer Erfahrungen ist nicht davon auszugehen, daß sie in jedem Falle negative Konsequenzen nach sich zögen:

Bei den nonsteroidalen Antirheumatika (NSAR) gibt es fast nie ein Complianceproblem: Wir leiten die Patienten in der Regel zu einer verständigen Selbstbehandlung an, da bisher nicht gezeigt werden konnte, daß eine konsequente NSAR-Behandlung irgendeinen Einfluß auf zentrale Krankheitsprozesse bzw. den Krankheitsverlauf ausübt. Vor Steroiden haben die meisten der erfahrenen Kranken so viel Angst, daß die zu befürchtende Mehr- bzw. Dauereinnahme eine seltene (dann aber unter Umständen gefährliche) Erscheinung ist. Nur bei der sog. Basistherapie mit langwirkenden Medikamenten sind negative Effekte von einer Mindereinnahme zu befürchten. Sie dürfte sich aber nur dann auf den Behandlungserfolg auswirken, wenn sie häufig und anhaltend auftritt. Eine gelegentliche Nichteinnahme wird auf die Blutspiegel dieser langwirkenden Medikamente mit Eliminationszeiten, die zum Teil im Bereich von Monaten liegen, keinen signifikanten Einfluß haben.

Eine paramedizinische Therapie beinhaltet vor allem drei – längst nicht immer realisierte – Gefährdungen: eine finanzielle, infolge der von den Krankenkassen nicht erstatteten Kosten, eine unmittelbare durch die eingesetzten Präparate (z.B. schlecht gereinigte Eiweißfraktionen) und eine mittelbare dann, wenn tatsächlich „alternativ" behandelt wird, wenn sich der Patient also gedrängt sieht, die von uns verordnete Therapie abzusetzen (forcierte Noncompliance). Gegen die oft übertriebenen Gefahren müssen mehrere positive Funktionen einer „heterodoxen" Therapie abgewogen werden:

Diese beinhaltet in experimentellen Aktivitäten und eigenen Kontrollbemühungen die immer wieder geforderte **Selbstbeteiligung** des chronisch Kranken; sie hilft gegen **Hoffnungslosigkeit** und gegen das **Schuldgefühl,** etwas versäumt zu haben; sie bewahrt den **sozialen Frieden:** Etwa die Hälfte der paramedizinischen Konsultationen sind das Ergebnis sozialen Druckes von Familienangehörigen, Freunden, Nachbarn, Kollegen (Raspe, 1989b).

45.1.3.3 Psychosoziale Probleme im Verlauf einer chronischen Polyarthritis

Dieses Thema überzeugt auch strikt somatisch Denkende. Es ist uns gewiß, daß eine anhaltend schmerzhafte, behindernde und gestaltverändernde Erkrankung soziale und seelische Probleme mit sich bringen muß. Diese Gewißheit ist ja der (unbewußte) Hintergrund für das Erstaunen über die Güte, die stille Freundlichkeit und die Geduld von cP-Patienten gewesen. Man erwartete Unruhe, Unglück und Unausgeglichenheit und fand Personen, die sich anscheinend oder scheinbar seelisch im Gleichgewicht befanden.

So hat die Diskussion der psychosozialen Implikationen und Folgen einer bereits eingetretenen cP in den letzten Jahren dominiert – nachdem die beiden zuerst skizzierten Forschungsschwerpunkte nicht zu unumstrittenen Ergebnissen geführt hatten. Hierzu haben auch innerrheumatologische Entwicklungen beigetragen: Vor allem in den anglo-amerikanischen Ländern ist ein Bedürfnis nach Instrumenten entstanden, mit denen sich der Gesundheitszustand („health status"; Meenan et al., 1980), die Lasten („impact") des Arthrikers und die Verlaufs- und Therapieergebnisse („patient outcome"; Fries et al., 1980) patientennah messen lassen. Entsprechend der von Feinstein (1987) inaugurierten „Klinimetrie" sind wichtige Dimensionen (z.B. Mißempfinden), ihre Komponenten (z.B. Schmerz), Variablen (z.B. Schmerzintensität) und Instrumente (z.B. numerische Ratingskala) entwickelt worden, um die subjektiven Angaben der Patienten in gültige, zuverlässige und objektive Meßwerte überführen zu können.

Solche Erhebungsinstrumente bedeuten ohne Zweifel einen erheblichen Gewinn für die klinische Verlaufsbeobachtung, Prognostik und Therapiebeurteilung von cP-Kranken (vgl. Wolfe et al., 1988; Raspe, 1989); ihre Ergebnisse ergänzen die bisher übliche rein biomedizinische Betrachtung. Sie helfen vor allem in der Beobachtung und Dokumentation der vielschichtigen **Chronifizierungsprozesse.** Diese können somatische Pathomechanismen und Krankheitsmanifestationen, subjektive Beschwerden und das Befinden, Krankheitsvorstellungen und -kognitionen, seelische Affekte und Leiden, das Krankheitsverhalten und schließlich auch die sozialen Folgen der Krankheit (z.B. Arbeitsunfähigkeit, -losigkeit) erfassen.

Dennoch stehen sie in ähnlicher Weise wie die erwähnten Persönlichkeitsinventare (s.o.) in einer spannungsvollen Beziehung zum Selbsterleben des Patienten, zur gemeinsamen Wirklichkeit von Arzt und Patient und zum Erleben des Arztes.

Krankheitsmanifestation, Lasten und Leiden

Prinzipiell ist kein Lebensbereich denkbar, der von einer cP nicht berührt werden könnte. Da es unmöglich ist, auf alle Störungen einzugehen, wollen wir uns auf Grundsätzliches beschränken:

Die körperlichen, biomedizinisch (d.h. klinisch-rheumatologisch, laborchemisch, röntgenologisch usw.) zu beschreibenden Krankheitsmanifestationen ziehen eine unabsehbare Zahl von materiellen und immateriellen Lasten und seelischen Leiden nach sich. Dabei betont der Begriff „Lasten" den sachlichen Kern der Probleme; der Begriff „Leiden" zielt auf bestimmte seelische Verfassungen der Betroffenen.

So sind z.B. die Gestaltveränderung des cP-Kranken und die Unvorhersehbarkeit des weiteren Krankheitsverlaufes gleichermaßen unbezweifelbare Tatsachen; und man wird bei verschiedenen Kranken ein Mehr oder Weniger an Gestaltveränderung und an prognostischer Unsicherheit unterscheiden können. Daraus läßt sich eine unterschiedlich starke Belastung ableiten – eine Belastung der Gleichgewichte, die ein Kranker bisher in seiner Beziehung zu sich selbst, zu seinen Mitmenschen und zu seinem Höchsten gefunden hat.

Ob diese Gleichgewichte nun aber „nur" belastet oder ernsthaft bedroht oder dauerhaft ausgelenkt oder womöglich sogar stabilisiert werden, ist nicht allein eine Funktion der Lasten. Es hängt auch von der „Bedeutungserteilung" der betroffenen Personen, ihrem adaptiven Potential und dem Erfolg ihrer „balancierenden Maßnahmen" ab.

Wird z.B. eine vergleichsweise geringe Gestaltveränderung als Kränkung eines heftig idealisierten Körperbildes erlebt und als Bedrohung sozialer Attraktivität gewertet, kommen abfällige Bemerkungen der Mitmenschen hinzu und schlagen Kaschierungsversuche (subjektiv) fehl, dann mag diese an sich leichte Last zum Ausgangspunkt einer anhaltenden und tiefgehenden Niedergeschlagenheit werden, die die Bezeichnung „Leiden" verdient.

Wenn wir uns im folgenden auf die Lasten und die Leiden von cP-Patienten konzentrieren und die eingewobenen und vermittelnden Prozesse der Bedeutungserteilung und der Problembewältigung übergehen, dann hat das weniger sachliche als forschungsgeschichtliche Gründe: Es fehlt an entsprechenden Untersuchungen – von wenigen Ausnahmen abgesehen (Fagerhough und Strauss, 1977; Felton et al., 1984).

Sehr viel genauer untersucht worden sind allerdings – beginnend mit der Studie von Edwards und Mitarbeitern (1964) – die **krankheits- und behand-**lungsbezogenen Vorstellungen** (beliefs) von cP-Patienten (vgl. Markson, 1971; Arluke, 1980). In ihnen werden bewußtseinsnahe Anteile jener Bedeutungserteilungen faßbar, die auch für das Krankheitsverhalten eine wesentliche Rolle spielen (vgl. Eraker et al., 1984; Affleck et al., 1987). Sie geben Antworten auf die Sinn- und Schuldfragen chronisch Kranker: Warum ich, warum jetzt, was habe ich, was haben andere falsch gemacht (vgl. Lipowski, 1970)?

Exkurs: Krankheits- und behandlungsbezogene Vorstellungen von cP-Patienten

Wir werden kaum einen Kranken antreffen, der sich nicht schon bestimmte Vorstellungen darüber gebildet hat, woher seine Krankheit kommt und wie sie sich weiter entwickeln und auf sein Leben auswirken wird.

Diese Vorstellungen sind unterschiedlich weitläufig, bestimmt und in sich geschlossen. Sie können sich auf die Krankheit wie auf die Behandlung beziehen. Unter den krankheitsbezogenen sind uns die ätiologischen, die pathomorphologischen und die prognostischen Überlegungen wichtig geworden:

Als Ursache ihrer Erkrankung nannten etwa zwei Drittel der von uns Befragten exogene Noxen (Raspe und Mattussek, 1985). Besonders häufig wurden Kälte und Feuchtigkeit angeführt. 6% machten ärztliche Maßnahmen (Medikamente, Operationen, Vakzinationen) verantwortlich; etwas seltener erwähnt wurden seelische Belastungen. Häufiger sollten erbliche Einflüsse und hormonelle Umstellungen (z.B. Geburten, Wechseljahre) eine Rolle gespielt haben.

Genauere Analysen zeigen, daß viele Patienten mehr als eine Ursache annehmen (Mosaiktheorie):

„Die Polyarthritis ergibt sich aus vielen Steinchen. Beruflich hatte ich in einer Offsetdruckerei viel mit Wasser zu tun. Dann hatte ich Streß in der Familie, ein Eheproblem. Es ist auch anlagebedingt. Großvater, Bruder und ein Cousin leiden auch unter Rheumaschüben. Dazu kam eine große körperliche Belastung beim Hausbau, wir haben dann gleich im feuchten Haus gewohnt."

Wenige Patienten denken in pathogenetischen Ketten:

„Schon seit längerer Zeit bestanden nach einer Nierentuberkulose Nierensteine – dadurch kam es zu einer Ausscheidungsstörung und schließlich zu Ablagerungen in den Gelenken."

Nicht selten sind Selbst- und vor allem Fremdvorwürfe zu ahnen.

Von diesen ätiologischen können wir bei fast allen Patienten pathomorphologische bzw. pathophysiologische Vorstellungen abgrenzen. Meist handelt es sich um humoralpathologische Ablagerungstheorien, vorzugsweise werden Harnsäurekristallablagerungen zur Erklärung der Gelenkveränderungen angeführt.

Die heutigen Laientheorien haben also eine Quelle in (medizingeschichtlich) sehr alten ärztlichen Theorien. Anderes scheint in volksmedizinischen Traditionen zu wurzeln, wie die Einschätzung physikalischer oder hormoneller Einflüsse.

Vollständige und (aus medizinischer Sicht) zutreffende Vorstellungen fanden wir bei maximal 3%. Zwischen 21 und 41% der Antworten mußten wir als definitiv falsch einordnen.

Aus ärztlicher Sicht sind auch die prognostischen Einschätzungen der Patienten problematisch: Rund 50% gehen von einem Fortschreiten ihrer Erkrankung aus, oft mit ganz unangemessenen Befürchtungen über die vor ihnen liegenden Behinderungen („Rollstuhl") und „Verkrüppelungen".

Bewertet man die Patientenvorstellungen allein vor dem Hintergrund medizinischen Wissens, dann sieht man vor allem Defizite (wie z.B. Meyers und Hall, 1977; Grennan et al., 1978).

Dies ist unzureichend. Man entwertet damit die sie tragenden individuellen und sozialen Leistungen und unterschätzt ihre verhaltens- und befindensprägende Kraft.

Patientenvorstellungen sind auch und vor allem ein Mittel der Krankheitsbewältigung. Sie sind adaptiv dadurch, daß sie dem undurchsichtigen, unvorhersehbaren und schwer zu kontrollierenden Krankheitsgeschehen eine – oft komplexe – kognitive Gestalt geben. Wir kennen Kranke, die bei schleichendem Krankheitsbeginn und nach diagnostischen Irrfahrten geradezu erleichtert waren, eine somatische Diagnose zu hören und sie zum Kristallisationspunkt eigener Vorstellungen machen zu können.

Zum zweiten erkären sie, woher „es" kommen könnte. Dabei haben sie oft eine rechtfertigende Funktion und beruhigen Sinn- und Schuldfragen. Diese sind nach Elders Untersuchung (Elder, 1973) besonders bei Unterschichtpatienten lebendig. Drittens stecken sie den Horizont der Hoffnungen und Befürchtungen ab und geben Hinweise darauf, wie man sich in der Krankheit verhalten soll. So tragen die Patientenvorstellungen nicht nur kognitive, sondern auch emotive und pragmatische Bedeutungen.

Es ist zu vermuten, daß sie auch auf die „Compliance" der Patienten Einfluß nehmen. Diese dürfte um so geringer sein, je weiter die Vorstellungen des Arztes und die seines Patienten auseinanderliegen (vgl. Jankowski et al., 1980; Lorig et al., 1984; Potts et al., 1984).

In allem nehmen sie Bezug auf die Biographie des Kranken, auf seine Herkunft, seine Familie, seinen Beruf. In ihnen verbinden sich (sub-)kulturell bereitliegende Deutungsmuster mit Familienromanen und eigenwilligen Bedeutungszuschreibungen. Mit dem medizinischen Wissen haben sie oft wenig gemeinsam. Im Regelfall bestätigen und bekräftigen sie die Verbindungen des Kranken mit seiner Gruppe, mit den Verhaltensmustern und dem Wissen seiner Mitmenschen. Dadurch distanzieren sie ihn nicht selten von der medizinischen Fachwelt.

Dieser sind die Patientenvorstellungen oft verborgen und schwer erreichbar. Dennoch oder gerade deshalb muß als weiteres Kriterium der ärztlichen Aufklärung das der **Verträglichkeit** gefordert werden: Ärztliche Mitteilungen müssen Anschluß an das gewinnen können, was der Patient – aus welchen Quellen auch immer – schon weiß.

Auch sehr bestimmte Patientenvorstellungen schließen ausgeprägte **Informations-** und **Aufklärungsbedürfnisse** nicht aus (Langer und Birth, 1987). Diese beziehen sich nach eigenen Untersuchungen bei etwa je 80% der Befragten auf die Ursachen und den weiteren Verlauf der Erkrankung und auf die medikamentöse Behandlung. Die Möglichkeiten einer psychologischen Behandlung/Betreuung finden ein relativ geringes Interesse, ebenso wie Fragen der Hilfsmittelversorgung oder der chirurgischen Therapie (bei 25–40% der Befragten) (vgl. Silvers et al., 1985). Damit ist eine schwierige Situation gegeben: Über zwei der besonders interessierenden Themen kann der Arzt sowenig Bestimmtes und Gesichertes sagen, daß Enttäuschungen schwer zu vermeiden sind. Auf der anderen Seite würden Rheumatologen mit ihren Patienten gerne über die in der Psychotherapie, Ergotherapie oder auch Chirurgie liegenden Behandlungs- und Betreuungsmöglichkeiten sprechen – über Themen also, die cP-Patienten deutlich weniger interessieren.

Lasten und Leiden von cP-Patienten

Die Lasten des Patienten mit einer chronischen Polyarthritis lassen sich meist einer der folgenden 3 Ebenen zuordnen (Tab. 45.1–2).

Die **Lasten des chronischen Polyarthritikers** weisen eine relativ hohe Krankheitsspezifität auf. Ein Patient mit einer Multiplen Sklerose, einer chronischen Bronchitis oder einer koronaren Herzerkrankung wird andere Primärsymptome erleben und berichten. Sie sind eng mit den pathophysiologischen und pathomorphologischen Prozessen der cP verbunden. Dennoch sind wir immer wieder von den Unterschieden überrascht, die sich in den Äußerungen von Patienten mit objektiv ähnlichen Krankheitsmanifestationen ergeben (vgl. de Haas et al., 1974). Entscheidend ist, welche individuelle Wirklichkeit die Schmerzen, die Kraftlosigkeit, die Bewegungsbehinderung und die Gestaltveränderung gewinnen.

Tabelle 45.1–3 gibt Auskunft über die Häufigkeit bzw. Schwere der (von uns operationalisierten) 4 Primärsymptome bei früher (mittlere Anamnesedauer 7 Monate) und langjähriger cP (mittlere Anamnesedauer über 10 Jahre).

Es fällt auf, daß sich rezent und seit Jahren Erkrankte im Ausmaß ihrer Beschwerden nur wenig unterscheiden – weniger, als man es in Kenntnis der

---

Tabelle 45.1–2. Lasten des Patienten mit einer chronischen Polyarthritis

---

Lasten des chronischen **Polyarthritikers** (z.B. Primärsymptome)

Lasten des **chronisch** Kranken (z. B. Glaubwürdigkeit, Zukunftsunsicherheit, soziale Isolation, beruflicher Abstieg)

Lasten des **Dauerpatienten** (z.B. anhaltende Kontroll- und Therapiebedürftigkeit, Nebenwirkungen, Aufklärungsdefizite)

Tabelle 45.1–3. Primärsymptome bei früher und langjähriger cP		
	frühe cP n = 90	langjährige cP n = 95
Schmerzintensität numerische Ratingskala 0 = keine bis 10 = unerträgliche Schmerzen mehr als 6 Punkte	5,2 $\bar{x}$ 41%	5,8 48%
Kraftlosigkeit innerhalb von 3 Stunden nach dem Aufstehen innerhalb von 6 Stunden	13% 28%	15% 54%
Behinderung bei Tätigkeiten des täglichen Lebens Funktionsfragebogen Hannover 0–100% Funktionskapazität(FK) weniger als 75% FK	79 $\bar{x}$ 35%	80 32%
ärztliches Urteil nach Steinbrocker Behinderung ≥ Klasse II	56%	55%
Subjektiv deutlich beeinträchtigtes Aussehen	23%	37%

jeweiligen Krankheitsmanifestationen vermuten sollte. Dies gilt für die Schmerzintensität, die ausgeprägte Kraftlosigkeit und für das Ausmaß der Behinderung. Möglicherweise werten krankheitserfahrene Patienten ihre Symptome geringer als gerade Erkrankte.

Das, was äußerlich als „Gewöhnung" imponiert, kann auf ganz unterschiedlichen Wegen zustande gekommen sein: durch Toleranzentwicklung oder langsame Erwartungseinschränkung, durch gezielte Autosuggestion und Ablenkung oder durch bewußte Dissimulation, durch aktives Ausgleichen bestimmter Funktionsdefizite oder durch gezielte medikamentöse Beeinflussung.

Im Gegensatz zu den Primärsymptomen scheinen die **Lasten des chronisch Kranken** (Tab. 45.1–4) wenig krankheitsspezifisch.

Auch MS-Kranke, chronische Bronchitiker oder koronar Herzkranke haben die immaterielle, aber objektive Last der prognostischen Unsicherheit zu tragen (für die cP vgl. Wiener, 1975). Auch sie müssen mit dem oben angeführten (Tab. 45.1–2) Problem der Glaubwürdigkeit umgehen. Dies ergibt sich besonders bei den Beschwerden, die sich der Fremdwahrnehmung nicht aufdrängen, wie beim Schmerz oder der Müdigkeit. Auch sie können in eine subjektive oder objektive Isolation geraten (Raspe et al., 1983),

auch sie haben bisher nicht erwähnte finanzielle Lasten zu tragen, für die nur schwer ein Kostenträger zu finden ist (Veränderungen der Garderobe oder der Wohnung, Transportkosten, Selbstbehandlungen u.a.).

Für die **Lasten des Dauerpatienten** kommen die jeweils betrachtete chronische Erkrankung, die für sie im Augenblick gültigen Behandlungsprogramme und die Vorsorgungssituation wieder stärker zum Tragen. Nach paralleler Untersuchung von 130 (nach Geschlecht, Alter, Krankheitsdauer und Behandlungsinstitution) vergleichbaren Patienten mit einem Typ-II-Diabetes oder einem erworbenen Herzklappenfehler erscheinen die von uns betreuten langjährig cP-Kranken stärker belastet (Tab. 45.1–5).

Zu den Lasten eines Patienten kann es auch gehören, verzögert oder womöglich gar nicht zu einer richtigen Diagnose und Therapie zu kommen. Im Falle unserer Patientin hatte es „nur" 7 Monate gedauert; immerhin wurde in dieser Zeit eine wenig sinnvolle Tonsillektomie durchgeführt. Auf der anderen Seite haben wir lernen müssen, daß auch qualifizierte Behandlungsangebote – unabhängig von ihrer medizinischen Indikation – zu einer Last werden können, sei es, daß sie an mühsam erreichte Gleichgewichte rühren (veränderte Rollenverteilung in der Familie bei stärkerer Behinderung der Hausfrau), sei es, daß sie die Unabhängigkeit des Patienten bedrohen (engmaschige Kontrollen bei einer Basistherapie) oder daß sie verbreitete Ängste mobilisieren (die Angst vor „Versteifung" nach Gelenkoperationen).

An die bisher erwähnten und an weitere Lasten können sich **seelische Leiden** knüpfen.

Wir kennen keinen Patienten, der im Verlaufe seiner Erkrankung nicht ängstlich oder besorgt, beschämt oder niedergeschlagen gewesen wäre. Von diesen kurzfristigen, meist reaktiven Verstimmungen sind anhaltende und tiefgreifende Störungen des seelischen Gleichgewichtes zu unterscheiden. Solche Störungen sind in psychometrischen Querschnittsstudien bisher immer nur bei einer Minderheit der untersuchten cP-Kranken nachgewiesen worden (vgl. Rimón, 1974; Zaphiropoulos und Burry, 1974; Robinson et al., 1977). Wir selbst fanden unter 228 Patienten mit einer langjährigen und 90 Patienten mit

Tabelle 45.1–4. Lasten des chronisch Kranken (bei langjähriger cP)		
prognostische Unsicherheit		etwa 90%
soziale Isolation	objektiv	max. 20%
	(subjektiv	60%)
berufliche Nachteile		30%
vorzeitige Invalidisierung (wegen cP)		25%

Tabelle 45.1–5. Lasten des Dauerpatienten	cP (%)	Diabetes (%)	Vitien (%)
anhaltende Behandlungsbedürftigkeit	90	90	90
Notwendigkeit engmaschiger ärztlicher Kontrollen	90	70	60
subjektive Medikamentennebenwirkungen			
im gesamten Krankheitsverlauf	70	18	38
in den letzten 4 Wochen	30	n.u.	n.u.
unbefriedigtes Informationsbedürfnis	60	29	20

(n.u. = nicht untersucht)

Tabelle 45.1–6. Klinisch relevante Depressivität und Ängstlichkeit bei früher und langjähriger cP

	frühe cP	langjährige cP
	n = 90	n = 228*
Depressivität (BDI)	23%	28%
Ängstlichkeit (STAI)	31%	30%

* Gepoolte Daten aus 3 Stichproben

Tabelle 45.1–7. Behinderung (nach Keitel et al., 1971) und Depressivität (BDI) bei langjähriger cP (n = 94)

Behinderung	Depressivität		
	gering	deutlich	
gering	25 (93%)	2 (7%)	27 FK 75–100%
mittelgradig	27 (63%)	16 (37%)	43 FK 50– 74%
stärker	13 (54%)	11 (46%)	24 FK 17– 49%
	65	29	94

$r = 0,41$, $p = 0,001$; $Chi^2 = 10,3$, $df = 2$, $p < 0,007$.
FK = Funktionskapazität

einer frühen cP (Tab. 45.1–6) nie mehr als 28% ausgeprägt Depressive ($\geq$ 8 Punkte im Beck-Depressions-Inventar; Beck et al., 1961) und 31% Ängstliche ($\geq$ 7 Standardwerte im State-Trait Anxiety Inventory; vgl. Laux et al., 1981). Eine bemerkenswert hohe Depressionsprävalenz von 42% (nach DMS-III-Kriterien) fanden Frank et al. (1988) mit Hilfe eines strukturierten psychiatrischen Interviews. So kann es sein, daß die Selbstausfüll-Instrumente wie das BDI zu einem „underreporting" führen.

Nach eigenen und fremden Längsschnittuntersuchungen sind Depressivität und Ängstlichkeit von cP-Kranken erstaunlich stabile Persönlichkeitsmerkmale. Werte, die bei den selben Patienten im Abstand von 24 Monaten erhoben werden, korrelieren wenigstens unter $r = 0,52$ miteinander (Raspe, 1987; vgl. Hawley und Wolfe, 1988). Im Vergleich zu epidemiologischen Erwartungswerten (z.B. Murrell et al., 1983) tragen cP-Kranke also ein wenigstens 1,5- bis 2fach höheres Depressionsrisiko.

Versuche, das Ausmaß dieser Form seelischen Leidens aus den genannten Lasten abzuleiten, waren begrenzt erfolgreich. Krankheitsmanifestationen und Primärsymptome klären in aller Regel um 20%, maximal 40% der Varianz der Depressionswerte. Dabei kommt der Bewegungsbehinderung das bei weitem größte Gewicht zu (Tab. 45.1–7): Mit abnehmender Funktionskapazität bzw. zunehmender Behinderung nimmt der Anteil des cP-Kranken mit einer klinisch relevanten Depressivität zu. Die Behinderung bildet die entscheidende Brücke zwischen den biomedizinischen Variablen und den seelischen Leiden. Damit ist ein Teil der Depressivität als reaktiv anzusehen (vgl. Mindham et al., 1981).

Fassen wir diesen Abschnitt zusammen, so lassen sich bei cP-Patienten eine Vielzahl von Lasten und Leiden mit jeweils unterschiedlicher Häufigkeit beschreiben. Wir haben uns hier auf diejenigen konzentriert, zu denen uns empirische Befunde vorliegen. Ausgespart blieben vor allem jene Probleme, die sich für den Kranken im Umgang mit anderen Menschen ergeben – in der Ehe, in der weiteren Familie, im Kollegen- und Freundeskreis, in der Gemeinde. Gerade die hier entstehenden Schwierigkeiten lassen sich nur noch mit Zwang einer der drei Ebenen der Tabelle 45.1–2 zuordnen. So können die eine Ehe möglicherweise belastenden Störungen der sexuellen Libido oder Potenz sowohl durch Krankheitsmanifestationen oder Primärsymptome als auch durch Medikamentennebenwirkungen oder depressive Vitalstörun-

gen ausgelöst werden (vgl. Raspe, 1982; Elst et al., 1984).

Beziehen wir solche Probleme mit ein, dann wird es ganz unwahrscheinlich, zwei Patienten mit einem gleichen „Problemgefüge" zu begegnen. Jeder einzelne Kranke repräsentiert einen neuen und unwiederholbaren Zusammenhang von Persönlichkeit, Krankheit und Lebenslage; jeder einzelne bedarf einer ebenso konzentrierten wie umfassenden Aufmerksamkeit. Wie unsere Krankengeschichte zeigt, können sich Problemgefüge plötzlich und unvorhersehbar ändern. Wir haben immer mit möglicherweise rasch wechselnden „Problemaktivitäten" zu rechnen (vgl. Hartmann, 1976).

Daher erfordern die Beobachtung verschiedener und die Begleitung einzelner Patienten neben dem rheumatologischen Sachverstand und der einfühlenden Aufmerksamkeit für die seelischen Leiden auch und vor allem eine kenntnisreiche Offenheit für die ganze Breite der materiellen wie immateriellen Lasten. Mit einer Verabsolutierung eines einzelnen Zugangs scheint dort wenig gewonnen, wo neben somatischen und seelischen Gleichgewichten auch interpersonelle, familiäre, wirtschaftliche, arbeits- und sozialrechtliche, ethnographische und bei wenigstens einem Drittel der Patienten auch religiöse Bezüge angesprochen werden.

Der Zwang zur mehrschichtigen Simultandiagnostik läßt sich schließlich auch damit begründen, daß die Krankheitsmanifestationen, die Primärsymptome, die weiteren Lasten und die seelischen Leiden untereinander keine einfachen gesetzmäßigen Zusammenhänge aufweisen (Moldofsky und Chester, 1970; Bishop et al., 1987; McFarlane und Brooks, 1988a).

Die chronische Polyarthritis ist damit als **multifokale Erkrankung** aufzufassen.

> Auf die Frage nach seinem zur Zeit größten Problem antwortete uns ein 50jähriger Angestellter mit einer langjährigen cP:
> „Das Knie, die Schmerzen ...
> Ich kann nicht schon wieder krankmachen ... Das Problem ist mit der Arbeit. Welche Firma kann sich das leisten, jemand angestellt zu haben, der immer krank ist ...
> Auch mit den Nerven, das Seelische ... man überlegt, warum muß einem das passieren?"

Nicht einmal aus erhebungsökonomischen Gründen bietet sich irgendwo ein Diagnoseverzicht an. Wo ein Patient in seiner Erkrankung steht, muß man sich zu jedem Zeitpunkt auf mehreren Dimensionen und mit vielen einzelnen Indikatoren vergegenwärtigen. Dies übersteigt die Fähigkeit jeder einzelnen Person oder Profession. Ein koordiniert-interdisziplinärer Zugang zum cP-Patienten ist unverzichtbar.

Das daher zu fordernde Behandlungsteam sollte neben einem internistischen Rheumatologen wenigstens eine Krankenschwester/Arzthelferin, eine Ergotherapeutin, eine Krankengymnastin und eine Sozialarbeiterin umfassen. Verfügt der Arzt über keine psychologische Schulung, wird ein entsprechend ausgebildeter Kollege/Psychologe unverzichtbar. Mit der lokalen Arbeitsgemeinschaft der Deutschen Rheumaliga sollte eine enge Zusammenarbeit gesucht werden.

Damit das, was durch das interdisziplinäre Vorgehen analytisch getrennt wurde, auch wieder zu einer Gesamtschau des Patienten zusammengeführt und in einigen wenigen tragenden Beziehungen (komprehensiv) realisiert werden kann, bedarf es eigener Vorkehrungen. Uns sind die stabile Zuordnung des Patienten zu einem Teammitglied und die regelmäßige Teambesprechung wichtig geworden.

Es ist zu betonen, daß ein solches Team am Wohnort des Patienten zur Verfügung stehen und – auf Dauer gesehen – ambulant und mobil arbeiten muß.

Anders sind die Grundforderungen an eine zeitgemäße rheumatologische und integriert-psychosomatische Betreuung von Rheumakranken nicht zu erfüllen.

Diese sei:
– rheumatologisch kompetent
– problemorientiert
– komprehensiv
– wohnortnah
– krankheitsbegleitend.

Der Arzt erleichtert sich seine Arbeit, wenn er sich in den ersten Kontakten mit einem ihm bisher fremden cP-Kranken die folgenden Fragen zur Beantwortung aufgibt (Raspe, 1989b):
– Warum kommt der Kranke – heute – zu mir?
– Wie präsentiert er sich in seiner Krankheit?
– Wie manifestiert sich die Krankheit?
– Was benennt der Kranke als sein zur Zeit größtes Problem?
– Wie beurteilt er seinen augenblicklichen Gesundheitszustand?
– Welche Vorstellungen hat er sich über die Krankheit und ihre Behandlung gebildet?
– Wie ist er bisher mit sich, seiner Krankheit und den Therapeuten umgegangen?
– Welche Prognosen stelle ich der Krankheit, diesem Kranken und unserer therapeutischen Beziehung?
– Auf die Veränderung welcher dieser drei Bereiche will ich in der Verlaufsbeobachtung achten?

45.2 Fibromyalgie

Jörg Michael Herrmann, Othmar W. Schonecke und *Werner Geigges*

45.2.1 Exemplarische Falldarstellung

Ein 21jähriger Betriebsschlosser wird von seinem Hausarzt mit der Diagnose BWS- und LWS-Syndrom bei Verdacht auf larvierte Depression zu einem stationären Heilverfahren in eine Rehabilitationsklinik für Innere Medizin/Psychosomatik überwiesen. Er leidet seit 2½ Jahren an starken Rücken- und Kreuzschmerzen, gürtelförmig, ohne Ausstrahlung in die Beine. Diese Schmerzen sind nicht streng belastungsabhängig. Eine eher geringere Schmerzintensität spürt er beim Laufen und Gehen. Seit 1½ Jahren trägt er ständig ein Stützkorsett, außerdem wurde er mit Massagen, Dehnungsbehandlungen und Wärmeanwendungen behandelt, sowie medikamentös mit Antirheumatika, dies alles ohne spürbare Besserung der Schmerzsymptomatik.

Der Patient schildert selbst seine Schmerzen: „Wie wenn einer mit der Latte auf den Buckel 'nauf schlägt". Diese Schmerzen sind ständig vorhanden, nur von kurzen, wenige Stunden andauernden Schmerzpausen unterbrochen. Am stärksten spürt er die Schmerzen während der Arbeit im Betrieb und vor allem am Montagmorgen beim Aufstehen, überhaupt in den Morgenstunden vermehrt, sowie ein Gefühl der Steifigkeit in den Gelenken.

An weiteren Beschwerden schildert der Patient ein mangelndes Konzentrationsvermögen, eine leichte Ermüdbarkeit sowie einen nicht erholsamen Schlaf. Außerdem bekommt er in geschlossenen Räumen ein Schwindelgefühl, es wird ihm dann schwarz vor den Augen, vor allem wenn viele Leute anwesend sind, z.B. in der Kirche.

Bei der körperlichen Untersuchung des 182 cm großen und 74 kg schweren Patienten findet sich ein diffuser Berührungsschmerz und Druckschmerz über dem gesamten Rücken ohne radikuläre oder dermatomtypische Begrenzung. Typische schmerzhafte Myogelosen lassen sich paravertebral im Bereich der HWS und des gesamten oberen Trapeziusrandes tasten, darüber hinaus besteht eine deutliche Klopf- und Druckschmerzhaftigkeit am thorako-lumbalen Übergang, ein diffuser Druckschmerz über den Intersakralgelenken sowie im Bereich der Ischiasdruckpunkte. Außerdem findet sich ein diffuser Klopf- und Druckschmerz über dem ganzen Abdomen bei deutlichem muskulärem Hartspann. Das Schober-Zeichen ist im Bereich der BWS und LWS unauffällig, der Finger-Boden-Abstand beträgt 2 cm. Bei Rechts- und Linksrotation berichtet er über ein diffuses Schmerzgefühl, das sich nicht eindeutig auf die Wirbelsäule projizieren läßt. Bereits vor der stationären Aufnahme war ambulant eine intensive orthopädische und radiologische Diagnostik erfolgt, bei der sich im Bereich der Wirbelsäule und Gelenke kein Anhalt für degenerative oder entzündliche Gelenkveränderungen fand.

Laborchemisch waren alle Werte normal, insbesondere die rheumabezogenen Laborwerte.

Der Patient ist als ältester von 5 Geschwistern (eine 3 und eine 4 Jahre jüngere Schwester, ein 9 Jahre jüngerer Bruder sowie eine 17 Jahre jüngere Schwester) bei seinen Eltern auf einem großen Hof in Süddeutschland aufgewachsen. Die Eltern kannten nur Arbeit, auch für die Kinder gab es keine Freizeit. Der Vater hat den Patienten immer zur Arbeit gezwungen, ihm wenig zugetraut und immer wieder zu ihm gesagt: „Du bist nichts und du wirst nie etwas werden."

Bereits zur Kindergartenzeit ist er durch Konzentrationsstörungen aufgefallen, nach der 4. Volksschulklasse mußte er auf die Sonderschule wechseln, darüber schämt er sich noch heute. Später schaffte er den Sprung auf die Hauptschule zurück mit erfolgreichem Hauptschulabschluß und erfolgreichem Abschluß zweier Facharbeiterprüfungen als Teilezurichter und Betriebsschlosser. Dennoch erlebt er auch heute im Betrieb immer wieder das Gefühl: „Manchmal fühle ich mich als der Dümmste dort".

Der Vater hat ihn immer dazu zwingen wollen, den Hof zu übernehmen, inzwischen hat er die Berufswünsche des Sohnes akzeptiert; er fühlt sich dennoch, verglichen mit seinen Geschwistern, von ihm nicht gleichwertig behandelt. Wenn er keine Rückenbeschwerden hat, wird er nach der Arbeit immer noch zur Mithilfe in der großen Landwirtschaft des Vaters herangezogen. Bereits als Kind hat er immer wieder versucht, dem Vater und seinen Ansprüchen auszuweichen. So hat er sich schon als Kind immer allein gefühlt. Die Beziehung zur Mutter war zwar besser, jedoch hat sie früh versucht, den Druck, der vom Vater auf die ganze Familie ausging, auf ihn zu übertragen. Der Patient wohnt noch zu Hause bei seinen Eltern, das Verhältnis zu den Geschwistern ist inzwischen „sehr gut", er hat keine Freundin: Eltern und Großeltern haben ihn stets ermahnt, aus religiös-moralischen Gründen großen Abstand zu Frauen zu wahren.

Weihnachten 1985/1986 traten erstmals nach dem Tod des Großvaters väterlicherseits (metastasierendes Nierenkarzinom) Rückenschmerzen auf. Früher fürchtete er diesen Großvater sehr, in den letzten Jahren bekam er aber einen sehr guten Kontakt zu ihm und zuletzt besuchte er ihn täglich. Bei Musterung und Nachmusterung wurde der Patient vom Wehrdienst wegen der Rückenbeschwerden zurückgestellt.

Im Rahmen der Gruppenpsychotherapie wurde die Selbstwert- und Autoritätsproblematik des Patienten deutlich: Sein Bedürfnis nach „väterlicher" Anerkennung und Stütze („Stützkorsett") führte zu entsprechenden Übertragungsbeziehungen, worunter die Schmerzsymptomatik zunächst deutlich abnahm. Sein Aggressionskonflikt trat bei Gruppenthemen wie „eheliche Gewalt" zum Teil abgewehrt, zum Teil im Sinne einer „Identifikation mit dem Aggressor" auf. Hierzu ist die – obengenannte – Eigenschilderung seiner Beschwerden von Bedeutung: „Wie wenn einer mit der Latte auf den Buckel 'nauf schlägt".

Nach der Abreise eines „väterlichen" Mitpatienten treten die Rückenschmerzen erneut in großer Intensität auf, damit ist eine starke Rückzugstendenz und psychomotorische Gehemmtheit sowie deutlicher Widerstand der Gruppentherapie gegenüber verbunden. Während die Gruppenteilnehmer erneut stützende Elternfunktion übernehmen und ihn ermuntern, erste Schritte in die Selbständigkeit zu gehen („eine eigene Wohnung suchen und eine Freundin finden"), bessert sich seine Schmerzsymptomatik erneut, gleichzeitig spürt er seine deutliche Ambivalenz in diesem typischen Abhängigkeits-Autonomie-Konflikt. Er spricht von seiner Angst vor dem Alleinsein („Du wirst es nie schaffen!") und davon, den Kontakt zu den Geschwistern zu verlieren, vor allem zu den beiden Jüngsten, und er schildert, daß er versucht, ihnen früh Selbstbewußtsein und eigene Kompetenz zu vermitteln, um sie vor den eigenen Schwierigkeiten zu bewahren (der Versuch, sie identifikatorisch neu „zu beeltern").

In dieser Ambivalenzphase zieht sich der Patient – sicher nicht ganz zufällig – eine Außenbandruptur des oberen Sprunggelenks zu, entscheidet sich jedoch für eine Gipsbehandlung, um weiter an der Gruppentherapie teilnehmen zu können. Zwar kommt es zwischenzeitlich, meist geleitet von seinem unbewußten Lebens-„Skript" des minderwertigen, ungeliebten Versagers, immer wieder zu kurzem innerem Rückzug und Beschwerdeprogredienz, insgesamt tritt jedoch eine deutlich anhaltende Abnahme der Schmerzsymptomatik ein, und der Patient findet Mut zu einer offeneren Auseinandersetzung mit den Gruppen-Eltern-Autoritäten und entdeckt erste realistische und konkretere Pläne der Ablösung vom Elternhaus.

45.2.2 Einleitung

Mitte August 1956 brach im Bergbau akut eine Epidemie aus, die zu einem erheblichen Anstieg der „Fehlschichtenzahlen", d.h. Arbeitsausfall führte. Die Kranken klagten über schwere Myalgien, vor allem im Bereich des Rückens. Bei der ärztlichen Untersuchung wurde ein echtes „rheumatisches" Syndrom diagnostiziert, das die angegebenen Beschwerden voll erklärte. Eine genaue Analyse der Zusammenhänge ergab, daß diese Epidemie im Anschluß an die Katastrophe in dem belgischen Bergwerk Marcinelle am 9. August 1956 ausgebrochen war. Es handelte sich also um „Weichteilrheumatismus", ausgelöst durch schwere Angstzustände, die durchaus einfühlbar waren und die sich als körperliche Symptome des Bewegungsapparates manifestierten. Die Angst saß den Bergleuten „im Nacken" und verursachte durch Hartspann und Verkrampfung der Muskulatur die schweren „epidemischen" Schmerzen (H. Sopp, 1958).

Das Fibromyalgie-Syndrom ist ein Sammelbegriff für eine Vielzahl von schmerzhaften Erkrankungen nichtentzündlicher Natur im Bereich der Weichteile von Körperstamm und Extremitäten.

Synonym werden Begriffe wie extraartikulärer Rheumatismus, Muskelrheumatismus, psychogener Rheumatismus, Fibrositis-Syndrom, arthritic neurosis, Myalgie, Weichteilrheumatismus oder „stiff shoulder" benutzt.

Tabelle 45.2–1. Klassifikation chronischer muskuloskelettärer Schmerzsyndrome (CMPS) ohne identifizierbare Ursache

Terminologie
– Primäres Fibromyalgie-Syndrom (PFS)
– Myofasziales Schmerzsyndrom (MPS)
– Temporomandibuläres Schmerz- und Dysfunktionssyndrom (TMPDS)

Nach Miehlke (1973, 1976) orientiert man sich am besten an den Gewebestrukturen des Bewegungsapparates. Dazu gehören das Subkutangewebe (Pannikulose), Sehnen (Tendinosen, Tendovaginosen, Tendoperiostosen, Tendomyosen), Faszien (Fasziose), Bänder, Muskeln (Myose), lockeres Bindegewebe (Fibrose) und Periost (W. Müller, 1971). Andere Autoren wie Mathies rechnen auch periphere Neuropathien, Angiopathien und Lymphangiopathien mit Manifestationen im Bereich des Bewegungsapparates zu den Weichteilerkrankungen.

Geringfügige organische Befunde am Bewegungsapparat werden häufig entweder als morphologisches Substrat eines Beschwerdebildes überbewertet oder als „rein psychische Störung" ignoriert. Beide Betrachtungsweisen werden dem Problem des sog. „Weichteilrheumatismus" nicht gerecht, da einerseits bereits minimale, mikroskopisch kaum erfaßbare Strukturänderungen von Sehnen, Muskulatur, Bändern oder Periost durch die Summation zu erheblichen Beschwerden und andererseits psychische Faktoren sowohl über die Formatio reticularis auf Willkürmotorik, wie auch über das limbische System und den Cortex (vgl. Kap. 36) erheblichen Einfluß auf Manifestation und Intensität von Beschwerden nehmen können. Dies erklärt sowohl die häufig nachweisbaren lokalisierten degenerativen Erscheinungen wie auch die Vielfalt und Lokalisation der Schmerzen, die für den Weichteilrheumatismus typisch sind.

Allerdings muß auch betont werden, daß im Bewegungsapparat die Variationsbreite von der Norm abweichender Befunde, die weder die Funktion beeinträchtigen, noch Beschwerden oder Beschwerdearmut zur Folge haben, sehr groß ist.

Inzwischen hat die Internationale Gesellschaft zum Studium des Schmerzes (IASP, 1986) eine Klassifizierung der Fibromyalgien vorgeschlagen (Tab. 45.2–1).

45.2.3 Epidemiologie

Angaben über die Häufigkeit des Fibromyalgie-Syndroms, das früher als „Weichteilrheumatismus" bezeichnet wurde, sind sehr unterschiedlich: Miehlke (1973) weist darauf hin, daß der Weichteilrheumatismus in den 50er Jahren und zu Beginn der 60er Jahre in den Statistiken des Bundesverbandes der Deutschen Ortskrankenkassen an der vierten Stelle aller Arbeitsunfähigkeit bedingenden Leiden stand.

Nach einer Untersuchung der Universitäts-Rheumaklinik und Poliklinik in Zürich wurde 1967 bei

23% der wegen einer rheumatischen Erkrankung untersuchten Patienten (n = 1525) die Diagnose „Weichteilrheumatismus" gestellt (Wagenhäuser, 1973). Nach Siegenthaler handelt es sich bei 55% aller Patienten mit einer rheumatischen Erkrankung um „Weichteilrheumatismus", während nach einer britischen militärmedizinischen Statistik sogar in 90% aller Rheumafälle ein Fibromyalgie-Syndrom vorliegt (Rausch, 1967).

45.2.4 Klinik und Symptomatologie

Durch die Entwicklung diagnostischer Kriterien (Yunus et al., 1981; Yunus, 1983) wird das Fibromyalgie-Syndrom folgendermaßen charakterisiert (Tabelle 45.2–2):

Für die Diagnose der Fibromyalgie sind nicht nur die obligaten Kriterien, sondern auch das Hauptkriterium und 3 Nebenkriterien notwendig. Fehlt das Hauptkriterium, müssen mindestens 5 Nebenkriterien vorhanden sein (Yunus et al., 1987).

Leitsymptom ist häufig ein ubiquitärer, schlecht lokalisierbarer Schmerz im Bewegungsapparat („Herr Doktor, es tut mir überall weh"). Prädilektionsstellen sind Lumbal- und Zervikalbereich. Grundsätzlich kann die Fibromyalgie aber jede Körperregion befallen. Neben dem Lumbal- und Zervikalbereich werden Schmerzen besonders häufig in den Schultern, der Serratusmuskulatur, den Innenseiten der Kniegelenke und im Bereich der Cristae iliacae angegeben. Besonders wichtig ist es, die Muskeln zu untersuchen, die häufig einen Hartspann aufweisen (Schmerzpunkte): Mitte des Oberrandes des M. trapezius, lateral des Oberrandes der 2. costosternalen Syndesmose, Ansatz des M. suboccipitalis, costoster-

nale Syndesmosen, Ansatz des M. supraspinatus, Ansätze des M. sternocleidomastoideus, medialer Rand der Scapula, medial und lateral des Ulnargelenkes, Ansätze des M. biceps brachii, hinterer Darmbeinkamm, Wirbelsäule (C4–C6 und L4–S1), Ileosakralgelenke, Trochanter major und medialer Bereich des Kniegelenkes (Trinkl, 1987).

Charakteristisch ist, daß die Beschwerden häufig während Freizeit, Ablenkung und Ferien deutlich geringer werden oder sogar verschwinden. Bei genauerer Analyse lassen sich oft Zusammenhänge zwischen Beginn der Beschwerden und psychischen Belastungen aufdecken (siehe Epidemie der Bergleute). Häufige Begleiterscheinungen sind psychovegetative Symptome wie verstärkter Dermographismus, Hyperhidrosis, funktionelle Magen-, Darm- und Herzbeschwerden. Bei der diagnostischen Bewertung muß das Nichtansprechen der Beschwerden auf Antirheumatika und eine Besserung nach Gabe von Psychopharmaka und Muskelrelaxantien kritisch gesehen werden, da auch bei organischen Beschwerden Psychopharmaka wirksam sind (vgl. Kap. 24).

An „psychischen Auffälligkeiten" fand R. Schild (1973) bei insgesamt 1400 Patienten folgende Charakteristika:

– Depressionssyndrom,
– inadäquater Befund, d. h. Zahl, Art und Intensität der Beschwerden stehen in einem deutlichen Mißverhältnis zu den Befunden der klinisch-somatischen Untersuchung,
– Therapieresistenz,
– Arztwechsel,
– Logorrhö,
– aggressive Haltung.

In all diesen Aspekten kommt – nach Schild – die Hilflosigkeit von Arzt und Patient dem Beschwerdebild gegenüber zum Ausdruck. „Logorrhö" steht hier als Bezeichnung für den Versuch des Patienten, affektive Bedürfnisse zu überspielen, die von der Umwelt nicht wahrgenommen und daher durch den Schmerz betont werden müssen. Die Aggression soll in den meisten Fällen als Versuch einer Bewältigung diffuser Angst vor schlecht faßbarer Bedrohung zu verstehen sein. Weiterbestehende Schmerzen könnten einen neuen Versuch des Patienten darstellen, den Konflikt schließlich durch eine Wendung der Aggression gegen sich selbst zu lösen.

45.2.5 Psychodynamik

Die bisher beschriebenen psychodynamischen Befunde sind sehr uncharakteristisch und lassen kein einheitliches Bild erkennen. Kontrollierte Studien haben inzwischen gezeigt, daß nur ein kleiner Teil der Patienten mit Fibromyalgie psychische Veränderungen aufweist, insbesondere lassen sich depressive Reaktionen bei Fibromyalgie-Patienten nicht häufiger finden als bei Patienten mit chronischen Schmerzen oder Polyarthritis (Ahles et al., 1987; Clark et al., 1985; Wolfe et al., 1984).

Tabelle 45.2–2. Diagnostische Kriterien des primären Fibromyalgie-Syndroms (Yunus et al., 1981; Yunus, 1983)

Obligate Kriterien
– Generalisierte Schmerzen oder Steifigkeitsgefühl zumindest in 3 anatomischen Regionen seit mindestens 3 Monaten
– Fehlen sekundärer Ursachen (z. B. traumatische Verletzungen, rheumatische Erkrankungen, infektiöse Arthropathien oder pathologische Laborwerte)
Hauptkriterium
5 oder mehr charakteristische Schmerzpunkte (s. u.)
Nebenkriterien
– Modulation der Symptome durch körperliche Aktivität
– Modulation der Symptome durch Wettereinflüsse
– Verschlimmerung der Symptome durch Angst und Streß
– Schlafstörung
– Allgemeine Ermüdbarkeit
– Angst
– Chronische Kopfschmerzen
– Funktionelle Oberbauchbeschwerden
– Gefühl der Gelenkschwellung
– Parästhesien, die weder Dermatomen noch radikulären Gesichtspunkten entsprechen

Es scheint so zu sein, daß der Patient mit Fibromyalgie seine Aggressionen offener zeigen kann als der Rheumapatient. Nach Weintraub (1977) sollen psychogene Kreuzschmerzen häufig bei Menschen vorkommen, die nicht imstande sind, sich mit einer Konfliktsituation auseinanderzusetzen und die unfähig sind, ihre psychischen Probleme zu verbalisieren.

Labhardt (1976) charakterisiert die Persönlichkeit von Patienten mit Fibromyalgie folgendermaßen:
– Äußerlich finden sich überwiegend beherrschte, zwanghaft perfektionistische Persönlichkeitstypen.
– Innerlich zeigt sich häufig ein Ambivalenzkonflikt zwischen Fremd- und Selbstbeherrschung einerseits und dienend-aufopfernder Haltung andererseits.
– Diese Ambivalenz führt zu chronisch gehemmter Aggressivität, die sich u.a. in gesteigertem Muskeltonus äußert, der als Dauerzustand das psychophysiologische Äquivalent der Fibromyalgie sein soll.
– Bei Frauen, die ihre weibliche Rolle nicht akzeptieren, finden sich häufig Züge des Beherrschenwollens.

Weintraub (1977) betont dagegen den Bedeutungsgehalt verschiedener, durch Fibromyalgie bedingter Schmerzzustände:
– Zervikalgie als Ausdruck einer emotional erschwerten Be-Haupt-ung, hartnäckiges Gesichtwahren.
– Dorsalgie als Ausdruck von Trauer, Verzweiflung, Mutlosigkeit und kompensatorischer zwanghaft aufrechter Haltung.
– Lumbalgie als Ausdruck psychischer Überlastung, Sprunghaftigkeit, Frustration, besonders bei gestörter Sexualität.
– Brachialgie als Ausdruck gehemmter Aggression: Wut, Zorn. Symbol: geballte Faust.

Egle et al. (1989) untersuchten 47 Patienten mit einem primären Fibromyalgie-Syndrom – als Kontrollgruppen dienten Patienten mit psychogenem Schmerzsyndrom und unselektierte Patienten ohne Schmerzen einer Allgemeinpraxis – und fanden erhöhte Werte für Aggression und Unfähigkeit, Aggressionen auszudrücken, eine gestörte emotionale Beziehungsfähigkeit sowie einen mangelnden Realitätsbezug. Die Autoren interpretieren diese Ergebnisse als Vorliegen unreifer Abwehrmechanismen, die die Möglichkeiten der Bewältigung schwieriger Lebenssituationen stark einschränken. Gleichzeitig werden zwischenmenschliche Konflikte verdrängt und durch eine ausgeprägte Fixierung auf eine somatische Schmerzgenese und hypochondrische Befürchtungen ersetzt. Psychodynamische Ursachen dafür sollen in einer besonderen Sauberkeitserziehung, einem strengen, legalistischen Familienstil und frühen Verlustängsten liegen.

45.2.6 Pathopsychophysiologie

Holmes und Wolff (1962) zeigten, daß aggressive Konflikte mit erhöhten Werten im Elektromyogramm einhergehen. Dagegen soll eine Verbalisierung von Konflikten den nach dem EMG erhöhten Muskeltonus deutlich herabsetzen (Shagaas und Malmo, 1954). Eine ebenfalls deutlich reduzierte EMG-Aktivität der Gesichtsmuskulatur konnte bei depressiven Patienten nachgewiesen werden. Schwartz und Mitarbeiter (1974) konnten mit Hilfe des EMG verschiedene affektive Zustände wie Depression, Fröhlichkeit oder Ärger unterscheiden.

Nach Fassbender (1973) finden sich elektronenoptisch Hinweise dafür, daß dem „Muskelrheumatismus" eine stufenweise Zerstörung der kontraktilen Substanz zugrunde liegt, deren Ursache in einem nerval bedingten Dauertonus und einer dadurch ausgelösten relativen Hypoxie zu suchen sei. Er betont, daß dem Weichteilrheumatismus mit Sicherheit weder in seiner muskulären noch in seiner bindegewebigen Manifestation ein entzündlicher Mechanismus zugrunde liegt (Abb. 45.2–1). Vielmehr soll es sich um morphologisch unterschiedliche Auswirkungen von Störungen der lokalen Sauerstoffversorgung handeln, deren Ursache verschiedenartig sein kann.

Bei Störungen im Bereich des subkutanen Fettgewebes spricht man von einer Pannikulose. Da hier Entzündungen nicht vorkommen, sollte der Begriff Pannikulitis nicht mehr benutzt werden. Pathophysiologisch nimmt man einen bisher noch unbekannten endokrinen Mechanismus an, der – siehe Abbildung 45.2–1 – über einen unspezifischen (mechanischen, psychischen, thermischen, traumatischen) Reiz zu einer gesteigerten subkutanen Wasseraufnahme und dadurch zu Schwellung und Schmerzen führen soll.

McCain et al. (1988) weisen darauf hin, daß bis heute keinerlei reproduzierbare und konsistente anatomische Veränderungen in der Struktur dieser Gewebe nachgewiesen werden konnten. Es bleibt weiter unklar, ob zu beobachtende strukturelle und metabolische Veränderungen Hinweise zur Ätiologie der Fibromyalgie liefern oder einfach eine Konsequenz der chronischen Schmerzen darstellen.

Smythe (1979) sieht im Fibromyalgie-Syndrom ein Syndrom selektiv erhöhter Schmerzempfindung durch quantitativ unveränderte physiologische Mechanismen bei unbeeinflußter allgemeiner Schmerzschwelle.

Interessant in diesem Zusammenhang sind auch die Schlafexperimente seiner Arbeitsgruppe: Sowohl

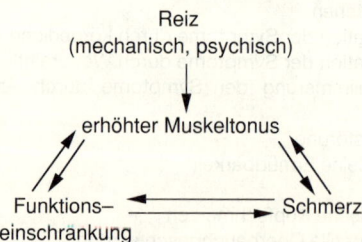

Abb. 45.2–1. Regulationsmechanismus des Muskeltonus und die Beziehung zwischen Nervensystem, Psyche und Muskulatur (modifiziert nach E. Neumayer, 1974).

bei Patienten mit Fibromyalgie-Syndrom wie auch bei gesunden Versuchspersonen konnte nach systematischer Störung des „non-rapid eye movement" (NREM)-Schlafes im EEG ein abnormer Alphawellenrhythmus registriert werden. Nach 3 Nächten gestörten NREM-Schlafes zeigten die Probanden die typische Fibromyalgie-Symptomatik, allerdings nur jene, die körperlich wenig aktiv waren.

45.2.7 Hinweise zur Differentialdiagnose

Die proteusartige Symptomatologie der Fibromyalgie birgt die besondere Gefahr, Schmerzzustände anderer Ätiologie zu übersehen. Um Fehldiagnosen zu vermeiden, ist es daher notwendig, alle Krankheitsbilder auszuschließen, die ähnliche Beschwerden auf anderer Basis verursachen. Von diesen sollen hier nur die folgenden genannt werden: Rücken- und Kreuzschmerzen können als Frühsymptom maligner Oberbauchtumoren (z.B. Pankreaskopf- und -schwanzkarzinom) oder auch bei gutartigen Oberbaucherkrankungen wie Cholelithiasis, Pankreatitis, Ulcus duodeni und ventriculi auftreten (Eppinger und Endsberger, 1975). Bei Pankreaskarzinomen kann die Differentialdiagnose schwierig sein, da hier oftmals Ängste, Depressionen und Persönlichkeitsveränderungen der Schmerzsymptomatik vorausgehen oder sie begleiten (Benos, 1974; Müller-Wieland, 1968).

Bei diffusen, nichtlokalisierbaren Schmerzen ist vor allem bei Frauen mittleren und höheren Alters an Knochenschmerzen bei Osteoporose, Osteomalazie oder Myelom zu denken.

Schmerzen in einer Extremität können durch einen benignen Glomustumor bedingt sein. Da bei dieser Erkrankung somatopsychisch häufig Angst und Depressionen bestehen, ist die Verwechslung besonders leicht möglich.

Ein besonderes Kapitel sind die Schmerzen bei Bandscheibenläsionen (Goldner, 1976). Hier ist die Differentialdiagnose besonders schwierig: Auf der einen Seite finden sich vor allem bei älteren Menschen fast regelmäßig anatomische Veränderungen der Wirbelsäule im Röntgenbild (Osteochondrose, Verschmälerung der Zwischenwirbelräume etc.), deren Ausmaß in keiner festen Beziehung zu Funktionseinschränkungen und Beschwerden steht. Auf der anderen Seite sind die anatomischen Veränderungen der Wirbelsäule fast immer selbst wieder die Folge von psychisch bedingten oder mitbedingten Haltungsanomalien. Hier steht daher vor allem die Frage im Vordergrund, ob und inwieweit ein anatomischer Defekt bereits zu einer selbständigen Krankheitsursache geworden ist.

Im übrigen sind alle im Zusammenhang mit dem Fibromyalgie-Syndrom besprochenen Vorgänge Faktoren, die in der Pathogenese der heute meist unter der Bezeichnung „Bandscheibenschäden" zusammengefaßten degenerativen Wirbelsäulenveränderungen eine Rolle spielen (vgl. Kap. 45.3).

45.2.8 Therapie

Die Therapie des primären Fibromyalgie-Syndroms umfaßt Psychotherapie, physikalische und medikamentöse Therapie.

Psychotherapeutische Techniken sind umstritten. Nach Seidel (1973) soll Psychotherapie bei psychogenen Bewegungsstörungen wenig Erfolg haben. Er betont die menschliche Führung durch den Hausarzt in Kombination mit gezielten symptomgerechten physikalischen Maßnahmen, die eine Besserung des organischen Befundes wie auch der psychischen Haltung erreichen sollen.

Bei der **Physiotherapie** müssen passive und aktive Maßnahmen unterschieden werden. Zu den passiven Maßnahmen zählen Hydro- und Thermotherapie, Elektro-, Ultraschall- und Balneotherapie. Zu den aktiven physiotherapeutischen Methoden gehören krankengymnastische Bewegungstherapie, Gymnastik, Schwimm- oder Turngruppen.

Bei der **Pharmakotherapie** steht die Schmerzbekämpfung der Muskelverspannungen mit Analgetika und Lokalanästhetika im Vordergrund. Zu vorübergehender Besserung führen einfache Analgetika wie Paracetamol oder Acetylsalicylsäure. Ergänzend können Psychopharmaka mit zentral myotonolytischem Effekt[1] eingesetzt werden. Bei depressiven Patienten mit Schlafstörungen empfiehlt sich die Gabe von 75 mg Saroten®[2] abends. Diese Dosierung kann um 25 mg/Woche bis zu einer Gesamtdosis von 200 mg/Tag erhöht werden. Bei depressiven Patienten ohne Schlafstörungen können äquivalente Dosen von Tofranil®[3] eingesetzt werden (Sternbach et al., 1973).

Nach Weintraub und Mitarbeitern (1975) ist die Unterscheidung zwischen somatopsychischen Begleiterscheinungen, d.h. Angst und Depression bei Schmerzsyndromen des Bewegungsapparates, strukturellen oder funktionellen Veränderungen oder psychosomatischen Schmerzzuständen im Sinne einer Konversion besonders wichtig. Konversionszustände sind zwar schwer zu diagnostizieren und zu beurteilen, einer tiefenpsychotherapeutischen Behandlung aber durchaus zugänglich. Wenn psychosoziale Probleme im Vordergrund stehen – und bei Störungen der Beweglichkeit oder der Bewegung werden die sozialen Kontakte, ja die aktive Auseinandersetzung mit der Umwelt immer wesentlich beeinflußt – ist die Mitarbeit von Sozialarbeitern oft unerläßlich. Bei konversionsneurotischen Zuständen wird die Verschickung zu einer Badekur kaum Erfolg haben. Auch wenn es den Patienten in der „Subkultur des Badeortes" besser gegangen war, stellen sich ihre Beschwerden nach Rückkehr in ihr gewohntes Milieu fast regelmäßig wieder ein.

Da muskuläre Verspannungen die häufigste Schmerzursache sind, kommt dem autogenen Training in der Therapie des Fibromyalgie-Sydroms eine wesentliche Bedeutung zu.

---

1 z.B. Diazepin-Derivate.
2 Amitriptylin.
3 Imipramin.

Für den Umgang mit diesen Patienten betont Goldner (1976), daß sie sich ständig nach neuen Behandlungsarten und Mitteln wie Akupunktur, neuen Medikamenten oder anderen speziellen Methoden, von denen „man gehört oder gelesen hat", erkundigen. Darin komme ihre Angst zum Ausdruck, an einer lebensbedrohlichen, z. B. malignen Krankheit zu leiden. Da sich dahinter häufig ein ungelöster Konflikt verschiedenster Art und ein weder vom Patienten noch vom Arzt verstandenes Bitten um Hilfe verberge, würden unkomplizierte Behandlungs- und Diagnostikmethoden die Angst dieser Patienten wesentlich verringern. Wenn der Arzt realisiert, daß die Angst des Patienten häufig die Ursache seiner dauernden Suche nach einer Antwort auf die Frage ist, woran er leide, würden sich beide schnell auf eine gemeinsame Richtung bei Diagnose und Therapie einigen.

Egle et al. (1989) betonen ebenfalls, wie das ausgeprägte Kontrollverhalten dieser Patienten, ihre fordernde Haltung und der von ihnen ausgehende Handlungsdruck auf der körperlichen Ebene die Arzt-Patient-Beziehung beeinflussen und die Gefahr einer iatrogenen Verstärkung der ohnehin schon vorhandenen Fixierung auf eine körperliche Krankheitsursache bedeuten.

Wichtig ist primär, dem betroffenen Patienten zu vermitteln, daß psychogene Schmerzen keine „eingebildeten" Schmerzen, sondern genau so real sind wie körperlich verursachte Schmerzen. Ein sich zunächst körperlich Ernst- und Angenommen-Fühlen ist die Voraussetzung, um später auch psychosoziale Zusammenhänge erarbeiten zu können.

45.3 Lumbago-Ischialgie-Syndrome

Mechthilde Kütemeyer und *Ulrich Schultz*

45.3.1.1 Geschichtliches

„Mit dem Rücken zur Wand", „Rückgrat haben", „einen breiten Rücken haben", „hartnäckig", „halsstarrig", „dem wurde das Kreuz gebrochen", „katzbuckeln", „zu Kreuze kriechen" und ähnliche Redewendungen zeugen von einem ursprünglichen Empfinden für die Bedeutung des Rückens. Nach altem Volksglauben, der auch in den medizinischen Sprachgebrauch Eingang gefunden hat, entsteht die akute Lumbago durch den „Hexen"- oder „Elfenschuß" (altengl.: haegtessan bzw. ylfsa gescot). Im Hexenschuß sah das Volk, wie der mittelalterliche Holzschnitt zeigt (Abb. 45.3–1), den stolzen Krieger durch weibliche zauberische Kräfte überraschend hinterrücks getroffen und in seiner aufrechten Haltung gebrochen. Sinnträchtig verbunden mit Machtkampf und Ohnmacht findet sich das Ischiassyndrom in einer jüdischen Urgeschichte:

Jakob – Sohn Isaaks und Rebekkas – stieß sich schon vor der Geburt im Leib seiner Mutter mit seinem Zwillingsbruder Esau. Und seiner Mutter wurde kundgetan, „zwei Leute werden sich scheiden aus deinem Leibe, und ein Volk wird dem anderen überlegen sein, und der Ältere wird dem Jüngeren dienen" (1. Mose 25,22). Der zweitgeborene Jakob, sanft, häuslich und Liebling der Mutter, aber mit der Verheißung der Überlegenheit über seinen wilden, kräftigen Bruder, den Jäger, versehen, verschafft sich mit List das Erstgeburtsrecht und den Segen des Vaters und muß vor der Rachsucht Esaus in die Fremde fliehen. In der Nacht vor der entscheidenden

Hexenschuß. Mittelalt. Holzschnitt

Abb. 45.3–1. Hexenschuß. Mittelalterlicher Holzschnitt (aus Beitl, 1933).

Wiederbegegnung mit seinem Bruder nach 20jähriger Trennung wird ihm, mit einem „Manne" ringend, das „Gelenk der Hüfte" verrenkt, so daß er am Morgen hinkt (1. Mose 32, 24–33). Siegend und gleichzeitig um Segen bittend, gewinnt der Patient Jakob in dieser Nacht – in der ihm auch der Name Israel verliehen wird – eine neue selbstbewußte, demütige und anerkennende Haltung dem Bruder gegenüber, die eine friedliche Begegnung ermöglicht.

Zur Erinnerung an dieses Ereignis verschonen die Israeliten beim Essen die „Spannader über der Hüfte", den „Hüftnerven", worunter einhellig der N. ischiadicus (griech.: ischios = zur Hüfte gehörend) verstanden wird (von Rad, 1976).

Cotugno (1779), Verfasser der ersten Monographie über die „Ischias nervosa", vermutet eine ödematöse Auftreibung („Hydrops") der Nervenscheide, welche zu einer Kompression oder durch die Schärfe der Ödemflüssigkeit zu einer Reizung führe. Anfang des 20. Jahrhunderts beginnt eine jahrzehntelange Diskussion, ob die idiopathische Ischias den Neuralgien oder den Neuritiden zuzuordnen sei. Gegenüber der Auffassung einer „Ischiasneuritis" oder „Neuritis des Plexus lumbosacralis" setzt sich Ende der vierziger Jahre gegen große Widerstände die Erkenntnis durch, daß es sich beim Lumbago-Ischialgie-Syndrom (LIS) primär um eine Wurzelkompression durch einen Diskusprolaps handelt (Mixter und Barr, 1934; Frowein und Firsching, 1984). Die Überbewertung dieser Entdeckung führt dazu, daß das LIS ausschließlich als mechanisch bedingte, operativ behandelbare Erkrankung angesehen wird. Einige warnen vor einer allzu operationsfreudigen „Diskushysterie" und machen auf die vegetativen und psychischen Momente der Krankheit aufmerksam. Inzwischen ist davon auszugehen, daß es sich beim LIS um ein ätiologisch vielfältiges Schmerzsyndrom handelt, das infektiös (z. B. Borreliose), psychisch (Schmerz als Konversion) und/oder degenerativ (z. B. Diskusvorfall) bedingt sein kann. Aus diesem Grunde halten wir es inzwischen für gerechtfertigt, von *den* Lumbago-Ischialgie-Syndromen zu sprechen.

45.3.1.2 Krankengeschichte

Ein 49jähriger Fahrlehrer leidet seit einem Jahr unter lumbalen Schmerzen, die in die Hinterseite des rechten Beines ausstrahlen, verbunden mit Taubheitsgefühl und Schwäche im rechten Fuß, so daß er seine Arbeit nur mühsam und mit Unterbrechung verrichten kann. Orthopädische Behandlungen mit paravertebralen Injektionen, Massagen und krankengymnastische Übungen haben bisher nur vorübergehende Besserung gebracht.

Bei der neurologischen Untersuchung fällt eine steilgestellte Wirbelsäule mit fehlender Brustkyphose und eine druckdolente paravertebral-lumbale Muskelverspannung auf. Bei positivem Lasègue finden sich Zeichen einer Kompression der Wurzel S1 rechts, eine Pronationsschwäche des rechten Fußes sowie eine bandförmige Hypalgesie am dorsolateralen Unterschenkel bis zum lateralen Fußrand; ASR rechts fehlend. Der Patient wirkt unruhig und gespannt, äußert Sorgen um sein Geschäft, drängt auf schnelle Behandlung und baldige Entlassung.

Trotz der gegebenen relativen Operationsindikation erlauben wir uns abzuwarten. Die biographische Anamnese ergibt, daß der Patient von Kindheit an um seine Anerkennung fürchten und um seine Stellung kämpfen mußte. Nach dem frühen Tod der Mutter – er war 12 Jahre alt – habe er „den Starken herausgekehrt". Wehmütige Gedanken an die Mutter habe er vor dem Vater verborgen. Der habe für „so etwas" keinen Sinn gehabt, habe immer zu viel verlangt und kleine Vergehen mit tagelangem Schweigen bestraft. Die Angst vor dem Vater sitze ihm bis heute „wie ein Stiefel im Nacken". In wechselnden Berufen als Gärtner, Busfahrer und Verkäufer fühlt er sich minderwertig und abhängig, später als selbständiger Fahrlehrer kann er seinem Bewegungs- und Tatendrang endlich Geltung verschaffen. Er richtet Filialen in mehreren Städten ein und beschäftigt 15 Angestellte. Seine Frau, die er als „ängstliche kleine Maus" geheiratet hat, besorgt die Büroarbeiten. Ihre Zärtlichkeiten kann er kaum erwidern, überhäuft sie statt dessen mit großzügigen Geschenken. Er läßt sich zweimal von ihr scheiden, um sie bald darauf wieder zu heiraten. Die Ehe ist erneut bedroht, als die Frau eine Stelle im Kaufhaus annimmt und schnell zur Abteilungsleiterin aufsteigt. Ein wiederholtes „Ziehen" im Rücken, das sich, nach längerem Autofahren und morgens beim Aufwachen, zu einem bohrenden Schmerz steigert, hindert ihn nicht, sein Arbeitspensum noch auszuweiten. Er fühlt sich durch die neue Selbständigkeit seiner Frau verunsichert und verraten, kann dies aber nicht zum Ausdruck bringen, „antwortet" statt dessen mit Impotenz, erleidet erstmalig morgens beim Aufstehen eine rechtsseitige Lumboischialgie, die sich nach einer Woche ohne Therapie bessert. Zu seiner 16jährigen Tochter entwickelt sich ein spielerisch-verliebtes Verhältnis, er geht häufig mit ihr zum Tanzen aus. Als sie sich einem „anderen Freund" zuwendet und das Elternhaus verläßt, beginnt er, enttäuscht und arbeitswütig seine Wohnung zu renovieren. Beim Anstreichen des Badezimmers treten, drei Jahre nach der ersten Attacke, erneut heftige radikuläre, diesmal persistierende Schmerzen auf.

Stationäre Bettruhe, Wärmeanwendungen und entspannende Medikamente bewirken über zwei Wochen kaum eine Besserung. Erst als der Patient etwas von seinen verborgenen Gefühlen, von seiner Trauer um die verlorene Mutter, seiner Angst vor dem Vater, seiner Befürchtung, nicht mehr gebraucht zu werden, mitteilen kann, tritt eine Entspannung und Schmerzlinderung ein, die Parese geht zurück, so daß auch die Neurochirurgen – trotz einer Abklemmung der Wurzeltasche S1 im Myelogramm – von einer Operation abraten.

45.3.2 Epidemiologie

45.3.2.1 Häufigkeit

Rückenschmerzen sind in den westlichen Industrienationen hinter grippalen Infekten der zweithäufigste Grund, einen Arzt aufzusuchen. Lumbago-Ischialgie-Syndrome (LIS) treten mindestens einmal im Leben, je nach Methodik der Erhebung, bei 14–80% der Bevölkerung auf – Punkt-Prävalenz: 15–39%, kumulative Lebenszeit-Prävalenz: 65% (Deyo und Tsui-Wu, 1987). Mehr als die Hälfte der Bevölkerung verspürt einmal im Jahr einen lumbalen Schmerz, ohne einen Arzt aufzusuchen oder sich arbeitsunfähig zu melden (Waddell, 1987).

Eine Lumbago-Ischialgie tritt bei zahlreichen Krankheiten als Hauptsymptom, bei einer Reihe weiterer Erkrankungen als Begleitsymptom auf. Die ätiologische Vielfalt der LIS hat bisher noch nicht zu einer international einheitlichen Klassifikation geführt (Merskey, 1986; Nachemson und Andersson, 1982), die auch für epidemiologische Studien als geeignet angesehen werden könnte. Dabei bleibt oft unberücksichtigt, daß 50% der über 40jährigen einen asymptomatischen Diskusvorfall haben (Wiesel et al., 1984). So ist anzunehmen, daß nicht-bandscheibenbedingte LIS häufiger einem Diskusvorfall zugeordnet werden, wobei die neuen bildgebenden Verfahren eine solche Fehlzuordnung begünstigen. Deshalb ist eine um so differenziertere Anamnese und neurologische Untersuchung zu fordern, will man nicht „Techniker" bekommen, die ständig radiologische Auffälligkeiten diagnostizieren (Morris et al., 1986). Dies gilt auch für lumbosakrale Übergangsanomalien (Lumbalisation, Sakralisation, Spina bifida occulta), bei denen LIS nicht häufiger auftreten, mit Ausnahme der Spondylolisthesis (Magora und Schwartz, 1980).

Unerklärlich bleibt die in den westlichen Industrienationen in den letzten 30 Jahren zu beobachtende drei- bis fünffache Zunahme LIS-bedingter Arbeitsunfähigkeit. „Dorsopathien" und „intervertebrale Diskopathien" sind der zweithäufigste Grund einer Arbeitsunfähigkeit, Krankenhausbehandlung oder Frühberentung. Schwerarbeiter sind wider Erwarten keinem höheren LIS-Risiko ausgesetzt. Jedoch scheinen Berufe mit überwiegend sitzender oder vornübergeneigter Tätigkeit und vermehrter Vibration (Traktor-, LKW-, Busfahrer, Piloten) ein signifikant erhöhtes Risiko für LIS darzustellen (Kelsey, 1975a, b), so daß etwa Bankangestellte häufiger erkranken als Industrie-Schwerarbeiter. Die trotz größerer Arbeitsbelastung geringere LIS-Morbidität von Landarbeitern im Vergleich zu Industriearbeitern wurde mit der Vielseitigkeit ihrer Arbeit gegenüber maschineller Eintönigkeit erklärt.

45.3.2.2 Verlauf

Trotz hoher Prävalenz- und Inzidenzraten enden 90% aller Rückenschmerzepisoden, vermutlich wegen ihres milden Verlaufs, ohne ärztliche Konsulta-

Tabelle 45.3–1. Einteilung von LIS-Verläufen

akut	0–6 Wochen oder 0–12 Wochen
subakut	6–12 Wochen oder 6–24 Wochen
chronisch	> 12 Wochen oder > ½ Jahr

tion. Nur 6% der Patienten mit milden Rückenschmerzen suchen einen Arzt auf (Horal, 1969). Unabhängig von der Art der Behandlung sind 70–90% der Lumbago- und 50% der Ischialgie-Patienten nach zwei Monaten beschwerdefrei, jedoch müssen 67 bis 90% der Patienten mit Rezidiven rechnen.

Obwohl der Verlauf von LIS bisher nicht einheitlich definiert ist (vgl. Tab. 45.3–1), sind gegenüber attacken- und episodenartig rezidivierenden die ab Krankheitsbeginn chronisch undulierenden Verläufe mit 0,5–5% aller LIS-Patienten selten (Weber, 1983). Dagegen ergibt sich bei Personen, die früher eine Lumbago oder Ischialgie von mindestens zwei Wochen erlebt haben, im späteren Stadium eine fast gleiche Verteilung (33%) von attacken- und episodenartig rezidivierenden, subakuten und chronischen Verläufen (Deyo und Tsui-Wu, 1987).

Nur 1–2% aller Lumbago-Patienten entwickeln eine Ischialgie oder einen Diskusvorfall (Frymoyer, 1988). Nicht mehr als 2,5% der bandscheibenbedingten LIS-Patienten sind von einem Cauda-equina-Syndrom bedroht, das der sofortigen operativen Behandlung bedarf.

45.3.3 Klinik

Bei den lumbalen Syndromen unterscheiden wir die lokalen, vorwiegend vom hinteren Längsband ausgehenden Schmerzen und die radikulär ausstrahlenden, die auf einer Wurzelirritation beruhen.

45.3.3.1 Prodromi

Den ersten heftigen Schmerzattacken geht gewöhnlich eine jahrelange „latente Phase" milder lumbaler Beschwerden voraus: gelegentliche morgendliche Steifigkeit, die nach Bewegung abnimmt, zeitweilige „Müdigkeit", Spannung oder Schwächegefühl im Rücken und in den Beinen. Diese Prodromi („low back insufficiency") werden von den meisten Patienten bagatellisiert und durch forciertes Tätigsein „verdrängt". Die differentialdiagnostische und prognostische Bedeutung der Prodromi und ihrer unterschiedlichen Verarbeitung ist bisher unklar; z.B. ist offen, ob sich diese Beschwerden bei degenerativ bedingten, entzündlichen oder psychogenen LIS gleichen.

45.3.3.2 Lokales Lumbalsyndrom

Als „Lumbago" (lat.: Lendenschmerz) bezeichnet man den heftigen, meist ziehenden Schmerz im Lumbalbereich, der plötzlich beim Bücken und Wiederaufrichten, bei Körperdrehung, beim Heben, in fast zwei Drittel der Fälle jedoch ohne eruierbaren Anlaß, z.B. morgens beim Aufwachen, erstmals verspürt wird und Stunden bis Wochen anhält. In der Regel findet sich eine mit lumbaler Muskelverspannung einhergehende Steifigkeit des Rückens. Dieses akute Ereignis wird von der langsam einsetzenden und persistierenden „Lumbalgie" unterschieden.

45.3.3.3 Lumbales Wurzelsyndrom

Unter „Lumboischialgie" („Wurzelreizsyndrom") versteht man den akut einsetzenden, aber auch chronischen, meist ziehenden Schmerz, der – zuweilen unter Abnahme der lumbalen Beschwerden – ins Bein, je nach betroffener Wurzel in die Dorsal-, Lateral- oder Ventralseite ausstrahlt und beim Pressen oder Husten zunimmt. Neben einem positiven Lasègue findet sich eine Druckdolenz sowohl im Lumbalbereich als auch über den Valleixschen Punkten. Bei zusätzlich neurologischen Ausfällen spricht man von „Wurzelkompressionssyndrom". Entsprechend der besonderen statischen Belastung des lumbosakralen Übergangs, des „Wetterwinkels" der Wirbelsäule, sind die Wurzeln L5 und S1, einzeln oder gemeinsam, in über 90% betroffen. Bezüglich der weiteren klinischen Symptomatik sei auf die dritte Auflage dieses Buches verwiesen.

45.3.3.4 Pathophysiologie

Den Prodromi liegen vermutlich Muskelverspannungen als Antwort auf Fehlhaltung und -belastung der Wirbelsäule, unter Umständen auch schon die ersten Gefügestörungen im Innern der Bandscheibe zugrunde.

Die Vielfalt der prodromalen Beschwerden kann auch vegetativ durch die enge Verbindung des dorsalen Spinalnerven mit dem sympathischen Grenzstrang bedingt sein. Vor allem der Ramus meningeus (N. sinuvertebralis) nimmt bei seinem Verlauf durch das Foramen intervertebrale ins Innere des Spinalkanals über den Ramus communicans albus vegetative Fasern auf, bevor er den Anulus fibrosus, das hintere Längsband, das dorsale Periost der Wirbelkörper und die Ventralseite der Dura mit Schmerzfasern versorgt. Die auch beim manifesten LIS beobachtbaren vegetativen Störungen (Schwere-, Kältegefühl und Muskelkrämpfe) mögen durch die Intimität zwischen dorsalem Spinalnerv und Sympathikus verursacht sein (Pette, 1953; Schiffter, 1985). Ein Teil der Beschwerden mag, wie thermographische Befunde nahelegen, Folge einer axonalen Ausschüttung vasoaktiver Substanzen sein. Dabei kommt vermutlich den dünnen, nicht-myelinisierten afferenten Schmerzfasern, dem häufigsten Fasertyp innerhalb der Hinterwurzel, eine besondere Bedeutung zu.

Beim lumbalen und beim Wurzelsyndrom sind jeweils alle Anteile eines „Bewegungssegments" beteiligt (Schmorl und Junghanns, 1968). Obwohl in der Praxis verschiedene Möglichkeiten – vom Prolaps ohne Wurzelkompression bis zur Wurzelkompression ohne Prolaps – vorkommen, kann die Lumbago als Stadium mit geringen, die Wurzelkompression als Stadium mit ausgeprägten Strukturveränderungen gelten. Die Lumbago wird als Ausdruck eines „dérangement interne", einer vom hinteren Längsband abgefangenen Protrusion der Bandscheibe bei noch intaktem äußerem Faserring angesehen. Das lumbale Wurzelsyndrom gilt als Hinweis auf eine Ruptur des Anulus fibrosus, wobei sich Diskusgewebe durch die paramediane, kaudal sich verbreiternde Längsbandlücke nach dorsolateral in Richtung Spinalwurzel verlagert. Eine dadurch bedingte Wurzelkompression führt zu veränderter radikulärer Mikrozirkulation mit Ischämie und intraneuralem Ödem und begünstigt, bei persistierender Kompression, die Entwicklung einer intra- und extraneuralen Fibrose, die eine chronische Entzündung unterhält (Rydevik et al., 1984). Bei zwei Drittel bandscheibenbedingter Kompressionssyndrome wird deshalb eine Wurzelschwellung, im CT nicht selten als Prolaps verkannt, beobachtet, die die alte Neuritis-/Radikulitis-Diskussion mit neuen Akzenten wiederbeleben könnte (Schultz et al., 1988).

In jüngster Zeit wird der vermehrten Ausschüttung von schmerzrelevanten Neuropeptiden, etwa der Substanz P und dem vasoaktiven intestinalen Polypeptid (VIP), mehr Aufmerksamkeit bei der Pathogenese radikulärer Schmerzen gewidmet (Weinstein et al., 1988), wobei die vasodilatatorische Wirkung dieser Neuropeptide gerade nicht erklärt, wie es bei bandscheibenbedingten LIS zu den – vermutlich vasokonstriktiv entstandenen – hypothermen „Dermatomen" bzw. Thermatomen kommt.

Die Bedeutung degenerativer Veränderungen für die Ursache eines lumbalen Schmerzsyndroms wird immer noch überschätzt. Im 5. Lebensjahrzehnt finden sich bei 70% der Bevölkerung degenerative Veränderungen ohne Schmerzen. Während eine Differenzierung in „organische" (somatogene) und „funktionelle" (psychogene) LIS für das therapeutische Procedere von Bedeutung ist, stellen grundsätzliche Erwägungen eine solche Einteilung auch wieder in Frage:
- Veränderungen der Wirbelsäule, etwa Torsionsskoliosen, können mit völliger Beschwerdefreiheit einhergehen.
- Strukturveränderungen der Wirbelsäule und der zugehörigen Weichteile sind häufig diskret und erklären nicht das Ausmaß der Beschwerden.
- Selbst ein Wurzelkompressionssyndrom kann zu den funktionellen LIS gerechnet werden, wenn trotz computertomographisch persistierendem Diskusvorfall Beschwerdefreiheit nach konservativer Therapie eintritt (Schultz et al., 1987).

45.3.3.5 Differentialdiagnose

Beim lokalen Lumbalsyndrom müssen ein Morbus Bechterew, eine Spondylitis, Diszitis, Affektionen des Sakroiliakalgelenks und gynäkologische Krankheiten (Schmerz sakral betont) ausgeschlossen werden. Bei progredientem Verlauf und stechendem, im Liegen

zunehmendem Schmerz oberhalb L4 muß an spinale Neurinome, Meningeome und extradurale Wirbeltumoren und -metastasen, auch im Retroperitonealraum und kleinen Becken gedacht werden, die radikuläre Schmerzen verursachen können. Beim lumbalen Wurzelsyndrom kommen Koxalgien (Schmerz verstärkt beim Gehen und bei Innenrotation des Hüftgelenks) und iatrogene Ischiadikusschädigungen sowie die diabetische Mononeuritis multiplex in Frage, bei der die reißenden Schmerzen bereits früh mit Atrophien und Paresen im Bereich des Oberschenkels einhergehen. Auch eine Borreliose, besonders häufig Anlaß von Fehldiagnosen, kann als Monoradikulitis eine Ischialgie erzeugen oder bei Gelenkbeteiligung eine solche imitieren. Tageszeitliche und lokomotorische Abhängigkeiten der Schmerzen sind zu beachten. Nachts zunehmende Schmerzen sind in der Regel (Ausnahme bakterielle Spondylitis mit morgendlichem Schmerzmaximum) entzündlicher und neoplastischer Genese oder Folge eines extrem lateralen Diskusvorfalls. Wenig bekannt sind die nächtlichen anfallsartigen Lumbalgien und/oder symmetrisch in beiden Beinen auf- oder absteigenden Schmerzen bei Angstneurose (vgl. 45.3.4.4). Alle LIS erfordern eine biopsychosoziale Anamnese. Psychogene LIS, die sich einer diskogenen Wurzelkompression zeitlich anschließen können, erfordern positive Hinweise bei der biographischen Anamnese und der neurologischen Untersuchung (vgl. 45.3.4.4).

45.3.4 Psychosomatik

Der Schriftsteller Hermann Hesse schreibt als Ischias-Patient über die Begegnung mit seinem Arzt: „...wir begrüßten einander, wie es gesitteten Boxern ziemt, vor dem Wettkampf mit herzlichem Händedruck. Vorsichtig begannen wir den Kampf, tasteten einander ab, probierten zögernd die ersten Schläge. Noch waren wir auf neutralem Gebiet, unser Disput ging um Stoffwechsel, Ernährung, Alter, frühere Krankheiten und troff von Harmlosigkeit, nur bei einzelnen Worten kreuzten sich unsere Blicke, klar zum Gefecht" (Hesse, 1975). Obwohl jeder Arzt bei dieser präzisen Schilderung an eigene Begegnungen mit LIS-Patienten und seine Wahrnehmung – „Kampf", forscher Blick und Händedruck – erinnert wird, kann er sich diese und die dabei provozierten, bewußten und unbewußten, eigenen Gefühle (Gegenübertragung) kaum eingestehen, was diagnostisch und therapeutisch jedoch von entscheidender Bedeutung wäre. Auch Hoff (1954) beschreibt bei LIS-Patienten eine „rebellische" Grundhaltung, die in bestimmten angstbesetzten Situationen zur Auslösung der Krankheit wesentlich beitrage.

45.3.4.1 Psychogene Lumbago-Ischialgie-Syndrome

In den Arbeiten zu „psychogenic backache", „low back pain" und Kreuzschmerzen werden fast ausnahmslos chronische Schmerz-Patienten untersucht, wobei vorausgesetzt wird, daß nur die „funktionellen" Lumbalgien ohne körperlichen Befund „psychogen" seien (Leavitt und Garon, 1979; Paul, 1950; Pongratz, 1980; Saul, 1941). Die Spaltung in eine Körpermedizin ohne Seele und eine Seelenheilkunde ohne Körper hat zur Folge, daß die bei diesen Patienten registrierbaren Muskelverspannungen, Druckschmerzen und Bewegungseinschränkungen nicht als körperliche Befunde gewertet und umgekehrt bei LIS-Patienten mit Diskusvorfall psychische Auffälligkeiten ignoriert werden.

In Frankreich und in den USA traten im Ersten und Zweiten Weltkrieg – im Unterschied zu den Kriegsneurosen deutscher Soldaten – epidemieartig Kreuzschmerzen auf, mit der Besonderheit einer steifen Verkrümmung und Vornüberneigung des Rückens (Abb. 45.3–2), die den üblichen Therapien trotzten. Dieses als „hysterical bent back" oder als „Camptocormia" bezeichnete Syndrom (griech.: champtein = sich beugen, krümmen; lat.: campter = Biegung; griech.: chormós = der Rumpf) wurde im Sinne einer Konversion als Ausdruck der Unterwerfung und gleichzeitig Verweigerung gegenüber militärischen Autoritäten gedeutet, in der sich eine ambivalente Vaterbeziehung widerspiegele (Sandler, 1947; Souques und Rosanoff-Saloff, 1915). Fehlte die Vornüberneigung, wurde eingeräumt, daß psychogene „Ischiaserkrankungen" in den „allermeisten Fällen nur schwer von den ‚echten' Erkrankungen zu unterscheiden" seien (Raether, 1917). Erstmals gerieten durch diese Epidemie psychosomatische Zusammenhänge von LIS ins wissenschaftliche Blickfeld.

Patienten mit „psychogenic low back pain" sollen – im Gegensatz zu denjenigen mit „organic pain", die einen klaren Beginn und umschriebene Lokalisation der Schmerzen angeben – an ihrem bunten Beschwerdebild erkannt werden. Neben einem dumpfen Dauerschmerz werde – ähnlich den prodromalen Beschwerden – ein Druck-, Spannungs-, Steifheits-, Müdigkeits- oder Engegefühl im Lumbalbereich angegeben; eine bildhafte Sprache werde bevorzugt: „als ob eine große Last drücke", „als ob ein Teil des Rückens fehle", der Rücken sich wie „durchgebrochen" anfühle oder „außer Kontrolle" sei, als ob da „ein Klumpen" sei, ein taubes, totes Gefühl. Die Ausstrahlung der Schmerzen in die Beine, aber auch nach oben bis in den Nacken und in andere Körperteile sei „unanatomisch" (vgl. Differenzierung zwischen psychogenem und somatogenem Schmerz in Kap. 36). Bei der Untersuchung werden inkonstante lumbale Muskelverspannungen und Bewegungseinschränkung, teilweise in Form „bizarrer" Gangstörungen, und diffuse Sensibilitätsstörungen beschrieben.

Die Patienten werden als aktive, ruhelose Menschen geschildert, die Müßiggang kaum ertragen. Innerlich unsicher und abhängig, seien sie extrem angewiesen auf Lob und Anerkennung, fordernd in ihren Versuchen, Sympathie und Unterstützung zu gewinnen, und feindselig, wenn Anerkennung ausbleibe, wobei in solchen Situationen Rückenschmerzen erstmals oder vermehrt auftreten (Holmes und Wolff,

1952). Psychodynamisch wird Angst vor passiver Abhängigkeit und Unterwerfung hervorgehoben, die bei Männern am häufigsten in Form von Autoritätsproblemen und Fluchttendenzen zum Ausdruck komme, bei Frauen in Form maskuliner Identifikation und masochistischem Altruismus. Die Verspannung der lumbalen Muskulatur sei – bei drohendem Zusammenbruch dieser angstbesetzten Abwehr – als „primitive" Ersatzabwehr zu sehen. Insofern wird der Organwahl („aufrechte Haltung") psychosomatische Bedeutung zugemessen. Dieses auch als „vertebral neurosis" bezeichnete Schmerzsyndrom wird, sofern der Ausdrucksgehalt der Krankheit betont wird, als Konversionsneurose, sofern die Beziehung von Angst, Muskelverspannung und Schmerz betont wird, als Organneurose angesehen (Fetterman, 1940).

Den funktionellen Rückenschmerzen – in den fünfziger Jahren in Deutschland dem rheumatischen Schmerzsyndrom zugerechnet – liege eine Störung in der analen Kleinkindphase zugrunde, in der durch Verweigerung, vor allem durch Zurückhaltung von Stuhl und Urin, Eigenständigkeit geübt wird. Werde diese Individuation, die sich u. a. körperlich in einer opisthotonen Versteifung des Rückens ausdrücken könne (Mahler et al., 1980), gebrochen, nehme das Kind in der Folge gegenüber der Welt eine die Eigenständigkeit trotzig betonende, gespannte Haltung ein, die über Muskelverspannung zum Schmerz führe.

Abb. 45.3–2. „Camptocormia" oder „hysterical bent back"; Rückenschmerz als Konversion.

Anhand biographischer Anamnesen von 50 Patienten mit „Wurzelreizerscheinungen" fand Fleck (1975) eine zwanghafte Helfereinstellung, übermäßigen Arbeitseifer und mangelnde Genußfähigkeit, Rivalisieren mit Geschwistern und Gleichaltrigen, hingegen ungehemmte Zuneigung zu Kindern und abhängigen Personen als Antwort auf ein Grundgefühl der Wertlosigkeit seit der Kindheit. Für die ersten Ischiasattacken, die später in persistierende Schmerzen übergingen, eruierte er Situationen äußersten Einsatzes für andere bei Ambivalenz zwischen Verpflichtung und unbewußtem Aufbegehren.

Holmes und Wolff (1952) belegten ähnliche Beobachtungen erstmals elektromyographisch durch eine Zunahme von Aktionspotentialen im M. erector trunci in charakteristischen Situationen. Bereits einfache Handlungen (Drücken der Hand des Untersuchers) führten bei Patienten mit funktionellem LIS zu prolongierter Mitbeteiligung weit entfernter, von gesunden Kontrollpersonen nicht beanspruchter Muskeln. Konfliktbesetzte Gespräche ließen lumbale und gluteäle Aktionspotentiale ansteigen. Die Verkürzung von Erholungszeiten der bei jeder Handlung überschießenden und prolongierten Muskelverspannungen erwies sich als schmerzauslösend. Die unangemessene Indienstnahme der Muskulatur, als Schutz gegenüber feindseliger Bedrohung gedeutet, wurde für die Genese und vor allem für die Chronifizierung des LIS verantwortlich gemacht.

45.3.4.2 Akute Lumbago-Ischialgie-Syndrome

Ein LIS verläuft ursprünglich in Episoden und Attacken; der persistierende Schmerz gehört – bis auf wenige Ausnahmen (vgl. Abschn. 45.3.2.2) – zum Spätstadium der Erkrankung. Das Verhalten vieler LIS-Patienten ist vor und im Frühstadium ihrer Erkrankung geprägt von Unruhe, hypomanischem Tatendrang und forcierter Selbstbehauptung, „Rückgrat-Beweisen auf Biegen und Brechen". Bei solchen Patienten läßt sich eine typische Vorgeschichte eruieren: In der Kindheit vorzeitig zu Verantwortung und harter Arbeit herangezogen, gleichzeitig durch Strenge und Entbehrung von den Eltern unmündig gehalten, entwickeln die späteren Patienten von der Pubertät an trotzige Eigenständigkeit, expansive Unternehmungslust und unermüdlichen Arbeitseifer. Regressive Bedürfnisse müssen dabei extrem verleugnet werden. Sie haben Angst vor Hingabe und neigen in ihren Beziehungen dazu, andere zu übertreffen und dominierend zu betreuen, etwa indem sie sich hilfsbedürftige Partner suchen. Sie selbst können Geschenke und Hilfe dagegen nur schwer annehmen. Bei den Frauen ist das expansive Verhalten weniger ausgeprägt, es findet sich eher ein überprotektives Helfen und „Bemuttern" bei phallischer Identifizierung (Kütemeyer und Schultz, 1988; Lewin, 1933).

Durch den Aktivismus dieser Patienten leidet der Schlaf („ich habe einen Apparat in mir, der hört nicht auf, sich zu drehen"), vor allem die körperliche Selbstwahrnehmung, das Empfinden für das Ausmaß der

eigenen Leistung, für schützende Ermüdungserscheinungen und für Prodromi als erste Dekompensationszeichen. Zeitlich manifestieren sich die Ischiasattacken in kritischen Situationen der Lebensgeschichte, in denen die „Überlegenheit" nicht mehr durchgehalten werden kann, am häufigsten, wenn Partner oder Kinder selbständig werden.

Der Fahrlehrer, dem als Kind Liebe und Anerkennung vom Vater nur zuteil wurde, wenn er sich stark und tüchtig zeigte, hatte sein Leben so eingerichtet, daß ihm in Familie und Beruf eine überlegene Position sicher war. Er hatte dabei störungslos enorme Leistungen vollbracht und Trennungen überstanden. Durch die neue Eigenständigkeit der Frau und das Erwachsenwerden der Tochter fühlte er sich „überflüssig". Die Möglichkeit, lange verleugnete Geborgenheitswünsche eindeutiger kundzutun, war ihm durch eine Identifizierung mit der abweisenden Einstellung seines Vaters gegen „so etwas" immer noch versperrt. Seine erste Lumboischialgie war im Anschluß an monatelange „Prodromi" nach dem beruflichen Aufstieg der Frau, die zweite schwerere Ischiasattacke nach dem Auszug der Tochter aufgetreten.

Die Kombination von Überanstrengung auf der einen, Bewegungsarmut und unphysiologischer Körperhaltung beim Autofahren auf der anderen Seite haben vermutlich einer Diskusdegeneration Vorschub geleistet, die sich vor und nach der erster Ischiasattacke durch Prodromi bemerkbar machte, bis das Renovieren der Wohnung – in hektischer Anspannung als kompensatorische Antwort auf innere Verunsicherung – die zweite Attacke mit Wurzelkompression auslöste. Erst die Kombination körperlich und psychisch entspannender Maßnahmen, unter Verzicht auf invasives Vorgehen, brachte die entscheidende Besserung.

Bei mehr als einem Drittel aller LIS finden sich in der Anamnese rezidivierende Gastritiden, Magen- oder Duodenalulzera. Der Konflikt zwischen Abhängigkeitsbedürfnis und Unabhängigkeitsbestreben ist bei Patienten mit LIS und Ulcus duodeni ähnlich. Bei LIS-Patienten steht die Neigung im Vordergrund, den Konflikt durch Hypermotorik und Muskelverspannung zu verarbeiten. Eine „Pseudo-Unabhängigkeit" findet sich häufiger als ein offen abhängiges Verhalten.

Psychodynamisch läßt sich das Verhalten dieser Patienten als Antwort auf ein Ungestilltbleiben narzißtischer und oraler Bedürfnisse in der Kindheit deuten, wobei sie zur Bewältigung ihrer späteren Lebensaufgaben überwiegend anal-aggressive Verhaltensweisen entwickeln, die der Eigenständigkeit und Ich-Abgrenzung dienen. Aus narzißtischer Ohnmacht und der Gefahr einer Depression flüchten sie gleichsam in narzißtische Omnipotenz (Größenphantasien, expansiver Tatendrang, Hochgefühl). Passive Bedürfnisse bleiben primärprozeßhaft chaotisch und bedrohlich; sie müssen, u.a. mit Hilfe eines erhöhten Muskeltonus, immer mehr aus dem Selbstbild ausgeblendet werden.

Bei der körperlichen Untersuchung spannen die Patienten ihre – nicht selten tätowierten – Extremitäten unwillkürlich übermäßig an, so daß die Reflexe zuweilen schwer beurteilbar sind; die Kraftprüfung wird zu einer Art „Kräftemessen" mit dem Arzt, das bereits mit dem forschen Händedruck bei der Begrüßung beginnt. Gleichzeitig fällt ein schmerzbedingtes Nachgeben, eine „Schwäche" auf, die leicht als radikuläre Parese mißdeutet wird, obwohl sie nie auf die Kennmuskeln beschränkt ist.

Die Ambivalenz von Regressions- und Unabhängigkeitswünschen setzt sich bis in die Therapie hinein fort. Dramatisch schildern die Patienten ihre Beschwerden, um im nächsten Moment wegen eiliger Geschäfte auf Entlassung zu drängen. Klagend und mitleiderregend der Schwester gegenüber, zeigen sie beim Arzt eine auf Imponieren bedachte Wortgewandtheit, empfindlich für alles ihr Ichgefühl Verletzende, beständig auf der Hut vor Demütigungen. Gerade noch vor Schmerzen zu keiner Bewegung fähig, steigen sie aus dem Bett, stützen einen gelähmten Mitpatienten oder helfen einem gestürzten Anfallskranken auf. Der Arzt soll keinen Beitrag zur Behandlung geliefert, sie wollen vielmehr alles selbst gemacht haben. Anstatt die Behandlung anzunehmen und dem Arzt persönlich näherzukommen, versetzen sie sich an seine Stelle; den Arzt mit eigenem medizinischem Wissen belehrend, werten sie seine Fähigkeiten ab. Andererseits sind die auf den Arzt gerichteten Wünsche besonders anspruchsvoll und leicht enttäuschbar, weshalb diese Patienten rasch mit Rückzug und wiederholtem Arztwechsel reagieren.

45.3.4.3 Psychometrische Befunde

Die von einigen Autoren anhand des MMPI eruierten erhöhten Hypochondrie-, Hysterie- und Depressionswerte sowie eine Zwanghaftigkeit wurden von anderen nicht bestätigt oder für funktionelle und organische LIS-Gruppen gleichermaßen gefunden (Sternbach et al., 1973 b). Patienten mit akuten LIS-Attakken und neurologischen Ausfällen zeigten im MMPI niedrigere Depressions- und höhere Maniewerte als die Gruppe mit chronischen Schmerzen ohne neurologische Ausfälle. Dies läßt sich als gerade noch kompensierte Abwehr (etwa einer Depression) oder Reaktionsbildung bei somatischer Dekompensation deuten, was sich durch Befunde im Gießen-Test – Selbstidealisierung und Omnipotenzphantasien als Abwehr von Ohnmachtsgefühlen – stützen läßt (Pongratz, 1980). LIS-Patienten mit persistierenden Schmerzen zeigten in verschiedenen psychometrischen Verfahren eine „Über-Normalität" mit hoher sozialer Kompetenz (Pfingsten et al., 1988). Bei der Selbsteinschätzung ihrer Behinderung jedoch sehen sie sich im Vergleich zu den objektiv schwerer behinderten Patienten mit rheumatoider Arthritis deutlich invalider und depressiver (Zung's Self Rating Depression Scale; Sternbach et al., 1973a). Die Arzt-Patient-Beziehung war von einem charakteristischen, Therapieversuche vereitelnden Versteckspiel („pain games" bzw. „Koryphäen-Killer-Syndrom") und von ausgeprägter Invalidenmentalität („dropping out style of life") bestimmt. Engel (1959) beschrieb eine durch Selbstbestrafungs-

neigung charakterisierte Schmerzpersönlichkeit, die „pain-prone patients", unter denen sich viele LIS-Patienten finden. Insofern wurde – nicht unwidersprochen – vermutet, daß der psychischen Schmerzbereitschaft und der latenten oder milden Depression eher ätiologische als schmerzreaktive Bedeutung zukommt (Maruta et al., 1976; Sternbach et al., 1973). Die Interpretationsschwierigkeit testpsychologischer Untersuchungen beruht wahrscheinlich darauf, daß die meisten Instrumente nicht für LIS-Patienten entwickelt wurden. Darüber hinaus wird zwischen zeitstabilen und situativ-reaktiven Merkmalen kaum unterschieden. Trotzdem könnten die widersprüchlichen Befunde die polaren Affekte von LIS-Patienten widerspiegeln. Es zeigen sich mit verschiedenen, zum Teil mit den selben Testmethoden offenbar einmal die Ergebnisse der hypomanischen Abwehr, ein andermal mehr die kritischen Affekte selbst (z. B. Angst, Depression). Die Versuche, mittels testpsychologischer Untersuchungen zwischen funktionellen (psychogenen) und organischen LIS zu trennen, scheitern bereits an der Fragwürdigkeit dieser klinischen Einteilung (vgl. Abschn. 45.3.3.4). Darüber hinaus werden bestimmte psychopathologische Befunde (häufiger bei organischen LIS) durch diese Instrumente offensichtlich nicht erfaßt. Das klinisch häufigere Auftreten einer gemischt funktionell/organischen Symptomatik, welche psychisch sowohl mit abgewehrten als auch offenen Elementen eines Autonomiekonflikts einhergeht, wird selten angemessen berücksichtigt.

Unter den Patienten mit persistierenden Schmerzen ließen sich mit dem MMPI die Subgruppen „Hypochondrie", „reaktive Depression", „Somatisierung" (psychogene Schmerzen ohne Depression) und „manipulative reaction" (alle Scores simulativ erhöht) differenzieren (Sternbach, 1974). Mehrfach erfolglos operierte Patienten unterschieden sich von erfolgreich oder einmal erfolglos operierten durch höhere Hypochondriewerte. Die Gruppe der mehrfach erfolglos Operierten sei aus psychischen Gründen derart auf ihre Beschwerden fixiert, daß jede weitere Operation ohne vorhergehende Psychotherapie kontraindiziert sei (Hehl et al., 1983).

In prospektiven MMPI-Untersuchungen zeigten LIS-Patienten mit schlechtem Therapieerfolg schon vor der Behandlung höhere Hypochondrie- und Hysteriewerte als Hinweis auf eine bereits prämorbid und im Frühstadium der Erkrankung vorliegende pathologische Schmerzverarbeitung. Mit dem Beckschen Depressionsinventar (BDI) konnte in 87% der Fälle der Behandlungserfolg richtig vorhergesagt werden; vor Behandlung erhöhte Depressionswerte korrelierten mit persistierendem Schmerz (Hasenbring und Ahrens, 1987). Dabei liegt der Mittelwert dieser Patientengruppe in der von Beck et al. (1961) entwickelten Kategorie „milde Depression", während psychiatrische Patienten mit einer endogenen oder neurotischen Depression einen deutlich höheren Mittelwert (>20) aufweisen. Insofern liegt bei chronifiziertem LIS keine klinisch manifeste „major depression" (DSM-III-R) vor.

Um psychometrische Verfahren differentialdiagnostisch und prognostisch nutzen zu können, sollten die Instrumente an LIS-Gruppen selbst entwickelt werden und sich an der klinischen Differenzierung organischer und psychogener LIS-Gruppen sowie ihrer Mischformen orientieren.

45.3.4.4 Bedingungen der Chronifizierung

Da das LIS, selbst bei chronischem Verlauf, keine lebenslange Erkrankung darstellt, bevorzugen wir den Terminus persistierend. Während prämorbid und im Akutstadium hypomanische Stimmung und Pseudo-Unabhängigkeit vorherrschen, tritt später, nicht selten zeitgleich mit der Schmerzpersistenz, die Kehrseite des Aktivismus, ein abhängiges Verhalten mit Initiativelosigkeit – eventuell identisch mit der von vielen Autoren eruierten „Depression" – in den Vordergrund. Die Patienten, die vorher dazu neigten, ihre Schmerzen zu verleugnen, bedrängen jetzt den Arzt mit drastischen Schmerzschilderungen, verbunden mit Versorgungs- und Rentenansprüchen. Sie zeigen dabei eine – für Depressive ungewöhnliche, aber aus ihren hypomanischen Zeiten bekannte – Widerständigkeit und Kampfbereitschaft, wie sie auch bei Patienten mit anderen persistierenden psychogenen Schmerzen anzutreffen sind.

Die Frage, wann und wodurch die Episoden und Attacken in den persistierenden Schmerz übergehen, läßt sich bisher nur hypothetisch beantworten. Das regressiv-fordernde Verhalten läßt daran denken, daß es sich nicht um die Folge persistierender Schmerzen, sondern um eine spezifische Bewältigungsform einer schon prämorbid und im Akutstadium vorhandenen, durch den Aktivismus nur verdeckten und kompensierten Neurose handelt (Spring et al., 1976).

Für den Übergang von Episoden und Attacken in den persistierenden Schmerz spielt, selbst bei operierten Patienten, die Schwere neurologischer Ausfälle während der Akutphase prognostisch keine (Cashion und Lynch, 1979), hingegen das Umschlagen in regressiv-ansprüchliches Verhalten eine entscheidende Rolle. Lange verleugnete Bedürfnisse zeigen sich jetzt massiv, aber nur indirekt in Form der Somatisierung. Durch die anhaltenden Schmerzen „entschuldigt" und vor Schamgefühlen geschützt, können die Patienten die mit ihrem Selbstbild unvereinbaren oralen Bedürfnisse konfliktfrei zulassen (primärer Krankheitsgewinn) und sich Zuwendung erzwingen (sekundärer Krankheitsgewinn) (Hoffmann, 1986). Der passagere Schmerz des Akutstadiums ist von der latenten Neurose gleichsam zur Somatisierung „benutzt" worden (Abb. 45.3–3).

Es zeichnen sich folgende, die Schmerzverarbeitung prägende biographische Bedingungen ab:

– Bei besonders rigidem Leistungsideal und Ich-synton erlebter Regressionsabwehr treten persistierende Schmerzen und Depression erst spät im Krankheitsverlauf, dann aber besonders hartnäckig auf.

– Bei sinnvoller, Entfaltung und Anerkennung versprechender Aktivität sind Chronifizierungen selten; durch Aktivierung der Patienten, etwa in Richtung beruflicher Selbständigkeit, können Besserungen sogar stabilisiert werden. Hingegen führt Belastung ohne befriedigende Perspektive häufig zu „Chronifizierung".

– Aus der Diskrepanz zwischen Verausgabung und vorenthaltener sozialer Anerkennung entwickelt sich bevorzugt eine Rückgrat-„Rentenneurose". Die gekränkte Selbstliebe des LIS-Kranken muß durch eine Rente entschädigt werden. Die Rentenverweigerung kann als erneute Entwertung erlebt werden und zu weiterer Eskalation der Schmerzen führen (Lieberz, 1989).

Persistierende LIS neigen zu psychogenen Ausgestaltungen: es finden sich nicht-radikuläre Schmerzen, etwa in beiden Beinen oder in höheren Wirbelsäulenabschnitten, nach der „Kleiderordnung" begrenzte Sensibilitätsstörungen oder/und funktionelle Gangstörungen, „Paresen", „Hyperkinesen" und psychogene Anfälle. Das LIS hat sich in eine „zweite Krankheit" (Weizsäcker, 1943), in eine Konversion verwandelt, in der sich ein psychosoziales Problem – und die lange kompensierte Neurose – darbietet und verbirgt.

Der Übergang vom „organischen" in den Konversionsschmerz läßt sich nicht nur an der abweichenden Ischiasanamnese und am antianatomischen Befund, sondern auch an der dramatisch-appellativen Schmerz- und Selbstdarstellung der Patienten erkennen (vgl. Kap. 36). Der Begriff Konversion wird hier im erweiterten Sinne der Umwandlung unbewußter psychischer Energien in Körpersymptome gebraucht. Insofern können alle neurotischen Modi der Konfliktverarbeitung mit episodischen LIS eine Verbindung eingehen und in persistierenden Schmerz konvertieren, so daß sich – insbesondere wenn der zugrundeliegende Affekt latent bleibt – vier „chronische" LIS-Gruppen unterscheiden lassen:

– LIS als konvertierte Depression (einschließlich Hypochondrie)

– LIS als konvertierte Angst (Freud, 1895)

– LIS als hysterisches Symptom (Konversion im engeren Sinne)

– LIS als somatisierte Zwangssymptomatik.

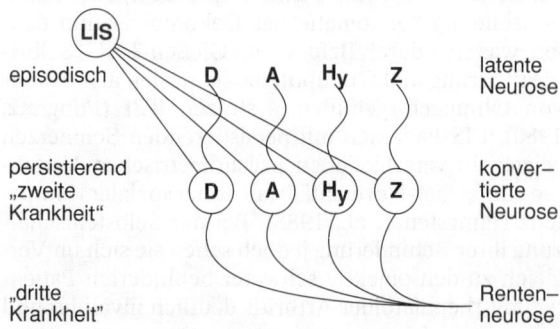

Abb. 45.3–3. Psychodynamisches Chronifizierungsmodell bei Lumbago-Ischialgie-Syndromen: ein LIS verbindet sich mit einem spezifisch neurotischen Konfliktverarbeitungsmodus (D = depressiv, A = angstneurotisch, Hy = hysterisch, Z = zwangsneurotisch). Dieser führt über eine Konversion zu einer „zweiten" und unter bestimmten Bedingungen zu einer „dritten" Krankheit (Rentenneurose).

Tabelle 45.3–2. Differenzierung psychogener Lumbago-Ischialgie-Syndrome

LIS	Konvertierte Depression	Konvertierte Angst	Hysterische Konversion	Konvertierter Zwang
Schmerz-charakter	brennend (drückend)	kribbelnd vibrierend	scharf dumpf phantomartig	„Druck"
Schmerz-verlauf	chronisch	anfallsweise (nachts)	undulierend	chronisch
Lokalisation	diffus	meist symmetrisch	oft einseitig	meist oben
Räumliche Ausbreitung	Tendenz zur kontinuierlichen Generalisierung	kontinuierlich aufsteigend absteigend und/oder in Sprüngen	nach der „Kleiderordnung"	nach oben
Globus Schluckstörung	+ +	+ + +	+	–
Schwindel	–	+ + +	(+)	–
Selbst-darstellung	monoton	dramatisch anklammernd	belle indifférence	betont sachlich latent aggressiv

Die Gruppen können ineinander übergehen oder kombiniert auftreten; insbesondere können sie, als gemeinsame Endstrecke, in die Rentenneurose – gleichsam eine „dritte Krankheit" – münden (Abb. 45.3–3).

Bereits aus der Schmerzschilderung lassen sich differenzierende Hinweise entnehmen (Tab. 45.3–2): Der (latent) depressive LIS-Patient gibt anhaltende diffuse, brennende und drückende Schmerzen mit Generalisierungstendenz (Ganzkörper-Schmerzsyndrom) an, der (latent) ängstliche Patient berichtet von kribbelnden, vibrierenden Schmerzen mit Parästhesien, die sich anfallsartig steigern, besonders nachts, und diskontinuierlich oder symmetrisch auf- und absteigend ausbreiten. Der hysterische Schmerz wird als messerscharf oder dumpf, in Richtung anorganisch beschrieben („wie ein Klumpen", „hölzern") mit Tendenz zur Ausweitung nach der „Kleiderordnung". Der persistierende Rückenschmerz des Patienten mit zwanghaften Anteilen neigt zur Verlagerung nach kranial, so daß zervikale und orofaziale Beschwerden überwiegen. Begleitbeschwerden wie Globusgefühl (Schluckstörung) werden in absteigender Häufigkeit vom ängstlichen, depressiven und hysterischen Patienten angegeben; ein diffuser Schwindel findet sich nicht bei der depressiven LIS-Gruppe, häufig hingegen in der Angst-Gruppe und zuweilen als Vorgefühl in der hysterischen LIS-Gruppe mit zusätzlich psychogenen Anfällen.

Durch die demonstrative Schilderung der persistierenden Schmerzen wird das nüchterne Interesse des Arztes gestört und durch Gefühle von Ohnmacht, Ärger, Ungeduld abgelenkt (vgl. Kap. 36). Die monotone, aggressionsgehemmte Schmerzdarstellung des de-pressiven und anankastischen Patienten und die dramatisch anklammernde des Patienten mit latenter Angst oder die „belle indifférence" des hysterischen Patienten erzeugt im Untersucher ein Gemisch aus hilflosem Helferwillen und aggressiver Abwehr. Diese – meist unbewußte – Ambivalenz kann sich in abweisenden Äußerungen des Arztes entladen („Sie haben nichts" – „Verschleiß" – „es könnte seelisch sein") und durch somatische Verlegenheitsdiagnosen (Fibromyalgie, Facettensyndrom) sowie durch invasive Maßnahmen neutralisiert werden, denen sich diese Patienten wegen ihrer Selbstbestrafungswünsche gerne unterwerfen. Der Effekt operativer Konversionsschmerzbehandlung ist vorübergehende Schmerzfreiheit oder aber drastische Eskalation der Schmerzen und funktionellen Ausgestaltung, nicht selten mit der Folge einer Operationssucht (Menninger, 1934).

45.3.4.5 Ansätze einer psychosomatischen Pathologie

Die Besonderheiten von LIS-Patienten spiegeln sich in einigen hervorstechenden körperlichen Befunden wider:

– Die in sagittaler Ebene häufig überstreckte Wirbelsäule, röntgenologisch als „Steilstellung" beschrieben, vermittelt bei der körperlichen Untersuchung einen Eindruck von der rückgratbetonten Haltung der Patienten. Lumbale und zervikale Lordose gehen beim gesunden Rücken aus der sakralen und thorakalen Kyphose hervor. Bei LIS-Patienten beobachten wir nicht nur eine abgeflachte Lendenlordose, sondern auch ein Fehlen der physiologischen Brustkyphose, so daß die sichtbaren Spuren der embryonal gebeugten Wirbelsäule verwischt werden. Die bei vielen LIS-Patienten schmerzbedingte vornübergebeugte Haltung kommt nicht lumbal, sondern im Hüftgelenk zustande.

– Die paravertebrale „brettharte" Muskelverspannung, die die akute Schmerzphase häufig lange überdauert, macht die nicht nur schmerzreaktive tonische-überaufrechte Haltung und Regressionsabwehr sichtbar und tastbar. Bei einigen Patienten finden sich beidseits lateral der LWS breite Muskelpakete, die von Ausläufern der Gesäßmuskeln, des M. latissimus dorsi und des M. obliquus abdominis externus (sog. Weichenwulst) gebildet werden – eine Art muskuläre „Lumbalisation".

– Der Rücken ist der Selbstwahrnehmung weitgehend entzogen. Die autochthone Muskulatur (M. erector trunci) nimmt anatomisch und funktionell eine Sonderstellung ein. Sie vollzieht die embryonale Wanderung der übrigen Muskeln nicht mit. Sie wird, anders als die übrige quergestreifte Muskulatur, vom Ramus *dorsalis* des N. spinalis versorgt, der segmental mit dem Grenzstrang des Sympathikus verbunden ist. So nehmen die autochthonen Rückenmuskeln eine Zwischenstellung zwischen der glatten, vegetativ versorgten und der Extremitätenmuskulatur ein. Wiewohl quergestreift, arbeiten sie vorwiegend unwillkürlich in Form unbewußter Mitbewegung bei Willkürbewegungen der Extremitäten. Sie erzeugen die Haltung des Rückens, die von innen kommt und der Willkür entzogen ist, etwa eine gedrückte oder eine überaufrechte, „stolze Haltung".

– Der Ort, an dem sich das LIS manifestiert, ist psychosomatisch bedeutsam. Am lumbosakralen Übergang findet phylo- und ontogenetisch die Aufrichtung statt. An der Grenze zwischen unbeweglichem, (embryonal) gebeugten und dem beweglichen aufgerichteten Wirbelsäulenabschnitt wirken sich nicht nur die Last des Körpers, sondern auch Konflikte zwischen Autonomie- und Abhängigkeitsbedürfnissen besonders aus.

– Die Bandscheibe, das größte nicht-vaskularisierte Gebilde im menschlichen Körper, unterhält ihren Stoffwechsel auf osmotischem Wege durch Druckschwankungen: Wasseraufnahme im entspannten Liegen, Abgabe von Wasser und Schlackenstoffen bei Belastungsdruck über 80 kp im Stehen und Sitzen. Die Versorgung der Bandscheibe leidet nicht durch vorübergehende Überlastung, sondern durch Daueranspannung, aber auch durch anhaltende Entlastung. Bei Unfällen – oder unter extremer Muskelkontraktion im epileptischen Anfall – kommt es eher zu einer Wirbelfraktur als zu einer Bandscheibenläsion. Durch die Biomechanik wird verständlich, daß bei fehlendem Rhythmus zwischen Anstrengung und Erholung die Bandscheibe vorzeitig austrocknet, degeneriert und, in Situationen vermehrter Anspannung, schließlich prolabiert und eine Wurzelkompression mit neurologischen Ausfällen verursacht. Von der Biomechanik der Bandscheibe her gesehen bedeutet es keinen Wi-

derspruch, daß sowohl die gespannte Hypermotorik der hypomanischen oder ängstlichen als auch die Bewegungsarmut der später depressiven LIS-Patienten lumbale Schmerzen hervorrufen.

45.3.5 Therapie

Die konservative Therapie, die für die meisten LIS-Patienten in Frage kommt, ist ein Stiefkind der Forschung geblieben (Deyo, 1983). Nach herkömmlicher Auffassung beginnt sie mit Bettruhe, Wärmeanwendungen und der Gabe von analgetischen (ASS/Paracetamol) und/oder muskelrelaxierenden Medikamenten (Diazepam). Anschließend entscheidet der Verlauf – häufig auch die Spezialität des behandelnden Arztes, nicht aber wissenschaftlich begründete Übereinkunft –, ob Massage und Krankengymnastik, Injektionen, Elektrotherapie, manuelle Therapie oder Extension zur Anwendung kommen. Angesichts der verwirrenden Fülle therapeutischer Angebote beeindruckt die Wirksamkeit der alleinigen Verordnung von strikter Bettruhe (Pearce und Moll, 1967).

45.3.5.1 Prinzipien psychosomatischer Therapie

Bereits die körperliche Untersuchung kann bei psychogenem LIS zu einer Behandlung werden.

> Eine 22jährige Verkäuferin mit Kreuzschmerzen entkleidet sich zur Untersuchung. Sie steht „vor mir, das Becken leicht nach vorn gekippt, den Rumpf, an dem sich die schlaffe Muskulatur kaum plastisch abzeichnet, leicht nach hintenüber geneigt. Ein haltungsschwacher Rücken, der rasch ermüdet und schmerzt. Während ich die Muskulatur mit meinen Händen untersuche, fühle ich, wie unter der Berührung der Rücken sich mehr rundet und meinen Händen anlegt. So, wie wenn er sich anlehnen wolle. Ich gebe meiner augenblicklichen Wahrnehmung und einer spontanen Deutungsphantasie nach und frage: ‚Sie sind anlehnungsbedürftig?' ‚Ja', und nach einer kurzen Schweigepause: ‚Lehnen Sie sich an Ihren Mann an?' Sie dreht sich überrascht um und antwortet: ‚Nein, er lehnt sich an mich an!' ‚Das geht aber nicht in der Ehe, Rücken an Rücken!' Sie sagt: ‚Das ging auch nicht mehr' und sie fährt fort: ‚Ich komme direkt von meiner Scheidung. Auf dem Weg zu Ihnen habe ich mir vorgenommen, selbständig zu werden'. Ich bin betroffen über die Situation. Sie bricht das Schweigen: ‚Wie können Sie das alles sehen und wissen? Das ist ja ein Wunder.' Ich bestärke sie in ihrem Entschluß, ihr Leben jetzt selbst in die Hand zu nehmen. Sie verläßt mich nachdenklich und erstaunt über das Erlebte. Wenige Monate später höre ich von ihr, daß es ihr gutgeht und sie schmerzfrei ist" (Heinl, 1986).

In der psychosomatischen LIS-Literatur finden sich therapeutische Anweisungen nur sporadisch. Einhellig wird eine vorsichtige und wenig aufdeckende psychotherapeutische Führung als sinnvoll angesehen. Wenig beachtet sind frühe Vorschläge, Physiotherapie und Psychotherapie bei der Akutbehandlung der Ischias zu kombinieren (von Weizsäcker, 1943). „Die Krankheit hat sozusagen zwei Gesichter, ein physisches und psychisches, gegen die man gleichzeitig vorgehen muß." Einem anfänglichen Stadium der Ruhe solle ein zweites Stadium des Trainings, der „allmählichen Wiedererziehung" folgen. Physio- und Psychotherapie sollten von Anfang an „Hand in Hand miteinander" gehen. Man begehe bei der Ischiasbehandlung oft den Fehler, mit den Verordnungen leicht zu wechseln, was dazu beitrage, den Patienten zu verwirren. „Es ist immer von Vorteil, wenn der Kranke den Eindruck gewinnt, daß man ihn planmäßig, nach einer bestimmten Methode behandelt" (Levy, 1913).

45.3.5.2 Psychosomatisches Therapieprogramm

Die lebensgeschichtlichen Bedingungen eines LIS berücksichtigend, haben wir für die stationäre Therapie ein dreistufiges psychosomatisches Programm entwickelt, bei dem physiotherapeutische Maßnahmen und Psychotherapie planmäßig ineinander greifen; zwei verschiedenen passiv-regressiven Phasen folgt eine kürzere aktive (Abb. 45.3–4). Die systematische Reihenfolge der Anwendungen und ihre konsequente Durchführung ist die Voraussetzung der therapeutischen Wirkung. Dafür ist ein therapeutisches Team erforderlich, das die heftigen Widerstände der Patienten gegenüber Regression verständnisvoll aufzufangen in der Lage ist, ohne das „Setting" zu verändern.

Das **dreistufige** Therapieprogramm hat sich auch zur präoperativen allgemeinen und lokalen Entspannung bewährt. Postoperativ sind hauptsächlich die Methoden der dritten Phase indiziert. Eine ambulante Behandlung kann, leicht abgewandelt, nach dem selben Muster erfolgen. Nach stationärer Therapie sollten, je nach hypomanischer, ängstlicher oder depressiver Grundstimmung, die entspannenden oder mehr die aktivierenden Maßnahmen eingesetzt werden.

Erste Phase: Allgemeine Entspannung

Das Ziel der ersten Phase ist eine körperliche und psychische allgemeine Entspannung. Durch Bettruhe (Stufenbett oder eine dem Patienten angenehme Lage), Wärme in Form von Schwitzpackungen (mit Lindenblütentee und Paracetamol) sowie muskelrelaxierende Medikamente (Diazepam) wird dem Patienten eine Regression ermöglicht. Nur wenige empfinden dies schon als lindernd und entlastend, da sie, ruhiggestellt, ihren abgewehrten, bedrohlichen Regressionsbedürfnissen vermehrt ausgesetzt sind. Sie drängen nach ihren gewohnten Aktivitäten, halten die verordnete Bettruhe nicht ein, wehren sich gegen die Einnahme von Diazepam (Angst vor „Abhängigkeit" und Kontrollverlust), beginnen sich als Helfer auf der Station zu betätigen und versuchen den Arzt zu aktiverem Vorgehen zu verführen. Dabei kommt es nicht selten vorübergehend zu Schmerzverstärkung (muskulärer – anankastischer Widerstand gegen Entspannung), obwohl der neurologische Befund schon im Begriff ist, sich zu bessern. Die eigentliche Leistung für Arzt und Pflegepersonal besteht darin, nicht mitzuagieren, sondern ein stabiles Gegenüber (Objekt) zu repräsentieren – was eine Immunität gegenüber der Hektik des Klinikalltags voraussetzt. Durch eine neurologische Untersuchung wird dem Patienten die Diskrepanz zwischen Schmerz und Befund demonstriert; am Therapieplan wird nichts geändert, dem Drängen des Patienten wird nicht nachgegeben, es wird zum Thema gemacht („Situationstherapie"). Die Konfrontation muß, soll sie nicht den Widerstand des Patienten verstärken, vorsichtig, fast beiläufig geschehen, etwa während der Visite: „Was macht Sie so unruhig, daß Sie nicht liegenbleiben können? – Was

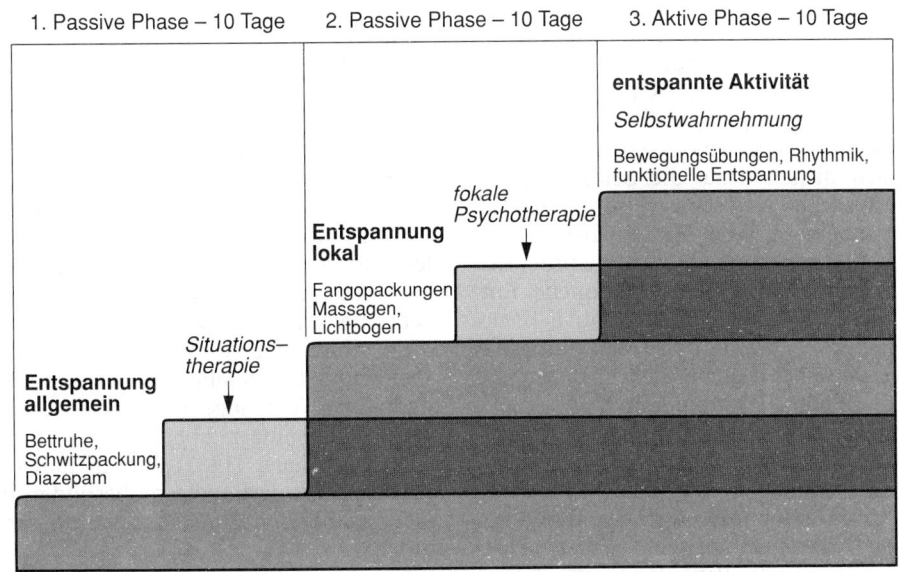

Abb. 45.3–4. Dreiphasiger Therapieplan zur konservativen Behandlung des Lumbago-Ischialgie-Syndroms.

treibt Sie so um? – Wer treibt Sie so an?" Solche Fragen vermitteln dem Patienten, daß er mit seiner Angst und Ambivalenz im Umgang mit seinen passiven Bedürfnissen verstanden und geschützt wird. Auf diese Weise kommt allmählich eine Distanzierung vom gewohnten Verhalten zustande. Durch die entspannenden Maßnahmen werden auch die Schlafstörungen behandelt. Beginnen die Patienten die verordnete Ruhe zu akzeptieren, was meist eine Entspannung und Besserung der Schmerzen zur Folge hat, kann – nach 7–10 Tagen – mit dem zweiten Abschnitt der passiven Phase begonnen werden.

Zweite Phase: Lokale Entspannung

Die zweite Phase hat die lokale Entspannung der lumbalen Muskulatur mit Fangopackungen und Massagen zum Ziel. Elemente der ersten Phase (Wärme in Form von Lichtbogen) werden beibehalten, vor allem die „Situationstherapie" ist weiterhin notwendig; denn die Massage mobilisiert erneut Widerstände. Der aktivistische Patient, gewohnt sich selbst zu helfen, muß mit der Bearbeitung seines Rückens körperliche Zuwendung hinnehmen. Er reagiert nicht selten erneut mit passagerer Schmerzzunahme, die wiederum die Objektkonstanz des Arztes und Pflegepersonals auf die Probe stellt. Häufig erst nach einer klärenden neurologischen Untersuchung und einem deutenden Ansprechen der Abwehr – „darf nicht auch einmal ein anderer etwas für Sie tun?" – kann die Massage als angenehm empfunden werden und lösend auf die lokale Muskelverspannung wirken. Zuweilen gelingt die Entspannung eher auf dem Wege der Unterwassermassage, die eine körperliche Berührung vermeidet.

Dritte Phase: Entspannte Aktivität – Selbstwahrnehmung

In der dritten Phase wird der Akzent auf Belastung und Aktivierung verschoben. Der Patient erlernt dabei Gelassenheit und einen veränderten Umgang mit seinem Körper. Die Krankengymnastin versucht beim Patienten mit isometrischen Übungen ein Gefühl für Verspannung und Erschöpfung, für seinen Körper überhaupt zu wecken. Auch die anschließenden Bewegungsübungen dienen dazu, die Selbstwahrnehmung, die bei der früheren Lebensweise gelitten hat, wieder zu entwickeln. In dieser Phase haben körperbezogene Psychotherapieformen, etwa die progressive Relaxation (Jacobson, 1938), funktionelle Entspannung (Fuchs, 1979) oder konzentrative Bewegungstherapie (Becker, 1981) ihren Platz. Autogenes Training ist wegen seiner leistungsmedizinischen und suggestiven Orientierung kontraindiziert.

Die Betonung der Selbstwahrnehmung bedeutet einen wesentlichen Unterschied zu dem in der Therapie bandscheibenbedingter Erkrankungen meist vertretenen Konzept der „Mobilisierung", „Stabilisierung" und „Kräftigung", das rein motorisch orientiert ist und die aktivistische Abwehr des Patienten verstärkt. Eine Differenzierung der Selbstwahrnehmung dagegen begünstigt – im Sinne der vorbeugenden Selbsthilfe – das rechtzeitige Erkennen und adäquate Umgehen mit späteren Rezidiven. Um die bereits erwähnten Faktoren der Chronifizierung und psychogenen Ausgestaltung zu vermeiden, müssen die das Auftreten des Wurzelkompressionssyndroms begünstigenden Lebensgewohnheiten und Beziehungsprobleme jetzt direkter angesprochen werden („fokale Psychotherapie"). Die Bedürftigkeit, die hinter dem Arbeitseifer, der Kampf- und Hilfsbereitschaft des Patienten steckt, darf jetzt Thema werden, wobei die wachsende Eigenständigkeit des Partners weniger bedrohlich, sondern entlastend und befreiend erlebt werden kann.

45.3.5.3 Therapieergebnisse

Mehr als die Hälfte der LIS-Patienten ist unabhängig von der Art der konservativen Therapie innerhalb von zwei bis drei Monaten beschwerdefrei. Deshalb sollte eine Nukleotomie vor Ablauf dieser Zeit – außer bei absoluter Operationsindikation – nicht durchgeführt werden (Bell und Rothman, 1984). Obwohl kurze Katamnesen die Vorzüge der Frühoperation zu belegen scheinen (Hakelius, 1970), ist im Langzeitverlauf (4–10 Jahre) die konservative Therapie der chirurgischen überlegen (Weber, 1983). Unter Berücksichtigung der Invalidisierten und Berenteten betragen die Kosten jeder Bandscheibenoperation etwa 100 000 DM (Dvorak et al., 1989a, b).

Hinsichtlich des Spontanverlaufs eines Diskusvorfalls läßt sich computertomographisch in einem Drittel der konservativ behandelten LIS-Patienten eine mäßige oder deutliche Verkleinerung beobachten (Fischer et al., 1988; Hackenbroch et al., 1984). Auch die übrigen zwei Drittel sind trotz gleichbleibenden Diskusvorfalls beschwerdefrei (Schultz et al., 1988; Köhler et al., 1989). Eine Verkleinerung prolabierten oder sequestrierten Diskusgewebes – besonders häufig bei mediolateraler Lokalisation – ist entweder Folge reparativer Vorgänge oder aber eines asymptomatischen Abgleitens in den terminalen Kaudabereich. Reparativen Vorgängen wurde noch wenig Beachtung geschenkt. Histopathologisch ließen sich bei 442 von 1000 nukleotomierten Patienten Gefäßeinsprossungen aus der Wirbelkörperspongiosa und durch das Foramen intervertebrale nachweisen, die möglicherweise eine Resorption begünstigen. Außerdem fand sich im rupturierten Anulus fibrosus einwachsendes zell- und kapillarreiches Granulationsgewebe (Becker, 1984). Zusätzlich gelingt dem Wirbelkörper durch die Ausbildung von Osteophyten die Abstützung der prolabierten Bandscheibe. Ob solche reparativen Vorgänge von spezifischen biographischen Bedingungen abhängen, ist ungeklärt, auch wenn Reischauer (1957) Selbstheilungen dieser Art nur bei „Leistungsfreudigen", besonders bei Patienten in freien Berufen ohne Versicherungsanspruch vermutet.

Allein eine 2–4wöchige strikte Bettruhe führt bei 68% der Patienten mit lumbalen Wurzelkompressionssyndromen (Katamnese: x̄ = 8 Jahre) zu Beschwerdefreiheit (Pearce und Moll, 1967). Trotzdem wird Bettruhe selten ausdrücklich empfohlen, aus Kostengründen sogar eine Liegezeitverkürzung angestrebt. Eine katamnestische Langzeitstudie des psychosomatischen Therapieprogramms ergab Beschwerdefreiheit bei 70% (einschließlich der Operierten) und 82% der ausschließlich konservativ behandelten LIS-Patienten (n = 37) (Kütemeyer und Schultz, 1983).

Ein Viertel bis gut die Hälfte nukleotomierter LIS-Patienten dagegen wird nicht beschwerdefrei (Schramm et al., 1978; Söderberg und Sjöberg, 1961; Spangfort, 1972). Die verschiedenen Beschwerden dieser heterogenen Gruppe werden unter dem Begriff „failed back syndrome" zusammengefaßt, wobei die Verfehlung („fail") projektiv dem Rücken und nicht dem Arzt zugeschrieben wird. Vermutlich finden sich in dieser Gruppe zahlreiche Patienten mit psychogenen LIS. Der Anteil der psychogenen LIS dürfte unter den 15–20% der reoperierten Patienten (Rish, 1984) besonders hoch sein, wobei der Erfolg nach der zweiten Operation – je nach Katamnesedauer und Dauer des postoperativen Intervalls – von 15% auf 0% absinkt. Handelt es sich anstelle eines Diskusprolaps lediglich um ein „bulging disc"-Phänomen (zirkuläre degenerative Vorwölbung der Bandscheibe über den Rand des Wirbelkörpers), so werden im Langzeitverlauf weniger als die Hälfte bandscheibenoperierter Patienten beschwerdefrei. Andererseits werden bei negativem Operationsbefund, d. h. fehlendem Diskusvorfall, 38% der LIS-Patienten schmerzfrei (Spang-

fort, 1972). Ob dies auf forcierter Muskelrelaxation während der Narkose oder auf der psychodynamischen Wirkung der Operation (Placebowirkung) beruht, ist unklar.

Im Gegensatz zur herrschenden Auffassung stellt eine radikuläre Parese, die dem Patienten meist nicht auffällt, eine zweifelhafte Operationsindikation dar, da der Spontanverlauf einer kompressionsbedingten Neuritis bisher unbekannt ist und Besserungen noch nach einem Jahr zu erwarten sind (Weber, 1975).

45.3.5.4 Fragwürdige Behandlungsmethoden

Eine psychosomatische Behandlung von LIS-Patienten wird häufig durch eine intramuskuläre Injektionsbehandlung ersetzt, die lebensgefährlich sein kann. Die chiropraktische Behandlung bandscheibenbedingter, insbesondere zervikaler Wurzelreiz- und -kompressionssyndrome ist Humbug, da degenerierte Wirbel nicht ausrenken und deshalb nicht eingerenkt werden können. Diese Methode kann zu einem radikulären Kompressionssyndrom und zu Insulten spinaler oder hirnversorgender Arterien führen. Die Indikation der Chemonukleolyse – unverletzter Anulus fibrosus – ist mit der Indikation einer konservativen Behandlung identisch und die Methode somit hinfällig. Hingegen stellt die transkutane Diskektomie bei strengster Indikation (therapieresistente Schmerzen bei medialer Protrusion) eine ernstzunehmende neurochirurgische Alternative dar.

46 Klinische Psychoneuroendokrinologie

Horst Lorenz Fehm und *Karlheinz Voigt*

Im Kapitel 10 „Psychoneuroendokrinologie" ist dargestellt, wie das Zentralnervensystem über neurosekretorische Neurone des Hypothalamus das endokrine System beeinflussen kann. Von daher ist es verständlich, ja zu erwarten, daß primär zentralnervöse Störungen zu Hormonstörungen führen, die schließlich auch klinische Relevanz gewinnen können. Im folgenden sollen diejenigen endokrinologischen Erkrankungen erörtert werden, bei deren Entstehung und Verlauf psychische Faktoren diskutiert werden. Es handelt sich um folgende:
– psychogener Zwergwuchs („maternal deprivation"-Syndrom)
– Hyperthyreose
– Cushing-Syndrom
– „psychogene Polydipsie"
Diese Auswahl von Diagnosen mutet sehr willkürlich an; dies darf nicht dahingehend fehlinterpretiert werden, daß für die hier nicht genannten endokrinologischen Erkrankungen psychische Faktoren keine Rolle spielten; sie besagt lediglich, daß es darüber heute noch keine oder zu wenig fundierte Untersuchungen gibt. Endokrine Störungen mit nachgeordneter klinischer Relevanz, wie z.B. das neuroendokrine Syndrom bei der endogenen Depression oder bei der Anorexia nervosa, wurden bereits im Kapitel 10 abgehandelt.

Wie das Zentralnervensystem das Endokrinium beeinflußt, ist der eine Aspekt der Psychoneuroendokrinologie. Der andere beschäftigt sich mit den Wirkungen peripher sezernierter Hormone auf zentralnervöse Funktionen. Daß es solche Wirkungen gibt, ist aus der klinischen Beobachtung lange bekannt: So gut wie alle endokrinologischen Erkrankungen können zu psychopathologischen Erscheinungen führen. Nicht selten sind die psychischen Symptome bei einer primär endokrinologischen Erkrankung schwerer und ernster als alle übrigen Manifestationen des gestörten hormonalen Gleichgewichts. In einem zweiten Abschnitt soll deswegen eine kurze Zusammenfassung über die psychopathologischen Erscheinungen bei endokrinologischen Erkrankungen gegeben werden. Bezüglich der Einzelheiten und der vollständigen Bibliographie können wir auf mehrere vorzügliche Übersichtsarbeiten verweisen, so Bleuler: Endokrinologische Psychiatrie, 1964; Smith et al.: Psychiatric Disturbance in Endocrinologic Disease, 1972; Beumont und Burrows: Handbook of Psychiatry and Endocrinology, 1982.

46.1 Die Bedeutung psychischer Faktoren für Entstehung und Verlauf endokrinologischer Erkrankungen

46.1.1 „Maternal deprivation"-Syndrom

Die Beobachtung, daß ein Mangel an emotionaler Zuwendung zu einer Wachstumsstörung bei Kindern führen kann, war an sich nicht neu, als 1967 Powell und Mitarbeiter 13 Kinder mit Minderwuchs als Folge einer „emotional deprivation" beschrieben. Dennoch wird der Begriff „maternal deprivation"-Syndrom mit Recht mit dem Namen Powell in Verbindung gebracht.

Diesem Autor und seinen Mitarbeitern waren bei minderwüchsigen Kindern, bei denen zunächst eine Hypophysenvorderlappeninsuffizienz vermutet worden war, schwere emotionelle Störungen aufgefallen, die auf die extreme häusliche Situation bezogen werden konnten. Fünf der elf Elternpaare waren geschieden oder getrennt. Schwere Eheprobleme gab es bei mindestens zwei weiteren Paaren. Fünf Väter waren exzessive Trinker, die meisten waren kaum zu Hause und keiner der Väter verbrachte viel Zeit mit den Kindern. Über die Mütter war weniger zu erfahren. Eine gestand offen, ihren Sohn zu hassen, eine andere war Alkoholikerin. Die Kinder selbst boten vielfältige Symptome wie Polydipsie und Polyphagie, hatten eigenartige Eßgewohnheiten, unruhigen Schlaf, fanden keinen Kontakt zu anderen Kindern und waren sprachverlangsamt.

Die Veränderung der psychosozialen Situation, nämlich Hospitalisierung mit psychologischer Behandlung und viel persönlicher Zuwendung, führte in allen Fällen zu einer erheblichen Wachstumsbeschleunigung, die die normale Wachstumsrate bei weitem überschritt (Abb. 46–1).

Interessant sind die bei diesen Kindern erhobenen endokrinologischen Befunde: Die meisten hatten erniedrigte STH-Spiegel und zeigten nicht den üblichen Anstieg des Wachstumshormons im Insulinhypoglykämietest. Bei neun der dreizehn untersuchten Kinder fand sich weiterhin eine verminderte Ausscheidung von 17-Hydroxykortikoiden im Urin und bei 12 Kindern ein fehlender Anstieg derselben im Metopirontest. Die Schilddrüsenwerte lagen im Normbereich.

Insgesamt würde man aufgrund dieser Befunde eine partielle Insuffizienz des Hypophysenvorderlap-

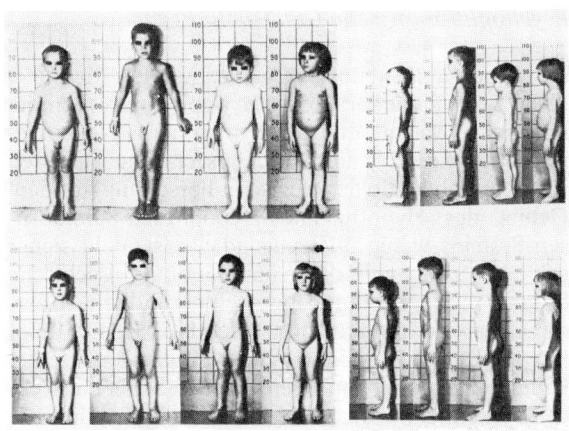

Abb. 46–1. Frontale und seitliche Ansicht von Kindern mit „maternal deprivation"-Syndrom vor (obere Reihe) und einige Zeit nach Hospitalisation (untere Reihe). Neben der Größenzunahme fällt die Rückbildung der auffälligen Vorwölbung des Abdomens auf.

pens diagnostizieren. Jedoch normalisierten sich die pathologischen endokrinologischen Befunde rasch während der Hospitalisierung, was einen hypophysären oder hypothalamischen Organdefekt unwahrscheinlich macht. Vielmehr muß man schließen, daß das Syndrom auf einer psychisch bedingten reversiblen Inhibierung der STH- und ACTH-Sekretion beruht. Unsere Kenntnisse über die Bedeutung zentralnervöser Neurotransmitter für die Regulation der STH- und ACTH-Sekretion legen die Hypothese nahe, daß die Verminderung beider Hormone beim Deprivationszwergwuchs primär auf einer veränderten Neurotransmitteraktivität im Hypothalamus beruht. Diese Hypothese wird durch eine Beobachtung an einem einzelnen Patienten mit diesem Syndrom gestützt (Imura et al., 1971). Bei diesem Patienten konnte der fehlende Anstieg des STH im Insulinhypoglykämietest durch eine Behandlung mit Propranolol (einem β-Rezeptorenblocker) wiederhergestellt werden. Daraus schlossen die Autoren, daß die Störung der STH-Sekretion primär auf einer gesteigerten β-adrenergen Aktivität hypothalamischer Zentren beruht.

In späteren Studien konnte gezeigt werden, daß auch die Somatomedine (Peptide, die unter dem Einfluß von Wachstumshormon in der Leber gebildet werden, die Effekte auf das Skelettwachstum vermitteln) bei Kindern mit dem „maternal deprivation"-Syndrom reversibel vermindert sind (D'Ercole et al., 1977). Erwähnenswert ist schließlich die Beobachtung von Frasier und Rallison (1972) an einer einzelnen Patientin mit „maternal deprivation"-Syndrom: das 5½ Jahre alte Mädchen zeigte eine Resistenz gegenüber der Behandlung mit menschlichem Wachstumshormon. Diese Beobachtung ist deswegen so interessant, weil 7 Jahre später eine tierexperimentelle Parallele gefunden wurde (Kuhn et al., 1979): Wenn man neugeborene Ratten von der Mutter trennte, kam es zu einer spezifischen Unterdrückung der Ansprechbarkeit verschiedener Gewebe auf das Wachs-

tumshormon. Als Parameter der STH-Wirkung wurde dabei die Aktivität der Ornithin-Decarboxylase im Gewebe gemessen.

Es ist verschiedentlich behauptet worden, daß man die endokrinen Störungen beim „maternal deprivation"-Syndrom auf Unterernährung zurückführen könne und psychische Faktoren damit allenfalls eine mittelbare Rolle spielten (Whitten et al., 1969; Krieger und Mellinger, 1971). Einerseits ist jedoch bei der Unterernährung das STH eher erhöht, andererseits ergeben sich bei den meisten publizierten Fällen keine Hinweise auf Unterernährung. Dennoch sollten solche Faktoren sorgfältig ausgeschlossen werden, ehe ein psychogener Zwergwuchs diagnostiziert wird.

Diese am Menschen gemachten Beobachtungen werden unterstützt durch zahlreiche tierexperimentelle Befunde, wobei ebenfalls eine Beeinflussung von Wachstum und Wachstumshormon durch „psychische" Faktoren gefunden wurde (z. B. Ruegamer et al., 1954). Schalch und Lee (1968) haben gezeigt, daß junge Ratten infolge systematischen „handlings" (in die Hand nehmen und streicheln) eine größere Wachstumsrate und eine größere Gewichtszunahme pro Einheit konsumierter Nahrung haben; gleichzeitig finden sich höhere Spiegel an STH. Dies macht wahrscheinlich, daß die beobachtete Wachstumsbeschleunigung auf eine vermehrte Sekretion von STH zurückgeführt werden kann. Eine ausführliche und kritische Übersicht über diese Arbeiten findet sich bei Brown und Reichlin (1972).

Insgesamt erscheint die Bedeutung psychischer Faktoren für die Entstehung klinisch relevanter Störungen der STH-Sekretion gesicherter als für irgendein anderes endokrines System.

46.1.2 Hyperthyreose

Die Hyperthyreose ist nach dem Diabetes mellitus die häufigste endokrine Krankheit. In der medizinischen Klinik hat durchschnittlich von 100 Patienten einer die Diagnose Hyperthyreose. Die Hyperthyreose ist eine Folge der Überschwemmung des Körpers mit Schilddrüsenhormon (Thyroxin = T_4 und Trijodthyronin = T_3), wobei das T_4 als Prohormon für das eigentlich biologisch aktive T_3 gelten kann.

Die häufigsten Symptome der Hyperthyreose sind Nervosität, emotionale Labilität, Schlaflosigkeit, feinschlägiger Tremor, übermäßiges Schwitzen und Wärmeintoleranz. Häufig findet sich auch ein Gewichtsverlust bei normalem oder sogar gesteigertem Appetit. Eine Myopathie mit Muskelschwäche und erhöhter Ermüdbarkeit kann vorkommen. Im allgemeinen wird bei jüngeren Patienten das klinische Bild von „nervösen" Symptomen beherrscht, während bei älteren Patienten oft nur kardiovaskuläre und/oder myopathische Symptome auftreten. Die Liste der Symptome zeigt, daß die Abgrenzung der Hyperthyreose von Zuständen psychovegetativer Labilität mit erhöhter Sympathikotonie schwierig sein kann. Früher waren dafür Begriffe wie „Thyreoneurose" oder „Pseudohyperthyreose" im Gebrauch. Heute ist es fast immer möglich, mit Hilfe der unmittelbaren Bestimmung von Thyroxin und Trijodthyronin im Blut eine Abgrenzung zwischen normaler und pathologischer Schilddrüsenfunktion zu treffen; als besonders wertvoll hat sich die Bestimmung des hypophysären TSH vor und nach Stimulation mit dem Thyreotropin-Releasing-Hormon (TRH) erwiesen.

Wenn eine Hyperthyreose festgestellt ist, muß differentialdiagnostisch zwischen einer großen Zahl von möglichen Ursachen unterschieden werden. Am häufigsten handelt es sich um einen Morbus Basedow oder um ein toxisches Adenom der Schilddrüse. Die diffuse mikrofollikuläre Autonomie der Schilddrüse ist im Einzelfall nicht sicher vom Morbus Basedow abgrenzbar. In seltenen Fällen muß an einen TSH-produzierenden Hypophysentumor, an die „Struma ovarii", an eine transitorische Hyperthyreose bei chronischer Thyreoiditis und anderes gedacht werden. Im folgenden beschäftigen wir uns nur mit dem Morbus Basedow.

Pathophysiologie des Morbus Basedow

Bereits vor 20 Jahren konnte gezeigt werden, daß im Serum von Basedow-Patienten ein Faktor vorhanden ist, der die Schilddrüse stimuliert. Wegen seiner im Vergleich zu TSH verlängerten Wirksamkeit wurde dieser Faktor „long acting thyroid stimulator" (LATS) genannt. Es stellte sich heraus, daß es sich beim LATS um ein Immunglobulin der Klasse IgG handelt, das von Lymphozyten dieser Patienten gebildet wird. Während mit der ursprünglichen In-vivo-Methode von McKenzie, die die Maus als Versuchstier benutzte, nur bei etwa der Hälfte der Patienten ein positiver Befund zu erhalten war, ist mit den neueren In-vitro-Methoden, die menschliches Schilddrüsengewebe benutzen, bei den meisten Patienten ein schilddrüsenstimulierender Faktor nachweisbar. Es handelt sich dabei um ein oder mehrere Immunglobuline, die gegen eine Komponente in der Plasmamembran von Thyreozyten als Antigen gerichtet sind. Es gibt inzwischen gute Hinweise, daß es sich bei dem Antigen um den TSH-Rezeptor selbst handelt. Durch die Verbindung des Antikörpers mit dem TSH-Rezeptor kommt es zu einer Reaktion, ähnlich als ob das TSH selbst mit seinem Rezeptor interagiert hätte.

Damit ist gleichzeitig gesagt, daß es sich beim Morbus Basedow um eine autoimmunologische Erkrankung handelt, die darauf beruht, daß schilddrüsenstimulierende Antikörper produziert werden. Es bestehen fließende Übergänge zur Hashimoto-Thyreoiditis, die ja auch durch Antikörper gegen Schilddrüsengewebe verursacht wird, bei der jedoch offenbar keine stimulierenden Antikörper gebildet werden. Man kann sich vorstellen, daß dem Morbus Basedow eine angeborene Störung der Immuntoleranz zugrunde liegt, so daß ein bestimmter Klon von Lymphozyten überleben kann; durch unbekannte auslösende Faktoren wird dieser Klon zur Proliferation und Sekretion von schilddrüsenstimulierenden Antikörpern gebracht.

Psychophysiologische Untersuchungen

Im Gegensatz zu unserer gut fundierten Kenntnis über die Stimulierbarkeit des Hypophysen-Nebennierenrinden-Systems und des Sympathikus-Nebennierenrindenmark-Systems durch psychische Faktoren ist die psychoendokrinologische Forschung auf dem Gebiet des Hypothalamus-Hypophysen-Schilddrüsen-Systems wenig überzeugend. Es ist nicht sicher geklärt, ob die Schilddrüse überhaupt auf psychische Stimuli reagieren kann. Eine sehr gründliche Übersicht über die Literatur findet sich bei Mason (1968) mit 96 Zitaten. Die Mehrzahl der Publikationen über Untersuchungen beim Menschen lassen eine Tendenz erkennen, wonach die Aktivität der Schilddrüse als Antwort auf einen emotionalen Stimulus ansteigt. Es gibt jedoch auch Autoren, die einen solchen Anstieg nicht beobachten konnten (z.B. Volpe et al., 1960). Wie auch immer, die Veränderungen sind gering und liegen innerhalb des Normbereiches. Dagegen scheint gesichert, daß beim Tier die Schilddrüsenhormone nach emotionaler Stimulation (z.B. Fesselung im Primatenstuhl) ansteigen (Mason und Mongey, 1972).

So schien lange Zeit ein Widerspruch zu bestehen zwischen der geringen Beeinflußbarkeit der Schilddrüse durch psychische Faktoren einerseits und der Bedeutung emotionaler Krisen in der Pathogenese der Hyperthyreose andererseits (s.u.). Dieser Widerspruch hat sich insofern geklärt, als wir heute wissen, daß das Hypothalamus-Hypophysen-System für die Ätiologie der Hyperthyreose keine Rolle spielt: Das TSH ist im Plasma und in der Hypophyse nicht vermehrt, sondern extrem erniedrigt. Bei der Hyperthyreose handelt es sich um eine autoimmunologische Erkrankung durch schilddrüsenstimulierende Antikörper. Damit ist klar, daß die Hyperthyreose nicht den neuroendokrinen Erkrankungen zugerechnet werden darf und daß die für neuroendokrine Krankheiten entwickelten psychophysiologischen Modelle sich nicht auf die Verhältnisse bei der Hyperthyreose übertragen lassen.

Wir wissen aber, daß auch das Immunsystem dem Einfluß psychischer Faktoren unterliegt. So konnte gezeigt werden, daß „emotionaler Streß" mit einer erhöhten Anfälligkeit gegenüber Infektionen und Neoplasien einhergehen kann, daß unter Streßbedingungen die Antikörperproduktion vermindert sein kann und Autoimmunerkrankungen häufiger auftreten etc. (Übersicht bei Solomon et al., 1974). Joasso und McKenzie (1976) haben gezeigt, daß bei Ratten unter dem Einfluß von Übervölkerung oder Isolation als „Stressoren" die Lymphozytenfunktion beeinträchtigt war.

So zeichnet sich ein neues Modell ab, das erklären könnte, wie psychische Faktoren die Schilddrüsenfunktion beeinflussen und wie die alten Beobachtungen über die Bedeutung emotionaler Belastungen für die Entstehung des Morbus Basedow und die derzeitigen Befunde zum Morbus Basedow als autoimmunologische Erkrankung in Einklang gebracht werden können.

Trauma-Theorie

Über die plötzliche Manifestation einer Hyperthyreose nach einer akuten seelischen Erschütterung ist oft berichtet worden, seit Parry 1825 zum ersten Mal ein solches Ereignis beschrieb. Seine Patientin war unmittelbar vor dem Auftreten der ersten Symptome aus einem Rollstuhl geschleudert worden, der rasch bergab fuhr. Von manchen Beobachtern wird die Häufigkeit solcher Ereignisse mit 100% angegeben; nach Oberdisse handelt es sich um ein sehr seltenes Ereignis (1964). Zu diesen Fragen hat Eickhoff (1949) wichtige tierexperimentelle Befunde beigesteuert. Es gelang ihm bei Wildkaninchen, die mit Frettchen gejagt wurden, einen Zustand zu erzeugen, der an eine Hyperthyreose denken läßt. Bei diesen Tieren tritt kurz nach dem Einfangen Adynamie, Tachykardie, Tremor und ein Exophthalmus auf. Histologisch kann man eine Verflüssigung und Ausschüttung des intrafollikulär gespeicherten Kolloids beobachten. Schilddrüsenexstirpation, Plummerung und Thiouracilgabe haben lebensverlängernde Wirkung. Im Radiojodtest findet sich eine vermehrte Speicherung und eine verminderte Ausscheidung des Radiojods im Urin. Da histologischer Befund, Jodidphase des Radiojodtestes und verminderte Ausscheidung des Radiojods im Urin nicht unbedingt für die Hyperthyreose charakteristisch sind, und sehr wohl auf eine unspezifische Alarmreaktion zu beziehen sein könnten, ist der letzte Beweis, daß es sich bei diesem Zustand um eine echte Hyperthyreose handelt, noch nicht gegeben. Dazu würde der Nachweis eines erhöhten intrathyreoidalen Jodumsatzes und ein Ansteigen der Schilddrüsenhormone im Serum gehören. Jedoch konnten Schäfer et al. (1965) beim Frettieren von Wildkaninchen kein Ansteigen des proteingebundenen Jods (PBJ) beobachten. Nach Oberdisse sollte man deswegen „aus diesen wichtigen und interessanten Beobachtungen von Eickhoff nicht auf einen experimentell erzeugten Zustand schließen, der der menschlichen Hyperthyreose entspricht. In jedem Fall sollte man bei der Beurteilung des menschlichen ‚Schreck-Basedow‘ zurückhaltend sein und Tierversuche nur mit Vorsicht heranziehen".

Psychodynamische Aspekte

Es gibt eine Reihe von Untersuchungen zu psychodynamischen Aspekten bei Patienten mit Hyperthyreose. Jeder Autor hat dabei seine eigene Version eines psychosomatischen Modells benutzt und es ist von daher sehr schwierig, die entstandenen Hypothesen auf einen gemeinsamen Nenner zu bringen.

Eine sorgfältige und oft zitierte Darstellung haben Ham und Mitarbeiter (1951) gegeben. Dort findet sich auch eine sehr gründliche Übersicht über die Literatur zum Thema emotionale Faktoren und Thyreotoxikose bis zu diesem Zeitpunkt. Die Autoren haben 24 Patienten exploriert und fanden dabei die folgenden charakteristischen Persönlichkeitsmerkmale:
– Vorzeitige Notwendigkeit, autark und reif zu sein durch frühe Verantwortung für sich selbst, für Ge-

schwister und Eltern. Die unzureichende elterliche Zuwendung in der Kindheit führt zu frustrierten Abhängigkeitswünschen. Dies ist die Folge eines oder mehrerer der folgenden Ereignisse:
> Tod, Scheidung oder Trennung der Eltern; andere Formen der Zurückweisung durch die Eltern; fortgesetzter oder übermäßiger ökonomischer Streß; oder anderer Ursachen wie
> mehrere Geschwister, wobei der Patient mit signifikant hoher Inzidenz das älteste Kind ist; traumatisierende Auseinandersetzung mit relevanten Sterbefällen.

– Unfähigkeit, Feindseligkeitsgefühle auszudrücken.
– Ein Kampf gegen Furcht durch Verleugnung der Unterdrückung, am charakteristischsten durch kontraphobische Verhaltensweisen.
– Ein lebenslanges Streben nach beruflichem Erfolg mit erschöpfender Arbeitsamkeit.
– Ein Bedürfnis, Kinder zu gebären, das signifikant größer ist als in einer Durchschnittspopulation.
– Häufige und affektgeladene Träume über Tod, Särge etc.

Aus diesen Beobachtungen wurde ein spezifisches dynamisches Modell rekonstruiert, wonach die Hyperthyreose dann auftritt, wenn die fortgesetzten Bemühungen um vorzeitige Autarkie versagen.

Diese von Ham und Alexander aufgestellten Hypothesen wurden 14 Jahre später von Hermann und Quarton (1965) einer Überprüfung unterzogen. Diese Autoren verglichen 24 hyperthyreote Patienten mit 15 euthyreoten und 11 hypothyreoten Personen. Die psychosoziale Situation der Patienten wurde mit Hilfe strukturierter Interviews evaluiert. Die Methoden, wie die Beurteilung zustande kam, werden dort im Detail berichtet. Dabei zeigte sich, daß keine der drei Gruppen von der anderen mit Hilfe eines der Kriterien der psychosomatischen Hypothese unterschieden werden kann. Der einzige signifikante Unterschied in Übereinstimmung mit der Ham-Alexander-Hypothese war, daß hyperthyreote Patienten mehr Kinder hatten. Die Schwierigkeiten, solche Hypothesen empirisch zu prüfen, wurden diskutiert. Darüber hinaus muß man allerdings fragen, ob es überhaupt sinnvoll ist, die „Objektivität" und Brauchbarkeit psychodynamischer Untersuchungen in dieser Weise zu prüfen oder ob hier nicht doch ganz verschiedene Verständnisebenen vorliegen.

Eine weitere gründliche Untersuchung der psychodynamischen Aspekte der Hyperthyreose stammt von Mandelbrote und Wittkower (1955). Diese Autoren berichten über 25 hyperthyreote Patienten, die von einem Team untersucht wurden, das aus einem Endokrinologen, einem Chirurgen, einem Nuklearmediziner, 2 Psychiatern und einem Psychologen bestand. Die emotionalen Aspekte der Thyreotoxikose wurden mit einer Kontrollgruppe verglichen, die aus Krankenhauspatienten bestand mit Leiden, bei denen psychologische Faktoren keine Rolle spielen sollten. Statistisch signifikante Unterschiede fanden sich für folgende Parameter:
– Nur 10% der hyperthyreoten Patienten gaben an,

eine „glückliche Kindheit" gehabt zu haben, im Vergleich zu 76% bei der Kontrollgruppe.

– Angepaßtes sexuelles Verhalten berichteten nur 8% der hyperthyreoten Patienten (52% der Kontrollgruppe) und 60% waren schwer gestört (impotent/frigide) im Vergleich zu 15% der Kontrollgruppe.

– Das Vorkommen von „neurotischen Tendenzen" war in der thyreotoxischen Gruppe signifikant häufiger, wobei die Betonung auf vermehrter Angst, geringer Selbstbehauptung und Depressivität lag.

Mandelbrote und Wittkower (1955) formulieren schließlich folgende Hypothese: Das Grundbedürfnis dieser Patienten ist das nach mehr Zuwendung, als sie bekommen können. Ihre Antwort auf diese Frustration ist aggressiver Natur und beinhaltet Todeswünsche. Die regressive Wiederbelebung dieser Phantasien und Impulse läßt Angst- und Schuldgefühle entstehen. Um diese Angst abzuwehren und in Übereinstimmung mit dem Grundbedürfnis, können sie in eine übermäßige Abhängigkeit von der Mutter geraten oder sich ostentativ von der Mutter abkehren und eine Lösung suchen im frühzeitigen Bemühen um Selbständigkeit. Wenn diese Abwehrmechanismen zusammenbrechen, entweder durch den Verlust der Mutter oder des Muttersubstitutes oder durch unwiderrufliche Zurückweisung oder Verlassenwerden, sind die Voraussetzungen zur Entstehung der Hyperthyreose gegeben.

In diesem Zusammenhang ist eine Untersuchung aus neuerer Zeit (Krüskemper und Krüskemper, 1970) interessant, wobei bei 14 Patienten mit dekompensierter diffuser Hyperthyreose vor Therapie und 3 Monate nach Erreichen einer euthyreoten Stoffwechsellage zahlreiche testpsychologische Untersuchungen durchgeführt wurden, die unter anderem die Persönlichkeitsfragebogen MPI (Maudsley Personality Inventory) und MMQ (Maudsley Medical Questionnaire) enthielten. Dabei zeigt sich in fast allen Fällen eine Verminderung der Stärke der neurotischen Tendenz. Die Lügenskala blieb konstant. Die Neuroseskala des MPI sank unter Behandlung signifikant ab. Diese deutliche Verschiebung zum Normalen unter der Therapie veranlaßte die Autoren zu der Meinung, daß die im aktiven Stadium der Hyperthyreose erhobenen Befunde im MPI und MMQ als Symptom der krankheitsspezifischen Stoffwechselstörung aufzufassen und somit einer Somato-Therapie zugänglich seien. Es ist auch der Versuch gemacht worden, die psychosomatische Spezifität der Persönlichkeitskonstellation, die für die Hyperthyreose postuliert wird, zu prüfen (Alexander et al., 1968).

Folgende Hypothese wurde geprüft:

Ist es möglich, zu einer korrekten Diagnose einer der folgenden sieben psychosomatischen Erkrankungen (Bronchialasthma, rheumatoide Arthritis, Colitis ulcerosa, essentielle Hypertonie, Neurodermitis, Hyperthyreose und Ulcus duodeni) zu kommen, ohne Hinweis auf irgendwelche organisch-medizinischen Symptome, ausschließlich auf der Basis der Formulierung der psychodynamischen und psychogenetischen

Muster, wie sie sich aus der Beurteilung der Protokolle von psychiatrisch-anamnestischen Interviews ergeben? Die psychologische Konfiguration, die für die Hyperthyreose typisch sein soll, wurde dabei wie folgt definiert:

„Das zentrale dynamische Ereignis bei der Thyreotoxikose ist der anhaltende Kampf gegen die Furcht, die sich auf die physische Integrität des Körpers und auf den biologischen Tod bezieht. Noch charakteristischer ist der Versuch, der Furcht Herr zu werden, indem man sie verleugnet und indem man kontraphobisch gefährliche Situationen herbeiführt und versucht, mit ihnen allein fertig zu werden. In der Anamnese finden sich häufige Auseinandersetzungen mit dem Tod naher Angehöriger und mit anderen dramatischen Ereignissen, die in der Kindheit als lebensbedrohlich erlebt wurden. Ebenso charakteristisch ist das frühreife Verhalten dieser Patienten. Träume von toten Personen in Särgen sind üblich. Die auslösende Situation ist oft eine Art von Lebensbedrohung, die die kontraphobische Abwehr zusammenbrechen läßt. Nicht selten folgt die Thyreotoxikose unmittelbar einem dramatischen Ereignis, wie z.B. einem Unfall (Schreck-Basedow). In anderen Fällen kann man nur mit sorgfältigem methodischem Vorgehen die bedrohlichen Ereignisse eruieren, die dem Ausbruch der Erkrankung unmittelbar vorausgehen. Ein solches Trauma wird jedoch eine Thyreotoxikose nur in prädisponierten Individuen hervorrufen – prädisponiert nicht nur durch Vererbung und Konstitution, sondern auch durch die Art der Lebensgeschichte, innerhalb derer die Bedrohung des reinen Überlebens wiederholt auftrat."

Das Ergebnis der Studie zeigt, daß der Prozentsatz richtiger Diagnosen größer war, als man hätte aufgrund der statistischen Wahrscheinlichkeit erwarten dürfen.

46.1.3 Das hypothalamo-hypophysäre Cushing-Syndrom

Unter einem Cushing-Syndrom wird ein Krankheitsbild mit typischer klinischer Symptomatik verstanden, dem immer ein Überangebot an endogenen oder exogenen Glukokortikoiden zugrunde liegt. Die Begriffe Cushing-Syndrom und Hyperkortisolämie werden synonym gebraucht. Eine Hyperkortisolämie kann verschiedene Ursachen haben, die differenziert werden müssen, weil sich daraus unterschiedliche therapeutische Konsequenzen ergeben. Es handelt sich um die folgenden vier Formen:

– das Cushing-Syndrom auf der Grundlage eines autonomen (d.h. ACTH-unabhängigen) Nebennierenrindentumors (Adenom oder Karzinom);

– das hypothalamo-hypophysäre Cushing-Syndrom, bei dem eine übermäßige ACTH-Sekretion der Hypophyse vorliegt;

– das ektopische ACTH-Syndrom, bei dem die Nebennieren durch ektopisch in Tumoren gebildetes ACTH übermäßig stimuliert werden (am häufigsten beim kleinzelligen Bronchialkarzinom);

– das iatrogene Cushing-Syndrom (häufigste Form) durch übermäßige Zufuhr von Glukokortikoiden oder ACTH.

Zur Differentialdiagnose stehen eine Reihe von Funktionstests zur Verfügung, wie der ACTH-Stimulationstest, der Dexamethasonhemmtest mit verschiedenen Dosen, der Lysin-Vasopressin-Test, der Metopirontest. Wertvolle Dienste leistet insbesondere auch die radioimmunologische Bestimmung von ACTH im Plasma, wie sie heute möglich ist.

Die wichtigsten Symptome der Hyperkortisolämie sind das rote, gedunsene Gesicht (Vollmondgesicht), die Stammfettsucht, Vorwölbung des Abdomens infolge Schwunds des elastischen Gewebes, durch Muskelschwund grazile Extremitäten und Striae rubrae. Weiterhin findet sich ein Diabetes mellitus, eine Hypertonie, eine Osteoporose und eine Amenorrhoe bzw. Hypogonadismus.

Bei Kindern vor der Pubertät führt das Cushing-Syndrom immer auch zu Minderwuchs. Über die psychischen Veränderungen, wie sie im Rahmen des Cushing-Syndroms auftreten, haben wir weiter vorn bereits berichtet.

Bewiesen wird das Vorliegen eines Cushing-Syndroms durch folgende Laborbefunde: Die Plasmakortisolspiegel sind erhöht und zeigen nicht die üblichen zirkadianen Schwankungen, sondern bleiben auch während der Nacht hoch. Die Ausscheidung an 17-Hydroxykortikoiden und 17-Ketosteroiden im Urin ist erhöht. Die erhöhten Steroidwerte lassen sich durch die Gabe von 2 mg Dexamethason pro Tag nicht supprimieren. Im Insulinhypoglykämietest findet sich nicht der übliche Anstieg des Plasmakortisols. Beim hypothalamo-hypophysären Cushing-Syndrom finden sich trotz der erhöhten Steroidspiegel hochnormale bis leicht erhöhte ACTH-Werte. Beim Cushing-Syndrom auf der Grundlage eines autonomen Nebennierenrindentumors ist ACTH dagegen nicht nachweisbar. Exzessiv erhöhte Spiegel sind für das ektopische ACTH-Syndrom charakteristisch; dabei werden häufig ACTH-ähnliche Peptide oder ACTH-Präkursoren sezerniert, die bei normaler radioimmunologischer Aktivität eine verminderte biologische Aktivität aufweisen.

Pathophysiologie des hypothalamo-hypophysären Cushing-Syndroms

Es ist interessant zu verfolgen, wie durch systematische Forschung der diesem Krankheitsbild zugrundeliegende pathophysiologische Mechanismus zunehmend klarer erkannt wurde (Fehm und Voigt, 1979). Dabei sind wir von einer endgültigen Klärung immer noch weit entfernt. Gerade zur Zeit wird wieder heftig diskutiert, ob es sich um eine primär hypophysäre oder primär hypothalamische Erkrankung handelt. Dabei wird oft übersehen, daß es auch gute Gründe für die Annahme gibt, daß die eigentliche Störung übergeordnete zentralnervöse Strukturen betrifft.

So hielt der Erstbeschreiber H. Cushing die nach ihm benannte Störung für eine primäre Erkrankung der Hypophyse, weil er bei seinen Patienten bei der Autopsie in der Hypophyse basophile Adenome fand. Die ACTH-Hypersekretion beim hypothalamo-hypophysären Cushing-Syndrom ist aber keineswegs autonom, sondern gehorcht durchaus den üblichen Regulationsprinzipien, nur eben auf einem anderen Niveau. So läßt sich durch höhere Mengen Dexamethason (8 mg/Tag) die ACTH- und Steroidsekretion doch supprimieren, und Lysin-Vasopressin stimuliert die ACTH-Sekretion. Auch mit dem neuerdings zur Verfügung stehenden Corticotropin-Releasing-Faktor (CRF) läßt sich die Kortisolsekretion – sogar überschießend – stimulieren. Es lag deswegen nahe, zu vermuten, daß die vermehrte Bildung dieses Faktors die eigentliche Ursache der ACTH-Hypersekretion ist, daß es sich also um eine primär hypothalamische Störung handelt.

Nun haben wir schon seit längerer Zeit Hinweise, daß Störungen zentralnervöser Funktionen eine Rolle beim hypothalamo-hypophysären Cushing-Syndrom spielen. Es sind dies die ja auch in der Diagnostik benützten Phänomene, daß der zirkadiane Rhythmus des Kortisols und des ACTH gestört ist und daß die insulininduzierte Hypoglykämie zu keiner ACTH-Ausschüttung führt. In jüngster Zeit konnten Befunde erhoben werden, die wahrscheinlich machen, daß die primäre Ursache der Erkrankung in Funktionsstörungen zentralnervöser Zentren gesucht werden muß, die dem Hypothalamus übergeordnet sind. Krieger und Glick (1974) fanden bei diesen Patienten eine erhebliche Verminderung der Schlaf-EEG-Stadien III und IV sowie des normalen nächtlichen Anstiegs des Wachstumshormons, der an das Auftreten dieser Schlafstadien geknüpft ist. Diese Störungen fanden sich auch bei Patienten, bei denen das Cushing-Syndrom durch totale Adrenalektomie therapiert worden war. Sie sind somit nicht einfach Folge der Hyperkortisolämie. Da die Abhängigkeit der Schlaf-EEG-Stadien und der STH-Konzentration vom Neurotransmittergehalt bestimmter ZNS-Strukturen bekannt ist, kann man annehmen, daß Veränderungen des zentralnervösen Neurotransmittermetabolismus auch für das hypothalamo-hypophysäre Cushing-Syndrom verantwortlich sind. Diese These konnte jetzt durch die gleichen Autoren in eindrucksvoller Weise bestätigt werden (Krieger et al., 1975). Sie konnten nämlich zeigen, daß man mit dem Serotoninantagonisten Cyproheptadin eine vollständige Remission bei diesen Patienten induzieren kann. Diese Substanz ist seit längerer Zeit als appetitförderndes Medikament in Gebrauch. Die seitherige Erfahrung zeigt, daß Cyproheptadin nur bei einem Teil der Patienten mit hypothalamo-hypophysärem Cushing-Syndrom wirksam ist. Wir selbst konnten bei vier mit dieser Substanz behandelten Patientinnen nur einmal eine Remission beobachten, wobei es zudem bereits nach 8 Wochen trotz Dosissteigerung zu einem Rezidiv kam (Kummer et al., 1979). Dies schmälert jedoch den heuristischen Wert der oben zitierten Befunde von Krieger nicht. Insgesamt ist von allen endokrinen Erkrankungen für das hypothalamo-hypophysäre Cushing-Syndrom eine zentralnervöse Ätiologie am überzeugendsten dokumentiert. Damit erscheinen aber auch die dabei beobachteten psychischen Veränderungen in einem anderen Licht und ebenso gewinnen psychosomatische Modelle weitaus mehr Überzeugungskraft.

Wir selbst haben uns bemüht, die Störung des Kortikosteroid-Feedback-Mechanismus, die für diese Patienten charakteristisch ist, eingehender zu analysieren. Wir fanden, daß der normalerweise negativ differentiale Feedback-Mechanismus umgekippt ist in ein positives Feedback (und damit einen Circulus vitiosus darstellt). Der integrale Feedback-Mechanismus funktioniert dagegen ungestört (Fehm et al., 1979). In weiteren Untersuchungen zeigte sich, daß ein solches Störungsmuster in der Feedback-Regulation durch Desipramin (trizyklisches Antidepressivum, das selektiv den Re-Uptake von Noradrenalin blockiert) ausgelöst werden kann. Ebenso war es möglich, die vorbestehende Störung der Feedback-Regulation von Cushing-Patienten durch Neuropharmaka zu beseitigen, die eine dem Desipramin entgegengesetzte Wirkung haben (Fehm et al., 1983). Diese Befunde sprechen dafür, daß die eigentliche Ursache des hypothalamo-hypophysären Cushing-Syndroms in einer Funktionsstörung des noradrenergen Systems im ZNS besteht. Auf der Basis dieser Befunde sollte es möglich sein, Neuropharmaka zu finden, die in der Therapie dieser Erkrankung mit Erfolg eingesetzt werden können (Fehm et al., 1980).

An dieser Stelle darf aber nicht unerwähnt bleiben, daß in den letzten Jahren die Hypophyse als Ort der primären Störung erneut in den Mittelpunkt des Interesses gerückt ist. Dies beruht auf den Erfolgen der neurochirurgischen Mikroadenomektomie, die sich seit 1974 weltweit als Therapie der Wahl beim hypothalamo-hypophysären Cushing-Syndrom durchsetzen konnte. Es zeigte sich, daß etwa 80 % dieser Patienten ein ACTH-produzierendes Mikroadenom in der Hypophyse aufweisen; die selektive Entfernung dieses Adenoms führt bei den meisten Patienten zu einer vollständigen Remission, häufig sogar mit vorübergehendem ACTH-Mangel. Eine endgültige Bewertung dieses Verfahrens wird erst möglich sein, wenn ausreichende Angaben über die Rezidivhäufigkeit nach erfolgreicher Adenomektomie vorliegen. Zusammenfassend muß festgestellt werden, daß sich die Frage nach der primären Ursache des hypothalamo-hypophysären Cushing-Syndroms derzeit nicht endgültig beantworten läßt. Es gibt gute Argumente für die eine wie die andere Hypothese.

Beziehung zwischen hypothalamo-hypophysärem Cushing-Syndrom und endogener Depression

In den letzten 15 Jahren haben mehrere Arbeitsgruppen immer wieder darauf hingewiesen, daß bei einem Teil der Patienten mit einer schweren endogenen Depression ausgeprägte Störungen der Kortisolsekretionsrate mit aufgehobener zirkadianer Rhythmik gefunden werden können (Carroll et al., 1976; Sachar, 1975). Bei ca. 40 % dieser Patienten fehlt die Hemmbarkeit der Kortisolsekretion im Dexamethasonhemmtest.

Daneben wurden auch Störungen der anderen Partialfunktionen des Hypophysenvorderlappens beschrieben. Insgesamt können die Störungen so ausgeprägt sein, daß eine Abgrenzung vom Cushing-Syndrom allein auf der Grundlage von endokrinologischen Befunden nicht möglich ist. Andererseits leiden die meisten Cushing-Patienten unter affektiven Störungen im Sinne einer Depression. Diese verblüffenden Ähnlichkeiten führten zu der faszinierenden Spekulation, daß es sich beim hypothalamo-hypophysären Cushing-Syndrom und bei der endogenen Depression um verschiedene Manifestationen der gleichen Störung handele. Unsere eigenen Untersuchungen zur Bedeutung des noradrenergen Systems in der Pathogenese des hypothalamo-hypophysären Cushing-Syndroms können so zwanglos mit der „Noradrenalin-Hypothese" der endogenen Depression zusammengebracht werden.

Durch die neuesten Ergebnisse gerät jedoch dieses ganze Hypothesengebäude bereits wieder ins Wanken. So fanden wir bei der testpsychologischen und psychiatrischen Nachuntersuchung von 17 Patienten nach erfolgreicher Behandlung eines hypothalamo-hypophysären Cushing-Syndroms (totale Adrenalektomie oder Mikroadenomektomie) eine vollständige Rückbildung der affektiven Störungen, obwohl die Störung der Kortikosteroid-Feedback-Regulation immer noch nachweisbar war (Voigt et al., 1984). Die Depression der Cushing-Patienten muß damit als eine Folge des Hyperkortisolismus aufgefaßt werden und nicht als eine unabhängige Manifestation der Grundkrankheit. Ebenso konnten wir zeigen, daß die Störung der Kortisolsekretion bei den depressiven Patienten nicht mit irgendeiner Dimension der Depression zu korrelieren ist, so daß es sich dabei möglicherweise um ein mehr oder weniger unbedeutendes, neuroendokrines Epiphänomen der endogenen Depression handelt (vgl. Kap. 10).

Psychodynamische Aspekte

Wir kennen nur eine Arbeit, die sich bemüht, ein psychosomatisches Konzept des Morbus Cushing zu erarbeiten (Gifford und Gunderson, 1970). Diese Autoren explorierten 16 Patienten mit nachgewiesenem hypothalamo-hypophysärem Cushing-Syndrom und fanden dabei die folgenden typischen Lebensläufe und Persönlichkeitsstrukturen:
- Störungen der frühen familiären Beziehungen,
- frühreife und ungewöhnliche Fähigkeiten,
- eine Neigung zu persönlichen Beziehungen von besonderer Intensität,
- während des ganzen Lebens Stimmungslabilität und Schwankungen der Aktivität,
- neurotische Konflikte mit sexuellen und aggressiven Impulsen,
- starke orale Persönlichkeitsmerkmale,
- psychische Traumatisierung durch Trennung oder Tod von wichtigen Bezugspersonen,
- Neigung zu Depressionen oder zu einer pathologischen Abwehr gegen Depressivität,
- rasche Besserung der akuten mentalen Symptome

nach Adrenalektomie, aber keine Veränderungen in der grundlegenden Persönlichkeitsstruktur. Insgesamt wird das hypothalamo-hypophysäre Cushing-Syndrom von diesen Autoren als psychophysiologische Reaktion auf emotionelle Verluste (bereavement) aufgefaßt.

Zusammenfassend muß festgestellt werden, daß man trotz aller Kenntnisse über die Ätiologie und Pathophysiologie des hypothalamo-hypophysären Cushing-Syndroms nur spekulieren kann, welche Bedeutung psychische Faktoren für Entstehung und Verlauf dieser Erkrankung haben können.

46.1.4 Psychogene Polydipsie und Diabetes insipidus

Physiologie

Die Bedeutung des Hypophysenhinterlappenhormons Vasopressin (= Adiuretin) bei der Regulation der Konzentration von Körperflüssigkeiten haben wir bereits im Kapitel 10 „Psychoneuroendokrinologie" kurz erwähnt. Der adäquate Reiz zur Freisetzung von Vasopressin ist die durch Wassermangel erhöhte Salzkonzentration im Blut. Die erhöhte Osmolarität wird über Osmorezeptoren wahrgenommen, die im Bereich des vorderen Hypothalamus nachgewiesen werden konnten. Vasopressin führt dann zu einer Antidiurese und wirkt so einer weiteren Wasserverarmung des Organismus entgegen. Selbstverständlich kann die Verminderung des renalen Wasserverlustes durch Vasopressin allein eine Dehydratation nicht verhindern. Der Organismus verliert ja ständig auch Wasser über die Lungen und die Haut. Deswegen muß die Wasseraufnahme – Trinkverhalten und Durstgefühl – in die Regulation integriert sein. Neurophysiologische Untersuchungen haben gezeigt, daß durch elektrische Stimulation bestimmter hypothalamischer Zentren („Durstzentrum") sehr prompt eine Polydipsie ausgelöst werden kann. Physiologischerweise wird Durstgefühl und Trinkverhalten durch eine lokale Hyperosmolarität ausgelöst, durch den gleichen Stimulus also, der auch zur Freisetzung von Vasopressin führt.

Vieles spricht dafür, daß die Osmorezeptoren, deren Stimulation letztlich zur Auslösung des Durstgefühls führt, identisch sind mit den obengenannten, die letztlich zur Freisetzung von Vasopressin führen (z.B. Andersson, 1971). Beide Phänomene werden durch cholinerge Neurone vermittelt: Durch lokale intrahypothalamische Injektionen cholinerger Substanzen wird sowohl das Trinkverhalten als auch die Antidiurese stimuliert. In diesem Zusammenhang ist von besonderem Interesse, daß sich beide Phänomene auch durch ein Peptidhormon, nämlich Angiotensin II, auslösen lassen. Wird diese Substanz intravenös beim Menschen appliziert oder bei Tieren in die Hirnventrikel injiziert, so wird das Trinkverhalten und die Freisetzung von Vasopressin ausgelöst (Lang et al., 1983). Der Effekt einer Injektion von hypertonischer Salzlösung in den 3. Ventrikel wird durch An-

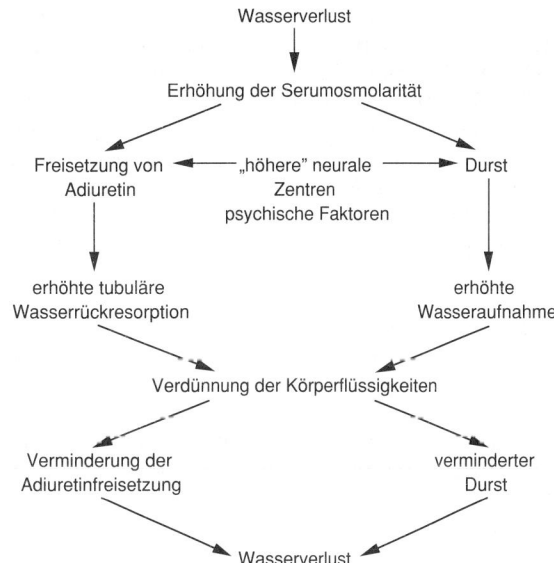

Abb. 46–2. Regulation des Wasserhaushaltes

giotensin II potenziert. Man nimmt deswegen an, daß dieses Hormon die Salzaufnahme der obengenannten Osmorezeptoren erhöht, wie es ja auch anderwärts die Salzaufnahme vaskulärer Strukturen fördert. In Situationen, wo Angiotensin II vermehrt ist, wie z.B. beim hypovolämischen Schock, ist die Flüssigkeitsaufnahme deutlich gesteigert.

Der Einfluß „höherer" neuraler Zentren auf die Vasopressinsekretion wird durch Beobachtungen wie die streßinduzierte Antidiurese bei Mensch und Tier belegt. Erfolgreiche Versuche, antidiuretische und diuretische Reaktionen zu konditionieren, sind mehrfach berichtet worden (Hofer und Hinkle, 1963). Weiterhin können Medikamente wie Morphin die Vasopressinfreisetzung stimulieren. Die Regulation des Wasserhaushaltes ist in Abbildung 46–2 schematisch dargestellt. Dieses Muster einer integrierten verhaltensmäßigen und neuroendokrinen Antwort auf einen Reiz ist typisch für viele homöostatische Regulationen, in die der Hypothalamus einbezogen ist.

Klinik

Bei einer Unterfunktion des Hypothalamus-Neurohypophysen-Systems müssen drei Krankheitsbilder differentialdiagnostisch erwogen werden, nämlich
– der „zentrale" Diabetes insipidus
– der nephrogene Diabetes insipidus
– die „psychogene Polydipsie".
Der zentrale Diabetes insipidus beruht auf einer Störung der Bildung oder Ausschüttung des Vasopressins (= Adiuretin). Vasopressin bewirkt eine Steigerung der Wasserrückresorption im distalen Teil des Nephrons. Sein Fehlen bedingt ein Unvermögen, den Urin zu konzentrieren, und damit eine Polyurie von 5 bis 20 Litern täglich mit entsprechender Polydipsie. Das spezifische Gewicht des Urins übersteigt selten 1005, die Plasmaosmolarität ist dagegen meist erhöht.

Zur Sicherung der Diagnose können der Durstversuch und der Vasopressintest herangezogen werden.

Der Hickey-Hare-Test (Infusion hyperosmolarer NaCl-Lösung) und der Nikotintest sind meist nicht nötig. Der Diabetes insipidus ist eine seltene Erkrankung. Männer sind etwas häufiger befallen als Frauen, ursächlich kommen verschiedene neoplastische, entzündliche und traumatische Schädigungen des Hypothalamus-Neurohypophysen-Systems in Frage. Bei etwa 50% der Patienten bleibt die Ursache unklar (idiopathischer Diabetes insipidus). Der noch seltenere **nephrogene Diabetes insipidus** beruht dagegen auf einer angeborenen Unfähigkeit des distalen Nephrons, auf endogenes oder exogenes Vasopressin anzusprechen. Die Urinmengen betragen bei Erwachsenen mit dem Vollbild stets mehr als 8 Liter am Tag. Die Abgrenzung des nephrogenen vom zentralen Diabetes insipidus erfolgt mit Hilfe des Vasopressintests, bei dem die Patienten mit nephrogenem Diabetes insipidus (im Gegensatz zum zentralen Diabetes insipidus) keine Reaktion zeigen.

Die **psychogene Polydipsie** („compulsive water drinking" der angloamerikanischen Literatur) wird aufgefaßt als ein durch psychische Faktoren ausgelöster Reizzustand des hypothalamischen Durstzentrums. Die Existenz eines solchen Zusammenhangs ist aufgrund unserer Einsicht in die Physiologie der Regulation des Wasserhaushaltes gut vorstellbar. Durch die exzessive Wasseraufnahme von bis zu 20 Litern täglich kommt es sekundär zu einer Hemmung der Vasopressinsekretion und zur Polyurie. Vier Fünftel der Patienten sind Frauen. Die Polydipsie entwickelt sich in der Regel allmählich über Wochen und Monate und kann intermittierend verlaufen (im Gegensatz zum Diabetes insipidus). Dieses Krankheitsbild ist für den Endokrinologen auch deswegen sehr wichtig, weil die Differentialdiagnose zum Diabetes insipidus sehr schwierig werden kann.

Zum einen kann durch eine längere Zeit bestehende psychogene Polydipsie die Fähigkeit der Neurohypophyse, Adiuretin freizusetzen, verlorengehen; zum anderen kann sich die Ansprechbarkeit der Niere gegenüber Adiuretin zunehmend verringern. So kann es sein, daß die üblicherweise zur Diagnose eines Diabetes insipidus durchgeführten Untersuchungen, nämlich Durstversuch und Vasopressintest, auch bei der psychogenen Polydipsie pathologisch ausfallen. Eingehende Untersuchungen der Pathophysiologie der Regulation des Wasserhaushaltes bei diesen Patienten werden dennoch in den meisten Fällen eine Unterscheidung ermöglichen. Darauf kann hier nicht näher eingegangen werden. Am ehesten schützt jedoch die Beachtung der Verhaltensstörungen und Persönlichkeitsveränderungen bei diesen Patienten vor einer Fehldiagnose. Leider finden sich dazu in der Literatur kaum Angaben. Barlow und De Wardener (1959) haben 9 Patienten mit psychogener Polydipsie untersucht und fanden bei 8 gravierende psychische Störungen, nämlich Wahnvorstellungen, Depression, Agitiertheit und hysterisches Verhalten. Nur 1 Patient schien „normal". Ähnliche psychische Störungen bestanden schon vor der Manifestation der Krankheit.

Dabei muß natürlich berücksichtigt werden, daß beim Diabetes insipidus Charakterveränderungen als Folgen des Leidens auftreten können. Dies ist von Angst (1959) eingehend untersucht worden. Danach ist an elementaren Trieben nicht der Durst allein bei Patienten mit Diabetes insipidus betroffen. Appetit und Schlafbedürfnis sind oft gesteigert, Sexualtrieb und Bewegungsbedürfnis vermindert. Die gesamte Antriebhaftigkeit soll leicht gedämpft sein. Es bestehe eine Neigung zu dysphorischen oder depressiv-apathischen Verstimmungen. Insgesamt lassen sich diese Störungen dem endokrinen Psychosyndrom Bleulers zuordnen.

46.2 Psychopathologische Erscheinungen bei endokrinologischen Erkrankungen

Nach Bleuler lassen sich die vielfältigen psychopathologischen Bilder, die bei Endokrinopathien beobachtet werden, leicht ordnen und in drei Grundformen psychischen Krankseins einteilen:

– in den akuten exogenen Reaktionstyp nach Bonhöffer,
– in das amnestische Syndrom und
– in das „endokrine Psychosyndrom".

Nach Schrappe (1975) müssen allerdings diese Grundformen um die syndromatischen Dimensionen aller jener Psychosen und Psychosyndrome erweitert werden, wie sie von Psychiatern sonst nur in Zusammenhang mit endogenen Psychosen, abnormen Persönlichkeitsstrukturen und Neurosen erörtert werden.

Die Erscheinungsbilder des **akuten Reaktionstyps** sind in leichten Fällen bloße Verstimmungen und Erregungszustände oder Müdigkeit, Apathie und Somnolenz. Diese Zustände können sich zum Koma oder zu Verwirrungen, Delirien oder Dämmerzuständen steigern. Andere Fälle verlaufen als Halluzinosen, wieder andere als akute Korsakow-Psychosen. Die Störung des Bewußtseins beherrscht viele schwere Fälle.

Zum **amnestischen Psychosyndrom** gehören neben der Störung der amnestischen Funktion (das Frischgedächtnis ist mehr betroffen als das Altgedächtnis) auch organische Denkstörungen mit Gedankenarmut, Perseverationstendenz und Kritikschwäche. Dazu kommen affektive Störungen, Verarmung des Gemütslebens und emotionelle Unbeherrschtheit.

Die häufigsten psychischen Veränderungen, die man bei endokrin Kranken findet, faßt Bleuler unter dem Begriff **„endokrines Psychosyndrom"** zusammen. Es handelt sich dabei um leichtere psychopathologische Erscheinungen im Sinne von Wesensveränderungen; ergriffen sind vor allem triebhafte Verhaltensweisen: der Nahrungstrieb, Durst, das Bedürfnis, Hitze oder Kälte auszuweichen, Schlafbedürfnis, der Bewegungstrieb, Aggressivität und der Sexualtrieb. Die gesamte Antriebhaftigkeit und emotionale Erregbarkeit kann gesteigert oder vermindert sein. So kommen einerseits erregte, erethische oder maniforme Zustände, andererseits apathische, indolente oder somnolente Zustände vor.

Stimmungsverschiebungen – andauernde und besonders vorübergehende – sind bei endokrin Kranken ebenfalls überaus häufig. Meist handelt es sich um Zustände von Zufriedenheit, ja Glücksgefühl, verbunden mit Trägheit und Tatenlosigkeit (sog. „Hypophysär-Stimmung"), dann reizbare, mürrische, verdrossene, weinerliche und ängstliche Verstimmungen in den verschiedensten Tönungen. Es muß gesagt werden, daß es kaum für eine bestimmte endokrine Erkrankung „spezifische" psychische Veränderungen gibt.

46.2.1 Cushing-Syndrom

Die Mehrzahl der Patienten mit Cushing-Syndrom, gleichgültig welcher Ätiologie, zeigt psychopathologische Erscheinungen; depressive Störungen sind dabei das häufigste Symptom. Jeffcoate und Mitarbeiter (1979) fanden eine Depression unterschiedlichen Schweregrades in 55% ihrer Patienten. 25 von 29 Patienten mit Cushing-Syndrom (86%) wurden von Cohen (1980) als depressiv diagnostiziert. Ein Teil dieser Patienten wiederum ist erheblich suizidal gefährdet. Andere häufig genannte Symptome beim Cushing-Syndrom sind: leichte Ermüdbarkeit, Adynamie, Reizbarkeit, Gedächtnisstörungen, Schlafstörungen, Konzentrationsstörungen und Angstzustände (Starkman et al., 1981). Die psychopathologischen Erscheinungen sind in ihrem Ausmaß mit der Höhe der Kortisolspiegel gut zu korrelieren. Paranoide Verkennungen, Halluzinationen und Schizophrenie-ähnliche Bilder sind nicht ungewöhnlich; ihre Häufigkeit wird mit 5–28% angegeben. Auf der anderen Seite ist es wichtig festzuhalten, daß sich die psychopathologischen Erscheinungen vollständig zurückbilden, wenn das Cushing-Syndrom erfolgreich behandelt wird.

Auch bei Patienten, die mit pharmakologischen Dosen von Glukokortikoiden behandelt werden, sind häufig psychopathologische Erscheinungen zu beobachten. Am geläufigsten ist die „vitalisierende" Wirkung der Glukokortikoide, die besonders bei alten und chronisch kranken Patienten zu beobachten ist. Diese Effekte werden gerne im Kontrast zur depressionauslösenden Wirkung der endogenen Steroide gesehen. Die Dauer dieser „euphorisierenden" Wirkung ist jedoch gewöhnlich auf die ersten Tage oder allenfalls Wochen der Glukokortikoidtherapie beschränkt und mit großer Wahrscheinlichkeit auf die Rückbildung der Beschwerden unter der Therapie zurückzuführen (v. Zerssen, 1976). Im übrigen entsprechen die unter der Steroidtherapie beobachteten Symptome vollständig denen bei den spontanen Formen des chronischen Hyperkortisolismus.

46.2.2 Nebennierenrindenunterfunktion

So gut wie alle Patienten mit nicht behandelter Nebennierenrindeninsuffizienz zeigen eine Wesensveränderung im Sinne eines endokrinen Psychosyndroms. Depressive und apathische Verstimmungen stehen dabei im Vordergrund. Es kommen jedoch euphorische Verstimmungen und Zustände von Hast und innerer Spannung und vieles andere vor. Die Addison-Krise wird häufig von einer Psychose vom akuten exogenen Reaktionstyp begleitet. Zusätzliche Hinweise auf eine gestörte Funktion des Zentralnervensystems gibt das EEG, das diffuse, langsame Aktivitäten mit hoher Amplitude zeigt. Bekannt ist weiterhin die generelle Überempfindlichkeit gegenüber Narkotika, insbesondere Barbituraten. Alle Veränderungen sind unter der Substitutionstherapie mit Glukokortikoiden reversibel, während Mineralokortikoide keinen Effekt haben.

46.2.3 Hypothyreose

Die Psychopathologie der Hypothyreose und des endogenen Kretinismus gehört zum alten psychiatrischen Wissen. Die psychopathologischen Erscheinungen stehen bei der Hypothyreose oft im Vordergrund der Symptomatologie.

Am leichtesten wird die Verlangsamung wahrgenommen, aber auch Stimmung und Affekte sind verändert; mit psychologischen Tests läßt sich oft ein leichtes Delir eruieren. Eine spezifische Heiterkeit, die der Witzelsucht von Patienten mit Frontallappentumoren ähnelt, ist häufig. Psychotische Bilder, die an die Schizophrenie und Zyklothymie erinnern, kommen vor. Bei der hypothyreoten Enzephalopathie finden sich EEG-Veränderungen, Krampfanfälle können auftreten. Bei einem gewissen Prozentsatz bessern sich die psychopathologischen Erscheinungen unter einer Substitutionsbehandlung mit Schilddrüsenhormon nicht. Dabei spielt eine permanente Hirnschädigung durch die Hypothyreose sicher eine Rolle.

Die deletäre Wirkung eines Schilddrüsenhormonmangels auf das sich entwickelnde Gehirn ist von den Patienten mit Kretinismus wohlbekannt.

46.2.4 Hyperthyreose

Die Autoren, die sich mit den psychischen Veränderungen bei der Hyperthyreose beschäftigen, interessieren sich vorwiegend für die mögliche ätiologische Bedeutung emotionaler Faktoren. Die Hyperthyreose wird von vielen Autoren zu den „klassischen" psychosomatischen Krankheiten gerechnet. Die Diskussion, ob die psychischen Faktoren Ursache oder Folge der endokrinen Störung sind, ist bei der Hyperthyreose besonders heftig. Wir verweisen dazu auch auf Abschnitt 46.1.2.

Hier sei nur festgehalten, daß bei der Hyperthyreose folgende Erscheinungen außerordentlich häufig sind: emotionale Labilität, Angst, Spannung, überschießende Reaktionen, Konzentrationsstörungen, Unruhe, Tremor, Schlafstörungen und manifeste Psychosen. Eine Minderheit von Patienten, besonders ältere, werden depressiv, zurückgezogen, apathisch und appetitlos. Zum Bild der thyreotoxischen Krise gehört das Delir und das Koma. Bei manchen Patienten bleibt nach der Behandlung eine organische Hirnschädigung zurück.

46.2.5 Hypopituitarismus

Die psychischen Veränderungen, wie sie beim Panhypopituitarismus beobachtet werden, entsprechen der Kombination der Erscheinungen, wie sie von der Hypothyreose, der Nebenniereninsuffizienz und dem Hypogonadismus bekannt sind, die ja alle im Rahmen der Hypophysenvorderlappeninsuffizienz auftreten. Entsprechend häufig finden sich psychische Phänomene beim Hypopituitarismus. In einer Studie waren von 78 untersuchten Patienten nur 6 frei von psychischen Störungen.

Werden die ausgefallenen Partialfunktionen der Hypophyse durch eine Substitutionstherapie mit Schilddrüsenhormon, Glukokortikoiden und Sexualhormonen ausgeglichen, so zeigt ein gewisser Teil der Patienten immer noch Apathie, Antriebsarmut und chronische Müdigkeit. Man kann vermuten, daß dies dem weiterbestehenden Mangel an Wachstumshormon und Prolaktin zuzuschreiben ist. Wir wissen jedoch so gut wie nichts über den Effekt von Wachstumshormonen auf psychologische Funktionen. Prolaktin hat zwar ein weites Spektrum von Wirkungen auf das Verhalten von niederen Tieren (Induktion des mütterlichen Verhaltens und des Wandertriebes), über eine ähnliche Wirkung beim Menschen gibt es jedoch keine gesicherten Erkenntnisse.

46.2.6 Akromegalie

Berichte über psychische Veränderungen bei der Akromegalie sind sehr spärlich; wahrscheinlich verursacht die Überproduktion an Wachstumshormon keine groben psychischen Störungen. Nach Bleuler treten psychotische Bilder nie auf. Sehr häufig ist jedoch eine Wesensänderung, die hauptsächlich in einem Mangel an Initiative und Spontaneität und in einer Stimmungsänderung in Richtung stiller und passiver Heiterkeit besteht. In einzelnen Fällen sollen bei Männern Zeichen einer „elementaren Mütterlichkeit" auftreten, die auf eine gleichzeitige Vermehrung von Prolaktin bezogen werden.

46.2.7 Nebenschilddrüsen

Die psychopathologischen Folgen des Hyperparathyreoidismus sind streng korreliert mit dem Ausmaß der Hyperkalzämie. Dies ist leicht verständlich, wenn man die Bedeutung des Kalziums bei vielen biologischen Prozessen bedenkt. Bei etwa ⅔ der Fälle werden leichte bis mittelschwere Veränderungen im Sinne des endokrinen Psychosyndroms beobachtet. Diese Verstimmungen und Triebänderungen können chronisch sein, sie können aber auch in charakteristischer Weise unvermittelt auftreten und wieder verschwinden. Die Stimmung ist depressiv, gelegentlich auch euphorisch oder reizbar. Der Kranke sieht seine Lebenstüchtigkeit dahinschwinden, verliert jeden Antrieb, gibt jeden Kampf auf und äußert den Wunsch zu sterben. Dabei finden sich alle Übergänge bis zur schweren Depression. Alle diese psychischen Abweichungen verschwinden wenige Wochen bis sechs Monate nach Exstirpation des Parathyreoideaadenoms (Petersen, 1968). Bei einem Serumkalziumwert von mehr als 5 mmol/l tritt der akute exogene Reaktionstyp auf. Seine Haupterscheinungen sind Störungen des Gedächtnisses, Bewußtseinstrübungen, Benommenheit, Desorientiertheit und eigentliche Verwirrungszustände. Auch diese Erscheinungen sind reversibel. Im wesentlichen die gleichen Erscheinungen, nämlich endokrines Psychosyndrom und die Erscheinungen des akuten exogenen Reaktionstyps, werden auch beim entgegengesetzten Krankheitsbild, dem Hypoparathyreoidismus beobachtet, der mit erniedrigten Kalziumspiegeln im Blut einhergeht. Dies unterstreicht, wie unspezifisch und uncharakteristisch die psychischen Veränderungen bei einer bestimmten endokrinen Störung sind.

47 Diabetes mellitus

Jörg Michael Herrmann, Wolfgang Beischer und *Christa Probst-Geigges*

47.1 Diabetes mellitus Typ I

Die folgende Falldarstellung ist insofern ein extremes Beispiel, weil beinahe alle für diese Patienten charakteristischen Komplikationen enthalten sind.

Bei der 38jährigen Patientin ist seit dem 4. Lebensjahr Diabetes mellitus bekannt. Während der letzten 6 Jahre treten starke Schwankungen der Blutzuckerwerte auf. Immer wieder kommt es während schwerer Entgleisungen des Blutzuckers zu stationären Klinikeinweisungen. Der behandelnde Arzt weist sie auf Diätfehler hin und mahnt zur Einhaltung eines strengen Diätplanes. Dies erlebt sie als Vorwurf. Die Zusammenarbeit zwischen ihr und dem Arzt gestaltet sich zunehmend schwieriger, bis es schließlich zum Arztwechsel kommt und zu einer stationären Aufnahme in eine Medizinische Klinik mit dem Ziel, die Diabetestherapie neu einzustellen.

Zu diesem Zeitpunkt existieren bereits diabetische Spätschäden mit einer ausgeprägten Polyneuropathie, einer diabetischen Nephropathie und einer geringen Retinopathie. Außerdem klagt die Patientin im Schulterbereich und in den Beinen über starke Schmerzen.

Sie erhält in der Medizinischen Klinik eine Insulinpumpe. Trotz längerer medikamentöser Behandlung klagt sie weiterhin über bestehende starke Schmerzen und wird für das Klinikpersonal zunehmend zur Problempatientin: Sie schränkt ihre täglichen Spaziergänge im Klinikgelände ein, zeigt kaum noch Eigenaktivität, zieht sich zunehmend in ihr Bett zurück, verweigert schließlich die Nahrungsaufnahme und nimmt rapide an Körpergewicht ab. Im Sinne einer malignen Regression geraten Patientin und Klinikpersonal zunehmend in einen nicht mehr auflösbaren Kampf um die Nahrungsaufnahme. Trotz hochkalorischer Sondenernährung stagniert ihr Körpergewicht bei 47 kg. Die behandelnden Ärzte vermuten, nach umfassender Diagnostik zum Ausschluß einer organischen Erkrankung, eine bewußte Manipulation der Nahrungsaufnahme. Wie in der Behandlung beim Hausarzt wiederholt sich in der Klinik das schwierige Interaktionsmuster zwischen Patientin und Therapeut um die Ernährungsfrage.

Zu diesem Zeitpunkt wird die Patientin in eine internistisch-psychosomatische Rehabilitationsklinik verlegt.

Gestützt von ihrem Ehemann erscheint sie zur Aufnahme. Ihr Körper ist abgemagert, sie bewegt sich sehr unsicher und langsam, die Beine erscheinen wie Gliedmaßen einer Marionette. Ihr Gesicht ist aufgeschwollen, sie wirkt entstellt, durch ihre Nase ist eine Magensonde eingeführt.

Im gemeinsamen Gespräch ist das Paar freundlich-reserviert. Die Patientin äußert sich bezüglich der Verlegung zunächst sehr hoffnungsvoll, ohne jedoch konkrete Vorstellungen einer für sie positiven Behandlung entwickeln zu können. Schließlich deutet sie sehr vage ihre Befürchtung an, daß sich auch jetzt Konflikte um ihre „adäquate" Nahrungsaufnahme wiederholen könnten. Während das Ehepaar im weiteren Gesprächsverlauf über die konflikthaften Erlebnisse um die Einhaltung einer diätetischen Ernährung berichtet, spricht vorwiegend der Ehemann. Die Patientin sitzt zusammengekauert auf ihrem Sessel, ihre Stimme wirkt zaghaft, bisweilen zeigt sie sich trotzig-schmollend, bisweilen lächelt sie verschmitzt und erinnert an ein schuldbewußtes kleines Mädchen.

Während der letzten drei Jahre haben Konflikte und Auseinandersetzungen um eine adäquate diätetische Nahrungsaufnahme zugenommen. In dieser Zeit zog sich das Ehepaar zunehmend aus sozialen Bezügen zurück, ging kaum noch aus, nahm kaum noch Einladungen von Bekannten wahr. Infolge von Diätfehlern kam es vermehrt zu Entgleisungen des Blutzuckers mit bedrohlichen körperlichen Zuständen, gelegentlich zu Klinikeinweisungen und zum raschen Fortschreiten diabetischer Spätschäden. Die Aufmerksamkeit und Sorge des Ehemannes zentrieren sich daraufhin verstärkt um die Nahrungsaufnahme seiner Frau, Spannungen und Konflikte kreisen um das Thema der richtigen Diät, Aggressionen und Gefühle der Hilflosigkeit finden Ausdruck in gegenseitigen Vorwürfen. Während sie sich von ihrem Mann bevormundet und kontrolliert fühlt und während der gemeinsamen Mahlzeiten kaum mehr etwas zu sich nimmt, erlebt er die Abgrenzungsversuche seiner Frau über ihre veränderten „seltsamen" Eßgewohnheiten zugleich als Aggression und Aufforderung zu verstärkter Sorge. Immer häufiger eskalieren die Spannungen zwischen dem Paar. Als schließlich polyneuropathische Schmerzen in Schulter und Beinen auftreten, scheint die maligne Auseinandersetzung zunächst beendet und ein neues labiles Gleichgewicht im Sinne eines komplementären Beziehungsmusters gefunden: Während sie sich hilfe- und trostsuchend an ihren Ehemann wendet, begegnet er ihr mit mütterlich-tröstender Zuwendung. Das Paar scheint über die Symptomatik eine Möglichkeit gefunden zu haben, regressive Bedürfnisse zu befriedigen und Nähe zu erleben. Diese „Lösung" forderte jedoch den Verzicht auf Autonomiebestrebungen und eine reife sexuelle Partnerschaft.

In den während der stationären Behandlung regelmäßig durchgeführten Paargesprächen wird deutlich, daß nach der Heirat vor allem bei ihm ein Kinderwunsch bestand. Nachdem der Hausarzt im Zusammenhang mit dem Diabetes mellitus auf mögliche Komplikationen während der Schwangerschaft hinwies, trat dieser Kinderwunsch zunächst in den Hintergrund. In der Folgezeit war die Patientin häufig depressiv. Über mehrere Jahre dachte das Ehepaar dann an die Möglichkeit einer Adoption, bei zunehmenden diabetischen Komplikationen wurde schließlich die Pflegschaft eines Kindes erwogen. Nachdem es während des letzten Jahres gehäuft zu Entgleisungen des Blutzuckers kam und die Patientin sich durch die diabetischen Spätschäden zunehmend kraftloser und auf Hilfe angewiesen fühlte, wurde der Kinderwunsch aufgegeben.

Das Ehepaar heiratete vor 10 Jahren, die Patientin war zu diesem Zeitpunkt 28 Jahre alt und als Lehrerin tätig, er war 31 Jahre alt, von Beruf Dipl.-Ingenieur und lebte bis zur Heirat bei den Eltern. Er hat bis heute zur 73jährigen Mutter ein sehr enges Verhältnis und war schon als Kind ihr Vertrauter. Der Kontakt zwischen dem Ehepaar und der Herkunftsfamilie des Ehemannes wird als eng und gut beschrieben. Die Patientin stammt aus einer ländlichen Gegend. Die Eltern, beide Anfang 70, betreiben bis heute eine große Landwirtschaft. Der jüngere 37jährige Bruder übernahm den elterlichen Hof. Die 33jährige Schwester lebt in unmittelbarer Nachbarschaft. Der Bruder leidet an einer Hämophilie und hat früh die elterliche Aufmerksamkeit ganz auf sich gelenkt. Er wurde in allem bevorzugt, ist das Lieblingskind beider Eltern und steht ständig im Mittelpunkt. Selbst als die Patientin mit 4 Jahren an einem Diabetes mellitus erkrankte, galt die elterliche Sorge unverändert dem Bruder. Für die Mutter war der Diabetes mellitus keine ernst zu nehmende Erkrankung, sondern Folge falscher Eßgewohnheiten. Die Mutter ermahnte sie regelmäßig zum „besseren" Essen, vom Vater wurde sie „gespritzt". Schon als Kind hat sie sich in der Familie als Außenseiterin und nicht richtig dazugehörig gefühlt. Während die Familie in der Landwirtschaft gemeinsame Arbeiten verrichtete, wurde sie aufgrund ihrer Erkrankung von körperlicher Betätigung ferngehalten. Zur Erlangung der Mittleren Reife lebte sie dann ab dem 10. Lebensjahr getrennt von der Familie in einem Internat. Obwohl sie gute schulische Leistungen erbrachte und die Weiterempfehlung für das Gymnasium erlangte, blieb die erhoffte elterliche Aufmerksamkeit und Anerkennung aus. Diese galt verstärkt dem Bruder, nachdem dieser seit dem 13. Lebensjahr nach einem komplizierten Beinbruch gehbehindert ist. Der Kontakt zur Familie gestaltete sich zunehmend spannungsreich, die eingetretene Entfremdung führte häufig zu gegenseitigen Mißverständnissen und Enttäuschungen. Der Graben zwischen der ländlichen Welt der Eltern und der „gebildeten" Welt der Tochter wurde mit ihrer Heirat noch größer. Der Schwiegersohn wurde sehr mißtrauisch aufgenommen, zumal er nicht der gleichen Konfession angehörte, was für die an strengen religiösen Normen orientierte Familie nur schwer akzeptabel war.

Die spannungsreiche Beziehung erreichte zu Weihnachten vor der stationären Klinikbehandlung ihren Höhepunkt, als das Ehepaar die Weihnachtsfeiertage bei der Herkunftsfamilie der Patientin verbrachte. Diese Tage verliefen für die Patientin sehr konfliktreich und enttäuschend, nachdem der Vater ihr einen zu geringen Erbanteil ausbezahlte. Zu einer Aussprache über das Erbe und die damit verbundene Enttäuschung und Wut kam es nicht. Vielmehr entzündeten sich zunehmend die Spannungen und Konflikte an ihrem Eßverhalten. Zu diesem Zeitpunkt hatte sie bereits begonnen, ihre Nahrungsaufnahme zu reduzieren in der Hoffnung, ihr durch Ödeme aufgeschwollenes Gesicht zu entschlacken. Vor allem die Mutter reagierte auf die Nahrungseinschränkung mit heftigen Vorwür-

fen, drängte die Tochter zum Essen. Als die Konflikte und Spannungen eskalierten und bei der Patientin massive Rückenschmerzen auftraten, die selbst durch starke Schmerzmittel unbeeinflußt blieben, geriet der Ehemann zunehmend unter Loyalitätsdruck gegenüber seiner Frau. Er versuchte die Spannungen dadurch zu schlichten, indem er um Verständnis warb und das Verhalten seiner Frau, ihre „Appetitlosigkeit" und ihre Schmerzen krankheitsbedingt als Folge der Polyneuropathie zu erklären versuchte. Als die Eltern einen Termin bei einem religiösen Heiler vermittelten, nahmen die gegenseitigen Kränkungen und Mißverständnisse zu. Die Patientin reagierte auf diesen Vorschlag empört und wütend, die Eltern fühlten sich in ihrem Bemühen ebenfalls gekränkt und mißverstanden, drohten aus ihrem Gefühl der Machtlosigkeit damit, daß eine Nichtkonsultation für die Tochter schlimme Folgen haben würde.

Wenige Tage später erfolgte die Aufnahme in die Medizinische Klinik. Im regressionsfördernden Milieu der Klinik entfaltete sich rasch die oben beschriebene depressive Problematik. Die Patientin griff auf familiäre Muster der Konfliktbewältigung zurück, indem sie über das Eßverhalten indirekt ihre Emotionen und Bedürfnisse zum Ausdruck brachte.

Nach der stationären Aufnahme in der Rehabilitationsklinik wird ein für die Patientin annehmbares Therapiekonzept erarbeitet: Mobilisierung und Wiedererlangung der Gehfähigkeit und selbständiger Umgang mit der Kontrolle der täglichen Insulinapplikation. Bezüglich der Störungen des Eßverhaltens wird der Patientin die Eigenverantwortung und Kontrolle über die Nahrungsaufnahme überlassen. Es wird vereinbart, daß die Patientin bei der für sie geeigneten individuellen Zusammenstellung eines Diätplanes und der Handhabung der Insulinpumpe unterstützt wird und nur bei Auftreten bedrohlicher körperlicher Zustände ärztlich eingegriffen wird. Das Experimentieren mit der Diät und der Insulinapplikation wird zum vorübergehenden Behandlungsziel, bei dem Schwankungen des Körpergewichtes und der Insulinwerte zu erwarten sind.

Während der ersten Wochen der Behandlung zeigen sich rasche Erfolge. Die Patientin wird durch die intensive krankengymnastische Behandlung zunehmend mobiler. Nach wenigen Tagen kann die Magensonde entfernt werden und sie nimmt ihre Mahlzeiten gemeinsam mit den anderen Patienten im Speisesaal ein. Während der Therapie und im Stationsalltag stehen anfangs die Ernährungsfrage und Insulineinstellung sowie die Handhabung der Pumpe im Vordergrund. Sie nimmt in dieser Zeit vorwiegend über das Thema ihrer richtigen Ernährung Kontakt auf, problematisiert Vorlieben bzw. die Möglichkeit der Zurückweisung ihr unliebsamer Speisen. Die Patientin „testet" das Team immer wieder darauf, inwieweit ihr die Eigenverantwortung bezüglich der Zusammenstellung ihres Diätplanes wirklich überlassen wird. Im Kontakt zu den Mitpatienten und dem Klinikpersonal fällt ihr zurückhaltendes und mißtrauisch wirkendes Verhalten auf, aktiv nimmt sie kaum Kontakt auf, sie wirkt verschlossen und ernst und gilt als Einzelgängerin. Regelmäßig bekommt sie Besuch von ihrem Mann, gelegentlich auch von Angehörigen. Nach Besuchen der Angehörigen wirkt sie häufig niedergeschlagen und traurig.

Nach einigen Wochen stationärer Behandlung klagt sie wieder vermehrt über Schmerzen in den Beinen, die Fortschritte in der krankengymnastischen Behandlung stagnieren. Der behandelnde Krankengymnast berichtet dem Team, daß er die Patientin plötzlich ängstlich, zögernd und der Behandlung gegenüber bremsend erlebt. Im Verlauf der krankengymnastischen Behandlung hatte die Patientin bis dahin Treppengehen und komplexe Bewegungsabläufe eingeübt. Mit Hilfe eines Gehstockes konnte sie sich zunehmend selbständiger bewegen. Die Stagnation der Fortschritte tritt zu einem Zeitpunkt ein, als Übungen des freien Gehens ohne Stock behandelt werden. Auf der Krankenstation wird sie als zunehmend klagsam und anhänglich erlebt. Das zu diesem Zeitpunkt geplante Paargespräch wird nur zögernd, eine Woche später als geplant, wahrgenommen: Der Ehemann ist beruflich sehr eingespannt und häufiger bei seinen Eltern zum Essen eingeladen, um von der Hausarbeit entlastet zu werden. Außerdem hat er die regelmäßigen Besuche bei seiner Frau am Wochenende eingeschränkt.

Sie hat sich während der letzten Zeit mit ihrer Entlassung auseinandergesetzt, hat sich zum Ziel gesetzt, die Arbeiten im Haushalt wieder voll zu übernehmen und beruflich wieder so „fit" wie vor einigen Jahren zu werden. Hinderlich beim Erreichen dieses Zieles erlebt sie ihre verstärkt auftretenden Schmerzen.

In der darauffolgenden Behandlungsphase setzt sie sich mit ihrer Leistungsproblematik und ihren Abhängigkeitswünschen auseinander. Sie bearbeitet ihre unrealistischen Erwartungen und Behandlungsziele. Es gelingt ihr zunehmend besser, ihre Autonomie- und Abhängigkeitswünsche zuzulassen und als nebeneinander bestehende, sich nicht gegenseitig ausschließende Bestrebungen anzunehmen. In der krankengymnastischen Behandlung erlebt sie die ihr überlassene Wahlmöglichkeit, das Tempo der Behandlungsschritte zu bestimmen, als sehr entlastend und befreiend. Indem sie die Übungen zum freien Gehen selbst bestimmt, forcieren bzw. auch einen Schritt zurückgehen kann, erfährt sie, daß sowohl ihre Autonomie- als auch ihre Abhängigkeitswünsche vom ganzen Team akzeptiert und getragen werden. So gelingt es ihr auch, den Stock, ihre „moralische Stütze", zunehmend bewußter zu handhaben, ihn wahlweise zu benutzen oder auf ihn zu verzichten.

Vom Klinikpersonal wird sie als offener und freundlicher erlebt. Im Gegensatz zum Behandlungsbeginn wird die Beziehung von fast allen Beteiligten als einfacher und lockerer beschrieben. Die Patientin sucht zu diesem Zeitpunkt häufig aktiv verschiedene Mitglieder des Teams auf, sucht Trost oder Unterstützung bei unterschiedlichen Fragen.

Im Paargespräch werden erstmals Ängste und Befürchtungen in bezug auf das gemeinsame Zusammenleben offen thematisiert und die bis vor einem Jahr häufig eskalierenden Krisen problematisiert. Eine Bearbeitung des symbiotischen Beziehungsmusters wird möglich und damit die Suche nach neuen Lösungen, Individuationsschritte einzuleiten.

Die räumliche Trennung durch den langen Klinikaufenthalt reaktivierte einerseits die ungelöste Ablösungsproblematik, ermöglichte andererseits dem einzelnen und dem Paar, für sich neue Individuationsschritte zu entdecken und gemeinsam einzuleiten. Bei der Entlassung aus der stationären Behandlung sind die Blutzuckerwerte stabil. Zwar ist sie noch untergewichtig, aber das Körpergewicht bleibt konstant.

Die Stabilität des körperlichen Zustandes bleibt auch nach der stationären Behandlung bestehen. Auch wenn die Suche nach einem neuen Gleichge-

wicht des Paares erst begonnen hat und die Konflikte mit den Herkunftsfamilien neuen Zündstoff zu Auseinandersetzungen liefern, erscheint das Paar in der begonnenen ambulanten Nachbehandlung im Umgang miteinander flexibler und wieder offener für außerfamiliäre Beziehungen.

47.1.1 Definition

Von Experten der European Study Group for Diabetes Epidemiology und der American Diabetes Association wurden 1979 neue Kriterien für die Diabetesdiagnostik vorgeschlagen, die von der WHO und der Deutschen Diabetesgesellschaft im wesentlichen übernommen wurden (National Diabetes Data Group, 1970; WHO-Expert Commitee on Diabetes Mellitus, 1980; Stellungnahme des Vorstands der Deutschen Diabetesgesellschaft, 1980):

Ein manifester Diabetes mellitus liegt vor, wenn die Konzentration der Blutglukose nüchtern 120 mg/dl (7,0 mmol/l) im Kapillarblut beträgt oder übersteigt. Ein manifester Diabetes mellitus besteht außerdem, wenn die Konzentration der Blutglukose zu einem beliebigen Zeitpunkt im Tagesprofil oder zwei Stunden nach oraler Belastung mit 75 g Glukose bzw. Oligosaccharidgemisch (bei Kindern 1,75 g/kg bis maximal 75 g) 200 mg/dl (11,0 mmol/l) im Kapillarblut erreicht oder übersteigt. Für die praktisch tätigen Ärzte ist besonders wichtig, daß sich ein Diabetes mellitus ausschließen läßt, wenn die Konzentration der Blutglukose nüchtern unter 100 mg/dl liegt und postprandial nicht über 140 mg/dl ansteigt.

Neben der manifesten Form der Erkrankung gibt es die sog. pathologische Glukosetoleranz. Eine pathologische Glukosetoleranz liegt vor, wenn zwei Stunden nach oraler Belastung mit 75 g Glukose Konzentrationen der Blutglukose zwischen 140 und 199 mg/dl (8,0–10,9 mmol/l) im Kapillarblut erreicht werden.

Bei dem nach den dargestellten Kriterien diagnostizierten Diabetes mellitus handelt es sich um ein Syndrom, dem unterschiedliche Krankheitsbilder zugeordnet werden müssen. Durch die Erkenntnisse jüngster Zeit imponiert das Krankheitsbild Diabetes mellitus als immer heterogener. Die in Tabelle 47–1 dargestellte Klassifikation des Diabetes mellitus und verwandter Stoffwechselstörungen entspricht ebenfalls internationalen Empfehlungen und berücksichtigt nur die heute allgemein anerkannten und genau definierten Heterogenitäten im Rahmen des diabetischen Syndroms.

Zahlreiche Charakteristika unterscheiden den Typ-I-Diabetes mellitus (juveniler Diabetes, IDDM $\triangleq$ insulin-dependent diabetes mellitus) vom Typ-II-Diabetes mellitus (Erwachsenendiabetes, NIDDM $\triangleq$ non-insulin-dependent diabetes mellitus). Schon lange sind die Unterschiede in der Klinik beider Krankheitsbilder bekannt:

Der Typ-I-Diabetes tritt meistens im jugendlichen Alter auf, die Betroffenen sind nur selten fettleibig, Symptome wie vermehrter Durst, vermehrtes Wasser-

Tabelle 47–1. Klassifikation des Diabetes mellitus und verwandter Stoffwechselstörungen (Stellungnahme des Vorstands der Deutschen Diabetesgesellschaft, 1980)

A) Klinische Nomenklatur
  1. Diabetes mellitus
    Typ I   insulinabhängiger Diabetes
    Typ II   insulinunabhängiger Diabetes
    Typ IIa ohne Adipositas
    Typ IIb mit Adipositas
Weitere Formen des Diabetes mellitus sind mit bestimmten Krankheiten oder Syndromen verknüpft:
    – Pankreaserkrankungen
    – endokrine Symptome
    – durch Medikamente oder Chemikalien ausgelöst
    – Störungen des Insulinrezeptors
    – bestimmte genetische Syndrome
    – andere Formen
  2. Pathologische Glukosetoleranz
    – mit Adipositas
    – ohne Adipositas
    – assoziiert mit den oben beschriebenen Krankheiten und Syndromen
  3. Gestationsdiabetes
B) Statistische Risikoklassen
  Pathologische Glukosetoleranz in der Vorgeschichte, erhöhtes Risiko für pathologische Glukosetoleranz

lassen, verminderte Leistungsfähigkeit, Müdigkeit, Gewichtsabnahme treten akut auf, es besteht die Bereitschaft zur ketoazidotischen Stoffwechselentgleisung, der nur durch Insulingabe vorgebeugt werden kann.

Nach neueren Erkenntnissen zeigt der Typ-I-Diabetes folgende weitere Charakteristika:

Die Restsekretion der Beta-Zellen der Langerhansschen Inseln, die heute mittels Bestimmung von C-Peptid erfaßt werden kann, ist beim Typ-I-Diabetes in der Regel von Beginn der Erkrankung an eingeschränkt und kommt nach wenigen Jahren der Erkrankung vollständig zum Erliegen. Morphologisch finden sich zum Zeitpunkt des Erkrankungsbeginns lymphozytäre Infiltrate in den Langerhansschen Inseln, im späteren Erkrankungsverlauf sind die Beta-Zellen mehr oder weniger vollständig verschwunden. Eine familiäre Belastung läßt sich beim Typ-I-Diabetes nur selten nachweisen, die Erkrankung tritt allerdings bei Trägern bestimmter HLA-Eigenschaften gehäuft auf. Schließlich ist der Typ-I-Diabetes mit zahlreichen Immunphänomenen assoziiert, u.a. kommt es zur Autoantikörperproduktion gegen Zellen der Langerhansschen Inseln. Die dargestellten zahlreichen Kriterien sind Mosaiksteine für die Zuordnung der Patienten zum richtigen Krankheitsbild, im abgehandelten Fall zum Typ-I-Diabetes.

47.1.2 Epidemiologie

Der Typ-I-Diabetes zeigt bei verschiedenen Rassen große Unterschiede in der Häufigkeit seines Auftretens. Die Verhältnisse bei der kaukasischen Rasse sind besonders gut untersucht. Bei Kaukasiern tritt der Diabetes mellitus des Typs I außerdem mit besonders hoher Häufigkeit auf. Nach Angaben aus England findet sich in der Altersgruppe zwischen 0 und 26 Jahren eine Prävalenz des Typ-I-Diabetes von 3,5/1000. Viel seltener und zum Teil auch gar nicht tritt der Typ-I-Diabetes bei Japanern, Indern, Chinesen, Philippinos, Indianern, Eskimos, Maltesern, Ceylonesen, südafrikanischen Negern, Polynesiern, Mikronesiern und Melanesiern auf (Zimmet, 1983).

Da die Diabetesfälle in der Bundesrepublik Deutschland nirgends registriert werden, sind genaue Aussagen über die Prävalenz der Erkrankung bei uns nicht möglich, sie dürfte ungefähr derjenigen in England entsprechen.

Die Inzidenz des Typ-I-Diabetes zeigt eine Altersabhängigkeit. Die meisten Erkrankungsfälle finden sich im Alter zwischen 11 und 12 Jahren, ein kleinerer Häufigkeitsgipfel tritt bereits zwischen 5 und 8 Jahren auf (Zimmet, 1983). Ob diese Häufigkeitsverteilung hinsichtlich des Erkrankungsalters genetische Gründe hat oder mit Umwelteinflüssen in Zusammenhang steht, ist nicht eindeutig geklärt.

Mehrere Studien berichten über eine Abhängigkeit der Erkrankungshäufigkeit auch von der Jahreszeit, mit einem Maximum im Winter und einem Minimum im Sommer (Zimmet, 1983). Umwelteinflüsse dürften bei dieser saisonalen Häufigkeitsverteilung des Auftretens des Typ-I-Diabetes eine entscheidende Rolle spielen, eine besondere Bedeutung könnte hierbei viralen Infekten zukommen.

47.1.3 Ätiologie und Pathogenese

Genetik

Bei eineiigen Zwillingen von Patienten mit Typ-I-Diabetes tritt in etwa 50% der Fälle ebenfalls ein Typ-I-Diabetes auf (Pyke, 1982). Hinsichtlich des Modus der Vererbung wurden im Laufe der vergangenen Jahre zahlreiche Hypothesen aufgestellt (Rotter et al., 1983). Der exakte Erbmodus ist nach wie vor unbekannt. Nach ausgedehnten Untersuchungen der vergangenen Jahre kann allerdings kein Zweifel daran bestehen, daß die HLA-Eigenschaften DR3, DR4 und die Kombination beider überdurchschnittlich häufig mit dem Vorkommen eines Typ-I-Diabetes assoziiert sind. Nach einer neueren Untersuchung kommt eine der DR-Eigenschaften oder die Kombination beider bei 112 von 123 Typ-I-Diabetikern (91%) vor. Fast 51% der 123 Patienten waren heterozygot für DR3 und DR4. Das relative Erkrankungsrisiko beträgt für Träger der Eigenschaft DR3 5,0, der Eigenschaft DR4 6,8 und der Kombination beider Eigenschaften (DR3/DR4) 14,3 (Wolf et al., 1983). Diese Ergebnisse sprechen dafür, daß zwei oder mehr Gen-Loci, die mit dem HLA/DR-Gen-Locus eng verbunden sind, eine besondere Empfänglichkeit für den Typ-I-Diabetes vermitteln (Köbberling und Illil, 1984).

Offensichtlich sind heterogene Krankheitsbilder mit der HLA-Eigenschaft DR3 bzw. DR4 assoziiert. Die Patienten mit Typ-I-Diabetes und der HLA-Eigenschaft DR3 erkranken bevorzugt an Autoimmunerkrankungen, vor allem anderer endokriner Drüsen, Antikörper gegen Inselzellen sind langfristig oder dauerhaft nachweisbar, Antikörper gegen Insulin treten kaum auf, der Erkrankungsbeginn ist nicht nur auf das jugendliche Alter beschränkt, und Frauen sind offensichtlich häufiger betroffen als Männer. Demgegenüber werden bei den Patienten mit Typ-I-Diabetes und der HLA-Eigenschaft DR4 kaum andere Autoimmunerkrankungen beobachtet, die Antikörper gegen Inselzellen treten nur sehr vorübergehend bei Beginn der Erkrankung auf, jedoch sind Antikörper gegen Insulin häufig, der Erkrankungsbeginn liegt fast ausschließlich im jugendlichen Alter, das männliche Geschlecht überwiegt (Goldmann, 1982; Rotter et al., 1983).

Von Bedeutung für die insulinabhängigen Diabetiker ist die Frage nach dem möglichen Risiko für die Nachkommen. Da ein definierter Erbgang bisher nicht bekannt ist (s.o.), sind genaue Angaben nicht möglich. Bei juvenilem Diabetes mellitus eines Elternteils liegt das statistische Risiko für die Kinder, im ersten Lebensjahrzehnt ebenfalls an einem Diabetes zu erkranken, bei 1–2% (Rotter et al., 1983).

Autoimmunphänomene

Die Beobachtung eines gehäuften Auftretens des Typ-I-Diabetes bei Patienten mit anderen Autoimmunerkrankungen, wie z.B. mit Immunthyreoiditis, wies zum ersten Mal auf die Möglichkeit hin, daß Autoimmunphänomene auch in der Pathogenese des Typ-I-Diabetes bedeutsam sein könnten. In der Folgezeit konnte der Nachweis von Antikörpern gegen Inselzellen bei Patienten mit Typ-I-Diabetes geführt werden. Inzwischen können mehrere Typen von Antikörpern unterschieden werden, die sich gegen das Zytoplasma der Zellen oder gegen die Zelloberfläche richten, zum Teil Komplement binden und mehr oder weniger selektiv gegen Beta-Zellen gerichtet sind (Brogren und Lernmark, 1982). Die Dauer des Auftretens der Antikörper gegen Inselzellen ist bei den einzelnen Erkrankungsfällen unterschiedlich, und hierin spiegeln sich wahrscheinlich Heterogenitäten bezüglich der Pathogenese des Typ-I-Diabetes wider.

Antikörper gegen Inselzellen ließen sich auch bei Verwandten ersten Grades von Typ-I-Diabetikern nachweisen. In einem Teil der Fälle trat Monate nach dem erfolgten Nachweis der Inselzellantikörper ein Typ-I-Diabetes auf, in anderen Fällen verschwanden die Antikörper wieder (Spencer et al., 1984).

Diese Beobachtungen sprechen für eine latente Phase des Typ-I-Diabetes, aus der sich entweder die manifeste Erkrankung mit den Zeichen des plötzlichen klinischen Beginns entwickeln kann oder aber eine Heilung resultiert.

Die biologische Rolle der Antikörper gegen Inselzellen im Zusammenhang mit der Pathogenese der Erkrankung bleibt zum jetzigen Zeitpunkt ungeklärt. Es kann sich um reine Epiphänomene der Erkrankung handeln, die Inselzellantikörper können im anderen Extrem aber auch zytotoxische und zytolyti-

sche Wirkungen ausüben; wahrscheinlich kommt unterschiedlichen Typen von Antikörpern eine unterschiedliche Bedeutung in der Pathogenese der Erkrankung zu (Brogren und Lernmark, 1982; Helmke et al., 1983).

In Verbindung mit dem Typ-I-Diabetes finden sich auch Störungen der zellulären Immunfunktion. Das Auftreten eines positiven Leukozytenmigrationshemmtestes wurde zuerst 1974 beschrieben und in der Folgezeit bestätigt (Irvine, 1982). In den letzten Jahren wurden Verschiebungen innerhalb der T-Zellpopulationen beschrieben, eine Verminderung der Suppressorzellen fiel dabei besonders auf (Buschard et al., 1983). Die Störungen der zellulären Immunität haben in bezug auf die Pathogenese des Typ-I-Diabetes in jüngster Zeit immer mehr Beachtung gefunden, ihnen könnte allein oder in Verbindung mit humoralen Immunphänomenen eine große pathogenetische Bedeutung zukommen. Abschließend sei darauf hingewiesen, daß die sog. BB-Ratte ein Tiermodell darstellt, das mannigfache Parallelen zum Typ-I-Diabetes zeigt und die weitere Erforschung des menschlichen Typ-I-Diabetes in den vergangenen Jahren sehr befruchtet hat (Like et al., 1982).

Da zelluläre und humorale Immunreaktionen durch psychische Faktoren, insbesondere Streß, beeinflußt werden können, ist anzunehmen, daß auch bei den beschriebenen Autoimmunphänomenen beim Typ-I-Diabetes psychosoziale Reize eine Rolle spielen (vgl. Kap. 11), obwohl gesicherte Erkenntnisse dazu bisher nicht vorliegen.

Viruserkrankungen

Zusammenhänge zwischen Viruserkrankungen und dem Auftreten eines Diabetes mellitus wurden bereits im letzten Jahrhundert vermutet. Bei Menschen wurden Fälle von Typ-I-Diabetes nach Infektion mit Mumpsviren, Coxsackie-B4-Viren und Masernviren beobachtet. Im Tierexperiment lassen sich durch Viren eine Insulitis und ein Diabetes mellitus erzeugen. An dem Zusammenhang zwischen Typ-I-Diabetes und Virusinfektion kann in Einzelfällen zwar kein Zweifel bestehen, bei Untersuchungen großer Kollektive ließ sich allerdings weder eine eindeutige Häufung des Diabetes nach den in Frage kommenden Virusinfektionen noch das Auftreten charakteristischer serologischer Hinweise auf die betreffenden Viren bei Patienten mit Typ-I-Diabetes nachweisen. Die Rolle der Viren bei der Pathogenese des Typ-I-Diabetes muß deshalb zum jetzigen Zeitpunkt als ungeklärt angesehen werden (Rayfield und Yoon, 1982; Kolb und Gries, 1982; Helmke et al., 1983).

Psychische Faktoren

Bis in die jüngste Zeit hinein werden immer wieder Untersuchungen publiziert, in denen der Frage nachgegangen wird, ob psychische Faktoren bei der Genese und Manifestation eines Diabetes mellitus vom Typ I eine Rolle spielen. Diese Untersuchungen konzentrieren sich im wesentlichen auf prämorbide Persönlichkeitsfaktoren (Swift et al., 1967; Fallstrom, 1974; Tavormina et al., 1976), die vor allem aus vermehrter Angst und Feindseligkeit sowie Problemen mit Geschwistern bestehen sollen. In anderen Untersuchungen konnten immer dann, wenn diabetische Kinder mit gesunden gleichaltrigen und gleichgeschlechtlichen Jugendlichen verglichen wurden, die beschriebenen Veränderungen nicht festgestellt werden (Galatzer et al., 1977; Kubany et al., 1956; McGraw und Tuma, 1977; Steinhauser et al., 1977).

Andere Untersucher interessierten sich weniger für prädisponierende Persönlichkeitsfaktoren als vielmehr für die Wirkung von „life-stress" auf die Entstehung des Diabetes mellitus vom Typ I, z.B. Objektverluste. Nach Slawson und Mitarbeitern (1963) sollen 56% der Jugendlichen mit Diabetes mellitus über Objektverlust einen Monat bis 4 Jahre vor Ausbruch ihrer Erkrankung berichtet haben, bei weiteren 24% der Patienten wurde von den Interviewern ein Objektverlust angenommen, der von den Patienten verleugnet, von den Interviewern aber als wichtig für die Erkrankung eingeschätzt wurde. Allerdings wurde bei diesen Untersuchungen keine Vergleichsgruppe untersucht. Weitere Untersuchungen (Stein und Charles, 1971 und 1975) zeigten ähnliche Ergebnisse, konnten aber letztendlich nie bestätigt werden (Bruch, 1959; Koch und Molnar, 1974; Koski, 1969).

Obwohl Streß möglicherweise eine wichtige Rolle bei der Genese des Diabetes mellitus vom Typ I spielt, fehlen bis heute prospektive Untersuchungen, die der individuellen Bedeutung von Streß bei der Entstehung eines Diabetes mellitus im Kontext mit genetischen, immunologischen und sozialen Faktoren nachgegangen wären.

47.1.4 Klinik

Ein akuter Beginn mit Polydipsie, Polyurie, verminderter Leistungsfähigkeit und Gewichtsabnahme ist typisch für den klinischen Verlauf. Meist führen diese Symptome den Patienten zum Arzt, häufig schon mit der richtigen Verdachtsdiagnose. Werden diese Symptome dagegen vom Patienten und/oder vom Arzt verkannt, so kommt es zum raschen Fortschreiten der Erkrankung bis zum diabetischen Koma.

Die Gabe von Insulin schafft entscheidende Abhilfe; vor der Einführung von Insulin in die Therapie sind diese insulinpflichtigen Patienten an der diabetischen Erkrankung verstorben.

Unter Behandlung mit Insulin und einer Diabetesdiät kommt es bei der Mehrzahl der Typ-I-Diabetiker in den ersten Tagen bis Wochen nach Beginn der Erkrankung zu einer mehr oder weniger ausgeprägten Erholung, der sog. Remissionsphase, in der eine Dosisreduktion des Insulins, ja sogar der völlige Verzicht auf die Insulingabe möglich ist. Die Remissionsphase ist durch eine teilweise Erholung der Restsekretion der Beta-Zellen der Langerhansschen Inseln gekennzeichnet. Die Ursachen für diese Remissionsphase sind bisher unbekannt.

Im weiteren Verlauf der Erkrankung entwickelt sich parallel mit einem neuerlichen Nachlassen und Versiegen der Restsekretion die jetzt zeitlebens fortbestehende Insulinpflichtigkeit.

Aus heutiger Sicht ergeben sich erst in dieser Phase die eigentlichen klinischen und medizinischen Probleme im Zusammenhang mit der Erkrankung des Typ-I-Diabetes. Diese Probleme bestehen in der Entwicklung der sog. Spätkomplikationen des Diabetes mellitus, die sich meistens erst nach 10 und mehr Jahren Dauer des Typ-I-Diabetes manifestieren. Unter diesen Spätkomplikationen werden vor allem die Folgekrankheiten im Bereich der kleinen und großen Gefäße des Organismus, die sog. Mikro- und Makroangiopathie, verstanden.

Bei der Mikroangiopathie handelt es sich um eine diabetesspezifische Erkrankung, die die Kapillaren in allen Organen des Organismus betrifft und an Niere und Auge zu klinisch besonders schwerwiegenden Folgezuständen, nämlich zur terminalen Niereninsuffizienz bzw. zur Erblindung, führt.

Bei der Makroangiopathie handelt es sich um eine nur durch wenige Besonderheiten charakterisierte Form der Arteriosklerose. Die Arteriosklerose entwickelt sich bei Diabetikern frühzeitiger und generalisierter und zeigt ein schnelleres Fortschreiten. An dieser Stelle muß auch auf die Neuropathie bei Diabetes mellitus hingewiesen werden, sie zeigt unter Einbeziehung des somatischen und des autonomen Nervensystems ein ausgesprochen vielfältiges Bild. Die Neuropathie ist zwar auch eine Folgeerkrankung des Diabetes mellitus, jedoch nicht im eigentlichen Sinne eine Spätkomplikation, da ihre Entstehung nicht in dem Ausmaß von der Erkrankungsdauer abhängig ist, wie dies für die Mikroangiopathie und die Makroangiopathie zutrifft.

Die Spätkomplikationen des Diabetes mellitus sind für die eingeschränkte Lebenserwartung der Diabetiker – sie beträgt etwa ⅔ der Lebenserwartung von Nichtdiabetikern, unabhängig vom Alter bei Erkrankungsbeginn (Petrides, 1977) –, für die 2- bis 20fache Übersterblichkeit der Diabetiker (Marble, 1974) – abhängig von der betrachteten Altersspanne – und für die erheblich beeinträchtigte Lebensqualität der Patienten mit Diabetes verantwortlich zu machen.

Wenn von dem letztlich natürlich erstrebenswerten Ziel einer vollständigen Heilung oder Vermeidung der Erkrankung abgesehen wird, so besteht die entscheidende klinische Aufgabe in bezug auf den Typ-I-Diabetes und den Diabetes insgesamt heute in der Suche nach Wegen und Möglichkeiten zur Vermeidung der Spätkomplikationen.

Hervorragende klinische und tierexperimentelle Untersuchungen sowie aktuelle Vorstellungen zur Pathogenese der Spätkomplikationen sprechen für eine entscheidende Rolle der Hyperglykämie bei deren Entstehung (Beischer und Pfeiffer, 1985). Damit stellt sich – zumindest vordergründig – das Ziel und die Aufgabe, die diabetische Stoffwechselführung möglichst normoglykämisch zu gestalten.

Mit dieser Zielvorstellung kommt neuen Ansätzen in der Stoffwechselkontrolle des Diabetes mellitus und neuen Entwicklungen in der Insulintherapie besondere Bedeutung zu. Neu an der Stoffwechselkontrolle ist vor allem die Einbeziehung des Patienten und zwar mit regelmäßigen Selbstkontrollen der Konzentration der Blutglukose. Hier stehen zahlreiche moderne Techniken zur Verfügung (Althoff et al., 1982; Beischer et al., 1985). Die Selbstkontrolle der Blutglukose ist nur ein Beitrag zur Erziehung des mündigen Patienten, der letztlich „Fachmann in eigener Sache" sein soll.

Die neuen Möglichkeiten der Therapie betreffen vor allem neue Möglichkeiten der Insulinanwendung. Die künstliche Beta-Zelle (auch künstliches Pankreas genannt) ermöglicht unter Einsatz eines Computers die feedbackgesteuerte Insulingabe, ähnlich wie im gesunden Organismus. Die Insulininfusionspumpen ahmen diese Gegebenheiten ohne Rückkopplung nach, indem eine meist subkutane Dauerinfusion von Altinsulin erfolgt und zu den Mahlzeiten zusätzlich Bolusinjektionen verabreicht werden (Pfeiffer und Kerner, 1984).

Mit der Kombination der neuen Möglichkeiten für Stoffwechselkontrolle und Therapie gelingt es nicht nur, deutliche Fortschritte in Richtung der normoglykämischen Diabetesführung zu erzielen, sondern auch die psychische Verfassung der Patienten ausgesprochen günstig zu beeinflussen (Dupuis, 1980).

47.1.5 Psychische Faktoren im Verlauf des juvenilen Diabetes mellitus

In zahlreichen Untersuchungen konnte bisher belegt werden, daß Stabilität oder Instabilität (sog. „brittle diabetes") der Krankheit im wesentlichen durch psychosoziale Faktoren determiniert sind (Johnson, 1980). Dabei haben sich drei Forschungsrichtungen herauskristallisiert:
– Untersuchung der psychosozialen Faktoren von jugendlichen Diabetikern in Abhängigkeit von der Compliance;
– Untersuchung der Familien diabetischer Kinder;
– Untersuchungen der Wirkung von Streß auf Blutzucker und freie Fettsäuren.

Psychosoziale Faktoren

Alexander (1971) konnte (an 2 Fällen) zeigen, daß diese Patienten ungewöhnlich starke rezeptive Tendenzen und Sehnsüchte nach Versorgtsein in sich trugen. „Diese Kranken behielten eine infantile, abhängige und fordernde Einstellung bei und litten an Versagung, weil ihre Forderungen nach Zuwendung und Liebe sich außerhalb jeder Möglichkeit der realen Situation eines Erwachsenen bewegten und infolgedessen niemals ausreichend befriedigt werden konnten. Auf diese Versagung reagierten die Patienten mit Feindseligkeit. Der Diabetes entstand, als diese infantilen Wünsche der Versagung anheim fielen."

Margolin (1953) berichtet von einem 16jährigen Diabetiker, der die Vorstellung hatte, daß seine Mutter ihm ihre Liebe versage und daß Liebe durch Essen

symbolisiert sei. Kennzeichnend für ihn war exzessiver Hunger und Durst, d.h. er forderte das, was seine Mutter ihm entzog. Er setzte Süßigkeiten mit Muttermilch gleich.

Stein und Charles (1975) verglichen 8 diabetische Kinder mit einer entsprechenden Kontrollgruppe in bezug auf frühkindliche Ernährungsgewohnheiten. Sie fanden bei den Diabetikerkindern ein gestörtes Nahrungsverhalten im Sinne oraler Abhängigkeit. Es fragt sich, ob es sich hier um eine reaktive oder primäre Eigenschaft handelt.

Koski (1969) stellte fest, daß diabetische Kinder vor allem Angstreaktionen, depressive Reaktionen, Suiziddrohungen und Aggressionen gegenüber Autoritätsfiguren zeigen.

Swift und Mitarbeiter (1967) fanden, daß diabetische Kinder emotional häufiger gestört waren, verglichen mit einer Kontrollgruppe zeigten sie mehr Schwierigkeiten in bezug auf Abhängigkeit/Unabhängigkeit, Selbstbewußtsein, manifeste und latente Ängste, sexuelle Identifizierung, Ausdruck von Feindseligkeit und orale Fixierung. Sie verglichen 50 diabetische und 50 gesunde Kinder. Sie fanden, daß junge Diabetiker gehäuft abnorme Vorstellungen ihres Körpers, latente Ängste, Dysphorie und abhängiges Verhalten zeigten.

Die diabetischen Kinder hatten auch Schwierigkeiten in der Krankheitsbewältigung, zu Hause und beim Spielen mit anderen Kindern. Diese Untersuchungen stehen im Gegensatz zu einer Arbeit von Davies und Mitarbeitern (1965), die 58 diabetische Kinder im Alter von 8 bis 15 Jahren mit Fragelisten bezüglich ihrer Einstellung dem Diabetes gegenüber untersuchten. Sie fanden, daß die diabetischen Kinder ihre Krankheit als einen normalen Bestandteil ihres Lebens betrachteten.

Der juvenile Diabetiker und seine Familie

Williams (1973) stellte nicht nur bei den Patienten, die an einer chronischen Erkrankung leiden, Reaktionen fest, sondern auch bei deren Familienmitgliedern. Er fand:
– Ärger und Gereiztheit gegenüber dem chronisch erkrankten Familienmitglied,
– Panikreaktionen beim Auftreten alarmierender Symptome,
– Hypochondrie bei Familienmitgliedern,
– überprotektive und unnötige Einschränkungen gegenüber dem erkrankten Familienmitglied,
– Angst, von derselben Krankheit betroffen zu werden oder dieselbe weiterzugeben.
Es fällt auf, daß bei Patienten und deren Familien ähnliche Reaktionsmuster gefunden werden.

Zu ähnlichen, allerdings unspezifischen Ergebnissen gelangte Minuchin (1986), der charakteristische Familienstrukturen bei Familien mit stoffwechsellabilen Kindern fand: Verstrickung, Überfürsorglichkeit, Starrheit und Konfliktvermeidung (vgl. Kap. 21).

Koski (1969) fand bei der Untersuchung von 60 diabetischen Kindern, daß vor allem die Familie Gefühle des Entsetzens, Schock, unbestimmte Furcht und Ängste, Depressionen und Schlafstörungen aufwies. Diese Feststellungen sind weitgehend in Übereinstimmung mit Hinkles Beschreibung der schwer einstellbaren Diabetiker (Hinkle et al., 1950). Es handelt sich hierbei um Patienten:

– mit äußerst labilem Diabetes mellitus,
– die durch ihr Betragen eine vernünftige Führung erschweren,
– die unfähig sind, eine Diät einzuhalten, und
– man findet gehäuft Familienprobleme oder Probleme zwischen diabetischen Kindern und ihren Familien.

Die primären Reaktionen der Eltern – zumeist der Mütter – auf die Diagnosemitteilung können in der Folge der psychischen Verarbeitung zu Verleugnung, Schuldgefühlen, Aggressionen oder Wunschdenken führen und so eine positive und konstruktive Adaptation des Kindes an die Krankheit erschweren oder sogar verhindern (La Hood, 1970).

Als Maßstab für die „emotionale Gesundheit" der Familie gilt die Fähigkeit, den Diabetes mellitus zu integrieren (Bruch, 1959). Die kindliche Anpassung an die Erkrankung gelingt nicht, wenn unspezifische prämorbide familiäre Verhaltensmuster durch die Manifestation der Krankheit bei einem Familienmitglied aktualisiert und fixiert werden, d.h., wenn es den Eltern nicht gelingt, die Krankheit ihres Kindes zu akzeptieren (A. Freud, 1952).

In testpsychologischen Untersuchungen wurden Mütter diabetischer Kinder folgendermaßen beschrieben: geringe Neigung zu Aggressivität, erhöhte Irritierbarkeit, Nachgiebigkeit und Verschlossenheit sowie mangelnde Extraversion und Durchsetzungsfähigkeit. In zweierlei Hinsicht sind diese Ergebnisse von Bedeutung:

Zum einen können diese Merkmale das Ausmaß emotionaler Belastung durch ein chronisch krankes Kind widerspiegeln, andererseits wird ein aktives krankheitsbezogenes Handeln, insbesondere bei der Bewältigung von Krisen im Verlauf der Krankheit erschwert (Steinhausen, 1977).

Aimez (1971) fand bei einer Untersuchung von 77 Diabetikern, daß die Insulintherapie häufig abgelehnt wird, daß die diätetischen Maßnahmen selten richtig befolgt werden, so daß diese Ablehnung via Schuldgefühle, Einstellungen gegenüber Autoritäten etc. alte Ängste mobilisieren kann. Eine gute Arzt-Patient-Beziehung half den Diabetikern, die vorgeschriebenen therapeutischen und diätetischen Maßnahmen zu befolgen. Wenige dieser untersuchten Diabetiker benötigten eine Psychotherapie. Auch Koski (1969) und Swift (1967) betonen, daß Verlauf und Einstellung des Diabetes mellitus vor allem davon abhängen, wie gut das Kind oder der Erwachsene die Diagnose annehmen und verarbeiten kann, aber auch davon, wieviel die Familie dem Diabetiker hilft, die Krankheit zu meistern.

Psychophysiologische Untersuchungen

Cannon und Mitarbeiter (1911) haben gezeigt, daß man bei Katzen, die man für kurze Zeit festbindet, eine emotionale Glukosurie auslösen kann. Die Menge der Zuckerausscheidung entsprach etwa der Wut der Tiere. Bei der Untersuchung einer Fußballmannschaft der Harvard-Universität nach einem Wettbewerbsspiel konnte er zeigen, daß emotionale Faktoren auch bei gesunden Menschen zu einer Glukosurie führen können: Von 25 Spielern hatten 12 Zucker im Urin; von diesen 12 waren 5 Ersatzspieler, die an dem Spiel gar nicht aktiv teilgenommen hatten. Auch bei Zuschauern, die sich aufgeregt hatten, fand sich eine Glukosurie (Cannon, 1929). Später gelang es Cannon nachzuweisen, daß psychische Erregung wie Wut, Hunger, Angst und Schmerz, die hauptsächlichen Emotionen („major emotions", wie er diese Gemütszustände nannte), zu einer erhöhten Adrenalinausschüttung führen. Eine erhöhte Katecholaminausschüttung führt zu einer Verminderung der Insulinsekretion, einem Anstieg der Glukagonsekretion und einer Zunahme der Glykogenolyse. Die dadurch bedingte Hyperglykämie führt, sobald die Nierenschwelle überschritten ist, zur Glukosurie. Die von Cannon erstmals nachgewiesenen Zusammenhänge zwischen emotionaler Erregung und endokrinen Reaktionen sind in der Folgezeit vor allem in der Streßforschung weiter untersucht worden (vgl. Kap. 10 „Psychoneuroendokrinologie"). Die Annahme, daß sie in der Ätiologie und Pathogenese des Diabetes mellitus von Bedeutung sein könnten, geht auf Weiss und English (1950), in gewisser Weise auch auf F. Dunbar und Mitarbeiter (1936) zurück:

Weiss und English gingen von der Feststellung Cannons aus, daß der Organismus auf Bedrohungen verschiedenster Art über eine Adrenalinausschüttung mit einer Bereitstellung zu Kampf oder Flucht reagiert, zu der auch eine Zuckermobilisation mit Hyperglykämie gehört. Sie stellten die Hypothese auf, daß in der frühen Kindheit erlittene psychische Schäden zu einer permanenten unbewußten Angst führen könnten, auf die der Organismus über Jahrzehnte hinweg so reagieren würde, als sei seine psychische und physische Sicherheit bedroht. Da bei einer unbewußt bleibenden Angst eine Lösung der psychophysischen Spannung durch Kampf oder Flucht mit entsprechender Affektabfuhr nicht möglich ist, sollte die Bereitstellung mit der sie begleitenden Hyperglykämie zu einem Dauerzustand werden können, aus dem dann schließlich über eine Erschöpfung des Inselapparates des Pankreas ein Diabetes mellitus resultieren würde. Auf diese Weise sollte eine an sich physiologische Bereitstellung – wie von Uexküll (vgl. Kap. 1) es formuliert hat – zu einer Bereitstellungskrankheit führen können. Die Hypothese deckt sich mit den klinischen Untersuchungen von Baker und Mitarbeitern (1969), die nachwiesen, daß diabetische Kinder nach einer Adrenalininjektion einen signifikant schnelleren Anstieg der Blutketonkörper haben als gesunde. Darüber hinaus fand er bei zwei adoleszenten Diabetikerinnen nach einem Streß-Interview einen deutlichen Anstieg der Blutzuckerwerte und der freien Fettsäuren, verbunden mit einer deutlichen Zunahme der Plasmasteroide und des Wachstumshormons, sowie eine gesteigerte Urinausscheidung von Adrenalin. Die Gabe von Betablockern vor dem Streß-Interview konnte die metabolischen Veränderungen blockieren, während die hormonalen Veränderungen unbeeinflußt blieben.

Hinkle und Wolf (1952) zeigten in detaillierten Einzelstudien, daß in den Lebensgeschichten von Diabetikern die Reaktionen auf Ereignisse im täglichen Leben mit der Feststellung übereinstimmen, daß die Lebensereignisse beim Ausbruch, Verlauf und bei Komplikationen der Krankheit eine wichtige Rolle spielen.

Sie beschreiben eine 17jährige Diabetikerin, die durch ihre häufigen Hospitalisationen wegen entgleistem Diabetes auffiel. Dieses ursprünglich unerwünschte Kind hatte Schwierigkeiten, mit seiner Mutter eine tragfähige Beziehung aufzubauen, zeitweise wurde es von einer Tante aufgezogen. Die Familie wechselte häufig den Wohnort, und für das Kind war es immer wieder schwierig, sich in eine neue Umgebung, Kindergruppen oder Schule einzuleben. Entsprechend reagierte die Patientin mit Rebellion, feindseligen Gefühlen gegenüber der Mutter und Angst. Nach einem erneuten Wohnungswechsel entdeckte man beim damals 10jährigen Kind einen Diabetes mellitus. Sowohl das Kind als auch die Mutter zeigten große Schwierigkeiten, die Krankheit zu akzeptieren. Furcht, Hoffnungslosigkeit und Dosierungsfehler waren häufig. In der Folge kam es beim Kind zu schwerwiegenden Entgleisungen des Diabetes mit notfallmäßigen Hospitalisationen, wann immer Schwierigkeiten zwischen den Eltern, neue Wohnungswechsel oder erzieherische Probleme auftauchten. So hatte das Mädchen innerhalb der letzten 5 Jahre 12 Hospitalisationen hinter sich, alle unmittelbar auf streßerfüllte Lebenssituationen folgend, ohne daß jemals Infekte oder andere Gründe zur Entgleisung des Diabetes vorlagen.

Das Mädchen war zum Zeitpunkt der Publikation der Arbeit 4 Jahre in enger Beobachtung und Behandlung (Hinkle und Wolf, 1952). In dieser Zeit wurde vor allem das Verhalten gegenüber der Mutter diskutiert und besprochen. Während dieser Periode hatte die Patientin keine Hospitalisation wegen einer ketoazidotischen Stoffwechselentgleisung. Gleichzeitig führte die Patientin ein Tagebuch, in dem sie sämtliche großen und kleinen Konflikte aufführte. Sie notierte ebenfalls die Resultate ihrer eigenhändig durchgeführten Urinanalysen. Beim Vergleich des Tagebuches mit den Urinanalysen stellte man einen ganz direkten Zusammenhang zwischen signifikanten Streßsituationen und dem Erscheinen einer Ketonurie, Durst bzw. Polyurie fest.

Hinkle und Mitarbeiter (1950, 1951) haben bei Diabetikern nach Streß-Interviews einen Anstieg des Blutzuckers, der Azetonämie und der Wasserausscheidung festgestellt. Die Autoren sehen daher in psychischem Streß einen wichtigen Faktor, der bei Diabetikern ein Coma diabeticum auslösen kann.

Nach Kemmer (1988) kommt es aber bei gesunden Probanden und Diabetikern unter Streßbedingungen (Kopfrechnen, freie Rede) zu einer Zunahme der inneren Erregung, der Herzfrequenz, des Blutdrucks, sowie der Hormone Adrenalin und Cortisol, nicht aber von Glukagon und STH. Trotz des Anstiegs von Cortisol und Adrenalin blieben die Serumkonzentrationen von Glukose, Ketonen und freien Fettsäuren unverändert. Dies könnte bedeuten, daß die im Alltagsleben (z.B. Autofahren im Berufsverkehr) nur leichten und kurzfristigen Anstiege von Adrenalin und Cortisol noch nicht zu Blutzuckeranstiegen führen, sondern daß erst chronischer Streß bei entsprechender Disposition zu einer Verschlechterung des Glukosestoffwechsels führt.

47.1.6 Therapeutische Hinweise

Die therapeutischen Grundlagen bilden Diät, Insulintherapie und körperliches Training. Eine wichtige Komplikation der Insulintherapie ist die Hypoglykämie. Aufgabe des Arztes ist es, den Patienten umfassend über die Symptome einer Hypoglykämie aufzuklären und ihn in die Lage zu versetzen, frühzeitig die richtigen Maßnahmen gegen eine Hypoglykämie zu ergreifen.

Die Angst des Arztes oder des Patienten vor einem hypoglykämischen Schock sind häufig Ursache einer hyperglykämischen Einstellung. Insulinallergien vom Sofort- und Spättyp mit Hautveränderungen, die als kosmetisches Problem, vor allem bei Mädchen, zu psychischen Reaktionen führen können, sind im Zeitalter der hochgereinigten Insuline und des menschlichen Insulins eine Seltenheit geworden.

Beim Kleinkind mit Diabetes mellitus ist es wichtig, mit den Eltern die Probleme der Diät, Insulindosierung, Zeichen einer Hypo- oder Hyperglykämie zu diskutieren. Genauso wichtig ist es, mit ihnen über ihre Gefühle, ein diabetisches Kind zu haben, zu sprechen; häufig sind die Eltern geplagt von Schuldgefühlen, daß irgendein Fehlverhalten ihrerseits den Ausbruch des Diabetes mellitus begünstigt habe. Ebenfalls die Vererbung des Diabetes mellitus muß – wie bereits erwähnt – mit den Eltern diskutiert werden, da diese oft voreilig beschließen, keine weiteren Kinder zu haben, aus Angst, diese könnten ebenfalls an Diabetes mellitus erkranken.

In der Schule ist das Kind oft verletzenden Fragen und unklaren Vorstellungen der Schulkameraden über seine Krankheit ausgesetzt. Man kann den kleinen Diabetikern helfen, indem man diese Fragen erörtert, falsche Vorstellungen korrigiert und sie auf ungeschickte Verhaltensweisen gegenüber ihren Schulkameraden aufmerksam macht. Schulkinder leiden oft unter Schuldgefühlen, sie hätten durch unmäßigen Genuß von Süßigkeiten ihren Diabetes mellitus selbst verursacht. In dieser Altersstufe ist es ebenfalls wichtig, dem Kind zu zeigen, daß die Einstellung des Diabetes eine gemeinsame Aufgabe des Kindes, der Eltern und des Arztes ist. Entsprechend muß man dem Kind auch eine altersgemäße Verantwortung und Entscheidungsmöglichkeit über die Behandlung und Lebensführung einräumen. Für die Eltern ist es oft schwierig, einen Teil dieser Verantwortung für die gute Einstellung des Diabetes mellitus dem Kind abzutreten. Daraus resultierende Gefühle wie Angst und Unsicherheit sollten vom Arzt mit den

Eltern besprochen und nicht stillschweigend übergangen werden. Die Erziehung eines diabetischen Kindes bietet den Eltern zusätzliche Schwierigkeiten: Geschwister des an Diabetes erkrankten Kindes können auf die zusätzliche Zuwendung zum diabetischen Kind mit Eifersucht reagieren, oder die Eltern können bewußt oder unbewußt vom diabetischen Kind weniger Mitarbeit in der Familie verlangen. Im Gespräch mit den Eltern soll der Arzt auf ungünstige Einstellungen dem diabetischen Kind gegenüber aufmerksam machen.

Der adoleszente Diabetiker gibt Eltern und Ärzten oft große Probleme auf. Diese Patienten wollen als unabhängige Menschen angesehen werden, und dementsprechend soll man sie möglichst viel an den Entscheidungen über ihre Behandlung teilhaben lassen. In ihrem Drang nach Unabhängigkeit haben diese Diabetiker oft Mühe, Insulin und diätetische Maßnahmen als notwendige Behandlung und nicht als Autoritätsausdruck des Arztes zu betrachten. Oft benutzen sie deshalb die Einstellung des Diabetes als Mittel, um ihre Unabhängigkeit gegenüber der Autorität oder dem Arzt zu demonstrieren. Bei unserer Patientin bestimmte ein Abhängigkeits-Autonomie-Konflikt über einen langen Zeitraum hinweg Krisen und Spannungen in der Arzt-Patient-Beziehung und führte zu erheblichen Compliance-Problemen.

Therapeutisch lohnt es sich, dem Unabhängigkeitsstreben dieser Patienten mit Verständnis zu begegnen, es in den Behandlungsplan einzubeziehen und auf ihre Renitenz nicht emotional gefärbt zu reagieren und sie nicht mit vermehrten Kontrollen, Strafen oder anderen Maßnahmen, die ihre Abhängigkeit aufdekken, zu behandeln. Es muß vielmehr versucht werden, dem Patienten Verständnis für sein Gefühl der verlorenen Unabhängigkeit entgegenzubringen. Kommen Diät- oder Insulindosierungsfehler vor, soll der Arzt den Patienten wissen lassen, daß er dies bemerkt hat, daß er sich aber vorstellen kann, daß verschiedene Gründe hierzu führen können. Viele Adoleszenten und Erwachsene zeigen eine bessere Kooperation, wenn die Einstellung des Diabetes als eine gemeinsame Aufgabe des Patienten und seines Arztes aufgefaßt wird, wobei der Arzt die Funktion des Beraters ausübt.

Adoleszente Diabetiker haben auch andere wichtige Fragen und Entscheidungen zu treffen, bei welchen ärztliche Beratung erwünscht ist: Der Adoleszent muß eine Berufswahl treffen, er stellt sich Fragen über Partnerwahl, Sexualität und Kinderwunsch. Für alle diese Fragen kann der Arzt Vermittler der entscheidenden Information sein.

Galatzer und Mitarbeiter (1982) berichten über ein Kriseninterventionsprogramm bei diabetischen Kindern. Die Ergebnisse dieser Studie belegen, daß ein multidisziplinäres Behandlungsprogramm den Familien und den Kindern helfen kann, besser mit dem traumatischen Ereignis der Diagnose umzugehen. Wichtig scheint eine schnelleinsetzende Therapie zu sein, dadurch werden die Chancen erhöht, die Compliance, die familiären Beziehungen und die soziale Anpassung zu verbessern. Der Aufwand, zu einem späteren Zeitpunkt durch ein multidisziplinäres Programm bei der Bewältigung des Traumas zu helfen, ist dreimal so hoch, wenn man dieselben Ergebnisse erzielen will wie bei dem Sofortprogramm. Ein Nebenbefund dieser Studie verweist darauf, daß in den sozial unteren Schichten ein Sofortprogramm, das in der Klinik durchgeführt wird, keine Verbesserung in den erwähnten Bereichen erbringt. Hier schlagen die Autoren vor, Programme zu entwickeln, bei denen die betroffenen Familien in ihrem häuslichen Milieu aufgesucht werden. Außerdem scheint es hier notwendig, bestimmte soziale Mängel zunächst z. B. durch finanzielle Hilfen zu beheben, bevor psychotherapeutische Maßnahmen überhaupt angenommen werden können.

Einen neuen Ansatz zur ganzheitlichen Betreuung von Diabetikern bildet das Genfer Modell eines integrierten psychosomatischen Zugangs zum Diabetespatienten (Gfeller und Assal, 1979): Auf einer Station innerhalb eines Hospitals von 2000 Betten werden in regelmäßigen Abständen 12 Diabetiker für 5 Tage aufgenommen. Diese Patienten bilden eine Behandlungsgruppe, deren Therapie in einem „halboffenen Milieu" stattfindet. Zwischen den Behandelnden und den Behandelten wird ein psychotherapeutischer Kontrakt abgeschlossen, der verbindlichen Charakter hat.

In der Zeit des Aufenthaltes wird Wert auf eine intensive und ausgiebige medizinische Behandlung und Diätberatung gelegt. So erhalten die Patienten in dieser Zeit 16 Stunden Unterricht in Medizin und täglich ist ein Diätbuffet für die Patienten aufgebaut. Das Genfer Behandlungsmodell zeichnet sich dadurch aus, daß hier die Patienten nicht als Notfälle aufgenommen, sondern zur Routinebehandlung eingewiesen werden. Die Gruppe der 12 Patienten bleibt geschlossen und kann so als Gruppe immer wieder aufgenommen werden. Von Fall zu Fall werden in diese Gruppe auch Ehepartner bzw. Eltern von juvenilen Diabetikern aufgenommen. Aus gruppendynamischen Gründen übersteigt die Größe der Gruppe nie 12 Patienten.

Während des Aufenthaltes wird mit den Patienten eine gruppentherapeutische Behandlung durchgeführt, deren Ziel es ist, dem einzelnen zu helfen, die notwendige Trauerarbeit zu leisten, die eine wesentliche Voraussetzung zur psychischen Bewältigung der Erkrankung darstellt.

Das medizinisch-therapeutische Personal bleibt konstant. Ärzte, Schwestern, Ernährungsberaterinnen, Psychologen etc. sind zu einem festen Team zusammengeschlossen. Die Patienten sollen zu den einzelnen Mitgliedern des Teams eine partnerschaftliche Beziehung aufbauen, um die sonst übliche Regression von Patienten im Krankenhaus zu vermeiden.

47.1.7 Besondere Probleme des Typ-I-Diabetikers

Sexualität

Diabetiker klagen häufig über sexuelle Impotenz bzw. Frigidität: So geben ca. 40% aller männlichen Diabetiker (Gfeller et al., 1981) sexuelle Schwierigkeiten an, und bei den diabetischen Frauen scheint ein ähnlich hoher Prozentsatz unter Störungen der Erlebnisfähigkeit zu leiden.

In der Nichtdiabetikerpopulation von Patienten mit sexuellen Schwierigkeiten liegen psychische Ursachen der Symptomatik zugrunde. Auch bei Diabetikern sind überwiegend psychische Ursachen für sexuelle Schwierigkeiten anzunehmen.

Die durch den Diabetes mellitus bedingte Schädigung der autonomen Nerven im Bereich S2–S4, die das Corpus cavernosum innervieren, ist beim Mann die organische Grundlage; bei der Frau kann diese neurogene Schädigung zu einer Verminderung der Sensibilität der Klitoris und Vagina führen.

So wird verständlich, daß eine organische Ursache für sexuelle Störungen hauptsächlich bei langjährigen Diabetikern gefunden wird. Bei diesen entwickelt sich – unter Beibehaltung der Libido – nach und nach eine Impotenz, während psychogen bedingte Störungen durch Nachlassen der Libido und rasches Auftreten der Symptome imponieren (vgl. Tab. 47–2).

Außerdem kann auch eine depressive Reaktion, z.B. bei der Diagnose einer diabetischen Organkomplikation, zu einem Libidoverlust führen (Köpp, 1989).

Spezifische Probleme in der Schwangerschaft

Schwangerschaft ist für jede Frau eine Zeit der Krise mit tiefen psychologischen und somatischen Veränderungen. Bibring (1959) spricht davon, daß eine solche Krise zu einem akuten psychischen Ungleichgewicht führt. Unter günstigen Umständen kann diese Krise zu einem spezifischen Reifungsprozeß und zur Übernahme neuer Funktionen führen. In der Mehrzahl der Fälle, in denen während der Schwangerschaft ein psychopathologischer Befund erhoben wird, zeigt sich jedoch, daß dieser meist von vorübergehender Natur ist, da er in der Regel Ausdruck von früheren Entwicklungskonflikten ist, auf die die Patientinnen während der Schwangerschaft regredieren.

Für eine an Diabetes erkrankte Frau, wo der Diabetes für sich allein genommen eine schwere Belastung für das psychische Gleichgewicht darstellt, bedeutet die Krise der Schwangerschaft eine Aktivierung der Konflikte aus der eigenen Beziehung zur Mutter.

Nichtgelöste Abhängigkeiten in der Mutter-Kind-Beziehung bekommen so ein zusätzliches Gewicht. Barglow und Mitarbeiter (1981) haben hundert schwangere diabetische Frauen untersucht und stellten bei ⅔ dieser Frauen schwere psychiatrische Symptome wie Depression, Angst und psychosenahe Zustände fest. Lediglich 35% ihrer Patientinnen konnten als normal angesehen werden. Leider fehlen vergleichende Untersuchungen.

Durch kooperative Betreuung der diabetischen Schwangeren und ihres Neugeborenen von Diabetologen, Frauenärzten und neonatologisch erfahrenen Kinderärzten in medizinischen Zentren hat sich die Prognose für die Mutter und vor allem auch für das Kind entscheidend gebessert. Die Häufigkeit von fetaler Morbidität und Mortalität entspricht fast derjenigen bei Nichtdiabetikerinnen, bei präkonzeptionellem Beginn der intensiven Betreuung normalisieren sich auch die früher erschreckend hohen Mißbildungsraten (Fuhrmann, 1982). Normoglykämische Führung der Blutglukose (60–120 mg/dl im Tagesprofil, HbA im Normbereich), regelmäßige Überwachung von Wachstum und Befinden des Feten mit modernsten Methoden der Gynäkologie und Geburtshilfe und neonatologische Intensivbetreuung sind die entscheidenden Säulen des Erfolgs (Beischer et al., 1985). Eine dauerhafte gute Motivation und Kooperation der Patientin ist einerseits Voraussetzung für den Erfolg und ergibt sich andererseits auch als dessen Folge (Law et al., 1980).

Abschließend sei darauf hingewiesen, daß bei der Mehrzahl der Patientinnen mit Schwangerschaft und Diabetes der Diabetes erst in der Schwangerschaft auftritt und meistens danach auch wieder verschwindet. Da auch die Kinder dieser Patientinnen gefährdet sind, stellt sich hier vor allem das Problem einer frühzeitigen Erkennung des Diabetes.

Proliferative diabetische Retinopathie

Eine proliferative diabetische Retinopathie (PDR) findet sich bei 60% der Patienten mit IDDM (insulin-dependent diabetes mellitus) nach 10 Jahren und bei 90% nach 20 Jahren. Das Vorkommen der PDR bei Patienten mit NIDDM (non-insulin-dependent diabetes mellitus) ist zwar geringer, aber immer noch hoch genug: So findet sich eine PDR bei 50% der Patienten nach 10jähriger und bei 70% der Patienten

Tabelle 47–2. Differentialdiagnose psychisch und somatisch bedingter Störungen der Sexualität (modifiziert nach Gfeller et al., 1981)

Ursache	Auftreten und Verlauf	Libido	Häufigkeit	Nächtl. Erektion	Behandlung	Prognose
psychisch	plötzlich	schwach	+++[1]	+++	Psychotherapie	gut
neurogen	fortschreitend	stark	+	±0	wenig befriedigend, Aufklärung über organische Ursachen, Paarbehandlung	ungünstig

[1] Vergleichbar der Population bei Nichtdiabetikern.

869

nach 20jähriger Krankheitsdauer. Insgesamt zeigen die Hälfte aller Patienten mit PDR bereits innerhalb von 5 Jahren eine signifikante Beeinträchtigung des Sehvermögens (Klein et al., 1984).

In einer kontrollierten Studie bei erwachsenen Patienten mit IDDM mit einer seit mindestens 5 Jahren bestehenden PDR fand Oehler (1980) bei 33% eine schwere Depression, bei 69% Probleme am Arbeitsplatz, bei 39% einen Verlust des Arbeitsplatzes, 73% der Patienten konnten nicht mehr Autofahren und 39% erhielten finanzielle Unterstützung durch öffentliche Sozialeinrichtungen.

Auch Jacobson et al. (1985) berichten über signifikant häufigere psychiatrische Symptome bei Patienten mit PDR im Vergleich zu diabetischen Patienten ohne Beeinträchtigung des Sehvermögens. Interessanterweise kommt es bei einer nur partiellen Verschlechterung des Visus zu ausgeprägteren psychischen Veränderungen als bei totaler Erblindung: So zeigen die Patienten mit partieller Beeinträchtigung des Sehvermögens mehr Ärger, Depression und Feindseligkeit als Blinde (Oehler, 1980).

Der Grund dafür ist unklar: Möglicherweise wird der Diabetiker mit der Visusverschlechterung in seinen noch bestehenden sozialen Bezügen stärker bedroht als der Blinde mit einer schon auf seine Erkrankung zugeschnittenen und damit stabilen psychosozialen Situation. Möglicherweise verleugnen auch die Patienten mit teilweiser Sehbeeinträchtigung sich selbst und anderen gegenüber ihre Erkrankung, was bei vollständigem Sehverlust nicht mehr möglich ist.

Die psychotherapeutische Unterstützung sollte nicht erst einsetzen, wenn der Patient bereits blind ist, sondern eine frühe psychosoziale Intervention könnte Angst und Furcht vor Erblindung und soziale Isolation vermindern.

Die von Oehler und Fitzgerald (1980) durchgeführte Gruppentherapie zeigte, daß entgegen den Befürchtungen von Ärzten, die Patienten würden durch die Gruppe erschreckt und in die Hoffnungslosigkeit gestürzt, die Gruppe, sofern sie qualifiziert geleitet wird, dem einzelnen Patienten helfen kann, den notwendigen Trauerprozeß zu bearbeiten. Allerdings kann eine diabetisch bedingte Organkomplikation wie Erblindung nicht nur ein Trauma sein, das es zu bewältigen gilt, sondern für den Patienten auch eine Schädigung darstellen, die im Sinne der Befriedigung unbewußter Strafbedürfnisse erlebt wird, wodurch die Verarbeitung der Erblindung möglicherweise auch behindert werden kann.

Einen weiteren wichtigen, bisher noch nicht untersuchten Aspekt betrifft die heute vielfach praktizierte Laserkoagulation bei PDR: Wie wird die Compliance des Diabetikers durch die Lasertherapie beeinflußt, wenn er plötzlich wieder besser sehen kann, oder wie sind die psychischen Auswirkungen, wenn ein Diabetiker durch die Lasertherapie erblindet (Wulsin et al., 1987)?

47.2 Typ-II-Diabetes

„James Möllendorpf, der älteste kaufmännische Senator, starb auf groteske und schauerliche Weise. Diesem diabetischen Greise waren die Selbsterhaltungsinstinkte so sehr abhanden gekommen, daß er in den letzten Jahren seines Lebens mehr und mehr einer Leidenschaft für Kuchen und Torten unterlegen war. Dr. Grabow, der auch bei Möllendorpfs Hausarzt war, hatte mit aller Energie, deren er fähig war, protestiert, und die besorgte Familie hatte ihrem Oberhaupte das süße Gebäck mit sanfter Gewalt entzogen. Was aber hatte der Senator getan? Geistig gebrochen, wie er war, hatte er sich irgendwo in einer unstandesgemäßen Straße, in der kleinen Gröpelgrube, An der Mauer oder Im Engelswisch, ein Zimmer gemietet, eine Kammer, ein wahres Loch, wohin er sich heimlich geschlichen hatte, um Torte zu essen ... und dort fand man auch den Entseelten, den Mund noch voll halbzerkauten Kuchens, dessen Reste seinen Rock befleckten und auf dem ärmlichen Tische umherlagen. Ein tödlicher Schlaganfall war der langsamen Auszehrung zuvorgekommen.

Die widerlichen Einzelheiten dieses Todesfalles wurden von der Familie nach Möglichkeit geheimgehalten; aber sie verbreiteten sich rasch in der Stadt und bildeten den Gesprächsstoff an der Börse, im Club, in der „Harmonie", in den Comptoirs, in der Bürgerschaft und auf den Bällen, Diners und Abendgesellschaften ..." (Thomas Mann „Buddenbrooks", 1960).

47.2.1 Definition

Auf die aktuellen Kriterien der Diabetesdefinition unabhängig vom Diabetestyp wurde im Abschnitt 47.1 „Typ-I-Diabetes" bereits eingegangen. Typische klinische, laborchemische, morphologische und genetische Merkmale unterscheiden den Typ-I-Diabetes vom Typ-II-Diabetes.

Zu den schon lange bekannten klinischen Merkmalen des Typ-II-Diabetes gehört, daß die Erkrankung meistens im Erwachsenenalter auftritt, die Mehrzahl der Betroffenen übergewichtig ist, die klinischen Symptome häufig spärlich sind und zum Teil bereits vorliegende Folgeerkrankungen des Diabetes betreffen, keine Bereitschaft zur ketoazidotischen Stoffwechselentgleisung besteht und die Behandlung mit Insulin nicht lebensnotwendig ist.

Zu den laborchemischen Merkmalen des Typ-II-Diabetes gehört eine während des gesamten Verlaufs der Erkrankung erhaltene Restsekretion der Beta-Zellen der Langerhansschen Inseln. Beim Typ-II-Diabetes können Autoantikörper gegen Zellen der Langerhansschen Inseln nicht häufiger als bei Nichtdiabetikern nachgewiesen werden, auch besteht keine überdurchschnittlich häufige Assoziation mit sonstigen Immunphänomenen oder Autoimmunerkrankungen. Ein besonderes Kennzeichen des Typ-II-Diabetes ist eine verminderte Insulinempfindlichkeit.

Morphologisch finden sich nur geringe und wenig spezifische Unterschiede zwischen den Langerhansschen Inseln von Patienten mit Typ-II-Diabetes und von Nichtdiabetikern. Bei den Patienten mit Typ-II-Diabetes liegt im Vergleich mit Nichtdiabetikern entsprechenden Alters eine etwas verstärkte Fibrose und

Hyalinose der Langerhansschen Inseln vor, die Anzahl der Beta-Zellen ist absolut und relativ zu den anderen endokrinen Zellen der Langerhansschen Inseln geringfügig vermindert, schließlich sprechen neueste Beobachtungen für eine etwas eingeschränkte funktionelle Aktivität der Beta-Zellen bei Patienten mit Typ-II-Diabetes (Gepts, 1984).

Genetisch besteht beim Typ-II-Diabetes eine höhere familiäre Belastung als beim Typ-I-Diabetes, eine Assoziation zu bestimmten Eigenschaften des HLA-Systems wurde bisher allerdings nicht beobachtet.

Bei Patienten mit Typ-II-Diabetes können individuelle Unterschiede bezüglich der aufgeführten klinischen, laborchemischen, morphologischen und genetischen Merkmale des Typ-II-Diabetes bestehen. Diese Tatsache weist darauf hin, daß auch der Typ-II-Diabetes mit Sicherheit ein Syndrom ist, das sich aus mehreren heterogenen Krankheitsbildern zusammensetzt.

Da die aufgezählten vielfältigen Unterscheidungsmerkmale zwischen Typ-I- und Typ-II-Diabetes keinesfalls in jedem Einzelfall vollständig erfüllt sein müssen, kann auch die Unterscheidung zwischen beiden Diabetestypen Schwierigkeiten machen.

47.2.2 Epidemiologie

Der Typ-II-Diabetes ist eine weltweit verbreitete Erkrankung. Allerdings werden für Patienten verschiedener Rasse und ethnischer Herkunft zum Teil große Unterschiede bezüglich des Vorkommens des Typ-II-Diabetes beobachtet. Diese Unterschiede finden sich auch für Patienten unterschiedlicher ethnischer Herkunft, die im selben Land unter ähnlichen Bedingungen leben. Andererseits zeigten epidemiologische Untersuchungen große Unterschiede für das Vorkommen des Typ-II-Diabetes bei Patienten gleicher ethnischer Herkunft in Abhängigkeit davon, ob sie auf dem Land oder in der Stadt leben. So fand sich bei einer epidemiologischen Untersuchung auf den Fidji-Inseln für die dort lebende Bevölkerung melanesischer Herkunft eine Diabeteshäufigkeit von 1,9% in ländlicher und 7,0% in städtischer Umgebung, während ebenfalls auf den Fidji-Inseln lebende Inder im ländlichen Bereich in 13,1% und im städtischen Bereich in 14,0% einen Diabetes zeigten.

Große Unterschiede in der Diabeteshäufigkeit fanden sich auch für ein und dieselbe ethnische Gruppe zwischen im Heimatland lebenden und in andere Gegenden der Welt ausgewanderten Menschen. So z.B. zwischen Japanern in Hiroshima und auf Hawaii.

Schließlich ergaben epidemiologische Untersuchungen zum Teil erhebliche Unterschiede bezüglich der Häufigkeit des Vorkommens des Typ-II-Diabetes in Abhängigkeit vom sozioökonomischen Status innerhalb derselben Bevölkerungsgruppe. Besonders eindrucksvoll konnte Himsworth (1949) für die Diabetesmortalität in England und Wales in der ersten Hälfte des 20. Jahrhunderts große Schwankungen in Abhängigkeit davon zeigen, ob Krieg oder Frieden herrschte und ob die Wirtschaft florierte oder in einer Krise steckte.

Die dargestellten Daten (West, 1978; Zimmet, 1983) belegen eindrucksvoll die Abhängigkeit des Typ-II-Diabetes von einerseits genetischen Faktoren und andererseits Einflüssen der Umwelt.

Nach mehreren Erhebungen der jüngeren Zeit, die von Mitarbeitern des Instituts für Sozialmedizin und Epidemiologie des Bundesgesundheitsamtes an Bevölkerungsgruppen in der Bundesrepublik durchgeführt wurden, dürften 3% der Gesamtbevölkerung an einem Typ-II-Diabetes erkrankt sein (Thefeld und Hoffmeister, 1982).

47.2.3 Ätiologie und Pathogenese

Bei der Entstehung des Typ-II-Diabetes wirken mehrere Faktoren zusammen. Eine schematische Darstellung dieser Faktoren ist in Abbildung 47–1 gegeben. Neben der Genetik sind es vor allem Umwelteinflüsse, die zur Manifestation des Typ-II-Diabetes beitragen. Im folgenden soll auf einzelne Punkte der Abbildung 47–1 eingegangen werden.

Genetik

Bei eineiigen Zwillingen von Patienten mit Typ-II-Diabetes tritt in nahezu 100% der Fälle ebenfalls ein Typ-II-Diabetes auf (Barnett et al., 1981a, b). Die hohe Konkordanz findet sich trotz zum Teil erheblicher Unterschiede in bezug auf die Lebensumstände der betroffenen Zwillingspartner. Dieses Ergebnis der Zwillingsstudien spricht für eine wichtige Rolle der Genetik bei der Entstehung des Typ-II-Diabetes.

Auch hier besteht kein einfacher Erbmodus. Offensichtlich besteht innerhalb der Gruppe des Typ-II-Diabetes Heterogenität. So scheint das Erbrisiko für Patientengruppen mit unterschiedlichem Gewicht ebenfalls unterschiedlich zu sein (Köbberling, 1976).

Für eine Untergruppe des Typ-II-Diabetes, den sog. Maturity Onset Diabetes of the Young (MODY), konnte inzwischen ein autosomal-dominanter Erbgang nachgewiesen werden (Tattersall, 1974). Die Patienten mit dieser Unterform des Typ-II-Diabetes erkranken bevorzugt bereits im jugendlichen Alter.

Es ist bekannt, daß sich das Insulin-Gen beim Menschen auf dem kurzen Arm des Chromosoms 11 befindet. Neuere Untersuchungen sprechen dafür, daß die Region des Insulin-Gens bei Patienten mit Typ-II-Diabetes Unterschiede im Vergleich mit Nichtdiabetikern zeigt (Owerbach und Nerup, 1982).

Störung der Insulinsekretion

Noch unter Anwendung biologischer Methoden zur Messung der Insulinkonzentration im Blut beobachteten Pfeiffer und Mitarbeiter (1959) eine „Starre der Insulinsekretion" bei Patienten mit Typ-II-Diabetes. Die radioimmunologische Insulinbestimmung ergab in der Folgezeit eine Störung der initialen Sekretion von Insulin vor allem nach Stimulierung mit Glukose

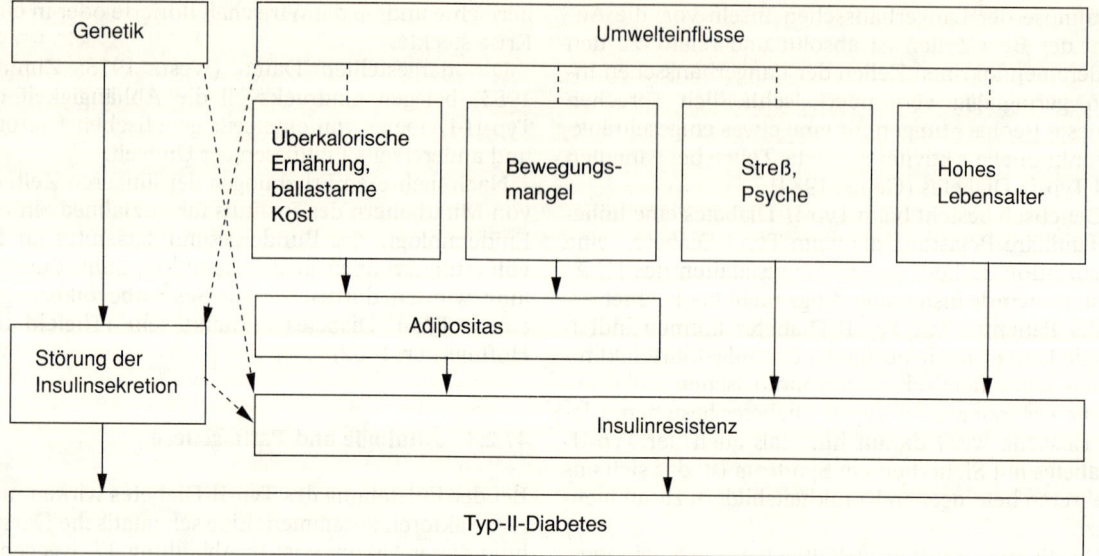

Abb. 47–1. Pathogenetische Faktoren beim Typ-II-Diabetes.

als besonderes Charakteristikum bei Typ-II-Diabetes (Kipnis, 1968; Cerasi und Luft, 1976). Die eingeschränkte initiale Sekretion der Beta-Zellen des Pankreas bestätigte sich auch bei Messung von C-Peptid anstelle von Insulin (Beischer, 1983; Beischer et al., 1984).

Auch hinsichtlich der gestörten Sekretion der Beta-Zellen zeigen Patienten mit Typ-II-Diabetes individuelle Unterschiede.

So konnten Fajans et al. bei nicht adipösen Patienten mit Typ-II-Diabetes zwischen Patienten mit insgesamt geringer Insulinantwort und mit ausgeprägter Insulinantwort unterscheiden.

Insulinresistenz

Als Insulinresistenz bezeichnet man einen Zustand der eingeschränkten Wirksamkeit von Insulin in vivo. Die in Frage kommenden Ursachen für eine Insulinresistenz sind in Tabelle 47–3 zusammengestellt.

Unter I. finden sich in der Tabelle Defekte am Erfolgsorgan. Voraussetzung für die biologische Wirkung von Insulin ist dessen Bindung an Rezeptoren der Zelloberfläche. Diese Rezeptoren wurden in den vergangenen Jahren ausgiebig untersucht und sind in ihrer Struktur teilweise aufgeklärt. Die auf die Bindung an den Rezeptor folgenden Stoffwechselvorgänge, die schließlich die bekannten Insulinwirkungen vermitteln, sind bisher nur wenig bekannt. Der Verlauf von Dosiswirkungskurven von Insulin in vivo gestattet eine Unterscheidung zwischen Defekten des Insulinrezeptors und Defekten im nachgeschalteten Stoffwechsel (Postrezeptordefekten).

In Gegenwart hoher Insulinkonzentrationen kommt es zur Verminderung der Rezeptorzahl auf der Zelloberfläche, es besteht also ein Insulinrezeptordefekt. Hohe Insulinkonzentrationen finden sich in der Regel bei überkalorischer Ernährung und bei

Adipositas. Im Gefolge davon findet sich demnach eine rezeptorbedingte Insulinresistenz (Olefsky und Kolterman, 1981; Kolterman et al., 1982). Die Mehrzahl der Patienten mit Typ-II-Diabetes sind übergewichtig. Die Insulinresistenz infolge des Übergewichts stellt bei diesen Patienten einen wichtigen Faktor in der Pathogenese des Diabetes dar.

Solange die Störung der Glukosetoleranz bei Normalgewichtigen oder Übergewichtigen nur geringfügig ist, ist die ebenfalls meist nur geringfügige Insulinresistenz in der Regel Folge eines Rezeptordefekts. Mit zunehmender Nüchternhyperglykämie nimmt auch die Insulinresistenz zu, Rezeptordefekt und Postrezeptordefekt bestehen jetzt nebeneinander, dem Postrezeptordefekt kommt allerdings mit Zunahme der Hyperglykämie eine immer größere Rolle zu. Durch bessere Einstellung der Glukose und Gabe von Insulin läßt sich der Postrezeptordefekt wieder beseitigen (Olefsky und Kolterman, 1981; Kolterman et al., 1982).

Bewegungsmangel begünstigt die Gewichtszunahme und kann damit indirekt zur Insulinresistenz bei-

---

Tabelle 47–3. Ursachen der Insulinresistenz (modifiziert nach Kolterman et al., 1982)

I. Störung am Erfolgsorgan
  – Insulinrezeptorstörung
  – Postrezeptorstörung
II. Abnormes Sekretionsprodukt der Beta-Zellen
  – abnormes Insulin
  – unvollständige Spaltung des Proinsulins zu Insulin
III. Antagonisten des Insulins im Blutkreislauf
  – Antikörper gegen den Insulinrezeptor
  – Antikörper gegen Insulin
  – erhöhte Spiegel gegenregulatorischer Hormone (z.B. Wachstumshormon, Cortisol, Glukagon, Katecholamine)

tragen. Andererseits wird in der Literatur auch über einen direkten begünstigenden Einfluß der körperlichen Bewegung auf die Bindung von Insulin an seine Rezeptoren in der Muskulatur berichtet (Koivisto et al., 1979).

Der Typ-II-Diabetes ist eine Erkrankung des Erwachsenen und insbesondere des älteren Menschen. In den letzten Jahren hat sich die Insulinresistenz nun auch als ein charakteristisches Merkmal des Alterns erwiesen (DeFronzo, 1979). Offensichtlich handelt es sich vor allem um einen Postrezeptordefekt. Dieser Postrezeptordefekt scheint – jedenfalls an isolierten Fettzellen – vor allem den Glukosetransport zu betreffen (Fink et al., 1984).

Eine Insulinresistenz kann bestehen, wenn die Beta-Zellen des Pankreas ein Insulin sezernieren, das biologisch nicht voll aktiv ist (vgl. Pkt. II, Tab. 47–3). Zwei derartige Zustände können heute unterschieden werden: die Sekretion eines abnormen Insulins und die unvollständige Spaltung des biologischen Insulinvorläufers Proinsulin.

In der Weltliteratur sind bisher drei Fälle bekanntgeworden, bei denen der Nachweis geführt werden konnte, daß das zirkulierende Insulin in seiner Struktur nicht dem üblichen menschlichen Insulin entspricht. Alle drei Fälle zeichneten sich durch Hyperglykämie, radioimmunologisch nachweisbaren Hyperinsulinismus und normale Empfindlichkeit gegenüber von außen zugeführtem Insulin aus. In einem Fall konnte gezeigt werden, daß die in Position 25 der B-Kette des menschlichen Insulins befindliche Aminosäure Phenylalanin gegen Leucin ausgetauscht war (Schoelson et al., 1983).

Bisher gibt es ebenfalls drei Berichte in der Weltliteratur über Familien, bei denen erhöhte Spiegel des Proinsulins im Blut gefunden wurden. Offensichtlich besteht eine Störung der Umwandlung des Proinsulins zu Insulin, die in den Beta-Zellen des Pankreas erfolgt. Bei zwei dieser Familien dürfte der Grund für die Hyperproinsulinämie in einer abnormen Struktur des Proinsulins liegen, im dritten Fall scheint bei normaler Struktur des Proinsulins die enzymatische Umwandlung des Prohormons in Insulin betroffen zu sein (Gruppuso et al., 1984).

Mit großer Wahrscheinlichkeit wird sich in weiteren Fällen ein abnormes Sekretionsprodukt der Beta-Zellen als Ursache für eine Insulinresistenz und in deren Gefolge für einen Diabetes des Typs II nachweisen lassen.

Der III. Punkt der Tabelle 47–3 betrifft Antagonisten des Insulins im Blutkreislauf als Ursache für eine Insulinresistenz. Antikörper gegen den Insulinrezeptor sind eine ausgesprochen seltene Ursache dieser Form der Insulinresistenz.

Eine wichtige Ursache für eine Insulinresistenz können Antikörper gegen Insulin sein. Sie treten in der Regel nur bei vorausgegangener Behandlung mit Insulin auf und finden sich deshalb vor allem bei Patienten mit Typ-I-Diabetes.

Diese Form der immunologischen Insulinresistenz wird bei Verwendung hochgereinigter Insuline und insbesondere hochgereinigter Humaninsuline in Zukunft eine immer geringere Rolle spielen.

Erhöhte Spiegel gegenregulatorischer Hormone führen bei endokrinen Erkrankungen mit Überproduktion dieser Hormone zum sekundären Diabetes, so z.B. bei der Akromegalie. Eine Überproduktion dieser Hormone – vor allem der Katecholamine – kann aber auch im Gefolge von Streß auftreten.

Psychische Faktoren

Patienten mit Übergewicht und Altersdiabetes sind – nach Cremerius (1978) ihr ganzes Leben hindurch an das Essen, das als etwas Zwanghaftes und Süchtiges von diesen Patienten erlebt wurde, fixiert. Essen dient dabei als Kompensation für frühkindlichen Mangel an Liebe und Fürsorge. Darüber hinaus kann ein bestimmtes Erziehungsverhalten im Sinne einer „overprotection" notwendige Schritte zur Individuation eines Kindes verhindern. Das Kind verzichtet dann auf diese Befriedigung und wählt statt dessen eine andere Befriedigungsform, nämlich die des Essens. Die Nahrungsaufnahme wird stark mit der Beziehung zur Mutter verknüpft, Essen bedeutet im übertragenen Sinne mütterliche Zuwendung und Liebe. Bleibt der Patient fixiert in einer Beziehung zum mütterlichen Objekt, wird die Angst vor dem Verlust der mütterlichen Zuwendung zum Trauma. Cremerius findet bei den Typ-II-Diabetikern eine seelisch bedingte Freßsucht, deren Ursache in einer neurotischen Grundstörung liegt, die durch starke depressive Strukturanteile charakterisiert wird. Im Gegensatz zu den Adipösen ohne Diabetes mellitus gelingt es den diabetischen übergewichtigen Patienten weit weniger, den neurotischen Grundkonflikt zu kompensieren, z.B. mit dem Symptom Adipositas. Folge dieser Instabilität sei nun frei flottierende Angst, die zu einer Hypersekretion von Adrenalin führt und eine langanhaltende Hyperglykämie bewirken kann. Cremerius versteht die bei diesen Patienten bereits prämorbid vorhandene „Freß-Fettsucht" als Körpersymptom, das dem Kranken helfen soll, die tieferliegende Depression zu bewältigen. Der Diabetes mellitus wird dann manifest, wenn es dem Eßtrieb nicht mehr gelingt, den neurotischen Grundkonflikt zu kompensieren: „Die Dekompensation des Freß-Fettsucht-Gleichgewichtes wird durch sympathikotone Angstzustände und paranoische Kampfstimmungen, welche durch bestimmte Konfliktsituationen aus der Latenz in die aktuelle Intensität überführt werden, verursacht" (Cremerius, 1978).

Aber auch andere Faktoren, z.B. Objektverluste, sollen in der Pathogenese des Typ-II-Diabetes eine Rolle spielen. So zeigten Slawson und Mitarbeiter (1963), daß bei 20 von 25 erwachsenen Diabetikern ein realer oder drohender Objektverlust dem Manifestwerden des Diabetes voranging. Bei 10 Patienten fanden die Autoren eine unvollendete Trauerreaktion, bei 14 eine emotionale Verwahrlosung. 5 Patienten waren stark depressiv.

Groen und de Loos (1973) beschreiben in einer Monographie, daß bei Diabetikern vor Ausbruch der Krankheit häufig der Verlust einer wichtigen Bezugsperson oder ein Liebesentzug mit anschließendem Gefühl von Einsamkeit, Traurigkeit und Nichtverstandensein gefunden werden.

Aimez und Mitarbeiter (1976) fanden bei Diabetikern konfliktgeladene Beziehungen zur Nahrungsaufnahme. Häufig hätten sie Schwierigkeiten, sich über ihre Abhängigkeits- und Unabhängigkeitsbedürfnisse Rechenschaft zu geben. Bei instabilen Diabetikern würden chronische Unterdrückung von Schuld, Angstgefühlen, Feindseligkeit sowie latente Depression dem Ausbruch des Diabetes mellitus vorangehen.

Bruni (1976) untersuchte und interviewte 1200 erwachsene Diabetiker. Er fand 139 Patienten, bei denen psychische Faktoren bei der Entstehung oder im Verlauf des Diabetes mellitus eine Rolle spielten.

47.2.4 Klinik

Der Beginn des Typ-II-Diabetes verläuft in der Regel schleichend. Polydipsie und Polyurie stehen nur selten im Vordergrund. Die Erkrankung kann durch Infektionen im Bereich von Haut und Schleimhaut, so vor allem durch Pilzinfektionen im Genitalbereich auffallen. Häufig führen erst bereits vorliegende Komplikationen des Diabetes zu Beschwerden. So kann sich hinter einer Potenzstörung eine diabetische Neuropathie verbergen, ein Nachlassen der Sehleistung kann Folge einer diabetischen Retinopathie sein oder der Typ-II-Diabetes wird anläßlich eines Herzinfarktes oder eines Schlaganfalls festgestellt. Nicht selten handelt es sich um einen Zufallsbefund anläßlich einer Routineuntersuchung.

Die entscheidende therapeutische Maßnahme ist die Verordnung und Einhaltung einer Diät, erforderlichenfalls einer Reduktionsdiät. Zwischen 60 und 80% der Patienten mit Typ-II-Diabetes sind übergewichtig (West, 1978).

Durch Gewichtsreduktion und damit einhergehender Beseitigung des erhöhten Insulinbedarfs und der Insulinresistenz gelingt es häufig, den Diabetes vollständig zum Verschwinden zu bringen, auch wenn sich eine gleichzeitig bestehende Störung der Insulinsekretion durch Gewichtsreduktion nicht beheben läßt. Erst wenn die Möglichkeiten der Gewichtsreduktion ausgeschöpft sind, oder wenn von vornherein kein Übergewicht bestand, können diätetische Maßnahmen durch eine medikamentöse Therapie ergänzt werden. Zur oralen Therapie bieten sich vor allem die Sulfonylharnstoffe an. Entscheidend für ihre Wirkung ist der stimulierende Einfluß auf die erhaltene Insulinsekretion (Pfeiffer, 1983). Die Glukoseaufnahme aus dem Darm kann auch durch Ballaststoffe und Quellstoffe verzögert werden (Huth und Bräuning, 1983). Solchen Maßnahmen kommt in der Regel allerdings nur additive Bedeutung zu.

Das Ausmaß der Insulinsekretion bei Diagnosestellung und unter Stimulierung vor allem mit Glukose ist für den Verlauf der Erkrankung von prognostischer Bedeutung. Patienten mit zwar verzögerter, aber reichlicher Insulinsekretion werden sehr selten insulinbedürftig. Patienten mit verzögerter und insgesamt eingeschränkter Insulinsekretion können nach einer Erkrankungsdauer von in der Regel mehreren Jahren durchaus insulinbedürftig werden. In besonderen Streßsituationen besteht für diese Patienten die Gefahr einer ketoazidotischen Stoffwechselentgleisung.

Das medizinisch größte Problem besteht beim Typ-II-Diabetes ebenso wie beim Typ-I-Diabetes in der Verhinderung der Spätkomplikationen. Hier spielt die Makroangiopathie, d.h. die Arteriosklerose mit ihren vielfältigen Folgeerkrankungen, eine besonders große Rolle.

Von großem Interesse ist, daß bereits bei pathologischer Glukosetoleranz ein erhöhtes Risiko für makroangiopathische Komplikationen besteht (Jarrett et al., 1982).

Bei den Patienten mit Typ-II-Diabetes kommen Hypertonie und Hyperlipidämie, zwei klassische Risikofaktoren für die Arteriosklerose, gleichzeitig überdurchschnittlich häufig vor. Dem gemeinsamen Vorkommen von Diabetes und Hypertonie wird eine potenzierende Auswirkung auf die Entstehung makroangiopathischer Komplikationen zugesprochen (Fuller et al., 1983).

Auch wenn die Beziehungen zwischen Hyperglykämie und Makroangiopathie weniger eindeutig sind als zwischen Hyperglykämie und Mikroangiopathie, so muß doch eine möglichst normoglykämische Diabetesführung auch im Interesse der Prophylaxe der Makroangiopathie gefordert werden (Beischer und Pfeiffer, 1985a, b). Der frühzeitigen Erkennung und Behandlung von Hypertonie und Hyperlipidämie kommt besondere ergänzende Bedeutung zu.

47.3 Allgemeine Hinweise

47.3.1 Krankheitsverlauf

Der Diabetes mellitus ist eine chronische Erkankung, bei der ungewiß ist, ob und wann Komplikationen auftreten. Je nach Persönlichkeitsstil des Patienten kann die Diagnose eines Diabetes mellitus ein mehr oder weniger schweres psychisches Trauma darstellen. Das Ziel einer erfolgreichen Diabetesbehandlung ist zunächst eine gute Abstimmung der körperlichen Tätigkeit, der Kalorienzufuhr und der Insulindosis. Schon dies ist nicht immer eine einfache Aufgabe, wie Loebert (1972) zeigte. Er fand, daß nur 17% der Diabetiker fähig waren, aufgrund ihrer Kenntnisse eine Diät einzuhalten. Nur 18% der Diabetiker konnten zwei oder mehr Symptome eines Coma diabeticum nennen, 24% konnten zwei Schocksymptome aufzählen und 97% waren fähig, sich ihrer zuletzt injizierten Insulindosis zu erinnern. Diese Zahlen lassen erkennen, wie wenigen Patienten eine Überwindung des Traumas, krank zu sein und eine realitätsgerechte Auseinandersetzung hiermit gelungen ist. Williams (1973) beschreibt folgende Reaktionsmuster bei chronisch erkrankten Patienten: Depression, Aggression, Ängstlichkeit, Abhängigkeit, Schuldgefühle und Hypochondrie.

Grant und Mitarbeiter (1974) untersuchten 37 erwachsene Diabetiker auf Zusammenhänge zwischen

„life-events" während der letzten 8 bis 18 Monate vor Änderungen in der Diabeteseinstellung. Er fand, daß vor allem unerwünschte „life-events" signifikant häufig zu einem Wechsel der Diabeteseinstellung führten.

47.3.2 Die Arzt-Patient-Beziehung

Die Grundlage für eine gute Arzt-Patient-Beziehung kann in einem patientorientierten Interview geschaffen werden. Dieses sollte offen geführt werden und dem Patienten die Möglichkeit geben, über seine Empfindungen, Gedanken zur Krankheit und Ängste zu sprechen (Herrmann und Schüffel, 1983; vgl. Kap. 12). Der Kliniker erhält so in einem Arbeitsgang Einsicht in psychologische, somatische und persönliche Daten des Patienten und deren Zusammenwirken. Die Art und Weise, wie einem Patienten mitgeteilt wird, daß er Diabetiker ist, kann darüber entscheiden, wie erfolgreich die Betreuung des Patienten sein wird. Auch hier lohnt es sich, dem Patienten immer wieder Gelegenheit zu geben, über seine Gefühle, Ängste und Befürchtungen zu sprechen; dabei sollte der Arzt vor allem der Hörer sein, der Patient der Sprecher. Man muß sich vor Augen halten, daß der Patient einige Zeit braucht, um die neue Tatsache, Diabetiker zu sein, zu akzeptieren. Während dieser Zeit stellt man mannigfaltige Reaktionen wie Verleugnung der neuen Diagnose, Aggression und Feindseligkeit oder auch blinde Unterwürfigkeit gegenüber dem Arzt fest. In dieser Phase ist es z.B. besser, vom Patienten einen Protest in Worten entgegenzunehmen, als später denselben in Form von Diätfehlern oder falschen Insulindosen zu erleben. In dieser Phase haben Patienten häufig starke Schuldgefühle gegenüber dem Arzt wegen ihrer Aggressionen. Man sollte nicht vergessen, daß es sich hier um normale psychische Reaktionen handelt, und daß es für den Patienten besser ist, wenn er seine Aggressionen gegenüber dem Arzt ausdrücken kann, als wenn er diese gegenüber seiner Familie oder in pathologischem Verhalten in der Therapie auslebt. Dabei ist es für den Arzt nicht leicht, solche Aggressionen zu ertragen; wenn er aber weiß, daß diese dem Schicksal gelten und nicht persönlich gemeint sind, lassen sie sich leichter aushalten. Spürt man, daß Patienten Aggressionen haben, lohnt es sich, diese für den Patienten in Worte zu fassen, um einer Aufstauung vorzubeugen.

Es muß auch darauf geachtet werden, wieviel ein Patient in einer Sitzung an theoretischem Wissen aufzunehmen fähig ist. Ängste, wie sie beim Bekanntwerden eines Diabetes meistens auftreten, können gewisse Patienten völlig unfähig machen, Wissen über die neue Krankheit aufzunehmen. Es bedarf des Gesprächs, dosierter Aufklärung und Zeit, um dem Patienten das für die Behandlung notwendige Wissen zu vermitteln. Es hat sich dabei immer wieder gezeigt, daß Lob und positive Kritik bessere Stimuli abgeben als Tadel und Strafe.

Ängste und Befürchtungen über den Diabetes, dessen Behandlung, Prognose und Komplikationen müssen diskutiert werden, sie können sonst neuen psychischen Störungen Vorschub leisten, die dann wiederum auf die Einstellung des Diabetes wirken. Nicht nur der Patient, auch der Arzt hat Ängste und Unsicherheiten in bezug auf Prognose und Komplikationen der Krankheit. Erst wenn der Arzt sich dessen bewußt wird und sich damit auseinandersetzt, wird er dem Patienten helfen können, mit diesen Befürchtungen zu leben, sie zu begreifen und zu meistern. Es lohnt sich, auch auf Phantasien der Patienten einzugehen – gleichgültig, ob diese vom Arzt aus gesehen falsche oder richtige Vorstellungen beinhalten – und diese nicht einfach mit rationalen Überlegungen abzutun. Für viele Patienten ist es wichtig, von Autoritätsfiguren immer wieder die Bestätigung zu erhalten, daß sie sich korrekt verhalten und gut bei der Therapie ihrer Krankheit mithelfen.

Es können aber auch unterschiedliche, für den Patienten selbst nicht vereinbare Wünsche vorhanden sein und in der Beziehung wirksam werden (wie bei unserer oben beschriebenen Patientin).

Für den Arzt ist es nicht immer leicht, diese Abhängigkeits- oder Kontrollbedürfnisse der Patienten zu ertragen. Eine weitere schwere Aufgabe für den Arzt besteht darin, auch dann zum Patienten zu stehen, wenn die Therapie schlechter geht. Nur zu oft tritt beim Arzt das Gefühl auf, er habe versagt und er sei unfähig, diesen Patienten erfolgreich zu behandeln, obwohl es sich beim Diabetes um eine chronische Erkrankung handelt, bei der kein großes Erfolgserlebnis für den Arzt möglich ist.

Beim erwachsenen Diabetiker (Typ I und II) treten vor allem Fragen über das Wann und Wo des Auftretens von Komplikationen des Diabetes mellitus in den Vordergrund, zusammen mit den sich daraus ergebenden Problemen in bezug auf Sexualität, berufliches Weiterkommen, Unabhängigkeit, Familienplanung. Man soll dem Patienten immer wieder die Möglichkeit geben, über diese Sorgen zu sprechen, wobei es ebenfalls wichtig ist, im Einverständnis mit dem Patienten auch seinem Ehepartner Gelegenheit zu geben, über seine Probleme um und mit dem Patienten zu sprechen.

Beim Altersdiabetiker entstehen häufig Schwierigkeiten, ihn über seine Krankheit genügend aufzuklären. Bei dieser Altersgruppe ist es speziell wichtig, die Informationen den geistigen Fähigkeiten des Patienten anzupassen. Hier muß vor allem die Umgebung des Patienten genügend über die Belange des Diabetes orientiert werden, aber auch über Art und Weise, wie man dem Patienten helfen kann, eine gute Diabeteseinstellung zu erreichen. Das fehlende Frischgedächtnis und der mangelnde Realitätssinn seniler Menschen machen sie häufig von ihrer Umgebung vollständig abhängig.

47.3.3 Methodische Probleme der psychosomatischen Forschung

Retrospektive Untersuchungen, die dazu dienen sollen, psychologische Faktoren zu finden, die bei der

Ätiologie des Diabetes von Bedeutung sein sollen, sind aus methodischen Gründen problematisch (vgl. Kap. 25). Zwar können retrospektive Studien zeigen, daß ein Zusammenhang zwischen psychischen Faktoren und der Diabeteserkrankung besteht, über das Maß der Erhöhung des Krankheitsrisikos allerdings kann durch diese Untersuchungen nichts ausgesagt werden. Nur wenn die psychischen Merkmale als unabhängige und die somatische Erkrankung als abhängige Variable erhoben werden, lassen sich letztlich gültige Aussagen über die Beteiligung psychischer und sozialer Faktoren am Diabetes treffen. Lediglich prospektive Untersuchungen sind geeignet, psychische Faktoren zu identifizieren, die an der Genese des Diabetes beteiligt sind. Daher müssen unabhängige Prädiktor-Variablen, die einen psychischen Risikofaktor darstellen, definiert werden. Gleichzeitig müssen die bekannten somatischen Risikofaktoren miterhoben werden. In einem derartigen Plan wäre

dann die Erkrankung die abhängige Variable und der „wirkliche" Einfluß der Prädiktor-Variablen ließe sich feststellen, vorausgesetzt die Prädiktor-Variablen können in ihrem Einfluß als konstant angesehen werden. Eine prospektive Untersuchung müßte epidemiologisch angelegt sein, d.h. an einer repräsentativen Stichprobe aus der Normalbevölkerung müßte die Koinzidenz von psychischen Faktoren und Diabetes geprüft werden. Im Bereich der psychosomatischen Diabetes-Forschung existiert keine derartige Studie.

Das Akzeptieren seiner chronischen Krankheit ist für jeden Diabetiker eine sehr schwierige psychische Aufgabe, bei der sowohl der Patient als auch seine Familie mit Depressionen, Aggressionen, Ängsten, Schuldgefühlen oder Hypochondrie reagieren können. Der Arzt kann dem Patienten entscheidend helfen, diese reaktiven Vorgänge zu überwinden.

48 Infektionskrankheiten

Jörg Michael Herrmann und *Werner Geigges*

48.1 Falldarstellung

Im folgenden wird eine Kasuistik vorgestellt, in der sich die Probleme eines Patienten mit einer Infektionskrankheit in besonderer Eindringlichkeit darstellen.

Ein untergewichtiger, 37jähriger Patient wird für ein Heilverfahren in eine Rehabilitationsklinik eingewiesen; durch Intervention des Hausarztes und der Rentenversicherung wurde der Patient im Eilverfahren aufgenommen, trotz üblicher Wartezeiten zwischen 4 bis 6 Monaten, und zwar unter der Einweisungsdiagnose „Depressiver Einbruch mit phobischer Komponente". Aufgrund dieser Diagnose war er bereits seit über 3 Monaten arbeitsunfähig krankgeschrieben.

Der Patient schildert seit einem Dreivierteljahr wiederholte Phasen von „Depressionen" und seit ca. 6 Monaten Appetitlosigkeit mit Gewichtsabnahme von 5 Kilogramm. Nach einer Kieferbehandlung wegen eines „Granuloms" vor einigen Monaten leidet er zunehmend unter innerer Unruhe und Angstzuständen mit Schweißausbrüchen. Er schildert sich als aggressiv gegen sich selbst, unbeherrscht, spontan explosiv und fast handgreiflich gegenüber den Arbeitskollegen; hinzu kommen Konzentrations- und Gedächtnisstörungen. Weiterhin klagt er über Schmerzen im Bereich des Rückens und gelegentlichen Schwindel, insbesondere bei Nikotinkonsum, sowie über leichten Reizhusten und häufige „Halsbeschwerden". Überhaupt traten in den letzten Monaten gehäuft Erkältungskrankheiten auf.

Seit dem 18. Lebensjahr leidet er an rezidivierenden Ulcera duodeni und Magenschleimhautentzündungen als Folge eines zunehmenden Alkoholabusus, den der Patient mit dem Tod der Mutter in Zusammenhang bringt. Sie starb, als er 15 Jahre alt war. Mit 27 Jahren war er stationär zum Alkoholentzug, seither ist er „trocken". Vom 30. bis 36. Lebensjahr trat immer wieder eine Hypothyreose bei Thyreoiditis auf. Vor einem Jahr wurde er wegen einer makrozytären Anämie, Vitamin-B$_{12}$-Resorptionsstörung und Verdacht auf eine Autoimmunerkrankung stationär behandelt.

Kurz zuvor wurde die Diagnose einer idiopathischen familiären Alopezie gestellt, weswegen er ein Toupet trägt.

Die Mutter des Patienten starb mit 44 Jahren an einem Hirntumor. Kurz darauf heiratete der heute 68-jährige Vater zum zweiten Mal. Wegen Auseinandersetzungen mit seiner Stiefmutter wurde der Patient mit 16 Jahren in einem Heim untergebracht. Damit begann für den Patienten offensichtlich eine fast 15jährige Alkoholkarriere und lange Phase sozialer Desintegration. Schon nach kurzer Zeit wurde der Patient aus dem Heim wieder entlassen, das Sozialamt übernahm seine Betreuung und er wurde bei einer alleinstehenden Frau untergebracht. In dieser Zeit begann der Patient eine Lehre als Bäcker und Konditor, die er aber wegen einer Mehlstauballergie abbrach. Danach arbeitete er als Hilfsarbeiter (Lagerist), bekam zunehmend Probleme wegen seines Alkoholismus und wurde schließlich entlassen. Nach der Alkoholentwöhnung begann er eine Ausbildung zum Krankenpfleger und Masseur, wurde aber nach einem dreivierteljährigen Praktikum vom Arbeitsamt nicht mehr weiter gefördert. Der Patient hat eine 7 Jahre ältere Schwester, zu der er eine gute Beziehung hat, sowie einen 8 Jahre jüngeren Bruder, der im vergangenen Jahr einige Monate bei ihm wohnte und ihn vor 9 Monaten wieder verließ. Der Patient schildert zusätzlich eine sehr gute Beziehung zum Hund des Bruders, der nach dessen Wegzug ins Tierheim mußte. Diese Verlusterlebnisse gehen einher mit dem Beginn der vom Patienten geschilderten Phasen von „Depression".

Bei der körperlichen Untersuchung findet sich bei dem 37jährigen Patienten ein reduzierter Allgemeinzustand (Größe 184 cm, Gewicht 64,7 kg). Am Rücken und an den Extremitäten finden sich akneartige Effloreszenzen, teilweise mit zentraler Verkrustung, die der Hausarzt in seinem Einweisungsgutachten als „abklingendes Ekzem bei Kratzeffekten infolge zeitweiligem Pruritus" beschreibt. Auch perinasal besteht eine intensive Rötung. 1–2 cm große Lymphknoten lassen sich supraklavikulär sowie inguinal beidseits tasten. Am Rachenring finden sich weißliche, nur zum Teil abstreifbare Beläge, die den Verdacht auf eine Candida-Infektion nahelegen.

Außer einer vermehrten Schweißneigung und Glanzaugen besteht sonst kein wesentlicher Befund.

Im Erstinterview wirkt der Patient ängstlich, unsicher sowie erheblich angespannt, mißtrauisch und gereizt. Weiterhin fällt auf, daß er sehr leise und monoton spricht und Mühe hat, Gefühle zu äußern.

In der 2. Woche des stationären Aufenthaltes wurde wegen der Candida-Infektion, der vergrößerten Lymphknoten, zwischenzeitlich aufgetretener subfebriler Temperaturen und eines pneumonischen Infiltrates im linken Lungen-Mittel- und -Unterfeld der Verdacht auf eine HIV-Infektion bzw. manifeste AIDS-Erkrankung geäußert. Eine HIV-Testung lehnte der Patient zunächst vehement ab. Nach zunehmender Temperaturerhöhung wollte er dann aber doch mit „offenen Karten" spielen und überreichte der Stationsärztin ein entsprechendes weiteres Schreiben des Hausarztes, in dem mitgeteilt wurde, daß der Patient seit einem Dreivierteljahr HIV-positiv sei.

Durch die Offenlegung der Diagnose verschwand die anfänglich ausgeprägte, mißtrauische Gespanntheit. Für den Patienten war „das Eis gebrochen". In weiteren Gesprächen konnte er berichten, daß ihn die

Erstmitteilung der Diagnose sehr deprimiert hat, daß er sich anschließend zurückzog und mit intensiven Grübeleien begann, insbesondere über das Wann, Wo und Wie der Infektion; er vermutet, daß er sich während der Zeit des Alkoholabusus, in der er zuletzt 24 Flaschen Bier bzw. 2 Flaschen Whisky trank, infiziert habe, kann es aber nicht mehr rekonstruieren.

In den Beziehungen zum Therapeutenteam kommen Bewältigungsmechanismen wie Wut und Enttäuschung, Vorwurfshaltungen unter dem Eindruck, zu kurz zu kommen – zum Teil auch als Abwehr eigener Schuldgefühle – und depressiver Rückzug sowie Angst vor Abhängigkeit abwechselnd zum Ausdruck und verunsichern die Therapeuten: Sie erleben sich zum Teil als störende Eindringlinge in die subjektive Welt des Patienten, stoßen auf starkes Mißtrauen („ich hatte mir vorgenommen, nichts über meine Diagnose zu verraten"), teilweise auf Verleugnung der Krankheit („ich möchte vor allem wieder kräftiger und leistungsfähiger werden") und eine starke Abwehrhaltung („ich möchte das nicht immer von neuem aufwühlen"). Ein Teil der Widerstände ist auf institutionelle Schwierigkeiten zurückzuführen, da eine manifeste AIDS-Erkrankung derzeit eine Ausschlußindikation für Rehabilitationsmaßnahmen des verantwortlichen Rentenversicherungsträgers darstellt.

Vor allem in den Gesprächen mit männlichen Therapeuten zeigen sich sehr schnell Gegenübertragungsprobleme: Die mißtrauische und gespannte Atmosphäre läßt die Therapeuten ungeduldig werden, der Patient wird unterbrochen, es werden Doppelfragen gestellt. Dies perpetuiert den Widerstand und die Abwehr des Patienten analog der früheren Beziehungen zum Vater bzw. zur Stiefmutter. Diese Interaktion findet sich nicht in den Gesprächen mit der Stationsärztin, die eher die Rolle der Mutter übernimmt, deren Verlust dem Patienten noch sehr nahe ist.

Trotz der zunehmenden körperlichen und seelischen Beeinträchtigungen durch die AIDS-Erkrankung und die beschriebenen Schwierigkeiten bei der psychischen Bewältigung von Diagnose und Prognose, hatte seit der Diagnosestellung die psychosoziale Umwelt des Patienten einen Grad an Stabilität erreicht, den der Patient seit seiner frühen „broken-home"-Jugend vermißte: Er lebte in einer eigenen Wohnung und genoß diesen Schutzraum („als Untermieter wurde es mir zu eng") und fand um sich herum einen kleinen beschützenden Kreis von Vertrauten, d.h. auch mit seiner Krankheit Vertrauten, wozu neben seiner älteren Schwester und deren Mann ein älterer Freund, der Hausarzt und eine Fachärztin („AIDS-Spezialistin") zählen. Zu diesen Personen unterhält der Patient enge Kontakte, ähnlich denen zu verläßlichen Eltern.

Im Therapieverlauf spiegelte sich im therapeutischen Milieu die Polarität zwischen beängstigender (stigmatisierender) und enttäuschender Außenwelt und der kleinen, fast verschworen anmutenden Welt sozialer Unterstützung zu Hause. Es kam zu einer vorsichtigen Annäherung der entsprechend dissoziierenden inneren Objektrepräsentanzen.

Konzentrations- und Gedächtnisstörungen sowie wechselnde Angaben zu verschiedenen Zeitpunkten, z.B. zum Alter von Bezugspersonen oder zeitlichen biographischen Abläufen, legen den Verdacht auf ein diskretes psychoorganisches Syndrom nahe, das entweder durch den Alkoholabusus vor 10 Jahren bedingt ist oder mit der manifesten AIDS-Erkrankung in Zusammenhang steht.

48.2   Einleitung

Diagnostik, Therapie und Epidemiologie der Infektionskrankheiten haben sich in den letzten Jahren rapide entwickelt, ohne daß der Frage nach Beziehungen zwischen psychischen Prozessen und Infektionen Aufmerksamkeit geschenkt wurde. Dabei ist dieser Zusammenhang seit langem bekannt.

Bei der Bundespost verläuft – nach H. Sopp (1958) – die jahreszeitliche Entwicklung der Krankenstandskurve fast spiegelbildlich zu der der Gesamtbevölkerung (Abb. 48–1).

Nach einem relativ hohen Krankenstand im Sommer fällt die Kurve im Herbst, um im Dezember ihren niedrigsten Jahresstand zu erreichen. Ende Januar/Februar steigt sie steil an und erreicht im Februar/März den Jahresgipfel. Wie kommt es, daß die Postbediensteten sich in ihrem Gesundheitszustand genau entgegengesetzt zu allen anderen Berufstätigen verhalten? Liegen bei den Postbediensteten Umstände vor, durch die sie im November und Dezember vor Erkrankungen, insbesondere grippalen Infekten, geschützt werden? Sopp vermutet, daß die Arbeitsbelastung vor und während der Weihnachtszeit einen Schutz darstellt und die Streßerholungsphase zu Infektionskrankheiten disponiert.

Zu der Überlegung, daß psychosoziale Faktoren in der Epidemiologie der Infektionskrankheiten eine Rolle spielen könnten, lieferte Selye (1946) mit seinem Konzept eines „General Adaptation Syndrome" (GAS) ein frühes Modell, das die Grundlage für weitere Forschungen bildete (vgl. Kap. 8).

Selye betont in seinem Streßkonzept sehr einseitig die Effekte der ACTH- bzw. der Hypophysen-Nebennieren-Aktivität, so daß dieses Modell, wie Selye selber einräumte (1973), nicht in der Lage ist, die komplexen Wechselwirkungen zwischen Mikroorganismen und dem Menschen abzubilden, die zu Infektion und Krankheit führen. Unser Wissen über die Mechanismen für Resistenz oder besondere Empfindlichkeit gegenüber Infektionserregern ist weiterhin sehr unvollständig. Subklinische Infektionen sind sehr viel häufiger als manifeste Krankheiten, eine Beobachtung, die die Aussage Viktor v. Weizsäckers untermauert, daß „die Gesundheit eines Menschen … überhaupt nur dort vorhanden (ist), wo sie in jedem

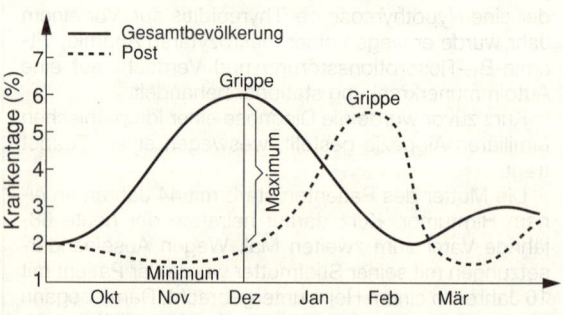

Abb. 48–1. Krankenstandskurven der Bundespostbediensteten und der Gesamtbevölkerung (nach H. Sopp, 1958).

Augenblick erzeugt wird. Wird sie nicht erzeugt, dann ist der Mensch bereits krank" (1955).

Von Uexküll (1988) beschreibt mit seinem Situationskreismodell Gesundheit und Krankheit als Ausdruck der Interaktion („Dialog") zwischen einem Individuum und seiner Umgebung auf einer biologischen, einer psychischen und einer sozialen Integrationsebene. Bezogen auf Infektionskrankheiten scheinen immunologische Prozesse eine wichtige Mediatorfunktion in der Beziehung zwischen psychosozialen Faktoren und Infektionsgeschehen im Sinne kreisförmiger Aufwärts- und Abwärtseffekte zu spielen, so daß für die zukünftige psychosomatische Erforschung dieser Zusammenhänge nach Elliott et al. (1982) alle Prozesse, die in einem „X-Y-Z-Modell" zusammenwirken, gleichzeitig studiert werden müssen, wobei sich X auf mögliche Aktivatoren oder Stressoren bezieht, Y auf kurzfristige physiologische Antworten wie immunologische Veränderungen und Z auf langfristige gesundheitliche Konsequenzen, wie eine Infektionskrankheit.

Die an den Anfang gestellte Fallgeschichte eines AIDS-Patienten zeigt – systemtheoretisch betrachtet – komplexe Aufwärtseffekte zwischen den verschiedenen biologischen, psychischen und sozialen Integrationsebenen auf, gleichzeitige Abwärtseffekte sind bisher nur zu vermuten, aber noch nicht sicher nachzuweisen. Kemeny et al. (1988) untersuchten die Beziehungen zwischen Verlust, Depression, Immunzustand und Immunverlauf bei männlichen Homosexuellen mit negativem bzw. positivem HIV-Testresultat und konnten zeigen, daß das Fehlen oder die Unterbrechung von sozialen Beziehungen, wie bei unserem AIDS-Patienten vielfach deutlich wurde, tiefgreifende psychobiologische Auswirkungen beim Menschen hat: Depressive HIV-infizierte Patienten hatten, verglichen mit nicht-depressiven HIV-positiven und HIV-negativen Patienten, weniger Helfer-/Effektor-T-Zellen, nicht-stimulierte Helfer-/Effektor-Zellen und mehr zytotoxische Suppressor-T-Zellen. Nach Weiner (1989) sind diese Immunveränderungen bei depressiven HIV-positiven Patienten nicht nur das Produkt der Virusinfektion: Depressionen sagen die schnellere Progredienz zur manifesten AIDS-Erkrankung voraus. Gegensätzliche Befunde werden im Kapitel 11 „Psychoimmunologie" zitiert.

48.3 Epidemiologie

Bei den oben erwähnten Untersuchungen von Postbeamten fand Sopp (1958), daß sie sich weder durch genetische, noch geographische Besonderheiten, noch durch Altersstruktur, noch besondere soziologische Merkmale, sondern einzig durch ihren Beruf und Arbeitsplatz von anderen Berufstätigen unterscheiden. Bemerkenswert ist, daß im Gegensatz zu anderen Berufen die Zeit des Krankheitsminimums mit der Zeit des Arbeitsmaximums zusammenfällt. Man könnte sagen, die Postbediensteten „können es sich wegen der besonderen Arbeitsanforderungen

nicht leisten", in dieser Zeit krank zu werden. Aber nur für einen Teil der Postler, nämlich die Päckchenzusteller und Briefträger, wäre wegen des Trinkgeldes in dieser Zeit eine lerntheoretische Erklärung im Sinne von Belohnung möglich.

Sopp konnte ferner nachweisen, daß die Erkrankungshäufigkeit an banalen Infekten vom Betriebsklima abhängig ist. Darüber hinaus erkranken Personen, die bereits länger in einem Betrieb arbeiten, seltener als solche, die in einem Betrieb erst kurze Zeit beschäftigt sind. Weiterhin erkranken ungelernte Arbeiter häufiger als gelernte. Sopp (1958) interpretierte diese Ergebnisse, zu denen – nach Pflanz (1962) – auch andere Autoren gekommen sind, mit der Bemerkung: „Nur wer sich wohl fühlt, ist zu qualifizierter Leistung imstande; nur wer sinnvoll arbeitet, fühlt sich wohl."

Hinkle und Wolff (1957) konnten in ausgedehnten epidemiologischen Langzeituntersuchungen nachweisen, daß Krankheiten immer dann gehäuft auftreten, wenn Situationen besondere Adaptationsleistungen erfordern.

Bei Untersuchungen an Studenten der Cornell University fanden Summerskill und Darling (1957), daß Studenten, die öfter als andere an diversen Erkrankungen, insbesondere psychoneurotischen Störungen litten, auch häufiger an sog. „banalen Infekten" erkrankten.

Greenfield und Mitarbeiter (1959) untersuchten 38 an infektiöser Mononukleose erkrankte Studenten mit dem Minnesota Multiphasic Personality Inventory (MMPI). Es fand sich eine signifikante inverse Beziehung zwischen psychischer Stabilität (z.B. Ich-Stärke) und Dauer der Rekonvaleszenz, die durch hämatologische Daten bestimmt wurde.

Vaillant (1979) beobachtete seit 1944 185 Männer über einen Zeitraum von etwa 40 Jahren. Nach 20 Jahren waren noch 100 Männer gesund, 54 hatten leichtere Krankheiten und 31 waren schwer krank bzw. gestorben. Zu Beginn der Untersuchung wurde eine Untergruppe mit Männern im Alter zwischen 21 und 46 Jahren gebildet. An psychischen Faktoren wurden die Kindheitsentwicklung (emotionale Probleme in der Kindheit, Mangel an familiärem Zusammenhalt und Beziehungen zu den Eltern, die nicht zu Vertrauen, Autonomie oder Eigeninitiative führten [Bewertung nach Vaillant, 1974]) und die psychosoziale Situation im Erwachsenenalter (psychische Gesundheit, finanzielle und berufliche Situation, Kontaktfähigkeit außerhalb der Familie, Schlafmittelverbrauch, psychiatrische Behandlungen, Krankheitshäufigkeit, eheliche Situation [Bewertung nach Vaillant, 1975]) herangezogen. Als psychisch gesund („best mental health") wurden 59 Männer eingestuft, von ihnen wurden nur zwei chronisch krank oder starben im Alter von 53 Jahren. Von den restlichen 48 psychisch instabilen Männern („worst mental health") wurden 18 chronisch krank oder starben. Die Unterschiede zwischen beiden Gruppen blieben auch statistisch signifikant, nachdem mit einer multifaktoriellen Regressionsanalyse die Auswirkungen von Alkohol- und Nikotingenuß, Adipositas und die Langlebigkeit von Vorfahren ausgeschlossen wurden, so daß andere als organische oder genetisch fixierte Faktoren eine Rolle spielen mußten.

Nach Spence und Mitarbeitern (1954) sind Infektionskrankheiten bei Kleinkindern um so häufiger, je niedriger die Sozialschicht der Eltern ist. Akute Bronchitiden, Pneumonien oder Keuchhusten kommen hier signifikant häufiger vor. Bei Kindern dieser Bevölkerungsschicht scheint die Mortalität an Infektionskrankheiten deshalb höher zu sein, weil das Erkrankungsalter früher liegt und damit die Gefährdung für den kindlichen Organismus größer ist. Das gilt vor allem für die Tuberkulose, bei der eine eindeutige Beziehung zwischen Erkrankungsalter, Unterernährung und Disposition sowie Verlauf auf der ganzen Welt nachweisbar ist. Es ist bekannt, daß mit Fortschreiten der Industrialisierung auch die Morbidität für Tuberkulose zugenommen hat. Hieran waren aber vor allem zu Beginn des 20. Jahrhunderts weniger die mit technischen Umstellungen verbundenen psychischen Belastungen als vielmehr die besonderen Durchführungsformen der Industrialisierung schuld, welche sowohl die Verbreitung der Bakterien wie auch die Disposition zur Erkrankung (Resistenzverminderung) förderten. Bereits Rudolf Virchow (1868) hat bei seiner Untersuchung der Typhusepidemie in Oberschlesien 1848 auf den Zusammenhang zwischen sozialen Lebensbedingungen und Krankheiten hingewiesen: „Denn daran läßt sich jetzt nicht mehr zweifeln, daß eine epidemische Verbreitung des Typhus nur unter solchen Lebensverhältnissen, wie sie Armut und Mangel an Kultur in Oberschlesien gesetzt hatten, möglich war. Man nehme diese Verhältnisse hinweg und ich bin überzeugt, daß der epidemische Typhus nicht wiederkehren würde."

Trotz dieser seuchenhygienisch wichtigen Zusammenhänge zwischen äußeren Lebensbedingungen und Infektionskrankheiten sind im einzelnen Fall auch psychische Faktoren für die Erkrankung, aber auch die Nichterkrankung, bedeutsam. Diese Faktoren werden in dem Maße wichtiger, in dem durch sozialmedizinische und andere Maßnahmen die äußeren Krankheitsfaktoren zurückgedrängt werden und die Häufigkeit der Infektionskrankheiten abnimmt. So standen die Infektionskrankheiten, inklusive Tuberkulose, Grippe und Pneumonie, in der Todesursachenstatistik 1924 mit 21% noch an der Spitze. 1961 sind diese Erkrankungen mit 6% erst weit hinter Herz- und Kreislaufkrankheiten (41,1%), bösartigen Tumoren (18,1%) oder unnatürlichen Todesarten (7,0%) zu finden (Schäfer und Blohmke, 1972). Diese Situation wird sich durch die Zunahme an AIDS-Erkrankungen verändern, so daß Infektionserkrankungen künftig auch in der Todesursachenstatistik wieder bedeutsamer werden (vgl. Kap. 49).

Klinischen Studien, die sich mit der Auswirkung von psychosozialen Faktoren auf eine Krankheit beschäftigen, wird vorgeworfen, sie hätten nur eine geringe Relevanz bei der Untersuchung spezifischer Infektionskrankheiten. Tatsächlich ist es bei den außerordentlich komplexen Zusammenhängen, die in der Pathogenese der meisten Infektionskrankheiten eine

Rolle spielen, nur selten möglich, psychische Belastungen von Umweltbedingungen wie Exposition, Ernährung, hygienischen, immunologischen Bedingungen und anderem mehr zu trennen. Aber immer präzisere Formulierungen von Vorhersagen bzw. Verhaltensvariablen erleichtern es, künftige Studien enger an speziellen Infektionskrankheiten zu orientieren, bzw. pathophysiologische oder immunologische Mechanismen zu untersuchen. Zusätzlich haben die Ergebnisse solcher Studien wiederholt gezeigt, daß weniger der Stressor selbst im weitesten Sinn als die Fähigkeit des Organismus, die Belastung zu bewältigen („coping"), maßgebend ist. Es ist also verständlich, daß unsere Kenntnis über die Rolle psychischer Faktoren bei Infektionskrankheiten nur lückenhaft und vielfach vorläufig ist.

48.4 Tierexperimentelle Studien

Ausgehend von den Versuchsbedingungen, mit denen auf die psychosoziale Situation Einfluß genommen wird, lassen sich drei große Gruppen von tierexperimentellen Studien unterscheiden:

- **Kontrollierte Stimulation:** Friedman und Mitarbeiter (1965) setzten Mäuse periodisch Licht- und Schüttelreizen aus, die die Tiere nicht vermeiden konnten. Nachdem die Tiere mit Coxsackie-B-Viren infiziert wurden, stellte sich heraus, daß sie deutlich an Gewicht verloren. Dagegen zeigten Kontrollgruppen von Mäusen, die nur Licht- und Schüttelreizen ausgesetzt wurden, keine Gewichtsveränderungen. Ohne Reize blieb das Gewicht beider Gruppen, ob infiziert oder nicht, gleich. D. h., daß erst das gemeinsame Auftreten der beiden Noxen den Widerstand gegen Infektionen verminderte.
- **Änderungen des sozialen Umfeldes:** Plaut und Mitarbeiter (1969) infizierten Mäuse mit Plasmodium berghei. Die postinfektiöse Überlebenszeit war umgekehrt proportional zur Anzahl der Tiere im Käfig.

  Friedman und Mitarbeiter (1970) setzten 1, 5, 10, 15 und 20 Mäuse in je einen Käfig und inokulierten alle mit Enzephalomyokarditis-Viren. Die Mortalität der alleinlebenden Mäuse war bedeutend höher als die der in Gruppen lebenden Mäuse.
- **Fortdauernde Einwirkungen von Stressoren:** Im Gegensatz zu den beiden ersten Gruppen beschäftigen sich diese Studien mit den Auswirkungen von Streß nach dessen Ende. Friedman und Mitarbeiter (1969) setzten zwei Gruppen von Mäusen zweimal täglich einem leichten elektrischen Schock aus. In der einen Gruppe waren Mäuse im Alter von 1–21 Tagen, in der anderen von 42–62 Tagen. Nach Abschluß der elektrischen Stimulation wurden beide Gruppen mit Enzephalomyokarditis-Viren infiziert. Die jüngere Gruppe zeigte daraufhin eine wesentlich höhere Mortalität als die ältere. Vergleichsgruppen gleicher Altersstufen, die nur zweimal täglich in die Hand genommen und gestrei-

chelt wurden, verhielten sich nach der Infektion wie die ältere elektrisch stimulierte Gruppe.

Tierexperimentelle Studien haben den Vorteil, daß mit genau definierten Populationen und bestimmbaren Versuchsbedingungen Untersuchungen über die Auswirkung verschiedener Stressoren einschließlich Erregern auf den Organismus möglich sind. Die hier zitierten Studien zeigen, daß auf diesem Weg wichtige pathogenetische Zusammenhänge erforscht werden können (vgl. Kap. 11).

48.5 Resistenz

Die ständige Präsenz einer großen Zahl von Erregern führt bei höher entwickelten Organismen zu einer Reihe von Abwehrmechanismen, die dem Schutz vor Infektionen dienen. Die Interaktionen zwischen Individuum und Erreger werden von einer großen Zahl endogener und exogener Faktoren beeinflußt. Eine schematische Darstellung dieses Prozesses ist in Abbildung 48–2 wiedergegeben.

Es besteht heute kein Zweifel daran, daß nur eine Minderheit der Bevölkerung, die mit einem potentiell pathogenen Erreger in Kontakt kommt, klinisch erkrankt. So wird beispielsweise geschätzt, daß nahezu 100% der Bevölkerung vorübergehend Träger von Meningokokken sind, aber nur ein Bruchteil infizierter Personen zeigen klinische Symptome (Shaw, 1965). Am Beispiel einer Coxsackie-B-Infektion läßt sich der sehr komplexe Vorgang der Wirt-Erreger-Interaktion veranschaulichen:

Das wechselnde Erscheinungsbild von Infektionskrankheiten während und in Abhängigkeit von Rei-

fung und Wachstum des Wirtsorganismus ist jedem Arzt bekannt. Coxsackie-B-Viren erzeugen bei älteren Kindern und Erwachsenen eine Reihe gewöhnlich gutartiger Symptome wie Pleurodynie und aseptische Meningitis. Die gleichen Viren können aber beim Kleinkind zu einem fulminanten Verlauf mit rapider Ausbreitung und multipler Organbeteiligung in Leber, Pankreas, Herz und Gehirn führen.

Ähnlich wie beim Menschen reagieren junge Mäuse – im Gegensatz zu alten Mäusen – äußerst empfindlich auf Coxsackie-B-Viren.

Einer der Faktoren, die bei der Resistenz gegenüber viralen Infektionen eine Rolle spielen, ist das Interferon. Heineberg und Mitarbeiter (1964) konnten zeigen, daß eine Coxsackie-B-Vermehrung bei jungen Mäusen zu einer nur geringen Stimulation der Interferonproduktion führte. Im Gegensatz dazu kam es nach Infektion mit diesem Virus bei älteren resistenten Mäusen zu einer charakteristischen eingeschränkten viralen Replikation, aber einer signifikant größeren Interferonantwort.

48.6 Immunologische Faktoren

Das Immunsystem ist an der Genese der meisten Krankheitsbilder direkt oder indirekt beteiligt. Es ist im lymphatischen System verankert und zusammen mit den Makrophagen für die biologische Individualität bestimmend. Zwei Arten von Lymphozyten repräsentieren die funktionale Trennung des Immunsystems: die T-Lymphozyten mit der Funktion der zellulären Reaktion und die B-Lymphozyten als Gedächtniszellen und Produktionsstätte der Immunglo-

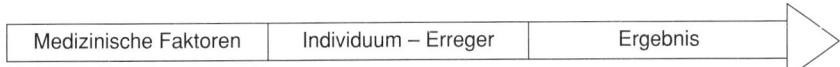

| Medizinische Faktoren | Individuum – Erreger | Ergebnis |

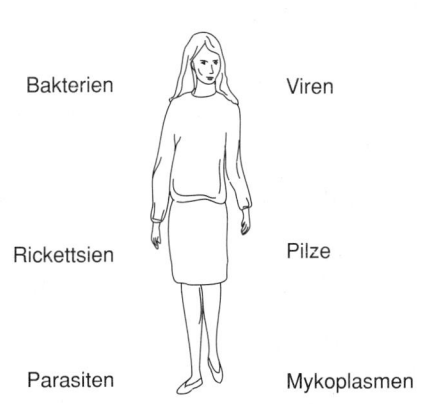

Endogene Faktoren:
Alter
Geschlecht
Schwangerschaft
Menstruation
Immunologische Abwehr
Frühere Auseinander-
  setzung mit Erregern
Bestehende Krankheiten
Hormoneller Status
„normale" Flora

Umwelt–Faktoren:
Infektionskrankheiten
  in der Umgebung
Psychischer Streß
Hygiene und Gesundheits–
  wesen
Medikamente
  (Hormone und Antibiotika)
Jahreszeit

Bakterien

Rickettsien

Parasiten

Viren

Pilze

Mykoplasmen

**Keine Infektion–
keine Krankheit**

Subklinische Infektion:
Erholung und Immunität
Vorübergehender oder
  Langzeit–Mikrobenträger

Akute Infektion:
Tod
Erholung und Immunität
Erholung als vorübergehender
  Mikrobenträger
Chronische Infektion

Chronische Infektion:
Mögliche maligne Entartung

Abb. 48–2. Schema endogener und exogener Faktoren, die das Individuum in seiner Interaktion mit den Erregern in seiner Umwelt beeinflussen können, sowie das mögliche Ergebnis derartiger Interaktionen (modifiziert nach Friedman und Glasgow, 1960).

buline für die humorale Reaktion. Die humoralen und zellulären Immunsysteme arbeiten nicht völlig unabhängig voneinander. Die Antigenerkennung und Antigenverarbeitung findet durch B-Lymphozyten, T-Lymphozyten und Makrophagen statt. Die Entwicklung von B-Lymphozyten zu antikörperproduzierenden Plasmazellen wird beeinflußt durch zwei Untergruppen von T-Lymphozyten, sog. „Helfer"-T-Zellen in unterstützender Funktion und „Suppressor"-T-Zellen in hemmender Funktion (Roitt, 1977). Die Steuerung der Antikörperproduktion ist demnach von mehreren Faktoren abhängig. Die immunologischen Reaktionsmuster sind entsprechend der an das Gesamtsystem gestellten Aufgaben vielfältig. Im wesentlichen spielen hier die anaphylaktischen, zytotoxischen, komplexvermittelten, zellvermittelten und durch Stimulation hervorgerufenen Reaktionen eine Rolle, die jeweils spezifische, zum Teil ineinandergreifende Aufgaben haben.

Die Kontrolle der Immunantwort liegt offenbar nicht nur auf zellulärer Ebene. Szentivanyi und Filipp (1958) untersuchten die Rolle des Hypothalamus bei der Anaphylaxie. An Meerschweinchen konnten sie zeigen, daß eine beidseitige, eng umgrenzte Zerstörung der Nuclei tuberales einem letalen anaphylaktischen Schock vorbeugt. Ähnliche Ergebnisse erzielten sie beim Kaninchen. Filipp und Szentivanyi (1958) berichteten über eine Verminderung von gewebefixierten Antikörpern gegen Rinderserum beim Meerschweinchen unter den oben genannten Bedingungen. Diese Befunde galten auch für eine passive Sensibilisierung sowohl mit homologem als auch heterologem (Kaninchen-)Serum.

Fessel und Forsyth (1963) konnten zeigen, daß mentaler Streß zu einer Erhöhung von 19-S-Globulinen führt und daß es bei Ratten nach Elektrostimulation des lateralen Hypothalamus zu deutlichen Veränderungen der Gammaglobulinkonzentration kommt. Darüber hinaus konnten Korneva und Khai (1963) nachweisen, daß die Läsion eines bestimmten Gebietes im dorsalen Hypothalamusanteil bei Kaninchen u. a. zu einer vollständigen Suppression der primären Antikörperantwort, einer verzögerten Retention von Antigenen aus dem Blut und einer fast vollständigen immunologischen Schwäche gegenüber einer streptokokkeninduzierten Myokarditis führte. Diese Autoren hatten Hinweise dafür, daß diese Effekte durch Wachstumshormon, das bei der Antikörpersynthese eine Rolle spielen soll, reversibel waren. Weitere Arbeiten zeigen die wichtige Rolle des autonomen Nervensystems einschließlich des Hypothalamus bei der Anaphylaxie und dem Asthma bronchiale (Widdicombe, 1963; Koller, 1968; Mills und Widdicombe, 1970; Gold et al., 1972; Parker, 1973).

Neben der oben angesprochenen humoralen Reaktionsebene bestehen offensichtlich auch zelluläre Regulationswege. So beschrieben Macris und Mitarbeiter (1970) beim Meerschweinchen mit Läsionen im Bereich des vorderen Hypothalamus eine Hypersensitivität der Haut u. a. auf Tuberkulin-Antigen.

Als Schaltstation zwischen höheren zentralnervösen Zentren und der Peripherie hat der Hypothala-

mus über Neurotransmitter und Neurohormone wesentlichen Einfluß auf viele Körperfunktionen. Eine große Zahl von tierexperimentellen Arbeiten weist auf seine Mediatorfunktionen bei der Immunantwort hin. Dabei fällt den Kortikosteroidspiegeln, aber auch Hypophysenhormonen eine wichtige Rolle zu. In einer Reihe von tierexperimentellen Untersuchungen konnte der Einfluß von Streß auf Immunvorgänge wiederholt nachgewiesen werden (Marsh et al., 1963; Spry, 1972; Yamada et al., 1964; Rasmussen et al., 1957; Johnson et al., 1963; Friedman et al., 1960, 1965). Auch wenn die tierexperimentellen Untersuchungen mit der entsprechenden Vorsicht zu bewerten sind und zirkadiane Rhythmen der Immunantwort meist nicht berücksichtigt sind, kann von einer übergreifenden zentralnervösen-peripheren Interaktion gesprochen werden.

Es mehren sich die Hinweise, daß Verbindungen zwischen Persönlichkeit, Streß und Reaktion des Immunsystems bestehen. In Abbildung 48–3 ist dargestellt, wie sich – bisher noch hypothetisch – die Beziehung zwischen Emotionen, Streß, nervalen und biochemischen Mechanismen und einer Dys- oder Unterfunktion des Immunsystems vorstellen läßt.

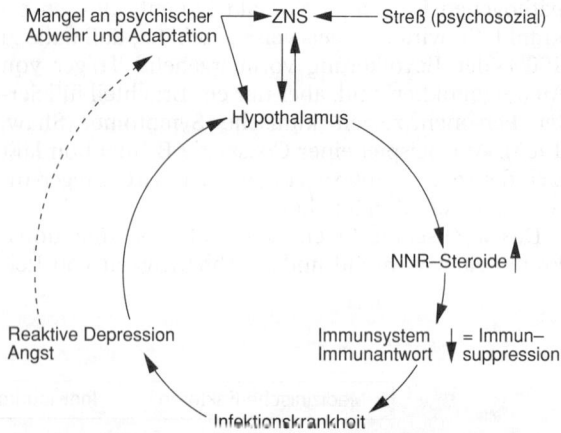

Abb. 48–3. Beziehung zwischen psychischen Faktoren und Immunsystem (modifiziert nach G. F. Solomon und R. H. Moos, 1964).

Was in dieser Abbildung als Immunsystem bezeichnet wird, muß, wie bereits oben erwähnt, nach unseren heutigen Vorstellungen weiter in zelluläre und humorale Immunität differenziert werden.

Personen, die schwierige Lebensumstände zu bewältigen haben, leiden häufiger an infektiösen und allergischen Erkrankungen, sind aber auch empfänglicher für kardiovaskuläre und psychiatrische Krankheiten. Das Ausmaß der Veränderungen des Immunsystems scheint mit dem Ausmaß an Streß, den die Lebensumstände hervorrufen, übereinzustimmen. Jemmot und Mitarbeiter (1983) untersuchten bei 64 Studenten der Zahnmedizin im Laufe eines Jahres fünfmal das über die Speicheldrüsen sezernierte Immunglobulin A. Die Meßperiode bestand aus einer anfangs streßarmen Zeit, gefolgt von drei streßreichen Abschnitten mit schweren Examina und einer abschließenden streßarmen Phase. Während der schweren Examina sanken die Immunglobulin-A-Sekretionsraten bei allen Probanden deutlich ab und normalisierten sich während der Entspannungsphasen wieder. Die Verbindung mit dem thematischen Apperzeptionstest ergab aber, daß Studenten mit besonders hoher Motivation und hohem Leistungsbewußtsein durchweg niedrigere Immunglobulin-A-Sekretionsraten hatten als ihre Kommilitonen. Das insgesamt einheitliche Bild wich aber in einem Punkt ab: Nach Abschluß der Examina lagen die Immunglobulinwerte der besonders hoch motivierten Studenten noch niedriger als während der Examina. Diese Studenten fühlten sich zwar nicht mehr gestreßt und hatten auch die gleichen Leistungen wie die anderen. Sie waren aber deutlich unzufriedener mit ihrem Abschneiden als ihre Kommilitonen. Die anhaltende Immunglobulin-A-Erniedrigung in der Gruppe der hoch motivierten und sehr leistungsbewußten Studenten läßt sich vielleicht damit erklären, daß für sie die Ruhephase der Semesterferien Streß durch Immobilisation bedeutet. McClelland und Burnham (1975) beschreiben in diesem Zusammenhang das „Inhibit Power Motive Syndrome" (IPS), ein Zusammentreffen von hoher Leistungsmotivation und starker Aktivitätseinschränkung, das besonders bei Führungspersönlichkeiten anzutreffen sei. Auch Frankenhaeuser und Mitarbeiter (1978) haben gezeigt, daß Nichtstun Streß bedeuten und einen erhöhten Ausstoß von Katecholaminen und Cortisol zur Folge haben kann. Sie beschreiben diese Veränderungen allerdings bei Menschen mit Typ-A-Verhalten nach Friedman, die aber eine ähnliche Persönlichkeitsstruktur haben sollen wie Patienten mit IPS. Entlastung kann bei bestimmten Individuen also auch als ein pathogenetischer Faktor angesehen werden (v. Uexküll und Pflanz, 1952). Ein weiterer Grund für die fehlende Immunglobulin-A-Normalisierung könnte auch darin liegen, daß diese Studenten bereits an den Streß des folgenden Ausbildungsjahres dachten.

Dorian et al. (1982) untersuchten die Belastung einer Prüfung bei Medizinstudenten im Vergleich zu einer Kontrollgruppe und ihren Einfluß auf die Immunfunktion. Sie fanden eine eingeschränkte zelluläre und humorale Immunfunktion bei den Prüflingen, sie stellten jedoch darüber hinaus einen reduzierten Cortisolspiegel und eine erhöhte Lymphozytenzahl fest, so daß anders als in den oben zitierten Untersuchungen von Frankenhaeuser et al. (1978) die eingeschränkte Immunfunktion demnach nicht auf die Cortisolkonzentration zurückgeführt werden kann. Schäfer et al. (1985) untersuchten Personen, die nahe dem Three-Mile-Island-Atomreaktor (Harrisburg) wohnten, und Personen, die nahe einer als gefährlich erachteten Mülldeponie ihren Wohnsitz hatten, und verglichen verschiedene immunologische Meßgrößen dieser Zufallsstichproben mit denen einer Kontrollgruppe. Sie fanden bei beiden Untersuchungsgruppen im Vergleich zur Kontrollgruppe eine starke Verminderung des sekretorischen Immunglobulins A. In der Three-Mile-Island-Gruppe war außerdem auch das

Verhältnis Helfer-/Suppressor-T-Zellen und die Anzahl der B- und T-Lymphozyten verringert.

Gonzáles (1983) beschrieb bei Patienten mit akuter nekrotisierender ulzerativer Gingivitis eine besondere psychosoziale Belastung in deren Vorgeschichte und eine verstärkte Neigung zu ängstlichen und depressiven Reaktionen sowie ausgeprägte emotionale Störungen. Gleichzeitig bestand eine Beeinträchtigung des Immunsystems, besonders der T-Lymphozyten. Daneben fand sich ein erhöhter Spiegel des freien Cortisols im Urin. Udelman (1982) fand eine Verminderung der T- und B-Lymphozytenzahlen, die zum Teil in zeitlichem Zusammenhang mit Änderungen der Lebenssituation standen. Interessanterweise schien eine antidepressive Behandlung wenigstens vorübergehend einen Anstieg der T-Lymphozyten auszulösen.

Bei der Untersuchung des Einflusses von Trennungserlebnissen bei Menschen auf deren Immunsystem fanden Bartrop et al. (1977) bei Verwitweten gegenüber einer Kontrollgruppe 8 Wochen nach dem Tod des Lebenspartners hochsignifikante Unterschiede in der Mitogenstimulierbarkeit. Andere immunologische Parameter, wie z. B. die Anzahl von T- und B-Zellen, Immunglobulintiter sowie die Hormonspiegel von Cortisol, Prolaktin und Thyroxin, zeigten hingegen keine richtungweisenden Veränderungen (vgl. Kap. 11). In einer vergleichbaren prospektiven Studie untersuchten Schleifer et al. (1983) die Mitogenstimulierbarkeit von T-Zellen vor und nach dem Tod des Ehepartners und führten eine Kontrollmessung nach einem Jahr durch. Sie fanden anders als Bartrop eine deutlich eingeschränkte T-Zellfunktion auch schon im ersten Monat nach dem Tod des Lebenspartners. Der Verlust des Lebenspartners ging gleichzeitig mit einem deutlich erhöhten Depressivitätsindex einher. In der Kontrolluntersuchung nach einem Jahr lagen dann die Werte für die T-Zellproliferation zwischen den Werten vor und nach dem Verlusterlebnis.

Totmann et al. (1980) untersuchten in einer prospektiven Studie an 52 freiwilligen Versuchspersonen den Einfluß des Faktors „psychosoziale Einsamkeit" auf den klinischen Manifestationsgrad einer experimentell induzierten Rhinovirusinfektion. Die Studie zeigte, daß die Symptomstärke der Erkrankung signifikant mit dem Persönlichkeitsmerkmal „Introversion" und ebenso mit dem durch Fragebogen und Interview ermittelten Ausmaß von Vereinsamung in den letzten 3 Monaten vor Experimentbeginn korrelierte. Ähnliche Beziehungen zwischen psychosozialer Vereinsamung und Abnahme der Immunkompetenz fanden auch andere Forscher, so z. B. Kiecolt-Glaser et al. (1987) bei reaktiv depressiven Patienten.

Nach v. Kerekjarto (1988) belegen zahlreiche empirische Daten, daß zwischen dem neuroendokrinen System und dem Immunsystem vielfältige Interaktionen stattfinden, und daß in beiden Systemen ähnliche Peptidsignale, Hormone und Rezeptoren die Kommunikation zwischen diesen beiden Systemen erleichtern. Diese Gemeinsamkeit im Signal- und Empfangsbereich scheint dem Organismus zu ermöglichen, mit Hilfe der sensorischen Funktion des Immunsystems außer kognitiven Stimuli auch Bakterien, Viren, Tumorzellen und sonstige Antigene in die zentralnervöse Verarbeitung einzubeziehen und damit die Gesamtantwort des Organismus zu optimieren.

48.7 Ausgewählte Krankheitsbilder

48.7.1 „Psychogene Angina"

Die folgenden Krankheitsbilder sollen zeigen, daß enge Beziehungen zwischen psychischen Faktoren und sehr häufigen Krankheitsbildern bestehen können, ohne daß man sie bereits schon einem bestimmten psychosomatischen Konzept zuordnen kann. V. v. Weizsäcker (1935) vermutete, daß solche Erkrankungen dazu beitragen können, Konflikte zu lösen, mit denen der Patient nicht fertig wurde. Als besonders eindrucksvolles Beispiel für die Bedeutung psychischer Faktoren bei infektiösen Erkrankungen gilt die Angina tonsillaris. Die Bezeichnung „psychogene Angina" (auch „Couvade-", „Männer-Kindbett-", „Hochzeits-", „Verlobungs-" oder „Junggesellen-"Angina) geht auf frühe Beschreibungen solcher Zusammenhänge durch v. Weizsäcker (1935) und Bilz (1936) zurück, die bei ihren Patienten besondere biographische Ereignisse in der Vorgeschichte der Krankheit fanden. Zwei Fallbeispiele von Viktor v. Weizsäcker illustrieren den Begriff der „Couvade-Angina":

> „Ein junges Mädchen wird mit starker Angina, kaum fähig, nur zu sprechen, in die Klinik eingeliefert. Ein junger Arzt äußert nach der Untersuchung: ‚Na, da haben Sie sich ja was Schönes geholt', worauf sie sagt: ‚Das ist immer noch besser, als ein Kind kriegen'. Später stellt sich heraus, daß sie am Vortage dem Drängen eines Verehrers, welches solche Folgen hätte haben können, widerstanden hat."
>
> „Einem jungverheirateten Mann schenkt seine Frau das erste Kind, einen Sohn. Er ist sehr bewegt, denn er hat sich ein Kind sehr gewünscht. Vielleicht war es ihm aber mit dem Eintritt der Geburt doch nicht so eilig gewesen, weil er in deren Vorbereitung recht lässig war. Tatsächlich wurde das Kind drei Wochen zu früh am Tag nach einer Szene geboren, in der er sich aufgeregt und in Gegenwart seiner Frau deren Sache nicht geschickt und kraftvoll geführt hatte. Einen Tag nach der Geburt wird er von heftiger Angina befallen. Damit hat er einen Grund, eine weite Reise nicht anzutreten. Sie hätte ihn zu einer Veranstaltung führen sollen, der er nicht gewogen war und auf der er ohne Begeisterung hätte öffentlich hervortreten müssen."

Psychodynamik

Wie aus den verschiedenen Bezeichnungen bereits deutlich wird, scheinen Liebesbeziehungen, Verlobung oder Ehe bei dieser Erkrankung eine große Rolle zu spielen. Die verschiedenen Fallbeispiele von v. Weizsäcker (1935) und Bilz (1936) zeigen, daß ein „erotischer Ansturm" zu einer – lebensgeschichtlichen – Krise führt, in deren Verlauf bzw. Zuspitzung die Krankheit ausbricht. V. v. Weizsäcker (1935) nannte das „Kränkung der erotischen Beziehung". Nach Überwindung der Krankheit und Rekonvaleszenz ist eine Entscheidung getroffen, eine neue Situation ist da. Nach v. Weizsäcker (1935): „Das Ganze ist wie eine historische Einheit: Wendung, kritische Unterbrechung, Wandlung."

Schellack (1957/1958), der Mitte der fünfziger Jahre neurosenpsychologische Faktoren in Ätiologie und Pathogenese der Tonsilliden untersuchte, fand Konflikte im Besitz- und Geltungsbereich, so etwa Ausbruch der Krankheit bei Geburts- und Weihnachtsfesten, Geburt von Geschwistern, bei Wettkämpfen und Examina und bei Liebesbeziehungen, in denen Angst vor der Aufgabe der eigenen Person besteht. Daneben beschreibt er bei diesen Patienten einen hohen Prozentsatz an depressiven Stimmungen (90%), Appetitstörungen (65%) und funktionellen Magen- oder Darmstörungen (je 30%). Die psychischen Untersuchungen weisen auf eine meist depressiv-zwangsneurotische Charakterstruktur hin. Frühe Entwicklungsstörungen führen zu einer Gehemmtheit im oralen und aggressiven Bereich.

48.7.2 Tuberkulose

Sehr eingehende Beobachtungen über psychische Zusammenhänge gibt es für die Tuberkulose. Sie stammen fast ausnahmslos aus der vorantibiotischen Ära. Der chronische, oft letale Verlauf und die damit verbundene langanhaltende Betreuung des Patienten durch den Arzt sowie die große Häufigkeit der Tuberkulose haben schon zu Beginn des Jahrhunderts auf psychische Faktoren bei dieser Infektionskrankheit aufmerksam gemacht. In der Literatur finden diese Beobachtungen, vor allem aber auch die krankheitsreaktiven Veränderungen, die zum Teil durch einen jahrelangen Sanatoriumsaufenthalt verursacht werden, in Thomas Manns „Zauberberg" (1924) ihren Niederschlag. Im wissenschaftlichen Schrifttum zeigen die oft widersprüchlichen Resultate, wie schwierig bei einem derart komplexen Krankheitsbild das Auffinden von psychischen Faktoren ist. Auch die Einführung spezifisch wirkender Antibiotika und die damit verbundene Änderung von Therapiegewohnheiten haben die weitere Präzisierung psychischer Faktoren erschwert.

Racamier (1950) hat 150 Patienten untersucht. Er beschreibt sie als beziehungslos, einsam und mit einem tiefen Gefühl von Unsicherheit („neurose d'abandon"). Ihm waren frühkindliche Störungen der affektiven Beziehungen der Patienten zur Mutter (90%) – und zwar bei zwei Dritteln im Sinne der Versagung, bei einem Drittel im Sinne der Verwöhnung – und/oder zum Vater (70%) aufgefallen.

In den fünfziger Jahren waren es dann vor allem Hübschmann (1952), Stern (1957/1958) und Bräutigam (1956/1957), die psychosomatische und somatopsychische Phänomene bei dieser Infektionskrankheit untersucht haben. Bräutigam (1956/1957) betont die Empfindlichkeit im Kontakt sowie die Labilität des Selbstwertgefühls dieser Patienten.

Dunbar (1948) findet als Persönlichkeitsprofil bei Tuberkulösen Entscheidungsschwäche, Selbstunsicherheit und masochistische Züge, sie nennt die Tuberkulose „einen Flirt mit dem Tode" (Bräutigam). Selbstzerstörerische Züge werden ebenfalls von Hübschmann (1952) beschrieben. Nach ihm haben diese für Entstehung und Verlauf der Erkrankung eine entscheidende Bedeutung. Der Tod wird im Grunde als eine letzte Geborgenheit ersehnt. Er beschreibt – die auch später verschiedentlich bestätigte Beobachtung –, daß Patienten, die sich gegen die strenge Disziplin in den Sanatorien auflehnten, vor allem die aus disziplinarischen Gründen entlassen wurden, eine bessere Prognose hatten als die fügsamen, „angenehmen" Patienten.

Wittkower (1949) glaubt nicht, daß bei diesen Kranken eine spezifische Persönlichkeitsstruktur zu finden sei, er hat versucht, verschiedene prämorbide bzw. prätuberkulöse Strukturen zu beschreiben. Er unterscheidet drei verschiedene Persönlichkeitstypen:
– den unsicheren Typ, den er noch unterteilt in übermäßig abhängige, sich anlehnende und ihre Unabhängigkeit kompensatorisch betonende Persönlichkeiten;
– den rebellierenden Persönlichkeitstyp;
– Menschen, die mit Konflikten beladen sind.
Allerdings fand er unter seinen Patienten auch eine Reihe von Kranken, die keinem dieser Persönlichkeitstypen zuzuordnen waren. Wittkower (1949) betont, daß die psychische Reaktion auf die Krankheit das Ergebnis des Zusammenwirkens der Umweltreize und der Persönlichkeitsstrukur ist.

Keine der bisher vorliegenden Untersuchungen, soweit sie über bloße Kasuistiken hinausgehen und methodisch ausgereift sind, gibt einen Anhalt dafür, daß die Tuberkulose einen bestimmten Persönlichkeitstyp bevorzugt oder einen bestimmten Persönlichkeitstyp ausbildet. Die Tuberkulose befällt Menschen aller Konstitutionen und Persönlichkeitsstrukturen und ruft ganz verschiedene psychische Reaktionen hervor (Stern, 1957/1958).

Schließlich seien noch die Ergebnisse psychologischer Untersuchungen erwähnt, die Melzer (1957) mit Hilfe des Rorschach-Tests bei Tuberkulösen durchgeführt hat. Danach finden sich bei diesen Kranken mangelnde Schaffenslust, geringe seelische Stabilität, erhöhte Reizbarkeit in bezug auf Affekte, Egozentrizität und Beherrschtwerden des Gefühlslebens durch verdrängte Konflikte. Die Kranken verlangen viel Rücksicht, ohne selbst dazu bereit zu sein. Darüber hinaus scheint eine tieferliegende Angst vorhanden zu sein, Fahrigkeit, Flüchtigkeit, Nonchalance und Nachlassen der Selbstdisziplin.

Psychosoziale Befunde und auslösende Situationen

Aus den oben erwähnten Arbeiten und epidemiologischen Untersuchungen (Pflanz, 1962) geht hervor, daß bei Morbidität und Mortalität der Tuberkulose exogene Faktoren bzw. Belastungssituationen wie Kriege, Seuchen, Unterernährung, Vertreibung größerer Bevölkerungsanteile (Flüchtlingsprobleme wie in Mitteleuropa nach dem I. und II. Weltkrieg oder später im Nahen Osten) eine große Rolle spielen.

Die Verteilung der Tuberkulose im Querschnitt der Bevölkerung ist nicht homogen. Besonders betroffen sind bei uns die untersten Sozialschichten, insbesondere Randgruppen wie etwa Wohnungslose. Ausländer sind eine eigene Risikogruppe, da die epidemiologische Situation in ihren Heimatländern meist ungünstiger ist als bei uns.

Sucht man nach Zusammenhängen zwischen biographischen Ereignissen und Krankheitsausbruch, z.B. Todesfälle von Angehörigen, Unfälle etc., so findet man – nach Bräutigam (1956/1957) – keinen Zusammenhang zwischen Lebenssituationen und Zeitpunkt der Erkrankung. Nach Hübschmann (1952) und Studt (1973) sollen vor allem Liebeskonflikte, wie Zeit des Kennenlernens, Verlobung, Entlobung, Heirat und verschiedene Ehe- und Familienkonflikte, besonders bei Frauen krankheitsauslösend wirken. Dagegen sollen bei Männern eher Probleme im Bereich der Berufs- oder Besitzsphäre eine Rolle spielen.

Verlauf und Therapie

Psychische Faktoren haben für den Verlauf der Krankheit eine besondere Bedeutung. Melzer (1957) berichtet detailliert über Reaktionen Tuberkulosekranker von der Zeit der stationären Aufnahme an über die lange Zeit der Kurbehandlung bis hin zu den Konsequenzen der Reintegration der Patienten in ihre soziale Umgebung.

Studt (1976) fand bei mehr als 70% der Patienten mit psychischen Belastungen einen therapeutischen Mißerfolg; nur 28% dieser Patienten erreichten einen Kavernenschwund gegenüber 87% der psychisch unauffälligen Patienten. Mehr als die Hälfte der Patienten hatten einen oder mehrere Kurabbrüche oder „disziplinarische Entlassungen" hinter sich. Sie hatten Heilverfahren oder Operationen verweigert oder Medikamente nicht eingenommen und Injektionen abgelehnt (Bräutigam und Christian, 1975). Sehr häufig fand sich bei diesen Patienten auch ein Alkoholabusus: Bei Alkoholikern sind Rückfälle doppelt so häufig wie bei Nichtalkoholikern.

Für die Therapie wird die große Bedeutung einer guten Führung durch den Arzt betont und die Wichtigkeit eines verständnisvollen Eingehens auf die Probleme des Patienten und auf Konflikte, die sich vor allem durch die Krankheitssituation ergeben. Die bei den Tuberkulösen oft anzutreffende Labilität und Ich-Schwäche soll eher eine supportive, stützende psychotherapeutische Technik erfordern als Aufdekken und Konfrontation (Bräutigam und Christian, 1975).

48.7.3 Akute Virushepatitis

Exemplarische Krankheitsverläufe (Häfner et al., 1955; Hagedorn, 1969; Hübschmann, 1977; Caliezi, 1980) lassen vermuten, daß der Ausbruch einer akuten Virushepatitis durch aktuelle Konflikte und Lebenskrisen begünstigt werden kann und der Gesundungsprozeß bei weiterbestehenden und neu auftretenden belastenden Lebenssituationen von Komplikationen begleitet ist.

Paar et al. (1987) fanden in einer Pilotstudie bei 18 Hepatitispatienten und einer Kontrollgruppe von 8 Unfallpatienten, daß sich eine gesteigerte psychosoziale Belastung vor und während der akuten Erkrankung vor allem auf den individuellen Krankheitsverlauf auswirkt: In der Gruppe der erschwert abheilenden Hepatitispatienten fanden sich mehr chronische Lebensbelastungen und neue belastende Lebensereignisse während ihres Genesungsprozesses gegenüber den komplikationslos gesundenden Hepatitispatienten bzw. den Unfallpatienten; außerdem wiesen jene Hepatitispatienten mit kompliziertem Verlauf zu Beginn und verstärkt gegen Ende ihres Krankenhausaufenthaltes höhere Depressionswerte auf als beide übrigen Versuchsgruppen. Diese stärkere depressive Verstimmung könnte die Immunkompetenz dieser Patienten zusätzlich beeinflußt haben (Gauss und Kubanek, 1984). Hieraus ergibt sich ein begründeter Ansatz für zusätzliche psychotherapeutische Angebote an diese Patienten.

48.7.4 Genitale Herpes-simplex-Virus-(HSV-)Infektionen

Die Bedeutung der HSV-Infektionen liegt darin, daß sie ein mögliches Modell für die Beziehungen zwischen Psyche und Verlauf von Infektionskrankheiten sein können.

Das HSV-Virus gelangt nach einer primären genitalen Herpesinfektion in die sensiblen sakralen Ganglien, wo es in latentem Zustand verbleibt. Eine große Anzahl von Faktoren werden als „Aktivatoren" für ein Herpesrezidiv vermutet: Sonnenlicht, Erkältung, Infektionserkrankungen, Hautverletzung, Menstruationszyklus, Nahrungsmittelallergien und nicht zuletzt emotionaler Streß. Die physiologischen Mechanismen, die es solchen Faktoren, wie Sonnenlicht oder emotionalem Streß ermöglichen, die Herpesviren zu reaktivieren, sind noch sehr unklar.

Kemeny et al. (1989) untersuchten über einen Zeitraum von 6 Monaten insgesamt 36 Individuen mit rezidivierenden genitalen HSV-Infektionen. HSV-Träger mit einem hohen Maß an akutem, chronischem und zukünftig erwartetem Lebensstreß zeigten eine Abnahme an T-Helfer-Zellen (T_H) und T-Suppressor-Zellen (T_S). Die Untersuchten mit einem hohen Maß an negativen Stimmungen (Angst, Depres-

sion oder Feindseligkeit) hatten weniger T_S-Zellen. Nur der Faktor Depressivität, gemittelt über die Studiendauer, korrelierte mit der HSV-Rezidivrate, und diese Korrelation erwies sich als unabhängig von Änderungen des Gesundheitsverhaltens, wie Alkoholkonsum, Schlaf oder Bewegung. Die Studie läßt vermuten, daß Depressivität zu einer Abnahme des T_S-Zell-Niveaus bzw. eines damit assoziierten anderen immunologischen Parameters führt und hiermit verbunden sich eine Änderung des biologischen Gleichgewichtes vollzieht, die ihrerseits die Virusreplikation und Ausbildung der typischen Infekt-Läsionen begünstigt. Durch die Abnahme des T_S-Zell-Niveaus stehen möglicherweise zu wenig zytotoxische Zellen zur Verfügung, um die virusinfizierten Zellen zu zerstören.

Die Ergebnisse zeigen auch, daß Streßexposition allein keinen geeigneten Auslöser für ein Herpesrezidiv darstellt. Anhaltende Stimmungen von Depressivität schaffen die spezifische Vulnerabilität für den Krankheitsprozeß. Darauf weisen auch die Untersuchungen von Katcher (1973) und Friedmann (1977) im Hinblick auf Herpes-labialis-Rezidive hin.

48.7.5 Psychogenes Fieber

Im Rahmen dieses Kapitels soll auch auf ein psychosomatisches Symptom eingegangen werden, das bisher nur selten untersucht wurde (vgl. Overbeck, 1973; R. Meyer und Beck, 1975). Es gehört zwar nicht zu den Infektionskrankheiten, zeigt aber, wie psychische Faktoren einerseits Reaktionen des Gesamtorganismus beeinflussen können und andererseits für die Differentialdiagnose bei fieberhaften Zuständen eine Rolle spielen.

Das Symptom Fieber ist für klinisch tätige Ärzte häufig ein Problem. Es werden die unterschiedlichsten Fieberverläufe bei Infekten, allergischen Reaktionen und Tumoren beobachtet. Die Temperaturunterschiede variieren von gering bis sehr hoch. Musher und Mitarbeiter (1979) zogen in einer prospektiven Studie den Schluß, daß die genaue Betrachtung von Fieberverlaufskurven für eine Diagnosestellung meist nur wenig hilfreich ist. Immerhin stellten sie fest, daß bei Infektionskrankheiten eher mit einem kontinuierlichen Fieberverlauf zu rechnen ist, der allerdings von zirkadianen Rhythmen beeinflußt ist. Überraschend war der Befund, daß etwa die Hälfte der nichtentzündlichen Fieberzustände nicht diesen täglichen Schwankungen unterworfen war. Musher und Mitarbeiter (1979) nehmen an, daß hier zentrale Temperaturregulationsmechanismen von höherstehenden Zentren zum Hypothalamus gestört sind.

Bei der Fieberentstehung ist in diesem Zusammenhang von Bedeutung, daß neben den bekannten Faktoren (Phagozytose; an einer Entzündungsreaktion beteiligte Substanzen wie Immunglobuline, chemotaktische Faktoren, Komplement; spontan endogenes Pyrogen erzeugende Zellen wie bei Malignomen) auch in ihrer Funktion gestörte B- und T-Lymphozyten über einen bisher noch nicht bekannten immunologischen Weg beteiligt sind.

Studt (1976) beschreibt, daß psychogene Temperaturerhöhungen meist im Bereich von 37,5–38,0 Grad Celsius bleiben und nur in seltenen Fällen bis 40 Grad Celsius ansteigen. Im Gegensatz zum Fieber besteht eine kühle und blasse Gesichtshaut und eine normale Puls- und Atemfrequenz, meist besteht auch kein Unterschied zwischen der Axillar- und Rektaltemperatur.

Winckelmann und Mitarbeiter (1982) berichten über 85 Patienten, bei denen über sechs Monate hinweg rezidivierendes Fieber über 38,5 Grad Celsius bestand. Bei 16 dieser Patienten blieb trotz invasiver Diagnostik die Ursache auch weiterhin ungeklärt. Reimann (1962) nannte diese Fälle periodisches Fieber, im englischen Sprachgebrauch auch als FUO („fever of unknown origin") bezeichnet. Im Gegensatz zu Reimann fand Winckelmann keine periodischen Fieberverläufe. Das Leiden begann zum Teil schon während der Kindheit und verlief über viele Jahre ohne erkennbare Organschädigung. Das einzige Zeichen einer Organbeteiligung war eine leichte Splenomegalie mit unauffälligem histologischem Befund. Mit Indometacin (200 mg/Tag) ließen sich bei rechtzeitiger Einnahme die Symptome weitgehend kupieren. Psychische Befunde wurden nicht miterhoben.

Meist unbemerkt bleiben die 3–4% aller Patienten, die neu in eine Klinik aufgenommen werden und, nach Studt (1976), eine kurzfristige, psychogene Temperaturerhöhung erleben; so ist z.B. Fieber bei funktionellen Syndromen bekannt. Ihr Anteil an der Gesamtzahl der Patienten ist in einer psychiatrischen Klinik etwa viermal größer als in einer medizinischen Klinik. Ein vorgetäuschtes Fieber, d.h. eine Manipulation des Patienten am Thermometer, wird nur bei 3% dieser Patienten gefunden.

Psychodynamik und psychosoziale Befunde

R. Meyer und D. Beck (1975) fanden bei 13 Patienten mit subfebrilen Temperaturen ohne organische Ursache schwere Störungen in den zwischenmenschlichen Beziehungen bzw. der Familienstruktur. Die Patienten zeichneten sich durch ein hochstehendes Ich-Ideal aus, worunter die Autoren moralisch bewertete Gedankeninhalte über richtungweisende und erstrebenswerte Ziele verstehen, die das Denken und Handeln eines Menschen bestimmen. Als psychodynamische Faktoren zeigten sich gehemmte Aggressionen sowie als deren Folge ein mangelhaft entwickeltes Durchsetzungsvermögen.

Gleichzeitig fanden die Autoren, daß diese Patienten darüber hinaus an verschiedenen, unspezifischen Symptomen litten. Overbeck (1973) beschreibt Fieberschübe, Palpitationen, Kopfschmerzen, Dämmer-, Angst- oder Erregungszustände, die mit Seh- und Hörstörungen, tonischer Muskelstarre, Beklemmungsgefühl und Hyperventilationstetanie einhergehen können. Erikson (1965) erwähnt diese Symptome bei „Kriegsneurotikern".

Auslösende Ursache

Nach R. Meyer und Beck (1975) tritt Fieber auf, wenn das hochstehende Ich-Ideal in Gefahr gerät, zusammenzubrechen, oder wenn symbiotische Abhängigkeitsbeziehungen auseinandergerissen werden. Damit in Verbindung könnten die von Studt (1976) beschriebenen Temperaturerhöhungen bei Klinikaufnahme gesehen werden, auch die von ihm beschriebenen Fieberschübe nach Ehe- und Familienkonflikten, die mit heftigem Ärger oder Streit einhergehen. Verschiedentlich wird beschrieben, daß Aufregung, Angst, Wut oder Ärger zu Temperatursteigerungen führen können. Dazu gehören: das leichte Fieber von Klein- und Schulkindern in bestimmten, wiederkehrenden Situationen oder bei Erwachsenen in Prüfungen (Medizinstudenten im Examen), unklares Fieber auf der Hochzeitsreise oder nach Antritt der ersten Stelle. Als krankheitsauslösende Ereignisse beschreibt Schellack (1957/58) mit heftigem Ärger und Streit einhergehende Ehe- und Familienkonflikte. Die Persönlichkeitsstruktur dieser Patienten sieht er als vorwiegend hysterisch.

Differentialdiagnose

Ausführlich beschreibt Hegglin (1964) den „Status febrilis" bei der „vegetativen Dystonie". Von einer konstitutionellen Hyperthermie (konstante Temperaturerhöhung bei Gesunden) sind Simulationsversuche („vorgetäuschtes Fieber") zu unterscheiden, die durch einen atypischen Verlauf und eine Diskrepanz zwischen Höhe der Temperatur und der Pulsfrequenz auffallen. Beim psychogenen Fieber zeigt das Blutbild, wie bei Tuberkulösen oder Patienten mit Thyreotoxikose, oft eine Lymphozytose bis 40%. Dagegen ist die Senkungsgeschwindigkeit der Blutkörperchen auffallend niedrig und überschreitet oft Werte von 1–2 mm in der ersten Stunde nicht. Nach Hegglin (1964) soll das Fehlen der Temperaturdifferenz bei rektaler und axillärer Messung von 0,5 Grad Celsius differentialdiagnostisch verwendet werden können. Als wichtiges differentialdiagnostisches Kriterium gilt die Beobachtung, daß psychogene Temperatursteigerungen nicht auf Antipyretika wie Pyrazolone oder Acetylsalicylsäure, aber auf Sedativa oder Hypnose reagieren.

48.7.6 Infektionsbedingte psychische Störungen

Anders als in den bisherigen Abschnitten geht es hier nicht um psychosoziale Faktoren und ihren Einfluß auf Genese und Krankheitsverlauf einer Infektionskrankheit, sondern um psychische Störungen als Hinweis auf eine körperliche Erkrankung. In der Zeit vor Einführung der Antibiotika waren bei Infektionen auftretende Psychosen oder psychische Störungen den Ärzten vertraut. Die infektionsbedingten psychischen Störungen besitzen jedoch auch in der Antibiotikaära ihre Aktualität und eigene Problematik, bei ihrer Diagnosestellung herrscht heute vielfach Unsicherheit.

Maneros et al. (1987) fanden bei 104 Patienten, die sich wegen infektionsbedingter psychischer Störungen in einer psychiatrischen Klinik befanden, daß fast die Hälfte wegen psychischer Auffälligkeiten direkt in die psychiatrische Klinik eingewiesen wurde. Die Infektionskrankheit wurde erst in der psychiatrischen Klinik entdeckt. Meist erschienen die psychischen Auffälligkeiten zunächst nicht körperlich bedingt. Die körperliche Untersuchung und einfache diagnostische Maßnahmen wie Fiebermessen wurden oft initial nicht durchgeführt. In der Mehrzahl der Fälle fanden sich im Anfangsstadium der infektionsbedingten psychischen Erkrankungen Störungen des Antriebes und der Psychomotorik, Affektstörungen und Sinnestäuschungen, Symptome also, die bei psychischen Erkrankungen oft zu finden sind. Drei Patienten wiesen als allererste psychische Auffälligkeiten eine akute Suizidalität auf. Die oft unspezifische und uncharakteristische Initialsymptomatik entwickelte sich während des stationären Aufenthaltes meist rasch zu einer für akute körperlich begründbare Psychosen typischen Symptomatik mit charakteristischer Bewußtseinstrübung (84% der Patienten). Die Hälfte der Patienten hatte eine Pneumonie, 18% ungeklärte fieberhafte Infekte, die übrigen Patienten Sepsis, Endokarditis, Typhus, Malaria, Scharlach oder andere Infektionskrankheiten. Bei zwei Drittel der Patienten reichte eine allgemeine internistische Behandlung, d.h. in den meisten Fällen eine antibiotische Therapie aus. Die dargestellten Zusammenhänge verdeutlichen eindrücklich die Notwendigkeit eines ganzheitlichen diagnostischen Zugangs zum Patienten, die Autoren empfehlen daher auch abschließend: „Der zunächst konsultierte Arzt darf sich nicht von psychischen Auffälligkeiten derart beeindrucken lassen, daß er bei solchen Patienten den Griff zu Stethoskop und Thermometer vergißt."

Während sich in der oben genannten Untersuchung die psychischen Störungen als Initialsymptomatik einer Infektionskrankheit manifestierten, weist White (1987) exemplarisch an zwei Fallgeschichten auf die Beziehung zwischen infektiöser Mononukleose und – mit einer Latenz von zum Teil 6 Monaten – schwerer depressiver Störung hin, die in beiden Fällen mit einer Wahnsymptomatik einhergingen. Zumindest in dem einen Fall fanden sich aufgrund der EEG- und Liquoruntersuchung diagnostische Hinweise für eine begleitende Enzephalopathie. Auch bei unserem Patienten zeigen sich u.a. durch die Konzentrations- und Gedächtnisstörungen Hinweise für ein diskretes psychoorganisches Syndrom.

48.8 Zusammenfassung

Die bisherigen psychosomatischen Überlegungen haben sich vor allem auf die Genese und den Verlauf von Erkrankungen bezogen. Sehr wenig beachtet wurden die sich hieraus ergebenden Konsequenzen für die Therapie. Die Berücksichtigung der Psychodynamik der einzelnen Patienten gibt dem behandelnden Arzt ein wichtiges Instrument in die Hand, mit dem er die sehr komplexen Vorgänge einer somatischen Erkrankung besser verstehen und auch besser darauf reagieren kann.

49 HIV-Infektion und AIDS

Sophinette Becker und *Ulrich Clement*

In den letzten Jahrzehnten hat sich die psychosomatische Medizin nur selten mit Infektionskrankheiten im allgemeinen (vgl. Kap. 48) und faktisch gar nicht mit sexuell übertragbaren Krankheiten beschäftigt. Dies lag vor allem daran, daß die Infektionskrankheiten von der Medizin als „besiegt" erklärt worden waren, zum einen durch Impfstoffe, zum andern durch Antibiotika. Krankheiten wie Syphilis oder Gonorrhoe hatten durch die Erfindung des Penicillins ihren Schrecken verloren.

Mit AIDS gibt es nach langer Zeit zum ersten Mal wieder eine Krankheit, die gleichzeitig sexuell übertragbar ist, gegen die es keinen Impfstoff gibt und die tödlich ist. AIDS hat die Aura von Krebs und Syphilis in einem. Wenn man bedenkt, daß schon Krebskranke oft von ihrer Umwelt so behandelt werden, als wären sie ansteckend, und sich auch selbst so erleben (vgl. Kap. 66), dann kann man sich das Ausmaß der sozialen Stigmatisierung von AIDS-Kranken vorstellen. Die soziale Einstellung gegenüber AIDS-Kranken wird darüber hinaus geprägt von der Tatsache, daß die hauptbetroffenen Gruppen nach wie vor homosexuelle Männer und i.v.-Drogenabhängige sind. Damit wird ein breites Assoziationsfeld über verbotene Sexualität, Promiskuität, Sucht, Kontrollverlust etc. angestoßen, das sich gesellschaftlich um so schärfer und verfolgender als Vorurteil gegen die Betroffenen richtet, je mehr eigene verpönte Impulse abgewehrt werden (Becker, 1985; Rühmann, 1985; Stevens und Muskin, 1987).

Auch wenn sich die öffentliche AIDS-Diskussion inzwischen mehr versachlicht hat, gibt es doch immer noch viele irrationale Ängste im Zusammenhang mit AIDS. Neben individuellen hypochondrischen und phobischen Reaktionen (vgl. Abschn. 49.8) äußern sie sich etwa im Festhalten an längst ausgeschlossenen Übertragungswegen (z.B. durch Mücken oder im Schwimmbad) oder in völliger Überschätzung der Infektionsgefahr. So passiert es z.B. immer wieder, daß Eltern wegen eines HIV-infizierten Kindes geschlossen ihre Kinder nicht mehr in den Kindergarten schicken, obwohl es weltweit nur einen Fall gibt, bei dem eine Übertragung von einem Kind auf ein anderes angenommen wird.[1] Wohl kaum eine Krankheit ist bisher in einem solchen Ausmaß vergesellschaftet worden wie AIDS (Sigusch, 1988). Zu dieser Vergesellschaftung gehören auch die Dramatisierung von AIDS als größter Gefahr für die Menschheit ohne jeden Bezug zu anderen gesellschaftlichen und individuellen gesundheitlichen Risiken (AIDS ist nach wie vor eine seltene Todesursache in der BRD), spekulative Hochrechnungen über die Ausbreitung von AIDS (vgl. Clement, 1987) sowie die Gleichsetzung von HIV-Infektion und AIDS.

Stand Krebs als Metapher für Lebensbedrohung durch Krankheit überhaupt (Sontag, 1978), so steht AIDS als Metapher für durch falsches, sündiges Leben selbstverschuldete Krankheit. Obwohl die Mehrheit der heute an AIDS Erkrankten sich zu einem Zeitpunkt infiziert hat, als kaum etwas über die Übertragungswege bekannt war, werden in der Öffentlichkeit nur die Bluter und die Transfusionsempfänger als „Opfer" des Virus wahrgenommen – und selbst sie haben häufig unter Diskriminierungen zu leiden.

Ärzte sind gegen gesellschaftliche Vorurteile nicht mehr gefeit als andere Mitglieder der Gesellschaft, und sie sind in sexuellen Fragen schlecht ausgebildet (Sigusch, 1979; vgl. Kap. 39). Unter diesen Umständen ist das Herstellen einer „gemeinsamen Wirklichkeit zwischen Arzt und Patient" (v. Uexküll und Wesiack, 1988) natürlich sehr erschwert. Dies kann bei HIV-Infizierten und AIDS-Kranken nur gelingen, wenn sich der behandelnde Arzt in besonderem Maße der Wahrnehmung und Reflexion seiner eigenen Gefühle, Ängste und Konflikte im Zusammenhang mit Homosexualität und Sexualität überhaupt stellt (Becker, 1988).

49.1 Die HIV-Infektion und ihre Verlaufsformen

Die auch unter Ärzten verbreiteten Sprachregelungen wie „AIDS-Infizierte" oder „AIDS-Vorfeld-Patienten" stehen für die falsche Vorstellung, eine HIV-Infektion sei gleichbedeutend mit AIDS. Und weil AIDS ein meist tödlich verlaufendes Krankheitsgeschehen ist, zeichnen solche Sprachregelungen die damit Bezeichneten ausnahmslos mit dem Tod. Diese unzutreffende Prognose reflektiert weder die offenen Fragen zum Verlauf der Infektion noch ihre Wirkung auf diejenigen, die wissen, daß sie mit dem Virus infiziert sind (Sigusch, 1987; Gschwind, 1988).

Prospektive Kohortenstudien mit Infizierten in den USA (z.B. Rutherford, 1987) kommen zu der Schätzung, daß nach sieben Jahren etwa 30 Prozent der Infizierten wahrscheinlich an AIDS erkranken wer-

---

[1] Es handelte sich um zwei Brüder; eine inzestuöse sexuelle Beziehung konnte nicht mit letzter Sicherheit ausgeschlossen werden.

den.[2] Ob überhaupt, und wenn, wie schnell die nach langen Jahren noch gesunden Infizierten erkranken werden, kann derzeit nicht vorhergesagt werden. Insbesondere ist bis heute ungeklärt, welche somato-psycho-somatischen Cofaktoren den Verlauf der HIV-Infektion beeinflussen bzw. dazu beitragen, daß es im Einzelfall sehr rasch, sehr langsam oder gar nicht zur manifesten Erkrankung AIDS kommt.

Wenn im folgenden die verschiedenen Verlaufsformen der HIV-Infektion, deren schwerste AIDS ist, beschrieben werden, ist es wichtig zu betonen, daß es sich dabei nicht um notwendig aufeinanderfolgende „Stadien" oder „Stufen" der Krankheit handelt.

Die Symptome und Erkrankungen, die im Verlauf einer HIV-Infektion auftreten können, sind mehrheitlich Ausdruck einer durch das HI-Virus verursachten Störung der körpereigenen Immunabwehr. Eine Ausnahme stellen die neurologischen Symptome dar, die zum Teil Ausdruck einer direkten Infektion des ZNS sind und deshalb auch ohne nachweisbaren Immundefekt auftreten können.

Eine HIV-Infektion ist in der Regel nachweisbar, wenn eine ausreichende Menge des Virus in den Körper gelangt ist und zur Bildung von Antikörpern geführt hat. Für die meisten Infizierten verläuft dieser Vorgang unbemerkt, nur in 10 bis 20 Prozent der Fälle kommt es wenige Tage bis drei Monate nach der Infektion zu einer akuten grippeartigen Erkrankung, deren Symptome in der Regel wieder abklingen. Eine Schwellung der Lymphknoten kann nach der akuten Krankheit bestehenbleiben oder unabhängig davon bei Infizierten auftreten. Bei bestimmter Dauer, Größe und Lokalisation dieser Schwellungen spricht man vom LAS (Lymphadenopathiesyndrom). Die prognostische Bedeutung des LAS für den weiteren Verlauf der HIV-Infektion ist bislang ungeklärt.

Eine schwerere Verlaufsform der HIV-Infektion ist der sog. AIDS-Related Complex (ARC), bei dem im Gegensatz zu den vorher beschriebenen Verlaufsformen der HIV-Infektion bereits ein Immundefekt und schwere klinische Symptome vorliegen. Nach der heute gültigen Definition spricht man von ARC, wenn mindestens zwei der in Tabelle 49-1 genannten Symptome und zusätzlich mindestens zwei der in dieser Tabelle genannten Laborbefunde vorliegen.

AIDS ist die schwerste Verlaufsform der HIV-Infektion. Aufgrund der geschwächten körpereigenen Abwehr (vor allem durch den massiven Verlust von T-Helfer-Zellen) kommt es zu malignen Erkrankungen (Kaposi-Sarkom und Lymphome) und/oder zu sog. opportunistischen Infektionen durch Erreger (Viren, Pilze, Bakterien, Protozoen), gegen die bei normaler Abwehrlage des Körpers eine natürliche Abwehr besteht. Die bei AIDS-Kranken verbreitetste opportunistische Infektion ist die Pneumocystis-carinii-Pneumonie.[3] Daneben können vor allem die Haut (inklusive Schleimhäute), der Darm und das Gehirn[4] betroffen sein.

Die Mortalität bei AIDS ist groß. Nach einer Untersuchung in New York (Rothenberg et al., 1987), die 5833 Patienten mit AIDS umfaßt, beträgt die Wahrscheinlichkeit zu überleben ein Jahr nach der AIDS-

Tabelle 49–1. Symptome des AIDS-Related Complex (zwei oder mehr der folgenden Symptome; aus Braun-Falco et al., 1987).

– Nachtschweiß
– Fieber oder Fieberschübe
– Gewichtsverlust von mehr als 10% des Körpergewichtes
– Durchfall, persistierend, ohne andere Ursache
– Persistierende Lymphknotenschwellungen wie bei LAS
– Haarleukoplakie

Laborbefunde
(zwei oder mehr zur Definition notwendig)

– Anämie oder Leukopenie oder Thrombopenie oder Lymphopenie
– Erhöhung der Gammaglobulinfraktion des Serum-Eiweißes
– Reduktion der T-Helfer-Lymphozyten T-Helfer-/T-Suppressor-Lymphozyten-Quotient unter 1,0
– Verminderung der Lymphozytenreaktion auf Mitogene
– Kutane Anergie auf Hauttest-Antigene
– Erhöhte Konzentration zirkulierender Immunkomplexe

Diagnose 49 Prozent, nach fünf Jahren 15 Prozent. Durch die wachsende Erfahrung mit der Krankheit AIDS sowie durch Fortschritte in der medizinischen Behandlung[5] kann bei einem Teil der AIDS-Kranken die Lebenszeit verlängert und die Lebensqualität verbessert werden. Darüber hinaus gibt es wohl auch bei AIDS-Kranken ungeklärte somato-psycho-somatische Prozesse, die den Verlauf der Erkrankung beeinflussen. So berichten Ärzte immer wieder einerseits von Patienten, die „eigentlich nach ihrem Immunstatus längst tot sein müßten", während andererseits Patienten mit „besserem" Immunstatus schon gestorben seien. Die Forschung über diese Fragen steckt noch in den Anfängen (vgl. Schiefer-Hofmann und Jäger, 1988).

49.2 Übertragungswege

Der Übertragungsmodus der HIV-Infektion ist bis heute nicht bekannt. Die Übertragungswege können aber epidemiologisch erschlossen werden. Nach heutigem Kenntnisstand wird die Infektion übertragen, indem infiziertes Körpersekret direkt in einen anderen Blutkreislauf gelangt. Die Infektionswahrscheinlichkeit scheint dabei von der Menge an infiziertem

---

2 Dabei ist zu beachten, daß diese Prognosen sich auf den Zeitraum nach der gestellten Diagnose, nicht nach der Infektion selbst beziehen, d.h. tendenziell günstiger liegen, wenn man vom Infektionszeitpunkt ausgeht.
3 Gegenwärtig erkranken AIDS-Patienten auch zunehmend an Tbc mit besonders schweren Verlaufsformen.
4 Zum Teil direkt, zum Teil durch Sekundärinfektion des ZNS (z.B. Toxoplasmose).
5 Derzeit vor allem Azidodeoxythymidin (AZT), das als „falsches" Nukleosid spezifisch das Thymidin-Nukleosid von der reversen Transkriptase des HI-Virus verdrängt.

Sekret, der Viruskonzentration in ihm und der Beschaffenheit der Eingangspforte abzuhängen. Die Empfänglichkeit für die Infektion (Vulnerabilität) scheint außerdem – bei gleicher Exposition – individuell und zwischen verschiedenen sozialen Gruppen/Schichten stark zu differieren. Solche Bedingungen der Infektion sind noch weitgehend unerforscht.

Blut und Ejakulat enthalten eine hohe Konzentration, eine etwas geringere die Vaginalflüssigkeit. In anderen Körpersekreten (Urin, Speichel, Tränen usw.) konnte das HIV zwar nachgewiesen werden, sie spielen aber für die Übertragung praktisch keine Rolle.

Eine Infektion ist im wesentlichen auf vier Wegen möglich:

– **auf sexuellem Weg:** Das wahrscheinlichste Übertragungsrisiko ist beim Analverkehr gegeben, wobei der rezeptive Partner das höhere Risiko trägt. Erheblich weniger wahrscheinlich ist eine Ansteckung beim Vaginalverkehr außerhalb der Menstruationszeit. Extrem unwahrscheinlich ist sie beim Oralverkehr, sofern nicht im Mundraum ejakuliert wird. Insgesamt ist das Übertragungsrisiko bei einem einmaligen Verkehr eines Infizierten mit einem Nichtinfizierten sehr gering; es gibt jedoch belegte Einzelfälle von Infektionen durch einmaligen Geschlechtsverkehr. Durch richtigen Gebrauch eines intakten Kondoms läßt sich eine Infektion verhindern;

– **durch kontaminierte Nadeln und Spritzen:** Wenn beim intravenösen Drogengebrauch gemeinsames Spritzbesteck benutzt wird, kann es zu einer Infektion kommen. Durch Benutzung von Einwegbesteck läßt sich dieser Übertragungsweg ausschließen;

– **durch Blut und Blutprodukte:** Es gab Infektionen durch Transfusionen, Gerinnungsfaktorpräparate und Transplantationen. Seitdem seit 1985 alle Blutkonserven in der BRD getestet werden, ist eine Übertragung auf diesem Weg praktisch ausgeschlossen, bei einer minimalen Restunsicherheit durch falsch-negative Testergebnisse oder durch Spenderblut frisch Infizierter, deren Antikörper noch nicht nachweisbar sind;

– **durch Mutter-Kind-Übertragung:** HIV-positive Mütter können vor oder während der Geburt, mit geringer Wahrscheinlichkeit auch beim Stillen, die Infektion an ihre Kinder weitergeben. Die Übertragung ist aber nicht zwangsläufig und wird gegenwärtig auf 25 bis 60 Prozent geschätzt.

Alle anderen Übertragungen (künstliche Befruchtung, bei der Arbeit im medizinischen Bereich) spielen quantitativ keine nennenswerte Rolle.

49.3 Epidemiologie

Aufgrund der Verbreitung und der schnell zunehmenden Kenntnisse sind epidemiologische Angaben rasch veraltet und daher nur in Verbindung mit dem Datum ihrer Gültigkeit sinnvoll.

Seit der Einführung der Bezeichnung „AIDS" 1982 hat die Zahl der Neuerkrankungen jedes Jahr zugenommen. In der BRD nennt das Bundesgesundheitsamt zum 31. 8. 1989 2668 Fälle, davon 1347 in den letzten 12 Monaten, wobei 56% bereits verstorben sind. Weltweit sind der WHO 167 000 Fälle gemeldet (Stand: 30. 6. 1989), wobei die tatsächliche Zahl auf das Zwei- bis Dreifache geschätzt wird. Diese kumulative Zahlenangabe, bei der die an der Krankheit Verstorbenen mitgerechnet werden, war zu Beginn der Epidemie durchaus sinnvoll, hat aber als epidemiologischer Parameter weniger Aussagekraft als die Prävalenz und Inzidenz. Während wegen der langen Inkubationszeit weiter mit einer Zunahme der AIDS-Inzidenz zu rechnen ist, gibt es aus prospektiven Kohortenstudien bei homosexuellen Männern Hinweise, daß die Neuinfektionen bei ihnen deutlich abnehmen (Kingsley et al., 1987), was sich auf die starken sexuellen Verhaltensänderungen zurückführen läßt.

Hauptbetroffenengruppen sind in Europa und den USA homosexuelle Männer, i.v.-Drogenabhängige und Hämophile. Der Anteil von auf vaginalen Geschlechtsverkehr zurückzuführenden AIDS-Erkrankungen ist in industrialisierten Ländern gering und steigt nur langsam.

Über Prävalenz und Inzidenz von HIV-Infektionen in der Bevölkerung lassen sich keine zuverlässigen Angaben machen. Seit Einführung der anonymen Laborberichtspflicht (1. 10. 1987) wurden beim Bundesgesundheitsamt etwa 18 000 positive Bestätigungstests gemeldet, was natürlich nur eine Untergrenze der tatsächlichen Infiziertenzahl darstellt. Ein „Eindringen" der HIV-Infektion in die allgemeine heterosexuelle Bevölkerung hat bisher nicht stattgefunden. Es kann höchstens von einem punktuellen „Einsickern" gesprochen werden.

49.4 Der HIV-Test

Für HIV-Infizierte, die keine klinischen Symptome bzw. „nur" ein LAS haben, gibt es derzeit keine medizinische Behandlungsmöglichkeit. Entsprechend erhalten sie auf ihre Frage, was sie denn tun könnten, damit die Krankheit AIDS bei ihnen nicht ausbreche, ärztlicherseits nur unspezifische Ratschläge im Sinne von „Leben Sie gesund, belasten Sie Ihr Immunsystem nicht!"[6, 7]

Gleichzeitig ist die Mitteilung eines HIV-positiv-Ergebnisses ein traumatischer Einschnitt im Leben eines Menschen mit erheblichen psychischen (und oft auch sozialen) Folgen, bis hin zur Suizidalität (vgl. Abschn. 49.6).

---

6 Als einzige spezifische Empfehlung wird die Vermeidung von Impfungen mit Lebendimpfstoffen wie BCG-, Salmonellen- und Gelbfieber-Impfung diskutiert, die jedoch ohnehin selten durchgeführt werden. Die WHO sieht keine Kontraindikation dieser Impfungen bei einer HIV-Infektion.
7 Die iatrogene Belastung des Immunsystems durch die Mitteilung des HIV-positiv-Ergebnisses wird dabei nicht bedacht.

Die Schere zwischen diagnostischen und therapeutischen Möglichkeiten der Medizin hat in den letzten Jahrzehnten erheblich zugenommen und wird mit der Entwicklung der Genom-Analyse zusätzlich völlig neue Dimensionen bekommen. „Es kommt zunehmend zur Produktion von medizinischen Damoklesschwertern" (Rosenbrock, 1988).

Dieses Problem wird in der Medizin international diskutiert. Vom Grundsatz des „nil nocere" ausgehend hat die WHO schon 1968 ein wesentliches Kriterium für die Indikation zur Früherkennungsdiagnostik aufgestellt: „It must be acceptable for the subject with regard to inconvenience, discomfort and risk of side effects ... Obviously there is no point in screening for a condition which cannot be treated" (Wilson und Jungner, 1968). Entsprechend wurde bei einer etwa zur gleichen Zeit wie der HIV-Test entwickelten (und ihm in bezug auf die Unsicherheit der Prognose bei gleichzeitiger Unmöglichkeit medizinischer Intervention sehr vergleichbaren) gentechnologischen Testmethode auf Veranlagung zu Chorea Huntington eine sehr sorgfältige Diskussion über ethische, psychische und soziale Implikationen ihrer Anwendung geführt (vgl. Crauford und Harris, 1986). Die Diskussion fand vor dem klinischen Einsatz dieses Gentests statt und verhinderte ihn mit der Begründung, daß die psychischen Folgen eines positiven Testergebnisses nicht abzuschätzen seien und zunächst weitere Erkenntnisse über das Verhältnis von Nutzen und Risiken dieser Testmethode erworben werden müßten (Crauford und Harris, 1986).

Eine ähnlich differenziert die Folgen abwägende Diskussion vor dem zum Teil massenhaften Einsatz des HIV-Antikörper-Tests hat nicht stattgefunden. Problematisierende Abhandlungen erschienen erst nach der Einführung des Tests, im Ausland auch von Ärzten (z.B. Miller et al., 1986), in der Bundesrepublik fast ausschließlich von Sozial- und Sexualwissenschaftlern (z.B. Rosenbrock, 1986, 1988; Becker und Clement, 1987). Anfangs wurden viele HIV-Tests ohne vorherige Aufklärung und Beratung durchgeführt, und oft wurde das HIV-positiv-Ergebnis den Betroffenen am Telefon mitgeteilt. Diese Bedingungen haben sich z.T. mittlerweile gebessert. Im Gegensatz zu anderen Ländern (z.B. Holland, Schweiz), in denen der HIV-Test nur bei sehr strenger, individueller Indikation angewandt wird, werden in der BRD nach wie vor massenhaft symptomlose Personen getestet, nach der Maxime „jeder sollte sich testen lassen, der glaubt, er könnte sich angesteckt haben".

Geht man den ärztlichen Motiven für diese Testungen nach (vgl. Rosenbrock, 1988), so geht es im seltensten Fall um die körperliche und seelische Gesundheit des Getesteten. Statt dessen überwiegen (abgesehen von nicht unerheblichen Forschungsinteressen) deutlich präventive Absichten, d.h. der Wunsch, die gesunde Bevölkerung vor den lebenslänglich ansteckenden HIV-Positiven zu schützen. In vermeintlicher Abwägung der Interessen des einzelnen gegen das Wohl der Allgemeinheit wird dabei die psychische Belastung, die das HIV-positiv-Ergebnis für den Betroffenen bedeutet, in Kauf genommen.

Das Problematische daran ist, daß diese aus präventiven Motiven handelnden Ärzte für Prävention gar nicht ausgebildet sind, da es anders als etwa in den anglo-amerikanischen Ländern bei uns keine „Public Health"-Tradition gibt. So vermischen sich dann ganz unärztliche Wünsche nach Erfassung und Kontrolle der Infizierten mit laienpsychologischen Begründungen für den HIV-Test als Mittel der Verhaltensbeeinflussung. Exemplarisch sei die verbreitetste davon genannt: „Nur wer seinen Immunstatus kennt, kann sich verantwortlich verhalten." Dieser Satz ist als generelle Aussage unhaltbar (vgl. die diesbezügliche Literaturübersicht über mehr als 40 Studien bei Michel, 1988). Sie kann sich darüber hinaus auch häufig ausgesprochen antipräventiv auswirken. Zum einen weil sie, indem sie den Test zur Prävention erklärt, vom eigentlichen Inhalt der Präventionsbotschaft ablenkt; zum andern, weil sie etwa bei jemandem mit einem negativen Testergebnis eine falsche Sicherheit erzeugen kann. Weltweit hat sich gezeigt, daß für den Erfolg von Prävention nicht der HIV-Test entscheidend ist, sondern die Qualität der allgemeinen und zielgruppenspezifischen Aufklärung und vor allem der persönlichen Beratung, in deren Rahmen dann auch individuell die Indikation des HIV-Tests erwogen werden kann.

Es ist nicht mehr wiedergutzumachen, daß durch den unkritischen massenhaften Einsatz des HIV-Tests iatrogen unnötiges Leid bei vielen erzeugt wurde, deren einziges, allerdings ihr Leben oft dramatisch veränderndes Symptom die Diagnose „HIV-positiv" ist[8] (vgl. Abschn. 49.6).

Aber es ist möglich, daß bei ärztlichen HIV-Beratungen in Zukunft nicht mehr der Test im Vordergrund steht, sondern die konkrete Lebenssituation des Ratsuchenden. Das setzt allerdings beim Arzt nicht nur voraus, daß er grundsätzlich erkannt hat, daß verantwortliches ärztliches Handeln auch das Unterlassen einer technisch möglichen diagnostischen Methode bedeuten kann. Er muß sich darüber hinaus sehr genau seiner Motive für die Anwendung des HIV-Tests bewußt sein und in jedem Einzelfall abwägen, ob er das, was er mit dem Test erreichen will, nicht auch ohne ihn erreichen könnte. Zu dieser Motivanalyse gehört auch die Reflexion seiner politischen und gesundheitspolitischen Einstellung. Das widerspricht zwar dem Selbstverständnis vieler Ärzte („Im Sprechzimmer hat Politik nichts verloren"), ist aber bei AIDS und im Umgang mit dem HIV-Test unerläßlich, damit nicht unbemerkt Gedanken wie „Man muß die vielen Gesunden vor den wenigen Kranken schützen" die Beziehung zum Patienten verunmöglichen.

Unstrittig ist die Anwendung des HIV-Tests zur Differential- und Ausschlußdiagnostik bei unklaren Krankheitsbildern, in der Prüfung von Blutprodukten, Transplantaten und Samenspenden. Ohne entsprechenden diagnostischen Anlaß ist die Testindikation nur im Einzelfall zu stellen.

---

8 Ein großer Teil der von uns befragten HIV-Positiven gibt an, wenn sie heute noch einmal vor der Wahl stünden, den Test nicht mehr machen zu lassen.

Ein in seinem Fach sehr erfolgreicher 24jähriger homosexueller Student, der kurz vor seinem Examen steht, kommt zur Beratung. Er ist bestens über AIDS informiert. Er hat seit seinem relativ späten Coming out mit 20 Jahren sehr promiskuitiv gelebt, hatte in manchen Phasen mehrere anonyme Sexualpartner pro Woche, mit denen er vorwiegend in Parks verkehrte. Einen festen Partner hat er gegenwärtig nicht. Er war zunächst zum Test beim Gesundheitsamt entschlossen gewesen, hatte sich dann aber von einem Freund davon abraten lassen. Er ist der Ansicht, daß er aufgrund seiner sexuellen Lebensweise in den letzten Jahren mit einer gewissen Wahrscheinlichkeit HIV-positiv sein könnte. Er meint jedoch, wenn er sicher wisse, daß er positiv sei, könne er das psychisch nicht verkraften. Er brauche das Bewußtsein eines gesunden Körpers, auch wenn das eine Illusion sein könnte. Er schildert hochambitionierte berufliche Zukunftspläne und meint, sein Lebensgefühl, im Vollbesitz seiner Kräfte zu sein, bräche zusammen, wenn er wisse, daß der Test positiv sei. Auch wenn er die Wahrscheinlichkeit für hoch halte, klammere er sich doch an die Chance, nicht infiziert zu sein. Das gelingt ihm unterschiedlich gut, er fällt zwischendurch immer wieder in Zweifel, ob er seine Entscheidung nicht doch auf eine Illusion aufbaut. Dennoch will er den Test nicht machen. Er kommt nun mit dem Problem, daß er zwar auf bestimmte Sexualpraktiken verzichtet, aber das Gefühl, sexuell getrieben zu sein und sexuelle Abenteuer suchen zu müssen, schwer beherrschen kann. Er sagt: „Ich werfe mir meine Geilheit vor, hasse diese sexuelle Unruhe auch, aber davon geht sie schließlich nicht weg." Nach einigen orientierenden Gesprächen entschließt sich der Patient zu einer ambulanten Psychotherapie, weil er seine sexuelle Getriebenheit besser verstehen will.

Bei diesem Patienten würde der HIV-Test das mögliche Risiko einer depressiven Dekompensation bergen, das durch keinen gleichwertigen positiven Effekt gerechtfertigt wäre; auch keinen präventiven, da der Patient bereits jetzt auf infektionsriskante Sexualpraktiken verzichtet. Ähnliches gilt für alle, die ihr sexuelles Verhalten ohnehin verändern, weil sie sich selbst vor einer Infektion schützen wollen; dies ist z. B. bei vielen promiskuitiv lebenden Homosexuellen der Fall. Natürlich ist die Veränderung des sexuellen Verhaltens (Verzicht auf bestimmte Praktiken oder Gebrauch von Präservativen) oft schwierig und stellt einen längeren Prozeß dar. Der HIV-Test als solcher fördert das Gelingen dieser Auseinandersetzung jedoch nicht, er kann sie sogar erschweren. Alle Erfahrungen zeigen, daß die meisten, die infiziert sein könnten, dies auch wissen (das ändert nichts an dem Schock, den das positive Ergebnis auslöst) und sich mit den Schutzmöglichkeiten auseinandersetzen. Dafür müssen genügend Hilfestellungen angeboten werden. Umgekehrt lassen viele den Test machen, um sich bestätigen zu lassen, daß sie negativ sind.

Auch wenn – bedingt durch die bisherige Testpolitik – die meisten schon mit dem Testwunsch zum Arzt kommen, muß dieser sich zunächst Zeit für ein ausführliches Gespräch nehmen. Dieses beinhaltet eine Risikoanamnese, eine Risikoaufklärung und eine Präventionsberatung, wobei alle drei sehr individuell

gehalten sein müssen. In diesem Gespräch kann der Arzt/Berater viele wichtige Hinweise auf Probleme des Patienten und auf möglicherweise notwendige Hilfen erfahren. Besteht im Anschluß immer noch ein Testwunsch seitens des Patienten, muß man mit ihm besprechen, was ein negatives, was ein positives und was die Ungewißheit über seinen Sero-Status für ihn ganz persönlich in seiner spezifischen Lebenssituation bedeuten würde. Schließlich muß der Arzt auf die auch bei optimaler Testhandhabung gegebene Möglichkeit eines falsch-negativen bzw. falsch-positiven Ergebnisses hinweisen.[9]

Die Entscheidung für oder gegen den Test muß beim Patienten liegen. Die Beratung kann ihm nur dabei helfen, die für ihn individuell richtige Entscheidung zu treffen. Eine Beratung, die generell zum Test auffordert oder generell davon abrät, stellt keine adäquate Hilfe dar.

Die Akzente liegen natürlich bei jeder Beratung anders. So sieht eine Beratung bei einem 15jährigen Mädchen, das nach dem ersten sexuellen Verkehr Angst hat, sich angesteckt zu haben, anders aus als bei einer ehemaligen Fixerin, die schwanger ist, wiederum anders bei einem Bluter mit Kinderwunsch und noch einmal anders bei einem homosexuellen Paar mit gelegentlichen sexuellen Außenkontakten, das innerhalb der festen Beziehung ohne Einschränkungen Sexualität haben will.

49.5 Prävention

49.5.1 Die Politik der Prävention

Solange AIDS und die HIV-Infektion nicht medizinisch behandelbar sind und solange nicht mit einem Impfstoff zu rechnen ist, kommt der Verhinderung neuer Infektionen eine zentrale Aufgabe zu. Über die präventiv orientierte Gesundheitspolitik gibt es in der BRD zwei konkurrierende Auffassungen. Beide sind sich hinsichtlich der Bedeutung von Aufklärung, Beratung und freiwilligem Selbstschutz einig, differieren aber in der Einschätzung der Notwendigkeit zusätzli-

---

9 Es handelt sich hierbei um ein statistisches Phänomen (vgl. Rosenbrock, 1988; Wittkowski, 1988): Je wahrscheinlicher das individuelle Risiko, desto geringer die Möglichkeit eines falsch-positiven Ergebnisses. Eine Studie aus dem Office of Technology Assessment (OTA) des US-Kongresses kommt zu dem Ergebnis, daß bei Ausschluß aller Laborfehler auf 100 000 – in welcher Gruppe auch immer – durchgeführte Tests fünf falsch-positive Ergebnisse kommen. Epidemiologisch ausgedrückt: In einer Population mit einer HIV-Prävalenz von 0,01% befinden sich neben den zehn positiven Testergebnissen auf 100 000 durchgeführte Tests auch fünf falsch-positive Befunde. Von den insgesamt 15 positiven Testergebnissen ist ein Drittel falsch-positiv. Es gibt außer der Langzeitbeobachtung keine Möglichkeit herauszufinden, welches Testergebnis zu diesem Drittel gehört.

Wird dagegen eine Population mit einer HIV-Prävalenz von 10% getestet, werden 10 000 positive Ergebnisse auf 100 000 durchgeführte Untersuchungen erwartet. Hinzu kommen – auch hier – fünf falsch-positive. Die Wahrscheinlichkeit, daß ein positives Testergebnis in dieser Gruppe auch zutreffend ist, beträgt 10 000 : 5, das sind 99,95%.

cher staatlicher Überwachungs- und Interventionsmaßnahmen. Die gegenwärtig mehrheitlich liberale
Position hält eine Regelung durch hoheitliche Mittel
nur in Ausnahmefällen eng definierter Verantwor-
tungslosigkeit für notwendig. Die konservative Position hält neben der Aufklärung einen Maßnahmen-
katalog für erforderlich, der sich am Bundesseuchenge-
setz orientiert und für sog. „Uneinsichtige" Untersu-
chungspflicht, Überwachung und gegebenenfalls seu-
chenpolizeiliches Einschreiten vorsieht. Unter psychologischen Gesichtspunkten ist dieser aufklärungs-
skeptischen Position entgegenzuhalten, daß durch
die Kontroll- und Strafandrohung vor allem der Eigenverantwortung Nichtinfizierter entgegengearbeitet wird. Sie erweckt nämlich den Eindruck, der Staat
könne den einzelnen vor einer Infektion schützen,
was ein falsches Sicherheitsgefühl schafft und zudem
Projektionen begünstigt, die Infizierte in die allein-
verantwortliche Täterrolle bringen.

49.5.2 Prävention durch Beratung

Während die allgemeine Aufklärung zwangsläufig
dieselbe einfache Botschaft für sehr viele unterschiedliche Menschen vermitteln muß, bezieht sich
die Beratung auf die individuelle Lage. Die Beratung,
die vor allem von Allgemeinärzten und Mitarbeitern
psychosozialer Einrichtungen geleistet wird, muß
sich sensibel auf den individuellen Lebensstil, insbesondere im Sexuellen, einstellen. Dabei sollte folgendes gelten:
- Die Ansteckungsmöglichkeiten müssen einfach
  dargestellt werden.
- Die relevanten sexuellen Sachverhalte müssen
  beim Namen genannt werden.
- Es muß vermittelt werden, daß die Ansteckungs-
  möglichkeit in bestimmten Situationen, nicht bei
  bestimmten Menschen liegt. Dadurch wird die irre-
  führende Empfehlung vermieden, man solle sich einen neuen Sexualpartner „genau ansehen", ehe man
  sich auf ihn einlasse. Ein sexuell zurückhaltender
  Mensch kann HIV-positiv sein, ein sexuell freizügiger Mensch kann HIV-negativ sein.
- Bei den sexuellen Verhaltensänderungen, die der
  einzelne vornehmen will, ist zu beachten, wie sie
  mit der Sexualmoral und seiner sexuellen Biographie im Einklang stehen. Veränderungen, die gegen die individuelle Moral oder sexuelle Triebrich-
  tung beabsichtigt werden, lassen sich nicht dauerhaft integrieren.
- Das Beratungsgespräch muß dem einzelnen auch
  die Konsequenz lassen, nichts an seinem sexuellen
  Leben zu verändern, selbst wenn dadurch Infek-
  tionsmöglichkeiten offenbleiben. Ein Berater, der
  diese Möglichkeit nicht zuläßt, riskiert, daß er von
  dem ratsuchenden Patienten nicht die Wahrheit erfährt oder daß dieser sich andere Berater sucht. Es
  liegt in der Verantwortung jedes Ratsuchenden,
  wie er mit einem geringen „Restrisiko" umgeht.
- Der Berater darf seine Vorstellungen über Sexualität und Normalität nicht zum Maßstab der Bera-

tung machen. Es ist nicht die Aufgabe von Beratungen, kollektiv und ohne Berücksichtigung der Person zu sexueller Treue oder zum Kondomgebrauch
aufzurufen.
Die zusätzlichen Probleme, die sich für getestete „Positive" ergeben, werden in Abschnitt 49.6 erörtert.

49.5.3 Folgen der Prävention

Es zeigen sich Verhaltensänderungen, vor allem bei
homosexuellen Männern und bei i.v.-Drogenabhän-
gigen, die in ihrem Ausmaß und ihrer Geschwindigkeit krankheitspräventives Verhalten in anderen Bereichen weit in den Schatten stellen.

Bei homosexuellen Männern hat die Häufigkeit des
Analverkehrs rapide abgenommen. Er wird außerdem
häufiger mit Kondom praktiziert; vor allem unter
nicht monogam lebenden hat die Partnerfluktuation
erheblich abgenommen (McKusick et al., 1985; Emmons et al., 1986; Martin, 1987; Winkelstein et al.,
1987; Literaturübersicht bei Becker und Joseph,
1988). Dabei ist die praktikbezogene Verhaltensänderung (seltener Analverkehr, und wenn, dann mit Kondom) bei Nicht-Partnergebundenen deutlich größer.
Partnergebundene Homosexuelle verändern ihre
Praktiken weniger. Ein häufiges neues Verhaltensmuster ist die beidseitige HIV-Testung und bei negativem
Ausgang eine Fortführung der gewohnten Sexualpraktiken, aber auch die Einhaltung differenzierter
Verhaltensregeln (innerhalb der Partnerschaft ohne,
außerhalb mit Kondom). Daß die soziale Nähe zum
Problem einen maßgeblichen Einfluß auf infektions-
präventives Verhalten hat, zeigt sich in einer Studie
von Calabrese et al. (1985), die in einer „low-inci-
dence-area" nur geringe Veränderungen im sexuellen
Verhalten fanden. Ähnliches gilt für heterosexuelle
Jugendliche (Strunin und Hingson, 1987; Kegeles,
1988), die nur eine geringe Bereitschaft zur Veränderung ihres sexuellen Verhaltens haben. Heterosexuelle Männer in Städten mit vielen AIDS-Fällen dagegen
berichten von einer erheblichen Reduktion der Zahl
ihrer Sexualpartner (Winkelstein et al., 1987). Auch
Alkohol- und Drogengenuß vermindern die Bereitschaft zu risikominimierendem Verhalten (Stall et al.,
1986).

Bei i.v.-Drogenabhängigen werden, entgegen dem
Stereotyp ihrer Uneinsichtigkeit, Verhaltensänderungen berichtet. Die Verwendung von Einwegspritzen
hat bei ihnen zugenommen (Friedmann et al., 1986;
Chiasson, 1987).

49.6 Die psychische Situation von HIV-Positiven und AIDS-Kranken

49.6.1 Die Mitteilung eines positiven Testergebnisses

Die Mitteilung eines positiven Testergebnisses erfordert größte beraterische Sorgfalt. In den meisten Fäl-

len erfährt der Betreffende das Ergebnis vom Hausarzt oder beim Gesundheitsamt. Der Arzt oder Berater muß sich darüber im klaren sein, daß er eine tief in das Leben eingreifende Mitteilung macht und daß mit dekompensatorischen Reaktionen bis hin zur Suizidalität zu rechnen ist, auch wenn der betreffende Patient im ersten Moment gefaßt erscheint.

Bei der Erörterung der Möglichkeit, in nicht vorhersehbarer Zeit an AIDS zu erkranken, ist die Nennung quantitativer Wahrscheinlichkeiten nicht hilfreich, ganz abgesehen davon, daß langfristige Verlaufsaussagen gegenwärtig nicht zuverlässig genug gemacht werden können. Dem Patienten muß vermittelt werden, daß er auch die Chance hat, gesund zu bleiben und nicht an AIDS zu erkranken. Dieser Aspekt der realistischen Hoffnung ist psychologisch von größter Bedeutung, weil er mitentscheidend dafür ist, ob der Patient die Mitteilung innerlich kompensieren kann.

Es empfiehlt sich, das erste Gespräch mit der Ergebnismitteilung nicht mit zuviel Information zu befrachten, sondern mindestens ein zweites Gespräch zu vereinbaren, nachdem der Patient die Mitteilung auch emotional ganz aufgenommen hat. In bezug auf die sexuellen Verhaltensempfehlungen muß der Arzt oder Berater sich bewußt sein, daß sie unterschiedlich leicht zu befolgen sind. Der Hinweis auf die Verantwortung des Patienten, die Infektion nicht weiterzugeben, ist richtig, aber für praktisch jeden HIV-Positiven ohnehin offensichtlich. Wichtiger als ein eindringlicher Appell ist das deutliche Angebot, bei „Rückfällen" und Schwierigkeiten bei den sexuellen Veränderungen zur Verfügung zu stehen. Dies ist nur möglich, wenn der Berater vermittelt, daß er nicht nur die Seite des verantwortlichen Gewissens, sondern auch die der beeinträchtigten Trieb- und Beziehungswünsche sieht. Ein Berater, der diese Ambivalenz nicht erträgt und einseitig mit dem öffentlichen Interesse der Nichtweitergabe von Infektionen identifiziert ist, wird von der psychosexuellen Lage eines HIV-Infizierten höchstens ausschnittweise etwas erfahren und verstehen. Erst wenn er auch das Recht eines Positiven auf gelebte Sexualität bejaht, kann ein Gespräch stattfinden, in dem auch Trauer über die Verluste und Schwierigkeiten bei der Kontrolle des sexuellen Verhaltens Platz haben.

49.6.2   Konfliktschwerpunkte bei HIV-Positiven

Der bedrohliche Mittelpunkt im Leben eines asymptomatisch HIV-Positiven ist das „Zeitbombengefühl", manifest gesund, potentiell aber todkrank zu sein.

Todesangst

Im subjektiven Erleben von HIV-Positiven kann die Angst vor dem Tod unterschiedliche Gestalt annehmen. Abhängig von der psychischen Struktur kann sie als Angst vor körperlichem Zerfall und Siechtum, also als narzißtische Katastrophe, als Angst vor Schwäche und Angewiesensein auf andere, also vor

dem Zusammenbruch der Autonomie, oder als Angst vor dem „sozialen Tod", vor Ausstoßung und Vernichtung durch die feindseligen Gesunden erlebt werden.

Ein 42jähriger Mann, dessen Freund ein halbes Jahr zuvor an AIDS verstorben ist, befindet sich selbst im Stadium WR 5 (AIDS-Related Complex). Er ist verheiratet, hat zwei Kinder. Seine Frau weiß erst seit wenigen Jahren von seiner Homosexualität, seit wenigen Monaten von seiner Infektion. Zu ihr besteht ein funktionales, aber kein emotional nahes Verhältnis. Der gutaussehende Patient ist in seinem kaufmännischen Beruf mit viel Publikumsverkehr sehr erfolgreich und beliebt. In seiner Jugend und seinem frühen Erwachsenenalter war er erstklassiger Tennisspieler, hatte in seinem Heimatort eine Art Star-Rolle. Nach einem späten Coming out mit fast dreißig Jahren versucht er immer wieder, seine Homosexualität vor sich selbst zu verleugnen, bis er seinen zehn Jahre jüngeren Freund kennenlernt. Diese Beziehung erlebt er in einer Intensität wie keine andere zuvor; der künstlerisch versierte Freund eröffnet ihm Welten, die ihm, der aus einem armen und wenig anregenden Elternhaus stammt, völlig neu sind. Er berichtet von einem nie gekannten Glanz und inneren Reichtum in dieser Lebensphase. Er zieht mit dem Freund in die benachbarte Stadt, kurz nachdem beide von ihrer HIV-Infektion erfahren, hält aber den Kontakt mit seiner Frau und den Kindern weiter aufrecht. Als der Freund ein halbes Jahr nach der Diagnose stirbt, gerät der Patient in eine suizidale Krise und wird kurzfristig stationär aufgenommen. Auch nach einigen Monaten ist er zutiefst hoffnungslos, kann sich ein Leben ohne den idealisierten Freund nicht vorstellen, lehnt auch ein psychotherapeutisches Angebot ab, das er als perspektivlos empfindet. Für ihn ist es unvorstellbar, bei vollem Ausbruch von AIDS sich von seiner Frau oder jemand anderem pflegen zu lassen, für ihn ist die Rolle des hilflos bedürftigen Kranken eine größere subjektive Katastrophe, als es der physische Tod wäre. Zudem beschäftigt ihn quälend die Phantasie, daß durch seine Krankheit, die er mit Gewißheit und unabwendbar erwartet, seine streng verheimlichte Homosexualität bekannt würde, was er damit verbindet, daß er der Verachtung seines ihn sonst so bewundernden sozialen Umfeldes ausgesetzt wäre. Er zieht noch einmal für kurze Zeit zu seiner Familie zurück und nimmt sich dann während seines Urlaubs das Leben.

Dieser Patient mit einer vorwiegend narzißtischen Charakterstruktur erlebt die Todesbedrohung vorwiegend als Verlust der Unversehrtheit seines Körpers und sich selbst als in körperlichem Verfall befindlichen Abhängigen, für den sogar der Abschied von seiner Familie und von unerledigten Lebensaufgaben in den Hintergrund tritt. Durch den Suizid rettet er sein glanzvolles Selbstbild vor dem erwarteten Siechtum, das für ihn schlimmer ist als der physische Tod. Was dieser Patient in die Tat umsetzte, ist für viele HIV-Positive zumindest eine passagere oder in Krisen wiederkehrende Phantasie: durch Selbstmord dem Leidensprozeß beim Ausbruch von AIDS zu entgehen.

Diskriminierungsangst

Eine Besonderheit gegenüber anderen Krankheiten liegt darin, daß eine hohe Erwartung feindseliger Reaktionen der näheren Umwelt viele HIV-Positive dazu bringt, ihren Immunstatus und auch Krankheitsanzeichen zu verbergen. Solche Erwartungen sind – bei der negativ besetzten Aura von AIDS – oft durchaus realistisch. Sie können aber auch das Ergebnis einer projektiven Abwehr einer schuldhaft erlebten sexuellen Lebensform sein. Die Unterscheidung von projektiver Abwehr und realer Ausstoßung und Feindseligkeit muß bei der Situation von HIV-Positiven besonders genau beachtet werden (Becker und Clement, 1989). Der Psychotherapeut oder der betreuende Arzt muß hier sensibel sein für ihre politisch-juristische Lage und das soziale Klima, in dem sie leben. Es gibt eine reale Diskriminierung (z. B. beruflich, versicherungsrechtlich), strafrechtliche Sanktionsandrohungen für bestimmte sexuelle Aktivitäten und ein trotz relativer Toleranz insgesamt homosexuellenfeindliches Gesamtklima. Erst die Kenntnis und Berücksichtigung dieser äußeren Realität erlaubt eine Einschätzung, inwieweit eine Wahrnehmung projektiven Charakter hat oder eine Verfolgungsangst als paranoid zu bezeichnen ist.

Sexualität und Partnerschaft

Die Sexualität HIV-Positiver steht in spezifischer Weise im Mittelpunkt ihres Erlebens. Die meisten haben sich auf sexuellem Weg infiziert, alle können ihre Infektion beim Geschlechtsverkehr weitergeben. Von ihnen wird erwartet, daß sie lebenslänglich ihr sexuelles Verhalten umstellen. Das bedeutet eine Einschränkung nicht nur des sexuellen Lusterlebens, viel erheblicher sind Beeinträchtigungen der Stabilisierungsfunktion, die Sexualität im psychischen Geschehen hat. Durch die starke innere und äußere Belastung sind HIV-Positive auf psychische und partnerschaftliche Kompensationsmöglichkeiten angewiesen, für die auch die Sexualität eine zentrale Rolle spielt. Die regressive Funktion der Sexualität (Hingabe, Anvertrauen) ist in der belasteten Lage meist von größerer Bedeutung als die „hedonistische" Lust. Das ist bei Schwierigkeiten mit der Umstellung des sexuellen Verhaltens (insbesondere die ausnahmslose Kondomverwendung beim Verkehr) zu beachten, die nicht – wie eine rationalistische Präventionslinie postuliert – auf Mangel an Einsicht zurückzuführen sind. Ein ungeschützter Geschlechtsverkehr kann ein unbewußter Selbststabilisierungsversuch in einer depressiven Krise sein; er kann verstehbar sein als manisch-animistische Abwehr der Todesangst („ich bin nicht ansteckend, also bin ich nicht infiziert, also sterbe ich nicht an AIDS"), aber auch als unbewußter Verschmelzungswunsch („ich infiziere den geliebten Partner und gehe mit ihm in den Tod").

Der erheblich häufigere Fall ist jedoch eine auch manifeste Beeinträchtigung der Sexualität im Sinne eines reduzierten sexuellen Verlangens, auch gelegentlicher Erektionsstörungen, zumindest in den er-

sten Monaten nach der Diagnosestellung. Verstehen läßt sich diese sexuelle Depression aus der Vermischung des sexuellen Verlangens mit der Angst, dem Partner etwas zuzufügen, und damit einer schuldhaften Überlagerung der sexuellen Wünsche. Auch bei konsequenter Kondombenutzung können solche sexuellen Schwierigkeiten auftreten. Dies liegt an der **Paradoxie des Kondom**gebrauchs, der zwar eine Neuinfektion verhindert, gleichzeitig aber an die Infektionsmöglichkeit erinnert und damit die sexuelle Phantasie durchdringt.

Von großer Bedeutung ist es, ob der Partner ebenfalls HIV-positiv ist. Psychisch ist diese Situation stabiler und weniger trennungsgefährdet, als wenn der Partner nicht infiziert ist. Bei ungleichem HIV-Status der Partner entstehen oft neue Kräfteverhältnisse in der Beziehung. Für viele HIV-Positive ist es eine sehr heikle Situation, wenn sie ihre Diagnose erfahren, dann die Stütze des Partners besonders brauchen, aber in Sorge sind, ob er sich nicht von ihnen trennt. Es erfordert ein hohes Maß an Stabilität beider Partner, um die partnerschaftliche Balance aufrechtzuerhalten. Unseren bisherigen Erfahrungen zufolge ist die Trennungsfrage bei heterosexuellen Paaren virulenter als bei homosexuellen Paaren.

Bei partnerlosen HIV-Positiven stellt sich fast immer ein Gefühl der Hoffnungslosigkeit ein, vielleicht nie wieder einen Partner zu finden. Manche versuchen, über eine gezielte Kontaktanzeige einen ebenfalls testpositiven Partner zu finden.

49.6.3 Psychische Bewältigungsmechanismen

Für die psychische Bewältigung ist es günstig, wenn HIV-Positive zumindest zeitweise ihre Diagnose und deren mögliche Implikationen verleugnen können, also in der Lage sind, den größten Teil ihres Lebens ohne Gedanken an die Todesbedrohung durch die Infektion zu verbringen. Diese stabilisierende Funktion der Verleugnung ist auch aus der Krebsforschung bekannt. Für HIV-Positive kann das heißen, daß sie die psychisch belastenden medizinischen Untersuchungen meiden.

Neben der **Verleugnung** spielen bei HIV-Positiven zwei charakteristische Bewältigungsformen eine große Rolle. **Altruismus**, d. h. anderen HIV-Positiven oder AIDS-Kranken zu helfen, kann die eigene Lage relativieren, kann vor allem die Angst vor sozialer Diskriminierung und vor dem Alleingelassenwerden binden. Die **aktive Auseinandersetzung** mit der Infektion, also das Gegenteil von Verleugnung, geschieht meist in Form bewußter Ernährung und eines Überdenkens der bisherigen Lebensweise. Die psychische Kompensierung der HIV-Diagnose scheint am ehesten von dem Bewußtsein abzuhängen, durch eigenes Verhalten den Erkrankungsbeginn verhindern oder hinauszögern zu können.

49.6.4 Zur besonderen Situation von AIDS-Kranken

AIDS-Kranke sind in vieler Hinsicht mit anderen unheilbar Kranken vergleichbar (vgl. Kap. 62 und 66). Ihre besondere Situation ist zum einen dadurch gekennzeichnet, daß sie relativ jung sind, die meisten Erkrankten sind 30 bis 40 Jahre alt, also in einem Alter, in dem Zukunftspläne und berufliche Ambitionen lebendig und unabgeschlossen sind. Zum anderen unterscheidet sich ihre Situation von der anderer Schwerkranker dadurch, daß sie eine ansteckende Krankheit haben und daß sie meistens einer sozial diskriminierten Minderheit angehören (homosexuelle Männer oder i.v.-Drogenabhängige), was erhebliche Folgen für ihr Selbsterleben und für den Umgang anderer mit ihnen hat. Ein besonderes Problem für AIDS-Patienten ist, daß ihr „sicherer" Tod so oft in der Öffentlichkeit beschworen wird, daß sie individuell weniger Raum für Hoffnung entwickeln können als z. B. Krebskranke. (Real hat sich durch die verbesserten medizinischen Behandlungsmöglichkeiten die Lebensdauer vieler AIDS-Kranker erheblich verlängert; auch ihre Lebensqualität hat sich zum Teil sehr verbessert, wobei der ambulanten Versorgung eine entscheidende Bedeutung zukommt.) Auch haben sie oft das Gefühl, daß ihr Tod gesellschaftlich erwünscht ist, was ihren persönlichen Trauerprozeß erschweren kann. Bei homosexuellen AIDS-Kranken kommt es oft zu schwierigen Konflikten zwischen der Primärfamilie (falls sie nicht den Kontakt abgebrochen hat) und dem Lebenspartner, die sich häufig am Krankenbett des Patienten zum ersten Mal begegnen. Hier kann ein Dritter (Arzt oder Psychotherapeut oder Sozialarbeiter) hilfreich sein, z. B. um eine Regelung zu finden, die dem Patienten die (nach allen Erfahrungen nur zu berechtigte) Sorge nimmt, nach seinem Tod würden seine Eltern dem Freund alles wegnehmen.

49.6.5 Die spezifische Situation der Betroffenengruppen

Homosexuelle Männer

Die größte Gruppe unter den HIV-Positiven und AIDS-Kranken sind nach wie vor homosexuelle Männer.[10] Bei vielen homosexuellen HIV-Positiven vermischt sich die Angst vor einer möglichen Erkrankung mit einer starken Verunsicherung ihrer homosexuellen Identität. Der Prozeß der Annahme der eigenen homosexuellen Orientierung, das sog. „Coming out", ist ein individuell ganz unterschiedlich verlaufender krisenhafter Prozeß, der auch häufig mißlingt und Beschädigungen hinterläßt (Becker, 1987). Im Idealfall wird die eigene Homosexualität entdeckt, zunächst zurückgedrängt, kann dann allmählich akzeptiert, in die eigene Identität integriert und schließlich auch nach außen hin vertreten werden. Auch das ideal verlaufende Coming out spielt sich immer im Spannungsfeld von drei Polen ab: dem eigenen Trieb-

schicksal, der sozialen Diskriminierung der Homosexuellen und der verinnerlichten eigenen Ablehnung der Homosexualität (Dannecker und Reiche, 1974). Der Kampf um die Annahme der eigenen Homosexualität ist mit dem Coming out nicht abgeschlossen. Bei vielen bleibt das Selbstwertgefühl als Homosexueller labil und leicht labilisierbar. Die Brüche im Coming out werden durch die Diagnose „HIV-positiv" um so heftiger reaktiviert, je konflikthafter die homosexuelle Orientierung für den einzelnen geblieben ist. Es kann zu massiven Schuldgefühlen und Selbstverurteilungen kommen, die sich auf die homosexuelle Orientierung und die Art beziehen, in der sie gelebt wurde (z. B. Promiskuität). Die gesellschaftliche Diskriminierung verstärkt die Selbstdiskriminierung.

> Ein 65jähriger Patient wird von der Hautklinik in die Psychosomatische Klinik überwiesen, nachdem er auf die Mitteilung seines HIV-positiv-Befundes hin in eine depressive Krise gerät, verbunden mit passagerer Suizidgedanken, Schlaflosigkeit und Angstzuständen. Ihm sei seine Infektion rätselhaft; er habe seit Jahren kaum noch Sexualpartner gehabt, über Einzelheiten läßt er sich nicht befragen. Seine homosexuelle Neigung verurteilt er mit aggressiver Heftigkeit, Homosexuelle seien „schlimmer als Huren und Zigeuner". Sein Coming out hat er in der nationalsozialistischen Zeit gehabt, mit deren Ideologie er affektiv noch sehr verbunden ist. Heute noch ist er, der als Kind ein massiv abwertendes Elternhaus erlebt hatte, stolz auf die Orden, die er als damals begeisterter Kriegsflieger erhalten hat. Die homosexuellenfeindliche Ideologie seiner prägenden Jugendzeit hat er sich bis hin zu Vernichtungsgedanken zu eigen gemacht: „Für meine Homosexualität hätte man mich erschlagen sollen." Andere Homosexuelle, die sich offen zeigen, verachtet er. Bis heute ist er mit den Homosexuellenverfolgern mehr identifiziert als mit den Verfolgten. Entsprechend fürchtet er, falls seine Infektion bekannt werden sollte, nicht mehr an seinem Wohnort leben zu können, hat eine panische und völlig unbeeinflußbare Angst, von den Nachbarn verstoßen und an den Pranger gestellt zu werden. Von seiner sozialen Vernichtungsangst ist er auch nicht durch positive Erfahrungen abzubringen, etwa nachdem er sich einer langjährig vertrauten Nachbarin geöffnet hatte und diese sich sehr loyal zeigte.

Homosexuelle AIDS-Kranke müssen sich oft mit der Erfahrung sozialer Isolation auseinandersetzen, wodurch frühere (durch die homosexuelle Entwicklung bedingte) Phasen der Isolierung schmerzlich wiederbelebt werden.

Manche Erkrankte sind durch die Erkrankung gezwungen, in ihre Herkunftsfamilie zurückzukehren, oft auch dann, wenn sie in den Jahren davor wenig Kontakt zu ihr hatten. AIDS-Patienten empfinden (anders als z. B. Krebspatienten) nicht nur eine innerliche Isolation, sondern hier rückt die Familie (oder Teile von ihr) oft real weg, häufig mit offenen und

---

10 Über 70% der AIDS-Kranken; für die HIV-Infizierten gibt es keine verläßlichen Zahlen.

versteckten Schuldzuweisungen.[11] Für viele Patienten stellt sich das zusätzliche Problem, ihrer Familie zugleich mit ihrer Krankheit auch ihre Homosexualität offenbaren zu müssen.

i.v.-Drogenabhängige und Ex-User

Ehemals drogenabhängige Frauen und Männer erleben die Diagnose „HIV-positiv" sehr unterschiedlich: Manche werden dadurch in ihrer Entwicklung sehr labilisiert, reagieren resignativ („wozu habe ich mir die ganze Mühe des Entzugs gemacht, wenn ich jetzt doch krank werde?") und sind rückfallgefährdet. Manche werden durch die Diagnose „HIV-positiv" regelrecht von ihrer Vergangenheit eingeholt, z.B. wenn sie eine neue Partnerschaft eingegangen sind und der Partner über die frühere Drogenabhängigkeit nichts weiß. Andere erleben durch die Diagnose ihren Lebenswillen wieder stärker, mobilisieren neue Energien in sich und machen zum Teil erstaunliche Entwicklungen aus dem Gefühl heraus: „Seitdem ich weiß, daß ich positiv bin, kämpfe ich um mein Leben." Ähnliche Reaktionen wurden auch bei Drogenabhängigen beobachtet, die während einer stationären Entwöhnungstherapie ihr HIV-positiv-Ergebnis erfuhren: Durch die Mitteilung der Diagnose „HIV-positiv" veränderten sich die Lebensziele (vor allem in bezug auf soziale Einbettung und Selbstverwirklichung), und die Motivation zur Drogenfreiheit stieg (Kochanowski-Wilmink und Belschner, 1988).

Im Gegensatz dazu ist die psychische Situation von HIV-positiven Drogenabhängigen, die nicht in Therapie sind bzw. schon zahllose gescheiterte Therapien hinter sich haben, meistens völlig desolat und vor allem von Resignation gekennzeichnet. Unter dem ständigen Beschaffungsdruck zur Finanzierung der Droge, der zu Kriminalität und Beschaffungsprostitution führt, empfinden viele die HIV-Infektion noch als ihr geringstes Problem. In anderen Ländern (z.B. Holland, Schweiz), in denen es für Heroinabhängige zum einen Substitutionsprogramme und zum anderen ein breites Angebot an niedrigschwelligen (also nicht Drogenfreiheit voraussetzenden) therapeutischen und sozialen Hilfsangeboten gibt, ist die psychische und soziale Situation der HIV-positiven Drogenabhängigen, die nicht zu einer Entwöhnungstherapie in der Lage sind, erheblich besser.

Bluter

Der gegenüber homosexuellen Männern und Drogenabhängigen in der Öffentlichkeit erhobene Vorwurf, sie bzw. ihr Lebensstil seien schuld an ihrer Infektion, wird gegenüber Blutern nicht gemacht. Sie haben aber auch oft unter sozialer Diskriminierung zu leiden; innerhalb ihrer Familien erfahren sie jedoch meistens keine Abgrenzung.

Gerade junge Bluter neigen oft dazu, ihre Grunderkrankung zu verleugnen und sich besonders waghalsig und wenig auf Verletzungen achtend zu verhalten. Diese Verleugnung bricht durch die HIV-Infektion zusammen. Manche Eltern infizierter Minderjähriger

verschweigen diesen die Infektion. „Doch meist spüren die Kinder sehr wohl die Angst und Belastung der Eltern, so daß sich ein gegenseitiges Verstecken unter dem Vorwand des Beschützenwollens entwickelt" (Pohlmann und Schramm, 1988).

Manche junge Bluter trifft die Diagnose „HIV-positiv" mitten in der pubertären Entwicklung. Einige erleben die Unsicherheiten bei der Familienplanung als zusätzliche Belastung.

49.7 Die psychotherapeutische und medizinische Versorgung

49.7.1 Medizinische Versorgung

Dem symptomlosen HIV-Infizierten kann keine medizinische Behandlung angeboten werden[12]. Gleichsam als Ersatz dafür beginnt mit dem positiven Testergebnis das sog. „Staging", d. h. umfassende diagnostische (vor allem Labor-, aber auch neurologische) Untersuchungen, die keine therapeutische Konsequenz haben, auch dann nicht, wenn gelegentlich spezifische und unspezifische Laborwerte verändert bzw. „pathologisch" sind. Meistens werden – auch geringfügige – Veränderungen der Laborwerte dem Patienten mitgeteilt, etwa leichte Schwankungen der T4/T8-Ratio, die auch bei Nichtinfizierten vorkommen können.

Auf seiten des Arztes spielt dabei (neben Forschungsinteressen) wohl vor allem das Motiv eine Rolle, die therapeutische Ohnmacht durch diagnostische Aktivität zu kompensieren. Die psychischen und psychosomatischen Folgen des „Staging" und des ständigen Mitteilens der Laborwerte für den Patienten werden allerdings viel zu wenig reflektiert (zur „Geburt des Leidens aus der Diagnose" vgl. Ohly, 1988).

Ein 35jähriger homosexueller Mann hatte zunächst für sich entschieden, den Test nicht zu machen, sich aber in der monogamen Beziehung zu seinem Freund an „safer sex" zu halten. Auf Drängen des Freundes, der die Ungewißheit nicht mehr aushielt und zu dekompensieren drohte, gingen beide zum Test und waren beide positiv. Der Patient, der in seiner Beziehung bisher immer der Starke, dem Freund Haltgebende gewesen war, und für dessen psychische Stabilität sein sportlich trainierter, unversehrter, intakter Körper immer eine große Rolle gespielt hatte, geriet durch das Testergebnis in eine massive Krise, die vor allem durch depressive Symptome, durch eine ständige ängstlich-hypochondrische Beobachtung und Kontrolle seines Körpers und durch Todesangst gekennzeichnet war. Da er seinen Freund nicht belasten wollte, wandte er sich an die Psychosomatische Klinik. Es gelang ihm,

---

[11] Bei Hämophilen und auch bei i.v.-drogenabhängigen AIDS-Kranken passiert dies sehr viel seltener.

[12] Die bisher vorliegenden Forschungsergebnisse über den Einsatz von ACT bei symptomlosen HIV-Infizierten haben an dieser Tatsache nichts geändert. Weitere Studien sind erforderlich (vgl. AIDS-Enquete-Kommission des Deutschen Bundestages, 60. Sitzung 11. 10. 89, Anhörung zu Stand und Entwicklung der AIDS-Forschung).

sich zu stabilisieren, die hypochondrische Körperselbstbeobachtung nahm ab zugunsten wieder vermehrter sportlicher Aktivitäten. Die anfangs ihn den ganzen Tag beschäftigende Angst vor der Erkrankung trat nur noch in bestimmten Situationen auf, er konnte sich zunehmend wieder mehr für andere Dinge als die Infektion interessieren. Außerdem beschäftigte er sich mit Vollwertkost und anderen Möglichkeiten, „gesund zu leben"; die Todesangst wich einem Gefühl, „ich tue etwas für meine Gesundheit und werde deshalb nicht erkranken". Bei einer Kontrolluntersuchung wurde ihm mitgeteilt, daß seine T-Helfer-Zellen etwas abgenommen hätten, aber noch weit über dem kritischen Wert lägen. Daraufhin dekompensierte der Patient erneut, entwickelte eine Vielzahl von Symptomen (Nachtschweiß, Lymphknotenschwellung etc.), die bei einer HIV-Infektion auftreten können, aber für diese spezifisch sind. Die Symptome verschwanden erst, nachdem ihm bei einer weiteren Untersuchung mitgeteilt wurde, die T-Helfer-Zellen seien wieder etwas gestiegen.

Mittlerweile hat sich der Patient von der Schulmedizin abgewandt zugunsten einer homöopathischen Behandlung, die ihn fast magisch mit Gesundheit erfüllt und ihm zu großer psychischer Stabilität verholfen hat.

Das Beispiel zeigt, wie wichtig ein sorgsamer Umgang mit Laborwerten ist. Die Erhebung von Laborwerten und ihre Mitteilung an den Patienten muß immer in Beziehung zu seiner psychischen Situation erfolgen, d.h. es muß immer mitbedacht werden, ob die Diagnostik bei dem HIV-Positiven stabilisierende Verleugnungsvorgänge erschüttert bzw. hypochondrische Befürchtungen mobilisiert. In vielen Fällen ist es für die Gesamtsituation eines HIV-Positiven besser, wenn er nicht regelmäßig einbestellt wird, sondern ihm nur angeboten wird, daß er jederzeit kommen kann, wenn er sich schlecht fühlt. Aus diesem Grund bevorzugen HIV-Positive oft die Versorgung durch ihren Hausarzt, weil sie sich bei ihm nicht als „Forschungsobjekt" fühlen.

Ein besonderes Problem der medizinischen Diagnostik bei HIV-Infektion ist die Erhebung neurologischer Befunde und ihre Bedeutung für die Arzt-Patient-Beziehung. Neurologische und neuropsychische Symptome können auch ohne einen nachweisbaren Immundefekt im Verlauf der Infektion auftreten. Schwere und Häufigkeit dieser Störungen nehmen jedoch mit der Schwere des Immundefekts zu. Die am häufigsten beobachtete neurologische Erkrankung ist die subakute Enzephalitis, eine langsam fortschreitende Gehirnentzündung. Frühe Symptome dieser Krankheit können u.a. sozialer Rückzug, Apathie, Depression, Merk- und Konzentrationsschwäche und verlangsamtes Denken sein. Diese Symptome sind jedoch nicht sehr spezifisch und können ebensosehr andere, z.B. psychische Ursachen haben. Es wird von Ärzten gern darauf hingewiesen, daß psychische Symptome bei HIV-Infizierten (vor allem depressive Verstimmungen) in Wirklichkeit oft neurologische seien. Nach unseren Erfahrungen liegt jedoch in der Praxis die viel größere Gefahr für die Arzt-Patient-Beziehung darin, daß der Arzt Gefühle des

Patienten wie Trauer und Wut „neurologisiert". Die Versuchung dazu ist groß, weil die Vorstellung, daß etwa Depression und sozialer Rückzug eines HIV-Infizierten organisch, also durch die Infektion bedingt sind, für den Arzt sehr entlastend sein kann, weil er ja dann nicht mehr (durch die Mitteilung der HIV-positiv-Diagnose) der „Verursacher" ist und sich zudem von der Notwendigkeit der Empathie entbunden fühlen kann.

Besonders kritisch ist die Erhebung neurologischer Befunde, ohne daß Symptome vorliegen. So können bei einem Teil der symptomlosen HIV-Infizierten diskrete Veränderungen im EEG nachgewiesen werden; diese Befunde haben keinerlei prognostischen Wert, können aber leicht zu einer Beziehungsabwehr des Arztes gegenüber dem Patienten führen. Bei AIDS-Patienten kommt es zum Teil zu schweren neurologischen und neuropsychischen Störungen wie Lähmungen, Krampfanfällen, Psychosen und Demenz. Diese Störungen und Ausfälle werden von den Patienten wahrgenommen und als sehr bedrohlich und kränkend erlebt; sie können das auch noch sehr lange vermitteln. Um so wichtiger ist, daß es nicht vorzeitig zu einem Beziehungsabbruch seitens der Ärzte und des Pflegepersonals kommt und der Patient nicht „abgeschrieben" wird, obwohl trotz der Einschränkungen seiner geistigen Fähigkeiten durchaus noch ein emotionaler Kontakt zu ihm möglich ist und er ihn auch braucht.

Grundsätzlich ist für alle (Ärzte, Psychotherapeuten, Pflegepersonal, Laienhelfer etc.), die mit HIV-Infizierten und AIDS-Kranken arbeiten, die Bewußtheit des eigenen AIDS-politischen Standpunktes von großer Wichtigkeit. Dies schließt auch das eigene Verhältnis zu den Hauptbetroffenengruppen mit ein: Jemand, der Homosexuelle oder Drogenabhängige wegen ihres Lebensstils oder ihrer subkulturellen Eigenheiten ablehnt, kann ebensowenig professionelle Hilfe bieten wie jemand, der damit überidentifiziert ist und keine therapeutische Distanz wahren kann. Die Auseinandersetzung mit den eigenen Phantasien über Tod, Omnipotenz, Schuld und Sexualität muß immer wieder erfolgen, gerade in besonders belastenden Situationen. So kann etwa ein bis dahin besonders angenehmer, „leicht zu führender" Patient bei Fortschreiten der Erkrankung fordernd, unbequem, „undankbar" werden, dem Arzt trotz dessen starken (aber vergeblichen) ärztlichen Engagements Vorwürfe machen. Das sind Situationen, in denen alte Vorurteile und Schuldzuweisungen leicht wieder durchbrechen können.

Supervision oder Balint-Gruppen auf AIDS-Stationen können hier für Ärzte und Pflegepersonal gute Möglichkeiten bieten, die eigenen Schwierigkeiten im Umgang mit AIDS-Patienten zu reflektieren, um besser damit umgehen zu können. Besonders vom Pflegepersonal wird oft der Wunsch nach solcher Unterstützung geäußert, weil es sich durch die Arbeit mit den AIDS-Kranken sehr belastet, oft auch überfordert fühlt. Ein wesentliches Moment der Belastung ist die Gleichaltrigkeit von Pflegepersonal und AIDS-Patienten, verbunden mit der Tatsache, daß für viele isolierte AIDS-Patienten „ihre Station" zu ihrem Zuhause wird, die Schwestern und Pfleger dann oft ihre Hauptbezugspersonen sind. Das führt häufig zu einem Dilemma, das eine Schwester so ausdrückte: „Gehe ich mehr auf Distanz, habe ich das Gefühl, dem Patienten zuwenig zu geben; lasse ich mich mehr ein und der Patient stirbt, stirbt ein Freund." Überfordert fühlen sich Pfleger und Schwestern z.B. dadurch, daß sie bei drogenabhängigen AIDS-Kranken die Verantwortung für den Umgang mit den Entzugserscheinungen haben, was sie nicht selten in Konflikt mit der Legalität bringt.

Die Notwendigkeit von Supervision und von in die Klinik integrierter psychotherapeutischer Behandlung wird bei AIDS-Kranken auch von solchen Medizinern gesehen, die sonst der Psychosomatik eher ablehnend gegenüberstehen.

49.7.2 Beratung und Psychotherapie

Ein Teil der HIV-Positiven braucht trotz der tief in das Leben eingreifenden Diagnose und trotz heftiger erster Reaktionen keine kontinuierliche Hilfe. Dies gilt vor allem für psychisch stabile, sozial und partnerschaftlich gefestigte Personen, solange sie symptomlos sind. Eine Unterstützung brauchen vor allem diejenigen, deren soziales Netz nicht tragfähig genug ist, und diejenigen, bei denen die HIV-positiv-Diagnose auf den Boden eines neurotischen Konflikts fällt und diesen aktiviert, z.B. bei einer schuldhaften Verarbeitung der Diagnose. Faßt man die HIV-positiv-Diagnose als Trauma auf (Weinel, 1989), kann man auch sagen, daß diejenigen psychotherapeutische Hilfe brauchen, bei denen das HIV-Trauma im Sinne einer kumulativen Traumatisierung (Khan, 1977) frühere Traumata wiederbelebt und bis dahin funktionierende Abwehr- und Kompensationsvorgänge jetzt nichts mehr nützen. (Es geht dabei z.B. um infantile Objektverluste oder um Brüche im Coming out, die bisher durch bestimmte Beziehungskonstellationen oder durch eine besondere Besetzung der narzißtischen Integrität des eigenen schönen und gesunden Körpers erfolgreich kompensiert wurden. Wenn aufgrund des HIV-Traumas diese Kompensation nicht mehr möglich ist, brechen gleichsam die alten Wunden wieder auf.) Während bei diesen Patienten langfristige Psychotherapie indiziert ist, brauchen andere HIV-Positive direkt nach dem Testergebnis eine Zeit lang psychotherapeutische Unterstützung und anschließend die Gewißheit, bei Bedarf kommen zu können, was sie dann in ganz unterschiedlichen Zeitabständen auch in Anspruch nehmen. Für manche HIV-Positive sind auch sog. Positiven-Gruppen (als Selbsthilfegruppe oder psychotherapeutisch geleitet) sinnvoll. Eine besondere Funktion dieser Gruppen ist die Möglichkeit, in der AIDS-Politik aktiv zu werden. Dieses politische Engagement (z.B. in der AIDS-Hilfe) bedeutet für manche HIV-Positive eine Möglichkeit, die eigene Lage und Zukunft mitzubeeinflussen, eine wichtige psychische Stabilisierung. Auf der anderen Seite entsteht in

solchen Gruppen auch manchmal ein großer normativer Druck – beispielsweise, daß sich ein HIV-Positiver mit dem Tod zu beschäftigen habe –, der den einzelnen in seinen Kompensationsbemühungen überfordern kann. Die Tatsache, daß die anderen Mitglieder der Gruppe auch „positiv" sind, wird oft als stabilisierend erlebt; Krisen in diesen Gruppen entstehen vor allem, wenn ein Mitglied der Gruppe erkrankt oder stirbt. Dies ist auch oft der Zeitpunkt, an dem sich Selbsthilfegruppen eine psychotherapeutische Leitung suchen.

Neben der bereits erwähnten notwendigen Reflexion der eigenen Phantasien über Tod, Omnipotenz, Sexualität und Schuld geht es im Umgang mit HIV-Positiven für den Berater vor allem darum, folgenden Widerspruch auszuhalten: Einerseits will der Berater, daß der HIV-Positive niemanden infiziert, andererseits hat dieser ein Recht auf gelebte Sexualität. Das Anerkennen dieses Dilemmas impliziert auch das Wissen darum, daß psychische Stabilität und „zeitstabile" sexuelle Verhaltensänderungen immer wieder in Widerspruch zueinander geraten können, eventuell sogar müssen. Weil dieses Dilemma nicht leicht auszuhalten ist, besteht die Gefahr, es in die eine oder andere Richtung aufzulösen (vgl. Becker und Clement, 1989): Wer HIV-Infizierte nur mit der Absicht berät, daß sie niemanden anstecken, wird keine Beziehung zu ihnen herstellen können, oder es kommt zur Empathieverweigerung und der therapeutische Dialog entgleist. Der Berater jedoch, der in sich selbst verleugnet, daß er auch möchte, daß der Infizierte niemanden ansteckt, verweigert eine wichtige Auseinandersetzung, was in der Folge dem Infizierten den Zugang zu inneren Konflikten (insbesondere mit Verantwortung, Schuld und Aggression) verunmöglichen kann.

Es stellt sich die brisante Frage der Verantwortung des Beraters, wenn er erfährt, daß ein HIV-Positiver die Infektion seines Sexualpartners in Kauf nimmt.[13] Angst des Beraters, durch Mitwissen mitschuldig an einer möglichen Neuinfektion zu werden, kann ihn zu einer „Flucht in die Eindeutigkeit" führen, etwa derart, daß er den Partner informiert, den Patienten selbst anzeigt oder ihn mit besonders eindringlichen Appellen direktiv „belehrt". Ein Berater, der seine Angst so umsetzt, entlastet sich zwar von moralischem Druck, vergibt auf lange Sicht aber therapeutische Möglichkeiten, da er danach vom Patienten nur noch als kontrollierende Instanz oder als moralischer Verfolger erlebt werden kann.

Ein Berater sollte deshalb grundsätzlich für sich den Bruch der Schweigepflicht ausschließen, auch wenn das im Einzelfall für ihn sehr belastend sein kann. Ein extremes Beispiel für eine solche Belastung ist etwa, wenn ein infizierter Ehemann seine Frau, die nichts von seinen außerehelichen Kontakten weiß, weder über seine Infektion informiert noch ein Kon-

---

13 Dieses schwierige Problem ist primär kein juristisches; nach herrschender Rechtsauffassung hat der Arzt das Recht, aber nicht die Pflicht, Dritte zu informieren, wenn Leben bedroht ist (Eberbach, 1988).

dom benutzt, um nicht ihren Verdacht zu erregen. Bei homosexuellen Patienten sind Berichte über ungeschützten sexuellen Verkehr für den Berater leichter zu ertragen, weil es eine vergleichbare „Ahnungslosigkeit" bei ihren Sexualpartnern nicht gibt.

Die Art der Bedrohung (das Virus kommt von außen und ist jetzt innen, es greift heimtückisch und diffus den ganzen Körper an) aktualisiert sehr frühe Phasen der psychischen Entwicklung (und die dazugehörigen Vernichtungsängste) und mobilisiert entsprechende frühe Abwehrmechanismen wie Spaltung, Projektion, projektive Identifikation, Verleugnung etc. Auf der Ebene der frühen Objektbeziehungen zu Partialobjekten kann man das so ausdrücken: „Das gute Objekt – sexuelle Erregung, Blut, Heroin –, von dem ich abhängig bin und mit dem ich gerade zum Zweck der Lebenserhaltung in Kontakt treten möchte, unternimmt gegen mich einen hinterhältigen Angriff, der mich in meinem innersten Kern (dem Autoimmunsystem als Repräsentanten der intakten Körper-Ich-Selbst-Imago) trifft und vernichten wird" (Reiche, 1988).

Zum Schutz des Ichs vor Desintegration durch Todes- und Vernichtungsangst kann nun das Zerstörerische, Verfolgende nach außen projiziert werden, z.B. auf den „bösen" Staat. Unauflösbar wird dieser Abwehrvorgang allerdings, wenn sich der Staat bzw. bestimmte gesellschaftliche Gruppen real verfolgend verhalten. Die Projektion des HIV-Infizierten wird dadurch zementiert, für ihn als solche unerkennbar; sie verliert darüber hinaus auch ihre protektive Funktion. Innere und äußere Realität sind hier oft schwer auseinanderzuhalten. Deshalb ist es wichtig, daß der Therapeut gegenüber dem Infizierten eine aktiv parteiliche Haltung in bezug auf öffentliche Kontrolle und Verfolgung einnimmt. Dabei ist es aber von entscheidender Bedeutung, wann und wie er diese Haltung ausdrückt, weil es umgekehrt auch die Gefahr gibt, daß sich der Therapeut (aus Überidentifikation mit dem Patienten oder aus Abwehr aggressiver Impulse) mit diesem gegen den Staat „verbündet" bzw. ihm diese Verbündung aufnötigt, was zu einem Harmoniezwang in der Therapie führt.

Eine sehr verbreitete Form der Verarbeitung von Vernichtungsangst ist die Introjektion eines allmächtigen guten Objekts, das das introjizierte Böse quasi magisch neutralisiert. Homöopathische Substanzen haben oft unbewußt diese Bedeutung für Infizierte (womit nichts gegen ihre sonstigen Wirkungen gesagt ist). Auch der Psychotherapeut kann gelegentlich für den Infizierten die Bedeutung eines solch allmächtigen gesunden guten Objekts bekommen. Es kommt dann zu einer idealisierenden Übertragung mit der Erwartung, daß durch die Psychotherapie der Ausbruch der Erkrankung verhindert werden könne – und gelegentlich zu entsprechenden Allmachtsphantasien in der Gegenübertragung. Die Verschmelzungswünsche und -phantasien mit diesem als Ideal phantasierten Objekt enthalten meist „nicht nur den Wunsch, an der Omnipotenz dieses gesunden Objekts zu partizipieren, sondern auch den, entweder mit diesem Objekt zu überleben oder aber mit ihm zu

sterben, es mit in den Tod zu nehmen". Gelingt es, „diese Verschmelzungsphantasien und -wünsche in ihren sehr unterschiedlichen Aspekten zu bearbeiten, dann wird dem Patienten auch zumeist ein Stück Individuierung, eine Auseinandersetzung mit seinem eigenen Tod möglich" (Weinel, 1988).

Die Angst vor Vernichtung und die idealisierende Übertragung bei HIV-Positiven erschweren das Bewußtmachen aggressiver und destruktiver Phantasien und Impulse, zu denen auch immer Ansteckungsängste und -wünsche gehören (Becker und Clement, 1989; Weinel, 1989).

Nach der bisherigen psychotherapeutischen Erfahrung mit HIV-Positiven „tritt die deutende Bearbeitung unbewußter neurotischer Konfliktanteile dabei meist zugunsten einer empathischen Begleitung in den Hintergrund, wobei aber trotzdem Übertragungs- und Gegenübertragungsreaktionen zu beachten sind" (Weinel, 1988).

49.8 AIDS-Phobie und AIDS-Hypochondrie

> Eine 30jährige Frau wird am Wochenende mit einer Urticaria stationär in die Hautklinik aufgenommen; allergische Ursachen lassen sich nicht finden. Im Gespräch stellt sich heraus, daß sie zwei Jahre lang jede nähere Beziehung zu einem Mann gemieden hat, nachdem sie sich aus einer langjährigen, sexuell unbefriedigenden, aber sehr bindenden Beziehung gelöst hatte. Vor ein paar Wochen hat sie einen Mann kennengelernt, zu dem sie sich sehr hingezogen fühlt. An diesem Wochenende war sie mit ihm verabredet, und es wäre wohl zum ersten sexuellen Verkehr gekommen, wenn sie nicht plötzlich die Urticaria entwickelt hätte und in die Hautklinik gekommen wäre. Die Ambivalenz (zwischen dem sexuellen Wunsch und der Angst vor Abhängigkeit) konnte mit der Patientin im Zusammenhang ihrer Biographie ansatzweise verstanden werden, die Urticaria ging zurück. Ein paar Wochen später kommt sie wieder: Sie habe mittlerweile mehrfach mit dem Mann geschlafen, es sei überwältigend gewesen, sie habe soviel erlebt wie nie zuvor. Leider habe sie die sexuelle Beziehung zu ihm jetzt beenden müssen, da sie panische Angst habe, sie könne sich bei ihm mit AIDS anstecken, er habe ja vor ihr schon drei langjährige Beziehungen zu Frauen gehabt. Die Angst gehe so weit, daß sie sich nicht einmal mehr von ihm anfassen lassen könne.

Bei dieser Patientin dominiert die subjektiv überhöhte Befürchtung, sich infizieren zu können, obwohl ein Risiko eher unwahrscheinlich ist; sie befürchtet aber nicht, bereits infiziert zu sein. Andere Patienten fürchten zum Teil ganz generalisiert jede direkte und indirekte körperliche Berührung. Andere leiden unter der oft quälenden subjektiven Gewißheit, bereits infiziert zu sein, obwohl sie kein nennenswertes Risiko eingegangen sind und oft meist bereits mehrfach – stets mit negativem Ergebnis – auf HIV getestet worden sind.

Für beide Arten neurotischer Angst vor AIDS hat sich der Begriff „AIDS-Phobie" durchgesetzt (Jäger, 1988). Richter (1987) schlägt dagegen vor, nur bei der ersten Gruppe von „AIDS-Phobie", bei der zweiten Gruppe dagegen von „AIDS-Hypochondrie" zu sprechen.[14] Gelegentlich kann eine AIDS-Phobie in eine AIDS-Hypochondrie übergehen.

Ärzte werden vorwiegend mit der AIDS-Hypochondrie konfrontiert, da sich AIDS-Phobiker ja nicht für infiziert halten. Patienten mit einer AIDS-Hypochondrie haben meistens eine phobische oder hypochondrische Vorgeschichte (Hualla und Jäger, 1988), die sich mit AIDS nur einen neuen Inhalt sucht.

Fast immer läßt sich eine sexuelle Auslösesituation finden. Dabei handelt es sich um (oft einmalige) sexuelle Episoden, die außerhalb des üblichen, sozial akzeptierten Rahmens einer bestehenden Liebesbeziehung stattfinden: Seitensprünge, Kontakte zu Prostituierten, homosexuelle Episoden heterosexueller Patienten, gelegentlich auch als bedrohlich erlebte Versuchungssituationen ohne realen sexuellen Kontakt.

Unabhängig davon, ob die konflikthafte sexuelle Situation real stattfand oder phantasiert wurde, geht es im wesentlichen um zwei Konfliktsituationen, die auch gemeinsam vorhanden sein können:

– Ein Trieb-Über-Ich-Konflikt kann aktualisiert sein, wenn sexuelle Wünsche schuldhaft verarbeitet werden; das Symptom hat dann einen Bestrafungscharakter, indem es Über-Ich-Motive befriedigt.

– Die sexuelle Auslösesituation kann für einen Bindungs-Autonomie-Konflikt stehen (Hirsch, 1988). Ähnlich wie bei der Herzneurose werden hier bei stark emotional gebundenen Patienten Autonomie- bzw. Trennungswünsche mit Hilfe des AIDS-hypochondrischen Symptoms zurückgedrängt, wodurch die Abhängigkeit von der Person wiederhergestellt werden kann, der die unbewußte Trennungsabsicht gilt. (Die andere Variante des Konflikts bei Herzneurotikern – das Symptom schützt vor einer die Autonomie bedrohenden Bindung – scheint eher AIDS-phobische Reaktionen zu begünstigen.)

Psychotherapeutisch gilt hier ähnliches wie für andere Phobien und Hypochondrien auch. Die Überweisung an einen Psychotherapeuten, ohne daß der Patient sich abgeschoben oder nicht ernstgenommen fühlt, gelingt nur, wenn der zugrundeliegende Konflikt von Arzt und Patient ansatzweise verstanden werden kann. Hilfreich dafür ist zum einen, wenn der Arzt nicht versucht, dem Patienten seine Befürchtungen auszureden, sondern sie ihn zu Ende denken läßt; zum anderen, wenn der Patient weiß, daß der Psychotherapeut, an den er überwiesen wird, auch HIV-Infizierte und AIDS-Kranke behandelt.

---

14 Wenn die phobischen AIDS-Ängste projektiv-aggressiv gegen Infizierte gerichtet werden, spricht Richter vom „AIDS-Paranoid". Dies scheint uns aber eher ein sozialpsychologisches als ein individuell-neurotisches Phänomen im klinischen Sinne zu beschreiben.

50 Psychische und soziale Faktoren in Entstehung und Verlauf maligner Erkrankungen*

Christoph Hürny

50.1 Psychosoziale Risikofaktoren für die Krebsentstehung

Krebs und Psyche haben auf den ersten Blick wenig miteinander zu tun. Als vorwiegend naturwissenschaftlich geschulte Ärzte sind wir es gewohnt, maligne neoplastische Prozesse biologisch zu betrachten, insbesondere da maligne Zellen mehr oder weniger autonom wuchern und sich charakteristischerweise den Steuerungsmechanismen des Organismus entziehen. Bei näherem Hinschauen sind jedoch vielfältige Wechselwirkungen zu erkennen. Diese sind ganz offensichtlich, wenn die Diagnose Krebs einmal gestellt ist, wenn es um die Information des Patienten geht, die Verarbeitung einer tödlichen Bedrohung, eines verstümmelnden Eingriffs, das Durchstehen einer eingreifenden Strahlen- oder Chemotherapie; oder wenn eine kurative Therapie nicht mehr in Frage kommt, wenn die Begleitung des zu Tode kranken Menschen unsere schwierige Aufgabe wird.

Doch bereits im unmittelbaren Vorfeld der Diagnosestellung können psychosoziale Faktoren eine entscheidende Rolle spielen. In einer retrospektiven Untersuchung von 200 Patienten mit kolorektalem Karzinom in England wurde eine Verzögerung von durchschnittlich 8,25 Monaten vom Auftreten der ersten Symptome bis zur Diagnosestellung gefunden. Ungefähr die Hälfte der Verzögerungszeit war dem Patienten zuzuschreiben, die andere Hälfte dem Hausarzt (Hollyday und Hardcastle, 1979). Die Gründe für dieses Verhalten von Patient und Arzt sind im einzelnen nicht klar. Unwissen, unbewußtes Vermeiden von bedrohlicher Information, Furcht vor Verstümmelung und Tod auf seiten des Patienten und des Arztes werden in der Literatur als mutmaßliche Gründe genannt (Holland, 1982).

Ich will jedoch noch weiter zurückgehen in der Biographie des Krebskranken und die Frage stellen, ob belastende Ereignisse und Entbehrungen oder Persönlichkeitsfaktoren, global ausgedrückt, ob psychosozialer Streß der Manifestation der Krebskrankheit vorausgehen und eventuell Genese und Ätiologie mitbeeinflussen kann. Die Fragestellung ist nicht neu. Bereits der römische Arzt Galen beobachtete, daß Frauen mit melancholischem Temperament gehäuft an Brustkrebs erkrankten. Hervorragende Ärzte der Neuzeit, z.B. Paget, Ewing und Leriche, bemerkten bei ihren Krebspatienten belastende Lebensumstände und besondere Charakterzüge (Hürny und Adler,

1985). 1893 untersuchte Snow am London Cancer Hospital 250 unausgewählte Krebspatienten. Er fand bei den meisten in zeitlichem Zusammenhang mit der Manifestation der Krebserkrankung schwere seelische Belastungen, Schwierigkeiten am Arbeitsplatz oder mechanische Traumen. Nur 19 Patienten wiesen keine Besonderheiten auf. Aufgrund seiner Erhebung fragte sich Snow, ob nicht in der Mehrheit der Fälle die Krebskrankheit eine „neurotische Ursache" habe (Le Shan, 1963; Snow, 1964).

> Frau M. wird 1944 als uneheliche Tochter einer Näherin geboren. Ihren Vater kennt sie nicht. 5 Jahre später heiratet die Mutter einen Alkoholiker. Da es der Mutter schlechtgeht, beginnt sie, die Tochter täglich zu schlagen. Mit 8 Jahren kommt das Mädchen zu einer alleinstehenden Pflegemutter, wo sie es gut hat, aber hart arbeiten muß. Sie macht eine Schneiderinnenlehre. In der Pubertät hat sie schwierige Auseinandersetzungen mit der Pflegemutter. Die Patientin geht frühe, multiple und unbefriedigende sexuelle Beziehungen ein. Mit 22 Jahren kommt es zur Mußheirat mit einem italienischen Maurer. Zwei Kinder werden geboren. Der Mann ist krankhaft eifersüchtig und schlägt die Patientin von Anfang an. Neben dem Haushalt arbeitet Frau M. hart als Putzfrau, da sonst zu wenig Geld vorhanden ist, muß aber ihren ganzen Lohn dem Mann abgeben. Nach 14 qualvollen Ehejahren ringt sich die Patientin zur Scheidung durch. Um sich und die beiden halbwüchsigen Kinder über Wasser zu halten, übernimmt sie eine Hauswartstelle und arbeitet als Putzfrau weiter. Sie fühlt sich allein, hilf- und hoffnungslos. Ein Jahr später beginnen Zwischenblutungen, nach weiteren 6 Monaten wird die Diagnose eines inoperablen Zervixkarzinoms gestellt. Bei der onkologischen Erstuntersuchung wirkt die Patientin, was ihr Gefühlsleben anbelangt, wie erstarrt und versteinert. Sie kann ihre Gefühle kaum ausdrücken, hat sich sozial isoliert und lebt zurückgezogen mit ihren beiden Kindern.

In dieser Lebenssituation und bei dieser Vorgeschichte der Patientin würde uns die Entwicklung einer schweren Depression, von Tabletten- oder Alkoholabusus, von psychogenen Schmerzen oder eines Ulcus-duodeni-Schubes nicht erstaunen. Hingegen fällt es schwer, die Manifestation eines Karzinoms aus ei-

---

* Eine englische Übersetzung dieses Kapitels wird in der Monographie „Psychosocial Aspects of Oncology" (J. Holland and R. Zittoun, eds.) der ESO, European School of Oncology, Springer, Berlin 1989 publiziert.

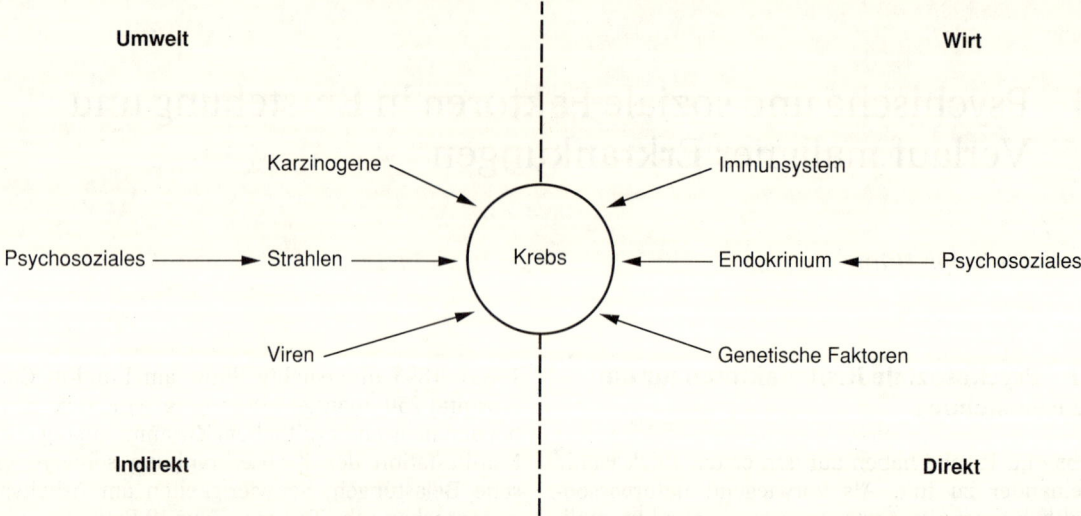

Abb. 50–1. Multifaktorielle Karzinogenese.

ner Lebenssituation heraus zu verstehen. Obwohl Krebs in unserem Inneren entsteht, erleben wir ihn in der Regel als von außen kommendes Unheil.

Die Ursachen maligner neoplastischer Prozesse sind heute trotz intensivster Forschung nur zu einem kleinen Teil bekannt, die meisten Krebstherapien sind empirisch. Es ist wenig wahrscheinlich, daß ein einzelner Faktor für die Erkrankung verantwortlich gemacht werden kann. Nach dem heutigen Stand des Wissens muß eine multifaktorielle Genese angenommen werden. In Abbildung 50–1 habe ich für einzelne Tumoren heute etablierte ätiologische Faktoren zusammengestellt und zwei mögliche Wirkungsweisen psychosozialer Faktoren angedeutet:

- **Indirekte psychosoziale Faktoren:** Ein bestimmtes, meist komplexes menschliches Verhalten führt zu vermehrter Karzinogenexposition, z.B. Rauchen → Lungenkrebs; Sonnenexposition → Melanom; frühes Alter beim ersten Geschlechtsverkehr, große Anzahl Sexualpartner → Zervixkarzinom; Alkoholgenuß → Leberzirrhose → Leberzellkarzinom; Alkohol plus Rauchen → Karzinom des oberen Verdauungstraktes.
- **Direkte psychosoziale Faktoren:** Ein psychosozialer Streß, z.B. der Verlust des Ehepartners, führt über psychische Prozesse, z.B. Trauer, zu somatischen Veränderungen, z.B. im Bereich des Immunsystems oder des endokrinen Systems. Eine Funktionsstörung, z.B. der Lymphozyten, begünstigt das Entstehen eines malignen neoplastischen Prozesses.

Im folgenden möchte ich mögliche Wirkungsweisen direkter psychosozialer Risikofaktoren etwas näher beleuchten. Um dies tun zu können, muß ich vorerst einige Bemerkungen zur Forschungsmethodik machen, die auch in diesem Gebiet zu wünschen übrigläßt.

50.1.1 Methodische Probleme

Die Frage nach direkten psychosozialen Risikofaktoren für die Krebskrankheit, also die Frage nach direkten kausalen Beziehungen zwischen psychosozialen Variablen und Krebs, ist im Prinzip ein epidemiologisches Problem. Epidemiologische Studien ermitteln Risikofaktoren, also ätiologische Teilerklärungen. Die gesamte Varianz wird praktisch nie durch einen einzelnen Faktor erklärt. In den sechziger Jahren glaubten einzelne Forscher, die Krebsentstehung durch eine psychosoziale Theorie hinreichend und umfassend erklären zu können (Bahnson und Bahnson, 1964). Die Idee, den Krebs als regressiven Regenerationsversuch auf biologischer Ebene bei Erschöpfung bzw. Blockierung der psychologischen Ausdrucksmöglichkeiten zu verstehen, ist einseitig und zeugt von mangelndem Verständnis für die biologische Komplexität des Problems.

Die folgenden allgemeinen epidemiologischen Kriterien gelten auch für die Beurteilung von Studien psychosozialer Risikofaktoren bei Krebs (Morrison und Pfaffenberger, 1981) (Tab. 50–1).

Tabelle 50–1. Epidemiologische Kriterien zur Beurteilung von Studien psychosozialer Risikofaktoren (nach Morrison und Paffenberger, 1981)

- Stichprobe
- Korrelationsstärke
- zeitlicher Ablauf
- Reproduzierbarkeit
- Persistenz
- Dosis-Wirkungs-Relation
- Spezifität
- biologische Plausibilität

Stichprobe:
Krebs ist wahrscheinlich nicht ein einziger Krankheitsprozeß, sondern ein Sammelbegriff für verschiedenste Krankheiten. Ein Basaliom der Haut ist kaum vergleichbar mit einer akuten myeloischen Leukämie, obwohl beiden ein histologisch definierter maligner neoplastischer Prozeß zugrunde liegt. In psychosozialen Studien sind häufig Patienten mit verschiedenen Primärtumoren und in verschiedenen Stadien zusammengefaßt. Weiter ist zu bedenken, daß der Begriff Krebs wie kaum eine andere Krankheit unheimliche Vorstellungen in uns weckt. Kontrollgruppen mit vergleichbaren Krankheiten sind schwierig zu finden. Das aktuelle Krankheitserleben kann vorbestehende psychosoziale Faktoren vortäuschen oder verändern. Eine Hypothese ist z.B., daß Krebspatienten eine Tendenz haben, Gefühle zu verleugnen und zu verdrängen, und daß diese Tendenz vorbestehend und kausal mit der Krebskrankheit zusammenhänge (Bahnson und Bahnson, 1964). Aufgrund auch einer gut kontrollierten retrospektiven Studie kann nicht entschieden werden, ob die Tendenz, zu verleugnen und zu verdrängen, eine spezifische Reaktion auf das Kranksein mit Krebs ist, oder ob sie schon vorher bestanden hat.

Korrelationsstärke:
Die Krankheit muß statistisch mit dem zur Diskussion stehenden Faktor assoziiert sein, d.h., der Faktor sollte in einer Gruppe mit der Krankheit substantiell häufiger oder seltener sein als in einer vergleichbaren Kontrollgruppe ohne die Krankheit.

Zeitlicher Ablauf:
Der zur Diskussion stehende kausale Faktor muß vor der Krankheit auftreten. Dies erscheint zunächst banal. Wenn man aber bedenkt, daß ein Neoplasma dann klinisch erkennbar wird, wenn es aus 10^9 Zellen besteht, muß man annehmen, daß ein maligner Tumor lange Zeit latent vorbesteht. Aus der Verdoppelungszeit eines Neoplasmas kann approximativ auf den Zeitpunkt des Beginns der malignen Proliferation zurückgeschlossen werden. Je nach Krebsart liegt dieser zwischen 3 und 10 Jahren vor der Krebsmanifestation (Fox, 1978). Kausale psychosoziale Faktoren müßten in der Regel also lange zurückliegen.

Reproduzierbarkeit oder Konsistenz:
Die gefundene Korrelation muß reproduzierbar sein, d.h. von verschiedenen Untersuchern an verschiedenen Patientenkollektiven, zu verschiedenen Zeiten und an verschiedenen Orten gefunden werden. Dies ist zum Teil problematisch, weil verschiedene Untersucher zur Erfassung psychosozialer Faktoren verschiedene, schwer vergleichbare Instrumente benutzt haben.

Persistenz:
Die Korrelation muß persistieren, wenn andere als Risikofaktoren bekannte Variablen eliminiert oder kontrolliert werden. Beim Lungenkrebs z.B. müßte ein gefundener direkter psychosozialer Risikofaktor dahin geprüft werden, ob er z.B. mit dem Rauchen zusammenhängt oder eigenständig ist.

Dosis-Wirkungs-Relation:
Das Krankheitsrisiko muß größer sein bei vermehrter „Exposition". Je ausgeprägter das psychosoziale Merkmal, desto höher sollte das Krebsrisiko sein.

Spezifität:
Der zur Diskussion stehende Risikofaktor muß für die untersuchte Krankheit spezifisch sein. In unserem Fall müßte er für Krebs oder sogar für eine bestimmte Krebskrankheit spezifisch sein und nicht auch bei der Auslösung anderer Krankheiten eine Rolle spielen.

Biologische Plausibilität:
Die hypothetische Korrelation muß biologisch plausibel sein im Licht des gegenwärtigen Wissensstandes über die Krankheit.

Wie aus den acht genannten Kriterien ersichtlich ist, sind die Ansprüche, die die epidemiologische Wissenschaft stellt, um einen kausalen Zusammenhang anzunehmen, sehr hoch. Bei den heute vorliegenden Studien sind diese Kriterien kaum je alle erfüllt. Methodische Mängel sind aber nicht nur in der Psychoonkologie anzutreffen, sondern leider ubiquitär. Methodische Mängel bedeuten auch nicht unbedingt, daß die Hypothesen falsch sind. Methodische Mängel gebieten Vorsicht bei der Interpretation der Daten, und sie rufen nach methodisch besseren Untersuchungen.

50.1.2 Hypothesen

Die Erforschung direkter psychosozialer Risikofaktoren bei der Krebserkrankung hat bisher hauptsächlich zwei Hypothesen hervorgebracht:
– **Verlust-Hypothese:** Schwere, persönliche Verluste führen über psychische Veränderungen zu körperlichen Funktionsstörungen, die Entstehen oder Manifestation einer Krebskrankheit begünstigen.
– **„Krebspersönlichkeit":** Bestimmte Persönlichkeitszüge prädisponieren zu Krebs.

Es kann nicht der Sinn dieses Kapitels sein, sämtliche Studien umfassend zu besprechen und kritisch zu analysieren. Wer sich dafür interessiert, kann dies an anderer Stelle nachlesen (Hürny und Adler, 1985). Zur „Krebspersönlichkeit" sei nur kurz folgendes bemerkt: Aufgrund der zahlreichen retrospektiven Studien, die verschiedene Persönlichkeitscharakteristika bei Krebspatienten beschreiben, läßt sich nicht entscheiden, ob diese vorbestehend oder als Reaktion auf die Krankheit zu verstehen sind. Offene Fragen bezüglich der Korrelationsstärke, des zeitlichen Ablaufes, der Reproduzierbarkeit, Persistenz, Spezifität und nicht zuletzt der biologischen Plausibilität zwingen zum Schluß, daß eine Krebspersönlichkeit als Risikofaktor heute nicht erwiesen ist (Bammer, 1981). Dies im Gegensatz zum Typ-A-Verhalten, einem Persönlichkeitsmuster, das als Risikofaktor der koronaren Herzkrankheit etabliert ist und den acht oben erwähnten epidemiologischen Kriterien standhält (Jenkins, 1978).

Nachfolgend sollen die Entwicklung der Verlust-Hypothese nachgezeichnet und einzelne typische Studien unter die Lupe genommen werden.

50.1.3 Verlust, Trauern und Erkrankung

„... laßt mich zu Euch jetzt von den Dingen reden, vor denen Ihr Euch hüten sollt; daß Ihr in Wut geratet und von Zeit zu Zeit Euch austobt, das gefällt mir, denn dies erhält die Hitze der Natur; was mir aber nicht gefällt, ist, wenn Ihr bekümmert seid und alle Dinge Euch zu sehr zu Herzen nehmt. Denn, wie die Gesamtheit der Physik uns lehrt, ist es dies vor allen Ursachen, das unserem Leib am meisten Schaden zufügt." Diese Empfehlung schreibt der italienische Arzt, Maestro Lorenzo Sassoli einem Patienten im Jahre 1402 (Zitat aus Le Shan und Worthington, 1956).

Hypothesen entstehen häufig aufgrund klinischer Beobachtungen. Am Anfang steht der anekdotische Fall. Ich habe zu Beginn von einer Patientin mit Zervixkarzinom, die ich im onkologischen Ambulatorium untersucht habe, berichtet. Dieser Fallbericht legt die Hypothese nahe, daß in der Vorgeschichte von Krebspatienten, strenggenommen Zervixkarzinom-Patientinnen, gehäuft Verluste auftreten. Will man wissen, ob gehäufte Verluste und später auftretender Krebs zufällig koinzidieren oder ob eine Korrelation besteht, so ist ein erster Schritt die Überprüfung der Beobachtung bei anderen Patienten mit derselben Erkrankung. Dies ist in extenso geschehen. Die Literatur ist abgesättigt mit Fallberichten. Le Shan z.B. hat bei der extensiven psychoanalytischen Behandlung von Krebspatienten mit verschiedener Lokalisation des Primärtumors in der Vorgeschichte häufig schwere Verluste mit tiefer und endgültiger Hoffnungslosigkeit gefunden (1958, 1963). Aufgrund weiterer, unkontrollierter Studien hat er ein seiner Meinung nach für Krebspatienten typisches Entwicklungsmuster beschrieben (1966). Das Drama beginnt mit dem Tod oder der definitiven Abwesenheit eines Elternteils in früher Kindheit. Das führt zum Gefühl von Verlassenheit oder Einsamkeit. Das Kind erlebt diesen Zustand zum Teil als selbstverschuldet. Es fühlt sich abgelehnt, und seine späteren Beziehungen sind oberflächlich und unstet. Durch Abtasten gelingt es später, eine dieser schwierigen Beziehungen etwas zu vertiefen und aufrechtzuerhalten. Ein zweiter, entscheidender Verlust dieser Bezugsperson reaktiviert die alten Gefühle von Hoffnungslosigkeit. Monate bis Jahre nach diesem zweiten Verlust manifestieren sich die ersten Krankheitszeichen von Krebs. Die vorhin geschilderte Lebensgeschichte der Patientin mit Zervixkarzinom paßt recht genau in dieses Muster.

Der nächste Schritt in der Überprüfung der Hypothese ist die retrospektive, kontrollierte klinische Studie. Als Beispiel sei die Untersuchung von Graham et al. (1971) angeführt. Sie untersuchten 447 Patientinnen mit Zervixkarzinom und 711 Kontrollpatienten in bezug auf traumatisierende Ereignisse, wie Tod einer Bezugsperson, Scheidung, Arbeitslosigkeit, finanzielle Entbehrungen und chronische Krankheit in der Familie in den 5 Jahren vor Stellung der Krebsdiagnose. Die Kontrollgruppe setzte sich aus Patienten mit einem anderen Primärtumor und anderen chronischen nicht-neoplastischen Krankheiten zusammen. Die genaue Verteilung der Diagnosen innerhalb der Kontrollgruppe wird nicht genannt. Die Zervixkarzinom-Patientinnen zeigten im Vergleich zu den Kontrollen keine gehäuften Verluste in den 5 Jahren vor Diagnosestellung. Diese Untersuchung sagt aus, daß Verluste für Zervixkarzinom-Patientinnen nicht spezifisch sind, schließt aber wegen der gemischten Kontrollgruppe (Krebs- und anders Kranke) Verlust als Risikofaktor für Krebs im allgemeinen nicht aus. Zudem wird nur die Tatsache des Verlustes und nicht seine psychischen Folgen, z.B. Hilf- und Hoffnungslosigkeit, berücksichtigt. Es wäre ja möglich, daß nicht der Verlust an sich, sondern bestimmte Auswirkungen des Verlustes krankmachend sind. Eventuell haben wir es mit einem krankheits-„spezifischen", aber nicht krebsspezifischen Faktor zu tun; also hätte eine Vergleichsgruppe von Gesunden dazugehört.

Schmale und Iker (1966, 1971) haben den Zusammenhang zwischen Hilf- und Hoffnungslosigkeit und der Manifestation eines Zervixkarzinoms in Unkenntnis des histologischen Befundes geprüft. Sie untersuchten 68 Frauen, die bei Routineuntersuchungen in der Zervixzytologie ein Stadium III nach Papanicolaou aufgewiesen hatten, sich sonst aber völlig gesund fühlten. Bei den Patientinnen, die in den vorangehenden 6 Monaten mit Hoffnungslosigkeit auf kürzliche Lebensereignisse reagiert hatten, wurde in der später durchzuführenden Konisationsbiopsie blind ein Karzinom vorausgesagt. Fehlte der Affekt von Hoffnungslosigkeit, wurde lediglich eine Dysplasie prognostiziert. Eine korrekte Klassifizierung war in 73,6% der Fälle möglich und somit nicht durch Zufall bedingt ($p < 0{,}001$). Karzinompatientinnen unterschieden sich von Dysplasie-Patientinnen in dieser Studie nicht in bezug auf tatsächlich durchgemachte Verluste, sondern lediglich in ihrer Neigung, auf drohende Verluste mit Hoffnungslosigkeit zu reagieren. Schmale faßt den Affekt der Hoffnungslosigkeit und die Neigung, mit Hoffnungslosigkeit zu reagieren, als permissive Bedingung auf, also als beitragenden, nicht als ätiologischen Faktor zur Manifestation eines Karzinoms bei Vorliegen einer Dysplasie.

Aufgrund der bisher zitierten, ausgewählten Arbeiten ist es schwierig zu entscheiden, ob Verluste und/oder deren psychische oder soziale Auswirkungen zum Risiko, an Krebs zu erkranken, beitragen können. Neben der Replikation retrospektiver Studien drängt sich die prospektive Überprüfung der Hypothese auf, d.h., daß man nicht nach Verlusten in der Vorgeschichte von bereits erkrankten Krebspatienten sucht, sondern gesunde Probanden, die einen schweren Verlust erleiden, über Jahre nachkontrolliert und prüft, ob im Vergleich zu Probanden, die keinen Verlust durchmachen, vermehrt Karzinome auftreten.

Große prospektive Studien sind an Menschen, die ihren Lebenspartner durch Tod verloren haben, also an Witwen und Witwern durchgeführt worden. Innerhalb einer Woche bis zu 10 Jahren nach dem Tod des Partners ist die Sterblichkeit im Vergleich zu altersentsprechenden Kontrollen ohne Verlust um das 2- bis 10fache erhöht (Jacobs und Ostfeld, 1977). Das erhöhte Risiko scheint zweigipflig zu verlaufen (Rogers, 1988). Kurz nach dem Verlust (1 Woche bis 1

Jahr) haben verwitwete Männer und Frauen ein erhöhtes Sterberisiko, vor allem durch akute Ereignisse bei koronarer Herzkrankheit. Langfristig, bis zu 10 Jahren nach dem Verlust des Ehepartners besteht nur für Männer ein erhöhtes Sterberisiko. Als Beispiel sei die große amerikanische Prospektivstudie in Washington-County MD von Helsing und Szklo genannt (1981). 4032 erwachsene Personen, die von 1963 bis 1974 ihren Ehepartner verloren, wurden bis 1975 nachkontrolliert und mit der gleichen Anzahl Verheirateter verglichen, die in derselben Zeit ihren Partner nicht verloren. Für Männer war die Sterblichkeit in den 12 Jahren in den verschiedenen Altersgruppen, abgesehen von den 75jährigen und älteren, um das 1,5- bis 4fache erhöht. Im Gegensatz zu früheren Studien ergab sich für Frauen kein Unterschied. Das starke Geschlecht scheint also in bezug auf das Verarbeiten von Verlusten schwächer zu sein. Bei den meisten Studien ist die Sterblichkeit nicht auf Konto einer bestimmten Krankheit erhöht, sondern die Todesursachen entsprechen in ihrer Verteilung denjenigen der Bevölkerung im allgemeinen (Helsing et al., 1981).

In bezug auf die Hypothese, daß Verlust als ätiologischer Faktor bei der Krebskrankheit eine Rolle spiele, können wir bis jetzt folgendes sagen: Das Risiko, in den Jahren nach dem Verlust des Ehepartners zu sterben, und zwar u.a. auch an Krebs, ist vor allem für Männer erhöht (Hürny und Holland, 1983). Verlust muß aber als spezifischer Risikofaktor für Krebs verworfen werden, d.h., es ist nach Verlust des Ehepartners nicht wahrscheinlicher, an Krebs als an etwas anderem zu sterben.

Es stellt sich nun die Frage, ob nicht der Verlust an sich, sondern die Art und Weise, wie jemand darauf reagiert, d.h. wie der Trauerprozeß abläuft, eine Rolle spielt. Auf diese Frage gibt vielleicht die Prospektivstudie von Shekelle et al. (1981) eine vorläufige Antwort. 2020 männliche Angestellte der Western Electric Co. machten 1958 einen auswertbaren MMPI (Minnesota Multiphasic Personality Inventory). Von denjenigen, welche damals D (Depressivität) als höchsten Wert in ihrem Persönlichkeitsprofil aufwiesen, waren 1975, also 17 Jahre später, 2,3mal mehr an Krebs gestorben als bei den Kontrollen (p < 0,001). Diese Korrelation blieb bestehen, wenn andere Risikofaktoren, wie Alter, Alkohol- und Tabakkonsum, Vorkommen von Krebs in der Familie und Arbeitssituation, statistisch eliminiert wurden. Eine Korrelation zwischen hohem D-Wert und der Mortalität an allen anderen Krankheiten wurde nicht gefunden. Beim Follow-up nach 20 Jahren war erneut eine signifikante Korrelation zwischen Depressivität und Krebsinzidenz und -mortalität zu erheben (Persky et al., 1987). In der prospektiven Walnut Creek Contraceptive Drug Study (Hahn und Petitti, 1988) füllten 8932 Frauen 1968–1969 einen auswertbaren MMPI aus. Bis 1982, also in den 13 Jahren nach der psychologischen Untersuchung erkrankten 117 Frauen an Brustkrebs. Die betroffenen Frauen zeigten im Vergleich zu den gesundgebliebenen außer einem kleinen, aber statistisch signifikanten Unterschied im

„Lügenscore" keine Unterschiede im initialen MMPI-Profil, insbesondere waren weder ein hoher D-Score im Profil, noch absolut erhöhte D-Werte vorhanden. Die Inzidenz anderer Neoplasien wurde nicht untersucht. Bei den Männern der Western Electric Study war Krebsmortalität stärker mit initial hohen D-Werten assoziiert als Krebsinzidenz. Bei der Walnut Creek Contraceptive Drug Study wurde die (Brust-) Krebsmortalität nicht untersucht. Nach diesen Untersuchungen scheint eine Neigung zu Depressivität, u.a. also zu Hilflosigkeit und Hoffnungslosigkeit, ein Risikofaktor für eine spätere Erkrankung an einem malignen Tumor zu sein, allerdings nur für Männer.

Verlust ist also wahrscheinlich nicht ein spezifischer Risikofaktor für Krebs, hingegen könnte die Neigung zu Depression, also eine mögliche Art, auf Verluste zu reagieren, kausal mit Krebs verbunden sein.

50.1.4 Psychobiologische Verbindungsglieder

Nimmt man an, daß Ereignisse und Zustände im psychosozialen Bereich, also z.B. Verluste, und die psychische Reaktion darauf, also Trauer, die Verletzbarkeit für körperliche Krankheiten erhöhen können, so stellt sich die Frage nach dem „Wie". Die ursächlichen Mechanismen des erhöhten Sterberisikos nach Verlust des Ehepartners z.B. sind nicht geklärt. Umweltfaktoren, die Verstorbene und Hinterbliebene in gleicher Weise treffen, sind möglich. Soziale Auswirkungen des Verlustes, z.B. Isolation, Vernachlässigung, mangelnde Pflege und Ernährung, könnten eine Rolle spielen. Ein wesentlicher Faktor ist wahrscheinlich auch die psychische Verarbeitung des Verlustes. In Abbildung 50–2 habe ich mögliche Verbindungswege zwischen direkten psychosozialen Risikofaktoren und biologischem Geschehen darzustellen versucht. Als Mediatoren zwischen Umwelt, Psyche und Körper werden heute hauptsächlich das Nervensystem, das hypophysär-endokrine System und das Immunsystem in Betracht gezogen. Sie stehen zudem unter gegenseitiger Wechselwirkung. Zahlreiche Hormone, vor allem Kortison, Adrenalin und Noradrenalin, Thyroxin, Wachstumshormon, Insulin und die verschiedenen Sexualhormone, spielen für die normale Entwicklung und Funktion der humoralen und besonders der zellgebundenen Immunabwehrprozesse eine wesentliche Rolle (Rogers et al., 1979). Zerstörung der Hypophyse führt zu Schädigung des Thymus und umgekehrt (Pierpaoli und Sorkin, 1973). Die Lymphozyten besitzen Rezeptoren für Kortikosteroide, Wachstumshormon, Insulin, Histamin, alpha-, beta-adrenerge und cholinerge Substanzen (Lippman et al., 1973; Lesniak et al., 1973; Krug et al., 1972; Bourne et al., 1974; Hadden et al., 1970). Die Wechselbeziehungen sind vielfältig und im einzelnen nicht genau bekannt. Trotzdem kann man sagen, daß komplizierte und komplexe Verbindungswege zwischen psychischem Erleben und Vorgängen auf zellulärer Ebene sich abzuzeichnen beginnen.

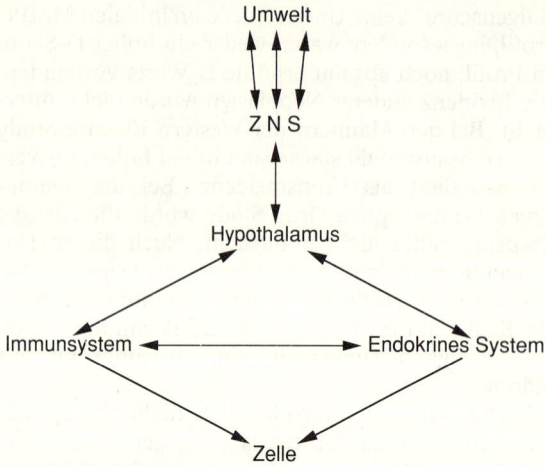

Abb. 50–2. Mögliche Wirkungsweisen direkter psychosozialer Risikofaktoren.

Abschließend sei als Beispiel die erste Studie zitiert, die mögliche Auswirkungen von Verlust auf das Immunsystem zeigt. 1977 hat die Gruppe von Bartrop (1977) in einer gut kontrollierten Studie bei 26 Witwern und Witwen eine substantielle Abschwächung der zellulären Immunantwort, also der T-Lymphozyten auf Mitogen-Stimulation sechs Wochen nach dem Tod des Partners gefunden. Die Abschwächung der T-Zell-Funktion war offensichtlich unabhängig von hormonalen Faktoren (Kortison, Prolaktin, STH und Thyroxin). Spätere Untersuchungen haben diesen Befund repliziert (Schleifer et al., 1983), und vielfältige Beziehungen zwischen psychosozialem Streß und Immunfunktion sind entdeckt worden (Locke, 1982; Kiecolt-Glaser et al., 1987 a, b). Es gilt allerdings zu betonen, daß es bis heute keine repräsentativen biologischen Indikatoren für subtile Erfassung der Funktion bzw. Dysfunktion des Immunsystems in vivo gibt. Sämtliche heute verfügbaren Lymphozytenuntersuchungen sind In-vitro-Tests und erfassen nur schwere Dysfunktionen des Immunsystems, wie z.B. AIDS.

50.1.5 Schlußfolgerung

Ausgehend von einem klinischen Fallbeispiel wird anhand von ausgewählten repräsentativen Untersuchungen aus der Literatur die Hypothese erörtert, ob schwere persönliche Verluste oder Entbehrungen als Risikofaktoren zu späterer Erkrankung an Krebs beitragen können. Es läßt sich der Schluß ziehen, daß schwere Verluste, z.B. der Tod des Ehepartners, vor allem bei Männern von erhöhtem Erkrankungs- und Sterberisiko (auch an Krebs!) gefolgt sind, daß aber der Verlust nicht ein spezifischer Risikofaktor für Krebs ist. Weiter kann vermutet werden, daß die Neigung zu Depression oder zu Hilf- und Hoffnungslosigkeit, also eine mögliche psychische Antwort auf Verluste, zu späterer Erkrankung an Krebs eventuell

spezifisch beitragen kann, allerdings auch nur bei Männern. Diese Vermutung muß aber durch weitere prospektive Untersuchungen erhärtet oder verworfen werden. Nach neueren Untersuchungen der Beziehung zwischen psychischem Erleben und Immunfunktion wäre die Hypothese jedoch biologisch plausibel.

Vor verfrühten Schlußfolgerungen, vor allem in der Presse, ist zu warnen. Dies kann unter Umständen unsere Patienten schwer verunsichern. Moderne Gesundheitsapostel haben die subjektive Einstellung zum Leben und zur Krebskrankheit in versimpelnder Weise mit Prognose gleichgesetzt (Cousins, 1979). Dadurch werden Patienten oft das Gefühl nicht los, daß wenn sie nicht „positiv" über ihre Krankheit denken können, diese rapide fortschreiten würde. Falsch verstandene Zusammenhänge zwischen Psyche und Krebs verhindern die adäquate Auseinandersetzung des Patienten mit seiner Krankheit. Dazu gehören vorübergehend Verzweiflung und Traurigsein.

„Wer glücklich ist, stirbt nicht an Krebs!" stand vor einiger Zeit in der Boulevard-Presse zu lesen. So einfach ist das Problem wohl kaum.

50.2 Psychosoziale Faktoren beim Verlauf maligner Erkrankungen

Es ist wenig wahrscheinlich, daß die weitere Entwicklung einer einmal manifesten bösartigen Geschwulst rein durch ihre zellkinetischen Eigenschaften bestimmt ist und somit vollständig unabhängig von der Reaktion ihres Trägers abläuft. Gegen die Annahme eines gänzlich autonomen Wucherns spricht die große Streubreite der Überlebenszeiten bei histologisch identischem Typ mit gleicher Lokalisation, gleichem Initialstadium und gleicher Therapie. Man kann einwenden, daß die klinischen und histologischen Untersuchungsmethoden zu grob seien, um einen individuellen Tumor genau zu erfassen. Einen Hinweis hierfür gibt das unterschiedliche Vorhandensein von verschiedenen Hormonrezeptoren bei Mammakarzinom von histologisch identischem Zelltyp. Gerade die Tatsache des hormonabhängigen Wachstums von Tumoren (Mamma-, Uterus-, Ovarial- und Prostatakarzinom) und der paraneoplastischen Hormonproduktion weist aber auch auf eine mögliche gegenseitige Wechselwirkung zwischen tumorspezifischen und psychoendokrinen Mechanismen hin (Abb. 50–3).

50.2.1 Spontanverlauf, Latenz und „Immunosurveillance"

Ein weiterer Hinweis dafür, daß der Krebs die Rechnung nicht ohne seinen Wirt machen kann, ist die Beobachtung, daß stark entdifferenzierte maligne Tumoren sich während des Krankheitsverlaufs wieder ausdifferenzieren und sich sogar in benigne Formen umwandeln können. Smithers hat 14 Fälle von Neuroblastomen, 5 Fälle von anaplastischen Hoden-

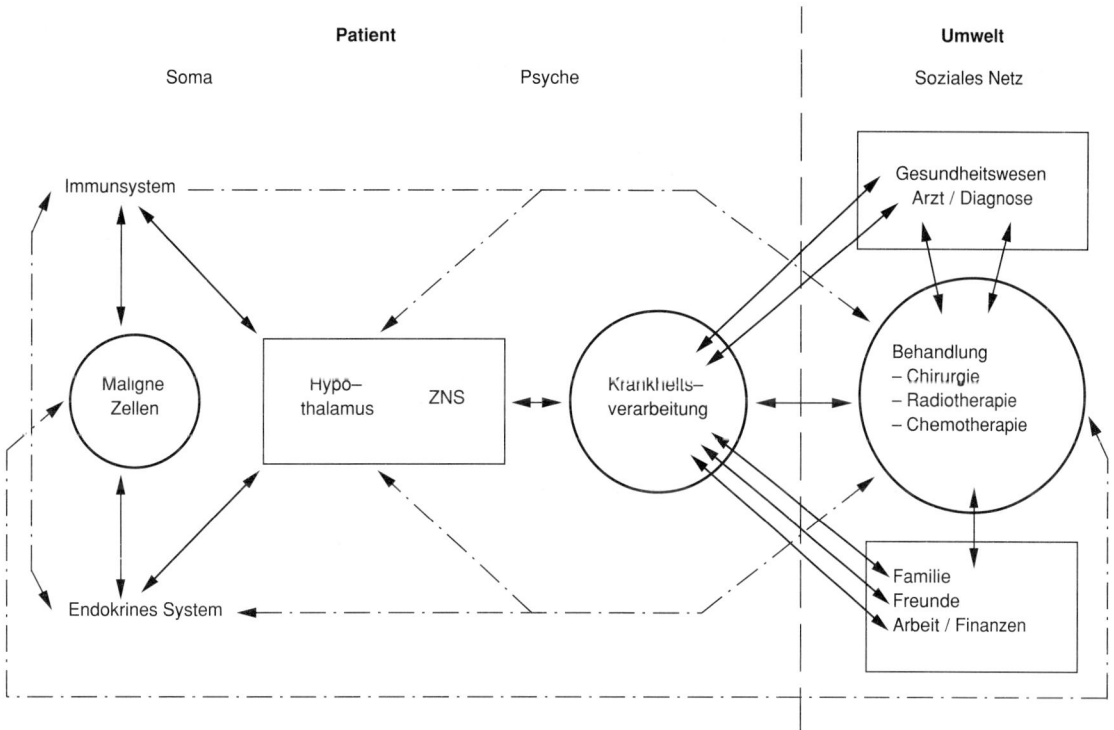

Abb. 50–3. Mögliche Wechselwirkungen zwischen Umwelt, Patient und Neoplasie.

tumoren und weitere von anderen Teratomen und embryonalen Tumoren zusammengestellt (1969). Als Ursache werden Wechselwirkungen zwischen genetischen, hormonellen, immunologischen und Umwelteinflüssen vermutet.

Boyd entwickelte 1966 am Beispiel von Patientinnen mit Carcinoma in situ der Cervix uteri, die verschiedenste Spontanverläufe zeigten, das Konzept der Latenz, d. h. der Möglichkeit, daß bei einer Vielzahl von Menschen latent Karzinome vorhanden sind, die klinisch gar nicht manifest werden oder wieder verschwinden. Er berichtet auch von vollständigen Remissionen bei Patienten mit anderer Tumorlokalisation (Boyd, 1966). Für das Konzept der Latenz spricht auch die Tatsache, daß bei genauer autoptischer Untersuchung der Prostata von 50- bis 59jährigen in etwa 10% der Fälle, von 60- bis 69jährigen in 36% und in 50% der Fälle bei über 80jährigen ein latentes Karzinom gefunden wird (Harbitz und Haugen, 1972). Das Konzept der Latenz und im Zusammenhang damit der „Immunosurveillance", d. h. der immunologischen Überwachung und Ausmerzung von ständig im Organismus durch Spontanmutationen entstehenden malignen Zellen, ist zwar biologisch plausibel, aber bis jetzt im Detail bei keiner spezifischen malignen Erkrankung nachgewiesen. Vermehrtes Auftreten von Neoplasien bei schweren angeborenen Immundefekten, bei langdauernder immunsuppressiver Therapie und bei AIDS ist zwar gut dokumentiert. Aber weder in der Vorgeschichte von Krebspatienten im allgemeinen noch bei der Diagnosestellung, noch im weiteren Verlauf der Erkrankung sind bis heute signifikante, labormethodisch oder kli-

nisch eindeutig faßbare Funktionsstörungen der zellulären oder humoralen Immunabwehr gefunden worden.

Die Chirurgen Everson und Cole haben 1966 aus über 700 Publikationen, die seit 1900 über das Phänomen der Spontanremission berichten, im Detail 130 histologisch verifizierte Fälle dokumentiert, zur Hauptsache Neuroblastome (N=38), hypernephroide Karzinome (N=21), Chorionkarzinome (N=13) und maligne Melanome (N=12). Häufigere Krebsarten wie das Mammakarzinom bilden sich seltener spontan zurück, spontaner Wachstumsstillstand über Jahre ist aber ein bekanntes klinisches Phänomen. Everson und Cole haben vergeblich nach biologischen Kriterien im Organismus und Umwelteinflüssen zur Erklärung der Spontanremission gesucht. Andere Autoren haben spontane Regressionen von Krebs im Zusammenhang mit verschiedenen Meditationspraktiken beobachtet (Meares, 1978; Ikemi et al., 1975). Systematische Untersuchungen psychosozialer Faktoren bei spontaner Tumorregression fehlen aber.

50.2.2 Soziale Schicht und Krankheitsverlauf

Neben bekannten biologischen und vermuteten psychologischen Faktoren sind Einflüsse der sozialen Umgebung auf Krankheitsentstehung und -verlauf gut dokumentiert. Praktisch jede Krankheit kommt in den unteren sozioökonomischen Schichten der Bevölkerung häufiger vor (Jenkins, 1983). Bei Krebspatienten mit verschiedenen Primärtumoren ist die Überlebenszeit nach der Diagnosestellung in gehobenen sozialen Schichten signifikant länger im Vergleich zu Krebspatienten mit niedrigem sozioökonomischem Status (Vägerö und Persson, 1987; Berg et al., 1977; Lipworth et al., 1970; Linden, 1969). Die Gründe hierfür sind nur zum Teil untersucht. Unterschiedliche Ernährung, unterschiedliche medizinische Versorgung infolge von Einkommen, Bildungsgrad und Wohnort (Region versus Zentrum), unterschiedliche berufliche und soziale Karzinogenexposition und Unterschiede in der Krankheitsverarbeitung werden vermutet.

50.2.3 Krankheitsverarbeitung und -verlauf

Im März 1987 kurz nach der vorzeitigen Pensionierung ihres Ehemannes entdeckt die 55jährige Frau W. einen Knoten in der rechten Brust. Innerhalb einer Woche muß wegen Karzinom eine Mastektomie durchgeführt werden. Der histologische Befund ergibt befallene axilläre Lymphknoten. Fünf Monate später beginnen ziehende Schmerzen in beiden Flanken. Nach langwierigen Abklärungen ohne Befund – Frau W. wird von verschiedenen Ärzten als psychisch krank betrachtet – tritt nach weiteren 2 Monaten eine unvollständige Querschnittslähmung auf. Die Patientin wird Mitte Oktober 1987 notfallmäßig laminektomiert, epidural und in mehreren Brustwirbelkörpern finden sich Metastasen. Unter Bestrahlung der Brustwirbelsäule gehen die Schmerzen gegen Ende des Jahres zurück. Im Januar 1988 tritt eine Hyperkalzämie bei generalisierter Skelettmetastasierung auf. Unter Chemo- und Hormontherapie bilden sich die akuten Symptome zurück. Die Patientin wird Anfang März zur Rehabilitation in unsere Klinik verlegt. Bei Eintritt ist die Patientin vollständig bettlägerig, es besteht eine rechtsbetonte Paraspastik mit Miktions-, Defäkations- und Sensibilitätsstörungen. Frau W. ringt mit dem Schicksal, sie versucht gegen ihre Krankheit zu kämpfen, ist oft dem Weinen nahe. Einen Zusammenhang zwischen dem Krebsleiden und der Lähmung schiebt sie aktiv weg. In harter Arbeit mit Krankenschwester, Physio- und Ergotherapeutin, unterstützt durch ihrem fürsorglichen Ehemann, erkämpft sich Frau W. in kleinen Teilschritten mehr Selbständigkeit. Sie kann im Rollstuhl mobilisiert werden und mit Hilfe der Schwester oder des Ehemannes vom Bett auf den Rollstuhl, vom Rollstuhl aufs WC transferiert werden. Nach 7 Monaten ist nach entsprechender Anpassung der Wohnung und mit einer Haushaltshilfe die Rückkehr nach Hause möglich. Nach unserer Ansicht stellt sich in bezug auf die Fortschritte ein Plateau ein und wir sind bei Fragen der Patientin in bezug auf den weiteren Verlauf skeptisch. Unter ambulanter Chemo-, Hormon- und Physiotherapie macht Frau W. unter großer Anstrengung weitere Fortschritte, so daß sie am Böckli wieder gehen lernt und hofft, daß sie doch noch einmal mit ihrem Mann in das geliebte Berghaus im Wallis ziehen kann. Vor Jahresfrist hatte sie kaum gewagt, diesen Wunsch zu äußern!

Solche Fallberichte sind in der Literatur zahlreich vorhanden. Die Verbindung von kämpferischer Haltung und guter Prognose entspricht aber möglicherweise unserem Wunschdenken. Wer wünschte sich im Fall des eigenen Betroffenseins nicht, daß eine mutige Haltung den Krebs in Schach halten könnte?

50.2.4 Methodische Probleme bei der Erfassung von Krankheitsverarbeitung

Seit den 50er Jahren wurde in zahlreichen, methodisch mehr oder weniger stichhaltigen, nur teilweise prospektiven Studien versucht, den Einfluß von unterschiedlicher Verarbeitung der Krebsdiagnose, u.a. einer hoffnungsvollen, kämpferischen Haltung bzw. des Aufgebens und der Hoffnungslosigkeit, auf den Krankheitsverlauf, gemessen an der Überlebenszeit, zu prüfen. Als methodischer Ansatz wurde bei den Studien der Vergleich von Patienten mit signifikant längerer Überlebenszeit und solchen mit signifikant kürzerer Überlebenszeit als statistisch erwartet herangezogen. Beim Durcharbeiten der Untersuchungen tauchen neben den im ersten Teil dieses Kapitels erwähnten allgemeinen methodischen Erwägungen entscheidende, bis heute nicht gelöste spezifische Fragen auf. Was ist Krankheitsverarbeitung und wie wird das sog. „coping" erfaßt? Welches sind die Wechselwirkungen zwischen Tumor, Patient und sozialer Umgebung über die Zeit? Welche Bedeutung hat soziale Unterstützung („social support") und wie wird sie erfaßt? Was sind die Merkmale einer adäquaten Krankheitsverarbeitung und welches die Kriterien für eine erfolgreiche Adaptation: Ist es das Befinden des Patienten? Ist es die Überlebenszeit? Ist es die soziale Integration? Falls eine adaptive vs. inadaptive Krankheitsverarbeitung definiert werden kann, gibt es Interventionen, die das Coping verbessern können?

Bei den meisten bisherigen Untersuchungen werden diese methodischen Fragen gar nicht gestellt. Die Überlebenszeit wird meist als zwangsläufiges und ausschließliches Maß der Krankheitsverarbeitung betrachtet. Als Indikator für die Krankheitsverarbeitung wird die Reaktion des Patienten auf die Diagnose zu einem bestimmten Zeitpunkt einmalig erfaßt und dabei vernachlässigt, daß Krankheitsverarbeitung als Prozeß zu verstehen ist und nicht als einmaliges Ereignis.

Die Diagnose „Krebs" ist eines der belastendsten Ereignisse, die einem Menschen widerfahren können. Die in Betracht kommenden Behandlungen, sei es ein möglicherweise verstümmelnder chirurgischer Eingriff (z.B. Mastektomie), sei es eine nebenwirkungsreiche Strahlen- und/oder Chemotherapie, sind einschneidend. Für die psychosoziale Anpassung ist die hervorragende Bedeutung der initialen Reaktion des Patienten und seiner Familie auf Diagnose und Behandlung evident. Die Krankheitsverarbeitung ist jedoch ein Prozeß, nicht ein einmaliges Ereignis. Verarbeitungsstrategien können sich mit der Zeit ändern. Die Krankheitsverarbeitung ist ein individuell sehr unterschiedlich ablaufender Prozeß. Dennoch können verschiedene Phasen, die Patienten in der Regel durchlaufen, definiert werden:

– Vorgeschichte: biographisch definierte Ängste und Phantasien über Krebs

- Erste Symptome: Wahrnehmung bzw. Vernachlässigung der ersten Symptome
- Erster Arztbesuch und anschließende Abklärung
- Diagnosestellung
- Erstbehandlung (häufig chirurgisch)
- Auftreten eines Rezidivs bzw. Metastasen
- Zweitbehandlung (häufig Chemo- bzw. Strahlentherapie)
- Heilung/chronische Krankheit/Sterben und Tod

Die verschiedenartigen Probleme, die der Krebspatient im Verlauf seiner Krankheit antrifft, lösen sehr wahrscheinlich eine ganze Reihe verschiedener Reaktionen in und um den Patienten aus, je nach dessen Erfahrungen mit Krebs und seinem Persönlichkeitsstil. Erst in den letzten Jahren ist eine differenzierte Betrachtung und Erfassung der Krankheitsverarbeitung im Patienten („coping") und seiner Umgebung („social support") in Angriff genommen worden. Die Gruppe von Heim (1988) hat in einer sehr differenzierten longitudinalen Studie von 60 Brustkrebspatientinnen die häufig vorkommenden Verarbeitungsstrategien definiert und konnte über die Zeit einerseits individuell zum Teil sehr wechselnde Krankheitsverarbeitungsmuster, aber auch gemeinsame Entwicklungen nachweisen.

Nach Herschbach und Henrich (1987) ist der Krankheitsverarbeitungsprozeß auf drei Ebenen definiert: Erstens durch die aktuellen Probleme, denen der Patient begegnet, zweitens durch das Ausmaß der Belastung, die der Patient durch jedes Problem erlebt, und drittens durch die Verarbeitungsstrategien, die er rigid oder flexibel braucht, um die Probleme zu lösen oder sich zu entlasten. In einer Querschnittsuntersuchung von über 300 Frauen mit Brust- oder Genitalkrebs haben die Autoren 7 Problembereiche und 15 mehr oder weniger häufig gebrauchte Verarbeitungsstrategien definiert (Tab. 50–2). Im Moment gibt es keine allgemein anerkannte Methodologie zur Erfassung von Krankheitsverarbeitung bei Krebspatienten. Dies wird durch die Vielfalt von angewendeten Instrumenten in den Krankheitsverlaufsstudien illustriert.

In der allgemeinen Coping-Forschung wird als Resultat der Krankheitsverarbeitung in der Regel die Adaptation des Patienten und seiner Umgebung in bezug auf Krankheit und Behandlung betrachtet. Das Ausmaß der geglückten Anpassung bestimmt im wesentlichen das physische, psychische und soziale Wohlbefinden des Patienten, anders gesagt, seine Lebensqualität.

In den Studien hingegen, die sich speziell mit der psychosozialen Reaktion des Patienten auf Krebs befassen, wird die Überlebenszeit als hauptsächlicher Parameter zur Auswertung der Krankheitsverarbeitung betrachtet und psychosoziales Wohlbefinden als Resultat der Adaptation kaum je als abhängige Variable untersucht. Zudem werden biologische Faktoren mit bekannter prognostischer Bedeutung, wie z. B. der axilläre Lymphknotenbefall bei Mammakarzinom, oft ignoriert. Nur wenige Untersuchungen (Funch und Mettlin, 1982) betrachten die psychoso-

Tabelle 50–2. Herschbach-Fragebogen zur Krankheitsverarbeitung
Selbsteinschätzung des Patienten auf 3 Ebenen:
7 Problembereiche
– Information/Behandlung
– Behandlungsfolgen
– Schmerzen
– Angst/seelische Belastung
– Leistungsfähigkeit
– Sozialverhalten
– Partnerschaft/Familie
Belastung durch die einzelnen Problembereiche
15 Verarbeitungsstrategien für jeden Bereich:
– Durchdenken von Gegenmaßnahmen
– Zähne zusammenbeißen
– Sich aussprechen bei jemandem
– Einsehen, daß man nichts ändern kann
– Ablenken durch Beschäftigung
– Sich trösten mit der Überlegung, daß es anderen noch schlechter geht
– Weinen und Verzweiflung
– Ausdrücken von Ärger und Wut
– Ratsuchen beim Fachmann
– Selbstkritik
– Einnehmen von Alkohol/Beruhigungstabletten
– „Es wird auch wieder aufwärts gehen"
– Abwarten von spontaner Problemlösung
– Vermeiden des Problems
– In Betracht ziehen der positiven Seiten des Problems

ziale Anpassung als untersuchungswert. In ähnlicher Weise werden in onkologischen Therapiestudien zur Evaluation von neuen chirurgischen, strahlen- oder chemotherapeutischen Behandlungen ausschließlich objektive Kriterien, das Verhalten des Tumors (Remission, Progression) und die Überlebenszeit, herangezogen. Psychosoziale Variablen, wie z.B. subjektives Wohlbefinden und Krankheitsverarbeitung, werden als weiche Daten betrachtet und selten berücksichtigt. Nach meinem Wissen gibt es bis heute keine publizierte Studie, die Krankheitsverarbeitung und die daraus resultierende psychosoziale Adaptation in die Evaluation von medizinischen Behandlungen miteinbezieht. Einen ersten Schritt in dieser Richtung hat die International Breast Cancer Study Group (IBCSG) unternommen (Hürny und Bernhard, 1989; Hürny, in press).

50.2.5 Krankheitsverarbeitung und Überlebenszeit

Was Krankheitsverarbeitung und deren Einfluß auf die Prognose anbetrifft, sind Patientinnen mit Brustkrebs weitaus am besten untersucht. Ich werde mich deshalb bei meinen weiteren Ausführungen auf diese Krankheit beschränken. Allgemeine kritische Übersichten können an anderer Stelle konsultiert werden (Hürny, 1985).

Wenn die Überlebenszeit als abhängige Variable der Krankheitsverarbeitung untersucht wird, müssen vorerst theoretisch mögliche Interaktionen zwischen Krankheitsverarbeitung und biologischem Verlauf

der Erkrankung geklärt werden. Grundsätzlich gibt es vier Möglichkeiten:
- Die Krankheitsverarbeitung kann sich auf die Compliance des Patienten mit der Behandlung auswirken und somit indirekt den Krankheitsverlauf beeinflussen.
- Die Krankheitsverarbeitung kann hauptsächlich durch den biologischen Verlauf bestimmt sein.
- Die Krankheitsverarbeitung kann den biologischen Verlauf der Krankheit direkt beeinflussen über psycho-neuro-endokrino-immunologische Mechanismen.
- Krankheitsverarbeitung und Krankheitsverlauf laufen unabhängig voneinander ab.

Wahrscheinlich sind die genannten Interaktionsmöglichkeiten meist miteinander vernetzt und im Einzelfall kaum auseinanderzuhalten. Wenn man den Einfluß von Krankheitsverarbeitung auf die Prognose in Betracht zieht, stellt sich konsequenterweise die Frage, ob dieser Einfluß durch adaptive bzw. maladaptive Verarbeitungsstrategien bedingt ist. In seiner Literaturübersicht versucht Heim die in den einzelnen Studien untersuchten psychosozialen Faktoren nach seinen Verarbeitungsstrategien, den „Berner Bewältigungsformen (BEFOS)" zu kategorisieren und definiert vorläufig adäquate versus inadäquate Krankheitsverarbeitung. Seine Schlußfolgerung ist, daß adäquates Coping die Überlebenszeit positiv, inadäquates Coping sie negativ beeinflußt (Heim et al., 1988). Dabei läßt er aber wesentliche methodische Mängel der verschiedenen Untersuchungen außer acht.

50.2.6 Methodische Kriterien zur Evaluation der Verlaufsstudien

Die in Tabelle 50–1 zusammengestellten Kriterien zur Beurteilung von Risikofaktoruntersuchungen gelten zum Teil auch bei dieser Fragestellung. Zusätzlich sind die in Tabelle 50–3 zusammengestellten Punkte besonders zu berücksichtigen.

Hypothesen:
Strenggenommen müssen die Hypothesen vor Beginn der Untersuchung festgelegt werden. Eine Studie, die eine vorher festgelegte Hypothese überprüft, ist unvergleichbar aussagekräftiger als das zufällige Finden von Zusammenhängen beim Auswerten der Daten. Bei den meisten publizierten Studien wird dieser Punkt nicht explizit erwähnt. Oft werden bei Reihenuntersuchungen aufgenommene psychosoziale Daten

---

Tabelle 50–3. Methodische Kriterien für die Auswertung von Untersuchungen des Einflusses psychosozialer Faktoren auf das Überleben bei Brustkrebspatientinnen

- Hypothesen
- Studiendesign
- Untersuchungszeitpunkte
- Untersuchungsinstrumente
- statistische Analyse
- Stichprobe

---

im nachhinein mit dem darauffolgenden Verlauf in Beziehung gebracht. So haben z. B. Funch und Marshall (1983) retrospektiv Daten analysiert, die 20 Jahre früher zu einem ganz anderen Zweck erhoben worden waren (Snell und Graham, 1971).

Studiendesign:
Bei einer kontrollierten retrospektiven oder nur halbwegs prospektiven Untersuchung, die Patientinnen mit günstigem Verlauf mit Patientinnen mit ungünstigem Verlauf vergleicht, kann nicht ausgeschlossen werden, daß der vermeintliche prognostische Faktor Ausdruck und nicht Ursache des günstigen bzw. ungünstigen Verlaufs ist. In der Untersuchung von Jensen z. B. wird ein hochselektioniertes, nicht repräsentatives Patientenkollektiv erfaßt, indem die einzelnen Patientinnen zu arbiträr gewählten Zeitpunkten nach der Diagnosestellung in die Studie aufgenommen und psychologisch untersucht werden. Bei der Auswertung wird dann die Zeit zwischen Diagnose und Studienbeginn nicht berücksichtigt (Jensen, 1987).

Untersuchungszeitpunkte:
Bei einem Großteil der Studien findet eine einmalige Untersuchung im Querschnitt statt (Funch und Marshall, 1983; Jensen, 1987; Derogatis et al., 1979; Greer et al., 1979; Pettingale et al., 1985; Morgenstern et al., 1984; Cassileth et al., 1985 a, 1988; Jamison et al., 1987; Levy et al., 1987). Da Krankheitsverarbeitung ein Prozeß über Zeit ist, sind Querschnittsuntersuchungen nicht repräsentativ, longitudinale Studien drängen sich auf. Eine mindestens zweimalige Erhebung zu genau definierten Zeitpunkten im Verlauf der Krankheit ist nötig.

Untersuchungsinstrumente:
Wie bereits erwähnt, gibt es im Moment keine Standardmethode zur Erfassung von Krankheitsverarbeitung. In den Studien werden die unterschiedlichsten Instrumente verwendet, was einen Vergleich der Untersuchungen praktisch unmöglich macht.

Statistische Analyse:
Die Anwendung nicht adäquater statistischer Verfahren ist ein häufiger Mangel. Die Untersuchung von zahlreichen Variablen bei einem kleinen Patientenkollektiv z. B. führt zu rein zufälligen statistisch signifikanten Korrelationen.

Stichprobe:
Die genaue biomedizinische und soziodemographische Beschreibung der Stichproben ist unabdingbar, weil ungleiche Verteilung bekannter soziodemographisch und biologisch prognostischer Faktoren einen psychologischen Faktor vortäuschen bzw. verwischen kann. Bei den bisherigen Untersuchungen sind die untersuchten Kollektive meist klein und nicht repräsentativ. In Tabelle 50–4 ist die minimal notwendige Charakterisierung einer zu untersuchenden Gruppe von Mammakarzinom-Patientinnen festgehalten.

Anstelle einer umfassenden Literaturübersicht sollen in der Folge die zwei meistzitierten Studien des Einflusses psychosozialer Faktoren auf den Krankheitsverlauf bei Brustkrebspatientinnen im Hinblick auf die obengenannten Kriterien exemplarisch und kritisch analysiert werden.

Tabelle 50–4. Minimal notwendige Charakterisierung der Stichprobe von Mammakarzinom-Patientinnen in biopsychosozialen Studien

Soziodemographische Charakteristika
– Alter
– Zivilstand
– Wohnsituation
– Beruf
– Berufliche Stellung
– Ausbildungsgrad
– Einkommen
Biomedizinische Charakteristika
– Menopausaler Status
– Primärtumorlokalisation und -größe
– Histologischer Differenzierungsgrad
– Hormonrezeptorstatus
– Krankheitsstadium
– Behandlungsmodalitäten

50.2.7 Exemplarische Analyse von zwei repräsentativen Untersuchungen des Einflusses psychosozialer Faktoren auf den Krankheitsverlauf bei Brustkrebspatientinnen

Die Gruppe von Greer am Kings-College-Hospital in London hat die erste psychosoziale Langzeitstudie bei Patientinnen mit Brustkrebs im Stadium I und II durchgeführt (Greer et al., 1979; Pettingale, 1985). 69 konsekutiv aufgenommene Patientinnen wurden präoperativ mit einem strukturierten Interview und verschiedenen psychologischen Tests (Hamilton Depression Scale, Hostility and Direction of Hostility Questionnaire, Eysenck Personality Inventory, Millhill Vocabulary Scale) untersucht. Drei Monate nach der Mastektomie führten die Autoren bei 57 Patientinnen ein längeres offenes Gespräch zur Erfassung der Reaktion auf die Diagnose durch. Sie teilten die Patientinnen zu diesem Zeitpunkt aufgrund von relativ groben klinischen Kriterien in 4 Gruppen:

– **Aktives Verleugnen** („es war nichts Ernsthaftes, die Amputation meiner Brust war nur eine Vorsichtsmaßnahme"); im Moment der Untersuchung keine offensichtlichen emotionalen Probleme.
– **Kämpferische Haltung** („ich kann Krebs bekämpfen und besiegen"); im Moment der Untersuchung keine offensichtlichen emotionalen Probleme.
– **Stoisches Akzeptieren** („ich weiß, daß es Krebs ist, aber ich muß einfach weiterleben, wie wenn nichts wäre"); postoperativ emotionale Schwierigkeiten, die zum Zeitpunkt der Untersuchung nicht mehr offensichtlich waren.
– **Hilf-** und **Hoffnungslosigkeit** („man kann nichts mehr machen, ich bin erledigt"); offensichtlich emotionale Probleme zum Zeitpunkt der Untersuchung.

Fünf Jahre nach der Mastektomie waren 28 Frauen ohne Lokalrezidiv oder Metastasen, 13 wiesen Metastasen auf und 16 waren am Tumor gestorben. Bei den 20 Frauen, die mit aktivem Verleugnen oder kämpferischer Haltung reagiert hatten, waren 15

nach 5 Jahren tumorfrei (75%), von 37 Frauen, die mit stoischem Akzeptieren oder Hilf- und Hoffnungslosigkeit reagiert hatten, nur 13 (35%). Bei den 16 Frauen, die in den 5 Jahren an Krebs gestorben waren, hatten 14 (88%) mit stoischem Akzeptieren oder Hilf- und Hoffnungslosigkeit reagiert, bei den 28 tumorfrei überlebenden nur 13 (46%, p < 0,025). Patientinnen, die zum Zeitpunkt der Diagnose allein lebten oder über Eheschwierigkeiten berichteten, hatten eine Tendenz weniger lang zu überleben (p < 0,09). Alle anderen untersuchten psychosozialen Variablen, insbesondere die psychologischen Tests waren nicht signifikant mit dem 5-Jahres-Überleben korreliert.

Nach einer Beobachtungszeit von 10 Jahren (19 Patientinnen leben ohne Rezidiv, 2 leben mit Metastasen und 36 sind gestorben) wurden diese Resultate bestätigt. In einer multivariaten Regressionsanalyse mit 8 prognostischen Faktoren war die psychologische Antwort auf die Diagnose der signifikanteste individuelle Faktor zur Voraussage des Todes durch eine beliebige Ursache (p < 0,003), durch Brustkrebs (p < 0,003) und für die Voraussage des ersten Rezidivs (p < 0,008) (Pettingale, 1985).

Cassileth und Mitarbeiter (1985a) am University of Pennsylvania Cancer Center untersuchten innerhalb von 2–8 Wochen nach der Diagnose 155 konsekutive Patienten (93 mit Brustkrebs Stadium II, 62 mit Melanom Stadium I („high or intermediate high risk") mit einem Fragebogen, der 7 psychosoziale Faktoren erfaßt, die in früheren Studien bei Krebspatienten die Prognose voraussagen konnten:

– soziales Netz,
– Befriedigung im Beruf,
– Gebrauch von Psychopharmaka,
– allgemeine Zufriedenheit mit dem Leben,
– subjektive Einschätzung der Gesundheit im Erwachsenenalter,
– Hoffnungslosigkeit,
– das subjektiv erlebte Ausmaß der notwendigen Anpassungsleistung an die Krankheit.

Jeder dieser 7 Faktoren wurde bei der Analyse der Daten als Subskala betrachtet. Durch Zusammenzählen der Subskalenwerte für jeden einzelnen Patienten wurde ein globaler psychosozialer Profilscore konstruiert. Eine zweite Gruppe von 204 Patienten mit weit fortgeschrittenen Krebsleiden wurde ebenfalls untersucht. Da die Resultate dieser Teiluntersuchung nicht direkt mit der Studie von Greer vergleichbar sind, gehe ich hier nicht weiter darauf ein. Mit Ausnahme der gut validierten Beck Hopelessness Scale handelt es sich bei der Erfassung der anderen 6 psychosozialen Faktoren nicht um psychometrisch geprüfte Instrumente, sondern um einfache Fragen. Nach einer medianen Zeit von 12,3 Monaten hatten 41 von den 155 Patienten ein Rezidiv erlitten. In drei verschiedenen Analysen (Kaplan/Meier-Überlebenszeitanalyse, Mantel-Cox-Modell und Varianzanalyse) hatten weder die individuellen Subskalenwerte noch die globalen Werte einen Einfluß auf die rezidivfreie Überlebenszeit. 3–8 Jahre nach der Diagnosestellung (85 Patienten rezidivfrei, 64 Patienten mit Rezidiv) wurden diese Resultate bestätigt. Positive Einstellung zum Leben und Hoffnung in bezug auf die Krankheit wurde mit gleicher Frequenz bei den Patienten gefunden, ob sie später ein Rezidiv erlitten oder in Remission blieben. Als eindeutige prognostische Faktoren wurde nur der Allgemeinzustand der Patienten und die Ausdehnung der Krankheit bei Diagnosestellung gefunden (Cassileth et al., 1988).

Die Resultate der beiden Studien sind diametral entgegengesetzt. Greer findet bei seinen Brustkrebspatientinnen, daß eine hoffnungsvolle oder kämpferische Haltung gegenüber der Krankheit die rezidivfreie und die gesamte Überlebenszeit verlängern kann, während Cassileth bei ihren Patientinnen mit Mammakarzinom und Patienten mit malignem Melanom keinen Einfluß einer positiven Einstellung auf das Überleben finden kann. Wie kommen solch unterschiedliche Resultate zustande? Meistens sind sie durch methodische Unzulänglichkeiten bedingt. Die Kriterien für die Evaluation solcher Untersuchungen sind in den Tabellen 50–3 und 50–4 zusammengestellt.

Beide Studien sind prospektiv. Die Hypothesen sind vor Studienbeginn klar definiert worden, die Patienten sind zu einem genau definierten Zeitpunkt im Verlaufe der Erkrankung untersucht worden.

In der englischen Studie ist der Follow-up von 10 Jahren deutlich länger als in der amerikanischen Studie (3–8 Jahre). Da Brustkrebs ein langsam wachsender Tumor ist, kommen signifikante Vorteile der adjuvanten Therapie in der Regel erst sehr spät im Verlauf der Krankheit zum Vorschein, meistens mehrere Jahre nach der Diagnose. In Analogie ist es möglich, daß frühe psychosoziale Faktoren erst spät Auswirkungen zeigen. Das Argument ist jedoch nicht sehr stichhaltig, da in der Greer-Studie bereits nach 5 Jahren deutliche Unterschiede vorhanden waren.

Von zentraler Bedeutung ist der Zeitpunkt der Untersuchung und die Wahl der Untersuchungsinstrumente. Da die Krankheitsverarbeitung sich über die Zeit ändert, sind mehrmalige Untersuchungen über die Zeit notwendig. In beiden Studien wurden die angenommenen prädiktiven Variablen nur einmal untersucht, innerhalb von 2–8 Wochen nach der Diagnose bei der Cassileth-Studie und 3 Monate nach Diagnosestellung in der Greer-Studie. Nach klinischer Erfahrung ist nach 3 Monaten die akute Krise, die durch die Krebsdiagnose ausgelöst wird, vorüber und die Situation hat sich etwas beruhigt. Es wäre möglich, daß zu diesem Zeitpunkt die üblichen, über längere Zeit bestehenden Verarbeitungsstrategien zum Vorschein kommen, während in der akuten Phase das Coping häufig wechselt und demnach schwieriger zu erfassen ist. Es ist also wahrscheinlicher, daß die Greer-Studie Verarbeitungsmuster erfaßt, die über längere Zeit wirksam sind und somit auch wahrscheinlicher den biologischen Verlauf beeinflussen könnten.

Wie weiter oben bereits angedeutet, sind zum heutigen Zeitpunkt keine generell akzeptierten Instrumente zur Erfassung der Krankheitsverarbeitung bei Brustkrebspatientinnen vorhanden. Instrumente, die für psychiatrische Patienten oder für die Allgemeinbevölkerung entwickelt wurden, sind in der Regel nicht empfindlich genug für spezifische Probleme von Brustkrebspatientinnen. In der Mehrzahl der Studien werden psychologische Tests, Fragebögen und strukturierte Interviews verwendet (Derogatis et al., 1979; Funch und Marshall, 1983; Cassileth et al., 1985a, 1988; Holland et al., 1986; Hislop et al., 1987; Jensen, 1987; Jamison et al., 1987; Levy et al., 1987). Ein Teil davon hat sich als valide, reliabel und spezifisch für Krebspatienten erwiesen, andere wurden in dieser Hinsicht gar nicht geprüft. Die Greer-Studie ist die einzige, die offene Gespräche zur Erfassung der Reaktion der Patientinnen auf die Krebsdiagnose benützt neben strukturierten Interviews und psychologischen Tests. Nach unserer Erfahrung besteht ein fundamentaler Unterschied zwischen Interviewdaten, die aus einer Interaktion mit einem erfahrenen psychologisch geschulten Arzt hervorgehen, und Daten, die von psychologischen Tests oder Fragebögen herstammen. Das Raster der objektiven Tests scheint uns viel gröber und oberflächlicher verglichen mit der Erfassung des Patienten im ärztlichen Gespräch. Auf der anderen Seite können Interviewdaten durch vorgefaßte Meinungen des Untersuchers verfälscht werden. In der Studie von Greer wird erwähnt, daß in einer Pilotuntersuchung zwei unabhängige Beobachter die Patientinnen mit hoher Übereinstimmung in die vier klinischen Kategorien einteilen konnten. Die vier Kategorien sind dann klinisch deskriptiv gut definiert, aber genauere Information über die Beobachterübereinstimmung („Interrater-Reliability") fehlt. Immerhin waren die Kategorien, was den weiteren Krankheitsverlauf anbelangt, prognostisch prädiktiv.

In der amerikanischen Studie wurde ein Fragebogen gebraucht, der mit 32 Fragen 7 komplexe psychosoziale Konstrukte zu erfassen sucht. Es ist fragwürdig, ob ein solch rudimentäres Instrument relevante Aspekte des Verarbeitungsprozesses erfassen kann. Mit Ausnahme der „Beck Hopelessness Scale" ist der Fragebogen im Hinblick auf seine psychometrischen Eigenschaften nicht untersucht worden.

In den meisten Studien läßt die biomedizinische und soziodemographische Beschreibung der untersuchten Patienten zu wünschen übrig, und der Einfluß dieser die Resultate möglicherweise verfälschenden Variablen wird nicht untersucht. In den hier zur Diskussion stehenden Studien wurden soziodemographische Charakteristika grob erfaßt, bei Cassileth als Antwort auf einen Brief von Funch (1985; Cassileth et al., 1985b), und bei Greer ist die Erfassung der sozialen Schicht nicht spezifisch beschrieben. Da der sozioökonomische Status möglicherweise einen prognostischen Einfluß hat (Vägerö und Persson, 1987; Berg et al., 1977; Lipworth et al., 1970; Linden, 1969), müßte er genauer untersucht werden.

In beiden Studien wurden möglicherweise verfälschende biomedizinische Faktoren nur zum Teil in Betracht gezogen. Bei Cassileth ist das Zusammenfassen von zwei Patientengruppen, die zwar, was das Krankheitsstadium anbelangt, vergleichbar sind, aber vollständig verschiedene Primärtumoren aufweisen, fragwürdig, obwohl statistisch die Lokalisation des Primärtumors die psychosozialen Werte nicht beeinflußte. Möglicherweise zeigt das einen Mangel an Spezifität und Empfindlichkeit des verwendeten Instrumentes an. Die Tatsache, daß die meisten Brustkrebspatientinnen adjuvante Chemotherapie erhielten, im Gegensatz zu den Melanompatienten, könnte auch von Bedeutung sein. Der menopausale Status der Patientinnen ist nicht erwähnt. Da das Alter der Patienten keinen Effekt auf die psychosozialen Werte hatte, ist jedoch ein wesentlicher Einfluß des menopausalen Status unwahrscheinlich. In der englischen Studie wurden eine ganze Anzahl von möglicherweise verfälschenden biomedizinischen Variablen getestet. Hingegen konnte einer der wichtigsten bekannten prognostischen Faktoren bei Brustkrebs, nämlich der histologische Befall von axillären Lymphknoten, nicht erhoben werden, da zur Zeit der Untersuchung die Axilla nicht chirurgisch exploriert wurde. Auch wenn behauptet wird, daß der Befall der axillären Lymphknoten für die 10-Jahres-Überlebensrate nicht von großer Bedeutung sei (Pettingale et al., 1985; Fentiman et al., 1984), kann nicht mit Sicherheit ausgeschlossen werden, daß die Unterschiede im Überleben nicht durch eine ungleiche Verteilung der lymphknotenpositiven Patientinnen in den vier Reaktionskategorien bedingt sind. Die bei dieser Untersuchung noch nicht erfaßbaren Hormonrezeptoren des Tumors könnten als bekannte prognostische Faktoren bei ungleicher Verteilung in den verschiedenen Reaktionsgruppen ebenfalls Unterschiede vortäuschen.

Die Zahl der untersuchten Patienten ist relativ klein in beiden Studien, aber größer bei Cassileth, was das Finden auch kleinerer Unterschiede etwas wahrscheinlicher macht. Es ist zu betonen, daß in der Greer-Studie mit Einschluß der Kontrollvariablen weit über 20 Variablen untersucht wurden. Bei der geringen Zahl von 57 untersuchten Patientinnen kann die Möglichkeit, statistisch signifikante Unterschiede durch Zufall zu finden, nicht vernachlässigt werden. Hingegen ist es unwahrscheinlich, alle zufälligen Signifikanzen beim selben Instrument, nämlich dem offenen Interview zu finden. Meines Erachtens ist das offene Interview eines der besten und empfindlichsten Instrumente zur Erforschung der Krankheits-

verarbeitung. Es ist aber sehr aufwendig und teuer, mit psychologisch geschulten Klinikern über Zeit wiederholte Gespräche bei einer großen Zahl von Patienten durchführen zu lassen.

Ich habe versucht, an zwei Studien mit gegenteiligen Resultaten aufzuzeigen, welche methodischen Mängel die Aussagekraft der Untersuchungen verringern können. Wenn man die Gesamtheit der vorliegenden Studien betrachtet, so kann zum heutigen Zeitpunkt nicht sicher entschieden werden, ob die Krankheitsverarbeitung einen Einfluß auf das Überleben nach Erstbehandlung bei Brustkrebs hat. Weitere sorgfältig geplante prospektive Untersuchungen sind nötig, um diese Frage weiter zu erhellen und zu differenzieren.

50.2.8 Zusammenfassung

Die Frage des Einflusses psychosozialer Faktoren auf den biologischen Verlauf der Krebskrankheit wird anhand von ausgewählten, repräsentativen Untersuchungen von Brustkrebspatientinnen erörtert und mit besonderer Berücksichtigung methodischer Aspekte kritisch durchleuchtet. Bisher wurde vor allem der Einfluß der Krankheitsverarbeitung auf den Krankheitsverlauf untersucht. Eine genaue Analyse der Studien zeigt, daß bis heute keine einheitliche, methodisch stichhaltige Erfassung der Krankheitsverarbeitung entwickelt worden ist. Wegen der Vielfalt der Untersuchungsmethoden sind diese Studien nur sehr schwer zu vergleichen. Somit ist auch eine abschließende Gesamtbeurteilung nicht möglich. Gesichert ist, daß die soziale Schichtzugehörigkeit einen Einfluß auf den Krankheitsverlauf hat und zwar nicht nur bei der Krebskrankheit.

Methodische Kriterien zur Evaluation der Verlaufsstudien werden aufgestellt und zur exemplarischen Analyse von zwei Untersuchungen mit gegenteiligen Resultaten herangezogen. Es zeigt sich, daß die gegensätzlichen Ergebnisse wahrscheinlich durch methodische Mängel der Studien bedingt sind. Generell ist also ein Einfluß der Krankheitsverarbeitung auf den Krankheitsverlauf nicht etabliert. Weitere methodisch bessere Prospektivuntersuchungen sind nötig und laufen gegenwärtig.

51 Anwendung psychosomatischer Konzepte in der Psychiatrie

Herbert Weiner
Übertragung aus dem Englischen: *Rolf H. Adler*

Anmerkung der Herausgeber: Wir bringen ein Kapitel über Psychiatrie in einem Lehrbuch der Psychosomatischen Medizin, weil wir der Meinung sind, daß psychosomatische Medizin Humanmedizin und Psychiatrie daher auch Teil der psychosomatischen Medizin ist. In dem Kapitel wird deutlich, welche neuen Probleme in einer Psychiatrie auftauchen, die von psychosomatischen Konzepten ausgeht und wie wenig fundiert gerade hier einseitig biochemisch orientierte Vorstellungen sind.

Die noch weitverbreitete Auffassung, psychosomatische Medizin beschränke sich auf die Untersuchung einer kleinen Zahl von Krankheiten oder die Rolle der Emotionen im Krankheitsgeschehen, ist vollständig veraltet und falsch. Die ursprünglichen Vorstellungen zweier früherer Generationen „psychosomatischer" Forscher sind auch häufig mißverstanden worden: Sie hatten nie behauptet, psychische und/oder soziale Faktoren würden für sich allein (ätiologische oder pathogenetische) Krankheitsursachen darstellen; sie betrachteten sie lediglich als mitentscheidende Faktoren. Spätestens seit 1950 (Alexander, 1950) wurden Krankheiten, im Gegensatz zur traditionellen Biomedizin, mit dem heiligen Gral des Glaubens an eine alleinige Ursache jeder Krankheit, als multideterminiert (multikausal) aufgefaßt. Dieser Unterschied der Auffassung stellt nur einen unter vielen dar, welche die traditionelle Biomedizin und psychosomatische Medizin unterscheiden (Ader, 1980; Coulehan, 1980, Engel, 1977, Weiner, 1977, 1978).

In den vergangenen vierzig Jahren haben die Erkenntnisse in der psychosomatischen Medizin eine rasche Ausweitung erfahren. Ihre Konzepte sind überarbeitet und ausgedehnt worden (Hofer, 1984; Taylor, 1987; von Uexküll, 1963; Weiner, 1977, 1982, 1987). Sie sind mit einer großen Vielfalt traditioneller Krankheiten in Beziehung gebracht worden, aber auch mit den sog. funktionellen Störungen wie beispielsweise der Hyperventilation (Lum, 1976; Magarian, 1982), dem irritablen Kolon (z.B. Weiner, 1988) und dem rheumatischen oder Fibromyositis-Syndrom (Moldofsky, 1989).

Meines Wissens sind die Vorstellungen der heutigen psychosomatischen Medizin noch nicht auf die Psychiatrie angewandt worden. Diese Auslassung ist merkwürdig, denn einer der Eckpfeiler der psychosomatischen Medizin stellt die Auffassung dar, daß nicht ein krankes Organ oder kranke Zellen im Zentrum des ärztlichen Interesses stehen, sondern Patienten. Eine Theorie der Humanmedizin (von Uex-

küll und Wesiack, 1988) integriert alle bedeutsamen Informationen über Patienten in ihren physikalischen und sozialen Umwelten: ihre genetische Ausstattung, ihre Erfahrungen während Reifung und Entwicklung, die Risikofaktoren, welche sie in sich tragen, die alle mit den mitbestimmenden Faktoren der Erfahrung von Krankheit und Leiden in Wechselbeziehung stehen. Die psychosomatische Medizin befaßt sich also mehr mit Ätiologie, unmittelbaren Vorläufern und Pathogenese von Krankheit und Leiden als mit deren Pathophysiologie. Sie untersucht die Auswirkungen von Leiden und Krankheit auf Patienten, deren Familien und deren Leben. Die Rolle der Qualität und Wechselfälle zwischenmenschlicher Beziehungen stehen im Brennpunkt ihres Interesses (Nemiah, im Druck; Taylor, 1987; Weiner, 1987). Wenn diese Inhalte nicht auch Kern und Substanz der Psychiatrie sind, sollten sie es werden!

Unglücklicherweise steht heute eine Psychiatrie seelenloser und entpersönlichter Gehirne am höchsten im Kurs (Reiser, 1988). Bemühungen um eine integrative Psychiatrie wirken anrüchig (Fink, 1988). Diese neue Psychiatrie preist sich selber als wissenschaftlich an, weil sie sich den Stempel der Biologie aufgeprägt hat. (Vergessen wir aber nicht, daß der Begriff Biologie, wie ihn diese Psychiater gebrauchen, auf Biochemie, Pharmakologie, abbildende Darstellungen des Gehirns und das Messen einzelner Hormone oder das von Neurotransmittern und ihren Stoffwechselprodukten im Blut, Liquor oder Urin beschränkt ist.) „Biologische" Psychiater scheinen die Tatsache vergessen zu haben, daß die ursprüngliche und andere Hauptsäule des Gebäudes der Biologie die Evolutionstheorie ist. Sie handelt von der Geschichte und Vielfalt der Organismen und ihrem Überleben bis zum heutigen Zeitpunkt der Geschichte. Sie befaßt sich auch mit den einmaligen und individuellen Charakteristika der Organismen, die selektiv wegen ihren größeren Anpassungsfähigkeiten ausgewählt wurden. Die Biologie fragt, wie sich Lebewesen unter dem Druck der Selektion verhalten und wie sie sich sozial benehmen. Evolutionsbiologie ist eine organismische Biologie (Mayer, 1982). Die präzise Analyse der Aktivität einzelner Neurone in Gewebsschnitten von Vogelgehirnen verrät uns nichts über die Rätsel und Präzision der Vogelwanderungen. Tote Vogelgehirne oder isolierte Neuronenzüge wandern nicht. Wissen um Einzelneurone ist notwendig, genügt aber nicht zum Verständnis komplexen Anpassungsverhaltens.

In diesem Kapitel werde ich mich bemühen zu zeigen, wie moderne psychosomatische Konzepte auf die Psychiatrie angewandt werden können.

51.1 Konzepte von Gesundheit, Leiden und Krankheit in der psychosomatischen Medizin

Psychosomatische Medizin ist eine Alternative zum traditionellen und im Westen vorherrschenden biomedizinischen, krankheitsorientierten Konzept. Sie ist eine patientorientierte Medizin. Sie versucht die Gefährdung eines Menschen zu Leiden und Krankheit umfassend darzustellen, ihre Auslösung, ihr Anhalten (Mirsky, 1958) und die Reaktionen des Patienten auf sie und die Behandlung. Ihr Blickfeld ist also weiter als das begrenztere des traditionellen Medizinmodells, das sich mit den nächstliegenden Mechanismen des Krankheitsgeschehens befaßt. Ihr ehrgeiziger Entwurf ist so umfassend, daß er unvollständig bleiben muß. Wir besitzen beispielsweise viele Einblicke in die sozialen Einflüsse, die manche Krankheiten fernhalten oder begünstigen – wie die ungleiche Verteilung vieler Krankheiten in den Bevölkerungen zeigt. Viele Beobachtungen sind über Bedingungen gesammelt worden, die den Ausbruch mancher Krankheiten begünstigen. Daß genetische Faktoren zur ätiologischen Varianz mancher Krankheiten beitragen, ist ebenfalls bekannt. In dieses Wissen sind die von der Biomedizin erarbeiteten Einsichten über Pathophysiologie und Anatomie der Krankheiten integriert worden. Was wir häufig noch sehr vermissen, sind tiefe Einsichten in die Pathogenese und die spezifischen Zusammenhänge zwischen den vorangehenden ätiologischen Faktoren einer Krankheit und ihrer Entstehung. Es müssen noch viele Wissenslücken geschlossen werden, bevor das Entstehen einer Krankheit vollständig beschrieben werden kann.

Die psychosomatische Medizin betrachtet es als eines ihrer Ziele, die Ätiologie von Krankheiten umfassend zu beschreiben. Faktoren liegen ihr besonders am Herzen, welche die Fähigkeit beschränken, Belastungen und Bedrohungen zu widerstehen, Faktoren, die, wie erwähnt, Folge der genetischen Ausstattung eines Individuums oder Folge einer Vielfalt von Erlebnissen mit der Umwelt während Reifung und Entwicklung oder Folge von beiden sein können. Die psychosomatische Medizin geht vom Axiom aus, daß die Kombination vieler Faktoren die Anpassungsmöglichkeiten eines Individuums steigern oder behindern kann. Mit anderen Worten vermag sie bei einigen Menschen die Gesundheit zu fördern oder andere dem Risiko zu erkranken auszusetzen, wenn sie Bedrohungen, Belastungen, Infektionen oder Katastrophen gegenüberstehen. Zu den herausragendsten vieler bedeutsamer, die Anpassung beeinflussender Faktoren gehört das Fehlen oder Bestehen und die Qualität einer zwischenmenschlichen Beziehung (House et al., 1988; Weiner, 1987).

Gleichzeitig berücksichtigt die psychosomatische Medizin die Beobachtung, daß das Verhalten und die Handlung eines Individuums es der Gefahr ernsthafter Erkrankungen aussetzen können. Gewisse sexuelle Verhaltensweisen oder intravenöser Drogenmißbrauch ebnen den Boden für das HIV-Retrovirus. Tabakrauchen und -kauen fördert die Gefahr von Atherosklerose, Lungenemphysem und -karzinom und Karzinomen von Mund und Kehlkopf. Schwangere Frauen, die rauchen oder Drogen und Alkohol mißbrauchen, gebären untergewichtige Kinder, die selbst medikamentenabhängig sind und das fetale Alkoholsyndrom erleiden. Alkoholmißbrauch kann zum Erkranken eines jeden Organs führen; er ist aber auch eindeutig mit Trauma, Invalidisierung oder Tod durch Autounfälle, mit körperlichen Gewaltverbrechen und sexuellem Fehlverhalten verbunden.

Die traditionelle Medizin widmet sich der biochemischen Pharmakologie und der Toxikologie von Alkohol, Nikotin und anderen Drogen. Sie versucht die Schäden zu beheben, die sich eingestellt haben, und kümmert sich weniger um den Schutz vor ihren sozialen, familiären und wirtschaftlichen Vorläufern: Alkohol- und Nikotinmißbrauch nimmt beispielsweise bei männlichen Arbeitern, die unverschuldet arbeitslos geworden sind, erheblich zu (Farrow, 1984).

Die psychosomatische Medizin fragt, warum Menschen zu einem bestimmten Zeitpunkt ihres Lebens krank werden, und warum sie an einer bestimmten Krankheit und nicht an einer andern erkranken. Antworten auf solche Fragen können erst gegeben werden, wenn die Umstände erfaßt werden, in denen Leiden und Krankheiten einsetzen, und durch die Untersuchung der vielfältigen Prädispositionen für eine bestimmte Krankheit. Die individuelle Wirklichkeit läßt einen bestimmten Menschen seine Umgebung als belastend und unkontrollierbar erleben; die Erfahrung kann nicht gemeistert ("bewältigt") werden. So erlebt das Individuum die belastende Erfahrung oft als unangenehm oder quälend, zum Teil wegen der Bedeutung und den Folgen für den betreffenden Menschen. Bestimmte Gefühle (z.B. Furcht oder Wut) werden geweckt, die Gefahr, Belastung und Widerwärtigkeit anzeigen: Kann die Situation abgewendet oder gemeistert werden, stellt sich Erleichterung ein; bei unsicherem Ausgang kann Hoffnung erlebt werden; sind Belastung, Bedrohung oder Verlust unkontrollierbar (wenn beispielsweise Machtlosigkeit der Situation gegenüber besteht), oder wenn sich die Niederlage einstellt, so werden Hoffnungs- und Hilflosigkeit erlebt und der Patient gibt auf (Engel, 1968).

Kontrollfähigkeit bedeutet in diesem Zusammenhang die Gabe eines Individuums, Antworten zu finden, die Belastung oder Bedrohung abwenden oder meistern. Diese Gabe hängt teilweise von den Kenntnissen ab, wie eine Situation anzugehen ist, und teilweise von der Fähigkeit, kognitive und emotionale Antworten zu steuern. Viele Stressoren, die mit Leidens- und Krankheitsbeginn in Zusammenhang gebracht werden, sind plötzlich und dramatisch, aber auch chronische Unlust, Irritation und Frustration führen zu Entmutigung, einem dauernd gedrückten

Zustand und einer Vielfalt von Krankheiten (Brown und Harris, 1978; Hinkle, 1974).

Die Bewältigung von Belastungen und Bedrohungen schwächt die Wirkung dieser Stressoren ab. Alter und Reifungszustand tragen entscheidend zur Fähigkeit bei, sie zu bewältigen (zu kontrollieren). Bewältigungsverhalten umfaßt die realistische Beurteilung der Streßsituation, das Einholen der nötigen Information, um mit ihr umzugehen, und das Entwerfen einer Handlung, die sie überwinden soll (Lazarus, 1966). Nicht immer können die gleichen Strategien benützt werden, um Bedeutung und emotionale Einwirkung eines Ereignisses zu beeinflussen. Bei der Bewältigung von Streß sind Hilfe, Unterstützung, Nähe und Beratung durch andere Menschen von entscheidender Bedeutung – in der Medizin handelt es sich oft um den Arzt, die Krankenschwester usw. So dämpfen soziale Verbundenheit, eine nahe persönliche Beziehung und nur allmähliche soziokulturelle Veränderungen die Auswirkung von traditionellen Risikofaktoren für die koronare Atherosklerose und den plötzlichen Tod oder Myokardinfarkt (wie stark fetthaltige Nahrung und Übergewicht) (Bruhn et al., 1966; Marmot und Syme, 1976).

Leider wissen wir noch nicht genug über die genauen Zusammenhänge zwischen dem Verlust der Kontrolle über eine Situation oder dem Versagen, sie zu bewältigen und den physiologischen Veränderungen, die Prozesse in Gang setzen, welche bei prädisponierten Menschen Krankheiten auslösen. Bei Mensch und Tier hängt der Kontrollverlust mit der Zunahme der Sekretion und dem Anstieg der Blutspiegel von Kortikosteroiden zusammen. Versagen und Niederlagen senken den Testosteronspiegel. Die Alarmierung durch Gefahren kann die Adrenalin- und Noradrenalinspiegel anheben (Ursin et al., 1978). Jedoch wissen wir noch sehr wenig über die Psychobiologie persönlich so aufwühlender Erfahrungen wie Trauern (Jacobs, 1987); dies ist um so bedauerlicher, als Trauern eine Kodeterminante des Ausbruchs vieler Krankheiten und veränderten Verhaltens ist (Weiner, 1987).

Die psychosomatische Medizin hat zum Verständnis beigetragen, daß es keine einheitlichen Krankheits-„Kategorien" gibt; die Regel ist vielmehr eine genetische (Rotter und Rimón, 1977) und psychobiologische Heterogenität (Julius und Esler, 1975; Rimón, 1969; Taylor et al., 1985; Vollhardt et al., 1982). Sie hat unser Wissen um die Verschlimmerungen und Rückbildungen von Krankheiten vermehrt (die gewöhnlich „spontane" genannt werden). Solche Beobachtungen entmystifizieren nicht nur solche Ereignisse, sondern erlauben auch genauere Einblicke in die Komplexität der Beziehung zwischen anatomischen Veränderungen und Symptomen. So weiß man heute, daß nach Einsetzen der Colitis ulcerosa die Schleimhaut des Kolons für immer histologisch abnorm bleibt, die Symptome der Krankheit aber kommen und gehen (Dick et al., 1966). Symptomverschlimmerungen stellen sich ein, wenn die Arzt-Patient-Beziehung auch nur kurz unterbrochen ist oder wenn der Patient aus dem Zustand von Aktivität und Kompetenz in einen Zustand von Hoffnungs- und Hilflosig-

keit gerät (Engel, 1956). Die Prognose nach einem Myokardinfarkt hängt zum Teil von der sozialen Situation des Patienten ab: Die Lebensspanne von Ledigen, Geschiedenen oder sozial isolierten Menschen kann verkürzt sein (House et al., 1988).

Es gibt Krankheit nicht als „Sache für sich". Sie kann von dem Kranken nicht getrennt werden: Die Persönlichkeit des Patienten mit chronischer Polyarthritis ist mitbestimmend für Prognose und Reaktion auf die Behandlung (McFarlane et al., 1987). Zudem reagieren Patienten auf ihre Krankheiten sowohl artspezifisch (z.B. mit Furcht), als auch ganz individuell. Dies hängt von ihren Glaubenssystemen und ihren Vorstellungen ab. Sie bestimmen zum Teil die Behandlungs-„Compliance"; sie tragen zu kürzeren oder längeren Spitalaufenthalten bei, zu Reaktionen auf Operationen und beziehen die Familie des Patienten (und sogar Anwälte) mit ein.

Diese Beispiele belegen, daß Menschen Krankheiten haben und daß es keine Krankheiten per se gibt (Krehl, 1932). Krankheiten sind reine Abstraktionen; sie können nicht wie die konkreten Objekte des Pflanzen- und Tierreichs eingeteilt werden.

Die psychosomatische Medizin ist über den Rahmen des einzelnen Menschen hinaus zu einer Medizin der zwischenmenschlichen Beziehungen geworden. Bis vor kurzem teilte die Psychiatrie diese Auffassung. Sie ist aber zu einem Krankheitsmodell zurückgekehrt, nach dem nicht Menschen, sondern Gehirne psychiatrische Störungen oder Krankheiten haben.

51.2 Begriffsprobleme in der Psychiatrie

Dieses Kapitel soll darlegen, daß die heutigen Entwicklungen der psychosomatischen Theorienbildung sich ebensogut auf die Psychiatrie anwenden lassen, wie auf alle anderen Zweige der Medizin (Engel, 1977; v. Uexküll et al., 1986; Weiner, 1977). Deshalb werde ich es vermeiden, ausführlich auf den traditionellen Begriffskonflikt in der Psychiatrie einzugehen, ob „physikalistische" (organische, biologische, neuropsychiatrische) oder psychologische und soziologische Erklärungen der Ätiologie und Pathogenese psychiatrischer Symptome und Leiden genügen. Wir brauchen nämlich bei psychisch Kranken eine integrative Auffassung über die Rolle all dieser Determinanten.

Die fruchtlose und brisante Kontroverse über entweder „physisch" oder „psychisch" geht in der Psychiatrie geschichtlich auf Bayles Beschreibung von chronischer Meningitis bei gewissen Geisteskranken zurück, auf die Beziehung zwischen einer bestimmten Form von Demenz und Pellagra und auf die Entdeckung von Treponema pallidum im Gehirn von Patienten mit paretischer Neurosyphilis. Bayles Beobachtung wurde verwendet, um Pinel und Esquirol zu widerlegen, die behaupteten, man könne psychiatrische Symptome – und Krankheit – weder verstehen, noch diagnostizieren, noch behandeln, wenn man den einzelnen Patienten nicht psychologisch voll erfasse mitsamt seiner persönlichen Geschichte (Goldstein, 1987): Verdickte Meningen genügen nicht zur Erklärung der Geisteskrankheit eines Menschen. Viele Geschichtsforscher haben belegt, daß der Kampf, der zwischen Bayle und Esquirol begann, nie beigelegt wurde. Die Auffassung, daß die Erklärung einzelner psychiatrischer Symptome oder Syndrome nur auf physikalischen Veränderungen des Gehirns beruhen könne, wird von dem traditionellen biomedizinischen Modell gestützt. Es beschränkt sich ausschließlich auf strukturelle, anatomisch umschriebene „Läsionen", welche Symptome durch isolierte ätiologische „Ursachen" oder die nächstliegenden molekularen Mechanismen erklären. Es vertraut vollständig auf pharmakologische Stoffe, Chirurgie und Strahlentherapie. Für diese Auffassung ist die paretische Neurosyphilis das Vorbild für alle folgenden „biologischen" und erklärenden Theorien in der Psychiatrie. Dieses grob vereinfachende Konzept möchte das weite Spektrum und die Vielfalt von Verhaltensweisen und Symptomen der Patienten mit Neurolues auf eine einzelne Ursache zurückführen: den Nachweis der Spirochäten im Gehirn. Die Symptome der paretischen Neurosyphilis sind jedoch individuell und idiosynkratisch; zum Teil sind sie Folgen des Versuchs des Patienten, seine kognitiven Defekte zu verdecken, die ihrerseits seine Fähigkeit für Problemlösungen und Arbeit einschränken und seine zwischenmenschlichen Beziehungen grundlegend verändern können (Goldstein, 1939; Schilder, 1951). Zudem können identische Verhaltensweisen und Symptome, die luetische Patienten aufweisen, auf verschiedenen Wegen als Folge anderer Krankheiten des Gehirns auftreten: Sie sind nicht spezifisch für die paretische Neurosyphilis und vermögen ohne Kenntnis der Anamnese des Patienten, seines bisherigen Kampfes mit den Schäden der Krankheit nicht verstanden werden.

Die Suche nach einem physikalischen Grund für jegliches Symptom setzte bereits mit Morgagni ein und wurde von Bichat, Laennec und ihren Geistesgenossen zu Beginn des 19. Jahrhunderts eifrig fortgesetzt. Sie hat seither angehalten. Im 20. Jahrhundert ist die materielle Ebene, auf der solche Erklärungen gesucht werden, immer mikroskopischer und eingeschränkter geworden – mit anderen Worten immer molekularer und physikochemischer. Reduktionismus und Materialismus haben sich verbrüdert und sind zum einzigen theoretischen Rahmen der Biomedizin geworden.

Das Fehlen von strukturellen Veränderungen zur „Erklärung" von Symptomen oder Symptombündeln und demzufolge von Krankheiten wird entweder auf heute noch fehlende Kenntnisse zurückgeführt (d. h. die strukturellen Veränderungen werden in der Zukunft nachgewiesen werden können), oder die Symptome werden als „eingebildete" erklärt – das wahre Forschungsgebiet der Psychiater! Diese Auffassung übersieht, daß es keine enge Beziehung zwischen Symptomen und strukturellen Veränderungen in Zellen, Geweben oder Organen gibt: Symptome (und Leiden) können ohne Läsionen vorkommen und diese (und Leiden und Krankheit) ohne Symptome. Placebos vermögen Symptome zu beheben, aber sie beeinflussen anatomische Veränderungen nicht (Peterson et al., 1977). Die Wirkung eines Medikaments braucht seiner, uns bekannten, pharmakologischen Wirkungsweise überhaupt nicht zu entsprechen (Luparello et al., 1968, 1970). Physiologische Funktionsänderungen führen ohne Erzeugung anatomischer Defekte zu Symptomen (und Leiden). Symptome von Leiden und Krankheit sind kulturgebunden, werden von Glaubenssystemen beeinflußt, von individuellen Interpretationen und ihnen untergeschobenen Bedeutungen (Fabrega, 1987; Kleinman, 1988). Kurz und gut: Menschen, nicht Organe haben Symptome.

Symptome sind Wahrnehmungen, und Wahrnehmungen besitzen Bedeutungen für die Menschen. Diese Einsicht führt unvermeidlich zu einer auf Menschen zentrierten Medizin. Sie bemüht sich, die Erkenntnisse der Biomedizin in einen gesetzmäßigen und umfassenden Rahmen zu integrieren und sie damit zu ergänzen und zu erweitern. Sie weiß, daß medizinische Phänomene – Symptome und Veränderungen von physiologischen Funktionen – nicht notwendigerweise die Folgen von Veränderungen der Struktur von Organen oder ihrer Teile zu sein brauchen. Das gleiche Symptom kann durch Veränderungen in der Funktion oder der Struktur oder durch Veränderungen in beiden entstehen. Diese Theorie ist umfassender und versucht Ergebnisse aus verschiedensten Forschungsgebieten zu integrieren. Natürlich muß sie sich noch mit ungelösten Fragen des Dualismus-Rätsels herumschlagen; aber es werden Anstrengungen unternommen, die sich mit diesem zentralen Problem befassen (Engel, 1977; von Uexküll, 1963, 1988; Weiner, 1986, 1989 b).

51.3 Krankheitstheorien und das Problem der Diagnose in der Psychiatrie

Die Aufgaben des Arztes bestehen im Diagnostizieren, Vorbeugen und/oder Behandeln von Krankheiten. Die Behandlung kann in einer Erleichterung von Symptomen, von Schmerzen und Leiden bestehen; am besten wird dieses Ziel durch die Heilung der Krankheit erreicht. Um dieses Ziel zu erreichen, muß der Arzt Konzepte über die Ursachen (Ätiologie und Pathogenese) der Krankheiten und nicht nur über ihre pathologische Anatomie und Physiologie besitzen. Für eine Prävention von Krankheit ist ein Wissen über die Ätiologie von Krankheiten besonders wichtig und Prävention ist der Behandlung nach Krankheitsbeginn vorzuziehen. Da unsere Kenntnisse über die Ätiologie und besonders die Pathogenese der meisten Krankheiten, außer bei bakteriellen, viralen, parasitären Infektionen und Unterernährung, beschränkt sind, sind Vorbeugung und Heilung eher die Ausnahme als die Regel. Da Ärzte zudem ihre Patienten gewöhnlich erst nach Ausbruch der Krankheit sehen, stützen sie ihre Diagnose nicht nur auf Symptome, sondern auch auf Störungen von Struktur und

Funktion und müssen herausfinden, wie diese Störungen entstanden sind.

Ob Ärzte es zur Kenntnis nehmen oder nicht, die Konzepte von Krankheiten (Kendell, 1976a; McHugh und Slavney, 1983) sind Angelpunkte der ärztlichen Arbeit, und nicht nur Diagnosen. Biomedizinische Modelle und Konzepte von Krankheit kommen meistens in zwei Formen vor – einer exogenen und einer endogenen Form (Copeland, 1977). Beide beeinflussen Behandlung und Diagnose, Hypothesen über Ätiologie und Pathogenese von Krankheiten und auch die Forschung.

Beflügelt von den erfolgreichen Klassifikationsversuchen für Pflanzen und Tiere von Linnaeus und Buffon, begann Sydenham im 17. Jahrhundert Krankheiten als natürliche (und nicht übernatürliche) Phänomene zu betrachten. Er teilte auch sie in umschriebene Klassen ein – in „definierte und gesicherte Arten" – und gründete sein Vorgehen auf Bündel von Symptomen, die er häufig willkürlich ordnete.

Die von Linnaeus und Buffon begonnene Einteilung in der Biologie hatte als weitreichende Folgen die Evolutionstheorie und die Genetik. Einteilungen in der Biologie sind Theorien, die gleichen Erscheinungen und identischen Eigenschaften gerecht werden, aber nicht identischen Ursachen; sie „schaffen und spiegeln die grundlegende Struktur von Wissenschaft". Arten sind nicht Wesenheiten: Gemeinsame Erscheinungen und gleiche Eigenschaften verdecken die Variation – das Rohmaterial der natürlichen Auslese. „Variation ist das primäre; Wesenheiten sind illusorisch" (Gould, 1985).

Biologen teilen ein, Ärzte diagnostizieren. Diese Unterscheidung wird selten getroffen. Biologen verfügen über ein einfaches Merkmal, um Mitglieder einer Art zu identifizieren: Die Nachkommen sich paarender Mitglieder der gleichen Art sind fruchtbar, diejenigen aus der Verbindung artfremder Lebewesen sind es nicht. Leider gibt es keine so einfachen Merkmale, Krankheiten zu unterscheiden. So können Mitglieder von zwei verschiedenen Tierarten die gleiche Krankheit aufweisen mit diagnostischen Kriterien, welche die Grenzen der Art überschreiten. Dann unterscheiden sich die Ziele von Diagnose und Klassifizierung: In der Medizin ist die Notwendigkeit nach Krankheiten oder ihren Unterformen einzuteilen ein dringendes Forschungsanliegen, denn gemischte Populationen verwirren die Resultate: Die Verwendung spezifischer HLA-Antigene oder von rheumatoiden Faktoren kann beispielsweise für die Einteilung der Unterformen der chronischen Polyarthritis und ihre Erforschung notwendig sein. Die chronische Polyarthritis kann jedoch klinisch ohne derart präzise Merkmale diagnostiziert werden.

Die Risikofaktoren (z.B. HLA-Antigene) für diese Krankheiten unterscheiden sich in verschiedenen ethnischen Gruppen, der phänotypische Ausdruck der Erkrankung ist aber sehr ähnlich.

Einteilung in Biologie und Diagnostik der Medizin unterscheiden sich auch sonst noch bedeutsam (Fabrega, 1987). Lebewesen wie Pflanzen und Tiere sind konkrete, ausgeformte Objekte, die man hierarchisch in Klassen, Gattungen und Arten einteilen kann. Krankheit und Leiden aber sind Abstraktionen, die in Lebewesen geschehen, die sich so oder so verhalten, ihre Symptome wahrnehmen und vorweisen, interpretieren, beschönigen, verschieben, übertreiben oder bagatellisieren. Sie teilen sie dem Arzt mit, der sie seinerseits interpretiert, bewertet und zu verifizieren sucht. Symptome neigen also dazu, sich im Verlaufe der Krankheit inhaltlich, in bezug auf die Dauer und Schwere zu verändern. Sie sind keine unveränderlichen Objekte.

Diese Bemerkungen treffen besonders auf die sog. „psychiatrischen" Leiden zu: Patienten mit Bulimia nervosa stellen beispielsweise ihre Freßanfälle im Spital ein und nehmen sie zu Hause wieder auf. Diese Patienten pflegen auch zu lügen, zu stehlen. Sie mißbrauchen Drogen und Alkohol, sind sexuell wahllos aktiv und schämen sich ihrer Verhaltensweisen. Die heutige Verlegung des Schwergewichts auf ihr „Eßverhalten" ist problematisch: eine Abstraktion. Es lenkt vom Bemühen ab, zu verstehen, was diesen Menschen wirklich fehlt.

Leiden und Krankheiten sind vom Menschen entworfene Begriffskategorien, die kranken Individuen übergeworfen werden; sie können angemessen sein, sind es aber nicht immer, weil kranke Menschen stark variieren. Symptome oder Verhaltensweisen können nicht vom Menschen abgelöst werden, mögen die Ärzte noch so sehr danach trachten. Verhalten ist keine konkrete und unveränderliche Sache wie eine bestimmte Pflanze oder ein bestimmtes Tier, es ist vielmehr eine Sammlung von Handlungen mit Zielen, Bedeutungen (Absichten) und Zwecken (Fabrega, 1987; von Uexküll, 1963). Sie sind alle durch soziokulturelle Gepflogenheiten gefärbt oder konditioniert und diese bestimmen ihrerseits, ob man sie als normales oder abnormales Verhalten beurteilt (wo hört normales Verhalten auf und wo beginnt abnormales?).

Wann betrachtet man ein Symptom als Manifestation eines Leidens und wann nicht? Bestimmte Formen von Sexualverhalten sind aus biologischer Perspektive atavistisch. Aus medizinischer (psychiatrischer) Sicht werden die gleichen Verhaltensweisen heute nicht mehr länger als Störungen oder Leiden betrachtet. Dadurch werden diejenigen, die (aus vielerlei Gründen) mit ihrem Verhalten nicht zu leben vermögen (mit seinem derzeitigen Risiko für HIV-Ansteckung), der ärztlichen Betreuung beraubt.

Die Grenzen zwischen Gesundheit, Störungen und Leiden sind also verwischt. Sie sind es auch aus anderen Gründen: Symptome (besonders wenn sie mit Bedeutungen, Glaubensvorstellungen über Leiden und Absichten durchtränkt sind) und Verhalten können nicht von Gepflogenheiten losgelöst betrachtet werden, die ihrerseits vielfältig sind und unter dem Einfluß religiöser, politischer, ökonomischer und sozialer Normen stehen. Normen, seien sie physiologische, immunologische, biochemische oder verhaltensmäßige, sind schwer zu fassen. Sogar diese „harten" („wissenschaftlichen") Merkmale wechseln je nach Geschlecht, ethnischer Zugehörigkeit, Alter, Tages- und

Jahreszeit. Handlungsweisen sind noch vielfältigeren Normen unterworfen.

Von wenigen Ausnahmen abgesehen sind klinische Zustände, die mit psychiatrischen Störungen oder Leiden gleichgesetzt werden, keine Krankheiten und hängen von historischen, kulturellen und sozialen Bedingungen ab (Fabrega, 1987; Kraupl-Taylor, 1980). Diese Feststellung wird von der Beobachtung gestützt, daß einige der Manifestationen der Schizophrenie (beispielsweise Katatonie und Hebephrenie) sich in unserem Jahrhundert verändert haben und ihre Prognose günstiger geworden ist (Hare, 1988).

Ein anderer Irrglauben liegt dem derzeitigen Begriff der psychiatrischen Diagnose zugrunde: Bestimmte Symptombündel (z.B. schwere Depressionen) werden als typischere Beispiele der Kategorie als andere Symptomgruppen der gleichen Kategorie betrachtet; z.B. die „atypischen" Störungen. Das logische Mißverständnis einer solchen Auffassung besteht darin, daß „wenn Kategorien lediglich durch die allen Mitgliedern gemeinsamen Eigenschaften definiert sind ... (beispielsweise bei depressiven Störungen) ..., keine Mitglieder typischere Beispiele bestimmter Kategorien sein sollten als andere" (Lakoff, 1987).

Diese Bemerkung möchte ich betonen: Der Auffassung, einzelne Krankheiten seien scharf abgegrenzte Einheiten, liegt eine problematische Behauptung über Wesenseigenschaften zugrunde. Diese entscheidende Einsicht haben medizinische Forscher erst in den letzten zwei Jahrzehnten gewonnen: Jede Krankheit ist variabel (d.h. heterogen) und diese Variabilität ist eine grundsätzliche Eigenschaft (Weiner, 1977, 1989a). In der Psychiatrie ist das Konzept der Heterogenität angeblich psychiatrischer Krankheitskategorien bis auf den heutigen Tag umstritten (und abgelehnt). Die Illusion, das identische Aussehen einer ihrer Krankheiten (etwa der Schizophrenie oder der bipolaren „Krankheit") sei Ausdruck ihrer gemeinsamen Ursache, belastet die Psychiatrie immer noch. Wird erst einmal die Auffassung der Heterogenität zugelassen, wird wahrscheinlich sogar eingesehen werden, daß auch strukturelle Läsionen durch eine Vielzahl von (pathogenetischen) Abläufen entstehen können – es gibt keine unveränderlichen Ursachen; und sie drücken sich nicht einmal in ihrer pathologischen Anatomie aus (Grossman, 1978).

Die Idealvorstellung von umschriebenen Krankheiten mit umschriebenen zellulären „Ursachen" spukt noch immer im Kopf der heutigen Psychiatrie, die leidenschaftlich versucht, Kategorien im Sinne von Linnaeus zu suchen und zu definieren und ihre anatomischen oder biochemischen („biologische Marker") Entsprechungen oder sogar ihre biologische Basis aufzustöbern. Die Suche nach Markern sollte viel bescheidener die Bestätigung für ein besonderes Bündel von Symptomen oder Verhaltensweisen zum Ziele haben: Man sollte ihnen nicht leichtfertig ätiologische oder pathogenetische Bedeutung unterstellen.

Da jede Krankheit oder Störung ein ausgesprochen variables Phänomen ist, kann sie sich in schwerer oder milder Form darbieten. Leidet der Patient mit chronischer Polyarthritis, der nur rheumatische Knoten aufweist, an der gleichen Krankheit wie derjenige mit zunehmenden Erosionen und Deformitäten der Gelenke? Haben Patienten mit milden und kurzdauernden Schüben von Schizophrenie (d.h. „schizophreniformen" Erscheinungen) die gleiche oder eine andere Krankheit als diejenigen mit einem Leiden, das mindestens sechs Monate anhält? Seit wann eröffnet die Dauer einer Krankheit wesentliche Einblicke in ihre Natur? Viele Faktoren tragen zur Prognose bei. Am anderen Ende der Skala stellt sich die Frage, wo Gesundheit aufhört und milde Krankheit und Leiden beginnen. Stellt ein Blutdruckwert von 140/90 mm Hg eine Krankheit dar in Anbetracht der Willkür, mit der diese Werte festgelegt wurden?

Viele Rätsel und Unvereinbarkeiten bestehen deshalb in unseren Vorstellungen von Diagnose und Krankheit fort. Bisher ist es der Psychiatrie mißlungen, Ätiologie und Pathogenese der meisten Krankheiten durch eine individuelle „Ursache" zu erklären und reine Krankheitskategorien aufzustellen. Sogar das Paradigma der Medizin – die Theorie der Infektionskrankheiten – ist eine grobe Vereinfachung, denn das infektiöse (externe) Agens ist für viele Symptome oder Krankheitszeichen weder notwendig, noch für sich alleine eine genügende Ursache. Die Reaktionsweise des Individuums ist für die Symptombildung und Erscheinungsform der Krankheit ebenso wichtig wie exogene und endogene „Ursachen".

Die alternative Ansicht, Leiden und Krankheit seien Folgen einer begrenzten Anpassungsfähigkeit des Menschen an belastende Erfahrungen (infektiöse Erreger eingeschlossen), ist von Copeland (1977) als endogenes Krankheitsmodell bezeichnet worden. Symptome stellen nicht nur die direkte Folge externer Ursache(n) oder das Endstadium von Organveränderungen dar. In ihnen kommen zusätzlich die Folgen der Anpassungsbemühungen des Individuums zum Ausdruck. Oft spiegeln sie Funktionsstörungen wider, die den strukturellen Veränderungen gewöhnlich zeitlich vorangehen. Eine vollständige Definition von Leiden und Krankheit setzt Kenntnisse über die Art der Beziehung zwischen Umgebung und Individuum voraus, und die Gesamtheit seiner ererbten und erworbenen Anpassungsfähigkeiten – Informationen, die viele Schichten der biologischen Organisation einbeziehen (Copeland, 1977). Dieses Modell führt den Arzt zu einer patient- und nicht zu einer krankheitsorientierten Medizin. Krankheit ist ein teilweises oder vollständiges Versagen der Anpassung: Da die Anpassungsmöglichkeiten des Menschen nicht unbeschränkt sind, bündeln sich die Manifestationen versagender oder mißlungener Anpassung in Krankheitskategorien, die individuell stark variieren.

Diese Auffassung ist mit dem derzeitigen diagnostischen Vorgehen in der Psychiatrie nur ein Stück weit vereinbar: Sie erfaßt nur zwei diagnostische Kategorien eines Konzepts der Mensch-Umwelt-Interaktion: die Anpassungsstörungen und das posttraumatische Streßsyndrom. Bei den Anpassungsstörungen ergeben sich Verhalten und Symptome aus den Bemühungen, sich anzupassen. (Es sei angemerkt, daß „adjustment" und „adaptation", also Angleichung resp.

Verarbeitung nicht synonyme Begriffe sind.) Beim traumatischen Streßsyndrom ist das Individuum bei der Verarbeitung von Erlebnissen gescheitert, die oft überwältigend und unausweichlich sind (wie Naturkatastrophen, Krieg, Folterung, Vergewaltigung usw.), mit kurz- und langzeitigen Folgen.

Andere diagnostische Kategorien – vor allem die zwei Hauptpsychosen und die Paniksyndrome – werden als „endogen" betrachtet. Dieses Adjektiv wird dabei aber in einem besonderen und anderen Sinn verwendet, als eben von uns. Es wird unterstellt, daß diese Störungen die Folge von im Körperinnern entstandenen biochemischen Produkten, etwa der Entstehung eines „Psychotogens", oder einer Stoffwechselstörung usw. sind. Der sozialen oder physischen Umwelt (ausgenommen als Quelle eines hypothetischen Virus) wird keine Bedeutung beigemessen. Das innere Agens soll die Symptome dieser „endogenen" Störung direkt verursachen. Die Rolle der Symptome bei der Einschränkung der Anpassungsmöglichkeiten des Patienten wird nicht gesehen. Physiologische Veränderungen (der Hormonsekretion, des Schlafmusters, der veränderten Rezeptor-„Empfindlichkeit" oder der Bindungseigenschaften), die bei diesen Störungen gefunden werden, werden als Vorläufer der Krankheit und nicht als mögliche Folgen oder Begleitzeichen von Anpassungsbemühung oder -versagen betrachtet.

Im Falle von Delir oder Demenz sollen die Symptome und Verhaltensweisen, wie schon betont, die direkte Folge infektiöser, metabolischer, physiologischer, pharmakologischer oder anatomischer Gehirnveränderungen sein und zur Erklärung genügen. Diese Hypothese ist falsch, weil Störungen der Kognition, Wahrnehmung, der Emotionen und des Verhaltens, welche Delir und Demenz begleiten, die Fähigkeiten des Patienten zur Anpassung einschränken und die zwischenmenschlichen Beziehungen stören, und so Furcht und Hilflosigkeit erzeugen.

51.4 Weitere Begriffsprobleme in der modernen Psychiatrie

Die Grenzen, Widersprüche, Einschränkungen und Unzulänglichkeiten usw. des biomedizinischen Modells (besonders in seiner exogenen Form) sind offensichtlich. Dennoch stützt sich die heutige Psychiatrie auf dieses Modell: So seien die klinischen psychiatrischen Störungen, also symptomatische Verhaltensweisen und andere Symptome (Halluzinationen, Wahnbildungen), direkter Ausdruck mutmaßlicher Hirnkrankheiten; sie seien (können? müssen? sollen?) in der Anatomie oder mindestens den physiologischen Prozessen des Gehirns verwurzelt (Scheller und Barchas, 1988). Diese Hirnkrankheiten seien die direkte Folge aberrierender Gene. Prägnant umschrieben, psychiatrische Symptome, auch Verhaltensweisen würden vom kranken Gehirn wie „Dampf aus einer Dampfmaschine" ausgestoßen, wie es T. H. Huxley ausdrückte. Die Behandlung besteht im Un-

terdrücken von Symptomen wie z.B. Gefühlen und Stimmungen (Angst mit „Anxiolytika", Depression mit Antidepressiva) und/oder Verhaltensweisen (beispielsweise „gewalttätige" oder „gefährliche") und verrückter Denkinhalte mit Neuroleptika.

Psychiatrie betrifft aber notwendigerweise menschliche Not, Gefühle, Gedanken, veränderte (wessen Realität ist die veränderte?) Wahrnehmungen, scheinbar unangepaßte Handlungen und verrückte Menschen. Beträfe sie diese nicht, hätte sie keine Existenzberechtigung. Aber Gehirne fühlen nicht; Menschen fühlen. Gehirne sind nicht verrückt, aber Menschen sind es. Und Verhaltensweisen werden aufgrund von sozialen Normen als abnormal beurteilt.

Das Neurologisieren von Seele und Geist scheitert an der Frage des Bewußtseins (Searle, 1984). Menschen sind sich bewußt, d.h. selbst-bewußt; Gehirne sind dies nicht. Ein zentrales Anliegen der Psychiatrie ist der veränderte Bewußtseinszustand (beispielsweise beim Delir und in Zuständen von Spaltungen und Depersonalisation).

Das Gehirn ist Grundlage von Verhalten (Handlungen). Handlungen sind zweckgerichtet. Auf ursprünglichster Ebene dienen sie dem Überleben und der Fortpflanzung; auch wenn sie aufgrund von Konventionen als abnormal oder symptomatisch bezeichnet werden, besitzen sie Bedeutung. Sie hätten sonst im natürlichen Ausleseprozeß keinen Bestand gehabt.

Gene sind auf vielfältige und unbekannte Art und Weise an der Entwicklung des Gehirns und an vielen psychiatrischen Störungen und Verhaltensweisen beteiligt. Deshalb müssen sie zu „abnormem", „schlechtangepaßtem" Verhalten (psychiatrischen Krankheiten) beitragen und tun das auch. Auch sie hätten nicht überlebt, hätten sie den Organismus nicht mit einem entschädigenden Vorteil ausgestattet (Rotter und Diamond, 1987).

„Unangepaßte" Verhaltensweisen sind kohärent, geplant oder organisiert. Diese Behauptung gilt sogar für einfachere Verhaltensebenen (wie z.B. dem Gehen). Die Koordination der Glieder und des Körpers beim Gehen sind das geplante Ergebnis von Millionen zusammenarbeitender Neuronen und von vielen Muskeln und Gelenken: Als Elemente in einem Gesamt betrachtet haben sie „Millionen von Freiheitsgraden. Aber das Gehmuster bringt sie in eine kohärente niedrig-dimensionale Form, gesteuert von einer niedrig-dimensionalen Kontrolle" (Garfinkel, 1983).

Die geläufige genetische Auffassung psychiatrischer Krankheiten stellt eine übermäßige Vereinfachung dar: Diese Konzepte beinhalten, daß abweichende Gene pleiotrop sind, was sie sein können, aber nicht zu sein brauchen. Gene sind nicht statische, sondern dynamische bewegliche Einheiten. Einige regulieren sich gegenseitig. Andere kodieren Proteine. Proteine sind nicht notwendigerweise das Endprodukt der Gentranslation; das Genprodukt wird oft weiterverarbeitet. Gene sind nicht dauernd aktiv, sondern werden aktiviert oder gehemmt durch die Stoffwechselprodukte der vom Gen induzierten Enzyme und auch durch Hormone aufgrund noch unbekannter Mechanismen. Mehrere Gene können beim Aufbau eines Stoffwechselprodukts zusammenwirken (z.B. bei einem Immunglobulin). Viele Gene sind nur während bestimmter Zeitabschnitte der Reifung und Entwicklung aktiv. Die von ihnen kodierten Strukturen (z.B. Gehirngebiete) sind ihrerseits nicht starr und unflexibel: Neurone können während der Entwicklung zugrunde gehen. Die Neuronenanatomie des visuellen Apparates wird nach der Geburt durch Erfahrungen umgeformt (Wiesel und Hubel, 1965). Erfahrungen vermögen sogar die Steuerung eines Enzyms durch ein Hormon zu beeinflussen (Schanberg et al., 1984). Sich entwickelnde Neurone (beispielsweise des autonomen Nervensystems) sind anfänglich „ungebunden"; sie können je nach den Signalen, die sie von ihrem Zielgewebe und der Umgebung erhalten, durch die sie wandern, adrenergisch oder cholinergisch werden (Patterson, 1978). Also bestimmt sogar auf der Ebene der Neuronen die Umgebung das Endprodukt, dessen Entstehung durch Gene in Gang gesetzt wurde.

Zu diesen Bedenken stößt noch die Tatsache, daß genetische Krankheitsmodelle nicht einheitlich sind. Es gibt:

- Mendels dominante, rezessive und X-gebundene Störungen. Sie sind in abnehmender Reihenfolge mit spezifischen Enzymmängeln gekoppelt (29% die rezessiven; 15% die X-gebundenen; 1,8% die dominanten);
- Polygenische Modelle der Krankheit, bei denen viele Gene mit unterschiedlichen Loci beteiligt sind. Ihre Auswirkungen sind additiv;
- Multifaktorielle Modelle versuchen zu erklären, wie multiple genetische und nicht-genetische Faktoren (beispielsweise der Erfahrung, der Ernährung, toxische) bei der Produktion des Phänotyps zusammenwirken. Das genetische Gut ist in der Bevölkerung normal verteilt, die Phänotypen sind es jedoch nicht (also herrscht Diskontinuität). Diese eigenartige Situation hängt von einem Schwelleneffekt ab: Unter dem Schwellenwert ist der Phänotyp normal, oberhalb ist er es nicht;
- Mitochondriale Vererbung (beispielsweise mitochondriale Myopathien, Myoklonus-Epilepsie);
- Metabolische Interferenz (zwei verschiedene Allele stören sich und produzieren einen abnormalen Phänotyp).

Es fällt nicht immer leicht zu entscheiden, ob eine Krankheit auf einem dominanten Gen mit verminderter Penetranz oder auf einer multifaktoriellen Vererbung beruht: Der Unterschied ist ein quantitativer.

Nur zwei psychiatrische Störungen haben einer Mendelschen Art der Vererbung zugeordnet werden können:

- Die X-gebundene Form der bipolaren affektiven Störung. Diese Patienten leiden manchmal auch an Farbenblindheit und Glukose-6-Phosphat-Dehydrogenasemangel und sind Sephardische Juden (Baron et al., 1987).
- Einige Formen des infantilen Autismus sind auch geschlechtschromosomgebunden, andere sind autosomal-rezessiv, weitere beruhen auf sporadi-

schen chromosomalen Anomalien (Smalley et al., 1988).

Bei bipolaren, unipolaren und schizoaffektiven Störungen in einer Amischen Bevölkerung* ist eine andere Erbform als die Mendelsche entdeckt worden (Egeland et al., 1987): Das Gen ist auf Chromosom 11 lokalisiert worden. Diese Beobachtung legt entweder ein dominantes Gen mit inkompletter Penetranz nahe oder einen einzelnen Hauptgenlocus mit multifaktorieller Vererbung. Schließlich findet sich bei Mitgliedern anderer Bevölkerungsgruppen die Lokalisation auf Chromosom 11 nicht (Hodgkinson et al., 1987). So zeichnet sich ab, daß die schweren affektiven Störungen und die Schizophrenie – bei der Chromosom 5 bei einer besonderen Familie als beteiligt beurteilt wurde (Sherrington et al., 1988) – genetisch heterogen sind (Kennedy et al., 1988). Bei einigen Formen der schweren affektiven Störungen scheint sich der gleiche Erbgang in verschiedenen Phänotypen ausdrücken zu können. Andererseits werden die meisten Schizophrenieformen wahrscheinlich auf multifaktorielle Art vererbt (wie eine große Zahl von anderen „medizinischen" Krankheiten).

Wir wissen noch nicht, wie sich die Gene auf dem kurzen Arm der Chromosomen 11 und 5 ausdrücken. Derzeit gibt es nur schwache Hinweise. Die Ausdehnung des ersteren, 3 Millionen Basenpaare lang, ist zwischen den Genen, die das Insulin und ein zellgebundenes Onkogen (cHarvey-ras-1) kodieren, und andererseits dem für Tyrosin-Hydroxylase kodierenden Gen (Craig et al., 1986) lokalisiert. Ein Gen von Chromosom 5 kodiert auch für den Glukokortikoidrezeptor, aber seine Nachbarschaft mit dem und die Bedeutung für das „Schizophrenie"-Gen muß noch erhellt werden.

51.4.1 „Biologische" und psychosoziale Theorien in der Psychiatrie: Die Notwendigkeit, sie zu integrieren

Derzeit herrschen sich gegenseitig ausschließende, ätiologische Theorien der psychiatrischen Störungen – die eine wird als „biologische", in einem eingeschränkten Sinne, und die andere als „psychosoziale" bezeichnet. Die Anhänger dieser „gegensätzlichen" Auffassungen stehen sich gegenüber, ohne den Versuch einer Synthese zu vollziehen. Die Folgen für die Betreuung und Behandlung der Patienten sind tragisch. Diese Lage bestand und besteht auch in der

---

* Nachkommen von Bauern aus dem Kanton Bern, Schweiz. Als Anabaptisten von der Obrigkeit geplagt, wanderten sie im 18. Jahrhundert ins Elsaß aus. Sie schlossen sich der strengen Lehre Jakob Ammanns, der aus Erlenbach im Berner Simmental stammt, an und wanderten im 19. Jahrhundert aus wirtschaftlicher Not nach Nordamerika aus, wo sie in 22 Staaten der USA und in Kanada leben. Ihre Zahl umfaßt etwa 90000. Die 3000 Amischen, die in der Umgebung des Städtchens Berne in Indiana (USA) leben, sprechen noch den alten Berner Dialekt und halten sich an die überlieferten Bräuche. Sie verzichten auf alle modernen Geräte, Autos, die Elektrizität und werden heute als Bio-Bauern von den Umweltschutz-Organisationen geschätzt (Anmerkungen des Übersetzers).

übrigen Medizin. Hier eine Synthese zu erzielen, ist das Ziel der modernen psychosomatischen Theorie.

Allgemein gesagt betonen die „biologischen" Theorien, daß einige der schweren psychiatrischen Syndrome (die affektiven, schizophrenen und Angstkrankheiten) Störungen sind, die auf „spontanen" (endogenen) Gehirnzuständen beruhen, die ihrerseits auf Veränderungen in der Konzentration von Neurotransmittern, Neuromodulatoren (z.B. Gehirnpeptide) und Hormonen zurückzuführen sind, auf Veränderungen in der elektrischen neuronalen Aktivität oder in der Gehirnstruktur (oder Gehirnregion). Diese Veränderungen werden als Folgen von Vererbungseinflüssen (Genen) betrachtet. Es wird aber nie klar, ob diese „Pathologien" die Syndrome oder Krankheiten erklären, oder alle oder einige ihrer Symptome und unangepaßten Verhaltensweisen. Was sie auch immer erklären mögen, es handelt sich um Entsprechungen zu den Syndromen oder Krankheiten, die als kausale, und nicht als zeitliche, dargestellt werden. Es gibt viele Gründe anzunehmen, daß solche Veränderungen selbst das Produkt einer weiteren Gruppe von Prozessen sind, von psychosozialen und solchen in der Umwelt: So ist die Frage gerechtfertigt, warum 11–30% aller Patienten mit schwerer Depression (vor allem der „nicht-endogenen" Form) auf Placebo ansprechen (Fairchild et al., 1986), oder warum diejenigen mit wiederholter saisonabhängiger Depression auf Lichtexposition reagieren (Lewy et al., 1985). Wären die vermuteten biochemischen oder physiologischen Störungen zeitlich vorangehende „Ursachen" und nicht Ergänzungen oder Folgen der Krankheiten, wie könnten sie dann durch Placebo verändert werden?

Die schweren depressiven Störungen sind durch veränderte vitale biologische Funktionen gekennzeichnet – des Atem- und Schlafrhythmus, der Temperaturregulation, der Nahrungseinnahme, durch den Verlust sexueller Bedürfnisse, Impotenz oder Amenorrhoe, der Herzfrequenz während des Schlafes, der Darmfunktion, -motilität – und bei einigen werden Abweichungen des zirkadianen Rhythmus in bezug auf Stimmung und Aktivität beobachtet. Wenn wir annehmen, daß die Nahrungsaufnahme durch eine Vielfalt von Neurotransmittern und Peptiden gesteuert wird, dann ist nicht zu erwarten, daß Veränderungen bei lediglich einem von ihnen die „Ursache" des Appetitverlusts bei einigen Formen darstellt oder die Ursache der Appetitsteigerung bei anderen Formen von schwerer Depression. (Vitale biologische Funktionen unterstehen immer der Kontrolle multipler regulierender, parallel angeordneter Prozesse.)

Die Untersuchung der Abweichung zirkadianer Vorgänge – besonders von Schlafrhythmen und Hormonsekretionsmustern (anstatt von Veränderungen ihrer Durchschnittsspiegel zu einem gegebenen Zeitpunkt) und der Stimmung und des Verhaltens – stellt eine ergiebigere Basis für das Erfassen von Zusammenhängen dar. Licht ist sicherlich ein wichtiger Synchronisator des Zirbeldrüsen-Melatonins und der Schlafrhythmen, soziale Beziehungen sind es aber auch (Hofer, 1984; Weiner, 1985) sowohl bei Tieren als auch beim Menschen. Soziale Beziehungen sind zum Erhalten des psychobiologischen Gleichgewichts junger Tiere und wahrscheinlich auch von Kindern besonders bedeutsam. Eine reiche Literatur legt nahe, daß Menschen mit Neigung zu depressiven Störungen besonders von Mitmenschen abhängig sind. Deshalb stellt der Zusammenbruch einer Beziehung, die für die abhängige Person lebenswichtig ist, einen Faktor im Vorfeld depressiver Störungen dar. Eine integrierte (psychosomatische) Hypothese wurde kürzlich vorgelegt (Ehlers et al., 1988): Sie verbindet Trennung mit dem Stören biologischer Rhythmen.

Nach diesen allgemeinen Bemerkungen sollen Daten besprochen werden, die für solch eine Integration bedeutsam sind. Eine begrenzte Zahl der wichtigsten psychiatrischen Syndrome soll den Anfang machen. (Einige davon, z.B. Konversionsreaktionen, Hypochondrie, die funktionellen (Somatisierungs-)Syndrome – kardiovaskuläre, gastrointestinale, Hyperventilationssyndrome, Schmerz, Kopfschmerz usw. – werden in anderen Kapiteln besprochen.)

51.5 Die schweren affektiven Störungen*

Sogar die anspruchsvollsten Experten in den USA verdammen den Dualismus, aber schreiben über die zwei Komponenten – „psychologische" (z.B. depressive Stimmungszustände, fehlendes Interesse an der Umwelt, Gefühle von Wertlosigkeit) und „biologische" (z.B. Veränderungen von Schlaf und Nahrungsaufnahme) – dieser Störungen, wie wenn es sich um zwei getrennte Gebiete innerhalb eines einzelnen Menschen handeln würde. Ihre Ambivalenz hinsichtlich der Ätiologie der schweren depressiven und affektiven Störungen kommt ebenfalls zum Ausdruck: Einerseits handelt es sich um eine „endogene" Störung, andererseits hängt sie mit Streß und dessen Neurobiologie zusammen (Gold et al., 1988).

Derzeit ist es gebräuchlich, die schweren affektiven Störungen in zwei Formen aufzuteilen: unipolare und bipolare. Bei der ersten kommt es zu wiederholten Depressionen; bei der zweiten treten zwischen den depressiven Schüben manische Episoden auf. Ihre klinischen Erscheinungen sind unterschiedlich. Den unipolaren und bipolaren Depressionen ist die melancholische Form gemeinsam, deren Validität fragwürdig ist (Young et al., 1986), deren klinische Merkmale aber recht eindeutig sind. Ihre Hauptzüge sind die Tiefe der Depression mit der größten Intensität am Morgen, die mit wahnhaftem Gefühl von Wertlosigkeit und Schuld einhergeht, der Wunsch Selbstmord zu begehen, die Arbeitsunfähigkeit, eine Verlangsamung von Handlungen und Denkvorgängen bis hin zum Mutismus, Konzentrationsschwierigkeiten, Hoffnungslosigkeit in bezug auf die Zukunft,

---

* In diesem Kapitel umgehe ich die großen Probleme, die ihre Definition aufwirft (vgl. Angold, 1988; Kendell, 1976b). Sie betreffen alle Ebenen ihrer Erforschung und stellen eine Hauptquelle der methodischen Fehler dar.

frühmorgendliches Aufwachen, Appetit- und Gewichtsverlust, Verstopfung, Abnahme der Libido, Impotenz beim Mann und Amenorrhoe bei der Frau. Einige Patienten sind aber ruhelos und ihr motorisches Verhalten ist gesteigert. Es überrascht deshalb nicht, daß die melancholische Form der Depression die in den Spitälern am häufigsten beobachtete ist (Klerman, 1948), denn der Patient ist schwer eingeschränkt und selbstmordgefährdet. Unklar bleibt aber, ob die Melancholie einen Extremzustand oder eine Spezialform einer depressiven Störung darstellt.

Zur oben erwähnten diagnostischen Verwirrung trägt die sog. „atypische" Form der Depression bei, die mit einer Zunahme des Schlafbedürfnisses, der Nahrungsaufnahme und des Gewichts einhergeht. Diese zweite Form kommt bei beiden Arten von schweren affektiven Störungen vor, hauptsächlich bei der bipolaren Störung (Casper et al., 1985).

Man könnte die schweren affektiven Störungen als Störungen der Regulation der Stimmung, des Selbstwertgefühls, des Schlafes, der Nahrungsaufnahme, der Fortpflanzung und der Ausscheidung auffassen, obwohl dies nicht gebräuchlich ist. Jede dieser vitalen Funktionen pendelt (unter Umständen zyklisch) in unterschiedlichen zeitlichen Phasen. Bei den schweren affektiven Störungen haben sich Phasenverschiebungen eingestellt (Garfinkel, 1983). So gesehen können diese Störungen, wie viele andere, als Phasenverschiebungs-Krankheiten betrachtet werden (Weiner, 1975, 1989b).

Obwohl ein Mensch während seines Lebens auch nur eine manische oder depressive Periode durchmachen kann, läuft die Störung gewöhnlich zyklisch ab. Mit zunehmendem Alter werden die zeitlichen Abstände zwischen den Schüben kürzer, und Schwere und Länge jeder Episode nehmen zu (außer sie würde durch Behandlung beendet) (Post et al., 1981). Es ist unklar, ob der Patient außerhalb der Schübe (innerhalb tolerabler Grenzen) je ganz von Stimmungsschwankungen oder von depressiven Verstimmungen frei ist: Es wird beispielsweise behauptet, daß 30% aller Patienten auch zwischen den Schüben phasenweise deprimiert sind (Goodwin und Jamison, 1984).

51.5.1 Ätiologie

Epidemiologie

Schwere affektive Störungen gibt es bei Kindern, Zahlen zu ihrer Prävalenz sind aber schwer zu finden, weil sie selten bei Bevölkerungsstichproben erhoben worden sind. Man nimmt an, daß ihre Prävalenz mit dem Alter zunimmt. Bei 2199 Kindern von 10–11 Jahren betrug sie 0,14%, und 1,5% bei 2303 Jugendlichen zwischen 14 und 15 Jahren (Rutter et al., 1976). Geschlechtsunterschiede bezüglich der Prävalenz fanden sich nicht. Mittels eines kurzen Frageinstruments stellten Kandel und Davies (1982) bei 4202 Jugendlichen zwischen 14 und 18 Jahren aber eine Prävalenz von 15–28% fest.

Bei Erwachsenen in einer städtischen Bevölkerung

traten die Episoden unipolarer Depression in 3–4% bei Männern und bei Frauen mindestens doppelt so häufig auf (Weissman und Myers, 1978). Dieser Geschlechtsunterschied bei Patienten mit unipolarer Depression hat noch nicht erklärt werden können. Bei bipolaren affektiven Störungen scheint er nicht vorzukommen. Bei diesen schwankt die Punktprävalenz zwischen 0,65–0,88% (Wing et al., 1978) oder ist geringer (Paykel, 1976). Die Punktprävalenz von Depression (verschieden definiert) liegt in einer städtischen Bevölkerung der USA jedoch zwischen 13 und 20% (Weissman und Myers, 1978). In der Primärversorgung (Barrett et al., 1988) oder bei Patienten von Polikliniken (Goldberg und Blackwell, 1970; Nielsen und Williams, 1980) liegt sie wohl noch höher. Bevölkerungsstichproben ergeben ein viel klareres Bild der Prävalenz depressiver Störungen, schwanken aber zwischen etwa 3,5% in Skandinavien (Sorensen und Strömgren, 1961) und 16% in Großbritannien (Brown et al., 1975). Diese Zahlen hängen von den diagnostischen Kriterien ab, die für diese Störungen verwendet werden.

Sozioökonomische und ethnische Faktoren

Bei Jugendlichen finden sich depressive Störungen häufiger in unteren sozioökonomischen Schichten (Kaplan et al., 1984; Schoenbach et al., 1984). Diese Verallgemeinerung gilt aber nicht für jede in den USA vorgenommene Bevölkerungsuntersuchung (Kandel und Davies, 1982). Die zuverlässigsten Untersuchungen in den USA und Großbritannien zeigen ein Überwiegen depressiver Störungen in den unteren sozioökonomischen Gruppen (Bebbington, 1978; Brown und Harries, 1986; Comstock und Helsing, 1976; Dohrenwend, 1975; Surtees et al., 1982).

Wiederholte Schübe affektiver Störungen schädigen eindeutig die Arbeitsleistung und erschweren eine ständige Anstellung. Bekanntlich haben aber begabte und kreative Dichter, Romanschriftsteller und Musiker an bipolaren affektiven Störungen gelitten.

Wie schon erwähnt, gibt es eine besondere Form der bipolaren Störung bei Sephardischen Juden, die an das X-Chromosom gebunden ist. In Untersuchungen an schwarzen und weißen amerikanischen Jugendlichen ist aber keine eindeutige Prävalenz depressiver Störungen festgestellt worden (Kandel und Davies, 1982; Schoenbach et al., 1984).

Genetische Aspekte

Zweifellos besteht eine familiäre Häufung von schweren affektiven Störungen, die Interpretation solcher Beobachtungen ist aber nicht einfach. Im großen und ganzen kann der Schluß gezogen werden, daß Kinder von Eltern mit einer Vielfalt von psychiatrischen Leiden mit einer größeren Wahrscheinlichkeit an verschiedenen psychiatrischen Störungen erkranken können, die aber nicht denjenigen der Eltern entsprechen (Rutter und Quinton, 1984). Eine Anamnese mit „Depression" bei einem Elternteil läßt Psychopathologie (Strober, 1984) bei etwa 40% seiner Kinder voraussagen (Beardslee et al., 1983). Die Verwandten (eingeschlossen ihre Kinder) von depressiven Menschen zeigen ein erhöhtes Risiko für schwere affektive Störungen (Weissman et al., 1984a, b). Die Verwandten ersten Grades von Menschen mit unipolarer Depression in ihrer Anamnese tragen ein erhöhtes Risiko (14–18%) für gerade diese Störung. Bei Verwandten von Patienten mit bipolaren affektiven Störungen besteht ein gleich starkes Risiko (7 oder 8 bis 10%) für beide Formen, die unipolare und die bipolare (Weissman et al., 1982).

Die Konkordanzrate für bipolare affektive Störungen beträgt bei monozygoten Zwillingen 65–70% und bei dizygoten 10–14% (Bertelson et al., 1977). Bedeutsam könnte das monozygote:dizygote Konkordanzverhältnis sein, das hier 6–7:1 beträgt und das, wenigstens bei bestimmten Familien, auf ein sehr hohes Vererbungsverhältnis hinweist. Wie schon bemerkt: Bipolare Störungen sind genetisch heterogen und können sich phänotypisch aus unbekannten Gründen in bipolarer und unipolarer Gestalt ausdrücken.

Die Bedeutung all dieser Werte ist überhaupt nicht klar. So kann die ätiologische Varianz bei bestimmten bipolaren Störungen fast, aber nicht ganz vollständig genetisch begründet sein, während bei anderen Formen das Übergewicht der Varianz nicht-genetische Kodeterminanten aufweist (beispielsweise bei den jahreszeitabhängigen affektiven Störungen). Auch wissen wir nicht, was vererbt wird: Ist es das Temperament, die Art zu reagieren, ein Syndrom, usw.? Die zyklische Natur dieser Störungen spricht für eine Reaktionsneigung, denn diese Störungen sind eindeutig nicht fixierte Züge. Trifft diese Hypothese zu, so sind die physiologischen Kovarianten dieser Tendenz noch zu erklären, die Auslösefaktoren und ihre Kodeterminanten in der Umgebung noch zu bestimmen.

Die Frage ist berechtigt: Was heißt es für ein Kind, bei Eltern aufzuwachsen, die entweder wiederholt depressiv sind oder zwischen depressiven und manischen Zuständen hin und her pendeln? (Wir wissen, daß manische Patienten auch Alkohol und Drogen mißbrauchen.) Depressive Menschen sind düster, pessimistisch, freudlos, klagsam, überkritisch, oft unnahbar oder reizbar, mit sich selbst mehr beschäftigt als mit anderen und strafen oft. Werden sie manisch, ist ihr Verhalten unvoraussagbar, reizbar, unkontrolliert, gewalttätig, gemein, sie lassen sich gehen und sind auf sich selbst bezogen. Die Scheidungsrate bei diesen Patienten ist hoch. Kein Wunder, daß solche Eltern ungenügende „elterliche Kenntnisse" besitzen (Angold, 1988; Weissman und Paykel, 1974). In einem Irrgarten von Zahlen und Fremdwörtern zerstreut finden sich eine Reihe von selbstverständlichen Beobachtungen, von denen einige für unser Verständnis dieser Störungen sehr bedeutsam sind (Cohen et al., 1954). Zu ihnen gehört die Tatsache, daß die Eltern in Familien mit bipolaren Patienten oft verfrühte Erwartungen von Leistung und Erfolg in ihre jungen Kinder setzen. In dieser Familienatmosphäre entwickelt sich ein Kind, das während des ganzen Lebens darauf erpicht ist, erfolgreich zu sein, aber dabei oft mit verheerenden Auswirkungen für sein Selbstvertrauen versagt. Das überwältigende Gefühl des Versagens und der Niederlage stellt einen Angelpunkt für unser Verständnis einiger Patienten mit schweren affektiven Störungen dar.

Die Erfahrung läßt weiter erkennen, daß Patienten mit depressiven Störungen Trennung von der Mutter oder deren Tod (dies gilt nicht für die Väter) vor dem 17. Lebensjahr erlitten haben, besonders Patienten mit niedrigerem sozioökonomischem Status (Brown und Harris, 1986). Es wird, wenn auch unrichtigerweise, behauptet, daß solch eine Erfahrung nur den „nicht-endogenen" Formen vorausgeht (Roy, 1987). Nur ganz vereinzelte Untersuchungen haben ihr Augenmerk auf das Familienleben solcher Patienten nach Erleiden des Verlustes gerichtet: Ist es ärmlich, stellt es einen bedeutenden Risikofaktor für Psychopathologie, besonders Depression dar (Breier et al., 1988).

Andere soziale Hintergrundfaktoren, welche Patienten einem Risiko für schwere affektive Störungen aussetzen, sind eine schlechte Ehe (definiert als Unvermögen, dem Ehepartner zu vertrauen) und eine Familienanamnese mit Depression, und zwar sowohl „endogene" als auch „nicht-endogene" Formen. Bei den letzteren finden sich zusätzlich die Verantwortung für drei oder mehr Kinder unter 14 Jahren und Arbeitslosigkeit (Brown und Harris, 1978, 1986; Roy, 1987).

Diese sog. Verletzlichkeitsfaktoren stützen sich auf große Stichproben. Offensichtlich spielen weitere Faktoren eine Rolle: Einer der wichtigsten ist das Vorliegen eines chronischen somatischen Leidens. Die Prävalenz von Depression bei Patienten mit Krebs (besonders des Pankreas) liegt über 50%. Sie ist auch bei Patienten mit primär chronischer Polyarthritis, systemischem Lupus erythematodes und besonders mit Colitis ulcerosa hoch (Übersicht bei Weiner, 1977, 1989a).

Die Untersuchung großer Stichproben vermag die feinen Einzelheiten individueller Merkmale bei Patienten zwischen den Schüben ihrer manischen oder depressiven Episoden natürlich nicht zu erfassen. Leider wird solchen Beobachtungen nicht das nötige Gewicht zugebilligt; sie wären aber von entscheidender Bedeutung. Sie zeigen, daß diese Patienten auffällig empfindlich für drohende oder tatsächliche Trennung von einem anderen Menschen sind, von dem sie sich fortwährend Liebe, Aufmerksamkeit, Lob, Unterstützung und Ermutigung erhoffen. Ihrerseits geben sie aber wenig zurück und sie scheinen gegenüber den Bedürfnissen und Wünschen anderer blind oder unempfindlich zu sein. Kritik und Vorwürfe verletzen ihre Gefühle tief. Versagen sie, fühlen sie sich wertlos. Werden sie verachtet oder zurückgewiesen, stürzen sie in eine Depression. Sie verlangen „Sonderbehandlung" und reagieren wütend, verletzt oder deprimiert, wenn sie ihnen nicht zuteil wird. Sie neigen zum Moralisieren, kritisieren andere und fühlen sich anderen gelegentlich (moralisch) überlegen (Cohen et al., 1954; Jacobson, 1971).

Auslösende Umstände und Verletzlichkeit

Nach psychosomatischer Theorie lebt der Mensch nicht im Vakuum. Lebenserfahrungen werden individuell bewertet und verarbeitet; Bedeutung wird ihnen zugemessen; und es wird auf sie reagiert oder nicht. Warum sollten sich Patienten, die zu depressiven Störungen neigen, anders verhalten? Wie oben angeführt sind sie durch besondere Verletzlichkeit ausgezeichnet, zu der sie auf unbekanntem Wege durch Gene, frühe Lebenserfahrung von Verlust, Armut und zueinander lieblosen oder sich in ihrer Ehe wenig stützenden Eltern gekommen sind. Diese Verwundbarkeit (oder Risikofaktoren) äußert sich wie eben dargestellt in besonderer Empfindsamkeit gegenüber Verlust der **externen Regler** ihrer Selbstachtung und den anderen eben geschilderten Charakterzügen.

Der Verlust kann real oder symbolisch sein. Die Hauptaufmerksamkeit kürzlicher Untersuchungen konzentrierte sich auf die Bedingungen, die einem Ausbruch des Leidens vorangehen (auslösende Kräfte) und die mit den besonderen Verletzbarkeiten der Patienten, d.h. bei anfälligen Menschen (Brown und Harris, 1986) in Wechselbeziehung treten. Gewisse Psychiater geben widerstrebend zu, daß solche Bedingungen beim Ausbruch affektiver Störungen einschließlich der melancholischen oder manischen eine Rolle spielen, aber sie sträuben sich dennoch dagegen: Sie führen an, solche Erlebnisse würden nur bei der Hälfte der Patienten (Patrick et al., 1978) gefunden oder nur beim Ausbruch der ersten Schübe, aber nicht bei späteren (Gold et al., 1988); oder nur bei milderen Formen der Depression und nicht bei unipolaren und bipolaren. Ein anderes, häufig angeführtes Argument lautet, daß diese Erlebnisse nicht „spezifisch" seien oder die Folgen und nicht die Vorläufer der Störungen darstellen würden, oder daß sie, sogar wenn sie lange anhalten, nur eine nebensächliche Rolle spielten (Breslau und Davis, 1986).

Aber die sorgfältigsten Untersuchungen auf diesem Gebiet erwecken schwere Zweifel an diesen Beteuerungen (Brown und Harris, 1986; Paykel, 1976). Nach ihnen spielen diese realen Erfahrungen der Patienten kausale Rollen, die zu den verschiedenen Formen depressiver Störungen führen. Solche Erlebnisse wiegen schwer. Sie umfassen Enttäuschungen und Verluste, die Mitmenschen, Rollen (als Frauen und Mütter, bei der Arbeit usw.) und Ideale betreffen. Oft pfropfen sich solche Erfahrungen auf gleichzeitige reale „Schwierigkeiten" auf (z.B. Anzahl der Kinder, finanzielle Schwierigkeiten, ärmliche Wohnbedingungen, Ehestreit usw.). Die verschiedenen Kategorien von Erlebnissen und Schwierigkeiten werden von den späteren Patienten als qualvoll erlebt. Beziehungen zu anderen Menschen, welche stützen, verständig sind oder Abhängigkeitsbedürfnisse stillen, können die Qual lindern (Costello, 1982).

Überwältigende Ereignisse – plötzliche Verluste – können unmittelbar auslösende Wirkung haben: Sie sind „schwer". Die Belastung durch andere Erlebnisse muß unter Umständen mehrere Tage anhalten. Sie werden als „milder" eingeschätzt. Selbstverständlich

können sich beide Kategorien (Ereignisse und Schwierigkeiten) im weiteren Verlauf kombinieren. Beachtet man alle diese Faktoren und ihr Zusammenwirken, so finden sie sich bei 89% aller Patienten vor dem Ausbruch der depressiven Störungen. Brown und Harris (1986) haben eine Anzahl von Studien zusammengefaßt, die zeigen, daß beide Kategorien einander in Art und Zahl sehr ähnlich sind und sowohl „endogenen" als auch „milderen" Depressionen vorangehen. Bei 73% der Depressionen ging der Beginn mit „schweren" Ereignissen oder einer bedeutsamen Schwierigkeit einher; kamen sie zusammen mit „milderen" vor, nahm der Prozentsatz noch weiter zu.

Diese und andere Daten (Bebbington et al., 1988) strafen die „endogene" Theorie der schweren affektiven Störungen Lüge. Die Vorstellung, eine Familienanamnese mit solchen Störungen genüge, um ihre Ätiologie zu erklären, oder daß es eine „reaktive" Form von schweren affektiven Störungen gebe, stimmt trotz mancher Studien, die diese Unterscheidung immer noch machen (z.B. Feinberg und Carroll, 1982), ebenfalls nicht (Bebbington et al., 1988; McGuffin et al., 1988; Patrick et al., 1978).

Die hier besprochenen Probleme sind aber noch viel komplexer. Die depressiven Störungen sind Prozesse, die sich über eine Zeit erstrecken. Ein Faktor, der zur Verletzbarkeit beiträgt, wie ein dauernder Mangel enger Beziehungen, kann sich im Verlauf einer sich verschlechternden Ehe verstärken: z.B. als Folge der chronischen Depression eines Ehepartners, die in ihrer Intensität schwankt. Viele Untersuchungen haben aber ergeben, daß Frauen mit Depression (unabhängig vom Schweregrad) häufiger als Männer beides, traumatische Ereignisse und chronische Schwierigkeiten, erlebt haben (Bebbington et al., 1988). In Anbetracht der Rollen, welche die britische Gesellschaft den Frauen zubilligt (besonders wenn sie in Armut leben), überrascht das nicht.

Die größte Gefahr (nicht die einzige) der depressiven Störungen ist der Selbstmord. Suizidversuche haben in einer Reihe von Ländern dramatisch zugenommen und stehen in enger Beziehung zu unverschuldeter Arbeitslosigkeit (Farrow, 1984; Platt und Kreitman, 1984) und sozialen Umständen (Weissman et al., 1973). Vollzogene Selbstmorde nehmen mit dem Alter an Häufigkeit zu und kommen vor allem bei Männern vor (Paykel, 1976). Menschen, die Selbstmordversuche begehen, hatten signifikant häufiger belastende Erfahrungen als Patienten, die später depressiv werden, und viel häufiger als Zufallsstichproben der Allgemeinbevölkerung. Die häufigsten Belastungen sind Erkrankung oder Tod eines Familienmitgliedes oder Trennung (durch Scheidung oder örtliche Trennung). Aber auch Streit und Dysharmonie mit dem Ehepartner, Arbeitslosigkeit, Gefängnis oder geschäftlicher Mißerfolg spielen eine Rolle. Weitere Faktoren vor einem Suizidversuch sind unbeeinflußbare Ereignisse, von denen ein großer Einfluß auf die Zukunft erwartet wird. Sie traten im Monat vor dem Versuch gehäuft auf (Paykel, 1976).

Über manische Episoden, die als „endogen" betrachtet werden, ist bisher nichts gesagt worden. Dem

Schub vorangehende Erlebnisse fanden sich in 66% von 50 untersuchten Patienten. Sie unterschieden sich z. T. von denjenigen, die der Depression vorangehen (obwohl Tod und Trennungen bei 13% dem Schub vorangingen): Bei 11 Patienten waren die Geburt eines Kindes und die Verlobung Auslösungsfaktoren (Ambelas, 1987). Diese Ereignisse wurden bei einigen Patienten von zunehmenden Schlafstörungen begleitet.

Die Rolle sozialer und psychischer Faktoren bei depressiven Störungen wird besonders deutlich, wenn ihr Einfluß auf den Verlauf und das Endergebnis dieser Leiden untersucht wird. Es handelt sich ja um chronische Leiden, die oft nach der Behandlung wieder auftreten und die bei 40–45% der Patienten über ein Jahr anhalten können (Brown und Harris, 1978; Brown et al., 1988; Keller et al., 1982). Die Entlastung von chronischen Lebensschwierigkeiten und/oder belastenden Ereignissen vermag einem depressiven Menschen ein Gefühl von Hoffnung zu vermitteln, das Leben neu beginnen zu können; gesellt sich dazu die Unterstützung durch einen Mitmenschen, so kann ein Schub abklingen und die Dauer der chronischen Depression sich verkürzen (Brown et al., 1988).

Zusammenfassend erbringen die besten und sorgfältigsten Untersuchungen genügend Belege für die Bedeutung einer vielfältigen Kombination von Ereignissen (besonders Verlust und Trennung) als auslösenden Faktoren. Oft sind sie mit schwierigen Lebenssituationen gekoppelt, die auf finanziellen, ehelichen und Familienproblemen beruhen. Diese Faktoren treten mit genetischen und entwicklungsbedingten Gegebenheiten und persönlichen Erfahrungen in Wechselbeziehungen. Vor allem aber hängt das psychische „Wohlergehen" des zukünftigen manischen und/oder depressiven Menschen von anderen ab, an die er sehr hohe innere Ansprüche und Erwartungen stellt. Tod und andere Trennungen bringen solche Beziehungen zu einem Ende, wenn Unterstützung und Verständnis nicht durch Ersatzpersonen geleistet werden. Widrige Umstände, die nicht beeinflußt werden können oder zu einem tiefen Empfinden von Versagen oder Niederlage führen, ziehen Depression nach sich. Tatsächliche Wiedervereinigung mit Bezugspersonen oder die Hoffnung darauf ebnet der Manie den Weg.

51.5.2 Physiologische und hormonelle Entsprechungen der schweren affektiven Störungen

Biogene Amine

Da Reserpin zentral zur Verarmung von Neuronen an biogenen Aminen führt (und bei 15% der Patienten, die das Medikament einnehmen, eine Depression auslöst) (Schildkraut, 1965), da die Monoaminooxidasehemmer den Abbau der Monoamine verlangsamen und antidepressiv wirken, und da die trizyklischen Antidepressiva die Wiederaufnahme der gleichen Amine hemmen (vor allem Noradrenalin), postulierte Schildkraut (1978), daß die schweren De-

pressionen auf einem Mangel an Noradrenalin in den zentralnervösen Synapsen beruhen.

Noradrenalin- und Adrenalinspiegel sind im Liquor und im Plasma depressiver Patienten, auch der melancholischen, entweder normal oder erhöht (Christensen et al., 1980; Roy et al., 1985). Die Spiegel des Hauptmetaboliten von Noradrenalin, 3-Methoxy-4-Hydroxy-Phenylglycol, sind in Urin und Liquor (Koslow et al., 1983) unbehandelter Patienten erhöht und fallen bei der Behandlung mit Antidepressiva (Linnoila et al., 1982). Aber seine Spiegel in Liquor und Blut sind in manischen Zuständen ebenfalls erhöht, und die Urinspiegel sind bei Patienten mit Angstzuständen hoch (Koslow et al., 1983; Sweeney et al., 1978).

Es ist aber ganz unklar, wie diese Veränderungen die Ausscheidung von Noradrenalin in die Synapse, seine Verteilung, Wiederaufnahme oder Abbau im Gehirn widerspiegeln. Die Hauptquelle von Noradrenalin im Plasma und Urin sind die peripheren sympathischen Nervenendigungen, während ein Teil des 3-Methoxy-4-Hydroxy-Phenylglycols im Liquor aus dem Plasma stammt. Man könnte also ruhig behaupten, daß bei schweren affektiven Störungen eine vermehrte sympathische Aktivität vorliegt. Derzeit wird aber angenommen, daß die beobachteten Veränderungen der Noradrenalin- und 3-Methoxy-4-Hydroxy-Phenylglycol-Spiegel eine gesteigerte Aktivität im Locus coeruleus widerspiegeln (Gold et al., 1988). Die Synthese von Noradrenalin und Adrenalin (im Nebennierenmark) ist bei depressiven Patienten tatsächlich gesteigert, und sie sezernieren auch mehr. Die 3-Methoxy-4-Hydroxy-Phenylglycol-Sekretion ist relativ zu diesen Steigerungen vermindert. Diese Unterschiede werden vor allem bei unipolaren und weniger bei bipolaren Depressionen gefunden (Maas et al., 1987).

Das Noradrenalin stellt aber nicht den einzigen Neurotransmitter dar, dem bei schweren affektiven Störungen Bedeutung zugeschrieben wird. Acetylcholinagonisten scheinen Depressionen induzieren zu können (Janowsky et al., 1980). Der Gammaaminobuttersäurespiegel im Liquor ist niedrig (Berrettini et al., 1982). Die Spiegel des Serotoninmetaboliten 5-Hydroxyindolessigsäure sind sowohl bei depressiven als auch nicht-depressiven suizidalen Patienten tief (Åsberg et al., 1976).

Die Hypothalamus-Hypophysen-Nebennierenrindenachse

Serumkortisol und Kortisolproduktion, Urin- und Liquorkortisolgehalt sind bei vielen, aber nicht allen Patienten mit schweren affektiven Störungen erhöht (Carroll et al., 1976; Rubin et al., 1987; Sachar et al., 1970). Die Serumkortisolspiegel sind altersabhängig; bei älteren Patienten liegen sie höher (Halbreich et al., 1984). Wird das Serumkortisol fortlaufend Tag und Nacht bestimmt, so finden sich gewöhnlich höhere als normale Spiegel, die am Abend nicht zur Ausgangsbasis zurückkehren. Amplitude und Dauer der einzelnen Kortisolsekretionsschübe sind nicht größer als die normalen bei einigen, aber nicht allen Patienten. Es tritt gewöhnlich keine Phasenvorverschiebung des Beginns des nächtlichen Rhythmus des Hormons ein. Im Gegensatz zu Patienten mit Anorexia nervosa und Morbus Cushing bleibt der zirkadiane Kortisolrhythmus erhalten (Rubin et al., 1987).

Eine aufregende Beobachtung wurde kürzlich gemacht. Bei erhöhten Kortisolspiegeln in depressiven Zuständen bildet sich eine Resistenz gegen das Kortisol durch „Inaktivierung" oder „Translokation" der Zytosol-Glukokortikoidrezeptoren auf Lymphozyten (Lowy et al., 1984; Whalley et al., 1986). Sollte dieses Phänomen auch im Hirn und der Hypophyse vorkommen, müßte eine Reihe von Hypothesen über die Aktivierung der Hypothalamus-Hypophysen-Nebennierenrindenachse bei schweren affektiven Störungen überdacht werden.

Dynamische Tests dieser Achse bei diesen Störungen legen nahe:

- Die Rezeptoren für ACTH in der Nebennierenrinde sind „sensitiver" oder in ihrer Zahl vermehrt (Amsterdam et al., 1983).
- Die Freisetzung und Sekretion von ACTH, aber nicht Kortisol, als Reaktion auf injiziertes Kortikotropin-Releasing-Hormon (CRH) ist vermindert (Holsboer et al., 1984), vermutlich durch die erhöhten Serumkortisolspiegel bedingt, die das hypophysäre Kortikotropin niedriger einstellen. Spannender ist die Beobachtung, daß die Frequenz der pulsierenden ACTH-Freisetzung bei depressiven Frauen verglichen mit normalen um 50% erhöht ist (Mortola et al., 1987). Dies könnte bedeuten, daß die CRH-Pulse schneller oszillieren und für höhere CRH-Spiegel im Liquor von Patienten mit schweren affektiven Störungen verantwortlich sind (Nemeroff et al., 1984).

Viele dieser Beobachtungen (z.B. Hyperkortisolämie) sind aber nicht spezifisch für die schweren affektiven Störungen (Christie et al., 1986; Taylor und Fishman, 1988). Diese Bemerkung gilt vor allem für den Dexamethasonsuppressionstest, der bei 70 oder mehr Prozent aller Patienten mit diesen Störungen positiv ist (Arana et al., 1985), aber auch im Verlauf einer Vielfalt von anderen Krankheiten positiv sein kann (Shapiro et al., 1983). Das Problem liegt weder in seiner fehlenden Spezifität für depressive Störungen, noch in der fehlenden Präzision für die Messung der Integrität der Hypothalamus-Hypophysen-Nebennierenrindenachse, sondern in der Bioverfügbarkeit oder der Kinetik des Dexamethasons: Beim Vorliegen hoher Serumkortisolspiegel fallen die Blutspiegel oral verabreichten Dexamethasons tief aus (Arana et al., 1984; Johnson et al., 1984) und es kann zu einem Fehlen der Hemmung des Kortisols kommen.

Was erklären diese Beobachtungen also? Erhöhte Plasmakortisolkonzentrationen kommen bei einer Vielfalt von Psychosen, bei Anorexia nervosa, Morbus Cushing und unter „Streß"-Bedingungen vor. Sie stehen mit dem Verlust der Kontrolle über Umgebungsbedingungen bei Tier und Mensch in Zusammenhang (Henry und Stephens, 1977).

Diese Beobachtungen können also keine Grundlage für eine Theorie der Pathogenese der schweren affektiven Störungen liefern, obwohl sie mit Kontrollverlust oder psychotischer Symptomatologie verbunden sein könnten (Christie et al., 1986). Anstrengungen wurden unternommen, die oben erwähnten Veränderungen der biogenen Amine mit der Aktivierung der Hypothalamus-Hypophysen-Nebennierenrindenachse in Zusammenhang zu bringen (Gold et al., 1988). Man weiß, daß stimulierte beta$_2$-adrenerge Rezeptoren der ACTH-enthaltenden Hypophysenzellen bei der Ratte zur ACTH-Sekretion führen (Mezey et al., 1983). Tatsächlich sind wie erwähnt Nordrenalin- und Adrenalinspiegel bei einer Anzahl von psychischen Störungen erhöht: So sind positive Korrelationen zwischen 3-Methoxy-4-Hydroxy-Phenylglycol-Urinspiegeln und Prä- und Post-Dexamethasontest-Konzentrationen von Plasmakortisol gefunden worden (Jimerson et al., 1983). Die Bedeutung solcher Korrelationen ist aber recht unklar. Vielleicht spiegeln sie einfach den nicht-spezifischen psychobiologischen Zustand eines psychotischen Menschen. Im Gegensatz dazu könnten sie auch auf eine häufigere oder gesteigerte Freisetzung von CRH in Neuronen des paraventrikulären Nucleus des Hypothalamus hinweisen (Sawchenko, 1989). Die Regulation von CRH- und ACTH-Sekretion ist äußerst komplex und umfaßt eine Vielzahl von Monoaminen und Peptiden (Reisine, 1989). Kurz und gut, wir wissen also noch lange nicht, warum die Oszillationen der CRH-Sekretion häufiger stattfinden.

Dazu kommt, daß CRH die Entladungen des sympathischen Nervensystems stark steigert. Es könnte also für die erhöhten Plasmaspiegel von Noradrenalin und Adrenalin (Brown und Fisher, 1985) und auch für die Hyperkortisolämie verantwortlich sein. Natürlich könnte der einmal erhöhte Katecholamin-Blutspiegel die ACTH-Sekretion weiter stimulieren.

Für die Beziehung zwischen häufigerer und stärkerer Freisetzung von CRH und damit auch Hyperkortisolämie und Ereignissen in der Umgebung, besonders Verlust und Trennung (Henry und Stephens, 1977; Mason, 1968), liegen reichlich Hinweise vor. Aber nur ganz wenige Untersuchungen haben diese Hypothese bei den schweren affektiven Störungen geprüft:

Dolan und Mitarbeiter (1985) haben nachgewiesen, daß sowohl auslösende Ereignisse wie auch Schwierigkeiten ohne Zweifel mit gesteigerten 24-Stunden-Mengen von freiem Kortisol (ein Maß für die Kortisolproduktion) verbunden waren. Diese Beziehung fand sich sowohl bei den milderen als auch bei den schwereren („endogenen") Depressionsformen.

Beta-Lipotropin und Beta-Endorphin

Beta-Lipotropin und -Endorphin entstehen wie auch das ACTH aus dem gleichen Vorläufermolekül. So würde man meinen, daß sie gleichzeitig mit ACTH sezerniert werden (wenigstens beim Tier). Aber bei depressiven Patienten scheint es nicht so zu sein. Die Reaktionen von Beta-Lipotropin und Beta-Endorphin auf Dexamethason korrelieren nicht notwendigerweise mit der Kortisolhemmung (Matthews et al., 1986). Die Bedeutung dieser Beobachtung ist nicht klar; die Ausschüttung dieser drei Hypophysenpeptide wird in depressiven Zuständen wahrscheinlich unterschiedlich gesteuert.

Somatostatin

Die Hauptaufgabe von Somatostatin besteht in der Hemmung des Wachstumshormons (STH); es hemmt aber ebenfalls die CRH-Sekretion (Heisler et al., 1982) und die Ausschüttung der Katecholamine. Somatostatinspiegel sind im Liquor von Patienten mit schweren affektiven Störungen erhöht (Agren et al., 1984; Rubinow et al., 1983) und korrelieren positiv mit dem negativen Dexamethasontest (Doran et al., 1986). Diese Untersuchung hat aber Schwächen: Die Validität des Dexamethasonhemmtestes ohne Messung der Serumkortisolspiegel ist verdächtig; und die Befunde sind für die Diagnose der schweren affektiven Störungen nicht spezifisch, weil sie auch bei schizophrenen Patienten, die „Nicht-Suppressoren" waren, gesehen wurden (Doran et al., 1986). Zudem wäre die Interpretation dieser Befunde einfacher, wenn STH-Blutspiegel oder zirkadiane Rhythmen auch gemessen worden wären.

Wachstumshormone

Die Steuerung der Wachstumshormonproduktion oder -ausschüttung ist komplex. Bei Adoleszenten wird das Wachstumshormon in Wellen während des Tages in den Kreislauf ausgeschüttet. Beim Erwachsenen wird es nur einmal sezerniert, und zwar lediglich während der ersten Slow-wave-Schlafperiode der Nacht. Mendlewicz et al. (1985) haben beim Erwachsenen mit schweren depressiven Störungen eine Umkehrung des Sekretionsmusters beschrieben. Man weiß nicht, ob diese Beobachtungen für die schweren depressiven Störungen spezifisch sind. Bei anderen Krankheiten (z.B. Anorexia nervosa) ist die verminderte Sekretion anderer Hypophysenhormone beschrieben worden. Dies bildet die Grundlage für eine neue Betrachtungsweise bestimmter Krankheiten.

Arginin-Vasopressin

Sowohl bei Depression als auch bei Manie sind verminderte Arginin-Vasopressin-Spiegel im Liquor beschrieben und bestätigt worden (Gold et al., 1981; Gjerris et al., 1985). Die Interpretation dieser Beobachtung fällt aus drei Gründen schwer:
- Aussagekräftiger wäre es gewesen, wenn das Verhältnis von Serum- zu Liquor-Arginin-Vasopressin vor und nach Salzinfusion gemessen worden wäre.
- Es ist gut bekannt, daß Arginin-Vasopressin mit CRH synergistisch die ACTH- und damit die Kortisolproduktion stimuliert; falls der verminderte Arginin-Vasopressin-Liquorgehalt eine zentrale Hemmung seiner Sekretion spiegelt, dann sollten die ACTH- und Kortisolspiegel tiefer als normal sein (sind es aber nicht).
- Die Arginin-Vasopressin-Verminderung sollte mit

einem partiellen Diabetes insipidus vergesellschaftet sein; er ist jedoch kein Merkmal der schweren affektiven Störungen.

Hypothalamus-Hypophysen-Schilddrüsenfunktionen

Dieses Gebiet ist verwirrend. Jahrelang haben Psychiater eine vielfältige Interaktion zwischen der Funktion der Schilddrüse und einer depressiven Stimmung vermutet (Whybrow und Prange, 1981). Tatsächlich sind mindestens 50% aller Basedow-Kranken ängstlich, agitiert und deprimiert (Übersicht bei Weiner, 1977). Viele hypothyreote oder myxödematöse Patienten erscheinen auch deprimiert.

Die Beurteilung wird dadurch erschwert, daß mit Appetit- oder Gewichtsverlust und/oder chronischer Krankheit der Schilddrüsenstoffwechsel eine nichtspezifische Anpassung erfährt. Thyroxin (T_4) wird in das physiologisch inaktive „inverse" Trijodthyronin (inverses T_3) anstatt in T_3 umgewandelt (Sullivan et al., 1973). Der Stoffwechsel etlicher Steroidhormone wird dadurch verändert (Smith et al., 1975). Appetit- und Gewichtsverlust sind herausragende Merkmale der schweren affektiven Störungen (besonders der Melancholie). Die Akten über das tiefe T_3-Syndrom als nicht-spezifische Folge von schweren affektiven Störungen sind noch nicht geschlossen. Die peripheren Schilddrüsenhormonspiegel (T_3, T_4) sind bei schweren affektiven Störungen gewöhnlich normal (Zach und Ackerman, 1988), aber nicht immer (Kirkegaard und Faber, 1981).

Bei ungefähr 25% der Patienten mit schweren affektiven Störungen findet sich eine verminderte Antwort des Thyreotropins (TSH) auf das Thyreotropin-Releasing-Hormon (TRH) (Loosen, 1985; Loosen und Prange, 1982). Dies ist wiederum nicht spezifisch für schwere affektive Störungen; die gleiche Veränderung findet sich bei betagten Gesunden, abstinenten Alkoholikern, bei Schizophrenie und Anorexia nervosa. Die Interpretation dieser Beobachtung liegt im dunkeln: Es kann sich um eine „Tieferstellung" der Regulation von Hypophysenhormonen handeln oder um eine Verminderung der hypophysären TSH-Reserve.

Kürzlich sind im Liquor von 14 schwer depressiven Patienten, 13 davon waren Frauen, erhöhte TRH-Spiegel gefunden worden. Es fanden sich aber keine Korrelationen zwischen dieser Veränderung und basalen Werten von T_3, T_4, TSH und der TSH-Antwort auf TRH (Banki et al., 1988). Da sich TRH an vielen Stellen im ZNS findet und die verschiedensten Aufgaben hat, kann die Quelle dieser Zunahme nicht bestimmt werden. Erhöhte Liquor-TRH-Spiegel lassen also Hypothesen über eine verminderte TSH-Reaktion auf TRH oder die Rolle von TRH bei schweren affektiven Störungen weder bestätigen noch verwerfen.

Schlaf- und Körpertemperatur-Rhythmen

Eines der auffälligsten Merkmale schwerer depressiver Störungen stellt das frühmorgendliche Aufwachen (z.B. zwischen 3–4 Uhr) dar, über das sich viele Patienten bitterlich beklagen, und das mit der „schwärzesten" Stimmung einhergeht. Die Gesamtschlafzeit ist bei schweren affektiven Störungen ebenfalls vermindert (in etwa 15% verlängert). Der depressive Patient hat Schwierigkeiten einzuschlafen und wacht oft auf.

Elektroenzephalographische (EEG-) Untersuchungen des Schlafs bei schweren depressiven Störungen haben ergeben:

- Die Latenz zwischen Einschlafen und erster REM-Phase ist verkürzt;
- die gewöhnlichen 4 REM-Perioden während der Nacht sind anders verteilt und zwar Richtung erste Nachthälfte;
- die durchschnittliche Dauer des Delta-slow-wave-Schlafs ist vermindert und der gewöhnliche Ablauf der vier Schlafstadien (besonders REM und Delta) ist verändert (Coble et al., 1976; Kupfer und Foster, 1972).

Die vier Hauptveränderungen im Schlaf-EEG treten nicht unbedingt zusammen auf; sie kommen auch bei der Manie vor, wahrscheinlich mit Ausnahme der Abnahme des Delta-Schlafs (Hudson et al., 1988). Die verkürzte REM-Latenz ist aber auch bei anderen Krankheiten beobachtet worden (z.B. Zwangsleiden).

Obwohl die Veränderungen im Schlaf-EEG als „Marker" für schwere affektive Störungen benützt werden, ist ihre Spezifität keineswegs gesichert. Verkürzte REM-Latenz kann lediglich eine Folge des frühmorgendlichen Erwachens sein, sie kann bei normalen Menschen (nicht-depressiven) erzeugt werden (Mullen et al., 1986) und hängt vermutlich auch von Alter und Gewicht ab. Gewichtsverlust geht mit verminderter Totalschlafzeit einher, mit Schlafunterbrechung und frühzeitigem Erwachen. Vermehrte Zufuhr von Kohlehydraten kann den REM-Schlaf steigern (Crisp, 1986).

Über die Gründe der Schlafstörungen bei schweren affektiven Störungen sind viele Vermutungen geäußert worden. Die faszinierendste bezieht chronobiologische Rhythmen und ihre Veränderungen ein. Bis jetzt kennt man zwei miteinander verbundene Schrittmacher im Hypothalamus, die den Organismus mit einigen dieser Rhythmen ausstatten, aber weitere Schrittmacher finden sich wahrscheinlich noch andernorts! Ein Oszillator kontrolliert hauptsächlich den REM-Schlaf, die Körperkerntemperatur, Plasmakortisol und andere Rhythmen. Der andere steuert Zyklen von Ruhe-Aktivität, Slow-wave-Schlaf, Plasma-STH und Schwankungen der Körpertemperatur (Moore-Ede et al., 1982).

Bei depressiven Störungen ist die Beziehung der zwei Schrittmacher verändert: Die zirkadiane Temperaturkurve ist flach und die Durchschnittstemperaturen liegen höher; die REM-Schlafstörungen bleiben dabei bestehen (Avery et al., 1982; Beersma et al., 1984). Die chronobiologischen Veränderungen bei schweren affektiven Störungen könnte man sich wie folgt vorstellen: Der erste Schrittmacher (ist gegenüber dem zweiten zeitlich vorangestellt worden) führt

zu früheren REM-Perioden und dazu, daß Kortisol und Kerntemperatur ihre Gipfelwerte nicht mehr erreichen (Wehr und Goodwin, 1981). Die Beobachtungen über STH passen aber nicht dazu: Sie legen eine „Regression" auf ein früheres Sekretionsmuster nahe.

Zirkadiane Rhythmen sind nicht nur endogen, sondern auch von äußeren Faktoren (Zeitgebern) gesteuert, die einen oder beide Schrittmacher beeinflussen oder die eine Funktion dem einen Schrittmacher und nicht dem anderen unterstellen können.

Licht, flugbedingte Zeitverschiebung, Schichtarbeit, sensorische und soziale Isolation und Medikamente vermögen eine Desynchronisation von Schrittmachern auszulösen (Wever, 1985, 1989). Unter Bedingungen konstanten Lichts kommt es zur Desynchronisation. Licht scheint den zirkadianen Schrittmacher für Körpertemperatur und Kortisol besonders zu beeinflussen (Czeisler et al., 1986).

Eine Unterform von schweren depressiven Störungen tritt jahreszeitlich gebunden auf und wird durch Lichteinfluß korrigiert (James et al., 1985; Lewy et al., 1985), aber bei anderen Formen ist dies nicht der Fall. Die Phasenvorverschiebung des Zubettgehens um sechs Stunden kann bei einigen Patienten die schweren depressiven Störungen beenden, und ein einmaliger nächtlicher Schlafentzug vermag, wenn er mit Antidepressiva kombiniert wird, diese Patienten aus dem depressiven Zustand zu holen (Moore-Ede et al., 1982).

Bei Tieren besteht wenig Zweifel darüber, daß Trennung die chronobiologischen Rhythmen verändert (Hofer, 1984). Bei Menschen kann die Anpassung an die Lebensgewohnheiten anderer Menschen und ihren Zeitplan in bezug auf das Zubettgehen, Aufstehen, Essen, körperliche Betätigung usw. den gleichen Einfluß haben (Ehlers et al., 1988); bei Frauen, die zusammenleben, gleichen sich sogar die menstruellen Zyklen einander an (McClintock, 1971).

51.6 Schizophrenie

Schizophrenie ist die Geisteskrankheit par excellence. Sie bringt Patienten und ihren Familien unermeßliches Leid und fügt der Gesellschaft durch Verlust von Produktivität riesigen Schaden zu. Sie ist das menschlichste aller Leiden, denn sie betrifft gerade die Eigenschaften, welche das menschliche am Menschen ausmachen: Denkprozesse und Sprache, die Fähigkeit zu nahen mitmenschlichen Beziehungen und das Selbstgefühl.

Da über die Neurobiologie dieser Kategorien geistigen Funktionierens nichts bekannt ist, ist die Behauptung verfrüht und geht in die falsche Richtung, daß es sich um eine progressive Erkrankung des Gehirns handelt. Dennoch suchen die Psychiater seit Griesinger weiter nach der Gehirnkrankheit, die der Schizophrenie zugrunde liegen soll, ohne sich auch nur die Frage zu stellen, was diese vermutete Störung denn an der Schizophrenie erklären soll.

Jeder Entwurf einer einheitlichen Theorie der Ätiologie und Pathogenese der Schizophrenie läuft Gefahr zu übersehen, daß sich hinter dieser diagnostischen Bezeichnung eine Vielfalt von Störungen verbirgt, deren Ursprünge und Ausbruchbedingungen voneinander abweichen. Um in Erscheinung zu treten, benötigt vermutlich jede einzelne Krankheitsform eine Kombination verschiedener Faktoren: Keine der Theorien wird allen Formen gerecht (solche Überlegungen sind in der Medizin schon öfters angestellt worden).

Die wahrscheinliche Heterogenität der Schizophrenie läßt die ätiologische Forschung scheitern. Noch schlimmer, die Meinungsverschiedenheiten über die diagnostischen Kriterien der Schizophrenie dauern an: Derzeit wird die Reliabilität der Diagnose betont und nicht ihre Bedeutung oder Validität. Es fehlen die Bemühungen, die Störungen der psychischen und Verhaltensfunktionen bei dieser Krankheit zu erfassen: Auf jeden Fall gelingt dies nicht durch Ausfüllen von Standardfragebogen, in welchen sie nicht enthalten sind! Ein Beispiel: Die Klage darüber, daß ein anderer Mensch oder eine Maschine die Gedanken des Patienten beeinflußt, blockiert, wegnimmt oder kontrolliert, kann Verschiedenes bedeuten – die Ablehnung, daß die eigenen Gedanken im Selbst entstehen, ein Versagen der Fähigkeit, die eigenen Gedanken von denen einer anderen Person zu unterscheiden, eine Verleugnung der eigenen Gedanken oder den Wunsch, von anderen kontrolliert zu werden.

Ein charakteristischer formaler Aspekt schizophrenen Denkens ist das Fehlen von Grenzen – eine Art von Begriffsvertauschung, bei der den Beziehungen zwischen einem Objekt und seiner symbolischen Vorstellung metaphorische Bedeutung zugeschrieben werden: Ein Teil eines Objektes vermag dann das ganze Objekt zu repräsentieren. Daraus folgt, daß der Schizophrene seine Welt nach eigenen Maßstäben einteilt. Der Kliniker aber scheint sich über diese Art und Weise, der Welt Bedeutung zu erteilen, hinwegzusetzen und behauptet einfach, daß der Patient „unlogisch", inkohärent oder bizarr denkt!

Noch bedeutsamer als das Fehlen von Grenzen ist die idiosynkratische und individuelle Art der Denkprozesse und der logischen Verknüpfungen des schizophrenen Menschen, die bei etwa 40% der erwachsenen Patienten dem offenen Ausbruch der Krankheit vorangeht oder nach ihrem Abklingen bestehenbleibt (Harrow und Marengo, 1986; Parnas und Schulsinger, 1986). Bei anderen Patienten (37%) kommt es nach der ersten Krankheitsperiode zu wiederholten Phasen idiosynkratischen Denkens. Das Anhalten solcher Denkstörungen (die zu einer unverständlichen persönlichen Sprache führen können) behindert offensichtlich die sprachliche Kommunikation zwischen dem Patienten und seiner Umgebung und auch seine Zusammenarbeit mit ihr (Marengo und Harrow, 1987).

Viele (vielleicht sogar der Großteil) der schizophrenen Patienten sind im wesentlichen asoziale Menschen („isoliert", „zurückgezogen" oder übermäßig scheu usw.) und viele von ihnen waren schon von früher Jugend an so. Diese Beobachtung legt nahe, daß die Fähigkeit, ein sozialer Mensch zu werden, sich nie entwickelte. Andere schizophrene Menschen geben (unbefriedigende) soziale Beziehungen aus einer Vielzahl von Gründen auf.

Entgegen derzeitigen Auffassungen gehören die Störungen der Fähigkeit zu sozialen Beziehungen zu den häufigsten diagnostischen Kernstörungen der Schizophrenie (Lehmann und Cancro, 1985). Sie stehen bei der paranoiden Form weniger im Vordergrund der Störung, die durch Verfolgungswahn, Größen- und/oder Eifersuchtsideen und/oder hypochondrische Befürchtungen mit oder ohne Halluzinationen gekennzeichnet ist. Die paranoide Schizophrenie kann abrupt oder allmählich auftreten und sich aus einem akuten undifferenzierten oder katatonen Beginn entwickeln. Häufig tritt sie später (im 4. Lebensjahrzehnt, besonders bei Frauen) auf. Patienten mit paranoider Schizophrenie leben in einer Welt, die mit Feinden, Konkurrenten und Rivalen bevölkert ist. Sie verdächtigen deshalb andere verwerflicher Motive. Männer vor allem befassen sich mit Fragen der Macht – von anderen Männern dominiert zu werden oder sie zu dominieren. Sie benehmen sich häufig feindselig in einer feindseligen Welt. Sie hüten sich vor ihr. Feindseligkeit bei anderen und Zurückweisung durch sie registrieren sie außerordentlich empfindsam.

Am Anfang der paranoiden Schizophrenie steht häufig der Vergiftungswahn oder die wahnhafte Furcht, ermordet zu werden usw., was meistens bedeutet, daß der Patient das Gefühl hat, eine wichtige Bezugsperson sei daran interessiert, ihn loszuwerden. Der Wahn repäsentiert häufig die Unfähigkeit des Patienten, anderen gegenüber zuzugeben, wie sehr der Patient nach einer Zurückweisung, einem Verlust oder einer Trennung den anderen vermißt.

Ob die paranoide Schizophrenie eine Untergruppe eines heterogenen Syndroms darstellt, ist unklar, obwohl einige Fachleute dies annehmen.

Die verschiedenen klinischen Formen von Schizophrenie (Kombinationen von Symptomen und Verhaltensweisen), ihr variabler natürlicher Verlauf und die verschiedenen Endzustände sind nicht nur das Produkt der Störung. Der Verlauf und die Prognose hängen auch vom sozioökonomischen Status des Patienten ab, der Haltung der Familienmitglieder, der Anwesenheit oder dem Fehlen von Familie und Freunden, Arbeitslosigkeit, Armut, sozialer Isolation, Hospitalisation, Vorliegen oder Fehlen von kompetenter Hilfe (sozialer, technischer, beruflicher Art) usw. (Gruenberg, 1967; Day et al., 1987).

So stellt sich folgende Frage: Wie kommt es zur Entwicklung eines Menschen, der nie soziale Beziehungen entwickelt oder ein asozialer oder andere verdächtigender Mensch wird, dessen Denken und dessen Logik eigenen persönlichen Regeln folgen und der unfähig ist, mit seinen Mitmenschen zu kommunizieren? Auf diese Frage kennen wir nur Teilantworten mit Erklärungen, die sich auf Genetik, frühkindliche Erfahrungen, Familienbedingungen, anatomische und physiologische Faktoren stützen.

51.6.1 Kindheitsanamnese und Prädispositionen

Die Kindheitsanamnese erwachsener Schizophrener

Es gibt keine Königsstraße zum Verständnis der Krankheit Schizophrenie. Verschiedene Routen werden vorangetrieben.

- Ungefähr 40% aller schizophrenen Patienten waren schon Jahre vor Krankheitsausbruch ruhige, einsame, gehorsame und nachgiebige („gute"), tagträumende, oft „passive" Kinder mit wenigen oder überhaupt keinen Freunden, besonders während der Adoleszenz. Sie leisten oft (nur) in einem Fach Überdurchschnittliches. Im Verlauf der Jahre nimmt ihre Schulleistung allmählich ab und eine Lücke klafft zwischen ihren Leistungen und ihren Begabungen. In der Adoleszenz vermeiden sie soziale Aktivitäten, lieben es zu lesen oder Musik zu hören und ziehen sich nach Möglichkeit auf sich selber zurück. Sie erscheinen der Umwelt desinteressiert und ohne Ziele.

  Sie sind auf Aufgaben, wie das Verlassen der Familie, ein Studium oder ein Handwerk zu erlernen oder zu heiraten, schlecht vorbereitet und oft erkranken sie, wenn diese Aufgaben sich ihnen stellen. (Das Haupterkrankungsalter an Schizophrenie ist etwa das 20. Lebensjahr.) Andere setzen ihr einsames, asoziales Leben einfach fort, bleiben häufig ohne Arbeit und sinken allmählich auf die tiefste sozioökonomische Stufe ab, leben am Rande der Gesellschaft als heimatlose Menschen, Vagabunden oder Prostituierte und treiben sich durch die nächtlichen Straßen. Solche Menschen sind also von früher Jugend an im Temperament anders oder werden später anders und bleiben lebenslang „schizoid" oder leiden an Schizophrenia simplex.
- Einige zukünftige schizophrene Patienten klammern sich als Kinder an ihre Mütter, schlafen im elterlichen Schlafzimmer bis in die späte Adoleszenz, sind Bettnässer und werden ängstlich, wenn sie von zu Hause und ihren Schlüsselfiguren entfernt sind. Die Trennung von ihren Müttern löst die Krankheit aus.
- Einige Kinder waren streitsüchtig, destruktiv, grausam, willensstark, rechthaberisch, ängstlich und reizbar. Sie schwänzten die Schule, benahmen sich asozial und wurden in ihrer frühen Adoleszenz wegen Vergehen oder Verbrechen ins Strafregister eingetragen (Hartmann et al., 1984; O'Neal and Robins, 1958). Werden diese Jugendlichen schizophren, dann in der paranoiden Form.
- Kinder, die in „schlechter Nachbarschaft" aufwachsen, Gefahren, Grausamkeiten und Verletzungen ausgesetzt sind, werden wachsam und beginnen, andere zu verdächtigen.
- Kinder, die von ihren Eltern die Motive andere zu verdächtigen übernehmen, lernen die Welt als bedrohlich und Mitmenschen als ihre natürlichen Feinde wahrzunehmen.

Die verschiedenen zu Schizophrenie führenden Wege können ganz unterschiedliche Ursprünge haben: Der gleiche schizophrene Phänotyp kann auf verschiedene Weise entstehen – einige Phänotypen mögen tatsächlich Phänokopien und andere Genotypen Genokopien sein.

Die Ätiologie der zu Schizophrenie führenden kindlichen Verhaltensmuster

Eine Vielfalt von Beobachtungen bezieht sich auf diese zentrale Frage. Sie verknüpfen nicht immer das Verhalten des Kindes (wie oben beschrieben) mit bestimmten Erfahrungen innerhalb der Familie (deren Mitglieder schizophren sein können), mit der sozialen Umwelt oder der genetischen Ausstattung.

Die Beobachtungen lassen sich in sechs Hauptgruppen gliedern:

- Wie erwähnt kann die soziale Umwelt, in der das Kind aufwächst, unvorhersagbar, bedrohlich oder entmutigend sein (Zubin et al., 1983).
- Die Familie kann zu viel fordern, kritisch und vorwurfsvoll sein. Das Kind findet sich ständigen Leistungsanforderungen in der Schule oder sozial usw. ausgesetzt. Diese Erwartungen stehen aber im Gegensatz zu den Befürchtungen, Ängsten und Fähigkeiten des Kindes (Hogarty, 1975; Katz und Lyerly, 1963).
- Die Familienmitglieder können offen feindselig, belästigend, zudringlich, rechthaberisch und kritisch sein (diesem Verhalten ist der Name „ausgedrückte Emotionen"! verliehen worden). „Ausgedrückte Emotionen" (Brown et al., 1972) sind mit Rückfällen und Wiederauftreten schizophrener Phasen in Zusammenhang gebracht worden (Koenigsberg und Handley, 1986; Leff und Vaughn, 1985). Es liegen keine Befunde vor, daß solche Belästigungen vor dem Ausbruch der Störung vorliegen? warum nicht? Es verlangt wenig Vorstellungskraft anzunehmen, daß das Verhalten von Familienmitgliedern über die Zeit gleichbleibt und sich nicht erst nach Ausbruch des Leidens ändert. Der zukünftige schizophrene Patient ist sehr wahrscheinlich gegenüber Feindseligkeit und Kritik auch anderer Menschen, und nicht nur seiner Familienmitglieder, sensibel gemacht worden. Tatsächlich sieht die klinische Erfahrung so aus: Die übliche Reaktion des schizophrenen Patienten besteht im Rückzug, in dem Gefühl belästigt und bekämpft zu werden, und/oder er beginnt zu halluzinieren (Beels und McFarlane, 1982).

  Häufige Kritik, die auch einen Wesenszug des Verhaltens der Familienmitglieder schwer depressiver Patienten darstellt, läßt den Rückfall voraussagen. Sie ist aber weniger intensiv (Leff und Vaughn, 1985) als in Familien schizophrener Patienten. Der Unterschied zwischen den beiden klinischen Gruppen liegt in der Art, wie sie verhaltensmäßig auf „ausgedrückte Emotionen" reagieren.
- Schizophrenie tritt in einigen Familien gehäuft auf. 3–15% der Verwandten ersten Grades (eingeschlossen die Geschwister) von schizophrenen Patienten sind auch schizophren. Sowohl Zwillings- als auch Adoptionsstudien stützen die Rolle geneti-

scher Faktoren in der Übermittlung und Häufigkeit von Schizophrenie. Diese Rolle ist aber nicht eindeutig: Die meisten zukünftigen schizophrenen Patienten sind nicht adoptierte Kinder. Diese Untersuchungen müssen sich deshalb mit der Möglichkeit abfinden, daß der zukünftige Patient mit einem schizophrenen Verwandten zusammengelebt hat.

Zusätzlich hat sich die Forschung vor allem mit Familien befaßt, in denen die Schizophrenie bei mehr als einem Mitglied vorkommt. Bei der überwiegenden Zahl von Familien aber, d.h. bei 75 bis 80%, ist nur ein Mitglied schizophren. So könnte man folgern, daß Schizophrenie meistens „sporadisch" auftritt; diese Schlußfolgerung kann, muß aber nicht berechtigt sein, weil ein (nicht-schizophrener) Elternteil eines Patienten das Gen (die Gene) für das Leiden weitergeben könnte (Gottesman, 1987).

In Familien von schizophrenen Patienten sind viele verschiedene Formen von Psychopathologie beobachtet worden. Es wird aber als unzulässig betrachtet, von einer Studie zur anderen zu verallgemeinern und von derzeitigen Beobachtungen in der Zeit rückwärts zu schließen.

- Zwischen den Veränderungen der formalen Aspekte von Denken und Sprache bei Eltern und Kindern besteht ein enger Zusammenhang: Spezifische Arten von Vagheit, bedeutungslose Bemerkungen, Sinnlosigkeit oder fehlende Fokussierung in Denken und Sprache kennzeichnen jede Familie oder Paare von Familienmitgliedern (Singer und Wynne, 1965). (Diese Beobachtungen wurden auf verschiedene Weise interpretiert.)
- Die Kinder schizophrener Mütter zeigten verglichen mit denjenigen nicht-schizophrener Mütter ganz eigene Antworten und eine schwächliche Konzeptbildung in kognitiven Tests. Die häusliche Atmosphäre war stürmisch. Die Mutter wurde vom Kind als unzuverlässig und immer zum Schimpfen bereit erlebt. Die Kinder fühlten sich einsam und zogen sich zurück, wenn sie durch das mütterliche Verhalten beunruhigt waren. Vor Aufgaben gestellt, strengten sie sich nicht an. Wurden die Mütter ins Spital gebracht, dekompensierten etwa 20% dieser Kinder.

In bezug auf diese Beobachtungen ergeben sich sofort eine Reihe konzeptueller Probleme: Welche charakteristischen Verhaltensweisen und Einstellungen des Kindes, das später zum schizophrenen Patienten wird, sollten sie erklären? Es gibt keine einheitlichen Merkmale. Wenn in einer Familie mit mehr als einem Kind nur eines krank wird, dann, so legen die Beobachtungen nahe, wurde es von den Eltern im Unterschied zu den anderen Kindern ungünstig behandelt. Diese Überlegung mag richtig sein. Es gibt aber keine Beweise dafür.

Psychophysiologische Prädispositionen

Eine Hypothese zur Prädisposition für Schizophrenie geht dahin, daß disponierte Menschen Information anders verarbeiten als diejenigen, die kein solches Risiko tragen: Bei Aufgaben, die ein beträchtliches Maß an fortwährender Aufmerksamkeit verlangen, schneiden Menschen mit hohem Schizophrenie-Risiko schlechter ab (Neuchterlein und Dawson, 1984). Nach einer zweiten Hypothese zeigen gewisse Kinder schizophrener Eltern intensivere und kürzere Latenz elektrodermaler Reaktionen (galvanic skin response) auf mäßig lauten Lärm als Kinder ohne Risiko (Mednick und Schulsinger, 1968). Diese Beobachtungen konnten in späteren Studien nur zum Teil bestätigt werden; mitunter wurde sogar Hyporeaktivität gemessen (Kugelmass et al., 1985).

Die Ergebnisse dieser Studien sind dahingehend interpretiert worden, daß Menschen mit Risiko für Schizophrenie „autonom" überreagieren. Elektrodermale Aktivität ist aber ein grobes und schlechtes Maß für allgemeine autonome Funktionen. Zudem müßte belegt werden, daß hyperreaktive Menschen später schizophrene Patienten werden; solche Langzeitstudien sind aber nicht durchgeführt worden. Die Tatsache, daß einige Risikopersonen hyper- und andere hyporeaktiv sind, weist lediglich auf Unterschiede hin, welche Menschen aufweisen, die ein Risiko für Schizophrenie in sich tragen: Es sind eben diese Unterschiede, welche die Erforschung dieser Krankheiten so sehr erschweren.

Extreme Reaktionsunterschiede charakterisieren aber auch akut schizophrene und schizophrene Patienten zwischen den Schüben. Dies gilt nicht nur für die galvanische Hautreaktion (Öhman, 1981), sondern auch für andere Meßwerte. Schizophrene Patienten sind in verschiedenen Beziehungen Außenseiter: Faßt man sie zu einer Gruppe zusammen, dann unterscheiden sich zwar ihre Durchschnittsreaktionen nicht von denjenigen normaler Vergleichsstichproben, wohl aber die Varianz der Reaktionen. Die wirkliche Bedeutung dieser Beobachtung ist nie vollständig geklärt worden. Sie könnte die Grundlage eines Forschungsprogramms abgeben, das autonome (und andere) Meßwerte mit jedem Aspekt der Vorläufer, des Verlaufs, der Verschlimmerung und auch der Erholung vom Leiden verbinden würde. Auf diese Weise könnten auch Unterschiede zwischen den heterogenen Unterformen herausgearbeitet werden (Dawson und Neuchterlein, 1987).

Ein weiteres psychophysiologisches Merkmal zwischen schizophrenen Patienten und ihren nicht-schizophrenen Familienmitgliedern stellt eine Störung der Kontinuität in der Abfolge der Augenbewegungen dar (Holzman et al., 1973): 52–86% aller Schizophrenie-Patienten folgen einem pendelförmig bewegten Gegenstand mit sakkadierten Bewegungen. Eine Minderzahl (7%) der Allgemeinbevölkerung zeigt die gleiche Störung. 35% der Eltern schizophrener Patienten zeigen sie auch. Beim monozygoten Zwilling findet sich eine 71% betragende Konkordanz für diese Abnormität. Bei dizygoten Zwillingen beträgt sie 54%. Diese Erscheinung ist aber nicht spezifisch für Schizophrenie, denn sie kommt in 15–40% bei schweren affektiven Störungen auch vor (Holzman et al., 1974; Solomon et al., 1987) und ebenfalls bei der

„schizotypischen" Persönlichkeitsstörung (Siever et al., 1984).

Die Funktionsstörung sakkadierter Augenbewegungen vererbt sich autosomal-dominant mit beträchtlicher Penetranz. Sie ist der phänotypische Ausdruck einer latenten Eigenschaft, die sich unabhängig von Schizophrenie oder mit ihr manifestiert (Matthysse et al., 1986).

Es ist ganz unklar, wie diese Funktionsstörung zur Schizophrenie und ihrer Psychopathologie beiträgt. Sie korreliert nicht signifikant mit formalen Denkstörungen (Solomon et al., 1987) und auch nicht mit der perzeptuellen Unaufmerksamkeit. Sie ist für die Krankheit nicht spezifisch, sie stellt höchstens einen „Marker" für die Risikopopulation bei Schizophrenie (und schweren affektiven Störungen) dar.

Im folgenden seien die Hauptpunkte dieser Übersicht zusammengefaßt: Die Schizophrenie ist ganz deutlich keine homogene Krankheit, und es führt kein einheitlicher Weg zu ihr. Viele Züge der klinischen Erkrankung – Passivität, eine formale Denkstörung, eingeschränkte soziale Beziehungen oder schlechte Steuerung aggressiver Impulse, die zu asozialem Verhalten führen – gehen der Krankheit voraus (Parnas, 1986). Diese Merkmale finden sich besonders bei Kindern, in deren Familien ein Mitglied (meistens die Mutter) schizophren ist. Es sei aber festgehalten, daß Kinder schizophrener Mütter nicht notwendigerweise alle die gleichen Verhaltensweisen und Eigenschaften aufweisen. Die Auswirkungen einer solchen Vererbung sind also nicht linear!

Kinder mit diesen Zügen sind für das Erwachsenenleben schlecht ausgerüstet. Diese Persönlichkeitszüge können genetisch verankert sein oder sich in der frühen Kindheit im Rahmen einer Vielfalt ungünstiger Umgebungen entwickeln. Sie weisen auf die einzigartige Verletzlichkeit und Anpassungsweise vieler schizophrener Patienten hin.

51.6.2 Ätiologie

Diese Beobachtungen bedürfen der Erklärung: Im Augenblick ist eine rein genetische Erklärung modern. Im besten Falle läßt sich diese einseitige Hypothese aber – und auch darum mit einigem Zögern – in eine Erklärung einbauen, welche die Prädisposition „Schizophrenie" auf Gene und den Ausbruch der Krankheit auf „Streß" zurückführt. Andere Erklärungen sind aber auch möglich.

Epidemiologische und soziale Faktoren

Viele Jahre lang galt die Prävalenzrate der Schizophrenie weltweit als konstant (etwa 1% der Bevölkerung). Diese Zahl fügte dem Rätsel dieser Störung ein weiteres hinzu. Wie Torrey (1987) es ausgedrückt hat: Keine ätiologische Theorie, wie sie auch gelautet haben möge, hätte diese Invarianz vorausgesagt, auch dann nicht, wenn man alle methodologischen Probleme solcher Studien vernachlässigt hätte. Die Tatsache bleibt bestehen, daß Prävalenzraten (Inzidenzraten

benehmen sich viel zufälliger) weiten Schwankungen unterworfen sind, und zwar bis zu einem Faktor von ungefähr 50! Bei den Amischen in Pennsylvania beträgt die Prävalenzrate für Schizophrenie 0,03% (Egeland und Hostetter, 1983), in einer abgeschiedenen Gemeinde im nördlichen Schweden 1,7% (Böök et al., 1978). In Gegenden, die so weit auseinanderliegen wie Taiwan (Rin und Lin, 1962) und Teile von Irland (Torrey et al., 1984), sind die Vorkommensraten außerordentlich unterschiedlich. Festgehalten sollte aber werden, daß die Prävalenzzahl in Taiwan die kulturelle Gruppe der Ureinwohner betrifft, die sich stark von der chinesischen Bevölkerung dieser Insel unterscheidet.

In diesen Daten sind wichtige Tatsachen mit tiefgreifender Bedeutung enthalten: Wählt man für epidemiologische Untersuchungen isolierte Populationen aus, bei denen Inzucht herrscht, findet man natürlich extreme Prävalenzen (die neuesten Daten über Molekulargenetik von schweren affektiven Störungen und Schizophrenie stammen aus Studien eben dieser Populationen). Inzucht vergrößert die Wahrscheinlichkeit der Erkrankung an diesen Störungen: Kommen ein bestimmtes Gen oder bestimmte Gene vor, so finden sie sich in solchen Bevölkerungen (Böök et al., 1978). Werden Studien über Morbidität und Prävalenz großer Populationen vorgenommen, gehen die feinen Unterschiede zwischen besonderen Bevölkerungsgruppen verloren: Neue amerikanische Studien über die Prävalenz nach einem Monat, nach sechs Monaten und während des ganzen Lebens bei insgesamt 18571 Menschen in fünf städtischen Populationen, mit 0,6%, 0,8% und 1,3% (Regier et al., 1988), beleuchten diesen Punkt.

Andere epidemiologische Studien konzentrieren sich auf die Verlaufsformen diagnostizierter Schizophrenien (Torrey, 1987) und auf die Häufigkeit, mit der die Störung bei Patienten aus verschiedenen sozioökonomischen Schichten vorkommt. Über letzteren Punkt ist keine Einigkeit erzielt worden. In den USA und Großbritannien (Freeman und Albert, 1986) ist die Prävalenzrate nur in den armen Schichten der mittelgroßen und großen Städte am höchsten, aber nicht der kleinen (Übersicht bei Weiner, 1985). In Indien trifft aber das Gegenteil zu: Das Vorkommen ist in höheren sozioökonomischen „Klassen" am höchsten (Saxena et al., 1972).

Genetische Theorien zur Ätiologie

Die intensivsten Forschungsbemühungen haben bisher die Aufdeckung einer erblichen Natur der Schizophrenie zum Ziel gehabt. Diese Art Forschung gliedert sich in drei Hauptkategorien: Blutsverwandtschafts-, Zwillings- und Adoptionsstudien. Die erste konzentriert sich auf das Vorkommen der Störung als Funktion des Verwandtschaftsgrades. Sie beträgt in verschiedenen Untersuchungen bei Verwandten ersten Grades (Eltern, Geschwister) 3–14%; ist ein Elternteil schizophren, steigt die Prävalenz auf 8–18%, sind beide Eltern betroffen, erreicht sie 15–55% (Rosenthal et al., 1972).

Die Konkordanzrate für Schizophrenie in 5 Zwillingsstudien betrug 35–58% (durchschnittlich 47%) bei monozygoten und 9–26% (15%) bei dizygoten Zwillingen (Gottesman und Shields, 1972). Diese Zahlen gestatten die Errechnung eines Vererbungsverhältnisses von ungefähr 3:1.

Adoptionsstudien liefern weitere gewichtige Argumente für die Heredität der Schizophrenie. Sie weisen darauf hin, daß es gleichgültig ist, ob man die Prävalenz der Schizophrenie bei Kindern untersucht, die nach der Geburt von einer schizophrenen Mutter abgegeben wurden, oder ob man die Prävalenz bei den Eltern untersucht, die ihre Kinder zur Adoption freigegeben haben, bevor die Krankheit bei den Eltern ausbrach – die Ergebnisse sind ähnlich (Kety et al., 1975; Rosenthal et al., 1972): Eine erhöhte Prävalenz für Schizophrenie (großzügig definiert) findet sich bei adoptierten Kindern. Diese Beobachtungen werfen die Frage nach dem Erblichkeitsmodus auf. Das wahrscheinlichste Modell ist ein multifaktoriell-genetisches (Schwellen-) Modell: Die Neigung für Schizophrenie wird genetisch vermittelt. In diesem Modell erklären die Gene 74% der ätiologischen Varianz, der Rest der Varianz beruht auf unidentifizierten (möglicherweise zufälligen) Umgebungsfaktoren (Gottesman und Shields, 1982; McGue et al., 1983). Die Umgebungsvariablen spielen eine größere Rolle, wenn die Diagnose Schizophrenie nicht ganz sicher ist.

Es ist eine Sache zu behaupten, Gene würden den Phänotyp bestimmen, es ist aber eine ganz andere zu behaupten, sie würden das Risiko oder die Wahrscheinlichkeit für die Krankheit erhöhen: Sogar monozygote schizophrene Vierlinge sind phänotypisch in keiner Beziehung gleich (DeLisi et al., 1984). Bei der Schizophrenie gibt es tatsächlich keinen bekannten Phänotyp (Cancro, 1985). Diese Feststellung wird durch die Beobachtung bekräftigt, daß 74% von 129 Müttern, deren Kinder zur Adoption freigegeben worden waren, 10 Jahre oder noch mehr danach als paranoid-schizophren diagnostiziert wurden. Die Wahrscheinlichkeit, daß ihre Nachkommen schizophren wurden, war aber beträchtlich kleiner als bei Kindern, deren Mütter andere schizophrene Störungen hatten (Jorgensen et al., 1987).

Wir wissen also im Grunde nicht, worin das Risiko eigentlich besteht. (Diese Feststellung entspricht derzeitigen psychosomatischen, theoretischen Vorstellungen.) Es braucht sich jedenfalls nicht phänotypisch auszudrücken: Der normale dizygote Zwilling eines schizophrenen Patienten gibt die Störung gleich häufig an seine Nachkommen weiter wie monozygote Probanden und ihr für Schizophrenie diskordanter Zwilling (Gottesman, 1987).

Einige Vorbedingungen für Schizophrenie sind oben erwähnt worden. Sie sind nicht einheitlich bei allen Patienten oder Stammbäumen vorhanden. Zusätzliche Faktoren könnten auf eine Verletzlichkeit gewisser Patienten hinweisen: Niedrige Plasmaspiegel des Enzyms Aminooxidase und verminderte Aktivität des Enzyms Monoaminooxidase in Blutplättchen (Baron, 1986).

Was diese peripheren Erscheinungen aber für das Gehirn bedeuten, ist unbekannt.

Zu den faszinierenden Möglichkeiten, die mit Vorbedingungen der Krankheit zusammenhängen mögen, gehören Entwicklungs- oder Reifungsdefekte des Gehirns (Weinberger, 1987), die nicht unbedingt genetisch verankert sein müssen, sondern mit intrauterinen Bedingungen, Hirnschäden bei der Geburt (Reveley und Murray, 1984) oder mit Lebenserfahrungen zusammenhängen mögen. Die erfolgversprechendste Forschungsstrategie widmet sich der Entwicklung und kombiniert anatomische und Persönlichkeitsaspekte. Die Suche nach einem einzelnen Faktor – sei er genetisch, enzymatisch, viral, hormonal usw. – wird die Rätsel der Schizophrenie kaum lösen können: Die zur Lösung des Rätsels vorgeschlagene Untersuchungsstrategie ist komplex.

Einige wenige Hinweise für einen Reifungs- oder Entwicklungsdefekt im Gehirn mindestens einiger schizophrener Patienten liegen vor: Er besteht in einer mangelhaften Organisation der Dendriten (vor allem ihrer apikalen Ausläufer) von Pyramidenzellen im Hippocampus. Typischerweise bilden diese Zellen einen Fächer, der senkrecht auf dem Ammonshorn steht; diese Anordnung entsteht im Laufe der Entwicklung. Die entsprechenden Befunde wurden von Kovelman und Scheibel (1986) und Conrad und Scheibel (1987) an einer kleinen Anzahl von Patienten gewonnen, von denen viele unter Medikamenten standen. Dennoch deuten solche Beobachtungen an, daß ein Entwicklungsmangel in der Anatomie eines der kritischen Hirnbezirke vorliegen könnte. Vermutlich besteht dieser Entwicklungsmangel in einer fehlerhaften neuronalen Wanderung und Ausrichtung in einer kritischen Phase der embryonalen Entwicklung. Dies mag auf einer Anzahl von Vorbedingungen beruhen.

Zu diesen Beobachtungen paßt die Feststellung einer verminderten Anzahl von Neuronen um den dritten und die Seitenventrikel sowie ihr Ersatz durch Glia bei schizophrenen Patienten (Kirch und Weinberger, 1986). Diese Beobachtung könnte sich als bedeutsam erweisen, weil aus dem Gebiet um die Ventrikel herum Neurone stammen, die im Verlaufe der fetalen Entwicklung in den Hippocampus und Cortex auswandern (Rakic, 1972).

Es gibt andere ähnliche Hypothesen über die bekannten Wirkungen von Genen und Umwelt auf die Hirnentwicklung (Haracz, 1984). Es klaffen aber noch beträchtliche Wissenslücken zwischen den anatomischen Beobachtungen und ihrer Bedeutung für die Verhaltensveränderungen von Kindern mit Risiko für Schizophrenie und Patienten, die bereits erkrankt sind.

51.6.3 Pathophysiologie und Pathobiochemie

Pathogenese der Schizophrenie

Wie bereits geschildert, sind schizophrene Patienten für die Bewältigung der Alltagsrealitäten des Erwachsenenlebens schlecht ausgestattet. Sie sind vor allem äußerst empfindsam für Trennungen und Verluste, Zurückweisungen durch andere (Brown und Harris, 1986), die Übernahme von Verantwortung, Verständnismangel und Feindseligkeit von Mitmenschen (Cancro, 1985).

Keine dieser Erfahrungen ist aber besonders spezifisch für Schizophrenie; sie finden sich auch bei vielen anderen Krankheiten und Störungen (Weiner, 1977, 1987). Sie erlangen für Schizophrene aber weitreichende und sehr persönliche Bedeutungen und lösen ganz spezifische Formen von Antworten aus – z.B. Rückzug oder den Glauben, daß andere Menschen versuchen würden, sie zu beseitigen.

Bei Prädisponierten kann die Störung durch viele zusätzliche Faktoren – Alkohol, LSD, Amphetamine, Cannabis – ausgelöst werden. Sie kann auch im Gefolge des Erlebens von Brutalität oder Vergewaltigung ausbrechen und dem betroffenen Menschen den Verdacht bestätigen, daß die Umwelt gefährlich ist und gemieden werden sollte.

Solche kasuistischen Beobachtungen bedürfen der Bestätigung. Der Sachverhalt ist aber deshalb so komplex, weil schizophrene Patienten sich voneinander unterscheiden und ihre Angaben oft schwer zu verstehen sind. Immerhin treten die Erstkrankheit oder der Rückfall meistens innerhalb weniger Tage oder Wochen nach einem besonders bedrohlichen Lebensereignis auf (Day, 1981; Leff und Vaughn, 1985).

Das Modell für Schizophrenie umfaßt also die Annahme eines verletzlichen Menschen, der mit Aufgaben konfrontiert wird, zu deren Bewältigung er nicht ausgerüstet ist, und der dazu neigt, auf Belastungen mit einem sehr eigenartigen Rückzug in eine ganz eigene Welt zu antworten oder reale oder phantasierte Feinde zu bekämpfen.

Pathoanatomie und Pathophysiologie der Schizophrenie

Unzählige Versuche sind unternommen worden, die Ätiologie und Pathogenese der Schizophrenie auf dem Boden einer Vielfalt einzelner Variablen zu erklären (Weiner, 1985). Keiner war erfolgreich. Solche Bemühungen verwechseln korrelierende Faktoren mit vorangehenden Ursachen; deshalb versagen sie als Erklärungen.

Anatomische In-vivo-Studien

Die Entwicklung bildgebender Verfahren im Bereich des Zentralnervensystems (Computertomographie [CT], magnetische Resonanz [MRI], Positronen-Emissionstomographie [PET] und regionale zerebrale Durchblutung) ermöglicht es Forschern, Veränderungen von Struktur und Funktion des Gehirns bei Schizophrenen und Patienten mit schweren affektiven Störungen im Vergleich zu Normalpersonen zu untersuchen.

Von den technischen Problemen beim Gebrauch von CT-Apparaten (Maser und Keith, 1983) und der Auswahl passender Kontrollpersonen (z.B. Weiner,

1985) abgesehen, weisen die Ergebnisse der meisten von etwa einem Dutzend Untersuchungen bis jetzt darauf hin, daß bei schweren affektiven Störungen und Schizophrenie ungefähr gleich viele Abnormitäten (10–30%) vorliegen dürften. In einer gewissen Zahl beider Patientengruppen ist das Verhältnis Ventrikel zur Gehirnmasse erhöht, die Hirnfurchen sind erweitert und/oder der Kleinhirnwurm ist geschrumpft (Jeste et al., 1988).

Die Bedeutung dieser Befunde ist unklar. Sicher sind sie nicht spezifisch für Schizophrenie. Sie können auch bei normalen Personen vorkommen (3,7%) (Nasrallah et al., 1982), besonders mit zunehmendem Alter.

Untersuchungen des Hirnglukosestoffwechsels bei schweren affektiven Störungen (Jeste et al., 1988) und Schizophrenie (Sheppard et al., 1983) mit PET sind ebenso erstaunlich wie widersprüchlich. Bei kleinen Gruppen akutkranker, unbehandelter schizophrener Patienten fanden sich keine metabolischen Unterschiede; Medikamentengabe veränderte den Glukosestoffwechsel vor allem in den frontalen Hirnabschnitten (Widen et al., 1983).

Mit der Weiterentwicklung der PET-Technik ist es möglich geworden, einige Neurotransmitter mit Positronen zu markieren und ihre Rezeptoren abzubilden (Sedvall et al., 1986). Bei schizophrenen Patienten ist besonders der Dopaminrezeptor untersucht worden, wegen der (kontroversen) Beobachtung erhöhter Dopaminrezeptordichte im Gehirn solcher Patienten bei der Autopsie (Seeman et al., 1984). Die Bedeutung dieses Befundes ist unklar: Es kann sich um ein Artefakt bei Neuroleptika-Medikation handeln (Mackay et al., 1982). Mittels PET ist eine 2- bis 3fach größere Dichte des Dopaminrezeptors in den Basalganglien unbehandelter Patienten festgestellt worden (Wong et al., 1986). Diese Beobachtung konnte von Farde et al. (1987) aber nicht bestätigt werden.

Wir wissen nicht, was diese kontroversen Befunde bedeuten. Selbst wenn sie bestätigt würden, bleibt die Frage unbeantwortet, was sie über die Schizophrenie auszusagen vermögen. Relationen zwischen positiven und negativen Symptomen sowie der Prognose und den zerebralen Ventrikelgrößen und der Durchblutung sind untersucht worden (Berman et al., 1987). Wie kann dieser Forschungsansatz aber in Anbetracht der Unspezifität der erweiterten Ventrikel für Schizophrenie fruchtbar sein? Was die regionale Durchblutung betrifft, so verändert sie sich in der Hirnrinde deutlich, je nach dem durch Reize ausgelösten sensorischen Impulszustrom und der Verarbeitung dieses Impulszustroms. Im Rahmen von Untersuchungen der regionalen Durchblutung werden solche Variablen gewöhnlich nicht kontrolliert, so daß sie zu widersprüchlichen Ergebnissen führen können (Jeste et al., 1988).

Pathophysiologie der Schizophrenie

Auf der Suche nach dem „pathogenen" Faktor der Schizophrenie ist jeder Winkel der Physiologie, Biochemie, Enzymologie, Neurochemie, Immunologie,

Toxikologie usw. ausgeleuchtet worden (Übersicht bei Weiner, 1985). Diese Suche war unfruchtbar, denn sie stützte sich entweder auf ein antiquiertes unikausales Modell (das sich auf Veränderung von einzelnen Variablen ausrichtete und nicht auf ganze Muster von Variablen) oder auf die Annahme, daß die bekannte pharmakologische Wirkung eines Medikamentes die Pathogenese der Krankheit enthüllen werde. (Sagt uns die Pharmakologie von Digitalis irgend etwas über die Ätiologie oder Pathogenese verschiedener Vorläufer der Herzinsuffizienz?)

Dopamine

Wie die schweren affektiven Störungen ihre Amin-Hypothese, so besitzt die Schizophrenie ihre Dopamin-Hypothese (Meltzer und Stahl, 1976). Sie lautet wie folgt: Phenothiazine und Butyrophenone blokkieren die Dopaminrezeptoren in vitro; ihre relative Fähigkeit, sich im Striatum des Tierhirns an solche Rezeptoren zu binden, steht in Beziehung zu ihrer klinischen Wirksamkeit bei schizophrenen Patienten (Creese et al., 1976). Darauf stützt sich die Annahme, daß bei der Schizophrenie eine übermäßige Produktion und/oder Sekretion von Dopamin in der präsynaptischen Nervenendigung stattfindet, oder eine erhöhte „Empfindlichkeit" (Affinität für Dopamin) oder eine erhöhte Zahl der postsynaptischen Dopaminrezeptoren vorliegt. (Diese zwei Hypothesen schließen sich vordergründig aus: Übermäßige Neurotransmittersekretion führt zu einer „Tieferstellung" der Regulation der Rezeptoren und umgekehrt.) Trotzdem wurde die Beobachtung, daß Methamphetamin-HCl und Methylphenidat-HCl (die zur Ausschüttung von Dopamin und Noradrenalin aus den Neuronen führen) bei einigen offen schizophrenen Patienten die Symptome intensivieren, nicht aber bei Patienten zwischen den Schüben, als indirekte Unterstützung dieser Hypothese gewertet. (Andere Patienten aber erholen sich unter diesen Medikamenten!)

Direkte Belege für diese Hypothese sind nicht eindeutig gefunden worden. Untersucht wurde die Frage mittels Hirngewebe von verstorbenen Schizophrenen und mit PET-Studien am lebenden Patienten. Die gesteigerte Produktion und Sekretion von Dopamin im Striatum und limbischen Bezirken des Gehirns bei Schizophrenen wurde mittels Erfassung eines der Hauptmetaboliten des Dopamins – der Homovanillinmandelsäure – im Liquor erfaßt. In einem Abschnitt von 10 Jahren haben neun verschiedene Untersuchungen (Übersicht z.B. bei Maas et al., 1988) keine gleichsinnigen Veränderungen der Spiegel von Homovanillinmandelsäure oder Dopaminmetaboliten im Liquor schizophrener Patienten gezeigt. (Wären niedrige Homovanillinmandelsäure-Spiegel gefunden worden, hätte dies der Hypothese der übermäßigen Dopaminsekretion widersprochen, ganz abgesehen von der Tatsache, daß diese Beobachtung nicht spezifisch für Schizophrenie sein muß [Bowers und Swigar, 1987; Davis et al., 1988].)

Diese widersprüchlichen Befunde sind damit erklärt worden, daß niedrige Homovanillinmandelsäure-Spiegel im Liquor nur bei einigen schizophrenen Patienten vorkommen – beispielsweise nur bei denjenigen mit erweiterten Ventrikeln im CT. Bei solchen Patienten wurden niedrige Spiegel anderer Homovanillinmandelsäure-Metaboliten – von Dihydroxyphenylacetsäure (DOPAS) und von konjugierter DOPAS – sowie von Homovanillinmandelsäure gefunden, während die Spiegel von Dopaminsulfat erhöht waren (van Kammen et al., 1986). Umgekehrt ausgedrückt: Sie spiegeln den Stand der Symptome und des Verhaltens des schizophrenen Patienten (Bowers, 1978). Zur Beurteilung der Homovanillinmandelsäure-Spiegel im Liquor oder Plasma müssen jedoch noch andere Faktoren berücksichtigt werden: Der Homovanillinmandelsäure-Liquor- oder Plasmaspiegel wird durch die vom Patienten eingenommene Diät (Kendler et al., 1983), die Tagesschwankungen und durch Geschlechtsunterschiede beeinflußt (Bowers, 1988).

Neuerdings wird Homovanillinmandelsäure im Plasma und Urin gemessen, obwohl bekannt ist, daß Dopamin auch aus anderen Quellen im Körper stammen kann als den sympathischen Neuronen und der Niere. Dennoch werden höhere Plasma-Homovanillinmandelsäure-Spiegel mit der „Intensität" der Symptome (Davidson und Davis, 1988) und mit dem Ansprechen auf Behandlung (Bowers, 1988) in Beziehung gebracht. Allen diesen einander widersprechenden Überlegungen zum Trotz, bleibt es jedoch völlig unklar, wie Plasma- und Urinkonzentration von Homovanillinmandelsäure unser Verständnis der Pathogenese der Schizophrenie fördern sollen. (Für weitere Argumente pro und contra die Dopamin-Hypothese vgl. Weiner, 1985.)

Noradrenalin

Der Schluß von der pharmakologischen Wirkung eines Medikamentes auf die Pathogenese einer Krankheit enthält Fallstricke, denn es gibt kein Medikament, das nur eine einzige Wirkung hat: Phenothiazine blockieren auch Noradrenalinrezeptoren. Dennoch sind Ergebnisse publiziert worden, die nahelegen, daß die Noradrenalinspiegel im Liquor und Hirnstamm von (paranoiden) schizophrenen Patienten erhöht sind (Hornykiewicz, 1982). Das Plasma-3-Methoxy-4-Hydroxy-Phenylglycol spiegelt aber das Ausmaß psychischer Störungen oder von Angst und nicht eine besondere Unterform der Schizophrenie (Ko et al., 1988).

Hypothalamus-Hypophysen-Nebennierenachse

In einer früheren, ausgezeichneten Untersuchung schwankten 17-Hydroxycorticosteroid und Adrenalin im Urin gemeinsam mit dem Verhalten und dem psychischen Zustand junger männlicher Schizophrener: Befanden sich die Patienten in einem stabilen Wahnzustand oder in einer Remission, waren die Werte niedrig. Waren sie ängstlich, in der Wahnstimmung bei Krankheitsausbruch oder depressiv in der Erholungsphase, lagen die Werte hoch (Sachar et al.,

1963). Die meisten Studien der letzten Zeit befassen sich mit dem Dexamethason-Suppressionstest. Die Kortisonspiegel können bei 20–30% der Patienten nicht supprimiert werden (Dewan et al., 1982; Targum, 1983). Wie schon erwähnt, ist die Bedeutung dieser Befunde fragwürdig.

Neuere Studien berücksichtigen nicht, daß das Serumkortisol wahrscheinlich mit dem klinischen Zustand des Patienten variiert. So werden die Kortisol- und ACTH-Serumspiegel und ihre Reaktion auf CRH alle als normal angegeben, beziehen sich aber auf die Diagnose und nicht auf den verhaltensmäßigen Zustand des Patienten (Roy et al., 1986).

Beta-Endorphin, Beta-Lipotropin usw.

Die Plasmaspiegel anderer Pro-Opiomelanocortin-(POMC-)Abkömmlinge bei Schizophrenie sind uneinheitlich. Die meisten Untersuchungen fanden keine Abweichungen von der Norm, eine einzige Untersuchung fand eine größere Variabilität der ACTH-, Beta-Endorphin- und Beta-Lipotropin-Spiegel als bei Vergleichspersonen. Die Spiegel der drei POMC-Abkömmlinge veränderten sich aber nicht parallel (Brambilla et al., 1984). In einigen Studien waren die Endorphinspiegel (chromatographisch erfaßt) im Liquor chronisch schizophrener Patienten erhöht, bei denen Neuroleptika abgesetzt worden waren (Lindström et al., 1986). Die hohen Endorphinspiegel waren mit tiefen Homovanillinmandelsäure-Spiegeln korreliert. Die Bedeutung solcher Korrelationen für eine Theorie der Schizophrenie bleibt aber schleierhaft.

Wachstumshormon (STH)

Die meisten Studien über STH-Sekretion bei Schizophrenie haben verschiedenartige Wege zur Stimulation benützt, sei es mit Dopaminagonisten (wie Apomorphin) oder Gammaaminobuttersäure-Agonisten (wie Natrium-Valproat). Die Resultate sind uneinheitlich. Unter Dopaminagonisten glichen sich die STH-Reaktionen bei schweren affektiven Störungen und Schizophrenie, übertrafen aber diejenigen bei freiwilligen Versuchspersonen. Bei Schizophrenie standen sie mit der Krankheitsdauer und den negativen Symptomen in Zusammenhang (Meltzer et al., 1984). Die Dopamin-Hypothese wurde dadurch nicht gestützt. In der zweiten Gruppe von Studien traten bei schizophrenen Patienten keine STH-Reaktionen ein, wohl aber bei Kontrollpersonen (Monteleone et al., 1986). Damit ist die Dopamin-Hypothese verworfen. Beide Untersuchungsmethoden sind schwer zu interpretieren, weil die STH-Sekretion von vielen Faktoren angeregt und gehemmt wird: Eine fehlende Reaktion könnte durch eine gesteigerte Hemmung oder verminderte Reizung zustande kommen (in dieser Untersuchung durch einen GABA-Agonisten).

Thyreotropin-Releasing-Hormon (TRH)

Die Zusammenfassung einer umfangreichen Literatur läuft darauf hinaus, daß der gleiche Prozentsatz (35 bis 45%) Schizophrener und Patienten mit schweren affektiven Störungen eine verminderte Reaktion des Thyreotropins auf TRH zeigt. Beziehungen zwischen diesem „verminderten" Ansprechen und der Antwort auf Behandlung sind gefunden worden (Langer et al., 1986), die Bedeutung dieser Befunde für das Verständnis von Ätiologie und Pathogenese dieser Störungen ist aber nicht analysiert worden.

Schlaf

Im Vergleich zu schweren affektiven Störungen sind sehr wenige chronobiologische Untersuchungen bei Schizophrenie unternommen worden, abgesehen von denjenigen des Schlafverlaufs. Viele akut schizophrene Patienten sind nachts wach und schlafen tagsüber oder klagen über schlechtes Schlafen, besonders in Zeiten von Angst. Über das Schlafverhalten chronisch schizophrener Patienten ist weniger bekannt. Medikamente, Alter der Patienten und Phase und Art des Leidens scheinen einen Einfluß auf das Schlafverhalten bei dieser Krankheit auszuüben.

Kupfer et al. (1970) stellten bei akut kranken Patienten gestörten Schlaf fest, und ihre Schlaf-EEG's waren durch verminderte REM- und Nicht-REM-Schlafstadien gekennzeichnet. Bei ruhigen Patienten oder während Remissionen zeigten sich diese Veränderungen nicht. Jüngere Studien berichten aber, daß sich bei akut und chronisch kranken Patienten verkürzte REM-Latenzzeiten finden, wie dies auch bei schizo-affektiven und Patienten mit schweren affektiven Störungen der Fall ist (Hiatt et al., 1985; Zarcone et al., 1987). Zusätzlich wurde bei 40–50% der schizophrenen Patienten ein vermindertes Schlafstadium 4 festgestellt und im allgemeinen weniger Nicht-REM-Schlaf (Hiatt et al., 1985). In einer anderen Untersuchung waren Schlafstadium 4 und die Zahl der langsamen Wellen pro Minute dieses Stadiums vermindert – eine Beobachtung, die bei schweren affektiven Störungen noch ausgeprägter ist. Eine kürzlich publizierte Untersuchung zeigte aber keine Abnormität der REM-Latenz und Dauer der ersten REM-Periode (Ganguli et al., 1987).

Zusammenfassend: Soweit die REM-Latenz verkürzt ist, handelt es sich um eine nicht-spezifische Erscheinung. Da wir die Ursache dieser Veränderung nicht kennen, bleibt ihre Bedeutung schleierhaft.

51.7 Abschließende Beurteilung

In diesem Beitrag ist versucht worden, die zwei Hauptgruppen schwerer psychiatrischer Störungen von einem psychosomatischen Standpunkt aus zu diskutieren. (Der beschränkte Raum erlaubte die Besprechung anderer Kategorien nicht.)

Es finden sich viele Analogien zwischen den schweren Depressionen, Schizophrenie und verschiedenen Kategorien von Krankheiten, die psychobiologisch untersucht worden sind (Weiner, 1977, 1989a).

Bei allen können ähnliche Verletzlichkeiten gegenüber Verlust, Trennung und Zurückweisung beobachtet werden. Das Angewiesensein auf Unterstützung durch Mitmenschen gehört zu den hervorstechenden Zügen der seelischen Organisation solcher Patienten. Angst und depressive Stimmungen finden sich als häufige Teilerscheinungen, ausgeprägt bei den Autoimmunkrankheiten. Bei Colitis ulcerosa kommen Schizophrenie-ähnliche Störungen vermehrt vor. Wo liegt dann der Unterschied? Selbstverständlich umfassen die somatischen Krankheiten eine Vielfalt von spezifischen Prädispositionen: Bei den Autoimmunkrankheiten steigern oder senken spezifische HLA-Antigene das Risiko, aber Geschlecht, ethnische und psychosoziale Faktoren spielen auch eine Rolle (Weiner, 1989a). Auch für sie brauchen wir ein komplexes multifaktorielles Krankheitsmodell.

Die einzigen gesicherten Prädispositionen für schwere affektive Störungen und Schizophrenie sind genetischer Natur. Wir wissen aber nicht, wie sich die verschiedenen Gene ausdrücken. Jüngere Ergebnisse legen nahe, daß es sich in beiden Fällen um genetisch-heterogene Krankheiten handelt. Außer bei der X-gebundenen Form von schweren affektiven Störungen entspricht ein multifaktorielles genetisches Modell, das einer Anzahl anderer und nicht genetischer Einflüsse auf die Prädisposition und Auslösung von schweren affektiven Störungen und Schizophrenie Rechnung trägt, am ehesten den Beobachtungen. Wie bei den meisten „somatischen" Krankheiten ist die Pathogenese von schweren affektiven Störungen und Schizophrenie unbekannt. Die bei ihnen bekannten ätiologischen Faktoren sind besser begründet. Bei der Erfassung der Pathophysiologie der schweren affektiven Störungen sind Fortschritte zu verzeichnen. Bei der Schizophrenie ist das pathophysiologische Wissen hingegen noch rudimentär. Einblicke in die Veränderungen des Gehirns bei Verlust- und Trennungserlebnissen (besonders in der Kindheit) würden unser Wissen vermehren (bei Tieren besteht eine solche Information). Das Fehlen eines konzeptuellen Rahmens von genügender Breite steht Fortschritten auf diesem Gebiet im Wege. Als Ausdruck eines veralteten medizinischen Modells wird die Suche nach dem unikausalen Agens fortgesetzt.

Das psychobiologische (psychosomatische) Modell erfaßt die bekannten Tatsachen von schweren affektiven Störungen und Schizophrenie besser. Bei vielen „somatischen" Krankheiten wird die Wahl der Krankheit durch Regulationsstörungen in bestimmten Organen entschieden (z.B. bronchiale Hyperreaktivität beim Asthma bronchiale; eine Vielfalt von Regulationsstörungen in der Magensäuresekretion bei den verschiedenen Formen des Ulcus pepticum duodeni; das Vorliegen von „stummer" koronarer Arteriosklerose beim Myokardinfarkt usw.). Mit diesen Zuständen treten Verlust- und Trennungserlebnisse in Wechselbeziehung. Verlust und Trennung sind für psychiatrische oder medizinische Krankheiten nicht spezifisch.

Sie interagieren beim Herbeiführen der Störungen mit einer vorbestehenden Prädisposition in Form von Regulationsstörungen funktioneller oder struktureller Genese. Bei schweren affektiven Störungen und Schizophrenie sind die Regulationsstörungen aber unbekannt.

Die Symptome einer Krankheit sind schlechte Kriterien für den Krankheitsprozeß: Sie können von diesem dissoziiert, oder Ausdruck von Anpassungsversuchen an den Krankheitsprozeß sein (z.B. überschießende [polyklonale] Antikörperbildung bei Autoimmunkrankheiten oder die übermäßige Freisetzung von Histamin bei den Hypersensitivitäts-Sofortreaktionen), oder sie entsprechen mißlungenen Bemühungen sich anzupassen (z.B. Clearance von Salz bei der „low renin"-Hypertension, oder von Antigen-Antikörper-Komplexen). Viele Symptome der psychiatrischen Störungen sind Ausdruck von (erfolglosen) Versuchen, sich an Verluste anzupassen oder sie zu bewältigen: ein Gefühl vollständigen Versagens beim depressiven Menschen; Halluzinationen des verlorenen oder entschwundenen Mitmenschen; oder die wahnhafte Vorstellung, daß solch ein Mitmensch den schizophrenen Patienten loswerden will (Helmsley, 1977). Viele der Faktoren, welche die Schizophrenie unterhalten, sind tatsächlich von den hypothetischen „Krankheits"-Prozessen unabhängig (Zubin und Spring, 1977).

Bei psychiatrischen Störungen ist das Gehirn das Zielorgan. Vermutlich liegen ihnen Regulationsstörungen im Gehirn zugrunde, die nicht näher bezeichnet werden können. Was in die Augen springt, sind ernsthafte Stillstände in der Persönlichkeitsentwicklung, die (wie eben besprochen) besonders bei der Schizophrenie die Prädisposition für die Erkrankung darstellen. Die Hirnprozesse, die asozialem Verhalten, „skurrilen" Denkprozessen, einer mangelnden Kontrolle von Aggressionen, asozialem, anklammerndem oder ängstlichem Verhalten zugrunde liegen, sind unbekannt. Sie scheinen aber dennoch Prädispositionen für einige Formen der Schizophrenie zu sein. Wahrscheinlich liegen bedeutende (nicht näher spezifizierbare) Reifungs- und Entwicklungsmängel des Gehirns vor, die über eine Vielzahl von epigenetischen und genetischen Abläufen (oder ihre Kombination) zustande kommen.

Dieses Modell ist ein interaktives (Turpin und Lader, 1986; Weiner, 1977): Spezifische Prädispositionen stehen mit seit langer Zeit vorliegenden (nicht krankheitsspezifischen) Schwierigkeiten und persönlichen Verwundbarkeiten in Wechselbeziehung. Wie bei einer Vielzahl anderer Krankheiten haben die Verfügbarkeit oder das Fehlen mitmenschlicher Beziehungen einen Einfluß auf auslösende Erlebnisse und den Verlauf der Krankheiten. Daneben spielen andere Faktoren (z.B. das Licht bei jahreszeitabhängigen affektiven Störungen) ebenfalls eine Rolle.

Abschließend kann nicht genug betont werden, daß Psychiatrie eine humane, patientzentrierte Medizin sein muß: Neurologisieren gefährdet den Patienten.

52 Gynäkologie und Geburtshilfe

Dietmar Richter und *Manfred Stauber*

52.1 Gynäkologie

In der Gynäkologie und Geburtshilfe ist die Berücksichtigung des psychosomatischen Aspektes in mehrfacher Hinsicht notwendig und hilfreich. Der Frauenarzt muß sich mit Zeitabschnitten im Leben der Frau beschäftigen, die mit enormen Umstellungsprozessen, mit körperlichen, hormonellen, psychischen und sozialen Reaktionen einhergehen, wie z.B. Pubertät, Adoleszenz, Schwangerschaft, Geburt, Wochenbett oder Klimakterium. Diese Lebensphasen, die von jeder Frau ganz persönlich gelöst werden müssen, was im Einzelfall mehr oder weniger gut gelingt, können auch Krisen darstellen, die zu Störungen und Krankheiten führen. Grundsätzlich gilt, daß besonders die Lebensphasen, die mit größeren, hormonellen Veränderungen einhergehen, immer auch eine gewisse psychische Labilisierung mit sich bringen können, auch wenn sich dies nach außen hin nicht immer bemerkbar machen muß.

Darüber hinaus untersucht bzw. behandelt der Frauenarzt die Geschlechtsorgane der Frau. Eine solche Untersuchung oder gar eine Operation im Bereich der Geschlechtsorgane bedeutet immer einen Eingriff in eine emotional stark besetzte Körperzone. Selbst rein organische Erkrankungen, wie entzündliche oder tumoröse Veränderungen an den Geschlechtsorganen, können daher zu erheblichen inneren, seelischen Reaktionen führen, um so mehr, wenn damit Teil- oder Radikaloperationen verbunden sind, welche das Körperbild und Körpererleben einer Frau zusätzlich stören können. Operationen an den Geschlechtsorganen – z.B. eine Ablatio mammae – sind also immer Eingriffe in einen sog. „Ich-nahen" Erlebensbereich. Funktionelle Sexualstörungen nehmen in der Frauenheilkunde einen breiten Raum ein. Dazu zählen nicht nur die sexuellen Dysfunktionen im engeren Sinne, wie Libido- oder Orgasmusstörungen, Dyspareunie oder Vaginismus, sondern gerade auch zahlreiche gynäkologisch-psychosomatische Symptome, wie Fluor, Pruritus, Miktionsstörungen und Unterbauchschmerzen, können Ausdruck einer gestörten Sexualität sein. Wir bezeichnen diese Störungen auch als maskierte, larvierte oder verleugnete Sexualstörungen. Wenn die Patientin überhaupt von sich aus auf Schwierigkeiten im sexuellen Bereich zu sprechen kommt, dann nicht selten dem Frauenarzt gegenüber, welcher aus der Sicht der Patientin noch am ehesten als kompetenter Berater in solchen Fragen betrachtet wird.

Aber nicht nur mit Sexualproblemen im engeren Sinne hat sich der Frauenarzt zu beschäftigen, sondern auch mit Partnerschaftsproblemen im weiteren Sinne. So kann z.B. die – meist unbewußte – Abwehr eines Partners sich somatisieren in zahlreichen psychosomatischen Symptomen wie Kopfschmerzen, Magenfunktionsstörungen, Unterbauchschmerzen, Miktionsstörungen, Fluor, Pruritus, Unverträglichkeitsreaktionen auf verschiedene kontrazeptive Methoden usw. Chronische, nicht organisch bedingte Unterbauchschmerzen, das sog. Pelipathiesyndrom, sind ein besonders deutliches Beispiel für die Somatisierung solcher Konflikte, ein unbewußter Protest einer sich tief enttäuscht, gekränkt und unverstanden fühlenden Frau.

Ein weiterer Aspekt für die Notwendigkeit einer psychosomatischen Betrachtungsweise in der Gynäkologie und Geburtshilfe liegt in der zweifelsohne schwierigen Rollenerwartung, die heute an die Frau gestellt wird, zumindest in unserer Gesellschaft. Es ist unmöglich, diesen Erwartungen jederzeit und adäquat gerecht zu werden. Andererseits gehört schon eine beträchtliche Autonomieentwicklung dazu, sich von solchen Erwartungszwängen als Frau unabhängig zu machen. Sicher kommt eine Frau – im Gegensatz zum Mann – während ihres Lebens eher in die Situation, ihren Lebensentwurf immer wieder neu zu überdenken, Lebensziele kurz- oder langfristig zu ändern oder gar ganz aufgeben zu müssen. Denken wir nur an eine nicht oder noch nicht geplante Schwangerschaft. Während ein Mann seiner beruflichen Karriere weiter nachgehen kann, bedeutet diese Situation für eine Frau viel mehr, unter Umständen Verzicht auf eine eben begonnene, befriedigende, erfolgreiche Berufstätigkeit.

Der Frauenarzt kann damit in die Rolle des Lebensberaters kommen. Manche Kollegin oder mancher Kollege mag sich vielleicht hier unwohl oder überfordert fühlen. Nicht wenige unserer Patientinnen haben aber trotzdem eine gewisse Erwartung an den Frauenarzt, als sei er tatsächlich Experte in Ehe- und Familienfragen, bei Partner- und Sexualproblemen. Faßt man die dargestellten Besonderheiten des Berufes des Frauenarztes zusammen, erfährt dieser Beruf eine beträchtliche Erweiterung hinsichtlich seiner Tätigkeitsmerkmale, was über Operieren, Entbinden, Erkennen und Behandeln von Frauenkrankheiten hinausgeht.

Die Notwendigkeit, sich auch diesen anderen Aufgaben zu stellen, ist einer der Gründe, warum das allgemeine Interesse an der gynäkologisch-geburtshilflichen Psychosomatik in jüngster Zeit stürmisch zugenommen hat. Vor allem der in einer Praxis niedergelassene und tätige Frauenarzt spürt rasch, daß er auf diese Fragen nicht vorbereitet, für diese Probleme nicht ausgebildet ist. Der Wunsch nach geeigneter psychosomatischer Fortbildung wird daher immer stärker. So verzeichnet die 1980 gegründete Deutsche Gesellschaft für psychosomatische Geburtshilfe und Gynäkologie eine rasch wachsende Mitgliederzahl und ist in kurzer Zeit mit über 800 Mitgliedern zu einer der größten Fachgesellschaften innerhalb der Deutschen Gesellschaft für Gynäkologie und Geburtshilfe geworden.

Der folgende Beitrag zur Psychosomatik in Geburtshilfe und Gynäkologie ist systematisch gegliedert. Nicht berücksichtigt wurden die Störungen im Bereich der gynäkologischen Urologie, die eine gesonderte Darstellung erfahren. Auch auf das umfangreiche Gebiet weiblicher Sexualstörungen wird aus gleichem Grund nur kurz eingegangen. Von der geburtshilflich-gynäkologischen Psychosomatik wurden in den letzten Jahren einige neue praxisorientierte Behandlungsmöglichkeiten erarbeitet. Am Beispiel von Amenorrhoe- und Unterbauchschmerz-Patientinnen werden einige dieser therapeutischen Möglichkeiten dargestellt.

52.1.1 Blutungs- und Zyklusstörungen

Der ovulatorische Menstruationszyklus der Frau ist abhängig von einem ungestörten Zusammenspiel des zentralen Nervensystems, des Hypothalamus, der Hypophyse und des Ovars. Unbestritten ist, daß psychische Faktoren diese funktionelle Einheit beeinflussen können.

Corpus-luteum-Insuffizienz, Anovulation, Oligomenorrhoe

Symptome einer Störung dieser Funktionsachse können sein: die Corpus-luteum-Insuffizienz, die Anovulation, die Oligomenorrhoe und – als auffälligstes Symptom – die sekundäre Amenorrhoe. Häufig werden bei der Entwicklung hypothalamischer Amenorrhoen und im Verlauf der Wiederaufnahme der normalen Ovarialfunktion Oligomenorrhoen, anovulatorische Zyklen sowie Corpus-luteum-Insuffizienz als Zwischenphasen in jeweils umgekehrter Richtung durchlaufen. In Übereinstimmung mit anderen Autoren (Leyendecker, 1979; Leyendecker et al., 1981) stellen diese graduell unterschiedlichen Zyklusstörungen im Rahmen der psychosomatisch bedingten suprahypothalamischen Ovarialinsuffizienz ein pathophysiologisches Kontinuum dar, in Abhängigkeit einer mehr oder weniger eingeschränkten hypothalamischen Sekretion von Gonadotropin-Releasing-Hormon (GnRH).

Während der Pubertät kommt es nach der Menarche häufig zunächst nur zu vereinzelten Blutungen, dann zu anovulatorischen Zyklen und Corpus-luteum-Insuffizienzen, bevor – als Ausdruck voller funktioneller Reife – normale biphasische Zyklen auftreten. Diese Ausreifung der Ovarialfunktion während der Pubertät hängt ab von der allmählich zunehmenden Stimulation der Hypophyse mit GnRH. In Analogie dazu kann der normale ovulatorische Zyklus jederzeit durch suprahypothalamische Beeinflussung (psychische Faktoren) in alle verschiedenen funktionellen Entwicklungsstadien bis hin zur Amenorrhoe zurückfallen.

Diese funktionellen Störungen – Corpus-luteum-Insuffizienz, Anovulation, Oligomenorrhoe und sekundäre Amenorrhoe – müssen daher als psychosomatisches Symptom aufgefaßt werden. Zu trennen sind davon die hyperprolaktinämische und die hyperandrogenämische Ovarialinsuffizienz sowie Ovarialinsuffizienzen infolge internistischer Erkrankungen, wie z.B. bei Hypothyreose, Morbus Cushing, Diabetes.

Sekundäres Amenorrhoe-Syndrom

Eine sekundäre Amenorrhoe kann auftreten im Verlaufe zahlreicher Organerkrankungen oder als Folge einer allgemein gravierenden Streßsituation. Eine sekundäre Amenorrhoe ist nicht selten Begleitsymptom psychiatrischer Erkrankungen.

Am häufigsten findet sich dieses Symptom bei psychosomatischer Ovarialinsuffizienz als Ausdruck einer neurotischen Persönlichkeitsproblematik. Eine als therapeutische Konsequenz heute noch immer überwiegend eingeleitete Hormonbehandlung bleibt unbefriedigend, da nur symptomatisch wirksam, wenn es darüber hinaus nicht gelingt, gemeinsam mit der Patientin den das psychosomatische Symptom „sekundäre Amenorrhoe" auslösenden Grundkonflikt zu lösen oder zumindest zu erhellen.

Hierbei hat sich die Synopsis von gynäkologisch-endokrinologischen und psychosomatischen Untersuchungsschritten bewährt (Richter et al., 1977; Peters et al., 1978; Richter, 1980, 1982).

Für die psychosomatisch bedingte Amenorrhoe existieren zahlreiche Synonyma: suprahypothalamische Amenorrhoe, funktionelle Amenorrhoe, psychische Amenorrhoe, psychoreaktive Amenorrhoe. Wir bezeichnen diese funktionelle Störung als „sekundäres Amenorrhoe-Syndrom", da es sich nicht nur um ein einziges Symptom, die Amenorrhoe, handelt, sondern es finden sich regelmäßig bei diesen Patientinnen eine ganze Reihe zusätzlicher körperlicher und/oder seelischer Symptome wie Magenfunktionsstörungen, Obstipation, Herz-Kreislauf-Störungen, Gewichtsschwankungen, anorektische Reaktion, rezidivierende Anginen, Myalgien, Kopfschmerzen, ausgeprägte Sexualstörungen, depressive Verstimmungen, Minderwertigkeitsgefühle, Konzentrations- und Arbeitsstörungen. Die psychosomatisch bedingte sekundäre Amenorrhoe (sekundäres Amenorrhoe-Syndrom) betrifft ca. 80% aller sekundären Amenorrhoen, die Prävalenzrate beträgt 1,4% (Petterson et al., 1973). Drei psychodynamisch wirksame Problembzw. Konfliktkreise führen zum Auftreten des sekundären Amenorrhoe-Syndroms (SAS).

Unbewußte Angst vor Verlust an Sicherheit, Geborgenheit und Wärme

Sicherheit, Geborgenheit und Wärme werden in erster Linie durch die Eltern oder einen Partner repräsentiert. Infolge frühkindlicher Erfahrungen spielen Sicherheits- und Geborgenheitswünsche im Erleben dieser Patientinnen eine übergroße Rolle. Andererseits wird das heranwachsende Mädchen in der Adoleszenz mit neuen Erlebnisdimensionen konfrontiert, wie Streben nach individueller und sozialer Unabhängigkeit. Das Ausprobieren und Einüben solcher Autonomiebestrebungen kann nicht angstfrei bewältigt werden, führt zu intrapsychischen Spannungen, zur Mobilisation unbewußter Trennungs- und Verlustängste. Vor diesem psychodynamischen Hintergrund erklären sich die auslösenden Konfliktsituationen, wie Trennung von den Eltern, Konflikte mit den Eltern, Ehekonflikte bzw. Scheidung der Eltern, Tod eines Elternteiles oder berufliche Existenzprobleme der Eltern. In modifizierter Weise liegt diese Problematik auch gewissen Partnerkonflikten zugrunde. Eine sekundäre Amenorrhoe tritt immer dann auf, wenn diese Partner die von ihnen unbewußt erwartete Rolle eines „ständigen Spenders von Nestwärme" nicht übernehmen können oder wollen oder gar ihrerseits Geborgenheitswünsche gegenüber ihren Partnerinnen zum Ausdruck bringen. Auch die von anderen Autoren wie Rosenkötter et al. (1968) als charakteristisch für das SAS behauptete „Ablehnung der Mutterschaft" gehört zu diesem Problemkreis. Bei diesen Frauen kollidieren die mit Schwangerschaft, Geburt und Kindererziehung verbundenen mütterlichen Aufgaben mit erheblichen eigenen, noch uneingestandenen Wünschen nach Versorgt- und Beschütztwerden.

Ablehnung oder Abwehr des Sexualtriebes

Die Ausbildung des reifen weiblichen Körpers bedeutet Manifestation sexueller Triebansprüche. Hat die vorausgegangene psychosexuelle Entwicklung bereits ungelöste Konflikte zurückgelassen, kann die Auseinandersetzung mit den eigenen andrängenden Triebimpulsen zu unlösbaren Spannungen führen. Sexuelle Triebbedürfnisse können generell als bedrohlich oder schuldhaft erlebt werden. Durch sexuelle Triebimpulse können aber auch Ängste mobilisiert werden, vor Verlust der eigenen Autonomie, vor Abhängigkeit vom Mann oder der eigenen Leidenschaft. Sexualität wird nicht als Bereicherung oder Erweiterung der eigenen psychosozialen Existenz, sondern als gefährliche Abhängigkeit erlebt. Ausgangspunkt für derartige Entwicklungen ist häufig eine prüde, asketische, allgemein triebfeindliche Familiensituation. Diese Triebfeindlichkeit steht im Kontrast zur sexuellen Liberalisierung unserer heutigen Zeit, so daß für die heranwachsende Frau Versuchung und Zwang zur Triebabwehr nebeneinander bestehen. Über die sekundäre Amenorrhoe hinaus können anorektische Reaktionen bevorzugt bei den Patientinnen auftreten, die in einer Familienatmosphäre aufwuchs, die neben einer allgemeinen Triebabwehr durch eine vorherrschende Leistungshaltung geprägt war. Die Auseinandersetzung mit den eigenen andrängenden sexuellen Impulsen bedeutet für diese Patientinnen die Zuspitzung eines Trieb-Geist-Konfliktes. Als Folge verstärkter Triebabwehr werden körperliche Reifezeichen, vor allem die sichtbaren wie Brust und Hüften, zurückgedrängt und ungeschehen gemacht. Bei depressiver Grundproblematik kann so die sexuelle Triebabwehr auf die orale Stufe transformiert werden.

Männliche Rollenidentifikation

Unter Vernachlässigung anderer Möglichkeiten weiblicher Selbstverwirklichung erleben sich diese Frauen vorrangig in ständiger Konkurrenz zum Mann. Alle Kräfte werden für das Ziel eingesetzt, „den Mann" zu erreichen bzw. zu überholen. Infolge kindheitsbedingter ödipaler Fixierungen stellen Eros, Sexualität und Mutterschaft weniger erstrebenswerte weibliche Lebensperspektiven dar. Überkompensatorisch findet sich bei diesen Frauen eine ausgeprägte allgemeine Ehrgeiz- und Leistungshaltung.

Entsprechend der individuellen Lebensgeschichte einer Patientin – unterschiedliche biologische, psychische, familiäre, gesellschaftliche, kulturelle Bedingungen – können einzelne wesentliche Aspekte weiblicher Selbstverwirklichung nicht entwickelt oder behindert sein. „Es gibt keinen Grund mehr für Amenorrhoe, wenn Weiblichkeit in ihren unterschiedlichen Aspekten wieder möglich und erstrebenswert wird" (De Senarclens und Fischer, 1978).

Bei der Zusammenschau endokrinologischer und psychosomatischer Befunde zeigt sich, daß bestimmte Hormonprofile mit gleichartigen psychologischen Merkmalen korrelieren. Es lassen sich zwei Patientinnengruppen deutlich voneinander unterscheiden: Die Patientinnen der einen Gruppe reagieren mit einem positiven Gestagentest. Sie zeigen im GnRH-Test eine noch deutliche Stimulierbarkeit der Hypophyse mit maximalen LH-Werten zwischen 10–40 ng/ml. Sie fühlen sich durch die Amenorrhoe in ihrem weiblichen Selbstwertgefühl beeinträchtigt und berichten relativ offen – sofern man danach fragt – über psychische Schwierigkeiten. Eine körperliche Begleitsymptomatik findet sich kaum oder tritt im Krankheitsgefühl deutlich hinter den seelischen Problemen zurück. Da diese Patientinnen einen Zusammenhang von seelischer Konfliktsituation und fehlender Menstruation oftmals ahnen, überrascht es sie nicht, wenn sie vom Arzt in diesem Sinne angesprochen werden. Auf die Frage, ob es sie stört, keine Menstruation zu haben, bekommt man etwa folgende Antwort: „. . . ja, es belastet mich, ich möchte, daß sich das wieder normalisiert, ich fühle mich nicht als vollwertige Frau . . ."

Der mehr situativ und nicht tiefer liegenden Problematik entspricht eine endokrine Störung leichteren Grades. Infolge des Leidensdruckes und des zumindest in Ansätzen vorhandenen Psychogenieverständnisses werden psychotherapeutische Hilfen von diesen Patientinnen bereitwillig angenommen. Der in formaler Psychotherapie ausgebildete Arzt oder Gynäkologe kann diese Patientinnen in Einzel- und Gruppentherapie erfolgreich behandeln.

Bei der Einzeltherapie genügen im allgemeinen 10 bis 30 Behandlungsstunden. In vielen Fällen stellt sich schon bald nach Behandlungsbeginn eine spontane Blutung ein. Das sollte nicht dazu verleiten, den psychotherapeutischen Prozeß vorzeitig zu beenden, da es sich hierbei um Übertragungserfolge handeln kann. Nur eine ausreichende Durcharbeitung der neurotischen Störung garantiert einen dauerhaften Erfolg. Wir haben mit der Einzel- wie auch mit der Gruppentherapie gute Erfahrungen gemacht (Richter et al., 1977). Diese psychotherapeutischen Behandlungsformen – so effektiv sie auch sein mögen – verlangen von Arzt und Patientin einen hohen Zeitaufwand. Darüber hinaus verfügen nur wenige Gynäkologen und Allgemeinärzte über eine zusätzliche psychotherapeutische Ausbildung und werden somit kaum in der Lage sein, SAS-Patientinnen auf diese Weise zu behandeln. Sie werden möglicherweise diese Patientinnen zu einem Psychiater oder Psychotherapeuten überweisen. Patientinnen, die bereits eine gewisse Introspektionsfähigkeit in die eigene Konflikthaftigkeit besitzen, mögen einer solchen Überweisung zustimmen. Nicht wenige Patientinnen erleben sich jedoch in einer solchen Situation zwischen einem „Fachmann für Hormone" und einem „Fachmann für die Psyche" aufgeteilt, wenn nicht sogar vom Arzt ihres Vertrauens irgendwie abgeschoben. Andererseits lehrt die Erfahrung – und auch wir konnten das zeigen –, daß manche dieser SAS-Patientinnen auch ohne jede Behandlung wieder eumenorrhoisch werden können, wenn situative Konflikt- oder Spannungszustände vorübergehen. Weitaus schwieriger gestaltet sich die Therapie bei den SAS-Patientinnen, welche eine endokrine Störung größeren Ausmaßes aufweisen. Im GnRH-Test finden sich niedrige LH-Antworten von deutlich unter 10 ng/ml. Der Gestagentest ist negativ oder eben noch positiv. Die niedrige Funktionsreserve der Hypophyse, als Folge ungenügender Stimulierbarkeit durch GnRH, muß hier als somatischer Ausdruck einer tief verdrängten psychischen Konfliktlage verstanden werden. Neben der Amenorrhoe finden sich regelhaft fast ausschließlich körperliche Beschwerden. Über psychische Beschwerden wird kaum geklagt. Diese Amenorrhoe-Patientinnen verhalten sich im Vergleich zur erstgenannten Gruppe in ihrer Begleitsymptomatik zur Amenorrhoe nahezu komplementär. Sie blocken Gefühle stark ab, können Gefühle nicht ausdrücken, seelische Affekte sind tief verdrängt, Konflikte werden nicht wahrgenommen. Das „Fehlen der Periode" stört nicht, sie wird allenfalls aus rationalen Gründen gewünscht. Infolge mangelnder Einsichtsfähigkeit in die eigene Konflikthaftigkeit fehlt ein Leidensgefühl fast völlig. Die Therapieerwartung an den Arzt besteht in einer hormonalen Behandlung. Ein direkter Hinweis auf die kausalen Zusammenhänge der Amenorrhoe und ein psychotherapeutisches Angebot wird von diesen Patientinnen meist verständnislos aufgenommen. Die Reaktionen reichen von ungläubi-

gen Fragen bis zur offenen oder empörten Ablehnung eines solchen „Ansinnens".

Bei diesen Patientinnen kann sich schon die psychosomatische Exploration als sehr schwierig erweisen, bzw. es kommt gar nicht dazu, weil sie infolge andersartiger Therapieerwartungen ein solches Vorgehen von seiten des Arztes abblocken. Andererseits ist aber gerade bei diesen Patientinnen eine psychosomatische Behandlung um so mehr angezeigt, weil auch auf Jahre hinaus mit einem Wiederauftreten der spontanen Regelblutung nicht zu rechnen ist, das Krankheitsbild des SAS sich vielmehr fixiert und chronifiziert, eine Situation, die viel zu wenig Beachtung findet. Von diesen Überlegungen ausgehend, haben wir ein kausaltherapeutisches Behandlungskonzept für SAS-Patientinnen entwickelt. Dabei kommt es zu einer schrittweisen Auflockerung des neurotischen Konfliktes und in über 90% zum Wiederauftreten ovulatorischer Zyklen (Richter, 1980). Zwei Aspekte sind in diesem Zusammenhang wichtig:

– Die Behandlung sollte möglichst von jedem Arzt durchgeführt werden können.
– Die Behandlung sollte im üblichen Sprechstunden-Setting stattfinden, ohne den zeitlichen Rahmen einer üblichen gynäkologischen Praxis zu sprengen.

Dazu ist es wichtig, auch die psychologische und soziale Dimension der Patientin in die meist überwertig somatisch orientierte Sprechstunde mit hineinzunehmen. Dies verlangt vom Arzt, von der gewohnten, eher aktiven und entschlußfreudigen Einstellung des „handelnden Arztes" auf eine mehr passive, rezeptive Haltung überzugehen. Während man sich mit der Anamnese, der gynäkologischen und hormonellen Diagnostik befaßt, werden Signale der Patientin, die mehr oder weniger indirekt und unbewußt gegeben werden, aufgegriffen. Im Verlaufe wiederholter Konsultationen kann sich so ein Gespräch entfalten, welches der Patientin ermöglicht, schrittweise ihre Problematik zu verbalisieren und bewußter werden zu lassen und neue Lösungen für ihre Schwierigkeiten zu finden. Durch den Arzt, gleichsam als Katalysator, kommt es zur schrittweisen Auseinandersetzung der Patientin mit sich selbst. Das Gespräch bleibt innerhalb der gynäkologischen Sprechstunde, es handelt sich nicht um formale Psychotherapie, sondern es wird ein Auseinanderklaffen zwischen gynäkologisch-endokrinologischer Behandlung einerseits und psychotherapeutischer Behandlung andererseits vermieden, es handelt sich um eine psychosomatische Vorgehensweise im echten Sinne des Wortes. Die Patientinnen erleben sich dabei nicht zwischen einem „Fachmann für Hormone" und einem „Fachmann für die Psyche" aufgeteilt, sie fühlen sich in ihrer physischen und psychischen Gesamtheit von einem Arzt umfassend angenommen.

Im einzelnen wird wie folgt vorgegangen (Tab. 52–1):

Bei der **1. Konsultation** wird die allgemeine und gynäkologische Anamnese erhoben, insbesondere Zeitpunkt und Umstände des Ausbleibens der Menstruation. Körperliche und seelische Begleitsymptome werden systematisch exploriert. Danach wird die Pa-

Tabelle 52–1. Psychosomatisch orientiertes Diagnostik- und Therapiekonzept bei Patientinnen mit sekundärem Amenorrhoe-Syndrom

1. Konsultation

allgemeine gynäkologische Anamnese	körperliche, seelische Begleitsymptomatik
gynäkologische Untersuchung	Frage: Stört Sie die fehlende Menstruation?
Zervixfaktor	
Funktionszytologie	(auf Affekte achten)
Blutentnahme zur Prolaktinbestimmung	Wie empfinde ich diese Patientin?
Erklärung der Physiologie des Menstruationszyklus (Ovar → Uterus, Östrogene, Gestagene)	
Rp. Farlutal Nr. XX (Op)	

2. Konsultation

Interpretation des Gestagentests	psychosoziale Situation der Patientin (Eltern, Partner, Beruf, Pläne usw.)
weitere Erklärung der Physiologie (Hypothalamus-Hypophyse-Ovar, FSH, LH, GnRH)	Wie empfinde ich die Patientin?
Rp. GnRH 25 μg (Op)	Wo liegt ihr Konflikt?

3. Konsultation

GnRH-Test: 25 μg i. v. Blutentnahmen −10, 0, 30 min	Während des Testablaufes Fortsetzung des Gesprächs über die aktuelle Lebenssituation
weitere Erklärung der Physiologie (ZNS/Psyche-Hypothalamus-Hypophyse)	Versuchsinterpretation des vermuteten Konflikts und beginnende Durcharbeitung, nur bei Psychogenieverständnis, sonst abwarten!

4. Konsultation

Interpretation des GnRH-Tests	Übergang von der organischen (Hormonstörung) zur psychosomatischen Selbstauffassung der Patientin
Hinlenkung der Vorstellung der Patientin auf die zentrale Blockierung als Ursache der sekundären Amenorrhoe	Einbeziehung der Begleitsymptomatik in die Gesamtbetrachtung
Gemeinsame Bearbeitung der Konfliktproblematik	

tientin gefragt, ob sie sich durch die fehlende Periode gestört fühlt, ob und warum sie die Regelblutung wiederbekommen möchte. Durch die Art und Weise, wie die Patientin auf diese Frage reagiert, gibt sie zu erkennen, ob sie durch die Amenorrhoe emotional irritiert ist oder ob eine Abwehrhaltung vorliegt. Die gynäkologische Untersuchung schließt die Prüfung der Zervixfaktoren und einen funktionszytologischen Abstrich ein. Nach Entnahme von 10 ml Nativblut zur Prolaktinbestimmung werden anhand einfach gestalteter Abbildungen die Wechselwirkungen der

Östrogene und Gestagene zwischen Ovar und Uterus und die Bedeutung des Gestagentests erläutert. Die erste Untersuchung endet mit der Verschreibung von Medroxy-Progesteronacetat als Gestagentest über 10 Tage.

Bei der **2. Konsultation** – 4 Wochen später – berichtet die Patientin, ob durch die Gestagene eine Blutung ausgelöst wurde oder nicht. Das Ergebnis des Gestagentests wird erläutert, die Kenntnisse über die Physiologie des Menstruationszyklus werden erweitert um die Erklärung der Wechselbeziehungen zwischen Hypothalamus-Hypophyse-Ovar. Mit der Besprechung der psychosozialen Situation der Patientin (Eltern, Partner, Beruf, Pläne usw.) wird die allgemeine Anamnese vervollständigt. Während man die Patientin auf sich wirken läßt, werden „mit dem inneren Auge" die drei hauptsächlichen symptomauslösenden Konfliktursachen von SAS-Patientinnen abgefragt: Hat die Patientin unbewußt Angst vor Verlust an Sicherheit, Geborgenheit und Wärme? Sind Sexualität und sexuelle Bedürfnisse für sie etwas Bedrohliches? Muß die Patientin mit der männlichen Rolle konkurrieren? Die zweite Untersuchung endet mit der Verschreibung von 25 μg GnRH zum GnRH-Test.

Im Mittelpunkt der **3. Konsultation** steht der GnRH-Test. Eine Blutentnahme erfolgt unmittelbar vor der intravenösen Injektion von 25 μg GnRH sowie 30 Minuten danach.

Während des Testablaufs erfolgt in einfachen Worten eine erneute erweiterte Erklärung der physiologischen Zusammenhänge von ZNS-Hypothalamus-Hypophyse-Ovar-Uterus. Das Gespräch über die aktuelle Lebenssituation wird weitergeführt. Hat man schon eine gewisse Sicherheit über den vermutlichen Konflikt der Patientin gewonnen und liegt keine Abwehrhaltung vor, kann der psychosomatische Zusammenhang angesprochen werden, im anderen Fall muß abgewartet werden.

Während der **4. Konsultation** wird das Ergebnis des GnRH-Tests erläutert und die Vorstellung der Patientin auf einen möglichen psychosomatischen Zusammenhang hingelenkt. Wenn – bei Patientinnen mit schwerer endokriner Blockierung und starrer Abwehrhaltung – der GnRH-Test die, aufgrund der bisherigen Befunde, erwartete niedrige LH-Antwort bestätigt, wird der Patientin gegenüber geäußert, man wisse jetzt, daß keine ernste organische Störung vorliege, daß lediglich die ihr schon bekannten Zwischenhirngebiete unzureichend Releasing-Hormon an die Hirnanhangsdrüse abgäben und damit eine Unterfunktion in den davon abhängigen Eierstöcken eingetreten sei. Man wisse auch, daß eine geringe Releasing-Hormonproduktion abhängig sei von nervösen Impulsen aus anderen Hirngebieten, welche wiederum eine Art Steuerungszentrale für im allgemeinen unbewußte Körpervorgänge, wie z.B. Magen- und Darmtätigkeit, darstellen. Diese vegetativen Steuerungszentralen seien durch alle möglichen Reize beeinflußbar, z.B. auch durch seelische Spannungszustände. Hier wird zum ersten Mal der Patientin gegenüber „die Psyche" im Sinne einer Reizdeutung erwähnt. Wir machen nun immer wieder die Erfahrung, daß jetzt die Patientin von sich aus die Frage stellt „... also könnten psychische Vorgänge bei mir eine Rolle spielen ...?" oder „... das hieße, meine fehlende Periode hätte seelische Ursachen ...?". Die Vorstellung der Patientin über die Ursache ihrer Störung wird damit auf die kausalen psychologischen Zusammenhänge hingelenkt. Wir bieten jetzt der Patientin an, gemeinsam herauszufinden, ob „Psychisches" eine Rolle spielen könnte. Diese Vorgehensweise kommt also zunächst der Vorstellung der Patientin von einer Hormonstörung entgegen und weist den Arzt als zuverlässigen, sachverständigen Partner bei der Abklärung der somatisch-hormonellen Situation aus. Nur auf dem Boden einer solcherart stabilisierten und vertrauensvollen Arzt-Patientin-Beziehung wird eine Umschaltung zu einer ganzheitlichen psychosomatischen Behandlung möglich. Dieser Übergang von der körperlichen zur psychosomatischen Selbstauffassung der Patientin beansprucht unterschiedliche Zeit. Manchmal geschieht dies innerhalb der ersten Konsultation, manchmal dauert es Monate oder sogar Jahre. Auch Patientinnen mit schwerer suprahypothalamischer Blockierung und zunächst ausgeprägter Abwehrhaltung können also im Rahmen der üblichen Sprechstundensituation weiterbehandelt werden. Unterschiedlich ist jedoch ein wesentlich aktiveres Vorgehen bei der Zuordnung bewußten Materials, bei der Strukturierung von Konflikten. Zu bearbeitende Problemkreise werden vom Arzt anfangs selbst umrissen, müssen aber auch flexibel geändert werden, wenn sie sich als falsch erweisen. Die Zeitdauer dieser Behandlungstechnik ist, je nach Schweregrad der Störung, mit einem halben bis zu zwei Jahren zu ver-

anschlagen, bei einem 4- bis 6wöchigen Sprechstundenintervall mit Konsultationszeiten von etwa 15 Minuten Dauer. In Abständen von ca. einem Vierteljahr werden Gestagen- bzw. GnRH-Tests durchgeführt, welche den psychosomatischen Therapieeffekt endokrinologisch dokumentieren durch kontinuierlichen Anstieg der Funktionsreserve der Hypophyse, bis wieder regelrechte ovulatorische Zyklen auftreten. Diese im Rahmen des üblichen Sprechstunden-Settings durchgeführte psychosomatische Behandlungstechnik hat – um es noch einmal zu unterstreichen – den großen Vorteil, daß Patientin und Arzt keine Aufspaltung, keinen Identitätsverlust erleiden. Die Patientin wird in ihrer psychischen wie physischen Gesamtheit vom Arzt umfassend angenommen.

Die Abbildungen 52–1 und 52–2 zeigen die endokrinen Veränderungen im Verlauf der Behandlung von 40 Patientinnen mit dem eben beschriebenen Behandlungskonzept. Die Gruppe I bilden 22 Patientinnen mit geringer hypophysärer Antwort im GnRH-Test. Die 30minütigen LH-Werte erreichten nicht 8 ng/ml. 14 Patientinnen hatten einen negativen, 8 Patientinnen einen noch positiven Gestagentest. Bei einem Durchschnittsalter von 22,7 Jahren betrug die Amenorrhoedauer bis zum Behandlungsbeginn 21,2 Monate. Bei den Patientinnen mit positivem Gestagentest trat nach durchschnittlich 4,7 Konsultationen, bei Patientinnen mit negativem Gestagentest nach 10 Konsultationen die erste spontane Menstruation auf. Die im Verlauf der Behandlung durchgeführten GnRH-Tests zeigen einen deutlichen Anstieg der maximalen LH-Antworten als Ausdruck der zunehmenden Funktionsreserve der Hypophyse infolge ei-

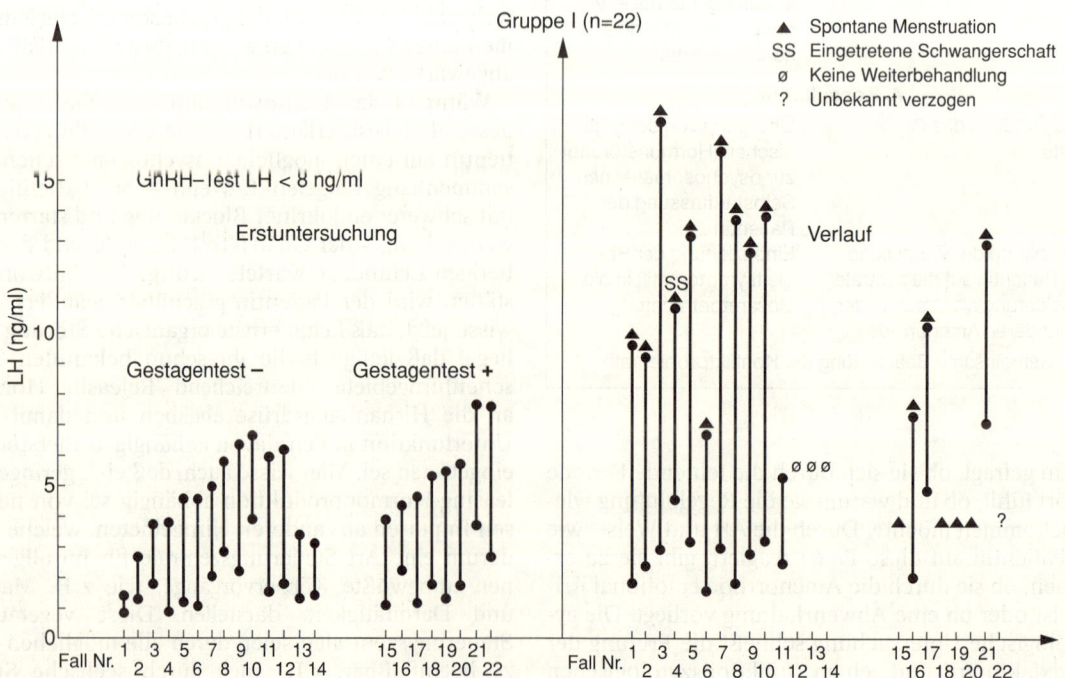

Abb. 52–1. SAS-Patientinnen-Gruppe I: GnRH-Tests bei der Erstuntersuchung und im Verlauf der psychosomatischen Behandlung vor Auftreten der spontanen Menstruation.

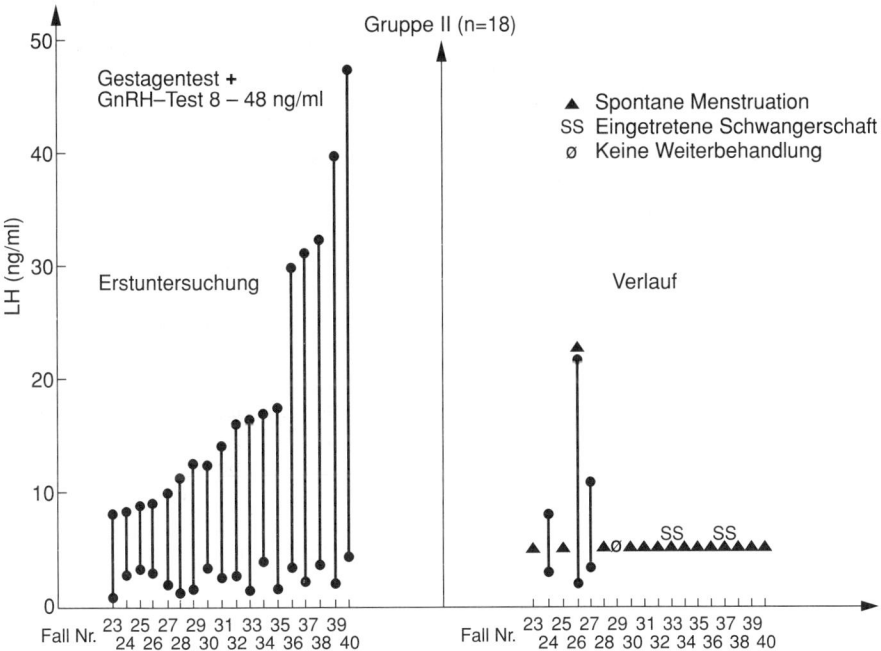

Abb. 52–2. SAS-Patientinnen-Gruppe II: GnRH-Tests bei der Erstuntersuchung und im Verlauf der psychosomatischen Behandlung vor Auftreten der spontanen Menstruation.

ner Abnahme der suprahypothalamischen Blockierung. Die Gruppe II bilden 18 Patientinnen mit einer Amenorrhoedauer von 16 Monaten bei einem Alter von 22,2 Jahren. Die GnRH-Tests zeigen bei der Erstuntersuchung eine deutliche Ansprechbarkeit der Hypophyse. Die 30minütigen LH-Werte erreichten Plasmaspiegel von 8–48 ng/ml. Da diese Patientinnen einen Zusammenhang von persönlichen Schwierigkeiten und der Amenorrhoe ahnten, konnte ihre Konfliktproblematik sofort angesprochen werden. Schon nach 2,7 Konsultationen bzw. 2,5 Behandlungsmonaten setzte bei 15 Patientinnen die spontane Regelblutung wieder ein, so daß sich weitere GnRH-Tests erübrigten. Zwei Patientinnen mit Kinderwunsch wurden schwanger. Diese Untersuchungen zeigen, daß der Schweregrad der psychosomatischen Störung, nämlich das sekundäre Amenorrhoe-Syndrom, sein endokrines Korrelat hat.

Anorexia nervosa

Vom sekundären Amenorrhoe-Syndrom – mit oder ohne anorektische Reaktion – muß die Anorexia nervosa als ein eigenes, wesentlich schwerwiegenderes Krankheitsbild abgegrenzt werden. Sie betrifft überwiegend junge Frauen im postpubertären Alter. In Anbetracht der häufig zu beobachtenden Chronifizierung dieses Leidens mit seinen negativen psychosozialen Folgen und einer Mortalitätsrate von 10–15% sollte die Diagnose so früh wie möglich gestellt werden und psychotherapeutische Hilfe einsetzen. Nicht selten geraten diese jungen Patientinnen in ein therapeutisches Vakuum zwischen Pädiater, Internist und Gynäkologe. Obwohl Begleitsymptomatik, Psychodynamik und Persönlichkeitsstruktur recht unter-

schiedlich sein können, ist eine Anorexia nervosa gekennzeichnet durch die phobische Vermeidung von Gewichtszunahme bei gestörter Wahrnehmung des Körperschemas und durch die Begleitamenorrhoe mit gravierender suprahypothalamischer Blockierung. Im GnRH-Test kommt es zu einer Umkehr der FSH/LH-Reaktion als Ausdruck einer „endokrinen Regression" auf die präpuberale Situation. Dieser „Rückfall" auf präpuberale endokrine Verhältnisse entspricht der psychisch-somatischen Regression. In Übereinstimmung mit den Erfahrungen beim sekundären Amenorrhoe-Syndrom kommt es auch im Verlauf der Behandlung von Anorexia-nervosa-Patientinnen nach der Umkehr der FSH/LH-Reaktion wieder zu ansteigenden LH-Spiegeln im GnRH-Test. Zur Psychopathologie und Behandlungsstrategie bei Anorexia nervosa siehe Kapitel 38.2.

Scheinschwangerschaft
(Pseudocyesis, grossesse nerveuse)

Diese selten gewordene Sonderform einer sekundären Amenorrhoe ist Ausdruck eines fast wahnhaft ausgelebten intensiven Wunsches nach einem Kind. Benedek (1952) deutete die Pseudocyesis als konversionshysterisches Symptom, auf der Triebebene mit dem Wunsch schwanger zu werden, auf der Abwehrebene mit der Furcht davor. Neben der Amenorrhoe kann es zur Gewichtszunahme, zur Zunahme des Leibesumfanges, zu Haut- und Pigmentveränderungen und zur Galaktorrhoe kommen, wie bei einer echten Schwangerschaft. Die gesamte psychische Energie dieser Frauen scheint auf das „erwartete Kind" ausgerichtet zu sein. Manche Frauen spüren Kindsbewegungen – es handelt sich um Darmperistaltik –, kaufen Babywäsche und richten ein Kinderzimmer ein.

Bei der Betreuung solcher Patientinnen ist ein einfühlsamer Umgang besonders wichtig. Als Einstieg empfiehlt sich die vorsichtige Konfrontation mit dem Leidensdruck, der durch den frustranen Kinderwunsch hervorgerufen wird. Es sollte angeboten werden, mit beiden Partnern die Möglichkeiten einer gezielten Kinderwunschbehandlung durchzusprechen.

Integrativ könnte während einer solchen Behandlung auch die Kinderwunschmotivation bearbeitet und der überwertige Kinderwunsch abgebaut werden (Stauber, 1977, 1979). Auch durch die Überlegung einer möglichen Adoption läßt sich der umschriebene Konflikt mildern.

Primäre Amenorrhoe

Die Menarche setzt in Mitteleuropa etwa um das 13. Lebensjahr ein. Wenn bis zum 16. Lebensjahr noch keine spontane Menstruation erfolgt ist, liegt eine primäre Amenorrhoe vor. Die sekundären Geschlechtsmerkmale sind noch gar nicht oder ungenügend entwickelt oder es finden sich äußere Stigmata für eine gravierende endokrine oder chromosomale Störung. Im Gegensatz zur sekundären Amenorrhoe ist die primäre Amenorrhoe in über 80% der Fälle Folge organischer Ursachen, anatomischer oder chromosomaler Anomalien. Heute kann durch gynäkologische Untersuchung, Hormonbestimmungen, Sonographie, eventuell Laparoskopie und Chromosomenanalyse die genaue Diagnose rasch gestellt werden. Die Mitteilung einer unheilbaren Störung kann eine junge Frau zutiefst erschüttern. Dies gilt besonders für die Anlagestörungen der Gebärmutter, wobei meist auch die Scheide ganz oder teilweise fehlt. Therapeutische Maßnahmen wie Hormonsubstitution oder operative Korrekturen anatomischer Anomalien bzw. Entfernung fehlentwickelter Gonaden müssen eingebettet werden in eine einfühlsame, kontinuierliche ärztliche Betreuung der Patientin, um ihr bei der mühsamen, neuen psychosexuellen Identitätsfindung zu helfen (Jürgensen, 1979). Trotz intensiver Bemühungen gelingt es bei einigen Patientinnen nicht oder nur schwer, eine Versöhnung mit dem gestörten Körperbild und eine realitätsgerechte Verarbeitung des narzißtischen Defekts zu erreichen (Kaplan, 1968).

Die psychosomatisch bedingte primäre Amenorrhoe ist Ausdruck einer Störung der psychosexuellen Entwicklung. Diese Mädchen kommen häufig aus Familien der ländlichen Bevölkerung, die in einer gewissen sozialen Isolierung leben (z.B. abgelegener Bergbauernhof). Die Mütter sind dominierend und blockieren durch zwanghafte Einengung die Entwicklung ihrer Töchter, insbesondere wenn andere weibliche Identifikationsmöglichkeiten für das heranwachsende Mädchen fehlen. Eine gynäkologische Untersuchung zum Ausschluß organischer Amenorrhoeursachen sollte um das 15. Lebensjahr herum erfolgen. Die Behandlung besteht in beruhigendem Abwarten. Mit zunehmenden sozialen Kontakten außerhalb der Familie kommt es zur allmählichen Nachrei-

fung, in Einzelfällen kann eine intensive psychotherapeutische Behandlung notwendig werden.

Dysmenorrhoe

Die schmerzhafte Regelblutung tritt überwiegend in der Adoleszenz auf und kann zu einer starken Beeinträchtigung des allgemeinen Wohlbefindens führen, besonders wenn sie von migräneartigen Kopfschmerzen, Übelkeit, Erbrechen und Kreislaufstörungen begleitet wird.

Früher wurde angenommen, daß ein kleiner hypoplastischer Uterus, durch Östrogene unzureichend aufgelockert, die dysmenorrhoischen Beschwerden verursache. Inzwischen gilt als gesichert, daß erhöhte Prostaglandin-$F_{2\alpha}$-Spiegel den dysmenorrhoischen Symptomenkomplex hervorrufen (Zahradnik et al., 1978). F-Prostaglandine sind die Leitsubstanz für die Schmerzentstehung im peripheren Gewebe. Die Uteruskontraktionen während der Regelblutung werden durch Prostaglandine gesteuert, die im Endometrium unter dem Einfluß von Östrogenen und Progesteron synthetisiert werden. Das Ausmaß der Prostaglandinsynthese bestimmt die Intensität der Uteruskontraktionen. Während der Proliferationsphase ist die Prostaglandinproduktion niedrig, sie steigt während der Sekretionsphase an, um im menstruellen Endometrium die höchsten Werte zu erreichen. Diese Prostaglandine können während der Menstruation in den Kreislauf gelangen und dort systemische Wirkungen entfalten. Die Dysmenorrhoe der jungen Mädchen muß also heute als Folge einer inadäquaten Ovarialfunktion betrachtet werden. Selbstverständlich können psychische Faktoren diese dysmenorrhoischen Schmerzen verstärken oder überhaupt auslösen, wobei es sich überwiegend um Rollenfindungskonflikte der jungen Frau handelt. Im Sinne einer Konditionierung ist dabei das Verhalten der eigenen Mutter von Bedeutung. Leidet diese selbst unter dysmenorrhoischen Beschwerden, hat sie ihre Tochter ungenügend auf die psychosexuellen Reifungsschritte vorbereitet oder reagiert sie sogar negativ auf deren Menstruationsblutung, wird die junge Frau ebenfalls zu Dysmenorrhoen neigen. Hertz und Molinski (1980) erklären die Dysmenorrhoe als Ausdruck des unbewußten Protests gegen eine als wenig erstrebenswert erlebte weibliche Rolle.

Die symptomorientierte Behandlung besteht in der Gabe von Spasmolytika bzw. Analgetika oder in der lokalen Applikation von Progesteron durch Einlegen eines Progesteron-T. Kontinuierliche Ovulationshemmereinnahme kann die Dysmenorrhoe deutlich bessern oder zum Verschwinden bringen. Diese Therapie sollte durch begleitende ärztliche Gespräche unterstützt werden, um der jungen Patientin zu helfen, eine positivere Einstellung und Neuorientierung gegenüber der als negativ erlebten Geschlechtsrolle zu finden.

Von dieser psychogen-funktionellen Dysmenorrhoe muß die organisch bedingte Dysmenorrhoe getrennt und durch entsprechende Diagnostik ausgeschlossen werden: Fehlbildungen des Genitales – z.B.

Hymenalatresie, Uterusmißbildungen, Zervixaplasie
– oder die Endometriose.

Prämenstruelles Syndrom

Beim prämenstruellen Syndrom treten periodisch in
einem Zeitraum von etwa 2–12 Tagen vor der Menstruation zahlreiche und verschiedenartige Beschwerden auf, die während der Regelblutung meist
rasch wieder verschwinden. Im körperlichen Bereich
kommt es zu einem allgemeinen Spannungsgefühl
mit besonderer Ausprägung in den Brüsten, im Unterleib und in den Beinen, zu Kopfschmerzen, zu einer
Gewichtszunahme aufgrund vermehrter Wassereinlagerung. Im psychischen Bereich läßt sich eine gestei
gerte nervöse Reizbarkeit neben depressiven Verstimmungen und Angstgefühlen beobachten. Diese zu unterschiedlichsten Lebensphasen der geschlechtsreifen
Frau einsetzenden Symptome können im Einzelfall
ein schweres Krankheitsgefühl darstellen. Ob primär
psychische oder endokrine Faktoren das prämenstruelle Syndrom auslösen, ist bisher unbekannt. Klinisch-experimentelle Studien konnten eine Unausge-
wogenheit der ovariellen Sexualsteroide nachweisen.
Aufgrund einer inadäquaten Lutealfunktion ist die
Progesteronproduktion des Corpus luteum unzureichend. Auf diese Weise kommt es zu einer relativen
Östrogendominanz mit entsprechenden generalisierten Auswirkungen (Sitruk-Ware et al., 1980). Wahrscheinlich ist eine passagere Hyperprolaktinämie am
Zustandekommen des Beschwerdekomplexes beteiligt (Benedek-Jaszmann, 1975; Carroll und Steiner,
1978). Frauen mit prämenstruellem Syndrom zeigen
ein geringes Selbstwertgefühl. Sie erleben die Menstruation negativ und neigen zu der Annahme, daß
ihr persönlicher Einfluß auf wichtige eigene Lebensereignisse gering ist (Spencer-Gardner et al.,
1983). Wie die Dysmenorrhoe ist für Hertz und Molinski (1980) das prämenstruelle Syndrom Korrelat
einer narzißtischen Problematik, Ausdruck einer
ganz bestimmten affektiven Haltung der Frau, die von
einem inneren Protest gegenüber der als minderwertig erlebten weiblichen Existenzmöglichkeit bestimmt
wird. Auf dem Boden eines geringen Selbstwertgefühls führen Kränkungen und Enttäuschungen in den
verschiedenen Lebensbereichen der Frau zu einer allgemeinen Unzufriedenheit, die in die Genitalsphäre
projiziert wird. Sie kompensiert dieses beeinträchtigte Selbstwertgefühl durch Aufmerksamkeitsverschie-
bung auf die zyklischen Menstruationsvorgänge, die
nun als Symbol allen weiblichen Eingeschränktseins
erlebt werden (Frick-Bruder, 1984). Beim prämenstruellen Syndrom richtet sich die Behandlung nach
den vorherrschenden Beschwerden. So können in einem Fall Gestagengaben, Ovulationshemmer oder
Bromocriptin, in einem anderen Fall eher Psychopharmaka hilfreich sein. Das begleitende ärztliche
Gespräch sollte versuchen – von jeweils aktuellen Lebenssituationen ausgehend – den Zusammenhang
zwischen der unzufriedenen ärgerlich-gespannten
Grundhaltung und der psychosomatischen Reaktion
aufzuhellen oder aufzudecken.

Psychosomatische Metrorrhagie
(Kontakt- oder Abwehrblutung)

Die psychosomatisch bedingte Metrorrhagie kann
auftreten als kurzzeitige Blutung während jeder Zy-
klusphase, als Tage oder Wochen dauernde leichte
Blutung, aber auch als plötzlich einsetzende starke
Blutung, welche ein operatives Eingreifen erforderlich macht. Häufig werden diese Blutungen während
oder nach dem Koitus beobachtet, ohne daß irgendwelche organische Veränderungen insbesondere an
der Zervix vorhanden sind. Da meist zunächst nicht
an psychosomatische Zusammenhänge gedacht wird,
finden sich in der Vorgeschichte dieser Patientinnen
eine Reihe von erfolglosen Behandlungsversuchen
mit hormonalen Pharmaka und operativen Maßnahmen wie Elektrokoagulation, Kryosation, Lasertherapie, Konisation und Abrasio. Sogar Hysterektomien
werden erwogen und durchgeführt. Für die Mehrzahl
dieser – wahrscheinlich durch neurovegetative Fehlsteuerungen vasomotorisch ausgelösten – Blutungen
werden starke Emotionen bzw. plötzliche Affekte
verantwortlich gemacht (Prill, 1960; Römer, 1969)
oder eine andauernde zwiespältige Einstellung gegenüber der Sexualität oder einem bestimmten Partner
(Mayer, 1944; Prill, 1960). Nach Heimann (1959)
können auch „ausgefallene Trauerreaktionen" bei larvierter Depression zu sog. „Trennungsblutungen" führen, z.B. bei Verlust einer geliebten Person. Wir
konnten einige Fälle mit zum Teil jahrelanger, zunächst therapieresistenter psychosomatischer Metrorrhagie beobachten bei tiefer, ablehnender Enttäuschung über einen Partner und gleichzeitiger Unfähigkeit, hieran etwas zu ändern, z.B. durch Trennung
oder das Durchsetzen eigener Wünsche in der Partnerbeziehung. Fremont-Smith und Meigs (1948) beschreiben eine 9 Monate therapieresistente Metrorrhagie als Ausdruck akuter Angst bzw. Ambivalenz
gegenüber einer Eheschließung. Uns ist in diesem Zusammenhang ein eindrucksvoller Fall bekannt.

Eine 24jährige Studentin wurde am Vorabend ihrer
standesamtlichen Heirat zunächst mit Kreislaufkollaps
und Hyperventilationstetanie notfallmäßig in die medizinische Universitätsklinik eingeliefert. Nach kurzfristiger Behandlung wurde sie wieder entlassen. Zwei
Stunden später kam sie mit derart starker uteriner Blutung in die Frauenklinik, daß noch in derselben Nacht
eine Abrasio notwendig wurde. Die Trauung kam nicht
zustande.

Konfliktzentrierte Gespräche können bei relativ offen
liegender Problematik ein rasches Verschwinden der
Blutungen bewirken. In anderen Fällen gelingt das
nicht, hier ist eine psychoanalytische Psychotherapie
angezeigt.

52.1.2 Zur sterilen Partnerschaft

Paare mit nicht erfülltem Kinderwunsch können unter großem Leidensdruck stehen. Vor allem die Frau
mit frustranem Kinderwunsch findet sich oft in ihrer

Wirkung auf die Umgebung unattraktiv, mißachtet und unbeliebt.

Weiterhin finden sich häufig reaktive Depressionen und vermehrte psychosomatische Symptome. Diese Problematik betrifft auch den Partner, wenn auch in abgeschwächter Form. Bei einer Untersuchung (Schulz-Ruthenberg, 1980) zeigten Paare mit unerfüllt gebliebenem Kinderwunsch gegenüber einem Vergleichskollektiv deutlich mehr psychosomatische Symptome, deutlich größere Schwierigkeiten im Berufsleben und in der Partnerbeziehung. Der nicht erfüllte Kinderwunsch führt zu einer erheblichen Störung im Selbstgefühl dieser Patientinnen und geht häufig mit den äußeren Zeichen einer depressiven Stimmungslage einher. Zur Abwehr der Kränkung werden vor allem zwei Mechanismen beobachtet; einmal die Verleugnung, die diesen Patientinnen trotz der mitgeteilten pathologischen Befunde völlig unrealistische Hoffnungen auf eine Erfüllung ihres Kinderwunsches beläßt und sie oft von einem Spezialisten zum anderen treibt. Der zweite beobachtete Abwehrmechanismus ist die Projektion. Diese Patientinnen schieben ihre innere Unzufriedenheit oft auf die für sie insuffizienten Ärzte oder den subfertilen Partner. Aufgrund groß angelegter Untersuchungsserien kann angenommen werden, daß etwa jede vierte Frau, die wegen Sterilität zur Behandlung kommt, aus psychischen Gründen steril ist (Stauber, 1979). Für die Praxis ergeben sich nach Ausschluß organischer Ursachen für diese Form der Sterilität folgende Hinweise: gehäuftes Auftreten psychischer und psychosomatischer Symptome, wie z. B. Amenorrhoe, Anovulation, Follikelinsuffizienz. Beim Mann imponieren meistens große Schwankungen in den Spermiogramm-Parametern (Spermienzahl, -motilität und -morphologie). Psychosozialer Streß im beruflichen und familiären Bereich korreliert hierbei mit subfertilen Werten. Die Interaktion in den psychogen-sterilen Partnerschaften zeigt häufig ein anklammernd-symbiotisches Beziehungsmuster. Diese Beziehungsform ist zwar stabil, sie beruht jedoch darauf, daß ein Partner das Verhalten des anderen determiniert, während sich der andere anpaßt.

Tabelle 52–2. Behandlungsaspekte bei unerfülltem Kinderwunsch

– **Sterile Paare mit „überwertigem" Kinderwunsch**
Leidensdruck + + + (anfallsweiser „Kinderhunger",
Spezialistensuche)
Agieren vorwiegend der Patientinnen (Ärzteverschleiß)
Erschwerte Arzt-Patient-Beziehung (psychologische
Führung!)

– **Sterile Paare mit „starkem" Kinderwunsch**
Leidensdruck + + (Drängen auf invasive medizinische
Eingriffe)
Depressive Reaktionen und negative soziale Resonanz
In vertrauensvoller Arzt-Patient-Beziehung gut führbar

– **Sterile Paare mit „gesundem" Kinderwunsch**
Leidensdruck + (Zögern gegenüber invasiven medizinischen Eingriffen)
Frustraner Kinderwunsch wird sozial untergebracht
Ausgewogene Arzt-Patient-Beziehung

Die Behandlung des Sterilitätsproblems sollte stets gleichzeitig bei beiden Partnern erfolgen. Besonders wichtig erscheint hierbei die Integration psychosomatischer Schritte in die Diagnostik und Therapie von Frau und Mann. Dabei ist die Wertigkeit des Kinderwunsches besonders in der Arzt-Patient-Beziehung zu berücksichtigen. Tabelle 52–2 vermag dem Arzt für die Behandlung steriler Paare einige Hinweise zu geben.

Das Vorgehen muß sowohl aus somatischen als auch aus psychischen Gründen für jede Patientin individuell gewählt werden. In einigen „Kinderwunschzentren" wurden in den letzten Jahren vermehrt psychosomatische Aspekte routinemäßig in die Behandlung integriert. Als Beispiel dient das Modell an der Universitäts-Frauenklinik Berlin-Charlottenburg, das in der folgenden Tabelle dargestellt ist (Tab. 52–3).

Ein wichtiges Ergebnis, das vor der Ära außerkörperlicher Befruchtungen gewonnen wurde, wird in Abbildung 52–3 dargestellt. Hieraus ist zu ersehen, daß von 1061 eingetretenen Schwangerschaften in der Kinderwunschsprechstunde der Universitäts-

---

Tabelle 52–3. Fertilitätssprechstunde für beide Partner (Modell UFK Berlin-Charlottenburg)

	Anamnese	Diagnostik	Therapie
♀	Kinderwunschdauer prim./sek. Sterilität Vorbehandlungen Zyklus	genitaler Befund BTK, Zervixfaktor Hormone, Genetik, US erweiterte Laparoskopie	Entzündungsbehandlung Ovulationsterminierung Insemination, Adoption Mikrochirurgie, IVF
♀♂	Leidensdruck durch KW Partnerbeziehung (stabil?) KW-Motivation Vita sexualis	psychosom. Symptome Persönlichkeitsstruktur Partnerinteraktion	psychische Führung (z. B. IVF) Behandlungspausen, Entspannung Psychotherapie (AT, Paartherapie) Kontaktangebot (Cave fix. KW)
♂	genitalspez. Erkrankungen Vorbehandlungen, OP Noxen (Nikotin, Medikam.)	genitale Befunde Spermiogramme (Streß?) Hormone, SH, Immunologie Hodenbiopsie	Entzündungsbehandlung Hormontherapie OP, Spermakonservierung für Insem., Adoption

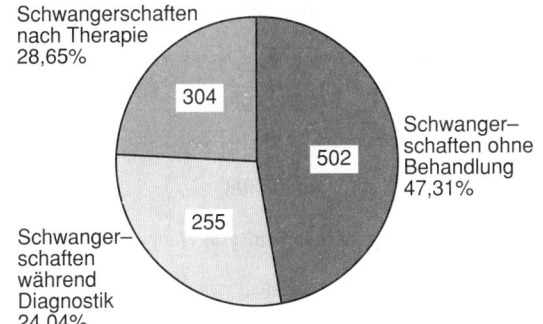

Abb. 52–3. Therapie und Konzeption bei sterilen Ehepaaren (n = 1061 eingetretene Schwangerschaften).

Frauenklinik Berlin-Charlottenburg nahezu die Hälfte ohne Behandlung eingetreten sind. Da diese „Erfolge" vorwiegend in der Wartezeit auf einen Behandlungsplatz oder auch in Behandlungspausen registriert wurden, muß man vor voreiligen invasiven Kinderwunschbehandlungen warnen (Stauber, 1988).

Durch die Möglichkeiten der extrakorporalen Fertilisierung sind die Hoffnungen steriler Paare auf die Erfüllung ihres Kinderwunsches deutlich erweitert worden. Obwohl man sich stets fragen muß, ob man den überwertigen Kinderwunsch nicht besser auf seinen tiefenpsychologischen Inhalt – häufig Kind als Substitut für die eigenen nicht erfüllten Wünsche – zurückführen sollte, besteht doch eine reale Möglichkeit, vor allem Frauen mit zerstörten Eileitern zu helfen. Bevor man dieses Verfahren anwendet, sollten beide Partner psychosomatisch untersucht werden. Kontraindikationen sind z.B. Psychosen oder starke neurotische Depressionen bei einem der Partner. Man sollte sich ebenfalls vor diesem Verfahren hüten, wenn ein Partner ambivalenten Kinderwunsch zeigt oder wenn allgemein das gewünschte Kind die Partnerschaft aufrechterhalten soll. Da allein die Durchführung dieses Verfahrens eine größere psychische als organische Belastung darstellt, ist eine vertrauensvolle Arzt-Patient-Beziehung erforderlich. Die Phantasien von Frau und Mann deuten auf ein ständiges Schwanken zwischen Hoffnung und Realität hin. Diese Paare bedürfen einer einfühlsamen Führung, wobei stets auf eine erhebliche psychische Belastung zu achten ist. Eine Bereitschaft zur Hilfe bei vergeblichem Kinderwunsch ist anzustreben.

52.1.3 Überlegungen zur modernen Reproduktionsmedizin

Allgemein ist noch darauf hinzuweisen, daß durch die moderne Reproduktionsmedizin und hier vor allem die außerkörperliche Befruchtung (In-vitro-Fertilisation, IVF) eine neue Dimension in die Medizin gekommen ist. Seit der Geburt des ersten extrakorporal gezeugten Kindes in England (1978), ist es möglich, die unmittelbare Entstehung des Menschen im Labor zu beobachten oder sogar an ihr zu manipulie-

ren. Dabei eröffnen sich Perspektiven, die vor allem dem psychosomatisch denkenden Arzt großes Unbehagen bereiten müssen. Folgende Fragen erscheinen für die Zukunft wichtig:

– Wo sind die Grenzen des technisch Machbaren?
– Ist ein Mißbrauch nicht schon vorprogrammiert?
– Laufen wir nicht Gefahr, daß die technische Entwicklung unserer geistigen Entwicklung davonläuft?
– Ist es nicht sinnvoll, die Wünsche der Patienten nach einer grenzenlosen Reproduktionsmedizin mit dem Vorschlag des Verzichts auf ein eigenes Kind zu beantworten?

Der mögliche Mißbrauch und die notwendige Grenzziehung in diesem Bereich der außerkörperlichen Befruchtung werden zur Zeit am häufigsten diskutiert. Sowohl in der Arbeitsgruppe an der Universitäts-Frauenklinik Berlin-Charlottenburg als auch in der Arbeitsgruppe an der Universitäts-Frauenklinik München wird das sog. „Berliner Modell" als Richtschnur anerkannt. Dieses Modell stellt eine Eigenbeschränkung dar und zeigt die neuralgischen Punkte auf, an denen die vielfältigen theoretischen Überlegungen zusammenlaufen. So werden die modernen Befruchtungstechnologien nach diesem „Berliner Modell" nur innerhalb der Familienstruktur vorgenommen. Es kommt also nicht zur Samenspende, zur Eispende oder zur Leihmutterschaft. Weiterhin werden keine verändernden Manipulationen am Embryo geduldet, also keine verbrauchenden Experimente, keine Embryoteilung, keine Embryofusion und keine Chimärenbildung. Durch eine maßvolle Stimulation bzw. Fertilisation entstehen somit auch keine überzähligen Embryonen. Schließlich sollen alle diese Befruchtungstechnologien nur nach strenger Indikation – auch von psychosomatischer Seite – vorgenommen werden.

Was die neuen Reproduktionstechniken betrifft, so müssen die Ärzte auch an die „Verformbarkeit ihres Gewissens" denken, da durch die routinemäßige Anwendung dieser Verfahren die Sensibilität für eine Grenzziehung verlorengehen kann. Ein Beispiel hierfür ist die Indikationsstellung für die In-vitro-Fertilisation, die sich bereits in den letzten fünf Jahren geändert hat. So hatten einmal alle Arbeitsgruppen damit begonnen, die In-vitro-Fertilisation streng an die Indikation irreparabel gestörter Eileiter zu binden. Als plausible Erklärung diente dazu die Feststellung: „Es geht darum, individuelles Leid nach Entzündungen und Operationen zu beseitigen". Die Indikation dehnte sich dann in vielen Arbeitsgruppen bereits auf „idiopathische" Sterilitäten aus, d.h. jedes Paar mit nicht erfülltem Kinderwunsch kann einer IVF unterzogen werden. Auf der Argumentationsebene: „Wir brauchen eine effektivere Kinderwunschbehandlung" wird meist die Tatsache vergessen, daß viele Paare in dieser idiopathischen Gruppe aus psychischen Gründen nicht zu einer Konzeption kommen.

Die nächste logische, wenn auch nicht psychologische Argumentationsebene besteht darin, den Patientenwunsch als einzig gültig hinzustellen. Dies hat dann natürlich zur Folge, daß man bereit sein muß zur Eispende, zur Samenspende, zum Embryokauf oder auch zur Surrogatmutterschaft. Die IVF als „Komforteingriff" mit kommerziellen Nebenerscheinungen wäre somit der nächste Schritt einer Revolution auf leisen Sohlen. Und in der Tat gibt es z.B. australische Arbeitsgruppen, die wenig Probleme in einer derartigen Fortentwicklung sehen und diese auch schon teilweise praktizieren.

Ein weiterer Schritt, der mit einer „Gewissensabschwächung" einhergeht, ist die Forschung an Embryonen. Durch das Argument: „Wir wollen doch den Fortschritt für die Methode der IVF, für die Krebsforschung, für genetisch gesunde Kinder" scheint der nächste Schritt dieser Entwicklung ohne großen inneren Widerstand getan. Auch hierfür gibt es bereits in ausländischen Arbeitsgruppen Stimmen und Taten.

Um diese langsame „Über-Ich-Abschwächung" zu einem möglichen Ende zu führen, soll die Frage aufgeworfen werden, ob nicht in Zukunft auch die Gefahr besteht, daß die In-vitro-Fertilisation nicht mehr nur als Heilungstechnik gebraucht wird, sondern als allgemeines Zeugungsverfahren, das nur noch „zu gesunden, guten, neuen Menschen" führt. Auch hier kann man wieder logisch argumentieren, daß es im Interesse der Medizin und natürlich auch der Gesellschaft stehen müßte, Erbkrankheiten gänzlich auszuschalten. Da der soziale Zwang ja auch heute schon in einigen Bereichen zu einer „Gewissenslaxierung" geführt hat (z.B. bei der Handhabung des § 218), könnte die hier aufgezeigte Revolution in kleinen Schritten eine Entwicklungshilfe in der modernen Reproduktionsmedizin darstellen.

Mit dem zuvor aufgezeigten Berliner Modell, das nach fünf Jahren Praxis nun auch von der Bundesärztekammer in fast gleichlautender Form in einem Rundbrief erschien, soll eine interdisziplinäre Diskussion erreicht werden, die einen möglichen Mißbrauch verhindern kann.

52.1.4 Kontrazeption

Wie noch zu keiner Zeit und in keiner Kultur hat der Mensch Macht und Kontrolle erlangt über seine Fruchtbarkeit. Neben sicherer Schwangerschaftsverhütung ermöglicht die moderne Endokrinologie auch Ovulationsauslösung durch computergesteuerte Hormonpumpen, Insemination und In-vitro-Fertilisation. Fast alles scheint machbar geworden. Andererseits wirft allein die Zahl von jährlich ca. 200000 an deut-

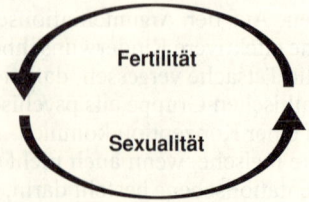

Abb. 52–4. Fruchtbarkeit und Sexualität.

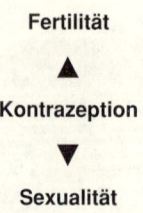

Abb. 52–5. Kontrazeption führt zur Trennung von Fruchtbarkeit und Sexualität.

schen Frauen vorgenommenen Schwangerschaftsabbrüchen die Frage auf, ob wir mit den Möglichkeiten sicherer Kontrazeption richtig umzugehen verstehen.

Fruchtbarkeit und Sexualität sind wesentliche Grundbedürfnisse des Menschen. Der Wunsch nach oder die Angst vor einem Kind auf der einen, das Bedürfnis nach sexueller Befriedigung auf der anderen Seite sind aber auch konflikthafte Bereiche, die tief in die seelisch-körperliche Befindlichkeit eingreifen. Die heute verfügbaren sicheren Kontrazeptiva ermöglichen eine absolute Loslösung der Sexualität von der Chance oder dem Risiko der Zeugung neuen Lebens (Abb. 52–4 und 52–5). Dieser bewußt gewollten Trennung können zahlreiche unbewußte Wünsche oder Befürchtungen entgegenwirken. Der bewußt gewollte und von den realen Lebensumständen her durchaus vernünftige Verzicht auf ein Kind schließt nicht aus, daß gleichzeitig ein starker unbewußter Wunsch nach einem Kind wirksam sein kann (Molinski, 1972; Hertz und Molinski, 1980). Diese Diskrepanz im bewußten und unbewußten Erleben erklärt die zahlreichen Probleme bei der Anwendung gerade sicherer kontrazeptiver Methoden. So sind beispielsweise die bei regelmäßiger Ovulationshemmereinnahme zu beobachtenden Nebenwirkungen nur selten spezifische Folge der Hormonwirkung der „Pille", sondern überwiegend psychosomatischer Ausdruck unbewußter Konflikte der Bereiche Sexualität und Fertilität. Daher führt das Verordnen einer hormonal anders zusammengesetzten „Pille" oder das Übergehen auf eine andere sichere kontrazeptive Methode, wie z.B. das Einlegen eines Intrauterinpessars, meist nicht zum erhofften Erfolg.

Die Symptome bleiben oder verschieben sich oder es treten neue Beschwerden auf.

Kontrazeption als Chance kann bedeuten: eine Frau muß keine Angst mehr vor ungewollter Schwangerschaft haben, emanzipatorische Bestrebungen, wie z.B. die Zuwendung zu einer als befriedigend erlebten Berufstätigkeit, können gefördert werden, die partnerschaftliche Sexualität kann freier, unbehinderter erlebt werden. Sichere Kontrazeption kann also unabhängiger, souveräner, glücklicher machen und das allgemeine Lebensgefühl sowie die weitere Lebensplanung positiv beeinflussen.

Kontrazeption als Risiko meint die Zwiespältigkeit, die vielschichtige Konflikthaftigkeit, die mit dem bewußten Abtrennen von Geschlechtlichkeit und Fruchtbarkeit von der Tiefe des Empfindens und Wahrnehmens verbunden sein kann (Petersen, 1980).

Folgende konflikthafte Konstellationen finden sich in der kontrazeptiven Beratung:

- Ablehnung von sicherer Kontrazeption, obwohl kein Schwangerschaftswunsch besteht:
  Es gibt Patientinnen, die absolut kein Kind (mehr) wollen, trotzdem aber keine Kontrazeption betreiben. Hier kann ein innerer Widerstand gegenüber der Trennung von Sexualität und Fertilität vorliegen. Sexualität kann nur zugelassen werden, wenn auch die Möglichkeit der Zeugung besteht. Das sichere Gefühl, nicht empfangen zu können, entwertet das sinnliche Sich-aufeinander-Einlassen. Hierdurch erklären sich manche Sexualstörungen wie Dispareunie, Libidoabnahme oder Libidoverlust bis zur völligen Ablehnung des Koitus, nachdem die gewünschte Kinderzahl erreicht ist. Auf der anderen Seite kann dann gerade in der Schwangerschaft der sexuelle Kontakt als besonders befriedigend erlebt werden. Daneben können sexuelle Lustgefühle von manchen Frauen als etwas Verbotenes erlebt werden. Auch sie können Sexualität nur unter der Bedingung zulassen, daß Empfängnis möglich ist. Kommt es zum Eintritt einer Schwangerschaft, wird dies von ihnen als eine Art Buße oder Strafe für das sinnliche Sich-Gehenlassen erlebt. Die Abwehr von lustvoll Sinnlichem hängt mit frühkindlichen triebfeindlichen Erziehungseinflüssen zusammen. Neurotische Schuldgefühle infolge einer als verboten erlebten Sexualität werden als psychosomatische Symptome in zahlreichen Nebenwirkungen der verschiedenen Kontrazeptionsmethoden sichtbar. Darüber hinaus kann die Empfehlung oder Verordnung sicherer Kontrazeptiva schon bestehende Sexualängste noch vermehren, wenn sich diese Frauen dann in verstärktem Maße der Verpflichtung zum sexuellen Kontakt ausgesetzt fühlen. Schließlich kann eine neurotische Angst vor Intimität und Hingabe zur Vermeidung zuverlässiger Kontrazeption führen. Die sexuelle Intimität gefährdet die eigene Identität. Die Hingabe an einen Partner kann nicht ohne Angst bewältigt werden. Nicht selten wird eine Schwangerschaftsangst vorgeschoben oder findet sich real tief verankert. In der Beratung werden dann gegen jede vorgestellte kontrazeptive Methode Einwände erhoben oder die Sicherheit selbst der zuverlässigsten Möglichkeiten wird bezweifelt.

- Unregelmäßige, unzuverlässige Kontrazeption:
  Hinter diesem häufig in der Kontrazeptionssprechstunde auftretenden Phänomen steht meist ein unbewußter Schwangerschaftswunsch. Die Patientinnen werden von ambivalenten Gefühlen hin und her bewegt, und zwar derart, daß weder eine Schwangerschaft noch eine dauerhaft zuverlässige Kontrazeption eine befriedigende Lösung für sie darstellt. Zeitweise kümmern sie sich um eine sichere Kontrazeption, dann gehen sie wieder phasenweise ein Schwangerschaftsrisiko ein. Unregelmäßiges oder unzuverlässiges Kontrazeptionsverhalten findet sich bei einem allgemeinen Mangel an Reife und Verantwortungsbewußtsein und bei Partnerkonflikten. Lehfeldt und Guze (1966)

konnten zeigen, daß sog. unerwünschte Schwangerschaften vermehrt bei neurotischen Paaren vorkommen, als Abwehr der Angst, den Partner zu verlieren. Fruchtbar zu sein, ist für manche Frauen ein so wesentlicher Bestandteil des Selbstwertgefühls, daß eine sichere Kontrazeption aus der Angst heraus vermieden wird, dieses hohe Gut zu verlieren. Andere wiederum fühlen sich gedrängt, sich selbst und ihre biologische Intaktheit immer wieder zu beweisen, indem sie schwanger werden, ohne wirklich ein Kind zu wollen. Uns ist in diesem Zusammenhang eine 35jährige ledige Patientin bekannt, die trotz zum Teil schrecklicher und schmerzhafter Abruptio-Erfahrungen in unterschiedlichsten Institutionen nicht zu einer konsequenten Kontrazeption zu bewegen war.

Orale Kontrazeption

Zahlreiche Untersuchungen existieren über die psychosomatischen Nebenwirkungen von oralen Kontrazeptiva (Übersicht bei Petersen, 1969; Nijs, 1972; Mall-Haefeli, 1974). In Abhängigkeit der methodischen Qualität der Untersuchungen schwanken die Angaben über psychische Nebenwirkungen erheblich. Grundsätzlich überspielen biographische und psychische Faktoren bei weitem die physische und endokrine Situation der Frau. Entscheidend, ob Nebenwirkungen auftreten oder nicht, ist die individuelle innere Situation der Frau.

Als positive seelische Veränderungen unter kontinuierlicher Ovulationshemmereinnahme lassen sich beobachten: eine Abnahme von Schwangerschaftsängsten, ein größeres seelisches Wohlbefinden mit Abnahme von ängstlicher Reizbarkeit, ein entspannteres sexuelles Erleben mit besserer psychosexueller Anpassung an den Partner.

Unerwünschte Nebenwirkungen im psychischen Bereich sind: depressive Verstimmungen, Antriebsschwäche, erhöhte nervöse Reizbarkeit, Libidoabnahme. Neben einer Veränderung der sexuellen Bedürfnisse findet sich eine vielfältige psychosomatische Symptomatologie wie unerwünschte Gewichtszunahme, Mastodynie, Kopfschmerzen, Schwitzen, Schwindelgefühle, Schlafstörungen, Erbrechen, gastrointestinale Beschwerden, Herzsensationen, Ödemneigung etc.

In Zusammenhang mit der Einnahme von oralen Kontrazeptiva lassen sich eine Reihe von Ängsten bzw. Befürchtungen beobachten.

- Selbstbeschädigungsängste:
  Von der Pille wird ein schädlicher Effekt auf den eigenen Körper erwartet. Die geäußerten Befürchtungen sind: Angst vor Unfruchtbarkeit, Angst vor früher einsetzenden Wechseljahren, Befürchtungen nachteiliger Folgen für ein Kind, das während oder nach der Pilleneinnahme empfangen wird. Angst vor Organerkrankungen, die als Folge der Ovulationshemmereinnahme auftreten können, wie Leberschädigung, Thrombose und Embolie, Zunahme von Krampfaderleiden, unkontrollierte Gewichtszunahme.

– Angst vor Veränderung der eigenen Emotionalität und der sexuellen Triebhaftigkeit:
Gefürchtet wird eine allgemeine Veränderung der eigenen Gefühlslage. Im sexuellen Erleben gibt es die Furcht vor Gefühlskälte z.B. bei der Vorstellung, daß die Verhinderung der Ovulation in Zusammenhang stehen könnte mit abnehmendem sexuellem Interesse. Andererseits können sich Ängste vor sexueller Aggressivität einstellen, der eigenen oder der des Partners, wenn die Angst vor einer Schwangerschaft wegfällt.

Intrauterine Kontrazeption

Nach den Ovulationshemmern kommt der intrauterinen Kontrazeption als nächst sicherer reversibler Methode steigende Bedeutung zu. Neben einer Gruppe von Frauen, die mit dieser Kontrazeptionsmethode zufrieden sind und sie ungestört tolerieren, scheint das IUP von Frauen bevorzugt zu werden, die sich durch die sichere generalisierte hormonelle Wirkung der Ovulationshemmer zu sehr beeinträchtigt fühlen (Jürgensen et al., 1979; Frick-Bruder, 1980). Angst vor Schmerzen bei der Einlage bzw. beim Entfernen des IUP, verstärkte, verlängerte Blutungen, Entzündungen und daraus resultierende Unfruchtbarkeit sind bewußte Vorbehalte gegenüber der intrauterinen Kontrazeption. Darüber hinaus finden sich Vorstellungen wie: die „Spirale" könnte wandern, verrutschen, einen selbst oder den Partner verletzen. Ein für die kontrazeptive Beratung wichtiger Zusammenhang zwischen emotionaler Einstellung des Arztes und der Akzeptanz bzw. Verträglichkeit des IUP wurde von Reading und Newton (1977) nachgewiesen. War der Arzt beim Einlegen des IUP emotional zugewandt, zuversichtlich und von der kontrazeptiven Methode überzeugt, so verringerten sich die Schmerzen beim Einlegen sowie Zahl und Ausmaß späterer Nebenwirkungen.

Definitive Kontrazeption (Sterilisation)

Die Einstellung gegenüber der Sterilisation hat sich erheblich verändert. Galt dieser Eingriff noch vor wenigen Jahren als sittenwidrige ärztliche Handlung, so wird diese definitive Kontrazeptionsmethode heute von der breiten Öffentlichkeit akzeptiert, von den Krankenkassen als Pflichtleistung übernommen. Zweifellos ist für die Frau im mittleren und höheren Fertilitätsalter diese Methode durch ihre Sicherheit, durch das Fehlen organischer Nebenwirkungen und das inzwischen risikoarme Routineverfahren der laparoskopischen Tubenkoagulation anderen kontrazeptiven Maßnahmen überlegen. In der Bundesrepublik Deutschland werden jährlich ca. 50 000 derartige Eingriffe vorgenommen. Um so erstaunlicher ist eine häufig zu beobachtende Unsicherheit und ein mangelhaftes faktisches Wissen gegenüber dieser Methode. Nach Wenderlein (1974) erwarten 40% eine Beeinträchtigung ihres Wohlbefindens, 25% eine Störung des sexuellen Erlebens und 10% Hormonstörungen. Daraus ergibt sich die Notwendigkeit einer umfassenden medizinisch-psychologischen Beratung unter Einschluß des Partners. Die Irreversibilität der Methode, der zukünftige bewußte Verzicht auf eigene Kinder verlangt eine gründliche Auseinandersetzung mit den seelischen Kräften des Fruchtbarkeitsbedürfnisses, nicht nur den eigenen, sondern auch denen des Partners, unter Einhaltung einer gewissen Bedenkzeit vor dem operativen Eingriff. Wesentliche Erkenntnisse über die psychosomatische Verarbeitung der Sterilisation konnten durch die Auswertung von 40 psychologisch-psychiatrischen Reihenkatamnesen von mehr als 1200 Frauen gewonnen werden (Petersen, 1977 und 1980). Nur 5% der Frauen sind später mit dem Eingriff nicht zufrieden. Nach Petersen zählen zu den günstigen Prognosebedingungen: eine ausreichende und qualifizierte Beratung, ein freier, wohlüberlegter, mit dem Partner durchdachter und von beiden Partnern getragener Entschluß, klare und eindeutige kontrazeptive Motivation, seelische und partnerschaftliche Ausgewogenheit und ein ausreichender Entscheidungszeitraum. Lebensalter und Kinderzahl – Faktoren, die von den meisten Ärzten als wesentlich für die Entscheidung berücksichtigt werden – haben kaum Einfluß auf die Verarbeitung der Operation. Dies ist für die Beratungspraxis wichtig, denn Lebensalter und Kinderzahl werden allzuleicht fälschlicherweise als die wesentlichen Merkmale für die Reife der Persönlichkeit, die Ausgewogenheit der Entscheidung und die Stabilität der Paarbeziehung unterstellt.

Nach der Operation kommt es zu einem phasenhaft ablaufenden psychischen Verarbeitungsprozeß von 1–4 Jahren Dauer (Barglow, 1964). Dieser Verarbeitungsprozeß betrifft auch den nichtoperierten Partner, so daß von einer psychosomatischen Reaktion des Paares ausgegangen werden muß. Verschiedene Verarbeitungsformen können auftreten: sensible, produktive, „dickhäutige" und neurotische.

Psychisch gesunde sensible Frauen zeigen unmittelbar nach dem Eingriff leichte depressive Verstimmungen, gefolgt von Tagphantasien und Träumen, die sich mit Schwangerschaft beschäftigen. In der ersten, ½ bis 1 Jahr dauernden Phase, die meist ohne somatische oder psychosomatische Symptombildung verläuft, kommt es zu Restitutionsphantasien etwa in Form zweifelnder Gedanken über die Sicherheit der Tubenkoagulation, oder die Frauen überprüfen ihre Unfruchtbarkeit, indem sie bewußt zum Zeitpunkt der Ovulation Geschlechtsverkehr suchen. Dagegen entwickeln extravertierte körperlich robuste Frauen („Dickhäuter") keinerlei psychosomatische Probleme. Bei bereits vor der Sterilisation neurotisch auffälligen Frauen können aufgrund einer chronischen seelischen Fehlverarbeitung zahlreiche psychische, psychosomatische oder sexuelle Störungen auftreten. Diese Problematik kann in einzelnen Fällen derart drängend werden, daß eine refertilisierende Operation erwogen werden muß.

Die früher häufiger durchgeführte Tubensterilisation unmittelbar post partum ist wegen der zahlreichen psychosomatischen Folgeerscheinungen weitgehend aufgegeben worden. Die Wöchnerin befindet sich – nicht nur nach einer schwierigen Schwangerschaft und/oder traumatischen Geburtserlebnissen – in einem Ausnahmezustand. Eine definitive Entscheidung sollte nicht vor Ablauf von drei Monaten nach der Geburt getroffen werden. Natürlich erleichtert eine aus medizinischen Gründen notwendige Sectio in Einzelfällen die Entscheidung zur gleichzeitigen Tubenligatur. Andererseits sind gerade die wegen zwingender medizinischer Indikation durchgeführten Sterilisationen besonders komplikationsreich hinsichtlich unerwünschter psychosomatischer Nebenwirkungen. Die Patientin kann den Eingriff als zusätzliches Trauma ihrer ohnehin schon verletzten psychophysischen Integrität fehlverarbeiten. Gleiches gilt für die mit einem Schwangerschaftsabbruch durchgeführte Sterilisation, um so mehr, wenn dies von der Patientin als Bedingung für die Abruptio erlebt wird.

52.1.5 Schwangerschaftsabbruch

Der Schwangerschaftsabbruch steht nach wie vor inmitten kontroverser weltanschaulich-politischer Diskussion. Für die einen ist er eine Forderung von Liberalität und Emanzipation, für die anderen ein Ärgernis, ein Zeichen zunehmenden moralischen Verfalls. Für diejenigen, die sich von Berufs wegen mit dem Problem befassen müssen, ist die Kenntnis wichtig, welche medizinischen und psychologischen Phänomene mit diesem Eingriff zusammenhängen. Nach der Reform des § 218 aus dem Jahre 1976 ist der Schwangerschaftsabbruch bei bestimmten Indikationen erlaubt. Man unterscheidet die medizinische, die eugenische, die ethische oder kriminologische und die Notlagenindikation. Die Zahl der aus Notlagenindikation vorgenommenen Schwangerschaftsabbrüche ist in den vergangenen Jahren stetig angestiegen. Heute erfolgen über ¾ aller legalen Abruptiones aus Notlagenindikation. Während bei der Beratung und Indikationsstellung bei medizinischer, eugenischer und kriminologischer Indikation im allgemeinen keine größeren Probleme auftauchen, kommt es im Entscheidungsprozeß zur Notlagenindikation in der Praxis zu Schwierigkeiten und Mißverständnissen. Viele Ärzte fühlen sich bei der Aufgabe überfordert, aufgrund der Angaben der Patientin über ihre persönlichen und familiären Verhältnisse zu erkennen, ob eine den Schwangerschaftsabbruch rechtfertigende Notlage vorliegt oder nicht. Andererseits wurde gerade der Schwangerschaftskonfliktberatung vom Gesetzgeber große Bedeutung zugemessen. Sie wurde bindend vorgeschrieben. Es wird verständlich, daß mancher Arzt vor solchen konflikthaften und zeitraubenden Gesprächen zurückschreckt und eine Mitwirkung beim Schwangerschaftsabbruch grundsätzlich ablehnt. Tatsächlich gehört Schwangerschaftskonfliktberatung zu den schwierigsten und verantwortlichsten Beratungsaufgaben, die vom Arzt im psychosozialen Bereich geleistet werden müssen. Die ratsuchende Frau muß zwischen zwei Alternativen entscheiden: Austragen des Kindes oder Schwanger-

schaftsabbruch. Zusätzlich besteht ein objektiver und subjektiver Zeitdruck. Hinzu kommt in vielen Fällen ein äußerer sozialer Druck durch Partner, Eltern, Freunde, Bekannte. Ziel der Schwangerschaftskonfliktberatung muß es sein, mit der Frau so zu sprechen, daß sie unter allmählicher Aufgabe aller zunächst vordergründigen Motivationen ihre wahren Tendenzen, Ambivalenzen, Wünsche und Interessen, Kollisionen mit sich selbst, mit dem Partner, mit der Familie, dem Beruf und der Gesellschaft klarer sehen kann. Erst dann wird sie ihre bestmögliche Entscheidung finden können. Eine solche Gesprächsführung setzt voraus, daß der beratende Arzt auf die Haltung verzichtet, daß diese oder jene Entscheidung die richtige Lösung für die Frau ist. Diese individuelle Gesprächsführung hat sich für die seelische Verarbeitung des Schwangerschaftsabbruchs als prognostisch besonders günstig erwiesen (Molinski, 1975; Poettgen, 1977).

In vielen Fällen stellt sich heraus, daß hinter den von der Frau vorgebrachten realen Gründen zum Schwangerschaftsabbruch viel grundsätzlichere und tiefere Konflikte liegen, so daß nicht zu erwarten ist, daß diese Konflikte durch einen Schwangerschaftsabbruch zu lösen sind. Jürgensen (1983) konnte zeigen, daß bei 26% der von ihr tiefenpsychologisch untersuchten Frauen die ungewollte Schwangerschaft einen unglücklichen Konfliktlösungsversuch bei Trennungstraumen darstellte. Die Schwangerschaft, die nach den realen äußeren Verhältnissen unerwünscht war, schien dennoch einem inneren Bedürfnis zu entsprechen, einem zwar unbewußten, aber das Verhalten bestimmenden Kinderwunsch. In die gleiche Richtung weisen die Nachuntersuchungen von Goebel (1984) an 125 Frauen. 39% führten auch kurz nach der Abruptio keine zuverlässige Kontrazeption durch, d.h. der tieferliegende Konflikt ging weiter. Der Schwangerschaftskonfliktberatung kommt somit eine entscheidende Weichenstellung in einer Lebenskrise zu, wenn sie in der Lage ist, auch die tiefenpsychologische Dimension der ratsuchenden Patientin zu erfassen.

52.1.6 Unterbauchschmerzen

Unterbauchschmerz-Patientinnen gehören zu den wirklichen Problempatientinnen in der gynäkologischen Sprechstunde. Zum einen ist bereits die diagnostische Abgrenzung schwierig, zum anderen bleiben therapeutische Maßnahmen häufig unbefriedigend oder gar erfolglos, weil psychosomatische Aspekte dieser Erkrankung nur ungenügend bekannt sind oder nicht berücksichtigt werden. Der Umgang mit Unterbauchschmerz-Patientinnen wird von Gefühlen der Hilflosigkeit, der Frustration oder der Ärgerlichkeit bestimmt. Das erklärt die häufig zu beobachtende aktiv operative Vorgehensweise von Gynäkologen, welche die Ursache der Unterbauchschmerzen gleichsam „an der Wurzel" packen soll. So begleiten nach erfolglosen Spasmolytika-, Analgetika-, Antibiotika-Behandlungen und physikalischen Maßnahmen nicht selten zahlreiche Operationen den Leidensweg dieser Patientinnen.

Pelipathie-Syndrom

Das Pelipathie-Syndrom galt lange Zeit als ein nur unscharf definiertes, schwer präzisierbares Krankheitsbild mit organisch nicht erklärbaren, meist chronischen Unterbauchschmerzen der Frau. Die Schwierigkeiten, mit diesem Problem zurecht zu kommen, zeigen sich eindrücklich in den zahlreichen Bezeichnungen und Interpretationen, welche in die Literatur Eingang gefunden haben, z.B.: Hysteralgie (Scanzoni, 1870), Krankheit mit den 20 Namen (Naujoks, 1920), Beckenneuralgie (Cotte, 1931), Parametropathia spastica (Martius, 1929), schwebende Pein (Sellheim, 1929), Zervikal-Syndrom (Young, 1938), Pelipathia vegetativa (Gauss, 1949; Prill, 1964), pelvic congestion (Taylor, 1949), Pseudoadnexitis spastica, Pelvipathie (Artner, 1982; Roemer, 1969), chronic pelvic pain (Sinclair, 1972; Renaer, 1981), chronische Schmerzzustände ohne deutlich organische Pathologie (Renaer, 1980), Unterleibsschmerzen ohne Organbefund (Molinski, 1982; Nijs, 1983), Pelipathie-Syndrom (Richter, 1979).

Nach Artner (1982) soll es inzwischen 150 Synonyma geben. Wir sprechen vom Pelipathie-Syndrom, weil es sich um ein polysymptomatisches Krankheitsbild mit einer ganzen Reihe zusätzlicher psychosomatischer Symptome handelt, wobei die Unterbauchschmerzen häufig das führende Leitsymptom sind, aber nicht sein müssen. In abnehmender Häufigkeit finden sich Sexualstörungen, Magenfunktionsstörungen, hypotone Kreislaufstörungen, Kopfschmerzen, Dysmenorrhoe, psychogener Fluor, Obstipation, Oligomenorrhoe, Schlafstörungen, prämenstruelles Syndrom. Der Gynäkologe kann drei unterschiedliche Palpationsbefunde unterscheiden: Druckschmerzhaftigkeit von Organen im kleinen Becken ohne Vorhandensein einer eigentlichen Organpathologie; schmerzhaft können sein: der Uterus, eine oder beide Adnexe, die Beckenwände, die Parametrien, die Sakrouterinligamente. Oder es findet sich ein „teigig gestauter" Tastbefund bei Vasokongestion im kleinen Becken. Schließlich kann die Palpation ganz unauffällig sein.

Organische Schmerzursachen können laparoskopisch ausgeschlossen werden, wie entzündliche Veränderungen, Adhäsionen im Genitalbereich, Varicosis pelvina, Endometriose, Allan-Masters-Syndrom. Allerdings darf bei Vorliegen eindeutiger pathologischer Organbefunde nicht ohne weiteres gefolgert werden, daß diese die eigentliche und alleinige Ursache für die Beschwerden sind, denn organische und psychosomatische Schmerzursachen können gleichzeitig wirksam sein. Zur weiteren Abklärung des Pelipathie-Syndroms gehört der Ausschluß extragenitaler organischer Schmerzursachen, z.B. orthopädisch bedingte Schmerzursachen, neurologische und gastrointestinale Schmerzursachen wie Colon irritabile, Divertikulose, Obstipation, chronische Appendizitis, Morbus Crohn.

Pathogenetisch handelt es sich beim Pelipathie-Syndrom um eine neurotische Reaktion auf ursächliche lebensgeschichtliche Zusammenhänge bzw. ungelöste, unbewußte Konflikte, die hauptsächlich folgende Problemkreise betreffen: innere Vereinsamung, Isolation, Auseinandersetzung mit sexuellen und Geborgenheitswünschen in der Partnerbeziehung. Bei überwiegend depressiver Persönlichkeitsstruktur gelingt es diesen Frauen kaum, im Alltagsleben eigene Wünsche und Erwartungen zu verwirklichen. So können Trennungs- und Verlustsituationen (Trennung oder Scheidung vom Partner) oder Verlust überwertiger sozialer Bezüge, wie z.B. Aufgabe oder Kündigung des Arbeitsplatzes, zum Auftreten der Unterbauchschmerzen führen. Hinter dem Pelipathie-Syndrom kann sich auch eine Abwehr sexueller Wünsche des Partners verstecken. Diese Frauen suchen in erster Linie Geborgenheit und Wärme und nehmen eigene sexuelle Wünsche kaum wahr, fühlen sich als Sexualobjekt mißbraucht oder vom Partner überfordert. Bei gestörter Entwicklung von Eros und Mütterlichkeit kann das Pelipathie-Syndrom einen Rückzug darstellen auf eine überwertig gelebte Beziehung zum eigenen Kind.

Zwei im Verhalten unterschiedliche Patientinnengruppen werden beobachtet. Ein Teil zeichnet sich aus durch eine allgemeine ärgerliche Befindlichkeit und versteckte Vorwurfshaltung. Diese Patientinnen äußern ihre Unzufriedenheit über den schleppenden Fortgang der diagnostischen Maßnahmen, kritisieren ärztliche Mit- und Voruntersucher. Durch ein destruktiv anmutendes Verhalten können sich Kollegen leicht provoziert fühlen und die Behandlung aufgeben oder sie lassen sich zu einer voreiligen Operation verleiten. Bei einer zweiten Gruppe von Pelipathie-Patientinnen handelt es sich um eine **verleugnete Depression** (Molinski, 1982). Im Vordergrund stehen zahlreiche körperliche Beschwerden, die psychischen Schwierigkeiten bleiben verdeckt. Daher suchen diese Patientinnen den Allgemeinarzt oder den Gynäkologen auf, nicht aber den Psychiater. Klagen über mangelnde Lebensfreude, allgemeine Lustlosigkeit und Antriebsschwäche, Appetitlosigkeit, Schlafstörungen, Abnahme des sexuellen Verlangens – depressive Symptome also – werden ausschließlich als Folge der Schmerzen erlebt. Diese Patientinnen verleugnen nicht nur ihre Depression, sondern psychische Schwierigkeiten überhaupt (Richter, 1979; Molinski, 1982). Diese Verleugnung der Depression und seelischer Schwierigkeiten macht einen therapeutischen Ansatz schwierig, weshalb bis vor kurzem dieses Krankheitsbild als nahezu unbehandelbar galt. Die psychosomatische Forschung in der Gynäkologie konnte allerdings in jüngster Zeit mit den Arbeitsgruppen Molinski (1978, 1982) und Dmoch (1979), Nijs (1981, 1982, 1983) und Renaer (1981), Richter (1979, 1983) und Strunk (1978) neue erfolgreiche Therapiekonzepte erarbeiten.

Erläuterung unseres Therapiekonzepts

Bei der ersten Konsultation werden die wichtigsten Daten aus der allgemeinen und gynäkologischen Anamnese erhoben, insbesondere Zeitpunkt und Umstände des Auftretens der Schmerzen. Nach Exploration der körperlichen und seelischen Begleitsymptomatik wird eine sorgfältige somatische Untersuchung vorgenommen. Der Patientin wird erläutert, daß es eine große Zahl möglicher Schmerzursachen im Unterbauch gäbe und daß man jetzt gemeinsam die Ursache dieser Schmerzen finden wolle. Dies werde aber unter Umständen einige Zeit in Anspruch nehmen und mache weitere Untersuchungsschritte erforderlich. Diese gründliche körperliche Untersuchung, sowie die gemeinsame systematische Planung weiterer diagnostischer Untersuchungsschritte, vertieft das Vertrauensverhältnis zwischen Patientin und Arzt, die Patientin fühlt sich ernst- und angenommen. Die Patientin wird nun in festen Zeitabschnitten einbestellt. Die Ergebnisse konsiliarärztlicher Ausschlußuntersuchungen werden in laienverständlicher Weise mitgeteilt und erläutert. Die allgemeine Anamnese wird weitergeführt mit der ganz selbstverständlichen Besprechung der psychosozialen Situation der Patientin (Partner, Kinder, Beruf, Pläne usw.). Gleichzeitig wird das in der Übertragung sich konstellierende emotionale Geschehen beobachtet, eventuell bereits aufgegriffen und angesprochen. Wichtig ist, daß die Behandlung ohne jede psychotherapeutische Etikettierung im Rahmen der üblichen gynäkologischen Sprechstunde durchgeführt wird. Ob, wann und wie Affekte angesprochen werden, hängt von der Erfahrung des Arztes, insbesondere aber von der Abwehrstruktur der Patientin ab. Sie allein bestimmt Zeitpunkt und Tempo des Überganges von der organischen zur psychosomatischen Selbsterfassung der Schmerzursachen. Für diesen Übergang von der organischen zur psychosomatischen Selbsterfassung der Patientin hat sich die diagnostische Laparoskopie im Rahmen eines etwa dreitägigen stationären Aufenthaltes als hilfreich herausgestellt. Sind extragenitale Schmerzursachen inzwischen ausgeschlossen, zentriert sich nämlich die endgültige Diagnoseerwartung der Patientin auf das „Hineinschauen in den Bauch". Hat die Laparoskopie die erwarteten unauffälligen Verhältnisse bestätigt, wird der Patientin gegenüber geäußert, man wisse jetzt, daß keine gravierende organische Erkrankung vorliegt. Vielmehr entstünden die Schmerzen durch Verkrampfungen und Stauungen im Beckenfüllgewebe, dort, wo zahlreiche Blutgefäße und Nerven verlaufen. Man wisse, daß solche Verkrampfungen und Spannungszustände sehr schmerzhaft sein könnten und durch Überanstrengung, Sorgen oder Streß jeglicher Art entstehen und unterhalten würden. Damit wird die Therapieerwartung der Patientin auf eine andere Ebene, in eine mehr psychologische Dimension verschoben und einer organischen Fixierung vorgebeugt. Wir machen jetzt das Angebot, herauszufinden, ob Streß, übermäßige Belastungen, Sorgen und Probleme beim Zustandekommen dieser Schmerzen eine Rolle spielen könnten. Hat man im Verlauf der bisherigen Behandlung schon gewisse Vorstellungen über einen möglichen psychodynamischen Fokus der Patientin entwickelt, können Versuchsinterpretationen aktueller Lebenssituationen der Patientin, falls sie von dieser einfühlbar sind, augenblicklich zu einer erheblichen Motivationssteigerung hinsichtlich einer jetzt mehr psychosomatisch ausgerichteten Behandlung führen. Aber auch Patientinnen, bei denen erst lediglich ein intellektuelles Interesse erwacht ist, herauszufinden, was wirklich mit ihnen los ist, sind bereit, ab jetzt einen gemeinsamen psychosomatischen Behandlungsweg einzuschlagen, wenn sie im Verlaufe der bisherigen Behandlung eine Beziehung zum Arzt aufbauen konnten, sein Engagement, seine Verläßlichkeit erfahren haben. Dieser Umgang mit der Patientin kommt also zunächst ihrer Vorstellung von einer somatischen Schmerzursache entgegen und stabilisiert die Arzt-Patientin-Beziehung. Parallel zur somatischen Diagnostik wird jedoch gleichzeitig auf das eingegangen, was die Pa-

tientin quasi zwischen den Zeilen an Erwartungen, Wünschen und Ängsten in bezug auf die Art der eingeschlagenen Behandlung und den Arzt mitteilt. Während Arzt und Patientin in einer scheinbar somatischen Nomenklatur miteinander umgehen, vollziehen sie die ersten Schritte von einer somatischen Auffassung der Krankheit der Patientin zur psychologischen Sicht ihrer Störung. Bei der nun beginnenden Bearbeitung der Konfliktbereiche muß man schrittweise vorgehen. Während einer einzelnen Konsultation von nicht mehr als 20 Minuten sollten nur immer einige Probleme angesprochen werden. Bei dieser Vorgehensweise kann die Patientin allmählich die eigenen Schwierigkeiten und Konflikte erkennen und sich ihnen stellen. Regelmäßig verabredete Sprechstundentermine geben der depressiven Patientin eine Art Zeitperspektive (Nijs, 1983). Die Aussicht auf eine andere Zukunft aber ist unentbehrlich für eine Änderung. Es gelingt so, die Patientin zu einem neuen Lebensmodus zu führen. Es kann gemeinsam überlegt werden, wie z. B. eine Überbelastung abgebaut oder wie ihre Belastungen auf andere, bessere Weise ertragen werden können. Bei diesen zögernden Schritten, neue Lebensformen auszuprobieren, muß die Patientin geduldig unterstützt werden. So lernt sie allmählich, das Leben für sich selbst zu genießen und den eigenen Wünschen mehr Platz einzuräumen.

Kreuzschmerzen

Kreuzschmerzen können mechanisch-statische oder emotionale Ursachen haben. Sie werden ähnlich dem Pelipathie-Syndrom diffus angegeben. Finden sich weitere psychosomatische oder psychische Symptome, besonders aus dem depressiven Formenkreis, so ist eine emotionale Pathogenese naheliegend. Der sog. Ermüdungskreuzschmerz kann Ausdruck einer Erschöpfungsdepression sein, wie sie bei psychisch, körperlich, arbeitsmäßig und beruflich oder in der partnerschaftlichen Beziehung überforderten Frauen beobachtet wird (Kielholz, 1971). Davon zu unterscheiden sind Kreuzschmerzen infolge Tonuserhöhung bzw. spastischer Verkrampfung der Muskulatur im Rücken und Lumbalbereich. Häufig bestehen neben dem Kreuzschmerz noch andere Myalgien (Condrau, 1969). Die Muskelverkrampfungen können als somatischer Ausdruck einer allgemeinen inneren Verkrampfung – „psychischen Rigidität" betrachtet werden. Diese Patientinnen wirken beherrscht und kühl oder zeigen ein eher männlich-dominierendes Verhalten. Sie haben Schwierigkeiten mit der weiblichen Rolle. Noch immer kommt es in der gynäkologischen Praxis vor, daß an sich weitgehend belanglose anatomische Veränderungen des Genitales für diese Kreuzschmerzen verantwortlich gemacht werden, wie ein retroflektierter Uterus, eine geringe Myomatose oder ein beginnender Descensus uteri et vaginae. Die früher zahlreichen, hinsichtlich der Schmerzursachen sinnlosen „lageverbessernden Operationen" sind heute erfreulicherweise wegen ihrer Erfolglosigkeit weitgehend verlassen worden.

Adnexitis

Therapeutische Maßnahmen bei der Adnexitis bleiben häufig unbefriedigend oder gar erfolglos. Vielfach notwendige Krankenhaus- und Kuraufenthalte können durch den wiederholten Ausfall der Frau im Haushalt oder am Arbeitsplatz zu sozialen Spannungen führen. Infertilität als Folge postentzündlicher Veränderungen des Tubarapparates kann jahrelange intensive medizinische Bemühungen in Gang setzen, einschließlich operativer Eingriffe, die meist nur geringen Erfolg haben. Ungewollte Kinderlosigkeit und sexuelle Schwierigkeiten infolge andauernder Schmerzzustände im kleinen Becken sind nicht selten Ursache von Partnerkonflikten, oder es droht schon in jungen Jahren die radikale verstümmelnde Operation als letzter Ausweg zur Sanierung ständig aufflackernder Adnexentzündungen.

Die übliche monokausale Adnexitistherapie bringt zwar die aktuelle Entzündung rasch zum Verschwinden, ändert aber nichts an der neurotischen Grundproblematik dieser Frauen (Lemnete et al., 1972; Richter, 1978, 1979, 1983). Daher ist die hohe Wiedererkrankungsrate nicht überraschend. Die Aszensionstheorie dominiert noch weitgehend in den pathogenetischen Überlegungen, obwohl Carol und Müller (1964) nur in 30% der Fälle einen Zusammenhang mit der Menstruation, nur in 6% mit gynäkologischen Eingriffen und in nur 7% mit Geburten und Fehlgeburten fanden. Bei Franke (1969) finden sich gar in nur 11,9% vorausgegangene Menstruationen, Geburten, Fehlgeburten oder gynäkologische Eingriffe. In einer großen Zahl von Fällen wird demnach kein Zusammenhang zwischen Eröffnung der Zervix und dem Auftreten von Adnexentzündungen festgestellt. Dies wird durch die Befunde von Parker und Jones (1966), Krebs und Schallenberg (1971) bestätigt, die bei vergleichenden bakteriologischen Untersuchungen von Adnex- bzw. Douglaspunktaten und Zervixschleim keine Übereinstimmung im kulturellen Erregerspektrum fanden. Schultz (1962) und Weidenbach (1966) diskutieren einen Infektionsweg vom Darm aus. Tatsächlich fanden Krebs (1972) und Parker und Jones (1966) bei der direkten Züchtung aus dem Entzündungsherd überwiegend anaerobe Keime und Keime der normalen Darmflora. Neuere, mittels verbesserter Abstrichmethoden durchgeführte Untersuchungen zeigen eine große Zahl gonorrhoischer Infektionen der Adnexe. Die Mehrzahl (60–80%) aller urogenitalen Gonorrhoe-Infektionen verläuft jedoch asymptomatisch (Ahmad und Parrish, 1974; Khoury und Linwood, 1974; Noonan, 1974). Es müssen also weitere Faktoren wirksam werden, wie allgemeine Abwehrlage oder lokale Resistenzschwäche infolge neurotischer Konflikte. Eigene tiefenpsychologische Untersuchungen an stationär behandelten Adnexitispatientinnen – Ersterkrankungen und Rezidive – ergaben überraschend häufig gleichartige Konfliktsituationen. Adnexitispatientinnen stehen zwischen zwei oder mehreren Partnern, ohne sich letztlich für einen allein entscheiden zu können. Zwei unterschiedliche Impulse werden bei verschiedenen männlichen Partnern ausgelebt. Bei einem rücksichtsvollen verläßlichen und immer verfügbaren „väterlichen" oder „kameradschaftlichen" Mann, der ihnen Beachtung und Zuwendung schenkt, finden sie Verständnis und Geborgenheit. Die Zuwendung zu diesem erweist sich

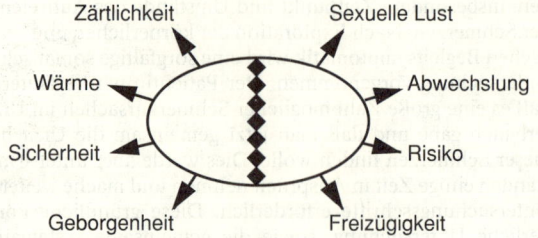

Abb. 52–6. Zwei zentrifugal desintegrierend erlebte Impulse bei Adnexitispatientinnen.

als ein ins Erwachsenenalter hineinreichendes Restbedürfnis der in der frühen Kindheit am eigenen Vater nicht ausreichend oder in falscher Weise befriedigten Geborgenheitswünsche. Auf der anderen Seite suchen sie jedoch auch Abwechslung, Ungebundenheit, eine Art erotische Spannung. Sie wenden sich daher weiteren Partnern zu, die in der Regel aktiv, dynamisch und kontaktfreudig, oft aber auch egoistisch, unzuverlässig und willkürlich sind. So sind diese Frauen ständig hin- und hergerissen. Das gleichzeitige Erleben von Geborgenheit und sexueller Lust mit einem einzigen Partner ist nicht möglich (Abb. 52–6).

Eine Patientin drückte diesen Konflikt einmal folgendermaßen aus: „... ich habe mich zu entscheiden für Freiheit und Risiko oder Geborgenheit und Liebe ..."

Anlehnungs- und Geborgenheitswünsche und sexuelle Sehnsüchte werden auf jeweils entgegengesetzte Partnerpersönlichkeiten projiziert. Bei Adnexitispatientinnen finden sich einmal Partnerschaften – meist von längerer Dauer –, in denen sich beide Partner in hohem Maße gegenseitig verfügbar halten, wobei aber auch oft eine eifersüchtige gegenseitige Einengung herrscht. Es bestehen sexuelle Schwierigkeiten mit Anorgasmie und Dyspareunie. Im Gegensatz dazu stehen impulsiv zustandegekommene, häufig aber auch nur impulsiv gewünschte Beziehungen zu auf diese Patientinnen faszinierend wirkenden Partnern, die über das Erotisch-Sexuelle nicht hinausgehen und die, falls sie verwirklicht werden, rasch scheitern. Für die Partnerbeziehungen finden sich überwiegend folgende zwei Möglichkeiten:

– Eine gleichzeitige oder gleichzeitig mögliche Beziehung zu einem zweiten Partner unmittelbar vor der Erkrankung (Richter, 1978).
– Die Patientin lebt die beiden, genau konträren Partnerschaftsmöglichkeiten in ihrer Lebensbiographie nacheinander (Bauer, 1982).

Die Adnexitispatientin vermißt in der symbiotischen „Anlehnungsbeziehung" die sexuelle Befriedigung, in der anderen Beziehung vermißt sie Geborgenheit und Stetigkeit. Die gerade bestehende Partnerschaft wird durch die jeweils latenten Sehnsüchte ge- und schließlich zerstört. Die Adnexitispatientin lebt in einem inneren Dauerkonflikt. Unmittelbar vor der Erkrankung finden sich Versuchungs- und Versagungssituationen, in welchen die Patientinnen auf ihre erotisch-sexuellen Sehnsüchte und Wünsche verzichtet hatten.

Warum Geborgenheit und Sexualität nur desintegrierend und nicht bei demselben Partner erlebt werden können, wird aus der Lebensbiographie verständlich. Häufig bestehen bis in die Gegenwart erheblich belastete Beziehungen zu den Müttern oder deren Ersatzpersonen (z.B. Heimmütter, erziehende Großmütter). Diese üben in auffallendem Maße Einfluß auf die Entscheidungen ihrer schon erwachsenen Töchter aus. Daraus entstanden bei den Patientinnen reaktive feindselige Gefühle, die sie ihren Müttern gegenüber aber ängstlich zurückhalten. Anbindung, Einengung durch große Strenge seitens der Mütter oder Bezugspersonen, verbunden mit einer unruhigen familiären Gesamtatmosphäre läßt sich bis in die Kindheit zurückverfolgen. Im Gegensatz zur Dominanz der Mütter fällt einheitlich eine Unverbindlichkeit der Vater-Tochter-Beziehung auf. Entweder waren die Väter ganz oder – z.B. berufsbedingt – häufig abwesend, oder sie waren schon älter (körperlich hinfällig), oder sie hielten sich generell in der Familie im Hintergrund. Damit konnten die Patientinnen am Vater kein realistisches Partnerbild entwickeln. In der späteren Wahl eines nachgiebigen, immer verfügbaren, jederzeit verständnisbereiten Partners erkennt man den Versuch, die von Kindheit an als gefährdet erlebte Geborgenheit zu finden und festzuhalten. Eine sich daraus ergebende Einengung wird dann aber als quälend empfunden, zumal sie gleichsam eine Neuauflage und Verstärkung der Einengung seitens der Mutter darstellt. Wie schon ihren Müttern gegenüber sind diese Frauen auch gegenüber ihren „Anlehnungspartnern" nicht in der Lage, sich adäquat durchzusetzen, Wünsche vorzutragen, Bedingungen für ein partnerschaftliches Miteinander festzulegen.

Psychosomatisch bedeutsam erscheint ein von Adnexitispatientinnen übereinstimmend und wiederholt geschildertes Verhaltensmuster, wie sie in einer Situation quälend erlebter Einengung mit angestauten Wutimpulsen verfahren. Sie entledigen sich ihres Ärgers, indem sie impulsiv unbewußt „aus Rache" eine erotische Situation suchen oder sich auf einen sexuellen Kontakt einlassen.

Einer 16jährigen Patientin war von der Heimmutter der Bikini weggenommen worden. Sie sagte ihr in ohnmächtigem Zorn: „... wenn du mir alles wegnimmst, lasse ich mir ein Kind machen, das kannst du mir dann nicht mehr wegnehmen ..." – Sie ging zu ihrem Freund, hatte erstmals mit ihm Geschlechtsverkehr.

Eine andere Patientin konnte ihre beabsichtigte Ehepartnerwahl gegenüber ihrer Mutter nicht durchsetzen und sagte deshalb ihrem Freund: „... komm, wir machen ein Kind, dann können wir heiraten ..."

Eine weitere Patientin war aus Verärgerung über ihren „Anlehnungspartner" allein ausgegangen und wollte aus Wut „voll Power flirten".

Psychodynamisch betrachtet wird die Sexualität zum Adressaten, psychosomatisch-topographisch wird der Genitalapparat zum Verarbeitungsorgan aggressiver Impulse. Adnexitispatientinnen zeigen diese spezielle Verarbeitungsweise von Wutimpulsen, welche die sexuelle Sphäre mißbräuchlich zur Kampfzone macht. Je mehr sie in den Anlehnungspartnerbeziehungen sich eingeengt empfinden, desto stärker verweigern sie unbewußt ihren Partnern die sexuelle Hingabe – Symptomatik Anorgasmie und Dyspareunie –, desto stärker aber spüren sie Sehnsüchte nach einer völlig anderen, überwiegend sexuellen Art von Partnerbeziehung. Die Patientinnen können diese Dynamik in sich nicht steuern. Solange die psychisch „eingespielte" Verarbeitung der Ärgerreaktionen mit Abreaktion in der sexuellen Sphäre gelingt, scheinen keine Adnexitiden aufzutreten. Das erklärt die völlig symptomfreien Phasen mancher Patientinnen bei an sexuelle Verwahrlosung grenzenden häufigen Partnerwechseln. Ein Adnexitisrezidiv scheint immer dann aufzutreten, wenn der stark erlebte Wunsch nach sexueller Entlastung nicht erfüllt werden kann, etwa wenn die Patientin aus Rücksicht auf ihre Anlehnungspartnerschaft verzichtet oder wenn ein angebahnter sexueller Kontakt aus anderen Gründen scheitert.

Wir sind überzeugt, daß angestaute, aggressive Impulse, die an der gewohnten Abfuhr in die Sexualsphäre gehindert werden, zu neuromuskulären Fehlinnervationen wie dem Tubenspasmus und zu anhaltenden unphysiologischen Kontraktionen der vegetativen Muskulatur im kleinen Becken führen können, mit den Sekundärfolgen Stase-Vorgänge und Hypoxie, Ödembildung in der Perisalpinx, Hypersekretion der Tubenepithelien. Damit wird jene lokale Resistenzminderung bewirkt, die über bereits vorhandene Keime zum Aufflackern einer neuen Entzündung führt.

Therapeutisch bewährt hat sich, diesen Basiskonflikt anzusprechen und zu bearbeiten, besonders in der Gruppenarbeit mit stationären Adnexitispatientinnen. Dadurch lassen sich über die Zeit der Hospitalisierung hinausgehende psychotherapeutische Impulse setzen und möglicherweise manche Fälle von chronischer Adnexitis vermeiden. Auch in der ambulanten Praxis sollte der Gynäkologe diesen Dauerkonflikt ansprechen und während wiederholter Konsultationen mit der Patientin bearbeiten.

52.1.7 Psychogener Fluor genitalis

Patientinnen mit Fluor genitalis spielen in der gynäkologischen Sprechstunde eine große Rolle. In der Annahme einer organischen Ursache wird zunächst überwiegend antimikrobiell behandelt. Obwohl sich in vielen Fällen dadurch am Symptom nichts ändert, wird kaum an psychosomatische Zusammenhänge gedacht. Dem psychogenen Fluor genitalis liegt eine übermäßige Sekretion der Vestibularis- und Zervixdrüsen und/oder eine vermehrte Transsudation der Vaginalwände zugrunde. Dieser Fluor kann eine eigenständige psychosomatische Störung darstellen oder als Begleitsymptom anderer psychosomatischer Erkrankungen auftreten.

Bei unerfülltem Wunsch nach sexueller Befriedigung wird der Fluor im Sinne eines permanenten „sexuellen Bereitstellungsreflexes" durch die eben erwähnte Sekretion der Vestibularis- und Zervixdrüsen und durch vermehrte Transsudation der Scheide infolge Vasokongestion im Parakolpium hervorgerufen. Perez-Gay (1983) unterscheidet den **Libidofluor,** als direkten Ausdruck unerfüllter sexueller Wünsche, vom **Abwehrfluor.** Hier wird eine mögliche sexuelle Befriedigung abgewehrt, weil die Befriedigung sexueller Wünsche grundsätzlich für das bewußte Erleben nicht akzeptabel ist oder weil vorangegangene sexuelle Erfahrungen so unbefriedigend waren, daß die Frau weiteren sexuellen Kontakten durch ausgeprägte Fluorentwicklung auszuweichen sucht. Beim sog. **Gewissensfluor** verhindert eine noch stärker wirksame Abwehr bzw. Über-Ich-Problematik eine konfliktfreie sexuelle Befriedigung. Der gleichsam „als Strafe" hervorgerufene Fluor kann dann begleitet sein von der Angst vor einer venerischen Infektion oder sogar karzinophobischen Phantasien.

Bei einer allgemeinen neurovegetativen Labilität können unspezifische Streßsituationen über Parasympathikusaktivierung zu zervikalem Fluor führen, der in diesen Fällen dann meist ein Begleitsymptom anderer psychosomatischer Störungen darstellt. So klagen Patientinnen mit psychosomatischen Magen-Darm-Störungen überwiegend über einen starken Fluor, oder der Fluor stellt sich ein bei Aufregungen, allgemeiner Anspannung oder Erschöpfung.

Nahezu pathognomonisch findet sich ein psychogener Fluor als Begleitsymptom beim Pelipathie-Syndrom und bei den Sekundäre-Amenorrhoe-Patientinnen, bei denen als auslösende Konfliktursache eine Abwehr des Sexualtriebs oder eine männliche Rollenidentifikation vorliegt. Bei diesem Fluorproblem ist – wie eingangs ausgeführt – eine gestörte Sexualphysiologie im Sinne eines permanenten Bereitstellungsreflexes bei unerfüllter oder abgewehrter sexueller Befriedigung die Ursache.

Bei einer vertrauensvollen Arzt-Patientin-Beziehung kann der psychosomatische Zusammenhang aufgedeckt werden.

52.1.8 Psychogener Pruritus vulvae

Der psychogene Pruritus vulvae kann akut und heftig, anfallsweise oder chronisch auftreten und sich als Jucken, Brennen, Stechen an der Vulva, in der Klitorisgegend oder am Perineum bemerkbar machen. Sekundär können Kratzeffekte wie Epithelläsionen, Ekzeme, Follikulitiden oder Pyodermien hinzukommen, insbesondere wenn die Patientin das zum Teil heftige Kratzbedürfnis nicht unterdrücken kann. Der Pruritus kann in jedem Alter auftreten, eine deutliche Zunahme ist jedoch während des Klimakteriums und in der frühen Postmenopause zu beobachten. Vorwiegend somatisch eingestellte Gynäkologen sehen daher im Pruritus ein Östrogenentzugssymptom. Die sehr bescheidenen Erfolge einer lokalen oder systemischen Östrogentherapie lassen aber daran zweifeln.

Psychodynamisch betrachtet liegen dem psychogenen Pruritus nahezu ausschließlich sexuelle Konflikte zugrunde (Stauber und Haupt, 1980).

Überwiegt der Abwehraspekt, findet sich bei diesen Frauen eine Abneigung gegen den Partner, eine Furcht vor Schwangerschaft, eine Aversion gegen sexuelle Praktiken, Angst vor Ansteckung oder Verbotenem. Andererseits können auch Frauen mit allgemeinen und sexuellen Kontaktwünschen an Pruritus leiden. Sie zeigen ein allgemeines Unbefriedigtsein in der Ehe oder unerfüllte Erwartungen an den Partner, etwa bei Impotenz des Mannes. Nach Kemper (1975) stellt der Pruritus die „Zerrform einer allgemeinen sexuellen Erregung" dar. Dieses Symptom sei „Ausdruck einer ständig abgewehrten Dauererregung des weiblichen Genitaltraktes, die mangels adäquater Befriedigung niemals zum Abklingen kommt". Einige Autoren bezeichnen den Pruritus vulvae direkt als masturbatorisches Äquivalent (Alexander, 1971; Ziolko, 1978). Prill (1964) sieht als Ursache für das Kratzen, Jucken und Kitzeln masochistische und im Bereich der Sexualorgane erotische Strebungen an. In der gynäkologischen Sprechstunde sollte nach Ausschluß exogener und endogener Pruritusursachen eine somatische Chronifizierung dieses Leidens verhindert werden, da diese eine kausale psychotherapeutische Behandlung später erschweren oder unmöglich machen kann. Konfliktzentrierte Gesprächspsychotherapie ist in vielen Fällen erfolgreich.

52.1.9 Klimakterium

Der Übergang der reproduktiven Lebensphase ins Senium wird als Klimakterium bezeichnet. In diesen Zeitraum, der etwa das 45. bis 55. Lebensjahr umfaßt, fällt als letzte Regelblutung die Menopause. Das durchschnittliche Menopausenalter liegt gegenwärtig bei ca. 51 Jahren. Das Klimakterium wird in eine prä- und postmenopausale Phase eingeteilt. Beginn und Dauer des Klimakteriums können höchst unterschiedlich sein, in Abhängigkeit des allmählichen Verlöschens der Ovarialfunktion. Die Zahl der im Ovar vorhandenen Primordialfollikel mit den sie umgebenden Granulosa- und Thekazellen bestimmt die Lebensdauer der generativen Funktion der Ovarien. Infolge degenerativer Vorgänge reagieren diese Zellverbände zunehmend refraktär gegenüber der endogenen Gonadotropinsekretion, die Bildung von Sexualhormonen nimmt immer mehr ab (Breckwoldt, 1983). Die Postmenopause ist endokrinologisch charakterisiert durch die hypergonadotrope Amenorrhoe.

Das Klimakterium kann sehr unterschiedlich verlaufen, völlig symptomlos oder mit einer ganzen Reihe charakteristischer körperlicher und seelischer Beschwerden. Diese sind im Einzelfall so gravierend, daß erhebliche psychische und soziale Beeinträchtigungen auftreten können.

An **körperlichen Beschwerden** finden sich vor allem Hitzewallungen, Schweißausbrüche, Schwindelzustände, Blutdruckanstieg, Kopfschmerzen, Schlafstörungen als Ausdruck eines erhöhten Sympathikotonus. Dieser ist Folge der abnehmenden Östrogenkonzentration im peripheren Blut – Östrogene wirken pharmakokinetisch im Sinne eines Parasympathikotonus. Der Östrogenmangel wirkt sich vor allem an den Zielorganen aus: Atrophien der Vaginalschleimhaut, der Vulva und der Blasenschleimhaut können zu Sexualstörungen wie Dyspareunie und Blasenstörungen wie der Urge-Harninkontinenz führen.

An **psychischen Symptomen** können auftreten: eine allgemeine Nervosität und innere Unruhe mit gesteigerter Reizbarkeit und depressiven Verstimmungen.

Nach Prill und Lauritzen (1970) zeigen ⅓ aller Frauen überhaupt keine, ⅓ leichte und ⅓ schwere Probleme im Klimakterium. Wenderlein (1977) fand keine oder nur schwächere Symptome bei Frauen oberer sozialer Schichten und bei Frauen mit höherem I.Q. und erklärte dies mit einem höheren Informationsgrad über die physiologischen Prozesse und einer größeren Bereitschaft zu einer Östrogensubstitutionstherapie. Maas und Kuypers (1974) und Lehr (1980) machen dafür allerdings zusätzliche Bedingungen verantwortlich, die man überwiegend bei dieser Frauengruppe antrifft, wie stärkere außerhäusliche Orientierung, geringere Familienzentriertheit und erweiterter allgemeiner Interessenradius. Auch van Keep (1976) fand bei Frauen niedriger sozialer Schicht mehr klimakterische Beschwerden im Vergleich zu Frauen höherer Schichtzugehörigkeit, da sie weniger in die Gesellschaft integriert und ausschließlich auf familiäre Rollen beschränkt waren.

Im 5. bis 6. Lebensjahrzehnt kommt es im allgemeinen zu einer Reihe von psychosozialen Veränderungen, zu einer Anhäufung von Belastungssituationen. Lehr (1961) spricht von einer Konfliktkumulation. Viele Frauen befinden sich in dieser Zeit in einem besonderen Spannungsfeld, im Schnittpunkt verschiedener Rollen mit zum Teil gegensätzlichen Rollenanforderungen und einer dadurch bedingten Phase der Verunsicherung zwischenmenschlicher Bezüge (Lehr, 1983). Konflikthaft können sein: die Auseinandersetzung mit den erwachsen werdenden und das Elternhaus verlassenden Kindern; die eigenen alternden, eventuell hilfsbedürftigen Eltern verlangen besondere Zuwendung und können auf einmal wieder an die Tochterrolle erinnern; der Partner – häufig auf dem Höhepunkt seiner beruflichen Entwicklung angelangt – kann zunehmend als fremdgeworden oder „entfernter" erlebt werden. Lebenslaufstudien haben widerlegt, daß die biologisch bedingten Veränderungen des Klimakteriums für eine Einbuße des Lebensgefühls und die zahlreichen vegetativen und psychischen Symptome verantwortlich zu machen sind. Die Konflikt- und Belastungssituationen während der Wechseljahre können manche Frau derart überfordern, daß nur noch eine Flucht in die Krankheit übrigbleibt. Die dann unter Umständen zahlreich auftretenden Beschwerden des klimakterischen Symptomenkomplexes können Appellationscharakter annehmen.

Nicht nur Laien, sondern auch viele Ärzte betrachten das Klimakterium als einen krisenhaften primär endogen bedingten Prozeß und glauben, etwaige Probleme im psychosozialen Bereich auf die körperlichen Umstellungsprozesse zurückführen zu müssen. Dabei können umgekehrt psychische Schwierigkeiten z.B. bei Partnerschaftskonflikten, Eltern-Kind-Konflikten oder auch berufliche Belastungssituationen selbst Ursache für körperliche Beschwerden werden (Troll und Turner, 1978). Zu warnen ist also vor einer grundsätzlichen Überbewertung biologisch-endokriner Vorgänge als Auslösermoment von Konflikt- und Belastungssituationen. Frauen, die in diesen Jahren mit irgendwelchen körperlichen oder seelischen Problemen den Arzt aufsuchen, sollten nicht einfach als „klimakterisch" eingestuft und sofort mit Hormonen und/oder Psychopharmaka behandelt werden. Der Arzt sollte sich vielmehr darum bemühen, die individuelle Situation der Patientin in ihrem psychosozialen Umfeld zu erfassen. Im Verlaufe wiederholter Konsultationen kann so eine Bestandsaufnahme der aktuellen Lebenssituation mit der Patientin erarbeitet und nach Wegen einer neuen zukünftigen Lebensgestaltung gesucht werden. Selbstverständlich hilft eine hormonelle Behandlung bei vegetativen Störungen, ein gezielter Einsatz von Antidepressiva bei depressiven Symptomen. Die alleinige Verordnung von Hormonen oder Psychopharmaka ohne ärztliches Gespräch über die aktuellen Schwierigkeiten der Patientin entspricht nicht einer psychosomatisch orientierten Begleitung durch eine für viele Frauen schwierige Lebensphase. Die ärztlich-therapeutische Begleitung während der Wechseljahre umfaßt demnach medizinisch-pharmakologische Hilfen, psychologische Betreuung und soziale Beratung unter Berücksichtigung der jeweiligen individuellen Situation der Frau.

52.1.10 Sexualstörungen

Funktionelle Sexualstörungen nehmen in der Gynäkologie einen breiten Raum ein. Dazu zählen nicht nur sexuelle Dysfunktionen wie Libido- oder Orgasmusstörungen, Dyspareunie und Vaginismus, sondern auch zahlreiche psychosomatisch-gynäkologische Symptome wie Fluor, Pruritus. Miktionsstörungen und Unterbauchschmerzen können Ausdruck einer gestörten Sexualität sein. Diese Störungen können als Einzelsymptom oder als Begleitsymptome zahlreicher anderer Krankheitsbilder auftreten. So sind beispielsweise sexuelle Schwierigkeiten beim sekundären Amenorrhoe-Syndrom oder beim Pelipathie-Syndrom fast pathognomonisch zu beobachten. Wenn die Patientin überhaupt von sich aus auf Schwierigkeiten im sexuellen Bereich zu sprechen kommt, dann nicht selten dem Gynäkologen gegenüber, der – aus Patientinnensicht – noch am ehesten als kompetenter Berater bei solchen Problemen betrachtet wird. Da eine sexualmedizinische Ausbildung der Ärzte an den Universitäten immer noch weitgehend fehlt, fühlen sich viele Ärzte bei der Behandlung von Sexualstörungen überfordert oder reagieren mit Abwehr. Auf eine systematische Darstellung der Sexualstörungen muß in diesem Zusammenhang verzichtet werden (vgl. Kap. 39). Hingewiesen sei allerdings darauf, daß gerade von gynäkologisch-psychosomatischer Seite aus in den letzten Jahren erfolgreiche neue Diagnostik- und Therapiekonzepte bei der Behandlung funktioneller Sexualstörungen erarbeitet wurden, welche die therapeutischen Möglichkeiten von Sexualstörungen wesentlich erweitert haben. Diese Therapiekonzepte, welche den verhaltenstherapeutischen Ansatz von Masters und Johnson (1970) um den tiefenpsychologischen Aspekt bereichert haben, erfahren zur Zeit eine größere Verbreitung (Molinski, 1976; Nijs, 1982; Höffken et al., 1982; Poettgen, 1983).

*Sekundäre Sexualstörungen als Folge
gynäkologischer Krankheitsbilder*

Organische Ursachen können über dyspareunische Beschwerden und nachfolgendes angstvolles Vermeiden der sexuellen Situation zu einer sekundären Libidostörung führen. Hierzu zählen bestimmte gynäkologische Krankheitsbilder. Organische Faktoren wirken sich jedoch nur dann als Sexualstörung aus, wenn sie auf eine besondere psychische Bereitschaft dazu von seiten der Frau treffen. Wiederholte Schmerzzustände während der Kohabitation können Schmerzängste konditionieren und schließlich zu Libidostörungen führen. Zu nennen sind: anatomische Besonderheiten wie Verwachsungen der Labia minora, ein rigides Hymen, Episiotomie- oder Dammrißnarben, chronisch-rezidivierende Entzündungszustände der Vulva oder Vagina bei Bakterien-, Pilz-, Trichomonaden-Infektionen oder ausgedehnter Condylomata-acuminata-Befall. Lage- und Halteanomalien der Genitalorgane, eine ausgeprägte Varikosis im kleinen Becken, die Endometriose, besonders bei Befall der Retrozervix und des Douglas-Raums, posttraumatische Zustände im Genitalbereich nach Kohabitations- oder Pfählungsverletzungen und das Allan-Masters-Syndrom können Ursache einer sekundären Sexualstörung sein. Eine ungenügende Lubrikation der Vagina, eine atrophische Kolpitis infolge postmenopausalen Östrogenmangels führen nicht selten zu einer zunehmenden sexuellen Abwehr der Frau bis hin zum völligen Libidoverlust. Hier muß gefragt werden, inwieweit nicht präformierte Erwartungshaltungen, „daß mit den Wechseljahren die Sexualität ohnehin nachläßt", die Abnahme der sexuellen Appetenz begünstigen. Bei der Libidostörung während Schwangerschaft und Wochenbett wirken organische, endokrine und psychische Faktoren zusammen, wobei wir den individuellen psychischen Bedingungen der einzelnen Frau die Hauptbeachtung zumessen (Richter, 1978). An organischen Störfaktoren nach operativen Eingriffen sind zu erwähnen: die Parametritis, ein schmerzhafter, zu kurzer Vaginalstumpf, peritoneale Reizerscheinungen, ein dicht über dem Vaginalabschluß fixiertes Ovar, ein zu enger Introitus nach zu hohem Dammaufbau.

52.1.11 Psychosomatische Probleme bei gynäkologischen Operationen und Genitalkarzinomen

Ungenügende Kenntnisse und falsche Vorstellungen über Lage, Struktur und Funktion der Genitalorgane sind in Laienkreisen weit mehr vorhanden, als selbst Ärzte sich dies immer wieder vergegenwärtigen. Auch bei gut informiert wirkenden Patientinnen stößt man nicht selten auf überraschende Wissenslücken oder erstaunliche Mißverständnisse. In noch größerem Umfang trifft dies zu für die Erwartungen über die Folgen von gynäkologischen Operationen. Am häufigsten werden Ängste vor dem Alt- und Dickerwerden geäußert und zwar unabhängig von dem geplanten Eingriff. Z.B. vermuten manche Patientinnen, obwohl sie wissen, daß die Hormone in den Eierstöcken gebildet werden, und sie diese – bei einfacher Hysterektomie – behalten werden, daß dem Organ Gebärmutter doch eine Bedeutung für die Erhaltung von Jugendlichkeit und schlanker Körperform zukomme. In Zusammenhang damit finden sich Vorstellungen, daß es im Körper zu – wahrscheinlich ungesunden – Stauungen kommen müsse, wenn das Menstrualblut vom Körper nicht mehr regelmäßig ausgeschieden werde. Das Ausbleiben der Regelblutung wird dann wieder mit Wechseljahren und beginnendem Alter gleichgesetzt. Weitverbreitet ist eine diffuse Vorstellung von einer allgemeinen Verminderung von Gefühlsempfindungen, besonders im sexuellen Bereich. So erwarten viele Frauen nach gynäkologischen Operationen eine zwangsläufig einsetzende Libidoverminderung oder befürchten für ihren Partner sexuell nicht mehr so interessant zu sein. Eine Operation an den Genitalorganen bedeutet immer einen Eingriff in eine emotional stark besetzte Körperzone. Es kann daher mit Recht vermutet werden, daß außer mangelnder Aufklärung auch tieferliegende Ängste oder Konflikte am Zustandekommen dieser Befürchtungen und falschen Vorstellungen beteiligt sind.

Dem präoperativen Gespräch kommt ganz entscheidende Bedeutung zu. Es genügt nicht, nur über die Indikation und die physiologischen Folgen der geplanten Operation aufzuklären, vielmehr muß das Gefühlsleben und Sexualverhalten offen angesprochen werden.

Psychosomatisch orientierte Patientenaufklärung

Um auch auf etwaige tieferliegende irrationale Ängste und Befürchtungen aufmerksam zu werden, hat es sich bewährt, nach der ärztlichen Erläuterung der Gründe für die Operation die Patientin direkt zu fragen, ob und wenn ja, welche Veränderungen sie danach im körperlichen, seelischen und sexuellen Bereich erwarte. Bei dieser Vorgehensweise kommt einerseits der Wissensstand der Patientin zum Ausdruck, falsche Vorstellungen können korrigiert werden. Andererseits kann der psychosomatisch orientierte Gynäkologe erkennen, ob sich die Patientin von der Operation vielleicht auch die Lösung partnerschaftlicher Probleme oder sexueller Schwierigkeiten erwartet. Aufgrund neurotischer Fehlwartungen sind es nämlich manche Patientinnen selbst, die zu einer Operation drängen können, da sie sich dadurch eine „Heilung" im umfassenden Sinne, auch von Schwierigkeiten im partnerschaftlichen Bereich erwarten. Werden solche Fehlerwartungen nicht erkannt und korrigiert, eine eventuell medizinisch nicht unbedingt notwendige Operation aufgeschoben oder verhindert, können psychische, psychosomatische oder sexuelle Störungen postoperativ auftreten. Die psychosomatisch orientierte präoperative Aufklärung durch einen darin erfahrenen Arzt ist eine äußerst wichtige präventiv-psychohygienische Maßnahme. Leider wird in der klinischen Alltagsroutine diese wichtige Aufgabe häufig nicht ernstgenommen, an einen unerfahrenen jungen Assistenzarzt delegiert oder gar weitgehend mit Hilfe von Aufklärungsbroschüren durchgeführt.

Ovarektomie

Bei einseitiger Ovarektomie sind keine Ausfallerscheinungen zu erwarten, da das verbliebene Ovar eine ausreichende Hormonproduktion sicherstellt. Treten dennoch vegetative, psychische oder sexuelle Beschwerden auf, kann auf eine seelische Fehlverarbeitung des Eingriffs durch die Patientin geschlossen werden, die sich möglicherweise seither in ihrer Weiblichkeit minderwertiger erlebt. Die Entfernung beider Ovarien führt zu einem Erlöschen des ovariellen und menstruellen Zyklus und zum Verlust der Fortpflanzungsfähigkeit. Durch die weiterbestehende Östrogenproduktion in der Nebenniere und im Fettgewebe treten auch in dieser Situation Ausfallerscheinungen nicht zwangsläufig auf. Bei Beschwerden und bei sehr jungen Frauen sollte eine zyklusgerechte Östrogen-Progesteron-Therapie erfolgen, um langfristigen pathologischen Veränderungen vor allem am Gefäß- und Skelettsystem vorzubeugen. Nach Dennerstein et al. (1977) erwarteten nach der Ovarektomie 42% eine Veränderung im Sexualbereich, 25% äußerten Befürchtungen vor Gewichtszunahme, 14% vor dem Verlust der Weiblichkeit, 9% vor Verminderung der geistigen Leistungsfähigkeit, 6% vor vorzeitigem Altern und vermehrter Körperbehaarung. Nach dem Eingriff waren bei 47% der Frauen diese Ängste weiterhin vorhanden. Besonders im sexuellen Bereich führten diese Ängste zu einer allgemeinen Verschlechterung der sexuellen Beziehungen mit Libidoabnahme und zunehmender Dyspareunie, während die Orgasmusfähigkeit nicht beeinflußt wurde. Zusammenfassend fand sich bei 29% der Patientinnen kein Unterschied im sexuellen Erleben im Vergleich zum Zeitraum vor der Operation. 37% gaben eine Verschlechterung, 34% allerdings eine Verbesserung des sexuellen Erlebens an. Eicher und Herms (1976) erklären das größere sexuelle Verlangen nach der Operation mit dem Wegfall von Schwangerschaftsängsten. Die Verminderung der Östrogene nach beidseitiger Ovarektomie kann demnach nicht für eine Verschlechterung der sexuellen Reaktionsfähigkeit verantwortlich gemacht werden. Johannson et al. (1975) untersuchten die Spätfolgen bei in jungen Jah-

ren beidseits ovarektomierten Frauen. Zu einer fortdauernden neurotischen Haltung hatte sich der Eingriff bei den Frauen ausgewirkt, deren wichtigster Lebensinhalt einmal die Mutterrolle und Kindererziehung gewesen war. Besser überwunden werden konnte die Operation von Frauen, die sich schon kurz danach in andere berufliche oder soziale Aktivitäten stürzten oder Kinder adoptierten. Auch Simon (1980) bestätigt, daß 81% der Patientinnen bald nach Ovarektomie wieder anfingen zu arbeiten, geistig oder körperlich sehr aktiv waren, um kastrationsbedingte Insuffizienzgefühle abzuwehren.

Hysterektomie

Weit mehr als die Ovarien beschäftigt der Uterus die Phantasie der Frau. Aufgrund der Lokalisation in der Mitte des Unterleibs, durch die Periodik der Menstruationsblutungen und durch seine Funktion als Fruchthalter nimmt er im Erleben der Frau einen hohen Stellenwert ein. Aus diesem Grund kann die Indikation zur Hysterektomie nicht sorgfältig genug gestellt werden. Tatsächlich gehört die Hysterektomie zu den am häufigsten durchgeführten gynäkologischen Operationen und nicht nur in der Laienöffentlichkeit wird die Frage gestellt, ob diese Operation in jedem Fall notwendig ist. Belanglose anatomische Veränderungen wie eine Retroflexio uteri, eine geringe Myomatose, Verwachsungen nach Voroperationen oder ein beginnender Deszensus sollten noch kein Grund für die Hysterektomie sein. Wenn allerdings Patientinnen mit solch minimalen anatomischen Befunden Beschwerden vorbringen, wie verstärkte oder unregelmäßige Blutungen, unklare Unterbauchschmerzen, sexuelle Schwierigkeiten, wird sich ein aktiv und operationsfreudig eingestellter Gynäkologe unter Umständen von der Patientin dazu „verführen" lassen, diese Probleme durch eine Operation lösen zu wollen. Eine solche in anatomisch-mechanistischer Denkweise allzu vordergründig gefällte Entscheidung zur Operation unter Mißachtung psychosomatischer Faktoren kann im Einzelfall gravierende psychosexuelle Folgen haben. Das Nichterkennen neurotischer Probleme im Zusammenhang mit einer Hysterektomie erklärt möglicherweise die von zahlreichen Autoren regelmäßig in 30–40% gefundenen sexuellen Störungen nach Hysterektomie (Richards, 1973; Eicher et al., 1975; Dennerstein et al., 1977; Martin et al., 1980; Zussmann et al., 1981). Neben depressiven Reaktionen (13%) fand Eicher in 25% allerdings auch eine positive Veränderung der sexuellen Erlebnisfähigkeit. Wie schon betont, kommt der psychosomatisch orientierten präoperativen Beratung – möglichst unter Einbeziehung des Partners – eine wichtige psychoprophylaktische Bedeutung zu. Richter et al. (1976) haben diesen Zusammenhang nachgewiesen. Bei guter präoperativer Aufklärung und Beratung unter Berücksichtigung emotionaler Aspekte zeigten Patientinnen nach Hysterektomie kaum nennenswerte psychische oder sexuelle Störungen.

Psychosomatisch orientierte Betreuung von Karzinompatientinnen

Bei der Behandlung von gynäkologischen Karzinomen sind große Fortschritte erzielt worden mit inzwischen beachtlichen Heilungs- und Überlebensraten. Trotzdem haften die Vorurteile der Unheilbarkeit sowie die Attribute „unheimlich, tückisch, unberechenbar" dieser Krankheit weiterhin an. In einer hemmungslosen Konsumgesellschaft, in der Jugend und Schönheit, Gesundheit und Leistungsfähigkeit über alles verherrlicht werden, müssen Gedanken an körperliches Siechtum, unheilbares Kranksein verdrängt, muß die Todesproblematik tabuisiert werden. Nicht zuletzt diese gesellschaftlichen Bedingungen sind es, welche die Krebsangst sowie eine fatalistische Einschätzung als Krankheit mit einem tödlichen Ausgang fördern. Daran haben auch umfassende Vorsorgeprogramme und Aufklärungskampagnen wenig geändert. Eingreifende therapeutische Maßnahmen wie radikale Operation, Strahlen- und Chemotherapie verändern und verunstalten den Körper und führen zu einer schweren Störung des Selbstwertgefühls. Krebspatientinnen empfinden sich als minderwertig, als sozial und emotional nicht mehr kommunikationsfähig, gesellschaftlich isoliert, als sexuell unattraktiv. Tatsächlich werden sie zwar bis zu einem gewissen Grad bemitleidet, oft jedoch gemieden am Arbeitsplatz, von Freunden, sogar von der eigenen Familie und dem geliebten Partner. Krebs zu haben bedeutet demnach nicht nur Furcht vor Ausgeliefertsein an eine tückische Krankheit, sondern insbesondere auch Angst vor sozialer Vereinsamung, vor Verlust von Anerkennung und Liebe. Über die individuelle psychische Störung der Patientin hinaus führt die Diagnose Krebs zu einer Störung der Familien- und Umweltsituation. Diese tiefgreifenden Veränderungen in der Patientin selbst und in bezug auf die Familiensituation müssen aufgefangen werden durch langzeitorientierte psychosoziale Interventionen. Man muß sich von der Einstellung freimachen, alles getan zu haben, wenn man operiert, bestrahlt oder eine Chemotherapie eingeleitet hat. Die Tatsache, der wir begegnen müssen, ist die individuelle Patientin, die an Krebs erkrankt ist. Diese hat außer dem Krebsleiden noch zahlreiche andere Eigenschaften. Sie hat ihre Familie, ihren Beruf, ihre Stellung in der Gesellschaft, ihren Charakter, ihre Überzeugungen und Leidenschaften. Sie hat eine eigene Lebensgeschichte. Die Lebensqualität einer Krebskranken wird zunächst weitgehend durch den Erfolg von Operation, Bestrahlung oder Chemotherapie bestimmt. Hinzukommen muß eine individuelle menschliche Zuwendung und kontinuierliche Betreuung.

Eine langfristige, psychosomatisch orientierte Betreuung von Krebspatientinnen müßte folgende Punkte umfassen: nach individueller Aufklärung über die Art der Erkrankung Abbau von Angst, Depression und Schuldgefühl der Patientin, Auseinandersetzung mit den therapiebedingten Veränderungen des Körpers im Sinne einer Stärkung des Selbstwertgefühls und des positiven Körpererlebens; sofortige Einbeziehung des Partners bzw. der Familie bei der Diagnosestellung; Verbesserung der Kommunikation in der Familie; Wiedereingliederung am Arbeitsplatz. Die Wiederaufnahme des Berufes trotz durch Krankheit und Operation bedingter Behinderungen kann viele von den Sorgen abhalten, als Krebskranke für immer abgestempelt zu bleiben. (Die automatische Gewährung einer Erwerbsunfähigkeitsrente ist ein grober psychologischer Fehler.) Ein solches psychosomatisch orientiertes Rehabilitationskonzept löst die Patientin aus dem Teufelskreis sozialer Isolation, schafft die Möglichkeit Teilziele, individuelle Lebensperspektiven zu entwickeln. Weil eine psychosomatisch orientierte Führung von Krebspatientinnen im Rahmen eines Gesamtbehandlungsplanes noch in den Anfängen steckt, muß ihre Notwendigkeit um so mehr gefordert werden (vgl. Kap. 50).

Psychosexuelle Störungen nach Karzinomtherapie

Ein besonderes Problem stellen psychosexuelle Störungen nach operativer und/oder Strahlentherapie von gynäkologischen Karzinomen dar. Nach Eicher (1979) erwartet die Hälfte von Mastektomiepatientinnen eine Beeinträchtigung der Partnerbeziehung nach der Operation und reagiert tatsächlich mit erheblichen sexuellen Störungen. Maguire et al. (1978) fanden bei mastektomierten Frauen signifikant häufiger Angstzustände und depressive Verstimmungen als bei einer Kontrollgruppe von Frauen mit benignen Brusterkrankungen. Vor der Entdeckung des Karzinoms hatten nur 8% der später operierten Frauen schwerwiegende sexuelle Probleme. Ein Jahr nach der Operation waren es 33%. Auch Nijs (1982) fand bei 20% eine Postmastektomiedepression im Verlauf des ersten Jahres nach der Operation.

Nach Wertheim-Operationen bleibt nur bei der Hälfte der Zervixkarzinompatientinnen im Zeitraum von 2–5 Jahren nach der Therapie die sexuelle Erlebnisfähigkeit unverändert erhalten (Erkrath und Randow, 1967; Henning und Schulz, 1975). Wir fanden ähnliche Verhältnisse (Pfleiderer et al., 1979). Etwas mehr als die Hälfte unserer Uteruskarzinompatientinnen berichteten von einer ausreichenden Libido. Geschlechtsverkehr, den 78% der Paare vor der Erkrankung noch ausübten, vollzogen 65% auch nach der Erkrankung. Eine Orgasmusfähigkeit, die vor dem Tumorbefall von 79% der Frauen angegeben wurde, bestand in 57% der Fälle auch nach der Behandlung noch. Wir haben festgestellt, daß die beste Prophylaxe einer Kohabitationsunfähigkeit in der frühzeitigen Wiederaufnahme des Geschlechtsverkehrs 2–6 Wochen nach dem stationären Behandlungsende besteht, natürlich mit entsprechender Rücksichtnahme des Partners. Das Paar muß darauf hingewiesen werden, daß ein blutiger postkoitaler Fluor durch gelöste Verklebungen hervorgerufen werden kann, was nicht schadet, sondern für die Erhaltung der Vagina als Kohabitationsorgan notwendig ist. Die krebsbehandelte Frau leidet an einem sexuellen Insuffizienzgefühl ihrem Partner gegenüber. Das vertrauensvolle ärztliche Gespräch – spätestens bei der Entlassungsuntersuchung und möglichst unter Einbeziehung des Partners – hilft negative Erwartungsängste abzubauen und verhindert in vielen Fällen sekundäre psychosexuelle Schwierigkeiten.

52.2 Geburtshilfe

In den Jahren von 1965 bis 1975 fanden bahnbrechende somatische Fortschritte Eingang in die Geburtshilfe. Der Zeitraum von der 28. Schwangerschaftswoche bis sieben Tage nach der Geburt wurde als besonders entscheidend für ärztliche Maßnahmen angesehen. Der Ausdruck „perinatale Medizin" wurde Symbol für einen neuen Schwerpunkt in den Forschungsarbeiten, und es wurden Gesellschaften für perinatale Medizin gegründet, die eine Wandlung von der Geburtshilfe hin zur Geburtsmedizin anstrebten. Die Mütter- und Säuglingssterblichkeit ließen sich durch die neugeschaffenen Überwachungsmethoden (Kardiotokographie, Ultraschall, Amnioskopie, Mikroblutuntersuchung usw.) und durch den vermehrten Einsatz operativer Maßnahmen entscheidend senken.

Obwohl diese sicherer gewordene Geburtsmedizin für die Mutter auch einen positiven emotionellen Aspekt im Sinne einer Angstreduktion bedeuten konnte, waren viele Schwangere sehr unzufrieden über die einseitige Betonung der organischen bzw. technischen Seite der Geburtshilfe. Die erste Kritik einer fehlenden emotionalen Ausgewogenheit beim Geburtserleben kam von den Frauen selbst.

Der an apparativen Techniken orientierten Geburtshilfe und Perinatalmedizin wurde vorgeworfen, daß sie auf wesentliche emotionale Werte der werdenden Mutter und des Vaters kaum Rücksicht nimmt, die Eltern ungenügend informiert, sie an medizinischen Entscheidungen nicht mitbeteiligt. Das eigene intime Geburtserlebnis als ein seltenes entscheidendes Lebensereignis sollte nicht einer kühlen Klinikorganisation zum Opfer fallen. So läßt sich, von den Frauen selbst ausgehend und unter einem gewissen Druck der Öffentlichkeit, eine erneute Veränderung der Geburtshilfe beobachten.

Psychosomatisch orientierte Geburtshelfer, Psychoanalytiker und klinische Psychologen, wie z.B. Conrad, Diederichs, Dmoch, Eicher, Fervers-Schorre, Frick, Molinski, Müller, Platz, Poettgen, Prill, Richter, Siedentopf, Stauber, Wenderlein hatten dieses Unbehagen der Mütter aufgegriffen und eine Integration psychosomatischer Aspekte in die Geburtshilfe gefordert. Von ihnen wurde auch die Deutsche Gesellschaft für psychosomatische Geburtshilfe und Gynäkologie ins Leben gerufen, die sich um den emotionalen Anteil von Schwangerschaft, Geburt und Wochenbett bemüht. Ergebnisse aus der Neurosenforschung und der Entwicklungspsychologie haben besonders zum Verständnis von Schwangerschaft, Geburt und Wochenbett beigetragen.

52.2.1 Zur Schwangerschaft

Während der Schwangerschaft erlebt die Frau eine Reihe von psychischen Veränderungen, die von Molinski (1972) ausführlich zusammengestellt wurden. Im Mittelpunkt dieses Erlebens fällt häufig ein psychischer Rückzug auf, so daß Hilfs- und Anlehnungsbedürftigkeit größer werden. Die körperliche Wahrnehmung ändert sich durch die physiologischen Prozesse in den verschiedenen Schwangerschaftsphasen und erinnert ständig an die zu erwartende neue Rolle. Der Weg von der Zweier- in die Dreierbeziehung wird in der Phantasie durchlebt, und es konstelliert sich ein Bild von der eigenen Mütterlichkeit. Dieses Bild bedeutet auch eine Auseinandersetzung mit der eigenen Mutter, was in einigen der später ausgeführten psychosomatischen Störungen in der Schwangerschaft deutlich wird.

Bereits in der Schwangerschaft besteht eine psychosomatische Wechselwirkung zwischen Mutter und Fötus. Untersuchungen haben bestätigt, daß psychisch belastete Schwangerschaften, z.B. durch sozialen Streß, wie er etwa bei ledigen Müttern beobachtet wurde, ein somatisches Risiko für Mutter und Kind darstellen (Davids und Rosengren, 1962; Benedek, 1971; Heinrichs, 1977; Lukesch, 1983; Weingart, 1983). Zu diesem Themenkreis gibt es auch eindrucksvolle Tierversuche von Bloch (1970), die zeigen, daß emotionelle Traumata, die stark angstbesetzt sind, negative somatische Einwirkungen auf Schwangerschaft und Geburt haben, z.B. in Form von Nidationsstörungen, Aborten, Mißbildungen und Totgeburten.

Vor allem die Vorstellung von der Geburt selbst geht bei jeder Schwangeren mit Phantasien und Ängsten einher. So beobachtet man gerade in der präpartalen Zeit bei vielen Schwangeren verstärkte Ängste, die mit der bevorstehenden Geburt zusammenhängen. Wenn man I-Gravidae nach ihren Ängsten in Zusammenhang mit der Geburt befragt, berichten sie von verschiedenartigsten Ängsten, die teilweise real, teilweise aber auch neurotisch und damit verzerrt

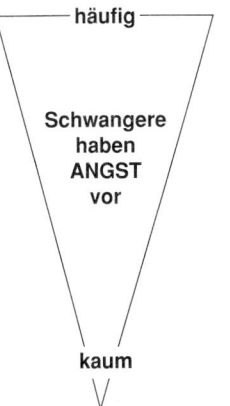

Abb. 52–7. Angsthierarchie bei Schwangeren (nach Perrez et al., 1978).

und schwer einfühlbar erscheinen. Ein Überblick über häufig vorkommende Ängste findet sich im Versuch der Erstellung einer „Angsthierarchie", die aus 60 verschiedenen Ängsten von Schwangeren ermittelt wurde (Perrez et al., 1978). Einen Auszug hieraus zeigt die Abbildung 52–7.

Es fallen hier Ängste auf, in die man sich gut einfühlen kann, allen voran die Angst vor einer Mißbildung beim Kind, die Angst vor Komplikationen, die Angst vor Schmerzen usw. Man kann diese Ängste als Realängste bezeichnen, wenn sie auch manchmal auf der Basis neurotischer Persönlichkeiten übersteigert empfunden werden. Solche realen Ängste – die teilweise auch auf falschen Vorstellungen beruhen – lassen sich abbauen durch eine fachgerechte Schwangerenberatung sowie durch sicherheitgebende Untersuchungen. So erscheint es sinnvoll, wenn man die Angst einer älteren Schwangeren vor einem mongoloiden Kind durch eine Amniozentese und eine zytogenetische Untersuchung beseitigt. Ähnlich kann man einer Frau durch wiederholte echographische Untersuchungen Sicherheit geben, wenn sie befürchtet, daß sich ihr Kind nicht termingerecht entwickelt. Allgemein sollten positive Auskünfte bei Schwangeren überwiegen, da die Frau in dieser Zeit besonders sensibel für Ängste ist.

Die zweite Gruppe von Ängsten ist eher zu den neurotischen Ängsten zu rechnen, so die Angst vor dem Verlust der Selbstkontrolle, die Angst vor dem Ausgeliefertsein oder die Angst vor dem eigenen Tod. Meist handelt es sich um Patientinnen, bei denen man Schwierigkeiten bei der Anpassung an die Mutterschaft findet. Eine ambivalente Einstellung zur Schwangerschaft kann sich in den beschriebenen neurotischen Ängsten zeigen. Mit zunehmendem Fortschreiten der Schwangerschaft und positiver Identifikation mit ihr lassen diese Ängste nach. Ebenso verschwinden mit zunehmender Schwangerschaftsdauer Symptome, die oft unter dem Stichwort „Impulsneurosen" subsumiert werden. Es handelt sich dabei meist um die Abfuhr oraler Bedürfnisse. Abnorme Gelüste, Hypersalivation, Heißhunger, Fettsucht und Stehlen spiegeln solche nur kurz dauernden Krisen im Erleben der Schwangerschaft wider. Was das Stehlen betrifft, so hat der Gesetzgeber in einigen Ländern bei Schwangeren das Strafmaß dieses „oralen Impulses" entsprechend gemindert (Weingart, 1983).

Hyperemesis gravidarum

Die Hyperemesis gravidarum ist das bekannteste psychosomatische Symptom in der Schwangerschaft, das vor allem im ersten Trimenon auftritt. Die Schwangere erbricht mehrmals täglich, mit der Gefahr der Elektrolytstörung und der Mangelernährung der Frucht. Der psychodynamische Hintergrund wurde umfassend von Molinski (1972) beschrieben. Er hat auf die Schwierigkeiten hingewiesen, in die eine Frau geraten kann, wenn sie mit der Rolle der Mutterschaft konfrontiert wird. Er fand vor allem Ängste bei den Frauen, die im Bereich des oralen und aggressiven Erlebens gehemmt sind. So können verdrängte orale und aggressive Impulse durch die Schwangerschaft aktualisiert werden und die Entwicklung einer befriedigenden Symbiose zwischen Mutter und Kind verhindern. Diese Frauen müssen deshalb den Fötus als oralen Konkurrenten, als „Mitesser" erleben. Die mobilisierten oralen und aggressiven Impulse können so zu einem verstärkten Schwangerschaftserbrechen als somatischem Korrelat führen. Ein Hinweis auf eine orale Störung dieser Schwangeren ist auch der immer wieder verblüffende therapeutische Effekt durch die alleinige stationäre Aufnahme. Bei den allermeisten Schwangeren, die wegen einer Hyperemesis gravidarum stationär aufgenommen werden, kommt es bereits unmittelbar nach der Aufnahme zu einer deutlichen Besserung oder einem Verschwinden des Erbrechens. Die Last ihres oralen Konkurrenten, „ihres Mitessers" wird durch die „Mutter" Klinik – sprich Ärzte und Schwestern – deutlich erleichtert – sie dürfen hier selbst wieder Kind sein, das versorgt wird. Das therapeutische Vorgehen bei Hyperemesis gravidarum besteht also primär in einer haltenden, unterstützenden Zuwendung, die innerhalb der geburtshilflichen Praxis in der Regel gut geleistet werden kann.

Psychogener und habitueller Abort

Das Abortgeschehen hat überwiegend organische Ursachen. Eine exakte sonographische, hormonelle und zytogenetische Begleitdiagnostik ist daher selbstverständlich. Trotzdem bleiben eine Reihe von Aborten ungeklärt und korrelieren mit psychischen Auffälligkeiten bei Frauen, die zumindest einen Risikofaktor für die Schwangerschaft aufweisen. Nach Prill (1967), Hertz und Molinski (1980) handelt es sich bei diesen Frauen meist um eine ambivalente Gefühlseinstellung. Einerseits wünschen sich solche Frauen zur Bestätigung ihrer Weiblichkeit auf der bewußten Ebene ein Kind, andererseits fühlen sie sich dieser Aufgabe nicht gewachsen. Als pathogenetischer Weg werden aufgrund chronifizierter Streßsituationen vegetative Fehlregulationen (Sympathikotonie) angenommen, die zu Uteruskontraktionen und schließlich zu einer Ablösung der Plazenta führen könnten. Eine vorangegangene Fehlgeburt kann auch eine ängstliche Erwartungshaltung bedingen, die diesen Mechanismus verstärkt. Therapeutisch bewährt hat sich hier – wie auch bei anderen psychosomatischen Störungen – kein aufdeckendes Verfahren, sondern mehr eine unterstützende, Ich-stärkende Arzt-Patientin-Beziehung in der Schwangerenvorsorge.

EPH-Gestose

Diese schwangerschaftsspezifische Erkrankung geht mit Symptomen der Ödembildung, der Proteinurie und der Hypertonie einher. Als pathogenetischer Mechanismus liegt ein generalisierter Arteriolenspasmus zugrunde. Obwohl man noch wenig über die Ätiologie dieses Symptomenkomplexes weiß, wird neben psychischen Faktoren in neuesten Arbeiten (Conradt, 1984) ein Magnesiummangel als somatische Ursache diskutiert. In der Klinik kommt man therapeutisch in den meisten Fällen gut mit Diät, Antihypertensiva und Sedativa zurecht. Es fällt aber auch hier auf, daß man durch Schaffung einer ausgewogenen emotionalen Situation dieses Leiden positiv beeinflussen kann. Jäschke und Dmoch (1984) haben ermittelt, daß das Selbstwertgefühl der schwangeren Frau mit EPH-Gestose durch die Konfrontation mit der Schwangerschaft in unterschiedlicher Weise in Frage gestellt wird. Diese Erschütterung des Narzißmus mobilisiert eine narzißtische Wut, die entweder auf zwanghafter oder depressiver Ebene abgewehrt wird. Der Bluthochdruck ist so im Sinne einer körperlichen Bereitstellung im Zusammenhang mit diesen aggressiven Impulsen zu sehen. Bei inkompletter, dauernd in Gang befindlicher Abwehr wird er eher als ein Korrelat zur Abwehr des Impulses angesehen. Die genannten Autoren haben bei den untersuchten 44 Patientinnen folgende sich wiederholende Konstellation beschrieben: das Spiel zwischen der seelische Kraft verzehrenden Abwehr und dem unvollkommen abgewehrten Impuls von Ärger und Affekten des Gekränktseins. Diese bringe die Gestosepatientin in eine charakteristische Spannung. Die „gestotische Beziehung", also die Konstellation von Affekt und Abwehr, könne sich so über humorale und vasomotorische Veränderungen auf das Kind auswirken.

Berger-Oser und Richter (1984) haben 10 Patientinnen mit EPH-Gestose nach psychoanalytischen Kriterien untersucht. In der Biographie fielen gehäuft Störungen im oralen Bereich auf (gravierendes Über- und Untergewicht). Bei allen Patientinnen fand sich eine „maligne" Symbiose zur Mutter. Die Mütter, die sich objektiv als eher gefühlskalt und desinteressiert erwiesen, hatten sich ihren Töchtern als „ideal" und aufopfernd dargestellt. Etwaige Störungen in der Beziehung zu ihren Müttern konnten die Mädchen daher nur als selbstverursacht ansehen und entwickelten massive Schuldgefühle. Wut auf die tatsächliche Gleichgültigkeit der Mütter mußten die abhängigen Töchter wegen der Gefahr einer unerträglichen Zerreißprobe im Keim ersticken. Zusammen mit dem entstehenden Kind, so hoffe die Schwangere, werde sie nun eine eigene Symbiose aufbauen, die endlich die Loslösung von der Mutter ermöglichen werde. Gleichzeitig treten aber Ängste davor auf, daß das Kind sie „ausbeuten" könne. In dieser Ambivalenz zwischen dem Wunsch nach Selbstverwirklichung und Angst vor Verlust der Mutter und Bedrohung durch das Kind liegt – nach Ansicht der letztgenannten Autoren – die Ursache dieser wichtigsten schwangerschaftsspezifischen Erkrankung.

Vorzeitige Wehen – Frühgeburt

In der geburtshilflichen Klinik gelingt es vermehrt, durch tokolytische Medikamente, Sedierung und Bettruhe eine drohende Fehlgeburt aufzuhalten oder doch zu verzögern. Obwohl diese therapeutischen Maßnahmen einen Fortschritt in der Herabsetzung der perinatalen Mortalität gebracht haben, ist doch hierdurch das Schwangerschaftserleben und eine günstige Vorbereitung auf die extrauterine Mutter-Kind-Beziehung beeinträchtigt. Psychosomatisch orientierte Geburtshelfer betonen hier immer wieder wehenauslösende emotionelle Faktoren. Ching und Newton (1980) haben die angloamerikanische Literatur zu diesem Thema aufgearbeitet und eine prospektive Studie über psychosoziale Faktoren durchgeführt. Sie haben auch die beschriebenen tiefenpsychologischen Hintergründe, wie innere Ablehnung der Schwangerschaft, Schwierigkeiten mit der weiblichen Rolle oder unreife Persönlichkeitsstrukturen, in ihre Untersuchung einbezogen. Obwohl eine große Zahl von Frauen (n = 335) untersucht wurde, fanden sie keinen signifikanten Hinweis für eine Beteiligung psychosozialer Faktoren an der Ursache der Frühgeburtlichkeit. In Arbeiten von Haldemann et al. (1976) und Hoyer und Thalhammer (1968) wurden allerdings gehäuft soziale Risikofaktoren, wie jugendliches Alter, unverheiratet sein und ein niedriger Sozialstatus, beschrieben. Auf tiefenpsychologischer Ebene werden unbewältigte Ängste und Ambivalenzen, die bis in das 3. Trimenon der Schwangerschaft persistieren, bei solchen Patientinnen beschrieben (Prill, 1983). Herms et al. (1982) konnten in einer prospektiven Studie zeigen, daß Berufstätige, besonders wenn es sich um qualifizierte Berufe handelte, eher zur Frühgeburt neigten als die familienorientierten Frauen. Frauen mit Frühgeburten hatten eine höhere Rate an psychosomatischen Symptomen wie Migräne, gastrointestinale Beschwerden und Schlaflosigkeit. Dmoch und Osorio (1984) haben vor allem depressive Persönlichkeitsstrukturen beschrieben. Petersen und Teichmann (1984), die eine noch laufende Studie zu diesem Thema durchführen, haben folgende entscheidenden Punkte hervorgehoben: Die Frühgeburtlichkeit ist eine sehr unspezifische Antwort auf eine allgemeine Überforderung – es gebe keinen spezifischen seelischen Konflikt. Die Frühgeburtlichkeit trage akzidentiellen Charakter, sei also nicht persönlichkeitsspezifisch. Therapeutisch wird von den meisten Autoren als wesentliches Ziel die Hilfe bei der Anpassung an die Schwangerschaft und deren Bewältigung angesehen. Das einfühlsame Visitengespräch steht dabei an erster Stelle. Je nach Indikationskriterien werden noch Einzelgespräche, autogenes Training, Hypnose und das respiratorische Biofeedback empfohlen.

Im Rahmen der Schwangerenberatung lassen sich meist alle hier angeführten Symptomgruppen durch eine intensivere Arzt-Patientin-Beziehung verdeutlichen und in Grenzen halten (Prill, 1976, 1977; Clyne, 1983; Stauber, 1976, 1979, 1983; Richter, 1980, 1982).

Psychosomatische Geburtsvorbereitung

Das Gelingen einer psychologischen Geburtserleichterung kann wesentlich durch eine psychosomatisch orientierte Betreuung der Schwangeren sowie ein psychohygienisches Geburtsvorbereitungsprogramm gefördert werden. Bei intensiver Zusammenarbeit von Geburtshelfer, Hebamme, Krankengymnastin, Kinderschwester und Kinderarzt entsteht die emotionale Atmosphäre, in welcher sich eine natürliche Geburtshilfe mit psychologischer Geburtserleichterung verwirklichen läßt. Die Einbeziehung möglichst aller an der Geburtshilfe beteiligten Personen in die Phase der Geburtsvorbereitung verhindert eine Diskrepanz zwischen den Erwartungen der Schwangeren und den späteren realen Kreißsaal- bzw. Geburtserfahrungen. Eine kontinuierliche Betreuung der Schwangeren während der Schwangerschaftsvorsorgeuntersuchungen durch denselben Arzt schafft Vertrauen und sollte nicht an organisatorischen Schwierigkeiten scheitern. Die psychosomatische Geburtsvorbereitung beginnt nämlich bereits mit einer gelungenen Arzt-Patientin-Beziehung in der Sprechstunde (Krebs, 1983; Perez-Gay, 1983; Prill, 1983; Richter, 1980, 1983; Stauber, 1979, 1983).

Die psychohygienische Geburtsvorbereitung geht heute weit über die Vermittlung biologisch-technischen Wissens oder krankengymnastisch eintrainierte Entspannungs- und Atemübungen hinaus. Auch Read- oder Lamaze-Kurse, die darauf abzielen, neben Vermittlung von Entspannungs- und Atemübungen Geburtsängste durch Aufklärung zu vermindern und Schmerzen zu reduzieren, entsprechen nicht mehr den aktuellen Bedürfnissen und Erfordernissen. Umfassende Geburtsvorbereitung muß nach den Problemen und Konflikten, nach Gefühlen, Ängsten, irrationalen Befürchtungen und Phantasien fragen, die in einem erweiterten Sinne mit Schwangerschaft und Geburt zu tun haben. Viele Eltern geraten unwissend und unvorbereitet in die Konflikt- und Problemkreise von Schwangerschaft, Geburt, Wochenbett und Elternrolle. Sie stehen nicht selten diesem Problem hilflos gegenüber. Andererseits sind Eltern während der Schwangerschaft für diese Fragen besonders aufgeschlossen, ja geradezu sensibilisiert. Es hat sich gezeigt, daß Eltern begierig diese Probleme aufgreifen und daß von solchen Geburtsvorbereitungsprogrammen, die Schwangerschaft und Geburt in einen größeren Zusammenhang menschlicher Grunderfahrung setzen, weitreichende präventiv-psychohygienische Impulse ausgehen können (Richter, 1980, 1982, 1983). Im Rahmen dieser Kurse hat sich die Einbeziehung des Partners sehr bewährt. Die Besichtigung des Kreißsaals, der Wochenstation und der Säuglingseinrichtung unter Kontaktaufnahme zu Hebammen und Schwestern wirkt angstmindernd. Praktische Übungen in Säuglingspflege geben der Schwangeren später mehr Sicherheit in der Versorgung des Neugeborenen. Von der Deutschen Gesellschaft für psychosomatische Geburtshilfe und Gynäkologie sind praktische Vorschläge für eine psychosomatisch orientierte Geburtsvorbereitung gemacht worden, die kursorisch in Tabelle 52–4 dargestellt sind.

52.2.2 Zur Geburt

Die Geburt ist nicht lediglich das physiologische Ende der Schwangerschaft, sondern ein psychosomatisches Ereignis – ein Erlebnis, das die Frau mit Leib und Seele erfaßt. Wohl kaum ein Ereignis im menschlichen Leben ist von so vielen Geheimnissen umgeben und mit einer solchen Vielfalt an Bedeutungsgehalten versehen worden. So spielt auch der Geburtsvorgang im Denken und Fühlen der Völker eine bedeutsame Rolle, die wiederum je nach weltanschaulicher, rassischer, kultureller und soziologischer Struktur verschieden ist.

In tiefenpsychologischen Arbeiten zur Geburt wird auf den Objektverlust der Frau hingewiesen, der individuell verschieden verarbeitet wird und postpartale Depressionen erklären kann. Die Mutter muß sich schließlich von dem einverleibten Kind trennen, was bei ihr eigene Trennungsängste aktualisieren kann.

Wenn wir uns mehr der geburtshilflichen Praxis zuwenden, so fällt auf, daß man dem Geburtsschmerz unter allen Phänomenen, die mit der Geburt zusammenhängen, von jeher die größte Beachtung schenkte. Zwischen folgenden beiden Extremen lagen die Ansichten:

Einmal – der Schmerz gehört wesensmäßig zur Geburt, nachdem es bereits in der Genesis heißt: „Du sollst dein Kind unter Schmerzen gebären".

Zum anderen – der Schmerz sei eine sinnlose und deshalb überflüssige Begleiterscheinung der Geburt.

Die zweite Ansicht hat sich in der Praxis mehr und mehr durchgesetzt – es haben sich Indikationen für die einzelnen geburtserleichternden Methoden herausgebildet, die sich an dem Risiko für Mutter und Kind orientieren.

Tabelle 52–4.	Psychosomatische Geburtsvorbereitung
Geburtshelfer	Paarweise Vorbereitung in Gruppen, Physiologie und Psychologie der Schwangerschaft, Noxen (Nikotin, Medikamente, Streß) Angstabbau durch Aufklärung über den natürlichen Geburtsablauf, dabei Vorstellung der apparativ-technischen Überwachungsmethoden lediglich als Sicherheit bringende Hilfsmittel, Operationen, Schmerzerleichterung, ambulante Geburt, Geburtserleben, Partneranwesenheit, Beziehung zum Kind, Besichtigung der für die Geburt ausgewählten Klinik Wochenbett: Mutter-Kind-Beziehung, Stillen, Signale und Entwicklungsschritte des Säuglings
Hebamme Physiotherapeutin	Körperarbeit mit Erfahrung der eigenen Leiblichkeit, individuelle Atmung, Entspannungsübungen, Gymnastik, Akzent auf „individueller Geburt", nicht auf Methoden Säuglingskurs, Körperpflege, Stillhilfen, soziale Hilfen, Mutterschutzgesetz
Kinderarzt	Körperliche und seelische Entwicklung des Kindes, Vorsorgeuntersuchungen, Impfungen, Ernährung des Säuglings und Kleinkindes

Geburtsschmerzen und Gebärstörungen

Der Geburtsschmerz setzt sich zusammen aus dem Wehenschmerz der uterinen Kontraktionen, dem Dehnungsschmerz der Zervix und dem Dehnungsschmerz des Beckenbodens. Weitere Schmerzen können vom Peritoneum und den Ligamenten im kleinen Becken ausgehen. Die Intensität und Dauer des Geburtsschmerzes können bei den Gebärenden individuell sehr verschieden sein.

Im allgemeinen empfindet der Mensch Schmerzen als Signal, als Alarmreaktion des Organismus: Etwas ist nicht in Ordnung. Schmerzen lösen daher Beunruhigung oder Angst aus. Für den Geburtsvorgang gilt diese kausale Verknüpfung nur bedingt; hier kann es weh tun, und trotzdem muß keine krankhafte Störung vorliegen. Diesen Zusammenhang gilt es Gebärenden zu erklären. Die Tatsache, daß es weh tut, kann zu einer erheblichen Angstquelle werden, kann zu der Vorstellung führen, daß man selbst oder das Kind in Gefahr ist. Im Gegensatz zum angloamerikanischen Begriff „labour" existiert im deutschen Sprachgebrauch der unglückliche Begriff „Wehe", welcher automatisch eine Schmerzassoziation auslöst. Man muß daher die Frau aufklären, daß die Wehe nicht eine sinnlose oder gefährliche Schmerzsituation, sondern ein physiologisch notwendiger und geburtsfördernder Vorgang ist. Phänomene des Schmerzes können als Ausdruck pathophysiologischer Abläufe des Organismus allein nicht ausreichend erklärt werden. Schmerzen sind immer ein komplexes sinnesphysiologisches und psychologisches Geschehen zugleich. Durch Schmerzen kann ein Mensch verändert werden, können Wesenszüge hervortreten, die ihm selbst und seiner Umgebung sonst gar nicht bekannt sind. Besonders dann, wenn – wie beim Geburtsschmerz – der Schmerz den ganzen Menschen erfaßt, beeinflussen seelische Bedingungen das Schmerzerlebnis und die Schmerzreaktion. Seelische Verfassung, Schmerzerleben und Schmerzreaktion wirken aufeinander ein, bedingen sich wechselseitig. Schmerzen unter der Geburt stellen also immer ein individuelles Problem dar. Daher müssen persönlichkeitsspezifische Phänomene mitberücksichtigt werden, die weit über das eigentliche Geburtsgeschehen hinausgehen. Wie eine Frau ihre Schwangerschaft akzeptiert hat, wie sie durch Wissen über den Geburtsvorgang Ängste abbaut, wie sie durch Vertrauen in ihre Leistungsfähigkeit in die Geburt geht, hat wesentlichen Anteil an der Bewältigung der Geburtsarbeit. Ihr seelisches und körperliches Wohlbefinden bestimmt die Toleranz des Geburtsschmerzes. Die Geburt darf deshalb nicht losgelöst vom Schwangerschaftsverlauf betrachtet werden, vielmehr bilden beide erlebnismäßig eine Einheit. Das bedeutet, daß positive wie negative Einflüsse während der Schwangerschaft Auswirkungen auf das Geburtsgeschehen und darüber hinaus auf die frühe Mutter-Kind-Beziehung haben (Prill, 1979).

Das „Angst-Spannungs-Schmerz-Syndrom", das durch Arbeiten von Platanov (1923), Read (1933), Lukas (1968, 1972), Molinski (1968) und anderen zu größerer Klarheit gekommen ist, kann zur Erklärung psychogener Gebärstörungen dienen.

So werden Wehen oft angstvoll erlebt, was dann mit vermehrter innerer Spannung verbunden ist. Diese Spannung führt auf muskulärem Weg zu einer Verkrampfung, auf vegetativem Weg zu Atmungsstörungen und Vasokonstriktion und affektiv zu einer gesteigerten Empfindlichkeit. Der dadurch verstärkt auftretende Schmerz bedingt eine verzögerte und damit oft komplizierte Geburt.

Zur Verminderung der Geburtsängste und damit der Schmerzen wurden in den letzten Jahrzehnten vermehrt psychologische Vorbereitungsmethoden angewandt. Im deutschsprachigen Raum hat vor allem die Tübinger Schule um Roemer (1977) und das autogene Training von J. H. Schultz (1970) – vertreten durch Arbeiten von Prill (1964, 1967, 1968) und Poettgen (1971, 1973) – großen Aufschwung erlebt. Zuvor hatte die englische Methode nach Dick-Read (1933), die russische Methode durch Velvolvski (1953) und die französische Schule nach Lamaze und Vellay (1952) bereits zahlreiche Anhänger gefunden. Diese Methoden, die, je nach Schwerpunkt, aufklärende, gymnastische, atemtechnische, lerntheoretische und autosuggestive Hilfen zur Geburtsvorbereitung geben, haben nach einer Sammelstatistik zwischen 75–96% Erfolge zu verzeichnen (Ruppin et al., 1977).

Neben diesen psychologischen geburtserleichternden Methoden haben die medikamentösen Verfahren vor allem in den 70er Jahren an Bedeutung gewonnen (Lenz, 1973; Beck und Potthoff, 1976). Aus psychohygienischer Sicht weist von diesen Verfahren vor allem die noch immer in mehreren Kliniken gehandhabte Allgemeinnarkose – die sog. Durchtrittsnarkose – einen großen Nachteil für die Mutter auf, da sie die Geburt selbst nicht bewußt miterleben kann. Dazu berichtet bereits H. Deutsch (1954) von Frauen, die nach einer prolongierten Entbindungsnarkose erklärten, daß das ihnen vorgestellte Kind nicht das ihrige, sondern vertauscht sei. Sie erklärt dieses Phänomen so, daß die ganze an den Geburtsvorgang geknüpfte, von der Außenwelt zurückgezogene psychische Energie im Moment der Entspannung dem Kinde zufließt. Das plötzliche Befreitsein von Schmerz und Angst, das Wissen, es geschafft zu haben, führt zu einem Gefühl des Triumphes und verleiht den ersten Momenten der Mutterschaft den Charakter der Ekstase. Da dieser Prozeß in Vollnarkose beeinträchtigt ist, appelliert sie an den Geburtshelfer, die Frau nicht ohne Grund um den „Lohn ihrer Arbeit" (engl. labour) zu bringen, d.h., um das triumphale Gefühl, es geschafft zu haben.

Wenn man die Vor- und Nachteile der psychologischen Vorbereitungsmethoden abwägt, so ist man geneigt, aufgrund der fehlenden Nebenwirkungen für Mutter und Kind sowie der positiven Auswirkungen auf das Geburtserleben und die spätere Mutter-Kind-Beziehung die lange Vorbereitungszeit in Kauf zu nehmen. Die medikamentösen Methoden sollten die psychologische Geburtsvorbereitung folglich nicht ersetzen, beide Methoden können sich jedoch in vie-

Tabelle 52–5. Geburtserleichternde Methoden im Vergleich

Analgesie in der Geburtshilfe	Medikamentöse Geburtserleichterung					Psychologische Geburts-erleichterung
+ + = günstig + = halbgünstig – = ungünstig	Allgemein-anästhesie	Regionalanästhesie			Analgetika Sedativa	(z. B. nach Read, Lamaze, AT)
Kriterien	i. v./Inhal.	Peridural-Kaudal-	PCB	Pudendus Damminfiltr.	Opiate Spasmolytika Tranquilizer	Entspannungs-übung Atemtechnik Gymnastik Vertrauensver-hältnis Arzt – Hebamme – Patientin
Ausdehnung I. **Analgesie** Wirkungsgrad	+ + + +	+ + + +	+ +	+ +	+ +	+ +
Eröffnung II. **Geburtsphase** Austritt	– + +	+ + + +	+ –	– +	+ –	+ +
Zur Vorbereitung III. **Zeit** Wirkungsdauer	+ + –	+ + +	+ +	+ +	+ +	– +
Mutter IV. **Nebenwirkungen** Kind	+ +	+ + +	+ +	+ + +	+ +	+ + + +
V. **Geburtserleben der Mutter**	–	+	+ +	+ +	+	+ +

len Fällen sinnvoll ergänzen. Tabelle 52–5 gibt einen kurzen Überblick über die derzeit angewandten Erleichterungsmethoden. Das Geburtserleben erscheint dabei aus psychosomatischer Sicht besonders wichtig, da es die Mutter-Kind-Beziehung positiv beeinflussen kann.

In der Praxis haben sich in den letzten Jahren einige Neuerungen durchgesetzt, die ihren Ursprung in psychohygienischen Überlegungen haben, so z.B. „der Vater bei der Geburt".

Sieht man den Ehemann nicht als Aufpasser, sondern als Helfer bei der Geburt, so kann er einmal seiner Frau bei der Beibehaltung der richtigen Atemtechnik und zur Verarbeitung der Wehen behilflich sein, er kann weiterhin seiner Frau Sicherheit und Geborgenheit geben, und schließlich kann der Mann als Informationsübermittler zum Personal dienen, vor allem bei gehemmten und ängstlichen Müttern.

Nach Prill (1976) bewirkt das gemeinsame Geburtserlebnis darüber hinaus eine Festigung des emotionalen Familiengefüges. Prill zeigt auch, daß keine sexuellen Funktionsstörungen nach einem negativen Geburtserlebnis beim Mann auftreten. Die Deutsche Gesellschaft für psychosomatische Geburtshilfe und Gynäkologie hat sich bei ihrer Tagung in Freiburg

(1982) die Frage gestellt: Welche psychosomatischen Forderungen lassen sich an das Geburtsgeschehen stellen? Der Grund für diese Suche nach wissenschaftlich begründeten psychosomatischen Forderungen waren die zahlreichen oft ideologischen Ansätze in der perinatalen Medizin über das Wie, Wo und Wann der Geburtsmethoden. In der Diskussion wurde deutlich, daß einseitig in die Waagschale geworfene Namen wie Read, Lamaze, Leboyer (1974) usw. dem individuellen Geburtserleben nicht gerecht werden. Eine starre Haltung mit reinem Methodendenken ist im psychosomatischen Sinne nicht vertretbar. Es wurden aus diesem Grunde auch nur einige essentielle psychosomatische Forderungen erhoben, die in Tabelle 52–6 zusammengefaßt sind.

Erste Untersuchungen belegen, daß die Einbeziehung psychosomatischer Verfahrensweisen im Sinne einer individuellen Geburtshilfe kein Sicherheitsrisiko darstellt. Kentenich (1983) konnte zeigen, daß die perinatale Morbidität und Mortalität durch eine individualisierte Geburtshilfe nicht negativ beeinflußt wird. Im Gegenteil, die Patientinnen beurteilten ihre Geburt hierdurch sehr positiv, was ihnen wieder eine bessere Motivation für eine gelungene Mutter-Kind-Beziehung gab.

Tabelle 52–6. Psychosomatische Forderungen an das Geburtsgeschehen (Basis: „die sichere Geburt")

Ziel: Sichere, angstfreie, schmerzarme, möglichst natürliche Geburt als individuelles Geburtserlebnis
- Einfühlsamer Umgang mit der Gebärenden (Angstreduktion)
  - durch die Hebamme (Akzent auf Zuwendung)
  - durch den Arzt (Akzent auf Sicherheit)
- Anwesenheit des Partners/vertrauter Bezugsperson
  - als Helfer bei der Verarbeitung von Wehen (spart Analgetika)
  - als Vermittler von Geborgenheit
- Individualisierte Schmerzerleichterung „Geburtserleben" der Gebärenden möglichst erhalten!
- Förderung des sofortigen Kontakts von Mutter und Kind durch intensiven Hautkontakt und frühes Anlegen des Kindes

52.2.3 Das Wochenbett

Aus psychosomatischer Sicht beginnt mit dem Wochenbett ein Prozeß, den der Psychoanalytiker Fornari (1970) „das zentrale Problem der gesamten Entwicklung des kindlichen psychischen Lebens" nennt: die Beziehung zwischen Mutter und Kind. Diesen Prozeß der frühen Mutter-Kind-Beziehung zu unterstützen, ist im Rahmen einer integrativen psychosomatischen Geburtshilfe die Aufgabe des Geburtshelfers, der Hebamme, des Pädiaters und des Pflegepersonals.

Die Erforschung der frühesten psychischen Entwicklung des Kindes ist eigentlich erst in den letzten Jahrzehnten erfolgt. Auf der einen Seite haben Spitz (1957, 1967), Ainsworth (1972), Bowlby (1952, 1972), Renggli (1974), Mueller-Braunschweig (1975) u.a. versucht, durch experimentelle Beobachtungen die Entwicklung des Säuglings zu verfolgen. Auf der anderen Seite wurden aus den Psychoanalysen von Kindern und Erwachsenen durch Freud (1964), Klein (1972), Ferenczi (1913) u.a. Material und Erkenntnisse zusammengetragen, die einen Einblick in die Genese neurotischer Störungen gaben.

Allen Autoren kommt es bei der Beschreibung der Mutter-Kind-Beziehung darauf an zu betonen, daß Mutter und Kind nach der Geburt noch eine Einheit bilden. So rechnet der Biologe Portmann (1963) das erste Lebensjahr des Menschen noch zur Embryonalzeit. Er stützt sich auf Untersuchungen von Lange (1903) und Scammon (1922) und belegt, daß der Mensch gemäß seiner Wuchsform im 1. Lebensjahr sowie seiner Gehirngröße ein Jahr zu früh auf die Welt kommt. Im Vergleich zur Tierwelt nennt er den Menschen einen Nesthocker, der noch der extrauterinen Nabelschnur bedarf. Für diesen zweiten postpartalen Uterus gelten beim Menschen in einem nur geringen Maße die erblich vorgegebenen instinktiven Ordnungen, die eine funktionierende Entwicklung garantieren.

Die Erforschung des besonderen Wahrnehmungsinstrumentes, welches die Mutter befähigt, die neonatale Situation ihres Kindes zu verstehen, liegt noch im dunkeln. Die Tatsache, daß dieses Wahrnehmungsinstrument sich kaum in Worte und schon gar nicht in verifizierbare Größen fassen läßt, spricht dafür, daß es in den tiefsten vorsprachlichen Schichten des menschlichen Gefühlslebens angesiedelt ist.

Dieses Phänomen der Beziehung der Mutter zu ihrem Säugling kann nur umschrieben werden. Wir sprechen von mütterlicher Intuition, von Empathie oder von einer gesteigerten Sensibilität. Freud (1964) nannte diese frühe Mutter-Kind-Beziehung eine „Masse zu zweit". Spitz (1967) spricht von einer Dyade, die er am besten charakterisiert sieht mit dem Dichterwort: „ein Egoismus zu zweit". Therese Benedek (1971) spricht von der „Mutter-Kind-Zweiheit". Winnicott (1974) sieht in der Abhängigkeit den Hauptzug des Säuglingsalters. Ein Säugling wird erst zum Säugling, sagt er, wenn er mit der mütterlichen Fürsorge zusammengebracht wird. Säugling und Mutterpflege bilden eine Einheit. Aus den Arbeiten von M. Klein (1972) dürfen wir folgern, daß der Säugling die Außenwelt, so z.B. die Brust der Mutter, als Teil von sich selbst phantasiert.

Wie funktioniert diese Symbiose zwischen Mutter und Kind? Nach Spitz (1967) laufen Mikrointeraktionen zwischen beiden ab. Winnicott (1974) sieht den Säugling als unreifes Wesen, das ständig am Rand unvorstellbarer Angst steht. Er spricht von der „holding function", die die Mutter einnehmen muß, um dem Kind Halt zu geben. Die Mutter sollte daher wie eine Hülle fungieren, die das Kind vor übermäßigen äußeren und damit auch inneren Spannungen beschützt. Die zunächst diffusen Gefühle beim Säugling wie Lust und Unlust können sich nur entwickeln, wenn die Gefühlsäußerungen jeweils von der Mutter angenommen und wiedergegeben werden. Der Mutter kommt bei diesem Prozeß eine Spiegelfunktion zu, wie Margret Mahler (1972, 1975) es genannt hat. Gelingt es der Mutter nicht, die Signale ihres Kindes zu verstehen, ist der „Dialog" zwischen ihr und dem Kind gestört, tritt beim Kind überstarke Unlust und Desorientiertheit auf. Wird das Kind immer wieder diesen negativen irritierenden Eindrücken ausgesetzt, kann es kein Urvertrauen (Erikson, 1961) entwickeln, das die notwendige Basis für eine weitere gesunde seelische Entwicklung darstellt. Statt dessen steht am Beginn seiner Entwicklung ein „kumulatives Trauma" (Kahn, 1964) mit der Folge oft unbeeinflußbarer neurotischer und psychotischer Krankheitsbilder. Nach Winnicott (1974) bedeutet „die genügend gute mütterliche Fürsorge" für den Säugling eine Ich-Unterstützung. Dadurch wird im Säugling eine Kontinuität des Seins aufgebaut, die die Grundlage der Ichstärke ist, die für eine spätere Bewältigung der einströmenden Belastungen und Versagungen notwendig ist. Versagt die mütterliche Fürsorge in den ersten Wochen und Monaten, dann kommt es zu einer Störung der Integrationsprozesse, die im Individuum ein Selbst aufbauen. Der Säugling kann diesen Ausfall an mütterlicher Qualität nicht selbst ausgleichen, weil er das Stadium der Ich-Strukturierung, die das ermöglichen würde, noch nicht erreicht hat.

Diese frühe Einheit von Mutter und Kind ist vielen Gefährdungen ausgesetzt. Der postpartale Uterus arbeitet nicht mit der gleichen Sicherheit wie der Mutterleib. Dies gilt heute erst recht, wo die ursprünglichen Formen des Familienlebens durch die moderne Arbeitswelt gelockert oder fast aufgehoben sind.

In diesem Zusammenhang soll noch auf die negativen Folgen hingewiesen werden, die eine Mutterentbehrung in frühester Kindheit haben kann. Bowlby (1952, 1972) hat gezeigt, daß die Folgen partieller Deprivation Angst, exzessive Liebesansprüche, starke Haßgefühle und Folgen der letzteren Schuld und Depression sind (Eggers, 1977; E. Freud, 1984). Die totale Deprivation beeinflußt tiefreichend die charakterliche Entwicklung und zerstört die Fähigkeit zum zwischenmenschlichen Kontakt. Einer breiteren Öffentlichkeit wurden diese Folgen der Mutterentbehrung durch die Untersuchungen weiterer genetisch denkender Autoren wie Spitz (1957, 1967), Goldfarb (1955), Ribble (1941, 1944), Anna Freud (1971) u.a. bekannt. Mueller-Braunschweig (1975) hat in einer umfangreichen Studie gezeigt, daß für die früheste Kindheit nur die sicherheitgebende kontinuierliche Betreuung durch eine Pflegeperson eine gelungene Ich-Entwicklung ermöglicht. Auch die Konstanz einer Gruppe, wie sie in der Erziehung in den Kibbuzim gegeben war, hat sich nach Bettelheim (1969) als problematisch erwiesen. Man sagt den Kibbuzkindern nach, daß sie in ihren zwischenmenschlichen Beziehungen eine emotionale Tiefe vermissen infolge der fehlenden Zuwendung durch **eine** Mutter. Inzwischen versucht man diese Mutterentbehrung wieder in einigen Kibbuzim rückgängig zu machen. Der Begriff der „maternal deprivation" hat mittlerweile die Kinderpsychologie und Kinderpsychiatrie stark beeinflußt, und zunehmend ergeben sich daraus auch Konsequenzen für die Sozial- und Gesundheitspolitik (so will z.B. das neue Adoptionsgesetz unterbinden, daß der Säugling zu lange von einer konstanten Pflegeperson getrennt wird).

Den Erkenntnissen über die Deprivation wird erst in den letzten Jahren mehr Verständnis in den geburtshilflichen Abteilungen entgegengebracht – dachte man doch bisher, daß die frühe Wochenbettzeit vorwiegend von somatischen und endokrinologischen Prozessen bestimmt sei.

Maas (1973, 1975) war es schließlich, der eine Bestandsaufnahme in der Deutschen Klinik für Diagnostik vorgenommen hat. Ausgehend von einer Statistik über die Häufigkeit des Auftretens psychoneurotischer und psychosomatischer Symptome, suchte er nach Gründen ihrer Entstehung im Hinblick auf präventivmedizinische Maßnahmen. Er zeigte auf, daß bis zu 60% der Patienten in der Deutschen Klinik für Diagnostik über psychoneurotische und psychosomatische Symptome klagen. Hinzu kommt noch, daß über ein Drittel der Patienten Beruhigungsmittel oder Schmerzmittel nimmt und außerdem die Sucht- und Selbstmordrate ständig zunimmt.

Jetzt ist zu fragen, ob man auf die Qualität der mütterlichen Fürsorge überhaupt Einfluß nehmen kann.

Obwohl wir wissen, daß die werdende Mutter aufgeschlossen ist, viel für eine gesunde seelische Entwicklung ihres Kindes zu tun, bleiben bei einigen Frauen diese Hilfen erfolglos. So gibt es neben der Deprivation andere krankmachende Mutter-Kind-Beziehungen, wie z.B. eine „liebevolle" Einstellung der Mutter, unter der sich eine unbewußt ablehnende verbirgt, oder wenn die Mutter selbst ein exzessives Verlangen nach Liebe und Geborgenheit hat.

Von größter Bedeutung ist die Erkenntnis, daß es nach der Geburt eine besonders sensitive Phase gibt, die für die emotionale Beziehung zwischen Mutter und Kind wesentlich ist (Klaus und Kenell, 1974). Den Verhaltensforschern ist eine solche Phase bei einer Reihe von Tieren bekannt. Bei Trennung von Muttertier und Jungen unmittelbar nach der Geburt reagieren diese mit einem abnormen Brutpflegeverhalten. So nimmt z.B. die Mutter das Junge nach einer postpartalen Trennung nicht mehr an. Erfolgt die Trennung jedoch erst am 5. Tag, dann nimmt die Mutter ihr arteigenes, schützendes und pflegendes Verhalten wieder auf.

Marshall et al. (1972) haben zwei Gruppen von Erstgebärenden verglichen. Während die Mütter der ersten Gruppe gleich nach der Geburt ihr Kind für eine Stunde behielten und es während des Klinikaufenthaltes jeweils zusätzlich zu den Stillzeiten fünf Stunden am Nachmittag bekamen, hatten die Mütter der Kontrollgruppe ihr Kind jeweils nur eine halbe Stunde zu den kliniküblichen Stillzeiten bei sich. Bei einer Nachuntersuchung einen Monat später zeigte sich, daß die Mütter mit dem intensiven Kontakt zu ihren Kindern gegenüber den Müttern der Kontrollgruppe deutlich liebevoller und engagierter mit ihren Kindern umgingen. Diese positive Einstellung war auch noch bei einer Nachuntersuchung, ein Jahr später, festzustellen.

Winter (1976) kam in einer weiteren Studie zu ähnlichen Ergebnissen. Vor allem in bezug auf das Stillen zeigte sich, daß die Mütter, die gleich nach der Geburt einen längeren Kontakt zu ihren Kindern hatten, gegenüber einer Kontrollgruppe dem Stillen positiver gegenüberstanden (vgl. Leboyer, 1974).

Über die Mutter-Kind-Situation bei Frühgeborenen hat Kenell (1976) eine Untersuchung durchgeführt. Einer Gruppe von Müttern wurde es erlaubt, sofort nach der Geburt in die Frühgeborenenstation zu kommen und so oft wie möglich bei dem Kind zu bleiben. Der anderen Gruppe wurde der erste Kontakt mit dem Kind erst nach 3 Wochen erlaubt. Nur bei der ersten Gruppe konnte der Autor später ein besonders inniges Verhältnis zwischen Mutter und Kind beobachten.

Es liegen noch weitere Untersuchungen an Frühgeborenen von Barnett (1974) und Sokoloff et al. (1974) vor. Alle diese Autoren kamen zu dem übereinstimmenden Urteil, daß Kinder in der monotonen Umgebung des Brutkastens zusätzlich emotionelle und taktile Zuwendung brauchen. Sie konnten zeigen, daß

Frühgeborene besser gedeihen, wenn man sie gleichzeitig in die Hände der Mutter gibt.

Die Tatsache, daß es nach der Geburt eine sensitive Phase gibt, die auf die Bindung zwischen Mutter und Kind von besonderem Einfluß ist, kann man auch aus Fällen ablesen, bei denen Frauen in der Klinik irrtümlicherweise nicht das eigene Kind versorgt haben. Prill (1976) zitiert hierzu einen Fall, der sich in Israel ereignet hat. Hier klärte sich die Verwechslung der Neugeborenen erst nach 14 Tagen auf. Die Mütter beider Kinder waren nur schwer bereit, das jeweils versorgte fremde Kind gegen ihr eigenes einzutauschen.

Aufgrund dieser Erkenntnisse muß es die Aufgabe des Perinatologen sein, Bedingungen zu schaffen, die die kostbare sensitive Phase zwischen Mutter und Kind fördern helfen. Damit ließen sich iatrogene Schäden abwenden, die zur Zeit noch meist aus Unwissenheit oder aufgrund zu großer Abwehr psychosomatischer Erkenntnisse entstehen.

Das „Rooming-in-System" – also das ganztägige Zusammenbringen von Mutter und Kind auf der Wochenbettstation – hat sich als Unterstützung beim Aufbau einer gelungenen Mutter-Kind-Beziehung bewährt. Die Einheit von Mutter und Kind wird dadurch erhalten. Zusätzlich entstehen Vorteile, die der Mutter mehr Sicherheit in dieser Zeit geben. So z.B.:
- die Entwicklung einer besseren pflegerischen Fähigkeit,
- die Entängstigung mancher Erstgebärenden gegenüber ihrem Kind und
- das schnellere Erkennen der normalen und individuellen Reaktionen des Kindes.

Selbstverständlich bewirkt das alleinige räumliche Zusammenbringen von Mutter und Kind noch keine gelungene Dyade, es stellt aber einen Nährboden hierfür dar. Deutsch (1954) glaubt, daß sich die Zahl der versagenden Mütter sehr verringern würde, wenn man die freie Entwicklung der mütterlichen Gefühle weniger reglementieren würde. Das trifft vor allem auf die Frage des Stillens zu, von dem Therese Benedek (1971) sagt, daß es gemeinsam mit dem Hautkontakt (vgl. Montagu, 1974) die extrauterine Nabelschnur zwischen Mutter und Kind darstelle.

Laktationsstörungen

Die Laktation gehört sicher zu jenen physiologischen Prozessen, die wie die Menstruation und alles, was mit Fortpflanzung zusammenhängt, starken psychischen Einflüssen unterliegen. Es war S. Freud 1892, der in einer seiner ersten Publikationen einen Fall von psychogener Agalaktie beschrieb, den er mit Hypnose erfolgreich behandelte (Freud, 1964).

Laktationsschwierigkeiten sind im Wochenbett besonders häufig zu beobachten. Auf tiefenpsychologischer Ebene läßt sich dieses Versagen oft als eine Flucht vor den Pflichten der Mutterschaft erkennen, die der Wöchnerin Angst einflößen. In der heutigen Situation kommt die junge Mutter mit ihrem beruflichen Engagement oft in Konflikt zwischen Ich-Interessen und Mutterschaft. Auf endokrinologischem

Wege erscheinen hierdurch Funktionsstörungen in der Laktation möglich.

Ein für Mutter und Kind befriedigendes Stillerlebnis gibt nach den Feststellungen vieler Autoren (Nitsch, 1975, 1977; Meves, 1976, 1977 u.a.) ein tragfähiges Fundament für eine genügend gute emotionelle Beziehung zwischen beiden. Man darf auch annehmen, daß das Stillen der Mutter ein besseres Verstehen der averbalen Signale des Kindes ermöglicht. Nach Winnicott (1974) erfährt die Mutter durch Anerkennung ihrer Leistung leichter eine positive Einstellung zum Muttersein. Es bewahrt sie auch vor der Enttäuschung, in ihren Pflichten als Mutter versagt zu haben und einer glücklichen Erfahrung beraubt zu sein, was die aggressiven Impulse zum eigenen Kind verstärken kann.

Das Kind erhält durch die Muttermilch nicht nur die optimale Nahrung in ernährungsbiologischer und immunologischer Hinsicht (Nitsch, 1975), der gleichzeitig vermittelte Hautkontakt übermittelt ihm zusätzlich das Gefühl der Wärme und Geborgenheit (vgl. Montagu, 1974). Daß diese wärmespendende Nähe zu den elementaren Bedürfnissen zählt, konnte Harlow (1959) im Tierversuch zeigen. Er bot kleinen Affen eine stoffbezogene Affensurrogatmutter an und eine andere aus Draht. Die Stoffmutter wurde von den jungen Affen bei weitem vorgezogen, auch dann, wenn nur die Drahtmutter Milch gab.

Wenn man das Stillen als einen Wegbereiter für eine gelungene Mutter-Kind-Beziehung ansieht, hat auch hier der Geburtshelfer eine präventivmedizinische Aufgabe. Wie eine Untersuchung in Berlin

---

Tabelle 52–7. Psychohygienische Ansatzpunkte in der perinatalen Medizin

– **Im Rahmen der Schwangerenberatung**
Vertrauensvolle Arzt-Patient-Beziehung („tender-loving-care")
Auf reale und neurotische Ängste eingehen (Ambivalenz erkennen)
Hilfestellung bei sozialen Problemen
Geburtsvorbereitung (Informationen, Säuglingskurs, Klinikbesichtigung, psychohygienische Aspekte)

– **Bei der Entbindung (Basis: „die sichere Geburt")**
Einfühlsamer Umgang mit der Gebärenden
{ Hebamme: Akzent auf Zuwendung
Arzt: Akzent auf Sicherheit
Möglichkeit zur Partneranwesenheit (Geborgenheit, Informationsübermittler)
Individualisierte Schmerzerleichterung („keine Ideologie")
Förderung des sofortigen Kontaktes von Mutter und Kind

– **Auf der Wochenbett- und Säuglingsstation**
Möglichkeit zum „Rooming-in" und „Self-Demand-Feeding"
Förderung einer gelungenen Mutter-Kind-Beziehung (individuelle Betreuung)
Ermutigende Unterstützung bei der Einbahnung des Stillens
Möglichkeit zur frühzeitigen Entlassung (ambulante Klinikgeburt)

(Goldstein, 1978) ergeben hat, sind 50,3% der Wöchnerinnen mit der Unterstützung beim Stillen und Abpumpen durch das Personal nicht zufrieden. In einigen Kliniken wird dieser Vorwurf sogar von ⅔ der Mütter vorgetragen. In der zitierten Untersuchung wurde auch der Einfluß des Klinikpersonals auf die Stillfrequenz besonders deutlich. Im Vergleich der Kliniken schwankte die Stillhäufigkeit zwischen 56 und 96%. Bei der Nahrungsverabreichung richteten sich nur 39,7% der Mütter nach dem Nahrungswunsch des Kindes. Dieses sog. „self-demand feeding" hat somit bei uns noch nicht die Verbreitung gefunden, die von vielen Autoren gefordert wird. Hier liegt die Vorstellung zugrunde, daß es dem Kind in der stark vulnerablen frühen Wochenbettphase erspart bleiben sollte, Vernichtungsängste durch zu große Spannungen auszuhalten.

Müller (1983) hat versucht, psychosomatische Forderungen für die Wochenbettstationen der geburtshilflichen Kliniken zu erstellen. Es ging ihm dabei vor allem um eine Umstrukturierung von Kinder- und Wochenbettstation durch die routinemäßige Einbeziehung psychosomatischer Aspekte.

Wenn man sich abschließend vergegenwärtigt, wie eminent wichtig die Förderung der sensiblen perinatalen Zeit für die spätere Persönlichkeitsentwicklung des Kindes ist, so kann man sich der hieraus erwachsenden präventivmedizinischen Verantwortung kaum entziehen.

Die für den Perinatologen leistbaren psychosomatischen Aufgaben sind nochmals in der Tabelle 52–7 zusammengefaßt.

53 Neurologie

Mechthilde Kütemeyer und *Ulrich Schultz*

53.1 Zur Geschichte der psychosomatischen Neurologie

Das erste neurologische Lehrbuch von Romberg (1851) unterteilte die Nervenkrankheiten in sog. „Sensibilitäts-" und „Motilitäts-Neurosen", wobei der Neurose-Begriff noch ganz im neurophysiologischen Sinne verwendet wurde. Eine den Bedeutungswandel der Neurose(n) berücksichtigende Begriffsgeschichte in Hinsicht auf das Verständnis neurologischer Krankheiten ist bisher nicht geschrieben. Eine selbständige, von der Inneren Medizin wie von der Psychiatrie unabhängige Neurologie gibt es erst seit der Gründung der Gesellschaft Deutscher Nervenärzte 1907. Obwohl sich bereits um die Jahrhundertwende eine Spaltung in eine seelenlose Neurologie und eine körperlose Psychologie abzeichnete (Sacks, 1987), war die Neurologie eine Schule der Psychosomatischen Medizin. Durch Charcot (1825 bis 1893) angeregt, stieß Freud (1856–1939) – anerkannter Neuropathologe und Neurologe in Wien – erstmals auf lebensgeschichtliche Zusammenhänge bei neurologischen Erkrankungen. Während die Neurologie noch um die Anerkennung als organmedizinisches Fach bemüht war, wurde sie mit dem epidemieartigen Auftreten der Kriegsneurosen konfrontiert. Auf der 8. Jahresversammlung Deutscher Nervenärzte 1916 in München kam es zu der denkwürdigen Auseinandersetzung zwischen Hermann Oppenheim und Max Nonne. Oppenheim ging bei den Kriegsneurosen von der mechanistischen Vorstellung einer „lokalen Commotion" des Gehirns durch das Trauma aus, die das Symptom hervorrufe. Nonne hingegen war von der Psychogenie der Kriegsneurosen überzeugt und propagierte eine suggestive Kurz-Psychotherapie. Sandor Ferenczi – ebenfalls ursprünglich Neurologe – bemerkte auf dem Kongreß „Zur Psychoanalyse der Kriegsneurosen" 1918 in Budapest ironisch: „Die Erfahrungen an Kriegsneurotikern führten allmählich weiter als zur Entdeckung der Seele – sie führten die Neurologen beinahe zur Entdeckung der Psychoanalyse". Zu einer ersten Integration der Psychoanalyse in die Neurologie kam es im deutschsprachigen Raum vor allem durch Abraham, Brun, Schilder, Simmel und von Weizsäcker, außerhalb durch Putnam, Jelliffe[1] und Cobb, die seit den zwanziger Jahren, über die Konversionssyndrome hinaus, neurologische Erkrankungen psychosomatisch erforschten (Trimble, 1989).

Die engen und vielfältigen Verbindungen zwischen Neurologie und Psychoanalyse waren jedoch auch getrübt, wobei es außerhalb von Deutschland und außereuropäisch durchaus unterschiedliche Entwicklungen gab. Berührungsängste mit der Neurologie äußerte Freud 1932 in einem Brief an v. Weizsäcker: „Von solchen Untersuchungen mußte ich die Analytiker aus erziehlichen Gründen fernhalten, denn Innervation, Gefäßerweiterung, Nervenbahnen wären zu gefährliche Versuchungen für sie gewesen, sie hatten zu lernen, sich auf psychologische Denkweisen zu beschränken" (Weizsäcker, 1947b). Die vermeintliche Gefährdung durch die Neurologie ging so weit, daß Freud, seine Herkunft verleugnend, die neurologischen Publikationen in den „Gesammelten Schriften" (1924b) und „Gesammelten Werken" (1948ff.) nicht aufnehmen ließ (Brun, 1936). Eine Ausnahme bildet die französische, nie übersetzte Arbeit über die Unterscheidung hysterischer und organischer Lähmungen (Freud, 1893).

Die bis ins 20. Jahrhundert reichende Kontroverse zwischen lokalisatorischen und gestaltpsychologisch beeinflußten antilokalisatorischen Vorstellungen über das zentrale Nervensystem ist bis heute nicht zugunsten einer Seite entschieden. Messungen mit der Positronen-Emissions-Tomographie (PET) zur Bestimmung des Hirnmetabolismus legen eine Modifikation der Lokalisationstheorie nahe: Auch bei sehr einfachen kognitiven Leistungen oder bei lokalen Erkrankungen werden mehr Hirnareale gleichzeitig aktiviert oder mitbetroffen, als dies nach der klassischen Lokalisationslehre vermutet wurde.

Der englische Neurologe und Royalist Hughlings Jackson (1835–1911) vertrat in dieser Auseinandersetzung eine eigene Position, wonach psychische Funktionen hierarchisch organisiert und vertikal repräsentiert seien. Genosse Goldstein (1878–1965)[2] dagegen beschäftigte das gleichzeitige oder alternierende Auftreten psychischer und somatischer Phänomene. Der Psychoanalyse aufgeschlossen, stellte er die These auf, daß bei lokalisierbaren neurologischen Störungen der Organismus immer als körperliche und psychische Einheit in veränderter Weise auf Anforderungen der Umwelt antwortet (Goldstein, 1934). Dies bedeutete den Beginn eines – auch von Viktor von Weizsäcker und anderen vertretenen – neuen

---

1 Oliver Sacks (1989) weist darauf hin, daß etwa Jelliffes Doppelqualifikation in Neurologie und Psychoanalyse nicht akzeptiert wurde: Für die Neurologen war Jelliffe Psychoanalytiker, für die Psychoanalytiker blieb er Neurologe.

2 „Prof. Genosse Kurt Goldstein" war Mitglied des Vereins Sozialistischer Ärzte in Berlin.

Konzepts von der Tätigkeit des Nervensystems, in dem die Lokalisation, ätiologisch bedeutsam, immer in psychosoziale Gegebenheiten eingebettet ist (Hallen, 1978).

Ein großer Teil der bahnbrechenden psychosomatisch-neurologischen Ansätze wurde nach der Vertreibung jüdischer Psychoanalytiker, die gleichzeitig Neurologen waren, in Deutschland kaum aufgegriffen (Kütemeyer und Schultz, 1984). Statt dessen stellte sich die Nervenheilkunde und die verbliebene Psychotherapie (etwa das Autogene Training von J. H. Schultz) als „Deutsche Seelenheilkunde" in den Dienst nationalsozialistischer Gesundheitsutopie. Die Folgen dieser Adaptation für Theorie und Klinik sind unermeßlich. Zur verdrängten Geschichte der psychosomatischen Neurologie gehört auch, daß Luria und Wygotski, anerkannte Neuropsychologen, als Begründer der russischen psychoanalytischen Bewegung unbeachtet geblieben sind. Statt dessen ist in Deutschland das Gestaltkreis-Modell (Weizsäcker, 1940) für die psychosomatische Neurologie konstitutiv geworden: Neurologische Störungen treten bei der Einordnung in die Umwelt immer als funktionelle Neubildungen in Erscheinung, bei denen nicht nur körperliche und psychische Phänomene, sondern auch Wahrnehmung und Bewegung eine Einheit bilden, d.h. bei scheinbar rein motorischen Störungen sind immer auch Änderungen der Wahrnehmung und des Leiberlebens zu beobachten und umgekehrt.

Die Entwicklung der Psychosomatik als Spezialdisziplin mit eigenen Institutionen nach 1950 erlaubte den Neurologen zunehmend die Abspaltung der biographischen Methode. Neue bildgebende Verfahren lenken den Blick wieder ganz auf die Lokalisierung neurologischer Krankheiten. Therapeutische Möglichkeiten, insbesondere bei Morbus Parkinson, den Epilepsien und entzündlichen Nervenkrankheiten, drängen psychosomatische Ansätze in den Hintergrund. Im Titel des 1928 gegründeten „Nervenarzt" wurde der Zusatz „mit besonderer Berücksichtigung der psychosomatischen Beziehungen" 1967 gestrichen. Während inzwischen fast alle Disziplinen der Medizin auf ihren Jahrestagungen psychosomatische Sektionen einrichten, fehlt eine solche bisher auf Tagungen der Deutschen Gesellschaft für Neurologie.

Goldsteins (1934) methodische Prinzipien, alle Erscheinungen, die ein Kranker bietet, ohne Vorrang zu berücksichtigen und immer in bezug auf den Organismus und die Situation, in der sie zur Beobachtung kommen, zu betrachten, könnten zur Öffnung verschlossener Türen in beide Richtungen beitragen. Die im selben Sinne erhobene Forderung Vogels (1953, 1956), mit dem ärztlichen Blick „in nystagtischer Bewegung" körperliche und psychische Erscheinungen gleichzeitig zu erfassen, wäre geeignet, die Neurologie zu ihrer ursprünglich psychosomatischen „Eigenart", aber auch die psychosomatische Forschung von der zunehmend psychologischen Spezialisierung und Isolierung zur klinischen Medizin zurückzuführen.

Bei der folgenden Darstellung beschränken wir uns auf wenige Aspekte einer psychosomatischen Neurologie. Die myatrophische Lateralsklerose und Querschnittssyndrome bleiben unberücksichtigt, der Morbus Menière wird in Kapitel 58 erwähnt. Es sei auf andere Darstellungen einer psychosomatischen Neurologie hingewiesen (Lamprecht, 1979; Kütemeyer und Masuhr, 1981). Kopfschmerzen sind in Kapitel 37, Lumbago-Ischialgie-Syndrome in Kapitel 45.3 behandelt. Von den in der Neurologie keineswegs seltenen klassischen Konversionssyndromen werden die hysterischen Anfälle besprochen. Für die Grundlage der Konversion verweisen wir auf Kapitel 30. In unserem Beitrag wird auf das verbreitete und wenig beachtete Phänomen konversionsneurotischer, depressiver, angstneurotischer und anankastischer Verarbeitung neurologischer Krankheiten eingegangen. Unter dem Begriff „sekundäre Symbolisierung" ist dieses Phänomen bekannt (Engel und Schmale, 1967): Viele neurologische Syndrome neigen zur funktionellen Ausgestaltung. Epileptische Anfälle können in hysterische übergehen oder sich mehr oder weniger inkognito unter diese mischen. Ein radikulärer Schmerz eines Depressiven kann sich zum lähmenden Ganzkörper-Schmerz ausweiten. Eine angstbedingte Verschlimmerung einer ataktischen oder parkinsonistischen Gangstörung kann den Patienten mehr behindern als die ursprüngliche Akinese oder Ataxie.

Ein bisher ungelöstes Problem stellt sich bei den neurologischen Krankheiten, die sich, im Unterschied zu internistischen Erkrankungen, an der Willkürmotorik manifestieren. Hier werden seit jeher „symbolische" Deutungen angeboten, wie sie sonst nur bei Konversionssyndromen „erlaubt" sind. Die theoretische Klärung dieser derzeit nicht allgemein anerkannten Deutungsmöglichkeit bleibt einer zukünftig differenzierenden psychosomatischen Neurologie vorbehalten.

53.2 Ausgewählte neuropsychologische Syndrome

Bei zerebralen Veränderungen (Infarkt, Tumor, Kontusion, Atrophie, Arteriosklerose) treten neben den bekannten hirnlokalen Psychosyndromen neuropsychologische Phänomene in Erscheinung, deren Kenntnis für den Umgang mit solchen Patienten wichtig ist.

53.2.1 Veränderungen des Gesamtverhaltens

Bei Hirnläsionen mit Beteiligung des Cortex sind über die lokale Schädigung hinaus allgemeine Beeinträchtigungen zu beobachten, die sich als Störung des „abstrakten" Verhaltens – Goldstein (1934) nennt es „kategoriales" Verhalten – zugunsten einer „konkreten", sachlichen Einstellung beschreiben lassen: Die Kranken sind unfähig, einen gedanklichen Entwurf für eine Handlung zu machen, Initiative zu ergreifen oder eine Situation als ganze zu erfassen. Sie haften an Einzelheiten und sind in besonderem Maße abhängig von äußeren Stimuli. Der Wechsel von Aufgaben fällt ihnen schwer. Infolgedessen wirken sie unbeweglich, zwanghaft, überordentlich und stereotyp. Goldstein (1928) hat dieser Einengung auch einen positiven Aspekt beigemessen. Die Patienten schaffen sich – unbewußt – ein überschaubares Umfeld, in dem ihre Störungen wenig in Erscheinung treten und ihr psychisches Gleichgewicht weniger bedroht ist. Wird von einem solchen Kranken ein abstraktes Verhalten, z.B. die freie Wahl zwischen zwei gestellten Aufgaben verlangt, kann es zu einer katastrophalen Reaktion in Form allgemeiner Unruhe, Zittern, Angst, zorniger Abwehr und Handlungsunfähigkeit kommen. Abrupte „inadäquate" Stimmungsänderungen können als solche „katastrophalen" Antworten auf eine Überforderung verstanden werden. So treten auch umschriebene neuropsychologische Störungen, etwa des Farbenunterscheidens bei der (amnestischen) Aphasie, in Verbindung mit einer konkreten Assoziation weniger hervor als bei abstrakter Prüfung. Solche Patienten können die Farbe „blau" nicht benennen, hingegen einen „blauen Bleistift", den „blauen Himmel" ohne Mühe identifizieren.

> Ein 68jähriger Psychotherapeut mit einer allgemeinen Hirnarteriosklerose benötigt wegen einer Orientierungs- und globalen Gedächtnisstörung im täglichen Leben die Hilfe seiner erheblich jüngeren Frau. Verwirrt und sprachlicher Kommunikation kaum fähig, kommt er auf die neurologische Station. Vom Arzt auf seine frühere Tätigkeit angesprochen, kann er plötzlich detailliert aus seiner ärztlichen Praxis erzählen. Seine Fähigkeiten sind also auf einen kleinen konkreten Ausschnitt seines beruflichen Lebens konzentriert geblieben, obwohl dieser intellektuell anspruchsvoller gewesen ist als sein übriger Alltag.

Zuweilen kann das Akzeptieren eines „inadäquaten" Affekts vorübergehende Besserung einer neuropsychologischen Störung zur Folge haben.

> Ein 67jähriger Patient mit mehreren flüchtigen Insulten und einer Pseudobulbärparalyse („Zwangsweinen", „Zwangslachen") entschuldigt sich während eines Weinkrampfes, daß er sich so gehenlasse. Er wird aufgefordert, dem Weinen freien Lauf zu lassen, es müsse doch seinen Grund haben. Unter heftigen Tränen bringt der Patient hervor, daß er nach dem frühen Tod seines älteren Bruders „innerlich" viel geweint, nach außen aber Haltung bewahrt habe. Daraufhin tritt das zwanghafte Weinen und Lachen nicht mehr auf.

Das Eingehen auf die Behinderung und Herausfinden verschiedener Möglichkeiten der „Einordnung" einer Störung im Umgang mit dem Kranken bei seinen täglichen Verrichtungen bietet mehr therapeutische Chancen als konstruierte, abstrakte Übungen zur Beseitigung eines „Defekts".

53.2.2 Anosognosie

Ein häufiges, von Ärzten wenig beachtetes Phänomen ist die fehlende Selbstwahrnehmung (Verleugnung) von Krankheitssymptomen. Die A-noso-gnosie (Babinski, 1918; Pick, 1922) wird am häufigsten bei zerebral bedingten Halbseitenlähmungen (Pötzl, 1924; Stengel und Steele, 1946), aber auch bei Blindheit (Anton, 1899), Hemianopsie (Hummel, 1981), Taubheit und fokalen Anfällen beobachtet (Beutler, 1976). Anosognosie im strengen Sinne ist ein vorübergehendes Phänomen unterschiedlicher Dauer zu Beginn einer zerebralen Erkrankung und muß von der anhaltenden Verleugnungstendenz vieler Patienten gegenüber schweren internistischen und anderen Krankheitssymptomen unterschieden werden (Huebschmann, 1952; Kütemeyer, 1956; Plügge, 1953; Schaeffer, 1961). Eine über die Anosognosie hinausgehende allgemeine Verleugnungstendenz wird aber auch bei zerebralen Erkrankungen beobachtet (Weinstein, 1968). Ob sich eine Anosognosie vor allem bei parietalen Läsionen der nicht-dominanten Hemisphäre zeigt, ist umstritten (Poeck, 1987).

Die Anosognosie kann in Schweregrade eingeteilt werden, die von einer leichten Fehlwahrnehmung über eine Indifferenz bis zum Leugnen auch auf drängendes Fragen hin reichen. Nicht selten wird der neurologische Ausfall bagatellisiert und mit Ausflüchten erklärt: „Ich habe keine Lust, den Arm zu bewegen", „die Schwester hat es mir verboten". Im Angelsächsischen wird zwischen „explicit denial" (Konfabulationen und Rationalisierung eines Defizits) und „implicit denial" (Verleugnung oder Nicht-Wahrnehmung) unterschieden (Weinstein und Kahn, 1953). Viele Kranke bewegen nach Aufforderung die gesunden Gliedmaßen statt der gelähmten oder heben den kranken Arm mit Hilfe des gesunden in die Höhe und täuschen sich so über die Behinderung hinweg.

Es wurde vermutet, daß Personen mit einem prämorbid narzißtisch besetzten Körperbild eine plötzliche Lähmung oder Blindheit schwerer in ihr Selbstbild integrieren können und deshalb für die Entwicklung einer Anosognosie prädisponiert sind (Critchley, 1955). Anosognosie wird als Schutz vor katastrophaler Erschütterung angesehen, der es dem Patienten erleichtere, verbliebene Leistungen zu nutzen.

In der Rückbildungsphase wird die Verleugnung zuweilen durch Erlebnisse der Fremdheit der gelähmten, meist linken Körperhälfte abgelöst. Der gelähmte Arm wird als „ein Klumpen", „eine Schlange" empfunden, wie eine andere Person mit „sie" oder „er", mit Kosenamen – „Toby", „silly billy", „floppy Joe", „Baby" – oder mit haßerfüllten Worten wie „delinquent", „communist" angesprochen, oder die ganze linke Rumpfseite fühlt sich „wie mit Brettern durchzogen" (Critchley, 1979; Ehrenwald, 1931). Diese „halbseitigen Depersonalisationserscheinungen" treten erst auf ausdrückliches Befragen und im intensiveren Umgang zutage, da die Patienten sich sonst adäquat verhalten – allenfalls durch Stimmungsschwankungen auffallen – und jetzt auch von der Lähmung Notiz nehmen. Diese „Misoplegia" ist Ausdruck beginnender Auseinandersetzung mit dem entfremdeten, verleugneten Körperteil, ein Vorgang, den der Arzt kennen muß, will er im Zuge der Besserung auftauchende Angstzustände, Depressionen und Widerstände gegen Übungsbehandlung verstehen und psychotherapeutisch nutzen.

53.2.3 Visuelle Halluzinationen im hemianopen Gesichtsfeld

Ähnlich vorübergehende, zwischen realer und Trugwahrnehmung sich abspielende, im klinischen Alltag häufig übersehene Phänomene stellen die visuellen Halluzinationen im hemianopen Gesichtsfeld dar (Kölmel, 1984). Gegenüber den einfachen Photopsien bei Hemianopsie (Muster, Nebel, „Blitze", Wellen) sind die komplexen szenischen visuellen Erscheinungen psychodynamisch von Bedeutung: Die Patienten berichten, meist erst auf ausdrückliches Befragen, von wunderlichen flüchtigen Bildern – Hände, Tiere (z.B. Schlangen), Blumensträuße, mehrere Menschen oder eine einzelne Gestalt –, die im gestörten Gesichtsfeld auftauchen und unbeweglich sind oder sich von der Peripherie zum Zentrum hin bewegen. Die visuellen Erscheinungen verschwinden nach Augenschluß oder Blickwendung und tauchen kurz darauf in derselben Form wieder auf. Die Patienten zweifeln nicht oder nur momentweise, daß die Bilder subjektiv, im eigenen Innern entstehen und nicht der Realität entsprechen. Deshalb bezeichnet man sie im Gegensatz zu den Halluzinationen, die für real gehalten werden, als visuelle Pseudohalluzinationen.

Bei halbseitig auftretenden komplexen Pseudohalluzinationen wird, im Unterschied zu den Photopsien, eine Störung nicht nur okzipital, sondern auch nach parietal und temporal reichend gefunden. In vielen Fällen läßt sich eine Verbindung der Halluzinationsinhalte mit der Biographie des Kranken herstellen; die komplexen Bilder lassen sich wie Träume deuten.

Eine 74jährige ehemalige Pädagogin, die einen Infarkt der A. cerebri posterior rechts mit homonymer Hemianopsie nach links erleidet, sieht in den ersten Tagen von links wiederholt mehrere Gestalten, Mönche in bräunlichen Felljacken mit faltigen Gesichtern und Bärten, auftauchen, die sich, wie in einer Prozession, in einer tibetanischen Landschaft mit Klöstern und schneebedeckten Bergen bewegen. Sobald sie versucht, die Szene genauer anzuschauen, ist die Erscheinung verschwunden, kehrt aber einige Tage lang immer wieder und bleibt dann ganz weg. Während sie von diesen Bildern nur zögernd berichtet, äußert sie die Erinnerungen im folgenden Gespräch bereitwillig: Sie hatte mit 28 Jahren Verbindung zu einer Gruppe junger Leute aufgenommen, die an einer Expedition nach Tibet beteiligt war. Begeistert hörte sie die Berichte und besorgte sich Bücher vom Himalaya, von Nepal und Tibet. Ihr sehnlicher Wunsch, selbst einmal dorthin reisen zu können, rückte durch ihre Heirat und durch den frühen Tod ihres Mannes als Soldat – sie mußte zwei Kinder ernähren – in unerreichbare Ferne.

Die Bilder im linken Gesichtsfeld der Patientin wecken die Erinnerung an einen besonders erfüllten Abschnitt ihres Lebens und an lange verlorene Entfaltungsmöglichkeiten.

Psychodynamisch erweisen sich die komplexen visuellen Erscheinungen im hemianopen Feld als halluzinatorische Wunscherfüllung und können als Stabilisationsversuche angesichts der Verunsicherung durch die Krankheit angesehen werden. Das verdunkelte Gesichtsfeld, das die Wahrnehmung der Außenwelt „aufgegeben" hat, wird zum Spiegel für Bilder aus dem Unbewußten.

Im Unterschied zum Traum wird aber hier die Abkehr von der Realität nicht von der ganzen Person vollzogen, weshalb die komplexen Pseudohalluzinationen als „partielles Träumen" bezeichnet werden (Gloning et al., 1955). Im Unterschied zum Traum scheinen die dazugehörigen Erinnerungen bewußtseinsnah. Sind die Halluzinationen als solche schwierig zu erfahren, kommen die entsprechenden Assoziationen spontan und mühelos. Ohne Schlaf, ohne Hypnose, ohne Bewußtseinstrübung ergibt sich hier die Möglichkeit, Einblick in die unbewußte Erlebniswelt eines körperlich Kranken zu bekommen.

53.3 Extrapyramidale Bewegungsstörungen

Extrapyramidale Bewegungsstörungen – im Angelsächsischen reduziert auf „motor disorders" – sind Erkrankungen der basalen Ganglien (Striatum, Pallidum, Nucleus subthalamicus und Substantia nigra), denen die automatische Ausführung erlernter motorischer Programme obliegt (Marsden, 1982). In jüngster Zeit wird ein Großteil den Dystonien zugeordnet, die topisch in fokale (Tic, Blepharospasmus, Meige-Syndrom, Schreibkrampf), segmentale (Tortikollis) und generalisierte (Tourette-Syndrom) eingeteilt werden. Es wurden vor allem der Tic – einschließlich der Sonderformen Blepharospasmus und Maladie de Gilles de la Tourette –, Tortikollis und Schreibkrampf psychosomatisch untersucht.

Bei der Chorea minor und major, beim Ballismus und der Athetose lassen nachweisbare Stammgangliendegenerationen den Impuls zu einem biographischen Zugang noch schwerer aufkommen. Eine Ausnahme macht das Parkinson-Syndrom, das trotz bekannter Schädigung der Substantia nigra psychosomatisch Beachtung gefunden hat.

Auch beim Tortikollis und Schreibkrampf werden, wiewohl niemals verifiziert, Stammganglienveränderungen vermutet. In der Vorgeschichte findet sich gelegentlich eine Enzephalitis. Das hyperkinetische Syndrom manifestiert sich aber oft erst Jahre nach der Hirnschädigung in biographisch kritischen Situationen, die, „wie ein Schlüssel ins Schloß der latenten extrapyramidalen Funktionsstörung passend, die Erkrankung in Gang bringen. Die alternative und kausale Zurückführung auf eine entweder organische oder psychogene Ursache wird dabei einer Kritik unterzogen" (Bräutigam, 1964). Die Auslösesituation ist dadurch gekennzeichnet, daß eine emotionale Bewegung nicht aufkommen will und so der Beweggrund für eine gezielte, willkürliche Antwort auf ein kritisches Ereignis ausbleibt. Statt dessen stellt sich die unwillkürliche motorische Stereotypie ein.

Extrapyramidale Bewegungsstörungen folgen zwar den Funktionsgesetzen extrapyramidaler Störungen, ein Ausdruckscharakter im Sinne einer Konversion ist aber nicht ausgeschlossen (Zacher, 1989).

53.3.1 Tic

Die plötzlichen unrhythmischen, in umschriebenen Muskelgruppen sich stereotyp wiederholenden Bewegungen, meist im Bereich des Gesichts, treten vornehmlich im Kindesalter und zwei- bis dreimal häufiger bei männlichen Patienten auf. Der Tic entwickelt sich häufig auf dem Boden einer allgemeinen motorischen Unruhe nach einem Schreckerlebnis.

Psychoanalytisch wird der Tic als präverbales Ausdrucksgeschehen gedeutet (Abraham, 1921; Ferenczi, 1921). Das Ich greift auf eine frühe Gebärdensprache zurück, wie sie dem Kleinkind vor der Sprachentwicklung zur Verfügung stand. Meist ist es ein feindseliger Impuls, der sich gegen den Willen des Patienten durchsetzt. Bei einer Patientin trat das heftige „verneinende" Kopfschütteln immer dann auf, wenn sie bei Begrüßung oder Abschiednehmen nach außen hin mehr Freundlichkeit zu zeigen suchte, als sie innerlich fühlte.

Der feindliche oder herabsetzende Impuls ist immer auf eine bestimmte Person gerichtet. So wollte Frau Emmy v. N. durch einen Schnalztic ihren kranken, liebevoll betreuten Vater unbewußt wecken, womit sich gegen sein Leben gerichtete Wünsche durchsetzten (Freud, 1895b). Die Objektbezogenheit des Tic wurde als ein „Konversionssymptom auf der sadistisch-analen Stufe" (Deutsch, 1925) oder als „prägenitale Konversion" (Fenichel, 1932) verstanden.

Entsprechend ihrer „analen Organisation" weisen Patienten mit Tic überwiegend eine zwanghaft-autoritätsabhängige Charakterstruktur auf. In den Analysen treten auch unterdrückte autoerotische Impulse zutage (Deutsch, 1925; Kulovesi, 1929; Reich, 1925). Das Vorherrschen einer Zwangsstruktur wird auch bei Patienten mit Tortikollis und Schreibkrampf gefunden.

53.3.2 Blepharospasmus

Beim Blepharospasmus kommt es zu einem beidseitigen unregelmäßigen tonischen Zukneifen und Aufreißen der Augenlider, das bei unterdrückter Erregung zunimmt. Zu Beginn der Erkrankung kann der Blepharospasmus durch Kunstgriffe, etwa Berühren der Stirn, unterdrückt werden. Tritt der Blepharospasmus schon bei Jugendlichen auf, wird er in der Regel dem psychogenen Tic zugeordnet. Die Progno-

se ist mit 95% Spontanheilung innerhalb von zwei Jahren günstig (Dichgans und Brinkmann, 1988).

Bei einem Patienten mit Blepharospasmus hat Schwöbel (1966) ein Wahren des Gesichts durch Verschweigen und Verheimlichen bei gleichzeitigem Schau- und Bekenntniszwang dargestellt. Das Auge als Medium der Auseinandersetzung – das neidische Auge, die stechenden Augen der anderen – spielt bei einer psychotherapierten Patientin mit Blinzeltic die entscheidende Rolle (Mitscherlich, 1973).

Vom doppelseitigen Gesichtstic ist differentialdiagnostisch der Hemispasmus facialis („Fazialistic") abzugrenzen, dem eine periphere – nicht selten diskrete oder unerkannte – Fazialisparese zugrunde liegt. Im Gegensatz zum Tic, Tortikollis und Schreibkrampf läßt sich dieser Hemispasmus facialis medikamentös mit Carbamazepin behandeln. Trotz der andersartigen Genese scheint die biographische Bedeutung dieses Syndroms eine ähnliche wie beim Tic zu sein.

> Ein Patient erhielt während einer Auseinandersetzung einen Schlag mit der Faust ins Gesicht. Er verteidigte sich nicht und erlebte keine affektive Reaktion. Zwei Tage später trat ein linksseitiger Fazialistic auf. Wenige Wochen vorher war seine Mutter gestorben, an der er sehr hing. Er sagte zu ihrem Tod: „Ich habe keine einzige Träne geweint. Meine Familie und ich sind kaltschnäuzig." Er erlebte nicht den Mangel an Emotion, sondern im Gegenteil, sein Verhalten erschien ihm als besonders wertvoll. In einem Betrieb, so sagt er, weint man nicht (Mitscherlich, 1973).

53.3.3 Schreibkrampf

Der Schreibkrampf wird zu den Beschäftigungsneurosen gerechnet (Rodenberg, 1962), d. h. zu den Störungen, die nur bei der Ausführung bestimmter – ambivalent erlebter – Tätigkeiten wie Schreiben, Violine- oder Klavierspielen auftreten. Ätiologisch und pathophysiologisch sind verschiedene Faktoren abzuklären (vgl. Abb. 53–1). Die Finger verkrampfen sich um das Schreibgerät, die Hand wird verdreht, meist supiniert, das Schreibgerät wird mit Gewalt gegen die Unterlage gedrückt, geradezu festgebohrt. Andere Verrichtungen mit der Hand, auch das Schreiben mit der Schreibmaschine, an der Tafel, in manchen Fällen sogar das Stenographieren, bleiben ungestört. Zusätzliche Verkrampfungen der übrigen Muskulatur, des Armes, der Schulter, und das Hinzukommen anderer Hyperkinesen (Tremor, Tic) werden beobachtet (Sheehy und Marsden, 1982).

Psychoanalytisch waren erfolgreiche Behandlungen für das Verständnis des Schreibkrampfes wegweisend (Jokl, 1922; Bergler und Eidelberg, 1933): „Wenn das Klavierspielen, Schreiben und selbst das Gehen neurotischen Hemmungen unterliegen, so zeigt uns die Analyse den Grund hierfür in einer starken Erotisierung der bei diesen Funktionen in Anspruch genommenen Organe, der Finger und der Füße ... Wenn das Schreiben, das darin besteht, aus einem

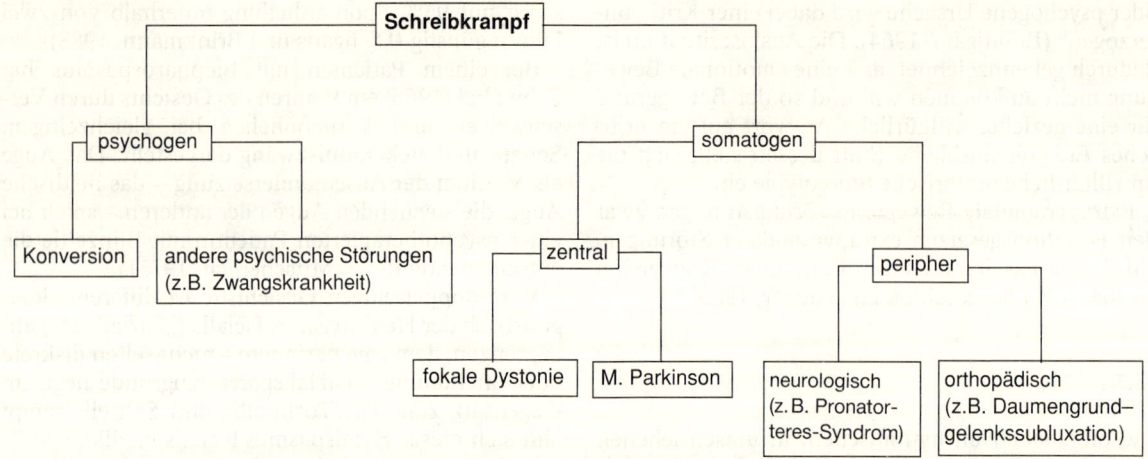

Abb. 53–1. Differentialdiagnose des Schreibkrampfs (nach Zacher, 1989)

Rohr Flüssigkeit auf ein Stück weißes Papier fließen zu lassen, die symbolische Form des Koitus angenommen hat, ... dann wird (das) Schreiben ... unterlassen, weil es so ist, als ob man die verbotene sexuelle Handlung ausführen würde. Das Ich verzichtet auf diese ihm zustehenden Funktionen, ... um einem Konflikt mit dem Es auszuweichen" (Freud, 1926).

> Bei einer Patientin tritt der Schreibkrampf erstmals auf, nachdem sie sich unter falschem Namen zu einem Rendezvous in einem Hotel eingetragen hat. Seither sind die Schreibschwierigkeiten am stärksten ausgeprägt, wenn sie etwas unterschreiben muß (Kemper, 1954).

> Eine Patientin entwickelt einen Schreibkrampf, als sie nach jahrelanger Freundschaft mit einem Mann, die nach dessen Heirat brieflich fortgesetzt wird, der Frage nach einer sexuellen Verbindlichkeit nicht mehr ausweichen kann (Stolze, 1953).

53.3.4 Torticollis spasmodicus

Der spastische Schiefhals wird als forme fruste der Torsionsdystonie angesehen (Meares, 1971; Mitscherlich, 1971). Er ist charakterisiert durch wiederholte unwiderstehliche, plötzliche und heftige oder allmähliche krampfartig gequält wirkende Drehung des Kopfes nach einer Seite, zuweilen gleichzeitige Neigung des Kopfes auf dieselbe oder zur Gegenseite, die vorwiegend durch einen „Spasmus" des vom N. accessorius innervierten M. sternocleidomastoideus und oberen Trapeziusrandes zustande kommt. Durch Anspannung antagonistischer Muskeln entsteht der Eindruck, als ob zwei Kräfte gegeneinander ankämpften und die eine die andere allmählich überwinde. Das Hinzutreten einer Retroversion mit blepharospastischen Elementen gilt als prognostisch ungünstig.

Hypothetische Erklärungsversuche und symptomatische Behandlungsmethoden – neuerdings mit nied-

rig dosiertem Botulinustoxin – belegen das Dilemma der Suche nach einer umfassenden Theorie. Der idiopathische Schiefhals sei klinisch meist nicht vom symptomatischen zu differenzieren (Rentrop und Straschill, 1986).

„Spontane" Remissionen – ein Drittel nach Psychotherapie, zwei Drittel ohne erkennbare Ursache – beweisen nicht eine Psychogenese des Tortikollis. Auf der anderen Seite sollten neurophysiologische Erklärungen nicht dazu führen, seelische Faktoren geringzuschätzen, wie das Beispiel einer kathartischen Ultrakurztherapie belegt (Drees und Schmidt, 1989).

Zur hypothetischen Striatumschädigung gesellen sich als auslösende Momente Traumen im Halsbereich, eingeengte Kopfbewegungen bei bestimmten Arbeiten und Konfliktsituationen, die durch visuelle Konfrontation mit einem feindlichen Menschen gekennzeichnet sind.

> Bei einem 64jährigen Schneidermeister tritt der Tortikollis im Laufe gerichtlicher Verhandlungen mit seinem Nachbarn auf. Obwohl er sich im Recht fühlt, kann er zu den Gerichtsverhandlungen nicht erscheinen. Er verläßt auch seine Wohnung kaum mehr, um seinem Prozeßgegner nicht begegnen zu müssen. „Anfang des Jahres habe ich dem noch frech in die Augen schauen können, wie der mir, jetzt kann ich ihn nicht mehr ansehen. Ich gehe ihm aus dem Weg, wo ich kann" (Bräutigam, 1964).

Psychoanalytisch wird der Tortikollis als präverbales Symbol im Rahmen einer frühen Störung angesehen und als fluchtartige Abwendung vom Gegner, gleichzeitig als kindliches Suchen nach der Mutterbrust gedeutet (Mitscherlich, 1971). Aus der Beobachtung, daß kleine Stützen – das sanfte Berühren des Kinns, ein zarter Druck auf die Nackenpartie oder das Anlehnen des Rückens beim Sitzen – den Tortikollis bessern, läßt sich das Angewiesensein der Patienten auf Unterstützung und „Rückendeckung" ermessen.

Die mangelnde Standfestigkeit in Auseinandersetzungen zeigt sich auch in einer bestimmten Defizienz

Tabelle 53–1. Verlaufsformen des Torticollis spasticus.

Patterson und Little (1943) (n = 103 Patienten)		Schulze und Gaebler (1988) (n = 113 Patienten)	
progredient	42%	chronisch	72%
rekurrierend	20%		
intermittierend	5%		
statisch	13%		
Remission	15%	Remission	28%

des Körperschemas. Zeichnungen der Patienten vom eigenen Körper haben keine Füße oder solche, auf denen ein Stehen nicht möglich ist (Mitscherlich, 1983).

53.3.5 Maladie de Gilles de la Tourette

Bereits ein Jahr nach Gilles de la Tourettes (1885) Erstbeschreibung der generalisierten Tic-Krankheit werden maniforme und hysterische Symptome als psychopathologische „Begleit"-Erscheinungen und wenig später „zahllose Phobien", „Ängste" und „vorübergehende melancholische Gedanken" beschrieben (Garcia, 1987; Shapiro et al., 1978). Die frühere Auffassung, daß es sich um eine neurologische Störung handle, deren somatischer Kern durch psychodynamische Kräfte aktiviert werde (Mahler und Rangell, 1943), findet erst neuerdings wieder Zustimmung (Sacks, 1987, 1989).

Die Symptomatik läßt gegensätzliche („anale") Impulse in Erscheinung treten: eine Hemmung autonomen Handelns und Sprechens in Form von Zwangshandlungen und Echolalie, Durchbruch aggressiver Impulse in Form ausfahrender Bewegungen und Koprolalie. Die enthemmten Elemente der Symptomatik werden subjektiv besonders scham- und schuldhaft, gleichzeitig aber als Befreiung erlebt. Eine Echopraxie kann sich zum virtuosen Imitationsdrang (-zwang) steigern. Tritt an die Stelle der Koprolalie wiederholtes kurzes Schreien, sollen die entfernt erlebten Mitmenschen erreicht, herbeigeholt und gleichzeitig vertrieben werden (Mitscherlich, 1973). Die Vielfalt und Ausdruckskraft der Symptomatik, insbesondere die maniforme Seite in Form überheller Wahrnehmungs- und Reaktionsfähigkeit, rasender Gedanken und witziger Einfälle wird selten gewürdigt. Die Problematik einer die unwillkürlichen Bewegungen und den Einfallsreichtum dämpfenden pharmakologischen Behandlung macht ein differenziertes psychosomatisches Behandlungskonzept erforderlich (Sacks, 1987; vgl. Kap. 13 und 17):

Der 24jährige Ray hat seine Erkrankung derart virtuos in sein Leben eingebaut, daß er die medikamentöse Einschränkung seiner Tics als schwere Einbuße erlebt und dazu übergeht, durch „Haldol-holidays" an den Wochenenden ein Stück seiner seelischen und motorischen Wildheit und Wendigkeit regelmäßig zuzulassen.

Zahlreiche Patienten unterliegen solch bizarren Ritualen, Zwängen und Zwangsvorstellungen, daß sie fälschlich als psychotisch angesehen werden.

53.3.6 Therapie

Die Behandlung der extrapyramidalen Bewegungsstörungen ist entsprechend der Stereotypie der Symptome im allgemeinen schwierig. Medikamente, z.B. Butyrophenone und Tiapridex, haben meist einen nur vorübergehenden oder unspezifischen (Placebo?-)Effekt, wobei unerwünschte Wirkungen – Gefühl der psychomotorischen „Blockade" und Gereiztheit – berücksichtigt werden müssen und oft zum (intermittierenden) Absetzen der Medikamente zwingen (Sacks, 1987). Bei lokalisierten Hyperkinesen, besonders beim isolierten Blepharospasmus und Hemispasmus facialis, werden Hoffnungen auf lokale, niedrig dosierte intramuskuläre Injektionen mit Botulinustoxin gesetzt. Stereotaktische Eingriffe (Mundinger und Riechert, 1963) oder Rhizotomie der Wurzeln C1–C4 und die Durchtrennung des N. accessorius beim Tortikollis (Mumenthaler, 1982) sind wegen der häufigen Rückfälle im Langzeitverlauf fragwürdig.

Die beim Schreibkrampf bewährte Kombination von Psychotherapie und Entspannungs- oder Bewegungsübungen scheint auch bei anderen extrapyramidalen Hyperkinesen erprobenswert (Freitag, 1921; Weizsäcker, 1941; Zacher, 1989). Bereits ein Berufswechsel oder Umstellung der Tätigkeit kann zur Symptomfreiheit führen.

Psychoanalytische Behandlungen wurden mit Erfolg beim Tic und beim Schreibkrampf durchgeführt (Klein, 1925; Kovács, 1925; Rinsley, 1986). Mitscherlich (1971, 1973 und 1983) erzielte durch körperbezogene Psychotherapie bei Patienten mit Tic, Tortikollis, Schreibkrampf und Maladie de Gilles de la Tourette anhaltende Besserungen. Bei mangelnder Verbalisierungsfähigkeit werden die Patienten auf der präverbalen, körperlichen Ebene angesprochen, indem Bewegungen, Mimik, Gestik und die zeichnerisch eruierbaren Körperschemastörungen besonders beachtet und in die Deutung einbezogen werden. Auf dem körperbezogenen Ansatz mögen krankengymnastische, nonverbale psychotherapeutische und verhaltenstherapeutische Erfolge beruhen. Die Ergebnisse von EMG-Biofeedback-Behandlungen sind wegen methodischer Mängel kritisch zu bewerten (Ince et al., 1986).

53.3.7 Morbus Parkinson

Epidemiologie

Der Morbus Parkinson gehört zu den häufigen neurologischen Erkrankungen. Die Prävalenz wird auf 1–2‰ geschätzt. In der Bundesrepublik leiden 250000 Menschen an dieser Erkrankung; seit über 50 Jahren erkranken unverändert jährlich 20 von 100000 Personen neu an Morbus Parkinson. Ungeklärt ist das seltenere Auftreten des Parkinson-Syndroms in Afrika und bei bestimmten ethnischen Gruppen (z.B. der farbigen Bevölkerung der USA). Das idiopathische, meist nach dem 60. Lebensjahr beginnende Parkinson-Syndrom wird von den symptomatischen – vor allem postenzephalitischen – Formen unterschieden. Eine arteriosklerotische Genese ist umstritten (Fischer, 1981). Neuerdings werden die Substantia nigra schädigende toxische Umweltfaktoren (z.B. MPTP, Mn oder Benzol) diskutiert. Bei experimenteller Provokation mit MPTP werden die psychischen Symptome des idiopathischen Parkinson vermißt.

Klinik

Die Hauptsymptome des Parkinson-Syndroms – Akinese, Rigor, Tremor und Bradyphrenie – erlauben schon vom optischen und akustischen Eindruck her die Diagnose. „Man braucht diese Kranken nur über die Schwelle des Sprechzimmers hereinkommen zu hören: kleinschrittig-schlurfend, unter Umständen trippelnd vorstoßend und dann wieder verhaltend. Man braucht sie nur zu hören, wie sie dann zögernd, leise zu sprechen anheben, eigentümlich monoton, modulationsarm, unartikuliert und immer weniger verständlich ... das übrige Gesicht unbewegt, fast zur Maske erstarrt, haben ausschließlich die ziemlich aufgerissenen Augen ihre Lebhaftigkeit behalten. Das Bild rundet sich, wenn man die Versteifung der Körperhaltung bis zur Statuenhaftigkeit auf sich wirken läßt" (Schulte, 1954).

Der Erkrankung liegt ein Dopaminmangel und ein relativer Acetylcholinüberschuß durch Zelluntergang in der Substantia nigra zugrunde. Weder histologisch noch biographisch kann bisher erklärt werden, wann und wodurch der Äquivalenz-Typ (annähernd gleiche Verteilung von Rigor, Tremor und Akinese) oder ein Tremordominanz- oder akinetisch-rigider Typ auftritt. Der Beginn der Symptome in biographisch bedeutsamen Situationen und ihre emotionale und situative Verlaufsabhängigkeit legen eine psychosomatische Betrachtung nahe. Verlangsamtes Auffassungs- und Reaktionsvermögen (Bradyphrenie) können eine intellektuelle Beeinträchtigung vortäuschen. Während Parkinson (1817) psychische Störungen verwarf („... the senses and the intellect being uninjured"), besteht heute eine Unsicherheit in der Einschätzung psychischer Behinderungen. Depressive Symptome werden zu Beginn der Erkrankung bei 25%, im späteren Stadium bei 30–90% der Patienten beobachtet. Angesichts dieses diagnostischen Dilemmas ist die Auffassung Goldsteins (1924) hilfreich, daß eine Störung der Einstellung zur Umwelt in Form mangelnder Spontaneität und Flexibilität vorliegt, die sich je nach Situation und je nach Blickwinkel des Untersuchers in spezifischen motorischen oder psychischen Phänomenen äußert. Die Behinderung des Parkinson-Kranken am Anfang einer Bewegung oder bei Änderung der Bewegungsrichtung und seine gehemmte Entschlußkraft und unmodulierte gedrückte Stimmung können demnach als verschiedene Ausdrucksformen derselben, die Kommunikation mit der Umwelt behindernden Grundstörung angesehen werden.

Parkinson-Patienten klagen über Vergeßlichkeit, Verlust an Interesse, Affektverflachung und Einbuße an Spontaneität (Fischer, 1982). Sie ziehen sich von sozialen Aktivitäten zurück, verlieren Selbstvertrauen und berichten von Ängsten bei geringsten Belastungen, aber auch von vorübergehend gesteigerter Geistesgegenwart und Leistungsfähigkeit. Diese situationsbedingten Schwankungen machen es schwer, neuropsychologisch durchgängige Störungen nachzuweisen (Todes, 1984; Lees, 1989). Dysphorie und zähflüssige Langsamkeit wechseln mit Affektdurchbrüchen. Nachts zunehmende, vermutlich rigor- und akinesebedingte Lumbalgien und andere Schmerzen („axiale Apraxie") sind häufig erster Anlaß, ärztliche Hilfe aufzusuchen. Die Schmerzen kommen auch durch die erzwungene Körperhaltung im Liegen mit angehobenem Kopf, das sog. imaginäre Kopfkissen („oreiller psychique") zustande.

Unabhängig von der Genese werden bei Parkinson-Patienten prämorbid übereinstimmende Wesenszüge gefunden (Bassyouni, 1987; Cohen-Booth, 1935; Booth, 1946, 1948; Mitscherlich, 1960; Kraus, 1964; Todes und Lees, 1985). In Anlehnung an den dominanten Elternteil entwickelt sich ein Ordnungsfanatismus, extremer Arbeitseifer und eine starke Identifikation mit rigiden sozialen Wertvorstellungen, die in einer gleichsam „religiösen Einstellung zum Erfolg" kulminieren. Bei der eingeengten, zunehmend der Spontaneität beraubten Existenz gelingt die Integration ursprünglich starker aggressiver und libidinöser Impulse immer weniger. Eine strenge Religiösität nährt auch extreme Selbstbeschuldigungs- und Bestrafungstendenzen (Kraus, 1964; Winnik und Bental, 1964). Parkinson-Symptome treten auf, wenn diese Einstellung nicht mehr zum erwarteten Erfolg führt.

Diese prämorbiden Wesenszüge wurden – nicht unwidersprochen (Ludin, 1988) – durch Einbeziehung der Angehörigen psychometrisch (Gießen-Test) im Vergleich zu Kontrollprobanden bestätigt (Poewe et al., 1983), wobei ähnlich zwanghafte und depressive Persönlichkeitszüge bei zahlreichen anderen körperlichen Erkrankungen (Colitis ulcerosa, Karzinom, Diabetes mellitus und Arthritis) gefunden wurden.

Ein 64jähriger Patient, viertes von neun Geschwistern, rühmt sich, als Miternährer der Familie schon als Kind 16 Pfennige pro Tag mehr als der Vater verdient zu haben. Der Vater, der neben Fabrikarbeit seinen gelernten Beruf als Bäcker bis in die Nacht hinein ausübte, versammelte dreimal am Tage die Familie kniend zum Gebet. „Wenn wir Äpfel gestohlen hätten, hätte er uns erschlagen." Der Patient brachte es in einer Fabrik zum Vorarbeiter; nebenbei war er Militärobmann, Feuerwehrkommandant und Volkssturmführer, bewirtschaftete nach Feierabend 26 Ar Ackerland und hielt sich 70 Hasen. Täglich ging er 900 abgezählte Schritte, hob jeden Abend dreißigmal die Hände hoch und kämmte sich dreißigmal, „damit ich weiß, was ich geschafft habe". Waren seine Leistungen vom Vater sehr anerkannt, so bekam er nach dessen Tod wegen seines Ehrgeizes das Mißtrauen des älteren Bruders zu spüren. Er erkrankt 52jährig an einer Enzephalitis mit Schluckstörung und Stimmbandlähmung, von der er sich nach einem halben Jahr erholt. Nach Rückkehr an seinen Arbeitsplatz richtet er Schichtarbeit ein und übernimmt selbst die Führung aller drei Schichten. Seine Untergebenen, die er erbarmungslos antreibt, begegnen ihm – wie sein Bruder – zunehmend mißtrauisch und widerständig, so daß er, vom Vorgesetzten deshalb gerügt, von zwei Seiten Sympathie und Anerkennung verliert. In dieser Konstellation treten, drei Jahre nach der Enzephalitis, die ersten Symptome in Form von Steifigkeit der Arme und des linken Beines auf, die sich rasch zum Vollbild eines postenzephalitischen Morbus Parkinson entwickeln und ihn für immer arbeitsunfähig machen (Kraus, 1964).

Mangelnde emotionale Entfaltung ist bei diesem Patienten in ungewöhnlichem Maße, anderen Parkinson-Kranken nicht unähnlich, durch Zwänge und Betriebsamkeit verdeckt. Es wurde eine Übereinstimmung mit dem „Typus melancholicus" von Tellenbach (1961) postuliert (Kraus, 1964), wobei sich die depressiven und zwanghaften Symptome nicht als krankheitsreaktiv, sondern – in Anlehnung an Goldstein (1924) – als der körperlichen Krankheit äquivalente, derselben Grundhaltung entspringende Störungen erweisen.

Phänomenologische Gegenüberstellungen von prämorbider Einstellung und späteren Krankheitssymptomen wirken deterministisch und gegenübertragungsbedingt (Booth, 1948): Durch Rigor und Akinese im affektiven (aggressiven) Handeln behindert, äußere sich die ziellose Aktivität in Form des Ruhetremors. Am maskenhaften Gesicht sei das jahrelange Zurückdrängen eigener Gefühle zugunsten sozialer Anpassung abzulesen. Das Erscheinungsbild der Patienten verändere sich gleichsinnig, so daß sich alle schwer Erkrankten ähnlich werden und wie verwandt wirken. Jelliffe (1933) versteht die vornübergeneigte Physiognomie der Kranken als Verteidigungshaltung, in der sich Feindseligkeit und Angst vereinigten. Die Akinese wird als Möglichkeit eines heilsamen Rückzugs von sinnentleerter Aktivität, der Rigor als Widerstand gegenüber überhöhten Anforderungen gedeutet (Korten und Ketterings, 1972).

Bei Kenntnis der psychischen Struktur lassen sich widersprüchliche Reaktionen besser verstehen. Bei Aufforderung und unter Beobachtung, in engen Gängen oder an einer Türschwelle nehmen Tremor und Akinese zu, während das Entgegensetzen eines Widerstandes oft die Bewegung erleichtert; der Gang, auf gerader Ebene gehemmt, kann beim Treppensteigen flüssiger werden. Vermutlich wirkt die Befriedi-

gung aggressiver Impulse oder eines Dominanzbedürfnisses symptombefreiend. Blockierende und stimulierende Einflüsse können nahe beieinander liegen. Die somatisch bisher nicht erklärbare Kinesia paradoxa (Babinski et al., 1921), die plötzliche Wiedergewinnung der Beweglichkeit bei emotionaler Erregung, und ihr Gegenstück, die Akinesia paradoxa („freezing"), sind möglicherweise an ähnlich verstehbare Situationen gebunden.

> Eine 50jährige Patientin erstarrt während einer Auseinandersetzung mit ihrer Mutter im Omnibus derart, daß sie förmlich auf ihrem Sitz festklebt und mit dem Bus in die neurologische Klinik transportiert werden muß. Im Bett liegend, akinetisch, behält sie noch stundenlang die angewinkelte „sitzende" Stellung bei.
>
> Ein Bischof, dessen Stimme im normalen Gespräch nur noch unverständlich murmeln kann, hält weiterhin über Jahre Predigten und Radioansprachen (Booth, 1948).

Die Bewegungsstörungen Parkinson-Kranker sind – entsprechend der „Einheit von Wahrnehmen und Bewegen" – mit Veränderungen der Wahrnehmung und des Leiberlebens verbunden (Kraus, 1974). Die Behinderung wird häufig in die Außenwelt projiziert, als Widerständigkeit der Umgebung wahrgenommen. Der Parkinson-Patient erlebt seinen Körper wie einen Gegenstand, den er durch angestrengten Willensakt in Bewegung oder zur Ruhe zu bringen hat. Diese Entfremdung des eigenen Leibes steigert sich, wenn er sich beobachtet und kontrolliert fühlt, und kann bei nachlassender Selbstkontrolle abnehmen.

Therapeutische Aspekte

Auf die heute zur Verfügung stehenden differenzierten medikamentösen Behandlungsmöglichkeiten mit Anticholinergika, L-Dopa, Dopaminagonisten und Amantadin kann hier nicht eingegangen werden, es muß auf die einschlägige Literatur verwiesen werden (Brandt et al., 1988; Ludin, 1988).

Aus psychosomatischer Sicht bleibt die medikamentöse Therapie trotz Verbesserung der Lebensqualität problematisch. Neben der nach ca. sechs Jahren zu erwartenden Wirkungslosigkeit der Antiparkinsonika und neben motorischen Nebenwirkungen (z.B. on-off-Phänomen) treten bei 13–60% der Behandelten, besonders unter höheren Dosen, unerwünschte psychische Wirkungen auf: Illusionen, Halluzinationen, Pseudohalluzinationen, Hypersexualität, Verwirrtheit, Agitiertheit, Angstträume und Schlaf-Wach-Umkehr (Presthus, 1980). Paranoide Ideen werden von den Patienten nur selten geäußert. Ungeklärt ist, welche Patienten bevorzugt psychotische Episoden erleben und in welcher Beziehung diese zur behandlungsbedingten Veränderung von Wahrnehmen und Bewegen stehen. Bevorzugt berichten bradyphrene Patienten von pseudohalluzinierten Personen, die nicht beunruhigen, sondern amüsieren (vgl. Abschn. 53.2.3); anderen sind sie eher lästig:

> Eine alleinstehende Frau wird in ihrer Wohnung von einem jungen Mann besucht, dem sie zunächst gut zuredet, um die Erscheinung schließlich mit dem Fliegenwedel zu vertreiben.
>
> Einer motorisch stark behinderten Patientin von großer geistiger Frische drängen sich, sobald sie die Augen schließt, Bilder von Personen auf, die ihr Bett umstehen und ihr Speisen und Getränke anbieten, aber so ungeschickt, daß sie nichts davon bekommt. Sobald sie die Augen öffnet, sind die Gestalten verschwunden (Fröhlich, 1984).

Krankengymnastik (ergänzt z. B. durch Konzentrative Bewegungstherapie) sollte deshalb weiterhin an erster Stelle stehen, aber angesichts der Problematik der Patienten weniger auf Übung, Leistung und Funktionieren als auf gesunde Selbstvergessenheit und Bewegungsfreude ausgerichtet sein (Spiele, rhythmische Bewegung, Tanz, Entspannung). Dabei wird die Selbstwahrnehmung und der Einfallsreichtum der Patienten im Herausfinden günstiger Bewegungskonstellationen gefördert (Neundörfer, 1978; Sacks, 1973).

Psychotherapie im engeren Sinne zielt auf Besänftigung des unmäßig fordernden Über-Ichs. Durch Identifikation mit der Autorität des Therapeuten gelingt es dem Patienten besser, die erzwungene Passivität ohne Schuldgefühle zu ertragen und verbliebene Fähigkeiten ohne ehrgeizige Überanstrengung zu nutzen und auszubauen. Bei der Schuldproblematik der Patienten ist besonders in Zeiten der Besserung auf selbstschädigendes Verhalten zu achten. „Toxische" Einflüsse von seiten der Umgebung müssen rechtzeitig erkannt werden. Die nächsten Angehörigen verhalten sich häufig überprotektiv und sadistisch, besonders wenn der Kranke sie in gesunden Tagen dominierte. Eine – vorübergehende – Distanzierung der Partner, z. B. durch rechtzeitige Krankenhausaufnahme, bleibt zuweilen die einzige Lösung, ehe psychosozial bedingte somatische Komplikationen eine Noteinweisung erzwingen.

53.4 Ausgewählte neurologische Schmerzsyndrome

53.4.1 Geschichtliches

Freud hat sich dem Thema Schmerz am ausdrücklichsten zu einer Zeit gewidmet, als er noch voller Hoffnung war, den „psychischen Apparat" in Begriffen der Hirnphysiologie darstellen zu können. Im „Entwurf einer Psychologie" (Freud, 1895a), für den er ursprünglich den Titel „Psychologie für Neurologen" vorgesehen hat, und in Briefen aus dieser Zeit finden sich zahlreiche Krankengeschichten, die das Vorherrschen von Schmerzen sowohl bei Angstneurose und Neurasthenie als auch bei der Hysterie und Melancholie belegen. Schmerzen spielen in den „Studien über Hysterie" (Freud, 1895b) eine herausragende

Rolle. Dem psychoanalytischen Schrifttum lassen sich folgende Hypothesen für eine neurologisch-psychosomatische Schmerzauffassung entnehmen:

Der Begriff „Schmerz" dient dem Patienten als Etikett für verschiedene Mißempfindungen von hyper- und hypästhetischem Charakter. Aus der Art des Schmerzes und seiner funktionellen Ausgestaltung lassen sich differentialdiagnostische Hinweise auf zugrundeliegende organische, aber auch psychische Erkrankungen entnehmen. Schmerz ist ein häufiges Konversionssymptom. Ein ursprünglich lokaler, passagerer (neurologischer) Schmerz wird von der (latenten) Neurose als „Erinnerungsspur" benutzt; der Körper dient dabei als „Landkarte von Erinnerungssymbolen". Für die Chronifizierung ist diese Verschmelzung von Akutschmerz und Neurose von entscheidender Bedeutung. Körper- und Seelenschmerz sind eine Antwort auf sich vollziehende Trennung libidinöser Organbesetzung/Objektbesetzung. Bei psychogenen Schmerzen haben Analgetika keine, chirurgische Eingriffe in der Regel verschlimmernde, dagegen stimulierende physio- und elektrotherapeutische Maßnahmen nicht selten eine bessernde Wirkung, die durch eine aufdeckende Psychotherapie unterstützt werden kann. Der Behandlungsbeginn chronischer (psychogener oder psychogen ausgestalteter) Schmerzen ist fast regelmäßig von vorübergehender Schmerzzunahme begleitet; die Kenntnis dieser Gesetzmäßigkeit schützt den Arzt vor unnötiger Diagnostik und Änderung des Behandlungsplans (Kütemeyer und Schultz, 1989; Hirsch, 1989).

53.4.2 Klinik

Kopfschmerzen und Lumbago-Ischialgie-Syndrome, die häufigsten neurologischen Schmerzsyndrome, sind gesondert behandelt (vgl. Kap. 37 und 45.3). Daneben kennt der Neurologe eine Fülle von Schmerzsyndromen, die trotz neurophysiologischer Erklärbarkeit therapieresistent bleiben. Die vielfältigen Schmerzen im Rahmen einer somatisierten Depression, Angst(neurose) oder Hypochondrie verweisen ihn, wo er „nichts Objektives" findet, auf das kranke Subjekt.

Neurologische Erkrankungen, die zu Bewegungseinschränkung führen (z. B. amyotrophische Lateralsklerose, Multiple Sklerose, Myasthenie, Morbus Parkinson und andere extrapyramidale Bewegungsstörungen), gehen nicht selten mit diffusen Schmerzen einher, die zuweilen zum ersten Anlaß eines Arztbesuchs werden. Ätiologisch kommen für diese Schmerzen neben radikulären Reizerscheinungen, Fehlhaltung und pathologischer Gelenkbelastung – Angst vor Kontrollverlust und spastisch-zwanghaftes „Gegenhalten" (Verspannung) in Frage.

Während Schmerz als Ausdruck der somatisierten Depression inzwischen geläufig ist, wird wenig beachtet, daß hysterische und anankastische Syndrome mit Schmerzen einhergehen, vor allem aber (unbewußte) Angst in Form von Schmerz in Erscheinung treten kann. Schmerz ist, neben Schwindel, Parästhesien und anderen neurologisch anmutenden Beschwerden, ein Hauptsymptom der „larvierten Angstneurose" (Freud, 1895c). Patienten mit unklaren „Myalgien" und „rheumatischen Beschwerden" ziehen bei der Medikamentenanamnese ein Päckchen Dociton® oder Dusodril® aus der Tasche, das ihnen der Hausarzt wegen „Herzrasen" oder gegen „Durchblutungsstörung" aufgrund von Schwindel und Kribbelgefühl in den Gliedern verordnete. Auf die gezielte Frage nach Angst wissen sie häufig von einem länger zurückliegenden nächtlichen Panikanfall mit Herzbeklemmung, Luftnot und Todesangst zu berichten, der sich selther nicht oder nur wenige Male wiederholt habe; seitdem hätten sich aber die Schmerzen und andere lästige Körperbeschwerden eingestellt.

Neurologische Schmerzen neigen zur Chronifizierung, aber auch, mehr als andere neurologische Erkrankungen, zur funktionellen Ausgestaltung: zur unanatomischen Ausstrahlung, verbunden mit bisweilen bizarren Bewegungs- und Sensibilitätsstörungen. Der „neurologische" Schmerz ist von einer latenten, zuvor kompensierten Neurose zur Somatisierung benutzt worden. Der Übergang vom „neurologischen" in den Konversionsschmerz läßt sich nicht nur an der abweichenden Schmerzanamnese und am antianatomischen Befund, sondern auch an der appellativ-bildhaften Schmerz- und Selbstdarstellung der Patienten in der Erstanamnese erkennen (vgl. Kap. 36). Aus der Anamnese lassen sich darüber hinaus Hinweise zur Unterscheidung von Schmerzen bei somatisierter Depression, larvierter Angst(neurose) und im Rahmen hysterischer oder anankastischer Syndrome entnehmen (Kap. 45.3).

53.4.3 Gesichtsschmerz

Die Beobachtung, daß bei atypischem Gesichtsschmerz eine vorausgehende Kränkung und erlittene Schläge ins Gesicht eine Rolle spielen (vgl. Kap. 36), scheint gelegentlich auch bei der Trigeminusneuralgie bedeutsam. Von vielen Patienten wird der Tic douloureux „wie ein Hieb ins Gesicht" beschrieben.

> Eine 75jährige Witwe leidet seit einem Jahr unter einer therapieresistenten idiopathischen Trigeminusneuralgie. Sie lebt in einem Altersheim, besucht aber täglich ihre jüngere verheiratete Schwester, von der sie finanziell unterstützt, gleichzeitig wie ein Dienstmädchen behandelt und beleidigt wird, so daß sie sich oft wie geohrfeigt fühlt. Nachdem sie sich unter Tränen über die Schwester hat beklagen können, gehen die Schmerzattacken bei unveränderter Medikation zurück.

Meist wirkt die Gabe von Carbamazepin prompt. Zuweilen wird erst mit hoher bis – kurzfristig – toxischer Dosierung Schmerzfreiheit erzielt. Die Patienten bedürfen in dieser Zeit intensiver Betreuung und wiederholter Information über die Reversibilität der unerwünschten Medikamentenwirkung. Dabei sollte ih-

nen die seelische Verarbeitung erlittener Kränkung ermöglicht werden. Operative Eingriffe kommen häufig einer autoaggressiven Haltung der Patienten entgegen und können infolgedessen nach vorübergehender Besserung fatale Verschlimmerungen und Rezidive zur Folge haben. Deshalb ist vor einer Operation Psychotherapie angezeigt.

Das sog. orofaziale Schmerz-Dysfunktions-Syndrom, das als Zeichen allgemeiner Verspannung oft mit Rückenschmerzen einhergeht, erweist sich meist als körperlicher Ausdruck einer larvierten Angst oder anankastischen Symptomatik (vgl. Kap. 59).

Im Gegensatz zu anderen chronischen Schmerzen, die zur Ausweitung bis zur Generalisierung (Ganzkörper-Schmerzsyndrom) neigen, zeichnet sich die zirkumskripte Hypochondrie, die vornehmlich alternde Personen trifft, durch eng umschriebene, bohrende und/oder brennende Schmerzen und unbestimmte Mißempfindungen (Fremdkörpergefühle) aus, besonders am Kopf, im Mundbereich, aber auch in anderen Körperregionen. Unerbittliches Operationsverlangen der Patienten bei Fehlen eines objektiven Befundes macht das Syndrom zur „crux medicorum". Ätiologisch werden psychoreaktive Momente einer endogenen Depression vorgezogen (Hallen, 1970).

53.4.4 Kokzygodynie

Die Kokzygodynie, die überwiegend bei Frauen jenseits des 30. Lebensjahres auftritt, äußert sich in quälenden ziehenden und brennenden Schmerzen in der Gegend der Steißbeinspitze, die beim Sitzen, Bücken, Laufen und im Liegen zunehmen, oft verbunden mit Druck- und Schweregefühl im Mastdarmbereich. Es findet sich ein umschriebener Druckschmerz an der Steißbeinspitze. Die vermuteten Entzündungen, Traumen oder „Mikrotraumen" (durch langes Sitzen) können nur selten eruiert werden oder erklären nicht die Persistenz der Beschwerden. Trotzdem darf die Auffassung, daß es sich, wie im folgenden Fallbeispiel, um eine Konversion oder Psychasthenie handelt (Oppenheim, 1923), nicht ungeprüft für alle Betroffenen gelten.

> Eine 31jährige Lehrerin hat wegen anhaltender Steißschmerzen sechs Monate lang bei Orthopäden vergeblich Hilfe gesucht, bevor sie – statt einer empfohlenen Operation – den Neurologen konsultiert. Die zierliche Person sitzt unbeweglich auf der Stuhlkante, gespannt, die Schultern hochgezogen, Kopf und Rücken gebeugt, und spricht mit leiser Stimme. Die Schmerzen haben begonnen, als sie erstmalig in einer Auseinandersetzung mit einem Kollegen bis zum Schluß auf ihrem Standpunkt beharrt habe. Früher habe sie immer „den Schwanz eingezogen", von da ab habe sie aufrechter, standfester sein wollen. Durch die Schmerzen fühle sie sich festgehalten, „wie eingeschweißt". Während der Behandlung mit funktioneller Entspannung (vgl. Kap. 22) wächst die Patientin in ihre neue Haltung hinein und verliert dabei ihre Schmerzen.

Von Therapieresistenz sollte erst nach einer solchen psychoanalytisch orientierten oder/und körperbezogenen Psychotherapie gesprochen werden.

53.4.5 Restless-legs-Syndrom

Ein mit quälenden Mißempfindungen und motorischer Unruhe in den Beinen einhergehendes Syndrom, bei den älteren Ärzten als „anxietas tibiarum" bekannt und von dem Neurologen Thomas Willis 1685 eingehend beschrieben, erhält durch Ekbom (1945) den heute gebräuchlichen Namen „restless legs". Diese unglückliche Bezeichnung unterschlägt die Fülle sensibler Mißempfindungen und den Angstcharakter des Syndroms, so daß lediglich die motorische Reaktion übrigbleibt. Die Patienten klagen über schwer definierbare Sensationen tief im Innern beider Unterschenkel, die – bei manchen mehr einem Wühlen oder Vibrieren, bei anderen mehr einem Kälte- oder Schwächegefühl ähnelnd, zuweilen auch den Charakter eines dumpfen Schmerzes annehmend – nur in Ruhe, überwiegend abends oder nachts auftreten. Die Patienten verspüren einen unwiderstehlichen Drang, die Beine zu bewegen, aufzustehen, umherzugehen und finden oft stundenlang keinen Schlaf. Von milden, kaum beachteten bis zu malignen Formen – „ein diabolisches Gefühl, es vergiftet mein ganzes Leben" – sind alle Schattierungen und Ausweitungen (restless arms) möglich. Eine Häufung bei Frauen während der Schwangerschaft, familiäres Auftreten und eine gleichzeitige Eisenmangelanämie werden beobachtet. Der neurologische und Gefäßbefund sind regelrecht. Die Ursache gilt als unbekannt. Unseres Erachtens handelt es sich um eine besondere Form der (larvierten) Angstneurose. Da die Patienten sich ihrer Beschwerden schämen und selten ärztliches Verständnis finden, führt erst eine andere Krankheit (Migräne, Raynaud-Syndrom, rheumatisches Fieber, Diskusprolaps, Ulcus ventriculi) – die psychodynamisch ebenfalls mit „Pseudounabhängigkeit" verbunden ist – zur Diagnose (Ekbom, 1960).

> Eine 75jährige Patientin, die wegen Lumboischialgie stationär behandelt wird, berichtet nebenbei von allnächtlicher Unruhe der Beine, unter der sie seit dem Tode ihres Mannes vor fünf Jahren leide. Sie fällt durch ihre jugendliche Erscheinung und forsches Reden auf. Sobald sie eine geringe Besserung ihrer Schmerzen verspürt, verläßt sie abrupt die Klinik und weist eine notwendige häusliche Hilfe barsch zurück.

Bei der „anxietas tibiarum" handelt es sich um das Äquivalent einer latenten Angst vor Regression, vor Kontrollverlust im Schlaf und zuweilen vor dem Tod.

> Bei zwei Patienten beginnt das Syndrom, als sie durch eine bösartige Erkrankung mit dem Tode konfrontiert werden. Während der eine, im Familienkreis betreut, dem nahenden Ende mehr und mehr ins Auge sehen kann und dabei die Unruhe der Beine verliert, eskalie-

ren bei dem anderen die quälenden Sensationen und ergreifen schließlich den ganzen Körper – er muß sich im Bett ständig hin und her werfen – als er, moribund, darauf besteht, nicht an seine Krankheit erinnert zu werden.

Bei der häufigen Therapieresistenz – Eisensubstitution, Vasodilatantien oder/und Phenobarbital (Luminaletten®) werden empfohlen, „Psychotherapie ist in vielen Fällen indiziert" (Schliack und Schiffter, 1976) – liegen nur vereinzelte Erfahrungen darüber vor, inwieweit ein Bearbeiten der Angst- und Todesproblematik die medikamentöse Wirkung unterstützen oder ersetzen kann.

53.4.6 Phantomschmerz

Vermutlich beschrieb der Chirurg Ambroise Paré (1552) als erster den Phantomschmerz, eine „wunderbar befremdende und ungeheuerliche Sache, die kaum zu glauben ist, es sei denn, man hat es mit eigenen Augen gesehen und mit eigenen Ohren gehört: Patienten, die sich noch viele Monate nach der Entfernung eines Beines über außerordentlich starke Schmerzen in diesem amputierten Körperglied beklagen" (Keynes, 1952).

Man versteht unter Phantomsensationen, die nicht schmerzhaft sein müssen und von etwa 85% amputierter Patienten berichtet werden, die anhaltende Wahrnehmung eines nichtexistierenden Körperteils; diese wird als Beweis für die Existenz eines Körperschemas angesehen. Phantomwahrnehmungen können schon bei Kleinkindern auftreten, selten jedoch vor dem 3.–4. Lebensjahr, da eine gewisse kognitive Reife und die Entwicklung eines Körper-Selbst Voraussetzung für das Phänomen seien (Poeck und Orgass, 1964; Poeck, 1969; Weinstein und Sersen, 1961).

Bei 3–4% der Patienten ist die Wahrnehmung des Phantomgliedes mit unerträglichen Dauerschmerzen verbunden. Periphere, zentrale und psychologische Erklärungen halten sich die Waage (Joraschky, 1983). Die „intensivere und mehr differenzierte Vertretung distaler Gliedabschnitte, die Verkürzung im Laufe der Jahre" – das Teleskopiephänomen – „und die überaus plastische, antizipierte Anpassung in wechselnden Situationen können als Effekt eines pathologischen Irritationszustandes im peripheren Nervensystem nicht zureichend gedeutet werden" (Poeck, 1963). Therapieresistente Phantomschmerzen können bei empathischem Umgang mit dem Patienten verschwinden.

> Bei einem Patienten, der im Ersten Weltkrieg beide Unterschenkel verloren hat, verwandeln sich harmlose Phantommißempfindungen in unerträgliche Schmerzen, die ihn zum Morphinisten machen, als sein Jugendtraum, das Hotel seines Vaters zu führen, wegen seiner Verstümmelung zunichte wird und er statt dessen eine Beamtentätigkeit übernehmen muß. Er träumt sich immer wieder im Besitz beider Beine, z.B. als Skifahrer, der er früher gewesen ist. Durch Psychothera-

pie wird sein Wunsch, einen anderen Beruf zu ergreifen, bestärkt, und er beginnt einen Großhandel mit Goldwaren, der ihm ein ungebundenes Leben ermöglicht. Durch diese Veränderung gelingt eine Entwöhnung und Behebung der Schmerzen, unter denen er mehr als 30 Jahre lang gelitten hat (Hallen, 1956).

Das Phantom ist Ausdruck des Unvermögens, auf die Integrität des eigenen Körpers zu verzichten (Schilder, 1923). Der Phantomschmerz repräsentiert einen Konflikt zwischen „archaisch angelegtem Streben nach Regeneration und endgültigem Verzicht auf das amputierte Glied" (Szasz, 1957). So ergibt sich ein therapeutischer Zugang über das Traumerleben, das von einer lebhaft andauernden Auseinandersetzung mit der Amputation geprägt ist.

Über gute Behandlungserfolge mit Psychotherapie und zusätzlich Hypnose wird berichtet. Dem hartnäckigen Operationsverlangen (Neuromentfernung, Sympathektomie, Chordotomie, postzentrale Gyrektomie) der Patienten sollte mit äußerster Zurückhaltung begegnet werden, weil es einer autoaggressiven, selbstverstümmelnden Einstellung der Patienten entspricht und mit der unbewußten Phantasie verbunden ist, den lästigen Regenerationswunsch „gewaltsam" abzuschneiden.

53.5 Epileptische Anfälle

53.5.1 Historische Aspekte

Epilepsie heißt Gepackt-, Ergriffen-, Überwältigtwerden und gibt den Eindruck wieder, den der Betroffene und der Beobachter von einem großen epileptischen Anfall haben. In der ersten Schrift „Über die heilige Krankheit" (430 v. Chr.) – Hippokrates' Autorenschaft ist umstritten – wird die volkstümliche dämonologische Auffassung durch eine naturwissenschaftliche Theorie ersetzt. „Heilig" bedeutet keine Auszeichnung; auch lag es antikem Bewußtsein fern, Epilepsiekranke als Heilige oder Propheten zu verehren (Temkin, 1971). Trotzdem gehören „ierós" und „sacer" zu jenen Urworten mit Doppelsinn, die heilig und verflucht oder beides in einem bedeuten konnten (Freud, 1910).

Die römische Namensschöpfung „morbus comitialis" – die Volksversammlungskrankheit – weist auf den Einfluß der Epilepsie bis in die Politik hin. Die Komitien – förmliche Versammlungen des römischen Volkes – waren sofort abzubrechen, sobald ein Sitzungsteilnehmer einen epileptischen Anfall erlitt, weil ein solcher als „Eingriff von oben", als göttlicher Einspruch gewertet wurde (Schneble, 1987). Die Kenntnis der Psychogenie kommt bereits in den Bezeichnungen „die verstellte fallende Sucht", „hysterische Epilepsie", „attaque simulée" oder „hystéro-épilepsie" zum Ausdruck. Psychosomatische Beobachtungen sind nicht erst in der Frühzeit der Psychoanalyse anzutreffen (Freud, 1948; Kulovesi, 1934; Men-

ninger, 1926; Ribble, 1936; Stegmann, 1913); zahlreich sind die feinen Hinweise bei den Neurologen des 18. und 19. Jahrhunderts, die sich mit der Phänomenologie und Ätiologie epileptischer Anfälle genauer befassen (Portal, 1828; Tissot, 1770).

Mit der Betonung erbbiologischer Aspekte in den zwanziger Jahren und der Etablierung des „Gesetzes zur Verhütung erbkranken Nachwuchses" 1933 wurden psychosomatische Ansätze vor allem in Deutschland verdrängt. Sie hielten nach 1945 mit der pharmakotherapeutischen Entwicklung nicht Schritt. „Vor den Erfolgsbannern der antikonvulsiven Therapie haben offenbar auch die Psychotherapeuten ihre bescheidenen Wimpel eingezogen, als ob im Felde der Epilepsie das therapeutische Gespräch endgültig zu einem Gespräch über die Tablette und die beste Art ihrer Einverleibung geworden wäre" (Vogel, 1961). Für Weizsäcker (1929) galt es als „noch nicht erwiesen, aber auch nicht zu widerlegen, daß das Einsetzen einer sehr frühzeitigen psychischen Behandlung den Verlauf der Krankheit und die Charaktergestaltung des Kranken entscheidend beeinflussen kann. Kein Arzt sollte den Mut haben, diese Möglichkeit a priori geringschätzig zu beurteilen". Trotzdem warnte er davor, „den Epileptiker zu analysieren", weil es so aussähe, „als sei die Gefahr, aus welcher diese Menschen in die Krankheit flüchteten, eine viel ernstere als bei den Hysterischen" (Weizsäcker, 1947b).

53.5.2 Klassifikation

Der verwirrenden Vielfalt epileptischer Anfallstypen und Verlaufsformen ist inzwischen durch eine mehrfach überarbeitete Internationale Klassifikation der Epilepsien Rechnung getragen worden (Wolf et al., 1987). Es verbietet sich jede Verallgemeinerung, insbesondere Diagnosen wie „zerebrales" oder „hirnorganisches Anfallsleiden".

Epilepsien werden in generalisierte und fokale eingeteilt. Bei generalisierten Anfällen weisen die ersten klinischen und elektroenzephalographischen (EEG-) Veränderungen auf eine initiale Beteiligung beider Hirnhemisphären hin; motorische Erscheinungen sind immer beidseitig. Bei fokalen Anfällen weisen die ersten klinischen und EEG-Veränderungen auf die initiale Aktivierung eines anatomischen und/oder funktionellen Neuronensystems hin, das auf einen Teil einer oder beider Hemisphären beschränkt ist. Ein fokaler Anfall kann sich jedoch, statt zu enden, zu einem generalisierten Anfall entwickeln (Janz, 1981).

Einer solchen Einteilung entspricht eine differente Selbstwahrnehmung: Personen mit fokalen Epilepsien (Jackson-Anfälle, isolierte Auren, komplex-fokale Anfälle) können fast immer Wahrnehmungen ihrer Anfälle berichten; Patienten mit generalisierten Epilepsien bemerken in der Regel nichts von ihren Anfällen. Dies könnte für die therapeutische Zugänglichkeit von Bedeutung sein. So brechen Epilepsiekranke mit generalisierten Anfällen die Therapie häufiger ab (Thorbecke, 1984).

53.5.3 Epidemiologie

Epilepsien zählen zu den häufigsten neurologischen Erkrankungen. Etwa 5% aller Menschen bekommen im Laufe ihres Lebens mindestens einmal einen epileptischen Anfall, meist in Form von Gelegenheitsanfällen. Eine aktive Epilepsie erleiden 6–7 von 1000 Personen. In der BRD leiden etwa 400000 Personen an einer „aktiven" Epilepsie; jährlich erkranken etwa 29000 Personen. Infolge der steigenden Unfallzahlen wird mit mindestens 5000–6000 Neuerkrankungen an einer traumatischen Epilepsie gerechnet (Janz, 1979; Penin, 1987).

53.5.4 Psychosomatische Aspekte

Epilepsiepatienten weisen ein vierfach höheres Suizidrisiko als die Durchschnittsbevölkerung auf. Wegen der Vielfalt der anfallsauslösenden bio-psycho-sozialen Faktoren und sozialen Folgen ist eine einseitig pharmakologische oder gar operative Behandlung ein Kunstfehler.

Erste Untersuchungen mittels Hypnose um die Jahrhundertwende versuchten die Annahme eines Unterschiedes zwischen epileptischen und hysterischen Dämmerzuständen zu widerlegen (Graeter, 1899; Muralt, 1900). Noch heute wird „in der Welt der Lehr- und Handbücher eine differentialdiagnostische Sicherheit vorgetäuscht" (Rabe, 1980). Auch ein EEG helfe nicht immer bei der Unterscheidung. Es gelang wiederholt, epileptische Amnesien mit Hypnose aufzuheben (Riklin, 1902, 1903; Schilder, 1925a, b; Ruffin, 1929).

Frühe Psychoanalysen von medikamentös unbehandelten Patienten mit epileptischen Anfällen unklarer Genese brachten eine Tendenz zur Spaltung der Persönlichkeit zutage, die sich nicht nur in Anfällen und Dämmerzuständen, sondern auch in Tagträumen mit geistiger Abwesenheit, Zerstreutheit und reicher Phantasietätigkeit äußerte, insbesondere in einer Neigung zu gewaltsamer Impulsivität, die durch hypertrophische moralische Hemmungen vom Bewußten abgedrängt würde (Stekel, 1924). Neben Psychotherapie wurden „Rückerziehung zur Arbeit" und andere Formen von Sozialtherapie empfohlen; z.B. sollte der Patient ohne Begleitung der Mutter oder einer anderen Person zur Therapie kommen.

Rasch wurde erkannt, daß Epilepsiekranke für eine Psychoanalyse „sprödes Material" sind: „Sie wollen sich an nichts erinnern und wehren sich gegen die freien Assoziationen, so daß man fassungslos vor den Toren ihrer Seele stehen würde, wenn sie nicht die Gabe hätten, reichlich zu träumen." Graven (1924) erreicht bei zehn Patienten durch Erarbeitung eines Sinnzusammenhangs zwischen Anfällen und unbewußten Mord- und Lustmordphantasien, Inzestwünschen, Suizid- und Wiedergeburtsträumen – Beobachtungen, wie sie auch von anderen berichtet werden (Menninger, 1926; Schilder, 1925a; Greenson, 1944; Heilbrunn, 1950) – eindrucksvolle Anfallsreduktionen.

Epileptische Anfälle werden auch als Regression bis zur intrauterinen Phase der paradiesischen Wunscherfüllung (Ferenczi, 1913), als Zeichen einer Triebentmischung zwischen Eros und Destruktion (Freud, 1923) und als pathologischer Orgasmus gedeutet, dessen Vollzugsorgan nicht das Genitale, sondern die Muskulatur sei (Reich, 1931). Freud (1948), der selbst offenbar keine Epilepsiepatienten behandelte, sieht die Anfälle Dostojewskis als Selbstbestrafung für den Todeswunsch gegen den gleichzeitig gehaßten und bewunderten Vater. Die Erstmanifestation epileptischer Anfälle in der Spätpubertät fällt häufig mit einer abnorm verzögerten Ablösung vom Elternhaus zusammen (Weizsäcker, 1929).

Beim Vergleich dreier Krankengeschichten (Epilepsie, Tetanie und Hysterie) ergibt sich eine im „Wut- und Anfallsgeschehen" verkörperte Affektkrise, die sich bei epileptischen Anfällen als brutale, bei tetanischen Anfällen als ohnmächtige und bei hysterischen Anfällen als masochistische Wut herausstellt (Janz, 1948/49).

In den fünfziger Jahren erfährt der vorurteilsbeladene Begriff „epileptische Wesensänderung" eine Differenzierung. Je nach Bindung großer Anfälle an den Schlaf-Wach-Rhythmus werden charakteristische Verhaltensweisen beobachtet (Janz, 1953). Patienten mit einer Aufwach-Epilepsie, die abends spät einschlafen und morgens nach dem Aufwachen – an chronischem Schlafdefizit leidend – Anfälle bekommen, gelten als alert und wendig, als leichtsinnig, suggestibel und leicht verstimmbar. „Ihre Verzweiflungen dauern nie lange an, ... ebensowenig ihre Reue." Demgegenüber wirken Patienten mit einer Schlaf-Epilepsie bedächtig, schwerfällig und zur Pedanterie neigend, aber auch zuverlässig und pflichtbewußt. Die Anfälle seien bei der Aufwachgruppe viel stärker in den kommunikativen Bereich eingebettet als bei der Schlafgruppe. Situative Anfallsbedingungen seien deshalb bei der Aufwach-Epilepsie besser aufhellbar als bei der Schlaf-Epilepsie. Beide Gruppen unterscheiden sich in ihren Formen der Abwehr und Bewältigung von Unlust und Konflikt. Gegenüber dem zwanghaften „Schlaf-Epileptiker" neigt der „Aufwach-Epileptiker" zur Verleugnung. Forciert optimistisch, gibt er sich so, als ob er frei von Schuldgefühlen und Angst, ausgeglichen und harmonisch wäre. Ganz im Gegensatz zu den Patienten der Schlafgruppe, die ihre im Grunde weit mehr verborgenen Anfälle genau notieren, „vergißt" der Aufwach-Epilepsiekranke seine Anfälle gern oder neigt dazu, sie auf die leichte Schulter zu nehmen oder ihre Entstehungsbedingungen zu verharmlosen. Es dominiert eine Entscheidungsschwäche, die sich von den alltäglichen Verrichtungen über die berufliche Existenz bis zu ethischen, erotischen oder religiösen Problemen erstreckt (Janz, 1968b, 1969).

Die enecetische (umständliche, haftende) Charakterveränderung des „Schlaf-Epileptikers" wird als Sicherung vor gesteigerten Haßtendenzen verstanden, wobei die Tagseite des „hypersozialen" Charakters die ihr zugehörige Nachtseite der Angriffslust und motorischen Expansivität nicht – oder nur in Form von epileptischen Anfällen – sichtbar werden lasse (Bräutigam, 1951/52).

Bei der psychopathologischen Differenzierung der Epilepsiekranken nach der tageszeitlichen Bindung großer Anfälle werden Patienten mit anderen epileptischen Anfällen (z. B. kleine fokale oder kleine generalisierte Anfälle) vernachlässigt, die u. a. durch unterschiedliche Selbstwahrnehmung der Anfälle charakterisiert sind. Patienten mit fokalen Epilepsien – psychopathologisch der Schlafgruppe zugeordnet – nehmen meist den Beginn ihrer Anfälle wahr. Dagegen bleiben generalisierte epileptische Anfälle (Absencen) vom Patienten in der Regel unbemerkt – mit Ausnahme des Impulsiv-Petit mal, das ohne Bewußtlosigkeit einhergeht. Die gestörte Selbstwahrnehmung ist vermutlich ein Grund für den Mangel autobiographischer Mitteilungen Epilepsiekranker, von denen uns nur eine bekannt ist (Cooks, 1987).

Fokale Anfälle

Bei fokal-sensorischen (Auren) und fokal-komplexen Anfällen (dreamy states, psychomotorische Anfälle) sind Fixierungen von Erlebnisinhalten bekannt, die in Analogie zu dem biologischen Phänomen des Tonusfangs von Jakob v. Uexküll (1912) als Erregungsfang bezeichnet werden.

> Pateisky (1957) berichtet den Fall eines 32jährigen Okzipitalhirnverletzten, der während seiner Aura in einem hellen Feld eine dunkle Person laufen sieht. Die inhaltliche Bedeutung dieser Aura, über die der Patient nichts anzugeben weiß, ergibt sich erst bei der Aussprache mit einem Kriegskameraden. Dieser berichtet, daß der Patient seine Hinterhauptsverletzung erlitten habe, als er, ein Maschinengewehr bedienend, im Zielfernrohr einem davoneilenden Soldaten nachschoß. In den späteren Auren wird das szenische Erlebnis vermutlich unter dem Druck eines starken Affekts „imprägniert".

An 50 Patienten mit Auren ließen sich Inhalt, Qualität und psychologische Bedeutung des unterschiedlichen Verdrängungsgrades ermitteln (Hill und Mitchell, 1953). Es fanden sich drei Kategorien, die sich nach Ausführlichkeit und Art der Erinnerung unterscheiden. Der ersten Kategorie wurden visuelle Erinnerungen zugeordnet, die meist als unangenehm, aber auch mit einem Gefühl der Vertrautheit empfunden wurden und sich durch ihre Unwiderstehlichkeit von normalen Erinnerungen unterschieden.

> Ein 39jähriger Mann hebt regelmäßig während einer epigastrischen Aura die Hand vor sein Gesicht, kniet nieder und ruft: „Schlag mich nicht, Vater, bitte, schlag mich nicht!" Er sagt, er sehe seinen Vater über sich, der ihn mit einem Feuerhaken bedrohe. Ein Angehöriger berichtet, daß ihn sein Vater kurz vor seinem ersten Anfall mit einem Feuerhaken geschlagen habe.

In der zweiten Kategorie drängten sich einzelne, in jeder Aura identische „parasitäre" Worte ins Bewußtsein; gelegentlich auftretende unbestimmte Szenen

hatten nicht diese Unmittelbarkeit wie bei der ersten Gruppe, sondern wurden als harmlos und bedeutungslos empfunden. Eine Beziehung zu biographischen Erfahrungen ließ sich nicht herstellen.

> Ein 31jähriger Mann beschreibt in der Aura eine wiederkehrende Wortkette, die mit dem Wort „esoterisch" beginnt und bis zu 20 Sekunden anhält, wobei die folgenden Worte in Vergessenheit geraten.

In der dritten Gruppe wird ein von Inhalten gereinigter dreamy state in Form eines unbestimmten, aber bedeutsamen Vertrautheitsgefühls, eines déjà-vu ohne inhaltliche Erinnerung erlebt.

Die Hypothese von drei verschiedenen Verdrängungsebenen wird mit der unterschiedlichen explorativen Zugänglichkeit belegt; auch sei während des Krankheitsverlaufes häufig ein Symptomwandel der Auren von Kategorie 1 zu Kategorie 3 zu beobachten. Zur Identifizierung der Auraerlebnisse und ihrer Korrelation mit biographischen Ereignissen sei in der ersten Kategorie keine vertiefte Exploration notwendig. Die zunächst als bedeutungslos eingeschätzten Aurainhalte der zweiten Gruppe erweisen sich bei eingehender Exploration als Deckerinnerungen. Die Auren der dritten Gruppe werden als unbewußte Residuen angesehen. Wurde hier trotzdem mit aktiver analytisch orientierter Psychotherapie eine Aufdeckung versucht, sei es wiederholt zu Suizidversuchen und zu Psychosen gekommen.

Unbewußte Wünsche oder Ängste können in visuellen Auren, Träumen und intermittierenden Psychosen identisch auftreten. Bei zwei Patientinnen mit psychomotorischen Anfällen änderten sich in der Psychotherapie die Auren- und Trauminhalte gleichsinnig (Epstein und Ervin, 1956).

Psychomotorische Anfälle zeichnen sich dadurch aus, daß der Patient nicht wie in allen anderen Anfällen Objekt der Anfallssymptomatik ist, sondern auch ein Handelnder. Der psychomotorische Anfall sei insofern ein Gestaltkreis (Weizsäcker, 1940), als er sich einmal als eine Wahrnehmung (in der Aura), ein andermal als Bewegung (Anfall nach dem Verlust des Bewußtseins mit Schmatzen, Kauen usw.) äußert. Solche Fragmente sinnlicher Wahrnehmung und Bewegung kommen in den Anfällen nicht zufällig zum Durchbruch, sondern stehen in enger Beziehung zur Vorgeschichte des Patienten (Hallen, 1970).

Generalisierte Anfälle

Unter generalisierten Anfällen werden nicht nur große Anfälle ohne Vorgefühl verstanden, sondern auch alle kleinen Anfälle, etwa die altersgebundenen pyknoleptischen (gehäuft auftretenden) Absencen, die im EEG mit regelmäßigen spitzen- und wellenförmigen Potentialen (spike-waves) in der durchschnittlichen Frequenz von 3/sec auftreten.

Um die Jahrhundertwende werden Absencen als hysterisch angesehen, da affektiv betonte Erlebnisse auf den Krankheitsverlauf gestaltend wirken, die damalige Bromtherapie (wie später Luminal) versagt, und keine intellektuellen Veränderungen zu beobachten sind (Fürstner, 1896).

In die Vor-EEG-Ära fällt der Versuch Kulovesis (1934), eine 22jährige Patientin mit – aus heutiger Sicht – Impulsiv-Petit mal psychoanalytisch zu behandeln:

> Die Übertragung in der Analyse sei meistens negativ, die Patientin äußert sich in dem gleichen trotzigen Schweigen wie sonst gegenüber der Mutter. In der Behandlung bestätigt sich die spätere Beobachtung, daß Beginn und Verlauf von Impulsiv-Petit-mal-Epilepsien in besonderer Weise von situativen Einflüssen abhängig sind (Janz und Christian, 1957). So erlebt die Patientin ihren ersten Anfall am Morgen ihrer Menarche, die für sie „eine völlige Überraschung" bedeutet. In der Zeit der Analyse stirbt derjenige ihrer Brüder, gegen den sie häufig Todeswünsche gehegt hat. Bei der Nachricht von seinem Tod erleidet sie einen epileptischen Anfall, ebenso wiederholt bei der Bearbeitung ihrer Träume während der analytischen Sitzungen.

Unter der Hypothese, daß epileptische Anfälle situativ verstehbar sein müßten, schildert Barker (1948) die Psychotherapie einer 23jährigen kinderlosen Hausfrau, die seit dem 11. Lebensjahr an einer primär generalisierten Epilepsie leidet.

> Da die Patientin ihre Lebensumstände, unter denen Absencen und große Anfälle zunehmend häufig auftreten, als unauffällig darstellt, versucht Barker über ein Tagebuch, in das die Patientin jedes situative und gedankliche Detail vor einer Absence eintragen soll, und durch freies Assoziieren die jeweilige „Bedeutung" des Anfalls zu rekonstruieren. Während Absencen anfänglich anstelle von Angst auftreten, entsprechen sie im weiteren Verlauf einem unbewußten Ressentiment und Wutgefühl gegenüber Mutter und Ehemann. Nachdem die Patientin zu Beginn etwa 7 bis 50 Absencen täglich und mindestens einen Grand mal wöchentlich erlitt, bleibt sie nach dieser modifizierten analytischen Behandlung mehrere Monate ohne Antiepileptika anfallsfrei und äußert statt dessen psychogene Zwerchfellkontraktionen.

Bei Patienten mit Aufwach-Epilepsie tritt der erste Anfall in Zeiten besonderer Überanstrengung mit Störung des Schlaf-Wach-Rhythmus, aber auch bei verzweifelten Ausbruchsversuchen aus der familiären oder sozialen Enge auf (Wenzl, 1965). Die Lösung einer konfliktgeladenen symbiotischen Beziehung zu den Eltern, insbesondere zur Mutter, die den Epilepsiekranken auf der Stufe des erwachsenen Kindes (Tellenbach, 1966) beläßt, kann zu einer Reduktion generalisierter epileptischer Anfälle führen.

Selbstinduzierte oder sensorisch ausgelöste epileptische Anfälle

Auf römischen Sklavenmärkten benutzte man, um den Kauf eines anfallskranken Sklaven auszuschließen, das Drehen eines Töpferrades in Augenhöhe in der Annahme, das dadurch ausgelöste Schwindelge-

Tabelle 53–2. Auslöser sensorischer oder selbstinduzierter Anfälle.

Einfache Auslöser	Komplexe Auslöser
Lichtreize (z. B. Fernsehen)	Musik
Einfache akustische Reize (z. B. ein lautes Geräusch)	Lesen
Somatosensorische Reize	Schreiben
Bewegungen	Rechnen
Schreck	Entscheidungssituationen
Muster	

fühl führe zu einem epileptischen Anfall (Temkin, 1971). Inzwischen sind eine Vielzahl von Reizen und Situationen bekannt, die mit epileptischen Anfällen beantwortet werden können. Am bekanntesten sind optische Stimuli, etwa Fahren durch eine Allee, Fernsehen, Computerspiele (Tab. 53–2). Epileptische Anfälle sind auch vom Patienten selbst auslösbar (Janz, 1968a), bei fokalen Epilepsien in Form musikogener Anfälle, bei generalisierten Epilepsien über eine Photosensibilität und Musterempfindlichkeit (Critchley, 1937). Photosensible Patienten wenden sich einer hellen Lichtquelle zu und klappern mit den Augenlidern oder bewegen die gespreizten Finger vor den Augen („Wischer", „Fächler"), was unter Kindern zu Spitznamen wie „Blinzelbiene" oder „Fuchtelliese" führt (Matthes, 1954). Über den Wiederholungszwang können solche Anfälle zur Sucht werden. Die Patienten wirken währenddessen wie in Trance und zeigen einen verklärten Gesichtsausdruck, woran sie sich später nicht erinnern. Es ist wenig darüber zu erfahren, was sie so „süchtig" nach Anfällen macht. Die naheliegende Vermutung, es handle sich um sexuelle Inhalte, hat sich nur in einem Fall bestätigt:

> Ein 32jähriger Patient gerät über das Fächeln und die nachfolgenden Absencen in einen sexuellen Erregungszustand, der zu Erektion und Masturbation führt, nicht selten aber auch mit aggressiven Entladungen in Form sadistischer Handlungen an Kindern endet (Ehret und Schneider, 1961).

In einem anderen Fall hat das Fächeln den Charakter einer Ersatzhandlung angenommen:

> Unvermittelt wendet sich ein junger Mann immer dann, wenn er sich über jemanden heftig geärgert hat, seinem Ärger aber schlecht Luft machen kann, zur Sonne und löst fächelnd einen Grand mal aus (Kammerer, 1963).

Daß nicht ausschließlich der „reflektorische" Vorgang die epileptische Erregung auslöst, zeigt ein 18jähriger Junge, der seine kortikalen Anfälle durch Reiben der rechten Wange, aber auch auf Aufforderung, sich für das Reiben bereitzuhalten oder sich dieses in Hypnose vorzustellen, auslösen kann (Goldie und Green, 1959).

Bash und Bash-Liechti (1959) konnten psychotherapeutisch die Bedingungen klären, unter denen eine bestimmte Musik zu einem anfallsauslösenden Reiz wird, weil diese den Charakter einer Deckerinnerung angenommen und sich in den Dienst einer neurotischen Verschiebung und Verdrängung gestellt hat. Inzwischen sind Versuche mit Verhaltenstherapie sensorisch ausgelöster Anfälle von größerem Interesse. Die guten Erfolge der Gruppe um Forster (1972) konnten von anderen nicht reproduziert werden (Jeavons und Harding, 1975). Möglicherweise hängt vieles vom Detail des Vorgehens ab. Übungen zur Erhaltung oder Verstärkung initialer Therapieerfolge sind über lange Zeit und täglich erforderlich (Fenwick, 1981).

Nicht-medikamentöse Anfallsunterbrechung

Nicht-medikamentöse Strategien zur Unterbrechung epileptischer Anfälle reichen von diätetischen bis zu zahlreichen Entspannungsverfahren (Yoga, progressive Relaxation und autogenes Training), ohne daß durch kontrollierte Studien der Wert dieser Methoden für die Reduktion der Anfälle evaluiert ist. Während Epilepsiepatienten solche Erfahrungen spontan selten, auf Befragen und insbesondere in Selbsthilfegruppen häufiger berichten, ist in der Literatur kaum etwas von den Bemühungen Epilepsiekranker zu finden, die Weiterentwicklung des begonnenen Anfalls zu verhindern.

Nach früheren – methodisch noch mangelhaften – Studien sollen 5% (nicht spezifiziert nach Art der Epilepsie) und 30% der Patienten mit fokalen Epilepsien (nach neueren Schätzungen 50–60%) in der Lage sein oder den Versuch unternehmen, ihren Anfall „zu kontrollieren, zu unterbrechen oder gar selbst auszulösen" (Paulson, 1963; Symonds, 1959).

Es handelt sich um mehr oder minder spezifische, nicht selten heimliche Methoden, die oft erstaunlich ausgeklügelt wirken. Patienten mit komplex-fokalen (psychomotorischen) Anfällen bevorzugen bei Ankündigung des Anfalls Strategien wie „sich rasch hinlegen", „Schokolade essen" (Zucker dagegen ist ohne Erfolg), „den Kopf schnell zwischen die Knie klemmen", „umherlaufen", „sich in die Wange zwicken". Bei Patienten mit einfach-fokalen (Jackson-)Anfällen unterscheiden sich die Praktiken danach, in welchem Gebiet des Körpers der Anfall eintritt und ob er mit sensiblen, sensorischen oder motorischen Symptomen beginnt.

Insofern lassen sich die Versuche der Anfallsunterbrechung in zwei Kategorien ordnen. Bei der ersten handelt es sich um spezifische Sinnesreize, die sich durch ihre Nähe zur anatomischen Region des subjektiv erlebten Anfalls auszeichnen. So werden motorische Anfälle mit motorischer Aktivität, sensible Anfälle mit sensiblen Reizen beantwortet. Die zweite Kategorie betrifft unspezifische, vorwiegend konzentrative oder entspannende Übungen, die die Patienten scheinbar planlos und unsystematisch den Anfallswahrnehmungen entgegenzusetzen versuchen.

Bei sieben von 71 Patienten mit komplex-fokalen Anfällen bestanden die Selbstunterbrechungsversuche aus je einzigartigen stereotypen somatischen und psychischen Elementen. Patienten, die ihre Anfälle zu unterbrechen versuchten, unterschieden sich von jenen, die das nicht angaben, durch einen höheren Bildungsgrad, bessere psychosoziale Integration und einen rechtsseitigen EEG-Fokus (Pritchard et al., 1985).

Die besondere Selbstwahrnehmung einer 41jährigen schwarzamerikanischen Sängerin gegenüber ihren Anfällen konnte für eine Dekonditionierung therapeutisch genutzt werden (Efron, 1956, 1957).

Die Patientin leidet seit 26 Jahren an Grand mal, die regelmäßig von einer Aura – in Form eines dreamy state mit Zwangsgedanken, einer sich anschließenden olfaktorischen, dann akustischen Aura mit Adversivbewegung des Kopfes – eingeleitet werden und trotz antiepileptischer Medikation mindestens zweimal monatlich auftreten. Es gelingt, mit einem spezifischen, rechtzeitig zu Beginn der Aura – vor dem von ihr als kritisch bezeichneten „Halbwegspunkt" – dargereichten unangenehmen „konträren" Geruch die weitere Entwicklung der Aura zu unterbrechen und damit den großen Anfall zu verhindern. Der olfaktorischen Aura kommt eine besondere Bedeutung zu, weil die Patientin bei ihrem ersten Anfall gerade dabei war, auf einem Feld Vergißmeinnicht zu pflücken. „Ich erinnere mich sehr gut, daß ich glaubte, diese Blumen zu riechen, obwohl ich wußte, daß sie keinen Geruch hatten. Etwa eine halbe Stunde fuhr ich fort, an ihnen zu schnuppern, weil ich gewiß war, daß sie gleich beginnen würden zu riechen . . ." In einem zweiten Therapieschritt koppelt Efron den konträren Geruchsreiz mit einem optischen Reiz in Form eines silbernen Armbandes, entsprechend einem anderen Anfallsfragment: Sie muß immer auf die Uhr blicken nach Art eines bedingten Reflexes. Die Patientin verspürt daraufhin bereits beim Anblick des Armbandes den unangenehmen Geruch, womit sie die Aura bereits zu Beginn unterbrechen kann. Schließlich genügt die gedankliche Vorstellung des Armbandes – was ihr beim Auftritt als Sängerin, wenn ein Anfall naht, zugute kommt –, um die Anfälle ohne Antiepileptika nicht wieder auftreten zu lassen (14monatige Katamnese).

Efron kommt zu folgenden Schlußfolgerungen:

- Der effektive sensorische Stimulus muß unangenehm oder schmerzhaft – man könnte auch sagen: konträr oder antidrom – sein.
- Der Stimulus muß der Art des Anfalls angemessen sein.
- Der Zeitpunkt des dargebotenen Gegenreizes ist von besonderer Wichtigkeit (je früher, desto effektiver).

Seit den sechziger Jahren werden, vor allem in den USA, verhaltenstherapeutische, speziell biofeedback-orientierte Ansätze und operante Konditionierungsmethoden angewandt. Deren Ziele sind:

- Die Identifizierung spezifischer (situationsgebundener, inter- oder intrapersonaler) „Trigger", die einem epileptischen Anfall (unbewußt oder bewußt) vorausgehen (Feldman und Paul, 1976).
- Dekonditionierung mit EEG- und Video-EEG-Techniken (Sterman und Macdonald, 1978; Sterman et al., 1974; Wyler et al., 1979).

Die bisherigen Ergebnisse sind nicht ermutigend: In keiner Studie hielt eine erzielte Anfallsreduktion länger an; eine Anfallsfreiheit wurde in keinem Fall erreicht. Die methodischen Mängel (heterogene Alters- und Anfallskollektive, fehlende Angaben zum medizinischen Behandlungsstand) überwiegen. Auch läßt sich der Verdacht nicht ausräumen, daß die Anfallsreduktion auf die intensive Zuwendung (bis zu zwei Stunden täglich) oder auf zuwendungsbedingte Vigilanzsteigerung zurückzuführen ist (Christian, 1986).

53.5.5 Psychotherapie bei Epilepsien?

Die Auffassungen über die Möglichkeiten einer aufdeckenden Psychotherapie bei Epilepsiekranken haben sich seit der Jahrhundertwende mehrmals geändert. Bis 1933 ist wegen der beschränkten medikamentösen Behandlungsmöglichkeit eine friedliche Koexistenz zwischen Pharmakotherapie und Psychotherapie zu beobachten. Im Dritten Reich werden beide von „erbbiologischen Therapien" überholt, die nicht wenigen Epilepsiekranken das Leben gekostet haben. In der Nachkriegszeit differenziert man psychopathologisch nach der tageszeitlichen Bindung großer Anfälle. Dabei ergibt sich, daß Patienten mit einer „Aufwach"-Epilepsie wegen ihrer Flüchtigkeit für eine Psychotherapie als ungeeignet angesehen werden. Obwohl Leder (1970) betont, daß biographisches Verstehen nicht den Anspruch einer Aufdeckung „hinter dem Anfallsdunkel nicht erinnerbare(r), unbewußte(r) Phantasien" erheben darf, wie es die frühen psychoanalytischen Autoren versucht haben, teilt er die erfolgreiche Psychotherapie einer Aufwach-Epilepsiekranken mit. Die Psychotherapie der stark verdrängenden Patienten mit Schlaf-Epilepsie wird wegen der möglichen Entfesselung einer nicht mehr steuerbaren Triebdynamik als bedenklich angesehen. Ungünstige Erfahrungen wurden aber nur bei jener Gruppe (Kategorie 3) mitgeteilt, deren Auren inhaltlos sind und in ihrer Bedeutung unbewußt bleiben (Hill und Mitchell, 1953). Die psychotherapeutischen Konsequenzen des Konzepts der tageszeitlichen Bindung großer Anfälle über ihre diagnostische Bedeutung hinaus sind noch nicht erschlossen. Nach Beauchesne (1980) ist eine auf Anfallsfreiheit zielende antiepileptische Therapie von Anfang an mit einer analytisch orientierten Psychotherapie zu verbinden. Er sieht in den Anfalls-„Krisen" eine sowohl die Ich-Integrität bedrohende als auch schützende, gefährliche Phantasien absorbierende Funktion. Eine vorsichtig aufdeckende Therapie sei deshalb mit averbalen, körperbezogenen Behandlungsformen (Entspannungsübungen, Psychodrama und malerische Selbstdarstellung) zu verbinden, die durch ihren spielerischen Charakter einen anderen Zugang zu den unbewußten, unaussprechlichen körperlichen „Krisen"-Erfahrungen ermöglichen. Verhaltenstherapie (Buddenberg, 1975) und Gruppentherapie scheinen sich auf das Befinden der Patienten und auf die Anfallshäufigkeit günstig auszuwirken (Brullemann, 1972; Rathmann-Kessel und Masuhr, 1984).

53.6 Pseudoepileptische (psychogene) Anfälle

53.6.1 Einfache pseudoepileptische (psychogene) Anfälle

Im Vergleich zum 19. Jahrhundert mag die Häufigkeit nichtepileptischer (hysterischer, pseudoepileptischer, psychogener, funktioneller) Anfälle abgenommen haben. Trotzdem können psychogene Anfälle noch heute große diagnostische und therapeutische Schwierigkeiten bereiten, weil Ärzte kaum integriert – epileptologisch und psychotherapeutisch – ausgebildet sind. Ein nach Grand mal erhöhter Serumprolaktinspiegel (Trimble, 1978) – unverändert nach nichtepileptischen Anfällen – findet sich bei tonisch-klonischen Anfällen nur in 80%, bei komplex-fokalen Anfällen in 43% und bei einfach-fokalen in 10% (Wyllie et al., 1984).

Pseudoepileptische Anfälle sind besonders schwer von epileptischen Anfällen des Frontallappens, von Synkopen (vgl. Kap. 36) und tetanischen Anfällen (vgl. Kap. 34) zu unterscheiden. Wegen der häufigen Verwechslung hysterischer Anfälle mit epileptischen halten wir den Begriff „pseudoepileptisch" für angebrachter (vgl. Kap. 30).

Pseudoepileptische Anfälle können isoliert oder mit epileptischen kombiniert auftreten. Etwa 1–5% der Epilepsiekranken haben, als psychogene Ausgestaltung der Epilepsie, pseudoepileptische Anfälle (Rabe, 1966, 1970). In der Regel folgen pseudoepileptische Anfälle einer bereits länger bestehenden Epilepsie, nicht selten alternativ bei medikamentös oder operativ erreichter Anfallsfreiheit, weil man, so eine Patientin, „Jahre der Deformität nicht einfach mit einem Messer abschneiden kann" (Ferguson und Rayport, 1956). Psychogene Anfälle verführen den Arzt, eine Pharmakoresistenz anzunehmen, die zu einer toxischen Höherdosierung der Antiepileptika verleitet. Pseudoepileptische Anfälle pflegen häufiger in Gegenwart von Publikum aufzutreten. Sie können zwischen „Bewegungssturm und Totstellreflex" (Kretschmer, 1946) alle Ausdrucksvarianten bieten. Da der Anfall szenisch-dramatisch, undulierend und meist länger als ein epileptischer Anfall verläuft, ist ein Mitagieren der Umwelt zu beobachten (Tab. 53–3). Die Augen sind meist geschlossen; ein Drittel der Patienten zeigt offene Augen mit starrem Blick. Geschlossene Lider sind im Gegensatz zu Synkopen nicht leicht zu öffnen und werden eher zugekniffen (Hallen, 1981).

Einnässen ist ebensowenig ein sicheres Unterscheidungsmerkmal wie ein Zungenbiß, jedoch dessen Lokalisation: epileptische Zungenbisse treten immer lateral, pseudoepileptische medial, multipel oder an der Spitze auf. Sollten einmal – bei sonst eindeutig pseudoepileptischem Anfall – lichtstarre mydriatische Pupillen irritieren, empfiehlt sich die Suche nach einem Fläschchen Mydriatikum, wie wir es bei einer unserer Patientinnen fanden – ein Beispiel für die latente Simulationstendenz bei länger bestehender Hysterie

Tabelle 53–3. Einige Merkmale zur Unterscheidung pseudoepileptischer (psychogener) Anfälle von generalisierten tonisch-klonischen Anfällen (Grand mal).

Merkmale	generalisierter tonisch-klonischer Anfall (Grand mal)	pseudoepileptischer (psychogener) Anfall
Beginn	plötzlich	allmählich – „einübend"
Verlauf	stereotyp	undulierend, regellos
Augen	meist geöffnet	meist geschlossen
Pupillen	lichtstarr	auf Licht reagierend
Zungenbiß	immer lateral	selten/Zungenmitte oder -spitze
Urin-/Stuhlabgang	häufig	selten
Zyanose	häufig	selten
Anfallsdauer	kurz	lang
Anfallsmodifikation durch		
a) Verhalten	nicht durch Ansprache	meist durch Ansprache/Nicht-Beachten
b) Antiepileptika	selten resistent	resistent/Verschlimmerung
Postiktal	Verwirrtheit/Nachschlaf	meist wach
Eindruck	furchterregend	szenisch-dramatisch mit Ausdruckscharakter

(Charcot, 1874). Simulierte Anfälle treten viel seltener auf als angenommen (Hammond, 1948). Auch beim Münchhausen-Syndrom sind sie eher ein zweitrangiges Symptom (Asher und Lond, 1951). Wird die Psychogenie eines Anfalls überhaupt erkannt, wird er häufig – unter Verleugnung der unbewußten Triebkräfte – fälschlich als simuliert angesehen.

53.6.2 Status pseudoepilepticus

Bei etwa 10% der Patienten mit pseudoepileptischen Anfällen häufen sich diese derart, daß sie als Status epilepticus verkannt werden (Riley und Roy, 1982). Niemals wurde ein Status pseudoepilepticus bei einem Epilepsiekranken beobachtet.

Während im Status epilepticus convulsivus mehrere große Anfälle nacheinander auftreten, ohne daß der Kranke zwischendurch zu Bewußtsein kommt, sprechen wir von einem Status pseudoepilepticus, wenn nichtepileptische Anfälle mehr als eine Stunde lang einander folgen, ohne daß der Kranke ansprechbar ist, obwohl er nicht bewußtlos ist (Schultz, 1986). Wegen der erheblichen Gefahren, denen die Patienten bei der fast regelmäßigen Fehldiagnose ausgesetzt sind, scheint es gerechtfertigt, diese von Charcot zwar beschriebene, aber nicht so genannte und später in der Literatur nur am Rande erwähnte Anfallsform als eigenständige, den Artefakt-Erkrankungen nahestehende psychogene Krankheit hervorzuheben – um so mehr, als diese in der Klassifikation von Anfallskrankheiten fehlt (Wolf et al., 1987). Der folgende Fall kann als typisch gelten:

Aus einer Intensivstation wird eine Patientin mit „Status epilepticus" überwiesen, nachdem die Anfälle bereits zehn Tage angedauert, an Häufigkeit zugenommen und zu Verletzungen, u. a. einer Radiusfraktur, geführt haben. Wir finden die Patientin nicht ansprechbar, mit Subclavia- und Blasenkatheter und Gipsverband am linken Arm, heftig und unregelmäßig an allen vier Extremitäten zuckend, Kopf und Blick sind extrem nach rechts gewendet. Zwischendurch streckt sich der Körper opisthoton. Speichelfluß, Augen geschlossen, durch minutenlanges Sistieren der Atmung zyanotisches Gesicht. Die Diagnose Status pseudoepilepticus – wegen der bunten Symptomatik erwogen – wird sicher, als die Patientin die Prüfung der Pupillenreaktion durch Zusammenkneifen der Lider vereitelt. Die Anfälle enden nach der Bemerkung des herbeigeholten Chefarztes: „Jetzt ist es genug. Wir haben Sie verstanden. Sie können jetzt aufhören." Die Patientin schlägt die Augen auf, lächelt und beginnt mit kindlicher Stimme von sich zu erzählen, indem sie wiederholt den Arm des Arztes an sich zieht und streichelt. Obwohl einzelne Anfälle von zwanzig Minuten Dauer in den nächsten Tagen auftreten, ist der Status pseudoepilepticus auf diese scheinbar banale Weise unterbrochen.

Der dramatische, appellative Charakter des einzelnen pseudoepileptischen Anfalls eskaliert im Status derart, daß nüchterne und genaue Beobachtung ausbleibt und ärztliches Eingreifen unmittelbar notwendig erscheint. Es bleibt offen, ob der Status pseudoepilepticus nicht erst die Folge einer aktivistischen, unreflektierten ärztlichen Haltung ist. Bei rechtzeitigem Erkennen (vgl. Tab. 53–3) läßt sich ein solcher Status in der Regel mit einfachen kommunikativen Mitteln unterbrechen, was aber noch keine kausale Therapie bedeutet.

53.6.3 „Bedeutung" und Verlauf

Die motorischen Erscheinungen des pseudoepileptischen Anfalls haben Ausdruckscharakter im Zusammenhang mit unbewußten, psychischen Konflikten. Den Inhalt des Anfalls bildet die halluzinatorische Reproduktion jenes mit Lebensgefahr verbundenen Ereignisses einschließlich der Gedankengänge und Sinneseindrücke, die „das bedrohte Individuum damals angesponnen" (Breuer und Freud, 1940).

Im Verlauf pseudoepileptischer Staten fällt eine Eskalation der Anfälle nach intensivmedizinischer und eine Anfallsreduktion nach kommunikativer Intervention auf. Trotz richtiger Diagnose werden diese Patienten, die häufig den medizinischen Hilfsberufen angehören, fast immer antiepileptisch weiterbehandelt. Das „Mimikry"-Bedürfnis der Patienten im Dienste des Selbstschutzes können Ärzte offenbar nur mitagierend beantworten, solange kein psychodynamisches Verständnis entsteht.

Ätiologisch werden inzestuöse Beziehungen und Mißhandlungen in der Kindheit für bedeutsam gehalten. Es gibt Hinweise, daß der Status pseudoepilepticus und die häufig hinzutretenden Selbstmißhandlungen eine Reinszenierung kindlicher Mißhandlung darstellen (Hirsch, 1986). So werden die chronische Suizidalität und andere psychiatrische Auffälligkeiten, auch in der Familie, verständlich. Eine zum Status pseudoepilepticus prädisponierende Sondergruppe stellen Patienten dar, deren Anfälle von den Müttern im Kindesalter, z.B. durch nächtliche Erstickungsversuche, Karotissinusdruck oder Medikamente ausgelöst wurden, oder die mit fingierten Anfallsschilderungen jahrelang den Ärzten als Epilepsiekranke präsentiert werden („fictitious epilepsy"). Einige dieser Kinder halten an einer fingierten Epilepsie auch später fest, was bis zur Invalidisierung führen kann (Meadow, 1984).

Nach dem Ende eines pseudoepileptischen Anfalls oder Status, das man ruhig abwarten oder kommunikativ herbeiführen sollte, ist eine psychotherapeutische Behandlung indiziert, die jedoch wegen des auf Spaltung und Verleugnung beruhenden weiteren Agierens von Arzt und Patient selten zustande kommt (Plassmann et al., 1985, 1986). Statt dessen dienen Krankenhäuser solchen Patienten weiterhin als Zuflucht, was sie zu bekannten „Krankenhauswanderern" macht (March, 1954).

53.7 Encephalomyelitis disseminata (Multiple Sklerose)

Die Multiple Sklerose (MS), auch Encephalomyelitis disseminata, Polysklerose oder nach ihrem Erstbeschreiber 1868 Morbus Charcot genannt, kann trotz intensiver Forschung als eine der neurologischen Krankheiten angesehen werden, bei der sich seit Charcots Zeiten die Situation für Patient und Arzt kaum verändert hat: Die Ätiologie der schubförmig oder chronisch progredient verlaufenden Krankheit ist bisher ungeklärt. Die Diagnose bereitet zuweilen Schwierigkeiten, weil sich die Symptome als „großer Imitator" neurologischer und funktioneller Krankheiten gebärden können, z.B. wenn, wie bei einer Angstneurose, flüchtige Parästhesien und Gangstörung berichtet werden.

53.7.1 Epidemiologie

Zwei Drittel der Ersterkrankungen liegen zwischen dem 20. und 40. Lebensjahr. Frauen erkranken etwa zweimal häufiger als Männer. Die MS ist vorzugsweise in gemäßigten Klimazonen anzutreffen und zeigt ein Nord-Süd-Gefälle mit höchsten Erkrankungsraten in Nordeuropa, Kanada und den nördlichen USA (Prävalenz: 50–100/100 000, in der Bundesrepublik 70/100 000 Einwohner). Afrikaner, Indianer, Eskimos, Japaner und Inder sind auffallend selten von MS betroffen. Um den Äquator ist das Risiko am niedrigsten; es scheint auf der südlichen Halbkugel, jenseits des 40. Breitengrades leicht anzusteigen, so daß Teile Australiens wieder ein mittleres Erkrankungsrisiko aufweisen (Poser und Ritter, 1980). Etwa doppelt so hoch wie die Zahl der diagnostizierten Patienten scheint der Anteil von Personen mit blandem MS-Verlauf zu sein.

53.7.2 Ätiologie und Prognose

Vermutlich handelt es sich um eine multifaktorielle Autoimmunerkrankung, die mit großflächigen, disseminiert verteilten Entmarkungsherden, sog. Plaques, einhergeht (Kesselring, 1989). Zu Spekulationen gaben die Ergebnisse verschiedener Migrationsstudien aus Regionen mit hohem in Länder mit niedrigem MS-Risiko und umgekehrt Anlaß: Wird das Alter zum Zeitpunkt der Migration berücksichtigt, so ergibt sich, daß das Risiko des Herkunftslandes mitgenommen bzw. beibehalten wird, wenn die Auswanderung im Erwachsenenalter erfolgt. Dagegen wird das Risiko des Gastlandes erworben, wenn die Migration im Kindesalter vor dem 15. Lebensjahr erfolgt. Diese Beobachtung wurde dahingehend interpretiert, daß ein pathogenetisch relevanter Faktor vor dem 15. Lebensjahr (Autoimmunvorgänge in Verbindung mit einer erworbenen slow-virus-Infektion?) wirksam sein muß (Kurtzke, 1980). Gegen eine solche Annahme spricht der chronisch-rezidivierende Verlauf. Eine kausale Therapie fehlt bislang, und prognostische Aussagen sind wegen der erheblichen Variation der Verläufe, besonders in den ersten Krankheitsjahren, unsicher (Wolter und Zimmermann, 1983).

53.7.3 Psychosomatische Aspekte

Die gegenwärtige Betonung epidemiologischer, virologischer und immunologischer Forschung hat psychosomatische Ansätze eher in eine Außenseiterrolle gebracht. Da hinsichtlich psychodynamischer Faktoren nur wenige systematische Untersuchungen vorliegen, dominieren in der psychosomatischen MS-Literatur Einzelbeobachtungen, die jedoch selten hinreichend detailliert sind (Mei-Tal et al., 1970). Zunehmendes Interesse findet neben psychopathologischen und neuropsychologischen Beobachtungen die Krankheitsverarbeitung („coping") (Dalos et al., 1983; Oberhoff-Looden, 1978; Seidler, 1978).

> Von Sir Augustus D'Este (1794–1848) stammt die erste eingehende Selbstschilderung. Sir D'Este bringt die ersten Symptome seiner MS (Retrobulbärneuritis) mit

28 Jahren in Zusammenhang mit dem Besuch bei einem nahen Angehörigen, zu dem er eine Beziehung wie die eines Sohnes hat. Bei seiner Ankunft stirbt dieser Mann. Kurz nach dessen Begräbnis, auf dem D'Este „dagegen ankämpfte, nicht zu weinen" (von ihm im Tagebuch unterstrichen), tritt eine Sehunschärfe auf (Firth, 1940, 1948). Der Verlust dieser Bezugsperson wird vermutlich deshalb traumatisch erlebt, weil D'Este nach Annullierung der elterlichen Ehe früh seinen Vater verlor.

Auch von späteren Autoren werden reale oder phantasierte Trennungen von elterlichen Schlüsselfiguren als Auslösesituation für eine MS beobachtet (Grinker et al., 1950; Paulley, 1976/77). Dabei scheinen chronische emotionale Störungen mit der Folge einer destabilisierenden Ängstlichkeit häufiger (87%) einer MS voranzugehen als akute seelische Belastungen (10%) (Philippopoulus et al., 1958). Letztere wurden in neueren Studien jedoch häufiger gefunden (Grant et al., 1989; Franklin et al., 1988; Warren et al., 1982). Trennungssituationen wirken vermutlich traumatisch, weil bei MS-Patienten seit frühester Kindheit eine besondere emotionale Abhängigkeit und psychische Unreife vorliege, die mit exzessivem Liebes- und Zuwendungsbedürfnis einhergehe (Inman, 1948; Jonez, 1951). Dieses früher nicht gestillte Bedürfnis nach Liebe und Zuwendung verberge sich hinter einer Maske von Gefügigkeit und „unschuldigem Lächeln", das einer belle indifférence gleichkomme (Paulley, 1975).

Die Patienten zeigen keine aggressiven Gefühle, vermeiden Auseinandersetzungen und verhalten sich im höchsten Grade konformistisch, worauf sie stolz seien. Weiterer Ausdruck ihrer Über-Normalität sei das Fehlen von neurotischen Störungen, wie Enuresis, Alpträume, Stimmungsschwankungen und Eßstörungen in der Kindheit. Eine Analogie zu Patienten mit urtikariellen Hauterkrankungen wurde postuliert. Wie die Haut entwickelt sich das Nervensystem aus dem Ektoderm; für eine frühe Traumatisierung sei von Bedeutung, daß die Myelinisierung des Gehirns und des Rückenmarks mit der Geburt noch nicht abgeschlossen ist (Jonez, 1951). Mit dem äußeren Harmonisierungsbedürfnis korrespondiert eine verborgene (Auto-)Aggression:

> Eine 26jährige MS-Patientin findet im Traum ein kleines Petrefakt, das das versteinerte Bild einer Spinne darstellt, die eine Fliege aussaugt.
> In einem anderen Traum bewundert sie aus der Ferne die Vitalität eines Mannes, der soeben ihre Freunde ermordet hat.
>
> Eine 27jährige MS-Patientin träumt, ihre Freundin habe sich selbst dadurch umgebracht, daß sie sich mit einem Beil Finger und Kopf abgehackt habe. Plötzlich äußert auch ihr Freund den Wunsch, sie solle ihm den Kopf abhacken, was sie nur unvollständig bewältigt.

Autoaggression kann sich auch hinter hypochondrischen Beschwerden verbergen:

> Ein 24jähriger MS-Patient äußert nach einer ACTH-Infusionstherapie wochenlang Befürchtungen, daß ein „Stück des Venenkatheters abgerissen und im Herzen hängengeblieben" sei, obwohl ihm die unversehrte Venenkatheterspitze gezeigt und über seine Ängste gesprochen wird.

Zwischen dem indifferenten, konformen Verhalten und der autoaggressiven Welt der Träume scheint – auch bei therapeutischem Umgang – kaum eine Verbindung herstellbar. Diese Dissoziation zwischen Primärprozeß im Unbewußten und Überangepaßtheit im Psychosozialen – von Marty und M'Uzan als „pensée opératoire" beschrieben – mag sich als Überlebensstrategie entwickelt haben: Kinder, bei denen nicht einzelne Bedürfnisse, sondern Bedürftigkeit an sich mißachtet wird – durch eine versagende oder eine überfürsorglich-einengende Mutter (mère calmante) – verleugnen ihren Anspruch auf Zuwendung und Aggression und erkaufen sich die Liebe der Eltern durch Fügsamkeit. Es entsteht ein „falsches Selbst" (Winnicott, 1974), das sich statt an den eigenen Bedürfnissen an den Forderungen anderer orientiert. Daneben lebt, verborgen und ohne bewußte Beziehung zu diesem sozial überangepaßten Selbst, das „Kind" weiter, jedoch unentwickelt, chaotisch, verletzbar und auf Anerkennung wartend. Groen et al. (1967) sprechen in diesem Zusammenhang von einer infantilen Persönlichkeitsstruktur, durch welche interpersonelle Konflikte bei mangelhaften Anpassungs- und Abwehrmechanismen nur als extrem traumatisch erlebt werden können. Der häufig zu beobachtende Wechsel zwischen vernünftig-genügsamem (erwachsenem) und kindlich-abhängigem Verhalten von MS-Patienten im therapeutischen Umgang könnte mit dieser Dissoziation in der Entwicklung zusammenhängen. Auch im Traum kann sich das Alternieren zwischen diesen beiden Polen dokumentieren:

> Eine 26jährige Patientin träumt sich wiederholt mit einem Kind auf dem Arm, das unter ihren Augen in Minutenschnelle heranwächst und schließlich älter ist als sie selbst.

Die Anpassungstendenz MS-Kranker kann anankastische Züge annehmen und sich in einem besonderen Pflichtbewußtsein und übertriebenem Arbeitsverhalten zeigen. Spätere MS-Kranke wählen oft Berufe mit mechanistischer Tätigkeit. Unter 120 MS-Patienten wurde keiner mit einem freien oder künstlerischen Beruf gefunden. „Alles steht in erhöhtem Maße im Banne von Zeit und Präzision, Maß, Zahl und Gesetz" (Wolff, 1949).

Die Unsicherheit eines überangepaßten Selbst im Hinblick auf die Wahrnehmung eigener körperlicher Bedürfnisse mag der diagnostischen Unsicherheit entsprechen, in die MS-Patienten den Arzt versetzen, wenn sie Konversionssymptome, angstneurotisch ausgestaltete organische Beschwerden und Symptome disseminierter Nervenläsionen gleichzeitig oder nacheinander bieten. Es ist zu vermuten, daß die

Gleichgültigkeit vieler MS-Patienten angesichts häufiger Sexualstörungen mit einer prämorbid bestehenden Selbstentfremdung zusammenhängt.

Affektive Besonderheiten

Psychopathologische Symptome während des Krankheitsverlaufs sind nicht ungewöhnlich, treten aber selten als Initialsymptome zutage. Demgegenüber eruierten Schiffer und Babigian (1984) bei 16% der Patienten psychiatrische Symptome als Erstmanifestation, weshalb bei psychiatrischen Krisen auch an eine MS zu denken ist.

Methodische Probleme ergeben sich, weil psychopathologische Studien meist auf neuropsychologische Daten, und neuropsychologische Studien auf eine psychopathologische Deskription verzichten. So wird etwa in der Hälfte der Fälle, bei denen sich in der neurologischen Untersuchung keine mentalen Störungen feststellen lassen, bei neuropsychologischer Untersuchung ein kognitives Defizit erfaßt (Peyser et al., 1980).

Umstritten bleibt, ob psychopathologische Auffälligkeiten bei MS-Kranken, insbesondere die Euphorie, als Folge eines demyelinisierenden Prozesses, als paradoxe Reaktion auf die Krankheit oder als Ausdruck der prämorbiden Persönlichkeit anzusehen sind (Groen et al., 1967; Paulley, 1976/77; Philippopoulos et al., 1958). Für ersteres spricht die als „organisches Psychosyndrom" imponierende optimistisch-verleugnende Einstellung, die überwiegend bei ausgeprägten neurologischen Ausfällen und nach langer Krankheitsdauer beobachtet wird. Euphorie scheint nicht häufiger zu sein als andere psychopathologische Auffälligkeiten (Tab. 53–4).

Auch abrupt wechselnde Affektäußerungen ohne erkennbare Verbindung zur inneren Stimmungslage dürften auf zerebralen Läsionen beruhen, wobei im biographischen Kontext die Affekte zuweilen verstehbar werden.

Krankheitsverarbeitung

Bereits mit der Aufklärung der Patienten beginnen die Probleme. Liegen schon zwischen den Erstsymptomen und der Diagnose 1–3 Jahre, so vergehen noch einmal 2–4 Jahre, bis der Patient die Diagnose erfährt. Dagegen scheinen Patienten bei Erstmanifestation nicht selten ihre Diagnose zu ahnen, bevor sich der Arzt sicher ist. Aus diesem Grunde empfiehlt sich eine frühzeitige, aber behutsame Aufklärung, wenn die Diagnose hinreichend abgesichert ist. Diese sollte jedoch nicht unerwähnt lassen, daß mehr als ein Drittel aller MS-Patienten über Jahrzehnte einen gutartigen Verlauf aufweisen und drei Viertel nur leicht behindert sind (Poser und Friedrich, 1983), wobei die blanden Fälle nicht berücksichtigt sind.

Die Unberechenbarkeit der Erkrankung führt bei MS-Patienten und Angehörigen gelegentlich zu magischen Krankheitsvorstellungen und beim Arzt zu umstrittenen Therapieverfahren wie Schweinehirnimplantation und Schlangengift (Bauer, 1983; Hallen, 1967; Görres et al., 1988). Der Nimbus der Unheilbarkeit läßt Suizidgedanken bei Patienten und Tötungsphantasien bei Angehörigen und Ärzten aufkommen (Kaufmann, 1951; Seidler, 1978). Das Suizidrisiko von MS-Patienten ist in den frühen Krankheitsstadien nach Bekanntwerden der Diagnose 14mal höher als in der Normalbevölkerung (Kahana et al., 1971). Propagandistisch wurde 1941 eine MS-Kranke in dem Film „Ich klage an" benutzt, um vordergründig Sterbehilfe, eigentlich die „Euthanasie" zu legitimieren: Die Patientin erhält von ihrem Mann eine tödliche Injektion, als dieser an der Erforschung eines MS-wirksamen Serums scheitert (Roth, 1985).

Hysterie und Angst(neurose) bei MS

Wegen bizarrer Beschwerden, die von MS-Patienten zu Beginn oder im Verlauf ihrer Erkrankung häufig angegeben werden, wird die MS nicht selten als Hysterie fehlgedeutet, worauf schon zur Jahrhundertwende hingewiesen wurde. Obwohl „die MS als ausgesprochen organische Nervenkrankheit und die Hysterie als echte Neurose in pathogenetischer Hinsicht Prototypen zweier gänzlich verschiedener Prozesse sind, können die klinischen Bilder derart zahlreiche Analogien zeigen, daß die Frage nach ihrer Unterscheidung einen der wichtigsten Abschnitte unserer differentialdiagnostischen Erörterung bilden muß. Die Ähnlichkeit beider Affektionen beruht ... darauf, daß Vielgestaltigkeit und Flüchtigkeit der Symptome ..., mehr oder minder plötzliches Einsetzen und Verschwinden von Empfindungs- und Bewegungsstörungen gemeinsame Merkmale sind; so kommt es zum Beispiel, daß Sehstörungen, welche sich im Gefolge

Tabelle 53–4. Psychische Auffälligkeiten bei Multipler Sklerose nach klinischem Eindruck (Mehrfachangaben).

psychische Auffälligkeiten	Cottrell u. Wilson (1926) n = 100	Surridge (1969) n = 108	Kahana u. Mitarb. (1971) n = 295	Payk (1973) n = 773	Poser (1980) n = 1572	Zusammen n = 2848
Unauffällig	0	15%	75%	47%	62%	56%
Euphorisch	63%	26%	5%	17%	11%	15%
Depressiv	10%	27%	6%	11%	8%	11%
Emotional (affekt-)labil	25%	–	8%	13%	18%	15%
Intellektuell beeinträchtigt	2%	61%	–	11%	9%	10%
Psychotisch	–	4%	6%	1%	0,1%	1%

der Sclerosis multiplex rapid entwickeln, an Intensität rasch und erheblich schwanken und ohne Residuen andauernd verschwinden können, ungemein leicht als ‚hysterische' aufgefaßt werden und zu schwerwiegenden Fehldiagnosen verleiten" (Müller, 1904). Wohl deshalb wurde die MS früher der Hysterie zugeordnet (Poser, 1984). Dieses Problem besteht klinisch auch heute noch, da flüchtige und bizarre MS-Beschwerden (Uthoffsches Phänomen), z.B. Sensibilitätsstörungen und Diplopien, bei der neurologischen Untersuchung nicht immer objektiviert werden können. Das Fluktuieren der Symptome kann nur zum Teil als Folge einer körpertemperaturabhängigen neuronalen Impulsleitungsstörung demyelinisierter Axone (Variationen der evozierten Potentiale) verstanden werden. Vielmehr zeigt sich nicht selten eine funktionelle Ausgestaltung der MS-Symptome: Eine Taubheit der Hand wird als Verlust des Gefühls im ganzen Arm empfunden.

> Eine Krankenschwester, bei der eine episodische Retrobulbärneuritis mit monokularer Farbenblindheit abgelaufen ist, beobachtet später während eines heißen Bades eine verminderte Sehschärfe und Farbenblindheit sowie eine Steifheit der rechten Hand, die sich mit visuell evozierten Potentialen unter Wärme nicht bestätigen läßt, so daß die Beschwerden als funktionell bewertet werden.

Diese diagnostische Unsicherheit nimmt zu bei angstneurotischer Ausgestaltung, da die somatisierten Angstäquivalente den MS-Symptomen oft täuschend ähnlich sein können (Freud, 1895 c). Die Differenzierung neurologischer und funktioneller MS-Symptome ist ebenso wichtig wie deren Betrachtung als verschiedene Ausdrucksformen derselben Krankheit.

Nicht weniger schwierig ist der Umgang mit jenen (angst)neurotischen Patienten, die über Sensibilitätsstörungen, Parästhesien und Gangstörungen klagen und ihrer Gewißheit Ausdruck geben, an einer MS erkrankt zu sein. Diese MS-Phobie kann spannungsbedingt kloniforme, vorübergehend sogar seitendifferente Muskeleigenreflexe in Erscheinung treten lassen und so das diagnostische Dilemma vergrößern.

53.7.4 Psychotherapeutische Aspekte

Wegen des wechselhaften Verlaufs und der Unkenntnis begünstigender Faktoren verbietet sich die Verallgemeinerung erfolgreich verlaufender psychosomatischer Behandlungen. Da solche zufällig mit der spontanen oder pharmakologischen Remission koinzidieren können, sollten sich die Kriterien weniger an der neurologischen als an der psychodynamischen Entwicklung orientieren. Psychotherapeutisch empfiehlt sich ein vorsichtig aufdeckendes Vorgehen, das, der frühen Störung entsprechend, die Konstanz einer sicheren und zugleich flexiblen Objektbeziehung gewährleistet. Ob der in einzelnen Fällen beobachteten schnellen und anhaltenden Remission im Rahmen einer stürmischen Übertragung, in der die auf den Therapeuten gerichteten Aggressionen nicht fehlen, eine allgemeinere Bedeutung zukommt, kann nur durch weitere psychotherapeutische Erfahrungen und prospektive psychosomatische Studien geklärt werden.

53.8 Myasthenia gravis

53.8.1 Klinische Symptomatik

Die Myasthenia gravis pseudoparalytica ist eine neuromuskuläre Autoimmunkrankheit, die sich an der motorischen Endplatte manifestiert. Ihre Inzidenz wird mit $1:20000$ angegeben. Vor dem 30. Lebensjahr sind von der Erkrankung mehr Frauen, in höherem Lebensalter mehr Männer betroffen. In sehr seltenen Fällen kommt die Erkrankung auch schon vor dem 10. Lebensjahr vor.

Die Myasthenia gravis äußert sich in einer abnormen Ermüdbarkeit der Willkürmuskulatur unter Belastung, die sich in Ruhe zurückbildet. Bevorzugt sind die kurzen (störanfälligeren) Augenmuskeln (Ptosis, Doppelbilder), die mimische Muskulatur (Kau-, Schluck- und Sprechstörung) und die proximale Extremitätenmuskulatur betroffen. Die gutartige okuläre Myasthenie kann in die faziopharyngeale und – in 80% der Fälle – in die generalisierte Form übergehen. Die Symptome können allmählich, aber auch plötzlich beginnen. Sie sind in der Regel abends deutlicher ausgeprägt als morgens; sie schwanken aber auch in Abhängigkeit von der Befindlichkeit der Patienten und von wechselnden Situationen. Durch die Facies myopathica machen die Patienten einen erschöpften Eindruck. Viele schildern eine allgemeine Mattigkeit und schnelle Ermüdbarkeit auch in nicht manifest geschwächten Muskeln.

Neben langsam progredienten Verläufen gibt es krisenhafte Verschlechterungen und Remissionen unterschiedlicher Dauer. Vital bedrohlich wird die Krankheit bei Schluck- und Atemlähmungen.

Pathophysiologisch werden nikotinerge Acetylcholinrezeptoren (ACH-R) durch spezifische IgG-Antikörper (AK) an der postsynaptischen Membran der Muskelendplatte blockiert, wodurch die neuromuskuläre Erregungsübertragung behindert wird. Ungeklärt sind die auslösenden Faktoren dieses „autoaggressiven" Vorgangs. Es wird angenommen, daß eine (virusbedingte?) Thymitis den Autoimmunprozeß anstößt, wobei der Thymus Myoidzellen mit ACH-R enthält. Im Rahmen einer Thymitis soll die ACH-R-AK-Produktion gegen diese Myoidzellen durch Thymuslymphozyten in Gang gesetzt werden (Wekerle und Ketelsen, 1977). 50% aller Myastheniepatienten weisen eine Thymushyperplasie oder ein Thymom auf, welche in 84% Antikörper gegen Skelettmuskel produzieren (Henze, 1988). Auffällig häufig ist die Myasthenie mit anderen immunologischen Erkrankungen (Schilddrüsenerkrankungen, PcP) assoziiert (Haas, 1988).

Wechselhaftigkeit und Situationsabhängigkeit der myasthenen Symptome und psychische Begleiterscheinungen bereiten, besonders im Anfangsstadium, differentialdiagnostische Probleme, so daß die Myasthenia gravis oft als Konversion, Psychasthenie oder „larvierte Depression" mißdeutet wird oder umgekehrt unklare Schwächezustände als Myasthenie behandelt werden. Neben der symptomatischen Behandlung mit Cholinesterasehemmern (Prostigmin) stehen „kausale" Therapieformen zur Verfügung, die am Immunsystem angreifen: Thymektomie, Cortison und immunsuppressiv wirkende Zytostatika (Azathioprin). Zunehmend hat sich die Kombination von Prostigmin (schnelle Wirkung) und Azathioprin (langsame Wirkung) bewährt. Durch eine Langzeittherapie mit Azathioprin konnte in vielen Fällen die Arbeitsfähigkeit wieder hergestellt und die Letalität der Myasthenie von 40% auf 5% gesenkt werden. Keine der genannten Therapien schützt jedoch vor Rezidiven; Prostigmin kann cholinergische Krisen hervorrufen.

53.8.2 Psychosomatische Aspekte

Eine Myasthenie manifestiert sich nicht selten nach psychischer Belastung, etwa nach Heirat, Schwangerschaft oder nach dem Tod von Angehörigen. Verschiedene Autoren diskutieren die Bedeutung emotionaler Belastungen für die Erstmanifestation und Exazerbation, wobei die Frage der ätiologischen Relevanz unterschiedlich bewertet wird (Oosterhuis und Wilde, 1964; Rabending und Röse, 1966). Die Art der psychischen Belastung wird meist nicht genau beschrieben, zuweilen entsteht der Eindruck, als ob sich durch Emotionen überhaupt – „sowohl freudige wie ängstliche Erregung" – die Myasthenie verschlechtere. Hillenbrand (1972) gelingt erstmals die Darstellung einer Psychodynamik der Myasthenie. Ihm fiel prämorbid eine Mutterabhängigkeit und aggressive Hemmung bei gleichzeitig forciertem Selbständigkeitsstreben und exzessivem Bewegungsdrang auf.

> Bei einer 17jährigen Patientin waren die ersten Symptome einer schweren generalisierten „therapieresistenten" Myasthenie in der Situation eines lähmenden Ausgeliefertseins an einen Freund aufgetreten, in der die bisherigen Verarbeitungsweisen versagten. Die myasthene Lähmung wurde als Ausdruck einer „Mischung von totaler Hingabe, Totstellreflex ... und völliger oraler Abhängigkeit" gedeutet. Die Patientin entwickelte in der Psychotherapie eine tänzerische Beweglichkeit, erlernte bisher unterdrückte Formen der Auseinandersetzung mit der Mutter und gesundete dabei.

Ein Patient Mitscherlichs schildert in der psychoanalytischen Behandlung ein Gefühl der Fremdbestimmtheit seines Körpers durch die Mutter. Mit der Myasthenie habe er sich „sadistisch" gegen seinen Körper gerichtet. Die Krankheit bedeute einen Heilungsversuch, d.h. eine Abwehr gegen diese Fremdherrschaft. Auf diese Weise werde die Myasthenie als Autoaggressionskrankheit in der Behandlung auch psychodynamisch verständlich (Mitscherlich, 1976).

Nimmt die Myasthenie nicht den üblichen Verlauf mit Beginn an den Augenmuskeln, dann sind, wie in der Psychotherapie deutlich wird, besonders „fremdbestimmte" Muskelfunktionen betroffen.

> Eine 23jährige kaufmännische Angestellte hat ein halbes Jahr vor Beginn der generalisierten Myasthenie, die sich vor allem in Sprech- und Schluckstörungen äußert, einen Steuerbeamten geheiratet, der ihre kreativen Bedürfnisse unterdrückt, sich mit ihrer fordernden Mutter verbündet und sie mit seiner Eloquenz überfordert. Die wechselnde Ausprägung der Dysarthrie zeigt sich von ihrer Fähigkeit abhängig, eine eigene Sprache zu sprechen. In lethargischer Stimmung, besonders zu Beginn der Therapiestunden, tritt schon nach wenigen Worten die Sprechstörung auf. Diese läßt nach, wenn sie sich zuweilen spielerisch phantasierend ihren Wunschträumen überläßt. Hingegen kann sie, besonders bei fortschreitender Behandlung, stundenlang ohne Störung sprechen, wenn sie Trauer und Aggression heftig zum Ausdruck bringt. Überraschende Besserungen sind auch nach ausgiebiger, lustbetonter körperlicher Anstrengung, beim Schwimmen und nächtelangen Tanzen zu beobachten. Die Patientin bringt nach Beendigung der einjährigen Psychotherapie – ohne Medikamente – zwei Kinder zur Welt und ist seit zwölf Jahren, bis auf seltene flüchtige myasthene Episoden bei Besuchsankündigungen der Mutter, symptomfrei (Kütemeyer, 1979).

> Bei einem 28jährigen indonesischen Ernährungswissenschaftler manifestiert sich die Muskelschwäche in den Beinen, als er sich, besonders auf Spaziergängen, von der Hektik seiner europäischen Frau bedrängt fühlt. Seine „streikenden" Beine, so lautet die Deutung, könnten ihm zu verstehen geben: „Wir wollen so langsam gehen, so besinnlich leben, wie wir es aus unserer östlichen Kultur gelernt haben." Nach wenigen Gesprächen fühlt er sich innerlich gestärkt, die Symptome sind ohne Medikamente schnell wieder rückläufig.

Körperbezogene Redewendungen erweisen sich in der Therapie von Myastheniepatienten als besonders wirksam. Zuweilen finden sich in Träumen Hinweise für die „Übersetzung" der Symptome:

> Die Beine sind auch bei einer 23jährigen Kürschnerin am deutlichsten betroffen. Zu Beginn der Psychotherapie ist sie mehrmals auf der Straße zusammengebrochen und kann sich nur mit fremder Hilfe wieder aufrichten. Im Initialtraum sieht sie ein Pferd, das mit seinen vier Beinen in der Dachluke eines Autos eingezwängt ist. Ähnlich empfindet sie sich durch ihren langjährigen Freund ihrer Bewegungsfreiheit beraubt. Nach erregenden Auseinandersetzungen, die zu einer räumlichen Trennung vom Freund führen, findet sie ihre vorläufige Standfestigkeit wieder – sie träumt sich auf einem Elefanten reitend – und entdeckt ihre Freude am Schwimmen und nächtelangen Tanzen, wobei sie keinerlei Ermüdung empfindet; ihre Beine zeigen nur noch sporadisch eine Schwäche.

Psychische Befunde im Sinne einer der Myasthenie vorausgehenden oder sie begleitenden „Neurasthenie" oder „Hysterie" sind seit Oppenheim (1901) bekannt.

Sie werden überwiegend unter dem Aspekt der Krankheitsverarbeitung betrachtet. Depression, phobisch-anankastisches Verhalten und hypochondrische Selbstbeobachtung werden als Reaktion auf die motorische Behinderung, auf die Unberechenbarkeit des Verlaufs und die „Unheilbarkeit" der Krankheit angesehen. Die ängstliche und depressive Verarbeitung der Myasthenie kann sich körperlich in Form von Adynamie, Schwäche und Gangstörung, neuropsychologisch in Form von Gedächtnisstörung (Gerber und Diener, 1989) äußern, die die Patienten unter Umständen mehr behindern als die Grunderkrankung. Die von etwa einem Drittel der Patienten geklagten Schulter-, Nacken- und Kopfschmerzen können als Ausdruck des Versuchs, die Kontrolle zu behalten, verstanden werden. Die myasthene Beeinträchtigung mobilisiert besonders Konflikte hinsichtlich Abhängigkeit (Brolley und Hollender, 1955; Chafetz, 1966). Den überaktiven, unabhängigen Patienten bereitet das Ausgeliefertsein an Arzt und Betreuer während der Intensivtherapie, den passiv-depressiven hingegen die Rehabilitation die größeren Schwierigkeiten.

Vereinzelt wird über psychotische Reaktionen berichtet, wobei die einzelnen myasthenen Symptome im Wahn eines Patienten eine positive Bedeutung erfahren können: Göttliches Eingreifen verhindert durch die Schwäche in den Armen den Genuß vergifteter Speisen; die Diplopie wird zum „Geschenk einer doppelten Sehkraft" (Hayman, 1941).

Wenn eine angstneurotische oder depressive Ausgestaltung der Myasthenie nicht rechtzeitig als psychogen erkannt und von der myasthenen Behinderung unterschieden wird, droht unnötige Höherdosierung der Cholinesterasehemmer.

> Eine 24jährige Kunststudentin, die seit zehn Jahren unter einer Myasthenie mit hohen AK-Titern leidet, erlebt nach der Thymektomie in Schwellensituationen (z.B. Abitur) mehrere myasthene Krisen mit anschließenden Krankenhausaufenthalten bis zu einem Jahr, so daß ihr strebsames Studentenleben schließlich einer Existenz im Rollstuhl bei den überfürsorglichen Eltern Platz macht. Wegen „Unwirksamkeit" von Mestinon und Imurek nimmt sie seit Jahren zusätzlich Prostigmin ein. Nach einer stationären psychosomatischen Therapie, in der sie ihre myasthenen Symptome von den angstneurotischen Körperstörungen zu unterscheiden, z.B. ihre vornehmlich am Morgen eskalierende „Schwäche" als Angst vor Autonomie verstehen lernt, kann sie bei verbliebenen milden myasthenen Beschwerden auf Rollstuhl und Prostigmin verzichten.

53.8.3 Psychotherapie bei Myasthenia gravis?

Es fehlen bisher kontrollierte Studien, die die Wirksamkeit psychotherapeutischer Intervention auf den Verlauf dieser Erkrankung belegen. Spontanremissionen treten bei 5–15% der Patienten auf (Grob et al., 1981; Emeryk et al., 1985). Trotzdem finden sich eine Reihe kasuistischer Mitteilungen, die die mancherorts vertretene Behauptung, Psychotherapie sei bei Myasthenie obsolet (Schumm, 1988), als voreilig erscheinen lassen.

Im Gegensatz zu oberflächlicher Befragung ohne psychotherapeutischen Umgang und zu einigen psychometrischen Befunden zeigt sich in der Psychotherapie: Seelische Erregung führt nicht zu Verstärkung der myasthenen Symptome, sondern umgekehrt, insbesondere wenn die Patienten sich verstanden fühlen, zu drastischer, zum Teil anhaltender Besserung. Symptome und seelische Spannungen scheinen sich gegenseitig zu vertreten. Ebenso hat, wie die Krankengeschichten zeigen, körperliche Anstrengung entgegen gängiger Auffassung nicht zwangsläufig einen ermüdenden, sondern unter „lustbringenden" Umständen einen deutlich mobilisierenden Effekt, wobei dem Tanzen immer wieder besondere Bedeutung zukommt.

Eine fast regelmäßig beobachtbare lebhafte Traumtätigkeit von Myastheniepatienten könnte – im Gegensatz zur Traumarmut der über-sthenischen Lumbago-Ischialgie-Patienten (vgl. Kap. 45.3) – psychosomatisch weiter aufschlußreich sein. Träume erleichtern die aufdeckende Psychotherapie auch bei anfänglicher Verschlossenheit der Patienten.

Eine symptomorientierte Psychotherapie sollte also eine medikamentöse Behandlung in jedem Fall begleiten, auch in myasthenen Krisen. In der neurologisch-psychosomatischen Praxis haben sich tägliche tabellarische Protokolle bewährt, in denen die Patienten ihre Beschwerden, ihre Stimmung, die Tagesereignisse und die eingenommenen Medikamente gegenüberstellend aufzeichnen. Ein solches Protokoll schärft die Wahrnehmung für myasthene Schwankungen sowie für den Unterschied myasthener und psychogener Symptome.

Die neurohumorale Erschöpfung ist offenbar nicht nur abhängig von Muskelanstrengung, sondern auch Ausdruck eines mangelnden Widerstandes gegen fremde Einflußnahme. Entwickelt sich unter der Behandlung (und mit Hilfe des Protokolls, das die Funktion eines Übergangsobjekts annehmen kann) die Eigeninitiative des Patienten, so werden die Muskelfunktionen wieder verfügbar. Auch die Cholinesterasehemmer können für den Umgang mit Autonomie und Abhängigkeit genutzt werden, insbesondere wenn der unbewußte Aspekt des Übergangsobjekt-Charakters der Tablette verstanden wird. Mestinon retard® ist dabei ungünstig. Die Patienten nehmen statt dessen ihre gewohnte Dosis oder einen Teil derselben in 10-mg-Portionen, die sie sich nach Bedarf und Symptomausprägung selbst einteilen. Dabei ist häufig eine eindrucksvolle Dosisreduktion und gesteigertes Wohlbefinden zu erzielen.

54 Pädiatrie

Dieter Bürgin und *Barbara Rost*

54.1 Allgemeine Gesichtspunkte

Psychosomatische Symptome können reine Nebengeräusche einer durchaus normalen Entwicklung sein, die belanglos sind und spontan verschwinden, oder aber die bestmögliche Konfliktlösung repräsentieren, die vom Kind oder Jugendlichen selbst ohne Hilfe in der jeweiligen Entstehungssituation gefunden wurde. Als eine Art Selbstheilungsversuch lassen sie sich als kreative Akte betrachten, die meist zwiespältige Tendenzen in sich vereinen.

Entsprechend den jeweils gebrauchten **Modellvorstellungen** ergeben sich verschiedene Verständniskonzepte für psychosomatische Störungen. Im folgenden wird hauptsächlich auf den Erfahrungsbereichen der psychoanalytischen Entwicklungspsychologie, der Psychoanalyse, der Kinderpsychiatrie und der Pädiatrie aufgebaut. Neben den neueren Ergebnissen der Arbeitsgruppen um Emde und Stern, welche vor allem für das Gebiet der Entwicklungspsychologie fruchtbar waren, haben im Bereich der klinischen Psychosomatik die Mitglieder der „Société psychanalytique de Paris" (P. Marty, M. Fain, C. David, M. de M'Uzan, L. Kreisler, M. Soulé und S. Lebovici), sowie für Kleinkinder und Kinder die Mitarbeiter des „Institut psychosomatique de Paris" viel beigetragen.

Jede **Belastung** des Ichs, welche seine momentane Verarbeitungsfähigkeit überschreitet, stellt eine Überforderungssituation dar. Es handelt sich dabei um einen relativen Begriff, der von der Art der Belastung ziemlich unabhängig ist, hingegen vom Alter des Patienten, von seiner Ich- (d.h. Verarbeitungs-) Stärke und von der puffernden Wirkung der Umwelt mitbestimmt wird. Dennoch gibt es relativ klassische Belastungssituationen, z.B.: jeder Entwicklungsschritt (z.B. der Eintritt in den Kindergarten, die Schule, die Manifestation der Pubertät oder auch ein Wechsel des Wohnorts bzw. der Sprache); jeder Verlust eines realen äußeren Objektes durch Tod, Trennung oder Scheidung der Eltern; viele Krankheiten; schwerer Schmerz; und schließlich auch die Entbehrung entwicklungsnotwendiger Kontinuität der emotionalen Fürsorge (emotionale Deprivation). Die Belastung selbst ist in ihrer Auswirkung außerordentlich abhängig von der Erwartungslage des Kindes, von der Möglichkeit einer vorwegnehmenden Verarbeitung (worry work) und auch von der stützenden Hilfsfunktion, die Erwachsene ausüben können. Eine Bewältigung im voraus kann wie eine Art psychologischer Immunisierung (Kliman, 1973) verstanden werden, bei der die traumatisierenden Belastungsfaktoren in zuträglichen Dosen mittels der Phantasie bearbeitet werden können. Wie eine potentiell traumatisierende Situation in der Phantasie vorstrukturiert werden kann, ist im psychosomatischen Modell des Situationskreises zusammengefaßt (vgl. Kap. 1). Je nach Art der Anpassungs- oder Abwehrleistungen, welche durch die Belastungen ausgelöst werden, sprechen wir von psychiatrischen, psychosomatischen Symptomen/Syndromen oder von Verhaltensstörungen.

54.1.1 Symptombeurteilung

Bei der Beurteilung von psychosomatischen Symptomen im Kindes- oder Jugendlichenalter ist immer davon auszugehen, daß die Betroffenen, infolge ihrer raschen seelischen Entwicklung, Zeiten spezifischer Verletzlichkeit durchlaufen. Folgende Punkte verdienen Beachtung, um die Wertigkeit eines Symptoms abschätzen zu können:

- Alter und Geschlecht des Patienten (z.B. hat es keinen Sinn, von einer Enuresis nocturna vor dem Ablauf des 4. Lebensjahres zu sprechen. Eine Trennungsangst ist bei einem 2jährigen durchaus angemessen, bei einem 10jährigen nicht mehr);
- Dauer, Häufigkeit und Intensität eines Symptoms;
- besondere Lebensumstände (z.B. die Geburt eines Geschwisters, der Schuleintritt eines Kindes etc.);
- Art des sozio-kulturellen Milieus, in welchem ein Kind aufwächst;
- Zahl der psychischen Bereiche, welche durch die Symptomatik betroffen sind;
- Verbindung eines Symptoms mit anderen psychischen Störungen (z.B. hat Nägelbeißen keinen signifikanten Hinweischarakter, hingegen sind Beziehungsstörungen sehr wohl häufig mit anderen Symptomen verknüpft) (Rutter, 1975);
- Verknüpfungen der vorliegenden psychosomatischen Symptome mit Verhaltensänderungen;
- situationsspezifisches Auftreten (z.B. eine Enuresis diurna nur zu Hause, nie im Kindergarten);
- Miteinbezug anderer psychischer, somatischer oder sozialer Funktionsbereiche (z.B. eine hysterische Beinlähmung, welche einen Schulbesuch unmöglich macht);
- subjektives Leiden des Kindes oder der Eltern unter der Symptomatik;
- soziale Restriktionen im Zusammenhang mit der psychosomatischen Symptomatik;
- Beeinträchtigung der psychischen Entwicklung des

Kindes und/oder der Familie durch die vorliegende Symptombildung;

- Beeinträchtigung der weiteren Umgebung durch die psychosomatischen Symptome (z.B. bei Enkopresis).
- Was soll das Symptom wem sagen? Wem dient es wozu?

Psychosomatische Erscheinungen beim Kind und Jugendlichen können nicht verstanden werden, wenn nicht der Entwicklungsstand der Emotionen, Affekte und Stimmungen und ihre Bedeutung in der Persönlichkeitsorganisation des Patienten berücksichtigt werden; wenn nicht ein Bild über die Rolle entsteht, welche die bedeutungsvollen Dialogpartner für den Ausdruck der Gefühle spielen; wenn nicht eine qualitative und quantitative Abschätzung der Bedeutung des emotionalen Dialoges im Hinblick auf die Ich-Funktionen erfolgen kann; wenn nicht deutlich werden kann, welche vor- oder frühzeitigen Phantasiebildungen über den Körper vom Kind aufgebaut worden sind.

54.1.2 Zur Definition/Klassifikation

Wird Psychosomatik definiert als ein Konzept, das psychische Faktoren und Konflikte in der Entstehung und/oder der Entwicklung organbezogener, läsioneller oder funktioneller physischer Krankheiten anerkennt und einbezieht, so ist eine anhaltende Oszillation zwischen der direkten Beobachtung des Patienten und dem progredienten Verständnis der klinischen Phänomene im Rahmen einer bestimmten Orientierung oder Theorie notwendig. Nur durch diese steten Bewegungen ist eine Dechiffrierung von Verhalten auf der einen Seite und auf der anderen die Vermeidung des Entstehens eines mythischen, d.h. ganz von der Theorie her gestalteten Patienten möglich. Hierdurch besteht aber auch die Gefahr eines epistemologischen Durcheinanders zwischen einer somatischen, einer psychiatrischen und einer sozialen Nosologie. Die Pathophysiologie gibt teilweise Antworten auf das „Wie", die Psychopathologie manchmal auf das „Warum" der Entstehung einer Störung.

Werden Konzeptualisierungen auf der Achse **narzißtische Besetzung/Objektbesetzung** vorgenommen, so bewegt man sich im Bereich eines dynamischen Gleichgewichtes der Besetzungsmodalitäten und ihrer Regressionen und gelangt zu Aussagen über Beziehungsqualitäten innerer oder äußerer Art. Konzeptualisierungen auf der Achse **Es/Ich/Über-Ich** bewegen sich vor allem im Bereich der klassischen Frühstrukturen und komplexer infantiler Phantasmen und umfassen das Modell der Konversion, welches sich auch auf prägenitale Konfliktkonfigurationen erstreckt, aber stets die relativ elaborierte Struktur eines neurotischen Ichs beansprucht. Konzeptionen auf der Achse des **Unbewußten/Vorbewußten/Bewußten** kümmern sich in erster Linie um Funktionen und Strukturdefekte im Vorbewußtsein und damit besonders um die ökonomischen Aspekte des psychosomatischen Geschehens. Das Vorbewußte als Drehscheibe psychischer Abläufe bestimmt mit seiner Struktur auch die Organisation der Abwehrvorgänge (Kreisler, 1981). Bei Borderline-Patienten sind die vorbewußten Strukturen fragil und durchlässig, so daß es zu wechselnden Ich-Funktionszuständen kommt. Größere Defekte und Funktionsstörungen im Vorbewußten, welche bis zur Isolierung unbewußter Vorgänge vom Bewußtsein führen, stellen die Basis der sog. „pensée opératoire" dar (Marty et al., 1963).

Die **Klassifikation** psychosomatischer Störungen beim kindlichen und jugendlichen Patienten ist noch bedeutend komplexer als beim Erwachsenen, weil sich die Patienten in rascher psychophysischer und sozialer Entwicklung mit Zeiten spezifischer Verletzlichkeit befinden und der Grad der Desomatisierung von Affekten/Konflikten entsprechend unterschiedlich ist. Aber auch, weil es für die Symptomatik des Kindes von eminenter Bedeutung ist, wie die Eltern auf körperliche Symptome reagieren und in welchem Maße sie selber seelische Befindlichkeit in körperlichen Ausdruck umsetzen.

54.1.3 Kinderwunsch, Schwangerschaft, Entwicklung des emotionalen Dialogs, primäre Mütterlichkeit

Die Erforschung psychosomatischer Störungen beim Kind und Jugendlichen hat auf früheste Wurzeln der Entwicklung zurückgeführt. Bereits bei der **Vorstellung eigener Elternschaft** treffen kulturelle, historische, familiäre und individuelle (psychische und somatische) Faktoren aufeinander. Bei den potentiellen Eltern mischen sich u.a. Phantasmen und Wünsche aus der frühkindlichen und postpubertären Sexualität mit Strebungen nach Unsterblichkeit oder Weitergabe eigener Werte und Erfahrungen, die zur Ausgestaltung eines inneren Bildes vom zukünftigen Kind beitragen. Im Laufe der Schwangerschaft entwickelt sich in beiden Eltern ein imaginiertes Kind (Ross, 1967), das nicht selten einem idealen Komplement des eigenen Selbst entspricht. Das Neugeborene trifft als Person mit eigener Realität mit diesen beiden imaginierten Kindern („Kind im Kopf", wie Soulé [1989] dies nennt) zusammen. Es muß diese Konfrontation bestehen oder wird nach dem einen oder anderen imaginären Modell, dem es nie entsprechen kann, umzuformen versucht. Im allgemeinen ist die Realität des Kindes stark genug, um die Phantasien der Eltern in den Hintergrund zu drängen. Ist im Extrem die Diskrepanz zwischen imaginiertem und realem Kind zur Zeit der Geburt aber noch groß, so sind Enttäuschung und Abweisung fast unvermeidbar; oder entspricht das reale Kind dem imaginierten allzusehr, so ist der Weg für eine übermäßige Idealisierung geebnet.

Beide Eltern müssen also bei der Geburt einen Besetzungsabzug von ihrem imaginierten Kind vornehmen und, neben der Freude über ihr Neugeborenes, eine Art Verlustarbeit leisten und das Ende einer Illusion anerkennen. Nur dann gelingt es ihnen, den **emotionalen Dialog** mit ihrem realen, in seiner Einzigartigkeit festgelegten Kind aufzubauen und zu ihrem neuen Status der „Elternschaft" zu stehen.

Die Aktivierung eines Zustandes, den Winnicott (1960) die **„primäre Mütterlichkeit"** nannte, der von außen gesehen fast einem pathologisch gesteigerten Einfühlungsvermögen gleichkommt und der mit einer Tendenz zu verstärktem, primärprozeßartigem Denken verknüpft ist (Condon, 1987), macht die Mutter für die Signale des realen Kindes besonders empfänglich. Wenn es sich aber z.B. um ein sehr aktives Kind handelt und die Mutter ein ruhiges Wesen erwartete, das sanft und gemächlich seine Umwelt erkundet, so fühlt sie sich durch seine Triebhaftigkeit, sein Schreien, seine Saug- und Gliederbewegungen bald invadiert und beeinträchtigt. Ist der reale Säugling bewegungsarm, still und wenig aktiv und hatte die Mutter, entsprechend ihrem imaginierten Kind, ein real sehr forderndes, oral zupackendes Kind erwartet, so läßt die Enttäuschung nicht lange auf sich warten. In beiden Extremfällen ist die Störung des emotionalen Dialoges rasch etabliert.

54.1.4 Das Neugeborene, der Säugling

Das Neugeborene zwingt durch seine Realität die Mutter zum Handeln, zu Neubesetzungen olfaktorischer, taktil-kinästhetischer, akustischer und visueller Art, und erreicht bestenfalls, daß bestimmte Anteile des realen emotionalen Dialoges mit der Mutter Ähnlichkeiten tragen mit dem zwischen dem idealisierten imaginierten Kind und der Mutter. In nicht wenigen Fällen erhält das reale Kind auf diese Weise ein Stück idealisierter Besetzung, die aus der Beziehung der Mutter zum imaginierten Kind stammt („mein Kind ist der schönste Säugling der Welt"). Dies gibt beiden Partnern Zeit, um die Idealisierung allmählich abzubauen und den realen Zügen mehr Platz einzuräumen, ein Vorgang, der oft bis zum Abschluß der Adoleszenz dauert!

Der Säugling kommt, von der Evolution vorangepaßt, mit einem beachtlichen Repertoire von Verhaltensweisen zur Welt, die im Kontext der Beziehung zu den primären Pflegepersonen aktiviert werden. Seine Entwicklung wird zwar durch genetische Faktoren organisiert, er selbst ist aber auch ein aktiv mitorganisierendes Wesen. Viele Entwicklungsabläufe in den ersten Jahren weisen auf angeborene und damit genetische Faktoren hin. Der Aktivitätszustand der Gene ist in der Entwicklung größtenteils genetisch festgelegt, wird aber auch durch Wechselwirkungen des Individuums mit der Umwelt mitgestaltet. Nicht die Muster eines kindlichen Verhaltens entwickeln sich in einer bleibenden Art, sondern die Muster der Beziehung zwischen dem Säugling und der primären Pflegeperson, welche später als Ganzes vom Kind internalisiert werden. Es werden also Beziehungsaspekte verinnerlicht, die während der gesamten Kindheit eine starke Auswirkung haben und über das ganze Leben hinweg in entsprechenden Beziehungskontexten reaktiviert werden können. Auf diese Art und Weise können Beziehungsformen zwischen Kleinkind und Mutter über mehrere Generationen hinweg weitergegeben werden.

54.1.5 Grundgegebenheiten der frühesten Kindheit

Zu den angeborenen, universal vorhandenen und zeitlebens wirksamen Grundgegebenheiten der frühesten Kindheit gehören
– Aktivität,
– die Fähigkeit zur Selbstregulation,
– die Bereitschaft für soziale Interaktion und
– die Gliederung der Gefühlserfahrungen (Emde, 1988).

Aktivität: Der Säugling wird mit einem Entwicklungsplan geboren, der sich systematisch entfaltet und zu immer höheren Integrationen, sowohl im neurophysiologischen als auch im Bereich des Verhaltens, führt. Er ist z.B. biologisch darauf vorbereitet, visuelle Reize zu suchen, um das Gehirn zu stimulieren, neuronale Verbindungen zu fördern und die Reifung des Zentralnervensystems zu beschleunigen. (Mit seinem Blick tastet er z.B. einen optischen Stimulus regelrecht ab.) Bereits ab der 7. Woche organisiert er seine optischen Eindrücke in eigentliche Gestaltungen oder Ganzheiten, erkennt Gesichter, richtet sein Hauptaugenmerk auf die Augen, besonders, wenn aus dem Gesicht noch eine Stimme spricht. Mit 2 Monaten, unabhängig von jeglicher Verstärkung durch die Außenwelt, beginnt er mit einer räumlichen und zeitlichen Gliederung seiner Welt, was sich in antizipatorischen Augenbewegungen zeigt, wenn attraktive Bilder links oder rechts, in einem bestimmten Rhythmus, gezeigt werden. Der Säugling sucht eine Gesetzlichkeit und Regelmäßigkeit solcher Abläufe zu erkennen, baut darauf eine vorwegnehmende Erwartung auf und richtet seine Aktivität dieser entsprechend aus. Aktivität entspricht also einem Bedürfnis, sensorisch-motorische Systeme anzuregen, sie zu üben.

Selbstregulation: Eine solche findet statt im Bereich der Physiologie, aber auch im Verhalten (z.B. Wachheit, Aufmerksamkeit, Schlaf-Wach-Zyklus, Wachstum, Entwicklung). Das Kind besitzt zudem eine Fähigkeit zur Selbstkorrektur, die sich auf wichtige Funktionen bezieht und die bei großen Mankos oder Herausforderungen aktiviert werden kann (z.B. kann eine schwere Deprivation in früher Kindheit unter günstigen Umständen später zum Teil noch korrigiert werden [Emde, 1982]).

Bereitschaft für soziale Interaktion: Das Kind kommt mit einer angelegten Bereitschaft für soziale Interaktionen zur Welt. Es besitzt eine organisierte Fähigkeit, eine Interaktion mit anderen menschlichen Wesen anzufangen, sie aufrechtzuerhalten oder sie zu beenden. Es kann schon kurz postnatal sequentielle Reize integrieren, eine Art Mittelwert bilden und sich auf diesen einstellen, und es vermag gewisse Charakteristika von Erfahrungen aus einer perzeptiven Modalität in eine andere zu transponieren sowie komplexe motorische Abläufe zu imitieren (z. B. Zunge herausstrecken). Da sich auch auf der Seite der primären Pflegeperson, insbesondere der Mutter, eine solche Bereitschaft verstärkt ausbildet (z. B. Augenkontakt unterstützen, Gesichtsausdruck und Töne des Säuglings imitieren), kann von einer eigentlichen Verhaltenssynchronizität gesprochen werden (Emde, 1988). Vom 6. Monat an folgt das Kind oft dem Blick der Mutter. Es besteht also auch ein angeborenes Vermögen für ein gemeinsames Betrachten einer optischen Realität.

Gliederung der Gefühlserfahrung: Auch die Fähigkeit, Erfahrungen nach ihrer affektiven Qualität von Lust oder Unlust zu gliedern, scheint hereditär angelegt zu sein. Für die Mutter sind die Gefühlsäußerungen des Kindes entscheidende Wegleiter ihrer Pflegeleistungen; für das Kind spielt in der Wahl seines Verhaltens zunehmend eine Rolle, ob die Mutter lust- oder unlustvoll auf seine Aktivität reagiert oder nicht (Bürgin, 1989b).

54.2 Psychodynamische Hypothesen zur Selbstentwicklung

Die Selbstrepräsentanz steht in einem fortgesetzten Entwicklungsprozeß, in dem, durch den ganzen Lebenszyklus hindurch, zunehmend komplexe Erfahrungs- und Erlebnisanteile synthetisiert und integriert werden müssen. Die Entfaltung angeborener Fähigkeiten durch Reifung und Entwicklung (optimale Stimulation, Aufbau von Erwartung und geringfügige Abweichung von dieser) läßt ein **Kernselbst** entstehen, das folgende Anteile umfaßt (Stern, 1985):

- Ein Gefühl von **Eigenaktivität,** das auf der Basis von Propriozeptivität dort entsteht, wo eigene Intentionalität erlebt, überschaubare Ursache-Wirkungs-Zusammenhänge erkennbar und die eigene Motorik zur Erweiterung des Erfahrungsbereiches gebraucht werden kann.
- Ein Gefühl von **Kohärenz** in psychischen, zeitlichen Bewegungs- und Intensitätsabläufen.
- Ein Gefühl von **Kontinuität,** das an die Gedächtnisentwicklung gebunden ist.
- Ein **affektives Signalsystem,** welches die Einschätzung von Wahrnehmung, die Gliederung der Verarbeitungswege von Information und die Auswahl von Aktivität entlang den Kraftlinien des Lust-Unlust-Prinzips mitgestaltet.

Es gibt eine Gruppe von grundlegenden Emotionen, die bereits im 1. Lebensjahr vorhanden, biologisch vorgegeben sowie ubiquitär zu beobachten sind und im Gesicht ausgedrückt und erkannt werden können (z. B. Freude, Wut, Traurigkeit, Ekel, Überraschung und Interesse). Sie bilden die Grundlage einer allgemein menschlichen Kommunikation, eines emotionalen Kontaktes mit dem anderen, und werden vom 1. Lebensjahr an bis zum Tod mit der Umwelt geteilt. Im 3. Monat ist der emotionale Ausdruck bereits entsprechend Lust/Unlust-Aspekten, Aktivität/Passivi-

tät und Innen- oder Außenwendung anhaltend organisiert. Diese Affekte geben dem Erleben Kontinuität und stellen emotionale Signale zwischen Säugling und primärer Pflegeperson dar. Sie sind Grundelemente für die Übermittlung von Bedürfnissen, Absichten und Befriedigungen. Positive Emotionen (Freude, Überraschung, Interesse) sind für die Entwicklung extrem wichtig und werden getrennt von den negativen organisiert. Sie können als Aktivatoren für soziale Interaktion, Exploration und Lernen bezeichnet werden und bedürfen eines umschriebenen Kontextes, um verstanden zu werden. Diese Grundemotionen sind für die „affektive Einstimmung" (affect attunement) und die „Erkundung im Sozialbezug" (social referencing) von zentraler Bedeutung (Bürgin, 1987).

54.2.1 Erkundung im Sozialbezug

Die Erkundung im Sozialbezug (Emde, 1988) ist ein emotionaler Prozeß, bei dem eine Person irgendwelchen Alters emotionale Information von einem bedeutungsvollen anderen sucht, um eine Gegebenheit oder Befindlichkeit zu verstehen, die unklar oder vieldeutig ist und über dem eigenen Klärungsvermögen liegt. Es besteht eine Situation von Unsicherheit, welche durch die eingeholte emotionale Information vom signifikanten anderen verringert werden soll. Dieses emotionale Kommunikationsphänomen ist wichtig über das ganze Leben hinweg. Mittels der Gefühle des anderen kommt es zu einer vikariierenden Lernerfahrung. Der affektive Kern des einen kommt in Kontakt mit dem des anderen. So entsteht Konstanz und Umformung zugleich in zwei sich verschränkenden affektiven Selbst. Die Erkundung im Sozialbezug ist in den ersten 6 Lebensmonaten eines Kindes in erster Linie eine, die von der Mutter ausgeht. Diese versucht sich zu orientieren, in welchem inneren Zustand, in welcher emotionalen Befindlichkeit sich der Säugling befindet. Vom 6. bis zum 18. Monat kehrt sich die Situation weitgehend um. Das Kind braucht emotionale Signale von der Mutter, um seine Unsicherheit zu reduzieren und sich zu orientieren. Wenn solche Situationen von Unsicherheit experimentell geschaffen werden, so zeigt sich, wie sehr der Ausdruck von Angst im Gesicht der Mutter prohibitiv, der von Freude und Interesse stimulativ für die weiteren exploratorischen Aktivitäten des Kindes ist. Wenn z. B. ein Kleinkind ein Zimmer exploriert und dabei auf eine Situation trifft, die ihm unvertraut ist, so schaut es zur primären Pflegeperson. Zeigt diese im Gesicht Angst oder Wut, so vermeidet das Kind die neue Situation. Zeigt sie Freude oder Interesse, so wagt sich das Kind weiter und untersucht die neue Situation. Unter bestimmten Bedingungen können auch Substitutspersonen vom Kind als „affektive Informanten" genutzt werden. Nicht nur der Gesichtsausdruck, sondern auch die Stimme kann prohibitiv oder encouragierend wirken. Wenn allerdings die Aussage des Gesichtes und die der Stimme widersprüchlich sind, so entsteht ein Dilemma für das Kind, das oft nur schlecht zu lösen ist und, wenn internalisiert, entwicklungsbeeinträchtigend wirken kann. Daß ein Kind beim Kontakt mit einer schwer depressiven oder psychotischen Mutter bezüglich seinen Erkundungen im Sozialbezug massiv eingeschränkt ist, liegt auf der Hand. Aktiv abweisende oder in hohem Ausmaß widersprüchliche emotionale Echosignale dürften bei der Entwicklung eines „falschen Selbst" eine nicht unwesentliche Rolle spielen. Zwischen dem 18. und 36. Monat wird das bei der Erkundung im Sozialbezug erhaltene emotionale Signal insbesondere vom Kind weiter auf seine Bedeutsamkeit und Intensität experimentell handelnd getestet. Es wird, gleichsam konflikthaft verhandelnd, vom Kind gefragt: Meinst du das wirklich? Je mehr Internalisierungen und Identifikationen stattfinden, desto mehr wird ein Teil dieser Erkundungen im Sozialbezug auch in den Innenraum des Kindes, in die inneren phantasmatischen Dialoge verlegt: die Selbstrepräsentanz orientiert sich in Unsicherheitssituationen am Über-Ich – ein lebenslanger Vorgang.

54.2.2 Affektive Einstimmung

Neben der Erkundung im Sozialbezug kommt der affektiven Einstimmung (Stern, 1984) für den Aufbau und die Konsolidierung der Selbstrepräsentanz eine weitere besondere Bedeutung zu. Damit der psychische Zustand eines Menschen für einen anderen erkennbar werden kann, ist es nötig, wenn der eine Partner ein Kleinkind ist, daß die emotionale Information auf nichtverbalem Wege übermittelt wird. Hierzu muß der seelische Zustand erst in Form eines offen sichtbaren Verhaltens manifest werden, und dann muß diese Art von Verhalten übersetzbar sein. Neben der „Empathie", der „phantasierten Interaktion" und dem „Spiegeln" dürfte das affektive Sich-Einstimmen ein weiterer Weg sein, um über eine Gefühlsqualität in einem Austauschdialog zu stehen. Vor dem 9. Lebensmonat, solange es in der Innenwelt des Kindes noch keine ausgeformten Repräsentanzen gibt und kein eindeutiges Gefühl für Getrenntheit besteht, existiert nur eine Art indirekte Identifikation des Säuglings mit den Gefühlen der Mutter und umgekehrt, eine reziproke affektive Bezogenheit zwischen beiden. Die affektive Einstimmung bezeichnet ein Geschehen, das die Qualität von Gefühlen eines gemeinsam erlebten Gefühlszustandes reflektiert, aber nicht mit einer identischen Wiedergabe des inneren Zustandes des einen durch den anderen Partner. Die Ausdrucksmodalität, welche die Mutter zum Reflektieren gebraucht, ist bezüglich gewisser Eigenschaften zwar derjenigen, die das Kind gebraucht, ähnlich, aber auch klar unterschieden davon. Sie entspricht nicht so sehr einer Reaktion auf das Verhalten des Kindes, als vielmehr auf gewisse Aspekte seines inneren Gefühlszustandes. Diese Einstimmung ist abzugrenzen von der Imitation (möglichst originalgetreue Reproduktion des Verhaltens von einem der Austauschpartner durch den anderen) und vom Spiegeln (völlige Gleichzeitigkeit der Geschehnisse). Sie stellt

eine besondere Form der **Intersubjektivität** dar. Die meisten Gefühlseinstimmungen der Mütter erfolgen in einer anderen sensorischen Modalität als der Ausdruck des Kindes. Wenn z.B. das Baby sich oral manifestiert, so reagieren die Mütter z.B. gestisch oder mimisch oder vice versa. Die Einstimmung erfolgt vor allem in bezug auf die Intensität des Gefühlsausdrucks, den Zeitverlauf oder das Profil. Hauptmotiv der Mütter für eine affektive Gefühlseinstimmung ist der Wunsch nach einer interpersonalen Gemeinsamkeit. Das gemeinsam Geteilte treibt die Entwicklung voran. Die transmodale Umformung des Gefühlsausdrucks des Kindes durch die Mutter gibt dem Kind ein Gefühl des emotionalen Wahrgenommenwerdens durch den anderen. Gefühlszustände eines Kleinkindes, auf die nie eine affektive Gefühlseinstimmung der Mutter erfolgte, bleiben eine Erfahrung, die nur allein gemacht werden konnte und die von einem interpersonalen Kontext ausgespart war. Das Kind besitzt ein gewisses Sensorium für das Ausmaß und die Güte einer geteilten emotionellen Erfahrung und kann zum Ausdruck bringen, daß die Störung der Gemeinsamkeit bedeutungsvoll ist.

54.2.3 Teilen/Wir-Gefühl

Vom 6. Lebensmonat des Kindes an lenken die Eltern mit Lob und Anerkennung einen Teil seiner Aktivitäten in der Weise, daß es Ziele und Erwartungen der Eltern erfüllen soll. Schließlich beherrscht das Kind die Verfolgung eigener Zielsetzungen und auch die der Eltern. Zwischen Beginn und Mitte des 2. Lebensjahrs ist ein Bedürfnis des Kindes zu beobachten, positive Affekte mit der primären Pflegeperson zu **teilen**. Das Teilen negativer Emotionen ist in Populationen, die unter Belastung oder Risiken stehen, bedeutend häufiger. Beide Interaktionsabläufe können internalisiert werden. Im Alter von 36 Monaten sind Verbote bereits so internalisiert, daß sie auch dann eingehalten werden, wenn die primären Pflegepersonen für kürzere Zeit abwesend sind oder wenn die Gebote durch andere Personen im spielerischen Kontext in Frage gestellt werden. In einer Situation, in der zwischen einer prosozialen Handlung und einer anders lautenden, verbindlichen Regel gewählt werden soll, entsteht ein moralisches Dilemma. Vielfach entscheidet sich das Kind dann für die prosoziale Handlung. Auch wenn ein von der primären Pflegeperson ausgesprochenes Verbot in Abwesenheit dieser in Frage gestellt wird, entsteht eine Konfliktsituation. Widersteht das Kind der Versuchung, so dokumentiert es ein **„Wir"-Gefühl** zwischen sich und der primären Pflegeperson, das ihm die Empfindung einer gesteigerten Beherrschung und Kontrolle vermittelt. Internalisierte Regeln schaffen somit eine zwischenmenschliche Welt mit **gemeinsam geteilten Bedeutungen**. Emde (1988) spricht von der Notwendigkeit, neben der Selbst- und der Ich-Psychologie auch eine „Wir"-Psychologie aufzubauen. Das Wir-Gefühl entspricht einer aktiven Erfahrung gemeinsam geteilter Wirklichkeit mit einem bedeutungsvollen anderen.

Sein Anfang liegt dort, wo sich eine Intersubjektivität zu entwickeln begonnen hat, also zwischen dem 7. und 9. Lebensmonat. Etwa ab dem 12. Monat verschränken sich zwei psychische Welten, Absichten und Zielsetzungen des einen beginnen sich mit denen des anderen abzugleichen, sofern ein Kontext gemeinsamer Aufmerksamkeit und gemeinsamen Fühlens etabliert worden ist. Emotionen haben nun nicht mehr nur Signalcharakter, sondern werden auch gebraucht, um entsprechende Antworten zu bekommen und um bestimmte Dinge mit dem bedeutungsvollen anderen auszuhandeln.

54.2.4 Wahres und falsches Selbst

Das Selbst des Säuglings ist im ersten Halbjahr seiner Existenz nur potentiell vorhanden, man könnte sagen, es sei mit dem der Mutter verschmolzen. Im emotionalen Dialog mit der primären Pflegeperson erwacht gleichsam das **wahre Selbst** zum Leben. „Das wahre Selbst erscheint, sobald es auch nur irgendeine psychische Organisation des Individuums gibt, und es bedeutet wenig mehr als die Gesamtheit der sensomotorischen Lebendigkeit" (Winnicott, 1974a). Im wahren Selbst wird nach Winnicott eine Kontinuität des Seins erlebt. Diese erst macht, in eigener Weise und Geschwindigkeit, den Erwerb einer personalen seelischen Realität und eines eigenen inneren Raumes möglich. Übergriffe bedrohen dieses Sein mit Vernichtung oder nötigen das Kind zum Reagieren und unterbrechen damit die Kontinuität des Seins. Die Hauptfunktion der haltenden Außenwelt besteht darin, Übergriffe auf ein Minimum zu reduzieren. Die Mutter verhilft dem Säugling durch ihre Fürsorge (z.B. angemessene affektive Einstimmung, Möglichkeit zu Konstanz in der Erkundung im Sozialbezug) zum Erleben der Illusion und der Omnipotenz. Störungen der Anpassung der Umwelt an die frühkindlichen Bedürfnisse des Kindes können zum Aufbau einer falschen Existenz führen. Ein falsches System von Beziehungen, das sich den Gegebenheiten der Umwelt übermäßig fügt, wird mit dem Ziel aufgebaut, das wahre Selbst zu schützen und zu verbergen. Auf diese Art und Weise ermöglicht das **falsche Selbst** dem wahren Selbst zu überleben und der Vernichtung zu entgehen. „Wenn die Umwelt sich nicht gut genug verhält, wird das Individuum zu Reaktionen auf Übergriffe veranlaßt, und die Prozesse des Selbst werden unterbrochen. ... Während das wahre Selbst geschützt wird, entwickelt sich ein falsches Selbst, das auf der Grundlage von Abwehr und Gefügigkeit, auf der Annahme der Reaktion auf Übergriffe aufgebaut ist. Die Entwicklung eines falschen Selbst ist eine der erfolgreichsten Abwehrorganisationen, die den Kern des wahren Selbst schützen soll, und ihr Vorhandensein ruft das Gefühl der Vergeblichkeit hervor" (Winnicott, 1976). „Wenn sich ein falsches Selbst in einem Individuum mit einem hohen intellektuellen Potential aufbaut, besteht eine starke Tendenz, daß der Intellekt der Ort des falschen Selbst wird, und in diesem Falle entwickelt sich eine Dissoziation zwischen intellektueller Aktivität und psychosomatischer Existenz" (Winnicott, 1974 b). Das daraus entstehende klinische Bild täuscht. Hohe schulische Erfolge können dann schweres reales Leiden kaschieren. Die defensive Verwendung eines erfolgreichen falschen Selbst „befähigt manche Kinder, so zu erscheinen, als seien sie vielversprechend, aber am Ende offenbart ein Zusammenbruch den Umstand, daß das wahre Selbst nicht vorhanden ist" (Winnicott, 1974 c). Die Funktion des falschen Selbst besteht darin, das wahre Selbst verborgen zu halten. Das Individuum existiert dadurch, daß es nicht gefunden wird. „Ein falsches Selbst kann sich gut in die Familienstruktur einfügen oder vielleicht zu einer Krankheit der Mutter passen, und man kann es sehr leicht mit Gesundheit verwechseln. Es trägt jedoch Instabilität und eine Neigung zu Zusammenbrüchen in sich" (Winnicott, 1978). Ein Zusammenbruch seelischer Funktionen kann dann ein gesundes Zeichen sein, ein Hinweis auf die Hoffnung, eine neuerlich verfügbare Umwelt benützen zu können, um eine Existenz auf einer Grundlage wieder aufzubauen, die sich real anfühlt.

54.2.5 Entwicklungspsychologie und Psychosomatik

Der Einbezug solcher entwicklungspsychologischer Aspekte ist insbesondere für die Beurteilung psychosomatischer Geschehnisse beim Säugling und Kleinkind unumgänglich. Da der Säugling in konstanter zwischenmenschlicher Interaktion mit der Mutter oder der sie ersetzenden primären Pflegeperson steht, gibt es eigentlich keine Psychosomatik des Säuglings, sondern nur eine solche der Dyade Mutter/Kind, die eine Art psychophysische Einheit bildet. Für die Entstehung einiger psychosomatischer Krankheiten müssen bestimmte strukturelle Voraussetzungen im seelischen Bereich eines Kleinkindes entwickelt worden sein (z.B. die Trennung der Repräsentanzen von Selbst und Nicht-Selbst, von Innen und von Außen).

Psychosomatische Krankheiten sind beim Kind verhältnismäßig häufig, einerseits wegen der Asymmetrie in der Eltern-Kind-Beziehung mit ungleicher Verteilung von Macht, Abhängigkeit und Formbarkeit sowie den größeren Möglichkeiten des Erwachsenen zur Manipulation. Andererseits aber auch, da die Desomatisierung, d.h. die Ablösung von seelischen Konflikten aus dem Bereich des Körperlichen, beim Kind noch nicht so weit fortgeschritten ist wie beim Erwachsenen. Und schließlich, weil Kinder durch Objektverluste, die immer einer seelischen Belastung gleichkommen, grundsätzlich verletzbarer sind als Erwachsene.

54.2.6 Selbstentwicklung in der Adoleszenz

Adoleszenz als psychische Entwicklungsphase beginnt ungefähr mit der physischen und sexuellen Reife, dem Anfang der Pubertät. Ihr Ende ist etwa dann erreicht, wenn eine sexuelle Identität etabliert und die

Art, mit Belastungen oder Angst umzugehen, weitgehend fixiert ist. Auch machen die psychologischen Einstellungen auf die physische Fähigkeit, Vater oder Mutter zu werden, einen nicht kleinen Teil dessen aus, um das es in der Adoleszenz geht. Zudem beginnt der Adoleszente sozial als unabhängige Person zu funktionieren.

Zu den **spezifischen Aufgaben,** die für den Adoleszenten typisch sind, gehören Umarbeitungsprozesse über viele Jahre auf dem Gebiet der Objektbeziehungen, die einerseits eine kritische Distanzierung von den primären Liebesobjekten (Mutter und Vater, bzw. ihren Substituten) und die allmähliche Neudefinition der Beziehungen zu ihnen umfassen (Prozeß, der üblicherweise als „Ablösung" bezeichnet wird), andererseits eine experimentierende Verbreiterung der Beziehungsfähigkeit mittels vielgestaltiger Kontakte zu Gleichaltrigen zur Folge haben. Die Beziehung zu den Eltern wird also von emotionaler Abhängigkeit zu Unabhängigkeit umgestaltet. Gedanken, Gefühle und Handlungen müssen notwendigerweise von einer Reaktion der Eltern unabhängig werden. Die Umgestaltung der Beziehungen zu bedeutungsvollen anderen ist auch mit einer Neuumschreibung sowohl des Ichs als auch des Selbst, einer Neueinstellung zu frisch gewonnenen aktuellen Werten (Über-Ich) und einer Modifikation in den Triebimpulsen verbunden. Ein Identitätswandel schließt die Vergewisserung ein, daß der sexuell reif gewordene Körper dem Adoleszenten selbst gehört, er auch dafür verantwortlich ist, und schafft die Voraussetzung für eine allmähliche Einfügung in den gesellschaftlichen Kontext (z. B. veränderte Art der Aggressionsverarbeitung).

Die Adoleszenz kann als eine Neuauflage der frühkindlichen Probleme, aber mit einem anderen Körper und einer anderen psychischen Organisation, angesehen werden. Bezüglich dieser Neuauflage ist es sehr wichtig, wie die früheren Entwicklungsphasen gemeistert worden sind. In vielen Kulturen gibt es Rituale, welche die Identitätsfindung und Eingliederung in den gesellschaftlichen Kontext, d.h. den Übergang vom Kind-Sein zum Erwachsen-Sein erleichtern. Bei uns fehlen sie zum größten Teil. Während der Adoleszenz kommt es üblicherweise zu einer verstärkten Ausgestaltung zentraler Konzepte wie z.B. solcher über Sexualität, Geburt, Leben, Krankheit, Tod und Religion (Bürgin, 1986b).

Mit dem Terminus **Narzißmus** werden Zustände des Selbstwertgefühls, der affektiven Einstellung eines Menschen zu sich selbst beschrieben. Ist diese realitätsgerecht, so spricht man von gesundem Narzißmus, ist sie es nicht, von einer narzißtischen Störung. Diese kann sich in einem übertriebenen Selbst- oder einem übermäßigen Minderwertigkeitsgefühl äußern. Ebenso wie es ein System der Triebregulation gibt, existiert ein narzißtisches Regulationssystem. Unter Regulation des Narzißmus wird die Aufrechterhaltung eines affektiven Gleichgewichtes bezüglich der Gefühle von innerer Sicherheit, Wohlbehagen, Selbstwertgefühl und Selbstsicherheit verstanden. Das narzißtische System ist sehr störanfällig, insbesondere in der Adoleszenz.

Der Adoleszente befindet sich in seiner Ablösungsentwicklung in einem außerordentlich verletzlichen Zustand, besonders dann, wenn er die Bindungen zu den primären Liebesobjekten, den Eltern, lockert, sich durch Erhöhung der eigenen Grandiosität zu stabilisieren versucht und tastend erste neue Beziehungen aufnimmt. Diese Beziehungen haben in der Regel eine narzißtische Qualität, d.h. der andere wird gesucht und geliebt, weil er in irgendwelchen Aspekten dem Bild der eigenen Person entspricht, man sich in ihm wiederfindet. Solche narzißtischen Beziehungen werden von den Beteiligten lange Zeit als ideal empfunden, zeichnen sie sich doch durch rasche Kontakte, Unkompliziertheit, gegenseitige Bestätigung und das Fehlen von größeren aggressiven Spannungen aus. Aber sie basieren zumeist auf einer Illusion, da der Partner in erster Linie eine Funktion für das eigene Selbstgefühl erfüllt und nicht als eigenständiger anderer Mensch erlebt wird. Bei Trennungen und Verlust des Beziehungspartners und beim Versagen der Kompensationsmechanismen läuft der Adoleszente Gefahr, daß es zu einem völligen Zusammenbruch seines narzißtischen Gleichgewichtes kommt, mit sehr heftigen Regressionen bis zum Verlust des Gefühls der eigenen Identität.

54.3 Psychosomatische Störungen und Familiendynamik

54.3.1 Rückkoppelungsprozesse

Kind, Eltern und Familie sind in unserer Gesellschaft nicht voneinander zu trennen. Die Familie ist ein gesellschaftliches Subsystem, das eine innere und erforschbare Gesetzmäßigkeit besitzt. Jedes Kind wird in ein spezifisches familiäres System hineingeboren und befindet sich vom Beginn seiner intra- und extrauterinen Existenz an in einem höchst komplizierten Kräftefeld. Die Familie ist ein System gegenseitiger zwischenmenschlicher Interaktionen, die in Form anhaltender **intrapsychischer** und **interpersoneller Rückkoppelungsprozesse** ablaufen. Jedes Mitglied beeinflußt die anderen und wird zugleich von den anderen beeinflußt (Stierlin, 1975). Es findet eine dauernde Entwicklung in Form von Kreisprozessen statt, die allerdings meist nach bestimmten gleichbleibenden intrafamiliären Mustern ablaufen. Letztere wiederum sind durch weitere äußere Felder, wie z.B. die soziale Schicht, die kulturellen Werte, die historischen Gegebenheiten etc., geprägt. Das übliche Ursache-Wirkungs-Denken erweist sich als willkürliche Abgrenzung, vergleichbar der Interpunktion bei sprachlichen Konventionen. Dies gilt sowohl für die intrapsychische Welt jedes Individuums wie auch für die interpersonalen Interaktionen zwischen Individuen, welche für einander bedeutungsvoll sind. Innerhalb eines solchen Interaktionsfeldes macht ein Kind viel mehr an seelischer Entwicklung (sowohl kognitiver wie emotionaler) durch als seine Erzieher, die Partner im familiären Gefüge sind ungleich. Eltern haben einen unaufholbaren zeitlichen sowie Macht- und Erfahrungsvorsprung. In unzähligen Lernschritten erwirbt das Kind, in konstanter kognitiver und emotionaler Interaktion und damit mitgeformt durch die Persönlichkeitsstruktur seiner Umgebungspersonen, diejenigen Kenntnisse und Erfahrungen, welche es für eine zunehmende Eigenständigkeit braucht.

54.3.2 Identifikationen

Alle Eltern/Erzieher haben bewußte und nicht bewußte Persönlichkeitsanteile. Als Erziehende stehen sie mit ihrer ganzen Person in Beziehung und Interaktion mit dem Kind. Gewollt oder ungewollt sind somit auch alle unbewußten Persönlichkeitsanteile der Eltern in diese kreisförmigen Beziehungsabläufe miteinbezogen. In den unabsehbar vielen verinnerlichten Interaktionen wird das Kind mit seiner gesamten, enormen Anpassungsfähigkeit sowohl durch die bewußten als auch durch die unbewußten Persönlichkeitsanteile der Eltern tief beeinflußt. (Es übernimmt z.B. Gefühle, Einstellungen, Ängste, Abwehren, Wertvorstellungen, Charakterhaltungen, Lebensziele oder Sinngebungen.) Mittels **partieller** oder **totaler Identifikation,** d.h. Vorgängen, durch welche ein Mensch eine Eigenschaft, ein Attribut oder sonst einen Aspekt eines anderen assimiliert und sich unbewußt diesem Vorbild im positiven oder negativen Sinne angleicht, übernimmt das Kind solche Anteile direkt oder ins Gegenteil verkehrt, oder es stößt sie zur Abgrenzung von sich weg. Solche Identifizierungsprozesse sind beim Kleinkind meist **global,** später sehr viel **selektiver** und verlaufen in den verschiedenen Abschnitten des kindlichen Lebens oft krisenhaft.

54.3.3 Intrapsychisch/intrafamiliär

Neben den Fragen der Heredität und der Auswirkung belastender äußerer Ereignisse ist eine zentrale Frage bei der Untersuchung von Kindern/Jugendlichen mit psychosomatischen Störungen die, ob eine vorliegende Symptomatik hauptsächlich Ausdruck eines in der Person des Kindes zentrierten, nicht anders verarbeitbaren, **interpersonellen,** d.h. familiären Konfliktes ist, oder ob es sich mehr um eine internalisierte, d.h. tief in der psychischen Struktur des Kindes verankerte, **intrapsychische** Störung handelt. Es kann sich bei dieser Frage nicht um ein Entweder/Oder handeln, sondern immer nur um ein mehr oder weniger gewichtetes Sowohl/Als-Auch, da intrapsychische und interpersonelle Abläufe wiederum nur idealtypische Konfigurationen innerhalb von Kreisprozessen darstellen. Bei der vorwiegend familiären Symptomatik kann der Konflikt von irgendeinem der Familienmitglieder ausgehen, die Symptomatik des Kindes aber auf eine andere Person des Familienverbandes zurückverweisen. Bei der vorwiegend intrapsychischen Problematik handelt es sich zumeist um eine Folgeerscheinung einer frühkindlich beeinträchtigten, in der

Innenwelt des Kindes strukturell fixierten, durch Affekte und Phantasmen mitausgeformten Interaktion, die vom Kind (Anlage, organische Störung) oder von den Erziehern (Deprivation, Störung der Separation/Individuation) oder von beiden (neurotische Beziehungsstörung) mitverursacht worden ist. Bei den meisten Störbildern handelt es sich um gemischte Formen.

54.3.4 Rollenzuschreibung, Parentifikation

Das Kind als der formbarste Teil eines familiären Systems wird durch pathogene Modalitäten der Beziehung oder der Rollenzuschreibung, infolge des unleugbaren Machtgefälles, am nachhaltigsten in seiner Entwicklung beeinträchtigt. Dies ist am deutlichsten, wenn es zum Sündenbock auserkoren worden ist, aber auch schon sehr klar erkennbar, wenn eine die Persönlichkeitsentwicklung des Kindes hemmende Parentifizierung vorliegt, d.h., daß das Kind Aufgaben und Funktionen für einen oder beide Elternteile übernehmen muß, die mit der Realität seiner Existenz nicht vereinbar sind (z.B. als Tochter eine gute Mutter für die Mutter zu sein). Es entwickelt dann oft ein „falsches Selbst" und wird auf Kosten seiner Eigenständigkeitsentwicklung zum Erzieher der Eltern, die ihre Entwicklung zu Autonomie und Identität noch nicht zu einem postadoleszenten Abschluß gebracht haben. Die Eltern/Erzieher mit ihren ungemein gefestigteren Persönlichkeitsstrukturen kennen wohl zumeist ihre bewußten Erziehungsziele und Interaktionsformen, sind definitionsgemäß aber blind für die **Übertragung** ihrer unbewußten Persönlichkeitsanteile auf die Kinder. (Als Übertragung wird ein unwillkürlicher, unbewußter Vorgang bezeichnet, durch welchen infantile Verhaltens- und Erziehungsmuster, Phantasien oder Wünsche innerhalb einer bestimmten Beziehung aktualisiert werden. Das früher Geschehene wird, auf die neue Beziehung transformiert, im gegenwärtigen Erleben wiederholt.)

54.3.5 Grenzen, Loyalitäten, Kräftegleichgewicht, Vermächtnisse

Jede Familie ist ein System, das sich in verschiedene Subsysteme unterteilt (z.B. jung/alt, Kinder/Eltern/Großeltern, Frauen/Männer etc.). Die Grenzen einer Familie können sowohl nach außen wie auch nach innen gegenüber den Subsystemen so geartet sein, daß sie eine entwicklungsungünstige Auswirkung haben, z.B. wenn sie durchlöchert oder fast aufgehoben sind (was gegen außen keine innere Kohärenz und gegen innen keine Individuation ermöglicht) oder wenn sie völlig rigide, undurchlässig, unveränderbar oder spezifisch verzerrt sind (was Sonderlingshaftigkeit, autistische Abkapselung oder mangelnde Anpassung an neue Situationen nach sich zieht). So geartete Systeme erlauben dem Kind keine altersgemäßen Entwicklungs- und Ablösungsschritte. Auch kann das intrafamiliäre, verbale oder averbale **Kommuni-**

kationssystem so beschaffen sein, daß jeweilige Bedürfnisse eines einzelnen, infolge mangelnder Authentizität, gar keinen angemessenen Ausdruck finden können.

In jeder Familie gibt es auch ein konstantes Kräftegleichgewicht, das aus **zentrifugalen** und **zentripetalen Kräften** zusammengesetzt ist. Dies gilt sowohl für die horizontale Ebene der jeweils gleichen Generation wie auch für die vertikale, welche eine Mehrgenerationenperspektive umfaßt. Überwiegen die zentripetalen Kräfte, so sind die Beziehungsmuster vor allem solche der gegenseitigen Bindung, was eine altersgemäße Ablösung erschweren bis verhindern kann. Infolge der **unsichtbaren Loyalitäten** (Boszormenyi-Nagy, 1980) entstehen so z.B. bei Adoleszenten übermäßige Ausbruchsschuldgefühle. Dominieren die zentrifugalen Kräfte, so müssen Ablösungen zu früh vollzogen werden, Loyalitäten bilden sich schwach aus, und es besteht eine Gefahr emotionaler Vernachlässigung. Oft können projektive Zuschreibungen oder Unterstellungen dazu führen, daß ein Kind von den Eltern zur Konfliktbewältigung gebraucht wird. Nicht selten wird eine solche Rollenzuschreibung sekundär als Krankheitsgewinn vom Kind ausgenützt, was dann bald zu einer schwer neurotischen, gegenseitigen Verstrickung führt. Bestimmte Rollenzuschreibungen und Beziehungsmodalitäten kennzeichnen jede Familie (Bauriedl, 1980). Es kommt vor allem dort zu Störungen, wo diese Geschehnisse zu stark, zu schwach oder verzerrt vorhanden sind. Dies zeigt sich beim Kind in einer Einschränkung seiner autonomen intrapsychischen Entwicklung. Denn es kommt zur Konfliktbildung zwischen den eigenständigen Wünschen des Kindes und den Zuschreibungsanforderungen durch die Eltern. **Transgenerationale Vermächtnisse** sowie **Verschleierungen** von Bedeutungen (Mystifikationen) können latente intrafamiliäre Konflikte weiterhin verschlimmern. Desgleichen **Familienmythen,** d.h. gemeinsame, oft nicht oder nur halb bewußte Phantasmen über familiäre Funktionen oder Geschehnisse, die nicht mehr hinterfragt werden dürfen.

54.3.6 Vulnerabilität, interaktionelle Spezifika

Interpersonale, familiäre Interaktionen beeinflussen die Psychophysiologie eines Kindes. Es sind aber nicht so sehr spezifische auslösende Situationen, sondern vielmehr bestimmte Interaktionsprozesse, die die Somatisierung oder den somatischen Ausdruck von Konflikten erleichtern. Wenn ernsthafte psychosomatische Störungen sich beim Kind oder Jugendlichen entwickeln, treffen häufig folgende Faktoren zusammen:

- Eine spezifische **physiologische Vulnerabilität** und/oder **organische Dysfunktion,** sei diese primär oder sekundär.
- Spezifische **interaktionelle Eigenheiten** innerhalb der Familie, insbesondere eine zu schwache oder zu starke Abgrenzung jedes einzelnen oder von familiären Subsystemen, ein übermäßiges gegenseiti-

ges Ausmaß von Besorgnis über das körperliche Wohlbefinden des anderen, eine ungenügende Fähigkeit zur Adaptation, d.h. die Tendenz, mittels starrer intrafamiliärer Strukturen einen Status quo zu erhalten, was in Zeiten wie z.B. der Adoleszenz, in welchen Veränderung und Wachstum unumgänglich sind, zu Schwierigkeiten führt, und die Tendenz, Konflikte in der Familie nicht auszuhandeln, sie nicht zu lösen, sondern sie zu vermeiden. Das Kind und der Jugendliche werden dann für die Vermeidung von elterlichen Konflikten oder als Zusammenhalt der Familie ge- (bzw. miß-)braucht (Minuchin, 1974, 1975, 1978; Kog et al., 1987). Die Symptomwahl erfolgt oft nach familiär vorgegebenen Mustern. Bildet sich eine ernsthafte psychosomatische Symptomatik als Ausdruck einer systemimmanenten Störung der Familie aus, so ist die Autonomieentwicklung des Kindes stets gestört, da die Familie, kaum hat sich das psychosomatische Symptom entwickelt, eine verstärkte Kontrolle über das kranke Kind ausübt und umgekehrt. Krankheit wird dann wie eine Art substantielle Währung in der gegenseitigen Interaktion gebraucht.

54.4 Das chronisch kranke (behinderte) Kind

Es erlebt sich selbst anders als die anderen Kinder, nimmt eine Sonderstellung ein und wird zu einer besonderen seelischen Arbeit genötigt, für die das gesunde Kind viele Jahre bis zur Pubertät und zum Abschluß der Adoleszenz Zeit hat. Unter dem Druck der Krankheit bzw. Behinderung muß es sich oft verfrüht mit den Grundphänomenen der menschlichen Existenz auseinandersetzen. Es bildet so bewußte und unbewußte Phantasien über die Zusammenhänge zwischen seinen Einschränkungen und den möglichen Ursachen aus. Krankheit wird oft als ein Geschehen erlebt, das einem durch einen äußeren Aggressor auferlegt wird. Normalerweise hat ein Kind genügend lange Zeit, seine Vorstellungen über Krankheit, über die Frage: „Was ist das Leben?" und über den Tod auszugestalten. Erst nach 14–16 Jahren, d.h. in der Adoleszenz, hat es seine entsprechenden Vorstellungen soweit entwickelt, daß sie denjenigen gleichen, die die Erwachsenen in seinem Kulturkreis aufweisen. Das chronisch kranke Kind hingegen ist genötigt, diesen Prozeß viel rascher zu vollziehen, ihn gleichsam im Eilzugtempo zu durchlaufen. Gelingt ihm das mit Hilfe seiner Eltern, dann kommt es zu einer Art inneren Frühreife.

Zu den Besonderheiten, die chronisch kranken Kindern auferlegt sind, gehört eine **längere Abhängigkeitszeit** als die, welche ein gesundes Kind zu durchlaufen hat. In der Adoleszenz muß es sich in der Regel aus engeren Bindungen lösen. Der Ablösungsprozeß wird somit für alle Beteiligten komplizierter, insbesondere aber für die chronisch kranken Jugendlichen, die sich ihre Eigenverantwortung zumeist hart erkämpfen müssen. Als weitere Belastung kommt ein zumeist **fluktuierender Gesundheitszustand** hinzu, welcher das Selbstwertgefühl konstanten Schwankungen und Einbrüchen unterwirft. Es ist auf diese Art und Weise viel schwieriger, sich eine innere Konstanz des Selbsterlebens aufzubauen. Verweigerung, Verleugnung, aggressives Verhalten, Regressionen, Reaktionsbildungen und Überkompensationen legen, als häufig erkennbare Abwehraktivitäten, Zeugnis von den entsprechenden Belastungen ab. Die Beeinträchtigungen lösen oft auch heftige Aufwallungen negativer Affekte aus (z.B. Angst, Wut, Ärger, Haß). Da diese auf niemanden zu richten sind, ihnen gleichsam eine Art **Schicksalsungerechtigkeit** zugrunde liegt, also kein Anspruch auf gegen eine Person gerichtete Rache besteht, werden sie leicht gegen das Selbst gewendet und manifestieren sich als Selbsthaß. Auch das erschütterte Vertrauen in die Verläßlichkeit des eigenen Körpers erschwert den Aufbau einer zuverlässigen Selbstrepräsentanz. Leicht entstehen Desorientierung, Konfusion, Resignation und Depression, dann aber auch wieder Erholung und Hoffnung. Diese Wechselhaftigkeit des Allgemeinzustandes macht die Entwicklung einer hohen Flexibilität und Elastizität im Psychischen notwendig.

Im weiteren können die Symptome selbst, die Erfordernisse der Therapie, die Sekundärerscheinungen der Krankheit, ihr wechselnder Verlauf und auch die Unsicherheiten im Hinblick auf die Zukunft (denn oft ist keine klare Voraussage möglich) besondere Belastungsmomente darstellen. Der gesunde Mensch geht mit mehr oder weniger Berechtigung und Wahrscheinlichkeit davon aus, daß seine Zukunft frei und offen sei. Das chronisch kranke Kind (bzw. Adoleszente) aber kann die Tatsache seiner behindernden Krankheit nie aus der Welt schaffen, was eine grundsätzlich andere Ausgangsposition darstellt. Diese muß allerdings nicht unbedingt ein Handicap sein, sondern kann auch zu einem konzentrierteren und intensiveren Erleben führen als bei jemandem, der sich noch nie damit auseinandersetzen mußte und alles fraglos akzeptierte. Probleme im Hinblick auf Berufstätigkeit, Partnerschaft (Ist ein Partner überhaupt zu finden? Was bedeutet die Behinderung für ihn? Aus welchen Motiven geht er eine solche Beziehung ein?), der Generativität (Frage der Weitergabe einer Krankheit; Problem, ob genügende Fürsorge für ein potentielles Kind möglich ist; mögliche Infertilität) und die Angst vor einer eventuellen Progression, zunehmender Abhängigkeit oder gar weiterer Einschränkungen mit verkürzter Lebenserwartung beschäftigen naturgemäß diese Jugendlichen. Wenn also eine Krankheit nicht direkt oder definitiv heilbar ist, so hat sich das Kind (bzw. der oder die Jugendliche) in jedem Fall mit dem Sinn seines Lebens und dem, was darin lebenswert ist, was eine optimale Lebensqualität ermöglicht, wenn der Körper nur eingeschränkte Möglichkeiten für Aktivität, Genuß und Lust vermittelt, auseinanderzusetzen. Es ist darauf angewiesen, kompensatorisch in jenen Bereichen eine besondere Entwicklung durchzumachen, in denen es durch seine Krankheit nicht eingeschränkt ist.

Während sich die psychosomatische Medizin bisher mit der Erfassung körperlich-seelischer Wechselwirkungen bei der Entstehung, im Verlauf und in der Behandlung von Krankheiten befaßte, erfährt sie heute durch die Entwicklung der modernen Medizin dahingehend eine Erweiterung, daß sie sich im Falle des Überlebens durch moderne Behandlungsmöglichkeiten mit den **Fragen des Wie-Überlebens** auseinandersetzen muß (Hoffmann, 1987); z.B. in der Onkologie (Bürgin, 1978), bei zystischer Fibrose, chronischen Nierenpatienten oder Unfallopfern (mit z.B. konsekutiver Para-/Tetraplegie). Eine wesentliche Aufgabe des jugendpsychiatrischen Spezialisten in diesem Kontext ist die Weiterbildung und Supervision des Behandlungs-/Rehabilitationsteams. Die direkte Arbeit mit dem Patienten und seiner Familie ist in den Situationen relevant, in denen, in der Regel durch vorbestehende psychosoziale Belastungsfaktoren, die Bewältigungs- und Anpassungsmöglichkeiten des Kindes und/oder seiner Familie unzureichend sind.

Für Kinder mit den klassischen, läsionellen psychosomatischen Krankheiten (z.B. Colitis ulcerosa, Morbus Crohn, Dermatitis atopica u.a.) gilt weitgehend das gleiche wie für das chronisch kranke Kind, da bis heute die ätiologische Rolle bestimmter Persönlichkeitsmerkmale/Konfliktkonstellationen offengelassen werden muß und sich die psychotherapeutische Begleitung dieser Kinder und ihrer Familien im Falle eines Versagens der Bewältigungs- und Adaptationsmechanismen außerordentlich bewährt hat.

54.5 Psychophysische Übergangsbereiche

Der **Spannungsabfuhr** über den Körper kommt bei Kindern und Jugendlichen eine große Bedeutung zu. Sie ist in Form von Störungen im Bereich der Motorik, bei den Dekompensationen mit regressiver Reaktivierung archaischer körperlicher Reaktionsmuster und beim Konversionsvorgang zu beobachten. Da sich bei den ersten zwei Formen wenig symbolerfüllte Phänomene zeigen, erscheinen viele psychosomatische Störungen im Hinblick auf ein vertieftes psychodynamisches Verständnis nicht sehr ergiebig.

54.5.1 Ausdruck über den Körper, Homöostase

Der unmittelbare Ausdruck über den Körper ist eine sehr ursprüngliche Form der Abfuhr, die zwar an Affekte gebunden ist, aber mit dem Körper verknüpft bleibt. Sie erfolgt mehr oder weniger automatisch. Hierzu sind die Zustände der Hyperaktivität oder Apathie zu rechnen, der autoerotischen oder autoaggressiven Aktivitäten (z.B. Masturbation und Mutilation), der rhythmischen Geschehnisse (Jaktationen) und der Stereotypien.

Die intrapsychische wie auch die interpersonale Homöostase ist durch die Neigung zur Wiederkehr des Verdrängten, bzw. den natürlichen Auftrieb von vorbewußten Beziehungskonflikten ungelöster Art ins bewußte Erleben, stets gefährdet. Die Entwicklung einer psychosomatischen Krankheit kann unter diesem Aspekt einen sinnvollen Versuch, Konflikte mit Hilfe einer körperlichen Erkrankung zu lösen, darstellen, im Sinne einer Anpassungsleistung (bei mehr dauerhafter Fehlverarbeitung) oder eines Selbstheilungsversuches (bei Überforderungen, akuten Konflikten oder Lebenskrisen).

54.5.2 Traumatisierende Faktoren, übermäßige Reizzufuhr, emotionales Manko, Strukturdefekte

Die Wirkung der **pathogenen, traumatisierenden Faktoren** ist immer abhängig von der Vulnerabilität der seelischen Strukturen und Dynamiken, auf die sie Einfluß nehmen. Gewisse Strukturen, d. h. psychische Funktionseinheiten mit größerer zeitlicher Persistenz, die im Verlaufe der Entwicklung eines Individuums, nämlich bevor die Bildung der erwachsenen Persönlichkeit durch die Beendigung der Adoleszenz abgeschlossen ist, allerdings noch verändert werden können, schaffen eine Art Disposition zur Traumatisierung, z. B. gewisse Frühgeburten, Patienten mit einer Hirnreifungsstörung, mit allergischer Atopie sowie Kinder mit sehr schwach ausgebildeten Regulationsmechanismen, niedriger Reizschwelle und ungenügender Ich-Integration. Je ausgewogener die Qualität der Abwehrorganisation, desto geringer die Dekompensation ins Somatische.

Zwei Gegebenheiten haben eine verhältnismäßig verbreitete traumatisierende Wirkung: Einerseits die **übermäßige Reizzufuhr,** welche beim Säugling und Kleinkind ein primäres Potential zur Entladung in den Körper enthält. Die Mutter oder ihr Substitut funktionieren in gewissen Fällen in den ersten beiden Lebensjahren, bevor die Psyche des Kindes autonom zu funktionieren gelernt hat, nicht als gute Abwehr und bewirken somit eine Überflutung des in Entwicklung befindlichen Ichs mit Reizen verschiedenster Art oder eine Inkohärenz der Reizzufuhr. Diese Form findet sich bei vielen Störungen (z. B. Schlafstörungen, Kopf- und Bauchschmerzen, der Drei-Monats-Kolik, dem psychogenen Erbrechen oder den Affektkrämpfen). Andererseits gehören die Formen des **emotionalen** (und/oder sensorischen) **Mankos** bzw. der Karenz mit ihren Unterformen des Ungenügens, der Dyskontinuität und der Verzerrung dazu. Auch sie sind bei vielen psychosomatischen Störungen nachzuweisen (z. B. schweren Eßstörungen, psychogenem Erbrechen, Rumination), aber auch bei Trennungen ohne Substitute in der frühen Kindheit, wechselnden Bezugspersonen, beim Hospitalismus und bei ungünstigen sozioökonomischen Verhältnissen.

Entsprechend der französischen Schule kann die Hypothese formuliert werden, daß solche traumatisierenden Faktoren dazu führen, daß die Strukturierung im vorbewußten Ich-Anteil defektuös oder auf primitiven Stufen stehenbleibt und daraus ein funktionelles Ungenügen der Abwehr resultiert, das beim Kind wegen seiner Beeinflußbarkeit aber noch rever-

sibel ist. Es charakterisiert sich durch ein Unvermögen zur phantasmatischen Elaboration. Unbewußte Triebimpulse fließen bei diesen Kindern nicht wie üblich in ihre Handlungen ein. Die gesamte Aktivität wird durch Tatsächliches bestimmt, durch die gerade gegenwärtige materielle Umgebung, durch die momentan vorhandene Situation oder durch die gegenwärtigen Personen. Hierdurch entsteht eine gewisse Kargheit der innerseelischen Repräsentanzen (Kreisler, 1985 a). Beim Erwachsenen zeigt sich dieses Syndrom als „pensée opératoire", welche permanent oder episodisch als Dekompensationszeichen auftreten kann (Marty und de M'Uzan, 1963).

Zustände mit wenig seelischer Ausstrukturierung finden wir auch bei den massiven Strukturdefiziten, welche fast sämtliche Bereiche der Entwicklung beeinträchtigen und z. B. bei massiver Privation, schweren Familienstörungen oder sozialen Defektentwicklungen oder auch beim psychosozialen Minderwuchs zu beobachten sind. Neben der Dekompensation mit psychosomatischen Krankheitserscheinungen kann sich der Zustand des Kindes noch verschlechtern, bis hin zu einer leeren, **depressiven Atonie,** in welcher die Kinder indifferent, ohne Angst oder wahrnehmbare Affekte erscheinen, wie mit eingefrorenen psychischen Funktionen ausgestattet. Dort zeigt sich ein völliges Absinken der Lebensantriebe, ein Libidorückzug vom Selbst und von den Objekten, mit zum Teil devitalisiertem, automatisiertem und „leerem" Verhalten als postregressive Organisationsstruktur.

54.5.3 Grundlinien psychosomatischer Krankheitsentstehung

Neben den Anschauungen der französischen Schule lassen sich zwei weitere Konzepte der psychosomatischen Krankheitsentstehung hervorheben:

Dekompensation und Regression

Die Dekompensation im Sinne einer regressiven Reaktivierung archaischer körperlicher Reaktionsmuster bei erheblichen frühen Persönlichkeitsstörungen, die sich in **Ich-Defekten** und **Störungen der Objektbeziehungen** ausdrücken: Unter länger anhaltenden oder schweren Belastungen kommt es zu unspezifischen Dekompensationen des psychischen Funktionierens mit Mobilisierung reflexhafter körperlicher Abwehrmuster, die habituell in Spannungssituationen auftreten und entweder durch psychosomatische Fixierungen in der frühen Kindheit, welche im Zusammenhang mit Krankheit entstehen, oder durch Abspaltung von Affekt- oder Erlebnisbereichen bei der Desomatisierung zustande kommen. Ist dieser „psychosomatische Funktionssektor" im Ich sehr groß, so genügen bereits kleinere Belastungen, um Reaktionen auf körperlicher Ebene, unter Umgehung einer psychischen Verarbeitung, auszulösen. Bei solchen Patienten ist die Repräsentanz des Körperselbst oft erheblich gestört (z. B. Beeinträchtigungen im Empfinden von Raum, Zeit, Temperatur oder Lateralität). Sie erscheinen im regressiven Zustand verarmt im Ausdruck von innerseelischen Vorgängen (z. B. der Symbolisierung, der Fähigkeit, Gefühle wahrzunehmen und der Bereitschaft, innerhalb einer Beziehung situationsadäquat zu kommunizieren). Wahrscheinlich liegen bei diesen Patienten frühe Ich-Defekte und archaische Abwehrformen vor. Diese Funktionsmodalitäten dürften in der frühen Separations-/Individuationsphase erworben worden sein. Sie sind immer verbunden mit einer narzißtischen Beziehungsform und der Unfähigkeit zu angemessener Aggressionsverarbeitung. Als interaktionales Verhaltensmuster, das sich in bestimmten Situationen einstellt, sind sie kein starres Persönlichkeitsmerkmal, sondern verändern sich im Laufe eines psychotherapeutischen Prozesses. Bei dieser narzißtischen Objektbeziehungsart wird der bedeutungsvolle andere in fast symbiotischer Form für die Regulation des eigenen Selbstwertgefühls, für Sicherheit und Wohlbefinden und für das narzißtische Gleichgewicht (z. B. als Spiegel des eigenen Selbst) in unentbehrlicher Art gebraucht. Der mögliche Verlust des Objektes mobilisiert heftigste regressive Trennungs- und Vernichtungsängste. Das Kind bzw. der Jugendliche sorgt deshalb manipulativ für die Realpräsenz solcher Selbstobjekte, da sonst Hilf- und Hoffnungslosigkeitsgefühle entstehen, und gebraucht demnach eine Art interpersonaler Abwehr. Kinder und Jugendliche, welche so geartete „psychosomatische Neurosen" entwickeln, stehen in ihrer Struktur den Patienten mit narzißtischen Neurosen oder Borderline-Syndrom recht nahe.

Konversion

Die neurotischen Konfliktlösungen auf der Körperebene bei relativ reifen Persönlichkeitsstrukturen nach dem Modell des Konversionsvorganges: Der Konversionsvorgang (vgl. Kap. 30) ist ein aktiver Prozeß in Richtung auf eine Symptombildung hin, bei welchem widersprüchliche Impulse intrapsychisch symbolisiert und mittels entsprechender Phantasmen in einer Art Körpersprache ausgedrückt werden. Es handelt sich hierbei um funktionelle Störungen ohne anatomisch-pathologisches Substrat, die sich in motorischen, sensorischen, somatoviszeralen und anderen Funktionsbereichen zeigen können. Konversionssymptome sind häufig vorübergehender Natur und können sich auf sehr unterschiedliche psychopathologische Strukturen aufpfropfen. Der Konversionsmechanismus erfolgt unbewußt und automatisiert und ist von der bewußten Simulation abzugrenzen. Unbewußte Vorgänge, die mittels der Konversion körperlich für andere wahrnehmbar werden, entsprechen unbewußten Bedürfnissen und ihrer Abwehr zugleich. Es handelt sich um eine Art Appell ohne Schrei, um einen Wunsch ohne Bitte. Bei etwa 2–3% aller Patienten im kinder- und jugendpsychiatrischen Krankengut handelt es sich um solche mit Konversionssymptomen. Während der Pubertät ist das Konversionssyndrom besonders häufig zu beobachten. In der Vorpubertät ist die Geschlechtsvertei-

lung ausgeglichen, während und nach der Pubertät überwiegen die Mädchen mit einem Verhältnis von etwa 4:1 (Bürgin, 1982 a).

Jeder Teil des Körpers kann zum Ort der Konversion ausgewählt werden. Für die Organwahl besteht möglicherweise ein somatisches Entgegenkommen (genetische Disposition, aktuelle Überbeanspruchung, frühkindliche Bahnung). Nicht selten werden auch durch familiäre Konstellationen Symptomtraditionen geschaffen (z. B. Kopfwehfamilien). Wurde früher die Konversion ganz der phallischen Phase zugeordnet, so ist in den letzten 20 Jahren zunehmend deutlicher geworden, daß es Konversionsmodalitäten auf allen Ebenen der seelischen Entwicklung gibt, d. h., daß Konflikte aus dem phallisch genitalen oder aus dem prägenitalen (d. h. oralen oder analen) Stadium mittels des Konversionsmechanismus zu lösen versucht werden (Rangell, 1969).

Ausgeprägte sensomotorische Funktionsausfälle im Rahmen von Konversionsneurosen (z. B. Mono-, Para- oder Tetraplegien, z. B. nach Bagatellunfällen im Verlaufe schwerer Adoleszenzkrisen) können in der Adoleszenzentwicklung die offenbar „beste Lösung" bei sonst unlösbar scheinenden, intrapsychischen oder interpersonellen Konflikten darstellen. Die Lähmungen ermöglichen den von massiven Schuldgefühlen gequälten Jugendlichen (z. B. wegen der Ablösung oder infolge inzestuöser Beziehungen), zu überleben, ohne im Selbstmord Sühne leisten zu müssen.

Simulation und Konversion können wie Extreme derselben Dimension betrachtet werden. Allerdings sind sie klar voneinander zu trennen, wenngleich es auch Übergänge gibt. Bei der Simulation handelt es sich um eine bewußte Vortäuschung (mit entsprechenden Schuldgefühlen), bei der Konversion um einen Verlust der bewußten Kontrolle und Automatisierung eines Ablaufs (oft mit einer „belle indifférence" verbunden, da vom Bewußten her das Ich ja keine Schuld trifft). Soll das Symbol aus der „Körpersprache" heraus wieder entschlüsselt werden, so kann dies nur entgegen dem ökonomischen Gefälle der bisherigen intrapsychischen Verarbeitung geschehen. Es wird somit Widerstand und Abwehrvorgänge mobilisieren und nur bei einer guten therapeutischen Allianz zwischen Patient und Therapeut möglich sein. Eine Symptombeseitigung gelingt im allgemeinen bei über zwei Drittel dieser Fälle recht gut, die Veränderung der neurotischen Fehlentwicklung allerdings ist bedeutend komplexer.

Zusammenbruch eines falschen Selbst, pathologische projektive Identifikation

Bei der Entwicklung ernsthafter psychosomatischer Krankheiten in der **Adoleszenz** sind neben komplexen familiären Faktoren folgende zwei psychologischen Konstellationen besonders häufig anzutreffen:
– Der Zusammenbruch eines „falschen Selbst". Dieser defensive fragile Überbau zerbricht unter dem Triebansturm. Das völlig überangepaßte Kind (Sonnenschein der Familie) regrediert auf frühe Entwicklungsstufen und reorganisiert sich, zumeist

unter Verwendung primitiver Abwehrmechanismen und nicht unbeträchtlichem sekundärem Krankheitsgewinn, neu auf dieser Ebene. Es geht dabei (notwendigerweise) das Risiko ein, daß aus funktionellen Störungen solche mit Läsionen werden.

– Der anhaltende Gebrauch von pathologischen projektiven Identifikationen: Beim Vorgang der Projektion werden, als intrapsychische Phantasie, bestimmten Objektrepräsentanzen abgespaltene Anteile der Selbstrepräsentanz zugesprochen (umgekehrt beim Vorgang der Introjektion). Das reale äußere Objekt ist nicht betroffen. Im therapeutischen Ablauf finden wir dieses Geschehen in der Phantasie über den Therapeuten. Es kann leicht vorkommen, daß sich Mitglieder eines therapeutischen Teams mit solchen Phantasien identifizieren. Wird nun vom Patienten, in der äußeren Welt, durch eine Handlung – oder auch durch ein „Tun des Nicht-Tuns" (Laotse) – etwas aktiv unternommen, um solche abgespaltenen Affekte oder andere Selbstanteile in die Person von anderen zu „injizieren", so daß sich diese mindestens partiell und temporär damit identifizieren müssen, so spricht man von einer Nötigung zu einer **projektiven Identifikation** (Ogden, 1982; Zwiebel, 1988; Sandler, 1987/88).

Dieser Mechanismus ist zugleich eine **Abwehroperation,** indem er unliebsame abgespaltene Selbstanteile in reale, bedeutungsvolle andere Personen verlegt (allerdings um den Preis einer dadurch erfolgenden Beziehungsstörung), wie auch eine **pathogene Kommunikationsform,** da er es dem Gegenüber ermöglicht, Erlebniskonfigurationen des Patienten direkt oder in modifizierter Form selbst erlebend wahrzunehmen. Oft erfolgen bei den Mitgliedern eines Behandlungsteams verschiedene projektive Identifikationen, die, statt agiert, in den regelmäßigen gemeinsamen Besprechungen reflektiert, relativiert und integriert werden können (Bürgin, 1989 a).

54.5.4 Bio-psycho-soziales Feld

Obwohl die Gesetze der biologischen, der psychologischen und der sozialen Welt ungleich und in weitgehend unbekannter Weise miteinander verknüpft sind, besteht eine Art Getrenntheit zwischen diesen Bereichen sowie ein sich gegenseitig beeinflussender Austausch. Dem bio-psycho-sozialen Feld mit seinen Wechselwirkungen muß in flexibel-dynamischer Weise Rechnung getragen werden, will man dem psychosomatisch kranken Kind und Adoleszenten als Arzt ganzheitlich entgegentreten. Analog zur Heisenbergschen Unschärferelation gibt es etwas Ähnliches bezüglich des ganzheitlichen Zugangs zum Patienten. Hat man sich auf einen der drei Aspekte Körper/Psyche/Umwelt, bzw. auf eine duale Interaktionsmodalität, genau eingestellt, so kann man die beiden anderen dualen Interaktionsmodalitäten nurmehr unscharf wahrnehmen, ist also genötigt, in einem Nacheinander die Perspektiven zu wechseln, muß aber das Ganze stets in einem labilen Gleichgewicht zu integrieren versuchen. Das Kind oder der Jugendliche hat ein Vorstellungsbild davon entwickelt, was Körper und Krankheit sind und sich ein Phantasma darüber aufgebaut, was die Seele oder das Psychische ist. Im allgemeinen hat es/er auch (ein zumeist unbewußtes) Konzept darüber entwickelt, wie Körper und Seele zur Umwelt in Beziehung stehen. Die Phantasmen des Arztes und des Patienten bzw. seines Umfeldes über das, was Krankheit bedeutet, brauchen ein Minimum geteilter gemeinsamer Wirklichkeit und somit eine gewisse gegenseitige Angleichung, wenn eine therapeutische Kooperation entstehen soll.

Zusammenarbeit von Spezialisten

Im Bereich der Psychosomatik von Säuglingen, Kleinkindern, Kindern und Jugendlichen arbeiten notwendigerweise meist mehrere Spezialisten verschiedener Fachgebiete zusammen. Im medizinisch-psychosomatischen Feld herrscht eine Denkweise mit räumlich-operativen Denkmodellen vor, welche durch Konzepte des Un- und Vorbewußten und durch zeitlich-systemische Perspektiven zu ergänzen sind. Im psycho-edukativen Feld steht eine Förder- und Forderpädagogik mit allgemeinen oder normierten Erziehungszielen im Vordergrund. Im psychosozialen Feld liegt der Schwerpunkt der Arbeit in der zum Teil aufsuchenden fürsorgerischen Hilfe im Sinne einer stützenden und führenden Verbesserung der Lebensumstände.

Pädiatrie und Kinder- bzw. Jugendpsychiatrie stehen in einem komplementär sich ergänzenden Verhältnis zueinander. Erst wenn die Diagnose über die Art und die Struktur von Funktionsstörungen oder Läsionen vom Somatischen und vom Psychischen her erfolgt ist, läßt sich in der bi- oder pluridisziplinären Kooperation, welche sich in einem kooperativen Kontext abspielen sollte, eine Gewichtung ätiologischer Faktoren und therapeutischer Wege erarbeiten. Die organischen und die psychischen Dysfunktionen können wie zwei Ufer des gleichen Lebensstromes betrachtet werden.

In derselben Krankheit finden sich verschiedenste psychische Strukturen und Konflikte widergespiegelt. Bei der gleichen psychischen Struktur und Konfliktart sind verschiedene somatische Krankheitsmanifestationen möglich (Bürgin, 1988). Obwohl es kein Standardprocedere der Abklärungsuntersuchung gibt, sollten in jedem Fall eine sorgfältige körperliche Untersuchung und ein oder mehrere Gespräche mit dem Kind, die Beobachtung und das Verstehen der Interaktionen zwischen Kind und Eltern, die Erfassung der persönlichen Charakteristika der Eltern (mit Hilfe von Gesprächen) und ihrer Beziehung zueinander und schließlich eine Evaluation des Gesamtfunktionierens der Familie erfolgen. Erst danach läßt sich das geeignete therapeutische Vorgehen festlegen. Bei allen funktionellen Störungen muß daran gedacht werden, daß sie sich auf Läsionen aufpfropfen können, die nicht erkannt worden sind. Auf der anderen Seite ist es unnötig, übermäßig lange auf der Suche nach Läsionen zu sein, wenn die primäre Funktionsstörung sehr offensichtlich ist.

Zeitliche Aspekte, projektive Faktoren

Den zeitlichen Aspekten ist besondere Bedeutung zuzumessen, haben doch z. B. Anorexien und Schlafstörungen bei Säuglingen, Kleinkindern, Kindern und Jugendlichen eine völlig unterschiedliche Bedeutung. Denn Entwicklung ist nicht nur eine Summierung im Sinne linearer Addition, sondern die anhaltende Umwandlung und Neuorganisation alter und neuer Erfahrungsinhalte, innerhalb welcher frühere Erlebnisinhalte in komplexere hierarchisch integriert werden. In Zeiten seelischer Belastung werden frühere Funktionsmodalitäten rasch wieder manifest. Die zuletzt integrierten Muster sind im Hinblick auf die regressive Auflösung am anfälligsten.

Bei Diskontinuitäten im Entwicklungsprozeß ist daran zu denken, daß diese entweder aufgrund von Reifungsprozessen, von interaktionsbedingten Erfahrungsprozessen oder von beiden zusammen herrühren können (Quinton et al., 1984.) Innerhalb der für die Entwicklung bedeutsamen Interaktionsprozesse sind nicht nur die Belastungs-, sondern auch die Schutzfaktoren abzuschätzen. Zu den gesicherten **protektiven Faktoren** können gezählt werden:
– ein seelisch gesunder primärer Beziehungspartner;
– Aufbau und Erhaltung einer guten Beziehung zu mindestens einem Elternteil;
– familiäre Harmonie;
– „easy temperament" (Rutter, 1985).

54.6 Spezifische Gesichtspunkte beim Säugling und Kleinkind

Auch der ausgeglichenste Säugling bringt mittels seines Körpers Unbehagen oder Konflikte zum Ausdruck. Entwickelt sich eine psychosomatische Symptomatik, so kann unterschieden werden zwischen der einfachen Funktionsstörung, bei welcher eine Konfliktdynamik mit Reizüberflutung vorliegt oder welche Ausdruck von Mangel und Frustration ist, der schweren funktionellen Störung, welche eine Tendenz zum Auslaufen in automatisierte Wiederholungszwänge hat, und schließlich der organischen Läsion, wie wir sie z. B. bei Infektionen, dem Asthma bronchiale oder beim psychosozialen Minderwuchs sehen. Wie beim Trauma wirken Belastungsfaktoren nur als Belastung, wenn die jeweilige Verarbeitungskapazität der vorbewußten und bewußten Ich-Anteile des Kindes überfordert ist. Psychosomatische Störungen beim Säugling und Kleinkind bringen im allgemeinen eine Überflutung des Ichs durch Triebimpulse, d. h. eine Überforderung der zur Verfügung stehenden Verarbeitungsfunktionen mit sich und enthalten zumeist kein symbolisches Äquivalent. Man kann in diesem Bereich also kaum von Konversionen sprechen, sondern vielmehr von **Folgen pathogener Interaktion.** Zu diesen gehören die bereits genannte

Reizüberlastung, welche beim Kind immer eine Tendenz zur psychosomatischen Dekompensation, zum Funktionszusammenbruch als Folge zu hohen Inputs fördert, aber auch Mangelzustände mit schweren Frustrationen (z. B. wiederholte Trennungen, wechselnde Bezugspersonen oder gespannte Familienverhältnisse).

Triebimpulse sind objektbezogene Begehren, welche durch ihre Ausrichtung auf das Objekt die seelischen Funktionen ordnen. Sie brauchen hierzu aber eine emotionale Resonanz, ein lebendiges, den Säugling selbst besetzendes Objekt. Wird das aufkeimende Selbst des Säuglings nur mangelhaft besetzt (sei dies anhaltend oder wechselnd), so kann sich der emotionale Dialog nicht etablieren, die Ich-Funktionen laufen infolge qualitativen und quantitativen Besetzungsmangels leer und münden im automatisierten Wiederholungszwang oder in autodestruktiven Abläufen. Denn der Säugling besitzt bereits eine reiche kommunikative Ausstattung auf allen sensorischen Kanälen (visuell, auditiv, olfaktorisch, kinästhetisch), er ist keinesfalls ein passiver Reizempfänger, sondern ein Individuum mit großen eigenen Kompetenzen. Er kann Signale der Mutter wahrnehmen, auf diese reagieren und selbst stimulierende Signale aussenden. Er ist somit ein initiatives Wesen mit der Fähigkeit und Neigung, innerhalb angeborener Programme das Objekt zu besetzen und eine Beziehung aufzubauen (Bürgin, 1982 b). Hierbei ist er auf die Mutter bzw. ihre Substitutsperson in hohem Maße angewiesen, insbesondere auf deren koordinative Fähigkeiten und auf das Besetztwerden durch sie, ohne welches seine eigenen Besetzungen im Dialogversuch ins Leere gingen. Er ist aber auch auf sie angewiesen, da sie für einen Großteil der Abwehr, welche später von seinem Ich übernommen wird, mittels der Gesamtheit ihrer mütterlichen Fürsorge aufkommt.

54.6.1 Frühgeborene und überansprechbare Säuglinge

Frühgeborene Kinder zeigen eine noch ungenügende Ausreifung im kommunikativen Bereich, mit einer gewissen Zurückhaltung, was das Bedürfnis nach Körperkontakt angeht, und mit der Tendenz, den Blick abzuwenden. Manchmal ist bereits bei der Geburt eine gewisse Dysharmonie in der Triebanlage und dadurch eine erhöhte Gefährdung des primären Narißmus zu beobachten. Hieraus resultiert eine große Verletzlichkeit. Durchschnittliche mütterliche Fürsorge reicht dann nicht, damit sich die angeborenen, interaktiven Muster und die Bewegungen von Reifung und Entwicklung ungestört entfalten können.

Bei nicht wenigen psychosomatisch auffälligen Säuglingen (z. B. beim Vorliegen von Schlafstörungen oder Drei-Monats-Koliken) finden sich Angaben über eine Schwangerschaft, die voll Angst und Spannung war, und es zeigen sich Charakteristika eines überansprechbaren Säuglings (Hypertonus, Hypervigilität, Überreizbarkeit). Die angeborenen Abwehrfähigkeiten sind geschmälert. Solche Kinder bedürfen übermäßigen Geschicks, um einen ausreichenden Schutz vor Reizüberlastung zu erfahren und den notwendigen emotionalen Dialog in Gang zu bringen. Baut sich auf dieser übermäßig verletzbaren seelischen Struktur noch ein konfliktuelles Beziehungsnetz auf, so entsteht bald eine entsprechende psychosomatische Dekompensation.

54.6.2 Privation/Deprivation

Besteht eine Privation, d.h. ein Ungenügen von seiten des Dialogpartners von Beginn ab, so entwickelt sich der Säugling kümmerlich, zeigt ein Bedürfnis nach konstanter motorischer Aktivität, eine Armut des emotionalen Ausdrucks (weint fast nie, ist ziemlich schmerzunempfindlich), baut keine spezifische Objektbeziehung auf, tritt zu jedem in Kontakt und zeigt auch keine Acht-Monats-Angst. Es findet keine dauerhafte Internalisierung einer Mutterrepräsentanz statt. Das psychosomatische Erkrankungsrisiko ist bei diesem mechanistischen seelischen Funktionieren hoch. Allerdings ist das Zustandsbild bei einem den Dialog fördernden Beziehungsangebot reversibel, da die Grundbedürfnisse nur wie verschüttet sind.

Bei schwerer emotionaler Deprivation, d.h. bei Zerreißung einer Objektbeziehung im Aufbau (z.B. durch Trennung oder durch depressive Dekompensation der Mutter) ohne Angebot einer Substitutsbeziehung, entsteht eine zentral ausgelöste Bremsung aller Aktivitäten mit Hypermotorik und verminderter Reagibilität/Kommunikativität. Nach der Phase des Protestes macht sich eine allgemeine Hemmung mit Besetzungsrückzug Platz, im Extrem bis zum Verlust des Interesses an der Außenwelt und der Entwicklung einer **anaklitischen Depression**. Das fehlende Objekt, der nicht mehr vorhandene Beziehungspartner, hat gleichsam alles Lebenswerte weggenommen, Schmerz und ein Loch im seelischen Erleben hinterlassen, das vom Säugling allein nicht ausgeglichen werden kann, sondern höchstens mittels einer depressiven Schutzbildung zu vernarben vermag. Kommt es zur anaklitischen Depression, so wird eine damit verknüpfte psychosomatische Dekompensation erleichtert (Kreisler, 1985 b).

Viele psychosomatische Erscheinungen beim Säugling und Kleinkind sind kurzfristig, d.h. es findet eine regressive Dekompensation mit einer restitutio ad integrum statt. Bei anderen kommt es zur Ausbildung einer unspezifischen Psychopathologie. Wiederum bei anderen kann sich eine psychosomatische Pathologie bis in die Kindheit hineinziehen. Ein geringerer Teil der Störungen (z.B. gewisse Formen von Asthma oder Adipositas) zieht sich bis ins Erwachsenenalter durch (Bürgin, 1987).

Die holothyme Abfuhr über den Körper ist beim Kleinkind eine häufige Form der Abwehr. Findet eine Fixierung statt, so erhält sich diese Abwehrform bis in weitere Entwicklungsphasen hinein, während der eigentlich günstigere Formen gefunden werden könnten. Eine Regression auf diese frühen Muster der Be-

ziehung und des Ich-Funktionierens ist von jedem Alter aus möglich.

Bei jeglicher derartiger Regression ist entscheidend, wieviel Ressourcen zur Progression in welcher Zeit mobilisiert werden können, wie die Gesamtreorganisation der Ich-Strukturen auf dem regressiven Niveau erfolgt und wie die Adaptationsvorgänge innerhalb der Familie an die regressive Dekompensation des betroffenen Familienmitglieds sind.

54.7 Krankheitsbilder

54.7.1 Schlafstörungen

Für das Verständnis von Schlafstörungen ist das komplizierte Zusammenspiel von psychischen und zentralnervösen Wirkungs- und Entwicklungsfaktoren von zentraler Bedeutung.

Der **Schlaf-Wach-Rhythmus** gehört zu den fundamentalen, hereditär angelegten, biologischen Rhythmen des menschlichen Organismus. Ein Ruhe-/Aktivitätswechsel entwickelt sich bereits intrauterin zwischen der 13. und 36. Schwangerschaftswoche. Von der 32. Schwangerschaftswoche an zeichnet sich immer deutlicher ein Schlaf-Wach-Rhythmus mit einer Periodizität von ca. 3½ Stunden ab. In den Schlafphasen sind von dieser Zeit ab auch zunehmend klassische REM-Phasen zu erkennen, welche beim Neugeborenen über 50% der Gesamtschlafzeit ausmachen (Bürgin, 1982).

Bis etwa zum 3. postnatalen Monat ist das Muster der Schlaf-Wach-Aktivität vor allem durch die biologische Reifung des Zentralnervensystems geregelt. Von frühester Zeit an und vor allem nach dem 3. Monat ist die Vernetzung von psychischen und physiologischen Reifungsschritten höchst intensiv. Um etwa den 3. Monat herum schlafen 70% der Säuglinge von Mitternacht bis zum frühen Morgen durch. Nach diesem Zeitpunkt spielen Umweltfaktoren wie Stimulation, Bedürfnisbefriedigung, Frustration, Aktivität und Fütterungsgewohnheiten eine zunehmend größere Rolle in der Gestaltung des Schlaf-Wach-Musters. Mit etwa 6 Monaten schlafen 83% der Kleinkinder bei Nacht (Moore und Ocko, 1957).

Über **Schlafstörungen** wird in der pädiatrischen Praxis häufig geklagt. Hierbei gibt es verschiedene entwicklungspsychologische Determinanten zu berücksichtigen:
- äußere Faktoren, welche das Gefühl von Sicherheit und Grundvertrauen bedrohen (interpersonale Störungen);
- traumatische Geschehnisse oder Erinnerungen;
- entwicklungspsychologische Belastungen im Zusammenhang mit Separation, Individuation, Autonomie- und Unabhängigkeitsentwicklung;
- Schwierigkeiten infolge von Triebkonflikten;
- verzögerte Entwicklung von Ich-Funktionen, vor allem der Realitätsprüfung und der kognitiven Entwicklung;
- unausgewogene Abwehrstrukturen, insbesondere

ein Überwiegen der oft eng mit dem Schlafzustand verbunden Regression.

1. Lebensjahr

Die Entwicklung eines individuellen Schlafmusters in den ersten 3 Lebensmonaten ist eng mit der Ich-Entwicklung verkoppelt. Sie ist das Produkt einer angeborenen biologischen Reifung und von Umweltfaktoren wie Stimulation, Befriedigung und Frustration. Schädigungen des Kindes schaffen eine Prädisposition für eine verzögerte Entwicklung des Schlaf-Wach-Rhythmus. Nach dem 3. Lebensmonat ist das Schlaf-EEG in seinem Grundmuster dem des Erwachsenen bereits angeglichen (klare Trennung in REM- und N-REM-Schlafstadien). In den ersten 3 Lebensmonaten ist der Schlaf-Wach-Rhythmus eng mit anderen Bedürfnissen, vor allem der Ernährung, verkoppelt. Erwachen wird durch Hunger provoziert, Einschlafen durch Sattheit. Zwei anfängliche Befindlichkeitszustände, nämlich Spannung und Entspannung, gliedern sich der Wachheit und dem Schlaf zunehmend an. Die Infiltration des Schlafes durch Triebimpulse ist somit eine vitale Unumgänglichkeit.

Die motorischen und psychischen Entwicklungen (z. B. Aufbau eines spezifischen Signalsystems mit der Mutter [Stern, 1979], das blickerwidernde Lächeln, die Konstituierung des Objektes um den 8. Monat herum, die symbiotische Phase mit der Mutter) des ersten Lebensjahres sind mit einer raschen Zunahme sensorischer Erfahrungen und Stimulationen aus der Außenwelt verbunden. Diese Außenfaktoren spielen eine zunehmend größere Rolle bei der Mitgestaltung des Schlaf-Wach-Musters. Trennungen von der Mutter oder deutliche Änderungen in ihrem Gefühlszustand (z. B. bei Spannungen in der Ehe, Scheidung, Verlust von bedeutungsvollen Personen, neuen Schwangerschaften, bei Depression oder Vernachlässigung) beeinflussen die Mutter-Kind-Beziehung in hohem Maße und können vorübergehende oder anhaltende Schlafstörungen verursachen.

Ein gut funktionierender Schlaf-Wach-Rhythmus entspricht einem mühelosen Rückzug der Libido von den Objekten auf das Selbst. Vom Psychologischen her sind zwei Arten von Schlaf zu unterscheiden: der erste, der nach einer Befriedigung auftritt und einen fast vollständigen Rückzug der objektgerichteten Libido zuläßt; der andere, der auf Frustration folgt und sich erst nach Erschöpfung einstellt. Eine übermäßige Spannung, welche keine Möglichkeit der Besänftigung gefunden hat, zieht ein totales Unvermögen zum Rückzug der Besetzung von den Objekten auf das Selbst nach sich. Sie erlaubt nicht, daß sich eine halluzinatorische Aktivität ausbildet, sondern hat eine autodestruktive, motorische Aktivität zur Folge (Fain, 1974). Der **Sättigungsschlaf** folgt nach einer „guten" Ernährung und entspricht einer ganz frühen und fast idealen Abwehraktivität. Die psychischen Prozesse lehnen sich an die physiologischen an. Unter günstigen Bedingungen wird der Schlaf rasch zu einem System der Autoregulation des Narzißmus. Beim **Erschöpfungsschlaf** ist das Kind oft gezwungen aufzu-

wachen, weil es durch die heftigen Frustrationen zum Träumen genötigt wird, zu einem Zeitpunkt, in dem es noch nicht vollständig dazu imstande ist. Bekommt es dann im Wachzustand Befriedigung, z. B. durch Wiegen, so kann es zur libidinösen Entspannung gelangen und daraufhin physiologische Ruhe finden.

Auf die innerseelischen Funktionsstrukturen, die sich unter dem Entwicklungsdruck dauernd verändern, wirken äußere Elemente ein, die den Schlaf verändern: der Tagesrhythmus und die Familienatmosphäre, die psychomotorischen Akquisitionen, die Ernährungsbedingungen und die Umstände und Qualitäten der zwischenmenschlichen Beziehungen.

Der Körper ist der Ort und das Mittel, über welches der Säugling sein Unbehagen ausdrückt. Er kann dies nur tun über Veränderungen in einer Grundfunktion, wie z. B. der Nahrungsaufnahme, der Ausscheidung, der Atmung oder des Schlaf-Wach-Rhythmus. Die libidinöse Besetzung des Kindes durch die Mutter erstreckt sich nicht nur auf den ganzen Körper und die erogenen Zonen, sondern auch auf funktionelle Mechanismen. Kann die Mutter nicht die genügenden Schutzmechanismen geben oder sind die angeborenen Abwehrmöglichkeiten besonders schwach, so bewirkt eine physiologische Belastung, daß ein frei flottierender Reiz bestehenbleibt, weil er noch nicht vom Kind selbst libidinös strukturiert werden kann. Die Mutter ist für das Kleinkind eine Hüterin des Schlafes. Das Kleinkind selbst dürfte den Schlaf als Äquivalent der primären Fusion mit der Mutter erleben.

Beim Übergang vom Wachsein zum Schlaf geschehen beim Kleinkind drei Dinge: Die Libido zieht sich von den Objekten zurück, ebenso die Interessen des Ichs von der Außenwelt, und das noch wenig differenzierte Ich funktioniert mehr in vorbewußten Strukturen. All diese Vorgänge erregen die Angst des Kindes (Freud, 1965). Autoerotische Betätigungen und der Gebrauch von Übergangsobjekten sind dem Kind eigene Hilfsmittel, um diesen Übergang zustande zu bringen.

Die Schlafstörungen sind wegen ihrer Häufigkeit an erster Stelle der psychosomatischen Krankheiten der frühen Kindheit zu nennen (Kreisler, 1985 c). Sie bewirken Störungen im Schlaf der Eltern mit heftigen Reaktionen (Eltern werden oft enorm wütend. Sie haben den Eindruck, als Eltern zu versagen und fühlen sich entmutigt) bis zur Erschöpfung und Dekompensation. Fast immer haben sie eine Auswirkung auf das Familienleben und ziehen die Väter bedeutend stärker ein als viele andere Störungen.

Die meisten Schlafstörungen treten erst einige Wochen nach der Geburt auf, am häufigsten im 2. und 3. Monat. Nicht selten stecken Ernährungsfehler dahinter oder die erwähnte Reizüberlastung (Ajurriaguerra, 1970). Zwischen dem 3. und 12. Monat werden Schlafstörungen oft dem Zahnen zugesprochen, stehen aber doch wohl häufiger mit dem affektiven Zustand der Mutter in Zusammenhang. Schwere Schlafstörungen, die wegen ihrer besonderen Dauer, Intensität oder starken motorischen Symptome wie Schreien, Agitation oder Apathie besonders auffallen, können manchmal erste Hinweise auf eine frühkindliche Psychose geben. Im Vergleich zum ersten Halbjahr sind die Schlafstörungen im zweiten Halbjahr des 1. Lebensjahres selten. (In jener Zeit finden sich bevorzugt Ernährungsstörungen, frühkindliche Anorexien, Rumination etc.).

Ein in dieser Zeit künstlich, d. h. mit medikamentöser Hilfe, induzierter Schlaf trägt ein großes Risiko in sich, nämlich, daß er auf eine Funktion ohne libidinösen Gehalt reduziert wird. Je mehr die primären Pflegepersonen also auf die körperlichen und seelischen Bedürfnisse des Kindes in befriedigender Art und Weise eingehen können, desto hilfreicher sind sie für die Behebung der Schlafstörung. Bereits Wiegen kann sehr beruhigend sein. Auch die Anerkennung der Belastung der Eltern und ihrer Ängste durch empathisches Zuhören bringt häufig eine so starke Entspannung mit sich, daß sie sich direkt auf die Kinder auswirkt. Gewisse Veränderungen in der Umgebung, wie z. B. die Verhinderung von Überstimulation beim Zurruhelegen des Kindes, haben oft eine günstige Wirkung.

2. und 3. Lebensjahr

Der monozyklische Schlaf-Wach-Wechsel des Erwachsenen wird allmählich, unter zeitweiliger Erhaltung eines Mittagsschlafes, erreicht.

Ätiologisch ist auch in dieser Zeit zu unterscheiden zwischen Gründen für Schlafstörungen, die in der Außenwelt liegen und solchen aus der Innenwelt des Kindes. Das Kleinkind schreitet in der Entwicklung seiner zwischenmenschlichen Beziehungen fort von einem Stadium der Bedürfnisbefriedigung zu einem der Objektkonstanz. Innerhalb des Separations-/Individuationsvorganges wird es nun durch Trennungen vom Primärobjekt sehr verwundbar. Es gilt die Trennungsangst zu ertragen. **Angst** selbst ist aber ein generalisierter Weckreiz mit sympathikotoner Reaktion des Vegetativums und somit ein dem Schlaf entgegengesetzter Stimulus. Je mehr sich die kognitive und emotionale Objektpermanenz etabliert, insbesondere in der zweiten Hälfte des 2. Lebensjahres, desto geringer wird die Trennungsangst. Überstimulation und erschreckende Erfahrungen in der Außenwelt (z. B. Unfälle, Operationen, verlängerte Trennungserlebnisse) erzeugen Angst und Spannung, welche in die Nacht hinein anhalten und im regressiven Ich-Zustand des Schlafes wiedererlebt werden können.

Zu den mehr internalisierten Konflikten gehören die der analen Phase, die mit Unterwerfung oder Trotz gegen verschiedene äußere Zwänge und Forderungen (z. B. Motorik, Sprache, Sauberkeit), aber auch mit der Autonomieentwicklung zu tun haben. Auch erzeugen eigene aggressive Impulse Angst. Oft besteht im regressiven Zustand des Schlafes Angst vor Kontrollverlust. Der Verlust des Vaters im Alter von 2 bis 4 Jahren durch Trennung oder Tod erzeugt, aufgrund des entwicklungspsychologisch bedingten, gesteigerten und gleichzeitig frustrierten Bedürfnisses nach dem Vater („Vaterhunger"), nicht selten Schlafstörungen (Herzog, 1980). Man könnte von einer freudigen Schlaflosigkeit des besonders lebhaften, von einer trotzigen des opponierenden und von einer ängstlichen des irritierten Kindes sprechen (Kreisler, 1976).

3. bis 6. Lebensjahr

Zu den Entwicklungsaufgaben des Vorschulkindes, das die Sprache schon recht gut meistert, individuiert ist und die entsprechende Ich-Entwicklung durchgemacht hat, gehören die triebbedingten Auseinandersetzungen mit heftigen Liebes- und Haßgefühlen, die im triangulären Beziehungsnetz dem Kind erste Lösungen des sog. ödipalen Konfliktes aufnötigen und es damit zur Auseinandersetzung drängen mit seinen Konzepten über Leben, Tod und besonders die Sexualität (d. h. die Frage, woher die Kinder kommen). Es erfolgen heftige Internalisierungsbewegungen von familiären und elterlichen Wert- und Rechtssystemen, was dem Kind zwar eine größere äußere Unabhängigkeit ermöglicht, dafür verstärkte innere Konflikte beschert. Milde und vorübergehende Schlafstörungen sind in diesem Lebensabschnitt in unserer Kultur häufig vorhanden (Sperling, 1955). Sie sind gleichsam Nebengeräusche der Konfliktbewältigungsversuche dieses Alters. Sind die Störungen aber tiefgehend oder länger anhaltend, führen sie zu starker Müdigkeit oder Angst, beeinträchtigen sie die Beziehung zu den Eltern, Geschwistern oder Gleichaltrigen und vermindern sie Aufmerksamkeit und Entwicklungsfortschritt, so ist eine psychiatrische Untersuchung unbedingt angezeigt (Nagera, 1966).

Entsprechend der Entwicklung des Todeskonzeptes und der Nähe von Vorstellungen über Tod und Schlaf wird Einschlafen in vorbewußten Ich-Anteilen oft mit dem Sterben und Nicht-mehr-Erwachen gleichgesetzt. Neugierde und konfliktuöse Konfrontation innerer Vorstellungen mit äußeren Realitäten stellen, zusammen mit traumatischen familiären Konfliktsituationen, weitere Ursachen für Ängste und damit von Schlafstörungen dar. Die Schlafstörungen dieses Zeitabschnittes sind im allgemeinen mit einigen therapeutischen Familien- oder Eltern-Kind-Gesprächen und nachfolgender Elternberatung gut zu beheben, es sei denn, das familiäre Beziehungsnetz sei sehr dysfunktional oder die intrapsychischen Konflikte des Kindes sehr tiefgehend, d. h. auf solchen aus vorherigen Entwicklungsabschnitten aufbauend.

Spezifische Formen

Einschlafängste und spezifische Einschlafrituale

Sie treten etwa zwischen 2 und 2½ Jahren auf, halten oft nicht sehr lange an, sind dann ein harmloses Begleitphänomen der seelischen Entwicklung und treten als Ängste vor dem Dunkeln (2–3jährige), vor Tieren im Bett (3–5jährige), vor Einbrechern oder bösen Menschen unter dem Bett (vor allem bei Mädchen von ca. 6 Jahren) oder vor Schatten, die seltsame Formen annehmen können, in Erscheinung. Komplexe

Rituale können dazu dienen, die in den Übergangsphasen zwischen Wachsein und Schlafen deutlicher hervortretenden Ängste magisch zu bannen. Sie haben oft Zwangscharakter (Uniformität, Repetition) und weisen manchmal auf eine leicht verzögerte Ich-Entwicklung hin.

Der Alptraum

Während beim Erwachsenen ein Druckgefühl auf der Brust mit Atemschwierigkeiten und Gefühlen von Hilflosigkeit und Lähmung im Vordergrund steht, gehört beim Kind vor allem eine ausgeprägte Angst zum Alptraum. Alpträume kommen insbesondere während der REM-Phasen des Schlafes vor. Sie sind meist mit starkem Anstieg der Atem- und Herzfrequenz verbunden, treten besonders häufig in der Vorschulzeit auf und werden nach dem Alter von 6 Jahren seltener (Ablon, 1979). Sie sind häufig mit anderen Schlafstörungen verknüpft.

Der Pavor nocturnus

Bei ungefähr 3% aller Kinder nachzuweisen, ist er bei Knaben etwas häufiger als bei Mädchen und kommt bevorzugt zwischen dem 5. und 7. Lebensjahr vor; gegen die Pubertät zu wird er praktisch nicht mehr gesehen. Der Pavor nocturnus ist charakterisiert durch heftige Angst, Schreien, oft Halbwachwerden, manchmal Schlafwandeln und immer durch eine retrograde Amnesie für die gesamte Episode. Sehr häufig sind auch Zeichen einer heftigen Mitbeteiligung des vegetativen Nervensystems vorhanden. Der klassische Pavor nocturnus tritt auf bei der Aufwachbewegung aus Stadium 4.

Die Kinder erwachen mit einem lauten, höchst erschreckten Schreien, stark erhöhter kardiorespiratorischer Aktivität, rufen nach Hilfe, sitzen auf dem Bett und starren mit weit geöffneten Augen wie auf etwas extrem Bedrohliches. Oft sprechen sie auf die Gesichter, die beruhigenden Worte oder die Umarmungen der Eltern zuerst nicht an. Sie können wie verzweifelt umhergehen, sich selbst verletzen und heftig sein. Es ist, als ob sie Illusionen oder Halluzinationen hätten. Die Herzfrequenz kann bis auf 180 Schläge pro Minute ansteigen.

Es können mehrere Pavor-nocturnus-Anfälle pro Nacht auftreten. Falls das Kind wirklich aufwachen kann, ist der Inhalt seines Erlebens meist nicht sehr komplex oder elaboriert, sondern eher von einem einzelnen, sehr lebhaften Bild, das mit einer physischen Erfahrung wie Fallen oder Erdrücktwerden verbunden ist, dominiert. Bedrohliche Gefühle werden in über 58% erinnert (Fisher, 1974). Die Auslösung eines Pavor nocturnus erfolgt häufig durch äußere Elemente wie Hospitalisierung, längere Trennung von der Mutter oder Todesfälle in der Familie.

Bei der diagnostischen Evaluation sind der Stand der Ich-Entwicklung, die Art aktueller traumatischer Geschehnisse sowie entsprechende intrapsychische oder intrafamiliäre Konflikte und Belastungen abzuschätzen. Erst dann kann entschieden werden, ob diese Schlafstörung mit einer Einschränkung der Entwicklung des Kindes verbunden und damit therapie-

bedürftig ist oder ob sie mittels der Ressourcen des Kindes oder seiner Umgebung überwunden werden kann.

Der Somnambulismus

Bei ca. 1 bis 15% aller Kinder ist irgendwann ein Schlafwandeln zu beobachten (Ablon, 1979). Der Somnambulismus ist oft verbunden mit Pavor nocturnus und Enuresis nocturna. Er beginnt meist mit einem Aufsitzen im Bett. Das Kind hat die Augen offen mit glasigem Blick, es steht auf und bewegt sich plump, stößt aber kaum je irgendwo an. Die Episode kann von einer Minute bis zu einer halben Stunde oder länger dauern. Sie endet meist damit, daß das Kind ins Bett zurückkehrt.

Manchmal werden einige kaum verständliche Worte gemurmelt. Nach dem Erwachen besteht eine Amnesie für die gesamte Periode. Auch der Somnambulismus tritt beim Aufwachen aus dem N-REM-Schlafstadium 3 und 4 auf, somit in den ersten Schlafstunden, und wird oft als eine Unreife des Zentralnervensystems verstanden. Vom Psychologischen her entspricht er etwa einem dissoziierten, hysterischen Symptom, dem Ausagieren eines Traumes. Auf jeden Fall ist er sehr oft mit Spannungen intrapsychischer oder intrafamiliärer Art verbunden.

Bei der Abklärungsuntersuchung sind Häufigkeit des Auftretens und Gefährlichkeit der motorischen Akte (de Villard, 1980), vergangene und gegenwärtige Entwicklungsschwierigkeiten, familiäre Probleme, aktuelle traumatische, intrapsychische und/oder intrafamiliäre Belastungen abzuschätzen.

Schulzeit und Adoleszenz

Das Latenzkind hat die ersten Stromschnellen seiner Trieborganisation in irgendeiner Art und Weise hinter sich gebracht. Mit dem Aufbau des Über-Ichs und damit des Gewissens ist es nach außen bedeutend unabhängiger geworden. Die Opposition beim Sichschlafenlegen, die vor allem im 2. und 3. Lebensjahr zu bemerken war, mit ihrer Weigerung, den Wutausbrüchen, dem Wiederaufstehen und dem Kampf gegen Passivität, ist überwunden. Das Kind hat gelernt zu akzeptieren, daß es die Besetzung von der Außenwelt abziehen und sich in den Schlaf gehen lassen kann, ohne daß Schlimmeres passiert. Trennungsangst, Angst vor den Bildern der eigenen Traumwelt, Gefühle des Ausgeschlossenseins vom Erwachsenenleben und Gefühle der Passivität bei nicht beherrschbaren Abläufen sind nicht mehr so angsterzeugend. Auch an die hypnagogen Phänomene beim Einschlafen hat es sich bereits ein Stück weit gewöhnt. Schlafrituale, welche vom 2. Lebensjahr ab zu beobachten sind und vor allem zwischen dem 4. und dem 6. eine Häufung zeigen, haben hierbei mitgeholfen. Die anscheinend unnützen, unverändert täglich wiederholten Abläufe bringen das Bedürfnis des Kindes zum Ausdruck, sich der Kontinuität seiner Umgebung zu versichern, und zwar in den Momenten, in welchen es zu einem Besetzungsabzug genötigt ist. Schlafrituale entsprechen einer Art Beschwörungsgesten bei bedrohlichen Phantasien. Auch die Schlafphobien, welche oft in der ödipalen Phase mit Dunkelangst, dem Wunsch nach offen gelassener Tür und einer Forderung des Neben-sich-ins-Bett-Legens verbunden sind, werden jetzt seltener.

Der **Traum** bindet Triebregungen und verhindert so die Überflutung des psychischen Apparates. Mittels der Abwehrvorgänge in der Traumarbeit wird das Begehren unkenntlich gemacht. Der Traum hilft somit, Angst zu desomatisieren, welche als Frucht eines Konfliktes zwischen dem Unbewußten, dem Sitz des Begehrens, und dem Vorbewußten, dem Sitz der Abwehr, notwendigerweise entsteht. Der Traum als Prozeß ist eine biologische Gegebenheit, der Traum als Erlebnis eine entwicklungspsychologische Tatsache, die Traumerzählung ein Hinweis dafür, wie sehr das Selbst in den Traumablauf verwoben ist (Houzel, 1985). Träumen kann nun klar von der äußeren Realität unterschieden werden. Die magischen Denkstrukturen weichen rationaleren Prozessen. Damit verändern sich auch die Inhalte möglicher Angstträume. Hauptsächlichste Gründe für die Nachtängste beim Kind sind die jedem Entwicklungsschritt innewohnenden Konflikte, traumatisierende Erlebnisse oder aber Beziehungskonflikte in der Familie. Wenn in dieser Zeit Schlafstörungen auftreten, die anhalten oder in Ausmaß und Tiefe zunehmen, so dürfen sie als Indikatoren eines tieferen pathologischen Geschehens intrapsychischer oder intrafamiliärer Art betrachtet werden. Sie weisen dann darauf hin, daß entweder das Kind oder die Familie als Ganzes vorhergehende Konflikte nur ungenügend gemeistert haben (Bürgin, 1986 a). Leichte Schlafstörungen waren bei einem Kollektiv von 9jährigen Schweizer Schülern und Schülerinnen in 45 bis 50% der Fälle vorhanden (Bettschart, 1978). In einer früheren Studie (Harnack, 1958) wurden bei Hamburger Schülern von 10 bis 11 Jahren in rund 15 bis 27% der Kinder Schlafstörungen diagnostiziert. Der Schule als äußerer Belastungssituation kommt für die Entstehung von Schlafstörungen in diesem Lebensabschnitt eine nicht geringe Bedeutung zu, da Kompetition, Kränkungen und Scheitern von Größenvorstellungen mit dem Gruppengeschehen und den pädagogischen Forderungen eng verbunden sind.

Die **Jactatio capitis** (bis zur Abschabung der Haare am Hinterkopf) und die **Jactatio corporis,** stereotype, streng rhythmisierte Bewegungen beim Einschlafen oder im Zustand des Alleinseins, die gelegentlich tranceartige oder hypnoide Zustände nach sich ziehen, werden bereits im 1. Lebensjahr gesehen. Ihre Häufigkeit zu Beginn des Schulalters liegt bei 2 bis 4%. Oft sind die Jaktationen mit Daumenlutschen oder Saugen an einem Tuch/Zipfel verbunden. Nach der Pubertät sind sie meist verschwunden oder nur selten noch festzustellen. Sie treten bei Knaben doppelt so häufig wie bei Mädchen auf. Für die Eltern gewinnen sie zumeist dann die Qualität von Symptomen, wenn die Begleitgeräusche für sie selbst oder für die Nachbarschaft störend werden. Die Jaktationen dienen nicht nur der einfachen motorischen Erregungsabfuhr, sondern auch der Selbstbeschaffung

von Reizen, der Selbstberuhigung (Ersatz von Wiegen) oder auch der Selbstbefriedigung. Es besteht eine deutliche Beziehung zur kindlichen Onanie. Beim Auftreten der Adoleszentenmasturbation hören sie auf. Die Häufigkeit, mit der sie bei Heimkindern angetroffen werden, weist darauf hin, wie sehr es sich um Ersatz- und Trostbefriedigungen für ein Manko an emotionaler Zuwendung handelt. Bei Kindern mit leichter Hirnreifungsstörung sind Jaktationen gehäuft zu verzeichnen. Werden sie als Signal für vermehrte Zuwendung emotionaler und körperlicher Art verstanden und können die familiären Situationen saniert werden, so ist die Prognose insgesamt gut.

Alpträume, Pavor nocturnus, Somnambulismus und Schlafstörungen mit Enuresis nocturna gibt es auch in der Adoleszenz, wenngleich viel weniger häufig. Meist begleiten Schlafstörungen in diesem Lebensabschnitt andere psychische oder psychosomatische Symptome. Insbesondere sind solche bei Jugendlichen mit Angstneurosen, Phobien oder hysterischen Neurosen zu beobachten. Manchmal treten Schlafstörungen auch infolge von Schuldgefühlen wegen Masturbation auf. Hypersomnien als regressive Flucht vor Konflikten sind für die Adoleszenz recht typisch. Schlafstörungen finden nicht selten ihre Ursache auch im Geschehen des Tages: „unvernünftiger" Lebensrhythmus mit Übermüdung, übermäßige Stimulation mit Überreizung (Tee, Kaffee, Tabak, Alkohol, Pharmaka, Drogen) oder exzessive Aktivität mit Überforderung (Kreisler, 1981). Sie können, sofern sie persistieren und durch ihre Schwere weitere psychische Funktionen beeinträchtigen, erste Signale eines neurotischen Geschehens, einer latenten depressiven oder einer psychotischen Erkrankung sein und bedürfen einer sorgfältigen fachärztlichen Evaluation, oft unter Einbezug der übrigen Familienmitglieder.

Die Narkolepsie mit den tagsüber unbezwingbaren, 10 bis 15 Minuten dauernden Schlafattacken, der Kataplexie, welche emotional ausgelöst sein kann, den Schlaflähmungen und den hypnagogen Halluzinationen kann bereits in der Mittadoleszenz beginnen.

Das **Klein-Lévin-Syndrom,** welches sich durch hypersomnische Perioden von 1 bis 3 Wochen mit extremem Hunger und zum Teil Adipositas, psychischer Labilität, depressiven oder Verwirrungszuständen äußert, die zwei- bis dreimal pro Jahr auftreten können, ist vor allem bei Knaben zu beobachten und kann auch schon in der Mitte der Adoleszenz auftreten.

Diagnostik und Therapie

Es gilt als erstes, die Schlafstörung genau zu bestimmen, ihre Intensität, Dauer, Verbindung mit weiteren psychischen Symptomen und die Folge- bzw. Sekundärerscheinungen abzuschätzen. Organische Gründe müssen sodann ausgeschlossen werden.

Die Mehrzahl von Kindern und Jugendlichen mit Schlafstörungen wird von Allgemeinpraktikern und Pädiatern gesehen. Der behandelnde Arzt muß den Weg suchen zwischen der Skylla, Organisches zu „psychologisieren", und der Charybdis, soziale oder psychologische Gegebenheiten zu „somatisieren". Einschlafstörungen weisen häufig auf Ängste, Durchschlafstörungen eher auf eine Problematik im Traumgeschehen hin, wogegen ein zu frühes Erwachen oft mit depressiven Phänomenen einhergeht.

Wenn immer möglich, sollte die Symptomatik ihre Auflösung im ärztlichen Gespräch und Dialog mit dem Kind, den Eltern oder der Familie finden. Manchmal müssen Veränderungen in der Umwelt und Lebensweise der Betroffenen vorgenommen werden. Ist auf eine Pharmakotherapie nicht zu verzichten, so muß man sich dennoch vergegenwärtigen, daß ein recht tiefgreifender Eingriff in die zerebrale Rhythmik und Physiologie von Schlaf und Traum vorgenommen wird. Es sollten deshalb in erster Linie Medikamente verwendet werden, mit denen auf Ängste und depressive Verstimmungen eingewirkt werden kann (Anxiolytika, Antidepressiva). Gegebenenfalls kann es auch dienlich sein, schlafanstoßende Effekte zu erzielen (Neuroleptika). Schlaferzwingende Medikamente (z.B. Barbiturate) sollten, wenn immer möglich, vermieden werden. Eine nur pharmakologische Behandlung von Schlafstörungen ohne eine gewisse, wenigstens minimale psychotherapeutische Betreuung ist zu vermeiden.

54.7.2 Die Drei-Monats-Kolik (Nabelkolik)

Etwa vom 15./20. Lebenstag an bis gegen Ende des 2. oder 3. Lebensmonates, in welchem plötzlich eine spontane Besserung auftritt, schreien diese Säuglinge vor allem am späten Nachmittag und nach Nahrungsaufnahme, wahrscheinlich infolge viszeraler Dysfunktion. Ihr Abdomen ist aufgebläht. Es handelt sich um sehr lebhafte, nervöse, extrem leicht stimulierbare Kinder. Ernährungsänderungen erbringen kaum eine Besserung. Oft ist die ganze Familie in das Krankheitsgeschehen miteinbezogen. Schnuller und Wiegen beruhigen solche Säuglinge.

Die Säuglinge selbst zeigen meist einen starken Hypertonus, leicht auslösbare primitive Reflexe und ein gieriges Saugen mit schnellem Schlucken. Es handelt sich wahrscheinlich um besonders vitale Kinder mit niedriger Reizschwelle (Spitz, 1967; Kreisler, 1976, 1981).

Auffällige Haltungen der Eltern, die in ihrer Fürsorge und Säuglingspflege eine etwas inkohärente, widersprüchliche und zum Teil überfürsorgliche Haltung einnehmen, sind nicht selten sekundär. Die Eltern sind erheblich dadurch verunsichert, daß es ihnen nicht gelingt, eine diffuse Überreizung des Säuglings abzumildern.

Wenn man davon ausgeht, daß die Regulationsvorgänge des Säuglings in diesem Alter noch außerordentlich schwach sind und den primären Pflegepersonen als äußeren Instanzen diejenigen regulativen Funktionen obliegen, die später durch Internalisierung vom Kleinkind selbst übernommen werden, so liegt auf der Hand, daß die Entwicklung des emotionalen Dialoges störanfällig ist. Das Wiegen oder die Saugbefriedigung haben zwar eine gewisse beruhi-

gende Wirkung, sind allein aber nicht ausreichend, um beginnende Schwierigkeiten im emotionalen Dialog zu überwinden.

Neben der sorgfältigen Abklärung und dem Ausschluß einer organischen Krankheit ist bei leichteren Fällen keine Behandlung notwendig, da um den 3. Lebensmonat herum, wenn es zum blickerwidernden Lächeln kommt, sich auch andere Abwehrmöglichkeiten im Innenleben des Säuglings entwickeln. Frühzeitig einsetzende klärende Gespräche mit der Mutter, mit dem Ziel der Information über die großen Anpassungsleistungen, die ihr Kind infolge des raschen Zuwachses an Wahrnehmungen leisten muß, sind meist ausreichend, tragen zur Verbesserung des emotionalen Dialoges bei und haben in der Regel eine rasche Beruhigung des Geschehens zur Folge. Bei schwereren Situationen mit beginnender Dystrophie kann eine Hospitalisation notwendig sein.

54.7.3 Psychogenes Erbrechen beim Säugling und Kleinkind

Sporadisch kann das Symptom bei verschiedenen affektiven Belastungen vorkommen. Habituell zeigt es sich im Rahmen der Anorexie des Kleinkindes und als funktionelles Erbrechen. Es wird digital provoziert und ist Ausdruck einer regressiven Zersetzung der Objektbeziehung. Die Nahrung wird zum schlechten Objekt, das auf diese Art und Weise auszustoßen versucht wird. Bei den schweren Formen des habituellen funktionellen Erbrechens findet man häufig nicht nur eine Reizüberlastung, sondern auch eine Reizinkohärenz, d.h. eine inkonstante Besetzung des Kindes durch die primäre Pflegeperson oder schwerer gestörte, bis ins Chaotische gehende Lebensformen (Kreisler, 1985f).

54.7.4 Rumination

Als Rumination wird die Regurgitation von Speisebrei mit partiellem oder totalem Wiederverschlucken nach kauähnlichen Bewegungen bezeichnet. Sie tritt vor allem in der zweiten Hälfte des ersten Lebensjahres auf, Knaben sind bevorzugt betroffen. Das Erbrechen kann durch Einführung eines Fingers in den Hals herbeigeführt werden, oder aber durch Muskelaktivitäten im Pharynx, Thorax, Abdomen und Zwerchfell. Das Kind spielt mit dem hochgestiegenen Mageninhalt und braucht, damit es nicht zum Erbrechen kommt, eine Fähigkeit, die regurgitierte Menge genau zu dosieren. Die anfängliche Mißstimmung weicht beim Ruminieren dem Ausdruck einer entspannten Zufriedenheit. Während des Ruminierens ist das Kind aber wie von der Außenwelt abgewendet, ganz auf sich bezogen, es erzeugt den Eindruck einer affektiven Leere. Bei Stimulation oder Präsenz von Menschen hört es sofort auf zu ruminieren. Fließt vielfach ein Teil der regurgitierten Nahrung aus dem Mund, kommt es zu Exsikkose, Dystrophie, vermehrter Infektanfälligkeit und selten auch zu Wachstumsrückstand. Die Kinder sind kontaktgierig, insbesondere mit den Augen, aber ihr Kontakthunger ist unspezifisch. Sie zeigen keine Acht-Monats-Angst, d.h. keine anhaltende Introjektion des Objektes. Nicht selten ist ein hohes motorisches Aktivitätsniveau mit hektischer oder zappeliger Gespanntheit zu beobachten. Rumination wird sehr oft bei emotionaler (De-)Privation, bei wechselnder Intensität der Besetzung durch die Mutter oder bei völligem Besetzungsentzug gesehen. Sie gehört zum großen Gebiet des psychogenen Erbrechens, wird nicht selten beobachtet nach einem gewaltsamen Unterbinden des Daumenlutschens, hat aber keine spezifisch auslösenden Determinanten. Die Mütter dieser Säuglinge sind häufig depressiv oder dem Kind gegenüber sehr ambivalent, können auf die neuen Bedürfnisse und Kontaktanforderungen ihrer Kinder nicht eingehen, sondern halten an der für sie einfacheren, frühen Beziehungsform fest (Kreisler, 1985 e).

Psychodynamisch kommt die Rumination einer zwanghaft repetitiven Regression auf eine orale Autoerotik zuungunsten der Objektbeziehung gleich. Die sensorischen Besetzungen werden aus ihrer primären Nachaußengerichtetheit auf die Sensationen aus dem Körperinneren zurückgezogen. Es handelt sich um die Kompensation einer narzißtischen Zufuhr, die üblicherweise durch die vielfältigen sensorischen und emotionalen Stimulationen der Mutter erfolgt, und um ein nachfolgendes, erotisiertes Spiel mit dem Nahrungsbrei. Diese perverse Handlung hat zum Ziel, das Manko nicht wahrzunehmen. Der regurgitierte Mageninhalt wird zum guten Objekt, das das Kind ernährt, ihm lustvolle Gefühle vermittelt, jederzeit verfügbar ist und damit scheinbar die Autonomie erhöht. Die Befriedigung bleibt aber unvollständig, hält nicht lange an. Wegen der steigenden Intensität der Bedürfnisse muß das Kind immer häufiger ruminieren. Die Rumination wird schließlich zu einer stereotypen Reaktion auf jede Mißempfindung, läuft aber zunehmend motorisch leer und vermittelt dann kaum mehr Funktionslust. Wird die Rumination zu unterdrücken versucht, bleibt nur noch das Abgleiten in eine depressive Dekompensation.

Gelingt es, eine entwicklungsgemäße Beziehung, d.h. einen lebendigen emotionalen Dialog mit affektiver Einstimmung herzustellen, so wird das autoerotische Objekt mit seinen oralen und muskulären Befriedigungen zugunsten der Objektbeziehung aufgegeben. Oft ist hierzu und zur Korrektur der metabolischen Lage eine Hospitalisation unumgänglich. Dort kann ein entsprechendes Beziehungsangebot gemacht und mit der Mutter an der Wiederherstellung der Beziehung und des emotionalen Dialogs, d.h. an der Reparatur des Mankos, gearbeitet werden. Über die Langzeitprognose dieser Kinder ist nur wenig bekannt.

54.7.5 Anorexie beim Säugling und Kleinkind

Nach Ausschluß einer organischen Krankheit ist die **essentielle** Säuglingsanorexie von den Anorexien der zweiten Hälfte des 1. Lebensjahres abzugrenzen. Die essentielle Form ist rar, besteht von Geburt an und entspricht wahrscheinlich einer speziellen Konstitutionsvariante. Es handelt sich um kleine, nervöse, außerordentlich schwache Kinder, die sehr passiv sind („trinkfaul") und bei denen sich eine eigentliche Opposition gegen die Nahrungsaufnahme erst nach einigen Wochen einstellt.

Bei den Anorexien, welche sich erstmals zwischen dem 6. und 12. Lebensmonat manifestieren, kann zwischen einer **inerten** Form mit großer Passivität, wenig direkt sichtbarer aktiver Beteiligung von seiten des Kindes und der **oppositionellen** Form unterschieden werden. Die erste ist oft mit Erbrechen verbunden. Statt der Ausbildung einer Art Acht-Monats-Angst projiziert das Kind die frustrierenden Anteile der Mutterrepräsentanz nicht auf den Dritten, den Fremden, sondern spaltet sie ab und projiziert sie auf die Nahrung, welche dann phobisch vermieden wird.

Bei der oppositionellen Form ist die aggressive Vermeidung sehr viel aktiver, die Kinder schreien, sind agitiert, wenden sich ab und entwickeln, insbesondere nach dem 1. Lebensjahr, ein trotzig-oppositionelles Verhalten gegenüber der Nahrung, das hungerstreikähnliche Züge trägt. In den schweren Kampfsituationen um das Essen bleibt das Kind meist Sieger. Die Reaktionen des Kindes sind Reaktionen auf Verhaltensweisen der Mutter oder umgekehrt. Hat sich der Circulus vitiosus einmal installiert, ist Kausalität nur noch in zirkulärem Sinne möglich.

Im allgemeinen sind mehr Mädchen als Knaben betroffen. Diese Kleinkinder sind überwach und treten extrem schnell in Kontakt. Anamnestisch wird über alle Formen der Verweigerung, der Verführung, der Gewalt und der Zwangsanwendung berichtet. Zumeist ist die ganze Familie in das Geschehen einbezogen. Das Trinken bleibt normal. Die auslösenden Faktoren sind unspezifisch (bei Kleinkindern oft Abstillen oder Trennung, bei größeren mehr Konflikte der Eifersucht, z.B. Geburt eines Geschwisters). Neben Faktoren der Beziehung gibt es noch endogene, z.B. eine vorzeitige oder verzögerte Ich-Entwicklung, die für die Ausbildung der Krankheit mitverantwortlich sein dürften.

Einfache Formen klingen rasch ab, da die besorgte Mutter üblicherweise schnell entsprechende Hilfe sucht. Je mehr die Störung sich organisiert und strukturiert, die Fronten sich verhärten, desto schwieriger ist es, ein Abklingen des Symptoms zu erreichen. So gibt es prolongierte Formen, bei denen anorektische Störungen bis in die Adoleszenz fortbestehen oder sich Symptomverschiebungen einstellen. Wie bei allen anderen frühen Störungen der Nahrungsaufnahme (z.B. den Kau-, Beiß- oder Schluckstörungen) sind hinter dem gleichen Symptom sehr unterschiedliche intrapsychische und intrafamiliäre Mechanismen am Werk. Prophylaktisch ist es immer günstig, wenn vermieden wird, Appetit zu erzwingen.

Schwere, komplexe Formen der frühkindlichen Anorexie sind kaum durch die üblichen Methoden zu verändern. Die Kinder verhalten sich so, als interessierten sie sich nicht für die Nahrung. Es scheint, als wäre das Hungergefühl direkt betroffen. Solche gravierenden Situationen sind bei dysharmonischen, präpsychotischen oder psychotischen Entwicklungen zu beobachten und zumeist mit Beiß- oder Kauhemmungen verbunden, aber auch bei akut phobischen Erscheinungen am Ende des 1. Lebensjahres zu sehen, begleitet von Schlafstörungen, anderen Phobien und der Verweigerung der Flüssigkeitsaufnahme bis hin zur Gefahr der Dehydrierung. Aber auch bei schweren depressiven Erscheinungen der frühen Kindheit mit massiver psychosomatischer Dekompensation sind anorektische Erscheinungen nicht selten. Bei diesen ernsthaften Formen besteht meist ein evidenter Konflikt zwischen Mutter und Kind. In der späteren Entwicklung dieser Kinder zeigen sich bevorzugt Symptome wie Enuresis, Enkopresis, Schlafstörungen, Abdominalspasmen, Verhaltensstörungen oder Störungen in der Charakterentwicklung (Kreisler, 1985 d).

Bezüglich der **Behandlung** ist bei den leichteren Formen eine Indifferenz gegenüber dem Symptom mit Überwachung des körperlichen Zustandes anzustreben. Das Kind sollte dekonditioniert und klar abgelehnte Nahrung ausgelassen werden. Bei der Umgebung ist eine Verhaltensänderung unumgänglich. Das anorektische Kleinkind darf nicht mit Tricks oder Zwängen überlistet werden. Die Nahrungsquantität sollte gering sein, so daß allmählich wieder ein Bedürfnis bzw. Hunger entstehen kann. Wenn immer möglich, ist auch eine Klärung der Beziehungsstrukturen zwischen Mutter und Kind anzustreben. Schwerere Fälle brauchen eine Hospitalisation und vereinzelt auch medikamentöse Unterstützung (Sedativa, Antihistaminika und eventuell auch Anabolika).

54.7.6 Psychosoziale Gedeihstörung und Minderwuchs

Dieses Syndrom kommt in allen Altersklassen bis zur Adoleszenz vor. Es ist gekennzeichnet durch einen „harmonischen" Wachstumsrückstand (mehr als 3 Standardabweichungen von der Norm) mit klarem Knick in der Wachstumskurve. Die Wachstumshormonsekretion erweist sich bei Stimulierung als blockiert. Sämtliche hypophysären Hormone sind verringert. Bei Trennung dieser Kinder von den Müttern zeigt sich bereits nach 10 Tagen eine Normalisierung der Hormonwerte, nach 4 Wochen eine solche des Wachstums.

Die Wachstumsblockierung hält so lange an, wie die Säuglinge oder Kleinkinder mit ihren Müttern zusammen sind. Die Störung ist oft mit Schlafstörungen, Erbrechen, verzögerter Sprachentwicklung und Sphinkterkontrolle, Jaktationen oder psychomotorischen Störungen kombiniert. Die Kinder sind gekennzeichnet durch eine enorme affektive Gier. Auf Trennung reagieren sie leicht mit depressiven Symptomen. In Substitutsbeziehungen, die aufgebaut werden, manifestieren sich als Übertragungsäquivalente bald große Aggressionen bis hin zu sadomasochistischen Interaktionen. Es besteht bei diesen Patienten ein großes Bedürfnis nach Kontrolle. Der Wachstumsrückstand läßt sich nur bei Säuglingen durch eine hypokalorische Situation erklären (Sibertin-Blanc, 1985).

In der Familie gibt es meist mehrere Kinder. Dasjenige mit dem Minderwuchs zieht als Sündenbock und Ursache steter Enttäuschung den Haß der Familie auf sich. Die Beziehungen zwischen Kind und Mutter oder Kind und Eltern sind verzerrt und oft durch emotionalen Mangel bestimmt. Die Eltern erwarten vom Kind diejenige Zuwendung und Befriedigung, dasjenige Verständnis, das sie in ihrer Kindheit selbst nicht bekommen haben. Die Patienten haben die Funktion von Stabilisatoren des Familiengleichgewichtes, da die Eltern selbst psychisch meist nur knapp kompensiert sind. Die Mütter sind oft depressiv, manchmal unterintelligent oder leiden an psychotischen Störungen.

Die Therapie besteht einerseits aus der Trennung des Kindes von der Familie bzw. der Mutter, andererseits aus einer intensiven therapeutischen Arbeit mit den Eltern/der Mutter. Wegen rigider Strukturen bei den Müttern entstehen in den Institutionen oft auch rigide Gegenübertragungsreaktionen.

Auch für andere Wachstumsstörungen unklarer Ätiologie können psychosomatische Zusammenhänge angenommen werden, vor allem wenn ein Knick in der Wachstumskurve und Verhaltens- oder Funktionsstörungen beim Kind auftreten, ohne daß entsprechende organische Ursachen dafür vorlägen.

Der psychosoziale Minderwuchs ist ein klassisches Beispiel für die vitale Rolle der Beziehung zwischen Kind und Umgebung für die somatische Entwicklung und weist auf die enorme Verflochtenheit des neuroendokrinen Systems mit den psychobiologischen Geschehnissen hin.

54.7.7 Respiratorische Affektkrämpfe

Sowohl die blasse wie auch die **zyanotische** Form des Anfalls treten vorwiegend zwischen dem 6. und 18. Lebensmonat auf. Der Affektkrampf wird durch einen Verdruß oder einen Vorwurf, der Wut erzeugte, ausgelöst. Das Kind weint laut, atmet immer schneller, blockiert schließlich die Respiration und hört auf zu atmen. Es wird blau, verliert das Bewußtsein und fällt schlaff dahin. Unter der bis zu einer Minute dauernden Atemlosigkeit können Krämpfe auftreten. Die **blasse** Form wird eher von Schmerz oder Angst ausgelöst. Bei ihr fällt das Kind nach einem kaum angedeuteten Schrei in Ohnmacht. Es manifestieren sich symmetrische Streckkrämpfe oder einseitige Zuckungen. Die Kinder kommen meist rasch wieder zu sich, sind nachher etwas niedergeschlagen und schlafen bald ein. Die Krämpfe sind Folge einer vorübergehenden zerebralen Anoxie, entweder infolge kurzdauernder Asystolie oder Apnoe. Bei länger dauernden Anfällen findet man nicht selten Urin- und Kotabgang. Das EEG ist in den Intervallen unauffällig.

Die respiratorische Funktion wird hier libidinös besetzt. Der asphyktische Zustand bewirkt einen veränderten Zustand des Bewußtseins, der als Verlockung erlebt werden kann und zugleich die Elimination eines unangenehmen Affektes ermöglicht. Die vitale Funktion des Atmens wird nach der Überreizung blockiert und stellt sich mit dieser Blockade in den Dienst der Verdrängung. Es gibt keinen Affektkrampf ohne Zuschauer. Dieses Geschehen hat eine Neigung zu raschem Sich-Einschleifen und zu Wiederholungen. Die Reaktion der Umwelt ist für solche Verstärkungen nicht unwichtig (Kreisler, 1976).

Üblicherweise verhindern die Abwehrmechanismen eine solche Gefährdung von Vitalfunktionen für die Spannungsabfuhr. Entwickelt sich eine derartige Fähigkeit zur abnormen Spannungsabfuhr, so zumeist aufgrund eines dissonanten emotionalen Dialogs und einer ungenügenden narzißtischen Unterstützung des Kindes, welche es ihm verunmöglicht, günstigere Auswege aus Kränkungs- und Machtkampfsituationen zu finden. Das dramatische Agieren der Allmacht hat erpresserischen Charakter und ist ein ungünstiger Weg, um die eigenen Bedürfnisse kundzutun. Diese Kinder entwickeln später nicht selten eine erhöhte seelische Labilität, eine Schwierigkeit beim Aufbau von Regulationsmechanismen (Neigung zu Jähzornanfällen) und wenig Konstanz im Gleichgewicht zwischen objektalen und narzißtischen Besetzungen.

Die Prognose für das Verschwinden der Symptomatik ist günstig, da die Anfälle meist im 4. Lebensjahr aufhören. Therapeutische Gespräche mit den Eltern, manchmal eine leichte medikamentöse Sedierung des Kindes und soziale Maßnahmen bei groben Fehlhaltungen der Eltern haben sich als günstig erwiesen.

54.7.8 Asthma bronchiale (vgl. Kap. 42)

Diese Reaktionsform des Bronchialsystems, welche durch rezidivierende, reversible Obstruktionen gekennzeichnet ist und durch unterschiedliche Ursachen und Auslösemechanismen in Gang gesetzt wird, gilt als eine der häufigsten chronischen Krankheiten im Kindesalter (ca. 1–2%) (v. d. Hardt et al., 1985). Bei 10jährigen Kindern sollen zwischen 4 und 10% an asthmoiden Beschwerden leiden. Vor der Pubertät sind die Jungen etwas häufiger betroffen als die Mädchen, bis zur Pubertät hin verliert sich die Symptomatik spontan bei etwa der Hälfte der Fälle. Bei 10% der Patienten dauert das Asthma bronchiale bis in die Adoleszenz hinein und weiter. Eine Kombination mit atopischer Dermatitis macht die Prognose ungünstiger.

Die klinische **Symptomatik** besteht in einer anfallsweisen Obstruktion der Bronchien, welche infolge eines Spasmus der Bronchialmuskulatur und der Produktion eines zähflüssigen Sekretes sowie durch eine ödematös-hyperämische Schwellung der Bronchialschleimhaut zustande kommt. Beim Säugling und Kleinkind manifestiert sich das Asthma meist im Rahmen einer infektiösen Lungenaffektion oder als asthmoide Bronchitis mit schleichendem Beginn. Nicht selten ist das Asthma mit den anderen Symptomen des Atopie-Syndroms (Dermatitis atopica, Rhinitis vasomotorica) verbunden, deren Erstmanifestation etwa zur gleichen Zeit wie die des Asthmas (zwischen dem 6. und 12. Lebensmonat) zu verzeichnen ist. Die exspiratorisch-obstruktiven Krisen erfolgen mit Vorliebe nachts. Während des Anfalls sind die Kinder infolge der Erstickungsangst unruhig. Die Angst selbst verstärkt in einem Kreisprozeß die Atemnot. Als leicht gelten bis zu 5, als mittelschwer bis zu 10 und als schwer bis zu 20 Anfälle pro Jahr (Hofman, 1983). Bei schwerem Asthma bestehen meist Infekte der Luftwege und eine respiratorische Insuffizienz.

Wahrscheinlich multifaktoriell vererbt, kann man sich fragen, ob das Asthma bronchiale wirklich eine Krankheitseinheit ist. Praktisch immer besteht eine **bronchiale** Übererregbarkeit, kombiniert mit einer **psychischen Übererregbarkeit.** Die Stimulationsbereitschaft der Mastzellen ist gesteigert. Zu den ursächlichen Faktoren gehören solche, die primär immunologische Reaktionen auslösen, und solche, welche mit unspezifischen Reaktionen verbunden sind. Vielfach gibt es auch gemischte Formen.

Das Asthma beim Säugling oder Kleinkind beginnt häufig zwischen dem 6. und 12. Lebensmonat und verschwindet im Verlaufe des 3. Lebensjahres. Diese Säuglinge zeigen zumeist eine auffällig leichte und gute Kontaktfähigkeit, auch zu Fremden, sind in ihrer übrigen Entwicklung unauffällig, aber durch eine fehlende Acht-Monats-Angst gekennzeichnet (Foliot, 1985).

Durch Kreisler (1974) sind bei diesen Kindern zwei Auffälligkeiten hervorgehoben worden:
– Die **Überlastung der Zweierbeziehung** durch die vorzeitige Einführung eines Dritten (z. B. bei wechselnden Pflegepersonen). Die Mutter leistet hier nicht den genügenden Schutz und die genügende Kontinuität. Das Kind regrediert auf die Dualbeziehung und fixiert sich dort.
– Die Mutter bietet eine sehr **exklusive Beziehung** mit besonders großem Einfühlungsvermögen und übersteigerter Fähigkeit zu affektiver Einstimmung an, die einen überbehütenden Charakter besitzt. Befriedigungen sollen nur im Kontakt zu ihr erlangt werden können. So wird das Kind von der Triangulation ferngehalten, progressive Tendenzen werden blockiert. Es erfolgt eine Verwöhnung durch übermäßige narzißtische Befriedigung, welche die Individuation und die Autonomieentwicklung behindert. Die Mütter dieser Kinder haben eine Tendenz, in der primären Mütterlichkeit zu verharren und besetzen ihre Identität als Partnerin des Mannes kaum mehr. Sie behalten ihr Kind auf dem Niveau eines Babys.

Es gibt keine besondere Persönlichkeitsstruktur asthmakranker Kinder. Dennoch lassen sich nach Kreisler (1974, 1985a) Unterschiede zwischen den Kindern **mit** und **ohne Allergien** festhalten. Das 1. Lebensjahr ist sowohl die kritische Zeit der Reifung des Immunsystems wie auch ein entscheidender Abschnitt für die Strukturierung der Objektbeziehung (blickerwiderndes Lächeln als erster und Acht-Monats-Angst als zweiter Organisator [Spitz, 1967]). Die Säuglinge und Kleinkinder mit Allergien zeigen ein enormes Kontaktbedürfnis, aber keine Wünsche nach längerdauernden und ausschließlichen Beziehungen. Sie wechseln und ersetzen die äußeren Objekte leicht mit neuen, zeigen keine Acht-Monats-Angst, verweilen etwa auf dem Niveau des ersten Organisators und vermeiden auf diese Art und Weise mittels einer generellen Blockierung teilweise den Separations-/Individuationsprozeß, die Triangulierung und die Elaboration aggressiver Impulse. Das Verharren auf diesem bezüglich Objektbeziehungen frühkindlichen Entwicklungsniveau wirkt wie eine Vermeidung, um nie in die depressive Position zu gelangen, in welcher das Objekt zugleich bekannt und unbekannt, geliebt und gehaßt wird. Auf diese Art und Weise erreichen diese Kinder scheinbar nie eine Entwicklungsebene, auf der Verluste erlebt werden konnten. In der Zeit zwischen der Entwicklung des ersten und des zweiten Organisators, d.h. rund um den 6. Monat, erfolgt meist der Versuch, die Realität der Frustration (z.B. die zeitweilige Absenz der Mutter) durch eine halluzinatorische Wunscherfüllung zu verleugnen. Dies gelingt den allergischen Säuglingen nicht gut. Die objektunspezifische Beziehungsart ermöglicht es ihnen, nie einen „Fremden" entstehen zu lassen. Statt einer autoerotischen Aktivität bildet sich infolge Triebblockierung auf dieser frühen Stufe eine Funktionsstörung (Atopie) aus. Der Preis für diese „Lösung" ist aber eine enorme Vulnerabilität, da später keine gut strukturierten Neurosen, sondern prägenitale Funktionsmuster entwickelt werden. Der Mechanismus der Verschiebung wird besonders häufig eingesetzt. Wenn der Aufbau einer Beziehung zum ganzen äußeren Objekt vermieden wird, so kann auch keine entsprechende Introjektion stattfinden. So ist für diese Kinder kein Verlaß auf ihre Fähigkeit, zu einem ganzen inneren Objekt in einer kontinuierlichen Beziehung zu stehen. Sie stützen sich deshalb auf die Außenwelt, auf Dinge, Situationen oder Personen, welche als Hilfs-Ich zu fungieren haben. Hieraus resultiert ihre große Verletzlichkeit und ihre Neigung, bei Trennungen von Dingen, Situationen oder funktionell gebrauchten Personen in schwere seelische Dekompensationen bis zur Depression zu geraten. Kreisler beobachtete, daß die Mütter dieser Kinder völlig unabsichtlich die schwachen integrativen Funktionen ihrer Kinder überforderten. Obwohl vielfach eine Störung der Mutter-Kind-Beziehung beobachtet werden kann (Biermann, 1977), gibt es keine spezifische Persönlichkeitsstruktur, d.h. keine „asthmatogene Mutter".

Allergische Kinder sind also schnell überfordert, versuchen, auf regressivem Niveau ein neues Funktionsgleichgewicht herzustellen und enden schließlich, wenn das nicht gelingt, in einer somatischen Dekompensation, die keine symbolische Dynamik aufweist. Affekte wie Wut und Angst, aber auch Familienkonflikte oder Schulprobleme können auslösend sein.

Tritt das Asthma nicht schon beim Säugling oder Kleinkind, sondern erst im Schulalter auf, so hält es oft bis zur Pubertät an. Regressionen auf eine frühe fusionelle Beziehungsmodalität mit dem Versuch einer primären Identifikation mit der Mutter sind immer wieder zu beobachten. Manifestiert sich das Asthma erst beim Erwachsenen, so ist auch dort beim Auftreten der Symptome eine Dekompensation in der Beziehungsmodalität festzustellen, wobei vorgängig das Objekt vor allem funktional gebraucht wurde und dann aus irgendeinem Grund in dieser Funktion ausfiel.

Therapeutisch ist es sinnvoll, zweigleisig vorzugehen: einerseits die Behandlung somatischer Art durchzuführen, andererseits eine genaue Psychodiagnostik vorzunehmen. Psychotherapeutische Maßnahmen, wie sie im Bereich der Kinder- und Jugendpsychiatrie allgemein angewandt werden (insbesondere auch familientherapeutische Verfahren [Steinhausen, 1981]), sollten aber rechtzeitig beginnen und nicht erst beim Vorliegen sekundärer somatischer Störungen mit Chronifizierung. Durch vorschnelle Anwendung und unkontrollierten Gebrauch von Tascheninhalatoren kann, vor allem in der Adoleszenz, eine Sucht und Abhängigkeit von Sympathikomimetika entstehen (bis zum Tod bei schwerem Abusus).

54.7.9 Dermatitis atopica

Neben einer hereditären Basis spielen bei der Entstehung der Neurodermitis allergische, infektiöse, klimatische, psychische und psychosoziale Einflüsse eine Rolle. Drei Viertel der Fälle manifestieren sich bereits im ersten Lebensjahr. Spezifische Persönlichkeitsmerkmale oder typische Mutter-Kind-Interaktionen können beim an Neurodermitis erkrankten Kind nicht nachgewiesen werden, wohl aber kann z.B. der Juckreiz, das Kardinalsymptom der Neurodermitis, für den Patienten eine Funktion im Dienste seiner zwischenmenschlichen Auseinandersetzung gewinnen, insbesondere bei Problemen der Autonomie, der Nähe- und Distanzregulation (Schleiffer, 1988). Chronischer quälender Juckreiz erschwert in hohem Maße die Entwicklung einer innerseelischen Eigenständigkeit, da er den Betroffenen das Erleben einer Abhängigkeit vom Symptom dauernd aufzwingt.

Psychotherapeutische Hilfe ist, wie bei vielen chronisch-läsionellen Krankheiten des psychosomatischen Formenkreises (z.B. den anderen chronischen Krankheiten der Haut, der Colitis ulcerosa, dem Ulkus, dem Morbus Crohn etc.), vor allem dann angezeigt, wenn Kinder, Jugendliche und/oder Eltern nicht in der Lage sind, die Krankheit psychisch zu verarbeiten und individuelle oder familiäre Probleme (z.B. depressive oder regressive Störungen) mit der somatischen Krankheit klar verkoppelt sind, was sich z.B. in ungenügender Kooperation bei der Behandlung, d.h. einem beeinträchtigten Gesundungswillen, zeigen kann.

54.7.10 Funktionelle Kopfschmerzen

Funktionelle Kopfschmerzen sind in der pädiatrischen Praxis recht häufig. Sie manifestieren sich bevorzugt in Übergangszeiten, die mit Neuintegration verbunden sind (Anfang von Schule, Mittelschule, Universität oder Berufseintritt) (Sirol, 1985). Die Diagnose kann erst nach Ausschluß einer organischen Ursache (z.B. einer subakuten oder chronischen Sinusitis, Konvergenz- oder Akkommodationsstörungen, eines erhöhten intrakraniellen Drucks bei Hirntumor) gestellt werden.

Migräne

Zumeist hereditär-familiär vorkommend (70% der Mütter und 20% der Väter sind auch betroffen). Es kann von einer weitgehend unklaren Ätiologie bei gut bekannter Pathophysiologie (Zusammenspiel von endokrinen, vaskulären und psychischen Faktoren) gesprochen werden. Auf 20 Schulkinder leidet eines zeitweilig an Migräne. Mehrfach ist auf eine atopisch-allergische Komponente im Krankheitsgeschehen hingewiesen worden. Der Kopfschmerz tritt verhältnismäßig plötzlich auf. Nach einigen Prodromalerscheinungen (z.B. Übelkeit, Sehstörungen bis zum Flimmerskotom) stellt sich ein pulsierendes Kopfweh mit Licht-, Lärm- und Bewegungsscheu und nicht selten auch Erbrechen und Bauchweh ein. Nach einem erholsamen Schlaf sind die gesamten Beschwerden oft vorbei. Die **abdominelle Form** kann sich wie ein akutes Abdomen manifestieren.

Die Kinder haben in ihrem Verhalten gewisse Ähnlichkeiten mit den Asthmatikern: Sie sind meist intelligent, lebendig, kontaktsuchend, gute Schüler, perfektionistisch, mit stark übersetzten intellektuellen Funktionen und wenig spürbarer Phantasieaktivität. Die Kapazität der Verdrängung ist offensichtlich überschritten. Diese Kinder vermeiden oft die Triangulierung und bringen mit ihrem Weh verschiedenste intrapsychische oder interpersonale Probleme zum Ausdruck. Die Familien sind sehr leistungsbetont.

Einfach-rezidivierende Kopfschmerzen

Die Schmerzen treten langsamer auf, sind viel diffuser, wechselnder, zeigen keinerlei Situationsspezifität, manifestieren sich aber vermehrt bei emotionalen oder intellektuellen Spannungszuständen. Im Rahmen einer **neurotischen Entwicklung** identifizieren sich Kinder manchmal mit den Kopfschmerzen ihrer Eltern. Beim **Dominieren von Hemmungen** werden meist auch die intellektuellen und emotionalen Abläufe gehemmt. Diese Formen sind dann mit Leistungsabfällen verknüpft. Funktionelle Kopfschmerzen werden auch bei beginnenden Depressionen, infolge massiver äußerer Zwänge und Einschränkungen (z.B. rigid-perfektionistische, strenge Eltern mit hohen Leistungsanforderungen und Überbehütung), bei Überbelastung schulischer oder körperlicher Art und bei Versagensangst (vor allem bei narzißtischen oder zwanghaft-perfektionistischen Jugendlichen) ge-

sehen. In diesen Fällen kapituliert das Ich gleichsam vor den Forderungen des Über-Ichs (Boudier, 1962; Sirol, 1985).

Alle schweren Kopfschmerzen bewirken eine schmerzliche Unterbrechung (Hemmung bis Blockierung) der intrapsychischen Funktionen. Als regressiver Ausdruck eines dann sich manifestierenden Strukturdefektes in vorbewußten Ich-Anteilen stehen sie im Dienste der Vermeidung und Verleugnung innerer Phantasieabläufe.

Bezüglich **Therapie** gelangen alle in der Kinder- und Jugendpsychiatrie typischen Behandlungs- und Therapieformen zur Anwendung.

54.7.11 Tic-Störungen

Als Tics werden monoton wiederkehrende, unwillkürliche, plötzlich einschießende Muskelzuckungen (ohne zentralnervöse oder muskuläre Krankheit) bezeichnet. Ihre Erstmanifestation in Form von einzelnen, multiplen, motorischen und/oder vokalen Tics liegt im Bereich von 2–15 Jahren. Tics zeigen oft große Schwankungen in ihrer Intensität, deren Ursachen weitgehend unklar sind, mit wiederholten stummen Perioden. Willkürlich können sie nur für kürzere Zeit unterdrückt werden, und sie wechseln leicht ihre Erscheinungsform.

Bei 10 bis 24% aller Kinder kommen irgendwann in der Entwicklung einmal Tics vor (Rothenberger, 1984/88). Unspezifische, chronische emotionale Belastungen (gekoppelt mit Unterdrückung aggressiver Strebungen) in Kombination mit einer organischen (hereditären?) Disposition werden ätiologisch für das Entstehen der Tic-Krankheiten verantwortlich gemacht. Immer wieder wurden auch bei der Tic-Entwicklung Konversionsmechanismen bei prägenitalen Konflikten beschrieben.

Rein phänomenologisch werden die **einfachen, vorübergehenden** von den **chronischen multiplen** Tics und dem **Gilles-de-la-Tourette-Syndrom** (multiple Tics mit Vokalisationen und z.T. Koprolalie und Echolalie) abgegrenzt. Zwischen diesen Formen gibt es fließende Übergänge, dennoch ist die Unterscheidung von Untergruppen der Tic-Störungen entlang eines Kontinuums sinnvoll, da dem rechtzeitigen Einsatz einer medikamentösen Therapie neben psychotherapeutischen und psychosozialen Hilfestellungen bei den chronischen multiplen Tics und beim Tourette-Syndrom große Bedeutung zukommt. Bei unzureichender, z.B. ausschließlich psychotherapeutischer Behandlung eines schweren Tic-Leidens besteht die Gefahr, daß sekundär psychoreaktive Störungen bei den betroffenen Kindern und ihren Familien gefördert werden.

54.7.12 Bauchschmerzen beim Kleinkind, in Latenz und Adoleszenz

Nach Ausschluß organischer Krankheiten (wie z.B. Appendizitis, nephrologisch-urologischen Störungen, Ovulationsproblemen bei jungen Mädchen oder Hirntumoren, vor allem der hinteren Schädelgrube) lassen sich die abdominelle Migräne, die funktionelle Obstipation und das funktionelle Bauchweh voneinander abgrenzen. Diese Funktionsstörungen können bei verschiedensten seelischen Krankheitsbildern gefunden werden und auch beim Megakolon und der Colitis ulcerosa.

Funktionelle Bauchschmerzen, die um das 3. Lebensjahr herum auftreten, haben oft eine Vorgeschichte von Nabelkoliken. Diejenigen, welche sich erst in der Latenz oder Präpubertät manifestieren, lassen zumeist keine solche Vorgeschichte erkennen. Die Gefühlsbindungen in den betroffenen Familien sind zumeist eng, oft überbehütend. Bauchweh kann unter solchen Aspekten als Versuch der Schaffung eines autonomen Bereiches verstanden werden. Zumeist besteht ein anal-sadistisches Organisationsniveau im Ich mit Problemen der Kontrolle/Unterwerfung, der Aktivität/Passivität. Angst vor ödipalen Positionen bewirkt eine Regression auf prägenitale Stadien. Das Bauchweh kann als Indikator von Angst und Unwohlsein fungieren (Sirol, 1985). Es bringt viel Ambivalenz zum Ausdruck und bewirkt oft die Einführung eines Arztes in die Szenerie des Familiendramas.

Die **abdominelle Migräne** ist meist hochaktiv, verläuft ohne Kopfschmerzen und tritt tagsüber in Erscheinung. Erst während der Pubertät erfolgt die Umwandlung in eigentliche Migräneattacken.

Bei der **funktionellen Obstipation** handelt es sich um wechselnde Schmerzen entweder bei einem Colon spasticum (dort plötzlich einsetzend; die Verstopfung ist nur von kurzer Dauer und es besteht kein Fieber) oder um eine chronische Obstipation im Wechsel mit Diarrhö (hier nicht selten Laxantien- oder Suppositorienabusus). Vielfach sind die entsprechenden Familien fast zwanghaft mit den Erscheinungen der Obstipation beschäftigt. Den Kindern wird oft durch minutiöse Defäkationskontrollen jegliche Intimität geraubt. Die Defäkation wird zum öffentlichen Akt und kann eine Defäkationsphobie mit sekundärem Krankheitsgewinn nach sich ziehen.

Funktionelle Bauchschmerzen, wie sie in der pädiatrischen Praxis häufig anzutreffen sind, dauern Stunden bis Tage (guter Allgemeinzustand). Sie sind mehrheitlich periumbilikal lokalisiert, oft mit Enuresis, Schlafstörungen, manchmal mit Anorexie, Abfall der Schulleistungen und äußeren Zwängen verknüpft. Bei den Eltern sind oft auch ambivalente Züge zu beobachten, z.B. sind sie bezüglich Zuwendung und Interesse am Kind wechselnd und widersprüchlich. Rigide Familienstrukturen überdecken oft nur mühsam die stark mit Sadismus durchsetzten Emotionen. Die Kinder selbst schwanken ambivalent zwischen Opposition und Anpassung, zwischen Abhängigkeit und Unabhängigkeit und weisen manchmal eine gewisse Entwicklungsverzögerung auf. Entsprechend ihrem verhaltenen Sadismus entwerten sie schnell die bedeutungsvollen Objekte. Als sekundären Krankheitsgewinn verfügen sie über die Kontrolle der Aufmerksamkeit und Zuwendung ihrer Eltern.

Funktionelle Bauchschmerzen sind oft Vorläufer späterer Beziehungskrisen in der Adoleszenz. Neben diesen mehr neurotischen Formen zeigen sich die gleichen Beschwerden auch bei Kindern mit Entwicklungsdysharmonien in der Präpubertät. Bei diesen besteht ein sehr labiles Selbstwertgefühl, sie sind eher passiv und schutzsuchend, vermeiden Rivalität, ihr Körperselbst ist sehr ambivalent besetzt, und sie klagen zumeist über diverse andere somatische Beschwerden, die nicht selten auf der Basis eines Konversionsmechanismus ablaufen und ein somatisches Sich-Sträuben gegen die beginnenden pubertären Umwandlungen zum Ausdruck bringen.

Die Symptome stellen teils Anpassungsphänomene an das familiäre Milieu dar, teils bringen verschiedene Konversionsmechanismen eine konflikthafte Vergangenheit in Form einer gestörten Gegenwart zum Ausdruck; z.B. wird auf diese Weise manchmal die Möglichkeit der Kastration verleugnet und die Angst, hoch besetzte Objekte zu zerstören, abgewehrt. Als Folge dieser Konversionsvorgänge kommt es gerade dort zur Regression auf anale und orale Ebenen in der Trieb- und manchmal auch in der Beziehungsorganisation, wo progressive Bewegungen sich einzustellen beginnen. Die Bauchschmerzen fungieren dann nicht nur als eine masochistische Form der Kontrolle über die Umwelt, sondern auch als ein Appell nach medizinischem Schutz und Hilfe.

Das therapeutische Vorgehen bedarf oft einer vertieften fachärztlichen Untersuchung als Grundlage und richtet sich nach den üblichen Methoden der Kinderpsychiatrie/-psychotherapie.

54.7.13 Enuresis

Als Enuresis wird eine nach der somatischen Funktionsreife primär oder sekundär auftretende, unkontrollierte, unwillkürliche, gehäuft im Schlaf vorkommende Miktion ohne organische Funktionsstörung bezeichnet. Abgegrenzt von der Inkontinenz, die eine organische Läsion beinhaltet, handelt es sich bei der Enuresis um eine funktionelle Störung. Die Retention wie auch die Expulsion von Harn wird sowohl durch glatte (Blase, glatter Sphinkter) als auch durch quergestreifte Muskulatur (Damm, quergestreifter Sphinkter) reguliert. Die Funktion der glatten Muskulatur kann durch unbewußte Phantasien beeinflußt werden, die der quergestreiften durch unbewußte oder bewußte Vorstellungen. Hierdurch entstehen unzählige Nuancen zwischen Absicht, Geschehenlassen, unabsichtlichem Tun und Tun, ohne davon wissen zu wollen (Schmit et al., 1985). Die Miktion, für deren harmonischen Ablauf es der Kontraktion der Blase, der Öffnung der Sphinkteren und der Entspannung des Beckenbodens bedarf, erfolgt erst reflexhaft und unwillkürlich, dann allmählich bewußt-aktiv und willkürlich. Die Sphinkterkontrolle installiert sich erst etwa am Ende des 1. Lebensjahres, automatisiert sich dann zunehmend und wird gegen das 3.–4. Lebensjahr autonom, d.h. unabhängig von Triebkonflikten.

Bei der Entwicklung der Miktionskontrolle ist zu unterscheiden zwischen der Reifung der somatisch-reflektorischen Funktionen, ihrer emotionalen Besetzung (Genuß, Beherrschung, Angst, Abwehr) und ihrer Bedeutung für die interpersonale Beziehung zur primären Pflegeperson. Bei der Sauberkeitsentwicklung verzichtet das Kind allmählich und um der Liebe des Objektes willen auf die mit der Miktion verbundene Lust. Dies erfolgt um so leichter, je mehr die Wünsche der Umwelt mit den Möglichkeiten des Kindes übereinstimmen, sich ihm anpassen. Daraus geht hervor, wie zentral die emotionale Bedeutung der Trockenheit bzw. Sauberkeit für die Mutter ist, eine wie kohärente oder widersprüchliche Haltung sie diesbezüglich einnehmen kann. Aber auch hereditäre Faktoren spielen eine Rolle, zeigen doch monozygote Zwillinge in ⅔ der Fälle eine Konkordanz bezüglich des Einnässens, was mit dem familiär gehäuften Auftreten der Enuresis gut in Übereinstimmung zu bringen ist. Erst nach Abschluß des 4. Lebensjahres haben ca. 90% der Kinder in unserem Kulturkreis tags und nachts Trockenheit erreicht. Von einer Enuresis kann vorher also nicht gesprochen werden. Die Häufigkeit der Störung (bei Knaben eindeutig häufiger als bei Mädchen) nimmt im Verlauf der Latenz ab. Nach der pubertären Reifung ist sie von etwa 10 auf ca. 1% abgesunken. Rund ⅕ sind **primäre** Enuretiker. Die Dauer des Intervalls bei der **sekundären** Enuresis ist sehr variabel. In etwa 80% handelt es sich um eine Enuresis **nocturna,** in etwa 15% um eine **gemischte** nocturna/diurna-Form und nur etwa in 5% um eine reine Enuresis **diurna.** Bezüglich Intensität kann die Enuresis episodisch bei Belastungen mit nachfolgender Regression (z.B. Familienkonfliken und Schulproblemen) auftreten. Eine Enuresis kann als schwer bezeichnet werden, wenn sie mehr als dreimal pro Woche auftritt.

Ein tieferer Schlaf als bei den Geschwistern konnte bei sorgfältiger Überprüfung nur in ⅔ der Fälle gefunden werden. Er dient oft als Vorwand und Entschuldigung. Bei der Schwierigkeit zu erwachen handelt es sich um einen aktiven Vorgang regressiver Art, der möglicherweise eine gemeinsame Wurzel mit der Enuresis selbst besitzt. Das EEG zeigt kaum einen Unterschied zu dem normaler Kinder. Höchstens besteht eine Tendenz zu einer längeren Dauer von Stadium 4. Die Miktion im Schlaf erfolgt in weitaus den meisten Fällen im Stadium 1b, folgend auf eine Tiefschlafphase. Selten ergibt sich auch eine Enuresis im REM-Schlaf. Es bleibt unklar, ob es sich hierbei um ein klinisch unterschiedliches Bild handelt oder nicht.

Hinter der gleichförmigen enuretischen Symptomatik treten höchst unterschiedliche psychische Strukturen in Erscheinung. Die wirklichen Reaktionen und Einstellungen eines Kindes zu seinem Symptom sind oft sehr schwer zu eruieren. Meistens übernimmt es die Haltung der Eltern. Es oszilliert zwischen regressiven Befriedigungen (Aufrechterhalten eines Reizes, autoerotische Funktion der Retention, symbolische Äquivalente eines Orgasmus: Laufenlassen, wenn es nicht mehr zu halten ist) und dem Wunsch nach

Funktionskontrolle. Die Kinder verhalten sich so, als könnten sie die Signale des Miktionsbedürfnisses nicht wahrnehmen, d.h. dieses und die mittels des Symptoms unbewußt erreichten Befriedigungen werden beide verleugnet. Der Schlaf erleichtert die Verleugnung. Enuretiker leeren ihre Blase vor dem Zubettgehen fast nie vollständig. Der Kampf gegen das Erwachen steht im Dienste der Verleugnung. Der primäre, unbewußte Krankheitsgewinn besteht im unmittelbaren, passiv-regressiven Genuß der Entlastung beim Laufenlassen des Harns und vorgängig in der hinausgezögerten masturbatorischen Reizung bei urethraler Fixierung. Bei dieser wird durch zunehmende Blasenfüllung die Spannung zwischen expulsiven und retentiven Tendenzen immer höher, bis sie sich durch das Einnässen orgastisch löst. Vorbewußt kann ein sekundärer Krankheitsgewinn aus der Auswirkung des Symptoms auf die Umgebung gezogen werden. Die Symptomatik wird gleichsam interpersonal genutzt, z.B. durch Vermeidung von Trennungen, durch engeren Kontakt zur Mutter, durch masochistische Befriedigungen beim familiären Ärger oder durch Vermeidung adoleszenter Reifungs- und Entwicklungsprozesse mit Fixierung in ödipaler Komplizenschaft. Das ambitendente Spiel des Kindes trägt zu einem gewissen Ausmaß perverse Züge. Bei der sekundären Enuresis findet man oft auslösende Ursachen wie Verluste, Familienkonflikte, Geschwisterrivalitäten, Schulprobleme, Krankheiten oder Trennungen.

Die betroffenen Kinder zeigen zwei Hauptauffälligkeiten:
– Eine **Tendenz zu emotionaler Abhängigkeit und depressiver Reaktion** mit übermäßiger Bindung an die Mutter, wenig Beziehung zu Gleichaltrigen, regressive Spiele und kindliche Gewohnheiten. Die Mütter haben eine Neigung, solche Kinder klein zu behalten und eine sehr enge Körperpflegebeziehung zu installieren. Das Kind versucht sich dann mittels seiner regressiv-depressiven Abhängigkeit soviel wie möglich vom „Vorenthaltenen" zu holen, ohne die Trauerarbeit um das Unwiederbringliche geleistet zu haben.
– **Probleme in der altersadäquaten Äußerung aggressiver Impulse** sind oft in verstärkter Opposition, offener Aggression oder aber vertuscht-aggressiven Verhaltensweisen erkennbar. Die Enuresis erlaubt als Symptom zugleich den getarnten Ausdruck aggressiver Impulse und die Demonstration regressiver Abhängigkeit.

Enuretische Knaben wirken eher ängstlich, passiv und abhängig, enuretische Mädchen eher ehrgeizig und nach Unabhängigkeit strebend. In vielen Fällen aber sind nur wenige direkt faßbare, psychische Phänomene zu erheben.

Auf einem **präödipalen** Niveau gibt die Enuresis dem Kind die Möglichkeit, sich passiv-regressive Befriedigungen zu erlauben, die Urethralerotik aktiv auszuprobieren, sadistisch-destruktive Impulse verdeckt zu agieren, Macht zu fühlen (Knaben) oder die Penislosigkeit zu verleugnen (Mädchen). Auf einem **ödipalen** Niveau bringt die Enuresis manchmal den

Zwiespalt zwischen dem Wunsch, groß zu werden und seine Genitalität zu genießen (was aber zu Kastrationsangst führt) und dem Wunsch, klein und damit enuretisch zu bleiben, aber von regressiven Befriedigungen zehren zu können, zum Ausdruck. In der **Präadoleszenz** dient sie immer wieder der Vermeidung genital-masturbatorischer Impulse durch Regression, was gleichzeitig die Abfuhr von Aggressionen erlaubt, die nicht in der Beziehung zum Ausdruck kommen dürfen. Es ist wie eine gespielte Autokastration mit gleichzeitiger Selbstversicherung der Nicht-Kastriertheit (mit „Wasser" das „Feuer der Erregung" löschen) (Freud).

Die Symptomwahl kommt wahrscheinlich durch eine organische Disposition und regressiv gelöste Konflikte zwischen dem 1. und 3. Lebensjahr zustande. Immer konfrontiert das Kind mit seinem enuretischen Symptom die Eltern mit ihren eigenen, verdrängten, präödipalen und regressiven Tendenzen. Viele ehemals enuretische Eltern bringen ihre Kinder gerade in dem Alter zur Konsultation, in welchem sie selbst ihre Enuresis loswurden. Kinder mit geistiger Behinderung, emotionaler Deprivation und auch minimaler zerebraler Dysfunktion sind etwas gehäuft unter den Enuretikern zu finden. Enuretische Kinder bleiben vulnerabel bezüglich ihrer Oppositionsneigung, den Konflikten zwischen Autonomie und Abhängigkeit, der Inbesitznahme des eigenen Körpers, der Entwicklung einer eigenen sexuellen Identität und der Beziehungsfähigkeit auf objektalem Niveau (Regressionsanfälligkeit).

Als **Therapieziel** ist das Aufgeben der regressiven Befriedigungen und die Veränderung der intrafamiliären Beziehungen anzustreben. Dies ist fast nur möglich, wenn ein Ersatz dafür angeboten wird, z.B. eine tragfähige therapeutische Beziehung. Gesundungstendenzen sowohl des Kindes wie auch der gesamten Familie sollten, wo immer möglich, unterstützt werden, um das Herausfinden aus der Ambivalenz zu erleichtern. Der Aufbau einer vertrauensvollen Beziehung, die Vermeidung von sadistischen oder unnütz-strengen Haltungen in der Gegenübertragung, die Aufklärung über normale Miktionsabläufe und die Vermeidung eines starren, unkreativen Vorgehens sind grundsätzlich wichtige Haltungen. Medikamentöse Hilfen (vor allem Imipramin) können zur Unterstützung, nicht aber zur alleinigen Symptombeseitigung eingesetzt werden. So appliziert fungieren sie manchmal als goldene Brücke für alle. Die zum Teil verbreitete Weckmatratze (Sorotzkin, 1984) (elektrisch ausgelöstes Wecksignal bei ersten Flüssigkeitstropfen) ist nur anzuwenden, wenn das Kind dafür motiviert ist und ihr Gebrauch die Autonomie des Kindes fördert. Alternativbefriedigungen wie sportliche Betätigungen können hilfreich sein. Grundsätzlich sollten unnötige Überreizung, Müdigkeit, Anstrengungen und Angst vermieden werden. Bei der Wahl des geeigneten psychotherapeutischen Vorgehens ist zu unterscheiden zwischen der Enuresis als Ausdruck einer allgemeinen Regression, als eines Symptoms schwerer Psychopathologie oder als Hinweis auf eine ungünstige psychosoziale Situation.

54.7.14 Enkopresis (Einkoten)

Als Enkopresis wird ein psychogen bedingtes, funktionelles Einkoten oder Einschmutzen nach dem 4. Lebensjahr bezeichnet. Das Symptom kann vom Beschmutzen bis zur völligen Stuhlentleerung in die Hosen variieren. Es ist etwa zehnmal seltener als die Enuresis (in der westlichen Welt ca. 1,5–3% der Primarschüler). Am häufigsten wird eine Enkopresis im Grundschulalter gesehen. Danach nimmt ihre Häufigkeit bis zum völligen Verschwinden in der Spätadoleszenz konstant ab. Knaben sind deutlich häufiger betroffen als Mädchen (ca. 3,5 . 1). Die Enkopresis, ein anhaltender „passage à l'acte" im Bereich der analen Körperfunktion, erfolgt meist tagsüber. Sie ist in $\frac{1}{3}$ bis $\frac{1}{2}$ der Fälle mit einer Enuresis verkoppelt. Gleichzeitig bestehen oft Schlafstörungen, aggressive Verhaltensweisen, Ängste und Eßprobleme. Den retentiven, häufiger vorkommenden Formen der Enkopresis liegt oft eine habituelle Obstipation oder ein funktionelles Megakolon zugrunde. Der Defäkationsmechanismus (Arhahn, 1974) besteht aus einer unwillkürlich-reflexhaften Kolon- und Sigmoidkontraktion, durch welche die Fäzes in die Ampulla recti gedrückt werden, und danach der Öffnung des äußeren analen Sphinkters mit Evakuation. Sowohl der unmittelbar motorische Vorgang als auch die Reizung der Mukosa und ebenso die Gefühle von Füllung und Entleerung können libidinös besetzt oder gegenbesetzt werden. In den ersten 3. Lebensmonaten gibt es keinen willentlichen Anteil bei der Defäkation. Erst danach entwickelt sich allmählich die Möglichkeit einer funktionellen Beherrschung.

Neben dem **organischen Megakolon** (Hirschsprungsche Krankheit), das durch eine sehr frühe, hartnäckige und schwere Verstopfung schon kurz nach der Geburt gekennzeichnet ist und auf einer Aplasie der nervösen Ganglienzellen im Plexus submucosus beruht, läßt sich das **funktionelle Megakolon,** eine erworbene Dysfunktion auf der Ebene der Defäkation, abgrenzen. Hier tritt die Verstopfung erst zwischen dem 6. und 12. Monat in allmählich progressiver Weise auf und bewirkt zuerst eine reflektorische Dilatation des Rektums und später des gesamten Kolons mit Ausbildung von sog. Kotsteinen. Durch eine aktive Kontraktion des Anus werden die Fäzes ins Sigmoid und ins Kolon zurückbefördert. Dieser gegenläufig zur üblichen Bewegung erfolgende Ablauf wird sekundär erotisiert. Nach unendlich wiederholten Bewegungen vor- und rückwärts, die einen gewissen masturbatorischen Charakter haben (heimlicher, unsichtbarer, innerer Ablauf, der erst noch erlaubt, die gesamte Familie zu manipulieren) und als Vorform eines perversen Vorganges bezeichnet werden können, kommt es schließlich zur zunehmenden Erschöpfung der Rektum- und Kolonmuskulatur. Parallel dazu erfolgt eine atonische Darmdilatation. Die Fäzes werden nun oft an den verhärteten Kotmassen vorbei vor- und zurückgedrückt. Die Defäkation oder das Schmieren in die Hose stellt dann gleichsam einen Betriebsunfall dar und erfolgt, so gesehen, bei diesen Fällen wirklich unwillkürlich (sog. Überlaufenkopresis).

In der Ätiologie und Pathogenese summieren sich in den meisten Fällen eine besondere Irritabilität des Ausscheidungstraktes, pathogene Effekte einer verfrühten, erzwungenen oder sonstwie unangemessenen Sauberkeitsentwicklung, zerebrale Reifungsverzögerungen und schädigende soziale Einflüsse. Schon daraus ist ersichtlich, daß die Enkopresis ein Indikator einer eher schweren Pathologie ist. Sehr oft finden sich schwerwiegende Eltern-Kind-Beziehungsstörungen, familiär belastende Verhältnisse und intrusive, in ihrem eigenen Selbstwertgefühl stark verunsicherte Mütter. Soulé et al. (1985) unterscheiden 4 Formen:

- Enkopresis als aktiver, willentlich-aggressiver Akt.
- Enkopresis als Folge eines Mankos bei ungünstigen Familienverhältnissen. Der Körper des Kindes wird durch die Mutter schlecht besetzt, Retention bedeutet gar nichts Lustvolles. Hingegen besteht ein Bedürfnis nach sofortiger fäkaler Entlastung. Diese Form ist oft kombiniert mit Mutismus oder Sprachverzögerung und großem Sammelbedürfnis. Die Kinder zeigen wenig Symbolisierungsfähigkeit und eine Schwierigkeit, Neues zu besetzen.
- Enkopresis als Spiel mit der Fäkalsäule, ein perverser Masturbationsersatz. Die Enkopresis ist hierbei ein Ausrutscher. Sie erfolgt aktiv und passiv zugleich, da die entsprechenden Patienten, im virtuosen Spiel mit ihren Omnipotenzgefühlen, nicht defäzieren wollen, um sich dem autoerotischen Vergnügen möglichst lange hingeben zu können.
- Enkopresis als Folge enteraler Krankheiten (z.B. einer längerdauernden Diarrhö oder operativer Eingriffe [Kreisler, 1974]), durch welche eine Erotisierung von retentiven und expulsiven Vorgängen stattfand. Diese Form ist meist sekundär und manifestiert sich als regressives Phänomen bei kleineren Belastungen (z.B. Geburt eines Geschwisters, Eintritt in den Kindergarten).

Enkopretiker sind oft unerwünschte, unehelich geborene, Scheidungs-, Pflege- oder Heimkinder, haben wenig Liebe und Geborgenheit oder eine inkonsequente Erziehung erfahren und oft die Rolle eines Sündenbocks erfüllt. Sie bewirken mit ihrem Symptom immer massive Reaktionen bei den Eltern (zumeist sind die Väter bei den Konsultationen anwesend, oft sind sie übermäßig beunruhigt durch die Verstopfung, verlangen zusätzliche Untersuchungen und brechen die Beziehung ab, wenn psychische Bereiche tangiert werden), welche oft eine sehr aggressive Qualität haben oder Verzweiflung und Hilflosigkeit zum Ausdruck bringen. Der sekundäre Krankheitsgewinn, ein enormes Machtgefühl, wird von den Kindern nur sehr ungern aufgegeben. Hinter dem Symptom können sich sehr unterschiedliche Persönlichkeitsstrukturen verbergen.

Angesichts der großen Sekundärschäden ist eine möglichst rasche Symptomheilung anzustreben. Innerhalb einer Einzelbehandlung besteht das Ziel darin, daß das Kind auf seine autoerotischen, analen Masturbationsvergnügen verzichtet und akzeptiert,

daß es sich um ein aktives, von ihm gemachtes Symptom handelt. Sobald die Kinder gesunden möchten, sollten sie akzeptieren können, am Morgen das Rektum zu entleeren, um der Masturbationsversuchung zu entgehen. Elternarbeit und ein familienorientiertes Vorgehen berücksichtigen die Tatsache, daß die Symptomatik zumeist auch ein Ausdruck innerfamiliärer Ereignisse, Spannungen und ungelöster interpersoneller Konflikte darstellt (Wille, 1984; Krisch, 1985). Physiotherapeutische Hilfen wie Bauchmassagen und organmedizinische Eingriffe bis zur digitalen Ausräumung aufgestauter, verhärteter Kotmassen können unterstützende Erleichterung schaffen. Das Symptom selbst aber wird nicht gerne aufgegeben, da es eine Autoerotik ohne Sublimation und Verdrängung ermöglicht, damit selbstfabrizierten Trost spendet und als sekundärer Gewinn eine große Machtausübung in der Familie zur Folge hat. Die Entwicklung verläuft äußerlich gesehen vielfach nicht sehr günstig ab (ca. $\frac{1}{3}$ der Patienten muß eine Klasse wiederholen, $\frac{1}{3}$ wird in eine Sonderklasse eingewiesen, die Hälfte „fremdplaziert". Bei $\frac{3}{5}$ traten während einer Katamnesedauer von 7 Jahren neue aggressive oder depressive Symptome auf, bei rund 15% persistierten Enuresis und Enkopresis [Wille, 1984]). Der Krankheitsverlauf kann einerseits in Richtung einer Normalisierung der Entwicklung gehen, andererseits aber auch in eine vermehrte Tendenz zum „passage à l'acte", mit Entwicklung entweder perverser Strukturen oder einer erhöhten Tendenz zu dissozial-delinquentem Verhalten.

54.7.15 Diabetes mellitus

Auch der Diabetes mellitus zählt zu den häufigen chronischen Krankheiten des Kindesalters (ca. ein Fall auf 600 bis 1200 Kinder unter 16 Jahren) und tritt bevorzugt zwischen dem 8. und 10. Lebensjahr auf. Die Manifestation erfolgt meist langsam zwischen dem 4. und 12. Lebensjahr, manchmal auch schon früher (zur Pathophysiologie vgl. Kap. 47).

Die Diagnose trifft die Eltern und Kinder meist aus heiterem Himmel. Wie bei jeder chronischen Krankheit hat sie eine narzißtische Verunsicherung, das Auftreten von Schuld-, Wut- oder heftigen Trauergefühlen, wenn nicht gar eine depressive Dekompensation bei Patient und Eltern zur Folge. Die betroffenen Kinder werden plötzlich zu Repräsentanten des Mangels, des Verlustes, des Versagens, verlieren kurz- oder längerfristig die Qualität, für die Eltern auch eine Quelle von Freude sein zu können. Sind sie bereits vor der Manifestation des Diabetes psychisch auffällig gewesen, so erleben sie ihr Anderssein und die mit der Therapie verbundenen Einschränkungen in der Lebensführung als noch belastender und können mit einer Intensivierung vorbestehender Symptome reagieren. Langzeitfolgen der Krankheit werden von der Familie in adaptiver Kollusion sowohl aus Angst vor wie auch zwecks Vermeidung von depressiven Zusammenbrüchen oft verleugnet. Nahrungsprobleme und Diätmaßnahmen können zu Kristallisationspunkten innerfamiliärer, oraler Kontrollkonflikte werden (Cramer, 1979, 1985). Tägliche Insulininjektionen und die mehrfach täglichen Urinkontrollen haben die Qualität auferlegter Zwänge. Schwere Hypoglykämien bzw. Komata werden von den Familien und den betroffenen Patienten oft wie Äquivalente des Todes erlebt und werden damit für die Umgebung zum Motiv für eine entsprechende Überbehütung.

In der Adoleszenz mit ihrem leicht störbaren seelischen Gleichgewicht wirft der Diabetes nicht nur Probleme des Selbstwertes, sondern auch solche der Triebregulation (Freßsucht) auf. Oppositionelle Tendenzen können zur Diätverweigerung und zum Nicht-Einhalten von Mahlzeitterminen führen, wenn die Behandlungsmaßnahmen dem Wunsch nach Selbstbestimmung entgegenlaufen.

Der Diabetes mellitus macht es dem Patienten schwer, sich ein angemessenes inneres Bild von seiner Krankheit zu machen. Er ruft deshalb unzählige Phantasmen bei den Patienten und ihren Eltern hervor. Vielfach besteht Unklarheit darüber, was von der Symptomatik somatisch und was psychisch bedingt ist. Es läßt sich keine typische Persönlichkeitsstruktur oder Konfliktkonfiguration des diabetischen Kindes oder Jugendlichen beschreiben. Stets besteht eine starke Abhängigkeit vom Körper, welche dem Kind durch die Krankheit auferlegt ist.

Knapp kompensierte Familiensysteme können unter dem Einfluß einer Zuckerkrankheit eines ihrer Kinder dekompensieren. Hierbei brechen schwere innerfamiläre Beziehungskonflikte auf mit heftigen Eifersuchts-, Protest-, Manipulations- und Kontrollmanifestationen. Ausgeprägte Verleugnungsmechanismen können die Lernfähigkeit des Kindes beeinträchtigen. Die Informationspolitik ist deshalb zentral. Repetitiv, einzeln und in Gruppen erfolgende Informationen reduzieren die Zahl der Hospitalisationen und verbessern das Selbstgefühl durch Förderung der Autonomie bei den Kindern. Die kollusive Vermeidung schwieriger Themen zwischen dem behandelnden Arzt und dem Kind bzw. seiner Familie dient oft der Aufrechterhaltung von Allmachtsphantasien, der Vermeidung von depressiven Zusammenbrüchen und dem Versuch, starken Übertragungsabhängigkeiten zu entgehen. Je mehr es dem Kind gelingt, Sorgfalt und Eigenverantwortlichkeit bezüglich Insulinsubstitution, Diät und körperlicher Aktivität zu übernehmen, und je mehr es ihm und der Familie möglich wird, mit der Krankheit zu leben, desto besser ist die Gesamtprognose.

54.7.16 Anorexia nervosa, Bulimarexie, Bulimia nervosa (vgl. Kap. 38)

Anorexia nervosa: Zu den Kardinalsymptomen dieses Syndroms, das viel häufiger bei Mädchen als bei Jungen auftritt (30 : 1), eine maximale Häufigkeit zwischen 14 und 18 Jahren aufweist, aber auch bereits zwischen dem 10. und 11. Lebensjahr und bis in die fünfte Lebensdekade hinein beobachtet werden kann, gehören die Anorexie (aktive Weigerung einer genügenden Kalorienaufnahme mit intensiver Angst vor dem Dickwerden), nachfolgender Gewichtsverlust von ca. 20 bis 25% ohne entsprechende somatische Erkrankung, sekundäre Amenorrhö (sofern bereits eine Menstruation bestand), Obstipation und motorische Überaktivität. Eine große Zahl sekundärer somatischer Erscheinungen (z.B. verstärkte Lanugobehaarung, Bradykardie, livide Akren und diverse Stoffwechselstörungen) sind je nach Hungerzustand zu beobachten. Andere psychiatrische Störungen (z.B. Depressionen, Schizophrenien, Zwangsneurosen oder somatische Krankheiten wie z.B. Hirntumor) müssen vor der Diagnosestellung einer Anorexie ausgeschlossen werden. Die Gewichtsabnahme wird entweder durch Beschränkung der Nahrungsaufnahme, Erbrechen oder Abusus von Laxantien bzw. Diuretika erzielt. Liegen beim anorektischen Syndrom zusätzlich noch Heißhunger-, d.h. Eß-Brech-Attacken vor, so spricht man von **Bulimarexie.** Mit einer Mortalitätsrate, über längere Zeit gesehen, zwischen 5 und 15% ist die Anorexie/Bulimarexie eine schwere psychosomatische Erkrankung mit Tendenz zur Chronifizierung, obwohl es ein breites Spektrum von kurzzeitigen, oft spontan heilenden anorektischen Reaktionen bis zu schwer beeinflußbaren, progredient verlaufenden Krankheitsbildern gibt. Dementsprechend kann hinter dem anorektischen Syndrom eine individuelle Psychopathologie stehen, die sich von psychotischen Strukturen über Borderline-Syndrome, narzißtische Neurosen und Symptomneurosen bis zur einfachen Adoleszentenkrise erstreckt.

Der Symptomatik, die immer mit einer Körperschemastörung einhergeht, liegt die Vorstellung zugrunde, daß sich der eigene Körper der aufgenommenen Nahrung bemächtigt, sie in seine Substanz integriert, sich mit ihrer Hilfe aufbläht, unkontrollierbar ausdehnt unter Verlust der Konturen und, in zu verabscheuender Art und Weise, durch diesen Prozeß Macht über das Ich gewinnt. Folge in diesem Erlebniszusammenhang ist, daß die Jugendlichen entweder keine Nahrung in den Körper hineinlassen (restriktiver Typ der Anorexia nervosa), oder der Nahrung den Durchtritt durch die Barriere der Magenschleimhaut verweigern (bulimischer Typ der Anorexia nervosa und Bulimia nervosa), oder die Nahrung, der sie die Passage in den Magen-Darm-Kanal gewährt haben, mit Vehemenz mit Hilfe von Laxantien, im Sinne von Reinigungsorgien, aus dem Körperinneren hinaustreiben. Immer geht es um das gleiche Grundanliegen: Der Körper darf die Nahrung nicht in sich aufnehmen, weil, in der Vorstellung der betroffenen Jugendlichen, der Körper dadurch einen Machtzuwachs gegenüber dem Ich erfahren würde und parallel dazu die Niederlage der geistig-seelischen Persönlichkeit befürchtet werden müßte.

Ohne daß bis heute eine vollständige Übersicht über die multikausalen Determinanten der gesamten psychophysischen Vernetzung und damit auch keine eindeutige Klarheit über Ätiologie und Pathogenese der Erkrankung besteht, kann man berechtigterweise von der Annahme ausgehen, daß es sich um eine primär psychogene Erkrankung handelt mit sekundär hypophysär-hypothalamischen Funktionsabweichungen. Die endokrinologischen Veränderungen tragen vermutlich wesentlich zur Unterhaltung und Chronifizierung des Krankheitsgeschehens bei und verlangen deswegen zwingend integrative somatopsychotherapeutische Behandlungsansätze.

Die Anorexie/Bulimarexie ist vorwiegend in industrialisierten Gesellschaften ohne Hunger und dort eher in der Mittel- und Oberschicht festzustellen. Es bestehen Hinweise auf einen Häufigkeitsanstieg in den vergangenen Dezennien. Da die Relevanz verschiedener, ätiologisch vermutlich wirksamer Einflußfaktoren in der Genese der Erkrankung nicht bekannt ist, muß ein komplexes Zusammenspiel individueller (psychophysische Prädisposition und persönliche Lebensgeschichte), familiärer und gesellschaftlicher Faktoren angenommen werden. Wie in verschiedenen Untersuchungen aufgezeigt wurde, sind viele junge Frauen unseres Kulturkreises mit ihrem Körper unzufrieden, wohl im Zusammenhang mit einem weiblichen Schönheitsideal, bei dem Schlankheit für Schönheit, Attraktivität, Dynamik und Erfolg steht (Gerlinghoff, 1988). Ob allerdings psychologische Parameter ausreichen, um die Geschlechtsverteilung bei der Anorexie zu erklären, muß offenbleiben. Die betroffenen Jugendlichen können sich nicht abfinden mit einer betont materialistischen Lebensauffassung, mit Orientierung an Modeströmungen und Konsumverhalten. Sie sind auf der Suche nach bleibenden geistigen Wertvorstellungen. Die Eßstörung wird hierbei auch zum Ausdruck einer gesellschaftskritischen Einstellung junger Menschen, die von Umwelterwartungen stark beeindruckbar sind und eigenständige, oft originelle Ideen und Impulse im Zusammenhang mit einer tiefen Lebensangst nicht zu realisieren wagen.

Die Eßsymptomatik ist auch ein Familiensymptom, ebenso wie der hohe Stellenwert des Leistungsstrebens. In den Familien, in denen sich bevorzugt Kommunikations- und Interaktionsstrukturen zeigen, die durch auffällig fehlende Eigen-, jedoch intensive Fremddefinition, Aggressionshemmung und heimliche Koalitionen gekennzeichnet sind und in denen traditionelle Rollenverteilungen vorherrschen, versuchen magersüchtige Töchter mit ihrer Eßstörung Anstoß zu Wandel und Veränderung zu geben. Bis zum Krankheitsausbruch hingegen waren sie besonders bemüht und meistens auch fähig, sich den Erwartungen ihrer Eltern entsprechend zu entwickeln. Die später magersüchtigen Töchter hatten von Anfang an oder haben im Verlaufe ihres Heranwachsens eine besondere Begabung entwickelt, die narzißtischen Bedürfnisse ihrer Eltern wahrzunehmen und ihren Eltern das Gefühl zu geben, gute Eltern zu sein – unter Verzicht auf oder Vermeidung altersadäquaten, eigenständigen Experimentierens. Folge ist, daß später magersüchtige Patientinnen eine überragende Fähigkeit haben, sich in gewisse Empfindungsbereiche des anderen einzufühlen, sie bleiben aber in bezug auf die Wahrnehmung eigener Bedürfnisse extrem unsicher und wenig fähig, diese in ausreichendem

Maße zu realisieren. In der Adoleszenz, wenn es darum geht, eigene Zielvorstellungen in Abgrenzung von den Eltern zu entwickeln und in Solidarisierung mit Gleichaltrigen neue Lebensformen zu erproben, sind diese Mädchen unvorbereitet, voller Lebensangst und schwerer Zweifel, ob sie je imstande sein werden, eigenständig denkende und handelnde Erwachsene zu werden.

Die intrapsychische Situation der Jugendlichen, die bis zum Eintritt der Pubertät durch ein falsches Selbst geschützt waren und mit Beginn der hormonellen Umstellung einen tiefgehenden Entwicklungszusammenbruch erfahren (Bürgin, 1988), ist charakterisiert durch eine höchstgradige Ambivalenz und Ambitendenz (Progression versus Regression, Loslösungsversus Bindungsstreben, Anpassung an Umwelterwartungen versus Eigenständigkeit). Anale Kontroll- und Manipulationsmechanismen und die Verleugnung der eigenen Befindlichkeit (Hunger, vitale Körperbedürfnisse, Sexualität) unter Zuhilfenahme projektiver Identifikationen sollen ein weiteres Absinken in depressive Hilflosigkeit verhindern. Hierbei besteht die Gefahr, daß die Jugendlichen ihre seelischen Energien in ambivalenten Absicherungskämpfen verzehren, wenn nicht möglichst frühzeitig intensive Hilfe angenommen werden kann (Gefahr der Chronifizierung).

Die **therapeutischen Bemühungen** zielen auf eine Vermeidung der invalidisierenden Chronifizierung (Restitution oder zumindest Stabilisierung des Gewichtes), von körperlichen und seelischen Schäden, insbesondere einer tödlichen Entwicklung (z. B. Suizid). Therapeutische Einflußnahme lassen die Mädchen und jungen Frauen nur dann zu, wenn sie sich in ihrer Kompetenz als kritische Beobachter familiärer Interaktionen in einem bestimmten gesellschaftlichen Kontext wahrgenommen und ernsthaft gehört fühlen. Es gelingt dann nicht selten, sie als eigenverantwortliche Partnerinnen für einen therapeutischen Vertrag in einem ambulanten Behandlungssetting zu gewinnen. Dieses Behandlungssetting beinhaltet neben der psychotherapeutischen Arbeit, immer unter Einbezug der Familie, die Akzeptanz eines Zielgewichtes, die Verpflichtung zu einer regelmäßigen Gewichtszunahme sowie zum Spitaleintritt, falls die vertraglich festgelegte Gewichtszunahme pro Zeiteinheit unter ambulanten Bedingungen nicht eingehalten werden kann. Die stationäre Behandlung, bei der in einzelnen Fällen eine zusätzliche antidepressive oder neuroleptische Pharmakotherapie notwendig werden kann, erfordert ein erfahrenes Behandlungsteam.

Bei der reinen **Eß-Brech-Sucht (Bulimia nervosa)** bleibt das Gewicht weitgehend konstant. Obwohl in der Spätadoleszenz häufig, bleibt sie oft lange verborgen, da die Schamschwelle sehr hoch ist und die Symptome nicht manifest werden, es sei denn durch Stehlen (sehr hohe Kosten für den Nahrungsmittelerwerb) oder panische Ängste, gepaart mit Selbstverachtung. Attacken gierigen Essens wechseln mit sofortigem Erbrechen. Die Beziehungsfähigkeit ist eingeschränkt. Bezüglich Pathogenese, Ätiologie und Therapie gilt Ähnliches wie bei der Anorexie und der Bulimarexie, zu denen es fließende Übergänge gibt.

54.7.17 Adipositas

Die Adipositas (Übergewicht von mehr als 20%, bezogen auf Alter und Geschlecht) hat in den industrialisierten Ländern stark zugenommen. Bei 50% beginnt sie in den ersten 2 Lebensjahren, bei den anderen in der Schulzeit oder Pubertät. Etwa 2 von 10 Schulkindern gelten als übergewichtig. Kinder aus unteren Sozialschichten sind bedeutend häufiger betroffen. Ätiologisch liegt auch hier eine Plurikausalität vor. Familiäre Dispositionen sind häufig (bei 80% fettsüchtiger Kinder sind auch Vater oder Mutter übergewichtig). Mögliche hereditäre Faktoren, Eigenheiten des Fettgewebes und die Regulation der Energiebilanz werden immer wieder als ätiologisch relevante Punkte genannt (Schmit, 1985). Aber auch familiäre Eßgewohnheiten mit einem kalorischen Überangebot spielen eine Rolle, ebenso wie eine gewisse konstitutionelle Neigung zur Passivität, die Verwendung von Nahrungsmitteln als Ersatz für emotionale Zuwendung, dominierendes Verhalten der Mutter, Submissivität des Vaters, Aggressionshemmung, Frustrationsintoleranz und eine deutliche Beeinträchtigung des Selbstwertgefühls der Kinder. Sekundäre Fehlverarbeitung der Folgen der Adipositas (z.B. körperliche Schwerfälligkeit, Bequemlichkeit, Gehänseltwerden, Kontaktstörungen) und sekundärer Krankheitsgewinn verschlimmern oft die primäre Pathologie. Die Prognose bezüglich des Gewichts ist nicht sehr günstig: 7–8 von 10 Kindern bleiben dick. Ein Therapieprogramm, das einen gewissen anhaltenden Erfolg erzielen will, muß sich über lange Zeit (mindestens 1–2 Jahre) erstrecken. Es muß Einfluß nehmen auf den körperlichen Bereich (Förderung der Bewegung), auf das Verhalten (Eß- und Kochsitten) und auf die Förderung der Selbstwahrnehmung, um zu einer Stabilisierung im Selbstgefühl beizutragen, damit dann erst, mittels individueller oder familiärer Therapie, neben der Diät Konflikte bearbeitet und weitere Einstellungsveränderungen erzielt werden können.

55 Dermatologie

Klaus Bosse

Im folgenden werden ausgewählte Aspekte der Psychosomatik in der Dermatologie besprochen. Die Auswahl spezieller modellhaft-repräsentativer Krankheitsbilder und für die Dermatologie allgemein bedeutsamer psychosomatischer Phänomene erfolgt nach der praktisch-ärztlichen Wichtigkeit und der Häufigkeit ihres Vorkommens. Subjektive Gewichtungen bei der Auswahl lassen sich nicht vermeiden, da eine allgemein anerkannte Bewertung der psychosomatischen Sicht in der Dermatologie noch aussteht.

55.1 Der psychosomatische Zugang zum Hautorgan

Die Möglichkeit einer sofortigen klinisch-morphologischen Blickdiagnostik ohne aufwendige Laboruntersuchungen sowie die der morphologisch orientierten Lokaltherapie erlaubt in vielen Fällen von Hauterkrankungen eine ausreichende ambulante somatische Beurteilung innerhalb weniger Minuten. Dies gilt nicht mehr, wenn der Hautarzt die „feste, aber für den außenstehenden Beobachter unsichtbare Schale der Umwelt" (Jakob von Uexküll, vgl. hierzu Th. v. Uexküll und Wesiack, Kap. 1) des Patienten, gleichsam seine „äußere Außenhaut", mit einbezieht. Noch schwieriger wird der diagnostische und therapeutische Zugang, wenn wir das komplizierte Bedingungsgefüge des Hautkranken mit seiner sozialen Umwelt auf dem Hintergrund seiner individuellen Entwicklung sehen (Medanski, 1980).

Aber gerade am Hautorgan offenbart sich die Notwendigkeit einer derartig orientierten Betrachtungs- und Behandlungsweise in ganz besonderem Maße (vgl. auch van Moffaert, 1982), indem an der Haut jeglicher emotionaler Input zu einem sofort sichtbaren Output in Form von Erröten, Erbleichen, Schwitzen oder sensorischen Mißempfindungen als Ausdruck der jeweiligen Gefühlslage führt. Die taktile und visuelle Kommunikation wird durch Hautkrankheiten nachdrücklich beeinträchtigt und schließlich veranlaßt die leichte Zugänglichkeit des Hautorganes den Patienten unbewußt, durch Sichberühren, Scheuern, Kratzen oder sonstige Selbstmanipulationen Besserung der Hauterscheinungen oder Entlastung von innerer Erregung zu suchen. Manchmal erinnert diese motorische (Ersatz-)Reaktion an eine Übersprunghandlung von Tieren.

Die Umgangssprache hat für den nahen Zusammenhang zwischen dem Hautorgan und der emotionalen Befindlichkeit des Individuums eine unendliche und bilderreiche Vielfalt von Ausdrucksweisen gefunden. Als pars pro toto sei hier lediglich an die „dünne Haut", das „dicke Fell", an das „sich die Haare raufen" eines Menschen oder das „unter die Haut gehen" einer Sache erinnert. Kein anderes Organ des Körpers – nicht einmal das Herz – spiegelt sich in der Sprache vergleichbar als Ausdrucksorgan des seelischen Lebens eines Menschen wider. Kaum ein anderes Organ des Körpers gewinnt im übertragenen Sinn des Funktionskreises J. v. Uexoülls eine derartige direkte Bedeutung für die ständig veränderliche Beziehung zwischen Organismus und Umgebung des Menschen: Die Haut als Merkmal (Farbe – Faltenbildung – Schweiß) erhält beim Gegenüber eine ständig registrierte (Merken) Bedeutung (Freude – Trauer – Angst) und bewirkt damit bei diesem ein Verhalten, das sich an der unterstellten Bedeutung des Merkmales orientiert (vgl. Th. v. Uexküll u. Wesiack, Kap. 1).

Der krankhaft veränderten Haut wird unter Umständen subjektiv ein erweiterter Bedeutungsgehalt zuerteilt (vgl. 55.3.3 „Paranoiatendenz des Hautkranken" und „Genügsamkeitsthese des Hautgesunden"). Der Funktionskreis wird damit unter Einschaltung von gegenseitigen Unterstellungen (Phantasie bzw. unterstelltes Fremdbild) zum Situationskreis (vgl. hierzu Th. v. Uexküll und Wesiack, Kap. 1, Abb. 1–2).

Zu den zahlreichen exogen naturwissenschaftlich, psychisch oder religiös-mythisch orientierten Krankheitskonzepten von hautkranken Patienten gehört die Vorstellung der „Reinigung des Körpers" von innen heraus: Zum „Ableiten der kranken Körpersäfte" muß nach Meinung vieler alter Unterschichtpatienten bestehenbleiben, die Abheilung wird deshalb sogar künstlich hinter dem Rücken des Arztes oder der Schwester verhindert. Auch für manche höhergebildete Patienten ist es schwer annehmbar, daß z.B. ein akutes Ekzem in der Regel keiner inneren Behandlung bedarf, denn äußere Unreinheit der Haut wird bereitwillig mit innerer Unreinheit gleichgesetzt; nicht nur vom Gesunden, sondern auch vom Erkrankten selbst. Als Hiob mit Gott zu hadern begann, erkrankte er an seiner Haut. Krankheit und Sünde stehen in seiner Umwelt in einem inneren Zusammenhang und deshalb hat auch zu seiner Zeit die Hautkrankheit bereits sozial schwerwiegende Konsequenzen. Menschen, die von Zaraath – einem Sammelbegriff für Hautkrankheiten im biblischen Schrifttum – befallen sind, dürfen nicht mehr an rituellen Kulthandlungen teilnehmen und gehören nicht mehr zur Dorfgemeinschaft (Falke, 1982). Auch die Bereitschaft des heutigen Menschen zu übertriebener, die Haut sogar schädigender Hygiene, bis hin zum Waschzwang, als Ursache ekzemähnlicher Störungen der Haut, erscheint auf diesem Hintergrund nicht verwunderlich. Die sozialen und intrapsychischen Dimensionen von Hautkrankheiten werden in der Umgangssprache der Straße, im theologischen Schrifttum und in der Geschichte des Mittelalters ebenso deutlich wie in der Sprechstunde des Dermatologen.

Die enge Beziehung des Hautorganes zum gefühlsmäßigen Erleben des Menschen geht somit weit über das Sprachliche hinaus und ist sicher nicht zufällig entstanden (Montagu, 1971).

Im Zustand der sozialen „Symbiose" zwischen Mutter und Säugling des ersten Lebenshalbjahres besteht für den letzteren nach Mahler (1978) eine Fusion zwischen Ich und Nicht-Ich, die alle Grenzen des Erlebens der äußeren Körperoberfläche verschwimmen läßt. Es handelt sich dabei nicht um eine Symbiose mit gegenseitiger Abhängigkeit im biologischen Sinn. Der erste, noch ausschließlich symbiotische Kontakt des Säuglings zur Mutter ist vorwiegend taktiler Natur. Die Haut vermittelt Wärme und Berührung, die taktile Kommunikation des Kindes über die Schleimhaut der Lippen und des Mundes setzt Unlustgefühle in solche der Befriedigung um. Zur Zeit der Aufnahme der visuellen Kommunikation mit der Mutter erhält die Haut neue Bedeutungen: Grenze und Trennfläche einerseits, Empfangsorgan für die Zärtlichkeit und Liebe der Mutter andererseits. Die frühesten und elementarsten Bedürfnisse des Säuglings nach zwischenmenschlicher Beziehung bedürfen der Haut und ihrer Vermittlung bis in das Leben des Erwachsenen hinein. Damit bekommen auch Störungen an diesem Vermittlerorgan ihren besonderen Krankheitswert: Angst vor Verlust zärtlicher Zuwendung wird ebenso wie die Einschränkung des Selbstwertgefühles vom Hautkranken gewichtig erlebt.

Schwieriger zu verstehen ist die Umkehrung dieses Prozesses, die in der täglichen Praxis des Dermatologen nicht selten evident wird. Nämlich die Tatsache, daß Verlust- und Trennungsängste oder andere innerseelische Konflikte zur Auslösung eines neuen Schubes einer Hauterkrankung führen können.

Schon vor den gewichtigen Hinweisen auf die Notwendigkeit der taktilen Kommunikation durch spätere vergleichende experimentelle Untersuchungen an verschiedenen Tierarten sowie durch die sorgfältigen Beobachtungen an Primaten in freier Wildbahn durch Ethologen und Verhaltensforscher, betonte die psychoanalytische Schule die Bedeutung affektiver und sensorischer Prozesse im Kindesalter für die spätere menschliche Entwicklung, insbesondere für die zukünftigen sozialen Fähigkeiten. Gegenstand dieser Forschungsrichtungen war überwiegend das in seinem Hautkontakt gestörte Individuum (Mensch/Tier).

Tierexperimentelle Untersuchungen an verschiedenen Spezies geben Hinweise auf die Bedeutung der kutanen Selbststimulation von weiblichen Tieren während der Trächtigkeit, für ihr Verhalten als Muttertier in der frühen post-partum-Periode gegenüber den Jungtieren sowie auf den Einfluß der taktilen Stimulation der Jungtiere für deren somatische Entwicklung und ihr späteres soziales Verhalten als Weibchen und Muttertiere (Übersicht bei Montagu, 1971). Die Beobachtungen werden ergänzt durch die Ergebnisse

der Verhaltensforschung bei Freilandbeobachtungen und wurden von Renggli (1976) auch unter psychoanalytischen Gesichtspunkten diskutiert. Besonders wichtige experimentelle, an Rhesusaffen durchgeführte Untersuchungen für diese Fragestellung wurden von Harlow (1958) publiziert (ausführliche Darstellung und Diskussion bei Montagu, 1971). In einer Versuchsanordnung, bei der Rhesusaffen-Jungtiere entweder von einer frotteebekleideten Muttergestalt oder einer Drahtmutter gestillt wurden, zeigte sich, daß die Berührungsbehaglichkeit bei der Frotteemutter für die Jungen wichtiger war als die milchspendende Eigenschaft der Drahtmutter. Er beschrieb außerdem die Beeinträchtigung des Zuwendungsverhaltens von Muttertieren zu ihren Jungen nach völligem Entzug der Mutterfigur bei fünf späteren „Versagern als Mütter".

Wenn psychosomatisch orientierte Dermatologen bei bestimmten Krankheitsgruppen (Selbstbeschädigungen, atopische Krankheiten u.a.) eine evidente Korrelation des zeitlichen Auftretens von Krankheitserscheinungen und typischen „life events" finden, so ist es naheliegend, psychologische Gesetzmäßigkeiten im aktuellen Sozialverhalten des Patienten zu suchen und dieses Verhalten in Analogie zu den geschilderten experimentellen Beobachtungen psychoanalytisch als Ergebnis der (früh)kindlichen Entwicklung zu begreifen.

Sowohl aus ethischen Gründen als auch wegen der Komplexität des Systems Mensch/Umwelt ist beim Menschen ein objektiver und reproduzierbarer experimenteller „Beweis" von Hypothesen aufgrund evidenter Beobachtungen des Analytikers (wie es der Kliniker in Form von Doppelblindstudien gerne sähe) selbst bei eineiigen Zwillingen nicht möglich, zumal es sich beim Erwachsenwerden um Zeiträume handelt, die in der Regel von einem einzelnen Beobachter nicht übersehen werden können.

Dies gilt auch für den Zusammenhang des Hautkontaktes zwischen Mutter und Kind einerseits und der späteren Entwicklung von sozialen Fähigkeiten und somatischen Reaktionen der Haut im Falle vorgegebener genetischer Disposition andererseits.

Besondere Bedeutung haben deshalb für die psychoanalytische Forschung und die psychosomatische Dermatologie die sorgfältigen kasuistischen Verlaufsstudien an Atopikern erlangt, wie sie z.B. von Schur (1955) und neuerdings von Thomä (1980) vorgelegt wurden, sowie die Beobachtungen zum endogenen Ekzem von R. Spitz (1957) an Heimkindern und schließlich analytisch orientierte retrospektive biographische Studien, wie sie auch von unserer Arbeitsgruppe seit Jahren durchgeführt werden (Heigl-Evers et al., 1976).

Versuche lernpsychologischer Interpretationen von Hautkrankheiten (endogenes Ekzem, Urtikaria u.a.) erfahren Unterstützung durch den Nachweis der Möglichkeit, die Histaminausschüttung beim Meerschweinchen als Lernprozeß durch Training zu aktivieren (Russell et al., 1984).

Der häufigsten Hauterkrankung, bei der psychische Faktoren eine wesentliche Mitursache für die Auslösung von Rezidiven spielen, dem endogenen Ekzem, soll deshalb ein eigener Abschnitt gewidmet werden.

Demgegenüber stellen sich individuell psychologische und soziale Konsequenzen von Hauterkrankungen am häufigsten anläßlich der Entstellung des äußeren Erscheinungsbildes des Patienten dar. Dieser Gesichtspunkt wird besonders deutlich im Falle der Psoriasis vulgaris, der Acne vulgaris und einiger anderer entstellender Hauterkrankungen.

Aus der Bedeutung der äußeren Entstellung durch Hauterkrankungen und der komplizierten damit einhergehenden Psychodynamik und Soziodynamik ergibt sich nicht nur die soziale Situation des Kranken als Phänomen interpersoneller Wahrnehmung, sondern auch die pathologische Angst mancher Menschen davor, entstellt zu sein bzw. entstellt zu werden. Diese Angst kommt als Begleitphänomen vorhandener Hautveränderungen wie auch als Ausdruck nicht vorhandener, aber befürchteter Veränderungen des Äußeren als Dysmorphophobie vor (vgl. Abschn. 55.3.4).

Der vermeintlich oder sichtbar Entstellte bedarf unbedingt „ursächlicher" ärztlicher Hilfe und/oder symptomatischer Selbstbehandlung mittels abdeckender oder auch provozierender Kosmetik. Aber nicht nur das objektiv mehr oder weniger veränderte Äußere als Zweck und Ziel der Behandlung, sondern auch und vielleicht gerade die Handlung selbst, das Berühren des eigenen Körpers an exponierten Stellen mit substituierenden erotischen und homoerotischen Zügen, weisen wieder auf die kommunikationsvermittelnde Funktion der Haut hin. Die verschiedenen Aspekte der Entstellung durch Veränderung des Äußeren werden im Abschnitt 55.3.3 dargestellt.

Unsere Erörterung von ausgewählten Gesichtspunkten der dermatologischen Psychosomatik führt schließlich zur Diskussion von besonderen Therapieformen, die sich neben fachpsychotherapeutischen Methoden als Ergänzung der somatischen Behandlung in der psychosomatischen Dermatologie anbieten (Abschn. 55.7).

Abschließend wird über fachspezifische Erfahrungen der Lehre und des Lernens von psychosomatischer Dermatologie berichtet. Es liegen Erfahrungen in großen Lehrveranstaltungen (bis zu 120 Studenten) und in studentischen Kleingruppen vor.

Die psychosomatische Orientierung eröffnet der Dermatologie eine bedeutende Erweiterung ihrer diagnostischen und therapeutischen Möglichkeiten. Hierauf wollte Sack 1926 hinweisen, als er zum Thema „Die Haut als Ausdrucksorgan" sagte, daß sich hier „... die experimentelle und theoretische Biologie einschließlich der Vererbungslehre ... und die Psychologie und Psychopathologie die Hände reichen zu einem eigenartigen, aber offenbar äußerst fruchtbaren Bunde". Er zitiert dabei ausdrücklich die Arbeiten von Driesch und von J. v. Uexküll wie auch die von Freud und seiner Schule.

In neuerer Zeit befaßten sich im deutschsprachigen Raum vor allem Borelli, Panse, Rechenberger, Hornstein, Musaph, v. Moffaert, Bosse und Hünecke mit Fragen der psychosomatischen Dermatologie. Eine sorgfältige und kritische monographische Bearbeitung erfolgte durch Whitlock, 1976 und 1980.

55.2 Das sog. endogene Ekzem als psychosomatische Hauterkrankung

(Synonyme: Atopische Dermatitis, Neurodermitis constitutionalis, Prurigo Besnier)

1881 beschrieben Broc und Jacquet diese chronisch juckende ekzemähnliche Hauterkrankung charakteristischerweise als „Neurodermitis" und erinnern damit noch heute an den emotionalen Anteil der Erkrankung.

Heute wird die Mitwirkung psychischer Abläufe neben den bekannten genetischen, immunologischen und allergologischen Faktoren bei der Auslösung von Rezidiven des sog. endogenen Ekzems wieder gesehen. Es wird jedoch darauf hingewiesen, daß der klinisch-dermatologischen, immunologischen und allergologischen Diagnostik sowie der symptomatischen dermatologischen Therapie eine aktuell entscheidende Rolle zukommt. Mittel- und langfristig erfährt diese durch Berücksichtigung psychosomatischer Gesichtspunkte eine wesentliche Erweiterung.

Das klinisch-morphologische Bild ist häufig beim Säugling bereits als Milchschorf, d.h. durch schuppende Beläge des Capillitiums, oder auch durch eine symmetrisch auftretende schuppende entzündliche Rötung beider Wangen charakterisiert. Kinder und Jugendliche leiden an einem rezidivierenden „Beugenekzem" (Kniekehlen, Armbeugen, Hand- und Fußgelenke, Hals) oder gelegentlich an einer nur rudimentären Symptomatik (Mundwinkelrhagaden, Ohrläppchenrhagaden, mykoseähnlich anmutende Veränderungen einzelner Finger oder Zehen). Der erwachsene endogene Ekzematiker zeigt neben den typischen, die Beugen betonenden Veränderungen der Haut ein pruriginöses, d.h. knotiges, juckendes Krankheitsbild mit einer häufig stark imponierenden Lymphadenitis. Der Übergang in eine Erythrodermie ist möglich (Rajka, 1975).

Der schubweise auftretende quälende Juckreiz zeigt eine tagesrhythmische Zunahme gegen Abend, eine Verstärkung in der Wärme und bei mechanischem Kontakt mit Wolle sowie einen individuellen Rhythmus, der meist durch Arbeitszeiten und arbeitsfreie Zeiten bestimmt wird. Entspannende, geglückte Urlaube führen regelmäßig zu einer Besserung der Hauterscheinungen, besonders im Nordsee- und im Höhenklima.

Therapie: Die lokale Anwendung von Steroiden oder deren innerliche Verabreichung führt meist schnell zu einer zeitweiligen symptomatischen Besserung. Begrenzt wird diese Möglichkeit durch die Nebenwirkungen der Kortisontherapie. Die kortisonfreie Lokalbehandlung (z.B. mit Teerderivaten und Harnstoff) ist schwierig und für den Patienten mühsam und belastend. Über die Art der Lokalbehandlung kommt es nicht selten zu Differenzen zwischen Arzt und Patient. Ständige Pflegebehandlung der Haut durch den Patienten verringert die sonst notwendige regelmäßige fachärztliche Betreuung. Die Einführung psychosomatischer Gesichtspunkte in die Therapie bringt nach eigenen Erfahrungen eine deutliche Reduktion der aktuell und langfristig notwendigen Lokaltherapie mit sich. Die günstige Wirkung von Nordsee- und Höhenklima dürfte u. a. auch auf der entspannenden Wirkung der Kuraufenthalte beruhen.

55.2.1 Exemplarischer Fall

A. G., männlich, 23 Jahre, drittes von drei Kindern

Bereits bei dem ersten ambulanten Konsil deutet der Patient psychische Zusammenhänge an („je mehr ich arbeite, desto schlechter ..., je mehr ich Blumen gieße und umtopfe, desto weniger Hauterscheinungen ...").

Der Patient ließ sich stationär aufnehmen, um sein Ekzem „auf Vordermann zu bringen". Er wollte in absehbarer Zeit ein akademisches Examen ablegen. Die Absicht, neben der Lokaltherapie den Patienten weitgehend aus dem Alltag herauszunehmen und anfänglich unter Neuroleptika entspannen und viel schlafen zu lassen, fand nicht seine Sympathie. Er wollte gleichzeitig sein Examen vorbereiten. Es gelang uns ohne Kortisonzusätze seine Hautsymptomatik zu bessern, um ihn – wie von ihm dringend gewünscht – alsbald entlassen zu können. Vorschläge des Arztes, möglicherweise sein Examen zu verschieben, um die Haut und seinen Allgemeinzustand zu stabilisieren, wurden unter dem Hinweis zurückgewiesen, daß er ja dann mehr vergesse. Auch eine intensive selbständige berufliche Nebenbeschäftigung wollte er nicht eingeschränkt wissen.

Unter der weiteren Examensvorbereitung verschlechterte sich sein Hautzustand wieder, während Versagensängste und Erfolgsgefühle ständig wechselten. Eine bisher bewährte Methode, dem frühsommerlichen Heuschnupfen und der damit verbundenen Hautverschlechterung durch einen Mittelmeeraufenthalt entgegenzuwirken, gelang diesmal nicht. Bei einem mehrtägigen Besuch zu einem Familienfest erlebte er einen massiven plötzlichen Ekzemschub. Er ließ sich von dem hinzugezogenen Arzt mit Kortison innerlich behandeln. Über sein daran anschließendes Verlangen nach uneingeschränktem Zugang zu Kortisonsalben kam es zur Auseinandersetzung mit dem bisher behandelnden Arzt (Patient: „Ich trage die Verantwortung für mich selbst!"; „Arzt: „Ich trage die Verantwortung für eine fachgerechte Behandlung der Patienten"), die dann zu einer beidseitigen Trennung führte. Ein Psychologe betreute den Patienten zunächst weiter. Im weiteren Verlauf konnte er sich anderweitig kortisonhaltige Salben in der von ihm gewünschten Menge verschreiben lassen und stand vor der Entscheidung, sich zum Examen zu melden. Er zeigte sich äußerst beunruhigt bei der Vorstellung, im Examen scheitern zu können, er wollte ein Prädikatsexamen ablegen. Er rang sich schließlich zur Entscheidung für das Examen durch. Danach wurde nicht, wie von ihm befürchtet, seine Haut schlechter und damit arbeitsstörender, sondern sie besserte sich sogar teilweise. Als er die dreitägigen Examensklausuren hinter sich gebracht hatte, kam es am Abend zu einer „noch nie erlebten Dreifachreaktion" (Patient). Es stellten sich sowohl Heuschnupfen als auch asthmatische Anfälle ein und schließlich reagierte die Haut mit außerordentlichen ekzematischen Erscheinungen. Zu seinem eigenen Erstaunen bildeten sich in den folgenden Tagen ohne besondere zusätzliche Maßnahmen alle Symptome wieder weitgehend zurück.

Bereits bei der ersten Vorstellung deutet der Patient seinen wichtigsten Grundkonflikt in einem eindrucksvollen Bild an: Er fühlt sich zur Leistung verpflichtet (der Vater ist in seinem Beruf besonders erfolgreich, die älteren Geschwister sind bereits etabliert). Er spürt aber, daß seine Haut sich bessert, sobald er dieser Verpflichtung nicht nachkommt. Er läßt sich zur stationären Behandlung aufnehmen, obwohl er sich mit der vorgesehenen entspannenden Behandlung nicht arrangieren kann: Bücher und Skripten auf dem Nachttisch als Absichtserklärung, die Besuche von Freunden und Studienkollegen als Unterbrechung demonstrieren seinen Wunsch, ein aktiver und beliebter Mensch zu sein. Der Konflikt zwischen Angst zu versagen und Leistungsanspruch wiederholt sich – jeweils gekoppelt mit Verschlechterung des Hautzustandes – bis zu dem gefürchteten Examen. Unter der Anspannung des letzten Examenstages kulminiert das Krankheitsgeschehen in einem vorher nie erlebten Ausmaß und ist ohne zusätzliche Therapie mit der Entspannung der darauffolgenden Tage wieder rückläufig. Eine ähnliche anfallsartige Verschlechterung erlebt der Patient während eines Familienfestes bei der Begegnung mit der Familie und Studienkollegen. In dem späteren Gespräch hierüber ist der Patient nicht in der Lage zu vermitteln, wo der Schwerpunkt seines Erlebens an diesem Tag lag (Familie und Geschwister? – Leistungsvergleich mit Studienkollegen?). Die vordergründig symptomauslösenden aktuellen Situationen sind eindeutig zeitlich zuzuordnen und werden vom Patienten gesehen. Seine Persönlichkeit und familiär tradierte Verhaltensweisen hindern ihn, alternative Problemlösungen zu suchen (Verzicht auf Leistungsanspruch und Erfolgsgefühl). Der Patient entzieht sich einer weiteren ärztlichen Betreuung, da eine ständige symptomlindernde Steroidbehandlung in eigener Regie des Patienten vom Arzt abgelehnt wird. Der Vorteil einer begleitenden psychologischen Betreuung wird in diesem Falle deutlich, indem der Patient auch nach dem Arztwechsel weiterhin sporadisch reflektierende Gespräche mit dem Psychologen wahrnimmt.

55.2.2 Aktuelle und chronische Konstellationen als Auslöser

Der endogene Ekzematiker kommt in der Regel somatisch orientiert – z. B. mit der Vorstellung eine Allergie zu haben – zum Hautarzt und erwartet eine dementsprechende Diagnostik und Behandlung. Wenn sich der behandelnde Arzt jedoch bei jedem Patienten regelmäßig die Mühe anamnestischer Erhebungen macht, so werden ihm sehr wohl die zeitlich mit dem Auftreten des Rezidivs korrelierten aktuellen Konfliktsituationen auffallen (Böddeker, 1976).

Aktuelle Verschlechterungen bei Kleinkindern treten meist gleichzeitig oder in der Folge von Verdauungsstörungen oder anderen Infekten, auch zur Zeit des Zahnwechsels, sowie bei Umstellungen in der Umgebung auf. Bei Kindern sind als typische Situationen die urlaubsbedingte oder sonstige Abwesenheit eines bzw. beider Elternteile, Spannungen zwischen den Eltern, Trennung der Eltern, aber auch Konkurrenzsituationen in der Geschwisterreihe (Geburt weiterer Geschwister, auf die der Patient nicht genügend vorbereitet war), sowie Auseinandersetzungen mit Altersgenossen, im Schulalter dann Klassenaufenthalte im Landheim oder sonstige Ferienaufenthalte in ungewohnter Umgebung, Schulwechsel oder Umzug zu beobachten. Beim Heranwachsenden sind es schwierige Konstellationen zu Zeiten der Berufsfindung, der Partnersuche oder drohende Prüfungen. Die rezidivierende Verschlechterung bei jüngeren Erwachsenen ist zur Zeit von Verlobung, Hochzeit, ehelichen Schwierigkeiten, Überlastungen durch Hausbau, durch verpflichtende Feste (Weihnachten für die Hausfrauen) oder beim Tod nahestehender Menschen zu finden.

Aktuelle Bezüge der beschriebenen Art sind meist sehr typisch und auffällig für den Arzt (Siegrist, 1980), wenn der Patient bereit ist, solche zeitlichen Korrelationen zu suchen und darüber zu sprechen. Eine wichtige Aufgabe des somatisch bzw. psychosomatisch orientierten Hautarztes ist es deshalb, den Patienten in dieser Richtung zu sensibilisieren und zur Selbstbeobachtung anzuregen.

Dagegen sind chronische Spannungszustände, die der Patient „akzeptiert" hat, sehr viel schwieriger therapeutisch zugänglich. Solche Konflikte und spannungsträchtige Umstände können bei gegebenen somatisch-genetischen Voraussetzungen als Dauerauslöser wirksam sein. Dies wird offensichtlich, wenn wir den Patienten aus seiner häuslichen und beruflichen Umgebung herausnehmen. Ohne Veränderung der Therapie, auch bei indifferenter Lokaltherapie und gezielter Anwendung von Tranquilizern oder Neuroleptika genügt dann oft der bloße Wechsel in die abschirmende Krankenhausatmosphäre und die Möglichkeit zur zeitweiligen Entspannung, um in solchen Fällen eine eindrucksvolle Besserung des Hautzustandes zu erreichen. Dasselbe gilt für Kleinkinder, die dann manchmal ihr Wohlbefinden nicht nur an der Haut, sondern auch durch ihr verändertes Verhalten auf der Station im Umgang mit Schwestern und anderen Kindern bekunden. Das auffallende Verhalten dieser Kinder und die schnelle Abheilung unter stationären Bedingungen legen die Interpretation als „umgekehrtes Hospitalisierungsphänomen" nahe. Unterstützt wird diese Deutung durch den oft kontaktgehemmten, wenig spontanen Umgang der Mutter mit ihrem Kind (vgl. 55.2.4 und 55.2.5) und auch durch das schnelle Rezidivieren der Hauterscheinungen nach der Entlassung in häusliche Bedingungen.

Den Modellfall einer sofortigen Reaktion mit Juckreiz können wir bei ambulanten Konsultationen beobachten: Führt der Arzt das Gespräch über die für den Patienten noch erträgliche Toleranzgrenze hinaus, oder frustriert er ihn in der experimentellen Situation, so beginnt letzterer sich zu kratzen. Kinder, deren Wunsch nach Beendigung der Konsultation oder andere Forderungen nicht erfüllt werden, beginnen sich ebenfalls zu kratzen (Hünecke und Bosse, 1981; Bosse und Hünecke, 1981).

55.2.3 Persönlichkeitsstrukturen und endogenes Ekzem

Im Verlauf der Beschäftigung mit dem endogenen Ekzem sind oft Versuche unternommen worden, eine Abhängigkeit des Krankheitsgeschehens von der Persönlichkeit nachzuweisen. Dementsprechende psychometrische Studien sind letztlich immer wieder gescheitert (vgl. die kritischen Stellungnahmen von Rostenberg, 1959; Whitlock, 1976; Rees, 1983). Auch die eigenen langjährigen Untersuchungen unserer Arbeitsgruppe (Hünecke, nicht publiziert) erbrachten noch keine schlüssigen Ergebnisse. Die Gründe für die Vergeblichkeit dieser Ansätze dürften im Methodischen liegen:

Selektiertheit/Unselektiertheit der Patienten

In sog. klinischen Stichproben mag unbemerkt das Faktum Einfluß nehmen, daß z.B. die Auffälligen (im Sinne von Neurosen) einer Testung zugeführt werden. Andererseits kann der Anteil derjenigen Patienten, bei denen eine überwiegend psychosomatische Genese vorliegt, zugedeckt und verwischt werden durch den Anteil an Patienten, deren Ekzemgenese sich mehr auf genetisch-dispositionelle Faktoren und/oder Umwelteinflüsse stützt.

Testwahl

Es ist zu unterstellen, daß bis heute keine verbindliche Begründung oder Übereinstimmung dahingehend besteht, weshalb mit dem einen oder anderen Testverfahren gearbeitet werden sollte. Weiterhin erscheint die Angemessenheit vieler Verfahren grundsätzlich fragwürdig, wenn man bedenkt, daß diese im Rahmen der Neurosenforschung entwickelt und normiert wurden.

Zeitpunkt der Untersuchung

Selten hat die Tatsache Berücksichtigung gefunden, daß mit großer Wahrscheinlichkeit Testantworten der Patienten durch die Akuität des Krankheitsgeschehens jeweils modifiziert werden, insbesondere durch ein symptomfreies Intervall. Bezeichnend ist, daß die meisten einschlägigen Erhebungen Querschnittserhebungen darstellen, in denen kurz- oder längerfristige Schwankungen nicht zur Darstellung kommen können. Auch die Frage der Dauer und der Intensität des Ekzems findet auf diese Weise kaum Eingang. Unter diesem Aspekt ist die langjährige Längsschnittuntersuchung von Rechardt (1970) wertvoll, der feststellte, daß die von ihm untersuchten Ekzematiker gleichzeitig mit der Besserung der Hautsymptomatik in zunehmendem Alter variable Persönlichkeitszüge zeigten (vgl. hierzu Thomä, 1980).

Testauswertung

Auch wenn mittlerweile Zurückhaltung gegenüber der möglichen subjektiven Auswertung von projektiven Tests geübt wird, bleibt doch bei den psychometrischen Testuntersuchungen die Auswertungsmethode problematisch, wenn nämlich Testbefunde nur über ihren „Mittelwert" wiedergegeben werden, da dieser in der Tat nur ein einziger und nicht zwingend der am besten geeignete Indikator für eine Verteilung von Testwerten sein muß.

Sicherlich sind die Bemühungen um eine objektive Erfassung der Persönlichkeitszüge bei endogenen Ekzematikern zu begrüßen, da auf diese Weise dem Wunsch nach einer nachvollziehbaren Psychodiagnostik entsprochen wird. Allerdings beginnt mit der Auswahl der Testinstrumente, der Anwendung und der Auswertung der Testergebnisse erst die vielfältige und entscheidende Arbeit, die eine Fülle von Fehlermöglichkeiten mit sich bringt.

So ist aus den Mittelwerten psychometrischer Daten wohl kaum eine Aussage über die Spezifität hinsichtlich eines allgemein gültigen Persönlichkeitstypus für den endogenen Ekzematiker zu erwarten. Dagegen können gerade Abweichungen vom Mittelwert im individuellen Fall wichtige Hinweise für die diagnostischen und therapeutischen psychosomatischen Ansätze ergeben. Wir verweisen in diesem Zusammenhang auch auf die Befunde von Schur (1955), der aus analytischer Sicht aufgrund seiner Erfahrung mit Atopikern betont, er habe „... in keinem Falle so etwas wie eine Spezifität hinsichtlich des Persönlichkeitstypus, des Konfliktes ..." gefunden.

55.2.4 Die Mutter und die Mutter-Kind-Beziehung bei Kindern mit endogenem Ekzem

Für dermatologisch wenig geübte Ärzte und Pfleger wie auch für langjährig im Umgang mit endogenen Ekzematikern geschulte, erscheint es immer wieder evident, daß die Mütter dieser kranken Kinder, die Kinder selbst und der Umgang miteinander auf eine eindrucksvolle Weise charakterisiert sind. „Feindseligkeit" und „Abweisung" der Mutter, „überprotektive" Mütter (Spitz, 1957), ein Umgang miteinander, der durch wenig Spontaneität und Mangel an Wärme auffällig ist, Jugendliche, die zurückgezogen ihren Hobbys leben, Kinder, die im Umgang mit anderen Kindern „schwierig" erscheinen. Diese Klischees drängen sich auch dem nicht belesenen, unvoreingenommenen Beobachter auf und wurden über Jahrzehnte anhand von Kasuistiken in der Literatur niedergelegt. Sie konnten dennoch nie in einer methodisch überzeugenden Weise abgesichert werden (Palos und Ring, 1983; Whitlock, 1980). Wir halten diese Beobachtungen dennoch für bedenkenswert und möchten ergänzen, daß die Mutter gegebenenfalls nicht per se, sondern stellvertretend für den in der Regel am meisten mit dem Kind beschäftigten Elternteil betroffen ist. Zur Illustration soll deshalb eine typische Konsultation wiedergegeben werden (Bosse und Hünecke,

1981), die von vielen Hautfachärzten ausdrücklich als besonders typisch bezeichnet wurde.

> Arzt zum ca. siebenjährigen Kind: „Weshalb kommst du denn? Was hast du?"
>
> Mutter: „Mein Kind hat eine Allergie." (Die Mutter faßt das zögernde Kind an den Schultern, dreht es und schiebt es zum Arzt. Das Kind läßt dies widerwillig mit sich geschehen.)
>
> Mutter: „Wir waren schon bei drei Ärzten und einem Hautfacharzt, aber es kommt immer wieder."
>
> Arzt (zum Kind): „Was kommt denn immer wieder?" (Der Arzt schaut das Kind fragend an, das Kind blickt zur Mutter.)
>
> Mutter (ungeduldig): „Nun, eben die Allergie, die festgestellt wurde." (Es folgt eine lange Ausführung über die vorbehandelnden Ärzte. Die Mutter endet schließlich ihre klagende und anklagende Leier, indem sie den Arm des Kindes anhebt.) „Sehen Sie, Herr Doktor, das arme Kind, es muß immer kratzen!" (Das bisher etwas widerwillige, unbeteiligt dastehende Kind beginnt bei diesem Stichwort wie auf Kommando mit dem rechten Handballen an der Ellenbeuge des linken, ausgestreckten Armes zu reiben.)
>
> Mutter: „Laß das, ich habe dir doch schon so oft gesagt, daß du nicht kratzen sollst!" (Zum Arzt gewendet) „Ach wenn Sie wüßten ..." (Beschreibung der Mutter, was sie alles für das Kind tut.) Ihr Finale: „Wir nehmen sie fast immer nachts zu uns ins Bett, wenn sie so kratzt. Dann wird sie ruhig und schläft."
>
> Vater (bisher ruhig und etwas abseits, fühlt sich angesprochen, mitverantwortlich): „Ja, das tun wir, aber Sie können sich ja vorstellen, was das für uns ..." (unterbricht sich, wendet sich ärgerlich zum Kind) „Laß das doch endlich!" (Nimmt die scheuernde Hand wie einen Gegenstand weg.)

Der Ablauf dieser Konsultation erinnert in seiner mechanisch anmutenden Motorik zwischen Mutter und Kind an ein Marionettenspiel, bei dem der Spieler weder durch sichtbare Mimik noch durch die Art der eigenen Bewegungen eine mehr als verbale Anteilnahme dokumentiert. Das Kind wird im Extremfall objekthaft wie eine Puppe gehandhabt. Selbst das plötzliche Bedürfnis des Kindes zu kratzen scheint von außen induziert zu sein, wie die Reaktion auf ein unausgesprochenes situativ wirksames Kommando. Die Etiketten „Feindseligkeit" und „Abweisung" für die Mutter schließen eine gleichzeitig vorliegende (pseudo-) „überprotektive" Haltung nicht aus und sind aus dem geschilderten Gespräch nachvollziehbar. Auch das als „schwierig" und „gehemmt" etikettierte Kind wird – als der schwächere Partner in dieser Dyade – durchaus verständlich. Wir können die genannten Charakterisierungen als Eindruck zwar nachvollziehen, aber es bleibt unseres Erachtens offen, ob es sich bei dem Verhalten mehr um die Folge übermäßiger Belastung der Mutter oder tatsächlich um eine ursächlich wirksame Persönlichkeitsstruktur des vornehmlich erziehenden Elternteiles handelt.

Deutlich wird in der Szene auch die eigenartige, wohl eingespielte Dynamik zwischen Eltern und Kind im Zusammenhang mit dem Kratzverhalten des Kindes (Bär, 1973). Wir unterscheiden psychosomatisch nach dem Ablauf einen Juck-Kratz-Zirkel mit primärem, das Kratzen auslösendem Jucken und den Kratz-Juck-Zirkel, bei dem – wie im obigen Gespräch – das konditionierte Kratzen sekundär zu einem crescendoartigen Jucken führt. Der geschilderte interaktionelle Ablauf zwischen Eltern und Kind ist zwingend und enthält Lerneffekte auf beiden Seiten. Manchmal drückt das Kind dies mit den Worten aus: „Wenn du nicht ..., dann kratze ich".

Wichtig erscheint uns auch die Angabe Rechardts (1970), daß 34% seiner Patienten eine Trennung der Eltern vor dem 16. Lebensjahr erlebt hatten. Besonders dann, wenn man berücksichtigt, daß der endgültigen Trennung eine jahrelange spannungsgeladene Zeit vorausgeht, die noch nicht einmal in allen Fällen zu den statistisch erfaßten Trennungen führt. Auch unsere eigenen Beobachtungen bestätigen diese Angaben. Sie legen den Schluß nahe, daß die gestörten Beziehungen der Eltern untereinander und die resultierende Störung der Eltern-Kind-Beziehung bei gegebener genetischer Veranlagung in entscheidender Weise den Verlauf des Krankheitsgeschehens beim Kind beeinflussen.

55.2.5 Die Auseinandersetzung des Patienten mit seiner Krankheit

Familiär tradierte Einstellungen zur Krankheit und die Reaktion der Umwelt auf die Einstellung des Patienten bestimmen weitgehend dessen Verhalten gegenüber dem Arzt.

> So begünstigte bei Frau A. die ständige Überbesorgtheit der Mutter („das arme Kind ...") bis ins Erwachsenenalter der Tochter hinein die Entwicklung eines hypochondrischen, überängstlichen und zeitweilig resignierten Wesens. Ständige Anforderungen an den Arzt, demonstratives Tragen von Verbänden an den Händen und am Hals, ausführliche verbale Darstellungen des eigenen Leidens mit der deutlichen Forderung nach Zuwendung kennzeichneten die Situation bei der ambulanten Behandlung. Auch der stationäre Aufenthalt war durch fortwährende, fast kindhafte Beanspruchung der Schwestern und Ärzte mit der Erwartung „versorgt" zu werden charakterisiert („es ist so schön, betüttelt zu werden").

Diese bei Frau A. im Extrem vorliegende Haltung ist in weniger ausgeprägtem Maße häufig bei erwachsenen Ekzematikern zu finden und wird bei der stationären Behandlung von uns für einen begrenzten Zeitraum bewußt im Sinne einer therapeutisch erwünschten Entspannung („Regression") respektiert. Droht jedoch die ärztliche und pflegerische Zuwendung unbegrenzt vom Kranken genutzt oder gar mißbraucht zu werden, so ist durch dosierte Forderungen die Eigeninitiative des Patienten zu aktivieren.

> Demgegenüber fiel Herr Y. durch unzureichende Durchführung der ambulanten Therapie und Ablehnung eines stationären Aufenthaltes auf. Das vom behandelnden Arzt wiederholt erbetene Protokoll über die intermittierend tageweise Anwendung steroidhaltiger Lokaltherapeutika wurde nicht geführt. Angebote zur Erleichterung der aus beruflichen Gründen schwierigen ambulanten Besuche in der Klinik wurden selten wahrgenommen. Eines Tages konnte der Patient dem Arzt seine Haltung verständlich machen, als er erzählte, daß das erbetene Protokoll über die Kortisonanwendung an der Haut und jeder Arztbesuch ihn jedesmal an seine Krankheit erinnerten. Diese Erinnerung wolle er aber nicht haben und seine Krankheit nicht anerkennen. Seine Mutter habe die Familie immer mit ihrer Krankheit in Atem gehalten und er konnte dies schon als Kind nicht ertragen. Außerdem schäme er sich seiner Krankheit, seit er sich davon überzeugt habe, daß die Verschlechterung der Haut immer dann eintrete, wenn er mit den Umständen nicht fertig werde. Jedermann könne dies sehen („alle merken, daß ich versage").

Solche bei allen chronisch Kranken vorkommenden Extreme der Krankheitsverarbeitung bekommen bei chronisch Hautkranken ihr besonderes Gewicht durch die Sichtbarkeit des Leidens: Als demonstrativ vorweisbarer Appell beim Hypochonder und Krankheitsgewinnler; beim Krankheitsverleugner dagegen durch die Tatsache, daß die ärztlichen Anweisungen bei der lokalen Behandlung der Haut vom Patienten vermeintlich eher umgangen werden können als bei peroraler Behandlung. Eine verringerte oder vermehrte Anwendung der vorgesehenen Lokaltherapeutika erscheint ihm weniger gefährlich als die Änderung der Dosierung von innerlich einzunehmenden Medikamenten.

Für analytisch orientierte Leser mag die Beobachtung von Pflegekräften bemerkenswert sein, daß Kinder und junge erwachsene endogene Ekzematiker gelegentlich auffallend häufig zu trinken verlangten, so daß sie im Verhalten gegenüber den Pflegekräften dem Verhalten von Säuglingen gegenüber der Mutter ähnlich wurden. Die Deutung und die Einordnung dieser Beobachtung als Rückkehr in den symbiotischen Funktionskreis bietet sich an.

55.2.6 Psychosomatische Kriterien bei der ambulanten und stationären Behandlung von endogenen Ekzematikern

Für den psychosomatisch orientierten (Haut-)Arzt besteht die Gefahr, ungenügend auf das somatische Konzept des Patienten einzugehen. Das unerwartete „Überstülpen" des eigenen, dem Patienten zunächst fremden Konzeptes ohne vorhergehende Klärung der Standpunkte induziert oder verstärkt den Widerstand des Patienten. Gerade der endogene Ekzematiker hat häufig Schwierigkeiten im Umgang mit sozialer Nähe. Er ist auch deshalb besonders empfindlich gegen ein vom Arzt unzeitgemäß eingeleitetes Gespräch, das über die diagnostischen und therapeutischen Erwartungen des Patienten hinausgeht. Die Phase der ausschließlich somatischen Orientierung kann über Wochen und Monate gehen (Rechenberger, 1980). Der behandelnde Facharzt kann diese Zeit zur somatischen Behandlung nutzen, wird aber jede Gelegenheit wahrnehmen, weitergehende Möglichkeiten anzubieten. Etwa mit dem einleitenden Satz: „Bei manchen Ekzematikern ist ein Zusammenhang mit seelischen Schwierigkeiten deutlich – vielleicht beobachten Sie einmal, ob …". Auf diese Weise kann der Patient zur Selbstbeobachtung angeregt werden und der Arzt bekommt die Möglichkeit, eine eingehende biographische Anamnese begleitend während der zunächst rein somatischen Behandlung zu erheben. Der im Verlaufe eines kürzeren oder längeren Zeitraumes sensibilisierte Patient kann schließlich zu einer seinen Möglichkeiten angepaßten Behandlung motiviert werden. Diese kann sich auf vertiefte Gespräche während der ambulanten Konsultation beschränken oder als begleitende Psychotherapie wahrgenommen werden. Ist das Bewußtsein des Patienten für den psychosomatischen Anteil seiner Erkrankung erst erwacht, so wird für ihn auch die Einsicht in den Sinn einer entspannenden stationären Behandlung geweckt.

Obwohl die Aufnahme zur stationären Behandlung in der Regel wegen des ausgeprägten morphologischen Hautbefundes erfolgt, wäre es unzulänglich, wollte man nicht auch die Chance nutzen, dem Patienten durch eine entsprechende Atmosphäre Gelegenheit zu zeitweiligem Rückzug zu geben. Für entsprechend sensibilisierte und motivierte Patienten sind die ersten zehn Tage des stationären Aufenthaltes von großer Wichtigkeit für den weiteren Verlauf der Gesundung. Ähnlich wie die Intensität der lokalen Maßnahmen gesteigert wird, empfiehlt sich daran anschließend der Versuch, mittels begleitender „diagnostischer" Gespräche das Problembewußtsein des Patienten für die eigene Situation zu verstärken. Gesprächspartner kann dabei – falls er genügend Vorerfahrungen und eine gute Beziehung zum Patienten hat – der behandelnde Arzt sein. In anderen Fällen empfiehlt es sich, einen außenstehenden, gleichsam neutralen Psychotherapeuten oder psychosomatisch geschulten Kollegen zuzuziehen. Dies gilt besonders dann, wenn eine weiterführende Psychotherapie vorgesehen ist. Der insgesamt für die dermatologische Therapie etwa drei bis fünf Wochen erfordernde Aufenthalt wird somit in seinem ersten Drittel zur Entspannung, im zweiten Drittel zum Bewußtwerden der individuellen psychosomatischen Zusammenhänge genutzt. Das letzte Drittel des Aufenthaltes dient vor allem der vorbereitenden Aktivierung des Patienten für die Entlassung. Während er lernt, seine Haut selbst zu behandeln und im Umgang mit einer begrenzten Anzahl von pflegenden und heilenden Maßnahmen unabhängiger vom Hautarzt zu werden, bereitet er sich innerlich auf eine neue Gewichtung seines täglichen beruflichen und familiären Lebens vor. Als vorbereitende Maßnahme für die Entlassung und die Rückkehr nach Hause ermuntern wir den Patienten, zunehmend Besuche zu haben oder am Wochenende zu einem Tagesaufenthalt nach Hause zu fahren. Die Reaktion der Haut am Tage nach dem Besuch ist dabei ein wichtiger Indikator für die erreichte Stabilität des Patienten. Die Kurzbesuche die-

nen der vorbereitenden Gewöhnung („Desensibilisierung") und helfen, die bekannten kurzfristigen Rezidive nach der Entlassung abzuschwächen oder gar zu vermeiden. In jedem Fall wäre es unseres Erachtens völlig unzulänglich, lediglich die momentan abgeheilte Haut als Kriterium für die Entlassung des Patienten zu bewerten („Drehtüreffekt").

Der beschriebene Ablauf der stationären Behandlung erhält zusätzliche Schwerpunkte, je nachdem, zu welcher Art von Krankheitsverarbeitung der Patient neigt. Die extremen Verhaltensweisen wurden bereits beschrieben.

Aus den besonderen Beziehungen zwischen Ekzemkind und der Familie (vgl. 55.2.4) ergibt sich, daß bei jeder Form von ärztlicher Behandlung oder Psychotherapie dieser Kinder und Jugendlichen die Familienstruktur und die familiäre Dynamik berücksichtigt werden müssen. Unsere langjährigen guten Erfahrungen in Zusammenarbeit mit Kindertherapeuten bestätigen dies (vgl. Slany, 1975; Wirsching und Stierlin, 1982).

Die Bedeutung dermatologischer, allergologischer, immunologischer und psychosomatischer Anteile für den Verlauf des endogenen Ekzems ist im Einzelfall sehr unterschiedlich und muß individuell in die Therapieform eingehen.

55.3 Haut und Entstellung

Der Hautkranke ist in seiner äußerlichen Attraktivität eingeschränkt, der Patient mit Dysmorphophobie befürchtet zwanghaft, äußerlich entstellt zu sein oder entstellt zu werden. Der Tätowierte versucht durch Veränderungen an seiner Haut bewußt sein Erscheinungsbild zu beeinflussen. Äußere Erscheinung, Kommunikation und soziale Einordnung sind eng gekoppelt und beeinflussen das individuelle Lebensgefühl des Menschen.

55.3.1 Die Psoriasis vulgaris (Schuppenflechte)

Die Psoriasis ist anlagebedingt und mit ca. 2% sichtbar Erkrankten der Gesamtbevölkerung weit verbreitet. Sie betrifft alle Altersstufen und beide Geschlechter. Der typische Herd ist gerötet, mit einer charakteristischen silbrig-weißen Schuppung bedeckt, scharf begrenzt, flächenhaft oder fleckförmig von unterschiedlicher Größe. Bevorzugte Lokalisationen sind Ellenbogen, Knie, Nacken, der behaarte Kopf sowie intertriginöse Regionen (submammär, Rima ani, periumbilikal, Bauchfalten bei Adipositas). Mykoseähnliche Veränderungen der Finger- und Fußnägel sowie mykoseähnliche hyperkeratotische Veränderungen der Handinnenflächen und Fußsohlen sind nicht selten. Schwere entzündliche Gelenkveränderungen können begleitend oder isoliert auftreten. Die Psoriasis verläuft schubartig, ausgelöst durch lokale Entzündung (Sonnenbrand), im Zusammenhang mit Infektionskrankheiten und Phasen psychischer Belastung. Die vielfältigen Möglichkeiten einer dermatologischen Behandlung erfordern eine intensive und zeitlich aufwendige Mitarbeit des Patienten. Die psychosomatische Betreuung betrifft vor allem die Einstellung des Patienten und die Verarbeitung der entstellenden Krankheit.

Exemplarische Fälle
(Auszüge aus Patientenberichten)

Pat. E, 50 J., w.: „Während der Schulzeit hat man mir zur Behandlung den Kopf kahlgeschoren und einen Verband angelegt. Darunter habe ich sehr gelitten und den Schulkameraden erklärt, es sei eine Verletzung am Kopf von einem Autounfall. Daraufhin wurde ich vom Vater geschimpft, weil ich nicht die Wahrheit gesagt habe. Später hat eine Schulfreundin, die auch eine Psoriasis hat, gesagt, ich hätte sie angesteckt."

Pat. M, 16 J., w.: „Vom Turnen bin ich immer befreit ... Manchmal werde ich mit dem Spitznamen ‚Lepra' angesprochen. Das kränkt mich. ... Zum Schwimmen gehe ich nicht, wenn die Schuppenflechte da ist. Jetzt mit 16 Jahren habe ich zum erstenmal mit einem Freund geschlafen, als die Schuppenflechte am schlimmsten war."

„Selbstmordversuch, ... das würde ich nie wieder machen." „Meine Schwester sagt: ‚Was Du immer hast' – Aber die Schwester ekelt sich vor den Hautveränderungen – ich komme mir manchmal wie ausgestoßen vor."

Pat. R, 33 J., w.: Aufnahme wegen Suizidgefährdung und reaktiver Depression. „Am Anfang dachte die Familie, es sei eine ansteckende Hautkrankheit" – ... „Wasch dir doch die Hände, sagt meine Schwiegermutter immer." „Zu meinem Mann sage ich manchmal: Du willst mich wohl nicht mehr – wenn du wüßtest, wie ich aussehe." – „Meine Schwiegermutter sagt: Jeder muß mit seiner Krankheit fertig werden."

„Ich stehe immer vor dem Spiegel und suche nach neuen Hauterscheinungen ... Ich geniere mich, mich vor meinem Mann auszuziehen ... Auch jetzt – nach der Behandlung – warte ich immer auf neue Hauterscheinungen ..."

Pat. M, 35 J., m.: „Mein Vater sagte: Junge, du mußt dich mehr waschen. – Ich war oft sehr verzweifelt und habe an Schlußmachen gedacht – ich bin schon ungefähr 15mal hier stationär behandelt worden und bin jetzt ein alter Hase mit meiner Krankheit. Ich kann jetzt ganz gut damit umgehen. – Wenn zu Hause die Polstergarnitur und der Fußboden mit Schuppen übersät sind, sagt meine Frau immer: Geh ins Krankenhaus."

Die spontanen biographischen Äußerungen der (im doppelten Sinn) betroffenen Patienten beziehen sich ausschließlich auf die schwerwiegenden psychosozialen **Folgen** der sichtbaren Entstellung. Die Auswirkungen sind altersabhängig und beziehen sich auf die Kommunikation in der Schule, beim Sport, später im Beruf, in der Ehe und betreffen schließlich die resignative und zeitweise suizidale oder aber verarbeitende Einstellung des alterfahrenen Psoriatikers. Der somatische Krankheitswert steht demgegenüber für den Patienten offensichtlich weit im Hintergrund: Er fühlt sich nicht primär physisch leidend, wenngleich sich in der eiligen Sprechstundensituation der gegenteilige Eindruck ergeben kann, wenn nämlich der Patient keine Chance hat, sein eigentliches Leiden zur Sprache zu bringen und er sich statt dessen mit einer Empfehlung für die Lokalbehandlung zufriedengibt. Spontane Äußerungen über die Auslösung eines Schubes in Abhängigkeit von der inneren Befindlichkeit des Patienten sind eine Ausnahme.

Die Besonderheiten des psychosomatischen Umganges mit dem Psoriatiker

Der erste Eindruck, den Psoriatiker beim behandelnden Arzt hinterlassen, ist der eines positiv gestimmten, sogar fröhlichen, kontaktfähigen Menschen, der Selbstvertrauen und eine gute Kommunikationsfähigkeit besitzt. In zahlreichen Untersuchungen (Übersicht bei Whitlock, 1980) wurde versucht, diesen klinischen Eindruck mit Hilfe psychometrischer Testverfahren zu bestätigen. Bojanovsky und Mitarbeiter (1981) fanden mittels Freiburger Persönlichkeitsinventar und weiterer Untersuchungsverfahren bei Psoriatikern eine „antineurotische Persönlichkeit", die sich signifikant von einer sorgfältig parallelisierten Gruppe „Neurotiker" und einer solchen von gesunden Personen unterschied. Die Autoren lassen offen, ob die gefundene Persönlichkeitsstruktur (Mit-)Ursache oder Folge der Erkrankung ist. Uns bietet sich die Vermutung an, daß es sich um eine übermäßige Adaptation handelt. Ähnlich kommt Rechenberger (1982) aufgrund tiefenpsychologischer Untersuchungen zu dem Schluß, daß diese „starke" Seite des Psoriatikers Ausdruck einer Reaktionsbildung bei einer besonders ausgeprägten seelischen Verletzlichkeit sei: Die Stärke sei Abwehr des Abgewehrten (der Verletzlichkeit). Alles in allem ist jedoch zu sagen, daß die bisherige Forschung keine allgemein überzeugenden Hinweise für einen einheitlichen krankheitstypischen Persönlichkeitscharakter der Psoriatiker erbracht hat. Die Mittelwertsbetrachtung psychometrischer Befunde ergab keine Abweichung vom psychisch normalen Standard. Aufgrund unserer Erfahrungen kommen wir zu dem Schluß, daß zeitliche Korrelationen bei der Auslösung von Psoriasisschüben und die massiven psychosozialen Folgen dennoch die psychosomatische Betrachtung dieses Krankheitsbildes ausreichend begründen. Die Mittelwertsbetrachtung psychosomatischer Daten ist jedoch unzureichend, da hierbei das Varianzspektrum nicht berücksichtigt wird (vgl. Bosse und Hünecke, 1984).

Stellen wir für den klinischen Umgang das oberflächlich manifeste Erscheinungsbild von Stabilität und Stärke des Psoriatikers in Frage, so entdecken wir den Patienten, der sich in den oben zitierten Äußerungen spiegelt. Die Kenntnis dieser Folgereaktion ist entscheidend für die Arzt-Patient-Beziehung und Voraussetzung für den psychosomatischen Anteil der Therapie, der die wichtige lokale Behandlung ergänzen sollte. Im Gespräch mit den Patienten bietet sich eine Kette von wichtigen psoriasistypischen Themen, ausgehend von der äußeren Entstellung und ihren Folgen, zur Bearbeitung an: Der tägliche Umgang mit anderen (vgl. 55.3.3), die resultierende Einschränkung des Selbstwertgefühles, die Belastung im Intimbereich, das Unverständnis des Betroffenen für abwehrende und distanzierende Reaktionen der gesunden Umwelt und der Lernprozeß des Psoriatikers und des Nichtpsoriatikers im Umgang mit der entstellenden Krankheit. Manchmal die „Schuld"-Übernahme der Eltern gegenüber den betroffenen Kindern. Die angestrebte Korrektur betrifft besonders die eigene Einstellung des Psoriatikers zu seinem Körper und die damit verbundenen tatsächlichen und befürchteten gegenseitigen Unterstellungen (Infektiosität, Unsauberkeit, Unverständnis für einander). Diese Themen können krankheitsbegleitend und in einem gezielten themenzentrierten Gespräch in eigens geplanten Konsultationen (vgl. 55.7.2) bearbeitet werden. Sie sind jedoch auch Inhalt des Austausches in den regionalen, sehr nützlichen Selbsthilfegruppen des Psoriasisbundes (vgl. Moeller, 1983, 1984). Gruppen unter psychologischer und/oder ärztlicher Leitung haben sich bereits therapeutisch bewährt (J. de Korte, 1982). Letzterer weist auf eine verminderte Rezidivneigung im Verlauf einer Gruppentherapie hin. Auch I. Rechenberger (1982) berichtet von psychotherapeutischen Erfolgen hinsichtlich der Rezidivbereitschaft.

Einen speziellen therapeutischen Zugang erfordern die überdurchschnittlich häufig bei Psoriatikern vorkommenden Phänomene Alkoholismus, Adipositas, Depression und Suizidneigung. Phänomene, die wir als unterschiedliche Versuche sehen, sich der bitteren Realität einer in hohem Maße stigmatisierenden chronischen Erkrankung zu entziehen.

55.3.2 Weitere Hauterkrankungen mit Entstellungswert

Zwar besitzt jede Hauterkrankung zum mindesten temporär einen größeren oder kleineren Entstellungswert, für solche des Gesichts, für Störungen des Haarwachstums und Veränderungen emotional stark besetzter Regionen (Augen/Mund, Brust, Genitalregion) gilt dies aber in besonderem Maße (Ehring, 1978).

Acne vulgaris und Acne conglobata

Die verschiedenen Formen der Akne betreffen überwiegend oder ausschließlich die Haut des Gesichts, ein bevorzugtes Organ menschlicher Verständigung. Die Akneerkrankungen sind in der Regel an die Zeit der hormonellen Umstellung in der Pubertät und Postpubertät gebunden, eine Zeit der Selbstfindung und besonderer Sensibilität für vermutete oder tatsächliche Ab- oder Aufwertung von außen. Bezüglich einer psychogenen Auslösung oder Verschlechterung der Akne geben Patienten häufig Hinweise, die auf

dem Boden psychoendokriner Mechanismen zu verstehen sind, z. B. Zyklusabhängigkeit (Amann, 1984; Whitlock, 1980). Rechenberger (1978) sieht als Gemeinsamkeit bei den verschiedenen Akneformen „eine Störung des Patienten zu sich selbst".

Unsere Untersuchungen mit Hilfe eines eigenen Verfahrens zur Erfassung der emotionalen Belastung Hautkranker zeigten:
- bei Aknepatienten ein verstärktes Krankheitsgefühl (Bosse et al., 1978);
- bei Aknepatienten läßt sich ein überzufällig großer Anteil von depressiv verstimmten Personen finden (Bosse und Hünecke, 1984).

Im Vordergrund des Erlebens stehen die psychosozialen Folgen, ähnlich der Psoriasis. Gloor und Mitarbeiter (1978) zeigten an 2009 18- bis 19jährigen Musterungskandidaten, daß der Krankheitswert der Acne vulgaris sich vorwiegend aus den Ressentiments der Mitmenschen und den beruflichen Folgen ergibt und offenbar um so größer wird, je geringer die Schulbildung der Aknepatienten ist. Zusätzlich zu der entscheidend wichtigen Lokalbehandlung halten wir eine entlastende Auseinandersetzung des Patienten mit der Entstellung und die aufklärende Motivierung zu ambulant überwachter Eigenbehandlung für wichtig, da die wirksame lokale Therapie (Vitamin-A-Säure oder Benzoylperoxid) mit erheblichen subjektiven Belastungen und zusätzlichen, zeitweilig entstellenden Veränderungen (Rötung, Schuppung) verbunden ist. Die Aknebehandlung durch eine dermatologisch erfahrene Kosmetikerin bringt dem Patienten neben der objektiven Besserung fast immer eine subjektive Erleichterung durch das Gefühl, etwas für sich getan zu haben (Wright et al., 1970). Auch die Erfahrung der direkten Zuwendung der Kosmetikerin in Form der manuellen Betätigung an der Haut dürfte gerade für junge Aknepatienten wichtig sein. Gleichzeitig bedeutet dies auch eine praktische Anleitung zur eigenen Aknetoilette. Hughes und Mitarbeiter (1983) berichten über eine deutliche Verminderung der Akne in randomisierten Gruppen im Vergleich zu den Kontrollgruppen unter der Behandlung mit verschiedenen entspannenden Therapieformen (Biofeedback u. a.).

Die sog. „Acne excoriée" gehört psychodynamisch gesehen nicht zum Formenkreis der Akneerkrankungen und wird deshalb im Abschnitt 55.4.2 behandelt.

Störungen des Haarwachstums

Mögen auch die Ursachen von Störungen des Haarwachstums sehr unterschiedlicher Natur sein, so besteht doch immer ein erheblicher Entstellungswert. Veränderungen der Haare stellen für den Betroffenen und den Beobachter einen Reiz dar, dessen subjektive und objektive Validität in Abhängigkeit von der Persönlichkeitsstruktur und der aktuellen Persönlichkeitskonstellation derer, die mit dem Reiz konfrontiert werden, weit divergieren können. Männer und Frauen zeigen durch ihr Krankheitsverhalten, in welch unterschiedlichem Ausmaß gerade das Aussehen des Haares mit dem Gefühl von Minderwertigkeit oder Gesundheit, gesellschaftlicher und erotischer Attraktivität assoziiert wird.

Anhand mehrerer Fallbeschreibungen zeigten wir (Bosse und Teichmann, 1973), wie die Projektion neurotischer Konfliktproblematik (persönliches Insuffizienzerleben) auf das Haar, bei minimalem objektivem Befund (Geheimratsecken), schwerwiegende Konsequenzen (Fehleinschätzung der Partner) im sozialen Bereich zur Folge hatte, im Falle einer Alopecia areata totalis aber die Kompensationsmöglichkeiten der Patientin (Integration im Umfeld) und die Reaktion des sozialen Umfeldes (Anerkennung als Klassensprecherin u.a.) eine psychoreaktive Fehlentwicklung vermeiden ließen. Für die psychosomatische Auslösung der Alopecia areata bieten sich auch eigene klinische Beobachtungen an. Die kritische Durchsicht der sehr widersprüchlichen Literatur hierzu (vgl. Puchalski und Szlendak, 1983; Whitlock, 1980) erlaubt uns jedoch bisher keine endgültige Stellungnahme in diesem Sinne. Neuere Untersuchungen weisen u.a. auf einen Autoimmunprozeß als Grundlage der Alopecia areata hin.

Besonders bei der langwierigen und häufig über Jahre fruchtlosen Therapie der Alopecia areata totalis stellt sich die Frage nach einer Perücke. Für Erwachsene bedeutet dies in der Regel die Frage nach dem Kostenersatz durch die Kasse. Diese wiederum fragt nach dem subjektiven Leidensdruck und dem Krankheitswert, um allenfalls einen Teilbetrag der Kosten zu übernehmen. Gelegentlich gibt es eine dramatische Zuspitzung, wenn ein Partner in einer jungen Beziehung dem anderen die Notwendigkeit, eine Perücke tragen zu müssen, vorenthalten hat. Häufiger treten Komplikationen bei Kindern auf, die selbst noch nicht den Entstellungswert realisiert haben und nunmehr durch die Aktivität der Eltern sich der Bedeutung ihrer Haarlosigkeit bewußt werden. Gibt der Arzt dem Bedürfnis der Eltern, die ihr ganz oder partiell haarloses Kind nicht ertragen können, nach, so verschiebt sich allzuschnell der Druck auf das bisher unbelastete Kind. Der Wechsel, im Unterricht, beim Schwimmen oder bei sonstigem Sport mit oder ohne Perücke zu erscheinen, läßt sich von Kindern und Klassenkameraden schwerer ertragen als ein eindeutiger Zustand, der einmalig mit der Klasse besprochen wurde. Die sorgfältige Abwägung der Sachlage mit Eltern und Lehrern unter Berücksichtigung aller Beteiligten ist erforderlich.

Die Trichotillomanie (Ausziehen von Haaren) (Bosse und Teichmann, 1973), das Auszupfen von Wimpern und Augenbrauen (Ziliotillomanie), das Zerren und Aufdrehen von Haaren mit nachfolgendem umschriebenem Effluvium können differentialdiagnostische Schwierigkeiten mit einer Alopecia areata bereiten. Gelegentlich wird der Entstellungswert der artifiziellen Kahlstellen vom Patienten genutzt (vgl. Krankheitsdynamik bei Artefakten, 55.4).

Die übermäßige Behaarung (Hypertrichosis – Hirsutismus) am Körper des Mannes (Brust, Rücken, Schultern) oder der Frau (Kinn, zwischen den Brüsten, in der Umgebung der Mamille, an den Extremitäten) führt immer zu starker Betroffenheit des einzelnen und erregt die Aufmerksamkeit der Allgemeinheit – unabhängig von der Frage, ob dieser Hypertrichose eine im Normbereich liegende konstitutionelle Veranlagung zugrunde liegt oder eine pathologische endokrine Abweichung (Überempfindlichkeit der Haarfollikel gegenüber Androgenen – Androgenüberproduktion durch Tumoren). Die Frage, ob das Ausmaß der Behaarung „objektiv" übermäßig ist oder nicht, ist für das subjektive Erleben des Patienten nicht entscheidend, jedoch wichtig für den therapeutischen Zugang. Bei objektiv vorliegender Hypertrichose ist durch somatische Behandlung (Elektroepilation, antiandrogene Therapie, Behandlung androgenbildender Tumoren) der Leidensdruck zu beeinflussen, während bei der konstitutionellen Hypertrichose im Normbereich, wie sie häufig bei Südländern, aber auch in unseren Regionen vorkommt, der dysmorphophobische Anteil und die Kränkung des Selbstwertgefühles im Vordergrund stehen. Diese letzteren Patienten – besonders Frauen – stellen sich oft im rasierten Zustand vor und verlangen ärztliche Sofortmaßnahmen, ohne die Möglichkeit einer Voruntersuchung, d.h. einer kritischen Abwägung oder gar Infragestellung ihres Anliegens durch den Arzt. Eine Wiedervorstellung des Patienten und Objektivierung etwaiger endokriner Störungen oder der nicht behandelten Hypertrichose werden unter Umständen sogar verzögert oder ganz abgelehnt, wenn sich der Patient in seinem Anliegen vom Arzt nicht akzeptiert fühlt.

Die psychosomatische Betreuung dieser Patienten erfolgt ähnlich wie diejenige der Patienten mit Dysmorphophobie (vgl. 55.3.4). Eine Abklärung der eigenen Einstellung und des aktuellen Verhaltens (soziale Nähe, Erotik/Sexualität) ist ebenso wichtig wie die medizinische Aufklärung. Wie immer bei Entstellungsproblemen ergibt sich die Frage, warum der Patient so „objektiv unangemessen" reagiert. Ob der analytische Zugang im Einverständnis mit dem Patienten gesucht werden kann, muß im Einzelfall entschieden werden. Die Zusammenarbeit mit dem Gynäkologen bzw. Endokrinologen ist in der Regel erforderlich.

Nävi (Muttermale)

Eine ähnliche Problematik, wie die im vorigen Abschnitt für die partielle oder totale Haarlosigkeit beschriebene, stellt sich im Umgang mit den zahlreichen angeborenen nävoiden Fehlbildungen. Auch hier sind zunächst ausschließlich die Eltern betroffen. Fast immer handelt es sich um Naevi pigmentosi et pilosi oder Hämangiome, und die Eltern befürchten, das Kind werde später in der Schule gehänselt. Leitlinien für den Umgang mit diesen Eltern sind:
- Die Klärung des Anteiles der Eltern bei dem Wunsch nach Therapie ist vorrangig.
- Die operative Entfernung eines Naevus pigmentosus et pilosus kann als Melanomprophylaxe indiziert sein.
- Das Schleifen eines Naevus pigmentosus ist nur während der ersten Lebenswochen aussichtsreich.
- Die meisten tuberösen Hämangiome bilden sich

spontan im Laufe der Jahre mit sehr gutem kosmetischem Ergebnis zurück, während beschleunigende Eingriffe (Operationen, Vereisung, Bestrahlung) Narben hinterlassen.

– Soweit eine kosmetisch indizierte Operation zur Diskussion steht, sollte das Kind vor weiteren Arztbesuchen bewahrt werden, um eine unnötige Sensibilisierung (Stigmatisierung) zu vermeiden. Die endgültige Entscheidung erfolgt bei uns erst in einem Alter, in dem das Kind eine kosmetische Notwendigkeit gegebenenfalls selbst empfindet. Die Motivation für Krankenhausaufenthalte und Narkose ist damit gesichert

Maligne Hauttumoren

Die Sichtbarkeit einer Hautveränderung bedeutet im Falle eines Malignoms an der Haut nicht nur eine ästhetische Abwertung (Entstellung). Der Prozeß stellt im Gegensatz zu dem unsichtbaren Malignom eines inneren Organes ein zusätzliches, täglich aktuelles Moment der lebensgefährlichen Bedrohung vom Tage der Entdeckung an dar. Die bekannten Stadien der Einstellung Krebskranker im Verlauf ihrer Erkrankung bedürfen deshalb im Falle von malignen Hauterkrankungen (Melanom, Karzinom, Basaliom, Mycosis fungoides) einer Ergänzung für die frühe Phase und die Zwischenphase. Eine Vielzahl von harmlosen, oft altersbedingt zunehmend auftretenden tumorösen Veränderungen der Haut verunsichern den zweifelnden Patienten zusätzlich. Palliative Behandlungen können deshalb auch unter dem Gesichtspunkt der sichtbaren Bedrohung sinnvoll sein.

55.3.3 Bedeutung und Verarbeitung der Entstellung

Die Besprechung des endogenen Ekzems und der Psoriasis zeigten modellhaft vereinfacht, wie psychische Faktoren entweder vorwiegend auslösend oder überwiegend als Folge bei Hautkrankheiten wirksam werden. Dabei wird allzu leicht übersehen, daß die psychischen Folgen des Krankheitserlebens, z.B. beim endogenen Ekzem, selbst zum Auslöser für eine somatische Verschlechterung im Sinne eines Circulus vitiosus werden können. Die ursächliche und als Konsequenz bedeutsame soziale Situation des Hautkranken als Phänomen interpersoneller Wahrnehmung wurde von Bosse und Mitarbeitern (1976) untersucht. Die Ergebnisse wurden von den Autoren kognitionstheoretisch, lernpsychologisch, psychoanalytisch und ärztlich-klinisch interpretiert. An 465 hautkranken und nicht hautkranken Probanden wurden u.a. die folgenden Hypothesen überprüft:

– Hautkranke leiden an einer paranoiden Überschätzung des Störungswertes, den ihrer Befürchtung nach das Hautstigma für ihre Partner darstellt (**„Paranoiatendenz der Hautkranken"**). In Bestätigung dieser Hypothese wurden Dissonanzen zwischen der eigenen Einstellung von Hautkranken und der von ihnen bei Hautgesunden vermuteten Einstellung gesichert.

– Hautgesunde neigen zu einer Einstellung, die dem Hautkranken eine verringerte Leistungsfähigkeit und damit eine geringere sozial gebilligte Berechtigung, Ansprüche zu stellen, zuweist. Dies betrifft berufliche und gegenseitige sozial-emotionale Bedürfnisbefriedigungen. Auch diese **„Genügsamkeitsthese"** des Hautgesunden konnte bestätigt werden. Sie beruht auf einer Dissonanz der eigenen Einstellung des Hautgesunden und der von ihm bei Hautkranken vermuteten Einstellung. Sie ergänzt gewissermaßen als Komplement die Paranoiatendenz des Hautkranken.

– Es wurde die Annahme geprüft, daß – aufgrund der Erfahrungen und Bedürfnisse bei der Anwendung von Kosmetika – gesunde Frauen im Vergleich zu hautkranken Frauen den Unterschied zwischen gesunder und kranker Haut deutlicher wahrnehmen als männliche gesunde und hautkranke Beurteiler. Die These über die geschlechtsrollenabhängige Bewertungstendenz gegenüber der Hautkrankheit konnte nicht bestätigt werden.

– Schließlich wurde überprüft und bestätigt, daß Hautkrankheit in unterschiedlichen Themenbereichen verschiedenartige Konsequenzen hat (Krankheitswert – Ästhetik – soziale Nähe – Erotik/Sexualität – Arbeits- und Leistungsbereich).

Im Sinne von Erkrankung wird eine Anormalität der Haut durchgehend anerkannt und von den Beurteilern bereitwillig mit der Attestierung von Arbeitsunfähigkeit beantwortet. Im gleichen extremen Maße (60–80% aller Befragten) wird auch ein Verlust des ästhetischen Erscheinungsbildes bescheinigt. Der persönlich-soziale Bereich (Intensität der Beziehungen) wird durch das Vorliegen einer Hautkrankheit unterschiedlich häufig als beeinträchtigt empfunden (Schwankungsbreite zwischen 25 und 75% aller Befragten, abhängig von deren Geschlechtszugehörigkeit, vom Vorliegen einer eigenen Hautkrankheit, vom Frageinhalt: „Sexuelle Zuwendung" bis „Café-tisch-Kontakt"). Im Arbeits- und Leistungsbereich wird der Hautkranke gegenüber dem Hautgesunden kaum noch oder gar nicht mehr negativ abgehoben.

Für den Arzt als Gesprächspartner ist es nützlich, das Selbstbild (Autostereotyp), das tatsächliche Fremdbild (Heterostereotyp) und die gegenseitig unterstellten Fremdbilder der Hautkranken und ihrer Umgebung nachvollziehen zu können. Weiterhin ist es für das Einfühlen in die komplexe Situation erforderlich, die Tatsache zu berücksichtigen, daß Hautkranke als **Jetzt-Kranke** (Selbstbild) und als **ehemals Hautgesunde** (vorgestelltes Fremdbild) unterschiedlich empfinden können. Weitere Gesichtspunkte zur Pathopsychologie der Entstellung durch Hautkrankheiten stellten Panse (1970), zur Kommunikation Hautkranker Teichmann und Bosse (1974), zum Erleben und zur Verarbeitung der Entstellung Hünecke und Bosse (1980) dar.

55.3.4 Dysmorphophobie

Patienten mit dysmorphophobischer Symptomatik sind häufig und bieten sich in der Sprechstunde des Dermatologen oder des Chirurgen zunächst in einer

Weise dar, die aktive therapeutische Maßnahmen zu Lasten diagnostischer Prozeduren zu forcieren sucht, oder sie werden erst dann auffällig, wenn trotz deutlicher Besserung minimale Restsymptome dem Patienten weiterhin Anlaß geben, objektiv unangemessene Behandlungen chirurgischer oder dermatologischer Art zu fordern. Sie befürchten, durch die vermeintliche Verunstaltung des Äußeren bei anderen Menschen Aufmerksamkeit oder Anstoß zu erregen. Häufig fixiert sich das negative subjektive Erleben auf eine schwer objektivierbare Äußerlichkeit: „Haarausfall" – „grobporige Haut" – „mißgeformte" Körperteile wie Nase, Ohren, Busen, Penis oder Falten im Gesicht. Lehnt der Arzt die somatische Behandlung ab, so ist eine lange Reihe von Arztwechseln die Folge. Hautschleifungen, Haartransplantationen, Nasen-, Ohren-, Busen-Korrekturen werden immer aufs neue gefordert. Die Folgen einer derartig erzwungenen Behandlung werden häufig selbst zum sichtbaren Anlaß für neue Klagen.

Mester (1983) unterscheidet drei Formen der dysmorphophobischen Symptomatik: Häufig als Ausdruck einer Adoleszenzkrise, gelegentlich vorkommend als auffälligstes Symptom einer depressiven Erkrankung und schließlich als umschriebene Wahnentwicklung. Zauner (1979) weist darauf hin, daß dieses häufige Syndrom in der Regel Ausdruck einer schweren Persönlichkeitsstörung ist und bei Chronifizierung zu weitgehender sozialer Isolierung bis zum Suizid führen kann. In den eigenen Fällen standen Unsicherheit und Ängste (mangelhaftes Körpervertrauen, Selbstwertzweifel, soziale Ängste und Rückzugstendenzen) sowie Anspruchshaltung und Zwänge (Geltungsstreben, Perfektion, Körperbesorgtheit) im Vordergrund (Hünecke und Bosse, 1985).

Der Versuch des schwer in seiner Identität gestörten Patienten, seine „Minderwertigkeit" auf sein vermeintlich defizitäres Äußeres zu projizieren, führt zwangsläufig zu der Forderung nach somatischer Therapie. Geht der angesprochene Arzt auf dieses Krankheitskonzept aktiv ein, so unterstützt er damit die schwerwiegende Fehlentwicklung des Patienten. Auch die bloße Konfrontation des Patienten mit seiner Fehleinschätzung des realen Befundes ist therapeutisch ineffektiv und führt zum Therapieabbruch oder Arztwechsel. Allenfalls der Versuch, dem Patienten eine (indifferente!) lokale somatische Mitbehandlung als Motivation in Aussicht zu stellen und ihn anzuleiten, seine Beschwerden als Ausdruck einer der Psychotherapie bedürftigen seelischen Erkrankung zu sehen, eröffnet eine begrenzte Erfolgsaussicht.

55.3.5 Tätowierungen

Tätowierungen haben mit der Entstellung gemeinsam, die körperliche Attraktivität zu beeinflussen. Gleichzeitig haben sie mit den im folgenden beschriebenen Selbstbeschädigungen an der Haut die Gemeinsamkeit, mittels ihres symbolischen oder konkreten Inhaltes sowohl etwas über die Persönlichkeit des Trägers auszusagen, als auch eine auffordernde Nachricht an die Umwelt zu vermitteln. Der Tätowierung als spezifischer und zusätzlicher averbaler Ausdrucksmöglichkeit in der Kommunikation von Außenseitergruppen stehen die artifiziellen Selbstbeschädigungen der Haut als bedrohliche Folge einer mehr oder weniger gehemmten verbalen Kommunikation gegenüber.

Tätowierungen kommen als schichttypisches und in der Regel an Gruppen gebundenes Phänomen besonders bei Jugendlichen in Erziehungs- und Haftanstalten vor. Sie werden jedoch auch als Zeichen der Zusammengehörigkeit und als Statussymbol in freiwilligen Zusammenschlüssen Jugendlicher sowie innerhalb bestimmter Berufsgruppen bevorzugt. Als äußerer, je nach den Umständen der Entstehung unterschiedlicher Anlaß werden in der Regel „Langeweile" – „Mutproben" – „Anpassung an die anderen", Geltungsbedürfnis oder „Protest" von den Tätowierten angegeben, manchmal auch das Bedürfnis, sich äußerlich attraktiver zu machen (Munkwitz und Neulandt, 1957).

Nach den Inhalten der Tätowierungen (Bosse und Teichmann, 1972) lassen sich diese in Anlehnung an die psychoanalytischen Entwicklungstheorien teils als vorwiegend schmückende, Ich-bezogene, narzißtische Tätowierungen und damit als Ausdruck der Phase autoerotischer Betätigung begreifen oder auch als partnerbezogener Hinweis auf Aggressions- und Liebesbeziehungen (Namen, Erinnerungsdaten) verstehen. Schließlich kommen eine Vielzahl von gruppenorientierten Tätowierungen (Symbole für Berufsgruppen und politische Gruppierungen) als Zeichen der Phase vermehrten sozialen Kontaktes zur Beobachtung. In der Gruppe der direkt partnerbezogenen Darstellungen nehmen solche mit sexuell orientierten Inhalten eine bevorzugte Stellung ein, während in der letztgenannten Gruppe vor allem die anklagenden und gegen die Gesellschaft revoltierenden Hinweise („traue eher einer Hure als der Justiz", Darstellungen vom Henker, Namen politischer Idole etc.) beeindrucken.

Die ärztliche und psychosomatische Aufgabe wird aktuell, wenn der Betroffene sich nicht mehr zu der bisherigen Aussage bekennen kann und/oder gesellschaftlichen Sanktionen unterliegt: Ablehnung des bisherigen Bekenntnisses durch einen neuen Partner, Verweigerung des Zuganges zu einer angestrebten Berufsgruppe (Polizei, Kindergärtnerin), Kennzeichnung als ehemaliger Häftling etc. Zahllose Briefe mit der Bitte um Entfernung von Tätowierungen – selbst um den Preis teilweise unvermeidbarer Narben – zeigen den Druck der Umwelt auf den Stigmatisierten. Die formaljuristische Frage der Krankenkasse nach dem Krankheitswert gibt dabei keine ausreichende Hilfe bei der ärztlichen Entscheidung für oder gegen eine operative Entfernung im Sinne einer langfristigen Resozialisierung des Betroffenen. Der Gesichtspunkt der Resozialisierung muß aber im Interesse des Betroffenen und der Allgemeinheit vorrangig vor jeder anderen Überlegung bei der Frage nach der Kostenübernahme Berücksichtigung finden. Technische, finanzielle und formale Überlegungen sind in aller Regel von sekundärer Wichtigkeit.

55.4 Formen der Selbstbeschädigung an der Haut

Zwischen den extremen Formen von Selbstbeschädigungen an der Haut bestehen vielfältige Übergänge bezüglich des somatischen Krankheitswertes, der mehr oder weniger bewußten Reflexion, der Motivation und Zielvorstellung, der angewandten Methoden, der Prognose und des erforderlichen ärztlichen Umganges mit den Patienten.

55.4.1 Der Artefakt im engeren Sinn

Der Artefakt (AF) wird in der Regel durch Kratzen mit den Fingernägeln, durch absichtliches Reiben, Scheuern, Bürsten, Verbrennen (Zigaretten), Anwendung von Laugen, Säuren oder mechanisch (Stricknadeln), seltener durch Abbinden oder Kneifen erzeugt. Dementsprechend ist die betroffene Lokalisation in der Regel der rechten Hand leicht zugänglich (linke Hand, linker Unterarm, rechte Körperseite). Exponierte Regionen (Gesicht, Handrücken, Unterarm, Brust) werden bevorzugt. Vorausgegangene Verletzungen oder Narben geben dem Patienten immer wieder Anlaß zu erneuten Manipulationen. Die Morphologie ist häufig charakteristisch für die Art der Auslösung: Flächenhafte Erosionen sind in atypischer Weise scharf begrenzt oder zeigen parallele Kratzspuren, der Zeitpunkt der „plötzlichen" Entstehung ist meist durch besondere, emotional belastende Umstände charakterisiert. Ein über Monate und Jahre andauernder Wechsel von Abheilung zu Zeiten stützender Zuwendung und Rezidivieren unter Belastung ist typisch.

Exemplarischer Fall

Weiblich, 15 J., Jüngste von fünf Kindern

Lokalbefund: Im Gesicht frische erosive, teils verkrustete Läsionen, in der Umgebung rot entzündlich. Scharfe Begrenzung, im Zentrum angedeutet „parallel-streifig", ca. 1-DM-Stück-groß; keinem dermatologischen Krankheitsbild zuzuordnen.

Subjektive Beschwerden: Patientin beklagt sich über Brennen und Schmerzen im Gesicht, gelegentliche Kopfschmerzen und Augenschmerzen. Augenärztliche Untersuchung o. B., neurologische Untersuchung o. B.

Verhalten: Die Patientin betont, daß sie nicht mit ihren Fingern ins Gesicht ginge – sie habe Angst, es könnten dann Narben verbleiben. Sie läßt sich von ihrem 6 Jahre älteren Freund „behandeln"; sie lehnt die Vermutung der Mutter ab, daß „es" (das Hautgeschehen) etwas mit einer vorausgegangenen Suizidandrohung zu tun haben könne.

Stationäre Aufnahme der Patientin: Die Läsionen bilden sich in wenigen Tagen vollständig zurück. In den täglichen Gesprächen wird anfänglich das Hautgeschehen nicht unmittelbar angesprochen. Es geht um die Person der Patientin: Ihren Lebensalltag, ihre Hobbys und ihre sozialen Kontakte (Familie, Freund, Freundin). In diesen Gesprächen wird eine tragfähige Beziehung zur Patientin entwickelt und Information zum besseren Problemverständnis gewonnen. Sie berichtet unter dem aktuellen Eindruck, das Krankenhaus kapsele sie von häuslichen Festen ab, und wie sie als 9jährige erreicht habe, nach einer Blinddarmoperation durch Essensverweigerung die ganze Verlobungsgesellschaft ihrer Schwester zu sich ins Krankenhaus zu bringen.

Insgesamt zeichnet sich ab:

Der Vater mußte wegen „ernsthafter" arterieller Durchblutungsstörung ins Krankenhaus. Die ganze Familie war aufgeregt, die Patientin geriet außer sich, drohte mit Suizid, wurde für eine Nacht in die Klinik eingewiesen, „normalisierte" sich, übernahm im Gespräch mit Mutter und Geschwistern die Meinung, daß sie in derartigen Situationen nicht nur an sich denken dürfe. Danach zeigten sich Läsionen am Unterarm. Der Freund unterstellt Verbrennung „. . . oder was sonst?" Daraufhin erste Läsionen im Gesicht. In diesen zeitlichen Rahmen fällt auch, daß der Freund der einzige war, der nach einem Grubenunglück unter Tage unverletzt blieb. Die Patientin geht schließlich im Gespräch auf ein unterstellendes Angebot ein, woher wohl „das in ihrem Gesicht kommen könne". Vielleicht käme es „von der Abbeize in der Schule, die beim Pinselreinigen so stark spritzt" und außerdem müsse sie sich auch weit über den großen Topf beugen. Tage später rückt sie davon wieder etwas ab. Sie beschwert sich, daß zu Hause die ganze Hautkrankheit nur im Zusammenhang mit ihren Nerven gesehen werde. Sie wird entlassen mit der Bemerkung, daß keine klare Ursache für die Symptome gefunden wurde. Aber es sei wichtig, daß eine erfolgreiche Therapie gefunden werde. Ihr wurde angeboten, sich vorzustellen, falls wieder Veränderungen auftreten sollten.

Eine Vielzahl von Hinweisen – das charakteristische klinische Bild und diffuse, wechselnde subjektive Beschwerden, Abwehr eines vom Arzt nicht verbal geäußerten Verdachtes, die vorausgegangene Androhung eines Suizids und schließlich die zeitliche Zuordnung der plötzlichen Entstehung von Hautläsionen in die aufregende Zeit der als ernsthaft erlebten Erkrankung des Vaters und des Grubenunglückes des Freundes – erwecken in ihrer Massierung den Verdacht auf das Vorliegen eines Artefaktes. Die rasche Abheilung unter stützender Zuwendung und indifferenter Wundbehandlung verstärkt diese Hinweise. Der im Gespräch anklingenden Einigung zwischen Arzt und Patientin über die Entstehungsweise durch äußere Einflüsse steht wiederholt das Bedürfnis der Patientin gegenüber, sich gegen eine – vom Therapeuten absichtlich nicht ausgesprochene – Unterstellung einer Selbstbeschädigung zu verwahren. Der Therapeut praktiziert damit eine für den Umgang mit AF-Patienten typische Gratwanderung zwischen diagnostischer Notwendigkeit und der Möglichkeit, eine tragfähige Arzt-Patient-Beziehung zu bewahren.

Die Persönlichkeit des Artefaktpatienten und die Bedeutung der Selbstbeschädigung

Bei der Konsultation fällt dem Arzt eine merkwürdige Konstellation auf: Die entstellende Hautveränderung und ein fast exhibitionistisch anmutendes Bedürfnis des Patienten. Nach den testpsychologischen Untersuchungen an 35 Frauen von Rauchfleisch und Mitarbeitern (1983) zeichnen sich die AF-Patientinnen durch Depressivität, aggressive Hemmungen und Autoaggressivität sowie Affektblockierung, geringe Frustrationstoleranz und wenig stabile Ich-Integration aus. Diese Befunde wurden als Ausdruck depressiv-schizoid-narzißtischer Strukturanteile der Persönlichkeit gedeutet. Janus (1972) weist auf die depressiv-hysterische Persönlichkeitsstruktur hin. Die psychiatrische Exploration ergibt in der Mehrzahl der Fälle belastende frühkindliche Entwicklungsbedingungen und aktuelle Konflikte zu Beginn des sichtbaren Krankheitsgeschehens (Rauchfleisch et al., 1983).

Die Bedeutung der Selbstbeschädigung für den Patienten ist nicht einheitlich: Die geplante Selbstbeschädigung mit dem Ziel sozialer Entlastung und/oder materiellen Gewinnes steht als Extremfall dem anderen Extrem eines AF als Ersatzhandlung für einen Suizid, einem „partiellen Suizid" (Teichmann et al., 1974) gegenüber. Am häufigsten dürften die Selbstbeschädigungen bei hysterischer Charakterstruktur und masochistischer Veranlagung sein (Übersicht bei Fischer, 1978). Letztere haben Appellcharakter mit dem Wunsch nach Zuwendung. Aus ärztlicher Sicht könnte man den AF der Haut als substituierenden Versuch averbaler Kommunikation einer im übrigen in ihrer Kommunikation behinderten Persönlichkeit verstehen.

Besonderheiten im Umgang mit Artefaktpatienten

Die typischen Hinweise, die zum Verdacht der AF-Diagnose führen, wurden bereits bei der Darstellung des exemplarischen Falles deutlich. Versucht der behandelnde Arzt einen konfrontativen Umgang mit dem Patienten in Form eines aufdeckenden Gespräches, wie es Nordmeyer und Mitarbeiter empfehlen (1983), so sind in der Regel, d.h. vornehmlich bei den depressiv-hysterischen Formen, die Ablehnung jeder weiteren Kooperation, Arztwechsel oder weitere Selbstbeschädigungen als „Beweis" gegen die These des Therapeuten die Folge. Selbst der objektive Nachweis eines AF (Salo et al., 1970) kann nur ausnahmsweise therapeutisch umgesetzt werden. Da eine psychotherapeutische Betreuung von diesen Patienten in der Regel nicht akzeptiert wird, praktizieren wir eine stützende Zuwendung bei der Wundbehandlung mit dem Ziel einer Arzt-Patient-Beziehung, die nicht über die wahre Entstehung des Leidens zu sprechen braucht, obwohl Arzt und Patient über die Krankheit als Artefakt im Bilde sind. Ein solches substituierendes Vertrauensverhältnis und das resultierende Interaktionsritual zwischen Arzt und Patient wird zur Basis einer zeitweilig erfolgreichen Führung mit gelegentlichen Rückfällen bis zur spontanen Remission

(Teichmann et al., 1974). Diese Regel gilt nicht für die geplanten Selbstbeschädigungen mit dem bewußten Ziel sozialer Entlastung (z.B. Militärdienst) oder materiellem Gewinn (Berentung). Der Versuch einer konfrontierenden Aussprache ist in diesen Fällen angezeigt. Favell und Mitarbeiter (1982) diskutieren verschiedene therapeutische Ansätze (medizinisch – psychodynamisch – verhaltensanalytisch) und kommen zu dem Schluß, daß der verhaltensanalytische Ansatz die Therapie der Wahl sei.

55.4.2 Acne excoriée des jeunes filles

Das bei oberflächlicher Betrachtung oft einer Acne vulgaris ähnliche Krankheitsbild wurde 1898 von Brocq beschrieben. Die Frage, ob diesen Hautveränderungen tatsächlich primär eine Acne vulgaris zugrunde liegen muß, wurde immer wieder aufgegriffen (Wrong, 1954). Alle Untersucher sind sich darüber einig, daß die eigenen Manipulationen und ihre Folgen das Bild charakterisieren: multiple Exkoriationen und flache schüsselförmige Närbchen, im Gesicht eher pigmentiert, an der Brust und in der Nacken/Schulter-Region meist unpigmentiert. Wie beim Artefakt (AF) im engeren Sinn sind fast ausschließlich Frauen betroffen, jedoch hier meist junge, differenzierte Frauen gehobener Bildungsstufe, z.B. Studentinnen. Die Manipulationen werden regelmäßig und zwanghaft, visuell (Spiegel) oder taktil gesteuert in Konfliktsituationen durchgeführt. Frau A. schreibt: „Heute haben wir uns zum ersten Mal in der Reisegruppe zerstritten. Jeder muffelt vor sich hin. Ich muffle nicht – ich kratze." Im Gegensatz zu den AF-Patienten sind die Acne-excoriée-Patienten regelmäßig schon bei den ersten Konsultationen auf ihren eigenen Anteil an den Hautveränderungen hin ansprechbar. Biographische Anamnesen verschiedener Autoren zeigten, daß häufig eine bestehende oder bereits abgeheilte Acne vulgaris – ähnlich der induzierenden Funktion von Narben beim AF – den Ort der zwanghaften Exkoriationen bestimmt. Das Verhalten bei der Konsultation weist ebenso wie die verhältnismäßig geringe somatische Beschädigung auf einen eher defensiven Charakter der Patienten hin, verglichen mit dem demonstrativ-appellativen Auftreten der AF-Patienten. Ein zwanghaftes Bedürfnis nach Perfektion, Sauberkeit und Ordnung fällt auf. Neuere kasuistische Darstellungen unter besonderer Berücksichtigung der psychodynamischen Entstehung der Zwangshandlung und der Entwicklung der Persönlichkeitsstruktur legten Vogel (1974) und unsere Arbeitsgruppe (Teichmann et al., 1974) vor. Nach unserer Auffassung liegt im Gegensatz zum „partiellen somatischen Suizid" des AF-Patienten bei der Acne excoriée die Bedeutung eher in einem „sozialen Suizid" als Folge der starken Entstellung sichtbar exponierter Regionen (Gesicht, Dekolleté, Schultern). Die differenzierten Patienten ermöglichen mittels konfrontierender Gesprächstechniken und einer Einübung in Selbstkontrolle (Tagebuchtechnik – Besprechung – Belohnung) meist einen erfolgreichen therapeutischen Zugang während der begleitenden dermatologischen Lokalbehandlung.

55.4.3 Sonstige Formen von Artefakten

Zahlreiche weitere, psychodynamisch und morphologisch charakteristische Formen von artifiziellen Veränderungen des äußeren Integumentes sind dem Dermatologen vom Aspekt her als wichtige differentialdiagnostische Erwägungen geläufig. Aktuell dramatisch, jedoch meist prognostisch günstig ist das Vorkommen sichtbarer Kahlstellen am behaarten Kopf durch Ausziehen von Haaren oder Drehen und Zerren der Haare unter dem Bilde der Trichotillomanie oder Ziliotillomanie. Konflikthafte Konstellationen durch Umschulung, Krankenhausaufenthalte oder Ferienaufenthalte bei Kindern, Examensnöte und Arbeitslosigkeit bei jugendlichen Erwachsenen erleben wir als auslösend. Eine Trichotillomanie mit deutlichen autoaggressiven Tendenzen als Ausdruck des Versuches, eine Konkurrenzsituation unter eifersüchtigen Geschwistern gegenüber dem Vater zu lösen, beschrieben wir 1973 (Teichmann, Bosse, Ahrens).

Lineare Nagelveränderungen bei gewohnheitsmäßig unkontrolliertem Zurückschieben des Nagelhäutchens lassen an eine Tinea unguium denken, periorale entzündliche Veränderungen durch hypermotorisches nervöses Lecken mit der Zunge erinnern an ein allergisches Kontaktgeschehen, Lichen-ruber-ähnliche oder leukoplakieartige Veränderungen durch Beißen und Saugen der Wangen und Lippenschleimhaut (Morsicatio) stellen relativ harmlose somatische Folgen von Gewohnheiten dar, die lediglich einer Aufklärung bedürfen. Übersichten über das gesamte Spektrum der Selbstbeschädigungen an der Haut stellten Waismann (1965) und Lyell (1976) zusammen.

55.5 Sonstige Formen von Verhaltensänderungen mit kutaner Manifestation

Die in diesem Abschnitt beschriebenen Krankheitsbilder sind nicht durchweg als psychosomatisch oder somatopsychisch aufzufassen. Die Besprechung erfolgt aus differentialdiagnostischen Gründen, da diese Patienten bevorzugt den Dermatologen zu Rate ziehen und die Abgrenzung gegenüber dermatologischen Krankheitsbildern häufig außerordentlich schwierig ist.

Es entspricht wohl einem allgemeinen Wandel in der Medizin, wenn um die Jahrhundertwende die verschiedensten dermatologischen Krankheitsbilder („bullöse Dermatitis", „neurotische Hautgangrän", „hysterisches Ödem", Blutungsphänomene, Exantheme, Urtikaria u.a.) bei Hysterikern beschrieben oder auf eine hysterische Entstehung zurückgeführt wurden (Voss, 1909). Heute finden wir dagegen in einer dermatologischen Ambulanz bei den unterschiedlichsten Hautsymptomen häufig Hinweise auf depressive Verstimmungen und phobieartige Zustandsbilder.

55.5.1 Depressive Zustände und ihre Äquivalente an der Haut

Funktionelle Syndrome – larvierte Depression

Rechenberger (1976) untersuchte 30 männliche und weibliche Urtikariapatienten unter tiefenpsychologischen Gesichtspunkten, nachdem die vollständige somatische Durchuntersuchung unter stationären Bedingungen keinen organisch-pathologischen oder allergologischen Befund erbracht hatte. Die Autorin kommt zu dem Schluß, daß in diesen Fällen von Urtikaria eine neurotisch depressive Charakterstruktur zugrunde liegt, und deutet das Krankheitsbild als funktionelles Syndrom. Biographisch konnte sie eine bewußtseinsnahe auslösende Situation ein bis zwei Tage vor dem Erstausbruch der Urtikaria erfassen. Andere charakteristische Beschwerdekomplexe bei Hautpatienten werden in der Dermatologie als Ausdruck einer „larvierten Depression" aufgefaßt. Die Patienten klagen über Haarausfall, „grobporige Haut", „schlechte Haut allgemein", Juckreiz oder Glossodynie (Hornstein, 1981). Die geklagten Beschwerden sind klinisch und mittels objektiver Untersuchungen kaum nachweisbar. Fühlt sich der Patient vom Arzt nicht ohne Vorbehalt mit seinem Syndrom akzeptiert, so folgt eine sich steigernde Liste weiterer Hautbeschwerden, die dann „ausgerechnet heute" nicht sichtbar sind. Die Nachfrage nach allgemeinen Symptomen erbringt wechselnde und schwer faßbare Klagen über Schlafstörungen, Arbeitsstörungen, Angst und Unzufriedenheiten aller Art sowie unter Umständen Informationen über die Behandlung anderer Organbeschwerden bei weiteren Fachärzten. Bei der Intensität der Klagen dieser Patienten ist der Hautarzt mangels nachweisbarer Symptome in Versuchung, den Ausweg in einer hautindifferenten Lokalbehandlung zu suchen, ohne dabei zu realisieren, daß er damit selbst zum Objekt in einem Kreislauf wird. Ergänzt er seine somatische Untersuchung durch eine erweiterte biographische Anamnese, so kann als Konsequenz derselben der unfruchtbare Kreislauf des Patienten von Arzt zu Arzt unterbrochen und der Patient mit seinem eigentlichen Leiden konfrontiert werden. Das schließt die Tatsache ein, daß keine Hautkrankheit nachweisbar ist, und die Frage, ob z.B. eine reaktive Depression oder eine neurotische Störung vorliegt. Die Abklärung wird im Regelfall der Hautarzt nicht vornehmen. Es bewährt sich jedoch als motivierend für den Patienten, eine begleitende hautpflegende Behandlung neben der Empfehlung einer psychiatrischen/psychotherapeutischen Behandlung anzubieten.

Reaktive Depression und Suizidalität bei Hautkranken

Reaktive Depressionen bis zur Suizidneigung als Folgesymptom werden bei extremen Verläufen der Psoriasis und der schweren Akne beobachtet. Depressive Zustandsbilder kommen in der Folge, wie auch als symptomverstärkende Voraussetzung, bei chronisch rezidivierenden endogenen Ekzemen und als häufige Symptomatik bei Dysmorphophobie- und Artefaktpatienten vor. Vorausgegangene Suizidversuche bei Artefaktpatienten sind nicht selten. Der Artefakt an der Haut wurde deshalb als „partieller Suizid" aufgefaßt.

Schließlich ist bei verhaltensauffälligen, eventuell depressiven Patienten mit Hauterscheinungen unklarer Genese in belichteten Regionen auch an eine Pellagra zu denken.

Differentialdiagnostisch ist zu beachten, daß bei der Behandlung chronischer Hautkrankheiten häufig Langzeittherapien mit Steroiden oder Chloroquin durchgeführt werden, die gelegentlich zu depressiven Zustandsbildern führen.

55.5.2 Halluzinosen

Die chronisch taktile Halluzinose (kutane Parasitophobie, Dermatozoenwahn)

Bei der kutanen Manifestation der Parasitophobie besteht ein klinisch eindeutiges Bild: Der Patient klagt über Befall mit Ungeziefer (Würmer, Läuse, „Drachen" u.a.) und bringt als „Beweis" eingetrocknete Hautabschilferungen, Wollfuseln und Schmutzpartikel, sorgfältig in Gläschen verpackt, zur Konsultation mit. Meist berichtet er über eine lange Reihe von Institutionen, die er jahrelang vergeblich konsultiert hat (Dermatologen, Mikrobiologen, Zoologen, Parasitologen, Veterinärmediziner, Gesundheitsämter, Regierungsinstitutionen und schließlich auch Psychiater). Er hat viel Geld für die fortlaufende Desinfektion seiner Wohnung, ständige Reinigung seiner Kleidung und Untersuchung der verdächtigen Partikel ausgegeben. Es besteht kein morphologischer Befund an der Haut, der Patient ist nicht von seiner Meinung abzubringen.

Ein Patient schreibt: „... Wir haben unser ganzes gespartes Geld aufgebraucht, um Mittel zu kaufen, damit wir diese Parasiten fortbekommen, aber nichts hilft, und die Hygiene streitet uns das ab und sagt, es wäre Einbildung und eine psychologische Sache. Ich habe bis zur Regierung hingeschrieben, aber von dort bekam ich die Antwort, daß es an Ort und Stelle erledigt wird ... Überall werden wir abgewiesen, wir seien bekloppt ... Bitte untersuchen Sie mir doch das Material bei Ihnen ... Meine Kleider sind alle von Parasiten besetzt und in der Wohnung kann man sich nicht aufhalten, ich habe Tag und Nacht keine Ruhe ... Es bleibt mir nur noch aus dem Leben zu scheiden, wenn keine Hilfe kommt ..."

Bei dem Bild der Parasitophobie handelt es sich offenbar nicht um eine geschlossene klinische Entität, jedoch oft um eine präsenile exogene, paranoid-halluzinatorische Psychose (Schulte und Tölle, 1975), von der fast ausschließlich Frauen betroffen sind. Häufig zeigen die Patientinnen eine depressiv-hypochondrische Einstellung, gelegentlich unterstellen sie Fremdbeeinflussung von seiten der Nachbarn etc. Die psychiatrische Abklärung ist im Einzelfall erforderlich (Mester, 1980). Die Prognose ist schlecht. Die Bedeutung der psychiatrischen Erkrankung liegt für

den Dermatologen vornehmlich in der notwendigen Abgrenzung von somatischen Hautleiden mit somatopsychischen Folgereaktionen. Therapeutisch ist ein Versuch mit Neuroleptika oder Thymoleptika neben einer indifferenten Lokalbehandlung gerechtfertigt.

Olfaktorische Halluzinose

Die olfaktorische Halluzinose bedarf einer psychiatrischen Abklärung. Die Nachdrücklichkeit der vorausgegangenen Krankheitsgeschichte und die Unfähigkeit der Patienten zur Einsicht sind ähnlich der im vorigen Abschnitt für die taktile Halluzinose geschilderten. Aufgabe des Dermatologen ist der Ausschluß einer somatisch verursachten chronischen Hyperhidrosis, die gegebenenfalls eine ursächliche Behandlung (z.B. Schilddrüse) oder symptomatische, operative Entfernung der axillären Schweißdrüsen oder lokale Behandlung erfordert. Duller und Gentry (1980) berichten über die Behandlung von 14 Erwachsenen mit „chronischer Hyperhidrose" mittels Biofeedback. Ihre Ergebnisse lassen vermuten, daß manche dieser Patienten nach Biofeedback-Training tatsächlich weniger stark schwitzen.

> Einer unserer Patienten (40 J., m.) führte sein nicht abgeschlossenes Studium, seine mißglückte berufliche Tätigkeit und seine mit schweren Problemen belastete Beziehung darauf zurück, daß er anfange zu riechen, wenn er sich in einem geschlossenen Raum aufhalte. Deswegen war er seit Jahren nicht mehr beruflich tätig. Er wünschte eine operative Entfernung der Schweißdrüsen, damit sein Leben in Ordnung käme. Die mangelnde Kooperation gegenüber der hinzugezogenen Psychologin begründete er mit der Furcht, sie wolle die von ihm geforderte Operation nicht befürworten. Aus demselben Grunde brach er kurzfristig die weiteren psychologischen und dermatologischen Konsultationen ab.

55.5.3 Phobien

Karzinophobie

Das besonders häufige Vorkommen einer übermäßigen Angst vor Krebs bei Patienten der dermatologischen Sprechstunde erklärt sich aus der bedrohenden Sichtbarkeit (selektive Wahrnehmung, mangelnde Kenntnis der Norm, Hoffnung auf Kontrollierbarkeit) der Hautveränderungen. Das Auftreten und Wachstum von Hauttumoren bei älteren Menschen, die farbliche Abweichung (Pigmentierung) und die dem Patienten nicht bekannte unterschiedliche Dignität sowie konkrete Erfahrungen im Familien- und Bekanntenkreis vermitteln zunächst ein Gefühl latenter Bedrohung, das dann plötzlich zu massiver Angst führt. Auffällig sind Patienten, die hypochondrisch wegen jeder Minimalveränderung eine histologische Kontrolle wünschen. Ebenso diejenigen, die trotz jahrelangen Bestehens der Veränderung den Arzt nicht aufgesucht haben und schließlich auf den Druck der

Familie hin, meist unter einem banalen dermatologischen Vorwand, scheinbar nebenher, mit einem fortgeschrittenen Prozeß zur Vorstellung kommen.

So fielen in der poliklinischen Sprechstunde an einem einzigen Tage eine ungewöhnlich große Anzahl der beschriebenen vorsichtigen oder auch hypochondrischen Patienten und drei Patienten mit fortgeschrittenem Melanom auf. Alle Patienten kannten ihre Veränderungen seit Jahren. Bei der Überprüfung zeigte sich, daß alle diese Patienten aus demselben Teil des Einzugsbereiches kamen. Zwei Tage vorher war in der dortigen Tageszeitung vom Melanom einer bekannten Filmschauspielerin berichtet worden. Eine im Ausmaß vergleichbare sachlich-informative Aktion über Melanom in einem anderen Teil des Einzugsbereiches hatte keine vergleichbare „Erfolgsquote".

Venerophobie

Heute selten, früher häufiger aufgrund von Ereignissen der Kriegs- und Nachkriegszeit, kommen Patienten mit einer durch nichts belegbaren und nicht behebbaren Sorge vor venerischen Infektionen, verbunden mit massiven Schuldgefühlen und Selbstvorwürfen, zur Vorstellung. Immer neue Untersuchungen werden gefordert, die Ergebnisse jedoch nicht akzeptiert. Die Patienten verlangen „Sicherheitskuren", die in der Hoffnung, den Patienten beruhigen zu können, vom Arzt verabreicht werden. Gerade letzteres bestätigt jedoch den zweifelnden Patienten in seinem Verdacht und führt ihn zum nächsten Venerologen. Die Venerophobie ist möglicherweise als besondere Variante der Bakteriophobie bzw. Parasitophobie aufzufassen oder als Ausdruck der hypochondrischen Angst des behandelten Luikers vor progressiver Paralyse zu sehen. Sie ist einer dermatologisch-venerologischen Therapie nicht zugänglich. Postgonorrhoische Schuldgefühle und Angst vor späteren Komplikationen drücken sich in ähnlicher Weise als schwer therapeutisch zugängliche unspezifische Urethriden aus. Ein sorgfältiger Ausschluß sonstiger Erreger (Trichomonaden, Mykoplasmen) ist erforderlich.

Der Umgang mit der Angst vor AIDS (vgl. Kap. 49) unterliegt heute einer ähnlichen Psychodynamik. Zusätzlich wird dabei die ungleich höhere somatische Bedrohung, der Mangel an therapeutischen Möglichkeiten und die Diskriminierung durch die Gesellschaft wirksam.

55.6 Psychosomatische Einflüsse auf sonstige Hauterkrankungen

Außer den in den vorausgehenden Abschnitten erwähnten modellhaften Krankheitsbildern gibt es weitere Hautkrankheiten mit psychosomatisch bedingten Einflüssen auf die Entstehung und den Verlauf. Eine gute Dokumentation liegt für die periorale Dermatitis vor (Hornstein, 1976; Hartung und Lehre, 1976; Thurn, 1976; Wilsch und Hornstein, 1976), klinisch beeindruckende Erfahrungen werden bei Herpes simplex und Verrucae vulgares, bei Urtikaria

(Schröpl, 1981), Lichen ruber, einzelnen Formen von Effluvium u.a.m. gemacht. Ein zeitgemäßes phobieartiges Krankheitsbild ist die Allergophobie. Eine gute Übersicht der Literatur stellte Whitlock (1980) zusammen.

55.7 Therapeutische Möglichkeiten des Facharztes in der dermatologischen Psychosomatik

In der Regel kommt der Hautkranke somatisch orientiert in die Sprechstunde und das konventionelle ärztliche Verhalten kommt dieser Einstellung entgegen. Zeitliche Gründe beim Arzt und die gewohnten Erwartungen des Patienten begünstigen die Einigung der beiden auf der somatischen Ebene. Abweichungen von dieser Position provozieren Widerstand. Es stellt sich die Frage: Was kann der psychosomatisch interessierte Dermatologe tun, um den Patienten während der Konsultation aus seinem somatisch begrenzten Konzept herauszuführen?

55.7.1 Das diagnostische Gespräch mit dem psychosomatisch/somatopsychisch Hautkranken

Übergang vom somatischen zum psychosomatischen Krankheitskonzept

Die Besonderheit der hautfachärztlichen Konsultation besteht darin, daß der psychosomatische Anteil einer Hautkrankheit schon sichtbar wird, ehe der Patient die Barriere der Verbalisierung überspringt: Das plötzliche Erblassen des Mund-Nasen-Dreieckes oder das unerwartete Auftreten von Juckreiz und Kratzen in schwierigen Gesprächssituationen bei dem endogenen Ekzematiker, das klinisch morphologische Bild des Artefaktes und der Acne excoriée sind hierfür beispielhaft. Der Arzt kann z.B. dem endogenen Ekzematiker in einer späteren Phase der Behandlung sogar anhand dieser auslösbaren Phänomene das Umsetzen von Emotion in körperliche Reaktion vor dem Spiegel demonstrieren und verständlich machen. Dennoch ist der Patient meist nicht sofort in der Lage, sein ausschließlich somatisches Krankheitskonzept („Allergie"–„Wasser"–„Luft"–„unnatürliche Ernährung") in Frage zu stellen und die Notwendigkeit einer biographischen Erhebung einzusehen. Die ersten Arzt-Patient-Kontakte werden sich deshalb so lange vorwiegend auf die somatischen Sachverhalte beschränken, bis sich das gegenseitige Vertrauen stabilisiert hat und weitergehende biographische Gespräche erlaubt. Die Beschränkung des Arztes auf die Frage nach den Umständen der letzten Verschlechterung bzw. dem Beginn der Hautveränderungen (Artefakt, endogenes Ekzem, periorale Dermatitis, Urtikaria, Acne excoriée) ergibt eventuell einen Hinweis, der als Ansatzpunkt für ein späteres Gespräch brauchbar ist („in der Schwangerschaft", „als XY starb",

„vor der Prüfung" ...). Der Einstieg in ein biographisches Gespräch mit dem Psoriatiker oder Aknepatienten ist demgegenüber anhand der sozialen Einschränkungen (Kleidung, Sport, Partnerkontakte ...) leichter möglich. Die Besonderheiten des „Rituals" anstelle eines Gespräches mit dem Artefaktpatienten wurden bereits besprochen.

Sensibilisierung und Motivierung als vortherapeutischer Prozeß

Hat sich im Verlauf der ersten Behandlungstermine die Bereitschaft des Patienten für ein weitergehendes biographisches Gespräch entwickelt, so kann er anhand seiner eigenen Beobachtungen und mittels schriftlicher aktueller Protokolle („Hausaufgaben") für den psychogenen Anteil seiner Erkrankung sensibilisiert werden und schließlich kann vom Arzt oder vom Patienten selbst auch die Frage nach etwaigen weitergehenden therapeutischen Konsequenzen aufgeworfen werden. Die diagnostischen und vortherapeutischen Gespräche können durch vorzeitige Einführung und ungeschickte Handhabung bei dem noch nicht innerlich vorbereiteten Patienten eine extreme Abwehrhaltung induzieren (Rechenberger, 1976; Hornstein, 1980).

Der Vorteil des gleichzeitig psychosomatisch und somatisch tätigen Facharztes besteht in der Möglichkeit, einen optimalen Zeitpunkt für die Aufnahme der psychosomatischen Thematik in Ruhe abwarten zu können. Gelegentlich zeigt sich, daß der Patient das vom Arzt angestrebte Gespräch zwar in der Sprechstunde abgelehnt, aber bei einem Psychotherapeuten aufgenommen hat.

Als unterstützender Faktor wirkt in der Phase der Sensibilisierung die ureigene Form der dermatologischen Behandlung: das Einsalben, Eincremen, Baden, das Erleben der Berührung und des Berührtwerdens bei der Selbstbehandlung und noch mehr bei der täglichen Versorgung durch den Pfleger während der stationären Behandlung. Ähnlich der Krankengymnastik, Kosmetik, der Akupunktur, dem Handauflegen oder der diagnostischen Palpation hat diese direkte Kontaktnahme eine emotionale therapeutische Bedeutung, die Patienten nicht selten ausdrücklich verbalisieren: „Versorgtwerden" – (evtl.) „sogar ohne Handschuhe, ist so schön". Diese institutionalisierte Berührung kann in jüngeren Altersgruppen (vgl. Abschn. „Acne vulgaris und Acne conglobata") ganz besonders bedeutsam werden und durch die entspannende Atmosphäre die Gesprächsbereitschaft des Patienten erheblich fördern. Die Stellung des Pflegers in der Dermatologie bekommt durch diese täglichen körperlichen Kontakte einen hohen und spezifischen Eigenwert.

55.7.2 Das psychotherapeutisch orientierte Gespräch in der Praxis des Hautarztes

Sehen wir davon ab, daß jedes über ein rein direktives, informatorisches Fragen und Antworten hinausgehende Ge-

spräch bereits die Arzt-Patient-Beziehung wesentlich beeinflußt und seinen eigenen therapeutischen Wert bekommt, so ist generell festzustellen, daß ein weiterführendes psychotherapeutisches Gespräch in der Praxis des Dermatologen mangels Vorbildung und aus organisatorischen Gründen die Ausnahme ist. Die fachpsychotherapeutische Behandlung ist deshalb nur als begleitende Mitbehandlung neben der dermatologischen Therapie möglich.

Möglich ist dagegen auch in der Praxis ein einzelnes, gezieltes, zeitlich und thematisch begrenztes Gespräch mit dem Patienten. Dies konnte unsere Arbeitsgruppe unter protokollierender Kontrolle mit Vor- bzw. Nachgesprächen überprüfen. Die ca. 30–35 Minuten dauernden Gespräche beschränkten sich auf drei Themenbereiche:

– Krankheitskonzept des Patienten und Wissen über die eigene Krankheit;
– Auswirkung der Krankheit und Abhängigkeit der Krankheit von der familiären und beruflichen Umwelt des Patienten;
– Konsequenzen für die Zukunft.

Eine Entspannung des Arzt-Patient-Verhältnisses, Klärung durch medizinische Information und die Fähigkeit, mögliche Änderungen im eigenen Leben und Verhalten in die Tat umzusetzen, werden schon durch ein einmaliges zeitlich geplantes Gespräch begünstigt. Neben diesen gezielten Gesprächen kann der Facharzt durch aufmerksames Zuhören und Beobachten auch bei den normalen Konsultationen psychosomatisch wichtige Information erhalten und vermitteln.

55.7.3 Koppelung dermatologischer und psychosomatisch-psychotherapeutischer Tätigkeit?

Die gegebenenfalls erwünschte oder erforderliche Weitervermittlung an den Psychiater, Psychosomatiker oder Psychotherapeuten scheitert oft an dem individuellen Krankheitskonzept des Patienten und seinen prinzipiellen Vorbehalten.

Eine wichtige Alternative in dieser Situation ist die Möglichkeit eines psychologischen, psychotherapeutischen oder psychiatrischen Dienstes in eigenen Räumen innerhalb der dermatologischen Klinik. Das diagnostische psychosomatische/psychologische Konsil in der Hautklinik erspart dem Patienten die Schwierigkeit, trotz seiner Abneigung und im Widerspruch zu seinem eigenen Krankheitskonzept in die „Nervenklinik" gehen zu müssen oder den Psychotherapeuten aufzusuchen. Diese Alternative hat sich in der Göttinger Hautklinik in wechselnden Formen seit mehr als zehn Jahren im Interesse der Patienten und der Ärzte bewährt. Auch die Möglichkeit der Zusammenarbeit und der gegenseitigen Information in der therapeutischen Phase ist innerhalb des Hauses wesentlich leichter als bei der Überweisung des Patienten von einer räumlich und personell getrennten Institution zur anderen. Als ähnlichen Versuch beschreiben Gould und Gragg (1983) eine dermatologisch/psychiatrische Verbundklinik für die Stanford University.

55.7.4   Co-Therapie von Arzt und Psychologe

Bei themenzentrierten Einzelgesprächen, Partnergesprächen und Gruppengesprächen mit Patienten der Hautklinik haben wir gute Erfahrungen gemacht, wenn der Psychologe die Gesprächsgestaltung, der Facharzt die Funktion des dermatologischen Informators und beide bei Bedarf die eines Moderators übernehmen. Wichtig ist dabei, die Balance der Führung aufgrund der Sachkompetenz zwischen Psychologe und Facharzt auch dann zu erhalten, wenn Patienten hierarchische Strukturen zugunsten des einen oder anderen erwarten. Vor- und Nachgespräche zwischen den beiden Leitern, gegebenenfalls auch über unterschiedliche Gesichtspunkte, kommen der Sache und allen Beteiligten zugute.

Erfahrungen liegen zu folgenden thematischen Anlässen vor:

– Entstellungsproblematik bei Dysmorphophobie und anderen Hautkrankheiten (Einzel- und Paargespräche);
– Konfliktbearbeitung bei Acne excoriée (Einzelgespräche);
– Impotenz (Einzel- und Paargespräche);
– Auseinandersetzungen mit Krankheit und Umwelt bei endogenen Ekzematikern, Psoriatikern und Acne-vulgaris-Patienten (Einzel- und Gruppengespräche).

55.7.5   Gruppenerfahrungen in der Dermatologie

Über die Entwicklung einer Gruppe von Psoriatikern unter psychologischer Leitung berichtet J. de Korte (1980). Er erwähnt, daß neben den sozialen Erfahrungen und einer Verminderung der Identifikation mit der kranken Haut auch das somatische Ausmaß der Erkrankung sich unter der Therapie gebessert habe. Leuteritz und Shimshoni (1982) berichten ebenfalls über positive Erfahrungen mit Gruppentherapie bei Psoriatikern unter psychologischer Leitung. Auch zahlreiche Selbsthilfegruppen, meist regional ausgehend vom Psoriasisbund (Sitz Hamburg) bewähren sich seit Jahren. Sie vermitteln den Teilnehmern Wissen über die Krankheit und die Möglichkeiten, mit der Krankheit zu leben (vgl. Moeller, 1983). Auch diese Gruppen sind eine wichtige Ergänzung zu der notwendigen fachärztlichen Versorgung.

Entsprechende eigene Erfahrungen liegen für Gruppen mit endogenen Ekzematikern bzw. mit den Eltern dieser Kinder vor. Unsere Zielvorstellungen bei der Gruppenarbeit sind vornehmlich eine Verbesserung der Fähigkeit zur Krankheitsbewältigung durch Ergänzung des Krankheitsverständnisses um den Sozialaspekt. Erwartungen der Patienten oder deren Eltern hinsichtlich einer kurzfristigen symptomatischen Besserung der Haut sind in Gruppen kaum zu realisieren. Eine Unterstützung der aktuellen dermatologischen symptomatischen Therapie ist nach unserer Erfahrung nur von vorsichtig konfrontierenden und stützenden Einzelgesprächen zu erwarten.

55.7.6   Führung und Selbständigkeit des Patienten in der dermatologischen Therapie

Der chronische Hautpatient (Psoriasis, endogenes Ekzem, bullöse Hauterkrankungen u.a.) hat in einem sehr hohen Maße und mehr als Patienten mit anderen Organerkrankungen die Chance, durch eigene Aktivität und bewußte Erfahrungen auch die lokale Therapie seiner Krankheit selbst zu handhaben. Die zunehmende Unabhängigkeit von Hausarzt und Facharzt durch den angemessenen Umgang des Patienten selbst mit drei bis fünf unterschiedlichen Zubereitungen für die Lokaltherapie kann den Patienten durchaus zugebilligt werden, sobald diese die Grenzen (z.B. Anwendung von Steroiden) ihrer eigenen Möglichkeiten kennen. Anstelle der oft erwünschten „Führung" eines passiven Patienten unter der väterlichen Autorität des Arztes kann dann mit zunehmender eigener Erfahrung des Patienten eine Aktivierung zur Selbständigkeit durch die Anleitung, Beratung und gemeinsame Entscheidung über die weitere Therapie treten. Jedoch ist darauf zu achten, daß sich erfahrungsgemäß gerade bei der lokalen Therapie für den Patienten wegen der scheinbaren Gefahrlosigkeit der Lokaltherapeutika ein unangemessener Gebrauch seiner Mündigkeit anbietet (vgl. Pierloot, 1983; Bosse, 1985).

55.7.7   Die Plazebowirkung in der Dermatologie

Es ist bekannt, daß die Plazebowirkung durch die Farbe und Größe der Verabreichung sowie durch unterschiedliche Einführung durch den Arzt variiert und gesteigert werden kann. Eine zusätzliche Steigerung der Plazebowirkung durch die lokale sichtbare Anwendung von Farbstofflösungen auf der Haut (Pyoktanin – Brillantgrün – Castellanische Lösung u.a.) sowie durch den Gebrauch riechender Zusätze (Teerprodukte) oder durch Scheinbestrahlungen ist möglich. Die individuelle Rezeptur und die Notwendigkeit deren Zubereitung durch den Apotheker eröffnen ebenfalls weitere Möglichkeiten der Intensivierung einer Plazebowirkung, zumal dann, wenn diese Rezeptur den gegenwärtigen Natürlichkeitsvorstellungen (z.B. „die Rezeptur enthält keine Konservierungszusätze") entgegenkommt (Bosse, 1985).

In der Regel beinhaltet jedes dermatologische Lokaltherapeutikum auch einen mehr oder weniger großen Anteil harmloser Plazebowirkung. Eine reine Plazebowirkung wird bei der suggestiven Scheinbestrahlung von Verrucae vulgares gelegentlich mit Erfolg benutzt.

56 Psychosomatische Aspekte in der Urologie

Ernst-Albrecht Günthert und *Peter Diederichs*

56.1 Einleitung

Die Beschreibung psychosomatisch bedingter Symptome und Erkrankungen des Urogenitaltrakts setzt das Wissen um die Komplexität dieses Organbereichs voraus. Erst die Berücksichtigung der drei ineinandergreifenden und voneinander abhängigen Funktionsaspekte als Produktions-, Reproduktions- und Lustorgan macht seine besondere Anfälligkeit für seelische Einflüsse verständlich. Die unmittelbare anatomische Nachbarschaft von Urogenitaltrakt und Enddarm bzw. Darmausgang ergibt darüber hinaus Wechselbeziehungen zwischen diesen beiden Organbereichen.

Indem die Psychoanalyse auf die Bedeutung des peripheren Harnapparates (Blase und Harnröhre) als Lust- und Triebzone hinwies, hat sie schon früh den dritten Funktionsbereich des Urogenitaltrakts als Lustorgan gewürdigt. Die Schleimhaut der Harnröhre besitzt ebenso wie die des Mundes, der Vagina und des Afters erogenen Charakter. Masturbationspraktiken an der Harnröhre sind daher keine Seltenheit. Aus dem Urinieren oder dem Zurückhalten des Urins zieht schon das Kind einen erotisch zu wertenden Lustgewinn (Libidotheorie). Freud (1905) weist als erster darauf hin, daß die Funktion des Urinierens im Dienst der infantilen Sexualität stehen kann. Sadger führt dann 1910 analog zur Analerotik den Begriff der Urethralerotik ein.

Ein die Libidotheorie erweiterndes Konzept der Urethralität entwickeln – offensichtlich unabhängig voneinander – Schultz-Hencke (1927) und Christoffel (1944). Das „urethrale Antriebserleben" ist für Schultz-Hencke kein Partialtrieb, sondern ein originäres biologisches Bedürfnis, das von Affekten begleitet wird (Gleichzeitigkeitskorrelation). Seine Befriedigung ist mit Lust und körperlicher Entspannung verbunden, seine Nichtbefriedigung mit einem Mangelzustand, einer Unlust, Spannung oder Protestreaktion (ausführlicher dargestellt ist das psychosomatische Konzept von Schultz-Hencke bei E. und W. Zander, 1988). Im urethralen Antriebserleben spiegelt sich zum einen die libidinöse Erfahrung des kleinen Kindes wider, sich im Miktionsakt vertrauensvoll zu verströmen und zum anderen, sich dabei unbekümmert um zeitliche und räumliche Zwänge seiner Willkür zu überlassen. Die von Schultz-Hencke und Christoffel herausgearbeitete aggressive Seite des Urethralen leitet sich aus der erfahrenen Willküreinschränkung durch die Sauberkeitserziehung her. Sie manifestiert sich auch in der „uropolemischen" All-tagssprache (jemanden „anpinkeln"). Die triebtheoretisch orientierten psychosomatischen Konzepte von Freud, Sadger, Schultz-Hencke oder Christoffel müssen durch die neuen Erkenntnisse der Selbst- und Objektbeziehungspsychologie erweitert werden. Die Mutter schränkt bei ihren Sauberkeitserziehungsmaßnahmen (in der Regel zwischen dem 2. und 3. Lebensjahr) nicht nur den urethralen bzw. analen Triebimpuls ein, sondern nimmt dabei auch Einfluß auf das sich bildende Selbst bzw. Körper-Selbst des Kindes. Die Beherrschung der Harn- und Stuhlausscheidung als die erste vom Kind aktiv geforderte Triebeinschränkung setzt gerade in der Phase des Kampfes um die Ich-Abgrenzung ein, also auf dem Höhepunkt des von Mahler (1978) beschriebenen normalen Entwicklungskonflikts zwischen Individuation und Separation. Vor diesem Hintergrund können auch die weiteren mit der Urethralität verbundenen Gefühlsbereiche der Hingabe und Geltung verstanden werden: Das kleine Mädchen uriniert im Sitzen und erlebt dabei Gefühle der Hingabe, des Fließenlassens und des Sich-Verströmens. Der Gefühlsbereich des „Sich-vertrauensvoll-Verströmenlassens" bildet dann später u. a. die Basis für ein Gefühl des „Sich-Hingebens und Hergebens". Der kleine Junge uriniert im Stehen und erlebt dabei Gefühle der Geltung, der Rivalität und der Selbstdarstellung. Das Urethrale liegt bei ihm in enger Nachbarschaft zum Phallischen („Wer kann den größten Bogen?"). In diesem Zusammenhang möchten wir darauf hinweisen, daß die Wahrnehmung des unterschiedlichen Miktionsverhaltens von Mädchen und Jungen für die psychosexuelle Entwicklung und Identitätsbildung von der Psychoanalyse vernachlässigt wurde. Nur Kestenberg (1988) macht in jüngerer Zeit in ihren Ausführungen über die Entwicklungsphasen der weiblichen Identität darauf aufmerksam, daß sich Mädchen benachteiligt fühlen, nicht wie die Jungen im Stehen urinieren und mit dem Anhalten und wieder Fließenlassen des Urins spielen zu können.

Zusammenfassend soll festgehalten werden, daß der unempathische Umgang der Eltern mit der urogenitalen Körperzone des Kindes zunächst bei der Körperpflege, später bei der Sauberkeitserziehung, negative Auswirkungen auf das urethrale Antriebserleben und die entsprechende Körper-Selbst-Repräsentanz haben kann. Das urogenitale Körper-Selbst ist ein wichtiger Baustein für die eigene Identität. Das Zusammentreffen von Sphinkterkontrolle und Ich- bzw. Selbst-Entwicklung weist auf die Bedeutung der Objektbeziehungen bei der Ätiopathogenese psychosomatisch bedingter Erkrankungen im Urogenital- und Analbereich hin. So verwundert es nicht, daß sowohl bei der Frau als auch beim Mann der Urogenitaltrakt mit Symptomen von Ausdruckscharakter auf Beziehungsstörungen reagiert.

In diesem Kapitel sollen neben einer kritischen Übersicht der bisher vorliegenden Forschungsergebnisse zu psychosomatischen Aspekten in der Urologie und Nephrologie (eine ausführliche Darstellung des gegenwärtigen Forschungsstandes findet sich bei Diederichs, 1983) die Erfahrungen des niedergelassenen Urologen in der täglichen Sprechstunde berücksichtigt werden. Besonders wichtig erscheint uns der Hinweis, daß schon bei der Erstbegegnung mit dem Patienten in der urologischen Sprechstunde die Möglichkeit psychosomatischen Geschehens erwogen werden muß, um iatrogene Fixierungen zu vermeiden. Dies zeigt auch der exemplarische Fall aus der Praxis.

Epidemiologische Daten sowohl über Inzidenz und Prävalenz als auch Schichtzugehörigkeit bei psychosomatisch bedingten urologischen Erkrankungen liegen mit Ausnahme einiger Untersuchungen zur Harnsteinbildung bisher nicht vor. Nach Breitwieser et al. (1981) und Günthert (1980) schwankt der Anteil an Patienten mit psychosomatisch bedingten Erkrankungen in der Praxis des niedergelassenen Urologen zwischen 30 und 50%.

Unter „psychosomatisch" verstehen wir ein multifaktorielles, ineinandergreifendes Geschehen im Sinne des bio-psycho-sozialen Modells (vgl. Kap. 1), wobei dem seelischen Faktor ein mehr oder minder großer Anteil in der Ätiopathogenese zukommt. Wir sprechen bewußt nur von „psychosomatischen Aspekten" in der Urologie, weil empirisch betrachtet das Ausmaß des seelischen Anteils bei urologischen Erkrankungen vielfach noch unbekannt ist. Historisch gesehen fällt auf, daß zu Beginn dieses Jahrhunderts mehr Interesse und Wissen um psychosomatisch bedingte Erkrankungen des Urogenitaltrakts bestand als heute; z.B. war der Herausgeber des ersten umfassenden Lehrbuchs der psychosomatischen Medizin, Oswald Schwarz (1925), ein Urologe. Sein damaliges Übersichtsreferat „Die psychogenen Miktionsstörungen" ist heute noch lesenswert.

56.2 Psychosomatisch bedingte Krankheiten des Urogenitaltrakts des Mannes

56.2.1 Ein exemplarischer Fall aus der Praxis

Ein 37jähriger Perser, der seit 12 Jahren in der Bundesrepublik lebt, kommt im Oktober 1981 erstmals in die Sprechstunde. Als Beschwerden werden angegeben: Brennen in der Harnröhre sowie neben Brennen auch ein Druckgefühl im Damm und im Analbereich. „Ich kann da unten nicht loslassen." Weiterhin besteht ein vermehrter Harndrang. Der Patient war drei Jahre wegen dieser Beschwerden in der Behandlung

eines Urologen, der eine Prostatitis feststellte. Nach Durchführung der großen urologischen Diagnostik (Ausscheidungsurogramm, Zystourethrographie, Zystourethroskopie, Blasendruckmessung, große Bakteriologie sowie Blutuntersuchungen) wurden Prostatamassagen und Harnröhreninstillationen durchgeführt („Ich kriege Angst und Schmerzen, wenn ich daran denke") sowie orale und parenterale antibakterielle Medikamente verabreicht und über lange Zeit auch Psychopharmaka verordnet. Eine Besserung der Beschwerden trat zu keinem Zeitpunkt ein. Ein einmal konsultierter Ordinarius für Urologie sagte nach eingehender Untersuchung, es sei alles psychisch. Wegen der vielen und hohen Arztrechnungen mit der Diagnose „chronische Prostatitis" muß der Patient einen Risikozuschlag an seine Privatkrankenversicherung zahlen. Die biographische Anamnese ergibt, daß der Patient als drittes Kind (2 Schwestern, 6 und 8 Jahre älter) geboren wurde und bis zu seinem 12. Lebensjahr in einem Dorf aufgewachsen ist. Der Vater mußte vor der Geburt des Patienten aus politischen Gründen ins Ausland fliehen. Der Patient hat den Vater dadurch nie kennengelernt. Die erziehenden Bezugspersonen waren die Mutter und die beiden Schwestern. Im Alter von 13 Jahren ist der Patient mit der Familie nach Teheran gezogen. Dort machte er eine Lehre als Teppichknüpfer. Der Lehrherr sei in ihn „verliebt" gewesen. Gezielte, jedoch behutsame Fragen, ob es zwischen dem Lehrherrn und ihm zu sexuellen Kontakten gekommen sei, werden ausdrücklich verneint. Der Patient steht heute noch unter dem Eindruck seiner streng religiösen islamischen Erziehung. Da die mitgebrachten Untersuchungsbefunde keine Abweichung von der Norm aufweisen, wird ein eingehendes therapeutisches Gespräch geführt. Die Mitteilung, daß es sich bei ihm nicht um eine organische Krankheit und schon gar nicht um ein chronisches Leiden handelt, bringt eine deutliche Entlastung, was auch im Hier und Jetzt zu einer spürbaren körperlichen Entspannung führt (siehe: der diagnostisch-therapeutische Zirkel, Kap. 1). Da bei der rektalen Untersuchung ein erhöhter Sphinktertonus auffällt, werden dem Patienten seine Beschwerden mit dem „Nicht-Loslassenkönnen" und den dadurch ausgelösten Verspannungen im Bereich des kleinen Beckens erklärt. Unter zusätzlicher Anwendung gezielter Wärme mittels heißer Sitzbäder sowie Ichthyol-Suppositorien verschwinden die Symptome nach etwa 10 Tagen völlig. In einem vom Patienten veranlaßten Brief an die Versicherung wird mitgeteilt, daß die bisher gestellten „somatischen Diagnosen" aufgrund der Untersuchungsergebnisse, auch der vorher konsultierten Ärzte, widerlegt sind, daß es sich hier um ein funktionelles Syndrom im Sinne einer Prostatopathie handelt und im Hinblick auf die Prostata kein Versicherungsrisiko besteht.

Zwei Jahre nach seinem ersten Besuch erscheint der Patient erneut in der Sprechstunde mit der Bitte um eine eingehende Kontrolluntersuchung. Die ursprünglichen Beschwerden sind nicht wieder aufgetreten. „Ich kann zwar immer noch nicht da unten loslassen, aber ein heißes Sitzbad hilft sofort." Grund für die jetzt gewünschte Kontrolluntersuchung ist die Forderung der Versicherungsgesellschaft von 1981. Falls innerhalb der nächsten 2 Jahre keine Behandlungen der Prostata notwendig werden und ein erneutes fachärztliches Gutachten einen pathologischen Befund der Prostata ausschließt, kann der Risikozuschlag erlassen werden. Die digitale Untersuchung der Prostata ergibt eine nicht schmerzhafte, dem Alter entsprechende Drüse. Im Prostataexprimat sowie im Ejakulat lassen sich keine pathologischen Keime nachweisen.

Der Patient gibt jetzt unaufgefordert an, daß er viel über das therapeutische Gespräch vor 2 Jahren nachgedacht habe. Tatsächlich hat er im Alter zwischen 4 und 12 Jahren mehrmals analen Verkehr gehabt, „aber das waren nur Kinderspielereien". Sein Lehrherr habe ihn jedoch massiv homosexuell belästigt, ihm immer wieder Arbeiten aufgetragen, die ihn in seine Privatwohnung führten. „Er wollte mich anal gebrauchen und vorn mit mir spielen. Es war gegen meinen Willen, er hat meine Ehre verletzt. Ich mußte immer zuzwikken, deshalb kann ich auch heute noch nicht loslassen."

Kommentar:

Im ersten therapeutischen Gespräch konnte der Patient unter Hinweis auf die physiologische Funktion der Prostata und die topographische Situation im kleinen Becken überzeugt werden, daß seine Symptome nicht durch eine krankhafte Organveränderung der Prostata, sondern durch die biographisch bedingten, nunmehr einsehbaren Verspannungen („Ich kann da unten nicht loslassen") ausgelöst wurden. Das zweite therapeutische Gespräch erbrachte wichtige biographische Einzelheiten, da der Patient in der Zwischenzeit seine Konfliktabwehr gegenüber der damaligen homosexuellen Verführung weitgehend aufgeben konnte. Darüber hinaus macht der Fall deutlich, daß dem Patienten durch den vorbehandelnden Urologen, der für die psychosomatischen Zusammenhänge keinen Blick hatte, eine Organerkrankung (Prostatitis) aufgedrängt wurde, die er nie hatte.

56.2.2 Die „Prostatitis" aus psychosomatischer Sicht

Bevor auf das vielschichtige und zugleich häufigste Krankheitsbild des Mannes in der urologischen Sprechstunde eingegangen wird, ist es notwendig, sich mit der mißverständlichen und irreführenden Diagnosebezeichnung „Prostatitis" auseinanderzusetzen. Klinisch muß zwischen der akuten bakteriellen Prostatitis und der chronischen Prostatitis unterschieden werden. Während die akute bakterielle Prostatitis aufgrund des dramatischen, meist hochfieberhaften Verlaufs, der typischen Symptome, der eindeutigen Organ- und Laborbefunde sowie des raschen Ansprechens auf gezielte antibakterielle Behandlung weder diagnostische noch therapeutische Probleme aufwirft, herrscht bei der sog. „chronischen Prostatitis" weltweit Uneinigkeit, nicht nur im Hinblick auf eine übereinstimmende und zutreffende nosologische Einordnung sowie auf eine definitive Diagnosebezeichnung, sondern auch im Hinblick auf den entsprechenden und übereinstimmenden therapeutischen Ansatz. So finden sich in der Fachliteratur zur sog. „chronischen Prostatitis" Begriffe wie „chronische bakterielle Prostatitis", „chronische abakterielle Prostatitis", „chronische unspezifische Prostatitis",

„Kongestionsprostatitis", „prostatisches Syndrom", „Prostatopathie", „Prostatodynie" und „vegetatives Urogenitalsyndrom". Alle diese Diagnosebezeichnungen stehen für weitgehend gleichartige Symptome: Druckgefühl im Dammbereich, Schmerzen in der Leistengegend, die bis in die Hoden ausstrahlen können, vermehrter Harndrang, Brennen oder Jucken in der distalen Harnröhre während der Miktion und auch unabhängig von ihr, Startverzögerung beim Wasserlassen, Kreuz- und Rückenschmerzen sowie Druckschmerz im Bereich des Schambeins („Blasenschmerzen"). Die Urologische Universitätsklinik Gießen hat von 1976–1979 eine umfassende Studie über die chronische Prostatitis durchgeführt (Weidner et al., 1980). Dabei wurden 267 Männer in einer eigens eingerichteten Prostatitis-Sprechstunde untersucht und über eine Mindestdauer von zwei Jahren beobachtet. Spätere Untersuchungen der gleichen Arbeitsgruppe zeigten entsprechend gleichlautende Ergebnisse (Weidner, 1984). Aus psychosomatischer Sicht ist bemerkenswert, daß von allen untersuchten Männern, die eines oder mehrere der oben aufgeführten Symptome hatten, bei 149 (56%) ein entzündliches oder bakterielles Geschehen in der Prostata ausgeschlossen werden konnte. Die Befunde in dieser Gruppe wurden von den Untersuchern zunächst als „vegetatives Urogenitalsyndrom", später als Prostatodynie (Junk-Overbeck et al., 1988; Pott et al., 1988) eingestuft. Berücksichtigt man, daß unter den verbleibenden „Prostatitis"-Patienten nur bei 9,4% ein echtes bakterielles Geschehen in der Prostata nachgewiesen werden konnte und bei 34,5% eine „abakterielle Prostatitis" (Nachweis von Chlamydien, Mykoplasmen oder ausschließlich Nachweis von Leukozyten) diagnostiziert wurde, gilt als bewiesen, daß bei weit mehr als der Hälfte der Patienten mit den Symptomen einer sog. „Prostatitis" ein entsprechendes entzündliches oder bakterielles Geschehen in der Prostata ausgeschlossen werden kann.

Seltene oder unregelmäßige Ejakulationen, Unterkühlung und Bewegungsarmut (vgl. Vahlensieck, 1985, 1988) können zur Prostatakongestion und damit zu Symptomen führen (Sexualanamnese!). Dabei kann es zu einer Erhöhung der Zellzahl (Leukozyten) im Prostataexprimat kommen, woraus bisher ein Hinweis auf entzündliches Geschehen in der Prostata abgeleitet wurde. Nach Weidner (1984) gilt die Wertigkeit dieser Befunde als umstritten. Eigene Beobachtungen zeigen, daß sich die Zellzahl des Prostataexprimats nach regelmäßiger Ejakulation im Verlauf von drei bis sechs Wochen normalisiert. Sieht man von der Prostatakongestion ab, muß die Frage gestellt werden, inwieweit die Prostata allein an der Entstehung der als „Prostatitis" bezeichneten Symptome beteiligt ist. Tatsächlich wird in Klinik und Praxis das oben beschriebene Symptombild, ungeachtet des oft fehlenden Organbefundes, noch viel zu oft als feststehende Diagnose „Prostatitis" vermittelt, ohne dem Patienten diese Diagnosebezeichnung näher zu erklären (s. exemplarischer Fall). Obwohl die Gießener Untersuchergruppen die Bezeichnung Prostatodynie (diese Diagnosebezeichnung kommt aus dem englisch-amerikanischen Sprachgebrauch) bevorzugen, bleiben wir bewußt bei der Diagnosebezeichnung „Prostatopathie", da diese dem psychosomatischen Charakter des Krankheitsbildes besser entspricht.

Die Prostatopathie ist die typische psychosomatisch bedingte urologische Erkrankung des Mannes (s. exemplarischer Fall). Transkulturell gesehen ist dieses Krankheitsbild weit verbreitet. Da Christoffel (1944) den ersten Fall in der „Weltliteratur" schon 1767 ausmachte – Rousseau beschrieb in seinen „Bekenntnissen" an sich die Symptome einer Prostatopathie – muß man annehmen, daß dieses Krankheitsbild über Jahrhunderte hinweg stabil geblieben ist. Aus psychosomatischer Sicht sind methodisch etwas anspruchsvollere Untersuchungen erst in letzter Zeit durchgeführt worden (Diederichs, 1983; Janssen et al., 1983; Riedell und Brähler, 1983). Psychometrische Befunde liegen von belgischen (Mendlewicz et al., 1971) und skandinavischen Autoren (Keltikangas-Järvinen et al., 1981; Nilsson et al., 1975) vor. Die gewonnenen Ergebnisse über Persönlichkeitsstruktur, Psychopathologie und Psychodynamik von Patienten mit Prostatopathie sind zum Teil unterschiedlich. Den Arbeiten gemeinsam sind Hinweise auf eine zentrale Selbstwertregulationsstörung und mangelnde positive Identifikation mit Vaterbildern (s. exemplarischer Fall). Janssen und Mitarbeiter (1983) haben ihre Patienten nach triebtheoretischen Kategorien diagnostiziert, wobei sie häufig eine psychische Störung auf zwangsneurotischem Niveau fanden. Auch wir beobachteten ein Überwiegen zwangsneurotischer Persönlichkeitsstrukturen (Diederichs, 1983).

Vor diesem psychodynamischen Hintergrund gewinnen Beobachtungen von Sinaki et al. (1977) aus der physiotherapeutischen Abteilung der Mayo-Clinic an Bedeutung, die als Ursache für das Zustandekommen der oben beschriebenen Symptome eine Myalgie auf der Basis einer Verspannung der Beckenmuskulatur, insbesondere der des Beckenbodens, annehmen. Entsprechende Beobachtungen machten Segura et al. (1979). Diener (1981) wies auf die Beckenkongestionierung zur Differentialdiagnose der Prostatitis hin (vgl. O'Shaughnesse und Parrino, 1956). Wilhelm (1985) untersuchte an der Urologischen Universitätsklinik Erlangen zwischen 1981 und 1982 97 Patienten mit den Symptomen einer sog. „Prostatitis" und konnte nur bei fünf Männern entzündliches bzw. bakterielles Geschehen in der Prostata nachweisen. Bei den verbleibenden 92 Patienten (89,2%) wurden die Beschwerden auf eine Beckenbodenmyalgie zurückgeführt. Die Diagnose basiert auf dem digitalen Rektalbefund. Dabei konnte eine vermehrte Druckdolenz der Prostata und des Levator ani auf der Seite der Beschwerden nachgewiesen werden. Diese Untersuchungen erklären die häufig beobachtete Einseitigkeit der „Prostatitis"-Beschwerden (vgl. McGivney et al., 1965). Wilhelm macht statische Ursachen für das Entstehen der Beckenbodenmyalgie verantwortlich. Aus psychosomatischer Sicht möchten wir diese Betrachtungsweise ergänzen: Die zwangsneurotische Persönlichkeitsstruktur geht entwicklungspsychologisch auf eine Störung in der analen und urethralen Phase, also der Zeit der Sauberkeitserziehung (Sphinkterkontrolle), zurück. Im Zentrum dieser Entwicklungsphase steht u.a. der Ambivalenzkonflikt von „Zurückhaltenwollen" und „Hergebenmüssen". Durch zu frühe und rigide Sauberkeitserziehung kann es im Enddarm-Becken-Bereich zu Störungen des biologischen Rhythmus von Festhalten und Loslassen kommen, die zu Verspannungen in dieser Region führen können. Erfahrungsgemäß leiden zwangsneurotisch strukturierte Menschen häufig an Krankheiten im unteren Darmabschnitt und am Muskel-Skelett-System. Dementsprechend fanden wir bei unseren Patienten mit den Symptomen einer Prostatopathie sehr oft einen erhöhten Analsphinktertonus, der auch vom mitbehandelnden Proktologen bestätigt wurde. Dieser Sphinkterhypertonus könnte nicht nur die bei Prostatopathie-Patienten häufig zu beobachtenden Analbeschwerden (anogenitales Syndrom) erklären, sondern würde auch für die von vielen Prostatopathie-Patienten beschriebene Schwierigkeit, den Urinstrahl zu starten („nicht aufmachen können" – „nicht loslassen können") verantwortlich zu machen sein.

Das Modell der myalgischen Schmerzen im Beckenbereich durch Verspannung des Muskelsystems infolge Retentivität läßt die bei über 50% der Prostatopathie-Patienten beobachteten funktionellen Sexualstörungen (Verlust der Libido, Erektionsstörungen, Schmerzen bei der Ejakulation, Ejaculatio praecox und retarda oder Anorgasmie) nun auch als Sekundärgeschehen verstehen. So ist z.B. der Ejakulationsschmerz das organische Korrelat der Angst, sich mit der Partnerin im Orgasmus vertrauensvoll zu „verschmelzen". Ergänzend sind auch eigene Beobachtungen (Günthert, 1986) an Homosexuellen zu erwähnen, die oft unter den Symptomen einer Prostatopathie leiden. Die Masturbationstechnik der Homosexuellen läßt die Lustphysiologie häufig nicht zur Entspannung kommen und kann auf diese Weise zu Verspannungen der Beckenmuskulatur, besonders des Beckenbodens, führen und so die Symptome einer Prostatopathie auslösen. Aus psychologischer Sicht muß dabei auch an unbewußt schuldhaft erlebte Homosexualität gedacht werden, die dann bei der Symptomentstehung verstärkend wirken kann.

Die qualitative Auswertung der Persönlichkeitsstruktur bei Männern mit den Symptomen einer Prostatopathie nach ichpsychologischen Aspekten zeigt jedoch, daß hinter der Zwangsstruktur große Unterschiede im Ich-Niveau, in der Stabilität des Selbst und der Qualität der zwischenmenschlichen Beziehungen zu finden sind. So ergaben sich bei zwei Drittel der von uns untersuchten Prostatopathie-Patienten Hinweise auf eine gröbere Störung des Ich oder Selbst sowie der Objektbeziehungen (ichstrukturelle Störungen). Die häufig beobachteten hypochondrischen Befürchtungen können das Ausmaß einer überwertigen Idee annehmen. Dabei stellt die Hypochondrie den Versuch des Patienten dar, die Kontrolle über die Gesamtheit seines Körper-Selbst zu behalten, indem die Aufmerksamkeit auf diejenigen Bereiche konzentriert wird, die sich ihm entfremdet haben (Kohut, 1973), beim Prostatopathie-Patienten der Urogenitaltrakt. Die Hypochondrie ist die Signalangst für eine Fragmentierungsgefahr des Selbst. Die Berücksichtigung narzißmustheoretischer bzw. selbstpsychologischer Aspekte bei psychosomatisch bedingten urologischen Erkrankungen (Diederichs, 1987) führt auch zu einem vertieften Verständnis der Psychodynamik der auslösenden Konfliktsituation bei Männern mit Prostatopathie. Sie liegt nach unserer Erfahrung im Geltungsbereich. Die narzißtische Seite des Urethralen und des Phallus prädestiniert diese Patienten, ihr schon primär labiles Körper-Selbst durch eine Kränkung offenbar werden zu lassen. So überrascht es nicht, daß viele Männer mit Beschwerden oder Erkrankungen im Urogenitaltrakt das Gefühl haben, im „Zentrum" getroffen zu sein (Grosch, 1958). In diesem Zusammenhang wird auch der Altersgipfel von 40 Jahren für Patienten mit den Beschwerden einer Prostatopathie verständlich. Die Schwellensituation besteht wohl darin, daß sich viele Männer in diesem Alter an der Grenze ihrer beruflichen und sexuellen Leistungsfähigkeit glauben, was auch ihr Körpererleben verändert.

Der Umgang mit dem Prostatopathie-Patienten, der sich häufig als unattraktiver, klagsamer Patient mit der Chronizität und der Therapieresistenz seines Leidens präsentiert, kann beim Arzt zu negativen Gegenübertragungsgefühlen führen, da er die Grenzen seines medizinischen Wissens und Könnens aufzeigt. Ungerechtfertigte Maßnahmen wie die Verordnung von antibakteriellen Medikamenten bei fehlendem Bakteriennachweis müssen strikt vermieden werden, um den Patienten nicht auf eine Organkrankheit zu fixieren, die er gar nicht hat. Auch ist es falsch, den Patienten mit der Bemerkung, seine Symptome seien „nur psychisch", allein zu lassen (s. exemplarischer Fall). Das bei vielen Prostatopathie-Patienten beobachtete häufige Wechseln des Urologen kann als ein Zeichen verstanden werden, daß sich der Patient nicht angenommen, ja vielleicht sogar geschädigt fühlt. Darüber hinaus kann der fehlgeleitete Patient den Arzt in die „Schädiger-Rolle" drängen (Janssen et al., 1983), indem sein neurotischer Wiederholungszwang ihn veranlaßt, den Urologen zu immer neuen diagnostischen und therapeutischen Aktivitäten zu provozieren. Andererseits – wenn auch deutlich seltener – können Patienten durch ungerechtfertigte Maßnahmen (z. B. Prostatamassagen) in ein Abhängigkeitsverhältnis zu ihrem Urologen geraten.

Aus eigener Erfahrung (Günthert, 1983) ist es wichtig, dem Patienten die pathophysiologischen und psychosomatischen Zusammenhänge in gleicher Weise verständlich zu machen (s. exemplarischer Fall). Erst dann ist es dem Patienten möglich, sich selbst als einen wichtigen Teil seines Krankheitsgeschehens verstehen zu lernen und die angebotenen flankierenden Maßnahmen (lokale Anwendung von Wärme; Ichthyol-Suppositorien) als einen sinnvollen Behandlungsvorschlag zur Verspannungslösung und Förderung der Durchblutung anzunehmen. Wenn es ihm darüber hinaus gelingt, die im therapeutischen Gespräch erarbeiteten psychosomatischen Zusammenhänge anzunehmen und zu verarbeiten, können die geklagten Beschwerden in den Hintergrund treten oder ganz überflüssig werden und damit verschwinden, wie der exemplarische Fall zeigt.

Ebenfalls aus eigener Erfahrung (Günthert, 1980, 1986) kann man auch bei der Mehrzahl der Patienten mit einer „chronischen bakteriellen Prostatitis" – Bakteriennachweis aus dem Prostataexprimat und dem Ejakulat – ohne antibakterielle Therapie auskommen. Diese Erfahrung findet ihre Bestätigung in der Tatsache, daß pathogene Erreger in pathogener Keimzahl auch bei Männern zu finden sind, die keinerlei Prostatabeschwerden haben (Fertilitätsuntersuchungen). Um so mehr muß die Frage nach der Kausalität des Nachweises von bakteriellem oder entzündlichem Geschehen in der Prostata gestellt werden. Tatsächlich sprechen diese Patienten ebenso wie die Patienten mit einer Prostatopathie in gleicher Weise auf das oben beschriebene Therapiekonzept an. Dies bestätigen auch die Befunde von Janssen et al. (1983) und Junk-Overbeck et al. (1988), die keine wesentlichen psychopathologischen Unterschiede zwischen Patienten mit Prostatodynie, Prostatopathie und Patienten mit einer Prostatitis gefunden haben, d.h., daß auch bei der „chronischen bakteriellen Prostatitis" psychosomatische Faktoren ätiopathogenetisch eine Rolle spielen können.

56.3 Psychosomatisch bedingte Symptome und Erkrankungen der Niere

Daß Affekte wie Zorn, Wut oder Angst die **Nierendurchblutung** beeinflussen können – bis zur Anurie –, ist seit langem bekannt und tierexperimentell schon 1926 von Dobreff nachgewiesen worden. Auch die experimentellen Untersuchungen am Menschen unter Hypnose (Bolland, 1957) zeigen die psychische Beeinflußbarkeit der Nierenfunktion (vgl. Blomstrand und Löfgren, 1956; Schwarz, 1928). In der Sprache drückt sich dieser psychosomatische Zusammenhang ebenfalls aus: „Es geht mir an die Nieren". Nach eigenen Beobachtungen (Günthert, 1986) können solche Affekte auch **polyurische Schübe** auslösen. Dabei werden im Zeitraum von 1 bis 4 Stunden große Mengen eines wasserklaren, hochverdünnten Urins ausgeschieden. Frauen sind in der eigenen Patientenklientel häufiger betroffen als Männer. „Ich weiß nicht, wo der viele Urin herkommt, ich habe doch gar nicht soviel getrunken. Ich nehme 1 bis 2 Kilo in wenigen Stunden ab" sind häufige Beschreibungen von Betroffenen. Während nephrologische Untersuchungen und Erklärungen dieses Phänomens bisher fehlen, spricht Hoff (1950) von „zentralnervös ausgelösten Polyurien". Neben Klessmann (1987), die einen Fall von psychogener Polyurie bei einer jungen Frau vor dem Hintergrund einer Konversionsneurose darstellt, beschreibt Benedek (1985) eine Steigerung der Diurese besonders bei Angstzuständen. Sie weist auch darauf hin, daß Polyurie von sich aus Angst auslösen kann in bezug auf die Beherrschung der Blasenfunktion, d.h. die Furcht „zu spät zu kommen", und auf diese Weise zu häufigem, vorsorglichem Wasserlassen führt. Unabhängig davon kann allein schon die rasche Füllung der Blase irritative Symptome auslösen: „**paradoxe Pollakisurie**". Dies erklärt, weshalb viel zu oft vom Arzt irrtümlich eine Blasenproblematik angenommen und dann therapeutisch am falschen Organ angesetzt wird. Die Beobachtung von Kleinsorge (1952), daß bei neurotischen Patienten vorwiegend ein **alkalischer Urin** als Folge einer erhöhten Phosphatausscheidung zu finden ist, wird in der eigenen Patientenklientel bestätigt.

Wichtige Hinweise auf die Beeinflußbarkeit der Nierenfunktion durch psychische Faktoren bietet die jüngere **Nierensteinforschung**. Nach Schneider (1986) stehen im Vordergrund der weltweiten Ursachenforschung für die Entstehung von Nierensteinen die chemischen Parameter. Dabei geht man davon aus, daß sich die Steinbildung in Krisen abspielt, in denen es zu Veränderungen der Konzentrations- und Lösungsverhältnisse des Urins kommt. Als eine der Ursachen solcher Steinkrisen wird nach Schneider (1973) „Streß" diskutiert. Streß führt u.a. zu einer erhöhten Ausschüttung der Nebennierenmarkhormone und damit über eine Beeinflussung der Nierenfunktion zur Veränderung der Harnzusammensetzung im Sinne eines Steinbildungsrisikos (Deetjen, 1979; Brundig et al., 1985). Auch Selye (1956) verweist schon in seinen frühen Streßuntersuchungen auf diesen Vorgang und führt den direkten Zusammenhang von Streßreaktion und Nephrokalzinose auf diesen Mechanismus zurück.

Epidemiologische Untersuchungen deuten darauf hin, daß Berufsgruppen mit vermehrtem Streßrisiko häufiger zur Steinbildung neigen (Schneider, 1973). Fliegendes Personal hat sehr viel häufiger Nierensteine als Bodenpersonal (Schmucki und Asper, 1977). Ebenso leiden Akademiker häufiger unter Nierensteinen als Handwerker (Joost et al., 1980). Schmucki et al. (1979) konnten in Tierversuchen an Ratten Veränderungen der Konzentration lithogener Substanzen unter Streßeinwirkungen nachweisen. Brundig et al. (1979, 1981) fanden Kalzium-Oxalat-Konzentrationserhöhungen bei Examenskandidaten und verglichen dabei die Werte von Steinpatienten und Gesunden. Eine vermehrte Kalziumausscheidung zeigten nur die Steinpatienten unter Streß, die Oxalatkonzentration war auch bei Gesunden unter dem Examensstreß erhöht, jedoch bei weitem nicht in dem Maße wie bei den Steinpatienten. Aus psychosomatischer Sicht galt bisher das besondere Interesse dem Zeitpunkt des Beginns der Steinbildung. Paar (1986) verweist zu Recht im Zusammenhang mit der extrakorporalen Stoßwellenlithotrypsie auf die damit neu gewonnene Möglichkeit, durch komplette Entsteinung den Zeitpunkt Null der Steingenese festzulegen. Damit wird für weitere Untersuchungen über die Bedeutung von Streßfaktoren bei der Steinbildung eine wichtige Grundlage geschaffen.

Systematische Untersuchungen über das Vorkommen von Nierenschmerzen und Nierenkoliken feh-

len, obgleich Schwarz schon 1928 auf psychische Einflüsse hingewiesen hat. Bei Patienten mit einer bekannten Solitärniere kann man häufig ein unbewußtes Schutzbedürfnis der Einzelniere beobachten, das im Sinne einer Schutzhaltung zu Verspannungen der nierenumgebenden Muskulatur und damit zu „Nierenschmerzen" führen kann.

56.4 Psychosomatisch bedingte Symptome und Erkrankungen der Blase

Für die Blaseninnervation (Festge, 1980) ist hervorzuheben, daß adrenerge Nervenendigungen im Bereich der gesamten Blase, zahlreicher aber am Blasenhals und im hinteren Harnröhrenbereich zu finden sind (Palmtag, 1981). Niedrige Konzentrationen von Adrenalin und Noradrenalin öffnen, hohe Dosen dieser Sympathikomimetika schließen dagegen den Blasenhals (Festge, 1980). Weiterhin werden Östrogeneinwirkungen beobachtet. Sie sollen die Ausflußbahn verengen. Gesichert ist ihr Einfluß auf die Schleimhäute von Harnröhre und Blase. Typische psychosomatisch bedingte Erkrankungen der Blase sind die sog. Reizblase und die chronisch rezidivierende Urethrozystitis. Beide Krankheitsbilder betreffen vorwiegend Frauen. Entsprechend ergeben sich hier Überschneidungen mit dem Praxisfeld des Gynäkologen.

Diederichs (in Vorbereitung) geht auf die Harnverhaltung, die Harninkontinenz (urge incontinence) und das Bettnässen ausführlicher ein, die ebenfalls zu den psychosomatisch bedingten Erkrankungen der Blase gehören.

56.4.1 Die Reizblase der Frau

Die sog. Reizblase (Synonyma: irritable bladder, urethral syndrome, vesical neuralgia oder cystalgies à urines claires), bei der ständiger Harndrang, Pollakisurie, gelegentlich auch Dysurie im Vordergrund stehen, gehört zu den häufigsten psychosomatisch bedingten urologischen Krankheitsbildern der Frau und gilt als funktionelles Syndrom, da typischerweise ein korrelierender organpathologischer Befund fehlt. Als pathophysiologische Grundlage der sog. Reizblase ist aus psychosomatischer Sicht eine durch Affekte (Enttäuschungswut und Angst) und Sexualstörungen ausgelöste Dyssynergie im gesamten Beckenbereich, besonders aber des Blasensphinkters und des M. detrusor zu erwägen.

Eine 35jährige, nicht verheiratete, beruflich erfolgreiche Frau wurde wegen häufigen Wasserlassens und quälender krampfartiger Schmerzen in der Blase in die Psychosomatische Ambulanz überwiesen. Sie war bis zu diesem Zeitpunkt nie ernsthaft krank gewesen. Anfangs traten diese intensiven Beschwerden nur während und nach der Miktion, zum Zeitpunkt der Überweisung auch unabhängig vom Wasserlassen auf. Sie beschrieb die Schmerzen: „... als ob mir jemand Säure in die Blase gekippt hätte; es brennt den ganzen Tag; egal was ich mache, ob ich ruhig bin oder mich mit Sockenstricken entspanne, ich bekomme mich nicht mehr in den Griff". Die Symptomatik beeinträchtigte ihr Allgemeinbefinden so stark – u.a. führte sie zu Schlafstörungen –, daß sie seit 2 Monaten krankgeschrieben war. Die zu diesem Zeitpunkt verzweifelt und resigniert wirkende Frau hatte die verschiedensten Fachärzte aufgesucht (Gynäkologen, Urologen, Dermatologen und Neurologen). Während der Weihnachtsfeiertage begab sie sich sogar in stationäre urologische Behandlung, ohne daß ein organpathologischer Befund erhoben werden konnte. Sowohl Bougieren der Harnröhre als auch die massive Gabe von Antibiotika und muskelrelaxierenden urologischen Medikamenten halfen ihr nicht weiter. Die erste psychologische Exploration ergab, daß die urologischen Symptome in zeitlichem Zusammenhang standen mit dem vor Monaten unternommenen halbherzigen Versuch, sich von ihrem langjährigen Partner zu trennen. Es handelt sich bei diesem um einen 8 Jahre älteren, geschiedenen, relativ erfolgreichen Geschäftsmann. Ihr Trennungswunsch war die Reaktion auf eine massive narzißtische Kränkung durch den Freund. Dieser hatte noch eine Beziehung zu seiner Sekretärin aufgenommen und konnte sich trotz des Drängens der Patientin nicht zwischen beiden Frauen entscheiden. Aus ihrer Kindheitsgeschichte ist eine enge Mutter-Bindung und ein idealisiertes Vater-Bild hervorzuheben. Der frühe Tod des Vaters (Patientin zweijährig) hat die Idealisierung begünstigt. In den urologischen Symptomen manifestierte sich zum einen ihre Enttäuschung (s. die aggressive Seite des Urethralen) und zum anderen ihre Hingabeproblematik bzw. ihre Schwierigkeit, sich vertrauensvoll „verströmen" zu lassen. Letztlich hatte die Patientin die früheren Angebote des Partners, ihn zu heiraten, abgeschlagen. Die Symptomatik besserte sich während der stationären Psychotherapie jeweils dann, wenn sie sowohl ihre Liebe und Bewunderung als auch ihren Haß auf diesen Mann erleben und verbalisieren konnte, insbesondere dann, wenn sie aus vollem Herzen über den Verlust des Partners weinte und sich Trost holte.

In einer eigenen psychometrischen und psychoanalytischen Untersuchung an 42 Frauen mit den Symptomen einer sog. Reizblase (Diederichs, 1983) fanden wir in allen Fällen eine Hingabeproblematik, wobei die Hingabe mit der Angst vor Selbstaufgabe einhergeht. Eine Patientin drückte das bei der Beschreibung ihrer Beschwerden folgendermaßen aus: „Ich ziehe dauernd da unten zusammen; es läuft ständig, aber ich habe das Gefühl, daß ich etwas nicht laufenlassen kann." Die Psychodynamik der Hingabeangst erklärt auch besser den von Molinski (1983) beschriebenen pathogenetischen Zusammenhang von Sexualstörung (Anorgasmie) und urologischen Symptomen: Vor dem Orgasmus kommt es bei der Frau zu einer Vasokongestion des kleinen Beckens, die auch den urethralen, den periurethralen und perivesikalen Bereich betrifft. Gleichzeitig entwickelt sich ein zunehmender Muskeltonus, der ebenfalls den gesamten urethrogenitalen Bereich erfaßt. Bei der Anorgasmie bleibt das Abklingen der Lustphysiologie aus, d.h. die Lösung des erhöhten Muskeltonus und die Rückbildung der Blutanreicherung finden nicht statt. Der Erregungszustand löst sich nur langsam und kann als genitale und urethrale Irritation im Sinne der Symptome einer Reizblase über mehrere Tage anhalten.

Natürlich stellen Frauen mit den Symptomen der sog. Reizblase – analog zu anderen psychosomatisch bedingten Erkrankungen – keine homogene Gruppe hinsichtlich Charakterstruktur, Psychodynamik und Ätiopathogenese dar, es deuten sich vielmehr drei Untergruppen an:

– Bei der einen kann die Reizblasensymptomatik wie oben beschrieben – als Korrelat einer verdeckten Sexual- bzw. Hingabestörung verstanden werden. Durch eine sorgfältige Sexualanamnese und genaue Beobachtung des zeitlichen Zusammenhangs beim Auftreten urologischer Symptome läßt sich dieser pathogenetische Mechanismus sehr oft rekonstruieren (Günthert, 1984).
– Bei einer anderen Untergruppe von Frauen mit den Symptomen der sog. Reizblase besteht dagegen der klinische Eindruck, daß diese Miktionsstörung sozusagen ein Ersatzsymptom für eine Agoraphobie ist, also ein Angstäquivalent, z.B. vor einer expansiven Versuchungssituation, darstellt. Diese Frauen berichten, daß sie ihre Wohnung nur noch in der Richtung verlassen können, in der sie eine Toilette finden. Ihr charakteristischer Ausspruch lautet: „Ich kenne alle Toiletten dieser Stadt".
– Unsere klinische Erfahrung (Diederichs, 1983) geht dahin, daß bei vielen Frauen mit dem urologischen Leitsymptom Reizblase die durch eine narzißtische Kränkung, insbesondere im Beziehungsbereich entstandene Enttäuschungswut abgewehrt wird (s. exemplarischer Fall).

Die Diagnosebezeichnung „Reizblase" weist schon auf abgewehrtes „Aggressives" hin. Hierzu paßt, daß Frauen mit Reizblasensymptomatik häufig über Begleitsymptome mit Spannungscharakter (migräneartige Kopfschmerzen oder Verspannungen im Schulter-Nacken-Bereich) klagen. In ihrer Partnerbeziehung neigen sie zu einer mehr „kämpferischen Kollusion", die aber vermutlich der Abwehr von regressiver Versuchung oder von Verschmelzungswünschen dient. Schließlich liegt hier auch eine Hingabestörung vor. Daraus wird ersichtlich, daß sich die Patientinnen in ihrer Psychodynamik nicht immer eindeutig abgrenzen lassen. Darüber hinaus findet sich die sog. Reizblase gelegentlich auch als psychosomatisches Begleitsymptom bei agitierten Depressionen.

56.4.2 Die Urethrozystitis (Blasenentzündung) der Frau

Bei dem häufigsten Krankheitsbild der Frau in der urologischen Sprechstunde handelt es sich um ein organisches Krankheitsgeschehen. Dabei stehen Charakteristika wie meist schlagartiger Beginn der Beschwerden, unerträglicher Harndrang, massive Dysurie, sehr oft auch makroskopische Hämaturie im Vordergrund. Es handelt sich um ein akutes, schweres, die Patientin massiv belastendes Krankheitsbild, das erfahrungsgemäß sehr rasch auf gezielte antibakterielle Behandlung anspricht und auf diese Weise Symptome und Organbefund der Urethrozystitis (entzündliche Veränderung der Blasenschleimhaut, Leukozyturie, Erythrozyturie und Bakteriurie) verschwinden läßt. Die signifikante Rezidivneigung bei vielen Frauen legt allerdings auch psychosomatische Zusammenhänge nahe. Unabhängig von dem von allen Untersuchern angenommenen Infektweg: Darmausgang – Damm – Scheide – Harnröhre – Blase werden verschiedene Auslöser, z.B. Unterkühlung diskutiert. Da viele Frauen den Beginn ihrer Zystitisbeschwerden in zeitlichem Zusammenhang mit Geschlechtsverkehr beschreiben (nach Kilmartin, 1982, bis zu 36 Stunden post coitum), vertreten Urologen und Gynäkologen (Hirsch, 1976; Kunin, 1978) die Theorie, daß es durch den Geschlechtsverkehr zu Mikrotraumen im Urethra-Trigonum-Bereich kommen kann, die dann das Entstehen einer Entzündung begünstigen. Dieser organische Erklärungsversuch ist aus psychosomatischer Sicht ergänzungsbedürftig, da nach unserer Erfahrung die Blasenentzündung – ähnlich wie andere Symptome oder Erkrankungen des Urogenitaltrakts – mit Konflikten im Beziehungsbereich einhergeht, z.B. konnte eine sich in psychotherapeutischer Behandlung (Diederichs, 1986) befindende Studentin angeben, daß die erste Blasenentzündung nach dem Entschluß, mit ihrem Partner zusammenzuziehen, aufgetreten ist. Bis zu diesem Zeitpunkt hatte sie eine sehr befriedigende Sexualität. Nach dem Zusammenzug häuften sich die Urethrozystitiden.

Beziehungsstörungen bedingen eine veränderte Lustphysiologie, die sich z.B. im Sinne einer Einschränkung der Orgasmusfähigkeit oder einer Scheidentrockenheit auswirken kann, wobei letztere die Entstehung der Mikrotraumen fördern würde. Wir beobachteten, daß Frauen mit rezidivierender Urethrozystitis (vgl. Illek, 1984) dazu neigen, den zwischenmenschlichen und sexuellen Bereich zu spalten. Während ihre Sexualität mit denjenigen Partnern, die sie lieben und von denen sie fasziniert sind, unbefriedigend bleibt, ist sie unkompliziert mit Männern, bei denen sie weniger emotional engagiert sind. Diese zunächst widersprüchlich wirkende Beobachtung wird verständlicher, wenn man bedenkt, daß bei dem geliebten Partner eine stärkere gefühlsmäßige Abhängigkeit droht. Sich auf der körperlich-sexuellen Ebene einzulassen, bedeutet dann, sich dem anderen hinzugeben und noch weitgehender auszuliefern (Diederichs, 1986).

56.4.3 Eine psychosomatische Theorie der chronisch rezidivierenden Urethrozystitis

Nicht alle Frauen mit einer Beziehungsstörung entwickeln eine Blasenentzündung. Es müssen also disponierende Faktoren von organischer Seite hinzukommen. Stamey (1985) hat in Untersuchungen über einen Zeitraum von 10 Jahren als einzigen biologischen Unterschied zwischen Frauen mit rezidivierender Urethrozystitis und Frauen ohne Urethrozystitis eine unterschiedliche Bakterienadhärenz nachweisen können. Um das Aufwandern von Bakterien aus dem Vulvabereich zu verhindern, empfiehlt er die prophylaktische abendliche Einnahme eines antibakteriellen Medikamentes in niedriger Dosis, ein Therapieangebot, das insgesamt enttäuscht und nicht befriedigen kann. Ebenso wie Stamey betonen auch Huland et al. (1984) im Hinblick auf die Anfälligkeit für die Urethrozystitis, daß hier weder ein anatomisches, urodynamisches noch ein mechanisches, also operativ zu behebendes Problem vorliegt, sondern daß es ein biologisch-immunologisches Geschehen ist. Wir würden hinzufügen, daß es sich um ein biologisch-psychoimmunologisches Problem handelt. Der Abwehrmechanismus der Blasenschleimhaut bzw. der Vaginalschleimhaut kann offenbar durch die oben aufgezeigte Beziehungsproblematik verändert werden, wobei die einzelnen pathophysiologischen intermediären Prozesse noch ungeklärt sind, ebenso das Problem der lokalen Immunschwäche. Zusammenfassend können bei der Pathogenese der rezidivierenden Urethrozystitis der Frau psychosomatische Faktoren eine Rolle spielen. Möglicherweise wirkt neben der organischen Disposition ein bestimmtes retentives Miktionsverhalten im Kindesalter als psychosomatische Fixierungsstelle determinierend. Anders (1984) fand bei einer Reihe von Mädchen mit Harnweginfekten funktionelle Blasenentleerungsstörungen. Als Ursache nimmt er an, daß Schmerzen bei einem ersten Harnweginfekt zur Miktionshemmung und auf diese Weise zur habituellen Retention führen können, die dann ihrerseits Ursache für neue Harnweginfekte sein kann. Darüber hinaus steht nach den Erfahrungen von Anders die Harnretention bei Mädchen über 8 Jahren in enger Beziehung zur Masturbation (vgl. Urethralerotik).

Es können vier Faktoren im Sinne des bio-psychosozialen Modells für die Ätiopathogenese der chronisch rezidivierenden Urethrozystitis herausgearbeitet werden:
- eine organische Disposition (erhöhte Bakterienadhärenz);
- eine frühe psychosomatische Fixierungsstelle (retentives Miktionsverhalten im Kindesalter);
- ein intrapsychischer Konflikt (Hingabeangst, Nähe-Distanz-Problem);
- geänderte gesellschaftliche Bedingungen (z.B. Entpolarisierung der Geschlechterrollen).

Bei Patientinnen mit chronisch rezidivierender Urethrozystitis sollte daher immer neben der notwendigen Organbehandlung eine eingehende Sexualanamnese ebenso wie eine sorgfältige biographische Anamnese im Hinblick auf psychosomatische Zusammenhänge erhoben werden. Die Sexualität dieser Frauen leidet meist erst infolge der wiederholt auftretenden urologischen Beschwerden. Im Sinne eines sekundären Krankheitsgewinnes kann die Blasenentzündung dann auch zur Vermeidung des sexuellen Kontaktes benützt werden. Die gemeinsame Erarbeitung der psychosomatischen Zusammenhänge vermittelt vielen Patientinnen ein besseres Verständnis ihres Krankheitsgeschehens. Die im therapeutischen Gespräch gewonnenen Einsichten ermöglichen ihnen unter Einhaltung aller vorbeugenden Maßnahmen (Sexualhygiene, Wasserlassen vor und nach dem Koitus), besser mit ihrem Kranksein umzugehen und damit der Rezidivneigung entgegenzuwirken (Günthert, 1984).

56.5 Rezidivierende Infektionen der Genitalschleimhäute

Die **unspezifische Urethritis** ist nur beim Mann von klinischer Bedeutung. **Herpes genitalis, Mykosen** und **Kondylome** betreffen beide Geschlechter, wobei jedoch die ätiopathogenetische Beteiligung psychischer Faktoren für die Kondylomatose und die unspezifische Urethritis von besonderem Interesse ist. Das Manifestwerden des Herpes genitalis unter Streß ist bekannt. Kondylome können vor dem Hintergrund einer larvierten Beziehungsstörung gesehen werden (Diederichs, 1988). Die Genitalschleimhäute sind sowohl beim Mann als auch bei der Frau das Kontaktorgan zum Partner, über das Nähe erfahren werden kann. Sie sind Symptomstätte für Beziehungsschwierigkeiten, bzw. erste Signale für „Grenzverletzungen" zwischen den Partnern (vgl. Kap. 55).

56.6 Die erektile Dysfunktion

Bis Anfang der siebziger Jahre galten neben den bis dahin bekannten organischen Auslösern (u. a. Gefäßveränderungen bei Diabetes mellitus, Nikotinabusus und neurogene Störungen) psychische Faktoren als häufigste Ursache der erektilen Dysfunktion. Nach Wershub (1959) sind 90% der Erektionsstörungen auf psychogene Ursachen zurückzuführen. Die 1952 von Goodwin und Scott erstmals vorgestellte Penisprothese blieb trotz technischer Verbesserungen umstritten. Rentrop (1983) warnt vor einer Anwendung dieser chirurgischen Behandlungsmethode bei Männern mit psychogener Erektionsstörung. Mit der Erforschung der Physiologie der Erektion (Lue et al., 1984) begann die euphorische Entwicklung neuer therapeutischer Möglichkeiten der Behandlung der erektilen Dysfunktion.

Der komplexe Vorgang der Erektion wird durch erotische Stimuli über zerebrale Impulse an die spinalen Erektionszentren eingeleitet. Endstation des nervalen Weges (sympathische und parasympathische Fasern) sind die Rezeptoren in den Corpora cavernosa, wobei als Neurotransmitter das vasoaktive Polypeptid VIP und das Prostaglandin E_1 eine wichtige Rolle spielen. Sie führen zur Relaxation der glatten Muskulatur der Schwellkörper und erlauben damit eine Steigerung der arteriellen Zufuhr. Durch nerval-muskuläre Aktivität kommt es dann zur venösen Abflußblockade, die über einen erhöhten intravasalen Druckanstieg im Venenbereich die Aufrechterhaltung einer ausreichend rigiden Erektion garantiert.

Mit seinem Selbstversuch hat Brindley (1983) bewiesen, daß die Injektion vasoaktiver Substanzen in die Schwellkörper zu einer Erektion führt. Heute hat die Schwellkörperautotherapie (SKAT) eine rasche Verbreitung gefunden (Stief et al., 1986). Diese Entwicklung kommt dem Bedürfnis vieler Patienten entgegen, ihre Sexualität wie ein Auto reparieren zu lassen. Dem entspricht auch die Tatsache, daß nach neuesten Veröffentlichungen rein psychogene Ursachen für die erektile Dysfunktion nur noch bei 15% der Patienten angenommen werden (Stief et al., 1987). Im Vergleich zu der Einschätzung von Wershub (1959) haben sich demnach die Ansichten im Hinblick auf die Ursachen der erektilen Dysfunktion ins Gegenteil verkehrt. Diese Zahlen erfordern aus unserer Sicht eine sehr zurückhaltende Beurteilung. Wir sehen uns durch Buvat et al. (1983) bestätigt, die psychotherapeutische Erfolge bei der Behandlung von erektiler Dysfunktion verzeichnen konnten, obwohl vaskuläre Defekte im Schwellkörper-Becken-Bereich vorlagen. Aus psychosomatischer Sicht kann die Erektionsstörung als körpersprachliches Signal für einen interpersonellen Konflikt verstanden werden. So war bei drei Männern, die uns nach einer SKAT aufsuchten, von den behandelnden Urologen eine Beziehungsstörung übersehen worden. Alle drei hatten eine normale Erektion bei der Masturbation und beim Vorspiel. Sie verloren die Erektion unmittelbar vor bzw. nach der Penetration. Obwohl in der urologischen Fachliteratur auf eine sorgfältige multidisziplinäre Abklärung vor Anwendung von SKAT hingewiesen wird, glauben wir, daß SKAT manchen Arzt dazu verleitet, diese noch nicht ausgereifte und mit vielen Nebenproblemen behaftete Methode übereilt und unreflektiert anzuwenden.

56.7 Die psychische und sexuelle Situation nach Genitaloperationen

Erkrankungen und Operationen des Urogenitaltrakts bedrohen die körperliche Integrität und damit das Selbstgefühl. Spengler (1988) weist darauf hin, daß im Gegensatz zu den psychosomatischen Erkenntnissen in der Gynäkologie, in der chirurgischen Urologie, abgesehen von den postoperativen sexuellen Funktionsstörungen, keine Forschungsergebnisse vorliegen. Er beobachtete an 32 Männern nach radikaler Prostatektomie bei 16% Suizidgedanken; 6 Patienten bedurften einer psychiatrischen Behandlung; 22% wirkten bei der postoperativen Untersuchung deutlich psychisch belastet. Die beobachteten auffälligen Erscheinungen nach urologischen Operationen reichten von Vermeidungsreaktionen (nicht ansehen, nicht berühren, Unterdrücken sexueller Zärtlichkeiten), Libidomangel und aversiven Reaktionen (ängstliche Selbstbeobachtung oder Ekel) bis zu den typischen sexuellen Funktionsstörungen der Erektion, des Orgasmuserlebens und des Ejakulationsablaufs.

Die transvesikale Adenektomie ebenso wie die transurethrale Resektion der Prostata mit Verlust des

Sphincter internus führt in der überwiegenden Mehrzahl der Fälle zur retrograden Ejakulation und damit zum trockenen Orgasmus. Die eingehende und ausführliche Aufklärung vor der Operation und die Führung des Patienten in den Monaten nach der Operation ist dabei von großer Bedeutung. Erfahrungsgemäß führt das Prostataadenom als solches nicht zur Einschränkung der Libido oder der Potentia coeundi. Mayer et al. (1983) haben das Sexualverhalten nach Prostataoperationen untersucht. Somatopsychische Reaktionen können auch im Zusammenhang mit der Harnableitung aus Niere und Blase auftreten (Stomaträger). Die Weiterentwicklung chirurgischer Techniken (Blasenersatz durch Darm bei voller Kontinenz) läßt jedoch hoffen, daß permanente Harnableitung bald eine Ausnahmeerscheinung sein wird.

56.8 Interaktionelle und therapeutische Aspekte

Auf die Psychodynamik der Arzt-Patient-Beziehung bei Männern mit Prostatopathie (Günthert, 1983; Janssen et al., 1983) wurde schon hingewiesen. Analoge Interaktionsprobleme sind auch bei Frauen mit urologischen Beschwerden, insbesondere mit Blasenbeschwerden zu vermuten. Schon Menninger (1941) hat auf den masochistischen oder selbstbestrafenden Aspekt urologischer Symptome hingewiesen. Deshalb empfiehlt sich hier eine besondere Zurückhaltung im Hinblick auf operative Eingriffe. Häufiges Bougieren der Harnröhre ebenso wie die Harnröhrenschlitzung können Ersatzbefriedigungscharakter bekommen. Bei therapieresistenten urologischen Symptomen muß auch an den sekundären Krankheitsgewinn gedacht werden, der – wie bei der rezidivierenden Urethrozystitis diskutiert wurde – in dem Vermeidenkönnen von Sexualität liegen kann.

Eine längerfristige frequente psychoanalytische, d.h. konfliktaufdeckende Psychotherapie ist – wie auch bei anderen psychosomatisch kranken Patienten – nur bei einem kleinen Teil indiziert, z.B. bei Frauen mit chronisch rezidivierender Blasenentzündung, da diese Patientinnen nach unserer Erfahrung introspektionsfähig sind und sich sprachlich differenziert ausdrücken können. Die Bearbeitung des Nähe-Distanz-Problems läßt sich nur in einem länger dauernden einzelpsychotherapeutischen Prozeß grundlegend aufarbeiten. Männer mit Prostatopathie sind dagegen schwer für eine analytische Einzel- oder Gruppentherapie zu motivieren. Wie jedoch der exemplarische Fall zeigt, kann der psychosomatisch orientierte Urologe im therapeutischen Gespräch dem Patienten wichtige Einsichten in Hintergrundkonflikte und psychosomatische Zusammenhänge vermitteln. Daneben können übende Verfahren (autogenes Training) bei diesen Patienten nicht nur muskuläre Entspannung herbeiführen, sondern darüber hinaus einen Schritt hin zu einer ersten Körper-Selbsterfahrung bedeuten.

Im Gegensatz zu Müller-Ehrenberg (1981) raten wir bei urologischen Erkrankungen zum zurückhaltenden therapeutischen Einsatz von Psychopharmaka.

57 Das Auge und seine Störungen aus psychosomatischer Sicht

Wolfgang Schultz-Zehden und *Friederike Bischof*

Die Einsicht in den engen Zusammenhang zwischen Auge und Psyche ist alt und in zahlreichen Zeugnissen aus Mythologie und Kunst festgehalten. An die Ödipussage und die biblische Blindenheilung muß man in dieser Hinsicht nicht erinnern; auch nicht an die zentrale Rolle des Auges in der bildenden Kunst als Vergegenwärtigung des Blicks. Seher oder Propheten werden vielfach als blind beschrieben, man spricht von einem „inneren" oder „geistigen" Auge.

Der Volksmund kennt viele Ausdrücke zur Umschreibung eines psychischen Zustandes durch die „Augensprache": Man kann liebäugeln, die Augen können einen festhalten, Blicke ineinander tauchen, man streift jemanden mit einem Blick. Augen können lachen, weinen, sprühen, funkeln, blitzen, drohen. Man ist ganz Auge, weint sich die Augen aus, hütet etwas wie seinen Augapfel. Liebe macht blind, Haß macht blind, es gehen einem die Augen auf etc.

Was heißt das alles? Lust und Leid, Kritik, Haß, Abneigung, Aggression lassen sich wortlos nur mit dem Blick der Augen ausdrücken. Die Augen spiegeln Gefühle wider und sind sozusagen mitverantwortlich für den mimischen Gesamtausdruck des Gesichts. Andererseits sind starke Emotionen in der Lage, das Sehen zu stören.

Wenn sich im medizinischen Denken die Erkenntnis durchsetzt, daß der Mensch eine Leib-Seele-Einheit ist, daß psychische Störungen somatische Veränderungen bewirken können und umgekehrt, so erscheint es eigentlich selbstverständlich, daß sich derartige Veränderungen ebenso wie an allen anderen Organen am Organ Auge manifestieren können. Trotzdem gibt es zu diesem Problem nur sehr wenig Literatur, die über die Beschreibung von Einzelfällen hinaus grundlegende Zusammenhänge zwischen Krankheit und psychischen Faktoren erkennen läßt.

Bis zur Etablierung der Psychosomatik als eigenständige medizinische Disziplin wurden psychogene Erkrankungen des Auges, vor allem von ophthalmologischer Seite, unter dem Sammelbegriff „Hysterie" erörtert. In den wenigsten Fällen geschah dies im Hinblick auf eine Konversionsproblematik. Um jedoch ein möglichst großes Spektrum verschiedenster Krankheitsbilder unter diesem Begriff vereinen zu können, mußte der umschriebene psychoanalytische Hysteriebegriff aufgegeben und von einem „Hysterietyp" oder „exogen bedingten hysteriformen Reaktionsweisen" gesprochen werden (Best, 1934). Die Vielfalt der Krankheitsbilder, die auf diese Weise entstand,

sprengte den bis dahin üblichen Rahmen, der im wesentlichen nur die hysterische Amaurose einschloß (zit. nach Niklewski, 1982).

57.1 Entwicklung, Funktion und Physiologie des Gesichtssinns

57.1.1 Die Entwicklung des Auges

Die Entwicklung des Auges zeigt den engen Zusammenhang zwischen dem Organ und psychischen Prozessen, denn es entsteht aus dem Gehirn. Schon beim 22 Tage alten Embryo treten zwei flache Furchen auf jeder Seite des noch nicht geschlossenen Vorderhirns auf. Mit dem Neuralrohrschluß werden diese Furchen zu Ausbuchtungen des Vorderhirns, die als Augenbläschen dem Oberflächenektoderm anliegen. Dann beginnt das Augenbläschen sich einzustülpen und bildet den doppelwandigen Augenbecher.

In der 7. Woche verschmelzen die Ränder der Augenbecherspalte und die Öffnung rundet sich zur Pupille ab. Im Laufe dieser Entwicklung bildet sich aus den Zellen des Oberflächenektoderms die Linsenplakode, die sich in den Augenbecher einstülpt und als Linsenbläschen die Verbindung zum Oberflächenektoderm verliert. Die äußere Schicht des Augenbechers bildet in der 5. Woche kleine Pigmentgranula und wird zum Pigmentepithel der Retina. Die hinteren vier Fünftel der inneren Schicht machen eine Reihe von Veränderungen analog der Wand der Gehirnbläschen durch und differenzieren sich in die verschiedenen Schichten der Netzhaut. Das vordere Fünftel beteiligt sich an der Bildung der Iris und des Ziliarkörpers.

Die lose Mesenchymschicht, die das Auge ab dem Ende der 5. Woche umgibt, bildet vorne die vordere Augenkammer mit dem Ziliarmuskel und den Mm. sphincter und dilatator pupillae. Ansonsten wandelt sie sich in eine innere Schicht (vergleichbar der Pia mater des Gehirns), die sich zur Choroidea entwickelt, und eine äußere Schicht (vergleichbar der Dura mater), die die Sklera bildet und sich in die Dura mater des N. opticus fortsetzt.

Anfänglich ist der Augenbecher mit dem Gehirn durch den Augenbecherstiel verbunden, der auch die Vasa hyaloidea trägt. Dieser wandelt sich im Laufe der Entwicklung zum N. opticus, der in seinem Zentrum die A. centralis retinae enthält (Langman, 1974).

Das Gehirn hat also sozusagen mit dem Augenbecher eine Antenne ausgefahren, die während der ganzen Individualentwicklung den engen Zusammenhang zum Gehirn beibehält. Auch die Weiterdifferenzierung ist gleichgeschaltet. Zwar sind Gehirn und Auge bei der Geburt weitgehend ausgebildet, jedoch brauchen sie das erste Lebensjahr, um endgültig Gestalt und Funktion anzunehmen, z. B. wird die Fixation in diesem Zeitraum erlernt.

57.1.2 Funktionen des Sehens

Die Grundform des Sehorgans entspricht einer Kugel mit einem Durchmesser von etwa 23 mm, einem Gewicht von 6,3 bis 8 g und einem Volumen von ca. 6 ccm. Man muß sich diese Größen klarmachen, um eine Vorstellung vom Auge, seinen Leistungen, aber auch von den Möglichkeiten, morphologische oder funktionelle Fehler chirurgisch zu korrigieren, zu erhalten.

An der vorderen, dem Licht zugewandten Fläche des Auges sind die bildentwerfenden Strukturen, hinten die bildempfangenden und -weiterleitenden Organe lokalisiert. Die Wand des Augapfels setzt sich aus einer äußeren, mittleren und inneren Augenhaut zusammen, die schalenartig übereinanderliegen: außen die Sklera (Lederhaut), die Schutzhülle des Auges, deren vorderer durchsichtiger Teil die Cornea (Hornhaut) ist, in der Mitte die dunkel pigmentierte Gefäßschicht (Choroidea oder Aderhaut), deren vorderer Abschnitt die Iris (Regenbogenhaut) ist, sowie innen die dünne Retina (Netzhaut).

Von diesen drei Hüllen ist die Netzhaut die weitaus wichtigste, sie erstreckt sich von der Papille, der Austrittsstelle des Sehnervs, bis zum Pupillarrand der Iris und wird unterteilt in die Pars optica und die Pars coeca. Die Pars optica der Netzhaut enthält den Empfangsapparat für den Lichtreiz, d. h. die das Tages- und Farbensehen vermittelnden Zapfen und die für das Sehen bei herabgesetzter Beleuchtung bestimmten Stäbchen, sowie ein Umschalt- und Leitungssystem, über das die durch den Lichtreiz ausgelöste Erregung den für den Gesichtssinn maßgebenden Teilen des Gehirns übermittelt wird. Die Pars optica ist schichtenartig gegliedert und in vivo vollkommen durchsichtig bzw. lichtdurchlässig, was notwendig ist, damit das Licht die Schicht der Sehzellen erreichen kann, die selbst wiederum dem Licht abgewandt sind.

Die Pars coeca ist der blinde Teil der Netzhaut am Übergang zum Ziliarapparat.

Die Fovea centralis empfängt Lichtreize von allen Objekten, die direkt angesehen, d. h. fixiert werden. Sie enthält nur Zapfen und ist die Stelle des schärfsten Sehens. Entsprechend der besonderen Funktion ist hier eine besondere Netzhautstruktur verwirklicht, vor allem enthält die Fovea nur Rezeptororgane, die sonst über den Sehzellen liegenden Schichten fehlen hier vollständig, so daß das einfallende Licht direkt auf das Zapfenlager fällt. Gegen die Netzhautperipherie hin nimmt die Zahl der Zapfen ab, das Verhältnis von Stäbchen zu Zapfen verschiebt sich mit zuneh-

mender Entfernung von der Fovea immer stärker zugunsten der Stäbchen (Heilman, 1979; Waldeyer, 1975).

Anders als mit der Kamera ist das Sehen mit den Augen kein einfacher und unmittelbarer Vorgang. Während bei der Kamera ein Abbild der Umwelt in das Gehäuse eintritt und auf dem Film sichtbar wird, bilden beim Menschen Auge und Gehirn zusammen ein System, das das aus der Außenwelt einströmende Datenmaterial analysiert, weiterleitet, verarbeitet und speichert. Sehen und Wahrnehmen sind an das Denken gekoppelt, wir denken oft wie wir sehen und umgekehrt. Die Ausscheidung unnötiger optischer Informationen ist ein wesentlicher Teil eines Ordnungsprozesses. Das vom Gehirn geleitete Auge sieht selektiv, es sieht subjektiv und bemerkt nur das, woran der Geist interessiert ist, und was es sehen möchte oder zu sehen gezwungen ist.

Im einzelnen verläuft der Akt des Sehens folgendermaßen: Lichtstrahlen fallen in die optischen Teile des Auges ein und bewirken auf der Netzhaut in bestimmten sensorischen Zellen (Zapfen, Stäbchen) einen physiologischen Impuls. Diesen ersten und primitiven Teil des Sehaktes nennen wir Empfindung.

Der physiologische Impuls reizt über retinale Neurone und Sehnervenfasern zelluläre Elemente der Hirnrinde (Sehcortex) und gleichzeitig Stammhirnteile (Formatio reticularis und limbisches System, das sog. „Gefühlshirn"). Dadurch werden emotionale Befindlichkeiten, Erinnerungen, Erfahrungen, Lern- und Denkprozesse in das Sehen einbezogen. Durch Querverbindungen zwischen Auge, Cortex und Stammhirn wird aus dem rein optischen, elektromagnetischen Reiz eine Funktion, die zum Erfassen, Sehen, Erkennen der Gegenstände und Situationen der Außenwelt führt. Diesen Vorgang bezeichnen wir als Wahrnehmung.

Die Vorgänge zur Gestaltwahrnehmung in allen Qualitäten (z. B. Farbe, Form, Kontrast, Tiefenschärfe, räumliches Bild) sind komplex und erfordern eine Rückkoppelung zwischen Auge und Gehirn. Nachdem ein optischer Reiz wahrgenommen und für sehenswert befunden worden ist, wird der optische Apparat auf das Bild eingestellt. Dazu gehören z. B. Blick auf das Objekt, so daß es am Fleck des schärfsten Sehens abgebildet wird, Fixierung, Akkommodation und Einstellen der Pupillenweite oder Detailbetrachtung durch koordinierte Augenbewegungen (Delay und Pichot, 1966; Grüsser, 1977; Schultz-Zehden, 1981).

Demnach wird nicht alles, was optisch auf die Netzhaut einfällt, auch wahrgenommen. Aus der riesigen Menge der Reize, die das Auge treffen, wird nur ein kleiner Teil ausgewählt. Diese Auswahl ist Wertschätzung oder Degradierung nach subjektiven Kriterien, entsprechend bisher gespeicherter Erfahrungen (Delay und Pichot, 1966; Grüsser, 1977; Schultz-Zehden, 1981).

Andererseits können Bilder, die vor dem geistigen Auge erscheinen, genauso plastisch sein wie reale Gegenstände. Beispiele hierfür sind der Traum und das katathyme Bilderleben. Hierbei wird wieder die enge

Verknüpfung zwischen Sehen und Erleben, Auge und Psyche deutlich. Diese Art der optischen Wahrnehmung ohne entsprechendes reales Korrelat wird als visionelles Sehen bezeichnet (Schultz-Zehden, 1981).

57.1.3 Das Auge als Träger nonverbaler Kommunikation

Ein alter Glaube ist, daß das Auge das Fenster zur Seele sei. Besonders die Pupille, die dieses Fenster öffnet und schließt, besitzt großen Symbolwert. Emotionen wie Furcht, Angst, Freude können die Pupille erweitern. Diese Reaktion kommt durch zentrale Hemmung des dritten Hirnnervs zustande, so daß an den Reaktionen der Pupille erkennbar wird, wie stark das sympathische bzw. parasympathische Nervensystem das Auge in seinen Funktionen steuert (Heilman, 1979).

Von den Muskeln des Gesichts sind die meisten Funktionseinheiten um die Augen gruppiert. Der mimische Ausdruck der Augen ist weniger bewußt kontrollierbar als z. B. der des Mundes. Daraus ergibt sich ein höherer Stellenwert der Augen als Mittel der nonverbalen Kommunikation. Störungen am Auge (Schielen, Blindheit) führen damit gleichzeitig zu Störungen der zwischenmenschlichen Kommunikation (Morris, 1978).

57.2 Exemplarische Krankengeschichte

Eine 22jährige ledige Krankenschwester kommt wegen Schielens in eine Augenarztpraxis, wohin sie ihr Hausarzt überwiesen hat, damit sie sich über die Möglichkeit, statt ihrer Brille Haftgläser zu tragen, informieren soll. Schon beim Erstinterview kommen die Hauptprobleme der Patientin zur Sprache: Minderwertigkeitsgefühle wegen des Schielens, Partnerprobleme aus dem gleichen Grund, Schlafstörungen und eine allgemeine Lebensangst.

Biographische und somatische Anamnese:
Das Mädchen zeigt von Geburt an Einwärtsschielen und hatte schon in frühester Kindheit unter Hänseleien deswegen zu leiden. Während der Schulzeit nahmen diese Belastungen noch zu, und alle Versuche der Mutter, die Äußerungen der Umwelt als Neid hinzustellen auf etwas, „das nicht jeder hat, einen kleinen Silberblick" halfen nicht, die Patientin vor einer schweren Krise zu bewahren. Mit fünf Jahren hatte die Patientin die erste augenärztliche Untersuchung und bekam eine Brille verordnet sowie eine Augenklappe. Sie konnte kaum etwas sehen und mußte trotzdem zum Trainieren „Fäden einfädeln und Knöpfe annähen". Nach 6 bis 8 Wochen wurden jedoch diese Übungen erfolglos abgebrochen.

Diesen ersten Arztbesuchen, um das Schielen behandeln zu lassen, stand der Vater der Patientin äußerst indifferent gegenüber und hielt sie für sinnlos, wie sich die Patientin erinnert. Das Verhältnis der Eltern zueinander war gestört, und die Ehe ging kurze Zeit später auseinander. Geschwister waren nicht vor-

handen. Nach der Enttäuschung über das Verhalten des Vaters hatte die Patientin in dem Augenarzt eine starke Vaterfigur gesehen, erlitt nun aber wieder eine Enttäuschung, als ihr der Arzt außer Brille, Augenklappe und Einfädelübungen nichts zu bieten hatte.

Weitere Ärzte wurden nicht konsultiert. Bis zu ihrem 14. Lebensjahr fand sich die Patientin damit ab, daß „man ihr eben nicht helfen könne". Später aber, als die ersten partnerschaftlichen Beziehungen gesucht wurden, litt die Patientin ganz erheblich unter ihrer Augenstörung. Sie berichtete, daß man es ihr sehr deutlich zu verstehen gegeben habe, sie stehe außerhalb der Norm und sei „irgendwie minderwertig".

Die Familie zog dann nach Berlin (aus einer Kleinstadt), und „hier ging es erst richtig los", wie sich die Patientin äußerte: „Vom Dorf und dann noch schielen!" Die Mutter versuchte wieder, der Patientin zu helfen, und ging mit ihr zum Augenarzt. Es wurde eine Operation vorgeschlagen und auch durchgeführt. Für die Patientin schien damit ein böser Traum vorüber zu sein.

Obwohl die Operation erfolgreich durchgeführt worden war, fand die Patientin nicht zu einer positiven Haltung und Bewertung ihrer selbst. Eine dauernde Angst verfolgte sie, es könne zu einem Rückfall kommen. Sie wagte deshalb auch nicht, Kontakte zu suchen bzw. zu halten, obwohl sie ein äußerst attraktives Mädchen war. „Wenn mir mit den Augen wieder etwas passiert, ist ja doch alles aus!" Damit stürzte sie sich in ihren Beruf und glaubte, damit ihr Leben allein ausfüllen zu müssen.

Als sich zu den Ängsten nun noch Schlafstörungen gesellten, und sie ihrem Hausarzt darüber berichtete, überwies sie dieser zum Augenarzt.

Verlauf der Behandlung:
Der jedem Hilfsangebot auf psychischem Gebiet gegenüber sehr aufgeschlossenen Patientin, die einerseits recht sicher auftrat, aber dann doch wieder ganz hilflos wirkte, war der Vorschlag, einen Kurs für autogenes Training zu versuchen, sehr willkommen. Sie machte gut mit und konnte in einer der späteren Sitzungen in der Gruppengesprächsrunde auch offen über ihre Ängste im zwischenmenschlichen Bereich sprechen, vor allem über die Angst, vom Partner wegen ihrer „Minderwertigkeit" verlassen zu werden. Sie hat von der Operation her noch ein erhebliches Trauma, empfindet die Brille als Belastung, als Verschandeln. In der Gruppe wurde dieses Verhalten mißbilligt und der Patientin auf verschiedene Weise geraten, ihre Beschwerden nicht überzubewerten. Aber erst in der Sitzung nach der Sonnengeflechtsübung ergab sich ein Ansatzpunkt fürs Umdenken: Eine andere Patientin erzählte, daß sie bei dieser Übung nichts spüre und auch nicht spüren könne, da sie ja keinen Uterus mehr habe. Sie aber habe sich damit abgefunden und fühle sich durchaus nicht als „halber Mensch" oder gar „ausgenommenes Huhn", wie man in der Klinik liebenswürdig die uterusamputierten Frauen als nun nicht mehr vollwertig betitelt habe.

Als der AT-Kurs beendet war, bat die Patientin den Augenarzt um ein Gespräch und fing sofort von der uterusamputierten Frau zu reden an und welch einen großen Eindruck diese noch relativ junge Frau auf sie gemacht habe: daß man also mit einem derartigen Fehler bereit und fähig sein könne zu leben. Seit diesem AT-Abend sehe sie das AT und ihre eigenen Probleme mit ganz anderen Augen an und beginne zu lernen, daß man auch mit Doppelbildern und mit dem bei Anstrengung wieder auftretenden Schielen leben könne.

In der Folgezeit wurden der Patientin Kontaktlinsen angeboten, die sie mit der Hilfe des AT und formelhafter Vorsatzgebung in sehr kurzer Zeit zu handhaben lernte.

Katamnese:
Die Patientin heiratete kurz nach der Operation einen Mann, an dem sie eigentlich „nichts Besonderes" fand, sie nahm ihn nur deshalb, weil er der erste Mann war, der sie überhaupt hatte heiraten wollen. Mit ihrer dauernden Furcht, er würde sie sofort verlassen, wenn ihr Schielen wieder manifest oder stärker würde, hat sie diesem Mann sicher erhebliche Probleme aufgeladen, die er nicht zu lösen imstande war, da jede kleinste Änderung in seinem Verhalten bereits als erster Schritt zur Trennung beargwöhnt wurde. Nach dem „flash" während des AT war der Patientin ein Wandel in der eigenen Persönlichkeit gelungen, so daß sie ein Jahr nach dem erwähnten entscheidenden Gespräch berichtete, das Leben mit ihrem Mann „ginge jetzt vorzüglich", sie habe erst in der vergangenen Zeit erkannt, was sie Gutes an ihm habe. Sie übt laufend AT und besucht in etwa vierteljährlichem Abstand Gesprächsabende der alten Gruppe (Schultz-Zehden, 1981).

57.3 Einzelne Augenstörungen und ihre Beziehung zur Psychosomatik

Im folgenden Abschnitt werden einzelne Krankheitsbilder aus psychosomatischer Sicht dargestellt. Zur Symptomatologie sei auf die speziellen Augenlehrbücher verwiesen (Hollwich, 1979).

Bei der Lokalisation somatischer Störungen, die sich aus verdrängten Konflikten und ähnlichem entwickeln, wird von „Organminderwertigkeit" (A. Adler, 1907) gesprochen oder ein locus minoris resistentiae diskutiert. Nach Jores hat „jedes Gefühl, jeder Affekt eine bestimmte organische Repräsentanz und ist – nicht immer, aber sehr häufig – regelhaft mit einem bestimmten Organ verbunden". Hinweise dafür finden sich in den volkstümlichen Sprachgewohnheiten.

Groddeck (1966) betrachtete Augenstörungen als den Versuch eines Patienten, bestimmte Dinge in seinem Leben nicht mehr sehen zu müssen. „Wo das Abwenden des Blickes, des Kopfes, des Körpers, das Schließen der Augenlider nicht ausreicht, störende Eindrücke der Außenwelt abzuwenden, tritt unter Umständen bei entsprechender Disposition eine Erkrankung hinzu, vom einfachen Gerstenkorn bis zur Erblindung." Die Erkrankung erhält damit einen „Sinn". Sie erscheint als „regulierendes Prinzip, das dem Leben des Patienten in seiner Ganzheit dienlich ist" (Knapp, 1981; Jores, 1981; Schultz-Zehden, 1981).

Bei einer Aufstellung psychosomatischer Augenstörungen korrespondieren häufig bestimmte, unterschiedlich tief lokalisierte Affektionen am Auge mit bestimmten Altersgruppen:

Bei Säuglingen und Kleinkindern finden sich oft psychische bzw. psychosoziale Hintergründe bei oberflächlichen Störungen des Auges und seines Halteapparates, also z.B. Schielen und Lidrandaffektionen.

Bei Jugendlichen zeigt sich öfter psychogenes Schielen, ferner präpubertäre Myopien und Hyperopien.

Im sogenannten „midlife" fallen psychosomatische Aspekte vor allem auf bei Störungen der Linsenumgebung, also in der „Mitte des Auges", und zwar Glaukome und damit verbundene Ängste wegen Sehverschlechterung bis hin zur Erblindungsangst; weniger häufig vorhanden sind auch entzündliche Lidrandaffektionen und pathologischer Tränenfluß.

Tief im Innenauge zeigen sich die häufigsten psychosomatischen Störungen des alten Menschen: Linsen-, Netzhaut- und Aderhautveränderungen (Schultz-Zehden, 1981).

57.3.1 Angeborene Störungen

Eine Fehlbildung am Auge kann für den Patienten nicht nur mit funktionellen Folgen verbunden sein, sondern auch in seine Persönlichkeit einbezogen werden, so daß sie zum Teil seiner Ich-Struktur wird. Die operative Korrektur einer angeborenen Ptosis oder des Strabismus sollte frühzeitig erfolgen, einerseits um die daraus entstehende Schwachsichtigkeit zu vermeiden, andererseits, um zu verhindern, daß eine derartige Fehlbildung zu einem festen Bestandteil des Körperbildes eines Menschen wird.

Eine weniger gefestigte Persönlichkeit kann einen entstellenden Fehler im Bereich der Augen als Charakterschutz verwenden oder sie auch vollkommen verleugnen.

Eine seit der Kindheit bestehende Fehlbildung, die die Möglichkeit zur Korrektur hatte, aber im Erwachsenenalter noch besteht, kann zu psychischen Symptomen führen. Gelegentlich kommt es vor, daß Eltern beim Kind eine Korrektur verhindert haben. Wenn in solchen Fällen der Patient später, wenn die elterliche Aufsicht nicht mehr besteht, eine Korrektur ablehnt, so möglicherweise deswegen, weil er nicht gegen den Wunsch der Eltern handeln und Schuldgefühle auf sich nehmen möchte (Heilman, 1979).

57.3.2 Reaktion des Patienten auf Augenkrankheiten

Langwierige Krankheiten und plötzliche Erkrankungen und Verletzungen des Auges rufen beim Patienten psychische Wirkungen hervor. Die Reaktionen auf Krankheit sind beim einzelnen unterschiedlich stark, sie machen sich entweder zeitweilig oder ständig bemerkbar, sind aber in irgendeiner Form immer vorhanden.

Gerade in der Ophthalmologie spielt die Angst eine ganz wesentliche Rolle, vermutlich eine größere als der Schmerz. Jede Einwirkung auf das Auge, ja schon allein die Phantasie einer Bedrohung des Sehvermögens löst in irgendeiner Form beim Patienten Angst aus. Da die Behinderungen des Sehakts praktisch immer einen mehr oder weniger starken Verlust an Kontakt zur Umwelt bedingen, ist der augenkranke Patient vermutlich immer ein Patient in einer besonderen emotionellen Situation.

Das Umgehen mit der Angst und deren Bewältigung steht im Vordergrund bei der psychischen Betreuung des Patienten. Dies gilt vor allem bei chirurgischen Eingriffen am Auge. Das aufklärende Gespräch des Arztes kann das Verhalten des Patienten entscheidend beeinflussen. Jede Augenoperation beinhaltet ein gewisses Risiko und Gefahren, deren Ausmaß nicht allein in der Hand des Chirurgen liegt: z.B. kann das falsche Verhalten des Patienten bei einem Eingriff in Lokalanästhesie den Erfolg einer Operation vollkommen zunichte machen. Postoperative Psychosen, wie sie häufig bei älteren Patienten auftreten, beruhen nicht immer nur auf den toxischen Wirkungen der verwendeten Sedativa und Analgetika, sie hängen auch davon ab, wie lange der Patient einen beidseitigen Augenverband tragen muß. Psychotische Symptome, die durch den mangelnden Kontakt zur Umwelt und die Angst vor der Erblindung verstärkt werden, lassen sich oft allein dadurch bessern, daß das nichtoperierte Auge freigegeben wird (Heilman, 1979).

57.3.3 Erblindung

Blindheit ist die gefürchtetste Behinderung des visuellen Systems. Die Angst vor der Dunkelheit ist eine Urangst des Menschen. Der spät erblindete Erwachsene befindet sich zumeist in einem Zustand psychologischer Unbeweglichkeit, der am besten mit dem Stadium des Schocks oder der Phase der „seelischen Anästhesie" beschrieben werden kann. Die Isolation von Gefühlen und Interessen erfolgt offenbar reflektorisch, um eine Desintegration der Persönlichkeitsstruktur zu verhindern. Der Grad der Rehabilitation hängt weitgehend von der Länge dieses Stadiums ab. In dieser Zeit muß der Augenarzt vermeiden, den Patienten in einem Zustand unberechtigter Hoffnungen zu belassen.

Auf das Stadium des Schocks folgt die Phase der Depression, in der Gefühle der Hoffnungslosigkeit, des Selbstbedauerns und des Mangels an Vertrauen vorherrschen und der Patient dem verlorenen Augenlicht nachweint. Die Bewältigung der Trauerarbeit ist für den Patienten nötig, um die Realität der Blindheit akzeptieren und Schritte der Rehabilitation einleiten zu können. Es ist erwähnenswert, daß bei der großen Anzahl der Menschen, die heute vor allem durch Verkehrsunfälle spät erblinden, die Zahl der Suizidversuche in dieser Phase äußerst gering ist (Heilman, 1979).

Ganz anders als die vollkommene Blindheit wirkt sich eine partielle aus. Es scheint schwieriger zu sein, sich an teilweise Blindheit zu gewöhnen als an den vollständigen Sehverlust. Offensichtlich ist Blindheit eine unverrückbare Tatsache, die nicht bezweifelt werden kann, während partielle Blindheit immer

noch eine gewisse Hoffnung auf spätere Besserung beinhaltet, so daß der Patient ständig in einem Zustand der Unsicherheit und der Angst vor völliger Erblindung lebt.

Beim Kind wirkt sich Blindheit anders aus als beim Erwachsenen. Hochgradig sehbehinderte oder vollständig blind geborene Kinder stellen oft mehr ein erzieherisches und soziales Problem dar, weniger ein medizinisches. Es hängt weitgehend von der Umwelt des blinden Kindes ab, ob aus ihm ein autonomer Erwachsener wird oder eine Person, die ständig auf die Hilfe anderer angewiesen ist. Die Aufgabe des Arztes ist hier, die genaue Diagnose zu stellen und die Eltern klar über Prognose, Behandlungsmöglichkeiten und Hilfseinrichtungen zu informieren (Heilman, 1979).

57.3.4 Entzündungen am äußeren und inneren Auge

Zur psychosomatischen Genese von Augenentzündungen liegt bisher noch wenig Material vor. Aufgrund von Einzelfallanalysen, die eine Koinzidenz von bestimmten Emotionen (depressiv-zwanghafte Struktur oder Objektverlustsituationen) mit Krankheitsschüben aufweisen, und bei denen durch Einbeziehung psychotherapeutischer Maßnahmen in die übliche Therapie die Erscheinungen zum Abklingen gebracht werden konnten, scheint ein psychosomatischer Zusammenhang wahrscheinlich (Bernhard und Huhn, 1983).

Die entzündlichen Veränderungen der Lidränder und der Bindehaut sind nicht immer allein durch Umwelteinflüsse bedingt, sondern auch durch eine psychische Reaktion darauf, im weitesten Sinne vergleichbar mit einer Allergie. Oft besteht dabei ein Zusammenhang mit Alkohol und Medikamenten, der jedoch vom Patienten nicht erkannt werden will.

Zu den entzündlichen Veränderungen des Innenauges konnten Bernhard und Huhn nachweisen, daß hier häufig Schwierigkeiten in Partnerschaft und Sexualität vorliegen. Z.B. konnte bei Chorioretinitis centralis serosa mit psychotherapeutischen Maßnahmen oft Besserung verzeichnet werden, wobei dem Patienten relativ schnell die Verbindung zwischen Augensymptom und eigenen Schwierigkeiten einsehbar wurde (Niklewski, 1982).

57.3.5 Muskelstörungen

Heterophorie oder latentes Schielen wird eine Abweichung der Gesichtslinien genannt, die nicht sehr groß ist und durch den Fusionszwang latent gehalten wird. Nach Jaensch und Schäfer (1938) weisen etwa 60 bis 80% der Bevölkerung eine Heterophorie auf, die aber keinerlei Beschwerden macht. Wird jedoch durch bestimmte Einwirkungen der Fusionszwang geschwächt oder fällt er ganz aus, so wird das Schielen zeitweilig manifest (periodisches Schielen) oder manifestiert sich für dauernd (permanentes Schielen).

Im Alltagsleben kommen solche den Fusionszwang beeinträchtigenden Umstände vor nach körperlichen und seelischen Anstrengungen, Störungen des Allgemeinbefindens, langem Lesen oder Näharbeit, Narkotika- und Alkoholeinwirkung. Die daraus resultierenden asthenopischen Beschwerden äußern sich in Kopf- und Augenschmerzen, Druckgefühl, Brennen wie bei Konjunktivitis und Doppeltsehen. Hier sind neben der orthodoxen augenärztlichen Therapie (Konvergenzübungen, Prismenbrillen, in schwierigeren Fällen Operation) auch psychotherapeutische Maßnahmen erfolgversprechend.

Aus der biographischen Anamnese und der Beobachtung der Bezugsperson erwachsener und jugendlicher Schielpatienten kann man vielfach feststellen, daß dem Betreffenden zumeist eine Doppelrolle zugeschrieben wird: Mit einem Auge sieht der Patient geradeaus, steht also auf dem Boden der gesellschaftlich erwünschten Tatsachen. Mit dem anderen Auge versucht er, den Gegebenheiten und Anforderungen seines persönlichen Lebens, denen er sich nicht gewachsen fühlt, auszuweichen.

Bei der Betrachtung des Kausalzusammenhangs von Schielen und gestörter Elternbeziehung wird immer deutlicher, daß in der frühkindlichen Phase Partnerschaftsprobleme und eheliche Schwierigkeiten der Eltern den Prozeß des Fixierenlernens behindern. Dabei ist es möglich, daß Muskelabweichungen manifest werden. Die Mutter-Kind-Beziehung ist problemhaft belastet, und das Kind leidet unter diesem emotionalen Defizit, so daß es die Kraft, die Fusion beider Augen zu bewerkstelligen, nicht aufbringen kann. Entsprechende therapeutische Maßnahmen beinhalten die große Schwierigkeit eines Gesprächs mit den Eltern des Schielkindes. Die betroffenen Eltern fühlen sich oft durch die „Anomalie" ihres Kindes bestraft, leiden darunter oder sind auf jeden Fall unsicher und unglücklich, so daß ihr Verhalten dem Kind gegenüber oft zur Verschlechterung des Krankheitsbildes beiträgt (Schultz-Zehden, 1981).

Die Augenlider sind von sympathischen Nervenfasern innerviert. Schon bei Ermüdung können die Lider nicht mehr ganz aufgehalten werden. Es gibt noch keine systematischen Untersuchungen über den Zusammenhang von Lidstörungen und psychischen Prozessen, jedoch ist anzunehmen, daß zumindest für die Ptosis der Mechanismus der Dekompensation ähnlich dem der Heterophorie ist.

57.3.6 Refraktionsanomalien und Asthenopie

Auch wenn hier ebenfalls eingehende Untersuchungen fehlen, besteht nach unseren Erfahrungen mit großer Wahrscheinlichkeit ein Zusammenhang zwischen psychischen Alterationen und asthenopischen Beschwerden, Konvergenz- und Fusionsschwächen, die sich unter Umständen zu einer Visusänderung auswachsen können. Besonders auffällig ist dieser Zusammenhang beim Krankheitsbild der präpubertären Myopie, das in den besonders schwierigen Jahren vor und zu Beginn der Pubertät gehäuft auftritt und

eindeutig mit Anpassungsschwierigkeiten, Verlassenheitsgefühlen und ähnlichem korreliert (Schultz-Zehden, 1981).

Durch die Einführung der Computertechnologie mit der Bildschirmarbeit hat der visuelle Streß durch die Dauerbeanspruchung der Augen erheblich zugenommen. Gleichzeitig kommt es zu einer geistig-seelischen Frustration, da der Arbeitsablauf durch die Maschine Bildschirm diktiert wird und soziale Kontakte während der Arbeit zu Kollegen weitgehend wegfallen. Durch die ständige Anspannung und Verspannung im visuellen System kommt es zu asthenopischen Folgezuständen, die therapeutischer Maßnahmen bedürfen.

Die meisten Menschen jedoch, die unscharf sehen, haben einen Brechungsfehler, d.h. zwischen der Brechkraft der brechenden Medien und der Augapfellänge besteht ein Mißverhältnis. Während bei der Myopie der Augapfel zu lang gebaut ist und die Bilder vor der Netzhaut zur Abbildung kommen, ist beim Hyperopen der Augapfel zu kurz, so daß es erst hinter der Netzhaut zur Abbildung kommt.

Solche Brechungsfehler können zwar durch optische Hilfsmittel (Brillen, Kontaktlinsen) korrigiert werden, jedoch ist der veränderte Bau des Auges keine unabhängige morphologische oder funktionelle Veränderung. Brechungsfehler können einen großen Einfluß auf die Entwicklung der Persönlichkeitsstruktur nehmen. Der intellektuelle, zurückgezogene Kurzsichtige und der extravertierte Weitsichtige sind bekannte Erscheinungen. Diese verallgemeinerungsfähigen Persönlichkeitsmerkmale sind jedoch durchaus erklärbar: Das kurzsichtige und das weitsichtige Kind unterscheiden sich vor Behandlung ihres Brechungsfehlers in ihrer Fähigkeit, in der Ferne bzw. in der Nähe scharf zu sehen und zurechtzukommen. Das kurzsichtige Kind findet sich in der Welt draußen nicht zurecht, weil es sie nicht erkennen kann, fühlt sich jedoch am Schreibtisch zuhause und zieht sich auf sich selbst zurück. Beim weitsichtigen ist es genau umgekehrt, es erkennt das Detail nicht und strebt nach außen (Heilman, 1979; Niklewski, 1982; Schultz-Zehden, 1981).

Die Presbyopie ist keine Krankheit, sondern ein normaler Alterungsprozeß, der jedoch auch für das normale Individuum häufig die Ursache von Sorgen werden kann. Die Lesebrille wird oft als Zeichen des herannahenden Alters empfunden und entsprechend emotional besetzt. Bei der Verordnung von Brillen (gleich aus welcher Indikation) müssen neben medizinisch-physikalischen auch psychologische Faktoren berücksichtigt werden. Viele Menschen lehnen Brillen ab, empfinden sie als kosmetisch störend oder als Hemmnis im täglichen Leben. Der Arzt sollte bei der Beratung auch diese Aspekte berücksichtigen (Heilman, 1979).

57.3.7 Katarakt

Der graue Star ist als eine degenerative Erkrankung der Linse zu verstehen, die mit der Zeit zu einer Trübung der an sich kristallklaren Struktur führt. Für viele Menschen ist die Diagnose gleichbedeutend mit unabwendbarer Blindheit und damit Quelle großer Angst. Die Sorgen beim Patienten stehen heute in keiner Relation zur Realität, da die Operation in der Hand des Erfahrenen ein Eingriff mit sehr guter Prognose ist. Fortgeschrittenes Alter ist keine Kontraindikation, vielmehr ist es erstaunlich, wie sehr die erfolgreiche Staroperation gerade den älteren Menschen tiefgreifend beeinflußt, ihm seine Selbständigkeit zurückgibt, so daß Lebensfreude wieder entstehen und neue Lebensinhalte gefunden werden können (Heilman, 1979).

57.3.8 Glaukom

Unter dem Begriff Glaukom oder grüner Star (Farbe der Iris) werden Krankheiten des Auges zusammengefaßt, deren gemeinsames Kardinalsymptom eine Steigerung des Augeninnendrucks ist, und in deren Folge morphologische Veränderungen am Sehnerven und Defekte im Gesichtsfeld auftreten. Der erhöhte Augeninnendruck beruht auf einem Mißverhältnis zwischen Zu- und Abfluß des Kammerwassers, auf einer Überproduktion desselben oder Behinderung der Zirkulation beim Durchfluß durch die Pupille oder beim Abfluß durch das Trabekelsystem in den Schlemmschen Kanal. Unterschieden werden müssen primäre Glaukomformen von sekundären, wobei für letztere eine vorausgegangene oder bestehende Erkrankung (z.B. Entzündungen, Verletzungen, Gefäßerkrankungen) des Auges als Ursache für die Augendrucksteigerung gefunden wird (Hollwich, 1979; Schultz-Zehden, 1981).

Das akute Glaukom ist eine sich innerhalb von Stunden entwickelnde anfallsweise Erhöhung des Augeninnendrucks auf das Drei- bis Fünffache der Norm. Hierbei werden neben der organischen Disposition (z.B. ein enger Kammerwinkel) psychische Traumen als auslösende Faktoren und eine psychophysische Labilität des Patienten als besondere Disposition zu Glaukomanfällen angegeben (Schultz-Zehden, 1981).

In der Literatur werden Glaukompatienten als psychisch-emotionell gestört, ängstlich, hypochondrisch, depressiv, introvertiert und an Minderwertigkeitskomplexen leidend beschrieben (Heilman, 1979; Niklewski, 1982).

Beim einfachen chronischen Glaukom ist der Abfluß des Kammerwassers trotz offenem Kammerwinkel erschwert. Es beginnt schleichend, geht in ein chronisches Stadium über und führt durch Entwicklung einer Druckexkavation der Papille mit Schädigung der Sehnervenfasern unauffällig, aber unaufhaltsam zu einem Gewebeschwund des Sehnerven und damit zur Erblindung, wenn nicht therapeutisch eingegriffen wird. Durch die Nebenwirkungen der Behandlung (Miosis und akkommodative Myopie), die zu einer Behinderung des Sehens und damit zur Beeinträchtigung des täglichen Lebens führen, und durch die Notwendigkeit, zeitlebens mehrmals täg-

lich Medikamente anwenden zu müssen, wird der Patient permanent an sein chronisches, nicht heilbares Leiden erinnert. Dies bedingt eine besondere psychische Situation, die damit auftretenden Probleme der Patientenbetreuung sind denen anderer chronischer Krankheiten wie Hypertonie oder Diabetes mellitus sehr ähnlich (Heilman, 1979; Schultz-Zehden, 1981).

Der Arzt muß in der Therapie bemüht sein, die Angst vor der Diagnose möglichst schnell zu kompensieren. Ein therapeutisches Bündnis zwischen Arzt und Patient ist für die Compliance enorm wichtig. Im Verlauf der Behandlung muß auch den obengenannten, mit der Therapie verbundenen Einschränkungen des täglichen Lebens Rechnung getragen werden.

57.3.9 Erkrankungen der Netzhaut

Auch zu diesem Thema gibt es nur vereinzelt Untersuchungen. Werry und Arends (1978) fanden bei Patienten mit Retinopathia centralis serosa signifikant höhere Werte auf der Hypochondrie- und Hysterieskala. Sie nehmen für diese Erkrankung eine neurotische Persönlichkeitsstruktur mit Neigung zur Konversion an und schlagen trotz der verfügbaren lichtchirurgischen Behandlungsmethoden (Laserung der entstandenen Netzhautlöcher) eine begleitende Psychotherapie vor (Niklewski, 1982).

Trichtel (1979) nahm einen Zusammenhang mit dem Licht-Maladaptationssyndrom (Licht-MAS) an, d.h. einer multifaktoriellen Funktionsstörung, die zu einer Beeinträchtigung der Anpassungsfähigkeit des Auges an seinen energetisch adäquaten Reiz führt. Bleibt der Zustand der Maladaptation längere Zeit bestehen und liegt die gleichzeitige Lichtbelastung über der Norm, so können je nach individuellem „Streßprofil" verschiedene chronische Augenleiden entstehen, deren wichtigste die Retinopathia centralis serosa und die nicht hereditären Formen der zentralen und peripheren Netzhautdegeneration sind. Die sich anläßlich eines Licht-MAS entwickelnden Augenerkrankungen sind psychosomatischer Natur, da zumeist psychosoziale Reize über psychophysiologische Reaktionen schließlich die organische Krankheit auslösen.

Mittels psychologischer Testverfahren und psychoanalytischer Interviews konnten bei R.c.s.-Erkrankten angstneurotische Züge festgestellt werden. Ferner waren die Patienten mit großer Regelmäßigkeit zuvor sowohl starkem Streß als auch einer übermäßigen Beleuchtung ausgesetzt (Trichtel, 1979, 1980 und 1983). Die häufig gemeinsam auftretenden zentralen und peripheren Netzhautdegenerationen sind zumeist in der zweiten Lebenshälfte auftretende Erkrankungen. Im Unterschied von der R.c.s. entstehen sie allmählich nach verschiedenen Prodromalsymptomen und nehmen einen chronischen, langsam progredienten Verlauf. Die Voraussetzungen zur Entstehung von Netzhautdegenerationen sind vielfältig. Sie beinhalten im wesentlichen neben mehr oder minder wirksam werdenden Erbfaktoren die langjährigen psychophysiologischen Folgen von Anpassungsstörungen (Licht-MAS) an die rasch wechselnden Störgrößen der technisierten Umwelt. Für diese Annahme sprechen die Umstände, daß Patienten mit degenerativen Netzhautveränderungen im Durchschnitt größere Pupillen aufweisen als Netzhautgesunde, und daß Trübungen der Linse mit dem hierdurch verminderten Lichteinfall in das Auge das Risiko der Makuladegeneration herabsetzen (Trichtel, 1983).

Klein und Moses (1974) untersuchten Fälle von Ablatio retinae prä- und postoperativ. Die Bedrohung völliger Blindheit ließ bei den Patienten typische Formen von Angst entstehen, die vor allem den drohenden Verlust der Unabhängigkeit, der Aktivität bis hin zu expliziten Kastrationsphantasien beinhalteten. Ein Teil der Patienten wurde durch die vorübergehende Erblindung zu einem stark regressiven Verhalten gebracht, ein Aspekt, der sicherlich bei vielen psychosomatischen Erkrankungen des Auges zu einem Krankheitsgewinn beiträgt, ist doch der Sehbehinderte massiv auf die Fürsorge seiner Umwelt angewiesen (Niklewski, 1982).

57.3.10 Hysterische Augenstörungen und Amaurose

Seit langem gilt die Hysterie als das klassische Beispiel einer seelischen Krankheit, die auch im Bereich des visuellen Systems zu einer Reihe von Symptomen führt, wie z.B. Einengung des Gesichtsfeldes, Auftreten von Gesichtsfeldausfällen, Änderung der Brechkraft, Funktionsstörung der Augenmuskeln, Schwachsichtigkeit bis hin zur hysterischen Amaurose (Niklewski, 1982).

Freud nahm an, daß die Bedeutung, die hinter dem hysterischen Verhalten „Ich kann nicht sehen" liegt, bedeutet: „Ich darf nicht sehen". Die Versuche Freuds, funktionelle Blindheit durch Hypnose hervorzurufen, haben klarwerden lassen, daß Blindheit hier zur Bestrafung wird (vgl. Kap. 30). Der Patient mit hysterischer Blindheit ist kein Augenkranker im eigentlichen Sinn, er ist eine kranke Gesamtpersönlichkeit und gehört in die Hände eines erfahrenen Psychotherapeuten (Freud, 1973; Niklewski, 1982).

57.3.11 Augensymptome als Begleiterscheinungen bei anderen Erkrankungen

Der Vollständigkeit halber sei an dieser Stelle auf sekundäre Augenerkrankungen hingewiesen, wie z.B. Visusverlust bei fortgeschrittenen Stadien der Hypertonie oder des Diabetes mellitus.

Schon beeinträchtigt durch die Grundkrankheit, werden die Patienten durch den Visusverlust in ihrem täglichen Leben noch mehr eingeschränkt. Oft wird die Augenkrankheit auch als eine Art Bestrafung für inkonsequentes Befolgen der Therapie (Diätfehler) angesehen, lassen sich doch die Spätfolgen durch konsequente Einhaltung von Diät und Medikation insbesondere beim Diabetes mellitus um ein paar Jahre hinausschieben.

57.4 Diagnostisch-therapeutisches Vorgehen bei psychosomatischen Augenstörungen

57.4.1 Diagnostik

Bei der psychosomatischen Diagnostik interessiert genauso wie bei der Beantwortung der Frage „Was hat der Patient?" das Problem „Wie ist der Patient?". Hierzu dient das sogenannte verstehende ärztliche Gespräch, das schließlich zu einer biographischen Anamnese (vgl. Kap. 12) führt. Die Rolle des Augenarztes bei der Diagnosestellung psychosomatischer Störungen ist insofern schwierig, als nicht mehr nur mit rein naturwissenschaftlichen Methoden faßbare Krankheitsbilder gesucht werden, sondern die Aufmerksamkeit mehr auf den Menschen und dessen subjektive Situation gerichtet sein muß. Es geht praktisch darum, die seelische Befindlichkeit des Patienten zu erforschen, oder – wörtlich genommen – das „innere Auge zu spiegeln".

Folgende Kriterien können als Anhaltspunkte dafür dienen, daß es sich bei dem Patienten um ein psychosomatisch aufzufassendes Augenleiden handelt (Schultz-Zehden, 1981):
- Neurotische Persönlichkeitsstruktur (labil, leicht erregbar, selbstunsicher, von schwankender Stimmung, klagend, hypochondrisch hinsichtlich der somatischen Beschwerden, völlig beherrscht bzw. angepaßt hinsichtlich psychischer Aspekte des Leidens usw.)
- Frühere psychosomatische Beschwerden (Symptomwandel)
- Frühere Augenstörungen (Organsprache, locus minoris resistentiae)
- Emotionale Krisen in zeitlichem Zusammenhang mit ersten Augenstörungen oder phasisches Auftreten der Symptome immer wieder bei emotionalen Krisen.

Hier spielt sich der wichtigste krankheitsdynamische Prozeß ab, diagnostisch und therapeutisch bereits während des sog. verstehenden ärztlichen Gesprächs: Es besteht ein verstehbarer Zusammenhang zwischen der körperlichen Störung, also etwa dem Glaukomanfall, und dem Erleben und Verhalten des Patienten, seinen Konflikten, Belastungen und entsprechenden Gefühlen.

57.4.2 Therapie

Neben den üblichen klassischen Therapieverfahren (vgl. Augenlehrbücher) empfehlen sich bei psychosomatischen Augenleiden psychotherapeutische Methoden. Besonders geeignet sind muskelrelaxierende Übungstherapien wie autogenes Training, funktionelle Entspannung, die vom Körperlichen ausgehen, beim Körper bleiben und auf Psychosoziales hinüberleiten, ferner das katathyme Bilderleben, seltener die Hypnose. Eine wichtige Art des Settings ist die dynamische Gruppentherapie (kombiniert mit AT oder KB). Gerade das AT in der Gruppe hat sich in Verbindung mit Gesprächstherapie als hervorragendes Mittel erwiesen, die Ängste von Glaukom- und Schielpatienten sowie die Probleme von Kontaktlinsenträgern zu bessern. Wichtig ist auch, die Umwelt des Patienten in die Therapie mit einzubeziehen, z.B. bei Schielkindern die Eltern und Geschwister. Zweck der Therapie ist eine supportive Wirkung, die den Patienten in die Lage versetzt, mit seiner Krankheit, seinen Ängsten, seiner Situation besser fertig zu werden (Schultz-Zehden, 1981 und 1983).

57.5 Zusammenfassung

Im „von Uexküllschen Situationskreis" (vgl. Kap. 1) wird von verschiedenen Integrationsebenen gesprochen, die durch Bedeutungskopplung miteinander verbunden sind. Im System Auge sind dies das äußere und das innere Sehen. Das Physische bewirkt die Organfunktion von außen, das Psychische das Reaktionsmuster von innen. Physikalische Reize lösen eine psychische Reaktion aus, die jedoch erst durch die Bedürfnisse des Menschen zu einer Wahrnehmung werden. Damit kommt eine Verbindung zwischen in-

nen und außen zustande. Das gesehene Bild ist das Ergebnis einer Synthese von äußerem Reiz und subjektiver Bedeutung. Ist die Verbindung zwischen dem äußeren Reiz und der inneren Bedeutung ungenügend, so ist hiermit die individuelle Wirklichkeit verzerrt, inneres und äußeres Sehen verlieren sich und werden unabhängig voneinander.

Schon Bodenheimer (1965) fordert, daß jede Augenerkrankung – ganz gleich welcher Entstehung – in erster Linie als Störung des Sehens aufzufassen ist, also „Augenerkrankung" mit „Sehstörung" zu identifizieren sei. Große Bindungsbereitschaft des sehgestörten Patienten schafft eine neue Art der Beziehungsebene zwischen Therapeut und Patient. Infolgedessen ist ein bestimmtes Maß an zurückhaltender Kommunikation zwischen dem Therapeuten und dem sehgestörten Patienten vonnöten. Von allergrößter Bedeutung ist dabei, daß der Patient mit dem dargebotenen Augensymptom nicht weiß, daß die Augenerkrankung für ihn individuelle biographische Bewandtnis hat, so daß der Arzt durch das Augensymptom einen Hinweis erhalten kann, woran der Mensch wirklich krank ist.

Die Hintergrundsbetrachtung von vordergründig dargebotenen Augensymptomen ist ein wichtiges Anliegen dieses Kapitels. Der Arzt – insbesondere der Augenarzt – muß lernen, das Auge von dieser Seite her zu betrachten.

58 HNO-Heilkunde

58.1 Störungen der Stimme

Hans H. Bauer

58.1.1 Funktionen und Entwicklung der menschlichen Stimme

Die Bildung einer normalen und leistungsfähigen Stimme setzt einen hochdifferenzierten und exakt koordinierten Ablauf vieler Einzelbewegungen eines morphologisch intakten Stimmapparates voraus. Aus didaktischen Gründen unterscheidet man physiologisch die Teilbereiche Atmung, Stimmgebung und Stimmansatz sowie die zugehörigen nervalen Strukturen. Die Erkrankungen, die mit morphologischen Veränderungen im Bereich des Kehlkopfes einhergehen, sind von der Hals-Nasen-Ohren-Heilkunde klinisch eingehend bearbeitet worden, für die funktionellen Syndrome gab es lange Zeit nur spärliche Ansätze.

Die Stimme ist Träger der menschlichen Sprache, Eigenklang des Individuums und für dieses spezifisch. Sie ist primärer Ausdruck und evolutionär älter als die Sprache, die erworben, nicht instinktiv und somit kulturell ist (Moses, 1956). Die Stimme ist der Spiegel der Persönlichkeit. Der Diskurs stimmlicher Kommunikation hat vor allem mit „Beziehung" zu tun, also mit Liebe, Haß, Angst, Wut, Respekt, Abhängigkeit usw. als affektiven Interaktionen zwischen Selbst und Umgebung. Die stimmliche Kommunikation ist Teil der Interaktion von Kind und Mutter, Basis der Sprachentwicklung und damit der Separations- und Individuationsprozesse.

In der Musik werden Gefühlszustände durch die Sprech- und vor allem die Gesangsstimme durch Komponisten, Interpreten und Schauspieler dargestellt. Sie korrelieren häufig mit musikalischen Elementen wie Tonfolgen, Klängen, Rhythmen, Betonung usw.

Besonders in der Spätklassik und Romantik werden die einzelnen Stimmgattungen bestimmten Persönlichkeitseigenschaften zugeordnet. So werden ältere bzw. würdevolle Personen, aber auch Intriganten von tieferen Stimmen dargestellt (Baß, Alt bzw. Mezzosopran), jüngere Frauen von hohen Stimmen (Sopran). Im Rivalitätskampf siegt meist der durchsetzungsfähigere Bariton über den Tenor usw.

Die Interpretation des Stimmausdrucks wird von Laien mehr von Assoziationen mit früheren Stimmerfahrungen des Hörers bestimmt. So wird die tiefe Stimme einer Frau am Fernsprecher bzw. auf Tonträgern ohne Sichtmöglichkeit in der Regel als typisch männliche Stimme bezeichnet, im Gespräch mit der Frau als weiblich tief. Das gleiche akustische Produkt wird somit je nach Assoziation mit visuellen Eindrücken anders beurteilt.

Moses (1956) verwendet für die Stimmdiagnostik ein System von meßbaren bzw. musikalisch nachvollziehbaren Kategorien: Atmung, Stimmumfang, Register, Resonanz, Rhythmus, ferner Melodie, Stimmstärke, Geschwindigkeit, Akzente, Emphase, auch Pathos, Genauigkeit, Pausen, Maniriertheit, Melismen (das sind diskrete Veränderungen der Sprechmelodie, des zeitlichen Ablaufs und/oder der Lautstärke; sie unterstreichen Willensäußerungen des Sprechers, um eine Reaktion beim Angeredeten zu erzielen).

Ostwald (1973) führte klanganalytische Untersuchungen bei emotionellen Störungen, psychopathologischen Veränderungen und Psychosen durch.

Die Phoniatrie/Pädaudiologie als Teilgebiet der HNO-Heilkunde bedient sich heute bei der klinischen Untersuchung von Stimmstörungen immer mehr objektiver Verfahren, die mit Hilfe einer technischen Apparatur reproduzierbare Meßdaten liefern, und „semiobjektiver" Methoden, bei denen die metrisch registrierbaren Ergebnisse der subjektiven Wertung unterliegen. So werden die gebräuchlichen und in vielen Bereichen noch diagnostisch überlegenen rein auditiven Verfahren zunehmend durch zum Teil computergestützte elektroakustische ergänzt (Kittel und Schürenberg, 1988).

58.1.2 Entwicklungsstörungen der Stimme

58.1.2.1 Definition und Vorkommen

Für das Wachstum des Kehlkopfes, der Resonanzhöhlen und die Beschaffenheit der Schleimhäute sind neben genetischen Faktoren vor allem hormonelle Einflüsse bestimmend, so das hypophysäre Wachstumshormon, die Hormone der Schilddrüse, der Nebennierenrinden und vor allem der Gonaden, besonders auffällig während der Pubertät. Die stimmliche Funktion kann auch von weiteren Komponenten beeinflußt werden. Im besonderen eine Dissoziation zwischen morphologischen Strukturen und willkürlicher bzw. unwillkürlicher Stimmgebung kann zu pathologischen Veränderungen der Stimmlage, des Klanges, des Stimmumfangs usw. sowie zu einer Einschränkung der Belastbarkeit der Stimme führen.

58.1.2.2 Pathogenetische Konzepte

Eine bedeutende Rolle spielen Vorbilder von Personen und deren Stimmen, die nachgeahmt werden. So werden besonders im Schlagergesang bisweilen abnorm hohe Männer- und Knabenstimmen von Männern, vielfach auch Männer- und tiefe Frauenstimmen von Frauen nachgeahmt. Verbreitet ist auch die Erhaltung der Kinderstimme durch Chorknaben während und nach dem Stimmwechsel. Hormonell Stimmgestörte akzeptieren häufig nicht ihre veränderte Stimme: Virilisierte Frauen (s.u.) versuchen oft durch erhöhte muskuläre Kontraktion die Sprechstimme zu erhöhen bzw. durch Engstellen des Ansatzrohres den Stimmklang aufzuhellen, was schließlich zu vorzeitiger Ermüdung der Stimme führt. Ein ähnliches funktionelles Fehlverhalten beobachtet man bei Männern, die sich nicht zur männlichen Rolle bekennen. So entsteht vor dem Hintergrund sexueller Identifikationsprobleme eine nervale Fehlsteuerung im Bereich des Stimmapparates, die daraus resultierende Stimmstörung erhält angesichts der mangelnden stimmlichen Leistungsfähigkeit Krankheitswert.

Bei Stimmwechsel-(Mutations-)Störungen muß stets auch an Unmusikalität gedacht werden, wobei z.B. die männliche Stimme als Vorbild nicht wahrgenommen wird und auch die auditive Kontrolle der eigenen Stimme eingeschränkt ist. Im Einzelfall muß stets abgeklärt werden, ob und wie weit die Mutationsstörung durch hormonelle Abweichungen verursacht wird oder ob sie funktionell bedingt ist.

58.1.2.3 Einzelne Störungen

Stimmwechselstörungen fallen durch die inadäquate Sprechstimmlage im Verhältnis zum meist nicht normalen Umfang der Stimme sowie durch veränderten Stimmklang auf. Je nach zeitlichem Verlauf der mittleren Sprechstimmlage unterscheidet man in der Phoniatrie verschiedene Formen (Abb. 58.1–1).

Mutationsfistelstimme bei Knaben

Obwohl der Kehlkopf bereits normale männliche Dimensionen angenommen hat, liegt die Sprechstimmlage oft noch höher als vor dem Stimmwechsel. Die Stimme klingt hell, ist wenig tragfähig und leicht ermüdbar. Nicht selten wechselt die Stimme bei Männern zwischen dem normalen Brustregister und der hohen Sprechtonlage („Stimmbruch"). Laryngoskopisch sieht man mitunter eine diffuse Rötung der normallangen Stimmlippen (sekundäre Hyperämie) so-

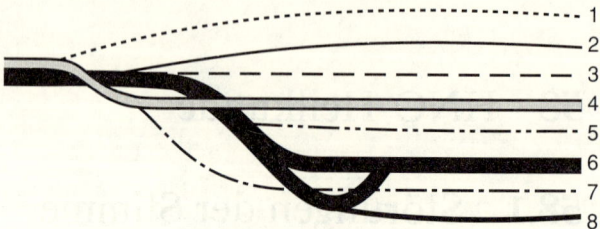

Abb. 58.1–1. Die mittlere Sprechstimmlage als Kardinalsymptom der Mutationsstörungen (aus Pascher, 1982).
1 Mutationsfistelstimme bei Mädchen
2 Mutationsfistelstimme bei Knaben
3 persistierende Kinderstimme (männlich und weiblich, endokrin bedingt)
4 normale Mutation bei Mädchen
5 unvollständige Mutation bei Knaben
6 normale Mutation bei Knaben
7 perverse Mutation bei Mädchen
8 Mutationsbaß beim männlichen Geschlecht

wie bisweilen einen dreieckigen Glottisspalt im hinteren Bereich.

Viele der heranwachsenden Männer werden von ihrer Mutter in die Sprechstunde begleitet. Nicht selten ist der Vater verstorben oder es bestehen Aggressionen gegen ihn, bzw. es liegt eine abnorme Mutterbindung vor. Die Identifikation mit der männlichen Rolle ist (noch) nicht erfolgt.

Die Therapie der Mutationsfistelstimme ist beispielhaft für eine phoniatrisch-logopädische Behandlung. Obgleich die Entstehung durch eine Reihe von Faktoren bewirkt werden kann, ist der therapeutische Ansatz ein **funktioneller.** Die formale Entstehung des Störungsbildes liegt in einer zu intensiven Kontraktion des M. cricothyroideus, wodurch die Stimmlippen zu stark gespannt werden. Durch Druck von vorn auf den Schildknorpel (Bresgenscher Handgriff) sinkt die Sprechstimmlage meist sofort um ca. 1 Oktave zur Norm ab. Stehen der Patient bzw. dessen Angehörige der normalen Stimme ablehnend gegenüber, müssen diese dahingehend beraten werden, daß die neue Stimme die adäquate darstellt und nur eine solche ausreichend belastbar ist. Die Behandlung kann durch eine Anästhesie des N. laryngeus superior unterstützt bzw. eingeleitet werden. Im Verlaufe der logopädischen Behandlung wird die tiefe Stimmlage allmählich fixiert.

Unvollständige Mutation

Leitsymptom ist der verspätete Beginn des Stimmwechsels, seine lange Dauer oder dessen ungenügender Erfolg, bzw. jener bleibt überhaupt aus. Der Kehlkopf ist meist ausgewachsen. Die Sprechstimmlage ist zu hoch und liegt bei den Männern oft zwischen der männlichen und weiblichen Sprechtonhöhe. Bei den Frauen ist die Sprechstimmlage meist ebenfalls zu hoch, der Stimmklang kindlich. Häufig besteht während der Zeit des Stimmwechsels eine Hyperfunktion, die ihre Ursache in anlagebedingten Mängeln des Stimmapparates, mangelnder Stimmschonung,

Hörstörungen oder Unmusikalität hat, nicht selten in seelischen Ursachen begründet ist. Ausgeschlossen werden muß stets ein primärer oder sekundärer Hypogenitalismus.

Die Behandlung erfolgt nach den gleichen Prinzipien wie bei der Mutationsfistelstimme, meist erstreckt sie sich über einen längeren Zeitraum. Ihr Erfolg hängt wesentlich davon ab, inwieweit der Stimmumfang nach unten erarbeitet werden kann.

Virilisierung der weiblichen Stimme

Androgene Hormone bewirken bei Frauen und Mädchen nach der Pubertät im Rahmen einer allgemeinen Virilisierung typische Veränderungen seitens der Stimme. Neben dem adrenogenitalen Syndrom findet man (selten) androgenbildende Tumoren im Bereich der Ovarien und der Nebennierenrinde. Eine Virilisierung als Folge androgener Drogen ist durch Testosteron und Anabolika möglich, ferner z.B. bei 19-Nortestosteron-Derivaten, die teilweise in Ovulationshemmern enthalten sind. Die Empfindlichkeit des weiblichen Organismus gegenüber diesen Substanzen ist individuell extrem verschieden.

Die Stimmstörungen sind anfangs nur subjektiv wahrnehmbar und lange Zeit unspezifisch. Später senkt sich der Stimmumfang nach unten, besonders im tieferen Bereich kommt ein männliches Timbre zustande. Vorwiegend musikalische Frauen versuchen lange Zeit, die weibliche Sprechtonhöhe und den weiblichen Klang zu erhalten, was häufig zu Überanstrengungen der Stimme führt.

Im Rahmen der allgemeinen Virilisierung treten Veränderungen der Stimme meistens zuerst auf, später folgen andere Symptome wie Behaarung im Gesicht und an den Beinen, Akne, Klitorishypertrophie und Amenorrhoe.

Die Veränderungen der Stimme sind praktisch irreversibel, die Therapie besteht daher vorrangig in der Einübung einer den neuen und veränderten anatomischen Verhältnissen entsprechenden tieferen Sprechstimmlage, wodurch eine relativ belastbare und tragfähige Stimme erreicht wird. Bei vielen Patientinnen treten psychische Erscheinungen auf, zumal sie insbesondere am Telefon als Männerstimmen verkannt werden. Ein wesentliches Ziel der Therapie liegt darin, daß die betroffenen Frauen angesichts der Notwendigkeit einer leistungsfähigen Stimme die neue Sprechstimmlage akzeptieren.

58.1.3 Funktionelle Stimmstörungen

58.1.3.1 Definition und Vorkommen

Unter dem Begriff „funktionelle Stimmstörungen" faßt man Störungsbilder mit uneinheitlicher Ätiologie zusammen, die mit Beeinträchtigung des Stimmklangs und/oder einer Einschränkung der stimmlichen Leistungsfähigkeit einhergehen, oft verbunden mit lokalen Mißempfindungen, wobei ursächliche organische Veränderungen im Bereich der an der Stimmbildung

beteiligten Organe nicht faßbar sind (vgl. z.B. Pascher und Bauer, 1984). Die stimmlichen Symptome können die Sprechstimme und die Gesangsstimme einzeln und gemeinsam betreffen. Ein häufiges Zeichen ist die Ermüdbarkeit der Stimme schon nach geringer Belastung. Die Qualität der Stimme wechselt oft rasch, was auf eine seelische (Mit-)Verursachung hinweist. Klinisch abgrenzbare Manifestationsformen sind die Taschenfaltenstimme (s.u.), die psychogene Aphonie und die spastische Dysphonie. Der Krankheitswert ist individuell sehr verschieden und hängt neben der Ausprägung und Prognose der Stimmstörung vom persönlichen Betroffensein im Berufsleben, in der privaten Sphäre bzw. auch von der kulturellen Bewertung der Stimme ab.

58.1.3.2 Epidemiologie

Der Anteil hyperfunktioneller Stimmstörungen (s.u.) ist relativ groß (70%), hypofunktionelle finden sich häufiger im mittleren und höheren Lebensalter und bei Frauen. Betroffen sind vielfach stimmintensive Berufe, z.B. Lehrer (bis 25%), die im Verlaufe des Unterrichts-Vormittags über Ermüdungserscheinungen seitens der Stimme, auch mit lokalen Mißempfindungen, klagen. Die Stimmstörung manifestiert sich häufig im Verlaufe eines Infekts oder bei emotionellen Belastungen (z.B. Autoritätsprobleme). Bei anlagebedingter Minderwertigkeit des Stimmapparates kann die Stimmstörung schon während der Referendarzeit auftreten und mangels alternativer Verwendungsmöglichkeit die berufliche Eignung in Frage stellen. Nicht selten sind Doppelbelastungen durch außerberufliches Engagement (Sänger, Chor, Nebenämter usw.), bei Frauen durch Beruf und Familie. Anfangs erholen sich die Stimmen bis zum folgenden Tag bzw. während der Ferien, später kommt es zu dauernder Heiserkeit.

Kindergärtnerinnen müssen sich stimmlich gegen den hohen Lärmpegel durchsetzen. Weitere stimmintensive Berufe sind: Schauspieler, Pfarrer, Soldaten, Verkäufer usw.

Sänger bemerken z.B., daß anfangs einzelne Töne insbesondere des oberen Stimmbereichs oder solche am jeweiligen Übergang der Register nur im Forte oder gar nicht gesungen werden können. Häufiger sind Tenöre und Soprane betroffen. Infekte, allgemeine Indisposition, ungeeignete Stimmtechniken und emotionelle Belastungen (z.B. viele und schwierige Rollen, Probleme in der beruflichen und privaten Sphäre) können zu kürzer oder länger dauernden Stimmschwierigkeiten führen („Krisen" der Sänger).

58.1.3.3 Pathogenetische Konzepte

Im Gegensatz zu den organisch bedingten Stimmstörungen stellen die funktionellen Dysphonien komplexe biologische Vorgänge dar, meist ein unübersehbares Flechtwerk von Korrelationen morphologischer, funktioneller, vegetativer, biochemischer, hormonel-

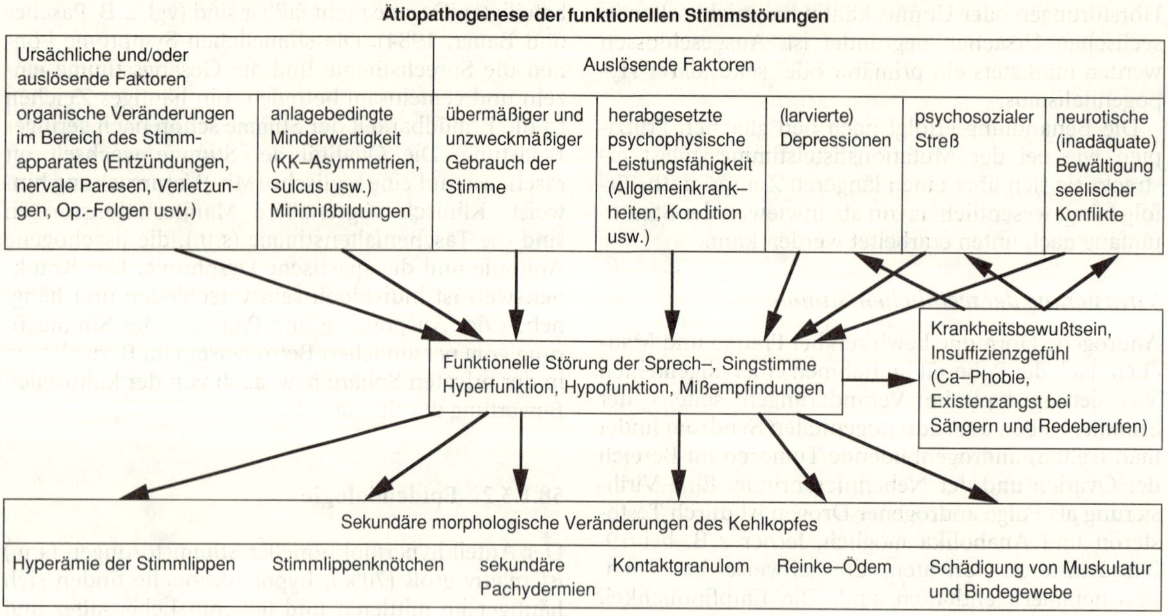

Ätiopathogenese der funktionellen Stimmstörungen

Ursächliche und/oder auslösende Faktoren		Auslösende Faktoren				
organische Veränderungen im Bereich des Stimmapparates (Entzündungen, nervale Paresen, Verletzungen, Op.-Folgen usw.)	anlagebedingte Minderwertigkeit (KK–Asymmetrien, Sulcus usw.) Minimißbildungen	übermäßiger und unzweckmäßiger Gebrauch der Stimme	herabgesetzte psychophysische Leistungsfähigkeit (Allgemeinkrankheiten, Kondition usw.)	(larvierte) Depressionen	psychosozialer Streß	neurotische (inadäquate) Bewältigung seelischer Konflikte

Funktionelle Störung der Sprech–, Singstimme Hyperfunktion, Hypofunktion, Mißempfindungen

Krankheitsbewußtsein, Insuffizienzgefühl (Ca–Phobie, Existenzangst bei Sängern und Redeberufen)

Sekundäre morphologische Veränderungen des Kehlkopfes					
Hyperämie der Stimmlippen	Stimmlippenknötchen	sekundäre Pachydermien	Kontaktgranulom	Reinke–Ödem	Schädigung von Muskulatur und Bindegewebe

Abb. 58.1–2. Psychosomatisches Konzept der Ätiopathogenese funktioneller Stimmstörungen (siehe Text).

ler und psychopathologischer Abweichungen (Bauer, 1961). Unter klinischen Aspekten lassen sich für die Ätiopathogenese funktioneller Stimmstörungen sieben Faktorenbündel nachweisen, die einzeln und kombiniert auftreten können (Abb. 58.1–2, obere Reihe).

Bei den stimmlichen Veränderungen unterscheidet man klinisch hyperfunktionelle und hypofunktionelle Zeichen im Bereich der Atmung, der Stimmgebung und des Stimmansatzes. Die Hyperfunktion kann dabei die kompensatorische Reaktion auf die muskuläre Schwäche, die Hypofunktion eine Schonhaltung nach hyperfunktioneller Überbeanspruchung sein, so daß man Zeichen beider Art beim Patienten gleichzeitig oder nacheinander sieht. Die Definition richtet sich nach der Art der primären Fehlfunktion.

Zeichen einer **Hyperfunktion** sind (Wendler und Seidner, 1987): hastige, zu häufige Atmung, ächzende Einatmung, Ausatmung unter Druck, unphysiologische schnüffelnde Einatmung durch die Nase während des Redens, harte bis knarrende Stimmeinsätze, Vorwölbung der Taschenfalten, teils bis zur Bildung einer Taschenfaltenstimme, Hochziehen des Kehlkopfes, gepreßter Tonansatz, Verlagerung der Resonanz in den Rachenraum, Steifhalten des Unterkiefers, Anspannen der Mundbodenmuskulatur, krampfhafte Lippenbewegungen und Stirnrunzeln. Stroboskopisch zeigen die Stimmlippen eingeschränkte Schwingungsamplituden, die auch bei Intensitätssteigerung nicht zunehmen. Die Schlußphase der Glottis ist scheinbar bei geringer Intensität verlängert. Perioden und Phasen sind oft irregulär.

Zeichen einer **Hypofunktion** sind: unzureichende oberflächliche Atmung, hauchiger Stimmeinsatz, leise bis flüsternde Stimmgebung, undeutliche Aussprache, ungenügendes Öffnen des Mundes. Der Stimm-

lippenschluß ist unvollkommen. Man sieht oft einen spindeligen Spalt oder ein offenes Dreieck im hinteren Bereich. Stroboskopisch finden sich erweiterte Schwingungsamplituden, die auch bei Intensitätsminderung nicht abnehmen. Die Schlußphase ist verkürzt. Amplituden und Phasen sind oft irregulär.

Neben den stimmlichen Symptomen klagen viele Patienten über **subjektive Mißempfindungen** mit Druckgefühl, Trockenheit oder vermehrtem Schleim, Jucken, Brennen, Kratzen, Fremdkörpergefühl, was zu Räusperzwang führt, eventuell sogar zu Schmerzen. Nicht selten stehen diese Symptome sogar im Vordergrund.

Anlagebedingte Minderwertigkeit des stimmgebenden Organsystems kann zu einer funktionellen Stimmstörung prädisponieren. Die Leistungsfähigkeit der menschlichen Stimme ist individuell sehr verschieden. Bei geringer Belastungsfähigkeit sind die Stimmen vielfach den Anforderungen in Beruf und privater Sphäre nicht gewachsen. Bei größeren Anforderungen, wie sie in vielen Berufen stimmlich und auch seelisch gestellt werden, treten funktionelle Fehlhaltungen auf. Der Funktionsablauf des Stimmapparates tritt aus seinem ökonomischen Arbeitsbereich heraus und löst damit nach Art des Circulus vitiosus weitere Fehlfunktionen aus.

Hinweise für eine konstitutionelle Schwäche sind Asymmetrien des Kehlkopfknorpelgerüsts, der Länge bzw. Höhe der Stimmlippen, ein Sulcus glottidis, Hypoplasien der Stimmlippenmuskulatur usw. Prophylaktische Stimmbelastungstests als Tauglichkeitsuntersuchungen für stimmintensive Berufe finden sich in der zusammenfassenden Darstellung von Heidelbach (1981). In der DDR ist das Bestehen derartiger Eignungsprüfungen vor dem Studium vieler Stimmberufe Voraussetzung.

Funktionelle Stimmstörungen werden insbesondere von Sprecherziehern und in der Gesangspädagogik als **stimmtechnische Fehler** bzw. als deren Folge aufgefaßt. Zu einer Überlastung der Stimme führen können u. a.: inadäquate Sprechstimmlage, Überschreiten des Stimmumfangs, Benutzung falscher Register, dauerndes zu lautes Sprechen, bei Sängern falsche Stimmtechniken und ungeeignete Übungs- und Ausbildungsmethoden.

Die Stimme wird von der individuellen Psychodynamik mitdeterminiert und ist normalerweise bereits allgemeinen konditionellen Schwankungen unterworfen. Erkrankungen und Zustände, die mit chronischer körperlicher Erschöpfung einhergehen, führen zu Hypofunktionen im Bereich der Muskulatur. Eine herabgesetzte psychophysische Leistungsbreite findet man häufig bei Kreislaufinsuffizienzen, -hypotonie, Anämie, chronischen Lebererkrankungen, während des Klimateriums der Frau (ca. 45–55 Jahre), des Involutionsalters des Mannes (ca. 55–60 Jahre), nach Gewichtsverlusten, bei Abmagerungskuren, als Folge von Genußgiften usw. Leitsymptom ist das Gefühl der Erschöpfung; die Patienten fühlen sich von einer Krankheit noch nicht genesen und allgemein wenig leistungsfähig.

58.1.3.4 Einzelne Störungsbilder

Funktionelle Stimmstörungen in Verbindung mit organischen Erkrankungen des Stimmapparates

Viele Organerkrankungen im Bereich des Stimmapparates können die Entstehung funktioneller Stimmstörungen begünstigen bzw. veranlassen, die sich so auf die primäre laryngeale Erkrankung aufpfropfen. Sie können zeitlich neben dieser vorhanden sein oder selbst nach Abklingen der organischen Veränderungen bestehenbleiben; z.B. führen entzündliche Erkrankungen der oberen und unteren Luftwege über passagere und bleibende Gewebsabweichungen zu Änderungen in den kinästhetischen Rückkoppelungsmechanismen. Hinweisendes Symptom ist wie bei allen funktionellen Stimmstörungen das Mißverhältnis zwischen den morphologischen Abweichungen einerseits sowie Art und Ausmaß der Stimmstörung andererseits. Die funktionelle Stimmstörung ist hier Teilerscheinung einer komplexen kompensatorischen Reaktion, oft auf das Bewußtwerden einer mangelnden stimmlichen Leistungsfähigkeit. Häufige Kehlkopferkrankungen, die funktionelle Stimmstörungen auslösen, sind: Laryngitiden aller Arten und Verlaufsformen, Kehlkopfnervenlähmungen, Verletzungen und Operationsfolgen im Bereich der Stimmlippen sowie hormonelle Stimmstörungen.

Eine Reihe von Erkrankungen im Bereich des Kehlkopfes sind Gewebsreaktionen auf laryngeale Fehlfunktionen, d.h. ihr Vorhandensein impliziert das Bestehen einer funktionellen Stimmstörung (vgl. Abb. 58.1–2, untere Reihe).

Rötung der Stimmlippen. Bereits eine normale intensive stimmliche Tätigkeit führt oft zu vermehrter Rötung der Stimmlippen, meist am freien Rand. Diese Hyperämie darf nicht mit entzündlichen Erscheinungen verwechselt werden.

Stimmlippenknötchen (Schrei- oder Sängerknötchen) sind Verdickungen am freien Rand der Stimmlippen, ein- oder beidseitig, meist im mittleren Drittel. Anfangs sind sie weich und können sich spontan bei Stimmruhe bzw. unter Stimmtherapie wieder zurückbilden. Später werden sie hart und müssen gegebenenfalls operativ entfernt werden.

Kontaktgranulom. Diese Krankheit tritt als Pachydermie, seltener als Kontaktulkus im Bereich der Processus vocales auf. Bevorzugt sind Männer, der Altersgipfel liegt bei 40–60 Jahren. Formalgenetisch wird ein Hammereffekt der aufeinanderschlagenden Processus vocales verbunden mit festem Stimmschluß angenommen. Miethe (1988) beschreibt eine charakteristische unspezifische Grundstruktur der Patienten: starre, ernste Miene mit fast ängstlichem Gesichtsausdruck, verantwortliche Berufsstellung. Die Körperspannung ist schlaff, aber kontrolliert. „Die Patienten können schlecht aus sich herausgehen, fressen viel in sich hinein, hoher Leistungsanspruch, Zug zur Pedanterie, Schuldgefühle, psychovegetative Symptome." Als Ursache der Granulome wird auch die Refluxösophagitis diskutiert.

Laryngoskopisch sieht man oft einen spindeligen Spalt der Glottis. Die Stroboskopie zeigt in Analogie zur Körperspannung das Zuviel und Zuwenig: im Bereich der vorderen zwei Drittel der Stimmlippen weite und wechselnde Amplituden, im hinteren Drittel keine oder minimale Schwingungsabläufe.

Weitere sekundäre morphologische Veränderungen, die als Folge übermäßiger und fehlerhafter Stimmfunktion auftreten können, sind Ödeme, Polypen sowie Schädigungen im Bereich muskulärer und/oder bindegewebiger Strukturen.

Funktionelle Stimmstörungen als unspezifische Reaktion auf emotionale Belastungen

Bei vielen funktionellen Stimmstörungen deckt die psychosoziale Anamnese Schwierigkeiten im emotionellen und im interaktionellen Bereich auf. **Erwachsene** schildern ihre persönliche Lebenssituation meist als unerträglich und aussichtslos. Sie geben Schwierigkeiten im Berufsleben oder in der zwischenmenschlichen Sphäre an: Ehekonflikte, Unzufriedenheit im Berufsleben, Spannungen mit Vorgesetzten und Mitarbeitern, verbunden mit dem Gefühl, nicht zu umgehenden Forderungen oder einem mit Ehrgeiz verfolgten Lebensziel nicht gewachsen zu sein. Bei **Kindern** erfährt man oft von intrafamiliären Spannungen, abnormer Verwöhnung, mangelnder Nestwärme, uneinheitlichem Erziehungsstil der Eltern, Geschwisterrivalität, Problemen im schulischen Bereich. Die Erhebung der Anamnese läßt oft deutlich werden, wie unmittelbar die Psychodynamik die Stimmfunktion beeinflußt: Trifft man im Gespräch den störungsauslösenden Komplex, so verändert sich die Stimme. Ihre Spannung nimmt zu, sobald die Lebensschwierigkeiten dem Patienten ins Bewußtsein treten; die Spannung löst sich wenigstens vorübergehend, wenn er seine aufgestauten Lebensprobleme vortragen konnte und die krankmachenden Zusammenhänge selbst erkennt.

Neben der Stimmstörung fallen bei den Patienten Anspannung, gereizte Stimmung und Angst auf. Die Prognose ist oft gut, wenn der Arzt das Problem mit dem Patienten durchspricht und nötigenfalls mit stützenden psychotherapeutischen Maßnahmen eingreift. Die Hintergründe müssen erkannt, Lösungsmöglichkeiten für deren Beseitigung bzw. Überwindung mit dem Patienten erörtert werden.

> Kaufmann, 46 J., seit 2 Jahren rezidivierend heiser, er vertrage das Klima am Wohnort seiner neuen Arbeitsstelle nicht. Sobald er den Urlaub in den Bergen verbringe, sei die Stimme „frei". Nach Rückkehr in den Wohnort sofort Verschlechterung. Die genaue Anamnese ergibt, daß die Stimme bereits während der Rückfahrt nachläßt. Es zeigt sich, daß der Patient in seinem neuen Arbeitsbereich völlig überfordert ist.

> Lehrer, 53 J., klagt seit 6 Monaten über wechselnde Stimmstörung. Seit dieser Zeit müsse er in einem Klassenraum mit schlechter Resonanz unterrichten. Die Anamnese deckt massive Autoritätsschwierigkeiten auf.

Funktionelle Stimmstörungen im Rahmen von Depressionen

Funktionelle Stimmstörungen können Symptome einer (larvierten) Depression sein. Typisch hierfür ist oft der Tagesverlauf der Symptomatik: Während bei den Berufsdysphonien die Stimmstörung im Laufe der Belastung zunimmt, folgt ihre Symptomatik hier der zirkadianen Periodik der Depression. Die alleinige logopädische Therapie ist oft wenig ergiebig.

Die Stimme wird von einer Lehrerin morgens als ausgesprochen schlecht geschildert, bessere sich aber im Verlaufe der ersten Unterrichtsstunden und sei nachmittags und abends relativ gut.

Ein Sänger benötigt oft relativ viel Zeit, um sich morgens einzusingen.

Funktionelle Stimmstörungen als Symptomatik neurotischer Erkrankungen

Eine Reihe funktioneller Stimmstörungen lassen sich nur unter **neurosepathogenetischen** Gesichtspunkten interpretieren. Die Aufdeckung der Ätiopathogenese bedarf von seiten des Arztes einer eingehenden biographischen Anamnese, in der Regel ist eine Behandlung durch einen Psychotherapeuten erforderlich, eventuell in Zusammenarbeit mit dem Logopäden.

Erfahrungsgemäß noch wenig beachtet wird, wie der Patient sich zu seiner Stimmstörung verhält. Dies geht als weiterer Faktor in die Krankheit ein und gestaltet deren Symptomatologie, Prognose und Verlauf. Schon das Bewußtwerden mangelnder stimmlicher Leistungsfähigkeit wird oft mit Hyperfunktionen

kompensiert, stimmliche Anstrengungen haben oft sekundäre Hypofunktionen zur Folge. Bekannt sind Patienten, die den äußeren und inneren krankmachenden Faktoren konsequent entgegentreten, sich vor Infekten schützen, Fehlbeanspruchungen der Stimme vermeiden, den Einsatz der Stimme auf die berufliche Tätigkeit konzentrieren, auch den Fehlhaltungen und den Fehlspannungen nach Kräften entgegentreten und für eine konsequente Stimmtherapie motiviert sind. Auf der anderen Seite stehen die Patienten, deren Stimmstörungen sich aus einem wenn auch oft unbewußten Streben nach Lebenssicherung und Krankheitsgewinn herleiten, d.h. wo diese zum Anlaß genommen werden, den alltäglichen Schwierigkeiten und den Anforderungen des Lebens auszuweichen, Rentenbegehren bzw. Schadensersatzforderungen zu unterstützen.

Das Bewußtwerden einer Stimmstörung löst bei vielen Patienten auch **sekundäre seelische Symptome** aus, die ihrerseits das Krankheitsgeschehen zusätzlich aufladen können. Verbreitet ist die Angst vor einer malignen Erkrankung, bei Sängern vor Stimmlippenknötchen. Bei vielen Berufsdysphonien ist das Gefühl des Krankheitsbewußtseins nachteilig, auch die Vorstellung, den stimmlichen Anforderungen nicht (mehr) gewachsen zu sein. Hier muß anhand einer eingehenden Diagnostik der Patient vom Nichtvorhandensein einer organischen Erkrankung überzeugt werden. Dieses ärztliche Vorgehen läßt viele aus Angst entstandene Stimmstörungen bald verschwinden; bei anderen Patienten ist es Voraussetzung, um eine eventuell notwendige Psychotherapie einleiten zu können.

Stimmstörungen mit klinisch abgrenzbaren Manifestationsformen

Funktionelle Aphonie

Das klinische Bild ist gekennzeichnet durch einen meist plötzlichen oder rezidivierenden praktisch vollständigen Verlust der Stimme. Die Kommunikation erfolgt ausschließlich flüsternd; hingegen ist der Hustenstoß in der Regel tönend. Oft wechselt die normale Stimme mit der Aphonie ab. Bei der **psychogenen Stummheit** erfolgt von seiten des Patienten keine sprachliche Kommunikation oder die tonlosen Artikulationsbewegungen geschehen demonstrativ verändert mit schwer verständlicher Aussprache. In allen Fällen fehlt ein ursächlicher pathologisch-anatomischer Befund, insbesondere im Bereich des Kehlkopfes. Die psychogene Ursache der funktionellen Aphonie ist schon lange bekannt. Sie wird als Manifestation einer Konversionsneurose angesehen (Aronson et al., 1968). Kinzl et al. (1988) fanden heterogene Persönlichkeitsstrukturen und unspezifische Konfliktsituationen. Für das Verständnis zu berücksichtigen ist – wie bei allen psychogenen Erkrankungen – die Bedeutung des primären (Symptombildung als psychische Entlastung) und des sekundären Krankheitsgewinns (positive soziale Folgen, wie vermehrte Beachtung und Entlastung von Aufgaben).

Sehr oft gelingt es dem HNO-Arzt, die normale Stimme in einer einzigen Sitzung wiederherzustellen. Hierfür wurden eine Reihe von Interventionen („Überrumpelungsversuche") bekannt (vgl. Zusammenfassung von Wirth, 1987). Ihre wesentlichen Elemente sind:

- die Überzeugung des Patienten, daß ein organisches Leiden nicht vorliegt,
- das reflektorische Erzwingen des Glottisschlusses mittels endolaryngealer Reizung,
- Ausschalten der audiophonatorischen Rückkopplung durch Vertäubung der Ohren, z.B. mit zwei Barany-Lärmtrommeln,
- eventuelles Ausschalten der kinästhetischen Eigenkontrolle mittels lokaler Anästhesie des Rachens und Kehlkopfes.

Sobald die tönende Stimme erreicht wurde, ist die Zwecktendenz von dem Symptom Stimmlosigkeit entkoppelt. Anschließend kann der Patient einer Psychotherapie zugeleitet werden.

Spastische Dysphonie

Hierbei handelt es sich um die extreme Form der hyperfunktionellen Stimmstörung. Die Stimme klingt gequält, stöhnend, ächzend, die Stimmeinsätze sind knarrend. Häufig werden die Vokale in mehrere Tonstöße geteilt. Der Stimmritzenkrampf tritt meist nur während der Phonation auf, während die nichtkommunikative Stimmgebung (Flüstern, Lachen, Singen) kaum betroffen ist. Auch die Artikulation ist mühsam, die Ausatmung erfolgt mit Hilfe der Bauchpresse, verbunden mit Spasmen der Thorax- und Abdominalmuskulatur.

Die Entstehung wird durch ein Überwiegen laryngealer Sphinktermechanismen gegenüber der Phonation erklärt. Meist geht der Erkrankung ein längerer Leidensweg seelischer, körperlicher oder sozialer Art voraus. Die spastische Dysphonie wird überwiegend als psychisch bedingte Stimmstörung (Konversionssymptom), die einen Lösungsversuch eines tiefgehenden emotionalen Konfliktes darstellt, angesehen. Manche Patienten können als Grenzfälle zur Psychose angesehen werden (Bloch, 1965). Vereinzelt werden auch neurologische Abweichungen und Symptome nachgewiesen und diese ursächlich in Betracht gezogen.

Die Behandlung ist in erster Linie psychotherapeutisch. Häufig kann die Auslösung der charakteristischen Stimmveränderungen auf ein schweres emotionales Trauma bezogen werden, bei manchen Patienten läßt sich jedoch ein psychischer Hintergrund der Krankheit nicht aufdecken. Zweckmäßig ist die Kombination mit einer konsequenten Atem- und Stimmbehandlung.

Neuerdings wurden Erfolge gesehen durch Resektion eines Teilstücks aus dem Nervus recurrens einer Seite, was eine Stimmlippenlähmung zur Folge hat, oder der Nerv wird durch Quetschung temporär ausgeschaltet. Angesichts dieser Behandlungserfolge wird für die Entstehung der spastischen Dysphonie auch eine Störung der propriozeptiven Eigenkontrolle der Stimme, möglicherweise auf dem Boden einer Virusinfektion, diskutiert.

Taschenfaltenstimme

Die Stimmbildung erfolgt hier nicht mittels der beiden Stimmlippen, sondern durch die oberhalb der Glottis zur Mitte vorgerückten Taschenfalten. Sie kann einen kompensatorischen Mechanismus darstellen, wenn die normale Funktion der Stimmlippen infolge Lähmung oder Defekten nicht möglich ist. Sie ist hier „erwünscht" und Ziel einer logopädischen Behandlung.

„Unerwünscht" ist die Taschenfaltenstimme hingegen, wenn die Stimmlippen funktionsfähig sind. Hier ist sie eine Erscheinung extremer laryngealer Hyperfunktion (s.o.), mitunter zum Zeitpunkt einer Kehlkopferkrankung entstanden und habituell fixiert, oder Folge seelischer Einflüsse. Für die Therapie ergeben sich mehrere Ansätze: logopädische Behandlung mittels Lockerungs- und Entspannungsübungen, eventuell Anästhesie der Nn. laryngei superiores oder der stylopharyngealen Muskulatur. Oft gelingt die funktionelle Umstellung auf die Stimmlippen durch Auseinanderhalten der Taschenfalten, z.B. mittels eines Endoskops, so daß diese sich nicht mehr berühren können, mitunter in einer Sitzung.

58.1.3.5 Richtlinien für die Therapie von Stimmstörungen

Die eingehende Anamnese und die vollständige Untersuchung stellen die Weichen für eine auch kausal orientierte, effiziente und rationale Behandlung.

Bei Organerkrankungen des Kehlkopfes kommt deren Therapie in der Regel das Primat zu. Die Diagnose „Laryngitis" wird oft zu leichtfertig gestellt und dabei werden dann hormonelle, funktionelle und seelische Ursachen übersehen. Auch der Stellenwert der stimmlichen Belastung wird oft als „Berufsdysphonie" fehleingeschätzt, häufig wurde die psychosoziale Anamnese nicht ausreichend erhoben.

Funktionelle Stimmstörungen haben bei stimmintensiven Berufen, aber auch bei außerberuflich stimmlich Tätigen in der Regel hohen Krankheitswert. Prinzip der logopädischen Behandlung ist die tonusregulierende Basistherapie. Die einzelnen Behandlungsphasen gliedern sich in

- Einführungsgespräch mit eventuell ergänzender Anamnese
- Wahrnehmungssensibilisierung für körperliche Spannungszustände zwecks Verbesserung der kinästhetischen, aber auch der akustischen Wahrnehmungsfähigkeit
- Atemregulierung
- Dynamische Bewegungsphonation
- Verbalisierung von psychischen Problemen.

Die logopädische Behandlung strebt die Schaffung einer ökonomischen Stimmfunktion, das Erreichen einer leistungsfähigen und klangvollen Stimme bei geringem Kraftaufwand an. Es gibt eine Reihe von Methoden in der phoniatrischen und logopädischen Literatur, im Einzelfall werden diese oft kombiniert, modifiziert und individuell auf den Patienten ausge-

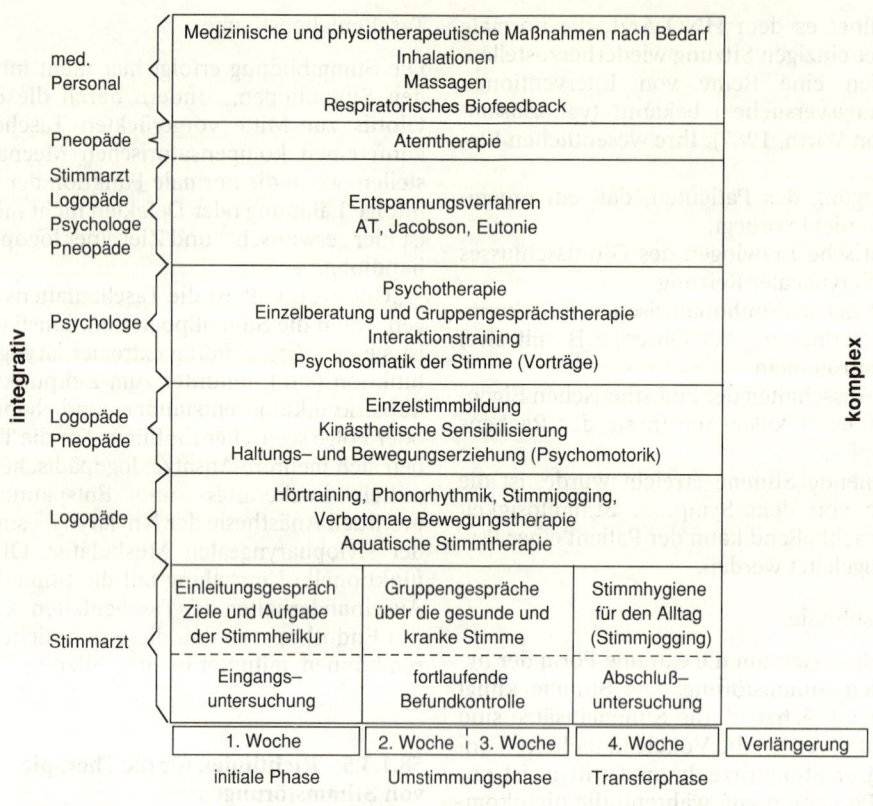

| med. Personal | Medizinische und physiotherapeutische Maßnahmen nach Bedarf Inhalationen Massagen Respiratorisches Biofeedback | | |

Pneopäde — Atemtherapie

Stimmarzt / Logopäde / Psychologe / Pneopäde — Entspannungsverfahren AT, Jacobson, Eutonie

Psychologe — Psychotherapie Einzelberatung und Gruppengesprächstherapie Interaktionstraining Psychosomatik der Stimme (Vorträge)

Logopäde / Pneopäde — Einzelstimmbildung Kinästhetische Sensibilisierung Haltungs– und Bewegungserziehung (Psychomotorik)

Logopäde — Hörtraining, Phonorhythmik, Stimmjogging, Verbotonale Bewegungstherapie Aquatische Stimmtherapie

Stimmarzt —

	Einleitungsgespräch Ziele und Aufgabe der Stimmheilkur	Gruppengespräche über die gesunde und kranke Stimme	Stimmhygiene für den Alltag (Stimmjogging)
	Eingangs– untersuchung	fortlaufende Befundkontrolle	Abschluß– untersuchung
	1. Woche	2. Woche ¦ 3. Woche	4. Woche	Verlängerung
	initiale Phase	Umstimmungsphase	Transferphase

intensiv

integrativ — *komplex*

Abb. 58.1–3. Therapeutischer Aufbau der Stimmheilkur im psychosomatischen Stimmheilzentrum (aus Gundermann, 1987).

richtet (vgl. Spiecker-Henke, 1982). Ziel der Therapie ist die Generalisierung der erlernten Übungsabschnitte.

Bei **Störungen der Gesangsstimme** muß oft ein Gesangspädagoge zugezogen werden.

Die **funktionelle Aphonie,** im besonderen die rezidivierende Form, bedarf in der Regel einer Psychotherapie, ebenso die spastische Dysphonie. Weitere Störungsbilder bedürfen alleiniger oder begleitender psychotherapeutischer Maßnahmen. Nicht selten stößt aber die Durchführung einer Psychotherapie bei funktionellen Störungen auf Schwierigkeiten. Vielen Patienten fehlt hierfür die Einsicht, sie entwickeln Widerstände allgemein gegen eine solche Behandlung, gegen ihre psychischen Auffälligkeiten oder gegen Psychotherapeuten, so daß die erforderliche Behandlung nicht zustande kommt. Ist die Stimme nicht nachhaltig gestört, fehlt auch vielfach der für eine solche Therapie erforderliche Leidensdruck. Aber selbst wenn die Psychotherapie zustande kommt, behindern nicht selten Gefühle des Versagens, der Scham und Schuld deren Fortgang (Behrendt, 1988).

Aus all diesen Überlegungen ergibt sich die Notwendigkeit der Integration der Psychosomatik und Psychotherapie in die HNO-Heilkunde. Nur der psychosomatisch geschulte Kliniker ist imstande, aus der Anamnese, dem sozialen Umfeld und den erhobenen Befunden eine integrierte ganzheitliche Betrachtung

zu ermöglichen, die wiederum die Weichen für eine auch kausal orientierte, nicht zuletzt effektive Behandlung stellt.

In der Phoniatrie wurden eine Reihe psychosomatisch orientierter Behandlungskonzepte entwickelt. Über den therapeutischen Aufbau solcher Stimmheilkuren siehe Gundermann, 1987 (Abb. 58.1–3).

58.1.4 Der kehlkopflose Patient

58.1.4.1 Anmerkungen zur Epidemiologie und Pathogenese des Kehlkopfkarzinoms

Das Kehlkopfkarzinom wird überwiegend bei Männern zwischen dem 55. und 70. Lebensjahr manifest. Die Häufigkeit wird im Mittel mit 3,9 auf 100 000 Einwohner jährlich angegeben. Die Relation männlicher zu weiblichen Patienten ist in Zunahme begriffen (früher: 20:1, neuere Angaben: 8:1), da offenbar immer mehr rauchende Frauen in das Krebsalter kommen. Kanzerogen sind die bei der Verbrennung des Tabaks entstehenden polyzyklischen, aromatischen Kohlenwasserstoffe. Die Indikation für die Totalexstirpation des Kehlkopfs ist das fortgeschrittene Larynxkarzinom, das eine Teilresektion nicht mehr möglich macht.

58.1.4.2 Die Bildung der Ersatzstimme

Grundsätzlich ist die Bildung der „Ösophagusstimme" anzustreben, die den Patienten von Zusatzgeräten unabhängig macht. Durch verschiedene Techniken erlernt der Kehlkopflose die Beförderung von Luft in den oberen Teil der Speiseröhre, wo diese willkürlich und dosiert ausgestoßen wird. Etwa in Höhe des Ösophagusmundes bildet sich eine Schleimhautfalte als vikariierende Glottis, die von der Luft in Vibrationen versetzt wird. Die Stimmbehandlung wird von Logopäden anfangs in Einzeltherapie, später in Gruppen durchgeführt (20–120 Sitzungen von 45–60 Minuten Dauer).

Die Anwendung eines Elektrolarynx als Tongenerator kommt in Betracht, wenn die Ösophagusstimme nicht den Anforderungen genügt oder die anatomischen Voraussetzungen nicht gegeben sind (Pharynxfistel usw.). Derartige batteriebetriebene Tongeneratoren werden meist außen an die Halshaut angesetzt und sind neben einer Reihe anderer Hilfsgeräte für Kehlkopflose im Handel.

58.1.4.3 Psychologische Probleme des kehlkopflosen Patienten

Vor dem operativen Eingriff stehen die Ängste im Vordergrund, die mit der Diagnose „Krebs", mit der Durchführung von Narkose und Operation, dem anfänglichen Verlust der Stimme, der eventuellen Bestrahlungsbehandlung oder Chemotherapie verbunden sind. Häufig kann selbst bei planmäßigem Behandlungsverlauf der Erkrankung der bisherige Beruf nicht mehr ausgeübt werden. Nach dem operativen Eingriff kann sich der Kehlkopflose zunächst nicht mitteilen, er kann seine aufgestauten Emotionen stimmlich nicht entladen. Es besteht Angst vor Ablehnung seitens der Umgebung und die Ungewißheit über den weiteren Krankheitsverlauf. Viele Patienten sind depressiv, der Alkoholkonsum nimmt zu.

58.1.4.4 Aufgaben der ärztlichen Betreuung

Vor dem chirurgischen Eingriff erhalten Arzt und Logopäde im Gespräch mit dem Patienten und dessen Angehörigen einen Einblick in die psychische, physische und soziale Situation des Betroffenen. Die unter Abschnitt 58.1.4.3 genannten psychologischen Probleme müssen eingehend besprochen werden. Einen breiten Raum nimmt die Aufklärung über die verschiedenen Möglichkeiten der stimmlichen Rehabilitation ein. Wichtig ist das persönliche Gespräch mit einem gut sprechenden Kehlkopflosen, wobei sich der Patient ein realistisches Bild von stimmlicher Rehabilitation und dem „Leben danach" machen kann. Durch Kontaktherstellung mit dem Sozialarbeiter müssen persönliche soziale Problemsituationen erörtert werden. Überdies beruhigt es den Patienten, wenn er bei bürokratischen Maßnahmen Unterstützung findet.

Zu den Aufgaben der Nachsorge gehören neben der regelmäßigen HNO-ärztlichen Untersuchung und der stimmlichen Rehabilitation die Zuführung zu regelmäßigen Gruppentreffen der Kehlkopflosen mit ihren Angehörigen, was in der Regel vielerorts monatlich stattfindet. Der Bundesverband der Kehlkopflosen organisiert in Zusammenarbeit mit HNO-Kliniken und Logopäden regionale Selbsthilfegruppen.

58.2 Funktionelle Störungen in der HNO-Heilkunde

Joseph Sopko

Die klassische Otorhinolaryngologie als Teilgebiet der Gesamtmedizin befaßt sich seit mehr als 100 Jahren vorwiegend mit den organisch-topographischen Gesichtspunkten der erwähnten Organe. Parallel mit dem allgemeinen Lebenswandel der heutigen industrialisierten Welt beobachtet man aber eine Verlagerung der Krankheiten auch unseres Fachgebietes. Die schweren, lebensbedrohlichen Krankheitsbilder, wie otogener Hirnabszeß, tonsillogene Sepsis, Sinus-cavernosus-Thrombose etc., sind dank dem medizinischen Fortschritt ausgesprochen selten geworden. In der Häufigkeit nehmen dafür die sog. Zivilisations- und Verbrauchskrankheiten immer mehr zu: Lärmschwerhörigkeit, Hörsturz, psychogene Aphonie, spastische Dysphonie etc. Psychische Konfliktsituationen können zu verschiedenen Krankheitsmanifestationen beitragen (beispielsweise sog. Examensangina).

In der Diagnostik der funktionellen Störungsbilder ist es ebenso wichtig, eine sorgfältige HNO-fachärztliche Untersuchung vorzunehmen, als auch die psychosomatischen Zusammenhänge zu erkennen. Eine gründliche Untersuchung beruhigt den Patienten und erweckt sein Vertrauen zum Arzt. Aber nur ein psychosomatisch ausgebildeter Kliniker vermag die relevanten somatischen, sozialen und psychischen Daten zu erheben und mit dem Befund der körperlichen Untersuchung in einem Arbeitsgang zu erfassen und ihre Bedeutung zu erkennen (Adler, 1986). Die moderne Otorhinolaryngologie sollte deshalb dem in ihrem Gebiet tätigen Facharzt zu einem integrativen somatopsychosozialen Krankheitsverständnis verhelfen und eine interdisziplinäre Zusammenarbeit nicht nur mit den anatomisch benachbarten Disziplinen der Neurologie und Ophthalmologie, sondern auch mit der sinngemäß ergänzenden Psychosomatik, Psychiatrie und Psychologie anstreben.

58.2.1 Schall, Klang und Lärm in ihrer Bedeutung für den Menschen

Die meisten Erscheinungen in der Natur sind mit Schall verbunden. Der Schall als physikalisches Phänomen entsteht durch die Schwingungen elastischer Körper. Die wohl geringste Intensität entsteht durch die Brownsche Molekularbewegung. Überdurchschnittlich empfindliche Menschen hören selbst dieses Molekularrauschen. In der Regel befindet sich die normale Hörschwelle 10 dB über diesem Molekular-

rauschen. Nur der sog. selektiven Frequenzempfindlichkeit des Gehörs (Békésy, 1970) ist es zu verdanken, daß der eigene tieffrequente Pulsschlag nicht immer gehört wird.

Wir sind aber ständig auf eine Schallumgebung angewiesen. So wie die Luft für das biologische Leben, ist der Schall für die geistige und körperliche Gesundheit des Menschen ein unentbehrliches Medium. Ähnlich dem ständigen Wechsel zwischen Inspirium und Exspirium brauchen wir die polaren Gegensätze von Stille und Schall, um seelisch gesund zu bleiben. In einer Camera silens – mit vollständiger Schallabsorption – eingeschlossen, entwickelt der Mensch bereits nach zwei Stunden psychotische Reaktionen von Verwirrtheit und Desorientierung. Unter der Stille verstehen wir aber keinen schalltoten Raum, sondern einen schallarmen Ort (Sopko, 1986), an dem das Individuum in ausgeglichenem Zustand sich ruhig erholen kann. Die verschiedenen Schallqualitäten wie Ton, Klang, Geräusch, Lärm und Knall können physikalisch gemessen werden. Die darin enthaltene Information und der emotionelle Inhalt können aber nur durch Bedeutungserteilungen (vgl. Kap. 1) psychologisch und linguistisch mit den bereits vorhandenen Engrammen verglichen werden.

Die Aufnahme und Transformation des Schalles in bioelektrische Impulse erfolgen im Hörorgan; die Analyse der Information hingegen findet in den Hörbahnen und -zentren (akustischer Analysator) statt. Die schädigende Wirkung von physikalisch definiertem Lärm und Knall auf das Hörorgan ist heutzutage genügend erforscht. Die weltweit anerkannten Grenzen der schädlichen Belastung bewegen sich für längerdauernde Lärmeinwirkung um 85 dB, bei kurzdauerndem Knall oder Explosion um 130 dB.

Viel bedeutsamer aber als an den heutigen Arbeitsplätzen ist die Lärmbelastung der breiten Bevölkerung durch den Luft- und Straßenverkehr sowie durch die ständig wachsende Industrialisierung.

Die **Schlafstörung** infolge Lärmeinwirkung ist hiervon wohl am bekanntesten. Individuelle Unterschiede in der Empfindlichkeit sind zwar groß und mit der Zeit tritt auch Gewöhnung (Habituation) auf einen gewissen konstanten Schallpegel ein. Doch verursacht bereits jedes vorüberfahrende Auto Veränderungen im Schlaf-EEG und an Flugzeuglärm kann man sich gar nicht gewöhnen (Jansen und Schulze, 1964).

Daß ein solcher Umgebungslärm – sei es im Schlaf- oder Wachzustand – potentiell gesundheitsschädigend ist, liegt auf der Hand. Die chronische Lärmbelastung kann für das Individuum einen ständigen Streß bedeuten (vgl. Streßkonzept in Kap. 8). Ganz besonders leiden Kinder unter Lärmeinwirkung. Gestört werden nicht nur Konzentrationsfähigkeit und Fleiß, sondern auch das Wachstum, die allgemeine Leistungsfähigkeit und die sozialen Beziehungen (Spoendlin, 1980).

Die **subjektive Lästigkeit** des Lärms ist von der **Beschäftigungsart** und der individuellen Einstellung abhängig. Einen intellektuell Arbeitenden kann bereits das leise Summen einer Fliege in seiner Tätigkeit wesentlich stören. Beim manuell Arbeitenden können hingegen die Ermüdungserscheinungen durch passenden Musikhintergrund positiv beeinflußt werden. Es ist dabei wiederum von Bedeutung, welche Art Musik gehört wird.

Durch Aufzeichnung von Atmungs-, Puls- und Blutdruckkurven konnte gezeigt werden, daß diese beim Anhören von moderner, atonaler, in der Lautstärke stark variierender Musik deutlich unregelmäßig werden. Hingegen soll das Anhören der klassischen Musik aus der europäischen Barockzeit die Regelmäßigkeit der Atmung, die Verlangsamung des Pulses und eine Normalisierung des Blutdruckes bewirken (Harrer, 1970). Auch das Fahrverhalten eines Lenkers im Straßenverkehr – ruhig oder aggressiv – kann durch die im Fahrzeug gehörte Musik beeinflußt werden.

Die gesamte **Psychomotorik** wird durch Lärm negativ beeinflußt. Die Reaktionen werden verlangsamt, Fehlleistungen nehmen zu, die Ermüdung stellt sich rascher ein.

Die **individuelle Einstellung** zum vorhandenen Lärm ist entscheidend über den Grad der Belästigung. Der Hundezüchter wird das Gebell seiner Vierbeiner nicht als lästig empfinden; sehr wohl aber sein Nachbar, der zu den Tieren keinerlei Zuneigung hegt.

Um die Lästigkeit des Lärmes zu erfassen, wurden von Kryter (1970) folgende Kriterien aufgestellt:
– statistisch faßbare physikalische Meßgrößen für Belästigung einer Gruppe von Leuten,
– zeitliche und spektrale Faktoren (z.B. Impulslärm oder ständiger Lärm),
– Informationsgehalt des Lärmes,
– Allgemeinzustand des Individuums.
Eine der störendsten Auswirkungen des Lärms ist die **Maskierung der Sprache.** Ab einem bestimmten Intensitätspegel kann weder der durch eine Menschenmenge verursachte, noch der von Maschinen produzierte Lärm mit Erhöhung der eigenen Stimmintensität übertönt werden. Arbeiter in lärmintensiven Betrieben erlernen intuitiv sehr rasch eine Ersatzkommunikation durch Hand- und Kopfzeichen. Stellt sich mit der Zeit eine Lärmschwerhörigkeit ein, so ändert sich konsekutiv auch die Artikulation. Weil die Hörschwelle und damit die Kontrolle wegfällt, werden die Zischlaute dumpf, die Artikulation im allgemeinen unpräzis, die Stimme zu laut.

Ist einmal dieser Zustand erreicht, so ist die Hörschädigung irreversibel. Deshalb lautet die erste Forderung in der Lärmbekämpfung – Prophylaxe. Der persönliche Hörschutz ist je nach Lärmintensität mit Gehörschutzkappen (Muscheln), Hörpfropfen oder Hörwatte möglich. Ärztliche Aufklärung in Betrieben soll zum Tragen von Hörschutzmitteln motivieren. Audiometrische Kontrollen sind bei Gefährdeten einmal jährlich angezeigt. Darüber hinaus sollte jeder zur Vermeidung von Lärm persönlich beitragen.

58.2.2 Taubheit, Gehörlosigkeit und Schwerhörigkeit

58.2.2.1 Definition

Als **Taubheit** wird ein völliges Fehlen von subjektiven Empfindungen und objektivierbaren Reaktionen der Cochlea auf akustische Reize bezeichnet. Die einzige Möglichkeit, diesen Patienten Höreindrücke zu vermitteln, besteht heutzutage in der Implantation von Elektroden (cochlear implants) in die Schnecke, falls der N. acusticus noch funktionsfähig ist.

Als **gehörlos** bezeichnet man Patienten, bei welchen zwar audiometrisch Hörreste nachweisbar sind, diese aber nicht für den Spracherwerb oder zur verbalen Kommunikation genügen. Diese Patienten benötigen hochleistungsfähige Hörgeräte und jahrelange spezielle audiopädagogische Schulung und Betreuung.

Schwerhörig sind Patienten, welche infolge Abschwächung oder Verzerrung des sprachlichen Signals nur bruchstückhaft die akustische Information ihrer Umgebung aufnehmen können. Mittels Hörgeräten ist es möglich, diese Schwäche ganz oder teilweise auszugleichen.

58.2.2.2 Ätiologie

Die **Ätiologie** der Taubheit und Gehörlosigkeit kann man in ca. 30 % der Fälle nicht eruieren. Von den bekanntesten Ursachen stehen Infektionen (Meningitis) an erster Stelle (40 %), gefolgt von angeborener oder frühkindlich erworbener Taubheit (11 %), Trauma (9 %), Wirkung ototoxischer Medikamente (5 %), Otosklerose und Akustikusneurinom (5 %) (Zusammenstellung von Laszig und Luetgebrune [1987] an 237 ertaubten Patienten). Bei ätiologischer Klärung der Schwerhörigkeit richtet man sich nach den allgemeinen otologischen Regeln.

58.2.2.3 Häufigkeit

Die Häufigkeit schätzt man auf 70–80 Taube pro 100 000 Einwohner (De Reynier, 1970). Diese Zahl hat sich seit dem letzten Jahrhundert weltweit kaum geändert. In der Schweiz konnte allerdings ein Rückgang von 245 im Jahre 1870 auf 154 im Jahre 1954 verzeichnet werden (Pfändler und Schnyder, 1959), infolge des Verschwindens des endogenen Kretinismus und der seltener werdenden Blutsverwandtschaft.

Hingegen steigt im allgemeinen die Zahl der Schwerhörigen parallel mit der Überalterung der Bevölkerung und der Zunahme der Lärmbelastung.

58.2.2.4 Symptomatologie

Wie wichtig das Gehör für die Integrität des Menschen ist, können sich Vollsinnige nur schwer vorstellen. Der Gehörsinn kann im Gegensatz zum Gesichtssinn nicht einmal im Schlaf ausgeschaltet werden. Bereits die Etymologie des Begriffes „Person" (per sonum = durch Klang) veranschaulicht die Bedeutung der akustischen Umwelt für die Persönlichkeit. Es sind sowohl die intellektuellen Fähigkeiten als auch Emotion und Gemüt, welche sich vorzugsweise über das Ohr entwickeln. Taubgeborene Kinder erreichen niemals die Persönlichkeitsstruktur Vollsinniger oder Blinder. Der Gehörlose oder Ertaubte kann also nicht bloß als normales Individuum ohne Hörfähigkeit aufgefaßt werden.

In den ersten sechs Lebensmonaten unterscheiden sich Normalhörende und Taubgeborene für den Beobachter nicht voneinander. Beide können selbst bei erhöhtem Lärmpegel – beispielsweise bei vorüberfahrenden Lastautos – schlafen. Beide spielen gleich mit ihren Extremitäten und Sprechorganen und erzeugen dabei Laute, die als Lallen bezeichnet werden. Doch nach dem sechsten Lebensmonat verläuft die sprachliche und geistige Entwicklung bei beiden verschieden. Wenn das normalhörende Kind stürmisch an der lautsprachlichen Entwicklung teilnimmt und die akustische Umwelt zum wichtigsten Faktor seiner geistigen Entwicklung wird, verstummt das gehörlose Kind allmählich, es bleibt an der Oberfläche der sichtbaren und greifbaren Gegenstände hängen. Das gleiche Schicksal erleiden auch Kinder, die zwischen dem vierten und siebten Lebensjahr – also noch vor der definitiven Sprachfestigung – ertauben. Das höchste Prinzip lautet dann, hörprothetisch und sprachtherapeutisch so rasch wie möglich einzugreifen, um den Zerfall der Sprache zu verhindern. Bei Taubgeborenen erlangen diese Maßnahmen im dritten Lebensquartal einen Dringlichkeitscharakter, denn nur durch die adäquate Reizung und Anregung können die Hörbahnen und Hörzentren überhaupt zur Information genutzt werden (je nach Ausmaß der Störung).

Ähnlich wie Gehör und Sprache zusammenhängen, bilden auch Sprache und Denken eine funktionelle Einheit. Die Entwicklung und das Niveau des Denkens sind mit demjenigen der Sprache aufs engste verbunden. Die Sprache ist somit nicht nur das vollkommenste Instrument des Denkens, sondern beide beeinflussen sich gegenseitig. Bei Gehörlosen und Tauben können abstrakte Vorstellungen nur mühsam und unvollständig vermittelt werden.

Der sprachliche Rückstand wird von der Umgebung nicht selten als Zeichen einer Debilität ausgelegt, in Wirklichkeit geht es aber nur um eine sekundäre Beeinträchtigung, die sich nach erfolgreicher Hörgeräteanpassung ausgleicht (Schlorhaufer, 1980).

Trotzdem konnten Baar (1957), Snijders (1976) et al. mit nichtsprachlichen Tests bei Gehörlosen ohne andere Behinderungen durchschnittlich nur einen IQ von 86–96 bestimmen. Wenn auch diese Kinder eine Lautsprache in den Gehörlosenschulen erlernen, so bleiben ihre Ohren doch für alle affektiven und emotionellen Feinheiten der Sprachmelodie, Dynamik und des Tempos verschlossen. Dies kann auch durch das sog. Vikariat des Gesichts- und Tastsinnes nicht ausgeglichen werden. Die Persönlichkeitsunterschiede zwischen Hörenden und Gehörlosen sind demzufolge nicht nur graduelle, sondern qualitative, wie das Kaiser (1962) dargelegt hat.

Die Folgen der Gehörlosigkeit führen häufig zu einer gestörten Eltern-Kind-Beziehung. Heute werden gehörlose Kinder rascher erfaßt und einer sog. „Früherziehung" zugeführt. Durch Einweisung in eine Sonderschule mit Internat kommt es zur Trennung von der Familie und zu Schuldgefühlen der Mutter. Für die Mutter-Kind-Beziehung, die gerade wegen der Behinderung des Kindes in der Regel besonders eng ist, bedeutet dies eine zusätzliche Belastung. Die Folge davon ist ein charakteristisches **Trennungsverhalten** mit anfänglicher Aggressivität und späterer Resignation und Depressivität.

All dies deutet darauf hin, daß das Problem der Gehörlosigkeit nicht allein durch Früherfassung und medizinische Frühbehandlung gelöst werden kann, sondern eine **interdisziplinäre Zusammenarbeit** und fachliche Betreuung durch den HNO-Arzt und Phoniater sowie Psychologen, Hörgeschädigten-Pädagogen und Hörgeräte-Akustiker braucht. Die Eltern sollten nach Möglichkeit im ganzen Rehabilitationsprozeß aktiv mitarbeiten. Trotz all dieser Bemühungen gelingt es kaum, Gehörlose in die normalhörende Gesellschaft voll zu integrieren. Es muß eine Eingliederung in die Schicksalsgemeinschaft der Gehörlosen angestrebt werden.

Wie tragisch der Hörverlust bei Erwachsenen sein kann, darüber sprechen Selbstzeugnisse berühmter Persönlichkeiten wie Beethoven, Schumann, Edison. Am Beispiel von Beethovens Ertaubung versucht Peyser (1942) vier Stadien nachzuweisen, die die Seele zu durchlaufen pflege, wenn der Mensch sein Gehör verliert: das Stadium der Verheimlichung, das Stadium der Verstimmung, das Stadium des Mißtrauens und das Stadium der Resignation (zit. nach Habermann, 1956).

Beethoven überwindet den für Musiker besonders schweren Schicksalsschlag, er stellt sich dem Daseinskampf mit gigantischer Energie und schafft so seine schönsten Kompositionen.

Wie viele schwerhörige Patienten aber vereinsamen allmählich, werden verbittert, mißtrauisch und depressiv – darüber gibt es keine Statistiken.

58.2.2.5 Hörapparate

Die Hörapparate können bei denjenigen Patienten eine Hilfe leisten, bei welchen die Schwerhörigkeit auf beiden Ohren einen Grad erreicht hat, der es ihnen nicht mehr ermöglicht, am beruflichen, gesellschaftlichen oder am Familienleben ausreichend teilzunehmen. Es wäre aber falsch anzunehmen, daß Schwerhörigkeit durch Hörapparate auszugleichen sei. Freilich kann das sprachliche Signal verstärkt werden. Den hauptsächlichen Nutzen davon hat der **Schallleitungsschwerhörige**. Aber bei **Innenohrschwerhörigen** ist das Hörfeld eingeschränkt. In diesen Fällen müssen Geräte mit Amplitudenbegrenzung (sog. peak-clipping) oder Dynamikkompression angepaßt werden. Am schlechtesten können Patienten mit sog. **Nerven-** oder **zentraler Schwerhörigkeit** hörprothetisch versorgt werden. Denn es besteht dabei außer der Hypofunktion (Schwerhörigkeit) auch eine Dysfunktion (Fehlhörigkeit) des Hörorgans. Das stereophone Hören kann nur durch eine beidseitige Versorgung gewährleistet werden. Aber selbst bei beidseitiger Hörgeräteanpassung haben viele alte Menschen Schwierigkeiten. Einerseits sind sie durch ein Überangebot an Information überbeansprucht, andererseits kann das Hörgerät eine Art Klaustrophobie (mit Druckgefühl in den Ohren, Schwindel und pulssynchronem Ohrensausen) verursachen. Schließlich scheuen sich einige Patienten vor einem Hörgerät, um nicht zum „alten Eisen" zu gehören.

Die Betreuung eines schwerhörigen Patienten beinhaltet nicht nur eine fachärztliche Behandlung und hörprothetische Versorgung, sondern auch eine Schulung (mit Ableseunterricht), welche oft durch Schwerhörigenvereine übernommen wird. Nur so kann das Optimum für die sprachliche Kommunikation der Schwerhörigen erreicht werden.

58.2.2.6 Cochlear-Implantate

Patienten mit Cochlear-Implantaten bilden eine ganz besondere Gruppe. Die Motivation für den operativen Eingriff und die postoperative Rehabilitation werden vorausgesetzt. Man muß die Erwartungen des Patienten dämpfen, denn das Hören mit Implantat muß über Monate erst wieder erlernt werden. Deshalb sind nicht nur umfassende somatische Abklärung und Behandlung, sondern auch eine psychologische Untersuchung und Betreuung unumgänglich.

Dort aber, wo es technisch nicht möglich ist, einem Ertaubten oder Schwerhörigen zu helfen, bleibt uns doch noch die Möglichkeit einer humanen, verständnisvollen Begleitung unserer Patienten. Wo dem Arzt das Heilen versagt bleibt, vermag er so weitgehend zu helfen (Habermann, 1956).

58.2.2.7 Die Seelentaubheit

Die Seelentaubheit (akustische Agnosie) muß differentialdiagnostisch von der echten Taubheit abgegrenzt werden. Wie schwierig das sein kann, ist schon daraus ersichtlich, daß viele Patienten in der Sprechstunde mit bereits angepaßten Hörgeräten erscheinen, von denen sie aber keinerlei Nutzen haben. Loebell (1944) beschrieb als erster dieses seltene Krankheitsbild als Unfähigkeit, Schallerscheinungen an ihrem Klang zu erkennen. Gestört sind die auditiven

Wahrnehmungsfunktionen, also auditive Aufmerksamkeit, auditive Merkfähigkeit, das Analysieren und Differenzieren von Klanggestalten, die Erfassung der richtigen Lautfolge, das Richtungshören und die Trennung von Nutz- und Hintergrundschall (Bauer, 1981). Das Schwellenaudiogramm kann sehr variabel oder normal sein. Nur die objektive Audiometrie (ERA, Impedanz) vermag zur näheren Klärung beizutragen.

Als Ursache kommen vor allem Geburtsschäden, Schädelhirntrauma und meningoenzephalitische Prozesse in Frage. Die Patienten sprechen zwar mechanisch das Vorgesprochene nach, wissen aber dem Gehörten keinen Sinn zuzuordnen. Sagt man einem normalen Kind: „Gib mir die Hand", so gibt es die Hand; sagt man es einem tauben oder idiotischen, so erfolgt gar nichts; sagt man es einem seelentauben Kind, so spricht es bestenfalls nach, ohne aber die Hand zu geben (Echosprache) (Berendes, 1971). Die klinische Diagnose läßt sich nur in einer Verlaufsbeobachtung stellen (Böhme, 1974).

Die Rehabilitationsmaßnahmen müssen jahrelang durchgeführt werden. Im Vordergrund stehen akustische und rhythmische Differenzierungsübungen, Musiktherapie und phonematische Übungen. Darüber hinaus werden auch andere seriale Modalitäten, wie die taktil-kinästhetische und visuelle, berücksichtigt (Affolter, 1972). Der Schulungsweg geht in der Regel durch eine Sonderklasse, aber nicht durch die Gehörlosenschule.

58.2.3 Tinnitus

58.2.3.1 Definition

Tinnitus ist eine vorübergehende oder dauerhafte, ein- oder doppelseitige Hörempfindung von Geräuschen/Tönen verschiedener Frequenz und Intensität ohne Einwirkung einer äußeren Schallquelle.

58.2.3.2 Häufigkeit

Tinnitus gehört ähnlich wie Juckreiz und Schmerz zu den häufigsten Mißempfindungen des Menschen. Besonders bei Streß oder nach Lärmbelästigung werden einige Minuten lang Ohrgeräusche wahrgenommen. Viele Ohrerkrankungen werden fast regelmäßig vom Tinnitus begleitet und dieser oft sogar noch störender als die Schwerhörigkeit selbst empfunden. Aber auch ohne eine erkennbare organische Ursache – monosymptomatisch – kommt Tinnitus relativ häufig vor.

Nach Angaben des National Center for Health Statistics (1980) sind 32 % der amerikanischen Bevölkerung von Tinnitus betroffen. Aber nur 7,2 % konsultieren den Hausarzt und 2,5 % kommen als Patienten direkt ins Spital (Smith und Coles, 1987). Nur bei dieser letzten Gruppe kann man vom sog. klinischen Tinnitus sprechen; allein dieser ist von praktischer medizinischer Bedeutung.

58.2.3.3 Symptomatologie und Ätiologie

Zur Beschreibung der Klangcharakteristik gebrauchen Patienten Vergleiche aus ihrer gewohnten Umgebung. Bei einem sind es Pfeiftöne, bei anderen Wasserrauschen, Wind, Motorengeräusche usw. Die ganze Tragik des Betroffenen ist auf geniale Art in „Caprichos" – Bild von Francisco Goya (1746–1828) – symbolisiert. In düsterer Atmosphäre bestürmen pfeifende, blasende und schreiende Dämonen den Künstler, der mit 46 Jahren ertaubt war und an quälendem Ohrensausen litt.

Friedrich Smetana (1824–1884), mit ähnlichem Schicksal, beschreibt seinen Tinnitus im e-Moll-Streichquartett (Aus meinem Leben, 1876) durch das schneidend schrille dreigestrichene e (1318 Hz) der ersten Violine – über sieben Takte durchgehalten.

Vincent van Gogh (1853–1890) schnitt sich ein Ohr ab, um sich vom quälenden Pfeifton zu befreien.

Bisweilen werden Patienten durch das Ohrensausen bis zum Suizidversuch getrieben.

Charakteristisch sind insbesondere die Intensitätsschwankungen. In ruhiger Umgebung wirken diese besonders störend und verursachen so Konzentrations- und Schlafstörungen.

Die durch Lautstärkevergleich ermittelte Intensität bewegt sich zwischen 5 und 15 dB (Opitz, 1980).

Tiefe, geräuschartige Frequenzen werden eher bei Erkrankungen des Mittelohres, hohe Töne fast immer nur bei Innenohrerkrankungen und unbestimmte Frequenzen bei zentralnervösen Erkrankungen angegeben. Doch dürfen diese Angaben nur als Hinweis und nicht als topische Zuordnung verstanden werden. Nach einer Zusammenstellung von Meikle und Mitarbeitern (1987) werden Geräusche am häufigsten in beiden Ohren zugleich angegeben (61 %), seltener links (12 %), rechts (9 %), im Kopfbereich (3 %), im Kopf- und Ohrbereich (10 %), mehr als drei Lokalisationen (4 %) und andere (2 %).

58.2.3.4 Befunde

Ein seltenes, **objektiv** nachweisbares, dabei meist pulsierendes Ohrensausen kann auch vom Untersucher mit einem ans Ohr des Patienten angelegten Stethoskop gehört werden (z. B. bei Aneurysma der Arteria carotis, Tumor des Glomus tympanicum oder jugulare). Die weitere Abklärung und Therapie richtet sich nach den chirurgischen Grundsätzen.

Das viel häufigere **subjektive** Ohrensausen kann nur annähernd genau durch Angaben des Patienten zu den vom Untersucher vergleichsweise dargebotenen Tönen und Geräuschen mit verschiedenen Intensitäten erfaßt werden. Zur Grunduntersuchung gehört hier auch die Bestimmung der Hörschwelle. Im allgemeinen wird eine normale Tonschwelle mit der normalen Cochlea gleichgesetzt. Nach Untersuchungen von Beck (1971) treten aber meßbare Auswirkungen auf die Tonschwelle erst dann auf, wenn mehr als 50 % der Hörelemente ausgefallen sind. Trotz einer normalen Hörschwelle kann also das Hörorgan bereits erheblich beeinträchtigt sein. Gerade diese geschädigten Zellen können aber aktiv – als ein zusätzlicher Geräuschgenerator – in Erscheinung treten und so Tinnitus verursachen (Meyer zum Gottesber-

ge, 1956; Lehnhardt, 1970). Zwischen den physiologischen spontanen Otoemissionen der gesunden Cochlea, einer aktiven Leistung des Innenohres, welche 1979 von Kemp entdeckt wurde, und dem Tinnitus konnte bis jetzt kein Zusammenhang gefunden werden (Hazell, 1984; Zwicker, 1987).

Außer in der erkrankten Cochlea kann Tinnitus in allen anatomischen Strukturen der Hörbahn – Kerne, aufsteigende Bahnen, Thalamus und Cortex – entstehen. Je höher die Läsion liegt, um so undifferenzierter wird das Geräusch empfunden. Aber selbst bei Durchschneidung des Nervus cochlearis bleibt der Tinnitus in 50 % weiterbestehen (Meyer zum Gottesberge, 1956). Die enge Verbindung der Hörbahn mit dem Thalamus weist geradezu auf die mögliche Beeinflussung durch psychische Reize hin. Bei einer primären Thalamusschwäche (Czernik, 1972), aber auch bei völlig ZNS-Gesunden kann Tinnitus als Antwort auf eine psychophysische Überforderung aufgefaßt werden (Böning, 1981).

In diesem mehrdimensionalen Bedingungsgefüge spielen demnach neben der speziellen Sinnesorgan- und Nervenläsion auch psychische Gegebenheiten eine entscheidende Rolle. Es sind dies Persönlichkeitsstruktur, individuelle Konfliktbewältigungsstrategien, konstitutionell vorgegebene oder erworbene zentralnervöse, biologische Strukturen mit thalamischer Reizschwelle und Filtertätigkeit für sensible Afferenzen. Dies alles entscheidet darüber, was letztlich störend empfunden wird. Dies bedeutet, daß die Patienten mit Ohrensausen ihren Arzt insbesondere zur Zeit einer erhöhten Streßbelastung aufsuchen. Entweder entsteht der Tinnitus erst im Augenblick der belastenden Situation, oder aber er wird dadurch akzentuiert. Die starke Verbindung mit dem emotionellen Hintergrund findet ihren Ausdruck auch in der umgangssprachlichen Form „es läutet mir in den Ohren". Die richtige Konsequenz bei Tinnitus-Abklärung besteht demnach im Aufnehmen einer Anamnese, welche nicht nur die akustischen Eigenschaften erfaßt, sondern auch das emotionelle Umfeld und die früheren Erlebnisse berücksichtigt. Der Tinnitus bedeutet meistens keine Organsprache, sondern gehört zu den psychophysiologischen Symptomen der Flucht-Kampf-Reaktion. Sekundär können Ohrgeräusche einen starken psychischen Einfluß ausüben und die Aufmerksamkeit des Patienten in einem übermäßigen Grad auf sich ziehen.

Bei einer 49jährigen Frau trat Tinnitus erstmals beim Tod ihrer Mutter auf. Bemerkenswert ist, daß die alte Mutter nicht nur pflegebedürftig, sondern auch schwerhörig war und demzufolge mußte die Patientin mit ihr sehr laut sprechen. Darüber hinaus geht es in der Familie mit fünf Kindern manchmal ohrenbetäubend zu. Die Patientin berichtet über zeitweise aufgetretene verhüllte Wut, die sie aber nie offen zeigte.

Der HNO-Befund und das Tonschwellenaudiogramm – 4 Tage nach Beginn der Beschwerden – waren völlig normal. Am linken Ohr konnte aber ein Tinnitus bei 6000 Hz und 30 dB bestimmt werden; er hatte einen pfeifenden Charakter.

Mit der Patientin wurde ein beruhigendes und aufklärendes Gespräch geführt. Die intelligente und einsichtige Patientin erkannte die Zusammenhänge mit der emotionellen Belastung, dem Zeitpunkt des Auftretens und der versteckten Wut. Die Patientin wurde mit Trental behandelt und wöchentlich kontrolliert. Bei den Gesprächen wurde stets die Harmlosigkeit des Symptoms betont. Bereits nach einer Woche besserte sich der Tinnitus und nach einem Monat war er gänzlich verschwunden. Wahrscheinlich war die verständnisvolle Betreuung und empathische Begleitung ebenso wirksam wie das Pentoxyphyllin.

58.2.3.5 Therapie

Außer der kausalen chirurgischen Therapie bei der kleinen Gruppe mit objektivem Tinnitus (infolge Aneurysma und Tumor), sind alle anderen therapeutischen Maßnahmen kompliziert, aufwendig und durch einen bescheidenen Heilungserfolg charakterisiert. Nach dem heutigen Wissensstand gibt es für Tinnitus keine spezifische Behandlungsform.

Die Mehrzahl der angewandten Methoden sucht eine bessere Durchblutung des Innenohrs zu erreichen. So werden Stellatumblockaden, Infusionen mit β-Sympathikotonika, Parasympathikolytika, Jontophorese, Physiotherapie, Akupunktur etc. verwendet.

Die symptomatischste aller Formen ist die Anwendung von sog. Tinnitus-Maskern (Feldmann, 1971), die auf der Erkenntnis beruht, daß Tinnitus beim Vorliegen von Umweltgeräuschen weniger störend wirkt. Im gleichen Sinne werden auch Walkmen verwendet, wobei man zum Einschlafen Radiogeräusche zwischen zwei Sendern auswählt. Solche Tricks werden in den Tinnitus-Selbsthilfegruppen erfunden und gegenseitig ausgetauscht.

Wahrscheinlich wird man Tinnitus, ähnlich wie Schmerz und Juckreiz, nie vollständig beherrschen können. Auch hier ist aber weder therapeutischer Nihilismus noch Polypragmasie angezeigt.

Es wäre schon viel erreicht, wenn man den Kranken aus dem Zustand eines „dekompensierten" in den eines „kompensierten" Tinnitus bringen könnte: Akzeptieren der Störung und Herabsetzen der Empfindlichkeitsschwelle. Hierzu eignen sich vor allem psychologische Behandlungsmethoden wie autogenes Training, Biofeedback-Verfahren, Suggestion und Hypnose sowie kognitive Verhaltenstherapie. Erste wohlkontrollierte Resultate sind recht ermutigend (Scott et al., 1985). Der HNO-Facharzt kann wohl in seiner andauernden Betreuung verhaltensändernd einwirken (Schneider, 1986).

Unabhängig von der angewandten therapeutischen Methode sollte man aber stets zwei Gesichtspunkte bedenken:
– dem Patienten gegenüber keine zu optimistischen Erwartungen in Aussicht stellen,
– die ungefährlichste und am wenigsten belastende Behandlungsart wählen.

58.2.4 Der Hörsturz

58.2.4.1 Definition

Der Hörsturz ist ein funktionelles Ohrsymptom, charakterisiert durch eine plötzliche, einseitige, periphere (ausschließlich sensorische) Hörverminderung ohne erkennbare Ursache.

58.2.4.2 Häufigkeit

Seit der Erstbeschreibung 1944 (de Kleyn) beobachtet man eine stetige Zunahme der Morbidität, welche nicht nur der verbesserten Diagnostik, sondern der Wirklichkeit entspricht. Diese Tendenz ist durchaus mit der Zunahme anderer Zivilisationskrankheiten infolge Streß, Bewegungsarmut, Überernährung, Genußmittelkonsum usw. vergleichbar. Heutzutage rechnet man pro Jahr 20 Fälle auf 100 000 Einwohner (Weinaug, 1984). Die Altersverteilung zeigt eine Häufung zwischen dem 40. und 60. Lebensjahr (Stange und Neveling, 1980), also in der beruflich und sozial aktivsten Zeit des Lebens. Männer und Frauen werden in etwa gleichem Maße betroffen.

58.2.4.3 Symptomatologie

Eine literarisch schöne und treffende Beschreibung stammt von Jean Jacques Rousseau, der 1736 im Alter von 24 Jahren einen Hörsturz erlitt (zit. nach Debain, 1957). In seinen Lebenserinnerungen schreibt Rousseau:

„Als ich eines Morgens, an dem es mir nicht schlechter ging als sonst, eine kleine Tischplatte auf ihrem Fuß richtete, fühlte ich in meinem ganzen Leibe einen plötzlichen, ganz unvorstellbaren Aufruhr. Ich kann es nur mit einer Art Sturm vergleichen, der sich in meinem Inneren erhob und im selben Augenblick durch alle Glieder tobte. Meine Arterien begannen derart heftig zu schlagen, daß ich das Klopfen nicht nur fühlte, sondern sogar hörte, vor allem die Kopfschlagadern. Dazu ein starkes Ohrensausen, so daß es wie ein drei- und vierfacher Lärm war, nämlich ein tiefes, dumpfes Sausen, ein helleres Rauschen wie von fließendem Wasser, ein schrilles Pfeifen und das geschilderte Pochen, dessen Schläge ich leicht zählen konnte, ohne mir den Puls zu fühlen oder meinen Körper mit den Händen zu berühren. Dieser innere Lärm war so groß, daß er mir mein bisher gutes Gehör raubte und mich zwar nicht ganz taub, aber so schwerhörig machte, wie ich es seitdem bin."

Diese stürmische Symptomatik des plötzlich aus voller Gesundheit (wie ein Blitz aus heiterem Himmel) aufgetretenen Ereignisses erschreckt den Patienten derart, daß er in der Regel notfallmäßig einen Arzt aufsucht. Außer einseitiger Hörverminderung mit Druckgefühl im Ohr (das Gefühl, als sei Watte im Ohr), ist ein lästiges Ohrensausen in 60–90 % ein häufiges Begleitsymptom (Neveling, 1967; Plester, 1978). Schließlich klagen ca. 30 % der Kranken über Unsicherheitsgefühl oder Schwankschwindel (Stange, 1969).

58.2.4.4 Befunde

Die einseitige Hörverminderung steht im Vordergrund der Befunde. Dies kann bereits durch Überprüfung der Flüster- und Umgangssprache wie auch durch Lateralisation des Stimmgabeltones ins gesunde Ohr (Prüfung nach Weber) festgestellt werden. Im **Reintonaudiogramm** wird am häufigsten (in 60 % nach Eichhorn und Martin, 1984) ein sog. pankochleärer Kurvenverlauf mit Hörverlust um 60 dB für alle Frequenzen angetroffen. Seltener wird der Verlust isoliert nur für die hohen, mittleren oder tiefen Frequenzen festgestellt. Aus otologischer Sicht bedeutet ein solcher Befund immer eine ernste Situation, nach dem Grundsatz: „Jede einseitige Innenohrschwerhörigkeit muß so lange auf Akustikusneurinom verdächtig bleiben, bis das Gegenteil bewiesen ist." Der Nachweis oder Ausschluß erfolgt mit der ERA (Electric Response Audiometry) und dem Computertomogramm oder der NMR (Nuklearmagnetische Resonanz) des Schädels. Der Hörsturz bleibt somit vorerst eine **Ausschlußdiagnose.** Außer dem Akustikusneurinom, in ca. 1% der Fälle (Steinert, 1986; Eichhorn und Martin, 1984), wird man insgesamt nur in ca. 10 % eine andere konkomittierende Krankheit finden können, wie Hyper- oder Hypotonie, degenerative Veränderungen der HWS, thromboembolische Krankheit oder Diabetes mellitus. Es gibt bereits über 100 verschiedene Ursachen, welche für einen Hörsturz verantwortlich gemacht werden können (Jaffe, 1973). Die Ansicht, daß Streß eine ausschlaggebende Rolle bei der Pathogenese des Hörsturzes hat, wird von fast allen Autoren geteilt. Greuel (1983, 1986) konnte aufgrund seiner psychologischen Untersuchungen Persönlichkeitsmerkmale des Hörsturzpatienten finden, die wie ein roter Faden bei der tiefenpsychologischen Anamneseerhebung erscheinen. Danach ist der Hörsturzpatient charakterisiert als Person, die in einem ständigen emotionalen Spannungszustand lebt, nicht zuletzt deshalb, weil sie in ihrem Leben selektiv nur ungelöste, nicht aber „erfolgreich" abgeschlossene Aufgaben wahrnimmt. Durch ihre eigene ständige Überforderung provoziert sie selbst belastende Lebensereignisse.

Wenn nun zu diesem Dauerstreßzustand zusätzlich ein schicksalhaftes Ereignis (Unfall, Todesfall oder Erkrankung) hinzutritt, kann eine „vegetative Detonation" erfolgen (Schultz van Treeck, 1956).

Wenn auch viele Faktoren zum Hörsturz beitragen können, so ist doch letztlich – nach heutiger Ansicht – der pathophysiologische Mechanismus auf eine Mikrozirkulationsstörung in der Cochlea und somit auf eine Hypoxie der Sinneshaarzellen zurückzuführen. Streß könnte dabei eine ausschlaggebende Rolle spielen. Das ist erklärlich aus der reichlichen vegetativen, vor allem adrenergen Nervenversorgung des Innenohres (Spoendlin und Lichtensteiger, 1967), aber auch aus der Tatsache, daß bei Streß die Thrombozytenadhäsivität steigt (Jacobi und Krüskemper, 1977). Somit verlangsamt sich die Mikrozirkulation bis zur völligen Stase – sog. Blood-Sludging (Fowler, 1950; Kellerhals, 1977) – mit konsekutiver

Hypoxie und Hörsturz. Hinzu kommt noch die ungünstige anatomische Gegebenheit, indem das Labyrinth von einer einzigen Endarterie (A. labyrinthi) versorgt wird, ohne die Möglichkeit einer Kollateralbildung.

58.2.4.5 Diagnose

Die Diagnose des Hörsturzes ist eine otologische: Bei normalem otoskopischem Befund ist eine einseitige Hörminderung objektivierbar. Die Vielzahl der möglichen konkomittierenden Krankheiten läßt die weitere Abklärung zu einer interdisziplinären Aufgabe werden.

58.2.4.6 Therapie

Bei der Vielzahl der angegebenen Therapieformen sind sich verschiedene Autoren nur in einem einzigen Punkt einig, nämlich, daß der Hörsturz notfallmäßig behandelt werden muß. Die Prognose ist unabhängig von der Behandlungsart um so besser, je früher die Therapie einsetzt. Die herrschende Polypragmasie ist nicht haltbar. Doch haben alle Methoden eine Verbesserung der Innenohrdurchblutung zum Ziel. So werden heutzutage hauptsächlich folgende Therapieformen angewandt: Dextran- und Rheomacrodex-Infusionen, Trental, Antikoagulation, CO_2-Inhalationen, Stellatumblockaden, Cortison, Acetylsalicylsäure etc. – um nur die wichtigsten zu nennen. Spektakulär erscheinen Erfolge, welche Greuel (1983) mit der Suggestiv- und Hypnosebehandlung erreichte. Nachdem er den Patienten das Schwere- und das Wärmegefühl mit sichtbarer Rötung der Ohrmuschel vermittelt hatte, erfolgte audiometrische Kontrolle, welche eine Besserung der Hörschwelle um 20–30 dB aufzeigte.

Noch verblüffender als die Suggestiv- und Hypnosebehandlung sind die Angaben von Weinaug (1984) über Spontanremissionen, welche in ca. 70 % eine vollständige Rückbildung, in ca. 90 % eine Besserung zeigen. Dies weist darauf hin, wie schwierig die einzelnen Therapieformen zu evaluieren sind.

Diese Angaben berechtigen aber ebensowenig einen therapeutischen Nihilismus wie ein starres Durchführen der Therapie nach einem bestimmten Schema. Das Ziel ist eine lückenlose Klärung der Ätiologie mit kausaler medikamentöser, chirurgischer oder psychologischer Therapie. Bis es erreicht ist, müssen wir für jeden Patienten die individuelle Therapieform finden. Die Erfassung psychischer Faktoren ist ein wesentlicher Bestandteil einer jeden Therapieform. Die Hospitalisation bietet gute Voraussetzungen, dem Patienten seine Problematik aufzuzeigen, damit er den Heilungsprozeß zur Umstrukturierung seiner Lebenssituation nützen kann.

Die **Prognose** des Hörsturzes ist im wesentlichen vom Alter und vom frühen Therapiebeginn abhängig. Rezidivneigung mit stufenförmiger Gehörverschlechterung kann bei ca. 8 % der Patienten erwartet werden (Eichhorn und Martin, 1984).

Im Gegensatz zum akuten Hörsturz ist eine **fluktuierende Hörschwelle** ein völlig unspezifisches Symptom, welches sowohl bei Menièrescher Erkrankung als auch bei Tumoren der hinteren Schädelgrube und bei Autoimmunerkrankungen auftreten kann (Pfaltz, 1988). Dies ist kein Notfall, verlangt aber zusätzlich nach einer gründlichen neurootologischen Untersuchung.

Eine 60jährige Patientin steht nach einer mikrochirurgisch behandelten Leukoplakie des Kehlkopfes in regelmäßiger fachärztlicher Kontrolle. Sie leidet sehr unter Karzinophobie, die sich insbesondere einige Tage vor den bevorstehenden Kontrollen bis zu psychotischen Reaktionen steigert. Diese Angst abzubauen gelang nicht, trotz intensiver Gespräche, Zeigen und Besprechen des Kehlkopfbefundes auf dem Video-Monitor. Die Patientin besuchte nun drei Tage vor der Kontrolle eine Flugschau – wider ihren Willen. Es war nur eine Woche nach der Tragödie in Ramstein (28. 8. 88), als eine Maschine der italienischen Staffel in die Zuschauer gestürzt war. Es kamen dabei zahlreiche Menschen ums Leben und Hunderte wurden verletzt. Die Patientin war – wie sie beschrieb – in einem seelisch aufgewühlten Zustand, voller Angst und Wut. Als die erste Flugzeugformation über die Zuschauer hinwegdonnerte, wurde ihr linkes Ohr wie zugedeckt. Sie versuchte, mit dem Kleinfinger im Gehörgang zu manipulieren, doch das Gefühl blieb bestehen. Gleichzeitig hörte sie links einen hohen Pfeifton und realisierte schließlich, daß sie links schlechter hörte. Das war ein Grund, sich von der Flugzeugschau zu entfernen und auf die Notfallstation zu kommen. Die audiometrisch aufgenommene Hörkurve ließ eine linksseitige Erhöhung der Hörschwelle um 30–40 dB ab 2000 Hz aufwärts erkennen. Die Patientin wurde hospitalisiert und erhielt insgesamt 10 Infusionen von Rheomacrodex. Die Hörschwelle normalisierte sich innerhalb einer Woche fast vollständig. An der Besserung der Innenohrdurchblutung und des Gesamtzustandes waren vermutlich auch die Bettruhe, die pflegerische Zuwendung und der Milieuwechsel mitbeteiligt.

58.2.5   Die psychogene Hörstörung

58.2.5.1   Definition

Unter psychogener Hörstörung wird eine unbewußte, zentrale, symmetrisch doppelseitige Schwerhörigkeit mittleren bis hohen Grades verstanden. Charakteristisch ist, daß die Störung nur während der Untersuchungssituation, nicht aber während ungezwungener Unterhaltung oder am Telefon auftritt.

58.2.5.2   Häufigkeit

Die Häufigkeit psychogener Hörstörungen ist sicher gering (Lüscher, 1957). Von allen Hörstörungen wird sie bei Erwachsenen in ca. 3 % (Doerfler, 1951) und bei Kindern in ca. 2 % (Leshin, 1960) angetroffen.

In Kriegszeiten steigt die Inzidenz auf 10–15 % (Johnson et al., 1956). Im neueren Schrifttum finden sich lediglich einzelne Falldarstellungen (Lehnhardt, 1974; Plentz, 1976; Veniar und Salston, 1983).

Vermutlich wird ein Teil der psychogenen Hörstörungen übersehen, weil sie in die Differentialdiagnose gar nicht einbezogen werden.

58.2.5.3   Symptomatologie

Die Untersuchung selbst wird erst auf Anraten von den Angehörigen, vom Betrieb oder von der Schule aus durchgeführt. Obwohl die Patienten in ihrer Umgebung als schwerhörig gelten, erfahren sie daraus doch keinerlei Nachteile – etwa in Form eines sozialen Abstiegs oder schulischen Versagens. Sie erzählen bereitwillig ihre Geschichte, aus der zwar lange Zeit zurückliegende Ohrenerkrankungen, Lärmexposition oder Kopfverletzungen, aber keine unmittelbare Ursache für die jetzige Gehörabnahme erkennbar ist. Erst eine erweiterte biopsychosoziale Anamnese erlaubt den Zusammenhang zwischen der schon lange zurückliegenden, kurzdauernden Schwerhörigkeit und der jetzigen unbewußten Abwendung vom Gehörten aufzudecken. Nur dann, wenn der Patient die Gelegenheit bekommt, allgemein über seine Probleme und nicht nur über „Schwerhörigkeit" zu erzählen, kommen auch seine Ängste, Befürchtungen und emotionellen Schwierigkeiten zum Vorschein.

Der Patient erlebt sich selbst als taub; er benutzt seine Taubheit, um das nicht zu hören, was er nicht ertragen kann. Es kann sich dabei um ein Verdrängen einer unannehmbaren Situation am Arbeitsplatz, in der Familie etc. handeln. Dieses unbewußte Verhalten kann Ausdruck einer Flucht aus einer belastenden Situation sein. Der fehlende Leidensdruck und die Beziehung zwischen Konflikt und Symptombildung weisen auf ein Konversionssymptom hin (vgl. Kap. 30).

58.2.5.4   Befunde

Bei allen psychoakustischen Untersuchungsmethoden (Hörweitenprüfung mit Flüster- und Umgangssprache, Tonschwellen- und Sprechaudiogramm, Békésy-Audiogramm) ist der Befund pathologisch, während das Unterhaltungsgehör völlig normal erscheint. Diese Diskrepanz ist charakteristisch für die psychogene Hörstörung. Es ist eindrücklich zu beobachten, wie der Patient, mit welchem sich der Untersucher vor kurzem noch völlig mühelos unterhalten konnte, plötzlich – sobald eine Prüfungssituation entsteht – hochgradig pathologische Befunde aufweist. So werden die Flüster- oder die Umgangssprache nicht oder kaum gehört. Die Reintonschwelle verläuft als flache Mulde in allen Frequenzen zwischen 50–90 dB, wobei die Knochenleitung der Luftleitung entspricht (Doerfler, 1951). Nur selten wird eine Schwankung der Hörschwelle zwischen den einzelnen Untersuchungen festgestellt (Lüscher, 1957), dafür aber eine Diskrepanz zu dem relativ besseren Resultat im Sprechaudiogramm.

Bei der automatischen Tonschwellenbestimmung im **Békésy-Audiogramm** sinken sowohl die Dauer als auch die Impulskurve ab. Das beschriebene Kriterium ist auch gegenüber Simulation das wichtigste Unterscheidungsmerkmal. Beim Simulanten liegt die Dauertonkurve über der Impulstonkurve, beim psychogenen Hörgestörten liegt sie – wie beim organisch Schwerhörigen – unter der Impulstonkurve und beide Kurven sinken parallel zueinander ab (Lehnhardt, 1974). Die **ERA** (Electric Response Audiometry) ergibt für psychogene Hörstörungen keine spezifischen Befunde, ermöglicht aber die tatsächliche Bestimmung der Hörschwelle bei Aggravation (Burian, 1969; Schmidt et al., 1983).

Von den objektiven audiometrischen Untersuchungsmethoden kann die **Stapediusreflexschwelle** normal ausgelöst werden. Das Vorhandensein des Reflexes bestätigt aber nicht, daß der Ton vom Cortex empfangen wurde, denn die Reflexbahnen reichen nur bis zum Niveau des Hirnstammes (Greiner et al., 1980).

58.2.5.5 Diagnose

Die Diagnose einer psychogenen Hörstörung kann nur aus der Gesamtheit aller Untersuchungsergebnisse gestellt werden. Positive diagnostische Kriterien kann die biopsychosoziale Anamnese ergeben. Zusätzliche Belege sind:
- Diskrepanz zwischen normalem Unterhaltungsgehör, aber schwer pathologischem Untersuchungsgehör,
- Diskrepanz zwischen hochgradig erhöhter Schwelle im Reintonaudiogramm und relativ guter Schwelle im Sprechaudiogramm,
- pathologisches Békésy-Audiogramm mit Absinken von Dauer- und Impulskurve,
- normale Stapediusreflexschwelle.

Simulation und Konversion können unterschieden werden (vgl. Kap. 30).

58.2.5.6 Therapie

Steht einmal die Diagnose oder zumindest der Verdacht einer psychogenen Hörstörung fest, so gilt an erster Stelle der Grundsatz: „primum nihil nocere".

Es ist sinnlos, dem Patienten gegenüber unverblümt aufzuzeigen, daß er keinen Hörschaden hat. Damit würde man ihn nur des primären und sekundären Gewinns seines Konversionssymptoms berauben. Vielmehr soll man versuchen zu erkennen, welche Notwendigkeit oder welche Rolle die Schwerhörigkeit in seinem Leben erfüllt.

Man hüte sich auch vor einer diagnostischen Etikettierung. Deshalb auch die behutsame Nomenklatur – psychogene Hörstörung – und nicht Schwerhörigkeit oder gar Taubheit. Dem Patienten gegenüber kann ruhig gesagt werden, daß seine Störung mit den zur Verfügung stehenden Untersuchungsmöglichkeiten zwar nicht faßbar ist, daß man ihn aber weiterhin längerfristig in Kontrolle behält. Anläßlich der Kontrolluntersuchungen kann man besser auf die Probleme des Patienten eingehen.

Im Rahmen der verschiedenen Kontakte mit dem Arzt vollzieht sich ein Prozeß der Bewußtwerdung, der es dem Patienten ermöglicht, die Zusammenhänge zwischen der Taubheit und den psychischen Problemen zu spüren. Das Konversionssymptom der psychogenen Taubheit kann sich auf diese Weise allmählich verflüchtigen. Obwohl bei Kindern meist eine schwerere psychische/psychosoziale Störung vorliegt, empfehlen einige Autoren (Veniar und Salston, 1983) eine suggestive Behandlungsmethode. Eine medikamentöse Therapie erübrigt sich in der Regel. Eine eigentliche Psychotherapie analytischer Art oder Hypnosebehandlung werden wohl nur selten notwendig sein. Die Prognose ist im allgemeinen gut.

58.2.6 Der Schwindel

58.2.6.1 Definition

Der Schwindel ist eine subjektive Empfindung von Gleichgewichtsstörung in physischer und/oder psychischer Hinsicht.

58.2.6.2 Erwünschter Schwindel

Erwünschter Schwindel wird vom Individuum willkürlich als bewußter Reiz der Labyrinthe herbeigeführt und als lustvolles, berauschendes Gefühl erlebt. Der adäquate Reiz für das Gleichgewichtsorgan kann durch Beschleunigungsbewegung, beispielsweise bei gewissen drehenden Tanzarten, auf dem Karussell etc., ausgelöst werden. Durch Alkohol und Drogenmißbrauch wird aber auch ein bewußter Rausch mit Schwindelgefühl und Distanz zur Realität gesucht.

58.2.6.3 Unerwünschter Schwindel

Unerwünschter Schwindel wird als lästige Antwort auf äußere Reize bei individuell empfindlichen Labyrinthen ausgelöst, beispielsweise bei der Seekrankheit. Bei Höhenschwindel und im Traumerleben kann ein unerwünschtes Schwindelgefühl auch ohne den äußeren Bewegungsreiz empfunden werden.

58.2.6.4 Pathologischer Schwindel

Pathologischer Schwindel hingegen ist ein krankhaftes Symptom bei verschiedensten Krankheiten, mit meistens seitenunterschiedlicher Reaktion der Labyrinthe, ohne äußere Krafteinwirkung.

58.2.6.5   Häufigkeit

Schwindel ist eine außerordentlich häufige Klage der Kranken. Der Allgemeinarzt wird deshalb von 2–5% seiner Patienten konsultiert (Hagnell, 1966; Rubin, 1976). Der HNO-Arzt und der Neurologe begegnen doppelt und dreifach so vielen Schwindelpatienten. Der Psychiater aber hört dieselbe Klage bei Patienten mit Neurosen und Psychosen, als Folge der Einnahme von Psychopharmaka.

58.2.6.6   Pathophysiologische Bemerkungen

Schwindel kann durch Bewegungsinformationen vom Labyrinth, von den Augen und von den Somatosensoren in Gelenken, Muskeln und Haut ausgehen. Diese Sinnesinformationen laufen im Hirnstamm zusammen und werden von der Formatio reticularis und dem Cerebellum beeinflußt. Über den Thalamus werden sie schließlich in die vestibulären kortikalen Gehirnareale projiziert.

58.2.6.7   Symptome

Die Palette der Mißempfindungen bei Schwindel ist groß. Aber bereits aus der subjektiven Beschreibung und der Art der Schilderung kann der Läsionsort vermutet werden.

Peripher-vestibulärer Schwindel

Bei peripher-vestibulärem Schwindel berichten die Patienten über eine Scheinbewegung der eigenen Person oder der Umgebung. Das ist die typische Vertigo (vertere, lat.: drehen) mit Drehschwindel, Lift- und Schwankschwindel. Falls noch gleichzeitig über Hörprobleme (Schwerhörigkeit, Ohrensausen) geklagt wird, besteht kaum noch Zweifel am sog. spezifisch otogenen Schwindel infolge einer Innenohrläsion.

Charakteristisch ist dabei auch eine starke Beteiligung des Vegetativums mit Übelkeit, Blässe, kaltem Schweiß und vor allem der Nausea mit Erbrechen (Patient mit der Brechschale). Bei ca. 30% aller Schwindelpatienten wird eine otogene Ursache gefunden.

Zentral bedingter Schwindel

Bei zentral bedingtem Schwindel werden die Beschwerden viel ungenauer beschrieben. Die Patienten sprechen von der Benommenheit, Taumeligkeit, Schweben im Raum, Verwirrtheit, Schwarzwerden oder Flimmern vor den Augen; im Schweizerdeutschen spricht man vom „Sturm" im Kopf u.ä. Allen diesen Erkrankungen ist aber das Gefühl der Desorientierung und der Unsicherheit der ganzen Person gemeinsam (Lüscher, 1959). Vegetative Begleiterscheinungen fehlen in der Regel. Dieser unspezifische Schwindel wird vor allem bei internistischen und neurologischen Erkrankungen vorkommen. Mit ca. 60% ist er die am häufigsten geklagte Schwindelart.

Psychogen bedingter Schwindel

Bei psychogen bedingtem Schwindel werden die Beschwerden noch unbestimmter und ungenauer formuliert. Oft geschieht dies sehr maniriert, mit wechselnden Angaben über die Art, Schwere und Dauer des Schwindels. Darüber hinaus sind diese Patienten suggestibel. In diese Gruppe gehören ca. 10% aller Schwindelpatienten. Häufig tritt Schwindel im Zusammenhang mit emotionaler Spannung, vor allem im Zusammenhang mit Angst, zum Teil zusammen mit Hyperventilation auf. Schwindel kann auch Konversionssymptom sein. Bereits die umgangssprachliche Formel: „Bei diesem Gedanken wird mir schwindlig" weist darauf hin (vgl. Kap. 30).

58.2.6.8   Befunde und Diagnose

Die diagnostischen Schwierigkeiten bei den Patienten mit Schwindel sind nicht nur durch die oft sehr ungenaue und unterschiedliche Schilderung der Beschwerden bedingt, sondern auch durch deren Flüchtigkeit. Meistens dauern die Schwindelanfälle nur einige Minuten bis Stunden. Demzufolge wird auch der erstuntersuchende Arzt nur gelegentlich einen Spontannystagmus nachweisen können. Der fehlende Nystagmus darf aber nicht eo ipso zur Annahme eines psychogenen Schwindels verleiten. Bei jedem Schwindelpatienten muß eine sorgfältige allgemeine und otoneurologische Untersuchung durchgeführt werden. Denn nur so können Innenohrerkrankungen, der Kleinhirnbrückenwinkeltumor, funktionelle HWS-Erkrankungen, Multiple Sklerose und andere Erkrankungen des ZNS frühzeitig erkannt werden. Selbst wenn ein Konversionssymptom bewiesen ist, kann der Kranke trotzdem noch zusätzlich somatisch krank sein.

58.2.6.9   Die Menièresche Erkrankung

Die Menièresche Erkrankung stellt eine peripher-vestibuläre Störung dar, bei der psychische Faktoren zum Anfall beitragen können. Ein erster massiver Anfall ist für den Patienten vernichtend, er steht Todesangst aus (Escher, 1985) und gerät in Panik, denn er empfindet dabei nicht nur Drehschwindel, sondern erlebt gleichzeitig starkes Ohrensausen, verbunden mit einer Höreinbuße und starker vegetativer Begleitsymptomatik mit Erbrechen, Schwitzen und Kollaps. Der Anfall dauert zwar nur einige Minuten, er kann sich jedoch in belastenden Lebenssituationen während Jahrzehnten immer wiederholen. Solche belastenden Lebenssituationen werden hauptsächlich mit Veränderungen am Arbeitsplatz – etwa durch Versetzung auf einen unbeliebten Posten – in Verbindung gebracht.

Zahlreiche Untersuchungsergebnisse (Groen, 1983; Lüscher, 1959) lassen bei Patienten mit diesem Leiden bestimmte Persönlichkeitszüge erkennen: Überdurchschnittliche Intelligenz, ausgeprägte Tendenz zu Zurückgezogenheit, Perfektionismus in Arbeit und Hobby; sie sind sehr ernsthaft, unfähig zu einer lockeren Unterhaltung, starr im Lebensstil und zeigen ein ausgeprägtes Über-Ich. Sie leben im Zustand einer ständigen Überforderung, unter Zeitdruck. Auffallend gleichförmig werden auch die frühkindlichen Erlebnisse geschildert. Aus sehr wohl geordneten Familienverhältnissen stammend, erlebten diese Kinder zwar eine kühle Korrektheit, niemals aber die notwendige „Nestwärme". In der Pubertät bleiben die üblichen Generationsprobleme aus. Als Musterschüler haben sie keine Konflikte mit den Lehrern gehabt. Diese Merkmale sollen illustrieren, daß es bei der Menièreschen Erkrankung psychische Besonderheiten gibt, welche beim Patienten in gewissen Lebenssituationen zur Auslösung eines Anfalls beitragen können.

58.2.6.10 Therapie

Die Therapie im Anfall muß sich auf die Symptome in der Notfallsituation konzentrieren: Bettruhe, Antivertiginosa, niedermolekulare i.v.-Dextraninfusionen. Im beschwerdefreien Intervall sind regelmäßige Kontrolluntersuchungen durch den psychosomatisch gebildeten Kliniker in engster Zusammenarbeit mit dem HNO-Spezialisten angezeigt. Ein ruhiges, verständnisvolles Zuhören und Aufklärung des Patienten über die Pathophysiologie seiner Erkrankung sowie eine taktvolle Anleitung zur Änderung seiner Lebensgewohnheiten sind entscheidend. Diese Bemühungen können medikamentös durch Betablocker und Kalziumantagonisten unterstützt werden. In jedem Fall sollten diese konservativen Behandlungsmöglichkeiten vor der ultima ratio – einer Neurektomie des Nervus vestibuli oder der Ausschaltung des Labyrinthes mittels lokaler Applikation von Gentamicin – ausgeschöpft werden. Patienten mit Berufen im öffentlichen Straßenverkehr oder bei welchen Absturzgefahr droht, müssen ihren Arbeitsplatz wechseln.

58.2.7 Psychosomatische Zusammenhänge bei Nasenerkrankungen

Die Nase kann im psychischen Geschehen des Menschen auf vielfältige Weise beteiligt sein. Bereits ihre äußere Form erfüllt bisweilen eine Symbolfunktion und vermag – mit oder ohne Deformitäten – Anlaß für krankhafte psychische Reaktionen sein. Der **Geruchssinn** als einer der Nahsinne erfüllt keine vitale Funktion; die verschiedenen Gerüche verursachen aber stark gefühlsbetonte Sinneseindrücke, deren lust- oder unlustbetonte Stimmung auch im Gesichtsausdruck, Schnüffeln, Abwehrreaktionen usw. ihren Ausdruck findet.

Zahlreiche aus der Nase auslösbare **Reflexe,** wie der nasopulmonale, der nasokardiale und der nasookulare, deuten auf die enge Verknüpfung mit anderen Organen, aber auch mit dem gesamten autonomen Nervensystem hin (z.B. K.o.-Boxschlag). Die Nasenschleimhaut und insbesondere die kavernösen Schwellkörper der Nasenmuscheln reagieren empfindlich sowohl auf äußere, atmosphärische wie auch auf die inneren, nervalen Reize. Ihre unmittelbare Auswirkung äußert sich dann in der unterschiedlichen Luftdurchgängigkeit. Der Sympathikotonus bewirkt nasale Vasokonstriktion mit Erweiterung der Nasenräume, der Parasympathikotonus bewirkt hingegen nasale Vasodilatation mit Verengung der Nasenräume. Verschiedene Nasenerkrankungen verursachen unspezifische pathologische, dabei funktionelle Symptome, wie Obstruktion, Rhinorrhoe, Niesreiz, Hyposmie und Näseln.

58.2.7.1 Hyperreaktivität der Nasenschleimhaut

Die Hyperreaktivität der Nasenschleimhaut steht unter dem starken Einfluß von Emotionen. Sie kann sowohl bei Allergikern wie auch bei Nichtallergikern gefunden werden (Zenner, 1987) und zeichnet sich dadurch aus, daß bereits gewöhnliche Stimuli zu einer Reaktion führen, welche beim Gesunden nur durch massive Reize entsteht. Der Unterschied besteht darin, daß bei der allergischen Rhinitis eine vorausgehende Immunisierung mit Allergenen (Pollen, Hausstaub, Tierhaare etc.) nachweisbar ist, wohingegen diese bei der vasomotorischen Rhinopathie vermißt wird. Beiden Geschehen ist jedoch die sog. biochemische Phase mit Freisetzung von Mediatoren (Histamin, Leukotriene und Substanz P) (Wolf, 1988) aus der Mastzelle gemeinsam. Diese bewirken dann in der Nasenschleimhaut Hyperämie mit Schwellung und Hypersekretion. Aufgrund dieser pathophysiologischen Erkenntnisse werden die zwei Krankheitsbilder auch als hyperergische und hyperreflektorische Rhinopathie bezeichnet (Terrahe, 1985).

Hyperergische (allergische) Rhinopathie

Über die Bedeutung der Psyche bei hyperergischer (allergischer) Rhinopathie bestehen in der Literatur gegenteilige Ansichten. Doch ist der Verlauf der allergischen Reaktion bei Patienten in Konfliktsituationen deutlich heftiger. Außer Nasenobstruktion, Rhinorrhoe und Kettenniesen ist das Krankheitsbild regelmäßig mit Konjunktivitis und Tränenfluß, weniger häufig auch mit Pharyngotracheitis verbunden. Die Patienten fühlen sich allgemein krank.

Hyperreflektorische Rhinopathie

Bei der hyperreflektorischen (vasomotorischen) Rhinopathie ist die Auslösung der Symptome durch psychische Reize offenbar. Oft genügt nur der Anblick einer unerwünschten Person und der Betreffende hat „die Nase voll", wäßrige Sekretion gesellt sich dazu (sog. Stundenschnupfen), weniger häufig folgt der Niesreiz. Augen- und Rachensymptome fehlen.

Meistens betrifft das Krankheitsbild besonders sensible, vasolabile und vegetativ gestörte Patienten. Die Symptome einer **Nasen-„Neurose"** äußern sich hauptsächlich in der rezidivierend verstopften Nase und dem zwanghaften Schniefen. Diese Patienten reagieren dann auf den emotionellen Streß mit gehäuft auftretendem Schnupfen. Einerseits handelt es sich dabei um eine erhöhte Infektanfälligkeit, andererseits aber auch um einen jahrein, jahraus ständig rezidivierenden, therapieresistenten Schnupfen.

58.2.7.2 Therapie

Die Therapie der hyperergischen (allergischen) Rhinitis besteht in der Desensibilisierung, Antihistaminika und Cortison allgemein und lokal. Bei der hyperreflektorischen (vasomotorischen) Rhinopathie erfüllen kortikosteroidhaltige Nasentropfen und -sprays auch eine stabilisierende Wirkung auf die Zellmembran der Nasenschleimhaut. Der therapeutische Akzent sollte jedoch viel mehr auf die Änderung der Lebensgewohnheiten gesetzt werden: Vermeidung von Genußgiften, Ausübung von Sport, Abhärtung und allgemeine körperliche und psychische Roborierung.

Der Therapieplan muß aber immer auch die lokalen rhinologisch-anatomischen Verhältnisse berücksichtigen. So werden allfällig vorhandene basale Cristae und Unregelmäßigkeiten des Nasenseptums korrigiert, denn diese unterhalten den Circulus vitiosus des Schwellungszustandes. Oft muß auch eine rhinologische Entwöhnungskur von den jahrelang als Selbsthilfe eingenommenen Nasentropfen durchgeführt werden (Schnieder, 1986). In solchen Fällen müssen dann gelegentlich auch die hyperplastischen Nasenmuscheln chirurgisch verkleinert werden.

Ein 17jähriger Jüngling erlitt vor sechs Jahren einen Verkehrsunfall mit Polytrauma und Bewußtlosigkeit. Er lebte mit seiner Mutter allein, nachdem der Vater von der Familie wegging. Während einer Woche mußte er durch eine pernasal eingelegte Magensonde ernährt werden. Seit der Entfernung der Sonde reagiert der Patient auf schulischen Streß mit verstopfter Nase, Nasenfluß und Schniefen. Sämtliche allergologischen Abklärungen waren negativ. Rhinoskopisch erscheinen die Nasenmuscheln geschwollen, aber blaß und mit wäßrigem Sekret gefüllt, sonst keine anatomischen Abnormitäten.

Dem Gymnasiasten wurden die Zusammenhänge seiner Nasenreaktion mit dem vor Jahren erlebten Unfall- und dem jetzigen Schulstreß aufgezeigt. Therapeutisch wurden „Kneippkuren der Nase" mit abwechslungsweisem Ein- und Ausspülen von warmem und kaltem Wasser durchgeführt. Durch diese „Abhärtung" besserten sich die Beschwerden, insbesondere was die Häufigkeit der vasomotorischen Reaktion betrifft, so daß er der Umgebung nicht mehr lästig auffällt und subjektiv nur gelegentlich gestört ist.

58.2.7.3 Atopie

Eine anlagemäßige, vererbte Bereitschaft zur Überempfindlichkeit mit einer bestimmten Überreaktion wird Atopie genannt. Außer der hyperreflektorischen (vasomotorischen) Rhinopathie gehören dazu auch Asthma bronchiale und die Neurodermitis in absteigender Häufigkeit (Schnyder, 1960). Die Rhinopathie und das Asthma sind parasympathikomimetisch stigmatisiert, dagegen haben die atopischen Hauterscheinungen einen sympathikomimetischen Charakter. Diese Unterschiede erklären auch, daß bei einem Patienten bei Verschlechterung seiner asthmatischen Beschwerden sich eine Besserung der Hauterscheinungen zeigt, und umgekehrt etwa eine Rhinopathie mit ihren Symptomen schwächer wird, wenn die Hauterscheinungen verstärkt auftreten (Ruppert und Rüdiger, 1982).

58.2.7.4 Nasennebenhöhlen

Alle lokalen Erkrankungen und Veränderungen der Nasenhaupthöhle stehen in engem Zusammenhang mit denjenigen der Nasennebenhöhlen. Dies ist bereits aus der gemeinsamen Bedeckung mit respiratorischer Schleimhaut einleuchtend. So wird man bei jeder Rhinitis, aber auch bei hypererger und hyperreflektorischer Rhinopathie eine klinisch vorerst irrelevante Begleitsinusitis oder Schleimhautschwellung beobachten können. Dieser Zustand erlangt erst dann praktische Bedeutung, wenn eine bakterielle Superinfektion hinzutritt. Dann wird der Verlauf ausgesprochen protrahiert; auf Antibiotikatherapie und Kieferhöhlenspülungen schlecht oder gar nicht ansprechbar. Oft entwickelt sich daraus eine Infektallergie. Die Schleimhaut verändert sich zusehends polypös. Die in der Nase sichtbaren Polypen sind nur „die Spitze des Eisbergs", dessen Basis versteckt in den Siebbeinzellen oder der Kieferhöhle liegt.

Durch diesen pathophysiologischen Prozeß chronifiziert die anfänglich nur anfallsweise auf emotionelle Belastung hin auftretende Erkrankung der hyperreflektorischen Rhinopathie. Die praktische therapeutische Konsequenz ist dann ein chirurgischer Eingriff. In der Nachbetreuung sollen jedoch auch die ursprünglich auslösenden seelischen Probleme psychotherapeutisch gelöst werden.

58.2.7.5 Näseln

Als Näseln wird eine pathologische Nasenresonanz beim Sprechen bezeichnet. Das geschlossene Näseln dämpft in der deutschen Sprache vor allem die drei nasalen Laute „m, n, ng". Es kommt bei allen obstruierenden Erkrankungen der Nase und des Nasenrachens vor, aber nur selten funktionell.

Das offene Näseln wird hingegen viel häufiger als funktionell bedingt angetroffen. Die Ursache liegt meistens in der unschönen Sprechgewohnheit, oft auch dialektbedingt, oder es ist Folge einer Schonhaltung des Gaumens bei schmerzhaften Rachenerkrankungen. Organisch ist es durch Unfähigkeit des Gaumenrachenverschlusses bedingt. Die wohl gravierendste Form des offenen Näselns entsteht bei Gaumenspalten. Die Betreuung der Lippen-Kiefer-Gaumenspaltenträger ist heute eine interdisziplinäre Aufgabe der Kieferchirurgie, Kieferorthopädie, Hals-Nasen-Ohren-Heilkunde, Phoniatrie und Logopädie. Die Leistungen der operativen Behandlung bedeuten einen echten Fortschritt in der Medizin. Genauso wichtig ist aber auch die jahrelange psychologische Führung der Patienten und ihrer Eltern. Auch beim besten operativ-ästhetischen Erfolg fühlen sich die Patienten in ihrem Körperschema betroffen. Wenn nun genau analysiert wird, wie die Einstellung der Mütter dieser Kinder ist, zeigt sich, daß Knaben insgesamt weniger betroffen sind und sich nur dann ungünstig entwickeln, wenn die Mütter sehr ängstlich sind. Bei Mädchen ist die ungünstige psychische Entwicklung als Folge der Mißbildung davon abhängig, ob die Mutter selbst stark auf die äußere Schönheit bedacht ist oder nicht. In beiden Fällen ist letztlich nicht der operative Erfolg, sondern die Verarbeitung der Krankheit und des noch festzustellenden Defektes dafür entscheidend, wie aus der Sicht des Patienten das Spätergebnis interpretiert wird (Heim, 1984).

58.2.8 Globus pharyngeus

58.2.8.1 Definition

Globus pharyngeus ist eine völlig unspezifische Mißempfindung im Hals, verursacht durch eine Reihe von organischen, funktionellen und psychischen Erkrankungen und Störungen. Bereits aus dieser Aufzählung der großen Gruppen von möglichen Ursachen ist ersichtlich, daß sich der Begriff „Globus hystericus" erübrigt, abgesehen von seinem abwertenden Beigeschmack (Schnieder, 1986). Hingegen ist die Differenzierung einer Dysphagie – als einer Schluckbehinderung – notwendig, aber nicht immer leicht, da beide Störungen gleichzeitig vorkommen können.

58.2.8.2 Häufigkeit

Globusbeschwerden sind außerordentlich häufig. In einer Reihenuntersuchung von 3176 Patienten zwecks Früherkennung von HNO-Karzinomen konnte Schnieder (1986) feststellen, daß über 50% der weiblichen und über 40% der männlichen Untersuchten schon einmal Globusbeschwerden hatten; hingegen fand er kein einziges Karzinom.

58.2.8.3 Symptome

Kloß im Hals, Fremdkörpergefühl, Kratzen, Brennen, Schmerzen etc.

58.2.8.4 Das organisch bedingte Globusgefühl

Das organisch bedingte Globusgefühl kann durch entzündliche oder tumoröse Prozesse verursacht sein. Entzündliche Veränderungen werden bereits durch einen chemischen Reiz, beispielsweise bei Konsum von scharfen Speisen, konzentriertem Alkohol oder Nikotin, verursacht und sind weitgehend von der individuellen Empfindlichkeit abhängig. Die mikrobielle Pharyngitis verursacht die gleiche Symptomatik von brennenden Schluckschmerzen. Beim Schlucken der Speisen nehmen die Schmerzen zu. Nach der Spray-Anästhesie der Rachenschleimhaut verschwinden sie, solange das Anästhetikum wirkt.

Das tumorös bedingte Globusgefühl bleibt trotz der Rachenanästhesie bestehen. Doch betrifft es meistens indolente Personen, welche sich im Verlauf von vielen Jahren an dieses Gefühl gewöhnt haben, und sie suchen den Arzt erst bei massiver Zunahme der Beschwerden auf. Zu diesen gehört vor allem die Irradiationsotalgie bei Larynx- und Hypopharynxkarzinom. Der negative otoskopische Befund mit Otalgie ist immer ein ernstes Symptom, welches einer Klärung bedarf. Im übrigen gilt nach wie vor der Grundsatz: Ein Globusgefühl bleibt so lange tumorverdächtig, bis das Gegenteil bewiesen wird. Dies verpflichtet immer zu einer sorgfältigen HNO-Untersuchung. Hingegen verursacht eine Struma oder gar ein Schilddrüsenkarzinom nur selten oder überhaupt kein Globusgefühl (Wey, 1977).

58.2.8.5 Das funktionell bedingte Globusgefühl

Das funktionell bedingte Globusgefühl entsteht durch Verspannung der Schluck- und Halsmuskulatur. Bekannt ist dies vor allem nach extremer körperlicher Belastung, beispielsweise bei Bodybuilding. Der gleiche Mechanismus kann aber auch durch extreme Reklination des Kopfes und die damit verbundene Überdehnung der Halsmuskulatur hervorgerufen werden (Seifert, 1988). Es bestehen auch engste Zusammenhänge mit der reflektorisch hervorgerufenen Verspannung der Halsmuskulatur bei den Erkrankungen der Halswirbelsäule (Seifert, 1988). Fast regelmäßig wird Globusgefühl auch bei funktionellen Stimmstörungen empfunden (Habermann, 1986); dies ist mit einem Muskelkater bei untrainierter, übermäßiger Muskelanstrengung vergleichbar.

58.2.8.6 Das psychisch bedingte Globusgefühl

Das psychisch bedingte Globusgefühl kann ein Ausdruck von unbewußten Konflikten mit deren Neutralisierung im Symptom sein, als psychophysiologisches Phänomen bei Depressionen, Überforderung und Streß auftreten. Für das psychogen bedingte Globusgefühl ist ein Verschwinden der Beschwerden beim Essen und Trinken charakteristisch. Es gibt aber auch Globusgefühl als monosymptomatische Form der Depression. Daß jemandem bei Aufregung und

Streß der Hals austrocknet, ist genauso verbreitet wie die kalten und feuchten Hände. Interessant ist dabei die negative Beeinflussung der lokalen Abwehrmechanismen mit konsekutiven Halsentzündungen und Angina – sog. Examensangina (Lüscher, 1959). Das „zuschnürende" Gefühl im Hals kann das erste Mal anläßlich einer organischen Halserkrankung auftreten, von welcher sich der Patient mit Räuspern und Hüsteln zu befreien versucht. Sehr rasch kann das letztere zur Gewohnheit werden und bleibt auch nach Abheilen der Grundkrankheit bestehen. Darüber hinaus werden die Symptome mit unterschiedlicher Qualität und an verschiedenen Lokalisationen wechselnd beschrieben. Diese Patienten sind auch sehr suggestibel.

58.2.8.7 Therapie

Bei organisch und funktionell bedingtem Globusgefühl ist eine ursächliche Therapie (chirurgische, logo-pädische, physiotherapeutische, manualmedizinische) unumgänglich. Beim psychogenen Globusgefühl kann bereits eine sorgfältige Untersuchung therapeutisch wirken. Doch ist es mit einer Konsultation nicht getan. Der Patient muß regelmäßig fachärztlich kontrolliert und begleitet werden. Die Rachenpinselung mit einer auffällig farbigen Flüssigkeit hat sich seit Generationen bewährt. Diese äußere Handlung ist gleichsam ein Zeichen der inneren Zuwendung des Arztes zu seinem Patienten. Wesentlich ist dabei die suggestive Beeinflussung, das Ernstnehmen und Anhalten zur Änderung von schädlichen Gewohnheiten. Nur so können unnötige chirurgische Eingriffe des in der Regel sehr operationsbereiten Patienten vermieden werden.

Medikamentöse Therapie mit Anxiolytika ist nur selten für eine kurze Zeit notwendig. Auch eine Psychotherapie bleibt nur für schwere Fälle ausnahmsweise vorbehalten.

Die Tendenz zur Spontanheilung ist erfreulich hoch.

59 Zahnheilkunde

Hans-Joachim Demmel und *Friedhelm Lamprecht*

Ein Patient kommt zum Zahnarzt, weil sein Zahnfleisch leicht blutet, er einen Zahn durch Lockerung zu verlieren befürchtet und Fehlstellungen durch Zahnwanderung in der Oberkieferfront bemerkt. Die intraorale Inspektion ergibt eine profunde marginale Parodontitis. Fünf Zähne sind gelockert; der Patient hat Zahnstein und subgingivale Konkremente. Diese Symptome führen in der Regel lediglich zur klassischen Parodontaltherapie. Die erweiterte Untersuchung zeigt markante Facetten auf den Zähnen vom Knirschen. Die Palpation der Kaumuskeln ergibt druckdolente Muskeln, die dem Bewegungsbild der Facetten auf den Zähnen entsprechen. Die erweiterte Anamnese ergibt: „dröhnende häufige Kopfschmerzen"; vor neun Monaten ein Schulter-Arm-Syndrom rechts mit Parästhesien; drei Kuren wegen „arthrotischer LWS-Veränderung"; häufige Magenkrämpfe, die über den Rücken ausstrahlen. Am Ende der ersten Exploration durch den Zahnarzt fragt der Patient von sich aus, ob für einen Teil seiner Leiden auch psychische Ursachen ausschlaggebend sein könnten. Andererseits sei er aber auch erschrocken bei dem Gedanken, „eine Macke zu haben". Er nimmt das Angebot der Überweisung an eine psychosomatische Fachklinik dankend an.

59.1 Die Bedeutung der Psychosomatik für die Zahnmedizin

Für den Zahnarzt kommt es nun darauf an, auch in seinem Bereich den „diagnostisch-therapeutischen Zirkel" nicht auf die organismischen Funktionen des Kauapparates zu beschränken, sondern über die Erhellung der Motivationskonflikte auf der interpersonellen Ebene das präsentierte Symptom im Situationskreis-Zusammenhang zu verstehen, wobei die diagnostisch-therapeutische Verquickung für Arzt und Patient um so deutlicher wird, je mehr der Zahnarzt seine gefühlsmäßige Resonanz auf den Patienten in seine diagnostischen Bemühungen mit einzubeziehen lernt. Die zahlreichen sich auf die Zähne oder den Beißvorgang beziehenden umgangssprachlichen Ausdrücke geben Hinweise darauf, in welchen emotionalen Bereichen die Motivationskonflikte zu suchen sind (Demmel et al., 1983). Ein dafür notwendiges Wahrnehmungs- und Explorationstraining sollte in das Curriculum der zahnärztlichen Ausbildung einbezogen werden, ähnlich wie es durch die Aufnahme der psychosozialen Fächergruppe in die medizinische Approbationsordnung geschehen ist. Auch soll-

te dem Zahnarzt eine qualifizierte psychosomatische Weiterbildung ermöglicht werden. Dies ist nötig, damit etwas, das verlorengegangenes ärztliches und damit auch zahnärztliches Allgemeingut darstellt, nicht abgespalten als Spezialdisziplin wiedererscheint, wie sich das in angelsächsischen und skandinavischen Ländern in dem Fach Psychodontie (Winnberg, 1969) ankündigt, und dann wieder mühsam integriert werden muß.

Man kann zwar durch Fragen Probleme einengen, aber auch ausgrenzen, oder, wie Balint sagt, auf Fragen bekommt man Antworten, sonst nichts. Läßt man den eingangs erwähnten Patienten zu Wort kommen durch offene Fragen, so wird der Patient als Mensch mit seinen Problemen spürbar und es ergeben sich andere Handlungsanweisungen.

Der 52 Jahre alte, sich betont jugendlich gebende, salopp gekleidete Patient berichtet zunächst über sein Zähneknirschen und fährt dann fort: „Ich verkrampfe die Muskulatur im Körper, Mutter hat immer gesagt, Zähne zusammenbeißen und durch". Bei Prüfungen hätte er immer Durchfälle und Magenbeschwerden bekommen. Wenn er jetzt Ärger mit seiner Frau habe, würde es ihm auf den Magen schlagen. Er hätte auch Schmerzen am After und häufig Blut im Stuhl. Verspannungen im Nacken, Kribbeln im rechten Arm, er würde sich da gern massieren lassen. Er müsse alles einigermaßen im Griff haben. Morgens früh würde er immer mit Zahnschmerzen aufwachen, er hätte auch Durchschlafstörungen. Seine Selbstbeschreibung lautet wie folgt: „Ich bin mit mir unzufrieden; ich habe nie ich sein können; ich fühle mich fremdbestimmt, bin ehrgeizig, fleißig, ordentlich und pünktlich; der, der ich war, kotzt mich an. Früher bin ich nur Funktion gewesen". Als Kind sei er sehr ängstlich und zurückhaltend gewesen. Seine Mutter habe die drei Söhne und den Vater total beherrscht, „ich habe immer gelernt, Contenance zu bewahren". In seiner Ehe sei dann alles umgekehrt, er würde alles machen, seine Frau wäre die Doofe, er hätte zu Hause die Patriarchen-Rolle. Mit 4–5 Jahren wäre er wegen einer Meningitis lange im Krankenhaus gewesen und er hätte seither häufiger unter Kopfschmerzen gelitten. Er hätte panische Angst vor brüllenden Menschen und auch Schwierigkeiten, insbesondere Frauen gegenüber, etwas zu entgegnen, wenn diese ihm gegenüber aggressiv sind. Häufig müsse er viel Süßigkeiten essen, sozusagen, um sich zu belohnen. Die früheste Erinnerung seines Lebens ist, wie er munter wird, aufwacht und eine Krankenschwester ihn fragt, ob sie erkennt, nachdem er drei Tage bewußtlos gewesen ist. Er bezeichnet seine Erziehung durch seine Mutter als formalisiert, ebenso die Gutenachtküsse. Er könne sich mit seiner Frau nicht streiten. „Ich habe noch nie zu meiner Frau etwas Abfälliges gesagt, wenn ich zu Hause ärgerlich werde, ziehe ich mich zurück." Er habe Schwierigkeiten, sich zu konzentrieren, sie würden zu Hause nebeneinanderher leben. Seine Bedürfnisse nach Zärtlichkeit befriedigt er in einer Außenbeziehung.

Die Kindheit des Patienten ist charakterisiert durch Ohnmacht, Liebesarmut und Einschränkung sich entwickelnder motorischer Expansivität mit Heranbildung eines sensiblen Wahrnehmungsvermögens und einer stark ausgeprägten rezeptorischen Sphäre ohne affektorische Einflußmöglichkeit, was ein Problembewußtsein schafft ohne Problemlösungsmöglichkeiten. Diese verletzliche Situation versucht der Patient durch ein Streben nach Unabhängigkeit, die meist eine Pseudounabhängigkeit ist, zu meistern, indem er sich kontraphobisch in Aktivitäten flüchtet und Leistung einen übermäßigen Stellenwert bekommt. Diese früher lebensnotwendige Haltung fordert ihren Tribut bei dem Patienten in Form von Schmerzen im Schulter- und LWS-Bereich, Magenkrämpfen und Zähneknirschen. Die beschriebene Reaktionsbildung hat zur Folge, daß der Patient jetzt in der Beziehung zu seiner Frau den Haustyrannen spielt, also die Einfluß- und Einwirkungsmöglichkeiten überbetont. Dabei ist aber die früher vorhandene sensible Wahrnehmung weitgehend ausgeschaltet, d. h., er hätte jetzt die Problemlösungsmöglichkeiten, erkennt aber sein Problem nicht mehr, was sich nach Anwendung des diagnostisch-therapeutischen Zirkels etwa wie folgt darstellt: Der Patient verhält sich gegenüber seiner Frau, wie er sich gegenüber seiner Mutter nicht verhalten konnte, und so, wie er sich seiner Frau gegenüber verhalten sollte, verhält er sich gegenüber seiner Freundin. Reaktionsbildung zur Mutter und Inzesttabu bestimmen also sein Verhältnis zu seiner Frau. Er selbst kann dann die Realisierung des Wunsches, in der Paarbeziehung glücklich zu werden, äußern, so daß die psychosomatische Konsultation mit der gemeinsam erarbeiteten Therapieempfehlung einer Paartherapie endet.

59.2 Die psychosomatische Besetzung der Mundregion

Die libidinöse Besetzung der Mundregion ist in einer frühen ontogenetischen Entwicklungsphase des Menschen so vorherrschend, daß diese Entwicklungsphase als orale bezeichnet wird (Abraham, 1982; Freud, 1968), in welcher die Mundregion den ersten Körperbereich darstellt, mit welchem zwischenmenschliche Beziehung hergestellt wird. Eine regressive Antwort auf eine verunsichernde Situation kann auch im späteren Leben zu einer Reaktivierung des ursprünglichen Bedeutungsgehaltes führen.

Die ausgedehnte zentralnervöse Repräsentanz von Mundschleimhäuten, Kaumuskulatur und Zunge, die Dichte der nervösen Versorgung und Rezeptorenbesetzung machen dieses Gebiet extrem empfindlich gegenüber Störgrößen, die aus umfassenderen Systembezügen einwirken können. Die Mundregion ist Ausführungsort einer Reihe von biologischen Akten (Viktor von Weizsäcker, 1947), wie z. B. des Kauens,

Beißens, Saugens, Schmeckens, Sprechens, Lächelns, Drohens, Küssens etc. In diesen biologischen Akten sind Stimmungen und Affekte unlösbar gekoppelt mit endokrinen, vegetativen und motorischen Bereitstellungsreaktionen, die wiederum durch die „Bedeutungserteilung" und durch die „Bedeutungsverwertung" über den Situationskreis die Verbindung und Kommunikation zu der individual-spezifischen Umwelt des Patienten ermöglichen. Erschwerend kommt hinzu, daß in der Mundregion phylogenetisch ableitbare außersprachliche Verhaltensresiduen zum Ausdruck kommen (Darwin, 1872; Every, 1965).

59.3 Das orofaziale Schmerz-Dysfunktions-Syndrom

Die besondere Bedeutung der orofazialen Region für den Ausdruck psychischer Situationen (Ekman, 1988) und den zwischenmenschlichen Kontakt läßt diesen Körperbereich geradezu prädestiniert erscheinen für psychophysiologische Syndrome und Konversionssymptome (vgl. Kap. 30). Differenzierte Untersuchungen über orofaziale Konversionssymptome fehlen bisher in der zahnärztlichen Literatur. Egle (1985) hat für die orofazialen Schmerzsyndrome eine erste Einordnung entsprechend dem DMS-III veröffentlicht.

In der ärztlichen oder zahnärztlichen Praxis erscheinen oft Patienten, ähnlich dem eingangs geschilderten Fallbeispiel, häufig mit der Verdachtsdiagnose „Trigeminusneuralgie", mit einem Symptombild, das nicht den klassischen Zeichen der Neuralgie entspricht. Sie klagen über Schmerzen im Kopfbereich, die ständig oder wellenförmig anhalten, nicht die Grenzen der Rami des Nervus trigeminus einhalten, oft auch beidseitig auftreten, mit Kopfschmerzen einhergehen und tageszeitlichen Schwankungen unterliegen. Häufig schmerzt das Kiefergelenk, die Mundöffnung kann behindert sein, und bei Bewegungen des Unterkiefers bemerkt der Patient ein Knacken im Gelenk. Nach neueren medizinischen Erkenntnissen muß der Begriff des Costen-Syndroms (Costen, 1934) als Bezeichnung für eine klinische Entität fallengelassen werden (Motsch, 1980; Siebert, 1981). Die fachärztliche, differentialdiagnostische Abklärung hat u. a. insbesondere zu berücksichtigen: Pulpitis, dentogene Ostitiden, Pericoronitis retinierter Weisheitszähne, Sinusitiden, Otitis media, Parotitis, Herpes zoster, akutes Glaukom, Tumoren, Erstsymptome der MS, Horten-Syndrom, andere Neuralgien im Kopfbereich, intramuskuläre Fibrositis, rheumatoide Arthritis (Blackwood, 1969; Cohen et al., 1979), entzündliche osteoarthrotische Veränderungen, Depressionsäquivalent, Konversionssyndrom (Engel, 1951), Phantombiß-Syndrom (Marbach, 1976). Es bleiben bei der zahnärztlichen Untersuchung als somatische Zeichen oft nur deutliche Schliffacetten auf den Zähnen, Störungen der Kiefergelenkbewegung (wie Seitenabweichung, eingeschränkte Mundöffnung und Gelenkgeräusche) und insbesondere Druckdolenz bestimmter Kaumuskeln (Krogh-Poulsen, 1980; Solberg, 1981). Eine genaue Funktionsanalyse des stomatognathen Systems zeigt fast immer eine Differenz zwischen der habituellen Unterkieferposition und der physiologisch günstigsten Position des Kiefergelenkes. Weiter sind Fehl- bzw. Frühkontakte einzelner Zähne zu finden (Krogh-Poulsen, 1967). Einer somatisch-mechanistischen Betrachtungsweise folgend, werden diese Patienten oft nur mittels Aufbißschienen mandibulär repositioniert und anschließend mittels Einschleiftherapie oder Onlays und Kronen aufwendig versorgt, d.h. durch spezielle Gestaltung der Höcker-Fossa-Beziehung und der Führungsflächen der Canini und Incisivi wird eine stabile maxilläre Abstützung erreicht (Bauer und Gutowski, 1975). Für eine ausführliche Darstellung dieser Okklusionskonzepte muß auf Slavicek (1982) verwiesen werden. Die monokausale, mechanistische, sich auf den Okklusionsvorgang beschränkende Betrachtungsweise des orofazialen Schmerz-Dysfunktions-Syndroms ist für die Diagnose und somit auch für die Therapie nicht ausreichend. Zum einen finden sich solche okklusalen Dysharmonien bei über 80% aller Menschen, ohne daß diese auch über Schmerzen im Sinne dieses Syndroms klagen, zum anderen wechseln die Beschwerden häufig oder verschwinden manchmal auch ganz ohne zahnärztliche Therapie (Bell, 1969; Desjatnikov et al., 1978). Darüber hinaus können die Schmerzen nach idealer funktionstherapeutischer, okklusaler Rekonstruktion in unveränderter Form weiterbestehen (Dworkin und Marbach, 1974). Erst ein mehrdimensionales Krankheitsverständnis schafft einen psychosomatischen Zugang zu diesen schwierigen Patienten (Kaban und Belfer, 1981; Meerwein, 1967; Schrenk, 1962), die oft gegen aufdeckende Psychotherapie feindselig eingestellt sind. Laskin (1969) begründete die psychophysiologische Theorie des orofazialen Schmerz-Dysfunktions-Syndroms. Zahlreiche Publikationen (u.a. Frei und Graber, 1977; Gold et al., 1974; Gold et al., 1975; Graber, 1978; Lefer, 1968; Lipton et al., 1974; Lupton, 1969; Pomp, 1974; Rugh und Solberg, 1976; Rugh, 1981; Violon, 1980), die diese Betrachtungsweise stützen, folgten. Schon einfache explorative Gespräche im Sinne Balints (Balint, 1980) zeigen beim orofazialen Schmerz-Dysfunktions-Syndrom immer einen psychisch auffallenden Hintergrund. Die Triggerfunktion der okklusalen Fehlkontakte und die Dyspositionierung der Mandibula begünstigen Bruxismus und Parafunktion (Knirschen und Pressen) bei psychischer Belastung (Heggendorn et al., 1979). Psychodiagnostische Untersuchungsverfahren zeigen Diskrepanzen zwischen Selbst- und Idealbild, diese Patienten leiden unter Minderwertigkeitsgefühlen und opfern sich oft für andere auf, ihre Wunschphantasien tendieren zur Norm, sie konkurrieren stärker, sind ungeduldig und häufig in Auseinandersetzungen verstrickt, dabei sind sie weniger durchsetzungsfähig und selbstkritischer. In der Auseinandersetzung mit sozial belastenden Situationen haben diese Patienten infantile Riesenansprüche an andere. Im Vergleich zu Patienten mit anderen psychosomatischen Störungen

beschreiben sie sich als vertrauensvoller im zwischenmenschlichen Kontakt. Sie sind depressiver als die Norm (Cathomen-Rötheli, 1979; Heiberg, 1980; Lamprecht et al., 1986; Staats und Graber, 1982). Eine weitere Differenzierung des orofazialen Schmerz-Dysfunktions-Syndroms erscheint möglich. Ätiologisch kann unterschieden werden zwischen dem durch okklusale Fehlkontakte induzierten Bruxismus und dem frontolateralen Bruxismus (Knirschen oder Pressen nur über die Front- und Eckzähne) ohne diese Fehlkontakte. Die Psychodynamik dieser zweiten Form wird als Angstäquivalent gedeutet (Demmel et al., 1988). Hier werden Ansätze für begleitende psychische Therapien für diese zahnärztlichen Problempatienten deutlich.

Wir haben uns (vgl. Einleitungsfall) aus dem Patientengut einen Mann herausgegriffen, da es häufig schwieriger ist, das zugrundeliegende Problem bei Männern zu entdecken. Sie kommen selten mit dem Leitsymptom in die Praxis, zeigen mehr Haltung und verbergen ihr Leiden. Entsprechend fallen sie im Sinne des orofazialen Schmerzsyndroms fast nur bei zahnärztlicher Routinekontrolle auf. Das zahlenmäßige Überwiegen der Frauen bei diesem Syndrom könnte sich teilweise dadurch erklären, daß die Männer erwartungsgemäß häufiger die Zähne zusammenbeißen müssen, obwohl sie „auf dem Zahnfleisch laufen" und sie dafür seltener zahnmedizinische Hilfe in Anspruch nehmen. Zum anderen ist auch bei der gesunden Frau die Muskelspannung im fazialen Bereich deutlich höher als beim Mann. Der erhöhte Tonus wird bei okklusaler Dysharmonie und Dyskinese eher zu Myogelose und Schmerz führen. Der Mann hat deutlich häufiger im lumbosakralen Bereich eine erhöhte Muskelspannung und folglich hier häufiger ein Schmerz-Dysfunktions-Syndrom (Sundsvold et al., 1981).

59.4 Streß und Parafunktion (Auswirkung auf das Parodont)

Maxilläre Dysharmonien durch okklusale Interferenzen können durch Abnutzung zur Adaptation führen und somit als Trigger ausgeschaltet werden (Graber et al., 1980). Solche autoadaptiven Vorgänge durch Zähneknirschen oder durch die mastikatorische Funktion gehören aber zu den Seltenheiten, da in der Regel hierdurch keine stabile zentrische Okklusion erreicht werden kann. Die Bewegungsbahn auf den Bruxofacetten kann als neuer Trigger dienen, soweit der Streß anhält. Wo die Erschöpfung der Adaptationsreserven des stomatognathen Systems zuerst auftritt und so zur chronischen Krankheit führt, hängt weitgehend von den Bedingungen des Individuums ab (Stallard, 1969). Die Mitbeteiligung psychischer Faktoren bei der Ätiologie der Dyskinesen bzw. Parafunktionen kann heute als gesichert angesehen werden (Belting und Gupta, 1960; Fallschlüssel, 1983; Glaros und Rao, 1977; Graber, 1971; Graber, 1978; Mikami, 1977; Schultz, 1961; Weinberg, 1977).

Der Kausalitätszusammenhang zwischen Parafunktion und atypischem Gesichtsschmerz ist begründet (Kaban und Belfer, 1981; Millstein-Prentky und Olson, 1979; Schulte et al., 1981; Schultz, 1961). Vieles spricht für einen Zusammenhang zwischen Parafunktion und Veränderung im parodontalen Gewebe, zumal Zahnwanderungen bei Zähne- und Zungenpressen beobachtet werden und ebenso Veränderungen im Sinne der Parodontitis (Ott und Wöhr, 1982). Morphologische Veränderungen der oralen Gewebe unter Streß sind beschrieben worden (Raetzke, 1985). Die Zahl der Untersuchungen über diese Zusammenhänge ist aber noch zu gering und lückenhaft, und so muß der Zusammenhang zwischen Parodontalerkrankung und Streß noch hypothetisch formuliert werden (Baker et al., 1961; Grant et al., 1979; Manhold, 1953; Milgram et al., 1983), da die Publikationen sich zum Teil widersprechen (Hanamura, 1987).

59.5 Zahnverlust als erstes Zeichen der Hinfälligkeit und der Erneuerungswunsch in der zahnärztlichen Prothetik

Lippenhaltung und Zahnstellung sind nonverbale Ausdrucksformen im sozialen Verhalten bei Mensch und Tier (Darwin, 1872; Every, 1965). Zahnverlust wird im Milchgebiß als sicheres Zeichen des Erwachsenwerdens begrüßt. Vom Erwachsenen wird der Zahnverlust aber als entwaffnend empfunden (Meerwein, 1967). Er wird als erstes Zeichen der Hinfälligkeit gewertet, und von der zahnärztlich-prothetischen Kunst wird die Wiederherstellung der Normalität erwartet; oft verbunden mit dem Wunsch, das Übel mit der Verwirklichung phantasierter Idealvorstellungen ins scheinbar Positive zu verkehren. In diesem Zusammenhang ist auch die sog. „psychogene Prothesenunverträglichkeit" von Bedeutung (Drost, 1978; Marxkors und Müller-Fahlbusch, 1976; Wupper, 1971). Festsitzender Brückenersatz wird noch eher toleriert, da er auch subjektiv als Wiederherstellung gefühlt werden kann. Herausnehmbarer Zahnersatz erneuert aber zumindest bei der täglichen Zahnpflege immer wieder das Bewußtsein des Verlustes der körperlichen Integrität. Das Erkennen der Problematik im Einzelfall und die psychische Betreuung dieser Patienten stellen große Anforderungen an den Behandler. Der Zahnverlust kann unbewußt als Sterben auf Raten empfunden werden und erinnert an die Endlichkeit des Körperlichen. Auf seinen symbolischen Bedeutungsgehalt als Kastrationsangst hat schon Freud (1968) hingewiesen.

59.6 Das Compliance-Problem bei Ernährung und Pflege (Karies)

Wie in vielen Bereichen der Medizin ist die Motivierbarkeit für ein Verhalten sehr schwierig, wenn der Erfolg des angestrebten Verhaltens im Nichteintreffen einer Krankheit liegt. Es ist bemerkenswert, daß trotz allgemeiner Kenntnis über die Wichtigkeit oraler Hygiene und Ernährung für die Vermeidung von Karies die überwiegende Mehrheit der Bevölkerung in nur ungenügendem Ausmaß die Zähne putzt und auch bei der Ernährung kariogene Substanzen nicht meidet. Diese Form der Non-Compliance gründet sich auf die allgemein unzureichende Motivierbarkeit für prophylaktische Maßnahmen bei nicht vorhandenem Leidensdruck. Die Compliance nach eingetretenem Zahnschaden ist trotz des dann vorhandenen Leidensdruckes ebenfalls häufig unzureichend, da die oralhygienischen Maßnahmen nur den weiteren Zahnverfall aufhalten, nicht aber den vorherigen Zustand wiederherstellen. Nur eine akzeptierte prothetische Arbeit wird u. U. wieder als erhaltenswert betrachtet werden können und nicht nur als Zwischenstadium zur Zahnlosigkeit. Nur ein erheblicher finanzieller und zeitlicher Aufwand in der Gesundheitsfürsorge kann die Compliance wesentlich verbessern.

59.7 Habits und kieferorthopädische Problematik

Seit Struwwelpeters Zeiten werden Daumenlutschen, Nägelbeißen, Lippensaugen und ähnliche Habits als kindische Unarten mit erzieherischer Macht bekämpft. Häufig muß auch der Zahnarzt als Autorität herhalten, der dem Kind erklären soll, wie schädlich Daumenlutschen ist. Obwohl das psychische Äquivalent für derartige Angewohnheiten nicht mehr unbekannt ist (Biermann, 1982; Fleischer-Peters und Zschiesche, 1980; Scholz, 1982), ist eine Kieferorthopädie, bei der psychoanalytische Gesichtspunkte mitberücksichtigt werden, noch eine extreme Ausnahme (Fleischer-Peters, 1985). Bei Zahn- und Kieferfehlstellungen, wie auch bei anderen Haltungsschäden des Körpers, muß mehr als bisher bedacht werden, daß sie auch psychisch bedingt bzw. verstärkt sein können (Sergl, 1967). Balters (1954) sieht die Kieferanomalie als psychophänomenologische Korrelation. So hat z. B. der Zahnengstand durch frühzeitigen Milchzahnverlust seine psychische Mitbedingtheit, wenn mit der Ursachenforschung nicht im Somatischen (floride Milchzahnkaries) haltgemacht wird, sondern die Gründe für mangelnde Selbstpflege und übermäßigen Zuckerkonsum im familiären Umfeld gesucht werden.

59.8 Angst in der Behandlungssituation

Das Gefühl der Angst bzw. Angstvermeidung ist ein wesentliches menschliches Motiv für Verhalten. Die ausgedehnte Repräsentanz sich von Zunge und Lippen herleitender Schmerzfaserprojektionen im Thalamus bildet ein physiologisches Korrelat für die Schmerzempfindlichkeit. Antizipierte Schmerzen fördern Angst bzw. Angstabwehrmechanismen. Diese werden nun ihrerseits durch zahnärztliches Verhalten verstärkt und tragen zur Schmerzintensivierung bei. Hauruck-Verfahren und billiger Trost „das tut nicht weh, das haben wir gleich" etc. sind hier weniger gefragt als ein Ernstnehmen des Patienten bei ihm zumutbarer größtmöglicher Mündigkeit. Bei der Betrachtung der Zahnbehandlungsangst ist zu vergegenwärtigen, daß es sich hier in der Regel um Eingriffe in die körperliche Integrität handelt. Es geht „an die Substanz". Die Angst vor Verlust der körperlichen Integrität kann beim Kind so weit gehen, daß es den Zahnarzt bei routinemäßigen Kontrollen nicht einmal in den Mund sehen lassen möchte. Die intime Sphäre der Mundhöhle bleibt dem Fremden verschlossen. Bei Erwachsenen kann es zur Projizierung tieferer Ängste vor Verlust und Trennung kommen. Die Verletzung der körperlichen Integrität kann als Hinweis auf die Endlichkeit des Lebens empfunden werden. So kann die empfundene Bedrohung des Daseins auch mit beitragen zu der häufigen Kollapsneigung in einer als ausweglos erlebten Situation. Eine Vielzahl von Patienten entspricht den Forderungen der Autorität des Zahnarztes im Sinne eines internalisierten autoritären Gewissens (Fromm, 1982) und ist „tapfer", fügt sich der vom Zahnarzt verordneten Therapie. Diese Patienten kommen regelmäßig zur Kontrolle, zeigen nach außen Einsicht in die Notwendigkeit der Zahnbehandlung, ohne Wunsch nach ausführlicher Erläuterung der Therapien. Die feuchte Hand, der Schweiß auf der Stirn, die veränderte Stimme und der Bewegungsablauf zeigen lediglich die psychophysiologischen Angstäquivalente. Aus dem Gefühl des Bedrohtwerdens, der Angst, die Haltung zu verlieren, sind auch aggressive Verhaltensweisen des Patienten verständlich. Oft sind diese auf das Hilfspersonal des Zahnarztes verschoben, denn der Zahnarzt wird als zu mächtig erlebt und könnte sich ja mit grober Behandlung „rächen". In die Angst fließen hier auch Schuldgefühl und Strafbedürfnis wegen mangelnder Zahnpflege mit ein. Nicht die Bagatellisierung der Angst, sondern das Ernstnehmen der Angst und das Gespräch darüber erleichtern den Umgang mit dem Patienten.

59.9 Therapeutische Konsequenzen

Die kartesianische Spaltung von Psyche und Soma ist in der Zahn-, Mund- und Kieferheilkunde noch vorherrschend. Es kann nicht bestritten werden, daß die rein somatisch begründeten Therapieansätze auch zu großen Erfolgen geführt haben. Es ist aber deutlich geworden, daß sie insbesondere im Bereich der Krankheitsvermeidung immer wieder an Grenzen

stoßen und auch zu Mißerfolgen führen müssen. Ob es sich z.B. um parodontale Probleme, Kiefergelenkschmerzen, Karies, den Ersatz von Zähnen oder um Zahnfehlstellungen handelt, erst die explorative Darstellung der Lebenssituation wird dem Zahnarzt Therapiekonzepte im Einzelfall ermöglichen, die eine Linderung der Leiden durch Einsicht des Patienten in seine Mitbeteiligung am Symptom sichern. Nicht die Psychotherapie für jeden Zahnpatienten ist die Forderung, sondern die Einbeziehung der Psyche des Menschen „Patient" in die Therapie des Zahnarztes, wie wir an unserem Beispiel gesehen haben.

Eine katamnestische Untersuchung, ein Jahr nach dem Explorationsgespräch, als der Patient zu einer Routinekontrolle in die Praxis kam, zeigte einen entspannten, sich wohlfühlenden Patienten, der keine Kopfschmerzen mehr hatte und auch im Zahnbereich schmerzfrei war. Die Untersuchung ergab eine wesentliche Verbesserung der Druckdolenz der Muskeln. Beiläufig erwähnt er, daß er seine Frau ganz anders wahrnehme und eine neue Qualität in die Beziehung gekommen sei. Der therapeutische Erfolg war hier sicher den explorativen Gesprächen und der Paartherapie zuzuschreiben, da die rein zahnärztlichen Maßnahmen (Einschleifen, Rekonstruktion) noch nicht durchgeführt waren.

Die Identität des Zahnarztes wird sich wesentlich ändern, wenn er vom bisher vorherrschenden bio-medizinischen zum bio-psycho-sozialen Krankheitskonzept übergeht. Erfolg für sich und Nutzen für den Patienten kann vom Zahnarzt mit bio-medizinischem Konzept nur in medizintechnischer und somatischer Perfektion gesehen werden. Aber „wer nur am Körper arbeitet, verfehlt die volle Hälfte der Wirklichkeit" (Weizsäcker, 1986). Das bio-psycho-soziale Konzept ermöglicht ihm zum einen, die Wirklichkeit des Kranken zu erfassen und ihm im weitesten Sinne ärztliche Hilfe anzubieten, und zum zweiten ermöglicht es dem Zahnarzt, seine Rolle als Therapeut – und damit auch seine eigene Wirklichkeit – neu zu definieren. Hierdurch kann der engagierte Zahnarzt auch seinem „burn-out" (Aronson et al., 1983) entgehen, denn er wird vom Patienten nicht mehr die fachlich-technische Anerkennung verlangen, die dieser ihm objektiv nicht geben kann, sondern die menschliche Anerkennung erhalten, wenn dieser sich verstanden fühlt. Schwierigkeiten hat insbesondere der niedergelassene Zahnarzt durch die depotenzierende Nichtanerkennung der psychosomatischen Arbeit durch die Kollegen und Institutionen und nicht zuletzt auch durch den Ausschluß psychosomatischer Positionen aus allen vom Zahnarzt anwendbaren Gebührenordnungen.

60 Psychosomatische Sicht des höheren Lebensalters

Hartmut Radebold

60.1 Definition, Kenntnis- und Forschungsstand

Erkrankungen über 60jähriger werden im ärztlichen Alltag noch immer weitgehend als altersbedingt und dazu aus einer defizitorientierten Perspektive angesehen. Der Prozeß des Alterns wird hierbei auf seine körperlichen Anteile reduziert und nicht mehr als einer verstanden, in dem körperliche, psychische und soziale Aspekte untrennbar miteinander verbunden sind. Faßt man darüber hinaus Altern als unabänderlich fortschreitenden (eben organisch bzw. hirnorganisch bedingten) Abbauprozeß auf, so entfallen in zunehmendem Maße ärztliche Behandlungsaufgaben, während Pflege, Versorgung, Bewahrung und unter Umständen Kontrolle in den Vordergrund treten. Gerade der über 60jährige Erkrankte benötigt gegenüber dieser einseitigen und defizitorientierten Sicht, daß psychosoziale Einflüsse – parallel zu physikalischen, chemischen und mikrobiologischen Faktoren – für die Entstehung, den Verlauf und die Endzustände seiner Erkrankungen gleichberechtigt einbezogen und gewichtet werden (v. Uexküll, 1979). Die hieraus resultierende psychosomatische Sicht verlangt die sich im höheren Lebensalter häufig verändernde Lebenssituation, insbesondere Auswirkungen und Bedeutung der vielfältigen Bedrohungen, Verluste und Kränkungen zu berücksichtigen.

Entgegen dieser Definition wird der Begriff psychosomatisch bisher fast ausschließlich für unterschiedliche Störungen und Erkrankungen jenseits des 60. Lebensjahres verwendet (Mittelmann, 1956; Kleemeier, 1958; Ernst, 1959; Condrau, 1966; Müller, 1967; Stenbeck, 1975; Gathmann, 1987); dazu zählen:

- Auftreten oder Verstärkung funktioneller Symptome oder Symptomgruppen, die psychosomatischen Charakter haben;
- Auftreten von Konversionssymptomen als Ausdruck neurotischer Erkrankungen;
- Auftreten körperlicher, meist funktioneller Symptome als Ausdruck und Begleiterscheinung depressiver, hypochondrischer oder paranoider Krankheitsbilder;
- Darstellung psychischer Konflikte und psychosozialer Schwierigkeiten durch Verstärkung bestehender organischer Krankheiten im Sinne der „Benutzung der Krankheit";
- seelische Auswirkung langfristiger schwerer oder chronischer organischer Erkrankungen.

Krakowski (1976) dient der Ausdruck psychosomatisch zur Beschreibung einer Methode der Integration und des Zuganges zu biologischen, psychologischen und sozialen Parametern des Alterns. Nowling und Busse (1977) verstehen unter psychosomatischen Problemen die Beschreibung von physiologischen und psychologischen Veränderungen, die den Alternsprozeß begleiten. Für Groen (1982) stellt Altern ein Paradigma des psychosomatischen Prozesses dar, in welchem körperliche, psychische und soziale Aspekte untrennbar verbunden sind.

Systematische, umfassende und längerfristige Forschungen zur Psychosomatischen Medizin im höheren und hohen Lebensalter fehlen fast völlig. Auf der 1. Konferenz des National Institute of Mental Health 1983 über den psychodynamischen Forschungsstand im Alter (Miller und Cohler, 1984) konnte bei der Diskussion über psychosomatische Krankheitsbilder lediglich der „überraschende Mangel" an Daten aus Klinik und Forschung konstatiert werden. Bis 1988 dürften maximal 300–400 weitgehend klinisch orientierte Publikationen zu dieser Thematik vorliegen, die größtenteils aus den USA stammen (Radebold, 1988, 1989).

Bisher haben die für die diesbezügliche psychosomatische Forschung relevanten Disziplinen (Psychoanalyse, Verhaltensforschung, Sozialpsychologie, Psychophysiologie usw.) die Determinanten „Altern" und „höheres und hohes Lebensalter" kaum berücksichtigt. Ebenso befaßt sich die Gerontopsychiatrie aufgrund ihrer weitgehend hirnorganisch ausgerichteten Forschung kaum mit neurotischen/reaktiven Erkrankungen.

Der von Bergener und Kark 1985 herausgegebene Sammelband „Psychosomatik in der Geriatrie" gibt einen unverändert gültigen Überblick über den derzeitigen deutschsprachigen Stand. Nur zögernd berücksichtigen deutschsprachige geriatrische Lehrbücher psychosomatische Aspekte (Lang, 1988).

60.2 Begegnung mit dem älteren Patienten

60.2.1 Der ältere Patient

Das nachfolgende Beispiel eines Erstgespräches verdeutlicht beispielhaft die Notwendigkeit einer psychosomatischen Sicht alternder Patienten in Praxis und Klinik. Gleichzeitig werden wichtige Aspekte der Symptomatik, des Verlaufs und der erfolgten Behandlung sichtbar:

Eine 76jährige Patientin wird nach 9wöchigem Aufenthalt auf der internistischen Station einer Universitätsklinik mit ihrem Einverständnis dem psychiatrischen Konsiliarius vorgestellt, da „kein nennenswerter organischer Befund besteht und die bisherige Behandlung nicht angeschlagen hat, sondern im Gegenteil zu einer erheblichen Verschlechterung führte".

Die Patientin wird im Rollstuhl hereingefahren. Sie wirkt einerseits müde-resignierend, andererseits aufgrund ihrer Sprache, ihres Verhaltens und Auftretens relativ aktiv und insgesamt jünger.

A.: Können Sie mir erzählen, was Sie für Probleme und Schwierigkeiten haben, oder was überhaupt mit Ihnen los ist?

P.: Ja, ich könnte nicht sagen, was weiter los ist. Allerdings, wir hatten einen Todesfall, aber das habe ich schon so weit klar, daß das nicht mehr ins Gewicht fallen könnte. Ich hab einfach keinen Appetit, gar keinen! Ein wenig Flüssigkeit, ja, aber weiteres kann ich gar nicht so gebrauchen. Mit einem Mal kam es. Ich hatte früher sonst immer einen guten Appetit.

A.: Und wann hat das angefangen mit dem schlechten Appetit?

P.: Seit ich da bin.

A.: Wie lange sind Sie jetzt hier?

P.: Neun Wochen.

A.: Neun Wochen. Und war das schon vor der Aufnahme mit dem schlechten Appetit?

P.: Da war's etwas besser, am Anfang, ja, da war's etwas besser! Aber einmal so zu Hause, da hatte ich ein Gefühl, ich kann die Butter nicht mehr sehen. Aber ich hab immer ordentlich gegessen. Fleisch, Wurst, alles kann ich gar nicht sehen.

A.: Ein Ekelgefühl davor?

P.: Ein Ekelgefühl! Und dann ist alles trocken und ich habe immer Durst. Bei mir ist in der Früh alles trokken, da kann ich am Anfang gar nicht reden, ich hab kein bißchen Speichel.

A.: Weswegen sind Sie denn hier eingewiesen worden?

P.: Wegen dem, weil ich zuerst immer hier Schmerzen hatte (zeigt auf den Leib). Dann hat mich der Herr Dr. B. behandelt, da war mir immer schwindlig oder halt immer übel geworden. Dann hat er damals, als er nach Karlsruhe ging (Therapiekongreß), gesagt, es wäre gut, wenn ich mal in die Klinik ginge, und hat mir 'ne Überweisung geschrieben, und so bin ich hierher gekommen. Er sagte mir noch beim Gehen, da liegt nichts Gefährliches vor, Sie sind in acht bis vierzehn Tagen wieder da. Und jetzt bin ich neun Wochen hier.

A.: Und was hat sich bisher an den Beschwerden geändert?

P.: Der Druck ist geblieben. Nun hatte ich am Anfang auch einen gebrochenen Arm, der hier behandelt wurde. Das war schon zu Hause, wo eigentlich der Arm gebrochen wurde. Acht Tage hatte ich den, und so mußte ich den eben hier ausheilen. Und der Gips kam vor acht Tagen weg.

A.: Und wie geht's Ihnen mit dem Arm?

P.: Gut! Ich geh immer in die Massage runter, da hab ich keine Beschwerden, gar nicht. Ich bin bloß recht müd und kein Appetit dazu, gar nichts.

A.: Wie geht's denn so mit dem Schlafen?

P.: Ja, mit Schlafmitteln ganz gut. Aber ohne hab ich hier noch nicht geschlafen.

A.: Und zu Hause, wie ging's da mit dem Schlafen?

P.: Da nehm' ich keine Schlafmittel.

A.: Da haben Sie gut geschlafen?

P.: Ja, ordentlich. Schon lang hab ich keine mehr genommen. Früher mal, wo ich die Grippe hatte. Ich hatte allerdings im Dezember 'ne schwere Grippe, und in dieser Zeit kam so vieles, da starb meine Tochter, noch dazu an Grippe. Dann ging es noch ganz ordentlich Januar/Februar durch, bis zum März, da merkte ich, ich bin einfach nicht in Ordnung. Immer das Gefühl, aber nicht Schmerzen. Keine Schmerzen.

A.: Sondern nur so ein Druck.

P.: Ja, ich hab auch jetzt keine Schmerzen. Bloß die Müdigkeit, ich möchte den ganzen Tag schlafen. Ich geh nicht raus, jeden Tag, so um die Klinik. Aber das Allerschlimmste ist, nichts zu essen sehen. Kann ich einfach nicht sehen. So ein bißchen bring ich schon zurecht, was sein muß.

A.: Kennen Sie solche Zustände von früher, daß Sie auch solchen Ekel hatten gegenüber Essen und so lustlos waren?

P.: Gegen Essen nie. Ich hatte immer guten Appetit und auch Stuhlgang ordentlich, immer, im Magen noch nie was gehabt. Mit Medikamenten hatte ich's nie. Ich hab von Herrn Dr. B. auch verschiedene Medikamente schon gehabt, aber hier hab ich halt so viele. Ich hab hier morgens sieben, und mittags glaub ich fünf oder sechs, und abends nochmal vier.

A.: Und Sie haben das Gefühl, daß das für Ihren Magen viel zuviel ist?

P.: Ich glaub das, weil's immer so, wie will ich sagen, so ein unguter Geschmack, also aufstößt und so sauer, einfach sauer, gar nicht so, daß ich immer denk, nehm ich dieses Medikament zuerst oder das, ist es vielleicht so besser, wenn ich dies erst nehme? Und es ist immer gleich. Ja, das kommt immer gleich hinterher, das Trockene im Mund, so daß ich manchmal kaum reden kann. In der Früh überhaupt ist es so schlimm.

A.: Daß Sie also auch Ihre ganzen Wünsche und Sorgen gar nicht richtig vorbringen können, weil Sie kaum reden können.

P.: Ja, überhaupt bin ich so, daß ich zu wenig spreche, das liegt mir nicht. Manche, die sagen das und das und das, und was will ich noch. Da sagen sie eben immer, ich sollte mich mehr regen. Aber das liegt mir einfach nicht.

A.: Sie lassen so alles über sich ergehen?

P.: Und was gefragt wird, ja, so ist das. Weiter kann ich nichts sagen.

A.: Daß Sie so wenig von sich erzählen können, worauf beruht denn das?

P.: Ach, das ist mir alles fremd. Und zweitens reden die anderen Insassen immer. Ich hab zuviel Angst. Vor was ich Angst habe, weiß ich auch nicht. So oft ich raus mußte, irgendwohin gebracht wurde, dann hatte ich Angst.

A.: Das passiert jetzt wieder?

P.: Ja. Jetzt passiert wieder etwas. Ich bin nicht empfindlich, Schmerz empfinde ich gar nicht. Mir kann man hier reinstechen, man kann's so machen oder so, das ist ganz egal. Das macht mir gar nichts aus. Die Infusionen oder sonst was, die Blutabnahme, das kann nochmal kommen, das stört mich nicht. Bei mir war nur lediglich das so schlimm, weil ich mit dem Gips bis hier oben immer nur einen Arm bewegen konnte.

A.: Aber Sie meinen doch, daß Sie jetzt ängstlich sind

gegenüber allem Neuen, was da kommt. Ist das zu Hause und draußen auch so?

P.: Also früher nicht. Aber jetzt, durch so viele Fälle. Wenn's bei mir zu Hause stramm klingelt, was ist jetzt passiert? Herzklopfen, dabei war gar nichts.

A.: Aber haben Sie jetzt so viel Verluste außer Ihrer Tochter erlitten? Denn der Postbote bringt ja nun diese beunruhigenden Nachrichten.

P.: Ja.

A.: Sie sagten, es sei soviel los gewesen zu Hause?

P.: Ja, es war so vieles. Mit den Autos, die Jungens, meine Söhne usw., aber außer diesem Fall, da lag ich selber schwerkrank an Weihnachten, wo meine Tochter starb – an einer Grippe, und kein Arzt kam, und doppelseitige Lungenentzündung hatte sie, sie war nicht mehr zu retten.

A.: Lag sie bei Ihnen?

P.: Nein, in S.

A.: In S.

P.: Und das ist gerade die gewesen, die am meisten für mich gesorgt hätte. Die waren in S. verheiratet und hatten keine Kinder, und sie kam so alle 14 Tage runter und ich rauf. Und dann sagte sie immer, so lange du lebst, hast du mich.

A.: Und das war auch Ihre Lieblingstochter?

P.: Meistens, weil sie mir am meisten an die Hand ging.

A.: Mit der Sie sich auch besser verstanden hatten?

P.: Ich versteh mich mit meinen Kindern, mit allen versteh ich mich. Aber sie hat sich am meisten um uns gekümmert, mein Mann ist jetzt drei Jahre tot und ich lebe hier allein. Aber ich kam gut zurecht, ich bin nicht unselbständig gewesen, nie. Aber gerade durch diese Sache hab ich mich etwas geändert.

A.: Sie sind also trauriger geworden und offenbar bedrückter und haben zu allem seitdem keine Lust mehr.

P.: Lust schon, wenn's etwas besser ginge. Ich dachte jeden Tag, wenn's morgen etwas leichter ist, vielleicht komme ich doch endlich so weit, daß ich zu Hause mich so ein bißchen fassen kann. Wie die Herren sagten, es lägen keine wesentlichen Krankheiten vor.

Die Patientin, die sich lebenslang als psychisch stabil und energisch erlebte, hatte den Verlust ihres Mannes vor 3 Jahren ohne ausgeprägte depressive Verstimmung überstanden. Ihre 2 Söhne lebten verheiratet in entfernter Nachbarschaft und hatten mehrfach angeboten, die Mutter zu sich zu nehmen. Sie selbst wollte aber in gewisser Distanz zu ihren Kindern leben.

Sie führte ihren Haushalt allein mit Hilfe einer Haushaltshilfe. Die verheiratete, in einer weitab liegenden Stadt lebende Tochter kam über das Wochenende nach Hause und versorgte die Mutter zusätzlich. Dabei bestand die stillschweigende Verabredung, daß die einzige Lieblingstochter bei einer Erkrankung der Mutter ihre Stelle wechseln und die Mutter bis zu ihrem Tode versorgen würde. Diese Tochter erkrankte akut mit 48 Jahren an einem Virusinfekt und verstarb innerhalb von 8 Tagen.

Zwei Monate später entwickelte sich die beschriebene Symptomatik, nachdem die Patientin zunächst jede Trauerreaktion verdrängt hatte. Zusätzlich fühlte sie sich von ihrem Hausarzt verlassen, der sie aufgrund eines Kongreßbesuches für „zehn Tage zur Durchuntersuchung" einwies.

Nach Ansprache der Trauerreaktion und Zulassen eines Teiles des Kummers konnte mit der Patientin eine Lösung erarbeitet werden, die ihr zu Hause eine größere Sicherheit und Verwöhnung bei Erhaltung der Selbständigkeit versprach. Dazu bestand die Möglichkeit, später in der Nähe der Söhne eine Wohnung zu nehmen. Unter dieser Lösung kam es kurzfristig zum Verschwinden der obigen Symptomatik und so weitgehender Wiederherstellung der Patientin, daß sie nach 14 Tagen wieder voll aktiv nach Hause entlassen werden konnte.

Ist diese 76jährige Patientin aufgrund ihres Alters, ihrer geklagten Symptomatik, ihres Verlustes und ihrer Erwartungen an den behandelnden Arzt für die allgemeinärztliche Praxis eine Ausnahme?

Bei der Mikrozensus-Erhebung 1978 gaben 9,07 Millionen Personen an, krank zu sein, gut 36% davon waren im Alter von 65 Jahren oder älter, 31% der Bevölkerung von 65 Jahren und älter standen in ständiger ärztlicher Behandlung. Entsprechend fand sich 1980 bei einer Untersuchung von 13 Allgemeinpraxen in Mannheim ein Anteil von 52,4% der Patienten von 55 Jahren und älter bei einer Schwankungsbreite von 33,9–72,4% bei Patienten dieser Altersgruppe (Zintl-Wiegand et al., 1980). Unter den Patienten überwogen die Frauen (teilweise über die prozentuale Geschlechtsverteilung in der Bevölkerung). Ebenso benannten bei einer zweiten Mannheimer Untersuchung (Cooper und Sosna, 1983) 90% der 343 über 65jährigen Probanden der Gesamtstichprobe einen Arzt für Allgemeinmedizin als Hausarzt, wobei 77,3% diesen innerhalb der letzten drei Monate konsultiert hatten.

Innerhalb der Alterskranken bilden die Patienten mit einer (zusätzlichen) psychiatrischen Diagnose eine große Teilgruppe (vgl. Abschn. 60.3.3); praktisch alle befinden sich in hausärztlicher Behandlung (Cooper und Sosna, 1983). Insgesamt konsultieren sie den Arzt häufiger als die nicht-psychisch Alterskranken (Zintl-Wiegand et al., 1980; Biron und Eder, 1980). Dabei besteht in der Regel ein langfristiges Arzt-Patient-Verhältnis (bei allen Altersgruppen) mit 1–5 Jahre bei 17,0%, mit 5–10 Jahren bei 16,5% und mit mehr als 10 Jahren bei 49,6%.

Könnte es sein, daß alternde Patienten aufgrund der zunehmenden Multimorbidität und der Somatisierungstendenz ihrem behandelnden Arzt nur noch über körperliche Beschwerden und Einschränkungen berichten können und diesen so „verführen", den Einfluß psychosozialer Faktoren als gering einzuschätzen?

Diese Annahme wird zunächst dadurch gestützt, daß offenbar rein psychische Symptome mit zunehmendem Alter abnehmen, es dagegen mit ansteigendem Alter zu einer Zunahme von „psychosomatischen" bzw. „psychophysiologischen" Symptomen kommt. So fand sich in der Sterling County Study (Katschnig und Strotzka, 1977) für diese Symptomgruppe ein praktisch kontinuierlicher Anstieg von etwa 30% für die Altersgruppe der unter 30jährigen bis etwa um 85% für die Altersgruppe der über 70jährigen. Bei der Wiener Gesundheitsstudie (Biron und Eder, 1980) zeigten die 60jährigen Männer und Frauen im Vergleich zu den 20- und 40jährigen häufiger psychische Symptome (die dazu noch im Zusammenhang mit entsprechenden Belastungen standen). Allerdings gaben diejenigen 60jährigen, die an zwei oder mehreren psychischen Beschwerden litten, gleichzeitig drei oder mehrere organische Beschwerden an. Insgesamt hatten in der Wiener Gesundheitsstudie 22% der 60jährigen psychische Beschwerden. Unter den über 65jährigen in Privathaushalten lebenden Mannheimern fanden Cooper und Sosna (1983) 26% mit mäßiger oder schwerer Gehbehinderung, 17% mit vergleichbarer Hör- und 13% mit Sehbehinderung. Insgesamt wurden 9,6% der Untersuchten als schwer und 25,4% als mäßig körperlich behindert eingeschätzt. Das bedeutet, daß 35% der über 65jährigen in Privathaushalten Lebenden mäßig bis erheblich körperlich behindert sind. Teilt man alle körperlich Behinderten, unabhängig von der Art, nur nach dem Ausmaß der Behinderung in drei Gruppen ein, so zeigt sich, daß 45,2% der mäßig bis stark und 23,1% der leicht Behinderten, gegenüber einem Zehntel (= 11,3%) der nicht Behinderten, als psychisch krank diagnostiziert wurden (Tab. 60–1). Mit der Schwere der Behinderung nimmt also die Häufigkeit psychischer Störungen deutlich zu. Dieser Zusammenhang ist sowohl für organische als auch für funktionelle psychische Störungen signifikant. Die in Heimen Lebenden weisen von vornherein einen höheren Anteil Behinderter auf. Damit scheinen mäßige bis schwere körperliche Behinderung – Sehbehinderung, Hörbehinderung, vor allem aber Bewegungsbehinderung – gewichtige Risikofaktoren für die psychiatrische Morbidität im Alter zu sein (Häfner, 1986).

Eine weitere Untersuchung (Zintl-Wiegand et al., 1980) belegt, daß die Ärzte für Allgemeinmedizin nur 59,5% der durch Psychiater als psychisch krank diagnostizierten Patienten aller Altersgruppen als solche identifizierten. Die Identifizierungsraten lagen in diesen Allgemeinpraxen zwischen 15 und 67,7%, d. h. sie resultierten in weit größerem Maße aus den Unterschieden, die die Praktiker selbst betrafen, als aus der unterschiedlichen Verteilung psychisch Kranker in Allgemeinpraxen. Der Verdacht, daß gerade über 60jährige psychisch Kranke in noch geringerem Um-

fang als unter 60jährige in der allgemeinärztlichen Praxis identifiziert werden, wird durch eine Untersuchung der überwiegend über 75jährigen Klientel von Sozialstationen in einer Großstadt (Prognos, 1987) gestützt; lediglich bei 31,5% aller Klienten mit einer psychischen Erkrankung bzw. Auffälligkeit wurde hausärztlicherseits eine psychiatrische Diagnose gestellt, wobei 18,4% fachpsychiatrisch (in irgendeiner Form) mitbehandelt wurden.

Diese wenigen zur Verfügung stehenden Daten – wenn auch aufgrund anderer Fragestellungen erhoben – belegen das häufige Nebeneinanderbestehen von psychischer und körperlicher Symptomatik, die Häufigkeit psychischer, organischer und funktioneller Erkrankungen in der allgemeinärztlichen Praxis, die Bedeutung von mäßigen bis schweren Behinderungen als gewichtigen Risikofaktoren für psychiatrische Morbidität im Alter und die auffallend geringe Fallidentifizierungsquote. Warum werden diese Patienten in der hausärztlichen Praxis nur in geringem Umfang als solche eindeutig erkannt?

60.2.2 Der Behandler: Seine Wahrnehmung und seine möglichen Reaktionen

Das wiedergegebene Erstinterview zeigt einige spezifische Interaktionsmuster der behandelnden Ärzte mit dieser Patientin:

Die Patientin steht seit vielen Jahren wegen verschiedener Bagatellerkrankungen bei dem gleichen Arzt in Behandlung, der auch den verstorbenen Ehemann betreut hat. Der Hausarzt bezieht offenbar die jetzige akute Erkrankung, die mit Angstzuständen und nachfolgenden funktionell erscheinenden Beschwerden beginnt, ursächlich nicht auf den völlig unerwarteten, traumatisch erlebten Verlust der Tochter. Er behandelt sie wegen einer nicht näher definierten körperlichen Erkrankung bei mitlaufendem Verdacht auf ein Krebsleiden. Gleichzeitig verordnet er Tranquilizer und Schlafmittel. Aufgrund der Verschlechterung wird sie nach zwei Monaten „zur diagnostischen Abklärung" in die Universitätsklinik eingewiesen. Die Patientin ver-

Tabelle 60–1. Körperliche Behinderung und psychische Krankheit (aus Häfner, 1986).

		körperliche Behinderung				
		nicht vorhanden (0)		leicht (1)		mäßig stark (2–4)
			%		%		%
psychisch gesund		118	88,7	73	76,8	46	54,8
psychisch krank		15	11,3	22	23,1	38	45,2
Summe		133	100,0	95	100,0	84	100,0
davon							
organisches Psychosyndrom			1,5		14,7		20,2
funktionelle psychische Erkrankung			9,8		8,4		25,0

mutet zu Recht, daß sie wegen „seines Kongreßbesuches" von ihrem Hausarzt „abgeschoben" wurde. Gleichzeitig verspricht er, daß sie in Kürze wieder zu Hause sei und er sich um sie kümmern werde. Beide Versprechen werden nicht eingehalten. Auch die Ärzte in der Klinik sind eindeutig auf eine „körperliche Ursache" der geklagten Beschwerden fixiert. Erst nach einer umfassenden diagnostischen Abklärung und mehreren vergeblichen medikamentösen Behandlungsversuchen wird schließlich ein Psychiater konsultiert, da „keine organische Ursache" feststellbar ist.

Auch der konsultierte Psychiater geht – wie das Erstinterview belegt – nicht auf das sofort erfolgende Angebot dieser Patientin ein, über den Tod ihrer Tochter zu sprechen. Er versucht zunächst, ihre Beschwerden zu sichten und ist erst nach dem dritten Hinweis bereit, ihre Angst und die abgewehrte Trauer anzusprechen.

Die Behandlungs- und Lebenssituation dieser Patientin stellt sich äußerlich als relativ günstig dar. Bisher in stabilen psychosozialen Lebensbedingungen und Beziehungen relativ selbständig lebend, erkrankt sie akut an Angstzuständen und funktionell erscheinenden Beschwerden. Der sie schon langjährig betreuende Hausarzt kennt ihre Lebenssituation, ihre bisherige Entwicklung mit dem Verlust des Ehemanns und hört von dem akuten Tod der Tochter. Weder der Hausarzt noch die behandelnden Klinikärzte können offenbar bei dieser 76jährigen Patientin die geklagten Symptome als funktionelle einordnen und in ursächliche Beziehung zu dem akuten Verlust der Tochter setzen, geschweige denn mit der Patientin die notwendige Trauerarbeit leisten. Im Gegenteil wird sie von Behandler zu Behandler „weiterverwiesen" und gemachte Versprechen werden nicht eingehalten.

Unterstellen wir die optimistische Annahme, daß eine 30- oder 50jährige Patientin mit dieser Vorgeschichte aufgrund adäquater Schlußfolgerungen entsprechend therapeutisch behandelt worden wäre: Könnte es das chronologische Alter von 76 Jahren sein, daß sich fast aufdrängende Zusammenhänge nicht gesehen und abgewehrt werden müssen (z.B. durch Weiterverweisungen mit Versprechungen)?

Damit stellt sich die Frage: Welche Voraussetzungen für die Behandlung älterer Patienten bringt ein Arzt mit, und wie reagiert er affektiv auf sie?

Geriatrische und sozialgerontologische Kenntnisse gehören aufgrund der bisherigen Approbationsordnung nicht zum Pflicht-Curriculum der medizinischen Ausbildung. Unverändert werden von den medizinischen Fakultäten nur in geringem Umfang (sozial)gerontologische, geriatrische und gerontopsychiatrische Lehrveranstaltungen angeboten (Pallenberg, 1983; Tokarski, 1989). Wie der Besuch entsprechender Fortbildungsveranstaltungen belegt, ist bisher die Bereitschaft zur geriatrischen Fortbildung auffallend gering. Die zur Behandlung alternder Patienten für erforderlich gehaltenen Erkenntnisse stützen sich weitgehend auf aus der Behandlung von Jüngeren übernommene Kenntnisse und ein langsam erworbenes Erfahrungswissen. Außerdem orientiert sich das vermittelte oder erworbene Wissen bisher fast ausschließlich an dem Defizitmodell des Alterns im Sinne eines unabänderlich fortschreitenden organischen, speziell hirnorganischen Abbaus mit der Konsequenz Versorgung, Pflege und unter Umständen Bewahrung anstatt einer Behandlung.

So stützt sich der Arzt in der Regel für die Interaktion mit und die Behandlung von älteren Patienten auf seinen „gesunden Menschenverstand" und seine eigenen Erfahrungen mit Älteren.

Zu diesen Älteren zählen zunächst Erwachsene, die er in seiner Kindheit aufgrund der bestehenden Altersdifferenz – nicht aufgrund ihres chronologischen Lebensalters – subjektiv als „alt" erlebt hat. Dazu gehören die eigenen Eltern (für ein 5jähriges Kind sind 35jährige Eltern sechsmal so alt und damit unvorstellbar alt, für einen 30jährigen Erwachsenen sind die 60jährigen Eltern nur doppelt so alt!), die Geschwister der Eltern, aber auch gleichaltrige oder ältere wichtige Bezugspersonen wie Kindergärtnerinnen, Lehrer, Krankenschwestern, Ärzte, Pfarrer und Hausbewohner. Dazu treten Erinnerungen an in Wirklichkeit – also chronologisch – alte Menschen wie Großeltern, alte Verwandte, aber auch Hausnachbarn. Diese Kindheitserinnerungen werden später durch die Erfahrungen mit den alternden Eltern und weiteren alternden wichtigen Bezugspersonen überlagert.

Die Wünsche und Erwartungen an diese Kindheitspersonen, die Vorstellungen und Enttäuschungen über sie, die Schwierigkeiten und Konflikte aus Kindheit und Adoleszenz mit diesen „Älteren" bleiben erhalten und sind – da größtenteils abgewehrt – unbewußt geworden. Mehr in Erinnerung geblieben sind die Altersveränderungen, Behinderungen, Krankheiten sowie Sterben und Tod der Großeltern und später der eigenen Eltern, unter Umständen auch ihre erlebte Pflege- und Hilfsbedürftigkeit.

Diese unbewußten Ängste, Befürchtungen, Wünsche, Sehnsüchte und Konflikte werden in jeder Interaktion mit Älteren reaktiviert und unbewußt gefühlsmäßig auf diese übertragen. Damit kommt es zu einer Umkehrung der klassischen Übertragungskonstellation. In ihr erlebt der jüngere oder gleichaltrige Patient den in der Realität oder Phantasie älteren Behandler in der unbewußten Wiederholung der Kind-Eltern-Beziehung. Jetzt begegnet der in der Regel sogar viel jüngere Behandler einem älteren Patienten und wiederholt damit eine Kind-Eltern- bzw. Enkelkind-Großeltern-Beziehung. Entsprechend sieht auch der Ältere den jüngeren Behandler als sein reales oder phantasiertes Kind an, auf das frühere Wünsche, Vorstellungen, Ängste und Konflikte unbewußt übertragen werden.

Damit befindet sich der jüngere Behandler nicht mehr in der Sicherheit, Anerkennung und Stabilität vermittelnden Position eines Elternteils, wie es seinen Vorstellungen über seine Position als Arzt entspricht (Radebold, 1979 und 1981).

Weiterhin sieht sich der Arzt durch seine alternden Patienten mit ihren zahlreichen, meist negativen und unübersehbaren Veränderungen konfrontiert. Dazu rechnen Kränkungen und Zurückweisungen durch die Umwelt und die Gesellschaft, Abnahme und Einschränkung physischer und psychischer Fähigkeiten, Zunahme an Krankheiten und Behinderungen, Minderung des sozialen Status, des Einkommens und eine Verschlechterung der Wohnsituation und zunehmende Verluste an wichtigen Bezugspersonen wie Partner, Verwandte, Freunde. Schließlich muß er seine alternden Patienten wegen langfristiger, oft schwerer und wenig beeinflußbarer Krankheiten behandeln, die letztlich zum Tode führen. Er erlebt durch seine alternden Patienten Kummer, Verzweiflung und Resignation, wobei er oft nur hilflos reagieren kann. Jeder Arzt merkt, daß seine therapeutischen Bemühungen – häufiger als bei der Behandlung Jüngerer – nur geringe Erfolge bringen und kurzzeitig anhalten. So kann er aufgrund seiner Allmachtsvorstellungen als „Helfer und Heiler" ausgeprägte narzißtische Kränkungen empfinden (Kastenbaum, 1963).

Schließlich fordert ihn die Wahrnehmung des Älterwerdens seiner Patienten heraus, sich seinen eigenen Vorstellungen, Phantasien, aber auch Ängsten bezüglich seines eigenen Alterns zu stellen. Der Umgang mit jüngeren oder gleichaltrigen Patienten erlaubt es dem Arzt mehr, sich mit ihrer Entwicklung, ihren Erfolgen und Fortschritten zu identifizieren und ermöglicht es ihm gleichzeitig, Fragen nach dem eigenen Älterwerden abzuwehren und zu verdrängen.

Damit sind die Schwierigkeiten benannt, denen der Arzt im Umgang mit seinen älteren Patienten begegnet. Ausgestattet mit geringen, dazu meist veralteten geriatrischen und gerontologischen Kenntnissen, sich auf Erfahrungswissen aus der Behandlung von Jüngeren stützend, orientiert an dem Leitbild des Defizitmodells des Alterns, konfrontiert mit Kummer, Sorgen und den erlebten Alternsveränderungen, gefragt bezüglich eigener Vorstellungen über sein Altern, soll er sich in eine gefühlsmäßige, teilweise intensive Arbeitsbeziehung einlassen, in der er affektiv die längst überwundene Position eines Kindes nacherlebt.

So werden die Reaktionen verständlich, die sich bei der Interaktion mit alternden Patienten in Klinik und Praxis beobachten lassen (Radebold, 1972):
– Eine Behandlung findet nicht statt, da jegliche Interaktion mit alternden Patienten vermieden wird.
– Findet eine Behandlung statt, so wird von vornherein rationalisierend (im Sinne einer Abwehr) argumentiert, daß Ältere nur „geringe Erfolgschancen" haben, „Jüngere bevorzugt zu behandeln seien", „die geringe Lebenserwartung eine intensive Behandlung verbiete" usw. Damit ist durch die unbewußt ablehnende Erwartung bereits der ungenügende Therapieerfolg vorausgesagt.
– Im Gegensatz zu Patienten im jüngeren und mittleren Lebensalter erscheinen die diagnostischen und therapeutischen Bemühungen oft verändert abzulaufen. Entweder wird sehr schnell, fast ungeduldig und überstürzt diagnostiziert und behandelt mit auffallend schnellen Entschlüssen zur Entlassung oder Verlegung ins Pflegeheim, oder Diagnostik und Behandlung wirken auffallend umfassend, strapaziös und eingreifend, ohne große Rücksichtnahme auf den körperlichen und seelischen Zustand des alternden Patienten.

– Die Behandlung ist weitgehend medikamentös ausgerichtet unter Zurückstellung aktivierender, rehabilitativer oder sogar sozio- oder psychotherapeutischer Maßnahmen. Rehabilitationsversuche, z.B. bei Schlaganfallpatienten, werden jenseits des 65./70. Lebensjahres trotz guter Erfolge in deutlich geringerem Umfang unternommen. Diese rein medikamentöse Behandlung erlaubt gleichzeitig dem Behandler eine indirekte Interaktion, da das Rezept häufig durch die Sprechstundenhilfe ausgehändigt wird und so der Arzt seinen Patienten lange Zeit nicht mehr zu sehen braucht (Balint und Norell, 1975).
– Die verbalen Äußerungen beinhalten oft sehr deutlich aggressive und ablehnende Züge. Massive Vorwürfe werden wegen „fehlender Mitarbeit", wegen zu „langsamen Verhaltens" oder wegen des „typischen Vergessens" erhoben. Dazu rechnen auch infantilisierende Umgangsformen mit Duzen, Anrede in der dritten Person, Erzählen entsprechender Witze und Anzüglichkeiten bei deutlichen Bemühungen um (Nach-)Erziehungsmaßnahmen. Nicht selten ist die Visitendauer im Vergleich zur Visite bei jüngeren Patienten verkürzt.
– Versprechungen bezüglich „neuer, großartiger Behandlungen" oder „phantastisch wirksamer Medikamente" sollen offenbar unbewußt die guten Absichten des Behandlers demonstrieren und gegenseitigen Enttäuschungen vorbeugen. Können die entsprechenden Erwartungen nicht befriedigt werden, liegt es häufig am „Alter" des betreffenden Patienten. Dazu werden diese Patienten häufig mit Versprechungen anstatt ausführlicher Informationen über die weitere Hilfestellung vertröstet, die dann nicht eingehalten werden.
– Die Schwierigkeiten eines adäquaten Umganges zwischen jüngeren und älteren Erwachsenen verdeutlicht auch die Über- oder Unterschätzung der Fähigkeiten und Möglichkeiten alternder Patienten. Eine Überschätzung mit der Unfähigkeit, deutliche hirnorganische Veränderungen, Abhängigkeit oder Hilflosigkeit wahrzunehmen, kann auf unterschiedliche Schwierigkeiten des Behandlers zurückgeführt werden: Entweder wünscht er sich noch „großartige und mächtige Eltern", die nicht abgebaut und altersverändert sein dürfen, oder seine Alterspatienten sollen seinen Idealvorstellungen vom eigenen Alter mit Weisheit, Abgeklärtheit und Selbständigkeit entsprechen. Die Unfähigkeit zur Wahrnehmung kann auch eine Verkehrung ins Gegenteil zur Abwehr eigener aggressiver Impulse bedeuten.
– Die Unterschätzung von Fähigkeiten und Möglichkeiten alternder Patienten spricht für eine vorweggenommene Abwertung.
– Die fast ausschließlich „organische Sicht" des Alters vermeidet die Notwendigkeit der Einbeziehung psychosozialer Aspekte. Häufiger treffen sich hierbei Arzt und Patient in einem unbewußten Bündnis. Der Patient kann sein Selbstbild eines unabhängigen, auch im Alter zurechtkommenden Menschen ohne Schwierigkeiten und Konflikte auf-

rechterhalten. Der Arzt vermeidet die Begegnung mit beunruhigenden, beängstigenden, bedrohlichen und bedrückenden Gefühlen und Ereignissen. Weder haben es die Jüngeren gelernt, mit ihren „Eltern", noch haben es die Älteren gelernt, mit ihren „Kindern" über Sorgen, Schwierigkeiten und Konflikte zu sprechen.

– Sehr bemühte, teilweise warmherzige liebevolle Interaktionsformen sprechen häufiger für den unbewußten Wunsch nach einer Wiedergutmachung oder für die Suche nach „neuen" (d.h. besseren und liebevolleren) „Eltern" und „Großeltern" oder für die Suche nach Wiederholung früherer befriedigender Beziehungen. Wenn die Patienten diese Wünsche nicht befriedigen können, kann es zu ausgeprägten Enttäuschungen mit Abbruch der Behandlungsbemühungen kommen.

Die sich in diesen Interaktionsformen widerspiegelnden gefühlsmäßigen Schwierigkeiten erschweren den Aufbau einer Beziehung zu älteren Patienten und die Einbeziehung und Verarbeitung von Konflikten und psychosozialen Schwierigkeiten. Im Falle der geschilderten Patientin hat sich der behandelnde Hausarzt offenbar gewünscht, daß sie – ähnlich wie beim Tod ihres Mannes – beherrscht, sicher und ohne allzu große Trauer über den Verlust ihrer Tochter hinwegkommt. Er und die nachfolgend behandelnden Ärzte einigten sich mit der Patientin unbewußt, in gemeinsamer Abwehr der Trauer über den Verlust, auf ein „organisches Leiden". Jeder vermied es, sie daraufhin anzusprechen, auf ihre eindeutigen Hinweise einzugehen und überwies sie an den nächsten Behandler.

60.3 Lebenssituation und mögliche Veränderungen des älteren Patienten

60.3.1 Demographische Daten

Der Anteil der über 65jährigen an der Gesamtbevölkerung der Bundesrepublik betrug 1986 insgesamt 9,127 Mio (= 15,2%; m. 10,8%, w. 19,2%); den gro-ßen Anstieg verdeutlichen die Zahlen für 1950 (9,4%; m. 9,0%, w. 9,7%) und für 1970 (13,2%; m. 10,7%, w. 15,4%). Diese Angaben belegen gleichzeitig die zunehmende Verschiebung zugunsten des Anteiles über 65jähriger Frauen (Statistisches Jahrbuch, 1988). Nach einer Modellrechnung der Bundesregierung (BMI, 1986) wird ihr Anteil im Jahre 2000 17,4% ausmachen; beträgt der Anteil der über 60jährigen heute 12,5 Mio, so ist nach einer Modellrechnung (BMI, 1986) im Jahre 2000 mit einer Anzahl von 14,8 Mio zu rechnen (Rückert, 1988).

Differenziert man nach Alter und Geschlecht, so werden die eingetretenen Veränderungen noch deutlicher: So stieg der Anteil der 60- bis 70jährigen von 1950 bis 1985 nur um 32%, aber der Anteil der über 90jährigen um mehr als 800% (Rückert, 1988). 1986 (Statistisches Jahrbuch, 1988) lebten in der Bundesrepublik 182 000 über 90jährige, im Jahre 2000 ist mit einem Anstieg auf eine Anzahl von 424 000 zu rechnen.

Ebenso unübersehbar ist eine weitere Veränderung, die sich anhand der Daten von 1950, 1970 und 1986 abzeichnet. Aufgrund der höheren Sterblichkeit der Männer überwiegt ab dem 55. Lebensjahr die Zahl der Frauen, und zwar mit steigendem Alter immer stärker (ab dem 80. Lebensjahr doppelt so hoch und ab dem 85. dreimal so hoch).

Die zunehmende Lebenserwartung (nicht nur die von Geburt an bestehende, sondern insbesondere die der 50-/60- und 70jährigen, Tab. 60–2) belegt die jetzt zur Verfügung stehende „sichere" Lebenszeit (Imhof, 1988). Damit stellt im Normalfall die Zeitspanne vom 50.–75. Lebensjahr die zweite Hälfte der Erwachsenenlebenszeit dar.

In Konsequenz verändert sich das Zahlenverhältnis der über 75jährigen zu den jüngeren Altersstufen dramatisch. Während im Jahre 1925, als die heute 60jährigen geboren wurden, auf einen Menschen im Alter von 75 und mehr Jahren insgesamt 67 Jüngere entfielen und 1970 noch 25, so waren es 1985 nur noch 13,4 (einschließlich Ausländer) und im Jahre 2000 werden es nur noch ca. 12,5 sein (Rückert, 1988).

Tabelle 60–2. Entwicklung der Lebenserwartung älterer Menschen (aus Rückert, 1988).

vollendetes Alter \ Jahr	1970/72 Bundesrepublik	1983/85 Bundesrepublik	Niederlande	1980/85 Schweiz	2000 Bundesrepublik	Schweiz
			Männer			
0	67,4	71,18	(72,4)	72,2	73,8	73,8
60	15,3	16,92	(11,0)		18,7	
70	9,4	10,42			11,6	
80	5,4	5,87			6,6	
90	2,8	3,55			4,1	
			Frauen			
0	73,8	77,79	(77,79)	79,8	81,2	81,3
60	19,1	21,36	(22,5)		23,8	
70	11,6	13,46	(14,5)		15,3	
80	6,2	7,26			8,6	
90	3,2	3,75			4,9	

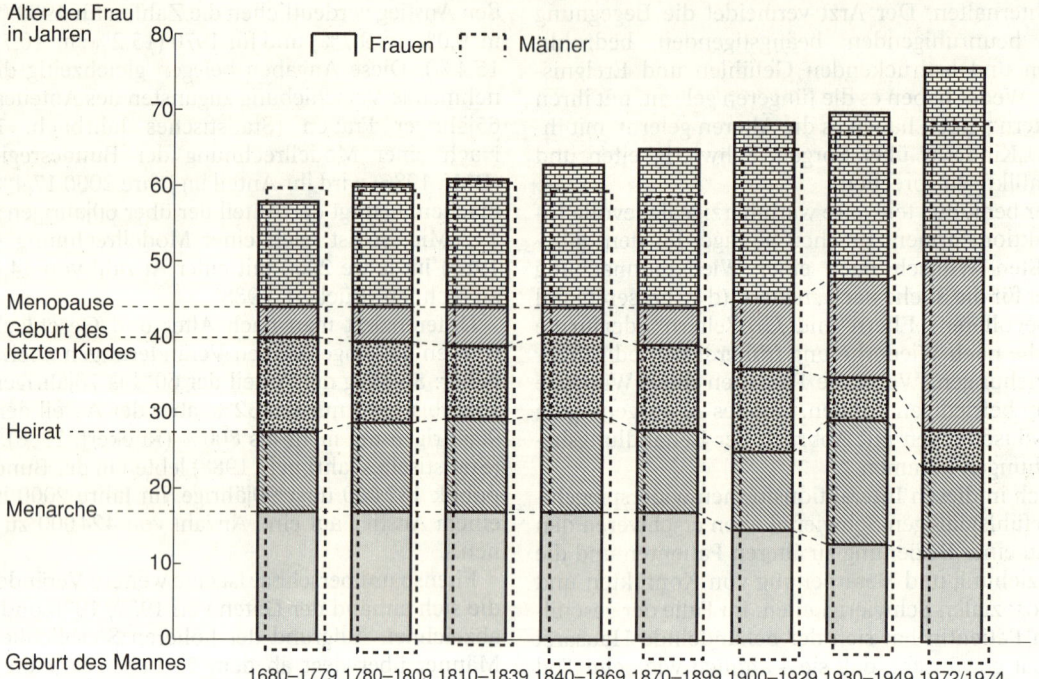

Abb. 60–1. Grundlegende Veränderungen in einzelnen Lebensphasen verheirateter Frauen während der letzten 300 Jahre (Angaben absolut in Jahren) (aus Imhof, 1981).

Die gesamte Lebensspanne ist von 58 auf über 76 Jahre angestiegen. Grundlegende Veränderungen ergaben sich:

– in Richtung auf eine Zunahme bei: Erwachsenenzeit (Menarche – Tod), fruchtbare Zeit (Menarche – Menopause); nicht „genutzte" Fruchtbarkeit (Geburt des letzten Kindes

– Menopause), „nachelterliche Gefährtenschaft" (Wegzug des letzten Kindes – Tod des Partners), Witwenschaftsdauer;

– in Richtung auf eine Abnahme bei: Kindheit (Geburt – Menarche), „genutzte" Fruchtbarkeit (Heirat – Geburt des letzten Kindes).

– Etwa gleich geblieben ist die „Jugendzeit" (Menarche – Heirat).

Gleichzeitig spiegelt die heutige Situation im historischen Vergleich auffallende Veränderungen im Lebensablauf wider (Abb. 60–1).

Während die Ehedauer über die Zeiten hinweg stets etwa die Hälfte des gesamten Lebens ausmachte und somit, relativ gesehen, ziemlich konstant blieb, nahmen die Phasen Kindheit (vor allem) und Jugend sowie die fruchtbare „genutzte" Zeit ab; neue wie Witwenschaft und besonders die nachelterliche Gefährtenschaft kamen hinzu.

Dabei betrug die Zunahme der „nachelterlichen Gefährtenschaft", also des Zeitraumes zwischen dem Ausscheiden des letztgeborenen Kindes aus der Familie (hier mit dem 20. Geburtstag gleichgesetzt) und dem Tod des Ehepartners 1972/74 20,9 Jahre, dadurch weitete sich gleichzeitig die sexuell aktive Phase auf 45,3 Jahre aus; ebenso verlängerte sich die Witwenzeit 1972/74 auf 8,3 Jahre (Imhof, 1981). So werden sie Anteile des heutigen normalen Alters.

Entsprechend ändert sich der Familienstand: Von den 65- bis 75jährigen Männern sind noch über 80%, von den 75jährigen noch über 60% verheiratet. Ganz anders sieht es bei den Frauen aus: Unter den 65- bis 75jährigen ist der Anteil der verheirateten mit 40% nur halb so groß wie bei den gleichaltrigen Männern

und unter den über 75jährigen Frauen sind nur noch 16% verheiratet, entsprechend hoch sind die Anteile der älteren Witwen (Rückert, 1988), die dann in Ein-Personen-Haushalten leben.

Die Generationen der heute über 60jährigen Frauen sind aufgrund ihrer früheren Erwerbstätigkeit und ihres Bildungsniveaus auch in ihrem Alter erneut benachteiligt (Abb. 60–2).

Eine den älteren Männern fast schon vergleichbare Einkommensposition erreichen die ledigen Frauen unter den „jüngeren Älteren". Aber schon die älteren ledigen Frauen fallen deutlich gegenüber den Männern der gleichen Altersgruppe in der Einkommensposition zurück. In der schlechtesten Einkommensposition der Alleinstehenden befinden sich die ältesten und die geschiedenen Frauen. 77,5% der geschiedenen Ältesten gaben 1982 an, ein Monatsnettoeinkommen unter DM 1200,– zu haben. Es waren insgesamt 1,64 Mio alleinstehender Frauen, die 1982 nach eigenen Angaben weniger als DM 1200,– Monatsnettoeinkommen zur Verfügung hatten. Von den alleinstehenden älteren Männern befinden sich nur 162 000 in dieser prekären finanziellen Lage. So ist auch nicht verwunderlich, daß unter den „Älteren" 1982 jeder Fünfte ein Sozialhilfeempfänger ist, aber auf einen älteren Mann, der Sozialhilfe beansprucht,

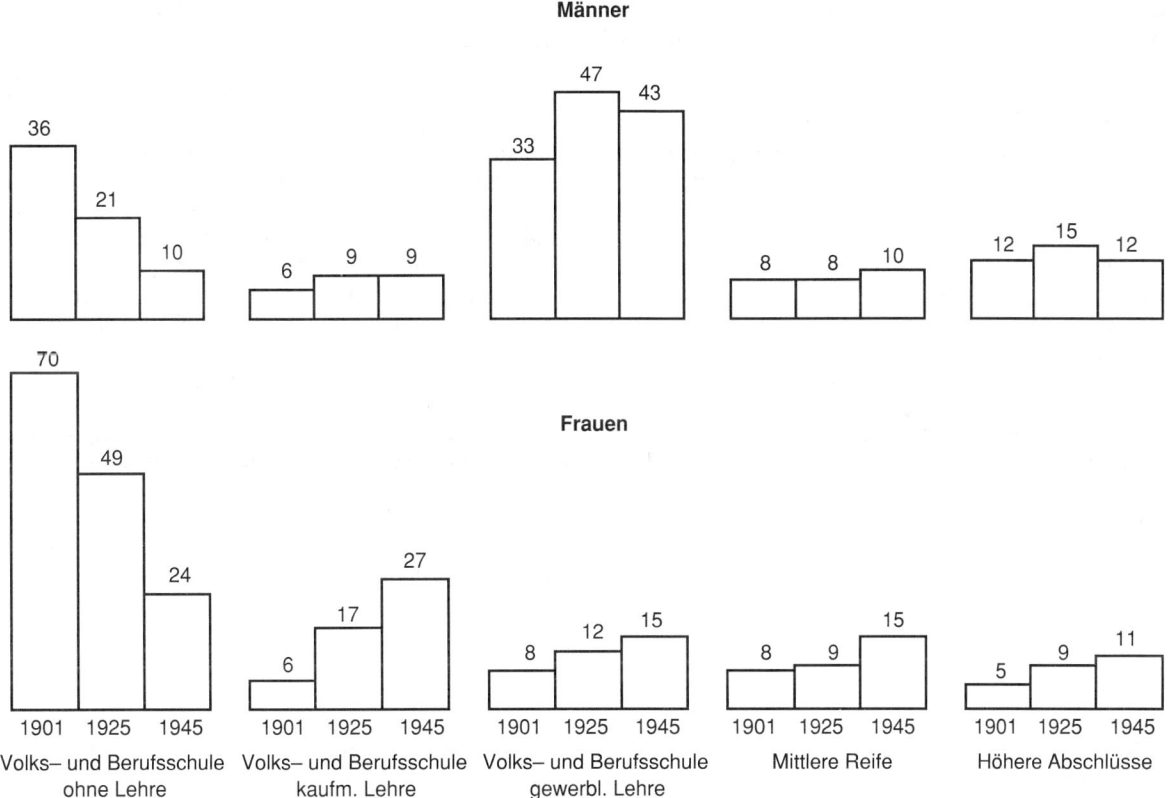

Männer

Frauen

	1901	1925	1945
Volks– und Berufsschule ohne Lehre			
Volks– und Berufsschule kaufm. Lehre			
Volks– und Berufsschule gewerbl. Lehre			
Mittlere Reife			
Höhere Abschlüsse			

Abb. 60–2. Schul- und Berufsabschlußquoten der Jahrgänge 1901, 1925 und 1945 in Prozent (aus dem 4. Familienbericht, 1986).

fast fünf Frauen gleichen Alters kommen (4. Familienbericht, 1986). In Konsequenz leben auch viele ältere Frauen unter ungenügenden, nicht altersgerechten Wohnbedingungen, insbesondere, was die Ausstattung ihrer Wohnung mit Innentoilette, Dusche/Bad und Zentralheizung betrifft.

Diese statistischen Angaben sollen dem ärztlichen Behandler helfen, sein gerontologisches Wissen zu überprüfen und zu korrigieren. Sie weisen nachdrücklich darauf hin, daß mit zunehmendem Alter in immer größerem Umfang Problemkumulationen (Grunow, 1978) möglich werden, die psychodynamisch gesehen Bedrohungen, Verluste und Kränkungen darstellen.

Erst eine differenzierende Sichtweise vermeidet einerseits Generalisierungen, erlaubt aber andererseits – aufgrund gezielter Nachfrage – das Ausmaß derartiger belastender Problemkumulationen im Einzelfall und gerade auch bei geriatrischen/gerontopsychiatrischen Patienten zu erkennen und zu würdigen. Diese differenzierende Sicht wird auch für die „neuen Alten" benötigt, d. h. die heute 50- bis 60jährigen. Sie umfassen einerseits Teilgruppen, die in guter körperlicher und psychischer Gesundheit bei befriedigenden bis guten sozialen Umständen mit entsprechender Versorgung selbständig ihr Leben gestalten (werden).

Daneben bestehen weitere Teilgruppen, die entweder als lebenslang benachteiligte und/oder chronisch Kranke in ihr Alter eintreten. Unbekannt bleibt auch, ob die „neuen Alten" nach ihrem 75. Lebensjahr unverändert aktiv, rüstig und autonom bleiben oder ob sie auch entscheidende Funktionsbeeinträchtigungen und Verschlechterungen aufgrund altersbedingter Gesundheitsveränderungen erleben. Man vergesse nicht, daß bisher die Zunahme an Lebensjahren eine Zunahme an „Behinderten-Jahren" darstellt (Schwartz, 1989). Dazu stellt langanhaltende Pflegebedürftigkeit wiederum einen Risikofaktor für psychosoziale Dekompensation, psychische Erkrankung oder Verschlechterung körperlicher Erkrankungen der nächsten Generation dar, wobei gleichzeitig die Chance erstmaliger oder erneuter Berufstätigkeit bzw. Verselbständigung entfällt.

60.3.2 Häufigkeit und Erstmanifestation klassischer „psychosomatischer" Störungen/ Erkrankungen

Widersprüchlich sind die Aussagen zur Erstmanifestation psychosomatischer Krankheiten. So findet Erfmann (1962) in 84% eine Erstmanifestation vor dem 40. Lebensjahr und jeweils nur 0,3% entfallen auf die Altersspanne vom 60.–64. und vom 65.–70. Lebensjahr. Vergleicht man diese Angaben mit den bei den einzelnen Krankheitsbildern gemachten, so findet sich doch eine erheblich größere Erstmanifestationsrate nach dem 60. Lebensjahr: Übereinstimmend sagen Fuchs (1947), Rowe und Rowe (1947), Rees (1956) sowie Paley und Luparello (1973) aus, daß eine erstmalige Manifestation des Bronchialasthmas[1] nach dem 60. Lebensjahr in 10–16% und nach dem 65. Lebensjahr in 2% zu beobachten ist. Ein Drittel der Ersterkrankungen an Magenulkus[1] fällt in die Zeit nach dem 60. Lebensjahr (Demole, 1975), dagegen bei Ulcus duodeni nur in 10% der Fälle. Ebenso wird von 10% Erstmanifestationen einer Colitis ulcerosa[1] zwischen dem 50. und 80. Lebensjahr ausgegangen (Law et al., 1961; Henning, 1975; Demole, 1975; Lang, 1976).

Schließlich zeigt sich auch eine deutliche Zunahme an Potenzstörungen, die für die 60jährigen mit 18 bis 25%, für die 65jährigen mit 25% und die 70jährigen mit 27–35% angegeben werden (Berezin, 1969 und 1976; Borelli, 1971).

Zusammenfassend läßt sich für psychosomatische Störungen und Erkrankungen aussagen, daß ihr Häufigkeitsgipfel vor dem Alterseintritt mit dem 60./65. Lebensjahr liegt. In einem gewissen Umfang sind Erstmanifestationen jenseits dieses Zeitpunktes zu beobachten. Auf jeden Fall bestehen während des Alterns in größerem Umfang derartige Krankheitsbilder fort.

Diese Untersuchungen weisen auf folgende Unterschiede zu Krankheiten im jüngeren und mittleren Lebensalter hin:
– Erneut oder neu auftretende Störungen und Erkrankungen beginnen offenbar schleichend mit einem eher langfristigen, in der Intensität abgeschwächten Verlauf bei geringerer Anzahl an Symptomen. Ein plötzlicher Beginn mit akutem Verlauf scheint sehr viel seltener zu sein.
– Bestimmte für frühere Lebensabschnitte typische Relationen, wie z. B. das Überwiegen der Männer bei Erkrankungen an Magenulkus, oder bestimmte Krankheitslokalisationen verändern sich ebenfalls mit dem Altern.
– Die funktionelle Störung wird zunehmend von der organischen Schädigung abgelöst.
– Vorhandene Krankheiten werden bei ansteigender Unspezifität zum Ausdruck von Konflikten und Schwierigkeiten benutzt.
Ätiologisch scheinen die für das jüngere und mittlere Lebensalter eher spezifischen innerpsychischen Konflikte zurückzutreten.

Funktionelle Syndrome lassen sich in großem Umfang bei alternden Patienten finden. Bekannt ist die zunehmende Rate an Schlafstörungen mit ansteigendem Alter (Regestein, 1980; Lund und Rüther, 1985). Bei Strauss und Wohlschläger (1971) klagten 60% der über 65jährigen über Einschlafschwierigkeiten und 95% über ein zu frühes Aufwachen. Interessant ist die Beobachtung von Hartleb (1966), daß unter den über 65jährigen Klinikpatienten mehr als 50% einen gestörten Schlaf angeben, objektiv aber nur

35% eine Schlafstörung aufwiesen. Belegt ist die umfangreiche Verordnung von Sedativa, Tranquilizern und Hypnotika, die sowohl auf den großen Umfang dieser Störungen als auch auf ihre innerpsychische Bedeutung für den Älteren hinweist.

Bekannt ist die Häufigkeit der Obstipation, an der allgemein 20–25%, bei gezielter Befragung jedoch 40% der über 65jährigen litten (Lang, 1976). Ebenso fanden sich funktionelle Störungen, insbesondere in Form von Sodbrennen, Aufstoßen, Magenbeschwerden, Übelkeit, Erbrechen, Durchfall, Verstopfung und Blähungen nach Sklar (1972) bei 60% seiner 300 über 65jährigen Patienten.

Auch die Schmerzzustände Alternder erweisen sich oft als psychosomatisch (Herzmann, 1985).

Die Ergebnisse der Wiener Gesundheitsstudie (Biron und Eder, 1980) weisen ebenfalls darauf hin, daß bestimmte funktionelle Störungen und psychosomatische Erkrankungen einerseits im 60. Lebensjahr häufig vorhanden sind und andererseits eine Zunahme zwischen dem 20. und 60. Lebensjahr zeigen (Tab. 60–3).

60.3.3 Gerontopsychiatrische Morbidität

Die inzwischen auch für die Bundesrepublik vorgelegten Prävalenzuntersuchungen zur gerontopsychiatrischen Morbidität (Dilling et al., 1984; Cooper und Sosna, 1983) bestätigen die Schätzungen der Psychiatrie-Enquete (1975), welche von einer Gesamtrate von 25% von psychischen Störungen/Erkrankungen im weitesten Sinne bei den über 65jährigen ausging.

Tabelle 60–3. Ärztliche Diagnose, angegeben durch den Arzt, nach Alter und Geschlecht (aus Biron et al., 1980)

	Männer (%)			Frauen (%)		
	60j.	40j.	25j.	60j.	40j.	25j.
Übergewicht						
„mittelgradig"	18	11	3	16	8	3
„stark"	10	4	2	25	7	2
Hypertonie						
„leicht"	22	9	6	21	5	2
„mittel"	12	4	4	13	4	–
„schwer"	2	4	2	2	3	–
psychovegetative Symptomatik und dgl.	7	11	15	16	22	17
Gastritis chronica	16	10	7	13	10	10
vermutl. „funktionelle" Herzbeschwerden	5	14	7	18	24	11
Hypotonie	0 (2)	4	3	2	10	17
Ulkuskrankheit	4	5	1	3	3	2
Bronchitis spastica, Asthma bronchiale	6	3	2	3	1	2
Verdacht auf chron. Alkoholismus	7	6	1	0 (2)	1	–
Debilität, Demenz	0 (2)	0 (2)	1	1	0 (2)	0 (1)
starkes Untergewicht	0 (1)	–	2	–	1	1

---

[1] Psychische und soziale Faktoren können beteiligt sein, sind aber nicht obligatorisch.

Die beiden Untersuchungen aus Oberbayern und Mannheim weisen weitgehend übereinstimmend auf eine Rate von 10,2–10,8% für die Gruppe der Neurosen und Persönlichkeitsstörungen hin (vgl. Kap. 4). Mit etwa 45% stellt diese Gruppe in beiden Untersuchungen eine auffallend große Gruppe dar und bestätigt gleichzeitig die Aussage, daß die „Alterspsychiatrie nicht die Psychiatrie der Demenz" (Lauter, 1974) ist. Diese bisher kaum erforschte Gruppe psychischer Erkrankungen über 60jähriger umfaßt aus ätiologischer Sicht sowohl von Kindheit/Jugendzeit an bestehende neurotische Erkrankungen (die sich unterschiedlich häufig im Lebenszyklus manifestieren) als auch erstmals nach dem 60. Lebensjahr auftretende reaktive Erkrankungen, insbesondere depressiver Ausprägung (Radebold, 1989a). Ihr zahlenmäßiger Anteil ist bisher unbekannt.

Katschnig und Strotzka (1977) weisen in ihrer Übersichtsarbeit darauf hin, daß die Ergebnisse über die Häufigkeit von Neurosen (und psychosomatischen Erkrankungen) in verschiedenen Altersgruppen widersprüchlicher als die über Geschlechtsunterschiede sind.

Allgemein wird angenommen, daß Neurosen im jüngeren und mittleren Lebensalter auftreten. Dem entspricht eine Aufstellung von Dohrenwend und Dohrenwend (1969), nach der der Altersgipfel für Neurosen in 7 epidemiologischen Untersuchungen vor dem 40. Lebensjahr lag. Hagnell (1970) fand bei Männern jenseits des 40. Lebensjahres und bei Frauen jenseits des 50. Lebensjahres ein deutliches Absinken der Inzidenz neurotischer Episoden. Demgegenüber weist die in Abschnitt 60.3.3 angeführte Aufstellung von Cooper und Sosna (1983) darauf hin, daß Neurosen und Persönlichkeitsstörungen bei ca. 10% (mit einer Schwankungsbreite je nach Untersuchung zwischen 6,8 und 12,6%) der über 65jährigen zu finden sind. Ein Teil dieser neurotischen Erkrankungen manifestiert sich wahrscheinlich erstmals nach dem 65. Lebensjahr im Sinne einer mißglückten Bewältigung der Alterssituation. Dementsprechend fanden Kay und Mitarbeiter (1955), daß die Hälfte der eindeutigen neurotischen Fälle (5–10%) zu den sich erst spät manifestierenden Neurosen zu rechnen sind.

Die bisherigen Befunde lassen sich so interpretieren, daß zwar der Häufigkeitsgipfel neurotischer Erkrankungen im mittleren Lebensalter (d.h. zwischen dem 30. und 50. Lebensjahr) liegt, neurotische Erkrankungen sich jedoch bei ca. 10% der über 65jährigen finden lassen. Ihre Erstmanifestation liegt in gewissem Umfang auch jenseits des 65. Lebensjahres.

Erschwert wird die Fragestellung noch durch die Befunde, die darauf hinweisen, daß einerseits Phobien und Ängste bei ansteigendem Alter abnehmen (Katschnig und Strotzka, 1977) und andererseits die Tendenz zur Somatisierung neurotischer Störungen mit steigendem Alter zunimmt (McDonald, 1966, 1973). In der Midtown Manhattan Study und in der Stirling County Study ergab sich außerdem eine positive Korrelation zwischen dem Alter und der Häufigkeit „psychosomatischer" bzw. „psychophysiologischer" Symptome. So findet sich in der Stirling County Study für diese Symptome ein praktisch kontinuierlicher Anstieg von etwa 30% für die Altersgruppe der unter 30jährigen bis etwa 85% für die 70jährigen (Katschnig und Strotzka, 1977). Bei der hohen Rate dieser Symptome ist zu diskutieren, ob psychische Störungen mit bestehenden körperlichen Beschwerden vermengt wurden oder ob die Zunahme derartiger Symptome mit dem Alter durch die gleichzeitige Zunahme somatischer Krankheiten bedingt war.

Als charakteristische psychische Alterserkrankungen sind die organischen Psychosyndrome/Demenzen anzusehen, deren Häufigkeit mit zunehmendem Lebensalter steil ansteigt.

Als die quantitativ und gesundheitspolitisch bedeutsamsten psychischen Erkrankungen im Alter sind die über einen langsamen Verlust von Gedächtnisleistungen und geistigen Fähigkeiten zur psychischen Behinderung und schließlich zur Pflegebedürftigkeit führenden Demenzen und die depressiven Erkrankungen anzusehen (Häfner, 1986). Dabei wird differentialdiagnostisch häufiger eine depressive Symptomatik zugunsten der Diagnose einer Demenz übersehen.

60.3.4 Spezifische Aspekte organischer Erkrankungen

Aus der Sicht der gerontologischen Grundlagenforschung wird Altern als ein nicht umkehrbarer Prozeß von einem Anfangspunkt (Zeugung) bis zu einem Endpunkt (Tod) angesehen. Dieser bald nach Abschluß der Reifungsperiode beginnende Prozeß be-

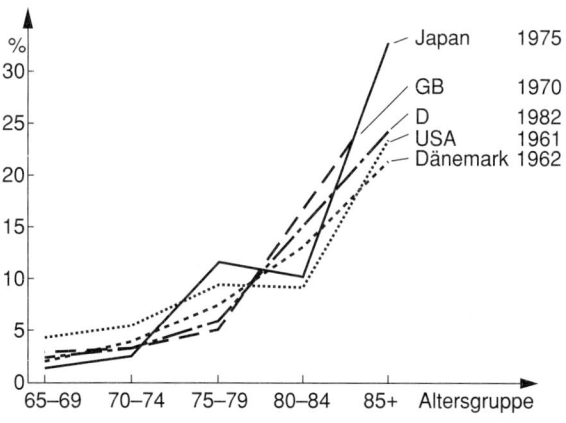

Abb. 60–3. Prävalenz organischer Psychosyndrome[1] in 5 Altersgruppen jenseits des 64. Lebensjahrs aus 5 Feldstudien[2] (Punktprävalenz per 100 Einwohner in jeder Altersgruppe) (aus Häfner, 1986).

[1] Alle schweren und mittelschweren Fälle von Demenz, Verwirrtheitszuständen und vergleichbaren exogenen Psychosen. Die N.Y.-State-Studie schließt einige Fälle funktioneller Psychosen ein.

[2] Die Daten sind den folgenden Studien entnommen: New York State Dept. of Mental Hygiene, 1961 (USA); Nielsen, 1962 (Dänemark); Kay et al., 1970 (UK); Kaneko, 1975 (Japan); Cooper & Sosna, 1983 (Bundesrepublik Deutschland).

trifft Zellen, Organe und Organsysteme, insbesondere aber zeigen sich molekular-biologische Veränderungen an den Bausteinen der Gewebe und Zellen. Aufgrund dieses lebenslangen „Alterns" ist es nicht möglich, von einem bestimmten Beginn des biologischen Alterns zu sprechen. Derartige, einen bestimmten Zeitpunkt betreffende Festsetzungen sind rein willkürlich und beziehen sich in der Regel auf den von dem Gesetzgeber festgelegten Termin der Berentung oder Pensionierung. Neben diesen Veränderungen zeigen sich auch Auswirkungen auf die Funktionen. „So kommt es zunehmend aufgrund von Einschränkungen und Störungen der Regulation zu einer Störung der Homöostase, d.h. des Zustandes, in welchem alle Funktionen, die Nahrungsaufnahme, die Funktionen des Magen-Darm-Kanals, die Ausscheidungsprozesse der Niere, die Kreislaufprozesse, die Regulationen durch Hormone und durch das Nervensystem in optimaler Harmonie ineinandergreifen … Weiterhin wird die Fähigkeit, sich an neue Umweltfaktoren anzupassen, mehr und mehr gestört. Die Adaptation z.B. an Kälte und Wärme oder an Sauerstoffmangel in der Höhe oder auch an physische Einflüsse einer sich verändernden Umwelt wird nun nicht mehr richtig oder nur langsam beantwortet. Während in der Jugend bei Gesunden so konstante Werte für die meisten Funktionen wie Blutdruck, Puls, Atmung, Reaktionszeit usw. bestehen, daß man in Lehrbüchern gut Mittelwerte angeben kann, streuen umgekehrt diese Werte mit eintretendem Alter mehr und mehr. Man kann alt werden mit niedrigem oder hohem Blutdruck, ebenso wie mit weißen Haaren oder jugendlicher Haarfarbe, mit Hautrunzeln oder ohne. Aber stets tritt als konstantes Alternszeichen die Verminderung der Adaptationsfähigkeit auf" (Verzàr, 1975).

Mit zunehmendem Lebensalter wird die Kluft zwischen Regenerationsvermögen und der Zunahme der degenerativen Prozesse immer größer, so daß die Adaptationsfähigkeit bereits aus diesem Grund eingeschränkt wird. Hinzu treten alterstypische Organveränderungen, die klinisch häufig nicht von pathologischen Vorgängen zu trennen sind (Klein, 1988). Dabei sind biologisches und kalendarisches bzw. chronologisches Alter eines Individuums nicht als identisch anzusehen. Die biologische Alterung verläuft zwar zeitabhängig, aber für die verschiedenen Zellen, Interzellularsubstanzen, Gewebe und Organe eines Individuums nicht synchron und nicht uniform. Deshalb ist das biologische Alter eines Individuums (= als Ganzes) nicht die Summe der Alterung seiner vorgenannten Teile. Die bisherigen Alterstheorien erklären noch nicht die Komplexität der sog. Altersursachen (Lindner, 1988).

Heute ist von der Annahme auszugehen, daß der überwiegende Teil von Organfunktionsstörungen im Alter nicht durch die morphologischen Altersvorgänge an sich, auch nicht durch die alterstypischen Regelmechanismen und die beschriebene verminderte Adaptationsfähigkeit des alternden Organismus verursacht wird, sondern vor allem durch krankhafte Prozesse. Sie werden als alternde Krankheiten bezeichnet, wenn ihr Verlauf chronisch ist, d.h., wenn sie den Menschen aus früheren Lebensabschnitten bis in das hohe Alter kontinuierlich begleiten.

Neben diesen alternden Krankheiten sind die primären Alterskrankheiten von großer Bedeutung. Sie treten erstmals im Alter auf und sind auch in ihrer Häufigkeitsverteilung eng an das höhere Lebensalter geknüpft (Lang und Diepgen, 1988; Abb. 60–4).

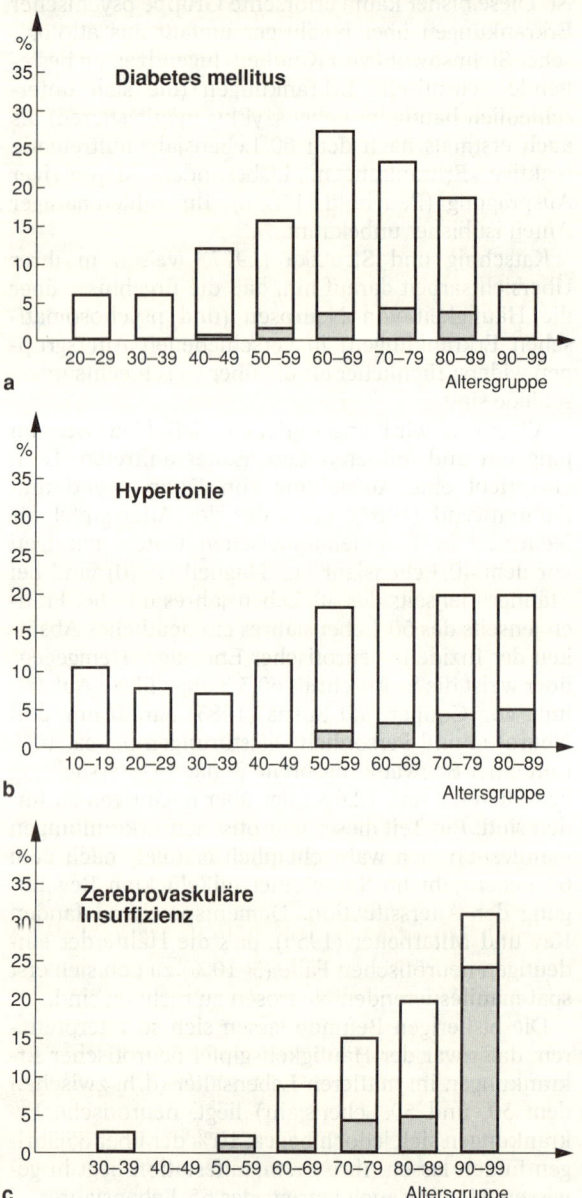

Abb. 60–4. Prozentualer Anteil verschiedener Erkrankungen an der Gesamtzahl der Patienten einer Altersgruppe in Abhängigkeit vom Lebensalter.
a) Diabetes mellitus und Alter: Die Häufigkeit nimmt bis zum 69. Lebensjahr zu; danach kommt es zu einer Abnahme.
b) Hypertonie und Alter: Zunahme der Häufigkeit bis zum 69. Lebensjahr; danach weitgehend unveränderte Häufigkeit.
c) Zerebrovaskuläre Insuffizienz und Alter: Ab dem 50. Lebensjahr bis ins höchste Alter läßt sich eine deutliche Zunahme der Häufigkeit erkennen (aus Lang und Diepgen, 1988).

Außer durch ihre Chronizität sind die Krankheiten der über 60jährigen durch Polypathie und Multimorbidität charakterisiert. Mit zunehmendem Lebensalter wächst aus biostatistischen Gründen die Wahrscheinlichkeit, immer mehr Funktionseinbußen und Krankheiten zu erleiden. Das gleichzeitige vielfache Kranksein ist geradezu ein spezifisches Merkmal des gealterten Organismus mit weitreichenden Konsequenzen und Anforderungen an Diagnostik und Therapie (Franke, 1983). Es ist dabei von großer praktischer Bedeutung, mehr oder weniger ruhende Leiden und Gebrechen („pathos") von akuten oder chronifizierenden aktiven Krankheiten („morbus") zu unterscheiden. Dementsprechend weisen die Begriffe Polypathie und Multimorbidität auf das gleichzeitige Nebeneinander mehrerer „pathoi" oder „morbi" hin (Schramm, 1988). Die Multimorbidität wird belegt mit dem etwa linearen Anstieg der Diagnosezahlen mit dem Alter (Abb. 60–5).

60.4 Psychosomatische Reaktion Älterer: unspezifische Nutzung bestehender körperlicher Symptomatik

Über 60jährige können während ihres Alterns in zunehmendem Maße und in großem Umfang (negative) Veränderungen erfahren. Diese umfassen aufgrund von Chronizität, Polypathie und Multimorbidität dauerhafte Einschränkungen physischer (Leistungsfähigkeit, Hören, Sehen, Beweglichkeit, Potenz) und psychischer (Orientierung, Gedächtnisleistungen, Kontrolle) Funktionen, die schließlich zu Hilfs- und Pflegebedürftigkeit sowie zunehmender Abhängigkeit von der Umwelt führen. Wichtige Beziehungen (Eltern, Partner/Partnerin, Geschwister, Freunde, auch schon Kinder und Enkelkinder) verändern oder verringern sich, insbesondere durch Umzüge, Krankheiten und Todesfälle. Schließlich kann sich auch die soziale Situation aufgrund geminderten Sozialstatus (Nicht-Rolle als Rentner/Pensionär), des eingeschränkten Einkommens (Rentner/-in, Situation als Verwitwete) bei ungenügenden Wohn- und Versorgungsverhältnissen entscheidend verschlechtern.

Psychodynamisch formuliert erleidet der alternde Erwachsene nach einer langen Phase psychosexueller und psychosozialer Identität und Integrität Schädigungen von Ich-Funktionen, nicht gewünschte Veränderungen und Verluste von wichtigen Objektbeziehungen sowie destabilisierende Einschränkungen seiner sozialen Umwelt. Dabei hat das Ich während des Alterns – genauso wie in den zurückliegenden Abschnitten des Lebenszyklus – zwischen libidinösen und aggressiven Triebimpulsen und den Ansprüchen des Über-Ichs zu vermitteln und gemäß dem Lust-Unlust-Prinzip unter Einbeziehung seiner Realität und der gegebenen Umwelt Befriedigungsmöglichkeiten zu suchen und anstehende psychosoziale Aufgaben zu lösen. Dabei entscheidet die bisherige innerpsychische Besetzung über die Bedeutung dieser Veränderungen und Verluste, die gleichzeitig als ausgeprägte (narzißtische) Kränkungen und Attacken (bei sich ablehnend verhaltender Umwelt) erlebt werden können.

Trennungen und Verluste werden als entscheidende Bedrohungen seelischer Gesundheit und damit als ernsteste Formen des psychogenen Stresses (Stenback, 1965; Bowlby, 1972; Schmale, 1972; Parkes, 1974; Joraschky und Köhle, 1981) gewertet. Mit Hilfe der Trauerarbeit (Lindemann, 1944; Bowlby, 1972; Parkes, 1974, 1980) gelingt es durch einen längeren phasenspezifischen Prozeß, drohende, stattfindende oder eingetretene Trennungen/Verluste innerpsychisch zu akzeptieren, zu bewältigen und zu verarbeiten. Pollock (1982) sieht den ständigen „mourning and liberation process" als charakteristisch für die zweite Hälfte des Erwachsenenlebens an, der gleichzeitig neue Beziehungen ermöglicht.

Nur vereinzelte Publikationen beschreiben altersgruppenspezifische Befunde bei der Verwitwung. Bereits 1951 fanden Stern und Mitarbeiter bei älteren Witwen einen „Mangel an sichtbaren mentalen Manifestationen von Gram". Nachfolgende Untersuchun-

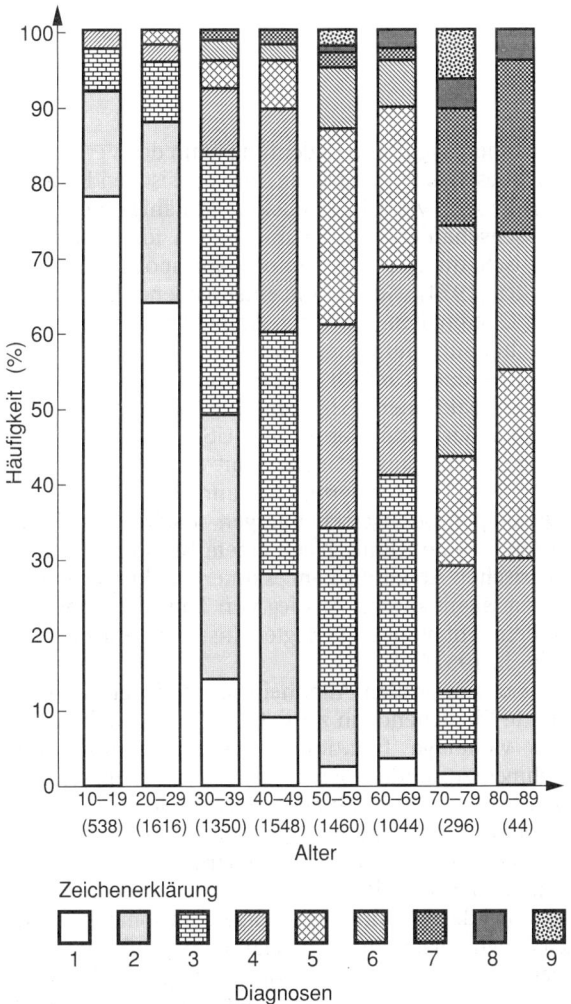

Abb. 60–5. Anzahl der Diagnosen in den einzelnen Altersstufen in einem ambulanten internistischen Krankengut (in Klammern die absolute Zahl der untersuchten Patienten, Gesamtzahl = 7896) (aus Schramm et al., 1982).

Tabelle 60–4. Prävalenz psychischer Störungen in der Bevölkerung über 65; berechnet aufgrund von Feldstudien (aus Cooper und Sosna, 1983).

Autoren	Untersuchungs-gebiet	Anzahl der Probanden	Schwere organische Psycho-syndrome (%)	Leichte organische Psycho-syndrome (%)	Funktionelle Psychosen (%)	Neurosen und Persön-lichkeits-störungen (%)	Ge-samt (%)
Sheldon, 1948	Wolverhampton, England (städtisch)	369	3,9	11,7	–	12,6	28-,2
Primrose, 1962	N. Schottland (ländlich)	222	4,5	–	1,4	12,6	–
Nielsen, 1962	Samsø (ländlich)	978	3,1	15,4	3,7	6,8	29,0
Kay et al., 1964	Newcastle, England (städtisch)	443	5,6	5,7	2,4	12,5	26,3
Parsons, 1965	Swansea, Wales (städtisch)	228	4,4	–	2,6	4,8	–
Dilling u. Weyerer, 1980	Oberbayern (halbländlich)	295	–	8,5	3,4	10,2	23,1
Cooper u. Sosna, 1983	Mannheim (städtisch)	519	6,0	5,4	2,2	10,8	24,4

gen stimmen weitgehend darin überein, daß über 60jährige Witwen im Vergleich zu jüngeren Frauen stärker zur Somatisierung neigen (Neugarten, 1968; Blau, 1973; Parkes, 1974; Ball, 1976; Cary, 1977). Weiterhin stellt sich eine erhebliche (vorübergehende) Verschlechterung des objektiven Gesundheitszustandes ein, und zwar insbesondere bei älteren verwitweten Männern (Marris, 1958; Clayton, 1974). Bei Stappen (1988) klagten 56% der Witwen (N = 30) über eine Verschlechterung ihres Gesundheitszustandes in den ersten 6 Monaten nach Partnerverlust. Epidemiologische Verlaufsanalysen (Young et al., 1963; Rees und Lutkins, 1967) belegen ein erhöhtes Sterblichkeitsrisiko Verwitweter, insbesondere für ältere Männer. In der Bundesrepublik findet sich bei älteren verwitweten Männern (vor allem bei über 75jährigen) im Vergleich zu den verheirateten eine deutlich erhöhte Sterblichkeit: Zur Zeit beträgt die Differenz zwischen verheirateten und verwitweten Männern dieser Altersgruppen 6 Jahre (Stappen, 1988).

Auch die erzwungene Beendigung von für die eigene Identität entscheidenden Funktionen/Aufgaben kann als Verlust erlebt werden. Neben Variablen wie sozioökonomischer Status und Bildungsniveau, Gesundheitszustand, zu erwartende Rentenhöhe, Einstellung des Ehepartners und der bestehenden Freizeitinteressen (Lehr, 1979; Tews, 1979) hat das subjektive Erleben entscheidenden Einfluß auf die „Ruhestands"-Situation. Schon früh wurden unter den Stichworten „Pensionierungsbankrott" und „Pensionierungstod" (Stauder, 1955; Jores und Puchta, 1959) entsprechende Krankheitsverläufe beschrieben, wobei der „Pensionierungstod" durch Zusammenwirken „der zur Pensionierung führenden Krankheit mit dem Trauma der Pensionierung" eintreten konnte. Ähnli-

che innerpsychische Bedeutung kann der Verlust von hochbesetzten psychischen oder physischen Funktionen haben, wie z.B. intellektuelle Fähigkeiten eines Wissenschaftlers, Hören und Sehen für geistig und musisch Interessierte, Motorik für sportlich Aktive. Parkes (1974) wies auf die entsprechende Bedeutung von Gliedmaßen, z.B. nach Amputationen, hin.

Schließlich führt auch der Verlust der bisherigen schützenden Umwelt zu entsprechenden somatischen Reaktionen. Friedsam (1962) beschrieb Ältere, die bei Wirbelstürmen in den USA im Vergleich zu Jüngeren fast ausschließlich mit Apathie und Hoffnungslosigkeit reagierten. Bekannt sind körperliche Dekompensationserscheinungen nach Aufnahmen Älterer in die Klinik. Nach einem Brand in einer geriatrischen Abteilung kam es innerhalb kurzer Zeit zu einem signifikanten Anstieg von Todesfällen der auf andere Abteilungen verlegten Insassen (Aleksandrowitsch, 1961).

In einem großen Altenheim zeigte jeder vierte bis fünfte Einziehende in zeitlichem Zusammenhang damit vielfältige funktionelle vegetative Störungen, Schmerzzustände ebenso wie Orientierungsstörungen und Unruhezustände (Solms-Wildenfels, 1985).

Diese Untersuchungen sprechen – vorsichtig interpretiert – dafür, daß über 60jährige bei Trennungs- und Verlustsituationen (darstellbar insbesondere am Beispiel der Verwitwung) weitgehend somatische Reaktionen zeigen, unter Wegfall emotionaler Reaktionen. Eine Teilgruppe – Männer häufiger als Frauen – weist in zeitlichem Zusammenhang damit schwere objektive Krankheitsbefunde mit ansteigender Sterblichkeit auf. Wahrscheinlich konnte in diesen Fällen die notwendige Trauerarbeit nicht geleistet werden und mußte daher abgewehrt werden.

Diese Tendenz zur Somatisierung zeigt sich auch bei anderen Krankheitsgruppen, so bei den neurotischen Erkrankungen. Ebenso wird als charakteristisch für die Symptomatik der Altersdepression die „somatogene" oder „larvierte" Depression angesehen (Oesterreich, 1981; Bergener, 1986); hier wird die psychische Symptomatik weitgehend durch funktionelle, vegetative oder körperliche Symptome ersetzt.

Schließlich wird diese Somatisierungstendenz auch bei einem Vergleich der Rangfolgen von Beschwerden zweier repräsentativer Populationen deutlich (Tab. 60–5). Bei den über 60jährigen verändert sich die Rangfolge zunehmend in Richtung „organischer" Symptomatik.

Unter Einbeziehung der in Abschnitt 60.3.2 und 60.3.3 zitierten Untersuchungsergebnisse lassen sich folgende Aussagen zur Manifestation psychosomatischer Störungen/Erkrankungen treffen:

– Erstmanifestationen psychosomatischer und neurotischer Erkrankungen zeigen sich in der Regel vor dem 50., eindeutig seltener zwischen dem 50. und 60./65. Lebensjahr. Diese Krankheiten im engeren Sinn sind charakterisiert durch spezifische Krankheitsbilder mit entsprechend strukturierter Symptomatik.
– Ihr zweiter Häufigkeitsgipfel liegt zwischen dem 45. und 60. Lebensjahr (bei depressiven Neurosen zwischen dem 50. und 70. Lebensjahr).
– Diese sich vor dem 60. Lebensjahr manifestierenden Krankheiten bestehen zu einem großen Teil nach dem 60. Lebensjahr fort. Dabei erscheint ihre Symptomatik diffuser, weniger gut abgrenzbar und

(aufgrund eines Bedeutungswandels?) unspezifischer.
– Nach dem 60. Lebensjahr manifestieren sich im Bedarfsfalle keine neuen spezifisch psychosomatischen (und auch neurotischen) Krankheitsbilder mit entsprechender abgrenzbarer Symptomatik. Bei traumatisierenden, ätiologisch wirksamen Einflüssen zeigen sich einerseits eine depressive Symptomatik (mit deutlicher Somatisierungstendenz) und andererseits in wohl noch größerem Umfang somatische Reaktionen. Diese reichen von neu auftretenden bzw. sich verstärkenden diffusen funktionellen/vegetativen Symptomen über eine „organisch" anmutende Symptomatik bis hin zu dekompensierenden organischen Erkrankungen mit teilweise nachfolgender Mortalität.

Diese beschriebenen Reaktionen weisen auf wichtige, bisher noch nicht geklärte Fragen hin:

Innerpsychische Konflikte, Trennungen/Verluste und traumatisierende familiäre Einflüsse treffen in Kindheit und Jugendzeit auf das sich in seiner psychosexuellen und psychosozialen Entwicklung befindliche Individuum. Sie werden mit Hilfe langfristiger Verarbeitungs-, Anpassungs- und Abwehrprozesse integriert und tragen so zur Bildung der psychosexuellen und psychosozialen Identität des zukünftigen Erwachsenen bei. Unbewältigt – also abgewehrt – führen sie zu neurotischen und psychosomatischen Erkrankungen, die sich dann unter spezifischen Voraussetzungen mehrfach im jüngeren und mittleren und auch später im höheren Lebensalter manifestieren können. Trennungen und Verluste über 60jähriger treffen – im idealtypischen Falle – Erwachsene mit reifer, langfristig erprobter, psychosexueller und psychosozialer Identität, die durch die Erfahrungen der Erwachsenenzeit zusätzlich geprägt und stabilisiert wurde. Außerdem hat der alternde Erwachsene in der Regel bereits mehrfach Trennungen und Verluste erlebt.

Bei den jetzt unverändert zu beobachtenden innerpsychischen wie auch intra- und intergenerativen Konflikten handelt es sich in der Regel nicht um grundsätzlich neue, sondern aus der Kindheit und Jugendzeit bekannte (unbewußte) zeitlose Konflikte, die bisher ungelöst in der oder durch die Alternssituation wiederbelebt wurden. Für den Umgang mit diesen Konflikten hat der Erwachsene aufgrund seiner Erfahrungen stabile Anpassungs- und Abwehrstrukturen entwickelt, die ihm auch während seines Alterns unverändert zur Verfügung stehen. Der grundsätzliche Unterschied liegt meines Erachtens darin, daß die Konflikte, Trennungen/Verluste und traumatisierenden (familiären) Einflüsse nicht mehr auf das sich entwickelnde Individuum treffen, sondern auf den reifen Erwachsenen mit seiner stabilen psychosexuellen und psychosozialen Identität. Sie können daher nicht mehr einen so schädigenden Einfluß ausüben.

Von denjenigen Älteren, die die beschriebenen Bedrohungen, Trennungen/Verluste, Kränkungen und Attacken erleiden, erkrankt nur ein (nicht eindeutig abschätzbarer, aber kleinerer) Teil. Eine stabile und

Tabelle 60–5. Rangfolgen der Beschwerden in einer repräsentativen ländlichen Alterspopulation (1987) und in einer repräsentativen städtischen Population zwischen 15 und 80 Jahren (1980) (die Zahlen in Klammern beziehen sich auf die Rangplätze, die die Beschwerden bei der Altenpopulation erzielten) (aus Jansen und Radebold, 1989).

1980	1987 (gerontologische Studie)
1. Erkältung, Schnupfen	1. Rückenschmerzen
2. Kopfschmerzen (10)	2. Wetterfühligkeit
3. Husten	3. Herzbeschwerden
4. Zahnschmerzen	4. Gliederschmerzen
5. Halsschmerzen	5. Gedächtnisstörungen
6. Kreuz-/Rücken- schmerzen (1)	6. Übergewicht
7. innere Unruhe, nervöse Unruhe (14)	7. Schlafstörungen
8. Fieber	8. Kälteempfindlichkeit
9. Bauchschmerzen/ Magenbeschwerden	9. Schwindel
10. Übergewicht (6)	10. Kopfschmerzen
11. Gliederschmerzen (4)	11. häufiges Wasserlassen
12. Schlafstörungen (7)	12. Unsicherheit im Gehen und Greifen
13. Schwäche/Ermü- dungszustände (15)	13. Sehstörungen
14. Blähungen	14. nervöse Unruhe
15. Herzbeschwerden (3)	15. Erschöpfung
16. Wetterfühligkeit (2)	16. Gleichgewichtsstörung

im Rückblick befriedigende bisherige Entwicklung, unbeeinträchtigt und ungeschädigt zur Verfügung stehende Ich-Funktionen (oder zum Ausgleich vorhandene Ich-Stärken), fortbestehende sichere und befriedigende Beziehungen, soziale Integration und Aktivitäten sowie die Übereinstimmung von Ich-Ideal mit der bisherigen Lebensverwirklichung helfen dem über 60jährigen, die zunehmende – teilweise auch mehrfach erforderliche – Trauerarbeit adäquat zu leisten.

Diese Aufgabe der Trauerarbeit erscheint unter besonderen Bedingungen und in spezifischen Situationen nur schwer oder gar nicht leistbar. Dazu zählen insbesondere:

– Die/der Ältere ist aufgrund defizitärer, eingeschränkter oder aufgrund von physischen/psychischen Krankheiten geschädigter Ich-Funktionen nicht mehr in der Lage, die für die Trauerarbeit notwendigen Verarbeitungs-, Anpassungs- und Abwehrprozesse zu leisten (These des eingeschränkten „Puffers").

Gerade neurotische, psychosomatische, aber auch psychotische Erkrankungen früherer Lebensjahrzehnte führen oft dazu, daß Ich-Funktionen nur eingeschränkt zur Verfügung stehen, keine Ich-Stärken und konfliktfreien Bereiche zum Ausgleich vorhanden sind und nur wenige und dazu noch gestörte Beziehungen bestehen. Organische und hirnorganische Erkrankungen können dann im Alter weitere Schädigungen bewirken.

– Die/der Ältere verliert die einzig wichtige hochbesetzte Beziehungsperson (sei es einen noch aus der Kindheit stammenden Elternteil oder eine/einen Schwester/Bruder, sei es die symbiotisch gestaltete Partnerbeziehung ohne weitere freundschaftliche Beziehungen); die Bewältigung wird noch dadurch erschwert, daß der Verlust ohne innere Vorbereitung plötzlich eintritt.

– Der Verlust erfolgt außerhalb des erwarteten Verlaufes (Neugarten, 1970), z.B. wenn das einzige oder wichtige hochbesetzte Kind anstelle des kranken Partners oder der alten Eltern stirbt.

– Bedrohungen, Verluste, Kränkungen treten innerhalb eines kurzen Zeitraumes in mehreren Bereichen auf: physische und psychische Funktionen, Beziehungen und soziale Sicherheit (These der Kumulation).

Können Trennungen/Verluste sowie Bedrohungen und Kränkungen aufgrund beschriebener Voraussetzungen und in besonderen Situationen mit Hilfe des Trauerprozesses nicht adäquat verarbeitet und bewältigt werden, so muß die/der Ältere sie verstärkt abwehren. Dazu werden zunächst die lebenslang benutzten Abwehrmechanismen und weitere zu Abwehrzwecken benutzte Verhaltensweisen eingesetzt. Konservatives Verhalten z.B. vermeidet neue Erfahrungen und Verführungen durch Rückgriff auf alte Erfahrungen. Weiterhin läßt sich eine Verschärfung, teilweise sogar Karikierung bestimmter Charakterzüge beobachten, die bisher eher hirnorganisch begründet angesehen wurde, wie z.B. der Übergang von sorgsamem Verhalten zu zwanghafter Pedanterie,

von Vorsicht zu Mißtrauen und von Sparsamkeit zu Geiz. Weiterhin können körperliche Veränderungen und Störungen zu Abwehrzwecken genutzt werden, so z.B. das Nicht-Hören, das Nicht-Sehen, das Nicht-Schreiben oder das Nicht-Bewegen.

Reichen diese Möglichkeiten der Abwehr nicht mehr aus, um fortbestehende oder sich verstärkende Bedrohungen und Trennungen zu bewältigen, zeigen sich regressive Phänomene, die dann als adaptive regressive Schritte im Dienste des Ichs zu verstehen sind. Sie werden von dem Individuum offenbar auch so erlebt und akzeptiert. Diese können unterschiedliche Aspekte des Es, des Ichs und des Über-Ichs betreffen. Dazu zählen folgende:

– das Auftreten „primitiverer", kindlicher Verhaltensweisen und Interaktionsformen der verschiedenen psychosexuellen Entwicklungsstufen mit stärkerem Hervortreten diesbezüglicher Phantasien, Gefühle, Träume;

– das Auftreten von Partialtriebbefriedigungen (entsprechend früherer Fixierungen) im genitalen, analen und oralen Bereich in Abhängigkeit von verschärftem oder eingeschränktem Über-Ich-Einfluß;

– der Rückzug von der Außenwelt mit stärkerer (im Sinne einer narzißtischen Regression; Levin, 1963) Besetzung der Körperfunktionen (oft im Zusammenhang mit den Partialtriebbefriedigungen);

– das Auftreten „primitiverer" Abwehrmechanismen wie Identifizierung mit dem Aggressor, Verleugnung und Projektion; sowie

– die Rückkehr zu früheren Gedächtnisinhalten, speziell aus Kindheit und Jugendzeit.

Regression wird hier definiert (Arlow und Brenner, 1976) als das Wiederauftreten seelischer Funktionsweisen, die für die psychische Tätigkeit des Individuums während früherer Lebensstadien charakteristisch waren. Vom deskriptiven Standpunkt aus kann man Regression als Primitivisierung einer Funktion verstehen. Diese Vorstellung läßt sich auf alle Teile des Seelenapparates anwenden, auf die Triebregungen des Es, die Funktionsweisen des Ichs und die Forderungen des Über-Ichs. Diese Definition betont jenen zuvor als „genetisch" ausgewiesenen Aspekt der Regression. Sie hebt die Bedeutung der Reifungs- und Entwicklungsprozesse bei der Bildung von Form und Funktion des seelischen Apparates hervor. Damit sind vier wesentliche charakteristische Merkmale der Regression gleichfalls beschrieben:

– Regression ist eine allgemeine Tendenz des seelischen Geschehens.

– Primitive Formen der seelischen Tätigkeit bestehen fort und können neben den reifen Formen des seelischen Geschehens vorkommen.

– Viele Formen der Regression, vielleicht sogar die meisten, treten vorübergehend auf und sind reversibel.

– In der Regel ist Regression weder ein globales noch ein einheitliches Geschehen. Sie wirkt sich gewöhnlich auf einzelne Aspekte des Trieblebens oder der Ich- oder Über-Ich-Funktion aus, nicht so sehr auf die Gesamtheit einer dieser Funktionen, und jene Funktionen, auf die sie tatsächlich einwirkt, sind in unterschiedlichem Maße davon betroffen.

Praktisch bis heute werden bestimmte als regressiv eingestufte Verhaltensweisen alternder Patienten als typisch, sogar als charakteristisch für das Altern angesehen. Dadurch erfolgt eine Gleichsetzung von regressiven Phänomenen mit (normalen) Alternsprozessen. Diese Annahme spiegelt sich auch in der Gleichsetzung von Altern mit einer zweiten, wenn auch umgekehrt verlaufenden Kindheit wider. Einerseits lassen sich überall (in der Familie, in der Klinik oder im Heim) bei sehr altersveränderten, hilflosen und abhängigen Patienten „kindliche" Verhaltensweisen beobachten. Andererseits sind auch frühere psychoanalytische Autoren von dieser Gleichsetzung von Alternsprozessen und Regression ausgegangen. So formuliert z. B. Deutsch (1925): „Der Anfang des Klimateriums (der Frau) geht Hand in Hand mit einer Regression physiologischer Funktionen . . ., damit beginnt eine Phase der Retrogression, (d.h.) einer Regression zu aufgegebenen infantilen libidinösen Positionen. Das Genitale verliert als Organ an Wert". Und ebenso führt Kaufmann (1940) für den Mann aus, daß die „Abnahme der Potenz, eingeführt wie bei der Frau, mit einem Verlust an Objektbeziehungen und Regressionen einhergeht". Diese Aussagen beruhten offenbar auf der Annahme einer Libidoinvolution, die bis zu Beginn der 60er Jahre vertreten wurde, ohne hinterfragt zu werden. So betonten Berezin (1963) und Zinberg (1963), daß es „aufgrund des Verlustes des Genitalprimats naturgemäß zu einer allmählichen Regression" kommt.

Nach dem 60. Lebensjahr treten eindeutig verstärkt regressive Prozesse, teilweise auch stärkerer Ausprägung, auf. Keinesfalls dürfen sie mit normalen Alternsprozessen gleichgesetzt werden, d.h. Regression stellt keinen Teilaspekt normalen Alterns dar. Diese regressiven Prozesse im Dienste des Ichs sind prinzipiell reversibel und können ebenso von progressiven Prozessen abgelöst werden, worauf auch der von Pollock (1982) als für das Altern charakteristisch angesehene „mourning-liberation process" hinweist.

Treten jetzt Bedrohungen, Trennungen/Verluste und Kränkungen in mehreren Bereichen gleichzeitig oder kurz nacheinander auf, stehen nur defizitäre, eingeschränkte oder geschädigte Ich-Funktionen zur Verfügung und betreffen die Verluste dazu noch die einzig wichtige Beziehungsperson, so tritt eine pathologische Regression auf (Weisman, 1970):

– Die Reaktionsweisen sind unangemessen, wirken undifferenziert im Vergleich mit früheren ausgewählten, sorgfältig ausgeführten Handlungen und Verhaltensweisen.
– Die Möglichkeiten zur Abwehr wirken geschwächt, mehr begrenzt und sind unflexibler, dazu bestehen geringe Möglichkeiten zum Erkennen und zum Umgang mit neuen Erfahrungen und Veränderungen.
– Die Realitätsprüfung ist eingeschränkt oder zeigt schnellere Dekompensation.
– Die Kommunikation zur Umwelt ist weniger symbolisch und mehr konkret. Ansichten werden generalisiert und sind nicht mehr differenziert. Konzepte und Ideen führen zu einer sofortigen sensomotorischen Bewältigung (Agieren). Die verringerte interne Verarbeitungsmöglichkeit ist ausgetauscht durch größere Externalisation von Gefühlen, Affekten und Motiven.
– Die Realitätsbeziehung beruht auf dem Hier-und-Jetzt-Prinzip bei jeglicher sich bietender Gelegenheit und bezieht sich nicht mehr auf die Wirklichkeit einer vergangenen und zukünftigen Zeitdimension.
– Die persönliche Autonomie ist deutlich stärker bezogen auf den gegenwärtigen Einfluß externer Personen, Dinge und Institutionen.

Klinisch manifestieren sich derartige pathologische Regressionen als Deprivationsphänomene, als akute hirnorganische Verwirrtheitszustände und vorübergehende wahnhafte Episoden alternder Patienten.

Manchmal läßt sich bei scheinbar selbständig und in äußerlich befriedigenden Lebenssituationen lebenden Menschen bei derartigen Bedrohungen und Verlusten als Reaktion eine Erstarrung mit völligem Rückzug beobachten, der schließlich über ein „Erlöschen" vielfältiger Funktionen zum relativ raschen Todeseintritt ohne organische Ursache führt, die bereits 1965 von Cath als Depletion beschrieben wurde.

Ungeklärt ist bisher, welches Ausmaß an Bedrohungen und Verlusten ein alternder Mensch ertragen kann. Möglicherweise verfügen alternde Menschen mit derartigen pathologischen Erscheinungen in Wirklichkeit nur über eine Scheinautonomie aufgrund stabiler symbiotischer Beziehungen.

Eine 66jährige Patientin hatte zusammen mit ihrem Mann nach seiner langersehnten Pensionierung eine komfortable Neubauwohnung fernab von dem bisherigen Bekanntenkreis bezogen.

Die seit 45 Jahren bestehende kinderlose Ehe wurde von Außenstehenden als unkompliziert, befriedigend, sehr eng und sich gegenseitig stützend beschrieben. Weitergehende Außenkontakte bestanden bis auf Bekanntschaften mit mehreren Ehepaaren nicht. Die neue, ebenfalls langerwünschte Wohnung sollte zur „Verwöhnung" in der Alterssituation beitragen.

Kurz nach dem Umzug erkrankte der gleichaltrige Mann an einem Herzinfarkt und verstarb innerhalb von 2 Tagen. Kurze Zeit darauf wurde bei der Patientin ein Genitalkarzinom diagnostiziert, welches vollständig operativ entfernt wurde. Die Patientin erholte sich jedoch nur sehr allmählich. Sie wurde apathisch und deutlich depressiv nach Hause entlassen, wobei sie aufgrund fortbestehender Beschwerden weiterhin fest davon überzeugt war, daß das Karzinom operativ nicht völlig entfernt worden sei und die Ärzte ihr nicht die Wahrheit sagten.

Nach Berichten aus ihrer Umgebung lebte sie während der nächsten 2 Monate in ihrer komfortablen, neu bezogenen Wohnung wie in einer „gläsernen Welt voller Spinngewebe", ohne daß es ihrem behandelnden Arzt, der betreuenden Sozialarbeiterin oder den Bekannten gelang, einen intensiven Kontakt aufrechtzuerhalten oder die Patientin zu neuen Interessen oder Aktivitäten zu bewegen. Dann wurde sie aufgrund all-

> gemeiner Apathie, zunehmender Resignation und starkem Gewichtsverlust erneut ins Krankenhaus eingewiesen, wo sich ihre Gedanken ausschließlich mit der als fortbestehend vermuteten Krebserkrankung und dem verstorbenen Mann beschäftigten. Nach weiteren 3 Wochen verstarb sie ohne Hinweis auf ein metastasierendes Karzinom oder eine andersartige organische Erkrankung (Obduktionsbefund).

Deutlich ist hier, wie eine 66jährige Patientin ohne anamnestische Hinweise auf ausgeprägte neurotische oder psychosomatische Vorerkrankungen bei einer sehr engen symbiotischen Partnerbeziehung aufgrund von Verlusten in mehreren Bereichen mit einer pathologischen Regression mit letalem Ausgang reagiert.

Die über die sich nach dem 60. Lebensjahr zeigende Somatisierungstendenz zitierten Untersuchungen geben in der Regel nur Querschnittsbefunde wieder und erlauben keine prozessuale Sicht. Allerdings wird mehrfach darauf hingewiesen, daß sich auch diejenigen Witwen/Witwer, die deutliche Verschlechterungen ihres subjektiven und objektiven Gesundheitszustandes aufwiesen, nach einem Zeitraum von 6 bis 24 Monaten weitgehend wieder erholten. Die beschriebenen Folgen wie fehlende emotionale Reaktionen, diffuse funktionelle/vegetative Symptomatik, mehr „organische" Symptomatik und dekompensierende organische Krankheiten mit teilweise nachfolgender Mortalität deuten offensichtlich auf unterschiedliche Ausprägungen regressiver Prozesse hin.

Warum wird der dargestellte Weg der Somatisierung gerade bei über 60jährigen häufiger genutzt und ist möglicherweise als für die Lebensabschnitte nach dem 60. Lebensjahr charakteristisch anzusehen? Bereits Müller (1967) hatte diskutiert (vgl. Abschn. 60.1), ob psychosomatische Krankheiten im Alter ihre relative Spezifität gerade deshalb verlieren, weil als Substrat vermehrt abnutzungsbedingte körperliche Erkrankungen zur Verfügung stehen. Polypathie und Multimorbidität stellen dafür ein breites Spektrum sowie unspezifische Möglichkeiten zur Verfügung. Nach Haag et al. (1989) haben sich im Alter organische Prozesse stärker verselbständigt, indem „die organischen Veränderungen zu einer biologischen Konstante" werden. Dazu führt die Regression im Dienste des Ichs zu einem Rückzug von der Außenwelt mit stärkerer (im Sinne einer narzißtischen) Besetzung (Levin, 1963) der Körperfunktionen, wodurch häufiger gleichzeitig entsprechende Partialtriebbefriedigungen zur Verfügung stehen. Wie bedrohlich muß sich die innerpsychische Situation gestalten, wenn der (dazu noch narzißtisch hoch besetzte) eigene Körper nicht mehr als „letzter verläßlicher Partner" zur Verfügung steht. Gleichzeitig bedeutet der Rückzug in die Innenwelt eine Rückkehr zu den verinnerlichten „guten" Objektbeziehungen (Kernberg, 1988). Schließlich wünschen viele Ältere – insbesondere die Männer – dem Ich-Ideal eines „ohne Gefühle alles bewältigenden Älteren" zu entsprechen; teilweise identifizieren sie sich auch mit den diesbezüglichen Normen und Ansichten ihrer Altersgruppe und Umwelt. Geklagte körperliche Symptome

stellen so unbewußt die noch einzig verbleibende Möglichkeit dar, auf sich aufmerksam zu machen und über diesen Weg Kontakte, Hilfe und Versorgung zu erhalten. Wahrscheinlich ist die für das Alter als charakteristisch angesehene Somatisierungstendenz auf mehrere Einflüsse, die im Einzelfall sehr unterschiedlich kombiniert sein können, zurückzuführen.

Unklar bleibt, ob die bei Bedrohung, Trennungen/Verlusten und Kränkungen zu beobachtende Somatisierungstendenz bei gleichzeitiger Unfähigkeit zu emotionaler Reaktion ein alterskohortenspezifisches Phänomen darstellt. Die heutigen über 65jährigen könnten aufgrund ihrer Erziehung mit fehlendem Austausch über Gefühle, gelernter Sprachlosigkeit, einem entsprechenden Ideal- und Normbild des Alterns verpflichtet, nicht fähig sein, sich während ihres Alterns anders zu verhalten und z.B. emotionale Trauerreaktionen zuzulassen. Handelt es sich dabei insbesondere um alternde, bereits früh an einer Alexithymie Erkrankte? Bisher über Jahrzehnte in einer einzigen stabilen symbiotischen Partnerschaft lebend, wird diese plötzlich durch Veränderungen bedroht oder durch den Tod des Partners aufgelöst; eine Identifizierung über den Beruf entfällt mit dem Ausscheiden aus dem Arbeitsprozeß; gleichzeitig fehlen die Norm- und Rollenvorstellungen über das Leben als Rentner/Pensionär. Im weiteren Ablauf zeigen gerade Witwen ein hohes Ausmaß an unauffälliger Angepaßtheit mit von der Umgebung erwünschten sozialen Verhaltensweisen (Fooken, 1979).

Damit bestünde eine gewisse Hoffnung, daß die nächsten Generationen von Älteren, insbesondere die „neuen Alten" (d.h. die heute 50–60jährigen) anders mit ihren Gefühlen umgehen, sie zulassen können und sich dadurch als gesünder erweisen. Aber auch sie werden trotz ihrer Aussicht auf eine „sichere Lebenszeit" (Imhof, 1988) aufgrund ihrer hohen Lebenserwartung genauso unverändert wie die heute Älteren Bedrohungen, Trennungen/Verlusten und Kränkungen begegnen.

Wie lassen sich nun Situationen verstehen, in denen eine akute und ausgedehnte zentralnervöse Schädigung (z.B. durch ein apoplektisches Syndrom) zu einem zunächst zentral bedingten Kontrollverlust der Ausscheidungsfunktionen mit weitgängiger Abhängigkeit und Hilflosigkeit bei zwischenzeitlich akuter Lebensbedrohung führt? Hierbei zu beobachtende Phänomene können erst sekundär als Regressionsphänomene angesehen werden. Konfrontiert mit dieser lebensbedrohlichen Schädigung seiner Ich-Funktionen kann das Ich dann mit Hilfe regressiver Phänomene reagieren und sich um eine neue psychische Stabilität bemühen. Auch hier entscheidet die Unterstützung der Selbständigkeit durch die nicht geschädigten Funktionen und Möglichkeiten des Ichs über erneute progressive Schritte und eine allmähliche Besserung.

Polypathie und Multimorbidität stellen dem über 60jährigen Erwachsenen ein großes Reservoir zu nutzender „körperlicher" Symptomatik zur Verfügung. Auf bereits gebahntem Wege kann diese im Sinne einer unspezifischen psychosomatischen Reaktion genutzt werden, um sowohl auf anstehende innerpsychische sowie intra- und intergenerative Konflikte als auch auf noch nicht durch einen Trauerprozeß bewältigte Bedrohungen, Verluste/Trennungen sowie Kränkungen hinzuweisen.

60.5 Hilfestellung und (psychotherapeutische) Behandlungsmöglichkeiten

60.5.1 Erforderliche Gesamtsicht

Unter Berücksichtigung geriatrischer Aspekte (Störmer, 1983) bedarf es bei über 60jährigen mit funktioneller, organischer oder psychischer Symptomatik einer biopsychosozialen Gesamtsicht, um den Einfluß physischer, psychischer und sozialer Faktoren und ihre Wechselwirkungen kennenzulernen und zu würdigen.

Der/die behandelnde Arzt/Ärztin für Allgemeinmedizin/Innere Medizin verfügt in der Regel aufgrund seiner/ihrer langen Behandlungstätigkeit über eine umfassende Kenntnis der Krankengeschichte seiner/ihrer über 60jährigen Patienten. Zusätzliche Informationen erbringen Hausbesuche, Partnerin/Partner und Familienangehörige sowie die weitere Umwelt (z.B. durch Mitarbeiter ambulanter Dienste oder von Alten- und Pflegeheimen).

In einem ersten Schritt wird es notwendig, die bisherige krankheitszentrierte in eine personzentrierte Sicht zu verändern und die eigene Interaktion mit dem Alterspatienten kritisch zu überprüfen. Kann ich mir als Behandler aufgrund meiner Kenntnisse ein umfassendes Bild über die Lebenssituation, die Beziehungen, die sozialen Umstände wie auch über mögliche Probleme/Konflikte meiner Alterspatienten machen oder kenne ich diese kaum oder nur teilweise? Wie ist meine Beziehung: mütterlich, väterlich, kindlich, distanziert, formalisiert oder routiniert, freundlich, mitleidend oder sogar liebevoll, d.h. was bedeutet sie/er für mich als behandelnden Arzt? Bewußt gestellt führen diese beiden Fragen bereits oft zu einer umfassenden Sicht, weisen auf bestehende Wissenslücken hin, verdeutlichen die affektive Beziehung und führen zu stärkerem Interesse gegenüber den Alterspatienten.

Sucht der Behandler so einen anderen Zugang, so kann er zunächst auf Abwehr stoßen: Alterspatienten "wissen" vorbewußt aufgrund langjähriger Erfahrungen, daß sich ihre Ärzte nur durch das "Angebot" von körperlichen Beschwerden ansprechen lassen: Neue und andersartige Beschwerden führen zu einem erneuten längeren Kontakt und weiteren Untersuchungen, bei langfristig schon bekannten Beschwerden wird häufig nur ein Wiederholungsrezept ausgehändigt. Außerdem fällt es über 60jährigen aufgrund ihres Selbstbildes schwer, über Belastungen, Krisen oder Schwierigkeiten zu sprechen. Für Menschen, die sich ihr Leben lang als aktiv, selbständig, unabhängig und

ihre Probleme meisternd erlebten, kann es eine tiefe Kränkung oder Beschämung bedeuten, sich jetzt eingestehen zu müssen, nicht mehr zurechtzukommen und Hilfe zu benötigen. Bisherige Generationen Älterer haben gelernt, „die Zähne zusammenzubeißen und schweigend selbst zurechtzukommen". So kann auch von seiten der Alterspatienten (und ihrer familiären Umwelt) das unbewußte Angebot erfolgen, die bestehende Symptomatik als rein „organisch" bedingt anzusehen und bestehende psychosoziale Einflüsse zu leugnen.

Die biopsychosoziale Gesamtsicht erfordert Kenntnisse aus folgenden Teilbereichen:

Welche (physischen, psychischen und sozialen) Einschränkungen ergeben sich aufgrund der jetzigen Krankheit? Welche Behinderungen bestehen bei körperlicher Leistungsfähigkeit, bei Beweglichkeit, beim Hören und Sehen? Wie hoch sind diese Einschränkungen innerpsychisch besetzt? Die Fragestellung führt von einer symptomorientierten zu einer funktionsorientierten Sicht. Das Ausmaß beeinträchtigter oder geschädigter Ich-Funktionen für die Lebens- und Alltagsbewältigung wird zugänglich und damit auch ihre innerpsychische Bedeutung.

Eine leichte Hemiplegie kann für eine Sportlerin den Verlust ihrer in der Beweglichkeit gesehenen Unabhängigkeit bedeuten und sich dadurch katastrophal auswirken. Ein Mann, der an Bewegung weniger interessiert ist und sich gern umsorgen läßt, wird dieser Hemiplegie nur geringe Bedeutung zumessen. Eine Arthrose im Handbereich bedeutet für einen musisch oder künstlerisch Tätigen den Verlust entscheidender Lebensinhalte; für eine Hausfrau dagegen ergibt sich die Möglichkeit, nun endlich die schon lange innerlich abgelehnte Hausarbeit aufzugeben. Schwerhörigkeit kann sich für kontaktfreudige Menschen als lebensbeeinträchtigend auswirken; Menschen, die Kontakt ablehnen, sich aus der Umwelt zurückziehen möchten oder unter einem hohen Geräuschpegel leiden, wird sie eher willkommen sein.

Welche Aktivitäten, Interessen und Fähigkeiten (Ich-Stärken) und welche konfliktfreien Bereiche bestehen zum Ausgleich? Diese Fragestellung beugt einer durch die Multimorbidität nahegelegten ausschließlich defizitorientierten Sicht vor.

Über welche Beziehungen verfügen die über 60jährigen (Partner, Familie, Verwandte, Freunde und Umgebung)? In welcher sozialen Situation (Status, Einkommen, Wohnung, Versorgung) leben sie zur Zeit? Problematische, konfliktträchtige oder sich reduzierende Beziehungen ebenso wie Armut, schlechte Wohn- und Versorgungsverhältnisse gestatten oft nicht, bestehende Autonomie zu erhalten und vorhandene Ich-Funktionen zu nutzen.

Drohen Veränderungen der Ich-Funktionen, in den Beziehungen oder bei den sozialen Lebensumständen bzw. sind diese bereits eingetreten? Handelt es sich um unvorhergesehene Ereignisse oder bestand eine längere Anpassungszeit? Treten die Bedrohungen/Verluste kumuliert auf? Wie erleben und beurteilen die über 60jährigen selbst ihre augenblickliche Situation? Wie stellen sie sich ihre weitere Entwicklung vor und welches Ziel streben sie mit einer (möglichen) Behandlung an? Wie sind sie früher mit ähnlichen Schwierigkeiten umgegangen? Einerseits lassen sich dadurch die „privaten" Ziele und die aus der Sicht der Patienten notwendige Reihenfolge von Behandlungsschritten kennenlernen und andererseits frühere Bewältigungsstrategien (Coping-Stile), die jetzt genutzt werden können.

In der Regel stehen diese Kenntnisse dem behandelnden Arzt für seine Gesamtsicht nicht zur Verfügung. Seine hierfür notwendigen Gespräche schaffen eine Atmosphäre des Interesses, bauen ein Arbeitsbündnis auf oder stabilisieren es. Damit wird gleichzeitig eine qualitativ andere Beziehung geschaffen, die sich – mehr oder weniger bewußt – auf folgende Annahmen und Erwartungen stützt:

- Die Beziehung besteht zwischen zwei psychosozial und psychosexuell Erwachsenen unterschiedlicher Entwicklungs- und Altersstufen mit unterschiedlichen Lebenserfahrungen.
- Die in der Regel langfristige Beziehung ist die zwischen jüngeren Behandlern und älteren Patienten, wodurch sich bestimmte Gefühlsübertragungskonstellationen (vgl. Abschn. 60.2.2) ergeben können.
- Die Behandlung wird von seiten des/der Älteren bis zum Lebensende (d.h. auch bei schweren Erkrankungen und in der Sterbesituation) gewünscht.
- Dem Behandler wird eine hohe psychosoziale Kompetenz zugewiesen. Gleichzeitig bestehen zur Zeit durch das psychotherapeutisch/psychosomatische und psychiatrische Behandlungssystem und auch in naher Zukunft kaum psychotherapeutische Behandlungsmöglichkeiten; d.h. eine diesbezügliche Hilfe kann nur durch den ärztlichen Behandler selbst erfolgen.

Erst eine Teilidentifizierung erlaubt, sich wirklich in die Krankheits- und Lebenssituation der/des Älteren hineinzufühlen. Wie empfindet man es, viele Tage bettlägerig und ständig auf fremde Hilfe angewiesen zu sein, eingeschränkt (Hören, Sehen) seine Umwelt wahrzunehmen oder dauernd Hilfsmittel (z.B. einen Stock oder einen Rollstuhl) benutzen zu müssen? Wie belastet fühlt man sich durch die große Anzahl verordneter Medikamente oder durch die durchgeführten Untersuchungen? Wie erlebt man ständige Verluste an wichtigen Beziehungspersonen?

60.5.2 Allgemeine Hilfestellung

Aufbauend auf das so geschaffene, anders strukturierte Arbeitsbündnis läßt sich auch in der Praxis der/des niedergelassenen Ärztin/Arztes ein breites Spektrum nachfolgend dargestellter Hilfsmöglichkeiten verwirklichen. Als gemeinsame Zielvorstellung gilt, daß die/der Ältere durch eine möglichst weitgehende, langanhaltende (und immer wieder neu zu schaffende!) Ich-Stabilität und Autonomie als ältere Erwachsene in der gewohnten sozialen Umgebung in die La-

ge versetzt werden, ihre Lebenssituation erneut zu bewältigen.

Stabilisierung von Ich-Funktionen

Unter einer funktionsorientierten Sicht erscheint es als entscheidend, eingeschränkte oder geschädigte körperliche (Leistungsfähigkeit, Hören, Sehen, Beweglichkeit, Blasen- und Darmkontrolle), psychische (Orientierung, Gedächtnisleistungen) und soziale (Wohnsituation, Versorgung, materielle Lebensumstände sowie Kontakte, Interessen etc.) Funktionen zu verbessern bzw. zu stabilisieren. Dies geschieht durch zusätzliche Behandlungen, Verordnungen von Hilfsmitteln, Training entsprechender Fertigkeiten, Nutzung von sozialen und weiteren ambulanten Diensten, Wohnberatung etc. Oft können auch Ich-Stärken (Interessen und Fähigkeiten) zum Ausgleich reaktiviert werden; Information, Beratung, Weiterverweisung und Vermittlung erweisen sich dabei als geeignete Instrumente. Ich-Funktionen können durch die in hohem Umfang verordneten Psychopharmaka, Schlaf- und Schmerzmittel negativ beeinflußt werden.

Vermeidung regressiver Einflüsse

Bereits die Interaktion mit der/dem Älteren kann hierarchisch, infantilisierend, pädagogisierend oder abhängig machend gestaltet werden. Einwände, Überlegungen und Wünsche der/des Älteren werden dabei kaum respektiert, eigene Meinungen nicht gefragt. Fördernde, aktivierende oder rehabilitative Maßnahmen unterbleiben, an ihrer Stelle werden pflegende, bewahrende oder sogar kontrollierende Vorgehensweisen (durch Partner, Familie und/oder professionelle Umwelt) unterstützt. Die/der Ältere wird nicht (mehr) als Gesprächspartner akzeptiert, sondern diagnostische Eingriffe, Behandlungsmaßnahmen werden eher mit den Familienangehörigen oder den professionellen Mitarbeitern „über den Kopf der Älteren hinweg" verabredet und durchgeführt. Gespräche erfolgen nicht in ihrer Gegenwart oder mit ihrem Einverständnis. Die (noch) bestehende Teil-Autonomie wird nicht respektiert.

Gezielte Ansprache

Oft wird es notwendig, ärztlicherseits bei bestehendem Arbeitsbündnis gesehene Probleme, Konflikte und Schwierigkeiten, drohende oder eingetretene Verluste ebenso wie soziale Notstände direkt und offen anzusprechen. Der/dem Älteren wird es dadurch möglich, in gewünschtem Umfang darauf einzugehen. Kummer, Trauer und Verzweiflung können zugelassen werden und gleichzeitig erfährt der Arzt, welche Verarbeitungs- und Bewältigungsmöglichkeiten bestehen. Oftmals werden alternative Lösungsmöglichkeiten gesehen, die dann im Gespräch mit dem Arzt abgeklärt und gegeneinander abgewogen werden können. So wird es der/dem Älteren möglich, Lösungen für anstehende Schwierigkeiten und Konflikte zu finden oder sich stärker auf einen Trauerprozeß einzulassen.

„Symbolische" Gabe

Gezeigtes Interesse, Ansprache, vermehrt zur Verfügung gestellte Gesprächszeit, Teilidentifizierung und Anerkennung der gezeigten Aktivitäten und rehabilitative Hilfe bei oft unerträglichen und schwierigen Lebens- und Krankheitssituationen stellen einen wichtigen Behandlungsbeitrag im Sinne einer „symbolischen" Gabe des Arztes dar, deren Bedeutung oft unterschatzt wird. Gerade bei vereinsamten Älteren kommt dem so gestalteten Kontakt zum behandelnden Arzt hohe Bedeutung zu.

Benötigte Zeit

Obwohl Alterspatienten in der Regel (Gesprächsdauer in der Praxis, Visitendauer etc.) eher zu wenig Zeit erhalten, benötigen sie im Gegenteil vermehrt Zeit. Der Arzt braucht Zeit, um ihre Biographie und ihre Lebenssituation kennenzulernen. Sie selbst benötigen Zeit, um über Probleme, Kümmernisse und Sorgen zu sprechen, sich an verändernde oder geänderte Lebensumstände zu gewöhnen, Konfliktlösungen im Gespräch zu finden oder einen Trauerprozeß zu durchleben.

Erforderliche Kenntnisse und Kooperation

Um dieses Spektrum der Hilfsmöglichkeiten zu realisieren, bedarf es umfassender Kenntnisse über lokale und regionale Hilfs- und Versorgungssysteme und weitreichender Kooperation. Der Arzt benötigt Wissen darüber, welche Möglichkeiten an Information, an Beratung (Altenberatungsstellen und spezifische Beratungsangebote), an ambulanten (Essen auf Rädern, Versorgungsdienste, Pflegeangebote) und stationären (Tagespflegeeinrichtungen, Alten- und Pflegeheime, Kurzzeitunterbringungsmöglichkeiten) Diensten der Altenhilfe sowie an Angeboten der Altenarbeit (Altenclubs, Altenzentren, Altenerholungsmaßnahmen, Volkshochschule und weitere Bildungsangebote etc.) lokal und regional bestehen. Ebenso bedarf es der Kooperation mit unterschiedlichen Berufsgruppen, so Altenpflege- und Krankenpflegekräften, Rehabilitationskräften, Sozialarbeitern und Sachbearbeitern unterschiedlicher Behörden/Ämter sowie ehrenamtlichen Mitarbeitern der unterschiedlichen Wohlfahrtsverbände/Kirchen. Leider verfügen Ärzte bisher nur in geringem Umfang über derartige Kenntnisse oder kooperieren entsprechend.

60.5.3 Spezifische Schwierigkeiten

Das geschaffene Arbeitsbündnis kann durch bestimmte Schwierigkeiten immer wieder in Frage gestellt werden:

Die beschriebene umgekehrte Übertragungskonstellation führt zu häufigen Beziehungsstörungen.

Wenn er es zuläßt, kann der Arzt häufiger unverständliche Gefühle und Handlungen gegenüber bestimmten Patienten beobachten. Diese werden mit Hilfe der eigenen Erinnerungen an die Kindheits- und Entwicklungsgeschichte zugänglicher. Welche Erwartungen, Befürchtungen, Vorstellungen und Wünsche habe ich bei diesem Patienten, welche Konflikte und Schwierigkeiten vermeide ich? An welche Kindheitspersonen erinnert mich dieser Patient? Umgekehrt verhilft eine genauere Kenntnis der Biographie dieses Patienten und seiner familiären Situation dazu, die unbewußten Übertragungen auf den Arzt und die jungeren Mitarbeiter als moglicherweise „letzte" (reale oder phantasierte) Kinder oder Enkelkinder besser zu verstehen.

Diese Übertragungskonstellation ändert sich bei fortschreitender Behandlung; teilweise sogar sehr schnell. Aufgrund der erfolgreichen Hilfestellung und der erlebten Kompetenz entwickelt der Patient trotz des großen Altersunterschiedes die bekannte klassische Übertragung im Sinne einer Kind-Eltern-Beziehung. Nur bei ausgeprägten Schwierigkeiten wird er den Arzt wieder als den „jungen Menschen" ansehen, „der doch nichts von den Alten versteht".

Regressive Patienten erwarten als Ausdruck der klassischen Übertragungskonstellation von ihrem Arzt die Handlungsweise eines mächtigen Elternteiles. Im „Hier und Jetzt" lebend, suchen sie eine tatkräftige Unterstützung, zum Teil auch gegen die Umwelt. In der Regel akzeptiert der Behandler dieses unbewußte Übertragungsangebot gern, da es seiner Vorstellung der Arzt-Patient-Beziehung entspricht. Die unbewußte Stabilisierung regressiver Tendenzen durch den Arzt kann in vielfältiger Form erfolgen: So z.B. durch eine langfristige Versorgung und Betreuung anstatt entsprechender Aktivierung, durch die Verlagerung von alltäglichen Entscheidungen von dem Patienten weg auf seine Umwelt, durch die Festlegung von Behandlungen und Verordnungen zusammen mit den Angehörigen ohne Berücksichtigung der Wünsche des alternden Patienten etc. Die häufig anzutreffende Ansicht vom Altern als „zweiter Kindheit" verstärkt noch diese Tendenz. Wenn alternde Patienten Kinder sind, dann müssen sie eben „versorgt", „betreut", „verwöhnt", aber auch „erzogen" und „beaufsichtigt" werden. Die unbewußte Identifizierung des Patienten mit diesen Ansichten schränkt die noch bestehenden Möglichkeiten einer zumindest teilweisen Verselbständigung völlig ein. Die zunächst so einleuchtende Gleichsetzung von Altern mit (umgekehrt verlaufender) zweiter Kindheit trifft auch deswegen nicht zu, weil bei Kindern die gegebene weitere Entwicklung zum Erwachsenen hilft, alle vorübergehenden Entwicklungsschwierigkeiten zu bewältigen. Dagegen führen dementielle Krankheitsbilder zu einer weiteren Verschlechterung, ausgeprägter Hilflosigkeit und schließlich zum Tode. Eine bewußte Annahme dieses Übertragungangebotes mit täglicher kurzfristiger Hilfestellung (Goldfarb, 1955, 1964, 1969) führt unter dem Schutz dieser Hilfestellung zu einer erneuten Verselbständigung mit der Möglichkeit einer mehr erwachsenengerechten Hilfe durch Gespräche.

Unverändert hat daher die Leitvorstellung zu gelten, daß auch alternde Patienten mit ausgeprägten physischen und psychischen Krankheitssymptomen und Ausfällen Erwachsene im höheren und hohen Lebensalter mit langjährigen Erfahrungen sind.

Jüngere, aber auch ältere Behandler reagieren häufig unsicher, hilflos oder ablehnend, wenn ihre alternden Patienten libidinöse oder aggressive Triebimpulse zeigen oder diesbezügliche Konflikte haben. Intensive Vorwürfe, Wut, Haß- und Neidgefühle gegen Jüngere, Gleichaltrige oder sogar noch Ältere, Rivalitätsempfindungen ebenso wie Wünsche nach Zärtlichkeit, Probleme mit Potenz und Selbstbefriedigung oder Wünsche nach neuen befriedigenden sexuellen Beziehungen entsprechen nicht dem Bild des „abgeklärten", „weisen" oder sich „jenseits von Gut und Böse" gewünschten alternden Patienten. Jüngere können und dürfen sich nicht vorstellen, daß ihre Eltern und Ältere ebenfalls gleichartige Konflikte haben oder sie selbst eines Tages mit derartigen Problemen konfrontiert sein könnten. Gleichaltrige vermeiden die Frage, ob sie selbst derartige Gefühle und Konflikte haben und bewältigen müssen (Radebold et al., 1973, 1981). Mögliche Reaktionen der Behandler sind in Abschnitt 60.2.2 beschrieben.

Häufig stützt sich die Behandlung der Patienten ausschließlich oder zumindest weitgehend auf Medikamente. Die bestehende Multimorbidität fordert oftmals eine medikamentöse Polypragmasie heraus, um den vielfältigen Krankheitssymptomen scheinbar gerecht zu werden. Oft ist der Nutzen anderer Behandlungsverfahren, wie krankengymnastische, ergotherapeutische, logopädische, diätetische, sozio- und psychotherapeutische Maßnahmen, sowie der aktivierenden Pflege entweder nicht bekannt oder wird nicht gesehen.

Verordnung, Änderung oder Absetzen eines Medikamentes bekommen gerade für den alternden Patienten und seinen Arzt eine wichtige zusätzliche innerpsychische Bedeutung. Das Rezept (Balint und Norell, 1975) erlaubt dem Behandler eine distanzierte indirekte Beziehung zu seinem Patienten und dient ihm gleichzeitig als Beweis seiner Bemühungen.

Die so häufig verordneten Hypnotika, Sedativa, Tranquilizer und Neuroleptika schützen einerseits den Patienten vor seinen Ängsten, Kümmernissen und seiner Traurigkeit und verhindern andererseits, daß diese an den Arzt herangetragen werden. Für beide Seiten bleibt damit das Bild eines ruhigen, abgeklärten, mit sich selbst zurechtkommenden Älteren erhalten.

Weiterhin erlebt der Patient die regelmäßige Verordnung als ständige Zuwendung und gleichzeitig als Beweis, daß der Arzt ihn noch nicht aufgegeben hat (unbewußt setzt er die Beendigung einer medikamentösen Behandlung ohne Ersatz mit der Aufgabe durch den behandelnden Arzt gleich, wobei die Beendigung Unheilbarkeit und Sterben bedeutet). Die Einnahme langjährig erprobter und vertrauter Medikamente – selbst bei ausgeprägter Placebowirkung – stützt offenbar bei sich regressiv verhaltenden Patienten als äußere strukturierende Maßnahme die innere Stabilität.

Gut bekannt ist, daß Patienten in großem Umfang nicht-rezeptpflichtige Medikamente und Hausmittel einnehmen: Einerseits glauben sie offensichtlich im Sinne einer magischen Vorstellung eine Stabilisierung, wenn nicht gar eine Verjüngung zu erreichen (dem entspricht auch die dazugehörige Reklame), und andererseits betonen sie damit ihre Selbständigkeit und Unabhängigkeit von der Umwelt. Diese vielfältige Bedeutung macht den oft so vehementen und intensiven Widerstand der Patienten gegen Veränderung oder sogar Absetzung eines Medikamentes verständlich. Erst eine durch andere Maßnahmen erreichte Sicherheit erlaubt es dem Patienten selbst, das Medikament zu reduzieren oder sogar abzusetzen.

Schwierigkeiten treten auch dadurch auf, daß über das anzustrebende Behandlungsziel unterschiedliche Vorstellungen bestehen. Die Familie wünscht sich einen ruhigen, anspruchslosen und sich fügenden Familienangehörigen; die Krankenschwester geht von dem Idealbild eines pflegebedürftigen, zu versorgenden und Dankbarkeit zeigenden Patienten aus; der Arzt oder andere Mitarbeiter wünschen sich einen aktiven, wieder selbständig werdenden (und damit in Konflikt mit seiner Umwelt kommenden) Patienten. Der Patient selbst spürt diese unterschiedlichen Anforderungen und wählt die ihm zusagende Möglichkeit.

In der Regel hat der Patient im Alter schon zahlreiche Verluste an wichtigen Bezugspersonen erlebt und kennt seine zunehmende Vereinsamung. So kommt jeder Unterbrechung einer Behandlung (z. B. durch den Urlaub oder Kongreßbesuch des behandelnden Arztes oder durch eine Krankenhauseinweisung) entscheidende Bedeutung zu. Gerade regressive Patienten setzen unbewußt die ihnen immer wieder zugemutete Trennung von einer wichtigen – möglicherweise „letzten" – Bezugsperson (Arzt und Pflegepersonal stellen oft die einzigen Kontakte dar) unbewußt mit dem Tod gleich. Ähnlich wie bei Kindern zeigen sie dann Deprivationserscheinungen, reagieren mit einer Verschlechterung ihrer Symptomatik, klammern sich an oder wenden sich resigniert ab. Daher muß eine Unterbrechung der Behandlung oder sogar eine manchmal erforderliche Trennung langfristig angekündigt und immer wieder angesprochen werden. Die Aufrechterhaltung der Kontakte durch entsprechende Besuche (auch z. B. des Arztes im Krankenhaus) oder die Fortführung der Behandlung zu einem bereits verabredeten festen Termin haben sich als hilfreich erwiesen.

60.5.4 Psychotherapeutische Behandlungsverfahren

Die im letzten Jahrzehnt vorgelegten deutschsprachigen Publikationen bzw. Handbucharticle (Petzold und Bubolz, 1979; Schlesinger-Kipp und Radebold, 1982; Radebold und Schlesinger-Kipp, 1983; Radebold, 1989 a, e) erlauben folgende Aussagen:

Spezifische Behandlungserfahrungen von psychosomatisch Alterskranken liegen bisher kaum vor (Radebold und Rassek, 1985). So berichteten Schwöbel (1965) über eine 60jährige Migränepatientin, Paley und Luparello (1973) über mehrere über 60jährige Asthmapatienten, Garfinkel (1980) über eine 72jährige Patientin mit Bronchialasthma und Wilke (1985) über eine 60jährige Patientin mit Colitis ulcerosa.

Die psychoanalytisch orientierte Einzelbehandlung hat sich unter Berücksichtigung üblicher Indikationskriterien bis mindestens zum 80. Lebensjahr bei neurotischen und reaktiven Erkrankungen als erfolgreich zur Symptomreduzierung, teilweise auch für eine Verhaltens- und Persönlichkeitsveränderung erwiesen: Als Krisenintervention (3–5 Behandlungen) dient sie der akuten Hilfestellung; bei pathologischen Trauerreaktionen haben sich 15–20 Einzelbehandlungen (im Verlauf mehrerer Monate) bewährt; erstmals auftretende reaktive und wiederauftretende neurotische Erkrankungen bedürfen einer längeren Behandlungszeit (2 Wochenstunden über mindestens 1–2 Jahre). Psychoanalysen (mehrjährige vierstündige Behandlung) wurden bisher kaum durchgeführt.

Über gesprächstherapeutische Einzelbehandlungen zur Hilfestellung bei Krisen und Schwierigkeiten im Rahmen der Rehabilitation körperlich, aber auch psychisch Erkrankter wurde bisher nur vereinzelt berichtet, ebenso wie z.B. über die Anwendung des autogenen Trainings, der Hypnose oder der Gestalttherapie.

Die Gruppe – hier als Gruppenpsychotherapie – bietet offensichtlich gerade für psychisch Alterskranke mit ihrer häufigen Vereinsamung und ihren Kontaktstörungen die Möglichkeit zu neuen Kontakten, zum Austausch von Erfahrungen, zur gegenseitigen Hilfestellung und Verständnisfindung. Sie bietet bei ausgeprägten regressiven Erscheinungen Schutz gegen ansteigende Angst oder Isolierung und läßt ein besseres Ertragen von Übertragung und Gegenübertragung im Sinne der Aufspaltung der aggressiven

oder libidinösen Impulse auf einzelne Gruppenmitglieder und den Therapeuten zu. Zudem fällt eine Identifizierung mit der Gruppe häufig leichter als mit dem einzelnen.

Das Spektrum psychoanalytisch orientierter Gruppenpsychotherapie zur Behandlung von neurotischen und reaktiven Alterskranken reicht von zweijährig laufenden Langzeitgruppen (einmal wöchentlich mit mindestens 90minütiger Dauer) über Tagesklinikgruppen (mehrmals wöchentlich bis zu 2–3 Monaten) bis zu Gruppen in Beratungseinrichtungen, in sozialpsychiatrischen Diensten sowie in Heimen.

Gesprächspsychotherapie in der Gruppe erfolgt im ambulanten Bereich, in Tageskliniken und Rehabilitationseinrichtungen zur Auseinandersetzung mit und Bewältigung von Krisen und Lebensschwierigkeiten. Lerntheoretische Programme, z.B. Orientierungs-, Selbst- oder Aktivierungstraining, werden in der Regel ebenfalls in Gruppen (von mehrwöchiger bis mehrmonatiger Dauer), in der Klinik oder im Heimbereich veranstaltet. Ebenso wurde über Gruppen, mit denen übende Verfahren oder eine Gestalttherapie durchgeführt wurden, berichtet.

Paar- und familientherapeutische Erfahrungen liegen bisher nur in geringem Umfang für den Altersbereich vor. Ältere werden in die Familientherapie meist zugunsten der jüngeren Kernfamilie einbezogen. Kaum ist die Familientherapie bisher auf die Hilfestellung für den Älteren unter Einbeziehung seiner Familie eingegangen.

Manchmal wird eine Veränderung des therapeutischen Ansatzes erforderlich, der jetzt zweiseitig, einerseits auf die Bearbeitung von aktuellen innerpsychischen Konflikten und andererseits auf Beratung, Behandlung und Hilfestellung im körperlichen und sozialen Bereich (Radebold et al., 1971, 1983), ausgerichtet sein muß.

61 Arbeit und Krankheit. Ein psychosomatisches Problem

Peter Novak

Im Zusammenhang mit bedeutenden Wandlungsprozessen der Arbeitsorganisation und der Produktionsmittel haben sich die Anforderungen an den arbeitenden Menschen und seine Reaktionen auf diese Anforderungen gewandelt (Mergner et al., 1975; Fürstenberg, 1977). Dabei haben psychosomatische Reaktionen, Störungen und Erkrankungen zunehmende Aufmerksamkeit und Bedeutung erfahren. Teilweise in Verbindung mit der Arbeitsmedizin sind in den letzten zwei Jahrzehnten erhebliche Fortschritte der sozialpsychologischen, sozialepidemiologischen und medizinsoziologischen Forschung über die Arbeitsbedingtheit psychosomatischer Krankheiten erzielt worden (Freese et al., 1978; Lühring und Seibel, 1981). Diese beziehen sich sowohl auf den theoretischen Problemzugang wie auch auf die empirische Analyse und die praktischen Konsequenzen. Die Differenziertheit und Heterogenität der vorliegenden Untersuchungen und ihrer Ergebnisse können in einem relativ knappen Beitrag allerdings nicht angemessen berücksichtigt werden. Dem Zweck, einen gewissen Überblick, zumindest jedoch einen Einblick in dieses Forschungs- und Praxisfeld zu vermitteln, soll daher eine exemplarische Darstellung dienen. Diese umfaßt zunächst einen historischen Rückblick, der verdeutlichen soll, daß die Problematik arbeitsbedingter psychosomatischer Störungen und Erkrankungen seit den Frühphasen der industriellen Produktionsprozesse bewußt ist. Es folgt die Darstellung und Analyse einer „frühen" theoretischen und empirischen Untersuchung. Der Hauptteil des Beitrags umfaßt theoretische Problemzugänge und empirische Analysen der sozialwissenschaftlichen Arbeitslosigkeits- und Schichtarbeitsforschung. Am Schluß steht die Repräsentation von Erkenntnissen über arbeitsbedingte psychosomatische Störungen und Erkrankungen in der ärztlichen Ausbildung mit einem Hinweis auf praktische Konsequenzen sowie eine Einschätzung der Beziehung der dargestellten Analysen des Verhältnisses von Arbeit und Krankheit zum Situationskreismodell.

61.1 Historische Bezüge heutiger Problematik

Die Arbeitsmedizin findet historische Zeugnisse ihrer heutigen Anliegen bereits in antiken medizinischen Schriften, zum Beispiel im berühmten Papyrus Ebers und in den hippokratischen Schriften. Sie spannt diesen Bogen historischer Selbstbesinnung weiter zu Paracelsus' 1534 entstandenem Buch über die Krankheiten der Bergarbeiter, zu Ramazzinis 1700 veröffentlichtem Werk „De morbis artificium diatriba" und so fort bis hin zur Einführung der Bezeichnung „Arbeitsmedizin" und zur gleichzeitigen öffentlichen Erklärung der Arbeitsmedizin zu einem selbständigen medizinischen Fachgebiet im Jahre 1929 (vgl. Koelsch, 1963).

Diese historisch bezeugten Anliegen der Arbeitsmedizin betreffen die durch Arbeit bedingten physikalischen und chemischen Belastungen des menschlichen Organismus und daraus resultierende Schädigungen und Krankheiten. Dabei tritt die psychische und soziale Organisationsebene menschlicher Existenz weder als unmittelbar selbst durch Arbeitsbedingungen betroffen in Erscheinung noch als Vermittlungsinstanz zwischen äußeren Einflüssen und Reaktionen des physischen Organismus. Das heißt, die historische Selbstbesinnung der Arbeitsmedizin auf ihre Anliegen läßt einen psychosomatischen Aspekt vermissen. Dies jedoch wird häufig oder gelegentlich zitierten historischen Zeugnissen mindestens seit dem 18. Jahrhundert nicht gerecht. Zwar sind die psychosomatischen Aspekte arbeitsbedingter Störungen und Krankheiten historisch nicht so lange und nicht so extensiv belegt wie beispielsweise bei den Krebserkrankungen, aber entsprechende Hinweise existieren immerhin seit den letzten Phasen der manufakturiellen Produktion, und sie nehmen laufend zu seit den Frühphasen der industriellen Produktion und seit der Entwicklung der mit ihnen im Bedingungsverhältnis stehenden besonderen gesellschaftlichen Ungleichheitsbeziehungen unter den Menschen. Exemplarisch seien einige dieser historischen Hinweise auf psychosomatische Aspekte arbeitsbedingter Störungen und Erkrankungen dargestellt; nicht, um perspektivische Verengungen der heutigen Spezialdisziplin „Arbeitsmedizin" zu demonstrieren, sondern vielmehr, um auf die Einsicht in die Notwendigkeit interdisziplinärer Zusammenarbeit in arbeitsmedizinischen Problemfeldern vorzubereiten.

61.1.1 Ramazzini: Fabrikarbeit, kompensatorisches Verhalten und Krankheit

In seinem vielzitierten Werk von 1700 beschreibt der Paduaner Arzt Bernardino Ramazzini durchaus nicht nur Staub, Dämpfe, Feuchtigkeit, schlechte Lüftung, unzweckmäßige Kleidung und ähnliches als Ursachen von Erkrankungen der Atemwege, von Hautinfektionen etc. vor allem bei Webereiarbeitern in Fabriken. Er weist auch darauf hin, daß Arbeitszeit und Geldlohn zu pathogenem Reproduktionsverhalten nötigen (Deppe und Regus, 1975). Bier und Branntwein werden einerseits nach Ramazzinis Beobachtungen häufig als nicht zubereitungsbedürftiger Nahrungsmittelersatz konsumiert, andererseits beschreibt er den Alkoholkonsum als Kompensationsmechanismus einer sonst unerträglichen sozialen Lage, die wesentlich von den Normsetzungen der Fabrikarbeit bestimmt wird. Besonders wichtig erscheint, daß Ramazzini die Fabrikarbeit als Sozialisationsinstanz für die Prägung pathogener Verhaltensformen betrachtete mit genereller Bedeutung für die gesamte Biographie. „Den meisten Handwerkern, die als Gesellen in ihrer Jugend in Fabriken gearbeitet haben, hängen die Gewohnheiten der Fabriken lebenslang an" (Deppe und Regus, 1975), „... weil die Fabriken auch selbst auf die Sitten und das moralische Verhalten derer, die sich in denselben aufhalten, meist einen sehr großen und für die ganze Lebenszeit junger Leute wichtigen Einfluß haben" (Deppe und Regus, 1975). In Ramazzinis These von der gesundheitsrelevanten „allgemeinen Unordnung in der allgemeinen Lebensordnung" (Deppe und Regus, 1975) können wir bereits die Vorform von Hallidays These der gesellschaftlichen Desintegration als Bedingung psychosomatischer Krankheit erkennen. Von diesem wichtigen Ansatz zur Theorie der Psychosomatik arbeitsbedingter Erkrankungen wird später die Rede sein.

61.1.2 Engels: Arbeitsbelastungen, soziale Lage und Krankheit

Eine besonders wichtige historische Quelle für die Analyse der Beziehungen zwischen Arbeitsbedingungen, Organisation des Alltagslebens, psychosozialen Störungen und Krankheit stellt die Untersuchung von Friedrich Engels über „Die Lage der arbeitenden Klasse in England" aus dem Jahre 1845 (1972) dar. Sie stützt sich auf englische Medizinal- und Verwaltungsberichte, die auch dem Parlament vorlagen, sowie auf eigene, fast zwei Jahre lang durchgeführte Beobachtungen. Eine infolge Arbeitsorganisation und Arbeitsbelastungen vollständige Desorganisation des Alltagslebens steht hier im Zusammenhang mit typischen psychosozialen Kompensationsversuchen, welche die Probleme nicht lösen können, dagegen aber Prozesse sozialer und gesundheitlicher Verelendung fördern. Hier bereits finden sich in einfacher, aber höchst eindrücklicher Darstellung die pathogenetischen Bedingungen und ihre Folgen, die auch heute Thema arbeitswissenschaftlicher, arbeitsmedizinischer und besonders sozialwissenschaftlicher Analyse sind.

Es wird berichtet von physischer und psychischer Erschöpfung durch Arbeitsbelastungen, ohne daß zeitliche und materielle Möglichkeiten im häuslichen und im familiären Bereich zur Entlastung bestehen. Verbunden mit der beständigen Angst allein um die

materielle Sicherung der Zukunft, wird besonders in Form des Alkoholkonsums ein psychosoziales Kompensationsverhalten entwickelt, das die Probleme hochbelastender, unbefriedigender und monotoner Arbeit nicht löst, sondern den ohnedies rapiden Gesundheitsverschleiß noch verstärkt (S. 165, 170).

Zum allgemein schlechten Gesundheitszustand der Industriearbeiter gehört besonders auch das Leiden an psychosomatischen und psychischen Störungen.

Es läßt sich geradezu ein „Demoralisierungssyndrom" der Arbeiter beschreiben, das vor allem auf fortschreitende Arbeitsteilung und Mechanisierung zurückgeführt wird. Jahrelange monotone repetitive Teilarbeit mit maximaler Beanspruchung von Aufmerksamkeitsleistungen ohne Möglichkeit zu selbständiger und kreativer Arbeitsgestaltung führe zu emotionaler Verödung und zur Verkümmerung körperlicher und geistiger Kräfte mit schwerwiegenden gesundheitlichen Folgen (S. 187, 244).

Der Rückgang schwerer körperlicher Arbeit durch beständige Erweiterung mechanisierter Arbeitsgänge bedinge ohne Sicherung der Arbeitsplätze Personalumschichtungen mit erheblichen Konsequenzen für traditionelle Geschlechtsrollenverteilungen im Alltagsleben, woraus sich schwerwiegende Störungen für die familiäre Beziehungsstruktur ergeben.

„In vielen Fällen wird die Familie durch das Arbeiten der Frau nicht ganz aufgelöst, sondern auf den Kopf gestellt. Die Frau ernährt die Familie, der Mann sitzt zu Hause, verwahrt die Kinder, kehrt die Stuben und kocht ... Man kann sich denken, welche gerechte Entrüstung diese tatsächliche Kastration bei den Arbeitern hervorruft und welche Umkehrung aller Verhältnisse der Familie, während doch die übrigen gesellschaftlichen Verhältnisse dieselben bleiben, dadurch entsteht" (S. 212 f).

61.1.3 Koelsch: Arbeitsbedingungen, physiologische Funktionsabläufe und psychische Verfassung

Zur „Hochzeit" der Sozialhygiene im Deutschen Reich erschien im Jahre 1913 das von M. Mosse und G. Tugendreich herausgegebene Sammelwerk „Krankheit und soziale Lage" – dessen Anliegen und Titel zum Vorbild für die unter dem gleichen Titel im Jahre 1976 von Heinz Harald Abholz herausgegebene Monographie wurde. Von besonderer Bedeutung für die Thematik unseres Artikels ist das Kapitel „Arbeit beziehungsweise Beruf in ihrem Einfluß auf Krankheit und Sterblichkeit" von Franz Koelsch, Kgl. Landesgewerbearzt in München, der 50 Jahre später das „Lehrbuch der Arbeitsmedizin" (1. Aufl. 1937 als „Lehrbuch der Arbeitshygiene") veröffentlichen sollte.

Aus dem Zusammenhang einer Wirtschaftspolitik, die den Konkurrenzdruck zwischen den Unternehmen verschärft, und einer fortschreitenden Mechanisierung der Produktion ergeben sich für Koelsch Arbeitsbedingungen, die zur Minderung psychischer und somatischer Resistenz führen, zu somatischen und psychischen Störungen, zu letztlich irreversibler Erschöpfung, schließlich zu vorzeitigem Gesundheitsverschleiß. Diese Bedingungen sind vor allem:

zunehmende Einschränkung der selbständigen Gestaltung der Arbeitsprozesse (Autonomie), Zeitdruck, lange Arbeitszeit, Monotonie, einseitige Belastungen, Mehrfachbelastungen, Nachtarbeit (S. 158).

Die arbeitsbedingte gesundheitliche Belastung erklärt Koelsch aus der Wechselwirkung physiologischer Funktionsabläufe, psychischer Verfassung und sozialer Lage (S. 158f). Besonderes Interesse wendet Koelsch Befunden zu, nach denen „psychoneurotische" und „organneurotische" Störungen häufiger bei körperlich als bei geistig Arbeitenden festgestellt wurden. Hinsichtlich der ursächlichen Bedingungen weist er wiederum auf psychische und materielle („toxische") Arbeitsbedingungen hin sowie auf die Verunsicherung und Bedrohtheit der sozialen Lage. Explizit geht er auf die Situation der Frau im Arbeitsleben ein.

„Bemerkenswert erscheint hierbei, daß bei den niederen Ständen, wo auch die Frau mitarbeiten und mitringen muß im Kampf ums Dasein, die Neurasthenie beim **weiblichen Geschlecht** rapid zunimmt ... Diese nervösen Störungen bleiben nicht auf die Psyche allein beschränkt, können vielmehr auch Organneurosen (Herzneurosen u. a.) auslösen (Petrén, Laehr, Tanzi u. a.)" (S. 161).

Schließlich stellt auch Koelsch schon die gesundheitlichen Folgen der Nachtarbeit dar, vor allem die damals bereits bekannte Tatsache, daß der notwendige Schlaf am Tage mit der Organisation des „normalen" Alltagslebens kollidiert, daß somit der Tagschlaf, der ohnehin den Nachtschlaf auf Dauer nicht ersetzen kann, zusätzlich gestört wird (S. 159).

61.1.4 Makrosoziale und mikrosoziale Zusammenhänge

Bemerkenswert an diesen Beispielen der Darstellung und Reflexion arbeitsbedingter Gesundheitsstörungen und Erkrankungen ist vor allem:
- Beeinträchtigende Arbeitsbedingungen sind in makrosoziale Zusammenhänge eingebunden. Ein solcher Ansatz ist in der heutigen Arbeitsmedizin längst nicht in dem Maße bis gar nicht entwickelt. Selbst die heutige empirische sozialwissenschaftliche Analyse verhält sich demgegenüber zum Teil noch zögerlich.
- Sowohl hinsichtlich der Beeinträchtigung durch Arbeitsbedingungen wie auch hinsichtlich der Bewältigungsmöglichkeiten von Arbeitsbelastungen werden die soziale, die psychische und die somatische Organisationsebene des Menschen in weitgehend gleichem Maße und miteinander integriert berücksichtigt; dies ist der heutigen Arbeitsmedizin noch nicht gelungen und auch die sozialwissenschaftliche empirische Analyse steht hier vor großen methodischen Schwierigkeiten.
- Es lassen sich bereits Ansätze erkennen, die jetzt besonders in den methodisch und theoretisch relativ weit entwickelten mikrosozialen Analysen der sozialwissenschaftlichen Streß- und Life-event-Forschung relevant geworden sind.

Eine frühe, sehr bedeutende Verbindung mikro- und makrosozialer Analyse arbeitsbedingter psychosomatischer Störungen und Erkrankungen hat James L. Halliday (1943) vorgelegt, die hier in der Absicht eines exemplarischen Hinweises, zugleich eines Hinweises auf eine auffällige Konvergenz zum Situationskreismodell von v. Uexküll und Wesiack (vgl. Kap. 1), skizziert sei.

61.2 Der Widerspruch von Bedürfnis und Anforderung und die Sanktionierung psychosomatischer Krankheit

Die Ansätze, die James L. Halliday bereits in den 40er Jahren zur sozialwissenschaftlichen Theoriebildung für die Erklärung der Arbeitsbedingtheit psychosomatischer Erkrankungen entwickelt hat (Halliday, 1943, 1949), sind von beispielgebender Bedeutung für heutige sozialwissenschaftliche Bemühungen um diesen Forschungsgegenstand in Verbindung mit streßtheoretischen Ansätzen. Eine kritische Analyse seines Beitrags zur psychosomatischen Theoriebildung lieferte Karola Brede (1972, S. 90ff).

Für das Thema dieses Artikels ist von besonderer Bedeutung Hallidays Untersuchung über psychosomatische Erkrankungen als Arbeitsunfähigkeitsursache bei Arbeitern im Untertagekohlebergbau. Die Ergebnisse seiner klinischen Fallanalysen und vergleichenden epidemiologischen Untersuchungen an dieser Population setzte Halliday in Beziehung zu Morbiditätsstatistiken über Industrie- und Landwirtschaftsarbeiter. Die Daten stammen aus Südwestschottland zwischen 1918 und 1939. Vor allem zwei epidemiologische Befunde waren Anlaß für seine Bemühungen um theoretische Erklärungsansätze: einmal die Zunahme der bei den Untertagearbeitern ohnehin überproportional vorkommenden psychosomatischen Erkrankungen und Störungen zwischen den beiden Weltkriegen, und zum anderen die Überrepräsentanz der jungen Bergbauarbeiter.

Anschließend an Beobachtungen und Hypothesen von Dickson (1936) stellte sich Halliday die Frage, ob und welche Beziehungen zwischen Morbiditätswandel einerseits und Wandlungen spezifischer Arbeitsbedingungen und spezifischen soziokulturellen Bedingungen andererseits bestehen könnten. Wichtig für diese Fragestellung waren folgende empirische Tatbestände: In der Periode zwischen den Weltkriegen wurde der Kohleabbau voll mechanisiert durch Einführung von Maschinen, welche die notwendigerweise individuelle Arbeitsgestaltung des Hauers ersetzten und diesen nötigten, sich dem alle gleichmachenden Rhythmus der Maschinen zu unterwerfen. Gleichzeitig brach die traditionsverwurzelte alte Subkultur der „miners" zusammen („changes in the nature of the miners as a group"; Halliday, 1943). Infolge Kriegseinsatzes mußten viele Bergbauarbeiter ihre traditionellen Siedlungen verlassen und kehrten nicht mehr dorthin zurück. Zugleich fanden jüngere Männer Zugang zum Bergbau, die zwar keinen traditionellen Bezug dazu hatten, aber eine Untertagearbeit dem Kriegsdienst vorzogen.

Die epidemiologischen und die sozialstrukturellen Befunde brachte Halliday in einen theoretischen Zusammenhang: In Gestalt der beruflichen Anforderungen durch Bergbauarbeit unter Tage tritt dem menschlichen Individuum gesellschaftliche Gewalt[1] in speziell strukturierten Formen entgegen. In gemeinsamer emotionaler, kognitiver und somatischer Auseinandersetzung mit diesen Formen gesellschaftlicher Gewalt gelingt es jedoch, spezifische soziokulturelle Verhaltens- und Wertmuster zu entwickeln, welche die belastenden, bedrohenden und entfremdenden Momente der besonderen Arbeitsanforderungen transformieren und so mit somatischen, emotionalen und kognitiven Bedürfnissen kompatibel machen. Die Entwicklung spezifischer Wert- und Verhaltensmuster gestattet es nicht nur, mit ungewöhnlichen Arbeitsbelastungen fertig zu werden, sondern die besonderen Formen von Arbeit werden auf diese Weise zum Kern soziokultureller Gruppenidentifikation, zum Kern der Bildung einer bestimmten Subkultur – um an die Terminologie von Halliday anzuknüpfen. Was ursprünglich Bedürfnis des Menschen war und ebenso, was ursprünglich gesellschaftliche Gewalt in Gestalt besonderer Arbeitsanforderungen war, ist in einem dialektischen Sozialisationsprozeß aufgehoben. Die Arbeitsanforderungen sind durch diesen Sozialisationsprozeß zum konstitutiven Bestandteil einer soziokulturell geformten Bedürfnisstruktur geworden. Eine ursprüngliche oder primäre Bedürfnisstruktur bzw. Verhaltensorganisation (Brede, 1972) bleibt ebenso unentdeckbar und ist daher nicht zu rekonstruieren wie auch die ursprüngliche Gestalt gesellschaftlicher Gewalt, solange die soziokulturell geformte Bedürfnisstruktur inneres Bedürfnis und äußere Anforderung im Gleichgewicht hält. Diese komplexe Bedürfnisstruktur vereint somatische, emotionale, kognitive und soziokulturelle Momente, darunter internalisierte Momente gesellschaftlicher Gewalt. Ihr korrespondieren spezifische Wert- und Verhaltensmuster. Bedürfnisstrukturen, Wert- und Verhaltensmuster bilden eine besondere Tradition, hier die Tradition der „coal miners", welche den reproduktiven und produktiven Lebensbereich bestimmt, d.h. Wohnen, Konsumverhalten, soziale Beziehungen, Belastungstoleranz und Verhaltensstile im Bereich Arbeit und so weiter. Diese Tradition wird vor allem in Prozessen der primären und der beruflichen Sozialisation überliefert.

Der in einer bestehenden – oder lebendigen – Tradition zwar stets vorhandene, aber als aufgelöster vorhandene Widerspruch zwischen innerem Bedürfnis und äußerem gesellschaftlichen Anspruch oder gesellschaftlicher Gewalt bricht zumindest tendenziell unter zwei Bedingungen auf:

1. Wenn der äußere gesellschaftliche Anspruch in einer Gestalt auftritt, die nicht mehr in die traditionsverwurzelte Bedürfnisstruktur zu integrieren ist.

2. Wenn keine soziokulturelle Tradition besteht, die über Sozialisationsprozesse inneres Bedürfnis und äußeren gesellschaftlichen Anspruch in einer Bedürfnisstruktur aufhebt.

Die epidemiologischen und klinischen Befunde über psychosomatische Erkrankungen bei Untertagearbeitern des Kohlebergbaus in einer bestimmten historischen Periode interpretiert Halliday auf der Folie seines theoretischen Ansatzes über die dialektische Beziehung zwischen innerem Bedürfnis und äußerem gesellschaftlichem Anspruch. Psychosomatische Störungen und Erkrankungen erscheinen dabei zunächst als Indikator für das Aufbrechen des Widerspruchs in dieser Beziehung, wobei in der sozialen Realität offenbar beide Bedingungen des Widerspruchs zwischen innerem Bedürfnis und äußerem Anspruch erfüllt sind.

1. Durch die vielfältigen Belastungen der Untertagebergbauarbeit tritt gesellschaftliche Gewalt den Menschen in Formen gegenüber, die nur vergleichsweise mangelhaft über gewachsene besondere Traditionen in Bedürfnisstrukturen zu integrieren sind. Zwei Tatbestände können im theoretischen Zusammenhang dieser These erklärt werden, wie sie umgekehrt den theoretischen Ansatz empirisch stützen. Einerseits treten bei Untertagearbeitern im Kohlebergbau häufiger psychosomatische Störungen und Erkrankungen auf als bei Landarbeitern und anderen Industriearbeitern, obwohl die „miners" über eine gewachsene, mit ihrer speziellen Arbeit verbundene Tradition verfügen. Andererseits nimmt die psychosomatische Morbidität der Untertagearbeiter sprunghaft zu, parallel zur Mechanisierung des Untertageabbaus von Kohle. Dieser Tatbestand weist darauf hin, daß zwar die bisherigen äußeren gesellschaftlichen Anforderungen an die Untertagebergbauarbeiter überhaupt durch Traditionsverwurzelung in gewachsene Bedürfnisstrukturen zu integrieren waren. Zugleich weist er aber auch darauf hin, daß die Tradition der „miners" die mit der Mechanisierung gegebenen gesellschaftlichen Anforderungen höchstens noch partiell in Bedürfnisstrukturen zu integrieren vermag.

2. Einen Erklärungswert erhält auch die zweite Bedingung für das Aufbrechen des Widerspruchs zwischen inneren Bedürfnissen und äußeren gesellschaftlichen Ansprüchen an empirischen Tatbeständen: Nach dem Ersten Weltkrieg war es nicht mehr möglich, von einer „nachwachsenden Generation" der

---

1 Mit dem Begriff „gesellschaftliche Gewalt" beschreiben und verstehen wir hier wie im folgenden Einwirkungen auf den Menschen, die seine intellektuellen, emotionalen und körperlichen Kräfte beanspruchen, ohne Rücksicht auf seine Belastbarkeit und ohne Rücksicht auf seine Interessen, seine familiären und anderen sozialen Beziehungen. Es sind Einflüsse, denen er um seines Überlebens willen nicht entgehen kann, denen er aufgrund der in der Gesellschaftsstruktur verankerten ungleichen Einfluß-, Einkommens- und Besitzverhältnisse sowie der dadurch bedingten ungleichen Lebenschancen ausgesetzt ist. Bei Ausübung und Einwirkung dieser Form von Gewalt spielen menschliche Individuen höchstens eine vermittelnde Rolle, zum Beispiel als Inhaber bestimmter sozialer Positionen oder Rollen, etwa als Unternehmensleiter, Chef etc. Dem Begriff „gesellschaftliche Gewalt", den wir hier mit besonderem Bezug zum Lebensbereich „Arbeit" verwenden, entspricht weitgehend die wichtige Analyse des Gewaltbegriffs, die der norwegische Friedens- und Konfliktforscher Johan Galtung (1975) vorgelegt hat, ohne allerdings umfassend Geschichte und Semantik des Begriffs zu berücksichtigen (vgl. hierzu Röttgers, 1974; ferner Rammstedt, 1973).

Untertagearbeiter im Kohlebergbau zu reden. Durch Abwanderung hatte sich die Population traditionsverwurzelter „miners" stark reduziert, so daß diese Tradition kaum noch in den Prozessen der primären und der beruflichen Sozialisation zu überliefern war. Zudem erwies sie sich angesichts der gleichzeitigen Mechanisierungsentwicklung im Kohleabbau als partiell brüchig. Im Vergleich zu den alten „miners" verfügten die jungen Untertagearbeiter nicht mehr über die soziokulturellen Möglichkeiten, äußere gesellschaftliche Gewalt in Gestalt der Arbeitsanforderungen in Bedürfnisstrukturen zu integrieren. Ebenfalls im Vergleich zu den alten „miners" traten bei ihnen Arbeitsunzufriedenheit und psychosomatische Morbidität in noch höherem Maße auf.

Der methodisch nachweisbare und theoretisch erklärbare Zusammenhang zwischen Arbeitsanforderung und psychosomatischer Krankheit ist für Halliday das empirische Paradigma für das Auseinandertreten von innerem Bedürfnis und äußerer gesellschaftlicher Anforderung in einen Widerspruch, der bis dahin in individuellen und gruppenspezifischen Bedürfnisstrukturen aufgelöst war, und zwar aufgelöst unter Vermittlung von Wert- und Verhaltensmustern, die in Sozialisationsprozessen tradiert werden konnten. Psychosomatische Erkrankung wird damit zum Ergebnis und zugleich zum Indikator von Ereignissen und Entwicklungen gesellschaftlicher Desintegration.

Halliday betrachtet das von ihm untersuchte psychosomatische Krankheitsgeschehen vor allem als kollektive Antwort einer Gruppe von Menschen auf zunehmenden Druck, der von schädigenden Faktoren ihrer spezifischen Umwelt im beruflichen, ökonomischen und sozialen Bereich ausgeht. Viel weniger dagegen vermag er dieses Geschehen als Manifestation einer Verschlechterung der materiell-biologischen menschlichen Natur dieser besonderen Gruppe zu deuten (Halliday, 1943):

„This phenomenon (i. e. psychoneurotic and psychosomatic affections and the incidence of these disorders in miners) may be regarded as a group response to an increase in the pressure of noxious factors in the miners' environment (occupational, economic and social), rather than as a manifestation of a deterioration in the nature of the human material comprising the mining community".

Unter kollektivem wie auch unter individuellem Aspekt „erhält Krankheit den Charakter einer primitiven emotionalen Antwort und dient damit der blinden Absicht, einer gefahrvollen Arbeitsumgebung zu entfliehen" (Halliday, 1943).

Den Beginn psychosomatischer Krankheitsentwicklung im Zusammenhang mit Arbeitsbedingungen leiten Ereignisse und Entwicklungen jener soziokulturellen Desintegration ein, die als Zerbrechen einer komplexen Bedürfnisstruktur beschrieben wurde, wodurch nun innerlich gewordene Bedürfnisse und äußerlich gewordene soziale Anforderungen selbständig und widersprüchlich auseinandertreten. Äußere Anforderungen als gesellschaftliche Gewalt bedrohen menschliche Individuen und Grup-

pen aktuell und vermuteterweise durch Gefahren der Verletzung und der Krankheit bis hin zur Vernichtung. Die virtuelle oder reale Wahrnehmung ernsthafter existentieller Bedrohung aktualisiert nun bloßgelegte fundamentale Strebungen nach Selbsterhaltung. Doch nicht allein die unmittelbaren Gefahren für Leib und Leben, die von den Bedingungen der Untertagearbeit ausgehen, lösen Strebungen nach purem Überleben aus. Hinzu treten, gleichsam als intervenierende Variablen, mit den Arbeitsbedingungen verknüpfte Schwierigkeiten in den beruflichen, ökonomischen und persönlichen Bereichen alltäglichen Lebens. Diesen unmittelbaren und mittelbaren Existenzbedrohungen beständig ausgesetzt zu sein, führt zu chronischen emotionalen Spannungszuständen, welche für psychoneurotische und psychosomatische Reaktionen prädisponieren.

Psychosomatische Erkrankung indiziert einmal Leidenszustände von Individuen und Gruppen an der Gesellschaft. Zum anderen aber fungiert sie als äußerstes Mittel der Selbsterhaltung durch Rückzug aus einer Bedürfnisstrukturen zerbrechenden Gesellschaft. Sie stellt einen Appell[2] an gesellschaftliche Praxis dar, das funktionale Gleichgewicht zwischen der Reproduktion des arbeitenden Menschen und der Produktion von Gütern wiederherzustellen. Die heutigen Forschungen und Maßnahmen zur Humanisierung des Arbeitslebens in der BRD können als exemplarische Resonanz auf einen Appell dieser Art verstanden werden. Auf der anderen Seite wird der soziale Rückzug auf dem Wege der Krankheit durch bestimmte gesellschaftliche Einrichtungen begünstigt, was Halliday ebenfalls schon erkannte. Er wird z. B. begünstigt durch sozialversicherungsrechtliche Regelungen, nach denen das Leiden an definierten und katalogisierten Krankheiten zu sozialen Leistungen berechtigt. Krankheiten, und darunter auch psychosomatische Krankheiten, werden also positiv sanktioniert. Sie verschaffen dem, der daran leidet, einen sekundären Krankheitsgewinn. Sozialrechtliche Regelungen für den definierten Krankheitsfall können so – auch dies eine These von Halliday – nicht nur Symptombildung, sondern auch Symptomwahl begünstigen (vgl. hierzu Brede, 1972).

Diese sozialrechtlichen Regelungen schützen andererseits vor dem Appell, krankheitsbegünstigende Arbeitsstrukturen zu verändern, indem sie belastete Individuen zur Krankheit motivieren oder sie über Prozesse der Diagnose als krank bzw. invalide etiket-

---

2 „Appell" verstehen wir als Signal, welches einerseits anzeigt, daß ein Individuum und/oder bestimmte Gruppen von Menschen bestimmte Probleme aufgrund ihrer Konstitution und aufgrund ihrer sozialen Lage nicht mehr lösen können, und welches zugleich andererseits anzeigt, daß ein Problemlösung auf einer anderen Ebene gefunden werden soll. Damit hat das Signal den Charakter einer Aufforderung. Signalisiert wird die Aufforderung, sich an eine andere Instanz zu wenden, die die Problemlösung übernimmt, oder die Aufforderung, daß eine andere Instanz sich der Problemlösung annehmen müßte (zum Beispiel das System der medizinischen Versorgung, das System der sozialen Sicherung, eine Organisation zur Wahrnehmung der Interessen von abhängig Beschäftigten etc.). Ob die „richtige" Instanz für die Problemlösung mitsignalisiert wird, bleibt unentschieden.

tieren und so aus dem Erwerbsleben zeitweise oder dauernd eliminieren.

Die positive Sanktionierung psychosomatischer Krankheit über sozialrechtliche Regelungen nimmt die affektiven Bedürfnisse arbeitender Menschen nicht nur in Schutz gegen die verselbständigte „technisch-instrumentelle Rationalität" beruflicher Anforderungen (Habermas, 1973), sondern auch umgekehrt: Sie ist ein gesellschaftlicher Kontrollmechanismus, der Kranke aus den Arbeitsprozessen herausselektiert, damit zur weiteren Verselbständigung und Immunisierung technisch-instrumenteller Rationalität gegen subjektive Ansprüche beiträgt und so die „Austrocknung kommunikativer Handlungszonen" weiter fördert (Habermas, 1973). Psychosomatische Krankheit als Antwort auf anders nicht zu bewältigende Arbeitsanforderungen stellt den Betroffenen vor die Wahl, entweder im Sinne der von Parsons analysierten sozialen Rolle des Kranken alles zu tun, um adaptiv wieder an der Erwerbstätigkeit teilzunehmen oder sich mit dem institutionell gesicherten – scheinbaren – sekundären Krankheitsgewinn dauerhaft zu begnügen. Im letzteren Fall allerdings muß er unter Umständen erhebliche soziale Nachteile in Kauf nehmen, wie z. B. ökonomische Einbußen, die seine bisherige Lebensführung und -planung tangieren, oder Stigmatisierungen („schafft nichts", „frühinvalide" etc.), die Perspektiven und soziale Beziehungen belasten und einschränken.

Psychosomatische Krankheit interpretiert Halliday zwar auch als Ergebnis eines motivierten individual- und gruppenspezifischen sozialen Rückzugs unter bestimmten sozialen Bedingungen, aber zugleich als Ergebnis komplexer sozialer Kontroll- und Ausschlußprozesse, welche die Dominanz technisch-instrumenteller Rationalität gegen kommunikative und affektive Bedürfnisse von Individuen und Gruppen durchsetzen. Was er am konkreten sozialhistorischen Beispiel zeigte und erklärte – Zerbrechen soziokulturell geformter Bedürfnisstrukturen als Prozeß gesellschaftlicher Desorganisation infolge von Traditionsverfall und Mechanisierung produktiver Arbeit – und was er bereits als Symptom allgemeiner sozialer Entwicklung deutete, das wurde in den 60er und 70er Jahren in der BRD Gegenstand gesamtgesellschaftlicher Analyse (vor allem Habermas, 1963, 1968 und 1973; Dreitzel, 1968).

Halliday eröffnete damit einen soziologischen Zugang zur Genese psychosomatischer Krankheit. Dieser relativiert gerade durch seinen besonderen Bezug zum Strukturwandel der Arbeitswelt die Bedeutung des Regressionsmodells als Grundlage für die Erklärung arbeitsbedingter psychosomatischer Krankheit, welches von der Notwendigkeit einer an die Primärprozeßebene gebundenen frühkindlich angelegten Krankheitsdisposition ausgeht (vgl. hierzu Ahrens und v. Gyldenfeldt, 1981). Neuere soziologisch orientierte theoretische Erklärungsansätze zur Genese psychosomatischer Krankheit mit Bezug zu Strukturen der Arbeitswelt sind in der Tat nicht auf das Regressionsmodell fixiert. Dementsprechend kann der Krankheitsprozeß einerseits das Ergebnis beruflicher Sozialisationserfahrungen auf der Basis zerbrochener soziokultureller Bedürfnisstrukturen sein. Andererseits können berufliche Sozialisationserfahrungen und Fragmente einer ehemals integrierten Bedürfnisstruktur in die primäre Sozialisation der nächsten Generation einfließen, um auf diesem Wege berufliche Sozialisationserfahrungen vorzuformen und psychosomatische Krankheitsprädisposition zu schaffen. Es handelt sich hier um sozialisationstheoretische Ansätze, deren konsistenteste Entwicklung Arbeiten von Alfred Lorenzer (insbesondere 1972 und 1976) repräsentieren (weitere wichtige Arbeiten: Leithäuser, 1976; B. Volmerg, 1976; U. Volmerg, 1978; Wakker, 1976, 1977 und 1981; Großkurth, 1979; Volpert, 1979; Lempert, 1979; Bruggemann, 1979; Ahrens und v. Gyldenfeldt, 1981). Auch der sozialisationstheoretisch zu vertiefende Ansatz des „pensée opératoire" (De M'Uzan, 1977; Stephanos, 1978) könnte zur Analyse arbeitsbedingter psychosomatischer Krankheit beitragen, wie ein Anwendungsversuch nahelegt (Friczewski, 1981). Allerdings: der bloße Bezug sozialisationstheoretischer Erklärungsansätze arbeitsbedingter psychosomatischer Erkrankungen zum Strukturwandel der Arbeitswelt stellt noch keineswegs eine Beziehung zur makrosoziologischen Analyse sozialhistorischer und interessenbedingter Wandlungen der Arbeitsprozesse dar (exemplarisch hierzu Deppe, 1973; Mergner et al., 1975; Fürstenberg, 1977; Braverman, 1977; Kühn und Hauss, 1978; Bispinck und Zwingmann, 1981; Konstanty, 1981).

61.3 Exemplarische Ergebnisse empirischer Forschung

61.3.1 Arbeitslosigkeit und psychosomatische Erkrankung

Aufgrund kulturhistorisch alter Erfahrung und Erkenntnis konstituiert Arbeit menschliche Identität und ist Bestandteil menschlicher Bedürfnisstrukturen. Dies gilt auch für die Industrialisierung gesellschaftlicher Produktion und blieb ebenso unangefochtene These der politischen Ökonomie von Marx und Engels wie der Psychoanalyse Freuds, obwohl und gerade weil Arbeit in Gestalt gesellschaftlicher Gewalt sich gegenüber kommunikativen und emotionalen Bedürfnissen verselbständigt hat.

„Keine andere Technik der Lebensführung bindet den einzelnen so fest an die Realität als die Betonung der Arbeit, die ihn wenigstens in ein Stück der Realität, in die menschliche Gemeinschaft, sicher einfügt" ... „Besondere Befriedigung vermittelt die Berufstätigkeit, wenn sie eine freigewählte ist, also bestehende Neigungen, fortgeführte oder konstitutionell verstärkte Triebregungen durch Sublimierung nutzbar zu machen gestattet" (Freud, 1974).

So sind Entzug von Arbeit, also Arbeitslosigkeit und ihre empirisch feststellbaren Folgen für Selbstwertgefühl, soziales Beziehungsgefüge sowie für seelische und körperliche Gesundheit wichtige Themen früher Arbeiten zur Lage der Industriearbeiterschaft (vgl. z. B. zum Spezialthema „Männliche Arbeitslosigkeit und Wandel familiärer Rollenstruktur" Engels, 1972).

Empirische Studien zur Arbeitslosigkeit und ihren Folgen gehören seit der Weltwirtschaftskrise der 20er Jahre zu den Klassikern der Soziologie und Sozialpsychologie (Marienthal-Studie: Jahoda et al., 1933; Warschauer Studie: Zawadski und Lazarsfeld, 1935). Ihre methodischen Ansätze und Ergebnisse sind immer noch maßgebend für Theoriebildung und Methodologie der zahlreichen Untersuchungen, welche die aktuelle Situation zum Gegenstand haben, wobei diese durch einen international wie in der BRD hohen und sich weiter erhöhenden Stand der Arbeitslosigkeit bestimmt ist.

Die umfangreichen sozialökologisch-epidemiologischen Untersuchungen von Brenner (1973) stellen differenzierte Beziehungen zwischen wirtschaftskonjunktureller Entwicklung, Arbeitsplatzunsicherheit und Arbeitslosigkeit sowie psychischen Erkrankungen und Einweisungen in psychiatrische Kliniken dar. Unter anderen theoretischen und methodischen Gesichtspunkten wichtige Übersichts- und Schwerpunktarbeiten zum Stand und zur Entwicklung der für die Psychosomatik relevanten Forschung über Arbeitslosigkeit hat Ali Wacker vorgelegt (1976, 1977 und 1981).

Als relativ unmittelbarer theoretischer Zugang zum Problem „Arbeitslosigkeit und psychosomatische Erkrankung" ist der Ansatz der Lebensereignisforschung („life event"-Forschung) betrachtet worden (Wacker, 1981; zur Life-event-Forschung vgl. insbesondere Holmes und Rahe, 1967; Dohrenwend und Dohrenwend, 1974; Theorell, 1976; Siegrist, 1977; Katschnig, 1980). Dementsprechend ist von folgenden Annahmen auszugehen:

– Arbeitslosigkeit ist ein Ereignis, welches routinemäßig eingespielte Erlebnisse, Erfahrungen, Erwartungen, Beziehungen, Handlungen etc. alltäglichen Lebens unterbricht und erhöhte Anpassungsleistungen an die neue Situation erfordert.
– Arbeitslosigkeit wird – wenn sie als unerwünscht erfahren wird – zugleich als unbeeinflußbar und als negative Folgen implizierend wahrgenommen.
– Arbeitslosigkeit selbst wie auch ihre Dauer und Häufigkeit sind für den Betroffenen so belastend, daß seine individuellen Bewältigungsmöglichkeiten (Coping-Strategien) und/oder jene, die ihm durch Familie und sonstige soziale Umgebung zur Verfügung stehen (social support systems), nicht mehr zureichen.
– Es treten infolgedessen emotionale Spannungszustände und physiologische Reaktionen auf, die insbesondere in Anwesenheit disponierender Faktoren zu Erkrankungen führen.

Vor allem hinsichtlich der letzten Annahme wird eine enge Beziehung zwischen Life-event-Forschung und der sozialwissenschaftlich wie biologisch orientierten Streßforschung deutlich sichtbar (Levine und Scotch, 1970; Solomon und Amkraut, 1974). Für diese Grundannahmen der Life-event-Forschung ebenso wie für Annahmen der Streßforschung konnten zwar in den Untersuchungen über Arbeitslosigkeit zahlreiche empirische Belege erbracht werden. Studien, die von einem konsistenten theoretischen Ansatz ausgehen und diesen in systematischer Weise empirisch überprüfen, liegen jedoch nicht vor.

Um die Konsistenz dieser beiden theoretischen Zugänge zur Untersuchung der Arbeitslosigkeit zu verstärken, werden auch Beziehungen zu sozialisations- und konflikttheoretischen Ansätzen hergestellt (Wacker, 1976 und 1981). Besonders die Bedeutung des sozialpsychologischen Faktors „Berufsorientiertheit" für individuelle Reaktionen auf Arbeitslosigkeit weist auf Wertorientierungen und Verhaltensmuster, die „Lebensroutine" wesentlich mitbestimmen, die aber gleichwohl keine Chance haben, sich in der gegebenen sozialen Realität zu bewähren. Die in den Prozessen primärer und sekundärer Sozialisation verinnerlichten Normen von Arbeitsfreude, Fleiß, Pünktlichkeit, Zuverlässigkeit etc. erweisen sich zwar im allgemeinen als gültig und werden positiv sanktioniert, im konkreten jedoch – d. h. in der realen Arbeitswelt – als abhängig von der wirtschaftlichen Konjunkturlage. Die im allgemeinen konforme Wertorientierung verliert unter Umständen, die für das Wirtschaftssystem typisch sind, gerade durch Realitätsbezug ihren Sinn und vermag realiter im Widerspruch zu ihrer Anerkennung im allgemeinen nicht vor sozialem Ausschluß zu schützen. Als die für das Wirtschaftssystem typischen Umstände sind Krisenzyklen mit konsekutiver Arbeitslosigkeit zu betrachten; als Realitätsbezug der individuellen Wertorientierung die Situation des Individuums am Arbeitsmarkt. Aufgrund sozialrechtlicher Regelungen und aufgrund ungeschriebener allgemein anerkannter Normen hat der Arbeitslose stets Arbeitswillen, Pünktlichkeit und Anpassungswillen im System der Lohnarbeit zu demonstrieren, bleibt aber hinsichtlich materieller Einschränkungen, Stigmatisierung und ungeklärter Lebensperspektiven auf Verbesserung der konjunkturellen Entwicklung angewiesen (Wacker, 1976).

Die These, daß Arbeit die soziokulturell dominante Basis der Lebenssicherung, aber auch eine zentrale Bedingung der Identitätsbildung darstellt, präformiert an sich die Frage nach der Verarbeitung der Erfahrung von Arbeitslosigkeit. Dennoch ist sie in empirischen Untersuchungen alternativ gestellt worden (Wacker, 1976): Führt die Erfahrung von Arbeitslosigkeit zur Überprüfung der „Realitätshaltigkeit" konformer Wertorientierungen oder zur verstärkten Anpassung an die gegebenen Anforderungen?

Obwohl „Umzentrierungen von Lebensinteressen" vor allem in Gestalt des Rückzugs auf die Familie und neue Aufgaben beobachtet wurden – weniger häufig bei Angehörigen der oberen Sozialschichten übrigens – hielten weitaus häufiger auch jugendliche und Langzeitarbeitslose an der Arbeit als zentralem Maßstab ihrer Selbstbewertung fest. Die überwiegende Zahl der Untersuchungen über Arbeitslosigkeit dokumentiert individuelles Scheitern am Widerspruch zwischen der Orientierung am zentralen Wert „Arbeit" und den realen Chancen, gemäß dieser Orientierung auch handeln zu können. Physische und psychische Verelendungserscheinungen nach zahlreichen vergeblichen Bemühungen um Wiedereingliederung ins Erwerbsleben werden zum empirischen Befund. Neben psychosomatischen Reaktionen und Erkrankungen treten Formen der Randschichtenexistenz auf, im besonderen auch Alkoholismus.

Aufgrund der bisherigen Befunde der empirischen Forschung ist davon auszugehen, daß sowohl aktuelle unfreiwillige Arbeitslosigkeit wie auch bevorstehende und befürchtete oder wie auch immer antizipierte Arbeitslosigkeit zu Veränderungen von Befindlichkeit und physiologischen Funktionen führen, die im ätiologischen Zusammenhang mit psychosomatischen Störungen stehen.

Besonders häufig treten auf: Stimmungslabilität, Angst, Gefühl innerer Leere, Selbstunsicherheit und Minderwertigkeitsgefühle, Schlafstörungen, Mattigkeit, gespannte Erschöpfung und leichte Ermüdbarkeit, allgemeine vegetative Beschwerden, Störungen des Herz-Kreislauf-Systems, Erkrankungen des Magen-Darm-Trakts, depressive Verstimmungen (Wacker, 1981).

Alter, Geschlechtszugehörigkeit, allgemeinbildender Schulabschluß, Dauer der Arbeitslosigkeit und die Ausprägung anderer sozialer Variablen sind offenbar von Bedeutung für den Zusammenhang zwischen Arbeitslosigkeit und psychosomatischen Störungen. Konsistente Ergebnisse hierzu liegen allerdings wegen unterschiedlicher Methodologie und Verfahrenstechnik relevanter Studien sowie wegen mangelnder Kontrollierbarkeit intervenierender situativer Variablen nicht vor. Dagegen läßt sich, wie erwähnt, für den Faktor „Berufsorientierung" eine konsistente Wirkung nachweisen: Stark an ihrer beruflichen Tätigkeit orientierte Männer reagieren häufiger auf Arbeitslosigkeit mit Beeinträchtigung ihres Selbstwertgefühls, mit Verschlechterung ihres gesundheitlichen Befindens, mit psychosomatischen Symptomen.

Die Bemühungen um wissenschaftlich-empirische Klärung der Zusammenhänge zwischen Arbeitslosigkeit und psychosomatischer Krankheit werden fortgesetzt, ohne daß es gelungen ist, über das bisher Erreichte hinauszukommen. Es bleibt dabei, daß z.B. bei aktuell Arbeitslosen häufiger schwerwiegende gesundheitliche Störungen nachweisbar sind als bei kontinuierlich Beschäftigten oder als bei zeitweilig Arbeitslosen, die aber zum Zeitpunkt der Untersuchung wieder in einem Arbeitsverhältnis standen. Auch scheint es nicht so zu sein, daß Arbeitslosigkeit an sich schon, also isoliert, als ein streßerzeugendes Lebensereignis betrachtet werden darf. Vielmehr steht sie im Kontext mit verschiedenen sozialen Bedingungen des Lebens, wie mit wirtschaftlicher Lage der Person und ihrer Familie, mit sozialer Unterstützung und zur Verfügung stehenden Bewältigungsmöglichkeiten materieller und psychischer Schwierigkeiten, mit Problemen der Kindererziehung und der Ausbildung, um nur einiges aus dem bisher untersuchten Alltagskontexten herauszugreifen (Ensminger und Celentano, 1988).

61.3.2 Schichtarbeit und Krankheit

Schichtarbeit mit und ohne Nachtarbeit und Nachtarbeit für sich genommen sind besondere Formen von Arbeitsbedingungen, unter denen in Industrieländern nach Ermittlungen der Internationalen Arbeitsorganisation (ILO, 1974) 15 bis 20 Prozent aller Erwerbstätigen standen; davon leistete die Hälfte Nachtarbeit. Die Tendenz wird als zunehmend beurteilt (Carpentier und Cazamian, 1981). In der BRD waren 1975 von mehr als 21 Millionen abhängig Erwerbstätigen ca. 17 Prozent in Schichtarbeit tätig, ca. 12 Prozent leisteten Nachtarbeit, ein Teil davon auch Schichtarbeit. Während hier die Schichtarbeit in den zum tertiären Wirtschaftssektor gehörenden Branchen Handel, Banken, Versicherungen und sonstige Dienstleistungen zwischen 1959 und 1975 zunahm, ging sie im öffentlichen Dienst (auch tertiärer Sektor) sowie in den Branchen Energiewirtschaft und Bergbau (sekundärer Sektor) zurück (Münstermann und Preiser, 1978).

Es gibt zahlreiche nach Zielsetzung, theoretischem Ansatz und Methodik unterschiedliche empirische Untersuchungen über Zusammenhänge zwischen Schichtarbeit und Erkrankungen, die konvergierende, aber auch widersprechende Ergebnisse produziert haben (zur Kritik theoretischer und methodischer Ansätze vergleiche Münstermann und Preiser, 1978, S. 87 ff). Vorwiegend physiologisch orientierte experimentelle und betriebliche Fallanalysen sind eine alte Domäne der Arbeitswissenschaft und Arbeitsmedizin (z.B. Menzel, 1950; Rutenfranz, 1967; vgl. auch Krell, 1980), aber es liegen auch sozialwissenschaftlich orientierte epidemiologische Studien und betriebliche Fallstudien vor (Rutenfranz und Werner, 1975; Volkholz, 1977; Bergmann et al., 1982). Die epidemiologischen Untersuchungen über die Auswirkungen der Nachtarbeit beziehen in ihr Design oder in die Diskussion ihrer Ergebnisse oft physiologische Experimentalstudien ein. Ihre Resultate sind in relativ hohem Maße vergleichbar. Übersichtsarbeiten über den Forschungsstand haben Carpentier und Cazamian 1975 dem Verwaltungsrat der ILO für die zugehörigen Industrieländer vorgelegt (1981) sowie Münstermann und Preiser dem Bundesministerium für Arbeit und Sozialordnung (BMAuS) für die BRD (1978). Als Auswirkungen der Nachtarbeit finden in diesen ausführlichen Berichten psychische und soziale Folgen sowie psychosomatische Störungen besonderes Interesse (vgl. auch Valentin et al., 1979; Werner et al., 1980; Clemens, 1981).

Es fällt dabei oft auf, daß sich Nacht- und Schichtarbeiter einerseits und Tagarbeiter andererseits hinsichtlich psychosomatischer Störungen gar nicht unterscheiden, daß manche Störungen bei Tagarbeitern entgegen der Erwartung sogar häufiger vorkommen. Vergleicht man mit diesen beiden Gruppen jedoch ehemalige Schichtarbeiter, so finden sich bei diesen höchst auffällig mehr Störungen. Dies weist im Sinne von Levi (1978) auf eine besondere Selektion von jungen, gesunden und hochleistungsfähigen aktuell tätigen Schichtarbeitern hin, wie auch auf Aussonderung älterer „verschlissener" Arbeitskräfte, die nun an „Schonarbeitsplätzen" tätig sind (Maschewsky, 1981).

Damit aber wird eine hohe Morbiditätsziffer der Schichtarbeiter verschleiert. Zum kleinen Teil hebt sich dieser Schleier durch Untersuchung der sogenannten Frühinvalidität, das heißt der Berentung wegen krankheitsbedingter Berufs- oder Erwerbsunfähigkeit vor Erreichen der Altersgrenze: „Abweichungen von der Normalarbeitszeit durch Schicht- und Nachtarbeit führen häufiger zur Frühberentung als Arbeit während der Tagesarbeitszeit, selbst wenn während der Tagesarbeitszeit Überstunden gemacht werden … Frauen scheinen bei der Schichtarbeit anfälliger für Gesundheitsschäden zu sein" (Specht, 1977; vgl. auch Fritsch et al., 1977). Zu den häufigsten gesundheitlichen Störungen ehemaliger Schichtarbeiter gehören: Magen-Darm-Beschwerden, Verdauungsstörungen, nervöse Störungen (Carpentier und Cazamian, 1981). Zu ähnlichen Befunden kommen andere Studien. Die häufigen Funktionsstörungen des Magen-Darm-Trakts werden überwiegend damit erklärt, daß vor allem regelmäßige Nachtarbeit die Abstimmung zwischen den an biologische zirkadiane Rhythmen gebundenen Stoffwechseldispositionen, den Zeiten der Einnahme von Mahlzeiten und den Tätigkeits- und Erholungsphasen unmöglich macht. Von den gewöhnlich eingenommenen beiden Tagesmahlzeiten unterbricht die eine in der Regel den durch Nachtarbeit notwendigen Tagesschlaf. Eine Nachtmahlzeit wird meistens kalt und ohne Appetit, aber zusätzlich gewürzt eingenommen, zusammen mit Mitteln zur Anregung wie Kaffee und Alkohol (Carpentier und Cazamian, 1981).

Nicht jedoch Nacht- und Schichtarbeit allein wirken sich belastend aus, sondern häufig treten andere ungünstige Arbeitsbedingungen als zusätzliche Belastungen hinzu, z.B. Hitze-, Lärm-, Staub-, Akkord- und Fließbandbedingungen sowie Arbeiten mit erhöhtem Unfallrisiko, körperliche Schwerarbeit, Überstundenarbeit, monotone Arbeiten und solche, die starke Anforderungen an die Aufmerksamkeit stellen. Für sich genommen haben diese Arbeitsbedingungen oft ähnliche gesundheitsschädliche Wirkungen wie Nacht- und Schichtarbeit (Müller-Seitz, 1979).

Wenn man vergleicht, unter welchen Arbeitsbedingungen Frührentner häufiger standen als Altersrentner, so waren es nicht nur Schicht- und Nachtarbeit, sondern auch jene Mehrfachbelastungen, die verhältnismäßig oft bei diesen vorkommen, nämlich Arbeit in Hitze und Kälte, Belastungen durch Staub, Geruch, Lärm. Im Vergleich zu Altersrentnern findet man bei Frührentnern sehr viel häufiger Berufs- und Arbeitsstellenwechsel, ein Hinweis auf eine Form gesundheitsbedingter beruflicher Mobilität bei älteren ehemaligen Schichtarbeitern, die oft mit Einkommens- und Statuseinbußen sowie mit Dequalifikation und Beeinträchtigung des Selbstwertgefühls verbunden ist (Carpentier und Cazamian, 1981; Maschewsky, 1981). Frührentner sind mit ihrem früheren Arbeitsleben häufiger unzufrieden als Altersrentner. Häufiger waren bei ihnen auch Krankenhausaufenthalte und Kuren. Unter ihnen fanden sich schließlich

häufiger Menschen, die während ihrer früheren Arbeit ohne warmes Essen auskommen mußten (Specht, 1977).

Übereinstimmend wird als unmittelbare Ursache der somatischen, sozialen und psychischen Folgen vor allem der Nachtarbeit der Eingriff in die zirkadianen Biorhythmen des Menschen betrachtet. Schlafstörungen, Unausgeschlafenheit, Müdigkeit, Zerschlagenheit sind in der Tat die häufigsten Befindensstörungen dieser Arbeitnehmer, die schon in den ersten Monaten oder erst nach 10 bis 20 Jahren der Nachtarbeit auftreten (Carpentier und Cazamian, 1981). Schlafstörungen gelten als Symptom, gelegentlich sogar als Ursache für grundlegende Stoffwechselstörungen (Bugard, 1974). Aufgrund zahlreicher Experimentalstudien lassen sich die zirkadianen Biorhythmen, zu denen wesentlich Dauer, Abfolge und zeitliche Plazierung qualitativ unterschiedlicher Phasen des Schlafs gehören, kaum durch äußere Eingriffe verändern, und ihre interindividuelle Varianz ist relativ gering.

Die Phasen des Tiefschlafs und des paradoxen Schlafs gelten als konstitutiv für ein Gleichgewicht des Stoffwechsels zwischen Anabolismus und Katabolismus. Der Tiefschlaf hat eine „anabolische" Funktion. In dieser Phase wird – bildlich ausgedrückt – der Energievorrat wieder aufgefüllt; muskuläre Ermüdung schwindet. Dagegen hat der paradoxe oder REM-Schlaf (rapid eye movement) „katabolische" Funktionen.

Diese Schlafphase scheint vor allem zur Überwindung „psychomentaler" Ermüdung wichtig zu sein. Hohe körperliche Belastung verlängert die Dauer des Tiefschlafs, hohe intellektuelle dagegen die des paradoxen Schlafs. Der nach der Nachtarbeit notwendige Schlaf am Tage beeinträchtigt nicht die Länge der Tiefschlafphase, während er die Phase des paradoxen Schlafs wesentlich verkürzt. Experimentelle Unterdrückung des paradoxen Schlafs hat zu Neurosen geführt. Daraus läßt sich der – empirisch überprüfte – Schluß ziehen, daß der körperlich tätige Arbeiter Nachtarbeit eher „verträgt" als der Arbeiter, dessen Tätigkeit „psychomentale" Anforderungen stellt, das heißt dessen Konzentrationsvermögen, Aufmerksamkeit, Wachsamkeit und so weiter beansprucht werden. Körperliche Ermüdung kann durch Tiefschlaf besser ausgeglichen werden, geistige Ermüdung schlechter (Carpentier und Cazamian, 1981). Die technologische Entwicklung tendiert im Zuge von Rationalisierungsmaßnahmen zur Zunahme der Automation, damit aber zur Zunahme „psychomentaler" und zum Rückgang körperlicher Anforderungen, ganz besonders bei Nachtarbeit.

Wenn der Schlaf sich nach den zirkadianen Rhythmen richten kann, so scheint der Tiefschlaf in zwei Hauptstadien abzulaufen, einmal zwischen dem Zubettgehen und zwei Uhr, zum anderen zwischen vier und fünf Uhr, während die Phase des paradoxen Schlafs besonders am Ende der Nacht auftritt. Sogenannte Tagmenschen scheinen das erste Tiefschlafstadium für ihren notwendigen Schlaf zu nützen und können relativ früh aufstehen im Gegensatz zu sogenannten Nachtmenschen, die das späte Tiefschlafstadium nützen und dann relativ spät aufstehen. Allerdings benötigen beide die Phase des paradoxen Schlafs.

Die Befunde über den Zusammenhang zwischen Nachtarbeit, „Abwärtsmobilität" und Frühinvalidität legen nahe, danach zu fragen, ob und welche Beziehungen bestehen zwischen dem spezifischen Symptom des Nachtarbeiters: Übermüdung und Prozessen des Alterns. Beim älterwerdenden Menschen sei festzustellen, daß er nachts häufiger erwache und daß sich die Tiefschlafphasen mit zunehmendem Alter verkürzen, während die Phase des paradoxen Schlafs sich nicht verändere. Den älteren Menschen belaste daher bei Nachtarbeit nicht nur die Tätigkeit während physiologischer Deaktivierung, sondern während des Schlafs bei Tage seien beide Formen des Schlafs gestört, was zur Beschleunigung der physiologischen Alterungsprozesse führe. Mit zunehmendem Alter ist aufgrund einiger empirischer Untersuchungen nicht etwa eine Gewöhnung an Nachtarbeit festzustellen, sondern im Gegenteil wachsende Unverträglichkeit, und je älter ein Arbeiter bei Übernahme der Nachtarbeit sei, um so schlechter passe er sich daran an (Carpentier und Cazamian, 1981).

Die Folgen des Eingriffs in die zirkadianen Rhythmen physiologischer Funktionen durch längerfristige Nachtarbeit lassen sich durch Ruhe während des Tages letzten Endes nicht ausgleichen. Diese Belastungen verstärken sich dadurch, daß die notwendige Ruhe des Nachtarbeiters mit der sozialen Organisation des Alltagslebens kollidiert: Einnahme der Mahlzeiten im Familienkreis, Verwandten- und Bekanntenbesuche, Schulzeiten der Kinder, kulturelles und Vereinsleben. Regelmäßige oder in Wechselschicht ausgeübte Nachtarbeit wirkt sich im Familienleben auf die praktische Organisation des Alltags und auf die Beziehungen der Familienmitglieder aus. Entweder paßt sich der Nachtarbeiter dem „normalen" Familienleben an und unterbricht zum Beispiel seinen Schlaf zum gemeinsamen Essen oder die Familie paßt sich seinem Zeitplan an oder er entwickelt einen von der Familie unabhängigen Zeitplan für sich. In allen Fällen bleiben psychosoziale und psychosomatische Störungen unvermeidbar. Herausgestellt hat sich, daß das psychische Gleichgewicht des Nachtarbeiters um so nachhaltiger gestört wird, je stärker die Ausübung seiner Rollen als Ehemann, Vater und Bürger beeinträchtigt ist. Bei verheirateten Nachtarbeitern erwies sich die Widerstandskraft gegen die berufsbedingten Belastungen als abhängig von der Einstellung der Ehefrau. Als günstig wirkte sich sowohl verständnisvolle Anpassung aus wie auch eine entschieden kompromißlose Haltung. Psychische Störungen fanden sich um so deutlicher ausgeprägt, je schlechter der Gesundheitszustand, je niedriger Arbeits- und Lebensalter, je höher der Bildungsstand und je jünger die Kinder waren.

Es gibt zahlreiche Befunde empirischer Studien, die auf Einschränkung und Störung informeller und formeller sozialer Beziehungen bei Nachtarbeitern hinweisen: Sie haben, verglichen mit Kontrollgruppen, häufiger das Gefühl, im sozialen Leben behindert zu sein und am Rande zu stehen. Ihr Freundeskreis ist kleiner. Nur selten nehmen sie leitende Stellungen in Vereinen, Gewerkschaften etc. ein, und sie nehmen in relativ stark eingeschränktem Maße an sportlichen, kulturellen und politischen Veranstaltungen teil (Carpentier und Cazamian, 1981).

Wie ist den negativen Folgen der Nachtarbeit zu begegnen? Übereinstimmung herrscht insoweit, als die Desynchronisation biologischer Rhythmen letztlich nicht organisatorisch zu bewältigen ist. Nachtarbeit ist die Schicht mit den meisten psychischen und sozialen Problemen. Ein einfacher regelmäßiger Schichtwechsel scheint hinsichtlich der Sicherung einer gewissen Regelhaftigkeit des familiären und übrigen sozialen Lebens noch am ehesten vorteilhaft zu sein. Als erstrebenswerte Neuerung wird generell eine Verringerung der Nachtarbeit durch weitere Perfektionierung automatisierter Produktion eingeschätzt. Um die schädlichen Auswirkungen und Nachteile für Dauernachtarbeiter zu vermindern, kommen noch folgende Maßnahmen in Betracht:
- Verringerung der wöchentlichen Arbeitszeit;
- kürzerer Einsatz im Nachtdienst;
- Senkung des Rentenalters;
- Ermöglichung von mehr beruflicher Mobilität durch wahrnehmbare Ausbildungsangebote;
- flexible Arbeitspläne mit gleitender Arbeitszeit, um Stabilisierung formeller und informeller sozialer Beziehungen zu erleichtern;
- Maßnahmen zur Verbesserung der Wohnverhältnisse, Einrichtung schneller Verbindungen zwischen Arbeitsplatz und Wohnung, Verbesserung der Mahlzeiten am Arbeitsplatz (Carpentier und Cazamian, 1981; vgl. auch Jansen et al., 1979; Ulich und Baitsch, 1979; Bielenski et al., 1980; Jansen et al., 1980).

Die wirksamste Empfehlung zur Minderung der psychosozialen und psychosomatischen Folgen der Nachtarbeit besteht ohne Zweifel darin, diese Arbeitsform abzubauen. Dem aber steht einerseits ein unabweisbarer gesellschaftlicher Bedarf nach Dienstleistungen entgegen, z.B. im Gesundheitswesen, im Bereich Transport, bei Polizei, Feuerwehr und so weiter. Andererseits besteht auf der Seite des Kapitals der Wunsch, stark mechanisierte und automatisierte technische Produktionsmittel, die teuer sind und schnell veralten, möglichst schnell zu amortisieren, was eine Tendenz zur Zunahme der Nachtarbeit begründet. Dieser Tendenz wären die Kosten der Nachtarbeit entgegenzustellen: die Bedrohung von Wohlbefinden und Gesundheit, die Störung von familiären und sonstigen sozialen Verantwortlichkeiten, die Bedrohung des sozialen Status, der private Aufwand für Erhaltung und Wiederherstellung von Gesundheit sowie für chronische Krankheit und Behinderung, schließlich auch der Anteil öffentlicher Kosten, z.B. Ausgaben für Gesundheitswesen und Sozialversicherung, Störung des gesellschaftlichen und politischen Lebens durch relativen Ausschluß einer Gruppe der daran Beteiligten (Carpentier und Cazamian, 1981).

61.3.3 Zur Selbsteinschätzung von Belastung und arbeitsbedingter Erkrankung

Höchst bedeutsam für die Einschätzung der Arbeitsbedingtheit psychosomatischer Störungen und Erkrankungen sind streßtheoretisch orientierte Ansätze hauptsächlich in den Sozialwissenschaften, aber gelegentlich auch schon in der Arbeitsmedizin, wonach die Arbeitenden selbst Art und Ausmaß der Belastungen ihres Arbeitsplatzes beurteilen, sowie auch Art und Ausmaß der eigenen beruflichen Beanspruchung einschließlich der Beurteilung psychosozialer und psychosomatischer Störungen (Wintersberger, 1976; Schienstock et al., 1979; v. Ferber und Standfest, 1981; Bamberg und Greif, 1982). Die Wirksamkeit dieser Ansätze in der italienischen Arbeitermedizin und ihre probeweise Einführung in wenige bundesdeutsche betriebsmedizinische Einrichtungen demonstriert deutlich auch für diesen Bereich die Relevanz des „diagnostisch-therapeutischen Zirkels" (vgl. Kap. 1).

Die eingesetzten Beurteilungsinstrumente erreichen ein relativ hohes Maß an Objektivität, Validität und Reliabilität. Auch unter formalem Gesichtspunkt sind diese Ansätze wichtig, denn das Arbeitssicherheitsgesetz der BRD von 1974 (ASiG § 3.3 c) hat zwar den Begriff „arbeitsbedingte Erkrankung" eingeführt, um den Begriff „Berufskrankheit" realitätsgerecht zu ergänzen, ohne allerdings den neu eingeführten Begriff „arbeitsbedingt" zu definieren.

61.4 „Arbeit und psychosomatische Probleme" in der ärztlichen Ausbildung

Angesichts der wissenschaftlichen und praktischen Bedeutung psychosozialer Belastungen durch Arbeit für Erkrankungen und Störungen stellt sich besonders im Rahmen eines Lehrbuchs die Frage, wie dieser Problemkomplex in der ärztlichen Ausbildung repräsentiert ist. Zu erwarten wäre eine Repräsentanz in den vorklinischen Ausbildungsfächern Medizinische Psychologie und Medizinische Soziologie sowie in den Fächern des zweiten klinischen Studienabschnitts Psychosomatik, Arbeitsmedizin und Sozialmedizin.

Die Darstellung des Prüfungsstoffs der Medizinischen Psychologie und Medizinischen Soziologie in der entsprechenden Anlage zur Approbationsordnung für Ärzte (AppOÄ Anl. 10 zu § 23, Abs. 2, Satz 2) enthält zum beschriebenen Problemkomplex nichts. Einen höchstens indirekten Hinweis bietet der vom Mainzer Institut für Medizinische und Pharmazeutische Prüfungsfragen (IMPP) herausgegebene Prüfungsstoffkatalog für diese Fächer unter dem Stichwort „Streß" (S. 182), wenn man dieses in Beziehung zum Stichwort „Arbeit" (S. 190) setzt. Der Examenskandidat soll zum Stichwort „Streß" unter anderem wissen: „Unterschied zwischen Stressoren und Streßreaktionen; Reizdeprivation, Reizüberflutung, Schmerz, Leistungsstressoren, soziale Stressoren;

Stadien der Streßreaktion; Zusammenhang zwischen Stressoren und physiologischen Veränderungen (z.B. Magenmotilität, Magensäuresekretion); Ansatzpunkte für funktionelle Störungen und psychosomatische Erkrankungen". Zum Stichwort „Arbeit" soll er wissen: „Merkmale der Industriearbeit; Arten der Arbeitsbelastung; Mechanisierung; Begriff und soziale Folgen der Automation".

In den Lehrbüchern der Medizinischen Psychologie wird der in diesem Artikel thematisierte Gegenstand nicht behandelt, wohl dagegen in denen der Medizinischen Soziologie. Siegrist (1977) stellt Zusammenhänge zwischen Arbeitslosigkeit und Krankheit dar unter Hinweis auf die bedeutende Untersuchung von M. H. Brenner (1973) über „Wirtschaftskrisen, Arbeitslosigkeit und Krankheit". Ferner behandelt er mit Hinblick auf vorwiegend sozialpsychologische Studien Beziehungen zwischen Arbeitssituation und psychischer sowie psychosomatischer Krankheit (Siegrist, 1977, 1988). P. Thoma diskutiert die Kausalnorm der gesetzlichen Unfallversicherung hinsichtlich des Zusammenhangs zwischen beruflicher Tätigkeit und psychischen beziehungsweise psychosomatischen Erkrankungen, welche maßgebend ist für die Anerkennung von Leistungsansprüchen gesundheitsgeschädigter Arbeitnehmer (Thoma, 1979; vgl. auch Borgers und Nemitz, 1978).

Die Darstellung des Prüfungsstoffs für Psychosomatik, Sozialmedizin und Arbeitsmedizin in den Anlagen zur AppOÄ (Anl. 16 zu § 29, Abs. 2, Satz 2) enthält allein schon aufgrund der sehr allgemeinen Formulierungen keinen Hinweis auf Arbeitsbedingungen und psychosomatische Krankheit. Explizite Hinweise auf diesen Gegenstand fehlen aber auch im Gegenstandskatalog des IMPP für „Psychosomatik" (GK III, S. 132–135). Spärliche, höchstens implizite Hinweise bietet immerhin der Gegenstandskatalog für „Sozialmedizin" (GK III, S. 342). Bezogen auf rheumatische Erkrankungen sollen folgende Kenntnisse erworben werden: „Probleme der epidemiologischen Feststellung, Prävalenz, Arbeitsunfähigkeit, Berufs- und Erwerbsunfähigkeit, mögliche psychosoziale Mitursachen, Rehabilitation".

In Lehrbüchern der Sozialmedizin ist das Thema „psychosoziale Belastungen, Krankheit und Arbeit" deutlicher präsent, zum Beispiel bei H. Schaefer und M. Blohmke (1978) im Kapitel „Krankheitsfaktoren beruflicher Art, die nicht Berufskrankheiten sind – Humanisierung am Arbeitsplatz", obwohl der Begriff „Psychosomatik" explizit in diesem Zusammenhang nicht vorkommt. Dagegen bietet der kurze, aber prägnante Beitrag „Arbeitswelt und Krankheit" von W. Dieckmann im „Lehrbuch Sozialmedizin" (1981) eine deutlich exponierte Übersicht über „psychosomatische Aspekte".

Der Gegenstandskatalog des IMPP für das Fach „Arbeitsmedizin" weist implizit mehrfach auf Arbeitsbelastungen und mögliche psychosoziale Folgen hin (GK III, S. 360f). „Arbeitsunfälle als wesentliche Teilursache für die Exazerbation bestehender chronischer Leiden" werden bezogen auf „akute und außergewöhnliche und/oder psychische Belastungen sowie

Traumen in ihrer Relevanz für die Auslösung von Herzinfarkt bei Koronararteriensklerose, Schlaganfall bei Hypertonie und/oder Veränderungen der Hirnarterien, Entgleisungen des Stoffwechsels bei Diabetes mellitus". Weitere Hinweise finden sich im Kapitel „Arbeitsphysiologie" unter den Stichworten „Belastung und Beanspruchung", „mentale Leistungsfähigkeit", „Arbeit und Ermüdung" sowie im Kapitel „Arbeitspsychologie" unter dem Stichwort „körperliche Auswirkungen mentaler Belastung". Deutlich repräsentiert sind psychosomatische Aspekte von Arbeitsbelastungen im Band I des Lehrbuchs „Arbeitsmedizin" von Valentin und Mitarbeitern in den Kapiteln „Arbeitsprobleme: Erkrankungen und Anpassungsstörungen bei Schichtarbeit" sowie „Herzinfarkt und Beruf" (Valentin et al., 1979).

61.5 Zu Forschungslage, Praxis und interdisziplinärer Zusammenarbeit

Die gegenwärtig schon große und weiterhin zunehmende sozialpolitische, wissenschaftliche, soziale und medizinische Bedeutung psychosomatischer Störungen und Krankheiten im Zusammenhang mit dem Arbeitsleben ist unstrittig. Demgegenüber müssen die Forschungslage, besonders auf dem Gebiet der Arbeitsmedizin, die betriebsärztliche und sonstige praktische medizinische Situation, schließlich die Ausbildungssituation als unbefriedigend bezeichnet werden (zur Arbeitsmedizin und Ergonomie vgl. Rutenfranz et al., 1980).

Die Forschungslage charakterisiert, daß die Analyse von Zusammenhängen zwischen Arbeit und psychosozialen Belastungen als Bedingungen für Erkrankungen multivariate Erklärungsmodelle und dementsprechende methodische Designs empirischer Untersuchungen erfordert. Der theoretische Problemzugang kann – in der Psychosomatik ist das Tradition – nur interdisziplinär eröffnet werden. Das bedeutet, für dieses Forschungsfeld ist psychosomatische, psychologische, soziologische, ergonomische und arbeitsmedizinische Expertise relevant. Abgesehen vom Mangel an interdisziplinärer Konsensfähigkeit – bei ohne Zweifel bestehender Verständigungsbereitschaft – ist aufgrund forschungsorganisatorischer und ressourcialer Bedingungen die Wahrscheinlichkeit sehr gering, in jeder relevanten Hinsicht problemadäquate hochkomplexe interdisziplinäre Forschungsvorhaben durchzuführen. Ein durchaus nicht gewisses Maß an „Einseitigkeit" von Forschungsansätzen und -ergebnissen bleibt daher unvermeidbar. Selbst wenn die üblichen methodischen Kriterien empirischer Forschung bei Verwendung statistischer Verfahren in maximal befriedigendem Umfang erfüllt sind, wird jene von der Arbeitsmedizin gestellte Forderung wohl kaum einzulösen sein, welche lautet: „Die exogene Verursachung von Gesundheitsschäden durch Arbeitswelt und Umwelt soll in der Fachwelt als wahr und richtig, sach- und fachgerecht akzeptiert worden sein und sollte Allgemeingeltung erlangt ha-

ben" (Valentin und Zober, 1982; vgl. auch Valentin et al., 1979). Außerdem steht das Postulat der Allgemeingeltung in Frage aufgrund Branchen-, Betriebs-, Arbeitsplatzspezifität und regionaler Spezifität gesundheitsgefährdender Arbeitsbedingungen sowie aufgrund des Einflusses unterschiedlicher Bewältigungsmöglichkeiten im betrieblichen und vor allem im außerbetrieblichen Bereich.

Es ist prinzipiell nicht möglich, die Verursachung von Krankheit durch arbeitsbedingte psychosoziale Belastungen nach der Kausalnorm der Berufskrankheiten, beispielsweise der Silikose, zu beurteilen, wo spezifische Agenzien spezifische funktionelle und strukturelle Veränderungen des menschlichen Organismus bewirken. Daher kann auch für eine versicherungsrechtliche Anerkennung und Entschädigung arbeitsbedingter psychosomatischer Erkrankungen das ätiologische Modell der Berufskrankheit nicht verwendet werden (zum Ansatz eines komplexen integrativen ätiologischen Modells vgl. Naschold und Novak, 1980). Hier stehen besonders Betriebsarzt (Florian, 1979) und Arbeitsmediziner, aber auch der behandelnde Arzt vor erheblichen Problemen.

Eine Erweiterung des versicherungsrechtlich leitenden Ätiologiemodells mit Hinblick auf psychosomatische Erkrankungen erscheint notwendig, welche im konkreten Beurteilungsfall dem erreichten Erkenntnisstand, der in der ärztlichen Aus- und Weiterbildung zu vermitteln ist, und der ärztlichen Praxiserfahrung hinreichend Rechnung trägt. Ebenso stellen arbeitsbedingte psychosomatische Erkrankungen den medizinischen Arbeitsschutz vor neue präventive Aufgaben, vor allem mit Hinblick auf Arbeitsorganisation und betriebliche Personalpolitik. Auch hier spielt das interdisziplinäre Zusammenwirken zwischen Ergonomie, Arbeitsmedizin, Arbeitspsychologie und Arbeitssoziologie eine bedeutende Rolle.

Unter Berücksichtigung der historischen Bezüge der heutigen wissenschaftlichen Problematik des Zusammenhangs von Arbeit und Krankheit läßt sich resümieren: Die sozialwissenschaftliche und teilweise auch die sozialwissenschaftlich orientierte arbeitsmedizinische Forschung hat Tendenzen entwickelt, Krankheit zu verstehen als Ergebnis einer dynamischen Beziehung zwischen arbeitendem Menschen und gegebenen Arbeitsbedingungen. Das bedeutet: nicht objektivierte Personenmerkmale für sich und nicht objektivierte Merkmale der Arbeitssituation für sich können hinreichende Begründungen dafür geben, daß Krankheit entsteht oder verhindert wird. Vielmehr resultiert die gesundheitsrelevante Wirksamkeit von Arbeitsbedingungen aus den selektierenden Wahrnehmungen und Interpretationen dieser Bedingungen durch die Menschen, die ihnen ausgesetzt sind. Diese ihrerseits machen „neue" Erfahrungen nur aufgrund ihrer spezifischen Konstitution, aufgrund individueller und kollektiver Vorerfahrungen, Vorurteile, Einstellungen, Wahrnehmungs- und Verarbeitungsmöglichkeiten. Die Beziehung Arbeitsumwelt – arbeitendes Individuum erscheint somit als individual- und sozialgeschichtlicher dialektischer Prozeß der Interaktion von Mensch und Umwelt.

Im Vorhergehenden ist an zahlreichen Stellen auf diese Tendenzen hingewiesen worden, besonders im Zusammenhang mit den Ansätzen der Lebensereignisforschung, der Streßforschung, der subjektiven Belastungseinschätzung. Die auf dem Wege empirisch-methodischer Hypothesenprüfung schließlich resultierenden Ansätze zu multivariaten Erklärungsmodellen für Zusammenhänge zwischen Arbeit und Krankheit vermögen allerdings nicht mehr den historischen Prozeßcharakter der Beziehung Arbeitsbedingungen – arbeitendes Individuum widerzuspiegeln. Eine nun notwendig werdende Vermittlung zwischen Ergebnissen empirischer Forschung und theoretischen Grundlagen mit dem Ziel, problemgerechte Handlungsstrategien zu entwickeln, ist bisher nicht über das Postulat von Levi (1978) hinausgekommen, die wissenschaftliche Analyse des Zusammenhangs zwischen Arbeitsbelastung und Krankheit müsse die gesamte Interaktion des Menschen mit seiner gesamten Umwelt berücksichtigen.

Dagegen überwindet der humanökologische Ansatz des Situationskreismodells, welches in diesem Lehrbuch die Grundlage für die Entwicklung und Darstellung der Psychosomatischen Medizin bildet, als integratives Modell von vornherein eine analytische Verkürzung der Mensch-Umwelt-Beziehung um die wesentliche Dimension der historisch prozeßhaften Interaktion. Die in diesem Kapitel dargestellten Analysen und Erklärungen des Verhältnisses von Arbeit und Krankheit als spezielle Form der Mensch-Umwelt-Beziehung zeigen daher bei kritischer Betrachtung erst der Tendenz nach Berührungspunkte mit dem Situationskreismodell. Dies weist aber zugleich in spezifische Richtungen ihrer notwendigen Weiterentwicklung in Verbindung mit Psychosomatischer Medizin.

Teil IV: Krankheitsverarbeitung bei körperlich Schwerkranken

62 Psychische Anpassungs- und Abwehrprozesse bei körperlichen Erkrankungen

Ekkehard Gaus und *Karl Köhle*

62.1 Anpassungs- und Abwehrvorgänge im Dienste der Krankheitsbewältigung (Coping)

62.1.1 Allgemeine Grundsätze über Anpassungs- und Abwehrvorgänge

Die Abschnitte 62.1 und 62.2 befassen sich mit den Möglichkeiten des Umgangs des menschlichen Organismus mit Streß und den verschiedenen Formen der Anpassung. Streß bedeutet dabei ein vorübergehendes oder dauerndes Ungleichgewicht zwischen individuellen Anpassungsfähigkeiten und der äußeren und inneren Realität. Streß setzt die Wahrnehmung einer bedrohlichen Situation, eines Stressors voraus, wobei die Antizipation einer solchen Wahrnehmung oft schwieriger zu bewältigen ist als die Konfrontation selber (Mechanic, 1962).

Die Wahrnehmung eines Reizes als Stressor spielt sich dabei in einem transaktionellen Prozeß zwischen Umgebungsreiz und dem betroffenen Individuum ab, in dessen Verlauf sich entscheidet, ob ein Reiz stressorische Bedeutung gewinnt (Temoshok, 1983). Im Situationskreiskonzept v. Uexkülls (vgl. Kap. 1) hängt dies davon ab, ob die zur Verfügung stehenden Deutungs- und Verhaltensprogramme den Anforderungen adäquat sind. In einem **allgemeinen** Sinn bezeichnet Anpassung und Situationsbewältigung („Coping") als „Bedeutungsverwertung" im Situationskreismodell eine Aktivität, der „Bedeutungserteilung" mit „Bedeutungsunterstellung" und Probehandeln vorausgehen. Störungen innerhalb der psychischen Entwicklung, aber auch pathogene Umgebungssituationen können dazu führen, daß die zur Verfügung stehenden Programme in der individuellen Wirklichkeit Erlebnissituationen aufbauen, die krankmachend sind. Krankheit entsteht also oft aus dem Versagen von Anpassungsprozessen (vgl. Weiner, 1983). Eine Intervention muß dann darauf zielen, dieses Versagen rückgängig zu machen oder den Grad der Anpassung zu verbessern.

In einem **spezifischeren** Sinn bezeichnet „Coping" Verhaltensweisen mit dem Ziel, durch **Krankheit** entstandene Probleme zu lösen. Wir wollen uns hier damit befassen, wie Individuen mit der tatsächlichen oder vermeintlichen Erfahrung einer schweren Erkrankung umzugehen pflegen. Eine solche Erkrankung stellt ein besonders eingreifendes Geschehen im Leben eines Menschen dar und bedeutet damit auch eine besondere Anforderung an sein Anpassungs- und Bewältigungsvermögen (Coping). Liegt eine seelische oder geistige Erkrankung allein oder zusätzlich vor, beeinflußt diese die adaptiven Fähigkeiten des Individuums in der Regel ungünstig. Wir wollen uns, obwohl der Umgang seelisch und geistig Behinderter mit Krankheit einen klinisch interessanten Aspekt darstellt, im wesentlichen auf die Problematik des Umgangs mit körperlichen Erkrankungen beschränken.[1] Es soll dargestellt werden, wie psychische Prozesse, insbesondere bei schwerkranken Patienten, dazu dienen, die Gefahr abzuwehren, eine Anpassung zu vollziehen und aktiv die Situation zu meistern. Das Ziel ist, einen Überblick über die vielfältigen Strategien von Menschen, mit einer schwer erträglichen Wirklichkeit fertig zu werden, zu gewinnen mit den daraus für die Therapie resultierenden Konsequenzen. Es interessiert dabei besonders die Frage, welche Bewertungskriterien für Anpassungs-, Bewältigungs- und Abwehrprozesse sich festlegen lassen, inwieweit einzelne Strategien psychische Symptome verhindern oder provozieren und welche Rolle sie insgesamt für den Krankheitsverlauf und Gesundungsprozeß spielen.

62.1.2 Verhältnis von Bewältigungs-, Anpassungs- und Abwehrvorgängen

Als Modellvorstellungen kann man bei der Reaktion auf eine Krankheit Bewältigungs- (in der angelsächsischen Literatur „coping mechanisms"), Anpassungs- und Abwehrmechanismen unterscheiden. Das am besten ausgearbeitete und am meisten diskutierte Co-

---

1 Schwere psychische Krankheiten schränken in der Regel die Fähigkeit, mit Belastungen umzugehen, stark ein und schaffen daher andere Voraussetzungen für die individuelle Belastbarkeit.

ping-Modell stammt von Lazarus und Mitarbeitern (Lazarus, 1966, 1974a, b, 1984; Cohen und Lazarus, 1980). Es betont den Prozeßcharakter im Sinne eines Kontinuums aufeinanderfolgender Prozesse von Bewertung und Neuanpassung und setzt eine aktive Auseinandersetzung voraus. Bewältigung versteht sich darin als Summe ständig sich verändernder kognitiver und verhaltensmäßiger Bemühungen, mit spezifischen äußeren und/oder inneren Ansprüchen fertig zu werden. Ausgehend von der Streßforschung, liegt dem ein Paradigmenwechsel zugrunde, nämlich die Wendung vom Fokus der Stressoren hin zur Fokussierung komplexer Bewertungs- und Verarbeitungsprozesse.

Lipowski (1970) definiert „Coping" im Hinblick auf körperliche Krankheiten als „... alle kognitiven und motorischen Aktivitäten einer kranken Person, um ihre körperliche und psychische Integrität zu wahren, reversibel geschädigte Funktionen wiederherzustellen und möglichst weitgehend jede irreversible Behinderung zu kompensieren ..."

Dabei kann zwischen **Bewältigungsstrategien,** die einzelne Techniken in Abhängigkeit von der jeweiligen Situation umfassen, und dem **Bewältigungsstil** eines Individuums unterschieden werden, der ein individuelles Repertoire an spezifischen Techniken voraussetzt (Coelho et al., 1974).

Bei **Abwehrprozessen** liegt, im Gegensatz zum Akzeptieren, die Betonung auf einer vollständigen oder teilweisen Zurückweisung einer Wirklichkeit oder deren Bedeutung für das Individuum. Im Verhalten schwerkranker Patienten findet man in der Regel sowohl Bewältigungs- als auch Abwehrprozesse. Beide haben Angst- und Spannungsverminderung angesichts der Bedrohung zum Ziel. Neuerdings wird insbesondere die Erhaltung des narzißtischen Gleichgewichts als Ziel betont. Dabei ist der Versuch einer strikten Trennung von Bewältigungs- und Abwehrvorgängen sicher illusorisch (Verwoerdt, 1972; A. Freud, 1965). Daß sie in der Literatur dennoch häufig getrennt betrachtet werden, rührt teils von

dem unterschiedlichen theoretischen Standort der Untersucher her, teils daher, daß hinsichtlich ihres Ablaufs und ihrer Auswirkungen Differenzierungen möglich sind (Haan, 1969, 1977). In der Literatur wird angenommen, daß mit der Größe der Bedrohung auch die Wahrscheinlichkeit des Auftretens von Abwehrprozessen wächst. Daraus ergeben sich oft schwierige klinische Probleme. Aus diesem Grund sollen die Abwehrprozesse hier genauer dargestellt und diskutiert werden.

Heim und Mitarbeiter (1982, 1983) haben ein **integriertes** Modell beschrieben, das durch ein System kybernetischer Rückkoppelung charakterisiert ist. Coping wird als komplexer, in einzelnen Schritten verlaufender und auf die jeweilige Krankheitsphase bezogener Vorgang beschrieben. Die einzelnen Schritte lassen sich folgendermaßen beschreiben: „Aus dem Zustand des Wohlbefindens heraus stellt der Patient erstmals gewisse Veränderungen in seiner Befindlichkeit fest. Daran schließt sich erst das Wahrnehmen der körperlichen Veränderung an, die nun beurteilt und auf ihre Konsequenzen hin geprüft werden muß. In einem letzten Schritt kommt es zur eigentlichen Krankheitsbewältigung, die auf verschiedenen Ebenen (handelnd, kognitiv, intrapsychisch) erfolgt und entsprechend vielfältige Bewältigungsformen einschließen kann" (Heim et al., 1983, 1986, 1988a). Die Abbildung 62–1 stellt diesen Vorgang vereinfacht dar. Wichtig ist dabei, daß bewußte Aktivitäten, die der Krankheitsbewältigung dienen, durch unbewußte Abwehrvorgänge mit beeinflußt werden. Damit ist eine strenge Dichotomie zwischen Bewältigungs- und Abwehrvorgängen aufgehoben. Steffens und Kächele (1988) machen in ihrer Übersicht den Versuch einer Integration von Abwehr und Bewältigung.

Die Frage nach Art und Funktion einer Strategie zur Situationsbewältigung hat praktische und theoretische Bedeutung: Für die therapeutische Praxis ist eine Kenntnis der beteiligten Prozesse Voraussetzung für eine sinnvolle Intervention. Theoretisch wäre es

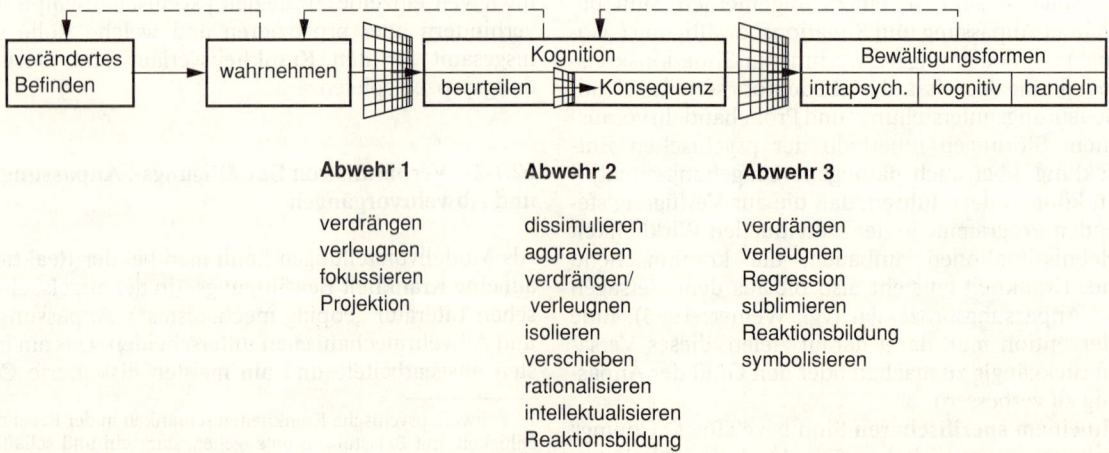

Abb. 62–1. Modell des „Coping"-Prozesses (aus Heim et al., 1983, S. 36).

interessant zu erfahren, wie solche Strategien mit Verläufen bestimmter Erkrankungen und der Anpassung an einzelne Therapieformen verbunden sind. D.h., es wäre zu untersuchen, wie die unterschiedliche individuelle psychische Belastbarkeit mit dem bevorzugten Stil, sich mit individueller Bedrohung auseinanderzusetzen, zusammenhängt (vgl. z.B. Heim et al., 1978, 1979 a, b, 1982).

62.1.3 Determinanten psychischer Steuerungsmöglichkeiten

Ich-Funktionen

Nach psychoanalytischen Konzepten sind Begriffe wie Anpassungsfähigkeit und Fähigkeit zur Situationsbewältigung sowie Abwehrprozesse eng verknüpft mit der Qualität der Ich-Funktionen.

Freud hat das Ich als theoretisches Konstrukt und Bestandteil des psychischen Apparates nach seinen Funktionen definiert. Ich-Stärke und Ich-Schwäche hängen dabei eng mit der Ausprägung, Ordnung und Rangskala einzelner Ich-Funktionen zusammen (Bellak et al., 1973). Hartmann (1960) unterschied bei den Ich-Funktionen zwischen autonomen Funktionen, Abwehrfunktionen und synthetischen Funktionen. Wichtig ist dabei, wie weit die autonomen Ich-Funktionen wie Wahrnehmung, Sprache, Motorik, Denken usw. einer Behinderung, beispielsweise durch Abwehrprozesse, widerstehen können. Besonders wichtig für die Anpassung sind die synthetischen Funktionen des Ichs (Spannungsverminderung, Konfliktbereinigung, Auflösung von Widersprüchen).

Die Zahl der Ich-Funktionen ist eine Frage der Konvention, der Praktikabilität und des theoretischen Konzepts. Es wurden verschiedene Versuche unternommen, Ich-Funktionen zu quantifizieren (Übersicht bei Bellak, 1973, S. 35–49).

Soziale Unterstützung in der Krankheitsbewältigung

Die isolierte Betrachtung eines Individuums kann nur einen unvollständigen Eindruck seiner Anpassungsmöglichkeiten geben. Soziokulturelle Variablen, beispielsweise die Familie und andere Kleingruppen, können ein Individuum unterstützen und Belastungen kompensieren helfen. Man kann auch typische familiäre Anpassungsmuster finden (Übersicht bei Croog et al., 1968). So verhalten sich beispielsweise geistig behinderte Kinder bei realistischer Reaktion der Familie weniger gestört (Freedman, 1957). Auch bei Kindern mit chronisch-terminaler Niereninsuffizienz wurde bei realistischer Einstellung der Eltern eine positivere Anpassung der Kinder an ihre Krankheit und an die Hämodialyse festgestellt (Steffen et al., 1974). Ähnliche Beobachtungen wurden bei jugendlichen Diabetikern, Apoplektikern (Robertson und Suinn, 1968; Overs und Belknap, 1967) und Patienten mit traumatischen Paresen gemacht (Adams und Lindemann, 1974).

Eine Verbesserung der Prognose bei familiärer Unterstützung konnte in der Literatur gezeigt werden (Wooley et al., 1978).

Aktivität und Selbsthilfevermögen fördern eine erfolgreiche Krankheitsbewältigung. Erfolglose Patienten tendieren dazu, sich vornehmlich mit der Fürsorge durch andere zu beschäftigen. Patienten, die in eine intakte Familie zurückgehen, sind erfolgreicher, umgekehrt scheinen – so die Schlußfolgerung der Autoren – Patienten ohne familiäre Unterstützung medizinische Institutionen gewissermaßen auch als Familienersatz aufzusuchen (Wooley et al., 1978). Görres et al. (1988) weisen auf die Bedeutung familiärer Bewältigungsstile für den Krankheitsverlauf bei Patienten mit multipler Sklerose hin (vgl. Parekh, 1988; Morgan et al., 1984 zum social support).

62.2 Anpassung bei körperlicher Krankheit

62.2.1 Bewertung von Anpassungs- und Bewältigungsstrategien

Eine gute und erfolgreiche Anpassung ist ein theoretisches Konstrukt, das ähnlich vage ist wie die Begriffe der Normalität oder Gesundheit. Üblicherweise wird die Anpassung körperlich Kranker aufgrund des Vorhandenseins oder Fehlens ausgeprägter Verstimmungszustände bewertet. Als Kriterien dienen zudem Kooperationsbereitschaft, Angepaßtheit des Verhaltens an soziale Normen und traditionelle Erwartungen im Krankenhaus, das Verhältnis von Willen und Fähigkeit zur Arbeit oder das Rehabilitationsergebnis, meist gemessen in Form der Wiederaufnahme beruflicher oder häuslicher Tätigkeit. Flexibilität, Rationalität und Effizienz sind wichtige Kriterien, um Anpassungsstrategien zu bewerten (Lazarus, 1974; Monat und Lazarus, 1977; Cohen und Lazarus, 1980).

Zur Bewertung der Anpassung kann man grundsätzlich ihr Ergebnis heranziehen. So kann man der Beurteilung das Auftreten psychopathologischer Phänomene, objektiv meßbare physiologische Indikatoren, die Hinweise auf eine besondere Belastung geben können, und somatische Reaktionen zugrunde legen. Man geht dabei von der theoretischen Annahme aus, daß eine Optimierung psychischer Funktionen mit einer Minimalisierung psychophysiologischer Reaktionen unter Belastung korreliere.

Haan (1969, 1977) geht von einem strukturalistischen Ich-Modell aus, das psychoanalytische und Vorstellungen von Piagets Lernmodell der Akkommodation und Assimilation einbezieht. Dabei ist Coping ein normatives Verhalten, das kontinuierlich in den rigideren Bereich der Abwehr und schließlich in den der psychopathologischen Vorgänge übergeht. „... the person will cope if he can, defend if he must and fragment if he is forced to do so ...“

Im Modell Haans wird vorgeschlagen, Ich-Prozesse unter dem Gesichtspunkt der Selbstbehauptung des Individuums in allgemeine Anpassungs- und Bewältigungsprozesse, Abwehrvorgänge und schließlich bruchstückhafte Prozesse als Ausdruck eines Ich-Versagens einzuteilen.

Anpassungs- und **Bewältigungsmechanismen** (**„Coping“**) werden durch ein flexibles, zweckgerichtetes Verhalten charakterisiert, das Wahlmöglichkeiten offenläßt und autonom organisiert ist. Dieses Verhalten ist zukunftsbezogen und berücksichtigt die Notwendigkeit der Gegenwart entsprechend den realen Bedürfnissen der jeweiligen Situation. Es ist weiterhin durch sekundärprozeßhaftes Denken charakterisiert, das bewußte und vorbewußte Momente miteinbezieht. Es richtet sich nach den körperlichen Bedürfnissen des Organismus und hat die Tendenz, störende Affekte in Grenzen zu halten. Die Befriedigung von Wünschen wird auf offene, regelrechte und angemessene Art ermöglicht.

Abwehrverhalten ist im Gegensatz dazu rigide, zwanghaft. Es ist hauptsächlich durch die Vergangenheit bestimmt und dabei wird die gegenwärtige Situation in einer verzerrten Weise wahrgenommen. Primärprozesse und unbewußte Elemente sind stärker vertreten. Die Annahme spielt eine Rolle, daß beunruhigende Affekte auf magischem Wege entfernt werden können. Typischerweise wird die Befriedigung von Wünschen zum Teil durch Flucht vor der Wirklichkeit erlaubt.

Ich-Versagen angesichts einer Bedrohung soll durch stereotypes, ritualisiertes und automatisiertes Verhalten gekennzeichnet sein, das sich nach privatistischen Annahmen richtet, das, unabhängig von der gegebenen Situation, primär durch instinktive Bedürfnisse bestimmt ist. Es herrscht eine Überschwemmung durch Affekte und Emotionen vor. Das Verhalten selbst erlaubt die unregulierte Erfüllung einzelner Wünsche. Im Unterschied zu Abwehrprozessen verkleinern oder negieren diese Prozesse, die einem Ich-Zusammenbruch entsprechen, nicht die Bedrohung. Auch setzen sie sich nicht mit der Bedrohung in der Reihenfolge von Bewertung – Entscheidung – Handlung auseinander, wie das bei reiferen Anpassungsprozessen der Fall ist (Haan, 1977).

Heim und Mitarbeiter (1983), die analog zum „Situationskreis" systemtheoretische Vorstellungen integrieren, verweisen mit Recht auf die Komplexität der Bewältigungsversuche mit ihren ständig wechselnden adaptiven und Abwehraspekten. So können dabei kognitive Vorgänge auf die Wahrnehmung zurückwirken und diese sekundär verändern. Auch in einer Reihe klinischer Untersuchungen, z.B. mit Hämophilen (Kipnowski, 1981) und Patienten mit medikamentös induzierter pulmonaler Hypertonie (Heim et al., 1978), erwies sich Coping als ein situationsspezifischer Prozeß, der allerdings stark in prämorbide Muster von Situationsbewältigung eingebunden ist (Davies-Osterkamp und Salm, 1980; Hamburg, 1974; Moos, 1977).

Cassileth und Mitarbeiter (1984) verglichen die Anpassungsprozesse bei fünf verschiedenen Gruppen von Patienten mit chronischen Erkrankungen (Rheuma, Diabetes, Krebs, Nierenerkrankung, Hauterkrankungen). Unabhängig von der Art der Erkrankung fanden sich bei der Auswertung Hinweise für eine mit der Dauer der Erkrankung zunehmend bessere Krankheitsbewältigung.

In Anlehnung an den Streßverarbeitungs-Fragebogen (SVF, Janke et al., 1978) wurde ein Instrumentarium entwickelt, um einzelne Bewältigungsformen zu unterscheiden

(Heim et al., 1982, 1983, 1988a). Dabei wurden etwa 20 verschiedene Formen getrennt.

Ein weiterer Versuch, Bewältigungs- und Anpassungsformen zu gliedern und für klinische Untersuchungen zu operationalisieren, richtet sich an der Einteilung in Prozesse mit vorwiegend internalisierenden und solchen mit vorwiegend externalisierenden Transaktionen aus. Wichtig ist dabei das Konzept von **Kontrollüberzeugung** bzw. **Kausalattribution** (vgl. Strickland, 1978). Sie beziehen sich auf generalisierte Erwartungen eines Individuums hinsichtlich der Frage, ob es durch eigenes Verhalten wichtige Ereignisse in seinem Leben verändern kann (internale Kontrolle) oder sich vorwiegend von außen bestimmt erlebt (externale Kontrolle). Maladaptives Krankheitsverhalten fand sich dabei in einer Reihe von Studien mit dem Vorherrschen externaler Kontrollüberzeugung korreliert. Von hier ergibt sich eine Beziehung zum Konzept der „erlernten Hilflosigkeit", die die Fähigkeit zur Streßbewältigung verringert und der eine Rolle in der Auslösung von Erkrankungen zugeschrieben wird (Abramson et al., 1978). Andererseits weist eine größere prospektive Studie mit Mammakarzinompatientinnen darauf hin, daß internale Ursachenzuschreibungen vermehrt mit einem depressiven Verarbeitungsmodus der Erkrankung korreliert sind (Riehl-Emde, 1989).

Untersuchungen der Bewältigungsstrategien von Patienten mit bestimmten Erkrankungen verwenden meist entweder nur globale Kriterien für Coping-Merkmale (wie gut – schlecht, Qualität sozialer Beziehungen), z.B. Weisman und Worden (1975), die die Überlebensdauer von Krebspatienten untersuchten und ein längeres Überleben mit „besserem Coping" korreliert fanden. Zum Teil werden auch nur indirekte Hinweise für Bewältigungsformen, z.B. das Auftreten psychopathologischer Merkmale wie Depression und Angst, mit der Überlebensdauer korreliert (Rogentine et al., 1979 bei Melanompatienten).

62.2.2 Faktoren, welche die Anpassungsprozesse bei körperlich Kranken bestimmen

Zur Antwort auf psychologischen Streß im Gefolge körperlicher Erkrankungen tragen viele Faktoren bei. Sie lassen sich drei Bereichen zuordnen:
– dem zugrundeliegenden körperlichen Leiden,
– Persönlichkeitsmerkmalen,
– der aktuellen Gesamtsituation des Kranken wie sozialer Hintergrund, verfügbare soziale Kontakte, Arzt-Patient-Beziehung, Bedeutung der Hospitalisierung, strukturelle Gegebenheiten des Krankenhauses.

Charakteristische Merkmale der körperlichen Erkrankung

Dazu gehören die Schwere der Krankheit, zeitliche Faktoren und betroffene Organsysteme.

Beispielsweise reagierten in einer Gruppe herzkranker Patienten alle nach Auftreten der Krankheit mit depressiven

Symptomen. Bei mildem Krankheitsverlauf bildeten sich diese rasch zurück, war das Leiden schwer, blieben sie bestehen (Dovenmühle und Verwoerdt, 1959). Malignompatienten mit schnell wachsenden Tumoren waren ängstlicher als solche mit langsam wachsenden Neoplasmen (Verwoerdt und Elmore, 1967; Verwoerdt, 1972).

Betroffene Organsysteme

Die Wertigkeit des betroffenen Organsystems im Erleben eines Patienten wird durch objektive und subjektive Merkmale bestimmt, z.B. Sichtbarkeit von außen (z.B. Haut), vitale Funktion (Herz, Lunge, Leber), ebenso durch seine symbolische Bedeutung. Bei der symbolischen Bedeutung eines Organs kann es sich um individualspezifische Vorstellungen oder um überindividuelle, beispielsweise kulturelle oder gesellschaftsspezifische Vorstellungen handeln (z.B. Herz). Die psychodynamische Bedeutung eines Organs oder Körperteils ist eng mit der Reaktion bei Schädigung oder Verlust verbunden. Organe oder Körperteile können dabei Quellen von Vergnügen, Stolz und Selbstvertrauen sein, Hilfen, um befriedigende zwischenmenschliche Beziehungen aufrechtzuerhalten, Mittel zur Verringerung intrapsychischer Konflikte, Unterstützung des Gefühls der persönlichen Identität, der Stabilität des Körperschemas und der erwünschten sozialen Rolle. Zur Illustration:

Die Reaktion kann beispielsweise übersteigert wirken bei einem Patienten mit einer Hirnerkrankung und einer Zwangsstruktur, weil intellektuelle Fähigkeiten dem Patienten Sicherheit und Selbstbewußtsein verleihen. Sie kann auch besonders stark sein bei einem Patienten mit einer Kolostomie, dem Sauberkeit und Ordnung „die halbe Welt" bedeuteten, ebenfalls bei solchen Patienten mit körperlichen Behinderungen, die höchsten Wert auf physische Aktivität und Fitness legen.

Bei radikalen Veränderungen des Körperschemas finden sich häufig Scham und Rückzug. Die tatsächliche oder antizipierte Reaktion der Bezugspersonen ist für diese Patienten, die nicht nur „krank", sondern durch ihre Einstellung „anders" geworden sind, besonders wichtig (Adams, 1974).

Bedeutung der Erkrankung

Patienten sehen die Bedeutung der Erkrankung im Licht ihrer Erfahrungen, ihres Wissens, ihrer Wertvorstellungen und Bedürfnisse. Die Bedeutung mag sich im Verlauf der Krankheit ändern, doch finden wir jeweils eine Vorstellung zu einem bestimmten Zeitpunkt vorherrschend. Wichtig erscheint dabei, ob eine internale oder externale Attribution vom jeweiligen Individuum bevorzugt wird. Daraus leiten sich bestimmte Reaktionsformen und Typen von Bewältigungsmustern ab. Lipowski (1983) unterscheidet unter den Bedeutungen vier Hauptkategorien: Herausforderung/Bedrohung, Verlust, Gewinn/Erleichterung, Bestrafung. In Tabelle 62–1 finden sich einzelne Bedeutungen und mögliche Folgen aufgelistet. Die Reaktion der Umgebung wirkt dabei auf die Beurteilung der Krankheit zurück, z.B. im Sinne einer Verstärkung beispielsweise des Verlusterlebens. Ge-

Tabelle 62–1. Auswirkungen der individuellen Bedeutung einer Krankheit für den Patienten

Bedeutung der Krankheit	Besonders häufige Folgen
Herausforderung	flexible und rationale Strategien
Feind	Angst, Feindseligkeit, projektives Denken
Bestrafung (ungerecht)	Aggression, Wut
Bestrafung (gerecht)	oft passive Hinnahme, gelegentlich psychische Entlactung
Schwäche	Scham, Kränkung, Verleugnung
irreparabler Verlust und Schaden	Depression, Feindseligkeit

schlechtserkrankungen, Epilepsie und Krebs haben häufig eine Konnotation einer sozialen Stigmatisierung.

Krankheit kann auch als Erleichterung oder Gewinn erlebt oder als Strategie eingesetzt werden. So kann es bei prämorbide bestehenden schweren Schuldgefühlen mit Krankheitseintritt zu einer paradoxen Besserung depressiver Verstimmungen kommen. Die körperliche Krankheit kann dabei unbewußt einem Reparationsversuch entsprechen, einen inneren Konflikt lösen helfen (Beck, 1981).

Als Strategie kann Krankheit beispielsweise von Eltern benutzt werden, um ihre Kinder um so fester an sich zu binden.

Persönlichkeitsmerkmale

Die Anpassung wird individuell beeinflußt von Faktoren wie Alter, Geschlecht, sozialer Status, Bildungsgrad, geistige Fähigkeiten, Persönlichkeitsentwicklung, beispielsweise frühe Objektbeziehungen, Persönlichkeitsstruktur, Körperschema und durch intrapsychische Konflikte, welche die psychische Flexibilität beeinträchtigen.

Mit dem Einfluß von Persönlichkeitsmerkmalen bei der Verarbeitung von Trennungserlebnissen, Lebensveränderungen und sozialen Krisen auf Krankheitsverlauf und Krankheitsverhalten befaßten sich zahlreiche Studien (z.B. Imboden et al., 1963: Trennungsangst; Rahe und Arthur, 1968: life change units; Schmale, 1958: Trennung, Depression).

Prämorbide Persönlichkeitsstrukturen können unter der Krankheitsbelastung deutlicher werden: Der Zwanghafte, der alle Dinge erklären und ordnen muß, um die Kontrolle nicht zu verlieren, kann noch zwanghafter reagieren. Krankheit braucht für passiv-abhängige oder masochistische Personen nicht unwillkommen zu sein; sie kann für eine überehrgeizige Person, die ihren Ambitionen nicht gerecht werden konnte, das kleinere Übel sein (Verwoerdt, 1972) im Sinne einer Entlastung.

Im Hinblick auf die Bewältigung belastender Situationen lassen sich neben anderen folgende Persönlichkeitstypen unterscheiden. Ihnen kommt prognostische Relevanz im

Hinblick auf Krankheitserleben und Krankheitsverhalten zu[2] (Kahana, 1972; Bibring, 1956):

- abhängige Personen mit besonderer Sensibilität gegen Verlusterlebnisse und einer ausgeprägten Tendenz zu sekundärem Krankheitsgewinn;
- pseudounabhängige Personen mit Zügen von Unreife und Abhängigkeit, mit begrenztem Verhaltensrepertoire und außerordentlicher Anfälligkeit für Fehlanpassungen;
- mütterlich-bemutternde (karitative) Personen, die im Krankheitsfall ihre Rolle aufgeben oder einschränken müssen;
- aktiv-aggressive, affektgeladene Personen mit der Tendenz, die Passivität der Krankenrolle als besondere Belastung zu empfinden, und der Tendenz zur Rebellion;
- passiv-aggressive Personen, die zu Problemen der Kooperation prädisponiert sind;
- querulatorische, argwöhnische (paranoide) Personen mit der Tendenz, in projektiver Form Fehler in der Umgebung zu suchen;
- unbeteiligt wirkende, abgesonderte, zurückgezogene (schizoide) Personen mit einer Tendenz zur Verniedlichung von Symptomen;
- dramatisierende, expressive (hysterische) Personen mit hauptsächlicher Kommunikation durch Emotionen und besonderer Anfälligkeit gegenüber einem Verlust an Attraktivität, Körperverletzung und sexuellem Versagen;
- ängstlich-gehemmte Personen mit dem ausgeprägten Bedürfnis nach Aufmunterung, die von seiten des Untersuchers besondere Rücksichtnahme erfordern;
- ordnungsliebende, kontrollierende, zwanghafte Personen mit besonderer Sensibilität gegenüber Gefühlen von Kontroll- und Autonomieverlusten;
- chronisch leidende, selbstquälerische, masochistische Personen mit der Tendenz zu sekundärem Krankheitsgewinn;
- überlegene, eitle (narzißtische) Personen mit besonderer Sensibilität gegenüber Verletzungen und Kränkungen;
- primitiv-magische Personen mit der Tendenz zu mangelnder Kooperation und häufiger Inanspruchnahme pseudomedizinischer Institutionen.

Groves (1978) beschreibt vier Stereotypen von Patienten mit einer Prädestination zu besonderen Anpassungsproblemen im klinischen Setting. Gemeinsam ist ihnen, daß Elemente von Zorn und Haß in der Arzt-Patient-Beziehung häufig vorherrschen und spezifische Gegenübertragungsreaktionen hervorrufen.

Genannt werden Patienten, die eine ausgeprägt abhängige Beziehung herstellen, sich wie „Kletten" an den Arzt heften und Aversionen hervorrufen, und solche Patienten, die sich durch ihre Erkrankung berechtigt fühlen, nach allen Seiten fordernd und erpresserisch aufzutreten. Sie provozieren oft aggressive Regungen beim Behandelnden. Drittens Patienten, die manipulativ jede Hilfe ablehnen oder frustran erscheinen lassen und oft Gefühle von Ohnmacht und Depression auslösen. Sie benutzen ihre Beschwerden wie einen Berechtigungsschein für eine Beziehung, der so lange gültig ist, wie das Symptom existiert. Verleugner mit der Tendenz zur Selbstzerstörung lösen oft Zorn und Ingrimm aus.

Patienten mit einer **Borderline-Persönlichkeitsstruktur** zeigen oft besonders maladaptive Formen des Umgangs mit Erkrankungen. Kommen sie mit medizinischen Institutionen in Berührung, führt dies nicht selten zu schweren Konflikten.

Dies hängt mit den vorwiegend primitiven Formen ihrer Abwehrmechanismen zusammen, die in ihrer Umgebung schnell Chaos, Frustration und Ärger erregen können. Spaltung, primitive Idealisierung, extensive Verleugnung und projektive Identifikation sind charakteristische Abwehrformen. Oft erleben diese Patienten die krankheitsbedingte Abhängigkeit als schwere Bedrohung und Wiederholung früherer traumatischer Erfahrungen. Da sie meist ein oberflächennahes Aggressionsreservoir aufweisen und die Neigung haben, Grenzen innerhalb von Beziehungen zu testen, werden sie nicht selten zum „enfant terrible" in klinischen Einrichtungen (Stoudemire und Thompson, 1982; Trimborn, 1983).

Situative Faktoren

Sie umfassen z.B. familiäre Beziehungen, die Reaktion der wichtigsten Bezugspersonen des Patienten auf die Erkrankung, mögliche Unterstützung innerhalb des sozialen Umfeldes, die wirtschaftliche Situation und ihre Veränderung durch die Erkrankung, die Einschätzung der Krankheitsrolle in der Gesellschaft. Dazu gehören auch Charakteristika des Aufenthaltsortes, beispielsweise des Krankenhauses mit seinem Zwang zu neuen Sozialbeziehungen, Abhängigkeit und Angewiesensein auf Ärzte und Schwestern. Am Beispiel der transluminalen Angioplastie wurde der Einfluß institutioneller Faktoren auf die Angemessenheit individueller Coping-Strategien beschrieben (Jordan, 1988).

62.2.3 Phänomenologie der Anpassung

Die Form der Auseinandersetzung mit einer Bedrohung ist häufig für ein bestimmtes Individuum spezifisch, so daß sich daraus typische Verhaltensmerkmale ableiten. Bei sorgfältiger Erhebung der Vorgeschichte läßt sich meist erkennen, wie ein Patient mit früheren Bedrohungen umgegangen ist und eine Hypothese dafür ableiten, wie er sich gegenüber der jetzigen Bedrohung verhalten wird. Phänomenologisch sind drei grundsätzlich verschiedene Formen der Auseinandersetzung mit einer Bedrohung zu unterscheiden.[3]

Kognitive Verarbeitungsweisen

Hier stehen sich als Extremformen das Prinzip der „Minimalisierung" von Bedrohung und auf der anderen Seite das Prinzip der übertriebenen Eigenbeobachtung mit maximaler Aufmerksamkeit für die Bedrohung gegenüber. Zu den kognitiven Prozessen werden auch einfache Erklärungsversuche für eine Erkrankung, eventuell mit Vorwürfen gegen sich selbst oder andere, gerechnet, ebenso wie pseudowissenschaftliche Hypothesen bis zu wahnhaften Vorstellungen.

---

2 vgl. Bibrings (1956) Beschreibung schwieriger „Patiententypen" in einem Allgemeinkrankenhaus.
3 vgl. Th. v. Uexküll, in Kap. 1, individuelle Programme mit Deutungs- und Handlungsanweisungen für das Individuum.

Affektive Verarbeitungsweisen

Hierunter faßt man das ganze Spektrum von Stimmungen, Affekten und Emotionen, von der normalen Angst- oder Trauerreaktion bis hin zu schwer pathologischen Zuständen zusammen. Auch den affektiven Stilarten sind unter teleologischem Aspekt adaptive Funktionen zuzuschreiben. So können Stimmungen, Emotionen und Affekte als „Anweisungen für Bereitstellungen" dienen (v. Uexküll, 1963) und „Handlungen" vorbereiten, die in „eine andere Stimmung hineinführen" (v. Uexküll, 1963). D.h., Stimmung kann hier Bestandteil eines zweckgerichteten Anpassungsversuches sein.

Verarbeitung durch bestimmte Handlungsmuster

Typische Möglichkeiten dieser Form der Auseinandersetzung sind Angriff, Kapitulation und Vermeidung. Vermeidung ist in der Regel von stark ausgeprägten kognitiven Prozessen wie Minimalisierungsversuchen unter Beteiligung psychischer Abwehrprozesse begleitet. Patienten, die ihre Kapitulation offen zeigen, werden nicht selten als psychiatrischer Hilfe bedürftig angesehen. Es besteht eine Ähnlichkeit zu dem, was als Giving-up-given-up-Komplex bezeichnet wird (Schmale, 1972).

Eine Stilart des Angriffs wählt z.B. der Arthritiker, der seine übertriebene sportliche Aktivität bewahrt und damit seine Gelenkschäden vergrößert. Kapitulation sollte nicht verwechselt werden mit dem Anteil passiven Akzeptierens, das bei jeder Anpassung an eine schwere Erkrankung erforderlich ist.

Die übertriebene Bevorzugung einer Stilart kann einem sozialisationsbedingten Mangel an alternativen Programmen entsprechen.

62.2.4 Krankheit und Trauerprozeß

Im Konzept der **Trauerarbeit** finden sich Merkmale des Copings und der Abwehr.

Krankheiten sind häufig mit dem Verlust von Objekten verbunden. Anpassung an eine Krankheit ist daher auch das Ergebnis der psychischen Vorgänge, die im Rahmen der Trauerarbeit ablaufen (vgl. Lindemann, 1944).

Trauer ist ein schmerzlicher Prozeß mit Kummer, Einschränkung von Interessen und Verlust an Liebesfähigkeit. Dieser Prozeß ist Ausdruck eines intrapsychischen Konflikts zwischen der realistischen Wahrnehmung des endgültigen Verlustes und dem Widerstand, ein libidinös besetztes Objekt aufzugeben.

Die Rücknahme der emotionalen Besetzung geht in kleinen Schritten vor sich und benötigt viel Zeit und Energie. Die initiale Reaktion auf den Verlust eines Liebesobjektes ist üblicherweise Verleugnung (Freud, 1917). Eine normale Zwischenstation des Trauerprozesses ist zeitweise Introjektion des Objektes, verbunden mit den Zeichen der Regression. Dieser Versuch, das Objekt zu erhalten, ist bereits ein erster Schritt zu seiner Aufgabe. Verzögerungen im Ablauf der Trauer-

arbeit oder Fixation auf einer bestimmten Stufe können zu pathologischen Erscheinungen führen. Die Trauerarbeit ist daher ein gutes Beispiel für einen adaptiven Prozeß, an dem intermittierend regressive Abwehrmechanismen beteiligt sind, deren Persistenz pathologisch sein kann.

Aufgezwungene gesellschaftliche Regeln, erneute Versagenssituationen, Selbstvorwürfe wegen eines wirklichen oder vermeintlichen Versäumnisses und das Unvermögen, sich realistisch mit dem Verlust auseinanderzusetzen (Meyer, 1977), können die Trauerarbeit ebenso stören wie Selbstvorwürfe körperlich Kranker das Ertragen ihrer Krankheit sehr erschweren können (Bard, 1956).

Eine Haltung der Versteinerung und Abkapselung, sozialer Rückzug und Aggression gegenüber der Umgebung, Ressentiments und Verbitterung, aber auch funktionelle Störungen und psychosomatische Erkrankungen können die Folge einer mißlungenen Trauerreaktion sein.

Lindemann selbst wurde auf die Bedeutung einer mißlungenen Trauerarbeit für die Pathogenese von Krankheiten aufmerksam durch die Beobachtung, daß nach einem Katastrophenereignis gehäuft Fälle von Colitis ulcerosa auftraten (1944).

Aus dem Konzept der „Trauerarbeit" scheinen uns folgende Punkte für den Umgang von Patienten mit körperlichen Erkrankungen besonders wichtig:
- Die Förderung eines bestimmten Maßes an Aktivität, das für die Patienten, besonders bei langfristiger Anpassung, notwendig wird.
- Die Beachtung, daß regelmäßig, aber zeitlich und nach Intensität begrenzt, psychische Symptome im normalen Prozeß auftreten.
- Die Notwendigkeit einer Intervention, die einen pathologisch verlaufenden Prozeß korrigieren kann. Voraussetzung dafür ist allerdings die Kenntnis des normalen Ablaufs und seiner Varianten.

Es ist nach Lindemann (1944) nicht notwendig, alle Gründe einer verfehlten Trauerarbeit aufzudecken. Auch bei der psychotherapeutischen Betreuung körperlich Kranker können Fehlentwicklungen in der Kindheit und deren Korrektur gegenüber aktuellen Hilfsmaßnahmen in den Hintergrund treten (Spiegel, 1973; Adams, 1974; Janis, 1958). Diese sollten vielmehr alle Hilfsmöglichkeiten der jeweiligen sozialen Umwelt des Patienten einbeziehen (Mechanic, 1974).

Zu berücksichtigen ist, daß sich eine pathologische „Trauerarbeit" nicht nur als individuelle Reaktion, sondern als gemeinsame Aufgabe auch im Rahmen eines Systems, z.B. der Familie, auffassen läßt. Dabei finden sich, wie auch bei der Betrachtung einzelner Individuen, typische familiäre Muster. Besonders wichtig scheint dies bei Erkrankungen von Kindern zu sein, wo Störungen der intrafamiliären Kommunikation besonders leicht zu pathologischen Verläufen führen können (vgl. Share, 1972).

Einzelne Studien, z.B. Kaplan (1971), untersuchten die Trauerreaktionen in einer Gruppe von Familien mit lebensbedrohlich erkrankten Mitgliedern (50 Familien mit Leukämikern).

Nicht selten werden akute Verlustereignisse – und diese können aus körperlichen Krankheiten resultieren – zum Auslöser der Manifestation gravierender seelischer oder körperlicher Beschwerden, meist auf dem Hintergrund entsprechender dispositioneller Merkmale, beispielsweise bei der Entstehung von Depressionen. Dementsprechend machen aufgrund frühkindlicher Erfahrungen verzerrte kognitive Strukturen (oder „Programme") es unmöglich, in Situationen mit gravierenden Verlusten die notwendige Trauerarbeit zu leisten. Dadurch werden die betroffenen Personen in besonderer Weise verletzlich. In diesem Fall sind weitergehende psychotherapeutische Maßnahmen angezeigt. In dem selben Maße, wie es Betreuungsprogramme bei Patienten mit Verlustreaktionen gibt (bereavement reactions, z.B. Horowitz et al., 1984), ist bei körperlich Schwerkranken auf Zeichen einer mißglückten „Trauerarbeit" zu achten.

62.3 Abwehrprozesse bei schwerkranken Patienten

62.3.1 Vorbemerkungen

Immer wieder erregen Gefaßtheit, Optimismus, manchmal sogar Heiterkeit von Patienten mit lebensbedrohlichen Erkrankungen beim Beobachter Erstaunen, wenn nicht gar Unverständnis. Theoretisch könnte es sich bei diesen Zuständen sowohl um eine Euphorie aufgrund toxischer oder sonstiger Einflüsse auf das Gehirn handeln als auch um psychodynamisch erklärbare Prozesse. Inwieweit die eine oder die andere dieser beiden Möglichkeiten beim jeweiligen Patienten vorherrscht, läßt sich häufig nicht mit letzter Sicherheit entscheiden. Da aber meist nur ganz bestimmte, häufig besonders belastende Aspekte der Wirklichkeit im Erleben der Patienten eine die günstigen Seiten betonende Verzerrung erfahren, kann daraus geschlossen werden, daß häufig psychologische Faktoren eine wesentliche Rolle spielen (vgl. Kap. 63).

Ein literarisches Beispiel für eine sehr differenzierte Form der Abwehr findet sich in Solschenizyns Roman „Krebsstation" in der Gestalt Wadims, eines 27jährigen Wissenschaftlers mit einem Melanom. Die reale Bedrohung wird von ihm nicht direkt geleugnet, aber in ein anspruchsvolles und ideales Selbstbild integriert, und das eigene Anspruchsniveau anfangs kompensatorisch gesteigert; erst allmählich erfolgt neben dem von Anfang an realistischen Akzeptieren der Diagnose auch ein Akzeptieren von Leiden und Schwäche, die mit der Krankheit verbunden sind.

Bei Infarktpatienten erstaunen z.B. immer wieder die Versuche, ihre Krankheit entweder nicht zur Kenntnis zu nehmen oder zu bagatellisieren, woraus bei oberflächlicher Betrachtung bei diesen vital gefährdeten Personen oft der Anschein extremen Leichtsinns und einer bei Erwachsenen schier unglaublichen Realitätsferne entsteht. Subjektiv entspricht das Verhalten der Patienten meist einer ihrer individuellen Wirklichkeit immanenten Logik.

Die folgenden Beispiele sollen die Auswirkungen von Abwehrvorgängen auf die initiale Reaktion, den Zeitpunkt der Inanspruchnahme ärztlicher Hilfe und das Krankheitsverhalten bei Infarktpatienten erläutern. Als erstes Beispiel sei ein Infarktpatient aus der Intensivstation angeführt:

> Der damals 64jährige Patient hatte schon zwei Jahre zuvor einen Herzinfarkt erlitten. Damals hatten sich an einem Samstag zur Mittagszeit, nachdem er morgens Überstunden im Betrieb gemacht hatte, typische Präkordialschmerzen eingestellt. Als die Schmerzen etwas nachließen, war der Patient in den Wald gegangen, um Teeblätter zu sammeln. Die pektanginösen Beschwerden mit Schmerzausstrahlung in beide Arme hatten sich während des ganzen Wochenendes wiederholt; trotzdem holte der Patient keine ärztliche Hilfe und wollte am Montag zur Arbeit. Eine erneute schwere Attacke verhinderte das und veranlaßte ihn schließlich, einen Arzt aufzusuchen. Mit der Diagnose „Angina pectoris" blieb der Patient dann zu Hause. Versuche, die Arbeit wieder aufzunehmen, scheiterten. Nach 6 Wochen wurde von einem Internisten ein abgelaufener Vorderwandinfarkt diagnostiziert.
>
> Jetzt, zwei Jahre später, als die Ehefrau des Patienten wegen einer Magenoperation im Krankenhaus war, erlitt der Patient wieder eine präkordiale Schmerzattacke mit Ausstrahlung in den linken Arm. Er öffnete das Fenster, versuchte zu schlafen und nahm Herzmittel seiner Frau ein. Weil er sich im Hausgang hätte erkälten können, habe er nicht versucht, Hilfe bei seinem Sohn, der im ersten Stock wohnte, zu holen. So wurde am nächsten Morgen gegen 8 Uhr derjenige Arzt verständigt, der zwei Jahre zuvor den ersten Infarkt nicht erkannt hatte, und der ihn auch dieses Mal wieder erst mit zehnstündiger Verspätung ins Krankenhaus schickte. In der Zwischenzeit hatte der Patient den Kontakt zu diesem Arzt abgebrochen gehabt.

Ähnlich berichteten Weisman und Hackett (1967) von einem Patienten mit Herzinfarkt, dessen Ehefrau zuvor an einer vom Hausarzt nicht erkannten Hirnblutung verstorben war. Der Patient hatte seitdem den Kontakt zu diesem Arzt über mehrere Jahre hinweg abgebrochen, verständigte aber, als er die Präkordialschmerzen spürte, eben diesen Arzt wieder nach langer Bedenkzeit.

62.3.2 Der Abwehrbegriff

Der Begriff „Abwehr" wurde von Freud (1894) verwendet, um zu illustrieren, wie unangenehme Vorstellungen und Affekte in bestimmten Fällen von Hysterien, Phobien, Zwängen und Halluzinationszuständen nicht wahrgenommen wurden. Der Begriff bezog sich primär auf die psychoanalytische Situation und wurde im wesentlichen aus Erfahrungen bei der psychoanalytischen Behandlung von Patienten abgeleitet und weiterentwickelt. Abwehr diene als

„... allgemeine Bezeichnung für die Techniken, deren sich das Ich in seinen eventuellen, zur Neurose führenden Konflikten bedient, während Verdrängung der Name einer bestimmten solchen Abwehrmethode bleibt, die uns infolge der Richtung unserer Untersuchungen zuerst besser bekannt geworden ist ..." (Freud, 1926, S. 186).

Abwehr bezeichnet auf dem Hintergrund des psychoanalytischen Instanzenmodells den Modus, mittels dessen sich das Ich mit unerträglichen Vorstellungen und Affekten, aber auch mit Aspekten der äußeren Wirklichkeit auseinandersetzt.

62.3.3 Abwehrmechanismen

Es handelt sich um die allen Arten von Abwehroperationen zugrundeliegenden Ich-Leistungen, die außerhalb des Bewußtseins ablaufen, aber analoge bewußte Derivate haben, z.B. als bewußte Entsprechung von Verdrängung Unterdrückung, als bewußtes Analogon zur Verleugnung Vermeidung. Die Operationen wirken dabei als Bestandteil einer Abwehrorganisation, die als „Substruktur des Ichs" bezeichnet werden kann. Sie erfüllen Notfalls- und Dringlichkeitsfunktionen.

Zu berücksichtigen ist ihre Verbindung mit den „autonomen" und „synthetischen" Funktionen des Ichs.

Abwehrfunktionen können die autonomen Prozesse beeinträchtigen, z.B., wenn Verdrängung das Gedächtnis, wenn Verleugnung die Wahrnehmungsfunktion oder Isolierung die synthetische Funktion vermindern.

62.3.4 Die Ursache von Abwehr

In der klassischen psychoanalytischen Theorie bezog sich Abwehr auf die Unterdrückung von Affekten, von sexuellen und aggressiven Triebgefahren. Sie vermögen gemeinsam Angst auszulösen. Angst als Gefahrensignal, das Kenntnis von einer äußeren oder inneren Bedrohung gibt, ist Auslöser für das Auftreten von Abwehrmechanismen. Die Gefahren können äußere, innere, verinnerlichte oder veräußerlichte Bedrohungen sein.

Ein Beispiel für eine äußere Gefahr wäre die Bedrohung durch einen operativen Eingriff, für eine innere Gefahr aggressive, vom Über-Ich mißbilligte Regungen. Eine verinnerlichte Gefahr kann am Beispiel eines Patienten illustriert werden, der vom Arzt während der Visite das ungünstige Ergebnis einer Lymphknotenbiopsie mitgeteilt erhielt. Einige Tage später, nachdem er diese Information erhalten, die schlimme Bedeutung aber abgewehrt hatte, äußerte er die maligne Diagnose als eigene vage Vermutung, womit er offensichtlich die bedrohliche Situation etwas zu entschärfen vermochte. Ein Beispiel für eine veräußerlichte Gefahr sind gefährliche innere, beispielsweise aggressive Regungen, die nach außen projiziert und dort im Rahmen der Objektbeziehungen bekämpft werden.

Ursprünglich ausgehend von trieb- und ichpsychologischen Konzepten ist Abwehr im Rahmen objekt- und selbstpsychologischer Theorien darauf ausgerichtet, Sicherheit und Wohlbefinden in der Beziehung zu den Objekten oder auch zum eigenen Selbst

zu bewahren (Joffe und Sandler, 1967). Lichtenberg und Slap (1972) betonen die Notwendigkeit der Abwehr schmerzhafter Affekte als Voraussetzung für die Ausbildung der Abwehrorganisation. Hoffmann (1987) plädiert dafür, der Abwehr narzißtischer Kränkungen und Beeinträchtigungen als Motiv zum Erhalt eines ausgeglichenen Selbstwertgefühls für die Entstehung der Abwehrorganisation eine entscheidende Bedeutung beizumessen.

62.3.5 Die Rolle der Abwehr

Der Abwehr kommt eine Doppelrolle zu: Abwehrmechanismen haben, besonders in der frühen Entwicklung und in der Latenzperiode, weitgehend Anpassungsfunktion und behalten auch im Erwachsenenalter adaptive Aspekte (Cremerius, 1968; Hartmann, 1960; A. Freud, 1965).

Allerdings muß nach dem Zeitpunkt des Auftretens in der psychischen Entwicklung, nach Intensität, Rigidität bzw. Flexibilität, Dauer und dem Bestehen einer gewissen Balance differenziert werden. Inwieweit die Mechanismen als pathologisch anzusehen sind, scheint neben dem zeitlichen Faktor und der Intensität einzelner Mechanismen für jeden Mechanismus unterschiedlich zu sein. Als relativ pathologisch bei Erwachsenen werden ein Vorwiegen von Verleugnung, Regression, Projektion und Introjektion angesehen (Bellak, 1973), wobei man zusätzlich nach dem Gesichtspunkt der die Abwehr auslösenden Bedrohung differenzieren muß. Einzelne Beurteilungskriterien sind:

– Erfolg oder Mißerfolg von Abwehroperationen, gemessen daran, ob die angstauslösende Ursache weiter abgewehrt werden muß;

– das Verhältnis des Anteils von Primär- und Sekundärprozessen, das Kriterium des „Angemessenseins" an die Situation;

– qualitative, auf das Verhalten bezogene Variablen, z.B. der Grad der Stereotypie und Rigidität des von der Abwehr bestimmten Verhaltens.

Bei körperlich Kranken sind Abwehrphänomene in solcher Häufigkeit zu beobachten, daß es absurd wäre, ihr Auftreten unter diesen Umständen von vornherein als pathologisch zu werten (Heim et al., 1978, 1979, 1983; Beutel, 1985, 1988a, b; Heim, 1986, 1988a).[4]

Für den klinischen Gebrauch dürfte es angebracht sein, zwischen bestimmten Abwehrphänomenen und dem psychoanalytischen Begriff der Abwehrmechanismen zu unterscheiden. Abwehrmechanismen dienen als theoretische Konstrukte zur Beschreibung einer psychischen Funktionsweise, um unerwünschte Impulse und Realitätsanteile abzuwenden oder in einem mehr akzeptablen Sinn zu verändern. Der globale Begriff Abwehr bezieht sich auf komplexe Erlebens- und Verhaltensweisen mit Abwehraspekt, an

---

4 Eine ausführliche Darstellung der Abwehrmechanismen mit vielen Beispielen für den klinischen Gebrauch findet sich bei White und Gilliland (1975).

denen Abwehrmechanismen beteiligt sind. Diese sind unbewußt und dienen in bestimmten Situationen als Ersatz für eine wirksamere Art, mit einer schwierigen Situation fertig zu werden. Zu einer echten Problemlösung tragen sie nicht bei. Stehen ihre Anwendung und das Resultat aber in einem angemessenen Verhältnis zum Problem oder zur Situation oder zu den individuellen Fähigkeiten, Probleme zu lösen, haben sie dennoch eine adaptive Funktion. Allerdings ist die Wahrscheinlichkeit eines situationsgerechten Verhaltens größer, wenn alle Aspekte des Problems wahrgenommen, alle Alternativen durchdacht und die passendste gewählt werden (White, 1974).

62.3.6 Einzelne Kategorien von Abwehrmechanismen

Eine Einteilung der verschiedenen Abwehrmechanismen kann nach **funktionalen** oder **genetischen** Gesichtspunkten vorgenommen werden. Mit der Bevorzugung einzelner Abwehrmechanismen findet man häufig einen bestimmten Verhaltensstil. So kann man bei Patienten, die, unter dem Gesichtspunkt der psychischen Entwicklung betrachtet, frühe Mechanismen bevorzugen, oft insgesamt ein unreifes Verhalten feststellen. Entsprechend kann man bei den Abwehrmechanismen solche reiferen wie die Verdrängung von einer anderen Gruppe regressiver Abwehrmechanismen mit schwacher Gegenbesetzungsfähigkeit (wie z.B. Identifizierung, Projektion) unterscheiden (Moser, 1964). Bei letzteren wird die Abwehr besonders über eine **Manipulation der Objektbeziehungen** vollzogen. Dies kann beispielsweise durch ein Oszillieren im Abstand zu einem Objekt, durch einen Wechsel von Nähe und Ferne geschehen.

Man kann dies bei regredierten schwerkranken Patienten häufig beobachten. Oft sind zunächst solche Objekte betroffen, von denen der Patient nicht vital abhängig ist, z.B. das medizinische Hilfspersonal.

Beispiel: Empfindet ein Patient Aggressionen gegenüber den behandelnden Ärzten, kann er diese z.B. abwehren, indem er eine Schwesternschülerin ständig ohne eigentlichen Grund ruft und unverrichteter Dinge wieder wegschickt (Manipulieren des Objektes).

Als reifer Mechanismus, der eine psychische Differenzierung voraussetzt, stehe die Verdrängung an der Spitze der Hierarchie der Abwehrmechanismen (Moser, 1964). Falls sie unzureichend wirksam sei, träten sog. sekundäre Mechanismen wie Rationalisierung, Intellektualisierung, Isolierung und Verleugnung auf, schließlich immer weniger differenzierte und ontogenetisch früher erworbene Mechanismen wie Projektion und Identifikation, die als regressive Abwehrmechanismen bezeichnet werden.

Eine Reihe von Klassifikationsschemata wurden entwickelt (Byrne, 1961) nach den Polen Repression versus Sensitization (Lazarus und Launier, 1978; Suppes und Warren, 1975). Letztere Autoren nehmen eine Einteilung nach Typen elementarer Transformationen zwischen Handlungsträger, Handlung und dem

Tabelle 62–2. Kurzdefinition von Abwehrmechanismen nach dem Subjekt-Objekt-Schema (nach Ehlers, 1983, S. 57).

- Wendung gegen die eigene Person: S wendet Aggressivität (D) gegen sich selbst (O)
- Introjektion: S verbindet sich mit ganzem Objekt (O), das abgewehrten Triebimpuls (D) repräsentiert
- Identifikation: S verbindet sich mit Teilfunktionen des Objektes (O), die Triebimpulse (D) repräsentieren
- Projektion: S verschiebt eigene abgewehrte Triebimpulse (D) auf das Objekt (O)
- Regression: S gibt ödipale Triebimpulse (D_1) auf und zieht sich zurück auf prägenitale Triebimpulse (D_2)
- Ungeschehenmachen: S nimmt einen verbotenen Triebimpuls (D) gegenüber einem Objekt (O) zurück
- Reaktionsbildung: S zeigt gegenüber Objekt (O) gegenteilige Einstellungen und Verhaltensweisen, als der verpönte Triebimpuls (D) erwarten ließe
- Verkehrung ins Gegenteil: S zeigt gegenüber Objekt (O) gegenteiligen Triebimpuls (D), als auslösende Situation erwarten ließe
- Isolierung: S trennt zusammengehörige Einfälle oder Handlungen sowie Vorstellung und Affekt, um Berührung des verpönten Triebimpulses (D) mit der eigenen Person (O) zu vermeiden
- Verzögerte Affektausbrüche: S verzögert Affektausbruch, um Zusammenhang mit bedrohlichem Triebimpuls (D) oder Objekt (O) zu vermeiden
- Affektäquivalente dominieren: S ist sich der affektiven Bedeutung einer den Affekt normalerweise begleitenden Körperreaktion nicht bewußt
- Verschiebung von Libido: S besetzt eine weniger bedrohliche Vorstellung oder Handlung gegenüber einem Objekt (O) mit Triebenergie (D)

Objekt der Handlung vor. Als Beispiel sei eine Zusammenstellung nach dem Subjekt-Objekt-Schema in Tabelle 62–2 aufgeführt.

Grundprobleme der Taxonomie und Bewertung der Abwehrmechanismen sind noch weit von einer Lösung entfernt (Beutel 1985, 1988a und b). Der Abwehrstil erwies sich in einer Untersuchung als unabhängig von DSM-III-Diagnosen (Bond, 1986).

62.3.7 Einzelne Abwehrmechanismen

Zu den wichtigsten Abwehrmechanismen, die unbewußt Verhalten und Befinden von körperlich Kranken beeinflussen, gehören (Laplanche und Pontalis, 1972):

Verdrängung

Sie bewirkt, daß nicht akzeptables Es-Material, meist mit einem Trieb zusammenhängende Vorstellungen (Gedanken, Bilder, Erinnerungen), sog. „Vorstellungsrepräsentanzen" des Triebs, aus dem Bewußtsein verbannt und durch eine Gegenbesetzung[5] am Wie-

---

5 Am häufigsten findet sich als Gegenbesetzung eine Reaktionsbildung, gelegentlich in Form einer Zwangshandlung, z.B. eines Waschzwanges.

dereintritt gehindert werden (vgl. Laplanche und Pontalis, 1972).

Beispiel: Ein Patient mit Herzinfarkt zeigt auf Station deutliche Anzeichen von Zorn und Aggressivität gegenüber dem Krankenhaus und der Behandlung; im Gruppengespräch, in dem Patienten ermuntert werden, über ihre Probleme zu reden, sagt er nichts über diese Gefühle, sondern äußert sich im Gegenteil außerordentlich lobend über seine Behandlung. Dahinter stehen aber Enttäuschung und Wut über seine Erkrankung. Im Verlauf der Sitzung berichtet er dann über einen Vorfall, bei dem Jugendliche eine Schaufensterscheibe zertrümmerten. Der Patient gerät darüber in einen ausgeprägten Wutzustand und redet sich seinen Zorn in einer heftigen Schimpfkanonade über „die Jugend" vom Leib.

Das Beispiel zeigt ein Zusammenwirken mehrerer Mechanismen. Dem ursprünglich verdrängten Affekt konnte durch Verschiebung auf ein ungefährlicheres Objekt, nämlich eine gesellschaftliche Gruppe, der häufig die „Sündenbockrolle" zufällt, Abfuhr verschafft werden, was insgesamt für den Patienten sehr viel weniger bedrohlich war. Der Arzt bleibt so verschont, da der Patient die Beziehung zu ihm benötigt.[6]

Rationalisierung, Intellektualisierung

Rationalisierung bedeutet einen Versuch zur logischen, verstandesmäßigen und moralischen Untermauerung eines alternativen und eher akzeptablen Vorgehens, Gedankens oder Gefühls. Es kann Denkweisen beinhalten, die tatsächliche Ursache-Wirkungs-Beziehungen ignorieren, unwichtige Aspekte einer Situation überproportional betonen und wichtige herunterspielen.

Intellektualisierung zeichnet sich durch das Übergewicht aus, das dem abstrakten Denken gegenüber auftauchenden Affekten und Phantasien gegeben wird. Sie ist stärker als die Rationalisierung darauf bedacht, Affekte in Distanz zu halten und zu neutralisieren.

Beispiel: Die bei Herzinfarktpatienten häufige Interpretation ihrer Beschwerden als Verdauungsstörungen und die „rationale" Erklärung durch die Einnahme bestimmter Speisen oder Getränke. Möglicherweise Überlegungen, wie häufig doch solche Verdauungsstörungen und wie wenig gefährlich sie in der Regel sind (Intellektualisierung).

Isolierung vom Affekt

Isolierung bedeutet, daß ein Erlebnis, ein Verhalten oder eine Vorstellung usw. nicht vergessen werden, aber ihren Gefühlsgehalt und/oder ihre assoziativen Verbindungen mit anderen Gedanken oder mit der übrigen Existenz des Individuums verlieren.

Beispiel: Der Patient, der seine schwere Krankheit detailliert kennt (z.B. durch die Lektüre von Fachbüchern) und mit dem Arzt detailliert, aber ohne Äußerung von Affekt, diskutiert. Bewußtes Analogon: Selbstbeherrschung („Ich weiß, wie schwer ich verletzt bin, und weiß, wie es passierte, aber es macht mir nichts aus!").

Reaktionsbildung

Es handelt sich um die Abwehr von Impulsen und Strebungen durch solche gegensätzlicher Bedeutung.

Häufiger erfolgt die Reaktionsbildung im Sinne der Gegenbesetzung zusammen mit Verdrängung.

Beispiel: Ein Patient bekämpft unbewußte aggressive Regungen gegenüber dem Krankenhauspersonal dadurch, daß er betont untertänig und freundlich auftritt.

Verschiebung

Darunter versteht man die ganze oder teilweise Verlagerung emotionaler oder sonstiger Komponenten einer Vorstellung, Situation oder eines Objektes auf eine andere, akzeptablere Vorstellung, Situation oder ein anderes Objekt, das mit dem ersten durch eine Assoziationskette verbunden ist. Die Gefühlskomponente bleibt dabei konstant.

Beispiel: Ein Koronarpatient verneint, von Herzschmerzen gepeinigt und beunruhigt zu sein, er beklagt sich statt dessen über das Kribbeln in den Zehen, d.h., die beängstigende Wahrnehmung der Schmerzen wird auf ein weniger bedrohliches Gebiet verschoben.

Identifizierung, Introjektion

Identifizierung bedeutet unbewußte Übernahme eines fremden Verhaltens bzw. fremder Werte, Einstellungen und Gefühle eines anderen. Introjektion ist Aufnahme eines Vorbildes im Sinne einer Einverleibung in das Ich, das es zum Gegenstand unbewußter Phantasien macht (Ferenczi, zit. nach Laplanche und Pontalis, 1972). Der seelische Vorgang kann als ein körperlicher erlebt und symbolisiert werden bis hin zu kannibalistischen Einverleibungsphantasien.

Beispiel: Der zwanghafte Patient, der sich nach Nierentransplantation mit rigiden Vorschriften bezüglich Infektionsprophylaxe und Medikation so stark identifiziert, daß er noch nach Monaten ein unnützes Ritual der Infektionsverhütung zelebriert, um Angst, beispielsweise vor Abstoßungskrisen, abzuwehren. Bewußtes Analogon: Imitation.

Projektion

Damit werden unerwünschte Vorstellungen, Impulse, Gefühle und Objekte nach außen einem anderen Objekt zugeschrieben, wobei es sich um eine Abwehr sehr archaischen Ursprungs handelt.

Beispiel: Der schwerkranke Patient, dessen größte Besorgnis es ist, ob seine Angehörigen seine Krankheit emotional verkraften können. Oder der Patient, der seine eigene Aversion und Aggression gegen die Behandlung auf Ärzte und Schwestern projiziert und in ihrem Verhalten Feindseligkeit empfindet.

Projektive Identifikation

Das auf M. Klein zurückgehende, insgesamt in der Literatur recht uneinheitlich definierte Konzept beinhaltet, daß nach Projektion unerwünschter Impulse

---

6 Dieses Beispiel zeigt das Zusammenwirken mehrerer Mechanismen (Verdrängung – Reaktionsbildung – Verschiebung).

und Vorstellungen auf eine andere Person der Patient sich mit dieser in dem Versuch identifiziert, in dieser Beziehung negative Selbstanteile unterzubringen. Die Person wird in der Regel nicht gemieden, da sich das Subjekt im Objektanteil weiter erlebt.

Ogden (1979) beschreibt drei Schritte des Prozesses der projektiven Identifikation:
- Die Phantasie, einen Selbstanteil bei einer anderen Person unterzubringen, der diese beeinflußt.
- Im Rahmen der Beziehung auf das Objekt ausgeübter Druck, gemäß der Projektion zu fühlen, zu denken und zu handeln.
- Reinternalisation der projizierten Gefühle. Typisch ist, im Unterschied zur einfachen Projektion, ein zumindest temporäres Verschwimmen von Selbst-Objekt-Grenzen bei der projektiven Identifikation. Das Subjekt fühlt sich aufs engste mit dem Objekt verbunden, empfindet so, als erlebe der Empfänger **seine** Gefühle.

Ungeschehenmachen

Bei diesem psychologischen Mechanismus ist „das Subjekt bemüht, so zu tun, als ob Gedanken, Worte, Gesten und abgelaufene Handlungen nicht geschehen wären". Es kann sich ausdrücken in Handlungen „irrational magischer Art", die die Vergangenheit aufheben. Solche Akte des Ungeschehenmachens erfolgen häufig im Rahmen der Verleugnung.

Beispiel: Der Infarktpatient, der sich nach Eintritt der Infarktzeichen physischen Belastungen aussetzt, z.B. eine größere Wanderung unternimmt, um sich selber Fitness zu beweisen.

Regression

Sie spielt bei körperlich schwerkranken Patienten eine besondere Rolle. Der Begriff enthält die Vorstellung einer Zurückentwicklung psychischer Vorgänge von einem bereits erreichten Zustand auf ein zuvor erreichtes Niveau. Es liegt das Bemühen zugrunde, auf frühere, in damaligen Situationen bewährte, einfachere Programme zurückzugreifen.

Entsprechend den verschiedenen Betrachtungsweisen der Psychoanalyse vollzieht sich dabei
„topisch" ein Zurückgehen „entlang einer Folge von psychischen Systemen" innerhalb des psychischen Apparates, „die von der Erregung normalerweise in einer vorgegebenen Richtung durchlaufen werden";
zeitlich eine Rückkehr zu Etappen, die in der Entwicklung bereits überschritten waren, und
im **formalen Sinn** der Übergang zu „Ausdrucksformen und Verhaltensweisen eines vom Standpunkt der Komplexität, der Strukturierung und der Differenzierung aus niedrigen Niveaus" (Laplanche und Pontalis, 1972). Das Ausmaß der Regression ist dabei in unterschiedlichen Bereichen verschieden (Greenson, 1973).

Körperfunktionen wie Essen, Schlafen und Ausscheidung erhalten Überwertigkeit. Die Selbstverständlichkeit ihres Funktionierens geht verloren. Damit kann sich eine Verstärkung der narzißtischen Besetzung des eigenen Körpers verbinden. Die Zurücknahme emotionaler Besetzungen auf den Bereich der ei-

genen Körpersphäre bedingt häufig einen sozialen Rückzug mit Verringerung von Objektbeziehungen, was zu Teilnahmslosigkeit und Apathie führen kann.

Beispiel: Der Patient, der das Krankenhaus-Setting als orale Verwöhnungssituation und die Ärzte als omnipotente Elternfiguren erlebt. Immer wieder trifft man Schwerkranke, deren Verhalten und Erscheinungsbild sich im Laufe ihrer Krankheit zunehmend kleinkindhaften Formen nähern (vgl. Patient B in Kap. 65).

62.4 Verleugnung als besonders wichtige Abwehrform bei lebensbedrohlich Erkrankten

Krankheit ist bei körperlich Kranken Teil ihrer äußeren und inneren Wirklichkeit. Eine partielle oder totale Zurückweisung dieser Wirklichkeit ist kennzeichnend für den Umgang Schwerkranker mit ihrem Schicksal. Um diese Beobachtungen genauer zu verstehen, ist eine Differenzierung zwischen den bisher beschriebenen Abwehrmechanismen und einer besonderen Art von Abwehr erforderlich, die als Verleugnung bezeichnet wird und, zumindest zu Beginn der Erkrankung, eine besondere Rolle spielt. Weil der Begriff Verleugnung zum Teil gleichsinnig mit Abwehr gebraucht wird und alle Vorgänge zur Abwehr einer bedrohlichen Wirklichkeit umfaßt, zum Teil im Sinne eines besonderen Abwehrmechanismus, der aus Beobachtungen in der psychoanalytischen Behandlung hervorgegangen ist, entstehen häufig Unklarheiten (Fine et al., 1969). Verständlich ist, daß Abwehrvorgänge bei der psychoanalytischen Behandlung von Neurotikern subtilerer Beobachtung zugänglich sind als bei der Behandlung schwerkranker Patienten im Krankenhaus.[7]

62.4.1 Definition, Entwicklung und Analyse des Begriffs

Der Begriff Verleugnung, von Freud (1923) im Zusammenhang mit der Haltung der Frau gegenüber ihrer „Kastration" und Penislosigkeit beschrieben, wurde ursprünglich für eine Abwehr der Wahrnehmung äußerer Wirklichkeitselemente geprägt. Im Laufe der Jahre wurde er jedoch auf andere Bereiche wie die innere Wirklichkeit erweitert (Lewin, 1950; Moore und Rubinfine, 1969; Jacobson, 1957; Waelder, 1951). Damit wurde er aber von der Verdrängung schwerer abgrenzbar, die ja ebenfalls innere Impulse abwehrt. Verleugnung als Abwehrform und Ich-Funktionen wie Realitätswahrnehmung stehen in einem gegenseitigen Spannungsverhältnis (Dorpat, 1983). Freud bezeichnete Verdrängung unter topischen Gesichtspunkten als intersystemischen Prozeß (Ich gegen Es), im Gegensatz zur Verleugnung als intrasystemischem

---

7 Zum Verständnis des Begriffs Verleugnung möchten wir trotzdem, von der ursprünglichen psychoanalytischen Bedeutung ausgehend, die aktuelle Sicht darstellen.

Prozeß (Ich gegen Ich) (1938 a, b), der sich gegen die Wahrnehmungsfunktion des Ichs richte.[8]

Neben den unterschiedlichen beteiligten Instanzen – nach der Terminologie des psychoanalytischen Strukturmodells – sollen abgewehrte Inhalte bei Verleugnung und Verdrängung unterschiedliche Schicksale haben; verdrängte Inhalte würden vom Bewußten ins Unbewußte verbannt, Verleugnung könne höchstens dazu dienen, Inhalte des Vorbewußten am Eintritt ins Bewußtsein zu hindern, was die Möglichkeit des Wiederauftauchens, verglichen mit verdrängten Inhalten, vergrößere.

Unter ontogenetischen Gesichtspunkten ist Verleugnung im Vergleich zur Verdrängung für A. Freud (1936) ein archaischerer Mechanismus, der für das unreife kindliche Abwehrverhalten adäquat sei, wo die Wahrnehmung der Wirklichkeit abrupt in Verleugnung und Ersatz durch Phantasiegebilde, Tagträume und symbolische Handlungen übergehen könne. Der Verleugnung folgten häufig andere Mechanismen wie Vermeidung oder Umkehr der Realität.

Im Laufe der ontogenetischen Entwicklung würde sich der Schwerpunkt der Verleugnung von einer Abwehr der Wahrnehmung der äußeren Wirklichkeit auf eine Verleugnung von Gefühlen verlagern, die durch die Wahrnehmung ausgelöst sind. Das entspräche dem Übergang vom Ich des Lustprinzips zu einem mehr am Realitätsprinzip orientierten Ich (Moore und Rubinfine, 1969).

Beziehungen zu anderen Abwehrmechanismen

Meist werden Verdrängung und Verleugnung einander gegenübergestellt. So wird Verleugnung als einmaliger, Verdrängung als kontinuierlicher Vorgang bezeichnet.

Allerdings kann es sich bei der emotionalen Besetzung von Wunscherfüllungsphantasien, die bei Verleugnungsprozessen auftreten können, analog zur Gegenbesetzung bei Verdrängung, um einen kontinuierlichen Prozeß handeln (Moore und Rubinfine, 1969), ebenso bei den die Verleugnung stützenden akzessorischen Mechanismen wie Rationalisierung usw.

Beispiel: Eine gelähmte Poliomyelitispatientin, die das Ausmaß der Krankheit verleugnet und taktile kinetische Halluzinationen hat. D.h., das Moment der Bewegung, das den Wunschvorstellungen der Patientin entspricht, taucht im halluzinatorischen Erleben auf.

Man nimmt bestimmte Gesetzmäßigkeiten des gemeinsamen Auftretens einzelner Mechanismen an. Mit Verleugnung gekoppelt findet man häufig primitive Formen von Introjektions- und Projektionsmechanismen, Verdrängung geht dagegen in der Regel mit differenzierten Formen von Projektion und Identifikation einher. Verleugnung, welche die gefühlsmäßige Bedeutung einer Vorstellung negiert, ist in ihrer Wirkung kaum von der Isolierung vom Affekt zu trennen. Ungeschehenmachen bedeutet letztlich Verleugnung durch Handlung.

Verleugnung und andere Persönlichkeitsvariablen

Verleugnung ist im Vergleich zu Rationalisierung, Intellektualisierung und Verschiebung ein relativ rigider Mechanismus. Sie wird mit engstirnigem, rechthaberischem und phantasielosem Verhalten verbunden. Dies entspricht der Funktion als Notfallmechanismus, bei dem ein großer Teil der Wirklichkeit ausgeblendet wird. Dabei wird die Problematik deutlich, wenn ein solcher Notfallmechanismus zu einem häufigen Verhaltensstil wird.

Hackett und Mitarbeiter (1968) berichten bei ihren Infarktpatienten von einem lebenslänglich auffälligen Verhaltensmuster, was sich in Erklärungen ausdrücken könne wie den folgenden: „Man nennt mich den Eisernen" oder „Ich habe das Glück gepachtet".

In einem kasuistischen Beitrag berichten Sullivan und Hackett (1963) von einem 53jährigen Patienten, der, als er von seiner Infarktdiagnose hörte, in rüder Weise das Krankenhauspersonal beschimpfte und nach einem Streit mit dem behandelnden Arzt eigenmächtig das Krankenhaus verließ. Der Patient starb, als er, nach Hause kommend, mehrere Treppen hochgestiegen und zusammengebrochen war. Es hatte sich um einen Patienten gehandelt, der im Waisenhaus großgeworden und viel herumgestoßen worden war und für den Tüchtigkeit und Arbeitsfähigkeit ganz oben auf der persönlichen Werteskala standen. Krankheit hatte er zeitlebens nie akzeptieren können. Es war für ihn unerträglich, ein „Krüppel" zu sein. Die Infarktdiagnose war mit seinem idealen Selbst nicht vereinbar.

62.4.2 Verleugnung als möglicher pathogener Faktor

Die Kollision mit anderen Funktionen des Ichs, die für die normalen Leistungen des Erwachsenen erforderlich sind und z.B. Realitätswahrnehmung und synthetische Funktion voraussetzen, führt dazu, eine fortgesetzte Verleugnung bei erwachsenen Personen als ungünstiges Zeichen zu werten. Exakte Realitätsprüfung charakterisiert Ich-Reifung. Umgekehrt besteht eine quantitative Beziehung zwischen dem Ausmaß von Verleugnung und dem Ausmaß der Ich-Regression. Da die Wiederkehr des verleugneten Inhalts bei ständiger Konfrontation mit der Realität unvermeidlich ist, wird, falls Verleugnung der einzige Abwehrmechanismus bleibt, bei langer Dauer eine noch

---

8 Daraus folgt die Ansicht S. Freuds, daß Verleugnung bei Erwachsenen ein pathologischer Mechanismus sei wegen der hervorgerufenen Ich-Spaltung. Demgegenüber kann darauf hingewiesen werden (Moore und Rubinfine, 1969), daß das Ich sich selbst aus unterschiedlichen Teilfunktionen zusammensetzt und eine selektive Spaltung der Abwehrfunktionen auf Kosten anderer nicht notwendigerweise eine durchgängige Ich-Spaltung bedeute. Allerdings sei anzunehmen, daß frühe und schwere Störungen der Objektbeziehungen zur Wahl der Verleugnung als bevorzugtem Abwehrmechanismus prädisponieren, beispielsweise bei einer schweren Störung der Mutter-Kind-Beziehung. Ich, Es und Über-Ich seien an der Verleugnung beteiligt, was darauf hinweise, daß die Ursprünge der Verleugnung in einem undifferenzierten Stadium des psychologischen Apparates liegen (Moore und Rubinfine, 1969, S. 33–35).

massivere Verleugnung mit einem regressiven Umbau der Ich-Struktur erzwungen (Modell, 1961).[9]

62.4.3 Verleugnung als sozialer Prozeß

Befaßt man sich mit dem Verhalten schwerkranker Patienten, findet man häufig das ganze Verhalten darauf ausgerichtet, die Erkrankung nicht zu akzeptieren. Dieser Aspekt der Verleugnung als komplexe Verhaltensstrategie in der Institution Krankenhaus wurde von Weisman und Hackett (1967) untersucht. Sie kritisieren die Vorstellung von Verleugnung als eines bloßen Mechanismus, dessen sich Individuen bedienen, um unliebsame Wirklichkeitsbestandteile vom Bewußtsein fernzuhalten und der immer wieder verfügbar sei wie ein Spinalreflex. Sie bezeichnen eine solche Vorstellung als „atomische" Betrachtungsweise. Es handle sich vielmehr um komplexe Verhaltensmuster, bei denen die Verleugnung das Resultat zahlloser Aktionen in einer Hierarchie von Einzelprozessen sei. Die Feststellung der Verleugnung und der Prozeß des Verleugnens verhielten sich wie Erinnerung zum Prozeß des Sich-Erinnerns mit seinen verschiedenen Hilfsprozessen und Untereinheiten wie Registrierung, Sichtung, Speicherung und Reproduktion des Gedächtnisinhaltes. Zu berücksichtigen sei der Zusammenhang, in dem sich Verleugnung abspiele: Verleugnungsobjekte, Zweck der Verleugnung, Bezug zur verleugneten Realität und das Schicksal nicht verleugneter Wirklichkeitsbestandteile müßten beachtet und es müsse gefragt werden, wer was wann zu welchem Zweck gegenüber wem in welcher Form verleugne, was er nicht verleugne und wodurch er den verleugneten Inhalt ersetze.

Gegenüber verschiedenen Bezugspersonen kann bei Patienten immer wieder ein unterschiedliches Maß von Verleugnung beobachtet werden:[10]

Als Beispiel sei ein Patient erwähnt, der nach Nierentransplantation dem Pflegepersonal eine weit pessimistischere Sicht seiner Situation präsentierte, als er das den Ärzten gegenüber tat, und dabei die Freiwilligkeit seines Entschlusses zur Transplantation verschiedenen Bezugspersonen gegenüber in unterschiedlichem Ausmaß in Zweifel zog.

Erwähnt sei auch eine Mitteilung von Weisman (1972) über eine Frau mit fortgeschrittenem Zervixkarzinom, die sich gegenüber Ärzten und Schwestern so verhielt, als wisse sie nichts über ihre Krankheit. Die Patientin zeigte sich hingegen bestürzt über ein Ulcus duodeni, das bei ihrem Ehemann festgestellt worden war. Einen Psychiater, der sie zur konsiliarischen Betreuung besuchte, fragte sie dann unvermittelt, ob es stimme, daß das Stadium 3 eines Zervixkarzinoms schlimmer sei als Stadium 1.

Dialysepatienten unterscheiden sich hinsichtlich ihrer Krankheitsverarbeitung in der Selbsteinschätzung deutlich von der Einschätzung durch ihre Ärzte, die ihnen häufiger verleugnende und depressive Verarbeitungsmodi zuschreiben, als sie es in der Selbstbeschreibung angeben (Muthny, 1988a; ähnliche Befunde Schüßler, 1989: nur geringer Zusammenhang zwischen Selbst- und Fremdeinschätzung der Krankheitsverarbeitung bei untersuchten Konsiliarpatienten).

Die Kenntnis der Dynamik von Verleugnungsvorgängen ist bei unheilbar Kranken von besonderer Bedeutung: Die Einstellung des Gesprächspartners bestimmt wesentlich das Thema, das mit todkranken Patienten besprochen wird. Wird ein Kranker sofort in einer unrealistischen Einstellung bestärkt, wird er nicht selten diese Gelegenheit schnell wahrnehmen. Ein wesentliches Motiv, den bevorstehenden Tod zu verleugnen, ist dabei die Furcht, Kontakt und Unterstützung von Angehörigen und Pflegepersonal zu verlieren, die der Wahrheit selber nicht ins Auge schauen können. D.h., der Patient wird aus einem anderen Motiv heraus als Todesangst, nämlich durch Sozialangst, für die Verleugnung sensibilisiert.

Verleugnung ist die Feststellung einer am Ereignis unbeteiligten Person. Da Wirklichkeit jedoch ein subjektives Phänomen ist,[11] wird mit einer solchen Feststellung die Grundfrage der Bewertung psychischer Funktionen überhaupt berührt: Das Verhalten einer Person mit einer anderen Hierarchie von Werten und Maßstäben kann dem beurteilenden Beobachter als skurrile Form von Wirklichkeitsverzerrung und Unvernunft imponieren, obwohl es in sich höchst logisch und folgerichtig ist.

Weisman und Hackett (1967) zitieren dazu einen Patienten mit Thoraxschmerzen, der einen Herzinfarkt hatte und nicht zum Arzt ging, in der Annahme, er habe Lungenkrebs. Angesichts der pessimistischen Version des Patienten erschiene es auf den ersten Blick unstatthaft, hier von Verleugnung zu sprechen. Bei genauerer Exploration stellte sich aber heraus, daß der Vater des Patienten an einem Herzinfarkt plötzlich verstorben war und der Patient es vorzog, unter dem Damoklesschwert einer chronisch-konsumierenden Erkrankung zu leben, als den für ihn schlimmeren Gedanken eines Infarktes mit der Assoziation „plötzlicher Tod" zu akzeptieren.

62.4.4 Klinische Untersuchungen zur Verleugnung

Bei klinischen Untersuchungen zur Verleugnung wurden meist unterschiedliche operationale Definitionen vorgenommen.[12]

---

9 Verschiedene Autoren haben eine Beziehung von Verleugnung zu verschiedenen psychischen Erkrankungen und Charakteranomalien betont: Deutsch (1933) zur Depression, Lewin (1950) zu manischen Zuständen, ebenso White und Gilliland (1975, S. 145), Fenichel (1974) zu Fetischismus, schizophrenen Reaktionen, Gedächtnisstörungen, Schlafstörungen, A. Freud (1936) zu Psychosen, Hamburg et al. (1953) zu Wahn- und Halluzinationszuständen, Jacobson (1957) zu Amnesien, Fehlhandlungen, Depersonalisationszuständen, Modell (1961) zur Auflösung von Ich-Grenzen.
10 Hingewiesen sei auf eine empirische Untersuchung von Miller und Rosenfeld (1975) über psychophysiologische Korrelate von Verleugnung bei Infarktpatienten. Es stellte sich heraus, daß Schwestern, die die Patienten über viele Stunden täglich betreuten, hinsichtlich der Einschätzung deren Verleugnung nur wenig mit den Einschätzungen der an der Studie beteiligten Ärzte übereinstimmten, die die Patienten nur einmal täglich aufsuchten.
11 In diesem Zusammenhang verweisen wir auf v. Uexküll Kap. 1, in dem gezeigt wird, daß es nur eine „individuelle Wirklichkeit", wie sie vom Individuum aufgebaut wird, gibt, und daß die „gemeinsame Wirklichkeit" eine soziale Konstruktion darstellt.
12 Meist wurden die Auswirkungen eines Konglomerats von Abwehrmechanismen im ursprünglichen psychoanalytischen Sinne untersucht.

Weinstein und Kahn (1955) untersuchten beispielsweise bei 104 Patienten mit neurologischen Erkrankungen solche Phänomene, die durch das Nichterkennen ihrer Erkrankung gekennzeichnet waren. Besonders ging es darum, zu untersuchen, inwieweit psychodynamische Vorgänge am Phänomen der „Anosognosie", d.h. am Nichterkennen größerer Funktionsausfälle des Zentralnervensystems durch den Patienten, beteiligt sein können. Dabei drängte sich in zahlreichen Fällen des Nichterkennens oder Vergessens der Verdacht auf unterstützende Abwehrstrategien der Patienten auf. Die Verkennung der Situation erstreckte sich häufig nicht nur auf Tatsachen und Affekte, die mit der Krankheit zusammenhingen, sondern zum Teil selektiv auch auf frühere schmerzliche Erfahrungen, beispielsweise die Verleugnung einer Fehlgeburt oder einer unglücklich verlaufenen Ehe. Organisches Defizit und individuelle Coping- und Abwehrstrategien verbinden sich, um den Patienten ein allmähliches Akzeptieren ihrer Behinderung zu ermöglichen (Säring, 1988; vgl. Kap. 63).

Beim „expliziten" Verleugnen wurden folgende klinische Erscheinungsformen unterschieden, die zur Einordnung von Beobachtungen herangezogen werden können:

- Vollständige Verleugnung; z.B.: Ein Blinder behauptet zu sehen;
- Verleugnung der größeren Behinderung, d.h. die schwere Störung wird verleugnet, triviale Aspekte werden betont; z.B.: Ein Halbseitengelähmter beklagt sich nur über Obstipation;
- Minimalisierung der Beschwerden; z.B.: Ein gelähmter Patient behauptet, er könne den Arm bewegen, er sei nur zu faul dazu. Ein linksseitig gelähmter Patient sagt, jeder habe im linken Arm weniger Kraft als im rechten;
- Projektion der Behinderung; z.B.: Ein Patient mit intrakranieller Blutung schreibt seine Kopfschmerzen den Angehörigen zu;
- Zeitliche Verschiebung der Behinderung; z.B.: Ein gelähmter Patient sagt, er habe schon früher nicht gehen können.[13]

Bracken und Bernstein (1980) stellten bei Patienten mit Rückenmarksverletzungen eine langfristig schlechtere Anpassung fest, wenn die Strategie ausgeprägter Verleugnung weit in die Phase der Rehabilitation hineinreicht.

Verleugnung und „Middle Knowledge"

Bei Beobachtungen an 350 schwerkranken Patienten, die konsiliarisch betreut wurden, konnte Weisman (1972) häufig ein Verhalten beobachten, das in der Mitte zwischen Wissen und Nichtwissen liegt. Er bezeichnet dieses Phänomen als „middle knowledge", ein Stadium „ungewisser Sicherheit" (S. 56–78).

Pragmatisch wird dabei eine Verleugnung ersten Grades, die das Faktum der Krankheit leugnet, Verleugnung zweiten Grades, d.h. Verleugnung von Auswirkungen der Erkrankung, und Verleugnung dritten Grades unterschieden, mit dem Unvermögen, sich bei voller Akzeptation der Diagnose und deren Bedeutung den eigenen Tod auszumalen. Weisman stellt einen bei Schwerkranken häufigen Verlauf dar, wobei die verschiedenen Verleugnungsformen in die korrespondierenden Formen der Akzeptation übergehen. In diesem Zusammenhang ist die klinische Erfahrung bedeutsam, daß Verleugnung um so rascher und vollständiger eintritt, wenn die Aufklärung völlig abrupt erfolgt ist (Bönisch und Meyer, 1975).

In den meisten Untersuchungen wird mit der Annahme eines umgekehrt proportionalen Zusammenhangs zwischen Angst und Verleugnung operiert. Diese Vorstellung leitet sich aus theoretischen Überlegungen über Angst als Auslöser von Abwehrprozessen her.

Weitere Untersuchungen

Katz und Mitarbeiter (1970) untersuchten psychische Reaktionen und Veränderungen endokriner Reaktionsmuster bei 30 Patientinnen, die auf eine Brustbiopsie warteten, und setzten diese mit den Abwehrstrategien der Patientinnen in Verbindung. Die Mechanismen der Verschiebung und insbesondere der Projektion erwiesen sich dabei als problematische Abwehrformen. Patientinnen mit einer ausgeprägten Verleugnungs- und Rationalisierungstendenz verzögerten die Diagnostik am meisten.

Derogatis und Mitarbeiter (1979) untersuchten 35 Brustkrebspatientinnen testpsychologisch. Patientinnen, die innerhalb eines Jahres verstarben, wiesen niedrigere Werte auf der Hostilitätsskala auf und insgesamt eine signifikant geringere psychiatrische Symptomatologie. Dabei wird ein Einfluß der Verleugnung diskutiert.

Rogentine und Mitarbeiter (1979) untersuchten prospektiv eine Gruppe von Melanompatienten. Dabei schien ein aktiveres und der Krankheit zugewandtes Coping-Verhalten mit einer besseren Prognose korreliert.

Es fand sich in einer Studie von Todd und Magarey (1978) bei Patientinnen vor Brustbiopsie eine Beziehung zwischen Verleugnung und Unterdrückung einerseits und der Verzögerung der Inanspruchnahme medizinischer Maßnahmen andererseits, ebenfalls bei verbal geäußerten Hinweisen für depressive Merkmale.

Brustkrebspatientinnen mit Verleugnung und aktiv kämpferischer Haltung wiesen nach 5 und 10 Jahren eine verlängerte Überlebenszeit auf (Greer, 1979; Pettingale, 1984).

62.4.5 Verleugnung und therapeutische Maßnahmen

Interessant ist die Frage nach der Stellung der Verleugnung in den verschiedenen Phasen des therapeutischen Prozesses bei unterschiedlichen Erkrankungen. Dazu sollen hier nur grundsätzliche Anmerkungen gemacht und bei den verschiedenen therapeutischen Situationen Funktion und Stellenwert der Verleugnung jeweils gesondert diskutiert werden. Entscheidend ist die Frage, ob Krankheit durch Verleugnung in positivem oder negativem Sinn beeinflußt werden kann. Und in welcher therapeutischen Situation soll man Verleugnung fördern und wann ist es angebracht, sie einzuschränken?

Krankheitsspezifität

Verleugnungsphänomene wurden bei verschiedenen Gruppen körperlich Kranker beobachtet, z.B. bei herzkranken Patienten, Tumorpatienten und Patienten mit neurologischen Störungen. Viele klinische Beobachtungen liegen bei Infarktpatienten (Hackett et al., 1969a, b; Sullivan und Hackett, 1963; Croog et al., 1968, 1971; Weisman, 1972; Gentry et al., 1972)

---

[13] Dabei wird deutlich, wie andere Abweisungsmechanismen an der Verleugnungsstrategie beteiligt sind, z.B. Verschiebung oder Rationalisierung.

und bei Malignompatienten vor (z.B. Shands et al., 1951; Weisman, 1972). Die meisten Arbeiten beziehen sich auf gefährliche Krankheiten und Malignome (Wolff et al., 1964a, b; Ziegler, 1984; Derogatis et al., 1979; Rogentine et al., 1979; Cassileth et al., 1984; Herschbach, 1987; Heim et al., 1988b). Becker (1986) beschreibt in seiner Studie einen ungünstigen Einfluß rigider und ineffizienter Abwehr auf den Erkrankungsverlauf bei onkologischen Patienten. Verleugnung bei harmlosen Erkrankungen wurde selten untersucht. Ausgehend von der theoretischen Vorstellung, daß mit der Angst die Abwehrtätigkeit wächst,[14] konzentrieren sich die wissenschaftlichen Fragestellungen auf solche Krankheiten, die man besonders mit der Auslösung von Angst in Verbindung brachte. Ein spezifischer Zusammenhang zwischen der Art der Erkrankung und der Tendenz zur Verleugnung ist nicht bekannt (vgl. Schüßler, 1988: kein für Arthritiker typisches Coping, ebenso Cassileth, 1984).

Therapiespezifität

Therapieverfahren, die ein größeres Maß an Kooperation und Einsicht vom Patienten verlangen, können durch Verleugnungsphänomene stärker behindert werden als andere, die weniger auf die Mitarbeit von Patienten angewiesen sind (z.B. Hämodialyse- versus Herzschrittmacherpatienten). Ebenso dann, wenn vom Patienten Umstellungen der Lebensweise erwartet werden, wie es bei vielen Herzinfarktpatienten der Fall ist oder bei Patienten, die langfristig Medikamente einnehmen müssen (z.B. Tuberkulosepatienten, Arthritiker, Hypertoniker, Patienten nach Pankreatektomie: vgl. Lang, 1989). Leider fehlen noch genaue Studien mit Kontrollgruppen, die Langzeitauswirkungen von Abwehrhaltungen von Patienten auf die Kooperation bei der Behandlung unterschiedlicher Erkrankungen untersuchen.

Phasenspezifität

Die Auswirkungen der Verleugnung sind bei Patienten mit unterschiedlichen Erkrankungen phasenspezifisch verschieden. Das Vorstadium des therapeutischen Prozesses umfaßt den Zeitraum bis zur Inanspruchnahme ärztlicher Hilfe. In diesem Stadium kann die Verleugnung von Beschwerden zur Verzögerung dieses Entschlusses führen, was bei malignen Prozessen (vgl. Shands et al., 1951) ebenso wie bei Infarktpatienten (vgl. Hackett et al., 1969a; Moss et al., 1969; Übersicht bei Doehrmann, 1977; vgl. auch Kap. 41.2) immer wieder beschrieben wurde und fatale Folgen haben kann. Allerdings sind diesbezüglich die Befunde nicht einheitlich (Beutel, 1988a, b).

In der Phase der stationären Behandlung, in welcher der Patient mit dem bedrohlichen Ereignis konfrontiert ist, muß die Rolle der Verleugnung anders eingeschätzt werden. Verleugnung kann hier zu einer erträglichen Einschätzung einer unerträglichen Situation führen.

Nicht selten trägt Verleugnung aber auch dazu bei, daß die Diagnose verzögert gestellt wird. Oft kann

man eine ausgeprägte Tendenz beobachten, körperliche Symptome als psychogen im Sinne eines Abwehrversuches zu präsentieren. Es kann geschehen, daß Patient und Arzt „an einem Strang ziehen" und gemeinsam ein Bündnis der Verleugnung eingegangen sind. Insgesamt ist festzustellen, daß dann, wenn die Verleugnung zu umfassend ist, dies zu einer Behinderung notwendiger Verarbeitungsprozesse, beispielsweise nach einem Verlust, führt und damit z.B. Trauerarbeit verhindert werden kann.

62.4.6 Verleugnung und Rehabilitation

Ebenfalls noch unklar ist die Rolle von Verleugnungsprozessen während der Rekonvaleszenz und im Rehabilitationsverlauf und ihre prognostischen Auswirkungen. Zum Teil wird Verleugnung auch hier als pathologisch beurteilt, da sie die Kooperation mit dem Arzt behindert, zum Teil aber auch als integrativer Faktor bezeichnet, der die Unverletzlichkeit des Körpers fingiert und mithilft, für den Patienten das Weiterleben erträglich zu machen (vgl. Adams und Lindemann, 1974 über querschnittsgelähmte Patienten). Berücksichtigt man, daß bei verschiedenen Untersuchungen Verleugnung in der Regel nicht als genügend klar definierte Variable untersucht wird, werden die unterschiedlichen Beurteilungen verständlich. Es besteht Einvernehmen, daß Verleugnung für die Entscheidung, ärztliche Hilfe in Anspruch zu nehmen, potentiell gefährlich ist, ebenso in der Phase der Rehabilitation, wenn die anfangs oft autoritäre Arzt-Patient-Beziehung verändert wird und der Arzt nur noch unterstützend, ermutigend und beratend zur Seite steht, vom Patienten aber aktive Mitarbeit erwartet wird. Grenzen der erwünschten Verleugnung sind in jedem Fall eine zu weitgehende Einschränkung der Realitätsprüfung und mangelnde Motivation zur therapeutischen Mitarbeit. Andererseits können übermäßige Angst, Hoffnungslosigkeit, Depression und regressives Verhalten Folge zu geringer Verleugnung sein. Jedoch ist zu berücksichtigen, daß Regression im Angesicht einer schweren Bedrohung sowohl aus zu geringer als auch aus übermäßiger Verleugnung und einer gestörten Umweltbeziehung infolge einer psychotisch anmutenden Verleugnung resultieren kann. Allerdings handelt es sich auch in diesen Fällen meist doch nur um einen partiell veränderten Umweltbezug, wobei der Patient streng selektiv vorgeht und das Bild der Umwelt so verändert, daß ihm die Bewältigung erleichtert wird. Aus diesem streng selektiven Vorgehen ergibt sich zwingend die Annahme eines sehr spezifischen geistigen Prozesses, der sich auch qualitativ von den Folgezuständen zerebraler Läsion unterscheidet, obwohl hier wie dort das Erkennen der Realität auffallend gestört sein kann. Denn was verleugnet wird, muß vorher zumindest

---

14 Diese Vorstellung wird durch Befunde von Levine und Ziegler (1975) gestützt, die die Verleugnung bei Patienten mit Schlaganfall, Lungenkrebs und Herzerkrankungen untersuchten.

teilweise wahrgenommen und bewertet worden sein. Auch findet bei der Verleugnung, die man bei schwerkranken Patienten sieht, in der Regel kein bedeutsamer Rückzug der emotionalen Besetzung von Objekten statt, wie es bei schweren psychopathologischen Zuständen angenommen wird; meist handelt es sich nur um Verlagerung von Schwerpunkten emotionaler Besetzung. Dies alles bedeutet, daß es für die Frage, ob der Arzt das Abwehrverhalten eines schwerkranken Patienten unterstützen oder den Patienten verstärkt mit der Wirklichkeit konfrontieren soll, keine allgemeinen Regeln gibt. Er muß in jedem Fall genau abwägen, was die Krankheit in ihrer jetzigen Phase für den einzelnen Patienten bedeutet, welche psychischen und sozialen Ressourcen ihm zur Verfügung stehen, welche Anpassungsleistungen der Patient in der Vergangenheit unter welchen Bedingungen vollzogen hat und welche zusätzlichen Beeinträchtigungen möglicherweise seine jetzige Anpassungsfähigkeit schmälern. Das erfordert nicht nur ein Gewinnen von Information, sondern auch die Fähigkeit zur Empathie.

63 Patienten mit körperlich begründbaren psychischen Störungen in der klinischen Praxis. Akute hirnorganische Psychosyndrome

Ekkehard Gaus und *Karl Köhle*

63.1 Fallgeschichte

Die 52jährige Patientin, die sich seit ungefähr 6 Jahren wegen einer chronischen dialysepflichtigen Niereninsuffizienz in ambulanter Dialysebehandlung befand, wurde stationär in eine psychiatrische Klinik eingewiesen. Zu Beginn der Erkrankung hatte als Ursache des Nierenversagens erst nachträglich ein Lupus erythematodes festgestellt werden können. Jetzt war bei ihr vor der Einweisung seit Monaten eine zunehmende ängstlich-depressive Verstimmung mit einem ausgeprägten sozialen Rückzug aufgefallen. Die Patientin, die früher den Haushalt alleine versorgt hatte, mußte jetzt durch ihren Ehemann und ihren Sohn abwechselnd betreut und beaufsichtigt werden und traute sich zuletzt nicht mehr, alleine in ihrer Wohnung zu bleiben. Etwa ein Jahr zuvor war sie wegen einer Herzschwäche ins Krankenhaus aufgenommen worden. Schon damals war sie sehr depressiv gewesen und hatte anschließend Antidepressiva verordnet bekommen, worauf sich die depressive Verstimmung nur kurzfristig besserte. Wegen ihrer Grunderkrankung (Lupus erythematodes) wurde sie mit Steroidhormonen behandelt. Etwa ein halbes Jahr zuvor war sie erneut wegen einer Zunahme ihrer depressiven Symptomatik und eines leicht erhöhten Serumkalziumspiegels in eine Klinik gekommen. Es wurde unter der Annahme eines tertiären Hyperparathyreoidismus eine Epithelkörperchen-Exstirpation durchgeführt. Nach einer einige Monate anhaltenden Besserung verschlechterte sich der Zustand der Patientin fortlaufend.

Bei der ersten Untersuchung war die Patientin bei klarem Bewußtsein, wirkte aber phasenweise substuporös, dabei ratlos, teils abweisend, teils anklammernd. Sie war zwar in allen Qualitäten orientiert, aber stark verlangsamt, erschien tief depressiv, verängstigt, im Antrieb und in ihrer Spontaneität erheblich vermindert, in ihrer Konzentrations- und Aufmerksamkeitsfähigkeit eingeschränkt. Im Gespräch zeigte sie sich schwerbesinnlich, im Denkablauf haftend, zäh. Für Halluzinationen fand sich kein Anhalt. Ihre Ängste schienen eine paranoide Färbung aufzuweisen. Insbesondere das Kurzzeitgedächtnis erschien stark gestört.

An Medikamenten hatte die Patientin über lange Zeit ein Cortisonpräparat in gleichbleibender Dosierung und ein Präparat zur Ulkusprophylaxe (Cimetidin) eingenommen. Vom Körperlichen her fanden sich Zeichen einer chronischen Nieren- und kompensierten Herzinsuffizienz, neurologisch Hinweise für eine beginnende Polyneuropathie. Das Ergebnis des kurz nach der Aufnahme durchgeführten Benton-Tests war hochpathologisch.

Insgesamt fiel neben der ängstlich-depressiven Verstimmung und dem Rückzugsverhalten der Patientin eine erhebliche Hirnleistungsschwäche auf. Eine Stoffwechseldekompensation, beispielsweise durch unzureichende Dialyse, oder eine Störung des Kalziumhaushalts konnten ausgeschlossen werden, ebenfalls fand sich kein Anhalt für eine Exazerbation der Autoimmunerkrankung. Im kranialen Computertomogramm zeigten sich keine über der Norm liegenden atrophischen Veränderungen.

Die Lebenssituation der Patientin war durch eine Kette von Verlusten und Entbehrungen geprägt. Nach Ausbruch ihrer schweren Nierenerkrankung vor 6 Jahren, die sie lange verleugnet hatte, mußte sie sehr rasch ins Hämodialyseprogramm aufgenommen werden. Während der letzten Jahre hatte es erhebliche Probleme mit dem Sohn der Patientin gegeben, der alkoholabhängig war und zuletzt seine Arbeit verloren hatte. Die Patientin hatte diesen Sohn jahrelang mit großer Aufopferung und Selbstverleugnung versorgt. Zuletzt war sie selbst völlig auf die Versorgung durch andere angewiesen. Im Gespräch zeigte sie sich darüber tief verzweifelt. Sie hatte sich bis zuletzt einer erneuten Aufnahme ins Krankenhaus widersetzt. Dort steigerte sich ihre Angst zunächst zur Panik; später, als sie darüber erzählen konnte, sagte sie, sie habe sich erst völlig hilflos und ausgeliefert erlebt und manches wie im Traum wahrgenommen. Sie habe immer Angst gehabt, irgendwohin gebracht zu werden, ohne zu wissen, wohin, und habe sich insbesondere vor der unvertrauten Umgebung in der neuen Dialysestation gefürchtet, wohin sie dreimal pro Woche gebracht wurde.

Es lag nicht fern, eine reaktive Komponente bei der depressiven Verstimmung zu vermuten. Insgesamt deutete aber das gesamte Bild mit Verlangsamung und, neben der affektiven Symptomatik, der im Vordergrund stehenden Beeinträchtigung ihrer geistigen Funktionen auf eine organische Entstehung hin. Dabei kamen aufgrund der Kollagenerkrankung, der Niereninsuffizienz mit begleitender Anämie und Einnahme verschiedener Pharmaka eine Reihe von Möglichkeiten in Betracht.

Da von Anfang an der Verdacht auf ein medikamenteninduziertes Psychosyndrom bestand, wurde zunächst das Ulkuspräparat abgesetzt und später die Cortisondosis schrittweise reduziert. Erst dann besserten sich innerhalb von Wochen Angst und Depression und die Auffassungsstörungen der Patientin. Der Benton-Test war bei späterer Wiederholung unauffällig. Im reduzierten Wechsler-Intelligenztest wies sie eine durchschnittliche intellektuelle Leistungsfähigkeit auf.

Nach Aufhellung der depressiv-ängstlichen Verstimmung kam es zu einer hypoman anmutenden Stimmungsschwankung der Patientin. Dabei wirkte sie überdreht, flirtend-kokettierend, neigte zur übermäßigen Flüssigkeitszufuhr, so daß in dieser Phase mehrfach eine Überwässerung auftrat.

Das akute organische Psychosyndrom, das vornehmlich als depressiver Verstimmungszustand aufgefallen war, der sich über viele Monate entwickelt hatte, hatte dazu geführt, daß die Patientin zu Hause vollständig auf die Hilfe anderer angewiesen und die Dialysebehandlung bei ihr erheblich erschwert, ja sogar gefährdet gewesen war. Da ein Suizidrisiko angenommen wurde, hatte das Dialysepersonal sich zuletzt nicht mehr getraut, die Patientin aus den Augen zu lassen, was den Arbeitsablauf auf der Dialysestation erheblich erschwert hatte. Der weitere Verlauf unter einer niedrigen Erhaltungsdosis mit Nebennierenrindenhormonen war komplikationslos.

Es war im folgenden möglich, mit ihr die konflikthafte familiäre Situation, ihre Unsicherheit, wie es mit dem Sohn weitergehen werde, auch unter Einbeziehung der Angehörigen, zu bearbeiten.

Der beschriebene Ablauf ist bezeichnend dafür, wie bei Entstehung und Ausprägung eines akuten organischen Psychosyndroms Schädigungen durch die Erkrankung, Medikamenteneinflüsse, Persönlichkeitseigenschaften und situationsspezifische Merkmale zu berücksichtigen sind und den Krankheitsverlauf insgesamt beeinflussen.

Das Beispiel zeigt, wie vielfältig die in Frage kommenden pathogenen Faktoren sind, die wir berücksichtigen müssen, wenn wir Patienten mit möglicherweise körperlich begründbaren psychischen Störungen begegnen.

63.2 Einführung und Problemstellung

Mit akuten organischen Psychosyndromen als körperlich begründbaren psychischen Störungen bei somatischen Erkrankungen befassen sich neben der Psychiatrie auch die jeweiligen organmedizinischen Fächer wie Innere Medizin, Chirurgie, Neurologie. Obwohl solche Störungen bei Patienten beispielsweise mit internen und chirurgischen Erkrankungen nicht selten auftreten, erfahren sie dort meist wenig Beachtung. In den Lehrbüchern dieser Disziplinen finden sich, wenn überhaupt, meist nur kurze Hinweise. Auch in der psychosomatischen Literatur werden sie, von wenigen Ausnahmen (z.B. Adler, 1986) abgesehen, wenig gewürdigt. Hingegen finden sie sich in den Lehr- und Handbüchern der Psychiatrie und psychiatrischen Übersichten und Monographien ausführlich abgehandelt (Bleuler et al., 1966; Bleuler, 1983; Huber, 1972, 1981, 1984; Schulte und Tölle, 1979; Burchard, 1980; Conrad, 1960; Battegay et al., 1984; Lipowski, 1980 a, b, c, 1989; Hunger, 1987).

Patienten mit hirnorganischen Psychosyndromen im Krankenhaus, die dem Psychiater vorgestellt oder überwiesen werden, zeigen meist besonders eindrückliche, oftmals dramatisch und bedrohlich erscheinende Symptome mit Unruhe und Agitiertheit, so daß die Behandlung der Grunderkrankung erschwert oder verunmöglicht wird. Dies hat dazu geführt, daß sich die Psychiatrie über lange Zeit vorwiegend mit diesen und weniger mit den leichten und unauffälligen Störungen psychischer Funktionen körperlich Kranker beschäftigt hat, die zahlenmäßig bei weitem überwiegen. Oft werden die Patienten, bei denen Stimmungsveränderungen und eine Verlangsamung der kognitiven Funktionen bestehen, auf Station als apathisch, uninteressiert und unkooperativ oder als leicht erregbar und aggressiv angesehen. Dabei wird nicht selten fälschlicherweise eine psychische Genese angenommen und das auffällige Verhalten dem Charakter zugeschrieben. Zum Teil werden die Störungen, insbesondere bei ruhigen und zurückgezogenen Patienten, übersehen. Dabei spielt auch die im Unterschied zu den meisten psychogenen Störungen feststellbare Tendenz der Patienten eine Rolle, das Ausmaß ihrer Einschränkungen zu überspielen und zu bagatellisieren. So kann man bei Visiten auf internen und chirurgischen Stationen immer wieder Patienten erleben, die ein verbindliches Lächeln und höfliche Floskeln mit dem Arzt austauschen, aber grobe Orientierungsstörungen erkennen lassen, wenn ihnen die Möglichkeit gegeben wird, sich zu äußern.

Mit diesem Kapitel in einem psychosomatischen Lehrbuch wollen wir einer Reihe von Gesichtspunkten Rechnung tragen. Patienten mit einer so gearteten Störung stellen einen nicht unerheblichen Teil derer dar, die im psychosomatischen Konsiliar- und Liaisondienst versorgt werden müssen (Krakowski, 1979). Neben reaktiv ausgelösten Angst- und Depressionszuständen handelt es sich um die häufigste Form psychischer Störungen bei körperlichen Erkrankungen (Schwab, 1982). Neben diesen praktischen Momenten stellt sich in der engen Verflechtung somatischer und seelischer bzw. erlebnisreaktiver Faktoren bei körperlich Erkrankten ein interessantes theoretisches Problem der Psychiatrie und klinischen Psychosomatik dar. Dabei hat der Psychosomatiker oder Liaison-Psychiater gerade bei den leichteren Formen oft die Aufgabe, dafür zu sorgen, daß sowohl die körperliche als auch die psychische Seite der Störung diagnostiziert und behandelt werden.

Das vorliegende Kapitel beschränkt sich auf eine überblickhafte Darstellung akuter hirnorganischer Psychosyndrome, wobei für die klinische Psychosomatik relevante Aspekte besonders betont werden.

Im Hinblick auf Einzelheiten der Psychopathologie, Ätiopathogenese und Systematik sei auf entsprechende Darstellungen im psychiatrischen Schrifttum verwiesen. Darin finden wir eine für den Außenstehenden erstaunliche und möglicherweise oftmals verwirrende nomenklatorische Vielfalt.

63.3 Definition, Begriffe

Zur Orientierung sollen einige häufig verwandte Begriffe mit Herkunft und Entstehung vorgestellt werden:

Wir verwenden im weiteren die Bezeichnung **akutes hirnorganisches Psychosyndrom** als Folge einer entweder primär, d.h. das Gehirn direkt, oder sekundär, d.h. beispielsweise bei einer körperlichen Allgemeinerkrankung, das Gehirn indirekt beeinträchtigenden Erkrankung oder Schädigung. Dieses ist zumindest potentiell reversibel.

Synonym, zum Teil ähnlich, mit einer gewissen Differenzierung, werden Begriffe wie „körperlich begründbare Psychosen" (K. Schneider, 1980), „organische Psychosen", „exogene Psychosen", „akute exogene Reaktionstypen" (nach Bonhoeffer, 1912), „symptomatische Psychosen" und „Funktionspsychosen" (Wieck, 1956, 1977), in der angelsächsischen Literatur auch „Delirium" i.w.S. (Lipowski, 1980a, b, c) verwandt.

Davon abgegrenzt werden **chronische organische Psychosyndrome** (auch organisches Defektsyndrom) als **irreversible** psychische Veränderungen aufgrund zerebraler Schädigung, zum Teil auch hirnlokale Psychosyndrome bei umschriebener lokalisierter Schädigung des Gehirns, die allerdings nur in Einzelfällen eine spezifische prägnante Differenzierung aufgrund von Veränderungen von Stimmung, Antrieb und kognitiven Leistungen erlauben, wie es auch für die endokrinen Psychosyndrome gilt (vgl. Kap. 10). Auch die genaue Abgrenzung von akut und chronisch bzw. reversibel und irreversibel im Querschnittsbild ist nicht in allen Fällen möglich.[1]

Alle Erkrankungen des Körpers und des Gehirns können grundsätzlich zu akuten organischen Psychosyndromen führen, wenn wir sie auch bei bestimmten Erkrankungen besonders häufig beobachten. Die heutigen Anschauungen leiten sich von den Erkenntnissen Bonhoeffers her (1912), der, anstelle der auf Kraepelin zurückgehenden Meinung, jede Hirnnoxe habe ein spezifisches Syndrom zur Folge, seine **fünf akuten exogenen Reaktionstypen** (mit den Möglichkeiten Delir, epileptiforme Erregung, Halluzination, Amentia und Dämmerzustände) als unspezifische Folgen verschiedener Schädigungsmöglichkeiten postulierte. Das bedeutet ein Abrücken von dem Versuch, unterschiedliche psychopathologische Bilder nach ätiologischen und pathogenetischen Gesichtspunkten zu klassifizieren.

Obwohl die einzelnen Syndrome Bonhoeffers verändert, zum Teil durch andere ersetzt wurden, orientieren sich spätere Vorstellungen an der Lehre vom exogenen Reaktionstyp. Zwar können bestimmte Färbungen auf spezifische Noxen bzw. Erkrankungen hinweisen (Medikamente, Drogen), doch betreffen die kennzeichnenden Züge, z.B. beim Delirium tremens, mehr die vegetative Begleitsymptomatik als das psychopathologische Syndrom (Huber, 1981). Typisch für das akute organische Psychosyndrom ist das Fließende des Zustandsbildes, an dem sich oft Gesetzmäßigkeiten des Ablaufs des zugrundeliegenden organischen Prozesses erkennen lassen (Conrad, 1960).

Als zentrales Symptom wird dabei die **Bewußtseinstrübung** dargestellt, die allerdings von Anfang an nicht als obligatorisch in jedem Fall angesehen wurde. Heute gilt die Bewußtseinstrübung als ein Leit- oder Achsensymptom beim akuten hirnorganischen Psychosyndrom, hingegen sind es Abbau und Veränderungen der Persönlichkeit und Intelligenz beim chronischen hirnorganischen Psychosyndrom. Bewußtseinsstörungen lassen sich in verschiedene Typen einteilen:

– nach dem Grad der Herabsetzung der Vigilanz
– nach der Einengung des Bewußtseinsfeldes
– nach der Lösung der Zusammenhänge seelischer Inhalte (vgl. Müller, 1973).

Dies entspricht einer idealtypischen Einteilung, wobei wir es häufig mit einer Mischung aus verminderter Vigilanz, Einengung des Bewußtseinsfeldes und gestörten Zusammenhängen zu tun haben.

Es werden auch quantitative Bewußtseinsstörungen mit der Übergangsreihe Koma-Sopor-Somnolenz-Verhangenheit und qualitative Bewußtseinsstörungen, z.B. mit produktiv-psychotischer Symptomatik, unterschieden. Den akuten Reaktionstypen gehen oft Störungen ohne wesentliche oder mit klinisch nicht mehr feststellbarer Bewußtseinstrübung voraus oder folgen ihnen nach. Wieck (1956, 1977), der den Begriff **Funktionspsychose** synonym mit den akuten exogenen Reaktionstypen verwandte, hat für diese leichteren Formen die Bezeichnung **Durchgangssyndrome** eingeführt. Er bezeichnet dabei die **Minussymptomatik** an kognitiven und intellektuellen Funktionen als gemeinsames Achsensymptom von Durchgangssyndromen und Funktionspsychosen. Im Situationskreismodell von v. Uexküll (Kap. 1) bedeutet dies, daß die gewohnten Programme, mit denen Probleme gelöst und Umweltbezüge hergestellt werden, nicht mehr oder nur noch teilweise zur Verfügung stehen und durch einfachere Programme ersetzt werden müssen.

Bei den Durchgangssyndromen handelt es sich um die Störungsformen ohne nachweisbare Bewußtseinstrübung. Im Verlauf einer progredienten primären Hirnerkrankung oder einer sekundär im Rahmen einer Körpererkrankung entstehenden Hirnfunktionsstörung nimmt ein Durchgangssyndrom häufig an Schwere zu und geht fließend in die Bewußtseinstrübung über. Nach Vergiftungen oder Hirntraumen finden wir meist eine umgekehrte Abfolge. Gering ausgeprägte Durchgangssyndrome als Zeichen einer körperlichen Erkrankung, die sich z.B. nur in ängstlicher oder depressiver Verstimmung, erhöhter Reizbarkeit oder Antriebsschwäche äußern, sind oft schwer zu erkennen, von psychoreaktiven Veränderungen kaum abzugrenzen und imponieren als pseudoneurasthenische Syndrome.

---

[1] Dies hat dazu geführt, daß im Diagnostic and Statistical Manual der American Psychiatric Association (DSM III; Köhler, 1984) nicht mehr streng zwischen akuten und chronischen Formen unterschieden und eine Einteilung in drei Gruppen vorgenommen wird, je nachdem, ob globale (Delir, Demenz) oder umschriebene kognitive Einschränkungen bzw. Abnormalitäten (akutes amnestisches Psychosyndrom, Halluzinose) oder aber organische Persönlichkeitsveränderungen bestehen.

63.4 Häufigkeit des Vorkommens akuter organischer Psychosyndrome

Häufigkeitsangaben über das Vorkommen akuter organischer Psychosyndrome in der Literatur sind sehr unterschiedlich. Dies rührt daher, daß sie sich oft auf nicht vergleichbare Gruppen untersuchter Patienten beziehen. So spielt beispielsweise das Alter für die Wahrscheinlichkeit des Auftretens eines organischen Psychosyndroms eine wesentliche Rolle, da mit steigendem Alter eine rapide Zunahme zu verzeichnen ist und insbesondere Patienten über 60 Jahre besonders gefährdet sind. Bestimmte Erkrankungen, z.B. mit metabolischen Veränderungen, und einzelne therapeutische Verfahren stellen ein besonderes Risiko dar. Schließlich sind Häufigkeitsangaben, die sich beispielsweise auf nichtpsychiatrische Populationen beziehen, oft durch diagnostische Unsicherheiten und Anwendung unterschiedlicher definitorischer Regeln zweifelhaft. So werden zum Teil nur eindrückliche Varianten mit produktiv-psychotischer Symptomatik oder psychomotorischer Unruhe berücksichtigt (z.B. bei Willi, 1966).

Im folgenden seien einige Literaturangaben zur Häufigkeit referiert: Nach Willi (1966) soll etwa ein Drittel der Bevölkerung zwischen 20 und 70 Jahren eine exogene Psychose durchmachen. Dabei handelt es sich sicherlich um die häufigste Form einer Psychose, der wir in der ärztlichen Praxis begegnen. Ihr Vorkommen bei nichtpsychiatrischen Krankenhauspatienten wird von Willi (1966) mit 5 bis 10% angegeben. In einer Literaturübersicht (Lipowski, 1980a,

S. 53–60) finden sich, abhängig von Institution und durchschnittlichem Patientenalter, Angaben von 5 bis 30%, bei über 60jährigen hingegen 40 bis 50% (Lipowski, 1983b). Im Konsiliardienst chirurgischer und internistischer Abteilungen bezogen sich bei über 60jährigen 65% der Konsiliaranforderungen auf diese Krankheitsgruppe (Krakowski, 1979). Die Häufigkeit organischer Psychosyndrome insgesamt unter Einschluß leichter reversibler und der irreversiblen Formen wird nach einer Untersuchung in Allgemeinpraxen in Mannheim mit 17% angegeben (Zintl-Wiegand, 1979), in einer anderen Untersuchung in einer süddeutschen Kleinstadt mit 13% (Dilling et al., 1975).

Während der letzten Jahrzehnte haben akute organische Psychosyndrome vermutlich zugenommen. Neben dem zunehmenden Durchschnittsalter der Bevölkerung hängt dies sicherlich auch mit veränderten Lebensgewohnheiten, z.B. steigendem Medikamenten-, Drogen- und Alkoholkonsum (vgl. Thompson et al., 1983), und der Einführung neuer invasiver therapeutischer Verfahren zusammen (vgl. Kap. 64). Schließlich dürfte auch eine verbesserte Diagnostik, z.B. durch zunehmende konsiliarische psychiatrische Betreuung der Patienten, und eine bessere Ausbildung während des Studiums zur häufigeren Diagnose beitragen.

63.5 Symptomatik

Entsprechend den unspezifischen Reaktionstypen (Bonhoeffer) lassen die einzelnen Formen bzw. Prägnanztypen akuter organischer Psychosyndrome meist keine Rückschlüsse auf Art und Lokalisation der körperlichen Erkrankung zu. Es gilt das Prinzip der **Unspezifität organischer Psychosyndrome.** Zwar sind akute und chronische, reversible und irreversible Formen vom Querschnittsbild her nicht mit absoluter Sicherheit zu unterscheiden, aber es finden sich zahlreiche differentialtypologische Kriterien, die dabei hilfreich sein können (vgl. Tab. 63–1).

Einzelne Prägnanztypen organischer Psychosyndrome zeigt Tabelle 63–2 (verändert und vereinfacht nach Huber, 1981). Hinsichtlich der näheren Beschreibung der psychopathologischen Begriffe sei auf psychiatrische Darstellungen verwiesen.

Phänomenologisch unterscheiden sich die verschiedenen Psychosyndrome neben dem Bewußtseinsgrad insbesondere nach dem Grad der psychomotorischen Erregung und dem Ausmaß und Aussehen produktiv-psychotischer Symptome. Vorwiegend nachts prägen sich diese häufig stärker aus und es kommt zu **deliranten Erscheinungsbildern,** die näher beschrieben werden sollen, mit ängstlich gefärbter psychomotorischer Unruhe, einer qualitativen Bewußtseinstörung mit abnormem Bedeutungserleben, erhöhter Suggestibilität, Wahrnehmungsstörungen und illusionären Verkennungen, verändertem Zeiterleben und einer Inkohärenz des Denkens. Dazu finden sich dann meist Merk- und Orientierungsstörungen. Bei der Untersuchung der Orientierungsfähigkeit erweist sich die zeitliche Orientierung als zuerst ge-

Tabelle 63–1. Symptome bei akuten und chronischen organischen Psychosyndromen (modifiziert nach Lipowski, 1980b, S. 1373).

	akute organische Psychosyndrome	chronische organische Psychosyndrome
Einsetzen	akut	gewöhnlich schleichend, falls akut, geht meist ein Koma oder delirantes Syndrom voraus
Dauer	gewöhnlich kürzer als 1 Monat	meist über Monate
Orientierung	meist gestört	in leichten Fällen ungestört
Denken	oft inkohärent	eher verarmt, z.T. inkohärent
Gedächtnis	vorwiegend Kurzzeitgedächtnis gestört	meist Kurz- und Langzeitgedächtnis gestört
Bewußtsein	oft getrübt, fluktuierend, meist nachts Verschlechterung	meist keine Bewußtseinstrübung
Wahrnehmung	oft gestört	seltener gestört
Schlaf/Wachrhythmus	immer gestört	gewöhnlich altersentsprechend

Tabelle 63–2. Einzelne Prägnanztypen akuter hirnorganischer Psychosyndrome (nach Huber, 1981).

Psychopathologische Syndrome der akuten (reversiblen) Form

Prägnanztypen

– **ohne Bewußtseinstrübung – Durchgangssyndrome**
  1. spontane, affektive (depressive, maniforme), pseudoneurasthenische, hysteriforme
  2. produktive: expansiv-konfabulatorische, paranoidhalluzinatorische, katatone
  3. Halluzinose: akustisch, optisch, haptisch
  4. akutes amnestisches Syndrom

– **mit Bewußtseinstrübung**
  1. mit **quantitativen** Bewußtseinsveränderungen: Bilder von (flüchtiger) Benommenheit (z.B. nach Commotio) bis Somnolenz über Sopor bis hin zu komatösen Zuständen (z.B. nach Contusio cerebri, schweren Vergiftungen oder hirntoxisch verlaufenden allgemeinen Erkrankungen, intrakraniellen Raumforderungen etc.)
  2. mit **qualitativ-produktiven** Bewußtseinsveränderungen:
     a) einfache Verwirrtheit
     b) amentielle Syndrome (Denkinkohärenz, Verwirrtheit, Ratlosigkeit, wahnhafte Verkennung der Umgebung, traumhaft-desorientierte Verfassungen)
     c) delirante Syndrome (motorische Unruhe, Desorientiertheit, illusionäre Verkennungen, vorwiegend optische Halluzinationen, szenisch-traumhafte Verwirrung, vegetative Stigmatisierung)
     d) Dämmerzustände (ängstlich-verhaltene oder auch zornig-gewalttätige Erregungszustände bei situativer Desorientiertheit und Umgebungsverkennung)

stört, am wenigsten die Orientierung zur Person. Patienten haben bei Orientierungsstörungen die Neigung, Nicht-Vertrautes zu Vertrautem zu verändern, indem sie beispielsweise behaupten, an ihrem Heimatort zu sein, unbekannte Personen als Angehörige verkennen.

Bei den **Halluzinationen** dominieren optische Phänomene, oft mit szenischem Charakter. Willi (1966) stellte in einer Untersuchung von 100 Patienten mit exogenen Psychosen bei 73% optische Halluzinationen fest, bei 23% akustische, beide Formen bei 17% der Untersuchten. Es treten auch osmische, gustatorische und haptische Halluzinationen auf.

Häufig kommen **Wahneinfälle** und **Wahnwahrnehmungen** vor. Die paranoiden Symptome sind meist fließend, zeigen wenig systematische Tendenzen und sind oft durch situationsspezifische Merkmale beeinflußt. Die **affektiven Störungen** können zu aggressiven Ausbrüchen und zur Verweigerung jeglicher pflegerischer oder ärztlicher Maßnahme oder, wie im Falle unserer Patientin, zur Gefährdung durch Suizid führen. Die einzelnen Symptome fluktuieren, zum Teil kommt es zu luziden Intervallen. Gelegentlich überlagern sich verschiedene Formen, z.B. ein akutes und chronisches hirnorganisches Psychosyndrom. Dies kann nach einer abendlichen Tranquilizergabe bei älteren Menschen passieren, wo es nachts zur funktionspsychotischen Dekompensation mit psychomotorischer Unruhe kommen kann, die dann

häufig als „paradoxe Reaktion" angesehen wird. Besteht eine Bewußtseinstrübung, findet sich später meist eine partielle oder komplette Amnesie. Interessant ist dabei die Frage, in welchem Maße psychische Abwehrvorgänge am Vergessen der meist unangenehmen Erfahrungen im Rahmen der exogenen Psychose beteiligt sind.

Willi (1966) untersuchte bei 100 Patienten mit exogenen Psychosen Erlebnisinhalt und Erscheinungsbild. Danach hat die aktuelle Lebenssituation für die Psychose oft eine „pathoplastische" Bedeutung, denn in ihrer Ausformung zeigt sich häufig die „Auseinandersetzung eines Schwerkranken mit der durch die Krankheit bestimmten Lebenssituation" (S. 57–72).

Dies unterstreicht den Anteil eines Verständnisansatzes, der psychodynamische Denkweisen einbezieht und zu einem tieferen Verständnis der Krankheitssituation des Patienten beitragen kann.

Dabei beeinflussen Persönlichkeit des Kranken mit den zur Verfügung stehenden Abwehr- und adaptiven („coping") Mechanismen, seine Biographie und seine aktuelle Lebenssituation die Ausgestaltung des akuten organischen Psychosyndroms. Diese Einflüsse können sich thematisch in psychotischen Symptomen, im vorherrschenden Affekt oder in der Krankheitsreaktion wiederfinden. So sind die Strategien von Patienten, mit Defiziten und Behinderungen umzugehen, sehr unterschiedlich (Säring, 1988 zum Problem der Anosognosie und Anosodiaphorie bei hirngeschädigten Patienten; Tölle, 1987 zur Wahnentwicklung bei körperlich Behinderten; Küchenhoff, 1986). Manche Patienten reagieren auf Fragen mit Scherzen, Sarkasmen und verstecken damit ihre kognitive Beeinträchtigung, andere mit Ausweichen oder mit explosivem Ärger. In Abhängigkeit von der vorherrschenden Stimmung und der affektiven Regulierung mischen sich organische, persönlichkeitstypische und situative Momente. Gelegentlich werden prämorbide Persönlichkeitszüge im organischen Psychosyndrom ins Groteske verzerrt.

63.6 Schweregrad und Verlauf

Zur Einschätzung des Schweregrades sind die Beeinträchtigung kognitiver Funktionen, affektive und produktiv-psychotische Störungen zu berücksichtigen. Diese lassen sich in der Regel im Gespräch mit dem Patienten erfassen. Umfangreiche Tests eignen sich nicht als diagnostische Maßnahmen, denn Patienten mit akuten organischen Psychosyndromen sind in der Regel zu krank, zu leicht ablenk- und ermüdbar, um einer längeren Testprozedur ausgesetzt zu werden. Einfache Tests sollen am Bett durchgeführt werden können und eine grobe Quantifizierung für den intra- und interindividuellen Vergleich erlauben.

Orientierung, Kurzzeitgedächtnis, Aufmerksamkeits- und Konzentrationsvermögen, abstrakte Denkfähigkeit und Geschwindigkeit und Dynamik von Denkvorgängen werden geprüft. Zur Quantifizierung kann z.B. der Benton-Test (1963) verwendet werden. Meist genügt das Gespräch mit dem Patienten, um das Ausmaß seiner zerebralen Beeinträchtigung festzustellen. Man läßt den Patienten z. B. Daten aus seinem Leben berechnen, kommt auf seine Angaben über Ort, Zeitpunkt und Art von Geschehnissen zurück und vergleicht sie.

Abhängig von der Schwere akuter organischer Psychosyndrome findet sich meist parallel eine diffuse Verlangsamung der Grundtätigkeit im EEG, die sich in der Regel mit klinischer Besserung normalisiert. Dabei ist in der Verlaufskontrolle der Grad der Verlangsamung wichtig.[2]

Akute organische Psychosyndrome können zu einer restitutio ad integrum führen. Möglich ist auch, daß sie irreversibel in ein chronisches hirnorganisches Psychosyndrom, bei dem verschiedene Prägnanztypen zu unterscheiden sind (Pseudoneurasthenie – organische Wesensänderung – Demenz), münden. In der Terminalphase einer schweren Erkrankung finden wir sie häufig als Zeichen des herannahenden Todes. Nahezu alle Patienten in der Agonie weisen ein hirnorganisches Psychosyndrom auf (Massie et al., 1983). Oft stellt das akute organische Psychosyndrom ein Durchgangsstadium dar, z. B. bei der langsamen Ausprägung einer sog. „Multiinfarktdemenz" durch wiederholte zerebral-ischämische Insulte, ebenfalls nach einem schweren Schädel-Hirn-Trauma zwischen Koma und Ausgang in einer chronischen hirnorganischen Beeinträchtigung.

63.7 Diagnose und Differentialdiagnose

63.7.1 Diagnosestellung

Die Diagnose eines akuten organischen Psychosyndroms stützt sich auf die psychopathologischen Auffälligkeiten und den Nachweis einer vorausgehenden oder gleichzeitigen körperlichen Funktionsstörung. Bilder mit ausgeprägter Bewußtseinstrübung haben in der Regel eine relativ eindeutige Zuordnung zu den Formen körperlich begründbarer psychischer Störungen. Bei den Durchgangssyndromen hingegen mit beispielsweise depressiven oder ängstlichen Verstimmungen, Antriebsstörungen oder produktiv-psychotischer, z. B. paranoid-halluzinatorischer, oder katatoner Symptomatik läßt sich von der psychopathologischen Erscheinungsform nicht auf die Zugehörigkeit schließen. Eine organische Genese wird durch folgende Kriterien wahrscheinlich:

- den Nachweis einer körperlichen Schädigung oder geeigneten Noxe, die der psychischen Störung vorausgeht oder parallel zu ihr verläuft;
- die psychische Störung variiert nach Grad und Schwere der Schädigung bzw. Noxe;
- mit erfolgreicher Behandlung der körperlichen Schädigung oder Abklingen der Noxe (z. B. bei Dosisreduzierung von Cortison; Übersicht bei Bräunig, 1988) oder erfolgtem Entzug kommt es zur Verbesserung der psychischen Symptomatik.

Diese kann allerdings das Vorhandensein der körperlichen Störung erheblich überdauern. Die klinische Erfahrung lehrt auch, daß die psychischen Veränderungen den feststellbaren körperlichen Störungen gelegentlich vorausgehen, so bei Avitaminosen, beim Pankreaskarzinom, beim Morbus Parkinson und der Chorea Huntington.

63.7.2 Differentialdiagnostische Überlegungen

Hinsichtlich der Differentialdiagnose lassen sich bei leichten Formen **psychoreaktive** Veränderungen auf die körperliche Erkrankung oft nicht mit Sicherheit abgrenzen. Hinzu kommt, daß die Wahrnehmung der Behinderung durch das organische Psychosyndrom beim Patienten zusätzlich zu reaktiven Veränderungen kognitiver Vorgänge, emotionaler und verhaltensmäßiger Äußerungen führt (vgl. Kap. 62). **Neurotische** Symptome können ähnlich aussehen. Man kann bei den affektiven Durchgangssyndromen, wie auch bei unserer Patientin, oft einen abrupten Wechsel, z. B. zwischen depressiven und hypoman erscheinenden Episoden beobachten, die gelegentlich an das Verhalten hysterischer Persönlichkeiten erinnern. Auch eine Angstsymptomatik kann im Vordergrund stehen mit der Notwendigkeit der Abgrenzung zu neurotisch bedingten Angstzuständen. Über Verlangsamung, Denkschwierigkeiten und Gedächtnisschwäche klagen oft auch Patienten mit einer **endogenen depressiven Erkrankung.**

Auf die Kriterien zur Abgrenzung gegenüber den chronischen hirnorganischen Psychosyndromen wurde bereits hingewiesen. Bedacht werden muß, daß Patienten mit einer beginnenden Demenz oft auch eine begleitende depressive Symptomatik aufweisen, bei der sich reaktive und somatogene Momente schlecht trennen lassen. Phänomenologisch fällt bei den chronischen hirnorganischen Psychosyndromen ein wesentlich geringeres Fluktuieren der Symptomatik auf. Es ist darauf zu achten, daß Patienten mit neurologischen Krankheitsbildern, die eine „Werkzeugstörung" aufweisen, z. B. eine sensorische oder motorische **Aphasie,** abgegrenzt werden, um Verwechslungen mit Verwirrtheitszuständen oder „negativistischem" Verhalten zu vermeiden. Funktionspsychosen können unter **schizophrenieähnlichen** Bildern verlaufen (Neumärker, 1989). Zur Differenzierung mag helfen, daß bei den exogenen Psychosen optische Halluzinationen, bei schizophrenen Psychosen akustische und leibliche häufiger sind.

Auch der Verlauf kann zur Differentialdiagnose beitragen: Primär durch Alkohol- oder durch Medikamentenabusus bedingte Syndrome lassen sich in der Regel medikamentös innerhalb weniger Tage zum Verschwinden bringen.

63.8 Ätiologie und Pathogenese

Das akute organische Psychosyndrom kann als gemeinsame Endstrecke einer Vielzahl pathophysiolo-

---

2 Beim Delirium tremens findet sich meistens eine schnelle Grundtätigkeit.

gischer Prozesse und Mechanismen angesehen werden. Viele Faktoren können Entstehung und Verlauf beeinflussen. Die folgende Abbildung 63–1 gibt dazu einen Überblick.

63.8.1 Somatische Faktoren

Einzelne Erkrankungen

Nahezu alle körperlichen Erkrankungen und Schädigungen können, indem sie direkt am ZNS ablaufen oder seine Funktion indirekt beeinträchtigen, zu organischen Psychosyndromen führen. Die nachfolgende Tabelle 63–3 soll einen orientierenden, wenn auch unvollständigen Überblick über einzelne organische Ursachen geben.

Statistisch läßt sich sagen, daß hirnorganische Psychosyndrome, die aufgrund extrakranieller Erkrankungen entstehen, meist solche vom reversiblen Typ sind und mit der Besserung der körperlichen Erkrankung verschwinden oder zumindest sich zurückbilden. Es gilt der Grundsatz einer **multifaktoriellen Genese.** Der schädlich einwirkende organische Faktor muß einen bestimmten kritischen, individuell variablen Grenzwert überschreiten, um ein Durchgangssyndrom oder eine Funktionspsychose zu erzeugen. Dieses hängt von zahlreichen körperlichen wie auch sonstigen Faktoren ab. So kann die Empfänglichkeit spezifisch gegen bestimmte Noxen erhöht sein, wie z.B. bei der Auslösung der Symptomatik einer akuten intermittierenden Porphyrie durch Medikamente, bei idiosynkratischen Reaktionen gegen bestimmte Stoffe, z.B. Acetylsalicylsäure, bei allergischen Prädispositionen, z.B. gegenüber Antibiotika.

Oft kommt das akute organische Psychosyndrom erst dann zustande, wenn mehrere Faktoren zusam-

Tabelle 63–3. Organische Ursachen akuter organischer Psychosyndrome (nach Lipowski, 1980b, S. 1369).

– Intoxikation durch Medikamente, Alkohol, Drogen, gewerbliche und sonstige Giftstoffe
– Alkohol-, Medikamenten- und Drogenentzug
– metabolische Störungen (Organinsuffizienz, Vitaminmangel, Störungen des endokrinen Systems, des Wasser- u. Elektrolythaushalts, Stoffwechseldefekte)
– Infektionskrankheiten (systemisch, intrakraniell)
– Schädel-Hirn-Verletzungen
– Epilepsie
– vaskuläre Erkrankungen (zerebro-, kardiovaskulär, Migräne)
– intrakranielle raumfordernde Prozesse (Tumoren, Abszesse, Hämatome, Parasiten)
– degenerative Erkrankungen mit Hirnbeteiligung
– Verletzungen durch physikalische Noxen (Verbrennungen, Erfrierungen, Strom)
– allergische Erkrankungen

menwirken, wie es häufig bei schweren Erkrankungen und in hohem Alter der Fall ist. Bei älteren Patienten finden wir oft, daß beispielsweise eine Herzerkrankung gemeinsam mit einer Infektion bei gleichzeitiger Störung des Flüssigkeits- und Elektrolythaushalts vorkommt. Zusätzlich bestehen dann meist auch noch vaskuläre Schädigungen. Es sollte immer versucht werden, die verschiedenen pathogenetisch wirksamen Faktoren zu identifizieren und sich nicht auf eine bekannte Erkrankung oder Noxe zu beschränken. Das Zusammenwirken verschiedener Noxen läßt sich beispielsweise nach ausgedehnten Verbrennungen zeigen. Andreasen und Mitarbeiter (1972) fanden in einer Untersuchung Häufigkeit und Schwere der Psychosyndrome mit dem Ausmaß der Verbrennungen korreliert. Delirien nach schweren Brandverletzungen weisen oft zwei Häufigkeitsgipfel auf, in der ersten Woche aufgrund der metabolischen Veränderungen, in der zweiten bis dritten Woche unter dem Einfluß von Infektionen. Häufig treten Psychosyndrome auf, wenn der Ausfall einer Organfunktion, z.B. der Nieren (vgl. Kap. 65) oder der Leber, zu multiplen Stoffwechselveränderungen und zur Beeinträchtigung anderer vitaler Organe führt (Übersicht bei Kaschkat, 1975). Die Regelhaftigkeit des Auftretens psychopathologischer Symptome bei einzelnen Störungen hat dazu geführt, daß man beispielsweise von einer urämischen oder hepatischen Enzephalopathie spricht. Es ist auch bekannt, daß bestimmte internistische Erkrankungen, wie beispielsweise der Lupus erythematodes, aber auch andere Kollagenosen, besonders häufig mit psychischen Symptomen verlaufen. Beim Lupus erythematodes sollen im Längsschnitt bei zwei Dritteln der Patienten neuropsychiatrische Symptome auftreten, die häufig uncharakteristisch sind, zum Teil der Diagnosestellung um Jahre vorausgehen können (Übersicht bei Krüger, 1984).

Sehr häufig finden wir schwere Psychosyndrome in der Behandlung von Patienten mit hämatologisch-onkologischen Erkrankungen, was durch die Viel-

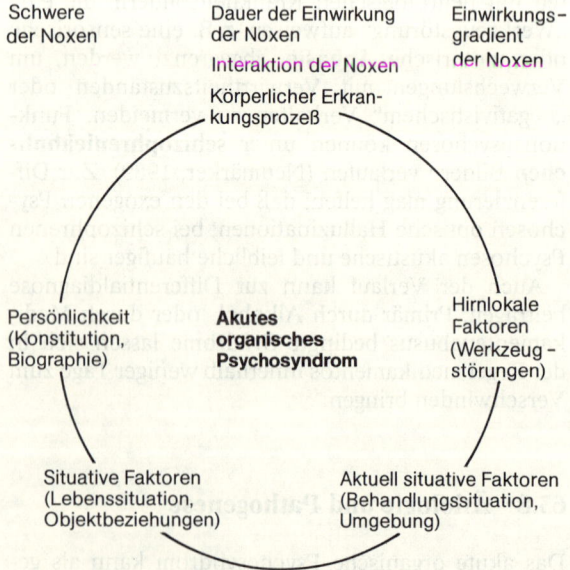

Schwere der Noxen

Dauer der Einwirkung der Noxen

Einwirkungsgradient der Noxen

Interaktion der Noxen

Körperlicher Erkrankungsprozeß

Persönlichkeit (Konstitution, Biographie)

Akutes organisches Psychosyndrom

Hirnlokale Faktoren (Werkzeugstörungen)

Situative Faktoren (Lebenssituation, Objektbeziehungen)

Aktuell situative Faktoren (Behandlungssituation, Umgebung)

Abb. 63–1. Ätiopathogenese und prägende Faktoren bei der Entstehung akuter organischer Psychosyndrome (in Anlehnung an H. H. Wieck, 1967).

zahl von Noxen und gestörter Funktionen erklärt werden kann. So treten bei vielen Zytostatika zentralnervöse Nebenwirkungen (Übersicht bei Strian und Maurach, 1980) gemeinsam mit den Folgen der Knochenmarksinsuffizienz und den Erscheinungen paraneoplastischer Syndrome auf.

Bei HIV-Infizierten erwiesen sich ca. 70% in den Frühstadien als kognitiv unauffällig, in Spätstadien nur ca. 40%. Allerdings zeigte sich das neuropsychologische Defizit nur bei 10–20% der Patienten als klinisch relevant. Die häufig zu beobachtenden ängstlichen und depressiven Beschwerden mit den subjektiven Einschränkungen der Konzentration dürften zu einem erheblichen Teil durch die Krankheitsbelastungen bedingt sein (Naber, 1989; Stieglitz, 1988).

Rolle von Medikamenten

Grundsätzlich muß man immer daran denken, daß die zur Behandlung der Grunderkrankung angewendeten Medikamente (z.B. Antihypertensiva, Antiarrhythmika, Herzglykoside, Diuretika, sedierende Pharmaka) die psychischen Störungen verursachen können, auch wenn sie im normalen therapeutischen Dosierungsbereich verwandt werden.

Dabei ist insbesondere bei Verwendung mehrerer Präparate zu berücksichtigen, daß gehäuft unerwünschte Wechselwirkungen auftreten können.

Insbesondere bei älteren Patienten muß damit gerechnet werden, daß Medikamente sehr viel häufiger organische Psychosyndrome verursachen. Sie weisen oft mehrere körperliche Erkrankungen gleichzeitig auf. Sie unterscheiden sich im Verteilungsmuster, der Eliminationsgeschwindigkeit und der Empfindlichkeit gegenüber Medikamenten stark von jüngeren Menschen (Greenblatt et al., 1982). Empfohlen wird, Dosierungen durchschnittlich auf 30–50% der Dosis, die für Erwachsene üblich ist, zu reduzieren. Bei psychotropen Medikamenten sind Nebenwirkungen doppelt so häufig wie bei anderen (Thompson et al., 1983). Das Ausmaß des Problems mag dadurch illustriert werden, daß beispielsweise in den USA die über 65jährigen 11% der Bevölkerung ausmachen, aber 30% der Verschreibungen erhalten (Thompson et al., 1983). Es ist wichtig, gerade bei älteren Patienten sich besonders streng an definierte Zielsymptome zu halten, die kleinstmögliche Dosis zu verschreiben und besonders häufig und sorgfältig eine Nutzen-Risiko-Abwägung vorzunehmen. Es sollte auch in Abständen ein Absetzversuch gemacht werden. Dieser empfiehlt sich überhaupt bei ungeklärten organischen Psychosyndromen, wie dies auch an unserem Fallbeispiel illustriert werden kann. Besonderes Augenmerk ist auf solche Medikamente zu richten, bei denen die Auslösung organischer Psychosyndrome relativ häufig ist (Zusammenstellung Hoffmann und Faust, 1983).

Bei bestimmten Grunderkrankungen, z.B. Lupus erythematodes oder Colitis ulcerosa, sollen Steroidpsychosen häufiger auftreten (Bräunig, 1988). Diese können zum Teil eine ungewöhnliche psychopathologische Ausprägung besitzen und beispielsweise bei längerem Verlauf und voller Reversibilität ohne produktiv-psychotische Symptomatik eine Demenz imitieren (Varney et al., 1984). Unter den Medikamenten, die besonders häufig zu Psychosyndromen führen, sind solche mit anticholinergen Nebenwirkungen (Übersicht bei Lipowski, 1989).

Prädestinierende Variablen

Zu den für die Auslösung in besonderem Maße begünstigenden Faktoren gehören der Literatur nach neben dem erwähnten höheren Alter Sucht, das Bestehen einer chronisch-konsumierenden Erkrankung und vorausgegangene zerebrale Schädigungen.

63.8.2 Sonstige Faktoren

Allgemeiner Streß

Klinische Beobachtungen deuten darauf hin, daß das Auftreten akuter organischer Psychosyndrome oft mit Veränderungen der Lebenssituation, der Umgebung und anderen aktuellen psychischen Streßsituationen zusammenfällt (Engel, 1959). So scheint der Wechsel in ein unvertrautes Milieu bei Abwesenheit von Bezugspersonen Häufigkeit und Schwere zu fördern. Dies deutet auf die Wichtigkeit interpersonaler Beziehungen hin. Patienten, die sich in der Dunkelheit schlechter zu orientieren vermögen oder deren Umweltbezug beispielsweise durch Erblindung oder Ertaubung schwer beeinträchtigt ist, sind besonders gefährdet (Cooper und Curry, 1976). Die Umgebung in Intensivbehandlungseinheiten kann dazu beitragen, den Schlaf-Wach-Rhythmus zu stören (vgl. Kap. 64).

Sensorische Deprivation

Viele Untersuchungen zur „sensorischen Deprivation" haben das Auftreten psychopathologischer Symptome bei einem Mangel an Sinnesreizen untersucht. Der Begriff, ursprünglich nur auf absolute sensorische Deprivation unter experimentellen Bedingungen angewandt, wurde später auch bei Zuständen partieller oder temporärer Reizverminderung, wie sie zahlreichen klinischen Situationen entspricht, z.B. bei Patienten in der „Eisernen Lunge" (Leiderman et al., 1958) oder im „Life Island", untersucht (vgl. Kap. 64).

Unter experimentellen Bedingungen wurden von zahlreichen Autoren Halluzinationen und Körperschemastörungen beschrieben (Zubek, 1969; Zuckerman, 1969a und b; Gross und Svab, 1969). Art und Ausmaß der Deprivation, der Persönlichkeitsmerkmale der Versuchspersonen und des Experimentators spielen als intervenierende Variablen eine Rolle. Visuelle Deprivation allein kann z.B. eine gesteigerte Sensibilität gegenüber Berührungsreizen, gesteigerte Schmerzempfindlichkeit, verstärktes Unterscheidungsvermögen für akustische, olfaktorische und gustatorische Reize schaffen. Insgesamt ist die Sensibilität gegenüber Deprivationsmaßnahmen individuell variabel und von bestimmten Persönlichkeitsmerkmalen abhängig. Probanden, die aufgrund testpsychologischer Ergebnisse zur **Verstärkung** von Sinnesreizen, z.B. zu verstärkten Schmerzwahrnehmungen neigen, sind gegenüber Reizentzug **widerstandsfähiger** als Probanden, die zur Gruppe derer gehören, die zur Reizverminderung tendieren (Petrie, 1967). Reizverminderer erhalten weniger Reize bei Sinnesentzug, Reizverstärker können bei Reizverarmung der Umgebung die wenigen Reize von außen und aus dem Körper noch ausnützen. Bei Überstimulation (z.B. Tieffluglärm) sind letztere mit ihren Schwierigkeiten der Abgrenzung vermehrt gefährdet (Küchenhoff,

1986). Empfindlichkeit gegenüber Schmerz und Empfindlichkeit gegenüber Reizentzug verhalten sich, wie mit Testpersonen in der „Eisernen Lunge" experimentell nachgewiesen werden konnte, umgekehrt proportional. Versuchspersonen, die im Embedded-Figures-Test (Witkin et al., 1962) „feldabhängiger" sind und primitivere Abwehrmechanismen benutzen als „feldunabhängige" Menschen, dekompensieren schneller in der Situation der sensorischen Deprivation.

Solche Untersuchungen vermögen, obwohl sie im einzelnen bruchstückhaft sind, doch Hinweise für den klinischen Umgang mit den Patienten zu geben. Zu berücksichtigen ist, daß ein ausgeprägtes akutes organisches Psychosyndrom schon eine sensorische Deprivation durch den Krankheitsvorgang als Ausdruck einer Informationsschwäche bedeutet (Burchard, 1980). Von daher wird es verständlich, daß eine durch die Behandlungssituation verstärkte Deprivation die Symptomatik akzentuieren kann, wie es der klinischen Beobachtung entspricht (Klein und Moses, 1974). Soziale Isolation (Linn und Kahn, 1956; Ziskind, 1958; Übersicht bei Lipowski, 1980a, S. 142; Jones, 1975) und eine längerdauernde Immobilisation (Levy, 1966) sollen den Ausbruch organischer Psychosyndrome fördern.

63.9 Therapie

An erster Stelle der Behandlung steht das Bemühen um **Beseitigung** der als **ursächlich** angesehenen **Faktoren** und die möglichst **kausale Behandlung** der zugrundeliegenden intra- oder extrakraniellen Erkrankung. Ergänzend dazu ist es wichtig, bei schweren Psychosyndromen eine ausreichende Ernährung und ausgeglichene Flüssigkeits- und Elektrolytbilanz sicherzustellen. Wichtig ist auch, für die Patienten **symptomatische Erleichterung** zu schaffen und Komplikationen wie beispielsweise Verletzungen bei unruhigen Patienten vorzubeugen. Optimal ist eine Sitzwache. Maßnahmen wie Immobilisieren oder die Verordnung sedierender Medikamente führen häufig zu einer Vermehrung der Unruhe. Es ist notwendig, die beschriebenen Störungen möglichst frühzeitig zu erkennen und zu behandeln. Bei agitierten, unruhigen und hochgradig verängstigten Patienten kann eine Sedierung und medikamentöse Behandlung notwendig sein. Beim Delirium tremens empfiehlt sich die Wahl von Clomethiazol (Distraneurin).

Immer bedacht werden muß, daß ein pharmakogenes Psychosyndrom vorliegen kann. Wenn beispielsweise Verdacht auf eine Delirauslösung durch Medikamente mit ausgeprägt anticholinergen Eigenschaften besteht, sollten keine anderen Pharmaka mit solchen Nebenwirkungen benutzt werden. Insbesondere bei älteren Patienten mit ausgeprägter Unruhe empfehlen sich hochpotente Neuroleptika. Niederpotente Neuroleptika können erhebliche Nebenwirkungen auf das Herz-Kreislauf-System haben und müssen in kleinen Dosen verabreicht werden. Die Gabe von Barbituraten zur Sedierung sollte vermieden werden. Die psychopathologischen Erlebnisinhalte sind häufig sehr angsterregend, weswegen stark anxiolytisch wirksame Medikamente wie Tranquilizer kurzzeitig angebracht sein können.

Hinsichtlich **allgemeiner supportiver Maßnahmen** sollte versucht werden, eine möglichst optimale Umgebung für die Patienten mit einem organischen Psychosyndrom zu gewährleisten. Dies schließt einen ruhigen, aber nicht zu grell beleuchteten Raum ebenso ein wie Kalender und Uhr in Sichtweite, um die Orientierung zu fördern. Schwestern und Pfleger sollten sich bemühen, mit den Patienten möglichst viel in Kontakt zu treten, sie notfalls wiederholt über alle Maßnahmen zu informieren und Orientierungshilfen zu geben. Verrichtungen wie Pulszählen oder Blutdruckmessen sollten erklärt werden. In der Literatur werden Versuche beschrieben, durch ständig wiederholtes Ansprechen und einfache Erklärungen die Orientierung im Sinne eines verhaltenstherapeutischen Trainings zu fördern (Garner, 1970). Es ist darauf zu achten, daß zu einem Zeitpunkt jeweils nur eine Information gegeben wird. Anweisungen sollten langsam und in ruhigem Ton vermittelt werden, damit sie der Patient besser aufnehmen kann (Hackett und Cassem, 1978). Gerade bei bewußtseinsgetrübten Patienten ist es wichtig, bei der Einteilung der Pflege darauf zu achten, daß die Pflegepersonen nicht zu häufig wechseln, was durch das Konzept der Zimmerpflege besser gewährleistet ist als durch die Funktionspflege. Eine vertraute Umgebung und vertraute Bezugspersonen können entlasten und zur Einsparung von sedierenden Medikamenten führen. Bei ängstlichen und depressiven Patienten ist es wichtig, ihnen emotionale Zuwendung anzubieten und sie gegebenenfalls mit stützendem psychotherapeutischem Vorgehen zu entlasten, wozu auch die Möglichkeit kathartischer Abreaktion gehört („Notfall-Psychotherapie"; Freyberger und Speidel, 1976). Ausmaß und Dauer einer Zwangsfixierung sollten auf das absolute Mindestmaß beschränkt und, wenn möglich, die Patienten fortlaufend durch Sichtkontakt überwacht werden. Die Anwesenheit von Angehörigen kann ein wichtiger Beitrag dazu sein, den Patienten die Orientierung zu erleichtern und sie emotional zu unterstützen. Gerade bei ausgeprägten Körperschemastörungen können korrigierende, geduldige Erklärungen für die Krankheitsverarbeitung hilfreich sein.

Im therapeutischen Vorgehen erweist sich die Berücksichtigung psychodynamischer Zusammenhänge in mehrfacher Hinsicht als sinnvoll und notwendig. Mit der Erfassung des Einflusses lebensgeschichtlicher und aktuell situativer Momente auf die Ausgestaltung der Symptomatik läßt sich so ein besseres Verständnis der Situation des Kranken gewinnen. Zudem stellt die durch die akute Erkrankung bedingte psychische Veränderung eine zusätzliche Belastung des Patienten dar, die ihn in seinen Möglichkeiten zur Krankheitsbewältigung einschränkt und zu weiteren psychoreaktiven Störungen, z.B. zu vermehrter Angst oder Depressivität, beeinträchtigtem Selbstwertgefühl, führen kann. Dies erläutert, warum häufig ein stützendes psychotherapeutisches Vorgehen angebracht ist.

64 Intensivmedizinische Behandlungsverfahren

Ekkehard Gaus und *Karl Köhle*

64.1 Vorbemerkungen

Bei Patienten in Intensivstationen bedürfen vitale Körperfunktionen infolge akuter Lebensgefahr einer dauernden Kontrolle und apparativer Überwachung. Neben der ständigen Kontrolle der Vitalfunktionen sind bilanzierte Infusions- und Transfusionstherapie, Korrektur metabolischer Abweichungen und besonders eingehende pflegerische Maßnahmen notwendig. Intensivtherapie stellt eine Extremsituation für die Patienten dar mit besonderen Anforderungen an Abwehr- und Anpassungsstrategien der betroffenen Patienten (vgl. Kap. 62).

Internistische Intensivstation und **interdisziplinäre Wachstation** unterscheiden sich zwar in vielerlei Hinsicht, aber der besondere Schweregrad der Erkrankung der Patienten, eine intensive, aufwendige Therapie und hohe Mortalitätsraten gehören zu den gemeinsamen Merkmalen, so daß eine zusammenfassende Darstellung der klinisch-psychosomatischen Gesichtspunkte berechtigt erscheint.

Die Mortalität in allgemeininternistischen Intensivstationen wird meist zwischen 15 und 35% angegeben. So lag in der Intensivstation der II. Medizinischen Universitätsklinik Mainz die Mortalitätsrate bei 4273 zwischen 1966 und 1975 behandelten Patienten bei 15%. Rechnet man die prognostisch günstigen Intoxikationsfälle ab, betrug sie 37%. Von den beatmeten Patienten starb etwa die Hälfte. Wenn zusätzlich zur Beatmung noch eine Hämodialyse erforderlich war, verstarben 80% der so behandelten Patienten (Schuster und Weilemann, 1981). Verfolgt man die aus Intensivstationen entlassenen Patienten weiter, so ergibt sich aus den Befunden, daß ein nicht geringer Teil der Behandelten innerhalb einer relativ kurzen Zeitspanne nachträglich verstirbt (Thibault et al., 1980: 25% binnen 15 Monaten).

Oft werden künstliche Beatmung durch Respiratoren, Herzstimulation durch temporäre oder permanente Herzschrittmacher und Maßnahmen wie Hämodialyse, Peritonealdialyse oder Hämoperfusion notwendig.

Diese Entwicklung der Medizin führt den Arzt in Grenzsituationen ärztlichen Handelns – etwa bei der Frage, in welchen Situationen und um welchen Preis es gerechtfertigt ist, durch lebensverlängernde Maßnahmen das Leiden von Patienten zu verlängern. Dies stellt extreme Anforderungen an den Arzt, auf die er durch seine Ausbildung meist nur ungenügend vorbereitet ist. Intensiveinheiten verkörpern dabei eine moderne und hochtechnisierte Medizin und ihnen wird innerhalb der Institution „Klinik" häufig ein elitärer Anspruch zuerkannt. Gleichzeitig wird an ihnen aber deutlich, wie ergänzungsbedürftig die Spitzen-

technologien in der Medizin sind. Interessant ist, daß es immer wieder Ärzte waren, die aus ihrer Erfahrung als Patienten heraus heftige Kritik gerade daran geübt haben, daß neben dem technisierten Ablauf die Versorgung emotionaler Patientenbedürfnisse leicht zu verkümmern droht (Robinson, 1975; Kautzky, 1976; Heinecker, 1980). Obwohl die körperlichen Probleme und die Gefährdung schwerstkranker, oft bewußtseinsgetrübter Patienten im Vordergrund stehen, findet sich auch auf der psychologischen Seite meist eine schwere Störung des seelischen Gleichgewichtes. So erleben viele Infarktpatienten mit der vitalen Bedrohung die Befürchtung schwerer sozialer Einbußen und des Verlustes von Kompensationsmöglichkeiten (vgl. Kap. 41.2). Noch evidenter wird die Bedeutung parallel verlaufender seelischer Störungen bei Patienten nach einem Suizidversuch (Böhme, 1981).

Reich und Gold (1983b) berichteten in einem kasuistischen Beitrag von einem Patienten, bei dem es auf der kardiologischen Wachstation zu einem Circulus vitiosus von Angst und schweren Rhythmusstörungen gekommen war, der mehrfache Defibrillationen erforderlich machte. Im psychiatrischen Interview und im Rahmen eines notfallpsychotherapeutischen Vorgehens konnte verstanden werden, daß zu diesem Teufelskreis der Umstand beitrug, daß der Patient die Behandlungssituation als Wiederholung eines biographisch bedeutsamen, schwer traumatisierenden Erlebnisses erfuhr. Erst die Bearbeitung dieses Zusammenhanges ermöglichte eine Entlastung und Unterbrechung des verhängnisvollen Geschehens.

Insgesamt können wir davon ausgehen, daß psychische Störungen bei lebensbedrohlichen Erkrankungen regelmäßig vorkommen und sich neben einer häufig festzustellenden Bewußtseinstrübung auch Rückzug und Depression finden können. Entsprechend kann man immer wieder eine überraschend schnelle Besserung des Zustandsbildes solcher Patienten bei ausreichender Zuwendung und verständnisvollem Umgang erleben. Daher ist es sicher nicht gerechtfertigt, bei der Intensivbehandlung einseitig die körperlichen Probleme zu berücksichtigen und eine Fülle diagnostischer und therapeutischer Verrichtungen durchzuführen, ohne die psychischen Bedürfnisse der Patienten zur Kenntnis zu nehmen. Dies gilt um so mehr, als Intensivtherapie einen hohen personellen Aufwand bedeutet und der Stellenschlüssel es ermöglichen würde, einzelnen Patienten mehr Zeit zu widmen. Wenn sich dennoch häufig eine Tendenz beobachten läßt, den Umgang bevorzugt auf technische Verrichtungen an Patienten zu beschränken, muß dies auch als Folge besonderer Distanzierungstechniken zur Entlastung von Ärzten und Pflegepersonal erklärt werden. In diesem Zusammenhang ist auch die hohe Fluktuationsrate, die man beim Personal von Intensivstationen findet, zu sehen.

Eine solche Distanzierung bedeutet notgedrungen auch eine Reduktion menschlicher Phänomene. Sie kann die affektive Belastung des untersuchenden und behandelnden Mediziners vermindern. Das Objekt, hier der Kranke, wird reduziert, um die Handlungsfähigkeit des Arztes zu verbessern. So sind in der Intensivmedizin diagnostische und therapeutische Maßnahmen leichter durchführbar, wenn die emotionale Seite im Umgang ausgeklammert bleiben kann. Eine auf die psychischen Bedürfnisse ihrer Patienten besonders eingehende Schwester reflektierte ihre eigene Abwehr: Am liebsten seien ihr die Bewußtlosen, die könnten wie ein Stück Holz behandelt werden, an ihnen seien all die nötigen Maßnahmen ohne weitere Rücksicht durchführbar, im Gegensatz etwa zu den Dialysepatienten, die sie gut kenne. Die verstorbenen Dialysepatienten würden sie oft noch monatelang in ihren Träumen beschäftigen.

Auch den Ärzten geht es so: Schon kleinere Eingriffe wie eine Lumbalpunktion fallen ihnen oft beim Bewußtlosen wesentlich leichter. Die Ausschaltung affektiver Beziehungen kann jedoch die Zusammenarbeit zwischen den Schwestern, Ärzten und Patienten beeinträchtigen. Der Patient, der seinen Schmerz und seine Ängste nicht schildern kann, kann z.B. zu wenig Schmerz- und Beruhigungsmittel erhalten, also nicht ausreichend behandelt werden. Die Krankenschwester, die bei der Reanimation eines Patienten eine Rippenfraktur verursacht hat, kann eventuell voller Schuldgefühle sein und Hemmungen haben, das Zimmer des Patienten wieder zu betreten. Sie ist damit in ihrer Fähigkeit, neue Symptome beim Patienten wahrzunehmen, erheblich beeinträchtigt.

Jede Arbeit mit oder an Menschen muß deren Antworten auf ein instrumentelles Vorgehen in Rechnung stellen. Innerhalb der Sozialwissenschaften wird dabei von „Gefühlsarbeit" (Strauss et al., 1980) gesprochen. Mit Fehlleistungen im Bereich der Gefühlsarbeit hängt es oft zusammen, wenn bei Patienten ein Gefühl der Erniedrigung, der Beleidigung, der verletzten Privatsphäre und der Eindruck, „wie ein Objekt" behandelt zu werden, aufkommt. Ein großer Teil der gegenwärtigen Klagen über eine entpersönliche Medizin in modernen Krankenhäusern bezieht sich auf derartige Fehlleistungen.

Indem die Aufmerksamkeit einseitig auf technologische Systeme verschoben wird, besteht die Gefahr, daß die psychosozialen Bedürfnisse des an Maschinen angeschlossenen Patienten aus dem Blickfeld pflegerischer und ärztlicher Aufmerksamkeit zu geraten drohen.

Grote-Janz und Weingarten (1983), die sich innerhalb eines Projektes „Patientenorientierte Intensivmedizin und medizinische Technologie" mit dieser Problematik befaßten und Analysen einzelner Interaktionen innerhalb einer Intensivstation vornahmen, beschreiben eine Szene mit einem neu eingelieferten Infarktpatienten, in der drei Schwestern gemeinsam die Routinehandlungen des Ankoppelns des Patienten an das zentrale intensivmedizinische Standardgerät, den EKG-Monitor, vornahmen: Die eine Schwester klebt Elektroden auf die Brust des Patienten, die andere kümmert sich um den Kabelsatz und die dritte versucht, das Monitorbild einzustellen. Diese Handlungen absorbieren die Aufmerksamkeit der drei Schwestern so, daß sie die Signale des Patienten, der fast weinend seine Angst äußert („jetzt werde ich meine Kinder nicht mehr sehen"), nicht aufgreifen. Abzumildern wäre diese belastende Situation durch empathische Gesten der Zuwendung.

Pannen, in diesem Fall der Umstand, daß sich über längere Zeit ein korrektes Monitorbild nicht einstellen läßt und die „Nullinie" auf dem Bildschirm verbleibt, verstärken die Aufmerksamkeitsfokussierung auf den technischen Vorgang, der zur mangelnden Wahrnehmung der Bedürfnisse des Patienten nach Trost, Aufklärung und Information über den Stellenwert der intensivmedizinischen Behandlung führt.

Das Dilemma der Aufmerksamkeitsfokussierung soll an einer Episode, die einer der Verfasser auf einer Überwachungsstation erlebte, verdeutlicht werden.

Mitten im Visitengespräch mit einem älteren Infarktpatienten kam es auf dem Bildschirm des Monitors innerhalb von wenigen Sekunden zur Abflachung der Kurvenausschläge bis zur Ausbildung einer Nullinie. Der überraschte Arzt fühlte sich für einige Augenblicke handlungsunfähig, indem er abwechselnd das Monitorbild mit der Nullinie und den Patienten, der ihn ruhig anschaute, beobachtete. Die Spannung löste sich, als der Arzt plötzlich wahrnahm, wie aus dem hinteren Teil des Monitors Rauch aufstieg. Dies erscheint uns als Beispiel dafür bezeichnend, daß zumindest für einen Augenblick die komplexe Wahrnehmung des klinischen Erscheinungsbildes des Patienten vom Ansatz her ausgeschaltet war zugunsten des Sogeffektes des Herzmonitors, der eine wichtige Vitalfunktion des Menschen sichtbar nach außen verlagert.

64.2 Psychische Störungen auf Intensivstationen – Übersicht

64.2.1 Häufigkeit

Die Zusammenfassung schwerkranker Patienten in eigens eingerichteten Stationen hatte zur Folge, daß man vermehrt auf eine große Zahl psychopathologischer Phänomene aufmerksam wurde, wie man sie bislang so gehäuft nur in psychiatrischen Institutionen gesehen hatte. Zunächst wurden vor allem krasse psychische Störungen wie delirante Bilder oder Formen mit einer schweren affektiven Symptomatik beschrieben. In der Anfangszeit der offenen Herzchirurgie wurde man besonders häufig auf Psychosen aufmerksam, die als „post-cardiotomy delirium" (Blachly und Starr, 1964), „cardiac psychosis" (Abram, 1965) und „postoperative psychosis" (Egerton und Kay, 1964) in der angelsächsischen Literatur beschrieben wurden. Dabei schwanken die Häufigkeitsangaben psychopathologischer Zustände nach Herzoperationen zwischen einigen wenigen und 100%. Trotz eines Rückgangs der Häufigkeit, insbesondere schwerster Formen (Layne und Yudofsky, 1971; Rabiner et al.,

1975), ist die Quote ungleich höher als bei allgemein-chirurgischen Eingriffen (1 : 1500 nach Knox, 1963).

Auch in internistischen Intensivstationen wurde man auf häufige psychopathologische Phänomene aufmerksam, was manche Autoren dazu veranlaßte, von einem **„Intensive-Care-Unit-Syndrome"** („ICU-Syndrom") zu sprechen (Meyer et al., 1961; Nahum, 1965; McKegney, 1966). Hier waren neben Funktionspsychosen mit produktiver Symptomatik und Bewußtseinstrübung vor allem schwere Angst- und Depressionszustände aufgefallen. Insbesondere wurde dies bei Infarktpatienten beobachtet (Hackett et al., 1968; Freyberger et al., 1969; Klapp und Freyberger, 1981). Die Häufigkeit von Psychosyndromen ist dabei von den auf Station angewandten Selektionskriterien abhängig.

64.2.2 Allgemeine Überlegungen zur Pathogenese

Im Rahmen von Überlegungen zur Pathogenese der beschriebenen psychischen Symptome stellte sich die Frage, in welchem Maß Grunderkrankung, therapeutische Maßnahmen und das gesamte Setting einer solchen Situation zur Entstehung und zur speziellen Ausformung organischer Psychosyndrome und psychoreaktiver Störungen bei Patienten in intensivmedizinischer Betreuung beitrugen. Es wurde kritisch gefragt, ob der verstärkte Aufwand apparativer Überwachung und Therapie und eine möglicherweise damit verbundene „Entfremdung" des Patienten eine pathogenetisch bedeutsame psychologische Noxe darstelle. Viele Untersuchungen belegen, daß die Intensivbehandlung keine Noxe per se darstellen muß, sondern im Gegenteil eine Möglichkeit für Patienten bietet, in emotionaler Hinsicht Sicherungs- und Haltefunktionen angeboten zu bekommen (Klapp et al., 1979, 1981, 1982, 1984).

64.2.3 Therapeutische Richtlinien im Umgang mit intensivmedizinisch betreuten Patienten

Eine psychotherapeutische Hilfe für Patienten in Intensivstationen muß mehrere Gesichtspunkte berücksichtigen: Der meist schlechte körperliche Zustand, eine vitale Gefährdung und die seelische Ausnahmesituation der Patienten erfordern Hilfe, die sich an den Regeln einer **stützenden** psychotherapeutischen Intervention orientiert (Freyberger und Speidel, 1977). Eine eingehende Erörterung stützender psychotherapeutischer Maßnahmen findet sich in Kapitel 16 über psychoanalytisch orientierte Therapieverfahren. Hier sei an einige Grundsätze erinnert: Neben den gemeinsamen Voraussetzungen stellen in der Intensivbehandlung unterschiedliche Behandlungsmaßnahmen auch unterschiedliche Belastungen dar, die zum Teil spezifisch sind (Gaus, 1975). Prophylaktische, diagnostische und therapeutische Maßnahmen müssen sich daher neben der Beachtung allgemeiner Grundsätze auch nach den spezifischen Anforderungen einzelner Verfahren richten, auf die

an entsprechender Stelle eingegangen werden soll. Konsiliarische Anforderungen werden meist gestellt, wenn Patienten auf Intensiveinheiten delirant oder so depressiv oder ängstlich werden, daß ihre Kooperation in Frage gestellt ist. Schließlich beziehen sich Anfragen auch oft darauf, die mögliche Psychogenese eines Symptoms oder die Bedeutung mitverursachender psychologischer Faktoren zu klären. Dadurch, daß psychiatrische und psychosomatische Abteilungen häufig institutionell und oft auch räumlich weit von Intensiveinheiten getrennt sind, wird die Betreuung durch einen Konsiliar- bzw. Liaisondienst in der Regel erschwert.

Schwere körperliche Beeinträchtigungen und Krisen, wie sie sich bei den intensivmedizinisch behandelten Patienten meist finden, überfordern häufig individuelle Fähigkeiten in der **Bewältigung und Anpassung** („**Coping**", vgl. Kap. 62). Nach den Vorstellungen der psychoanalytisch orientierten Persönlichkeitstheorie bedingen sie oft eine Schwächung der **„Ich-Funktionen"** (vgl. Bellak et al., 1973; Janis, 1958). Psychotherapeutische Behandlungsmaßnahmen müssen daher anstreben, den Aufbau einer stabilen Objektbeziehung und effizientere Abwehrstrategien angesichts der vitalen Bedrohung zu fördern und individuelle Fähigkeiten im Sinne des „Coping" zu stärken. Ein wichtiger Schritt im therapeutischen Vorgehen ist das Angebot einer Haltefunktion im Sinne Winnicotts (1974, 1976).

Bei einer „anaklitischen" oder **„Notfallpsychotherapie"** (Freyberger et al., 1969, 1977) kann der Therapeut mehrfach täglich kurze Besuche am Krankenbett eines besonders bedrohten Patienten machen, ihn ermutigen, ihn seines Beistands versichern und ihm Gelegenheit zur Abreaktion von Gefühlen geben. Dabei kann der Therapeut als Elternersatz erlebt werden. Eine Identifikation des Patienten mit dem Therapeuten sollte in der akuten Notsituation gefördert werden, ebenso können vorübergehend Omnipotenzvorstellungen des Patienten über den Arzt ebenso akzeptiert werden wie eine regressive Beziehung zum Therapeuten nach dem Vorbild des Mutter-Kind-Verhältnisses. Wichtig ist neben einer möglichst ausreichenden Verfügbarkeit eine große Frustrationstoleranz des Behandelnden (Freyberger und Speidel, 1977). Der der Krankheit gegenüber oft hilflose und hoffnungslose Patient muß sich dem Therapeuten in der Krise anvertrauen können. **Äußere Hilfsmaßnahmen,** z.B. durch gezielte Orientierungshilfen bei bewußtseinsgetrübten Patienten, die es ihnen erleichtern, ihre Umgebung zu strukturieren, können zur Stützung von Ich-Funktionen beitragen, und es kann auch der Versuch gemacht werden, innere Reserven bei den Patienten zu mobilisieren, z.B. dadurch, daß an Erinnerungen angeknüpft wird, in denen die Patienten erfolgreich Konflikte und Schwierigkeiten bewältigt haben.

Der Zustand schwerkranker Patienten in Intensivstationen steht häufig auch in krassem Widerspruch zu ihrem „idealen Selbst", d.h. der Wunschvorstellung, die sie von sich selbst haben. Je weiter diese Wunschvorstellung von der Wirklichkeit entfernt ist, desto größer ist die Kränkung und Beeinträchtigung ihres Selbstwertgefühls. Eine Möglichkeit, die Kränkung zu verringern, besteht darin, positive Aspekte des momentanen und künftigen Selbst zu betonen, z.B. bei tracheotomierten Patienten, die nicht sprechen, sich aber durch Gesten und Bilder und eventuell schriftlich ausdrücken können, verbliebene Ausdrucksmöglichkeiten zu fördern und mit Rehabilitationsvorstellungen zu verknüpfen, ihnen Interesse und Mitgefühl zu zeigen, Anerkennung zu spenden und Gelegenheit zu geben, ihre Gefühle auszudrücken.

Alle diese therapeutischen Bemühungen setzen voraus, daß man über die medizinische, soziale und psychologische Vorgeschichte eines Patienten informiert ist. Zu berücksichtigen ist, daß Schwestern, Pfleger und Ärzte in der Regel zu dem Vorurteil neigen, daß Schwerkranke in erster Linie in Ruhe gelassen werden sollten. Demgegenüber wissen wir, daß gerade Schwerkranke im Vergleich zu Gesunden verstärkt emotionelle Bedürfnisse haben. Wichtig ist, daß sich die Kontakte nicht nur auf diagnostische und therapeutische Verrichtungen beschränken, die die Ruhe des Patienten stören oder Schmerzen verursachen, sondern daß dem Kranken Interesse an seiner Person vermittelt wird.

Untersuchungen über die prozentuale Häufigkeit der Präsenz des Personals am Krankenbett von Patienten auf Intensivstationen ergaben, daß ca. ein Drittel der Zeit direkt mit Verrichtungen beim Patienten verbracht wird. Der Kontakt beschränkt sich aber meist auf kurze Interaktionen, nämlich die Durchführung pflegerischer Maßnahmen, denen gegenüber Gespräche selten sind (Hannich et al., 1983 a, b).

Ein besonders wichtiger Punkt betrifft die Notwendigkeit, **Angehörige** in die **Betreuung** auch auf der Intensivstation einzubeziehen. So schlägt Schara (1981) vor, Schilder wie „Besuche verboten" in „Besuche erbeten" umzuändern.

64.3 Internistische Intensivstationen

64.3.1 Vorbemerkungen

1958 wurden die ersten internistischen Intensivstationen in den USA eingerichtet. Im Laufe der 60er Jahre wurde diese Entwicklung in der Bundesrepublik in großem Stil in die Regelversorgung übernommen. Parallel dazu setzte sich in der Intensivmedizin die Einführung der externen Herzmassage, der Kardioversion und der externen Schrittmacherbehandlung mit Elektrodenkatheter durch. Das Zusammentreffen der Belastungen und Einschränkungen der Intensivbehandlung mit den Persönlichkeits- und Konfliktmerkmalen, wie sie für Herzinfarktpatienten als typisch beschrieben wurden, ist im Hinblick auf psychosomatische Fragestellungen von besonderem Interesse (vgl. Kap. 41.1 und Kap. 41.2).

Eine 50jährige Patientin litt nach einer Reanimation noch unter einem akuten organischen Psychosyndrom. Innerhalb von drei Wochen war sie in drei verschiedenen Krankenhäusern behandelt worden. Beim Erstgespräch begann sie auf die Frage, wie es ihr ginge, gleich zu weinen. Sie ließ die Hand der Schwester nicht mehr los. Die Patientin brauchte längere Zeit, bis sie darüber sprechen konnte, wie schlimm diese Verlegungen für sie gewesen waren. Die Verlegungen seien immer plötzlich angeordnet worden; über die Gründe und die beabsichtigte Zeitdauer der Verlegungen sei sie nie informiert worden. Sie hatte ohnehin Schwierigkeiten, sich in ihrer Umgebung zurechtzufinden, wußte durch die Verlegungen jetzt nicht mehr, wo sie war. Die Patientin hatte zunächst die Befürchtung, an einem Ort zu sein, wo sie weder ihr Sohn noch ihre Bekannten finden konnten. Die häufigen Verlegungen hatten zu ihrer Vorstellung beigetragen, daß man ihr mit ihrer Krankheit ohnehin nicht mehr helfen könne. Diese Befürchtungen der Patientin ließen sich durch die Information über die günstige Prognose der Erkrankung abbauen. Das Entstehen einer tragfähigen Beziehung mit ihr wurde dadurch wesentlich erleichtert.

64.3.2 Das „ICU-Syndrom"

Autoren aus der Zeit der „Pionierphase" der Einrichtung von Intensivstationen, die von den dort zu beobachtenden häufigen psychischen Störungen wie Funktionspsychosen und Durchgangssyndromen beeindruckt waren, gingen so weit, von einem „Intensive-Care-Unit-Syndrome" als nosologischer Einheit (McKegney, 1966; Nahum, 1965, „ICU-Syndrom") zu sprechen, wobei sie einen eigenen pathogenen Einfluß des Milieus dieser Intensiveinheiten postulierten. Systematische Untersuchungen erfolgten dabei im Hinblick auf Infarktpatienten.

In psychologischer Hinsicht besonders pathogen wurden sensorische Monotonie, Schlafdeprivation, die Vielzahl technischer Apparaturen, die Nähe zu anderen schwerkranken Patienten und das Miterleben von Todesfällen und Verschlechterung von Mitpatienten angesehen (Hackett et al., 1968; Kornfeld et al., 1965 und 1969a; Parker und Hodge, 1967). So wurden zahlreiche Vorschläge gemacht, die das Milieu der Intensivstation für die Patienten weniger bedrohlich machen sollten, z.B. die Einrichtung von separaten Kabinen, Auflockerung durch Bilder, Radio, möglichst geringe Störung der Nachtruhe, Berieselung mit beruhigender Musik.

Das Miterleben des Todes von Mitpatienten und von Reanimationsmaßnahmen ist, insbesondere bei Infarktpatienten, als schädlich anzusehen, was gegen die Einrichtung offener Räume ohne Trennwände spricht. Auch wenn Abwehrvorgänge wie Verleugnung die Bedrohlichkeit solcher Wahrnehmungen verringern und, was häufig der Fall ist, Patienten einen Todesfall überhaupt nicht wahrgenommen haben wollen oder aber jeglichen Bezug zum eigenen Zustand verneinen, konnte doch ein signifikanter Anstieg von Puls und systolischem Blutdruck bei Koronarpatienten in Intensivstationen nach dem Tod eines Mitpatienten festgestellt werden (Bruhn et al., 1970).

Sensorische Deprivation – das haben theoretische und klinische Studien gezeigt – kann Sensibilität, Schmerzempfindung, das Entstehen von produktiven Symptomen wie Halluzinationen und primärprozeßhaftes Denken fördern (vgl. Kap. 63). So treten bei Patienten mit Augenoperationen, die zusätzlich schwerhörig sind, Funktionspsychosen besonders häufig auf.

Vergleichende Studien (Leigh et al., 1972; Holland et al., 1973) konnten testpsychologisch keine sicheren Auswirkungen des unterschiedlichen Milieus verschiedener Stationen feststellen. Auch andere Autoren bestätigen, daß das Personal von Intensivstationen die Behandlungsumgebung eher als angsterregend einschätzt als die Patienten selbst (Klapp et al., 1980). Für die Patienten spielt offensichtlich das durch die intensive Überwachung vermittelte Moment zusätzlicher Sicherheit eine wesentliche Rolle. Befragungen lieferten Anhaltspunkte dafür, daß es die meisten Patienten vorziehen, nicht in vollständig abgetrennten Kabinen zu sein (Leigh et al., 1972; Klein und Kellner, 1979). Technische Apparaturen wie Monitoren können bei entsprechender Information und Motivierung der Patienten als „mechanischer Schutzengel und Leibwächter" und so auch gefühlsmäßig als Sicherungsmoment empfunden werden, ebenso wie die in Intensivstationen bessere personelle Ausstattung und die technische Kompetenz des Personals. Dies ergibt sich aus eigenen Erfahrungsbeispielen der Autoren aus der Intensivbetreuung von Patienten, aber auch aus Patienteninformationen, die durch semistrukturierte Interviews gewonnen wurden.

Im Rahmen der wissenschaftlichen Überprüfung dieser Hypothesen untersuchten Cay und Mitarbeiter (1972) retrospektiv an 208 Patienten, überwiegend Infarktpatienten, Reaktionen und Einstellungen zur Intensivbehandlung und fanden eine überwiegend positive Meinung bei diesen untersuchten Patienten (ebenso Dominian und Dobson, 1969; Jelen et al., 1979; Hannich und Wendt, 1983 a, b; Greiner et al., 1981). Allerdings ist gegenüber retrospektiv erhobenen Fragebogenuntersuchungen hinsichtlich der Validität der Resultate äußerste Skepsis angebracht.

Die Bostoner Arbeitsgruppe (Hackett et al., 1968, 1977; Cassem und Hackett, 1971) ist bei ihren Untersuchungen zur Intensivbehandlung von der Prämisse ausgegangen, daß in der klinischen Situation das Fehlen von Angst oder Depression Ausdruck von Abwehrmechanismen wie Verleugnung sei. Andere Untersucher (Klapp, 1984) haben diese Annahme kritisiert, da die klinischen Zeichen von Angst oder Depression und Abwehr sich umgekehrt proportional verhielten.

64.3.3 Prophylaktische und psychotherapeutische Maßnahmen

Als therapeutische Konsequenz können wir ableiten, gerade im Umgang mit Schwerkranken, Apparate und Eingriffe jeweils genau, dem krankheitsbedingten Zustand entsprechend, zu erklären und möglichst

viel Kontakt zu den Patienten aufzunehmen. Beim Umgang mit den für die Intensivstation spezifischen Problemen sollten ähnliche Regeln und Grundsätze gelten, wie sie für Infarktkranke während der Intensivbehandlungsphase dargestellt sind (vgl. Kap. 41.2). Dazu gehören z. B. Entängstigung, Depressionsbearbeitung, Errichtung eines Arbeitsbündnisses und emotionale Stützung, insbesondere auch bei der Verlegung von der Intensivstation. Gelegentlich erfolgen Verlegungen nachts infolge eines Bettenmangels und werden dann, falls keine Vorbereitung des Patienten erfolgte, auch als besonders traumatisch erlebt. Kimball (1976) hat in diesem Zusammenhang von „Verlegungsarrhythmien" gesprochen. Ohne rücksichtsvolle Aufklärung der Patienten stellen solche Verlegungen eine akute Gefährdung der Patienten dar. Es empfiehlt sich, mit den Patienten gezielt eine Art von „Verlegungsgespräch" zu führen. Die Sicherheit vermittelnden Funktionen von Intensivstationen machen es verständlich, daß die Verlegung auf eine Allgemeinstation bei manchen Patienten Ängste und Befürchtungen auslöst.

In einer Studie (Toth, 1980) wurden zwei Gruppen von Infarktpatienten verglichen, von denen die erste unstrukturiert auf die Verlegung vorbereitet, die zweite Gruppe speziell über die Verlegungspraxis informiert wurde. Bei der Kontrolle physiologischer Angstindizes fanden sich diese im Verlauf von 24 Stunden am Verlegungstag in der ersten Gruppe signifikant erhöht. In einer anderen Untersuchung an Herzinfarktpatienten waren die testpsychologisch ermittelten Angstparameter nach Verlegung von der Intensivstation am stärksten ausgeprägt (Philip und Cay, 1979).

Befragungen von Patienten auf Intensivstationen ergaben ein großes Informationsbedürfnis (Hannich und Wendt, 1981). Informations- und Kommunikationsdefizite gehören zu den wichtigsten Streßfaktoren im Intensivmilieu (Klapp, 1984). Wichtig ist, eine ständige Präsenz des Personals auch auf einer emotionalen Ebene zu gewährleisten. Heinecker (1980) spricht anhand eigener Erfahrungen von der Gefahr einer „Empathiebarriere" beim Personal.

Wichtig sind auch vorwegnehmende Erklärungen über den Fehlalarm von Monitoren bei Bewegungen des Patienten, wobei man auf die Demonstration dieses Vorganges besonderen Wert legen sollte. Die Angst, die bestimmte Ereignisse, z. B. große Visiten usw., auslösen können, kann durch vorherige Information vermindert werden (Wise, 1985). So wird verhindert, daß diese Vorgänge im Patientenerleben eine ungewöhnliche, bedrohliche Bedeutung erlangen. Vermögen schon medizinische Dispute, die am Krankenbett die bisherigen Therapiemaßnahmen skeptisch diskutieren, bei vielen Patienten Ängste und Zweifel auszulösen, so können sie bei Intensivpatienten zu Katastrophenreaktionen und akuten Todesängsten führen.

Auch eine falsche Vorstellung des Patienten über seine Erkrankung kann zu ausgeprägten Beschwerden beitragen und die Rehabilitation erschweren:

So hatte ein Infarktpatient, der mehrfach reanimiert worden war und über eine schwere Gedächtnisstörung klagte, bei sorgfältiger psychologisch-psychiatrischer Testung aber keine Anzeichen eines organischen Psychosyndroms aufwies, die Vorstellung, daß bei ihm das Verbindungsgefäß zwischen Herz und Gehirn verschlossen sei. Diese Vorstellung ängstigte ihn sehr und machte ihn hoffnungslos. Nachdem ihm der tatsächliche Befund erklärt worden war, besserten sich sowohl die Angst als auch die „Gedächtnisschwäche" (Stein et al., 1969).

Oberste Maxime aller psychotherapeutischen Maßnahmen muß daher sein, alle Vorgänge zu fördern, die einer intrapsychischen Verarbeitung der Erkrankung und ihrer Folgen dienlich sind und äußere Hilfs- und Unterstützungsressourcen zu mobilisieren. Dies setzt Einfühlungsvermögen und die Kenntnis der Anpassungs- und Abwehrstrategien der Patienten voraus, die durch Anamnese und Verhaltensbeobachtung herauszufinden sind. Wichtig ist, daß Informationen über Patienten innerhalb des Behandlungsteams ausgetauscht werden und daß es möglich ist, für Problempatienten, z.B. solche, deren Verhalten unkooperativ erscheint, gemeinsam Strategien zu entwickeln (vgl. Koumans, 1965; Köhle et al., 1973). Dabei sollten alle Interaktionen in den Therapieplan miteinbezogen werden. Den organisatorischen Rahmen können Fallbesprechungen des Teams, die der Balint-Gruppenarbeit nachempfunden sind, herstellen. Die Wirksamkeit eines so auf den Patienten zugeschnittenen Vorgehens zeigt das folgende kasuistische Beispiel. Aus ihm geht hervor, wie Wahrnehmungsänderungen beim Personal durch sorgfältige Betrachtung des Patientenverhaltens selbst in der Lage sind, eine Änderung des Patientenverhaltens zu induzieren.

> Zwei Schwestern der Intensivstation klagten darüber, daß ein Patient, der seit Wochen wegen einer aufsteigenden Lähmung beatmet worden war und der erst am Vortag extubiert werden konnte, immer noch nicht auf eine normale Station verlegt worden sei. Auf die fortbestehende Gefährdung des Patienten aufmerksam gemacht, begannen sie über sein ständiges Läuten und Quengeln zu klagen. Der Patient hätte weiter paranoide Befürchtungen geäußert und u.a. behauptet, es würde ihm von den Schwestern Gift in die Infusion getan. Schließlich habe er darüber geklagt, daß er der am schlechtesten versorgte Patient der Station sei. Man spürte, daß die Schwestern den Patienten verlegen wollten, weil er für sie nur noch schwer zu ertragen war. Die extreme Abhängigkeit des Patienten und seine Versuche, damit umzugehen, wurden dann mit den Schwestern durchgesprochen. Erstaunlich war, daß der Patient den Schwestern während der nächsten Schicht „wie verwandelt" erschien – ohne daß sich ihr Verhalten dem Patienten gegenüber für sie selbst erkennbar geändert hatte. Alle Klagen und paranoiden Befürchtungen waren verschwunden. Er bedankte sich dafür, daß er der bestversorgte Patient auf der Station sei. Etwas am Verhalten des Stationspersonals hatte sich als Folge der Besprechung verändert und dem Patienten gezeigt, daß er sich verstanden fühlen konnte. Das Beispiel belegt, wie Patienten veränderte Einstellungen des therapeutischen Teams wahrzunehmen und darauf zu reagieren vermögen.

64.4 Psychosyndrome nach Herzoperationen

64.4.1 Phänomenologie und Verlauf

Psychische Störungen treten nach Operationen am offenen Herzen in einer besonderen Häufung und Intensität auf. Dies hat, obwohl Herzoperationen, verglichen mit allgemeinchirurgischen Eingriffen, quantitativ wenig ins Gewicht fallen, dazu geführt, daß an vielen Zentren Gruppen mit psychiatrischen und psychosomatischen Fragestellungen sich mit herzoperierten Patienten befaßt haben. Abram (1965), Blachly und Starr (1964, 1967), Egerton und Kay (1964) und Kornfeld und Mitarbeiter (1965) gehörten zu den ersten Autoren, die auf die besondere Häufigkeit und Schwere von Funktionspsychosen nach Herzoperationen hinwiesen. Nach ihren Angaben traten die Symptome in typischen Fällen erst einige Tage nach der Herzoperation auf. In einer Beschreibung Pieringers und Reisners (1970) wird das Krankheitsbild aufgrund der Beobachtung von 52 Patienten folgendermaßen beschrieben, wobei illusionäre Verkennungen und paranoide Inhalte im Vordergrund der Symptomatik stehen:

> „... die Patienten bauten sich eine Welt illusionärer Verkennungen mit Wahnbildern auf, in der sie sich dann selbst nicht mehr zurechtfanden. Sie waren charakterisiert durch realistische, meist ängstliche Inhalte, denen das Unheimliche und schwer Einfühlbare der schizophrenen Psychose fehlte. So wurden die Ärzte zu Ganoven, die Schwestern zu einer verschworenen Gesellschaft, die Wachstation zum Konzentrationslager, der EKG-Schreiber zum Fernsehapparat, die Urinflasche zum Knüppel, mit dem man sich gegen Feinde zur Wehr setzen müßte. Das Sauerstoffzelt war bei einem Kranken ein Auto, in welchem er ohne Führerschein in Moskau rasch über eine Kreuzung fuhr. Der Infusionsständer imponierte als Galgen, die Infusionsleitung als Strick, mit dem einer erhängt werden sollte ..." (S. 249).

Die Befunde in der Literatur belegen, daß sich demgegenüber ein breites Spektrum psychopathologischer Auffälligkeiten finden läßt, für die in den zahlreichen Untersuchungen sehr unterschiedliche Klassifikationsversuche unternommen worden sind (vgl. Kap. 63).

Nach den von der Hamburger Arbeitsgruppe im Rahmen eines Sonderforschungsbereiches der DFG durchgeführten Untersuchungen über psychopathologische Auffälligkeiten nach Herzoperationen, bei denen die Datenerhebung nach dem AMP-System erfolgte (Scharfetter, 1972), wurden Symptombilder unterschieden, die durch folgende Faktoren gekennzeichnet sind:

– Emotionale Störung (Verstimmung)
– Bewußtseins- und Orientierungsstörung
– Paranoid-halluzinatorische Störungen (Produktivsyndrom) (Speidel et al., 1979 b).

Aus den Resultaten geht hervor, daß nach den Rhythmusstörungen Psychosyndrome an zweiter Stelle der postoperativen Komplikationen bei Herzoperierten stehen (Goetze, 1980). Dabei wurden in einer Übergangsreihe vom leichten Durchgangssyndrom bis zu schwer deliranten Zuständen mit Bewußtseinstrübungen einschließlich Koma alle Formen beobachtet.

In älteren, vorwiegend aus den USA stammenden Studien wurde berichtet, daß ausgeprägte Psychosyndrome häufig erst nach einem freien Intervall zu einem Zeitpunkt auftreten, da sich die körperlichen Funktionsparameter wieder der Norm nähern. So trat bei den von Kornfeld und Mitarbeitern (1965) beobachteten Patienten die organische Psychose durchschnittlich 4,2 Tage nach der Operation auf.

Nach den Ergebnissen der Hamburger Arbeitsgruppe können psychische Störungen nach Herzoperationen meist schon sehr frühzeitig ohne freies postoperatives Intervall beobachtet werden (Speidel et al., 1979 b).

Dabei steht in einem frühen Stadium meist eine Minussymptomatik im Vordergrund, nach 5 bis 7 Tagen überwiegt oft die affektive Symptomatik mit einem zurückgezogen-depressiven Verhalten. Paranoid-halluzinatorische Bilder treten meist später als Orientierungsstörungen auf. Bei drei Vierteln der Patienten, die postoperativ auffällig werden, waren schon am ersten postoperativen Tag Störungen vorhanden. Bei der Hälfte der auffälligen Patienten dauerten die Symptome ein bis zwei Tage, bei einem Viertel 3 bis 4 Tage, bei 12% 10 Tage und länger, wie anhand der Beobachtung von 209 Fällen festgestellt werden konnte (Speidel et al., 1979a). Am häufigsten fanden sich die Symptome zwischen dem 2. und 7. Tag. Der Schweregrad der Störungen wies folgendes Verteilungsmuster auf: unauffällig = 59%, leicht psychopathologisch auffällig = 21%, mittel = 11%, schwer = 9% (Dahme et al., 1977).

64.4.2 Exemplarische Fallgeschichte

Als Beispiel sei eine unserer Patientinnen angeführt, die eine durch ihre prämorbide Struktur deutlich beeinflußte Ausgestaltung einzelner Wahrnehmungen im postoperativen Verlauf nach der Operation eines Mitralvitiums zeigte.

> Postoperativ berichtete die Patientin als Eindruck aus dem Wachraum u. a. von einer ihr endlos scheinenden Prozession junger und gutaussehender Ärzte, die sich stundenlang um ihr Bett bewegt habe. Das habe ihr einerseits Genugtuung und Sicherheit verschafft, sie aber irgendwie auch geängstigt.
>
> Die Patientin vermittelte schon präoperativ den Eindruck einer hysterisch strukturierten Persönlichkeit, die sich beispielsweise in kokettem und verführerischem Verhalten gegenüber den Ärzten und in der Abwertung des eigenen Ehemannes kundtat, was in krassem Widerspruch stand zu ihrem eigenen miserablen körperlichen Zustand. Unsere psychodynamische Erklärung dafür war, daß die Koketterie der Patientin ihre Minderwertigkeitsgefühle angesichts ihres kranken Körpers kompensieren helfen sollte. Die postoperativen Wahrnehmungen der Patientin deuteten wir als in diesem Sinn konsequentes Phantasieerleben, das die ärztlichen Visiten in einer ihren Persönlichkeitseigentümlichkeiten gemäßen Weise umgestaltete.

64.4.3 Häufigkeit

Tabelle 64–1 soll einen Überblick über die Häufigkeit psychopathologisch auffälliger Zustände nach Herzoperationen in einer Reihe von Untersuchungen geben (nach Speidel et al., 1979b, gekürzt).

Anzunehmen ist, daß die Häufigkeitsangaben in starkem Maß durch Art und Differenziertheit des jeweiligen methodischen Ansatzes bestimmt sind. Bei Patientengruppen, die systematisch auf psychische Störungen untersucht und regelmäßig vom Psychiater betreut werden, werden auch dezentere Störungen mit geringer Einschränkung der psychischen Funktion erfaßt. Es wurde gezeigt, daß sich bei sorgfältiger Untersuchungstechnik bei einem hohen Prozentsatz aller Patienten während der postoperativen Phase episodisch mehr oder minder ausgeprägte Anzeichen einer Funktionspsychose oder eines Durchgangssyndroms fanden. So sind die vereinzelt genannten sehr niedrigen Prozentraten sicherlich auf unzureichende Untersuchungsmethoden zurückzuführen. Die angegebenen Raten bewegen sich meist zwischen 30 und 60%. Die Angaben sind schwer vergleichbar im Hinblick auf den unterschiedlichen Entwicklungsstand der Herzchirurgie, unterschiedlich erfaßte Diagnosegruppen und Klassifikationsmerkmale der psychischen Auffälligkeiten.

Tabelle 64–1. Angaben über psychopathologische Komplikationsraten nach Herzoperationen (modifiziert nach Speidel et al., 1979b).

Autoren	Jahr	N	%
Bolton und Bailey	1956	1500	3
Gilberstadt und Sako	1967	75	13
Layne und Yudofski	1971	58	14
Kaplan	1956	18	17
Fox et al.	1954	32	19
Rimon et al.	1968	92	23
Heller et al.	1970	100	24
Abram	1965	15	30
Rubinstein und Thomas	1969	36	31
Gilman	1965	35	33
Bliss et al.	1955	37	33
Kornfeld et al.	1965	119	38
Henrichs et al.	1971	110	40
Huse-Kleinstoll et al.	1976	105	40
Dahme et al.	1977	209	41
Tufo et al.	1970	100	43
Jakubik	1972	60	43
Freyhan et al.	1971	150	51
Egerton und Kay	1964	60	51
Blachly und Starr	1964	139	57
Kimball	1970	76	57
Meyendorf	1976	150	60
Morgan	1971	72	72
Pieringer und Reisner	1970	52	90
Lehmann et al.	1969	15	100

64.4.4 Ätiologie und Pathogenese

Weitgehend übereinstimmend werden in der Literatur vorher bestehende **zerebrale Organschädigungen** und **neurologische Auffälligkeiten** als Risikofaktoren für das Auftreten postoperativer Psychosyndrome angenommen (Egerton und Kay, 1964; Hazan, 1966; Kennedy und Bakst, 1966; Rubinstein und Thomas, 1969; Tufo et al., 1970; Freyhan et al., 1971; Kimball, 1976; Speidel et al., 1978, 1979 a, b, 1980), ebenfalls der **präoperative kardiale Status** (Blachly und Starr, 1964; Heller et al., 1970; Freyman et al., 1971, Kornfeld et al., 1965, 1967; Quinlan et al., 1974; Huse-Kleinstoll, 1976). Mitralklappenersatz prädisponiert im Vergleich zu anderen operativen Eingriffen (Egerton und Kay, 1964; Hazan, 1966). Als weitere Risikofaktoren werden genannt: die extrakorporale Zirkulation und ihre Dauer und eine zu schnelle postoperative Dehydratation.

Bei vielen Herzkranken sind schon präoperativ psychische Störungen anzutreffen (Kampman et al., 1979). Patienten, die bereits präoperativ psychopathologische Auffälligkeiten zeigten, hatten in der Hamburger Studie eine höhere Rate postoperativer Störungen. Danach waren (Götze, 1980) nur 30% der präoperativ Unauffälligen, hingegen drei Viertel der präoperativ Auffälligen auch postoperativ gestört. Auffallend ist, daß den meisten Untersuchungen psychopathologischer Auffälligkeiten nach Herzoperationen die Vorstellung eines Krankheitsmodells zugrunde liegt, das einseitig die psychopathologischen Folgen hirnorganischer Funktionsstörungen berücksichtigt.

64.4.5 Bedeutung psychischer Faktoren

Ausgehend von den Beobachtungen über mutmaßliche pathogene Einflüsse der Wachraumatmosphäre, wurden psychologische Gesichtspunkte und atmosphärische Merkmale in einer Reihe von Arbeiten untersucht. Ihnen lag die hypothetische Vorstellung zugrunde, daß, ohne daß die einzelnen organischen Veränderungen als Wegbereiter geleugnet würden, die psychischen Symptome auch als besondere Form der Antwort auf die Herzoperation, beispielsweise als „Katastrophenreaktion", erklärbar wären (Blacher, 1972; Abram, 1965; Kornfeld et al., 1965, 1967, 1969 b, 1972; Kimball, 1969).

Besonders erwähnenswert sind Untersuchungen von Kimball und Mitarbeitern (1969, 1972, 1973, 1976) an der Yale-Universität, bei denen von 1968 bis 1971 180 Patienten mit Herzoperationen einer bis zu 30 Monate dauernden Verlaufsbeobachtung und ausführlicher psychiatrischer Exploration und psychologischer Testung unterzogen wurden.

Nach dem psychiatrischen Interview wurden 109 Patienten nach dem Modus der Verarbeitung ihrer Erkrankung in vier Gruppen eingeteilt (angepaßt-symbiotisch-verleugnend-deprimiert/hoffnungslos). In der Untersuchung erwiesen sich für das kurzfristige Überleben somatische Faktoren und Operationsmerkmale als entscheidend, allerdings war die nach psychologischen Gesichtspunkten vorgenommene Gruppeneinteilung auch hier von prädiktivem Wert bezüglich des Auftretens postoperativer Komplikationen. Für das längerfristige Überleben und die spätere Rehabilitation erwies sich der Aussagewert psychischer Faktoren als sehr groß. Die 36 von den insgesamt 109 Patienten, die präopera-

tiv aufgrund des Interviews als depressiv und ohne Hoffnung eingestuft worden waren, hatten, was die psychiatrische und sonstige postoperative Prognose und ihre Rehabilitationsaussichten betraf, eindeutig die schlechteste Prognose.

Einen analogen Zusammenhang zwischen Depression und Hoffnungslosigkeit auf der einen Seite und vermehrter Mortalität auf der anderen Seite fand Morgan (1971) bei vergleichbaren Ausgangsdaten bei der Untersuchung von 72 herzoperierten Patienten.

In der Hamburger Untersuchung wurden drei Patientengruppen mit offenen Herzoperationen untersucht: Patienten mit Mitral-, Aortenklappenersatz und mit aortokoronarem Bypass. Bei der varianzanalytischen Untersuchung der dispositionellen Merkmale zeigten sich für die einzelnen Gruppen unterschiedliche Merkmale hinsichtlich der anteiligen psychischen und somatischen Komponenten. Die diagnosespezifische Auswertung ergab wesentliche Unterschiede der Risikofaktoren für postoperative psychopathologische Auffälligkeiten, so daß eine allgemeine Aussage über Prädiktoren zweifelhaft ist und ein differenzierterer Ansatz notwendig wird (Speidel, 1981). Aus den erhobenen Befunden seien im folgenden einige wesentliche genannt:

– Bei koronarchirurgischen Eingriffen haben psychosoziale Faktoren einen besonders großen Einfluß auf postoperative psychische Störungen (Meffert et al., 1983).
– Patienten mit Partnerschafts- und Familienkonflikten zeigen unabhängig von der Diagnosegruppe häufiger postoperative Psychosyndrome (Speidel et al., 1982).
– Extremformen von sensitivem und repressivem Umgang mit präoperativer Angst zeigten keinen Einfluß auf die Häufigkeit, aber auf den Schweregrad der psychischen Störungen.
– Ähnlich fanden hinsichtlich der Formen der Angstbewältigung Davies-Osterkamp und Mitarbeiter (1978, 1980) in einer anderen Untersuchung, daß solche Patienten nach Herzoperationen besonders durch Psychosyndrome gefährdet sind, die bereits präoperativ Bewältigungsstile bevorzugen, bei denen die Auseinandersetzung mit der Bedrohung vermieden wird.
– Es findet sich der schon beschriebene Zusammenhang vorher bestehender präoperativer und postoperativer psychischer Gestörtheit.
– Im zeitlichen Verlauf nimmt die stimmungsmäßige Labilität der Patienten oft zu.

Wenn auch einzelne Befunde in verschiedenen Arbeiten unterschiedlich, teils sogar widersprüchlich sind und in der faktorenanalytischen Untersuchung bei der Komplexität der Anordnung mit der Zahl der Faktoren auch verstärkt methodenkritische Einwände berücksichtigt werden müssen (Dahme et al., 1982), scheint der Einfluß psychosozialer Faktoren aus den vorliegenden Ergebnissen doch ausreichend belegt. Diese Faktoren scheinen auch auf das Langzeitergebnis der Operation einen wesentlichen Einfluß zu haben. Schon Frank und Mitarbeiter (1979) stellten in einer Nachuntersuchung von 800 herzoperierten Patienten fest, daß der Langzeiterfolg oft durch psychische Probleme gefährdet ist. Ähnlich fanden Heller und Mitarbeiter (1974) ein Jahr nach der Herzoperation bei 90% der Operierten eine somatische Besserung, aber häufig eine psychische Verschlechterung. Auch Blacher (1978) und Lützenkirchen und Mitarbeiter (1980) stellten im weiteren postoperativen Verlauf oft depressive Verstimmungen fest. Diese Befunde unterstreichen die Notwendigkeit einer psychiatrischen Evaluation und im Bedarfsfall die Notwendigkeit eines Angebots psychotherapeutischer Hilfestellung nach der Operation. Die Bedeutungen psychiatrischer Auffälligkeiten für die Langzeitprognose werden durch die Befunde Rabiners und Willers (1979) unterstrichen, die einen Zusammenhang zwischen der Häufigkeit psychiatrischer Komplikationen vor Entlassung und einer schlechteren Fünf-Jahres-Überlebensrate bei Herzoperierten fanden.

64.4.6 Allgemeine Bemerkungen zur Psychologie operativer Eingriffe

Versucht man, am Modellfall von Herzoperationen psychisch besonders belastende Momente zu isolieren, sind zunächst Erkenntnisse über die psychodynamische Interpretation des Operationserlebnisses als solches zu berücksichtigen.

Schon H. Deutsch (1942) hat dazu in ihrem Artikel „Some psychoanalytic observations in surgery" den individuellen Stellenwert von Krankheit und Operation und Besonderheiten der Persönlichkeitsstruktur des Patienten, beispielsweise neurotische Anteile, als wichtige Hinweise für die postoperative Reaktion bezeichnet. Narkoseerlebnis und operativer Eingriff selbst seien dabei gesondert zu bewerten. Anästhesie entspreche häufig unbewußt einem Konfrontationserlebnis mit dem Tod und werde als Trennung erlebt; der Eingriff selber könne Kastrationsängste wiederbeleben als Folge der Bedrohung eines Organs und entsprechend psychodynamischen Konstellationen beispielsweise als Strafe empfunden werden für ambivalente Impulse.

Janis (1958) stellte in einer Monographie über Operationsstreß fest, daß Patienten mit einem „vigilanten", d. h. auf Aufmerksamkeit fokussierten Bewältigungsstil eher postoperativ durch psychiatrische Komplikationen gefährdet sind. Möglicherweise hängt dies damit zusammen, daß diesen Patienten durch die postoperative Situation gewohnte Coping-Stile vorenthalten sind, was zu einer vermehrten Vulnerabilität führt (vgl. Cohen und Lazarus, 1973; Sime, 1976; Davis-Osterkamp, 1978). Konträre Anpassungsmuster können bei unterschiedlichen Patientengruppen zu einer gleichermaßen befriedigenden Emotionsregulierung führen (Borgert, 1988; Götze, 1988).

64.4.7 Prophylaxe und Therapie

Für den psychologischen Umgang mit Patienten, die operiert werden, ist es besonders wichtig, möglichst exakt über ihr momentanes psychisches Befinden orientiert zu sein. Die tägliche Einschätzung dieses Befindens sollte einen ähnlichen Stellenwert einnehmen wie beispielsweise die Bestimmung biophysikalischer und biochemischer Parameter, z.B. Blutdruck, Elektrolyte usw. Ein wichtiges Problem besteht darin, das Personal so zu schulen, daß es Patientenverhalten adäquater zu beobachten, bewerten und die Beobachtungen zu dokumentieren versteht. Aus diesen Beobachtungen sollen für die Patienten therapeutische Maßnahmen abgeleitet werden. Für die Prophylaxe psychischer Störungen sind alle als pathogenetisch relevant angesehenen Faktoren in die Überlegungen einzubeziehen, wie z.B. Schlafdeprivation oder Monotonie der Umgebung. Hilfe bei der Orientierung, Angebot von Kontakten, emotionale Unterstützung, eine möglichst große Konstanz der Bezugspersonen aus dem Personal sind notwendige Forderungen. Folgende Hinweise sprechen dafür, daß sich die postoperative Häufigkeit von Durchgangssyndromen und Funktionspsychosen durch einfache **psychoprophylaktische Maßnahmen** günstig beeinflussen läßt:

Nach präoperativen Interviews und sorgfältiger postoperativer psychischer Betreuung wurden nach Herzoperationen weit weniger Funktionspsychosen beobachtet als in einer vergleichbaren Kontrollgruppe (Lazarus und Hagens, 1968). Auch Layne und Yudofsky (1971) konnten beim Vergleich zweier Gruppen herzoperierter Patienten mit und ohne präoperativ geführte psychiatrische Interviews analoge Feststellungen machen. Charakteristisch ist dabei, daß die Schwestern regelrecht überredet werden mußten, mit den schwerkranken Patienten nach der Operation zu sprechen.

Information allein vermöge die Angstindizes operierter Patienten nicht zu senken, wohl aber eine Kombination von Information und Stützung (Andrew, 1970). Information und Zuwendung konnten auch den postoperativen Verbrauch schmerzstillender Medikamente verringern und die Entlassung aus dem Krankenhaus beschleunigen (nach Egbert et al., 1964). Sie berichteten über einen Vergleich randomisierter Gruppen bei 97 Patienten mit intraabdominalen Operationen, daß sich durch Aufklärung und postoperative Führung der durchschnittliche postoperative Morphiumverbrauch um die Hälfte reduzieren ließ.

Aus der Praxis ihrer Konsiliartätigkeit geben Hackett und Weisman (1960) in ihrem Bericht über 400 konsiliarisch betreute Patienten auf allgemeinchirurgischen Stationen ihre Erfahrungen wieder, daß plötzlich auftretende emotionale Störungen nach der Operation häufig daraus resultieren, daß chronische Konflikte durch das psychische Trauma der Operation neu entfacht und ihre Lösung erschwert worden sei, beispielsweise dadurch, daß Fähigkeiten, Motivation oder wichtige Teile des Ich-Ideals betroffen wurden. Die Autoren fanden auch einen auffallenden Zusammenhang zwischen präoperativ antizipierten Vorstellungen, beispielsweise über Beeinträchtigung der Wahrnehmung bei Augenoperationen, der Geschicklichkeit bei Amputation, der Kommunikation bei Kehlkopfoperationen und der symbolischen Bedeutung postoperativer psychischer Reaktionen. Nach Erfahrungen aus anderen chirurgischen Bereichen scheinen die emotionale Besetzung des operierten Organs, der Grad der Verstümmelung, das Ausmaß der Beeinträchtigung von Sinneswahrnehmungen als Operationsfolge (z.B. bei Augenoperationen, insbesondere nach Ablatio retinae (Jackson, 1969; Klein und Moses, 1974)), im Hinblick auf die postoperative Reaktion prognostisch relevant zu sein. Bei Patienten mit Kolostomien (Golden und Nahum, 1964; Dlin et al., 1969, 1973) und mit schweren Verbrennungen (Hamburg et al., 1953a, b) wurden ausgeprägte psychopathologische Reaktionen in Gestalt schwerer emotionaler

Verstimmungszustände und Veränderungen von Sinneswahrnehmungen beschrieben.

Zu diskutieren ist, ob die emotionale Sonderstellung, die das Herz im menschlichen Erleben einnimmt, zur besonderen Häufung psychopathologischer Reaktionen nach Herzoperationen beiträgt (Blacher, 1983). Die Reaktion der Öffentlichkeit auf die ersten Herztransplantationen gab beredtes Zeugnis für diese Sonderstellung. Wichtige Informationen für Patienten mit Herzklappenoperationen betreffen mechanische Vorstellungen über Wirkung und Lokalisation der Klappen und die Art des verwendeten Materials.

So kann die Vorstellung, daß die schadhafte Klappe sich im Innern des Herzens befindet, besonders beängstigend sein (Blacher, 1975).

Interessanterweise wurden bei Kindern Funktionspsychosen nach Herzoperationen extrem selten beschrieben. In welchem Ausmaß dies von körperlichen Faktoren abhängt, läßt sich noch nicht abschließend übersehen. Egerton und Kay (1964) fanden einen Fall unter 36, Kornfeld und Mitarbeiter (1965) keinen unter 20, Danilovicz und Gabriel (1971) keinen unter 67 Patienten, allerdings häufig ausgeprägte Angstreaktionen, Aggressionen und absolute Fügsamkeit im Sinne eines Hospitalismussyndroms mit Apathie und Rückzug. Erwähnenswert ist dabei, daß ja bei Kindern auch keine psychogenen, auf das Herz bezogene Erkrankungen zu beobachten sind. Bekannt ist dazu, daß die Erfahrung der eigenen Leiblichkeit entwicklungsabhängig ist und eine umschriebene Herzempfindung bei Kindern im Rahmen der Entwicklung des Körperbildes erst etwa mit der Pubertät auftreten soll (Plügge und Mappes, 1962).

Hinsichtlich der **Organisation** der Zusammenarbeit und **Institutionalisierung** psychosomatischer Ansätze in der Herzchirurgie empfiehlt sich die direkte Mitarbeit eines Psychosomatikers oder Psychiaters im Team. Nur so erscheint es möglich, frühzeitig, eventuell präventiv zu intervenieren und psychiatrische Komplikationen sowohl direkt im postoperativen als auch im späteren Verlauf rasch zu behandeln. Einzelne Erfahrungsberichte belegen die Möglichkeiten, aber auch Schwierigkeiten der Kooperation (Jordan et al., 1983; Speidel et al., 1981; Freyberger, 1975).

64.5 Patienten nach Reanimation

64.5.1 Vorbemerkungen

Externe Herzmassage, die Anwendung transvenöser Schrittmacher zur Herzstimulation und elektrische Defibrillation haben zusammen mit der Monitorüberwachung die Reanimation bei Kreislaufstillstand zu einer wirksamen Methode der Lebensverlängerung gemacht.

Nach einer Untersuchung in einem amerikanischen Lehrkrankenhaus wurden von 294 Reanimierten 14% entlassen. Die übrigen Patienten verstarben im Krankenhaus. Nach 6 Monaten waren davon noch drei Viertel am Leben. Unter ihnen waren 93% nach 6 Monaten nicht hirnorganisch beeinträchtigt. Bei Entlassung waren fast alle Patienten depressiv (Bedell et al., 1983). Auch Eisenberg (1982) und Mitarbeiter stellten in einer Untersuchung über einen längeren Zeitraum eine gute Langzeitprognose von außerhalb von Krankenhäusern erfolgreich reanimierten und dann hospitalisierten Patienten fest. Nach 4 Jahren war von den entlassenen Patienten fast die Hälfte noch am Leben.

Studien über psychologisch-psychiatrische Auffälligkeiten bei reanimierten Patienten befassen sich mit den akuten Syndromen nach der Reanimation und der Beeinflussung der Langzeitrehabilitation dieser Patienten (Dlin et al., 1966; Druss und Kornfeld, 1967; Falicki und Sepp-Kowalk, 1969; Dupont et al., 1969; Minuck und Perkins, 1970; Dobson et al., 1971; White und Liddon, 1972; Dlin et al., 1974; Krakowski und Krakowski, 1975; Drühe, 1989).

> Herr M., ein etwa 50jähriger Patient mit kombiniertem Aortenvitium, befand sich seit längerem an der Grenze zur kardialen Dekompensation. Im täglichen Leben und im Beruf war der Patient bestrebt, von seiner Krankheit möglichst wenig Notiz zu nehmen. Statt dessen belastete er sich immer wieder bis zur Erschöpfung, indem er sich z.B. als Pannenhelfer betätigte. Er hatte während einer Herzkatheteruntersuchung, die wegen seines Vitiums erfolgte, eine Episode von Kammerflimmern erlitten. Dabei war der Blutdruck nie unter die kritische Grenze abgefallen. Nach dem Ereignis folgte auf der Intensivstation eine Phase, in der Patient mit geschlossenen Augen dalag und schwer bewußtseinsgetrübt schien.
>
> Auffallend war, daß er nur auf laute Anrufe reagierte, wobei, je nachdem, wer ihn ansprach, seine Aufmerksamkeit variabel erschien. Darauf folgte eine Phase häufiger Tagträume, in der der Patient sehr unruhig war und oft aufschrie. Während der folgenden Nacht war er delirant, riß die Infusionen heraus und erlebte sich im Schaufenster einer Schlächterei, wobei er die Infusionsflasche als Teil der Geschäftsauslagen ansah. Am nächsten Tag berichtete er über zahlreiche Träume, die sich meist mit Kriegserlebnissen beschäftigten und beunruhigende Gewaltszenen enthielten. Erlebnisse mit Inhalt eigener Schuld, zum Beispiel als Angehöriger der deutschen Armee in Italien während des 2. Weltkrieges, erschienen dem Patienten als Alpträume schlimmsten Ausmaßes. Er war in dieser Phase sehr erregt, berichtete über seine Empfindungen während der Reanimationsmaßnahmen, die er vage als Gefühl des „Pumpens" erlebt hatte. Als ein Nachbarpatient im selben Zimmer starb, stellte sich Herr M. schlafend und äußerte, von dem Geschehen nichts wahrgenommen zu haben, als er von Mitpatienten später darauf angesprochen wurde. Er döste vor sich hin. Er begann dann aber, über den Tag mehrfach zu erbrechen und konnte schließlich in einem Gespräch seine ganzen Befürchtungen, die durch den Todesfall ausgelöst waren, zum Ausdruck bringen. Dabei wurde deutlich, daß sein erneuter Rückzug Teil seines Abwehrverhaltens war gegenüber den schweren Todesängsten.
>
> In ähnlicher Weise kann man bei vielen körperlich kranken Patienten ein Mischbild organisch und psychisch bedingter Rückzugszustände finden.

64.5.2 Akute Reaktionen

Inwieweit Patienten die Reanimationsmaßnahmen selbst empfinden, wird in der Literatur widersprüchlich beurteilt. Nach Angaben von Dlin und Mitarbeitern (1974) scheinen Patienten Reanimationsmaßnahmen häufig wenigstens akustisch oder taktil mitzuerleben, ohne daß über Schmerzen berichtet wird. Meist wird im Anschluß daran eine initiale Schockreaktion beschrieben mit ausgeprägter Störung von Wahrnehmungs-, Aufmerksamkeits- und Denkfunktionen (Druss und Kornfeld, 1967; Dlin et al., 1966, 1974; Hackett et al., 1968). Zur Erklärung dient zum einen die Annahme einer Funktionspsychose, zum Teil wird auch zusätzlich eine primär vorliegende emotionale Schockreaktion mit einem dem „Totstellreflex" analogen Verhalten angenommen (Dlin et al., 1966; Biörck und Edhag, 1973). Auffallend ist, daß diese Reaktion auch in Fällen auftritt, in denen der Kreislaufstillstand innerhalb von kürzester Zeit behoben werden kann. Viele Patienten geben in der ersten Phase nach dem Herzstillstand den Eindruck an, „tot" zu sein. Sie empfinden auch nachher keine Schmerzen und äußern keine Gefühle, was als wirksamer Selbstschutzmechanismus im Augenblick höchster Gefahr interpretiert werden kann. Vorhandene Erinnerungsfragmente der Patienten an das Ereignis können in ihrer Inkohärenz und mit ihren erschreckenden Einzelheiten jedoch auch später Angst auslösen und zu psychopathologischen Reaktionen führen. Dabei ist die Rolle organischer und psychoreaktiver Momente kaum voneinander zu trennen.

Schwere Alpträume mit aggressiven und brutalen Trauminhalten, wie auch bei unserem Patienten Herrn M., sind häufig in der Literatur beschrieben (Druss und Kornfeld, 1967; Dobson et al., 1971). Die Alpträume der Patienten sind auch als Affektabfuhr aufgefaßt worden, indem sie den Patienten erlauben sollen, sich von Erinnerungen zu trennen, d.h. als Bewältigungsversuche des Ichs interpretiert werden könnten. Die Erinnerungen der Patienten sind durch Abwehrvorgänge beeinflußt. Auch Verleugnungsvorgänge der Ärzte verhindern, daß über das bedrohliche Erlebnis gesprochen wird.

So erfuhr ein Infarktpatient in der Intensivstation nach erfolgreicher Reanimation nur durch Andeutungen, daß etwas Besonderes vorgefallen sein mußte. Als er sich bei der Visite schlafend stellte, wurde ihm aus der Unterhaltung der Ärzte am Krankenbett das Geschehene klar.

64.5.3 Auswirkungen auf die Rehabilitation

Auch in den späteren Rehabilitationsphasen scheint das Vorkommen eines Herzstillstandes das Auftreten psychischer Symptome zu fördern. Als besonders häufig werden aufgrund lange fortbestehender Ängste ausgeprägte Abhängigkeits- und Versorgungswünsche beschrieben (Dobson et al., 1971), die durch überprotektives Verhalten von Partnern unterstützt werden (White und Liddon, 1972).

Ebenso wird eine radikale Veränderung von Lebensgewohnheiten im Sinne einer vermehrten Schonhaltung beschrieben, die weit über das allgemein übliche und erforderliche Maß hinausgehen kann (Druss und Kornfeld, 1967).

Auch Bedell und Mitarbeiter (1983) stellten fest, daß das Ausmaß der langfristigen funktionellen Einschränkung nach der Reanimation mehr durch das Angstniveau bestimmt war als durch physiologische Parameter. Hingegen konnten Krakowski und Krakowski (1975) in einer vergleichenden Untersuchung von 14 Patienten mit Herzstillstand und einer Kontrollgruppe von 14 Infarktpatienten ohne Herzstillstand keinen eindeutigen Einfluß der Reanimation auf die spätere Rehabilitation nachweisen.

Reich und Mitarbeiter (1983a) berichteten in einer Längsschnittuntersuchung von 6 erfolgreich Reanimierten nach Herzstillstand. Sie stellten bei sorgfältiger Nachuntersuchung einschließlich testpsychologischer Methoden Veränderungen im Sinne einer vermehrten Reizbarkeit, Depressivität, Apathie, eine Störung der Impulskontrolle, der Einsichts- und Urteilsfähigkeit fest, die sich deutlich von der Primärpersönlichkeit abhob und wohl zumindest teilweise Ausdruck einer leichten hirnorganischen Beeinträchtigung sein dürfte. Ebenso wird die oft beschriebene fehlende emotionale Betroffenheit über die Gedächtnisbeeinträchtigung auch auf die hypoxischen Schädigungen zurückgeführt (Drühe, 1989).

64.5.4 Therapeutische Gesichtspunkte

Ärzte sollten Patienten auf Fragen, die den Herzstillstand betreffen, offen antworten und im Gespräch die Abreaktion all der Gefühle ermöglichen, die mit dem Ereignis verbunden werden. Hinweise auf eine Gefährdung der Patienten durch solche Gespräche ließen sich nicht finden (Dlin et al., 1974). In der Fachwelt und in der Laienpresse sind Berichte erschienen über Empfindungen von Patienten, die wiederbelebt werden mußten. Sehr häufig werden dabei von den Patienten angenehme Gefühle und Bilder angegeben (Moody, 1977, „Todesnäheerlebnisse"). Es empfiehlt sich, in der akuten Phase sehr stützend vorzugehen. Im weiteren Verlauf sollte frühzeitig die nachstationäre Behandlung geklärt und Einzelheiten einer etwaigen beruflichen Rehabilitation angesprochen werden. Dabei muß man sich auch beispielsweise mit hypochondrischen Befürchtungen von Patienten, die durch das Reanimationserlebnis ausgelöst wurden, auseinandersetzen. Zur eingehenderen Erörterung möchten wir auf das Kapitel 66 „Zum Umgang mit unheilbar Kranken" verweisen.

Als Beispiel sei einer unserer Patienten angeführt, der nach einem Infarkt reanimiert werden mußte und bei dem sich bei einer Nachuntersuchung ergab, daß sich nach dem Ereignis eine Impotenz entwickelt hatte, die den Patienten und seine Beziehung zu seiner wesentlich jüngeren Frau belastete. Ein klärendes Gespräch mit dem Patienten mit dem Ziel, seine Ängste auf ein realistisches Maß zu reduzieren, hatte den Erfolg, daß der Patient sein lästiges Symptom verlor.

64.6 Patienten mit Herzschrittmacher

64.6.1 Vorbemerkungen

Die ersten Herzschrittmacher wurden 1959 implantiert. In den meisten psychologisch orientierten Arbeiten wird über leichte Angst- und Depressionszustände nach der Implantation berichtet, wie sie für Kranke mit der Erfahrung der Unzuverlässigkeit einer Organfunktion typisch sind.

Herr A., ein 70jähriger Patient, erhielt wegen eines progredienten Blockbildes im EKG bei koronarer Herzerkrankung mit schweren Angina-pectoris-Anfällen einen permanenten Schrittmacher. Schon vor Auftreten gelegentlicher bradykarder Rhythmusstörungen hatte der Patient unter häufigen Schwindelattacken gelitten, die als Folge der allgemeinen Gefäßerkrankung aufgefaßt wurden. Er war zeitlebens sehr ordnungsbewußt gewesen. Die Ehefrau des Patienten war seit etlichen Jahren selber Schrittmacherpatientin gewesen und hatte eine Reihe von Schrittmacherkomplikationen erlitten, die sie äußerst verunsichert hatten, so daß sie selten wagte, ohne ihren Mann etwas zu unternehmen, beispielsweise kleine Besorgungen zu machen. Unser Patient war daher von Anfang an dem Schrittmacher gegenüber sehr skeptisch eingestellt. Er begründete dies mit den Erfahrungen seiner Ehefrau und seiner allgemeinen Lebenserfahrung, daß er nie ein Glückspilz gewesen und es häufig noch schlimmer gekommen sei, als er vorhergesehen hatte. Vor und nach der Operation war der Patient stark verängstigt und verunsicherte die betreuenden Ärzte durch zahlreiche, ständig wechselnde Beschwerden, die er teils triumphierend, teils vorwurfsvoll hervorbrachte.

Postoperativ wiederum gehäuft auftretende Angina-pectoris-Anfälle, eine Venenentzündung mit septischer Streuung und kurzfristig hohe Temperaturen bestätigten ihn in seinen Befürchtungen. Obwohl diese Komplikationen rasch behoben werden konnten, war die Rekonvaleszenz verzögert, da er nun immer häufiger über Schwindelgefühle klagte, die er auf die Narkose zurückführte. Die Beschwerden traten vor den jeweils vorgesehenen Entlassungsterminen verstärkt auf. Im Gespräch ergab sich, daß der Patient völlig verunsichert und verzweifelt darüber war, wie er nun mit seiner Frau, die jetzt zusätzlich Beschwerden im Genitaltrakt äußerte, und ihren weitgehenden Versorgungswünschen umgehen sollte, da er sich als Schrittmacherträger selber als hochgradig verletzlich und gefährdet erlebte. Diese Angst zeigte sich überspitzt in der Frage, wer denn nun wen beim Spaziergang stützen solle. In dieser Situation war deutlich die Konkurrenz der Partner um Zuwendung und Unterstützung zu empfinden. Erst nach mehreren Gesprächen mit beiden Partnern konnte eine Entlastung des Patienten erreicht werden, ohne daß in den folgenden Monaten größere Schwierigkeiten auftraten. Trotz der unzweifelhaft weiterbestehenden hirnorganischen Beeinträchtigung bei Zeichen einer zerebrovaskulären Insuffizienz zeigen der Verlauf der Symptomatik und die Reaktion beider Partner auf, wie die Veränderung der familiären Rollen und die Beschwerden des Patienten von seiner Unsicherheit und Angst vor seiner neuen Situation gefördert wurden.

64.6.2 Anpassungsprobleme

Eine Reihe psychologischer Untersuchungen ergab, daß Patienten nach anfänglicher Skepsis ihren Herzschrittmacher relativ gut zu tolerieren pflegen (Speidel et al., 1969; Galdston und Gamble, 1969; Greene und Moss, 1969; Kortmann, 1974; Doenecke et al., 1974; Hesse, 1975; Alt et al., 1983; Friedrich et al., 1983). Während der akuten Phase kurz nach der Implantation finden sich bei vielen Patienten Angst- und Depressionszustände, gelegentlich auch Panik mit Ablehnung des Schrittmachers. Im späteren Verlauf traten bei ca. einem Viertel der Patienten in den ersten Monaten depressive Verstimmungen, zum Teil als Folge von Enttäuschungsreaktionen, ambivalenten Einstellungen und Ablehnungsgefühlen auf, insbesondere bei Auftreten von Komplikationen. Bei den meisten Patienten erfolgte später eine relativ komplikationslose Integration ins Körperschema. Gelegentlich kann es zu einer ausgesprochenen Fehladaptation kommen. Schumacher (1965) beschreibt dies ausführlich bei einem 28jährigen Patienten, dem nach Diphtherie-Myokarditis ein Schrittmacher eingesetzt werden mußte.

Der Prozeß der Integration der Prothese ins Körperbild scheint bei den meisten Patienten in einer Weise zu erfolgen, daß das internalisierte Objekt Hilfe, Stütze und Sicherheit gewährt und vom Träger idealisiert wird. Auch der Arzt erhält oft die Rolle eines allmächtigen Objektes. Fast alle Patienten erscheinen sehr zuverlässig zu ihren Kontrolluntersuchungen in der Ambulanz. Erst Schrittmacherkomplikationen vermögen die mehr abstrakt erlebte Abhängigkeit in eine als konkret und belastend erlebte zu verwandeln. Aus diesem Grund werden in früheren Arbeiten (z. B. Becker et al., 1967), als Schrittmacherkomplikationen noch wesentlich häufiger waren, psychosoziale Anpassungsstörungen öfter angegeben. In der genannten Arbeit fanden sich bei 31% von 97 untersuchten Patienten Zeichen einer Fehlanpassung. Neben dem Auftreten von Komplikationen sind auch prämorbide Persönlichkeitszüge und Anpassungsmuster, die interpersonalen Beziehungen und das Alter der Patienten für die Gewöhnung an den Schrittmacher prognostisch bedeutsam (Greene und Moss, 1969; Hesse, 1975). Bei jungen Patienten treten postoperative Ängste und Depressionszustände häufiger auf, bei älteren Patienten, die die Prothese passiver ertragen, traten erwartungsgemäß mehr Durchgangssyndrome und Funktionspsychosen auf (Hesse, 1975). Kinder tolerieren Schrittmacher offenbar relativ problemlos (Galdston und Gamble, 1969). Als problematisch muß sicherlich die weitverbreitete Tendenz bezeichnet werden, auch jüngeren Patienten mit einem Herzschrittmacher Schwerbehindertenstatus zuzuerkennen.

64.6.3 Therapeutische Gesichtspunkte

Schon allein die sorgfältige Information der Schrittmacherpatienten und ihrer Familien über die Funktion des Schrittmachers, Komplikationen und Möglichkeiten der Vorbeugung und Behandlung können die Anpassung der Patienten erleichtern. Oft macht sich die Angst der Patienten an der Vorstellung der Begrenztheit der Lebensdauer der Batterie fest. Auch sollten Patienten allgemein über die Herzfunktion, über den Begriff der „Blockbilder", über die Indikation zur Schrittmacherbehandlung im jeweiligen Fall, über Einzelheiten der Schrittmacherfunktion, die medikamentöse Behandlung und die Pulsmessung Bescheid wissen. Es gibt Patienten, die mit dem Begriff „Herzblock", den sie im Laufe ihrer Behandlung hören, häufig etwas verbinden, was der umgangssprachlichen Bedeutung, beispielsweise dem mechanischen Bild des Versiegens des Blutflusses, entspricht und verständliche Ängste auslöst. Solchen Fehlvorstellungen sollte durch Erklärung vorgebeugt werden. Entsprechendes gilt für Begriffe wie z. B. „Betablocker".

Zu beachten ist das hohe Durchschnittsalter der häufig multimorbiden Schrittmacherpatienten, bei denen oft auch die geistige Auffassungsgabe beeinträchtigt ist; d. h., daß die Unterrichtung besondere Ausführlichkeit, Geduld und Zuwendung erfordert, da gerade bei geriatrischen Patienten mit Schrittmachern die medizinische Betreuung häufig Lebenshilfe und Führung ist, deren Güte die Prognose wesentlich mitbestimmt (Witt, 1972). Unbedingt muß dem Schrittmacherpatienten auch vermittelt werden, daß der Schrittmacher zwar eine Hilfe, aber kein neues Herz bedeutet, d. h., daß sich auch nach der Implantation eines Schrittmachers Symptome einer Herzinsuffizienz ausbilden können und ein Abbruch der medikamentösen Behandlung nicht erfolgen darf. Bei passageren Schrittmachern ist darauf zu achten, daß die Wegnahme des Schrittmachers ohne entsprechende Rückversicherung und Stützung des Patienten zu psychologischen Komplikationen führen kann.

64.7 Psychologische Gesichtspunkte zur Situation der künstlichen Beatmung

64.7.1 Vorbemerkungen

Künstliche Beatmung bedeutet totale Abhängigkeit, Hilflosigkeit, ständige Bedrohung durch Pannen und ausgeprägte Kommunikationsbehinderung. Panikartige Angstzustände, die sich aus teilweise geringfügigen Störungen des technischen Ablaufes ergeben, psychische Labilität mit einer deutlichen Tendenz zu schwerer depressiver Verstimmung und eine emotionale Besetzung des lebensnotwendigen Apparates finden sich häufig bei beatmeten Patienten (Freyberger, 1975).

64.7.2 Exemplarische Fallgeschichte

Frau M., eine 70jährige Patientin mit Cor pulmonale bei Lungenemphysem, mußte wegen einer akuten Verschlechterung ihrer chronischen Bronchitis mit Ausbildung einer Lungenentzündung in der Intensivstation künstlich beatmet werden. Die Frau hatte, während sich ihr Gesundheitszustand ständig verschlechterte, alleine in einem großen Mietshaus gelebt, dessen Wohnungen nach und nach in Gastarbeiterapparte-

ments umgewandelt wurden. Ihre Kontakte beschränkten sich auf gelegentliche Besuche von Sohn und Schwiegertochter und Kontakte mit ihrer „Telefon-Freundin", die kurz vor dem jetzigen stationären Aufenthalt der Patientin erkrankt war. Sie hatte das Gefühl, daß niemand, auch der Hausarzt nicht, ihre ständige gesundheitliche Verschlechterung bemerkt hatte. Auf der Intensivstation erholte sich die Patientin nur ganz langsam. Der Zustand blieb äußerst labil, und sie litt an einem schweren organischen Psychosyndrom. Es hatte zunächst Uneinigkeit zwischen den Ärzten bestanden, ob die Prognose nicht überhaupt infaust sei.

Nachdem die Patientin in einem noch sehr schlechten Zustand auf die Allgemeinstation verlegt worden war, wobei man mit ihrem baldigen Sterben rechnete, konnte nach Abklingen des Psychosyndroms eine langsame Stabilisierung ihres Zustandes erreicht werden. Anfangs bestanden große Schwierigkeiten bei der assistierten Beatmung; allmählich lernte die Patientin, mit dem Gerät umzugehen. Da sie über längere Zeit Tag und Nacht intermittierend einen Beatmungsapparat benutzen mußte, fanden mit Schwestern und Nachtwachen intensive Gespräche statt, und es entwickelte sich eine tragfähige Beziehung. Nach einiger Zeit war die Patientin vom Respirator fast unabhängig geworden und konnte kleinere Spaziergänge unternehmen. Zu dieser Zeit verschlechterte sich das Befinden ihrer Mitpatientin, die an einem metastasierenden Karzinom mit Querschnittslähmung litt, und es entwickelte sich ein langsames, quälendes Sterben. Da Bettenmangel herrschte, konnte die Patientin erst während der unmittelbaren Sterbephase an einem Wochenende in ein anderes Zimmer verlegt werden. Auffallend war, daß sich unsere Patientin M. in dieser Zeit, während mit der Sozialarbeiterin schon Pläne für einen Umzug in ein Altersheim vorbereitet wurden, fast völlig aus der Kommunikation zurückzog und wie zu Anfang ihres Aufenthaltes an der Station wieder verlernte, mit dem Respirator umzugehen. In ihrem körperlichen Befinden kam es zu einer Verschlechterung, die, parallel zum Sterben der Mitpatientin, in einen plötzlichen dramatischen Rückfall mündete.[1] Im Verlauf eines Tages nach dem Tod der Mitpatientin mußte sie erneut auf die Intensivstation verlegt werden, wo sie innerhalb von Stunden verstarb.

An dieser Fallgeschichte fällt auf, daß sich die langsame und ständige Verschlechterung des Befindens der Patientin nahezu unbemerkt durch den Hausarzt entwickelt hatte, bis ein Zustand erreicht war, der fast hoffnungslos schien. Es bietet sich die Hypothese an, daß es der Patientin an Möglichkeiten gefehlt hat, nachdrücklich genug auf ihre Bedürfnisse und ihren Zustand aufmerksam zu machen. Das schließlich erreichte labile Gleichgewicht wurde dann wieder empfindlich gestört durch das für die Patientin erschreckende Erlebnis des quälenden Sterbens der Mitpatientin, der eigenen Hilflosigkeit und der der Ärzte und Schwestern. Auch jetzt wurde zunächst der Bedrohung des emotionalen Gleichgewichtes der Patientin, schließlich den erneuten Zeichen somatischer Verschlechterung erst verhältnismäßig spät Aufmerksamkeit gewidmet. Parallel mit dem depres-

---

1 Erinnert sei an das Konzept von „Conservation-Withdrawal".

siven Rückzug kam es zu verstärkter Atemnot, die das Beatmungsgerät wieder erforderlich machte und schließlich sogar zum Verlust der Fähigkeit, mit dem Gerät umzugehen, führte. Auch jetzt hatte es der Rückzug der Patientin erschwert, rasch auf ihre Gefährdung und die Bedrohung ihrer Bedürfnisse aufmerksam zu machen.

64.7.3 Psychische Probleme beatmeter Patienten

Da beatmete Patienten in der Regel nicht frei kommunizieren können und der Arzt auf averbale Ausdrucksweisen angewiesen ist, wird gerade bei diesen Patienten das Ausmaß von Angst und Depression meist unterschätzt. Besonders die Patienten, die infolge einer allgemeinen Lähmung beatmet werden müssen, bedürfen infolge der Kommunikationsbarrieren besonderer Zuwendung. Patienten, die tracheotomiert wurden, haben meist die Angst, nie mehr richtig sprechen zu können. Durch ihr schlechtes Befinden leidet auch ihre Aufnahmefähigkeit, so daß sie häufiger Versicherung bedürfen, daß sie nicht dazu verurteilt sind, stumm zu bleiben (Blacher, 1975).

Viele psychologische Erkenntnisse bei beatmeten Patienten wurden während der großen Polioepidemien der 50er Jahre in den USA gewonnen. Dabei waren Menschen von medizinischen Apparaten völlig abhängig. Aus den Veröffentlichungen geht hervor (Prugh und Tagluri, 1954; Holland und Coles, 1957; Mendelson et al., 1958), daß Patienten, die am schlimmsten durch Lähmungen betroffen waren, ihren Zustand am meisten verleugneten. Bei vielen Patienten beobachtete man eine sehr starke Gewöhnung an die „Eiserne Lunge", die die Entwöhnung langwierig machte und viel Geduld und Eingehen auf die Patienten erforderte.

Bei Patienten mit chronischem Cor pulmonale, die dauernd oder intermittierend eine assistierende Respiratorbehandlung erfahren, fand man, daß die subjektive Atemnot von bestimmten emotionalen Zuständen wie Depression, Wut usw. abhängig war. Danach scheinen sowohl Aktivierungszustände wie Angst, Wut als auch Zustände des Rückzugs wie Depression, Verzweiflung subjektiv die Empfindung Dyspnoe auslösen und verstärken zu können (Dudley et al., 1968). Diese Patienten entwickeln ähnlich wie Schrittmacherpatienten ein ausgeprägtes Gefühl von Abhängigkeit gegenüber der technischen Apparatur. Insgesamt sind sie aber sehr viel stärker als jene in ihrer Lebensqualität beeinträchtigt (McSweeney, 1982).

64.7.4 Möglichkeiten der Prophylaxe und Therapie

Als therapeutische Konsequenz im Umgang mit beatmeten Patienten sollten folgende Gesichtspunkte berücksichtigt werden:
- Gerade bei diesen Patienten, die teilweise oder (bei generalisierten Lähmungen) vollständig ihrer Ausdrucksmöglichkeiten beraubt sind, besonders auf diskrete Zeichen psychischer Störungen zu achten (Blick, vegetative Zeichen etc.).
- Entsprechend den regressiven Bedürfnissen der beatmeten Patienten für eine Atmosphäre zu sorgen, die durch Stützung, Ermunterung und Hilfsbereitschaft gekennzeichnet ist.
- Zu beachten, daß auch kleinere und unbedeutend erscheinende technische Pannen von den Patienten in ihrer Hilflosigkeit als maximale Lebensbedrohung erlebt werden.
- Die Patienten häufig anzusprechen und ausreichenden Sozialkontakt zu ermöglichen (z.B. Verwandtenbesuche).
- Die verbliebenen Kommunikationsmöglichkeiten zu nutzen. Das ist besonders wichtig bei Patienten, die beispielsweise bei paralytischen Zuständen unterschiedlicher Ursache beatmet werden und neben der Sprache auch anderer Ausdrucksmöglichkeiten, z.B. der Gestik, beraubt sind. Eine besondere Grenzsituation stellt das Locked-in-Syndrom dar als Zustand völliger Bewegungslosigkeit und Sprechunfähigkeit des nur noch zu vertikalen Blick- und Blinzelbewegungen fähigen Patienten mit erhaltenem Bewußtsein.
- Für die technische Möglichkeit zu sorgen, daß die Patienten jederzeit auf sich aufmerksam machen können (z.B. Glocke, Schrifttafel). Dadurch können die Ängste vor Pannen in Grenzen gehalten werden.
- Den Patienten alle Handlungen zu gestatten, die so viel wie möglich Unabhängigkeit demonstrieren, um die persönliche Abhängigkeitsproblematik zu entschärfen.
- Verständnis dafür zu empfinden, daß die Patienten vieler Möglichkeiten zur Abreaktion von Spannungen beraubt sind und Zeichen von Aggressivität Ausdruck sowohl dieser Spannungen als auch ihres Selbstbehauptungswillens sein können.
- Die Patienten **schrittweise** vom Beatmungsgerät zu entwöhnen, sie auf die Trennung vorzubereiten und zu demonstrieren, daß im Bedarfsfall jederzeit erneut eine apparative Beatmung möglich ist.
- Auch bei konfliktbearbeitendem psychotherapeutischem Vorgehen sollten die supportiven Momente nicht vernachlässigt werden. So kann beispielsweise eine zu tiefgehende Mobilisierung von Affekten gefährlich sein, da die damit verbundenen physiologischen Begleiterscheinungen eine Dekompensation hervorrufen könnten.

64.8 Psychologische Gesichtspunkte zur Arbeitssituation des Personals von Intensivstationen

64.8.1 Vorbemerkungen

Es gibt viele Hinweise auf die besonderen Belastungen für das Personal von Intensivstationen. Eine Reihe von Autoren hat sich in den letzten Jahren mit diesen Problemen befaßt (Cassem et al., 1970, 1972;

Freyberger et al., 1972; Vreeland und Ellis, 1969; Hay und Oken, 1972; Klapp et al., 1981, 1982, 1984; Caldwell, 1981; Weiner, 1981; Koran et al., 1983; Bernhard, 1983 und 1984).

Ähnlich wie im Umgang mit onkologischen Patienten ergeben sich aus der Arbeitssituation besondere Schwierigkeiten und Probleme (vgl. Kap. 66).

64.8.2 Belastende Faktoren der Arbeitssituation auf Intensivstationen

Als besonders belastende Faktoren werden genannt:
- Die **hohe Mortalitätsrate** mit den Folgen sich ständig wiederholender Objektverluste, die das Gefühl therapeutischen Erfolgs schmälern. Schließlich wird meist der Tod eines Patienten, entsprechend dem berufsethischen Grundsatz der Verpflichtung zum Heilen, als Niederlage oder Mißerfolg empfunden. Wichtig ist dabei, daß es, vor allem bei jungen Patienten, zu Identifikationsvorgängen des zumeist jungen Personals in Intensivstationen kommt. Bei älteren Patienten finden wir Identifikationen auf der Ebene der Elternbeziehung.
- Die ständige **Konfrontation mit schwerkranken Patienten,** d.h. der Zwang, in einer Atmosphäre bedrückenden menschlichen Leidens zu arbeiten. Dazu gehören u.a.:
- Der hohe Prozentsatz bewußtloser und bewußtseinsgetrübter Patienten, mit denen der Kontakt behindert oder nicht möglich ist.
- Die häufige Verstümmelung und Verunstaltung schwerkranker Patienten („Schläuche aus jeder Körperöffnung"), die den Vorstellungen von der Integrität des menschlichen Körpers zuwiderläuft.
- Die ständige Kontrolle der Körperfunktionen, die als lähmende Routine erlebt werden kann und der Notwendigkeit der „Gefühlsarbeit" entgegengesetzt ist.
- Die Notwendigkeit für das Pflegepersonal, bei akuten Noteinweisungen zunächst, wie es die Situation erfordert, weitgehend autonom zu handeln, was im Widerspruch zu einem hierarchischen Rollenverständnis steht.
- Die Angst, durch Fehler gravierende Folgen für die schwerkranken Patienten zu verursachen. Sie provoziert Unsicherheit und kann im Rahmen eines „Circulus vitiosus" zu neuen Fehlern führen.
- Die Tatsache, daß Reanimationsmaßnahmen nicht selten Verletzungen von Patienten zur Folge haben, was zu Hemmungen und nachfolgenden Schuldgefühlen führen kann.
- Die Schwierigkeiten im Umgang mit den Todesängsten von Patienten (Campbell, 1980) führen zu charakteristischen, häufig zu beobachtenden **Abwehrmanövern.** Dazu gehören die Flucht in eine **kontraphobische Hyperaktivität** ebenso wie der häufig anzutreffende rauhe Ton, der einer **Abwehr** auf der **affektiven Ebene** entspricht, oder die Distanzierung. Dies wiederum kann dazu führen, daß sich das Team als „emotional verödet" (Klapp und Scheer, 1982) erlebt und noch mehr unter dem Ein-

druck von Schuldgefühlen leidet. Die Hektik der Aufnahmesituation bedeutet dabei einen erheblichen Rollenschutz; mit dem Tod eines Patienten bricht dieser Rollenschutz zusammen.
- Nicht selten entstehen komplizierte Übertragungs/Gegenübertragungsverstrickungen. So können im Umgang mit heranwachsenden Suizidpatienten Probleme der eigenen Adoleszenz aktiviert werden. Die hohe Fluktuationsrate stellt ein besonderes Problem von Intensivstationen dar. Es gibt Hinweise dafür, daß unter dem Personal solcher Stationen Beziehungsprobleme überproportional häufig auftreten.

  Bernhard (1984) hat dazu die Hypothese geäußert, daß die Arbeitssituation der Intensivstation die Möglichkeit bietet, neurotische Ängste auf dem Weg einer institutionalisierten Abwehr zu verwandeln und als somatisch faßbare Realangst zu bekämpfen.
- Die anfänglich bei fast allen Patienten bestehende Hilflosigkeit und völlige Abhängigkeit fordern überprotektives Verhalten des Pflegepersonals geradezu heraus. Bei Besserung des Zustandes der Patienten ergeben sich daraus immer wieder Konflikte zwischen den wiedererwachenden gesunden Autonomiebestrebungen der Patienten und den überfürsorglichen Tendenzen ihrer Betreuer, die häufig aggressiv wirken.
- Die meist rasche Trennung von genesenden Patienten, d.h. der Verzicht darauf, den Heilungsverlauf als für Heilberufe typische Befriedigungsmöglichkeit, z.B. in Form des Dankes von Patienten und Angehörigen, zu erleben. Dies wird verstärkt durch die Beobachtung, daß diese Patienten häufig die Intensivstation im späteren Genesungsverlauf meiden.
- Der Umgang mit meist schwer beunruhigten und verstörten Angehörigen, wobei deren Dank teilweise makabre Qualität erlangt: „Sie taten alles, was Sie konnten".
- Ähnlich wie in onkologischen Stationen sind innerhalb des Personals in Zeiten besonderer Belastung und Anspannung Reaktionen zu beobachten, die auf onkologischen Stationen als „burn-out"-Syndrom beschrieben worden sind (Meerwein, 1984) und die zu erheblichen Störungen auf der Gefühlsebene und der Effizienz der Arbeit führen können.
- Das im Sinne einer institutionalisierten Abwehr zu verstehende gesteigerte „Elitebewußtsein", das oft auf solchen Stationen anzutreffen ist, verstärkt die Abschottung, die durch eine restriktive Handhabung der Besuchsregelung und des Zugangs häufig noch weiter verstärkt wird.

64.8.3 Präventive und therapeutische Möglichkeiten

Die Auseinandersetzung mit besonders ausgeprägten Streßfaktoren, wie sie für Intensivstationen als typisch angenommen werden, kann zu verschiedenen

Symptombildern führen. Als häufigste werden depressive Verstimmungszustände, Arbeitsstörungen und besonders als inadäquat empfundene Verhaltensweisen genannt, wie das schon beschriebene laute, polternde, burschikose Handeln, welches das Resultat einer Abwehrform ist, aber der Umgebung, z. B. den Angehörigen der Patienten, unverständlich erscheint, schließlich auch der Rückzug auf eine rein technische Beziehung im Umgang mit schwerkranken Patienten. Auch der besondere Arbeitsenthusiasmus, der sich häufig findet, ist teilweise eine Folge von Abwehrfunktionen. Das Gruppenverhalten nimmt oft besonders kohärente Formen an. Schließlich wird auch die schon erwähnte Form exaltierter Fröhlichkeit, Schnoddrigkeit und gespielter Ober-

flächlichkeit beobachtet, die bei Außenstehenden zu Unverständnis und zu Konflikten führen kann (z.B. wird vom Sterben als „Löffel-Wegschmeißen" gesprochen) (vgl. Schors, 1979).

Als therapeutische Möglichkeiten bieten sich organisatorische Lösungsversuche an im Sinne veränderter Stationskonzepte, die beispielsweise teamorientierte Konsiliarpsychiatrie und regelmäßige Gruppensitzungen einschließen, wie sie allerdings bislang meist nur im Rahmen von Forschungsvorhaben realisiert werden konnten. Bei der Mitarbeiterauswahl sollten auch Gesichtspunkte der Befähigung zur Teamarbeit und der psychischen Belastbarkeit eine Rolle spielen.

65 Psychosomatische Gesichtspunkte beim künstlichen Organersatz und der Transplantation. Beispiel: Die Behandlung der chronischen terminalen Niereninsuffizienz

Ekkehard Gaus, Karl Köhle, Uwe Koch, Manfred Beutel und *Fritz A. Muthny*

65.1 Vorbemerkungen

Der alte Wunschtraum, funktionsuntüchtige Organe durch eine apparative Prothese bzw. durch die Übertragung von funktionstüchtigen Zellen, Geweben und Organen zu ersetzen, ist in den letzten zwei Jahrzehnten in den Bereich des Möglichen gerückt. Dabei hat der apparative Ersatz der Nierenfunktion durch verschiedene Verfahren ebenso wie die Nierentransplantation bahnbrechend gewirkt und ist seit langem zum klinischen Routineverfahren geworden, dessen psychosomatische Aspekte ausführlich untersucht wurden. Aus diesem Grund werden diese Gesichtspunkte in der Behandlung der chronischen terminalen Niereninsuffizienz exemplarisch dargestellt. Dies kann aber nicht darüber hinwegtäuschen, daß andere Verfahren wie die Transplantation des Herzens, der Leber, des Knochenmarkes mittlerweile dem Stadium einer experimentellen Therapie entwachsen sind. Neben den Gemeinsamkeiten im Erleben der Patienten führen die besonderen Umstände der jeweiligen klinischen Situation bei der Transplantation unterschiedlicher Organe auch zur spezifischen Ausgestaltung von Belastungen, Reaktionen, Anpassungsmustern und Komplikationen. Sie resultieren aus der Verschiedenartigkeit der Grunderkrankung und der angewandten therapeutischen Verfahren. Wir möchten daher nach der als Beispiel vorgesehenen und ausführlichen Darstellung von Dialyse und Nierentransplantation wenigstens einige Besonderheiten weiterer Transplantationsverfahren im Anschluß skizzieren.

65.1.1 Häufigkeit, Behandlungsformen und Prognose der chronischen Niereninsuffizienz

1986 wurden in der BRD nach der EDTA[1]-Statistik ca. 20000 an chronischem Nierenversagen leidende Patienten mit einer Nierenersatztherapie behandelt (ca. 30/100000 Einwohner).

Die terminale chronische Niereninsuffizienz entwickelt sich je nach Ätiologie sehr unterschiedlich rasch und läßt so auch dem Patienten unterschiedlich viel Zeit, sich auf den lebensbedrohlichen irreversiblen Krankheitszustand einzustellen. So kann ein akut innerhalb von Stunden einsetzendes Nierenversagen nach einem Unfall mit Kreislaufschock oder nach einer schweren Verbrennung weiter andauern und zur „Dialysepflichtigkeit" führen. Andererseits stehen unter den Ursachen des chronischen Nierenversagens Erkrankungen im Vordergrund, die unter Umständen erst nach einem über Jahre dauernden pathogenetischen Prozeß zur chronischen terminalen Niereninsuffizienz führen können: Immunologische Krankheitsprozesse (hauptsächlich Glomerulonephritiden), entzündliche (z.B. Pyelonephritiden) und mechanische Ursachen (Nierensteine), Mißbildungen (z.B. familiäre Zystennieren), nierentoxische Ursachen (z.B. langdauernder Schmerzmittelabusus) sowie pathologische Gefäßprozesse als Diabetes-Spätfolge.

Mit dem Absinken der Nierenfunktion unter einen kritischen Wert kommt es zum Anstieg harnpflichtiger Substanzen im Blut (Urämie) und zur Flüssigkeitsretention, so daß schließlich eine Nierenersatztherapie erforderlich wird. Die wichtigsten Behandlungsverfahren sollen kurz dargestellt werden.

Bei der **Hämodialyse,** der häufigsten Behandlungsform, wird Blut aus einer in der Regel am Arm angelegten arteriovenösen Fistel, dem sog. „Shunt", entnommen und durch die Dialysemaschine einem Dialysator mit einer selektiv-permeablen künstlichen Membran zugeleitet. Dort werden dem Blut durch Diffusionsvorgänge gegen eine Dialyseflüssigkeit harnpflichtige Substanzen entzogen und Elektrolyte ausgeglichen, das „gereinigte" Blut wird über den Shunt wieder dem Körper zugeführt. Die technische Durchführung jeder Dialysesitzung (3mal wöchentlich 4 bis 7 Stunden) kann in Form der Zentrumsdialyse vom Klinikpersonal durchgeführt werden, in Form der sog. Limited-Care-Dialyse auch den geeigneten Patienten stärker einbeziehen oder nach entsprechendem Training von Patient und Partner auch zu Hause erfolgen (Heimdialyse-Anteil in der BRD ca. 17%).

Bei der **Peritonealdialyse** (EDTA-Statistik, 1986: 2% der dialysepflichtigen Patienten) wird das Bauchfell als Austauschfläche benutzt, d.h., es wird über

---

1 EDTA = European Dialysis and Transplant Association.

einen Dauerkatheter eine Dialyseflüssigkeit in den Bauchraum eingefüllt, die hier über eine bestimmte Verweilzeit verbleibt und durch Diffusion über die Peritonealgefäße dem Blut harnpflichtige Stoffe entzieht bzw. einen Ausgleich der Elektrolyte herbeiführt.

Bei der **Nierentransplantation,** die bei uns die zweithäufigste Therapieform der chronischen Niereninsuffizienz darstellt (in der BRD bis 31. 12. 1985 8606 Transplantationen; ungefähr 15 bis 20% der Patienten mit terminaler Niereninsuffizienz verfügen über ein funktionierendes Transplantat), übernimmt das Transplantat, das in der Regel von einem gewebstypisch weitgehend übereinstimmenden toten Spender übertragen wird, die Nierenfunktion. Übertragungen von einem verwandten **Lebendspender** sind in der BRD mit insgesamt 3% der Transplantationen selten. Eine Abstoßung des Organs muß durch eine Dauermedikation von Immunsuppressiva verhindert werden (Cortison, Azathioprin, Ciclosporin A). Die Ein-Jahres-Funktionsrate bei transplantierten Leichennieren ist auf über 85% angestiegen. 10-Jahres-Patientenüberlebensraten von 80% sind möglich geworden. Mit der Überwindung vieler technischer Probleme in Dialyse und Transplantation sind die Überlebensmöglichkeiten des chronisch niereninsuffizienten Patienten deutlich gestiegen, bzw. die Behandlungsindikation konnte beträchtlich erweitert werden in Richtung auf Patienten mit höherem Alter und größeren medizinischen Risiken.

Zur ausführlichen Darstellung der Dialyseverfahren vergleiche Franz (1981), zu medizinischen Aspekten der Nierentransplantation sei auf Chatterjee (1979) verwiesen.

Andererseits hat diese Lebensverlängerung psychosoziale Gesichtspunkte, vor allem im Hinblick auf die Belastung von Patient und sozialem Umfeld durch Erkrankung und Behandlungsverfahren, in den Vordergrund rücken lassen. Die psychischen Probleme, die bei diesen Patienten je nach Krankheitsverlauf, individuellen Voraussetzungen und sozialer Unterstützung auftreten können und in vielen Fällen psychotherapeutische Hilfen notwendig machen, sind zum einen typisch für chronisch Kranke im allgemeinen, zum Teil aber auch spezifisch für die jeweiligen Behandlungsformen und medizinischen Komplikationen. Nach einer Übersicht zum Stand des psychosozialen Betreuung in der BRD (Pommer, 1985) auf der Basis einer Befragung von Ärzten und Schwestern in Dialysezentren bilden mehr als ein Drittel der terminal Nierenkranken eine psychosoziale Risikogruppe.

65.1.2 Methodische Probleme der psychosomatischen Forschung bei Dialyse- und Transplantationspatienten

Die Gruppe der Dialysepatienten stellt sicher neben den Herzinfarkt- und Krebspatienten die am besten untersuchte Population chronisch Kranker dar. Lange Zeit galt die Dialysesituation als geradezu paradig-

matisch für psychische Beeinträchtigungen als Folge der Fortschritte der technologischen Medizin.

Das Themenspektrum der psychonephrologischen Forschung reicht von Beiträgen zu psychischen und sozialen Faktoren in der Ätiologie der Krankheit über psychopathologische Beeinträchtigung bis hin zu Analysen des Einflusses einer langjährigen Berufstätigkeit in einem Dialysezentrum auf die Befindlichkeit des Personals.

Diese Forschung wird von Personen mit einem sehr unterschiedlichen wissenschaftstheoretischen Hintergrund betrieben. In der Literatur finden sich wenige gut belegte Kasuistiken ebenso wie experimentelle Studien. Das Methodenspektrum weist ebenfalls eine erhebliche Variation auf. Man findet klassische Persönlichkeitsfragebögen wie den MMPI, selbstentwickelte Coping-Fragebögen und projektive Verfahren als Grundlage der empirischen Beiträge.

Aus der Vielzahl der Publikationen darf aber keineswegs geschlossen werden, daß die Untersuchungsmöglichkeiten bei Patienten mit chronischer Niereninsuffizienz besonders günstig seien, oder daß ein besonders breiter Fundus gesicherten Wissens inzwischen gebildet werden konnte. Es ist vielmehr so, daß Forschung im Dialyse- und Transplantationsbereich durch eine ganze Reihe von Faktoren erheblich erschwert wird:

- Auch wenn zur Zeit in der Bundesrepublik über 20000 Patienten mit chronischer Niereninsuffizienz behandelt werden, so handelt es sich doch letztlich um eine seltene Erkrankung. Dies hat für die Forschung verschiedene Konsequenzen: Zum einen müssen die meisten Untersuchungen mit kleinen Fallzahlen auskommen, zum anderen sind wünschenswerte prospektive Untersuchungen, z. B. zum Einfluß psychosozialer Faktoren auf Krankheitsentstehung und Verlauf, kaum realisierbar. Das Fehlen prospektiver Untersuchungen führt dazu, daß z. B. Fragestellungen zur prämorbiden Persönlichkeitsstruktur nur retrospektiv betrachtet werden können. Damit läßt sich aber nicht klären, ob eine Auswirkung im psychischen Bereich Ursache oder Folge der Erkrankung ist.

- Da es sich bei Dialysepatienten um Schwerkranke handelt, sind schon aus forschungsethischen Gründen den Untersuchungsmöglichkeiten enge Grenzen gesetzt. Unterschiedlichkeit der Betreuungsbedingungen in verschiedenen Zentren erschwerten oder machten gar die Bildung von Kontrollgruppen unmöglich.

- In der Praxis begegnet der Forscher zahlreichen Kooperationsproblemen mit Ärzten und Dialysepersonal. Skepsis gegenüber der psychosomatischen Forschung bzw. dem Forscher selbst ist häufig. Die Forschung wird dabei leicht als Störfaktor erlebt: Sie ist oft nur realisierbar, wenn von seiten einer psychosozialen Forschungsgruppe gleichzeitig ein Betreuungsangebot erbracht wird.

Vergleiche zwischen verschiedenen Behandlungsformen der Dialyse, z.B. Zentrumsdialyse, Limited-Care- und Heimdialyse sind von einer schwierigen Selektionsproblematik überlagert.

Bei den Forschungsinstrumenten zeigt sich, daß die klassischen psychometrischen Verfahren, wie die Persönlichkeitsfragebögen, häufig für die Gruppen nicht geeicht und für die aufgeworfenen psychonephrologischen Fragestellungen nur von begrenztem Wert sind. Für die nach einzelnen Fragestellungen ad hoc entwickelten Forschungsinstrumente fehlen dagegen meist Belege ihrer Objektivität, Reliabilität und Validität (vgl. Kap. 14). Eine andere Schwierigkeit der psychosomatischen Forschung bei Niereninsuffizienten liegt in der Trennung zwischen den psychischen und somatischen Einflußquellen. So ist z.B. bei psychopathologischen Auffälligkeiten schwer zu entscheiden, ob dies die Folge von hirnorganischen Veränderungen durch Urämie ist oder eine psychische Reaktion auf die lebensbedrohlichen Belastungen unter der Dialyse darstellt (vgl. auch Kap. 63).

65.1.3 Exemplarische Krankengeschichten

An zwei Beispielen sollen typische Probleme dieser Patienten dargestellt werden.

Patientin A

Die 27jährige litt seit etwa 7 Jahren unter wechselnden Allgemeinbeschwerden. Später waren zusätzlich anfallsweise Gelenkschmerzen und Fieberschübe aufgetreten. Sie hatte zahlreiche Ärzte erfolglos konsultiert; schließlich wurde ein Lupus erythematodes mit Nierenbeteiligung diagnostiziert. Während der Betreuung der Patientin durch verschiedene Abteilungen der Klinik waren immer wieder Rivalitätskonflikte um die Patientin aufgetreten. Derartige Rivalitäten wiederholten sich auch während der weiteren Behandlung auf der internistisch-psychosomatischen Krankenstation (vgl. Kap. 28), vor allem im Verhältnis zwischen den Stationsärzten und konsiliarisch tätigen Ärzten. Wegen des medikamentenbedingten „Vollmondgesichts" neigte die Patientin dazu, ihre Cortisondosis eigenmächtig zu verändern. Schließlich entwickelte sich bei fortschreitender Grunderkrankung eine zunehmende Niereninsuffizienz.

Bei der stationären Aufnahme bestanden massive Schwellungen und Ergüsse; die Abklärung des vorausgegangenen Behandlungsverlaufs ergab, daß die späte Aufnahme und der Zustand der Patientin auch im Zusammenhang mit der bisherigen Arzt-Patient-Beziehung gesehen werden mußten: Der betreuende Ambulanzarzt war in Urlaub gegangen, die Patientin hatte – vermutlich aus einem Gefühl von Verlassenheit und Trotz – ihre Diätvorschriften nicht eingehalten und die vorhandenen Möglichkeiten zur ärztlichen Hilfe nicht genutzt. Schon früher hatte sich ihr Befinden mehrfach rapid verschlechtert, wenn der jeweils betreuende Arzt nicht erreichbar war.

Die Patientin wurde ins chronische Hämodialyseprogramm aufgenommen. Der Dialyseverlauf wurde durch häufige Diätfehler kompliziert. Zusätzlich klagte die Patientin in wechselndem Ausmaß über Bauchbeschwerden. Nach Monaten entwickelte sich eine akute Entzündung der Bauchspeicheldrüse mit mehrfachen Rezidiven, hinzu kam eine Lungenembolie. Der körperliche Zustand stabilisierte sich erst nach langer, intensiver Behandlung. Während dieser Phasen machten Diätverstöße häufig Notdialysen, zusätzlich zu den drei wöchentlichen Dialysebehandlungen, erforderlich. Die Patientin weigerte sich, die Krankenhauskost zu essen und nahm nur von der Mutter eigens für sie zubereitete Speisen zu sich. Nachts wurde sie mehrfach beim Trinken größerer Flüssigkeitsmengen beobachtet. Sie litt sehr unter ihrem entstellten Aussehen (Vollmondgesicht, Haarausfall, zunehmende Kachexie).

Der Wunsch der Patientin nach einer Übernahme in die Ambulanz der Zentrumsdialyse ließ sich nicht realisieren; aus probeweisen Beurlaubungen über das Wochenende kam sie mehrfach in überwässertem Zustand zurück, so daß eine Notdialyse durchgeführt werden mußte. Die Patientin verstarb schließlich an einem erneuten septischen Geschehen.

Überlegungen zur Psychodynamik

Die Patientin und die Mitglieder ihrer Familie, insbesondere ihre Mutter und ihr Bruder, haben von Anfang an die Mitarbeiter der Station in ungewöhnlich starkem Maße beschäftigt; besonders auffallend war ihr aggressives und vorwurfsvolles Verhalten: Sie versuchten die Schwestern herumzukommandieren, alle Mitarbeiter gegeneinander auszuspielen und zogen sich immer wieder beleidigt zurück. Die Patientin geriet immer wieder mit anderen Kranken in Konflikte; dabei versuchte sie, die Mitarbeiter der Station als Bundesgenossen zu gewinnen. Während der Patientin anfangs die besondere Aufmerksamkeit des Personals galt, löste sie im Verlauf bei allen Beteiligten zunehmend Ärger und Ablehnung aus. Trotz aller Bemühungen gelang es nicht, in zufriedenstellendem Maße tragfähige emotionale Beziehungen zu ihr aufzubauen. Entsprechende Versuche wurden auch durch die Angehörigen der Patientin entscheidend behindert.

Anamnestisch ist bemerkenswert, daß die Patientin ihre Arbeitsstelle 11mal gewechselt hatte, weil keine dieser Stellen ihren Vorstellungen entsprochen hätte. Früher war sie besonders stolz auf ihr Aussehen gewesen und hatte den Wunsch gehabt, Mannequin zu werden. Die jetzige Entstellung ihres Aussehens schien für sie die größte Kränkung zu sein. Die Krankheit hatte so ihr narzißtisches Gleichgewicht wie der Einbruch einer Katastrophe gestört.

Während der Behandlung wurden nun die Beziehungen innerhalb ihrer Herkunftsfamilie sichtbar; sie hatte offenbar bisher nur flüchtige Bekanntschaften mit älteren Männern gehabt. Sie fühlte sich durch die Mutter in ihrer Entfaltung sehr behindert. Die Familie schien sich von der Umgebung weitgehend zu isolieren. Auffallend war die Neigung der Mutter, sich mit den Leistungen der Kinder zu identifizieren. Verständlicherweise bedeutete damit die jetzige Erkrankung der Patientin auch für die Mutter selbst eine herbe Enttäuschung.

Die Beziehung zwischen Mutter und Tochter wurde von Ärzten und Schwestern als schwer gestört erlebt. Die Mutter wirkte ausgesprochen starr, mißtrauisch und überfürsorglich. Sie schirmte die Tochter von ihrer Umgebung ab, wurde als ständig „über dem Bett liegend" wahrgenommen. Mit zunehmender Krankheitsdauer war es kaum mehr möglich, unmittelbar mit der Patientin zu kommunizieren, ihre Kontakte hatten über die Mutter zu verlaufen. Alle Maßnahmen, die diese Beziehung störten, führten zu ärgerlich-aggressiven Reaktionen. Gleichzeitig wurde die Patientin in ihrem Verhalten zunehmend infantiler. Sie wirkte manchmal wie ein kleines trotziges Kind, das mit seinen Bedürfnissen zu kurz kommt – beispielsweise, wenn sie sich nachts am Milchtopf in der Küche zu schaffen machte.

Der Vater der Patientin wirkte still und ängstlich; er trat in diesem Ablauf völlig zurück. Er blieb dem Krankenbett meist fern und wartete oft stundenlang im Auto vor dem Krankenhaus, weil er sich am Bett „nicht beherrschen" könne. Der jüngere Bruder wurde von der Patientin idealisiert; er fiel durch eine rationalisierende Form der Abwehr auf, die es ihm jedoch gleichzeitig erschwerte, sich gefühlsmäßig mit der Krankheit seiner Schwester auseinanderzusetzen. Er hatte für diese Zeit sein Technikstudium nahezu völlig unterbrochen und sich detaillierte Informationen über das Krankheitsbild angelesen. In kontrollierender Form versuchte er, sich immer wieder in die Behandlung einzumischen. Die Komplikation der Erkrankung führte zu wachsender aggressiver Verbitterung innerhalb der Familie mit der Neigung, Enttäuschung und Ärger projektiv zu verarbeiten. Mit nahezu allen medizinischen Institutionen, mit denen die Familie in Berührung kam, entstanden Konflikte.

Während die prämorbide Persönlichkeitsentwicklung der Patientin neurotische Merkmale mit Hinweisen auf eine mißglückte ödipale Konfliktlösung und ausgeprägte narzißtische Verwundbarkeit zeigte, traten mit Verschlechterung der Krankheit und der hiermit verbundenen Lebensbedrohung zunehmend regressive Tendenzen auf: Vor allem ein symbiotisch-anklammerndes Verhalten gegenüber der Mutter und die Wiederbelebung früherer Abwehrformen.

Die starren Interaktionsmuster innerhalb der Familie erschwerten für alle Beteiligten einen flexibleren Umgang mit dem Unglück des einen Familienmitglieds. Emotionaler Austausch und damit gemeinsame Trauerarbeit war nicht möglich; jedes Familienmitglied versuchte verzweifelt, für sich und seiner Struktur entsprechend mit der Erkrankung umzugehen. Der Vater zog sich verzweifelt-stumm zurück, die Mutter verhielt sich überfürsorglich und rivalisierend, der Bruder rationalisierend und kontrollierend-einmischend. Der Gefahr der Auflösung der Familie wirkte die Projektion der aggressiven Reaktionen nach außen entgegen. Diese Verlagerung des Ursprungs der Aggression nach außen trug entscheidend mit dazu bei, daß es nicht gelang, mit der Patientin eine tragfähige therapeutische Beziehung herzustellen oder die Familie für ein gemeinsames therapeutisches Vorgehen zu gewinnen. So war es auch nicht möglich, der Patientin zu helfen, ihre autoaggressiven Verhaltensweisen zu modifizieren, über die sie wesentlich zum ungünstigen Krankheitsverlauf beitrug.

Patient B

Kriseninterventionen bei einem extrem zurückgezogenen Heimdialysepatienten

Der 23jährige Germanistikstudent wurde uns auf eigenen Wunsch von seinem Nephrologen zur Mitbehandlung überwiesen. Seit dem 15. Lebensjahr leidet der Patient an einer terminalen Niereninsuffizienz auf dem Boden einer chronischen Glomerulonephritis; seit dem 18. Lebensjahr bedarf er einer chronischen Heimdialysebehandlung, die seit seinem 19. Lebensjahr zu Hause durch seine Mutter durchgeführt wird. Zwei Jahre vor der Überweisung traten zunehmend Allgemeinbeschwerden auf, die im Verlauf eines Jahres in ein schweres Krankheitsbild mit Fieberschüben, Perimyokardbeteiligung und peripherer Neuropathie mündeten. Die Ätiologie dieses Zustandsbildes konnte trotz intensiver stationärer und ambulanter Diagnostik nicht endgültig abgeklärt werden. Retrospektiv könnte es

am ehesten als Folge einer Zytomegalie-Virus-Infektion angesehen werden. Auffallend war der zeitliche Zusammenhang der Krankheitserstmanifestation mit einer bevorstehenden Zwischenprüfung. Die Allgemeinbeschwerden des Patienten, insbesondere seine Appetitlosigkeit, eine chronische Neigung zu erbrechen, allgemeine Schwäche und Kraftlosigkeit, aber auch ein pathogenetisch ungeklärter rezidivierender Aszites bildeten sich nach Abklingen des akuten schweren Krankheitszustandes nicht zurück. Der Wunsch nach zusätzlicher psychosomatischer Diagnostik und Therapie entstand, als der Gesamtzustand sich nach Jahresfrist wiederum vor dem neu angesetzten Zwischenprüfungstermin erheblich verschlechterte, und der Patient sein Studium wieder unterbrechen mußte. Inzwischen waren auch die Spannungen in der Familie eskaliert. Auch die engagiert-fürsorgliche Mutter fühlte sich durch die ständigen Forderungen ihres einzigen Kindes zunehmend überlastet und in ihren eigenen Entfaltungsmöglichkeiten ebenso wie der Vater unerträglich eingeschränkt. Die Indikation zu einer Behandlung mit psychoanalytisch orientierten Gesprächen und — parallel dazu — mit funktioneller Entspannung nach M. Fuchs (1976) (vgl. Kap. 22) stellten wir aufgrund des zeitlichen Zusammenhangs zwischen Auftreten und Verschlechterung des Zustandsbildes mit den Prüfungsterminen, vor allem aber aufgrund der vom Patienten berichteten extremen Abhängigkeitsproblematik gegenüber seinen Eltern in Verbindung mit einer totalen Isolation gegenüber seiner übrigen Umwelt. Dabei war nicht klar zu trennen, inwieweit die Problematik bei dem leidgeprüften Patienten erst als Folge der schweren körperlichen Komplikation aufgetreten war oder inwieweit die bevorstehende Zwischenprüfung eine Krise im Konflikt zwischen seinen Wünschen nach Verselbständigung einerseits und seinen realen Abhängigkeitswünschen andererseits hat entstehen lassen, der dann die Entwicklung der körperlichen Krankheit förderte.

Die Behandlung hatte etwa in gleichem Maße eine Ermutigung des Patienten zu eigener Aktivität und die vorsichtige Bearbeitung seiner Abhängigkeitsproblematik zum Ziel. Es wurde bald deutlich, in welchem Ausmaß sich die reale versorgungsbedingte Abhängigkeit, aber auch die eigenen regressiven Abhängigkeitswünsche des Patienten sowie die in dieser Situation notwendig gewordene Unterdrückung heftigster aggressiver Regungen, insbesondere gegenüber dem Vater, gegenseitig verstärkten. Diese Prozesse blockierten jede Initiative und Eigenaktivität des Patienten, der sich zudem ja in einem sehr schlechten Allgemeinzustand befand. In psychologischer Hinsicht hatte er sich in diesem Konflikt extrem narzißtisch zurückgezogen („als Weltraumpilot hinter dem Mond"), die körperliche Schwäche wurde durch extremen Bewegungsmangel perpetuiert. Im Verlauf von etwa 15 Gesprächen gelang es, diese Situation in einem ersten Ansatz zu bearbeiten und den Patienten zu ermutigen, gleichzeitig ein systematisches körperliches Trainingsprogramm zu beginnen. Erste Erfolge stützten das verständlicherweise tief beeinträchtigte Selbstgefühl und entschärften die häusliche Situation. Am Studienort fand der Patient unerwartet hilfreiche Unterstützung von Kommilitonen und Lehrern. Es gelang ihm schließlich auch, die erhebliche Diskrepanz zwischen den tatsächlichen Studienanforderungen und seinen diese verharmlosenden Phantasien bzw. Erwartungen durch konsequente und intensive Prüfungsvorbereitung ausreichend zu überbrücken. Die psychotherapeutische

Begleitung konnte am Studienort fortgesetzt werden. Der Patient bestand die Zwischenprüfung und konnte das Hauptstudium erfolgreich abschließen. Das Entgegenkommen von Kommilitonen und Lehrern verhalf dem Patienten aus seiner Isolation und Anonymität heraus und unterstützte ihn dabei auch, in Problem- und Konfliktsituationen mit anderen vermehrt zu kommunizieren. Der günstige Rahmen in diesem Studienabschnitt hat dazu beigetragen, daß der Patient aus seinem depressiven Rückzugsstand wieder herausfand und auch einen Teil seiner familiären Ablösungsproblematik lösen konnte. Parallel dazu haben sich der körperliche Gesamtzustand und die Allgemeinbeschwerden – soweit dies bei der Grunderkrankung möglich ist – weitgehend gebessert. Daraufhin brach der Patient die psychotherapeutische Begleitung ab. Er hatte inzwischen alle Examensprüfungen bestanden, lediglich die schriftliche Examensarbeit war noch anzufertigen; jetzt traten die als überwunden und gelöst geglaubten Konflikte allerdings wiederum in Form einer depressiven Rückzugstendenz, mit starkem Rückgang seiner Eigenaktivität und vermehrter Wahrnehmung der Abhängigkeitskonflikte, erneut auf. Neben der bevorstehenden endgültigen Beendigung des Studiums hatte aus der Sicht des Patienten auch eine Verschlechterung des Behandlungsklimas in der betreuenden Dialyseklinik zu dieser Entwicklung beigetragen. Das Resultat war eine erhebliche Arbeitshemmung, hinzu kamen die bereits bekannten körperlichen Beschwerden wie Appetitlosigkeit, Erbrechen und Müdigkeit. Jetzt konnte sich der Patient zu einer psychoanalytischen Behandlung mit dem Ziel entschließen, seine Abhängigkeits- und Verselbständigungsprobleme vertieft zu verstehen.

65.2 Die Hämodialyse

65.2.1 Belastungen für Dialysepatienten

Mit den vielfältigen psychosozialen Belastungen von Patienten, die an einer chronischen terminalen Niereninsuffizienz erkrankt sind, haben sich im Laufe der letzten 20 Jahre zahlreiche Untersuchungen, vorwiegend aus den Gebieten der Psychosomatik und Psychotherapie und der Psychiatrie, beschäftigt. Konzentrierten sich diese anfänglich fast ausschließlich auf die Patienten, sind mittlerweile zunehmend auch Probleme der betroffenen Partner und des Personals von Dialysestationen mit den komplexen Interaktionsschwierigkeiten in den Blickpunkt des Interesses gerückt. Neben den spezifischen belastenden Faktoren für Dialysepatienten sind ihre Möglichkeiten der Verarbeitung dieser Belastungen und die Reaktionen der Umgebung (Familie, Personal) von Bedeutung. Einzelne Streßfaktoren lassen sich nach verschiedenen Ursprungsbereichen gliedern.

Verlust der Körperfunktionen und der körperlichseelischen Integrität vor Dialyseaufnahme

Patienten mit einer chronischen terminalen Niereninsuffizienz haben oft eine lange Zeit chronischer Krankheit hinter sich, in der sie unter vielen Symptomen zu leiden hatten, die ihren Lebensalltag beeinträchtigten. Zu diesen gehören körperliche Schwäche, Schmerzen, Appetitlosigkeit, Erbrechen, Juckreiz, Schlaf- und Konzentrationsstörungen. Hinzu kommen diätetische Einschränkungen und häufig die Notwendigkeit, Medikamente einzunehmen. Die Intoxikation durch retinierte Substanzen führt zu vielfältigen und anfänglich uncharakteristischen Störungen unterschiedlicher Körperfunktionen mit zunehmender Intensität. Im Stadium der kompensierten Retention läßt sich bei 90% der Patienten eine hirnorganische Beeinträchtigung, meist in Form einer Minderung von Aufmerksamkeit, Konzentration, der visuell-motorischen Koordination und des verbalen Abstraktionsvermögens feststellen (Wanick et al., 1977). Mit der Anlage eines Shunts wird der Organschaden für den Patienten äußerlich sichtbar.

Verlust der Körperfunktionen und der körperlichseelischen Integrität nach Dialyseaufnahme

1. Nach Aufnahme ins chronische Hämodialyseprogramm in der Klinik oder einer Praxis muß sich der Patient an die neuen **Einschränkungen seiner Lebenssituation** gewöhnen: meist dreimal pro Woche ist er für mehrere Stunden an den Dialyseapparat gebunden, was zusammen mit den Wegzeiten gut dem zeitlichen Aufwand einer Halbtagsbeschäftigung entspricht und eine grundlegende Veränderung seiner bisherigen Lebensweise erfordert. Zahlreiche **körperliche Beschwerden** können als Folge der biochemischen Veränderungen und als Reaktion auf die mit der Dialyseprozedur verbundenen körperlichen und seelischen Belastungen auftreten. Dazu zählen beispielsweise Appetitstörungen, Schwindel, Erbrechen, Kopfschmerzen, Schlafstörungen, Wadenkrämpfe.

Obwohl die Dialyseprozedur vergleichsweise sicher ist – binnen 10 Jahren soll das Mortalitätsrisiko aufgrund technischer Defekte nur 2% betragen (Friedman und Lundin, 1982) – erleben viele Patienten neben physischem Unwohlsein während der Dialyse, insbesondere zur Zeit des Anschlusses an den Apparat, angstvolle Momente.

Trotz der gestiegenen Überlebensraten sind die Patienten von zahlreichen Risiken und **Komplikationsmöglichkeiten** bedroht, die eine Quelle zusätzlicher seelischer Belastungen sind. Dazu gehören Infekte, Shunt-Komplikationen (z.B. Verschluß), Hochdruck und dessen Folgen wie zerebrale Insulte, Rhythmusstörungen des Herzens, Herzbeutelerguß, Lungenödem, Blutungen, Hämolyse, Blutverluste durch Leckbildung, Schockzustände, Muskelkrämpfe, Ablagerung von Plastikpartikeln in den inneren Organen, Schädigungen durch Spurenelemente, atherogene Stoffwechselveränderungen (Nestel, 1982), außerdem psychiatrische Symptome, wie akute und chronische organische Psychosyndrome, und Medikamentennebenwirkungen. Osteopathie, Neuropathie und Anämie komplizieren oft den Verlauf. Insbesondere das Risiko arteriosklerosebedingter Komplikationen ist hoch. Eine besondere Risikogruppe stellen niereninsuffiziente Diabetiker dar. Der Tod von Mitpatienten, mit denen sie oft eine enge und langwährende „Leidensgenossenschaft" verbindet, führt den Patienten ihre Bedrohung immer wieder neu vor Augen.

Alle Dialysepatienten sind vom Verlust ihrer körperlichen Integrität betroffen und von weiteren Verlusten bedroht.

Mit der Situation chronischen Krankseins und der Dialyse verbinden sich oft **Veränderungen des Körperschemas,** die auf äußere Merkmale wie beispielsweise Narben, Shunt, fahlgraues Hautkolorit und Vollmondgesicht (vgl. Pat. A) zurückgehen. Der Verlust der Fähigkeit zu urinieren kann im Sinne einer Frustration urethraler libidinöser Wünsche (Kaplan De-Nour, 1969) das Körperschema verändern (vgl. Kap. 56). So wurden bei Patienten Phänomene im Sinne eines „Phantom-Urinierens" beobachtet, indem Patienten mit Anurie über entsprechende Sensationen des Urinierens berichteten. Die Einbeziehung des Dialyseapparates vermag eine Erweiterung des Körperschemas zu bedingen (Abram, 1968, 1969; Kaplan De-Nour, 1969; Short und Wilson, 1969).

Zeichentests wiesen auf eine Beeinträchtigung geschlechtlicher Differenzierung, eventuell auch als Hinweis auf eine Störung der Rolle als Sexualpartner hin (Cazzullo et al., 1973), aber auch auf gravierende desintegrative und regressive Phänomene (Basch et al., 1981). Das psychische Leben vieler Dialysepatienten ist durch eine Neigung zum Konkretismus und eine zunehmende Verarmung an Phantasie gekennzeichnet (Castelnuovo-Tedesco, 1973). Es ist anzunehmen, daß diese Phänomene zumindest teilweise Folge seelischer Abwehr- und Anpassungsvorgänge an eine physisch und psychisch äußerst belastende Situation sind (Blodgett, 1981/1982).

2. Verlust von sozialen Bindungen, Triebbefriedigungsmöglichkeiten und Frustration von Triebbedürfnissen.

Mit der Dialysebehandlung gehen meist die Einbuße oder ausgeprägte **Veränderung von Rollen** und ein **Rückzug aus sozialen Bindungen** einher (Vollrath et al., 1976; Kaplan De-Nour, 1980 a, b, 1982, 1983; Koch, 1984). Viele Patienten wie auch ihre Partner weisen, wie sich in Fragebogenuntersuchungen herausstellte, ein ausgeprägtes Gefühl mangelnder sozialer Potenz auf. Angst vor weiteren Verlusten und Komplikationen beeinflußt ihr Leben. Dabei findet sich die globale Angst häufig auf dem Weg der Verschiebung in eine Angst vor einzelnen behebbaren Störungen umgewandelt. Auffallend ist, daß die Partner von Patienten mehr Ängste um ihre Gesundheit und Komplikationsmöglichkeiten zugeben als die Patienten selber (Koch, 1984). Dies entspricht einem Modus der Delegation von Ängsten (Balck et al., 1978). Oft befürchten die Angehörigen, daß die Patienten eine Verschlechterung verschweigen. Sie nehmen mehr Rücksicht auf die Patienten als umgekehrt, neigen dazu, Konflikten auszuweichen und sich bei Auseinandersetzungen zurückzuziehen. Dieses Schonverhalten hat einen verstärkten sekundären Krankheitsgewinn der Patienten zur Folge.

Das Ausmaß sozialer Einbußen variiert dabei in Abhängigkeit vom Dialyse-Setting. Patienten in der Praxisdialyse und ihre Partner geben beispielsweise mehr Verluste an als Patienten und Angehörige in der Heimdialyse (Speidel, 1981; Koch et al., 1982, 1984).

Verlust oder Einschränkung von Hobbies behindern Patienten in ihrer Autonomie, auch wenn durch Dialysemöglichkeiten an Urlaubsorten, transportable Dialysesysteme und gemeinsame Urlaubsveranstaltungen von Dialysepatienten (DÄB 4/1981) erhebliche Verbesserungen im Vergleich zu früher erreicht werden konnten. Dialyseintervalle können im Sprachgebrauch der Patienten die Qualität eines neuen Zeitmaßes gewinnen.

Eine Minderung des **sozialen Status** der Familie führt in der Regel auch zu **Veränderungen** innerhalb des **Familiengefüges,** z. B. zu einem Wechsel der Position in der Rangordnung einzelner Familienmitglieder. Systemisch betrachtet, ändern sich sowohl die extra- als auch die intrasystemischen Grenzen der Familie. In der Literatur zu psychologischen Aspekten von Dialyse und Transplantation wird der Berücksichtigung dieser systemischen Sicht zunehmend mehr Bedeutung eingeräumt (Balck et al., 1978, 1982, 1988; Gerhardt und Heberling, 1982). Zunächst bedingt die terminale Niereninsuffizienz eines Familienmitgliedes mit der Abhängigkeit von der Maschine und vom medizinischen Personal eine Einschränkung der Möglichkeit, sich abzugrenzen und damit eine erhöhte Durchlässigkeit der Familiengrenzen. So beeinflußt das Verhalten von Partnern, je nachdem, ob es mehr identifikatorisch oder antagonistisch geprägt ist, die Stellung des Dialysepatienten in der Familie. Wichtig ist dabei die prämorbide Rollenverteilung, beispielsweise die Frage, ob es sich bei dem Erkrankten um den Brotverdiener gehandelt hat, ob die Flexibilität der ehelichen Beziehungen überhaupt eine neue Definition von Rollen zulassen kann (Gerhardt und Heberling, 1982), oder ob das erkrankte Familienmitglied wie im Falle unserer Patientin A die narzißtischen Bedürfnisse eines Elternteils in starkem Maße verkörpert hatte. Ist die Beziehung schon zuvor hauptsächlich durch einen Dissens über die Rollenverteilung geprägt, wird die Anpassung komplizierter. Obwohl die Patienten ihre sozialen Aktivitäten erhalten wollen, kontrastiert dies mit deren tatsächlicher Abnahme bei Patienten und Partnern, wie viele Autoren feststellen konnten (Kaplan De-Nour, 1982; Speidel et al., 1981; Koch, 1982, 1984).

Eine überprotektive Haltung der Familie kann die Regressionsneigung eines Patienten verstärken, sein Selbstwertgefühl und seine Autorität innerhalb der Familie schmälern, das familiäre Gleichgewicht bedrohen und zu erheblichen intrafamiliären Spannungen beitragen. Viele betroffene Familien versuchen, die Krankheit und die damit verbundenen Probleme so weit wie möglich aus dem Alltag herauszuhalten, Spannungen durch eine erhöhte Kongruenz der Meinungen und die damit verbundenen Gefühle wie Trauer, Ärger und Unzufriedenheit zu unterdrücken (Czaczkes und Kaplan De-Nour, 1978). Insbesondere Kindern wird oft möglichst viel Information vorenthalten (Maurin und Schenkel, 1976). Immer wieder wird auch beobachtet, wie eine Trauerreaktion von

Angehörigen bei Verschlechterungen und Komplikationen vorweggenommen wird und zur gegenseitigen Distanzierung beiträgt. Eine solche Form der antizipierten Trauer fand sich auch beim Vater unserer Patientin A.

Nicht selten reagiert die Familie als Ganzes auf den Schock der Erkrankung, was sich auch in den **psychiatrischen Morbiditätsziffern** ihrer Mitglieder niederschlägt. Für Dialysepatienten werden sehr unterschiedliche Häufigkeiten für das Vorkommen **depressiver Syndrome** angegeben. Die Zahlen, in unterschiedlichen Kollektiven mit unterschiedlichen Methoden gewonnen, differieren stark. In älteren Arbeiten finden sich bei bis zu 60% der Patienten depressive Syndrome, zum Teil finden sich Angaben, die der Häufigkeit psychiatrischer Symptome in der Allgemeinpraxis entsprechen (Livesley, 1982; Übersicht bei Kaplan De-Nour, 1983; Shea et al., 1965; Farmer et al., 1979: 31%; Lowrie, 1979: 22%).

Bei Kindern von Heimdialysepatienten wurde eine hohe Rate depressiver Symptome beschrieben (Tsaltas, 1976). Von der untersuchten Gruppe mußte bei zwei Dritteln der Kinder der Schulpsychologe in Anspruch genommen werden. Beobachtet wurde auch ein häufiger emotionaler Rückzug der Kinder von den Patienten (Speidel et al., 1978; Friedlander und Viederman, 1983).

In der psychoanalytischen Literatur wird auf den Einfluß chronisch kranker und körperlich schwerbehinderter Väter auf Entwicklungsdefizite und pathologische Entwicklungsverläufe bei Kindern hingewiesen (vgl. Lussier, 1980; Yorke, 1980; Castelnuovo-Tedesco, 1981). Angenommen wird, daß unter dem Druck verstärkter Kastrations- und Desintegrationsängste, insbesondere auch während der Adoleszenz, die Identitätsbildung als phasenspezifische Aufgabe erschwert ist (Blos, 1978; Krejci und Bohleber, 1982).

3. Sexuelle Probleme. Libido, Potenz und sexuelle Erlebnisfähigkeit sind bei einem Großteil der Patienten, aber auch in einem geringeren Maß bei ihren Partnern gemindert (Steele et al., 1976). Mit diesem Problem befassen sich eine Reihe von Übersichtsarbeiten (Levy, 1974, 1981, 1984; Abram et al., 1973; Milne et al., 1978; Czaczkes, 1978; Speidel et al., 1983; Degen et al., 1983; Kaplan De-Nour, 1978; Lim und Fang, 1975).

Levy hat 1974 in einer Fragebogenuntersuchung bei 429 Hämodialyse- und transplantierten Patienten bei 70% der Befragten eine Störung der sexuellen Erregung und Erlebnisfähigkeit festgestellt, die während des Dialysezeitraumes noch zunehme, was als Hinweis auf eine starke Mitbeteiligung psychosozialer Faktoren gewertet wurde. Viele Untersuchungen schon in der Anfangsphase der Hämodialyse befaßten sich mit diesem Thema (Abram et al., 1973; Harari et al., 1971; Freyberger und Bauditz, 1969; Cazzullo und Invernizzi, 1973; Speidel et al., 1970; Friedman et al., 1970). Testpsychologisch konnten Hinweise auf eine sexuelle Versagensangst gefunden werden (Vollrath et al., 1976). Einzelne Studien wiesen auf einen Zusammenhang von ehelichen Problemen, Depression und Störungen sexueller Funktionen hin (Steele et al., 1976; Finkelstein und Steele, 1978). Dabei blieb aber die Frage der ursächlichen Beziehungen und der Gewichtung einzelner Faktoren für gestörte Teilfunktionen wie Libido, Orgasmus- und Erektionsfähigkeit noch weitgehend offen.

Als psychologisch bedeutsam werden die mit der Dialysesituation und den Komplikationen verbundenen Streßerlebnisse, Angst und Depression, Selbstgefühlstörungen, Abhängigkeits- und Hilflosigkeitsgefühle, regressive Tendenzen mit einem steigenden Bedürfnis nach Zärtlichkeit und abnehmendem Interesse an genitaler Sexualität, Störungen der Partnerbeziehung, z. B. bei wachsender Feindseligkeit, genannt. Oft tritt an Stelle der Attraktivität des Partners eine betont fürsorgliche Haltung, die den Patienten in seinem Selbstwerterleben verletzt. Gelegentlich läßt sich die Angst von Patienten und Partnern feststellen, den Shunt beim Verkehr zu beschädigen. Es sollte auch nicht übersehen werden, daß unter den niereninsuffizienten Patienten und ihren Partnern sich schon primär solche mit neurotischen Störungen und Beeinträchtigungen der Sexualität finden, die gelegentlich durch die Erkrankung und durch virulent werdende Partnerschaftskonflikte eine Aktualisierung erleben (Watts, 1983).

An möglichen **organischen Faktoren** für eine Störung sexueller Funktionen bei den Dialysepatienten werden neben Medikamenten, z. B. Antihypertensiva, insbesondere hormonelle Veränderungen (Sexualhormone, Hyperprolaktinämie, Hyperparathyreoidismus), Anämie und Neuropathie genannt. In letzter Zeit wird der Bedeutung organischer Faktoren in der Genese sexueller Störungen bei Dialysepatienten zunehmend mehr Gewicht eingeräumt (Levy, 1981, 1984). So wird von Weizman und Mitarbeitern (1983) und Procci und Mitarbeitern (1983) auf die Rolle des erhöhten Prolaktinspiegels und anderer hormoneller Veränderungen als mögliche Ursache hingewiesen, wobei bei Männern eine Störung der Erektionsfähigkeit im Vordergrund steht.

Zur Differenzierung organischer und psychischer Faktoren kann die automatische Messung nächtlicher Erektionen, die bei Männern in 80% während der ersten Phase des REM-Schlafes auftreten (NPT, Nocturnal Penile Tumescence) und bei Dialysepatienten viel seltener sind (Karacan et al., 1978; Procci et al., 1983), benutzt werden.

Die Erfahrung einer Störung oder Erschwerung der Sexualfunktionen kann zu einer sekundären Impotenz führen. Therapeutisch können bei solchen Fällen auch sexualtherapeutische Techniken nach Masters und Johnson (1966) versucht werden (Berkman, 1978).

Viele Möglichkeiten, Angst, Aggression und sonstige Spannungen in einer sozial konformen Weise abzureagieren (Rauchen, Trinken, Sport usw.), sind den Dialysepatienten entweder versperrt oder ihnen wird davon abgeraten. Ein bei Dialysepatienten häufig zu beobachtendes Masturbieren kann beispielsweise als Ersatz und Versuch einer Abreaktion von Spannungen gesehen werden.

4. Einschränkung von Flüssigkeit, Diät. Die meisten Dialysepatienten leiden unter **Einschränkung von Flüssigkeit und Diäterfordernissen** (Cramond et al., 1967; Shea et al., 1965; Friedman et al., 1970; Kaplan De-Nour und Czaczkes, 1972a, 1976; Gardner, 1981). Diese Erfordernisse kollidieren mit zumeist verstärkten oralen Bedürfnissen dieser Patienten, wie sie sich aus den regressiven Tendenzen im Zusammenhang mit der Erkrankung und der Behandlungssituation ergeben. In der Literatur wird ein suchthaftes Nahrungs- und Trinkverlangen beschrieben, was immer wieder tödliche Komplikationen zur Folge hat. Nach solchen Exzessen („food kleptomania", Schreiner, 1966) folgt teilweise eine schlagartige Besserung der Depression (Villard, 1969). Bei manchen Patienten drückt sich der übermächtige Essens- und Trinkwunsch in Träumen, beispielsweise mit Banketten in luxuriöser Umgebung, aus (Cramond et al., 1967, 1968), ähnlich wie das auch bei Kriegsgefangenen und bei Hungerexperimenten beobachtet wurde. Zwischen 25 und 60% der Patienten halten sich nicht an diese Empfehlungen, wie eine Übersicht der Literatur zeigt. Dabei ist es unabdingbar, einzelne Bereiche der Compliance gesondert zu betrachten, z.B. Diät- und Medikamentencompliance. Bei Diätverordnungen sollte darauf geachtet werden, diese spezifisch auf den einzelnen Patienten zugeschnitten zu gestalten, um unnütze Restriktionen zu vermeiden (Gardner, 1981).

5. Beziehungen zum Dialysepersonal. Die langdauernde Beziehung und der ständige Umgang des Personals mit den Dialysepatienten führen notwendigerweise zu intensiven gefühlsmäßigen Bindungen und Reaktionen. Das Ausmaß der Bedrohung der Patienten und der Komplikationen belasten Schwestern und Pfleger. Kaplan De-Nour (1983) nennt als wichtigste **Streßfaktoren** im Umgang mit Dialysepatienten: Unkooperatives, feindseliges Verhalten, frustrierte Erwartungen, beispielsweise durch psychische und physische Komplikationen, Zweifel über den Sinn der Behandlung, Aggression und Spannung aus einer unbefriedigenden Interaktion mit dem Patienten.

Die Reaktionen des Personals entsprechen den vorherrschenden Abwehr- und Anpassungsprozessen der einzelnen Mitglieder und der Gruppe. Sie können sich sowohl in übertriebener Fürsorge, Betroffenheit wie in Ärger und Zurückweisung der unkooperativ erscheinenden Patienten, in einer Indifferenz als Ausdruck eines gefühlsmäßigen Rückzugs äußern. Dieser Rückzug entspricht der Abwehr eigener Gefühle von Hilf- und Hoffnungslosigkeit (vgl. Abram, 1972). Fehlverhalten und emotionale Fehlreaktionen resultieren häufig aus Mißverständnissen. Die Kooperation eines Patienten, ebenso wie seine aggressiven Verhaltensweisen, kann als persönliche Kränkung und Frustration des engagierten Personals erlebt und so verarbeitet werden, daß die Beziehung zum Patienten gestört wird oder aber die eigene Belastung bei der Arbeit unerträglich wird. Beispiele sind eine Verleugnung der Aggressivität und der fehlenden Compliance des Patienten. Führt die Enttäuschung zum Rückzug, zur Distanzierung und Abwendung, bleibt der Patient in einer kritischen Situation alleingelassen. Das verstärkt seine maladaptiven Verhaltensweisen in der Regel. Oft läßt sich dann eine noch gesteigerte Fürsorglichkeit und Opferbereitschaft des Per-

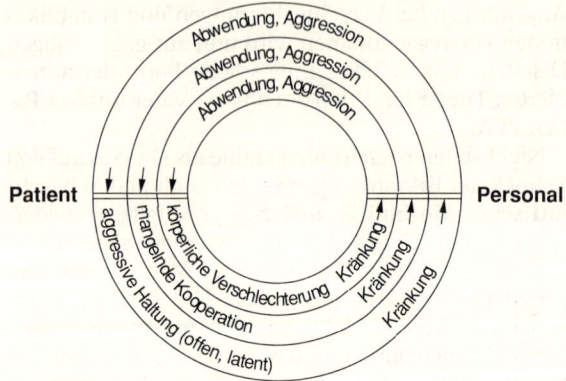

Abb. 65–1. Einige Möglichkeiten pathologischer Interaktion von Dialysepatienten und Stationspersonal.

sonals als Reaktionsbildung auf eigene aggressive Regungen beobachten, die zu weiterer Belastung und latenter Unzufriedenheit führen muß (Abb. 65–1).

In einer Fragebogenuntersuchung, die sich an das Personal von Dialysestationen wandte, wurde die überwiegende Einstellung des Dialyseteams als freundlich-akzeptierend und mütterlich-protektiv befunden, was unter Berücksichtigung der schwierigen und oft durch Konflikte komplizierten Behandlungssituation als weiterer Hinweis für die Beteiligung von Abwehrvorgängen angesehen wurde (Balck et al., 1983, 1984). Häufig ist auch eine Verschiebung oder projektive Verarbeitung aktivierter Aggressionsgefühle zu beobachten, was zu vordergründig unverständlichen Reaktionen, Spannungen und Auseinandersetzungen innerhalb des Teams Anlaß geben kann. Die Abwehr aggressiver Impulse durch Wendung gegen sich selbst kann zu depressiven und psychosomatischen Reaktionen beim Personal führen. Die Belastung kann sich auch in häufigen Angstträumen ausdrükken (Foy, 1970).

Eine häufige Sequenz einzelner Reaktionen sei im folgenden beschrieben: Ist die Verleugnung maladaptiven Krankheitsverhaltens des Patienten nicht aufrechtzuerhalten und mißlingt es, Ärger und Enttäuschung zu unterdrücken, resultieren Schuldgefühle, auf die das Personal häufig mit Vermeidung des Patienten antwortet. Alleingelassen, wird dieser um so feindseliger und vorwurfsvoller, was Wut und Enttäuschung der Behandelnden verstärkt. Dabei kommt es oft zu Spaltungsvorgängen. Psychogenetisch spricht man von Spaltung, wenn primitive Aggression es verhindert, daß ausschließlich gute und böse Selbst- und Objektrepräsentanzen in ein Konzept des Selbst und der Objekte mit guten und bösen Teilrepräsentanzen integriert werden. Die Identifikation mit den Patienten und ihren gespaltenen Selbst- und Objektrepräsentanzen, Schuldgefühle wegen eigener aggressiver Regungen oder Versagen kann beim Personal dazu führen, die Patienten in „gute" und „böse" aufzuteilen. Derartige Beziehungskonflikte können für die Versorgung der Patienten und den weiteren Erkrankungsverlauf folgenreich sein.

Foster und Mitarbeiter (1973) wiesen in einer prospektiv angelegten Untersuchung darauf hin, wie die Einschätzung zweier ansonsten vergleichbarer Gruppen von Dialysepatienten als „gut" oder „schlecht" durch das Team mit der späteren Todesrate korrelierte. In der einen Gruppe starben 7 Patienten in 2 Jahren, in der anderen Gruppe kein Patient.

Selbstgefährdende Verhaltensweisen der Patienten vermögen infolge der engen emotionalen Verflechtung mit dem Personal bei diesem zu ausgeprägten Ohnmachtsgefühlen zu führen und latente Todesängste zu aktivieren. Die Patienten können Frustration und Ärger, die beispielsweise ihrer Familie, der Erkrankung oder der Behandlungssituation gelten, auf das Personal richten, wobei häufiger Schwestern oder Pfleger als die Ärzte betroffen sind. Auf die Ambivalenz eines Patienten, die Dialysebehandlung fortzusetzen, reagieren Schwestern oder Pfleger nicht selten mit panischem Rückzug und Ärger, um der Diskussion über den Wunsch eines Patienten, zu sterben, aus dem Weg zu gehen (Selvin, 1983). Dabei erlauben professionelle Distanzierungstechniken es den Ärzten eher, den Todeswunsch eines Patienten zu verleugnen. Nicht selten führt das zu Spannungen innerhalb des Teams, wenn der bedrohlich erlebte Patientenwunsch nicht offen besprochen, sondern dieser zum Anlaß heftiger medizinischer Kontroversen über Detailprobleme genommen wird. Eine Verschlechterung des Behandlungsklimas wirkt sich oft negativ auf die Kooperation und das Befinden der Patienten aus (vgl. Pat. B).

Die Belastungen von Dialyseschwestern und -pflegern finden u. a. auch in ihren Träumen Ausdruck, z. B. Träumen, daß bereits verstorbene, früher von ihnen betreute Dialysepatienten wieder aus ihren Gräbern auferstehen würden.

Zur gegenseitigen Unzufriedenheit mag beitragen, daß oft **unterschiedliche Vorstellungen und Erwartungen** auf seiten der Patienten und des Personals existieren, die pathologisch überdeterminiert sind. Eine Reihe von Untersuchungen haben die Verläßlichkeit von Einschätzungen des Teams über das Befinden von Dialysepatienten problematisiert (Kaplan De-Nour und Czaczkes, 1971, 1972, 1974). Insbesondere das subjektive Empfinden von Patienten wurde vom Team häufig im Sinne einer Tendenz zur Abwehr und euphemistischen Beurteilung der Lage der Patienten falsch eingeschätzt.

Besteht eine ausgeprägte Inkongruenz der Meinungen des Teams über Patienten, soll dies einen negativen Einfluß darauf haben, wie die Patienten mit der Dialysebehandlung zurechtkommen (Kaplan De-Nour et al., 1972b). Umgekehrt soll eine homogene Einschätzung der Behandelnden das Anpassungsverhalten von Dialysepatienten günstig beeinflussen.

Ärzte sollen häufiger im Team zu Rückzug aus Schuldgefühlen, Schwestern und Pfleger zu Überprotektion von Patienten, zum Besitzdenken neigen (Kaplan De-Nour und Czaczkes, 1968a). Infolge der engen Bindung und Abhängigkeit werden immer wieder heftige Trennungsreaktionen beim Weggang eines Arztes oder einer betreuenden Schwester mit Depression und Selbstaufgabetendenzen beobachtet. So führte der Urlaub des betreuenden Ambulanzarztes der Patientin A zu einer Verschlechterung ihrer Mit-

arbeit. Cramond (1970) beschrieb die Apathie, die auf vielen Dialysestationen zu beobachten ist, und bezog sie auf die Situation einer unterdrückten Rebellion. Zu bedenken ist, daß unkooperatives Verhalten, aber auch somatische Symptome wie Juckreiz, Schlaflosigkeit bei Dialysepatienten Folge von Schwierigkeiten im Umgang mit dem Personal sein und appellativen Charakter haben können.

Da sich solche langfristigen und intensiven Beziehungen zwischen Personal und Patient nur in wenigen anderen Bereichen der Medizin finden, verdienen die hier auftretenden Trennungsprobleme und Trauerreaktionen besondere Beachtung.

Calland (1972) und Ellis (1974), beide Ärzte in Dialysebehandlung, haben in engagierter Form kritisiert, wie psychosoziale Probleme bei Dialysepatienten außer acht gelassen werden.

Das Ausmaß und die Komplexität der Probleme in Dialysestationen lassen, ebenso wie die intensiven emotionalen Beziehungen, es geraten erscheinen, für das Personal ein Angebot in Form von Balint-Gruppen oder in ähnlicher Weise konzipierten Teamkonferenzen zu schaffen (Drees, 1976; Leonard, 1981).

65.2.2 Besondere Konfliktbereiche

Abhängigkeit – Unabhängigkeit, Aggression – Unterwürfigkeit

Der Verlust einer Organfunktion führt, je nach Ersatz, zu unterschiedlichen Formen der **Abhängigkeit.** Auf einer unbewußten Ebene kann die „Prothese" von ihrem Träger nach dem Modell früher Objektbeziehungen und deren innerer psychischer Repräsentanz erlebt werden. Die Situation der Hämodialysebehandlung wird dabei von einer Beziehung zu einem intermittierend angewandten äußeren Objekt bestimmt, die für den Patienten Zwang bedeutet und mit Gefühlen von Ohnmacht und Ambivalenz besetzt ist. Abhängigkeit von der Maschine oder der Verlust des Lebens ist zunächst die brutale Alternative. Von dieser Abhängigkeit kann sich der Patient nicht lösen. Die Abhängigkeit bezieht sich dabei sowohl auf die Dialyseapparatur, die Prozedur, als auch auf die an der Behandlung beteiligten Personen.

Zunächst wurde von vielen Autoren Abhängigkeit als das zentrale Problem für Hämodialysepatienten bezeichnet (Kaplan De-Nour, 1968b; Czaczkes und Kaplan De-Nour, 1978). Dies wurde mit zunehmender Erfahrung dahingehend modifiziert, daß Abhängigkeit, je nach Persönlichkeitsstruktur, sehr unterschiedlich erlebt werden und beispielsweise Patienten mit ausgeprägten Abhängigkeitswünschen sehr entgegenkommen kann.

Zur Aktualisierung solcher Wünsche kommt es häufig im Zuge der regressiven Entwicklung, die ein schweres chronisches Organleiden (vgl. Viederman, 1974) begleitet. Besonders eindrucksvoll sind solche Phänomene beispielsweise in der Adoleszenzphase zu beobachten (Ganowsky et al., 1983), wo wir ohnehin häufig einen raschen Wechsel zwischen progressiven und regressiven Vorgängen beobachten können (vgl. Pat. B).

Viele Schrittmacherpatienten vermögen Ängste abzuwehren, indem sie die „Prothese" wie ein idealisiertes inneres Objekt behandeln, analog einer idealisierten Elternfigur beim kleinen Kind, das auf diese Weise Ohnmacht in Allmacht zu verwandeln vermag. Nicht selten ist auch bei Dialysepatienten ein Beziehungsmuster zu beobachten, das den Charakter einer symbiotischen Bindung hat. Kemph (1966) berichtet von einem Dialysepatienten, dessen Angst proportional zur Entfernung vom Dialyseapparat anwuchs. Der Patient blieb, obwohl er nur wenige Häuserblocks vom Dialyseort entfernt wohnte, so lange wie möglich im Krankenhaus. Es läßt sich hier eine Beziehung zur Entwicklung des Trennungsverhaltens kleiner Kinder herstellen, die sich nur kurzzeitig von der Mutter entfernen können und häufig zur Rückversicherung zurückkehren müssen, was als eine Form „emotionalen Auftankens" interpretiert wurde (Mahler et al., 1980).

Auf einen besonderen Aspekt der Abhängigkeits-/Unabhängigkeitsproblematik ist unter Anwendung der **Double-Bind-Hypothese** hingewiesen worden (Bateson, 1983). Danach sind sowohl Erwartungshaltungen als auch Handlungsanweisungen des Dialyseteams oft widersprüchlich (Alexander, 1976), zum Teil lassen sich die direkten Widersprüche auch im gleichzeitigen verbalen und averbalen Verhalten aufzeigen. Vereinfacht ausgedrückt, fordern die Behandelnden vom Patienten, er solle möglichst selbständig, unabhängig sein, sich normal fühlen, dankbar sein, daß ihm durch die Dialyse der Tod im urämischen Koma erspart wurde; gleichzeitig findet sich, meist latent, die Erwartung, daß der Patient Abhängigkeit akzeptiert, gefügig ist, die Annahme, daß er alles andere als wie ein normaler Mensch leben kann und Opfer einer schrecklichen Erkrankung ist. Daraus resultieren paradoxe Erwartungen und Handlungsanweisungen, denen sich der Patient, der ja an die Situation gewissermaßen gekettet ist, nicht entziehen kann und die zu entsprechenden pathogen wirksamen Lernstörungen führen können.

Reichsman und Levy (1974) beobachteten bei Patienten, die mit der Forderung nach Wiederaufnahme einer beruflichen Tätigkeit konfrontiert wurden, heftige Konfliktreaktionen, die sie auch auf den Ambivalenzkonflikt zwischen passiver Hingabe und Abhängigkeit, wie sie dem Dialyseverfahren entspricht, und den Erwartungen der Umgebung, im täglichen Leben eine aktive und autonome Rolle zu spielen, zurückführten.

Die **Auswirkungen** des **Konfliktes** um **Abhängigkeit** und **Unabhängigkeit** sind, je nach der Behandlungsphase, dabei unterschiedlich: Verallgemeinert ist anzunehmen, daß Patienten mit einer besonderen Tendenz zum Ausleben von Unabhängigkeitsstrebungen am Anfang der Dialysebehandlung die größten Probleme haben, Patienten mit Abhängigkeitstendenzen eher in der Phase der Rehabilitation (Blodgett, 1981/82). Patienten in der Heimdialyse oder Patienten, die mittels der chronischen ambulanten Peritonealdialyse (CAPD) behandelt werden, sind weniger durch den Verlust ihrer Autonomiebedürfnisse bedroht (Baum et al., 1982).

Immer wieder ist in der Literatur (Kaplan De-Nour und Shanan, 1980; Freyberger, 1984) auf die Bedeutung der **Aggression** für das Entstehen vieler Probleme von Dialysepatienten aufmerksam gemacht worden. Die vielfältigen Abhängigkeitsbeziehungen stellen dabei eine wichtige Quelle aggressiver Gefühle dar.

Klinisch führt die projektive Verarbeitung der Aggression eher zu Schwierigkeiten mit der Umgebung. Stehen die introjektiven Mechanismen der Verarbeitung im Vordergrund, so ist eine Neigung zu depressiven Erlebnisweisen anzunehmen. Ein hohes Aggressionsniveau führt zu einer Zunahme psychischer Abwehrprozesse mit der Tendenz zur entsprechenden Nivellierung von Affekten (Glassman und Siegel, 1970). Dies dürfte zum beobachteten Phänomen der „sekundären Alexithymie" (Freyberger, 1984) beitragen.

Koch et al. (1984) stellten fest, daß Patienten mit längerer Dialysebehandlung depressiver, ungeselliger, zurückgezogener sind und ein geringeres Selbstvertrauen haben. Die Hypothese einer Beziehung dieses Phänomens zur depressiv verarbeiteten Dialysesituation ließ sich dabei testpsychologisch stützen: Patienten mit einer längeren Dialysedauer wenden die Aggression eher gegen die eigene Person (standardisierter Aggressionsfragebogen, SAF) und nicht gegen äußere Objekte. Sie sind depressiver und klagen mehr über psychosomatische Beschwerden.

Compliance von Dialysepatienten

Darunter verstehen wir die Fähigkeit und Bereitschaft von Patienten, sich an einen Therapieplan zu halten und aktiv an der Wiederherstellung der Gesundheit mitzuarbeiten. Zur Compliance gehören das Befolgen ärztlicher Anordnungen bezüglich Medikamenten und Diät, aber auch der allgemeinen Lebensführung, das Vermeiden von schädlichen Aktivitäten und allgemein die Kooperation bei der Behandlung, z.B. auch das pünktliche Erscheinen zur Dialyse.

Die Compliance ist dabei, wie sich aus vielen Untersuchungen zeigen läßt, in ihren unterschiedlichen Dimensionen häufig sehr verschieden ausgeprägt und jeweils von anderen Bedingungsfaktoren beeinflußt. Am häufigsten beschäftigten sich die Untersucher mit der medikamentösen und Diät-Compliance. Deren gravierende Mängel lassen oft Behandlungen scheitern und beeinflussen das Überleben auch von Dialysepatienten ungünstig (Calsyn et al., 1978).

Die Situation der Hämodialyse ermöglicht es, die Kooperation von Patienten exakt zu beobachten (strikte Bindung an ein Zentrum, Objektivierbarkeit von Fehlern durch Gewichts- und Elektrolytschwankungen).

Kaplan De-Nour und Mitarbeiter (1972a und 1974b) haben in einer der wenigen multizentrischen Studien zu diesem Problem die Kooperation von Dialysepatienten und ihre Beziehung zu einzelnen Persönlichkeitsmerkmalen untersucht (Abb. 65–2).

Bei folgenden Persönlichkeitszügen und situativen Merkmalen waren Diätfehler signifikant häufiger:
- niedrige Frustrationstoleranz mit prämorbiden Verhaltensmustern von Ungeduld, Zuspätkommen, schnellem Mißmut;
- Ausagieren aggressiver Tendenzen im Sinne einer mangelnden Fähigkeit zur Kontrolle aggressiver Impulse, z.B. Diätfehler während des Urlaubs des betreuenden Arztes;

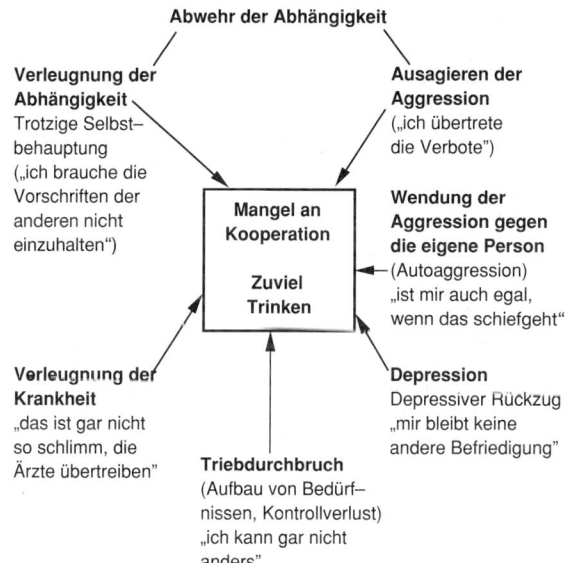

Abb. 65–2. Wichtige Anpassungsvorgänge, die zu gestörtem Krankheitsverhalten (zuviel trinken) beitragen können (vgl. Kaplan De-Nour und Czaczkes, 1974 b).

– Wendung aggressiver Gefühle gegen die eigene Person, beispielsweise im Rahmen depressiver und suizidaler Impulse;
– Ausagieren von Unabhängigkeitsstrebungen im Sinne: „Ich lasse mir nicht alles vorschreiben";
– sekundärer Krankheitsgewinn.

Auch eine bestimmte Form der Verleugnung (vgl. Yanagida und Streltzer, 1979) wird als prädisponierender Faktor genannt, ebenso wie depressive Verstimmungen (Kaplan De-Nour und Czaczkes, 1976; Procci, 1981), das Bestehen intrafamiliärer Konflikte und Spannungen im Team (Kaplan De-Nour, 1983).

Oft sind Compliancefehler nur aus der komplizierten individuellen Familienpathologie verständlich, wie bei Patienten, die nach erfolgreicher Transplantation Immunsuppressiva absetzten, gezeigt wurde (Armstrong und Weiner, 1981/82).

Dialysepatienten spüren beispielsweise im Unterschied zu Hypertoniepatienten meist die negativen Folgen ihrer mangelnden Kooperation direkt. Patientin A beantwortete Streßsituationen, Komplikationen und erlittene Kränkungen mit Verweigerung der Kooperation. Sie zeigte dabei zunehmend ein infantil-trotziges Verhalten. Ihr Eindruck, vom Schicksal ungerecht behandelt und geschlagen zu sein, ihre eingeschränkte Frustrationstoleranz und die im Verlauf der regressiven Entwicklung verstärkten oralen Bedürfnisse kollidierten mit den Mindestanforderungen einer Diätbehandlung (Abb. 65-2).

Wiederholte gravierende Compliancefehler machen bei Dialysepatienten eine psychotherapeutische Intervention dringend erforderlich. Sie müssen auch die Planung des geeigneten Therapieverfahrens bei terminaler Niereninsuffizienz beeinflussen. Wichtig ist, daß, je nachdem welche Dimension der Compliance gestört ist, unterschiedliche psychodynamische Faktoren zu berücksichtigen und andersgeartete therapeutische Strategien zu verfolgen sind (Diät, Flüssig-keit, Medikamente, Verhalten bei den Dialysen, Pünktlichkeit etc.). Compliancefehler können Ausdruck des Bedürfnisses von Patienten sein, die der depressiven Stimmung zugrundeliegenden aggressiven Gefühle zu äußern (Procci, 1981). Eine wesentliche Rolle spielen auch Krankheitsphase und Alter der Patienten.

Stoudemire und Thompson (1983) empfehlen bei Compliancemängeln ein diagnostisches und therapeutisches Stufenprogramm, das eine psychiatrische Evaluation des Patienten und seiner Umgebung und entsprechende individuum-, familien- und teamzentrierte Interventionen ebenso einschließt wie eine sorgfältige Verhaltensanalyse des Patienten hinsichtlich seiner Fehler.

Suizidale Handlungen

Gelegentlich werden in der Literatur gravierende Diätfehler generell als Ausdruck suizidaler Impulse bei Dialysepatienten interpretiert. Dies ist sicherlich nicht gerechtfertigt. Zum Teil dürften schwere Diätfehler, die bei depressiven Dialysepatienten häufiger sind, Ausdruck suizidaler Wünsche sein, doch sollten zur diagnostischen Klärung eine Reihe weiterer Kriterien wie Interessenverarmung, Aufgabe von Zukunftsplänen und -gedanken, auch von Beziehungen, berücksichtigt werden. Zu bedenken ist, daß Dialysepatienten der Suizid besonders leicht zugänglich ist.

65.2.3 Psychische Anpassungsprozesse bei Dialysepatienten

Zahlreiche Autoren gehen aufgrund von klinischer Erfahrung, Fallstudien und klinischem Rating davon aus, daß Dialysepatienten, u.a. aufgrund ihrer hohen krankheitsbezogenen Belastung, in besonderem Maße Abwehrmechanismen einsetzen (Armstrong, 1978; Beard und Sampson, 1981; Freyberger, 1980; Norton, 1969; Reichsman und Levy, 1977; Wilson et al., 1974; vgl. Kap. 62). Darin nimmt **Verleugnung** einen zentralen Stellenwert ein, oft als einzige untersuchte Strategie der Adaptation (zusammenfassend Blodgett, 1981/82). Testpsychologische Untersuchungen ergaben zum Teil widersprüchliche Ergebnisse.

Häufig wird in derartigen Untersuchungen auf die Anwesenheit von Verleugnung geschlossen, wenn Reaktionen von Patienten auf Persönlichkeitsinventarien sich nicht von der Normalbevölkerung unterscheiden, die Autoren jedoch der Ansicht sind, die Patienten lebten unter schwerwiegendem Streß und daher sei das Auftreten von psychopathologischen Reaktionen zu erwarten (vgl. Blodgett, 1981/82). Dies ist kein hinreichender Beweis für das Vorliegen von Verleugnung, da Verleugnung nicht direkt gemessen wurde und die Gültigkeit standardisierter Persönlichkeitsinventarien für chronisch Kranke umstritten ist (Yanagida und Streltzer, 1979). Als weitere Abwehrmechanismen werden u.a. **Verdrängung, Reaktionsbildung, Projektion** und **Regression** genannt (Adler, 1972; Viederman, 1974).

Einige neuere Arbeiten betonen die Bedeutung von **aktiven, realitätsorientierten Bewältigungsbemühungen (Coping).** Der Sprachgebrauch ist jedoch uneinheitlich. Viele Autoren (z. B. Adler, 1972; Yanagida et al., 1981) setzen Abwehrmechanismen mit „Coping-Strategien" oder „Coping-Mechanismen" gleich. Wenige Untersuchungen erfassen tatsächlich Coping-Strategien bei Dialyse- und Nierentransplantationspatienten.

Shanan und Mitarbeiter (1976) verglichen eine Stichprobe aus 38 männlichen und 21 weiblichen Patienten mit terminalem Nierenversagen mit einer in bezug auf Alter, Geschlecht, Ausbildung, Herkunft und ehelichem Status „gematchten" Vergleichsgruppe hinsichtlich aktivem und passivem Coping, gemessen mit der Shanan Sentence Completion Technique (SSCT). Sie fanden, daß Nierenpatienten im Unterschied zu Vergleichsgruppen signifikant weniger aktive Problemlösungsbemühungen zeigten. Sie konzentrierten sich stärker auf negative Aspekte des Selbst und in geringerem Maße auf konstruktive interpersonale Beziehungen und instrumentelle Aktivitäten.

In den vergangenen Jahren haben die Konzepte der Kontrollüberzeugung bzw. Kausalattribution zunehmend Anwendung gefunden (vgl. Strickland, 1978; Armstrong und Woods, 1983). Kontrollüberzeugungen beziehen sich auf generalisierte Erwartungen eines Individuums, ob es durch sein eigenes Verhalten wichtige Ereignisse in seinem Leben beeinflussen kann (internale Kontrolle) oder nicht (externale Kontrolle). Es zeigte sich, daß Dialysepatienten eher externale Kontrollüberzeugung haben als gesunde Vergleichsgruppen (Goldstein, 1976; Wenerowicz et al., 1978; zusammenfassend Strickland, 1978). Limited-Care- und Heimdialysepatienten berichteten über ein höheres Ausmaß an internaler Kontrolle als Zentrumsdialysepatienten (Devins et al., 1981). Es zeigte sich, daß internale Kontrollüberzeugung einen positiven Zusammenhang zur Compliance aufweist (Poll und Kaplan De-Nour, 1980; Wenerowicz et al., 1978; Strickland, 1978).

Trotz inhaltlich-konzeptueller und methodischer Mängel zeigt die gegenwärtige Forschung, daß Dialysepatienten ein breites Spektrum von Abwehr- und Bewältigungsstrategien anwenden, um mit den beträchtlichen Belastungen umzugehen, die mit ihrer Krankheit verbunden sind.

Gemessen an dem gegenwärtigen Stand der Theoriebildung des Coping-Konzepts (vgl. Cohen und Lazarus, 1980), ist der Forschungsstand in diesem Bereich jedoch als wenig fortgeschritten zu bezeichnen.

65.2.4 Determinanten der Adaptation

In der Literatur werden eine Reihe von **psychischen Determinanten** von Adaptationsprozessen diskutiert. Diese sind teils globaler Natur und versuchen, Anpassung im allgemeinen vorherzusagen (z. B. Viederman, 1974), teils wird versucht, in quasi-prospektiven Untersuchungen Anpassungen in spezifischen Bereichen vorherzusagen (z. B. Kaplan De-Nour,

1981; vgl. Kap. 62). Als wichtige Faktoren der Adaptation werden genannt:

– **Frühkindliche Sozialisation des Patienten** (z. B. Carey, 1976; Viederman, 1974). Viederman führt in Kasuistiken Adaptationsprozesse auf befriedigende bzw. unbefriedigende Mutter-Kind-Beziehungen zurück.

– Faktoren der **prämorbiden** oder **aktuellen Persönlichkeit:**

Beard und Sampson (1981) nennen als günstige Voraussetzungen der Adaptation die Fähigkeit zu befriedigenden, tiefen Objektbeziehungen, Levy und Wyndbrandt (1975) gute Objektbeziehungen und Sublimierungsfähigkeiten. Malmquist (1972) betont die Bedeutung des Fehlens neurotischer Beschwerden vor Dialysebeginn. Gegen die Annahme, daß die Adaptation an die Dialysebehandlung überwiegend durch die sog. prämorbide Persönlichkeitsstruktur determiniert wird, spricht aber u. a. die Untersuchung von Streltzer und Mitarbeitern (1977). Die Autoren fanden, daß selbst bei schweren vorbestehenden psychiatrischen Erkrankungen gute Langzeitdialyseresultate erzielt werden konnten. In einer quasi-prospektiven Untersuchung fanden Hagberg und Malmquist (1974) als positive Prädiktoren der Adaptation: stabile Persönlichkeitsstruktur, Konstanz des Lebensstils, regelmäßige soziale Außenkontakte, das Überwiegen reiferer Abwehrmechanismen (z. B. Verdrängung anstelle von Isolierung), als negative Prädiktoren den Mangel an Sozialkontakten, eine negative Reaktion auf die Nierenerkrankung sowie schlechte berufliche Chancen. **Intelligenz** korrelierte nur mit der Rehabilitationsgeschwindigkeit, nicht jedoch mit dem erreichten Rehabilitationsniveau. Auch in anderen Untersuchungen hat Intelligenz nur einen geringen prädiktiven Wert (zusammenfassend Kaplan De-Nour, 1984).

– Von zahlreichen Autoren wird die Bedeutung der **Vorerfahrung** in der **Bewältigung von belastenden Lebensereignissen** und **Krankheit** betont (Hagberg und Malmquist, 1974; Carey, 1976; Anderson, 1975; Joel und Vieder, 1973).

– Als wesentliche Determinante der Adaptation erweist sich zunehmend die **Unterstützung** des Patienten **durch sein soziales Umfeld** (Maurin und Schenkel, 1976; Palmer et al., 1983), vor allem durch eine tragfähige Partnerbeziehung (Speidel et al., 1979).

65.2.5 Sozio- und psychotherapeutische Maßnahmen bei Dialysepatienten

Die Vielfalt psychologischer, sozialer und psychiatrischer Probleme bei Patienten mit chronischer terminaler Niereninsuffizienz macht es notwendig, ein Angebot an psychiatrischen, psycho- und soziotherapeutischen Hilfen zur Verfügung zu stellen. Psychosoziale Hilfsangebote sind nur partiell und ansatzweise in Dialysezentren verwirklicht. In der Regel sind systematische psycho- und soziotherapeutische Betreuungsangebote ausschließlich im Rahmen oder im Zusammenhang mit universitären Einrichtungen und Forschungsvorhaben realisiert. Die Hindernisse, die einer weiteren Verbreitung und systematischen Integration im Wege stehen, sind vielfältig, so z.B. beschränkt der wachsende Kostendruck personell aufwendige Maßnahmen.

Strain (1981) nennt einige wesentliche Hemmnisse für eine effizientere psychologische Betreuung:

- die technologische Orientierung in den Dialyseeinheiten;
- eine häufig mangelnde psychosoziale Kompetenz der Nephrologen;
- eine oft einseitige psychologische, biologische Phänomene wenig berücksichtigende Orientierung von Psychologen und Psychotherapeuten;
- eine Aversion vieler Patienten gegen psychologische Hilfsangebote.

Organisationsformen

Als Organisationsformen bieten sich ein Konsultationsmodell und ein am Liaison-Service orientiertes Modell an (vgl. Kap. 28). Im ersten Fall wird der Psychosomatiker, Psychiater oder Psychologe konsiliarisch im Bedarfsfall zugezogen, im zweiten Fall arbeitet ein psychologisch geschulter Kollege als Voll- oder Teilzeitmitarbeiter direkt auf der Station mit. Dabei bietet ein Liaisonansatz den Vorteil kontinuierlicher Arbeit auf einer Dialyseeinheit, die Möglichkeit, Patienten frühzeitig, bevor sie eine besondere psychische Auffälligkeit entwickeln, zu untersuchen und von Anfang an einen Kontakt zu ihnen herzustellen (Famularo und Kimball, 1983; Newman, 1983; Bloom, 1981; Freyberger et al., 1979). Dies erleichtert den Aufbau einer therapeutischen Beziehung bei Komplikationen und Krisen und vermindert die Traumatisierung des Patienten, die dann entsteht, wenn bei Schwierigkeiten ein ihm unbekannter Spezialist für psychologisch-psychiatrische Probleme hinzugezogen wird. Es ermöglicht schon präventiv therapeutisch tätig zu werden. Der Liaisonansatz bietet bessere Möglichkeiten, dem Dialyseteam den Blick für psychologische Zusammenhänge zu schärfen. Besonders bewährt haben sich dabei regelmäßige Teamkonferenzen. Angesichts der Vielfalt sozialer Probleme ist die Mitarbeit eines Sozialarbeiters dringend notwendig.

Psychotherapie bei Dialysepatienten

Form und Zeitpunkt psychotherapeutischer Interventionen bei Dialysepatienten werden von der Zielsetzung, z.B. der Frage, ob es sich um eine akute Belastungs- oder Konfliktreaktion handelt oder um das Resultat einer Entwicklungsstörung, von der Krankheitsphase und von spezifischen Indikationen bestimmt. So können beispielsweise Zeichen einer psychoneurotischen Störung, die den Krankheitsverlauf ungünstig beeinflußt, Suchtverhalten, suizidale Tendenzen, Komplikationen oder sonstige Behandlungskrisen Anlaß zu therapeutischen Interventionen geben, ebenso wie Störungen des Krankheitsverhaltens auf dem Hintergrund ungelöster Konflikte (Unkontrolliertheit versus Kontrolle, Autonomie versus Anklammerung, Emotionalität versus rationalistische Tendenzen). In den letzten Jahren wird die Rolle des Schmerzmittelmißbrauchs in der Entstehung der

chronischen Niereninsuffizienz zunehmend mehr Beachtung geschenkt. Personen mit chronischen Schmerzsyndromen und entsprechendem Abusus zeigen schon prämorbid ein auffälliges Persönlichkeitsprofil mit einer Neigung zu maladaptiven Verhaltensweisen. Es ist davon auszugehen, daß sich daraus vielerlei Komplikationen in der späteren Behandlung herleiten. Psychotherapeutische Ansätze müssen der Suchttendenz dieser Patienten Rechnung tragen.

Grob schematisch läßt sich hinsichtlich der wichtigsten **Problembereiche** folgende Einteilung **psychotherapeutischer Indikationsbereiche** treffen:

- psychoreaktive Störungen im Gefolge der Nierenerkrankung
- Beziehungsprobleme mit Angehörigen
- Beziehungsprobleme mit dem Personal
- Noncompliance.

Es existiert eine große Anzahl von Berichten über psychotherapeutische Behandlungsversuche mit Dialysepatienten. Es handelt sich jedoch meist um kleine Fallzahlen ohne randomisierte Kontrollgruppe, so daß allenfalls tendenzielle Wertungen angebracht sind. Nach Kaplan De-Nour und Czaczkes (1976) beispielsweise ließen sich in einer Einheit mit ständiger psychiatrischer Betreuung psychologische Komplikationen präziser vorhersagen als in einer Dialysestation ohne konsiliarische Versorgung.

In einem supportiven therapeutischen Ansatz kommt neben der Möglichkeit zur kathartischen Abreaktion der Präsenz des Therapeuten eine besondere Funktion zu. Er kann sich als Objekt, von dem der Patient nicht vital abhängig ist, für die narzißtischen Bedürfnisse der Patienten anbieten, sie zu gefühlhaft differenzierten Äußerungen anregen. Beispielsweise kann dazu gehören, offensichtlich maladaptives Verhalten zu problematisieren und nach seiner psychodynamischen Bedeutung zu fragen, ebenso Fehlvorstellungen hinsichtlich des Körperschemas und unbegründete Ängste, z.B. bezüglich Vererbbarkeit von Nierenkrankheiten, zu korrigieren und Abwehrfunktionen der Patienten einzuschätzen.

Im Falle eines ausreichenden Konfliktbewußtseins und genügender Fähigkeiten zur Selbstreflexion sollte angestrebt werden, oberflächen- und bewußtseinsnahe Konflikte mit dem Patienten zu bearbeiten. Dabei sollte der besonders in der Anfangsphase häufig starke Widerstand von Patienten berücksichtigt und gegebenenfalls vorübergehend auf eine weitere Konfliktbearbeitung verzichtet werden (Kaplan De-Nour, 1968b; Hollon, 1972). Möglicherweise ist die Abneigung vieler Patienten, psychotherapeutische Angebote anzunehmen, eine Folge der krankheitsimmanenten regressiven Entwicklung. Insgesamt sind die Voraussetzungen für eine konfliktbearbeitende Psychotherapie häufig nicht günstig, da die chronische körperliche Erkrankung und die Behandlungssituation wenig Freiheitsgrade für die Umstrukturierung der Persönlichkeit bieten, so daß Stützung, Ermutigung und direktes Eingreifen im Vordergrund stehen. Allerdings ist auch bei diesen Patienten die flexible Handhabung des Wechsels zwischen der stützenden oder mehr Einsicht vermittelnden Intervention vorzuziehen. Dieser Wechsel muß sich insbesondere auch an den einzelnen Krankheitsphasen und eintretenden Krisen und Komplikationen orientieren.

Bei der psychoanalytischen Behandlung von Patienten mit einer chronischen und letztlich meist zum Tode führenden Erkrankung wie der terminalen Niereninsuffizienz treten spezifische Gegenübertragungsreaktionen auf. Die zumeist ausgesprochenen Kontraindikationen dürften unbewußt auch der Verlustangst der Therapeuten entsprechen. Andererseits mag der besonders starke Wunsch, einen solchen Patienten in analytische Behandlung zu nehmen, auch Rettungsphantasien und narzißtische Wünsche einschließen.

Eine besonders intensive Betreuung, die auch die Eltern einzubeziehen hat, ist bei der psychotherapeutischen Behandlung von **Kindern** und **Jugendlichen** mit terminaler Niereninsuffizienz nötig (Drotar und Ganovsky, 1976, 1981). Analog zu den Formen der Kinderpsychotherapie sind besondere Modifikationen der psychotherapeutischen Techniken vonnöten, die den spielerischen Umgang einbeziehen (Sampson, 1981). Auf dem Hintergrund der Ohnmacht der Eltern sind Schuld- und Enttäuschungsreaktionen der Eltern besonders häufig (Wolff et al., 1984). Dabei ist, analog zu anderen Erkrankungen und Gebrechen, die Einstellung der Eltern zum Defekt besonders wichtig für das kindliche Erleben und die Entwicklung der Persönlichkeit (Castelnuovo-Tedesco, 1981). Psychotherapeutische Interventionen bei adoleszenten niereninsuffizienten Patienten müssen den besonderen phasenspezifischen Konflikten und Problemen dieser Altersgruppe Rechnung tragen (Ganowsky et al., 1983).

Familien- und paartherapeutische Ansätze

Die Erkrankung eines Patienten an einer Niereninsuffizienz verändert meist die Homöostase im Familiensystem (vgl. Kap. 21). Auch eine therapeutische Intervention kann als bedrohlich erlebt und von den Angehörigen mit Abwehr und Ablehnung beantwortet werden. Wie groß die Widerstände sein können, zeigte sich auch am Beispiel unserer Patientin A, bei der Versuche, ihr zu etwas mehr Autonomie zu verhelfen, mit der Mutter-Tochter-Dyade in Konflikt gerieten und den Widerstand der Familie herausforderten. Neben anderem spricht dieser Umstand für eine systematische Einbeziehung der Familie in die therapeutischen Aktivitäten (Balck, 1982, 1988).

Gruppentherapeutische Ansätze

Über gruppentherapeutische Behandlungsversuche gibt es zahlreiche Berichte. Sie werden, vor allem im Hinblick auf den Gruppeneffekt und den Erfahrungsaustausch, häufig relativ günstig beurteilt (Wijsenbeck und Munitz, 1970; Hollon, 1972; Buchanan, 1975, 1981; Sorensen, 1972; Steinglass et al., 1982; Campbell und Sinha, 1980; Bolm et al., 1979). Dabei ist es sicher nicht unproblematisch, die emotionale Nähe von Patienten mit begrenzter Lebenserwartung zu fördern. Da Patienten häufig mit der Erkrankung

und deren Behandlung thematisch befaßt sind, empfiehlt sich die Teilnahme eines Arztes. Buchanan (1981) verweist darauf, daß mit Gruppen mit mehr pädagogischer Zielrichtung weniger Schwierigkeiten auftraten als mit Gruppen, deren Ziel primär an der Konfliktbearbeitung ausgerichtet war. Allerdings sollten gruppendynamische Prozesse, z. B. wenn aggressive Affekte einzelner Patienten gegen das Personal Autoritätsängste in der Gruppe auslösen, nicht vernachlässigt werden.

Externe Supervisionsmöglichkeiten und eine langfristige Etablierung der Gruppen sind von Wichtigkeit. Als therapeutisch fruchtbar erwies es sich (Buchanan, 1981), in einer ersten Phase die Informationsvermittlung und die Förderung der Kommunikation, auch unter Einbeziehung der Angehörigen, in den Mittelpunkt zu stellen, in einer späteren Phase mit motivierten Patienten der Gruppendynamik mehr Aufmerksamkeit zu widmen (vgl. auch Steinglass et al., 1981).

Andere Therapieformen

Als weitere therapeutische Möglichkeiten sind das autogene Training (Lohmann, 1973) und die Hypnose (Scott, 1973; Surman und Tolkoff-Rubin, 1984) zu nennen (vgl. Kap. 22). Bei Sexualstörungen wurden spezifische, an Masters und Johnson ausgerichtete Techniken (Berkman, 1978) erprobt. Bei Kopfschmerzen, Verspannungen und Schlafstörungen können auch verhaltenstherapeutische Techniken angewendet werden (Basler et al., 1979). Zu erwähnen sind dabei insbesondere Selbstkontrollansätze, vergleichbar der Verhaltenstherapie bei Adipösen und Alkoholikern.

65.2.6 Besonderheiten bei Heimdialysepatienten

Statistisch ist die Überlebensquote von Heimdialysepatienten etwas höher als bei Patienten in der Zentrums- und Praxisdialyse, was sicherlich auch Folge der Selektion ist. In somatischer Hinsicht bietet die Heimdialyse eine Reihe von Vorteilen. Körperliche Komplikationen erweisen sich in der Regel nicht als wesentliches Hindernis. Die berufliche Rehabilitationsquote von Heimdialysepatienten wird als im Durchschnitt besser bezeichnet als die von Patienten in der Zentrumsdialyse (Asaba et al., 1971; Schoeppe et al., 1971).

In psychologischer Hinsicht verlangt die Heimdialyse durch den Verbleib im häuslichen Milieu weniger Anpassungsleistungen, ermöglicht den Patienten mehr Autonomie, liberalere Diätregeln, bürdet ihnen andererseits, wie auch den Partnern, eine größere Verantwortung auf. Der Patient ist geradezu gezwungen, beim Wechsel von der Zentrums- zur Heimdialyse eine duldend-passive Haltung zugunsten eines möglichst autonomen Umgangs mit der Krankheit und der Behandlungssituation aufzugeben, was im Gegensatz zu den zumeist aktivierten regressiven Bedürfnissen steht (Kessel, 1971; Castro et al., 1973; Albert et al., 1974). Patienten mit langer Praxis in der Zentrumsdialyse scheinen für den Wechsel weniger geeignet. Zum Teil setzen Patienten und Angehörige diesem Wechsel erheblichen Widerstand entgegen und erzwingen eine Rückkehr in das gewohnte Behandlungssetting. Häufige Fehler von Patienten und Partnern können Ausdruck verborgener Widerstände sein (Shimizu und Richardson, 1981). Schon während der Trainingsphase sollte den Patienten möglichst viel Verantwortung übertragen werden. Es konnte beobachtet werden, daß in der Dreiecksbeziehung beim Heimdialysetraining (Patient – Partner – Dialyseschwester/Pfleger) leicht Rivalitätssituationen entstehen können (Speidel et al., 1978). An das betreuende Personal wird die Forderung gestellt, sich anstatt in einer zentralen Rolle als entfernte Berater zu erleben.

Allgemein formuliert, prädisponiert die Situation der Heimdialyse eher zum Auftreten von **familiären Konflikten,** während sich in der Zentrums- und Praxisdialyse häufiger Konflikte mit den Vertretern der jeweiligen Institution abspielen. Die Heimdialysesituation stellt besondere Anforderungen an die Partnerbeziehung. Auch in medizinischer Hinsicht wird der Partner zur wichtigsten Bezugsperson. Verschiebungen familiärer Rollenverteilungen können zu verstärkten Konflikten in der Familie (Swanson und Affliti, 1974) und zur Überlastung des Systems Familie führen (Stewart und Johansen, 1976). So können beispielsweise Partner, die früher vom Patienten sehr abhängig waren, die geforderte Rollenumkehr oft nicht ertragen (Streltzer et al., 1976). Auffallend ist, wie gerade zuvor äußerlich „unauffällig" wirkende Dialysepaare in der Heimdialyse dekompensieren (Levenberg et al., 1978).

Shambaugh und Mitarbeiter (1969) haben bei den Partnern häufiger Angst, Unsicherheit und Depression festgestellt als bei den Patienten selbst. Mit 18 Ehepaaren wurde eine Gruppentherapie durchgeführt. Im Laufe der Behandlung kam es zu einer zunehmenden Verbalisierung der Konflikte, aber auch zum Gewahrwerden eines wachsenden emotionalen Abstandes zum Patienten bis hin zu latenten oder offenen Todeswünschen. Aus solchen Wünschen resultieren nicht selten aggressive Impulse, die paradoxerweise darin münden können, daß sich der Partner auf dem Boden einer Reaktionsbildung besonders aufopferungsvoll und mit akribischer Genauigkeit an der Dialysebehandlung beteiligt.

Allerdings wurden in einer Fragebogenuntersuchung bei Partnern von Heimdialysepatienten weniger Ängste (Koch et al., 1982) festgestellt als bei Partnern von Patienten in anderen Therapiesettings, ebenfalls niedrigere Werte bei den Faktoren Objektverlust, Belastung und depressive Reaktion und eine geringer ausgeprägte Verschlechterung der Sexualbeziehungen (Koch, 1984a, b). Heimdialysepatienten werden als aktiver, optimistischer beschrieben, sie wagten eher körperliche Anstrengungen und interessierten sich aktiver für den Kontakt zur Umwelt. Hinsichtlich der Persönlichkeitsstruktur erwiesen sich im FPI (Freiburger Persönlichkeitsinventar) Patienten und Partner in der Heimdialyse als weniger psychosomatisch gestört, weniger aggressiv, weniger dominant und extravertiert, verglichen mit Patienten und Partnern der Zentrums- und Praxisdialyse (Koch et al., 1982).

Obwohl die Heimdialyse gegenüber der Zentrumsdialyse eine erhebliche Kostenersparnis bedeutet, ist der prozentuale Anteil der Heimdialyse im letzten Jahrzehnt gefallen (Romeis et al., 1983). Fraglich ist, inwieweit daran auch finanzielle Interessen beteiligt sind.

Da zwischen den verschiedenen Dialysesettings und der Transplantation keine wesentlichen prognostischen Unterschiede bestehen, sollte die Wahl der Behandlungsmethode von der Evaluation des Lebensstils, den Rehabilitationsmöglichkeiten, den zu erwartenden Komplikationen und Merkmalen der Familie abhängen (Vollmer et al., 1983). Die Auswahl für die Heimdialyse sollte eine sorgfältige Beurteilung der familiären Situation einschließen. Wenig tragfähige Partnerbeziehungen sind in der Regel als eine Kontraindikation für die Heimdialyse zu werten. Therapieangebote sollten die Partner einbeziehen (Conley et al., 1981).

65.3 Peritonealdialyse

An dieser Stelle soll auch auf die Behandlungsform der Peritonealdialyse eingegangen werden, zum einen, weil dieses Verfahren in den letzten Jahren weltweit zahlenmäßig an Bedeutung gewonnen hat (in Großbritannien bei 20% aller dialysepflichtigen Patienten CAPD, in der BRD mit 2% gering repräsentiert; EDTA-Statistik, 1986), andererseits, weil von einigen Autoren psychologische Vorzüge des Verfahrens besonders betont werden (z.B. Oreopoulos, 1981), und auch besondere Anforderungen an die psychischen Voraussetzungen des Patienten gestellt werden (z.B. Gonsalves-Ebrahim et al., 1982).

Zum Verfahren selbst: Die Möglichkeit, das Bauchfell als Austauschfläche und natürliche Dialysemembran zu verwenden, war erst mit Einführung der Antibiotikatherapie in den 50er Jahren möglich geworden. Trotzdem blieb dieses Verfahren weitgehend beschränkt auf die Indikation der Notfalldialyse bzw. Ausweichdialyse, wenn Gefäßprobleme, Shuntprobleme oder eine Gegenindikation für Heparin bestanden, die die Hämodialyse unmöglich machten oder einschränkten (z.B. bei spätdiabetischem Syndrom). Die Peritonealdialyse wird in zwei Formen durchgeführt: Früher ausschließlich in Form der sog. **intermittierenden Peritonealdialyse (IPD),** wobei der Patient in der Regel stationär oder ambulant jeden 2. Tag 10 bis 15 Stunden mit einem Dialysegerät dialysiert wurde, das Ein- und Auslauf der Dialyseflüssigkeit regelte. Eine weitere Indikation wurde durch den Tenckoff-Verweilkatheter (der die Peritonitisrate senkte) und durch die Entwicklung der **chronisch ambulanten Peritonealdialyse (CAPD)** durch Popovich möglich, bei der der Patient in einer Art Heimdialyse die Dialyseflüssigkeit aus einem sterilen Beutel (2 l Flüssigkeit) einlaufen läßt und alle 4 bis 6 Stunden den Beutel wechselt, so daß über 24 Stunden eine relativ kontinuierliche Dialyse erreicht und damit einer der Hauptnachteile, die geringe Entgiftungseffizienz des Verfahrens, durch erhöhte Dialysezeiten ausgeglichen wird.

Den medizinischen Vorteilen der CAPD (gute Kontrolle der Hypertonie und Anämie, kontinuierliche Dialyse, gute Entgiftung vor allem im Bereich der sog. „Mittelmoleküle" und geringe Kosten) wird auf der anderen Seite die hohe Inzidenzrate gefährlicher Peritonitiden gegenübergestellt, die immer noch relativ häufig einen längeren Klinikaufenthalt, unter Umständen auch einen Abbruch des Verfahrens erforderlich machen können. So stellen auch neuere größere Untersuchungen im amerikanischen und europäischen Raum weitgehend übereinstimmend eine Peritonitis auf ca. 9 Patienten-Behandlungsmonate fest (Oreopoulos, 1981; EDTA-Statistik, 1983).

Weitere medizinische Komplikationen, die auftreten können (nach Oreopoulos, 1980, 1981), bestehen in erniedrigtem Blutdruck, der die periphere Durchblutung verschlechtern kann, und gastrointestinalen Komplikationen bzw. abdominalen Krämpfen infolge des erhöhten intraabdominellen Drucks durch das Dialysat sowie Rückenschmerzen bei vorgeschädigter Wirbelsäule. Die Verbreitung der CAPD als Therapiemethode zeigt regional starke Differenzen und Tendenzen: So werden beispielsweise in Toronto 55% der Heimdialysepatienten mit CAPD behandelt (Oreopoulos, 1981). Dagegen wird die Methode in Europa bei allerdings ansteigender Tendenz seltener eingesetzt.

Psychologische Vorteile der CAPD oder gar eine „signifikante psychologische Besserung" durch das Verfahren (Oreopoulos, 1981) sahen die Vertreter dieser Behandlungsmethode vor allem in der gewonnenen Unabhängigkeit von einer Maschine, der größeren Mobilität und Reisemöglichkeiten sowie in den geringeren diätetischen Einschränkungen im Vergleich zur Hämodialyse.

Psychische Voraussetzungen des Patienten werden vor allem im Hinblick auf die kognitiven Fähigkeiten, die Kooperationsfähigkeit und Compliance gemacht, da das Peritonitis-Risiko als Hauptbedrohung nur durch ein konsequent hygienisches Vorgehen bei Beutel- und Systemwechsel in akzeptablen Grenzen gehalten werden kann. Hier werden unter Umständen sogar zwanghafte Persönlichkeitszüge für vorteilhaft gehalten. Eine eingehende psychologische Voruntersuchung vor allem im Hinblick auf bisheriges Complianceverhalten und kognitive Fähigkeiten wird empfohlen (Demarco, 1980).

Gonsalves-Ebrahim und Mitarbeiter (1982) arbeiten sechs psychologische Bereiche der Selektion heraus: Grundstimmung und Realitätskontakt (Ausschluß von schwer Depressiven und Psychotikern), kognitive Fähigkeiten, Körperbild und entsprechende Einstellungen zur CAPD, frühere Fähigkeiten, mit Krisen umzugehen, Streßbelastbarkeit, Persönlichkeitsstruktur und soziale Unterstützung durch Angehörige bzw. deren Akzeptanz der CAPD.

Neben den für viele niereninsuffiziente Patienten ähnlichen Belastungen durch die Erkrankung werden von einigen Autoren **CAPD-spezifische Belastungen und Probleme** angeführt und zum Teil auch empirisch untersucht, so vor allem die Beeinflussung des Körperschemas durch den Verweilkatheter sowie die Auswirkungen der Behandlungsprozedur auf die Partner- und sexuelle Beziehung. Die Ergebnisse sind hier widersprüchlich (z.B. Lindsay et al., 1980; Singh et al., 1980). Zusätzliche Probleme werden vor allem bei Diabetespatienten unter CAPD beschrieben. Diese Patienten wählen zum Teil die CAPD als Behandlungsform, um damit eventuell ein rasches Fortschreiten der Erblindung aufzuhalten, sind aber insgesamt

besonders belastet durch die oft stark eingeschränkte Sehfähigkeit (die die Durchführung der CAPD nur mit einem verläßlichen Partner sinnvoll erscheinen läßt) und durch die Angst vor einer Verschlechterung derselben. Sie unterliegen einer Reihe zusätzlicher Complianceanforderungen, wie Einhalten einer Diabetesdiät und Durchführung der Insulintherapie (die bei der CAPD auch über das Dialysat erfolgen kann). Die Diabetespatienten unter CAPD gelten so als eine besondere Risikogruppe unter medizinischem und auch psychosozialem Aspekt (z.B. Comty et al., 1974). Ergebnisse von Vergleichsstudien zwischen Hämodialyse und CAPD sind häufig durch den höheren Anteil von besonders risikoträchtigen Diabetespatienten in der CAPD-Stichprobe nur eingeschränkt interpretierbar.

Lindsay und Mitarbeiter (1980) finden keine Unterschiede bezüglich Sexualstörungen, Befürchtungen und Persönlichkeitsvariablen (Depression, Verleugnung, Angst, Selbstwertschätzung und Introversion), stellen aber in ihrer Vergleichsstudie an 150 Hämodialyse- und 89 CAPD-Patienten fest, daß die CAPD als Behandlungsverfahren zwar als weniger belastend erlebt wird, aber unter Umständen stärker mit der Berufsausübung interferiert.

Da das Wissen des Patienten über das Verfahren, seine Erwartungen an bzw. seine Widerstände gegen die Methode wie auch die Qualität des Trainingsprogramms beträchtlichen Einfluß auf den Anpassungsprozeß und die erzielte Behandlungsqualität haben dürften (vgl. auch Jeffrey et al., 1982; Burton et al., 1983), erscheinen die psychologische Vorbereitung des Patienten und die Unterstützung des trainierenden Personals durch psychosoziale Fortbildung als wichtige flankierende Maßnahmen zur Gewährleistung einer guten Adaptation und Compliance. Die Einschätzung des Verfahrens durch die behandelnden Ärzte und die Indikationsstellung hängen stark von deren persönlicher Erfahrung und Vertrautheit mit der CAPD als Behandlungsmethode ab (Muthny et al., 1988). Psychosozialen Faktoren wird bei der Abwägung von Nutzen und Risiko ein hoher Stellenwert eingeräumt, wie sich bei einer Umfrage bei 77 Dialyseärzten ergab.

65.4 Nierentransplantation

65.4.1 Einführung

In der Bundesrepublik Deutschland sind nur 15–20% der Patienten mit terminaler Niereninsuffizienz mit einem funktionierenden Transplantat versorgt. Mit diesem vergleichsweise geringen Anteil der Nierentransplantation an der Behandlung chronisch Niereninsuffizienter liegt die BRD in der europäischen Statistik im unteren Drittel; in Großbritannien sind es 46%, in Norwegen sogar über 66% der Niereninsuffizienten, die mit einem Nierentransplantat leben.

Ein Vergleich der Häufigkeit von Totnierentransplantation mit der von Lebend-(Verwandten-)Transplantation zeigt ebenfalls starke regionale Differenzen schon in der europäischen Statistik. So lag der Prozentsatz der Lebendnierenspenden bei den fast 700 der 1981 in der BRD durchgeführten Nierentransplantationen bei etwa 1%, während beispielsweise in Schweden 23% der Nierentransplantierten das Organ von einem lebenden Verwandtenspender erhalten hatten. Obwohl die Lebendnierenspende – betrachtet man die Transplantatüberlebenszeit – nur für den Fall entscheidende medizinische Vorteile gegenüber der Totnierenspende bietet, in dem Empfänger und Spender eineiige Zwillinge sind, so hat sie doch auch noch den Vorteil der Planbarkeit der Operation (bedeutsam vor allem in den Fällen, wo andere Behandlungsverfahren der chronischen Niereninsuffizienz nicht mehr möglich sind). Andererseits führt die Lebendnierentransplantation zu einer Gefährdung und Belastung einer weiteren Person, was viele Chirurgen vor allem in der BRD zur Zurückhaltung veranlaßt.

Bei der Totnierentransplantation muß eine im Gewebstypus (HLA-System) möglichst weitgehend ähnliche Spenderniere für den Empfänger ausgesucht werden. Dies geschieht über einen zentralen Computer in der EDTA-Zentrale in Leyden/Holland, wo jede neu gemeldete Spenderniere bzw. ihr Gewebstypus mit den Daten der Patienten auf der Warteliste verglichen und der optimale Empfänger ermittelt wird. Wenn abgeklärt ist, ob der potentielle Empfänger erreichbar und transplantabel ist, werden Niere und Empfänger zu einem Transplantationszentrum transportiert, in dem in der Regel innerhalb 24 Stunden nach Entnahme des Organs die Transplantation erfolgt.

65.4.2 Psychische und psychosomatische Aspekte über den Verlauf der Transplantation

Eine Übersichtsdarstellung hierzu findet sich in Tabelle 65–1. (Da phasenhafte Einteilungen [z. B. Chambers, 1982 und Freyberger, 1983] viele Überschneidungen aufweisen und praktisch manchmal schwierig handhabbar sind, wurden hier nur wichtigste Ereignisse des Verlaufs und psychische Reaktionen und Probleme gegenübergestellt.)

Patienten, die zur Nierentransplantation kommen, haben in der Regel eine mehrjährige traumatische Erfahrung mit der Entwicklung der Erkrankung und der Dialysebehandlung hinter sich, selten auch bereits eine vorangegangene Nierentransplantation (1981 waren in der BRD ca. 10% der Transplantationen Zweittransplantationen; EDTA, 1983).

Viele Patienten verbinden mit dem Gedanken an die Transplantation große Hoffnungen und Erwartungen, die sich auf den Rückgang der physischen Belastung, Unabhängigkeit von der Maschine, das Wegfallen vieler Zwänge bezüglich Zeit, Ort und Ernährungsweise beziehen und hoffen, nach der Transplantation wieder ein aktiveres Leben (berufliche Möglichkeiten, familiäre Aktivitäten, Freizeitgestaltung, Reisemöglichkeiten usw.) führen zu können. Für diese Patienten ist die Wartezeit oft sehr belastend, zumal sie mit der Anspannung des ständig Verfügbar-Sein-Müssens verbunden ist.

Für andere Patienten, die sich zwar für die Transplantation entschieden haben, aber gleichzeitig Angst vor der Operation bzw. den damit verbundenen Schmerzen haben, kann eine ausgesprochen ambiva-

Tabelle 65–1. Psychische Probleme im Umfeld der Transplantation.

Ergebnisse im Verlauf	Mögliche psychische Reaktionen und Probleme
Entscheidung zur Transplantation	Entscheidungskonflikt
Transplantationsvorbereitung	Angst vor erforderlichen Operationen bzw. psychische Belastung durch Eingriffe wie z.B. Nephrektomie
Wartezeit	Anspannung durch ständiges Verfügbar-Sein-Müssen
Transplantation (mögl. med. Komplikationen: – Nierenarterienstenose – Infektionen – Abstoßungsreaktionen – Nebenwirkungen der immunsuppressiven Medikamente)	Angst vor der Operation und Narkose Angst vor den Schmerzen Unrealistische Erwartungshaltung Banges Erwarten des Funktionsbeginns Belastung durch postoperative Komplikationen Psychische Belastung durch Abstoßungsreaktionen Angst vor Verlust der Niere Psychische Belastung durch Morbus Cushing Compliance-Problematik (spez. bezogen auf Immunsuppressiva) Fremdkörpererlebnisse Integrationsprobleme
Endgültige Abstoßung Explantation Rückkehr zur Hämodialyse	Trauer, Depression, Hoffnungslosigkeit, „giving up", Suizidalität Psychische Belastung durch erneute Operation Suche der „Schuld" bei anderen oder bei sich selbst

lente Gefühlslage entstehen. In einigen Fällen ist der Entscheidungskonflikt so ausgeprägt, daß psychotherapeutische Intervention notwendig wird.

Im folgenden sollen Ergebnisse einer Längsschnittstudie zur Erfassung der emotionalen Befindlichkeit der Patienten in der Phase der Transplantation dargestellt werden (Muthny, 1983). Das Projekt „Psychische Probleme im Umfeld der Nierentransplantation und Möglichkeiten psychotherapeutischer Interventionen" hatte sowohl eine Betreuungsaufgabe (Kriseninterventionen, psychologische Beratung, längerfristige Psychotherapien und Personalfortbildung) als auch einen Forschungsauftrag (Querschnittstudie zur Ausgangssituation vor der Transplantation, Längsschnittstudie zum Verlauf der Transplantation und über die mittelfristige Rehabilitation).

Die folgende Kurzdarstellung des psychologischen Verlaufsaspekts bezieht sich im wesentlichen auf die Ergebnisse von klinischen Ratings an 33 Patienten (ausführliche Darstellung bei Koch et al., 1983; Muthny, 1984).

In der **Situation, in der der Patient das Nierenangebot erhält** (in der Regel telefonische Nachricht durch den Dialysearzt), äußerten fast die Hälfte der Patienten Erschrecken im ersten Augenblick. Ca. 40% der Patienten äußerten Angst, zwei Drittel zeigten Unruhe, ca. 20% konnten sich nicht sofort entscheiden. Trotz dieser häufigen initialen Schreckreaktion äußerten 80% der Patienten noch in der präoperativen Phase auch Gefühle der Freude.

Nach dem **Aufwachen aus der Narkose** äußerten 60% der Patienten Glücksgefühle und Dankbarkeit, aber 55% gaben auch Gefühle von Angst, Furcht und Pessimismus für diese Phase an.

In den ersten 5 postoperativen Tagen war Hoffnung die dominierende Gefühlsäußerung der Patienten. Ca. 20% verleugneten mögliche Risiken und Komplikationen und etwa derselbe Prozentsatz von Patienten reagierte regressiv in dieser Zeit. 40% gaben an, von Partner und Familie die bedeutendste emotionale Unterstützung zu erhalten, ca. 20% erwähnten Ärzte und Personal als wichtigste psychosoziale Hilfe.

Während des stationären Aufenthalts nach der Transplantation (Dauer 18–50 Tage) ergaben sich sehr unterschiedliche Verläufe in Abhängigkeit von der präoperativen Ausgangssituation, der Zeitspanne bis zum Funktionsbeginn der Niere, dem Auftreten von Abstoßungskrisen und dem Gesamterfolg der Transplantation.

3 Verläufe sollen im folgenden exemplarisch skizziert werden:

> Eine 49jährige Patientin zeigte einen besonders günstigen medizinischen Verlauf mit extrem frühem Ausscheidungsbeginn (unmittelbar postoperativ) und einer raschen somatischen Besserung nach der Operation. Die bereits im Erstinterview bei der Dialyse ausgesprochene optimistische Einstellung der Patientin zeigte sich auch nach der Nierentransplantation in ausgeprägten Hoffnungs- und Freudegefühlen (die Patientin sprach von „Wiedergeburt") und war durch 2 Abstoßungskrisen nicht wesentlich erschüttert. Obwohl die Patientin zu keinem Zeitpunkt einer Krisenintervention ernsthaft bedurft hätte, nahm sie doch bereitwillig unser Gesprächsangebot wahr, und der Therapeut sah im wesentlichen seine Aufgabe darin, sicherzustellen, daß die Hochstimmung der Patientin keine Gefahr für die Medikamentencompliance (Immunsuppressiva) nach der Entlassung darstellte, und sie sich der Alarmsignale einer Abstoßungskrise bewußt blieb und adäquat handeln konnte.

> Ein 46jähriger alleinstehender Patient war demgegenüber durch einen besonders ungünstigen medizinischen Verlauf psychisch sehr belastet. Zwar kam die Urinausscheidung recht früh in Gang, der Patient wurde jedoch nie dialysefrei und mußte schließlich nach 34 Tagen im Anschluß an eine finale starke Abstoßungskrise explantiert werden. Der in seiner Grundhaltung eher skeptische und mißtrauische Patient schöpfte nach Beginn der Ausscheidung etwas Hoffnung, diese wurde aber im ungünstigen weiteren somatischen Verlauf schnell zunichte. Da der Patient aufgrund seiner verschlossenen und ängstlichen Art wenig Kontakt zu Mitpatienten und Personal fand, war hier ein intensiver Betreuungsaufwand erforderlich, um dem Patienten über die psychischen Krisen, die stark durch Angst und Depressivität bestimmt waren, hin-

wegzuhelfen. Bei diesem Patienten war bereits beim Erstinterview ein geringes Coping-Repertoire eingeschätzt worden. Nach der besonders angstbesetzten Explantation reagierte der Patient eher erleichtert. Er kehrte in ein für ihn im Gegensatz zur Transplantationsstation gewohntes soziales Umfeld (Mutter, Schwester) und in eine eher vertraute Dialysesituation zurück.

> Ein 38jähriger Patient steht für einen ebenfalls häufigen dritten Verlaufstypus, den man als „medizinisch und psychisch problematisch" bezeichnen könnte. Nach anfänglichem Optimismus und beträchtlichen Ausgangshoffnungen reagierte der Patient zunehmend ungeduldig und ängstlich gespannt auf das Ausbleiben der Nierenfunktion, zum Teil wohl auch aus den traumatischen Erfahrungen der ersten fehlgeschlagenen Transplantation heraus, als die finale Abstoßung und die Trennung von der Ehefrau ihn während des stationären Aufenthalts extrem belastet hatten. Die Spannungen des Patienten wirkten sich auch auf die Patient-Personal-Interaktion aus und erforderten zusätzlich zur intensiven Betreuung des Patienten (26 Gespräche über den stationären Aufenthalt hinweg) auch Gespräche mit dem Personal. In der Entlassungssituation wurde noch einmal ein beträchtliches Ausmaß an Angst und Spannung sichtbar, obwohl der Patient mit guter Nierenfunktion entlassen werden konnte. Aus diesem Grund fanden weitere Nachgespräche statt, bis der Patient Vertrauen in das transplantierte Organ und die neue Lebenssituation gewonnen hatte.

Trotz der individuell sehr unterschiedlichen Reaktionsweisen unterscheidet Chambers (1982) 5 Phasen, denen sie jeweils bestimmte Probleme und Bewältigungsstrategien zuordnet:

– **Präoperative Phase** (3 Tage vor bis 12 Stunden nach der Operation): Die Patienten werden in dieser Phase als optimistisch und zuversichtlich beschrieben, von der belastend erlebten Dialyse wegzukommen.
– **Unmittelbare postoperative Phase** (1 bis 5 Tage): Diese wird charakterisiert durch Immobilisierung, apparative Überwachung mit Gefühlen von Angst, Ärger und gelegentlich steroidinduzierten Psychosen. Als spezifisch für Nierentransplantierte werden Euphorie sowie Gefühle der Wiedergeburt angesehen.
– Die **frühe postoperative Phase** (5–14 Tage) gibt häufig Hinweise auf das Funktionieren des Transplantats und wird charakterisiert durch die Bewußtheit, daß die Niere dennoch versagen kann, andererseits auch durch eine psychische Entspannung, vermehrtes Wohlbefinden, verminderte Angst und relative Ruhe. Bei Auftreten von Abstoßungskrisen oder Komplikationen kann sich dies rasch in das Gefühl verändern, „auf einem Pulverfaß zu sitzen", verbunden mit Depression, Angst, Agitiertheit und vermindertem Selbstwertgefühl.
– **Mittlere postoperative Phase** (2.–6. Woche): Es findet allmählich eine Anpassung an die neu erworbene Nierenfunktion, die Medikation und ihre Nebenwirkungen sowie geringfügige medizinische Komplikationen statt. Mit der Funktion des Organs stellt sich zunehmend auch die Frage nach Gefühlen gegenüber dem Nierenspender. Neben Gefühlen der Erleichterung und Zufriedenheit werden auch aggressive, ärgerliche Gefühle beobachtet.
– **Späte postoperative Phase** (nach Entlassung): Es wird eine Entfremdung von der Umgebung, die die Nieren-

transplantation und ihre Folgen nicht nachvollziehen kann, beobachtet. Dem Patienten werden zahlreiche Adaptationsleistungen abverlangt, während er sich zugleich der Nierenfunktion noch unsicher ist.

Zusammengefaßt beobachtet die Autorin bei praktisch allen Patienten Episoden von Angst bzw. Depression. Diese sind meist vorübergehend und medizinischen Komplikationen zuzuordnen.

Das praktizierte Liaisonkonzept sieht u.a. vor, daß die Autorin in gemeinsamen Konferenzen mit Schwestern bzw. Chirurgen Behandlungspläne aufstellt, die die besonderen psychischen Probleme der Patienten berücksichtigen. In der Nachsorge wird auf personelle Kontinuität der Betreuung geachtet.

65.4.3 Besondere Probleme im Umfeld der Nierentransplantation

Probleme der Organintegration

Im Gegensatz zu anderen Operationen, bei denen ein krankes Organ entfernt wird, handelt es sich bei der Transplantation um eine lebenserweiternde („life-extending") Operation, d.h. ein neues Organ kommt hinzu (Castelnuovo-Tedesco, 1981). Dies erfordert eine Integration des neuen Organs (das zudem der Selbstuntersuchung des Patienten zugänglich ist) in das Körperbild. Der Prozeß der Integration wird von verschiedenen Autoren an Kasuistiken im Rahmen psychotherapeutischer Betreuung untersucht (Basch, 1973; Lefebvre et al., 1973; Viederman, 1974). Übereinstimmend gehen die Autoren davon aus, daß die psychische Repräsentanz des Fremdorgans als ein vom Selbst getrenntes Objekt zunächst mit Objektlibido besetzt und schrittweise in das Körperbild integriert wird. Dieser Prozeß kann zahlreichen Störungen unterworfen sein, die sich u.a. in hypochondrischer Beobachtung des Organs, besonderer Ängstlichkeit nach Bagatelltraumen sowie Mißempfindungen im Bereich des Organs äußern können (Cramond, 1971).

Viederman (1974) unterscheidet als Bedeutungen des Introjekts:
– Feindseliges Introjekt, verbunden mit Angst vor Abstoßung oder Verlust der Niere;
– gutwilliges Introjekt, verbunden mit Selbstwertsteigerung;
– Abwesenheit von Konflikt bei wahrgenommener Ähnlichkeit von Selbst und Objekt;
– Identifikation mit dem Objekt.

Besondere Probleme werden bei gegengeschlechtlichen Spendern berichtet (Cramond, 1971). Bei Kadavernierenempfängern werden Schuldgefühle berichtet und Phantasien – insbesondere bei Kindernierentransplantaten –, dem Spender ein vitales Teil gestohlen zu haben, so daß jener verletzt oder getötet wurde (Castelnuovo-Tedesco, 1981). Bei der Lebendnierenspende wird dem Verhältnis zum Spender zum Teil große Bedeutung beigemessen (z.B. Viederman, 1974). Systematische Untersuchungen an größeren Stichproben fehlen bislang allerdings zum Problem der Organintegration.

Belastungen durch Medikation, medizinische Komplikationen und Compliance

Neben nahezu obligaten Abstoßungskrisen (s.u.) sind Nierentransplantierte in vielen Fällen einer Reihe medizinischer Komplikationen ausgesetzt. Als Frühkomplikationen werden Blutungen, Urinfisteln, Magen-Darm-Ulzera sowie Zytomegalie-Virus-Infektionen beschrieben. Spätkomplikationen umfassen chronische Abstoßungsreaktionen, Nierenarterien- und Ureterstenose sowie die Aktivierung präexistenter Infektionsherde (Tbc, Hepatitis; zusammenfassend Wilms, 1984). Die Patienten reagieren auf derartige Komplikationen zum Teil sehr heftig mit Depressionen, Angst, Agitiertheit bzw. Ärger (Chambers, 1982; McKegney et al., 1981; Muthny et al., 1984).

Voraussetzung für das Überleben des Transplantats ist die regelmäßige Einnahme von Immunsuppressiva über den gesamten Zeitraum der Transplantatfunktion, die mit zahlreichen Nebenwirkungen behaftet ist. Insbesondere die zahlreichen Kortikosteroidnebenwirkungen (u.a. cushingoide Fazies, gastroduodenale Ulzera, Osteoporosen, Katarakt, Hautveränderungen sowie steroidinduzierte Psychosen, vgl. Wilms, 1984) werden von Patienten sehr belastend erlebt. Vor allem bei jüngeren Patienten können Symptome wie Stammfettsucht und Vollmondgesicht zu beträchtlichen Störungen von Körperbild und Selbstwertgefühl beitragen (Muthny et al., 1984). Die Noncompliance – nach erfolgter Nierentransplantation – wird niedrig eingeschätzt (nach Armstrong und Weiner, 1981/82 auf 1–4%). Die Autoren verweisen auf die besondere Bedeutung der Beachtung von Noncompliance bei Kindern und Jugendlichen. In der Literatur werden bislang lediglich Einzelfälle von Transplantatabstoßung aufgrund von Noncompliance berichtet (z.B. Basch, 1980). Mit der Verwendung von Cyclosporin als alleinigem oder in Kombination verwendetem Immunsuppressivum ist langfristig eine leichte Erhöhung der Transplantat- und Patientenüberlebensrate möglich. Durch Verzicht auf Steroide oder deren Dosisreduktion lassen sich viele unangenehme Nebenwirkungen vermeiden, allerdings ist Cyclosporin selbst nephrotoxisch und hat eine nur geringe therapeutische Breite.

Abstoßungsreaktionen, Rückkehr zur Dialyse

Im ersten Jahr nach der Transplantation ist mit ein bis zwei akuten Abstoßungskrisen zu rechnen, die für den Patienten zum Teil massive Bedrohungen darstellen. Wichtig ist auch hier die Kooperation des Patienten zur rechtzeitigen Diagnosestellung und hochdosierten immunsuppressiven Therapie, die wiederum mit erheblichen und belastenden Nebenwirkungen verbunden sein kann.

Die Gewißheit der endgültigen Abstoßung bzw. das Ende der Hoffnung auf den Funktionsbeginn nach zum Teil wochenlangem Warten bei funktionsloser Niere wird häufig als Schock erlebt. Schwere Depressionen und Gefühle von Verzweiflung bis hin zur Selbstaufgabe werden beschrieben (Milne, 1977; Freebury, 1974). Der Versuch des Patienten, Erklärungen für dieses Ereignis zu finden, kann zur Überzeugung führen, selbst etwas falsch gemacht zu haben oder falsch behandelt worden zu sein und zu vorwurfsvollen Reaktionen gegenüber dem Personal Anlaß geben. Die Aussicht auf Rückkehr zur Dialyse und das damit verbundene Scheitern der Erwartungen und Hoffnungen an die Transplantation kann schwere psychische Reaktionen bis hin zur Selbstaufgabe und zu erhöhter Suizidalität nach sich ziehen (zusammenfassend Abram und Buchanan, 1976).

Lebendspendenproblematik, Spendersuche und -selektion

Untersuchungen zur Nierenspende lebender Verwandter stammen überwiegend aus den USA, da diese in Europa weit weniger durchgeführt werden, und sind daher auch nur begrenzt übertragbar. Die Spendersuche erfolgt durch eine komplexe familiäre Interaktion, in deren Verlauf die Ausübung von sozialem Druck auf potentielle Spender beschrieben wird (Lefebvre et al., 1973). In diesem Prozeß werden häufig Verwandte als Vermittler eingeschaltet. Der Spender nimmt in der Familie eher eine Randposition ein (vgl. Abram und Buchanan, 1976; Milne, 1977). Die Nierenspende von der Mutter an ihr Kind wird als besonders reibungslos dargestellt, da sie sozialen Erwartungen entspricht (Kemph, 1966). Versuche, das ärztliche Selektionsproblem, ob die Nierenspende akzeptiert werden soll (insbesondere Frage der Freiwilligkeit), durch Meßverfahren zu lösen (Milne, 1977), erscheinen unbefriedigend.

Reaktionen des Spenders

Frühere Arbeiten berichten über Konflikte zwischen Altruismus und unbewußter Feindseligkeit (Kemph, 1970) sowie dem postoperativen Auftreten von Depressionen und Ängsten beim Spender. Eine breit angelegte Untersuchung von Bennett und Harrison (1974) an 300 Nierenspendern fand jedoch bei 295 keine gravierenden psychischen oder physischen Veränderungen durch die Spende. Simmons (1983) fand in ihrer Langzeitstudie an 135 Lebendnierenspendern ein erhöhtes Selbstwertgefühl, eine positive Einstellung und wenig Bedauern über ihre Spende. Erfolglose Spender berichteten dabei mehr Bedauern, Schwierigkeiten in der Beziehung zum Empfänger und weniger Nähe zu dem Empfänger.

Reaktionen des Empfängers

Psychische Symptome werden häufiger bei Lebend- als bei Kadavernierenempfängern berichtet. Die Abstoßung des Lebendtransplantats geht nach Abram und Buchanan (1976) gehäuft mit Schuldgefühlen und Selbstvorwürfen einher.

Die dargestellten Ergebnisse weisen auf die Bedeutung der Beachtung der Familiendynamik bei der Spendersuche sowie intensiver Vor- und Nachbetreuung von Spender und Empfänger hin.

65.5 Psychosoziale und berufliche Rehabilitation nach Dialysebehandlung und Nierentransplantation

Vergleichsuntersuchungen zur Rehabilitation nach Hämodialyse und Transplantationsbehandlung sind methodisch nicht unproblematisch, weil die Ergebnisse häufig überlagert sind von spezifischen Selektionsprozessen bei der Zuweisung zu der einen oder anderen Behandlung. Darüber hinaus ist festzustellen, daß prospektiv angelegte Längsschnittuntersuchungen zum Rehabilitationsverlauf fehlen und sich die bisherigen Aussagen im wesentlichen auf retrospektive Untersuchungen stützen.

Bereits Beard (1971) verweist darauf, daß seelische Belastungen nach der Transplantation bei zuvor hämodialysierten Patienten nicht prinzipiell verschwunden sind. Seiner Untersuchung zufolge läßt sich die während der Hämodialyse verlorengegangene Lebensqualität durch Transplantation nur bedingt wiederherstellen, die Patienten fühlen sich auch als Transplantierte chronisch krank, behindert und minderwertig.

Kaplan De-Nour und Shanan (1980) finden, daß zwei Drittel der transplantierten Patienten im Gegensatz zu nur einem Drittel der Dialysepatienten frei von psychiatrischen Komplikationen waren. Auch nach der Studie von Burdett (1978) weisen Nierentransplantationspatienten weniger Einbußen im sozialen Leben auf, sie fühlen sich seltener isoliert. Procci (1978) stellt darüber hinaus Verbesserungen der Sexualfunktion gegenüber der Zeit der Hämodialyse fest, wobei Männer größere Fortschritte als Frauen aufweisen. Katschnig und Koniecza (1982) kommen in ihrer sorgfältig angelegten Vergleichsstudie (Zweipunkt-Befragung) zu folgenden Resultaten:

– 80% der Nierentransplantationspatienten bezeichnen ihren körperlichen Zustand als sehr gut oder mindestens gut, während dies bei Dialysepatienten nur bei 33% der Fall ist.
– Ihren seelischen Zustand beschreiben 60% der Nierentransplantationspatienten als sehr gut oder gut; von den Dialysepatienten stellen dies nur 30% fest.
– Wie Dialysepatienten berichten auch Transplantationspatienten über psychische Beeinträchtigungen wie Reizbarkeit (72 versus 66%), Sorgen und Grübeln (73 versus 59%), Deprimiertheit (76 versus 43%), fehlende Zukunftsorientierung (70 versus 46%).

Interessant ist auch, wie nach der Untersuchung von Katschnig und Koniecza (1982) die Patienten selbst ihre gegenwärtige Behandlungsform einschätzen. So äußern 45% der Dialysepatienten, daß eine Transplantation ihre Lebensqualität entscheidend verbessern würde, und 51% würden sich prinzipiell einer Transplantation unterziehen. Fast die Hälfte der Dialysanden hält die Transplantation für die bessere Behandlungsform, dies, obwohl nur ein kleiner Teil von ihnen auf der Transplantationsliste steht. Die zur Zeit mit einem Transplantat lebenden Patienten bewerten dagegen ihre jetzige Situation deutlich positiver. So berichten 97% von einer entscheidenden Verbesserung der Lebensqualität gegenüber der Dialysezeit. 100% halten sogar die Nierentransplantation für die bessere Behandlungsform, 94% würden sich auch dann einer erneuten Transplantation unterziehen, wenn das Transplantat abgestoßen würde. Der Unterschied zwischen dem Leben mit der transplantierten

Niere und dem Leben, das der Patient als Gesunder führen würde, wird von 56% als eher gering eingeschätzt, und 74% beurteilen die Belastungen durch das Leben mit der transplantierten Niere als eher gering. Nierentransplantierte sollen im Vergleich zu Dialysepatienten eine höhere allgemeine Lebenszufriedenheit und ein besseres emotionales Wohlbefinden aufweisen (Evans et al., 1985); ähnliche Befunde erhoben Muthny et al. (1989) durch eine Befragung von insgesamt 1119 Patienten über verschiedene Behandlungsverfahren (Transplantation, Hämodialyse, CAPD). Dabei nahmen die Angaben der CAPD-Patienten hinsichtlich der Lebensqualität eine Mittelstellung ein zwischen der Gruppe der Transplantierten und der Patienten an der Dialyse. Möglicherweise tragen aber auch Selektionseffekte dazu bei, daß CAPD-Patienten ihre Lebenssituation besser einschätzen als Dialysepatienten. Kalman et al. (1983) hatten im Gegensatz zu den erwähnten Befunden festgestellt, daß sich Transplantatempfänger und Dialysepatienten nicht wesentlich in der Häufigkeit psychischer Probleme unterscheiden.

Berufliche Rehabilitation

Die vorliegenden Untersuchungen zur beruflichen Rehabilitation von Dialyse- und Transplantationspatienten deuten insgesamt darauf hin, daß die Häufigkeit der Wiederaufnahme der Berufstätigkeit bei Transplantationspatienten erheblich höher ist als bei Dialysepatienten (Evans et al., 1985; Muthny, 1989).

Ahlmen und Olander (1973) finden allerdings, daß diejenigen transplantierten Patienten, die vor der Transplantation bereits berentet waren, in keinem Fall ihre Berufstätigkeit wieder aufnahmen, obwohl sie überwiegend berufsfähig waren. Nicht berentete Dialysepatienten nahmen immerhin in 60% der Fälle nach der Transplantation wieder ihre berufliche Tätigkeit auf. Graf und Mitarbeiter (1979) konnten zeigen, daß der Grad der beruflichen Rehabilitation relativ unabhängig von der Transplantatüberlebenszeit, der Gewebeverträglichkeit, der Dialysezeit vor Transplantation und dem postoperativen Krankenhausaufenthalt ist. Kaplan De-Nour und Shanan (1980) berichten in ihrer in Israel durchgeführten Untersuchung, daß die berufliche Rehabilitation bei Transplantierten weniger als bei Dialysepatienten von Ausbildung und Berufsprestige vor der Nierenerkrankung abhängig ist.

Eine Analyse der EDTA-Statistik (1980) zeigt, daß die Erwartungswerte (bezogen auf den Anteil der Bevölkerung, der in der entsprechenden Lebenssituation überhaupt berufstätig wäre) bei 52% für Dialysepatienten und 70% für Transplantationspatienten liegen. Hierin spiegelt sich im wesentlichen eine unterschiedliche Alterszusammensetzung der Dialyse- und Transplantationspopulation wider, ein Faktor, der bei den meisten oben zitierten Untersuchungen vernachlässigt wurde. Die weitere Analyse zeigt, daß der Anteil Arbeitsunfähiger in der Dialysegruppe bei 13,7%, in der Transplantationsgruppe bei 6,2% liegt. Ebenfalls höher ist der Anteil Arbeitsloser bei den Dialysepatienten (10,1%) gegenüber den Transplantationspatienten (6,1%). 82,3% der Transplantationspatienten sind ganz- oder halbtags beschäftigt (oder in Ausbildung oder voll in der Haushaltstätigkeit stehend) gegenüber 65% bei den Dialysepatienten.

Vergleicht man den Grad der beruflichen Rehabilitation nach klinischer Hämodialyse sowie nach Nierentransplantation über verschiedene Industrienationen, so zeigt sich eine große Schwankungsbreite. Der Prozentsatz Rehabilitierter liegt in Schweden mit 39,2% für Hämodialyse am niedrigsten, in Italien mit 80,4% am höchsten, die Bundesrepublik Deutschland liegt mit 46,4% an zweitletzter Stelle. Bei der Transplantation hat Dänemark mit 62% die niedrigste, Italien wiederum mit 94% die höchste Quote, die Bundesrepublik bewegt sich hier wiederum eher im unteren Drittel. Unterschiedliche Systeme sozialer Sicherung, unterschiedliche Strukturen des Arbeitsplatzangebotes und unterschiedlich gut mit der Arbeitszeit abgestimmte Dialysepläne dürften hier eine Rolle spielen. Auffällig ist bei diesem Vergleich auch, in welchem Ausmaß das Teilzeitarbeitsangebot Auswirkungen auf den beruflichen Rehabilitationsstatus zeigt. Hier ist ein Vergleich zwischen der Bundesrepublik Deutschland und der Deutschen Demokratischen Republik interessant. Sowohl bei der klinischen Hämodialyse als auch bei der Transplantationsbehandlung ist die Quote der beruflichen Rehabilitation in der DDR wesentlich höher; dies scheint zum Teil die Folge eines sehr viel flexibleren Angebots an Teilzeitbeschäftigung zu sein.

65.6 Weitere Transplantationsverfahren (Herz, Leber und Knochenmark)

Auch Patienten, die einer **Herztransplantation** bedürfen, haben meist eine längere Zeit chronischen Krankseins mit reduziertem Leistungsvermögen und vielfältigen Einschränkungen, vergleichbar mit niereninsuffizienten Patienten, hinter sich. In dieser Zeit ist es bei vielen zu seelischen Veränderungen im Sinne reaktiver Ängste und Depressionen, aber auch zu hirnorganisch bedingten Störungen und Fehlanpassungen gekommen (vgl. auch Kap. 41.1, 62 und 63). Ähnlich wie bei Dialysepatienten führt die Zeit des Wartens auf ein geeignetes Transplantat oft zu einer großen Spannung und Belastung. Bei der Transplantation des Herzens imponiert insbesondere die Bedeutung des Organs als „Sitz des Lebens", „Verkörperung der Gefühle", die sich in vielerlei Redewendungen und Sprichwörtern ausdrückt. Dies erklärt sicherlich zum Teil den sensationellen Aspekt der ersten durchgeführten Herztransplantationen.

In der psychosomatischen Forschung konzentrierte sich das Interesse auf die Evaluation präoperativer Reaktionen, z.B. von Angst und Depressivität, bei Transplantationskandidaten (Freeman et al., 1984) und auf die Feststellung postoperativer psychischer Komplikationen, insbesondere deliranter Zustände (vgl. Hotsen et al., 1976; vgl. auch Kap. 64). Über den psychiatrischen Verlauf des Patienten Barney Clark, der ein künstliches Herz erhalten hatte, wurde eine detaillierte psychiatrische Krankengeschichte publiziert (Berenson, 1984). An zahlreichen Transplantationszentren, die Herztransplantationen durchführen, wurden psychosomatische bzw. psychiatrische Konsiliar- und Liaisondienste eingerichtet (vgl. Mai, 1985, 1986a; Künsebeck, 1987) mit dem Ziel, durch präoperative Diagnostik und Begleitung postoperative Komplikationen zu verhindern bzw. zu mildern.

In der Literatur wird angegeben, daß sich viele Transplantatempfänger intensiv mit dem Schicksal des Spenders befassen. Darüber kommen auch Gefühle von Angst, Schuld und ambivalente Einstellungen zutage. Verleugnung tritt dabei als häufig vorkommender Abwehrmechanismus in Erscheinung (Mai, 1986b).

Die Ein-Jahres-Überlebensrate herztransplantierter Patienten liegt mittlerweile bei gut 80%. Hinsichtlich der psychosozialen Situation im weiteren Verlauf sind viele Patienten, insbesondere, wenn sich die Komplikationen in Grenzen halten, mit ihrer körperlichen und geistigen Leistungsfähigkeit zufrieden. Hingegen wird eine Tendenz zur Unzufriedenheit in finanzieller und beruflicher Hinsicht erwähnt (Künsebeck, 1987). Möglicherweise könnte dies aber auch Ausdruck einer Abwehr der Bedrohung sein, im Sinne einer Verschiebung, da für die Patienten die Gefahr einer Abstoßungsreaktion doch allgegenwärtig ist und sie sich immer wieder eingreifenden, beispielsweise bioptischen Untersuchungen unterziehen müssen.

Patienten, bei denen eine **Lebertransplantation** durchgeführt wird, sind bereits präoperativ oft erheblich psychoorganisch beeinträchtigt und durch schwere Komplikationen wie Blutung bedroht. Das Ausmaß der psychoorganischen Veränderungen beeinflußt die postoperative Anpassung ebenso wie die Art der Grunderkrankung, z.B. die Frage, ob ein Malignom vorliegt (vgl. Kap. 66), ebenso wie die Frage der Verantwortlichkeit des Patienten oder anderer für den Krankheitsprozeß, z.B. bei einer toxischen Genese.

Die **Knochenmarktransplantation** hat bei einer Reihe von Erkrankungen, z.B. aplastischen Anämien verschiedener Genese, Immunglobulinmangelsyndromen, Stoffwechseldefekten sowie insbesondere akuten und chronischen Leukämien, eine zunehmende Bedeutung gewonnen. Die Übertragung von Knochenmark bietet im Unterschied zur Transplantation von Organen keine operationstechnischen Probleme. Knochenmark wird durch die mehrfache Aspiration aus dem Beckenkamm des Spenders gewonnen und nach Zubereitung intravenös infundiert. Das Risiko für den Spender ist sehr gering. Nach ca. 2 Wochen beginnen die ins Knochenmark gewanderten Stammzellen mit der hämatopoetischen Differenzierung. Vor der Transplantation wird der Patient hochdosiert mit Chemo- und Radiotherapie behandelt („Konditionierung"). Danach ist er existentiell davon abhängig, Spendermark zu erhalten. Es gibt kein Zurück mehr. Zur Verminderung der Infektionsgefahr ist der Patient in der Regel in einer sterilen Einheit. Zur Wiederherstellung des erythropoetischen und megakaryozytären Systems werden 4–6 Wochen benötigt, dabei bedürfen die Patienten intensiver supportiver

Maßnahmen. Die Zeit bis zur Wiederherstellung der vollständigen Immunkompetenz beträgt im allgemeinen 3–4 Monate. In dieser Zeit ist der Patient durch vielfältige Komplikationen, insbesondere infektiöser Art, z.B. durch Herpes zoster, Zytomegalie-Virus etc. bedroht, ebenso durch eine akute und chronische GVH-Reaktion (graft versus host) auch im weiteren Verlauf.

Patienten, denen wegen einer Leukämie eine Knochenmarktransplantation vorgeschlagen wird, mußten sich meist plötzlich mit der Realität der lebensbedrohlichen Erkrankung auseinandersetzen. Die Entscheidung zur Transplantation wird oft schnell getroffen, sie sollte bei den Leukämien möglichst in der ersten Remissionsphase erfolgen. Meist ist auch die Bereitschaft der Familienmitglieder zur Spende groß (Hörner et al., 1987c). Die Phase der Konditionierung wird von den Patienten meist als schlimmste Zeit erlebt (Hörner et al., 1987a, c), wobei körperliche Beschwerden wie Erbrechen und Übelkeit im Vordergrund stehen. Im weiteren Verlauf kommt es bei Komplikationen wie Infektionen und GVH-Erkrankungen oft zu Krisen. Neusser et al. (1988) fanden bei 20% der Patienten klinisch relevante affektive Störungen, überwiegend im Zusammenhang mit medizinischen Komplikationen, zum Teil sind auch hirnorganische Psychosyndrome zu beobachten (Hörner et al., 1987b). Bei der Untersuchung der Spender fiel auf, in welchem Maß sie sich mit den Wechselfällen der Behandlung der Patienten identifizieren (Hörner, 1987c). Eine besondere Risikogruppe hinsichtlich psychischer Probleme, aber auch Complianceschwierigkeiten, stellen Adoleszente dar (Alby et al., 1987), was sowohl den Einbezug der Familie als auch eine spezielle Betreuung oft notwendig macht.

Nach einem Bericht der Leukaemia Working Party 1987, der Daten über 2224 in 52 europäischen Zentren zwischen 1979 und 1986 transplantierte Patienten umfaßt (Gratwohl, 1987), ist die chronische myeloische Leukämie (CML) mittlerweile die häufigste Indikation zur Knochenmarktransplantation geworden. Ca. 45% der Patienten mit akuter Leukämie und 75% der Patienten mit aplastischer Anämie können nach der Knochenmarktransplantation mit langfristiger Remission bzw. als wahrscheinlich geheilt entlassen werden.

Insgesamt ist die Knochenmarktransplantation nicht nur eine „ultima ratio" bei Patienten mit aplastischer Anämie und Leukämie, sondern bei geeigneten Patienten eine „Frühtherapie der Wahl" (Übersicht bei Thiele, 1985). Allerdings ist sie immer noch ein quoad vitam mit hohem Risiko behafteter Eingriff. Insbesondere ist die im Vergleich zu anderen therapeutischen Verfahren erhöhte Frühsterblichkeit problematisch, ebenso wie die Massivität der möglichen Komplikationen. Die Anwendung supraletaler Dosen bei der Konditionierung, die kein Zurück mehr erlaubt und eine existentielle Abhängigkeit vom Knochenmarkspender schafft, ist eine Grenzsituation des Lebens für Patienten und Behandelnde, die alle Beteiligten auf eine harte Probe stellt.

66 Zum Umgang mit unheilbar Kranken

Karl Köhle, Claudia Simons und *Bernhard Kubanek.* Mit einem Beitrag von *Jutta Zenz*

„Dieses ausgezeichnete Hôtel ist sehr alt, schon zu König Chlodwigs Zeiten starb man darin in einigen Betten. Jetzt wird in 559 Betten gestorben. Natürlich fabrikmäßig. Bei so enormer Produktion ist der einzelne Tod nicht so gut ausgeführt, aber darauf kommt es auch nicht an. Die Masse macht es. Wer gibt heute noch etwas für einen ausgearbeiteten Tod? Niemand. Sogar die Reichen, die es sich doch leisten könnten, ausführlich zu sterben, fangen an, nachlässig und gleichgültig zu werden; der Wunsch, einen eigenen Tod zu haben, wird immer seltener. Eine Weile noch, und er wird ebenso selten sein wie ein eigenes Leben. Gott, das ist alles da. Man kommt, man findet ein Leben, fertig, man hat es nur anzuziehen. Man will gehen, oder man ist dazu gezwungen; nun, keine Anstrengung. Voilà, votre mort monsieur. Man stirbt, wie es gerade kommt; man stirbt den Tod, der zu der Krankheit gehört, die man hat (denn seit man alle Krankheiten kennt, weiß man auch, daß die verschiedenen letalen Abschlüsse zu den Krankheiten gehören und nicht zu dem Menschen, und der Kranke hat sozusagen nichts zu tun).

In den Sanatorien, wo also gern und mit so viel Dankbarkeit gegen Ärzte und Schwestern gestorben wird, stirbt man einen von den an der Anstalt angestellten Toden; das wird gerne gesehen. Wenn man aber zu Hause stirbt, ist es natürlich, jenen höflichen Tod der guten Kreise zu wählen, mit dem gleichsam das Begräbnis erster Klasse schon anfängt und die ganze Folge seiner wunderschönen Gebräuche. Da stehen dann die Armen vor so einem Haus und sehen sich satt. Ihr Tod ist natürlich banal, ohne alle Umstände. Sie sind froh, wenn sie einen finden, der ungefähr paßt. Zu weit darf er sein; man wächst immer noch ein bißchen. Nur wenn er nicht zugeht über der Brust oder würgt, dann hat es seine Not.

Wenn ich nach Hause denke, wo nun niemand mehr ist, dann glaube ich, das muß früher anders gewesen sein. Früher wußte man (oder vielleicht, man ahnte es), daß man den Tod *in* sich hatte wie die Frucht den Kern. Die Kinder hatten einen kleinen in sich und die Erwachsenen einen großen. Die Frauen hatten ihn im Schoß und die Männer in der Brust. Den *hatte* man, und das gab einem eine eigentümliche Würde, einen stillen Stolz."

Rainer Maria Rilke: „Die Aufzeichnungen des Malte Laurids Brigge", 1909.

„Saluti et solatio aegrorum"
Josef II.   Inschrift über dem Tor des Allgemeinen Krankenhauses der Stadt Wien.

66.1 Umgang mit unheilbar Kranken: zur Sonderstellung einer Aufgabe innerhalb der klinischen Medizin

In diesem Kapitel stellen wir vor allem die Probleme im Umgang mit unheilbar **Krebs**kranken dar, da das Problembewußtsein hier am stärksten ausgeprägt ist.

Dies folgt nicht nur aus der Häufigkeit von Krebserkrankungen als Todesursache, sondern aus den (Laien-)Vorstellungen zur Krankheit „Krebs" (s. u.). Wie irrational diese Vorstellungen zum Teil sind, wird deutlich, wenn man die Überlebenszeiten nach Manifestation der verschiedenen Malignomerkrankungen mit denjenigen bei anderen akut bedrohlicheren Erkrankungen wie Ösophagusvarizenblutung bei Leberzirrhose, Herzinfarkt oder zerebraler Ischämie vergleicht. Ähnliche Probleme im Umgang finden sich auch bei akut lebensbedrohlich Kranken; bei Krebskranken stellen sie sich u.a. auch deshalb besonders ausgeprägt dar, da die Erkrankung selten plötzlich und unerwartet zum Tod führt.

Wir konzentrieren uns hier auf die Probleme im Umgang mit **unheilbar** Krebskranken, wie sie dem onkologisch arbeitenden Kliniker begegnen. Damit geht es in diesem Kapitel vor allem um die psychosoziale Betreuung zum Tode Kranker und Sterbender. Wir können hier nicht auf die ebenso wichtigen psychosozialen Probleme in der Rehabilitation von Patienten eingehen, deren Krebserkrankung heilbar ist oder doch wenigstens mit Phasen länger dauernder Remission verläuft (vgl. Heyde, 1983 und Herschbach 1983, 1987a, b).

In der klinischen Praxis und in der Forschung nehmen die Probleme im Umgang mit unheilbar Krebskranken eine Sonderstellung ein. Neben der Häufigkeit maligner Neubildung hat hierzu vor allem die soziale Einstellung gegenüber an Krebs Erkrankten beigetragen.

„Krebs" steht heute als Metapher für Lebensbedrohung durch Krankheit überhaupt (Sontag, 1978). Im Erleben der eigenen Bedrohung und – nach dem Muster der Infektionskrankheiten – der Mitbedrohung anderer hat „Krebs" die Nachfolge von Pest und Tuberkulose angetreten. Die damit verbundene soziale Stigmatisierung kann der Leser gut nachvollziehen, wenn er sich selbst die von J. Dornheim (1983) den Bewohnern eines Dorfes auf der Schwäbischen

Alb gestellte Frage vorlegt: „Würden Sie mit einem Krebskranken aus einem gemeinsamen Glas Wasser trinken?" Zum Vergleich kann dieselbe Situation etwa mit einem Herzinfarktkranken vorgestellt werden.

66.1.1 Zur Kooperation zwischen Onkologen und Psychosomatikern

Die Entwicklung von Forschungsansätzen in diesem Bereich wurde durch das zunehmende Problembewußtsein führender Onkologen und ihre Bereitschaft zur Kooperation mit Psychosomatikern gefördert. Diese Onkologen initiierten, ausgehend von den sie bedrängenden Problemen der Patienten, vielfach konkrete Kooperationsansätze in der klinischen Arbeit, aus denen sich dann auch wissenschaftliche Projekte entwickeln konnten.

Solche interdisziplinären Arbeitsgruppen bestehen bzw. bestanden in den USA seit vielen Jahren, u.a. in Boston (A. Weisman) und Buffalo bzw. New York (J. C. Holland zusammen mit ihrem Ehemann, dem Onkologen J. F. Holland), in der Schweiz in Zürich (F. Meerwein zusammen mit G. Martz; Honsalek, 1983). In der Bundesrepublik entwickelte sich eine derartige Kooperation seit 1968 u.a. in Ulm (ausgehend von der Schwerpunktgruppe Hämatologie, vgl. Köhle und Kubanek, 1981; Schreml et al., 1981), in Hamburg und in Nürnberg (Gallmeier, 1984; Herty, 1984; Pontzen et al., 1988). Erst die engagierte Förderung der Stiftung Deutsche Krebshilfe ermöglichte entsprechende Arbeitsansätze in München bei H. Begemann (Bettex, 1982), Heidelberg (Herfarth und Sellschopp [1981, 1984]) und an den Tumorzentren der Universitäten Essen und Köln sowie eine Intensivierung der Arbeit der Hamburger Gruppe um von Kerekjarto. Während es anfangs darum gegangen war, die Zusammenarbeit zwischen Onkologen und Psychosomatikern zu institutionalisieren, geht es heute darum, differenziertere Ansätze empirischer Forschung zu verwirklichen. Diese Ansätze werden zur Zeit in der BRD vor allem vom Bundesministerium für Forschung und Technologie („Rehabilitation von Krebskranken") unterstützt (u.a. an der TU München, an den Universitäten Köln und Hamburg).

Vorgeschichte und Verlauf dieser Kooperationsansätze lassen in der Regel die Notwendigkeit eines besonders guten Vertrauensverhältnisses zwischen den beteiligten Onkologen und Psychosomatikern als notwendige Voraussetzung für ein Gelingen erkennen. Ist dieses Vertrauensverhältnis nicht aufgrund persönlicher Beziehungen bereits vorhanden, benötigt die Entwicklung eines solchen „Liaison-Ansatzes" (vgl. Kap. 28) meist nicht nur längere Zeit, sondern auch die Möglichkeit, wirklich gemeinsame Erfahrungen in der Krankenversorgung zu machen (vgl. Köhle und Kubanek, 1981; Köhle, 1988).

Es scheint, daß klinische Onkologen in besonderem Maße ein Problembewußtsein für die psychologischen Aspekte im Umgang mit ihren Patienten entwickeln. Dies dürfte auch damit zusammenhängen, daß klinische Onkologen nicht nur in besonderem Maße mit den Grenzen der therapeutischen Möglichkeiten konfrontiert sind, sondern die sich daraus ergebenden Konsequenzen auch eingehend reflektieren; beitragen könnte zu dieser Einstellung auch, daß Onkologen im allgemeinen und Hämatologen im besonderen systemorientiert denken, was eine ganzheitliche Betrachtung gegenüber einer einseitigen Organorientiertheit fördert (Schonecke, 1984).

Die meist aus der konkreten klinischen Arbeit entstandene interdisziplinäre Kooperation hat sich mittlerweile auch in der Namensgebung für diesen Arbeitsbereich niedergeschlagen: Weisman (1979) spricht von „psychosocial cancerology", im deutschsprachigen Bereich hat sich nach ersten Ansätzen („Psycho-Hämatologie", Heimpel, 1970) inzwischen „Psycho-Onkologie" (Meerwein, 1981) eingebürgert.

„Psycho-Onkologie" vermeidet die Einschränkung dieses Arbeitsbereiches durch die in der Sache ungerechtfertigte Assoziation von Krebs mit Tod. Gerade die Erforschung und Bearbeitung der für die Rehabilitation von Krebskranken entscheidenden psychosozialen Probleme könnte in diesem Rahmen durchgeführt werden. „Thanatologie" bzw. „klinische Thanatologie" (Roswell Park, 1912) erscheinen als Bezeichnungen für diesen Arbeitsbereich innerhalb der klinischen Medizin aus dem genannten Grund weniger günstig; sie haben sich – wohl auch wegen der mit dem Selbstverständnis ärztlicher Tätigkeit nicht zu vereinbarenden Betonung des Todes – trotz ihrer viel früheren Einführung nicht durchgesetzt.

66.1.2 Die Situation von Ärzten und Pflegepersonal

Die Aufgabe, Krebskranke zu behandeln und zu betreuen, stellt für Ärzte und Pflegepersonal in quantitativer und qualitativer Hinsicht eine noch zunehmende Herausforderung dar. In der Bundesrepublik erkranken jährlich etwa 250 000 Menschen an Krebs (Inzidenz), die Zahl der insgesamt an Krebs Erkrankten (Prävalenz) beträgt mindestens 700 000 (Gerdes, 1984).

Krebserkrankungen bilden in der Bundesrepublik die zweithäufigste Todesursache.

Häufig begleiten Ärzte und Schwestern unheilbar Kranke über längere Zeit; Fortschritte der Therapie verlangsamen den Krankheitsprozeß oder ermöglichen Remissionen. Während der oft langen Krankenhausaufenthalte werden Ärzte und Schwestern in zunehmendem Maße für die Patienten zu bedeutsamen Bezugspersonen, zu Partnern bei der Bewältigung dieser schwierigen und für die Bewertung des ganzen Lebens mitentscheidenden Situation. Belastungen ergeben sich zunächst daraus, daß die in den Heilberufen Tätigen für diese Arbeit nicht ausgebildet sind und keine ausreichende Hilfestellung für die Ausbildung einer professionalen Gestaltung dieser Beziehungen erhalten; das Engagement in diesen Beziehungen verläuft meist analog zum Engagement in privaten Beziehungen, führt jedoch häufig zu einer chronischen Überlastung. Hinzu kommt, daß Ärzte und Krankenpflegepersonal gerade in den intensiven Beziehungen zu den von ihnen betreuten Patienten durch die Folgen ihres eigenen professionellen Tuns belastet werden: Die von ihnen durchgeführte Therapie beeinträchtigt nicht selten die Lebensqualität der Patienten erheblich, kann zu zusätzlichem Leid, zusätzlicher Gefährdung und auch zum Tode führen.

Informell diskutieren vor allem die beteiligten Schwestern und Pfleger, aber auch die Ärzte nicht selten darüber, ob sie die von ihnen verordneten und durchgeführten Therapiemaßnahmen auch im Falle einer eigenen entsprechenden Erkrankung angewandt wissen wollten. Das Ausmaß der eigenen Belastungen durch die Folgen der durchgeführten Therapie wird an der Heftigkeit deutlich, mit der diese Diskussionen geführt werden, und an der Entschiedenheit, mit der die angewandten therapeutischen Ansätze – jedenfalls in diesen Diskussionen – für die eigene Person abgelehnt werden.

All diese Belastungen haben Auswirkungen auf den Umgang mit den Kranken: Eine hämatologisch-onkologische Spezialstation mit 16 Betten ist ganz überwiegend mit jüngeren Patienten belegt, die aller Voraussicht nach während des gegenwärtigen Aufenthalts oder doch im Verlauf der nächsten ein bis zwei Jahre werden sterben müssen. Die Schwestern dieser Station engagieren sich in hohem Maße für eine patientorientierte, ganzheitliche Pflege. Während der regelmäßigen wöchentlichen Stationskonferenz betonen sie jedoch, daß sie bei einem solchen Belegungsgrad mit zum Tode Kranken sich nicht mehr in der Lage fühlen, auch auf die psychischen Bedürfnisse der von ihnen Betreuten einzugehen; um wenigstens den Routineaufgaben nachgehen zu können, müßten sie „ihre Ohren nach hinten klappen". Solche Abwehrvorgänge zum Schutze des eigenen psychischen Gleichgewichts und zur Erhaltung der Arbeitsfähigkeit finden sich im klinischen Bereich häufig und müssen respektiert werden – es sei denn, es gelingt Arbeitsformen einzuführen, die die Reflexion und Modifikation dieser Belastungen erlauben.

Das Sterben wurde in Zusammenhang mit den Fortschritten der klinischen Medizin immer mehr ins Krankenhaus verlegt. Während 1900 in Deutschland etwa 10 % der Sterbenden in Krankenhäusern versorgt wurden, sind es in der Bundesrepublik heute etwa 60 %, in Großstädten bis zu 80 % (Schied, 1979). Internistische Stationen in Schwerpunktkrankenhäusern sind oft zu einem Drittel mit Patienten belegt, deren Lebenserwartung im Durchschnitt weniger als zwei Jahre beträgt, jeder 10. Patient stirbt während des akuten Aufenthaltes. Auf Intensivstationen, hämatologischen bzw. onkologischen Spezialstationen verschärft sich diese Situation noch erheblich (vgl. Kap. 64; Hannich und Wendt, 1984; Klapp, 1984; Grote-Janz und Weingarten, 1983).

Die Frage, ob neben Diagnostik und Therapie auch die mitmenschliche Begleitung unheilbar Kranker zu den professionellen Aufgaben der in der Medizin Tätigen gehört, wird zu selten ausdrücklich gestellt und zu klären versucht. Diese Frage wurde jedenfalls nicht in allen Perioden der abendländischen Medizin positiv beantwortet (Schadewaldt, 1969). Ein Teil der auftretenden Schwierigkeiten dürfte auch daher rühren, daß wir in der christlichen Tradition diese Begleitung in einer konkreten Arbeitssituation übernehmen, ohne jedoch zu versuchen, diese Aufgabe im Zusammenhang mit unserer übrigen Tätigkeit und der Medizin als Wissenschaft systematisch zu klären.

Für Ärzte und Pflegepersonal kommt das Dilemma dadurch zustande, daß Patienten Bedürfnisse an uns richten, für die die Medizin als Wissenschaft keine Verständnis- bzw. Deutungsmöglichkeiten bietet. In der Medizin als Wissenschaft ist der Tod nur Endpunkt von Krankheit, nicht Bezugspunkt in einem Verständnissystem. Diese Form wissenschaftlicher

Medizin vermag keine Deutung des Todes zu vermitteln; damit besteht die Gefahr, daß das Sterben und der sterbende Mensch in ihr bedeutungslos werden (Scheytt, 1984). Die Konsequenzen werden in der klinischen Praxis deutlich: Tod und Lebensprobleme unheilbar Kranker und Sterbender werden weitestgehend ausgeklammert, häufig total verleugnet (vgl. Sudnow, 1973; Glaser und Strauss, 1974, 1968; Übersicht bei Christian und Widmaier, 1984).

Frau Kübler-Ross (1971) berichtete, daß noch Ende der 60er Jahre die ärztlichen Kollegen einer großen amerikanischen Universitätsklinik auf ihre Bitte hin, mit sterbenden Patienten im Rahmen von Seminaren sprechen zu können, entgegen der täglich zu beobachtenden Evidenz behaupteten, an ihrer Klinik gäbe es überhaupt keine sterbenden Kranken. Ansätze zur Erweiterung des medizinischen Verständnissystems um die psychosomatische Betrachtungsweise haben seit etwa 1965 verstärkt auch zur systematischen Untersuchung der Probleme unheilbar Kranker und Sterbender, zur Entwicklung von Betreuungskonzepten und entsprechenden Ansätzen in der Fort- und Weiterbildung geführt. Heute stellen sich in diesem Bereich vordringlich zwei miteinander verbundene Aufgaben: Die bisher entwickelten Konzepte sollten in die klinische Praxis umgesetzt werden; gleichzeitig und in Verbindung mit den Erfahrungen der klinischen Arbeit sollten breiter angelegte Forschungsansätze eine Weiterentwicklung dieser Konzepte ermöglichen.

Für die Umsetzung in der klinischen Praxis ist es wichtig, klar zu sehen, daß sich eine solche Erweiterung der praktischen Tätigkeit zunächst aus sozialwissenschaftlichen Verständnisansätzen ableitet, die noch nicht in eine entsprechende Erweiterung der Theorie der Heilkunde integriert sind.

Versuche, Leiden und Tod in den wissenschaftlichen Ansatz einer anthropologischen Medizin als systematische Bezugskategorien einzuführen, blieben bisher für die Praxis der Medizin bedeutungslos. Aufgabe einer solchen Medizin wäre es, „die Teilhabe des Todes am Leben" wissenschaftlich zu erforschen. Der Tod würde dann nicht mehr isoliert gesehen: „Der Tod aber ist nicht ein Ereignis. Er ist eine umfassende Ordnung, und sein Abglanz ruht auf jedem Wandel, jedem Untergang, jedem Schlaf und jedem Abschied. Er, als Gesetz, bestimmt auch die Farbe des Erlebens des Lebenden – es ist die Farbe des Leidens" (Viktor von Weizsäcker, 1947). In einer solchen Wissenschaft würde gelten: „Der Tod ist nicht der Gegensatz zum Leben, sondern der Gegenspieler der Zeugung und Geburt; Geburt und Tod verhalten sich wie Rückseite und Vorderseite des Lebens, nicht wie logisch einander ausschließende Gegensätze. Leben ist: Geburt **und** Tod. Das ist eigentlich unser Thema" (v. Weizsäcker, 1950).

Bei der Versorgung unheilbar Kranker wird in besonderem Maße deutlich, daß Krankenversorgung immer neben instrumenteller Tätigkeit – die sich aus dem naturwissenschaftlichen Verständnisansatz des Krankheitsgeschehens ableitet – auch „Gefühlsarbeit" (A. Strauss et al., 1980) enthält.

Zunächst soll ein Beispiel aus der täglichen Arbeit die ärztliche Aufgabe veranschaulichen, instrumentelle Tätigkeit und „Gefühlsarbeit" zu verbinden.

Eine 24jährige Patientin, die an Morbus Hodgkin im Terminalstadium leidet, klagte während der täglichen Visite über andauernde Müdigkeit und sagte in diesem Zusammenhang: „Ich möchte doch nicht immer schlafen". Der visiteführende Arzt steht nun vor der Aufgabe, diese Klagen und im Zusammenhang der Klagen auch den Satz „Ich möchte doch nicht immer schlafen" auf mehreren miteinander verflochtenen Ebenen zu klären und entsprechend Hilfestellungen für die Kranke zu entwickeln:

- Die Müdigkeit kann Folge des Fortschreitens des Krankheitsprozesses sein. Die Patientin kann Information hierüber wünschen, sie kann aber auch ihre Enttäuschung darüber andeuten, daß ihr die Medizin nicht besser zu helfen vermag; entsprechend spürt der visiteführende Arzt eventuell auch einen gegen ihn gerichteten Vorwurf.
- Die Müdigkeit kann Folge bzw. Nebenwirkung der Chemotherapie und damit des ärztlichen Tuns sein. Auch hier wird es zunächst um Klärung und Information gehen; soweit die Klage der Patientin Anklage enthält, wird aber auch eine Klärung der Beziehung erforderlich.
- Die Müdigkeit der Patientin kann Ausdruck einer depressiven Reaktion im Rahmen ihrer Krankheitsverarbeitung sein und würde dann entsprechende ärztlich-psychotherapeutische Unterstützung verlangen.
- Der Satz „Ich möchte doch nicht immer schlafen" kann auch die Todesangst der Patientin andeuten.

Die Patientin wendet sich jedenfalls mit ihren Fragen, ihren Klagen, ihren emotionalen Reaktionen bis hin zu ihrer Todesangst an den visiteführenden Arzt. Sie spricht ihn an und fordert ihn zur Hilfestellung auf; so konfrontiert kann er die Probleme zumindest in der Regel erst einmal nicht einfach an Spezialisten weiterdelegieren. Es geht für ihn vielmehr darum, die kommunikativen Aufgaben in den Gesamtzusammenhang ärztlicher Tätigkeit einzubeziehen.

Was benötigt die Patientin von ihrem Arzt, was kann der Arzt für die Patientin tun?

Zunächst benötigt die Patientin in ihrer neuen Situation Orientierungshilfe. Auf den ersten beiden Verständnisebenen geht es vor allem um Information zum Krankheitsverlauf, zur Therapie, zu deren Wirkungen und Nebenwirkungen. Der Arzt wird versuchen, die eigenen Vorstellungen und Phantasien der Patientin zu erfahren, um dann mit seiner Information – eventuell korrigierend – hieran anzuknüpfen. In einer von Verständnis getragenen Beziehung wird er sich dabei von den versteckten oder offenen Vorwürfen der Patientin nicht irritieren lassen; diese Vorwürfe rühren vor allem aus der Enttäuschung darüber, daß sich die Krankheit nicht heilen läßt, und gelten in der Regel nicht der Person des Arztes. In dieser Situation sollte der Arzt eigene Tendenzen zum Ausweichen, etwa vor den Fragen von Patienten, sorgfältig beachten und reflektieren; diese Gefahr nimmt mit Verschlechterung der Prognose der Patienten nach allen vorliegenden Untersuchungen zu. Häufig bestimmen emotionale Reaktionen, vor allem depressive Verstimmung und Angst, die Situation des Patienten. In unserem Beispiel benötigt die Patientin in diesem Bereich vor allem Unterstützung in ihrer krankheitsbedingten Selbstwertproblematik. Ihre abnehmende körperliche Leistungsfähigkeit erlebt sie als drohende Wertlosigkeit. Allein die Tatsache, daß der Arzt hierüber mit ihr spricht, zeigt ihr, daß sie noch nicht abgeschrieben ist, aus der Ge-

meinschaft der Lebenden noch nicht ausgestoßen ist. Die Beziehung zum Arzt kann so dazu beitragen, die Bedrängnis zu vermindern, Hoffnung aufrechtzuerhalten und so eventuell auch eine Sinnfindung zu ermöglichen. Auch Todesangst wird der Arzt dadurch zu mindern vermögen, daß er ihr nicht ausweicht, ihr standhält, daß er bereit ist, sich etwa die Suizidphantasien der Patientin anzuhören und daß er im einzelnen auf ihre Befürchtungen eingeht, die ja meist nicht dem Tod gelten, sondern die Art des Sterbens betreffen; die Furcht vor unerträglichen Schmerzen, vor dem Ersticken, vor dem Alleingelassenwerden. Im Verlauf unseres als Beispiel gewählten Visitengesprächs informiert der visiteführende Arzt die Patientin ausführlich; gleichzeitig versucht er geduldig, auf ihre depressiven Gefühle und die angedeuteten Ängste einzugehen. Die Patientin scheint die bereits längere Zeit bestehende Beziehung zu ihrem Arzt jetzt wieder aufs neue daraufhin zu prüfen, ob sie sich in ihr aufgehoben fühlen kann, danach kann sie ihre Verleugnung zurücknehmen. Sie äußert jetzt offen ihre Befürchtungen und spricht über ihre depressive Verstimmung. Jetzt erst kann der Arzt sie angemessen in ihrer Trauer unterstützen; daraufhin kommt die Patientin auf ihre gegenwärtige Situation und drohende Verluste zu sprechen: Es geht ihr darum, zu klären, was sie noch selbst in ihrem Haushalt tun kann und wo sie auf die Hilfe anderer angewiesen sein muß. Sie kann die anfängliche Verleugnung von Krankheit, Bedrohung und Angst etwas zurücknehmen, sich wieder mit ihrer konkreten Situation auseinandersetzen und Teilaspekte dieser Situation akzeptieren.

Kommunikation, „Gefühlsarbeit" mit unheilbar Kranken hat zwei dialektisch aufeinander bezogene **Ziele:** Ärzte und Pflegekräfte versuchen den Kranken zu helfen, in die Gemeinschaft der Lebenden integriert zu bleiben; im Falle eines ungünstigen Verlaufes bieten sie dem Patienten jedoch auch Unterstützung beim Verlassen dieser Gemeinschaft an.

Eine solche Erweiterung der Aufgaben der Krankenversorgung führt zu einer erheblichen Mehrbelastung aller Beteiligten. Jetzt gehört auch **„Gefühlsarbeit"** zur professionellen Tätigkeit; dies hat zur Voraussetzung, daß sich Ärzte und Pflegepersonal in den Umgang auch mit unheilbar Kranken einlassen und ihre Beziehung zu dem Patienten als Mittel für Verständnis und Unterstützung systematisch benützen. Die Reflexion der eigenen Gefühle von Arzt und Schwester in dieser Beziehung ermöglicht ein Verständnis der psychischen Prozesse, die im Kranken ablaufen. Arzt und Pflegepersonal lassen sich vom Kranken bewegen, um über ihre eigene Mitbewegung etwas über den Kranken zu erfahren, um Zugang zu seiner jeweils „individuellen Wirklichkeit" zu finden. Eine solche Mitbewegung bildet die Voraussetzung für den Versuch, mit dem Patienten jeweils eine „gemeinsame Wirklichkeit" aufzubauen. Dieser Versuch kann unterschiedlich beschrieben werden: als probeweiser Rollentausch (Mead, 1968); als ein Ablauf von Handlungen, in dem die Partner sich gegenseitig explizit vermitteln, daß jeder seinen nächsten Schritt von der Antwort des anderen abhängig macht („intelligente", d.h. „verständnisvolle Geste", Mead, 1968;

nach v. Uexküll, 1982); als flexibel kontrollierter und reversibler Prozeß von Identifikations- und Projektionsvorgängen; als Übertragungs- und Gegenübertragungsgeschehen. Gemeinsames Merkmal dieser Konzepte ist es, die Gegenseitigkeit in der Beziehung der beiden Interaktionspartner ernst zu nehmen. Für die klinische Arbeit ergibt sich hieraus die Forderung nach der Bereitschaft, sich vom jeweils anderen mitbewegen, ja sich von ihm „verändern" zu lassen (v. Weizsäcker, 1928).

Im Umgang mit unheilbar Kranken fällt es uns besonders schwer, eine solche Gegenseitigkeit, eine solche Bereitschaft zur Mitbewegung zu verwirklichen. Der Eintritt in die Wirklichkeit des Patienten kann von uns als bedrohlich erlebt werden; er kann uns mit unserer eigenen Einstellung zum Tod, mit unseren eigenen bisher mehr oder weniger bewußten Todesängsten konfrontieren; dies löst häufig Abwehrvorgänge in uns selbst aus, die unsere Bereitschaft und Fähigkeit, sensibel auf den Kranken einzugehen, blockieren können. Die Bereitschaft zur Mitbewegung setzt voraus, „Passivität", Bewegtwerden durch den anderen, ertragen zu können. Es ist schwierig, uns von unheilbar Kranken bewegen zu lassen, ohne – im Extrem – emotional mitsterben oder uns zu weit distanzieren zu müssen.

Mehrere Schwestern einer onkologischen Abteilung heirateten terminal Kranke und sind bald darauf Witwen.

Ein Ordinarius aus einem operativen Fach beteuert im Anschluß an einen Vortrag über die Arbeit in einem Londoner „Hospice" (Klinik für terminal Kranke): Ängste von Krebskranken und Sterbenden könne er nicht mehr als ein Problem ansehen; mit der Entdeckung der Endorphine sei dieses Problem für ihn gelöst.

Schwestern berichten in einer Balint-Gruppe davon, daß sie auf ihrer Station die Zuständigkeit für neuaufgenommene schwerkranke Patienten durch ein Losverfahren entscheiden.

Wie schwer die Verbindung von Gefühlsarbeit und instrumenteller Tätigkeit sein kann, zeigt das Beispiel einer Visite bei einem jungen Leukämiekranken.

Der Patient hat seit mehreren Tagen ungeklärtes Fieber, am ehesten kommt als Ursache eine Infektion am Subclaviakatheter in Frage. Der Arzt fragt den Patienten, wie es ihm gehe. Der Patient antwortet: „Ganz gut". Er sei nur beunruhigt durch das Fieber. Wahrscheinlich weiß er, an welche Ursache die Ärzte denken. Der Arzt wiederholt sehr einfühlsam: „Beunruhigt?" und fordert dadurch den Patienten auf, mehr von seiner Beunruhigung zu sprechen. Gleichzeitig nimmt der Arzt jedoch seinen Spatel aus dem Kittel, führt den Spatel in den Mund des Patienten ein und bittet ihn „ah" zu sagen und verhindert so ein Weitersprechen des Patienten. Das Thema Beunruhigung wird danach nicht wieder aufgegriffen.

Der Umgang mit unheilbar Kranken kann für uns oft auch zu einer Kränkung führen. Den Motiven unserer Berufswahl, den Zielen unserer Ausbildung und unseren eigenen Bedürfnissen entspricht es in der Regel mehr, Sieger denn Verlierer im Kampf gegen die Krankheit zu sein. Kränkung und Verunsicherung sind dann besonders ausgeprägt, wenn wir das Gefühl eigener Omnipotenz im Beruf zur Bewältigung der situationsbedingten Unsicherheit oder generell zur Stabilisierung unseres psychischen Gleichgewichtes benötigen.

Häufig wird diese Kränkung indirekt über Kompensationsversuche sichtbar: Im Anschluß an eine Visite hält ein Chefarzt alle Beteiligten 20 Minuten lang mit Erzählungen aus seiner Schulzeit auf; er berichtet von seinem Mut und seinem Erfolg während des Turnunterrichts.

Es wäre sicher günstig, die Rolle von Todesängsten bei Ärzten in der Ausbildung berücksichtigen und während der späteren Tätigkeit reflektieren zu können. Allgemeiner sollten Fragen im Zusammenhang mit berufsbedingter Unsicherheit und die Möglichkeit zum Erwerb sinnvoller Bewältigungsstrategien in den Unterricht einbezogen werden.

Bei Ärzten spielt die Abwehr eigener Todesängste eine bedeutsame Rolle: Feifel (1959, 1965) fand testpsychologisch, daß Ärzte in der Regel den Tod mehr fürchten als eine Vergleichsgruppe von Patienten. Kaspar (1965) wies darauf hin, daß starke, jedoch verdrängte Todesangst als unbewußtes Motiv bei der Berufswahl von Ärzten eine wichtige Rolle spielt.

Die Berufstätigkeit kann der Abwehr solcher persönlicher Ängste dienen. Die Konfrontation mit unheilbar Kranken droht in diesem Fall die ursprünglichen Ängste zu aktivieren und mobilisiert dann in verstärktem Maße die gegen diese Ängste gerichteten Abwehrprozesse. In Verbindung mit dem Bemühen um die Stabilisierung des eigenen Selbstgefühls kann ein Arzttyp entstehen, der ein ausgesprochenes Omnipotenzideal anstrebt. Dieses Ideal wird jedoch im Umgang mit unheilbar Kranken und Sterbenden ebenso in Frage gestellt, wie die Abwehr der eigenen Todesängste. Als Schutzreaktionen können dann beispielsweise die von Medizinsoziologen immer wieder beobachtete distanzierte Haltung gegenüber unheilbar Kranken und Sterbenden, aber auch ein übertriebener therapeutischer Aktivismus in aussichtslosen Situationen beobachtet werden (Glaser und Strauss, 1974; Siegrist, 1978; Sudnow, 1973; vgl. Kap. 18, Kap. 67).

Andererseits wird das Ausmaß der im Umgang mit unheilbar Kranken für Ärzte und Krankenschwestern auftretenden Belastungen bisher meist noch ganz erheblich unterschätzt. Diese Unterschätzung hängt auch damit zusammen, daß die kommunikative Tätigkeit, die „Gefühlsarbeit", aus der professionellen Tätigkeit ausgegliedert bleibt und in den Privatbereich, analog zu nicht beruflichen zwischenmenschlichen Beziehungen, verwiesen wird. Von den Helfern wird allenfalls „mehr Menschlichkeit" gefordert, mit den Belastungen jedoch werden sie alleine gelassen. Wird kommunikative Tätigkeit, „Gefühlsarbeit", dagegen als Bestandteil der professionellen Tätigkeit anerkannt, so wird es auch möglich, die hierfür erforderlichen Rahmenbedingungen und den mit dieser Tätigkeit verbundenen Aufwand und die erforderlichen Ausbzw. Fortbildungsmaßnahmen im einzelnen zu klären.

Bei einer sorgfältigen Betrachtung dieser Belastungen, etwa im Verlauf von Balint-Gruppen (Drees, 1981, 1984; Köhle et al., 1972; Naujoks et al., 1985), zeigt sich erst, in welchem Maße Ärzte und ganz besonders Krankenschwestern bei aller gegebenen Bereitschaft noch emotionale Arbeit leisten müssen, wollen sie einerseits sich auf das Schicksal und die Gefühle ihrer Patienten einlassen, andererseits professionelle Helfer bleiben. Es geht darum, eine Haltung im Sinne der „detached compassion" (Pattison, 1981) zu entwickeln. Wird Fassung und Gleichmut nicht über eine die Beziehung blockierende Abwehr aufrechterhalten, so erfordert dies eine ganz erhebliche integrative Leistung. A. Strauss spricht von „Identitätsarbeit", die Arzt und Pflegepersonal für sich zu leisten haben; zu dieser Identitätsarbeit gehört es beispielsweise, auch die für die Einfühlung in den Patienten notwendige Projektion eigener Bedürfnisse zurücknehmen zu können und für diese Bedürfnisse außerhalb der professionellen Beziehung Befriedigungsmöglichkeiten zu finden. Nicht selten projizieren wir ja eigene Bedürfnisse, z.B. eigene intensive Versorgungswünsche, auf Patienten und befriedigen unsere Wünsche bei der Versorgung dieser Patienten mit, indem wir uns mit den Kranken identifizieren und in dieser Form an ihrer Versorgung partizipieren. Dieser Aspekt des „Helfersyndroms" (Schmidbauer, 1977, 1983) erklärt eine weitere Facette der Verletzlichkeit der Helfer: Eine solche indirekte Befriedigung eigener Bedürfnisse ist nur dann ausreichend möglich, wenn die Patienten sich hierauf sensibel mit einstellen; weichen die Patientenbedürfnisse selbst oder die gewünschten Modalitäten ihrer Befriedigung zu sehr von den Erwartungen der Helfer ab, führt dies zu einer Irritation so motivierter Helfer.

Die individuelle Belastung der Helfer im Umgang mit unheilbar Kranken und der für ihre eigene Stabilisierung nötige psychische Aufwand können so groß werden, daß kaum mehr Kraft zur Reflexion und für die Suche nach Entlastungs- oder Hilfsmöglichkeiten bleibt. Eine Folge dieser oft extremen Belastung kann das Gefühl des „Ausgebranntseins" bei den Helfern selbst sein, das dann oft zum vorzeitigen Ausscheiden aus dieser Tätigkeit oder dem Beruf überhaupt führt (Aronson et al., 1983).

Teamkonflikte auf Krankenstationen sind in solchen Situationen häufig und können oft nicht mehr auf einer nur rationalen Ebene verstanden und gelöst werden. Dies hängt auch damit zusammen, daß zur geschilderten Verunsicherung und zur Frustration eigener Bedürfnisse oft noch Spannungen kommen, die durch Patienten induziert wurden: Klagen und Vorwürfe der Patienten stellen das Bemühen der Helfer in Frage; innere Konflikte zwischen guten und bösen Anteilen, die die Kranken selbst nicht mehr überbrücken können, werden durch Aufspaltung dieser Anteile und Projektion auf verschiedene Teammitglieder zu lösen versucht; dies führt dazu, daß die ursprünglich inneren Konflikte des Patienten oft externalisiert unter Teammitgliedern ausgetragen werden. Hilfsangebote von außen, etwa von konsiliarischen Diensten, können die Teammitglieder häufig nicht mehr annehmen; mit der Annahme des Hilfsangebotes wäre die Wahrnehmung der eigenen Hilfsbedürftigkeit verbunden; dies widerspricht jedoch dem eigenen Ideal und wird als persönliche Insuffizienz erlebt. Es mag paradox erscheinen, doch gerade hier liegt eine Quelle für die Entwertung sinnvoller Kooperationsangebote.

Wird diese komplexe Situation erkannt, so werden auch die zu fordernden Konsequenzen deutlich: Ärzte und Pflegepersonal benötigen während ihrer Ausbildung und während ihrer praktischen Tätigkeit vielfältige Unterstützung, von der Vermittlung von Verständniskonzepten bis zur Möglichkeit, eigene emotionale Reaktionen mit anderen besprechen und klären zu können. Auch für diesen Teil der professionellen Tätigkeit sind angemessene Organisationsformen zu schaffen; beispielsweise wird es nötig, auf Krankenstationen Konferenzen für alle Beteiligten einzurichten, die einen systematischen Informationsaustausch und eine Klärung des eigenen Verhaltens im Umgang mit dem Patienten gewährleisten (Köhle et al., 1980; Gaus et al., 1980; vgl. Kap. 28).

Gerade wenn Ärzte und das Krankenpflegepersonal ihre Patienten ganzheitlich behandeln und betreuen wollen, gerade wenn die an sie herangetragenen Bedürfnisse der Patienten nicht delegiert werden können, erscheint es dringend notwendig, Organisationsmodelle in der Krankenversorgung zu entwickeln, die die Kooperation und die gegenseitige Unterstützung aller an der Betreuung dieser Patienten Beteiligten – Ärzte, Schwestern, Seelsorger, Sozialarbeiter und eventuell hinzugezogene Fachpsychotherapeuten – gewährleisten. Bei jedem zu entwickelnden Lösungsansatz bleibt es deshalb zentral wichtig, die psychosoziale Betreuung im engsten Zusammenhang mit der medizinischen Behandlung durchzuführen, da die intensive Kommunikation auch mit unheilbar Kranken Voraussetzung jeder rationalen Planung von Diagnostik, Therapie und – soweit möglich – Rehabilitation ist. Noch nicht einmal Analgetika können ohne Kenntnis der emotionalen Reaktion des Patienten angemessen verordnet werden; Fragen der Compliance des Patienten bei verschiedenen Therapiemaßnahmen lassen sich nur unter Berücksichtigung der psychischen Situation und der Beziehung zum Arzt diskutieren. Eine ausschließliche Delegation der psychologischen Betreuung würde den Arzt darüber hinaus zum reinen Exekutivorgan von Therapieprotokollen reduzieren.

66.1.3 Die Situation der Kranken

Die Situation unheilbar Kranker, insbesondere unheilbar Krebskranker, wird vor allem von der Gefahr der sozialen Isolation und den Belastungen des Selbstgefühls bestimmt.

- Unheilbar Kranke befinden sich immer in Gefahr, sozial isoliert zu werden, ihre Bezugsgruppe zu verlieren, ja, ausgestoßen zu werden.
- Unheilbar Kranke sind in ihrem Selbstgefühl, insbesondere in ihrem Selbstwerterleben, häufig extrem labilisiert; mit dem Verlust der körperlichen Integrität und den Einbußen an Funktionsfähigkeit und anderen Krankheitsfolgen werden die psychischen Anpassungsmöglichkeiten oft überfordert.

– Unheilbar Kranke haben ein großes Bedürfnis nach Kommunikation und nach Orientierungsmöglichkeiten über ihre Situation.

– Unheilbar Kranke sind in hohem Maße für das Verhalten ihrer Umwelt und damit auch für unser Verhalten als Ärzte und Pflegepersonen sensibilisiert.

Viele von Sozialwissenschaftlern durchgeführte empirische Untersuchungen wiesen übereinstimmend auf die ausgeprägte Diskrepanz zwischen dem großen Kommunikationsbedürfnis der Patienten und dem zu geringen Kommunikationsangebot sowohl bei niedergelassenen Ärzten als auch im Krankenhaus hin. Zwar findet sich diese Diskrepanz im Krankenhaus bei allen Patienten, bei unheilbar Kranken jedoch in besonders hohem Ausmaß (Waitzkin und Stöckle, 1972; McIntosh, 1974, 1976; Hartmann, 1976; Ley, 1977; Glaser und Strauss, 1974, 1968; Sudnow, 1973; Engelhardt, 1973; Rohde, 1974; Siegrist, 1978; Raspe, 1982 a; zu Einzelheiten dieser Untersuchungen vgl. Kap. 18).

Dieses Kommunikationsdefizit führt nach Hartmann (1976) nicht selten zu einem ausgeprägten Deprivationssyndrom: er spricht von einem „psychischen Hospitalismus" auch bei Erwachsenen, der sich z.B. in Form eines depressiven Rückzugsverhaltens darstellen kann.

Berücksichtigen Ärzte und Schwestern die kommunikativen Bedürfnisse unheilbar Kranker nicht ausreichend, wehren sie deren Bedürfnisse ab und weichen sie den Kranken aus, so fühlen diese sich abgelehnt, mißachtet, gekränkt, ohnmächtig, erniedrigt. Sie können über ihre Angst, ihr Gefühl von Hilflosigkeit und die Minderung ihres Selbstwerterlebens nicht sprechen; enttäuscht ziehen sie sich selbst noch weiter aus den Beziehungen zurück. Ihre depressiven Gefühle sind dabei mit aggressiven Regungen gegen Ärzte und Pflegepersonen vermischt; sie müssen diese aggressiven Regungen aufgrund ihrer Abhängigkeit jedoch unterdrücken. Es entsteht schließlich ein Zustand „feindseliger Abhängigkeit".

Es ist wichtig, die Heftigkeit der unausgesprochenen aggressiven Regungen und die damit verbundenen Belastungen auf beiden Seiten, bei den Kranken und bei den Helfern, zu kennen, um die Gefahren dieser Situation nicht zu unterschätzen. Wir versuchen, diese Affekte durch Zitate aus Thomas Bernhards „Der Atem" (1981) zu illustrieren. Dieser Ausschnitt aus der Autobiographie des Dichters beschreibt, wie er als 18jähriger mit den Folgen einer Lungentuberkulose im „Sterbezimmer" liegt und unter dem Mangel an Kommunikationsmöglichkeiten leidet.

„Eine durch das Verhalten der Gesellschaft wahrscheinlich gerechtfertigte Verlegenheitslösung war diese Visite immer gewesen, die täglich die Ärzte, an jedem Freitag an ihrer Spitze auch den Primarius, in das Sterbezimmer geführt hatte ..." (es folgt die Schilderung der Wahrnehmung der Schwestern durch den Autor). „Die Visite hatte mir jedesmal die in weiß daherkommende Machtlosigkeit der Medizin gezeigt. Ihr Auftritt hatte immer nur Eiseskälte und mit dieser Eiseskälte die Zweifel an ihrer Kunst und an ihrem Recht

hinterlassen. Einzig und allein vor meinem Bett waren sie aus der Fassung geraten, weil sie es, immer wieder unvermutet und urplötzlich, jetzt hier im Sterbezimmer mit einem Lebenden und mit keinem Toten zu tun hatten. Hier waren sie, wenn auch nur untereinander, gesprächig und diskussionsbereit, wenn sie mir da auch immer unverständlich geblieben sind. Es war niemals möglich gewesen, mit ihnen einen tatsächlichen Kontakt aufzunehmen. Jeder Versuch in dieser Richtung war von ihnen gleich durch ein rüdes Zurück- und Zurechtweisen meiner Person abgebrochen worden. Sie wollten sich der Außenwelt, wie es den Anschein hatte, um keinen Preis, nicht einmal um den Preis einer ganz einfachen, ganz kurzen Unterhaltung, um den Preis eines auch nur angedeuteten Übermuts öffnen. Sie waren immer nur die an jedem Tag auf einmal und mit der gleichen Rücksichtslosigkeit vor meinem Bett aufgestellte weiße Mauer geblieben, in welcher kein menschlicher Zug zu entdecken war. Dem Jüngling waren die Ärzte immer als Schreckensbotschafter erschienen, an die ihn seine Krankheiten erbarmungslos ausgeliefert hatten. Er hatte zu den Ärzten immer nur eine Schreckensbeziehung haben können. Sie waren ihm niemals und in keinem Augenblick vertrauenserweckend gewesen ..."

Nach einer Ausdehnung des Angriffs gegen die Ärzte und ihre Kunst fährt Bernhard fort:

„Ich hatte ununterbrochen den Wunsch gehabt, mit meinen Ärzten zu sprechen, aber ausnahmslos haben sie niemals mit mir gesprochen, nicht die geringste Unterhaltung mit mir geführt. Meine Natur verlangte immer schon nach Erklärung, besser noch Aufklärung und ich wäre vor allem, was meine Ärzte betrifft, für Erklärung und Aufklärung dankbar gewesen. Mit den Ärzten war aber nicht zu sprechen gewesen. Sie haben sich in die Unbequemlichkeit einer Unterhaltung mit mir von vornherein nicht eingelassen. Immer hatte ich das Gefühl, daß sie vor Erklärung und Aufklärung Angst hatten. Und es ist ja Tatsache, daß die Kranken, die den Ärzten ausgeliefert sind in den Krankenhäusern, niemals mit Ärzten in Kontakt kommen, geschweige denn zu Erklärung und Aufklärung kommen. Die Ärzte schirmen sich ab, errichten die, wenn nicht natürliche, so doch künstliche Mauer der Ungewißheit zwischen dem Patienten und sich. Die Ärzte sind ununterbrochen hinter dieser von ihnen als Mauer aufgerichteten Ungewißheit verschanzt. Ja, sie operieren mit der Ungewißheit. Wahrscheinlich sind sie sich ihrer eigenen Unfähigkeit und also Machtlosigkeit bewußt und wissen, daß der Patient allein die Initiative zu ergreifen hat, will er seinen Krankheitszustand eindämmen oder aus seinem Krankheitszustand wieder herauskommen. Die wenigsten Ärzte geben zu, daß sie beinahe nichts wissen und ebenso beinahe nichts tun können. Die Ärzte, die hier im Sterbezimmer Visite machten, hatten ihre Patienten niemals aufgeklärt und hatten alle diese Patienten im Stich gelassen. Im medizinischen und im moralischen Sinn. Ihre Medizin war naturgemäß machtlos, ihre Moral wäre ihnen ein zu hoher Einsatz gewesen. Hier notiere ich, was im Kopf des Jünglings vorgegangen ist, der ich damals gewesen bin, nichts weiter. Später mag alles in einem anderen Licht erschienen sein, damals nicht" (S. 52–56).

Ärzte und Schwestern spüren in der klinischen Situation auch die abgewehrten aggressiven Impulse, reagieren darauf ihrerseits gereizt oder ziehen sich noch weiter zurück; dieser sich negativ verstärkende Prozeß kann zu einer zunehmenden Isolation der Patienten und zu weiteren negativen Reaktionen auf seiten der Ärzte und Schwestern führen. In diesem Prozeß kommt es oft auch zu einer Fehl- oder doch Überinterpretation des Rückzugsverhaltens des Patienten als

Folge vermeintlicher organischer Ursachen, wie zerebralsklerotischer Veränderungen oder Hirnmetastasen.

Über solche Erklärungsansätze wird es dann wieder möglich, eigene aggressive Gereiztheit bzw. Schuldgefühle durch Mitleid für den „armen" Kranken zu ersetzen, der ja aus organischen Gründen nicht mehr zur Kommunikation fähig ist. Der Arzt und die anderen Mitarbeiter einer Krankenstation werden in einer solchen Situation aber auch mit der eigenen Berufstätigkeit unzufriedener, sie stellen gelegentlich die eigene Kompetenz oder die von Kollegen in Frage. Forscher, die solche Situationen beobachteten (u.a. A. Strauss et al., 1980), beschrieben auch, daß in solchen Beziehungskrisen – und meist erst dann – „Spezialisten" wie Psychiater, Psychosomatiker oder Seelsorger zugezogen bzw. die „schwierigen" Patienten diesen Spezialisten zugewiesen werden. Die Überweisung des Patienten bildet jetzt die Alternative zur eigenen Inanspruchnahme fachkompetenter Hilfe, die eine engere Zusammenarbeit, etwa nach dem Liaison-Modell (vgl. Kap. 28), voraussetzen würde.

Im folgenden Beispiel berichten wir über eine solche ungünstige Entwicklung, einen Interventionsversuch und den anschließenden Verlauf.

Eine 45jährige Frau wurde wegen metastasierendem Mammakarzinom aufgenommen. Sie erschreckte die Schwestern anfangs durch unerklärliche Verhaltensweisen: so legte sie sich z.B. quer ins Bett, zog sich völlig nackt aus oder spuckte den Schwestern beim Füttern die Tabletten ins Gesicht. Dies und ihr völlig unselbständiges Verhalten – sie ließ sich füttern und waschen – tolerierten die Schwestern aufgrund des Nachweises von Hirnmetastasen und Hirndruckzeichen. Sie rechneten damit, daß es mit der Patientin bald zu Ende gehe und bemühten sich zumindest nicht zusätzlich um eine Intensivierung der Kommunikation. Die Kranke lag schließlich im Einzelzimmer der Station, das am weitesten vom Arbeitszimmer der Schwestern entfernt war; sie war völlig verstummt, auch bei Fragen war ihr Vokabular auf „ja" und „nein" zusammengeschrumpft, meist reagierte sie nur mit Kopfschütteln oder abwehrenden Handbewegungen. Dabei hatte ihr Aussehen – die schwarzen Haare fielen wirr in ihr bleiches, ungepflegtes Gesicht, aus dem die dunklen Augen deutlich hervortraten – für die Schwestern etwas Furchterregendes an sich.

Während der Visite fiel allerdings auf, daß die Patientin auf direkte Anrede nicht reagierte und die Gespräche über sich hinweggehen ließ, aber dennoch aktiv beteiligt, wie in einer bemühten Abwehrhaltung wirkte. Auf die psychosomatisch ausgebildete Schwester machte sie den Eindruck, als habe sie ihre eigenen Gedanken zu dem Schauspiel der Visite, als wüßte sie im Grunde mehr über ihren Zustand als die Ärzte, die sich mit ihrer Kurve beschäftigten. Beim Verlassen des Zimmers schaute die Kranke der Visite nach, ohne dabei den Kopf zu bewegen. Auch den Schwestern fiel auf, daß die Patientin ihnen, ohne den Kopf zu bewegen, beim Verlassen des Zimmers nachsah, bis die Tür ganz geschlossen war. Eine Schwester erlebte dies als versteckte Aufforderung, die Tür noch einmal zu öffnen und die Patientin aus der Distanz zu fragen, ob sie noch etwas brauche oder vergessen habe.

Diese und weitere Beobachtungen sprachen eher dafür, daß der extreme Rückzug der Patientin als Folge von Depression, ja Verzweiflung angesehen werden sollte, als daß er allein durch die hirnorganischen Veränderungen erklärt werden könnte.

Über das Verhalten der Kranken wurde in der Stationskonferenz gesprochen. Danach begann eine Schwester, sich intensiver mit ihr zu beschäftigen. Nach einer Woche wagte die Patientin sich in kleinen Schritten aus ihrer Zurückgezogenheit heraus. Sie begann sich wieder verbal zu äußern, auch wenn zunächst heftige Ablehnung und Verneinung im Vordergrund standen. Allmählich lernte sie wieder, manches für sich selbst zu tun, begann zu fragen, Wünsche zu äußern, und schließlich auch von ihrer Angst zu sprechen. Die Kommunikation intensivierte sich immer mehr, allmählich war die Kranke fast unbemerkt zum Mittelpunkt der Station geworden. Nach zwei Wochen fragte die Patientin völlig unerwartet und unvermittelt während der Visite: „Was hab' ich denn eigentlich jetzt für eine Krankheit?" Als der Stationsarzt sie zu informieren begann, äußerte sie selbst die Vermutung, daß sie jetzt auch im Gehirn Krebsgeschwülste habe.

Durch das konsequente Bemühen der Schwestern und das neue Verständnis des Arztes war wieder Kommunikation möglich geworden, die Patientin hatte aus ihrer Rückzugshaltung herausgefunden. Die Zuwendung ihrer Mitwelt ermöglichte ihr, nun wieder Interesse an ihrer Umwelt zu haben. Stück für Stück konnte sie mit Hilfe der Schwestern ihren Aktionsraum erweitern, parallel dazu wuchs ihr Selbstwertgefühl. Gespräche mit der Familie zeigten, daß sie keineswegs – wie sie es erlebt hatte – aufgrund ihrer Leistungsunfähigkeit von Mann und Tochter als wertlos erlebt wurde und „abgeschrieben" war. Trotz fortbestehender Hirnmetastasen und leichter Hirndruckzeichen konnte die Patientin im weiteren Verlauf noch einmal für mehrere Wochen nach Hause entlassen werden, wo sie sich selbst versorgte und zum Teil auch den Haushalt führen konnte.

66.1.4 Zielvorstellungen für den Umgang mit unheilbar Kranken

Aus der dargestellten Problemanalyse ergeben sich für uns folgende Zielvorstellungen für eine in die klinische Versorgung integrierte psychosomatische Betreuung unheilbar Kranker:

Die Therapie des Grundleidens, palliative Maßnahmen – insbesondere Schmerztherapie – und das Eingehen auf die kommunikativen Bedürfnisse des Patienten sollten sorgfältig aufeinander abgestimmt werden. Der Patient sollte dabei mit uns als informierter und weitestmöglich selbständiger Partner in der Behandlung seiner Erkrankung kooperieren können, ausreichend Hilfe bei der Verarbeitung seiner Krankheit, der aus ihr resultierenden Behinderungen sowie in seiner Trauerarbeit erhalten. Wir sollten den Kranken darin unterstützen, seine psychischen und physischen Kräfte soweit wie möglich zu erhalten oder wieder zu mobilisieren, um die verbleibende Lebenszeit gemäß seiner Persönlichkeit optimal nutzen zu können. Ziel unserer Hilfestellung sollte auch sein, die Integrität seiner Person ebenso wie seine zwischenmenschlichen Beziehungen möglichst weitgehend aufrechtzuerhalten oder wiederherzustellen. Unser Arbeitsbündnis mit dem Patienten sollte zu seiner emotionalen Stabilisierung beitragen und es ermöglichen, drohende gravierende emotionale Komplikationen rechtzeitig zu erkennen und nach Möglichkeit zu verhindern.

Wir versuchen in diesem Beitrag, dem Verlauf der Erkrankung bzw. dem Verlauf der Beziehung zwischen Helfern und unheilbar Kranken zu folgen. Am Beginn der Beziehung zwischen Arzt und unheilbar Krankem steht in der Regel die Frage, ob der Arzt sich in eine offene Kommunikation mit dem Patienten über das Wesen seiner Erkrankung, auf eine Kommunikation über die Diagnose einlassen will. Die längerfristige Entwicklung des Umgangs mit unheilbar Kranken ist ebenso sehr von den Prozessen der Krankheitsverarbeitung auf seiten des Patienten wie von den Reaktionen auf seiten seiner Umwelt abhängig. Diese Reaktionen auf seiten von Ärzten, Schwestern und aller anderen an der Versorgung unheilbar Kranker Beteiligter werden wir ebenfalls ausführlicher darstellen, da ihre Kenntnis und Berücksichti-

gung eine längerfristige Betreuung Krebskranker erst ermöglicht. Ebenso bedeutsam erscheinen uns auch die Probleme im Umgang mit Angehörigen unheilbar Kranker; ihre Kenntnis ermöglicht oft Interventionen, die die so nötige Unterstützung des Patienten durch seine soziale Umwelt – heute oft unter dem Begriff „social support" diskutiert – entscheidend verbessern können. Schmerzen können die Lebensqualität unheilbar Kranker erheblich beeinträchtigen; hier hat der Arzt praktisch immer die Möglichkeit zu wirksamer Hilfe; empirische Untersuchungen haben – so überraschend es sein mag – jedoch gezeigt, daß diese Hilfsmöglichkeit bei weitem nicht im möglichen Ausmaß und nicht konsequent entsprechend den pathophysiologischen und pharmakologischen Gegebenheiten angewandt wird (Angell, 1982; Schreml et al., 1981). Wir stellen deshalb auch die Schmerztherapie in diesem Kapitel ausführlicher dar. Am Beispiel der längerfristigen Betreuung eines jugendlichen Krebskranken schließlich versuchen wir, sowohl die Betreuungsmöglichkeiten als auch die für den Betreuer hiermit verbundenen Belastungen zu veranschaulichen.

66.2 Die Kommunikation über die Diagnose

„Die Hauptqual für Iwan Iljitsch war die Lüge – jene aus irgendeinem Grunde von allen anerkannte Lüge, daß er nur krank sei, nicht aber sterbe, und daß er sich nur ruhighalten und die Kur durchführen müsse, damit alles wieder sehr gut werde. Er aber wußte: sie konnten tun, was sie wollten, es würde doch nichts mehr herauskommen als noch qualvollere Leiden und der Tod. Und ihn quälte diese Lüge, es quälte ihn, daß man nicht eingestehen wollte, was alle wußten, und was auch er wußte, und daß man ihn über seine entsetzliche Lage belügen und ihn zwingen wollte, an dieser Lüge teilzunehmen. Die Lüge, die Lüge, dieser an ihm am Vorabend seines Todes verübte Betrug, die Lüge, welche dieses schreckliche, feierliche Ereignis seines Todes auf das Niveau aller ihrer Besuche und Gardinen sowie des Störs zum Mittagessen herabdrücken sollte ... das war schrecklich, qualvoll für Iwan Iljitsch. Und seltsam! Er war viele Male, während sie mit ihm alle diese ihre törichten Dinge anstellten, um ein Haar nahe daran, sie anzuschreien: So hört doch auf zu lügen, ihr wißt es, und ich weiß es, daß ich sterbe, so hört doch wenigstens auf zu lügen! Aber er hatte nie den Mut, dies zu tun. Der schreckliche, entsetzliche Vorgang seines Sterbens – er sah es – wurde von den Seinigen auf die Stufe einer zufälligen Unannehmlichkeit, zum Teil sogar einer Unschicklichkeit herabgesetzt (in der Art, wie man mit einem Menschen umgeht, der beim Betreten des Salons einen üblen Geruch um sich verbreitet); man verfuhr gemäß derselben ‚Schicklichkeit', der er sein ganzes Leben lang gedient hatte; er sah, daß niemand mit ihm Mitleid hatte, weil niemand seine Lage auch nur verstehen wollte."

„Außer oder infolge dieser Lüge war für Iwan Iljitsch am quälendsten, daß ihn niemand so bemitleidete, wie er sich wünschte, bemitleidet zu werden: in manchen Augenblicken, nach langen Schmerzen, wünschte Iwan Iljitsch mehr als alles, so sehr er sich schämte, es einzugestehen, daß ihn jemand wie ein krankes Kind bemitleide. Er wollte, daß man zu ihm zärtlich sei, ihn küsse und über ihn weine, wie man Kinder liebkost und tröstet. Er wußte, daß er ein hoher Gerichtsbeamter war, daß sein Bart schon zu ergrauen anfing und daß infolgedessen dies alles unmöglich war; und dennoch verlangte ihn danach." ... „Diese Lüge rings um ihn und in ihm selbst vergiftete am meisten die letzten Lebenstage Iwan Iljitschs."

„... daß er einen berühmten Arzt aufsuche. Er fuhr hin. Alles war, wie er erwartet hatte. Alles war so, wie es immer gemacht wird. Auch die Erwartung war dieselbe, die er bei sich im Gericht kannte und das Beklopfen und Behorchen und die Fragen, die wohl im voraus bestimmte und darum unnötige Antworten verlangten, und die bedeutsame Miene, die zu verstehen gab: ‚Sie müssen sich nur uns überantworten, und wir werden es schon machen. – Wir wissen, und daran ist nicht zu zweifeln, wie alles gemacht werden muß, alles auf die eine Art bei jedem Menschen, bei wem Sie wollen' ... Es war alles genau wie beim Gericht. Dieselbe wichtige Miene, die er im Gericht den Angeklagten zeigte – hier wurde sie von dem berühmten Arzt ihm selber gezeigt. Der Arzt sagte: ‚Das und das weist darauf hin, daß in Ihrem Innern das und das vorhanden ist; wenn aber das nach den Untersuchungen von dem und dem sich nicht bestätigt, dann wird man bei Ihnen das und das annehmen. Wenn man aber das und das annimmt ..., dann ...' Für Iwan Iljitsch war nur die eine

Frage wichtig: ob sein Zustand gefährlich sei oder nicht. Der Arzt ignorierte diese unangebrachte Frage. Vom Standpunkt des Arztes war es eine müßige Frage, die nicht zur Erörterung stand; für ihn gab es nur das Abwägen der Wahrscheinlichkeiten, ob es eine Wanderniere, ein chronischer Darmkatarrh oder eine Erkrankung des Blinddarms war. Für ihn gab es keine Frage nach dem Leben Iwan Iljitschs, sondern es gab nur einen Streit zwischen der Wanderniere und dem Blinddarm. Und diesen Streit entschied der Doktor vor den Augen Iwan Iljitschs aufs glänzendste, zugunsten des Blinddarms, unter dem Vorbehalt, daß die Harnuntersuchung neue Indizien ergeben könne und das Urteil revidiert werden müsse. Alles das war ganz genau dasselbe, was Iwan Iljitsch selbst tausendmal an den Angeklagten in so glänzender Weise vollbracht hatte. Ebenso glänzend machte der Arzt sein Resümée und sah dabei triumphierend, sogar fröhlich über die Brille hinweg den Angeklagten an. Aus dem Resümée des Doktors folgerte Iwan Iljitsch, daß es um ihn schlecht stehe, daß dies aber ihm, dem Doktor, und vielleicht auch allen anderen gleichgültig sei, er aber leiden müsse. Und diese Schlußfolgerung traf Iwan Iljitsch schmerzlich, indem sie in ihm das Gefühl des großen Mitleids mit sich selbst und der großen Wut gegen diesen Doktor, dem eine so wichtige Frage gleichgültig war, erregte.

Er sagte aber nichts, sondern stand auf, legte das Geld auf den Tisch und sagte mit einem Seufzer: ‚Wir Kranke richten wohl oft unangebrachte Fragen an Sie ... überhaupt ist es eine gefährliche Krankheit, oder nicht? ...' Der Arzt sah ihn streng mit einem Auge über die Brille hinweg an, als wolle er gleichsam sagen: ‚Angeklagter, wenn Sie sich nicht in den Grenzen der an Sie gerichteten Fragen halten wollen, werde ich gezwungen sein, anzuordnen, daß man Sie aus dem Sitzungssaal entfernt.'

‚Ich habe Ihnen bereits gesagt, was ich zu sagen für notwendig und passend hielt', sagte der Doktor. ‚Das weitere wird die Untersuchung ergeben'. Und der Doktor verbeugte sich."

Leo Tolstoi: „Der Tod des Iwan Iljitsch"

66.2.1 Zur Entwicklung der Einstellung von Ärzten in dieser Frage

Über die Frage der sog. „Diagnosemitteilung" auch an unheilbar Kranke wird zwar auch heute noch unter Ärzten gelegentlich kontrovers diskutiert. Insgesamt hat sich die Einstellung der Ärzte während der letzten 30 Jahre im angloamerikanischen, aber diesem folgend auch im deutschen Sprachraum eindrucksvoll gewandelt: Während früher die Mehrzahl der Ärzte eine Information von Malignomkranken über ihre Diagnose ablehnte, wird dieses Vorgehen heute von der Mehrzahl befürwortet. Dabei muß allerdings offenbleiben, inwieweit dieser Einstellungsänderung bereits eine entsprechend weitgehende und auch qualitativ ausreichende Veränderung des durchschnittlichen ärztlichen Handelns entspricht.

Historisch gesehen spielte in dieser Diskussion die Darstellung von Einzelerfahrungen eine wesentliche Rolle. Auffallend ist dabei, daß die hieraus abgeleiteten Konsequenzen über so lange Zeit nicht systematisch mit wissenschaftlichen Methoden überprüft wurden (Simpson, 1982). Dies könnte damit zu tun haben, daß es hier weniger um die Abklärung einer wissenschaftlichen Fragestellung als um die Aufrechterhaltung einer dem Arzt Sicherheit und Schutz bietenden Ideologie ging (Jonasch und Sellschopp, 1984).

Die Ablehnung einer offenen Kommunikation mit dem Patienten wurde in der Regel mit einer möglichen Gefährdung des Patienten begründet, oder, wie es Hufeland in apodiktischer Kürze formulierte: „Den Tod verkündigen, heißt den Tod geben" (nach Schadewaldt, 1969).

„Im Jahre 1868 kam ein wegen seiner Tapferkeit im Kriege mehrfach ausgezeichneter Oberst in Uniform zu Billroth an die Klinik und erbat sich von ihm die volle Wahrheit über seine Erkrankung. Er habe im Felde dem Tod oft ins Auge gesehen und sei auf das Schlimmste gefaßt. Billroth klärte ihn daraufhin nach gründlicher Untersuchung über die krebsige Natur seines Zungenleidens auf. Der Kranke empfahl sich unter aufrichtigen Danksagungen, verließ das Zimmer und stürzte sich sofort vom Gangfenster des ersten Stockes herab, wobei er sich tödlich verletzte und beinahe einen Assistenten der Klinik erschlagen hätte. Dieser tragische Ausgang machte einen großen Eindruck auf Billroth und alle anwesenden Ärzte. Der Meister berichtete oft über dieses Erlebnis in der Vorlesung" (von Eiselsberg, 1941 nach Raspe, 1982 b).

Ein Zitat aus Bismarcks „Gedanken und Erinnerungen" zeigt, vor welchem Hintergrund diese Erfahrung, aber auch die inzwischen abgelaufene Entwicklung zu sehen ist (vgl. P. Meerwein, 1980):

„Die behandelnden Ärzte waren am 20. Mai 1887 im Begriff, den Kronprinzen bewußtlos zu machen und die Exstirpation des Kehlkopfes auszuführen, ohne ihm ihre Absicht angekündigt zu haben. Ich erhob Einspruch, verlangte, daß nicht ohne die Einwilligung vorgegangen und, da es sich um den Thronfolger handle, auch die Zustimmung des Familienhauptes eingeholt werde. Der Kaiser, durch mich unterrichtet, verbot, die Operation ohne Einwilligung seines Sohnes vorzunehmen."

Felix Deutsch, Psychoanalytiker und Freuds Hausarzt, hatte 1923 Freud über den malignen Charakter seiner Gaumenerkrankung nicht offen informiert. 1956, anläßlich Freuds 100. Geburtstag, stellt Deutsch noch einmal seine seinerzeitigen Überlegungen dar, berichtet aber auch über einen späteren, an ihn gerichteten Brief Freuds, in dem dieser erwähnt, wie er voller Anklage darüber gewesen sei, daß ihm damals die richtige Diagnose vorenthalten wurde und daß er sich getäuscht gefühlt hatte:

„Ich konnte mich immer jeder Art von Realität anpassen und sogar Unsicherheit ertragen, sofern sie durch die Realität erzwungen war. Aber alleingelassen sein mit der eigenen Unsicherheit, ohne die Stütze und das Kissen der Ananke, der unerbittlichen und unausweichlichen Notwendigkeit, bedeutet, die Beute der erbärmlichen menschlichen Feigheit und ein unwürdiges Schauspiel für andere zu werden" (Deutsch, 1956).

Die Voraussetzung dieser früheren, kontroversen Diskussion bestand in der Annahme, unheilbar Kranke seien vollständig unfähig, ihre Situation selbst zu interpretieren und folglich ganz auf die Information und Interpretation durch ihre Ärzte angewiesen (Begemann-Deppe, 1976). Alle vorliegenden empirischen Untersuchungen widersprechen jedoch der Gültigkeit dieser Annahme ebenso wie der früher befürchteten zusätzlichen Gefährdung der Patienten durch offene Kommunikation über das Wesen ihres Leidens.

66.2.2 Das Ziel: Arzt und Patient teilen eine gemeinsame Wirklichkeit
Kommunikation versus „Aufklärung"

Alle empirischen Untersuchungen zeigen, daß die dargestellte Kontroverse tatsächlich von falschen Voraussetzungen ausgeht, daß die dichotomisierende Fragestellung „Aufklärung oder nicht?" falsch gestellt ist. Praktisch alle Patienten bringen nämlich – zumindest in der Klinik – bereits ein Vorwissen um die mögliche Lebensbedrohlichkeit ihrer Erkrankung mit. Sie sind dabei keineswegs ausschließlich auf explizite, verbale Mitteilungen ihrer Ärzte angewiesen, um sich in ihrer neuen Lebenssituation orientieren zu können. So zeigen mehrere Untersuchungen, daß mindestens 90% aller Malignompatienten im Verlauf der Erkrankung auch ohne solche „Aufklärung" ihre Diagnose in Erfahrung bringen (Begemann-Deppe, 1976; Oken, 1961; Bahnson, 1975; Kübler-Ross, 1972).

Zunächst möchten wir die Art dieses Vorwissens und mögliche Informationsquellen der Patienten an Beispielen illustrieren:

Eine 53jährige, äußerlich undifferenziert wirkende Geschirrspülerin beklagt sich im Erstgespräch darüber, daß sie vom Hausarzt und von den Ärzten auswärtiger Krankenhäuser mit ihren Fragen nach dem Wesen der vorliegenden Erkrankung nur abgewiesen worden sei. Ich spreche sie darauf an, daß ihr doch sicher selbst viele Gedanken durch den Kopf gegangen seien. Darauf meint die Patientin: „Wissen Sie, ich bin halt immer blutärmer geworden. Da ich nach außen kein Blut verloren habe, habe ich mir gedacht, das kann nur innerlich von einer Art Krebs aufgefressen werden." Die Mitteilung der Diagnose einer akuten Leukämie konnte bei dieser Patientin ohne weiteres an ihr eigenes Vorwissen anknüpfen.

Vor der Tür einer 19jährigen, erst seit wenigen Tagen erkrankten Patientin unterhielten sich Ärzte und Angehörige darüber, ob und wie man der Patientin die Diagnose einer akuten Leukämie mitteilen sollte. Als ich die Patientin im ersten Gespräch fragte, was sie sich denn selbst für Gedanken über ihre Erkrankung ge-

---

macht habe, meinte sie: „Ich kenne die Diagnose schon." Auf der Toilette stand ein Urinkrug, an dem ein Zettel mit dem Namen der Patientin und der Diagnose „Verdacht auf akute Leukämie" angebracht war.

Eine 17jährige, ebenfalls an akuter Leukämie erkrankte Patientin wurde gerade noch rechtzeitig aufgeklärt, bevor sie im Verlauf einer Schulführung durch ein Forschungszentrum alle Einzelheiten über Symptome und Prognose der Leukämie dargestellt bekam.

Einer anderen Kranken mit akuter Leukämie wurde von Ärzten und Schwestern eines auswärtigen Krankenhauses keine Information über ihre Erkrankung gegeben. Vor der Verlegung in die Ulmer Klinik empfiehlt ihr jedoch die Stationsschwester, besser noch zu beichten und vermittelt ihr so die krankheitsbedingte Bedrohung.

Bei diesen Beispielen handelt es sich nicht um Einzelfälle, sondern um typische Vorkommnisse.

Für die Orientierung der Kranken in ihrer neuen Lebenssituation spielt das gesamte Verhalten ihrer Bezugspersonen, nicht nur deren ausdrückliche verbale Mitteilungen, eine große Rolle. Die Patienten sind für das Verhalten der anderen außerordentlich sensibilisiert und nehmen auch Veränderungen im Ausdruck ihrer Bezugspersonen und andere Formen averbaler Kommunikation in erhöhtem Maße wahr. Sie spüren deutlich, wenn Angehörige oder Ärzte nicht offen mit ihnen kommunizieren, ihnen im Gespräch ausweichen, bestimmte Themen vermeiden oder ihnen ungerechtfertigt Hoffnung machen.

Patienten nehmen selbstverständlich die Betroffenheit ihrer Angehörigen wahr, auch wenn diese zur eigenen Beruhigung zunächst ein offenes Gespräch über den Ernst der Situation vermeiden. Nicht selten kommen solche Angehörige weinend aus dem Krankenzimmer und beteuern dem Arzt gegenüber, der Kranke wisse nichts von seiner Lebensgefährdung.

Die Sensibilisierung des Patienten resultiert zunächst aus seinem Gefühl der Gefährdung. Wird die Kommunikation hierüber mit ihm eingeschränkt, so ist er in seiner Verunsicherung verstärkt auf indirekte Zeichen angewiesen und wird sich an ihnen zu orientieren versuchen. Dies gilt auch für das Erleben des ärztlichen Verhaltens.

B. Harker (1972), eine amerikanische Sozialwissenschaftlerin, die an einem Mammakarzinom erkrankte, beschrieb ihre Erfahrung: „Die erste intensive emotionale Reaktion trat bei mir genau in dem Moment ein, als der Arzt bei der Untersuchung eines Knotens meiner Brust innehielt und dann die Untersuchung fortsetzte".

Kleine Verhaltensänderungen des Arztes erhalten Signalcharakter. So meinte ein Patient: „Als der Professor sich bei mir aufs Bett setzte, habe ich gewußt, jetzt muß ich sterben".

Ein ca. 60jähriger Patient trägt wegen Herzrhythmusstörungen seit Jahren einen Schrittmacher; er kommt regelmäßig einmal im Jahr zu einer kurzen stationären Kontrolluntersuchung. Über seine eben gemachten Er-

---

fahrungen berichtet er während des Interviews im Psychosomatischen Praktikum: Er begegnete seinem Stationsarzt auf dem Flur der Krankenstation. Der Arzt sagt zu ihm: „Kann ich Sie sprechen?" Der Patient erschrickt, bisher hatte es immer nur eine kurze Abschlußbesprechung gegeben. Der Arzt bittet den Patienten in sein Zimmer und bietet ihm einen Stuhl an. Der Patient erlebt dies zum ersten Mal. Er sagt zum Arzt: „Muß ich sterben?" Der Arzt teilt dem Patienten mit, daß der dringende Verdacht auf Lungenkrebs besteht.

Eine zum Infektionsschutz in einem Isolierbettsystem behandelte jugendliche Leukämiekranke berichtete, daß sie am Abend nach der Knochenmarkspunktion das Ergebnis vorhersagen könne: „Einen guten Befund erfahre ich immer am selben Tag bis 17 Uhr, einen schlechten erst am nächsten Tag." Dieselbe Patientin beobachtete, daß der behandelnde Arzt bei einem guten Befund näher an ihr Bett trat, bei einem schlechteren größere Distanz hielt.

Für den Arzt, der nicht offen mit seinen Patienten kommuniziert, ist ihr Vorwissen oft nur schwer beurteilbar, da die Patienten sich auf ihre Ärzte angewiesen fühlen und auf deren Schutz- und Abwehrhaltungen Rücksicht nehmen.

So meinte ein Leukämiepatient, mit dem in Ulm offen über seine unheilbare Erkrankung gesprochen worden war, vor seiner Rückkehr an eine andere Universitätsklinik, an der er auf seine diesbezüglichen Fragen früher keine Antworten erhalten hatte: Der dortige Arzt sei so nett und väterlich zu ihm gewesen, daß er **ihm** die Konfrontation mit seinem jetzigen Wissen nicht zumuten wolle; deshalb werde er dort offene Gespräche über seine Erkrankung wie früher vermeiden.

Ein Arzt berichtet, ein Patient mit Kolonkarzinom habe ihn bei der Visite gefragt: „Herr Doktor, ich habe doch …" Der Arzt erwartete mit Schrecken, jetzt werde der Patient sagen: „Krebs?", doch der Patient fuhr nach kurzer Pause fort: „… nichts Bösartiges?" Der Arzt meinte in der Diskussion, der Patient habe seine Abwehr gespürt und versucht, ihm die Konfrontation mit der tödlichen Erkrankung zu ersparen.

Zur Diskussion steht nach Kenntnis dieser Befunde für uns nicht mehr die Frage: „Aufklären oder nicht?", nicht die Frage nach der „Mitteilung der Diagnose", sondern die Frage nach der Art des Umganges mit dem Kranken, der bereits über sein eigenes, jeweils individuelles Vorwissen verfügt. Dieses Vorwissen – der amerikanische Ausdruck „middle knowledge" trifft dies gut – kann als Resultat von Informationsmöglichkeiten einerseits und Abwehrvorgängen gegen das Bewußtwerden der Bedrohung andererseits aufgefaßt werden. Dabei ist zu berücksichtigen, daß nicht nur die aktiven Bemühungen um Orientierung des Patienten, sondern auch seine innerseelischen Abwehrvorgänge im Zusammenhang mit seiner sozialen Situation betrachtet werden müssen. Auch die Verleugnung ist in diesem Sinne ein „sozialer Akt" (Weisman und Hackett, 1967).

Dies illustriert das Verhalten eines 42jährigen Patienten mit Kolonkarzinom. Der Patient hatte zwei Jahre vor der eigenen Erkrankung seinen Vater und seinen älteren Bruder unter der gleichen Symptomatik an derselben Krankheit verloren. Ärzten gegenüber äußerte er niemals auch nur die leichtesten Beschwerden, gegenüber den Schwestern klagte er über heftigste Leibschmerzen, das Hausmädchen fragte er: „Gell, ich habe Krebs?" Es ist klar, daß die Mitteilung des Hausmädchens wieder leichter aus dem Bewußtsein entfernt werden kann als eine etwaige Bestätigung durch den Arzt. Angemerkt sei, daß derselbe Patient, der sich den Ärzten gegenüber gesund darstellte, in hochdramatischer Weise vom Weltuntergang träumte und sich zunächst mit seinen angstvollen Phantasien alleine herumschlagen mußte, da die Ärzte dieser Station eine offene Kommunikation ablehnten.

Aufgabe des Arztes ist es, die Kommunikation mit dem unheilbar Kranken in Dialogform zu führen. Der Arzt gewinnt in diesem Dialog Zugang zum Vorwissen des Kranken, zu seinen existentiellen Ängsten, zu Phantasien über die Erkrankung und Abwehrvorgängen und knüpft dann, im einzelnen darauf abgestimmt, mit seinen Informationen daran an. Der Arzt versucht also die individuelle Wirklichkeit (v. Uexküll) seines Patienten erst einmal kennenzulernen, bevor er ihm aus seiner Sicht die zu erwartenden krankheitsbedingten Veränderungen nahezubringen versucht. Dabei ist es wichtig, daß der Arzt aufgrund seiner Information über den Patienten berücksichtigt, welche **individuelle Bedeutung** die Erkrankung und ihre Folgen für den jeweiligen Patienten haben. Zuerst sollte also immer der Patient zu Wort kommen und all die Gedanken und Phantasien äußern können, die ihn beschäftigen; erst dann blendet sich der Arzt ein. Durch ein solches Vorgehen ist die Mitteilung der Diagnose gleichsam „ex cathedra" im Sinne der sog. „Aufklärung" von vornherein ausgeschlossen. In diesem Vorgehen wird vielmehr eine Vermittlung zwischen der Sicht des Arztes und der jeweiligen individuellen Wirklichkeit des Patienten angestrebt, so daß es im Verlauf allmählich gelingt, daß Arzt und Patient eine gemeinsame Wirklichkeit (v. Uexküll; Weisman, 1979) teilen. In einer solchen Entwicklung wird der Arzt sich einerseits immer wieder rückversichern, wie der Patient die Information aufnimmt, und dem Kranken andererseits immer wieder die Ziele des Vorgehens erläutern.

Der Arzt bzw. alle Mitarbeiter einer Krankenstation stehen somit vor der Aufgabe, die durch die Krankheit gegebene neue Situation mit dem Patienten über längere Zeit durchzuarbeiten. Während heute im klinischen Alltag die „Diagnose" häufiger als früher mitgeteilt wird, ergeben sich – bedingt vor allem durch Mangel an Zeit, Raum und Ausbildung – häufig große Defizite in der weiteren Kommunikation mit der Folge zum Teil grotesker Mißverständnisse bzw. Mängel bei der Informationsaufnahme und -verarbeitung durch den Patienten, wie sie aus Untersuchungen über die Einholung eines „informed consent" bei experimentellen Therapiemaßnahmen bekannt sind (Köhle et al., 1982).

Aus dem prozeßhaften Charakter solcher Kommunikation über die Krankheit ergibt sich die Forderung, nach Möglichkeit mit dem Patienten über längere Zeit in Kontakt zu bleiben. Ist eine solche längere Betreuung abzusehen, dann bewährt es sich, bereits bei der Erstuntersuchung des Patienten darauf hinzuweisen, daß er über alle erhobenen Befunde informiert werden wird und ihn anschließend zu fragen, ob er mit einem solchen Vorgehen einverstanden sei.

Dieses Vorgehen empfiehlt sich unseres Erachtens allerdings bei allen Kranken. Erst im Falle der Diagnose einer malignen Erkrankung mit dem Patienten die Kommunikation zu beginnen, ist sehr viel schwieriger, weil der Patient, wie in den dargestellten Beispielen gezeigt, die Verhaltensänderung des Arztes als Zeichen der Bedrohung interpretiert. Es erweist sich überhaupt als günstig, das Problem von Information und Kommunikation aus der Reduktion auf die Situation des Arztes „vor den Kranken mit infauster Prognose", des Arztes „vor den letzten Dingen" wieder auf die Gesamttätigkeit des Arztes auszuweiten (Raspe, 1982 a, b); durch eine solche „Dramatisierung" (Raspe, 1982) wird nämlich klärende Forschung und die systematische Entwicklung entsprechender Versorgungsansätze eher blockiert als gefördert.

66.2.3 Patienten wünschen eine offene Kommunikation mit dem Arzt über ihre Krankheit

In unserer eigenen Untersuchung hatten von 100 Leukämiepatienten von Anfang an 96 die Mitteilung der erst zu stellenden Diagnose gewünscht. 4 Kranke wollten nicht informiert werden.

Krankenhauspatienten, die nach ihrem Informationsbedürfnis im Falle einer unheilbaren Krankheit gefragt wurden, äußerten zum ganz überwiegenden Teil den Wunsch nach offener, rückhaltloser Information (Raspe, 1976, 1977, 1982; Jonasch und Sellschopp, 1984). Der Arzt kann davon ausgehen, daß mindestens 80% der Krankenhauspatienten eine offene Information über Diagnose und Prognose wünschen, lediglich 3–5% nicht über die Art der Erkrankung, die übrigen nicht über deren Unheilbarkeit informiert werden wollen. Das Informationsbedürfnis der Patienten steht dabei in Wechselbeziehung zur Informationsbereitschaft der Ärzte. Die meisten empirischen Untersuchungen wurden in Institutionen durchgeführt, in denen die Patienten im allgemeinen informiert werden.

Werden die Patienten nicht informiert, können sie kaum nach ihren Informationsbedürfnissen befragt werden. McIntosh (1976) arbeitete als teilnehmender Beobachter auf einer Station für Krebskranke, auf der die Patienten nicht über ihre Diagnose informiert wurden. Auch hier vermuteten 63,5% der Kranken ihre Diagnose richtig, jedoch wollten hiervon nur 32% ausdrücklich ihre Diagnose bestätigt wissen. Patienten, denen Information vorenthalten wird, äußern – wie sich auch in empirischen Untersuchungen von Visitengesprächen zeigt (Siegrist, 1978; Nordmeyer, 1978; Raspe, 1982 c) – ihr Informationsbedürfnis gegenüber dem Arzt nur noch in geringem Maß; sie äußern etwa während

der Visite kaum noch Fragen. Diese Passivität von Patienten kann von einseitig aktiven Ärzten fehlinterpretiert und als Argument für die Annahme benützt werden, ihre Patienten hätten gar kein weitergehendes Informationsbedürfnis. Gesprächsanalytische Untersuchungen haben im einzelnen gezeigt, wie die aktive Beteiligung des Patienten etwa während des Visitengesprächs behindert werden kann, aber auch gefördert werden kann (vgl. Kap. 18).

Werden informierte Malignompatienten im weiteren Verlauf befragt, so begrüßen 90% dieser Kranken auch im Rückblick die offene Information (Kelly und Friesen, 1950; Oken, 1961; Jonasch und Sellschopp, 1984).

Für den Arzt ist es wichtig zu wissen, daß – zumindest in der durchschnittlichen Behandlungssituation – etwa die Hälfte der Patienten sich mit Fragen, mit Sorgen und mit der Bitte um Rat nicht aktiv von sich aus an ihn wenden. Damit besteht die Gefahr, daß ein großer Teil der Informations- und Kommunikationsbedürfnisse verborgen bleibt. Es gehört zur Aufgabe des Arztes, die Äußerung dieser latenten Bedürfnisse zu ermöglichen.

66.2.4 Unheilbar Kranke benötigen die fachkompetente Hilfe ihres Arztes bei der Orientierung über ihre Situation

Auch Krankenhauspatienten, die über ihre Krankheit informiert wurden, haben in einem hohen Prozentsatz das Bedürfnis, den Informationsaustausch mit dem Arzt über die Zeit des Krankenhausaufenthaltes fortzuführen (Pender, 1974; Raspe, 1977, 1982 a).

Während von Ärzten informierte Patienten nur zu einem geringeren Anteil sich noch über andere Quellen informieren (Gilbertsen und Wangensteen, 1962; Achté und Vauhkonen, 1971; Krant und Johnston, 1976; Pender, 1974), versuchten die nicht von Ärzten informierten Kranken sich über andere Quellen zu informieren.

Die von McIntosh (1976) untersuchten Malignomkranken benutzten vor allem drei Möglichkeiten:
– Wichtige Anhaltspunkte bot ihnen die durchgeführte Therapie. Vor allem Bestrahlungen wurden von den Patienten regelmäßig als Beweis für das Vorliegen einer Krebserkrankung angesehen, wie dies heute dem Laienverständnis entspricht (Bestrahlung = Krebs).
– Die Tatsache, daß den Patienten die Diagnose nicht ausdrücklich und verständlich mitgeteilt wurde, bestätigte sie in ihrem Verdacht, an Krebs zu leiden.
– Viele Patienten erschlossen ihre Diagnose indirekt aus den verbalen und averbalen Mitteilungen von Arzt und Stationspersonal; darüber hinaus befragten sie „erfahrene" Mitpatienten und erfuhren die Diagnosen gelegentlich auch von den Angehörigen.

Unserer Erfahrung nach haben Patienten im Krankenhaus reichlich Gelegenheit, sich über Laboraufträge, Befundzettel und auch Krankengeschichten zu informieren. Sehr häufig suchen sie dann die Fachtermini in meist veralteten Gesundheitslexika zu klären, deren Angaben sie häufig noch mehr als unbedingt nötig erschrecken. Diese Erfahrungen werden durch die Untersuchungen von Dubach und v. Rechenberg (1977) sowie von Raspe (1977, 1982 a) bestätigt.

Die Möglichkeit zur kontinuierlichen Kommunikation mit dem Arzt benötigen die Kranken vor allem auch, um die Bedeutung von Diagnosen und Befunden, die Wertigkeit ihrer Beschwerden zutreffend einstufen und damit ihre Ängste und Befürchtungen in Bezug zur äußeren Realität setzen zu können; sie benötigen den Arzt als fachkompetenten Berater. Im Verlauf des weiteren Umgangs wird dieser Prozeß von Orientierung und gegenseitiger Abstimmung ständig fortgeführt. Ein solcher Austausch scheint angesichts von Untersuchungen über Verständnis und Verarbeitung medizinischer Information durch Patienten besonders wichtig: Fachtermini und Angaben zur Therapie werden nur von einem wesentlich geringeren Prozentsatz der Patienten verstanden als die eigentliche Diagnose (Samora et al., 1976; Boyle, 1970; Kane und Deuschle, 1967; Dubach und v. Rechenberg, 1977). Die Annahmen der behandelnden Ärzte über das Wissen ihrer Patienten treffen in einem hohen Prozentsatz nicht zu (Dubach und v. Rechenberg, 1977). Untersuchungen zur Aufnahme und Verarbeitung von Information, die Patienten nach dem Prinzip des informed consent vor experimentellen Therapiestudien gegeben wurde, haben diese Befunde bestätigt (Köhle et al., 1982).

66.2.5 Der rational therapierende Arzt benötigt seinerseits die Kommunikation mit dem Kranken

Der Arzt kann nur dann die gesamte Situation seines Patienten angemessen beurteilen, wenn er mit ihm kommuniziert. Das offene Gespräch ist Voraussetzung für die Beurteilung von Symptomen und Beschwerden, für die Einschätzung emotionaler Reaktionen – etwa auch einer Suizidgefährdung – und die Indikationsstellung für unterstützende psychotherapeutische Maßnahmen. Ohne ein solches Gespräch kann der Arzt schon das Wissen seiner Patienten nicht zutreffend einschätzen (McKinlay et al., 1975; Kane und Deuschle, 1967; Dubach und v. Rechenberg, 1977); ihre Sorgen, Befürchtungen und auch Hoffnungen bleiben ihm erst recht verborgen, sie werden jedoch dann durch irrationale Einflüsse mitbestimmt, was nicht selten zu einer Behinderung der Behandlungsmaßnahmen führt.

Dabei ist es wichtig, neben dem Angebot von Information sich auch immer wieder über das Verständnis des Patienten rückzuversichern und dabei die Verarbeitung des vermittelten Inhalts einschließlich der hierdurch angeregten Phantasien zu erfragen:

> Ein mehrfach zu Radioisotopenuntersuchungen geschickter Patient fragte während der Visite stereotyp immer wieder nach dem Zweck dieser Untersuchung; dieser wurde ihm immer wieder geduldig erklärt. Der behandelnde Arzt versuchte insoweit durchaus auf die offen geäußerten Patientenbedürfnisse einzugehen, er verhielt sich „symmetrisch" (Siegrist, 1978). Erst nach längerer Zeit sprach der Arzt den Patienten darauf an, warum er denn immer wieder dieselbe Frage stelle. Jetzt berichtete der Kranke, daß auf der Tür des Untersuchungsraumes ein Totenkopf-Zeichen („Vorsicht Radioaktivität") angebracht gewesen sei und er deshalb mit der Durchführung der Untersuchung angstvolle Vorstellungen verbinde.

Offene Kommunikation ist auch für die Klärung von Compliance-Problemen unabdingbar. Patienten berichten in einer offen gestalteten Beziehung auch über irrationale Rettungsphantasien und entsprechende Behandlungsversuche, etwa über den Besuch bei Heilpraktikern. Dem Arzt wird es so eher möglich, auch in Krisensituationen Berater des Patienten zu bleiben und bei wirklicher Gefahr für die Zusammenarbeit rechtzeitig zu intervenieren.

66.2.6 Die Kommunikationsbereitschaft der Ärzte und die institutionellen Voraussetzungen entsprechen noch nicht dem Kommunikationsbedürfnis unheilbar Kranker

Umfragen bei Ärzten in den USA mittels Fragebogen ergaben während der letzten 25 Jahre eine deutliche Veränderung der **Einstellung.** So gaben noch während der 50er Jahre bis zu 90% der Klinikärzte an, Malignomkranke im allgemeinen nicht zu informieren (Fitts und Ravdin, 1953; Oken, 1961), obwohl die überwiegende Mehrzahl der Ärzte selbst im Falle einer Erkrankung offen informiert werden wollte (Oken, 1961; Burton, 1965). Demgegenüber meinten 93% der von Carey und Posovac (1978) befragten Ärzte, der Arzt sollte Malignompatienten umfassend informieren, und 87%, Sterbende hätten ein Anrecht, über ihren Zustand informiert zu werden. Rea und Mitarbeiter (1976) ermittelten dieselbe Einstellung gegenüber terminal Kranken bei über 60% der 151 an der Befragung kooperierenden Ärzte.

In den genannten Untersuchungen tendieren vor allem jüngere Ärzte zu einer offenen Information ihrer Patienten.

In der Bundesrepublik führte Raspe (1977) Interviews mit 53 in Kreiskrankenhäusern tätigen Ärzten über ihr Informationsverhalten gegenüber körperlich Kranken (ohne spezielle Berücksichtigung ihrer Prognose). 53% dieser Ärzte wollen über die Diagnose, 83% über die Prognose nicht vollständig, sondern eher mit Einschränkungen informieren.

Das **tatsächliche Informationsverhalten** von Ärzten gegenüber Patienten wurde nur selten direkt beobachtet. Klinische Beobachtungen ergaben zunächst, daß Sterbende seltener von Ärzten und Krankenschwestern aufgesucht werden, ihnen am Krankenbett weniger Zeit gewidmet wird (Sudnow, 1973; Glaser und Strauss, 1974, 1968; Hackett und Weisman, 1969).

Teilnehmende Beobachter beschrieben ausweichendes Verhalten von Ärzten gegenüber unheilbar Kranken (Buckingham et al., 1976; McIntosh, 1976).

Siegrist fand im Rahmen seiner Untersuchungen zur verbalen Kommunikation während der ärztlichen Visite, daß Ärzte auf die Bitte nach Information zum Krankheitsbild gegenüber Schwerkranken mit ungünstiger Prognose signifikant häufiger ausweichen und nicht informieren als gegenüber Leichtkranken mit günstiger Prognose. Solche ausweichenden, „asymmetrischen" Reaktionen fanden sich bei der ersten Patientengruppe in 91% all der Fälle, in denen

Patienten während der Visite Fragen zu ihrem Krankheitsbild stellten! (Siegrist, 1972, 1978).

Von Ärzten, die die Information ihrer Patienten ablehnen, wird die **Ineffizienz** ihres Vorgehens, häufig „schonendes Betrügen" genannt, nicht berücksichtigt bzw. verleugnet. In einigen Studien wurde versucht, das Ergebnis von Information demjenigen von Nicht-Information gegenüberzustellen.

Gerle und Mitarbeiter (1960) begleiteten insgesamt 101 Malignompatienten mit inoperablen Tumoren; 92% der Kranken hatten eine Überlebenszeit von weniger als einem Jahr. Nach vorausgegangenem psychiatrischem Interview wurden 38 zufällig ausgewählte Patienten durch den Chirurgen informiert, 43 Patienten wurden nicht informiert. 20 der „nichtaufgeklärten" Kranken informierten sich anderweitig über ihre Diagnose.
Hackett und Weisman (1969) berichten von 20 Malignomkranken, von denen nur 7 „aufgeklärt" worden waren; 15 wußten über ihre wahre Diagnose und Prognose weitgehend Bescheid.
65 (88%) der 74 von McIntosh (1976) in einer Klinik beobachteten nichtinformierten Patienten wußten oder vermuteten, daß sie an Krebs litten.

Hess (1975) beobachtete, daß 70% der von ihm eher zurückhaltend informierten Patienten innerhalb weniger Monate durch Nachbarn, Mitpatienten oder unbeteiligte Dritte über ihre Diagnose informiert wurden.
Nachdem sich auch im deutschsprachigen Raum die Einstellung der Ärzte in Richtung einer größeren Bereitschaft zur offenen Kommunikation verändert hat, kommt es jetzt vermehrt darauf an, die konkrete Umsetzung dieser Einstellung in die alltägliche Praxis und die bei einem solchen Vorgehen auftretenden Schwierigkeiten zu diskutieren; gleichzeitig wird es wichtig, ausreichende Unterstützungsmöglichkeiten für Ärzte und Pflegepersonen bereitzustellen, die sich in einem solchen intensiveren Umgang mit unheilbar Kranken vermehrt auch eigenen Belastungen aussetzen.

66.2.7 Folgen offener Kommunikation: Entlastung der Situation; Verbesserung der Kooperation; keine Minderung der Hoffnung; keine Vergrößerung der Suizidgefahr

Nach unserer eigenen Erfahrung entlastet die offene Kommunikation die Patienten, ihre Familienangehörigen, Ärzte und Pflegepersonal sowie die Beziehung zwischen allen Beteiligten. Information über Diagnose, erforderliche Untersuchungen, Untersuchungsergebnisse, therapeutische Maßnahmen und prognostische Möglichkeiten vermindern Angst, da die Befürchtungen des Patienten in Beziehung zur Realität gesetzt werden. Im Dialog lernt der Arzt die emotionalen Reaktionen des Patienten, der Patient die Überlegungen des Arztes kennen. Der Arzt kann den Patienten so gezielter unterstützen, der Patient kann seine Erkrankung vorübergehend auch aus der Distanz der ärztlichen Überlegungen mitbeurteilen; seine Autonomie bleibt so größer, seine Abhängigkeit geringer. Die Kommunikation mit dem Arzt stärkt das Selbstwerterleben des Kranken; er fühlt sich trotz seiner Erkrankung ernstgenommen und akzeptiert, als für die eigene Zukunft verantwortlich anerkannt. Dies trägt unseres Erachtens bereits wesentlich zur Prophylaxe von schweren Depressionszuständen und Suizidtendenzen bei. Systematische Nachuntersuchungen von informierten Malignomkranken bestätigen diese Erfahrungen.

In Nachbefragungen äußerte sich die überwiegende Mehrheit (70–93%) der über ihren Zustand informierten Malignompatienten positiv über das Vorgehen der sie behandelnden Ärzte (Aitken-Swan und Easson, 1959; Gilbertsen und Wangensteen, 1962; Achté und Vauhkonen, 1971; Jonasch und Sellschopp, 1984; Weisman, 1979). Dies gilt sowohl für Malignompatienten mit günstiger, als auch für solche mit ungünstiger Prognose. Von den Patienten werden als besonders vorteilhaft hervorgehoben: die Fähigkeit, die eigene Krankheit verstehen zu können; die Möglichkeit zur aktiven Planung der eigenen Zukunft und zur aktiven Beteiligung an der Behandlung; die Chance, der Bedrohung gefaßter gegenüberstehen zu können (Gilbertsen und Wangensteen, 1962).
Gerle und Mitarbeiter (1960) heben unter den Ergebnissen ihrer kontrollierten Untersuchung hervor, daß die überwiegende Mehrheit der Patienten positiv auf die Information reagierte; die Kranken hätten nicht nur ihre Situation gefaßt akzeptiert, sondern auch ihren bisherigen Lebensstil aufrechterhalten, soweit sie nicht durch die Krankheitssymptomatik hierin beeinträchtigt worden seien. Die Kommunikation mit den Familienangehörigen sei bei den informierten Patienten freier gewesen und hätte den Angehörigen mehr Chancen geboten, den Kranken zu helfen. Eindeutig läßt sich aus dieser Untersuchung ableiten, daß sich für die informierten Patienten nicht mehr emotionale Belastungen ergaben als für die nicht informierten.

Für die Patienten schädliche Auswirkungen der offenen Kommunikation beobachteten wir in unserer eigenen Arbeit nicht. Stärkere emotionale Reaktionen der Patienten auf die Information sind zu erwarten und einfühlbar. Ihr Verlauf wird in Abschnitt 66.3 dargestellt. Auffallend, „krankhaft", wäre ein Ausbleiben solcher Reaktionen in einer bedrohlichen Lebenssituation.
Häufig wird die Befürchtung geäußert, die Information auch unheilbar Kranker über das Wesen ihres Leidens würde ihnen
1. die Hoffnung nehmen und
2. die Suizidgefahr vergrößern.
Die erste Befürchtung beruht unseres Erachtens auf einer einseitigen Auffassung von Hoffnung, die zweite hält empirischer Nachprüfung nicht stand.
1. Zur Frage der **Hoffnung:** Hoffnung wird häufig lediglich als Hoffnung auf Besserung der Symptomatik oder auf Verlängerung der Lebenserwartung verstanden. Hoffnung aufrechterhalten bedeutet dann, die auf therapeutische Maßnahmen gerichteten Erwartungen auch über jedes aus der Sicht des Arztes berechtigte Maß hinaus bis zuletzt zu unterstützen. Hoffnung bezieht sich hier ausschließlich auf Möglichkeiten, trotz aller Bedrohung zu überleben. Hoffnung wäre dann gleichzusetzen mit Verleugnung der Realität.

Selbstverständlich bleibt es wichtig, alle auf das Überleben bezogenen Erwartungen, soweit sie sich aus der Sicht des Arztes rechtfertigen lassen, zu unterstützen. Der Arzt muß nur wissen und erkennen, daß diese Form der Hoffnung bei ungünstigem Krankheitsverlauf zwangsläufig abnehmen wird.

Die Hoffnung des Kranken bezieht sich – vor allem mit fortschreitender Erkrankung – nicht ausschließlich aufs Überleben. Für den Kranken ist die Entwicklung seiner sozialen Beziehungen, die Aufrechterhaltung der Integrität seiner Person zusammen mit dem Erleben des eigenen Wertes entscheidend. Sein Selbstwertgefühl ist dabei auch von den Möglichkeiten abhängig, die es ihm erlauben, sich in seiner jetzt gegebenen Situation – auch und gerade angesichts des Todes – selbst verwirklichen und mit seinen Bezugspersonen kommunizieren zu können. Der Theologe Böckle spricht vom Bedürfnis, nicht Fragment zu bleiben, von der Suche nach Vollendung und dem Bemühen um Gültigkeit der eigenen Person. Die Antithese zur Hoffnung ist für den Kranken die Verzweiflung. Das Erleben von Wertlosigkeit und Alleingelassenwerden führt zur Verzweiflung. Der Kranke fürchtet den sozialen Tod, der dem körperlichen oft genug vorausgeht, mehr als den physischen Tod.

Auch v. Uexküll (1973) weist darauf hin, daß bei der Angst vor dem Sterben „die Furcht vor dem Ausgeschlossenwerden aus der Gruppe der Mitmenschen gegenüber allen anderen Elementen – wie Furcht vor Schmerzen, körperliche Beeinträchtigung oder Nicht-mehr-Sein, unter dem man sich nichts vorstellen kann – bei weitem überwiegt. Ausgeschlossensein wird fast immer als Zusammenbruch des Selbstwerterlebens, als Verlust der Achtung vor sich selbst erfahren und wird dann schwerer ertragen als jedes andere Schicksal".

Im Gegensatz zur Hoffnung, die sich, die Realität verleugnend, allein auf das Überleben bezieht, kann die Hoffnung, die mit den sozialen Beziehungen und dem eigenen Wert, mit der Selbstverwirklichung verbunden ist, im Krankheitsverlauf zunehmen. Hierzu können vor allem die Beziehungen zu den Angehörigen, aber auch der Umgang des Arztes und der Pflegenden mit dem Patienten beitragen. Die Bereitschaft zur Kommunikation, das Eingehen auf die Bedürfnisse des Kranken, die Unterstützung seiner Planungen für die Zukunft kann die sozialen Stigmatisierungsprozesse beim Krebskranken zumindest abmildern und dazu beitragen, daß die oft zumindest vorübergehend gegebenen Möglichkeiten zur Rehabilitation besser genützt werden.

In der Einstellung zu sich selbst, in den Beziehungen zu seinen Angehörigen, aber auch in den Beziehungen zu seinen professionellen Helfern kann es dem Kranken in der verbleibenden Zeit gelingen, Neues zu verwirklichen. Le Shan (1969) betont, daß es bei der psychotherapeutischen Begleitung Sterbender nicht um eine „extension of time" gehe, sondern um eine „extension of values". Auf diese Möglichkeit, im Vergleich zum bisherigen Leben neue Werte zu verwirklichen, bezieht sich auch die zweite Form der Hoffnung. Dabei geht es vor allem darum, eine Ein-

stellung zur gegebenen Situation zu bekommen, die eine Entwertung der eigenen Person in Beziehungen zu anderen verhindert, vielmehr die Beziehung zum eigenen Selbst und zu den anderen angstfrei und reicher werden läßt („Einstellungswerte", Frankl, 1959; vgl. Spiegel-Rösing, 1980). Der Arzt unterstützt den Patienten bei der Aufrechterhaltung dieser Hoffnung vor allem durch die Aufrechterhaltung seiner eigenen positiven Beziehung zu ihm. Innerhalb einer solchen Beziehung kann er dann auch dazu beitragen, daß der Kranke während der Auseinandersetzung mit seinem Schicksal sich selbst und seine Bezugspersonen aus verständlicher Enttäuschung heraus nicht total entwertet, sondern die verinnerlichten guten Abbilder des eigenen Selbst und der Beziehungen zu anderen aufrechterhalten oder wiederbeleben kann (Meerwein, 1981).

Ein 52jähriger Patient mit chronischer myeloblastischer Leukämie wirkte zunächst sehr zurückgezogen. Vor allem die Beziehung zu seiner Ehefrau sei schon vor der Erkrankung distanziert gewesen, jetzt sei zwischen ihnen „ein Bambusvorhang". Sie hätten sich nichts mehr zu sagen. Ihre Beziehung bleibe nur wegen der beiden Kinder (6 und 9 Jahre) aufrechterhalten.

Im Verlaufe zahlreicher Gespräche gelang es dem Patienten, sein Rückzugsverhalten im Zusammenhang mit seiner heftigen Enttäuschung zu verstehen: In seinem Erleben hatte er alle eigenen Interessen seiner Berufstätigkeit und seiner Familie geopfert; wodurch habe er die Krankheit verdient? Nachdem er sich zunächst gegenüber dem Arzt hatte öffnen können, gelang es ihm allmählich, auch die Beziehung zu seiner Frau wieder offener zu gestalten, mit ihr über seine Sorgen und Ängste zu sprechen. Schließlich traf er mit ihr zusammen umfassende Vorbereitungen für das Leben seiner Familie nach seinem Tod. Nach vielen Bedenken und einigem Zögern verstand er, daß es zum Wertvollsten gehören könnte, was er seinen Kindern für ihre Entwicklung mitgab, wenn er sie in angemessener Weise an seiner Auseinandersetzung mit Krankheit und Tod teilnehmen ließe. Er bereitete seine Kinder behutsam auf die Trennung vor und sorgte gleichzeitig für ihre Zukunft, vor allem für ihre Ausbildung. Emotional war der Kranke nach dieser Einstellungsänderung aufgelebt. Er selbst sprach wiederholt davon, daß er das Gefühl habe, aus der ihm verbliebenen Zeit für sich selbst und für seine Familie etwas Wertvolles gemacht zu haben.

Der Arzt kann zur Hoffnung des Patienten vor allem beitragen, indem er dessen eigenes Bemühen um Integration in die Gemeinschaft der Lebenden unterstützt und so seine Identität stärkt (Scheytt, 1984). Die Arbeitsbereiche von Ärzten, Pflegenden und Seelsorgern berühren sich in diesem Bereich besonders eng; hier bietet sich für diese Berufsgruppen eine fruchtbare Möglichkeit zur Zusammenarbeit (Scheytt, 1984).

2. Suizid: Wir beobachteten bei den über ihre Diagnose informierten und von uns im Rahmen des Liaison-Dienstes bzw. auf der internistisch-psychosomatischen Krankenstation betreuten Patienten im Zeitraum von 1968–1984 keinen Suizidversuch.

Eine erfolgreiche Suizidhandlung – ein Stich mit dem Messer ins Herz – beging in diesem Zeitraum ein Privatpatient, dessen behandelnder Arzt eine offene Information nach eingehender Diskussion mit dem mitbehandelnden Onkologen abgelehnt hatte.

Die relativ wenigen systematischen empirischen Studien über Suizidtendenzen, Suizidversuche und erfolgreiche Suizidhandlungen unheilbar Kranker (Achté und Vauhkonen, 1971; Campbell, 1966; Reich und Kelly, 1976; Farberow et al., 1971; Laxenaire et al., 1972; Oken, 1961; Fitts und Ravdin, 1953; Weisman, 1976, 1979; Plumb und Holland, 1981; Fox et al., 1983) ergeben zusammengefaßt zunächst, daß die Suizidhäufigkeit unter unheilbar Kranken nicht ungewöhnlich hoch ist. Vergleichende Untersuchungen zwischen informierten und nicht informierten Patienten liegen unseres Wissens nicht vor.

Die Berichte erfahrener Kliniker sprechen dafür, daß Suizidhandlungen bei Krebskranken nur selten, nicht häufiger als bei der Durchschnittspopulation vorkommen (Weisman, 1979; Jonasch und Sellschopp, 1984). Litin (1960) betont die extreme Seltenheit einer Suizidhandlung als Folge einer Diagnosemitteilung; er habe einen solchen Zusammenhang in der Region von Rochester innerhalb von 10 Jahren nur einmal beobachtet.

Die vorhandenen empirischen Untersuchungen berichten überwiegend nur von kleinen Zahlen und weisen verschiedene methodische Probleme auf (Übersicht bei Fox et al., 1983). Fox und Mitarbeiter haben für Connecticut im Zeitraum von 1940–1973 bei insgesamt 144 530 Krebskranken für 192 Patienten Selbstmord als primäre Todesursache gefunden. Verglichen mit der Gesamtsuizidrate in der Bevölkerung war diese Häufigkeit nur bei Männern statistisch erhöht; bei Männern war die Suizidrate in der Zeit nach der Diagnosestellung am häufigsten, sie nahm mit dem zeitlichen Abstand zur Diagnosestellung ab.

Suizidphantasien beobachteten wir bei den von uns betreuten Patienten häufig. Die Patienten berichten derartige Phantasien allerdings nur dann, wenn sie ihre Beziehung zum Arzt als tragfähig erleben. Weisman fand solche Suizidphantasien bei etwa jedem 10. (1979), Plumb und Holland (1981) bei jedem 7. der von ihnen untersuchten Patienten mit fortgeschrittener Krebserkrankung.

Malignompatienten, die Suizidversuche unternehmen, handeln meist aufgrund einer tiefgehenden Verzweiflung. Weisman (1976, 1979) führt solche Suizidversuche auf die Verletzlichkeit dieser Patienten als Folge ihrer Erkrankung zurück. Diese Verletzlichkeit lasse sich an Äußerungen der Kranken erkennen, so an Hinweisen auf ihre Isoliertheit, Wertlosigkeit, Hilflosigkeit, Erschöpfung und Angst. Therapeutisch empfiehlt Weisman, diese Verletzlichkeit im Umgang zu berücksichtigen und den Kranken vor allem Unterstützung bei der Aufrechterhaltung ihrer Autonomie zu vermitteln und ihnen Gesprächsmöglichkeiten mit dem Ziel emotionaler Entlastung und Stützung des Selbstwerterlebens anzubieten.

Zusammenfassend ist zu sagen, daß ohne offene Kommunikation mit dem Kranken weder das Suizidrisiko einschätzbar ist, noch gezielt Maßnahmen zu einer prophylaktisch wirksamen Unterstützung des Patienten durchgeführt werden können. Eine intensive Kommunikation zwischen Arzt und Patient erhöht das Suizidrisiko nicht, sondern ermöglicht erst ein spezifisches Eingehen auf die Nöte des Kranken.

66.2.8 Mißverständnisse und Berichte über negative Erfahrungen mit offener Kommunikation

Von erfahrenen Klinikern werden neben den geschilderten positiven aber auch negative Erfahrungen mit der offenen Kommunikation mit Krebskranken berichtet.

So wird beobachtet und kritisiert, daß „Aufklärung" – oft von jungen Kollegen – zu forsch und konfrontativ an die Patienten herangetragen werde und dann zu einer Verunsicherung der Patienten führen könne. Ein solches forciertes Bemühen um Information kann aufgrund von zwei Mißverständnissen zustande kommen.

Dieses Verhalten kann zunächst ebenso durch ein persönliches Schutzbedürfnis des Arztes mitbestimmt werden wie das Zurückhalten von Information. In beiden Fällen wird eine intensivere Beziehung zum Patienten und damit eine zusätzliche Belastung vermieden. „Aufklärung für alle Kranken" kann ebenso ideologische Funktion bekommen wie die grundsätzliche Ablehnung von Informationsvermittlung. Angesichts der Möglichkeit zu einem derartigen Mißverständnis möchten wir noch einmal mit Nachdruck auf das Ausmaß der emotionalen Belastungen hinweisen, denen insbesondere onkologisch tätige Ärzte und Krankenschwestern bzw. -pfleger ausgesetzt sind. Hier ist ein dringender Bedarf für eingehende Untersuchungen zu diesen Belastungen und über den Bedarf an Unterstützungsmöglichkeiten.

Ein zweites Mißverständnis erscheint uns ebenso bedeutsam. „Aufklärung" wird in Analogie zu diagnostischen und therapeutischen Maßnahmen dem Patienten sozusagen appliziert, ohne daß der Arzt eine intensivere Beziehung aufnimmt und sein Vorgehen wechselseitig mit den Bedürfnissen und Reaktionen des Patienten abstimmt. Diese Form des Mißverständnisses ist grundsätzlicher Natur. Information wird hier nicht im Rahmen einer kommunikativen Tätigkeit, einer Gegenseitigkeit implizierenden Beziehung verstanden, sondern nach dem Muster instrumenteller Tätigkeit aufgefaßt und zu organisieren versucht. Das Mißverständnis beruht auf einem verkürzten Verständnis ärztlichen Tuns, bei dem der ganze Bereich der „Gefühlsarbeit" unberücksichtigt bleibt.

In diesem Zusammenhang können auch die Klagen über eine Verunsicherung der Patienten durch zu differenzierte Information diskutiert werden. Die beklagte Verunsicherung erfolgt nicht durch die vermittelte Information, sondern durch den Mangel an gleichzeitiger Unterstützung bei der Verarbeitung dieser Information. Der Patient behält oder gewinnt Sicherheit nicht, indem ihm Information vorenthalten wird, sondern indem der Arzt dem Patienten Sicherheit in der Gemeinsamkeit der Beziehung anbietet. Erst in dieser Beziehung kann er dem Patienten helfen, Krankheit und Krankheitsfolgen in seine Biographie einzuordnen, trotz krankheitsbedingter Einschränkungen und den damit verbundenen Veränderungen seiner sozialen Rolle die eigene Identität zu bewahren, sein Selbstwertgefühl zu stabilisieren und über die unabänderlichen Verluste zu trauern.

66.2.9 Schwierigkeiten bei dem Versuch, offen mit unheilbar Kranken zu kommunizieren

Patienten lehnen die Information ab

Der kleinen Gruppe von Kranken – unter den von uns Betreuten ca. 3–4%, in der Literatur bis zu maximal 10% –, die von vornherein wünschen, nicht über ihre Krankheit informiert zu werden, soll die Information selbstverständlich nicht aufgedrängt werden. Andererseits sollte sich der Arzt gerade bei diesen Patienten intensiv um die Möglichkeit zur Kommunikation bemühen. Unserer Erfahrung nach handelt es sich bei diesen Patienten nicht selten um sehr einsame Menschen, bei denen das Bedürfnis nach Autonomie bzw. Unabhängigkeit besonders ausgeprägt ist. Auch diese Patienten wissen im Grunde um die Bedrohlichkeit ihrer Erkrankung – dieses Wissen ist ja auch Voraussetzung verleugnender Abwehr –, sie möchten sich jedoch die Möglichkeit, die Bedrohung selbst vom Bewußtsein abzuwehren, erhalten und sich diese Möglichkeit nicht fremdbestimmt durch Information nehmen lassen. Im Umgang kommt es darauf an, diesen Patienten zu erkennen zu geben, daß man an ihnen interessiert, um sie bemüht bleibt, jedoch bereit ist, die von ihnen vorgegebene Distanz einzuhalten, sie nicht mit Information zu bedrängen. Das wiederholte Angebot von Zeichen einer solchen unaufdringlichen Kommunikationsbereitschaft ist gerade für diese Kranken oft besonders wichtig und kann dazu beitragen, emotionale Komplikationen zu verhindern oder doch abzumildern.

Sekundäre Abwehr von Information: „Wiederverleugnung" („re-denial") – „middle knowledge"

Auch um Information bemühte Ärzte beobachten immer wieder, daß Patienten, die bereits mehrfach und offen über ihre Erkrankung und auch die Prognose informiert wurden, sich so verhalten, als hätten sie keinerlei Wissen von der Bedrohlichkeit ihrer Situation. Dieses Phänomen wird als „Wiederverleugnung" („re-denial") beschrieben (Meerwein, 1980, 1981). Der Arzt nimmt solche Wiederverleugnung entweder indirekt über Compliance-Probleme und selbstgefährdende Verhaltensweisen wahr, oder er wird unmittelbar von entsprechenden Äußerungen der von ihm betreuten Kranken überrascht.

> So rief mir nach einem langen Gespräch über ihre eingeschränkte Lebensperspektive eine 40jährige Kranke mit Wirbel- und Gehirnmetastasen bei Bronchialkarzinom nach der Verabschiedung noch nach: „Ich bin doch überhaupt nicht krank", und drängte auf Bestätigung, als ich schon in der Zimmertür stand.

Dieses Hin- und Herbewegen auf der Skala zwischen Wissen und Nichtwissen – oft erscheint es so, daß Patienten gleichzeitig wissen und nicht wissen – hat Weisman (1979) als „middle knowledge" beschrieben. Weisman betont auch immer wieder, in welchem Ausmaß der Bewußtseinszustand des Patienten, seine „individuelle Wirklichkeit", von seinen sozialen Beziehungen und damit auch vom Arzt-Patient-Verhältnis abhängt.

Selbstverständlich kann niemand ständig im vollen Bewußtsein tödlicher Bedrohung leben; immer wird diese Bedrohung auch wieder verleugnet. Für den Arzt ist es dabei jedoch wichtig zu beobachten, wann und unter welchen Bedingungen die Verleugnung wieder eintritt. Er hat zu beurteilen, ob und inwieweit verleugnende Abwehr den Patienten zusätzlich schädigen könnte – etwa bei Unterbrechung lebenswichtiger Therapiemaßnahmen – oder ob sich die Indikation zur Intervention stellt. Er hat zu klären, ob diese Verleugnungsvorgänge für den Patienten nicht auch einen sinnvollen Schutz darstellen könnten, der ihm für die verbleibende Zeit sogar kreative Möglichkeiten eröffnet; ein solcher Schutz sollte dann nicht angetastet werden.

> Bei dem erwähnten Beispiel tritt die Wiederverleugnung bei der Verabschiedung, bei der Trennung von mir ein. Der appellative Charakter der Mitteilung weist darauf hin, daß die Patientin die Konfrontation mit der krankheitsbedingten Bedrohung allenfalls innerhalb einer Beziehung – hier zu mir, aber, wie sie berichtete, auch zu den Ärzten der Station – vorübergehend zulassen kann und dann wieder Schutz in einer Illusion sucht.

Meerwein (1980) hat ein entwicklungspsychologisches Konzept zum Verständnis der psychischen Vorgänge bei der Wiederverleugnung erarbeitet und nachdrücklich auf die darin enthaltene kreative Leistung der bedrohten Patienten hingewiesen. Im Bemühen um Sicherheit schafft sich der Mensch die Illusion von Unsterblichkeit, die Illusion einer Welt ohne Todesbedrohung. Meerwein betrachtet den Vorgang der Wiederverleugnung analog zum Auftreten des sog. „Übergangsobjektes" (Winnicott, 1969) beim Kind (vgl. auch L. Schacht, Kap. 6 „Die früheste Kindheitsentwicklung und ihre Störungen aus der Sicht Winnicotts"), das sich während des Trennungsprozesses von der Mutter ein Surrogat für die abwesende Mutter „erschafft" und dabei einen vorhandenen Gegenstand, wie den Zipfel einer Decke, oder eine angebotene Stoffpuppe benutzt. Das Kind versichert sich über das von ihm geschaffene Surrogat der abwesenden Mutter im Sinne einer Illusion.

> Die Patientin im dargestellten Beispiel schafft sich seit langem solche illusionären Welten als zentrales Hilfsmittel fürs Überleben. So hat sie als wichtigsten Gesprächs-„Partner" seit vielen Jahren einen Stoffhasen, der auch im Krankenhaus unter ihrem Kopfkissen liegt; ihm vertraut sie ihre persönlichen Gedanken und Sorgen an, nicht wirklichen Bezugspersonen. Sie ist

als Künstlerin am Theater tätig; die kreative Gestaltung illusionärer Welten für andere, aber auch für sich selbst, macht ihr Leben aus. Sie meint, außerhalb des Berufes nicht existieren zu können.

Wird diese Funktion der Wiederverleugnung als schützende Illusion erkannt, so verbietet sich ein forciertes Konfrontieren der Patienten mit der Realität von selbst. Gerade bei diesen Kranken gilt noch vermehrt: die **Aufrechterhaltung der Beziehung** über den Krankheitsverlauf hat Vorrang vor der Informationsvermittlung.

A. E. Meyer und von Kerekjarto berichten über die Reaktion des Dichters Theodor Storm auf seine tödliche Erkrankung. Man kann diesen Bericht als Anstoß zu kritischem Nachdenken über unsere Fähigkeit aufnehmen, die Belastungsfähigkeit von Kranken zu beurteilen; vielleicht kann dieser Bericht aber auch im Zusammenhang mit den Überlegungen Meerweins zur Bedeutung der Illusion als kreativer Leistung angesichts der Bedrohung durch tödliche Krankheit verstanden werden.

„Theodor Storm erkrankte im Frühjahr 1827 – in seinem 69. Lebensjahr – an Magenbeschwerden. Um diese Zeit schrieb er sein erschütterndes Gedicht ‚Beginn des Endes‘:

Ein Punkt nur ist es, kaum ein Schmerz,
Nur ein Gefühl, empfunden eben;
Und dennoch spricht es stets darein,
Und dennoch stört es Dich zu leben.

Wenn Du es anderen klagen willst,
So kannst Du's nicht in Worte fassen,
Du sagst Dir selber: „Es ist nichts!"
Und dennoch will es Dich nicht lassen.

So seltsam fremd wird Dir die Welt
Und leis verläßt Dich alles Hoffen,
Bis Du es endlich, endlich weißt,
Daß Dich des Todes Pfeil getroffen.

Da es sich so offensichtlich um einen Patienten handelt, der um seiner Krankheit zum Tode wußte, sah Theodor Storms Hausarzt keine Gründe, ihm die Diagnose eines Magenkrebses zu verschweigen. Wider Erwarten versank Theodor Storm in eine Depression, und er brach seine Arbeit am ‚Schimmelreiter‘ ab und wurde für sich und seine Angehörigen eine Last. Schließlich wußte sich die Familie nicht anders zu helfen, als mit einem ‚frommen Betrug‘. Sie baten einen anderen Arzt, Storm zu versichern, daß seine Krankheit völlig harmlos sei. Theodor Storm gewann seine gute Laune wieder. Er beendete den ‚Schimmelreiter‘, den viele für seine schönste Leistung halten. Und er feierte seinen 70. Geburtstag unter Anteilnahme der ganzen Bevölkerung. Einige Monate danach starb er" (A. E. Meyer und v. Kerekjarto, 1980; P. Meerwein, 1980).

Wünsche von Angehörigen, Patienten
Information vorzuenthalten

Der Unterstützung unheilbar Kranker durch ihre Familienangehörigen kommt aus ärztlicher Sicht, aber auch im Erleben der Kranken selbst (Jonasch und Sellschopp, 1984), ganz vorrangige Bedeutung zu. Die Einbeziehung der Familienmitglieder in den Informationsprozeß sollte möglichst frühzeitig erfolgen.

Probleme können dadurch entstehen, daß sich Angehörige mit dem Wunsch an den Arzt wenden, dem Kranken die Diagnose zu verschweigen. Dieser Wunsch sollte für den Arzt keine Kontraindikation für die Information des Patienten darstellen. Ein solcher Vorschlag zu „schonendem Betrügen" entsteht vielfach aus eigener Unsicherheit der Angehörigen, aus Schuldgefühlen oder aus dem Wunsch, schon länger schwelende familiäre Konflikte auch weiterhin nicht auszutragen.

Wir versuchen mit dem Patienten selbst die möglichst frühzeitige Einbeziehung seiner Angehörigen in das Gespräch über die Krankheit zu planen. So bleibt der Patient auch für die Entwicklung seiner künftigen Stellung in der eigenen Familie verantwortlich. Werden – wie dies häufig geschieht – dagegen zunächst die Angehörigen informiert, so nimmt man gewissermaßen eine Entmündigung des Patienten vorweg. Zudem wird in diesem Falle meist die Belastbarkeit der Angehörigen überschätzt. Nach Zustimmung des Patienten empfiehlt es sich, möglichst frühzeitig und mit Nachdruck einen konkreten Gesprächstermin mit den Angehörigen, eventuell zusammen mit dem Patienten zu vereinbaren, sonst scheitert diese Absicht an den Alltagsschwierigkeiten der Ärzte und der Angehörigen. Dabei ist es sinnvoll, den Angehörigen deutlich genug Verständnis für ihre belastende Situation auszudrücken, die Zusammenarbeit in der Betreuung des Patienten mit ihnen für den weiteren Verlauf zu planen und auch ihnen in diesem Zusammenhang die Möglichkeit zu Gesprächen anzubieten.

66.3 Die längerfristige Betreuung unheilbar Kranker

66.3.1 Emotionale Reaktionen der Patienten während der Auseinandersetzung mit der Erkrankung

Der Umgang mit unheilbar Kranken wird verständlicherweise immer wieder von zum Teil heftigen Emotionen mitgeprägt. Die emotionalen Reaktionen der Patienten stehen dabei im Zusammenhang mit ihrer Auseinandersetzung mit dem krankheitsbedingten Schicksal. Die Entwicklung dieser Auseinandersetzung ist abhängig vom Verlauf der körperlichen Erkrankung, von den seelischen Bewältigungsmöglichkeiten des Patienten und von der Unterstützung, die er in seinen sozialen Beziehungen erhält.

Für den Kliniker ist es wichtig, charakteristische und häufig auftretende emotionale Reaktionen und Verhaltensweisen unheilbar Kranker zu kennen; dies erleichtert den Versuch, diese emotionalen Reaktionen und Verhaltensweisen im Zusammenhang der Anpassungs- und Abwehrvorgänge des Patienten, also im Zusammenhang mit dem Bemühen des Patienten um eine Bewältigung seiner Situation zu verstehen.

Wir verzichten hier auf eine ausführliche Darstellung der Bewältigungsstrategien unheilbar Kranker; diese finden sich im Kapitel 62. Für den Kliniker ist es entscheidend wichtig, auf das Wechselspiel zwischen äußeren Belastungen und inneren Bewältigungsmöglichkeiten zu achten und jeweils individuell ein Verständnis für die besondere Verletzlichkeit des Patienten in einzelnen Teilbereichen zu gewinnen (Weisman, 1979). Es bewährt sich dabei immer, die emotionalen Reaktionen und Verhaltensweisen von Patienten im Zusammenhang mit ihrem Bemühen um eine Bewältigung ihrer Situation zu verstehen und dieses Bemühen zunächst einmal als sinnvolle aktive Leistung zu würdigen. Ein solches Vorgehen führt auch zu einer Verminderung der Belastung bei Ärzten und Pflegekräften; insbesondere aggressive Reaktionen des Patienten werden dann nicht von vornherein als gegen Ärzte oder Pflegepersonen gerichtet aufgefaßt, sondern im Zusammenhang mit der Auseinandersetzung des Patienten mit seiner Krankheit verstanden.

Wir folgen Frau Kübler-Ross (1969) bei der Auswahl und Darstellung der wichtigsten emotionalen Reaktionen, nicht jedoch in ihrer Auffassung eines mehr oder weniger regelmäßigen phasenhaften Verlaufes dieser Reaktionen.

Wir stellen diese emotionalen Reaktionen als mögliche Reaktionen vor, auf die der klinisch Arbeitende häufiger treffen wird. Diese Reaktionen sind dabei keineswegs spezifisch für unheilbar Kranke. Sie finden sich auch sonst als Reaktionen auf schwere Krankheit oder andere als bedrohlich wahrgenommene Lebensereignisse. Es bewährt sich nicht, davon auszugehen, daß diese Reaktionen in Form von „Stadien" oder „Phasen" mit großer Regelmäßigkeit von der Wahrnehmung der tödlichen Erkrankung bis zum Tode ablaufen, wie dies Frau Kübler-Ross dargestellt hatte. Alle bekanntgewordenen empirischen Untersuchungen sprechen ebenfalls gegen eine solche Auffassung (Simpson, 1982). Die einzelnen Reaktionen können isoliert auftreten. Sie können sich, vor allem auch im Zusammenhang mit Remissions- und Rezidivphasen der Erkrankung, wiederholen und auch miteinander vermischen.

Auftreten und Verlauf dieser Reaktionen werden entscheidend von den Bezugspersonen und der übrigen sozialen Umwelt der Patienten mitbestimmt (Begemann-Deppe, 1976). Der Prozeß der Krankheitsverarbeitung, der auch als Trauerprozeß aufgefaßt werden kann, kann durch das Verhalten der Umwelt sowohl gefördert als blockiert werden. Ein solcher Einfluß kommt nicht nur dem Verhalten der natürlichen Bezugspersonen, sondern auch den Bezugspersonen im medizinischen Versorgungssystem zu.

Wir haben solche fördernden und blockierenden Einflüsse auf die Auseinandersetzung unheilbar Kranker mit ihrem Schicksal vor allem auf unserer internistisch-psychosomatischen Krankenstation (vgl. Kap. 28) immer wieder beobachtet. So verschwanden beispielsweise Blockaden im Trauerprozeß von Patienten nicht selten dann, wenn es Mitarbeitern der Station – etwa nach einer Stationskonferenz – gelang, eigene emotionale Schwierigkeiten zu lösen und sich wieder verstärkt auf den Patienten einzulassen.

Den Behandlungsbedingungen kommt so für die weitere Verarbeitung der Erkrankung große Bedeutung zu: Hier gemachte Erfahrungen können das spätere Erleben und Verhalten der Patienten mitbestimmen. Die Erfahrung, von den Behandelnden im Krankenhaus gemieden zu werden, dürfte schnell dazu beitragen, später verschärft sensibel auf stigmatisierende Tendenzen anderer zu reagieren und sich sozial zurückzuziehen. Einbeziehung in die Kommunikation und möglichst aktive Beteiligung an der Planung und Durchführung der Behandlung könnten dagegen auch die spätere aktive Auseinandersetzung mit der Erkrankung und die Aufrechterhaltung der sozialen Beziehungen fördern.

Während gelegentlich eine Stadieneinteilung des psychologischen Anpassungsprozesses im Verlauf der unheilbaren Erkrankung ganz abgelehnt wird (Simpson, 1982), plädiert Weisman (1979) aufgrund seiner großen Erfahrung sowohl als Kliniker als auch als Forscher dafür, bei allen Krebskranken analog zum Vorgehen in der somatischen Diagnostik und Therapie auch eine psychosoziale Stadieneinteilung („psychosocial staging") vorzunehmen. Ein solches Vorgehen könnte die Integration des psychosomatischen Verständnisansatzes in die onkologische Routinearbeit, etwa im Rahmen eines Liaison-Dienstes oder auf einer entsprechend integrativ arbeitenden Krankenstation, unseres Erachtens durchaus unterstützen. Es würde eine erste Konzeptualisierung der für alle Beteiligten zugänglichen Dokumentation der Diagnostik im psychosozialen Bereich fördern; alle Beteiligten müßten sich für die Kommunikation untereinander festlegen, wo der Patient aktuell in seiner Entwicklung steht.

Weisman (1979) unterscheidet vier Stadien:
– Existential plight
– Mitigation and accommodation
– Decline and deterioration
– Preterminality and terminality
Weisman hat seine Stadieneinteilung entsprechend den Auswirkungen der Erkrankung im psychosozialen Bereich getroffen; er betont dabei die Analogie zur somatischen Stadieneinteilung auf der Grundlage der Invasivität, der Ausbreitung und der zellulären Differenzierung der Krebserkrankung.

Die Berücksichtigung der unterschiedlichen Verlaufsformen der Erkrankung bei der Beurteilung der psychosozialen Probleme hat sich vor allem auch für die Rehabilitation von Krebskranken als bedeutsam erwiesen (Sellschopp, 1984).

Schock und Verleugnung

Die Konfrontation mit der bedrohlichen Erkrankung führt zunächst zu Unruhe und Angst, vielfach zur Lähmung der eigenen Orientierungsmöglichkeiten und nach außen gerichteten Aktivitäten. Die massive Bedrohung kann das Gefühl von Hilflosigkeit auslösen und regressive Prozesse in Gang bringen. Die extreme Belastung kann auch die kognitiven Fähigkeiten beeinträchtigen: Denken und Erleben werden weniger differenziert. Die Entdifferenzierung der kognitiven Prozesse und das Gefühl der Ohnmacht gegenüber den als bedrohlich unkontrollierbar erlebten körperlichen Vorgängen können sich im Sinne eines Circulus vitiosus gegenseitig steigern. Es können dann stark überschießende Reaktionen auch bezüglich der Bewertung der zu erwartenden Einbußen und des gesamten eigenen Lebens auftreten (Schonecke, 1984).

Häufig wird zunächst versucht, die Bedrohung durch Verleugnung zu vermindern; die Bedrohung wird damit vom Bewußtsein ferngehalten, die Augen werden vor ihr verschlossen. Der Patient sucht zu formulieren: „Das kann doch bei mir nicht möglich sein", oder kurz: „Nicht ich".

Als physiologischer Abwehrmechanismus erlaubt die Verleugnung dem Ich des Patienten, seine auf die Realität bezogenen Funktionen wieder aufzunehmen. Dagegen kann nach Qualität und Quantität pathologische Verleugnung zu unrealistischen Verhaltensweisen, wie zur Verzögerung von Diagnosestellung und Behandlungsbeginn, zur Nichteinnahme von Medikamenten und anderen Formen der Selbstschädigung, zum Abbruch der Therapie und zum Aufsuchen von Heilpraktikern führen. Den letztgenannten Versuch, der oft den Versuch beinhaltet, einen Verbündeten im Verleugnungsprozeß zu finden, beobachteten wir bei fast allen von uns untersuchten Leukämiekranken.

Verleugnungsvorgänge werden von der Umgebung oft deshalb unterstützt, weil sie auch die Umgebung vor der Konfrontation mit der bedrohlichen Situation des Kranken verschonen. Dies gilt auch für Ärzte: sie sind oft erleichtert, wenn die Patienten nicht mehr von der Bedrohung sprechen, sondern wieder „Zuversicht" und „Hoffnung" gewinnen; der Abwehrcharakter solcher Veränderungen im Verhalten der Patienten wird dann oft übersehen. Oft wird von Ärzten und Pflegepersonal sogar der Rückzug von Kranken aus der Kommunikation noch im Sinne einer Erleichterung wahrgenommen.

Ärztliche Aufgabe in dieser Situation ist es, dem Patienten dabei zu helfen, sich in der neuen Realität wieder zu orientieren; es gilt vor allem dazu beizutragen, **pathologische Verleugnungsvorgänge,** die die Behandlung der Erkrankung und die Planung der Rehabilitation behindern, abzubauen. Es geht um den **Aufbau eines Arbeitsbündnisses** zwischen Arzt und Patient, dessen Ziele die rationale Langzeitbehandlung der Krankheit und eine auf die Krankheit abgestimmte sinnvolle Umstellung der Lebensweise sind. Grundlage dieses Arbeitsbündnisses bildet die offene gegenseitige Information zwischen Arzt und Patient.

Geduldige Information auch „intellektuell nicht differenzierter" Patienten ist Vorbedingung für die Herstellung eines derartigen Arbeitsbündnisses. Dabei geht es nicht nur um die Sachinformation, sondern auch um die damit verbundene emotionale Zuwendung unter Berücksichtigung der Abwehrvorgänge.

> So droht ein an akuter Leukämie leidender älterer Landwirt nach der 20. Knochenmarkspunktion: „Wenn das jetzt immer noch nichts geholfen hat, lasse ich das nicht mehr machen." Er hatte bis dahin angenommen, daß der Knochenmarksaspiration therapeutische Funktion zukomme; von ihrer diagnostischen Bedeutung hatte er nichts gewußt.

Eine Belastung für den Arzt stellen bei Vorherrschen dieser Reaktion die **übergroßen Erwartungen** der Patienten dar. Die bedrohliche Situation führt zu Hilflo-sigkeit und belebt kindliche Erwartungen und Verhaltensweisen wieder, wie sie einst gegenüber den Eltern bestanden. Wie das Kind allmächtige Eltern, so wünscht sich der Patient einen omnipotenten Arzt, von dem er in der eigenen Hilflosigkeit Heilung und Heil erwartet. Schon um die sonst unvermeidlichen späteren Enttäuschungen und einen entsprechenden Rückzug aus der Kommunikation vermeiden zu helfen, muß der Arzt versuchen, die Autonomie des Patienten so weit wie nur irgend möglich aufrechtzuerhalten bzw. wiederherzustellen und unrealistische Erwartungen mit ihm durchzusprechen. Zu einem solchen Gespräch kann auch der Versuch gehören, den Patienten vorsichtig mit dem Ernst der Erkrankung und den gegebenen beschränkten Hilfsmöglichkeiten des Arztes zu konfrontieren und zu beginnen, ihn im notwendigen Trauerprozeß zu unterstützen. In diesen Gesprächen kann der Arzt auch dazu beitragen, die kognitiven Verarbeitungsmöglichkeiten des Patienten zu unterstützen, indem er ihn mit der neuen Situation vertrauter macht und Bewältigungsmöglichkeiten mit ihm durchspricht.

Zu den Aufgaben des Arztes gehört es auch, bei der Unterstützung der Krankheitsverarbeitung die familiären Beziehungen zu berücksichtigen. Je besser die Kommunikation zwischen Patient und Familie ist, desto weniger muß der Patient die Bedrohung durch die Erkrankung in pathologischer Form aus seinem Bewußtsein fernhalten.

Die charakteristischen Reaktionen bei der Verarbeitung der Erkrankung möchten wir am Beispiel eines 35jährigen Patienten mit akuter Leukämie veranschaulichen.

> Beim ersten Gespräch wirkt der Patient auf mich überstark anklammernd; wie ein total abhängiges Kind ist er an jeder meiner Bewegungen orientiert. Er ist nicht Gesprächspartner, steht mir nicht mit einer gewissen Distanz gegenüber, sondern versucht mit allen Mitteln, an mir einen Halt zu finden. Total verunsichert, will er zunächst nichts von dem, was ich ihm mitteilen könnte, erfahren. Seine übergroßen, auf mich gerichteten Erwartungen wirken auf mich erschreckend und ebenfalls verunsichernd. Der Patient vermeidet, an seine Zukunft zu denken: „Da darf man nicht an sich selbst denken, das wird zu gefährlich".
>
> Der Patient war von einem Onkologen über die Diagnose und auch die Prognose – er hatte „etwas von einer mittleren Lebenserwartung von 3 Monaten" verstanden – informiert worden.
>
> Am ersten Tag war er fast panikartig damit beschäftigt, über die weitere Versorgung seiner Familie nachzudenken. Diese Sorge schien ihn völlig zu beherrschen. Erst später konnte er allmählich von seinem Betroffensein, seiner Angst, seiner inneren Unruhe sprechen. Als die Sprache endlich auf ihn selbst kommt, wird deutlich, daß in die übersteigerte Sorge um seine Familie auch die Abwehr seiner Wut darüber eingeflossen war, daß gerade ihn die Krankheit betroffen hatte: Er meinte, „da darf man nicht an sich selbst denken, das wird gefährlich". Dabei fiel ihm eine Episode ein, die er vor zwei Jahren mit einem Freund erlebt hatte: Nach reichlichem Alkoholgenuß sei das Gespräch darauf gekommen, was sie wohl im Falle einer unheilbaren Erkrankung tun würden. Der Freund

> habe zu ihm gesagt, er würde Amok laufen, würde alle anderen, seine ganze Familie erschießen, und schließlich auch sich selbst. Der Patient hatte daraufhin ungewöhnlich heftig ablehnend reagiert, er hat jede Beziehung zu seinem Freund abgebrochen. Auch weitere Sorgen des Patienten entsprechen einer Reaktionsbildung gegen die abgewehrte Aggression: Er bedrängt uns z.B. mit der Frage, ob Leukämie vererblich oder ansteckend sei. Das wäre das Schlimmste, wenn seine Kinder diese Krankheit von ihm bekommen könnten.
>
> Während der ersten Schockreaktion mußte er seine heftigen negativen Affekte gegenüber der Familie vollständig abwehren, später war es ihm möglich, über solche Regungen auch offen zu sprechen.

Schock, Beeinträchtigung der eigenen Orientierungsmöglichkeiten und verleugnende Abwehr sind in der Regel am stärksten während der ersten Zeit der Konfrontation mit der tödlichen Erkrankung ausgeprägt. Weisman (1979) gibt zur groben Orientierung als Dauer dieser Form der Auseinandersetzung mit der existentiellen Betroffenheit einen Zeitraum von 3–4 Monaten bzw. von ca. 100 Tagen an.

Für den Arzt ist es während dieser Zeit oft nicht ganz einfach, das notwendige Verständnis für die adaptive Funktion auch ausgeprägter Verleugnungsvorgänge zu entwickeln; dies gilt insbesondere dann, wenn diese Verleugnungsvorgänge trotz aller Bemühungen um Information immer wieder neu auftreten und zumindest zeitweise zu einer illusionären Verkennung der Realität führen.

Der Beginn der Auseinandersetzung findet häufig ganz oder doch zu einem großen Teil im Rahmen der stationären Behandlung statt. Unseres Erachtens sind die Erfahrungen, die die Patienten während dieser Zeit der psychischen Überlastung machen, mit prägend für ihr späteres Verhalten gegenüber ihrer Erkrankung. Sie haben zu Beginn der diagnostischen und therapeutischen Maßnahmen noch keine Erfahrungen im Umgang mit der Erkrankung; werden sie von Ärzten und Pflegepersonal auf der Station als Krebskranke mit möglicherweise ungünstiger Prognose eher gemieden, so erleben sie hier erstmals Zeichen einer sozialen Stigmatisierung, dies kann dazu führen, daß die Kranken sich im weiteren Verlauf auch selbst immer stärker sozial isolieren. Werden Patienten dagegen soweit wie möglich informiert und aktiv an der Planung und Durchführung der stationären Behandlung beteiligt, so kann damit die Grundlage für eine aktive und selbstverantwortliche Auseinandersetzung mit der Krankheit gelegt werden und von vornherein Angst, Depression, Hoffnungslosigkeit und Resignation entgegengearbeitet werden (Schonecke, 1984).

Zorn, Wut, Enttäuschung und Vorwürfe

Die Frage, „warum gerade ich?" steht im Zentrum des Erlebens. Mit ihr verbindet sich oft die Frage nach einer Eigenschuld, die Frage des Hiob. Der Patient ist von Gott und der Welt enttäuscht, innerlich wütend gegen seine Bezugspersonen. Wird die Aggressivität offen geäußert, ist die Situation leichter überschaubar. Häufig vermeidet der von seinem Arzt abhängige Patient jedoch die Äußerung seiner aggressiven Impulse, um weiter akzeptiert zu werden. Es entsteht dann ein Zustand „feindseliger Abhängigkeit", in dem

sich der Patient aus der Kommunikation zurückzieht. Offene und abgewehrte Aggressivität des Kranken können auf seiten des Arztes zu Gereiztheit führen.

Der Arzt fühlt sich in seinem Hilfsangebot zurückgewiesen, der Patient erscheint ihm als „undankbar". Die Reaktion des Arztes hierauf besteht vielfach ebenfalls in einem enttäuschten Rückzug vom Patienten, so daß nun die Kommunikation von beiden Seiten her gefährdet wird. Als Ergebnis eines solchen Kommunikationsabbruches finden sich immer wieder Patienten, deren Rückzug so weit geht, daß ihr auffälliges Verhalten fälschlicherweise auf somatische Veränderungen, wie z.B. Hirnmetastasen, zurückgeführt wird (vgl. Bsp. in Abschn. 66.1.3). In solchen Krisensituationen werden Patienten häufig an Psychosomatiker oder Psychiater überwiesen. Die Überweisung erst in einer solchen Krise ist in aller Regel jedoch ungünstig: Der Patient erlebt sich selbst in seinem Bemühen um Bewältigung der Situation ebenso gescheitert wie seinen Arzt; er fühlt sich jetzt zu einem Spezialisten für „psychische Störungen" abgeschoben. Dagegen vermag ein kontinuierliches, geduldiges Angebot zur Kommunikation auch dann, wenn der Patient aggressive Impulse erkennen läßt, das Verhalten solcher Kranken oft in erstaunlichem Ausmaß und in kurzer Zeit zu verändern. Voraussetzung hierfür ist, daß es dem Arzt selbst gelingt, diese Reaktion des Patienten nicht als auf die eigene Person bezogen zu erleben und entsprechend abwehrend zu reagieren; die Enttäuschung des Patienten bezieht sich nicht auf den einzelnen Arzt oder die einzelne Schwester, sondern letztlich immer auf die Unheilbarkeit der Erkrankung. Der Arzt sollte auf eine vorausgegangene Idealisierung der Medizin und einzelner Ärzte und damit im Zusammenhang stehende Erwartungen achten und die aktuelle Enttäuschung im Zusammenhang damit verstehen. Es wird ihm dann oft möglich, die Aggressionsproblematik in Ruhe mit dem Patienten zu besprechen und ihm zu zeigen, daß das einmal geschlossene Arbeitsbündnis auch hierdurch nicht gefährdet wird.

> Der oben geschilderte Leukämiekranke sprach in dieser Phase beunruhigt von eigenen aggressiven Phantasien gegen die Mitglieder seiner Familie, z.B. von Phantasien, die Familienmitglieder, für die er bisher in übertrieben aufopfernder Weise gesorgt hatte, zu erschießen. Aber auch die Beziehung zum Arzt wird von aggressiven Impulsen gefärbt: Im Gespräch über die weitere Zusammenarbeit beteuert der Patient ganz unvermittelt: „Ja, man muß sich doch vertragen, zum Raufen darf es nicht kommen".

Aggressive Regungen gegenüber Ärzten, Krankenschwestern oder -pflegern sind häufig, allerdings werden sie nur selten direkt, öfter dagegen in versteckter Form geäußert: Die Kritik gilt dem Essen in der Klinik, Verhaltensweisen anderer, weniger angepaßter Patienten oder dem Verhalten von Mitarbeitern, die in der Klinikhierarchie niedriger stehen. Im folgenden Beispiel wurden die aggressiven Regungen auf Mitpatienten verschoben:

> Eine unheilbar Krebskranke wird auf einer Station etwa 25mal in ein anderes Zimmer verlegt, da sie aufgrund eigener Erfahrungen neuen Mitpatientinnen durch die Schilderung der Nebenwirkungen immer wieder Schrecken vor der geplanten Chemotherapie einzujagen versucht.

Patienten versuchen nicht selten, sich vom eigenen Schreck über die Erkrankung durch das Erschrecken anderer zu entlasten. Das Ausmaß der in dieser Reaktion enthaltenen Aggressivität wird oft erst an der Heftigkeit der Reaktion der Betroffenen deutlich.

> Ein junger Leukämiekranker, der zum Infektionsschutz während der Chemotherapie in einem Isolierbettsystem behandelt wird, begrüßt eines Morgens die ihn betreuende Krankenschwester mit dem Hinweis: „Sie sehen aber heute auch sehr blaß aus". Die in der Betreuung onkologischer Patienten sehr erfahrene Krankenschwester befürchtet daraufhin tagelang, selbst an Leukämie erkrankt zu sein; sie beobachtet sich nicht nur genau im Spiegel, sondern läßt eine Reihe von Laboruntersuchungen durchführen.
>
> Ein 40jähriger Patient mit Morbus Hodgkin berichtet, daß ihn ein alter Bekannter auf der Straße gefragt habe, wie es ihm gehe. Er habe geantwortet: „Mir geht es gut. Ich weiß, daß ich Krebs habe". Daraufhin sei der Bekannte erschrocken und habe erstaunt nachgefragt. Er habe ihm dann erklärt: „Sehen Sie, ich weiß wie ich dran bin. Aber wissen Sie, wenn Sie jetzt gleich über die Straße gehen, ob Sie nicht von einem Auto überfahren werden?" Der Patient berichtet weiter, wie er sich gefreut habe, als er „bemerkte, wie der andere erschrak und bei ihm der Groschen fiel".

Depression

Bei der depressiven Reaktion lautet die Frage: „Was bin ich jetzt, als Kranker, noch wert?" Die Beschädigung des Körpers, die Funktionseinbußen, der Verlust von Befriedigungsmöglichkeiten und die Veränderung der Rolle in der Familie beeinträchtigen das Selbstgefühl, insbesondere das Selbstwerterleben des Patienten.

Die diffusen, oft unbeeinflußbaren Klagen des depressiv verstimmten Kranken belasten und verunsichern den Arzt. Häufig kann der Depressive Hilfe nicht wirklich annehmen, obwohl er ständig Hilfe zu fordern scheint. In seiner Enttäuschung weist er nicht selten Bemühungen der Helfer, ihm näherzukommen, zurück; dies kann über einen gekränkten Rückzug der Helfer rasch in Form eines Circulus vitiosus zu einer weitgehenden Isolation führen. Die Gesamtsituation und solches abwehrendes Verhalten aktiviert bei Ärzten und Schwestern gelegentlich eigene depressive Tendenzen, deren Abwehr wiederum Energie verzehrt; hinzu kommt die Angst vor der Möglichkeit eines Suizidversuchs und den Folgen innerhalb der Institution.

In der depressiven Reaktion nimmt auch die Verletzlichkeit des Patienten enorm zu. Die Reaktion auf die situationsbedingte Verunsicherung und Hilflosig-

keit kann sich zu einer Art „existentieller Verzweiflung" (Weisman, 1979) steigern.

Faktorenanalytisch ließen sich in einer Untersuchung um einen Kern von Depression und Machtlosigkeit Variablen in vier Bereiche von Bedrohung gliedern:
- Vernichtung (Hoffnungslosigkeit, Angst, geschlossene Zeitperspektive),
- Entfremdung (auch Verlassenheit, Isolation, Zurückweisung, Wertlosigkeit),
- Überwältigtsein von Gefahr und Belastung und
- Verleugnung.

Hieraus ergibt sich für den Arzt die Bedeutung seines Bemühens um kontinuierliche Aufrechterhaltung einer tragfähigen Beziehung zum Patienten und um Unterstützung seines Selbstwerterlebens. Im Umgang mit dem Arzt soll der Patient zunächst spüren können, daß auch seine depressive Reaktion als Reaktion auf die Krankheit verstanden und akzeptiert wird. Im Bemühen um die Aufrechterhaltung der Beziehung wird der Arzt Versuche des Patienten, ihn abzulehnen und die Beziehung zu entwerten, als eine Reaktion des Kranken im Zusammenhang mit seiner Enttäuschung einerseits und seinem verzweifelten Bemühen um Autonomie und Aufrechterhaltung des eigenen Wertes andererseits zu verstehen suchen.

In der Begleitung unheilbar Kranker kommt der Unterstützung ihres Selbstgefühls, insbesondere ihres Selbstwerterlebens, zentrale Bedeutung zu. Die Aufrechterhaltung der Beziehung selbst vermag hierzu entscheidend beizutragen; daneben kann das Selbstwerterleben verstärkt werden, indem die Patienten auf den Wert vergangener, positiver Aspekte ihres Lebens, auf eigene Leistungen, aber auch auf neue Möglichkeiten zum Wertvollsein auch als Kranke hingewiesen werden. Dabei stellt sich auch immer wieder die Frage, was der Kranke in seiner Umgebung („in unserer Gesellschaft") noch wert ist. Es lohnt sich, äußerste Sorgfalt und Geduld darauf zu verwenden, mit dem Patienten gemeinsam jede Möglichkeit zur Aufrechterhaltung seiner alten oder zum Auffinden einer entsprechend bedeutsamen neuen Rolle innerhalb seiner Familie und seiner beruflichen Tätigkeit zu prüfen. Dies kann ein wesentlicher Beitrag zu einer – wenn auch oft nur vorübergehend während Remissionsphasen möglichen – Rehabilitation des Patienten sein (vgl. Diskussion zur Frage der Hoffnung in Abschnitt 66.2.7).

In konsequenter Anwendung der Konzepte der psychoanalytischen Objektbeziehungspsychologie haben Meerwein und Mitarbeiter (1976, 1981) darauf aufmerksam gemacht, wie wichtig es ist, zusammen mit unheilbar Kranken die positiven Anteile früherer Beziehungen zu erarbeiten; es geht dabei darum, den Kranken die Wahrnehmung der inneren Abbilder dieser Beziehungen wieder zugänglich zu machen und diese „inneren Objekte" vor der Entwertung zu schützen, die aus der aktuellen Enttäuschung rührt. Gerade beim Auftreten heftigster aggressiver Impulse kommt es oft zu inneren Spaltungsvorgängen: Der Patient kann eigene gute und böse Anteile in sich – diese Anteile beinhalten Abbilder der eigenen Person, der Bezugspersonen und der Beziehung zwischen den Beteiligten – nicht mehr integrieren. Schuldgefühle und schließlich auch Entwertung der eigenen Person können die Folgen sein, die ihrerseits große Destruktionskraft entfalten. Der Arzt muß darauf gefaßt sein, daß er in diese Entwertungstendenzen mit einbezogen wird und daß sich die Projektion abgespaltener negativer Anteile solcher inneren Abbilder auf ihn richten kann.

> Der erwähnte Leukämiepatient zeigte im Verlauf eine Tendenz zum Rückzug aus der Kommunikation. Er fühlte sich erleichtert, als er auch akzeptiert wurde, als er „den Moralischen" bekam und längere Zeit weinte. Daraufhin vermochte er auch über das Gefühl seiner Wertlosigkeit zu klagen, und schließlich konnte er zum ersten Mal über seine Impotenz, die mit der Krankheit aufgetreten war, sprechen. Es gelang, ihm zu zeigen, daß seine Potenz in dieser Situation nicht ausschließlich im sexuellen Bereich, sondern in einem umfassenderen Sinn, nämlich in den gemeinsamen Planungen mit seiner Familie für die Zeit nach seinem Tod gesucht werden könne. Diese Gespräche vermochten die Depression aufzuhellen; später sprachen beide Ehepartner davon, daß das letzte Jahr des gemeinsamen Lebens zu ihrer schönsten Zeit gehört habe.

Die Depression kann durch die krankheitsbedingte Einschränkung aller Befriedigungsmöglichkeiten mitbedingt oder doch verstärkt werden. Arzt und Pflegepersonal sollten deshalb zusammen mit der Familie immer wieder bis ins einzelne gehend überlegen, welche Befriedigungsmöglichkeiten dem Patienten noch erhalten oder wieder zugänglich gemacht werden können.

Von dieser erstgenannten, mehr vergangenheitsbezogenen Depressionsform ist eine zweite zu unterscheiden, die am besten als **vorwegnehmende Trauer** bezeichnet wird. Diese Trauer betrifft den zukünftigen, durch den Tod bestimmten Abschied von allen Bezugspersonen. Sie vollzieht sich eher still, im günstigsten Fall gemeinsam mit den Angehörigen; der Arzt sollte diese Kommunikation mit den Angehörigen nach Möglichkeit fördern.

Argumente gegen die offene Kommunikation über die Diagnose beziehen sich häufig auf die depressive Reaktion von Patienten; es wird angenommen, diese lasse sich durch „schonendes Betrügen" vermeiden. Während es sich vom Kranken her bei dieser depressiven Reaktion um eine „gesunde", einfühlbare Reaktion in seiner bedrohlichen Lebenssituation handelt, geht es in dieser Argumentation dem Arzt oft darum, die Konfrontation mit dem traurigen Patienten, von dem wir uns als zur Hilfestellung bereite Ärzte in Frage gestellt erleben, zu vermeiden. Ein klagender oder gar weinender Patient löst dann übertriebene ärztliche Aktivität aus, während es dem Kranken oft mehr helfen könnte, ihn in seiner spezifischen Wertproblematik in der nun einmal gegebenen realen Situation zu verstehen.

„Feilschen" oder „Handeln"

Der Patient hat jetzt im Prinzip die Unheilbarkeit seiner Erkrankung anerkannt; er versucht jedoch, Aufschub zu erreichen. Das Thema heißt: „Noch nicht jetzt". Das Feilschen um Aufschub, das große Informationsbedürfnis und ständiges Fragen nach neuen Behandlungsmethoden können für den Arzt recht lästig werden. Verletzt fühlt er sich, wenn der Patient trotz aller Bemühungen beim Aufkommen neuer Verleugnungstendenzen vorübergehend Zuflucht bei anderen Ratgebern, wie z. B. Heilpraktikern, sucht. Im allgemeinen ist jedoch gerade die aufrichtige Information des Patienten auch in dieser Phase in der Lage, das Arbeitsbündnis aufrechtzuerhalten und weiter zu stärken.

Versuche des Patienten, die Behandlung selbst zu kontrollieren, sollten im Rahmen der Förderung der Autonomie unterstützt werden, so z. B. die Tendenz vieler Leukämiekranker, genau Buch über die Werte ihrer Blutbilder zu führen. Ärzte und Schwestern erleben solche Versuche oft als unangenehm, die Patienten als mißtrauisch.

> Unseren Leukämiekranken beschäftigte über lange Zeit ein Traum, in dem er Schach spielte, wobei seine Partner sowohl die Ärzte als auch der Tod waren. Als ich mit dem Patienten über diesen Traum sprach, konnte er nach langem Zögern erstmals offen mit seinen Todesvorstellungen verbundene Ängste äußern. Insbesondere beunruhigte ihn auch die Tatsache, daß er auf den Krankheitsverlauf selbst so wenig Einfluß nehmen konnte.

Wie intensiv solche Wünsche nach Macht und Kontrolle in der Arzt-Patient-Beziehung sein können, illustriert der Traum eines 40jährigen Leukämiekranken:

> Im Traum erhielt er im Lotto den Hauptgewinn. Er baute mit diesem Geld eine riesige Spezialklinik für Leukämiekranke mit einer Reihe neuer spezieller Behandlungsmöglichkeiten. Den ihn behandelnden Arzt stellte er als Chefarzt ein, er selbst behielt sich jedoch die Gesamtleitung der Klinik als Direktor vor.
>
> Angesichts der Bedrohung laufen beim Patienten offensichtlich auch regressive Prozesse mit einer Wiederbelebung von Omnipotenzphantasien ab: Er stattet seinen Arzt mit ungeheurer Potenz aus, gleichzeitig kann er ihn jedoch noch kontrollieren. Dies hilft ihm dabei, sowohl seine reale Abhängigkeit vom Arzt als auch sein tatsächliches Ausgeliefertsein an die tödliche Krankheit wenigstens etwas zu kompensieren.

Aufmerksamkeit bedarf die Tendenz der Patienten in dieser Phase, sich Aufschub über zu große materielle oder die Verarbeitung der Erkrankung blockierende ideelle, pseudoreligiöse Opfer zu erkaufen.

Die Situation Sterbender

Die Konfrontation mit dem Sterben als letzter Grenzsituation zwingt den Arzt zur Klärung seiner Werte und der damit verbundenen Zielvorstellungen für den Umgang mit unheilbar Kranken.

Die in der Literatur vertretenen Ziele bewegen sich zwischen zwei Polen (Spiegel-Rösing, 1980, 1984 a, b, c): einige in der Psychotherapie Sterbender erfahrene Autoren betonen die Möglichkeit, Prozesse der Selbstfindung und Selbsterweiterung im Sterben zu unterstützen (Le Shan, 1969), andere verweisen demgegenüber auf die oft ganz unerträgliche Wirklichkeit des Sterbens und all die Kräfte, die die Person des Patienten zu zerstören drohen (Bleeker, 1978). Dem optimistischen bis euphorischen Anspruch, jedem Patienten das Akzeptieren seines Todes zu ermöglichen (Kübler-Ross), steht die Forderung gegenüber, die krankheitsbedingten Begrenzungen psychotherapeutischen Bemühens, die „Häßlichkeit des Sterbens" anzuerkennen, auch wenn man sich darum bemühe, dem Patienten einen „angemessenen Tod" zu ermöglichen.

Weisman (1974, 1976) hat die Ziele im Umgang mit Sterbenden im Rahmen seines Konzeptes eines angemessenen Todes („appropriate death") dargestellt. Ein angemessener Tod ist ein Tod in Übereinstimmung mit dem eigenen Selbstverständnis und Ich-Ideal. Kernkonzepte sind: „awareness, acceptability, resolution, relief".

Hilfestellung zur Ermöglichung eines „angemessenen Todes" hat dabei folgende Ziele: relative Schmerzfreiheit, Verminderung des Leidens, Minimalisierung emotionaler und sozialer Beeinträchtigung. Der Betroffene sollte innerhalb der Grenzen seiner Behinderung auf einem möglichst hohen Niveau sein Leben gestalten können, auch wenn ihm nur Teile früherer Befriedigungsmöglichkeiten geboten werden können. Es sollte ihm möglich sein oder wieder möglich werden, verbleibende Konflikte zu erkennen und nach Möglichkeit zu lösen; es sollte versucht werden, alle verbliebenen Wünsche – im Zusammenhang mit seiner gegenwärtigen Belastung, aber auch im Zusammenhang mit seinen Idealvorstellungen – wenigstens annäherungsweise zu befriedigen. Der Kranke sollte die Möglichkeit haben, die für ihn bedeutsamen Schlüsselpersonen aufzusuchen oder sich auch von ihnen zu trennen; dabei sollte ihm auch gestattet sein, über diejenigen Kontrolle auszuüben, zu denen er Vertrauen hat (Weisman, 1974, 1976).

Es herrscht heute Übereinstimmung darüber, daß die Wahrnehmung der Bedrohung und die Auseinandersetzung mit ihr – soweit möglich – Bestandteil eines würdigen, menschlichen Sterbens ist (J. E. Meyer, 1973, 1979, 1983; Strotzka, 1984). Dagegen gehen die Meinungen darüber auseinander, in welchem Ausmaß es Patienten möglich sein soll, ihren Tod bewußt „anzunehmen". Gegenüber einer in der Konsequenz den Tod auch heroisierenden optimistischen Darstellung, vor allem bei Elisabeth Kübler-Ross und einigen anderen amerikanischen Autoren, erscheint uns aus klinischer Erfahrung eine zurückhaltendere und differenzierende Betrachtungsweise angezeigt.

Überschwengliche, das Akzeptieren des Todes geradezu fordernde Darstellungen können im Bereich der Krankenversorgung zu ganz unrealistisch übertriebenen Erwartungen an die Patienten führen, ihren Tod geradezu heroisch und bewußt „auf einer Art privatem Kalvarienberg" anzunehmen, nachdem sie die „Kreuzwegstationen" des entsprechenden Phasenschemas durchlaufen haben (Weisman, 1979).

Forderungen nach einer so weitgehenden Annahme des Todes führen leicht auch zu einer Überforderung der Mitarbeiter im ärztlichen und pflegerischen Dienst in der Begleitung zum Tode Kranker.

Die Annahme des Todes geschieht nur selten in vollem bewußtem Einverständnis. Letztlich bleibt der Tod häufig erschreckend und oft unannehmbar, wie es ein Kranker ausdrückte: „Der Tod kommt immer zu früh". Seine Annahme erfolgt viel häufiger im Verlaufe eines stillen, mehr oder weniger resignierenden Nachgebens, eines Sich Anpassens an das sich nähernde Ende (vgl. auch J. Holland, 1982). Dabei verkürzt sich allmählich die Zeitperspektive, weiterreichende Zielvorstellungen und Pläne nehmen an Bedeutung ab. Das Bedürfnis nach Nähe, ebenso in den familiären Beziehungen wie in den Beziehungen zu Ärzten und zum Pflegepersonal, nimmt zu (J. Holland, 1982). Alltäglichen Ereignissen in diesen Beziehungen und der Befriedigung von Basisbedürfnissen und einfachen Wünschen kommt größere Bedeutung zu. Dabei kann es für den Schwerkranken wichtig sein, daß ihm die Regression in Abhängigkeit und Unselbständigkeit gestattet wird (J. E. Meyer, 1974). Eissler spricht von dem Bedürfnis, sich in eine sublimierte Art von Liebe gleichsam eingehüllt zu fühlen, ohne zu Gegenleistungen aufgerufen zu sein. Im Rahmen der Regression könne möglicherweise die Sehnsucht nach einem früheren Zustand der Lust, in dem gleichzeitig auch das Gefühl höchster Sicherheit herrschen konnte, geweckt oder wiedererlebt werden (Eissler, 1955, am Beispiel einer Patientin).

Vom Arzt wird in dieser Situation besonders viel verlangt. Er muß in seiner Person die sich zu widersprechen scheinenden Aufgaben des Therapeuten und des Begleiters, kurative und palliative Ziele integrieren. Zunächst besteht in dieser Situation die Gefahr, daß die Kommunikation von seiten des Arztes abgebrochen wird, da die medizinische Behandlung häufig sinnlos wird, der Arzt sich als professioneller Helfer überflüssig fühlt und ihn zudem der Ausgang des Krankheitsverlaufes emotional sehr belastet. Oft kommt es in dieser Situation darauf an, ob der Arzt selbst sich mit der Wahrheit auseinanderzusetzen vermag, die den Mythos des Arztes als Sieger über den Tod gefährden könnte (J. E. Meyer, 1973).

Manche Krankenschwestern reagieren irritiert, insbesondere wenn männliche Patienten im Zuge regressiver Verhaltensweisen „schwach" erscheinen, „sich gehenlassen". In der Diskussion zeigt sich in der Regel, daß dieses Verhalten nicht mit den Idealvorstellungen der betreffenden Schwester von „Männlichkeit" oder auch anerzogenen Idealvorstellungen von der eigenen Person kompatibel ist.

Sozialwissenschaftliche Untersuchungen haben wiederholt darauf aufmerksam gemacht, daß Ärzte und Schwestern dazu neigen, sich aus Beziehungen zu Patienten mit ungünstiger Prognose zurückzuziehen. So nimmt die Informationsbereitschaft von Ärzten gegenüber solchen Patienten stark ab (Raspe, 1979, 1982a); diese Kranken haben auch eine wesentlich geringere Chance, während der Visite an den Arzt gerichtete Fragen angemessen beantwortet zu bekommen (Siegrist, 1972); auch auf Ängste und andere während der Visite geäußerte Affekte gehen die Ärzte bei Schwerkranken seltener ein als bei Leichtkranken (Koch et al., 1982). Schwestern reagieren auf das Läuten solcher Kranken verzögert: Le Shan und Gassman (1958) fanden auf einer Station eine Korrelation zwischen der Zeit, die vom Läuten des Patienten bis zum Erscheinen der Krankenschwester verging, und der Prognose der Patienten; je aussichtsloser die Prognose, desto länger die Reaktionszeit der Schwestern. Diese Befunde sprechen dafür, daß Ärzte und Schwestern entgegen den Normen ihrer Rollen und entgegen ihrem beruflichen Selbstverständnis einem intensiveren Umgang mit solchen Patienten, die sie subjektiv belasten, eher ausweichen (vgl. auch Glaser und Strauss, 1974).

Gelegentlich wird angesichts der Todesbedrohung der Patienten aus dem Ausweichen offene Flucht:

> Die oben erwähnte querschnittsgelähmte Künstlerin mit metastasierendem Bronchialkarzinom berichtete, daß sie während der Visite ihren Stationsarzt gefragt habe, ob sie auf seiner Station – und sie meinte positiv: in seiner Obhut – auch sterben könne. Unmittelbar auf diese Frage hin habe der Arzt ohne Antwort abrupt das Zimmer verlassen; am Abend habe er sie dann noch einmal besucht und ihr ausdrücklich gesagt, daß er über dieses Thema nicht mit ihr sprechen wolle; mit Sterben und Tod wolle er nichts zu tun haben, dieses Thema möchte er vermeiden.

Die Möglichkeit zu intensiver verbaler Kommunikation nimmt zwar mit Verschlechterung des Zustandsbildes und angesichts des drohenden Todes ab; dennoch können Arzt und Pflegepersonal während dieser Zeit auch viel für den Patienten tun. Nach Eissler (1978) kommt es neben der allgemeinen Hilfsbereitschaft und der antizipierenden Befriedigung von Bedürfnissen auf die aus einer sublimierten Liebe resultierenden Gefühle von unaufdringlicher Sorge und Mitleid an, die auch in dieser Situation Vertrauen, Mut und Trost schaffen und Kummer und Verzweiflung entgegenwirken können.

Die Sorge für den Patienten vermittelt sich dabei vor allem auch über die Qualität der Pflege und über ein sorgfältiges Hinhören und Eingehen auf seine oft nur indirekt geäußerten Bedürfnisse. Besondere Bedeutung kommt der sorgfältigen, soweit irgend möglich gemeinsam mit dem Patienten geplanten Gabe von Analgetika und – soweit nötig – Sedativa zu.

Die Möglichkeiten palliativer Therapie wurden in den letzten Jahren erheblich verbessert. Ihre Umsetzung im klinischen Alltag scheitert allerdings häufig daran, daß die Konflikte im Spannungsfeld zwischen kurativem und palliativem Ansatz in der Medizin im allgemeinen und in der Onkologie im speziellen nur selten systematisch reflektiert werden (Petzold und Frehen, 1987).

Für den Patienten ist es hilfreich, die Bereitschaft von Ärzten und Pflegepersonal zu spüren, trotz des ungünstigen Verlaufes und trotz der eigenen therapeutischen Ohnmacht bei ihm zu bleiben, die Beziehung aufrechtzuerhalten, die Begleitung nicht abzubrechen. In einer solchen Beziehung bringen die Kranken ihr Erleben vom bevorstehenden Ende und – im engsten Zusammenhang hiermit – häufig eine rück-

blickende Bewertung des eigenen Lebens – oft in einer symbolischen Sprache zum Ausdruck, die es erlaubt, diese Inhalte anzudeuten, ohne sie bewußt auszusprechen. Die Kranken sprechen dann z.B. davon „eine Reise zu planen", „nach Hause zu gehen" oder davon, daß „die Kohle bzw. das Geld nicht mehr lange reicht". Auch intensive Schuldgefühle können in einer solchen Sprache symbolisch angedeutet werden (Piper, 1977; Prest, 1970; Kübler-Ross, 1982).

Es lohnt sich auf bildhafte Redeweisen von Patienten individuell zu achten. Bliesener ist in unserer Kölner Arbeitsgruppe dabei solchen bildhaften Stellen in Interviews mit Leukämiepatienten näher nachgegangen (1987). Hier ein Beispiel:

„Die Therapien bilden für mich meine Heilungschance ... Ich bin in einen Sumpf gefallen und habe irgendwo eine Leiter gefunden. Diese Leiter ist die Therapie. Und diese Leiter kletter' ich hoch, ... bis daß ich wieder oben bin, aber im Moment sehe ich wieder Tageslicht." Bliesener kommentiert: Die „offizielle Botschaft" dieses Vergleichs ist sicher, daß die in regelmäßigen Intervallen angesetzte Chemotherapie, analog den Sprossen einer Leiter, aus der Krankheit heraus in die Gesundheit führt. Bleibt man aber im Bilde und phantasiert es konsequent zu Ende, so findet man einen ganz anders akzentuierten, latenten Gehalt: Die Leiter droht beim Besteigen in den Sumpf mit abzusinken (Rückfall und Tod); sie führt andererseits wie die biblische Jakobsleiter nach oben ins Licht (Himmel und wiederum Tod). Bliesener merkt an, daß die zur Gesundung gedachte Therapie ja nicht selten die Gefahr enthält, den Tod mit herbeizuführen (wie es auch bei diesem Patienten war).

Der Theologe Michael Nüchtern (1981) hat in einer Betrachtung über Emil Noldes Bild „Der Arzt, der Kranke, der Tod und der Teufel" (Abb. 66–1) eindrucksvoll herausgearbeitet, wie wichtig es für den Todkranken sein kann, daß der Arzt auch in dieser Situation bei ihm bleibt. Wir geben diese Betrachtung hier verkürzt wieder:

„Ganz vorne auf dem Tischchen steht eine Arznei. Die braucht der alte Mann jetzt nicht mehr. Ganz ruhig blickt er den Tod an. Auf seiner Decke schiebt er ihm die Hand entgegen. Die Knochenhand des Todes winkt: Komm. Der Arzt, der so eigentümlich dazwischen steht, sieht den Tod nicht an. Er blickt auf den Sterbenden, die Nase, den Mund, die Augen und weiß, weiß mit Trauer fast, wie es um den alten Mann steht. Er kann nichts mehr machen, er tritt einen Schritt zurück – und bleibt doch da. Wozu soll der Arzt, wo er doch nichts mehr machen kann, noch dabeibleiben? Hat er nicht seine Schuldigkeit getan? Ich versuche mir vorzustellen, wie das Bild aussehen würde, wäre der Arzt nicht mit auf dem Bild.

Der Sterbende wäre allein mit dem Tod, und allein vor allem ausgesetzt der teuflischen Fratze, die im Hintergrund gemalt ist. Ich glaube, wer allein mit dem Tod ist, den schützt niemand vor dem Teufel, das heißt vor Schrecken und Verzweiflung. Wo der Kranke die Hand nach dem Tod ausstreckt, greift der Arzt nicht ein, er bleibt stehen zwischen dem Sterbenden und dem Teufel. Der Arzt scheint gar nicht zu wissen, was er durch sein Dasein für den Sterbenden tut: er hält ihm den Teufel vom Leibe. Er trennt Tod und Teufel für den Sterbenden, damit dieser ja sagen kann zu seinem Tod. Sterbehilfe heißt – im mythischen Bilde – Tod und Teufel trennen für den, der stirbt. ...

Abb. 66–1. Emil Nolde: „Kranker, Arzt, Tod und Teufel" (1911), Stiftung Seebüll.

Der Arzt sieht den Sterbenden an. Im Gesicht des Kranken sieht er den Tod wie in einem Spiegel. Im Gesicht des Kranken sieht er nichts vom Teufel, nicht Schrecken, nicht Verzweiflung. Nur weil der Arzt dableibt, nicht davonläuft vor dem Sterben und somit Tod und Teufel für den Sterbenden trennt, kann er auch selbst etwas von dem Sterbenden lernen, was er nicht erfahren könnte, würde er nicht dabeibleiben: daß ein Mensch in Frieden sterben kann und seinen Tod bejaht. Der Arzt hilft dem Kranken, und der Sterbende hilft dem Arzt. Die dämonische Fratze hat das Spiel verloren."

An den Grenzen ihrer therapeutischen Möglichkeiten kann so der offene Umgang mit dem Patienten Arzt und Pflegepersonal helfen, diese Begrenztheit bewußter wahrzunehmen und zu verarbeiten. Gelingt Ärzten und Schwestern eine solche Auseinandersetzung, so ist es immer wieder erstaunlich, in welchem Maße Patienten von ihrem bevorstehenden Sterben zu sprechen beginnen; dann wird es auch möglich, gemeinsam die Umstände des Sterbens zu planen.

Für den klinisch tätigen Arzt ist dabei wichtig, zwischen der Angst vor dem Tod und der Furcht vor den mit dem Sterben verbundenen Belastungen zu unterscheiden (u. a. Heimpel, 1984). Bei vielen Patienten sind mit der terminalen Phase tiefgehende Befürchtungen verknüpft, insbesondere in bezug auf die dann zu ertragenden Schmerzen. Es kann Patienten entscheidend helfen, wenn sie ihre Befürchtungen und die damit verknüpften Phantasien (z.B. auch Verblutungs- oder Erstickungsängste, Vorstellungen von der Intensivstation) ihrem Arzt mitteilen und mit ihm besprechen können. Wichtig ist, daß der Arzt nach dem Kennenlernen dieser Vorstellungen dem Patienten

glaubhaft erläutert und versichert, daß präterminale Schmerzen und auch präterminale Erstickungszustände heute durch medikamentöse Therapie verhindert werden können (vgl. Kap. 36).

Andererseits sollten Ärzte und Pflegepersonal sich die Ziele im Umgang mit unheilbar Kranken auch nicht zu hoch stecken. Oft ist schon viel gelungen, wenn ein „schlechter Tod" (Weisman, 1979) in Verzweiflung und Isolation vermieden und die Andeutung einer Versöhnung erreicht werden kann.

> Während der Stationskonferenz auf einer hämatologisch-onkologischen Station wird ein 20jähriger Leukämiekranker als extrem zurückgezogen erlebt, jeden Näherungsversuch hat er bisher zurückgewiesen. Bei den Schwestern hat diese Zurückweisung heftigen Ärger und Ablehnung hervorgerufen. Während der Konferenz gelingt es, das Verhalten des Patienten wenigstens zum Teil als Folge eines Konfliktes zwischen spätadoleszenten Verselbständigungswünschen und regressiven Versorgungswünschen in der Folge des krankheitsbedingten Zustandes zu verstehen. Danach ergab sich folgende Entwicklung: Diejenige Schwester, die sich besonders um den Patienten zu kümmern versuchte, wollte sich am Wochenende von ihm bis zum folgenden Montag verabschieden. Der bis dahin schroff abweisende Patient meinte darauf: „Da sehen wir uns nicht wieder"; zur Überraschung der Schwester gab er ihr die Hand zum Abschied und bedankte sich für die gute Versorgung. Als die Schwester am Montag zurückkkam, war der Patient verstorben.

Soweit es irgend möglich ist, sollten die Familienmitglieder in die Versorgung ihrer sterbenden Angehörigen einbezogen und in dieser Versorgung unterstützt werden. Die Beteiligung an der terminalen Versorgung eröffnet den Angehörigen nicht selten noch die Möglichkeit zu intensiven Gesprächen und zu einem ausdrücklichen Abschiednehmen. Dies ist zugleich auch eine wirkungsvolle Prophylaxe gegen pathologische Trauerreaktionen. Hat der Patient den Wunsch, zu Hause zu sterben, so sollten alle Möglichkeiten ausgeschöpft werden, um diesen Wunsch zu erfüllen.

In dem dialektischen Spannungsverhältnis unserer Aufgaben im Umgang mit unheilbar Kranken – dem Bemühen, ihm möglichst lange ein qualitätsvolles Leben in der Gemeinschaft zu erhalten und dem Bemühen, ihn beim Verlassen dieser Gemeinschaft zu unterstützen – ist es wichtig, die zur Trennung führenden Entscheidungen des Patienten zu respektieren. Wir sollten ihn im richtigen Augenblick gehen lassen können, unsererseits die Trennung akzeptieren und ihn nicht unnötig aus unseren Bedürfnissen heraus festzuhalten versuchen. Die Tendenz Angehöriger, den Patienten festzuhalten, kann dem Kranken Abschiednehmen und Sterben sehr erschweren. Die bekannte Psychologin A.-M. Tausch, selbst an einer fortgeschrittenen Krebserkrankung leidend, hat dies in Vorträgen – bezogen auf ihre Töchter – wiederholt hervorgehoben (vgl. Tausch, 1981).

In diesem Zusammenhang sei auch noch auf die Bedeutung des Abschiednehmens vom verstorbenen, aufgebahrten Patienten hingewiesen. Ist dies möglich, kommt es seltener zu überschießenden pathologischen Reaktionen während der Trauerzeit; es treten vor allem seltener Illusionen im Sinne hartnäckiger Vorstellungen auf, der Angehörige sei in Wirklichkeit gar nicht gestorben (u. a. Parkes, 1974).

> Der mehrfach genannte Leukämiekranke war während seiner letzten Lebensstunden bewußtlos. Wir Ärzte zogen uns zurück, die Familie blieb bei ihm. Unmittelbar vor seinem Tod kam der Patient noch einmal zu sich, winkte jeden der Angehörigen zu sich heran, verabschiedete sich von jedem mit einem Händedruck; dann sank er zurück und verstarb nach einigen Atemzügen.

Versuchen Ärzte und Pflegepersonal in der beschriebenen Weise (vgl. auch Kap. 28) dazu beizutragen, unheilbar Kranken ein angemessenes Sterben zu ermöglichen, dann nimmt auch die Gefahr ab, den Umgang mit dem Tod entweder zu heroisieren oder zu profanisieren, indem der Tod auf das „Natürliche" des Vorgangs reduziert wird (J. E. Meyer, 1979). Neben der Sorge um die aktuellen Bedürfnisse Sterbender können dann auch immer wieder Gedanken, die sich auf das im Sterben Kommende richten, geäußert werden (J. E. Meyer, 1979). Die Sensibilität von auf die Zukunft gerichteten Äußerungen, von Äußerungen der Neugier, von der Ernst Bloch (1959) in „Das Prinzip Hoffnung" („Forschende Reise in den Tod") spricht, könnte sich als mindestens ebenso wichtig erweisen wie eine Vergrößerung des noch dürftigen psychologischen Wissens um die Bedingungen der Fähigkeit, den Tod zu „akzeptieren" (Übersicht bei Wittkowski, 1978).

66.3.2 Hinweise auf die psychische Belastungs-fähigkeit von Patienten

Die Auseinandersetzung mit der unheilbaren Erkrankung wird durch den Krankheitsverlauf selbst, durch die Fähigkeit des Patienten zur psychischen und sozialen Anpassung („coping") sowie durch die Unterstützung, die er aus seiner Umgebung erfährt („social support"), beeinflußt. Nur bei einem offenen Umgang mit dem Kranken erhält der Arzt ausreichend Einblick in diesen Prozeß und vermag eventuell notwendige zusätzliche Unterstützungsmaßnahmen rechtzeitig zu planen. Auch hier geht es darum, Zugang zur individuellen Wirklichkeit des Patienten zu bekommen und so die subjektiv erlebte Belastung, ihre Verarbeitung und die wahrgenommenen sozialen Unterstützungsmöglichkeiten kennenzulernen.

Auf die Psychologie der Anpassungs- und Abwehrprozesse bei körperlichen Erkrankungen gehen wir in diesem Lehrbuch in einem eigenen Kapitel (Kap. 62) ausführlich ein. Hier möchten wir die Reaktion zweier Patienten auf die extreme Belastung durch die Erkrankung (akute Leukämie) und deren Behandlung in einem Isolierbettsystem illustrieren. Beide Patienten verarbeiteten diese Situation auch in ihren Träumen.

> Der Patient Z. (40 Jahre) träumte vor Behandlungsbeginn seine „Lebenssituation" in einem dreiteiligen Spiegel. In der Mitte sah er das Isolierbett, hinter ihm die Werkskapelle, die nach seiner Genesung zu seinem Empfang bei Wiederaufnahme der Arbeit spielte; links und rechts des Isolators bewegten sich Leichenzüge, die, wie der Patient selbst bemerkte, den Tod symbolisierten. Dem Ich des Patienten schien es zu gelingen, die belastende Behandlungssituation als für seine Genesung notwendig zu akzeptieren. Während des gesamten Behandlungsverlaufs war die Kooperation mit diesem Patienten ausgezeichnet.
>
> Patient W. (36 Jahre) träumte – ebenfalls vor Beginn der Behandlung – von einem Leuchtschirm, der sich über seinem Körper bewegte. Auf dem Leuchtschirm kamen Zahlen zur Darstellung, die hin und her flackerten, der Patient mühte sich vergebens ab, diese Zahlen zu einer sinnvollen Information zu verarbeiten; er erlebte es als ausgesprochen kränkend, daß ihm dies nicht gelang. Das Ich des Patienten fühlte sich nicht imstande, sich in der neuen Situation zurechtzufinden. Dieser Kranke hatte auch in der Realität sein seelisches Gleichgewicht schon vor der Erkrankung nur mühsam aufrechterhalten können. Seine Ehe war von einem beständigen Kampf zwischen Abhängigkeit und Unabhängigkeit, zwischen Überlegenheit und Unterlegenheit gekennzeichnet. Das Gleichgewicht konnte schon seit Monaten nur über gegenseitige Selbstmorddrohungen aufrechterhalten werden. Schon kurz nach Behandlungsbeginn zeigte dieser der Dekompensation von Anfang an nahe Patient Zeichen einer psychischen Dekompensation, z.B. Depersonalisationserlebnisse und panische Angstzustände.

Im Krankenhaus hängt die Belastbarkeit des Patienten ganz wesentlich von der hier in den Beziehungen zu Ärzten und Schwestern angebotenen Unterstützung ab. Die Beurteilung der Belastbarkeit des Patienten hat damit den Umgang zwischen den Helfern und dem Patienten zu berücksichtigen.

66.3.3 Reaktionen von Ärzten und Pflegepersonal

Der Umgang mit unheilbar Kranken und die damit verbundenen Belastungen können zu charakteristischen Schutz- und Abwehrhaltungen von Ärzten und Pflegepersonal führen. Die Kenntnis dieser Reaktionen ist wichtig, will man eine Arbeitssituation schaffen, in der die professionellen Helfer sich längerfristig in den Umgang mit unheilbar Kranken einlassen, ohne selbst überlastet zu werden und „auszubrennen" (Aronson et al., 1983). Neben der eigentlichen Belastung ist bei einer solchen Entwicklung natürlich auch immer die Motivation der Helfer für die Wahl dieser Arbeitssituation zu berücksichtigen (Schmidbauer, 1977, 1983).

Die hier beschriebenen Schutz- und Abwehrhaltungen verhindern Offenheit und Gegenseitigkeit in den Beziehungen zu dem Kranken; sie können sowohl zu einer Vergrößerung der Distanz als auch zu einer Übersteigerung der eigenen Aktivität, die den Patienten einseitig zu bestimmen droht, beitragen.

Vermeidung

Unheilbar Kranke bemerken es, wenn wir ihnen ausweichen; so meinte ein Malignomkranker zu einer Schwester: „Sie können mir ruhig alles sagen; ich habe ganz bestimmt Krebs, das nehme ich jedenfalls an!" Die Schwester fragte ihn, wie er darauf komme. „Weil Sie mir dauernd ausweichen. Es hat mir keiner etwas gesagt. Sie haben aber immer versucht, ein anderes Gesprächsthema zu finden oder sind mit irgendwelchen Worten über meine Frage hinweggegangen."

Vermeidendes Verhalten tritt gehäuft in Krisensituationen auf. Für den Patienten zeigt solches Verhalten oft eine Information über negative Entwicklungen der Krankheiten an, wie dies etwa die Wahrnehmung der isolierten Leukämiekranken über den Zusammenhang zwischen dem Zeitpunkt der Mitteilung des Knochenmarkbefundes und der Art des Befundes illustrierte (s. o.).

Verleugnung

Der seelische Abwehrvorgang der Verleugnung ermöglicht es, eine drohende Gefahr von der bewußten Wahrnehmung auszuschließen; die Augen werden sozusagen vor der beängstigenden Realität verschlossen. Häufig verleugnen Patienten und Personal gemeinsam.

So wurde vom Stationsarzt über mehrere Wochen gemeinsam mit einer an einem metastasierenden Pankreaskarzinom leidenden älteren Kollegin die Illusion einer Hilfe durch einen operativen Eingriff aufrechterhalten.

In einer „Balint-Gruppe" berichtet eine Krankenschwester davon, daß es sie jedesmal erleichtere, wenn sie einer zytostatisch behandelten Malignomkranken, die sich angstvoll mit Fragen an sie klammere, sage: „So sicher ist das mit Ihrer Diagnose ja noch gar nicht." Als die Patientin schließlich vom Arzt ausführlich informiert wurde, fragte sie die Schwester vorwurfsvoll, warum sie ihr denn immer wieder Hoffnung gemacht habe. Die Schwester sagte der Patientin, wie schwer es ihr selbst gefallen sei, ihr Wissen wegzuschieben. Nach diesem Gespräch hatte die Schwester zum erstenmal das Gefühl einer wirklichen Verständigung mit dieser Kranken.

Die Motive für die Verleugnung der Schwester konnten wir in der „Balint-Gruppe" verstehen. Sie hatte sich in der Gruppe zunächst beinahe darüber empört, daß die Frau als Malignomkranke therapiert werde, obwohl die Diagnose noch nicht hundertprozentig gesichert sei. Gleichzeitig berichtete sie, daß sie sich, im Unterschied zu anderen Kranken, im Umgang mit der Malignomkranken jedesmal besonders hilflos fühle und wohl deshalb versuche, die Diagnose in Frage zu stellen. Die Schwester wünschte sich eine heilbare Krankheit der Patientin, um sich selbst nicht so hilflos zu erleben. Dieses Gefühl der Hilflosigkeit reichte bei der Schwester weit zurück: Sie hatte als Kind früh ihre eigene Mutter infolge einer Karzinomerkrankung verloren.

Flucht in die Überaktivität

Hilflosigkeit gegenüber unheilbar Kranken und ihrer Angst ist besonders schwer zu ertragen. Oft wird von uns und vom Pflegepersonal diese Hilflosigkeit durch ein Übermaß an Aktivität, gelegentlich geradezu durch eine Flucht in Überaktivität, kompensiert. Besonders deutlich wird diese Haltung, wenn sie zugleich aus der Unsicherheit zu Beginn der Berufstätigkeit resultiert: Ein Kollege und ich (K. K.) verbrauchten als Medizinalassistenten an einem kleinen Kreiskrankenhaus auch aus diesem Grund so viel Medikamente, daß der Etat zwei Monate früher als sonst aufgezehrt war. Überaktivität von Ärzten und Schwestern lassen die Beziehung zum Kranken in besonderem Maße asymmetrisch werden; Ärzte und Schwestern versuchen ständig etwas für den Patienten zu tun, die Möglichkeiten des Patienten, sich zu äußern oder aktiv zu sein, werden dadurch eingeschränkt. So hört der Arzt z. B. weniger auf die Klagen des Patienten, läßt den Patienten nicht ausführlich berichten, sondern tendiert dazu, rascher Analgetika und Sedativa zu verordnen und unter Umständen auch ein „aggressiveres" therapeutisches Regime zu planen.

Krankenschwestern berichten, daß sie Schlafmittel oft mindestens so sehr zur eigenen Beruhigung wie zur Beruhigung der Patienten austeilen. Zu berücksichtigen ist, daß Patienten für die Gestaltung der Beziehung, für diese Einengung des Bewegungsraumes, sehr sensibel sein können.

Ein Beispiel aus dem Pflegebereich: Ein Patient wird gebettet, die Schwester begleitet ihre Handlungen mit sanften Worten: „So, jetzt schütteln wir eben noch das Kissen, und die Bettflasche legen wir ihm unter die Füße – so, und dann ist alles recht". Der Patient, der, wie es schien, bisher teilnahmslos im Bett gelegen hatte, setzt den Kommentar der Schwester ebenso sanft im gleichen Tonfall fort: „Dann bringen wir ihm noch einen Sarg".

Die Beziehung wird eindeutig von den Aktivitäten der Schwester beherrscht, der Kranke hat in ihr keinen Raum mehr, von sich aus Gefühle mitzuteilen. Er kann der Schwester aber noch sarkastisch das Spiegelbild ihrer eigenen Handlungen vor Augen führen und damit ausdrücken, daß er sich „zu Tode gepflegt" fühlt.

Entmündigung, Verkindlichung, Versachlichung

Wie schon bei der Flucht in die Überaktivität wird die Beziehung immer einseitiger gestaltet. Ärzte und Schwestern drängen von ihrer überlegenen Position aus den Patienten immer mehr zurück, engen seinen Raum im Interaktionsfeld ein, statt ihn bei der Entfaltung eigener Aktivität und bei der Artikulation seiner Bedürfnisse zu unterstützen. Für den Patienten kann es dann sinnlos werden, eigene Wünsche und Sorgen zu äußern. Er wird in die Rolle eines Kindes gedrängt, das von den Schwestern, wie früher von der Mutter, gepflegt, versorgt, gefüttert, gewärmt wird. Schwestern, aber auch Ärzte, leiten aus ihrer langjährigen Erfahrung den Anspruch ab, sehr genau zu wissen, was der Patient braucht und was er nicht braucht, um sich wohlzufühlen. Hierdurch wird jedoch allenfalls die Befriedigung von Schwestern und Ärzten sichergestellt.

Während in der Pflege die Verkindlichung des Patienten überwiegt, wird der Kranke bei einer solchen Entwicklung bei den Ärzten immer mehr zum reinen „Objekt" rein naturwissenschaftlich mißverstandener Medizin. Die Kommunikation während der Visite findet zunehmend nicht mehr mit ihm, sondern über ihn statt, es wird immer mehr über die Wirkung von Zytostatika, die Veränderung chemischer Parameter und anderes mehr gesprochen. Die Situation des Patienten wird verharmlost, er wird mit aufmunterndem Schulterklopfen oder ähnlichem getröstet. Ein solches Verhalten hat wenig mit der Freundlichkeit oder Unfreundlichkeit einzelner Ärzte, mit „Menschlichkeit" oder „Unmenschlichkeit" einzelner Schwestern zu tun; es entspricht vielmehr professionellen Schutzhaltungen, ohne die die extremen Belastungen im Umgang mit unheilbar Kranken oft zumindest so lange nicht zu ertragen sind, wie Ärzten und Schwestern nicht eine entsprechende Aus- und Weiterbildung für den Umgang mit diesen Patienten angeboten wird.

Isoliert betrachtet, nehmen manche Situationen groteske Gestalt an:

> Zwei Ärzte untersuchen das Ohr eines todkranken Leukämiepatienten, das phlegmonös entzündet ist und wiederholt geblutet hat. Einer meint etwas jovial zum Patienten: „Ja, Meister, am besten schnitten wir das Ohr doch wohl ab!" Darauf der andere Arzt: „Aber nein, wir wollen es doch heilen!" Und der Patient: „Von mir aus, schneiden Sie's halt ab, darauf kommt es auch nicht mehr an!"
>
> Die Situation ist um so grotesker, als der Patient an einem Bein bereits zweimal amputiert werden mußte und jetzt das Ohr wie ein Gegenstand betrachtet wird, der eigentlich schon keine Verbindung mehr mit einem lebendigen Menschen hat. Es scheint, als fühlten sich die beteiligten Ärzte und Schwestern so sehr durch die Krankheit und den Patienten bedroht, daß sie in dieser Situation nur dann stark bleiben können, wenn sie den Patienten kleiner machen.

Überidentifikation

Ein gewisses Maß an Einfühlung in die Situation des Patienten ist notwendig, um ihn verstehen zu können. Zur Einfühlung benützen wir Identifikations- und Projektionsvorgänge. Um Kranke ärztlich versorgen zu können, ist die Fähigkeit zur flexiblen Rücknahme solcher Projektions- und Identifikationsvorgänge nötig. Setzt man sich innerlich zu sehr an die Stelle des Patienten – „ich an seiner Stelle" –, so kann der Patient mit seinen Bedürfnissen, seinen Sorgen und Ängsten nicht mehr zu Wort kommen. Es wird dann unmöglich, vom subjektiven Erleben des Kranken auszugehen, das sich oft weitgehend vom Erleben des Arztes unterscheidet.

Identifikationsvorgänge mit unheilbar Kranken treten leichter bei Patienten auf, die eine solche Identifikation fördern, die im selben Alter sind oder denselben Beruf ausüben wie Arzt oder Schwester. Die Gleichsetzung im Erleben kann auch die eigene Beziehung zu engen Angehörigen betreffen:

> Eine ältere Krankenschwester, Mutter von drei Kindern, hatte ihre Berufstätigkeit nach langer Unterbrechung wieder aufgenommen, nachdem die Tochter einer befreundeten Familie an Leukämie verstorben war. Wiederholt sprach sie in der „Balint-Gruppe" davon, daß dieses Mädchen einer ihrer eigenen Töchter sehr ähnlich war. Ihr war die magische Vorstellung gekommen, durch ihre eigene Berufstätigkeit eine ähnliche Erkrankung von ihren Töchtern fernzuhalten. Im Umgang mit jungen Leukämiepatientinnen hatte sie regelmäßig besondere Schwierigkeiten. Sie brachte es trotz allen Bemühens nicht fertig, sich mit diesen Kranken in ein Gespräch einzulassen. Sie erlebte sich derart verunsichert, daß sie jede Gelegenheit benutzte, um diese Patientinnen zu meiden.

Unsere Abwehr gilt dabei dem Mitbetroffensein durch die dem Patienten geltende Bedrohung und der dadurch bedingten Bedrohung unserer eigenen Rolle als Helfer. Wir fühlen uns um so intensiver bedroht, je mehr wir uns mit dem Patienten identifizieren und uns projektiv in ihn hineinversetzen. Diese Prozesse können durch eine Art Sog verstärkt werden, der vom Patienten ausgeht: Die Bedrohung und oft auch die Einschränkung der Funktionsmöglichkeiten führen zu einer Regression in den Objektbeziehungen, oft zu einem intensiven Wunsch nach einer bergenden Zweierbeziehung nach dem Modell der symbiotischen Mutter-Kind-Beziehung. Gleichzeitig damit auftretende innerpsychische Spaltungsvorgänge können zu weiteren Konflikten führen, wenn etwa auf der Krankenstation vom Patienten ein Partner idealisiert wird, während die bösen eigenen Anteile auf einen anderen Partner projiziert werden. Es kann dann sowohl zu Konflikten zwischen dem Patienten und dem negativ besetzten Partner, als auch – über eine Externalisierung des inneren Konfliktes des Patienten – zum Konflikt zwischen zwei verschiedenen Teammitgliedern kommen. Die letztgenannte Möglichkeit zeigt, wie intensiv und weitreichend die Wechselwirkungen zwischen dem Patienten bzw. seinen emotionalen Reaktionen und der Gruppe der therapeutisch Tätigen in der Institution sind.

Die Intensität der wechselseitigen Einflüsse zwischen Patienten und Helfern auf einer Krankenstation kann kaum überschätzt werden. Dies zeigt auch eine Untersuchung, die unter anderem die Zusammenhänge zwischen dem Betriebsklima auf verschiedenen Stationen und dem Einschlafverhalten der Patienten aufdeckte: Die Patienten gaben um so häufiger Einschlafschwierigkeiten an, je schlechter die Schwestern das Betriebsklima einstuften, je weniger qualifiziert nach Ansicht der Schwestern die Vorgesetztenposition besetzt war und je seltener die Arbeitsschritte aus ihrer Sicht (als Kennzeichen von Bürokratie) vorhersehbar waren, je mehr der Kranke abgewertet wurde und je engmaschiger das mündliche Kontaktnetz unter den Schwestern war, das den Patienten unvermeidlich ausschließt (de Ridder, 1980).

Resignation, Abbruch der Therapie

Nicht selten kommt es beim Umgang mit unheilbar Kranken auch zu resignativen Reaktionen bei Ärzten und Schwestern. Es wird dann diskutiert, ob die Fortsetzung der Therapie noch sinnvoll ist, ob die Patienten hierdurch nicht nur gequält werden. Neben fachlichen Überlegungen spielt in diesen Diskussionen oft die eigene Enttäuschung über die therapeutischen Möglichkeiten eine wesentliche Rolle; diese Enttäuschung behindert dann die Wahrnehmung der Bedürfnisse und Gefühle des Patienten.

> Wir berichteten im Kapitel 28 über einen 58jährigen Maurer mit akuter Leukämie, der am rechten Bein zweimal amputiert werden mußte, dabei einen Herzstillstand erlitt und sich zunächst nicht an den Rehabilitationsmaßnahmen beteiligte. Als die Amputationswunden und eine eitrige Otitis media nicht heilten und die Grunderkrankung keine positive Entwicklung mehr erkennen ließ, kam bei Schwestern und Ärzten der Station immer wieder die Diskussion über den Sinn ihres Tuns, über den Sinn einer weiteren „supportiven" Therapie mit Bluttransfusionen und Antibiotika auf: „Soll man nicht endlich die Therapie abbrechen?" – „Nun haben wir wieder angefangen, jetzt müssen wir auch weitermachen!" – „Welchen Sinn hat so ein Leben, wenn einer nur noch im Bett existieren kann?" – „Soll man ewig so weiterbehandeln?" – „Wir können aber doch nicht einfach gar nichts mehr tun!" – „Wir dürfen doch die Hoffnung nicht aufgeben. Wer kann schon entscheiden, wann es Zeit ist zu sterben?"
>
> In der Stationsbesprechung wurde allmählich klar, daß die Tendenz bei einem Teil der Ärzte und Schwestern zur Resignation, zum Abbruch der Therapie, auch eine Reaktion auf die intensive emotionale Belastung durch diesen Patienten und sein Schicksal darstellte. Diese Schutzreaktion erschwerte es ihnen jedoch, auf die Bedürfnisse des Patienten einzugehen. Das Verhalten des Patienten veränderte sich grundlegend, nachdem er in die Entscheidung über das weitere Vorgehen einbezogen wurde. Er entschied sich fürs Weiterleben und begann von diesem Zeitpunkt ab, sich intensiv an seiner Rehabilitation zu beteiligen.

66.3.4 Rückzugsverhalten von Patienten als Reaktion auf abwehrendes Verhalten von Ärzten und Schwestern – Euthanasiewünsche

Lassen sich Ärzte und Schwestern intensiver in den Umgang mit unheilbar Kranken ein, so werden an sie von den Patienten nicht selten Euthanasiewünsche herangetragen. Die unter ihren Schmerzen leidenden, nicht selten von Angst gequälten Patienten bitten in ihrer Verzweiflung Ärzte oder Schwestern, ihnen eine „erlösende" Injektion zu geben oder ähnliches.

In dieser Situation versuchen wir immer besonders sorgfältig, die Situation des Patienten zu verstehen. Bisher wurde uns dabei jedesmal deutlich, daß es sich um einen verzweifelten Befreiungswunsch in einer Situation besonders großer Unfreiheit handelt, keineswegs – wie oft unzutreffend angenommen wird – um

eine besonders freie Entscheidung des Patienten. Wir bemühen uns darum, die Situation des Kranken zu verbessern, Bedürfnisse zu entdecken, deren Befriedigung noch möglich ist, seine Beziehungen zu seiner Umgebung zu verbessern, Beeinträchtigungen seines Selbstwertgefühls auszugleichen. Gelingt dies, äußern die Kranken keine Euthanasiewünsche mehr. Wir sind auf solche Euthanasiewünsche nie eingegangen. Selbstmordversuche von Kranken sind in dieser Situation nie aufgetreten.

Euthanasiewünsche stellen für uns jedoch immer eine alarmierende Mitteilung, einen Notruf dar!

66.4 Probleme im Umgang mit Angehörigen unheilbar Kranker

Krebskranke erwarten und benötigen in hohem Maße Unterstützung durch ihre Familienangehörigen (Sellschopp, 1984). Das Verhalten der Angehörigen beeinflußt die Verarbeitung der unheilbaren Erkrankung durch die Patienten oft wesentlich mit. Die Angehörigen ihrerseits sind durch die Krebserkrankung eines Familienmitglieds oft erheblich mitbelastet; auch sie benötigen nicht selten Unterstützung, sollen sie in dieser Situation die emotionalen Belastungen ertragen und die Beziehungen zu den Kranken aufrechterhalten können (Richter, 1981; Buddeberg, 1984a, b; Dreifuss, 1982; Stierlin et al., 1983, 1984; Wirsching und Stierlin, 1982). Sowohl in der Forschung als auch in der Praxis besteht bezüglich der Einbeziehung der Familie unheilbar Kranker ein ausgesprochenes Defizit (Buddeberg, 1984a, b). Wir beschreiben hier einige Reaktionen, die dem Kliniker im Umgang mit den Angehörigen unheilbar Kranker häufiger begegnen.

Schon bei der Krankenhausaufnahme ergibt sich nicht selten ein Spannungsverhältnis zwischen den Angehörigen und Ärzten oder Schwestern. Häufig versuchen Angehörige, die Aufnahme zu beschleunigen – und später die Entlassung zu verzögern – aus ihrer Sorge, zu Hause nicht genügend für den Patienten tun zu können. Ärzte und Schwestern mißverstehen diesen Versuch nicht selten als einen Versuch, den Kranken „abzuschieben". Spannungen zwischen Ärzten und Pflegepersonal einerseits und den Angehörigen der Kranken andererseits werden jedoch vor allem durch Schuldgefühle und Enttäuschungsreaktionen der Angehörigen mitverursacht.

66.4.1 Schuldgefühle

Schuldgefühle für Ärzte und Schwestern ergeben sich aus der oft unbefriedigenden Arbeitssituation, insbesondere dem Ausmaß des Arbeitsanfalls und dem sich hieraus ergebenden Zeitdruck und aus Interaktionsproblemen mit dem Patienten.

Klinikärzte und niedergelassene Kollegen klagen darüber, daß sie todkranke Patienten nicht so behandeln können wie sie möchten, da ihnen die hierfür nötige Zeit nicht zur Verfügung steht.

Ärzte und Schwestern werden oft durch überkritische, anklagende und vorwurfsvolle Angehörige von unheilbar Kranken irritiert. Diese scheinen ihnen vorzuwerfen, nicht alles heute Mögliche für den Patienten zu tun. Ärzte und Schwestern können hierauf aggressiv-gereizt reagieren oder sich gekränkt zurückziehen.

In der Regel bewährt es sich, solches Verhalten von Angehörigen nicht im Sinne persönlich gegen Ärzte und Schwestern gerichteter Vorwürfe aufzufassen, sondern zu versuchen, die Belastungen der Angehörigen zu eruieren und ihr Verhalten als Reaktion auf diese Belastungen zu verstehen. Die Angehörigen sind durch die lebensbedrohliche Krankheit meist selbst emotional stark belastet, sie fühlen sich dem Geschehen hilflos ausgeliefert, haben in ihrer Vorstellung Mitschuld an der Krankheit oder meinen, nicht genügend für ihr Familienmitglied getan zu haben. Solche Schuldgefühle entstehen häufig ganz unabhängig von jeder realen Schuld: In allen Beziehungen spielen ambivalente Einstellungen eine Rolle; im Falle einer Erkrankung des Beziehungspartners werden negative Wünsche und Phantasien bewußt oder unbewußt mit dem Auftreten der Erkrankung oder dem ungünstigen Verlauf verknüpft, dies kann zu Schuldgefühlen und hierdurch motivierten Reaktionen führen.

Der Ehemann einer Malignomkranken erkundigte sich unmittelbar nach der stationären Aufnahme bei der Schwester nach den verantwortlichen Ärzten und wie er sie telefonisch erreichen könne. Die Schwester hatte dabei den Eindruck, der Ehemann versuche sie einzuschüchtern, um eine möglichst vollkommene Versorgung seiner Frau sicherzustellen. Dies und das weitere Verhalten des Mannes irritierte die Schwester und sie spürte, daß sie begann, die Patientin abzulehnen.

Von der Patientin erfährt sie dann, daß sie zusammen mit dem Ehemann unter Aufnahme eines hohen Kredits ein Mietshaus erworben habe. Der Ehemann hatte die Patientin gedrängt, im Erdgeschoß ein Einzelhandelsgeschäft einzurichten und so zur Schuldentilgung beizutragen. Die Frau war mit diesen Plänen zunächst nicht einverstanden, sie hatte gehofft, sich nach dem Hauserwerb ganz der Erziehung der Kinder widmen zu können. Schließlich hat sie sich jedoch überreden lassen. Seinerzeit bei der Entscheidung habe der Mann geäußert: Alles wäre zu realisieren, „wenn sie innerhalb der nächsten Jahre nicht krank würde".

Es ist einfühlbar, daß der Ehemann nun gegenüber seiner Frau Schuldgefühle empfindet und seinerseits versucht, Schwestern und Ärzte unter Kontrolle zu bringen. Nachdem der Schwester diese Zusammenhänge deutlich wurden, empfand sie der Patientin gegenüber keine Ablehnung mehr; sie sah jetzt, daß diese Ablehnung durch das Verhalten des Ehemanns ausgelöst worden war. Sie konnte mit der Kranken freier sprechen und ließ sich vom Verhalten des Ehemanns nicht mehr irritieren.

Auch ein Überengagement von Angehörigen in der Mitbetreuung von Patienten kann durch solche Schuldgefühle motiviert sein. Diese Angehörigen beteiligen sich dann ohne Rücksicht auf die eigenen Kräfte z. B. an der Pflege. Sie sollten darin unterstützt werden, auch an ihre eigenen Bedürfnisse zu denken, schon aus Rücksicht auf die langfristige Versorgung des Patienten.

Auch Ärzte und Schwestern können sich bei der Mitbetreuung ihrer Angehörigen überfordern. So versuchte ein jüngerer Kollege, dessen Mutter mit einer unheilbaren Malignomerkrankung auf unserer Station lag, sich in die neueste Literatur über die Behandlung dieser Erkrankung einzuarbeiten und sich so an der Therapie zu beteiligen. Er sprach jedoch nicht mit den erfahrenen Onkologen der Klinik, obwohl er selbst in einem anderen Fachgebiet arbeitete. So überbeanspruchte er sich offensichtlich in dem Wunsch, eigenständig zur Behandlung seiner Mutter beizutragen.

66.4.2 Enttäuschung

Krankheit und Tod einer Bezugsperson hinterlassen bei den Angehörigen Schmerz, Trauer, oft auch große Enttäuschung. Solche Enttäuschung kann auch das Verhalten der Angehörigen gegenüber den Mitarbeitern im Krankenhaus beeinflussen.

Die Frau eines Malignompatienten hatte bis zu seiner Erkrankung gehofft, sich nach erfülltem gemeinsamem Leben jetzt bald zur Ruhe setzen zu können. Sie hatte sich in ihrem Erleben bisher für die Erziehung ihrer Kinder „aufgeopfert" und meinte, daß sie gewissermaßen jetzt erst hätte beginnen können, ihr eigenes Leben zu leben. Zusammen mit ihrem Mann hatte sie sich ein Haus gebaut, die Kinder waren nun erwachsen, ein friedlicher Lebensabend schien greifbar nahe. Nun erkrankte der Mann, in der Klinik wurde ein Karzinom diagnostiziert. Die Frau sagte: „Jetzt stirbt er. Warum muß ausgerechnet uns das passieren?" Die Chemotherapie brachte nicht den erhofften Erfolg, der Mann starb tatsächlich. Die Ehefrau war nun maßlos enttäuscht, daß er sie so „im Stich gelassen hatte" und versuchte in ihrer ohnmächtigen Wut, die Klinik für den Tod ihres Mannes verantwortlich zu machen.

66.4.3 Vorwegnehmende Trauerreaktionen

Die Trauer von Angehörigen beginnt mit der Wahrnehmung der Unheilbarkeit einer Erkrankung. Bei längerverlaufenden Krankheiten schließen Angehörige ihren Trauerprozeß nicht selten vor dem Tod des Patienten ab. Sie haben sich dann innerlich bereits vom Patienten getrennt, warten darauf, neue Beziehungen eingehen zu können oder gehen solche tatsächlich ein. Vorwegnehmende Trauer führt leichter in eine solche Entwicklung, wenn die Beziehung bereits vor der Erkrankung instabil war oder wenn ihre Qualität in besonderem Ausmaß von mit Gesundheit verbundenen Eigenschaften abhing.

Haben Angehörige ihren Trauerprozeß in einer gespannten Beziehung einmal abgeschlossen, so bleibt von der Beziehung oft nur grausame gegenseitige Quälerei übrig.

So meinte der Ehemann einer Dialysepatientin, als sie in ihrem Garten Beerensträucher pflanzte: das sei doch sinnlos, sie würde die Früchte doch nicht mehr ernten können. Derselbe Ehemann beklagte sich nach Aussagen der Patientin am Telefon bei ihrer eigenen Mutter darüber, daß sie nun schon länger lebe, als die Ärzte vorausgesagt hätten.

66.4.4 Die Betreuung von Angehörigen ist oft über den Tod des Familienmitglieds hinaus erforderlich

Der Verlust eines Familienmitglieds, insbesondere eines Ehepartners, bedeutet immer eine einschneidende Lebenskrise. Auf die sich hieraus ergebenden Belastungen, die Verarbeitung des Verlustes im Rahmen der Trauer, sowie die möglichen therapeutischen Interventionen können wir hier nicht im einzelnen eingehen (vgl. Kast, 1982; Parkes, 1974; Pincus, 1977; Spiegel, 1973; Worden, 1987).

> Der wiederholt geschilderte Leukämiepatient hatte vor seinem Tod immer wieder Sorge darüber geäußert, wie seine Frau die Trennung überstehen würde. Die Ehefrau hatte früher bei anderen Belastungen mit starken Depressionen reagiert. Nach dem Tod des Patienten bat uns die Ehefrau um ein Gespräch. Ihr selbst ginge es einigermaßen gut, aber die 13jährige Tochter sei krank, sie leide unter starken Konzentrationsstörungen und Halluzinationen, vor allem abends berichte sie davon, daß sie den Vater leibhaftig in der Wohnung stehen sehe. Es zeigte sich, daß die Frau nach dem Verlust des Ehemanns die Tochter als ihren Partner noch enger an sich zu binden versucht hatte. Bei der Tochter waren im Zuge der pubertären Entwicklung ödipale Wünsche wiederbelebt worden, die sie in der jetzigen Situation vermehrt abwehren mußte. Es genügte, der Mutter die Entwicklungsprobleme der Tochter in der Pubertät aufzuzeigen. Die Mutter konnte die Notwendigkeit einer unabhängigen Entwicklung der Tochter akzeptieren, worauf sich die Symptomatik bei der Tochter vollständig zurückbildete. Ein Jahr später allerdings kam die Mutter selbst und suchte wegen neurotischer Ängste psychotherapeutische Hilfe.

Hauptsächlich bei jüngeren Frauen beobachteten wir nach dem Tod des Ehemanns wiederholt den Wunsch, selbst in einem sozialen Beruf, z.B. als Krankenschwester, zu arbeiten. Da solche Wünsche so unmittelbar nach dem Tod häufig auch aufgrund von Schuldgefühlen zustande kommen, raten wir diesen Angehörigen, mit solchen Entscheidungen zunächst einige Zeit zu warten.

Aufgabe des Umgangs mit Malignomkranken ist es, die Angehörigen frühzeitig so weit in die Betreuung einzubeziehen, daß solchen Entwicklungen nach Möglichkeit vorgebeugt werden kann, solange Versöhnung möglich ist.

In der Praxis klinischer Arbeit wird oft versäumt, frühzeitig nach der Aufnahme des Patienten mit den Angehörigen einen konkreten Gesprächstermin zu vereinbaren; hierdurch wird oft wertvolle Zeit verloren oder die Möglichkeit zur Einbeziehung der Angehörigen ganz versäumt. Es empfiehlt sich, gleich bei der Aufnahme des Patienten einen festen Gesprächstermin mit den nächsten Angehörigen – nach Mög-

lichkeit im Beisein des Patienten – zu vereinbaren. Angesichts der bedrohlichen Situation machen sich die Angehörigen dann praktisch immer – entgegen den Erwartungen von Ärzten und Schwestern – von anderen Verpflichtungen frei.

66.5 Die Behandlung von Schmerzen unheilbar Kranker

66.5.1 Allgemeines

Wenige Aufgaben des Arztes bei unheilbar Kranken sind wichtiger als die Linderung von Schmerzen. Die Schmerzbehandlung wird jedoch häufig unsystematisch und unzulänglich durchgeführt. Ihr sollte beim Krebskranken von vornherein ein klar definierter Stellenwert im therapeutischen Gesamtkonzept zukommen – schon weil bei diesen Patienten die Furcht, mit unbehandelten Schmerzen alleingelassen zu werden, von Anfang an eine große Rolle spielt (Catalano, 1976).

Der Schmerz des Krebspatienten im fortgeschrittenen Stadium ist in der Regel chronisch; er ist ursächlich oft nicht beeinflußbar und gilt häufig noch als therapieresistent. Er unterscheidet sich damit von akuten Schmerzzuständen, deren Ursachen, wie Trauma oder Operationsfolgen, meistens zeitlich begrenzt sind und ausheilen. Das Problem „Schmerz" beim Karzinompatienten ist durch die Vielfalt der Symptomatik bei den fast unbegrenzten Möglichkeiten von Infiltration und Destruktion von Organstrukturen durch den Tumor, aber auch durch die Einflüsse der psychischen und sozialen Situation komplex. In der Endphase der Erkrankung leiden etwa 60% aller Patienten an chronischen Schmerzen, die durch eine spezifische antineoplastische Therapie nicht mehr zu beeinflussen sind und daher einer symptomatischen medikamentösen Therapie bedürfen (Bonica, 1979).

Chronische Schmerzen beim Karzinompatienten bessern sich ohne intensive Behandlung selten. Die Schmerztherapie beim Malignomkranken sollte nach standardisierten Richtlinien durchgeführt werden, die aufgrund pharmakologischer, pathophysiologischer und psychologischer Erkenntnisse aufgestellt werden (Twycross, 1979). Eine angemessene Therapie ist für den individuellen Patienten nur dann möglich, wenn eine exakte Diagnose gestellt wurde und folgende Faktoren der Schmerzentstehung und -therapie bedacht werden: die Lokalisation des Schmerzes; der Mechanismus der Schmerzentstehung; die psychische und soziale Situation des Patienten; die verfügbaren Möglichkeiten und die praktische Durchführbarkeit der Schmerztherapie. Die Therapiemöglichkeiten reichen von neurochirurgischen Eingriffen über lokale Bestrahlung bis zur symptomatischen medikamentösen Schmerztherapie oder bedürfen der Kombination verschiedener Therapiemodalitäten. Die Festlegung einer angemessenen individuellen Schmerztherapie sollte daher interdisziplinär unter

Einbeziehung der verschiedenen Fachrichtungen z. B. in einer Schmerzambulanz erfolgen. Die Durchführung der Therapie sollte dann allerdings, soweit irgendwie möglich, vom „behandelnden" Arzt geleitet werden, um die für den Krebspatienten so wichtige dauerhafte Arzt-Patient-Beziehung zu gewährleisten.

66.5.2 Wechselwirkungen von somatischen und psychischen Faktoren bei der Schmerzentstehung

Die Neurophysiologie des Schmerzes bei Krebspatienten ist nicht verschieden von der allgemeinen Pathophysiologie des Schmerzes (Adler, 1983; vgl. Kap. 36). Sie wird aber mehr als bei anderen Patienten durch die besondere Situation des Krebspatienten beeinflußt. Schmerz ist eine subjektive, emotionale Erfahrung, die sehr individuell ertragen wird, aber auch von dem gleichen Patienten in verschiedenen Situationen und zu verschiedenen Zeiten unterschiedlich stark empfunden werden kann (Melzack, 1973). Emotionale Faktoren wie Angst, Depression und Unsicherheit sensibilisieren die Schmerzempfindung; zwischen Schmerz und emotionalen Reaktionen gibt es die Möglichkeit gegenseitiger Verstärkung, aber auch einer gegenseitigen Stellvertretung.

> Eine 17jährige Leukämiekranke mit ungünstiger Prognose kommt während eines Gesprächs mit Schwestern im Stationszimmer auf ihren bevorstehenden Geburtstag zu sprechen; traurige Gefühle klingen an; bald danach läutet die Patientin von ihrem Zimmer aus und klagt über heftigste Schmerzen.

Bei Kranken mit heftigen, auch mit hohen Analgetikadosen nicht beherrschbaren Schmerzen und bei Patienten mit stark wechselnden Schmerzen kann die sorgfältige Klärung der psychischen und sozialen Situation zur Lösung des Problems beitragen. Verzweifelte Patienten, die sich nach längerer Krankheitsdauer bereits aus ihrem Berufsfeld und/oder aus der Gemeinschaft ihrer Angehörigen ausgeschlossen fühlen und keine sinnvolle Perspektive für ihr weiteres Leben sehen, haben nicht selten nur noch den körperlichen Schmerz, um sich selbst lebendig zu spüren und anderen ihre Not mitteilen zu können. Kommt den Schmerzen diese Funktion zu, so lassen sie sich häufig erst dann lindern, wenn der Umgang mit Ärzten und Pflegepersonal den Patienten aus ihrer verzweifelten Isolation herausgeholfen hat; dies gelingt oft erst nach Teambesprechungen nach dem Modell von Balint-Gruppen (vgl. Kap. 28); die Minderung der Schmerzen bei gleichzeitiger Abnahme des Analgetikabedarfs nach gezielten Gesprächen mit dem Patienten kann sehr eindrucksvoll sein.

Langanhaltende Schmerzen können das emotionale Verhalten von Patienten stark beeinflussen und insbesondere zur Entwicklung einer depressiven Verstimmung beitragen, die ihrerseits wieder eine negative Auswirkung auf das Schmerzerleben hat.

Die von Engel (1969) gegebene Definition des Schmerzes wird am ehesten all diesen Phänomenen gerecht: „Schmerz ist eine grundlegend unangenehme Empfindung, die dem Körper zugeschrieben wird und dem Leiden entspricht, das durch die psychische Wahrnehmung einer realen, drohenden oder phantasierten Verletzung hervorgerufen wird."

Schmerz kann aber auch den Charakter einer Mitteilung haben (z. B. als Ausdruck einer Kränkung), die verstanden werden muß, um die Schmerztherapie effektiv zu gestalten. Bei der Schmerzentstehung spielen auch die individuellen Phantasien und Ängste des Patienten bezüglich seiner Erkrankung, der erforderlichen therapeutischen Maßnahmen und der mit beiden verbundenen Schmerzen und Verluste („Verstümmelungsproblematik") eine Rolle.

Wesentliche Faktoren für den Erfolg der Schmerztherapie sind die Zuwendung und ein sicheres und systematisches Therapiekonzept des behandelnden Arztes. Ärztliche Hilflosigkeit und Ängstlichkeit wird vom Patienten erspürt und verstärkt dessen Furcht und Hoffnungslosigkeit. Eine Schmerztherapie bei Krebskranken ist zum Scheitern verurteilt, wenn die emotionalen Bedürfnisse des Patienten und seine Krankheitsvorstellungen nicht in das Behandlungskonzept einbezogen werden und sich der Arzt nicht ständig über die Wirksamkeit der Therapie rückversichert. Die Sicherheit eines solchen therapeutischen Gesamtkonzepts verstärkt den therapeutischen Effekt der Medikamente und verhindert das Abwandern des Patienten zu alternativen Heilmethoden.

Eine positive Einstellung des Pflegepersonals und der Ärzte hat unabhängig von der pharmakologischen Wirkung von Analgetika einen therapeutischen Effekt („Plazeboeffekt"). Ein solcher zusätzlicher Plazeboeffekt ist bei jeder Schmerztherapie in mehr oder minder großem Ausmaß zu erwarten und sollte genutzt werden.

Andererseits ist die Gabe von „Plazebos" anstelle pharmakologisch wirksamer Analgetika unseres Erachtens bei allen Patienten und besonders bei Krebskranken strikt zu vermeiden. Solche Plazebogaben werden nach einer Studie von Goodwin und Mitarbeitern (1979) besonders häufig bei „Problempatienten" angewandt, die durch „übertriebene Anforderungen" beim Pflegepersonal unbeliebt sind. Die vorkommende positive Wirkung solcher Plazebogaben darf auch keineswegs als Indiz für eine nicht organische Verursachung der Schmerzen fehlinterpretiert werden. Der Karzinompatient mit chronischen Schmerzen ist für solche Versuche mit einer Plazebotherapie denkbar ungeeignet; er ist in besonderem Maße auf das Verständnis der Pflegenden angewiesen und darf unter keinen Umständen als „Patient, der Schmerzen aggraviert", gebrandmarkt werden.

66.5.3 Diagnostik und Bemessung des Schmerzes

Schmerz ist nur ein Symptom, dessen Pathogenese und Lokalisation immer exakt abgeklärt werden sollten, da seine Ursachen sehr vielfältig sein können (Bonica, 1979).

Kreuzschmerzen bei einem malignen Tumor können durch eine Knochenmetastase mit Infiltration des Endosts oder Periosts bedingt sein; sie können aber auch erstes Symptom einer drohenden Querschnittslähmung, verursacht durch eine epidurale Metastase, sein.

Zusätzlich zu den somatischen Ursachen sind die psychischen Faktoren zu berücksichtigen, die Schmerzen verursachen und verstärken können. Individuelle Vorstellungen, hervorgerufen durch die Symptomatik der Schmerzen, müssen in die Diagnose einbezogen werden (z.B. können starke Schmerzen aufgrund von Ulzerationen im Pharynx zu einer Erstickungsfurcht führen). Solche Befürchtungen können, werden sie nicht individuell abgeklärt, nicht selten eine relative Therapieresistenz bedingen.

Voraussetzung für die Wahl und die Beurteilung einer adäquaten Schmerzlinderung durch eine analgetische Therapie ist eine Bemessung und Dokumentation der Intensität ("keine Schmerzen" bis "kaum auszuhalten") und der Dauer ("nie" bis "die ganze Zeit"). Die Beurteilung der Schmerzen eines individuellen Patienten vor und während einer Therapie kann zwar nicht in "objektiven" Meßparametern ausgedrückt werden, zur Verfügung stehen jedoch sehr wohl beurteilbare und validierte Methoden zur subjektiven Erfassung von Schmerz und Schmerzerleben, die praktisch und wissenschaftlich auswertbar sind. Intensität und Häufigkeit des Auftretens von Schmerzen sollten subjektiv durch den Patienten und durch die Pflegenden bewertet werden (Schreml et al., 1983). Eine tägliche Bewertung von Intensität, Dauer und Häufigkeit der Schmerzen durch die Patienten auf visuellen Analogskalen hat sich zusammen mit der Beurteilung durch den Stationsarzt auf

einer Stufenskala als praktikabel erwiesen und ermöglicht erst, die Wirksamkeit der Schmerztherapie zu überprüfen und längerfristig zu dokumentieren (Schreml et al., 1983a, b).

66.5.4 Anleitung zur Schmerztherapie

Ziel der Schmerztherapie sind Schmerzfreiheit und Erhaltung der größtmöglichen Aktivität sowie Unabhängigkeit bei der geringstmöglichen Störung des Sensoriums und des affektiven Verhaltens.

Das Vorgehen in der Schmerztherapie beim Malignompatienten kann in zwei prinzipiell verschiedene Maßnahmen aufgeteilt werden:
- Spezifische Therapiemodalitäten, die die schmerzverursachende Tumorläsion beseitigen oder zurückdrängen;
- symptomatische medikamentöse Therapie durch eine periphere oder zentralnervöse Beeinflussung der Schmerzempfindung oder – seltener – durch eine Unterbrechung und Manipulation der Schmerzleitung (Quindlen, 1982).

Auch in der Terminalphase sollte immer erwogen werden, ob durch palliative chirurgische Eingriffe, Bestrahlung oder zytostatische Therapie der schmerzverursachende Tumor angegangen werden kann, bevor eine ausschließlich symptomatische analgetische Therapie angewandt wird. Dabei sind die Nebenwirkungen der palliativen Therapie gegen den zu erwartenden Nutzen für den Patienten sehr sorgfältig abzuwägen. Tumorspezifische Schmerztherapie, falls erforderlich, kombiniert mit einer symptomatischen analgetischen Behandlung, kann auch hoffnungslosen, von Schmerzen geplagten Patienten noch zu einer Phase sinnvollen Weiterlebens verhelfen.

Hier kann nicht im einzelnen auf spezifische palliative Maßnahmen und spezielle Ansätze der Schmerztherapie, wie z.B. Leitungsanästhesie oder Nervenstimulation, eingegangen werden (vgl. Quindlen, 1982). Im folgenden stellen wir lediglich allgemeine Gesichtspunkte der medikamentösen Schmerzbehandlung bei Tumorkranken mit chronischen Schmerzzuständen dar.

Analgetika

"Analgetika" ist ein Oberbegriff für mehrere heterogene Pharmaka, die an verschiedenen Strukturen des Systems ansetzen, welches die Schmerzwahrnehmung und Schmerzleitung steuert (Burton, 1982). Die gebräuchlichsten Medikamente sind:
- Analgetika, die eine periphere antiinflammatorische und antipyretische Wirkung haben, wie z.B. die Acetylsalicylsäure (ASS) oder das Paracetamol (Tab. 66–1);
- Opiate und Opioide, welche an den Opiatrezeptoren des zentral absteigenden nozizeptiven Systems eingreifen (Tab. 66–2);
- lokale Anästhetika, die eine membranstabilisierende Wirkung haben und damit die Schmerzleitung im afferenten Schenkel blockieren.

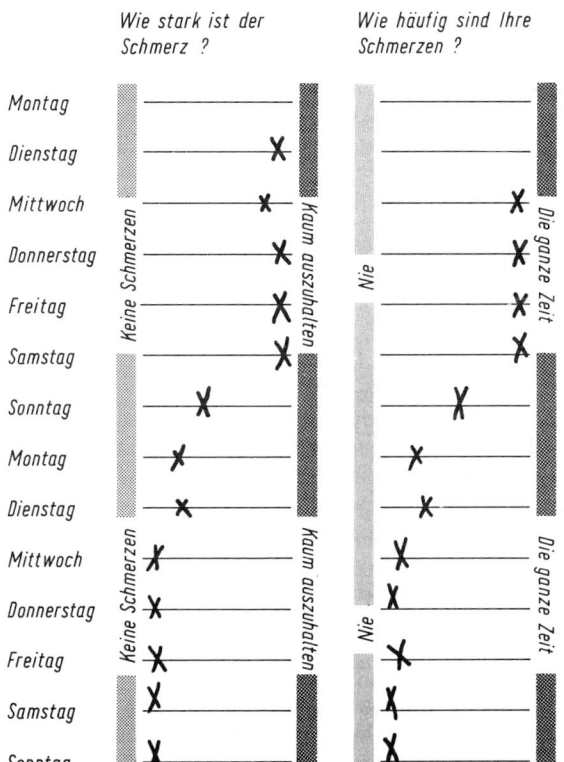

Abb. 66–2. Beispiel einer Beurteilung der Schmerzempfindung durch den Patienten auf einer visuellen Analogskala (freundlicherweise überlassen von Prof. Schreml, Ulm).

Tabelle 66–1. Analgetika (antiinflammatorische)

Medikament	Analgeti-scher Effekt	Antipyreti-scher Effekt	Anti-inflamma-torischer Effekt	Einzel-dosis (mg)	Dosie-rungs-intervall (in Std.)	Nebenwirkungen
Acetylsalicylsäure (ASS) (Aspirin®) (Colfarit®)	++	++	+++	500–1250	3	Magenirritation, Übelkeit, Erbrechen, gastrointestinale Blutungen, Störung der Thrombozytenfunktion
Paracetamol (Benuron®) (Enelfa®)	++	++	0	500–1000	2–3	Selten! Kopfschmerz, Hautallergien, hämolytische Anämien, Nierenschäden
Metamizol (Novalgin®) (Novamin Sulfon®)	+++	+++	+++	500–1000	4	Hautallergien, Agranulozytose (1 : 200 000)!

Die antiinflammatorischen, also aspirinähnlichen Medikamente gelten als schwache, die Morphiumderivate als starke Analgetika (Catalano, 1976; Reuler et al., 1980). Zusätzlich sollten unterstützend Psychopharmaka als „schmerzmitteleinsparende Medikamente" eingesetzt werden (vgl. Kap. 36 und Kap. 24).

Acetylsalicylsäure (ASS) ist nach den klassischen Studien von Moertel und Mitarbeitern (1972) die wohl beste analgetische Einzelsubstanz der peripher eingreifenden Analgetika, wenn sie ausreichend dosiert wird (Tab. 66–1). Tumorschmerzen, die durch begleitende entzündliche Prozesse verursacht werden, sprechen durch die starke antiphlogistische Wirkungskomponente besonders gut an. Paracetamol kann alternativ zu ASS eingesetzt werden, besonders wenn Nebenwirkungen auf den Gastrointestinaltrakt und auf die Thrombozytenfunktion zu beachten sind (Gerbershagen, 1979).

Der Einsatz von Kortikoiden ist beim Kompressionsschmerz von Nervenstrukturen durch Tumormassen sinnvoll. Dexamethason reduziert die Entzündung um und im Tumor und führt zu einer Verminderung der Kompression besonders in geschlossenen anatomischen Strukturen. Dexamethason (bis zu 6 × 4 mg/Tag) ist daher indiziert bei intrakraniellen Tumoren, aber auch bei Schmerzen durch Druck auf periphere Nerven.

Narkotika (Opiate und Opioide) sind die wirkungsvollsten Analgetika zur Behandlung chronischer Schmerzen, insbesondere bei terminalen Patienten (Twycross, 1979; Shimm et al., 1979). Narkotika beeinflussen nicht nur die Schmerzempfindung günstig, sie können auch eine euphorisierende Indifferenz gegenüber den Ängsten und Anspannungen der Patienten bewirken. Diese euphorisierende Wirkung der Narkotika wird auch bei Patienten mit chronischen, therapieresistenten Schmerzen ungerechtfertigterweise mit dem Problem Sucht und Abhängigkeit in Verbindung gebracht. Sucht- und Gewöhnungsprobleme sind bei diesen Patienten jedoch von sehr untergeordneter Bedeutung. Die Erfahrung zeigt, daß sie selten

(0,1%) eine echte Abhängigkeit entwickeln und auch nach Absetzen der Narkotika ohne diese Medikamente auskommen (Miller, 1976; Porter und Jick, 1980). Zudem handelt es sich häufig um terminale Patienten mit geringer Lebenserwartung, die unter unerträglichen Schmerzen leiden.

Kriterien für die Wahl eines bestimmten Narkotikums sind: die erwünschte Zeit-Wirkungs-Kurve, die günstigste Verabreichungsart und die geringsten Nebenwirkungen (Tab. 66–2). Vereinfachend ist festzustellen, daß die Nebenwirkungen verschiedener Opiate und Opioide etwa vergleichbar sind, wenn äquianalgetische Dosen, bezogen auf 10 mg Morphiumhydrochlorid, gegeben werden (Beaver, 1980). Bei älteren Patienten ist eine Akkumulation von Opiaten zu beachten. Die atemdepressorischen Nebenwirkungen, vor denen immer wieder gewarnt wird, treten bei Beachtung gravierender Kontraindikationen und oraler Applikation nur selten (1%) auf (Miller, 1976). Wegen der fast regelmäßigen Obstipation bei länger andauernder Opioidtherapie sollte zusätzlich routinemäßig ein Laxativum verabreicht werden. Opioidantagonisten (z.B. Pentazocin, Buprenorphin) dürfen nicht gleichzeitig mit einem Opioidagonisten (Morphin, Codein) verabreicht werden, da sie seine Wirkung antagonisieren können!

Bei Patienten mit therapieresistenten chronischen Schmerzen ist häufig im Laufe der Zeit eine schmerzadaptierte Dosissteigerung aufgrund einer Toleranzentwicklung notwendig. Opiate sollten so lange wie möglich **oral** gegeben werden, da die Dosisanpassung einfacher und die Wirkungsdauer günstiger ist als bei der parenteralen Gabe; gleichzeitig wird hier die Unabhängigkeit des Patienten gefördert.

Der rasche Wirkungseintritt – wesentlicher Vorteil der parenteralen Gabe – ist bei der Behandlung von Patienten mit chronischen Tumorschmerzen ohnehin unwichtig, weil ja eine Schmerzprophylaxe angestrebt wird. Die nicht ganz ungefährliche intravenöse Verabreichung von Opiaten ist bei chronischen Schmerzzuständen im Verlauf einer wohlgeplanten Therapie nur in den seltensten Fällen notwendig. Die kontinuierliche, patientengesteuerte Infusion von Opiaten subkutan oder über ein Portsystem erlaubt eine vermin-

Tabelle 66–2. Analgetika (Opiate und Opioide)

Analgetikum	Äquianalgetische Dosis mit 10 mg Morphium in mg Parenteral	Oral	Wirkungsdauer verglichen mit Morphium	Wirkungsbeginn (Min.)	Verabreichungsformen Dosis
Morphium HCl (Amphiolen®) (Morphin-Thilo®)	10	60	4–5 Std.	15–20	Amp. 10 mg, 20 mg
Morphinsulfat-Pentahydrat (MST®)	–	60	12 Std.	30–120	Retard-Tabletten 10 mg, 30 mg, 60 mg, 100 mg
Hydromorphon (Dilaudid®)	1,5	0,5	4–6 Std.	15–30	Amp. 2 mg
L-Methadon (L-Polamidon®)	5	10	4–6 Std.	30–60	Amp. 5 mg Tabl. 2,5 mg
Oxycodon (Eukodal®)	15	30	2–3 Std.	10–15	Amp. 10 mg Tabl. 5 mg
Buprenorphin (Temgesic®)	0,3	0,8	6–8 Std.	30–60	Amp. 0,3 mg Tabl. 0,2 mg
Codein	in Deutschland nur als Tablette verfügbar	120	4 Std.	15–30	Tabl. 30 mg, 50 mg

derte Dosis bei geringeren Nebenwirkungen. Ähnliches gilt für die epidurale Applikation von Opioiden. Doch sollten solche invasiven Verfahren nur nach Versagen der systemischen, vorzugsweisen oralen Opioidtherapie angewendet werden.

Für die Auswahl unter den zahllos angebotenen Analgetikapräparaten ist es wichtig, sich auf wenige Medikamente der einzelnen Wirkungsgruppen zu beschränken, um möglichst gute Detailkenntnisse und eigene Erfahrungen über Wirksamkeit, Wirkungsdauer, Dosierung, Nebenwirkungen und Interaktion der verwendeten Medikamente mit anderen Medikamenten zu erreichen.

Planung der Schmerztherapie

Um die Schmerztherapie effektiv, aber auch für alle Beteiligten durchsichtig und verständlich zu gestalten, sollte ein standardisierter, gestufter Therapieplan (Tab. 66–3) aufgestellt werden. Ziel dieses Stufenplans ist eine weitgehende und andauernde Schmerzfreiheit, d.h. eine Prävention des Schmerzes. Die Anordnung der Analgetikagaben sollte regelmäßig und in ausreichender Dosierung so erfolgen, daß Schmerzen nicht mehr auftreten. Analgetika sollten nicht „nach Bedarf" verordnet werden, also nicht nach dem Motto „Wenn Sie Schmerzen haben, rühren Sie sich". Eine individuelle optimale Dosierung wird durch die Berücksichtigung des Schmerzerlebens des Patienten bei der Titrierung des Analgetikums erreicht.

Die Intensität und die Qualität des Schmerzes und nicht die geschätzte Prognose der Erkrankung bestimmen, ob schwache oder starke Analgetika verwendet werden. Schwache Analgetika sollten nicht in jedem Fall abgesetzt werden, da sie aufgrund der unterschiedlichen pharmakologischen Wirkungsmecha-

nismen zusammen mit Morphinderivaten einen therapeutischen Kombinationseffekt bewirken können.

Mit der üblichen Schmerztherapie – die Medikamente werden „nach Bedarf" angeordnet und nach Abpackungsmengen dosiert – ist die erwünschte Schmerzprävention meist nicht zu erzielen. Häufig wird nicht nur eine zu niedrige Dosierung angesetzt, sondern es werden auch gedankenlos („3 × 1"!) fixierte Zeiten gewählt, ohne die individuelle Schmerzintensität und die Wirkungsdauer des verordneten Analgetikums zu berücksichtigen. Bei einem solchen Vorgehen wird vom Patienten erwartet, daß er den Zeitraum bis zur nächsten Medikamentengabe abwartet, ungeachtet der von ihm erlittenen Schmerzen. Dadurch kann sich beim Patienten aus Angst vor dem Schmerz und aus der Sorge, die nächste Medikation nicht rechtzeitig zu erhalten, der Schmerz bis ins Unerträgliche steigern. Erfolgt die Schmerztherapie nach einem solchen Schema, wird dem Patienten gleichzeitig Wohlverhalten und Disziplin verordnet – einem Patienten, der häufig wegen unerträglicher Schmerzen die Minuten bis zur nächsten Medikation zählt. Hiermit wird lediglich die Abhängigkeit des Patienten von Analgetika und Pflegenden gefördert, er wird in die demütigende Rolle eines Bittstellers gedrängt.

Bei einer stufenweisen präventiven Schmerzbehandlung wird dem mitbestimmenden Patienten die Angst vor den Schmerzen genommen. Er wird der Sorge enthoben, rechtzeitig die nächste Medikation zu erhalten und es wird so eine „operante Konditionierung" verhindert.

Ein Problem einer sehr konsequenten präventiven Schmerzbehandlung ist die Ausschaltung des Symptoms Schmerz als Signal für ernsthafte Komplikationen (z.B. Ileus). Dies spielt nur in der wirklich terminalen Phase der Erkrankung keine Rolle mehr. Außerdem wird bei Anwendung eines solch starren Schemas unserer Ansicht nach auch der Entscheidungsspielraum des Patienten eingeengt.

Tabelle 66–3. Stufenplan der Schmerztherapie

Graduierung des Schmerzes analog der Schmerzskala	Stufen	Verordnungs-schema	Medikament	Dosierung[1]
gelegentlich, auszuhalten	I	bei Bedarf	Acetylsalicylsäure (ASS) oder Paracetamol	max. 4 × 500 mg tägl. p.o.
	II	regelmäßig	ASS oder Paracetamol oder Supp.	6 × 500 mg p.o.
	III		ASS oder Paracetamol oder Supp.	6 × 500 mg p.o.
			+ Morphinsulfat retard[2] (MST®)	2 × 20–30 mg
	IV		Morphinsulfat retard (MST®)	2 × 30 mg
			+ Haloperidol (Haldol®) bei agitierten Patienten bzw.	3 × 1–2 mg p.o.
			+ Amitriptylin (Laroxyl®) bei depressiven Patienten	2–3 × 25 mg p.o.
Dauerschmerz, kaum auszuhalten	V		Morphinsulfat (MST®) + Haloperidol	2 × 30–120 mg
			oder Hydromorphon	4 × 2–4 mg s.c. oder als Supp.

[1] Für starke Analgetika initialer Dosisbereich, weitere Dosis bedürfnisentsprechend titriert.
[2] Anstatt MST® kann das Opioid Buprenorphin (Temgesic®) 2–3 × 0,2–0,4 mg sublingual gegeben werden oder ein oraler Morphin-Cocktail.

Oraler Morphin-Cocktail (Rp.-Beispiel für 18 Einzelgaben à 15 mg): Das Volumen des Cocktails bleibt mit 5 ml konstant. Die Anfangseinzeldosis Morphin kann dann entsprechend der Bedürfnisse austitriert werden.
Morph. hydrochl. 0,27 g
4,5 ml Haldol-Janssen 0,009 g
Aqua purif. ad 90,0
Divide in partes aequales
No. XVIII à 5 g, alle 4 h eine Dosis

Am sinnvollsten erscheint uns ein Schema, das zur weitgehenden Schmerzfreiheit die Schmerzempfindung des Patienten noch stärker berücksichtigt und ihm mehr Eigenbestimmung ermöglicht. Dabei wird eine initial optimal erscheinende Dosierung von Analgetika mit fixierten Zeiten verschrieben. Zusätzlich wird der Patient vor jeder Änderung des Therapieplans befragt, ob und in welchem Ausmaß er eine Schmerzlinderung benötige, um danach die Analgetikadosis optimal zu titrieren. Damit ist eine weitgehende Schmerzfreiheit des Patienten garantiert, gleichzeitig wird ihm die in dieser Situation notwendige Entscheidungsmöglichkeit noch belassen. Korrekturen und Fehlverhalten können im Gespräch geklärt werden. Es ist selbstverständlich, daß bei einem solchen Schema eine genaue Protokollierung des Analgetikaverbrauchs, z.B. im Kardex-System, erforderlich ist.

In der Endphase einer Krebserkrankung kann das Gespräch über die Schmerzbehandlung auch ein wichtiger Anknüpfungspunkt für die Fortsetzung der Kommunikation sein und dem Patienten das Gefühl vermitteln, einerseits nicht verlassen zu werden und andererseits doch auch selbst Kontrollmöglichkeiten zu behalten. Für den Pflegenden ergibt sich bei diesem Vorgehen die Notwendigkeit, das Gespräch mit dem Patienten nie abbrechen zu lassen; dabei kann er dann auch Komplikationen im emotionalen Bereich, wie etwa eine stärkere depressive Verstimmung, rechtzeitig erkennen.

Widerstände gegen eine wirksame Schmerzkontrolle

Die systematische und rationale Therapie der chronischen Tumorschmerzen wird häufig durch die nihilistische Einstellung des Arztes gegenüber dem chronischen und unheilbar Kranken erschwert. Diese Einstellung kann als Reaktion auf das Unvermögen, solche Kranke zu heilen, entstanden sein. Eine Überforderung des Pflegepersonals durch diese Hilflosigkeit kann sowohl die Tendenz zu einer unzulänglichen Schmerztherapie durch Plazebogaben (s.o.) fördern, als auch zu einer Überdosierung von Analgetika und Psychopharmaka führen. Der Patient wird ruhiggestellt, weil sein Leiden für die Umgebung unerträglich ist; er wird dann – wie es im Krankenhausjargon oft ausgedrückt wird – „abgeschossen". Hinzu kommt, daß vielen Pflegenden unglücklicherweise die pharmakologischen Grundlagen und die neueren Konzepte der Schmerztherapie noch unbekannt sind.

Eine Untersuchung an einem Lehrkrankenhaus in den USA zeigte, daß 32% der chronisch kranken Patienten trotz Behandlung mit Opiaten noch an ausgeprägten Schmerzen litten und 41% dieser Patienten nicht schmerzfrei waren (Marks und Sachar, 1973). Die Auswertung der Krankengeschichten ergab eine eindeutige Unterdosierung der Analgetika. Widerstände gegen eine ausreichende Schmerzbehandlung von Tumorpatienten nach dem Prinzip der Schmerzprophylaxe durch regelmäßige Analgetikagabe beschreiben Schreml und Mitarbeiter (1983). Diese Autoren fanden eine anfangs deutlich reservierte Einstellung des betreuenden Pflegepersonals gegenüber einer standardisierten Schmerztherapie mit praventlvem Ziel.

Die bürokratischen Hürden bei der Verschreibung von „Betäubungsmitteln" sind eine häufige Ursache einer nicht ausreichenden Schmerzbehandlung von Tumorpatienten.

Patienten können eine systematische Schmerzprophylaxe ablehnen, weil sie Angst vor Sucht und Gewöhnung haben, aber auch weil bei ihnen die Vorstellung besteht, Schmerzen ertragen zu müssen, um Schuld abzubauen (Angell, 1982). Nicht selten findet sich auch die Furcht, daß eine zu früh einsetzende Unterdrückung von Schmerzen durch starke Medikamente bei starken terminalen Schmerzen keine ausreichende Therapie mehr zulasse. Die Ablehnung der Schmerztherapie aus solchen Gründen muß mit jedem Patienten im einzelnen geklärt werden. Nicht selten ergeben sich in einem solchen Gespräch Hinweise auf weitergehende Ängste oder Schwierigkeiten in den sozialen Beziehungen.

66.6 Zusammenfassende Empfehlungen für den Umgang mit unheilbar Kranken

Es war in diesem Beitrag unser Ziel, den Leser auf die Probleme unheilbar Kranker und Möglichkeiten zur Unterstützung dieser Patienten aufmerksam zu machen. Wir verzichten darauf, ins einzelne gehende Vorschläge für den Umgang mit unheilbar Kranken zu geben. Hier setzt die Aufgabe konkreter Angebote in Aus-, Fort- und Weiterbildung ein.

Die Gestaltung der Betreuung unheilbar Kranker hängt immer von den jeweiligen persönlichen kommunikativen Fähigkeiten und der Ausbildung der beteiligten Experten sowie von den institutionellen Gegebenheiten ab. Wir können hier nur einige allgemein gehaltene Empfehlungen geben.

66.6.1 Vorgehen bei der Kommunikation mit unheilbar Kranken

Wir empfehlen eine **offene Kommunikation** auch mit unheilbar Kranken über das Wesen ihrer Erkrankung und über die sich hieraus ergebenden Probleme. Offene Information sollte die Regel darstellen, Ausnahmen von dieser Regel sollten begründet werden. Vorbedingungen für ein solches Vorgehen sind:
– eine **tragfähige Beziehung** zum Patienten, die langfristig aufrechterhalten werden kann;

– ausreichende **Information** des Arztes über den Patienten in folgenden Bereichen:
– Vorwissen über die Erkrankung
– Krankheitserleben einschließlich der Vorstellungen über die krankheitsbedingten körperlichen Veränderungen
– Vorerfahrungen mit Ärzten und Krankenhäusern
– frühere belastende Lebenssituationen und ihre Verarbeitung
– wichtige biographische Daten und
– Lebensumstände zur Zeit der Krankheitsmanifestation.

Zunächst sollte im Gespräch geklärt werden, ob der Patient die offene Information wünscht. Ist dies der Fall, wird der Patient gebeten, sein Vorwissen und seine Phantasien zur Krankheit darzustellen; der Arzt fördert diese Darstellung durch offenes Nachfragen (vgl. Kap. 12) und knüpft dann mit seinen Informationen erläuternd an das Vorwissen des Patienten an. Er vermittelt die erforderliche Information schrittweise, im Verlauf mehrerer Gespräche.

66.6.2 Vorgehen im weiteren Verlauf der Betreuung

Der Arzt bemüht sich im weiteren Verlauf darum, den Patienten in seiner jeweiligen emotionalen Verfassung zu erreichen und soweit nötig und möglich zu unterstützen. Dabei empfiehlt es sich, die Entwicklung der Krankheitsverarbeitung sorgfältig zu beobachten und zu dokumentieren. Gemeinsam mit dem Patienten und seinen Familienangehörigen sollte bereits möglichst früh während der stationären Behandlung ein Konzept für die anschließende Rehabilitation zusammen mit der weiteren Behandlung und Betreuung entwickelt werden. Die Rolle des Arztes und der anderen beteiligten Experten ist es vor allem, Patienten und ihre Angehörigen so ausreichend zu informieren, daß sie selbst die für ihre Zukunft wichtigen Entscheidungen treffen können.

66.6.3 Integration der psychosomatischen Betrachtungsweise in die klinische Arbeit

Eine konsequente Berücksichtigung psychosozialer Gesichtspunkte in der Betreuung unheilbar Kranker hat bestimmte Formen der Arbeitsorganisation zur Voraussetzung. Die Probleme lassen sich nur in einem Team lösen, dem nach Möglichkeit neben Ärzten und Schwestern auch in diesem Bereich erfahrene Psychotherapeuten, Sozialarbeiter und Seelsorger angehören. Die Einstellung der Teammitglieder und die Organisation des Arbeitsablaufes müssen die Weitergabe der von allen benötigten Informationen ebenso sicherstellen, wie eine der jeweiligen Aufgabe und Kompetenz entsprechende Teilung der Verantwortung. Der einzelne Mitarbeiter benötigt das Team auch, um die sich im Umgang mit unheilbar Kranken ergebenden emotionalen Belastungen reflektieren und verarbeiten zu können. Einzelheiten der sich aus einem solchen Arbeitsansatz ergebenden Anforderungen an die Organisation der stationären Versorgung haben wir im Kapitel 28 dargestellt.

66.6.4 Spezielle Ansätze: Palliativstationen, Hospize (sog. „Sterbekliniken"); Betreuung der Patienten zu Hause

Die von England ausgehende Hospiz-Bewegung hat der Verbesserung der institutionellen Versorgungsansätze für unheilbar Kranke wesentliche Impulse gegeben (Übersicht bei Wilkes, 1982). Der Vorzug solcher Einrichtungen liegt in der Möglichkeit, modellhaft Verbesserungen in der Betreuung unheilbar Kranker entwickeln und erproben zu können. Wir würden eine Betreuung auch unheilbar Kranker im Rahmen der allgemeinen Krankenversorgung bevorzugen, wie wir sie z.B. im Rahmen unseres Konzeptes einer klinisch-psychosomatischen Krankenstation (vgl. Kap. 28) beschrieben haben; die weitere Ausgliederung zum Tode Kranker aus der Medizin ließe sich so vermeiden (vgl. auch Gallmeier und Bruntsch, 1988; Kappauf et al., 1988). Die rasche Weiterentwicklung der instrumentell orientierten Krankenversorgung und die damit zunehmenden Spannungen zwischen kurativem und palliativem Ansatz (Petzold und Frehen, 1987) schränken jedoch den Spielraum für innovative Ansätze im Bereich der Routineversorgung zunehmend ein. Wir glauben deshalb, daß es berechtigt und wichtig ist, auch im deutschen Sprachraum Erfahrungen mit solchen modellhaften Spezialeinheiten zu gewinnen.

In der Bundesrepublik arbeiteten im Herbst 1988 vier palliative Einrichtungen (Hospize) nach unterschiedlichen Konzepten: die Station für palliative Therapie der Chirurgischen Universitätsklinik Köln, das Hospiz „Haus Hörn" in Aachen, das Hospiz zum Heiligen Franziskus in Recklinghausen und das Christopherus-Haus in Frankfurt. In den USA existieren rund 1680 Hospizprogramme, in denen 1987 172000 Personen, 90% davon Krebspatienten, betreut wurden. In England existieren etwa 150 Hospizprogramme.

Aus eigener Anschauung kenne ich (K. K.) die Arbeit der Palliativstation der Chirurgischen Universitätsklinik Köln. Von der Stiftung Krebshilfe gefördert, wird hier eine kleine Station mit fünf Betten betrieben (eine Erweiterung der Station ist geplant), der ein Hausbetreuungsdienst und das „Bildungsforum Chirurgie" angegliedert sind. Die Station wird von einer Ärztin und sechs examinierten Krankenpflegekräften geführt, auf Teilzeitbasis arbeiten ein Anästhesist und die Sozialarbeiterin, die gleichzeitig den Hausbetreuungsdienst leitet, sowie die Klinikseelsorger beider Konfessionen mit. Zusätzlich unterstützen ehrenamtliche Helferinnen das Team.

Ziele der Arbeit sind die lindernde Behandlung körperlicher Beschwerden („Symptomkontrolle") sowie die psychosoziale Unterstützung der Kranken.

Von April 1983 bis Ende 1987 wurden 218 Patienten insgesamt 281mal (51 Patienten mehrfach) aufgenommen. Davon waren vorher 144 Patienten der Chirurgischen Klinik, 34 Patienten anderer Kölner Universitätskliniken, 40 kamen von auswärtigen Krankenhäusern. Als Aufnahmegrund dominierten (bei 81,9%) Schmerzen. Bei je 22% spielten Ernährungsschwierigkeiten, psychische Probleme und/oder die soziale Situation eine wesentliche Rolle. Die Aufenthaltsdauer der Patienten auf der Station betrug im Mittel 24,2 Tage. Patienten, die mit bestmöglich gelinderten Beschwerden entlassen werden, können durch den Hausbetreuungsdienst in Zusammenarbeit mit dem Hausarzt und eventuell anderen ambulanten Diensten weiterbetreut werden. Erfahrungen aus der Arbeit der Station und der Hausbetreuung in der Behandlung, Pflege und Betreuung schwerstkranker Krebspatienten werden im „Bildungsforum Chirurgie" an Ärzte, Pflegekräfte, Sozialarbeiter und interessierte Laien weitergegeben.

Die Hälfte der neu aufgenommenen Patienten verstarb während des ersten Aufenthaltes (Thielemann und Pichlmaier, 1988).

Ich habe zunächst mit einer sehr skeptischen Einstellung an der Arbeit dieser Station im Rahmen einer wöchentlichen Stationskonferenz über knapp drei Jahre teilgenommen. Ich lernte, daß es dem Stationsteam in modellhafter Weise gelungen ist, einen Raum zu schaffen, in dem eine Integration instrumenteller und kommunikativer Arbeitsansätze gelingen kann. So erfolgt sorgfältigste Schmerzdiagnostik und -therapie eingebettet in ein Verständnis der gesamten Lebenssituation des Patienten. Nicht selten gelingt nämlich Schmerztherapie erst dann befriedigend, wenn Selbstwert- oder Beziehungsprobleme des Patienten abgeklärt oder stellvertretend über Beziehungen zu Mitarbeitern der Station gemildert werden können. Mein Eindruck ist, daß neben der Grundeinstellung aller Beteiligten und ihrem großen Engagement die Beschränkung auf Palliation diesen Raum eröffnen half. Die Arbeit dieser Station setzt Maßstäbe für eine Kultur der Beschwerdelinderung und der Kranken- bzw. Sterbebegleitung; diese Kultur kann auf andere Krankenstationen einer Klinik ausstrahlen. Ich halte es für außerordentlich wichtig, daß gerade angesichts der von den Beteiligten meist selbst sehr schmerzlich empfundenen Mängel der Krankenversorgung im Spannungsfeld zwischen kurativen und palliativen Ansätzen in der Behandlung Krebskranker eine derartig exemplarische Betreuung zum Tode Kranker realisiert werden kann. Auf dieser Station ist möglich, was in der Hektik des therapeutischen Alltags oft nur sehr schwer zu realisieren ist: ein Angebot zum Gespräch, zur Beziehung – nach dem Bedarf der Kranken und immer eingebettet in die medizinische Versorgung. Nicht selten führt das Gespräch von den Schmerzen zu den noch offenen Lebensproblemen, zu den unerledigten Aufgaben; oft kann im Zusammensein auf der Station der Wert des Lebens erst bestimmt, Abschied und Trauer vollzogen, Versöhnung noch erreicht werden. Nicht selten sprechen Patienten erleichtert davon, daß sie sich nicht hatten vorstellen können, daß „Sterben so schön sein kann", oder Mitpatienten davon, daß sie auch „so ruhig sterben" möchten.

Insgesamt dürfte die Möglichkeit, auch unheilbar Krebskranke in ihrer häuslichen Umgebung zu behandeln, bisher unterschätzt worden sein. Neue Arbeitsansätze engagierter Onkologen in der Klinik (Tagesklinik an der TU München, Fink et al.; Mobile ambulante Nachbehandlung, Schlag et al., 1988) und in der Praxis (z.B. Kleeberg, Erdmann et al. in Hamburg; Feyen et al. in Friedrichshafen) zeigen, daß bei einer entsprechenden Zusammenarbeit mit hierfür ausgebildeten Krankenschwestern weitreichende Veränderungen in der Versorgung möglich sind.

66.6.5 Selbsthilfegruppen

Selbsthilfegruppen Krebskranker unterscheiden sich von anderen Selbsthilfegruppen – etwa von Adipösen und Alkoholikern – dadurch, daß ihre Leiter und Organisatoren ebenso wie ihre Mitglieder meist in größerer Abhängigkeit vom medizinischen Versorgungssystem bleiben. Auch bei Krebskranken bewährt sich das Selbsthilfekonzept: Viele Patienten erfahren in den Gruppen entscheidende emotionale Unterstützung und finden hier die Möglichkeit, viele Fragen zu ihrer Krankheit, ihrer Behandlung und ihrer künftigen Lebensführung zu klären. Die Gesprächsmöglichkeit in den Gruppen hilft den Kranken oft entscheidend dabei, zu kritischen und selbständigen Partnern auch der professionellen Helfer im medizinischen Versorgungssystem zu werden. Eine unaufdringliche Unterstützung der Selbsthilfegruppen in der Zusammenarbeit mit Onkologen scheint sich insgesamt zu bewähren. Eigene Erfahrungen und Berichte in der Literatur lassen jedoch auch Probleme der Arbeit von Selbsthilfegruppen erkennen: Die zum Teil extremen emotionalen Belastungen Krebskranker können zu erheblichen aggressiven Spannungen innerhalb der Gruppe und zu zusätzlicher emotionaler Belastung einzelner Gruppenmitglieder führen; in der Gruppe können – oft von den anderen unbemerkt – zusätzliche Stigmatisierungsprozesse ablaufen (Schafft, 1983). Die teilweise Professionalisierung der Gruppenleiter im Zusammenhang mit der öffentlichen Förderung und mit Zentralisierungstendenzen von Selbsthilfeorganisationen kann zu zusätzlichen Problemen führen (Schafft, 1983). Insgesamt scheint uns eine mit den Betroffenen gemeinsam vorgenommene sorgfältige Analyse ihrer Bedürfnisse heute vordringlicher als eine überstürzte Expansion „organisierter Selbsthilfe".

66.6.6 Aus- und Fortbildung

Bei Studenten besteht nach unseren Erfahrungen sehr großes Interesse an thanatologischen Fragen. Ein entsprechendes Angebot während des Studiums könnte auch dem späteren „Praxisschock" bei der Konfrontation mit sterbenden Patienten vorbeugen.

Vor allem für Ärzte und Krankenpflegepersonal wurden zahlreiche Unterrichtsmodelle für die Verbesserung der Kommunikation auch mit unheilbar Krebskranken ausgearbeitet (Übersicht bei Koch und Schmeling, 1982; Strain, 1982). Die Effektivität intensiver Trainingskurse konnte auch objektiviert werden (Koch und Schmeling, 1978, 1982, 1984; Huck und Petzold, 1984). Schwieriger ist es, die organisatorischen Vorbedingungen für eine Verbesserung der Kommunikation aller Mitarbeiter im Krankenhaus zu schaffen.

Bewährt hat sich die Verbindung von Fortbildung und psychosomatischem Liaison-Dienst für onkologische Abteilungen (Hales und Borus, 1982), die mehrjährige Teilnahme von Hämatologen bzw. Onkologen an Balint-Gruppen innerhalb der Institution (Drees, 1981, 1984) sowie die systematische und kontinuierliche Einbeziehung bereits von Studenten in die psychologische Betreuung Schwerstkranker (Thoma et al., 1979).

66.7 Bericht über die Betreuung eines jugendlichen Krebskranken

66.7.1 Zur Vorgeschichte

Der folgende Bericht von Jutta Zenz über die Betreuung des 16jährigen krebskranken Stefan illustriert viele der dargestellten Probleme im Umgang mit unheilbar Kranken. Frau Zenz ist Krankenschwester; sie hat sich über viele Jahre im Bereich der Psychosomatischen Medizin intensiv fortgebildet und später den einjährigen Weiterbildungskurs in patientzentrierter Pflege/Psychosomatischer Medizin an der Universität Ulm geleitet. Sie betreute Stefan während mehrerer stationärer Aufenthalte im Verlauf eines Jahres. Diese Betreuung erfolgte im Rahmen der internistisch-psychosomatischen Krankenstation der Ulmer Universitätsklinik, auf der der Versuch unternommen wurde, den psychosomatischen Verständnisansatz möglichst weitgehend in die internistische Krankenversorgung zu integrieren (vgl. Kap. 28). Die Organisation der Arbeit auf dieser Krankenstation war entsprechend dem Gesamtkonzept modifiziert worden, ebenso die Weiterbildung der Mitarbeiter; die Ärzte befanden sich gleichzeitig in internistischer und psychotherapeutischer Weiterbildung, die Krankenschwestern hatten alle an dem im Zusammenhang mit dem Stationskonzept durchgeführten einjährigen Vollzeitweiterbildungskurs teilgenommen (Köhle et al., 1980, 1983a, b, 1984).

> Stefan erkrankte im Alter von knapp 16 Jahren an einem Osteosarkom im Bereich des linken Oberschenkels. In den Monaten vor der Diagnosestellung waren zunehmende Schmerzen im Bein aufgetreten; als begeisterter Fußballspieler hatte Stefan zunächst angenommen, diese Beschwerden seien Folge sportbedingter Überbeanspruchung oder Verletzung.
>
> Auf die Diagnosemitteilung reagierten Patient und Eltern panikartig, sie wandten sich massiv gegen die vorgeschlagene Amputation des Beines. Zunächst wurde eine lokale Tumorresektion ohne Amputation mit Metallimplantat vorgenommen. Stefan wurde zur zytostatischen Nachbehandlung auf die internistisch-psychosomatische Station des Zentrums für Innere Medizin verlegt.
>
> Im Rahmen einer zytostatischen Adjuvansbehandlung wurde Stefan alternierend mit unterschiedlichen Zytostatika behandelt. Unter anderem wurde er mit hohen Dosen eines zytostatischen Medikamentes und anschließender Antidotgabe therapiert; dabei wäre die

Dosis des Zytostatikums ohne Antidotgabe tödlich gewesen. Ziel dieser Therapie war die Prophylaxe eines Rezidivs bzw. die Prophylaxe von Lungenmetastasen. Etwa ein halbes Jahr nach Diagnosestellung der Malignomerkrankung kam es dennoch zu einer Neuausbreitung des Tumors. Jetzt wurde eine Oberschenkelamputation durchgeführt.

Stefans Eltern betreiben eine kleine Landwirtschaft. Seiner Mutter gelang es während des schwierigen Behandlungsverlaufs, seine Probleme und seine Not immer wieder zu verstehen; sie vermochte ihn gefühlsmäßig in besonderem Maße zu unterstützen; der Mutter gelang es auch bei krisenhafter Zuspitzung von Konflikten zwischen Stefan und dem Behandlungsteam zu vermitteln. Der Vater war chronischer Alkoholiker; er beging einige Jahre nach Stefans Tod Selbstmord.

Stefan war in der Familie gegenüber seinem älteren Bruder benachteiligt; da der Bruder an der Landwirtschaft kein Interesse hatte, sollte Stefan sie später von den Eltern übernehmen. Nach Schulabschluß begann er keine Ausbildung, sondern blieb gleich zu Hause.

Stefan setzte der zytostatischen Behandlung über lange Zeit einen meist untergründigen Widerstand entgegen. Seine verständliche massive Enttäuschung über den Mißerfolg der eingreifenden Therapie und ihre zeitweise – im Stadium der Knochenmarksaplasie – lebensgefährdenden Folgen äußerte er zum Teil auch in heftigen Vorwürfen, insbesondere gegenüber den Schwestern und Pflegern. Auf der Station kam es im Zusammenhang der Behandlung von Stefan immer wieder zu heftigen Konflikten; dabei identifizierten sich einige Teammitglieder immer wieder mit den den Ernst der Erkrankung verleugnenden Anteilen des Patienten. Die Konflikte im Umgang mit Stefan wurden mehrmals in der regelmäßigen Stationskonferenz diskutiert. Dabei wurde allmählich deutlich, daß verschiedene Teammitglieder sich mit unterschiedlichen Anteilen des Patienten identifiziert hatten und jetzt gewissermaßen externalisiert im Team die heftigen inneren Konflikte des Patienten ausgetragen worden waren. Gefördert wurden diese Konflikte durch die empörten Angriffe Stefans gegenüber Schwestern und Ärzten bei fast jeder Behandlungsmaßnahme und jeder auftretenden Nebenwirkung, trotz aller vorausgegangenen Bemühungen um Information und Vorabsprachen. Allen Beteiligten gelang es nur sehr allmählich, den naturgemäß langsamen Entwicklungsprozeß von Stefan zu akzeptieren, in dessen Verlauf er begann, die Bedrohung seines Lebens wahrzunehmen und sich mit dieser Bedrohung auseinanderzusetzen (vgl. Bliesener, 1982, 1986; Gaus und Köhle, 1982).

66.7.2 Die Betreuung von Stefan
(J. Zenz)

Irgendwann sprach er mich mal auf dem Flur der Station an: „Was machen Sie hier eigentlich? Laufen immer ohne Kittel rum, dürfen trotzdem in jedes Zimmer?"

Er lief immer mit Krücken und war mir auch aufgefallen. Ich habe mir Zeit genommen und mich zu ihm gesetzt. Ich habe erklärt, was ich tue und es schien ihm zu gefallen.

„Und was machen Sie hier?" – „Ich? Ich habe einen Tumor im Fuß gehabt, jetzt ist er raus. Man hat nach Information von Ärzten Knochenkalk eingesetzt." – „Was für ein Tumor

war es denn?" – „Ein Osteosarkom." – „Aha, tut das noch weh?" – „Wieso?" – „Weil Sie Krücken brauchen?" – „Ja, es tut noch weh. So hat es auch angefangen. Ich konnte nur noch schlecht Fußball spielen. Dann auch beim Laufen Beschwerden. Da hat mich mein Hausarzt mal her überwiesen. Da hat man dann alles festgestellt." – „Ach, Sie spielen gerne Fußball?" – „Ja, ich schaue auch gerne zu, ich bin ein Fan. Ich bin auch gerne mit meinen Kumpels zusammen zum Fußball und Schafkopfen, so am Wochenende auf ein Bier. Die nehmen mich so, wie ich bin. Leider kann ich nicht mehr so wie die." – „So, ich muß jetzt weiter!" – „Kommen Sie mal öfter hier vorbei?" – „Ja, Sie können mich auch gerne wieder ansprechen, wenn Sie mich sehen." – „Tschüß."

Vielleicht habe ich noch zwei- oder dreimal mit ihm gesprochen, ich weiß es nicht mehr so genau. Dabei ging es meist um seinen Fuß und daß er immer noch schmerzt. Ängstlich wirkt er und meint, da ist doch was nicht in Ordnung. Die Ärzte sagen ihm, daß der Knochenkalk locker ist, das verursache diese Schmerzen. Aber er ist mißtrauisch.

„Meine Schmerzen sind zu stark und es tut so, als ob da wieder ein Tumor sei."

Er besteht auf Klärung, auch wenn's Bein ab muß! „Aber vielleicht ist es doch nichts? Was soll ich ohne Bein machen?"

Soviel Energie und soviel Angst vor der Konsequenz. Ich finde ihn mutig, wie er sich so um sich selbst kümmert, trotz der Angst, die er dabei hat, daß man vielleicht dabei was findet!

„Was denn?" – „Na einen neuen Tumor! Und dabei bin ich so jung und habe von meinem Leben noch nichts gehabt."

Bestimmtheit und Enttäuschung, als ob die Diagnose schon von ihm gestellt worden wäre. Das ist sogar mir ein wenig unheimlich.

„Sie gehen also davon aus, daß man was findet bei Ihnen?" – „Ja, was soll ich denn sagen, ich habe doch nun mal die Schmerzen! Solche Schmerzen hat einfach kein normaler Mensch!"

Beim nächsten Treffen ist er schon auf einer chirurgischen Station. Ich suche sein Zimmer, er ist nicht da.

„Hallo, schön daß Sie kommen."

Er sitzt mit einem Rollstuhl auf dem Flur. Der linke Unterschenkel fehlt. Er sieht so schlecht aus und die Augen voller Angst. Er faßt mich an und zieht mich auf einen Stuhl. Er spricht nichts und ich sage:

„Sie sehen so angstvoll aus. Was ist denn mit Ihnen?" – „Oh, hier ist alles so schrecklich. Alle sind immer wieder so schnell von meinem Bett weg, als hätt' ich Aussatz. Und als ich aufgewacht bin, war mein Bein weg. Das wußte ich ja, aber es war doch sehr schlimm." – „Und was sagen die Ärzte?" – „Ja, wie ich gesagt habe, ich habe einen neuen Tumor. Jetzt hofft man, daß alles weg ist."

Er faßt mich wieder an. Das berührt mich, er war sonst so zurückhaltend und oft sogar abweisend. Ich weiß nicht recht, was ich sagen soll. Da hilft er mir auf die Sprünge.

„Wie ist es denn oben auf der Station, ist alles beim alten?" – „Ja, warum? Haben Sie ein bißchen Sehnsucht nach uns?" – „Ja!" – „Soll ich denn oben mal fragen, ob wir Sie nicht übernehmen können?"

Es scheint das Richtige gewesen zu sein. Er lacht.

„Aber das wird schlecht gehen, die Wunde, die Fäden?" – „Ach, das können wir oben auch, sonst holen wir einen Chirurgen."

Nach der Verabschiedung auf dem Weg zur Station wird mir heiß, meine Güte, was hab' ich versprochen? Ich bin weder der Stationsarzt, der ihn aufnimmt, noch die Schwester, die ihn pflegt, und habe doch so getan, als könnte ich alles entscheiden und habe Hoffnung gemacht. Mein Herz klopft, als ich in der Konferenz meinen Wunsch vorbringe.

„Na klar, das geht, warum nicht?"

So kann es einem auch mal gehen, denke ich und freue mich. „Wir rufen mal unten an." – „Nein, dafür bin ich nicht." Stefan soll das selbst bei den Ärzten vorbringen, ich finde nicht, daß man ihn behandeln soll wie ein Kind, das nicht für sich selbst sorgen kann. Er fragt und kommt nach oben.

Beim nächsten Treffen scheint er viel ausgeglichener. Ich komme ins Zimmer und frage, ob ich mich aufs Bett setzen darf. Ein bißchen fühle ich mich wie eine Mutter, die es geschafft hat, daß er sich ein wenig besser fühlt – Ich? – na ja.

Er zieht die Decke weg und präsentiert mir sein Bein. Eingewickelt in Verbände, stupst er immer wieder mit dem Finger dagegen und fragt:

„Wie finden Sie das denn? Einfach weg, einfach weg, man hat mir schon eine Prothese angeboten, aber ich laufe doch nicht damit, ich bin meine Krücken gewöhnt."

Er will gar keine Antwort, er fragt mehr sich selbst, dann spricht er weiter.

„Was werden meine Kumpels sagen, die wissen das sicher schon, aber nur zwei waren da, warum wohl?" Dann die Erklärung: „Na ja, die haben wenig Zeit, die haben ja auch ihren Job. Aber die Langeweile hier im Krankenhaus."

Ich sag so an ihn hin: „Ja, Sie können jetzt nicht mehr so weglaufen, wie Sie das vor der OP noch konnten, und hier ist alles so mühsam und das Krankenhaus deprimiert Sie auch?"

Er schaut mich an und antwortet, im ersten Augenblick für mich nicht klar.

„Sie können mich ruhig duzen, das machen alle, da fühle ich mich wohler. Ich bin ja noch keine 17. Ich wollte Sie noch um etwas bitten, könnten Sie nicht mal fragen, ob ich am Wochenende immer nach Hause darf? Da sehe ich mal was anderes."

Ich will seinen Wunsch gerne auf Station unterstützen, aber fragen soll er doch lieber selbst, ich komme mir sonst zu mütterlich vor und eigentlich wirkt er nicht besonders unselbständig. Das leuchtet ihm ein, er probiert es und es klappt. Das war ein wichtiger Schritt. Wieder nach draußen. Er hat mir doch eine gute Antwort gegeben, er hat einen Kumpel mehr gewonnen und ist öfter raus aus der deprimierenden Umgebung. Immer wieder sprech' ich mit ihm, dabei hab' ich das Gefühl, daß ich auf ihn angewiesen bin.

Er bringt durch seine Spontaneität und Offenheit Klarheit in unseren Kontakt, er setzt mich instand, Schwerkranke besser zu verstehen. Zwischenzeitlich wird er immer mal wieder entlassen. Aber auch eine Chemotherapie wird angesetzt. Es geht ihm häufig sehr schlecht – alles will er hinschmeißen – weg – nach Hause – zu seiner Mutter. Die versteht ihn. Aber immer ist da auch der Vater, den er haßt. Er ist immer betrunken und schlägt ihn noch immer, der sieht nicht, wie krank er schon ist. Ein Konflikt, wo er besser aufgehoben ist. Überall sind schlechte und gute Anteile. Ausschlaggebend sind meist seine Freunde, nach denen er sich sehnt.

Manchmal, wenn ich bei ihm und seinen Infusionen sitze, weint er, manchmal ist er aggressiv und nicht zugänglich. Dann macht er mir wegen Kleinigkeiten Vorwürfe. Er habe die Nase voll, ich würde mich mit allen verbünden, nur nicht mit ihm. Am anderen Tag eine andere Einstellung. Plötzlich sagt er, er hätte mal richtig mit dem Arzt gesprochen.

„Über was?" – „Was die schlimmsten Folgen meiner Krankheit wären, er hat mir alles so erklärt, daß ich es verstehen konnte." – „Was hat er denn gesagt?" – „Daß der Tumor Metastasen machen kann, die zum Schluß auch in die Lunge gehen können, deswegen will ich auch therapiert werden,

ich will nicht ersticken! Wissen Sie, ich habe solche Angst vor den Schmerzen. Angst, daß ich ersticke. Ich will soviel Medikamente, daß ich keine Schmerzen mehr spüre. Sie müssen mir dann helfen, daß ich soviel Schmerzmittel bekomme, wie ich will. Außerdem will ich bis zum Schluß immer nach Hause! Man darf mich aber nie in einem Krankenwagen fahren, die Nachbarn sollen nicht erfahren, wie schlecht es um mich steht, die zerreißen sich sonst das Maul."

Dabei hat er schon kein Haar mehr auf dem Kopf!

Die Fußballweltmeisterschaft will er noch sehen: „Meinen Sie, ich schaffe das?"

Bis jetzt sei doch noch alles drin, warum er mich fragt, ob er keine Hoffnung mehr habe? „Nein." – „Warum nicht?" „Ich habe wieder solche Schmerzen an meinem Stumpf. Wissen Sie, das macht mir schon die ganze Zeit Angst, aber ich konnte das nicht sagen." – „Warum nicht?" – „Weil die Ärzte immer gemeint hätten, daß die Knubbel an der OP-Naht auch Blutergüsse sein könnten, das habe ich gehofft, aber die Schmerzen, die kenne ich, das sind Tumorschmerzen."

Ich bekomme Angst, er wird recht haben, er hat bis jetzt immer gut beurteilen können, wie es um ihn steht. Ich will alles beschönigen, aber ich lasse es lieber. Ich will keine falsche Aussage machen, weil ich weiß, daß Schwerkranke noch schneller als andere Patienten nicht ertragen können, wenn man sie aus Angst vor der Wahrheit belügt.

Der Kontakt zu mir wäre schnell beendet und das will ich nicht. So verabschiede ich mich und sehe zu, daß ich mich fange. Ein Profi verleugnet eben manchmal auch gerne mit, denke ich und ärgere mich über mich selbst.

Nach Untersuchungen und langem Hin und Her muß das Bein ab. Stefan hadert und zankt. Man hätte das doch gleich machen sollen, so scheibchenweise, das sei doch schrecklich, gleich von Anfang an hätte man das richtig machen sollen. Er will seine Befunde nicht mehr von einem Arzt mitgeteilt bekommen. Ich soll das tun, ihm alles erklären.

Vielleicht soll ich Filter sein; die Angst ist bei mir die gleiche, aber vielleicht darf er bei mir eher mal die Fassung verlieren und muß nicht immer seine Männlichkeit und seine Fassung zeigen.

Ich weiß nicht – ich tu's mit Einverständnis des Arztes. Es sind Metastasen im Bein, die Histologie hat's gezeigt, doch zuvor gehen schrecklicherweise die Befunde verloren. Er glaubt, daß wir ihn belügen, er hat mehr Angst vor der Ungewißheit als vor einem schlechten Befund, da weiß man wenigstens, wo man dran ist. Das Vertrauen scheint zu verschwinden, ich kämpfe drum, die andern sicher auch, aber mich strengt das an und ich frage mich, ob ich nicht auch gegen die Gleichgültigkeit eines anonymen Krankenhauses kämpfe. Keiner fühlt sich für den Befund verantwortlich, keiner hat ihn gesehen, der Pathologe, der ihn erhoben hat, ist nicht da. „Wie, die Gewebeprobe ist noch gar nicht bei uns gelandet, da kann man verrückt werden!" Der Patient glaubt es doch.

Er kommt zurück mit einer Amputation bis zum Becken. Beim ersten Besuch ist er fast nicht zu sehen. So weiß, so still und so klein scheint er geworden zu sein. Er spricht kaum, fast ist der Kontakt zu ihm abgebrochen.

Doch er schlägt die Bettdecke zurück und wie beim ersten Mal schnippt er wieder mit den Fingern an seiner verbundenen OP-Narbe und sagt:

„Stecken Sie mir doch mal mein Schlafanzugbein hoch, das Bein brauch' ich ja jetzt nicht mehr, ich hab' nichts mehr zum reintun." Er lacht.

Wieder eine lange Zeit die gleichen Fragen und die Angst, die Übelkeit, das Erbrechen und die Schmerzen. Zwischendurch immer die Hoffnung auf Heilung. Mit Zytostatika muß behandelt werden, weil sonst Metastasen in die Lunge kommen. Bluttransfusionen verweigert er, da er aber blutet, braucht er sie. Was ist los?

Mit viel Geduld bekomme ich heraus, daß er Angst hat, daß das frische Blut nicht nur ihn wieder belebt, sondern auch das Wachstum des Tumors beeinflussen könnte. Die Phantomschmerzen nehmen zu. Er spürt wieder seinen Fuß. Er braucht viel Medikamente. Die Gesprächsthemen bestimmt er. Manchmal ist die Angst vor Schmerzen, der Tod, das Sterben und das Alleinsein im Mittelpunkt und manchmal geht es um Fußball, von dem ich Gott sei Dank was verstehe, und manchmal um Mädchen und Heirat. Er findet eine Schwester auf der Station sei ein Klasseweib! Daran spielt sich eine ganze Auseinandersetzung mit seiner Sexualität ab. Er macht einen Prozeß der Verliebtheit bis hin zum resignativen Verzicht durch und ist oft deprimiert. Und plötzlich spricht mich seine Mutter gehäuft an. Sie braucht jemand, der sie versteht. Sie traut sich nicht heran an die Doktors und überhaupt sei sie nur eine Bauersfrau und verstünde nicht viel, aber ihr Stefan, der spricht so viel über mich, da wollte sie doch mal sehen, wer das ist. Was er mir denn so erzählen würde?

„Ja wissen Sie, ich glaube, das sollten Sie Ihren Sohn selbst fragen, ich weiß nicht, ob er will, daß ich es Ihnen erzähle. Es hat meist mit seiner Krankheit zu tun und seiner Angst vor'm Sterben." Davon will sie nichts wissen, sie erschrickt, das soll er ruhig mit mir besprechen, sie würde das nicht aushalten. Sie will so tun, als sei nichts, die Fassung behalten. Ihr Sohn braucht so eine starke Unterstützung, die könnte sie ihm dann nicht mehr geben. Sie will so tun, als ob nichts sei, sie will die Fassung vor ihrem Sohn nicht verlieren. Der braucht doch jemand, der ihn gegenüber dem Vater in Schutz nimmt, der schon oft im Delir seinen Sohn, der so krank sei, verprügelt hätte. Ob ich verstünde, daß das mit dem Mann die Strafe für ihre Sünde sei? Ich fragte, was sie meint. Ja, daß sie ihn eigentlich nie geliebt habe, und daß sie sich oft gewünscht hat, daß sie kein Kind von diesem Mann hätte. Das sei eine Sünde und vielleicht sei das jetzt die Strafe Gottes, ihr den Sohn zu nehmen, wo sie ihn liebt und auf keinen Fall hergeben möchte. Sie weint hemmungslos und ich versuche zu erfahren, was sie denn so quält. Da sagt sie leise, daß sie sich manchmal gewünscht hätte, früher, daß der Stefan tot sein sollte, weil sie das Leben zu Hause nicht mehr ertragen habe, weil sie kein Kind von so einem Mann haben wollte. Ihr ältester Sohn, der sei von einem anderen Mann, der sei doch auch so krank, der habe Asthma, sie hätte schon gehört, daß das auch was mit der Psyche zu tun habe, das stimme auch, immer wenn er sich aufregt, oder wieder ein Kind kommt, sei das Asthma besonders schlimm. Ich habe das Gefühl, sie schüttet ihre Seele aus, und nehme sie einfach mal in meine Arme. Was ich denn zu ihren Verrücktheiten sagen würde, fragt sie. Ich meine, daß wohl jede Mutter mal das Gefühl hat, daß sie ihr Kind nicht mag, daß sie aber deswegen noch keine schlechte Mutter sei, sondern nur ehrlich. Sie strahlt mich an und schildert ihren Lebenslauf. Sie meint, daß ihr Vater, obwohl schon lange tot, noch immer ihre Stütze sei, sie sei wie er, gut und ruhig. Er habe immer vermittelt. Die Mutter sei eher hektisch gewesen, und zudem war sie noch leidend und habe damit die ganze Familie schikaniert. Der Vater sei ein ruhiger Pol gewesen, von ihm habe sie die Kraft, das alles durchzustehen. Sie hat auch Fragen, die ich gar nicht beantworten muß.

„Warum will mein Sohn, daß ich mich von meinem Mann scheiden lassen soll, das geht doch nicht; er aber quält und quält mich damit. Sie sind doch studiert, was meinen Sie denn? Ich habe ihm erklärt, daß ich ihn nach so langer Ehe und nach dem, was er schon alles durchgemacht hat, auch als junger Mann, nicht im Stich lassen kann! Eigentlich hat er das verstanden, aber er drängt mich immer wieder, mei-

nen Sie, er will nicht, daß ich all meine Liebe nach seinem Tod meinem Mann zuwende und für ihn dann wie eine Mutter bin?" – Ich frage: „Warum meinen Sie das?" – „Ja er sieht doch schon häufig, daß ich mich intensiv um meinen Mann kümmern muß, wenn er ein Delir hat und auf dem Ofen sitzt und weint, wenn er Tiere durchs Zimmer krabbeln sieht, oder wenn er alles unter sich läßt. Vielleicht denkt der Stefan dann, ich hätte für meinen Mann mehr Zeit als für ihn und ist eifersüchtig." Ihre Augen strahlen mich an, als ich nicke und meine: „Da können Sie recht haben."

Vor lauter Freude erzählt sie mir dann noch die Geschichte, wie Stefan seinem Vater den Arm in der Luft festgehalten hat, als dieser ihn in seinem kranken Zustand schlagen wollte, er hätte den Vater angeblitzt und gesagt, du schlägst mich nie wieder! Und das alles, obwohl er auf Krücken gehen mußte. Da hätte ihr Mann aber mal geschluckt und hätte seit der Zeit den Bub nie mehr angegriffen. Das gleiche erzählt mir Stefan, für ihn ist es noch wichtiger, er freut sich wie ein Mann nach einer Schlägerei, die er gewonnen hat, so meint er. Trotz objektiver Schwäche jetzt stärker zu sein als früher, bedeutet für ihn Hoffnung, Dinge in seinem Leben noch bewältigen zu können. Er darf nach Hause. Er kommt glücklich zurück. Es war so schön, seine Mutter hat ihn verwöhnt und ihn bekocht bis zum Gehtnichtmehr. Jetzt hat er Angst vor dem Einheitsbrei und den Befunden! In kurzen Abständen wird die Lunge geröntgt. Er bittet mich, mit ihm die Befunde zu besprechen, ich tue es, allerdings nach Absprache mit dem Stationsarzt und nicht ohne ihn zu fragen, warum er mich dazu braucht. Seine Antwort ist ganz schnell da und klar:

„Warum eigentlich nicht? Sie betreuen mich doch." – Mich packt die Furcht vor meiner eigenen Courage. Möglichst locker meine ich zu ihm: „Ich komme dann, da können wir ja drüber sprechen."

10 Sekunden später stürzt er mit seinen Krücken auf den Flur; er humpelt hinter mir her, mit kahlem Kopf voller Grinde und aufgebrochenen Geschwüren, einem ‚abben‘ Bein und einem Zungenkrampf. Wie eine ‚erstickende Leiche‘ kommt er mir vor, und ich fürchte mich vor ihm, ich will für Sekunden so tun, als ob ich ihn nicht gesehen habe. Er folgt mir keuchend ins Stationszimmer. Mit angstvoll aufgesperrten Augen und wild gestikulierend grunzt er mich an: „Angst – gelähmt –". Er zeigt auf seinen Hals und der Kopf ist nach hinten gebogen; er weint und ich ekle mich; ich glaube, das dauert nur Sekunden, ich empfinde aber eine Ewigkeit. Dann habe ich Mitleid und nehme seinen grindigen Kopf in die Hände und spreche beruhigend auf ihn ein; ich versuche sein Kinn nach unten zu drücken und die Halsmuskeln zu entspannen; ich rede und rede, er brauche keine Angst zu haben, er sei doch da und glaube fast, ich sage es zu mir selbst. Ich merke, wie alle fluchtartig den Raum verlassen und sogar die Türe hinter sich zumachen; ich komme mir alleingelassen vor; ich lasse ihn nicht los und streiche so lange über seinen Kopf, bis die Spannung im Mund nachläßt und er wieder sprechen kann, er ist völlig erschöpft und ich auch. Tapfer meine ich: „Na, sehen Sie, es geht ja wieder." – Er schaut mich an und meint: „Ja was war das?" – „Ich weiß es nicht, was meinen Sie?" – „Ja, ich bin so erschrocken, daß ich mir vorstellte, daß Sie mir was von neuen Metastasen sagen könnten, da hab' ich plötzlich keine Luft mehr bekommen."

Er lacht und ich frage warum, ich verstünde jetzt seine Reaktion nicht. Ich bin auch etwas befreit, als er sagt, ja vielleicht hat es mir vor Angst die Sprache verschlagen, aber damit hatte er sicher recht. Ein paar Minuten später das gleiche noch einmal, aber so kurz und dann nie mehr. Dann sind immer diese Schmerzen da. Stationsarzt, Oberarzt, Schwestern und ich, wir geben uns doch sehr Mühe, Stefan einigermaßen beschwerdefrei zu halten, zumal er sich das in einem früheren Gespräch gewünscht hat. Doch es gelingt einfach nicht. Ja, es wird sogar zu einem Problem. Eine Schwester ärgert sich bei mir: „Der schikaniert mich ja. Kaum hat er den Doktors versprochen, ein bestimmtes Medikamentenschema einzuhalten, schreit er nach mir und will eine Spritze. Was soll das? Gebe ich sie nicht, wird er aggressiv."

Ich versuche zu trösten, mir geht es genauso, kaum setze ich einen Fuß über die Schwelle, will er ein Medikament oder er lehnt ab, was ich bringe. Das hilft der Schwester aber nichts, sie will die Sache geklärt haben, und zwar sofort. Wir gehen also zu ihm. Er sagt zu allem ja, will morgen bei der Visite das Problem besprechen. Morgen ist dasselbe in Grün. Also wieder hin. Wir sind beide gegen ihn sehr aufgebracht. Wir beruhigen uns, auch wenn jemand so schwer krank ist, hat er doch nicht das Recht, uns auf der Nase rumzutanzen. Zu zweit stehen wir am Bett und reden auf ihn ein. So geht das einfach nicht, Stefan! Einmal hü, einmal hott und so weiter und so weiter; wir sind zu zweit in Fahrt, wie mag er sich da fühlen?

Ich werde laut, als er widerborstig widerspricht: „Wir sind doch hier nicht in einem Kindergarten, wenn du nicht die Medikamente so nehmen willst, wie sie angeordnet werden, dann besprich das gefälligst auch mit dem Doktor und dem Oberarzt."

Es ist eine harte Auseinandersetzung wegen einer Kleinigkeit. Sie endet mit einer Demutsgeste von Stefan, und die Schimpfe wird mir fad im Mund. Kann man einen Schwerkranken so fertigmachen, warum tun wir das? Diese Gedanken gehen mir durch den Kopf, als ich vor der Türe stehe, und ich fühle mich mies. Drei Tage mache ich ein Gesprächsangebot, und er nimmt nicht an. Da steht doch was. Am 4. Tag mache ich einen Sprung vorwärts und lasse mich nicht mehr abweisen, ich spreche die Sache noch mal an. Und siehe da, ich bin erleichtert, daß ich nicht verletzt habe, sondern eine offenere Situation entstanden ist.

Er sagt gelassen: „Ja, ich wollte mal sehen, wie Sie auf mich reagieren, wie weit ich bei euch gehen kann!" – „Warum? Hast du uns also bewußt geärgert? Das kann ich kaum glauben." – „Na ja, ein wenig anders ist es schon, wissen Sie, ich muß eigentlich immer fühlen, wo meine Schmerzen sind in meinem Körper, wieviel Schmerz ich noch aushalten kann; aber wenn es dann eben zu stark wird, dann brauche ich sofort was dagegen."

Ich verstehe plötzlich sein Handeln, er will sich empfinden, er will nicht jeden Tag gleichförmig erleben, er will sich spüren, auch wenn es Schmerzen macht, auch wenn er spürt, hier sind meine Metastasen. Wie einleuchtend, aber auch wie schrecklich, Schmerzen als etwas Positives zu erleben, was Lebendiges. Ich habe so was theoretisch im Kopf wie sekundärer Krankheitsgewinn oder andere schöne Bezeichnungen, aber in der Realität ist es für mich doch neu.

An einem anderen Tag überrascht er mich mit dem Ausspruch: „Wenn eins kommt, muß eins gehen." Ich frage ihn, was er damit meint. „Eine Bauernweisheit, wissen Sie das nicht. Mein Bruder hat zwei Kinder; als das erste kam, ist seine Schwiegermutter gestorben, jetzt ist das zweite da, jetzt sterbe ich."

Ich kann dazu nichts sagen, weil ich glaube, daß er sich Brücken baut, um zu verstehen, daß er jetzt bald sterben muß, und ich wüßte nicht, ob ich das so gut könnte. Er sieht es vielleicht so, daß er eine Chance hat, in dem neuen Kind weiterzuleben. Er weiß nicht, was einem so alles durch den Kopf geht, aber kann man überhaupt konkreter sehen, daß man stirbt? Kann man nicht!

Er hustet Blut, er kann nichts essen, er hat zunehmend Schmerzen. Er fragt mich, woher das Blut kommt, aus dem Magen oder aus der Lunge? Ich zögere; wenn ich sage, aus der Lunge, weiß er, wie schnell er stirbt. Ich habe mehr

Angst, ihm alles zu sagen, als er, alles zu erfahren. Er herrscht mich an: „Ich warte auf Antwort!" Ich greife hilfesuchend zu meiner Gesprächsführung – ziehe mich zurück und fühle mich schlecht dabei: „Was meinst du denn?" – „Ich glaube, es kommt aus der Lunge!" – Ich meine zaghaft: „Ich glaube das auch, aber vielleicht kannst du den Doktor nochmal fragen?" – „Ach der, der zeichnet doch schon Striche auf meinen Rücken, wo die Metastasen sitzen. Wo der Erguß ist, das weiß ich selbst. Aber gut, ich frage bei der Visite, wenn Sie dabei sind!"

Er fordert mehr Nähe als früher, zwischendrin auch Abstand, wo er Tage nicht mit mir sprechen will. Er will dann schlafen, aber ich soll ruhig sitzen bleiben an seinem Bett. Manchmal möchte er aufsitzen, er kann das aber nicht mehr allein, weil er so starke Schmerzen hat. Ich soll ihm eine Stütze geben. Ich schlage ihm vor, eine breite Binde um seinen Brustkorb zu wickeln, damit er ein festes Gefühl hat. Das geht gut und er will, daß ich das öfter mit ihm mache, die Krankengymnasten hätten nicht so viel Zeit und verstünden nicht, wenn er meckert und ein bestimmtes Verhalten fordert. Ich habe das Gefühl, ich lebe für seine Wünsche, ich bin ausführendes Organ. Telefonisch versuchen wir im Winter Zitroneneis zu bekommen. Das geht so weit, daß eine Dame aus unserer Küche ihren Eismacher mitbringt von zu Hause; denn Zitroneneis ist das einzige, was er bei sich behält.

Er schreit häufig seine Mutter an, ohne Grund, einfach so. Sie kommt aus dem Zimmer, ganz bleich und zermürbt. „Ich habe das Gefühl, Stefan rückt immer weiter von mir weg. Das macht mir zu schaffen, und zwar nicht, weil Stefan sich zurückzieht, sondern, und jetzt muß ich Ihnen mal was ganz Verrücktes erzählen, weil alle Bekannten im Dorf sich von uns zurückziehen! Wissen Sie, man gewöhnt sich an alles, auch daß einem das Kind stirbt, aber daß man selbst so vergessen wird von den anderen, das ist schlimmer. Im Dorf fragt uns keiner mehr, ob wir bei der Ernte helfen. Sie wissen ja, daß wir jede freie Minute bei Stefan sind. Sie haben vergessen, mich für meine 35jährige Mitgliedschaft bei den Landfrauen zu ehren! Kein Gruß und nichts, das macht mich fix und fertig! Ist das nicht verrückt?" – „Nein das finde ich nicht." Ich finde, das ist ein sozialer Tod, der vielleicht nicht mehr aufzuholen ist von den Angehörigen. Sie weint, weil ich sie verstehe.

Ich gehe in Urlaub und sage Stefan, daß er mich auch telefonisch erreichen kann. Das will er aber nicht, was ich auch verstehe, es klingt ja fast so, als ob ich Angst hätte, daß er sterben könnte.

Nach Weihnachten kommt er wieder mit sehr viel Angst. Er will, daß ich ihm die Haare bürste, das kann er aber eigentlich selbst, warum also ich? Ja also, er ekelt sich so vor seinen eigenen Haaren, die ihm büschelweise ausgehen, er kann die Situation besser meistern, wenn er nicht allein ist. „Eine Perücke, ja das geht in Ordnung!"

Immer die gleiche Angst, vor jedem Befund, immer wieder Angst. Einmal sagt er trotzdem: „Ich müßte mich doch schon dran gewöhnt haben!" Manchmal scheint das so. Er sagt, es brodle bei ihm in der Lunge, er würde hören, wo, und würde seinen Wasserstand messen. Ich soll oft bei ihm sein, wenn er schläft, ich könne dann aufpassen, daß er nicht einfach stirbt. Ich denke, er hat gar nicht so unrecht. Einmal, als ich bei ihm sitze, schreckt er nach einer Weile auf und meint: „Ist das denn bei normalen Menschen auch so, daß sie alle Geräusche so intensiv wahrnehmen und sich damit beschäftigen?"

Ein langes Gespräch kommt in Gang über Kranke und Krankenhaus. Der muß doch Sinne ausbilden, wie Taube oder Blinde, weil sie so abgeschnitten sind von der Außenwelt. Häufig läßt er mich nur vom Bett weg, wenn ich ihm verspreche, daß ich morgen wiederkomme. Als ob ich die Garantie sei, daß er bis dahin nicht stirbt.

Eines Morgens erzählt er mir verschmitzt, daß er und sein Zimmerkollege einen Test gemacht hätten. „Was für einen?" – „Ja, mit den Pfarrern" – „Wie war das?" – „Ja, der evangelische hat gewonnen." – „Wieso?" – „Der hat die Flasche Sekt mitgebracht, die wir uns gewünscht hatten. Der katholische Geistliche hat nur etwas konsterniert geschaut. Wie finden Sie das? Der Pfarrer ist doch ein pfundiger Kerl, bringt Kranken einfach einen Sekt mit." – Ich finde das insgeheim auch. Sage aber nicht, daß ich denke, wie wichtig es für die Patienten sei, noch als vollwertige Trinker gesehen zu werden. Auch das ist ein Stück normales Leben, auch wenn man den Sekt hinterher nicht mehr trinken kann!

Er entdeckt jetzt jeden Tag neu, wie man die Welt aus seinem Bett heraus entdecken kann. Ich glaube, das füllt ihn richtig aus. Manchmal so stark, daß die Schwestern sich dagegen wehren. Das erträgt er dann aber ganz gut, wenn auch mit lauthalsem Protest. Sein Lieblingswort ist Scheiße. Er ist sehr empfindlich, wenn sich Menschen von ihm zurückziehen. So schimpft er auf den Hämatologen, der sich vor ihm drückt! Immer wieder zeigt er seine starken Seiten. Er versetzt Ärzte, wenn sie mit einer Medikation nicht pünktlich an seinem Bett erscheinen. Wenn seine Mutter nicht kommt, läßt er sie am nächsten Tag buchstäblich an seinem Bett verhungern und spricht kein Wort mit ihr. Auch mich schickt er mit den Worten weg: „Ich habe heute keinen Sprechtag!"

Er wirkt nicht mehr passiv. Er motiviert seine Mitpatienten, ihn aus der Klinik zu begleiten, ins Kino zu humpeln und vieles mehr. Er wirkt so, als ob er die Station führt und geht oft zu weit. Eines Tages kann er dann nicht mehr, seine Kraft ist zu Ende. Er beschimpft noch Mitpatienten, die zuviel rauchen, und verbannt sie ins Treppenhaus oder gibt anderen Patienten von seinen Medikamenten ab, aber das dauert nur kurze Zeit.

Er kann immer schlechter atmen, aber Sauerstoff will er auf keinen Fall, dann weiß er, daß er stirbt. Er will lieber ein Spray, keine Asthmapatienten, keine Sauerstoffsonde, so wie die alten Männer, die schon neben ihm gestorben sind. Aber manches Mal braucht er sie doch, was für eine Niederlage für ihn! Er weint deswegen oft. Wieder und wieder hält er mich an seinem Bett fest. Er versucht, sich am Leben zu halten. „Alles kommt jetzt so viel zu schnell", sagt er wütend. Oft kann er nicht mehr gelassen sein, wo er es will, er hat Angst und Schmerzen, daß er hemmungslos weint: „Ich will doch noch so viel!" Zwischendurch denkt er nach und meint, er wisse gar nicht, warum er jetzt so aufgelöst sei, er hätte ja schon lange gewußt, wie schwer seine Erkrankung ist und daß er sterben müsse.

Dann will er wieder nach Hause, alles mit den Augen mitnehmen, wie er sagt. Am liebsten will er zu Hause sterben. Die Mutter am Bett sieht fassungslos zu mir und meint: „Aber Junge, wie soll ich das denn schaffen?" Das ist ihm jetzt egal; er will, daß alle für ihn da sind, er schreit: „Du kannst mich nicht leiden, du läßt mich ja im Stich. Du mußt eben lernen zu spritzen! Ich will das so." Die Mutter ist fast am Ende, sie weint auf dem Flur und meint zu mir hin: „Aber sagen Sie doch mal, ich kann das doch wirklich nicht! Ich habe Angst! Das schaffe ich nicht, das Pflegen und so!"

Die Mutter ist weg und Stefan holt mich. Er hat Angst, wenn er nach Hause muß, daß er dann nicht so gut versorgt ist wie hier, wenn er plötzlich blutet, wenn er erbricht und erstickt, keiner ist dann da. Ich frage ihn, warum er dann nach Hause möchte, er gibt mir die richtige Antwort: „Weil ich dahin gehöre!"

In der Nacht tobt er und schlägt auf den Arzt ein, der von den Schwestern geholt wird. Am nächsten Tag bekommt er das über seine Mitpatienten mit und ist völlig verstört. Er kann sich an nichts mehr erinnern als an eine furchtbare Angst. Davor hat er jetzt die meiste Furcht. Er will in der Nacht nicht mehr allein bleiben, weil er Angst vor solchen Zuständen hat. Die Station ermöglicht, daß er ein paar Tage eine Sitzwache erhält.

Plötzlich sagt er zu mir, ich solle das Fell aus seinem Bett wegnehmen, das sei zu weich, er könne sich nicht mehr spüren. Ich tue das, obwohl er dann eher einen Dekubitus bekommen kann.

Ich war einen Tag nicht da, und ausgerechnet in dieser Zeit sollte sein Bettnachbar verlegt werden. Er hatte das Gefühl, daß ich das angeordnet hätte, und stellt mich am nächsten Tag zur Rede. Ich weiß von den anderen, daß der Bettnachbar nicht mehr zusehen kann, wie Stefan stirbt, und seinen Auszug selbst mit dem Arzt besprochen hat. Ich versichere Stefan, daß ich damit nichts zu tun habe, meine aber, daß das gar nicht das Wichtigste an dieser Situation sei, sondern vielleicht, daß er sich vorkommt, als ob er im Stich gelassen würde. Er weint und weint und kann sich gar nicht beruhigen, die Zeiten an seinem Bett werden immer länger.

Ich freue mich, daß er die Beziehung nicht abgebrochen hat, sondern weiter den Kontakt mit mir sucht. Ich habe das Gefühl, er nimmt Abschied in allen möglichen Formen. Er erlebt jede noch so kleine Trennung. Er macht Pläne, mehr als früher – ich mache sie mit ihm – so wie er kann, so mach' ich es auch. Er will ein Spanferkelessen organisieren, ehe er stirbt; einen Leichenschmaus, wo er noch dabeisein kann, wie er sagt. Er fragt mich, ob ich glaube, daß er das noch schaffen kann? „Ich weiß das genauso gut oder schlecht wie du; ich weiß doch über dich nicht besser Bescheid als du." – „Ja, das stimmt! Selbst die Ärzte mit ihrer ganzen Diagnostik können keine genauen Zeitangaben über meinen Tod machen. Ja, da haben Sie recht, das weiß ich besser."

Immer wieder will er nicht allein sein. Die Tür bleibt häufig offen, damit er die Geräusche der Station hören kann, was ihn manchmal beruhigt. Sein Halt ist eine 1-Liter-Cola-Flasche, die er täglich kauft. Ich habe das Gefühl, daß er denkt, „solang ich die nicht ausgetrunken habe, kann ich nicht sterben". Er reagiert erbost, als seine Mutter einmal zu ihm sagt: „Aber Bub, so viel kannst du doch sowieso nicht trinken."

Jetzt tut ihm sein Herz so furchtbar weh. Das muß zuviel leisten. Ich soll bleiben und zählen, wie oft er hustet, auch wenn er dabei schläft. „Warum?" – „Damit ich weiß, ob ich mich verschlechtere." Ich schau wohl ein wenig betroffen, denn er sagt, er hätte keine Angst mehr vorm Sterben, denn jetzt sei es doch bald so weit? Er schont sich für den Kontakt mit seinen Eltern, hält seinen Kontakt zu mir auf Sparflamme. Nur nicht stören, da sein soll ich schon.

Er will nach Hause, alle versuchen, das mit Mühe zu organisieren, denn es ist demnächst Ostern. Dann ist alles geregelt: Eine Schwester aus einem nahe gelegenen Krankenhaus soll kommen, um ihn zu spritzen. Er ist sehr beruhigt.

Die Unruhe nimmt von Tag zu Tag zu. Am Nachmittag vom 24. 3. sitze ich mit seiner Mutter am Bett, und er schläft; wir unterhalten uns leise. Ich will ihm den Schweiß abwischen, der ihm von der Stirn rinnt. Da schreckt er auf und schreit: „Lassen Sie mich in Ruhe, das kann ich nicht brauchen." Ich habe ihn wohl beim Sterben gestört. Ich bin ganz verstört. Ich kann nicht mehr helfen. Wie ungewohnt, daß Stefan alles selbst macht.

Meine Zuwendung schlägt um in Haß, als ich Stunde für Stunde sehen muß, wie er nicht sterben kann. Ich will, daß er endlich tot ist.

Sein Vater kommt betrunken und faßt zum erstenmal zärtlich seine Hand an, die schon ganz kalt ist. Da läuft er hinaus und holt sich ein Bier, wie seine Frau sagt.

Am nächsten Tag ist Stefan tot.

Die Mutter erinnert mich in einem zweistündigen Gespräch immer an Stefan. Ich gehe nach Hause und werde erst einmal krank. Ich bin so heiser, daß ich glaube, mir hat es die Sprache verschlagen.

Teil V: Aus- und Weiterbildung, berufspolitische Situationen

67 Die Ausbildung zum Arzt[1]

Wolfram Schüffel

67.1 Vorbemerkung

Der Autor hat die Hoffnung, daß konzeptgeleitete Überlegungen zum ärztlichen Ausbildungsprozeß verstehen helfen, wie berufliche Werte internalisiert werden und ärztliche Einstellungen entstehen.

Die Ausbildung zum Arzt wird in der vorliegenden Darstellung als ein sechsjähriger Prozeß gesehen, während dessen sich der Student grundlegende berufliche Werte aneignet. Es wird angenommen, daß die hier erworbenen oder die mitgebrachten und dann modifizierten Werte seine spätere Tätigkeit als Arzt nachhaltig beeinflussen.

Die medizinische Ausbildung wird somit als ein Entwicklungs- und Prägungsprozeß verstanden. In der Fachsprache wird dieser als „sekundäre Sozialisation" bezeichnet; die „primäre Sozialisation" ist der Erwerb derjenigen Werte, Kenntnisse und Fertigkeiten, die das Individuum zu einem Mitglied seiner Familie oder engsten Bezugsgruppe werden lassen.

Für die sekundäre Sozialisation, also Arztwerdung, ist aus psychosomatischer Sicht die entscheidende Aufgabe, die krankheitsbezogene Orientierung einer übergeordneten personenbezogenen ärztlichen Vorgehensweise unter- bzw. einzuordnen. Dabei wird es sich um Erhalt und Gewinn von Gesundheit, erst an weiterer Stelle um Kampf gegen Krankheit handeln. Es geht um den Gewinn „einer Perspektive des Systems Gesundheit" (Pauli, 1988). Die Frage ist: Wie wird sich der Student und der junge Arzt auf diese Aufgabe vorbereiten können?

Die Aufgabe kann mit einer Frage umschrieben werden, die sich der psychosomatisch orientierte Student und Arzt bei der Behandlung jedes Patienten stellt. Sie lautet in einer Formulierung von F. Hartmann, der sich seinerseits auf V. von Weizsäcker bezieht: „Warum kommt **dieser** Mensch zu **diesem** Zeitpunkt mit **dieser** Krankheit zu **mir**?" (F. Hartmann, 1983).

Es soll versucht werden, die Umstände solcher Internalisierungsprozesse zu beschreiben, die dem Studenten helfen, als Arzt späterhin eine derartige Frage durchgehend zu stellen. Der rote Faden dieses Buchabschnittes wird die Frage sein: Wie kann sich der Student eine personenbezogene **Einstellung** aneignen, die er berufsbegleitend und stets in der Zusammenarbeit mit anderen weiterentwickelt?

Im Text wird häufig auf sog. **Anamnesegruppen** verwiesen. Hierzu kurz eine Erläuterung: Bei Anamnesegruppen handelt es sich um zehnköpfige Gruppen von Medizinstudenten, die jeweils zur Hälfte aus Vorklinikern und Klinikern, Studentinnen und Studenten zusammengesetzt sind. Die Teilnehmer treffen sich wöchentlich für zwei Stunden mit dem Ziel, daß bei jedem Treffen mit einem körperlich Kranken ein Erstgespräch geführt, d.h. eine Anamnese erhoben wird. Das Ziel des Erstgespräches ist zweifach: Zunächst soll die Situation des Patienten, ausgehend von dessen Beschwerden erfaßt werden. Diese Situation beinhaltet: die Beschwerden, die Benennung des Krankheitsbildes (sofern im traditionellen Diagnosenschema einzuordnen), den Umgang des Patienten mit seinen Beschwerden, die Beziehungen des Patienten zu seinen Nächststehenden einschließlich seines Arztes. Dann soll der Eindruck geschildert werden, den der Patient auf den Studenten gemacht hat, der das Gespräch führte. Die Beobachtungen der Mitstudenten sollen als Beitrag zur Gesamtsituation des Patienten und seiner Umwelt gewertet werden. Dabei gibt es kein „falsch" oder „richtig"; vielmehr geht es immer um die Frage, **wie** ist der Beitrag zu verstehen. Die Studenten diskutieren ihre Beobachtungen vor dem Hintergrund ihrer eigenen Interaktionserfahrung, unter Anleitung eines studentischen Tutors. Der Tutor hat seinerseits in der Regel eine Anamnesegruppe durchlaufen, ein Training als Gruppenleiter mitgemacht und beteiligt sich an einer **begleitenden Supervision**. Die Gruppenarbeit läuft über zwei Semester, umfaßt ca. 50 Stunden (25 Doppelstunden) und wird in der Regel durch ein bis zwei gemeinsam verbrachte Wochenenden unterstützt. Eine ausführliche Beschreibung des Lernmodells der Anamnesegruppe und der darin gemachten Erfahrungen wurde vorgelegt (Schüffel et al., 1983).

Im folgenden wird eine **konzeptgeleitete** Betrachtungsweise der verfügbaren Kenntnisse des sekundären Sozialisationsprozesses entworfen. Die Konzepte entstammen sozialpsychologischer, soziologischer (67.3.1, 67.3.2, 67.3.3) und psychoanalytischer bzw.

---

[1] Herrn Prof. J.J. Groen (Leyden) nachträglich zu seinem 80. Geburtstag gewidmet. Er hat uns allen die Arbeit ermöglicht, indem er als Arzt und Ratgeber versicherte, daß wir nichts Unmögliches anstreben.

gruppenpsychologischer (67.3.4) Betrachtungsweise. Nach einer kurzen Vergegenwärtigung der vorgegebenen Studienbedingungen, d.h. Form des Curriculums, Dozenten und Prüfungsbedingungen (67.4), werden Verbindungen zum Modell des Situationskreises hergestellt (67.5). Schließlich wird überlegt, wie in konkreter Form das Bemühen von Studenten und Dozenten gefördert werden kann, patientenbezogene Einstellungen zu erwerben bzw. zu vermitteln (67.6).

Einen besonderen Dank möchte ich aussprechen: All den Studenten, die seit 1969 in Ulm beginnend zu Fragen der Sozialisation mit mir gearbeitet und schließlich mit Ärzten und Psychologen gemeinsam die **Anamnesegruppe** als eine Lernmethode im sozioaffektiven Bereich erarbeitet haben; U. Egle, Mainz, der langjährig mitgearbeitet und dieses Manuskript kritisch beurteilt hat; H. Pauli, Bern, der mir mit größter Mühe, Sorgfalt und reicher Anregung geholfen hat, die Gedanken zu sortieren und die Ansprüche in diesem so schwierigen Bereich immer wieder mit einer realistischen Elle zu messen; Th. von Uexküll (Ulm/Freiburg), der dieses Manuskript mit mir durcharbeitete.

67.2 Betroffenheit als unreflektierte emotionale Erfahrung

Bei vielen Studenten gibt es ein Schlüsselerlebnis für die Wahl eines Berufes, bei dem man lernen kann, medizinisch und menschlich mit Krankheit und Tod umzugehen. Entscheidende Erfahrungen während des Studiums sind dann Momente der Betroffenheit, wie sie etwa bei der Konfrontation mit der Leiche im Präpariersaal, mit Verstümmelten, mit Schwerkranken oder mit Sterbenden während der klinischen Semester ablaufen. Es sind auch diejenigen Erfahrungen, die der Student macht, wenn er Professoren und Assistenzärzte im Umgang mit ihren Kranken beobachtet.

Mit seiner Betroffenheit bleibt aber der einzelne zumeist allein. Im Unterricht kommen Probleme dieser Art nicht zur Sprache, Dozenten und Pflegepersonal erwecken den Eindruck, es sei anstößig, Gefühle zu haben. So entsteht eine Atmosphäre, in der man auch mit den Mitstudenten nicht über seine Erfahrungen sprechen kann.

Ein westdeutscher Medizinstudent kommentiert seine Studienerfahrungen stellvertretend für viele Kommilitonen (Spiegel, 1983):

„Nach sechs Jahren Studium würden wir nicht einmal einem Minimalkatalog von ärztlichen Fähigkeiten genügen, wenn es den gäbe. Theoretisch hatten wir einen Wust zusammenhangloser Fakten im Kopf, praktisch handwerklich sind wir Dilettanten. Was wir menschlich sind, bleibt dem Zufall überlassen. ... Wie wichtig sind menschliche Qualitäten und wieweit sind sie ausbildbar? Muß ein Arzt eine Persönlichkeit sein? Dieser Aspekt fällt während des Studiums völlig unter den Tisch. Grundsatzprobleme sind nicht gefragt. Bei uns gibt es nur zwei Veranstaltungen, die sich mit elementaren Fragen der Medizin befassen: Übungen zum Arzt-Patient-Verhältnis und ein Seminar zu Fragen medizinischer Ethik, angeboten für alle Semester. Teilnehmerzahl bei letz-

terer Veranstaltung: sechs (an einer Fakultät von 5000 Studenten). ... Am Ende eines langen Medizinstudiums an einer ehrwürdigen Universität muß ich erkennen, daß ich nicht gelernt habe, meine eigenen innerpsychischen Abläufe kritisch zu erkennen. Ich weiß sehr wenig darüber, welche Wirkungen die „Droge Arzt" auf den Krankheitsverlauf hat und wie man erkennen kann, ob Besserungen auf mein Rollenverhalten oder auf spontane Heilungen zurückzuführen sind. ... Angesichts der Zustände versinken viele von uns in Resignation, Frust und Zynismus. Und wir haben Ängste, weil wir spüren, daß wir auf das, was auf uns zukommt, nicht genügend vorbereitet sind."

67.3 Ergebnisse medizinischer Sozialisationsforschung

Orientierungspunkt dieses Abschnittes ist die Frage, wie Medizinstudenten Werte einer personenbezogenen Medizin internalisieren und hierdurch gefördert werden können, sich bestimmte Haltungen oder **Einstellungen** anzueignen. Zunächst werden Beobachtungen referiert, die sich mit der Abwehr gegenüber einer Wahrnehmung zwischenmenschlicher Prozesse und deren Bearbeitung während der Ausbildung auseinandersetzen. Es wird dann die Frage aufgeworfen, ob die resultierenden Einstellungen vorübergehende Haltungen oder eine anhaltende Persönlichkeitsveränderung darstellen. Zur Erläuterung dieses Problems werden Beobachtungen zum sozialen Verhalten von Medizinstudenten und Ärzten einander gegenübergestellt. Dann wird der Frage nachgegangen, wie umschriebene, sozialisierende Kräfte auf die Studenten einwirken, und wie die Einwirkung dieser sozialisierenden Kräfte psychodynamisch zu verstehen ist.

67.3.1 Einstellung als zentraler Begriff

Als zentraler Begriff wird der Begriff der Einstellung gesehen. Hierunter wird verstanden (Rezler, 1972):

„Als Einstellung versteht man eine relativ überdauernde Organisation von Überzeugungen, die sich auf ein Objekt, ein Subjekt oder ein Konzept beziehen. Sie veranlassen dessen Träger zu bevorzugten Handlungsweisen. – Einstellungen werden erlernt und sind anders als bloße Gedanken mit Emotionen verbunden. Eine Einstellung umfaßt drei Komponenten: a) ein kognitives oder Wissenselement; b) ein affektives oder gefühlsmäßiges Element; und c) eine Neigung zum Handeln." Nachfolgend geht es um Einstellungen gegenüber dem Patienten und Einstellungen gegenüber dem Mitarbeiter, auf den der Arzt der heutigen spezialisierten Medizin mehr denn je angewiesen ist.

Hilfsbereitschaft gegenüber dem Patienten, die in Zynismus umschlägt?

In der Mitte der 50er Jahre erschienen in den USA als Ergebnis eines in der Nachkriegszeit stimulierten Interesses an Fragen der Entwicklung zum Arzt eine Reihe heute noch äußerst aufschlußreicher Untersuchungen und Beobachtungen zur Sozialisation des Arztes (Becker, 1958; Merton et al., 1957; Fox, 1957; Ham, 1956). Ihnen gemeinsam ist, daß sie den Medizinstudenten als **Subjekt** ihrer Betrachtungsweise einführen, d. h. nach den persönlichen Reaktionsweisen des Medizinstudenten und dessen Entwicklung fragen.

Unter den damaligen Forschern haben Eron und Mitarbeiter besonderes Interesse gefunden (1955, 1958). Sie beobachteten, daß Medizinstudenten im Umgang mit Patienten und ihren Problemen zu Beginn und zum Abschluß ihres letzten Jahres (4. Jahr der Medical School in den USA) einen Unterschied erkennen ließen, den sie als Entwicklung einer „zynischen" Einstellung bezeichneten. Da Erons Untersuchungsergebnisse für die Sozialisation des Medizinstudenten von zentraler Bedeutung sind, sollen sie hier ausführlicher dargestellt werden (Eron, 1955).

Untersucht wurden drei Einstellungen, die sie als Zynismus, Humanität und Angst bezeichneten, mit Hilfe eigens entwickelter Einstellungsskalen. Items für Zynismus waren z. B.: „Die meisten lügen, wenn es um ihr Wohl geht"; „die meisten schließen aus Aspekten der Nützlichkeit Freundschaft"; „wenn Du nicht für Dich sorgst, keiner wird's sonst machen". Items für Humanität waren: „Wenn ich vom Leiden anderer höre, drängt es mich zu helfen"; „bei der Berufswahl ist die Möglichkeit zum Helfen mit das Wichtigste"; „Todesstrafe sollte wegen ihrer Inhumanität abgeschafft werden"; „bei gegebener Möglichkeit sollte man selbst unter persönlichen Opfern helfen". – Die Hypothese von Eron und Mitarbeitern war, daß Zynismus und Angst zunehmen, Humanität abnimmt. Verglichen wurden die Angehörigen der Eingangs- und der Ausgangsklassen von Yale.

Ergebnisse: Es fand sich eine signifikante Zunahme der Zynismuswerte. Die Werte für Angst und Humanität blieben etwa gleich. – Über diese Ergebnisse hinaus fanden Eron und seine Mitarbeiter, daß sich eine zuvor nicht bestehende Korrelation zwischen den drei Variablen hergestellt hatte: Eine negative Korrelation zwischen Zynismus und Humanität einerseits und eine positive Korrelation zwischen Zynismus und Angst andererseits. Anders formuliert: Zynismuswerte waren signifikant weniger mit Humanitätswerten gekoppelt, wie dies auch zu erwarten ist. Hohe Angstwerte – und das ist aufschlußreich – waren dagegen mit hohen Zynismuswerten gekoppelt. Zwischen Angst- und Humanitätswerten ließen sich keine signifikanten Korrelationen herstellen.

Nach ihren Berufswünschen gefragt, ergaben sich zwischen den Vertretern der Pädiatrie und der Psychiatrie die größten Unterschiede: Studenten, die Pädiater werden wollten, hatten weniger Zynismus- und wesentlich weniger Angstwerte als Studenten, die Psychiater werden wollten. Den Zynismuswerten der Wunsch-Psychiater lagen diejenigen der Wunsch-Chirurgen nahe, während die Angstwerte der ersten die der zweiten bei weitem überstiegen.

Eron und Mitarbeiter faßten die Ergebnisse ihrer Untersuchungen dergestalt zusammen, daß sie mit zunehmender Ausbildungslänge einen Zusammenhang zwischen Zynismus und Angst postulierten, der um so ausgeprägter ist, je stärker die Angst ist. „Die zunehmende Homogenität und Einheitlichkeit der Befunde der im vierten Jahr Studierenden scheint in der Tat das Ergebnis einer Exposition gegenüber derselben Art von Erfahrung zu sein" (1955, S. 564).

Weiterführende Untersuchungen zum Thema „Zynismus" und kritische Stellungnahmen

Diese Untersuchungen lösten lebhafte Diskussionen, zum Teil auch neue Untersuchungen aus. Hierunter findet sich eine der wenigen deutschsprachigen Sozialisationsarbeiten im Medizinbereich. Deren Autoren fanden die Ergebnisse von Eron und Mitarbeitern bestätigt (Beckmann et al., 1972). Sie stellten fest:

„Der durchschnittliche, zum Medizinstudium entschlossene Abiturient glaubt noch, daß er sich eher häufig Sorgen um andere Menschen mache, … aber während des Medizinstudiums verschwinden diese vorsorglichen Empfindungen mehr und mehr, und am Ende stellt sich der Mediziner als jemand dar, der sich so selten wie niemand sonst unter den Studentengruppen sorgenvolle Gedanken um andere Menschen macht."

In einer Übersichtsarbeit der Forschungsergebnisse aus dem nordamerikanischen Bereich zwischen 1970 und 1980 fand Rezler die Grundaussage der Ergebnisse von Eron dem Inhalt nach bestätigt (1982): „Mit fortschreitender Ausbildung bleiben der Bedarf nach Zuwendung zum Patienten niedrig, der Bedarf nach Autonomie, Erfolg und theoretischer Orientierung hoch, Bedarf nach Dominanz, Durchhalten-Wollen verstärkt sich, und es nehmen Spontaneität, spielerisches Vermögen – allerdings auch Dogmatismus und Zwanghaftigkeit – ab" (Burstein, 1980; Degenais et al., 1975; Donovan, 1970; Juan, 1973; Kelley, 1981; Parlow, 1971; Rothmann et al., 1973). Ein neues und liberales Curriculum ermöglicht zwar den Erhalt humanitärer Einstellungen, kann dennoch nicht die Entwicklung zynischer Komponenten verhindern (Weber, 1972). Auch die Teilnahme an einem Programm über Familienmedizin ändert die Entwicklung ebensowenig (Canning et al., 1973) wie die Teilnahme an einem Seminar über Vertretung von Patientenbedürfnissen (Donovan, 1976; Canning et al., 1973). Diese teilweise vor 30 und mehr Jahren gemachten Lehr- und Lernerfahrungen können unverändert auch heute gemacht werden; es wird von einer „mirror like familiarity of 30 year old accounts of medical student life" (Bloom und Speedling, 1989) gesprochen.

Diese Veränderung in eine negative Richtung, und hier besonders bezogen auf Krebskranke, Sterbende, Abhängige, vollzieht sich in umschriebener Weise zu Beginn der klinischen Erfahrungen (Baldigo, 1975; Johnson, 1980; Kutner, 1978). Es findet eine Anpassung ursprünglich liberaler Einstellungen an die konservativen der klinischen Lehrer statt (Coe, 1977). – In dieser trüben Situation läßt sich als einziger Lichtblick ausmachen: Bei einem Vergleich unter Ärzten vier Jahre nach ihrer Approbation finden sich bei Berufsangehörigen mit sog. „hoher Interaktion" (z. B. Familienärzte) deutlich geringfügigere Anzeichen von Zynismus als bei Berufsangehörigen, die eine „niedrige Interaktionsrate" (z. B. Chirurgen) aufweisen

(Reinhardt und Gray, 1972). Lediglich aus Großbritannien wird ein deutlich zu beobachtender Wandel in Richtung vermehrter Offenheit der Abgänger britischer Medical Schools berichtet; er wird auf den Einfluß der psychosozialen Fächer und den der Allgemeinmedizin zurückgeführt (Crisp, 1984).

Das Verhalten zum Mitstudenten und das Einüben von Kollegialität

Hier formuliert Rezler (1982):

„Die Mehrheit der Studenten ist nicht bereit, mit anderen Studenten aus den Bereichen der Pflege, verwandter Gesundheitszweige, Sozialarbeit zusammenzuarbeiten. Infolgedessen sind sie auch zu Beginn ihrer Praxis nicht darauf vorbereitet, mit Angehörigen des Gesundheitswesens zusammenzuarbeiten." – Weiterhin:

„Autoritäre, dogmatische und zynische Studenten finden nicht zur Familienmedizin" (Canning, 1973). Vertreter der Familienmedizin beurteilen Werte wie Freundschaft, Hilfsbereitschaft höher als internistische und chirurgische Assistenzärzte. Allerdings fand sich kein Unterschied im Dogmatismus (Canning, 1973). Vergleichbare Ergebnisse fanden andere Autoren (Collings, 1975; Holzmann, 1979; Plovnik, 1979; West, 1982).

Speirer und Weidelt befragten 491 deutsche Medizinstudenten und -studentinnen im Zeitraum Sommersemester 1981 bis Wintersemester 1982/83 nach ihrem Selbstbild und 172 nach ihrem Fremdbild, d. h. dem Bild der Kommilitonen (1984). Hochsignifikant (p = 0,001, t-Test) wurden die Mitstudenten in abnehmender Größenordnung als weniger vertrauenswürdig, weniger feinfühlig, weniger verständnisvoll, weniger freundlich, weniger demokratisch, weniger zuverlässig, weniger selbstlos, weniger sympathisch, weniger fähig, weniger aktiv, weniger offen, weniger fortschrittlich wahrgenommen. Sie wurden als reicher und selbstbewußter eingeschätzt, auf gleicher Stufe stehend hinsichtlich der Eigenschaften gründlich, machtlos, idealistisch und sicher. Speirer und Weidelt kommentierten ihre Ergebnisse als „Hinweise auf eine positiv-idealisierende Tendenz der Selbsteinschätzung" und fanden, daß „schon die Medizinstudenten ihre Kommilitonen als wenig solidarisch, kooperationsbereit und menschlich attraktiv wahrnehmen, daß sie einander mißtrauen und miteinander wenig freundlich umgehen". Sie nehmen als Ursachen ein Überlegenheitsgefühl bzw. Überlegenheitsbedürfnis an und die Tendenz, eigene Schwächen und Probleme vor anderen zu verbergen und sie bevorzugt projektiv bei anderen wahrzunehmen.

In eigenen Untersuchungen im Rahmen der Arbeit mit Anamnesegruppen forderten wir Studenten auf, 25 Beziehungen aus dem professionellen, dem studentischen und dem privaten Bereich auf 25 Eigenschaften hin zu beurteilen (Schüffel et al., 1977). Als Methode setzten wir den repertory-grid-Test (G. Kelly, 1955) ein. Wir erhielten eine Matrix von Einschätzungswerten, die faktorenanalytisch (Slater, 1977) untersucht wurden. Es ergaben sich drei Hauptkom-

ponenten, die ca. 80% der beobachteten Varianz aufklärten (Egle, 1982). Bei der ersten Hauptkomponente (knapp 50% der Varianz) handelte es sich um die Dimension Ferne/Nähe, bei der Hauptkomponente II (ca. 20%) um die Inhalte Helfen/Ablehnen und bei der Hauptkomponente III (ca. 10%) um die Dimension Bindung/Bindungslosigkeit. Bei einem Vergleich der Hauptkomponenten bzw. Faktoren und der Ladungen, welche die einzelnen Beziehungen aufwiesen, ergaben sich für Studienanfänger und höhersemestrige Studenten unterschiedliche Einschätzungen.

Studienanfänger wollten eine umfassende Hilfe leisten. Sie strebten eine Nähe zum Patienten an, die der Nähe zum Partner oder Freund gleich war oder diese sogar übertraf. In ihrer eigenen Beziehung zum Patienten grenzten sie sich deutlich von der real beobachteten Arzt-Patient-Beziehung ab, die sie geradezu als angst- und schreckenerregend ablehnten.

Höhersemestrige Studenten (4. klinisches Semester) hatten das Gefühl, dem Patienten nicht ausreichend zur Seite zu stehen. Sie spürten aber auch Abneigung gegenüber dem Patienten und fühlten sich entmutigt und eingeschüchtert. Sie meinten, sich bestimmend gegenüber dem Patienten verhalten zu müssen und dieses Verhalten auch ausbauen zu sollen. Erst in zweiter Linie kam es ihnen auf ein offeneres und einfühlsameres Verhalten an. Gleichzeitig spürten sie ein hohes Maß an Verpflichtung gegenüber dem Patienten. – Kontrastierend zu den Anfängern beruhte ihre Einschätzung der idealen Arzt-Patient-Beziehung weniger auf Vergleichen mit dem Privatbereich als auf solchen mit real beobachteten Arzt-Patient-Beziehungen. In einer widersprüchlichen Weise sahen sie diese realen Beziehungen weiterhin als wenig erstrebenswert an. Somit befanden sie sich in einem inneren Konflikt: Einerseits suchten sie Vorbilder, also Rollenträger, um deren Verhalten zu übernehmen; andererseits lehnten sie diese Vorbilder ab (Schüffel et al., 1983).

Von den 25 Beziehungen wiesen die reziproken Beziehungspaare Arzt/Patient und Patient/Arzt die beiden größten Varianzanteile auf (Schüffel et al., 1977). Man konnte auch sagen, daß der Student diese beiden Beziehungen als die kritischsten ansah. Wir interpretierten diese Beobachtung derart, daß wir sagten, dem Studenten gehe es um gefühlsmäßige Zustände, die in der Realität einer medizinischen Fakultät praktisch nicht zum Gegenstand einer bewußt herbeigeführten Auseinandersetzung werden. Sie sind aber vorhanden, werden konflikthaft erlebt und beanspruchen dementsprechend die Aufmerksamkeit des Studenten. – Als wichtigste Eigenschaften bzw. Konstrukte der vom Studenten beobachteten Beziehungen fanden wir (Schneider, 1982): sich einfühlen können, sich hingezogen fühlen, Gefühle äußern, sich dankbar fühlen, sich als Hilfe sehen, eingeschüchtert sein, verärgert sein, entmutigt werden, abhängig sein, bestimmend sein, benötigt werden, sich verpflichtet fühlen.

Die Schutzbedürftigkeit des angehenden Arztes

Die Untersuchungsergebnisse zum Umgang mit dem Patienten und mit dem Kommilitonen bzw. dem Arzt stimmen nachdenklich. Statt sie verurteilend oder moralisierend zu werten, müssen die Phänomene als Ergebnisse langfristig ablaufender Prozesse gesehen und analysiert werden.

Besonders der Begriff des Zynismus wird erst im Rahmen der ablaufenden Sozialisationsprozesse, d. h. als ein Schutzmechanismus verständlich.

Das englische Wort cynicism oder cynic wird im Oxford Dictionary folgendermaßen definiert (1976): „Ancient Greek philosopher of sect founded by Antisthenes, marked by ostentatious contempt for ease and pleasure; one who sarcastically doubts human sincerity and merit". Für den Begriff cynical heißt es: „like a cynic, incredulous of human goodness; sneeking". „Sneeking" ist mit „höhnisch, spöttisch" zu übersetzen (Cassel's, 1978).

Wir müssen fragen, worauf sich das Urteil „Zynismus" und „zynisches Verhalten" bezieht. Wenn wir es in Zusammenhang mit den ebenfalls durchgehend festgehaltenen Beobachtungen bringen, daß Medizinstudenten helfen wollen, ja geradezu von einem „Helfersyndrom" gesprochen wurde (Schmidtbauer, 1977), so bietet sich die Interpretation an, daß junge Medizinstudenten angesichts kaum vorstellbaren Leidens vor ihren eigenen Hilfsimpulsen Schutz suchen.

Vorübergehende Anpassung oder anhaltende Panzerung?

Es besteht Unsicherheit, inwieweit die geschilderten Einstellungsveränderungen überdauernder oder vorübergehender Natur sind (Bloom, 1980). So wird sowohl die Meinung geäußert, es handele sich um eine Anpassung an die Ausbildungssituation, wie die, es handele sich um eine charakterliche Entwicklung im Sinne einer Panzerung gegen Gefühle.

Überdauernde Veränderung

Eine beträchtliche Gruppe unter den Sozialisationsforschern sieht in diesen Ergebnissen den Niederschlag überdauernder Veränderungen. Sie interpretieren die Befunde als eine Entwicklung weg vom Idealismus des Anfängers und dessen Bestreben zum Helfen, hin zu einem distanzierenden Behandeln, das von der Person des Patienten absieht und überindividuell festgelegte Eingriffe entsprechend dem Kanon medizinischer Diagnostik und Therapie vornimmt (Eron, 1955, 1957; Christie und Merton, 1958; Gordon und Mensh, 1962). – Begleitend hierzu wurde ein kognitiv definierter Entwicklungsprozeß beschrieben (Fox, 1957), der als „detached concern" bezeichnet wurde. Die hierzu führende Lernmethode sei ein „training for uncertainty". Mit „detached concern" ist gemeint, daß vom Arzt ein sorgfältiges Abwägen der zur Verfügung stehenden Mittel und ein Absehen vom eigenen Gefühl erwartet wird. Mit der formulierten Ungewißheit ist gemeint: Ungewißheit hinsichtlich des eigenen, persönlichen Wissens und Ungewißheit hinsichtlich des allgemeinen medizinischen Wissens, das sich in einer ständigen Fortentwicklung befindet.

Vorübergehende Veränderung

Mit der Formulierung „vorübergehender Natur", die von einer anderen Gruppe Sozialisationsforscher vertreten wird, ist gemeint, daß der Student zunächst wenig ausdifferenzierte Erwartungen von der Rolle des Arztes hat bzw. sie in Verhaltensweisen umsetzen kann (Becker und Geer, 1958; Nathamson, 1958). Der angehende Arzt wird sich, so die Hypothese, dem Dozenten-Arzt zunächst anpassen und erst allmählich, gewöhnlich nach dem Abschluß des Studiums, in seinen eigenen Stil hineinfinden. Das einstellungsmäßige Lernen wird hier als ein voranschreitender Reifungs- bzw. Anpassungsprozeß und nicht wie im Falle der ersten Interpretation als ein Prägungsprozeß interpretiert.

Belastung und Belastbarkeit entscheiden; wie das Studium vorübergehend und dauernd prägt

Die Kontroverse setzt sich über drei Jahrzehnte Sozialisationsforschung fort und ist auch bei der Diskussion der Ergebnisse zu verfolgen, die Shuval aufgrund ihrer Beobachtungen an Medizinstudenten vom ersten bis sechsten Jahrgang der Universitäten von Tel-Aviv und Jerusalem vorlegte (1980). Sie stellte als eines ihrer wesentlichen Ergebnisse fest, daß mit zunehmender Studiendauer der Konsensus der Studenten hinsichtlich personen-, status- und wissenschaftsbezogener Rollenkomponenten des Arztes abnimmt. Ein Tiefpunkt wird mit dem Abschluß des Studiums erreicht. Sie interpretiert ihre Ergebnisse:

„Über die Zeit entwickelt sich eine Vielzahl von Einstellungen, was möglicherweise das Ergebnis unterschiedlicher Reaktionsweisen auf unbeständige Modelle oder auch Ausdruck von wachsender Reife und Individualismus ist. Sozialisation führt also offenbar nicht zu einem einzigen, klar umschriebenen und strukturierten Modell normativer Praxis, sondern vielmehr zur Einschätzung des Berufes als eines Ortes alternativer Praxisstile. Ein derartiges Muster läßt vermuten, daß die formale Sozialisation in eine gewisse **Offenheit** gegenüber Möglichkeiten der Innovation in der Praxis einmündet; denn alternative Ansichten werden von verschiedenen Mitgliedern der Gruppe (gemeint ist der jeweilige Jahrgang; Anmerkung des Autors) vertreten, und es gibt keine Vorherrschaft eines einzigen normativen Praxisstiles. *Andererseits können wir nicht die Möglichkeit außer acht lassen, und das trifft besonders für den Zeitpunkt vor Beginn der Praxis zu, daß niedrige Übereinstimmung eine unstrukturierte Konfusion hinsichtlich professioneller Normen widerspiegelt"* (Hervorhebung durch Autor).

Eigene Erfahrungen, die aus der Unterrichtszeit zwischen 1969 und 1989 an zwei Universitäten (Ulm/Donau und Marburg/Lahn), vorzugsweise aus dem sog. Pflichtunterricht und dort aus Gruppenarbeit im 6. und 8. Semester stammen, zwingen zur Zurückhaltung bei der Interpretation der Shuvalschen Forschungsergebnisse im Sinne der von ihr beschriebenen Offenheit. Wir fanden, daß in der Tat mit wachsender Semesterzahl der Konsensus unter den Studenten abnahm. Dafür beobachteten wir häufig eine ausgeprägte Gruppenbildung, die in Extremfällen Cliquencharakter hatte. Diese Cliquen zeichneten sich im wesentlichen dadurch aus, daß in der einen Hauptgruppierung die ganzheitlich, in der anderen Hauptgruppierung die spezialistisch orientierten Studenten zu finden waren. Beide Gruppierungen hatten oft kaum Kontakt miteinander. Wurden beide in einer Kleingruppe zusammengebracht, so lehnten sie einander häufig in einer heftigen, mitunter rigiden Weise ab. So konnten wir beobachten, daß sich die Mitglieder derartiger Subgruppen zu Beginn nicht mit Namen bekannt machen wollten, die Untergruppen sich im Block zusammenscharten und sich gegenüber der Gesamtgruppe abschirmten, sich nicht auf eine gemeinsame Patientenbetreuung einigen wollten usw. Wir interpretierten die Phänomene damit, daß die Studenten durch diese Gruppenbildung mühsam eine Identität aufrechtzuerhalten suchten.

In abgeschwächter Form stellten wir verwandte Phänomene auch in Anamnesegruppen fest. Dort ließen sich aufgabenzentrierte von selbsterfahrungsorientierten Mitgliedern unterscheiden (Schneider und Wenzel, 1983). Wir nannten sie, einem Vorschlag Schneiders folgend, etwas salopp „Techniker" und „Emotionalisten" (Schüffel et al., 1983).

Im Gegensatz zu Shuval fanden wir, daß die Verhaltensweisen der einzelnen Studenten recht starr waren. Informelle Befragungen zeigten uns, daß die Kleingruppenbildung ein langdauerndes, semesterüberdauerndes Geschehen darstellt. Respektierten die Dozenten die spontane Gruppenbildung, so erleichterte das die Aufgabe, den Mitgliedern der verschiedenen Untergruppierungen die Inhalte des Psychosomatikpraktikums nahezubringen.

Berücksichtigt man diese Überlegungen, dann könnten Shuvals Beobachtungen auf die zitierten vorausgehenden Diskussionen „passager versus überdauernd" zurückgeführt und interpretiert werden: Vom Studenten werden, so die hier vertretene Hypothese, persönliche Grundwerte mitgebracht. Sie entstammen der primären Sozialisation. Befragt man junge Medizinstudenten nach Einstellungen gegenüber bestimmten medizinischen Fragen, so wird sich zunächst kein größerer Dissenz finden, da diese Werte noch nicht auf medizinische Bereiche angewendet wurden. Erst im Laufe des Studiums ergibt sich aus einer Art „psychologischen Entgegenkommens" eine berufsbezogene Ausprägung der mitgebrachten Werte und hierdurch entsteht zum Studienende der erwähnte Dissenz.

Beispiel: Medizinstudenten werden heute nahezu einhellig sagen, daß es die Pflicht des Arztes sei, die Persönlichkeit des Patienten zu berücksichtigen. Hierbei müsse man die eigene persönliche Reaktionsweise in seine Vorgehensweise als Arzt einbeziehen. Diese Äußerung würde gleichermaßen von jungen Medizinstudenten bestätigt, die aus einem aufgeschlossenen und Unsicherheit ertragenden Elternhaus kommen, in dem gleichzeitig effektive Bewältigungsstrategien für psychosoziale Probleme vermittelt wurden, wie auch von solchen Studenten, die aus

autoritären und angstbestimmten Elternhäusern stammen. Im Laufe des Studiums werden sich aber die Vertreter beider Gruppierungen unterschiedlich entwickeln. Die eine Gruppe wird sich mehr durch Offenheit, die andere mehr durch Dominanzverhalten und Rigidität auszeichnen.

Im Rahmen einer solchen Hypothese wird auch wahrscheinlicher, daß die zweite von Shuval aufgegriffene Interpretation[2] zutrifft: Es findet sich unter den Abgängern medizinischer Schulen eine höhere Heterogenität der Einstellungen; diese sind aber, anders als entsprechend der ersten Interpretation von ihr vermutet, sehr viel fester ausgeprägt und wahrscheinlich durch Persönlichkeitseigenschaften mitbestimmt, wie sie u.a. von Eron beschrieben wurden. Hier ließen sich zwanglos Erons Beobachtungen einfügen, daß Zynismus- und Angstwerte in signifikanter Weise positiv miteinander korrelieren.

Die eingangs gestellte Frage, ob der beobachtete Entwicklungsprozeß eine anhaltende Panzerung oder eine vorübergehende Anpassung darstelle, ist wahrscheinlich eine Scheinfrage. Vielmehr ist anzunehmen, daß es je nach Belastungsfähigkeit des einzelnen Studenten zu vorübergehenden oder überdauernden Persönlichkeitsentwicklungen während des Medizinstudiums kommt. Es wird wesentlich von der Art der Belastung und der Art der angebotenen Bewältigungsmittel abhängen, auf welche Reaktionsweisen und schließlich Einstellungen der Student und beginnende Arzt zurückgreifen wird. – Eine ähnliche Auffassungsweise von der Beeinflußbarkeit ärztlicher Einstellungen vertrat M. Balint, der annahm, daß praktische Ärzte zunächst Verhaltensweisen verlernen müssen, ehe sie neue Verhaltensweisen einüben können (1957).

Andererseits finden sich eindrucksvolle Hinweise darauf, daß vielfach nicht „verlernt", sondern in extremer Weise Gelerntes vertieft wird. Dieser Vorgang läßt sich insbesondere bei jungen Ärzten beobachten, die gezwungen sind, die bisherige Studentenrolle aufzugeben und die Arztrolle zu übernehmen. Gerade in diesen kritischen Jahren sind zwei grundlegende Beobachtungen zu machen, die bezeichnet werden könnten als:

67.3.2 Zwei Seiten derselben Medaille: „Strong Service Ethic" und die Häufung von Psychopathologie unter Ärzten

Die Zahl psychosomatischer Zusammenbrüche unter Studenten und Ärzten scheint erheblich zu sein. So wird in den USA angenommen, daß Medizinstudenten zumindest in 10–25% auf ärztliche oder psychotherapeutische Hilfe zurückgreifen müssen (Nadelson et al., 1983). Sie sehen ihre Umgebung als „fordernd und bedrohlich", sehen sich in einem „Krieg mit der Fakultät". Die Probleme werden als Tren-

---

2 Sie spricht von „unstrukturierter Konfusion", wahrscheinlich ein statistisches Artefakt, das aus der Nichtbeachtung der Subgruppen herrührt.

nungsängste, homosexuelle Ängste, Depressionen, Depersonalisation und Versagensängste beschrieben (Nadelson et al., 1983, S. 75 ff.). Die Dunkelziffern der Dekompensationen bzw. Erkrankungen sind ausgesprochen groß, da anscheinend informelle Hilfsmöglichkeiten für diese Population zur Verfügung stehen.

Wie sieht die Situation nach der Approbation aus? Wiederum nordamerikanischen Erhebungen zufolge, die in Ermangelung deutscher Erhebungen herangezogen werden müssen, findet sich unter Ärzten ein hohes Maß an Psychopathologie. Die quantitativen wie qualitativen Hauptprobleme sind: Medikamenten- und Alkoholabhängigkeit, Depressionen und Eheprobleme (Langsley, 1983). Etwa 7–8% aller nordamerikanischen Ärzte sind Alkoholiker oder werden sich hierzu entwickeln; die Suizidrate ist unter Ärzten (und Sozialarbeitern) zweimal höher als in der Normalbevölkerung. In der Gruppe der 25- bis 39jährigen stellen Suizide 26% aller Todesfälle dar. Man hat in den USA davon gesprochen, daß ein ganzer Jahrgang einer Medical School des Landes rein rechnerisch dafür eingesetzt werden muß, die jährlich durch Suizid aus dem Leben geschiedenen Ärzte zu ersetzen. Die Suchtrate ist 30–100mal höher als in der Normalbevölkerung. 15% aller bekannten Süchtigen sind Ärzte, von denen 9% wiederum Suizid begehen.

Immer wieder wird von Medizinstudenten als attraktivster Aspekt der Medizin gesehen, anderen helfen zu können. Dies wird explizit ausgeführt (Ewan et al., 1981; Beckmann et al., 1972) bzw. ist implizit in den Berichten enthalten (Eron, 1955, 1958; Shuval, 1980). Ausdrücklich wird von einer „multipotentiality" gesprochen (Ewan, 1981). Egle beobachtete mit Hilfe eines Einschätzverfahrens und faktorenanalytischer Berechnungen ein extremes Ausmaß des Bedürfnisses unter jungen Studenten, Patienten beizustehen (1980). Hierbei wollen sie, wie bereits oben ausgeführt, in ihrer Beziehung zum Patienten eine Nähe entwickeln, die selbst die Nähe zum persönlichen Freund übertrifft.

Auf die Erwartungen der Studenten hinsichtlich ihrer eigenen Person oder hinsichtlich ihrer Kommilitonen wurde bereits oben eingegangen. Sie sind zusammenfassend so zu interpretieren, daß Studenten sich sehr häufig als Einzelgänger und/oder ohne Schutz fühlen. Gleichzeitig neigen sie, das sei nochmals hervorgehoben, zu einer idealisierenden Selbsteinschätzung, die als Schutz für das bedrohte positive Selbstwerterleben gedeutet werden kann.

Wie der Schlüssel zum Schlüsselloch paßt die Beobachtung, daß seitens der Gesellschaft von Medizinstudenten überdurchschnittliche Leistungen erwartet werden. Hier, wie insgesamt in der westlichen Welt, gehören Medizinstudenten zu einer gesellschaftlichen Führungsgruppe. Bei weitem überwiegt die Nachfrage das Angebot vorhandener Studienplätze. In der Folge findet ein rigoroses Filterverfahren statt, das in der letzten Zeit in der BRD zu einer extremen Anforderung führte: Nur Schüler mit „Superzensuren" zwischen 1,0 und 1,2 im Notendurchschnitt erhielten den Zugang. Andererseits gab es die „Warter", die mit

Berufserfahrung oder als „reine Warter" auf die Hochschule kamen. Heute, d.h. nach der Einführung eines kombinierten Verfahrens mit Test, durchschnittlicher Abiturnote und Los, wird die Lage noch komplizierter. Häufiges Thema unter Studenten ist die Frage: „Wie hast Du den Platz bekommen?".

Interpretiert man die überdurchschnittlichen Dekompensationen unter Studenten und Ärzten, so ist die wahrscheinlichste Erklärung, daß sie einerseits hohe Anforderungen an sich stellen, bzw. diese auch realiter da sind, und daß sie sich gleichzeitig überfordert und ohne zwischenmenschlichen Rückhalt fühlen. Besonders gravierend scheint die Rolle für Studentinnen zu werden. Nicht selten hören sie sinngemäß Äußerungen wie die eines Standesvertreters zu einem männlichen Medizinstudenten: „Sie als Medizinstudent sollten etwas Gutes tun und eine Medizinstudentin heiraten; dann hätten Sie zwei Plätze versorgt". – Eine Studentin sagt: „Im Studium trifft man fast so viele Frauen wie Männer; nachher im Krankenhaus merkt man, daß es plötzlich viel weniger Frauen gibt, daß es anders geworden ist."

Der Sozialisationsprozeß der Medizinstudenten und jungen Ärzte vollzieht sich offensichtlich unter größtem Druck. Die resultierenden Verhaltensweisen schwanken dann häufig zwischen zwei Extremen. Auf der einen Seite sind es höchste Anforderungen an das Leistungsvermögen des einzelnen, die sich in einer im Angelsächsischen so bezeichneten „strong service ethic" ausdrücken; sehr häufig kommt es aber zur Dekompensation, und die Kehrseite dieser Medaille sind dann schwerste pathologische Zustände und psychosomatische Krankheitsbilder. Eindrucksvoll, in Art eines Tagebuches, beschreibt ein junger, soeben graduierter Arzt seine Erlebnisse im „second best hospital" der USA und führt dem Leser das ungeheure Ausmaß psychischer (und physischer) Belastung vor Augen, dem der junge Arzt ausgesetzt ist und die hier einstellungsprägend ist (Shem, 1980).

67.3.3 Sozialisierende Kräfte: Patient, Arzt, Student

Um die beobachteten Phänomene der Sozialisation zu erklären, untersucht die soziologisch orientierte Sozialforschung sog. sozialisierende Kräfte (socializing agents). Entscheidende sozialisierende Kräfte sind für den Medizinstudenten der Patient, der Arzt-Dozent, der Kommilitone und das Krankenhauspersonal.

Von ihnen wird angenommen, daß sie eine Entwicklung des Studenten beeinflussen, die sich durch die Bereitschaft ausdrückt, entweder stärker zur Ausübung der studentischen oder stärker zur Ausübung der ärztlichen Rolle zu tendieren. Je stärker die Bereitschaft der Fakultät, so die Annahme, dem Studenten Verantwortung zuzutrauen, um so mehr wird dieser zur Übernahme der Arztrolle tendieren. Je stärker aber inaktivierende Tendenzen in der Fakultät zu finden sind, um so mehr wird der Student in einer studentischen Rolle verharren. Nachfolgend soll auf die ersten drei der vier sozialisierenden Kräfte eingegangen werden. Auf eine Diskussion der Rolle des Personals müssen wir bedauerlicherweise wegen der Klinikferne des deutschen Unterrichtes im Vergleich zum angelsächsischen verzichten.

Der überfordernde Patient

Den Studenten geht es immer wieder darum, Patienten umfassend helfen zu können. Dieser Anspruch auf Hilfeleistung wird zunächst in einer Weise empfunden, die als nahezu total zu bezeichnen ist.

Interessante Einblicke in das Ausmaß dieser Ansprüche lassen Erfahrungen aus Zusammentreffen mit Patienten zu, bei denen derartige Versorgungsansprüche in der Tat ausgeprägt sind, wie z. B. Colitisulcerosa-Patienten mit ihren symbiotischen Bedürfnissen. Hier kann geradezu paradigmatisch die Reaktion der Studenten studiert werden, wie auch in studentischen Selbsterfahrungsberichten nachzulesen ist (Bregulla et al., 1982).

> Eine 35jährige Kolitispatientin beschreibt ihre Lebensgeschichte, die durch umschriebene traumatische Trennungserlebnisse gekennzeichnet ist. Zunächst hat sie ihre ältere, als Mutterersatz erlebte Schwester durch einen Tumor verloren, dann den als Mutterersatz empfundenen Ehemann durch Scheidung. Ihre Situation in der Ursprungsfamilie beschreibt sie als wenig haltgebend und gleichzeitig in einer verwirrenden Weise, in der sie keine Hinweise auf Grenzen zwischen einzelnen Mitgliedern und Generationen der Familie zu finden scheint. Die Studenten spüren sowohl die Isolation der Patientin wie ihren Versuch, in den nun vor ihr liegenden Lebensabschnitten allmählich tragende familiäre Strukturen erwachsen zu lassen. Sie spüren auch die symbiotischen Anlehnungsbedürfnisse der Patientin. – Die Studenten sind zunächst betroffen, dann sind sie verwirrt und schließlich rebellieren sie. Sie geben an, nicht zu wissen, wo sie in der Patientengeschichte aufhören sollen. Infolgedessen wollten sie auch gar nicht anfangen. Sie schließen: Diese Patientin sei genauso wie andere Patienten in ihrer eigenen „Box"; sie selbst, d. h. die Studenten, seien in ihrer „Box". Die studentische Tutorin der Praktikumsgruppe faßt die Reaktion der Kommilitonen treffend zusammen: „Sie demonstrieren **ihre** totale Verweigerung, die der totalen Beanspruchung seitens der Patientin gegenübersteht".

Diese Beobachtung wurde im Gruppenunterricht des Psychosomatikpraktikums gemacht. Abgesehen von diesem Unterricht bleiben derartige Prozesse unbemerkt, von einer Bearbeitung ganz zu schweigen. Die Folge ist, daß Studenten mit wachsender Semesterzahl psychosoziale Prozesse, insbesondere intrapsychische Beobachtungen zurückdrängen oder ausklammern und sich auf technisch zu bewältigende Abläufe konzentrieren. Dies manifestiert sich u. a. in einer technologisch orientierten Sprache, die mit wachsender Semesterzahl zunimmt. Entsprechende Phänomene machen sich bereits in der kurzen Zeit zwischen Anfang und Ende des Klinischen Untersu-

chungskurses im 6. Semester bemerkbar. Während am Anfang die Situation des Patienten etwa folgendermaßen geschildert wird: „Der Patient schwitzte und er erschien mir ängstlich; er preßte seine Hand auf die linke Brust und wies auf Schmerzen in dieser Region hin, die in den linken Arm ausstrahlten", so kann es bereits 8–10 Wochen später heißen: „Der Patient war kollapsig, klagte über substernale Schmerzen; im EKG fand sich eine ST-Senkung".

Andererseits ist es möglich, mit den Studenten in eine Diskussion darüber zu treten, daß jeder nur teilweise in seiner „Box" ist. Wiederum am Beispiel eines Kolitispatienten, der im Klinischen Untersuchungskurs gesehen wurde:

> Der 56jährige, doppelberuflich als Landwirt und Tiefbauarbeiter tätige Mann klagt über schmerzlose Blutungen, die zu einem Hb-Abfall auf 6 g% geführt haben. Der Patient berichtet, wie von ihm erwartet wird, daß er den Hof für die Familie erhält, während gleichzeitig die erwachsenen Kinder hinausgehen und offensichtlich das alternde Elternpaar alleine zurückbleibt. Der Patient möchte, daß Transfusionen und Medikamente gegeben werden. – Die Studenten stimmen ihm sofort zu. Sie überlegen nicht, daß dies über die Jahre wiederholt geschehen ist, keine Änderung eintrat und auch jetzt der Patient anscheinend keinerlei körperliche Symptome wahrnahm. Der extreme Hb-Abfall auf 6 g% wird nicht registriert. Erst allmählich erkennen sie ihre Betroffenheit und sie rebellieren zugleich: Man könne ja doch nichts machen. Dann läßt sich allmählich erarbeiten, wie eine eventuell langfristige Behandlungsstrategie aussieht, die ganz wesentlich vom Hausarzt getragen wird. Es gelingt, das „Boxen-Gefühl" zu erkennen, auch als **Notwendigkeit** anzuerkennen. Zwar geht von dem Patienten ein Drang zum Helfen, geradezu zum körperlichen Verschmelzen aus. Einzelne Studenten tragen zusammen, wie sie den Drang, aber auch das Abgrenzen beobachtet haben. Beide Positionen werden zunächst mit Absolutheit vorgetragen; beide können sie als vom Patienten ausgehend schließlich akzeptiert werden. Bis diese Einsicht gelingt, sind freilich sowohl für den Dozenten wie für die Studenten erhebliche Aggressionen durchzustehen, die an Hostilität grenzen.

Mit Hilfe der erwähnten Grid-Untersuchungen konnte in unserer Arbeitsgruppe nachgewiesen werden, daß insbesondere die höhersemestrigen Studenten den Patienten in der Regel überaus ambivalent einschätzen (Egle, 1982). Sie verspüren dem Patienten gegenüber geradezu Abneigung, sie haben das Gefühl, sich nicht genügend für ihn einzusetzen, fühlen sich entmutigt und auch eingeschüchtert. Gleichzeitig streben sie an, ihrer Verpflichtung gegenüber dem Patienten in einem stärkeren Maße als bisher nachzukommen. Kurz: Der Patient wird als überfordernd empfunden.

Der distanzierende und abgelehnte Arzt-Dozent

In der beschriebenen Situation des Kolitispatienten, ebenso wie in vielfältigen anderen patienten- und kliniknah geführten Diskussionen, wird vom ärztlichen Dozenten erwartet, daß er rezeptartig Hilfen verschreibt. In der Tat ist ja die Klinik, aber auch die Praxis an derartigen Rezeptverschreibungen orientiert.

Die Studenten durchschauen aber sehr schnell, daß derartige Rezepturen nur in einem marginalen Bereich der multimorbiden Krankheitssituation helfen. Sie beobachten die krankheitsauslösenden und -unterhaltenden psychosozialen Faktoren und sehen gleichzeitig, daß auf diese nur peripher eingegangen wird. Sie äußern entsprechende Kritik oder behalten sie in Vorwegnahme einer Zurückweisung für sich. Es kommt in der Folge zu einer Störung der Kommunikation zwischen Studenten und Fakultät, besonders ausgeprägt im klinischen Bereich. Bloom beschreibt in seinen minutiösen Darstellungen eindrucksvoll, wie in einer nordamerikanischen Medical Faculty, die stellvertretend für Medical Schools allgemein gesehen wird, heftige und teilweise irrationale, aggressiv-hostile Spannungen zwischen Studentenschaft und Dozentenschaft bestehen (1973).

In unseren eigenen Untersuchungen fanden wir, daß Studenten verschiedenster Semester den Arzt fast ausnahmslos in einer derart negativen Form einschätzten, daß sich geradezu der Ausdruck des **Antimodelles** ärztlicher Tätigkeit aufdrängte (Schüffel et al., 1979). Wir hatten bei der Darstellung dieser Ergebnisse die Formulierung eines „training in anger" geprägt, um hiermit zweierlei auszudrücken: Die Ausbildung vollzieht sich in einem Feld des Ärgers, und sie vermittelt gleichzeitig Fähigkeiten zu mannigfaltiger, zum Teil sublimer Entäußerung von Ärger.

Bloom weist darauf hin, daß der Student widersprüchliche Signale seitens der Fakultät zu dem aufnimmt, was von ihm erwartet wird. Es resultiert hieraus Unsicherheit und nachfolgend Angst; der einzelne Student verhält sich nach dem Motto „bloß überleben" (don't make waves). Dabei werden Dozenten als uninteressiert am Schicksal des Studenten wahrgenommen. Selbst bei den für deutsche Verhältnisse geringen Studentenzahlen einer amerikanischen Fakultät klagen die Studienabgänger darüber, in der Fakultät unbekannt geblieben zu sein. Wachsend werden die Fakultätsvertreter als Vertreter einer Mühle, eines „tough grind" empfunden; die negativen Gefühle ihnen gegenüber erreichen den Höhepunkt vor dem Schlußexamen. Während die vorklinischen Fakultätsmitglieder die Studenten als zu klagend, zu ängstlich und zu angespannt, zu unsauber und unhöflich empfinden, beurteilen die klinischen Fakultätsmitglieder die Studenten in erster Linie als unreif, bar jedes Idealismus, aber auch als nicht (!) zynisch. Bloom faßt zusammen (1973, S. 136):

- Die Fakultät findet, daß ihre Vorstellungen und diejenigen der Studenten unvereinbar sind („basically opposed").
- Studentenwerte sieht die Fakultät als durch „praktische" Motive bestimmt, während sie ihre eigenen als „humanistisch-akademisch" beurteilt.
- Vollzeitfakultätsmitglieder, klinisch wie vorklinisch, nehmen die schärfsten Beurteilungen der

Studenten vor; der weitere Kreis der klinischen Fakultät[3] neigt eher dazu, grundlegende Wertvorstellungen und Motive mit den Studenten gemeinsam zu haben.

Der zurückweisende und dennoch geschätzte Kommilitone: peer learning

Wir müssen fragen, wie unter derart schwierigen Ausbildungsbedingungen zumindest ein Teil der Studentenschaft Wertvorstellungen im Sinne einer personenbezogenen Medizin erhält und weiterentwickelt.

Es scheint, daß dem peer learning, also dem Lernen durch und mit Gleichgestellten, im Bereich des einstellungsmäßigen Lernens eine große Bedeutung zukommt. Daher zunächst einige Anmerkungen zu dieser Form des Lernens im Medizinstudium. G. L. Miller, einer der prominentesten nordamerikanischen Sozialisationsforscher, weist darauf hin, daß die Möglichkeiten des peer learning innerhalb der medizinischen Ausbildung nicht im mindesten berücksichtigt sind (1973). Dem stimmt Shuval sinngemäß zu und entwickelt weiterführende Gedanken (1980). Sie weist auf die in diesem Abschnitt eingangs erwähnten vier sozialisierenden Kräfte (agents) hin, nämlich Patient, Arzt, Kommilitone und Krankenhauspersonal; in der englischen Sprache mit vier P's griffig ausgedrückt: Patient, Physician, Peer, Personnel. Ihrer Ansicht nach sind die Einflußmöglichkeiten des peer und der peer group möglicherweise die nachhaltigsten:

„Die Rolle des peer ist ebenso bedeutsam oder, in mancherlei Hinsicht, bedeutsamer als die des offiziellen ärztlichen Ausbilders. Die lange kollektive Gruppenexistenz und deren intimer Stil kritischer Beurteilung ihrer Mitglieder sensibilisieren das Individuum gegenüber Gruppennormen wie gegenüber Reaktionen auf sein Verhalten. Auch wenn Studenten in professionelle Rollen schlüpfen wollen, die Spielregeln begrenzen die Geschwindigkeit und Vollständigkeit, mit der eine Rollenübernahme erwartet werden kann. Es gibt sehr gut bekannte und zu beachtende Fahrpläne. Die Geschwindigkeit wird kontrolliert durch soziale Kontrollmechanismen wie Herablassung, Scham, Wertschätzung und Ironie" (Shuval, 1980, S. 156f.).

Wir fanden bei unseren Untersuchungen zum peer learning in Anamnesegruppen, daß sich die Kommilitonen untereinander ausgesprochen distanziert einschätzten (Schüffel, 1983). In den von uns identifizierten drei Hauptkomponenten eines Wertesystems von Studenten zur Beurteilung ihrer Ausbildungssituation fanden wir im Hinblick auf die Einschätzung des Kommilitonen erhebliche Veränderungen. Während der Kommilitone bei einer Einschätzung der Beziehung selbst/Kommilitone bzw. reziprok Kommilitone/selbst zunächst als fernstehend gesehen wurde, war er zum Schluß der Gruppenarbeit außerordentlich nahe gerückt. Zu Beginn der Gruppenarbeit wur-

---

3 Hierunter sind zu verstehen: an nichtuniversitären Einrichtungen praktizierende Ärzte und Lehrbeauftragte. – Blooms Buch ist ein Standardbuch im Bereich medizinischer Sozialisation und eine wahre Fundgrube für Informationen wie weiterführende Fragen.

de eine Bindungslosigkeit, d.h. der Gegensatz von Solidarität empfunden. Das war zum Schluß der Gruppenarbeit nicht mehr der Fall.

In der Beziehung zu Patienten wurde parallel hierzu eine geringere Hilfsbedürftigkeit und eine größere Bindungsfähigkeit im Sinne einer Zunahme der Solidarität mit dem Patienten wahrgenommen.

Diese Veränderungen wurden nach einem Jahr Gruppenarbeit in Anamnesegruppen beobachtet. Gleichzeitig konnte ein psychomotorischer Lernzuwachs verzeichnet werden: Zum Schluß der Gruppenarbeit wiesen die Gruppenmitglieder eine erhebliche Verbesserung in ihrer Fertigkeit auf, Anamnesen zu erheben (Buchinger und Schüffel, 1983).

Die vorliegenden Forschungsergebnisse lassen sich wie folgt zusammenfassen: Medizinstudenten weisen zunächst ein sehr hohes Maß an Hilfsbereitschaft gegenüber der Gesamtproblematik von Patienten auf. Sehr schnell kommt es jedoch zu einer Distanzierung vom Patienten im psychologischen Bereich, die für den außenstehenden Beobachter unangemessen erscheint und von einer Anzahl von Sozialisationsforschern als „zynisch" bezeichnet wurde. Die Distanzierung nimmt unterschiedliche Ausmaße an und kann je nach individueller Belastung und Anpassungsleistung vorübergehend oder langdauernd sein. Die individuelle Reaktion macht sich einerseits in hoher professioneller Leistungsbereitschaft wie andererseits überdurchschnittlich häufigen psychopathologischen Reaktionen bemerkbar. Die berufliche Entwicklung wird im allgemeinen nachhaltig durch vier sozialisierende Kräfte bestimmt, nämlich durch Patienten, Ärzte, Kommilitonen und das Personal. Die Bedeutung des peer scheint in einigen Bereichen sehr einflußreich zu sein; die Bedeutung des Personals ist im deutschen Ausbildungswesen praktisch auszuklammern.

Mit der Beschreibung sozialisierender Kräfte gelingt es, Einflußgrößen zu bestimmen. Dieses Konzept kommt dann an die Grenze seiner Aussagemöglichkeit, wenn gefragt wird, welche Kräfte beim einzelnen Studenten nun darüber entscheiden, wie die genannten sozialisierenden Kräfte ihre Wirkung entfalten. Psychoanalytische bzw. psychodynamisch orientierte Konzepte führen hier weiter.

67.3.4 Versuche zur psychodynamischen Erklärung

Es wird davon ausgegangen, daß der Student in ununterbrochener Reihenfolge Momenten der Betroffenheit, also besonderen Belastungen ausgesetzt ist, welche der innerpsychischen Bearbeitung bedürfen. Es scheint möglich, bei dieser Bearbeitung regelhaft auftretende Abschnitte zu unterscheiden. Es wird über zwei konzeptionelle Ansätze dieser Art berichtet.

Die Bewahrung des Selbstkonzeptes

Es wurde vorgeschlagen (Kahn et al., 1981), die in einzelnen Studienabschnitten ablaufenden und durch eine Belastung ausgelösten intrapsychischen Bearbeitungsprozesse als das Bemühen des Individuums zu verstehen, sein Konzept vom eigenen Selbst zu erhalten. Kahn und Mitarbeiter beobachteten bei Studenten, die zu Beginn der klinischen Ausbildung an einem Untersuchungskurs teilnahmen, folgende vier Abschnitte:

- Die Studenten waren verwirrt (confusion);
- die Studenten leugneten eine Bedrohung ihres Selbstkonzeptes (denial);
- die Studenten äußerten Ärger und versanken gleichzeitig in Depression (anger, depression);
- die Studenten lernten Problemlösungen oder Reintegration des Selbstkonzeptes durch neue Verhaltensstrategien (resolution or reintegration).

Kahn und Mitarbeiter zogen das mehrstufige Modell von Kübler-Ross heran, das beim Krebs- und Sterbenskranken die klinischen Phänomene der initialen Ablehnung und des allmählichen Annehmens der Diagnose erklärt. Ausgangspunkt der Überlegung war, daß die bisherige Sicht des Selbst nicht mehr aufrechterhalten werden kann.

Persönlicher und professioneller Reifungsprozeß während einer Gruppenarbeit

E. Krejci beschrieb bei Studenten, die eine Anamnesegruppe durchliefen, eine siebenphasige Entwicklung, um zu einem erweiterten Verständnis persönlichen und professionellen Verhaltens zu kommen (1983):

- Die Studenten stießen auf tabuisierte Bereiche und fühlten sich gehemmt;
- sie verbündeten sich mit der Abwehr der Patienten;
- sie konnten sich selbst nur schwierig wahrnehmen;
- sie konnten nur mit Schwierigkeiten zwischen Hypothese und Realität unterscheiden, d.h. „Bedeutungserteilung" und „Überprüfung" der Bedeutungserteilung im Situationskreis (vgl. 67.5);
- sie erlebten Selbsterfahrung als Selbstkonfrontation;
- sie hatten Angst, den Patienten auszunutzen;
- sie mußten sich zwischen patientenbezogener und gruppenbezogener Selbsterfahrung entscheiden.

Krejcis Einschätzungen wurden vor dem Hintergrund einer psychoanalytisch orientierten Beurteilung der Gruppenarbeit einer Anamnesegruppe formuliert. Krejci ging von der Überlegung aus, daß die „Verlagerung des Interesses" von einer organbezogenen zu einer personenbezogenen Medizin „eine tiefgreifende Veränderung in dem Rollenverständnis des Arztes" zur Folge hat (1983, S. 190). Obwohl die Lernsituation und teilweise die konzeptionelle Orientierung unterschiedlich sind, scheinen dennoch Entsprechungen zwischen den Ergebnissen der Arbeiten von Kahn und Krejci aufzeigbar. Krejci legt größeres Gewicht auf die Beschreibung der Lerninhalte, Kahn betont stärker die Abwehrmechanismen bzw. die Bewältigungsformen.

- Nach Krejci können sich Studenten zunächst nur schwierig selbst wahrnehmen, da sie angesichts tabuisierter Bereiche gehemmt sind und sich nach-

folgend mit der Abwehr der Patienten verbünden. Kahn beobachtet in diesem Abschnitt Verwirrung.
- Krejci beobachtet bei Studenten die Schwierigkeit, zwischen Hypothese und Realität zu unterscheiden; Kahn betont hier die Abwehr der resultierenden Bedrohung und spricht von einer geleugneten Bedrohung des Selbstkonzeptes.
- Krejci beobachtet ängstlich erlebte Selbstkonfrontation und schuldhaft erlebtes „Ausnutzen" des Patienten; Kahn beschreibt Ärger und Depression.
- Krejci beschreibt neue Entscheidungsmöglichkeiten zwischen patientenbezogener und gruppenbezogener Selbsterfahrung; Kahn spricht vom Erlernen neuer Verhaltensstrategien.

Krejci führte ihre Gruppenarbeit zusammen mit Th. von Uexküll weiter. Den Mitgliedern der Anamnesegruppe wurde nach Abschluß ihrer Arbeit nahegelegt, ihre Erfahrungen jüngeren Studenten weiterzuvermitteln, d. h. Tutoren neu zu bildender Anamnesegruppen zu werden. Auf diese Weise wurde die bisherige Anamnesegruppe in eine Gruppe von Tutoren übergeführt (Krejci, 1983). Diese Gruppe arbeitete ca. eineinhalb Jahre zusammen. Erneut konnte Krejci sieben Abschnitte der Gruppenarbeit studentischer Tutoren von Anamnesegruppen benennen:
- Die Tutoren erlebten in der Anamnesegruppe ein Protestreservoir;
- sie erkannten die Kollision der Interessen einer beziehungsorientierten mit den Interessen einer organorientierten Medizin;
- sie erkannten die Kollision ihrer eigenen Probleme mit denen der Patienten;
- sie führten eine innere Auseinandersetzung, ihre Position als Gruppenleiter zu akzeptieren;
- sie wurden in ihrer Autorität als Gruppenleiter in Frage gestellt;
- sie erkannten, daß Solidarität innerhalb des Selbsthilfekonzeptes erforderlich ist, um unrealistisch überhöhte Erwartungen abarbeiten zu können;
- sie erkannten und anerkannten bisher als selbstverständlich angenommene Gruppenbeziehungen.

An diesen Arbeitsergebnissen erscheint wesentlich, daß wiederum phasenartige Entwicklungen beschreibbar und abgrenzbar wurden – diesmal auf einer neuen und höheren Integrationsebene. Während es sich bei den sieben Phasen der Anamnesegruppe um die Entwicklung des Selbstkonzeptes im Hinblick auf den Patienten handelte, ging es bei den Phasen der Arbeit in der Tutorengruppe um das Selbstkonzept als Gruppenleiter inmitten der medizinischen Institution.

Mit diesen Arbeitsergebnissen hat Krejci die Anregung geliefert, sich nacheinandergeschaltete Entwicklungsphasen vorzustellen, in denen der Student zwei Integrationsebenen unterschiedlicher Komplexität durchläuft. Er beginnt mit einer biologisch-spezialistisch orientierten Ausbildungswelt; inselartig führt er in der Anamnesegruppe psychosoziale Gesichtspunkte in seine Konzepte ein, die sich in eine personenbezogene, holistische Richtung zu entwickeln beginnen. Auf der zweiten Integrationsebene und weniger inselförmig, d. h. stärker auf den Ausbil-

dungsplatz bezogen, fragt sich der Student, wie eine personenbezogene und holistische Ausbildung und die entsprechenden medizinischen Versorgungsansätze zu realisieren sind.

Nachfolgend sollen Merkmale des Ausbildungsplatzes, genauer des Curriculums, beschrieben werden, wie sich diese gegenwärtig an den medizinischen Fakultäten und Fachbereichen der BRD wiederfinden lassen.

67.4 Merkmale des Curriculums

Zu diesen Merkmalen werden hier gezählt: das Curriculum in seinem inneren und äußeren Aufbau, die Dozenten und die von ihnen vertretenen Konzepte, die Prüfungsbedingungen.

67.4.1 Das Curriculum in seinem äußeren und inneren Aufbau

Die Approbationsordnung gibt einen vierstufigen Rahmenplan für das Medizinstudium vor. Eine Zielbestimmung liegt bisher nicht vor. Im Prinzip wurde der Basisarzt angestrebt (vgl. Ulmer Curriculum, Schüffel, 1983, S. 360 ff.).

Die Inhalte der einzelnen Studienabschnitte:
1. Stufe, 1.–4. Semester (Vorklinik): Anatomie, Biochemie, Physiologie, Soziologie;
2. Stufe, 5. und 6. Semester (erster klinischer Abschnitt): Untersuchungsmethoden, Pathologie und Propädeutik der Medizin als Hauptveranstaltungen;
3. Stufe, 7.–10. Semester (zweiter klinischer Abschnitt): Innere Medizin, Chirurgie als Hauptfächer; es kommen hinzu: ökologische Fächer, Psychiatrie, Psychosomatik und Psychotherapie;
4. Stufe, 6. Jahr mit drei Tertialen (Praktisches Jahr): Innere Medizin und Chirurgie als Hauptfächer, ein Wahlfach.

Es wurde gesetzlich festgelegt, daß eine zweijährige, nach dem Medizinstudium abzuleistende Tätigkeit als „Arzt im Praktikum" eingeführt wird, die „die Voraussetzung zur Erteilung der zur selbständigen Ausübung des Berufes berechtigenden Approbation als Arzt" sein wird.

Der vierstufige Ausbildungsrahmen folgt einem an sich logischen Gedankengang, der pilotartig zunächst in Gießen (von Uexküll) erprobt wurde. Die erste Stufe dient der Unterrichtung in den Basisfächern, zu denen mit der Approbationsordnung im Gegensatz zur früher bestehenden Bestallungsordnung die psychosozialen Fächer hinzukamen. Fakultativ waren Hausbesuchsprogramme mit Allgemeinärzten vorgesehen; in der zweiten Stufe ein erster und patientennaher, in kleinen Gruppen durchzuführender Kontakt mit klinisch-stationär Kranken. Sowohl die Propädeutik wie die Pathologie sind als flankierende Maßnahmen für die Auseinandersetzung mit klinischen Phänomenen zu sehen. In der dritten Stufe erfolgt einerseits die Vertiefung, andererseits mit den Spezialfächern die Differenzierung des vorher Gelernten. In der vierten Stufe wird auf einige wenige Fächer fokussiert; im Wahlfach können exemplarisch Probleme erarbeitet werden, und der Student übernimmt zum ersten Mal eine umschriebene ärztliche Verantwortung im Sinne des Basisarztes.

Zum inneren Aufbau des Curriculums: Nach wie vor weist das Medizinische Curriculum der BRD zwar einen in sich logischen Aufbau auf, der aber für die Entwicklung personenbezogener Einstellungsweisen in der Medizin verhängnisvoll ist. Im Prinzip ist in diesem Curriculum das Lernziel enthalten, am Modell der unbelebten, bestenfalls der animalischen Welt Konzepte der Medizin zu verstehen und sie anzuwenden. Der Unterricht beginnt mit der unbelebten Welt, d.h. mit der Physik und der Chemie, geht über zur Anatomie und läßt allmählich die belebte Welt in Form der Biologie zu, um sich der Biochemie und der Physiologie zuzuwenden. Zwar wurden Psychologie und Soziologie als Unterrichtsfächer eingeführt; dies vollzog sich jedoch in einer additiven Weise, d.h. die Einführung geschah nicht in einer problemorientierten integrativen Vorgehensweise. In einer verhängnisvollen Weise dominiert das von Bloom (1988) international beobachtete Prinzip des passiven Lernens statt eines aktiven, selbsterfahrungsbetonten Lernens.

Ein vergleichbarer Aufbau findet sich in der Klinik. Dort kommen mit der Pathophysiologie und der Pathochemie neue Fächer hinzu, welche vorklinische Erklärungsmechanismen zwar differenzieren, die grundlegenden biologisch orientierten Mechanismen aber beibehalten. Pathophysiologie und Pathologie werden zunehmend zu Wegweisern durch die Klinik, und eine Wissenschaft vom menschlichen Verhalten und Einrichtungen der menschlichen Kultur wird in einer zu vernachlässigenden Weise abgehandelt. Am Beispiel der Pathologie ist besonders zu zeigen, daß ausschließlich deren Maßstäbe auch weiterhin zur Qualitätsbeurteilung herangezogen werden. Mir ist keine medizinische Fakultät der BRD bekannt, die etwa eine Fallkonferenz unter Hinzuziehung des Hausarztes sowie der Sozialarbeiterin oder der Gemeindeschwester regelmäßig anböte, um dort den Studenten Qualitätsmerkmale der Patientenversorgung nahezubringen.

Entscheidend an dem skizzierten Aufbau des Curriculums ist, daß zunehmend komplexe Erklärungsmuster für medizinische Probleme zusammengetragen und dem Studenten vermittelt werden. Sie beruhen auf Daten, die aus der anorganischen Umwelt, bestenfalls aus der Umwelt von Säugetieren stammen. Die Gesetzmäßigkeiten der Muskelphysiologie können etwa am Muskel des Frosches, die Kreislauffunktion am Herz-Kreislauf-System des Säugers erläutert werden. Es ist das offensichtliche Ziel eines derart aufgebauten Curriculums, ein sich zunehmend verfeinerndes Erklärungsraster auf biologisch-somatischer Ebene zu liefern, um hiermit die klinische Vielfalt der Patientenprobleme anzugehen und den Patienten

und seine Krankheit beurteilen zu können. Es handelt sich um ein weltweites Problem. Für die USA wird gesagt (Bloom, 1989):

„We are not training tomorrow's physicians for the real needs of the population. ... We believe that a narrow reductionistic positivism has prevailed in medical education in spite of the prevalence of functionalism in the social science of medicine."

Hinzu kommt der Umfang. Zur Qualität kommt die erdrückende Quantität der Lernanforderungen. Unterrichtsstunden bzw. Dozentenzahlen, die im Bereich der psychosozialen Fächer eingebracht werden, sind demgegenüber verschwindend klein.

Durch dieses Studium geht der Student immer wieder beunruhigt oder sogar betroffen hindurch. Das Curriculum ist aber so angelegt, als gäbe es derartige Momente der Beunruhigung oder Betroffenheit nicht. Das Problem der unbearbeiteten Betroffenheit, das die Studenten vor Beginn des Studiums, während des Präparierens, beim klinischen Untersuchungskurs, in der Begegnung mit Sterbenskranken, im Praktischen Jahr und schließlich als junge Ärzte angesichts von Fragen der Sexualität, des Sterbens haben – hinzuzufügen sind Abhängigkeit und Sucht – alles das wird verschwiegen. Es bestehen kaum Angebote, diese Betroffenheit zu bearbeiten. Statt dessen werden ungeheure Forderungen aufgebaut, die nun neuerdings mit dem erwähnten Ausbildungsziel auch explizit formuliert werden. Wie soll der Student eine dem einzelnen Menschen und der Allgemeinheit verpflichtete ärztliche Einstellung erwerben, wenn er während seines gesamten Studienganges kein einziges Mal ärztliche Verantwortung einüben lernt? Studenten versichern heute glaubhaft, daß sie das gesamte Studium von 6 Jahren durchlaufen können, ohne jemals mit der Aufgabe konfrontiert worden zu sein, den gesamten Untersuchungsbefund eines (!) Patienten zu erheben – geschweige denn, dessen Therapie zu entwerfen und die therapeutischen Maßnahmen durchzuführen. Es drängt sich geradezu der Eindruck einer „beinahe gespenstischen Realitätsferne unserer Fakultäten" (Th. von Uexküll) auf. Allerdings geht es in diesem Fall nicht allein um die Realitätsferne der Fakultäten, sondern auch um diejenige der zuständigen politischen Gremien. Hierzu wird der Wissenschaftsrat in seinen „Empfehlungen zur Verbesserung der Ausbildungsqualität in der Medizin" vom 12. 12. 1988 zitiert (Th. von Uexküll, 1989):

„Daß derzeit auch der Anschein nicht mehr aufrechterhalten werden kann, bei der bestehenden Situation könne eine angemessene, den gesetzlichen Anforderungen entsprechende Ausbildung vermittelt werden."

67.4.2 Die Dozenten

Vor die Anforderungen der Approbationsordnung gestellt, sehen sich Dozenten vor zwei Problemen. Sie sollen zum einen patientennah und zum anderen in kleinen Gruppen unterrichten. Von diesen Dozenten wird in der Regel ein Wissenschaftskonzept vertreten, für das psychosoziale Belange, insbesondere die des intrapsychischen Bereiches als unwissenschaftlich erscheinen. Sie haben ihrerseits eine Sozialisation durchlaufen, während der sie lernten, sich auf isolierte Fragen aus dem biologischen Bereich zu konzentrieren und dieses Vorgehen als beispielhaft für das anzustrebende naturwissenschaftliche Vorgehen in der Medizin anzusehen. Sie haben in der Regel aber nicht gelernt, „Gesundheit als ein Gleichgewicht miteinander verbundener Systeme auf der biologischen, sozialen und sozialkulturellen Ebene" zu sehen (Pauli, 1983).

Für den Beobachter aus der BRD wurde in einer beneidenswerten Art und Weise die Studienreform in der Schweiz vorangetrieben (Pauli, 1978). Auch hier stand im Vordergrund, zu einer größeren Patientennähe zu kommen. Eine im Zuge dieser Reform durchgeführte Begleituntersuchung zum Gruppenunterricht erbrachte, daß dieser nicht nur von Studenten, sondern auch von ärztlichen Dozenten außerordentlich geschätzt wurde (Bangerter und Noack, 1983). Die Einschätzung des Gruppenunterrichtes durch die Studenten und Dozenten weist jedoch eine Diskrepanz auf: Von den Arzt-Dozenten sahen 53% (eine der höchsten Einschätzungen) den größten Vorteil des Kleingruppenunterrichtes in der Interaktion und in der Kommunikation der Gruppen, während nur 31% den Vorteil in der Vermittlung von Kenntnissen, Fähigkeiten und Fertigkeiten sahen. Als die Studenten die **gleiche** Gruppenarbeit einschätzten, sahen aber nur 29% von ihnen als **tatsächlich** erlebten Vorteil die Vermittlung von Interaktion und Kommunikation. Dagegen meinten 57% (!), die Bedeutung des Unterrichtes habe in der Vermittlung von Kenntnissen, Fähigkeiten und Fertigkeiten gelegen. Es zeigte sich also geradezu das Spiegelbild der Einschätzung der tatsächlich abgelaufenen Gruppenarbeit. Dementsprechend bezogen sich die meisten Gruppenvorschläge der Studenten hinsichtlich einer Verbesserung des Gruppenunterrichtes auf das Thema „Auswahl, Anleitung und Entlastung der Arzt-Tutoren". – Ihrerseits forderten die Dozenten am häufigsten, die Studenten sollten besser theoretische Voraussetzungen in die Gruppenarbeit einbringen. Für den außenstehenden Beobachter drängen sich mit dieser unterschiedlichen Einschätzung des Unterrichtes Assoziationen an den von Bloom geschilderten Kleinkrieg zwischen Dozentenschaft und Studentenschaft auf (s. o.).

Das Problem der patientenbezogenen Kommunikation in der kleinen Gruppe, d.h. letztlich das Problem des Austauches von Gleichrangigen oder von peers und damit das Problem der späteren Kollegialität, zieht sich wie ein roter Faden durch die Ausbildung hindurch.

Dieses allgemein zugrundeliegende und ständig zu beobachtende Doppelproblem einer Kommunikation mit dem Patienten und einer Kommunikation mit dem Kollegen wird in der deutschen Ausbildungssituation durch folgende Faktoren verschärft: Ein sprunghaftes Ansteigen der Studentenzahlen auf fast das Doppelte innerhalb von zehn Jahren; eine teutonische Reglementierungswut, welche die Fakultäten zwischen Kiel und München zwingt, die Approbationsordnung exerziermäßig durchzuführen, statt ihnen Freiheit zur Entwicklung eigener Curricula zu lassen; die praktisch nicht existente Bewertung unter-

richtsmäßiger Leistungen des Dozenten für seine berufliche Entwicklung.

67.4.3 Die Prüfungsbedingungen

Studenten lernen das, wer könnte es ihnen verübeln, was die Prüfungen abverlangen. Bei den Prüfungen handelt es sich um Wissensstoff, d.h. die Aufmerksamkeit des Studenten wird ausschließlich auf den kognitiven Lernbereich gelenkt.

Hier hilft auch die erwähnte Novellierung der Approbationsordnung nichts, die eine gewisse Verschiebung des Schwergewichtes zugunsten klinischer Fertigkeiten dadurch bewirken soll, daß während des Staatsexamens mündlich geprüft wird. Vielmehr werden differenzierte Prüfungsverfahren erforderlich. Hierzu würde sicherlich weiterhin das multiple choice gehören, aber ebenso notwendig schriftliche Abhandlungen in freier Form („essay" der Angelsachsen) und vor allem mündlich-praktische Überprüfungen durch Prüfer, die untereinander in einem Austausch stehen. – Diese Möglichkeiten können allerdings erst dann langfristig wirksam werden, hierauf weist Novak hin, wenn innerhalb der Medizin in Zusammenarbeit mit der Medizinsoziologie Räume geschaffen werden, die eine Analyse der strukturell bedingten Grenzen ärztlichen Handelns ermöglichen (1983, 1987).

67.5 Wissenschaftstheoretische Überlegungen und ein konkretes Beispiel dermatologischer Ausbildung

Buchborn fordert die „komplementäre Anwendung von Natur-, Verhaltens- und Sozialwissenschaften...", um ein mehrdimensionales Krankheitsverständnis zu begründen" (1984). Anschütz sieht gleichermaßen in den Naturwissenschaften wie in den Verhaltenswissenschaften diejenigen Wissenschaftsbereiche, aus denen wissenschaftliche Konzepte und Modellvorstellungen der Medizin kommen (1984). In der Bearbeitung der Kooperationsprobleme zwischen Generalisten und Spezialisten sieht Siegenthaler geradezu das Grundproblem der modernen Medizin (1984).

Der Student wird während seiner Ausbildung mit Befunden aus den verschiedensten Wissenschaftsbereichen konfrontiert und begegnet deren Vertretern in Person. Die biologisch orientierten Wissenschaften dominieren hierbei. Die Aufgabe, aus den resultierenden heterogenen Erfahrungen ein integrierendes Konzept für den Umgang mit Kranken zu machen, wird dem einzelnen Studenten überlassen. In den Eingangskapiteln dieses Buches wird allgemein, im Kapitel über Ulcus duodeni speziell und exemplarisch auf die wichtigsten ungelösten wissenschaftstheoretischen Fragen für ein solches Konzept hingewiesen. Diese sind:

– Der Problemkreis der Beziehungen zwischen Organismus und Umgebung. Zu dessen Lösung wur-

de das Modell der Umwelt bzw. der individuellen Wirklichkeit entwickelt.
– Der Problemkreis des Zusammenhanges zwischen physiologischen, psychologischen und sozialen Faktoren. Zur Beantwortung dieser Fragen wurde das Konzept der Beziehungen zwischen Subsystemen, Systemen und Suprasystemen eingeführt.

Die damit strukturierte Betrachtungsweise läßt sich nicht nur auf den Patienten, sondern auch auf den Studenten anwenden. Sie hat den Vorteil, daß sie erlaubt, die Entwicklung, die er während seiner Ausbildung durchläuft, als Weiterentwicklung seiner individuellen Wirklichkeit darzustellen. So läßt sich beschreiben, wie die individuelle Wirklichkeit des Studenten an seinem Ausbildungsplatz durch die zur Verfügung gestellten Programme für Problemlösungen zu erweitern ist. Der Student bemüht sich, die Probleme des individuellen Patienten zu erfassen und auf sie einzugehen. Nach dem Modell des Situationskreises besteht also eine Problemkonstellation, die durch Bedeutungserteilung aufgebaut und nachfolgend gelöst werden muß, wobei die Richtigkeit der Bedeutungserteilung geprüft wird. Im herkömmlichen Curriculum wird die Problemkonstellation biologisch definiert, dementsprechend wird die Bedeutungserteilung biologisch überprüft.

Im Situationskreis überprüft der Student seine Informationen aber auch immer aufgrund mitgebrachter Programme. Es wird von seinem Selbstwertgefühl abhängen, ob und in welchem Maße er auch Programme heranziehen kann, die ihm nicht während des regulären Unterrichtes vermittelt werden; denn diese werden automatisch zu Spannungen zwischen ihm und dem Dozenten führen.

Diese Aussage hört sich zunächst banal an. Sie hat jedoch unmittelbare praktische Relevanz, wenn z. B. an die vielfältigen Bemühungen gedacht wird, Balints Gruppenarbeit mit niedergelassenen Ärzten auf Studenten im Praktischen Jahr zu übertragen. Solche Versuche mißlingen gewöhnlich, weil unterschiedliche Wertvorstellungen zwischen dem Balint-Gruppenleiter und der Fakultät zu Spannungen führen können, die für den einzelnen Studenten unerträglich werden. Dies hatte Balint bereits frühzeitig erkannt. Seine Frau, E. Balint, hat dies in einfacher und treffender Weise ausgedrückt: „Michael felt that the professors were between himself and the students" (1982).

Krejcis Ergebnisse lassen sich zwanglos im Rahmen des Situationskreismodelles verstehen: Der Student wird in die Lage versetzt, zunächst abseits vom offiziellen Wertsystem der Fakultät mit dem Patienten und seinen Kommilitonen zusammen eine subjektive Situation aufzubauen. Er nimmt jetzt die Informationen des Patienten wahr, ohne diese sofort in eine biologisch vorgegebene Form pressen zu müssen. So hören die Studenten im Falle des oben erwähnten Bauern und Tiefbauarbeiters mit Kolitis dessen Schwäche und spüren ihre eigenen Befürchtungen, in dem allgemeinen Strudel der Hilflosigkeit unterzugehen. Sie können nun Strukturen entwickeln und diese in

Form von Fragen und Mitteilungen an den Patienten zurückvermitteln. Das gibt dem Patienten die Möglichkeit, über die drohende Auflösung der Familie und seine Ängste zu berichten, mit der Frau allein gelassen zu werden. Diesen Situationskreis, der sich in deutlicher Abhebung von einem rein biologisch orientierten Situationskreis entwickelt, hatten wir als den persönlichen Situationskreis des Medizinstudenten bezeichnet (Schüffel et al., 1983).

Nicht nur die mitgebrachten, sondern auch die heterogenen Programme der einzelnen Fächer müssen in Verbindung gebracht werden. Sie bieten dem Studenten Erklärungsmodelle für die Symptome und für das Verhalten eines Patienten an, d. h. für Phänomene, die auf verschiedenen Integrationsebenen angesiedelt sind. Der Medizinstudent muß lernen, die verschiedenen Integrationsebenen in Verbindung zu bringen. Das Modell des Situationskreises verweist hier auf den Begriff der Bedeutungskoppelung. Im Kapitel 43 „Ulcus duodeni" war ausgeführt worden, daß wir einerseits sagen können, „einer angebotenen Hypersekretion des Magens entspricht eine erhöhte Appetenz des Säuglings, Umgebungsreize an gastrointestinale Zeichen zu koppeln. Damit beschreiben wir das Entstehen einer Umwelt, in der alles und jedes eine Fütterungsbedeutung, gewissermaßen einen ‚basalen Freßton' erhält". – Unter Bezugnahme auf die Freudsche Terminologie können wir auch sagen: „Eine angeborene Hyperpepsinogenämie wirkt als somatische Triebquelle. Diese wird in den psychischen Drang übersetzt, Umgebungsreize für die Mund- und Zungenschleimhaut als nahrungsspendende (oder nahrungsverweigernde) Umwelt zu interpretieren, und mit ihrer Hilfe (die Befriedigung des Dranges) zu erreichen. Über Einverleibung von Umwelt (des Triebobjektes Milch) soll es zum Abstellen der somatischen Triebquelle kommen."

Entscheidend ist bei diesem Denkmodell, daß wir bei der Verknüpfung der verschiedenen Integrationsebenen Verklammerungen im Sinne von „Aufwärts- und Abwärts-Effekten" in ihrer biographischen Zeitgestalt einüben. So kann, um ein Beispiel zu bringen, eine Ablehnung des Individuums im sozialen Bereich Konsequenzen auf verschiedenen Ebenen haben: Sozial kommt es zu Einschränkungen des Bekanntenkreises, zu Abbruch von Freizeitaktivitäten; psychisch kommt es zu Selbstvorwürfen und Zweifeln am Selbstwert; physiologisch kommt es zu einer blassen Mukosa und einem Verschluß des Pylorus.

Der Student muß nun seinerseits lernen, zunächst die verschiedenen Integrationsebenen zu identifizieren und dort nach Symptomen zu fahnden. Dann muß er versuchen, zu Verknüpfungen zu kommen, indem er „Aufwärts- und Abwärts-Effekte" in Rechnung stellt. Ferner muß er berücksichtigen, daß Bedeutungskoppelungen auch autonom werden können. Das kann bedeuten, daß ein dem Beobachter unauffällig erscheinendes soziales Ereignis vom Patienten als eine soziale Zurückweisung gedeutet wird. Das – scheinbar neutrale – Ereignis könnte dann automatisch mit einer blassen Mukosa und Pylorusverschluß bei entsprechender klinischer Symptomatik verbunden sein.

Auch hier finden sich bei Krejci Hinweise, wie Lernprozesse ablaufen, die Ereignisse verschiedener Integrationsstufen verknüpfen helfen. Die Studenten lernen in der Tutorengruppe, ihre Mittutoren als die Verfechter unterschiedlicher Integrationsebenen wahrzunehmen. Indem sie aber eine gemeinsame Aufgabe haben, also vor der gleichen Problemkonstellation stehen, lernen sie die unterschiedlichen Äußerungen als Ausdruck unterschiedlicher Betrachtungsweisen der gleichen Situation wahrzunehmen.

Das wesentlichste Geschehnis für die Studenten ist jedoch die Erfahrung, daß ihre eigenen Reaktionen, d.h. ihre Gefühle und Stimmungen berechtigt sind. Dies gilt sowohl für ihre Reaktionen gegenüber dem Patienten wie gegenüber der Fakultät. Diese Feststellung mag wiederum banal erscheinen. Aber: Ein systemorientiertes Denken, wie es hier vertreten wird, ist weithin als Notwendigkeit akzeptiert worden. Hierauf stützt sich auch in der Regel die Unterscheidung zwischen Generalisten und Spezialisten bzw. von Holisten und Reduktionisten (Pauli, 1984). Der psychosomatische Zugang läßt sich jedoch noch nicht allein mit Hilfe von Modellen beschreiben, die nur die individuelle Wirklichkeit des Studenten oder des Arztes im Auge haben. Was den psychosomatischen Zugang vielmehr ausmacht, ist die Fokussierung auf die Interaktion bzw. das Beziehungsgefüge in einem interindividuellen (Supra-)System von in der Regel zwei Personen, dann derjenigen von Kleingruppen. Die Annahme ist, daß die individuelle Wirklichkeit des Patienten nur dann erfaßt werden kann, wenn es Arzt und Patient gelingt, eine gemeinsame Wirklichkeit aufzubauen.

In einem Bericht zum Thema „Dermatologische Psychosomatik im studentischen Unterricht" zeigt Bosse (Göttingen), daß die Herstellung dieser gemeinsamen Wirklichkeit keine theoretische Forderung bleiben muß, sondern zu einer Realität werden kann. Er greift hierbei die oben erwähnte Betroffenheit der Studenten auf. Bosse formuliert:

„Anhand von Diapositiven entstellter Hautkranker werden die eigenen Ängste der Studierenden bewußtgemacht und im gemeinsamen Gespräch verarbeitet. Wir gehen davon aus, daß das Verständnis für die Gefühle des Patienten die Reflexion der eigenen Emotionen voraussetzt. Die aktive verbalisierende Teilnahme einer großen Zahl von Studenten zeigt, daß diese Reflexion auch im großen Hörerkreis möglich ist. Die späteren Rückmeldungen zeigen, daß auch eine beträchtliche Zahl aktiv Zuhörender, wenngleich nicht diskutierender Teilnehmer das Bedürfnis zum emotionalen Lernen hat.

In der Diskussion werden grundlegende Probleme der Interaktion mit hautkranken Menschen zur Sprache gebracht und Möglichkeiten zur Lösung derselben durchdacht: Die eigene Betroffenheit, Verdrängung derselben, Mitleid mit dem Patienten; Selbstbild und Reaktion (Rückzug, Mißtrauen des Patienten), wie würde der Lernende mit der Hautkrankheit umgehen? Angst vor Ansteckung, Schuld und Strafe. Die Problematisierung dieser Fragen soll die subjektiven Normen des künftigen Arztes reflektieren und relativieren sowie schließlich die Fähigkeit, sich in den hautkranken Patienten zu versetzen, entwickeln.

Diagnostische und therapeutische Erwägungen werden in diesen Gesprächen allenfalls zweitrangig berücksichtigt.

Diese Art von Lernveranstaltungen korrigiert die traditionelle Leitvorstellung des professionellen Arztbildes, indem es die Möglichkeit eröffnet, Angst vor Krankheit, Hilflosigkeit, Ekel vor Hautkrankheit und ähnliche Gefühle zwischen den Studierenden unter Anleitung zur Sprache zu bringen."

67.6 Wie können patientenbezogene Einstellungen erworben werden?

67.6.1 Eine Zusammenfassung des Bisherigen

Wir waren davon ausgegangen, daß sich der Medizinstudent während seiner Ausbildung als Leitsatz seines Handelns die Frage zu eigen machen sollte: „Warum kommt **dieser** Mensch zu **diesem** Zeitpunkt mit **dieser** Krankheit zu **mir?**" – Die Sozialisationsforschung hat nachgewiesen, daß während der medizinischen Ausbildung ständig personenbezogenes von objektivistisch-reduktionistischem Lernen verdrängt zu werden droht. Die Folge ist, daß der Student gegenüber dem Patienten zu einer distanzierenden Haltung gedrängt wird, zu einer Einstellung, welche von verschiedenen Untersuchern mit dem Terminus „Zynismus" umschrieben worden ist. Statt dessen wird es notwendig, die oben (67.2) erwähnte Betroffenheit aufzugreifen. Der Student muß ständig die Möglichkeit geboten bekommen, seine Betroffenheit reflektieren zu lernen und sie in einer Zusammenarbeit mit Experten einsetzen zu können (Th. von Uexküll, 1989). – Momente derartiger Betroffenheit sind festgehalten und liegen gedruckt vor (Loew, 1989). Damit sind auch Kräfte verbunden, die der Student bereits in das Studium einzubringen scheint: Mißtrauen gegenüber den Wünschen, Beziehungen aufzunehmen. Sie äußern sich als Neigung zum Einzelkämpfertum. Folgen dieser teilweise übermäßigen Abwehrleistungen können sein: eine langanhaltende professionell-charakterliche Panzerung bis hin zu extremer Leistungsbetontheit oder Dekompensation. Die Kräfte, die diesen Entwicklungsprozeß beeinflussen, gehen aus von Patient, Dozent, Student und sonstigen Mitarbeitern im Gesundheitswesen. Die innerpsychischen Veränderungen sind am besten mit Hilfe eines psychoanalytisch fundierten Konzeptes zu verstehen, das die Entwicklung eines professionellen Selbstkonzeptes bzw. dessen Erhalt unter extremen Belastungen zu erklären sucht. Zur konzeptionellen Orientierung scheint ein zweistufiges Modell hilfreich zu sein, welches die Entstehung von zwei Situationskreisen beschreibt: die Entwicklung des Selbst in der Student-Patient-Interaktion, d.h. der individuelle Situationskreis des Studenten; die Entwicklung des Selbst angesichts eines Austausches mit Kommilitonen und Fakultät, d.h. der professionelle Situationskreis des Studenten. – Derartige Entwicklungsprozesse werden nachhaltig durch die äußeren Rahmenbedingungen bestimmt, d.h. durch den Aufbau des Curriculums, das biologisch-spezialistisch orientiert ist; durch die Dozenten mit einem überwiegend reduktionistischen Wissenschaftskonzept und ihrer Unerfahrenheit in Gruppenprozessen sowie durch die Prüfungsbedingungen in ihrer ausschließlich kognitiven Orientierung. Die Probleme, die einer psychosomatisch orientierten Ausbildung zum Arzt entgegenstehen, sind somit gewaltig. Bei der Verwirklichung psychosomatischer Ziele in diesem Bereich muß man daher mit großen Zeitabschnitten rechnen.

67.6.2 Eine kurze Liste von Empfehlungen für einen langen Weg

Als eine Art Destillat lassen sich wesentliche Punkte herausarbeiten, welche die angestrebte Internalisierung von Werten für berufliches Handeln im Rahmen einer psychosomatisch orientierten Ausbildung zu fördern scheinen:

1. Patientenkontakte sollten möglichst frühzeitig, d.h. schon im ersten vorklinischen Semester erfolgen.
2. Kleingruppenarbeit sollte mit jedem Patientenkontakt parallel gehen.
3. Konzepte sind zu vermitteln, die von problemzentrierten Patient-Student-Interaktionen ausgehen und diese Interaktionen in einen Systembezug bringen.
4. Patientenarbeit sollte ständig von Supervision begleitet sein.
5. Eine Ausbildungsstelle, die Teil der Fakultät ist, sollte überwachen, daß an Studenten in der Arbeit mit dem Patienten **Verantwortung** delegiert und die Bildung studentischer **Neigungsgruppen** gefördert wird.
6. Eine Clearing-Stelle sollte in ständiger Rückkopplung mit allen Beteiligten die Qualität der Krankenversorgung und den Austausch zwischen Student und Fakultät verfolgen, beurteilen und konzeptionell weiterbearbeiten.
7. Ein überregionaler Austausch sollte für Studenten, Leiter der Ausbildungsstelle und Supervisor gewährleistet sein.

Einzelne Anmerkungen zur Kurzliste

Zu 1. Patientenkontakt: Ein Arzt und Hochschullehrer erzählte mir, daß er während seines vorklinischen Studiums in Heidelberg bei einem Assistenten von V. von Weizsäcker vorsichtig anfragte, ob er bei einer so zentralen Sache wie der Psychosomatischen Medizin bereits als Vorkliniker teilnehmen dürfe. – Die Antwort war recht lapidar: „Natürlich, je zentraler die Sache, um so früher sollten Sie daran teilnehmen". – Dementsprechend ist für Dozenten und Studenten der möglichst frühzeitige Patientenkontakt von größter Wichtigkeit. Ebenso wichtig bei diesen Kontakten ist die Berücksichtigung der mehrfach beschriebenen Phasen der Betroffenheit des Studenten, d.h. es müßten in unterschiedlichen Phasen unterschiedliche Themen angesprochen werden. Man muß wissen, daß ein Student häufig in Belastungssituationen scheinbar unzuverlässig mitarbeitet oder unre-

gelmäßig erscheinen wird. Dieser letzte Punkt der scheinbaren studentischen Unberechenbarkeit macht Dozenten immer wieder außerordentlich zu schaffen.

Zu 2. Kleingruppenarbeit: Die Gruppen sollten sich als Neigungsgruppen finden. Mit Nachdruck sollte man sich allen Bemühungen widersetzen, Gruppen durch Computer zusammenzustellen. Wünschenswert ist, daß eine ausgewogene Beteiligung von Männern und Frauen und teilweise auch von Angehörigen verschiedener Jahrgänge stattfindet. Die Gruppenbildungen sollten im Auge haben, daß sich die Teilnehmer für bestimmte Arbeitsziele verpflichten, initial für eine regelmäßige Zusammenarbeit mit dem Patienten. Ihr Selbsthilfebedürfnis sollte von Beginn an unterstützt, ja die Gruppen sollten geradezu unter dem Aspekt der „angeleiteten Selbsthilfe" (guided self help) begründet werden. – Das Zeitproblem scheint das größte Problem der Gruppenarbeit zu sein. Beziehungsaufnahme und Beziehungsentwicklung sind ein Prozeß, der Zeit erfordert. Unsere Erfahrungen mit Anamnesegruppen haben gezeigt, daß Angaben zum Zeitbedarf möglich sind (Schüffel, 1983). Wir beobachteten, daß der Student durchschnittlich ein Semester benötigt, um eine Gruppe als vertrauenswürdig zu empfinden. Ein weiteres Semester wird benötigt, zu einem Austausch über wechselseitige Erwartungen zu kommen. Erst dann bildet sich in der Sprache des Situationskreismodells der individuelle Situationskreis des Medizinstudenten. Zwei weitere Semester sind für Tutoren notwendig, um von ihren Omnipotenzvorstellungen Abstand zu nehmen. Eine Heidelberger Tutorin von Anamnesegruppen beschrieb diese Entwicklung folgendermaßen: „Aus zu jungen Eltern durften ältere Geschwister (der Mitglieder der Anamnesegruppen; Anmerkung des Autors) werden" (Bregulla, 1983). Hier hatte sich – nach dem zweiten Jahr der Arbeit – der professionelle Situationskreis zu bilden begonnen. Erst danach setzt eine konzeptionelle Bearbeitung ein, die dem Studenten persönlich erlaubt, die eigene Entwicklung zwischen den scheinbaren Antipoden einer studentischen und einer ärztlichen Rolle bei gleichzeitig einwirkenden sozialisierenden Kräften wahrzunehmen. Wir schätzen, daß die zur Bildung des individuellen und des professionellen Situationskreises erforderliche Zeit mindestens zweieinhalb Jahre beträgt.

Zu 3. Konzeptvermittlung: Studenten wehren sich zunächst durchgehend gegen Konzepte und können diese erst im Verlaufe längerer Gruppenarbeit übernehmen. Dennoch sollte versucht werden, recht früh auf die Notwendigkeit von Konzeptübernahmen einzugehen. Keineswegs sollte gesagt werden, daß sich Konzepte der Psychosomatischen Medizin in erster Linie durch ihren Systembezug auszeichnen; vielmehr sollte immer wieder betont werden, daß die Psychosomatische Medizin von der Arzt-Patient-Interaktion ausgeht und dann zum Systembezug vorstößt. Hilfreich ist, auf Einsichten der Psychotherapieforschung hinzuweisen (Strupp, 1984). Heute steht fest, daß jede Psychotherapie auf den aktuellen zwischenmenschlichen Beziehungen von Patient und Therapeut basiert. Entscheidend ist, ob es dem Therapeuten und dem Patienten gelingt, eine „konstruktive Beziehung" (Strupp) zu entwickeln. Keine Schule konnte gegenüber einer anderen ihre Überlegenheit beweisen. Andererseits ist, darauf muß immer wieder verwiesen werden, nur dann von Psychotherapie zu sprechen, wenn das zwischenmenschliche Agieren vor einem konzeptionellen Hintergrund geschieht und auf diesen bezogen wird.

Es wurden im deutschsprachigen Raum spezielle Konzepte entwickelt, um Studenten in eine patientenzentrierte Medizin einzuführen. Diese können hier nur global aufgeführt werden. Auf weiterführende Literatur wird in diesem Zusammenhang verwiesen. Freyberger entwickelte ein klinisches Programm im Rahmen des psychosomatischen Konsiliardienstes in Hannover (1983). Unter intensiver Gruppen-supervision betätigten sich Studenten des Praktischen Jahres als Hilfstherapeuten bei körperlich Schwerkranken und erfüllten ihre Aufgabe nachweislich erfolgreich. – Bräutigam, Knaus und Wolf führten in Heidelberg in Zusammenarbeit mit dem Londoner University College das Heidelberger Psychotherapie-Ausbildungsprojekt ein, in dem Patienten der dortigen Psychosomatischen Klinik ambulant über größere Zeiträume von Studenten therapeutisch betreut wurden. – Jork schlug ein von ihm so bezeichnetes „Praxisorientiertes Ausbildungsmodell Frankfurt" vor (1980). Hier werden also systematisch aufeinander aufgebaute Unterrichtsangebote vom 1. bis 10. Semester vorgestellt, die zum Kontakt mit dem Patienten des Allgemeinarztes anleiten.

Auf die von uns entwickelten Ausbildungs- und Lernmethoden der Anamnesegruppe ist durchgehend Bezug genommen worden. Sie scheinen besonders für die ersten fünf der sechs Ausbildungsjahre geeignet zu sein. Grundsätzlich läßt sich die Anamnesegruppe als das Muster einer problemzentrierten Kleingruppenarbeit mit Selbsterfahrungscharakter verstehen, in der die verschiedensten Themen der medizinischen Ausbildung, und hier insbesondere solche in Phasen schwerer Betroffenheit, abgehandelt werden können (Egle, 1983). Eine ausführliche Darstellung des Konzeptes und der Ergebnisse sowie der mit dieser Methode verbundenen Probleme ist an anderer Stelle erfolgt (Schüffel, 1983). – B. Luban-Plozza hat „Junior-Balint-Gruppen" beschrieben (Luban-Plozza, 1978, 1982), die Motivationsgruppen darstellen, in denen Studenten erstmals ihre Fähigkeiten zu einem personenbezogenen Interagieren wahrnehmen können. Er hat diese Form des motivierenden Arbeitens bei den jährlichen Balint-Wochen in Ascona (Schweiz) international bekannt gemacht.

Zu 4. **Supervision:** Wesentliche Inhalte der Supervision sind im Hinblick auf den Patienten die Ängste des Studenten, einen hilflosen Patienten mit höchsten Ansprüchen vorzufinden. Die hieraus resultierende Überforderung verdichtet sich unserer Erfahrung nach in drei Themenbereichen, die wir als die „drei S" bezeichnet haben: Sexualität, Sucht und Sterben. Hier scheint es sinnvoller, die Bewältigungsmöglichkeiten der Probleme mit den Studenten zu erarbeiten, als in einer konfliktzentrierten Art auf psychodynamische Zusammenhänge einzugehen. Die Supervision sollte aber auch immer die Interaktion mit dem Kommilitonen im Auge haben. Auch hier sollte es weniger um die Bearbeitung der Ursprünge der zu beobachtenden Rivalität und der Aggressionen gehen als vielmehr um Formen konstruktiver Zusammenarbeit.

Eine wichtige Anmerkung scheint angebracht: Immer werden während der Supervision institutionsbezogene Konflikte abgehandelt. Infolgedessen erscheint es sinnvoll, einen externen Supervisor heranzuziehen, der zur fraglichen Institution in keinem Abhängigkeitsverhältnis steht. Die positiven Auswirkungen einer solchen Supervision können kaum hoch genug eingeschätzt werden, wie wir es selbst im Falle einer Zusammenarbeit mit den Herren B. Klapp und M. Wirsching (Gießen) erfahren konnten.

Zu 5. **Ausbildungsstelle:** Hier geht es um das Problem der Delegation von Verantwortung für Patienten an Studenten. Fakultätsmitglieder fürchten häufig, der Student könne bei selbständig betriebener Kleingruppenarbeit dem Patienten schaden. Dies hat u.a. zur Folge, daß vorklinischen Studenten wegen angenommener Unkenntnis die Teilnahme an Anamnesegruppen verwehrt wird. Diese Befürchtungen sind durch vielfältige Erfahrungen widerlegt worden (Schüffel et al., 1983). Im Interesse der Studenten besteht die Hauptaufgabe darin, die spontane Gruppenbildung entsprechend deren Neigungen zu fördern. Hierbei besteht die

Schwierigkeit, daß die Gruppenzusammensetzung von Fach zu Fach variiert. Eine durchgehende Gruppenbildung wird in den offiziellen Curricula der Fakultäten kaum in Betracht gezogen.

Zu 6. **Clearing-Stelle:** Sie hätte die Aufgabe, die Qualität der geleisteten Patientenarbeit in Rückkopplung mit den beteiligten Studenten zu beurteilen („monitoring"). Jeder Patientenkontakt sollte dokumentiert und seine Ergebnisse sollten dem behandelnden Arzt zugänglich gemacht werden. Arzt und Student sollten die erhobenen Befunde besprechen. – Aus dem Miteinander von Arzt und Student, das ja in der Regel im Rahmen der Fakultät stattfindet, entwickeln sich weiterführende Fragen. So wurden längerfristige Krankenbetreuungen durchgeführt und Stationsbesprechungen eingerichtet (Bhandari et al., 1983). Es gibt Dozenten, die einer derartigen Arbeit skeptisch-ablehnend, andere, die ihr positiv-fördernd gegenüberstehen. Da es sich hier um die genannten „sozialisierenden Kräfte" handelt, ist es von besonderer Bedeutung, unter Mitwirken einer Clearing-Stelle zu einem Austausch zu kommen. Dies könnte unter Umständen der Lehr- und Studienausschuß oder eine informelle, aus Studenten und Ärzten zusammengesetzte Arbeitsgruppe sein.

Zu 7. **Überregionaler Austausch:** Aufgrund ihres anregenden und gleichzeitig informativen Charakters haben sich überregionale Treffen zwischen den Studenten als überaus förderlich erwiesen. Hierzu gehört das jährliche von B. und W. Luban-Plozza eingerichtete Balint-Treffen in Ascona, neuerdings (1990) in Szeged (Ungarn), auf dem Studenten aus verschiedenen Fakultäten der BRD, der Schweiz, Österreichs, aus Frankreich und Italien nicht nur über Erfahrungen auf dem Gebiet der personenbezogenen Ausbildung berichten, sondern auch die Möglichkeit haben, aktiv am Erfahrungsaustausch von Ärzten teilzunehmen und sich schließlich an einem jährlichen Preisausschreiben für Studenten auf dem Gebiet der patientenzentrierten Ausbildung zu beteiligen. – Zur Förderung der Arbeit in Anamnesegruppen wurde ein spezielles Arbeitstreffen in Form des sog. Maitreffens entwickelt, das jährlich in Marburg stattfindet.

Derartige überregionale Treffen sind auch für die Mitarbeiter der entsprechenden Ausbildungsstellen und für die Supervisoren von großer Hilfe. Sind doch auch die Dozenten medizinischer Fakultäten dem Druck verschiedener sozialisierender Kräfte ausgesetzt, der sie in Richtung eines spezialistisch-reduktionistischen Verhaltens drängt. Ein selbsterfahrungsbezogener Austausch kann derartigen Entwicklungen entgegenwirken. Die Belastungen der Dozenten sind bisher mit ganz wenigen Ausnahmen nicht der Gegenstand spezieller Überlegungen gewesen (Schüffel, 1983; Woodmansey, 1983). Hier zeichnet sich eine erfreuliche Wende ab, die durch zwei Entwicklungen zu belegen ist. In Europa gewinnt die Association for Medical Education in Europe (AMEE) langsam aber stetig Gewicht. Innerhalb des Deutschen Kollegiums für Psychosomatische Medizin (DKPM) wurde 1982 eine Arbeitsgruppe für Hochschullehrer/Hochschulfragen eingerichtet.[4]

67.6.3 Ein langer Weg in Einzelschritten

Aufgrund seiner Erfahrungen an der Berner Universität kommt E. Heim unter dem Hinweis auf das fortschrittliche Curriculum von McMaster zu dem

---

4 Prof. Dr. P. Hahn, L. Krehl-Klinik, 6900 Heidelberg; oder Sekretariat des DKPM, Abteilung Psychosomatik, Zentrum für Innere Medizin der Universität, Baldingerstr., 3550 Marburg.

Schluß, daß wir in der Regel nicht wie Reformuniversitäten völlig neue Curricula einführen können, sondern in einzelnen Schritten vorgehen müssen (1984). Gleichwohl kann, wie das Berner Beispiel zeigt, ein integriertes 6jähriges Curriculum für psychosoziale Medizin entstehen, das nunmehr durch eine „responsive Evaluation" abgesichert, ausgebaut und entwickelt werden sollte (Heim, 1989). Ermutigend sind auch die Erfahrungen aus den Anamnesegruppen, die sich jetzt an ca. 25 deutschsprachigen Universitäten finden (Schüffel, 1988). Beschränken wir uns auf klar definierte Ausbildungsziele, auf umschriebene Lernabschnitte und berücksichtigen eine problemzentrier-

te Rückkopplung zwischen Lehrenden und Lernenden, so lassen sich sehr wohl Fortschritte erzielen, die dem immer wieder zu begegnenden Pessimismus entgegengehalten werden können. Der Dekan einer nordamerikanischen Universität beschrieb den Weg zu dem Ziel einer patientenorientierten Ausbildung einem Korrespondenten des TIME-Magazins folgendermaßen (23. 5. 83): „Wenn wir unsere Studenten mitfühlender erziehen wollen, so müssen wir als Fakultät und Verwaltung gleichfalls mitfühlend sein". („If we want our students to be compassionate, we as faculty and administrators have to be compassionate too".)

68 Fort- und Weiterbildung in der Psychosomatischen Medizin

Wolfgang Wesiack, Karl Köhle und *Othmar W. Schonecke*

Im Laufe der letzten Jahre hat die Diskussion um die Wünsche und Möglichkeiten einer Integration der psychosomatischen Betrachtungsweise in die Medizin eine gewisse Klärung gebracht. Für die Fort- und Weiterbildung läßt sich heute als sinnvolles Ziel ein zweigliedriges Modell vertreten:

- **Fortbildung** für Ärzte aller Fachrichtungen – unserer Ansicht nach sollte sie obligatorisch sein: Curriculum, das den Erwerb psychosozialer Kompetenz vermittelt, wie es Voraussetzung für die kassenrechtliche Zulassung zur „psychosomatischen Grundversorgung" ist (Helmich et al., 1990).
- **Weiterbildung** für Ärzte und Psychologen, die in ihrer Berufsarbeit eine psychosomatische Betrachtungs- und Behandlungsweise vertieft integrieren wollen: ihnen wird das Curriculum empfohlen, das vom Deutschen Kollegium für Psychosomatische Medizin (DKPM) entwickelt worden ist. Es soll Voraussetzungen schaffen, die dem Erwerb eines Zusatztitels entsprechen, wie er für Psychotherapie (für Ärzte) bereits existiert. Die „Lehr-Psychosomatiker" sollen sich aus der Gruppe dieser Ärzte und Psychologen rekrutieren.

Gebietsbezeichnung „Psychosomatik" für Ärzte?

Psychosomatische Medizin im Sinne dieses Lehrbuches ist gleichzeitig ein Grundlagenfach der klinischen Medizin und eine abgegrenzte Subdisziplin mit spezifischen Aufgaben. Da „Psychosomatik" als Grundlagenfach eine Form des Zugangs zum kranken Menschen bezeichnet, die in vielen, wenn nicht allen klinischen Fächern benötigt wird, erscheint uns eine eigene Gebietsbezeichnung für dieses Fach problematisch.

68.1 Fortbildung: „Psychosomatische Grundversorgung"

In der BRD ist am 1. 10. 1987 die Neufassung der „Psychotherapie-Richtlinien" (Richtlinien des Bundesausschusses der Ärzte und Krankenkassen über die Durchführung der Psychotherapie in der Kassenärztlichen Versorgung vom 3. 7. 1987) in Kraft getreten. In diese Neufassung ist der Begriff der **„Psychosomatischen Grundversorgung"** eingeführt worden. Damit besteht die Chance, ärztliche Leistungen in die kassenärztliche psychotherapeutische Versorgung

einzubinden, die bisher nur unzureichend in der Weiterbildungs- und Gebührenordnung berücksichtigt waren. Psychotherapeutische Leistungen des Primärarztes könnten so einen neuen Stellenwert bekommen.

Wir geben hier zunächst den Text „Psychosomatische Grundversorgung" aus den Psychotherapie-Richtlinien wieder (nach Faber und Haarstrick, 1989, S. 107–108).

68.1.1 Text der Psychotherapie-Richtlinien

„Die Psychosomatische Grundversorgung kann nur im Rahmen einer übergeordneten somato-psychischen Behandlungsstrategie Anwendung finden. Voraussetzung ist, daß der Arzt die ursächliche Beteiligung psychischer Faktoren an einem komplexen Krankheitsgeschehen festgestellt hat oder aufgrund seiner ärztlichen Erfahrung diese als wahrscheinlich annehmen muß. Ziel der Psychosomatischen Grundversorgung ist eine möglichst frühzeitige differentialdiagnostische Klärung komplexer Krankheitsbilder, eine verbale oder übende Basistherapie psychischer, funktioneller und psychosomatischer Erkrankungen durch den primär somatisch orientierten Arzt und ggf. die Indikationsstellung zur Einleitung einer ätiologisch orientierten Psychotherapie durch einen psychoanalytisch oder verhaltenstherapeutisch behandelnden Arzt.

Die begrenzte Zielsetzung der Psychosomatischen Grundversorgung strebt eine an der aktuellen Krankheitssituation orientierte seelische Krankenbehandlung an; sie kann während der Behandlung von somatischen, funktionellen und psychischen Störungen von Krankheitswert als verbale Intervention oder als Anwendung übender Verfahren vom behandelnden Arzt durchgeführt werden."

Verbale Intervention

„Die verbalen Interventionen orientieren sich in der Psychosomatischen Grundversorgung an der jeweils aktuellen Krankheitssituation; sie fußen auf einer systematischen, die Introspektion fördernden Gesprächsführung und suchen Einsichten in psychosomatische Zusammenhänge des Krankheitsgeschehens und in die Bedeutung pathogener Beziehungen zu vermitteln. Der Arzt berücksichtigt und nutzt dabei die krankheitsspezifischen Interaktionen zwischen Patient und Therapeut, in denen die seelische Krankheit sich darstellt. Darüber hinaus wird angestrebt, Bewältigungsfähigkeiten des Kranken, evtl. unter Einschaltung der Beziehungsperson und Personen aus dem engeren Umfeld, aufzubauen.

Die verbalen Interventionen können nur in Einzelbehandlungen durchgeführt und nicht mit suggestiven und übenden Techniken in derselben Sitzung kombiniert werden; sie können in begrenztem Umfang sowohl über einen kürzeren Zeitraum als auch im Verlauf chronischer Erkrankungen über einen längeren Zeitraum niederfrequent Anwendung finden, wenn eine ätiologisch orientierte Psychotherapie" (psychoanalytisch begründete Verfahren und Verhaltenstherapie) „nicht indiziert ist." Wird eine solche Psychotherapie begonnen, ist die Psychosomatische Grundversorgung nicht weiter abrechenbar.

Übende und suggestive Techniken

Hier werden aufgeführt:
- Autogenes Training als Einzel- oder Gruppenbehandlung (Unterstufe)
- Jacobsonsche Relaxationstherapie als Einzel- oder Gruppenbehandlung
- Hypnose in Einzelbehandlung

In die Anwendung dieser Techniken sind Instruktionen und die Bearbeitung therapeutisch bedeutsamer Phänomene (auch abrechnungstechnisch) eingeschlossen. Während einer tiefenpsychologisch fundierten und analytischen Psychotherapie dürfen diese Techniken grundsätzlich nicht angewandt werden, d.h. nicht abgerechnet werden.

68.1.2 Kommentar

Zielsetzung

Die Zielsetzung entspricht zunächst unseren in diesem Buch dargestellten Vorstellungen: Der Arzt sollte möglichst frühzeitig eine „Gesamtdiagnose" im Sinne Balints stellen, danach die Indikation für eine verbale oder übende Basistherapie, die er selbst durchführt, bzw. die Indikation für eine Überweisung zum Fachpsychotherapeuten prüfen.

„Basistherapie" wird ganzheitlich verstanden. Sie umfaßt somatische Therapie und seelische Krankenbehandlung. Letztere erfolgt mit begrenzter Zielsetzung: Sie orientiert sich an der aktuellen Krankheitssituation; diese kann durch akute seelische Krisen, aber auch durch chronische Krankheiten und Behinderungen bestimmt werden. Angestrebt wird Symptombeseitigung sowie Einsicht des Patienten in pathogene Zusammenhänge und in die Notwendigkeit einer prophylaktischen Umorientierung.

Definition – Abgrenzung zu „Beratung" und „Erörterung"

„Die ärztliche **Beratung** stellt Information und Empfehlung durch den Arzt in den Vordergrund. Das Beratungsgespräch hat vorwiegend monologischen Charakter" (Faber und Haarstrick, 1989).

„In der **Erörterung** dagegen findet zwischen Arzt und Patient oder zwischen Arzt und Bezugsperson ein Dialog statt." „Es muß ein persönlicher Kontakt zwischen Patient und Arzt in der direkten Begegnung zustande kommen. Der Arzt muß die Reaktion des Gesprächspartners Patient wahrnehmen, seine Stimmungslage und Aufnahmebereitschaft beobachten und diese Beobachtung in die Planung gezielter therapeutischer Maßnahmen einbeziehen" (Faber und Haarstrick, 1989). Die Psychosomatische Grundversorgung unterscheidet sich „von den genannten ärztlichen Tätigkeiten qualitativ..." Sie stellt „noch wesentlich höhere Anforderungen" an den Arzt.

68.1.3 Behandlungsmethode

Die „Basistherapie" der Richtlinien enthält verbale Interventionen sowie übende und suggestive Verfahren.

Die verbale Intervention „stellt eine besondere Form der ärztlichen Gesprächsführung dar, die das Ziel verfolgt,
- eine „Innenschau" (Introspektion) des Patienten anzuregen,
- Einsichten in die psychosomatischen Zusammenhänge seines Krankheitsgeschehens zu vermitteln und
- die Bedeutung gegebenenfalls krankmachender persönlicher Konflikte des Patienten für ihn erkennbar zu machen" (Faber und Haarstrick, 1989).

68.1.4 Durchführungsbestimmungen und Wirtschaftlichkeitsprüfung

Verbale Interventionen können nur in einer Einzelbehandlung durchgeführt werden. In derselben Sitzung können sie nicht mit übenden oder suggestiven Techniken kombiniert werden. Die Dauer der Sitzung beträgt mindestens zwanzig Minuten.

Die verbale Intervention unterliegt im Rahmen der kassenärztlichen Versorgung der allgemeinen Wirtschaftlichkeitsprüfung. Der Umfang dieser Leistung wurde nicht verbindlich festgelegt. „In akuten seelischen Krisen sollte verbale Intervention über einen ‚kürzeren Zeitraum' Anwendung finden ..." „In der Regel dürften dafür vier bis sechs Wochen ausreichen." Im Verlaufe chronischer Krankheiten und Behinderungen kann verbale Intervention auch über einen „längeren Zeitraum" zur Anwendung kommen, d.h. aber in jenem „begrenzten Umfang", der sich aus der „jeweils aktuellen Krankheitssituation" ergibt.

Die „Wirtschaftlichkeitsprüfung" wird für die verbale Intervention „bei einer hochfrequenten, täglich oder mehrfach am Tage oder über längere Wochen durchgeführten Anwendung" „besondere Bedeutung haben" (Faber und Haarstrick, 1989). Wird sie über einen längeren Zeitraum angewandt, „ist auch zu prüfen, ob nicht eine ätiologisch orientierte Psychotherapie" indiziert ist.

„Zur Entlastung der Prüfverfahren und der Therapeuten wurde in den Richtlinien festgelegt, daß im Behandlungsfall in der Regel zwölf Sitzungen in jeder der drei zugelassenen Techniken" (verbale Intervention, übende und suggestive Techniken) „zur Verfügung stehen." „Die verbalen Interventionen unterliegen" jedoch „keiner von vornherein eingrenzenden Bestimmung der Richtlinien, sondern nur der Allgemeinen Wirtschaftlichkeitsprüfung durch die KV" (Faber und Haarstrick, 1989).

68.1.5 Erforderliche Qualifikation – Fortbildung

Die **Psychotherapie-Richtlinien** legen fest: „Die Teilnahme des Arztes an der Psychosomatischen Grundversorgung setzt mehrjährige Erfahrung in selbständiger ärztlicher Tätigkeit, Kenntnisse in der Theorie einer psychosomatisch orientierten Krankheitslehre und reflektierte Erfahrungen über die therapeutische Bedeutung der Arzt-Patient-Beziehung voraus."

Die **Psychotherapie-Vereinbarungen** fordern darüber hinaus, daß der Arzt Maßnahmen in der Psychosomatischen Grundversorgung entsprechend den Nummern 850 (Diagnostik) und 851 (Verbale Intervention) BMÄ/E-GO nur mit der Einwilligung der für seinen Kassenarztsitz zuständigen kassenärztlichen Vereinigung ausführt. Diese Einwilligung setzt folgende Nachweise voraus:
- Eine mindestens dreijährige Erfahrung in selbstverantwortlicher ärztlicher Tätigkeit
- Den Erwerb von Kenntnissen in einer psychosomatisch orientierten Krankheitslehre
- Reflektierte Erfahrungen über die psychodynamische und therapeutische Bedeutung der Arzt-Patient-Beziehung (Faber und Haarstrick, 1989).

Wir verstehen Psychotherapie vor allem als Beziehungstherapie. Hierbei kommt der Fähigkeit des Arztes zur Selbstwahrnehmung in der Beziehung zum Patienten besondere Bedeutung zu. Für den psychosomatisch tätigen Arzt empfiehlt sich deshalb ein Mindestmaß an Selbsterfahrung. Ein solches Mindestmaß ist für die Zulassung zur Psychosomatischen Grundversorgung allerdings nicht vorgeschrieben – im Gegensatz zur Weiterbildung zum Psychotherapeuten. Ärzten, die eine solche **Selbsterfahrung** anstreben, empfehlen wir eine Einzelselbsterfahrung, eine „Lehr-Therapie", da sie selbst mit Patienten ja auch in der Zweierbeziehung arbeiten. Gruppenselbsterfahrung kann als Ergänzung sinnvoll sein.

Die **theoretische Fortbildung** sollte wenigstens zwanzig Doppelstunden umfassen. Für die inhaltliche Ausgestaltung bieten Helmich et al. (1990) einen Leitfaden; eine Beteiligung der Fortzubildenden an der Auswahl der Themen verbessert den Lernerfolg („Lernen auf Verlangen").

Eine Einführung in **psychosomatisch-biographische Anamneseerhebung** sollte wenigstens zehn Doppelstunden umfassen.

Die Fähigkeit zur **Reflexion der Arzt-Patient-Beziehung** wird über eine kontinuierliche Teilnahme an einer Balint-Gruppe über wenigstens vierzig Doppelstunden vermittelt. Diese Balint-Gruppe sollte von einem psychosomatisch erfahrenen Arzt mit dem Zusatztitel „Psychotherapie" geleitet werden. In Ausnahmefällen kann eine Einzelsupervision über wenigstens vierzig Stunden bei einem entsprechend ausgebildeten Arzt als gleichwertig angesehen werden.

68.1.6 Weiterbildungsrichtlinien für Österreich

In Österreich ist bis zum gegenwärtigen Zeitpunkt die Weiterbildung in Psychosomatischer Medizin und Psychotherapie noch nicht gesetzlich geregelt. Zur Zeit wird um ein Psychotherapiegesetz gerungen.

1989 hat jedoch die Österreichische Bundesärztekammer ein Dreistufenmodell beschlossen, das in Zukunft verbindlich für die Weiterbildung sein soll:

Modul 1: Erhöhung der psychosozialen Kompetenz der Ärzte.

Zunächst werden diese Weiterbildungsveranstaltungen, die etwa der „psychosomatischen Grundversorgung" in der BRD entsprechen, auf freiwilliger Basis durchgeführt. Es ist aber vorgesehen, diese zu einem späteren Zeitpunkt in die dreijährige Turnusausbildung der Ärzte als Pflichtweiterbildung zu integrieren.

Modul 2: Weiterbildung in psychosomatischer Medizin.

In Anlehnung an die auf Seite 1263 veröffentlichten Rahmenrichtlinien für Ärzte wurde ein Weiterbildungscurriculum entworfen, das sich über einen Zeitraum von 2 Jahren erstreckt. Es umfaßt etwa 150 Stunden Theorie und praktische Übungen sowie je 100 Stunden Selbsterfahrung und Balint-Gruppenarbeit.

Im Juli 1989 wurde im Bereich der Landesärztekammer Tirol für 45 Ärzte und Ärztinnen das erste zweijährige Curriculum abgeschlossen. Nach Einreichen eines Fallberichtes und nach erfolgreich bestandenem Prüfungsgespräch erhalten diese Ärzte und Ärztinnen von der Ärztekammer ein Diplom „Psychosomatische Medizin".

Ab Herbst 1989 werden ähnliche Weiterbildungsveranstaltungen auch von anderen österreichischen Landesärztekammern durchgeführt.

Modul 3: Weiterbildung zum Fachpsychotherapeuten.

Diese Weiterbildung erfolgt nach den Richtlinien der verschiedenen psychotherapeutischen Fachgesellschaften. Die Ärztekammer wird den so weitergebildeten Ärzten ein Diplom „Psychotherapie" verleihen.

Die Weiterbildung der Psychologen soll in einem Psychologengesetz geregelt werden, das in Vorbereitung ist.

68.1.7 Anspruch und Wirklichkeit

Wir wollen nicht verschweigen, daß unsere Zielvorstellung, Psychotherapie in die primärärztliche Versorgung zu integrieren, einen hohen Anspruch an die Aufnahmebereitschaft, die emotionale Tragfähigkeit und die fachliche Kompetenz des Arztes und d. h. auch einen hohen Anspruch an seine Fortbildungsbereitschaft stellt. Die Vielfalt der Problematik, die Verwobenheit körperbezogener Klagen bzw. körperlicher Erkrankungen mit psychischen und sozialen Problemen stellt an den primärversorgenden Arzt in seiner Arbeitssituation zum Teil höhere Anforderungen, als sie sich für den spezialisierten Fachpsychotherapeuten ergeben; dieser kann mit wenigen, nach ihrer Motivation ausgewählten Patienten unter klar definierten Rahmenbedingungen arbeiten. Die Einbeziehung der Psychotherapie in die Tätigkeit des primärversorgenden Arztes erscheint uns jedoch von so grundsätzlicher Bedeutung für das Gesundheitssystem, daß wir meinen, daß ein großer Aufwand hier in besonderem Maße gerechtfertigt ist. Die Finanzierung „verbaler Interventionen" im Rahmen der „psychosomatischen Grundversorgung" durch die Krankenkassen ist ein erster Schritt in einer anzustrebenden Entwicklung. Die Honorierung dieser anspruchsvollen Tätigkeit erscheint allerdings auch in diesem Ansatz noch weitgehend unangemessen; es handelt sich um eine persönliche Leistung des Arztes im engsten Sinne des Wortes, die zudem eine kontinuierliche, mit nicht unerheblichen Kosten verbundene Fortbildung erfordert.

Für die Beurteilung der Situation ist es wichtig, sich zu verdeutlichen, daß nach den vorliegenden epidemiologischen Untersuchungen zwischen 11,3% (minimal) und 38% (maximal) aller Patienten in unserem Gesundheitswesen an Störungen leiden, bei denen eine psychosomatische Abklärung und – in der Regel – eine psychotherapeutische Behandlung erforderlich ist. Bisher wird für diesen Anteil der Kranken weniger als 1% der Kosten unseres Gesundheitswesens aufgewendet (Meyer, 1988). Wichtig erscheint uns, daß die Versorgung dieser Patienten innerhalb des ärztlichen Versorgungssystems stattfinden kann und nicht aus diesem ausgegliedert wird.

68.2 Weiterbildung

Der Weiterbildungsausschuß des Deutschen Kollegiums für Psychosomatische Medizin bemühte sich bei der Erarbeitung dieser Rahmenrichtlinien um eine die einzelnen Schulrichtungen übergreifende Konzeption. Dabei kam es vor allem auch darauf an, die jeweils grundlegenden Psychotherapieformen, einerseits die aus der Psychoanalyse abgeleiteten, andererseits die aus der Lerntheorie abgeleiteten, jeweils ausreichend zu berücksichtigen.

Mediziner und Psychologen sind aufgrund verschiedener Grundstudien in einer entsprechend unterschiedlichen Ausgangslage für die psychosomatische Weiterbildung. Während das Medizinstudium eine große Menge an Wissen über organische Bedingungen und deren Zusammenhänge vermittelt, sowie Regeln für dessen Anwendung auf den konkreten

Einzelfall, ist das Studium der Psychologie, wie es an den deutschen Hochschulen durchgeführt wird, mehr theoretisch und methodisch orientiert, auch wenn in den letzten Jahren praxisbezogene Lehrinhalte vor allem im zweiten Studienabschnitt deutlich mehr in den Vordergrund getreten sind.

Die Inhalte des Psychologiestudiums beziehen sich auf menschliches Verhalten und dessen Bedingungen, wobei die organischen Bedingungen des Verhaltens einen eher geringeren Raum einnehmen. Das Studium der Medizin beinhaltet andererseits nur in geringem Umfang Wissensinhalte, die den menschlichen Organismus als einen sich in einer Umgebung oder „Umwelt", vor allem auch sozial verhaltenden verständlich machen. Absolventen eines Medizinoder Psychologiestudiums sind also entsprechend komplementär nur eingeschränkt darauf vorbereitet, sich mit Fragen eines Verständniszuganges zu befassen, der z. B. auch Kranksein im menschlichen Leben als einen Vorgang versteht, der nicht nur organischphysikalische Relevanz besitzt – diese allerdings auch –, sondern das gesamte Leben in einer bestimmten Umgebung betrifft und von diesem Leben in seinem Verhältnis von Organismus und Umgebung auch bedingt ist.

Ziel einer psychosomatischen Weiterbildung ist es, diesen allgemeinen Verständnisansatz spezifisch und konkret auszufüllen, um ein wissenschaftlich begründetes Handeln zu ermöglichen.

Für Psychologen kann und soll die psychosomatische Weiterbildung nicht ein Studium der Medizin ersetzen, wie dies auch umgekehrt nicht möglich ist, sie soll vielmehr die spezifischen Kenntnisse von Psychologen für die Psychosomatik erweitern und anwendbar machen.

Dem interdisziplinären Charakter der Psychosomatik entsprechend, soll die Weiterbildung die notwendige Verständigung zwischen Ärzten und Psychologen in der Versorgung von Patienten, in der Forschung und gegebenenfalls in der Weiterbildung von Kollegen ermöglichen und weiterentwickeln. Dazu ist es auch notwendig, Grundlagen medizinischen und psychologischen Wissens zu erwerben.

Wir gehen davon aus, daß die Weiterbildung im Bereich der Psychosomatischen Medizin entsprechend den tatsächlichen Arbeitsanforderungen interdisziplinär angelegt sein muß, und daß an einer solchen interdisziplinären Zusammenarbeit alle in der Krankenversorgung tätigen Personen beteiligt sein sollten. Wir legen hier zwar lediglich die Rahmenrichtlinien für die Weiterbildung der Ärzte und Psychologen vor, berücksichtigen dabei jedoch die gestellten Aufgaben zur interdisziplinären Kooperation. Entsprechend der Bedeutung der Beziehung zwischen Arzt und Patient in psychosomatischer Diagnostik und Therapie kommt der Selbsterfahrung und der Supervision der diagnostischen und therapeutischen Tätigkeit in den Rahmenrichtlinien des Deutschen Kollegiums für Psychosomatische Medizin besondere Bedeutung zu. Die angeführten Ziele des Selbsterfahrungsprozesses schließen dabei natürlich nicht die grundsätzliche Überzeugung aus, daß sich jeder Therapeut in einem ständigen Prozeß der Entwicklung, Reifung und Erweiterung seiner Persönlichkeit befinden sollte. Dies kann jedoch nicht ein Weiterbildungsziel im engeren Sinne sein. Eine persönliche Psychoanalyse z.B. wird demnach nicht gefordert, wohl aber empfohlen.

Die Zusammenstellung der erforderlichen theoretischen und praktischen Kenntnisse entspricht Minimalforderungen und erlaubt mehr eine allgemeine Orientierung. Der Weiterbildungsausschuß war sich hier im klaren, daß ein Vielfaches an Zeit zur Vermittlung dieses Lehrstoffes erforderlich wäre und deshalb ein sehr intensives Literaturstudium erwartet werden muß. Wir hoffen, daß das aus diesen Weiterbildungslinien abgeleitete Curriculum dazu beitragen wird, daß die psychosomatischen Gesichtspunkte bei vielen Patienten mit funktionellen Störungen und körperlichen Erkrankungen früher und häufiger erkannt werden können, als dies heute der Fall ist; hierdurch ließen sich schwerwiegende Chronifizierungsprozesse vermeiden, wie sie heute noch die Regel sind.

Vom Deutschen Kollegium für Psychosomatische Medizin wurde eine Kommission, bestehend aus Psychologen und Ärzten, beauftragt, Richtlinien für die psychosomatische Weiterbildung von Ärzten und Diplompsychologen zu erarbeiten, die nun im folgenden wiedergegeben werden, um spezifische Inhalte der Weiterbildung zu kennzeichnen.

In Abschnitt 68.3 bringen wir die Rahmenrichtlinien zur Weiterbildung für Ärzte und in Abschnitt 68.4 die Rahmenrichtlinien zur Weiterbildung für Psychologen. Da die Balint-Gruppenarbeit einen besonders wichtigen Anteil an allen Bemühungen um Weiter- und Fortbildung im Bereich der Psychosomatischen Medizin hat, wurde ihr Abschnitt 68.5 gewidmet.

68.3 Rahmenrichtlinien für die Weiterbildung der Ärzte

68.3.1 Eingangsbedingungen

Die Weiterbildung eines jeden Arztes beginnt nach der Approbation. Frühestens nach zwei Jahren klinisch-organmedizinischer Weiterbildung unter Leitung eines dazu ermächtigten Facharztes und nach einem Jahr Tätigkeit in der Psychiatrie bei einem mindestens zur zweijährigen Weiterbildung in Psychiatrie ermächtigten Arzt, kann der Arzt sich um Aufnahme in eine Institution bemühen, die eine psychosomatische Weiterbildung im engeren Sinne vermittelt. Während dieser Zeit soll der Arzt mindestens ein Jahr lang (ca. 35 Doppelstunden) kontinuierlich an einer Balint-Gruppe teilgenommen haben. Der Leiter muß von der für ihn zuständigen Ärztekammer zur Weiterbildung in Balint-Gruppenarbeit ermächtigt sein.

68.3.2 Psychosomatische Weiterbildung

Die psychosomatische Weiterbildung im engeren Sinne sollte vor allem während einer dreijährigen klinischen und/oder poliklinischen Tätigkeit in einer Institution erfolgen, in der organische, psychologische und psychosoziale Gesichtspunkte in Diagnostik und Therapie integrativ einbezogen werden. Wegen der Kontinuität der Weiterbildung erscheint es uns wichtig, daß von diesen drei Jahren mindestens zwei Jahre zusammenhängend in einer Institution abgeleistet werden. Diese Institutionen müssen vom DKPM als Ausbildungsinstitution anerkannt worden sein. Während dieser dreijährigen klinischen und/oder poliklinischen Weiterbildungstätigkeit sollte kontinuierlich ein Selbsterfahrungs- und ein Supervisionsprozeß ablaufen.

68.3.3 Selbsterfahrung

Sowohl die Teilnahme an Selbsterfahrungsgruppen als auch die Teilnahme an Balint-Gruppen zielt darauf ab,

– das Erleben von therapeutischen oder Interventionsprozessen sowie die eigenen Grenzen tiefer zu erkennen und – wenn möglich – zu erweitern,
– sich in einer psychotherapeutischen Methode ausreichende Kenntnisse und Erfahrungen anzueignen und diese Methode selbst zu erleben,
– die Interaktionen zwischen Patient und Arzt, Arzt und Behandlungsteam, Arzt und Institution zu erleben und zu reflektieren;
– die Selbsterfahrung sollte das Erleben und Erlernen einer körperbezogenen Methode beinhalten.

Im einzelnen werden mindestens 70 Doppelstunden in einer kontinuierlich laufenden Selbsterfahrungsgruppe gefordert. Je nach Weiterbildungsschwerpunkt muß die Selbsterfahrung psychoanalytisch oder verhaltenstherapeutisch orientiert sein. Bei psychoanalytischer Orientierung sind nur 35 Doppelstunden Gruppenselbsterfahrung erforderlich, wenn eine selbstgewählte weiterbildende Lehranalyse von mindestens 100 Stunden vorliegt. Während der psychosomatischen Weiterbildungszeit sind mindestens weitere 35 Doppelstunden Balint-Gruppenarbeit erforderlich.

Die Weiterbildung der körperbezogenen Selbsterfahrung sollte mindestens 40 Stunden umfassen.

68.3.4 Supervision

Der Supervisionsprozeß sollte folgende Zielrichtungen umfassen:

– **Problemfokussierung.** Inwieweit vermag der Kandidat eine umfassende Diagnose und entsprechende therapeutische Indikationen zu stellen? Dabei beinhaltet die „umfassende" oder „Gesamtdiagnose" die Synthese der somatischen und psychosozialen Teilaspekte unter Berücksichtigung der ätiologischen, pathogenetischen und krankheitsreaktiven Faktoren.

– **Therapiemethode.** Inwieweit beherrscht der Kandidat die Anwendung verschiedener therapeutischer Verfahren und kennt deren Reichweite? Ist er in der Lage, auch die psychologische Bedeutung der verschiedenen somatischen diagnostischen und therapeutischen Methoden zu beurteilen?

Die Supervision sollte als Einzelsupervision und als Gruppen- bzw. Teamsupervision erfolgen.

Nachzuweisen sind mindestens 10 selbstausgeführte und abgeschlossene Behandlungen mit insgesamt mindestens 400 Behandlungsstunden und mindestens 100 Supervisionsstunden. Die Fälle sind eingehend zu dokumentieren. Empfehlenswert ist zusätzlich noch eine Supervision der Institution.

68.3.5 Wissenschaftliche Grundlagen

Die Vermittlung theoretischer Kenntnisse hat vor allem zum Ziel:

– Für die angewandten diagnostischen und therapeutischen Verfahren wichtige Grundlagen zu vermitteln.
– Den Arzt zur interdisziplinären Zusammenarbeit zu befähigen.

Das theoretische Angebot baut auf den in der ärztlichen Ausbildung vermittelten Kenntnissen in den Fächern medizinische Psychologie, medizinische Soziologie, Psychiatrie, Psychotherapie und Psychosomatik auf.

(1) Theoretische und praktische Kenntnisse in allgemeiner und spezieller Psychosomatischer Medizin:

– diagnostisch-therapeutische Strategien (Methodik und Praxis der psychosomatischen Konsultationen, Technik der psychosomatisch orientierten Gesprächsführung)
– psychoanalytische, lerntheoretische, psychophysiologische und sozialwissenschaftliche Modelle der Psychosomatik
– spezielle Krankheitslehre
– hauptsächliche wissenschaftstheoretische Auffassungen, die für das Gebiet der Psychosomatik wesentlich sind.

(2) Wissenschaftliche Grundkenntnisse:

– psychoanalytische Persönlichkeitstheorie
– Theorie der Emotionen, Streßtheorien
– Theorie des Lernens, Verhaltenstheorien
– Theorie der sozialen Interaktion und Kommunikation
– biologische Grundlagen des Verhaltens.

(3) Einführende Kenntnisse (Überblick und allgemeine Prinzipien):

– Entwicklungspsychologie (Sozialisation, Verhaltensgenetik)
– Persönlichkeitspsychologie (differentielle Psychologie, ausgewählte Persönlichkeitstheorien, Konstitutionsforschung)
– Sozialpsychologie (Einstellungen, Gruppenprozesse, Familie, soziologische Grundbegriffe)
– psychologische Diagnostik (Technik der Erstuntersuchung, Verhaltensbeobachtung, Tests, diagnostische Strategien)
– Psychopathologie, allgemeine und spezielle Neurosenlehre, Theorie abweichenden Verhaltens
– Indikation und Evaluation von Therapien (Methodik und ausgewählte Forschungsergebnisse)
– Systemanalyse und Organisationsstruktur in therapeutischen Einrichtungen.

Die Vermittlung dieser Kenntnisse sollte nicht so sehr durch Vorlesungen, sondern vor allem in Form von Seminaren, Arbeitsgruppen usw. erfolgen und insgesamt nachweislich mindestens 250 Stunden umfassen.

68.4 Rahmenrichtlinien für die Weiterbildung der Psychologen

68.4.1 Eingangsbedingungen

Die Weiterbildung eines jeden Psychologen beginnt nach dem Diplom. Frühestens nach zwei Jahren praktischer klinisch-psychologischer Tätigkeit (davon mindestens sechs Monate im Bereich der psychiatrischen Krankenversorgung) kann der Psychologe mit der Weiterbildung in Psychosomatischer Medizin im engeren Sinne in einer hierfür vom DKPM anerkannten Institution beginnen. Ferner ist die mindestens einjährige (ca. 70 Stunden) kontinuierliche Teilnahme an einer fallorientierten Gruppensupervision unter besonderer Berücksichtigung der Erlebens- und Beziehungsaspekte in der Therapeut-Patient-Beziehung nachzuweisen.

68.4.2 Psychosomatische Weiterbildung

Die psychosomatische Weiterbildung im engeren Sinne sollte vor allem während einer dreijährigen klinischen und/oder poliklinischen Tätigkeit in einer Institution berufsbegleitend erfolgen, in der organische, psychologische und psychosoziale Gesichtspunkte in Diagnostik und Therapie integrativ einbezogen werden. Zur Wahrung der Kontinuität der Weiterbildung sollten von diesen drei Jahren mindestens zwei Jahre zusammenhängend in einer Institution abgeleistet werden. Diese Institutionen müssen vom DKPM als Ausbildungsinstitutionen anerkannt worden sein.

68.4.3 Selbsterfahrung

Während der Weiterbildungstätigkeit muß der Nachweis über die kontinuierliche Teilnahme an einem Selbsterfahrungsangebot erbracht werden. Dazu gehört auch das Selbsterleben der jeweiligen Therapiemethoden in exemplarischer Form sowie das Erleben und Reflektieren der Interaktion zwischen Patient und Psychologe, Psychologe und Behandlungsteam, Therapeut und Institution.

Diese Selbsterfahrung sollte schwerpunktmäßig Methoden berücksichtigen, die im Rahmen psychosomatischer Therapien wichtig sind.

Die spezifischen Inhalte der Selbsterfahrung werden in Ausführungsbestimmungen gesondert geregelt; der erforderliche Zeitaufwand darf 200 Stunden nicht unterschreiten.

68.4.4 Supervision

Der Supervisionsprozeß muß übergreifend folgende Zielrichtungen umfassen:
- **Problemfokussierung.** Der Kandidat soll in der Lage sein, eine umfassende Diagnose und entsprechende therapeutische Indikationen zu stellen. Dabei beinhaltet die „umfassende" oder „Gesamtdiagnose" die interdisziplinäre Synthese der somatischen und psychosozialen Teilaspekte unter Berücksichtigung der ätiologischen, pathogenetischen und krankheitsreaktiven Faktoren.
- **Therapiemethode** und **therapeutische Beziehung.** Der Kandidat soll die Anwendung verschiedener therapeutischer Verfahren beherrschen, deren Reichweite kennen und sie fallspezifisch zur Anwendung bringen können. Er soll in der Lage sein, die psychologische Bedeutung und die medizinischen Folgen der verschiedenen diagnostischen und therapeutischen Methoden zu beurteilen.
- **Interdisziplinäre psychosoziale Kompetenz.** Der Kandidat soll psychosoziale Kompetenz im eigenen Verhalten in dem Sinne erwerben, daß er zur Beurteilung und Beeinflussung gruppendynamischer Prozesse und der Interaktion von Ärzten, Psychologen, Pflegepersonal und anderen im Bereich der Psychosomatik tätigen Berufsgruppen fähig ist.

Die Supervision sollte als Einzelsupervision und als Gruppen- bzw. Teamsupervision erfolgen. Nachzuweisen sind mindestens 10 selbstausgeführte und abgeschlossene Behandlungen mit insgesamt mindestens 400 Behandlungsstunden sowie mindestens 100 Supervisionsstunden. Die Fälle sind eingehend zu dokumentieren.

Empfehlenswert ist zusätzlich noch eine Supervision der Institution. Die spezifischen Inhalte des Supervisionsprozesses sowie der im einzelnen dafür erforderliche Zeitaufwand werden in Ausführungsbestimmungen gesondert geregelt.

68.4.5 Wissenschaftliche Grundlagen

Die Vermittlung theoretischer Kenntnisse hat vor allem zum Ziel:
- Die für die angewandten diagnostischen und therapeutischen Verfahren wichtigen medizinischen Grundlagen zu vermitteln.

- Den Psychologen zu interdisziplinärer Zusammenarbeit zu befähigen.

Das theoretische Angebot baut auf den im Diplomstudiengang Psychologie erworbenen und während der klinisch-psychologischen Weiterbildung vertieften Kenntnissen in Biologie, Psychologie, Psychophysiologie, Psychopathologie, Psychotherapie, der biologischen Grundlagen psychiatrischer und psychosomatischer Erkrankungen sowie der wissenschaftstheoretischen Begründung der Psychosomatik auf.

Insbesondere sollen folgende theoretische Lehrinhalte unter besonderer Berücksichtigung der Psychosomatik im Rahmen der Weiterbildung vermittelt werden:
- Grundkenntnisse in Pathophysiologie, Biochemie und Pharmakologie
- Krankheitslehre (innere, gynäkologische, neurologische, pädiatrische, orthopädische, HNO-, dermatologische Erkrankungen usw.)
- Indikation, Durchführung und Konsequenzen von Operationen
- Ökologische Medizin (Epidemiologie, Arbeits- und Sozialmedizin)
- Institutionslehre, Kenntnis der RVO und Sozialgesetzgebung
- Systemanalyse und Organisationsstruktur therapeutischer Einrichtungen und professionelle Kooperationsformen in Klinik und Praxis
- Gesundheitserziehung und -training
- Didaktik der Fortbildung und Schulung
- Methoden psychosomatischer Diagnostik (z.B. psychophysiologische Untersuchungsmethoden, vegetative Funktionsprüfung, Erstinterviewtechnik, Verhaltensbeobachtung, Persönlichkeits- und Leistungstests, Sozialanamnese)
- Differentialdiagnostik in der Psychosomatik unter Berücksichtigung psychischer, biologischer und sozialer Bedingungen und Folgen von Krankheit.

Die Vermittlung dieser Kenntnisse sollte nicht so sehr durch Vorlesungen, sondern vor allem in Form von Seminaren, Arbeitsgruppen usw. erfolgen und insgesamt nachweislich mindestens 250 Stunden umfassen.

68.5 Die Bedeutung der Balint-Gruppenarbeit für die Aus-, Weiter- und Fortbildung in der Psychosomatischen Medizin

Michael Balint hatte die, man kann schon sagen, geniale Idee, die ärztlich-naturwissenschaftliche und psychoanalytische Kompetenz in den heute nach ihm benannten Gruppen zur Zusammenarbeit anzuregen. Er versammelte praktizierende Ärzte um sich, ließ sie über ihre Problempatienten und Problemsituationen berichten und half ihnen als Psychoanalytiker, den psychodynamischen und psychosozialen Aspekt des Krankheitsgeschehens herauszuarbeiten. Dabei zeigt sich ein hochinteressantes Phänomen, auf dem letztlich die ganze Fruchtbarkeit und Tiefe der Balint-

Gruppenarbeit beruht: Es zeigt sich nämlich, daß sich die gesamte Beziehungsproblematik des Patienten in der konkreten und daher meist für den Arzt problematischen Interaktion zwischen ihm und seinem Patienten wiederfindet, wenn auch in einer durch die Persönlichkeit des Arztes mehr oder minder stark modifizierten Form. Indem nun der Arzt darüber in der Gruppe berichtet, vollzieht sich zwischen ihm und der Gruppe ähnliches. Das im Erleben und im Bericht des referierenden Arztes sich darstellende Beziehungsproblem des Patienten induziert einen Gruppenprozeß, der wiederum die Beziehungsproblematik des Patienten darstellt. Jetzt allerdings gebrochen und widergespiegelt durch das Erleben der einzelnen Gruppenteilnehmer, wodurch das Problem gewissermaßen prismatisch in viele Facetten der ursprünglichen Problematik aufgelöst und dadurch einerseits verbreitert und vertieft, andererseits aber auch durch zusätzliche Faktoren, nämlich die individuelle Psychodynamik der einzelnen Gruppenmitglieder und durch den unabhängig vom referierten Fall ablaufenden Gruppenprozeß kompliziert wird.

Für den Balint-Gruppenleiter stellt sich daher das Beziehungsproblem des Patienten auf mehreren Ebenen dar:

Durch den referierenden Arzt bekommt er zunächst eine Schilderung der Symptomatik und der „realen" Beziehungsproblematik des Patienten. In den Schwierigkeiten, die der Arzt mit seinem Patienten hat, erscheint das Problem in neuer Verarbeitung wieder, um schließlich im Gruppenprozeß nochmals und vervielfältigt aufgegriffen werden zu können.

Diese gewiß sehr verkürzte und vereinfachte Darstellung dessen, was sich nach unserer Erfahrung (W.) in einer Balint-Gruppe abspielt, wollen wir noch durch die Skizze einer Gruppensitzung zu veranschaulichen suchen, die natürlich zwangsläufig wiederum stark verkürzt und vereinfacht hier wiedergegeben wird.

Eine sehr erfahrene Ärztin für Allgemeinmedizin berichtet von der Behandlung einer Bauernfamilie, die ihr große Schwierigkeiten bereitet.

Die Familie besteht aus dem, wie sie sich ausdrückt, „Jungbauern", der aber schon Anfang fünfzig ist, der Bäuerin, die Mitte vierzig ist, und der Altbäuerin, die um die achtzig und die Mutter des Bauern ist. Der Altbauer, der eine schwache Figur gewesen sein soll, ist bereits vor mehreren Jahren gestorben. Außerdem sind in der Familie noch mehrere Kinder, die aber für die referierende Ärztin nur eine untergeordnete Bedeutung haben, weil vor allem die eben geschilderten drei Personen wechselseitig ihre Patienten sind. Die Altbäuerin hat arthrotische und kreislaufbedingte Altersbeschwerden, der Bauer leidet an chronischen Magenbeschwerden und Zwölffingerdarmgeschwüren und die Bäuerin an Depressionen, die schon wiederholt stationäre Behandlung erforderten. Was die referierende Ärztin so irritiert und zur Verzweiflung bringt, ist die Tatsache, daß jedesmal, wenn sie einen der drei „wiederhergestellt und aus dem Sumpf gezogen" habe, ein anderes Familienmitglied „in einem Spiel ohne Ende erkrankt". Da ihr das Ziel vorschwebt, eine glückliche und zufriedene Familie vor sich zu haben, in der jeder seinen Lebensraum hat, ist sie ob ihrer vermeintlichen ärztlichen Insuffizienz verzweifelt, zumal sie sich um

jedes einzelne Familienmitglied große Sorgen macht, wenn es, was meist der Fall ist, bedrohlich dekompensiert. Die Resonanz der Gruppenmitglieder ist nun keineswegs depressiv und besorgt, entsprechend der Stimmungslage der Referentin, sondern teilweise verärgert, aggressiv mit gelegentlich boshaften Bemerkungen. Schließlich wird der Gruppe und der Referentin klar, daß hier drei zutiefst unzufriedene und narzißtisch verunsicherte Menschen aneinandergekettet sind, die mit verteilten Rollen einerseits sich bekämpfen und miteinander rivalisieren, andererseits in depressiver Hoffnungslosigkeit versinken. In diesem triangulären Spannungs- und Kampffeld, das seit zwanzig Jahren besteht, muß die Ärztin ihre ursprüngliche narzißtische Größenphantasie, die große Friedensstifterin und Heilerin sein zu wollen, aufgeben und sich auf das Mögliche beschränken. Es besteht darin, dem System und allen Gliedern desselben Überlebenschancen zu sichern und den einzelnen Familienmitgliedern zu helfen, sich Lebensräume zu schaffen, in denen sie sich einigermaßen entfalten können. Dies wird jedoch nur möglich sein, wenn auch verbale Äußerungen von Aggressivität und Rivalisieren bzw. das Abgrenzen des eigenen Lebensbereiches stärker zugelassen werden als bisher. In der Familie herrschte nämlich unausgesprochen die Ideologie „wir wollen eine gute und friedfertige Familie sein und verabscheuen deshalb Streit und Rivalisieren".

In unbewußter, aber sehr folgenschwerer Weise war es zu diesem Zusammenspiel – einer Kollusion im Sinne von Willi – zwischen den Wunschvorstellungen der Ärztin und ihren Patienten gekommen, das, solange es weiterbestand, die therapeutische Potenz der Ärztin vollständig paralysierte. Anstatt verstehen, nachdenken und dann sinnvoll handeln zu können, fühlte sich die Ärztin so sehr in die Psychodynamik dieser Familie hineingezogen, daß sie nur noch ihre eigene Zerrissenheit, Unzufriedenheit, Wut, Verzweiflung und Hilflosigkeit spürte. Erst im Verlauf der Gruppensitzung wurde den Gruppenmitgliedern und der Referentin klar, daß diese Gefühle sog. „Gegenübertragungsphänomene" sind, die wir einerseits zunächst benötigen, um überhaupt die Psychodynamik unserer Patienten zu verstehen, von denen wir uns dann aber befreien und distanzieren müssen, um rational handeln zu können.

Am Beispiel dieser Balint-Gruppensitzung, von der wir allerdings hier nur eine Haupteinsicht herausarbeiten konnten und all die vielen auch wichtigen Nebenlinien, zu denen uns die emotionalen Reaktionen und die kognitiven Einfälle der Gruppenmitglieder geführt haben, vernachlässigen mußten, wollen wir nun ohne Anspruch auf Vollständigkeit die wichtigsten Lernziele und Lerninhalte der Balint-Gruppenarbeit festhalten.

Während der angehende Arzt in seiner naturwissenschaftlich-klinischen Ausbildung lernt, den Patienten gewissermaßen als Objekt zu beobachten, um klinische Diagnosen stellen zu können – bei der Altbäuerin „degenerative Gelenkveränderungen und altersbedingte Herz- und Kreislaufschäden", beim Bauern ein „Zwölffingerdarmgeschwürsleiden" und bei der Bäuerin eine „Depression" –, führt ihn die Balint-Gruppenarbeit über die klinische Diagnose hinaus zu einer umfassenden Diagnose oder Gesamtdiagnose, die die Psychodynamik des Patienten und seine Beziehungspathologie miteinschließt. Jetzt wird uns nachfühlbar einsichtig, daß die Altbäuerin, der Bauer und die Bäuerin nicht nur verschiedene Erkrankungen „haben", sondern auch, wie es dazu gekommen ist, was sie für sie bedeuten, warum immer wieder abwechselnd ein Mitglied dieser Trias erkrankt und wie sie damit umgehen.

Um diese emotionale Seite des Krankheitsgeschehens in die Gesamtdiagnose zu integrieren, genügt jedoch keineswegs der Gebrauch der fünf Sinne und des Verstandes. Es muß ein Nacherleben und Einfühlen in die Situation der Patienten hinzukommen, wenn wir nicht bei der objektivierenden klinischen Diagnose stehenbleiben wollen. Erst das bewußte Wahrnehmen und diagnostische Verarbeiten der emotionalen Resonanz des Gruppenleiters, der referierenden Ärztin und der Gruppenmitglieder, eröffnete den Zugang zu dieser Dimension des Krankseins unserer Patienten.

Der Arzt lernt also gewissermaßen den Gebrauch eines sechsten Sinnes, seiner affektiven Resonanz oder – psychoanalytisch gesprochen – seiner Gegenübertragung, ohne die er nicht zu einer Gesamtdiagnose kommen kann.

Durch Einbeziehen seiner Emotionalität lernt er das, was man seit Balint treffend „Beziehungsdiagnose und Beziehungstherapie" genannt hat. Erst durch das Verstehen der Beziehungspathologie vermag er sinnvoll und gezielt in diese einzugreifen.

Er vermag jedoch die Beziehungspathologie seiner Patienten nur zu erfassen, wenn er sich – zumindest zunächst – in das Beziehungsgeflecht einbeziehen läßt. Er macht die Erfahrung, daß er zu einem wichtigen Element des Systems „Patient und seine Umwelt" geworden ist und daß ihn gerade dieses Elementsein, wenn es reflektiert wird, zu besonderer Wirkung befähigt. Dadurch wird die Tätigkeit des einzelnen Arztes sehr aufgewertet. Er ist nicht mehr, wie in der objektivierend-naturwissenschaftlichen Medizin, jederzeit durch einen beliebigen anderen Arzt austauschbar.

Dieses Einbeziehen der Emotionalität des Arztes und die Neubestimmung seiner Position in der Arzt-Patient-Beziehung vollzieht sich nicht ohne Selbsterfahrung. Obwohl diese in der Balint-Gruppe nicht direkt angesprochen wird, und die Psychodynamik des Patienten und seine Beziehungspathologie und nicht die persönliche Sphäre des Referenten Ziel der Balint-Gruppenarbeit sind, gewinnt jeder Teilnehmer durch seine emotionale Beteiligung auch ganz erheblich an Selbsterfahrung. „Warum reagiere ich so auf diesen Patienten und auf diese Situation?" Das sind Fragen, die sich fast jedes Gruppenmitglied immer wieder selbst insgeheim stellt und die es indirekt – wir betonen indirekt – auch beantwortet bekommt.

Durch diese allmähliche Änderung des Arbeitsstiles und des Erlebens des Arztes ändert sich auch zwangsläufig seine Theorie. Das erlernte linear-monokausale Denken, das meist mit einer apostolisch-autoritativen Haltung des Arztes einhergeht, wird allmählich durch ein relationales Denken in Beziehungssystemen ersetzt. Der Patient wird nicht nur in der Praxis, sondern auch in der Theorie zu einem Partner, mit dem gemeinsam nach den Gründen seines Krankseins und den Lösungen seiner Lebensprobleme gesucht wird. Objektiv identifizierbare Befunde müssen, so gesehen, erst in eine Gesamtdiagnose integriert werden. Gewissermaßen nebenher erlernt der Arzt die sog. „Technik" des guten diagnostisch-therapeutischen ärztlichen Gesprächs. Außerdem lernt er viele komplizierte Arzt-Patient-Situationen kennen, wodurch sich sein ärztlicher Gesichtswinkel und seine Möglichkeiten stark erweitern.

Schließlich – last but not least – erfährt der Arzt in der Balint-Gruppe auch eine emotionale Entlastung und Stützung.

Die referierende Ärztin in unserem Fallbeispiel war zunächst verzweifelt und fühlte sich recht insuffizient. Nachdem sie diese Gefühle als ihre adäquate emotionale Resonanz auf die Situation ihrer Patienten erkannte und die Zusammenhänge besser durchschauen konnte, fühlte sie sich entlastet und fähig, diesen Patienten im Rahmen des Möglichen zu helfen. Die ursprünglichen Größen- und Kleinheitsphantasien wichen einer realistischen Beurteilung der Situation. In einer solchen emotionalen Entlastung, die fast immer in einer einigermaßen zufriedenstellend verlaufenden Balint-Gruppensitzung stattfindet, ist bereits eine ganz wesentliche emotionale Stützung enthalten, denn es wird deutlich, was der referierende Arzt, der sich subjektiv so insuffizient fühlt, doch bereits alles geleistet hat. Eine weitere wichtige Stützung besteht in der Erfahrung, daß die anderen Kollegen die gleichen oder zumindest sehr ähnliche Schwierigkeiten haben und daß die Gruppe einen emotionalen Rückhalt gegen die viele Frustrationen des ärztlichen Alltags bietet, denen gerade all jene Ärzte ausgesetzt sind, die Tag für Tag – und nicht selten auch nachts – ihre schwere Pflicht tun, ohne jene Kompensation zu erhalten, die manchen Ärzten zuteil wird, die als sog. Koryphäen im Rampenlicht der Öffentlichkeit stehen. Es ist gut möglich, ja sogar wahrscheinlich, daß wir einige Lernziele und Lerninhalte, die durch die Balint-Gruppenarbeit vermittelt werden, nicht aufgezählt haben. Allein die aufgezählten, so meinen wir, rechtfertigen jedoch das Urteil, daß Balint-Gruppenarbeit für jeden, insbesondere aber für den niedergelassenen Arzt und für den Arzt für Allgemeinmedizin durch keine andere Aus-, Weiter- und Fortbildungsmethode ersetzt werden kann und sich daher mit Recht zunehmender Beliebtheit erfreut. Sie vermag wie keine andere Methode den Arzt darauf vorzubereiten, seine Patienten im Sinne einer integrativen Psychosomatik zu beurteilen und zu behandeln.

69 Die Einführung der psychosomatischen Betrachtungsweise als wissenschaftstheoretische und berufspolitische Aufgabe
Gedanken zum Problem der ärztlichen Verantwortung

Thure von Uexküll

69.1 Die Utopie einer Humanmedizin und die Wirklichkeit der Heilkunde

Wir verknüpfen mit dem Begriff Utopie gewöhnlich die Vorstellung von etwas Wirklichkeitsfremdem. Um die Undurchführbarkeit eines Planes oder eines Vorschlags zu unterstreichen, sagen wir, sie seien utopisch. Damit verdecken wir aber den wirklichen Inhalt des Begriffs; denn Utopie und Wirklichkeit definieren und – was noch wichtiger ist – korrigieren sich gegenseitig. Da diese Wechselwirkung die Welt verändert, haben sie eine gemeinsame Geschichte, in deren Verlauf Utopien wirklichkeitsnäher und Wirklichkeiten utopienäher werden können.

Wenn wir die Einführung der psychosomatischen Betrachtungsweise in die Heilkunde als Geschichte der Utopie einer Humanmedizin sehen, können wir einen Weg erkennen, auf dem sich sowohl die psychosomatische Betrachtungsweise wie die Wirklichkeit der Heilkunde verändert haben und auch in Zukunft verändern werden.

Die Gründungsväter der Psychosomatischen Medizin, oder die Wiederentdecker einer psychosomatischen Betrachtungsweise in einer inzwischen sehr technisch gewordenen Heilkunde, waren noch Bilderstürmer. Sie wollten die Medizin durch die Psychoanalyse, d.h. die naturwissenschaftliche Theorie der Krankheitsursachen durch die psychoanalytische Lehre der unbewußten Krankheitsmotive ersetzen. Einführung der psychosomatischen Betrachtungsweise bedeutete für sie, Krankheit als Konversionsgeschehen zu deuten. Sie sahen noch nicht die Gefahren unkritischen Symboldeutens, obgleich sie bei Georg Groddeck (1866–1934) schon offenkundig waren.

In der nächsten Generation begann sich die psychosomatische Betrachtungsweise zu differenzieren. Mit der Einsicht in die Eigengesetzlichkeit körperlicher Vorgänge gegenüber dem psychischen Geschehen wurde die Utopie ein Stück wirklichkeitsnäher. Das drückte sich, wenn auch auf verschiedene Weise, in den Entwürfen aus, mit denen Franz Alexander (1951) und Viktor von Weizsäcker (1949) die psychosomatische Betrachtungsweise in die Medizin einführen wollten.

Aber die Konfrontation mit den Problemen des Körpers wirkte auch als ein Schock. Sie hinterließ das Dualismus-Trauma. Damit begann eine Epoche, die nicht nur erkenntnistheoretisch, sondern auch in der Organisation des Gesundheitswesens von der Leib-Seele-Problematik überschattet war – und weitgehend noch heute ist. Während man innerhalb der Psychosomatischen Medizin bemüht war, den Dualismus durch Konzepte zu überwinden, die eine Brücke zwischen zwei heterogenen Seinsbereichen schlagen sollten, schrieb die gesundheitspolitische Wirklichkeit den Dualismus als ein System der Krankenversorgung fest – allerdings mit einem sehr einseitigen Übergewicht der somatischen Seite.

Dabei spielte sich im einzelnen folgendes ab: In der somatischen Medizin führte die rasch fortschreitende Spezialisierung zu einem Zerfall der großen Fächer, vor allem der Inneren Medizin. In ihr hatte noch bei Klinikern wie L. von Krehl, G. von Bergmann, R. Siebeck, A. Jores, L. Heilmeyer und anderen nicht nur Interesse, sondern auch die Bereitschaft für eine Einführung der psychosomatischen Betrachtungsweise bestanden (Th. v. Uexküll, 1984). Das hörte mit der Entstehung der Subdisziplinen auf. Die Spezialisierung auf ein Organ verführt zu einseitiger somatischer Betrachtungsweise und zwingt zu Konzentration auf immer komplizierter und technischer werdende Spezialmethoden eines immer begrenzteren Gebietes.

In der Psychosomatischen Medizin provozierte diese Entwicklung zwei gegensätzliche Tendenzen. Man könnte sie die Tendenz zum Spezialisten und zum Generalisten nennen. Die Spezialisten sehen das Ziel in einer elitären Weiterentwicklung psychotherapeutischer Methoden zur Behandlung neurotischer Patienten. Die Bezeichnung „psychosomatisch" wird von ihnen nur noch aus taktischen Gründen beibehalten. Die Einführung psychosomatischer Betrachtungsweise in die somatische Medizin ist für sie weder möglich noch erstrebenswert. Bei den Generalisten führte die Trennung von den somatischen Kliniken und ihren Problemen zu einer Tendenz, Modelle von hoher Subtilität für den psychischen Bereich, aber relativ primitive, oft naive Vorstellungen für das Körpergeschehen zu entwickeln. Auch das verhindert letzten Endes die Einführung der psychosomatischen Betrachtungsweise in die Heilkunde. Wer glaubt, mit Hilfe des Konzepts der Resomatisierung von Max Schur (1955) oder der zweiphasigen Verdrängung von Alexander Mitscherlich (1966) psychosomatische Betrachtungsweise in die Medizin einführen zu können, erfährt, daß die Wirklichkeit der Heilkunde dieser Utopie davongelaufen ist.

Die Utopie einer Humanmedizin muß sich wiederum verändern; denn Hämatologen, Kardiologen, Gastroenterologen oder Endokrinologen wissen sehr viel mehr über den menschlichen Körper als Psychosomatiker, die sich außerhalb der Organkliniken angesiedelt haben. Wenn sie von dort aus psychosomatische Probleme untersuchen wollen, müssen sie sich bei den Organspezialisten über Immunvorgänge, Herz- und Kreislaufprozesse und Abläufe im Magen-Darm-Trakt informieren lassen, ohne zu den einseitigen Körpervorstellungen der Organspezialisten kritisch Stellung nehmen zu können. Das aber wäre für eine Integration psychosomatischer Betrachtungsweise unerläßlich; denn die somatischen Spezialisten haben im Zuge der zunehmenden Fraktionierung der Heilkunde das Mosaik, das den kranken Menschen darstellen soll, über den Einzelheiten ihrer Facetten aus den Augen verloren. So ist Heilkunde heute zu einem Milchstraßensystem von Spezialdisziplinen geworden, in dem sich nicht nur Patienten, sondern auch Ärzte verirren können. Diese Wirklichkeit der Heilkunde ruft nach einer anderen Utopie: Sie muß Modelle zur Lösung der Frage anbieten, wie Teile in einem Ganzen integriert sind. Das sind andere Modelle als die, welche versuchen, Körper und Seele aus zwei heterogenen Seinsbereichen über „Zwischenstufen" zu verbinden, in denen psychische Vorgänge auf undurchsichtige Weise in „Mechanismen" umgeformt werden, die dann in das Räderwerk der Körpermaschine eingreifen sollen. Die Aufgabe, die sich einer dritten Generation von Psychosomatikern stellt, lautet nicht mehr Einführung der psychosomatischen Betrachtungsweise in eine mechanistische Medizin, sondern Entwicklung der Theorie und Praxis einer Heilkunde, die psychische und somatische Abläufe als interdependente und interaktive Funktionen (Weiss und English, 1949) eines einheitlichen Systems begreifen kann.

69.2 Paradigmawechsel als Begriff und als praktische Konsequenz

69.2.1 Die ethische Orientierungslosigkeit der modernen Medizin

Wir haben in Kapitel 1 ausgeführt, daß diese Theorie einen Paradigmawechsel in der Medizin voraussetzt, und haben versucht, die erkenntnis- und wissenschaftstheoretischen Konsequenzen darzustellen. Nachdem in Teil III die praktischen Konsequenzen einer Einführung der psychosomatischen Betrachtungsweise für die pathogenetischen Konzepte, für die diagnostischen und therapeutischen Verfahren, für die Organisationsformen der Krankenversorgung und in den folgenden Teilen für die Interpretation der verschiedensten Krankheitsbilder aufgezeigt wurden, ist es die Aufgabe eines Schlußkapitels, die Konsequenzen aufzuzeigen, die sich für die berufspolitischen Probleme und das Problem der Verantwortung des Arztes ergeben.

Wer die Situation der Medizin unvoreingenommen betrachtet, stellt überrascht und beunruhigt fest, daß sie für das Problem der ärztlichen Ethik einen blinden Fleck entwickelt hat. Weder aus Anatomie, noch aus Physiologie, noch aus Pathologie, noch aus Genetik oder Molekularbiologie lassen sich Richtlinien für ärztliche Entscheidungen ableiten, die über technische Anweisungen hinausgehen und an denen sich der Arzt in Situationen orientieren kann, in denen es um die Zulässigkeit von technisch Möglichem geht. Das ist in einer Zeit, in der die Medizin über nie dagewesene Möglichkeiten verfügt, in menschliches Leben einzugreifen, eine erschreckende Feststellung. Der Grund für diesen Defekt ist unschwer auszumachen: Die Medizin verfügt über kein verbindliches Menschenbild. Sie schwankt zwischen zwei einander ausschließenden Konzepten, die zwei verschiedenen Paradigmen entsprechen. In Kapitel 1 haben wir vereinfachend von einem strukturellen und einem funktionellen Ansatz gesprochen, um diese beiden Paradigmen zu umschreiben.

Die Geschichte der Medizin lehrt uns, daß sich die beiden einander scheinbar entgegengesetzten Ansätze zu allen Zeiten im ärztlichen Denken ergänzten, daß aber die Art und Weise, wie dies geschah, in verschiedenen Epochen sehr verschieden war. Die Lehre von dem Primat der Struktur konnte erst seit dem 17. Jahrhundert konsequent an Boden gewinnen. Damit entstand das Paradigma der Maschine als Erklärungsmodell für Lebensvorgänge (vgl. Anmerkung 1 am Ende des Kapitels), und das Primat der Struktur war zum Prinzip erhoben: Funktionsstörungen waren eine Folge von Strukturschäden.

So entstand das Menschenbild, dem der französische Philosoph Julien Offray de Lamettrie (1709 bis 1751) schon früh die griffige Formel des „L'homme machine" gegeben hatte. Der Glaube an dieses Menschenbild fand durch den Bau automatischer Maschinen eine neue Bestätigung. Die Vision des Maschinenmenschen wandelte sich zur Vision eines Roboters, den ein künstliches Gehirn zur Produktion einer „künstlichen Intelligenz" befähigen soll, die der „natürlichen Intelligenz" des heutigen Menschen weit überlegen sein wird. Damit beginnen die Grenzen zwischen Wissenschaft und Science-fiction bedenklich zu verschwimmen (vgl. Weizenbaum, 1977; vgl. Anmerkung 2 am Ende des Kapitels).

Demgegenüber machten Ärzte, die von dem Bild des Maschinenmenschen nicht geblendet waren, schon immer darauf aufmerksam, daß Maschinen keine Gefühle haben, daß sie weder Hoffnung noch Verzweiflung kennen und keine Schmerzen empfinden. Sie warnten vor der gefährlichen Verkürzung der Probleme des Kranken, wenn man die Bedeutung des Gefühlslebens für Gesundheit und Krankheit ignoriert. Diese Warnungen bekamen durch die Entdeckungen Sigmund Freuds über die Bedeutung seelischer Konflikte für die Entstehung und den Verlauf neurotischer Erkrankungen besonderes Gewicht. Seine Methode, mit Kranken zu sprechen, erweiterte die diagnostischen und therapeutischen Möglichkeiten der Medizin um eine neue Dimension. Aber Freud scheute sich, die Einsichten, die er an psychisch Kranken gewonnen hatte, auch für organische Leiden nutzbar zu machen.

Diese Zurückhaltung war auch ein Grund dafür, daß die Medizin lange Zeit die revolutionäre Konsequenz nicht sah, welche die Entdeckung Freuds für ihr Menschenbild bedeutete: Statt der Einsicht zum Durchbruch zu verhelfen, daß das Bild des Maschinenmenschen als Gesundheits- und Krankheitsmodell versagt, verirrte sie sich in der Sackgasse des dualistischen Denkens, und das Paradigma der Maschine blieb das offizielle Dogma der Berufs- und Wissenschaftspolitiker.

Dabei hatte schon 1932 Gustav von Bergmann gezeigt, daß am Anfang der Krankheit nicht die lädierte Struktur steht, die der Pathologe postmortal in Leichen findet, sondern die gestörte Funktion, die früher oder später zu einem Strukturschaden führen kann, aber nicht muß. Er hatte an zahlreichen klinischen Beispielen und eindrucksvollen Tierversuchen eine „Funktionelle Pathologie" (1932), eine Pathologie der Funktionen entwickelt, welche die einseitige Pathologie der Strukturen ablösen sollte.

Seine These, daß Krankheit mit einer Funktionsstörung beginnt, wurde dann durch Viktor von Weizsäcker einen Schritt weitergeführt. Er stellte die Frage: Wenn am Anfang der Krankheit die Funktionsstörung steht, was steht dann am Anfang der Funktionsstörung? Er gab die Antwort: Die Person des Kranken mit ihren psychischen und sozialen Konflikten, und forderte als Konsequenz die Einführung des Menschen als Subjekt in die Medizin. Er hatte schon früh erkannt, daß die Entdeckungen der Psychoanalyse dazu führen müßten, das einseitig von der Pathologie geprägte Menschenbild durch ein neues zu ersetzen, in dem die mechanistischen Vorstellungen vom menschlichen Körper neu interpretiert sind. Schon 1925 schrieb er:

„Wenn sich die Entdeckungen der Psychoanalyse in Forschung, in Wissenschaft und Lehre mit dem naturwissenschaftlichen Bestande in einer organischen und sinnvollen Weise zusammenschließen" sollen, sei „eine medizinische Anthropologie, eine allgemeine Lehre vom Menschen als Grundwissenschaft der Heilkunde" vonnöten.

Nur, die Frage, wie dieser organische und sinnvolle Zusammenschluß aussehen sollte, blieb unbeantwortet; und so fand auch die Einführung des Menschen als Subjekt in die Medizin bis heute nicht statt. Solange die Medizin an einem mechanistischen Körperbild festhält, bleibt sie in der dualistischen Vorstellung von zwei einander ausschließenden Seinsweisen gefangen. Ein seelisches Subjekt läßt sich durch keinen Kunstgriff in eine Maschine installieren. Entweder man beginnt mit einer Interpretation der körperli-

chen Vorgänge nach einem anderen Modell – oder das Menschenbild der Medizin bleibt zwiespältig, in sich widerspruchsvoll und unverbindlich – wie Spötter gesagt haben: das Bild von einem Gespenst in einer Maschine.

Für das Problem der ethischen Maßstäbe ist das Menschenbild von zentraler Bedeutung; denn von ihm hängt es ab, wie ich mich selbst und den anderen sehe. Die ethische Orientierungsfunktion, die ein Menschenbild für das Zusammenleben von Menschen hat, wird in der Medizin besonders deutlich, weil die Ziele, die sie erreichen will, und die Verfahren, die sie dafür verwendet, der Gesundheit und dem Leben von Menschen gelten, die sich einem Arzt – im Vertrauen auf zwei Dinge – ausliefern:

– Im Vertrauen auf seine Fähigkeit, die Probleme des Kranken so zu verstehen, als ob es seine eigenen wären, d.h. seine Sensibilität und seine Bereitschaft, sich in die Lage des Kranken zu versetzen, und

– im Vertrauen auf seine Fähigkeit, durch medizinisches Wissen und Können zu helfen.

Die Bedeutung des ersten Punktes ist heute so deutlich geworden wie nie zuvor; denn mit den Möglichkeiten, durch medizinisches Wissen und Können zu helfen, haben sich auch die Gefahren dieser Hilfe vervielfacht.

Das Fehlen eines verpflichtenden Menschenbildes, an dem sich der Arzt für seine Entscheidungen zwischen technisch möglichen und ethisch verantwortbaren Maßnahmen orientieren kann, führt in der täglichen Berufspraxis immer wieder zu bedrückenden Situationen. Das gilt besonders für die Intensivstationen (vgl. Kap. 64). Seit es möglich geworden ist, Funktionen lebenswichtiger Organe wie Herz, Lunge oder Niere durch elektronisch gesteuerte Apparate zu ersetzen, ist die Grenze zwischen Leben und Tod verschiebbar geworden, verschiebbar durch Ärzte, welche die Entscheidung treffen müssen, ob diese Apparate eingesetzt werden sollen, und wenn sie einmal eingesetzt sind, ob und wann sie abgestellt werden. Damit hört der Tod auf, persönliches Schicksal des Sterbenden zu sein, und wird zu einem ärztlich zugelassenen oder vorenthaltenen Betriebsunfall. Nur wenn ein lebensbedrohlich Erkrankter begründete Hoffnung auf die Wiederherstellung eines menschenwürdigen Lebens hat, ist diese Situation ethisch konfliktfrei. Aber schon die Frage, was unter einem menschenwürdigen Leben zu verstehen ist, stellt den Arzt vor ethische Probleme.

Auch der unheilbar Kranke braucht den Arzt als Vertrauensperson und Ratgeber, ja er braucht ihn sogar besonders dringend (vgl. Kap. 66). Aber welchen Rat soll ein Arzt geben, der kein anderes Menschenbild kennt als das einer Medizin, von der Paul Martini vor ca. 40 Jahren sagte:

„Wer eine kontinuierliche Reihe vom Atom bis zum Menschen annimmt, wird, indem er im Experiment vom Leblosen zum Lebendigen schließlich bis zum Menschen fortschreitet, keine unübersteigbaren Schranken, weil keine grundsätzlichen Unterschiede, finden können" (1948).

Für den Arzt, der sich an einem solchen Menschenbild orientiert, bleiben Begriffe wie Seele, Bewußtsein, Schicksal und Sinn leere Worthülsen.

Geradezu gespenstisch wird die Situation in der Forschung. Der Internist Diehl (1984) schildert, wie die Visionen, die Aldous Huxley 1936 in seinem Zukunftsroman „Brave New World" dargestellt hat, durch die heutige Wirklichkeit längst überholt sind. Er schreibt:

„Embryotransfer, Genklonierung mit Reproduktion identischer Lebewesen beliebiger Zahl, Produktion von Embryonen für ‚organverbrauchende Experimente', bei denen Föten regelrecht ‚geschlachtet' werden, um einzelne Organe nach Belieben zu gewinnen, Samen-, Ei- und Embryonenbanken, Herstellung von Menschen-Tier-Hybriden, künstliche Geschlechtswahl und Vermarktung von Samen- und Eizellen sowie von Embryonen sind nicht utopische Horrorbilder, sondern Realität unseres Lebens geworden. ... Noch an keiner Stelle medizinisch-wissenschaftlicher Forschung ist der einzelne Wissenschaftler und der beteiligte Arzt in seiner ethischen Verantwortung so unmittelbar gefordert worden."

„Drohendes Unheil können wir in fünf Jahren nicht als ‚ungewollten Schaden' apostrophieren, den wir hätten verhindern können. Unsere Aufgabe ist es, unsere ethischen Normen zu überprüfen, angesichts dieser naturwissenschaftlichen Entwicklung, die einigen betroffenen Menschen nützt, aber unabsehbaren Schaden für die gesamte Menschheit bringen kann."

69.2.2 Der Wandel in den Voraussetzungen der Naturwissenschaften und der Nachholbedarf der Medizin

Das Menschenbild, das sich von dem Maschinenparadigma ableitet, kennt keinen ins Gewicht fallenden Unterschied zwischen Humanmedizin und Veterinärmedizin (vgl. Anmerkung 3 am Ende des Kapitels). Es gibt dem Arzt keine Richtlinien, an denen er sich in kritischen Situationen, z. B. für oder gegen die Weiterführung einer Therapie bei einem unheilbar Kranken, oder in Fragen der Zulässigkeit oder Unzulässigkeit biologischer oder medizinischer Versuche orientieren könnte. Ich sprach von einem blinden Fleck der Medizin für ethische Probleme. Man versucht in letzter Zeit, dieses Vakuum durch sog. Ethikkommissionen auszufüllen, die der Medizin das substituieren sollen, was ihr offensichtlich fehlt.

Die Medizin ist dem Menschen gegenüber in das gleiche Dilemma geraten wie die Naturwissenschaften gegenüber der Natur. In beiden Fällen haben die technischen Auswirkungen der Forschungsergebnisse zu Problemen geführt, über die wir die Kontrolle verloren haben. Man hat der Medizin den Vorwurf gemacht, sie sei zu sehr Naturwissenschaft geworden, daher habe sie den Menschen aus dem Blick verloren. Wir haben aber schon in Kapitel 1 betont, daß dieser Vorwurf an dem Kern des Problems vorbeigeht, weil die Medizin nicht zu viel, sondern zu wenig Naturwissenschaft, oder genauer gesagt, im 20. Jahrhundert eine Naturwissenschaft des 19. Jahrhunderts geblieben sei. Sie habe einen Nachholbedarf aufzuarbeiten, da die Naturwissenschaften die Voraussetzungen, von denen sie an die Natur herantreten, radikal geändert hätten, ohne daß die Medizin die notwendigen Konsequenzen gezogen habe.

Die radikale Änderung der Voraussetzungen zeigt sich in dem scharfen Einschnitt, der die moderne Physik von der klassischen Physik trennt; denn er symbolisiert die Erkenntnis, „daß wir niemals mit dem Objekt unserer Vorstellung in direkte Beziehung treten, sondern daß wir es immer nur mit der Vorstellung dieses Objekts zu tun haben" (J. v. Uexküll, 1936).

Was bedeutet diese Änderung in einem scheinbar weit von den praktischen Problemen des ärztlichen Berufs entfernten Bereich abstrakter Theorienbildung für das Problem der ärztlichen Verantwortung? Zunächst scheinbar gar nichts. Dann aber etwas entscheidend Wichtiges: Solange der Blick des Forschers magnetisch auf das Objekt der Forschung fixiert ist, solange er von dort alle Antworten auf seine Probleme erwartet, bleibt die kritische Reflexion seines Erklärungsmodells auf die Frage eingeengt, wie gut oder wie schlecht es geeignet ist, die verborgene Realität des Objekts zu enthüllen. Nur die Verifikations- und Falsifikationsmöglichkeiten des Erklärungsmodells sind von Interesse (Popper, 1973). Die Frage, was es für den Forscher, für seine Art zu fragen und für seinen Umgang mit sich selbst, seinen Mitmenschen und den Objekten seiner Forschung bedeutet, ist durch die zwanghafte Fixierung des Blicks blockiert.

Die Bedeutung des Paradigmawechsels in der Physik für das Problem der menschlichen Verantwortung – auch für das der Verantwortung des Arztes – liegt also in der Befreiung unseres Blickes aus einer magischen Gefangenschaft. Wir beginnen Zusammenhänge zu sehen, die durch eine Fata Morgana von Objekten verborgen waren, die sich unserem Zugriff entzogen, sobald wir ihnen ganz nahe gekommen zu sein schienen. Nicht das Objekt, sondern die Vorstellung, die wir von ihm haben, entscheidet über unser Verhalten ihm gegenüber. Die Verantwortung für unser Verhalten verschiebt sich von den Objekten auf unsere Vorstellung und damit auf die Erklärungsmodelle, die unsere Vorstellung von den Objekten prägen. Damit tauchen lange verschüttete Inhalte eines Wissens um Verantwortung des Menschen sich selbst, seinen Mitmenschen und der Natur gegenüber wieder auf. Es ist kein Zufall, daß der Paradigmawechsel in den Naturwissenschaften und die Entdeckung der Ökologie in das gleiche Jahrhundert fallen.

Sobald der Arzt die Abhängigkeit von den Erklärungsmodellen zu durchschauen beginnt, in die er mit ihrer Übernahme gerät, zeigen sich ihm die drei Aspekte menschlicher Verantwortung in zugespitzter Form: als Verantwortung für den Kranken, für die eigene individuelle Wirklichkeit und für die Medizin als gesellschaftliche Institution. Gleichzeitig beginnt er aber auch die Gegenkräfte zu spüren, die in jedem gesellschaftlichen System die Änderung traditioneller Erklärungsmodelle zu verhindern trachten, weil jede Neubewertung der Zusammenhänge auch eine Neuverteilung von Einflußbereichen und Machtpositionen bedeutet. Die gleichen Probleme zeigen sich auch in den Naturwissenschaften – und vor allem in der mit ihnen eng verbundenen Technik. Der enge Zusammenhang zwischen Erkenntnis- bzw. Wissenschaftstheorie auf der einen und Wissenschafts- und Berufspolitik auf der anderen Seite wird sichtbar.

69.3 Das Paradigma der Medizin und die Spielregeln für das ärztliche Handeln

Jedes Paradigma in dem hier verwendeten Sinn hat einen doppelten Aspekt: Als wissenschaftliches Erklärungsmodell strukturiert es die Auseinandersetzung des Wissenschaftlers mit dem Objekt seiner Forschung und in der Medizin die Interaktion des Arztes mit dem Patienten; als Organisationsprinzip für die Zusammenarbeit mit anderen Wissenschaftlern bestimmt es die Form der Wissenschaft als Institution und in der Medizin die Struktur des Gesundheitssystems. Den ersten Aspekt kann man den „mikrostrukturellen", den zweiten den „makrostrukturellen" Aspekt nennen.

Ich will zunächst von dem mikrostrukturellen Aspekt als Ausdruck der strukturierenden Macht sprechen, die das Paradigma als Erklärungsmodell auf das Handeln des Arztes und seine Interaktion mit dem Kranken ausübt. Dieser Aspekt müßte – so würde man meinen – jedem Arzt aus seiner täglichen Berufspraxis geläufig sein. Das ist aber nicht der Fall; denn die Selbstverständlichkeit beruflicher Routine verhindert die Reflexion der Spielregeln, die das ärztliche Handeln bestimmen.

Erst die Möglichkeit, das traditionelle Paradigma durch das Erklärungsprinzip der psychosomatischen Betrachtungsweise zu ersetzen, gibt dem Arzt die Möglichkeit einer vergleichenden Reflexion seines Tuns. Dabei erfährt er zu seinem Erstaunen, in welchem Ausmaß seine Interaktion mit dem Kranken, dessen Angehörigen und dem Pflegepersonal bis in Einzelheiten nicht nur der Diagnostik und Therapie, sondern auch der Kommunikation zwischen den Beteiligten, durch das Paradigma der Medizin, die er ausübt, vorprogrammiert ist. Damit wird die abstrakte These von einem Paradigmawechsel in der Medizin konkret.

M. Balint hat dem Buch, in dem er über seine gemeinsame Forschung mit niedergelassenen Ärzten über deren Beziehungen zu ihren Patienten berichtet, den Titel: „Der Arzt, sein Patient und die Krankheit" gegeben (1957). Damit hat er zum Ausdruck gebracht, daß es sich bei der Interaktion zwischen Arzt und Patient um einen Zusammenhang zwischen drei Akteuren handelt, den man als ein System bezeichnen kann, weil das Verhalten eines jeden durch das Verhalten der anderen bestimmt wird. Dabei spielt die Krankheit – dem Objekt in der modernen Physik vergleichbar – eine besondere Rolle, weil sie sowohl den Arzt wie den Patienten zwingt, sich eine Vorstellung von dem rätselhaften Geschehen zu machen, das sich hinter dem Namen „Krankheit" verbirgt. Krankheit ist eine unserer direkten Beobachtung verschlos-

sene Realität, von der uns nur das Symptom Kunde gibt.

Ähnlich hat es Foucault (1973) formuliert:

„Das Symptom", schreibt er, „ist die Form, in der sich die Krankheit präsentiert. Daher seine wichtige Rolle. Von allem Sichtbaren ist es dem Wesentlichen am nächsten; es ist die erste Umschreibung der unzugänglichen Natur der Krankheit. Husten, Fieber, Seitenschmerz und Atembeschwerden machen nicht selber die Brustfellentzündung aus – diese ist nämlich den Sinnen niemals zugänglich, sondern ‚entdeckt sich nur der Verstandestätigkeit'."

Die Bilder, welche die Medizin für die „den Sinnen niemals zugänglichen" Krankheiten entworfen hat, nennen wir Diagnosen. Sie sind Produkte der Verstandestätigkeit scharfsinniger Ärzte, die sich – wie alle Wissenschaftler – auf die Zielsetzung, Fragestellung und Methodik ihrer Beobachtungen und, was nicht minder wichtig ist, auf eine gemeinsame Sprache geeinigt haben. Daher ergeben Diagnosen intersubjektiv übereinstimmende Bilder, über die sich Ärzte untereinander verständigen können. Diagnosen begründen so eine gemeinsame Wirklichkeit der Medizin, an der sich die individuellen Beobachtungen des einzelnen Arztes orientieren können. Diagnosen sind Mittel, mit deren Hilfe der Arzt eine undurchsichtige Wirklichkeit strukturieren kann.

Kranke erwarten von ihrem Arzt Hilfe bei der Deutung ihrer für sie unheimlich und undurchsichtig gewordenen individuellen Wirklichkeit. Sie wünschen sich von dem Arzt eine Diagnose, welche die Vorstellungen bestätigt oder korrigiert, die sie sich selbst von ihrer Krankheit machen; denn auch Kranke stellen Diagnosen, welche die unzugängliche Realität ihrer Krankheit deuten. So betont Raspe (1985):

„Alle Untersuchungen belegen, daß wir kaum einen Kranken antreffen werden, der sich nicht schon bestimmte Vorstellungen darüber gebildet hat, woher seine Krankheit kommt und wie sie sich weiter entwickeln und auf sein Leben auswirken wird. ... Diese Vorstellungen sind unterschiedlich weitläufig, bestimmt und in sich geschlossen."

Diagnosen sind daher nicht nur Deutungen für die dem Arzt unzugängliche Realität der Krankheit, sondern auch Schlüssel, die ihm die individuelle Wirklichkeit eines Kranken erschließen können. Wenn wir die Diagnose kennen, mit der ein Kranker seine Krankheit deutet, können wir die individuelle Wirklichkeit, in der er lebt und die sein Erleben und sein Handeln bestimmt, ein Stück weit rekonstruieren. Deshalb ist es wichtig, daß Ärzte sich nach den subjektiven Vorstellungen erkundigen, die sich Patienten über ihre Krankheiten gebildet haben.

Erklärungsmodelle erlauben uns aus zwei Gründen eine Orientierung in undurchsichtigen Situationen: Sie geben uns **Deutungsanweisungen,** die uns sagen, wie wir undurchsichtige Phänomene ordnen und in eine Beziehung zueinander und zu uns setzen sollen. Gleichzeitig geben sie uns **Handlungsanweisungen,** wie wir aktiv oder passiv mit ihnen umgehen sollen. Daraus erhellt die Wichtigkeit, die Erklärungsmodelle für Patienten haben, deren individuelle Wirklichkeit durch die Krankheit in einer für sie undurchsichtigen und unheimlichen Weise verändert ist.

Für den Arzt sind seine medizinischen Diagnosen die Erklärungsmodelle, und da er unter einem mikrostrukturellen Aspekt in dem System „Arzt-Patient-Krankheit" andere Aufgaben hat als der Patient, unterscheiden sich die Deutungs- und Handlungsanweisungen medizinischer Erklärungsmodelle von denen der Patienten.

In dem Beispiel Foucaults gibt das Erklärungsmodell „Brustfellentzündung" dem Arzt die Deutungsanweisung, Husten, Fieber, Seitenschmerz und Atembeschwerden als „Symptome" zu identifizieren und zu einem Krankheitsbild zu vereinigen. Aus dieser Deutungsanweisung ergibt sich für ihn dann die Handlungsanweisung der einzuschlagenden Therapie. Für das Erklärungsmodell, mit dem der Patient seine Beschwerden als Symptome einer Krankheit deutet, interessiert sich Foucault weniger. Es genügt, daß es ihn veranlaßt einen Arzt aufzusuchen.

Auf diese Weise strukturiert das Erklärungsmodell des Arztes die Struktur des sozialen Mikrosystems „Arzt-Patient-Krankheit". Immerhin ist es wichtig, daß das ärztliche Erklärungsmodell mit dem des Patienten zusammenstimmt oder sich wenigstens mit ihm verträgt (Kleinman, 1980).

Hinter den Erklärungsmodellen der Diagnosen, die scharfsinnige Ärzte erfunden haben, steht als gemeinsamer Bezugsrahmen eine allgemeine Gesundheits- und Krankheits-Theorie oder – wie wir sagten – ein bestimmtes Paradigma. Für das traditionelle Paradigma, das den Organismus des Kranken als eine Maschine auffaßt, die aus vielen kleinen Maschinen zusammengesetzt ist, bedeutet „Gesundheit" den einwandfreien Zustand der Körpermaschine und „Krankheit" eine Störung ihres Mechanismus. Die Konsequenzen, die sich daraus für die Struktur des sozialen Mikrosystems „Arzt-Patient-Krankheit" ergeben, habe ich dargestellt. Auch die Konsequenzen für die dualistische Struktur des sozialen Makrosystems der „Gesundheitsversorgung unserer Gesellschaft" mit seiner hochentwickelten und hochdotierten Körpermedizin neben einer unterprivilegierten Seelen-Heilkunde werden unter dem Aspekt dieses Paradigmas verständlicher.

Wie sieht demgegenüber das allgemeine Erklärungsmodell oder Paradigma aus, das die psychosomatische Betrachtungsweise ihren Diagnosen zugrunde legt? Welche Konsequenzen ergeben sich aus ihm für das soziale Mikrosystem „Arzt-Patient-Krankheit" und für das soziale Makrosystem „Gesundheitsversorgung der Gesellschaft"?

Betrachten wir zunächst dieses Erklärungsmodell etwas genauer, so sehen wir, daß der Paradigmawechsel zwei Aspekte hat:

- Lebewesen werden nicht als Maschinen (mit mehr oder weniger fertigen Strukturen), sondern als „autopoietische", d.h. sich ständig selbst aufbauende und selbst entwickelnde Systeme (Maturana, 1976) gedeutet. Damit läßt sich „Gesundheit" nicht als ein (mehr oder weniger fertiger) Zustand oder Besitz verstehen, sondern muß als Prozeß der Erzeugung

von Gesundheit (Salutogenese; Antonowsky, 1987) aufgefaßt werden. Erzeugung von Gesundheit entspricht dem Prozeß, in dem das System sich selbst erzeugt.

„Pathogenese" ist dann nicht mehr ein Prozeß, der eine vorgefundene Gesundheit abbaut oder zerstört, sondern eine Blockierung der Gesundheitserzeugung, die mehr oder weniger ausgedehnt, mehr oder weniger langdauernd sein kann und dementsprechend mehr oder weniger schwerwiegende Konsequenzen hat.

- Lebewesen sind nicht nur autopoietische, sondern auch hierarchisch gegliederte Systeme, d.h. der autopoietische Prozeß spielt sich gleichzeitig auf einer biotischen, einer psychischen und einer sozialen Integrationsebene ab. Auf jeder dieser Ebenen treten „emergent" neue Phänomene (Eigenschaften und Fähigkeiten) auf (vgl. Anmerkung 4 am Ende des Kapitels).

Die Modelle Funktionskreis und Situationskreis beschreiben Gesundheitserzeugung (Salutogenese = Autopoiese) als Prozesse, in denen der Organismus auf der biotischen und das Individuum auf der psychischen und sozialen Integrationsebene bestimmten Ausschnitten der Umgebung eine (ihren Bedürfnissen entsprechende) Bedeutung erteilen und diese Ausschnitte (durch ein bedeutungsverwertendes Verhalten) als Teile des Systems „assimilieren".

Die verschiedenen Ebenen sind ständig durch somato-psycho-soziale „Aufwärts-Effekte" (von der Zelle zum sozialen System) und gegenläufige soziopsycho-somatische „Abwärts-Effekte" (vom sozialen System zur Zelle) miteinander verbunden.

69.4 Ärztliches Handeln und das Problem der Ethik

Das Paradigma der Maschine und die von ihm abgeleiteten Diagnosen als Erklärungsmodelle für Krankheiten geben dem Arzt, wie wir gesehen haben, keine ethischen Orientierungshilfen. Es kennt nur eine „Ethik des Kunstfehlers". Ich sprach von dem Skotom der Medizin für die ethischen Probleme unserer Zeit und habe Diehl (1984) zitiert, um die Gefahren deutlich zu machen, die uns durch dieses Defizit drohen.

Kann das Paradigma des hierarchisch gegliederten, lebenden Systems diese Lücke füllen? Ehe wir in dieser entscheidenden Frage ein Urteil fällen, müssen wir klarstellen, was in diesem Zusammenhang mit dem Begriff „Ethik" gemeint ist. Wir können auf die Feststellung Kants zurückgreifen, daß für den Menschen die Pflichten der Ethik auf die „für uns begreiflichen" moralischen Verhältnisse „des Menschen gegen den Menschen" beschränkt sind (Ritter, 1972).

Für den Arzt geht es bei „den moralischen Verhältnissen des Menschen gegen den Menschen" um die Verpflichtung, durch seine Handlungen die Gesundheit des anderen zu vermehren und nicht zu mindern.

Dieser Satz klingt banal. Auch der Arzt, der sich mit Hilfe des Maschinen-Paradigmas orientiert, will die Gesundheit seiner Patienten vermehren. Der Satz bekommt aber eine andere Bedeutung, wenn wir ihn im Zusammenhang mit dem Konzept der Salutogenese betrachten. Nach diesem Konzept sind Gesundheit und Krankheit – trotz der Betonung, die das Modell der Autopoiese auf Autonomie und Selbstverantwortung legt – nicht Angelegenheit des Organismus bzw. des Individuums allein. Gesundheit und Krankheit entstehen auf der biotischen, psychischen und sozialen Integrationsebene aus der Interaktion zwischen dem Individuum und seiner Umgebung. Auf jeder dieser Ebenen setzen die Leistungen des Organismus und die Rollen des Individuums Gegenleistungen und Gegenrollen der Umgebung voraus.

Allgemein gilt, daß jede Leistung einer Gegenleistung und jede Rolle einer Gegenrolle bedarf, wenn Interaktion und damit Erzeugung von Gesundheit zustande kommen soll. So bedürfen z.B. auf einer biotischen Integrationsebene die Leistungen „Atmung" und „Stoffwechsel" des Organismus der Gegenleistungen „Luftzufuhr" und „Nahrungsangebot" der Umgebung. Auf einer psychischen und sozialen Integrationsebene bedürfen Rollen, wie z.B. „Sprechen" und „Nehmen" des Individuums, der Gegenrollen „Zuhören" und „Geben" von Interaktionspartnern der Umgebung.

Ein entscheidendes Kriterium für unser Erleben von Gesundheit ist das Gefühl von Autonomie. Dieses Gefühl ist jedoch auf „selbstverborgene Weise" zugleich ein soziales Regulativ. Um diesen Zusammenhang deutlich zu machen, muß ich etwas weiter ausholen:

Christian und Haas (1949) haben ein Modell der „Bipersonalität" entwickelt, das die Verknüpfung somatischer, psychischer und sozialer Vorgänge in der Salutogenese durchsichtiger macht. Durch eine subtile Analyse des Verhaltens und des Erlebens von zwei Menschen während des gemeinsamen Gebrauchs einer Säge konnten sie feststellen, daß ein subjektives Erleben von Autonomie objektiv die harmonische Ergänzung ihrer Leistung durch die Gegenleistung des Partners voraussetzt. Sie schreiben:

„Wenn beide Partner sich auf dem Höhepunkt der Zusammenarbeit höchst selbständig erleben, zeigt die Analyse (der objektiven Anteile des Arbeitsgangs), daß beide objektiv in strenger Gegenseitigkeit der Abläufe verbunden sind. Das Erlebnis freier Selbständigkeit wird also nur gewonnen, wenn die Gegenseitigkeit des Tuns objektiv erreicht ist." Sie fügen hinzu: „Diese freie Selbständigkeit des einzelnen beruht auf einer (dem einzelnen) verborgenen komplementären Beziehung."

Umgekehrt wird jede Disharmonie, jede Störung der Qualität „zusammen" als Störung der autonomen Verfügbarkeit über die eigenen Kräfte – das Ausbleiben einer erwarteten Gegenleistung unter Umständen sogar als Lähmung – erlebt.

Nach dem Modell, das sich daraus ableitet, ist Gesundheit, die als ein Gefühl der Autonomie und Leistungsfähigkeit erlebt wird, gleichzeitig ein Indikator für soziale Integration. Soziale Integration wiederum

meint einen Zustand, in dem man sich, ohne darüber nachdenken zu müssen, auf die Gegenleistungen und Gegenrollen der Umgebung verlassen kann.

Diese Definitionen lassen uns verstehen, wie es möglich ist, daß Kranke und chronisch Behinderte bei entsprechenden Gegenleistungen der Umgebung trotz ihrer krankheitsbedingten Einschränkungen Autonomie und Gesundheit erleben können. Das macht die ethische Bedeutung des Grundsatzes der Solidargemeinschaft mit Kranken und Hilfsbedürftigen sichtbar.

Für das Problem Ethik und ärztliches Handeln folgt daraus die Forderung, den Begriff „Solidarität" genauer zu fassen und die Gegenleistungen und Gegenrollen zu definieren, die erforderlich sind, um die Gesundheitserzeugung eines Patienten zu unterstützen. Die Aufgaben, die sich hier stellen, lassen sich unter dem Begriff „Rehabilitation" zusammenfassen; denn Rehabilitation besagt: Leistungsdefekte durch Gegenleistungen der Umgebung kompensieren und dadurch eine blockierte Salutogenese wieder in Gang bringen. Umgekehrt wissen wir aus vielen eindrucksvollen Untersuchungen, daß gestörte soziale Integration durch den Ausfall wichtiger Gegenleistungen der Umgebung – z.B. bei Objektverlust (Tod des Ehepartners, Arbeitslosigkeit, sozialer Abstieg usw.) – die Salutogenese Gesunder blockieren und zu erhöhter Morbidität und Mortalität führen kann.

Diese Zusammenhänge werden auf dem Hintergrund des neuen Paradigmas noch deutlicher, wenn wir uns klarmachen, daß Gesundheit und Krankheit keine Gegensätze in dem Sinne sind, daß Gesundheit Krankheit und Krankheit Gesundheit ausschließt. Sie bezeichnen Endpunkte eines Kontinuums, auf dem sich jeder Mensch während seines Lebens hin und her bewegt. Dazu stellt Antonovsky (1987) fest:

„Zu jedem Zeitpunkt kann mindestens ein Drittel, möglicherweise mehr als die Hälfte der Bevölkerung jeder Industrienation aufgrund des einleuchtenden Parameters eines pathologischen Merkmals als krank bezeichnet werden. Das zeigt, daß Krankheit keine relativ seltene Abweichung von irgendeiner Norm (sondern ein ubiquitäres Phänomen) ist. Man kann diese Tatsache unter zwei Gesichtspunkten betrachten: Die pathogenetische Betrachtungsweise muß erklären, warum Menschen krank wurden, warum sie eines der festgestellten Krankheitsmerkmale erworben haben. Die salutogenetische Betrachtung konzentriert sich auf die Entstehung von Gesundheit und stellt die radikal andere Frage: Warum befinden sich Menschen auf der positiven Seite des ‚Gesundheits/Krankheits-Kontinuums'? Oder warum bewegen sie sich in Richtung auf den Gesundheits-Pol, wo immer ihre Position sich zu einem gegebenen Zeitpunkt auch befunden haben mag?"

Diese Frage zwingt uns den Begriff „Rehabilitation" grundsätzlich zu überdenken und neu zu definieren. Das hat theoretische und praktische Konsequenzen: Wenn auch der unheilbar Kranke am äußersten Ende dieses Kontinuums noch über salutogenetische Potenzen verfügt, die es zu fördern gilt, solange es möglich ist, so definiert der Begriff „Rehabilitation" theoretisch den ethischen Hintergrund, auf dem sich jede ärztliche Handlung orientieren muß. Praktisch gewinnt der Begriff in einer Gesellschaft mit einer ständig wachsenden Zahl geriatrisch und chronisch Kranker eine immer größere Bedeutung.

In diesem Zusammenhang sind die Feststellungen eindrucksvoll, die zwei Sozialarbeiterinnen im Rahmen ihrer Betreuungsberichte über Patienten einer onkologischen Krankenstation gemacht haben (Diehl und Köhle, 1989). Darin werden die theoretischen und praktischen Konsequenzen am Beispiel des Krankheitsverlaufs der 27jährigen, unheilbar krebskranken Studentin Ute deutlich gemacht:

„Der spezifische Verlauf von Utes Erkrankung veranlaßt uns auch, den Begriff Rehabilitation in bezug auf Patienten mit palliativem Behandlungsschema zu überdenken. Während bei kurativ zu behandelnden Patienten die Zielsetzung der Rehabilitation darin besteht, eine dauerhafte Beeinträchtigung der persönlichen, sozialen und beruflichen Lebensumstände durch die Erkrankung zu verhindern, mußten wir uns am Beispiel Ute auf den stufenweise fortschreitenden Verschlechterungsprozeß einstellen. Die Interventionen waren immer darauf abgezielt, ihren Wunsch nach Autonomie bestmöglich zu unterstützen. ... Die Realisation der Rehabilitationsschritte im Sinne unseres Verständnisansatzes möchten wir ... insofern als gelungen bezeichnen, als die subjektiven und objektiven Bedürfnisse der Patientin erkannt werden konnten; eine gemeinsame Suche nach Möglichkeiten der Realisation stattfand; Unterstützung bei der Verwirklichung gewährt wurde; ...

Durch die Realisation dieser (und anderer genau definierter) Rehabilitationsziele war es möglich, depressive Phasen während der Krankheitsverschlechterung bei Ute so gut es ging aufzufangen, um ihr während der abschließenden Phase ihres Lebens noch ein individuell höchstmögliches Maß an Lebensqualität zu erhalten."

Die Frage, was die Position eines Menschen zur positiven oder negativen Seite des Gesundheits/Krankheits-Kontinuums verschiebt, scheint im konkreten Umgang mit Patienten aufgrund der unmittelbaren Erfahrung relativ einfach beantwortbar zu sein. Unverhältnismäßig schwieriger wird diese Antwort jedoch, wenn es darum geht, epidemiologische Daten zu ihrer Erhärtung vorzulegen. Das ist nicht verwunderlich, wenn man bedenkt, daß auf jeder Integrationsebene jede einzelne Leistung zu ihrer optimalen Realisierung eine oder mehrere Gegenleistungen erfordert, und daß darüber hinaus die verschiedenen Integrationsebenen ständig durch ein Netz von Aufwärts- und Abwärts-Effekten verknüpft sind. Beides erklärt, warum der statistisch gesicherte Nachweis eines positiven oder negativen Einflusses bestimmter Faktoren auf die Entstehung und den Verlauf von Krankheiten so kompliziert und aufwendig zu sein pflegt. Beides macht aber gleichzeitig deutlich, daß Untersuchungen, die sich nur auf eine – und bevorzugterweise die biotische – Integrationsebene beschränken, unzureichend und letztlich unwissenschaftlich sind. Exemplarisch ist das in Kapitel 50.2 über die Rolle psychischer und sozialer Faktoren in der Entstehung und im Verlauf maligner Erkrankungen dargelegt.

Wenn ich versuche, diese Überlegungen zu dem neuen Paradigma und seinem Gesundheitsbegriff in einem Satz zusammenzufassen, so würde er etwa so lauten: Die Erzeugung von Gesundheit, Autonomie und Selbstverwirklichung eines Menschen ist (auf eine ihm verborgene Weise) zugleich Erzeugung von Gesundheit, Autonomie und Selbstverwirklichung des anderen.

Für das Problem der Ethik folgt aus dieser Feststellung ganz allgemein ein Prinzip gegenseitiger Verantwortung für unser Handeln, das in einer realistischen, d.h. durchaus unsentimentalen Weise, dem „Tat twam asi" (Das bist Du) der Upanischaden und dem „Liebe Deinen Nächsten wie Dich selbst" der christlichen Lehre entspricht.

Für das Problem der Ethik in der Medizin als Wissenschaft der Salutogenese des Menschen folgt aus den Thesen die Forderung, daß Störungen in den Beziehungen des Individuums zu seiner psychosozialen Umgebung ebenso ernstgenommen werden müssen wie Störungen in den Beziehungen des Organismus zu seiner biotischen Umgebung.

Der Arzt hat die Aufgabe, über Gegenleistungen und Gegenrollen der Umgebung eines Kranken, als eines Menschen, zu wachen, dessen Leistungen und Rollen den salutogenetischen Prozeß nicht mehr ohne zusätzliche Unterstützung aufrechterhalten können. Als Teil der Umgebung des Kranken muß der Arzt diese zusätzliche Unterstützung auf der biotischen, der psychischen und sozialen Ebene sicherzustellen suchen, und auf der psychischen und sozialen Ebene unter Umständen deren Gegenleistungen und Gegenrollen zeitweise sogar selbst übernehmen.

Das verbietet ihm, den diagnostischen und therapeutischen Prozeß in der Weise zu trennen, wie es die objektivistisch-biotechnische Medizin verlangt. Ihre Forderung, am Objekt Patient müsse zuerst der „Krankheitsprozeß" diagnostiziert sein, ehe die Therapie beginnen dürfe, ist nur für solche Bereiche des Menschen sinnvoll und durchführbar, die sich tatsächlich weitgehend objektivieren lassen. So kann z.B. ein Karzinom erst operativ entfernt werden, nachdem es genau diagnostiziert worden ist. Sobald es jedoch nicht vollständig entfernt werden kann und allmählich weiterwächst, müssen Diagnose und Therapie als diagnostisch-therapeutischer Zirkel unter Umständen bis zum Lebensende des Patienten rückgekoppelt bleiben. Der Arzt muß sich fragen: Welche Organe sind bereits, und wie stark, von der Krankheit (durch Metastasenbildung) befallen? Wie widerstandsfähig ist der Patient noch? Welche diagnostischen und therapeutischen Eingriffe kann er ungefährdet ertragen? Wie können sein Leben, seine Leistungsfähigkeit verlängert und sein Leiden vermindert werden? Was bedeutet sein Leiden für ihn, wie kann er es sinnvoll verarbeiten, d.h. doch noch in seinen salutogenetischen Prozeß einbeziehen?

All das sind diagnostische Fragen, d.h. Fragen des Erklärungsmodells, das dem Arzt Deutungsanweisungen geben muß, um die undurchsichtige Situation der Erkrankung eines Menschen so zu strukturieren, daß sich daraus Handlungsanweisungen ergeben. Bricht der Arzt den diagnostischen Prozeß etwa mit der Feststellung „inoperables Karzinom" ab, so bricht er auch den therapeutischen Prozeß ab und überläßt den Patienten sich selber, dem Schicksal und irgendwelchen mehr oder weniger mitfühlenden und in der Regel wenig kompetenten Pflegepersonen. Dieses Zerbrechen des diagnostisch-therapeutischen Zirkels und seine Folgen für den Kranken kann man in manchen Kliniken und Facharztpraxen sehr genau beobachten. Die Konsequenzen, die sich daraus für die Verpflichtung der Gesellschaft auch chronisch und unheilbar Kranken gegenüber ergeben, liegen auf der Hand.

69.5 Die individuelle und die soziale Wirklichkeit

Die aufgezeigte Interdependenz zwischen individuellem Handeln, ethischer Verantwortung und gesellschaftlichen Belangen wird noch deutlicher, wenn man sich klarmacht, daß jeder Mensch in einer, letztlich nur ihm selbst zugänglichen, individuellen Wirklichkeit lebt. Wollen wir etwas über einen anderen Menschen erfahren oder ihm etwas von uns mitteilen, so müssen wir mit ihm, wenigstens partiell, eine gemeinsame Wirklichkeit aufbauen. Diese Forderung gilt in besonderer Weise für den Arzt. Nur wenn der Arzt bereit und in der Lage ist, mit dem Kranken eine gemeinsame Wirklichkeit aufzubauen, kann er entscheiden, welche Krankheitsfaktoren auf Störungen der individuellen Wirklichkeit des Kranken beruhen, ob diese (über Abwärts-Effekte) psychosomatisch zu körperlichen Störungen geführt haben, und/oder ob und wieweit die individuelle Wirklichkeit des Kranken aufgrund primär somatischer Prozesse (durch Aufwärts-Effekte) somatopsychisch verändert ist. Seine Entscheidungen zwischen verschiedenen diagnostischen Möglichkeiten oder sein Entschluß, dem Patienten zu oder gegen einen diagnostischen oder therapeutischen Eingriff zu raten, setzen Informationen über das Wirklichkeitserleben des Kranken voraus.

Die Kenntnis der individuellen Wirklichkeit des Kranken ist auch eine unabdingbare Voraussetzung für eine, dem Fassungsvermögen und der emotionalen Belastbarkeit des Kranken angepaßte, möglichst vollständige Aufklärung. Erst sie setzt den Patienten in die Lage, Entscheidungen, die sein Leben und sein Wohl betreffen, selber zu fällen. Nur so wird es möglich, einen, auch juristisch geforderten, „informed consent" und nicht den „confused consent" eines verwirrten und verängstigten Patienten zu erlangen.

Diese Überlegungen zeigen die Notwendigkeit, den Begriff des „Scharlatan" neu zu definieren. Bisher verstand man unter einem „Scharlatan" einen Arzt, der sich nicht an die Regeln der biotechnischen Medizin hielt. Heute müssen wir einen Arzt Scharlatan nennen, der ohne Kenntnis der individuellen Wirklichkeit seiner Patienten und ohne ausreichende Informationen über eventuelle Folgen medizinischer Maßnahmen für deren individuelle Wirklichkeiten „wissenschaftlich anerkannte" Heil- und Operationsverfahren zur Anwendung bringt.*

69.6 Psychosomatische Medizin als Humanmedizin

Wir haben das Konzept der individuellen Wirklichkeit in den Mittelpunkt unserer Überlegungen über die Aufgaben einer psychosomatisch orientierten Heilkunde gestellt. Der Grund dafür ist leicht einzusehen; denn der Mensch verwirklicht sich nicht nur durch die Entwicklung seines Körpers, sondern durch seine kreative Leistung, die vorgefundene Umgebung in eine individuelle Wirklichkeit zu transponieren. Bei seinen Überlegungen über die Möglichkeit einer medizinischen Anthropologie stellt Viktor von Weizsäcker (1956) fest:

„Was mir dabei den größten Eindruck macht, ist das, daß trotz ihrer anatomischen Ähnlichkeit verschiedene Menschen so ungeheuer verschieden sind".

Aber durch die Entdeckung der individuellen Wirklichkeit oder des Menschen als Subjekt (v. Weizsäcker) allein, wird Medizin noch nicht Humanmedizin. Solange sie nur somatopsychische Aufwärts-Effekte erkrankter Zellen und Organe auf die individuelle Wirklichkeit eines Menschen zur Kenntnis nimmt und von den Problemen psychosomatischer Abwärts-Effekte nichts wissen will, bleibt die Bezeichnung „human" etwas Äußerliches. Man verwechselt dann den Begriff mit einer „Humanität", von der man meint, sie sei für die Medizin eine zwar wünschenswerte Beigabe, auf die man aber zur Not auch verzichten könne, ohne daß die Effektivität der Medizin darunter leiden würde. Für diese Auffassung ist das Quantum an Humanität, das man Kranken in der Praxis oder in der Klinik bietet, nur eine Frage des Pflegesatzes.

Medizin wird erst Humanmedizin, wenn sie beides sieht: die Bedeutung körperlicher Krankheiten für die individuelle Wirklichkeit des Menschen und die Bedeutung seiner individuellen Wirklichkeit für die Gesundheit oder Krankheit seines Körpers und dessen Organe.

Mit der Entdeckung der individuellen Wirklichkeit als medizinisch relevantes „Organ" entsteht ein neues Menschenbild. Mit ihm tritt die Heilkunde aus dem ethischen Niemandsland in einen Bereich, in sich für die Entscheidungsprobleme des Arztes Orientierungspunkte zeigen – in dem aber auch neue Aufgaben und auch neue Probleme sichtbar werden. Die Einsicht, daß die individuelle Wirklichkeit eines Menschen Voraussetzung für seine Fähigkeit zum Handeln – für seine Autonomie – ist, besagt, daß ärztliche Hilfe Hilfe zur Selbsthilfe sein muß, daß salopp formuliert, die Krankheit dem Kranken, aber nicht dem Arzt gehört, und daß der Kranke über existentielle Fragen, die ihn betreffen, selbst entscheiden muß.

Der Arzt kann bei solchen Entscheidungen nur Berater sein. Er darf Entscheidungen nur abnehmen,

---

* Die Ausführungen dieses Abschnitts folgen Gedanken, die in Uexküll und Wesiack „Theorie der Humanmedizin" genauer dargelegt sind.

wenn der Kranke entscheidungsunfähig ist oder wenn er einen Kranken vor sich selbst schützen muß. Beides erfordert eine sehr kritische Prüfung der Frage, ob, und wenn ja wo, der Autonomie des Kranken durch seine Krankheit Grenzen gezogen sind. Das scheinen für den Außenstehenden Selbstverständlichkeiten zu sein. Sie sind es aber nicht für den, der in kritischen Situationen handeln muß und der die Hilflosigkeit, aber auch die Überheblichkeit von Ärzten in solchen Situationen erlebt hat.

Verantwortung dem Kranken gegenüber ist aber nur ein Aspekt des ethischen Problems in der Medizin, wenn auch ein besonders wichtiger. Zu dem Komplex der Verantwortungsproblematik gehört auch die Verantwortung des Arztes der Medizin gegenüber, die er ausübt – und nicht zuletzt die Verantwortung für sich selbst, für seine eigene individuelle Wirklichkeit.

69.7 Das Paradigma der Medizin und die Struktur des Gesundheitswesens

Wir kommen jetzt zu der Frage nach den Konsequenzen des Paradigmawechsels für das „Makrosystem" der Gesundheitsversorgung einer Gesellschaft.

Wir haben festgestellt, daß ein Paradigma in der Heilkunde nicht nur die Spielregeln für ärztliches Handeln festlegt, sondern als Organisationsprinzip für die Zusammenarbeit zwischen Forschern und Ärzten (sowie zwischen Ärzten und Politikern) auch die Struktur der Medizin als Institution, und damit des Gesundheitswesens, bestimmt.

Damit stehen wir vor der Frage nach den Konsequenzen, die ein Paradigmawechsel, wie ihn die Einführung der psychosomatischen Betrachtungsweise bedeutet, für die Struktur unseres Gesundheitswesens haben muß. Einzelheiten solcher Konsequenzen sind in dem Kapitel über die „Institutionalisierung der Psychosomatischen Medizin im klinischen Bereich" dargestellt. Sie behandeln vor allem organisatorische Fragen, die sich für psychosomatische Abteilungen und ihre Beziehungen zu den traditionellen großen Kliniken stellen. Ich will im folgenden die Frage diskutieren, wie sich ein Paradigmawechsel in dem hier verstandenen Sinn auf die Medizin als Ganzes, einschließlich der traditionellen Großkliniken, auswirken würde. Ich will, mit anderen Worten, den Gedanken der Utopie einer Humanmedizin ernst nehmen.

Dafür gibt es realistische Gründe, die damit zusammenhängen, daß die strukturellen Konsequenzen des Paradigmas der Maschine, die aus immer zahlreicheren kleinen Maschinen zusammengesetzt ist, an organisatorische, finanzielle und psychologische Grenzen stoßen.

Nach einem Paradigma, für das ein Ganzes nur aus einer Summe von Einzelteilen besteht, muß sich die Medizin auf die Organe konzentrieren und kann den Menschen als deren zufälligen Träger vernachlässigen. Die Aufteilung der Heilkunde in immer zahl-

reichere, hochtechnisierte Organkliniken, die nur noch aus verwaltungstechnischen Gründen zusammengefaßt sind, ist die logische Konsequenz, die sich im modernen Krankenhausbau am sinnfälligsten zeigt: Wer ein modernes Großklinikum betritt, findet das Paradigma der Maschine, die aus vielen kleinen Maschinen zusammengesetzt ist, ins Monumentale transponiert. Er betritt eine Maschinenhalle, deren gigantische Ausmaße an die Gefängnisse erinnern, die Piranesi dargestellt hat. Der Eindruck nackter Grausamkeit dieser Landschaft aus Stahl und Beton wird durch Hinweisschilder unterstrichen, deren Aufschriften („Nicre rechts", „Herz links", „Lunge im ersten Obergeschoß") an einen labyrinthischen Schlachthof erinnern.

Man erlebt, daß hier eine unaufhaltsame Dynamik der Zerstückelung am Werk ist, eine spezifische Tendenz zur Fraktionierung. Sie drängt dazu, Organkliniken wieder in Spezialkliniken für bestimmte Gewebe, diese wieder in Kliniken für bestimmte Zellarten und schließlich für Zellbestandteile aufzugliedern. Am konsequentesten wird dieser Anspruch bereits in der Soziobiologie von Dawkins (1982) vertreten. Danach ist das Individuum nur die Replikationsmaschine selbstsüchtiger Gene, seine Person und sein persönliches Schicksal sind völlig gleichgültig. Die Forderung nach Kliniken für Gene wäre die folgerichtige Konsequenz.

Wir haben an Beispielen konkreter Modelle in der Praxis niedergelassener Ärzte und klinischer Einrichtungen Veränderungen beschrieben, die sich mit der Einführung der psychosomatischen Betrachtungsweise in der Wirklichkeit der medizinischen Versorgung ergeben. Dabei zeigt sich, daß dieser Paradigmawechsel keine „alternative Medizin" bedeutet, wohl aber einen alternativen, d.h. menschengerechteren Gebrauch der Möglichkeiten, welche die moderne Medizin für kranke Menschen gebracht hat. In diesem Zusammenhang wurden auch die ökonomischen Konsequenzen untersucht, die sich aus dieser Veränderung des Konzepts für den Arzt und für das Gesundheitswesen ergeben. Dabei stellte sich heraus, daß Ärzte als Preis für eine Form der Berufsausübung, die sie persönlich als befriedigender erlebten, auf etwa ein Drittel der Einkünfte ihrer konventionell arbeitenden Kollegen verzichten mußten. Auf der anderen Seite belegen die wiedergegebenen Erfahrungen eindrucksvoll, welche Kosteneinsparung eine Krankenversorgung erreichen kann, die durch rechtzeitige Berücksichtigung psychosozialer Komplikationen Chronifizierungen verhindert, nutzlose, unter Umständen schädliche und immer von neuem wiederholte somatische Untersuchungen unnötig macht und den Medikamentenverbrauch auf ein sinnvolles Maß einschränkt (Th. v. Uexküll, 1981). Eine Heilkunde, die sich an einem derartigen Modell orientiert, ist auch keine Rückkehr zu einer Medizin ohne Spezialisten. Die Vorstellung von Supraordinarien, die alle Spezialmethoden der somatischen und psychologischen Medizin beherrschen, entspringt Omnipotenzphantasien, die schon lange den Kontakt mit der Wirklichkeit der Heilkunde verloren haben.

Eine an einem integrativen Modell orientierte Heilkunde müßte – allgemein formuliert – das additive Nebeneinander von Organfächern und psychotherapeutischen Disziplinen durch ein System klar gegliederter Integrationsebenen ersetzen, in dem Spezialisten noch insoweit Generalisten sind, daß sie die psychosozialen Zusammenhänge sehen und deren Bedeutung für die Probleme ihres Fachgebietes, sowie vor allem die Bedeutung der Probleme ihres Fachgebietes für die psychosozialen Zusammenhänge ihres Patienten beurteilen können. Sie müßten, anders formuliert, soweit Allgemeinmediziner sein, daß sie den diagnostisch-therapeutischen Zirkel (Wesiack, 1984) von der Ebene der Interaktionen zwischen Patient und Arzt bis hinunter zu den Ebenen der Organe, Gewebe und Zellen der Spezialität, die sie vertreten, verfolgen können. Auf diese Weise wären sie in der Lage, in großen Zügen die Aufwärts- und Abwärts-Effekte zu übersehen und zu beurteilen, die für das Krankheitsbild eines Patienten entscheidend sind.

Damit ist das Problem einer „Allgemeinmedizin" angesprochen, unter der mehr als nur eine Gruppe standespolitisch definierter Ärzte, die eine Art Alibifunktion für den Rest nur noch spezialisierter Ärzte übernehmen, gemeint ist. Das Problem der Integration kann in einem arbeitsteiligen Gesundheitswesen nur gelöst werden, wenn alle Ärzte, die kranke Menschen alleinverantwortlich betreuen, neben ihren spezialistischen Kenntnissen auch allgemein-medizinisches, d.h. psychosomatisches oder bio-psycho-soziales Wissen haben (Pauli, 1984).

Die Notwendigkeit, ein neues Gleichgewicht zwischen spezialistischen und „generalistischen" Kenntnissen und Fertigkeiten zu finden, wird heute von allen einsichtigen Ärzten und Gesundheitspolitikern gesehen. Siegenthaler hat diese Notwendigkeit 1984 auf dem Deutschen Internistenkongreß eindringlich betont. Aber solche Proklamationen bleiben Rhetorik, solange man meint, jeder gute Arzt sei von Natur aus schon Generalist und könne, wenn man ihm nur die dafür aufgewendete Zeit vergütet, mit Freundlichkeit und gesundem Menschenverstand auch die komplexen bio-psycho-sozialen Probleme seiner Patienten lösen.

Auch die oft geäußerte Meinung, welche die Technisierung der Medizin für das Schwinden an Humanität verantwortlich machen will, geht an dem Kern des Problems vorbei. Sie unterstellt, die Konzentration auf technische Verfahren würde den Ärzten und dem Pflegepersonal die Zeit rauben, die für Zuwendung, Wärme und menschliches Verstehen notwendig ist. In dem Kapitel über „Psychosomatische Aspekte intensivmedizinischer Behandlungsverfahren" wird deutlich, daß der Zeitfaktor zwar eine limitierende, aber keineswegs entscheidende Rolle spielt, wenn technische Verfahren in der Medizin inhuman werden.

Ahnefeld (1984) hat zweifellos recht, wenn er die Technisierung der Medizin mit dem Hinweis verteidigt, daß technische Verfahren nicht per se inhuman seien, daß im Gegenteil neue technische Errungenschaften, angefangen vom Beatmungsgerät über den Herzschrittmacher bis zur Hüftprothese, Fortschritte für die Rettung von Leben und die Minderung von Leiden gebracht haben, die vorher undenkbar waren; wenn er betont, daß neue technische Entwicklungen, wie die Computertomographie, dem Kranken unendlich viel an Schmerzen und Komplikationen erspart und bessere und frühzeitigere Diagnosen ermöglicht haben. Überdies bürden neue technische Verfahren Ärzten und Pflegepersonal nicht nur neue Aufgaben auf, sondern entlasten sie auch von früheren, die oft noch zeitraubender und kaum weniger technisch waren.

Der entscheidende Punkt liegt woanders: Ahnefeld formuliert ihn, indem er feststellt, daß ein Mehr an Technik nicht ein Weniger, sondern ein Mehr an Arzt erfordere. Dieses Mehr ist schon deshalb kein bloßes Zeitproblem, weil Zuwendung, Wärme und menschliches Verstehen keine spezifisch ärztlichen Fähigkeiten sind. Ärzte werden darin oft genug von den Angehörigen der Kranken übertroffen. Das „Mehr an Arzt" ist seine Bereitschaft und seine Fähigkeit, sich selbst als diagnostisches und therapeutisches Instrument einzusetzen und der Versuchung zu widerstehen, sich in diesen Funktionen durch technische Verfahren ersetzen zu lassen.

Wenn wir versuchen, diese abstrakten Überlegungen in ein konkretes Modell umzusetzen, stellt sich auf der ärztlichen Ebene die Forderung, daß die Organspezialisten mehr von der psychosozialen Wirklichkeit des Menschen und ihren Wechselwirkungen mit seinem Körper wissen müssen, als es heute der Fall ist. Kardiologen, Gastroenterologen, Endokrinologen und die anderen Organspezialisten müssen lernen, den Einfluß der Vorgänge, die sich in den individuellen Wirklichkeiten ihrer Patienten ereignen, auf die Organe, die sie behandeln, in Rechnung zu stellen. Sie müssen die zahlreichen Untersuchungen zur Kenntnis nehmen, die nachweisen, daß Schicksalsschläge, welche die individuelle Wirklichkeit eines Menschen erschüttern, mit einer erhöhten Rate von Organkrankheiten einhergehen, daß hoher Blutdruck, Herzinfarkt, Magengeschwüre, Erkranken der Schilddrüse und andere körperliche Leiden mit dem Tod naher Angehöriger, dem Scheitern von Lebensplänen, Arbeitslosigkeit, Isolierung und anderen psychosozialen Belastungen zu tun haben. Dies setzt voraus, daß sie bereit sind, die Bilder zu ändern, die sie sich nach dem herrschenden Paradigma von der verschlossenen Realität der Krankheit gemacht haben.

Dazu kommt die Forderung, daß der Spezialist die Grenzen und die Ergänzungsbedürftigkeit seines Fachs im Rahmen eines Gesundheitssystems erkennen muß, in dem Kranke integriert psychosomatisch betreut werden sollen. Das erfordert die Bereitschaft zur Kooperation, nicht nur mit anderen Spezialisten, sondern auch mit anderen Berufsgruppen, Psychologen, Sozialarbeitern usw., eine Kooperation, die damit beginnen muß, eine gemeinsame Sprache für den Aufbau einer gemeinsamen Wirklichkeit auch unter den Mitarbeitern z.B. einer Krankenstation zu entwickeln. Mit der Erfüllung dieser Forderung steht und fällt jeder Versuch einer Integration als Gegengewicht und Ergänzung der notwendigen Spezialisierung.

Wir wissen noch nicht, welche Organisationsform diesem Modell am besten entspricht. In der Praxis erprobte Versuche beweisen jedoch, wie oben erwähnt, daß dieses Modell in einem kleineren oder größeren Rahmen verwirklicht werden kann (Th. v. Uexküll, 1981). Der Erfolg oder Mißerfolg solcher Modelle hängt von einer Reihe verschiedener, zum Teil innerer, zum Teil äußerer Faktoren ab, zu denen vor allem der Widerstand der „Normalmedizin" (vgl. Anmerkung 5 am Ende des Kapitels) gehört, der von Situation zu Situation, von Landschaft zu Landschaft wechseln kann.

Ein in der praktischen Erprobung sieben Jahre erfolgreicher Großversuch wurde mit einem Departmentsystem in Ulm gemacht. Er beruhte auf dem Konzept, daß alle Kranken auf Allgemeinstationen betreut wurden, deren Chefärzte Kliniker waren, die neben ausreichenden Kenntnissen als Allgemeininternisten (oder Allgemeinchirurgen) über Kenntnisse auf einem Spezialgebiet verfügten. Den reinen Spezialisten standen demgegenüber nur die Einrichtungen zur Verfügung, die für ihre Funktion als spezialistische Konsiliarien mit diagnostischen und therapeutischen Aufgaben notwendig waren. Sie hatten keine eigenen Bettenabteilungen.

Anstelle der so oft geübten „Aufteilung der Verantwortung" (Balint) unter verschiedene Spezialisten, mit der Gefahr der Entstehung eines „Verantwortungs-Vakuums", sollte das Prinzip eingeübt werden, daß jeder Kranke während seines Aufenthaltes im Klinikum seinen persönlichen Arzt hat, der für ihn verantwortlich ist. Dieser Arzt – gewöhnlich der Stationsarzt – sollte unter Leitung des zuständigen Oberarztes und Chefarztes lernen, wann ein Spezialist zugezogen werden muß, und wie man ihn zuzieht, ohne die Verantwortung für den Kranken zu delegieren. Mit dieser Regelung sollte auch die heute von Ärzten und Laien meist kritiklos hingenommene Vorstellung einer Prestigeskala abgebaut werden, an deren Spitze der Spezialist steht.

Dieses Departmentsystem – und die mit ihm geplante Einführung der psychosomatischen Betrachtungsweise durch das Beispiel einer Modellstation (Köhle et al., 1977) – scheiterte schließlich an den Gegenkräften, die in jedem gesellschaftlichen System Änderungen der traditionellen Erklärungsmodelle – und der ihnen entsprechenden Organisationsformen – zu verhindern trachten (Th. v. Uexküll, 1977). Andere Versuche mit weniger ehrgeizigen Zielsetzungen konnten sich dagegen trotz größerer oder geringerer Schwierigkeiten behaupten.

Inzwischen wächst bei einsichtigen Internisten das Bewußtsein der Notwendigkeit, „das psychosomatische Versorgungsangebot in der inneren Medizin zu erhöhen und die psychosomatische Betrachtungsweise sowohl wissenschaftlich-psychophysiologisch weiter zu fundieren, wie praktisch als ärztliche Anthropologie zu verwirklichen" (Buchborn, 1984). Zur Frage, wie dieses Programm organisatorisch gelöst werden kann, meint Buchborn:

„Das kann zwar im Hinblick auf die immer noch wachsenden Ansprüche an spezialistisches Wissen und Können nicht bedeuten, daß der Internist auch noch die Methoden der Psychodiagnostik und der Psychotherapie erlernen und beherrschen soll. Ideal wäre vielmehr die **gemeinsame** und **gleichzeitige** Betreuung psychosomatischer Kranker in kombiniert internistisch-psychotherapeutischen Behandlungseinheiten, wie sie modellhaft an einigen Medizinischen Universitätskliniken (z.B. Hannover, Heidelberg, Ulm) erprobt wurde."

Diese Feststellung kann durch den Hinweis auf andere universitäre (und kommunale) Kliniken ergänzt werden, in denen psychosomatische Betreuung organisch Kranker auch von Internisten durchgeführt wird, die zusätzlich psychotherapeutische Kompetenzen erworben haben, z.B. Bern (vgl. Kap. 27) und Lübeck (Feiereis, 1982).

Zu der Frage nach den Konsequenzen des Paradigmawechsels für die Struktur des sozialen Makrosystems „Gesundheitswesen" gehört auch – und zwar mit großem Gewicht – die Frage nach den Konsequenzen für die Struktur medizinischer Fakultäten und ihren Zusammenhang mit den Problemen der Ausbildung zum Arzt. Da diese Probleme schon in Kapitel 67 behandelt sind, genügt hier der Hinweis, daß als Konsequenz des Paradigmawechsels die Fakultäten in ihrer derzeitigen Struktur und Aufgabenstellung die Position als Vorbild und Muster für das Gesundheitssystem unserer Gesellschaft verlieren werden und verlieren müssen. Durch ihre Entwicklung zu Zentren hochqualifizierter biotechnischer Forschung und der Maximalversorgung einer relativ kleinen, hochselektierten Gruppe von Behandlungsbedürftigen haben sie weltweit die Fähigkeit eingebüßt, Ärzte auszubilden, die in der Lage sind, die Aufgaben der Primärversorgung einer Bevölkerung wahrzunehmen (Bloom, 1988).

69.8 Die Heilkunde als Teilbereich der Industriegesellschaft

Ich sprach von der Verantwortung, die der Arzt der Medizin gegenüber hat, die er ausübt und der er als Mitglied einer gesellschaftlichen Institution mit einer Jahrtausende alten Geschichte angehört. Das Bewußtsein einer Verantwortung dieser Institution gegenüber, die keine nationalen, politischen oder religiösen Grenzen kennt, hat sich in Bekenntnissen und Eiden niedergeschlagen, die von den Hindu-Ärzten bis in die Gegenwart reichen (Kleeberg, 1979). Wir können dieser Verantwortung nur genügen, wenn wir bereit sind, die Einrichtungen unserer zeitgenössischen Medizin und deren Entwicklungen unter dem Aspekt einer Humanmedizin zu bewerten.

Dazu gehört auch eine illusionslose Einschätzung der Kräfte, die Veränderungen der bestehenden Verhältnisse zu verhindern trachten und die sich vor allem gegen eine Entwicklung stemmen, die zu einem Paradigmawechsel in dem hier definierten Sinne führt.

Die Medizin ist ein Teilgebiet der Gesamtgesellschaft und besitzt daher nur eine begrenzte Autonomie. Aber sie ist ein wichtiges Teilgebiet, von dem auch Anstöße für die Entwicklung der zeitgenössischen Gesellschaft ausgehen. Ferber (1971) hat auf die enge Verflechtung der modernen Medizin mit der Industriekultur hingewiesen und gezeigt, wie Grundprinzipien für die Deutung von Krankheit als ein Geschehen, das von der Person des Kranken abgelöst im Tierversuch und im Laboratorium reproduziert werden kann, den Grundprinzipien der technischen Produktion entsprechen.

Zu den strukturbildenden Faktoren unserer Industriegesellschaft gehören auch wirtschaftliche Interessen, die in kapitalistischen und sozialistischen Staatsformen einen Machtfaktor bilden, der auch die Entwicklung der Medizin beeinflußt. Dieser Punkt ist historisch und berufspolitisch von besonderem Interesse. Eine Schrift der WHO (1984) stellt diesen Zusammenhang folgendermaßen dar:

„Als Virchow 1849 schrieb, ‚Medizin ist eine soziale Wissenschaft und Politik ist Medizin im Großen‘, nahm er den Weitblick auf das Problem der Gesundheit voraus, wie es heute gesehen werden muß. Durch die Arbeiten von Pasteur, Koch und anderen, die zur Theorie der Verursachung von Krankheit durch Erreger führten, ging in der zweiten Hälfte des 19. Jahrhunderts viel von dieser weiten Sicht verloren. Man konzentrierte sich auf die konsequente Verfolgung neuer medizinischer Entdeckungen und die Entwicklung der technischen Möglichkeiten, spezifische Krankheiten zu bekämpfen. Der krönende Erfolg dieser Strategie war vielleicht die kürzlich gelungene Auslöschung der Pocken als Gefahr für die Menschheit.

Im Zuge dieser Entwicklung wurden Kliniken und Laboratorien zu den Zentren der Medizin und des wissenschaftlichen Fortschritts. Schließlich sah man in ihnen das Herzstück des Gesundheitswesens überhaupt. In vielen nationalen Systemen wurden sie darüber hinaus Mittel für Kapitalinvestition zum Erzielen von finanziellem Gewinn, ohne Rücksicht darauf, ob die angewandten Techniken nachweislich notwendig oder effektiv waren. Weltweit konnte das gleiche profitorientierte Denken hinter der Werbung für Medikamente und technische Entwicklungen beobachtet werden, die nicht immer von sozialem Nutzen waren.

Die hochtechnisierte Medizin scheint den Menschen aus der Hand zu gleiten und sich in die falsche Richtung zu entwickeln: weg von dem Ziel einer Gesundheit für viele und hin zu einer kostspieligen Behandlung für die wenigen, die es sich leisten können."

Hier wird ein anderer Aspekt der Technisierung der Medizin angesprochen als der, den Ahnefeld im Auge hat. Seine Gefahr liegt in einer Entwicklung, für die der biomedizinische Bereich unter dem Gesichtspunkt seiner wirtschaftlichen Verwertbarkeit besonderes Gewicht erhält, während der psychosoziale uninteressant oder sogar nur negativ bewertet wird. Diese Entwicklung führt nicht zu dem „Mehr an Arzt", das Ahnefeld für die Anwendung technischer Verfahren fordert, sondern zu einem immer „Weniger an Arzt", das die technischen Verfahren inhuman werden läßt.

Pauli (1984) stellt fest, daß diese Sicht auch die Gewichtung der Fächer in Lehre und Forschung bestimmt:

„Die Hindernisse gegenüber einer Neuorientierung von Lehre und Forschung sind politisch-struktureller Natur. Es haben sich machtvolle Institutionen im biomedizinischen Bereich gebildet, die sowohl die vorhandenen Mittel binden, als auch eine Entwicklung in den übrigen Anteilen des Handlungsfeldes im Gesundheitsbereich blockieren."

Hier wird die begrenzte Autonomie der Medizin als Teilbereich einer Gesellschaft augenfällig, der es noch nicht gelungen ist, ein Gleichgewicht zwischen ökonomischen Forderungen und ökologischer Verantwortung vor der Natur – auch der Natur des Menschen – zu finden.

69.9 Die individuelle Wirklichkeit des Arztes und die erkenntnistheoretischen und berufspolitischen Entscheidungen

Auf dem Hintergrund dieser Situation der Medizin in dem Kräftefeld gesellschaftlicher Entwicklungen wird das Problem der ärztlichen Verantwortung in seinen drei Aspekten deutlich: dem Patienten, der Medizin als gesellschaftlicher Institution und sich selbst gegenüber. In dem letzten Aspekt konzentrieren sich die Widerstände gegen einen Paradigmawechsel wie in einem Brennpunkt: daß die derzeitigen Formen der Aus- und Weiterbildung und die „finanziellen Entlohnungs- und sozialen Anreizsysteme dem Bestreben einer humanen Krankenversorgung entgehenstehen" (Infas, 1983); daß die standespolitischen Strukturen die Integration einer psychosomatischen Betrachtungsweise erschweren, und daß kognitive und emotionale Widerstände überwunden werden müssen, die im einzelnen selbst liegen.

Wer der Meinung ist, die Übernahme der psychosomatischen Betrachtungsweise sei nur eine Frage der Überzeugungskraft der vorgebrachten Argumente, übersieht nicht nur die gesellschaftlichen Gegenkräfte, sondern auch die Bedeutung der Zugehörigkeit zu einer etablierten Gruppe für die Identitätsfindung des einzelnen und die Effektivität von Repressalien gegen Abtrünnige; er unterschätzt auch die Bedeutung eines herrschenden Paradigmas für die kognitiven und affektiven Strukturen des Arztes. Diesen letzten Punkt hat Thomas Kuhn (1973) sehr deutlich gemacht:

„Die Arbeit im Zeichen des Paradigmas kann auf keine andere Weise durchgeführt werden, und das Paradigma im Stich lassen, hieße die Wissenschaft, die es definiert, nicht mehr ausüben."

Trotzdem hat es, wie er sich ausdrückt, immer wieder „Deserteure" gegeben, die ein herrschendes Paradigma im Stich gelassen und das Abenteuer eines Paradigmawechsels auf sich genommen haben. Sie sind, wie er schreibt, „die Angelpunkte, um die wissenschaftliche Revolutionen sich drehen". Damit ist der Spielraum für individuelle Entscheidungen angesprochen, die, wenn auch gegen innere und äußere Widerstände und unter der ständigen Gefahr des Scheiterns, bestehende Strukturen ändern können. Wissenschafts- und berufspolitische Entscheidungen sind keine Automatismen. Man kann bestehenden Institutionen Immobilismus vorwerfen. Man kann die Forderung aufstellen, die Medizin müsse „ihr reduktionistisches multidisziplinäres Modell durch ein interdisziplinäres anthropologisches ersetzen" (Pauli, 1984); nur – „die Medizin" sind letztlich wir selbst – die Ärzte als Studenten von gestern und die Studenten als die Ärzte von morgen.

Wissenschafts- und berufspolitische Entscheidungen sind nicht nur anonyme Systemzwänge. Sie sind auch Sache des einzelnen. Jeder von uns trägt Verantwortung für die Medizin, die wir haben, und für das, was morgen aus ihr wird. Jeder muß sich entscheiden, ob er das Abenteuer des Paradigmawechsels auf sich nehmen oder dem herrschenden Paradigma weiterhin Gefolgschaft leisten will.

Mit dem Entschluß, die psychosomatische Betrachtungsweise in die Medizin einzuführen, handelt sich jeder Arzt und bereits jeder Student Schwierigkeiten ein, die es nötig machen, daß er sich frühzeitig klarwird, wofür er sie auf sich nimmt. Hier ist die Einsicht entscheidend, daß es nicht nur um eine mehr oder weniger abstrakte Verantwortung für „die Medizin", sondern auch um die sehr konkrete Verantwortung für sich selbst geht, für die Entwicklung der eigenen individuellen Wirklichkeit und der Fähigkeit, gemeinsame Wirklichkeiten aufzubauen, in denen der Arzt mit Kranken kommunizieren kann.

Damit ist eine Fähigkeit angesprochen, ohne die weder Ausbildung noch Weiterbildung noch Forschung die Utopie einer Humanmedizin jemals verwirklichen können.

Der Historiker Carlo Ginzburg (1983) nennt diese Fähigkeit „moralische Vorstellungskraft" und schreibt, sie sei die „grundlegende Sache" für junge Menschen, die Geschichte studieren wollen. Er rät ihnen, viele Romane zu lesen, weil dadurch die Fähigkeit entwickelt werde, „sein Leben zu vervielfachen und beispielsweise der Fürst Andrej in Tolstois Krieg und Frieden oder der Mörder der alten Wucherin in Dostojewskis Schuld und Sühne zu sein". „Moralische Vorstellungskraft", fügt er hinzu, „hat nichts mit Träumerei zu tun". Sie sei vielmehr ein unentbehrliches Mittel gegen die weitverbreitete Neigung, „sich andere Menschen als alter ego – und das heißt, als äußerst langweilige Personen vorzustellen".

Was für angehende Historiker gilt, das gilt in einem noch höheren Maße für angehende Ärzte. Auch ein junger Mensch, der Medizin studieren will, sollte Romane lesen und prüfen, ob seine moralische Vorstellungskraft ausreicht, um beispielsweise Hans Castorp in Thomas Manns Zauberberg, Wadim, Dr. Donnavan oder Schulubin in Solschenizyns Krebsstation, der Fürst Myschkin in Dostojewskis Idiot oder der Kranke in Tolstois Tod des Ivan Iljitsch zu sein.

Im Unterschied zu dem angehenden Historiker muß sich der angehende Arzt, der bei der Schilderung menschlicher Schicksale die Erfahrung macht, ein anderer Mensch zu sein und eine ihm vorher unbekannte Wirklichkeit zu betreten, aber zwei Fragen vorlegen: Was geschieht mit mir, während ich das erlebe, und was fange ich in der ärztlichen Situation mit diesem Erleben an?

Wenn er diesen Fragen nachgeht, wird er feststellen, daß es in ihm eine Instanz gibt, die seine Empfindungen in einer Weise ordnet, daß sie für ihn lesbar werden. Er entdeckt, mit anderen Worten, daß sich diese Instanz als Übersetzer betätigt und bekannte Zeichen in dem Kontext einer neuen Stimmung interpretiert: Die Stimmung, die bisher seine Motivationen und Einstellungen zu sich selbst und zu seiner Umgebung getragen hatte, schlägt um und damit verändern sich auch die Bilder, die seine Phantasie von ihm selbst in seiner individuellen Wirklichkeit entwirft.

Die geheimnisvolle Verzauberung, welche die Grundstimmung jeder echten Erfahrung ausmacht, hat Winnicott (1973) beim kleinen Kind als Subjekt-Objekt-Identität beschrieben und dargestellt, wie die Fähigkeit, den anderen als vom eigenen Selbst abgelöste Person zu erleben, eine lange, für jeden Menschen verschieden verlaufende Entwicklung voraussetzt. Es gibt überzeugende Hinweise, daß jede neue Erfahrung wieder mit dem Erleben einer Subjekt-Objekt-Identität beginnt und dann in großen Zügen den Entwicklungsgang der Trennung in Subjekt und Objekt wiederholen muß. Nach diesen Hinweisen ist jeder, den das Schicksal eines anderen bewegt, zunächst der andere. Je stärker und unmittelbarer ein angehender Arzt diese Identität erlebt, um so wichtiger ist für ihn die kritische Reflexion, die ihn befähigt, sich in dem Zustand der Identifikation mit einem Kranken, seiner Verzweiflung, seiner Depression oder seiner Euphorie als Instrument zu beobachten, das im Mitschwingen die Stimmung des anderen registriert.

Aus diesem kritischen Abstand kann er seine emotionale Betroffenheit auf dem Hintergrund der ärztlichen Aufgabe betrachten und erkennen, was er bisher nur mehr oder weniger dunkel geahnt hatte: daß emotionale Betroffenheit rationales Urteilen und Handeln verhindern kann. Damit sieht er ein allgemeines Problem, das jede Ausbildung zum Arzt auf die eine oder andere Weise lösen muß, und das die Persönlichkeitsbildung jedes jungen Menschen berührt, der sich auf diese Ausbildung einläßt.

Die meisten beginnen das Medizinstudium, ohne dies Problem zu sehen und ohne sich Gedanken zu machen, wie sie es lösen könnten. Sie haben auch kaum eine Vorstellung, wie dies Problem im Rahmen der Ausbildung, die sie beginnen, gelöst werden wird. Die Tatsache, daß sie ein konsequent aufgebautes Desensibilisierungsprogramm erwartet, gehört zu den streng tabuierten Themen unserer Industriekultur.

So beginnen viele ihr Studium mit dem Wunsch nach Nähe zum Patienten (vgl. Kap. 67) und wissen nicht, wie es geschieht, daß ihnen dieser Wunsch mehr und mehr abhanden kommt. Sie durchschauen nicht, daß eine Ausbildung, die zur Vorbereitung auf den Umgang mit kranken Menschen mit Physik, Chemie und dem Sektionskurs an der Leiche beginnt, die im klinischen Teil die psychosozialen Probleme der Kranken weitgehend ignoriert (Adler, 1981), in der über die Frage, was Sexualität, Abhängigkeit, Sucht und Sterben bedeuten, nicht gesprochen wird, und die statt dessen das Ideal des unbeteiligten wissenschaftlichen Beobachters einübt, den Betroffenheit für den ärztlichen Beruf disqualifiziert – daß eine solche Ausbildung zu einer merkwürdigen Schizophrenie führt: Sie erreicht einen weitgehenden Abbau der Tabuschranken vor der Intimsphäre des Körpers bis zur Indifferenz gegenüber verletzenden und eindringenden Verfahren, die mit einer irrationalen Scheu vor einem Eindringen in die psychische Intimsphäre eines Kranken gekoppelt ist.

Im Verlauf dieses Desensibilisierungsprozesses lösen sich wissenschafts- und berufspolitische Probleme wie von selbst und die Entwicklung der Persönlichkeit des zukünftigen Arztes verläuft nach dem uniformen Muster, das die Rollen vorprogrammiert, die er später als Arzt in der Makrostruktur des Gesundheitssystems spielen wird.

In Kapitel 67 werden Versuche dargestellt, in denen Studenten das Problem ihrer Sozialisation zum Arzt nicht durch passive Desensibilisierung zu lösen suchen, sondern sich bemühen, das Problem ihrer Persönlichkeitsbildung selbst in die Hand zu nehmen. Es wird gezeigt, wie man die Fähigkeit erwerben kann, mit seiner emotionalen Betroffenheit kritisch umzugehen, wie emotionales Beteiligtsein am Erleben des Kranken den Arzt nicht urteils- und handlungsunfähig zu machen braucht, sondern im Gegenteil Informationen über den Patienten und sein Kranksein vermitteln kann, die für Urteilsfindung und Handlungsentscheidung unerläßlich sein können.

Der Arzt muß lernen, in der professionellen Beziehung zum Patienten zwischen Nähe und Distanz zu „pendeln". Zunächst muß er sich in den Patienten einfühlen, ja sich mit ihm identifizieren, um dessen individuelle Wirklichkeit kennenzulernen. Um dieses Erleben aber reflektiert zum Nutzen des Patienten verwenden zu können, benötigt er danach wieder Distanz; d.h., er muß in der Lage sein, seine Identifikation flexibel zurückzunehmen, um die Konsequenzen seiner Erfahrung für die Diagnostik und Therapie erfassen und abwägen zu können. Daher unterscheidet sich die professionelle ärztliche Beziehung von den Alltagsbeziehungen zwischen Menschen, z.B. einer Freundschaft, durch die Fähigkeit, reflektiert zwischen Nähe und Distanz zu wechseln. Damit werden vom Arzt neben einer unspezifischen Kompetenz zur Kommunikation spezifische Lernschritte einer Aus- und Weiterbildung gefordert.

Wie schwer es den Studierenden gemacht wird, diese Lernschritte schon während ihrer Ausbildung zurückzulegen, ist in Kapitel 67 dargestellt worden. Aber wie schwer es ist, das herrschende Paradigma auch während der Berufsausübung zu verlassen und sich an dem Paradigma zu orientieren, das mit der Einführung der psychosomatischen Betrachtungsweise verbunden ist, erfahren Studierende und junge Ärztinnen und Ärzte, die in Balint-Gruppen für die Beziehungen zwischen Patient und Arzt sensibilisiert wurden, an dem Widerstand der Ärzte und des Pflegepersonals, mit denen sie zusammenarbeiten müssen und von denen sie zum Teil abhängig sind. Wenn sie erleben, wie Schwestern und Kollegen die Vermittlung der Stimmung eines Kranken feindselig abwehren oder ihre Versuche einer Zuwendung zu Kranken boykottieren oder lächerlich machen, erfahren sie die wissenschafts- und berufspolitischen Probleme der Integration psychosomatischer Betrachtungsweise am eigenen Leibe.

Auch auf dem Hintergrund dieser Erfahrungen muß man die kritische Frage nach der Validität einer Ausbildung zum Arzt stellen, die, wie Pauli (1984) betont, etwa zehn Jahre zur Vermittlung spezialistischer Kenntnisse beansprucht, aber für „eine synthetisch und systemisch orientierte Sicht umfassender Probleme kaum Raum und Zeit läßt".

Um zu verstehen, was die Forderung bedeutet, daß der Arzt auch Verantwortung für sich selbst und seine eigene individuelle Wirklichkeit zu übernehmen hat, muß man sich klarmachen, daß nur ein zufriedener, ausgeglichener und in sich gefestigter Arzt wirklich in gleichberechtigter Zusammenarbeit mit anderen Ärzten und Angehörigen anderer Gesundheitsberufe seinen Patienten helfen kann. Die Forderung, in Selbstverleugnung nur für andere zu leben und zu handeln, die so oft erhoben wird, entspricht einem einseitigen Zerrbild ärztlicher Tätigkeit.

Anders formuliert: Um sich selbst als diagnostisches und therapeutisches Werkzeug in die Interaktion mit dem Patienten einbringen und mit ihm eine gemeinsame Wirklichkeit aufbauen zu können, trägt der Arzt auch Verantwortung für die Entwicklung und die Effektivität dieses Werkzeugs.

Anmerkungen

Anmerkung 1: Struktur

„Anfänge dieses Modells (oder dieser Art von Modellen) gehen zweifellos auf sehr frühe Epochen der Heilkunde zurück und sind wahrscheinlich als Bestandteile aller Heilbehandlungssysteme zu finden; denn es interpretiert den Körper als räumliches Orientierungsschema für die Möglichkeiten von ‚Eingriffen' der menschlichen Hand. Das Körpermodell der chinesischen Volksmedizin für Akupunktur ist ein Beispiel für solch ein räumliches Orientierungsschema für Handgriffe von Heilbehandlern. Solche Körpermodelle waren früher in Vorstellungen über die Bedeutung von Körperteilen, Organen oder Säften für kosmische, religiöse oder magische Zusammenhänge eingebettet. In den westlichen Kulturen hat sich das räumliche Orientierungsschema für manuelle Eingriffe in den Körper eines Kranken seit etwa 150 Jahren zu ‚der wissenschaftlichen Theorie' der Kausalzusammenhänge von Strukturen und Prozessen in einem hochkomplexen Mechanismus verselbständigt. ...

Von Ferber (1971) hat auf den inneren Zusammenhang dieser Entwicklung mit dem Entstehen der Industriekultur hingewiesen. Dieser Aspekt läßt uns besser verstehen, wie es möglich war, den Grundsatz der räumlichen Orientierung für manuelle Eingriffe der menschlichen Hand im Verlauf der stürmischen Entwicklung zu dem hochkomplexen Theoriengebäude der modernen Medizin konsequent durchzuhalten. Die zunehmende Verfeinerung der Möglichkeiten für direkte Eingriffe der menschlichen Hand durch technische Apparaturen und für indirekte Eingriffe durch Pharmaka erzwang eine fortschreitende Differenzierung des Körpermodells, das umgekehrt wieder die Verfeinerung der Technik für Eingriffe vorantrieb. So entstand das imponierende Gebäude der modernen Medizin, das den menschlichen Körper nach dem Paradigma einer hochkomplexen physikalisch-chemischen Maschine interpretiert. Krankheit ist danach eine räumlich lokalisierbare Störung in einem technischen Betrieb, der zwar eine sehr komplizierte, aber mit Hilfe des technischen Vorbilds doch überschaubare Struktur besitzt. Von diesem allgemeinen Paradigma lassen sich die Diagnosen für konkrete Krankheiten als spezielle Spielregeln für den Umgang mit Kurzschlüssen, Rohrbrüchen, Transportproblemen oder ähnlichen technischen Fragen ableiten. Wie ein Techniker auf der Basis eines Schaltplans den Betriebsschaden eines Autos, eines Fernsehers oder Computers lokalisieren und danach die Reparatur planen kann, so kann der Arzt eine Krankheit, die als Betriebsschaden im menschlichen Körper – als Klappenfehler im Herzen, als Geschwür im Magen oder als Enzymdefekt in einem Gewebe oder Transportsystem – interpretiert ist, durch entsprechende technische Eingriffe reparieren" (Th. v. Uexküll, 1986).

Anmerkung 2: L'homme machine

Zu der Frage, ob und wieweit Computerprogramme mit Programmen menschlichen Denkens und Fühlens gleichgesetzt werden können, sagt Grant Johnson:

„... und wenn er (der Computer) Witze macht, dann kann man sicher sein, es sind nicht seine Witze. (Es gibt natürlich auch Menschen, die unentwegt Witze erzählen, die nicht die ihrigen sind, doch dies sagt etwas über sie aus. Die Tatsache, daß ein Rechner programmiert werden kann, um dasselbe zu tun, sagt über ihn überhaupt nichts aus.)"

Bateson (1985) formuliert das Problem folgendermaßen:

„Wir wollen nun für einen Augenblick die Frage erwägen, ob ein Computer denkt. Ich würde sagen, daß er das nicht tut. Was ‚denkt' und sich auf ‚Versuch und Irrtum' einläßt, ist der Mensch **plus** Computer **plus** Umgebung. Und die Grenzen zwischen Mensch, Computer und Umgebung sind rein künstliche, rein fiktive Linien. Sie sind Linien durch die Wege, auf denen Informationen oder Unterschiede übertragen werden. Sie sind keine Grenzen des Denksystems. Was denkt, ist das Gesamtsystem, das sich auf Versuch und Irrtum einläßt, nämlich der Mensch plus Umgebung."
(Vgl. auch Dreyfuss, 1985.)

Anmerkung 3: Humanmedizin versus Veterinärmedizin

Wieweit die Aufhebung eines Unterschiedes zwischen Human- und Veterinärmedizin tatsächlich als Programm einer wissenschaftlichen Entwicklung aufgefaßt wird, zeigt der folgende Passus aus der Eröffnungsrede des Kongresses der Deutschen Gesellschaft für innere Medizin von Franz Volhard 1930:

„Man kann es geradezu als Kriterium und höchste Leistung der rationellen Therapie bezeichnen, daß sie in einer Gruppe von Fällen ohne Rücksicht auf den individuellen Kranken, seine Persönlichkeit, seine seelische Verfassung, seine Konstitution mit seiner Krankheit fertig wird. ... Das bitter gemeinte Wort aus der Zeit vor der Wiederentdeckung der Seele: ‚Zwischen Medizin und Tierheilkunde besteht nur noch ein Unterschied bezüglich der Kundschaft', trifft heute im Gegensatz zu früher tatsächlich für eine ganze Reihe von Krankheiten zu, bei denen aufgrund wissenschaftlicher Erkenntnis Heilung sozusagen garantiert werden kann, unabhängig von der Individualität der Kranken und der Persönlichkeit des Arztes. Das Ziel der Forschung kann nur sein, die Zahl dieser rationell angreifbaren Krankheitszustände zu vergrößern."

Die Medizin hat es noch nicht zur Kenntnis genommen, daß die Hoffnung, die Volhard auf das Maschinenmodell setzte, nicht in Erfüllung gegangen ist. Zwar gelang es bei einigen akuten, vor allem infektiösen Krankheiten mit diesem Modell noch beachtliche Erfolge zu erringen und durch den Einsatz von Geräten zum Ersatz biologischer Funktionen Leben zu verlängern. Aber die Zunahme der Krankheiten durch selbstverschuldete Schäden wie Alkohol und Drogen, Übergewicht, Bewegungsmangel, Straßenverkehrsunfälle u. a. m. machte die Erfolge bald mehr als wett, und der Einsatz lebensverlängernder Geräte bei chronisch Kranken macht das Ungenügen eines Menschenbildes, aus dem die persönlichen Faktoren und das Vorhandensein von Gefühlen ausgeklammert sind, besonders offensichtlich.

Anmerkung 4: Emergenz/Dialektik

Die Entdeckung, daß in der Beziehung zwischen einem Ganzen und seinen Teilen eine Fülle theoretischer und praktischer Probleme steckt, ist ebensowenig neu, wie der Versuch, diese Probleme begrifflich zu fassen. Dafür spielt der Begriff der „Dialektik" von Plato über Aristoteles, das Mittelalter bis zu Kant und schließlich zu Hegel und Marx eine wichtige, oft genug widersprüchliche Rolle. In dem historischen Wörterbuch der Philosophie (1972) sind ihm 60 Seiten gewidmet!

Der Grund, warum der Begriff „Emergenz" dem der „Dialektik" vorzuziehen ist, hat zunächst mit dieser schillernden Geschichtlichkeit und den verschiedenartigen und oft gegensätzlichen Konnotationen zu tun, die mit dem Dialektik-Begriff verknüpft sind. Der Begriff Emergenz erlaubt Definitionen, die das Gemeinte schärfer fassen und die Beziehungen zwischen Element und System (Teil und Ganzem) genauer definieren. Vor allem beschreibt er den wesentlichen

Punkt: das sprunghafte (emergente) Auftreten neuer Phänomene, deren Eigenschaften sich nicht auf ihre Teilkomponenten reduzieren lassen.

Für das Verständnis dieser Zusammenhänge ist der von Medawar und Medawar (1977) gegebene Hinweis auf die Bedeutung der **Restriktionen,** denen die Möglichkeiten der Elemente im Rahmen eines Systems unterliegen, besonders fruchtbar. Sie machen darauf aufmerksam, daß die Restriktion der Aktionsmöglichkeiten, die isolierte Gebilde haben, Voraussetzung für die Integration zu einem einheitlichen Gebilde und für das Auftreten neuer Eigenschaften ist. Damit werden zwei Phänomene verständlich:

- daß jede Integration von Elementen zu einer **Reduktion von Komplexität** als Ausdruck einer neuen Ordnung auf einer höheren Stufe führt;
- daß ein Versagen der Restriktionen und die Freisetzung von Aktionsmöglichkeiten, welche Elemente auf der Elementar-Ebene haben, zu einer Verletzung der höheren Ordnung führt und ein Grundvorgang jeder Pathologie ist.

Beispiele gibt es auf jeder Integrationsebene: Auf der Ebene sozialer Gebilde finden wir Gesetze und moralische Normen als Restriktionen der Aktionsmöglichkeiten der Individuen, auf der Ebene der Organismen sehen wir die erwähnten Beispiele für die Restriktion der Zellteilungen frei lebender Einzeller im Verband eines Gewebes, Organs oder Organismus oder Restriktion der Bewegungsmöglichkeiten einzelner Muskeln im Rahmen koordinierter Bewegungen der Gliedmaßen.

Die ursprünglichste und einfachste Form dieser neuen Ordnung ist auf jeder Ebene die **Stimmung.** Sie manifestiert sich als eine Struktur der **Entsprechungen,** die mehr oder weniger differenziert, mehr oder weniger flüchtig als das Sich-Ergänzen von Leistung und Gegenleistung, Rolle und Gegenrolle, von Umwelt und „Nische" (z. B. bei Flügel und Luft, Flosse und Wasser, Fuß und Boden) in Erscheinung tritt. Die Feststellung der Entsprechungen ist für J. v. Uexküll die eigentliche Aufgabe der Biologie, deren Ziel die Aufstellung einer „Kompositionslehre der Natur" (1940, 1980) sein muß.

Plessner (1976) spricht von einer „Entsprechungsregel": „Nach dem Prinzip, das v. Uexküll einmal so formuliert hat: ‚Wo ein Fuß ist, da ist auch ein Weg, wo ein Mund ist, da ist auch Nahrung, wo eine Waffe ist, da ist auch ein Feind',

sollten sich diese Entsprechungen (auch für den Menschen) leicht und eindeutig angeben lassen."

Warum das weder leicht noch eindeutig möglich ist, stellt das Problem der Conditio humana dar.
(Vgl. auch Uexküll, J. v., 1940 und 1980.)

Anmerkung 5: „Normalmedizin"

Der Begriff „Normalmedizin" wird hier in dem gleichen Sinn verwendet wie der Begriff „Normalwissenschaft", den Th. Kuhn (1973) geprägt hat, und an dem er den Widerstand einer etablierten Wissenschaft gegen jeden Paradigmawechsel erläutert. Er führt aus, daß ein Paradigma seinen Status durch die Tatsache erlangt, daß es bei der Lösung einiger Probleme, welche ein Kreis von Fachleuten als brennend definiert hat, erfolgreicher war als andere, mit ihm konkurrierende Paradigmata.

Der Erfolg eines Paradigmas, schreibt er dann, sei am Anfang weitgehend eine Verheißung auf Erfolg, bei ausgesuchten und noch unvollkommen definierten Beispielen. Er fährt dann fort:

„Die normale Wissenschaft besteht in der Verwirklichung jener Verheißung, einer Verwirklichung, die durch Erweiterung der Kenntnis der vom Paradigma als besonders aufschlußreich offenbarten Fakten, durch Verbesserung des Zusammenspiels dieser Fakten mit den Voraussagen des Paradigmas sowie durch weitere Präzisierung des Paradigmas selbst herbeigeführt wird.

Von denen, die nicht tatsächlich Fachleute in einer ausgereiften Wissenschaft sind, erkennen nur wenige, wieviel ‚Aufräumarbeit' solcher Art ein Paradigma übrigläßt und wie faszinierend diese Arbeit tatsächlich sein kann. Das aber gilt es zu verstehen. Aufräumarbeiten sind das, was die meisten Wissenschaftler während ihrer gesamten Laufbahn beschäftigt, und sie machen das aus, was ich hier normale Wissenschaft nenne. Bei näherer Untersuchung … erscheint dies Unternehmen als Versuch, die Natur in die vorgeformte und relativ starre Schublade, welche das Paradigma darstellt, hineinzuzwängen. In keiner Weise ist es das Ziel der normalen Wissenschaft, neue Phänomene zu finden; und tatsächlich werden die nicht in die Schublade hineinpassenden oft überhaupt nicht gesehen. Normalerweise erheben die Wissenschaftler auch nicht den Anspruch, neue Theorien zu finden, und oft genug sind sie intolerant gegenüber den von anderen gefundenen."

Abb. 28–4.   Zeichnung einer 25jährigen Patientin mit Colitis ulcerosa: „Alle Männer werden in ein Höllenfeuer geworfen."

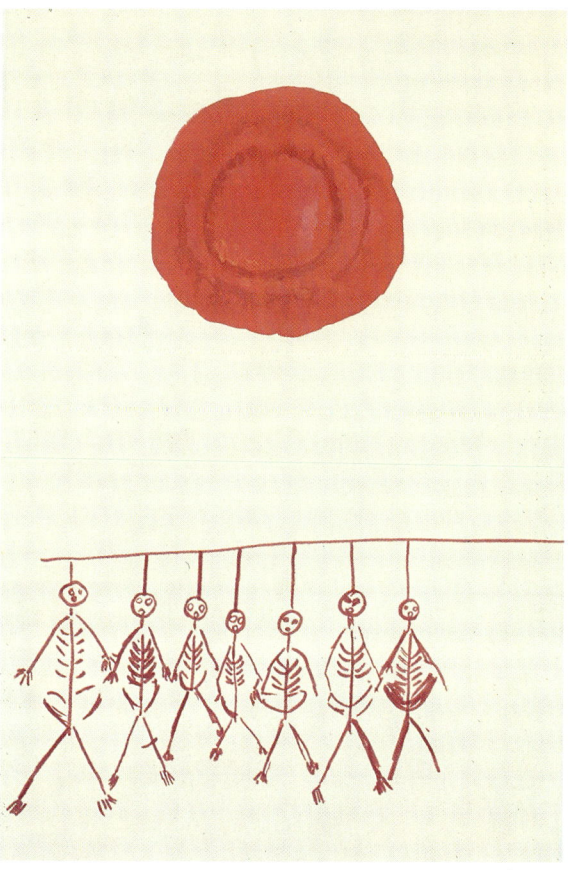

Abb. 28–5.   Zeichnung einer 25jährigen Patientin mit Colitis ulcerosa: Auf den Gerippen der im Höllenfeuer verkohlten Leichen möchte die Kranke Xylophon spielen. Zum Therapeuten: „Das größte Gerippe links sind Sie". Im oberen Bildteil: „Brennende Zigarre" (vgl. Text).

Abb. 28–6.   Zeichnung einer 25jährigen Patientin mit Colitis ulcerosa: Die Patientin läßt die Personen ihrer Umgebung – den Arzt eingeschlossen – als Marionetten um sich herumtanzen.

Abb. 28–7.   Zeichnung einer 25jährigen Patientin mit Colitis ulcerosa: Die Kranke stellt ihre eigenen bisher abgewehrten Versorgungswünsche dar.

Abb. 38.3–8. 19j. Patientin F. L.: „Das Tier in mir, das mich umschlingt und immer wieder zum ‚Fressen' treibt, das diesen wahnsinnigen Hunger in mir verursacht und das ich nicht zur Ruhe bringen kann. Ich würde gern seine tierische Energie umpolen." (7. 11. 85)

Abb. 38.3–9. 25j. Patientin N. M.: „Ohne Text." (12. 2. 87)

Abb. 38.3–10. 17j. Patientin D. U.: „Ein Traum, den ich immer wieder träume." (4. 6. 87)

Abb. 38.3–11. 23j. Patientin H. L.: „Ich gehe übermorgen für ein Test-Wochenende nach Hause. Die Weltkugel ist die Klinik, der Schirm ist der Halt durch die Ursachenerkenntnis meiner Krankheit. Das Wochenende ist für mich sehr entscheidend, aber eben ein Sprung ins Ungewisse (Weltall)." (23. 10. 86)

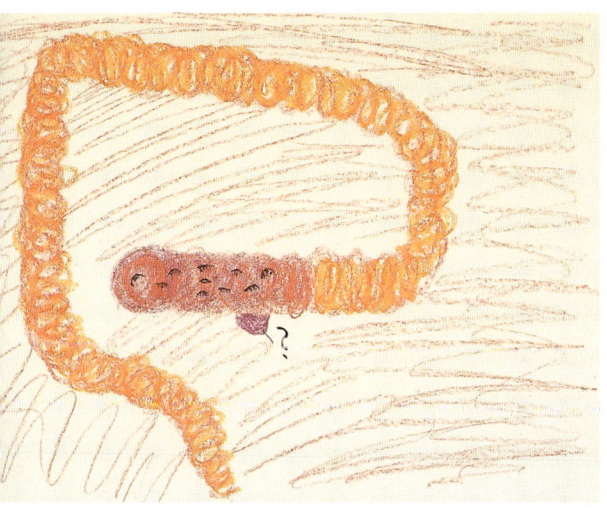

Abb. 44.1–1. 31jährige Patientin M. E. mit Colitis ulcerosa: „Mein Darm. Mit vielen nagenden ‚Ameisen' drin." (8. 12. 85)

Abb. 44.1–2 50jährige Patientin J. L. mit Colitis ulcerosa: „Zukunft." (26. 9. 85)

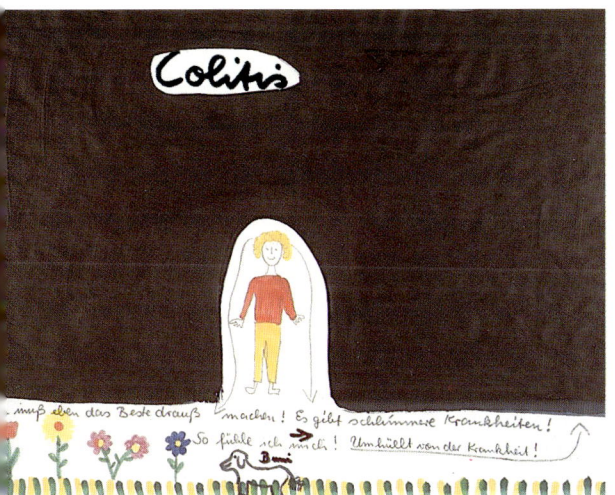

Abb. 44.1–3. 34jährige Patientin C. M. mit Colitis ulcerosa: „Colitis – man muß eben das Beste draus machen! Es gibt schlimmere Krankheiten! So fühle ich mich! Umhüllt von der Krankheit!" (15. 2. 88)

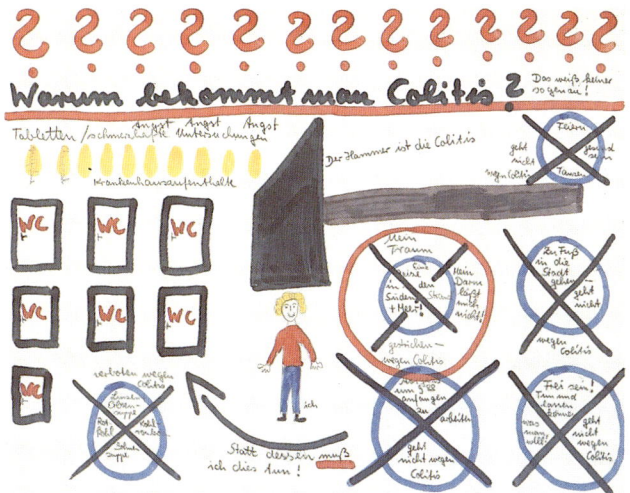

Abb. 44.1–4. Dieselbe Patientin wie Abb. 3.: „Warum bekommt man Colitis? Das weiß keiner so genau! Angst, Angst, Angst, Tabletten (Azulfidine, Claversal)/schmerzhafte Untersuchungen, Krankenhausaufenthalte. Der Hammer ist die Colitis. Feiern, Gesundsein, Tanzen – geht nicht wegen Colitis. Linsen- und Erbsensuppe, Rotkohl, Kohlrouladen, Bohnensuppe – verboten wegen Colitis. Mein Traum: Eine Reise in den Süden, Strand und Meer! Mein Darm läßt mich nicht! Gestrichen wegen Colitis. Zu Fuß in die Stadt gehn – geht nicht wegen Colitis. Morgens um 9.00 Uhr anfangen zu arbeiten – geht nicht wegen Colitis. Freisein! Tun und lassen können, was man will, geht nicht wegen Colitis. Statt dessen muß ich dies tun!" (WC!). (15. 2. 88)

Abb. 44.2–3. 26j. Patientin B. T. mit Morbus Crohn: „Die Fisteln nehmen unerbittlich ihren Gang!" (14. 3. 88)

Abb. 44.2–5. 41j. Patient J. T. mit Morbus Crohn: „Zerbrochene Brücke, suche einen Ausweg, um über den Fluß zu kommen, durch den dunklen Wald ins Sonnenlicht, was noch nicht zu erkennen ist, weil der Wald so endlos tief ist." (25. 6. 87)

Abb. 44.2–6. 34j. Patientin L. S. mit Morbus Crohn: „Die Krankheit läßt mich nicht mehr denken und fühlen, teils bin ich gestorben, teils zerrinnt das verbleibende Leben." (25. 7. 85)

Abb. 44.2–7. 25j. Patientin J. L. mit Morbus Crohn: „Rot ist für mich eine aggressive Farbe, ebenso wie ich auch das Feuer als aggressiv und bedrohlich, ja sogar lebensgefährlich empfinde. Die Wolke stellt die Therapie dar, die versucht, das Feuer zu löschen. Noch hat der Regen keine Chance. Die beiden Blätter bedeuten meine Hoffnung, auch wenn die Blätter noch immer in Gefahr sind, verbrannt zu werden. Das Feuer steht für alles, was Gefühle vernichten oder überdecken will, so wie z. B. die Umwelt oder auch meine innere Sperre. Der Baum, dessen Gefühle (natürliche Farbe) verbrennt, das bin ich." (13. 8. 87)

Literatur

Abelson, H. J., P. M. Fishbourne, J. Cisin: National Survey on Drug Abuse 1977 – A National Survey – Youth, Young Adults and Older People, Vol. 1. National Institute on Drug Abuse, U. S. Department of Health, Education and Welfare, Washington/D. C. 1977.

Abholz, H. H. (Hrsg.): Krankheit und soziale Lage. Befunde der Sozialepidemiologie. Campus, Frankfurt-New York 1976.

Ablon, S. L., J. E. Mack: Sleep disorders. In: Noshpitz, J. (ed.): Basic Handbook of Child Psychiatry, Vol. 2, pp. 643–660. Basic Books, New York 1979.

Abplanalp, J. M., R. M. Rose, A. F. Donelly, L. Livingston-Vaughan: Psychoendocrinology of the menstrual cycle: II. The relationship between enjoyment of activities, moods, and reproductive hormones. Psychosom. Med. 41 (1979) 605–615.

Abraham, A. D., G. Bug: ³H-Testosterone distribution and binding in rat thymus cell in vivo. Molec. cell. Biochem. 13 (1976) 157–163.

Abraham, K.: Beitrag zur Tic-Diskussion. Int. Z. Psychoanal. 7 (1921) 393–395.

Abraham, K.: Beiträge der Oralerotik zur Charakterbildung (1925). Gesammelte Schriften, Bd. 2. Fischer, Frankfurt 1982.

Abraham, K.: Versuch einer Entwicklungsgeschichte der Libido. In: Psychoanalytische Studien, I. Fischer, Frankfurt 1969.

Abraham, S. F., P. J. V. Beumont: How patients describe bulimia or binge eating. Psychol. Med. 12 (1982) 625–635.

Abraham, S. F., M. Mira, D. Llewellyn-Jones: Bulimia: a study of outcome. Int. J. Eat. Dis. 2 (1983) 175–180.

Abrahams, V. C., S. M. Hilton, A. Zbrozyna: Active muscle vasodilatation produced by stimulation of the brain stem: its significance in the defense reaction. J. Physiol. (Lond.) 154 (1960) 491–513.

Abram, H. S.: Adaptation to open heart surgery: psychiatric study of response to threat of death. Amer. J. Psychiat. 122 (1965) 659–667.

Abram, H. S.: The psychiatrist, the treatment of chronic renal failure and the prolongation of life. Amer. J. Psychiat. 124 (1968) 1351–1358; 126 (1969) 157–167; 128 (1972) 1534–1539.

Abram, H. S., L. R. Hester, A. Epstein: Sexual activity and renal failure. Proc. 5th Int. Congr. Nephrol. Karger, Basel 1973.

Abrams, R. D.: The patient with cancer – his changing pattern of communication. New Engl. J. Med. 274 (1966) 317–322.

Abramson, L. Y., M. E. P. Seligman, J. D. Teasdale: Learned helplessness in humans: critique and reformulation. J. abn. Psychol. 87 (1978) 49–74.

Abteilung Innere Medizin III: Richtlinien zur Diagnose und Therapie hämatologischer und onkologischer Erkrankungen. Klinikum der Universität Ulm, 5. Aufl. Ulm 1981.

Achté, K. A., M.-L. Vauhkonen: Cancer and the psyche. Omega 2 (1971) 46–56.

Achté, K. A., M.-L. Vauhkonen: Suicides committed in general hospitals. Psychiatria Fennica, 1971. Yearbook of the Psychiatric Clinic, Helsinki 1971.

Acker, J. E.: Socio-economic factors affected by an in-hospital cardiac rehabilitation program. In: Stocksmeier, U. (ed.): Psychological Approach to the Rehabilitation of Coronary Patients, pp. 96–100. Springer, Berlin-Heidelberg-New York 1976.

Ackerman, S.: Early life events and peptic ulcer susceptibility: an experimental model. Brain Res. Bull. 1 (1980) 43–49.

Ackerman, S. H.: Premature weaning, thermoregulation and the occurrence of gastric pathology. In: Weiner, H., M. A. Hofer, A. J. Stunkard (eds.): Brain, Behaviour and Bodily Disease. Raven, New York 1981.

Ackerman, S., H. Weiner: Peptic ulcer disease: some considerations for psychosomatic research. In: Hill, O. W. (ed.): Modern Trends in Psychosomatic Medicine. London 1976 und Weiner, H. (ed.): Psychobiology and Human Disease. Elsevier, New York 1977.

Ackerman, S. H., M. A. Hofer, H. Weiner: Early maternal separation increases gastric ulcer risk in rats by producing a latent thermoregulatory disturbance. Science 201 (1978).

Adam, K., L. Adamson, V. Brezinova, I. Oswald: Do placebos alter sleep? Brit. med. J. 1 (1976) 195–196.

Adams, D., J. P. Flynn: Transference of an escape response from tail shock to brain stimulated attack behavior. J. exp. Anal. Behav. 9 (1966) 401–408.

Adams, D. B., A. R. Gold, A. D. Burt: Rise in female-initiated sexual activity at ovulation and its suppression by oral contraceptives. New Engl. J. Med. 299 (1978) 1145–1150.

Adams, H. E., M. Feuerstein, J. L. Fowler: Migraine headache: review of parameters, etiology, and intervention. Psychol. Bull. 87 (1980) 217–237.

Adams, J. E., E. Lindenmann: Coping with long-term disability. In: Coelho, G. V., D. A. Hamburg, J. E. Adams (eds.): Coping and Adaptation. Basic Books, New York 1974.

Adams, M. R., J. R. Kaplan, T. B. Clarkson, D. R. Koritnik: Ovarectomy, social status, and atherosclerosis in cynomolgus monkeys. Arteriosclerosis 5 (1985) 192.

Adelson, L., W. Hoffmann: Sudden death from coronary disease related to lethal mechanism arising independently of vascular occlusion of myocardial damage. J. Amer. med. Ass. 176 (1961) 129.

Ader, R.: Letter to the editor. Psychosom. Med. 36 (1974) 183–184.

Ader, R.: Conditioned adrenocortical steroid elevations in the rat. J. comp. Physiol. Psychol. 90 (1976) 1156–1163.

Ader, R.: Psychosomatic and psychoimmunologic research. Psychosom. Med. 42 (1980) 307–322.

Ader, R.: Behavioral influences on immune responses. In: Weiss, S. M., J. A. Herd, B. H. Fox (eds.): Perspective on Behavioral Medicine. Acad. Press, New York 1981.

Ader, R. (ed.): Psychoneuroimmunology. Acad. Press, New York 1981 a.

Ader, R.: A historical account of conditioned immunologic responses. In: Ader, R. (ed.): Psychoneuroimmunology, pp. 321–352. Acad. Press, New York 1981 b.

Ader, R.: Conditioned taste aversions and immunopharmacology. Ann. N. Y. Acad. Sci. 443 (1985 a) 293–307.

Ader, R.: Conditioned immunopharmacological effects in animals: implications for a conditioning model of pharmacotherapy. In: White, L., B. Tursky, G. E. Schwartz (eds.): Placebo: Theory, Research and Mechanisms, pp. 306–323. Guilford, New York 1985 b.

Ader, R., N. Cohen: Behaviorally conditioned immunosuppression. Psychosom. Med. 37 (1975) 333–340.

Literatur

Ader, R., N. Cohen: Conditioned immunopharmacologic responses. In: Ader, R. (ed.): Psychoneuroimmunology. Acad. Press, New York 1981.

Ader, R., N. Cohen: Behaviorally conditioned immunosuppression and murine systemic lupus erythematosus. Science 215 (1982) 1534–1536.

Ader, R., N. Cohen: CNS-immune system interactions: conditioning phenomena. Behav. Brain Sci. 8 (1985) 379–394.

Ader, R., N. Cohen: Conditioned responses are indeed conditioned. Behav. Brain Sci. 9 (1986) 760–763.

Ader, R., N. Cohen, L. J. Grota: Adrenal involvement in conditioned immunosuppression. Int. J. Immunopharmacol. 1 (1979) 141–145.

Ader, R., N. Cohen, D. L. Felten: Brain, behavior, and immunity (editorial). Brain, Behavior, and Immunity 1 (1987) 1–6.

Ader, R., L. J. Grota, N. Cohen: Conditioning phenomena and immune function. Ann. N. Y. Acad. Sci. 496 (1987) 532–544.

Adler, A.: Studie über die Minderwertigkeit von Organen. Urban & Schwarzenberg, Berlin-Wien 1907.

Adler, A.: Soziale Beschaffenheit des Seelenlebens. In: Adler, A.: Menschenkenntnis. Fischer, Frankfurt 1966.

Adler, H. M., V. B. O. Hammett: The doctor-patient relationship revisited. An analysis of the placebo effect. Ann. intern. Med. 78 (1979) 595–598.

Adler, M. L.: Kidney transplantation and coping mechanisms. Psychosomatics 13 (1972) 337–341.

Adler, R.: Therapieresistente Schmerzen. Medikamentöse Therapie des Karzinomschmerzes. Schweiz. med. Wschr. 108 (1978) 456–461.

Adler, R. H.: Die Mißachtung der Gefühle – ein Hindernis für die Entwicklung einer Patient-orientierten Medizin. Schweiz. med. Wschr. 111 (1981) 1245–1249.

Adler, R. H.: The differentiation of organic and psychogenic pain. Pain 10 (1981) 249–252.

Adler, R.: Schmerzen bei Tumorpatienten – Diagnostik und medikamentöse Therapie. Schweiz. Rundsch. Med. 72 (1983) 1301–1306.

Adler, R.: Der Kliniker als Psychosomatiker. In: Uexküll, Th. v., R. Adler, J. M. Herrmann, K. Köhle, O. W. Schonecke, W. Wesiack (Hrsg.): Psychosomatische Medizin, 3. Aufl. Urban & Schwarzenberg, München 1986.

Adler, R. H., W. Hemmeler: Praxis und Theorie der Anamnese, 2. Aufl. Fischer, Stuttgart 1988.

Adler, R., F. Lomazzi: Perceptual style and pain tolerance, I. The influence of certain psychological factors. J. psychosom. Res. 17 (1973) 369–379.

Adler, R., F. Lomazzi: Die Bedeutung der individuellen Schmerzempfindlichkeit für die Beurteilung von Schmerzzuständen. Schweiz. med. Wschr. 104 (1973) 1192–1195.

Adler, R., F. Lomazzi: Mild analgesics evaluated with the „Submaximum Effort Tourniquet Method". I. The influence of psychological factors on their effect. Psychopharmacologia 38 (1974) 351–356.

Adler, R. H., Th. v. Uexküll: Individuelle Psychologie als Zukunftsaufgabe der Medizin. Schweiz. Rundschau f. Medizin Praxis 76 (1987) 1275–1280.

Adler, R. H., K. Macritchie, G. L. Engel: Psychologic processes and ischemic stroke (occlusive cerebrovascular disease). I. Observations on 32 men with 35 strokes. Psychosom. Med. 33 (1971) 1–29.

Adler, R., A. Gervasi, B. Holzer: Perceptual style and pain tolerance, II. The influence of an anxiolytic agent. J. psychosom. Res. 17 (1973) 381–387.

Adler, R. H., S. Zlot, Ch. Hürny, Ch. Minder: Engel's psychogenic pain and pain-prone patient: a controlled, retrospective clinical study. Psychosom. Med. 51 (1989) 87–101.

Adlersberg, D., O. Porges: Die neurotische Atmungstetanie, eine neue klinische Tetanieform. Wien. Arch. Inn. Med. 8 (1924) 185–238.

Adlersberg, D., L. Schaefer, S. R. Drachman: Development of hypercholesteremia during cortisone and ACTH therapy. J. Amer. med. Ass 144 (1950) 909–914.

Adorno, T. W.: Zum Verhältnis von Psychoanalyse und Gesellschaftstheorie. Psyche 6 (1952) 1–18.

Adsett, A., J. G. Bruhn: Short-term group psychotherapy for post-myocardial infarction patients and their wives. Canad. med. Ass. J. 99 (1968) 577–584.

Affleck, G., C. Pfeiffer, H. Tennen, J. Fifield: Attributional processes in rheumatoid arthritis. Arthr. and Rheum. 30 (1987) 927–931.

Affolter, F.: Diagnostische und therapeutische Probleme bei frühkindlicher Hirnschädigung aus pädaudiologischer Sicht. In: Städeli, H. (Hrsg.): Die leichte frühkindliche Hirnschädigung. Huber, Bern 1972.

Agnew, D. C., H. Merskey: Words of chronic pain. Pain 2 (1976) 73–81.

Ago, Y. et al.: A comparative study on somatic treatment and comprehensive treatment of bronchial asthma. J. Asthma Res. 14 (1976) 37–43.

Ago, Y. et al.: Psychosocial factors influencing intractability of bronchial asthma, and psychosomatic approach. Shinshin Igaku 20 (1980) 403–409.

Agras, S. W., D. W. Barlow, H. N. Chapin: Behavior modification of anorexia nervosa. Arch. gen. Psychiat. 30 (1974) 279–286.

Agras, W. S., B. Dorian, B. G. Kirkley, B. Arnow, J. Bachman: Imipramine in the treatment of bulimia: a double-blind controlled study. Int. J. Eat. Dis. 6 (1987) 29–38.

Agras, W. S., C. B. Taylor, H. C. Kraemer, M. A. Southam, J. A. Schneider: Relaxation training for essential hypertension at the worksite: II. The poorly controlled hypertensive. Psychosom. Med. 49 (1987) 264–273.

Agren, H., G. Lundqvist: Low levels of somatostatin in human CSF mark depressive episodes. Psychoneuroendocrinology 9 (1984) 233–248.

Ahles, T. A., A. King, J. E. Martin: EMG biofeedback during dynamic movement as a treatment for tension headache. Headache 24 (1984) 41–44.

Ahles, T. A., M. B. Yunus, A. T. Masi: Is chronic pain a variant of depressive disease? Pain 29 (1987) 105–111.

Ahlmen, J., R. Olander: The influence of preoperative tension on vocational rehabilitation following renal transplantation. Acta med. scand. 194 (1973) 13–16.

Ahmad, M., V. W. Parish: Study of the occurrence of gonorrhea in postpartum women. Amer. J. Obstet. Gynec. 118 (1974) 368.

Ahnefeld, F. W.: Ansprache zur Eröffnung des Anästhesisten-Kongresses, 1984.

Ahrens, S.: Experimentelle Untersuchungen kognitiver Funktionen bei Ulcus-Patienten. In: Zander, W. (Hrsg.): Experimentelle Forschungsergebnisse in der Psychosomatischen Medizin. Vandenhoeck & Ruprecht, Göttingen 1981.

Ahrens, S.: Konsultationsverhalten psychosomatischer Patienten. Z. psychosom. Med. Psychoanal. 28 (1982a) 242–254.

Ahrens, S.: Empirische Ergebnisse zum Konsultationsverhalten neurotischer, psychosomatischer und somatisch kranker Patienten II. Z. psychosom. Med 28 (1982b) 335–346.

Ahrens, S.: Alexithymie und kein Ende? Z. psychosom. Med. 33 (1987) 201–220.

Ahrens, S.: Die instrumentelle Forschung am instrumentellen Objekt. Psyche 3 (1988) 225–241.

Ahrens, S., H. v. Gyldenfeldt: Instrumentelle Orientierung und Affektregulation bei psychosomatisch Kranken. In: Deppe, H.-U., U. Gerhardt, P. Novak (Hrsg.): Medizinische Soziologie, Jahrbuch 1, S. 152–174. Campus, Frankfurt-New York 1981.

Ahrens, S., G. Deffner: Zur Affektverarbeitung bei psychosomatischen Patienten II. Z. psychosom. Med. 30 (1984) 357–376.

Ahrens, S., G. Deffner: Alexithymie – Ergebnisse und Methodik eines Forschungsbereichs der Psychosomatik. Psychother. Med. Psychol. 35 (1985) 147–159.

Ahrens, S., G. Deffner, H. Feiereis: Zur Differenzierung von Colitis- und Morbus Crohn-Kranken anhand psychosozialer Variablen. Z. psychosom. Med Psychoanal. 32 (1986) 301–315.

Aimez, P., P. Ferrari, B. Guy-Grand: Facteurs psycho-émotionnels dans le diabète sucré. Diabète et Métab. 2 (1976) 73–79.

Aimez, P., M. Tutin, B. Guy-Grand, F. Desme, H. Bour: Étude psycho-sociologique des contraintes thérapeutiques du diabète sucré. La Presse Médicale 79 (1971) 1149–1152.

Ainsworth, M.: Weitere Untersuchungen über die schädlichen Folgen der Mutterentbehrung. In: Bowlby, J. (Hrsg.): Mutterliebe und kindliche Entwicklung. Reinhardt, München-Basel 1972.

Aitken, C.: Medical leadership in rehabilitation medicine. Proc. 3rd Congr. Int. College Psychosom. Med., Rome 1975.

Aitken-Swan, J., E. C. Easson: Reactions of cancer patients on being told their diagnosis. Brit. med. J. 288 (1959) 779–783.

Ajuriaguerra, J.: Manuel de la psychiatrie de l'enfant, p. 181. Masson, Paris 1970.

Akert, K., P. Hummel: Anatomie und Physiologie des limbischen Systems, 2. Aufl. Hoffmann-La Roche, Basel 1968.

Albers, L.: Untersuchungen zum „kulturgebundenen Syndrom" Naeng. Dissertation, Heidelberg 1988.

Albert, F. W. et al.: Organisation, Probleme und Erfolg der Hämodialyse. Med. Klinik 69 (1974) 146.

Albutt, C.: Disease of the heart, including angina pectoris, II. MacMillan, London 1915.

Alby, N., A. Devergie, E. Vilmer et al.: Adolescents: a population at risk for psychopathological complications after BMT: 46 cases. Bone Marrow Transplant. 2 (Suppl. 1) (1987) 248.

Aleksandrowicz, D. R.: Fire and its aftermath on a geriatric ward. Bull. Menninger Clin. 25 (1961) 23–32.

Alexander, F.: Functional disturbances. J. Amer. med. Ass. 100 (1933) 469–473.

Alexander, F.: The influence of psychologic factors upon gastrointestinal disturbances: a symposion. General principles, objectives and preliminary results. Psychoanal. Quart. 3 (1934) 501–539.

Alexander, F.: Emotional factors in essential hypertension. Psychosom. Med. (1939) 173–179.

Alexander, F.: Fundamental concepts of psychosomatic research. Psychosom. Med. 5 (1943) 205–210. Deutsch in: Overbeck, G., A. Overbeck (Hrsg.): Seelischer Konflikt – körperliches Leiden, S. 46–55. Hamburg 1978.

Alexander, F.: Fundamentals of psychoanalysis. London 1949.

Alexander, F.: Psychosomatic medicine (1950). Deutsch: Psychosomatische Medizin, 3. Aufl. De Gruyter, Berlin-New York 1977.

Alexander, F., Th. M. French, G. H. Pollock: Psychosomatic specificity. Univ. of Chicago Press, Chicago-London 1968.

Alexander, L.: The double-bind theory and hemodialysis. Arch. gen. Psychiat. 33 (1976) 1353–1356.

Alexander, N., L. B. Hinshaw, D. R. Drury: Development of a strain of spontaneously hypertensive rabbits. Proc. Soc. exp. biol. Med. 86 (1954) 855.

Alexander, R. W., J. N. Davis, R. J. Lefkowitz: Direct identification and characterization of beta-adrenergic receptors in rat brain. Nature (Lond.) 258 (1975) 437–440.

Alexander, T.: An objective study of psychological factors in ulcerative colitis in children. Lancet 85 (1965) 22–24.

Alexander-Williams, J., P. Buchmann: Perianale Komplikationen beim Morbus Crohn. In: Ottenjann, R., H. Fahrländer (Hrsg.): Entzündliche Erkrankungen des Dickdarms. Springer, Berlin 1983.

Alexander-Williams, J., T. A. Betts, S. Pidd: Psychiatric disturbance and the effect of gastric operations. Clin. Gastroenterol. 6 (1977) 694–699.

Allyon, T., N. H. Azrin: The measurement and reinforcement of behavior of psychotics. J. exper. Anal. Behav. 8 (1965) 357–383.

Almay, B. G. L., F. Johannson, L. v. Knorring, L. Terenius, A. Wahlström: Endorphines in chronic pain, I. Differences in LFS endorphine levels between organic and psychogenic pain syndromes. Pain 5 (1968) 153–162.

Almy, T. P.: The gastrointestinal tract in man under stress. In: Sleisenger, M. H., J. S. Fordtran (eds.): Gastrointestinal Disease. Saunders, Philadelphia-London-Toronto 1973.

Almy, T. P.: Wrestling with the irritable colon. Med. Clin. N. Amer. 62 (1978) 203–210.

Almy, T. P.: Epilogue and future. In: Chey, W. Y. (ed.): Functional Disorders of the Digestive Tract, pp. 333–334. Raven, New York 1983.

Almy, T. P.: Clinical features and diagnosis of functional GI disorders. In: Chey, W. Y. (ed.): Functional Disorders of the Digestive Tract, pp. 7–11. Raven, New York 1983.

Almy, T. P., P. Sherlock: Genetic aspects of ulcerative colitis and regional enteritis. Gastroenterol. 51 (1966) 757–761.

Almy, T. P., M. Tulin: Alterations in colonic function in man under stress: I. Experimental production of changes simulating the „irritable colon". Gastroenterol. 8 (1947) 616.

Almy, T. P., F. K. Abbot, L. Hinkle jr.: Alterations in colonic function in man under stress. Gastroenterology 15 (1950) 95–103.

Almy, T. P., F. Kern, F. K. Abbott: Constipation or diarrhea as reactions to life stress. Res. Publ. Ass. Res. nerv. ment. Dis. 29 (1949) 724–730.

Almy, T. P., F. Kern, M. Tulin: Alterations in colonic function in man under stress: II. Experimental production of sigmoid spasm in healthy persons. Gastroenterol. 12 (1949) 425.

Almy, T. P., L. E. Hinkle, B. Berle, K. Kern: Alterations in colonic function in man under stress: III. Experimental production of sigmoid spasm in patients with spastic constipation. Gastroenterol. 12 (1949) 437.

Alonzo, A., A. Simon, M. Feinleib: Prodromata of myocardial infarction and sudden death. Circulation 52 (1975) 1056–1072.

Alt, E., H. Müller, A. Wirtzfeld: Schrittmachertherapie aus der Sicht des Patienten. Med. Klinik 78 (1983) 598–600.

Althoff, P.-H., C. Rosak, K. Schöffling: Die Selbstkontrolle des Diabetes mellitus durch den Patienten. Dtsch. Ärztebl. 22 (1982) 31–45.

Althusser, L.: Das Kapital lesen, Bd. 1. Rowohlt, Reinbek 1972.

Altmann-Herz, U., A. Reindell, E. Petzold, H. Ferner: Zur psychosomatischen Differenzierung von Patienten nach Herzinfarkt. Psychosom. Med. Psychoanal. 29 (1983) 234–252.

Alun Jones, V., M. Shorthouse, P. McLaughlan, E. Workman, J. Hunter: Food intolerance: a major factor in the pathogenesis of irritable bowel syndrome. Lancet II (1982) 1115–1117.

Amann, W.: Ist die Acne vulgaris eine psychosomatische Erkrankung? Ärztl. Kosmetologie 14 (1984) 162–170.

Ambelas, A.: Life events and mania: a special relationship? Brit. J. Psychiat. 150 (1987) 235–240.

Ambrose, C. T.: The requirement for hydrocortisone in antibody-forming tissue cultured in serum-free medium. J. exp. Med. 119 (1964) 1027–1049.

Amecke, F., R. Rost: The prognostic significance of an overshooting exercise blood pressure as an indicator for subsequent manifestation of hypertension. Sportwissenschaftlicher Verlag, Tübingen 1983.

American Psychiatric Association: Diagnostic and statistical manual of mental disorders, 2nd ed. Washington, D. C. 1968.

American Psychiatric Association: Diagnostic and statistical manual of mental disorders, 3rd ed. Washington, D. C. 1980.

American Psychiatric Association: Diagnostic and statistical manual of mental disorders, 3rd ed., revised. Washington, D. C. 1987.

American Psychological Association: Technical recommendations for psychological tests and diagnostic techniques. Washington 1954.

Amir, S., Z. Amir: The pituitary gland mediates acute and chronic pain responsiveness in stressed and nonstressed rats. Life Sci. 24 (1978) 439–448.

Amkraut, A. A., G. F. Solomon, H. C. Kraemer: Stress, early experience and adjuvant-induced arthritis in the rat. Psychosom. Med. 33 (1971) 203–214.

Amsel, A.: Frustrative nonreward in partial reinforcement and discrimination learning: some recent history and theoretical extensions. Psychol. Rev. 69 (1962) 306–328.

Amsterdam, J. D., A. Winokur, E. Abelman, I. Lucki, K. Rickels: Cosyntropin (ACTHa 1–24) stimulation test in depressed patients and healthy subjects. Amer. J. Psychiat. 140 (1983) 907–909.

Ananth, J.: Side effects on fetus and infant of psychotropic drug use during pregnancy. Int. Pharmacopsychiat. 11 (1976) 246–260.

Anders, D.: Mädchen mit rekurrierenden Harnweginfekten. Therapiewoche 34 (1984) 907.

Anders, H. J.: Die Gesundheits-Panel-Methode. Ergebnisse, Vergleiche Deutschland/Frankreich. Med. Inform. 10 (1979) 416–440.

Andersen, A. E.: Practical comprehensive treatment of anorexia nervosa and bulimia. Johns Hopkins Univ. Press, Baltimore 1985.

Andersen, A. E., J. E. Hedblom, F. A. Hubbard: A multidisciplinary team-treatment for patients with anorexia nervosa and their families. Int. J. Eat. Dis. 2 (1983) 181–192.

Anderson, D. E.: Cardiovascular adaptation to behavioral stress mediated by salt intake. In: Schmidt, T., T. Dembroski, G. Blümchen (eds.): Biobehavioral Factors in Coronary Heart Disease. Springer, Berlin-Heidelberg-New York 1984.

Anderson, E. E., J. V. Brady: Prolonged pre-avoidance effects upon blood pressure and heart rate in the dog. Psychosom. Med. 35 (1973) 4–12.

Anderson, K.: The psychological aspects of chronic hemodialysis. Canad. psychiat. Ass. J. 20 (1975) 385–391.

Anderson, K. O., L. A. Bradley, L. D. Young, L. K. McDaniel, C. M. Wise: Rheumatoid arthritis: review of psychological factors related to etiology, effects, and treatment. Psychol. Bull. 98 (1985) 358–387.

Anderson, R. W.: The relation of life situations, personality features and reactions to the migraine syndrome. In: Dalessio, D. J. (ed.): Wolff's Headache and Other Head Pain,

pp. 403–417. 4th ed. Oxford Univ. Press, New York-Oxford 1980.

Andersson, B.: Thirst and brain control of water balance. Amer. Sci. 59 (1971) 408.

Andolfi, M.: Familientherapie – Das systemische Modell und seine Anwendung. Lambertus, Freiburg 1982.

Andrasik, F.: Biofeedback applications for headache. In: Bischoff, C., H. C. Traue, H. Zenz (eds.): Clinical Perspectives on Headache and Low Back Pain, pp. 181–200. Hogrefe & Huber, Toronto 1989.

Andrasik, F., K. A. Holroyd, F. Abell: Prevalence of headache within a college student population: a preliminary analysis. Headache 19 (1979) 384–387.

Andrasik, F., E. B. Blanchard, J. G. Arena, S. J. Teders, R. C. Teevan, L. D. Rodichok: Psychological functioning in headache sufferers. Psychosom. Med. 44 (1982) 171–182.

Andrasik, F., E. B. Blanchard, D. F. Neff: Biofeedback and relaxation training for chronic headache: a controlled comparison of booster treatments and regular contacts for long-term maintenance. J. consult. clin. Psychol. 52 (1984) 609–615.

Andreasen, N. J., N. Russel, C. E. Hartford: Factors influencing adjustment of burn patients during hospitalization. Psychosom. Med. 34 (1972) 517–525.

Andresen, B.: Differentielle Psychophysiologie valenzkonträrer Aktivierungsdimensionen. Lang, Frankfurt 1987.

Andrew, J. M.: Recovery from surgery, with and without preparatory instruction, for three coping styles. J. Pers. soc. Psychol. 15 (1970) 223–226.

Andrew, J. M.: Coping style and declining verbal activities. J. Geront. 28 (1973) 183–197.

Andrews, G., M. S. Amstrong, H. Brodaty, D. Hadzi-Pavlovic, W. Hall, P. R. Harvey, D. J. Sansom, C. C. Tennant, P. Weeks, J. Grigor, B. J. Hughson, G. Johnson, L. G. Kiloh: A treatment outline for depressive disorder: The Quality Assurance Project. Aust. N. Z. J. Psychiat. 17 (1983) 129–146.

André, C., L. Descos, P. Landais, J. Fermanian: Laboratory supplementation of Crohn's disease activity index. Lancet II (1980) 594–595.

Angell, M.: The quality of mercy. New Engl. J. Med. 306 (1982) 98–99.

Angell, M.: Disease as a reflection of the psyche. New Engl. J. Med. 312 (1985) 1570–1572.

Angermann, I.: Psychodynamik und Psychotherapie der gestörten Sexualität. In: Eicher W. (Hrsg.): Sexualmedizin in der Praxis, S. 389–436. Fischer, Stuttgart-New York 1980.

Angermeier, W. F., M. Peters: Bedingte Reaktionen. Grundlagen: Beziehungen zur Psychosomatik und Verhaltensmodifikation. Springer, Heidelberg 1973.

Angleitner, A.: Methodische und theoretische Probleme bei Persönlichkeitsfragebogen unter besonderer Berücksichtigung neuerer deutschsprachiger Fragebogen. Habilitationsschrift, unpubliziert. Bonn 1976.

Angold, A.: Childhood and adolescent depression. I. Epidemiological and etiological aspects. Brit. J. Psychiat. 152 (1988) 601–607.

Angst, J.: Die Psychiatrie des Diabetes insipidus. Arch. Psychiat. Nervenkr. 199 (1959) 633.

Anochin, P. K.: Das funktionelle System als Grundlage der physiologischen Architektur des Verhaltensaktes. In: Abh. Ges. Hirnforsch. Verhaltensphysiol. Bd. 1. Jena 1967.

Anschütz, F.: Die körperliche Untersuchung. Springer, Berlin-Heidelberg-New York 1973.

Anschütz, F.: Das psychosoziale Defizit. Die Neue Ärztliche 247 (1988) 2.

Anstee, B. H.: The pattern of psychiatric referrals in a general hospital. Brit. J. Psychiat. 120 (1972) 631–634.

Anton, G.: Über die Selbstwahrnehmung der Herderkrankungen des Gehirns durch den Kranken bei Rindenblindheit und Rindentaubheit. Arch. Psychiat. Nervenkr. 32 (1899) 86–127.

Antonowsky, A.: The salutogenetic perspective: toward a new view of health and illness. Advances 4 (1987) 47–55.

Antonowsky, A.: Unraveling the mystery of health. How people manage stress and stay well. Jossey-Bass, San Francisco 1987.

Apley, J.: The child with recurrent abdominal pain. Pediat. Clin. N. Amer. 14 (1967) 64–72.

Apley, J., B. Hale: Children with recurrent abdominal pain: how do they grow up? Brit. med. J. 3 (1973) 7–9.

Appels, A.: Myocardial infarction and depression. A cross-validation of Dreyfuss' findings. Act. nerv. sup. 21 (1979) 65–66.

Appels, A.: Vitale Erschöpfung und Depression als Vorboten des Herzinfarkts. In: Langosch, W. (Hrsg.): Psychosoziale Probleme und psychosoziale Interventionsmöglichkeiten bei Herzinfarktpatienten, S. 33–46. Minerva, München 1980.

Appels, A.: The syndrome of vital exhaustion and depression and its relationship to coronary heart disease. In: Siegrist, J., M. J. Halhuber (eds.): Myocardial Infarction and Psychosocial Risks, pp. 116–119. Springer, Berlin-Heidelberg-New York 1981.

Appels, A.: The year before myocardial infarction. In: Dembroski, T. M., T. H. Schmidt, G. Blümchen (eds.): Biobehavioral Bases of Coronary Heart Disease, pp. 19–37. Karger, Basel-München-Paris 1983.

Appels, A., J. Poole, J. Lubsen, E. van der Does: Psychological prodromata of myocardial infarction. J. psychosom. Res. 23 (1979) 405–421.

Appels, A., P. Mulder, M. van't Hof: The predictive power of the A/B typology in Holland. Results of a 10-year follow-up study. Paper presented at the conference „Biobehavioral factors in coronary heart disease", Winterscheid 1984.

Appels, A., P. Mulder, M. van't Hof, C. Jenkins, J. van Houten, F. Tan: The predictive power of the A/B typology in Holland. Results of a 10-year follow-up study. In: Schmidt, T., T. Dembroski, G. Blümchen (eds.): Biological and Psychological Factors in Cardiovascular Disease, pp. 56–62. Springer, Berlin-Heidelberg-New York 1986.

Appels, A., P. Höppener, P. Mulder: A questionnaire to assess premonitory symptoms of myocardial infarction. Int. J. Cardiol. 17 (1987) 15–24.

Arana, G. W., R. J. Workman, R. J. Baldessarini: Association between low plasma levels of dexamethasone and elevated levels of cortisol in psychiatric patients given dexamethasone. Amer. J. Psychiat. 141 (1984) 1619–1620.

Arana, G. W., R. J. Baldessarini, M. Ornsteen: The dexamethasone suppression test for diagnosis and prognosis: commentary and review. Arch. gen. Psychiat. 42 (1985) 1193–1204.

Araujo, G. da, P. O. van Arsdel, T. H. Holmes, D. L. Dudley: Life change, coping ability and chronic intrinsic asthma. J. psychosom. Res. 17 (1973) 359–363.

Arbeitsgruppe Fachbericht über Probleme des Alterns (Hrsg.): Altwerden in der Bundesrepublik Deutschland: Geschichte – Situationen – Perspektiven, Bd. I–III. Dtsch. Zentrum für Altersfragen, Berlin 1982.

Arbib, M. A.: Memory limitations of stimulus response models. Psychol. Rev. 75 (1969) 507–510.

Arentewicz, G., G. Schmidt: Sexuell gestörte Beziehungen. Springer, Berlin-Heidelberg-New York 1980.

Aresin, L.: Über Korrelationen zwischen Persönlichkeit, Lebensgeschichte und Herzkrankheit. VEB Fischer, Leipzig 1960.

Argelander, H.: Die Analyse psychischer Prozesse in der Gruppe. Psyche 17 (1963/64) 450–470; 481–515.

Argelander, H.: Gruppenanalyse unter Anwendung des Strukturmodells. Psyche 22 (1968) 913–933.

Argelander, H.: Die szenische Funktion des Ichs. Psyche 24 (1970) 325.

Argelander, H.: Gruppenprozesse – Wege zur Anwendung der Psychoanalyse in Behandlung, Lehre und Forschung. Rowohlt, Reinbek 1972.

Argelander, H.: Der Flieger. Suhrkamp, Frankfurt 1972.

Argyle, M.: Körpersprache und Kommunikation. Junfermann, Paderborn 1979.

Arhahn, P.: L'exploration fonctionelle de la motricité du rectum et de l'anus. Rev. pédiat. 10 (1974) 281–290.

Arlow, J. A.: Identification mechanisms in coronary occlusion. Psychosom. Med. 7 (1945) 195–209.

Arlow, J. A., C. Brenner: Psychoanalytic concepts and the structural theory. Int. Univ. Press, New York 1964. Deutsch: Grundbegriffe der Psychoanalyse. Rowohlt, Reinbek 1976.

Arluke, A.: Judging drugs: patient's conceptions of therapeutic efficacy in the treatment of arthritis. Human Organization 39 (1980) 84–88.

Armario, A., J. M. Castellanos, J. Balasch: Dissociation between corticosterone and growth hormone adaption to chronic stress in the rat. Horm. metab. Res. 16 (1984) 142–145.

Armentrout, D. P.: The impact of chronic pain on the self concept. J. clin. Psychol. 35 (1979) 517–521.

Armstrong, S. H.: Psychological maladjustment in renal dialysis patients. Psychosomatics 19 (1978) 169–171.

Armstrong, S. H., M. F. Weiner: Noncompliance with post-transplant immunosuppression. Int. J. Psychiat. Med. 11 (1981/82) 89–95.

Armstrong, S. H., A. Woods: Patient self-reported adjustment and health beliefs in compliant vs. non-compliant hemodialysis patients. In: Levy, N. B. (ed.): Psychonephrology 2, pp. 79–92. Plenum, New York 1983.

Arndt, H. J.: Stimmstörungen. In: Biesalski, P., F. Frank (Hrsg.): Phoniatrie-Pädaudiologie, S. 227. Thieme, Stuttgart 1982.

Arnett, F. C., S. M. Edworthy, D. A. Bloch, D. J. McShane, J. F. Fries, N. S. Cooper, L. A. Healey, S. R. Kaplan, M. H. Liang, H. S. Luthra, R. A. Medsger, D. M. Mitchell, D. H. Neustadt, R. S. Pinals, J. G. Schaller, J. T. Sharp, R. L. Wilder, G. G. Hunder: The American Rheumatism Association 1987 revised criteria for the classification of rheumatoid arthritis. Arthr. and Rheum. 31 (1988) 315–324.

Arnold, K., E. Lang: Informationsmedizin – Aufgabe für den Arzt. Herz u. Gefäße 4 (1984) 511.

Arnold, M. B.: Stress and emotion. In: Appley, Th. H., R. Trumbull (ed.): Psychological Stress: Issues in Research. Appleton Century Crofts, New York 1967.

Arnold, M. B.: Brain function in emotion: a phenomenological analysis. In: Black, P. (ed.): Physiological Correlates of Emotion. Acad. Press, New York 1970.

Arnold, R.: Ulkus-Therapie im Wandel? Dtsch. Ärztebl. 38 (1981) 1767–1773.

Arnold, R.: Pathogenese des Ulcus duodeni. In: Blum, A. L., J. R. Siewert (Hrsg.): Ulcus-Therapie, S. 47–71. 2. Aufl. Springer, Heidelberg 1982.

Arnold, W.: Der Pauli Test. Anweisung zur sachgemäßen Durchführung, Auswertung und Anwendung des Kraepelinschen Arbeitsversuchs. 4. Aufl. Barth, München 1970.

Aronson, A. E., J. R. Brown, E. M. Litin, J. S. Pearson: Spastic dysphonia. I. Voice, neurologic and psychiatric aspects. J. Speech Hear. Dis. 33 (1968) 203.

Aronson, E., A. M. Pynes, D. Kafry: Ausgebrannt. Vom Überdruß zur Selbstentfaltung. Klett-Cotta, Stuttgart 1983.

Arrenbrecht, S.: Specific binding of growth hormone to thymocytes. Nature 252 (1974) 255–257.

Literatur

Arrowood, M., K. Uhlrich, C. Gomillion et al.: New markers of coronary-prone behavior in a rural population. Psychosom. Med. (Abstr.) 44 (1982) 119.

Artner, J.: Funktionelle Unterleibsbeschwerden der Frau. Med. Klinik 77 (1982) 683.

Arznei-Telegramm: Im Westen nichts Neues? Alte und neue Benzodiazepine – Tranquilizer im Vergleich. Arznei-Telegramm (1981) 77–82.

Asaba, H. et al.: Hemodialysis in the home – 20 months experience. Scand. J. Urol. Nephrol. 5 (1971) 71–76.

Åsberg, M., L. Träksman, P. Thorén: 5-HIAA in cerebrospinal fluid: a biochemical suicide predictor. Arch. gen. Psychiat. 33 (1976) 1193–1197.

Aschoff, J.: Circadiane Rhythmen im endocrinen System. Klin. Wschr. 56 (1978) 425–435.

Asher, R., M. D. Lond: Munchhausen's syndrome. Lancet (1951) 339–341.

Asken, M.: Psychoemotional aspects of mastectomy: a review of recent literature. Amer. J. Psychiat. 132 (1975) 56–59.

Askevold, F.: Measuring body image. Psychother. and Psychosom. 26 (1975) 71–77.

Askevold, F.: What are the helpful factors in psychotherapy? Int. J. Eat. Dis. 2 (1983a) 39–43.

Askevold, F.: The diagnosis of anorexia nervosa. Int. J. Eat. Dis. 2 (1983b) 193–197.

Association for American Medical Colleges (AAMC): Physicians for the 21st century. The GPEP Report. AAMC, Washington 1984.

Attas, H. J., M. T. Gutierrez, S. Bellet et al.: Effect of stimulation of hypothalamus and reticular activating system on production of cardiac arrhythmia. Circulat. Res. 12 (1963) 14.

Auer, I. O.: Immunology in Crohn's disease. Z. Gastroenterol. 17 (Suppl.) (1979) 83–93.

Avery, D. H., G. Wildschitz, O. J. Rafaelson: Nocturnal temperature in affective disorder. J. affective Disord. 4 (1982) 61–71.

Ax, A. F.: The physiologic differentiation between fear and anger in humans. Psychosom. Med. 15 (1953) 433–442.

Axelrod, S., M. Noonen, B. Atanacio: On the laterality of psychogenic somatic symptoms. J. nerv. ment. Dis. 168 (1980) 517–525.

Ayllon, T., N. H. Azrin: Reinforcement and instruction with mental patients. J. exp. Anal. Behav. 7 (1964) 327–331.

Azima, H.: Changes in organization of mood as a therapeutic and research problem in psychopharmacology. Neuropsychopharmacology (1959) 491.

Baar, E.: Sprachfreie Entwicklungsteste für taube, schwerhörige und sprachlich speziell gestörte Kinder im Alter von 1–7 Jahren. Karger, Basel 1957.

Babinski, J.: Anosognosie. Rev. neurol. 34 (1918) 365–367.

Babinski, J., B. Jarkowski, V. Plichet: Kinésie paradoxale, mutisme parkinsonien. Rev. neurol. 37 (1921) 1266–1270.

Bach, J. F., T. B. Strom: Corticosteroids. In: Turk, J. L. (ed.): The Mode of Action of Immunosuppressive Agents, pp. 21–104. Elsevier, Amsterdam 1985.

Bachrach, H. M., L. A. Leaff: „Analyzability". A systematic review of the clinical and quantitative literature. J. Amer. psychoanal. Ass. 26 (1978) 881–920.

Badura, B.: Krankheitsbedingte Belastungen und Unterstützungen: Das Beispiel Herzinfarkt. In: Badura, B. (Hrsg.): Soziale Unterstützung und chronische Erkrankung. Zum Stand epidemiologischer Forschung, S. 168–182. Suhrkamp, Frankfurt 1981.

Badura, B., C. v. Ferber (Hrsg.): Selbsthilfe und Selbstorganisation im Gesundheitswesen. Oldenbourg, Wien 1981.

Baer, L. H. J., B. R. M. Kuypers: Behavior therapy in dermatological practice. Brit. J. Derm. 88 (1973) 591–598.

Bär, P. E., J. Reed, P. C. Bartlett, J. P. Vincent, B. J. Williams, G. G. Bourianoff: Studies of gaze during induced conflicts in families with a hypertensive father. Psychosom. Med. 45 (1983) 233–242.

Baer, U., K.-J. Bauknecht, R. Souchon, C. Viebahn: Strahlenspätschäden am Darm. Münch. med. Wschr. 123 (1981) 503–507.

Baerwolf, H.: Katamnestische Ergebnisse stationärer analytischer Psychotherapie. Z. psychosom. Med. 5 (1958) 80–91.

Bagby, R. M., G. J. Taylor, L. Atkinson: Alexithymia: a comparative study of three self-report measures. J. psychosom. Res. 32 (1988) 107–116.

Bahnson, C. B.: Psychologic and emotional issues in cancer: the psychotherapeutic care of the cancer patient. Semin. Oncol. 2 (1975) 293–309.

Bahnson, C. B.: Stress psychologique familial et problème des antécédents psycho-affectifs du cancer, p.13. Masson, Paris 1978.

Bahnson, C. B.: Stress and cancer: the state of the art. Part 1. Psychosomatics 21 (1980) 975–981.

Bahnson, C. B.: Stress and cancer: the state of the art. Part 2. Psychosomatics 22 (1981) 207–220.

Bahnson, C. B., M. B. Bahnson: Cancer as an alternative to psychosis: a theoretical model of somatic and psychologic regression. In: Kissen, D. M., L. L. Le Shan (eds.): Psychosomatic Aspects of Neoplastic Disease, pp. 184–202. Pitman, New York 1964.

Bahnson, C. B., M. B. Bahnson: Ego defenses in cancer patients. Ann. N. Y. Acad. Sci. 164 (1969) 546–559.

Baider, L.: The silent message: communication in a family with a dying patient. Journ. Marriage and Fam. Counseling 3 (1977) 23.

Baird, I. M., J. T. Silverstone, J. I. Grimshaw et al.: Prevalence of obesity in a London borough. Practitioner 212 (1974) 706–714.

Bakal, D. A.: The psychobiology of chronic headache. Springer, New York 1982.

Bakal, D. A., J. A. Kaganov: Symptom characteristics of chronic and non-chronic headache sufferers. Headache 19 (1979) 285–289.

Baker, E. G., G. H. Crook, E. D. Schwabacher: Personality correlates of periodontal disease. J. dent. Res. 40 (1961) 390.

Baker, G. H. B.: Life events before the onset of rheumatoid arthritis. Psychother. and Psychosom. 38 (1982) 173–177.

Baker, G. H., D. A. Brewerton: Rheumatoid arthritis: a psychiatric assessment. Brit. med. J. 282 (1981) 2014.

Baker, L., A. Barcai, R. Kaye, N. Haque: Beta-adrenergic blockade and juvenile diabetes: Acute studies and long-term therapeutic trial. J. Pediatr. 75 (1969) 19–25.

Balck, F. B.: Zufriedenheit in der Zweierbeziehung. Diss., Hamburg 1982.

Balck, F. B.: Hämodialyse und Partnerschaft. Thieme, Stuttgart 1988.

Balck, F. B., U. Koch, H. Speidel (Hrsg.): Psychonephrologie. Springer, Heidelberg 1984.

Balck, F. B., J. Kniess, U. Koch, H. Speidel: Probleme in der Partnerbeziehung bei Dauerdialysepatienten. Verh. Dtsch. Ges. Inn. Med. 84 (1978).

Balck, F. B., M. Dvorak, H. Speidel et al.: Staff's problems and staff's affective reactions to dialysis patients' problems. In: Levy, N. B.(ed.): Psychonephrology 2, pp. 15–30. Plenum, New York 1983.

Baldigo, J. et al.: Impression formation in the patient interview: a training experience with first year medical students. AAMC-RIME Conference 1975, 206.

Balint, E.: Persönliche Mitteilung. Ascona 1982.

Balint, E., J. S. Norell: Six minutes for the patient. London 1973. Deutsch: Fünf Minuten pro Patient. Suhrkamp, Frankfurt 1975.

Balint, M.: The basic fault. Therapeutic aspects of regression. London 1952. Deutsch: Therapeutische Aspekte der Regression. Die Theorie der Grundstörung. Stuttgart 1970.

Balint, M.: Primary love and psychoanalytic technique. London 1952. Deutsch: Die Urformen der Liebe und die Technik der Psychoanalyse. Bern-Stuttgart 1965.

Balint, M.: Der Arzt, der Patient und die Krankheit. Klett-Cotta, Stuttgart 1983.

Balint, M., D. H. Ball, M. L. Hare: Unterricht von Studenten in patientenzentrierter Medizin. Psyche 23 (1969) 532–546.

Balint, M., P. H. Ornstein, E. Balint: Focal psychotherapy. London 1972. Deutsch: Fokaltherapie. Frankfurt 1972.

Balint, M., J Hunt, D. Joyce, J. Marinker, J. Woodcock: Das Wiederholungsrezept – Behandlung oder Diagnose. Klett, Stuttgart 1975.

Ball, J. F.: Widows' grief: the impact of age and mode of death. Omega: J. Death Dying 7 (1976/77) 307–333.

Ballenger, J. C.: Pharmacotherapy of the panic disorders. J. clin. Psychiat. 47 (Suppl.) (1986) 27–32.

Ballenger, J. C., G. D. Burrows, R. L. DuPont et al.: Alprazolam in panic disorder and agoraphobia: results from a multicenter trial. I. Efficacy in short-term treatment. Arch. gen. Psychiat. 45 (1988) 413–422.

Balter, M. B., J. Levine, D. J. Mannheimer: Cross-national study of the extent of anti-anxiety sedative drug use. New Engl. J. Med. 290 (1974) 1179–1196.

Balters, W.: Zur Kunst der Menschenbehandlung. In: Deutscher Zahnärztekalender, S. 165. Hanser, München 1954.

Balzer, K., S. Förster, H. Goebell, V. Seifert, I. Köcker: Demographische und soziale Charakteristik von Patienten mit Morbus Crohn in einer Großstadtregion. Eine Studie mit Nachbarschafts- und Krankenhauskontrollen. Z. Gastroenterol. 23 (1985) 347–354.

Balzer, W.: Comparative interaction process analysis of psychosomatic and psychoneurotic patients in an inpatient group-psychotherapy. In: Koptagel-Ilal, G. (ed.): Proc. 13th Europ. Conf. Psychosom. Res. Istanbul 1981.

Bamberg, E., S. Greif: Streß: Bedrohung der Gesundheit oder subjektiver Begriff. Psychosozial 5 (1/1982) 8–28.

Bammer, K.: Krebs und Psychosomatik. Kohlhammer, Stuttgart-Berlin-Köln 1981.

Bancroft, J.: The relationship between hormones and sexual behaviors in humans. In: Hutchison, J. B. (ed.): Biological Determinants of Sexual Behavior, pp. 493–519. Wiley, New York 1978.

Bancroft, J.: Grundlagen und Probleme menschlicher Sexualität. Enke, Stuttgart 1985.

Bancroft, J., N. S. Skakkebaek: Androgens and human sexual behavior. In: Porter, R., J. Whelan (eds.): Sex, Hormones and Behavior. Ciba Foundation Symposium 62, pp. 209–226. Excerpta Medica, Amsterdam 1979.

Bandura, A.: Principles of behavior modification. Holt, Rinehart and Winston, New York 1969.

Bandura, A.: Self-efficacy: toward a unifying theory of behavioral change. Psychol. Rev. 84 (1977) 191.

Bandura, A., F. L. Menlove: Factors determining vicarious extinction of avoidance behavior through symbolic modeling. J. pers. Soc. Psychol. 8 (1968) 99–108.

Bangerter, Ch., H. Noack: Der klinische Gruppenunterricht im Urteil von Studenten und Tutoren. Inst. f. Ausbildungs- und Examensforschung. Med. Fakultät der Univ. Bern, 2. Aufl. Bern 1983.

Banki, C. M., G. Bissette, M. Arato, C. B. Nemeroff: Elevation of immunoreactive CSF-TRH in depressed patients. Amer. J. Psychiat. 145 (1988) 1526–1531.

Banks, B. M., B. I. Korelitz, L. Zetzel: The course of nonspecific ulcerative colitis: review of 20 years' experience and late results. Gastroenterology 32 (1957) 983–1012.

Barcai, A.: Family therapy in the treatment of anorexia nervosa. Amer. J. Psychiat. 128 (1971) 286–290.

Barchilon, J.: Analysis of a woman with incipient rheumatoid arthritis. Int. J. Psychoanal. 44 (1963) 163–177.

Barchilon, J., G. L. Engel: Dermatitis: An hysterical conversion symptom in a young woman. Psychosom. Med. 14 (1952) 295–305.

Barcroft, H., O. G. Edholm, J. McMichael, E. P. Sharpey-Schafer: Post hemorrhagic fainting. Study by cardiac output and forearm flow. Lancet 15 (1944) 489.

Bard, M., R. Dyk: The psychodynamic significance of beliefs regarding the cause of serious illness. Psychoanal. Rev. 43 (1956) 146–162.

Bard, P.: Emotion: I. The neurohumoral basis of emotional reactions. In: Murchinson, C. (ed.): Handbook of General Experimental Psychology. Clard Univ. Press, Worcester/Mass. 1934.

Bardin, C. W., J. F. Catterall: Testosterone: a major determinant of extragenital sexual dimorphism. Science 211 (1981) 1285–1294.

Barefoot, J. C., W. G. Dahlstrom, R. B. Williams: Hostility, CHD incidence and the total mortality: a 25-year follow-up of 255 physicians. Psychosom. Med. 45 (1983) 59–63.

Bargen, J. A.: The management of colitis. National Medical Monographs, National Medical Book Comp., New York 1935; zitiert nach: White et al., 1939.

Bargen, J. A., W. G. Sauer, W. P. Sloan, R. P. Gage: The development of cancer in chronic ulcerative colitis. Gastroenterology 26 (1954) 32–37.

Barglow, P.: Pseudocyesis and psychiatric sequelae of sterilization. Arch. gen. Psychiat. 2 (1964) 571.

Barglow, P., R. Hatcher, J. Wolston, R. Phelps, W. Burns, R. Depp: Psychiatric risk factors in the pregnant diabetic patient. Amer. J. Obstet. Gynec. 140 (1981) 46–52.

Barker, W.: Studies on epilepsy: the petit mal attack as a response within the central nervous system to distress in organism-environment integration. Psychosom. Med. 10 (1948) 73–94.

Barlow, E. D., H. E. de Wardener: Compulsive water drinking. Quart. J. Med. 28 (1959) 235.

Barnert, C.: Conversion reactions and psychophysiologic disorders: A comparative study. Psychiatry in Medicine 2 (1971) 205–220.

Barnett, A. H., C. Eff, R. D. G. Leslie, D. A. Pyke: Diabetes in identical twins. A study of 200 pairs. Diabetologia 20 (1981a) 87–93.

Barnett, A. H., A. J. Spiliopoulos, D. A. Pyke, W. A. Stubbs, J. Burrin, K. G. M. M. Alberti: Metabolic studies in unaffected co-twins of non-insulin dependent diabetics. Brit. med. J. 282 (1981b) 1656–1658.

Barnett, C. R.: zit. nach: Montagu, A.: Körperkontakt. Klett, Stuttgart 1974.

Baroldi, G., G. Falzi, F. Mariani: Significance of morphological changes in sudden coronary death. In: Manninon, Halonen: Sudden Coronary Death. Advanc. Cardiol. 25 (1978) 82–95.

Baroldi, G., G. Falzi, F. Mariani: Sudden coronary death. A postmortem study in 208 selected cases compared to 97 'controlled' subjects. Amer. Heart J. 98 (1979) 20–31.

Barolin, G. S.: Migräne. Das angiocephale Attackensyndrom: Diagnostik, Ätiologie und Therapie. Facultas, Wien 1969.

Baron, M.: Genetics of schizophrenia: II. Vulnerability traits and gene markers. Biol. Psychiat. 21 (1986) 1189–1211.

Baron, M., N. Risch, R. Hamburger, B. Mandel, S. Kuschner, M. Newman, D. Drumer, R. Bellmaker: Genetic linkage between X-chromosome markers and bipolar affective illness. Nature (Lond.) 326 (1987) 289–292.

Barr, R., V. Abernethy: Single case study. Conversion reaction. Differential diagnosis in the light of biofeedback research. J. nerv. ment. Dis. 164 (1977) 287–292.

Barry, R. J.: „Primary bradycardia" and „vagal inhibition" as two manifestations of a trivial delay function. J. Psychophysiol. 4 (1987) 375–379.

Barsky, A. J.: The paradox of health. New Engl. J. Med. 318 (1988) 414–418.

Bárta, K., L. Benýšek, B. Rotrekl: Zirkulierende antinukleäre Globuline bei der Colitis ulcerosa. Gastroenterologia 102 (1964) 16–22.

Bartels, E.-J., H. Feiereis, J. Horn, Th. Mansky, G. A. Martini, M. Otte, O. Samland, H.-M. Schuchardt: Morbus Crohn. Dtsch. Ärztebl. 86 (1989) 1772–1779.

Bartemeier, L. H.: A historical note on the psychoanalytic hospitals. Psychiat. J. Univ. Ottawa 3 (1978) 77–79.

Bartle, S. H., L. F. Bishop: Psychological study of patients with coronary heart disease with unexpectedly long survival and high level of function. Psychosomatics 15 (1974) 68–69.

Bartling, G., L. Echelmeyer, M. Engberding, R. Krause: Problemanalyse und Planung des therapeutischen Veränderungsprozesses – ein Leitfaden. (Im Druck.)

Barton, D., M. T. Kelso: The nurse as a psychiatric consultation team member. Int. J. Psychiat. Med. 2 (1971) 108–115.

Bartrop, R. W., E. Luckhurst, L. Lazarus, L. G. Kiloh, R. Penny: Depressed lymphocyte function after bereavement. Lancet I (1977) 834–836.

Basch, S. H.: The intrapsychic integration of a new organ. Psychoanal. Quart. 42 (1973) 364–384.

Basch, S. H.: Emotional dehiscence after successful renal transplantation. Kidney Int. 17 (1980) 388–396.

Basch, S. H., F. Brown, W. Cantor: Observations on body image in renal patients. In: Levy, N. B. (ed.): Psychonephrology 1, pp. 93–102. Plenum, New York 1981.

Bash, K. W., J. Bash-Liechti: Die Psychotherapie eines Falles von musikogener Epilepsie. Schweiz. Arch. Neurol. Psychiat. 83 (1959) 196–221.

Basler, H. D., H. Otte, T. Schneller, D. Schwoon: Verhaltenstherapie bei psychosomatischen Erkrankungen. Kohlhammer, Stuttgart 1979.

Bassyouni, J.: Welche Rolle spielt die prämorbide Persönlichkeitsstruktur in der Ätiologie, formalen Pathogenese und Therapie des Parkinson-Syndroms? Med. Diss., Homburg/Saar 1987.

Bastiaans, J.: Emotiogene Aspekte der essentiellen Hypertonie. Verh. Dtsch. Ges. Inn. Med. 69 (1963) 7.

Bastiaans, J.: Psychoanalytic investigations on the psychic aspects of acute myocardial infarction. Psychother. and Psychosom. 16 (1968) 202–209.

Bastine, R.: Forschungsmethoden in der klinischen Psychologie. In: Schraml, W. J., U. Baumann (Hrsg.): Klinische Psychologie I. Huber, Bern 1975.

Bateson, G.: Geist und Natur. Eine notwendige Einheit. Suhrkamp, Frankfurt 1982.

Bateson, G.: Ökologie des Geistes. Suhrkamp, Frankfurt 1985.

Bateson, G. et al.: Schizophrenie und Familie. Suhrkamp, Frankfurt 1969.

Batra, K. V., R. Schrek: Effect of age and species on sensitivity of lymphocytes to prednisolone and phytohemagglutinin. Proc. soc. Biol. Med. 125 (1967) 871–874.

Battegay, R.: Die Hungerkrankheiten. Unersättlichkeit als krankhaftes Phänomen. Huber, Bern 1982.

Battegay, R., J. Glatzel, W. Pöldinger, U. Rauchfleisch: Handwörterbuch der Psychiatrie, S. 340–344. Enke, Stuttgart 1984.

Baudouin, C.: Suggestion und Autosuggestion. Sibyllen-Verlag, Dresden 1924.

Bauer, A., A. Gutowski: Gnathologie. Quintessenz, Berlin 1975.

Bauer, B., M. Bergmann: Psychologische Befunde bei Duodenalulzera. Dtsch. Z. Verdau.- u. Stoffwechselkrkt. 6 (1981) 288–294.

Bauer, H.: Ätiologie und Pathogenese der funktionellen Stimmstörungen im Blickfeld der Erb- und Konstitutionsforschung. Dtsch. med. J. 12 (1961) 599.

Bauer, H.: Klanganalytische Untersuchungen fortlaufender sprachlicher Signale bei den verschiedenen Formen des Näselns. Fortschr. Med. 83 (1965) 409.

Bauer, H.: Klinik der Stimmstörungen. In: Biesalski, P. (Hrsg.): Phoniatrie und Pädaudiologie, S. 104. Thieme, Stuttgart 1973.

Bauer, H.: Die Bedeutung der Anamnese für die Therapie von funktionellen Stimmstörungen. Sprache-Stimme-Gehör 4 (1980) 93.

Bauer, H.: Störungen der Stimme, der Sprache und des Sprechens. In: Spiel, W. (Hrsg.): Die Psychologie des 20. Jahrhunderts, Bd. XII, S. 270. Kindler, Zürich 1980.

Bauer, H.: Das Gehör – ein mehrdimensionales Phänomen vom Aspekt der Pädaudiologie. Sprache-Stimme-Gehör 5 (1981) 6.

Bauer, H.: Konservativ ärztliche Behandlung von Stimmstörungen. In: Kittel, G. (Hrsg.): Phoniatrie und Pädaudiologie. Kursbuch, S. 60. Dtsch. Ärzte-Verlag, Köln 1989.

Bauer, H. J.: Umstrittene MS-Therapie. Nervenarzt 54 (1983) 400–405.

Bauer, J.: Psychosomatik der Adnexitis. Gynäk. Praxis 6 (1982) 725.

Baum, M.: Extinction of avoidance responding through response prevention (flooding). Psychol. Bull. 74 (1970) 276–284.

Baum, M., D. Powell et al.: Continuous ambulatory peritoneal dialysis in children. New Engl. J. Med. 307 (1982) 1538–1542.

Baumann, U. (Hrsg.): Indikation zur Psychotherapie. Perspektiven für Praxis und Forschung. Fortschritte der klinischen Psychologie 25. Urban & Schwarzenberg, München-Wien-Baltimore 1981.

Baumann, U.: Differentielle Therapiestudien und Indikation. In: Baumann, U. (Hrsg.): Indikation zur Psychotherapie. Perspektiven für Praxis und Forschung. Fortschritte der klinischen Psychologie 25. Urban & Schwarzenberg, München-Wien-Baltimore 1981.

Baumann, U., B. Wedel: Stellenwert der Indikationsfrage im Psychotherapiebereich. In: Baumann, U. (Hrsg.): Indikation zur Psychotherapie. Perspektiven für Praxis und Forschung. Fortschritte der klinischen Psychologie 25. Urban & Schwarzenberg, München-Wien-Baltimore 1981.

Baumann, U., R. D. Stieglitz: Psychotherapieforschung – Schwerpunkt psychologische Methoden. In: Häfner, H. (Hrsg.): Forschung für die seelische Gesundheit. Eine Bestandsaufnahme der psychiatrischen, psychotherapeutischen und psychosomatischen Forschung und ihrer Probleme in der BRD. Springer, Berlin-Heidelberg 1983.

Baumeyer, F.: Der psychogene akute Herzanfall. Psychosom. Med. 12 (1966) 1.

Bauriedl, T.: Beziehungsanalyse. Suhrkamp, Frankfurt 1980.

Beaglehole, R., C. E. Salmond, A. Hooper, J. Huntsman, J. M. Cassel, J. A. Prior: Blood pressure and social interaction in Tokelauan migrants in New Zealand. J. chron. Dis. 30 (1977) 803–812.

Beahrs, J. O.: The hypnotic psychotherapy of Milton H. Erickson. Amer. J. clin. Hypnosis 14 (1971) 73–90.

Beard, B. H.: The quality of life before and after renal transplantation. Dis. nerv. Syst. 32 (1971) 24–31.

Beard, B. H., T. F. Sampson: Denial and objectivity in hemodialysis patients. In: Levy, N. B.(ed.): Psychonephrology 1, pp. 169–176. Plenum, New York 1981.

Beardslee, W. R., J. Bemporad, M. B. Keller, G. L. Klerman: Children of parents with major affective disorder: a review. Amer. J. Psychiat. 140 (1983) 825–832.

Beattie, J., G. R. Brow, C. N. H. Long: Physiological and anatomical evidence for the existence of nerve tracts connecting hypothalamic with spinal sympathetic centers. Proc. roy. Soc. (B) 106 (1930) 253.

Beauchesne, H.: L'épileptique. Dunod, Paris 1980.

Beaumont, G.: The use of psychotropic drugs in other painful conditions. J. int. med. Res. (Suppl. 2) 2 (1976) 56–57.

Beaumont, W · Experiments and observations on the gastric juice and the physiology of digestion. Allen, Plattsburgh 1833.

Beaver, W. T.: Management of cancer pain with parenteral medication. J. Amer. med. Ass. 244 (1980) 2653–2657.

Bebbington, P. E.: The epidemiology of depressive disorder. Culture, Medicine, and Psychiatry 2 (1978) 297–341.

Bebbington, P. E., T. Brugha, B. MacCarthy, J. Potter, E. Sturt, T. Wykes, R. Katz, P. McGuffin: The Camberwell Collaborative Depression Study I. Depressed probands: adversity and the form of depression. Brit. J. Psychiat. 152 (1988) 754–756.

Becher, H. K.: Relationship of significance of wound to pain experienced. J. Amer. med. Ass. 161 (1956) 1609–1613.

Becher, H. K.: Pain, placebos and physicians. Practitioner 189 (1962) 141–155.

Bechler, B., A. Cogoli, D. Mesland: Lymphozyten sind schwerkraftempfindlich. Naturwissenschaften 73 (1986) 400–403.

Bechterew, W.v.: Was ist Suggestion? J. Psychol. Neurol. 3 (1904) 110–111.

Beck, A. T.: Cognitive therapy and the emotional disorders. Int. Univ. Press, New York 1976.

Beck, A. T., C. H. Ward, M. Mendelson, J. Mock, J. Erbaugh: An inventory for measuring depression. Arch. gen. Psychiat. 4 (1961) 561–571.

Beck, C.: Funktionsstörungen des Innenohrs aus morphologischer Sicht. Z. Hörgeräte-Akustik, Sonderheft 1971.

Beck, D.: Die Kurzpsychotherapie. Bern 1974.

Beck, D.: Das Koryphäen-Killer-Syndrom. Dtsch. med. Wschr. 102 (1977) 303–307.

Beck, D.: Krankheit als Selbstheilung. Insel, Frankfurt 1981.

Beck, J. C., K. Brochner-Mortensen: Observations on the prognosis in anorexia nervosa. Acta med. scand. 149 (1954) 409–430.

Beck, L., S. Potthoff: Zusammenfassende Übersicht über die praktische Anwendung der medikamentösen Analgesie bei der Geburt. Gynäkologe 9 (1976) 223.

Becker, E.: Histopathologie degenerativer Bandscheibenerkrankungen. Med. Diss., Berlin 1984.

Becker, E. P.: Persönlichkeit und Neurosen in der Zwillingsforschung. Ein historischer Überblick. In: Heigl-Evers, A., H. Schepank (Hrsg.): Ursprünge seelisch bedingter Krankheiten, Bd. 1, S. 177–181. Vandenhoeck & Ruprecht, Göttingen 1982.

Becker, H.: Die Vater-Tochter-Beziehung in der Familiendynamik bei Anorexia-nervosa-Patientinnen. Nervenarzt 51 (1980) 568.

Becker, H.: Compliance und subjektive Krankheitstheorie des Patienten. Dtsch. Ärztebl. 80 (1983) 25–28.

Becker, H.: Psychoonkologie. Springer, Berlin 1986.

Becker, H.: Psychoanalyse, Handlung und Körper. Prax. Psychother. Psychosom. 32 (1987) 170–177.

Becker, H.: Konzentrative Bewegungstherapie. Ein nonverbales Psychotherapieverfahren zur Erweiterung der Indikation. In: Stolze, H. (Hrsg.): Konzentrative Bewegungstherapie, S.187. 2. Aufl. Springer, Berlin-Heidelberg-New York 1989.

Becker, H.: Konzentrative Bewegungstherapie. Integrationsversuch von Körperlichkeit und Handeln in den psychoanalytischen Prozeß. 2. Aufl. Thieme, Stuttgart-New York 1989.

Becker, H., H. Lüdeke: Erfahrungen mit der stationären Anwendung psychoanalytischer Therapie. Psyche 32 (1978) 1.

Becker, H., P. Körner, A. Stöffler: Psychodynamics and therapeutic aspects of anorexia nervosa. A study of family dynamics and prognosis. Psychother. and Psychosom. 36 (1981) 8–16.

Becker, H. S., B. Geer: The fate of idealism in medical school. Amer. soc. Rev. 23 (1958) 50–56.

Becker, M. C., I. R. Zucker, V. Parsonnet, L. Gilbert: Rehabilitation of the patient with a permanent pacemaker. Geriatrics 22 (1967) 106–111.

Becker, M. H., J. G. Joseph: AIDS and behavioral change to reduce risk: a review. Amer. J. publ. Health 78 (1988) 394–410.

Becker, S.: AIDS – die Krankheit zur Wende? Psychologie heute 11 (1985) 60–65.

Becker, S.: Die Annahme der eigenen Homosexualität. In: Jäger, H. (Hrsg.): AIDS – Psychosoziale Betreuung von AIDS- und AIDS-Vorfeldpatienten, S. 68–80. Thieme, Stuttgart 1987.

Becker, S.: Die Krankheit AIDS in der Medizin. Über den Umgang mit Angst und Tabu. In: Sigusch, V., S. Fliegel (Hrsg.): AIDS. DGVT, Tübingen 1988.

Becker, S., U. Clement: HIV-Infektion und Sexualität. Dtsch. Ärztebl. 84 (1987) 1980–1984.

Becker, S., U. Clement: HIV-Infektion – Psychische Verarbeitung und politische Realität. Psyche 43 (1989) 698–709.

Becker-Carus, C.: Grundriß der physiologischen Psychologie. Quelle und Meyer, Heidelberg 1981.

Beckmann, D., H. E. Richter: Giessen-Test. Bern-Stuttgart-Wien 1972.

Beckmann, D., J. Scheer, H. Zenz: Methodenprobleme in der Psychotherapieforschung. In: Pongratz, L. (Hrsg.): Handbuch der Psychologie, Bd. 8/2: Klinische Psychologie. Hogrefe, Göttingen 1978.

Beckmann, O., M. L. Moeller, E. Richter, J. W. Scheer: Studenten-Urteile über sich selbst, über ihre Arbeit und über die Universität. Aspekte, Frankfurt 1972.

Bedell, S. E., T. L. Delbanco, E. F. Cook, F. H. Epstein: Survival after cardiopulmonary resuscitation in the hospital. New Engl. J. Med. 309 (1983) 569–576.

Beecher, H. K.: The powerful placebo. J. Amer. med. Ass. 159 (1955) 1602–1606.

Beeken, W. L.: Evidence of virus infection as a cause of Crohn's disease. Z. Gastroenterol. 17 (Suppl.) (1979) 101–104.

Beels, C. C., W. R. McFarlane: Family treatment of schizophrenia: background and state of the art. Hosp. Community Psychiat. 33 (1982) 541–550.

Beere, P. A., S. Glagov, C. K. Zarins: Retarding effect of lowered heart rate on coronary atherosclerosis. Science 226 (1984) 180–182.

Beersma, D. G. M., S. Daan, R. H. van den Hoofdakker: Distribution of REM latencies and other sleep phenomena in depression as explained by a single ultradian rhythm disturbance. Sleep 7 (1984) 126–136.

Begemann-Deppe, M.: Im Krankenhaus sterben: Das Problem der Wissenskonstitution in einer besonderen Situation. In: Begemann, H. (Hrsg.): Patient und Krankenhaus. Urban & Schwarzenberg, München 1976.

Begemann-Deppe, M.: Sprechverhalten und Thematisierung

Literatur

von Krankheitsinformation im Rahmen von Stationsvisiten. Phil. Dissertation, Universität Freiburg 1978.

Behrendt, W. S.: Gestalt- und integrative Therapie bei der Behandlung von Stimmstörungen mit psychogenen Aspekten. Fol. phoniat. 33 (1981) 358.

Behrendt, W. S.: Motivation und Widerstand des dysphonischen Patienten vor dem Hintergrund seiner Psyche. Vortrag, II. Kommunikationsmedizinische Tage, 15.–17. 4. 1988, Bad Rappenau.

Beigler, J. S. et al.: Report on liaison psychiatry at Michael Reese Hospital 1950–1958. AMA Arch. Neurol. 81 (1959) 733–737.

Beischer, W.: Regulation der Insulinsekretion. In: Bachmann, W., H. Mehnert (Hrsg.): Kombinationstherapie Insulin/Sulfonylharnstoff, S. 31–42. Karger, Basel 1983.

Beischer, W., E. F. Pfeiffer: Zur Prognose des Diabetes mellitus. Teil I: Bisherige Prognose und neue Möglichkeiten für Stoffwechselkontrolle und Therapie. Fortschr. Med. 103 (1985a) 501.

Beischer, W., E. F. Pfeiffer: Zur Prognose des Diabetes mellitus. Teil II: Zukünftige Prognose, Chancen, Realität und Hoffnungen. Fortschr. Med. 103 (1985b) 506.

Beischer, W., E. Dittus, M. Pfeiffer, B. Beischer, E. F. Pfeiffer: Therapeutic and prognostic relevance of fast beta-cell stimulation capacity. In: Melchionda, N., D. L. Horwitz, D. S. Schade (eds.): Recent Advances in Obesity and Diabetes Research, pp. 1–18. Raven, New York 1984.

Beischer, W., R. Brachmann, H. Zier, R. Koberstein, C. Schomann-Ziesenböck, M. Pfeiffer, W. Kerner, U. Loos: Selbstkontrolle der Blutglukose mittels visueller und reflektometrischer Auswertung von Haemo-Glukotest® 20–800: Objektive Befunde und Urteil der Patienten. Münch. med. Wschr. 19 (1985) 481.

Beischer, W., J. J. Hoet, E. F. Pfeiffer: Diabetes und Schwangerschaft. Dtsch. Ärztebl. 82 (1985) 727.

Beisel, W. R., J. M. Talbot: The effects of space flight on immunocompetence. Immunol. Today 8 (1987) 197–200.

Beitl, R.: Deutsche Volkskunde, S. 452. Dtsch. Buch-Gemeinschaft, Berlin 1933.

Bell, G. R., R. H. Rothman: The conservative treatment of sciatica. Spine 9 (1984) 54–56.

Bell, H. W.: Nonsurgical management of the pain-dysfunction syndrome. J. Amer. dent. Ass. 79 (1969) 161.

Bell, S. M., M. D. S. Ainsworth: Infant crying and maternal responsiveness. Child Dev. 43 (1972) 1171–1190.

Bellak, L., L. Small: Kurzpsychotherapie und Notfallpsychotherapie. Suhrkamp, Frankfurt 1972.

Bellak, L., M. Hurvich, H. K. Gediman: Ego functions in schizophrenics, neurotics and normals. Wiley, New York 1973.

Bellak, L., J. B. Chassan, H. K. Gediman, M. Hurvich: Ego function assessment of analytic psychotherapy combined with drug therapy. J. nerv. ment. Dis. 157 (1973) 465–469.

Belting, Ch.M., P. Gupta: Incidence of periodontal disease among persons with neuropsychiatric disorders. J. dent. Res. 39 (1960) 744.

Bender, S. W.: Crohn's disease in children. Z. Gastroenterol. 17 (Suppl.) (1979) 164–170.

Benedek, Th.: The functions of the sexual apparatus and their disturbances. In: Alexander, F. (ed.): Psychosomatic Medicine. Allen & Unwin, London 1952. Deutsch: Die Funktionen des Sexualapparates und ihre Störungen. In: Alexander, F. (Hrsg.): Psychosomatische Medizin. De Gruyter, Berlin-New York 1971.

Benedek-Jaszmann, L. J.: Treatment of the premenstrual syndrome with bromocriptine. Acta endocr. (Kbh.) 78 (Suppl.) 193 (1975) 29.

Benedetti, G.: Behandlung anorektischer Kinder durch Psychoanalyse der Mütter. Helv. paediat. Acta 6 (1956) 539–561.

Benedetti, G.: Beitrag zum Problem der Alexithymie. Nervenarzt 51 (1980) 534–541.

Bengel, J., U. Koch: Evaluationsforschung im Gesundheitswesen. In: Koch, U., G. Lucius-Hoene, R. Stegie (Hrsg.): Handbuch der Rehabilitationspsychologie. Springer, Berlin-Heidelberg 1988.

Bengtsson, C.: Ischaemic heart disease in women. Acta med. scand. (Suppl.) 5 (1973) 549.

Bennet, A. H., J. H. Harrison: Experience with living familial renal donors. Surg. Gynecol. Obstet. 139 (1974) 894–989.

Bennet, W. B., I. Gurin: The dieter's dilemma: eating less and weighing more. Basic Books, New York 1982.

Benos, J.: Psychische Störungen, ein Frühkriterium des Pankreaskarzinoms. Med. Welt 25 (1974) 952–953.

Benson, H.: The relaxation response: physiologic bases, history, and clinical usefulness. In: Dembroski, T. M., T. H. Schmidt, G. Blümchen (eds.): Biobehavioral Bases of Coronary Heart Disease. Karger, Basel-München-Paris 1983.

Benson, H., S. Alexander, C. L. Feldman: Decreased premature ventricular contractions through the use of the relaxation response in patients with stable ischemic heart disease. Lancet 2 (1975) 380.

Benson, H., B. A. Rosner, B. R. Marzetta, H. M. Klemchuk: Decreased blood pressure in borderline hypertensive subjects who practiced meditation. J. chron. Dis. 27 (1974) 163–169.

Benson, H., B. A. Rosner, B. R. Marzetta, H. M. Klemchuk: Decreased blood pressure in pharmacologically treated hypertensive patients who regularly elicited the relaxation response. Lancet 23 (1974) 289.

Benson, H., D. B. Shapiro, G. E. Schwartz: Decreased systolic blood pressure through operant conditioning techniques in patients with essential hypertension. Science 173 (1971) 740–742.

Bentley, S., D. J. Pearson, K. J. Rix: Food hypersensitivity in irritable bowel syndrome. Lancet I (1983) 295–297.

Benton, A. L.: The visual retention test: clinical and experimental explications. 3rd ed. Psychological Corporation, New York 1963.

Bepperling, W., M. Klotz: Analytische Psychotherapie und Funktionelle Entspannung als kombinierte Behandlungsmethode. Hippokrates, Stuttgart 1978.

Berczi, I. (ed.): Pituitary function and immunity. CRC, Boca Raton/FL 1986.

Berczi, I., K. Kovacs (eds.): Hormones and immunity. MTP, Lancaster 1987.

Berendes, J.: Einführung in die Sprachheilkunde, 9. Aufl. Barth, München 1971.

Berendes, J.: Psychologisches in der HNO-Praxis. Z. Laryng. Rhinol. 51 (1972) 1.

Berenson, C., B. Grosser: Total artificial heart transplantation. Arch. gen. Psychiat. 41 (1984) 910–916.

Berezin, M. A.: Some intrapsychic aspects of aging. In: Zinberg, N. E., I. Kaufman (eds.): Normal Psychology of the Aging Process, pp. 93–117. Int. Univ. Press, New York 1963.

Berezin, M. A.: Sex and old age: a review of the literature. J. geriat. Psychiat. 2 (1969) 131–149.

Berezin, M. A.: Sex and old age: a further review of the literature. J. geriat. Psychiat. 9 (1976) 189–220.

Berg, J. W., R. Ross, H. B. Latourette: Economic status and survival of cancer patients. Cancer 39 (1977) 467–477.

Bergener, M.: Depressionen im Alter: Entstehungsbedingungen, Symptomatologie, Diagnostik und Differentialdiagnostik. In: Bergener, M. (Hrsg.): Depressionen im Alter. Steinkopff, Darmstadt 1986.

Bergener, M., B. Kark (Hrsg.): Psychosomatik in der Geriatrie. Steinkopff, Darmstadt 1985.

Berger, F., R. Rauskolb, M. Schütz, S. Stephanos: Gestosis

and the psychosomatic phenomenon – an empirical investigation. Psychother. and Psychosom. 27 (1976/77) 154–158.

Berger, P., Th. Luckmann: Die gesellschaftliche Konstruktion der Wirklichkeit. Fischer, Frankfurt 1969.

Berger, W.: Organisches und Funktionelles bei Stimmstörungen. Z. Laryngol. 19 (1930) 80.

Berger-Oser, R., D. Richter: Zur Psychosomatik der EPH-Gestose. In: Jürgensen, O., D. Richter (Hrsg.): Psychosomatische Probleme in der Gynäkologie und Geburtshilfe. Springer, Berlin 1985.

Bergin, A. E., M. L. Lambert: The evaluation of therapeutic outcome. In: Bergin, A. E., S. L. Garfield (eds.): Handbook of Psychotherapy and Behavioral Change: An Empirical Analysis. Wiley, New York 1978.

Bergler, E., L. Eidelberg: Der Mammakomplex des Mannes. Int. J. Psychoanal. 19 (1933) 547–583.

Berglund, G., B. Larsson, O. Andersson, O. Larsson, K. Suärdsudd, P. Björntorp, L. Wilhelmsen: Body composition and glucose metabolism in hypertensive middle-aged males. Acta med. scand. 200 (1976) 163–169.

Bergmann, E., W. Bolm, B. Seitz, S. Bartholomeyczik: Schichtarbeit als Gesundheitsrisiko. Mehrfachbelastungen und Beanspruchungen bei Schichtarbeitern. Campus, Frankfurt-New York 1982.

Bergmann, G.v.: Das vegetative Nervensystem und seine Störungen. In: Bergmann, G.v., B. Staehelin (Hrsg.): Handbuch der Inneren Medizin. Springer, Berlin 1926.

Bergmann, G.v.: Funktionelle Pathologie. Springer, Berlin 1932.

Bergmann, J. F.: Die Theorie des sozialen Systems von Talcott Parsons. Frankfurt 1967.

Bergold, J., D. Kallinke: Lerntheoretische Überlegungen zur Psychosomatischen Medizin. Fortbildungskurse der Schweiz. Ges. f. Psychol. 6 (1973) 78.

Bergolt, H., K. J. Ebscher, H. Malchow: Der niedergelassene Arzt und die Patienten-Compliance. Fortschr. Med. 99 (1981) 1784–1790.

Bericht zur Lage der Psychiatrie in der Bundesrepublik Deutschland – Zur psychiatrischen und psychotherapeutisch-psychosomatischen Versorgung der Bevölkerung. Dtsch. Bundestag, 7. Wahlperiode, Drucksache 7/4200, 1975.

Berkenbosch, F., J. van Oers, A. Del Rey, F. Tilders, H. Besedovsky: Corticotropin-releasing factor-producing neurons in the rat activated by interleukin-1. Science 238 (1987) 524–526.

Berkman, H. A.: Sex counseling with hemodialysis patients. Dial. Transplant. 7 (1978) 924–927.

Berkman, L. F., S. L. Syme: Social network, host illness and mortality. A nine year follow-up study of Almeda county residents. Amer. J. Epidem. 109 (1979) 186–204.

Berkowitz, L.: Roots of aggression – a reexamination of the frustration-aggression hypothesis. New York 1969.

Berlyne, D. E.: Novelty and curiosity as determinants of exploratory behavior. Brit. J. Psychol. 41 (1950) 68–80.

Berlyne, D. E.: The motivational significance of collative variables and conflicts. In: Abelson, R. P. et al. (eds.): Theories of Cognitive Consistency. Rand McNalley, Chicago 1968.

Berman, K. F., D. R. Weinberger, R. C. Shelton, R. F. Zec: A relationship between anatomical and physiological brain pathology in schizophrenia: lateral cerebral ventricular size predicts cortical blood flow. Amer. J. Psychiat. 144 (1987) 1277–1282.

Berman, P. M., J. B. Kirsner: The aging gut: II. Diseases of the colon, pancreas, liver and gallbladder, functional bowel disease and iatrogenic disease. Geriatrics 27 (1972) 117–124.

Bernfeld, S.: Vom Gemeinschaftsleben der Jugend. Quellenschriften zur seelischen Entwicklung. Psychoanalytischer Verlag, Leipzig-Wien-Zürich 1922.

Bernhard, P.: Rollengebundene Angstbewältigung von Patienten und Personal auf der medizinischen Intensivstation. Prax. Psychother. Psychosom. 28 (1983) 181–189.

Bernhard, P.: Angst und Angstbewältigung von Patient und Personal einer medizinischen Intensivstation. Psychother. med. Psychol. 34 (1984) 50–55.

Bernhard, P., W. Huhn: Psychosomatische Aspekte bei Uveitis. In: Studt, H. H. (Hrsg.): Psychosomatik in Forschung und Praxis, S. 521–535. Urban & Schwarzenberg, München-Wien-Baltimore 1983.

Bernhard, Th.: Der Atem. Eine Entscheidung. Deutscher Taschenbuch Verlag, München 1981.

Bernheim, H.: Die Suggestion und ihre Heilwirkung. (Übers. S. Freud). Deuticke, Leipzig-Wien 1888.

Bernreuter, R. G.: The Personality Inventory. Consulting Psychologists Press, Palo Alto/Ca. 1931.

Bernstein, D. A., T. D. Borkovec: Entspannungs-Training. Handbuch der progressiven Muskelentspannung. 3. Aufl. Pfeiffer, München 1982.

Bernstein, I. L.: Learned taste aversions in children receiving chemotherapy. Science 200 (1978) 1302–1303.

Bernstein, J. G.: Drug interactions. In: Cassem, N. H., T. P. Hackett (eds.): Massachusetts General Hospital Handbook of General Hospital Psychiatry, 2nd ed. PSG, Littleton 1987.

Bernstein, L. H., M. S. Frank, L. J. Brandt, S. F. Boley: Healing of perineal Crohn's disease with metronidazole. Gastroenterology 79 (1980) 357–365.

Berrettini, W. H., J. I. Nurnberger jr., T. A. Hare, E. S. Gershon, R. M. Post: Plasma and CSF-GABA in affective illness. Brit. J. Psychiat. 141 (1982) 483–487.

Berry, D. T. R., W. B. Webb: Mood and sleep in aging women. J. Pers. soc. Psychol. 49 (1985) 1724–1727.

Bertalanffy, L.v.: General systems theory. Braziller, New York 1968.

Bertel, O., R. Stauber, U. C. Dubach: Diagnostische Abklärung und Verlauf bei 105 Patienten mit Synkopen. Schweiz. med. Wschr. 115 (1985) 439–441.

Bertelson, A., B. Harvald, M. Hauge: A Danish twin study of manic-depressive disorder. Brit. J. Psychiat. 130 (1977) 330–351.

Besedovsky, H. O., A. Del Rey: Immune-neuroendocrine network. Progr. Immunol. 6 (1986) 578–587.

Besedovsky, H. O., A. Del Rey: Neuroendocrine and metabolic responses induced by interleukin-1. J. Neurosci. Res. 18 (1987) 172–178.

Besedovsky, H. O., E. Sorkin, D. Felix, H. Haas: Hypothalamic changes during the immune response. Eur. J. Immunol. 7 (1977) 323–325.

Besedovsky, H. O., A. Del Rey, E. Sorkin, C. A. Dinarello: Immunoregulatory feedback between interleukin-1 and glucocorticoid hormones. Science 233 (1986) 652–654.

Besedovsky, H. O., A. Del Rey, E. Sorkin, W. Lotz, U. Schwulera: Lymphoid cells produce an immunoregulatory glucocorticoid increasing factor (GIF) acting through the pituitary gland. Clin. exp. Immunol. 59 (1985) 622–628.

Best, F.: Die Augenstörungen bei Hysterie. Zbl. ges. Ophthalm., 30. Bd. (1934) 321–400.

Best, W. R.: Design and analysis of clinical trials in Crohn's disease. Z. Gastroenterol. 17 (Suppl.) (1979) 29–36.

Best, W. R., J. M. Becktel, J. W. Singleton, F. Kern jr.: Development of a Crohn's disease activity index. Gastroenterology 70 (1976) 439–444.

Beta-Blocker Heart Attack Trial Research Group: A randomized trial of propranolol in patients with acute myocardial infarction: I. Mortality results. J. Amer. med. Ass. 247 (1982) 1707–1714.

Bettelheim, B.: Die Kinder der Zukunft. Molden, Wien 1969.

Bettelheim, B.: Der Weg aus dem Labyrinth. Deutsche Verlagsanstalt, Stuttgart 1975.

Bettex, M.: Erste Arbeitsergebnisse aus dem psychoonkologischen Dienst. Med. Psychol. 8 (1982) 141–151.

Bettschart, W. et al.: L'enfant de 9 ans. Séries paedopsychiat., Fasc. 5. Schwabe, Basel 1978.

Beumont, P. J. V., J. Russell: Anorexia nervosa. In: Beumont, P. J. V., G. D. Burrows (eds.): Handbook of Psychiatry and Endocrinology. Elsevier, Amsterdam 1982.

Beumont, P. J. V., C. J. Bearwood, G. F. M. Russell: The occurrence of the syndrome of anorexia nervosa in male subjects. Psychol. Med. 2 (1972) 216–231.

Beumont, P. J. V., M. O'Connor, W. Lennerts, S. W. Touyz: Ernährungsberatung in der Behandlung der Bulimia. In: Fichter, M. M. (Hrsg.): Bulimia nervosa. Enke, Stuttgart 1989.

Beutel, M.: Zur Erforschung der Verarbeitung chronischer Krankheit: Konzeptualisierung, Operationalisierung und Adaptivität von Abwehrprozessen am Beispiel von Verleugnung. Psychother., Psychosom., med. Psychol. 35 (1985) 295–302.

Beutel, M.: Verarbeitung chronischer Krankheit. Chemie-Verlag, Weinheim 1988a.

Beutel, M.: Spezifische und generelle Aspekte der Verarbeitung chronischer Erkrankungen. In: Kächele, H., W. Steffens (Hrsg.): Bewältigung und Abwehr. Springer, Berlin 1988b.

Beutel, M.: Bewältigungsprozesse bei chronischen Erkrankungen. Edition Medizin, Weinheim 1988c.

Beutler, J.: Anosognosie. Med. Diss., Aachen 1976.

Bibring, G. L.: Psychiatry and medical practice in a general hospital. New Engl. J. Med. 254 (1956) 366–372.

Bibring, G.: Some considerations of the psychological processes in pregnancy. Psychoanal. Study Child 14 (1959) 113–121.

Biebl, W.: Über Selbsthilfegruppen. IMA-Kongreßhefte 1980–1981.

Biebl, W., T. Platz, J. Kinzl, G. Judmaier: Psychosomatische Untersuchung bei Patienten mit Colitis ulcerosa und Morbus Crohn. Prax. Psychother. Psychosom. 29 (1984) 184–190.

Bielenski, H., W. Streich, V. Volkholz: Praktizierte Maßnahmen zur Schichtarbeit. Humanisierung des Arbeitslebens (hrsg. vom Bundesminister für Arbeit und Sozialordnung), Bd. 38, Bonn 1980.

Biermann, B.: Der Hypnotismus im Lichte der Lehre von den bedingten Reflexen. J. Psychol. Neurol. 38 (1929) 265.

Biermann, B.: Autogenes Training mit Kindern und Jugendlichen. Reinhardt, München-Basel 1975.

Biermann, G.: Das kindliche Bronchialasthma aus psychosomatischem Blickwinkel. Diagnostik 10 (1977) 16–20.

Biermann, G.: Die Mundwelt des Kindes. Fortschr. Kieferorthop. 43 (1982) 91.

Biggo, G., E. Costa (eds.): Advances in biochemical psychopharmacology, Vol. 38. Raven, New York 1983.

Bilodeau, C. B., T. B. Hackett: Issues raised in a group setting by patients recovering from myocardial infarction. Amer. J. Psychiat. 128 (1971) 73–78.

Bilz, R.: Psychogene Angina. Epikritische Betrachtungen über eine Mandelentzündung und ihre Psychopathologie. 1. Beitr. Zbl. Psychother. Hirzel, Leipzig 1936.

Bilz, R.: Der Vagus-Tod. Eine anthropologische Erörterung über die Situation der Ausweglosigkeit. Med.Welt 17 (1966) 117–122; 163–170.

Binder, H.: Seminar über Gruppentherapie mit dem autogenen Training. Lehmann, München 1964.

Binder, V.: Epidemiology, course and socio-economic influence of inflammatory bowel disease. Schweiz. med. Wschr. 118 (1988) 738–742.

Binet, A., T. Simon: Le développement de l'intelligence chez l'enfant. Ann. Psychol. 14 (1908).

Binger, C. A. L., N. W. Ackerman, A. E. Cohn et al.: Personality in arterial hypertension. Psychosom. Med. Monogr. Brunner, New York 1945.

Biniek, E. M.: Psychotherapie mit gestalterischen Mitteln. Wiss. Buchgesellschaft, Darmstadt 1982.

Biörck, G., O. Edhag: Loss of consciousness from arrhythmia: the patient's experience. Acta med. scand. 193 (1973) 201–205.

Bion, W. R.: Learning from experience. Heinemann Medical Books, London 1962.

Bion, W. R.: Erfahrungen in Gruppen und andere Schriften. Klett, Stuttgart 1971.

Biondi, M., P. Pancheri: Mind and immunity. A review of methodology in human research. Advanc. psychosom. Med. 17 (1987) 243–251.

Birbaumer, N.: (Hrsg.): Neuropsychologie der Angst. Urban & Schwarzenberg, Berlin 1973.

Birbaumer, N.: Physiologische Psychologie. Springer, Berlin 1975.

Birbaumer, N.: Biofeedback. In: Pongratz, L. J. (Hrsg.): Handbuch der Psychologie, Bd. 8/2: Klinische Psychologie. Hogrefe, Göttingen 1977.

Birbaumer, N.: Psychophysiologische Ansätze. In: Euler, H. A., H. Mandel (Hrsg.): Emotionspsychologie, S. 54. Urban & Schwarzenberg, München-Wien-Baltimore 1983.

Birbaumer, N.: Psychologische Analyse und Behandlung von Schmerzzuständen. In: Zimmermann, M., H. O. Handwerker (Hrsg.): Schmerz. Springer, Berlin 1984.

Birbaumer, N.: Psychophysiologische Grundlagen. In: Miltner, W., N. Birbaumer, W.-D. Gerber (Hrsg.): Verhaltensmedizin, S. 61. Springer, Berlin-Heidelberg-New York-Tokyo 1986.

Birnbaum, D.: Peptic ulcer and the central nervous system – aetiology and management. Clin. Gastroenterol. 2 (1973) 245–257.

Biron, F., A. Eder, M. Frass, I. Frassine, I. Grunmiller, E. Herndl, H. Klima, E. Kremeier, R. Kuzmits, P. Lorant, M. M. Müller, E. Trombik: Ergebnisse der Wiener Gesundheitsstudie 1979. Inst. f. Stadtforschung, Wien 1980.

Bischoff, C.: Wahrnehmung der Muskelspannung. Hogrefe, Göttingen 1989.

Bischoff, C., K.-J. Müller: Portable EMG-biofeedback – a single case study with a muscle contraction headache sufferer. In: Bischoff, C., H. C. Traue, H. Zenz (eds.): Clinical Perspectives on Headache and Low Back Pain, pp. 201–218. Hogrefe & Huber, Toronto 1989.

Bischoff, C., G. Sauermann: Nicht-instrumentelles motorisches Verhalten von Personen mit und ohne Spannungskopfschmerz während und nach kognitiver Belastung. In: Wittchen, H.-U., J. C. Brengelmann (Hrsg.): Experimentelle Befunde der psychologischen Therapieforschung bei akuten und chronischen Schmerzzuständen, S. 93–111. Springer, Berlin 1985.

Bischoff, C., H. C. Traue: Myogenic headache. In: Holroyd, K. A., B. Schlote, H. Zenz (eds.): Perspectives in Research on Headache, pp. 66–90. Hogrefe, Lewiston-New York 1983.

Bischoff, C., H. C. Traue, H. Zenz: Muskelspannung und Schmerzerleben von Personen mit und ohne Spannungskopfschmerzen bei experimentell gesetzter aversiver Reizung. Z. exp. angew. Psychol. 29 (1982) 357–385.

Bischoff, C., H. C. Traue, H. Zenz (eds.): Clinical perspectives on headache and low back pain. Hogrefe & Huber, Toronto 1989.

Bishop, D., A. Green, S. Cantor, W. Torresin: Depression, anxiety and rheumatoid arthritis. Clin. exp. Rheum. 5 (1987) 147–150.

Bishop, E. R.: Monosymptomatic hypochondriasis. Psychosomatics 21 (1980) 731–741.

Bismarck, O.v.: Gedanken und Erinnerungen. Reden und Briefe. Safari-Verlag, Berlin 1951.

Bispinck, R., B. Zwingmann: Aspekte der gewerkschaftlichen Humanisierungspolitik und ihrer infrastrukturellen Voraussetzungen. WSI-Mitteilungen 34 (1981) 62–73.

Björkerud, S.: Effect of adrenocortical hormones on the integrity of the rat aortic endothelium. In: Schettler, G., Weizel: Atherosclerosis III. Proc. 3rd Int. Symp., pp. 245–249. Springer, New York 1974.

Blacher, R. S.: The hidden psychosis of open-heart surgery. J. Amer. med. Ass. 222 (1972) 305–308.

Blacher, R. S.: The meaning of heart valve surgery to the patient. Int. J. Psychiat. Med. 6 (1975) 517–521.

Blacher, R. S.: Paradoxical depression after heart surgery: a form of survivor syndrome. Psychoanal. Quart. 47 (1978) 267–283.

Blacher, R. S.: Death resurrection and rebirth. Observations in cardiac surgery. Psychoanal. Quart. 52 (1983) 56–72.

Blachly, P. H.: Open-heart surgery. In: Abram, H. S. (ed.): Psychological Aspects of Surgery. Int. Psychiat. Clinics, Boston. Vol. 4, No. 2 (1967) 133–153.

Blachly, P. H., A. Starr: Post-cardiotomy delirium. Amer. J. Psychiat. 121 (1964) 371–375.

Black, S., J. H. Humphrey, J. S. Niven: Inhibition of Mantoux reaction by direct suggestion under hypnosis. Brit. med. J. 1 (1963) 1649–1652.

Blackwell, B.: Patient compliance. New Engl. J. Med. 289 (1973) 249–252.

Blackwell, R. E., R. Guillemin: Hypothalamic control of adenohypophyseal secretion. Ann. Rev. Physiol. 35 (1973) 357–390.

Blackwood, H. J. J.: Pathology of the temporomandibular joint. J. Amer. dent. Ass. 79 (1969) 118.

Blagg, C. R.: Home dialysis: six years' experience. New Engl. J. Med. 233 (1970) 1121.

Blalock, S. J., B. McEvoy DeVellis, R. F. DeVellis, S. H. van Sauter: Self-evaluation processes and adjustment to rheumatoid arthritis. Arthr. and Rheum. 31 (1988) 1245–1251.

Blanchard, E. B., F. Andrasik: Psychological assessment and treatment of headache: recent development and emergent issues. J. consult. clin. Psychol. 50 (1982) 859–879.

Blanchard, E. B., S. T. Miller: Psychological treatment of cardiovascular disease. Arch. gen. Psychiat. 34 (1977) 1402–1413.

Blanchard, E. B., F. Andrasik, T. A. Ahles, S. J. Teders, D. O'Keefe: Migraine and tension headache: a meta-analytic review. Behav. Ther. 11 (1980) 613–631.

Blanchard, E. B., K. A. Appelbaum, P. Guarnieri, B. Morrill, M. P. Dentinger: Five-year prospective follow-up on the treatment of chronic headache with biofeedback and/or relaxation. Headache 27 (1987) 580–583.

Blanchard, E. B., S. T. Miller, G. G. Abal, M. R. Haynen, R. Wicker: Evaluation of biofeedback in the treatment of borderline essential hypertension. J. appl. Behav. Anal. 12 (1979) 99–109.

Blanchard, E. B., D. O'Keefe, D. Neff, S. Jurish, F. Andrasik: Interdisciplinary agreement in the diagnosis of headache type. J. Behav. Assess. 3/1 (1981) 5–9.

Blankenburg, W.: Der Leib als Partner. Psychother. med. Psychol. 33 (1983) 206.

Blaser, A.: Der Urteilsprozeß bei der Indikationsstellung zur Psychotherapie. Huber, Bern 1977.

Blatt, C. M., S. H. Rabinowitz, B. Lown: Central serotonergic agents raise the repetitive extrasystole threshold of the vulnerable period of the canine ventricular myocardium. Circulat. Res. 44 (1979) 723–730.

Blatteis, C. M.: The neurobiology of endogenous pyrogens. In: Bligh, J., K. Voigt (eds.): Thermoreception and Temperature Control. Springer, Berlin 1989.

Blau, D.: On widowhood. J. geriat. Psychiat. 8 (1975) 29–40.

Bleecker, E. R., B. T. Engel: Learned control of ventricular rate in patients with atrial fibrillation. Psychosom. Med. 35 (1973) 161.

Bleecker, J. A. C.: Brief psychotherapy with lung cancer patients. Psychother. and Psychosom. 29 (1978) 282–287.

Bleuler, E. (Hrsg.): Lehrbuch der Psychiatrie, 15. Aufl. Springer, Berlin 1983.

Bleuler, M., J. Willi, H. P. Bühler: Akute psychische Begleiterscheinungen körperlicher Krankheiten. Thieme, Stuttgart 1966.

Bleuler, R.: Endokrinologische Psychiatrie. In: Gruhle, H. W., R. Jung, W. Mayer-Groß, M. Müller (Hrsg.): Psychiatrie der Gegenwart, Bd. I/1B, S. 161. Springer, Berlin 1964.

Bliesener, Th.: Strategien der Abweisung von Initiativen. Zur Kommunikation zwischen Krankenhauspersonal und Patient in der Visite. Psychol. Dipl.-Arbeit, Universität Bonn 1978.

Bliesener, Th.: Erzählen unerwünscht. Erzählversuche von Patienten in der Visite. In: Ehlich, K. (Hrsg.): Erzählen im Alltag, S. 27–36. Suhrkamp, Frankfurt 1980a.

Bliesener, Th.: Wie kann man als Patient in der Visite zu Wort kommen? In: Tschauder, G., E. Weigand (Hrsg.): Perspektive textextern. Akten des 14. linguistischen Kolloquiums Bochum, Bd. 2, S. 27–36. Niemeyer, Tübingen 1980b.

Bliesener, Th.: Die Visite – ein verhinderter Dialog. Initiativen von Patienten und Abweisungen durch das Personal, S. 162. Narr, Tübingen 1982a.

Bliesener, Th.: Konfliktaustragung in einer schwierigen „therapeutischen Visite". In: Köhle, K., H.-H. Raspe (Hrsg.): Das Gespräch während der ärztlichen Visite, S. 249–268. Urban & Schwarzenberg, München 1982b.

Bliesener, Th.: Die Therapie ist meine Leiter nach oben. Bildhafte Redeweisen von Leukämiekranken als Zugang zu ihrer Lebenswelt. In: Spillner, B. (Hrsg.): Perspektiven der angewandten Linguistik, Bd. 12, S. 60–61. Narr, Tübingen 1987.

Bliesener, Th., J. Siegrist: Greasing the wheels. J. Pragmatics 5 (1981) 181–204.

Bliesener, Th., K. Köhle (Hrsg.): Die ärztliche Visite – Chance zum Gespräch? Westdeutscher Verlag, Wiesbaden 1986.

Bliss, E. L., W. R. Rumel, C. H. Branch: Psychiatric complications of mitral surgery. Arch. Neurol. Psychiat. 74 (1955) 249–252.

Bliwise, N. G., D. Bliwise, W. Dement: Age and MMPI scores in insomnia. Sleep Res. 14 (1985) 126.

Bloch, D. A.: Family systems medicine: The field and the journal. Family Systems Medicine 1 (1983) 3.

Bloch, E.: Das Prinzip Hoffnung. Suhrkamp, Frankfurt 1959.

Bloch, S.: Beobachtungen über den Einfluß einiger vor der Gravidität applizierter emotioneller Traumata auf die nachfolgende Trächtigkeit und die postnatale Entwick-

lung der Jungen der auf das Trauma folgenden Trächtigkeit bei der Laboratoriums-Maus. Z. psychosom. Med.
16 (1970) 360.

Block, J.: Parents of schizophrenic, neurotic, asthmatic
and congenitally ill children. Arch. gen. Psychiat. 20
(1969) 659.

Block, J., P. H. Jenning, E. Harvey, E. Simpson: Interaction
between allergic potential and psychopathology in
childhood asthma. Psychosom. Med. 26 (1964) 307–320.

Blodgett, C.: A selected review of the literature of adjustment to hemodialysis. Int. J. Psychiat. Med. 11 (1981/
82) 97–124.

Blohmke, M., J. Neipp: Chronisch entzündlicher Gelenk-
rheumatismus und Krankheiten der Knochen und Gelenke aus epidemiologischer und sozio-ökonomischer
Sicht. Arbeitsmedizin Sozialmedizin Präventivmedizin
16 (1981) 1–6.

Blohmke, M., B. Koschorrek, O. Stelzer: Häufigkeit von
Risikofaktoren der koronaren Herzkrankheit in ver-
schiedenen Altersgruppen und sozialen Schichten bei
Männern. Z. Gerontol. 3 (1970) 201–209.

Blohmke, M., B. Depner, B. Koschorrek, O. Stelzer: Soziale Faktoren und Krankheit bei Arbeitnehmern. Gentner,
Stuttgart 1975.

Blomstrand, F., F. Löfgren: Influence of emotional stress
on the renal circulation. Psychosom. Med. 18 (1956)
420.

Bloom, F. E.: The endorphins: a growing family of pharmacologically pertinent peptides. Ann. Rev. Pharmacol.
23 (1983) 151–170.

Bloom, S. W.: Power and dissent in the medical school.
Free Press, New York 1973.

Bloom, S. W.: The process of becoming a physician and
the context of medical education. In: Noack, H. (ed.):
Medical Education and Primary Health Care, pp. 144–160.
London 1980.

Bloom, S. W.: Structure and ideology in medical education: an analysis of resistance to change. J. Health and
soc. Beh. 29 (1988) 294–306.

Bloom, S. W., E. J. Speedling: The education of physicians:
training for what? In: Saladin, P., H. J. Schaufelberger,
P. Schlappi (Hrsg.): Festschrift zur Emeritierung von
Prof. H. G. Pauli. Bern 1989.

Bloom, V.: Functions of a liaison psychiatrist in a kidney
center. Dial. Transplant. 10 (1981) 51–56.

Blos, P.: Adoleszenz. Klett-Cotta, Stuttgart 1978.

Bluestone, H.: DSM-III und die Psychoanalyse. Forum
Psychoanal. 1 (1985) 157–160.

Blum, A. L.: Der Einfluß von Tagamet auf Diagnose und
Therapie des Ulcus. Gespräch. Swiss Med. 4 (1982) 7–9.

Blumenthal, J. A., R. B. Williams, Y. Kong, S. M. Shanberg,
L. W. Thompson: Type A behavior pattern and coronary
atherosclerosis. Circulation 58 (1978) 634–639.

Blumenthal, J. A., R. S. Williams, R. B. Williams et al.: Effects of exercise on the Type A (coronary-prone) behavior pattern. Psychosom. Med. 42 (1980) 289–320.

Bockus, H. L.: Present status of chronic regional or cica-
trizing enteritis. J. Amer. med. Ass. 127 (1945) 449–456.

Bockus, H. L., J. Bank, A. A. Wilkinson: Neurogenic mucous colitis. Amer. J. M. Soc. 176 (1928) 813–829.

Bockus, H. L., J. L. A. Roth, E. Buchman, M. Kalser, W. R.
Staub, A. Finkelstein, A. Valdes-Dapena: Life history of
non-specific ulcerative colitis: relation of prognosis to
anatomical and clinical varieties. Gastroenterologia 86
(1956) 549–581.

Bodamer, J.: Sexualität und Liebe. Furche, Hamburg 1970.

Bodenheimer, A. R.: Gibt es eine ophthalmologische Psychosomatik? Brauchen wir eine? Ophthalmologica 149
(1965) 424–437.

Bodman, F.: The psychologic background of colitis. Amer.
J. med. Sci. 190 (1935) 535–545.

Böddecker, K.-W., M. Böddecker: Verhaltenstherapeutische Ansätze bei der Behandlung des endogenen Ekzems unter besonderer Berücksichtigung des zwanghaften Kratzens. In: Bosse, K., P. Hünecke (Hrsg.): Psychodynamik und Soziodynamik bei Hautkranken. Vandenhoeck & Ruprecht, Göttingen 1976.

Böhler, U.: Gestaltungstherapie. In: Schepank, H., W.
Tress (Hrsg.): Die stationäre Psychotherapie und ihr
Rahmen. Springer, Berlin 1988.

Böhme, G.: Stimm-, Sprech- und Sprachstörungen. Fischer, Stuttgart 1974.

Böhme, G. (Hrsg.): Therapie der Sprach-, Sprech- und
Stimmstörungen. Fischer, Stutgart 1980.

Böhme, K.: Psychiatrische Aufgaben und Erfahrungen auf
einer internistischen Intensivstation. Internist 22 (1981)
32–38.

Böhme-Bloem, C., M. J. Schulte: Bulimie: unterschiedliche
Psychogenese, Symptomwahl und Therapie. In: Speidel,
H., B. Strauß (Hrsg.): Zukunftsaufgaben der psychoso-
matischen Medizin. Springer, Berlin 1989.

Böning, J.: Klinik und Psychopathologie von Ohrgeräu-
schen aus psychiatrischer Sicht. Z. Laryng. Rhinol. 60
(1981) 101.

Bönisch, E., G. E. Meyer: Medizinische Extremsituationen
und der sterbende Patient. In: Kisker, K. P., J.-E. Meyer,
C. Müller, E. Strömgren (Hrsg.): Psychiatrie der Gegenwart, 2. Aufl. Springer, Berlin 1972.

Böök, J. A., L. Wetterberg, K. Modrzewka: Schizophrenia
in a North Swedish geographical isolate 1900–1977:
Epidemiology, genetics and biochemistry. Clin. Genetics
14 (1978) 373–394.

Boethins, J., U. Lindblom, B. A. Meyerson, L. Widen: Effects of multifocal brain stimulation on pain and soma-
tosensory functions. In: Zottermann, Y. (ed.): Sensory
Functions of the Skin. pp. 531–548. Pergamon Press,
Oxford 1976.

Bötticher, H. R.: Langzeitprognose von M. Crohn und
Colitis ulcerosa. Fortschr. Med. 95 (1977) 1623–1624.

Bötticher, H. R., C. Köhle, H. Matthies, C. Wunderlich:
Personale und interpersonale Momente bei psychoso-
matischen Krankheiten, dargestellt an den Beispielen
Ulcus duodeni und Colitis ulcerosa. In: Probleme und
Ergebnisse der Psychologie 75, S. 59–78. Jena 1980.

Bojanovsky, A., B. Schöninger, H. G. Kugler, J. Bojanovsky: Persönlichkeitsstruktur und „life events" bei Psoriatikern. Akt. Derm. 7 (1981) 17–19.

Bolland, G.: Experimentelle Untersuchung über die psychische Beeinflußbarkeit der Nierenfunktion in Hypnose. Z. Psychother. 7 (1957) 109.

Bolm, G., H. Jaekel, U. Holtschoppen: Gruppenpsychotherapie von Dialysepatienten. Psychother. med. Psychol.
29 (1979) 105–112.

Bolton, H. E., C. P. Bailey: Psychosomatic aspects of surgery. Grune & Stratton, New York 1956.

Bond, A. J., D. C. James, M. H. Lader: Physiological and
psychological measures in anxious patients. Psychol.
Med. 4 (1974) 364–373.

Bond, M. P., J. S. Vaillant: An empirical study of the relationship between diagnosis and defense style. Arch. gen.
Psychiat. 43 (1986) 285–288.

Bondoulas, H., H. S. Schmidt, R. W. Clark, P. Geleris, S. F.
Schaal, R. P. Lewis: Anthropometric characteristics, cardiac abnormalities and adrenergic activity in patients
with primary disorders of sleep. J. Med. 14 (1983)
223–238.

Bongartz, B., W. Bongartz: Hypnose: Wie sie wirkt und
wem sie hilft. Kreuz, Zürich 1988.

Bongartz, W.: Abnahme von Plasmacortisol und weißen Blutzellen nach Hypnose. Eine Pilotstudie. Exp. klin. Hypnose 2 (1986) 101–108.

Bonhoeffer, K.: Die Psychosen im Gefolge von akuten Infektionen, Allgemeinerkrankungen und innerer Erkrankungen. In: Aschaffenburg, G. (Hrsg.): Handbuch der Psychiatrie, Bd. 8. Deuticke, Leipzig 1912.

Bonica, J. J.: Importance of the problem. In: Bonica, J. J., V. Ventafridda (eds.): Advances in Pain Research and Therapy, Vol. 2. Raven, New York 1979.

Bonica, J. J.: Introduction to management of pain of advanced cancer. In: Bonica, J. J., V. Ventafridda (eds.): Advances in Pain Research and Therapy, Vol. 2, pp. 115–130. Raven, New York 1979.

Bonnet, M. H.: Performance and sleepiness as a function of frequency and placement of sleep disruption. Psychophysiol. 23 (1986a) 263–271.

Bonnet, M. H.: Performance and sleepiness following moderate sleep disruption and slow wave sleep deprivation. Psychol. Beh. 37 (1986b) 915–918.

Bonshey, H. A.: Neural mechanisms in asthma. In: Weiner, H., M. A. Hofer, A. J. Stunkard (eds.): Brain, Behaviour and Bodily Disease, pp. 27–44. Raven, New York 1981.

Booth, G.: Organ function and form perception. Use of the Rorschach method with cases of chronic arthritis, parkinsonism and arterial hypertension. Psychosom. Med. 8 (1946) 367–385.

Booth, G.: Psychodynamics in parkinsonism. Psychosom. Med. 10 (1948) 1–14.

Booth, G. C.: Personality and chronic arthritis. J. nerv. ment. Dis. 85 (1937) 637–662.

Booth, I. W., J. T. Harries: Inflammatory bowel disease in childhood. Gut 25 (1984) 188–202.

Booth-Kewley, S., H. S. Friedman: Psychological predictors of heart disease: a quantitative review. Psychol. Bull. 101 (1987) 343–362.

Bootzin, R. R., P. M. Nicassio: Behavioral treatments for insomnia. In: Hersen, M., R. M. Eisler, P. M. Miller (eds.): Progress in Behavioral Modification, Vol. 6, pp. 1–45. Acad. Press, New York 1978.

Boranic, M., D. Pericic, M. Radacic, M. Poljak-Blazi, V. Sverko, G. Miljenovic: Immunological and neuroendocrine responses of rats to prolonged or repeated stress. Biomed. 36 (1982) 23–28.

Borel, J. F., C. Feurer, H. U. Gubler, H. Stähelin: Biological effects of cyclosporin A: a new antilymphocytic agent. Agents Actions 6 (1976) 468–475.

Borelli, S.: Psyche und Haut. In: Jadassohn, J.: Handbuch der Haut- und Geschlechtskrankheiten, Bd. VIII. Hrsg. von H. A. Gottron. Springer, Berlin-Heidelberg-New York 1967.

Borelli, S.: Potenz und Potenzstörungen des Mannes. Hartmann, Berlin 1971.

Borelli, S., U. W. Schnyder: Neurodermitis constitutionalis sive atopica. In: Jadassohn, J.: Handbuch der Haut- und Geschlechtskrankheiten, Bd. II/1. Hrsg. von A. Marchionini. Springer, Berlin-Heidelberg-Göttingen 1962.

Borgers, N., B. Nemitz: Bedingungen werksärztlicher Tätigkeit und das Arbeitssicherheitsgesetz. In: Haug, W. F. (Hrsg.): Jahrbuch für kritische Medizin 3, S. 116–130. Argument Verlag, Berlin 1978.

Borgert, A., L. R. Schmidt: Präoperative Bewältigungsprozesse bei Hysterektomie-Patientinnen. Psychother. med. Psychol. 38 (1988) 288–293.

Bortner, R. W.: A short rating scale as a potential measure of pattern A behavior. J. chron. Dis. 22 (1969) 87–91.

Bortner, R. W., R. H. Rosenman: The measurement of pattern A behavior. J. chron. Dis. 20 (1967) 525–533.

Boss, M.: Die Blutdruckkrankheit als menschliches Problem. Psyche 2 (1949) 499–517.

Bosse, K.: Therapeutischer Erfolg – post hoc oder propter hoc? In: Mahrle, S., H. Ippen (Hrsg.): Dermatologische Therapie. Perimed, Erlangen 1985.

Bosse, K., A. T. Teichmann: Die Tätowierung – Motiv und Motivation. Kosmetologie 5 (1972) 174–177.

Bosse, K., A. T. Teichmann: Haar und Persönlichkeit. Kosmetologie 4 (1973) 132–135.

Bosse, K., P. Hünecke: Der Juckreiz des endogenen Ekzematikers. Münch. med. Wschr. 123 (1981) 1013–1016.

Bosse, K., P. Hünecke: Psychosomatic aspects of psoriasis. In: Current Research Problems in Psoriasis. Grosse, Berlin 1984.

Bosse, K., P. Hünecke, R. Nordhausen: Zum „Krankheitsgefühl" Hautkranker. Ärztl. Kosmetologie 8 (1978) 228–238.

Bosse, K., P. Faßheber, P. Hünecke, A. T. Teichmann, J. Zauner: Zur sozialen Situation des Hautkranken als Phänomen interpersoneller Wahrnehmung. Z. psychosom. Med. Psychoanal. 21 (1976) 3–61.

Boszormenyi-Nagy I., G. M. Spark: Invisible loyalties. Harper & Row, New York 1973. Deutsch: Unsichtbare Bindungen. Klett-Cotta, Stuttgart 1981.

Boudier, P.: La céphalée de l'enfant ou contribution à l'étude des états prémorbides de l'enfance. Rev. fr. Psychanal. 26 (1962) 633–654.

Bourne, H. R., L. M. Lichtenstein, K. M. Melmon: Modulation of inflammation and immunity by cyclic AMP. Science 184 (1974) 9–28.

Bovbjerg, D.: Interleukin-1 induces taste aversion as well as decreased food and water consumption in mice. Psychosom. Med. 50 (1988) 196.

Bovbjerg, D., N. Cohen, R. Ader: The central nervous system and learning – a strategy for immune regulation. Immunol. Today 3 (1982) 287–291.

Bovbjerg, D., R. Ader, N. Cohen: Acquisition and extinction of conditioned suppression of a graft-vs.-host response in the rat. J. Immunol. 132 (1984) 111–113.

Bovbjerg, D., N. Cohen, R. Ader: Behaviorally conditioned enhancement of delayed-type hypersensitivity in the mouse. Brain, Behavior, and Immunity 1 (1987a) 64–71.

Bovbjerg, D., Y. T. Kim, G. W. Siskind, M. E. Weksler: Conditioned suppression of plaque-forming cell responses with cyclophosphamide: the role of taste aversion. Ann. N. Y. Acad. Sci. 496 (1987b) 588–594.

Bovensiepen, R., R. Oesterreich, K. Wilhelm, M. Arndt: Die elterliche Erziehungseinstellung als Ausdruck der Familiendynamik bei Kindern mit Asthma bronchiale. Prax. Kinderpsych. Kinderpsychiatr. 29 (1980) 163.

Bowers, K. S.: Situationism in psychology: an analysis and a critique. Psychol. Rev. 80 (1973) 307–336.

Bowers, M., E. N. Jackson, J. A. Knight, L. LeShan: Wie können wir Sterbenden beistehen? Kaiser, München 1971.

Bowers jr., M. B.: CSF acid monoamine metabolites in psychotic syndromes: what might they signify? Biol. Psychiat. 13 (1978) 375–383.

Bowers jr., M. B.: Plasma monoamine metabolites in psychotic disorders. Arch. gen. Psychiat. 45 (1988) 595–596.

Bowers jr., M. B., M. E. Swigar: Acute psychosis and plasma catecholamine metabolites. Arch. gen. Psychiat. 44 (1987) 190.

Bowlby, J.: Maternal care and mental health. WHO, Geneva 1952.

Bowlby, J.: Separation. Penguin, Harmondsworth 1972.

Bowlby, J. (Hrsg.): Mutterliebe und kindliche Entwicklung. Reinhardt, München-Basel 1972.

Bowlder, J., B. Kossmann: Drug therapy and chronic headache. In: Holroyd, K. A., B. Schlote, H. Zenz (eds.): Perspectives in Research on Headache, pp. 115–125. Hogrefe, Lewiston-New York 1983.

Literatur

Boyar, R., J. Finkelstein, H. Roffwarg, S. Kapen, E. Weitzman, L. Hellman: Synchronization of augmented luteinizing hormone secretion with sleep during puberty. New Engl. J. Med. 287 (1972) 582–586.

Boyd, W.: The spontaneous regression of cancer, Vol. III. Thomas, Springfield/Ill. 1966.

Boyer, W., B. Chernow, C. R. Lake: Psychopharmacology in the intensive care unit. Psychiat. Clin. N. Amer. 7 (1984) 1–7.

Boyle, C. M.: Differences between doctor's and patient's interpretations of some common medical terms. Brit. med. J. 2 (1970) 286–289.

Bracken, M. B., M. Bernstein: Adaptation to and coping with disability one year after spinal cord injury: an epidemiological study. Soc. Psychiat. 15 (1980) 33–41.

Bradley, J. J.: Severe localized pain associated with the depressive syndrome. Brit. J. Psychiat. 109 (1963) 741–745.

Brady, J. V.: Ulcer in „executive" monkeys. Sci. Amer. 199 (1958) 95–100.

Brady, J. V.: Psychophysiology of emotional behavior. In: Bachrach, A. J. (ed.): Experimental Foundations of Clinical Psychology. Basic Books, New York 1962.

Brady, J. V.: Towards a behavioral biology of emotion. In: Levi, L. (ed.): Emotions: Their Parameters and Measurement. Raven, New York 1975.

Brady, J. V.: Conditioning and emotion. In: Levi, L. (ed.): Emotions: Their Parameters and Measurement. Raven, New York 1975.

Brady, J. V., D. V. Kelly, C. Plumcee: Autonomic and behavioral responses of the rhesus monkey to emotional conditioning. Ann. N. Y. Acad. Sci. 150 (1969) 959–975.

Brady, J. V., R. W. Porter, D. G. Conrad, J. W. Mason: Avoidance behavior and the development of gastrointestinal ulcers. J. exp. Anal. Behav. 1 (1958) 69–73.

Brähler, E., J. Scheer: Der Gießener Beschwerdebogen (GBB). Huber, Bern 1983.

Brähler, E., K. Möhlen: Psychodiagnostische Prädiktoren für die postoperative Prognose des Zwölffingerdarmgeschwürs. Psychother. med. Psychol. (1988) 153–158.

Bräunig, P., J. Bleistein: Kortisoninduzierte Psychosen. Nervenarzt 59 (1988) 596–602.

Bräutigam, W.: Zur epileptischen Wesensänderung. Psyche 5 (1951/52) 523–544.

Bräutigam, W.: Beitrag zur Psychosomatik der Lungentuberkulosen. Fortschr. Tuberk.-Forsch. 7 (1956) 184.

Bräutigam, W.: Beobachtungen zur Erkrankungssituation und zur Psychotherapie bei Lungentuberkulosen. Z. Psychother. med. Psychol. 7 (1957) 104.

Bräutigam, W.: Grundlagen und Erscheinungsweisen des Torticollis spasticus. Nervenarzt 25 (1964) 451–462.

Bräutigam, W.: Typus, Psychodynamik und Psychotherapie herzphobischer Zustände. Z. psychosom. Med. 10 (1964) 276–285.

Bräutigam, W.: Sexualmedizin im Grundriß. Thieme, Stuttgart 1977.

Bräutigam, W., P. Christian: Psychosomatische Medizin. Thieme, Stuttgart 1975.

Bräutigam W., M. v. Rad, K. Engel: Erfolgs- und Therapieforschung bei psychoanalytischen Behandlungen. Psychosom. Med. Psychoanal. 26 (1980) 101–118.

Braid, J.: Neurohypnology or the rationale of nervous sleep considered in relation with animal magnetism. Churchill, London 1843.

Brambilla, F., F. Facchinetti, F. Petraglia, L. Vanzulli, A. R. Genazzani: Secretion pattern of endogenous opioids in chronic schizophrenia. Amer. J. Psychiat. 141 (1984) 1183–1189.

Brammel, H. L., S. A. Niccoli: A physiological approach to cardiac rehabilitation. Nurs. Clinics N. Amer. 11 (1976) 223–236.

Brand, R.: Coronary-prone behavior as an independent risk for coronary heart disease. In: Dembroski, T. M., S. M. Weiss, J. L. Shields, S. G. Haynes, M. Feinleib (eds.): Coronary-Prone Behavior. Springer, New York 1978.

Brand, R. (1982): Eutonie und KBT – ein Methodenvergleich. In: Stolze, H. (Hrsg.): Konzentrative Bewegungstherapie, S. 197. 2. Aufl. Springer, Berlin-Heidelberg-New York 1989.

Brand, R. J., R. H. Rosenman, R. J. Sholtz, M. Friedman: Multivariate prediction of coronary heart disease in the Western Collaborative Group Study to the findings of the Framingham Study. Circulation 53 (1976) 355–438.

Brand-Jacobi, I.: Die Klassifikation von Anorexia nervosa und Bulimia nervosa als Syndrome gestörten Eßverhaltens. Akt. Ernähr. 9 (1984) 18–22.

Brand-Jacobi, I., Th. Paul, V. Pudel: Symptomatologie der Bulimia nervosa: Eine Untersuchung an 500 betroffenen Frauen. Verh. Dtsch. Ges. Inn. Med. 90 (1984) 1080–1083.

Brandes, J.-W., F. Eulenburg: Der lange Weg zur Diagnose Morbus Crohn. Z. Gastroenterol. 14 (1976) 400–406.

Brandlmeier, P.: Die Allgemeinpraxis. Springer, Berlin-Heidelberg-New York 1974.

Brandt, L. J., L. H. Bernstein, S. J. Boley, M. S. Frank: Metronidazole therapy for perineal Crohn's disease: a follow-up study. Gastroenterology 83 (1982) 383–387.

Brandt, Th., J. Dichgans, H. C. Diener (Hrsg.): Therapie und Verlauf neurologischer Erkrankungen. Kohlhammer, Stuttgart 1988.

Brandtzaeg, P., K. Baklien: Immunopathology of the intestinal lesion in Crohn's disease. Z. Gastroenterol. 17 (Suppl.) (1979) 77–82.

Brandtzaeg, P., I. Fjellanger, S. Gjeruldsen: Human secretory immunoglobulins. 1. Salivary secretions from individuals with normal or low levels of serum immunoglobulins. Scand. J. Hematol. 12 (1970) 3–83.

Brantigan, C. O., T. A. Brantigan, N. Joseph: Effect of beta-blockade and beta-stimulation on stage fright. Amer. J. Med. 72 (1982) 88–94.

Brashear, R. E.: Hyperventilation syndrome. Lung 161 (1983) 257–273.

Braun, K. H.: Kritik des Freudo-Marxismus. Köln 1979.

Braun, R. N.: Die Allgemeinpraxis als Zeitfaktor. Dtsch. med. Wschr. 88 (1965) 2084.

Braun, R. N.: Lehrbuch der ärztlichen Allgemeinpraxis. Urban & Schwarzenberg, München 1970.

Braun-Falco, O., F. Deinhardt, D. Goebel (Hrsg.): AIDS. Leitlinien für die Praxis. MMV Medizin-Verlag, München 1987.

Braus, H.: Anatomie des Menschen (fortgeführt von C. Elze), Bd. I, S. 15–16. 3. Aufl. Springer, Berlin 1954.

Braverman, H.: Die Arbeit im modernen Produktionsprozeß. Campus, Frankfurt-New York 1977.

Breckwoldt, M.: Klimakterium: Biologische Grundlagen und klinische Aspekte. In: Richter, D., M. Stauber (Hrsg.): Psychosomatische Probleme in Geburtshilfe und Gynäkologie. Kehrer, Freiburg 1983.

Brede, K.: Sozioanalyse psychosomatischer Störungen. Athenäum, Frankfurt 1972.

Brede, K.: Sozialpsychologische Modellvorstellungen. In: Balmer, H. (Hrsg.): Psychologie des 20. Jahrhunderts, Bd. IX: Ergebnisse für die Medizin 1, S. 274–291. Zürich 1979.

Bregulla, S.: Asconauten und Ackerbauern. In: Pöldinger, H. (Hrsg.): Beziehungdiagnostik und Beziehungtherapie. Springer, Heidelberg 1983.

Breier, A., J. P. Kelsoe jr., P. D. Kirwin, S. A. Beller, O. M. Wolkowitz, D. Pickar: Early parental loss and development of psychopathology. Arch. gen. Psychiat. 45 (1988) 987–993.

Breitwieser, P., O. Sareyka: Häufigkeit psychosomatischer Fälle in der urologischen Praxis. Urologe B 21 (1981) 14.

Brener, J.: A general model of voluntary control applied to the phenomena of learned cardiovascular change. In: Obrist, P. A., A. H. Black, J. Brener, L. V. Dicara (eds.): Cardiovascular Psychophysiology. Aldine, Chicago 1974.

Brenner, C., A. P. Friedman, S. Carter: Psychologic factors in the etiology and treatment of chronic headache. Psychosom. Med. 11 (1949) 53–56.

Brenner, H.: Themenzentrierte Gespräche und Verhaltensübungen zur Einstellungs- und Verhaltensänderung von Herzinfarktpatienten. In: Langosch, W. (Hrsg.): Psychosoziale Probleme und psychosoziale Interventionsmöglichkeiten bei Herzinfarktpatienten, S. 201–226. Minerva, München 1980.

Brenner, H.: Herzinfarkt und Selbstverantwortung. Präventionen 4 (1981) 13–18.

Brenner, H. D.: Zur Effizienz von Psychotherapie. Münch. med. Wschr. (1978) 78–86.

Brenner, M. H., A. Mooney, T. J. Nagy: Assessing the contributions of the social sciences to health. AAAS Symposium. Westview, Boulder/Colorado 1980.

Breslau, N., G. C. Davis: Chronic stress and major depression. Arch. gen. Psychiat. 43 (1986) 309–314.

Breuer, J., S. Freud: Studien über Hysterie. Deuticke, Leipzig-Wien 1895.

Breuer, J., S. Freud: Zur Theorie des hysterischen Anfalles. Int. Z. Psychoanal. 25 (1940) 107–110.

Brickenkamp, R. (Hrsg.): Handbuch psychologischer und pädagogischer Tests. Hogrefe, Göttingen 1975.

Bridges, K. W., D. P. Goldberg: Somatic presentation of DSM-III psychiatric disorders in primary care. J. psychosom. Res. 29/6 (1985) 563–569.

Brindley, G. S.: Cavernosal alpha-blockade. Brit. J. Psychiat. 143 (1981) 332.

Broadbent, D. E.: Perception and communication. Pergamon, New York 1958.

Broadbent, D. E.: Decision and stress. Acad. Press, New York 1971.

Broca, P.: Anatomie comparée des circonvolutions cérébrales. Le grand lobe limbique et la scissure limbique dans la série des mammifères. Rev. anthrop. 1 (1878) 385.

Brod, J., V. Fencl, A. Hejl, J. Jirka: Circulatory changes underlying blood pressure elevation during acute emotional stress in normotensive and hypertensive subjects. Clin. Sci. 18 (1959) 269–279.

Brodie, D. A., H. M. Hanson: A study of factors involved in the production of gastric ulcers by the restraint technique. Gastroenterol. 38 (1960) 353–360.

Brodsky, M., D. Wu, P. Denes et al.: Arrhythmias documented by 24-hour continuous electrocardiographic monitoring in 50 male medical students. Amer. J. Cardiol. 39 (1977) 390.

Brody, H.: Placebos and the philosophy of medicine. Univ. of Chicago Press, Chicago 1977.

Brogren, C. H., A. Lernmark: Islet cell antibodies in diabetes. In: Johnston, D. G., K. G. M. M. Alberti (eds.): Clinics in Endocrinology and Metabolism 11 (1982) 409–430.

Brolley, M., M. H. Hollender: Psychological problems of patients with myasthenia gravis. J. nerv. ment. Dis. 122 (1955) 178–184.

Brooke, S. T., B. C. Long: Efficiency of coping with a real life stressor: a multimodal comparison of aerobic fitness. Psychophysiol. 24 (1987) 173–180.

Brooks, S. M., C. F. Sanborn, B. H. Albrecht, W. W. Wagner: Diet in athletic amenorrhea. Lancet I (1984) 559.

Broström, O.: The role of cancer surveillance in long-term prognosis of ulcerative colitis. Scand. J. Gastroenterol. 88 (Suppl.) (1983) 40–42.

Brotman, A. W., D. B. Herzog, P. Hamburg: Long-term course in 14 bulimic patients treated with psychotherapy. J. clin. Psychiat. 49 (1988) 157–160.

Brow, G. R., C. N. H. Long, J. Beattie: Irregularities of the heart under chloroform: their dependence on the sympathetic nervous system. J. Amer. med. Ass. 95 (1930) 715.

Brown, A. R.: Non-gangrenous ischemic colitis. Brit. J. Surg. 59 (1972) 463–473.

Brown, G. M.: Endocrine alterations in anorexia nervosa. In: Darby, P. L., P. E. Garfinkel, D. M. Garner, D. V. Coscina (eds.): Anorexia nervosa. Recent Developments in Research, pp. 231–247. Liss, New York 1983.

Brown, G. M., S. Reichlin: Psychologic and neural regulation of growth hormone secretion. Psychosom. Med. 34 (1972) 45.

Brown, G. W., T. Harris: Social origins of depression. Tavistock, London 1978.

Brown, G. W., T. Harris: Establishing causal links: the Bedford College studies of depression. In: Katschnig, H. (ed.): Life Events and Psychiatric Disorders: Controversal Issues. Cambridge Univ. Press, Cambridge 1986.

Brown, G. W., J. L. T. Birley, J. K. Wing: Influence of family life on the course of schizophrenic disorder, a replication. Brit. J. Psychiat. 121 (1972) 241–258.

Brown, G. W., M. Ní Bhrolcháin, T. O. Harris: Social class and psychiatric disturbance among women in an urban population. Sociol. 9 (1975) 225–254.

Brown, G. W., Z. Adler, A. Bifulco: Life events, difficulties and recovery from chronic depression. Brit. J. Psychiat. 152 (1988) 487–498.

Brown, J. T., C. D. Mulrow, A. G. Stoudemire: The anxiety disorders. Ann. intern. Med. 100 (1984) 558–564.

Brown, M., L. Fisher: Brain peptides as intercellular messengers. J. Amer. med. Ass. 251 (1984) 1310–1315.

Brown, M., L. Fisher: Corticotropin-releasing factor: effects on the autonomic nervous system and visceral systems. Fed. Proc. 44 (1985) 243–248.

Brown, M., J. Rivier, W. Vale: Actions of bombesin, thyrotropin-releasing factor, prostaglandin E2 and naloxone on thermoregulation in the rat. Life Sci. 20 (1977) 1681–1687.

Brownell, K. B., J. H. Kelman, A. J. Stunkard: Treatment of obese children with and without their mothers: changes in weight and blood pressure. Pediatrics 71 (1983) 515–523.

Brownell, K. D., J. P. Foreyt (eds.): Handbook of eating disorders. Basic Books, New York 1986.

Browning, C. H., S. I. Miller: Anorexia nervosa: a study in prognosis and management. Amer. J. Psychiat. 124 (1968) 1128–1132.

Bruce, E. H., M. K. Edwards, R. Frederick, R. A. Bruce, T. H. Homes: Is coping with life stresses enhanced by cardiac rehabilitation programs? In: Stocksmeier, U. (ed.): Psychological Approach to the Rehabilitation of Coronary Patients, pp. 75–84. Springer, Berlin-Heidelberg-New York 1976.

Bruce, T.: Emotional sequelae of chronic inflammatory bowel disease in children and adolescents. Clin. Gastroenterol. 15 (1986) 89–104.

Bruch, H.: Psychological aspects of diabetic children. In: Oberdisse, K., K. Jahnke (eds.): Diabetes mellitus. Thieme, Stuttgart 1959.

Bruch, H.: Death in anorexia nervosa. Psychosom. Med. 33 (1971) 135–144.

Bruch, H.: Eating disorders: obesity, anorexia nervosa and the patient within. Basic Books, New York 1973.

Bruch, H.: Perils of behavior modification in treatment of anorexia nervosa. J. Amer. med. Ass. 230 (1974) 1419–1422.

Bruch, H.: Der goldene Käfig. Das Rätsel der Magersucht. Fischer, Frankfurt 1980.

Literatur

Bruch, H.: Anorexia nervosa: therapy and theory. Amer. J. Psychiat. 139 (1982) 1531–1538.

Bruch, H.: Four decades of eating disorders. In: Garner, D. M., P. E. Garfinkel (eds.): Handbook of Psychotherapy for Anorexia Nervosa and Bulimia. Guilford, New York 1985.

Bruggemann, A.: Erfahrungen mit wichtigen Variablen und einigen Effekten beruflicher Sozialisation in einem Projekt zur Humanisierung des Arbeitslebens. In: Groskurth, P. (Hrsg.): Arbeit und Persönlichkeit, S. 146–175. Rowohlt, Reinbek 1979.

Bruhn, J. G.: Psychological predictors of sudden death in myocardial infarction. J. psychosom. Res. 18 (1974) 187–191.

Bruhn, J. G., B. C. Chandler, S. Wolf: A psychological study of survivors and nonsurvivors of myocardial infarction. Psychosom. Med. 31 (1969) 8–19.

Bruhn, J. G., A. E. Thurman, B. C. Chandler: Patients' reactions to death in a coronary care unit. J. psychosom. Res. 14 (1970) 65–70.

Bruhn, J. G., A. Paredes, C. A. Adsett, S. Wolf: Psychological predictors of sudden death in myocardial infarction. J. psychosom. Res. 18 (1974) 187–191.

Bruhn, J. G., B. Chandler, M. C. Miller, J. Wolf, T. N. Lynn: Social aspects of coronary heart disease in two adjacent, ethnically different communities. Amer. J. publ. Health 56 (1966) 1493–1506.

Brullemann, C. H.: Group-therapy with epileptic patients at the „Instituut voor Epilepsiebestrijding". Epilepsia 13 (1972) 225–231.

Brun, R.: Sigmund Freuds Leistungen auf dem Gebiete der organischen Neurologie. Schweiz. Arch. Neurol. Psychiat. 37 (1936) 200–207.

Brun, R.: Über Freuds Hypothese vom Todestrieb. Psyche 7 (1953/54) 81–111.

Brundig, P., W. Berg, H.-J. Schneider: Streß und Harnsteinbildungsrisiko. II. Der Einfluß von Streß auf litholytische Harnsubstanzen. Urol. int. 34 (1979) 265.

Brundig, P., W. Berg, H.-J. Schneider: Untersuchungen zum Bildungsrisiko von Kalzium-Oxalat-Harnsteinen unter besonderer Berücksichtigung von Streßmomenten. Urol. int. 34 (1979) 105.

Bruner, J. S.: Über kognitive Entwicklung. In: Bruner, J. S., R. R. Olver, P. M. Greenfield (Hrsg.): Studien zur kognitiven Entwicklung, S. 33. Klett, Stuttgart 1971.

Bruni, B.: Diabetes and psychic factors: Proceedings of the 6th Congress of the Intern. Diabetes Federation, Stockholm 1967.

Brunson, B. J., K. A. Matthews: The Type A coronary-prone behavior pattern and reactions to uncontrollable events: an analysis of learned helplessness. J. Pers. soc. Psychol. 40 (1981) 906–918.

Bruntsch, U., W. M. Gallmeier: Schmerztherapie im fortgeschrittenen Krebsstadium. Münch. med. Wschr. 112 (1980) 7–9.

Brusis, O. A., H. Weber-Falkensammler (Hrsg.): Handbuch der Koronargruppenbetreuung. Perimed, Erlangen 1986.

Buchanan, D. C.: Group therapy for kidney transplant patients. Int. J. Psychiat. Med. 6 (1975) 523–531.

Buchanan, D. C.: Psychotherapeutic intervention in the kidney transplant service. In: Levy, N. B.(ed.): Psychonephrology 1, pp. 265–278. Plenum, New York 1981.

Buchborn, E.: Die Medizin und die Wissenschaften vom Menschen, Eröffnungsansprache. In: Lasch, H. G., B. Schlegel (Hrsg.): Hundert Jahre Deutsche Gesellschaft für Innere Medizin, S.957–971. Bergmann, München 1982.

Buchborn, E.: Ergebnisse der Psychotherapieforschung bei psychosomatischen Erkrankungen. Der Internist 25 (1984) 674–681.

Buchinger, B., W. Schüffel: Entwicklung eines Untersuchungsinstrumentes zur Beurteilung des ärztlichen Gespräches. Verh. Dtsch. Ges. Inn. Med. Bergmann, München 1983.

Buchmann, P., J. Alexander-Williams: Classification of perianal Crohn's disease. Clin. Gastroenterol. 9 (1980) 323–330.

Buckingham, R. W., S. A. Lack, B. M. Mount, L. D. McLean, J. T. Collins: Living with the dying: use of the technique of participant observation. Canad. med. Ass. J. 115/12 (1976) 1211–1215.

Buddeberg, B., C. Buddeberg: Familientherapie bei Anorexia nervosa. Prax. Kinderpsych. Kinderpsychiatr. 28 (1979) 37.

Buddeberg, C.: Möglichkeiten und Grenzen der Verhaltenstherapie in der Behandlung schwer wesensveränderter Epileptiker. Nervenarzt 46 (1975) 447–452.

Buddeberg, C.: Ehen krebskranker Frauen. Eine prospektive Untersuchung über familiäre Auswirkungen eines Mammakarzinoms. Urban & Schwarzenberg, München 1984.

Buddeberg, C.: Sexuelle Probleme von Krebspatienten. Münch. med. Wschr. 126 (1984) 225–226.

Buddeberg, C.: Sexualberatung. Enke, Stuttgart 1987.

Budzynski, T., J. Stoyva, C. Adler: Feedback-induced muscle relaxation: application to tension headache. J. behav. exper. Psychiat. 1 (1970) 205–211.

Budzynski, T. H., J. M. Stoyva, C. S. Adler, D. M. Mullaney: EMG biofeedback and tension headache: A controlled outcome study. Psychosom. Med. 35 (1973) 484–496.

Bühler, F. R., A. S. De Lèche, G. Schüler, F. Gutzwiller, F. Baumann, W. Schweizer: Das Hypertonieproblem in der Schweiz. Schweiz. med. Wschr. 106 (1976) 99–107.

Bürgin, D.: Das Kind, die lebensbedrohende Krankheit und der Tod. Huber, Bern 1978.

Bürgin, D.: Konversionsneurosen. Kassenarzt 22/44 (1982a) 5066–5078.

Bürgin, D.: Über einige Aspekte der pränatalen Entwicklung. In: Nissen, G. (Hrsg.): Psychiatrie des Säuglings- und frühen Kindesalters. Huber, Bern 1982b.

Bürgin, D.: Schlafstörungen in Kindheit und Adoleszenz. Münch. med. Wschr. 128/8 (1986a) 127–132.

Bürgin, D.: Entwicklungsstörungen in der Adoleszenz. Zbl. Jugendrecht 73 (1986b) 128–137.

Bürgin, D.: Die Bedeutung der affektiven Austauschvorgänge für den Aufbau des Selbst in der Kindheit. In: Rauchfleisch, U. (Hrsg.): Allmacht und Ohnmacht. Huber, Bern 1987.

Bürgin, D.: Psychische Störungen bei Kindern und Jugendlichen. Vorübergehende Episoden oder Fatum? In: Nissen, G. (Hrsg.): Prognose psychischer Krankheiten im Kindes- und Jugendalter. Huber, Bern 1987.

Bürgin, D.: Beziehungskrisen in der Adoleszenz. Huber, Bern 1988.

Bürgin, D.: Liaisonpsychiatrische Aspekte im Bereich der Kinder- und Jugendpsychiatrie. Schweiz. Rundsch. Med. 1990a (in press).

Bürgin, D.: Die Bedeutung der prä- und postnatalen Bewegungsentwicklung für die Etablierung des Mutter-Kind-Dialogs. In: Strassburg, H. M., M. Pachler (Hrsg.): Der unruhige Säugling. Fortschr. Soz. Päd. Hanse'sches Verlagskontor, Lübeck 1990b.

Bürke, H., V. Irrgang, E. Rüther: Psychopharmakotherapie der Angst. In: Strian, F. (Hrsg.): Angst – Grundlagen und Kritik. Springer, Berlin 1983.

Bugard, P.: Stress, fatigue, dépression. Vol. 2. Doin, Paris 1974.

Bulkley, B. H., W. C. Roberts: The heart in systemic lupus erythematosus and the changes included in it by corticosteroid therapy. A study of 36 necropsy patients. Amer. J. Med. 58 (1975) 243–264.

Bullinger, M., D. C. Turk: Selbstkontrolle: Strategien zur Schmerzbewältigung. In: Keeser, W., E. Pöppel, P. Mitterhusen (Hrsg.): Schmerz. Urban & Schwarzenberg, München 1982.

Bulloch, K.: Neuroanatomy of lymphoid tissue – a review. In: Guillemin, R., M. Cohn, T. Melnechuk (eds.): Neural Modulation of Immunity, pp. 111–142. Raven, New York 1985.

Bumke, O.: Lehrbuch der Geisteskrankheiten, 4. Aufl. Bergmann, München 1936.

Bundesfamilienministerium: Vierter Familienbericht: Die Situation der älteren Menschen in der Familie. Dtsch. Bundestag, 10. Wahlperiode, Drucksache 10/6145. BMJFFG, Bonn 1986.

Bundesminister für Arbeit und Sozialordnung (Hrsg.): Selbstbehandlung und Selbstmedikation medizinischer Laien. Bd. 67, 1981.

Bundesminister für Arbeit und Sozialordnung: Forschungsbericht zur Humanität im Gesundheitswesen. Bonn, 1983.

Bundesminister für Forschung und Technologie (Hrsg.): Wohnortnahe Versorgung von Rheumakranken. Bonn 1988.

Burack, B.: The hypersomnia-sleep apnea syndrome: its recognition in clinical cardiology. Amer. Heart J. 107 (1984) 543–548.

Burch, P.: Ischemic heart disease: epidemiology, risk factors and causes. Cardiovasc. Res. 14 (1980) 307–338.

Burchard, J. M.: Lehrbuch der systematischen Psychopathologie. Schattauer, Stuttgart 1980.

Burdett, C.: Personal and social gains and losses during treatment by dialysis and transplantation. Unpublished Diss., 1978.

Burish, T. G., M. P. Carey: Conditioned aversive responses in cancer chemotherapy patients: theoretical and developmental analysis. J. consult. clin. Psychol. 54 (1986) 593–600.

Burnam, M. A., J. W. Pennebaker, D. C. Glass: Time consciousness, achievement striving and the Type A coronary-prone behavior pattern. J. abn. Psychol. 84 (1975) 76–79.

Burnet, F. M.: Immunological surveillance. Pergamon, Oxford 1970.

Burnstock, G., T. Hökfelt (eds.): Non-adrenergic, non-cholinergic autonomic transmission mechanism. Neurosci. Res. Program Bull. 17/3 (1979).

Burrow, T.: The group method of analysis. Psychoanal. Rev. 14 (1926) 268.

Burstein, A. G. et al.: A longitudinal study of personality characteristics of medical students J. med. Educ. 55 (1980) 786–787.

Burton, A. F., J. M. Storr, W. L. Dunn: Cytolytic action of corticosteroids on thymus and lymphoma cells in vitro. Canad. J. Biochem. 45 (1967) 289–297.

Burton, C. V., C. D. Ray: Neurostimulation. In: De Vita jr., V. T., S. Hellman, S. A. Rosenberg (eds.): Cancer. Principles and Practice of Oncology, pp. 1670–1676. Lippincott, Philadelphia 1982.

Burton, G.: Nurse and patient. The influence of human relationships. Tavistock, London 1965.

Burton, H. J., L. Canzona, L. Way, R. R. Holden, J. Conley, R. M. Lindsay: Determinants for successful adaptation of patients on CAPD. In: Levy, N. B. (ed.): Psychonephrology 2. Plenum, New York 1983.

Buschard, K., C. Röpke, S. Madsbad, J. Mehlsen, J. Rygaard: T-lymphocyte subsets in patients with newly diagnosed type 1 (insulin-dependent) diabetes: a prospective study. Diabetologia 25 (1983) 247–251.

Bushfield, B. L., P. Schneller, D. Capra: Depressive symptom or side effect? A comparative study of symptoms during pre-treatment and treatment periods of patients on three antidepressant medications. J. nerv. ment. Dis. 134 (1962) 339–345.

Butollo, W.: Chronische Angst. Theorie und Praxis der Konfrontationstherapie. Lernpsychologische Konzepte zur Angst. Urban & Schwarzenberg, München 1979.

Buvat, J., L. Dehaene, A. Lemaire, M. Buvat-Herbaut: Arteriell bedingte erektile Impotenz. Sexualmedizin 12 (1983) 248.

Buvat, J., A. Lemaire, M. Buvat-Herbaut, P. Lepretre, J. C. Fourlinnie: Psychoneuroendocrine investigations: 115 cases of female anorexia nervosa at the time of their maximum emaciation. Int. J. Eat. Dis. 2 (1983) 117–128.

Byrne, D.: The repression-sensitization scale. J. Personal. 29 (1961) 334–339.

Byrne, D.: Repression-sensitization as a dimension of personality. In: Maher, B. (ed.): Progress in Experimental Personality Research. Acad. Press, New York 1964.

Byrne, D. G.: Psychological responses to illness and outcome after survived myocardial infarction. A follow-up. J. psychosom. Res. 26 (1975) 105–112.

Byrne, D. G.: Neuroticism, illness behavior and coronary heart disease. Psychother. and Psychosom. 26 (1975) 317–321.

Byrne, D. G.: Personal determinants of life event stress and myocardial infarction. Psychother. and Psychosom. 40 (1983) 106–114.

Byrne, D. G., H. M. Whyte: Life events and myocardial infarction revisited: the role of measures of individual impact. Psychosom. Med. 42 (1980) 1–10.

Cabot, R. C.: Social service and the art of healing. Moffart, Yard & Co., New York 1909.

Cade, B.: Strategic therapy. J. Fam. Therapy 2 (1980) 89.

Caffey, J., R. Silbey: Regrowth and overgrowth of the thymus after atrophy induced by the oral administration of adrenocorticosteroids to human infants: Benjamin Knox Rachford lecture. Pediatrics 26 (1960) 762–770.

Cahoon, D. D., A. J. Turner: Three hypotheses concerning the establishment and maintenance of psychosomatic processes. Beh. Ther. 2 (1971) 97–100.

Cake, M. H., G. Litwak: The glucocorticoid receptors. In: Litwak, G. (ed.): Biochemical Actions of Hormones, Vol. 3, pp. 317–390. Acad. Press, New York 1975.

Calabrese, L. H., M. R. Proffitt, S. Rehm, M. Lederman, J. T. Carey, H. B. Houser, K. Edmonds, J. J. Ellner: Lack of correlation between promiscuity and seropositivity to HTLV-III from a low-incidence area for AIDS. New Engl. J. Med. 312 (1985) 1256–1257.

Caldwell, T., M. F. Weiner: Stresses and coping in ICU-nursing, I. A review. Gen. Hosp. Psychiat. 3 (1981) 119–127.

Callahan, L. F., R. H. Brooks, T. Pincus: Further analysis of learned helplessness in rheumatoid arthritis using a „rheumatology attitudes index". J. Rheum. 15 (1988) 418–426.

Calland, C. H.: Iatrogenic problems in end-stage renal failure. New Engl. J. Med. 287 (1972) 334–336.

Callaway, E., R. S. Layne: Interaction between the visual evoked response and two spontaneous biological rhythms: the EEG-alpha cycle and the cardiac arousal cycle. Ann. N. Y. Acad. Sci. 112 (1964) 421–431.

Calsyn, D. A., J. Louks, C. W. Freeman: The use of the MMPI with low back patients with a mixed diagnosis. J. clin. Psychol. 32 (1976) 532–536.

Calsyn, D. A., D. J. Sherrad, C. W. Freeman: Vocational adjustment, psychological assessment and survival on hemodialysis. Trans. Amer. Soc. artif. intern. Org. 24 (1978) 125–126.

Campbell, D. R., B. K. Sinha: Brief group psychotherapy

with chronic hemodialysis patients. Amer. J. Psychiat. 137 (1980) 1234–1237.

Campbell, H. J.: Irrtum der Seele. Scherz, Bern 1973.

Campbell, P. C.: Suicides among cancer patients. Connecticut Health Bulletin 80 (1966) 207–212.

Campbell, T. L.: Family's impact on health: A critical review. Family Systems Medicine 4 (1986) 135–191.

Campbell, T. W.: Death anxiety on a coronary care unit. Psychosomatics 21 (1980) 127–136.

Cancro, R.: History and overview of schizophrenia. In: Kaplan, H. I., B. J. Sadock (eds.): Comprehensive Textbook of Psychiatry IV. Chapt. 15.1. Williams & Wilkins, Baltimore 1985.

Canning, C., R. L. Kane, R. Gray: Curricular change and medical student attitudes. AAMC-RIME Conference 1973, 24.

Cannon, W. B.: The James Lange theory of emotions: a critical examination and alternation. Amer. J. Psychol. 39 (1927) 106–124.

Cannon, W. B.: The mechanism of emotional disturbance of bodily functions. New Engl. J. Med. 198 (1928) 877–884.

Cannon, W. B.: Again the James Lange and the thalamic theories of emotion. Psychol. Rev. 38 (1931) 281–295.

Cannon, W. B.: The wisdom of the body. London 1939.

Cannon, W. B.: Bodily changes in pain, hunger, fear and rage. Branford, Boston 1953. Deutsch: Urban & Schwarzenberg, München 1975.

Cannon, W. B.: „Voodoo" death. Psychosom. Med. 19 (1957) 182.

Cannon, W. B., D. De la Paz: Emotional stimulation of adrenal secretion. Amer. J. Physiol. 27 (1911) 64.

Cannon, W. B., S. W. Britton: Studies on the conditions of activity in endocrine glands. The influence of motion and emotion on medulloadrenal secretion. Amer. J. Physiol. 79 (1926) 433–465.

Cannon, W. B., A. T. Shol, W. S. Wright: Emotional glycosuria. Amer. J. Physiol. 29 (1911) 280–287.

Canter, A.: Changes in mood during incubation of acute febrile disease and the effects of pre-exposure psychological status. Psychosom. Med. 34 (1972) 424–425.

Capponi, R., M. E. Kawada, C. Varela, L. Vargas: Diabetes mellitus by repeated stress in rats bearing chemical diabetes. Horm. metab. Res. 12 (1980) 411–412.

Captan, R. L., Th. Nadelson: The Oklahoma complex. A common form of conversion hysteria. Arch. intern. Med. 140 (1980) 185–186.

Carey, M. E.: A child's struggle for independence following kidney transplantation. Maternal Child Nurs. J. 5 (1976) 45–55.

Carey, M. P., T. G. Burish: Etiology and treatment of the psychological side effects associated with cancer chemotherapy: a critical review and discussion. Psychol. Bull. 104 (1988) 307–325.

Carey, R. G.: The widowed: a year later. J. couns. Psychol. 24 (1977) 125–131.

Carey, R. G., E. J. Posovac: Attitudes of physicians on disclosing information to and maintaining life for terminal patients. Omega 9 (1978) 67–77.

Carey, R. M., R. A. Reid, C. R. Ayers, S. S. Lunch, W. L. McLain, E. D. Vaughan: The Charlottesville blood pressure survey: value of repeated blood pressure measurements. J. Amer. med. Ass. 236 (1976) 847–851.

Carey, W. B.: Maternal anxiety and infantile colic. Is there a relationship? Clin. Pediat. 7 (1968) 590–595.

Carl, A., J. Fischer-Antze, H. Gaedtke, S. O. Hoffmann, W. Wendler (1982): Vergleichende Darstellung gruppendynamischer Prozesse bei Konzentrativer Bewegungstherapie und Analytischer Gruppentherapie. – Zugleich ein Versuch zur formalen Beschreibung dieser Prozesse. In:

Stolze, H. (Hrsg.): Konzentrative Bewegungstherapie, S.167. 2. Aufl. Springer, Berlin-Heidelberg-New York 1989.

Carleson, R., B. Fristedt, J. Philipson: Ulcerative colitis. A follow-up investigation of a 20-year primary material. Acta med. scand. 172 (1962) 647–656.

Carmelli, D., G. E. Swan, R. H. Rosenman: The relationship between wives' social and psychologic status and their husbands' coronary heart disease. Amer. J. Epidemiol. 122 (1985) 90–100.

Carol, W., W. Müller: Der akut entzündliche Adnexprozeß, seine Differentialdiagnose und Therapie. Dtsch. Gesundh.- Wesen 19 (1964) 854.

Carpentier, J., P. Cazamian: Nachtarbeit – ihre Auswirkungen auf Gesundheit und Wohlbefinden (1977). Rationalisierungs-Kuratorium der deutschen Wirtschaft (RKW), Eschborn 1981.

Carrobles, J. A. I., A. Cardona, J. Santacreu: Shaping and generalization procedures in the EMG-biofeedback treatment of tension headaches. Brit. J. clin. Psychol. 20 (1981) 49–56.

Carroll, B. J., M. Steiner: The psychobiology of premenstrual dysphoria: the role of prolactin. Psychoneuroendocrinology 4 (1978) 171.

Carroll, B. J., G. C. Curtis, J. Mendels: Neuroendocrine regulation in depression, I. Limbic system-adrenocortical dysfunction. Arch. gen. Psychiat. 33 (1976) 1039.

Carroll, B. J., G. C. Curtis, J. Mendels: Cerebrospinal fluid and plasma free cortisol concentrations in depression. Psychol. Med. 6 (1976) 235–244.

Carter, A. B.: Prognosis of certain hysterical symptoms. Brit. med. J. 1 (1949) 1076–1079.

Carter, W. R., J. Herrman, K. Stokes, D. J. Cox: Promotion of diabetes onset by stress in the BB rat. Diabetol. 30 (1987) 674–675.

Cartwright, A., L. Hockey, J. L. Anderson: Life before death. Routledge and Kegan Paul, London-Boston 1973.

Carver, C. S., D. C. Glass: Coronary-prone behavior pattern and interpersonal aggression. J. Pers. soc. Psychol. 36 (1978) 361–366.

Carver, C. S., A. E. Coleman, D. C. Glass: The coronary-prone behavior pattern and the suppression of fatigue on a treadmill test. J. Pers. soc. Psychol. 33 (1976) 460–466.

Carver, C. S., E. de Gregorio, R. Gillis: Challenge and Type A behavior pattern among intercollegiate football players. Sport Psychol. 3 (1981) 140–148.

Case, R. B., S. S. Heller: Letter to the editor – reply. New Engl. J. Med. 313 (1985) 451.

Case, R. B., S. S. Heller, N. B. Case, A. J. Moss: Multicenter Post-Infarction Research Group: Type A behavior and survival after acute myocardial infarction. New Engl. J. Med. 313 (1985) 737–741.

Casey, K. L.: The neurophysiologic basis of pain. Postgrad. Med. 53 (1973) 58–63.

Cashion, E. L., W. J. Lynch: Personality factors and results of lumbar disc surgery. Neurosurgery 4 (1979) 141–145.

Casper, R. C.: On the emergency of bulimia nervosa as a syndrome. A historical view. Int. J. Eat. Dis. 2 (1983) 3–16.

Casper, R. C.: Some provisional ideas concerning the psychologic structure in anorexia nervosa and bulimia. In: Darby, P. L., P. E. Garfinkel, D. M. Garner, D. V. Coscina (eds.): Anorexia Nervosa. Recent Developments in Research, pp. 387–392. Liss, New York 1983.

Casper, R. C., E. Redmond jr., M. N. Katz, C. B. Schaffer, V. M. Davis, S. H. Koslow: Somatic symptoms in primary affective disorder: presence and relationship to the classification of depression. Arch. gen. Psychiat. 42 (1985) 1098–1104.

Cassano, G. B., G. Perugi, D. M. McNair: Panic disorder:

review of the empirical and rational basis of pharmacological treatment. Pharmacopsychiat. 21 (1988) 157–165.

Cassel, J.: Physical illness in response to stress. In: Levine, S., A. Scotch (eds.): Social Stress, pp. 189–209. Chicago 1970.

Cassel, J.: Studies of hypertension in migrants. In: Oglesby, P. (ed.): Epidemiology and Control of Hypertension, pp. 41–48. Stratton, New York 1975.

Cassel's German Dictionary. Macmillan, London 1978.

Cassell, E. J.: Changing ideas of causality in medicine. Soc. Res. 46 (1979) 728–743.

Cassem, N. H., T. P. Hackett: Psychiatric consultation in a coronary care unit. Ann. intern. Med. 75 (1971) 9–14.

Cassem, N. H., T. P. Hackett: Sources of tension for the CCU nurse. Amer. J. Nurs. 72 (1972) 1426–1430.

Cassem, N. H., T. P. Hackett: The setting of intensive care. In: Cassem, N. H., T. P. Hackett (eds.): Massachusetts General Hospital Handbook of General Hospital Psychiatry, 2nd ed. PSG, Littleton 1987.

Cassem, N. H., T. P. Hackett, C. Buscom: Reactions of coronary patients to the CCU nurse. Amer. J. Nurs. 70 (1970) 319–324.

Cassileth, B. R., E. J. Lusk, T. B. Strouse, D. S. Miller et al.: Psychosocial status in chronic illness: a comparative analysis of six diagnostic groups. New Engl. J. Med. 311 (1984) 506–510.

Cassileth, B. R., E. J. Lusk, D. S. Miller, L. L. Brown, C. Miller: Psychosocial correlates of survival in advanced malignant disease? New Engl. J. Med. 312 (1985) 1551–1555.

Cassileth, B. R., D. S. Miller, C. Miller, E. J. Lusk, L. Brown: Reply to letters to the editor. New Engl. J. Med. 313 (1985) 1356.

Cassileth, B. R., W. P. Walsh, E. J. Lusk: Psychosocial correlates of cancer survival: a subsequent report 3 to 8 years after cancer diagnosis. J. clin. Oncol. 6 (1988) 1753–1759.

Castelnuovo-Tedesco, P.: Organ transplant, body image, psychosis. Psychoanal. Quart. 42 (1973) 349–363.

Castelnuovo-Tedesco, P.: Psychological consequences of physical defects. A psychoanalytic perspective. Int. Rev. Psychoanal. 8 (1981) 145–154.

Castelnuovo-Tedesco, P.: Transplantation: psychological implications of changes in body image. In: Levy, N. B. (ed.): Psychonephrology 1, pp. 219–226. Plenum, New York 1981.

Castillo-Ferrando, J. R., M. Garcia, J. Carmona: Digoxin levels and diazepam. Lancet II (1980) 368.

Castro, L., G. Gahl, M. Kessel: Ergebnisse eines Heimdialysetrainings- und Behandlungsprogramms. Dtsch. med. Wschr. 98 (1973) 641–646.

Catalano, R. B.: The medical approach to management of pain caused by cancer. Semin. Oncol. 2 (1975) 379–392.

Cath, S.: Some dynamics of middle and later years: a study in depletion and restitution. In: Berezin, M., S. Cath (eds.): Geriatric Psychiatry. Int. Univ. Press, New York 1965.

Cathomen-Rötheli, M.: Untersuchungen über die Persönlichkeitsstruktur an Myoarthropathie erkrankter Patienten. Schweiz. Mschr. Zahnheilk. 86 (1979) 29.

Cattell, J. McK.: Mental tests and measurement. Mind 15 (1890).

Cattell, R. B., H. W. Eber: The Sixteen Personality Factors Questionnaire. Inst. for Personality and Ability Testing, Champaign/Ill. 1964.

Caul, W. F., D. C. Buchnan, R. C. Hays: Effects of unpredictability of shock on incidence of gastric lesions and heart rate in immobilized rats. Physiol.Beh. 8 (1972) 669–672.

Cay, E. L.: Psychological problems in patients after a myocardial infarction. Advanc. Cardiol. 24 (1982) 108–112.

Cay, E. L., P. Dugard, A. E. Philip: Return to work after a heart attack. J. psychosom. Res. 17 (1973) 1–13.

Cay, E. L., A. E. Philip, C. Aitken: Psychological aspects of cardiac rehabilitation. In: Hill, O. W. (ed.): Modern Trends in Psychosomatic Medicine. Butterworth, London 1976.

Cay, E. L., H. Vetter, A. E. Philip: Psychological reactions to a CCU. J. psychosom. Res. 16 (1972) 437–447.

Cay, E. L., A. V. Gardner, N. Vetter, A. E. Philip: Rehabilitation after a heart attack; the team approach. Proc. 3rd Congr. Int. College Psychosom. Med., Rome 1975.

Cay, E. L., N. Vetter, A. E. Philip, P. Dugard: Psychological status during recovery from an acute heart attack. J. psychosom. Res. 16 (1972) 425–435.

Cazzullo, C., G. Invernizzi, R. Ventura et al.: Psychosomatic implications in chronic hemodialysis. Psychother. and Psychosom. 22 (1973) 341–346.

CDC: Human immunodeficiency virus infection in the United States: a review of current knowledge. MMWR 36 (1987) 1–48.

Cebelin, M. S., C. S. Hirsch: Human stress cardiomyopathy. Hum. Pathol. 11 (1980) 123–132.

Cerasi, E., R. Luft: Insulin secretion and the development of diabetes mellitus in the adult. Acta med. scand. (Suppl.) 601 (1976) 109–148.

Chafetz, M. E.: Psychological disturbances in myasthenia gravis. Ann. N. Y. Acad. Sci. 135 (1966) 424–427.

Chakraborty, R., W. J. Schull, E. Harburg, M. A. Schork, P. Roeper: Hereditary stress and blood pressure. A family set method. V. Heritability estimates. J. chron. Dis. 30 (1977) 683–699.

Chambers, M.: Psychological aspects of renal transplantation. Int. J. Psychiat. Med. 12 (1982) 229–236.

Chambers, W. N., M. Rosenbaum: Ulcerative colitis. Psychosom. Med. 15 (1953) 523–532.

Chang, K. J.: Opioid peptides have actions on the immune system. Trends in Neurosci. 7 (1984) 234–235.

Chang, Y. H., C. M. Pearson, C. Abe: Adjuvant polyarthritis. IV. Induction by a synthetic adjuvant: immunologic, histopathologic, and other studies. Arthr. and Rheum. 23 (1980) 62–71.

Charcot, J. M.: Klinische Vorträge über Krankheiten des Nervensystems. Verlag der Metzlerschen Buchhandlung, Stuttgart 1874.

Charcot, J. M.: Poliklinische Vorträge. (Übers. S. Freud). Deuticke, Leipzig-Wien 1889.

Charvat, J., P. Dell, B. Folkow: Mental factors and cardiovascular diseases. Cardiologia 44 (1964) 124.

Chatterjee, S. N. (ed.): Manual of renal transplantation. Springer, New York-Heidelberg-Berlin 1979.

Chaudhary, N., S. C. Truelove: Human colonic motility: a comparative study of normal subjects, patients with ulcerative colitis, and patients with the irritable colon syndrome. Gastroenterology 40 (1961) 1–17, 18–26, 27–36.

Chaudhary, N. A., S. C. Truelove: The irritable colon syndrome. A study of the clinical features, predisposing causes, and prognosis in 130 cases. Quart. J. Med. 31 (1962) 307–322.

Chazan, J. A., W. Winkelstein: Household aggregation of hypertension. J. chron. Dis. 17 (1964) 9–18.

Chesney, M. A., J. R. Eagleston, R. H. Rosenman: Type A behavior: assessment and intervention. In: Prokop, C. K., L. A. Bradley (eds.): Medical Psychology: Contributions to Behavioral Medicine. Acad. Press, New York 1981.

Chesney, M. A., G. W. Black, J. H. Chadwick et al.: Psychological correlates of the coronary-prone behavior pattern. J. behav. Med. 4 (1981) 217–229.

Chesney, M. A., G. Sevelius, G. W. Black, M. M. Ward, G. E. Swan, R. H. Rosenman: Work environment, Type A behavior and coronary heart disease risk factors. J. occup. Med. 23 (1981) 551–555.

Chester, R.: Health and marital breakdown. J. psychosom. Res. 17 (1973) 317.

Chiang, B. W., L. V. Perlman, M. Fulton et al.: Predisposing factors in sudden cardiac death in Tecumseh, Michigan. Circulation 41 (1970) 31.

Chiasson, R. E., A. R. Moss, R. Onishi, D. Osmond, J. R. Carlson: Human immunodeficiency virus infection in heterosexual intravenous drug users in San Francisco. Amer. J. publ. Health 77 (1987) 169–172.

Ching, J., N. Newton: A prospective study of psychological and social factors in pregnancy related to preterm and low-birthweight deliveries. Vortr. 6. int. Kongr. psychosom. Geburtshilfe u. Gynäkologie, Berlin Sept. 1980.

Chiodo, J., P. R. Latimer: Hunger perception and satiety responses among normal-weight bulimics and normals to a high-caloric, carbohydrate-rich food. Psychol. Med. 16 (1986) 343–349.

Chodoff, P., S. B. Friedman, D. A. Hamburg: Stress defenses and coping behavior: observations on parents of children with malignant disease. Amer. J. Psychiat. 120 (1964) 743–749.

Christensen, J.: In: Johnson, L. R. (ed.): Physiology of the Gastrointestinal Tract, p. 686. Raven, New York 1987.

Christensen, N. J., P. Vestergaard, T. Sørensen, O. J. Rafaelson: Cerebrospinal fluid adrenaline and noradrenaline in depressed patients. Acta psychiat. scand. 61 (1980) 178–182.

Christian, P.: Das Personenverständnis im modernen medizinischen Denken. Tübingen 1952.

Christian, P.: Die Atembewegung als Verhaltensweise. Nervenarzt 28 (1957) 243–247.

Christian, P.: Psychohygiene der Zivilisationskrankheiten. Ärztl. Mitteil. 2 (1960) 2406.

Christian, P., R. Haas: Wesen und Formen der Bipersonalität. Grundlagen für eine medizinische Soziologie. Beiträge aus der Allgemeinen Medizin. Bd. 7, Enke, Stuttgart 1949.

Christian, P., R. Kropf, H. Kurth: Eine Faktorenanalyse der subjektiven Symptomatik vegetativer Herz- und Kreislaufstörungen. Arch. Kreislaufforschung 45 (1965) 171–194.

Christian, P., K. Fink-Eitel, W. Huber: Verlaufsbeobachtungen über 10 Jahre bei 100 Patienten mit vegetativen Kreislaufstörungen. Z. Kreislaufforschung 55 (1966) 342–357.

Christian, P., P. Mohr, M. Schrenk, W. Ulmer: Zur Phänomenologie der abnormen Atmung beim sogenannten „Nervösen Atmungssyndrom". Nervenarzt 26 (1955) 191–197.

Christian, W.: Der Einfluß des Vigilitätsgrades auf die Auslösung epileptischer Anfälle. Nervenarzt 57 (1986) 257–262.

Christian-Widmaier, P.: Der institutionelle Rahmen thanato-therapeutischer Arbeit. In: Spiegel-Rösing, J., H. Petzold (Hrsg.): Die Begleitung Sterbender, S. 183–236. Junfermann, Paderborn 1984.

Christie, J. E., L. J. Whalley, H. Dick, D. H. R. Blackwood, I. M. Blackburn, G. Fink: Raised plasma cortisol concentrations: a feature of drug-free psychotics and not specific for depression. Brit. J. Psychiat. 148 (1986) 58–65.

Christie, R., R. K. Merton: Procedures for the sociological study of the values climate in medical schools. J. med. Educ. 33 (1958) 125–153; zit. n. Bloom, S. W.: Power and dissent in the medical school. Free Press, New York 1973.

Christoffel, H.: Trieb und Kultur. Zur Sozialpsychologie, Physiologie und Psychohygiene der Harntriebhaftigkeit mit besonderer Berücksichtigung der Enuresis. Schwabe, Basel 1944.

Cinciripini, P. A., D. A. Williamson, L. H. Epstein: Behavioral treatment of migraine headaches. In: Ferguson, J. M., C. B. Taylor (eds.): The Comprehensive Handbook of Behavioral Medicine, Vol. 2: Syndromes and Special Areas, pp. 207–227. MTP Press, Lancaster 1981.

Ciompi, L.: Affektlogik. Klett-Cotta, Stuttgart 1982.

Claessens, D.: Familie und Wertsystem. Eine Studie zur „zweiten, sozio-kulturellen Geburt des Menschen". Berlin 1962.

Claessens, D.: Das Konkrete und das Abstrakte. Suhrkamp, Frankfurt 1980.

Claesson, M. H., C. Ropke: Quantitative studies on cortisol-induced decay of lymphoid cells in the thymolymphatic system. Acta path. microbiol. scand. 76 (1969) 376–382.

Claman, H. N.: Corticosteroids – immunologic and anti-inflammatory effects. In: Berczi, I., K. Kovacs (eds.): Hormones and Immunity, pp. 38–42. MTP, Lancaster 1987.

Clark, D. C.: Oral complications of anorexia nervosa and/or bulimia: with a review of the literature. J. oral. Med. 40 (1985) 134.

Clark, N. M.: The effectiveness of education for family management of asthma in children: a preliminary report. Health Educ. Quart. 8 (1981) 166–174.

Clark, P., S. P. Glasser, E. Spoto: Arrhythmias detected by ambulatory monitoring. Lack of correlation with symptoms of dizziness and syncope. Chest 77 (1980) 722–725.

Clark, S., S. M. Campbell, M. E. Forehand, E. A. Tindall, R. M. Bennett: Clinical characteristics of fibrositis in a „blinded" controlled study using standard psychological tests. Arthr. and Rheum. 28 (1985) 132–136.

Clarkson, T.: Implications for atherogenesis. In: Weiss, S. M., K. A. Matthews, T. Detre, J. A. Graeff (eds.): Stress, Reactivity, and Cardiovascular Disease. NIH Publ. 84–2698 (1984) 35–39.

Clarkson, T., J. R. Kaplan, M. R. Adams, S. B. Manuck: Psychosocial influences on the pathogenesis of atherosclerosis among nonhuman primates. Circulation 76 (Suppl. 1) (1987) 29–40.

Clarkson, T., K. Weingand, J. R. Kaplan, M. R. Adams: Mechanisms of atherogenesis. Circulation 76 (Suppl. 1) (1987) 20–28.

Classen, M., H. G. Damman, W. Domschke et al.: Kurzzeittherapie des Ulcus duodeni mit Omeprazol und Ranitidin. Ergebnisse einer deutschen Multizenterstudie. Dtsch. med. Wschr. 110 (1985) 210.

Clauser, G.: Das Anorexia nervosa-Problem unter besonderer Berücksichtigung der Pubertätsmagersucht und ihrer klinischen Bedeutung. Ergeb. inn. Med. Kinderheilk. 21 (1964) 97–164.

Clayton, P. J.: Mortality and morbidity in the first year of widowhood. Arch. gen. Psychiat. 30 (1974) 747–750.

Cleghorn, J. M.: Organization of psychosocial care in a teaching hospital. Int. Congr. Psychosom. Med., Amsterdam 1973 (Manuskript).

Clemens, W.: Arbeitsbelastungen und gesundheitliche Auswirkungen bei Schichtarbeit. Sozialer Fortschritt 30 (1981) 25–29.

Clement, U.: Höhenrausch. In: Sigusch, V. (Hrsg.): AIDS als Risiko. Konkret Literatur Verlag, Hamburg 1987.

Clemente, C. D., M. H. Chase: Neurological substrates of aggressive behavior. Ann. Rev. Physiol. 35 (1973) 329–356.

Clyne, M.: Das ärztliche Erstgespräch. Sexualmed. 10 (1975).

Clyne, M.: Änderungen des ärztlichen Umgangs mit Patienten. Balint-Gruppen. In: Prill, H. J., D. Langen (Hrsg.): Der psychosomatische Weg zur gynäkologischen Praxis. Schattauer, Stuttgart 1983.

Cobb, L. A., J. A. Werner, G. B. Trobaugh: Sudden cardiac death: I. A decade's experience with out-of-hospital resuscitation. Modern Concepts Cardiovasc. Dis. 49 (1980) 31–36.

Cobb, L. A., R. S. Braun, H. Alvarez, W. A. Schaffer: Resuscitation from out-of-hospital ventricular fibrillation: 4-year follow-up. Circulation 51/52 (Suppl. 3) (1975) 223–228.

Cobb, S.: Borderlands of psychiatry. Harvard Univ. Press, Cambridge/Mass. 1943.

Cobb, S.: Contained hostility in rheumatoid arthritis. Arthr. and Rheum. 2 (1959) 419–425.

Cobb, S., R. M. Rose: Hypertension, peptic ulcer, and diabetes in air traffic controllers. J. Amer. med. Ass. 224 (1973) 489–492.

Cobb, S., W. Bauer, I. Whiting: Environmental factors in rheumatoid arthritis. J. Amer. med. Ass. 113 (1939) 668–670.

Coble, P., F. G. Foster, D. J. Kupfer: Electroencephalographic sleep diagnosis of primary depression. Arch. gen. Psychiat. 33 (1976) 1124–1127.

Cochrane, R.: Neuroticism and the discovery of high blood pressure. J. psychosom. Res. 13 (1969) 21–25.

Cochrane, R.: Hostility and neuroticism among unselected essential hypertensives. J. psychosom. Res. 17 (1973) 215–218.

Coe, R. M., M. Pepper, M. Mattis: The „new" medical student: another view. J. med. Educ. 52 (1977) 89–98.

Coelho, G. V., D. A. Hamburg, J. E. Adams (eds.): Coping and adaptation. Basic Books, New York 1974.

Cogoli, A., A. Tschopp: Lymphocyte reactivity during spaceflight. Immunol. Today 6 (1985) 1–4.

Cohen, E., R. Hillis: The use of hypnosis in treating the temporomandibular joint pain dysfunction syndrome. Oral Surg. 48 (1979) 193.

Cohen, F., R. S. Lazarus: Active coping processes, coping dispositions and recovery from surgery. Psychosom. Med. 35 (1973) 375–389.

Cohen, F., R. S. Lazarus: Coping with the stresses of illness. In: Stone, G. C., R. Cohen, N. E. Adler et al. (eds.): Health Psychology, pp. 217–257. Jossey Bass, San Francisco 1980.

Cohen, H. D. et al.: The effects of stress on components of the respiration cycle. Psychophysiol. 12 (1975) 377–380.

Cohen, J.: Statistical power analysis for the behavioral sciences. Acad. Press, New York 1977.

Cohen, M. B., G. Baker, R. A. Cohen, F. Fromm-Reichman, E. V. Weigert: An intensive study of 12 cases of manic depressive psychosis. Psychiatry 17 (1954) 103–107.

Cohen, M. E., D. W. Badal, A. Kilpatrick, E. A. Reed, P. D. White: The high familial prevalence of neurocirculatory asthenia (anxiety neurosis, effort syndrome). Amer. J. hum. Genet. 3 (1951) 126–158.

Cohen, M. E., F. Conzolazzio, R. E. Johnson: Blood lactate response during moderate exercise in neurocirculatory asthenia, anxiety neurosis, or effort syndrome. J. clin. Invest. 26 (1947) 339–342.

Cohen, M. E., P. D. White: Life situations, emotions and neurocirculatory asthenia. Psychosom. Med. 13 (1951) 335–357.

Cohen, M. E., P. D. White, R. E. Johnson: Neurocirculatory asthenia, anxiety neurosis, or the effort syndrome. Arch. intern. Med. 81 (1948) 260–281.

Cohen, N., R. Ader, N. Green, D. Bovbjerg: Conditioned suppression of thymus-independent antibody response. Psychosom. Med. 41 (1979) 487–491.

Cohen, S., G. W. Evans, D. S. Krantz, D. Stokols: Physiological, motivational and cognitive effects of aircraft noise on children. Amer. Psychol. 35 (1980) 231–243.

Cohen, S. I.: Cushing's syndrome: a psychiatric study of 29 patients. Brit. J. Psych. 136 (1980) 120.

Cohen-Booth, G.: Paralysis agitans. Entstehungsbedingungen und Beeinflussungsmöglichkeiten. Nervenarzt 8 (1935) 69–83.

Cohn, E. M., I. I. Lederman, E. Shore: Regional enteritis and its relation to emotional disorders. Amer. J. Gastroent. 54 (1970) 378–387.

Cohn, R.: Von der Psychoanalyse zur themenzentrierten Interaktion. Von der Behandlung einzelner zu einer Pädagogik für alle. Klett, Stuttgart 1975; 6. Aufl. 1984.

Cohn, R. C.: Themenzentrierte Interaktion. In: Heigl-Evers, A. (Hrsg.): Die Psychologie des 20. Jahrhunderts, Bd. VIII. Kindler, Zürich 1979.

Colgan, S. M., E. B. Faragher, P. J. Whorwell: Controlled trial of hypnotherapy in relapse prevention of duodenal ulceration. Lancet, June 11 (1988) 1299–1300.

Collet, L., J. Cottraux, C. Juenet: GSR feedback and Schultz relaxation in tension headaches: a comparative study. Pain 25 (1986) 205–213.

Collings, F., R. Roessler: Intellectual and attitudinal characteristics of medical students selecting family practice. J. Fam. Pract. 2 (1975) 431–432.

Collins, A., P. Eneroth, B. M. Landgren: Psychoendocrine stress responses and mood as related to menstrual cycle. Psychosom. Med. 47 (1985) 512–527.

Collins, F. H., P. E. Baer, G. G. Bourianoff: Orienting behavior of children with hypertensive fathers. Soc. psychophysiol. Res., Vancouver 1980.

Combrinck-Graham, L.: Structural family therapy in psychosomatic illness: Treatment of anorexia nervosa and asthma. Clin. Pediat. 13 (1974) 827.

Compernolle, T., K. Hoogduin, L. Joele: Diagnosis and treatment of the hyperventilation syndrome. Psychosomatics 20 (1979) 612–625.

Comroe, J. H.: Some theories of the mechanisms of dyspnea. In: Howell, J. B. L., E. J. M. Campbell (eds.): Breathlessness. Blackwell, Oxford 1966.

Comstock, G. W., K. J. Helsing: Symptoms of depression in two communities. Psychol. Med. 6 (1976) 551–563.

Comty, C. M., A. Leonard, F. L. Shapiro: Psychosocial problems in dialysed diabetic patients. Kidney Int. 6 (Suppl. 1) (1974) 144.

Condon, J. T.: Altered cognitive functioning in pregnant women: a shift towards primary process thinking. Brit. J. med. Psychol. 60 (1987) 329–334.

Condrau, G.: Zur Psychosomatik des alternden Menschen. Ther. Umschau 23 (1966) 458–467.

Condrau, G.: Psychosomatik der Frauenheilkunde. Huber, Bonn 1969.

Conen, D., D. Frey: Colon irritabile – ja oder nein? Ist die Anamnese eine Entscheidungshilfe? Schweiz. med. Wschr. 15 (1982) 531–534.

Conley, J. A., H. J. Burton, A. Kaplan-De Nour et al.: Support systems for patients and spouses on home dialysis. Int. J. Fam. Psychiat. 2 (1981) 45–54.

Connolly, J.: Life events before myocardial infarction. J. hum. Stress 2 (1976) 3–17.

Conrad, A. J., A. B. Scheibel: Schizophrenia and the hippocampus: the embryological hypothesis extended. Schizophrenia Bull. 13 (1987) 577–587.

Conrad, K.: Die symptomatischen Psychosen. In: Gruhle, H. W., R. Jung, W. Mayer-Groß, M. Müller (Hrsg.): Psychiatrie der Gegenwart, Bd. II, S. 369–436. Springer, Berlin 1960.

Conradt, A.: Neuere Modellvorstellungen zur Pathogenese der Gestose unter besonderer Berücksichtigung eines Magnesium-Mangels. Z. Geburtsh. Perinatol. 188 (1984) 49.

Conron, G., K. J. Hardy: Psychological factors as a prediction of success in duodenal ulcer surgery. Aust. N. Z. J. Psychiat. 10 (1976) 151–155.

Cook, W., D. Medley: Proposed hostility and pharisaic-virtue scales for the MMPI. J. appl. Psychol. 38 (1954) 414–418.

Coombs, D. W., R. L. Saunders, M. Gaylor, M. G. Pagean:

Epidural narcotic infusion reservoir: implantation technique and efficacy. Anaesthesiol. 56 (1982) 469–473.

Cooper, A. F., A. R. Curry: The pathology of deafness in the paranoid and affective psychoses of later life. J. psychosom. Res. 20 (1976) 97–105.

Cooper, B., H. G. Morgan: Epidemiologische Psychiatrie. Urban & Schwarzenberg, München-Wien-Baltimore 1977.

Cooper, B., U. Sosna: Psychische Erkrankungen in der Altenbevölkerung. Nervenarzt 54 (1983) 239–249.

Cooper, C. L.: Stress and breast cancer. Wiley, Chichester 1988.

Cooper, C. L., R. F. Davis Cooper, E. B. Faragher: A prospective study of the relationship between breast cancer and life events, type A behavior, social support and coping skills. Stress Medicine 2 (1986) 271–277.

Cooper, D.: Psychiatrie und Anti-Psychiatrie. Suhrkamp, Frankfurt 1971.

Cooper, D. E., R. W. Holmstrom: Relationship between alexithymia and somatic complaints in a normal sample. Psychother. and Psychosom. 41 (1984) 20–24.

Cooper, P. E., J. B. Martin: Neuroendocrinology and brain peptides. Trends in Neurosci. 5 (1982) 186.

Cooper, P. J., C. G. Fairburn: Binge-eating and self-induced vomiting in the community. Brit. J. Psychiat. 142 (1983) 139–144.

Cooper, P. J., C. G. Fairburn: The depressive symptoms of bulimia nervosa. Brit. J. Psychiat. 148 (1986) 268–274.

Cooper, P. J., M. J. Taylor: Body image disturbance in bulimia nervosa. Brit. J. Psychiat. (in press).

Cooper, P. J., D. J. Charnock, M. J. Taylor: The prevalence of bulimia nervosa. A replication study. Brit. J. Psychiat. 151 (1987) 684–686.

Cooper, T., T. Detre, S. M. Weiss (eds.): Coronary-prone behavior and coronary heart disease: a critical review. Circulation 63 (1981) 1199–1215.

Cooperstock, R.: Sex differences in psychotropic drug use. Soc. Sci. Med. 12 (1978) 179–186.

Copeland, D. D.: Concepts of disease and diagnosis. Perspect. Biol. Med. 20 (1977) 528–538.

Corbalan, R., R. L. Verrier, B. Lown: Psychologic stress and ventricular arrhythmia during myocardial infarction in the conscious dog. Amer. J. Cardiol. 34 (1974) 692–696.

Corley, K. C., H. P. Mauk, F. O. M. Shiel: Cardiac responses associated with „yoked-chair" shock avoidance in squirrel monkeys. Psychophysiol. 12 (1975) 439–444.

Cormier, B., E. Wittkower, V. Marcotte, F. Forget: Psychological aspects of rheumatoid arthritis. Canad. med. Ass. J. 7 (1957) 533–541.

Cornell, W.: Pädagogische Verhaltenspsychologie. München 1969.

Cosgriff, S. W., A. J. Diefenbach, W. Vogt jr.: Hypercoagulability of the blood associated with ACTH and cortisone therapy. Amer. J. Med. 9 (1950) 752–756.

Costa, P. T.: Is neuroticism a risk factor for CAD? Is Type A a measure of neuroticism? In: Schmidt, T., T. Dembroski, G. Blümchen (eds.): Biological and Psychological Factors in Cardiovascular Disease, pp. 85–95. Springer, Berlin-Heidelberg-New York 1986.

Costa, P. T., R. R. McCrae: Hypochondriasis, neuroticism, and aging: When are somatic complaints unfounded? Amer. Psychol. 40 (1985) 19.

Costa, P. T., A. B. Zonderman, B. T. Engel, W. F. Baile, D. L. Brimlow, J. Brinker: The relation of chest pain symptoms to angiographic findings of coronary artery stenosis and neuroticism. Psychosom. Med. 47 (1985) 285.

Costa, P. T., D. S. Krantz, J. A. Blumenthal, C. D. Furberg, R. H. Rosenman, R. B. Shekelle: Task force 2: psychological risk factors in coronary artery disease. Circulation 76 (Suppl. 1) (1987) 145–149.

Costello, C. G.: Social factors associated with depression: a retrospective community study. Psychol. Med. 12 (1982) 329–339.

Costen, J. B.: Syndrome of ear and sinus symptoms dependent on disturbed function of the temporomandibular joint. Ann. Otol. 43 (1934) 1.

Cotte, G., J. Dechaume: Les plexalgies hypogastriques. Documents histopathologiques, considérations pathogéniques. Presse méd. 39 (1931) 373.

Cottier, C., R. Adler, H. Vorkauf, R. Gerber, T. Hefer, C. Hürny: Pressured pattern or Type A behavior in patients with peripheral arteriovascular disease: controlled retrospective exploratory study. Psychosom. Med. 45 (1983) 187.

Cottington, E. M., K. A. Matthews, E. Talbott, L. H. Kuller: Environmental events preceding sudden death in women. Psychosom. Med. 42 (1980) 567–574.

Cottrell, S. S., S. A. K. Wilson: The affective symptomatology of disseminated sclerosis. A study of 100 cases. J. Neurol. Psychopath. 7 (1926) 1–30.

Cotugno, D.: De Ischiade nervosa commentarius novis curis auctior. Secunda ed. Simoniana, Napoli 1779.

Coulehan, J. L.: Human illness: cases, models, paradigms. The Pharos 43 (1980) 2–8.

Coumel, P., J. Fidelle, V. Lucet et al.: Catecholamine-induced severe ventricular arrhythmias with Adams-Stokes syndrome in children. Report of four cases. Brit. Heart J. 40 (Suppl.) (1978) 28–37.

Cousins, N.: Anatomy of an illness (as perceived by the patient). New Engl. J. Med. 26 (1976) 1458–1463.

Cousins, N.: Anatomy of an illness. Reflections on healing and regeneration. Bantam, Toronto 1979.

Coué, E.: Die Selbstbemeisterung durch bewußte Autosuggestion. Schwabe, Basel-Stuttgart 1966.

Cowan, W. K., G. D. Sorenson: Electron microscopic observations of acute thymic involution produced by hydrocortisone. Lab. Invest. 13 (1964) 353–370.

Cowley, D. S., T. S. Hyde, S. R. Dager, D. L. Dunner: Lactate infusions: the role of baseline anxiety. Psychiat. Res. 21 (1987) 169–179.

Cox, B. M., K. E. Ophein, H. Teschemacher, A. Goldstein: A peptide-like substance from pituitary that acts like morphine. Life Sci. 16 (1975) 1777–1782.

Cox, G. B., C. R. Chapman, R. G. Black: The diagnosis of psychogenic pain. J. Behav. Med. 1 (1978) 437–443.

Craig, S. P., V. J. Buckle, A. Lamouroux, J. Mallet, I. Craig: Localization of the human tyrosine hydroxylase gene to 11p15: gene duplication and evolution. Cytogenet. cell. Genet. 42 (1986) 29–32.

Cram, J. R.: EMG biofeedback and the treatment of tension headaches: a systematic analysis of treatment components. Behav. Ther. 11 (1980) 699–710.

Cramer, B., F. Feihl, F. Palacio Espasa: Le diabète juvenile, maladie difficile à vivre et à penser. Etude psychiatrique multifocale d'enfants diabétiques. Psychiat. de l'Enfant 12 (1979) 5–66.

Cramond, W.: The psychological problems of renal dialysis and transplantation. Modern Trends in Psychosom. Med. 2 (1970) 278–297.

Cramond, W.: Renal transplantation: experiences with recipients and donors. Sem. Psychiat. 3 (1971) 116–132.

Cramond, W., P. Knight, J. Lawrence: Psychological aspects of the management of chronic renal failure. Brit. med. J. (1968) 539–543.

Cramond, W., P. Knight, J. Lawrence: The psychiatric contribution to a renal unit undertaking chronic hemodialysis and renal homotransplantation. Brit. J. Psychiat. 113 (1967) 1201–1212.

Crary, B., M. Borysenko, D. C. Sutherland, I. Kutz, J. Z. Borysenko, H. Benson: Decrease in mitogen responsiveness of mononuclear cells from peripheral blood after epinephrine administration in humans. J. Immunol. 130 (1983a) 694–697.

Crary, B., S. L. Hauser, M. Borysenko, I. Kutz, C. Hoban, K. A. Ault, H. L. Weiner, H. Benson: Epinephrine induced changes in the distribution of lymphocyte subsets in the peripheral blood of humans. J. Immunol. 131 (1983b) 1178–1181.

Crauford, D. I. O., R. Harris: Ethics of predictive testing for Huntington's chorea: the need for more information. Brit. med. J. 293 (1986) 249.

Creese, J., D. R. Burt, S. H. Snyder: Dopamine receptor binding predicts clinical and pharmacological potencies of anti-schizophrenic drugs. Science 192 (1976) 481–483.

Creese, J., A. P. Feinberg, S. H. Snyder: Butyrophenone influences on opiate receptors. Eur. J. Pharmacol. 36 (1976) 231–235.

Cremerius, J.: Rheumatische Muskel- und Gelenkerkrankungen als funktionelles Geschehen. Psychosom. Med. 3 (1955) 173–181.

Cremerius, J.: Freuds Konzept über die Entstehung psychogener Körpersymptome. Psyche 9 (1957) 125.

Cremerius, J.: Die Beurteilung des Behandlungserfolges in der Psychotherapie. Springer, Berlin 1962.

Cremerius, J.: Diskussionsbemerkung. In: Meyer, J. E., H. Feldmann (Hrsg.): Anorexia nervosa, S. 68–69. Thieme, Stuttgart 1965a.

Cremerius, J.: Zur Prognose der Anorexia nervosa. Arch. Psychiat. Z. ges. Neurol. 207 (1965b) 378–393.

Cremerius, J.: Zur Frage der nosologischen Einordnung funktioneller Syndrome. Med. Welt 19 (1968) 689–692.

Cremerius, J.: Abriß der psychoanalytischen Abwehrtheorie. Z. Psychother. med. Psychol. 18 (1968) 1–14.

Cremerius, J.: Die Prognose funktioneller Syndrome. Enke, Stuttgart 1968.

Cremerius, J.: Zur Dynamik des Krankenhausaufenthaltes von Ulcuskranken. Z. psychosom. Med. 17 (1971) 282–293.

Cremerius, J.: Ist die „psychosomatische Struktur" der französischen Schule krankheitsspezifisch? Psyche 31 (1977) 293–317.

Cremerius, J.: Zur Prognose der Anorexia nervosa (11 sechsundzwanzig- bis neunundzwanzigjährige Katamnesen psychotherapeutisch unbehandelter Fälle). Z. psychosom. Med. Psychoanal. 24 (1978) 56–69.

Cremerius, J.: Zur Theorie und Praxis der Psychosomatischen Medizin. Suhrkamp, Frankfurt 1978.

Cremerius, J., S. O. Hoffmann, W. Hoffmeister, W. Trimborn: Die manipulierten Objekte. Psyche 33 (1979) 801–828.

Crespi, L. P.: Quantitative variation of incentive and performance in the white rat. Amer. J. Psychol. 55 (1942) 467–517.

Crismer, R., C. Drèze: La recto-colite ulcéro-hémorragique. Étude statistique et clinique de 120 observations. Acta gastro-enterol. belg. 24 (1961) 476–506.

Crisp, A.: Social factors in disease: the education of doctors in the UK and in Europe. 15th European Conf. Psychosom. Res., London 1984.

Crisp, A. H.: A treatment for anorexia nervosa. Brit. J. Psychiat. 112 (1965) 505–512.

Crisp, A. H.: Some aspects of the evolution, presentation and follow-up of anorexia nervosa. Proc. roy. Soc. Med. 58 (1965) 814–820.

Crisp, A. H.: The prevalence of anorexia nervosa and some of its associations in the general population. Adv. psychosom. Med. Vol. 9 (1977) 38–47.

Crisp, A. H.: Early recognition and prevention of anorexia nervosa. Dev. Med. Child Neurol. 21 (1979) 393–395.

Crisp, A. H.: Let me be. Grune & Stratton, New York 1980.

Crisp, A. H.: Some aspects of the psychopathology of anorexia nervosa. In: Darby, P. L., P. E. Garfinkel, D. M. Garner, D. V. Coscina (eds.): Anorexia nervosa. Recent Developments in Research, pp. 15–28. Liss, New York 1983.

Crisp, A. H.: „Biological" depression: because sleep fails. Postgrad. Med. J. 62 (1986) 179–185.

Crisp, A. H., T. Burns: The clinical presentation of anorexia nervosa in the male. Int. J. Eat. Dis. 2 (1983).

Crisp, A. H., B. McGuiness: Jolly fat: relation between obesity and neurosis in general population. Brit. med. J. 1 (1975) 7–9.

Crisp, A. H., R. S. Kalucy: Aspects of the perceptual disorder in anorexia nervosa. Brit. J. med. Psychol. 47 (1974) 349–361.

Crisp, A. H., D. A. Toms: Primary anorexia nervosa and weight phobia in the male. Brit. med. J. 47 (1972) 334–338.

Crisp, A. H., B. Harding, B. McGuiness: Anorexia nervosa. Psychoneurotic characteristics of parents: relationship to prognosis. J. psychosom. Res. 18 (1974) 167–173.

Crisp, A. H., R. L. Palmer, R. S. Kalucy: How common is anorexia nervosa? A prevalence study. Brit. J. Psychiat. 128 (1976) 549–554.

Crisp, A. H., C. Cohen, P. C. B. McKinnon, C. S. Corker: Observations of gonadotropic and ovarian hormone activity during recovery from anorexia nervosa. Postgrad. med. J. 49 (1973) 584–590.

Crisp, A. H., R. S. Kalucy, J. H. Lacey, B. Harding: The long-term prognosis in anorexia nervosa. In: Vigersky, R. A. (ed.): Anorexia nervosa. Raven, New York 1977.

Critchley, M.: Musicogenic epilepsy. Brain 60 (1937) 13–27.

Critchley, M.: Personification of paralysed limbs in hemiplegics. Brit. med. J. II (1955) 284–286.

Critchley, M.: Misoplegia or hatred of hemiplegia. In: Critchley, M. (ed.): The Divine Banquet of the Brain, pp. 115–120. Raven, New York 1979.

Crohn, B. B.: Granulomatous diseases of the small and large bowel. A historical survey. Gastroenterology 52 (1967) 767–772.

Crohn, B. B., L. Ginzburg, G. D. Oppenheimer: Regional ileitis. A pathologic and clinical entity. J. Amer. med. Ass. 99 (1932) 1323–1328.

Crohn, B. B., H. Yarnis: Regional ileitis, 2nd ed. Grune & Stratton, New York 1966.

Croiset, G., H. D. Veldhuis, R. E. Ballieux, D. deWied, C. J. Heijnen: The impact of mild emotional stress induced by the passive avoidance procedure on immune reactivity. Ann. N. Y. Acad. Sci. 496 (1987) 477–484.

Cronbach, L. J., P. E. Meehl: Construct validity in psychological tests. Psychol. Bull. 52 (1955) 281.

Croog, S. H., D. S. Shapiro, S. Levine: The heart patient and the recovery process. A review of the directions of research on social and psychological factors. Soc. Sci. Med. 2 (1968) 111–164.

Croog, S. H., D. S. Shapiro, S. Levine: Denial among male heart patients. Psychosom. Med. 33 (1971) 383–397.

Crown, J. M., S. Crown: The relationship between personality and the presence of rheumatoid factor in early rheumatoid disease. Scand. J. Rheum. 2 (1973a) 123–126.

Crown, J. M., S. Crown: Personality in early rheumatoid disease. J. psychosom. Res. 17 (1973b) 189–196.

Crown, J. M., S. Crown, A. Fleming: Aspects of the psychology and epidemiology of rheumatoid disease. Psychol. Med. 5 (1975) 291–299.

Cruz-Coke, R.: Environmental influences and arterial blood pressure. Lancet 2 (1960) 885–886.

Csaba, G.: Ontogeny and phylogeny of hormon receptors, Vol. 15. In: Monographs in Developmental Biology. Karger, Basel 1984.

Literatur

Cull, R. E.: Barometric pressure and other factors in migraine. Headache 21 (1981) 102–104.

Cullen, J. W., B. H. Fox, R. N. Isom: Cancer: the behavioral dimensions. Raven, New York 1976.

Cumes-Rayner, D. P., J. Price: Blood pressure reactivity: pitfalls in methodology. J. psychosom. Res. 32 (1988) 181–190.

Cumin, R., E. P. Bonetti, R. Scherschlicht, W. E. Haefely: Use of the specific benzodiazepine antagonist, Ro 15–1788, in studies of physiological dependence on benzodiazepines. Experientia (Basel) 38 (1982) 833–834.

Cunnick, J. E., D. T. Lysle, A. Armfield, B. S. Rabin: Shock-induced modulation of lymphocyte responsiveness and natural killer activity: differential mechanisms of induction. Brain, Behavior, and Immunity 2 (1988) 102–113.

Cunningham, A. J.: Conditioned immunosuppression: an important but probably nonspecific phenomenon. Behav. Brain Sci. 8 (1985) 397.

Curtius, F.: Individuum und Krankheit. Springer, Berlin 1959.

Curtius, F.: Die Colitis ulcerosa und ihre konservative Behandlung. Springer, Berlin 1962.

Curtius, F.: Moderne Asthmabehandlung. Springer, Berlin 1965.

Curtius, F.: Zur kombinierten pharmako- und psychotherapeutischen Behandlungsmethode der Colitis ulcerosa. Prax. Psychother. 13 (1968) 81–83.

Curtius, F., H. G. Rohrmoser: Zur Psychotherapie der Colitis ulcerosa. Dtsch. med. Wschr. 80 (1955) 105–108.

Czaczkes, J. W., A. Kaplan De-Nour: Chronic hemodialysis as a way of life. Brunner/Mazel, New York 1978.

Czeisler, C. A., J. S. Allan, S. H. Strogatz, J. M. Ronda, R. Sánchez, C. D. Ríos, W. O. Freitag, G. S. Richardson, R. E. Kronauer: Bright light resets the human circadian pacemaker independent of the timing of the sleep-wake cycle. Science 233 (1986) 667–671.

Czembirek, H., R. Pötzi, D. Tscholakoff, E. Salomonowitz, G. Wittich: Radiologisch-endoskopische Korrelation bei Crohnscher Erkrankung des Kolons. Fortschr. Röntgenstr. 138 (1983) 519–525.

Czernik, A.: Zur Psychopathologie und Persönlichkeitsstruktur der „primären Thalamusschwäche". Arch. Psychiat. 216 (1972) 101.

D'Atri, D. A., A. M. Ostfeld: Crowding: its effects on the elevation of blood pressure in a prison setting. Prev. Med. 4 (1975) 550–566.

D'Ercole, A. J., L. E. Underwood, J. J. van Wyk: Serum somatomedin-C in hypopituitarism and in other disorders of growth. J. Pediat. 90 (1977) 375.

Da Costa, J. M.: Membranous enteritis. Am. J. M. So. 62 (1871) 321–335..

Da Costa, J. M.: On irritable heart; a clinical study of a form of functional cardiac disorder and its consequences. Amer. Heart J. Jan. 1871.

Dager, S. R., A. Khan, K. A. Comess, V. Raisy, D. L. Dunner: Mitral valve abnormalities and catecholamine activity in anxious patients. Psychiat. Res. 20 (1987) 13–18.

Dahl, L. K.: Der mögliche Einfluß der Salzzufuhr auf die Entwicklung der essentiellen Hypertonie. In: Bock, K. P., P. Cottier (Hrsg.): Essentielle Hypertonie. Springer, Berlin 1960.

Dahlem, N. W., R. A. Kinsman, D. J. Horton: Panic-fear in asthma: requests for as-needed (PRN) medications in relation to pulmonary function measurements. J. Allergy clin. Immunol. 60 (1977) 295.

Dahlstrom, W. G., G. S. Welsh, L. E. Dahlstrom: An MMPI handbook. Vol. II: Research applications. A revised edition. Univ. of Minnesota Press, Minneapolis 1975.

Dahme, B.: Psychophysiologische Funktionsdiagnostik – Methodische Überlegungen am Beispiel der Herzneurose und des Asthma bronchiale. In: Tewes, U. (Hrsg.): Angewandte Medizinische Psychologie. Klotz, Frankfurt 1984.

Dahme, B., R. Richter: Einige Anforderungen an psychosomatische Aktivationsdiagnostik. Med. Psychol. 6 (1980) 103–122.

Dahme, B., R. Richter: Zur Psychophysiologie des Asthma bronchiale. Therapiewoche 31 (1985) 935–942.

Dahme, B., I. Achilles, B. Flemming, P. Götze, H. J. Meffert, G. Huse-Kleinstoll, M. J. Polonius, H. Speidel: Klassifikation psychopathologischer Auffälligkeiten nach Herzoperationen. Thoraxchir. 25 (1977) 345–349.

Dahme, B., B. Flemming, P. Götze, H. J. Meffert, G. Huse-Kleinstoll, H. Speidel: Psycho-Somatik der Herzchirurgie. In: Beckmann, D., S. Davies-Osterkamp, J. W. Scheer (Hrsg.): Medizinische Psychologie. Springer, Heidelberg 1982.

Dahmer, H.: Psychoanalyse und historischer Materialismus. Psyche 22 (1968) 172–177.

Dahmer, H.: Libido und Gesellschaft. Frankfurt 1982.

Dahmer, J.: Anamnese und Befund. Die systematische ärztliche Untersuchung. 5. Aufl. Thieme, Stuttgart-New York 1984.

Dalessio, D. J.: Mechanisms and biochemistry of headache. Postgrad. Med. 56 (1974) 55–62.

Dalessio, D. J.: Mechanisms of headache. Med. Clin. N. Amer. 62 (1978) 429–442.

Dalessio, D. J. (ed.): Wolff's headache and other head pain. 4th ed. Oxford Univ. Press, New York-Oxford 1980.

Dally, P.: Anorexia nervosa. Heinemann, London 1969.

Dally, P., J. Gomez, A. J. Isaacs: Anorexia nervosa. Heinemann, London 1979.

Dalos, N. P., P. V. Rabins, B. R. Brooks, P. O'Donnell: Disease activity and emotional state in multiple sclerosis. Ann. Neurol. 13 (1983) 573–577.

Danckwardt, J. F.: Zur Interaktion von Psychotherapie und Psychopharmakotherapie. Psyche 32 (1978) 111–154.

Danckwardt, J. F.: Psychopharmaka – ein Problem für Psychotherapeuten. Prax. Psychother. Psychosom. 25 (1980) 99–113.

Daniels, G. E., J. F. O'Connor, A. Karush, L. Moses, C. A. Flood, M. Lepore: Three decades in the observation and treatment of ulcerative colitis. Psychosom. Med. 24 (1962) 85–93.

Danilovicz, D. A., H. P. Gabriel: Postoperative reactions in children. Amer. J. Psychiat. 128 (1971) 185–188.

Dannecker, M., R. Reiche: Der gewöhnliche Homosexuelle. Fischer, Frankfurt 1974.

Dantzer, R., E. Satinoff, K. W. Kelley: Cyclosporine and alpha-interferon do not attenuate morphine withdrawal in rats but do impair thermoregulation. Physiol. Behav. 39 (1987) 593–598.

Darby, P. L., P. E. Garfinkel, D. M. Garner, D. V. Coscina (eds.): Anorexia nervosa. Recent developments in research. Liss, New York 1983.

Darke, S. G., A. G. Parks, J. L. Grogono, D. J. Pollock: Adenocarcinoma and Crohn's disease. A report of 2 cases and analysis of the literature. Brit. J. Surg. 60 (1973) 169–175.

Darwin, C.: Der Ausdruck der Gemütsbewegungen bei den Menschen und den Tieren. Schweizerbart, Stuttgart 1872.

Dattore, P. J., F. C. Shantz, L. Coyne: Premorbid personality differentiation of cancer and noncancer groups: a test of the hypothesis of cancer proneness. J. consult. clin. Psychol. 48 (1980) 388–394.

Davids, P. D., W. R. Rosengren: Social stability and psychological adjustment during pregnancy. Amer. psychosom. Med. 24 (1962) 579.

Davidson, D. M., M. A. Winchester, C. B. Taylor et al.: Effects of relaxation therapy on cardiac performance and sympathetic activity in patients with organic heart disease. Psychosom. Med. 41 (1979) 303–309.

Davidson, M., K. L. Davis: A comparison of plasma homovanillic acid concentrations in schizophrenic patients and normal controls. Arch. gen. Psychiat. 45 (1988) 561–563.

Davies, J. O.: The control of renin release. Amer. J. Med. 55 (1973) 333–350.

Davies, R. J.: Allergy and the skin. In: Lessof, M. H. (ed.): Immunological and Clinical Aspects of Allergy, pp. 217–295. MTP, Lancaster 1981.

Davies-Osterkamp, S.: Angst und Angstbewältigung bei chirurgischen Patienten. Therapiewoche 28 (1978a) 8253–8261.

Davies-Osterkamp, S., K. Möhlen: Postoperative Genesungsverläufe bei Patienten der Herzchirurgie in Abhängigkeit von präoperativer Angst und Angstbewältigung. Med. Psychol. 4 (1978b) 247–260.

Davies-Osterkamp, S., A. Salm: Ansätze zur Erfassung psychischer Adaptationsprozesse in medizinischen Belastungssituationen. Med. Psychol. 6 (1980) 66–80.

Davies-Osterkamp, S., K. Möhlen, H. R. Lademann, H. Scheld: Postoperative reactions in open-heart surgery patients. In: Speidel, H., G. Rodewald (eds.): Psychic and Neurological Dysfunctions after Open-Heart Surgery. Thieme, Stuttgart 1980.

Davis, D. M., J. C. Schipp, E. G. Pattishall: Attitudes of diabetic boys and girls toward diabetes. Diabetes 14 (1965) 106–109.

Davis, F.: Uncertainty in medical prognosis, clinical and functional. In: Scott, W. R., E. H. Volkart (eds.): Medical Care, pp. 311–321. Wiley, New York 1966.

Davis, J. M., S. H. Koslow, R. D. Gibbons, J. W. Maas, C. L. Bowden, R. Casper, I. Harin, J. I. Javaid, S. S. Chang, P. E. Stokes: Cerebrospinal fluid and urinary biogenic amines in depressed patients and healthy controls. Arch. gen. Psychiat. 45 (1988) 705–717.

Davis, M.: Variations in patients' compliance with doctors' order. J. med. Educ. 41 (1966) 1037–1048.

Davis, M.: Variations in patients' compliance with doctors. Psychiat. Med. 2 (1971) 31–54.

Davis, P. M., E. Sherwood-Jones: A service for the adult asthmatic. Thorax 35 (1980) 111–113.

Davison, A. M. (ed.): EDTA statistics. Proc. Europ. Dialysis and Transplant Ass. Pitman, London 1983.

Davison, G. C.: Systematic desensitization as a counter-conditioning process. J. abn. Psychol. 73 (1968) 91–99.

Davison, G. C., G. T. Wilson: Critique of „Desensitization: social and cognitive factors underlying the effectiveness of Wolpe's procedure". Psychol. Bull. 78 (1972) 28–31.

Davy, J.-P., J.-C. Andersson, E. Poilpre: Devenir et éléments de pronostic dans l'anorexie mentale. Ann. méd.-psychol. 2 (1976) 464–480.

Dawkins, R.: The extended phenotype: the gene as the unit of selection. Freeman, Oxford 1982.

Dawson, M. E., K. H. Neuchterlein: The role of autonomic dysfunctions within a vulnerability-stress model of schizophrenic disorders. In: Magnusson, D., A. Öhman (eds.): Psychopathology: An Interactional Perspective. Acad. Press, Orlando/Fl. 1987.

Day, R.: Life events and schizophrenia: the „triggering hypothesis". Acta psychiat. scand. 64 (1981) 97–112.

Day, R., J. Zubin, S. R. Steinhauer: Psychosocial factors in schizophrenia in light of vulnerability theory. In: Magnusson, D., A. Öhman (eds.): Psychopathology: An Interactional Perspective. Acad. Press, Orlando/Fl. 1987.

Day, S. C., E. F. Cook, H. Funkenstein: Evaluation of emergency room patients with transient loss of consciousness. Amer. J. Med. 73 (1982) 15–23.

Debain, J. J.: Les surdités brusques. Zit. nach: Stange, G., R. Neveling in: Berendes, J., R. Link, F. Zöllner (Hrsg.): Hals-Nasen-Ohrenheilkunde in Praxis und Klinik, Bd. 6/Ohr II. Thieme, Stuttgart-New York 1980.

DeBakey, M., A. Gotto: The living heart. Charter, New York 1977.

DeBold, A. J.: Atrial natriuretic factor: a hormone produced by the heart. Science 230 (1985) 267–269.

DeBoor, C.: Die Colitis ulcerosa als psychosomatisches Syndrom. Psyche 18 (1964) 107–119.

DeBoor, C.: Strukturunterschiede unbewußter Phantasien bei Neurosen und psychosomatischen Krankheiten. Psyche 18 (1964) 664–673.

DeBoor, C.: Zur Psychosomatik der Allergie, insbesondere des Asthma bronchiale. Huber-Klett, Bern-Stuttgart 1965.

DeBoor, C.: Psychosomatische Symptome und delinquentes Verhalten. Psyche 30 (1976) 625–641.

DeBoor, C., A. Mitscherlich: Verstehende Psychosomatik: Ein Stiefkind der Medizin. Psyche 27 (1973) 1–20.

Decastro, F. J., R. Biesbrock, C. Erikson, P. Farrell, W. Leong, D. Murphy, R. Green: Hypertension in adolescents. Clin. Pediat. 15 (1976) 24–26.

Dedo, H. H.: Recurrent laryngeal nerve section for spastic dysphonia. Ann. Otol. (St. Louis) 85 (1976) 451.

Deetjen, P.: Diskussionsbemerkung in: Gasser, G., W. Vahlensieck (Hrsg.): Pathogenese und Klinik der Harnsteine VII, S. 168. Steinkopff, Darmstadt 1979.

DeFaire, U.: Life change pattern prior to death in ischemic heart disease. A study on death-discordant twins. J. psychosom. Res. 19 (1975) 237–278.

Defourmy, M., P. Hubin, D. Luminet: Alexithymia, 'pensée opératoire' and predisposition to coronopathy. Pattern 'A' of Friedman and Rosenman. Psychother. and Psychosom. 27 (1976/77) 106–114.

DeFrance, J. F. (ed.): The septal nuclei. Plenum, New York 1976.

DeFronzo, R. A.: Glucose intolerance and aging. Evidence for tissue insensitivity to insulin. Diabetes 28 (1979) 1095–1101.

Degen, K., J. J. Strain, B. Zumoff: Biopsychosocial evaluation of sexual function in end-stage renal disease. In: Levy, N. B. (ed.): Psychonephrology 2, pp. 223–234. Plenum, New York 1983.

Degenais, F., E. Rosinski: Social class level, performance and change in medical education. AAMC-RIME Conference 1975, 199.

Degkwitz, R.: Zum umstrittenen psychiatrischen Krankheitsbegriff. Urban & Schwarzenberg, München 1981.

Degkwitz, R., H. Helmchen, G. Kockott, W. Mombour (Hrsg.): Diagnosenschlüssel und Glossar psychiatrischer Krankheiten. 5. Aufl, korrigiert nach der 9. Revision der ICD. Springer, Berlin 1979.

DeHaas, W. H. D., W. de Boer, F. Griffioen, P. Oosten-Elst: Rheumatoid arthritis and the robust reaction type. Ann. rheum. Dis. 33 (1974) 81–85.

Dekker, E., H. Pelser, J. Groen: Conditioning as a cause of asthmatic attacks. A laboratory study. J. psychosom. Res. 2 (1957) 97–108.

DeKlein, A.: Sudden complete or partial loss of function of the octavus-system in apparently normal persons. Acta oto-laryng. (Stockholm) 32 (1944) 407.

DeKorte, J.: Psychotherapeutische Möglichkeiten bei Psoriasis. Akt. Derm. 8 (1982) 160–162.

Delay, J., P. Pichot (Hrsg.): Medizinische Psychologie, 2. Aufl, S. 27–43. Thieme, Stuttgart 1966.

Delgado, J. M.: Effect of brain stimulation on task-free situations. In: Hernandez-Peon, R. (ed.): The Physiological Basis of Mental Activity. Elsevier, Amsterdam 1963.

Delgado, J. M.: Aggression and defense under cerebral ratio

control. In: Clemente, C. D., D. B. Lindsley (eds.): Aggression and Defense. Univ. of California Press, Los Angeles 1967.

Delgado, J. M.: Gehirnschrittmacher. Ullstein, Frankfurt 1971.

DeLisi, L. E., A. F. Mirsky, M. S. Buchsbaum, D. P. van Kammen, K. F. Berman, C. Caton, M. S. Kafka, P. T. Ninan, B. H. Phelps, F. Karoum, G. N. Ko, E. R. Korpi, M. Linnoila, M. Sheinan, R. J. Wyatt: The genain quadruplets 25 years later: a diagnostic and biochemical follow-up. J. psychiat. Res. 13 (1984) 59–76.

Delius, L.: Psychosomatische Aspekte bei Herz-Kreislaufstörungen. Psychosom. Med. 10 (1964) 242.

Delius, L., J. Fahrenberg: Ein kritischer Beitrag zur Psychosomatik der essentiellen Hypertonie. Med. Klinik 27 (1963) 1102–1107.

Delius, L., J. Fahrenberg: Psychovegetative Syndrome. Thieme, Stuttgart 1966.

Delius, L., J. Fahrenberg: Funktionelle kardiovaskuläre Störungen. Internist 13 (1972) 1–6.

Delius, L., P. Christian, H. Enke, A. Jores, H. P. Koepchen, C. Kulenkampf, F. Labhardt, O. Meier, H. Schaefer, Th.v. Uexküll: Ordnung und Störung der Herz-Kreislaufregulation im Zusammenhang mit emotionalen Vorgängen. Verh. Dtsch. Ges. Inn. Med. 70 (1964) 255.

Dell, P. F., H. A. Goolishian: Ordnung durch Fluktuation: Eine evolutionäre Epistemologie für menschliche Systeme. Familiendynamik 6 (1981) 104–122.

Demarco, V.: Psychiatric aspects of diabetes on CAPD. Periton. Dialys. Bull. 1980.

Dembroski, T. M.: Cardiovascular reactivity in type A coronary-prone subjects. In: Oborne, D. J., M. M. Gruneberg, J. R. Eiser (eds.): Research in Psychology and Medicine, Vol I: Physical Aspects: Pain, Stress, Diagnosis and Organic Damage. Acad. Press, New York 1979.

Dembroski, T. M.: Zusammenhang zwischen Psychophysiologie und Verhalten bei Typ-A-Personen. In: Dembroski, T. M., M. J. Halhuber (Hrsg.): Psychosozialer „Stress" und koronare Herzkrankheit: 3. Verhalten und koronare Herzkrankheit, S. 47–82. Springer, Berlin-Heidelberg-New York 1981.

Dembroski, T. M., P. T. Costa: Coronary prone behavior: components of the type A pattern and hostility. J. Pers. 55 (1987) 211–236.

Dembroski, T. M., J. M. MacDougall: Stress effects on affiliation preferences among subjects possessing the Type A coronary-prone behavior pattern. J. Pers. soc. Psychol. 36 (1978a) 23–33.

Dembroski, T. M., J. M. MacDougall: Behavioral and psychophysiological perspectives on coronary-prone behavior. In: Dembroski, T. M., T. H. Schmidt, G. Blümchen (eds.): Biobehavioral Bases of Coronary Heart Disease. Karger, Basel-München-Paris 1983.

Dembroski, T. M., J. M. MacDougall: Beyond global Type A: relationships of paralinguistic attributes, hostility, and anger – in relationship to coronary heart disease. In: Field, T., P. MacCabe, N. Schneiderman (eds.): Stress and Coping, pp. 223–242. Erlbaum, Hillsdale 1985a.

Dembroski, T. M., J. M. MacDougall, R. Lushene: Interpersonal interaction and cardiovascular response in Type A subjects and coronary patients. J. hum. Stress 5 (Dec. 1979a) 28–36.

Dembroski, T. M., J. M. MacDougall, J. L. Shields et al.: Components of Type A coronary-prone behavior pattern and cardiovascular responses to psychomotor performance challenge. J. behav. Med. 2 (1978c) 159–176.

Dembroski, T. M., J. M. MacDougall, R. B. Williams et al.: Components of Type A, hostility and anger – in relationship to angiographic findings. Psychosom. Med. 47 (1985b) 219–233.

Dembroski, T. M., J. M. MacDougall, P. T. Costa, G. A. Grandits: Antagonistic hostility as a predictor of coronary heart disease in the Multiple Risk Factor Intervention Trial. Psychosom. Med. (1989).

Dembroski, T. M., J. M. MacDougall, J. A. Herd, J. L. Shields: Effect of level of challenge on pressor and heart responses in Type A and B subjects. J. appl. soc. Psychol. 9 (1979b) 208–228.

Dembroski, T. M., J. M. MacDougall, J. A. Herd, J. L. Shields: Die Erforschung des Verhaltensmusters Typ A zur koronaren Herzkrankheit. Eine problemgeschichtliche Literaturübersicht. In: Dembroski, T. M., M. J. Halhuber (Hrsg.): Psychosozialer „Stress" und koronare Herzkrankheit: 3. Verhalten und koronare Herzkrankheit, S. 194–264. Springer, Berlin-Heidelberg-New York 1981.

Dembroski, T. M., S. M. Weiss, J. L. Shields, S. G. Haynes, M. Feinleib: Coronary-prone behavior. Springer, Berlin-Heidelberg-New York 1978b.

Demling, L., G. Hegemann, M. Classen, J. v. d. Emde: Die Prognose der Colitis ulcerosa. Dtsch. med. Wschr. 94 (1969) 247–253.

Demmel, H.-J., F. Lamprecht, A. Riehl: Psychosomatic aspects of orofacial pain dysfunction syndrome. Abstr. VIIth World Congress of the International College of Psychosomatic Medicine, p. 28. Hamburg 1983.

Demmel, H.-J., R. Neubauer: Chronische Myalgie bei frontolateralem Bruxismus. Abstr. Int. Arbeitstagung d. Dtsch. Kollegiums f. Psychosom. Med. S. 28. Innsbruck 1988.

Demole, M.: Krankheiten des Verdauungsapparates. In: Martin, E., J.-P. Junod (Hrsg.): Ein kurzes Lehrbuch der Geriatrie. Huber, Bern 1975.

DeMolina, A. F., R. W. Hunsperger: Organization of the subcortical system governing defense and flight reactions in the cat. J. Physiol. 160 (1962) 200–213.

Denford, J., J. Schachter, N. Temple, P. Kind, R. Rosser: Selection and outcome in in-patient psychotherapy. Brit. J. med. Psychol. 56 (1983) 225–243.

DeNike, L. D.: The temporal relationship between awareness and performance in verbal conditioning. J. exp. Psychol. 68 (1964) 521–529.

Denker, P. G.: Results of treatment of psychoneuroses by the general practitioner – a follow-up of 500 cases. N. Y. St. J. Med. 46 (1946) 2164–2166.

Dennerstein, L., S. F. Abraham: Affective changes and the menstrual cycle. In: Beumont, P. J. V., G. D. Burrows (eds.): Handbook of Psychiatry and Endocrinology, pp. 367–400. Elsevier, Amsterdam 1982.

Dennerstein, L., C. Wood, G. D. Durrows: Sexual response following hysterectomy and oophorectomy. Obstet. Gynec. 49 (1977) 92.

Dennerstein, L., C. Spencer-Gardner, G. Burrows: The premenstrual syndrome – a holistic approach. In: Keep, P. A. van, W. H. Utiar (eds.): The Premenstrual Syndrome. MTP Press, Lancaster 1981.

Denton, D.: The hunger for salt: an anthropological, physiological and medical analysis. Springer, New York 1982.

Deppe, H.-U.: Industriearbeit und Medizin. Zur Soziologie medizinischer Institutionen. Athenäum, Frankfurt 1973.

Deppe, H.-U., M. Regus (Hrsg.): Seminar: Medizin, Gesellschaft, Geschichte. Suhrkamp, Frankfurt 1975.

DeReynier, J. P.: Deafness in the world today. WHO Chron. 24 (1970) 3.

DeRidder, P.: Patient und Krankenhaus. Personenbezogener Dienst auf der Station. Bd. 1: Die Trauer des Leibes. Enke, Stuttgart 1980.

Der Spiegel: Wegweiser durch den Pillendschungel. Der Spiegel 37 (1983) 221–235.

Derogatis, L. R., M. D. Abeloff, N. Melisaratos: Psychological coping mechanisms and survival time in metastatic breast cancer. J. Amer. med. Ass. 242 (1979) 1504–1508.

Dersee, Th. (Hrsg.): Der große Wie-lebst-Du-denn? Das Buch für Selbsthilfe, Selbstorganisation und Patientenrecht. Verlagsgemeinschaft Gesundheit, Berlin 1982.

Descartes, R.: Les passions de l'âme. Paris 1649.

Descotes, J., R. Tedone, J. C. Evreux: Different effects of psychotropic drugs on delayed hypersensitivity responses in mice. J. Neuroimmunol. 9 (1985) 81–85.

DeSilva, R. A.: Central nervous system risk factors for sudden cardiac death. In: Greenberg, H. M., E. M. Dwyer (eds.): Sudden Coronary Death, pp. 143–160. N. Y. Acad. Sci., 1982.

DeSilva, R. A.: Psychological stress and sudden cardiac death. In: Schmidt, T., T. Dembroski, G. Blümchen (eds.): Biological and Psychological Factors in Cardiovascular Disease. Springer, Berlin-Heidelberg-New York 1986.

DeSilva, R. A., B. Lown: Ventricular premature beats, stress and sudden death. Psychosomatics 19 (1978) 649–659.

DeSilva, R. A., R. L. Verrier, B. Lown: Protective effect of the vagotonic effect of morphine sulfate on vulnerability to ventricular fibrillation. Cardiovascul. Res. 12 (1976) 161–172.

DeSilva, R. A., R. L. Verrier, B. Lown: Effects of psychologic stress and vagal stimulation on vulnerability to ventricular fibrillation in the conscious dog. Amer. Heart J. 95 (1978) 197–203.

DeSilva, R. A., Q. R. Regestein, B. Lown: Unpublished observations.

Desjatnikov, V. F., T. V. Nikitina, N. I. Chartulari, I. I. Pavlova: Pain in dental patients as an expression of depression. Quintessence Int. 9 (1978) 81.

DeSwaan, A.: Zur Soziogenese des psychoanalytischen „Settings". Psyche 32 (1978) 793–826.

Deter, H. C.: Zur Methodik von katamnestischen Untersuchungen bei psychosomatischen Patienten am Beispiel einer Gruppe von 31 Anorexie-Patienten. Z. Psychother. med. Psychol. 31 (1981) 48–52.

Deter, H. C.: Psychosomatische Behandlung des Asthma bronchiale. Springer, Berlin-Heidelberg-New York 1986.

Deter, H. C.: Die krankheitsorientierte Gruppentherapie im Rahmen der psychosomatischen Behandlung des Asthma bronchiale. In: Deter, H. C., W. Schüffel (Hrsg.): Gruppen mit körperlich Kranken, S. 67–83. Springer, Berlin-Heidelberg-New York 1988.

Deter, H. C., G. Allert: Group therapy for asthma patients; a concept for the psychosomatic treatment of patients in a medical clinic. Psychother. and Psychosom. 40 (1983) 95–105.

Deter, H. C., C. Heintze-Hook: Möglichkeiten der Einbeziehung körpertherapeutischer Verfahren in die tiefenpsychologisch fundierte, krankheitsorientierte Gruppentherapie von Asthmapatienten. In: Brähler, E. (Hrsg.): Körpererleben, S.90. Springer, Berlin-Heidelberg-New York 1986.

Deter, H. C., E. Petzold, R. Hengst-Theis, U. Breiden, G. Lanzinger-Rossnagel: Katamnestische Ergebnisse einer klinisch-psychosomatischen Behandlung von 103 Patienten mit Anorexia nervosa aus internistischer Sicht unter besonderer Berücksichtigung der Mortalität. Inn. Med. 1 (1983) 3–12.

Deucher, F.: Die Colitis ulcerosa. Ergeb. Chir. Orthop. 39 (1955) 69–197.

Deutsch, F.: Applied psycho-analysis. New York 1949.

Deutsch, F.: Reflections on Freud's one hundredth birthday. Psychosom. Med. 18 (1956) 279–283.

Deutsch, F.: Symbolization as a formative stage in the conversion process. In: On the Mysterious Leap from the Mind to the Body. New York 1959.

Deutsch, H.: Psychoanalyse der weiblichen Sexualfunktionen. Int. psychoanal. Verlag, Wien 1925.

Deutsch, H.: Zur Psychogenese eines Ticfalles. Int. Z. Psychoanal. 11 (1925) 325–332.

Deutsch, H.: Zur Psychologie der manisch-depressiven Zustände, insbesondere der chronischen Hypomanie. Int. Z. Psychoanal. 9 (1933).

Deutsch, H.: Some psychoanalytic observations in surgery. Psychosom. Med. 4 (1942) 105–115.

Deutsch, H.: Psychologie der Frau, 2. Bd. Huber, Bern-Stuttgart 1954.

Deutsche Diabetesgesellschaft: Stellungnahme des Vorstands zu den internationalen Vorschlägen für die Diagnostik des Diabetes mellitus. Diabetologie-Information 2 (1980) 6–10.

Deutsche Liga zur Bekämpfung der Atemwegserkrankungen e.V. Dtsch. med. Wschr. 105 (1980).

Devereux, R. B.: Mitral valve prolaps. Primary Care 12 (1985) 39–54.

Devereux, R. B., T. G. Pickering, G. A. Harshfield, H. D. Kleinert, L. Denby, L. Clark, D. Pregibon, M. Jason, B. Kleiner, J. S. Borer, J. H. Laragh: Left ventricular hypertrophy in patients with hypertension: importance of blood pressure response to regularly recurring stress. Circulation 3 (1983) 470–476.

Devins, G. M., Y. M. Binik, D. B. Hollomby, P. E. Barré, R. D. Guttman: Helplessness and depression in end-stage renal disease. J. abn. Psychol. 90 (1981) 531–545.

Devor, D., R. D. Knauft: Exploratory laparotomy for abdominal pain of unknown etiology. Diagnosis, management, and follow-up of 40 cases. Arch. Surg. 96 (1968) 836–839.

Devroede, G. J., W. F. Taylor, W. G. Sauer, R. J. Jackmann, G. B. Strickler: Cancer risk and life expectancy of children with ulcerative colitis. New Engl. J. Med. 285 (1971) 17–21.

Dewan, M. J., A. K. Pandurangi, M. L. Boucher, B. F. Levy, L. F. Major: Abnormal dexamethasone suppression test results in chronic schizophrenic patients. Amer. J. Psychiat. 139 (1982) 1501–1503.

DeWied, D., W. H. Gipsen: Behavioral effects of peptides. In: Grainer, H. (ed.): Peptides in Neurobiology, pp. 397–448. Plenum, New York 1977.

Deyo, R. A.: Conservative therapy for low back pain. Distinguishing useful from useless therapy. J. Amer. med. Ass. 250 (1983) 1057–1062.

Deyo, R. A., Y.-J. Tsui-Wu: Descriptive epidemiology of low back pain and its related medical care in the United States. Spine 12 (1987) 264–268.

Deyo, R. A., Th. S. Inui, B. Sullivan: Noncompliance with arthritis drugs. J. Rheum. 8 (1981) 931–936.

Deyo, R. A., A. K. Diehl, M. Rosenthal: How many days of bed rest for acute low back pain? A randomized clinical trial. New Engl. J. Med. 315 (1986) 1064–1070.

Diagnostisches und statistisches Manual psychischer Störungen DSM-III. Beltz, Weinheim 1984.

Diagnostisches und statistisches Manual psychischer Störungen DSM-III-R. Beltz, Weinheim 1989.

Diamond, E. L.: The role of anger and hostility in essential hypertension and coronary heart disease. Psychol. Bull. 92 (1982) 410–433.

Diamond, E. L., C. S. Carver: Sensory processing, cardiovascular reactivity, and the Type A coronary-prone behavior pattern. Biol. Psychol. 10 (1980) 265–275.

Diamond, S.: Depression and headache. Headache 23 (1983) 122–126.

Diamond, S., D. Montrose: The value of biofeedback in the treatment of chronic headache: a 4-year retrospective study. Headache 24 (1984) 5–18.

Dicara, L. V.: Learning in the autonomic nervous system. In: Barber, T. X. et al. (eds.): Biofeedback and self-control. Aldine, Chicago 1971.

Literatur

Dicara, L. V., N. E. Miller: Instrumental learning of systolic blood pressure responses by curarized rats: dissociation of cardiac and vascular changes. Psychosom. Med. 30 (1968) 489–494.

Dichgans, J., A. Brinkmann: Dystonien und Athetosen. In: Brandt, Th., J. Dichgans, H. C. Diener (Hrsg.): Therapie und Verlauf neurologischer Erkrankungen, S. 728–740. Kohlhammer, Stuttgart 1988.

Dichgans, J., H. C. H. Diener, W. D. Gerber, E. J. Verspohl, H. Kukiolka, M. Kluck: Analgetika-induzierter Dauerkopfschmerz. Dtsch. med. Wschr. 109 (1984) 369–373.

Dick, A. P., L. P. Holt, E. R. Dalton: Persistence of mucosal abnormality in ulcerative colitis. Gut 7 (1966) 355–360.

Dick-Read, G.: Natural childbirth. Heinemann, London 1933; deutsch: Die natürliche Geburt. Mutterwerden ohne Schmerz. Hoffmann & Campe, Hamburg 1971.

Dickes, R. A.: Brief therapy of conversion reactions: An inhospital technique. Amer. J. Psychiat. 131/5 (1974) 584–586.

Dickson, D. E.: Morbid miners. Edinburgh Medical Journal 43 (1936) 696.

Dieckmann, H.: Mutterbindung und Herzneurose. Psychosom. Med. 12 (1966) 25.

Dieckmann, W.: Arbeitswelt und Krankheit. In: Viefhues, H. (Hrsg.): Lehrbuch Sozialmedizin, S. 60–64. Kohlhammer, Stuttgart-Berlin-Köln-Mainz 1981.

Diederichs, P.: Urologische Psychosomatik. Springer, Berlin 1983.

Diederichs, P.: Zur Psychosomatik der Miktionsstörungen. Habilitationsschrift, Berlin 1983.

Diederichs, P.: Körpererleben von Männern mit Prostatopathie. In: Brähler, E. (Hrsg.): Körpererleben. Springer, Berlin 1986.

Diederichs, P.: Sexualität und Miktionsstörung. Gynäkologe 19 (1986) 37.

Diederichs, P.: Zur Relevanz narzißmustheoretischer Aspekte in der psychosomatischen Medizin. In: Rudolf, G., U. Rüger, H. H. Studt (Hrsg.): Psychoanalyse der Gegenwart, S. 223. Vandenhoeck & Ruprecht, Göttingen 1987.

Diederichs, P.: Psychosomatische Störungen des männlichen Urogenitaltrakts. In: Brähler, E., A. Meyer (Hrsg.): Partnerschaft, Sexualität und Fruchtbarkeit, S. 207. Springer, Berlin 1988.

Diehl, V., A. Diehl: Medizin zwischen Heil und Unheil. Antrittsvorlesung, Köln 1984.

Diehl, V., K. Köhle: Antrag an die Robert-Bosch-Stiftung auf Verlängerung des Projekts: Integration des psychosomatischen Arbeitsansatzes in die I. Med. Klinik der Universität zu Köln. 1989.

Diener, W.: Beckenkongestionierung: Zur Differentialdiagnose der Prostatitis. Urologe B 21 (1981) 11.

Dilling, H.: Epidemiologie psychischer Störungen und psychiatrische Versorgung. Fortschr. Med. 96 (1978) 1870–1874.

Dilling, H.: Epidemiologie. In: Häfner, H. (Hrsg.): Forschung für die seelische Gesundheit, S. 201–213. Springer, Berlin-Heidelberg-New York 1983.

Dilling, H., S. Weyerer, H. Lisson: Zur ambulanten psychiatrischen Versorgung durch niedergelassene Nervenärzte. Soz. Psych.10 (1975) 111–131.

Dilling, H., S. Weyerer, I. Enders: Patienten mit psychischen Störungen in der Allgemeinpraxis und ihre psychiatrische Überweisungsbedürftigkeit. In: Häfner, H. (Hrsg.): Psychiatrische Epidemiologie, S. 135–160. Springer, Berlin-Heidelberg-New York 1978.

Dilling, H., S. Weyerer, R. Castell: Psychische Erkrankungen in der Bevölkerung. Enke, Stuttgart 1984.

Dillon, K. M., B. Minchoff, K. H. Baker: Positive emotional states and enhancement of the immune system. Int. J. Psychiat. Med. 15 (1985/86) 13–17.

Dimberg, U., A. Oehman: The effects of directional facial cues on electrodermal conditioning to facial stimuli. Psychophysiol. 20 (1983) 160–167.

Dimsdale, J. E., T. P. Hackett, P. C. Block et al.: Emotional correlates of Type A behavior pattern. Psychosom. Med. 40 (1978) 580–583.

Dimsdale, J. E., T. P. Hackett, A. M. Hutter et al.: Type A personality and extent of coronary atherosclerosis. Amer. J. Cardiol. 42 (1978) 583–586.

Dimsdale, J. E., T. P. Hackett, A. M. Hutter et al.: Type A behavior and angiographic findings. J. psychosom. Res. 23 (1979) 273–276.

Dimsdale, J. E., R. P. Newton, Th. Joist: Neuropsychological side effects of beta-blockers. Arch. intern. Med. 149 (1989) 514–525.

Dippel, B., E. Schnabel, S. Bossert, J.-C. Krieg, M. Berger: Vom Lernprozeß im Umgang mit bulimischen Patienten. Prax. Psychother. Psychosom. 33 (1988) 21–34.

Dirks, J. F., A. Paley, K. H. Fross: Panic-fear research in asthma and the nuclear conflict theory of asthma: similarities, differences and clinical implications. Brit. J. med. Psychol. 52 (1979) 71–76.

Dirks, J. F., K. R. Sharon, P. N. Moore: The prediction of psychomaintenance in chronic asthma. Psychother. and psychosom. 36 (1981) 105–115.

Dirks, J. F., J. C. Shraa, E. L. Brown, R. A. Kinsman: Psychomaintenance in asthma: hospitalization rates and financial impact. Brit. J. med. Psychol. 53 (1980) 349–354.

Dirks, J. F., D. J. Horton, R. A. Kinsman, K. H. Fross, N. F. Jones: Patient and physician characteristics influencing medical decisions in asthma. J. Asthma Res. 15 (1978) 171–178.

Dirks, J. F., R. A. Kinsman, N. F. Jones, S. L. Spector, P. T. Davidson, N. W. Evans: Panic-fear: a personality dimension related to length of hospitalization in respiratory illness. J. Asthma Res. 14 (1977) 61–71.

Dirnagl, K., J. Kugler: Wetter und Kopfschmerz. In: Barolin, G. S. (Hrsg.): Kopfschmerz 1981/1, S. 55–66. Braun, Karlsruhe 1981.

Ditschuneit, H., H. H. Ditschuneit, J. Wechsler: Adipositasbehandlung – Nulldiät oder kalorienreduzierte Diät. Internist 20 (1979) 151–158.

Ditto, B.: The application of the twin design to the study of individual differences in psychophysiological measures. Psychophysiol. 25 (1988) 423.

Dixhoorn, J. van, H. J. Duivenvorden, J. A. Staal, J. Pool, V. Verhage: Cardiac events after myocardial infarction: possible effect of relaxation therapy. Eur. Heart J. 8 (11) (1987) 1210–1214.

Dlin, B. M.: Emotional aspects of ileostomy and colostomy. In: Linder (ed.): Emotional Factors in Gastrointestinal Illness. Excerpta Medica, Amsterdam 1973.

Dlin, B. M., W. Winters, K. Fischer, P. Koch: Psychological adaptation to pacemaker following cardiac arrest. Psychosomatics 7 (1966) 73–80.

Dlin, B. M., A. Perlman, R. Ringold: Psychosexual response to ileostomy and colostomy. Amer. J. Psychiat. 126 (1969) 122.

Dlin, B. M., A. Stern, S. Poliakoff: Survivors of cardiac arrest. Psychosomatics 15 (1974) 61–67.

Dmoch, W.: Zur Therapie der Unterleibsschmerzen ohne Organbefund. Vortr. 1. Seminar Univ. Frauenklinik ü. Psychosomatik in Geburtshilfe und Gynäkologie. Düsseldorf 1979.

Dmoch, W., C. Osario: Untersuchungen zur Psychodynamik und Persönlichkeitsstruktur bei Frauen mit vorzeitigen Wehen. In: Frick-Bruder, V., P. Platz (Hrsg.): Psychosomatische Probleme in der Gynäkologie und Geburtshilfe. Springer, Berlin 1984.

Dobreff, M.: Experimenteller Beitrag über den Einfluß von Affekten und Muskelarbeit auf die Urinausscheidung. Arch. ges. Physiol. 213 (1926) 511.

Dobson, M., H. Tattersfield, M. Adler: Attitudes and long-term adjustment of patients surviving cardiac arrest. Brit. med. J. 24 (1971) 207–212.

Dockray, G. J.: The physiology of cholecystokinin in brain and gut. Brit. med. Bull. 38 (1982) 253–258.

Dodge, D. L., W. T. Martin: Social stress and chronic illness. London 1970.

Doehrman, S. R.: Psychosocial aspects of recovery from coronary heart disease – a review. Soc. Sci. Med. 11 (1977) 199–218.

Dölle,W.: Funktionelle Syndrome (Reizmagen, Colon irritabile, Obstipation). Konsequenzen und praktisches Vorgehen. In: Goebell, H., J. Hotz, E. H. Farthmann (Hrsg.): Der chronisch Kranke in der Gastroenterologie. Springer, Berlin-Heidelberg-New York-Tokyo 1984.

Dölle, W., K. H. Wiedmann: Das Colon irritabile. Excerpta Medica, Amsterdam 1978.

Doenecke, P., R. Föthner, G. Harbauer, L. Bethe: Medizinische und soziale Rehabilitation von Schrittmacherträgern. Münch. med. Wschr. 116 (1974) 983–986.

Doerfler, L. G.: Psychogenic deafness and its detection. Ann. Otol. 60 (1951) 1045.

Doering, C. H., B. C. McAdoo, H. C. Kramer, H. K. H. Brodie, N. J. Dessert, D. A. Hamburg: Psychological effects of gonadotropin-releasing hormone in the adult male. In: Usdin, E., D. A. Hamburg, J. D. Barchas (eds.): Neuroregulators and Psychiatric Disorders, pp. 267–275. Oxford Univ. Press, New York 1977.

Dörner, G.: Sex hormone dependent brain differentiation and sexual behavior. In: Wuttke, W., R. Horowski (eds.): Gonadal Steroids and Brain Function. Exp. Brain Res., Suppl. 3, pp. 238. Springer, Berlin 1981.

Doherty, W. J., M. A. Baird: Family therapy and family medicine. Guilford, New York-London 1983.

Dohrenwend, B. P.: Sociocultural and social-psychological factors in the genesis of mental disorders. J. Health soc. Behav. 16 (1975) 365–392.

Dohrenwend, B. P., B. S. Dohrenwend: Social status and psychological disorder: a causal inquiry. Wiley, New York 1969.

Dohrenwend, B. P., B. S. Dohrenwend: Sex differences and psychiatric disorders. Amer. J. Sociol. 81 (1976) 1447–1459.

Dohrenwend, B. S., B. P. Dohrenwend (eds.): Stressful life events: their nature and effects. Wiley, New York 1974.

Dohrenwend, B. S., B. P. Dohrenwend: Some issues in research on stressful life events. J. nerv. ment. Dis. 166 (1978) 7–15.

Dolan, R. J., S. P. Calloway, P. Fonagy, F. V. A. De Souza, A. Wakeling: Life events, depression and hypothalamic-pituitary-adrenal axis function. Brit. J. Psychiat. 147 (1985) 429–433.

Dollard, J., N. E. Miller: Personality and psychotherapy: an analysis in terms of learning, thinking and culture. McGraw-Hill, New York 1950.

Dominian, J., M. Dobson: Study of patients' psychological attitudes to a coronary care unit. Brit. med. J. 4 (1969) 795–798.

Doms, R., L. Hupfauf, D. Langen: Psychosomatische Aspekte bei funktionellen Kiefergelenksbeschwerden. Dtsch. zahnärztl. Z. 24 (1969) 337.

Donald, A., M. D. Kristt, P. T. Engel: Learned control of blood pressure in patients with high blood pressure. Circulation 51 (1957) 370.

Dongier, M.: Psychosomatic aspects in myocardial infarction in comparison with angina pectoris. Psychother. and Psychosom. 23 (1974) 123–131.

Donnison, C. P.: Blood pressure in the African native. Its bearing on the etiology of hypertension and arteriosclerosis. Lancet 1 (1929) 6–7.

Donovan, J. C., L. F. Salzman, P. Z. Allen: The Rochester study: noncognitive data. AAMC-RIME Conference 1970, 207.

Donovan, J. C., L. F. Salzman, P. Z. Allen: Studies in medical education: the role of cognitive and psychological characteristics as career choice correlates. Amer. J. Obstet. Gynec. 114 (1972) 461–468.

Doran, A. R., D. R. Rubinow, A. Roy, D. Pickar: CSF somatostatin and abnormal response to dexamethasone administration in schizophrenic and depressed patients. Arch. gen. Psychiat. 43 (1986) 305–373.

Dorian, B., P. Garfinkel, E. Keystone, R. Gorczyinski, P. Darby, D. Garner: Occupational stress and immunity. Psychosom. Med. 47 (1985) 77.

Dornheim, J.: Kranksein im dörflichen Alltag. Tübinger Vereinigung für Volkskunde e.V., 1983.

Dorossiev, D., V. Paskova, Z. Zachariev: Psychological problems of cardiac rehabilitation. In: Stocksmeier, U. (ed.): Psychological Approach to the Rehabilitation of Coronary Patients, pp. 26–31. Springer, Berlin-Heidelberg-New York 1976.

Dorpat, T. L.: The cognitive arrest hypothesis of denial. Int. J. Psychoanal. 64 (1983) 47–55.

Douglas, D., H. Anisman: Helplessness or expectancy incongruency: effects of aversive stimulation on subsequent performance. J. exp. Psychol. (Hum. Percept. Perform.) 1 (1975) 411–417.

Dovenmühle, R., A. Verwoerdt: Physical illness and depressive symptomatology, I. Incidence of depressive symptoms in hospitalized cardiac patients. J. Amer. med. Ass. 170 (1959) 932–947.

Doyle, J. T., W. B. Kannel, P. McNamara et al.: Factors related to sudden death from coronary disease: combined Albany-Framingham studies. Amer. J. Cardiol. 37 (1976) 1073.

Drayer, J. I. M., M. A. Weber, E. R. Chard: Non-invasive automated blood pressure monitoring in ambulatory normotensive men. In: Weber, M. A., J. I. M. Drayer (eds.): Ambulatory Blood Pressure Monitoring. Steinkopff, Darmstadt 1984.

Drees, A.: Konfliktbearbeitung, Beratung und Psychotherapie in Dialyse-Einrichtungen. Gr. Ther. Gr. Dy. 11 (1976) 150–157.

Drees, A.: Psychosomatische Aspekte der Colitis ulcerosa. Therapie Gegenw. 116 (1977) 1330–1346.

Drees, A.: Balintgruppen in Institutionen. Gruppenpsychother. Gruppendynamik 20 (1984) 76–86.

Drees, A., F. J. Schmidt: Entkopplung oder Ultrakurztherapie des Torticollis. Z. psychosom. Med. 35 (1989) 38–47.

Dreifuss, E.: Der Krebspatient und seine Familie – Erfahrungen aus der Klinik. Schweiz. Rundschau Med. 71 (1982) 1927–1934.

Dreitzel, H. P.: Die gesellschaftlichen Leiden und das Leiden an der Gesellschaft (1968). Taschenbuch-Ausgabe, Enke, Stuttgart 1972.

Drewes, C., J. v. Wietersheim: Themenzentrierte Gruppentherapie. In: Feiereis, H. (Hrsg): Diagnostik und Therapie der Magersucht und Bulimie. Marseille, München 1989.

Dreyfuss, F., H. Dasberg, M. I. Assael: The relationship of myocardial infarction to depressive illness. Psychother. and Psychosom. 17 (1969) 73–81.

Dreyfuss, H. L.: Die Grenzen künstlicher Intelligenz. Athenäum, Frankfurt 1985.

Driscoll, R. H., S. C. Meredith, M. Sitrin, I. H. Rosenberg: Vitamin D deficiency and bone disease in patients with Crohn's disease. Gastroenterology 83 (1982) 1252–1258.

Drossman, D. A.: Psychogenic abdominal pain. In: Chey, W. Y. (ed.): Functional Disorders of the Digestive Tract, pp. 251–258. Raven, New York 1983.

Drost, R.: Sogenannte Prothesenunverträglichkeit. Zahn-ärztl.Welt 87 (1978) 848, 907.

Drotar, D., M. A. Ganowsky: Mental health intervention with children and adolescents with end-stage renal failure. Int. J. Psychiat. Med. 7 (1976/77) 179–192.

Drotar, D., M. A. Ganowsky: A family oriented supportive approach to dialysis and renal transplantation in children. In: Levy, N. B. (ed.): Psychonephrology 1, pp. 79–92. Plenum, New York 1981.

Drühe, C., W. Hartje: Hypoxische Amnesie nach Herzstillstand. Nervenarzt 60 (1989) 280–283.

Drummond, P. D.: Vascular responses in headache-prone subjects during stress. Biol. Psychol. 21 (1985) 11–25.

Druss, R. G., D. A. Kornfeld: The survivors of cardiac arrest. J. Amer. med. Ass. 201 (1967) 291–296.

Dtsch. Ärztebl.: Medizinische Erfahrungen während einer Kreuzfahrt mit Dialysepatienten. 4 (1981) 132.

Dubach, U. C.: Maßvolle Medizin. Medica 4 (1984) 199–204.

Dubach, U. C., K. N. v. Rechenberg: Krankheitsverständnis und Patienten-Arzt-Beziehung in der Ambulanz. Dtsch. med. Wschr. 102/35 (1977) 1239–1244.

Dubin, W. R., K. J. Weiss, J. M. Dorn: Pharmacotherapy of psychiatric emergencies. J. clin. Psychopharmacol. 6 (1986) 210–222.

Dubois, P.: Die Psychoneurosen und ihre seelische Behandlung, 2. Aufl. Francke, Bern 1910.

Duden Band 7, Etymologie, Herkunftswörterbuch der deutschen Sprache. Bibliographisches Institut, Mannheim-Wien-Zürich 1963.

Dudley, D. L., T. H. Holmes, C. J. Martin, H. S. Ripley: Changes in respiration associated with hypnotically induced emotion, pain, and exercise. Psychosom. Med. 26 (1964) 46–57.

Dudley, D. L., C. J. Martin, T. H. Holmes: Dyspnea: psychological and physiologic observations. J. psychosom. Res. 11 (1968) 325–339.

Dührssen, A.: Die Überprüfung prognostischer Urteile bei psychogenen Erkrankungen. Z. Psychother. 2 (1952) 174.

Dührssen, A.: Katamnestische Untersuchungen bei Patienten nach analytischer Psychotherapie. Z. Psychother. 3 (1953) 167–170.

Dührssen, A.: Die Beurteilung des Behandlungserfolges in der Psychotherapie. Z. psychosom. Med. 3 (1956/57) 201.

Dührssen, A.: Katamnestische Ergebnisse bei 1004 Patienten nach analytischer Psychotherapie. Z. psychosom. Med. 8 (1962) 94–113.

Dührssen, A.: Katamnestische Untersuchungen zur Gruppentherapie. Ergebnisse bei 270 behandelten Patienten 5 Jahre nach Abschluß der Therapie. Z. psychosom. Med. 10 (1964) 120.

Dührssen, A: Psychogene Erkrankungen bei Kindern und Jugendlichen. Göttingen 1974.

Dührssen, A., E. Jorswiek: Eine empirisch-statistische Untersuchung zur Leistungsfähigkeit psychoanalytischer Behandlungen. Nervenarzt 36 (1965) 166.

Duffy, E.: Activation. In: Greenfield, N. S., R. A. Sternbach (eds.): Handbook of Psychophysiology. Holt, Rinehart and Winston, New York 1972.

Duller, P., W. Doyle Gentry: Use of biofeedback in treating chronic hyperhidrosis: a preliminary report. Brit. J. Derm. 103 (1980) 143–146.

Dunbar, F.: Emotions and bodily changes. New York 1938.

Dunbar, F.: Psychoanalytic notes relating to syndromes of asthma and hay fever. Psychoanal. Quart.(1938) 25.

Dunbar, F.: Mind and body. Random House, New York 1948.

Dunbar, F.: Psychosomatic diagnosis. Hoeber, New York 1948.

Dunbar, F.: Psychiatry in the medical specialities. Hoeber, New York 1959.

Dunbar, H. F., T. P. Wolfe, J. M. Rioch: Psychiatric aspects of medical problems: the psychic component of the disease process (including convalescence) in cardiac, diabetic and fracture patients. Amer. J. Psychiatr. 93 (1936) 649–679.

Dupont, B., E. Flenstedt-Jensen, E. Sandoe: The long-term prognosis for patients after cardiac arrest. Amer. Heart J. 78 (1969) 444–449.

Dupuis, A.: Assessment of the psychological factors and responses in self-managed patients. Diabetes Care 3 (1980) 117–120.

Durkheim, E.: Erziehung und Soziologie (1922). Düsseldorf 1972.

Dusch, T. v.: Lehrbuch der Herzkrankheiten, S. 334. Engelmann, Leipzig 1868.

Dustan, H. P.: Obesitas und hypertension. In: Lauer, R. M., R. B. Shekelle (eds.): Childhood Prevention of Atherosclerosis and Hypertension, pp. 305–312. Raven, New York 1980.

Dvorak, J., M. H. Gauchat, L. Valach: The outcome of operation for lumbar disc herniation. A 4–17 years follow-up with emphasis on the somatic aspects. Part I. Spine 13 (1988a) 1418–1422.

Dvorak, J., L. Valach, P. Fuhrimann, E. Heim: The outcome of operation for lumbar disc herniation. A 4–17 years follow-up with emphasis on psychosocial aspects. Part II. Spine 13 (1988b) 1423–1427.

Dworkin, B. R., R. J. Filewich, N. E. Miller, N. Craigmyle: Baroreceptor activation reduces reactivity to noxious stimulation: implications for hypertension. Science 205 (1979) 1299–1301.

Dworkin, B. R., T. Pickering, N. E. Miller, L. Eisenberg, B. Brucker: Continuous noninvasive recording of blood pressure changes during attempted muscular contraction in patients with severe muscular paralysis. Personal communication 1975 (zit. nach: Obrist, P. A.: Cardiovascular Psychophysiology. Plenum, New York 1981).

Dworkin, S., J. Marbach: Group therapy with chronic MPD patients. J. dent. Res. (spec. issue) 53 (1974) 126.

Dyck, D. G., S. M. Driedger, R. Nemeth, T. A. G. Osachuk: Conditioned tolerance to drug-induced (Poly I:C) natural killer cell activation: effects of drug-dosage and context-specific parameters. Brain, Behavior, and Immunity 1 (1987) 251–266.

Dyck, D. G., A. H. Greenberg, T. A. G. Osachuk: Tolerance to drug-induced (Poly I:C) natural killer cell activation: congruence with a pavlovian conditioning model. J. exp. Psychol. (Anim. Behav.) 12 (1986) 25–31.

Dyck, D. G., T. A. G. Osachuk, A. H. Greenberg: Drug-induced (Poly I:C) pyrexic responses. Congruence with a compensatory conditioning analysis. Psychobiol. 17 (1989) 171–178.

Dyck, D. G., T. A. G. Osachuk, L. Janz, A. H. Greenberg: Mechanisms of conditioned tolerance to immunostimulatory drugs. Paper presented at the XXIV. Int. Congress of Psychology, Sydney, Australia, August 1988.

Eagle, K.: Evaluation of patients with syncope. New Engl. J. Med. 29 (1983) 1630.

Eaker, E. D., M. Feinleib: Psychological factors and the 10-year incidence of cerebrovascular accident in the Framingham Heart Study. Psychosom. Med. 45 (1981) 84.

Eaker, E. D., S. G. Haynes, M. Feinleib: Spouse behavior

and coronary heart disease in men: prospective results from the Framingham Heart Study. Amer. J. Epidemiol. 118 (1983) 23–41.

East, T.: The story of heart disease. Dawson, London 1957.

Easterbrook, J. A.: The effect of emotion on cue utilization and the organization of behavior. Psychol. Rev. 66 (1959) 183–201.

Eastman, J., G. B. Mesibov: Family intervention in a private pediatric practice. J. Marital and Fam. Therapy 7 (1981) 461.

Eaton, J. W. et al.: Resistance to psychiatry in a general hospital. Ment. Hosp. 16 (1965) 156.

Ebbinghaus, H.: Über eine neue Methode zur Prüfung geistiger Fähigkeiten und ihrer Anwendung bei Schulkindern. Z. Psychol. 13 (1897).

Eberbach, W. H.: Die ärztliche Schweigepflicht. In: Jäger, H. (Hrsg.): Aids und die HIV-Infektionen. ecomed, Landsberg 1988.

Eberhard, G.: Peptic ulcer in twins – a study in personality, heredity, and environment. Acta psychiat. scand. (Suppl.) 205 (1968) 44.

Eberspächer, H. E.: Über die Integration von Psyche und Soma im methodischen Vorgehen der Funktionellen Entspannung. In: Lamprecht, F. (Hrsg.): Spezialisierung und Integration in Psychosomatik und Psychotherapie, S. 160. Springer, Berlin-Heidelberg-New York 1987.

Eccles, J. C.: Hirn und Bewußtsein. In: Ditfurth, H. v. (Hrsg.): mannheimer forum. Boehringer Mannheim 1977/78.

Eckensberger, D., G. Overbeck, E. Wolff: Ein objektivierendes Verfahren zur diagnostischen Untergruppenbildung von Ulkuskranken. Z. psychosom. Med. Psychoanal. 23 (1977) 371–386.

Ecker, A.: Emotional stress before strokes: a preliminary report of 20 cases. Ann. intern. Med. 40 (1954) 49–56.

Eckert, E. D.: Behavior modification in anorexia nervosa: a comparison of two reinforcement schedules. In: Darby, P. L., P. E. Garfinkel, D. M. Garner, D. V. Coscina (eds.): Anorexia nervosa. Recent Developments in Research, pp. 377–385. Liss, New York 1983.

Edholm, O. C.: Emotion and stress. In: Assemacher, J., D. S. Farner (eds.): Environmental Endocrinology, pp. 2–6. Springer, Berlin 1978.

Edinger, J. D., A. L. Stout, T. J. Hoelscher: Cluster analysis of insomniacs' MMPI profiles: relation of subtypes to sleep history and treatment outcome. Psychosom. Med. 50 (1988) 77–87.

Editorial: Benzodiazepine: kein Mittel gegen Alltagsärger. Dtsch. Ärztebl. (1981) 2229.

Editorial: Placebos. Brit. med. J. 1952, zit. n. Brody, H.: Placebos and the philosophy of medicine. Univ. of Chicago Press, Chicago 1977.

Editorial: The humble humbug. Lancet 267 (1954) 321.

Edmonds, E. P.: Psychosomatic non-articular rheumatism. Ann. rheum. Dis. 6 (1947) 36–49.

Edwards, E. A., R. H. Rahe, P. M. Stephens, I. P. Henry: Antibody response to bovine serum albumin in mice: the effects of psychosocial environmental change. Proc. Soc. exp. Biol. (N. Y.) 164 (1980) 478–481.

Edwards, F. C., S. C. Truelove: The course and prognosis of ulcerative colitis. Gut 4 (1963) 299–308; 309–315. Gut 5 (1964) 1–22.

Edwards, F. C., N. F. Coghill: Clinical manifestations in patients with chronic atrophic gastritis, gastric ulcer, and duodenal ulcer. Quart. J. Med. 37 (1968) 337–360.

Edwards, M. H., J. J. Calabro, M. E. Wied: Patients' attitudes and knowledge concerning arthritis. Arthr. and Rheum. 7 (1964) 425–435.

Efron, R.: The effect of olfactory stimulus in arresting uncinate fits. Brain 79 (1956) 267–281.

Efron, R.: The conditioned inhibition of uncinate fits. Brain 80 (1957) 251–262.

Egbert, L. D., S. E. Battit, C. E. Welch, M. K. Bartlett: Reduction of postoperative pain by encouragement and instructions of patients. A study of doctor-patient-rapport. New Engl. J. Med. 270 (1964) 825–827.

Egeland, J. A., A. M. Hostetter: Amish study, I. Affective disorders among the Amish, 1976–1980. Amer. J. Psychiat. 140 (1983) 56–61.

Egeland, J. A., D. S. Gerhard, D. L. Pauls, J. N. Sussex, K. K. Kidd, C. R. Allen, A. M. Hostetter, D. A. Housman: Bipolar affective disorders linked to DNA markers on chromosome 11. Nature (Lond.) 325 (1987) 783–787.

Egeren, L. F. van: Cardiovascular changes during social competition in a mixed motive game. J. Pers. soc. Psychol. 37 (1979) 858–864.

Egerton, N., J. H. Kay: Psychological disturbances associated with open-heart surgery. Brit. J. Psychiat. 110 (1964) 433–439.

Eggers, C.: Depressive Kleinkinder. Welche Rolle spielt die Mutter-Kind-Beziehung? Medical Tribune 16 (1977) 54.

Egle, U.: Die Arzt-Patient-Beziehung als affektives Lernziel im Medizinstudium – Konzept und Evaluation der Anamnesegruppe. Dissertation, Marburg 1982.

Egle, U.: Patientenorientierte Medizinerausbildung – Reader zur studentischen Selbsthilfe. Marburg 1982 (unveröffentlicht).

Egle, U. T.: Auf der Suche nach den Wurzeln psychogen bedingter Mund-Krankheiten. Zahnärztl. Mitt. 75 (1985) 2413.

Egle, U. T., M. L. Rudolf, S. O. Hoffmann, K. König, M. Schöfer, R. Schwab, H. v. Wilmowsky: Persönlichkeitsmerkmale, Abwehrverhalten und Krankheitserleben bei Patienten mit primärer Fibromyalgie. Z. Rheum. 48 (1989) 73–78.

Ehlers, A., J. Margraf, W. T. Roth, C. B. Taylor, R. J. Maddock, J. Sheikh, M. L. Kopell, K. L. McClenahan, D. Gossard, G. H. Blowers, W. S. Agras, G. H. Kopell: Lactate infusion and panic attacks: do patients and controls respond differently? Psychiat. Res. 17 (1986) 295–308.

Ehlers, C. L., E. Frank, D. J. Kupfer: Social zeitgebers and biological rhythms. An unified approach to understanding the etiology of depression. Arch. gen. Psychiat. 45 (1988) 948–952.

Ehlers, W.: Die Abwehrmechanismen: Definitionen und Beispiele. Prax. Psychother. Psychosom. 28 (1983) 55–66.

Ehlich, K.: Zur Struktur der psychoanalytischen Deutung. Unveröffentl. Aufsatz, Universität Tilburg 1981.

Ehrenfels, Ch. v.: Historisches Wörterbuch der Philosophie, Bd. 3, S. 546 u. 550. Schwabe & Co., Basel-Stuttgart 1969.

Ehrenwald, H.: Anosognosie und Depersonalisation. Nervenarzt 4 (1931) 681–688.

Ehret, R., E. Schneider: Photogene Epilepsie mit suchtartiger Selbstauslösung kleiner Anfälle und wiederholten Sexualdelikten. Arch. Psychiat. Nervenkr. 202 (1961) 75–94.

Ehrhardt, A. A., H. F. L. Meyer-Bahlburg: Effects of prenatal sex hormones on gender-related behavior. Science 211 (1981) 1312–1318.

Ehring, F.: Die Nachsorge bei Haut- und Gesichtskrebs. GBK-Mitteilungsdienst 20 (1978).

Ehrström, M. C.: Psychogene Blutdrucksteigerung. Kriegshypertonien. Acta med. scand. 122 (1945) 546–570.

Eibach, H.: Die Psychodynamik einer chronischen Herzneurose im Lichte des Katathymen Bilderlebens – Behandlung und zugleich ein Beitrag zur 'endlichen Analyse'. In: Leuner, H., O. Lang (Hrsg.): Psychotherapie mit dem Tagtraum, S. 203–228. Huber, Bern 1982.

Literatur

Eibl-Eibesfeldt, I.: Die Biologie des menschlichen Verhaltens. Grundriß der Humanbiologie. Piper, München 1984.

Eicher, W.: Die sexuelle Erlebnisfähigkeit und die Sexualstörungen der Frau. Fischer, Stuttgart 1975.

Eicher, W. (Hrsg.): Sexualmedizin in der Praxis. Fischer, Stuttgart-New York 1980.

Eicher, W., V. Herms: Sexualverhalten nach gynäkologischen Operationen und Carcinomtherapie. Sexualmed. 5 (1976) 36.

Eicher, W., V. Herms: Zur Epidemiologie des Mamma-Carcinoms. Fortschr. Med. 97 (1979) 1683.

Eicher, W., V. Herms, C. Repschläger, F. Kubli: Psychosomatik der Hysterektomie. Sexualmed. 4 (1975) 351.

Eichhorn, T., G. Martin: Verlauf und Prognose beim Hörsturz. HNO 32 (1984) 341.

Eicke, D.: Der Körper als Partner – Plädoyer für eine psychosomatische Krankheitslehre. München 1973.

Eickhoff, W.: Schilddrüse und Basedow. Thieme, Stuttgart 1949.

Eiff, A. W. v., C. Piekarski: Stress reactions of normotensives and hypertensives and the influence of female sex hormones on blood pressure regulation. In: De Jong, W., A. P. Provoost (eds.): Progress in Brain Research, Vol. 47: Hypertension and Brain Mechanisms, pp. 289–299. Elsevier, Amsterdam-Oxford-New York 1977.

Eiff, A. W. v., H. Neus: Verkehrslärm und Hypertonie-Risiko. Münch. med. Wschr. 122 (1980) 894–896.

Eiff, A. W. v., G. Kloska, H. Quint: Essentielle Hypertonie. Klinik, Psychophysiologie und Psychopathologie. Thieme, Stuttgart 1967.

Eikelboom, R., J. Stewart: Conditioning of drug-induced physiological responses. Psychol. Rev. 89 (1982) 507–528.

Einstein, A., L. Infeld: The evolution of physics. Simon & Schuster, New York 1938.

Eisenberg, L.: Science in medicine: too much or too little and too limited in scope? Amer. J. Med. 84 (1988) 483–491.

Eisenberg, M. S., A. Hallstrom, L. Bergner: Long-term survival after out-of-hospital cardiac arrest. New Engl. J. Med. 306 (1982) 1340–1344.

Eissler, K. R.: The effect of the structure of the ego on psychoanalytic technique. J. Amer. psychoanal. Assoc. 1 (1953) 104–143.

Eissler, K. R.: The psychiatrist and the dying patient. Int. Univ. Press, New York 1955. Deutsch: Der sterbende Patient: Zur Psychologie des Todes. Frommann-Holzboog, Stuttgart 1978.

Ekbom, K. A.: Restless legs. Acta med. scand. (Suppl.) 158 (1945) 1–123.

Ekbom, K. A.: Restless legs syndrome. Neurology 10 (1960) 868–873.

Ekman, P.: Universal and cultural differences in facial expression of emotion. In: Nebraska Symposium on Motivation 1971, pp. 207–283.

Ekman, P.: Gesichtsausdruck und Gefühl. Junfermann, Paderborn 1988.

Elbert, T., B. Rockstroh, W. Lutzenberger, M. Kessler, R. Pietrowsky, N. Birbaumer: Baroreceptor stimulation alters pain sensation depending upon tonic blood pressure. Psychophysiol. 25 (1988) 25–29.

Elder, R. G.: Social class and lay explanations of the etiology of arthritis. J. Health Social Behavior 14 (1973) 28–38.

Elder, S. T., Z. R. Ruiz, H. L. Deabler, R. L. Dillenkofer: Instrumental conditioning of diastolic blood pressure in essential hypertensive patients. J. appl. anal. Behav. 6 (1973) 377–382.

Eliasson, K., P. Hjemdahl, T. Kahan: Circulatory and sympatho-adrenal responses to stress in borderline and established hypertension. J. Hypertension 1 (1983) 131–139.

Eliasson, S., B. Folkow, P. Lindgren, B. Uvnäs: Activation of sympathetic vasodilatator nerves to the skeletal muscles in the cat by hypothalamic stimulation. Acta physiol. scand. 23 (1951) 333–351.

Elkin, I., P. A. Pilkonis, J. P. Docherty, S. M. Satsky: Conceptual and methodological issues in comparative studies of psychotherapy and pharmacotherapy, I: Active ingredients and mechanisms of change. Amer. J. Psychiat. 145 (1988) 909–917.

Elkin, T. E., M. Hersen, R. Eissler et al.: Modification of caloric intake in anorexia nervosa: an experimental analysis. Psychol. Rep. 32 (1973) 75–78.

Ellertsen, B., H. Nordby, D. Hammerborg, S. Thorlacius: Psychophysiologic response patterns in migraine before and after temperature biofeedback. Prediction of treatment outcome. Cephalalgia 7 (1987) 109–124.

Elliott, G. R., C. Eisdorfer: Stress and human health. Springer, New York 1982.

Elliott, R.: The significance of heart rate for behavior: a critique of Lacey's hypothesis. J. Pers. soc. Psychol. 22 (1972) 398–409.

Elliott, R.: The motivational significance of heart rate. In: Obrist, P. A., A. H. Black, J. Brener, L. V. Dicara (eds.): Cardiovascular Psychophysiology. Aldine, Chicago 1974.

Elliott, R.: Further comment on the Lacey hypothesis. J. Pers. soc. Psychol. 30 (1974) 19–23.

Elliott, R. S., J. C. Buell: Role of the central nervous system in sudden cardiac death. In: Dembroski, T. M., T. H. Schmidt, G. Blümchen (eds.): Biobehavioral Bases of Coronary Heart Disease, pp. 257–270. Karger, Basel-München-Paris 1983.

Elliott, R. S., A. D. Forker, R. J. Robertson: Aerobic exercises as a therapeutic modality in the relief of stress. Advanc. Cardiol. 18 (1976) 231–242.

Elliott, R. S., G. L. Todd, F. C. Clayton, G. M. Pieper: Experimental catecholamine-induced acute myocardial necrosis. In: Manninon, Halonen: Sudden Coronary Death. Advanc. Cardiol. 25 (1978) 107–118.

Ellis, J. P.: A patient's commentary. In: Levy, N. B. (ed.): Living or Dying, pp. 57–61. Thomas, Springfield/Ill. 1974.

Elst, P., T. Sybesma, J. van der Stadt, A. P. A. Prins, W. Hissink Muller, A. den Butter: Sexual problems in rheumatoid arthritis and ankylosing spondylitis. Arthr. and Rheum. 27 (1984) 217–220.

Ely, D. L., J. P. Henry: Ethological and physiological theories. In: Kutash, I. L., L. B. Schlesinger et al. (eds.): Handbook on Stress and Anxiety. Jossey Bass, San Francisco 1980.

Emde, R.: Anaclitic depression. A follow-up from infancy to puberty. Psychoanal. Study Child 37 (1982) 67–94.

Emde, R.: Development terminable and interminable. I. Innate and motivational factors from infancy. Int. J. Psychoanal. 69/1 (1988) 23–42. II. Recent psychoanalytic theory and therapeutic considerations. Int. J. Psychoanal. 69/2 (1988) 283–296.

Emeryk, B., K. Rowinska, T. Nowak-Michalska: Do true remissions in myasthenia gravis exist? J. Neurol. 231 (1985) 331–335.

Emmelkamp, P. M. G.: Treatment of obsessive-compulsive patients: the contribution of self-instructional training to the effectiveness of exposure. Behav. Res. Ther. 18 (1980) 61–66.

Emmett, S. W. (ed.): Theory and treatment of anorexia and bulimia. Brunner/Mazel, New York 1985.

Emmons, C. A., J. G. Joseph, R. C. Kessler, C. B. Wortman, S. B. Montgomery, D. G. Ostrow: Psychosocial predictors of reported behavior change in homosexual men at risk for AIDS. Health Educ. Quart. 13 (1986) 331–345.

Emrich, H. M., D. v. Zerssen: Beta-Rezeptoren-Blocker:

Grundlagen und Therapie. In: Langer, G., H. Heimann (Hrsg.): Psychopharmaka – Grundlagen und Therapie. Springer, Wien 1983.

Endler, N. S., J. McV. Hunt, A. J. Rosenstein: An S-R-inventory of anxiousness. Psychol. Monographs 76 (1962) 1.

Engel, B. T.: Clinical applications of operant conditioning techniques in the control of cardiac arrhythmias. Semin. in Psychiat. 5 (1973) 433.

Engel, B. T., A. F. Bickford: Response specifity, stimulus-response and individual-response specifity in essential hypertension. Arch. gen. Psychiat. 5 (1961) 478–489.

Engel, B. T., K. R. Gaarder, M. S. Glasgow: Behavioral treatment of high blood pressure: I. Analysis of intra- and interdaily variations of blood pressure during a 1-month baseline period. Psychosom. Med. 43 (1981) 255–270.

Engel, B. T., K. R. Gaarder, M. S. Glasgow: Behavioral treatment of high blood pressure: III. Follow-up results and treatment recommendations. Psychosom. Med. 45 (1983) 23–29.

Engel, G. L.: Primary atypical facial neuralgia. An hysterical conversion symptom. Psychosom. Med. 13 (1951) 375–396.

Engel, G. L.: Selection of clinical material in psychosomatic medicine. Psychosom. Med. 16 (1954) 368–373.

Engel, G. L.: Studies of ulcerative colitis, III. The nature of psychological processes. Amer. J. Med. 19 (1955) 231–256.

Engel, G. L.: Studies of ulcerative colitis. IV. The significance of headaches. Psychosom. Med. 18 (1956) 334–346.

Engel, G. L.: Studies of ulcerative colitis V. Psychological aspects and their implications for treatment. Amer. J. digest. Dis. 3 (1958) 315–337.

Engel, G. L.: Psychogenic pain and the pain-prone patient. Amer. J. Med. 26 (1959) 899–918.

Engel, G. L.: Biologic and psychologic features of the ulcerative colitis patient. Gastroenterology 40 (1961) 313–322.

Engel, G. L.: Is grief a disease? A challenge for medical research. Psychosom. Med. 23 (1961) 18.

Engel, G. L.: Fainting. 2nd ed. Thomas, Springfield 1962.

Engel, G. L.: A life setting conductive to illness. The giving up – given up complex. Ann. intern. Med. 69 (1968) 293–300.

Engel, G. L.: Psychological development in health and disease. Philadelphia 1962. Deutsch: Psychisches Verhalten in Gesundheit und Krankheit. Huber, Bern-Stuttgart-Wien 1969.

Engel, G. L.: Psychological factors in ulcerative colitis in man and gibbon. Gastroenterology 57 (1969) 362–365.

Engel, G. L.: Psychological processes and gastrointestinal disorders. In: Paulson, M. (ed.): Gastroenterologic Medicine. Lea & Febiger, Philadelphia 1969.

Engel, G. L.: Psychological aspects of illness. Unpublished manuscript 1970.

Engel, G. L.: In: MacBryde, C. M., R. S. Blacklow (eds.): Signs and Symptoms: Applied Physiology and Clinical Interpretation. Chap. 30, 5th ed., Lippincott, 1970.

Engel, G.L: The education of the physician for clinical observation. The role of the psychosomatic (liaison) teacher. J. nerv. ment. Dis. 154 (1972) 159–164.

Engel, G. L.: Is psychiatry failing in his responsibilities to medicine? Amer. J. Psychiat. 128 (1972) 1561–1563.

Engel, G. L.: Diskussionsbeitrag. In: Weiss, J. M. (ed.): Influence of psychological variables on stress-induced pathology. In: Ciba Foundation Symposium 8: Physiology, Emotion and Psychosomatic Illness. Elsevier, North Holland-London-Amsterdam 1972.

Engel, G. L.: Identification, inspiration and learning. Arch. intern. Med. 135 (1975) 1981–1983.

Engel, G. L.: Psychisches Verhalten in Gesundheit und Krankheit. 2. Aufl. Huber, Bern-Stuttgart-Wien 1976.

Engel, G. L.: The need for a new medical model: a challenge for biomedicine. Science 196 (1977) 129–136.

Engel, G. L.: Psychologic stress, vasodepressor (vasovagal) syncope, and sudden death. Ann. intern. Med. 89 (1978) 403.

Engel, G. L.: Colitis ulcerosa. In: Uexküll, Th.v. (Hrsg.): Lehrbuch der Psychosomatischen Medizin. Urban & Schwarzenberg, München-Wien-Baltimore 1979.

Engel, G. L.: The biopsychosocial model and medical education. New Engl. J. Med. 306 (1982) 802–805.

Engel, G. L.: How much longer must medicine's science be bound by a 17th century world view? In: White, K. L. (ed.): The Task of Medicine, Dialogue at Wickenburg. The Henry J. Kaiser Family Foundation, Menlo Park CA 1988.

Engel, G. L., J. Romano: Delirium, a syndrome of cerebral insufficiency. J. chron. Dis. 9 (1959) 260–277.

Engel, G. L., A. H. Schmale: Eine psychoanalytische Theorie der somatischen Störung. Psyche 23 (1969) 241–261.

Engel, G. L., A. H. Schmale: Conservation – withdrawal: a primary regulatory process for organismic homeostasis. Ciba Found. Symp. 8 (1972) 52–75.

Engel, G. L., A. H. Schmale jr.: Psychoanalytic theory of somatic disorder. J. Amer. psychoanal. Ass. 15 (1967) 344–365. Deutsch in: Overbeck, G., A.Overbeck (Hrsg.): Seelischer Konflikt – körperliches Leiden, S. 246–268. Hamburg 1978.

Engel, G. L., E. B. Ferris, M. Logan: Hyperventilation: analysis of clinical symptomatology. Ann. intern. Med. 27 (1947) 683–704.

Engel, G. L., F. Reichsman, H. L. Segal: A study of an infant with a gastric fistula: I. Behavior and the rate of total HCl secretion. Psychosom. Med. 18 (1956) 374.

Engel, G. L., W. W. Hamburger, M. Reiser, J. Plunkett: Electroencephalographic and psychological studies of a case of migraine with severe preheadache phenomena. Psychosom. Med. 15 (1953) 337–348.

Engel, K.: Testing cooperation in parents with children destined for home dialysis. Psychother. and Psychosom. 30 (1978) 28.

Engel, K., A.-E. Meyer: Theorie und Empirie einer mehrfaktoriellen stationären Anorexie-Therapie für schwerkranke Patienten. Med. Welt 33 (1982) 1812–1816.

Engel, K., C. Krüger, A.-E. Meyer: Elemente einer mehrfaktoriellen stationären Anorexie-Behandlung für schwererkrankte Patienten. Med. Welt 38 (1987) 1441–1449.

Engel, K., E. Haas, M. v. Rad, W. Senf, H. Becker: Heidelberger Rating. Med. Psychol. 5 (1979) 253–268.

Engel, K., M.v. Rad, H. Becker, W. Bräutigam: Das Heidelberger Katamneseprojekt. Med. Psychol. 5 (1979) 124–137.

Engelhardt, K.: Der Patient in seiner Krankheit. Thieme, Stuttgart 1971.

Engelhardt, K.: Kranke im Krankenhaus. Enke, Stuttgart 1973.

Engelhardt, K.: Morbus Crohn. Schlesw.-Holst. Ärztebl. 35 (1982) 734–741; 838–843.

Engels, F.: Die Lage der arbeitenden Klasse in England (1845), 5. Aufl. Dietz, Berlin 1972.

Enke, H., G. Gercken: Der seelische Befund bei essentiellen Hypertonikern. Psychodiagnostisch-statistische Untersuchungen. Klin. Wschr. 33 (1955) 551.

Enna, S. J.: The role of neurotransmitters in the pharmacologic actions of benzodiazepines. In: Mathew, R. J. (ed.): The Biology of Anxiety. Brunner & Mazel, New York 1983.

Ennulat, A.: Zur sozialen Integration von Patientinnen mit Anorexie und Bulimie. Med. Diss., Lübeck 1988.

Ensminger, M. E., D. D. Celentano: Unemployment and psychiatric distress: social resources and coping. Soc. Sci. Med. 27 (1988) 239–247.

Entwurf einer 5. Verordnung zur Änderung der Approbationsordnung für Ärzte, Bonn, 9.9.83.

Eppinger, G., G. Endsberger: Kreuzschmerzen, Hinweis und Ablenkung von Grundleiden. Z. Allgemeinmed. 31 (1975) 1423–1425.

Eppinger, H., L. Hess: Die Vagotonie. Springer, Berlin 1910.

Epstein, A. N.: The lateral hypothalamic syndrome. In: Stellar, E., J. M. Sprague (eds.): Progress in Physiological Psychology, pp. 263–317. Acad. Press, New York 1971.

Epstein, A. N.: The neuroendocrinology of thirst and salt appetite. In: Ganong, W. F. (ed.): Frontiers in Neuroendocrinology, Vol. 5, pp. 101–134. Oxford Univ. Press, New York 1978.

Epstein, A. W., F. Ervin: Psychodynamic significance of seizure content in psychomotor epilepsy. Psychosom. Med. 18 (1956) 43–55.

Epstein, F. H.: Die Epidemiologie des Hochdrucks. Verh. Dtsch. Ges. Inn. Med. 80 (1974) 36–42.

Epstein, S.: Toward an unified theory of anxiety. In: Maher, B. (ed.): Progress in Experimental Personality Research. Acad. Press, New York 1967.

Epstein, S., W. D. Fenz: Steepness of approach and avoidance gradients in humans as a function of experience. Theory and experiment. J. exp. Psychol. 70 (1965) 1–12.

Eraker, St.A., J. P. Kirscht, M. H. Becker: Understanding and improving patient compliance. Ann. intern. Med. 100 (1984) 258–268.

Erdmann, G., W. Janke: Interaction between physiological and cognitive determinants of emotional state: experimental studies on Schachter's theory of emotion. Biol. Psychol. 6 (1978) 61–74.

Erdmann, H., H. G. Overrath, W. Adam, Th.v. Uexküll: Organisationsprobleme der ärztlichen Krankenversorgung. Dtsch. Ärztebl. 71 (1974) 3421–3426.

Erfmann, I.: Age and manifestation of psychosomatic disorders. Vita humana 5 (1962) 161–166.

Erickson, M. H.: The collected papers of M. H. Erickson on hypnosis, Vol. I–IV. Edited by E. L. Rossi. Irvington, New York 1980.

Erickson, M. H., E. L. Rossi: Autohypnotic experiences of Milton H. Erickson. Amer. J. clin. Hypnosis 20 (1977) 36–54.

Erickson, M. H., E. L. Rossi: Hypnose. Induktion, psychotherapeutische Anwendung, Beispiele. Pfeiffer, München 1978.

Erickson, M. H., E. L. Rossi: Hypnotherapie. Aufbau, Beispiele, Forschungen. Pfeiffer, München 1981.

Erikson, E. H.: Childhood and society. New York 1950. Deutsch: Kindheit und Gesellschaft. Klett, Stuttgart 1965.

Erkrath, F. A., H. Randow: Vita sexualis nach Karzinombehandlung im Vergleich zu nicht-krebskranken berufstätigen Frauen. Zbl. Gynäk. 33 (1967) 1209.

Ermann, M.: Psychotherapeutische Katamnestik: Zielsetzung und Perspektiven. Psychother. and Psychosom. 24 (1974) 359–367.

Ermann, M.: Die Grundstörungen bei depressiven Neurosen und psychosomatischen Störungen. Z. psychosom. Med. 26 (1980) 316–328.

Ermann, M.: Zum ichpsychologischen Verständnis der vegetativen psychosomatischen Störungen. In: Studt, H. H. (Hrsg.): Neue Ergebnisse der Psychosomatik. Urban & Schwarzenberg, München-Wien-Baltimore 1983.

Ernst, K.: Die Prognose der Neurosen. Springer, Berlin-Göttingen-Heidelberg 1959.

Ernst, K., G. A. Schoenenberger: DSIP: basic findings in human beings. In: Inoué, S., D. Schneider-Helmert (eds.): Sleep Peptides: Basic and Clinical Approaches, pp. 131–173. Japan Sci. Soc. Press, Tokyo/ Springer, Berlin 1988.

Ernst, S., W. Klosterhalfen, S. Klosterhalfen: Examination stress and salivary immunoglobulin A. J. Psychophysiol. 1 (1987) 297.

Eron, L. D.: Effect of medical education on medical students' attitudes. J. med. Educ. 10 (1955) 559–566.

Eron, L. D.: The effect of medical education on attitudes: a follow-up study. J. med. Educ. 33 (1958) 25–33.

Erskine-Milliss, J., M. Schonell: Relaxation therapy in asthma: a critical review. Psychosom. Med. 43 (1981) 365–372.

Escher, F.: Psychosomatik in der ORL-Praxis. In: Kellerhals, B. et al. (Hrsg.): Aktuelle Probleme der Otorhinolaryngologie 8 (1984) 322. Huber, Bern 1984.

Esler, M., S. Julius, O. Randall, V. DeQuattro, A. Zweifler: High-renin essential hypertension: adrenergic cardiovascular correlates. Clin. Sci. Mol. Med. 51 (Suppl. 3) (1976) 181s-184s.

Espie, C. A., W. R. Lindsay, D. N. Brooks, E. M. Hood, T. Turvey: A controlled comparative investigation of psychological treatments for chronic sleep-onset insomnia. Behav. Res. Ther. 27 (1989) 79–88.

Eßer, G.: Colitis ulcerosa und Morbus Crohn aus chirurgischer Sicht. Med. Welt 32 (1981) 606–610.

Estes, W. K., B. F. Skinner: Some quantitative properties of anxiety. J. exp. Psychol. 29 (1941) 390–400.

Ettl, T.: Bulimia nervosa – die heimliche unheimliche Aggression. Z. psychoanal. Theorie u. Praxis 3 (1988) 48–76.

Etzel, B. C., J. L. Gewirtz: Experimental modification of caretaker-maintained high-rate operant crying in a 6 and a 20 week old infant (Infants tyrannotearus): extinction of crying with reinforcement of eye contact and smiling. J. exp. Child Psychol. 5 (1967) 303–317.

Evans, F. J., M. R. Cook, H. D. Cohen, E. C. Orne, M. T. Orne: Appetitive and replacement naps: EEG and behavior. Science 197 (1977) 687–689.

Evans, J. G., E. D. Acheson: An epidemiological study of ulcerative colitis and regional enteritis in the Oxford area. Gut 6 (1965) 311–324.

Evans, M. B., G. L. Paul: Effects of hypnotically suggested analgesia on physiological and subjective response to cold stress. J. consult. Psychol. 35 (1970) 362–371.

Evans, R. W.: Cyclosporine in cadaveric renal transplantation. New Engl. J. Med. 311 (1984) 127.

Evans, R. W., D. L. Manninen, L. P. Garrison et al.: The quality of life of patients with end-stage renal disease. New Engl. J. Med. 312 (1985) 553–559.

Everhart, D. L., B. Klapper, W. H. Carter, S. Moos: Evaluation of dental caries experience and salivary IgA in children age 3–7. Caries Res. 11 (1977) 211–215.

Every, R. G.: The teeth as weapons – their influence on behavior. Lancet 7387 (1965) 685.

Ewan, C. E., M. J. Bennet: Medicine in prospect – the first year student's view. Med. Educ. 15 (1981) 287–293.

Export, V.: Das Reale und sein Double: Der Körper. Benteli, Bern 1987.

Eysenck, H. J.: Dimensions of personality. London 1947.

Eysenck, H. J.: The effects of psychotherapy: an evaluation. J. consult. Psychol. 16 (1952) 319–324.

Eysenck, H. J.: The dynamics of anxiety and hysteria. London 1957.

Eysenck, H. J.: The scientific study of personality. London 1958.

Eysenck, H. J.: The biological basis of personality. Thomas, Springfield 1967.

Eysenck, H. J.: A theory of the incubation of anxiety/fear responses. Beh. Res. Ther. 6 (1968) 309–321.

Eysenck, H. J.: The measurement of emotions: psychological parameters and methods. In: Levi, L. (ed.): Emotions: Their Parameters and Measurement. Raven, New York 1975.

Eysenck, H. J.: Personality as predictor of cancer and cardiovascular disease, and the application of behavior therapy in prophylaxis. Europ. J. Psychiat. 1 (1987) 29–41.

Eysenck, S. B. G., H. J. Eysenck: Physiological reactivity to sensory stimulation as a measure of personality. Psychol. Rep. 20 (1967) 45–46.

Ezra, J.: Social and economic effects on families of patients with myocardial infarction. Univ. of Denver, 1961.

Faber, F. R., R. Haarstrick: Kommentar Psychotherapie-Richtlinien. Jungjohann, Neckarsulm-München 1989.

Fabrega, H.: Disease and social behaviour: An interdisciplinary perspective. MIT Press, Cambridge/Mass. 1974.

Fabrega jr., H.: Psychiatric diagnosis. A cultural perspective. J. nerv. ment. Dis. 175 (1987) 383–394.

Fagerhaugh, Sh. Y., A. Strauss: Politics of pain management. Addison-Wesley, Menlo Park 1977.

Fahrenberg, J.: Das Komplementaritätsprinzip in der psychophysiologischen Forschung und psychosomatischen Medizin. Z. klin. Psych. Psychother. 27 (1979) 151–167.

Fahrenberg, J., L. Delius: Eine Faktorenanalyse psychischer und vegetativer Regulationsdaten. Nervenarzt 34 (1963) 437–443.

Fahrenberg, J., R. Hampel, H. Selg: Das Freiburger Persönlichkeitsinventar. Handanweisung, 4., revidierte Aufl. Hogrefe, Göttingen 1984.

Fahrenberg, J., P. Walschburger, F. Foerster, M. Myrtek, W. Müller: Psychophysiologische Aktivierungsforschung – Ein Beitrag zu den Grundlagen der multivariaten Emotions- und Streßtheorie. Minerva, München 1979.

Fahrenberg, J., F. Foerster, W. Müller, M. Myrtek, H. J. Schneider: Aktivierungsforschung im Labor-Feld-Vergleich. Zur Vorhersage von Intensität und Mustern psychophysischer Aktivierungsprozesse während wiederholter psychischer und körperlicher Belastung. Minerva, München 1984.

Fahrenberg, J., F. Foerster, W. Müller, M. Myrtek, H. J. Schneider: Adequate scaling of heart-rate reactions – a comparative study based on resting levels, measures of basal (sleep) state, vita maxima, and individual range. In: Orlebeke, J. F., W. Mulder, L. van Doornen (eds.): Psychophysiology of Cardiovascular Control. Modells, Methods and Data, pp. 479–490. Plenum, London 1985.

Fahrländer, H.: Funktionelle Beschwerden des Magen-Darm-Kanals. Dtsch. Med. J. 23 (1972) 162–167.

Fahrländer, H.: Die chronisch entzündlichen Darmkrankheiten: Colitis ulcerosa und Enterocolitis regionalis Crohn. Verh. Dtsch. Ges. Inn. Med. 85 (1979) 52–65.

Fahrländer, H.: Morbus Crohn und Colitis ulcerosa. Dtsch. med. Wschr. 108 (1983) 1491–1492.

Fahrländer, H.: Die Langzeitbehandlung des Reizdarms. In: Goebell, H. et al. (Hrsg.): Der chronisch Kranke in der Gastroenterologie, S. 485–491. Springer, Berlin 1984.

Fain, M.: Régression et psychosomatique. Rev. franç. Psychanal. 30 (1966) 451–456.

Fain, M.: Réflexions sur la structure allergique. Rev. franç. Psychanal. 33 (1969) 201–226.

Fain, M.: Prélude à la vie fantasmatique. Rev. franç. Psychanal. 35 (1971) 291–294.

Fain, M.: In: Kreisler, L., M. Fain, M. Soulé: L'Enfant et son Corps. PUF, Paris 1974.

Fain, M., P. Marty: A propos du narcissisme et sa genèse. Rev. franç. Psychanal. 29 (1965) 561–572.

Fain, M., L. Kreisler: Discussion sur la genèse des fonctions représentatives. Rev. franç. Psychanal. 34 (1970) 285–306.

Fairburn, C. G.: Binge eating and its management. Brit. J. Psychiat. 141 (1982) 631–633.

Fairburn, C. G.: The management of bulimia nervosa. J. psychiat. Res. 19 (1983) 465–472.

Fairburn, C. G., P. J. Cooper: Self-induced vomiting and bulimia nervosa: an undetected problem. Brit. med. J. 284 (1982) 1153–1155.

Fairburn, C. G., P. J. Cooper: The epidemiology of bulimia nervosa. Int. J. Eat. Dis. 2 (1983) 61–67.

Fairburn, C. G., P. J. Cooper: The clinical features of bulimia nervosa. Brit. J. Psychiat. 144 (1984) 238–246.

Fairburn, C. G.: Cognitive-behavioral treatment for bulimia. In: Garner, D. M., P. E. Garfinkel (eds.): Handbook of Psychotherapy for Anorexia Nervosa and Bulimia. Gullford, New York 1985.

Fairburn, C. G., J. Kirk, M. O'Connor, P. J. Cooper: A comparison of two psychological treatments for bulimia nervosa. Behav. Res. Ther. 24 (1986) 629–643.

Fairburn, C. G., J. Steere, P. J. Cooper: Die Diagnose der spezifischen Psychopathologie bei Bulimia nervosa. In: Fichter, M. M. (Hrsg.): Bulimia nervosa. Enke, Stuttgart 1989.

Fairchild, C. J., J. Rush, N. Vasavada, D. E. Giles, M. Khatami: Which depressions respond to placebos? J. psychiat. Res. 18 (1986) 217–226.

Fajans, St. S., M. C. Cloutier, R. L. Crowther: Clinical and etiologic heterogeneity of idiopathic diabetes mellitus. Diabetes 27 (1978) 1112–1125.

Falger, P., A. Appels: Psychological risk factors over the life course of myocardial infarction patients. Advanc. Cardiol. 29 (1982) 132–139.

Falicki, Z., B. Sep-Kowalk: Psychic disturbances as a result of cardiac arrest. Polish med. J. 8 (1969) 200–206.

Falke, G.: Probleme des Psoriatikers in der Gesellschaft aus der Sicht des Patienten. Akt. Derm. 8 (1982) 167–169.

Falkner, B., H. Kushner, G. Onesti, E. T. Angelakos: Cardiovascular characteristics in adolescents who develop essential hypertension. Hypertension 3 (1981) 521–527.

Falkner, B., G. Onesti, E. T. Angelakos, M. Fernandes, C. Langman: Cardiovascular response to mental stress in normal adolescents with hypertensive parents. Hypertension 1 (1979) 23–30.

Fallenbacher, B.: Autogenes Training und progressive Muskelrelaxation: psychophysiologische Befunde bei psychosomatischen Krankheiten. Med. Diss., Lübeck 1989.

Fallik, A., M. Sigal: Hysteria – the choice of symptoms. A review of 40 cases of conversion hysteria. Psychother. and Psychosom. 19 (1971) 310–318.

Fallschlüssel, G.: Persönlichkeitsprofil und Persönlichkeitsentwicklung von Patienten mit Funktionsstörungen im Kausystem. Dtsch. zahnärztl. Z. 38 (1983) 670.

Fallstrom, K.: On the personality structure in diabetic schoolchildren aged 7–15 years. Acta paediat. scand. (Suppl.) 251 (1974) 1–71.

Famularo, R., C. P. Kimball: Liaison psychiatry considerations in renal hemodialysis patients with acute organic cerebral disorders. In: Levy, N. B. (ed.): Psychonephrology 2, pp. 71–78. Plenum, New York 1983.

Farberow, N. L., S. Ganzler, F. Cutter, D. Reynolds: An eight-year survey of hospital suicides. Life-Threatening Behavior 1 (1971) 184–202.

Farde, L., F.-A. Wiesel, H. Hall, C. Halldin, S. Stone-Elander, G. Sedvall: Letter to the editors: no D2 receptor increase in PET study of schizophrenia. Arch. gen. Psychiat. 44 (1987) 671–672.

Farmer, C. J., S. A. Snowden, U. Parsons: The prevalence of psychiatric illness among patients on home dialysis. Psychol. Med. 9 (1979) 509–514.

Farquharson, R. F., H. H. Hyland: Anorexia nervosa. The

course of 15 patients treated from 20 to 30 years previously. Canad. med. Ass. J. 94 (1966) 411–419.

Farrow, S. C.: Unemployment and health: a review of methodology. In: Cullen, J., J. Siegrist (eds.): Breakdown in Human Adaptation to Stress. Vol. 1. Nijhoff, Boston 1984.

Fassbender, H. G.: Pathologie des Weichteilrheumatismus. Ärztl. Praxis 51 (1973) 2497–2501.

Fauci, A. S., D. C. Dale: The effect of in vivo hydrocortisone on subpopulations of human lymphocytes. J. clin. Invest. 53 (1974) 240–246.

Fauler, I., P. Safian: Visitengespräche. Internistische und psychosomatische Krankenstation. Ein methodischer und inhaltlicher Vergleich. Abschlußbericht 2, Teilprojekt B 5/SFB 129, Universität Ulm 1983.

Fauler, I., P. Safian, U. Koch: Analyse der von Ärzten und Patienten zweier klinischer Populationen verbalisierten Affekte mittels des Gottschalk-Gleser-Verfahrens. In: Koch, U., G.Schöfer (Hrsg.): Sprachinhaltsanalyse in der psychiatrischen und psychosomatischen Forschung. Psychologie Verlags Union, Weinheim 1986.

Fava, G. A., L. Pavan: Large bowel disorders, II. Psychopathology and alexithymia. Psychother. and Psychosom. 27 (1976/77) 100–105.

Fava, G. A., N. Sonino, Th.N. Wise: Management of depression in medical patients. Psychother. and Psychosom. 49 (1988) 81–102.

Favell, J. E., N. H. Azrin, A. A. Baumeister, E. G. Carr, M. F. Dorsey, R. Forehand, R. M. Foxx, O. J. Lovaas, A. Rincover, T. R. Risley, R. G. Romanczyk, D. C. Russo, S. R. Schroeder, J. V. Solnick: The treatment of self-injurious behavior. Beh. Therapy 13 (1982) 529–554.

Featherstone, H. J., B. D. Beitman: Diabetic hyperglycemia and glycosuria as a manifestation of bulimia. Sth. med. J. 77 (1984) 936–937.

Federn, P.: Ich-Psychologie und die Psychosen, S. 107–151 (1933). Suhrkamp, Frankfurt 1978.

Fehlenberg, D.: Die empirische Analyse der Visitenkommunikation: Institutionskritik und Ansätze für eine reflektierte Veränderung institutioneller Praxis. Osnabrücker Beiträge zur Sprachtheorie 24 (1983) 29–56.

Fehlenberg, D., C. Simons, K. Köhle: Ansätze zur quantitativen Untersuchung ärztlicher Interventionen im Visitengespräch. In: Köhle, K., H.-H. Raspe (Hrsg.): Das Gespräch während der ärztlichen Visite, S. 232–248. Urban & Schwarzenberg, München 1982 a.

Fehlenberg, D., C. Simons, K. Köhle: Doctor-patient communication during ward rounds: II. Investigation of psychotherapeutic intervention. Paper read at the 14 th Europ. Conference on Psychosomatic Research, Noordwijkerhout (NL) 1982b.

Fehm, H. L., K. H. Voigt: Pathobiology of Cushing's disease. In: Joachim, H. L. (ed.): Pathobiology Annual 1979. Raven, New York 1979.

Fehm, H. L., K. H. Voigt, E. F. Pfeiffer: Die Bedeutung des Zentralnervensystems in der Ätiologie des Morbus Cushing. Verh. Dtsch. Ges. Neurol. 1 (1980) 308–315.

Fehm, H. L., E. Klein, R. Holl, K. H. Voigt: Evidence for extrapituitary mechanisms mediating the morning peak of plasma cortisol in man. J. clin. Endocr. 58 (1984 a) 410–414.

Fehm, H. L., E. Klein, R. Holl, K. H. Voigt: Evidence for ACTH-unrelated mechanisms in the regulation of cortisol secretion in man. Klin. Wschr. 62 (1984 b) 19–24.

Fehm, H. L., K. H. Voigt, G. W. Kummer, E. F. Pfeiffer: Positive rate-sensitive corticosteroid feedback mechanism of ACTH secretion in Cushing's disease. J. clin. Invest. 64 (1979) 102–108.

Fehm, H. L., R. Steck, J. Hohnloser, K. H. Voigt, E. F. Pfeiffer: Influence of neuroactive drugs on corticosteroid feedback regulation of ACTH secretion in man. Horm. Metab. Res. 15 (1983) 29.

Feiereis, H.: Beobachtungen zur psychischen Struktur bei Colitis-Kranken. Verh. Dtsch. Ges. Inn. Med. (1967) 701.

Feiereis, H.: Klinik und Therapie der Colitis ulcerosa. Marseille, München 1970.

Feiereis, H.: Klinik und Prognose der Colitis ulcerosa. Lebensversicherungsmedizin 27 (1975) 62–69.

Feiereis, H.: Psychodynamik des Konfliktes und psychosomatische Therapie bei Colitis ulcerosa. Schlesw.-Holst. Ärztebl. 30 (1977) 658–667.

Feiereis, H.: Langzeitverlauf und Prognose der Colitis ulcerosa unter kombinierter konservativer Therapie. Verh. Dtsch. Ges. Inn. Med. 85 (1979) 208–211.

Feiereis, H.: Scheinlösung Krankheit – Der somatisierte Konflikt. In: Feiereis, H., H.-J. Thilo (Hrsg.): Basiswissen Psychotherapie. Vandenhoeck & Ruprecht, Göttingen 1980.

Feiereis, H.: Integrierte psychosomatische Diagnostik und Therapie am Beispiel der inneren Medizin. Schleswig-Holstein. Ärztebl. 10 (1982) 823.

Feiereis, H.: Zur Psychotherapie des M. Crohn. Langenbecks Arch. Chir. 364 (1984) 407–411.

Feiereis, H.: Das Gespräch mit somatisch und psychosomatisch Kranken. In: Reimer, C. (Hrsg.): Ärztliche Gesprächsführung. Springer, Berlin 1985 a.

Feiereis, H.: Colitis ulcerosa – Morbus Crohn. Kombinierte Langzeitführung mit Medikation und Psychotherapie. Therapiewoche 35 (1985 b) 3075–3083.

Feiereis, H.: Morbus Crohn. In: Feiereis, H., H.-J. Kabelitz (Hrsg.): Internistische Pharmakotherapie. Marseille, München 1985 c.

Feiereis, H.: Diabetes mellitus Typ I und Bulimie – eine bedrohliche Doppelkrankheit. Dtsch. med. Wschr. 113 (1988) 1876–1878.

Feiereis, H.: Bauchschmerzen aus psychosomatischer Sicht. Therapiewoche 38 (1988) 1452–1460.

Feiereis, H. (Hrsg.): Diagnostik und Therapie der Magersucht und Bulimie. Marseille, München 1989.

Feiereis, H., M. Otte: Der endoskopische Befund bei Enteritis granulomatosa regionalis Crohn. internist. prax. 23 (1983 a) 65–73.

Feiereis, H., M. Otte: Der endoskopische Befund bei entzündlichen Dickdarmkrankheiten. internist. prax. 23 (1983 b) 651–662.

Feiereis, H., M. Wetzel: Ergebnisse einer Langzeitbeobachtung von 279 Patienten mit schwerer Colitis ulcerosa, besonders unter sozialmedizinischem Aspekt. Kassenarzt 29 (1989) 44–48.

Feiereis, H., K. Kamrowski, H. G. Rohrmoser: Colitis ulcerosa und Psyche. Arch. Psychiat. Z. ges. Neurol. 202 (1962) 657–677.

Feiereis, H., F. Janshen, V. Sudau: Assoziative Maltherapie. In: Feiereis, H. (Hrsg.): Diagnostik und Therapie der Magersucht und Bulimie. Marseille, München 1989.

Feifel, H.: Attitudes toward death in some normal and mentally ill populations. In: Feifel, H.: The Meaning of Death, p. 114. McGraw-Hill, New York 1959.

Feifel, H.: The function of attitudes toward death. Death and dying: Attitudes of patient and doctor, Vol. 5. Symposium Nr. 11. Group for the Advancement of Psychiatry, New York 1965.

Feighner, J., E. Robins, S. Guze: Diagnostic criteria for use in psychiatric research. Arch. gen. Psychiat. 26 (1972) 57–63.

Feinberg, M., B. J. Carroll: Separation of subtypes of depression using discriminant analysis: I. Separation of unipolar endogenous depression from non-endogenous depression. Brit. J. Psychiat. 140 (1982) 384–391.

Feinleib, M., R. Garrison, N. Borhan, R. Rosenman, J. Christian: Studies of hypertension in twins. In: Oglesby, P. (ed.): Epidemiology and Control of Hypertension. Stratton, New York 1975.

Feinstein, A. R.: Zit. nach: Lewis, A. J.: Diagnosenschlüssel und Glossar psychiatrischer Krankheiten. 5. Aufl, korrigiert nach der 9. Revision der ICD. Springer, Berlin 1980. Vorwort der englischen Ausgabe (1974).

Feinstein, A. R.: Clinimetrics. Yale Univ. Press, New Haven 1987.

Feldenkrais, M.: Abenteuer im Dschungel des Gehirns. Der Fall Doris. Suhrkamp, Frankfurt 1977.

Feldenkrais, M.: Bewußtheit durch Bewegung. Suhrkamp, Frankfurt 1978.

Feldman, F., D. Cantor, S. Soll, W. Bachrach: Psychiatric study of a consecutive series of 19 patients with regional ileitis. Brit. med. J. IV (1967) 711–714.

Feldman, R. G., N. L. Paul: Identity of emotional triggers in epilepsy. J. nerv. ment. Dis. 162 (1976) 345–352.

Felten, D. L., S. Y. Felten, D. L. Bellinger, S. L. Carlson, K. D. Ackerman, K. S. Madden, J. A. Olschowski, S. Livant: Noradrenergic sympathetic neural interactions with the immune system: structure and function. Immunol. Rev. 100 (1987) 225–260.

Feltkamp, H., K. A. Meurer, E. Godehardt: Tryptophan-induced lowering of blood pressure and changes of serotonin uptake by platelets in patients with essential hypertension. Klin. Wschr. 62 (1984) 1115–1119.

Felton, B. J., T. A. Revenson, G. H. Hinrichsen: Stress and coping in the explanation of psychosocial adjustment among chronically ill adults. Soc. Sci. Med. 18 (1984) 889–898.

Fenichel, O.: Prägenitale Konversionsneurosen. In: Spezielle psychoanalytische Neurosenlehre. Wien 1932. Und in: Psychoanalytische Neurosenlehre, Bd. II, S. 168–183. Walter, Olten 1975.

Fenichel, O.: Zur Kritik des Todestriebes. Imago 21 (1935) 458–471.

Fenichel, O.: The psychoanalytic theory of neurosis. Norton, New York 1945. Deutsch: Psychoanalytische Neurosenlehre. Walter, Olten 1974.

Fentiman, I. S., J. Cuzick, R. R. Millis, J. L. Hayward: Which patients are cured of breast cancer? Brit. med. J. 289 (1984) 1108–1111.

Fenwick, P.: Precipitation and inhibition of seizures. In: Reynolds, E. H., M. Trimble (eds.): Epilepsy and Psychiatry, pp. 306–321. Churchill Livingstone, Edinburgh-London-Melbourne 1981.

Fenz, W. D.: Strategies for coping with stress. Paper presented at the conference „Dimensions of Anxiety and Stress", Oslo 1975.

Ferber, C. v.: Der Tod, ein unbewältigtes Problem für Mediziner und Soziologen. Kölner Zeitschr. f. Soziologie und Sozialpsychologie 22 (1970) 237–250.

Ferber, C. v.: Gesundheit und Gesellschaft. Kohlhammer, Stuttgart 1971.

Ferber, C. v.: Medizinkultur und Laienkultur nebeneinander – gegeneinander – miteinander. In: Markuard, O., E. Seidler, O. Staudinger (Hrsg.): Medizinische Ethik und soziale Verantwortung. Schöningh, Paderborn 1989.

Ferber, C. v., E. Standfest: Gesundheitsvorsorge gegen arbeitsbedingte Krankheiten. In: Badura, B., C. v. Ferber (Hrsg.): Selbsthilfe und Selbstorganisation im Gesundheitswesen. S. 165–183. Oldenbourg, Wien 1981.

Ferenczi, S.: Die Psychoanalyse der Hypnose und Suggestion. Gyógyászat, Budapest 1910. Ref. Zb. ges. Neurol. 30 (1911) 734.

Ferenczi, S.: Entwicklungsstufen des Wirklichkeitssinnes. Int. J. Psychoanal. 1 (1913).

Ferenczi, S.: Die Psychoanalyse der Kriegsneurosen. In: Internationale Psychoanalytische Bibliothek, Bd. I, S. 9–30. Int. Psychoanal. Verlag, Wien 1919.

Ferenczi, S.: Psychoanalytische Betrachtungen über den Tic. Int. Z. Psychoanal. 7 (1921) 33–62.

Ferenczi, S.: Über forcierte Phantasien (Aktivität in der Assoziationstechnik). Z. Psychoanal. 10 (1924) 6–16.

Ferenczi, S.: Über den Anfall der Epileptiker. Beobachtungen und Überlegungen. In: Bausteine zur Psychoanalyse, Bd. III, S. 170–179. Int. Psychoanal. Verlag, Wien 1938. 3. Aufl. Huber, Bern 1984.

Ferenczi, S.: Weiterer Ausbau der aktiven Technik in der Psychoanalyse. In: Ferenczi, S. (Hrsg.): Bausteine der Psychoanalyse, Bd. II. Bern 1939.

Ferguson, J. M.: Treatment of an anorexia nervosa patient with fluoxetine. Amer. J. Psychiat. 144 (1987) 1239.

Ferguson, S. M., M. Rayport: The adjustment to living without epilepsy. J. nerv. ment. Dis. 140 (1965) 26–37.

Ferner, H., A. Reindell: Familientherapie bei Colitis ulcerosa und Morbus Crohn. Therapiewoche 29 (1979) 6314–6319.

Ferstl, R., W. Müller-Ruchholtz: Psychoneuroimmunologie – ihre Forschungsgebiete und ihre konzeptuellen Probleme. Z. f. klin. Psychol. 16 (1987) 199–204.

Fessel, W. J., R. P. Forsyth: Hypothalamic role in control of gamma globulin levels. Arthr. and Rheum. 6 (1963) 771.

Festge, O. A.: Neue Aspekte der Harnblasenphysiologie. Urol. int. 35 (1980) 28.

Fetterman, J. L.: Vertebral neuroses. Psychosom. Med. 11 (1940) 265–275.

Feuerstein, M., H. E. Adams, J. Beiman: Cephalic vasomotor and electromyographic feedback in the treatment of combined muscle contraction and migraine headaches in a geriatric case. Headache 16 (1976) 232–237.

Feuerstein, M., C. Bush, R. Corbisiero: Stress and chronic headache: a psychophysiological analysis of mechanisms. J. psychosom. Res. 26 (1982) 167–182.

Feurle, G. E.: Berücksichtigung epidemiologischer Erkenntnisse bei der Therapie des Morbus Crohn. Dtsch. med. Wschr. 111 (1986) 835–838.

Feurle, G. E., O. Keller, K. Hassels, H. J. Jesdinsky: Soziale Auswirkungen des Morbus Crohn. Dtsch. med. Wschr. 108 (1983) 971–975.

Feurle, G. E., W. Kruschitz, J. Küchenhoff, D. Normann: Colitis – Morbus Crohn. In: Bräutigam, W. (Hrsg.): Kooperationsformen somatischer und psychosomatischer Medizin. Springer, Berlin 1988.

Fichter, M. M.: Strukturiertes Interview zur Anorexia nervosa. Expertenbeurteilung; Erhebungsbogen und Manual (unveröffentlicht). München 1980.

Fichter, M. M.: Epidemiologie der Anorexia nervosa. Akt. Ernähr.-Med. 1 (1984) 8–13.

Fichter, M. M.: Magersucht und Bulimia. Springer, Berlin 1985.

Fichter, M. M.: Psychologische Therapien bei Bulimia. In: Fichter, M. M. (Hrsg.): Bulimia nervosa. Enke, Stuttgart 1989a.

Fichter, M. M.: Bulimia nervosa und bulimisches Verhalten. In: Fichter, M. M. (Hrsg.): Bulimia nervosa. Enke, Stuttgart 1989b.

Fichter, M. M., C. Chlond: Hypertrophe Osteoarthropathie bei Bulimia nervosa mit chronischer Intoxikation mit Laxantien. Nervenarzt 59 (1988) 244–247.

Fichter, M. M., R. Hoffmann: Bulimia beim Mann. In: Fichter, M. M. (Hrsg.): Bulimia nervosa. Enke, Stuttgart 1989.

Fichter, M. M., W. Keeser: Das Anorexia nervosa-Inventar zur Selbstbeurteilung (ANIS). Arch. Psychiat. Nervenkr. 228 (1980) 67–89.

Fichter, M. M., R. Meermann: Zur Psychopathometrie der

Anorexia nervosa. In: Meermann, R. (Hrsg.): Anorexia nervosa., S. 17–81. Enke, Stuttgart 1981.

Fichter, M. M., K. M. Pirke: Somatische Befunde bei Anorexia nervosa und ihre differentialdiagnostische Wertigkeit. Nervenarzt 53 (1983) 635–643.

Fichter, M. M., K. Pirke: Hormonelle Dysfunktionen bei Bulimia. In: Fichter, M. M. (Hrsg.): Bulimia nervosa. Enke, Stuttgart 1989.

Fichter, M. M., P. Doerr, K. M. Pirke, R. Lund: Behavior, attitude, nutrition and endocrinology in anorexia nervosa. Acta psychiat. scand. 66 (1982) 429–444.

Fichter, M. M., S. Weyerer, S. Kellnar, S. Dilling: Zur Epidemiologie des Alkoholismus. Med. Welt 37 (1986) 752–757.

Fichter, M. M., S. Weyerer, L. Sourdi, Z. Sourdi: The epidemiology of anorexia nervosa: a comparison of greek adolescents living in Germany and greek adolescents living in Greece. In: Darby, P. L., P. E. Garfinkel, D. M. Garner, D. V. Coscina (eds.): Anorexia nervosa. Recent Developments in Research, pp. 95–105. Liss, New York 1983.

Fiedler, P. A.: Diagnostische und therapeutische Verwendbarkeit kognitiver Verhaltensanteile. In: Hoffmann, N. (Hrsg.): Grundlage kognitiver Therapie. Huber, Bern 1979.

Fielding, R.: Behavioral treatment in the rehabilitation of myocardial infarction patients. In: Osborne, D. J., M. M. Gruneberg, J. Eiser (eds.): Research in Psychology and Medicine, pp. 176–182. Acad. Press, London-New York-Toronto 1979.

Fielding, R.: A note on behavioral treatment in the rehabilitation of myocardial infarction patients. Brit. J. soc. clin. Psychol. 19 (1980) 157.

Fields, H. L., J. D. Levine: Biology of placebo analgesia. Amer. J. Med. 70 (1981) 745.

Filipowski, D. M., R. S. Jorgensen, A. W. Langer, P. D. Gelling: Cardiovascular reactivity in cognitive challenge: the moderating influence of anger expression. Psychophysiol. 25 (1988) 445.

Filipp, G., A. Szentivanyi: Anaphylaxis and the nervous system, Part III. Ann. Allergy 16 (1958) 306–311.

Filler, D., K. Schwemmle: Die Langzeitprognose bei der Colitis ulcerosa. Lebensversicherungsmedizin 28 (1976) 47–51.

Filter, P. M., P. Wesemann, P. Bayerl, M. Demmering, J. Franz, W. Kettner, H. Korenberg, W. Reim, S. Schmidt: Patientenzentrierte Medizin an einer internistischen Fachklinik. Med. Klinik 74 (1979) 1505–1510.

Fimmel, C. J., F. Sabbatini, F. Pace, P. Caradonna, A. L. Blum: Neue Ulcustherapeutika – ist Cimetidin überholt? Ther. Umschau 39 (1982) 841–851.

Finch, S. M., J. H. Hess: Ulcerative colitis in children. Amer. J. Psychiat. 118 (1962) 819.

Fine, F. D. et al. (eds.): The mechanism of denial. Monograph III. Int. Univ. Press, New York 1969.

Fink, P. J.: Response to the presidential address: is „biopsychosocial" the psychiatric shibboleth? Amer. J. Psychiat. 145 (1988) 1061–1067.

Fink, R. I., O. G. Kolterman, M. Kao, J. M. Olefsky: The role of the glucose transport system in the postreceptor defect in insulin action associated with human aging. J. clin. Endocr. 58 (1984) 721–725.

Finkelstein, F. O., T. E. Steele: Sexual dysfunction and chronic renal failure. Dialys. Transpl. 7 (1978) 877–878.

Finnerty, F. A., E. C. Mattie, F. A. Finnerty III: Hypertension in the inner city, I.: Analysis of clinical drop-outs. Circulation 47 (1973) 73–75.

Firth, D.: The case of Augustus d'Este (1794–1848): the first account of disseminated sclerosis. Proc. roy. Soc. Med. 34 (1940) 381–384.

Firth, D.: The case of Augustus d'Este. Cambridge Univ. Press, Cambridge 1948.

Fischer, E. G., N. E. Falke: β-endorphin modulates immune functions. Psychother. and Psychosom. 42 (1984) 195–204.

Fischer, G. H.: Einführung in die Theorie psychologischer Tests. Grundlagen und Anwendungen. Huber, Bern-Stuttgart-Wien 1974.

Fischer, G. H.: Neuere Testtheorie. In: Feger, H., J. Bredenkamp (Hrsg.): Messen und Testen. Enzyklopädie der Psychologie, Themenbereich B, Serie I, Forschungsmethoden der Psychologie, Bd. 3. Hogrefe, Göttingen 1983.

Fischer, H.: Selbstbeschädigung und Selbstverstümmelung. Krankenhausarzt 51 (1978) 93–96.

Fischer, H. K., B. M. Dlin: The dynamics of placebo therapy: a clinical study. Amer. J. med. Sci. 232 (1956) 504–512.

Fischer, P. A.: Parkinson-Syndrom. Diagnose – Verlauf – Therapie. Sandoz, Nürnberg 1981.

Fischer, P. A.: (Hrsg.): Psychopathologie des Parkinson-Syndroms. Editiones Roche, Basel 1982.

Fischer, R., M. Schumacher, U. Thoden: Verlauf nicht operierter lumbaler Bandscheibenvorfälle. Radikuläre Störungen und computertomographische Befunde. Schmerz 2 (1988) 26–32.

Fischman, H. K., D. D. Kelly: Sister chromatid exchanges induced by behavioral stress. Ann. N. Y. Acad. Sci. 496 (1987) 426–435.

Fisher, C. et al.: A psychophysiological study of night mares and night terrors. J. nerv. ment. Dis. 158 (1974) 174.

Fisher, C. M.: Clinical syndromes in cerebral hemorrhage, pathogenesis and treatment of cerebrovascular disease. In: Fields, W. W. (ed.): Clinical Syndromes in Cerebral Hemorrhage. Pathogenesis and Treatment of Cerebrovascular Disease. Thomas, Springfield/Ill. 1961.

Fisher, S.: Body boundary and perceptual vividness. J. abn. Psychol. 73 (1968) 392.

Fisher, S.: Institutional authority and the structure of discourse. Discourse Processes 7 (1984) 201–224.

Fiske, D., L. Luborsky, M. Parlow, H. Hunt, M. Orne, M. Reiser, A. Tuma: The planning of research on effectiveness of psychotherapy. Arch. gen. Psychiat. 22 (1970) 22–32.

Fitts, W. T., J. S. Ravdin: What Philadelphia physicians tell patients with cancer. J. Amer. med. Ass. 153 (1953) 901–904.

Flader, D., W. D. Grodzicki: Die psychoanalytische Deutung – eine diskursanalytische Fallstudie. In: Flader, D., W. D. Grodzicki, K. Schröter (Hrsg.): Psychoanalyse als Gespräch, S. 139–193. Suhrkamp, Frankfurt 1982.

Flannery, J. G.: Alexithymia, II. The association with unexplained physical distress. Psychother. and Psychosom. 30 (1978) 193–197.

Fleck, H. C.: Über psychodynamische Faktoren bei Wurzelreizerscheinungen. Z. psychosom. Med. 21 (1975) 118–128.

Fleck, L., J. Lange, H. Thomä: Verschiedene Typen von Anorexia nervosa und ihre psychoanalytische Behandlung. In: Meyer, J. E., F. Feldmann (Hrsg.): Anorexia nervosa, S. 87. Thieme, Stuttgart 1965.

Fleck, S.: A general systems approach to severe family pathology. Amer. J. Psychiat. 133 (1976) 669.

Fleischer-Peters, A.: Psychologie und Psychosomatik in der Kieferorthopädie. Hanser, München-Wien 1985.

Fleischer-Peters, A., S. Zschiesche: Ist Lutschen wirklich schädlich? Fortschr. Kieferorthop. 41 (1980) 563.

Fleischhacker, W. W., C. Barnas, C. Stuppäck: Benzodiazepines: utilization and patterns of use in a university hospital. Pharmacopsychiat. 22 (1989) 111–114.

Flor, H.: Die Rolle psychologischer Faktoren bei der Entstehung und Behandlung chronischer Wirbelsäulensyndrome. Psychother. med. Psychol. 37 (1987) 424–429, 4765.

Flor, H., D. C. Turk: Chronic back pain and rheumatoid arthritis: predicting pain and disability from cognitive variables. J. Behav. Med. 11 (1988) 251–265.

Flor, H., R. D. Kerns, D. C. Turk: The role of spouse reinforcement, perceived pain, and activity levels of chronic pain patients. J. psychosom. Res. 31 (1987) 251–259.

Florian, F.: Aufgaben und Stellung des Betriebsarztes nach dem Arbeitssicherstellungsgesetz (ASiG). Münch. med. Wschr. 121 (1979) 261–262.

Flynn, J. T., M. A. K. Kennedy, S. Wolf: Essential hypertension in one of identical twins. Res. publ. Ass. nerv. ment. Dis. 29 (1949) 954–961.

Folgering, H., A. Cox: Beta-blocker therapy with metoprolol in the hyperventilation syndrome. Respiration 41 (1981) 33–39.

Foliot, C.: L'asthme de l'enfant. In: Lebovici, S., R. Diatkine, M. Soulé (eds.): Traité de psychiatrie de l'enfant et de l'adolescent, Vol. II., pp. 571–578. PUF, Paris 1985.

Folkins, C. H., A. Amsterdam: Control and modification of stress emotions through chronic exercise. In: Amsterdam, A., J. H. Wilmore, A. N. DeMaria (eds.): Exercise in Cardiovascular Health and Disease. Yorks, New York 1977.

Folkow, B.: Nervous control of the blood vessels. Physiol. Rev. 35 (1955) 629.

Folkow, B.: Vascular changes in hypertension – review and recent animal studies. In: Berglund, G., L. Hansson, L. Werkö (eds.): Pathophysiology and Management of Arterial Hypertension, pp. 95–113. Lindgren & Söner, Möndal (Sweden) 1975.

Folkow, B.: Physiological aspects of primary hypertension. Physiol. Rev. 62 (1982) 347–503.

Folkow, B., E. Neil: Circulation. Oxford Univ. Press, London 1971.

Folkow, B., E. H. Rubinstein: Cardiovascular effect of acute and chronic stimulations of the hypothalamic defense area in the rat. Acta physiol. scand. 28 (1966) 48–57.

Folkow, B., B. Uvnas: Discussion of sympathetic vasodilator system and blood flow. Physiol. Rev. 40 (Suppl. 4) (1966) 77.

Folkow, B., G. Göthberg, S. Lundin, S. W. Ricksten: Structural „resetting" of the renal vascular bed in spontaneously hypertensive rats (SHR). Acta physiol. scand. 100 (1977) 270–272.

Folkow, B., P. Hedner, B. Lisander, E. H. Rubinstein: Release of cortisol upon stimulation of the hypothalamic defense area in cats. Försvarsmed. 3 (Suppl. 2) (1967) 114–119.

Folkow, B., M. Hallbäck, Y. Lundgren, R. Siverston, L. Weiss: Importance of adaptive changes in vascular design for establishment of primary hypertension. Studies in man and in spontaneously hypertensive rats. Circulat. Res. (Suppl. 1) 32/33 (1973).

Fooken, I.: Situation von Witwen/Witwern unter psychologischen Aspekten. Z. Geront. 12 (1979) 266.

Foppa, K.: Lernen, Gedächtnis, Verhalten. Ergebnisse und Probleme der Lernpsychologie. Kiepenheuer und Witsch, Köln 1968.

Ford, C. S., F. A. Beach: Formen der Sexualität. Rowohlt, Reinbek 1971.

Ford, C. V.: The somatizing disorders. Elsevier, New York 1983.

Ford, C. V., G. A. Glober, P. Castelnuovo-Tedesco: A psychiatric study of patients with regional enteritis. J. Amer. med. Ass. 208 (1969) 311–315.

Fordtran, J. S.: The psychosomatic theory of peptic ulcer: In: Sleisenger, M. H., J. S. Fordtran (eds.): Gastrointestinal Disease. Saunders, Philadelphia-London-Toronto 1973.

Fordyce, W.: Behavioral methods for chronic pain and illness. Mosby, St. Louis 1976.

Fordyce, W. E.: Behavioral concepts in chronic pain and illness. Mosby, St. Louis 1976.

Fordyce, W. E.: Learning processes in pain. In: Sternbach, R. A. (ed.): The psychology of pain. Raven, New York 1978.

Fordyce, W. E., J. C. Steger: Chronischer Schmerz. In: Keeser, W., E. Pöppel, P. Mitterhusen (Hrsg.): Schmerz. Urban & Schwarzenberg, München 1982.

Fordyce, W. E., F. S. Brena, R. J. Holcomb, B. J. DeLateur, J. D. Loeser: Relationship of patient semantic pain descriptions to physician diagnostic judgements, activity level measures and MMPI. Pain 5 (1978) 2293–303.

Fornari, F.: Psychoanalyse des ersten Lebensjahres. Fischer, Frankfurt 1970.

Forster, F. M.: The classification and conditioning treatment of the reflex epilepsies. Int. J. Neurol. 9 (1972) 73–86.

Forsyth, R. P., R. E. Harris: Circulatory changes during stressful stimuli in rhesus monkeys. Circul. Res. Suppl. 26/27 (1970) 13–20.

Foss, L., K. Rothenberg: The second medical revolution. From biomedicine to infomedicine. New Science Library, Boston 1987.

Foster, F. G., G. L. Cohn, F. P. McKegney: Psychobiologic factors and individual survival on chronic renal hemodialysis. A 2-year follow-up. Psychosom. Med. 35 (1973) 64–82.

Foster, G. M.: Concepts of health and disease in cultural perspective. In: Nizetic, B. Z., H. G. Pauli, P.-G. Svensson (eds.): Scientific Approaches to Health and Health Care. WHO, Copenhagen 1986.

Fothergill, J.: Complete collection of the medical and philosophical works. London 1781.

Foucault, M.: Die Geburt der Klinik. Eine Archäologie des ärztlichen Blicks. Hanser, München 1973.

Foudraine, J.: Wer ist aus Holz? Piper, München 1973.

Foulkes, S. H.: Gruppenanalytische Psychotherapie. Reihe „Geist und Psyche", Bd. 2130. Kindler, München 1974.

Fournier, D. v.: Tumorwachstum als Kriterium der Malignität. In: Frommhold, W., P. Gerhardt (Hrsg.): Das Mammakarzinom, S. 68–79. Thieme, Stuttgart 1982.

Fowler, E. P.: Sudden deafness. Ann. Otol. (St. Louis) 59 (1950) 980.

Fowles, D. C.: The three arousal model: implications of Gray's two-factor learning theory for heart rate, electrodermal activity, and psychopathy. Psychophysiol. 17 (1980) 87–104.

Fowles, D. C.: Psychophysiology and psychopathology: a motivational approach. Psychophysiol. 25 (1988) 373–391.

Fox, B. H.: Premorbid psychological factors as related to cancer incidence. J. Behav. Med. 1 (1978) 45–133.

Fox, B. H.: Psychosocial factors and the immune system in human cancer. In: Ader, R. (ed.): Psychoneuroimmunology, pp. 103–157. Acad. Press, New York 1981.

Fox, B. H.: Current theory of psychogenic effects on cancer incidence and prognosis. J. psychosoc. Oncol. 1 (1983) 17–31.

Fox, B. H., E. J. Stanek, S. C. Boyd: Suicide rates among cancer patients in Connecticut. Manuscript 1983 (National Cancer Institute, Bethesda).

Fox, B. H., D. R. Ragland, R. J. Brand, R. H. Rosenman: Type A behavior and cancer mortality: theoretical considerations and preliminary data. Ann. N. Y. Acad. Sci. 496 (1987) 620–627.

Fox, R.: Training for uncertainty. In: Merton, R., G. C. Reader, P. L. Kendell (eds.): The Student Physician, pp. 207–241. Harvard Univ. Press, Cambridge/Mass. 1957.

Foy, A. L.: Hemodialysis, dreams of patients and staff. Amer. J. Nurs. 70 (1970) 80.

Frahm, H.: Beschreibung und Ergebnisse einer somatisch orientierten Behandlung von Kranken mit Anorexia nervosa. Med. Welt 38 (1966) 2004–2011; 2068–2072.

Frahm, H.: Anorexia nervosa. In: Hornbostel, H., W. Kaufmann, W. Siegenthaler (Hrsg.): Innere Medizin in Praxis und Klinik, Bd. 4, Kap. 16, S. 13–19. Thieme, Stuttgart 1973.

Frank, C., G. Harrer, G. Ladurner: Locked-in Syndrom – Erlebnisdimensionen und Möglichkeiten eines erweiterten Kommunikationssystems. Nervenarzt 59 (1988) 337–343.

Frank, E., C. Anderson, D. Rubinstein: Frequency of sexual dysfunction in „normal" couples. N. Engl. J. Med. 299 (1978) 111–115.

Frank, J.: Persuasion and healing. Johns Hopkins Univ. Press, Baltimore 1961.

Frank, J. D.: Problems of controls in psychotherapy. In: Rubinstein, E. A., M. B. Parlow (eds.): Research in Psychotherapy. APA 1958.

Frank, J. D.: Therapeutic components of psychotherapy. A 25-year progress report of research. J. nerv. ment. Dis. 159 (1974) 325–342.

Frank, J. D.: The present status of outcome studies. J. consult. clin. Psychol. 47 (1979) 310–316.

Frank, K. A., S. Heller, D. S. Kornfeld: Psychological intervention in coronary heart disease. Gen. Hosp. Psychiat. 1 (1979) 18–23.

Frank, K. A., S. S. Heller, D. S. Kornfeld, A. A. Sporn, M. B. Weiss: Type A behavior pattern and coronary angiographic findings. J. Amer. med. Ass. 240 (1978) 761–763.

Frank, L. K.: Projective methods. Springfield/Ill. 1948.

Frank, R. G., N. C. Beck, J. C. Parker, J. H. Kashani, T. R. Elliott, A. E. Haut, E. Smith, C. Atwood, M. Brownlee-Duffeck, D. R. Kay: Depression in rheumatoid arthritis. J. Rheum. 15 (1988) 920–925.

Franke, H.: Zur Rehabilitation akut entzündlicher Adnexerkrankungen mit Glukokortikoiden. Dtsch. Ges.-Wesen 24 (1969) 1015.

Franke, H.: Wesen und Bedeutung der Polypathie und der Multimorbidität in der Altersheilkunde. In: Platt, D. (Hrsg.): Handbuch der Gerontologie, Bd. 1: Innere Medizin, S. 449. Fischer, Stuttgart 1983.

Franke, H., H. Bracharz, H. Laas, E. Moll: Studie an 148 Hundertjährigen. Dtsch. med. Wschr. 31 (1970) 1950f.

Frankel, B. L., R. D. Cousey, R. Buchbinder, F. Snyder: Recorded and reported sleep in chronic primary insomnia. Arch. gen. Psychiat. 33 (1976) 615–623.

Frankel, R. M., H. B. Beckman: Impact: an interaction-based method for preserving and analysing clinical transactions. In: Pettigrew, L. (ed.): Explorations in Provider and Patient Interactions. Humana, Nashville 1982.

Frankenhaeuser, M.: Experimental approaches to the study of human behavior as related to neuroendocrine functions. In: Levi, L. (ed.): Society, Stress, and Disease, Vol. I. Oxford Univ. Press, London 1971.

Frankenhaeuser, M.: Experimental approaches to the study of catecholamines and emotion. In: Levi, L. (ed.): Emotions: Their Parameters and Measurement. Raven, New York 1975.

Frankenhaeuser, M.: Sympathetic-adrenomedullar activity, behavior and the psychosocial environment. In: Venables, P. H., M. J. Christie (eds.): Research in Psychophysiology. Wiley, London 1975.

Frankenhaeuser, M.: Psychoneuroendocrine approaches to the study of emotion as related to stress and coping. Arnold, W. I. (ed.): Nebraska Symposium on Motivation, pp. 123–161. Univ. of Nebraska Press, Lincoln 1978.

Frankenhaeuser, M.: The sympathetic-adrenal and pituitary-adrenal response to challenge. Comparison between the sexes. In: Dembroski, T. M., T. H. Schmidt, G. Blümchen (eds.): Biobehavioral Bases of Coronary Heart Disease, pp. 91–105. Karger, Basel-München-Paris 1983.

Frankenhaeuser, M., A. Rissler: Effects of punishment on catecholamine release and efficiency of performance. Psychopharmacol. 17 (1970) 378–390.

Frankenhaeuser, M., L. L. Lundberg, L. Forsman: Note on arousing type-A persons by depriving them of work. Reports from the Department of Psychology, Univ. of Stockholm, No. 539, 1978.

Frankenstein sen., R.: Installation eines Meßplatzes zur Abschreckung vor psychophysiologischer Forschung. Brain and Psychophysics 703 (1897) 1–13.

Frankl, V. E.: Grundriß der Existenzanalyse und Logotherapie. In: Frankl, V. E., V.v. Gebsattel, J. H. Schultz (Hrsg.): Handbuch der Neurosenlehre und Psychotherapie. München-Berlin 1959.

Frankl, V. E.: Der Wille zum Sinn. Ausgewählte Vorträge über Logotherapie. Huber, Bern-Stuttgart-Wien 1972.

Franklin, G. M., L. M. Nelson, R. K. Heaton, J. S. Burks, D. S. Thompson: Stress and its relationship to acute exacerbations in multiple sclerosis. J. neurol. Rehab. 2 (1988) 7–11.

Fransella, F., A. H. Crisp: Comparisons of weight concepts in a group of (1) neurotic, (2) „normal" and (3) anorexic females. Unpublished, 1975.

Franz, H. E. (Hrsg.): Blutreinigungsverfahren – Technik und Klinik. Thieme, Stuttgart-New York 1981.

Franz, I. W.: Belastungsblutdruck bei Hochdruckkranken. Springer, Berlin-Heidelberg-New York 1981.

Franzke, E.: Der Mensch und sein Gestaltungserleben. Huber, Bern 1977.

Frasier, S. D., M. L. Rallison: Growth retardation and emotional deprivation: relative resistance to treatment with human growth hormone. J. Pediat. 80 (1972) 603.

Frazier, S. H.: Anorexia nervosa. Dis. nerv. Sys. 26 (1965) 155–159.

Frederikson, M.: Orienting and defensive reactions to phobic and conditioned fear stimuli in phobics and normals. Psychophysiol. 18 (1981) 456–465.

Fredrickson, M.: Behavioral aspects of cardiovascular reactivity in essential hypertension. In: Schmidt, T., T. Dembroski, G. Blümchen (eds.): Biological and Psychological Factors in Cardiovascular Disease, pp. 418–446. Springer, Berlin-Heidelberg-New York 1986.

Fredrickson, R. C. A., H. C. Hendrie, J. N. Hintgen, M. H. Aprison (eds.): Neuroregulation of autonomic, endocrine and immune systems. Nijhoff, Boston 1986.

Freebury, D. R.: The psychological implications of organ transplantation. A selective review. Canad. Psychiat. Ass. J. 19 (1974) 593–597.

Freedman, A. M.: Psychiatric aspects of familial dysautonomia. Amer. J. Orthopsychiat. 27 (1957) 96.

Freedman, A. M.: Psychopharmacology and psychotherapy in the treatment of anxiety. Pharmacopsychiat. 13 (1980) 277–289.

Freedman, A. M., H. I. Kaplan, B. J. Sadock (eds.): Modern synopsis of psychiatry II, Chap. 13: Classification. Williams & Wilkins, Baltimore 1978.

Freedman, L. Z., A. B. Hollingshead: Neurosis and social class. Amer. J. Psychiat. 113 (1956/57) 769–775.

Freeman, A., D. Watts, R. Karp: Evaluation of cardiac transplant candidates. Psychosomatics 25 (1984) 197–207.

Freeman, C., D. Calsyn, J. Louks: The use of the Minnesota Multiphasic Personality Inventory with low back pain patients. J. clin. Psychol. 32 (1976) 295–298.

Freeman, H., M. Alpert: Prevalence of schizophrenia in an urban population. Brit. J. Psychiat. 149 (1986) 603–611.

Freeman, R. I., C. D. Thomas, L. Solyom, I. E. Miles: Body image disturbances in anorexia nervosa: a reexamination and a new technique. In: Darby, P. L., P. E. Garfinkel, D. M. Garner, D. V. Coscina (eds.): Anorexia nervosa. Recent Developments in Research, pp. 117–126. Liss, New York 1983.

Frei, P., G. Graber: Der Kaumuskelsynchronisator. Schweiz. Mschr. Zahnheilk. 86 (1977) 1195.

Freidson, E.: Patients' view of medical practice. Russel Sage, New York 1961.

Freis, E. D.: Effects of treatment on morbidity in hypertension. Results in patients with diastolic blood pressures averaging 115 through 129 mm Hg. J. Amer. med. Ass. 202/11 (1976) 116.

Freis, E. D.: Salt, volume and the prevention of hypertension. Circulation 53 (1976) 589–594.

Freitag, A.: Zur Pathologie und Therapie der funktionellen Schreibstörungen. Arch. Psychiat. 63 (1921) 574–590.

Freiwald, M., R. Liedtke, S. Zepf: Die Imagination des erkrankten Organs von Patienten mit Colitis ulcerosa und funktionellen Herzbeschwerden im experimentellen katathymen Bilderleben. Psychother. med. Psychol. 25 (1975) 15–24.

Fremont-Smith, A. Meigs: Amer. J. Obstet. 55 (1948) 1042.

French-Belgian Collaborative Group: Ischemic heart and psychological patterns. Advanc. Cardiol. 29 (1982) 25–31.

Frese, M., S. Greif, N. Semmer (Hrsg.): Industrielle Psychopathologie. Huber, Bern-Stuttgart-Wien 1978.

Freud, A.: Das Ich und die Abwehrmechanismen. London 1936.

Freud, A.: Role of bodily illness in the mental life of children. In: Eissler, R., A. Freud, H. Hartmann, E. Kris (eds.): Psychoanalytic Study of the Child, Vol. 7. Univ. Press, New York 1952.

Freud, A.: Die Schriften der Anna Freud, Bd. VIII, S. 2274. Kindler, München 1965.

Freud, A.: Normality and pathology in childhood. Int. Univ. Press, New York 1965.

Freud, A.: Observations on child development. In: The Writings of Anna Freud, Vol. 5. New York 1969.

Freud, A.: Wege und Irrwege der Kinderentwicklung. Klett, Stuttgart 1971.

Freud, E.: Mutter-Kind-Beziehung bei Frühgeburten. Pers. Mitteilung. 1984.

Freud, S.: Ein Fall von hypnotischer Heilung (1892). Ges. Werke, Bd. I, S. 3–17. Fischer, Frankfurt 1964.

Freud, S.: Quelques considérations pour une étude comparative des paralysies motrices organiques et hystériques (1893). Ges. Werke, Bd. I, S. 37–45.

Freud, S.: Die Abwehr-Neuropsychosen (1894). Ges. Werke, Bd. I. Imago, London 1950.

Freud, S.: Entwurf einer Psychologie (1895). In: An den Anfängen der Psychoanalyse. Fischer, Frankfurt 1975.

Freud, S.: Über die Berechtigung, von der Neurasthenie einen bestimmten Symptomenkomplex als „Angstneurose" abzutrennen (1895). Ges. Werke, Bd. I, S. 341. Frankfurt 1952.

Freud, S.: Die Traumdeutung (1900). Ges. Werke, Bd. II/III, S. 541. Frankfurt 1969.

Freud, S.: Drei Abhandlungen zur Sexualtheorie (1905). In: Freud, S.: Sexualleben. Studienausgabe Bd. V, S. 37. Fischer, Frankfurt 1972.

Freud, S.: Bruchstücke einer Hysterie-Analyse (1905). Ges. Werke, Bd. V, S. 281. Fischer, Frankfurt 1972.

Freud, S.: Über den Gegensinn der Urworte (1910). Ges. Werke, Bd. VIII, S. 214–221.

Freud, S.: Formulierungen über zwei Prinzipien des psychischen Geschehens (1911). Ges. Werke, Bd. VIII, 3. Aufl. Fischer, Frankfurt 1969.

Freud, S.: Zur Dynamik der Übertragung (1912). Ges. Werke, Bd. VII, S. 364–374. Fischer, Frankfurt 1976.

Freud, S.: Erinnern, Wiederholen und Durcharbeiten (1914). Ges. Werke, Bd. X, 3. Aufl. Fischer, Frankfurt 1969.

Freud, S.: Das Unbewußte (1915). Ges. Werke, Bd. X, 3. Aufl. Fischer, Frankfurt 1969.

Freud, S.: Zur Frage der Laienanalyse (1916). Ges. Werke, Bd. XIV, S. 278. Fischer, Frankfurt 1976.

Freud, S.: Trauer und Melancholie (1917). Ges. Werke, Bd. X. Imago, London 1950.

Freud, S.: Vorlesungen zur Einführung in die Psychoanalyse (1917). Ges. Werke, Bd. XI. Frankfurt 1969.

Freud, S.: Jenseits des Lustprinzips (1920). Ges. Werke, Bd. XIII. Frankfurt 1969.

Freud, S.: Massenpsychologie und Ich-Analyse (1921). Ges. Werke, Bd. XIII, S. 72–161. Frankfurt 1946.

Freud, S.: Das Ich und das Es (1923). Ges. Werke, Bd. XIII. Imago, London 1950.

Freud, S.: Neurose und Psychose (1924a). Ges. Werke, Bd. XIII, S. 387–391.

Freud, S.: Hemmung, Symptom und Angst (1926). Ges. Werke, Bd. XIV, S. 192. Fischer, Frankfurt 1976.

Freud, S.: Dostojewski und die Vatertötung (1928). Ges. Werke, Bd. XIV.

Freud, S.: Das Unbehagen in der Kultur (1930). Studienausgabe Bd. IX. Fischer, Frankfurt 1974.

Freud, S.: (1932): Zit. nach Cremerius, J.: Freuds Konzept über die Entstehung psychogener Körpersymptome. Psyche 11 (1957/58) 125–139.

Freud, S.: Neue Folge der Vorlesungen zur Einführung in die Psychoanalyse (1933). Ges. Werke, Bd. XV. Frankfurt 1967.

Freud, S.: Die endliche und die unendliche Analyse. Ges. Werke, Bd. XVI. Fischer, Frankfurt 1937.

Freud, S.: Die Ichspaltung im Abwehrvorgang (1938). Ges. Werke, Bd. XVII. Imago, London 1950.

Freud, S.: Brief an V. v. Weizsäcker (3. XI. 1932). In: Weizsäcker, V. v.: Körpergeschehen und Neurose. Klett, Stuttgart 1947.

Freud, S.: Zit. nach: Uexküll, Th. v.: Psychologische Aspekte der essentiellen Hypertonie. Verh. Dtsch. Ges. Inn. Med. 69 (1963) 496.

Freud, S.: Gesammelte Werke. Fischer, Frankfurt 1968.

Freud, S.: Zur Einführung des Narzißmus. Ges. Werke, Bd. X. Fischer, Frankfurt 1969.

Freud, S.: Die zukünftigen Chancen der Psychotherapie. Ges. Werke, Bd. VIII, S. 104–115. 5. Aufl. Fischer, Frankfurt 1969.

Freud, S.: Studien über Hysterie. Ges. Werke, Bd. I, S. 197 und 278. Fischer, Frankfurt, 4. Aufl. 1972.

Freud, S.: Die psychogene Sehstörung in psychoanalytischer Auffassung. Ges. Werke, Bd. VIII, S. 94–102. Fischer, Frankfurt 1973.

Freud, S.: Über Psychoanalyse. Ges. Werke, Bd. VIII. Fischer, Frankfurt 1973.

Freyberger, H.: Psychosomatik und Psychotherapie – Colitis ulcerosa. In: Krauspe, C., K. Müller-Wieland, F. Stelzner (Hrsg.): Colitis ulcerosa und granulomatosa. Urban & Schwarzenberg, München 1972.

Freyberger, H.: Psychosomatik. In: Lawin, P. (Hrsg.): Praxis der Intensivbehandlung. Thieme, Stuttgart 1975.

Freyberger, H.: Topic: intensive care unit. Psychother. and Psychosom. 26 (1975) 337–343.

Freyberger, H.: Psychosomatic aspects of an intensive care unit. In: Havells, J. G. (ed.): Modern Perspectives in the Psychiatric Aspects of Surgery. Brunner/Mazel, New York 1976.

Freyberger, H.: Psychosomatische Aspekte bei Durchfallerkrankungen. Klinikarzt 5 (1976) 971–976.

Freyberger, H.: Psychosomatik der Colitis ulcerosa und des Morbus Crohn.Therapiewoche 27 (1977) 6675–6682.

Freyberger, H.: Psychosomatik des Erwachsenen. In: Bock, H. E., W. Gerok, F. Hartmann (Hrsg.): Klinik der Gegenwart, Bd. IX, S. 613–675. Urban & Schwarzenberg, München 1977.

Freyberger, H.: Supportive psychotherapeutic techniques in primary and secondary alexithymia. In: Bräutigam, W., M.v. Rad (eds.): Toward a Theory of Psychosomatic Disorders, pp. 337–348. Basel 1977.

Freyberger, H.: Klinisch-psychosomatische Praxis. Krankenhausarzt 51 (1978) 645–658.

Freyberger, H.: Beiträge der Psychosomatik zur Pathophysiologie und Klinik von Kolonerkrankungen. In: Müller-Wieland, K. (Hrsg.): Handbuch der inneren Medizin, Bd. III/4: Dickdarm. 5. Aufl. Springer, Berlin 1982.

Freyberger, H.: The renal transplant patient: three-stage model and psychotherapeutic strategies. In: Levy, N. B. (ed.): Psychonephrology 2. Plenum, New York 1983.

Freyberger, H.: Psychotherapeutische Ansätze bei Dialysepatienten und ihren Partnern. In: Balck, F. B., U. Koch, H. Speidel (Hrsg.): Psychonephrologie. Springer, Heidelberg 1984.

Freyberger, H. et al.: Consultation-liaison-psychiatry activities in a renal transplant unit. Psychother. and Psychosom. 32 (1979) 157–163.

Freyberger, H., K. Müller-Wieland: Kombinierter internistisch-psychosomatischer Therapieansatz bei Colitis ulcerosa. Med. Klinik 61 (1966) 228–230.

Freyberger, H., W. Bauditz: Psychosyndrome und somatische Reaktionen bei chronisch Nierenkranken im Hämodialysedauerprogramm. Verh. Dtsch. Ges. Inn. Med. 75 (1969) 971–977.

Freyberger, H., H. Speidel: Die supportive Therapie in der klinischen Medizin. Bibl. Psychiat. Neurol. (Basel) 152 (1976) 141–169.

Freyberger, H., D. Haan, K. Müller-Wieland: Psychosomatische Aufgabenbereiche auf Intensivbehandlungsstationen. Internist 10 (1969) 240–243.

Freyberger, H., B. Porschek, H. Bödeke et al.: Das Berufsbild der Intensivschwester und des Intensivpflegers. Z. prakt. Anästh. 7 (1972) 123–140.

Freyberger, H., R. Liedtke, W. Wellmann: Möglichkeiten und Grenzen der Psychotherapie bei Colitis ulcerosa und Morbus Crohn. Dtsch. Ärztebl. 77 (1980) 2731–2734.

Freyberger, H., W. Wellmann, H. Ziegler: Psychosomatische Aspekte der entzündlichen Erkrankungen des Darmkanals. In: Gall, F. P., H. Groitl (Hrsg.): Entzündliche Erkrankungen des Dünn- und Dickdarms, 2. Aufl. perimed, Erlangen 1983.

Freyberger, H., J. Nordmeyer, W. Lempa, H. W. Künzebeck, H. J. Avenarius, W. Wellmann, R. Schöl, R. Liedtke: Clinical and educational activities of a psychosomatic division. Advanc. psychosom. Med. 11 (1983) 166–175.

Freyhan, F. A., S. Gianelli, R. A. O'Connell, J. A. Mayo: Psychiatric complications following open heart surgery. Comprehens. Psychiat. 12 (1971) 181–195.

Freyschuss, V.: Cardiovascular adjustment to somatomotor activation: the elicitation of increments in heart rate, aortic pressure and vasomotor tone with the initiation of muscle contraction. Acta Physiol. scand. Suppl. 342 (1970) 1–63.

Friar, L. R., J. Beatty: Migraine: management by trained control of vasoconstriction. J. consult. clin. Psychol. 44 (1976) 46–53.

Frick-Bruder, V.: Intrauterinpessar. Psychologische Aspekte des Für und Wider. Gynäk. Praxis 4 (1980) 103.

Frick-Bruder, V.: Die Psychosomatik der Gynäkologie. In: Jores, A. (Hrsg.): Praktische Psychosomatik, S. 317–330. Huber, Bern-Stuttgart-Wien 1981.

Frick-Bruder, V.: Das prämenstruelle Syndrom. Sexualmed. 3 (1984) 153.

Friczewski, F.: Arbeitswissenschaft und Psychosomatik. Internationales Institut für Vergleichende Gesellschaftsforschung, Wissenschaftszentrum Berlin, reprint 81–211, 1981.

Friczewski, F., R. Thorbecke: Arbeitssituationen und koronare Herzkrankheiten. In: Haug, W. F. (Hrsg.): Lohnarbeit – Staat – Gesundheitswesen. Argumente für eine soziale Medizin VII (AS 12), S. 190–220. Argument-Verlag, Berlin 1976.

Friedberg, C. K.: Erkrankungen des Herzens, Bd. 1, S. 899–900. Thieme, Stuttgart 1972.

Friedenwald, J., M. Feldman, L. J. Rosenthal: Mucous colitis, observations in 500 cases. Ann. intern. Med. 3 (1929) 521–545.

Friedlander, H. J., M. Viederman: Children of dialysis patients. In: Levy, N. B. (ed.): Psychonephrology 2, pp. 93–105. Plenum, New York 1983.

Friedman, A. P.: Ad hoc committee on classification of headache. J. Amer. med. Ass. 179 (1962) 717–718.

Friedman, A. P.: Characteristics of tension headache: a profile of 1420 cases. Psychosomatics 20 (1979) 451–461.

Friedman, A. P., H. H. Merritt: Headache: diagnosis and treatment. Davis, Philadelphia 1959.

Friedman, E. A., A. P. Lundin: Environmental and iatrogenic obstacles to long life on hemodialysis. New Engl. J. Med. 306 (1982) 167–169.

Friedman, E., N. Goodwin, L. Chandhry: Psychosocial adjustment to maintenance hemodialysis. N. Y. St. J. Med. I (1970) 629–637; II (1970) 767–774.

Friedman, E., A. Katcher, V. Brightman: Incidence of recurrent herpes labialis and upper respiratory infection: a prospective study of the influence of biologic, social and psychologic predictors. Oral Surg. 43 (1977) 873–878.

Friedman, G. D., H. K. Ury, A. L. Klatsky, M. S. Siegelaub: A psychological questionnaire predictive of myocardial infarction. Psychosom. Med. 36 (1974) 327–343.

Friedman, H. S., S. Booth-Kewley: Validity of the Type A construct: a reprise. Psychol. Bull. 104 (1988) 284–381.

Friedman, M.: Type A behavior and myocardial infarction: reply. Letter to the editor. Amer. Heart J. 111 (1986) 1216.

Friedman, M., M. A. Brown, R. H. Rosenman: Voice analysis test for detection of behavior pattern. J. Amer. med. Ass. 208 (1969) 828–836.

Friedman, M., S. O. Byers, J. Diamant, R. H. Rosenman: Plasma catecholamine response of coronary prone subjects (Type A) to a specific challenge. Metabolism 24 (1975) 205–210.

Friedman, M., J. S. Kadanin: Hypertension in only one of identical twins. Report of a case with consideration of psychosomatic factors. Arch. intern. Med. 72 (1943) 767–774.

Friedman, M., J. H. Manwarning, R. H. Rosenman et al.: Instantaneous and sudden deaths. Clinical and pathological differentiation in coronary artery disease. J. Amer. med. Ass. 225 (1973) 1319–1328.

Friedman, M., R. H. Rosenman: Type A behavior and your heart. Knopf, New York 1974.

Friedman, M., C. E. Thoresen, J. J. Gill, L. H. Powell, D. Ulmer, L. Thompson: Feasibility of altering Type A behavior pattern after myocardial infarction. Recurrent Coronary Prevention Project Study. Circulation 66 (1982) 83.

Friedman, M., C. E. Thoresen, J. J. Gill, L. H. Powell, D. Ulmer, L. Thompson, V. A. Price, D. D. Rabin, W. S. Breall, T. Dixon, R. Levy, E. Bourg: Alteration of Type A behavior and reduction in cardiac recurrences in postmyocardial infarction patients. Amer. Heart J. 108 (1984) 237–248.

Friedman, M., C. E. Thoresen, J. J. Gill, L. H. Powell, D. Ulmer, L. Thompson, V. A. Price, D. D. Rabin, W. S. Breall, T. Dixon, R. Levy, E. Bourg, B. Brown: Alteration of Type A behavior and its effect on cardiac recurrences in post-myocardial infarction patients. Summary results of the Recurrent Coronary Prevention Project. Amer. Heart J. 112 (1986) 653–665.

Friedman, R., L. K. Dahl: Psychic and genetic factors in the etiology of hypertension. In: Wheatley, D. (ed.): Stress and the Heart. Raven, New York 1977.

Friedman, S. B.: The concept of marginality applied to psychosomatic medicine. Psychosom. Med. 50 (1988) 447–453.

Friedman, S. B., L. A. Glasgow: Psychologic factor and resistance to infectious disease. Pediat. Clin. N. Amer. 13 (1960) 315.

Friedman, S. B., J. W. Mason, D. A. Hamburg: Urinary 17-hydroxy corticosteroid levels in parents of children with neoplastic disease. Psychosom. Med. 25 (1963) 364.

Friedman, S. B., R. Ader, L. A. Glasgow: Effects of psychological stress in adult mice inoculated with Coxsackie B viruses. Psychosom. Med. 27 (1965) 361–368.

Friedman, S. B., L. A. Glasgow, R. Ader: Psychosocial factors modifying host resistance to experimental infections. Ann. N. Y. Acad. Sci. 164 (1969) 381–393.

Friedman, S. B., R. Ader, L. A. Glasgow: Differential susceptibility to a viral agent in mice housed alone or in groups. Psychosom. Med. 32 (1970) 285–299.

Friedman, S. R., D. C. Des Jarlais, J. L. Sotheran: AIDS health education for intravenous drug users. Health Educ. Quart. 13 (1986) 383–393.

Friedrich, R., J. Jordan, C. Schlienf, G. Overbeck, W. Dehe: Psychische Adaptation an einen implantierten Herzschrittmacher. Münch. med. Wschr. 125 (1983) 193–196.

Friedsam, H.: Reactions of older persons to disaster-caused losses – a hypothesis of relative deprivation. In: Tibbits, C., W. Donahue (eds.): Social and Psychological Aspects of Aging. Columbia Univ. Press, New York 1962.

Fries, H.: Secondary amenorrhea, self-induced weight reduction and anorexia nervosa. Acta psychiat. scand. 248 (Suppl.) (1974).

Fries, J. F., P. Spitz, R. G. Kraines, H. R. Holman: Measurement of patient outcome in arthritis. Arthr. and Rheum. 23 (1980) 137–145.

Frisch, R. E., R. Revelle, S. Cook: Height, weight and age at menarche and the „critical weight" hypothesis. Science 174 (1972) 1148–1149.

Fritsch, W., G. Enderlein, S. Kruschwitz, J. Lossow, W. Moszler, B. Weichert: Nachtschichtarbeit und Gesundheitszustand werktätiger Frauen. Ztschr. f. d. gesamte Hygiene und ihre Grenzgebiete 23 (1977) 885–888.

Fritze, E. (Hrsg.): Lehrbuch der Anamneseerhebung und allgemeinen Körperuntersuchung. 3. Aufl. Verlag Chemie, Weinheim 1983.

Fröhlich, F.: Psychische Veränderungen bei Patienten mit behandelter Parkinsonkrankheit. Med. Diss., Bern 1984.

Froelich, R. E., F. M. Bishop: Die Gesprächsführung des Arztes. Springer, Heidelberg 1973.

Froelicher, V. T.: Exercise and the prevention of the atherosclerotic heart disease. Wanger, N. K. (ed.) F. A. Davis, Philadelphia 1978.

Froese, A., T. P. Hackett, N. H. Cassem, E. L. Silverberg: Trajectories of anxiety and depression in denying and non-denying acute myocardial infarction patients during hospitalization. J. psychosom. Res. 18 (1974) 413–420.

Froese, A., T. P. Hackett, N. H. Cassem, E. L. Silverberg: Galvanic skin potential as a predictor of mental status, anxiety, depression and denial in acute coronary patients. J. psychosom. Res. 19 (1975) 1–9.

Fromm, E.: Über den Ungehorsam. Deutsche Verlagsanstalt, Stuttgart 1982.

Frowein, R. A., R. Firsching: Von der Ischiasneuritis zum vertebragenen Wurzelkompressionssyndrom. In: Hohmann, D., B. Kügelgen, K. Liebig, M. Schirmer (Hrsg.): Neuroorthopädie 2: Lendenwirbelsäulenerkrankungen mit Beteiligung des Nervensystems, S. 319–330. Springer, Berlin-Heidelberg-New York 1984.

Frymoyer, J. W.: Back pain and sciatica. New Engl. J. Med. 318 (1988) 291–300.

Fuchs, A.: Allergy problems in elderly persons. Geriatrics 2 (1947) 235.

Fuchs, M.: Funktionelle Entspannung. Stuttgart 1974.

Fuchs, M.: Einführung in die Funktionelle Entspannung. In: Fuchs, M. (Hrsg.): Funktionelle Entspannung in der Kinderpsychotherapie, S. 11. Reinhardt, München-Basel 1985.

Fuchs, M.: Das leibliche und seelische Unbewußte, die Funktionelle Entspannung und das therapeutische Gespräch. Prax. Psychother. Psychosom. 33 (1988) 120.

Fuchs, M.: Funktionelle Entspannung. Hippokrates, Stuttgart 1989.

Fürmaier, A. M.: Zur Psychodynamik des Morbus Crohn. Diss., Tübingen 1979.

Fürstenau, P.: Probleme der vergleichenden Psychotherapieforschung. Psyche 26 (1976) 423.

Fürstenau, P., E. Mahler, H. Morgenstern, H. Müller-Braunschweig, H. E. Richter: Untersuchungen über Herzneurose. Psyche 3 (1964) 177.

Fürstenberg, F.: Einführung in die Arbeitssoziologie. Wissenschaftliche Buchgesellschaft, Darmstadt 1977.

Fürstner, K.: Zur Pathologie gewisser Krampfanfälle (hysterische Anfälle bei Kindern, Spätepilepsie). Arch. Psychiat. Nervenkr. 28 (1896) 494–509.

Fuhrmann, K.: Diabetic control and outcome in the pregnant patient. In: Peterson, C. M. (ed.): Diabetes management in the '80s, pp. 66–79. Praeger, New York 1982.

Fuller, J. H., M. J. Shipley, G. Rose, R. J. Jarrett, H. Keen: Mortality from coronary heart disease and stroke in relation to degree of glycaemia: Whitehall Study. Brit. med. J. 287 (1983) 867–870.

Funch, D. P.: Psychosocial variables and the course of cancer. New Engl. J. Med. 313 (1985) 1354.

Funch, D. P., C. Mettlin: The role of support in relation to recovery from breast cancer. Soc. Sci. Med. 16 (1982) 91–98.

Funch, D. P., J. Marshall: The role of stress, social support and age in survival from breast cancer. J. psychosom. Res. 27 (1983) 77–83.

Funk, G. A., M. M. Jensen: Influence of stress on granuloma formation. Proc. Soc. exp. Biol. (N. Y.) 124 (1967) 653–655.

Funkenstein, D. P.: Nor-epinephrine-like and epinephrine-like substances in relation to human behavior. J. nerv. ment. Dis. 124 (1956) 58–68.

Furth, P.: Nachträgliche Warnung vor dem Rollenbegriff. Argument 66 (1971) 494–522.

Gaddini, R.: Early psychosomatic symptom and the tendency towards integration. Psychother. and Psychosom. 23 (1974) 26–34.

Gaddini, R.: The pathology of the self as a basis of psychosomatic disorders. Psychother. and Psychosom. 28 (1977) 260–271.

Gaillard, M., F. Wulliemier: Etude catamnestique comparative de deux traitements de l'anorexie nerveuse. J. psychosom. Res. 26 (1982) 113–121.

Galatzer, A., M. Frish, Z. Laron: Changes in self-concept and feelings toward diabetic adolescents. In: Laron, Z. (ed.): Psychological Aspects of Balance of Diabetes in Juveniles. Pediatr. Adolesc. Endocrinol., Vol. 3, 1977.

Galatzer, A., S. Amir, R. Gil, M. Korp, A. Leron: Crisis intervention program in newly diagnosed diabetic children. Diabetes Care 5 (1982) 414–419.

Galdston, R., W. J. Gamble: On borrowed time. Observations on children with implanted cardiac pacemakers and their families. Amer. J. Psychiat. 126 (1969) 104.

Gall, F. P., E. Mühe, B. Angermann: Resultate operativer Therapie des Morbus Crohn. In: Ottenjann, R., H. Fahrländer (Hrsg.): Entzündliche Erkrankungen des Dickdarms. Springer, Berlin 1983.

Gallistel, C. R.: Self-stimulation: the neurophysiology of reward and motivation. In: Deutsch, J. A. (ed.): The Physiological Basis of Memory. Acad. Press, New York 1973.

Gallmeier, W. M.: Psycho-Onkologie. Münch. med. Wschr. 126 (1984) 211–213.

Gallmeier, W. M., U. Bruntsch: Hospize, Palliativstationen etc. – Ausdruck der Krise der Medizin? Münch. med. Wschr. 130 (1988) 275–277.

Galton, F.: Inquiries into human faculty and its development. London 1883.

Galtung, J.: Strukturelle Gewalt. Rowohlt, Reinbek 1975.

Gambert, S. R., T. L. Garthwaite, C. H. Pontzer, T. C. Hagen: Fasting associated with decrease in hypothalamic β-endorphin. Science 210 (1980) 1271–1272.

Gamzu, E., G. Vincent, N. Tare, W. Benjamin, J. Farrar, A. C. Sullivan: Effects of stress on immune function in mice and rats. Abstract of poster presented at Society for Neuroscience, Anaheim 1984.

Gandras, G.: Konzentrative Bewegungstherapie. In: Feiereis, H. (Hrsg.): Diagnostik und Therapie der Magersucht und Bulimie. Marseille, München 1989a.

Gandras, G.: Progressive Relaxation. In: Feiereis, H. (Hrsg.): Diagnostik und Therapie der Magersucht und Bulimie. Marseille, München 1989b.

Ganguli, R., C. F. Reynolds II, D. J. Kupfer: Electroencephalographic sleep in young, never-medicated schizophrenics. Arch. gen. Psychiat. 44 (1987) 36–44.

Ganong, W. F.: Neurophysiologic basis of instinctual behavior and emotions. In: Ganong, W. F. (ed.): Review of Medical Physiology, pp. 173–185. Lange, Los Altos/Ca. 1971.

Ganong, W. F. (ed.): Nervous system. Lange, Los Altos/Ca. 1977.

Ganowsky, M. A., D. Drotar, S. Makker: Growing up with renal failure. In: Levy, N. B. (ed.): Psychonephrology 2, pp. 195–206. Plenum, New York 1983.

Ganten, D., K. Fuxe, M. I. Philips, J. F. E. Mann, U. Ganten: The brain isorenin-angiotensin system: biochemistry, localization, and possible role in drinking and blood pressure regulation. In: Ganong, W. F. (ed.): Frontiers in Neuroendocrinology, Vol. 5, pp. 61–99. Oxford Univ. Press, New York 1978.

Garcia, C.: Über die klinischen Grenzen des Gilles de la Tourette-Syndroms. Nervenarzt 58 (1987) 748–753.

Garcia, J.: Autogenes Training und Biokybernetik. Hippokrates, Stuttgart 1983.

Garcia, J., R. A. Koelling: Relation of cue to consequence in avoidance learning. In: Seligman, M. E. P., J. L. Hager (eds.): Biological Boundaries of Learning. Appleton Century Crofts, New York 1972.

Garcia, J., F. R. Ervin, R. A. Koelling: Very long CS-UCS intervals in taste-aversion learning. In: Seligman, M. E. P., J. L. Hager (eds.): Biological Boundaries of Learning. Appleton Century Crofts, New York 1972.

Garcia, J., W. G. Hankins, K. W. Rusiniak: Behavioral regulation of the milieu interne in man and rat. Science 185 (1974) 824–831.

Gardiner, B. M.: Psychological aspects of rheumatoid arthritis. Psychol. Med. 10 (1980) 159–163.

Gardner, J. L.: 'Hyperdietism' – its prevention, control and relation to compliance in dialysis and transplant patients. Dial. Transplant. 10 (1981) 57–60.

Garfield, S. L., A. E. Bergin (eds.): Handbook of Psychotherapy and Behavioral Change: An Empirical Analysis. Wiley, New York 1978.

Garfinkel, A.: A mathematics for physiology. Amer. J. Physiol. 245 (1983) R 455 – R 466.

Garfinkel, P. E., D. M. Garner: The multidetermined nature of anorexia nervosa. In: Darby, P. L., P. E. Garfinkel, D. M. Garner, D. V. Coscina (eds.): Anorexia nervosa. Recent Developments in Research, pp. 3–14. Liss, New York 1983.

Garfinkel, P. E., H. Modolfsky, D. M. Garner: The outcome of anorexia nervosa: significance of clinical features, body image and behavior modification. In: Vigersky, R. A. (ed.): Anorexia nervosa, pp. 315–329. Raven, New York 1977.

Garfinkel, R.: Treatment of a psychosomatic ailment in an elderly woman. Psychosomatics 21 (1980) 1015–1016.

Garner, D. M.: Psychotherapy outcome research with bulimia nervosa. Psychother. and Psychosom. 48 (1987) 129–140.

Garner, D. M., P. E. Garfinkel: The eating attitude test: an index of the symptoms of anorexia nervosa. Psychol. Med. 9 (1979) 273.

Garner, D. M., P. E. Garfinkel (eds.): Handbook of psychotherapy for anorexia nervosa and bulimia. Guilford, New York 1985.

Garner, D. M., P. E. Garfinkel, H. Modolfsky: Perceptual experiences in anorexia nervosa and obesity. Canad. psychiat. Ass. J. 23 (1978) 249–263.

Garner, D. M., P. E. Garfinkel, M. P. Olmsted: Does anorexia nervosa occur on a continuum? Int. J. Eat. Dis. 2 (1983).

Garner, D. M., P. E. Garfinkel, M. P. Olmsted: An overview of sociocultural factors in the development of anorexia nervosa. In: Darby, P. L., P. E. Garfinkel, D. M. Garner, D. V. Coscina (eds.): Anorexia nervosa. Recent Developments in Research, pp. 65–82. Liss, New York 1983a.

Garner, D. M., M. P. Olmsted, J. Polivy: The Eating Disorder Inventory: a measure of cognitive-behavioral dimensions of anorexia nervosa and bulimia. In: Darby, P. L., P. E. Garfinkel, D. M. Garner, D. V. Coscina (eds.): Anorexia nervosa. Recent Developments in Research, pp. 173–184. Liss, New York 1983b.

Garner, D. M., W. Rockert, M. P. Olmsted, C. Johnson, D. V. Coscina: Psychoeducational principles in the treatment of bulimia and anorexia nervosa. In: Garner, D. M., P. E. Garfinkel (eds.): Handbook of Psychotherapy for Anorexia Nervosa and Bulimia. Guilford, New York 1985.

Garner, H. H.: Confrontation technique. Ill. med. J. (1970).

Gaskin, R. J., A. Gad, D. S. Barros, S. N. Joffe, J. H. Baron: Natural history and morphology of secretagogue-induced duodenal ulcers in rats. Gastroenterol. 69 (1975) 903–910.

Gastorf, J. W.: Physiologic reactions of Type A's to objective and subjective challenge. J. hum. Stress (1981) 16–20.

Gatfield, P. D., S. B. Guze: The prognosis and differential diagnosis of conversion reactions: A follow-up study. Dis. nerv. Syst. 23 (1963) 623–631.

Gathmann, P.: Das pathologische psychosomatische Reaktionsmuster beim Alternden: Epidemiologische, diagnostische, präventive und therapeutische Bemerkungen. Z. Geront. 20 (1987) 210–218.

Gathmann, P., L. Linzmayer, J. Grünberger: Colitis ulcerosa – eine Studie zur Objektivierung der Persönlichkeitsmerkmale des Kolitispatienten. Wien. med. Wschr. 131 (1981) 421–425.

Gaus, E.: Zur Psychologie therapeutischer Extremsituationen in der Medizin. Diss., Ulm 1975.

Gaus, E., K. Bechter, K.-H. Dreyer, W. Merkle, A. Rein: Untersuchungen zur Bedeutung und Rolle von Medikamenten. In: Quint, H., P. L. Janssen (Hrsg.): Psychotherapie in der psychosomatischen Medizin. Springer, Berlin 1987.

Gaus, E., M. Klingenberg, K. Köhle: Psychosomatische Gesichtspunkte in der Behandlung von Hypertoniepatienten – Möglichkeiten eines integrierten internistisch-psychosomatischen Ambulanzkonzeptes. Psychother. med. Psychol. 33 (1983) 53–60.

Gaus, E., M. Klingenburg, C. Simons, C. Westphale, K. Köhle: Die Funktion von Stationskonferenzen im Rahmen eines internistisch-psychosomatischen Stationskonzeptes. Verh. Dt. Ges. Inn. Med. 8 (1980) 1487–1491.

Gaus, E., K. Köhle: Ängste des Patienten – Ängste des Arztes. Anmerkungen zur Konfliktaustragung in einer schwierigen Visite bei einem Todkranken. In: Köhle, K., H.-H. Raspe (Hrsg.): Das Gespräch während der ärztlichen Visite. Urban & Schwarzenberg, München 1982.

Gaus, E., B. Kubanek: Psychosoziale Faktoren und Immunkompetenz. Internist 25 (1984) 667–673.

Gausch, K.: Schmerzpatient, Schmerzprojektion und initiale Funktionstherapie. Dtsch. zahnärztl. Z. 35 (1980) 587.

Gauss, C. J.: Eine häufig vorkommende, mehrfach beschriebene, meist verkannte und oft operativ umsonst angegangene Erkrankung: die Pelipathia vegetativa. Dtsch. med. Wschr. 74 (1949) 1288.

Gazzard, B. G., H. L. Price, G. W. Libby, A. M. Dawson: The social toll of Crohn's disease. Brit. med. J. II (1978) 1117–1119.

Geist, W., H. Urban, K. Köhle: Der Nachtdienst in der Krankenpflege aus der Sicht patientenzentrierter Medizin. In: Psychosoziale Probleme im Krankenhaus. Urban & Schwarzenberg, München 1976.

Gente, H. P.: Marxismus, Psychoanalyse, Sexpol. Bd. 1 und 2. Frankfurt 1972.

Gentry, W. D., S. Foster, T. Haney: Denial as a determinant of anxiety and perceived health status in the coronary care unit. Psychosom. Med. 34 (1972) 39–44.

Gepts, W.: Pathomorphology of the pancreas in diabetes. In: Federlin, K., J. Schiltholt (eds.): The Importance of Islets of Langerhans for Modern Endocrinology, pp. 93–110. Raven, New York 1984.

Gerber, W. D.: Verhaltensmedizin der Migräne. edition medizin, Weinheim 1986.

Gerber, W. D.: Uni- und multidimensionale Therapieansätze. In: Gerber, W. D., G. Haag (Hrsg.): Migräne. Springer, Berlin-Heidelberg 1982.

Gerber, W. D., G. Haag (Hrsg.): Migräne. Springer, Berlin 1982.

Gerber, W. D., H. C. Diener: Klinisch-psychologische und neuropsychologische Untersuchungen zur Myasthenia gravis. In: Jacobi, P. (Hrsg.): Jahrbuch der medizinischen Psychologie 2: Psychologie in der Neurologie, S. 185–199. Springer, Berlin-Heidelberg-New York 1989.

Gerber, W. D., W. Miltner, N. Birbaumer, W. Lutzenberger: Cephalic vasomotor feedback therapy: a controlled study of migrainors and normals. In: Holroyd, K. A., B. Schlote, H. Zenz (eds.): Perspectives in Research on Headache, pp. 163–170. Hogrefe, Lewiston-New York 1983.

Gerber, W. D., W. Miltner, N. Birbaumer, G. Haag: Konkordanztherapie. Röttger, München 1989.

Gerber, W. D., W. Miltner, H. Gabler, E. Hildenbrand, W. Larbig: Bewegungs- und Sporttherapie bei chronischen Dauerkopfschmerzen. In: Gerber, W. D., W. Miltner, K. Meyer (Hrsg.): Verhaltensmedizin: Ergebnisse und Perspektiven interdisziplinärer Forschung, S. 55–66. edition medizin, Weinheim 1987.

Gerbershagen, H. U.: Nonnarcotic analgetics. In: Bonica, J. J., V. Ventafridda (eds.): Advances in Pain Research and Therapy, Vol. 2, pp. 255–262. Raven, New York 1979.

Gerbert, B.: Psychological aspects of Crohn's disease. J. behav. Med. 3 (1980) 41–58.

Gerdes, K.: Rehabilitation von Krebskranken. Ausschreibung des Bundesministers für Forschung und Technologie. Gesellschaft für Strahlen- und Umweltforschung, München 1984.

Gerhards, F.: Emotionsausdruck und emotionales Erleben bei psychosomatisch Kranken. Beiträge zur psychologischen Forschung 14. Opladen 1988.

Gerhards, S. F., J. Rojahn, K. Boxan, C. Gnade, M. Petrik, I. Florin: Biofeedback vs. cognitive stress-coping therapy in migraine headache patients. In: Holroyd, K. A., B. Schlote, H. Zenz (eds.): Perspectives in Research on Headache, pp. 171–182. Hogrefe, Lewiston-New York 1983.

Gerhardt, U., R. Heberling: Rehabilitation und Familie bei terminaler Niereninsuffizienz. In: Angermeyer, M. C., H. Freyberger (Hrsg.): Chronisch kranke Erwachsene in der Familie, S. 95–107. Enke, Stuttgart 1982.

Gerich, L.: Ein Beitrag zur psychosomatischen Betrachtung der Enteritis regionalis. Z. klin. Psychol. Psychother. 28 (1980) 350–369.

Gerle, B., G. Lunden, P. Sandblom: The patient with inoperable cancer from the psychiatric and social standpoints. Cancer 13 (1960) 1206–1217.

Gerlinghoff, M.: Magersucht. Psychologie Verlagsunion, München 1988.

Gershon, E. S., J. L. Schreiber, J. R. Hamovit, E. D. Dibble, W. Kaye, J. I. Nurnberger, A. E. Andersen, M. Ebert: Clinical findings in patients with anorexia nervosa and affective illness in their relatives. Amer. J. Psychiat. 141 (1984) 1419–1422.

Gesundheitsreformgesetz: Stand 17. Dezember 1988.

Gewirtz, J. L.: Soziales Lernen. In: Zeier, H. (Hrsg.): Pawlow und die Folgen. Psychologie des 20. Jahrhunderts, Bd. IV. Kindler, Zürich 1977.

Gfeller, R., J.-Ph. Assal: Une expérience-pilote en diabétologie clinique et en psychologie médicale: l'unité de traitement et d'enseignement pour malades diabétiques de l'Hôpital cantonal de Genève. Médecine et Hygiène (Genève) 1346 (1979) 2966–2970.

Gfeller, R., J.-Ph. Assal, J. U. Ekoe: Les problèmes sexuels des diabétiques. Therapeutische Umschau/ Revue thérapeutique 38 (1981) 1069–1074.

Ghanta, V., N. S. Hiramoto, H. B. Solvason, S. K. Tyring, N. H. Spector, R. N. Hiramoto: Conditioned enhancement of natural killer cell activity, but not interferon, with camphor or saccharin-LiCl conditioned stimulus. J. Neurosci. Res. 18 (1987 a) 10–15.

Ghanta, V., R. N. Hiramoto, B. Solvason, N. H. Spector: Influence of conditioned natural immunity on tumor growth. Ann. N. Y. Acad. Sci. 496 (1987 b) 637–646.

Gianturco, O. T., M. S. Breslin, A. Heyman, W. D. Gentry, C. D. Jenkins, B. Kaplan: Personality pattern and life stress in ischemic cerebrovascular disease, I. Psychiatric findings. Stroke 5 (1974) 453–460.

Gifford, S., J. G. Gunderson: Cushing's disease as a psychosomatic disorder. Medicine 49 (1970) 397.

Giger, M., H. J. Nüesch, U. Seefeld, M. Jaeger, W. Wüst, R. Siebenmann, P. Deyhle, A. L. Blum: Die Salmonellen-Kolitis. Schweiz. med. Wschr. 109 (1979) 1309–1313.

Gilberstadt, H., Y. Sako: Intellectual and personality change following open-heart surgery. Arch. gen. Psychiat. 16 (1967) 210–214.

Gilbertsen, M. D., O. H. Wagensteen: Should the doctor tell the patient that the disease is cancer? Cancer 12 (1962) 82–86.

Giles, T.: A team approach: bulimia. Diab. Educ. 12 (1986) 69–70.

Giligan, I., L. Fung, D. Piper, C. Tennant: Life event stress and chronic difficulties in duodenal ulcer: a case control study. J. psychosom. Res. 31 (1987) 117–123.

Gill, J. J., V. A. Price, M. Friedman, C. E. Thoresen, L. H. Powell, D. Ulmer, B. Brown, F. R. Drews: Reduction of Type A behavior in healthy middle-aged American military officers. Amer. Heart J. 110 (1985) 503–514.

Gilles de la Tourette, G.: Etude sur une affection nerveuse caractérisée par l'incoordination motrice accompagnée d'écholalie et de coprolalie. Arch. Neurol. 9 (1885) 19–42; 158–200.

Gillette, S., R. W. Gillette: Changes in thymic estrogen receptor expression following orchidectomy. Cell. Immunol. 42 (1979) 194–196.

Gilman, S.: Cerebral disorders after open-heart operations. New Engl. J. Med. 272 (1965) 489–498.

Ginzburg, C.: Spurensicherungen. Wagenbach, Berlin 1983.

Giordani, B., S. B. Manuck, J. K. Farmer: Stability of behaviorally-induced heart rate changes in children after one week. Child Dev. 52 (1981) 533–537.

Giron, L. T., K. A. Crutcher, J. N. Davis: Lymph nodes – a possible site for sympathetic neuronal regulation of immune response. Ann. Neurol. 8 (1979) 520–525.

Gitelson, N.: The first phase in psychoanalysis. Int. J. Psychoanal. (1962).

Gitnick, G. L., M. H. Arthur, I. Shibata: Cultivation of viral agents from Crohn's disease. Lancet II (1976) 215–217.

Gjerris, A., M. Hammer, P. Vendsborg, N. J. Christiensen, O. J. Rafaelson: Cerebrospinal fluid vasopressin – changes in depression Brit. J. Psychiat. 147 (1985) 696–701.

Glaros, A. G., S. M. Rao: Bruxism: a critical review. Psychol. Bull. 84 (1977) 767.

Glaser, B. C., A. L. Strauss: Awareness of dying. Aldine, Chicago 1965. Deutsch: Interaktion mit Sterbenden. Beobachtungen für Ärzte, Schwestern, Seelsorger und Angehörige. Vandenhoeck & Ruprecht, Göttingen 1974.

Glaser, B. C., A. L. Strauss: Time for dying. Aldine, Chicago 1968.

Glaser, R., J. K. Kiecolt-Glaser, C. E. Speicher, J. E. Holliday: Stress, loneliness, and changes in herpesvirus latency. J. Behav. Med. 8 (1985) 249–260.

Glasgow, M. S., K. R. Gaarder, B. T. Engel: Behavioral treatment of high blood pressure, II: Acute and sustained effects of relaxation and systolic blood pressure biofeedback. Psychosom. Med. 44 (1982) 155–170.

Glasmacher, A., B. Krusenotto, R. Gugler: Selbsteinschätzung und Elternbild bei Patienten mit Morbus Crohn: Eine kontrollierte Untersuchung mit dem Gießen-Test. Klin. Wschr. 66 (Suppl. 13) (1988) 66–67.

Glass, D. C.: Behavior patterns, stress and coronary heart disease. Erlbaum, Hillsdale 1977.

Glass, D. C., M. L. Snyder, J. F. Hollis: Time urgency and the Type A coronary-prone behavior pattern. J. appl. soc. Psychol. 4 (1974) 125–140.

Glass, D. C., D. T. Ross, W. Isecke, R. H. Rosenman: Relative importance of speech characteristics and content of answers in the assessment of behavior pattern A by the structured interview. Basic and Applied Social Psychology 23 (1982) 161–168.

Glassman, A. H., S. P. Roose, E.-G. V. Giardina, J. Th. Bigger jr.: Cardiovascular effects of tricyclic antidepressants. In: Meltzer, H. J. (ed.): Psychopharmacology: The Third Generation of Progress. Raven, New York 1987.

Gloning, I., K. Gloning, K. Weingarten: Über optische Halluzinationen. Wien. Z. Nervenheilk. Grenzgeb. 10 (1955) 58–66.

Gloor, M., C. Eicher, H. Wiebelt, G. Moser: Soziologische Untersuchungen bei der Acne vulgaris. Z. Hautkr. 53 (1978) 871–880.

Gloor, P.: Diskussionsbeitrag zu B. Kaada: Brain mechanisms related to aggressive behavior. In: Clemente, C. D., D. B. Lindsley (eds.): Aggression and Defense. Univ. of California Press, Los Angeles 1967.

Gnirss, F., D. Schneider-Helmert, J. Schenker, V. Winkler: Schlafstörungen bei psychisch Kranken. Nervenarzt 49 (1978) 394–401.

Godelier, M.: Ökonomische Anthropologie. Untersuchungen zum Begriff der sozialen Struktur primitiver Gesellschaften. Reinbek 1973.

Godow, A. G.: Human Sexuality. Mosby, St. Louis-Toronto-London 1982.

Goebel, P.: Ein Vergleich der psychosozialen Situation von 125 Interruptio-Patientinnen vor und nach einem Schwangerschaftsabbruch. Z. psychosom. Med. 30 (1984) 270.

Göllner, R., W. Volk, M. Ermann: Analyse von Behandlungsergebnissen eines 10jährigen Katamneseprogrammes. In: Beese, F. (Hrsg.): Stationäre Psychotherapie. Vandenhoeck & Ruprecht, Göttingen 1978.

Görres, H. J., G. Ziegeler, H. Friedrich, G. Lücke: Krankheit und Bedrohung. Formen psychosozialer Bewältigung der Multiplen Sklerose. Z. psychosom. Med. 34 (1988) 274–290.

Goethe, J. W. v.: Kampagne in Frankreich (1792).

Goetz, P. L., R. A. Succop, J. B. Reinhart, A. Miller: Anorexia nervosa in children: a follow-up study. Amer. J. Orthopsychiat. 47 (1977) 597–603.

Goetz, S., R. H. Adler, R. Weber, C. Minder: Psychological processes and ischemic stroke (occlusive cerebrovascular disease) II. A controlled, retrospective clinical study on 19 women with strokes, 19 with non-vascular disease and 19 healthy hospital volunteers. (in preparation).

Götze, P.: Psychopathologie der Herzoperierten. Enke, Stuttgart 1980.

Götze, P., G. Huse-Kleinstoll: Präoperative Angst und Angstbewältigung: Psychodiagnostische Probleme und therapeutische Implikationen aus psychoanalytischer Sicht. Psychother. med. Psychol. 38 (1988) 232–239.

Goffman, E.: Stigma. Über Techniken zur Bewältigung beschädigter Identität. Frankfurt 1970.

Gold, P. W., F. K. Goodwin, G. P. Chrousos: Clinical and biochemical manifestations of depression. Relation to the neurobiology of stress. New Engl. J. Med. 319 (1988) 348–353, 413–420.

Gold, P. W., F. K. Goodwin, R. M. Post, G. L. Robertson: Vasopressin function in depression and mania. Psychopharmacol. Bull. 17 (1981) 7–9.

Gold, S., J. Lipton, J. Marbach, S. Dworkin, B. Gurion: Sites of psychophysiologic complaints in MPD patients: I. The orofacial region. J. dent. Res. 53 (spec. issue) (1974) 127.

Gold, S., J. Lipton, J. Marbach, S. Dworkin, B. Gurion: Sites of psychophysiologic complaints in MPD patients: II. Areas remote from orofacial region. J. dent. Res. 54A (1975) 165.

Gold, W. M., G. F. Kessler, D. Y. C. Yu: Role of vagus nerves in experimental asthma in allergic dogs. J. appl. Physiol. 33 (1972) 719–725.

Goldband, S.: Stimulus specifity of physiological response to stress and the Type A coronary-prone behavior pattern. J. pers. soc. Psychol. 39 (1980) 670–679.

Goldberg, D.: A psychiatric study of patients with diseases of the small intestine. Gut 11 (1970) 459–465.

Goldberg, D. P., P. Blackwell: Psychiatric illness in general practice. Brit. J. Psychiat. 2 (1970) 439–443.

Goldberg, D. P., K. Bridges: Invited review somatic presentations of psychiatric illness in primary care setting. J. psychosom. Res. 32 (1988) 137–144.

Goldberg, D. P., B. Cooper, M. R. Eastwood, H. B. Kedward, M. Shepherd: A standardized psychiatric interview for use in community surveys. Brit. J. prevent. soc. Med. 24 (1970) 18–23.

Goldberg, E. M.: Family influences in psychosomatic illness. London 1958.

Goldberg, J. K., A. H. Kutscher, B. Schoenberg, H. Gralnick, H. Kutscher: Psychopharmacologic and analgesic agents employed in the terminal care of 100 cancer patients. J. Thanatol. 2 (1972) 635–643.

Goldberg, M.: Über meine Therapie-Formel in der konzentrativen Bewegungstherapie. Prax. Psychother. XIX (1974) 327.

Goldberg, R. J., L. O. Cullen: Use of psychotropics in cancer patients. Psychosomatics 27 (1986) 687–700.

Goldberg, R. J., H. Leigh, D. Quinlan: The current status of placebo in hospital praxis. Gen. Hosp. Psychiat. 1 (1979) 196–201.

Golden, J. S.: The surgeon and the psychiatrist: special problems in psychiatric liaison. In: Pasnau, R. O. (ed.): Consultation-Liaison Psychiatry, pp.123–133. Grune & Stratton, New York 1975.

Golden, J. S., H. Nahum: Emotional reactions to mutilating surgery. In: Wahl, C. W. (ed.): New Dimensions in Psychosomatic Medicine. Little, Brown & Co., Boston 1964.

Goldenberg, M., C. E. Sluzki: Setting up a psychiatric service in a general hospital. Ment. Hyg. 55 (1971) 85–90.

Goldfarb, A. I.: Psychotherapy of aged persons, IV. One aspect of the psychodynamics of the therapeutic situation with aged patients. Psychoanal. Rev. 42 (1955) 180–187.

Goldfarb, A. I.: Patient-doctor-relationship in treatment of aged persons. Geriatrics 12 (1964) 18–23.

Goldfarb, A. I.: The psychodynamics of dependency and the search for aid. In: Kalish, R. (ed.): The Dependencies of Old People. Univ. of Michigan, Ann Arbor 1969.

Goldfarb, W.: Emotional and intellectual consequences of psychologic deprivation in infancy. In: Hoch, P. H., J. Zubin (eds.): Psychopathology of Childhood. Grune & Stratton, New York 1955.

Goldfried, M. R., R. N. Kent: Herkömmliche gegenüber verhaltenstheoretischer Persönlichkeitsdiagnostik. In: Schulte, D. (Hrsg.): Diagnostik in der Verhaltenstherapie. Urban & Schwarzenberg, München 1976.

Goldie, L., J. M. Green: A study of the psychological factors in a case of sensory reflex epilepsy. Brain 82 (1959) 505–524.

Goldmann, S. F.: Das Haupt-Histokompatibilitätssystem HLA und die Genetik der Zuckerkrankheit. Dtsch. Ärztebl. 79 (1982) 45–48.

Goldner, J. L.: Musculoskeletal aspects of emotional problems. Sth. med. J. (Bgham., Ala.) 69 (1976) 6–8.

Goldschmidt, O.: Vorgeschichte und Entwicklung. Psyche 27 (1973) 1022.

Goldstein, A., L. I. Lowney, B. K. Pal: Stereospecific and nonspecific interactions of the morphine narcotic congener levorphanol in subcellular fractions of mouse brain. Proc. nat. Acad. Sci. (Wash.) 68 (1971) 1742–1747.

Goldstein, A. M.: Denial and external locus of control as mechanisms of adjustment in chronic medical illness. Essence 1 (1976) 5–22.

Goldstein, A. M.: The subjective experience of denial in an objective investigation of chronically ill patients. Psychosomatics 13 (1972) 20–27.

Goldstein, A. P., N. Stein: Maßgeschneiderte Psychotherapie. Springer, Berlin-Heidelberg 1980.

Goldstein, F. J., P. Mojaverian, M. H. Ossipov, B. M. Swanson: Evaluation in analgesic effect and plasma levels of morphine by desipramine in the rat. Pain 14 (1982) 279–282.

Goldstein, J.: Console and classify: the french psychiatric profession in the nineteenth century. Cambridge Univ. Press, Cambridge 1987.

Goldstein, K.: Über den Einfluß motorischer Störungen auf die Psyche. Dtsch. Z. Nervenheilk. 83 (1925) 119–133.

Goldstein, K.: Beobachtungen über die Veränderungen des Gesamtverhaltens bei Gehirnschädigung. Mschr. Psychiat. Neurol. 68 (1928) 217–242.

Goldstein, K.: Der Aufbau des Organismus. Einführung in die Biologie unter besonderer Berücksichtigung der Erfahrungen am kranken Menschen. Nijhoff, Den Haag 1934. Reprint 1963.

Goldstein, K.: The organism. American Book Company, New York 1939.

Goldstein, M.: Untersuchung über die Häufigkeit und Dauer des Stillens und den Einfluß psychosozialer Faktoren in West-Berlin. Diss., FU Berlin 1978.

Goldstein, S., A. J. Moss, W. Greene: Sudden death in acute myocardial infarction; relationship to factors affecting delay in hospitalization. Arch. intern. Med. 129 (1972) 720–724.

Goligher, J. C., F. T. de Dombal, J. McK. Watts, G. Watkinson: Ulcerative colitis. Baillière, Tindall & Cassell, London 1968.

Gonsalves-Ebrahim, L., A. D. Gulledge, S. Miga: Continuous ambulatory peritoneal dialysis: psychological factors. Psychosomatics 23 (1982) 944–949.

González, E. R.: Stressed whites especially prone to 'trench mouth'; study finds. J. Amer. med. Ass. 249 (1983) 157–158.

Gonzalez, R. G.: Bulimia and adolescence: individual psychoanalytic treatment. In: Schwartz, H. J. (ed.): Bulimia: Psychoanalytic Treatment and Theory. Int. Univ. Press, Madison 1988.

Good, B. J.: The heart of what's the matter. The semantics of illness in Iran. Culture, Medicine and Psychiatry 1 (1977) 25–58.

Goodman, L. S., A. Gilman: The pharmacological basis of therapeutics. Macmillan, New York 1985.

Goodwin, F. K., K. R. Jamison: The natural course of manic depressive illness. In: Post, R. M., J. L. Ballenger (eds.): Neurobiology of Mood Disorders. Williams & Wilkins, Baltimore 1984.

Goodwin, J. S., J. M. Goodwin, A. V. Vogel: Knowledge and use of placebos by house officers and nurses. Ann. intern. Med. 91 (1979) 106–110.

Gorczynski, R. M., S. MacRae, M. Kennedy: Conditioned immune response associated with allogeneic skin grafts in mice. J. Immunol. 129 (1982) 704–709.

Gorczynski, R. M., S. MacRae, M. Kennedy: Factors involved in the classical conditioning of antibody responses in mice. In: Ballieux, R. E. (ed.): Breakdown in Human Adaptation to „Stress". Vol. II, Part 3: Psychoneuroimmunology and Breakdown in Adaptation: Interaction within the Central Nervous System, the Immune and Endocrine Systems, pp. 704–712. Nijhoff, Boston 1984.

Gordon, J. V., J. H. Mench: Values of medical students at different levels of training. J. educ. Psychol. 53 (1962) 48–51; zit. n. Bloom, S. W.: Power and dissent in the medical school. Free Press, New York 1973.

Gorman, J. M.: Panic disorders. In: Ban, T. A., P. Pichot, W. Pöldinger (eds.): Modern Problems of Pharmacopsychiatry, Vol. 22. Karger, Basel 1987.

Gorman, J. M., K. Shear, R. B. Devereux, D. L. King, D. F. Klein: Prevalence of mitral valve prolaps in panic disorder: effect of echocardiographic criteria. Psychosom. Med. 48 (1986) 167–171.

Gorski, R. A., J. H. Gordon, J. E. Shryne, A. M. Southam: Evidence for a morphological sex difference within the medial preoptic area of the rat brain. Brain Res. 148 (1978) 333–346.

Gotlieb-Jensen, K.: Peptic ulcer: genetic and epidemiologic

aspects based on twin studies. Munksgaard, Kopenhagen 1972.

Gottesman, I. I.: The psychotic hinterlands or the fringes of lunacy. Brit. med. Bull. 43 (1987) 557–569.

Gottesman, I. I., J. Shields: Schizophrenia and genetics: a twin study vantage point. Acad. Press, New York 1972.

Gottesman, I. I., J. Shields: Schizophrenia: the epigenetic puzzle. Cambridge Univ. Press, Cambridge 1982.

Gottschalch, W., M. Neumann-Schönwetter, G. Soukup: Sozialisationsforschung. Frankfurt 1971.

Gottschalk, L. A.: Some problems in the evaluation of the use of psychoactive drugs in the treatment of non-psychotic personality disorders. In: Efron, D. H., J. O. Cole, J. Levine, J. R. Wittenborn (eds.): Psychopharmacology: A Review of Progress 1957–1967. Public Health Service Publication No. 1836. U. S. Government Printing Office, Washington 1968.

Gottschalk, L. A.: Analysis of speech samples to determine effect of lorazepam on anxiety. Clin. Pharmacol. Ther. 13 (1972) 323–328.

Gottschalk, L. A., G. C. Gleser, H. W. Wylie, S. M. Kaplan: Effects of imipramine on anxiety and hostility levels derived from verbal communication. Psychopharmacologia 7 (1965) 303–310.

Goudie, A. J.: Aversive stimulus properties of drugs: the conditioned taste aversion paradigm. In: Greenshaw, A. J., C. T. Dourish (eds.): Experimental Psychopharmacology, pp. 341–391. Humana, Crescent Manor 1987.

Gould, S. J.: The flamingo's smile. Norton, New York 1985.

Gould, W. M., T. M. Gragg: A dermatology-psychiatry liaison clinic. J. Amer. Acad. Derm. 9 (1983) 73–77.

Goy, R. W., B. S. McEwen: Sexual differentiation of the brain. MIT Press, Cambridge/Mass. 1980.

Goyeche, J. R., Y. Ago, Y. Ikemi: Breathing and psychosomatic medicine. Asian med. J. 21 (1978) 53–60.

Goyeche, J. R., Y. Ago, Y. Ikemi: Asthma: the Yoga perspective. Part I: The somatopsychic imbalance in asthma: towards a holistic therapy. J. Asthma Res. 17 (1980) 111–121.

Graber, G.: Psychosomatik und Okklusion. In: Singer, F., F. Schön (Hrsg.): Europäische Prothetik heute, S. 169. Quintessenz, Berlin 1978.

Graber, G., H. P. Vogt, W. Müller, J. Bahous: Weichteilrheumatismus und Myoarthropathien des Kiefer- und Gesichtsbereiches. Schweiz. Mschr. Zahnheilk. 90 (1980) 609.

Grace, W. J.: Life stress and regional enteritis. Gastroenterology 23 (1953) 542–553.

Grace, W. J., S. Wolf, H. G. Wolff: Life situations, emotions and colonic function. Gastroenterology 14 (1950a) 93–108.

Grace, W. J., S. Wolf, H. G. Wolff: Life situations, emotions and ulcerative colitis. J. Amer. med. Ass. 142 (1950b) 1044–1048.

Grace, W., S. Wolf, H. G. Wolff: The human colon. Hoeber, New York 1951.

Grace, W. J., C. W. Holman, S. Wolf, H. G. Wolff: Action of various pharmacologic and other agents on the colon of man. Arch. Surg. 61 (1950) 1036.

Gräff, C.: Konzentrative Bewegungstherapie in der Praxis. Hippokrates, Stuttgart 1983.

Graeter, C.: Ein Fall von epileptischer Amnesie durch hypnotische Hypermnesie beseitigt. Z. Hypnotismus, Suggestionsther., S.-Lehre, verw. Forsch. 8 (1899) 128–140.

Graf, H., A. Wolf, H. K. Stummvoll, J. Kovarik, W. Pinggera: Rehabilitation nach Nierentransplantation. In: Mortalität und Morbidität nach Nierentransplantation, S. 396–400. Egermann, Linz 1979.

Graham, J. P. D.: High blood pressure after battle. Lancet 1 (1945) 239–240.

Graham, N. M. H., R. M. Douglas, P. Ryan: Stress and acute respiratory infection. Amer. J. Epidemiol. 24 (1986) 389–401.

Graham, S., I. M. Snell, J. B. Graham, L. Ford: Social trauma in the epidemiology of cancer of the cervix. J. chron. Dis. 24 (1971) 711–725.

Grandison, L., A. Guidotti: Stimulation of food intake by muscimol and beta-endorphin. Neuropharmacology 16 (1977) 533–536.

Grant, D. A.: A preliminary model for processing information conveyed by verbal conditioned stimuli in classical conditioning, In: Black, A. H., W. F. Prokasy (eds.): Classical Conditioning II. Appleton Century Crofts, New York 1972.

Grant, D. A., I. B. Stern, F. G. Everett: Periodontics. Mosby, St.Louis 1979.

Grant, I., G. C. Kyle, A. Teichman, J. Mendels: Recent life events and diabetes in adults. Psychosom. Med. 36 (1974) 121–128.

Grant, I., G. W. Brown, T. Harris, W. I. McDonald, T. Patterson, M. R. Trimble: Severely threatening events and marked life difficulties preceding onset or exacerbation of multiple sclerosis. J. Neurol. Neurosurg. Psychiat. 52 (1989) 8–13.

Grant, R., P. Lindgren, A. Rosen, B. Uvnäs: The release of catecholamines from the adrenal medulla on activation of the sympathetic vasodilatator nerves to the skeletal muscles in the cat by hypothalamic stimulation. Acta physiol. scand. 43 (1958) 135–154.

Gratwohl, A., J. Hermans, A. Lyklema, F. E. Zwaan: Bone marrow transplantation for leukemia in Europe. Bone Marrow Transplant. 2 (Suppl. 1) (1987) 15–18.

Graven, P.: Die aktive analytische Behandlung der Epilepsie. Fortschr. Sexualwiss. Psychoanal. 1 (1924) 58–169.

Graves, P. L., M. Phil, L. A. Mead, T. A. Pearson: The Rorschach interaction scale as a potential predictor of cancer. Psychosom. Med. 48 (1986) 549–563.

Grawe, K.: Differentielle Psychotherapie, I. Indikation und spezifische Wirkung von Verhaltenstherapie und Gesprächstherapie. Eine Untersuchung an phobischen Patienten. Huber, Bern 1976.

Grawe, K.: Indikation in der Psychotherapie. In: Pongratz, L. (Hrsg.): Handbuch der Psychologie, Bd. 8: Klinische Psychologie, 2. Halbband, S. 1849–1883. Hogrefe, Göttingen 1978.

Grawe, K.: Der gegenwärtige Stand in der Indikationsfrage in der Psychotherapie. In: Schulz, W., H. Hautzinger (Hrsg.): Bericht ü. d. Kongr. Klinische Psychologie u. Psychotherapie Berlin 1980, Bd. 2.

Grawe, K.: Überlegungen zu möglichen Strategien der Indikationsforschung. In: Baumann, U. (Hrsg.): Indikation zur Psychotherapie. Perspektiven für Praxis und Forschung. Urban & Schwarzenberg, München-Wien-Baltimore 1981.

Grawe, K.: Vergleichende Psychotherapieforschung. In: Minsel, W., R. Scheller (Hrsg.): Brennpunkte der klinischen Psychologie, Bd. 1: Psychotherapie. Kösel, München 1981.

Gray, G., P. Flynn: A survey of placebo use in a general hospital. Gen. Hosp. Psychiat. 3 (1981) 199–203.

Gray, J. A.: The psychology of fear and stress. McGraw-Hill, New York 1971.

Gray, J. A.: Angst und Streß. Kindler, München 1971.

Gray, J. A.: The structure of the emotions and the limbic system. In: Ciba Foundation Symposium 8: Physiology, Emotion and Psychosomatic Illness. Elsevier, North Holland-London-Amsterdam 1972.

Gray, J. J., K. Ford, L. M. Kelly: The prevalence of bulimia in a black college population. Int. J. Eat. Dis. 6 (1987) 733–740.

Greco, R. S., R. A. Pittenger: One man's practice. Tavistock, London 1966.

Greely, A.: Erotische Kultur, Partnerschaft und Intimität. Styria, Graz-Wien-Köln 1977.

Green, A. W.: Sexual activity and the postmyocardial infarction patient. Amer. Heart J. 89 (1975) 246–252.

Green, J. P., S. Maayani: Tricyclic antidepressant drugs block histamine H2 receptor in brain. Nature (Lond.) 269 (1977) 163–165.

Green, M.: Diagnosis and treatment: psychogenic, recurrent, abdominal pain. Pediatrics 40 (1967) 84–89.

Green, R.: Sex-dimorphic behavior development in the human: prenatal hormone administration and postnatal socialization. In: Porter, R., J. Whelan (eds.): Sex, Hormones and Behavior. Ciba Foundation Symposium 62, pp. 59–80. Excerpta Medica, Amsterdam 1979.

Greenblatt, D., E. M. Sellers, R. J. Shader: Drug disposition in old age. New Engl. J. Med. 306 (1982) 1081–1087.

Greenblatt, D. J., R. J. Shader, D. R. Abernethy: Current status of benzodiazepines I, II. New Engl. J. Med. 309 (1983) 354–358; 410–416.

Greenblatt, D. J., R. J. Shader, J. S. Harmatz, K. Franke, J. Koch-Weser: Absorption rate, blood concentrations, and early response to oral chlordiazepoxide. Amer. J. Psychiat. 134 (1977) 559–562.

Greene, W. A., A. J. Moss: Psychosocial factors in the adjustment of patients with permanently implanted cardiac pacemakers. Ann. intern. Med. 70 (1969) 897–902.

Greene, W. A., S. Goldstein, A. J. Moss: Psychological aspects of sudden death. Arch. intern. Med. 129 (1972) 725–731.

Greene, W. A., G. Conron, D. S. Schalch, B. F. Schreiner: Psychologic correlates of growth hormone and adrenal secretory responses in patients undergoing cardiac catheterization. Psychosom. Med. 32 (1970) 599–614.

Greenfield, N. S., R. Roessler, A. P. Crosley: Ego strength and length of recovery from infectious mononucleosis. J. nerv. ment. Dis. 128 (1959) 125–128.

Greenson, R.: Das Arbeitsbündnis und die Übertragungsneurose. Psyche 20 (1966) 81–103.

Greenson, R.: Technik und Praxis der Psychoanalyse. Klett, Stuttgart 1967.

Greenson, R.: The technique and praxis of psychoanalysis. New York 1967.

Greenson, R., M. Wexler: Die übertragungsfreie Beziehung in der analytischen Situation (1969). In: Greenson, R. (Hrsg.): Psychoanalytische Erkundungen, S. 308–335, Klett-Cotta, Stuttgart 1982.

Greenson, R. R.: On genuine epilepsy. Psychoanal. Quart. 13 (1944) 139–159.

Greer, S., T. Morris: Psychological attributes of women who develop breast cancer: a controlled study. J. psychosom. Res. 19 (1975) 147–153.

Greer, S., M. Watson: Towards a psychobiological model of cancer: psychological considerations. Soc. Sci. Med. 20 (1985) 773–777.

Greer, S., T. Morris, K. W. Pettingale: Psychological response to breast cancer: effects of outcome. Lancet 13 (1979) 785–787.

Greiner, G. F., C. Conraux, P. Feblot: Zentrale und psychogene Hörstörungen. In: Berendes, J., R. Link, F. Zöllner (Hrsg.): Hals-Nasen-Ohrenheilkunde in Praxis und Klinik, Bd. 6/ Ohr II. Thieme, Stuttgart-New York 1980.

Greiner, R., S. Lehrl, H. Grohmann: Zur psychischen Situation von Herzinfarktpatienten auf der Intensivstation. Med. Welt 32 (1981) 961–963.

Grennan, D. M., S. Taylor, D. G. Palmer: Doctor-patient communication in patients with arthritis. N. Z. med. J. 87 (1978) 431–434.

Greuel, H.: Suggestivbehandlung beim Hörsturz. HNO 31 (1983) 136.

Greuel, H.: Persönlichkeitsmerkmale als Hörsturzrisiko. HNO 34 (1986) 146.

Griffiths, R. R., Ch. A. Sannerud: Abuse of and dependence on benzodiazepines and other anxiolytic/sedative drugs. In: Meltzer, H. J. (ed.): Psychopharmacology: The Third Generation of Progress. Raven, New York 1987.

Grinker, R.: Some current trends and hypothesis of psychosomatic research. In: Deutsch, F. (ed.): The Psychosomatic Concept of Psychoanalysis, pp. 37–62. New York 1953.

Grinker, R.: Die Physiologie der Affekte. Psyche 15 (1961) 38–58.

Grinker, R. R., G. C. Ham, F. P. Robins: Some psychodynamic factors in multiple sclerosis. Amer. Res. nerv. Dis. 28 (1950) 456–459.

Grob, D., N. G. Brunner, T. Namba: The natural course of myasthenia gravis and the effect of various therapeutic measures. Ann. N. Y. Acad. Sci. 377 (1981) 652–669.

Groddeck, G.: Psychische Bedingtheit und psychoanalytische Behandlung. Hirzel, Berlin 1917.

Groddeck, G.: Psychosomatische Forschung als Erforschung des Es. (Abdruck des Vortrags: Das Es und die Psychoanalyse, 1925). Psyche 4 (1950/51) 481–487.

Groddeck, G.: Psychoanalytische Schriften zur Psychosomatik. Herausgegeben von G. Clauser. Limes, Wiesbaden 1966.

Groddeck, G., S. Freud: Briefe über das Es. Kindler, München 1974.

Groeben, N., B. Scheele: Argumente für eine Psychologie des reflexiven Subjekts. Steinkopff, Darmstadt 1977.

Groen, J. J.: Psychogenesis and psychotherapy of ulcerative colitis. Psychosom. Med. 9 (1947) 151–174.

Groen, J. J.: Psychosomatic aspects of ischemic (coronary) heart disease. In: Hill, O. W. (ed.): Modern Trends in Psychosomatic Medicine 3, p. 976. Butterworth, London 1976.

Groen, J. J.: Psychosomatic aspects of aging. In: Groen, J. J. (ed.): Clinical Research in Psychosomatic Medicine. Van Gorkum, Assen 1982.

Groen, J. J.: Psychosomatic aspects of Menière's disease. Acta oto-laryng. (Stockholm) 95 (1983) 407.

Groen, J. J.: From clinical experience to tested hypothesis: the role of psychological factors in coronary heart disease. In: Schmidt, T., T. Dembroski, G. Blümchen (eds.): Biological and Psychological Factors in Cardiovascular Disease. Springer, Berlin-Heidelberg-New York 1986.

Groen, J. J., J. M. van der Valk: Psychosomatic aspects of ulcerative colitis. Gastroenterologia 86 (1956) 591–608.

Groen, J. J., H. E. Pelser: Experiences with, and results of group psychotherapy in patients with bronchial asthma. J. psychosom. Res. 4 (1960) 191–205.

Groen, J. J., Z. Feldman-Toledano: Educative treatment of patients and parents in anorexia nervosa. Brit. J. Psychiat. 112 (1966) 671–681.

Groen, J. J., W. S. de Loos: Psychosomatische aspecten van diabetes mellitus. De Erven Bohn BV, Amsterdam 1973.

Groen, J. J., J. J. G. Prick, J. Bastiaans: De Betekenis van Persoonlijkheid en Conflictsituaties voor het Onstaan, het Belogs en de Behandeling van Multiple Sklerose. De Erven Bohn, Haarlem, The Netherlands 1967.

Groen, J. J., J. H. Medalie, H. Neufeld, E. Rijs: On epidemiological investigation of hypertension and ischemic heart disease in Israel. Israel J. med. Sci. 4 (1968) 775.

Groen, J. J., J. M. van der Valk, A. Wellner, D. Ben-Ishay: Psychobiological factors in the pathogenesis of essential hypertension. Psychother. and Psychosom. 19 (1971) 1–26.

Groen, J. J., B. Hansen, J. M. Herrmann, H. Schäfer, T. H. Schmidt, K. H. Selbmann, Th.v. Uexküll, P. Weckmann: Effects of experimental emotional stress and physical exercise on the circulation in hypertensive patients and control subjects. J. psychosom. Res. 26 (1982) 141–154.

Grol, R. P. T. M.: Die Prävention somatischer Fixierung. Springer, Heidelberg 1985.

Grollnick, L.: A family perspective of psychosomatic factors in illness; a review of the literature. Family Process 11 (1972) 457–486.

Gromotka, R., N. Henning: Vegetatives Nervensystem und Krankheiten der Verdauungsorgane. Bibl. psychiat. neurol. (Basel) 130/4 (1966) 16–97.

Gromus, B., W. Kahlke, U. Koch: Möglichkeiten einer Gruppentherapie durch interdisziplinäre Kooperation von Ernährungsberatern, Internisten und Psychologen bei Übergewichtigen ohne und mit weiteren ernährungsabhängigen Risikofaktoren. Kohlhammer, München 1984.

Grosch, M.: Über Hypochondrie. Z. psychosom. Med. 4 (1958) 195.

Groskurth, P.: Berufliche Sozialisation als entscheidende Grundlage der Persönlichkeitsbildung. In: Groskurth, P. (Hrsg.): Arbeit und Persönlichkeit, S. 7–19. Rowohlt, Reinbek 1979.

Gross, J., L. Svab: Die experimentelle sensorische Deprivation als Modellsituation der psychotherapeutischen Beziehung. Nervenarzt 40 (1969) 21–25.

Gross, W. M.: Mental health survey in a rural area. Eugenics Rev. 40 (1948) 140.

Grossarth-Maticek, R.: Seelischer Druck kann Krebs fördern. Von der Gefahr, es allen recht zu machen. Bild d. Wissenschaft 1 (1988) 88–93.

Grossarth-Maticek, R., J. Bastiaans, D. T. Kanazir: Psychological factors as strong predictors of mortality from cancer, ischaemic heart disease and stroke – the Yugoslav prospective study. J. psychosom. Res. 29 (1985) 167–176.

Grossman, M. I.: Abnormalities of acid secretion in patients with duodenal ulcer. Gastroenterol. 75 (1978) 524–526.

Grossman, S. P.: Behavioral effects of chemical stimulation of the ventral amygdala. J. comp. physiol. Psychol. 57 (1964) 29–36.

Großpietzsch, R., S. M. Großpietzsch: Die Wahrheitsfrage in der sozialmedizinischen Begutachtung. Öff. Gesundh.-Wesen 48 (1986) 277–280.

Großpietzsch, R., M. Ihmann: Medizinmanagement – ein neues Anforderungsprofil an den Medizinischen Dienst der Krankenversicherung. Öff. Gesundh.-Wesen 51 (1989) 34–36.

Grota, L. J., R. Ader, N. Cohen: Taste aversion learning in autoimmune Mrl-lpr/lpr and Mrl +/+ mice. Brain, Behavior and Immunity 1 (1987) 238–250.

Grote-Janz, C., E. Weingarten: Technikgebundene Handlungsabläufe auf der Intensivstation: Zum Zusammenhang von medizinischer Technologie und therapeutischer Beziehung. Z. Soziologie 12 (1983) 328–340.

Groves, J. E.: Taking care of the hateful patient. New Engl. J. Med. 298 (1978) 883–887.

Gruen, W.: Effects of brief psychotherapy during the hospitalization period on the recovery process in heart attacks. J. consult. clin. Psychol. 43 (1975) 223–232.

Gruenberg, E.: The social breakdown syndrome: some origins. Amer. J. Psychiat. 123 (1967) 1481–1489.

Grünberg, H. W. von: Das psychotherapeutische Gespräch in der Sprechstunde des Hausarztes. Dtsch. Ärztebl. 83 (1985) 666–670.

Grüsser, O.-J.: Gesichtssinn und Okulomotorik. In: Schmidt, R. F., G. Thews (Hrsg.): Physiologie des Men-

schen, 19. Aufl., S. 226–262. Springer, Berlin-Heidelberg-New York 1977.

Grunberger, B.: Le narcissisme. Paris 1971. Deutsch: Vom Narzißmus zum Objekt. Frankfurt 1976.

Grunert, J. (Hrsg.): Körperbild und Selbstverständnis. Psychoanalytische Beiträge zur Leib-Seele-Einheit. Kindler, München 1977.

Grunhaus, L., S. Gloger, A. Rein, B. S. Lewis: Mitral valve prolaps and panic attacks. Israel J. med. Sci. 18 (1982) 221–223.

Grunow, D.: Problemsyndrome älterer Menschen und die Selektivität organisierter Hilfe der örtlichen Sozialverwaltung. In: Dieck, M., G. Naegele (Hrsg.): Sozialpolitik für ältere Menschen. Quelle & Meyer, Heidelberg 1978.

Gruppuso, P. A., P. Gorden, R. Kahn, M. Cornblath, W. P. Zeller, R. Schwartz: Familial hyperproinsulinemia due to a proposed defect in conversion of proinsulin to insulin. New Engl. J. Med. 31 (1984) 629–634.

Gschwind, H.: Die HIV-Infektion und ihre Verlaufsformen. In: Sigusch, V., S. Fliegel (Hrsg.): AIDS. DGVT, Tübingen 1988.

Gück, J., E. Matt, E. Weingarten: Zur alltagssprachlichen Repräsentation intensivmedizinischer Behandlungsroutinen. Zwischenbericht zum Forschungsvorhaben „Die sprachliche Herstellung und Aufrechterhaltung von Normalität in intensivmedizinischen Extremsituationen." FU Berlin, Fachbereich Medizin. Grundlagenfächer 1981.

Gück, J., E. Matt, E. Weingarten: Zur interaktiven Ausgestaltung der Arzt-Patient-Beziehung in der Visite. In: Gerhardt, U. et al. (Hrsg.): Jahrbuch Medizinische Soziologie 3. Campus, Frankfurt 1983a.

Gück, J., E. Matt, E. Weingarten: Sprachliche Realisierung von hierarchischen Kontexten – Eine konversationsanalytische Untersuchung intensivmedizinischer Visitenkommunikation. In: Soeffner, H. G. (Hrsg.): Ansätze und Materialien zu einer Soziologie der Interaktion. Campus, Frankfurt 1983b.

Günthert, E.-A.: Psychosomatische Aspekte der männlichen Adnexerkrankungen. In: Verhandlungsbericht der Deutschen Gesellschaft für Urologie, S. 57. Springer, Berlin 1980.

Günthert, E.-A.: Psychosomatische Probleme in der urologischen Sprechstunde. Erfahrungen aus der Tätigkeit des niedergelassenen Urologen. Urologe A 19 (1980) 232.

Günthert, E.-A.: Prostatitis aus psychosomatischer Sicht. In: Brunner, H., W. Krause, C. F. Rothauge, W. Weidner (Hrsg.): Chronische Prostatitis, S. 255. Schattauer, Stuttgart 1983.

Günthert, E.-A.: Psychosomatische Aspekte der Zystitis der Frau aus der Sicht des niedergelassenen Urologen. In: Verhandlungsbericht der Deutschen Gesellschaft für Urologie, S. 323. Springer, Berlin 1984.

Günthert, E.-A.: Der Problemfall in der urologischen Sprechstunde: „Symptome der sogenannten Reizblase der Frau". Urologe A 25 (1986) 82.

Günthert, E.-A.: Die urologische Anamnese aus psychosomatischer Sicht. In: Sitzungsbericht: Psychosomatische Aspekte in der Urologie. Würzburg 1986.

Günthert, E.-A.: Psychosomatic aspects of prostatitis. In: Weidner, W., H. Brunner, W. Krause, C. F. Rothauge (eds.): Therapy of Prostatitis, p. 161. Zuckerschwerdt, München 1986.

Guillemin, R.: The brain as an endocrine organ. Neurosci. Res. Program Bull. 16 (Suppl. Aug. 1978). MIT Press, Boston 1978.

Guillemin, R., M. Cohn, T. Melnechuk (eds.): Neural modulation of immunity. Raven, New York 1985.

Guilleminault, G., E. Lugaresi (eds.): Sleep/wake disorders: natural history, epidemiology and long-term evaluation. Raven, New York 1983.

Guilleminault, G., S. J. Connolly, R. A. Winkle: Cardiac arrhythmia and conduction disturbance during sleep in 400 patients with sleep apnoe syndrome. Amer. J. Cardiol. 52 (1983) 490–494.

Gull, W.: Anorexia nervosa. Lancet I (1888a) 516.

Gull, W.: The address in medicine. Lancet II (1888b) 171.

Gundermann, H.: Die kommunikative Stimmtherapie. In: Gundermann, H. (Hrsg.): Aktuelle Probleme der Stimmtherapie, S. 61. Fischer, Stuttgart 1987.

Gundert-Remy, U., V. Möntmann, E. Weber: Studien zur Regelmäßigkeit der Einnahme verordneter Medikamente bei stationären Patienten. Inn. Med. 2 (1978) 78–83.

Gurman, A. S., D. P. Kniskern: Handbook of family therapy. Brunner/Mazel, New York 1981.

Gurtner, H. P.: Kardiovaskuläre Synkopen. Schweiz. med. Wschr. 144 (1984) 1514–1525.

Guthy, E.: Ätiologie des Morbus Crohn. Dtsch. med. Wschr. 108 (1983) 1729–1733.

Guyton, A. C., D. B. Young, J. W. Declue, J. D. Ferguson, R. E. McCaa, A. Cevese, N. C. Trippodo, J. E. Hall: The role of the kidney in hypertension. In: Berglund, G., L. Hansson, L. Werkö (eds.): Pathophysiology and Management of Arterial Hypertension. Lindgren & Söner, Mölndal (Sweden) 1975.

Guze, S. B.: The role of follow-up studies: Their contribution to diagnostic classification as applied to hysteria. Seminars in Psychiatry 2 (1970) 392–402.

Haag, A., U. Stuhr, T. Wiencke: Psychosomatische Aspekte bei alten Menschen in der stationären Versorgung. In: Speidel, H., B. Strauss (Hrsg.): Zukunftsaufgaben der psychosomatischen Medizin. Springer, Berlin-Heidelberg 1989.

Haag, G., W. D. Gerber, N. Birbaumer, K. Mayer, W. Lutzenberger, G. Schroth: Differentielle Indikation zur Psychotherapie der Migräne. In: Huber, H. P. (Hrsg.): Migräne, S. 205–232. Urban & Schwarzenberg, München-Wien-Baltimore 1982.

Haan, N.: A tripartite model of ego functioning. J. nerv. ment. Dis. 148 (1969) 14–30.

Haan, N.: Coping and defending – processes of self-environment organization. Acad. Press, New York 1977.

Haas, J.: Myasthenia gravis. Aktuelle Therapie unter pathophysiologischen Aspekten. Dtsch. Ärztebl. 85 (B) (1988) 126–130.

Habener, J. F.: Principles of peptide-hormone biosynthesis. In: Martin, J. B., S. Reichlin, J. L. Bick (eds.): Neurosecretion and brain peptides, pp. 21–34. Raven, New York 1981.

Haberland, H. F. O.: Infektion und Nervensystem. Münch. med. Wschr. 34 (1926) 1389–1393.

Habermann, G.: Physiologie und Phonetik des lauthaften Lachens. In: Loebell, H., W. Tonndorf (Hrsg.): Heft 10 der Hals-Nasen-Ohrenheilkunde, zwanglose Schriftenreihe. Barth, Leipzig 1955.

Habermann, G.: Zur Psychologie und Soziologie des Schwerhörigen. Wiss. Z. Univ. Halle 5 (1956) 1089.

Habermann, G.: Funktionelle Stimmstörungen und ihre Behandlung. Arch. Ohr.-Nas.-Kehlk.-hk. 227 (1980) 171.

Habermann, G.: Stimme und Sprache, 2. Aufl. Thieme, Stuttgart-New York 1986.

Habermas, J.: Technik und Wissenschaft als Ideologie. Suhrkamp, Frankfurt 1968.

Habermas, J.: Thesen zur Theorie der Sozialisation. MS-Druck, Frankfurt 1968.

Habermas, J.: Theorie und Praxis, 4. Aufl. Suhrkamp, Frankfurt 1971.

Habermas, J.: Legitimationsprobleme im Spätkapitalismus. Suhrkamp, Frankfurt 1973.

Habermas, J.: Bemerkungen zu Alexander Mitscherlichs analytischer Sozialpsychologie. Psyche 37 (1983) 352–363.

Hackenbroch, M. H., B. Waldecker, C. Prömper: Verlaufsbeobachtungen bei konservativ behandelten computertomographisch diagnostizierten Bandscheibenvorfällen. Orthop. Prax. 20 (1984) 298–303.

Hackethal, K. H.: Das durch Magenresektion behandelte Magen- und Zwölffingerdarmgeschwür. In: Linneweh, F. (Hrsg.): Die Prognose chronischer Erkrankungen. Springer, Berlin 1960.

Hackett, T. P.: Psychological assistance for the dying patient and his family. Ann. Rev. Med. 27 (1976) 371–378.

Hackett, T. P.: Myocardinfarkt: Emotionale Faktoren, die die Prognose verschlechtern (Interview). Tempo Medical 1 (1977) 7–12.

Hackett, T. P., A. D. Weisman: Psychiatric management of operative syndromes. Psychosom. Med. 22 (1960) 267–282; 356–372.

Hackett, T. P., A. D. Weisman: The treatment of dying. In: Massermann, J. H. (ed.): Current Psychiatric Therapies, Vol. 2, pp. 121–126. Grune & Stratton, New York 1962.

Hackett, T. P., A. D. Weisman: Denial as a factor in patients with heart disease and cancer. Ann. N. Y. Acad. Sci. 164 (1969) 802–817.

Hackett, T. P., N. H. Cassem: Factors contributing to delay in responding to the signs and symptoms of myocardial infarction. Amer. J. Cardiol. 24 (1969) 651–658.

Hackett, T. P., N. H. Cassem: Development of a quantitative rating scale to assess denial. J. psychosom. Res. 18 (1974) 413–420.

Hackett, T. P., N. H. Cassem: White-collar and blue-collar responses to heart attack. J. psychosom. Res. 20 (1976) 85–95.

Hackett, T. P., N. H. Cassem: The psychology of intensive care: problems and their management. In: Usdin, G. (ed.): Psychiatric Medicine. Brunner/Mazel, New York 1977.

Hackett, T. P., N. H. Cassem (eds.): Massachussetts General Hospital Handbook of General Hospital Psychiatry. Mosby, St. Louis 1978.

Hackett, T. P., N. H. Cassem, L. A. Wishnie: The coronary care unit – an appraisal of its psychological hazards. New Engl. J. Med. 279 (1968) 1365.

Hackett, T. P., N. H. Cassem, L. A. Wishnie: Detection and treatment of anxiety in the coronary care. Amer. Heart J. 78 (1969) 727–730.

Hadden, J. W., E. M. Hadden, E. Middleton jr.: Lymphocyte blast transformation: I. Demonstration of adrenergic receptors in human peripheral lymphocytes. J. cell. Immunol. 1 (1970) 583–595.

Hadfield, J. A.: Treatment by suggestion and hypnoanalysis. In: Miller, E. (ed.): Neurosis in war. Macmillan, New York 1940.

Haeberle, E. J., A. Bedürftig: AIDS – Beratung, Betreuung, Vorbeugung. Anleitungen für die Praxis. De Gruyter, Berlin-New York 1987.

Häfner, H. (Hrsg.): Psychiatrische Epidemiologie. Springer, Berlin-Heidelberg-New York 1978.

Häfner, H.: Psychische Gesundheit im Alter. Fischer, Stuttgart 1986.

Häfner, H., H. Freyberger: Ikterus als psychosomatisches Krankheitsbild. Beitrag zur speziellen Psychosomatik bei Lebererkrankungen. Psychother. med. Psychol. 5 (1955) 107–116.

Hänsel, D.: Eßstörungen. Die Bedeutung des Problems, Übersicht zu den Erscheinungsbildern. In: Brakhoff, J. (Hrsg.): Eßstörungen. 2. Aufl. Lambertus, Freiburg 1987.

Häußler, S.: Der praktische Arzt heute und morgen. Gentner, Stuttgart 1967.

Hagberg, B.: A prospective study of patients in chronic hemodialysis III. J. psychosom. Res. 18 (1974) 151–160.

Hagberg, B., A. Malmquist: A prospective study of patients in chronic hemodialysis: IV. Pretreatment psychiatric and psychological variables predicting outcome. J. psychosom. Res. 18 (1974) 315–319.

Hagedorn, E.: Psychosomatische Aspekte bei Funktionsstörungen und Erkrankungen der Leber. Z. psychosom. Med. Psychoanal. 15 (1969) 1–31.

Hagnell, O.: The premorbid personality of persons who develop cancer in a total population investigated in 1947 and 1957. Ann. N. Y. Acad. Sci. 125 (1966) 846–855.

Hagnell, O.: A prospective study of the incidence of mental disorder. Scandinavian Univ. Books, Stockholm 1966.

Hagnell, O.: The incidence and duration of episodes of mental illness in a total population. In: Hare, E. H., J. K. Wing (eds.): Psychiatric Epidemiology. Oxford Univ. Press, London 1970.

Hagnell, O.: zit. bei: Modestin, J.: Schwindel als psychosomatisches Phänomen. Psychother. med. Psychol. 33 (1983) 77.

Hahn, P.: Zur Analyse der auslösenden Situation bei der sog. „Herzphobie". Psychosom. Med. 11 (1965) 264.

Hahn, P.: Der Herzinfarkt in psychosomatischer Sicht. Vandenhoeck & Ruprecht, Göttingen 1971.

Hahn, P.: Die Bedeutung des „somatischen" Entgegenkommens für die Symptombildung bei der Herzneurose. Therapiewoche 26 (1976) 963–969.

Hahn, P., A. Reindell: Psychosomatik des Herzinfarkts. In: Reindell, H., H. Roskamm (Hrsg.): Herzkrankheiten. Springer, Berlin 1977.

Hahn, P., P. Vollrath, E. Petzold: Aus der Arbeit einer klinisch-psychosomatischen Station. Prax. Psychother. 20 (1975) 66–77.

Hahn, R. C., D. B. Petitti: Minnesota Multiphasic Personality Inventory-rated depression and the incidence of breast cancer. Cancer 61 (1988) 845–848.

Hakelius, A.: Prognosis in sciatica. A clinical follow-up of surgical and non-surgical treatment. Acta orthop. scand. (Suppl.) 129 (1970) 3–76.

Haland-Wirth, I., H. Wirth: Über die familientherapeutische Behandlung eines 13jährigen asthmakranken Jungen und seiner Familie. Familiendynamik 6 (1981) 275.

Halbreich, U., D. Kas: Variations in the Taylor MAS of women with pre-menstrual syndrome. J. psychosom. Res. 21 (1977) 391–393.

Halbreich, U., G. M. Asnis, D. Zumoff, R. S. Nathan, R. Schindeldecker: Effect of age and sex on cortisol secretion in depressives and normals. J. psychiat. Res. 13 (1984) 221–229.

Haldemann, R., U. Gigon, B. Baur, E. Pusterla, D. Sidiropoulos: Statistische Auswertung bei einem Frühgeburtenkollektiv von 245 Fällen. Zbl. Gynäk. 98 (1976) 468.

Halder-Sinn, P.: Effektivität psychotherapeutischer Interventionen. In: Wittling, W. (Hrsg.): Handbuch der klinischen Psychologie, Bd. 6, S. 92–115. Hoffmann & Campe, Hamburg 1980.

Hales, R. E., J. F. Borns: Teaching psychological issues to medical house staff: a liaison program on oncology service. Gen. Hosp. Psychiat. 4 (1982) 1–6.

Haley, J.: Research on family patterns: an instrument measurement. Family Process 3 (1964) 48.

Haley, J.: Die Psychotherapie Milton H. Ericksons. Pfeiffer, München 1978.

Haley, J.: Milton Ericksons Beitrag zur Psychotherapie. Hypnose und Kognition 5 (1988) 19–33.

Halhuber, M. J.: Rehabilitation nach Herzinfarkt. Dtsch. med. Wschr. 98 (1973) 1570–1572.

Halhuber, M. J.: Vortrag. 25. Dtsch. Kongr. Ärztl. Fortbildung. Berlin, Juni 1976.

Halhuber, M. J.: (Hrsg.): Psychosozialer Streß und koronare Herzkrankheiten. Springer, Berlin 1977.

Halhuber, M., M. Altpeter-Becüwe, H. Angster, R. Beckmann, K. A. Bungeroth, I. Gehring et al.: Rehabilitation des Koronarkranken. Perimed, Erlangen 1982.

Hall, A., A. H. Crisp: Brief psychotherapy in the treatment of anorexia nervosa: preliminary findings. In: Darby, P. L., P. E. Garfinkel, D. M. Garner, D. V. Coscina (eds.): Anorexia nervosa. Recent Developments in Research, pp. 427–439. Liss, New York 1983.

Hallen, O.: Zur biographischen Genese des Phantomschmerzes. Psychother. med. Psychol. 6 (1956) 3–7.

Hallen, O.: Über die Behandlung der Multiplen Sklerose. Therapiewoche 17 (1967) 635–639.

Hallen, O.: Über circumscripte Hypochondrien. Nervenarzt 41 (1970) 421–425.

Hallen, O.: Der Einfluß psychologischer Lehren auf naturwissenschaftliche Theoriebildung (dargestellt am Beispiel der cerebralen Lokalisationslehre). Nervenarzt 49 (1978) 734–736.

Hallen, O.: Die Unterscheidung epileptischer und nichtepileptischer Anfälle. Therapiewoche 31 (1981) 3608–3610.

Halliday, J. L.: The concept of psychosomatic rheumatism. Ann. intern. Med. 15 (1941) 666–677.

Halliday, J. L.: Psychological aspects of rheumatoid arthritis. Proc. roy. Soc. Med. 35 (1942) 455–457.

Halliday, J. L.: Dangerous occupation, psychosomatic illness, and morale. Psychosom. Med. 5 (1943) 71–84.

Halliday, J. L.: Psycho-social medicine. New York 1943.

Halliday, J. L.: Psychosocial medicine. New York 1948.

Halliday, J. L.: Psychosocial medicine. A study of the sick society. Heinemann, London 1949. Deutsche Teilübersetzung in: Mitscherlich, A., T. Brocher, O.v. Mering, K. Horn (Hrsg.): Der Kranke in der modernen Gesellschaft, S. 159–171. Kiepenheuer & Witsch, Köln-Berlin 1967.

Halmi, K. A.: Anorexia nervosa: demographic and clinical features in 94 cases. Psychosom. Med. 36 (1974) 18–26.

Halmi, K. A.: Anorexia nervosa: recent investigations. Ann. Rev. Med. 29 (1978) 149–162.

Halmi, K. A.: The diagnosis and treatment of anorexia nervosa. In: Zales, M. R.: (ed.): Eating, Sleeping and Sexuality, pp. 43–58. Brunner/Mazel, New York 1982.

Halmi, K. A.: Classification of eating disorders. Int. J. Eat. Dis. 2 (1983) 21–26.

Halmi, K. A.: Classification of eating disorders. J. psychiat. Res. 19 (1985) 113–119.

Halmi, K. A.: Die Wahrnehmung von Sättigung bei Bulimia. In: Fichter, M. M. (Hrsg.): Bulimia nervosa. Enke, Stuttgart 1989.

Halmi, K. A., J. R. Falk: Behavioral and dietary discriminators of menstrual function in anorexia nervosa. In: Darby, P. L., P. E. Garfinkel, D. M. Garner, D. V. Coscina (eds.): Anorexia nervosa. Recent Developments in Research, pp. 323–329. Liss, New York 1983.

Halmi, K. A., G. Broadland, J. Loney: Prognosis in anorexia nervosa. Ann. intern. Med. 78 (1973) 707–709.

Halmi, K. A., G. Broadland, C. Rigas: A follow-up study of 79 patients with anorexia nervosa: an evaluation of prognostic factors and diagnostic criteria. Life Hist. Res. Psychopathol. 4 (1975) 290–301.

Halmi, K. A., P. Powers, S. Cunningham: Treatment of anorexia nervosa with behavior modification. Effectiveness of formula feeding and isolation. Arch. gen. Psychiat. 31 (1975a) 93–96.

Halmi, K. A., J. R. Falk, E. Schwarz: Binge eating and vomiting: a survey of a college poulation. Psychol. Med. 11 (1981) 697–706.

Halsted, J. A., R. Schwarz, S. R. Rosen, H. Weinberg, S. M. Wyman: Correlated gastroscopic and psychiatric studies of soldiers with chronic non-ulcerative dyspepsia. Gastroenterol. 7 (1946) 177–190.

Halter, F.: Spontanverlauf des peptischen Ulkus. Schweiz. Rundschau Med. 67 (1978) 1869–1870.

Halter, F.: Der Einfluß von Cimetidin auf die Ulkusforschung. Swiss. Med. 4 (1982) 20–22.

Ham, G. C., F. Alexander, H. T. Carmichael: A psychosomatic theory of thyreotoxicosis. Psychosom. Med. 13 (1951) 18.

Ham, T. H.: Methods in development and revision of a program of medical education. J. med. Educ. 31 (1956) 519–521.

Hambling, J.: Emotions and symptoms in essential hypertension. Brit. J. med. Psychol. 24 (1951) 242–253.

Hamburg, D. A.: Coping behavior in life-threatening circumstances. Psychother. and Psychosom. 23 (1974) 13–25.

Hamburg, D., B. Hamburg, S. de Goza: Adaptive problems and mechanisms in severely burned patients. Psychiat. 16 (1953a) 1–20.

Hamburg, D., C. Artz, E. Reiss: Clinical importance of emotional problems in the care of patients with burns. New Engl. J. Med. 248 (1953b) 355–359.

Hamburg, D., G. Bibring, C. Fisher, A. Stenton, R. Wallenstein, H. Weinstock, E. Haggard: Report of the ad hoc committee on control fact gathering data of the American Psychoanalytic Association. J. Amer. Psychoanal. Ass. 15 (1967) 841–861.

Hamm jr., T. E., J. R. Kaplan, T. B. Clarkson, B. C. Bullock: Effects of gender and social behavior on the development of coronary artery atherosclerosis in cynomolgus macaques. Atherosclerosis 48 (1983) 221.

Hammar, J. A.: Konstitutionsanatomische Studien über die Neurotisierung des Menschenembryos. IV. Über die Innervationsverhältnisse der Inkretorgane und des Thymus bis in den 4. Fötalmonat. Z. mikrosk. anat. Forsch. 38 (1935) 253–293.

Hammer, B.: Assoziierte Erkrankungen bei Patienten mit Colitis ulcerosa und deren Verwandten ersten Grades. Diss., Kiel 1968.

Hammond, R. D.: Simulated epilepsy: report of a case. Arch. Neurol. Psychiat. 60 (1948) 327–328.

Hampel, P.: Cyclosporin A und Adjuvans-Arthritis. Unveröffentlichte Diplomarbeit, Univ. Düsseldorf 1986.

Han, J. S., L. Terenius: Neurochemical basis of acupuncture analgesia. Ann. Rev. Pharmacol. 22 (1982) 193–220.

Hanamura, H.: Periodontal status and bruxism. J. Periodont. 58 (1987) 173.

Handwerker, H. O., M. Zimmermann: Schmerz und vegetatives Nervensystem. In: Sturm, A., W. Birkemayer (Hrsg.): Klinische Pathologie des vegetativen Nervensystems, Bd. 1, S. 468–497. Fischer, Stuttgart 1976.

Hannich, C., M. Wendt: Der Schwerkranke und Sterbende auf der Intensivstation. Eine Beurteilung seiner Situation und der des Behandlungsteams. In: Howe, J., R. Ochsmann (Hrsg.): Tod – Sterben – Trauer, S. 157–161. Fachbuchhandlung für Psychologie, Frankfurt 1984.

Hannich, J. H., M. Wendt, P. Bertlich: Streßerleben und seelische Anpassungsprozesse bei traumatologischen und postoperativen Intensivpatienten. In: Tewes, U. (Hrsg.): Angewandte Medizin-Psychologie. Klotz, Frankfurt 1983a.

Hannich, J. H., M. Wendt, P. Lawin: Psychosomatik der Intensivmedizin. Thieme, Stuttgart 1983b.

Hansson, L., R. Sivertson: Rückbildung struktureller Gefäßveränderungen nach antihypertensiver Therapie. In: Dietz, R., D. Ganten, K. G. Hofbauer, J. B. Lüth (Hrsg.): Essentieller Hochdruck und seine Behandlung. Schattauer, Stuttgart-New York 1977.

Hanvik, L. J.: Profiles in patients with low back pain. J. consult. Psychol. 15 (1951) 350–353.

Happich, C.: Das Bildbewußtsein als Ansatzstelle psychischer Behandlung. Zbl. Psychother. 5 (1932) 633.

Haracz, J. L.: A neural plasticity hypothesis of schizophrenia. Neurosci. Biobehav. Rev. 8 (1984) 55–71.

Harari, A., H. Munitz, H. Wijsenbeck et al.: Psychological aspects of chronic hemodialysis. Psychiat. Neurol. Neurochir. 74 (1971) 219–223.

Harbitz, T. B., O. A. Haugen: Histology of the prostate in elderly men. Analysis in an autopsy series. Acta path. microbiol. scand. 80 A (1972) 756–768.

Harburg, E., J. C. Erfurt, L. S. Hauenstein, C. Chape, W. J. Schull, M. A. Shork: Socio-ecological stress, suppressed hostility, skin color, and black-white male blood pressure: Detroit. Psychosom. Med. 35 (1973) 276–296.

Harding, H. E.: A notable source of error in the diagnosis of appendicitis. Brit. med. J. 2 (1962) 1028–1029.

Hare, E.: Schizophrenia as a recent disease. Brit. J. Psychiat. 153 (1988) 521–531.

Hare-Mustin, R. T.: Family therapy following the death of a child. J. Marital and Fam. Therapy 5 (1979) 51.

Haring, C.: Lehrbuch des autogenen Trainings. Enke, Stuttgart 1979.

Harker, B. L.: Cancer and communication problems: a personal experience. Psychiatry in Medicine 3 (1972) 163–171.

Harlow, H. F.: Learning and satiation of response in intrinsically motivated complex puzzle performance by monkeys. J. compar. physiol. Psychol. 43 (1950) 289–294.

Harlow, H. F.: Basic social capacity of primates. Hum. Biol. 31 (1959) 40.

Harlow, H. F., R. R. Zimmermann: Affectional responses in infant monkey. Science 130 (1959) 421.

Harnack, G. A.: Nervöse Verhaltensstörungen beim Schulkind. Thieme, Stuttgart 1958.

Harrer, G.: Somatische Aspekte des Musikerlebens. Med. Mon. Sp. 6/70 (1970) 124.

Harris, A. H., J. V. Brady: Animal learning – visceral and autonomic conditioning. In: Rosenzweig, M. R., L. W. Porter (eds.): Annual Review of Psychology. Annual Reviews, Palo Alto 1974.

Harris, A. S., H. Otero, A. Bocage: The induction of arrhythmias by sympathetic activity before and after occlusion of a coronary artery in the canine heart. J. Electrocardiol. 4 (1971) 34.

Harris, C. M.: Tiredness and headaches. In: Munro, A. (ed.): Psychosomatic Medicine, pp. 40–45. Churchill Livingstone, Edinburgh-London 1973.

Harris, I. D.: Relation of resentment and anger to functional gastric complaints. Psychosom. Med. 8 (1946) 211–215.

Harrison, W. J.: Autoantibodies against intestinal and gastric mucous cells in ulcerative colitis. Lancet I (1965a) 1346–1350.

Harrison, W. J.: Thyroid, gastric (parietal-cell), and nuclear antibodies in ulcerative colitis. Lancet I (1965b) 1350–1352.

Harrow, M., J. T. Marengo: Schizophrenic thought disorder at follow-up: its persistence and prognostic significance. Schizophrenia Bull. 12 (1986) 373–393.

Harrow, M., J. T. Marengo: Schizophrenic thought disorder at follow-up. Arch. gen. Psychiat. 44 (1987) 651–659.

Hartig, M.: Probleme und Methoden der Psychotherapieforschung. Urban & Schwarzenberg, München-Berlin-Wien 1975.

Hartkamp, N.: Die psychosomatische Theorie von Georg L. Engel und Arthur H. Schmale. In: Zepf, S. (Hrsg.): Tatort Körper – Spurensicherung, S. 45–47. Springer, Heidelberg 1986.

Hartkamp, N.: Zur Kritik an Max Schurs Konzept der Desomatisierung und Resomatisierung. In: Zepf, S. (Hrsg.):

Tatort Körper – Spurensicherung, S. 27–44. Springer, Heidelberg 1986.

Hartleb, O.: Behandlung von Schlafstörungen im Alter – aus internistischer Sicht. In: Kaiser, H. (Hrsg.): Schlafstörungen im Alter und ihre Behandlung. Thieme, Stuttgart 1966.

Hartmann, E., J. Cravens: The effect of long-term administration of psychotropic drugs on human sleep: I. Methodology and the effects of placebo. Psychopharmacol. 33 (1973) 153–167.

Hartmann, E., E. Milofsky, G. Vaillant, M. Oldfield, R. Falke, C. Ducey: Vulnerability to schizophrenia: prediction of adult schizophrenia using childhood information. Arch. gen. Psychiat. 41 (1984) 1050–1056.

Hartmann, F.: Kranksein im Krankenhaus. Vortrag auf der 109. Versammlung der Ges. Deutscher Naturforscher und Ärzte am 23. 9.1976 in Stuttgart.

Hartmann, F.: Von der Diagnose zum problemoffenen Krankenblatt. Therapiewoche 26 (1976) 916–920.

Hartmann, H.: Ich-Psychologie und Anpassungsprobleme. Int. Univ. Press, New York 1939. Psyche 14 (1960) 81–164.

Hartmann, H.: Ich-Psychologie und Anpassungsprobleme. Klett, Stuttgart 1970.

Hartmann, H.: Wie prägt den Arzt seine Wissenschaft? – Marburg, 21.11.83 (unveröffentlichter Vortrag).

Hartmann, H., E. Kris, R. M. Loewenstein: Notes on the theory of aggression. Psychoanal. Stud. Child 3 (1949) 9–36.

Hartmann, N.: Neue Wege der Ontologie. Zit. n.: Historisches Wörterbuch der Philosophie. Schwabe u. Co., Basel-Stuttgart 1969.

Hartung, M. L., S. Lehre: Psychologische Befunde bei einer Gruppe von Patientinnen mit perioraler Dermatitis. In: Bosse, K., P. Hünecke (Hrsg.): Psychodynamik und Soziodynamik bei Hautkranken. Vandenhoeck & Ruprecht, Göttingen 1976.

Hartwich, P., E. Steinmeyer: Strukturmodell zur Darstellung krankheitserschwerender Faktoren der Anorexia nervosa mittels der Pfadanalyse. Arch. Psychiat. Nervenkr. 219 (1974) 297–312.

Hase, H. D., L. R. Goldberg: The comparative validity of different strategies of driving personality inventory scales. Psychol. Bull. 67 (1967) 231–248.

Hasenbring, M., S. Ahrens: Depressivität, Schmerzwahrnehmung und Schmerzerleben bei Patienten mit lumbalem Bandscheibenvorfall. Psychother. med. Psychol. 37 (1987) 149–155.

Hassler, R.: Zur funktionellen Anatomie des limbischen Systems. Nervenarzt 35 (1964) 386–396.

Hathaway, S. R., J. C. McKinley: Minnesota Multiphasic Personality Inventory. Psychological Corporation, New York 1967.

Hattingberg, J.: Psychologische Methoden bei der Rehabilitation nach Herzinfarkt. Med. Klinik 64 (1969) 1907.

Haug, F.: Kritik der Rollentheorie. Frankfurt 1972.

Hauri, P.: Treating psychophysiologic insomnia with biofeedback. Sleep Res. 13 (1984) 146.

Hauss, W. H., J. Lindner: Wesen des Alters und der Krankheiten im Alter. In: Hauss, W., W. Oberwittler (Hrsg.): Geriatrie in der Praxis. Springer, Berlin 1975.

Hautzinger, M.: Verhaltenstraining bei Übergewicht. Müller, Salzburg 1978.

Havik, O. E., J. G. Maeland: Verbal denial and outcome in myocardial infarction patients. J. psychosom. Res. 32 (1988) 145–157.

Haviland, M. G., J. P.MacMurray, M. A. Cummings: The relationship between alexithymia and depressive symptoms in a sample of newly abstinent alcoholic inpatients. Psychother. and Psychosom. 49 (1988) 37–40.

Havlik, R. J., R. J. Garrison, M. Feinleib, S. Padgett, W. P. Castelli, P. M. McNamara: Evidence for additional blood pressure correlates in adults 20–56 years old. Circulation 61 (1980) 710–715.

Hawkins, J. R.: Clinical experience with beta-blockers in consultant psychiatric practice. Scot. med. J. 20 (1975) 294–297.

Hawley, D. J., F. Wolfe: Anxiety and depression in patients with rheumatoid arthritis: a prospective study of 400 patients. J. Rheum. 15 (1988) 932–941.

Hay, D., D. Oken: The psychological stresses of intensive care unit nursing. Psychosom. Med. 34 (1972) 109–118.

Hayes, D. M.: The impact of the health care system on physician attitudes and behaviors. In: Cullen, J. W., B. H. Fox, R. N. Isom (eds.): Cancer. The Behavioral Dimension. Raven, New York 1976.

Haymaker, W. E., E. Anderson, W. H. J. Nauta (eds.): The hypothalamus. Thomas, Springfield/Ill. 1969.

Hayman, M.: Myasthenia gravis and psychosis. Report of a case with observations on its psychosomatic implications. Psychosom. Med. 3 (1941) 120–137.

Hayn, H.: Das Hyperventilationssyndrom. Der Prakt. Arzt 1 (1974) 34–36.

Haynes, S. G., M. Feinleib: Type A behavior and the incidence of coronary heart disease in the Framingham Heart Study. Advanc. Cardiol. 29 (1982) 85–95.

Haynes, S. G., S. Levine, N. Scotch et al.: The relationship of psychosocial factors to coronary heart disease in the Framingham Study, I. Methods and risk factors. Amer. J. Epidemiol. 107 (1978a) 362–383.

Haynes, S. G., M. Feinleib, W. B. Kannel et al.: The relationship of psychosocial factors to coronary heart disease in the Framingham Study, II. Prevalence of coronary heart disease. Amer. J. Epidemiol. 107 (1978b) 384–402.

Haynes, S. G., M. Feinleib, W. B. Kannel et al.: The relationship of psychosocial factors to coronary heart disease in the Framingham Study, III. 8-year incidence of coronary heart disease. Amer. J. Epidemiol. 111 (1980) 37–58.

Haynes, S. G., M. Feinleib, E. Eaker: Type A behavior and the 10-year incidence of coronary heart disease in the Framingham Heart Study. In: Rosenman, R. H. (ed.): Psychosomatic Risk Factors and Coronary Heart Disease: Indications for Specific Preventive Therapy. Huber, Bern 1983.

Haynes, S. N.: Muscle contraction in headache: a psychophysiological perspective of etiology and treatment. In: Haynes, S. N., L. Gannon (eds.): Psychosomatic Disorders. A Psychophysiological Approach to Etiology and Treatment, pp. 447–483. Praeger, New York 1982.

Hazan, S. J.: Psychiatric complications following cardiac surgery. J. thorac. cardiovasc. Surg. 51 (1966) 307.

Hazell, J. W. P.: Spontaneous cochlear acoustic emissions and tinnitus. Clinical experience on the tinnitus patient. J. Laryngol. Suppl. 9 (1984) 106.

Hazum, E., K. J. Chang, P. Cuatrecasas: Specific nonopiate receptors for beta-endorphin. Science 205 (1979) 1033–1035.

Heath, R.: Pleasure and brain activity in man. J. nerv. ment. Dis. 154 (1972) 3–18.

Heaton, K. W., J. R. Thornton, P. M. Emmet: Dietary factors in Crohn's disease. Z. Gastroenterol. 17 (Suppl.) (1979) 140–144.

Heberden, W.: Some account of a disorder of the breast. Med. Transactions roy. Coll. Physicians 2 (1772) 59.

Hecker, M., M. Chesney, G. Black, N. Frautschi: Components of type A behavior and coronary heart disease. Psychosom. Med. 50 (1988) 153–164.

Heckers, H., F. W. Melcher, W. Kamenich, K. Henneking: Morbus Crohn und Fettverzehr. Verh. Dtsch. Ges. Inn. Med. 90 (1984) 568–571.

Heckhausen, H.: Motivation und Handeln. Springer, Berlin 1980.

Heefner, J. D., R. M. Wilder, J. D. Wilson: Irritable colon and depression. Psychosomatics 19 (1978) 540–547.

Hees, P. A. M. van: An index of inflammatory activity in patients with Crohn's disease. Acta gastro-enter. belg. 47 (1984) 282–288.

Hees, P. A. M. van, P. van Elteren, H. J. J. van Lier, J. H. M. van Tongeren: An index of inflammatory activity in patients with Crohn's disease. Gut 21 (1980) 279–286.

Hefferline, R. F., L. J. Bruno: The psychophysiology of private events. In: Shapiro, D., T. X. Barber, L. V. DiCara, J. Kamiya, N. E. Miller, J. Stoyva (eds.): Biofeedback and Self-Control, p. 72. Aldine, Chicago 1973.

Hefti, M. L.: Risiko- und Invaliditätsbeurteilung bei Morbus Crohn. Lebensversicherungsmedizin 33 (1981) 106–112.

Heggendorn, H., H. P. Voigt, G. Graber: Experimentelle Untersuchungen über die orale Hyperaktivität bei psychischer Belastung, im Besonderen bei Aggression. Schweiz. Mschr. Zahnheilk. 89 (1979) 1148.

Hegglin, R.: Differentialdiagnose innerer Krankheiten, 12. Aufl. Thieme, Stuttgart 1972.

Hehl, F. J., U. Makowka, R. Schleberger: Zur Psychosomatik des Operationserfolges bei Bandscheibengeschädigten. Z. klin. Psychol. Psychopath. Psychother. 31 (1983) 53–66.

Heiberg, A. N.: Alexithymic characteristics and somatic illness. Psychother. and Psychosom. 34 (1980) 261.

Heiberg, A. N., A. Heiberg: A possible genetic contribution to the alexithymia trait. Psychother. and Psychosom. 30 (1978) 205–210.

Heiberg, A. N., B. Helöe, B. S. Krogstad: The myofacial pain dysfunction: dental symptoms and psychological and muscular function. Psychother. and Psychosom. 30 (1978) 81–97.

Heidelbach, J. G.: Über unsere Erfahrungen bei der Eignungsbeurteilung für den Sängerberuf und ihre weitere Bedeutung aus stimmärztlicher Sicht. In: Spitzer, L. (Hrsg.): Probleme der Sängerausbildung, S. 27. Mechitharisten-Druckerei, Wien 1981.

Heigl, F.: Indikation und Prognose in Psychoanalyse und Psychotherapie. Vandenhoeck & Ruprecht, Göttingen 1972.

Heigl-Evers, A., F. Heigl: Gesichtspunkte zur Indikation für die kombinierte Einzel- und Gruppenpsychotherapie. Gruppenpsychotherapie und Gruppendynamik, Bd. 4, Heft 1 (1970) 82–99.

Heigl-Evers, A., F. Heigl: Für wen ist Gruppenpsychotherapie geeignet? In: Heigl-Evers, A. (Hrsg.): Psychoanalyse und Gruppe. Vandenhoeck & Ruprecht, Göttingen 1971.

Heigl-Evers, A., R. Schneider, K. Bosse: Biographische Daten von endogenen Ekzematikern. Z. psychosom. Med. Psychoanal. 21 (1976) 75–84.

Heigl-Evers, A., U. Streek: Psychoanalytisch-interaktionelle Therapie. Psychother. med. Psychol. 35 (1985) 176.

Heijnen, C. J., R. E. Ballieux: Influence of opioid peptides on the immune system. Advances 3 (1986) 114–121.

Heijningen, H. van, N. Treurniet: Psychodynamic factors in acute myocardial infarction. Int. J. Psychoanal. 47 (1966) 370–374.

Heilbrunn, G.: Psychodynamic aspects of epilepsy. Psychoanal. Quart. 19 (1950) 145–157.

Heilman, K.: Augenheilkunde. In: Hahn, P. (Hrsg.): Die Psychologie des 20. Jahrhunderts, Band Psychosomatik, S. 744–756. Kindler, München 1979.

Heim, E.: Krankheit als Krise und Chance. Kreuz, Zürich 1979a.

Heim, E.: Coping oder Anpassungsvorgänge in der psychosomatischen Medizin. Z. psychosom. Med. 25 (1979b) 251–262.

Heim, E.: Konsequenzen für die Praxis aus der Psychotherapieforschung der letzten Jahre. Schweiz. Arch. Neurol. Neurochir. Psychiat. 128 (1981) 211–226.

Heim, E.: Models, methods and lecturers in psychosocial medicine. Paper read at the Réunion des Enseignants de Psychologie Médicale des Universités Européennes, 11.–14. 4. 1984, Delphi, Greece.

Heim, E.: Psychosoziale Aspekte der otorhinolaryngologischen Medizin. In: Kellerhals, B. et al. (Hrsg.): Aktuelle Probleme der Otorhinolaryngologie 8 (1984) 332. Huber, Bern 1984.

Heim, E.: Coping und Adaptivität: Gibt es ein geeignetes oder ungeeignetes coping? Psychother., Psychosom., med. Psychol. 38 (1988a) 8–18.

Heim, E., J. Willi (Hrsg.): Psychosoziale Medizin, Bd. 2: Klinik und Praxis, S. 343–390. Springer, Berlin 1986.

Heim, E., M. Thommen: Curriculum psychosozialer Medizin – Entwicklung und Evaluation. In: Saladin, P., H. J. Schaufelberger, P. Schlappi (Hrsg.): Festschrift zur Emeritierung von Prof. H. G. Pauli. Bern 1989.

Heim, E. et al.: Dyspnea: psychophysiologic relationships. Psychosom. Med. 34 (1972) 405–423.

Heim, E., A. Moser, R. Adler: Defense mechanisms and coping behavior in terminal illness. Psychother. and Psychosom. 30 (1978) 1–17.

Heim, E., R. Adler, A. Moser: Beeinträchtigung der psychosozialen Anpassung durch terminale Krankheit. Z. psychosom. Med. 28 (1982) 347–362.

Heim, E., K. Augustiny, A. Blaser: Krankheitsbewältigung (Coping) – ein integratives Modell. Psychother. med. Psychol. 33 (1983) 35–40.

Heim, E., K. Augustiny, A. Blaser, C. Bürki et al.: Bewältigung von Brustkrebs – eine longitudinale Studie. In: Kächele, H., W. Steffens (Hrsg.): Bewältigung und Abwehr, S. 133–161. Springer, Berlin 1988b.

Heim, E., K. Augustiny, A. Blaser, C. Bürki, D. Kühne, M. Rothenbühler, L. Schaffner, L. Valach: Coping with breast cancer – longitudinal prospective study. Psychother. and Psychosom. 48 (1988) 44–59.

Heiman, M.: Separation from a love object as an etiological factor in functional uterine bleeding. J. Mt. Sinai Hosp. 26 (1959) 56.

Heimann, P.: On countertransference. Int. J. Psychoanal. 31 (1950) 81–84.

Heimann, P.: Bemerkungen zur Gegenübertragung. Psyche 9 (1964) 483–493.

Heimpel, H.: Persönliche Mitteilung 1970.

Hein, H.: Sekretorisches Immunglobulin A: Aufbau und Bedeutung. Fortschr. Med. 93 (1975) 865–875.

Heineberg, H., E. Gold, F. C. Robbins: Differences in interferon content in tissues of mice of various ages infected with Coxsackie B 1 virus. Proc. Soc. exp. Biol. (N. Y.) 115 (1964) 947.

Heinecker, R.: Erfahrungen als Patient einer Intensivstation und Vorschläge zur Humanisierung einer solchen Station. Dtsch. med. Wschr. 12 (1980) 417.

Heinemann, M.: Entwicklungsstörungen der Stimme. In: Pascher, W., H. Bauer (Hrsg.): Differentialdiagnose von Sprach-, Stimm- und Hörstörungen, S. 31. Thieme, Stuttgart 1984.

Heinl, H.: Groddeck und die Integrative Leibtherapie, S. 179–185. In: Siefert, H., F. Kern, B. Schuh, H. Grosch (Hrsg.): Groddeck-Almanach. Stroemfeld/Roter Stern, Frankfurt 1986.

Heintze-Hook, C.: Analytische Psychotherapie und Funktionelle Entspannung bei Asthmapatienten. Vortrag auf der Tagung „Asthma – eine Beziehungsstörung", Bad Orb 7.–9. 11. 1986, Kurzreferat von C. Heuer in: A. F. E. INTERN 4 (1986) 15.

Literatur

Heirichs, O.: Die Relevanz psychosozialer Faktoren für die Schwangerschaft und die perinatale Periode bei ledigen und geschiedenen Müttern. Diss., FU Berlin 1977.

Heisel, J. S., S. E. Locke, L. J. Kraus, R. M. Williams: Natural killer cell activity and MMPI scores of a cohort of college students. Amer. J. Psychiat. 143 (1986) 1382–1386.

Heisler, S., T. D. Reisine, V. Y. H. Hook, J. Axelrod: Somatostatin inhibits multireceptor stimulation of cyclic AMP formation and corticotropin secretion in mouse pituitary tumor cells. Proc. Natl. Acad. Sci. (Wash.) 79 (1982) 6502–6506.

Helgason, T.: Prevalence and incidence of mental disorders. Acta psychiat. scand. 58 (1978) 256–266.

Heller, A.: Theorie der Gefühle. Hamburg 1981.

Heller, S., K. A. Frank, I. R. Malin et al.: Psychiatric complications of open heart surgery. A reexamination. New Engl. J. Med. 283 (1970) 1015–1020.

Heller, S., K. A. Frank, D. S. Kornfeld, J. R. Malm, F. O. Bowman: Psychological outcome following open-heart surgery. Arch. intern. Med. 134 (1974) 908–914.

Hellerstein, H. K., E. H. Friedman: Sexual activity and the postcoronary patient. Med. Aspects hum. Sexuality 3 (1969) 70–76.

Hellhammer, D., I. Florin, H. Weiner (eds.): Neurobiological approaches to human disease. Huber, Toronto 1988.

Hellmuth, G. A., W. J. Johannsen, T. Sorauf: Psychological factors in cardiac patients. Arch. Environm. Health 12 (1966) 771–775.

Helmchen, H., U. Rüger: Neurosen und psychosomatische Erkrankungen als klassifikatorisches und diagnostisches Problem. Z. psychosom. Med. 26 (1980) 205–216.

Helmich, P., E. Hesse, K. Köhle, H. J. Mattern, Th. v. Uexküll, W. Wesiack: Psychosoziale Kompetenz in der Primärversorgung. Springer, Heidelberg 1990.

Helmke, K., M. Seitz, R. Brockhaus, R. Weimer, A. Otten, K. Federlin: Autoimmunphänomene beim Diabetes mellitus. Zur pathogenetischen und diagnostischen Bedeutung. Immun. Infekt. 11 (1983) 199–208.

Helsing, K. J., M. Szklo: Mortality after bereavement. Amer. J. Epidemiol. 114 (1981) 41–52.

Helsing, K. J., G. W. Comstock, M. Szklo: Causes of death in a widowed population. Amer. J. Epidemiol. 116 (1981) 524–532.

Helzer, J. E., S. Chammas, C. C. Norland, W. A. Stillings, D. H. Alpers: A study of the association between Crohn's disease and psychiatric illness. Gastroenterology 86 (1984) 324–330.

Hemsley, D. R.: What have cognitive defects to do with schizophrenic symptoms? Brit. J. Psychiat. 130 (1977) 167–173.

Hendrickx, M.: Quantitative Untersuchung der Immunglobuline IgA und IgG im Speichel von 7- bis 8jährigen Kindern unter besonderer Berücksichtigung des jahreszeitlichen Verlaufs und ökologischer Faktoren. Dissertation (med.dent.), Univ. Düsseldorf 1981.

Hendrie, H. C., F. Paraskevas, F. D. Barager, J. D. Adamson: Stress, immunoglobulin levels and early polyarthritis. J. psychosom. Res. 15 (1971) 337–342.

Hendriksen, C., V. Binder: Social prognosis in patients with ulcerative colitis. Brit. med. J. 281 (1980) 581–583.

Hendriksen, C., S. Kreiner, V. Binder: Long-term prognosis in ulcerative colitis. Gut 26 (1985) 158–163.

Hengst, J. C. D., R. A. Kempf: Immunomodulation by cyclophosphamide. Clin. Immunol. Allergy 4 (1984) 199–216.

Hennig, J., A. Buske: Klassische Konditionierung eines Immunparameters beim Menschen. Unveröffentlichte Diplomarbeit, Universität Münster 1988.

Henning, G., S. Schulz: Sexualverhalten von Patientinnen nach der Therapie von Zervixkarzinomen. Zbl. Gynäk. 97 (1975) 1562.

Henning, N.: Gastrointestinaltrakt. In: Hauss, W., W. Oberwittler (Hrsg.): Geriatrie in der Praxis. Springer, Berlin 1975.

Henrichs, T. F., I. W. Mackenzie, C. H. Almond: Psychological adjustment and psychiatric complications following open-heart surgery. J. nerv. ment. Dis. 152 (1972) 332–345.

Henry, J.: The relation of social to biological processes in disease. Soc. Sci. Med. 16 (1982) 369–380.

Henry, J. P.: Mechanisms of psychosomatic disease in animals. In: Brandly, C. A., C. E. Cornelius, W. I. B. Beveridge (eds.): Advances in Veterinary Science and Comparative Medicine 20 (1976) 115–145.

Henry, J. P.: Present concept of stress theory. In: Usdin, E. et al. (eds.): Catecholamines and Stress: Recent Advances. Elsevier, New York 1980.

Henry, J. P.: Coronary heart disease and arousal of the adrenal cortical axis. In: Dembroski, T. M., T. H. Schmidt, G. Blümchen (eds.): Biobehavioral Bases of Coronary Heart Disease, pp. 365–381. Karger, Basel-München-Paris 1983.

Henry, J. P., J. C. Cassel: Psychosocial factors in essential hypertension: recent epidemiologic and animal experimental evidence. Amer. J. Epidemiol. 90 (1969) 171–200.

Henry, J. P., P. M. Stephens: Stress, health, and the social environment. A socio-biologic approach to medicine. Springer, New York 1977.

Henry, J. P., P. M. Stephens: The social environment and essential hypertension in mice: possible role of the innervation of the adrenal cortex. In: De Jong, W., A. P. Provoost (eds.): Progress in Brain Research, Vol. 47: Hypertension and Brain Mechanisms, pp. 263–276. Elsevier, Amsterdam-Oxford-New York 1977.

Henry, J. P., J. P. Meehan, P. M. Stephens: The use of psychosocial stimuli to induce prolonged systolic hypertension in mice. Psychosom. Med. 29 (1967) 408–432.

Henry, J. P., D. L. Ely, P. M. Stephens: Changes in catecholamine-controlling enzymes in response to psychosocial activation of defense and alarm reactions. In: Physiology, Emotion and Psychosomatic Illness. Excerpta Medica, Amsterdam 1972.

Henry, J. P., P. M. Stephens, J. Axelbrod, R. A. Mueller: Effect of psychosocial stimulation on the enzymes involved in the biosynthesis and metabolism of noradrenaline and adrenaline. Psychosom. Med. 33 (1971) 227–237.

Henry, J. P., D. L. Ely, P. M. Stephens, H. L. Ratcliffe, G. A. Santisteban, A. P. Shapiro: The role of psychosocial factors in the development of arteriosclerosis in CBA mice: observations on the heart, kidney, and aorta. Atherosclerosis 14 (1971) 203–218.

Henryk-Gutt, R., W. L. Rees: Psychological aspects of migraine. J. psychosom. Res. 17 (1973) 141–153.

Henseler, H.: Narzißtische Krisen. Zur Psychodynamik des Selbstmordes. Rowohlt, Hamburg 1974.

Hentschel, E.: Medikamentöse Langzeitbehandlung des Ulcus pepticum. Dtsch. Ärztebl. 1 (1984) 25–30.

Henze, T.: Labordiagnostik bei Myasthenia gravis. In: Holzgrafe, M., H. Reiber, K. Felgenhauer (Hrsg.): Labordiagnostik von Erkrankungen des Nervensystems. Perimed, Erlangen 1988.

Herbart, F.: Psychologie als Wissenschaft, neu gegründet auf Erfahrung, Metaphysik und Mathematik (2 Teile). Königsberg 1824. (Reprint Amsterdam 1968).

Herberman, R. B., H. T. Holden: Natural killer cells as antitumor effector cells. J. Natl. Cancer Inst. 62 (1979) 441–445.

Herd, J. A.: Behavioral factors in the psychology mechanism of cardiovascular disease. In: Weiss, S. M., J. A. Herd, B. H. Fox (eds.): Perspectives on Behavioral Medicine, pp. 55–66. Acad. Press, New York 1981.

Herlitz, J., D. Elmfeld, S. Holmberg, I. Malek, G. Nyberg, K. Pennert, L. Ryden, K. Swedenberg, A. Vedin, F. Waagstein, J. Waldenström et al.: Göteborg Metoprolol Trial: mortality and cause of death. Amer. J. Cardiol. 53 (1984) 9D-14D.

Herman, C. P., J. Polivy: A boundary model for the regulation of eating. In: Stunkard, A. J., E. Stellar (eds.): Eating and Its Disorders. Raven, New York 1984.

Hermanek, P., J. Tonak: Ischämische Kolitis. Klinikarzt 5 (1976) 88–90.

Hermann, E., D. Schneider-Helmert, G. A Schoenenberger. One personality pattern for all insomniacs is a myth. Sleep Res. 17 (1988) 125.

Hermann, H. T., G. C. Quarton: Psychological changes and psychogenesis in thyroid hormone disorders. J. clin. Endocr. 25 (1965) 327.

Hermann-Maurer, E. K.: Psychotherapeutischer Umgang mit Insomniepatienten. Schweiz. Arch. Neurol. Psychiat. 138 (1989) 262–272.

Hermann-Maurer, E. K., D. Schneider-Helmert, G. A. Schoenenberger: Diagnostic principle at the Medical Center Mariastein (Sleep Disorders Center). In: European Federation of Professional Psychologists Association (ed.): Benefits in Psychology: Improving the Quality of Hospital Care. Proc. of the congress 1988, Rome (in press).

Herms, V., J. Gabelmann, F. Kubli: Psychosomatic aspects of premature labor. In: Prill, H. J., M. Stauber (eds.): Advances in Psychosomatic Obstetrics and Gynecology. Springer, Berlin 1982.

Herrmann, J. M.: Psychosomatische Therapie bei Hypertonie. Münch. med. Wschr. 128 (1986) 869–872.

Herrmann, J. M., W. Schüffel: Das ärztliche Interview. Rocom, Basel 1983.

Herschbach, P.: Einige Überlegungen zur psychosozialen Rehabilitation von Krebskranken. Die Rehabilitation 22 (1983) 33–35.

Herschbach, P.: Stationäre onkologische Rehabilitation – eine Bestandsaufnahme. Z. f. personenzentrierte Psychologie und Psychotherapie 6 (1987) 15–29.

Herschbach, P.: Stationäre onkologische Rehabilitation – eine Bedarfsanalyse. Z. f. personenzentrierte Psychologie und Psychotherapie 6 (1987) 31–45.

Herschbach, P., G. Henrich: Probleme und Problembewältigung von Tumorpatienten in der stationären Nachsorge. Psychother. med. Psychol. 37 (1987) 185–192.

Herschbach, P. A., M. Rosbund, J. Brengelmann: Probleme von Krebspatientinnen und Formen ihrer Bewältigung. Onkol. 8 (1985) 219–231.

Herty, H.: Psychosoziale Arbeit auf einer onkologischen Station. Münch. med. Wschr. 126 (1984) 223–224.

Hertz, D. G., H. Molinski: Psychosomatik der Frau. Springer, Berlin 1980.

Herz, E., E. H. Glaser: Spasmodic torticollis II. Clinical evaluation. Arch. Neurol. Psychiat. 61 (1949) 227–239.

Herzmann, C. E.: Die psychosomatische Problematik des Schmerzes bei alten Menschen. In: Bergener, M., B. Kark (Hrsg.): Psychosomatik in der Geriatrie. Steinkopff, Darmstadt 1985.

Herzog, D. B.: Are anorexic and bulimic patients depressed? Amer. J. Psychiat. 141 (1984) 1594–1597.

Herzog, D. B., M. B. Keller, P. W. Lavori, I. S. Bradburn, L. Ott: Ergebnisse zum Krankheitsverlauf der Bulimia nervosa. In: Fichter, M. M. (Hrsg.): Bulimia nervosa. Enke, Stuttgart 1989.

Herzog, D. B., M. B. Keller, P. W. Lavori: Outcome in anorexia nervosa and bulimia nervosa. J. nerv. ment. Dis. 176 (1988) 131–143.

Herzog, J. M.: Sleep disturbances and father hunger in 18 to 28 months old boys. Psychoanal. Study Child 35 (1980) 219–233.

Hess, G.: Psychologische Korrelate langfristiger Operationseffekte bei Ulcuspatienten. Psychol. Diplomarbeit, Marburg 1983.

Hess, R.: Inhalt und Grenzen der ärztlichen Aufklärungspflicht. Prakt. Arzt 12 (1975) 2120–2122.

Hess, W. R.: Die funktionelle Organisation des vegetativen Nervensystems. Schwabe, Basel 1948.

Hess, W. R.: Das Zwischenhirn. Basel 1949.

Hess, W. R.: Beziehungen zwischen psychischen Vorgängen und Organisation des Gehirns. Studium Generale 9 (1956) 467 und 10 (1957) 327.

Hesse, E.: Hilfe zur Selbsthilfe durch den Hausarzt. Therapiewoche 30 (1980) 3951–3961.

Hesse, H.: Kurgast und die „Aufzeichnung bei einer Kur in Baden". Suhrkamp, Frankfurt 1975.

Hesse, K. A. F.: Meeting the psychological needs of pacemaker patients. Int. J. Psychiat. Med. 6 (1975) 359–372.

Heyck, H.: Der Kopfschmerz. 4. Aufl. Thieme, Stuttgart 1975.

Heyde, W.: Aspekte onkologischer Rehabilitation und Nachsorge. Die Rehabilitation 22 (1983) 1–17.

Heyer, G.: Psychogene Funktionsstörungen des Verdauungstraktes. In: Schwarz, O. (Hrsg.): Psychogenese und Psychotherapie körperlicher Symptome, S. 228–258. Deuticke, Wien 1925.

Heyer, G. R. (1925): Die Atmung. In: Heyer-Grote, L. (Hrsg.): Atemschulung als Element der Psychotherapie, S.54. Wiss. Buchgesellschaft, Darmstadt 1970.

Heymans, C., E. Neil: Reflexogenic areas of the cardiovascular system. Little, Brown & Co., Boston 1958.

Hiatt, J. F., T. C. Floyd, P. H. Katz, I. Feinberg: Further evidence of abnormal non-rapid-eye-movement sleep in schizophrenia. Arch. gen. Psychiat. 42 (1985) 797–802.

Hiebsch, H.: Sozialpsychologische Grundlagen der Persönlichkeitsforschung. Berlin 1971.

Hiebsch, H., W. Vorwerg: Einführung in die marxistische Sozialpsychologie. Berlin 1971.

Hightower jr., N. C., A. C. Broders jr., R. D. Haines, J. F. McKenney, A. W. Sommer: Chronic ulcerative colitis. Amer. J. digest. Dis. 3 (1958) 722–733, 861–876, 931–941.

Hilgard, E. R.: A quantitative study of pain and its reduction through hypnotic suggestion. Proc. nat. Acad. Sci. (Wash.) 57 (1967) 1581–1586.

Hilgard, E. R.: The alleviation of pain by hypnosis. Pain 1 (1975) 213–231.

Hilgard, E. R.: Hypnosis and pain. In: Sternbach, R. A. (ed.): The Psychology of Pain. Raven, New York 1978.

Hilgard, E. R., G. H. Bower: Theories of learning. Appleton Century Crofts, New York 1966.

Hill, D., W. Mitchell: Epileptic anamnesis. Fol. psychiat. neerl. 56 (1953) 718–725.

Hill, D. L.: A review of cyclophosphamide. Thomas, Springfield/Ill. 1975.

Hill, O. W.: Anorexia nervosa. Modern Trends in Psychosomatic Medicine 3 (1976) 382–403.

Hill, O. W.: Epidemiological aspects to anorexia nervosa. Advanc. psychosom. Med. Vol. 9 (1977) 48–62.

Hill, O. W., L. Blendis: Physical and psychological evaluation of „non-organic" abdominal pain. Gut 8 (1967) 221–229.

Hillard, J. R., P. J. Hillard: Bulimia, anorexia nervosa and diabetes. Deadly combinations. Psychiat. Clin. N. Amer. 7 (1984) 367–379.

Literatur

Hillard, J. R., M. C. Lobo, R. P. Keeling: Bulimia and diabetes: a potentially life-threatening combination. Psychosomatics 24 (1983) 292–295.

Hillenbrand, D.: Psychosomatische Aspekte der Myasthenia gravis. Psychother. med. Psychol. 22 (1972) 69–76.

Hilpert, H.: Über den Einfluß familiärer Faktoren auf die Epidemiologie der Appendektomie. Nervenarzt 51 (1980) 417.

Himsworth, H. P.: Diet in the aetiology of human diabetes. Proc. roy. Soc. Med 43 (1949) 323.

Hinkle, G.: The role of freudianism in American sociology. Diss., Univ. of Wisconsin, Wisconsin 1951.

Hinkle jr., L. E.: Ecological observations of the relation of physical illness, mental illness, and the social environment. Psychosom. Med. 23 (1961) 289–296.

Hinkle jr., L. E.: The effect of exposure to culture change, social change and changes in interpersonal relationships on health. In: Dohrenwend, B. S., B. P. Dohrenwend (eds.): Stressful Life Events: Their Nature and Effects. Wiley, New York 1974.

Hinkle jr., L. E., B. Benjamin, W. N. Christenson: Coronary heart disease: 30-year experience of 1160 men. Arch. Environm. Health 13 (1966) 312–321.

Hinkle jr., L. E., H. T. Thaler: The clinical classification of cardiac deaths. Circulation 65 (1982) 457–464.

Hinkle jr., L. E., S. Wolf: Importance of life stress in course and management of diabetes mellitus. J. Amer. med. Ass. 148 (1952) 513–520.

Hinkle jr., L. E., H. G. Wolff: The nature of man's adaptation to his total environment and the relation of this to illness. Arch. intern. Med. 99 (1957) 442–460.

Hinkle jr., L. E., H. G. Wolff: Health and the social environment: experimental investigations. In: Leighton, A. H., J. E. Clausen, R. N. Wilson (eds.): Explorations in Social Psychiatry, pp. 105 ff. New York 1957.

Hinkle jr., L. E., H. G. Wolff: Ecological investigations of the relation between illness, life experiences and social environment. Ann. intern. Med. 49 (1958) 1373–1388.

Hinkle jr., L. E., C. Conger, S. Wolf: Studies on diabetes mellitus: relation of stressful life situations to the concentration of ketone bodies in the blood of diabetic and non-diabetic humans. J. clin. Invest. 29 (1950) 513–520.

Hinkle jr., L. E., F. M. Evans, S. Wolf: Studies in diabetes mellitus III. Psychosom. Med. 13 (1951) 160–183.

Hinkle jr., L. E., W. N. Christenson, F. D. Kane, A. Ostfield, W. Thetford, H. G. Wolff: An investigation of the relation between life experience, personality characteristics, and general susceptibility to illness. Psychosom. Med. 20 (1958) 278–295.

Hinze, E., H. Krüger: Das Herzangstsyndrom bei alten Patienten. Z. Gerontol. 14 (1981) 34–39.

Hirsch, E. Z., J. A. Maksem, D. Gagen: Effects of stress and propranolol on the aortic intima of rats. Arteriosclerosis 4 (1984) 526.

Hirsch, H. A.: Bakterielle und mykotische Erkrankungen der ableitenden Harnwege bei der Frau. In: Verhandlungsbericht der Deutschen Gesellschaft für Urologie, S. 310. Springer, Berlin 1976.

Hirsch, M.: Realer Inzest. Psychodynamik des sexuellen Mißbrauchs in der Familie. Springer, Berlin-Heidelberg-New York 1986.

Hirsch, M.: Psychogener Schmerz als Repräsentant des Mutterobjektes. Psychother. med. Psychol. 39 (1989) 202–208.

Hirsch, M., J. M. Herrmann: Hypochondrie und Objektbeziehungstheorie am Beispiel der AIDS-Phobie. In: Schüffel, W. (Hrsg.): Sich gesund fühlen im Jahre 2000, S. 191–198. Thure von Uexküll zum 80. Geburtstag gewidmet. Springer, Berlin 1988.

Hislop, I. G.: Psychological significance of the irritable colon syndrome. Gut 12 (1971) 452–457.

Hislop, I. G.: Onset setting in inflammatory bowel disease. Med. J. Aust. I (1974) 981–984.

Hislop, I. G.: The effect of very brief psychotherapy on the irritable colon syndrome. Med. J. Aust. 2 (1980) 620–622.

Hislop, T. G., N. E. Waxler, A. J. Coldman, J. M. Elwood, L. Kan: The prognostic significance of psychosocial factors in women with breast cancer. J. chron. Dis. 40 (1987) 729–735.

Hochberg, M. C.: Adult and juvenile rheumatoid arthritis: current epidemiologic concepts. Epidem. Rev. 3 (1981) 27–44.

Hockman, C. H., H. P. Mauck, E. C. Hoff: ECG changes resulting from cerebral stimulation, II. A spectrum of arrhythmias of sympathetic origin. Amer. Heart J. 71 (1966) 695.

Hockman, C. H., H. P. Mauck, N. S. Chu: ECG changes resulting from cerebral stimulation, III. Action of diphenylhydantoin on arrhythmias. Amer. Heart J. 74 (1967) 256.

Hodapp, V., G. Weyer: Zur Streß-Hypothese der essentiellen Hypertonie. In: Vaitl, D. (Hrsg.): Essentielle Hypertonie. Springer, Berlin-Heidelberg 1982.

Hodges, H., J. J. Kline, G. Barbero, R. Flanery: Life events occuring in families of children with recurrent abdominal pain. J. psychosom.Res. 28/3 (1984) 185–196.

Hodgins, E.: Episode. A report on an accident which occurred inside my head. Atheneum 1964.

Hodgkinson, S., R. Sherrington, H. Gurling, R. Marchbanks, S. Reeders, J. Mallet, M. McInnis, H. Petursson, J. Brynjolfsson: Molecular genetic evidence for heterogeneity in manic depression. Nature (Lond.) 325 (1987) 805–806.

Höffken, K. D., L. Beusen, W. Dmoch, H. Molinski, P. Nijs: Modifizierte Paar- Therapie. Die tiefenpsychologische Variante der Masters- und Johnson-Therapie. Sexualmed. 11 (1982) 501.

Höfler, W.: Amöbiasis. Int. Welt 8 (1982) 258–267.

Hoehn-Saric, R.: Neurotransmitters in anxiety. Arch. gen. Psychiat. 39 (1982) 735–742.

Höllt, V.: Multiple endogenous opioid peptides. Trends in Neurosci. 6 (1983) 24–26.

Hoelscher, T. J., K. L. Lichstein, T. L. Rosenthal: Home relaxation practice in hypertension treatment: objective assessment and compliance induction. J. consult. clin. Psychol. 54 (1986) 217–221.

Hölzl, R., W. E. Whitehead: Psychophysiology of the gastrointestinal tract. Plenum, New York 1983.

Hönmann, H.: Zur Psychosomatik der Colitis ulcerosa. Therapiewoche 32 (1982) 2740–2746.

Hönmann, H.: Epidemiologie. In: Uexküll, Th. v., R. Adler, J. M. Herrmann, K. Köhle, O. W. Schonecke, W. Wesiack (Hrsg.): Psychosomatische Medizin, S. 378–388. 3. Aufl. Urban & Schwarzenberg, München 1986.

Hönmann, H., H. Schepank, P. Riedel: Beschwerden bei psychisch Gesunden und psychisch Kranken in der Allgemeinbevölkerung. In: Studt, H. H. (Hrsg.): Psychosomatik in Forschung und Praxis, S. 3–22. Urban & Schwarzenberg, München-Wien-Baltimore 1983.

Hörner, W., B. Osen, E. Hannemann et al.: Psychosoziale Situation erwachsener Knochenmarkempfänger. (Abstr.) Psycho 13 (1987a) 760.

Hörner, W., E. Hannemann, M. Haen et al.: Organic brain syndrome in BMT: reversible or irreversible deterioration of intellectual function? Bone Marrow Transplant. 2 (Suppl. 1) (1987b) 257.

Hörner, W., S. Ammann, K. Foerster: Zur psychosozialen Situation erwachsener Knochenmarkspender, deren Empfänger rezidivfrei leben. Vortr. 26. Tag. Dtsch. Koll. Psychosom. Med., Bad Dürkheim 1987c.

Hörner, W., B. Osen, G. Ehninger, K. Foerster: Life-situation of adult bone marrow transplant recipients. J. Canc. Res. clin. Oncol. 114 (Suppl.) (1988) 115.

Hofer, M. A.: The organisation of sleep and wakefullness after maternal separation in young rats. Dev. Psychobiol. 9 (1976) 189–206.

Hofer, M. A.: The roots of behavior. Freeman, San Francisco 1981.

Hofer, M. A.: On the relationship between attachment and separation process in infancy. In: Plutchik, R. (ed.): Emotion, Theory, Research and Experience: Emotion in Early Development. Acad. Press, New York 1982.

Hofer, M. A.: Relationships as regulators. Psychosom. Med. 44 (1984) 183–187.

Hofer, M. A., H. Weiner: Mechanism for nutritional regulation of automatic cardiac control in early development. Psychosom. Medicine 37 (1972).

Hoff, F.: Klinische Physiologie und Pathologie. Thieme, Stuttgart 1950.

Hoff, H.: Der psychische Faktor bei Schmerzen und Veränderungen der Wirbelsäule. Wien. klin. Wschr. 66 (1954) 632–635.

Hoff, H., R. Clotten, W. Thurner: Zur Frage des Hyperventilationssyndroms. Wien. med. Wschr. 46 (1952) 917.

Hoffman, A. L.: Psychological factors associated with rheumatoid arthritis. Nurs. Res. 23 (1974) 218–234.

Hoffmann, B.: Handbuch des autogenen Trainings. 3. Aufl. dtv, München 1981.

Hoffmann, C., V. Faust: Psychische Störungen durch Arzneimittel. Thieme, Stuttgart 1983.

Hoffmann, L.: Foundations of family therapy. A conceptual framework of systems change. Basic Books, New York 1981. Deutsch: Grundlagen der Familientherapie. Isko, Hamburg 1982.

Hoffmann, S. O.: Können wir mit dem DSM-III leben? Forum Psychoanal. 1 (1985) 320–323.

Hoffmann, S. O.: Das psychogene Schmerzsyndrom – eine psychosomatische Krankheit. In: Studt, H. H. (Hrsg.): Psychosomatik in der inneren Medizin, Bd. 1, S. 93–101. Springer, Berlin-Heidelberg-New York 1986.

Hoffmann, S. O.: Einführung in die Neurosenlehre und psychosomatische Medizin. Schattauer, Stuttgart 1987.

Hoffmann, S. O.: Forschungstendenzen im Bereich von Psychotherapie und Neurosenlehre in den letzten 15 Jahren – ein persönlicher Eindruck. Psychother. med. Psychol. 37 (1987) 10–14.

Hoffmann, S. O.: Die psychoanalytische Abwehrlehre – aktuell, antiquiert oder obsolet? Forum Psychoanal. 3 (1988) 22–39.

Hofman, D.: Die Klinik des Asthma bronchiale im Kindesalter. Mschr. Kinderheilk. 131 (1983) 125–127.

Hogarty, G.: Informant ratings of community adjustment. In: Waskow, I., M. Parloff (eds.): Psychotherapy Change Measures. NIMH, Rockville MD 1975.

Holland, J., S. Sgroi, S. Marwit, N. Solkoff: The ICU syndrome. Fact or fancy? Int. J. Psychiat. Med. 4 (1973) 241.

Holland, J. C.: Psychologic aspects of cancer. In: Holland, J. F., E. Frei (eds.): Cancer Medicine, p. 1176. Lea & Febiger, Philadelphia 1982.

Holland, J. C., A. H. Korzun, S. Tross, D. F. Cella, L. Norton, W. Wood: Psychosocial factors and disease free survival (DFS) in stage II breast carcinoma. ASCO Proc. 5 (1986) 237.

Holland, J. C. B., M. R. Coles: Neuropsychiatric aspects of acute poliomyelitis. Amer. J. Psychiat. 114 (1957) 54–63.

Hollatz, F., H. U. Ziolko: Zur Differentialdiagnose der Anorexia nervosa. Koinzidenz von somatischer Erkrankung und psychogener Magersucht. Münch. med. Wschr. 118 (1976) 263–266.

Hollenberg, M. D., P. Cuatrecasas: Hormone receptors and membrane glycoproteins during in vitro transformation of lymphocytes. In: Clarkson, B., R. Baserga (eds.): Control of proliferation of animal cells, pp. 423–434. Cold Spring Harbor, New York 1974.

Hollister, L. E.: Tricyclic antidepressants. New Engl. J. Med. 299 (1978) 1106–1109.

Hollister, L. E.: Pharmacotherapeutic considerations in anxiety disorders. J. clin. Psychiat. 47 (1986) 33–36 (Suppl.).

Hollmann, W., R. Rost, B. Dufaux, H. Liesen: Prävention und Rehabilitation von Herz- und Kreislaufkrankheiten durch körperliches Training. Hippokrates, Stuttgart 1983.

Hollon, T. H.: Modified group therapy in the treatment of patients on chronic hemodialysis. Amer. J. Psychother. 26 (1972) 501–510.

Hollwich, F.: Augenheilkunde. 9. Aufl. Thieme, Stuttgart 1979.

Hollyday, H. W., J. D. Hardcastle: Delay in diagnosis and treatment of symptomatic colorectal cancer. Lancet I (1979) 309–311.

Holm, J. E., K. A. Holroyd, K. G. Hursey, D. B. Penzien: The role of stress in recurrent tension headache. Headache 26 (1986) 160–167.

Holme, J., A. Helgeland, J. Hjierman, P. Leren: Socio-economic status as a coronary risk factor: the Oslo Study. Acta med. scand. (Suppl.) 660 (1982) 147–151.

Holmes, T. H., H. G. Wolff: Life situations, emotions, and backache. Psychosom. Med. 14 (1952) 18–33.

Holmes, T. H., R. H. Rahe: The social readjustment rating scale. J. psychosom. Res. 11 (1967) 213–218.

Holroyd, K. A., D. Penzien: EMG biofeedback with tension headache: therapeutic mechanisms. In: Holroyd, K. A., B. Schlote, H. Zenz (eds.): Perspectives in Research on Headache, pp. 147–162. Hogrefe, Lewiston-New York 1983.

Holroyd, K. A., D. B. Penzien: Client variables and the behavioral treatment of recurrent headache: a meta-analytic review. Unpublished, 1985.

Holroyd, K. A., F. Andrasik, T. Westbrook: Cognitive control of tension headache. Cogn. Ther. Res. 1 (1977) 121–133.

Holroyd, K. A., D. B. Penzien, K. G. Hursey, D. L. Tobin, L. Rogers, J. E. Holm, P. J. Marcille, J. R. Hall, A. G. Chila: Change mechanisms in EMG-biofeedback training: cognitive changes underlying improvements in tension headache. J. consult. clin. Psychol. 52 (1984) 1039–1053.

Holsboer, F., U. v. Bardesleben, A. Gerken, G. K. Stalla, O. A. Müller: Blunted corticotropin and normal cortisol response to human corticotropin-releasing factor in depression. New Engl. J. Med. 311 (1984) 1127.

Holst, D.v.: Psychosocial stress and its pathophysiological effects in tree shrews (tupaia belangeri). In: Schmidt, T., T. Dembroski, G. Blümchen (eds.): Biological and Psychological Factors in Cardiovascular Disease, pp. 476–490. Springer, Berlin-Heidelberg-New York 1986.

Holst, E. V., H. Mittelstaedt: Das Reafferenzprinzip. Naturwissenschaften 37 (1950) 469–476.

Holste, Th., L. Joanni: Untersuchungen über okklusale Schliffacetten und pathologische Befunde im Kausystem jugendlicher Patienten. Dtsch. zahnärztl. Z. 37 (1982) 173.

Holt, R. R.: Imagery: the return of the ostracized. Amer. Psychologist 19 (1964) 254.

Holtermüller, K.-H.: Natürlicher Verlauf der Ulcuskrankheit. In: Blum, A. L., J. R. Siewert (Hrsg.): Ulcus-Therapie, S. 123–137. 2. Aufl. Springer, Heidelberg 1982.

Holtermüller, K.-H.: Was ist gesichert in der konservativen Ulcustherapie? Internist 23 (1982) 653–679.

Holtzman, J. M., C. H. Toewe, J. D. Beck: Specialty preference and attitudes toward the aged. J. Fam. Pract. 9 (1979) 667–672.

Holzgreve, H.: Die Kooperation des Patienten bei der Hochdrucktherapie. Münch. med. Wschr. 122 (1980) 267.

Holzman, P. S., L. R. Proctor, D. W. Hughes: Eye-tracking patterns in schizophrenia. Science 181 (1973) 179–180.

Holzman, P. S., L. R. Proctor, D. L. Levy, N. J. Yasillo, H. Y. Meltzer, S. W. Hurt: Eye-tracking dysfunctions in schizophrenic patients and their relatives. Arch. gen. Psychiat. 31 (1974) 143–151.

Hommer, D. W., P. Skolnick, S. M. Paul: The benzodiazepine/GABA receptor complex and anxiety. In: Meltzer, H. J. (ed.): Psychopharmacology: The Third Generation of Progress. Raven, New York 1987.

Homo-Delarche, F., D. Duval: Glucocorticoid receptors in lymphoid tissue. In: Berczi, I., K. Kovacs (eds.): Hormones and Immunity, pp. 1–19. MTP, Lancaster 1987.

Honigfeld, G.: Non-specific factors in treatment I, II. Dis. nerv. Syst. 25 (1964) 145–156; 225–239.

Honsalek, J.: Das Behandlungsteam des Karzinom- und des Leukämiepatienten. Schweiz. Rundschau Med. 72 (1983) 44–48.

Hontschik, B.: Indikation zur Appendektomie – in der Praxis zu wenig restriktiv? Chir. Praxis 40 (1989) 221–227.

Hoppe, K. D.: Über den Einfluß der Übergangsobjekte und -phänomene auf die Symptombildung. Jahrb. Psychoanal. 3 (1964) 86–115.

Hoppe, K. D. (ed.): Hemispheric specialization. Psychiatric Clinics of North America 11. Philadelphia 1988.

Hoppe, R.: Anorexia nervosa. Eine klärende Diskussion des derzeitigen Wissensstandes. Selbstverlag, Berlin 1982.

Horal, J.: The clinical appearance of low back disorders in the city of Gothenburg, Sweden. Acta orthop. scand. (Suppl.) 118 (1969) 8–108.

Horder, J., E. Horder: Illness in general practice. Practitioner 173 (1954) 177–187.

Horkheimer, M.: Geschichte und Psychologie (1932). In: Dahmer, H. (Hrsg.): Analytische Sozialpsychologie, Bd. 1, S. 158–178. Frankfurt 1980.

Horlick, L., R. Cameron, W. Firor, U. Bhalerao, R. Baltzahn: The effects of education and group discussion in the post myocardial infarction patient. J. psychosom. Res. 28 (1984) 485–492.

Horn, J., C. Herfarth: Das Gastarbeiterulcus. Med. Klin. 73 (1978) 1417–1421.

Horne, J.: Why we sleep. The functions of sleep in humans and other mammals. Oxford Univ. Press, Oxford-New York-Tokyo 1988.

Horney, B.: Neue Wege in der Psychoanalyse. Stuttgart 1951.

Hornstein, O. P.: Die Entwicklung des psychosomatischen Konzepts von der perioralen Dermatitis. In: Bosse, K., P. Hünecke (Hrsg.): Psychodynamik und Soziodynamik bei Hautkranken. Vandenhoeck & Ruprecht, Göttingen 1976.

Hornstein, O. P.: Was kann die Dermatologie von der Psychotherapie erwarten? – Ein Plädoyer. Z. Hautkr. 55 (1980) 913–927.

Hornstein, O. P.: Leitsymptom Zungenbrennen. Dtsch. Derm. 29 (1981) 143–149.

Hornykiewicz, O.: Brain catecholamines in schizophrenia: a good case for noradrenaline. Nature (Lond.) 299 (1982) 484–486.

Horowitz, M. J., C. Marmar, D. Weiss, K. Newitt, R. Rosenbaum: Brief psychotherapy of bereavement reactions. Arch. gen. Psychiat. 41 (1984) 438–448.

Horrobin, D. F., J. P. Mtabaji, R. A. Karmali, M. S. Manku,

B. A. Nassar: Prolactin and mental illness. Postgrad. Med. J. (Suppl. 3) 52 (1976) 79–85.

Hotson, J., T. Pedley: The neurological complications of cardiac transplantation. Brain 99 (1976) 673–694.

Houde, R. W.: Systemic analgesics and related drugs: narcotic analgesics. In: Bonica, J. J., V. Ventafridda (eds.): Advances in Pain Research and Therapy, Vol. 2. Raven, New York 1979.

House, J.: The relationship of intrinsic and extrinsic work motivations to occupational stress and coronary heart disease risk. PhD thesis, Ann Arbor/Michigan 1972.

House, J.: Occupational stress and coronary heart disease: a review and theoretical integration. J. Health soc. Behav. 15 (1974) 12–27.

House, J.: Social relations and health: theory, evidence and implications for public health policy. Referat anläßlich des Kongresses „Zukunftsaufgabe Gesundheitsförderung", Berlin 29. 4. 1989. (Manuskript).

House, J., C. Robbins, H. L. Metzner: The association of social relationships and activities with mortality: prospective evidence from the Tecumseh Community Health Study. Amer. J. Epidemiol. 116 (1982) 123–140.

House, J., K. R. Landis, D. Umberson: Social relationships and health. Science 241 (1988) 540–545.

Houser, B.: An investigation of the correlation between hormonal levels in males and mood, behavior and physical discomfort. Horm. Behav. 12 (1979) 185–197.

Houston, B. K.: Viability of coping strategies, denial and response to stress. J. Pers. 41 (1973) 50–58.

Houston, B. K.: Psychological variables and cardiovascular and neuroendocrine reactivity. In: Matthews, K. A., S. M. Weiss, T. Detre, T. M. Dembroski, B. Falkner, S. B. Manuck, R. B. Williams jr. (eds.): Handbook of Stress, Reactivity, and Cardiovascular Disease, pp. 207–229. Wiley, New York 1986.

Houzel, D.: Les troubles du sommeil de l'enfant et de l'adolescent. In: Lebovici, S., R. Diatkine, M. Soulé (eds.): Traité de psychiatrie de l'enfant et de l'adolescent, Vol. II., pp. 445–465. PUF, Paris 1985.

Howard, J. H., D. A. Cunningham, P. A. Rechnitzer: Health patterns associated with Type A behavior: a managerial population. J. hum. Stress 2 (1976) 24–32.

Howland, E. W., A. W. Siegman: Toward the automated measurement of the Type A behavior pattern. J. behav. Med. 5 (1982) 37–53.

Hoyer, D.: Colitis ulcerosa. Diss., Essen 1983.

Hoyer, H., O. Thalhammer: Geburtshilfliche und sozioökonomische Faktoren in der Genese der Frühgeburt. Geburtsh. Frauenheilk. 28 (1968) 709.

Hoyt, C. J.: The test reliability obtained by analysis of variance. Psychometrica 6 (1941) 153.

Hsu, L.: Outcome of anorexia nervosa. A review of the literature. Arch. gen. Psychiat. 37 (1980) 1041–1046.

Hsu, L., D. Holder: Bulimia nervosa: treatment and short-term outcome. Psychol. Med. 16 (1986) 65–70.

Hsu, L., A. H. Crisp, B. Harding: Outcome of anorexia nervosa. Lancet 1 (1979) 61–65.

Hualla, T., H. Jäger: Die Schwabinger AIDS-Phobiker-Studie. In: Jäger, H. (Hrsg.): AIDS-Phobie. Thieme, Stuttgart 1988.

Huang, S. W., S. M. Plaut, G. A. Taylor, L. E. Wareheim: Effect of stressful stimulation on the incidence of streptozotocin-induced diabetes in mice. Psychosom. Med. 43 (1981) 431–437.

Huber, G.: Klinik und Psychopathologie der organischen Psychosen. In: Kisker, K. P., J.-E. Meyer, C. Müller, E. Strömgren (Hrsg.): Psychiatrie der Gegenwart, Forschung und Praxis, Bd. II, 2. Aufl. Springer, Berlin 1972.

Huber, G.: Psychiatrie. Schattauer, Stuttgart 1981.

Huber, G.: Organische und symptomatische Psychosen. In: Battegay, R., J. Glatzel, W. Pöldinger, U. Rauchfleisch: Handwörterbuch der Psychiatrie. Enke, Stuttgart 1984.

Huck, K., H. Petzold: Death education, Thanatogogik – Modelle und Konzepte. In: Spiegel-Rösing, J., H. Petzold (Hrsg.): Die Begleitung Sterbender, S. 501–576. Junfermann, Paderborn 1984.

Hudson, J. I., H. G. Pope Jr.: Psychopharmakologische Behandlung der Bulimia. In: Fichter, M. M. (Hrsg.): Bulimia nervosa. Enke, Stuttgart 1989.

Hudson, J. I., H. G. Pope, J. M. Jonas, D. Yurgelun-Todd: Phenomenologic relationship of eating disorders to major affective disorder. Psychiat. Res. 9 (1983 a) 345–354.

Hudson, J. I., H. G. Pope, J. Jonas, D. Yurgelun-Todd: Family history study of anorexia nervosa and bulimia. Brit. J. Psychiat. 142 (1983 b) 133–138.

Hudson, M. S., S. M. Wentworth, J. I. Hudson: Self-induced glykosuria. A novel method of purging in bulimia. J. Amer. med. Ass. 249 (1983 c) 2501.

Hudson, J. I., M. S. Hudson, S. M. Wentworth: Bulimia and diabetes. New Engl. J. Med. 309 (1983 d) 431–432.

Hudson, J. I., J. F. Lipinski, F. R. Frankenburg, V. J. Grococinski, D. J. Kupfer: Electroencephalic sleep in mania. Arch. gen. Psychiat. 45 (1988) 267–273.

Hudzinski, L. G.: The significance of muscle discrimination training in the treatment of muscle contraction headache. Headache 24 (1984) 203–210.

Huebschmann, H.: Psyche und Tuberkulose. In: Weizsäcker, V.v. (Hrsg.): Beiträge aus der Allgemeinen Medizin, H. 8. Enke, Stuttgart 1952.

Huebschmann, H.: Über Angst- und Zwangssymptome bei Patienten mit Herzinfarkt. Med. Klinik 59 (1964) 893–894.

Huebschmann, H.: Vom Leiden organisch Herzkranker. Landarzt 42 (1966) 677–682.

Huebschmann, H.: Zur Psychopathologie von Patienten mit Herzinfarkt. Landarzt 43 (1967) 1152–1157.

Huebschmann, H.: Krankheit – ein Körperstreik. Herder, Freiburg 1974.

Huebschmann, H.: Bewußte und unbewußte Erwartungen der Patienten am Beispiel zweier Hepatitiskranker. Niedergelass. Arzt (1977) 1–6.

Hünecke, P., K. Bosse: Entstellung – Erleben und Verarbeitung der äußeren Erscheinung. In: Whitlock, F. A.: Psychophysiologische Aspekte bei Hautkrankheiten. Perimed, Erlangen 1980.

Hünecke, P., K. Bosse: Kratzen. Methodischer Zugang und Objektivierung von symptomverstärkenden Situationen bei endogenen Ekzematikern. Münch. med. Wschr. 123 (1981) 992–995.

Hünecke, P., K. Bosse: Dysmorphophobie als casus pro diagnosi. Z. Hautkr. 60 (1985).

Hünecke, P., K. Bosse: Über die Persönlichkeitsstruktur des Psoriatikers. Z. psychosom. Med. 31 (1985) 105–117.

Huenemann, R., L. Shapiro, M. Hampton, B. Mitchell: A longitudinal study of gross body composition and body conformation and their association with food and activity in a teenage population. Amer. J. clin. Nutr. 18 (1966) 325–338.

Hürny, C.: Critical review of quality of life: psychosocial aspects of adjuvant therapy in breast cancer. In: Senn, H. J., A. Goldhirsch, B. Osterwalder (eds.): Recent Results in Cancer Research, Vol. 115: Adjuvant Therapy of Primary Breast Cancer, pp. 279–282. Springer, Berlin-Heidelberg-New York 1989.

Hürny, C., J. C. Holland: Letter to the editor. Gen. Hosp. Psychiat. 5 (1983) 301–303.

Hürny, C., R. Adler: Psychoonkologische Forschung. In: Meerwein, F. (Hrsg.): Einführung in die Psycho-Onkologie, 3. Aufl. Huber, Bern-Stuttgart-Wien 1985.

Hürny, C., J. Bernhard: Coping and survival in patients with primary breast cancer: a critical analysis of current research strategies and proposal of a new approach integrating biomedical, psychological and social variables. In: Senn, H. J., A. Goldhirsch, B. Osterwalder (eds.): Recent Results in Cancer Research, Vol. 115: Adjuvant Therapy of Primary Breast Cancer, pp. 255–271. Springer, Berlin-Heidelberg-New York 1989.

Hughes, H., B. W. Brown, G. F. Lawlis, E. Fulton: Treatment of acne vulgaris by biofeedback relaxation and cognitive imagery. J. psychosom. Res. 27 (1983) 185–191.

Hughes, J., T. W. Smith, H. W. Kosterlitz, C. A. Fothergill, B. A. Morgan, H. R. Morris: Identification of two related pentapeptides from the brain with potent agonistic activity. Nature (Lond.) 258 (1975) 577–579.

Hughes, P. L., L. A. Wells, C. J. Cunningham, D. M. Ilstrup: Treating bulimia with desipramine. Arch. gen. Psychiat. 43 (1986) 182–186.

Huland, H., R. Busch, H. Klosterhalfen: Über die Ätiologie von Harnwegsinfekten. Dtsch. med. Wschr. 109 (1984) 1370.

Hull, C. L.: A behavior system. Yale Univ. Press, New Haven 1952.

Hulse, S. S., H. Fowler, W. K. Honig (eds.): Cognitive processes in animal behavior. Erlbaum, Hillsdale 1978.

Hummel, M.: Das Antonsche Phänomen bei homonymer Hemianopsie. Med. Diss., Berlin 1981.

Hunger, J., B. Leplow, J. Keim: Zur Struktur des hirnorganischen Psychosyndroms. Nervenarzt 58 (1987) 603–609.

Hunt, J.: Intrinsic motivation: information and circumstance. In: Schroder, H. M., P. Suedfeld (eds.): Personality theory and information processing. Ronald, New York 1971.

Hurst, P. S., J. H. Lacey, A. H. Crisp: Teeth, vomiting and diet: a study of the dental characteristics of seventeen anorexia nervosa patients. Postgrad. med. J. 53 (1977) 298–305.

Husband, A. J., M. G. King, R. Brown: Behaviorally conditioned modification of T cell subset ratios in rats. Immunol. Lett. 14 (1986) 91–94.

Huse-Kleinstoll, G., B. Dahme, B. Flemming, A. Haag, H. J. Meffert, M. J. Polonius, G. Rodewald, H. Speidel: Einige somatische und psychologische Prädiktoren bei psychopathologischen Auffälligkeiten nach Herzoperationen. Thoraxchir. 24 (1976) 386–389.

Husslein, H.: Psychosomatik in Gynäkologie und Geburtshilfe – Probleme und Schwierigkeiten. Klin. Wschr. 89 (Suppl.) (1977) 10–13.

Huth, K., Ch. Bräuning (Hrsg.): Pflanzenfasern – Neue Wege in der Stoffwechsel-Therapie. Karger, Basel 1983.

Huygen, F. J. A., A. J. A. Smits: Family therapy, family somatics and family medicine. Family Systems Medicine 1 (1983) 23.

Ibrahim, M. A., J. G. Feldman, H. A. Sultz: Management after myocardial infarction: a controlled trial of the effect of group therapy. Int. J. Psychiat. Med. 5 (1974) 253–269.

Ibrahim, M. A., J. G. Feldman, H. A. Sultz, M. G. Staiman, L. Z. Young, D. Dean: The management of myocardial infarction: a controlled trial of the effect of group psychotherapy. Int. J. Psychiat. Med. 5 (1974) 523.

Ignelzy, R. J., R. A. Sternbach, G. Timmermans: The pain ward follow-up analyses. Pain 3 (1977) 277–280.

Ikemi, Y., T. Nagakawa, M. Sutiga: Psychosomatic considerations on cancer patients who have made a narrow escape from death. Dynam. Psychiat. 8 (1975) 77–93.

Illek, S.: ... auf die Blase geschlagen? Empirische Untersuchung zum Zusammenhang zwischen Beziehungserleben und rezidivierenden Harnweginfekten bei Frauen. Unveröffentlichte Diplomarbeit. Psychologisches Institut der FU, Berlin 1984.

Illig, H.: Symptomatologie der funktionellen Syndrome des Gastrointestinaltraktes. Münch. Med. Wschr. 43 (1961) 2082–2086.

Imboden, J. B., A. Canter, L. Cluff: Separation experiences and health records in a group of normal adults. Psychosom. Med. 25 (1963) 433–440.

Imhof, A. E.: Die gewonnenen Jahre. Beck, München 1981.

Imhof, A. E.: Die Lebenszeit. Beck, München 1988.

Imperato-McGinley, J., L. Guerrero, T. Gautier, R. E. Peterson: Steroid 5 α-reductase deficiency in man: an inherited form of male pseudohermaphroditism. Science 186 (1974) 1213–1215.

Imperato-McGinley, J., R. E. Peterson, T. Gautier, E. Sturla: Androgens and the evolution of male gender identity among male pseudohermaphrodites with 5 α-reductase deficiency. New Engl. J. Med. 300 (1979) 1233–1237.

Imura, H., T. Yoshimi, K. Ikekubo: Growth hormone secretion in a patient with deprivation dwarfism. Endocr. Jap. 15 (1971) 301.

Ince, L. P., M. S. Leon, D. Christidis: EMG biofeedback for handwriting disabilities: a critical examination of the literature. J. behav. Ther. exp. Psychiat. 17 (1986) 95–100.

Indefrey, P., F. Alexander: Der Mensch als komplizierte Maschine. In: Zepf, S. (Hrsg.): Tatort Körper – Spurensicherung, S. 15–26. Springer, Heidelberg 1986.

Ingram, P. W., G. Evans: Right iliac fossa pain in young women. Brit. med. J. 2 (1965) 149–151.

Inman, W. S.: Can emotional conflict induce disseminated sclerosis? J. med. Psychol. 39 (1948) 135–154.

Insel, Th.R., J. Zohar: Psychopharmacologic approaches of obsessive-compulsive disorder. In: Meltzer, H.J. (ed.): Psychopharmacology: The Third Generation of Progress. Raven, New York 1987.

Ireland, C. E., P. H. Wilson, J. P. Tonkin, S. Platt-Hepworth: An evaluation of relaxation training in the treatment of tinnitus. Behav. Res. Ther. 23 (1985) 423–431.

Irvine, W. J.: Immunological aspects of diabetes. In: Brodoff, B. N., S. J. Bleicher (eds.): Diabetes mellitus and Obesity, pp. 355–363. Williams & Wilkins, Baltimore-London 1982.

Irwin, M., M. Daniels, S. C. Risch, E. Bloom, H. Weiner: Plasma cortisol and natural killer cell activity during bereavement. Biol Psychiat. 24 (1988) 173–178.

Irwin, M., M. Daniels, T. L. Smith, E. Bloom, H. Weiner: Impaired natural killer cell activity during bereavement. Brain, Behavior, and Immunity 1 (1987) 98–104.

Israel, S. L.: Premenstrual tension. J. Amer. med. Ass. 110 (1938) 1721–1723.

Iversen, G.: Erfahrungen mit dem autogenen Training in der Gruppenarbeit. Dtsch. Ärztebl. 66 (1969) 1924–1929.

Iversen, S. D.: Neuropeptides. Do they integrate body and brain? Nature 291 (1981) 454.

Izard, C., S. Buechler: Aspects of consciousness and personality in terms of differential emotion theory. In: Plutchik, R., H. Kellerman (eds.): Emotion – Theory, Research and Experience, Vol.I: Theories of Emotion. Acad. Press, New York 1980.

Jackson, C. W.: Clinical sensory deprivation. A review of hospitalized eye-surgery patients. In: Zubek, J. P. (ed.): Sensory Deprivation, pp. 332–373. Appleton, New York 1969.

Jackson, D. D., J. Yalom: Family research on the problem of ulcerative colitis. Arch. gen. Psychiat. 15 (1966) 410. Deutsch: Familiale Interaktionsmuster und Colitis ulcerosa. In: Brede, K. (Hrsg.): Einführung in die psychosomatische Medizin, S.242. Fischer, Frankfurt 1974.

Jackson, M.: Psychopathology and pseudo-normality in ulcerative colitis. Psychother. and Psychosom. 28 (1977) 179–186.

Jacob, D. L., H. Robinson, A. T. Masi: A controlled home interview study of factors associated with early rheumatoid arthritis. Amer. J. publ. Health 62 (1972) 1532–1537.

Jacobi, E., G. Krüskemper: Thrombozyten-Adhäsivität und thrombozytäre CAMD unter Streß. Med. Welt (Stuttgart) 28 (1977) 888.

Jacobs, S. C.: Psychoendocrine aspects of bereavement. In: Zisook, S. (ed.): Biopsychosocial Aspects of Bereavement, Chapt. 9. American Psychiatric Press, Washington D. C. 1987.

Jacobs, S. C., A. Ostfeld: An epidemiological review of the mortality of bereavement. Psychosom. Med. 39 (1977) 344–357.

Jacobson, A. M., L. I. Rand, S. T. Hauser: Psychologic stress and glycemic control: a comparison of patients with and without proliferative diabetic retinopathy. Psychosom. Med. 47 (1985) 372–381.

Jacobson, E.: Progressive relaxation. Univ. of Chicago Press, Chicago 1938.

Jacobson, E.: Denial and repression. J. Amer. psychoanal. Ass. 5 (1957) 61–92.

Jacobson, E.: The self and the object world. New York 1964.

Jacobson, E.: Depression. Int. Univ. Press, New York 1971.

Jacobson, R.: Language in relation to other communication-systems. In: Collected Writings II. Paris 1971.

Jaeger, A., J. Wüst, R. Lüthy, H. J. Nüesch, J. Munzinger: Milder Spontanverlauf einer pseudomembranösen Kolitis. Schweiz. med. Wschr. 111 (1981) 350–355.

Jäger, A. O.: Dimensionen der Intelligenz, 3. Aufl. Hogrefe, Göttingen 1973.

Jäger, H. (Hrsg.): AIDS-Phobie. Thieme, Stuttgart 1988.

Jährig, C., U. Koch: Die Arzt-Patient-Interaktion in der internistischen Visite eines Akutkrankenhauses – Eine empirische Untersuchung. In: Köhle, K., H.-H. Raspe (Hrsg.): Das Gespräch während der ärztlichen Visite. Urban & Schwarzenberg, München 1982.

Jaensch, P. A., F. Schäfer: Das Schielen und seine Behandlung. Bücherei des Augenarztes, Heft 4. Enke, Stuttgart 1938.

Järnerot, G., K. Lantorp: Antibodies to EB virus in cases of Crohn's disease. New Engl. J. Med. 286 (1972) 1215–1216.

Järnerot, G., I. Järnmark, K. Nilsson: Consumption of refined sugar by patients with Crohn's disease, ulcerative colitis, or irritable bowel syndrome. Scand. J. Gastroent. 18 (1983) 999–1002.

Jäschke, B., W. Dmoch: Der psychische Befund bei Frauen mit verschiedenen Formen der EPH-Gestose. Pers. Mitteilung 1984.

Jaffe, B. F.: Clinical studies in sudden deafness. Fortschr. Hals-Nasen-Ohrenheilk. 20 (1973) 220.

Jahoda, M., P. F. Lazarsfeld, H. Zeisel: Die Arbeitslosen von Marienthal. Ein soziographischer Versuch (1933). Suhrkamp, Frankfurt 1975.

Jakubik, A.: Mental disorders in patients after cardiosurgical operations. Acta med. pol. 13 (1972) 103–111.

Jalan, K. N., R. J. Prescott, W. Sircus, W. I. Card, J. P. A. McManus, C. W. A. Falconer, W. P. Small, A. N. Smith, J. Bruce: An experience of ulcerative colitis III. Long term outcome. Gastroenterol. 59 (1970) 598–609.

James, J. M., R. M. Pearson, D. N. W. Griffith, P. Newbury: Effect of oxprenolol on stage-fright in musicians. Lancet II (1977) 952–954.

James, S. P., T. A. Wehr, D. A. Sack, B. L. Parry, N. E. Rosenthal: Treatment of seasonal affective disorder with light in the evening. Brit. J. Psychiat. 147 (1985) 424–428.

James, W.: What is emotion? Mind 9 (1884) 188–205.

James, W.: Psychologie. Leipzig 1909.

Jamison, R. N., T. G. Burish, K. A. Wallston: Psychogenic factors in predicting survival of breast cancer patients. J. clin. Oncol. 5 (1987) 768–772.

Jamison, R. N., D. A. Matt, W. C. V. Parris: Effects of time-limited vs. unlimited compensation on pain behavior and treatment outcome in low back pain patients. J. psychosom. Res. 32 (1988) 277–283.

Janis, I. L.: Psychological stress. Psychoanalytic and behavioral studies of surgical patients. Wiley, New York 1958.

Janke, W.: Psychophysiologische Grundlagen des Verhaltens. In: Kerekjarto, M v (Hrsg.): Medizinische Psychologie. Springer, Berlin 1974.

Janke, W., G. Erdmann, W. Boucsein: Der Streßverarbeitungsfragebogen (SVF). Ärztl. Prax. 30 (1978) 1208–1210.

Janke, W., G. Erdmann, W. Boucsein: Streßverarbeitungsfragebogen (SVF). Hogrefe, Göttingen 1985.

Jankovic, B. D., B. M. Marcovic, N. H. Spector (eds.): Neuroimmune interactions: Proceedings of the Second International Workshop on Neuroimmunomodulation. Ann. N. Y. Acad. Sci. 496 (1987) whole volume.

Jankowski, R. F., E. Robinson, S. Paterson, W. C. Dick: Patient's expectations of health care: fulfilled and unfulfilled. Practitioner 224 (1980) 351–353.

Janowsky, D. S., S. C. Risch, D. Parker, L. Huey, L. Judd: Increased vulnerability to cholinergic stimulation in affective disorder patients. Psychopharmacol. Bull. 16 (1980) 29–31.

Jansen, B., H. Radebold: Beschwerden einer repräsentativen Stichprobe über 60jähriger in einer ländlichen Region. In: Speidel, H., B. Strauss (Hrsg.): Zukunftsaufgaben der psychosomatischen Medizin. Springer, Berlin-Heidelberg 1989.

Jansen, G., J. Schulze: Beispiele für Schlafstörungen durch Geräusche. Klin. Wschr. 3 (1964) 3.

Jansen, R., J. Münstermann, K. Preiser: Schichtpläne bei unterschiedlichen Wochenarbeitszeiten. Humanisierung des Arbeitslebens (hrsg. vom Bundesminister für Arbeit und Sozialordnung), Bd. 37, Bonn 1979.

Jansen, R., M. Möllenstedt, K. Preiser: Schichtarbeit im öffentlichen Dienst. Humanisierung des Arbeitslebens (hrsg. vom Bundesminister für Arbeit und Sozialordnung), Bd. 41, Bonn 1980.

Janssen, P. L.: Psychoanalytische Therapie in der Klinik. Klett-Cotta, Stuttgart 1987.

Janssen, P. L., R. Kukahn, K.-H. Spieler, L. Weißbach: Psychosomatische Untersuchungen zur chronischen Prostatitis. Z. psychosom. Med. 29 (1983) 253.

Jantsch, E.: Die Selbstorganisation des Universums. Vom Urknall zum menschlichen Geist. Deutscher Taschenbuch Verlag, München 1982.

Jantschek, G., I. Jantschek: Familientherapie. In: Feiereis, H. (Hrsg.): Diagnostik und Therapie der Magersucht und Bulimie. Marseille, München 1989.

Jantschek, G., I. Jantschek, J.v. Wietersheim: Einzel- und Familientherapie bei Patientinnen mit Eßstörungen. In: Lamprecht, F. (Hrsg.): Spezialisierung und Integration in Psychosomatik und Psychotherapie. Springer, Berlin 1987.

Jantschek, G., I. Jantschek, J.v. Wietersheim, C. Drewes, U. Drossard, F. Kröger, E. Petzold, S. Becker: Familienuntersuchungen bei chronisch entzündlichen Darmkrankheiten. In: Speidel, H., B. Strauß (Hrsg.): Zukunftsaufgaben der psychosomatischen Medizin. Springer, Berlin 1989.

Janus, L.: Persönlichkeitsstruktur und Psychodynamik bei dermatologischen Artefakten. Z. psychosom. Med. Psychoanal. 18 (1972) 21–28.

Janus, L.: Psychoanalytisch-psychophysiologische Untersuchungen bei Patienten mit funktionellem Cervikalsyndrom. Z. psychosom. Med. Psychoanal. 24 (1978) 101–115.

Janz, D.: Wut und Anfallsgeschehen. Psyche 2 (1948/49) 97–120.

Janz, D.: „Aufwach"-Epilepsien (als Ausdruck einer den „Nacht"- oder „Schlaf"-Epilepsien gegenüberzustellenden Verlaufsform epileptischer Erkrankungen). Arch. Psychiat. Nervenkr. 191 (1953) 73–98.

Janz, D.: Soziale Aspekte der Epilepsie. Psychiat. Neurol. Neurochir. 66 (1963) 240–248.

Janz, D.: Leitbilder der Epilepsie bei Hippokrates und Paracelsus. Jb. Psychol., Psychother., med. Anthropol. 14 (1966) 2–16.

Janz, D.: Über das Suchtmoment in der Epilepsie. Nervenarzt 39 (1968a) 350–355.

Janz, D.: Zur Abgrenzung verschiedener Psychosyndrome bei Epilepsie. Hippokrates 39 (1968b) 402–407.

Janz, D.: Die Epilepsien. Spezielle Pathologie und Therapie. Thieme, Stuttgart 1969.

Janz, D.: Epidemiologie und Klassifikation von Epilepsien und epileptischen Anfällen. Akt. Neurol. 6 (1979) 189–196.

Janz, D.: Epilepsien. In: Hopf, H. C., K. Poeck, H. Schliack (Hrsg.): Neurologie in Praxis und Klinik, Bd. II/6. Thieme, Stuttgart 1981.

Janz, D., W. Christian: Impulsiv-Petit Mal. Dtsch. Z. Nervenheilk. 177 (1957) 346–386.

Jarrett, R. J., P. McCartney, H. Keen: The Bedford survey: ten year mortality rates in newly diagnosed diabetics, borderline diabetics and normoglycaemic controls and risk indices for coronary heart disease in borderline diabetics. Diabetologia 22 (1982) 79–84.

Jasnoski, M. L., J. Kugler: Relaxation, imagery, and neuroimmunomodulation. Ann. N. Y. Acad. Sci. 496 (1987) 722–730.

Jaspers, K.: Allgemeine Psychopathologie, 8. Aufl. Springer, Berlin-Heidelberg-New York 1965.

Jeavons, P. M., G. F. A. Harding: Photosensitive epilepsy. Heinemann, London 1975.

Jeffcoate, W. J., J. T. Silverstone, C. R. W. Edwards, G. M. Besser: Psychiatric manifestations of Cushing's syndrome: response to the lowering of plasma cortisol. Quart. J. Med. 191 (1979) 465.

Jefferson, G. E.: The social complex. J. Rehab. 32 (1966) 59–60.

Jeffrey, J. E., H. J. Burton, A. P. Heidenheim, R. M. Lindsay: A comparison of home training and problems encountered with initial home dialysis. Hemodialysis vs. CAPD. AANNT J. 9 (1982) 56–62.

Jelen, S., E. Kolb, G. Temper: Intensivbehandlung im Erleben des Patienten. In: Schara, J. (Hrsg.): Humane Intensivmedizin, S. 134–143. Perimed, Erlangen 1979.

Jelliffe, S. E.: Die Parkinsonsche Körperhaltung. Einige Betrachtungen über unbewußte Feindseligkeit. Int. Z. Psychoanal. 19 (1933) 485–498.

Jemmott, J. B., S. E. Locke: Psychosocial factors, immunologic mediation, and human susceptibility to infectious diseases: how much do we now? Psychol. Bull. 95 (1984) 78–108.

Jemmott, J. B., K. Magloire: Academic stress, social support, and secretory immunoglobulin A. J. Pers. soc. Psychol. 55 (1988) 803–810.

Jemmott, J. B., J. Z. Borysenko, M. Borysenko, D. C. McClelland, R. Chapman, D. Meyer, H. Benson: Academic stress, power motivation, and decrease in salivatory secretory immunoglobulin A secretion rate. Lancet I (1983) 1400–1402.

Jenkins, C. D.: Psychologic and social precursors of coronary disease. New Engl. J. Med. 284 (1971) 244–255; 307–317.

Jenkins, C. D.: Recent evidence supporting psychologic and social risk factors for coronary disease. New Engl. J. Med. 294 (1976) 987–994; 1033–1038.

Jenkins, C. D.: Epidemiological studies of the psychosomatic aspects of coronary heart disease. Advanc. psychosom. Med. 9 (1977) 1–19.

Jenkins, C. D.: Behavioral risk factors in coronary artery disease. Ann. Rev. Med. 29 (1978) 543–562.

Jenkins, C. D.: Kritische Betrachtungen des Zusammenhangs zwischen Typ-A-Verhalten und verschiedenen Manifestationen koronarer Herzkrankheiten. In: Dembroski, T. M., M. J. Halhuber (Hrsg.): Psychosozialer „Stress" und koronare Herzkrankheit: 3. Verhalten und koronare Herzkrankheit, S. 83–111. Springer, Berlin-Heidelberg-New York 1981.

Jenkins, C. D.: Psychosocial risk factors for coronary heart disease. Acta med. scand. (Suppl.) 660 (1982) 123–136.

Jenkins, C. D.: Social environment and cancer mortality in men. New Engl. J. Med. 308 (1983) 395–398.

Jenkins, C. D., S. J. Zysanski: Behavioral risk factors and coronary heart disease. Psychother. and Psychosom. 34 (1980) 149.

Jenkins, C. D., R. H. Rosenman, M. Friedman: Replicability of rating the coronary-prone behavior pattern. Brit. J. prev. soc. Med. 22 (1968) 16–22.

Jenkins, C. D., R. H. Rosenman, S. J. Zysanski: Prediction of clinical coronary heart disease by a test for the coronary-prone behavior pattern. New Engl. J. Med. 290 (1974) 1271–1275.

Jenkins, C. D., R. H. Rosenman, S. J. Zysanski: Coronary-prone behavior: one pattern or several? Psychosom. Med. 40 (1978) 25–43.

Jenkins, C. D., R. H. Rosenman, S. J. Zysanski: Jenkins activity survey. Psychol. Corp., New York 1979.

Jenkins, C. D., B. A. Slanton, M. D. Klein, I. A. Raragean, D. E. Harken: Correlates of angina pectoris among men awaiting coronary by-pass surgery. Psychosom. Med. 45 (1983) 141–153.

Jenkins, P. E., R. A. Chadwick, J. A. Nevin: Classically conditioned enhancement of antibody production. Bull. Psychonom. Soc. 21 (1983) 485–487.

Jennings, J. R., C. C. Wood: Cardiac cycle time effects on performance, phasic cardiac responses, and their intercorrelation in choice reaction time. Psychophysiol. 14 (1977) 297–307.

Jennings, J. R., M. W. Van der Molen, C. Terezis: Primary bradycardia and vagal inhibition as two manifestations of the timing of vagal influence on the heart beat. J. Psychophysiol. 4 (1987) 361–374.

Jenny, S., P. Deyhle: Ulcus duodeni. Endoskopiebefund und psychosozialer Status. Z. Gastroenterol. 14 (1976) 728.

Jensen, M. R.: Psychobiological factors predicting the course of breast cancer. J. Pers. 55 (1987) 317–342.

Jenzer, H. R.: Das Mitralklappenprolapssyndrom: Merkmal oder Krankheit? Schweiz. Rundsch. Med. 70 (1981) 1572–1582.

Jessup, B. A., R. W. J. Neufeld, H. Merskey: Biofeedback therapy for headache and other pain: An evaluative review. Pain 7 (1979) 225–270.

Jeste, D. V., J. B. Lohr, F. K. Goodwin: Neuroanatomical studies of major affective illness: a review and suggestions for further research. Brit. J. Psychiat. 153 (1988) 444–459.

Jette, A. M.: Improving patient co-operation with arthritis treatment regimens. Arthr. and Rheum. 25 (1982) 447–453.

Jimerson, D. C., T. R. Insel, V. I. Reus, I. J. Kopin: Increased plasma MHPG related to dexamethasone-resistant depressed patients. Arch. gen. Psychiat. 40 (1983) 173–176.

Joasso, A., J. M. McKenzie: Stress and the immune response in rats. Int. Arch. Allerg. Appl. Immunol. 50 (1976) 659.

Joel, E., S. M. Wieder: Factors involved in adaptation to the stress of hemodialysis. Smith Coll. Stud. soc. Work 43 (1973) 193–205.

Jörgens, H., K. Dieckhöfer: Internistische und psychopathologische Aspekte zur Colitis ulcerosa (mit kasuistischen Beiträgen). Z. psychosom. Med. 18 (1972) 305–323.

Jörgens, V., M. Pavlovic, P. Greßnich, M. Berger, H. Zimmermann: Katamnestische Untersuchungen an 138 Patientinnen mit Anorexia nervosa. Verh. Dtsch. Ges. Inn. Med. 86 (1980).

Jørgensen, Å., T. W. Teasdale, J. Parnas, F. Schulsinger, H. Schulsinger, S. A. Mednick: The Copenhagen high-risk project. The diagnosis of maternal schizophrenia and its relation to offspring diagnosis. Brit. J. Psychiat. 151 (1987) 753–757.

Joesoet, M. R., S. F. Wetterhalt, F. De Stefano, N. E. Stroup, A. Tronck: The association of peripheral arterial disease with hostility in a young healthy veteran population. Psychosom. Med. 51 (1989) 285–289.

Joffe, W. G., J. Sandler: Kommentare zur psychoanalytischen Anpassungspsychologie mit besonderem Bezug zur Rolle der Affekte und der Repräsentanzenwelt. Psyche 21 (1967) 728–744.

Johannsen, A.: Persönlichkeit und Körperschema von Patienten mit Störungen im gastrointestinalen und kardiovaskulären Bereich. Diagnose-Instrumentarium und Kurzzeit-Therapie. Dissertation, Hamburg 1984.

Johannson, B. W., L. Kaij, S. Kullander: On some late effects of bilateral oophorectomy in the age range 15–30 years. Acta obstet. gynec. scand. 54 (1975) 449.

Johansson, F.: Differences in serum cortisol concentrations in organic and psychogenic chronic pain syndromes. J. psychosom. Res. 26 (1982) 351–358.

Johnen, R.: Funktionelle Entspannung im stationären Setting. Erste Ergebnisse einer explorativen Vorstudie. In: Lamprecht, F. (Hrsg.): Spezialisierung und Integration in Psychosomatik und Psychotherapie, S. 167. Springer, Berlin-Heidelberg-New York 1987.

Johnen, R., H. Müller-Braunschweig: Psychoanalytische Aspekte der Funktionellen Entspannung. In: Fuchs, M. (Hrsg.): Funktionelle Entspannung, 4. Aufl. Hippokrates, Stuttgart 1989.

Johnsen, B. H., K. Hugdahl: Preparedness and electrodermal fear-conditioning: an old problem revisited. Psychophysiol. 25 (1988) 457.

Johnson, A., L. B. Shapiro, F. Alexander: Preliminary report on a psychosomatic study of rheumatoid arthritis. Psychosom. Med. 9 (1947) 230–295.

Johnson, C. L., D. J. Berndt: Preliminary investigation of bulimia and life adjustment. Amer. J. Psychiat. 140 (1983) 774–777.

Johnson, C. L., S. Q. Love: Bulimia: multivariate predictors of life impairment. J. psychiat. Res. 19 (1985) 343–347.

Johnson, C., C. Lewis, S. Love, L. Lewis, M. Stuckey: Incidence and correlates of bulimic behavior in a female high school population. J. Youth Adolesc. 13 (1984) 15–26.

Johnson, C. L., M. K. Stuckey, L. D. Lewis, D. M.

Schwartz: A survey of 509 cases of self-reported bulimia. In: Darby, P. L., P. E. Garfinkel, D. M. Garner, D. V. Coscina (eds.): Anorexia nervosa. Recent Developments in Research, pp. 159–171. Liss, New York 1983.

Johnson, G. F., G. Hunt, K. Kerr, I. Caterson: Dexamethasone suppression test (DST) and plasma dexamethasone. J. psychiat. Res. 13 (1984) 305–313.

Johnson, Grant: ...und wenn er Witze macht, sind es nicht die seinen. Dialog mit dem Computer. Kursbuch 75 (1984) 38–56.

Johnson, J. H., I. G. Sarason: Moderator variables in life stress research. In: Sarason, I. G., C. D. Spielberger (eds.): Stress and Anxiety, Vol. 6. Wiley, New York 1979.

Johnson, K. O., W. Work, G. Maccoy: Functional deafness. Ann. Otol. 65 (1956) 154.

Johnson, L. C., A. Lubin: On planning psychophysiological experiments: design, measurement, and analysis. In: Greenfield, N. S., R. A. Sternbach (eds.): Handbook of Psychophysiology. Holt, Rinehart and Winston, New York 1972.

Johnson, S. B.: Psychosocial factors in juvenile diabetes: a review. J. Behav. Med. 3 (1980) 95–116.

Johnson, T., J. F. Lavender, E. Hultin, A. F. Rasmussen: The influence of avoidance-learning stress on resistance to Coxsackie B virus in mice. J. Immunol. 91 (1963) 569.

Johnson, W., K. Hoffman: Medical students' attitudes towards patient's physical, psychological and health state characteristics. AAMC-RIME Conference 1980, 269.

Johnston, D. W.: How does relaxation training reduce blood pressure in primary hypertension? In: Schmidt, T., T. Dembroski, G. Blümchen (eds.): Biobehavioral Factors in Coronary Heart Disease. Springer, Berlin-Heidelberg-New York 1984.

Jokl, R. H.: Zur Psychogenese des Schreibkrampfes. Int. Z. Psychoanal. 8 (1922) 168–190.

Jonasch, K., A. Sellschopp: Der Prozeß der Aufklärung als psychosozialer Streß für Patienten und das Behandlungsteam einer chirurgisch-onkologischen Station. Verh. Dt. Ges. Inn. Med. 90 (1984).

Jones, D. J., M. M. Fox, H. M. Babigan, H. E. Hutton: Epidemiology of anorexia nervosa in Monroe County, New York: 1960–1976. Psychosom. Med. 42 (1980) 551–558.

Jones, E.: Papers of psychoanalysis. London 1936.

Jones, E.: Report of the clinic work 1926–1936. London Clinic of Psychoanalysis, decenniae report. London 1936.

Jones, E.: Sigmund Freud, Bd.III. Bern 1962.

Jones, M.: Social psychiatry. Tavistock, London 1952.

Jones, M.: Prinzipien der therapeutischen Gemeinschaft. Huber, Bern-Stuttgart 1976a.

Jones, M.: Maturation of the therapeutic community. Human Sciences Press, New York 1976b.

Jones, R. D.: Delirium. In: Pasnau, R. O. (ed.): Consultation-Liaison Psychiatry, p. 219–226. Grune & Stratton, New York 1975.

Jonez, H. D.: Psychotherapy in multiple sclerosis. Ann. Allergy 9 (1951) 653–659.

Joost, J., G. Egger, G. Hohlbrugger, H. Marberger: Epidemiologie des Nierensteinleidens in Tirol. Oest. Ärztetg. 35 (1980) 1016.

Joraschky, P.: Das Körperschema und das Körper-Selbst als Regulationsprinzipien der Organismus-Umwelt-Interaktion. Minerva, München 1983.

Joraschky, P.: Das Körperschema und das Körper-Selbst. In: Brähler, E. (Hrsg.): Körpererleben, S. 34. Springer, Berlin-Heidelberg-New York-Tokyo 1986.

Joraschky, P., K. Köhle: Maladaptation und Krankheitsmanifestation. Das Streßkonzept in der psychosomatischen Medizin. In: Uexküll, Th. v. (Hrsg.): Lehrbuch der Psy-

chosomatischen Medizin. Urban & Schwarzenberg, München-Wien-Baltimore 1981.

Jordan, J., M. Schmidt: Rehabilitationspsychologische Aspekte bei Erkrankungen des rheumatischen Formenkreises. In: Koch, U., G. Lucius-Hoene, R. Stegie (Hrsg.): Handbuch der Rehabilitationspsychologie, S. 455–478. Springer, Heidelberg 1988.

Jordan, J., G. Overbeck, W. Joos: Psychische Bewältigungsmechanismen bei offenen Herzoperationen in Abhängigkeit von der Persönlichkeitsstruktur des Patienten. Z. psychosom. Med. 29 (1983) 380–403.

Jordan, S. M., E. D. Kiefer: The irritable colon. J. Amer. med. Ass. 93 (1929) 592–595.

Jores, A.: Vom kranken Menschen. Thieme, Stuttgart 1960.

Jores, A.: Zivilisationskrankheiten Französisch-Westafrikas. Med. Klinik 48 (1960) 2145.

Jores, A.: Der Tod des Menschen in psychologischer Sicht. In: Sborowith, A. (Hrsg.): Der leidende Mensch. Personale Psychotherapie in anthropologischer Sicht. Ein Sammelbuch, S. 417–428. Wiss. Buchgesellschaft, Darmstadt 1960.

Jores, A.: Ulcus ventriculi. In: Linneweh, F. (Hrsg.): Die Prognose der Neurosen. Springer, Berlin 1960.

Jores, A.: Die Medizin in der Krise unserer Zeit. 3. Aufl. Bern 1961.

Jores, A.: Der Mensch und seine Krankheit, 2. Aufl. Stuttgart 1962.

Jores, A.: Der Kranke mit psychovegetativen Störungen. Vandenhoeck & Ruprecht, Göttingen 1973.

Jores, A.: Praktische Psychosomatik. 2. Aufl. Huber, Bern-Stuttgart-Wien 1981.

Jores, A., H. Puchta: Der Pensionierungstod. Med. Klinik 54 (1959) 1158–1164.

Jores, A., M. v. Kerekjarto: Der Asthmatiker. Ätiologie und Therapie des Asthma bronchiale in psychologischer Sicht. Huber, Bern 1967.

Jorswieck, E., J. Katwan: Neurotische Symptome. Eine Statistik über Art und Auftreten in den Jahren 1947, 1956 und 1965. Z. psychosom. Med. 13 (1967) 12.

Joseph, F.: Transference and countertransference in the case of a dying patient. Psychoanalysis and the Psychoanalytic Review 49 (1962) 21–34.

Josten, J.: Emotional adaptation of cardiac patients. Scand. J. rehab. Med. 2 (1970) 49–52.

Jouve, A., M. Dongier, M. Delage, R. Mayaud: Recherches psychosomatiques en cardiologie. Presse Médicale 69 (1961) 2545–2548.

Jovanovic, U. J.: Methodik und Theorie der Hypnose. Fischer, Stuttgart-New York 1988.

Juan, I., R. Paiva, H. B. Haley: High and low levels of dogmatism in relation to personal characteristics of medical students: a follow-up study. AAMC-RIME Conference 1981, 31.

Jürgensen, O.: Gynäkologische Endokrinologie. In: Die Psychologie des 20. Jahrhunderts, Bd. 9. Kindler, Zürich 1979.

Jürgensen, O.: Schwangerschaftskonfliktberatung. Abtreibung als wiederholter Trennungsversuch. Sexualmed. 12 (1983) 15, 26.

Jürgensen, O., W. Klein, H. G. Siedentopf: Psychologie der intrauterinen Kontrazeption. Sexualmed. 8 (1979) 49.

Juli, D., H. Brenner: Ein verhaltenstherapeutisch orientiertes Modell für die Gruppenarbeit mit Herzinfarktpatienten. Herz/Kreisl. 9 (1977) 661–669.

Julius, S.: The psychophysiology of borderline hypertension. In: Weiner, H., M. A. Hofer, A. J. Stunkard (eds.): Brain, Behaviour and Bodily Disease. Raven, New York 1981.

Julius, S., J. Conway: Hemodynamic studies in patients with

borderline blood pressure elevation. Circulation 38 (1968) 282–288.

Julius, S., M. A. Schork: Borderline hypertension – a critical review. J. chron. Dis. 23 (1971) 723–754.

Julius, S., M. D. Esler: Autonomic nervous cardiovascular regulation in borderline hypertension. Amer. J. Cardiol. 36 (1975) 672–685.

Julius, S., C. Cottier: Behavior and hypertension. In: Dembroski, T. M., T. H. Schmidt, G. Blümchen (eds.): Biobehavioral Bases of Coronary Heart Disease, pp. 271–289. Karger, Basel-München-Paris 1983.

Julius, S., A. V. Paskual, R. London: Role of parasympathetic inhibition in the hyperkinetic type of borderline hypertension. Circulation 44 (1971) 413–418.

Julius, S., A. B. Weder, A. L. Hinderliter: Does behaviorally induced blood pressure variability lead to hypertension? In: Matthews, K. A., S. M. Weiss, T. Detre, T. M. Dembroski, B. Falkner, S. B. Manuck, R. B. Williams jr. (eds.): Handbook of Stress, Reactivity, and Cardiovascular Disease, pp. 71–82. Wiley, New York 1986.

Jung, C. G.: Psychoanalyse und Assoziationsexperiment. In: Jung, C. G. (Hrsg.): Diagnostische Assoziationsstudien 1 (1906) 258–281.

Jung, C. G.: Über psychische Energetik und das Wesen der Träume. Rascher, Zürich 1948.

Junk-Overbeck, M., W. Pott, U. Pauli: Empirische Untersuchungen zur Psychosomatik der chronischen Prostatitis. In: Brähler, E., A. Meyer (Hrsg.): Partnerschaft, Sexualität und Fruchtbarkeit, S. 217. Springer, Berlin 1988.

Justice, A.: Review of the effects of stress on cancer in laboratory animals: importance of time of stress application and type of tumor. Psychol. Bull. 98 (1985) 108–138.

Kaada, B.: Brain mechanisms related to aggressive behavior. In: Clemente, C. D., D. B. Lindsley (eds.): Aggression and Defense. Univ. of California Press, Los Angeles 1967.

Kaban, L. B., M. L. Belfer: Temporomandibular joint dysfunction: An occasional manifestation of serious psychopathology. J. oral Surg. 39 (1981) 742.

Kabiersch, A., A. Del Rey, C. G. Honegger, H. O. Besedovsky: Interleukin-1 induces changes in norepinephrine metabolism in the rat brain. Brain, Behavior and Immunity 2 (1988) 267–274.

Kächele, H.: Zum Begriff „psychogener Tod" in der medizinischen Literatur. Z. psychosom. Med. 16 (1970) 105–222.

Kächele, H.: Die Beurteilung des Behandlungserfolges in der Psychotherapie. Materialien zur Psychoanalyse u. analyt. orient. Psychotherapie, Sekt. G. C 12 (1975) 3–38.

Kächele, H.: Aktuelle Trends der Ergebnisforschung in der Psychotherapie und deren Bedeutung für die Psychosomatik. Psychother. med. Psychol. 36 (1986) 307–312.

Kächele, H., R. Schors: Ansätze und Ergebnisse psychoanalytischer Psychotherapieforschung. In: Baumann, U., H. Berbalk, G. Seidenstücker (Hrsg.): Klinische Psychologie, Trends in Forschung und Praxis. Huber, Bern-Stuttgart-Wien 1981.

Kächele, H., I. Fiedler: Ist der Erfolg einer psychotherapeutischen Behandlung vorhersagbar? Erfahrungen aus dem Penn Psychotherapy Project. Psychother. med. Psychol. 35 (1985) 201.

Kächele, H., W. Steffens (Hrsg.): Bewältigung und Abwehr. Springer, Berlin 1988.

Kagaminori, S.: Occupational life tables for cerebrovascular disease and ischemic heart disease in Japan compared with England and Wales. Jap. Circulat. J. 45 (1981) 195.

Kagan, A. R., L. Levi: Health and environment – psychosocial stimuli. A review. In: Levi, L. (ed.): Society, Stress and Disease 2, pp. 241–260. Oxford Univ. Press, London 1975.

Kahana, E., U. Leibowitz, M. Alter: Cerebral multiple sclerosis. Neurology 21 (1971) 1179–1185.

Kahana, R. J.: Personality and response to physical illness. In: Lipowski, Z. J. (ed.): Psychosocial Aspects of Physical Illness. Advanc. Psychosom. Med. Vol. 8. Basel 1972.

Kahana, R. J., G. L. Bibring: Personality types in medical management. In: Zinberg, N. E. (ed.): Psychiatry and Medical Practice in a General Hospital, pp. 108–123. Int. Univ. Press, New York 1964.

Kahn, E., S. L. Lass, R. Hartley, H. K. Kornreich: Affective learning in medical education. J. med. Educ. 56 (1981) 646–652.

Kahn, M. R.: Ego distortion. Cumulative trauma and the role of the reconstruction in the analytic situation. Int. J. Psychoanal. 45 (1964) 272.

Kahn, R. J., D. M. McNair, R. S. Lipman, L. Covi, K. Rickels, R. Downing, S. Fisher, L. M. Frankenthaler: Imipramin and chlordiazepoxide in depressive and anxiety disorders. Arch. gen. Psychiat. 43 (1986) 79–85.

Kaibara, N., T. Hotokebuchi, K. Takagishi, I. Katsuki, M. Morinaga, C. Arita, S. Jingushi: Pathogenetic difference between collagen arthritis and adjuvant arthritis. J. exp. Med. 159 (1984) 1388–1396.

Kaiser, E.: Über die Bedeutung des Hörens für die Entwicklung der Persönlichkeit. N. Bl. Taubst.-Bild. 6/7 (1962) 169.

Kalbak, K.: Incidence of atherosclerosis in patients with rheumatoid arthritis receiving long-term corticosteroid therapy. Ann. rheum. Dis. 31 (1972) 196–200.

Kallestad Laboratories: Use of Kallestad low level IgA rid plates in determination of secretory IgA (product information 1984).

Kalman, T. P., P. G. Wilson, C. M. Kalman: Wie häufig sind psychische Veränderungen bei Nierentransplantat-Empfängern und Patienten unter Langzeitdialysebehandlung? J. Amer. med. Ass. (1983) 771–775.

Kaluza, K., H. Lehnert, H. Losse, K. Dorst: Langzeitwirkung und prognostische Kriterien eines verhaltenstherapeutischen Programms bei essentieller Hypertonie. Psychother. med. Psychol. 36 (1986) 179–186.

Kamin, L. J.: Attention-like processes in classical conditioning. In: Jones, M. R. (ed.): Miami symposium on the prediction of behavior: aversive stimulation. Univ. of Miami Press, 1968.

Kaminer, Y., M. Feingold, K. Lyons: Bulimia in a pair of monozygotic twins. J. nerv. ment. Dis. 176 (1988) 246–247.

Kaminski, G.: Diskussionsbeitrag zum Symposion. In: Hörmann, H. et al. (Hrsg.): Die Beziehung zwischen psychologischer Diagnostik und Grundlagenforschung. Ber. 25. Kongr. Dt. Ges. Psychol. Hogrefe, Göttingen 1968.

Kammerer, T.: Süchtiges Verhalten bei Epilepsien. Photogene Epilepsie mit selbstinduzierten Anfällen. Dtsch. Z. Nervenheilk. 185 (1963) 319–330.

Kampman, R., R. Hirvenoja, A. Juolasmaa et al.: Psychic complications following open-heart surgery. In: Speidel, H., G. Rodewald (eds.): Psychic and Neurological Dysfunctions after Open-Heart Surgery. Thieme, Stuttgart 1980.

Kandel, D. B., M. Davies: Epidemiology of depressive mood in adolescents: an empirical study. Arch. gen. Psychiat. 39 (1982) 1205–1212.

Kane, R. L., K. W. Deuschle: Problems in patient-doctor communications. Medical Care 5 (1967) 260–271.

Kanfer, F. H., G. Saslow: Behavioral analysis. Arch. gen. Psychiat. 12 (1965) 529–538.

Kanfer, F. H., G. Saslow: Behavioral diagnosis. In: Francks, C. M. (ed.): Behavioral Therapy: Appraisal and Status. McGraw-Hill, New York 1969.

Kanfer, F. H., G. Saslow: Verhaltenstheoretische Diagnostik. In: Schulte, D. (Hrsg.): Diagnostik in der Verhaltenstherapie. Urban & Schwarzenberg, München 1976.

Kannel, W. B.: Assessment of hypertension as a predictor of cardiovascular disease: the Framingham study. Symposium Malta 1974, pp. 69–86. Ciba, Horsham (England) 1975.

Kannel, W. B., T. R. Dawber: Hypertensive cardiovascular disease: the Framingham study. In: Onesti, G., E. Kim, J. H. Moyer (eds.): Hypertension: Mechanisms and Management, pp. 93–110. New York 1973.

Kannel, W. B., T. R. Dawber, M. E. Cohen: The electrocardiogram in neurocirculatory asthenia (anxiety neurosis or neurasthenia): a study of 203 neurocirculatory asthenia patients and 757 healthy controls in the Framingham study. Ann. intern. Med. 49 (1958) 1351–1360.

Kaplan, A. S.: Anticonvulsant treatment of eating disorders. In: Garfinkel, P. E., D. M. Garner (eds.): The Role of Drug Treatments for Eating Disorders. Brunner/Mazel, New York 1987.

Kaplan, E. H.: Congenital absence of the vagina. Psychiatric aspects of diagnosis and management. N. Y. St. J. Med. (1968) 1937.

Kaplan, G. A., P. Reynolds: Depression and cancer mortality and morbidity: prospective evidence from the Alameda County study. J. Behav. Med 11 (1988) 1–13.

Kaplan, H. S.: The new sex therapy. Brunner & Mazel, New York 1974.

Kaplan, H. S., D. Langer: Sexualtherapie. Ein neuer Weg für die Praxis. Enke, Stuttgart 1979.

Kaplan, J. R., S. B. Manuck, T. B. Clarkson et al.: Social stress and atherosclerosis in normocholesteronemic monkeys. Science 13 (1983) 733–735.

Kaplan, J. R., S. B. Manuck, T. B. Clarkson, F. M. Lusso, D. B. Taub: Social status, environment, and atherosclerosis in cynomolgus monkeys. Arteriosclerosis 2 (1982) 359–368.

Kaplan, J. R., M. R. Adams, T. E. Hamm jr., T. B. Clarkson: Psychosocial phenomena and female „protection" from coronary artery atherosclerosis in cynomolgus macaques (macaca fascicularis). In: Beamish, R. E., P. K. Singal, N. S. Dhalla (eds.): Stress and Heart Disease, pp. 250–261. Nijhoff, Boston 1985.

Kaplan, J. R., S. B. Manuck, M. R. Adams, K. W. Weingand, T. B. Clarkson: Inhibition of coronary atherosclerosis by propranolol in behaviorally predisposed monkeys fed an atherogenic diet. Circulation 76 (1987) 1364–1372.

Kaplan, S., F. Kozin: A controlled study of group counseling in rheumatoid arthritis. J. Rheum. 8 (1981) 91–99.

Kaplan, S. L., G. Hong, C. Weinhold: Epidemiology of depressive mood in adolescents: an empirical study. J. Amer. Acad. Child Psychiat. 23 (1984) 91–98.

Kaplan De-Nour, A.: Some notes on the psychological significance of urination. J. nerv. ment. Dis. 148 (1969) 615–623.

Kaplan De-Nour, A.: Hemodialysis: sexual functioning. Psychosomatics 19 (1978) 229–235.

Kaplan De-Nour, A.: The hemodialysis unit. Advanc. psychosom. Med. 10 (1980a) 132–150.

Kaplan De-Nour, A.: Maintenance hemodialysis. In: Freyberger, H. (ed.): Psychotherapeutic Intervention in Life Threatening Illness. Advanc. psychosom. Med. 10, pp. 132–150. Karger, Basel 1980b.

Kaplan De-Nour, A.: Prediction of adjustment to chronic hemodialysis. In: Levy, N. B. (ed.): Psychonephrology 1, pp. 117–132. Plenum, New York 1981.

Kaplan De-Nour, A.: Psychosocial Adjustment to Illness Scale (PAIS): a study of chronic hemodialysis patients. J. psychosom. Res. 26 (1982) 11–22.

Kaplan De-Nour, A.: An overview of psychological problems in hemodialysis patients. In: Levy, N. B. (ed.): Psychonephrology 2, pp. 3–14. Plenum, New York 1983.

Kaplan De-Nour, A.: Persönlichkeitsfaktoren und Adaptation. In: Balck, F. B., U. Koch, H. Speidel (Hrsg.): Psychonephrologie. Springer, Heidelberg 1984.

Kaplan De-Nour, A., J. Czaczkes: Emotional reactions and problems of the medical team in a chronic hemodialysis unit. Lancet 2 (1968a) 987–991.

Kaplan De-Nour, A., J. Czaczkes: Professional team opinion and personal bias – a study of a chronic hemodialysis unit team. J. chron. Dis. 24 (1971) 533–541.

Kaplan De-Nour, A., J. Czaczkes: Personality factors in chronic hemodialysis patients causing noncompliance with medical regimen. Psychosom. Med. 34 (1972a) 333–344.

Kaplan De-Nour, A., J. Czaczkes: Bias in assessment of patients on chronic dialysis. J. psychosom. Res. 18 (1974a) 217–221.

Kaplan De-Nour, A., J. Czaczkes: Personality and adjustment to chronic hemodialysis. In: Levy, N. B. (ed.): Living or Dying. Thomas, Springfield/Ill. 1974b.

Kaplan De-Nour, A., J. Czaczkes: Personality factors influencing vocational rehabilitation. Arch. gen. Psychiat. 32 (1975) 573.

Kaplan De-Nour, A., J. Czaczkes: The influence of patient's personality on adjustment to chronic dialysis. J. nerv. ment. Dis. 162 (1976) 323–333.

Kaplan De-Nour, A., J. Schaltiei, J. Czaczkes: Emotional reactions of patients on chronic hemodialysis. Psychosom. Med. 30 (1968b) 521–533.

Kaplan De-Nour, A., J. Czaczkes, P. Lilos: A study of chronic hemodialysis teams – differences in opinions and expectations. J. chron. Dis. 25 (1972b) 441–448.

Kaplan De-Nour, A., J. Shanan: Quality of life of dialysis and transplant patients. Nephron 25 (1980) 117–120.

Kapoor, W., M. Karpf, S. Wieand, J. Peterson, G. Levey: A prospective evaluation and follow-up of patients with syncope. New Engl. J. Med. 309 (1983) 197–204.

Kapoor, W. N., M. Karpf, Y. Maher, R. A. Miller, G. S. Levey: Syncope of unknown origin. The need for a more cost-effective approach to its diagnostic evaluation. J. Amer. med. Ass. 247 (1983) 2687–2691.

Kapp, F. T., M. Rosenbaum, J. Romano: Psychological factors in men with peptic ulcers. Amer. J. Psychiat. 103 (1947) 700.

Kappauf, H., R. Dietz, W. Pontzen, W. M. Gallmeier: Sterben im Krankenhaus – ein Betriebsunfall? Münch. med. Wschr. 130 (1988) 292–294.

Kappeler, M., F. Halter: Differentialdiagnose entzündlicher Dickdarmkrankheiten im endoskopischen Bild. internist. prax. 25 (1985) 279–287.

Kappus, W.: Ich nehme ab. Deutsche Gesellschaft für Ernährung (Hrsg.). Brönner, Frankfurt 1982.

Karacan, L., A. Dervent, G. Cunningham et al.: Assessment of NPT as an objective method of evaluating sexual functioning in ESRD patients. Dial. Transplant. 7 (1978) 872–877.

Karasek, R. A., R. S. Russell, T. Theorell: Physiology of stress and regeneration in job-related cardiovascular illness. J. hum. Stress 1 (1982) 29–42.

Karasek, R. A., T. Theorell, J. Schwartz, C. Pieper, L. Alfredson: Job, psychological factors and coronary heart disease. Advanc. Cardiol. 29 (1982) 62–67.

Karasu, T. B.: Psychotherapy of medically ill. Amer. J. Psychiat. 136 (1979) 1–11.

Karasu, T. B.: Psychotherapy and pharmacotherapy: toward an integrative model. Amer. J. Psychiat. 139 (1982) 1102–1113.

Kardener, S. H., M. Fuller, T. N. Mensh: A survey of physicians attitudes and nonerotic contact with patients. Amer. J. Psychiat. 130 (1973) 1077–1088.

Karmaus, W., V. Müller, G. Schienstock (Hrsg.): Streß in der Arbeitswelt. Bund, Köln 1979.

Karstens, R., K. Köhle, D. Ohlmeier, S. Weidlich: A multidisciplinary approach for the assessment of psychodynamic factors in young adults with acute myocardial infarctions. Psychother. and Psychosom. 18 (1970) 281–285.

Karush, A., G. Daniels: Colitis ulcerosa. Psychoanalyse zweier Fälle. Psyche 7 (1953) 401–452.

Karush, A., R. B. Hiatt, G. E. Daniels: Psychophysiological correlations in ulcerative colitis. Psychosom. Med. 17 (1955) 36–56.

Karush, A., G. Daniels, J. O'Connor, L. Stern: The response to psychotherapy in chronic ulcerative colitis. Psychosom. Med. 30 (1968) 255–276; Psychosom. Med. 31 (1969) 201–226.

Karush, A., G. E. Daniels, Ch. Flood, J. F. O'Connor: Psychotherapy in chronic ulcerative colitis. Saunders, Philadelphia 1978.

Kaschkat, G.: Psychiatrische Syndrome bei Anämien, Kollagenerkrankungen, Mineralhaushaltstörungen und postinfektiösen Zuständen. Internist 16 (1975) 15–19.

Kasl, S. V., S. Cobb: Blood pressure changes in men undergoing job loss: a preliminary report. Psychosom. Med. 32 (1970) 19–38.

Kasl, S. V., A. S. Evans, J. C. Niederman: Psychosocial risk factors in the development of infectious mononucleosis. Psychosom. Med. 41 (1979) 445–466.

Kaspar, A. M.: The doctor and death (1959). In: Feifel, H.: The Meaning of Death, pp. 218–233. McGraw-Hill, New York 1965.

Kasper, H.: Enteritis regionalis (Morbus Crohn). In: Ritter, U., M. Claassen (Hrsg.): Ergebnisse der Gastroenterologie 1976. Demeter, Gräfelfing 1977.

Kasper, H., H. Sommer: Dietary fiber and nutrient intake in Crohn's disease. Amer. J. clin. Nutr. 32 (1979) 1898–1901.

Kast, V.: Trauern. Phasen und Chancen des psychischen Prozesses. Kreuz-Verlag, Stuttgart 1982.

Kastenbaum, R.: The reluctant therapist. Geriatrics 18 (1963) 296–301.

Kastenbaum, R.: Death, society and human experience. Mosby, Saint Louis 1977.

Kastenbaum, R., R. Alsenberg: The psychology of death. Springer, New York 1972.

Kastenbaum, R., P. T. Costa: Psychological perspectives on death. Ann. Rev. Psychol. 28 (1977) 225–249.

Katcher, A., A. Honori, V. Brightman, L. Luvorsky, I. Ship: Prediction of the incidence of recurrent herpes labialis and systemic illness from psychological measurements. J. dent. Res. 52 (1973) 49–58.

Katkin, E. S., E. N. Murray: Instrumental conditioning of autonomically mediated behavior: theoretical and methodological issues. Psychol. Bull. 70 (1968) 52–68.

Katon, W., A. Kleinman, G. Rosen: Depression and somatization: a review I, II. Amer. J. Med. 72 (1982) 127–135; 241–247.

Katschnig, H. (Hrsg.): Sozialer Streß und psychische Erkrankung. Urban & Schwarzenberg, München-Wien-Baltimore 1980.

Katschnig, H., H. Strotzka: Epidemiologie der Neurosen und psychosomatischen Störungen. In: Blohmke, M., C. Ferber, K. Kisker, H. Schäfer (Hrsg.): Handbuch der Sozialmedizin, Bd. II. Stuttgart 1977.

Katschnig, H., T. Konieczna: Die psychosoziale Situation chronisch hämodialysierter und nierentransplantierter Patienten sowie ihrer Angehörigen. Unveröffentl. Forschungsbericht, Ludwig Boltzmann-Institut, Wien 1982.

Katz, J. L., H. Weiner: A functional, anterior hypothalamic defect in primary anorexia nervosa? Psychosom. Med. 37 (1975) 103–105.

Katz, J. L., H. Weiner, T. F. Gallagher, L. Hellman: Stress, distress, and ego defenses. Psychoendocrine responses to impending breast tumor biopsy. Arch. gen. Psychiat. 23 (1970) 131–142.

Katz, M., S. Lyerly: Methods of measuring adjustment and social behavior in the community. Psychol. Rep. 13 (1963) 237–243.

Kaufman, M. R.: Old age and aging: the psychoanalytic point of view. Amer. J. Orthopsychiat. 10 (1940) 73–79.

Kaufman, M. R. (ed.): The psychiatric unit in a general hospital. Int. Univ. Press, New York 1965.

Kaufman, M. R., S. G. Margolin: Theory and practice of psychosomatic medicine in a general hospital. Med. Clin. N. Amer. 32 (1948) 611–616.

Kaufman, M. R., M. Heiman: Evolution of psychosomatic concepts: anorexia nervosa. A paradigm. Int. Univ. Press, New York 1964.

Kaufman, M. R., L. Roose: The function of a liaison service in medical education. Psychotherapeutic implications for the non-psychiatrist. Psychother. and Psychosom. 24 (1972) 220–221.

Kaufman, R. A.: A psychiatric unit in a general hospital. J. Mt. Sinai Hosp. 24 (1957) 572–579.

Kaufman, W.: Discussion of Jonez, H. D.: Psychotherapy in multiple sclerosis. Ann. Allergy 9 (1951) 658.

Kaufman, W.: Pathophysiologische und klinische Aspekte des Renin-Angiotensin-Aldosteron-Systems. Nauheimer Fortbildungslehrgänge, Bd. 37 (Hypertonie), S. 10–17. Steinkopff, Darmstadt 1973.

Kautzky, R. (Hrsg.): Sterben im Krankenhaus. Herder, Freiburg 1976.

Kay, D.: Anorexia nervosa: a study in prognosis. Proc. roy. Soc. Med. 46 (1953) 669–674.

Kay, D., K. Shapiro: The prognosis in anorexia nervosa. In: Meyer, J. E., F. Feldmann (Hrsg.): Anorexia nervosa. Thieme, Stuttgart 1965.

Kay, D. W., M. Roth, B. Hopkins: Aetiological factors in the causation of affective disorders in old age. J. ment. Sci. 101 (1955) 302.

Kayser, H.: Die verschiedenen Formen der therapeutischen Gemeinschaft und ihre Indikation für die Praxis. Psychother. med. Psychol. 24 (1974) 80–94.

Kayser, H., H. Krüger, W. Mävers, P. Petersen, M. Rohde, H.-K. Rose, A. Veltin, V. Zumpe: Gruppenarbeit in der Psychiatrie. Thieme, Stuttgart 1973.

Keele, K. D., M. D. Lond: Pain sensitivity tests. The pressure algometer. Lancet I (1954) 636–639.

Keitel, W., H. Hoffmann, G. Weber, U. Krieger: Ermittlung der prozentualen Funktionsminderung der Gelenke durch einen Bewegungsfunktionstest in der Rheumatologie. Dt. Gesundheitswesen 26 (1971) 1901–1903.

Keller, K.: Psychosomatik. Eine Bestätigung der Allgemeinmedizin. Z. f. Allgemeinmedizin 14 (1975).

Keller, M. B., R. W. Shapiro, P. W. Lavori, N. Wolfe: Recovery in major depressive disorder – analysis with the life table and regression models. Arch. gen. Psychiat. 39 (1982) 905–910.

Keller, S. E., J. M. Weiss, S. J. Schleifer, N. E. Miller, M. Stein: Suppression of immunity by stress: effect of a graded series of stressors on lymphocyte stimulation in the rat. Science 213 (1981) 1397–1400.

Keller, S. E., J. M. Weiss, S. J. Schleifer, N. E. Miller, M. Stein: Stress-induced suppression of immunity in adrenalectomized rats. Science 221 (1983) 1301–1304.

Keller, S. E., S. J. Schleifer, A. S. Liotta, R. N. Bond, N. Farhoody, M. Stein: Stress-induced alterations of immunity in hypophysectomized rats. Proc. Natl. Acad. Sci. (Wash.) 85 (1988) 9297–9301.

Kellerhals, B.: Die Behandlung der akuten Innenohrschwerhörigkeit (Hörsturz) und akustisches Trauma. Z. Laryng. Rhinol. 56 (1977) 357.

Kelley, K., D. Harris: Medical students' personality changes from freshmen to seniors. AAMC-RIME Conference 1981, 55.

Kelley, K. W., R. Dantzer: Is conditioned immunosuppression truly conditioned? Behav. Brain Sci. 9 (1986) 758–760.

Kelley, K. W., R. Dantzer, P. Mormede, H. Salmon, J. M. Aynaud: Conditioned taste aversion induces immunosuppression in the absence of an immunosuppressive drug. C. R. Acad. Sci. (Paris), série III, 299 (1984) 123–126.

Kelley, K. W., R. Dantzer, P. Mormede, H. Salmon, J. M. Aynaud: Conditioned taste aversion suppresses induction of delayed-type hypersensitivity immune reactions. Physiol. Behav. 34 (1985) 189–193.

Kellner, R.: Family ill health. An investigation in general practice. Tavistock, London 1963.

Kelly, G. A.: The psychology of personal constructs, Vol. I/ II. Norton, New York 1955.

Kelly, W. D., S. R. Friesen: Do cancer patients want to be told? Surgery 27 (1950) 822–826.

Kelsey, J. L.: An epidemiological study of acute herniated lumbar intervertebral discs. Rheumatol. Rehab. 14 (1975a) 144–159.

Kelsey, J. L.: An epidemiological study of the relationship between occupation and acute herniated lumbar intervertebral disc. Int. J. Epidemiol. 4 (1975b) 197–205.

Keltikangas-Järvinen, L.: Concept of alexithymia II: The consistency of alexithymia. Psychother. and Psychosom. 47 (1987) 113–120.

Keltikangas-Järvinen, L., H. Järvinen, T. Lehtonen: Psychic disturbances in patients with chronic prostatitis. Ann. clin. Res. 13 (1981) 45.

Kemeny, M. E., F. Cohen, L. S. Zegans, M. A. Conant: Psychological and immunological predictors of genital herpes recurrence. Psychosom. Med. 51 (1989) 195–208.

Kemeny, M. E., J. L. Fahey, S. Schneider, H. Weiner, S. Taylor, B. Visscher: Bereavement-associated alterations in phenotypes of lymphocytes in HIV+ and HIV- homosexual men. Proc. IV. Int. Conf. on AIDS. Stockholm, June 12–16, 1988.

Kemmer, F. W.: Einflüsse von Stresshormonen und psychischen Belastungen auf die diabetische Stoffwechsellage. Urban & Schwarzenberg, München 1988.

Kemp, D. T.: Physiologically active cochlear micromechanic – one source of tinnitus. In: Tinnitus. Ciba Foundation Symposium 85 (1981) 54.

Kemper, T. D.: The socio-bio-social chain: essays in social structure and testosterone. Unpublished, 1989.

Kemper, W.: „Organwahl" und psychosomatische Medizin. Psychother. med. Psychol. 4 (1954) 101–113.

Kemper, W.: Die Störungen der Liebesfähigkeit beim Weibe. Wiss. Buchgesellschaft, Darmstadt 1975.

Kemph, J. P.: The role of the psychiatrist on the kidney transplant team. Int. Congr. Acad. Psychosom. Med. 1966.

Kemph, J. P.: Observations of the effects of kidney transplant on donors and recipients. Dis. nerv. Syst. 31 (1970) 323–325.

Kendall, R., D. Hall, A. Hailey, H. Babigan: The epidemiology of anorexia nervosa. Psychol. Med. 3 (1973) 200–203.

Kendell, R. E.: The concept of disease and its implication for psychiatry. Brit. J. Psychiat. 128 (1976a) 588–594.

Kendell, R. E.: The classification of depressions: a review of contemporary confusion. Brit. J. Psychiat. 129 (1976b) 15–28.

Kendler, H. H., T. S. Kendler: Developmental processes in discrimination learning. Hum. Develop. 13 (1970) 65–89.

Kendler, K. S., R. C. Mohs, K. L. Davis: The effects of diet and physical activity on plasma homovanillic acid in normal subjects. J. psychiat. Res. 8 (1983) 215–224.

Kenell, J. H.: zit. nach Prill, H. J.: Neuere Erkenntnisse der Mutter-Kind-Beziehung nach der Geburt. Vortr. 41. Tag. Dtsch. Ges. Gynäkologie Geburtshilfe, Hamburg 1976.

Kennedy, J. A., H. Bakst: The influence of emotion on the outcome of cardiac surgery: a predictive study. Bull. N. Y. Acad. Med. 42 (1966) 811–845.

Kennedy, J. L., L. A. Giuffra, H. W. Moises, L. L. Cavalli-Sforza, A. J. Pakstis, J. R. Kidd, C. M. Castiglione, B. Sjögren, L. Wetterberg, K. K. Kidd: Evidence against linkage of schizophrenia to markers on chromosome 5 in a north Swedish pedigree. Nature (Lond.) 336 (1988) 167–170.

Kentenich, H: „Natürliche Geburt" in der Klinik. Diss., FU Berlin 1983.

Kenyon, F. E.: Review-Article: „Hypochondrial States". Brit. J. Psychiat. 129 (1976) 1–14.

Kepes, E. R., P. S. Thomas: Continuous intravenous (i. v.) morphine to evaluate oral methadone requirement in cancer pain. Pain (Suppl. 1) Abstract No. 116. Abstracts of the 3rd World Congr. on Pain. Edinburgh, Scotland, Sept. 4–11, 1981.

Kerber, G. W., M. Greenberg, J. M. Rubin: Computed tomography evaluation of local and extraintestinal complications of Crohn's disease. Gastrointest. Radiol. 9 (1984) 143–148.

Kerekjarto, M. v.: Psychoneuroimmunology. Plenary Lecture, 17th Europ. Conf. on Psychosom. Res., Marburg, 4th-9th September 1988.

Kerekjarto, M. v., B. Dahme, O. Hansen, T. Küchler, D. Phillip-Dormston, R. Richter, K. Schulz, F. Wistuba: Vergleichende klinische Studien zu psychosomatischen Erkrankungen – insbesondere zum Asthma bronchiale. Abschlußbericht über das DFG-Forschungsprojekt A6 im Sonderforschungsbereich 115, Hamburg 1981.

Kernberg, O. F.: Factors in the psychoanalytic treatment of narcicisstic personalities. J. Amer. psychoanal. Ass. 18 (1970) 51–85.

Kernberg, O.: Summary and conclusions of „Psychotherapy and Psychoanalysis: Final report of the Menninger Foundation Psychotherapy Research Project." Int. J. Psychiat. 11 (1973) 62–77.

Kernberg, O.: Some methodological and strategic issues in psychotherapy research: research implications of the Menninger Foundation's psychotherapy research project. In: Spitzer, R., D. Klein (eds.): Evaluation of Psychological Psychotherapies. Hopkins Univ. Press, Baltimore 1976.

Kernberg, O.: Borderline-Störungen und pathologischer Narzißmus. Suhrkamp, Frankfurt 1978.

Kernberg, O.: Objektbeziehungen und Praxis der Psychoanalyse. Klett-Cotta, Stuttgart 1981.

Kernberg, O.: Innere Welt und äußere Realität. Verlag Internationale Psychoanalyse, München-Wien 1988.

Kernberg, O., E. Bernstein, L. Coyne, A. Appelbaum, C. Horwitz, H. Voth: Psychotherapy and psychoanalysis: final report of the Menninger Foundation Psychotherapy Research Project. Bull. Menninger Clin. 36 (1972) 3–275.

Kerr, W. J., J. W. Dalton, P. A. Gliebe: Some physical phenomena associated with the anxiety states and their relation to hyperventilation. Ann. intern. Med. 11 (1937) 961–992.

Kessel, M.: Heimdialyse als familiäres Problem. Med. Welt 22 (1971) 1031.

Kesselring, J. (Hrsg.): Multiple Sklerose. Kohlhammer, Stuttgart 1989.

Kestenberg, J. S.: Entwicklungsphasen weiblicher Identität. Psyche 42 (1988) 349.

Kety, S. S., D. Rosenthal, P. H. Wender, F. Schulsinger, B. Jacobsen: Mental illness in the biological and adoptive families of adopted individuals who have become schizophrenic. In: Fieve, R. R., D. Rosenthal, H. Brill (eds.): Genetic Research in Psychiatry. Johns Hopkins Univ. Press, Baltimore 1975.

Keupp, H., M. Pflanz, J. Siegrist (Hrsg.): Medizin und Sozialwissenschaften. Bd. 5: Wirtschaftskrisen, Arbeitslosigkeit und psychische Erkrankungen, hrsg. von M. H. Brenner. Urban & Schwarzenberg, München-Wien-Baltimore 1979.

Keynes, G.: The apology treaties of Ambroise Paré. Chicago Univ. Press, Chicago 1952.

Keys, A.: Coronary heart disease in 7 countries. Circulation (Suppl. 1) 41 (1970) 138–144.

Keys, A., H. L. Taylor, H. Blackburn, J. Brozek, J. T. Anderson, E. Simonson: Mortality and coronary heart disease among men studied for 23 years. Arch. intern. Med. 128 (1971) 201–214.

Khan, M. M. R.: Das kumulative Trauma. In: Khan, M. M. R. (Hrsg.): Selbsterfahrung in der Therapie, S. 50–70. Kindler, München 1977.

Khoury, S. A., C. J. Linwood: Administration problems and solutions in screening for gonorrhea. Health Serr. Res. 89 (1974) 286.

Kickbusch, I., A. Trojan (Hrsg.): Gemeinsam sind wir stärker. Selbsthilfegruppen und Gesundheit. Fischer, Frankfurt 1981.

Kiecolt-Glaser, J. K., R. Glaser: Methodological issues in behavioral immunology research with humans. Brain, Behavior, and Immunity 2 (1988) 67–78.

Kiecolt-Glaser, J. K., R. Glaser, E. C. Shuttleworth: Chronic stress and immunity in family care givers of Alzheimer's disease victims. Psychosom. Med. 49 (1987) 523–535.

Kiecolt-Glaser, J. K., D. Ricker, J. George: Urinary cortisol levels, cellular immunocompetency, and loneliness in psychiatric inpatients. Psychosom. Med. 46 (1984 b) 15–23.

Kiecolt-Glaser, J. K., L. D. Fisher, P. Ogrocki, J. C. Stout, C. E. Speicher, R. Glaser: Marital quality, marital disruption, and immune function. Psychosom. Med. 49 (1987) 13–34.

Kiecolt-Glaser, J. K., W. Garner, C. Speicher, G. M. Penn, J. Holliday, R. Glaser: Psychosocial modifiers of immunocompetence in medical students. Psychosom. Med. 46 (1984 a) 7–14.

Kiecolt-Glaser, J. K., R. Glaser, D. Williger, J. Stout, G. Messick, S. Sheppard, D. Ricker, S. C. Romisher, W. Briner, G. Bonnell, R. Donnerberg: Psychosocial enhancement of immunocompetence in a geriatric population. Health Psychol. 4 (1985) 25–41.

Kielholz, P.: Diagnose und Therapie der Depressionen für den Praktiker. Lohmann, München 1971.

Kielholz, P.: Die larvierte Depression. Huber, Bern 1973.

Kiener, F.: Untersuchungen zum Körperbild (body image). Teil 1: Z. klin. Psychol. Psychother. 21 (1973) 335. Teil 2: Z. klin. Psychol. Psychother. 22 (1974) 45.

Kiesler, D. J.: Some myths of psychotherapy research and the search for a paradigm. Psychol. Bull. 65 (1966) 110–136.

Kilmartin, A.: Blasenentzündung. Zystitis – Urethritis. Ehrenwirth, München 1982.

Kim, C., H. Choi, J. K. Kim, M. S. Kim, H. J. Park, B. T. Ahn, S. H. Kang: Influence of hippocampectomy on gastric ulcer in rats. Brain Res. 109 (1976) 245–254.

Kimball, C. P.: Psychological responses to the experience of open-heart surgery. Amer. J. Psychiat. 126 (1969) 96–107.

Kimball, C. P.: Conceptual developments of psychosomatic medicine 1939–1969. Ann. intern. Med. 73 (1970) 307–316.

Kimball, C. P.: The experience of open-heart surgery. III. Toward a definition and understanding of postcardiotomy delirium. Arch. gen. Psychiat. 27 (1972) 57–63.

Kimball, C. P.: Medical psychotherapy. Psychother. and Psychosom. 25 (1975) 193–200.

Kimball, C. P.: The experience of cardiac surgery and cardiac transplant. In: Howells, J. G. (ed.): Modern Perspectives in the Psychiatric Aspects of Surgery. Brunner/Mazel, New York 1976.

Kimball, C. P., D. Quinlan, F. Osborne, B. Woodward: The experience of cardiac surgery. V. Psychological patterns and prediction of outcome. Psychother. and Psychosom. 22 (1973) 310–319.

Kimmel, H. D.: Instrumental conditioning of autonomically mediated responses in human beings. Amer. Psychologist 29 (1974) 325–335.

Kind, H.: Kritisches zu körperbezogenen Psychotherapieformen. Psychother. med. Psychol. 35 (1985) 167.

Kindt, W.: Zur interaktiven Behandlung von Deutungen in Therapiegesprächen. J. Pragmatics 8 (1984) 731–751.

King, M. G., A. J. Husband, A. W. Kusnecov: Behaviorally conditioned immunosuppression using anti-lymphocyte serum: duration of effect and role of corticosteroids. Med. Sci. Res. (Biochem.) 15 (1987) 407–408.

King, S. H.: Psychosocial factors associated with rheumatoid arthritis. J. chron. Dis. 2 (1955) 287–302.

Kingsley, L. A., R. Karlow, C. R. Rinaldo: Risk factors for seroconversion to human immunodeficiency virus among male homosexuals. Lancet I (1987) 345–349.

Kinsman, R. A. et al.: Observations on patterns of subjective symptomatology of acute asthma. Psychosom. Med. 36 (1974) 129–143.

Kinsman, R. A., J. F. Dirks, J. C. Schraa: Psychomaintenance in asthma. Personal styles affecting medical management. Respir. Therapy 4 (1981) 39–46.

Kinzl, J., W. Biebl, H. Rauchegger: Functional aphonia: psychosomatic aspects of diagnosis and therapy. Fol. phoniat. 40 (1988) 131.

Kipnis, D. M.: Insulin secretion in diabetes mellitus. Ann. intern. Med. 69 (1968) 891–901.

Kipnowski, A., J. Kipnowski: Zum Problem der Krankheitsverarbeitung. Med. Welt 32 (1981) 1023.

Kipnowski, H. J.: Biographische und testpsychologische Untersuchungen an Colitis-ulcerosa-Patienten. Diss., Bonn 1978.

Kipnowski, J., A. Kipnowski: Biographische und testpsychologische Ergebnisse bei Patienten mit chronisch rezidivierender Colitis ulcerosa. Psychother. med. Psychol. 32 (1982) 31–34.

Kipnowski, J., A. Kipnowski: Psychosomatischer Beitrag zur Ätiopathogenese der Colitis ulcerosa. Z. psychosom. Med. 27 (1981) 372–380.

Kipnowski, J., Ch. Schmidt, S. Miederer, A. Kipnowski: Konservativ-medikamentöse und psychosomatische Aspekte in der Behandlung der Colitis ulcerosa – ein integratives Therapiekonzept. Med. Welt 39 (1988) 182–186.

Kirch, D. G., D. R. Weinberger: Post-mortem histopathological findings in schizophrenia. In: Nasrallah, H. A., D. R. Weinberger (eds.): The Neurology of Schizophrenia. Elsevier, New York 1986.

Kiresuk, T. J., R. E. Sherman: Goal attainment scaling: a general method for evaluating comprehensive Community Mental Health Programs. Community ment. Health J. 4 (1968) 443–453.

Kirkegaard, C., J. Faber: Altered serum levels of thyroxine, triiodothyronines, diiodothyronines in endogenous depression. Acta endocr. (Kbh.) 96 (1981) 199–207.

Kirschbaum, D. S., P. M. Stalonas, T. R. Zastowny, A. J. Tomarken: Behavioral treatment of adult obesity: attentional controls and a 2-year follow-up. Behav. Res. Ther. 23 (1985) 675–682.

Kirsner, J. B.: Experimental hypersensitivity reactions in the colon and the problem of ulcerative colitis. Amer. J. digest. Dis. 5 (1960) 868–879.

Kirsner, J. B.: Experimental „colitis" with particular reference to hypersensitivity reactions in the colon. Gastroenterology 40 (1961) 307–312.

Kirsner, J. B.: The immunologic response of the colon. J. Amer. med. Ass. 191 (1965) 809–814.

Kirsner, J. B.: Clinical challenge of functional gastrointestinal disorders. Postgrad. Med. 39 (1966) 565–575.

Kirsner, J. B.: Drug therapy in ulcerative colitis. Mod. Med. 34 (1966) 115–117.

Kirsner, J. B.: Genetic aspects of inflammatory bowel disease. Clin. Gastroenterol. 2 (1973) 557–575.

Kirsner, J. B.: Inflammatory bowel disease. Considerations of etiology and pathogenesis. Amer. J. Gastroent. 69 (1978) 253–271.

Kirsner, J. B.: The irritable bowel syndrome. Arch. intern. Med. 141 (1981) 635–639.

Kirsner, J. B., R. G. Shorter: Inflammatory bowel disease, 2nd ed. Lea & Febiger, Philadelphia 1980.

Kish, G. B.: Studies of sensory reinforcement. In: Honig, W. K. (ed.): Operant behavior: Areas of Research and Application. Appleton Century Crofts, New York 1966.

Kissen, D. M.: The influence of some environmental factors on personality scores in psychosomatic research. J. psychosom. Res. 8 (1964) 145–149.

Kissileff, H. R., B. T. Walsh, J. G. Krall, S. M. Cassidy: Laboratory studies of eating behavior in women with bulimia. Physiol. Behav. 38 (1986) 563–570.

Kitagawa, E. M., P. M. Hauser: Differential mortality in the United States: a study in socio-economic epidemiology. Harvard Univ. Press, Cambridge/Mass. 1973.

Kittel, F.: Type A and other psychological factors in the relation to CHD. In: Schmidt, T., T. Dembroski, G. Blümchen (eds.): Biological and Psychological Factors in Cardiovascular Disease, pp. 61–84. Springer, Berlin-Heidelberg-New York 1986.

Kittel, G., B. Schürenberg: Objektive und semiobjektive Untersuchungsmöglichkeiten von Stimme, Sprache und Gehör. Proc. XV. Congr. UEP Erlangen, 14.–18. 9. 1988. Dtsch. Ärzte-Verlag, Köln 1988.

Kiviniemi, P., L. Lyytikäinen: The rheumatoid factor and some psychic characteristics in rheumatoid arthritis. Arthr. and Rheum. 25 (1982) 47.

Kläger, J.: Biofeedback von Darmgeräuschen zur Therapie des Colon irritabile: Experiment zur Prüfung der Trainierbarkeit von Darmgeräuschen. Unveröffentlichte Dissertation, Tübingen 1984.

Klagsbrun, S. C.: Cancer, emotions, and nurses. Amer. J. Psychiat. 126 (1970) 1237–1244.

Klapp, B.: Psychosoziale Intensivmedizin. Springer, Berlin 1984.

Klapp, B., H. Freyberger: Psychosomatik der Intensivmedizin. Dtsch. med. Wschr. 106 (1981) 227–229.

Klapp, B. F., J. W. Scheer: Psychologische Aspekte der intensivmedizinischen Betreuung. In: Beckmann, D., S. Davies-Osterkamp, J. W. Scheer (Hrsg.): Medizinische Psychologie. Springer, Heidelberg 1982.

Klapp, B. F., J. W. Scheer, E. Glaser: Die internistische Intensivbehandlung in der Einschätzung der Patienten. Ein Vergleich kardialer mit nicht-kardialen Patienten, unter besonderer Berücksichtigung der Infarktpatienten. Intensivmed. 16 (1979) 153–158.

Klapp, B. F., W. Laubach, J. W. Scheer: Die Intensivbehandlung als psychosomatisches Aufgabengebiet – Probleme und Konfliktmomente im Behandlungsteam. Verh. Dtsch. Ges. Inn. Med. 86 (1980).

Klaus, H. M., J. H. Kenell: Auswirkungen früher Kontakte zwischen Mutter und Neugeborenem auf die spätere Mutter-Kind-Beziehung. In: Biermann, G. (Hrsg.): Jahrbuch der Psychohygiene. Reinhardt, München 1974.

Kleeberg, J.: Eide und Bekenntnisse in der Medizin. Karger, Basel 1979.

Kleemeier, R. W.: Somatopsychological effects of illness in the aged person. Geriatrics 13 (1958) 441–449.

Kleiger, J. H., N. F. Jones: Characteristics of alexithymic patients in a chronic respiratory illness population. J. nerv. ment. Dis. 168 (1980) 465–470.

Klein, D. F.: Delineation of two drug-responsive anxiety syndromes. Psychopharmacology 5 (1964) 397–408.

Klein, D. F.: Anxiety. In: Ban, T. A., P. Pichot, W. Pöldinger (eds.): Modern Problems of Pharmacopsychiatry, Vol. 22. Karger, Basel 1987.

Klein, G.: Adaptationsfähigkeit des alternden Organismus. In: Lang, E. (Hrsg.): Praktische Geriatrie. Enke, Stuttgart 1988.

Klein, H., R. Moses: Psychological reaction to sensory deprivation in patients with ablatio retinae. Psychother. and Psychosom. 24 (1974) 41–52.

Klein, M.: Zur Genese des Tics. Int. Z. Psychoanal. 11 (1925) 332–350.

Klein, M.: Neid und Dankbarkeit. In: Das Seelenleben des Kleinkindes. Klett, Stuttgart 1962.

Klein, M.: Das Seelenleben des Kleinkindes. Klett-Cotta, Stuttgart 1983.

Klein, M., K. Kellner: Interaktion zwischen Intensivstationspatienten. Prakt. Anästh. 14 (1979) 406–411.

Klein, P.: Tanztherapie, eine einführende Betrachtung im Vergleich mit Konzentrativer und Integrativer Bewegungstherapie. Pro Janus, Suderburg 1983.

Klein, R., B. E. K. Klein, S. E. Moss, M. D. Davis, D. L. Demets: The Wisconsin epidemiologic study of diabetic retinopathy. II. Prevalence and risk of diabetic retinopathy when age at diagnosis is less than 30 years. Arch. Ophthalmol. 102 (1984) 520–526.

Klein, R. F., V. A. Kleiner, D. S. Zipes: Transfer from a CCU – some adverse reactions. Arch. intern. Med. 122 (1968) 104–108.

Klein, R. F., T. F. Garrity, J. Gelein: Emotional adjustment and catecholamine excretion during early recovery from myocardial infarction. J. psychosom. Res. 18 (1974) 425–433.

Kleinman, A. M.: Medicine's symbolic reality: on a central problem in the philosophy of medicine. Inquiry (Oslo) 16 (1973) 206–213.

Kleinman, A. M.: Patients and healers in the context of culture. An exploration of the borderland between anthropology, medicine and psychiatry. Univ. of California Press, Berkeley-Los Angeles-London 1980.

Kleinman, A. M.: The illness narratives. Suffering, healing and the human condition. Basic Books, New York 1988.

Kleinsorge, H.: Phosphaturie und Neurose. Ärztl. Wschr. 7 (1952) 133.

Kleinsorge, H.: Hypnose. Methodik und Indikation. Fischer, Stuttgart-New York 1986.

Kleinsorge, H.: Selbstentspannung und gezieltes Organtraining. Trainingsheft für das autogene Training, 7. Aufl. Fischer, Stuttgart-New York 1988.

Klerman, G. L.: History and development of modern concepts of affective illness. In: Post, R. M., J. L. Ballenger (eds.): Neurobiology of Mood Disorders. Williams & Wilkins, Baltimore 1984.

Klerman, G. L.: Drugs and psychotherapy. In: Garfield, S. L., A. E. Bergin (eds.): Handbook of Psychotherapy and Behavior Change, 3rd ed. Wiley, New York 1986.

Klerman, G. L., M. M. Weissman, B. J. Rounsaville, E. S. Chevron: Interpersonal psychotherapy of depression. Basic Books, New York 1984.

Literatur

Klessmann, E.: Psychogene Polyurie – ein anachronistisches Konversionssyndrom? Psychother. med. Psychol. 37 (1987) 205.

Klessmann, E., H. A. Klessmann: Ambulante Psychotherapie der Anorexia nervosa unter Anwendung des Katathymen Bilderlebens. In: Leuner, H., G. Horn, E. Klessmann (Hrsg.): Katathymes Bilderleben mit Kindern und Jugendlichen. Reinhardt, München 1978.

Klessmann, E., H. A. Klessmann: Heiliges Fasten – Heilloses Fressen. Huber, Stuttgart 1988.

Klicks, B. R., M. J. Burgess, J. A. Abildskov: Influence of sympathetic tone on ventricular fibrillation threshold during experimental coronary occlusion. Amer. J. Cardiol. 36 (1975) 45.

Kliman, G.: Notfälle in der pädiatrischen Praxis. Hippokrates, Stuttgart 1973.

Klosterhalfen, W.: Experimenteller Streß und Adjuvans-Arthritis: Ein Beitrag zur Psychoimmunologie. Athenäum, Frankfurt 1987.

Klosterhalfen, W.: Experimental stress and adjuvant arthritis: a contribution to psychoimmunology. German J. Psychol. 12 (1988) 62–63.

Klosterhalfen, W.: A promising new strategy for studying conditioned immunomodulation. Behav. Brain Sci. 12 (1989) 150.

Klosterhalfen, S., W. Klosterhalfen: Conditioned immunopharmacologic effects and adjuvant arthritis: further results. In: Spector, N. H. (ed.): Proceedings of the First International Workshop on Neuroimmunomodulation, pp. 183–187. IWGN, Bethesda, MD 1985 a.

Klosterhalfen, S., W. Klosterhalfen: Conditioned taste aversion and traditional learning. Psychol. Res. 47 (1985 b) 71–94.

Klosterhalfen, S., W. Klosterhalfen: Classically conditioned effects of cyclophosphamide on white blood cell counts in rats. Ann. N. Y. Acad. Sci. 496 (1987a) 569–577.

Klosterhalfen, S., W. Klosterhalfen: Potentiation of drug-induced leukopenia in rats by conditioning. J. Psychophysiol. 1 (1987 b) 299.

Klosterhalfen, W., S. Klosterhalfen: Pavlovian conditioning of immunosuppression modifies adjuvant arthritis in rats. Behav. Neurosci. 97 (1983) 663–666.

Klosterhalfen, W., S. Klosterhalfen: On demonstrating that conditioned immunomodulation is conditioned. Behav. Brain Sci. 8 (1985c) 404–405.

Klosterhalfen, W., S. Klosterhalfen: Psychoimmunologie. In: Amelang, M. (Hrsg.): Bericht über den 35. Kongreß der Dtsch. Ges. f. Psychol. in Heidelberg 1986, Bd. 2, S. 59–69. Hogrefe, Göttingen 1987.

Klosterhalfen, W., S. Klosterhalfen: Classically conditioned cyclosporin A effects – a model to study brain-immune system interactions. Immunobiol. 178 (1988 a) 155–156.

Klosterhalfen, W., S. Klosterhalfen: Effects of restraint on adjuvant arthritis in two strains of rats. In: Hellhammer, D., I. Florin, H. Weiner (eds.): Neurobiological Approaches to Human Disease, pp. 392–396. Huber, Toronto 1988 b.

Klosterhalfen, W., S. Klosterhalfen: Psychologische Faktoren, Immunität und Krankheit. In: Speidel, H., B. Strauß (Hrsg.): Zukunftsaufgaben der Psychosomatischen Medizin, S. 133–156. Springer, Heidelberg 1989.

Klosterhalfen, W., J. Kugler, S. Klosterhalfen: Film-induced mood changes and salivary immunoglobulin A. J. Psychophysiol. 1 (1987) 302.

Klosterhalfen, S., U. Stockhorst, H.-J. Steingrüber: Are anticipatory nausea and vomiting in chemotherapy learned responses? Psychophysiol. 3 (1989) 325.

Klotz, U., I. Reimann: Influence of cimetidine on the pharmacokinetics of desethyldiazepam and oxazepam. Eur. J. clin. Pharmacol. 18 (1980) 517–520.

Klotz, U., V. I. Auttila, I. Reimann: Cimetidine-diazepam interaction. Lancet II (1979) 699.

Klüver, H.: The temporal lobe syndrome produced by bilateral ablations. In: Wolstenholm, E. E., C. M. O'Connor (eds.): Neurological Basis of Behavior, pp. 175–182. Churchill, London 1958.

Klüver, H., P. C. Bucy: Preliminary analysis of the temporal lobes in monkeys. Arch. Neurol. Psychiat. (Chic.) 42 (1939) 979.

Klumbies, G.: Psychotherapie in der Inneren und Allgemeinmedizin. Hirzel, Leipzig 1983.

Klußmann, R.: Psychosomatische Aspekte kolektomierter Patienten. Fortschr. Med. 97 (1979) 318–320.

Knapp, M.: Renal failure – dilemmas and development. Brit. med. J. 284 (1982).

Knapp, P. H.: Free association as a biopsychosocial probe. Psychosom. Med. 1 (1980) 197–219.

Knapp, P. H., C. Mushatt, S. J. Nemetz: Asthma, melancholia, and death: I. Psychoanalytic considerations. Psychosom. Med. 28 (1966 a) 114.

Knapp, P. H., E. C. Herman, C. Mushatt, S. J. Nemetz: Asthma, melancholia, and death. Psychosom. Med. 28 (1966 b) 134–154.

Knapp, P. H., S. Levin, R. H. McCarter, H. Werner, G. Zetzel: Suitability for psychoanalysis. A review of a hundred supervised analytic cases. Psychoanal. Quart. 29 (1960) 459–477.

Knapp, S. A. In: Jores, A. (Hrsg.): Praktische Psychosomatik. 2. Aufl., S. 371–395. Huber, Bern-Stuttgart-Wien 1981.

Knapp, T. J., L. A. Wells: Behavior therapy for asthma: a review. Behav. Res. Ther. 16 (1978) 103–115.

Knapp, T. W.: Ein „Kognitiv-Behaviorales Streßbewältigungstraining" (KBST) zur Behandlung von Migräne: Eine kontrollierte verhaltenstherapeutische Fallstudie. Z. klin. Psychol. Psychother. 29 (1981) 238–246.

Knapp, T. W.: Migräne 1: Symptomatologie und Ätiologie. Beltz, Weinheim 1983a.

Knapp, T. W.: Migräne 2: Psychologische Therapie. Beltz, Weinheim 1983 b.

Knauth, P., W. Rohmert, J. Rutenfranz: Systematic selection of shift plans for continuous production with the aid of work-physiological criteria. Applied Ergonomics 10 (1979) 9–15.

Knight, R. P.: Evaluation of the results of psychoanalytic therapy. Amer. J. Psychiat. 98 (1941) 434–446.

Knipschild, P.: Medical effects of aircraft noise. Int. Arch. occup. environm. Health 40 (1977) 185–190.

Knölker, U.: Psychotherapie bei Colitis ulcerosa in der Adoleszenz. Prax. Kinderpsychol. Kinderpsychiat. 35 (1986) 8–16.

Knoflach, P.: Ätiologie und Pathogenese von Morbus Crohn und Colitis ulcerosa. Wien. Klin. Wschr. 98 (1986) 754–758.

Knop, J., A. Fischer: Duodenal ulcer, suicide, psychopathology and alcoholism. Acta psychiat. scand. 63 (1981) 346–355.

Knox, C. J.: Psychiatric aspects of mitral valvotomy. Brit. J. Psychiat. 109 (1963) 656–668.

Ko, G. N., D. C. Jimerson, R. J. Wyatt, L. B. Bigelow: Plasma 3-methoxy-4-hydroxyphenylglycol changes associated with clinical state and schizophrenic subtype. Arch. gen. Psychiat. 45 (1988) 842–846.

Koch, L.: Psychosomatische Erkrankungen von Frauen und Körperpsychotherapie (Konzentrative Bewegungstherapie). Vortrag auf der Internat. Arbeitstagung des DKPM, Innsbruck 1988.

Koch, M. F., G. D. Molnar: Psychiatric aspects of patients

with unstable diabetes mellitus. Psychosom. Med. 36 (1974) 57–68.

Koch, U.: Erleben der Dialysesituation. In: Balck, F. B., U. Koch, H. Speidel (Hrsg.): Psychonephrologie. Springer, Heidelberg 1984 a.

Koch, U.: Selbst- und Fremdbildveränderungen unter der Dialyse. In: Balck, F. B., U. Koch, H. Speidel (Hrsg.): Psychonephrologie. Springer, Heidelberg 1984 b.

Koch, U., C. Schmeling: Umgang mit Sterbenden – ein Lernprogramm für Ärzte, Medizinstudenten und Krankenschwestern. Med. Psychol. 1 (1978).

Koch, U., C. Schmeling: Umgang mit Schwer- und Todkranken. Urban & Schwarzenberg, München 1982.

Koch, U., R. Schmid: Abschlußbericht. Evaluation des Modellprogramms „Psychosoziale Betreuung krebskranker Kinder und Jugendlicher". (1988, unveröffentlichtes Manuskript).

Koch, U., B. Siegrist: Psychosomatische Dienste in medizinischen Kliniken – die Kooperationsfrage unter forscherischer Perspektive. In: Bräutigam, W. (Hrsg.): Kooperationsformen somatischer und psychosomatischer Medizin. Springer, Berlin-Heidelberg 1988.

Koch, U., H. Speidel, F. Balck: Psychische Problem von Hämodialysepatienten und ihren Partnern. In: Beckmann, D., S. Davies-Osterkamp, J. W. Scheer (Hrsg.): Medizinische Psychologie. Springer, Heidelberg 1982.

Koch, U., M. Beutel, M. Broda, F. A. Muthny: Psychische Probleme vor und nach einer Nierentransplantation und Möglichkeiten psychologischer Interventionen. Projektzwischenbericht, Abt. Rehabilitationspsychologie der Univ. Freiburg 1983.

Koch, U., J. Fauler, P. Safian, C. Jährig: Affekte bei Ärzten und Patienten während der Visite: eine Analyse verbalisierter Affekte mit dem Gottschalk-Gleser-Verfahren an Hamburger und Ulmer Visitengesprächen. In: Köhle, K., H.-H. Raspe (Hrsg.): Das Gespräch während der ärztlichen Visite, S. 196–209. Urban & Schwarzenberg, München 1982.

Koch, W., E. Kriener: Langzeitverlauf beim Morbus Crohn. Therapiewoche 31 (1981) 8159–8166.

Kochanowski-Wilmink, J., W. Belschner: Lebensperspektiven Drogenabhängiger nach einer HIV-Infektion. In: Sigusch, V., S. Fliegel (Hrsg.): AIDS. DGVT, Tübingen 1988.

Kocher, R.: Psychopharmakotherapie bei Schmerzzuständen. In: Langer, E., H. Heimann (Hrsg.): Psychopharmaka – Grundlagen und Therapie. Springer, Wien 1983.

Kocher, R.: The use of psychotropic drugs in the treatment of cancer pain.. In: Zimmermann, A., P. Drings, G. Wagner (eds.): Recent Results in Cancer Research, Vol. 89. Springer, Berlin 1984.

Kocher, R., J. Schär: Zur Behandlung schwerer Schmerzzustände mit einer Kombination von Tofranil (Imipramin) und Nozinan (Laevopromazin). Praxis 57 (1968) 1459–1464.

Kockott, G.: Verhaltenstherapie bei sexuellen Störungen. In: Eicher, W. (Hrsg.): Sexualmedizin in der Praxis, S. 453–494. Fischer, Stuttgart-New York 1980.

Kockott, G.: Psychiatrische Aspekte bei der Entstehung und Behandlung chronischer Schmerzzustände. Nervenarzt 53 (1983) 365–376.

Koe, B. K., A. Weissman: P-chlorophenylalanine: a specific depletor of brain serotonin. J. Pharmacol. exp. Ther. 154 (1966) 499–516.

Köbberling, J.: Genetic heterogeneities within idiopathic diabetes. In: Creutzfeldt, W., J. Köbberling, J. V. Neel (eds.): The Genetics of Diabetes mellitus, pp. 79–87. Springer 1976.

Köbberling, J., H. T. Illil: Genetik des Diabetes mellitus. Münch. med. Wschr. 126 (1984) 727–730.

Köhle, K.: Psychosomatische Untersuchungen an Patienten mit Gliedmaßenarterienverschlüssen. Diss., München 1969.

Köhle, K.: Ein Konzept zur Bearbeitung von psychologischen Problemen auf Schwerkrankenstationen. In: Bönisch, E., J. E. Meyer (Hrsg.): Psychosomatik in der klinischen Medizin, S. 118–139. Springer, Berlin 1983.

Köhle, K.: Probleme im Umgang mit der Angst körperlich Schwerkranker. In: Götze, P. (Hrsg.): Leitsymptom Angst, S. 67–75. Springer, Heidelberg 1984.

Köhle, K.: Zur psychischen Führung von Patienten mit Reizmagen und Colon irritabile. In: Goebell, H. et al. (Hrsg.): Der chronisch Kranke in der Gastroenterologie, S. 474–480. Springer, Berlin 1984.

Köhle, K., C. Simons: Psychodynamic aspects in young patients suffering from peripheral vascular occlusions. Psychother. and Psychosom. 18 (1970) 313–320.

Köhle, K., B. Kubanek: Zur Zusammenarbeit von Psychosomatikern und Internisten. Erfahrungen aus zwölf Jahren. In: Uexküll, Th.v. (Hrsg.): Integrierte Psychosomatik, S. 17–54. Schattauer, Stuttgart 1981.

Köhle, K., H.-H. Raspe (Hrsg.): Das Gespräch während der ärztlichen Visite. Urban & Schwarzenberg, München 1982.

Köhle, K., H. Mall: Follow-up study of 36 anorexia nervosa patients treated on an integrated internistic-psychosomatic ward. Int. J. Eat. Dis. 2 (1983) 215–219.

Köhle, K. et al.: Klinische Psychosomatik. In: Begemann, H. (Hrsg.): Patient und Krankenhaus, S. 91–135. Urban & Schwarzenberg, München 1976.

Köhle, K., D. Böck, A. Grauhan: Die internistisch-psychosomatische Krankenstation. Editiones Roche, Basel 1977.

Köhle, K., B. Kubanek, C. Simons: Informed consent – psychologische Gesichtspunkte. Der Internist 23 (1982) 209–217.

Köhle, K., A. Erath-Vogt, D. Böck: Das Erstgespräch mit Patienten. Ein Lehrprogramm für die Krankenpflege, 6. Aufl. Rocom, Basel 1983.

Köhle, K., E. Gaus, R. Karstens, D. Ohlmeier: Ärztliche Psychotherapie bei Herzinfarktkranken während der Intensivbehandlungsphase. Therapiewoche 22 (1972) 4379–4382.

Köhle, K., C. Simons, B. Scholich, N. Schäfer: Critical theses concerning the future development of integrated psychosomatic departments. Psychother. and Psychosom. 22 (1973) 200–204.

Köhle, K., K. H. Schultheis, C. Simons, B. Scholich: Bedingungen und Möglichkeiten psychosomatischer Krankenbehandlung auf internistischen Stationen. Verh. Dtsch. Ges. Inn. Med. 79 (1973) 1444–1447.

Köhle, K., C. Simons, D. Böck, A. Grauhan (Hrsg.): Angewandte Psychosomatik. Die internistisch-psychosomatische Krankenstation – Ein Werkstattbericht. Rocom, Basel 1980.

Köhle, K., H. Kächele, H. Franz, H. Urban, W. Geist, H. Bosch: Integration der psychosomatischen Medizin in die Klinik: Die Funktion einer Schwesternarbeitsgruppe „patientenzentrierte Medizin". Med. Klin. 67 (1978) 1611–1615; 1644–1648.

Köhler, D., M. Langer, U. Schultz, T. Weiss, A. Stäbler: Klinische und computertomographische Verlaufskontrollen konservativ behandelter Diskusvorfälle. Röntgen-Bl. 8 (1989) 346–351.

Koehler, K., H. Saß (Hrsg.): Diagnostisches und Statistisches Manual Psychischer Störungen (Deutsche Bearbeitung und Übersetzung). Beltz, Weinheim 1984.

Koehler, T.: Psychosomatische Krankheiten. Kohlhammer, Stuttgart 1989.

Koehler, T.: Stress and rheumatoid arthritis: a survey of empirical evidence in human and animal studies. J. psychosom. Res. 29 (1985) 655–663.

Köhler, T.: Zur Psychogenese der rheumatoiden Arthritis – was wissen wir wirklich? Z. Rheum. 46 (1987) 183–188.

Koehler, W.: Nachweis einfacher Strukturfunktionen beim Schimpansen und beim Haushuhn: über eine neue Methode zur Untersuchung des bunten Farbsystems. Abh. Preuss. Akad. d. Wiss. Berlin (1929) 51–101.

Köhnken, G., G. Seidenstücker, U. Baumann: Zur Systematisierung von Methodenkriterien für Psychotherapiestudien. In: Baumann, U., H. Berbalk, G. Seidenstücker (Hrsg.): Klinische Psychologie, Trends in Forschung und Praxis. Huber, Bern-Stuttgart-Wien 1979.

Koella, W. P.: Die Physiologie des Schlafes. Fischer, Stuttgart-New York 1988.

Kölmel, H. W.: Visuelle Halluzinationen im hemianopen Feld bei homonymer Hemianopsie. Springer, Berlin-Heidelberg-New York 1984.

Koelsch, F.: Arbeit bzw. Beruf in ihrem Einfluß auf Krankheit und Sterblichkeit. In: Mosse, M., G. Tugendreich (Hrsg.): Krankheit und soziale Lage, S. 154–232. Lehmann, München 1913.

Koelsch, F.: Lehrbuch der Arbeitsmedizin, Bd. 1. 4. Aufl. Enke, Stuttgart 1963. (1. Aufl. Lehrbuch der Arbeitshygiene, Enke, Stuttgart 1937).

König, R.: Psychophysiologische Untersuchungen zur interozeptiven Wahrnehmung von extern zugeführten Atemwegswiderständen mit Hilfe der Signal-Detektions-Theorie. Psycholog. Diplomarbeit, Hamburg 1981.

Koenigsberg, H. W., R. Handley: Expressed emotion: from predictive index to clinical construct. Amer. J. Psychiat. 143 (1986) 1361–1373.

Köpp, W.: Bedeutung seelischer Faktoren bei der Behandlung des Diabetes mellitus. Münch. med. Wschr. 131 (1989) 58–62.

Koerfer, A., C. Neumann: Alltagsdiskurs und psychoanalytischer Diskurs. In: Flader, D., W. D. Grodzicki, K. Schröter (Hrsg.): Psychoanalyse als Gespräch, S. 97–137. Suhrkamp, Frankfurt 1982.

Körner, P.: Zur Prognose der Anorexia nervosa: Verlaufsstudien mit 38 Patienten. Med. Diss., Heidelberg 1978.

Kog, E., R. Pierloot, W. Vandereycken: Methodical considerations of family research in anorexia nervosa. Int. J. Eat. Dis. 2 (1983) 79–84.

Kog, E., H. Vertommen, W. Vandereycken: Minuchin's psychosomatic family model revised: a concept-validation study using a multitrait-multimethod approach. Fam. Process 26 (1987) 235–253.

Kohlenberg, R. J.: Tyramine sensitivity in dietary migraine: a critical review. Headache 22 (1982) 30–34.

Kohut, H.: The analysis of the self. New York 1971.

Kohut, H.: Narzißmus. Eine Theorie der psychoanalytischen Behandlung narzißtischer Persönlichkeitsstörungen. Suhrkamp, Frankfurt 1974.

Kohut, H.: The restoration of the self. New York 1977. Deutsch: Die Heilung des Selbst. Frankfurt 1979.

Koivisto, V. A., V. Soman, P. Conrad, R. Hendler, E. Nadel, Ph. Felig: Insulin binding to normocytes in trained athletes. J. clin. Invest. 64 (1979) 1011–1015.

Kolata, G.: Puberty mystery solved. Science 223 (1984) 272.

Kolb, H., F. A. Gries: Viruserkrankungen, Autoimmunität und Insulinmangeldiabetes. Dtsch. Ärztebl. 79 (1982) 18–39.

Kollar, E. J., D. T. Fullerton, R. Di Censo, C. F. Agler: Stress specificity in ulcerative colitis. Comprehens. Psychiat. 5 (1964) 101–112.

Koller, E. A.: Atmung und Kreislauf im anaphylaktischen Asthma bronchiale des Meerschweinchens, III. Die Lungenveränderungen im Asthmaanfall und die inspiratori-

sche Reaktion. Helv. Physiol. Pharmacol. Acta 26 (1968) 153–170.

Kollman, B. S., R. L. Verrier, B. Lown: The effect of vagal nerve stimulation upon vulnerability of the canine ventricle: role of sympathetic-parasympathetic interventions. Circulation 52 (1975) 578–585.

Kolodny, R. C.: Evaluating sex therapy: Process and outcome at the Masters & Johnson Institute. J. Sex. Res. 17 (1981) 301–318.

Kolodny, R. C., W. H. Masters, V. E. Johnson: Textbook of sexual medicine. Little, Brown and Co., Boston 1979.

Kolter, M., B. Tabatznik, M. M. Mower et al.: Prognostic significance of ventricular ectopic beats with respect to sudden death in the late post-infarction period. Circulation 47 (1973) 959.

Kolterman, O. G., J. A. Scarlett, J. M. Olefsky: Insulin resistance in non-insulin dependent type II diabetes mellitus. Clin. Endocr. Metabol. 11 (1982) 363–388.

Konstanty, R.: Thesen für ein Arbeitsschutzgesetz. WSI-Mitteilungen 34 (1981) 105–112.

Koolhaas, J. M., D. S. Fokkema, B. Bohus, G. A. van Oortmerssen: Individual differences in blood pressure reactivity and behavior of male rats. In: Schmidt, T., T. Dembroski, G. Blümchen (eds.): Biological and Psychological Factors in Cardiovascular Disease, pp. 517–526. Springer, Berlin-Heidelberg-New York 1986.

Koop, H.: Neuere Aspekte in der Therapie von Gastritis und peptischen Erkrankungen des Ösophagus und Magens. In: Mayer, K. P., W. Gerok (Hrsg.): Trends in der Gastroenterologie. Urban & Schwarzenberg, München 1990.

Koran, L. M., R. H. Moos, B. Moos, M. Zasslow: Changing hospital work environments: an example of a burn unit. Gen. Hosp. Psychiat. 5 (1983) 7–13.

Kordy, H.: Probleme der Gruppenstatistik und Einzelfallforschung. Z. different. diagnost. Psychol. 3 (1982) 231–239.

Kordy, H.: Bemerkungen zur empirischen Erforschung von Einflußgrößen bei Indikationsentscheidungen. Z. personenzentr. Psychol. Psychother. (1983) 119–131.

Kordy, H., D. Scheibler: Individuumsorientierte Erfolgsforschung: Erfassung und Bewertung von Therapieeffekten anhand individueller Behandlungsziele. Z. klin. Psycholog., Psychopathol. u. Psychother. 32 (1984) 218–233.

Kordy, H., W. Senf: Überlegungen zur Evaluation psychotherapeutischer Behandlungen. Psychother. med. Psychol. 35 (1985).

Kordy, H., M.v. Rad, W. Senf: Success and failure in psychotherapy: hypotheses and results from the Heidelberg follow-up project. Psychother. and Psychosom. 40 (1983) 1–4, 211–227.

Korelitz, B., D. Gribetz, I. Danziger: The prognosis of ulcerative colitis with onset in childhood. I. The pre-steroid era. Ann. intern. Med. 57 (1962) 582–591.

Koriat, A., R. Melkman, J. R. Averill, R. S. Lazarus: The self-control of emotional reactions to stressful film. J. Pers. 40 (1972) 601–619.

Korneva, E. A., L. M. Khai: Effect of destruction of hypothalamic areas on immunogenesis. Fizio Zh SSSR Sechenov 49 (1963) 42.

Kornfeld, D. S.: Psychiatric complications of cardiac surgery. Int. Psychiat. Clinics, Boston, Vol. 4, No. 2 (1967) 115–131.

Kornfeld, D. S.: Psychiatric aspects of patient care in the operating suite and special areas. J. Anaesthesiol. 31 (1969) 166–171.

Kornfeld, D. S.: Psychiatric view of the intensive care unit. Brit. J. Med. 1 (1969a) 108–110.

Kornfeld, D. S.: The hospital environment: its impact on the patient. Advanc. psychosom. Med. 8 (1972) 252–270.

Kornfeld, D. S., S. Zimberg, J. R. Malm: Psychiatric complications of open-heart surgery. New Engl. J. Med. 273 (1965) 287–293.

Kornitzer, M., F. Kittel, M. Dramaix, G. DeBacker: Job stress and coronary heart disease. Advanc. Cardiol. 29 (1982) 56–61.

Korsch, B. M., V. Negrete: Doctor-patient communication. Sci. Amer. 227 (1972) 227–266.

Korsch, B. M., E. K. Gozzi, V. Francis: Gaps in doctor-patient communication. Pediatrics 42 (1968) 855.

Korten, J., K. Ketterings: Anthropologische Aspekte der Parkinsonschen Krankheit. Nervenarzt 43 (1972) 201–205.

Kortmann, R.: Beobachtungen zur Rehabilitation von Schrittmacherpatienten. Med. Welt 25 (1974) 579–582.

Kortweg, G. C., J. T. Boelles, J. Ten Cate: Influences of stimulation of some subcortical areas on the electrocardiogram. J. Neurophysiol. 20 (1957) 100.

Kosbab, F. B.: Symbolismus, Selbsterfahrung und die didaktische Anwendung des Katathymen Bilderlebens in der psychiatrischen Ausbildung. Z. Psychother. med. Psychol. 22 (1972) 210–224.

Koskenvuo, M., J. Caprio, A. Kesaniemi, S. Sarna: Differences in mortality from ischemic heart disease by marital status and social class. J. chron. Dis. 33 (1980) 95.

Koski, M. L.: The coping processes in childhood diabetes. Acta paediat. scand. (Suppl.) 198 (1969) 1–52.

Koslow, J. H., J. W. Maas, C. L. Bowden, J. M. Davis, I. Hanin, J. Javaid: CSF and urinary biogenic amines and metabolites in depression and mania: a controlled, univariate analysis. Arch. gen. Psychiat. 40 (1983) 999–1014.

Kossmann, B., I. Bowlder: Alternative treatments in chronic headache. In: Holroyd, K. A., B. Schlote, H. Zenz (eds.): Perspectives in Research on Headache, pp. 126–136. Hogrefe, Lewiston-New York 1983.

Kost, U. (1979): Vom Erkennen der Erlebnisstörung in der Konzentrativen Bewegungstherapie. In: Stolze, H. (Hrsg.): Konzentrative Bewegungstherapie, S.460. 2. Aufl. Springer, Berlin-Heidelberg-New York 1989.

Kotses, H. et al.: Operant muscular relaxation and peak expiratory flow rate in asthmatic children. J. psychosom. Res. 22 (1978) 17–23.

Kottje-Birnbacher, L.: Paartherapie mit dem Katathymen Bilderleben – eine Falldarstellung. Familiendynamik 6 (1981) 260.

Koumans, A. J.: Psychiatric consultation in an intensive care unit. J. Amer. med. Ass. 194 (1965) 633–637.

Kovelman, J. A., A. B. Scheibel: Biological substates of schizophrenia. Acta neurol. scand. 73 (1986) 1–32.

Kovács, V.: Analyse eines Falles von „Tic convulsif". Int. Z. Psychoanal. 11 (1925) 318–325.

Kraepelin, E.: Der psychologische Versuch in der Psychiatrie. Psychologische Arbeiten 1, 1895.

Kraepelin, E.: Psychiatrie: Ein Lehrbuch für Studierende und Ärzte, 5. Aufl. Barth, Leipzig 1896.

Kraft, H.: Autogenes Training. Methodik und Didaktik. Hippokrates, Stuttgart 1989.

Krakowski, A.: Psychosomatic aspects of aging. Psychiat. J. Univ. Ottawa 1 (1976) 151–157.

Krakowski, A. J.: Psychiatric consultations for the geriatric population in the general hospital. Bibl. Psychiat. 159 (1979) 163–185.

Krakowski, A. J., A. J. Krakowski: Long-range psychosomatic effects on cardiac arrest survivors. 3 rd Congr. Int. Coll. Psychosom. Med., Rome 1975.

Krank, M. D., M. G. MacQueen: Conditioned compensatory responses elicited by environmental signals for cyclophosphamide-induced suppression of antibody production in mice. Psychobiol. 16 (1988) 229–235.

Krant, M. J., L. Johnston: Communication and the late-stage cancer patient (meeting abstract). Proc. Amer. Soc. clin. Oncol. 17 (1976) 251.

Krantz, D. S., S. B. Manuck: Acute psychophysiologic reactivity and risk of cardiovascular disease: a review and methodologic critique. Psychol. Bull. 96 (1984) 435–464.

Krantz, D. S., D. C. Glass, M. L. Snyder: Helplessness, stress level and the coronary-prone behavior pattern. J. exp. soc. Psychol. 10 (1974) 284–300.

Krantz, D. S., M. A. Schaeffer, J. E. Davis et al.: Extent of coronary atherosclerosis in men. Psychophysiology 18 (1981) 654–664.

Krantz, D. S., M. J. Sanmarco, R. H. Selvester, K. A. Matthews: Psychological correlates of progression of atherosclerosis in men. Psychosom. Med. 41 (1979) 467–476.

Krapf, G.: Autogenes Training aus der Praxis, 2. Aufl. Springer, Berlin-Heidelberg-New York 1976.

Krasner, L.: The therapist as a social reinforcement machine. In: Strupp, H. H., L. Luborski (eds.): Research in Psychotherapy, Vol. II, p. 61–94. APA 1962.

Kraupl-Taylor, F.: The concepts of disease. Psychol. Med. 10 (1980) 419–424.

Kraus, A.: Biographische und verhaltenspsychologische Untersuchungen beim postenzephalitischen Parkinsonismus. Ein Beitrag zur Psychosomatik extrapyramidaler Erkrankungen. Med. Diss., Heidelberg 1964.

Kraus, A.: Störungen der Wahrnehmung und des Leiberlebens beim Parkinsonismus. Nervenarzt 45 (1974) 639–646.

Krause, R.: Zur Onto- und Phylogenese des Affektsystems und ihrer Beziehung zu psychischen Störungen. Psyche 37 (1983) 1016.

Krause, R.: Eine Taxonomie der Affekte und ihre Anwendung. Psychother. med. Psychol. 138 (1988) 77.

Krause, R.: Emotionsstörungen. In: Scherer, K. U. (Hrsg.): Psychologie der Emotion, Bd. C/IV/3. Enzyklopädie der Psychologie. Hogrefe, Göttingen 1988 a.

Krause, W. H.: Integrierte psychosomatische stationäre Rehabilitation. Öff. Gesundh.-Wes. 47 (1985) 392.

Krauspe, C., K. Müller-Wieland, F. Stelzner (Hrsg.): Colitis ulcerosa und granulomatosa. Urban & Schwarzenberg, München 1972.

Krebs, D.: Bakteriologische Probleme bei der Adnexentzündung. Gynäk. 5 (1972) 214.

Krebs, D., W. Schallenberg: Bakteriologische Befunde bei gynäkologischen Erkrankungen unter besonderer Berücksichtigung anaerober Keime. Arch. Gynäk. 102 (1971).

Krebs, G.: Die Geburtsvorbereitung nach G. Dick-Read und ihre Weiterentwicklung bis in die Gegenwart. In: Prill, H. J., D. Langen (Hrsg.): Der psychosomatische Weg zur gynäkologischen Praxis. Schattauer, Stuttgart 1983.

Kreeger, L. (Hrsg.): Die Großgruppe. Klett, Stuttgart 1977.

Krehl, L.v.: Entstehung, Erkennung und Behandlung innerer Krankheiten. Vogel, Berlin 1932.

Kreisler, L.: L'enfant psychosomatique. PUF, Paris 1976.

Kreisler, L.: L'enfant du désordre psychosomatique. Privat, Toulouse 1981.

Kreisler, L.: Conduite à tenir devant l'insomnie d'un adolescent. Der informierte Arzt 1 (1981) 55–58.

Kreisler, L.: (a) La pathologie psychosomatique, pp. 423–443. (b) La clinique psychosomatique du nourrisson, pp. 695–712 (c) L'insomnie du nourrisson, pp. 713–722. (d) L'anorexie mentale du nourrisson, pp. 723–732. (e) La rumination ou mérycisme, pp. 733–739. (f) Les vomissements psychogènes, pp. 741–743. In: Lebovici, S., R. Diatkine, M. Soulé (eds.): Traité de psychiatrie de l'enfant et de l'adolescent, Vol. II. PUF, Paris 1985.

Kreisler, L., M. Fain, M. Soulé: L'enfant et son corps. PUF, Paris 1974.

Krejci, E.: Colitis haemorrhagica und Colitis ulcerosa in der psychosomatischen Klinik. Diss., Freiburg 1962.

Krejci, E.: Anamneseerhebung als Gespräch: Lernprozesse auf dem Weg zum „Psychosomatischen Arzt" – Erfahrungen mit einer freiwilligen Studentengruppe in Freiburg/ Breisgau. In: Schüffel, W. (Hrsg.): Sprechen mit Kranken – Erfahrungen studentischer Anamnesegruppen. Urban & Schwarzenberg, München 1983.

Krejci, E., W. Bohleber: Spätadoleszente Konflikte. Vandenhoeck & Ruprecht, Göttingen 1982.

Krell, U.: Arbeitswissenschaftliche Beurteilung von Schicht- und Nachtarbeit. Zbl. für Arbeitsmedizin, Arbeitsschutz, Prophylaxe und Ergonomie 30 (1980) 148–158.

Kretschmer, E.: Hysterie, Reflex und Instinkt. 4. Aufl. Thieme, Leipzig 1946.

Kretschmer, E.: Psychotherapeutische Studien. Thieme, Stuttgart 1949.

Kretschmer, E.: Gestufte Aktivhypnose – Zweigleisige Standardmethode. In: Frankl, V. E., V.v. Gebsattel, J. H. Schultz (Hrsg.): Handbuch der Neurosenlehre und Psychotherapie, Bd. IV, S. 130–141. Urban & Schwarzenberg, München-Berlin 1959.

Kretschmer, E.: Protreptik. In: Frankl, V. E., V.v. Gebsattel, J. H. Schultz (Hrsg.): Handbuch der Neurosenlehre und Psychotherapie. Bd. IV, S. 122–129. Urban & Schwarzenberg, München-Berlin 1959.

Kretschmer, E.: Medizinische Psychologie, 13. Aufl. Thieme, Stuttgart 1971.

Kreuz, L. E., R. M. Rose: Assessment of aggressive behavior and plasma testosterone in a young criminal population. Psychosom. Med. 34 (1972) 321.

Krieg. In: Rohen, J. W. (Hrsg.): Funktionelle Anatomie des Nervensystems. Schattauer, Stuttgart 1975.

Krieg, H., H. Brünner: Der rezidivierende Dünndarmileus beim Morbus Crohn. Therapiewoche 31 (1981) 5587–5594.

Krieg, J. C., H. Backmund, K. M. Pirke: Cranial computed tomography findings in bulimia. Acta psychiat. scand. 75 (1987) 144–149.

Krieg, J. C., C. Lauer, G. Leinsinger, W. Schreiber, K. Pirke: Hirnstruktur und Hirnfunktion bei Anorexia nervosa: eine computertomographische Untersuchung. Psychother. med. Psychol. 39 (1989) 256–259.

Krieger, D. T.: Brain peptides: what, where, and why? Science 222 (1983) 975–985.

Krieger, D. T., S. Glick: Sleep EEG stages and plasma growth hormone concentration in stages of endogenous and exogenous hypercortisolemia or ACTH elevation. J. clin. Endocr. 39 (1974) 980.

Krieger, D. T., L. Amorosa, F. Linick: Cyproheptadine-induced remission of Cushing's disease. New Engl. J. Med. 293 (1975) 893.

Krieger, J., R. C. Mellinger: Pituitary function in the deprivation syndrome. J. Pediat. 79 (1971) 216.

Kris, E.: On preconscious mental processes (1952). In: Kris, E. (ed.): Psychoanalytic Explorations in Art. New York 1962.

Krisch, K.: Enkopresis. Huber, Bern 1985.

Kröner, B.: The empirical validity of clinical headache classification. In: Holroyd, K. A., B. Schlote, H. Zenz (eds.): Perspectives in Research on Headache, pp. 56–65. Hogrefe, Lewiston-New York 1983.

Kröner-Herwig, B., K. W. Weich: Untersuchungen zur Prädiktion des Erfolgs verhaltensmedizinischer Interventionen bei chronischem Kopfschmerz. Z. klin. Psychol. 17 (1988) 55–69.

Krönig, B.: Blutdruckvariabilität bei Hochdruckkranken; Ergebnisse telemetrischer Langzeitmessung. Hüthig, Heidelberg 1976.

Krogh-Poulsen, W.: Zusammenhänge zwischen Lokalisation von Abrasionsfacetten und Schmerzen in der Kaumuskulatur und deren Bedeutung für Diagnostik und Behandlung. Öst. Z. Stomat. 64 (1967) 402.

Krogh-Poulsen, W.: Orthofunktion und Pathofunktion des mastikatorischen Systems unter Berücksichtigung der beteiligten Muskelgruppen. In: Drücke, W., B. Klempt (Hrsg.): Kiefergelenk und Okklusion, S. 13. Quintessenz, Berlin 1980.

Kronenfeld, J. J., C. Wasner: The use of unorthodox therapies and marginal practitioners. Soc. Sci. Med. 16 (1982) 1119–1125.

Krüger, K. W.: Lupus erythematodes und Zentralnervensystem. Nervenarzt 55 (1984) 165–172.

Krüskemper, G.: Patienten mit rheumatischen Beschwerden. In: Basler, H.-D., I. Florin (Hrsg.): Klinische Psychologie und körperliche Krankheit, S. 146–161. Kohlhammer, Stuttgart 1985.

Krüskemper, G., H. L. Krüskemper: Neurotische Tendenzen und Extraversion bei Hyperthyreose. Z. psychosom. Med. 16 (1970) 178.

Krüskemper, G., H. Zeidler: Testpsychologische Untersuchungen an Patienten mit chronischer Polyarthritis. Dtsch. med. Wschr. 37 (1975) 1833–1837.

Krüskemper, G., P. Schejbal: Kranksein als Streß. In: Eiff, A. W. von (Hrsg.): Streß, S. 154–164. Thieme, Stuttgart 1980.

Krug, H., F. Krug, P. Cuatrecasas: Emergence of insulin receptors on human lymphocytes during in vitro transformation. Proc. Natl. Acad. Sci. (Wash.) 69 (1972) 2604–2608.

Kruse, W.: Entspannung. Autogenes Training für Kinder. Dtsch. Ärzteverlag, Köln 1974.

Kruse, W.: Einführung in das autogene Training mit Kindern, 4. Aufl. Dtsch. Ärzteverlag, Köln 1984.

Krystal, H.: The genetic development of affects and affect repression. Ann. psychoanal. 2 (1974) 98–126.

Krystal, H.: Self-representation and the capacity of self-care. Ann. psychoanal. 6 (1978) 206–246.

Krystal, H.: Trauma and affects. Psychoanal. Study Child 33 (1978) 81.

Krystal, J.: Assessing alexithymia. In: Krystal, H. (ed.): Integration and Self-healing. Affect-Trauma-Alexithymia. Ann Arbor 1988.

Krystal, J., E. L. Giller, D. V. Cicchetti: Assessment of alexithymia in post-traumatic stress disorder and somatic illness: introduction of a reliable measure. Psychosom. Med. 48 (1986) 84–94.

Kubanek, B., H. Heimpel, G. Paar, A. Schoengen: Hämatologische Veränderungen bei der Anorexia nervosa. Blut 35 (1977) 115–124.

Kubany, A. J., T. S. Danowski, C. Moses: The personality and intelligence of diabetics. Diabetes 5 (1956) 462–467.

Kubie, L. S. A.: A psychoanalytic approach to the pharmacology of psychological processes. In: Uhr, L., J. G. Miller (eds.): Drugs and Behavior. Wiley, New York 1960.

Kubitz, K. A., B. S. Peavey, B. S. Moore: The effect of daily hassles on humoral immunity. An interaction moderated by locus of control. Biofeed. Self Regul. 11 (1986) 115–123.

Kuder, G., M. W. Richardson: The theory of estimation of test reliability. Psychometrica 2 (1937) 151.

Kübler-Ross, E.: On death and dying. What the dying have to teach doctors, nurses, clergy and their own families. McMillan, New York 1969. Deutsch: Interviews mit Sterbenden. Kreuz-Verlag, Stuttgart 1973.

Kübler-Ross, E.: On the use of psychopharmacologic agents for the dying patient and the bereaved. J. Thanatol. 2 (1972) 563–567.

Kübler-Ross, E.: Questions and answers on death and dying. Macmillan, New York 1974.

Kübler-Ross, E.: Was können wir noch tun? Kreuz-Verlag, Stuttgart 1974.

Kübler-Ross, E.: Vortrag. Thanatologie-Symposion, Kopenhagen, Dezember 1976.

Kübler-Ross, E.: Death. The final stage of growth. Prentice-Hall, Englewood Cliffs/New Jersey 1975. Deutsch: Reif werden zum Tode. Kreuz-Verlag, Stuttgart-Berlin 1976.

Kübler-Ross, E.: Die geheime Sprache sterbender Kinder. Deutsche Ärzteblatt 79 (1982) 55–67.

Küchenhoff, J.: Oneiroides Erleben bei intensivbehandelten panplegischen Polyradikulitis-Patienten. Nervenarzt 58 (1987) 524.

Küchenhoff, J., H. Kordy, W. Kruschitz, W. Senf: Persönlichkeit, Krankheitsverarbeitung und Krankheitsverlauf bei Morbus Crohn. In: Speidel, H., B. Strauß (Hrsg.): Zukunftsaufgaben der psychosomatischen Medizin. Springer, Berlin 1989.

Kügler, E.: Empirisch-psychologische Untersuchung von rheumatoider Arthritis im Vergleich zum chronischen Lumbalsyndrom unter besonderer Berücksichtigung der Aggressivitätsproblematik. Unveröffentlichte Diplomarbeit, Psychologisches Institut der Univ. Düsseldorf 1980.

Kühn, H., F. Hauss: Entwicklungstendenzen im medizinischen Arbeitsschutz. In: Haug, W. F. (Hrsg.): Jahrbuch für kritische Medizin 3, S. 96–115. Argument-Verlag, Berlin 1978.

Kühn, H. A., E. Nägele: Colitis ulcerosa. Ergeb. inn. Med. Kinderheilk. 25 (1967).

Künsebeck, H. W.: Psychosoziale Situation und Lebenszufriedenheit bei Herztransplantierten. Psycho 13 (1987) 760.

Künsebeck, H. W., W. Lempa, H. Freyberger: Kurz- und Langzeiteffekte ergänzender Psychotherapie bei Patienten mit Morbus Crohn. In: Lamprecht, F. (Hrsg.): Spezialisierung und Integration in Psychosomatik und Psychotherapie. Springer, Berlin 1987.

Küster, W., W. Lenz: Morbus Crohn und Colitis ulcerosa. Häufigkeit, familiäres Vorkommen und Schwangerschaftsverlauf. Ergeb. inn. Med. Kinderheilk. 53 (1984) 103–132.

Küster, W., J. Purrmann, S. Funk, G. Strohmeyer: Zur Genetik des Morbus Crohn. Med. Klinik 82 (1987) 679–682.

Kütemeyer, M.: Anthropologische Medizin in der inneren Klinik. In: Vogel, P. (Hrsg.): Victor von Weizsäcker: Arzt im Irrsal der Zeit. Eine Freundesgabe zum 70. Geburtstag am 21.4.1956, S. 243–265. Vandenhoeck & Ruprecht, Göttingen 1956.

Kütemeyer, M.: Symptom changes during the psychotherapy of patients with myasthenia gravis. Psychother. and Psychosom. 32 (1979) 279–286.

Kütemeyer, M.: Versuch der Integration psycho-somatischer Medizin in eine neurologische Universitätsklinik. In: Uexküll, Th. v. (Hrsg.): Integrierte Psychosomatische Medizin, S. 187–226. Schattauer, Stuttgart-New York 1981.

Kütemeyer, M., K. F. Masuhr: Psychosomatische Aspekte in der Neurologie. In: Jores, A. (Hrsg.): Praktische Psychosomatik, S. 353–370. 2. Aufl. Huber, Bern-Stuttgart-Wien 1981.

Kütemeyer, M., U. Schultz: Verlauf des akuten Wurzelkompressionssyndroms nach dreistufiger konservativer Therapie. In: Seitz, D., P. Vogel (Hrsg.): Hämoblastosen. Zentrale Motorik. Iatrogene Schäden. Myositiden, S. 699–701. Springer, Berlin-Heidelberg-New York 1983.

Kütemeyer, M., U. Schultz: Kurt Goldstein (1878–1965): Begründer einer psychosomatischen Neurologie? In: Pross, C., R. Winau (Hrsg.): Nicht mißhandeln. Das Krankenhaus Moabit 1920–1945, S. 133–139. Edition Hentrich, Berlin 1984.

Kütemeyer, M., U. Schultz: Psychosomatik des Lumbago-Ischias-Syndroms. In: Uexküll, Th.v. (Hrsg.): Lehrbuch der Psychosomatischen Medizin, 3. Aufl. S. 835–848. Urban & Schwarzenberg, München-Wien-Baltimore 1986.

Kütemeyer, M., U. Schultz: Psychosomatische Überlegungen zum funktionellen Kreuzschmerz der Frau. In: Prill, H. J., M. Stauber, A. Teichmann (Hrsg.): Psychosomatische Gynäkologie und Geburtshilfe 1987, S. 51–62. Springer, Berlin-Heidelberg 1988.

Kütemeyer, M., U. Schultz: Frühe psychoanalytische Schmerzauffassung. Psychother. med. Psychol. 39 (1989) 185–192.

Kütemeyer, W.: Körpergeschehen und Psychose. Stuttgart 1953.

Kütemeyer, W.: Die Krankheit in ihrer Menschlichkeit. Vandenhoeck & Ruprecht, Göttingen 1963.

Kütemeyer, W.: Anthropologische Medizin als klinisches Forschungsprinzip an einem Fall von primär chronischer Polyarthritis erläutert. Unveröffentlichte Antrittsvorlesung, Heidelberg 1967.

Kugelmass, S., J. Marcus, J. Schmueli: Psychophysiological reactivity in high-risk children. Schizophrenia Bull. 11 (1985) 66–73.

Kuhn, C. M., G. Evoniuk, S. M. Schanberg: Loss of tissue sensitivity to growth hormone during maternal deprivation in rats. Life Sci. 25 (1979) 2089.

Kuhn, H. et al.: Psychophysiologische Untersuchungen beim Asthma bronchiale. In: Zander, W. (Hrsg.): Experimentelle Forschungsergebnisse in der psychosomatischen Medizin, S. 129–140. Vandenhoeck & Ruprecht, Göttingen 1981.

Kuhn, Th.: Die Struktur wissenschaftlicher Revolutionen. Suhrkamp, Frankfurt 1973.

Kuiper, P. C.: Die seelischen Krankheiten des Menschen. Huber, Bern-Stuttgart-Wien – Klett-Cotta, Stuttgart 1968.

Kulenkampff, C., A. Bauer: Über das Syndrom der Herzphobie. Nervenarzt 31 (1960) 443.

Kuller, L., M. Cooper, J. Perper: Epidemiology of sudden death. Arch. intern. Med. 129 (1972) 714–719.

Kulovesi, Y.: Zur Entstehung des Tics. Int. Z. Psychoanal. 15 (1929) 82–95.

Kulovesi, Y.: Ein Beitrag zur Psychoanalyse des epileptischen Anfalls. Int. Z. Psychoanal. 20 (1934) 542–549.

Kummer, G. W., H. L. Fehm, K. H. Voigt, E. F. Pfeiffer: Transient cyproheptadine-induced remissions in Cushing's disease. Acta endocr. (Kbh.) (Suppl.) 225 (1979) 62.

Kunin, C. M.: Sexual intercourse and urinary infections. New Engl. J. Med 298 (1978) 336.

Kupfer, D. J., F. G. Foster: Interval between onset of sleep and rapid-eye-movement sleep as an indicator of depression. Lancet II (1972) 684–686.

Kupfer, D. J., R. J. Wyatt, J. Scott, F. Snyder: Sleep disturbance in acute schizophrenic patients. Amer. J. Psychiat. 126 (1970) 1213–1223.

Kurtzke, J. F.: Multiple sclerosis: an overview. In: Rose, C. F. (ed.): Clinical Neuroepidemiology. Pitman, Kent 1980.

Kuschinsky, G.: Wirkungen und Indikationen von Placebo. Dtsch. Ärztebl. (1975) 663–667.

Kusnecov, A. W., A. J. Husband, M. G. King: Behaviorally conditioned suppression of mitogen-induced proliferation and immunoglobulin production: effect of time span between conditioning and re-exposure to the conditioned stimulus. Brain, Behavior, and Immunity 2 (1988) 198–211.

Kusnecov, A. W., M. Sivyer, M. G. King, A. J. Husband, A. W. Cripps, R. L. Clancy: Behaviorally conditioned suppression of the immune response by antilymphocyte serum. J. Immunol. 130 (1983) 2117–2120.

Kutner, N. G.: Medical students' orientation toward the chronically ill. J. med. Educ. 53 (1978) 111–118.

Literatur

Kutter, P.: Elemente der Gruppentherapie. Vandenhoeck & Ruprecht, Göttingen 1976.

Kutter, P.: Psychoanalytische Ansätze. In: Euler, H. A., H. Mandel (Hrsg.): Emotionspsychologie, S. 52. Urban & Schwarzenberg, München-Wien-Baltimore 1983.

Kutter, P.: Methoden und Theorien der Gruppenpsychotherapie. Frommann-Holzboog, Stuttgart 1985.

Kutzner, J., R. Wanitschke, K. Ewe, W. Goldhofer, H. Gabert: Strahlenproktitis als Folge onkologischer Radiotherapie. Proktol. 2 (1979) 14–19.

Kvist, N., O. Jacobsen, P. Nörgaard, H. H. Ockelmann, H. K. Kvist, G. Schou, S. Jarnum: Malignancy in Crohn's disease. Scand. J. Gastroenterol. 21 (1986) 82–86.

L'Abate, L.: The handbook of family psychology and therapy. Vol. I. Dorsey, Chicago 1985.

Labbé, E. L., D. A. Williamson: Treatment of childhood migraine using autogenic feedback training. J. consult. clin. Psychol. 52 (1984) 968–976.

Laberke, J. A.: Klinische Erfahrungen mit dem Autogenen Training bei Herz- und Kreislauferkrankungen. In: Luthe, W. (Hrsg.): Autogenes Training. Thieme, Stuttgart 1965.

Labhardt, A.: Psychosomatische und psychodynamische Aspekte weichteilrheumatischer Erkrankungen. Referat, 17. Tagung d. Dtsch. Ges. Rheumatologie, Regensburg 1976.

Lacey, J. H., D. E. Batemann, R. v. Lehn: Autonomic response specifity. Psychosom. Med. 15 (1953) 8.

Lacey, J. I.: The evaluation of autonomic responses: toward a general solution. Ann. N. Y. Acad. Sci. 67 (1956) 123–164.

Lacey, J. I.: Psychophysiological approaches to the evaluation of psychogenetic process and outcome. In: Rubinstein, E. A., M. B. Parloff (eds.): Research in Psychotherapy. APA, Washington 1962.

Lacey, J. I.: Somatic response patterning and stress: some revisions of activation theory. In: Appley, M. A., R. Trumbell (eds.): Psychological Stress: Issues in Research, pp. 14–42. Appleton, New York 1967.

Lacey, J. I., B. Lacey: Verification and extension of the principle of autonomic response-stereotypes. Amer. J. Psychol. 71 (1958) 50–61.

Lacey, J. I., B. Lacey: Some autonomic-central nervous system interrelationships. In: Black, P. (ed.): Physiological Correlates of Emotion. Acad. Press, New York 1970.

Lacey, J. I., B. Lacey: On heart rate responses and behavior. A reply to Elliott. J. pers. soc. Psychol. 30 (1974) 1–18.

Lacey, J. I., B. Lacey: Studies of heart rate and other bodily processes in sensomotor behavior. In: Obrist, P. A., A. H. Black, J. Brener, L. V. Dicara (eds.): Cardiovascular Psychophysiology. Aldine, Chicago 1974.

Lacey, J. I., J. Kagan, B. Lacey, H. A. Moss: The visceral level in situational determinants and behavioral correlates of autonomic response patterns. In: Knapp, P. H. (ed.): Expressions of Emotions in Man. Int. Univ. Press, New York 1963.

Lader, M. H.: Palmar skin conductance measures in anxiety and phobic states. J. psychosom. Res. 11 (1967) 271–281.

Lader, M. H., L. Wing: Physiological measures, sedative drugs, and morbid anxiety. Oxford Univ. Press, London 1966.

Lader, M. H., N. Sartorius: Anxiety in patients with hysterical conversion symptoms. J. Neurol. Neurosurg. Psychiat. 31 (1968) 490–495.

Ladouceur, R., Y. Gros-Louis: Paradoxical intention vs. stimulus control in the treatment of severe insomnia. J. Behav. Ther. Exp. Psychiat. 17 (1986) 267–269.

Laessle, R. G.: Eßstörungen und Depression. Psychobiologische Studien bei Anorexia nervosa und Bulimie. Lang, Frankfurt 1987.

Laessle, R. G.: Affektive Störungen und bulimische Syndrome. In: Fichter, M. M. (Hrsg.): Bulimia nervosa. Enke, Stuttgart 1989.

Laessle, R. G., U. Schweiger, U. Daute-Herold, M. Schweiger, M. M. Fichter, K. M. Pirke: Nutritional knowledge in patients with eating disorders. Int. J. Eat. Dis. 7 (1988) 63–73.

Lagercrantz, R. H. C.: Ulcerative colitis. In: Linneweh, F. (Hrsg.): Prognose chronischer Erkrankungen. Springer, Berlin 1960.

LaHood, B. J.: Parental attitudes and their influences on the medical management of diabetic adolescents. Clin. Pediat. 9 (1970) 468–471.

Lain Entralgo, P.: La historia clinica, historia, theoria del relato patografico. Consejo Superior de Investigaciones Cientificas. Madrid 1950.

Lain Entralgo, P.: Heilkunde in geschichtlicher Entscheidung. Müller, Salzburg 1956.

Laing, R. D., H. Phillipson, A. R. Lee: Interpersonelle Wahrnehmung. Suhrkamp, Frankfurt 1971.

Lakoff, G.: Women, fire and dangerous things. Univ. of Chicago Press, Chicago 1987.

Lamaze, F., P. Vellay: L'accouchement sans douleur par la méthode psychophysique. Premiers résultats portant sur 500 cas. Gaz. méd. Fr. 59 (1952) 1445.

Lambert, M. J.: Introduction to assessment of psychotherapy outcome: historical perspective and current issues. In: Lambert, M. J., E. R. Christensen, St. De Julio (eds.): The Assessment of Psychotherapy Outcome. Wiley, New York-Chichester-Brisbane 1981.

Lambert, M. J., E. R. Christensen, St. De Julio (eds.): The assessment of psychotherapy outcome. Wiley, New York-Chichester-Brisbane 1981.

Lammers, C. A., B. D. Naliboff, A. J. Straatmeyer: The effects of progressive relaxation on stress and diabetic control. Behav. Res. Ther. 22 (1984) 641–651.

Lamprecht, F., H.-J. Demmel, A. Riehl: Psychosomatische Befunde bei orofacialem Schmerzdysfunktionssyndrom. Z. Psychosom. Med. 32 (1986) 382.

Lamprecht, F. v.: Neurologie. In: Hahn, P. (Hrsg.): Psychologie des 20. Jahrhunderts, Bd. IX. Kindler, Zürich 1979.

Lancet (1979) 17–18: Sexual behavior and the sex hormones. Editorial.

Landau, E.: Sterbehilfe mit dem Katathymen Bilderleben. In: Leuner, H. (Hrsg.): Katathymes Bilderleben, S. 255–262. Huber, Bern 1980.

Landmann, J. T., R. M. Davis: Psychotherapy outcome: Smith's and Glass' conclusions stand under scrutiny. Amer. Psychol. 37 (1982) 504.

Lang, E. (Hrsg.): Praktische Geriatrie. Enke, Stuttgart 1988.

Lang, E., T. Diepgen: Altern und Krankheit. In: Lang, E. (Hrsg.): Praktische Geriatrie. Enke, Stuttgart 1988.

Lang, H., H. Faller, S. Schilling: Krankheitsverarbeitung aus psychosomatisch-psychotherapeutischer Sicht am Beispiel pankreatektomierter Patienten. Psychother., Psychosom., med. Psychol. 39 (1989) 239–247.

Lang, P. J.: The application of psychophysiological methods in the study of psychotherapy and behavior change. In: Bergin, A. E., G. L. Garfield (eds.): Handbook of Psychotherapy and Behavior Change: An Empirical Analysis. Wiley, New York 1971.

Lang, P. J.: Die Anwendung psychophysiologischer Methoden in Psychotherapie und Verhaltensmodifikation. In: Birbaumer, N. (Hrsg.): Psychophysiologie der Angst. Urban & Schwarzenberg, München 1977.

Lang, P. J., D. E. Rice, R. A. Sternbach: The psychophysiology of emotion. In: Greenfield, N. S., R. A. Sternbach (eds.): Handbook of Psychophysiology. Holt, Rinehart and Winston, New York 1972.

Lang, R. E., T. Unger, W. Rascher, D. Ganten: Brain angiotensin. In: Iverson, L. L., S. D. Iversen, S. H. Snyder (eds.): Handbook of Psychopharmacology, Vol. 16, pp. 307–349. Plenum, New York 1983.

Langbein, K., H.-P. Martin, P. Sichrowsky, H. Weiss: Bittere Pillen. Kiepenheuer & Witsch, Köln 1983.

Lange (1903): zit. nach Portmann, A. in: Flanagan, G. L. (Hrsg.): Die ersten neun Monate des Lebens. Rowohlt, Reinbek 1963.

Lange, C.-G.: Om sindsbevaegelser. Et psyko. fysiol. studie. Krmar, Kopenhagen 1885.

Langen, D.: Gestufte Aktivhypnose, 3. Aufl. Thieme, Stuttgart 1969.

Langen, D.: Psychodiagnostik, Psychotherapie. Thieme, Stuttgart 1969.

Langen, D.: Kompendium der medizinischen Hypnose, 3. Aufl. Karger, Basel 1972.

Langer, E., H. Heimann (Hrsg.): Psychopharmaka – Grundlagen und Therapie. Springer, Wien 1983.

Langer, G., G. Koinig, R. Hatzinger, G. Schönbeck, F. Resch, H. Aschauer, M. S. Keshaven, W. Sieghart: Response of thyrotropin to thyrotropin-releasing hormone as predictor of treatment outcome: prediction of recovery and relapse in treatment with antidepressants and neuroleptics. Arch. gen. Psychiat. 43 (1986) 861–868.

Langer, H. E., U. Birth: Probleme und Interessenschwerpunkte von Rheumapatienten und Planung von Patienteninformation. Rheuma 7 (1987) 7–16.

Langman, J.: Medizinische Embryologie, 3. Aufl., S. 322–331. Thieme, Stuttgart 1974.

Langner, T. S., S. T. Michael: Life stress and mental health. The Midtown Manhattan Study. In: Thomas, A. C. (ed.): Rennie Series in Social Psychiatry, Vol. II. Glencoe-Collier-Macmillan, London 1963.

Langosch, W.: Psychosoziale Aspekte des Herzinfarktes. In: Roskamm, H. (Hrsg.): Handbuch der inneren Medizin IX/3, S. 238–248. Springer, Berlin-Heidelberg-New York 1984.

Langosch, W., P. Seer, G. Brodner, D. Kallinke: Stationäre Verhaltenstherapie mit Koronarkranken: Ergebnisse einer vergleichenden Untersuchung. In: Brengelmann, J. C., G. Bühringer (Hrsg.): Therapieforschung für die Praxis 3, S. 141–166. Röttger, München 1982a.

Langosch, W., P. G. Hahn, A. Reindel: Psychosomatik des Herzinfarktes. In: Roskamm, H., H. Reindel (Hrsg.): Herzkrankheiten. Pathophysiologie, Diagnostik, Therapie, S. 936–944. Springer, Berlin-Heidelberg-New York 1982b.

Langosch, W., P. Seer, G. Brodner, D. Kallinke, B. Kulick, P. Heim: Behavior therapy with coronary heart disease patients below age 40. Act. nerv. sup. (Suppl. 3) (1982c) 157–162.

Langosch, W., G. Brodner, E. Farinelli: Type A behavior pattern in a German sample of postinfarction patients below age 40. Act. nerv. sup. (Suppl. 3) (1982d) 157–162.

Laplanche, J., J.-B. Pontalis: Das Vokabular der Psychoanalyse, Bd. I und II. 5. Aufl. Suhrkamp, Frankfurt 1982.

Laragh, J.: Conquering the quiet killer. Time, January (1975) 30.

Larbig, W.: Schmerz. Grundlagen-Forschung-Therapie. Kohlhammer, Stuttgart-Berlin-Köln-Mainz 1982.

Larbig, W.: Psychoanalytische Therapieansätze. In: Gerber, W. D., G. Haag (Hrsg.): Migräne, S. 223–233. Springer, Berlin-Heidelberg 1982b.

Larbig, W., T. Elbert, B. Rockstroh, W. Lutzenberger, N. Birbaumer: Elevated blood pressure and reduction in pain

sensitivity. In: Orlebeke, J., G. Mulder, L. van Doornen (eds.): Psychophysiology of Cardiovascular Control. Plenum, New York 1985.

Lary, D., N. Goldschlager: Electrocardiographic changes during hyperventilation resembling myocardial ischemia in patients with normal coronary arteriograms. Amer. Heart J. 87 (1974) 383–390.

Lasagna, L.: Placebos. Sci. Amer. 193 (1955) 68–71.

Lascelles, P. T., P. R. Evans, H. Merskey, M. A. Sabur: Plasma cortisol in psychiatric and neurologic patients with pain. Brain (1974) 533–538.

Lasègne, E. C.: De l'anorexie hystérique. Arch. gén. méd. 21 (1873) 385 u. Med. Times Gaz. II (1873) 265; 367.

Lask, B., M. Kirk: Childhood asthma; Family therapy as an adjunct to routine management. Journal Fam. Therapy 1 (1979) 33.

Lask, W.: Psychological aspects of inflammatory bowel disease. Wien. Klin. Wschr. 98 (1986) 544–547.

Laskin, D.: Etiology of the pain dysfunction syndrome. J. Amer. dent. Ass. 79 (1969) 147.

Lasser, R. B., J. H. Bond, M. D. Levitt: The role of intestinal gas in functional abdominal pain. New Engl. J. Med. 293 (1975) 524–526.

Laszig, R., T. Luetgebrune: Klinische Topodiagnostik der Ertaubung. In: Lehnhardt, E., M. S. Hirshorn (ed.): Cochlear Implantat. Springer, Berlin-Heidelberg-New York 1987.

Latimer, P. R.: Crohn's disease: a review of the psychological and social outcome. Psychol. Med. 8 (1978) 649–656.

Laudenslager, M. L., S. M. Ryan, R. C. Drugan, R. L. Hyson, S. F. Maier: Coping and immunosuppression: inescapable but not escapable shock suppresses lymphocyte proliferation. Science 221 (1983) 568–570.

Laudenslager, M. L., M. Fleshner, P. Hofstadter, P. E. Held, L. Simons, S. F. Maier: Suppression of specific antibody production by inescapable shock: stability under varying conditions. Brain, Behavior, and Immunity 2 (1988) 92–102.

Lauren, K.: 31 patients with anorexia nervosa treated in a pediatric clinic with family therapy and integrative body psychotherapy. Conf. Psychosom. Res., Istanbul 1980, p. 275.

Lausberg, H., J. v. Wietersheim, E. Wilke, H. Feiereis: Bewegungsbeschreibung psychosomatischer Patienten in der Tanztherapie. Psychother. med. Psychol. 38 (1988) 259–264.

Lauter, H.: Epidemiologische Aspekte alterspsychiatrischer Erkrankungen. Nervenarzt 45 (1974) 277–288.

Laux, G.: Tranquilizer – Abhängigkeit als neues Suchtproblem der klinischen Psychiatrie. In: Laux, G., H. Reimer (Hrsg.): Klinische Psychiatrie. Hippokrates, Stuttgart 1982.

Laux, L., P. Glanzmann, P. Schaffner, C. D. Spielberger: Das State-Trait-Angstinventar. Beltz, Weinheim 1981.

Lavie, P.: Ultrashort sleep-waking schedule. III. 'Gates' and 'forbidden zones' for sleep. Electroenceph. clin. Neurophysiol. 63 (1986) 414–425.

Law, D., H. Steinberg, M. Sleisenger: Ulcerative colitis with onset after the age of 50. Gastroenterology 41 (1961) 457–464.

Law, S. W., W. Gibbons, A. N. Poindexter: Patient co-operation. A determinant of perinatal outcome in the pregnant diabetic. J. reprod. Med. 24 (1980) 197–201.

Laxenaire, M., L. Bentz, C. Chardot: Abord psychologique du malade cancéreux. Ann. Médico-Psychologiques 1 (1972) 195–207.

Layne, O. L., S. C. Yudowsky: Postoperative psychosis in cardiotomy patients. The role of organic and psychic factors. New Engl. J. Med. 284 (1971) 518–520.

Lazarus, A. A.: Behavior therapy and beyond. McGraw-Hill, New York 1971.

Lazarus, A. A.: Multidimensional behavior therapy. Springer, New York 1976.

Lazarus, H. R., J. H. Hagens: Prevention of psychosis following open-heart surgery. Amer. J. Psychiat. 124 (1968) 1190–1195.

Lazarus, R. S.: Psychological stress and the coping process. McGraw-Hill, New York 1966.

Lazarus, R. S.: The concept of stress and disease. In: Levi, L. (ed.): Society, Stress and Disease. Oxford Univ. Press, London-New York-Toronto 1971.

Lazarus, R. S.: Psychological stress and coping in adaptation and illness. Psychiat. Med. 5/4 (1974a) 321–333.

Lazarus, R. S.: The self-regulation of emotion. In: Levi, L. (ed.): Emotions: Their Parameters and Measurement. Raven, New York 1975.

Lazarus, R. S., E. Alfest: Short circuiting on threat by experimentally altering cognitive appraisal. J. abnorm. soc. Psychol. 69 (1964) 195–205.

Lazarus, R. S., R. Launier: Stress related transactions between persons and environment. In: Lewis, M., L. A. Pervin (eds.): Perspectives in Interactional Psychology. Plenum, New York 1978.

Lazarus, R. S., S. Folkman: Stress, appraisal and coping. Springer, New York 1984.

Lazarus, R. S., J. R. Averill, E. M. Opton: Toward a cognitive theory of emotion. In: Levi, L. (ed.): Society, Stress and Disease, Vol I. Oxford Univ. Press, London-New York-Toronto 1971.

Lazarus, R. S., G. R. Averill, E. M. Opton: The psychology of coping: issues of research and assessment. In: Coelho, G. V., D. A. Hamburg, J. E. Adams: Coping and Adaptation. Basic Books, New York 1974b.

Lazarus, S., A. D. Kanner, S. Folkman: Emotions: a cognitive-phenomenological analysis. In: Plutchik, R., H. Kellerman (eds.): Emotion – Theory, Research and Experience, Vol.I: Theories of Emotion. Acad. Press, New York 1980.

Leavitt, F., D. C. Garron: The detection of psychological disturbance in patients with low back pain. J. psychosom. Res. 23 (1979) 149–154.

Leavitt, F., D. C. Garron, C. M. D'Angelo, T. W. McNeil: Low back pain in patients with and without demonstrable organic disease. Pain 6 (1979) 191–200.

Lebovici, S., R. Diatkine, M. Soulé (eds.): Traité de psychiatrie de l'enfant et de l'adolescent, Vol. II. PUF, Paris 1985.

Lebovits, B. Z., R. B. Shekelle, A. M. Ostfeld, P. Oglesby: Prospective and retrospective psychological studies in coronary heart disease. Psychosom. Med. 29 (1967) 265–272.

Leboyer, F.: Der sanfte Weg ins Leben. Desch, München 1974.

Leder, A.: Zur Psychopathologie der Schlaf- und Aufwach-Epilepsie. Nervenarzt 38 (1967) 434–442.

Leder, A.: Zur Psychotherapie bei Epilepsie. Therapiewoche 20 (1970) 674–678.

Lee, S. G., G. M. Cartairs, M. J. Pickersgill: Essential hypertension and the recall of motives. J. psychosom. Res. 15 (1971) 95–105.

Lees, A. J.: Neuropsychologische Störungen beim Morbus Parkinson. Beziehungen zur psychomotorischen Hemmung und Zwangskrankheit. Nervenarzt 60 (1989) 71–79.

LeFave, M. K., R. W. J. Neufeld: Anticipatory threat and physical danger trait anxiety: a signal-detection analysis of effects on autonomic responding. J. Res. Personality 14 (1980) 283–306.

Lefebvre, P.: The narcissistic impass as a determinant of psychosomatic disorders. Psychiat. J. Univ. Ottawa 5 (1980) 5–11.

Lefebvre, P., J. C. Crombez et al.: Psychological dimension and psychopathological potential of acquiring a kidney. Canad. psychiat. Ass. J. 18 (1973) 495–500.

Lefebvre, P., R. Leroux, J. C. Crombez: Object-relations in the dermatologic patient. A contribution to the psychoanalytic theory of psychosomatic disorders. Psychiat. J. Univ. Ottawa 5 (1980) 17–20.

Lefer, J.: Fusion and rheumatoid arthritis. Contemp. Psychol. 9 (1972) 63–78.

Lefer, L.: The patient with the temporomandibular joint pain dysfunction syndrome. Fortschr. Psychoanal. 3 (1968) 69.

Leff, J. P., C. E. Vaughn: Expressed emotions in families. Guilford, New York 1985.

Lehfeldt, H., H. Guze: Psychological factors in contraceptive failure. Fertil. and Steril. 17 (1966) 110.

Lehmann, H. E., R. Cancro: Schizophrenia: clinical features. In: Kaplan, H. I., B. J. Sadock (eds.): Comprehensive Textbook of Psychiatry, IV, Vol. 1, Chapt. 15.4. Williams & Wilkins, Baltimore 1985.

Lehmann, J. H., H. Gramnann, K. Hauss, G. W. Rodewald, T. Schmitz: Akute organische Psychosyndrome nach Herzoperationen. Nervenarzt 39 (1968) 529–536.

Lehnert, H., K. Kaluza, H. Vetter, H. Losse, K. Dorst: Longterm effects of a complex behavioral treatment of essential hypertension. Psychosom. Med. 49 (1987) 422–430.

Lehnhardt, E.: Zur Abgrenzung der psychogenen Hörstörung von der aggravierten Schwerhörigkeit. HNO 22 (1974) 134.

Lehnhardt, E.: Praxis der Audiometrie, 6. Aufl. Thieme, Stuttgart 1987.

Lehr, U.: Veränderungen der Daseinsthematik der Frau im Erwachsenenalter. Vita hum. 4 (1961) 193.

Lehr, U.: Psychologie des Alterns. Quelle & Meyer, Heidelberg 1979.

Lehr, U.: Altersstereotypen und Altersnormen – das Bild des alten Menschen in unserer Gesellschaft. In: Hartmann, K. D., K. F. Köppler (Hrsg.): Fortschritte der Marktpsychologie. Fachbuchhandl. Psychologie, Frankfurt 1980.

Lehr, U.: Klimakterium – sozialpsychologische Aspekte. In: Richter, D., M. Stauber (Hrsg.): Psychosomatische Probleme in Geburtshilfe und Gynäkologie. Kehrer, Freiburg 1983.

Leibig, T., E. Wilke, H. Feiereis: Zur Persönlichkeitsstruktur von Patienten mit Colitis ulcerosa und Morbus Crohn, eine testpsychologische Untersuchung während der Krankheitsremission. Z. psychosom. Med. Psychoanal. 31 (1985) 380–392.

Leiderman, P. H. et al.: Sensory deprivation: clinical aspects. Arch. Med. 101 (1958) 289–290.

Leigh, H., M. F. Reiser: A general system taxonomy for psychological defense mechanisms. J. psychosom. Res. 26 (1982) 77–81.

Leigh, H., M. A. Hofer, J. Cooper: A psychological comparison of patients in „open" and „closed" CCU. J. psychosom. Res. 16 (1972) 449–457.

Leighton, D. C., J. S. Harding, D. B. McLin, A. M. Macmillan, A. H. Leighton: The character of danger: psychiatric symptoms in selected communities. Vol. 3 of the Stirling County Study of psychiatric disorder and sociocultural environment. Basic Books, New York 1963.

Leighton, D. C., J. S. Harding, D. B. McLin, C. C. Hughes, A. H. Leighton: Psychiatric findings of the Stirling County Study. Amer. J. Psychiat. 119 (1963) 1021–1026.

Leithäuser, T.: Kapitalistische Produktion und Vergesellschaftung des Alltags. In: Leithäuser, T., W. R. Heinz (Hrsg.): Produktion, Arbeit, Sozialisation, S. 48–68. Suhrkamp, Frankfurt 1976.

Lemnete, E., V. Valeanu, C. Daniel: Psychosomatic disorders commitant with pelvic inflammation in women. Psychosomatic Medicine in Obstetrics and Gynecology, 3rd Int. Congr. 1971. Karger, Basel 1972.

Lempa, W., W. Wellmann, H.-W. Künsebeck, H. Freyberger: Ergebnisse der kombinierten internistisch-psychosomatischen Behandlung bei M. Crohn-Patienten (kontrollierte Studie). Verh. Dtsch. Ges. Inn. Med. 90 (1984) 1086–1089.

Lempa, W., H. W. Künsebeck, J. Nordmeyer, U. Darin, L. Witzig, H. Freyberger: Empirische Belege zur Effektivität des studentischen Psychosomatik-Wahlfach-Modells der Medizinischen Hochschule Hannover. Psychotherapie, Psychosomatik, Medizin, Psychologie 1985.

Lempert, W.: Zur theoretischen und empirischen Analyse der Beziehungen zwischen Arbeit und Lernen. In: Groskurth, P. (Hrsg.): Arbeit und Persönlichkeit, S. 87–111. Rowohlt, Reinbek 1979.

Lennard-Jones, J. F.: Bacteria and Crohn's disease. Z. Gastroenterol. 17 (Suppl.) (1979) 105–108.

Lennard-Jones, J. E.: The clinical outcome of ulcerative colitis depends on how much of the colonic mucosa is involved. Scand. J. Gastroenterol. 88 (Suppl.) (1983) 48–53.

Lentz, R.-J.: Psychogene Hörstörungen im Kindesalter. Fortschr. Med. 94 (1976) 989.

Lenz, D.: Die Bedeutung des Plazentasitzes für das Auftreten einer fetalen Bradycardie nach Paracervicalanästhesie. Diss., FU Berlin 1973.

Leonard, M. O.: Professional stress and the response of nurses caring for patients with chronic renal failure. In: Levy, N. B. (ed.): Psychonephrology 1, pp. 35–42. Plenum, New York 1981.

Leplow, B., C. Kramer, B. Dahme, R. Richter: Biofeedback-Training des Atemwiderstandes. In: Sill, V. (Hrsg.): Kongreßbericht der 19. Tagung der Norddtsch. Ges. f. Lungen- und Bronchialheilkunde, S. 183–190. Universimed, Frankfurt 1985.

Lepper, M.: Die Psychotherapie an einer Rehabilitationsklinik für Herzkranke. In: Halhuber, M. J., H. P. Milz (Hrsg.): Praktische Präventivkardiologie. Urban & Schwarzenberg, München 1972.

Lesch, O. M., S. Lentner, R. Mader, M. Musalek, A. Nimmerrichter: Medikamentengebrauch und Verkehrssicherheit (eine für Österreich repräsentative Studie). Zschr. Verkehrsrecht 31 (1986) 161–166.

Le Shan, L. L.: A basic psychological orientation apparently associated with malignant disease. Psychiat. Quart. 35 (1961) 314–330.

Le Shan, L. L.: Untersuchungen zur Persönlichkeit der Krebskranken. Z. psychosom. Med. Psychoanal. 9 (1963) 246–253.

Le Shan, L. L.: An emotional life-history pattern associated with neoplastic disease. Ann. N. Y. Acad. Sci. 125 (1966) 780–793.

Le Shan, L. L.: Mobilizing the life force. Ann. N. Y. Acad. Sci. 164 (1969) 847–861.

Le Shan, L. L., R. E. Worthington: Loss of cathexes as a common psychodynamic characteristic of cancer patients. An attempt of a clinical hypothesis. Psychol. Rep. 2 (1956) 183–193.

Le Shan, L. L., M. L. Gassman: Some observations on psychotherapy with patients suffering from neoplastic disease. Amer. J. Psychother. 12 (1958) 723–734.

Leshin, G. J.: Childhood non-organic hearing loss. J. Speech Dis. 25 (1960) 290.

Lesniak, M. A., J. Roth, P. Gordon, J. R. Gavin: III. Human growth hormone radioreceptor assay using cultured human lymphocytes. Nature (New Biol.) 241 (1973) 20–22.

Lesse, S.: The masked depression syndrome. Amer. J. Psychother. 37 (1983) 456–475.

Lesser, L. I., B. J. Ashenden, M. Debuskey, L. Eisenberg: Anorexia nervosa in children. Amer. J. Orthopsychiat. 30 (1960) 572–580.

Leuner, H.: Kontrolle der Symbolinterpretation im experimentellen Verfahren. Z. Psychother. med. Psychol. 4 (1954) 201.

Leuner, H.: Symbolkonfrontation – ein nicht-interpretierendes Vorgehen in der Psychotherapie. Arch. Neurol. Psychiat. 76 (1955) 23–49.

Leuner, H.: Die experimentelle Psychose. Springer, Berlin-Göttingen-Heidelberg 1962.

Leuner, H.: Das assoziative Vorgehen im Symboldrama. Z. Psychother. med. Psychol. 14 (1964) 196–211.

Leuner, H.: Basic principles and therapeutic efficacy of guided affective imagery (GAI). In: Singer, J. L., K. S. Pope (eds.): The Power of Human Imagination. Plenum, New York 1978.

Leuner, H.: Grundlinien des Katathymen Bilderlebens aus neuerer Sicht. In: Leuner, H. (Hrsg.): Katathymes Bilderleben. Huber, Bern 1980.

Leuner, H.: Katathymes Bilderleben – Grundstufe, 2. Aufl. Thieme, Stuttgart 1981.

Leuner, H.: Katathymes Bilderleben. Unterstufe. Thieme, Stuttgart 1981.

Leuner, H., E. Schroeter: Indikation und spezifische Applikation der Hypnosebehandlung. Huber, Bern-Stuttgart-Wien 1975.

Leuteritz, G., R. Shimshoni: Psychotherapie bei Psoriasis – Results at the Dead Sea. Z. Hautkr. 57 (1982) 1612–1615.

Leuzinger, M.: Kognitive Prozesse bei der Indikationsstellung. In: Baumann, U. (Hrsg.): Indikation zur Psychotherapie. Perspektiven für Praxis und Forschung. Urban & Schwarzenberg, München-Wien-Baltimore 1981.

Levander-Lindgren, M.: Studies in neurocirculatory asthenia (Da Costa's syndrome), I. Variations with regard to symptoms and some pathophysiological signs. Acta med. scand. 172 (1962) 665–676.

Levenberg, S. B., C. Jenkins, D. Wendorf: Studies in family-oriented crisis intervention with hemodialysis patients. Int. J. Psychiat. Med. 9 (1978) 83–92.

Levenson, R. W.: Effects of thematically relevant and general stressors on specifity of responding in asthmatic and non-asthmatic subjects. Psychosom. Med. 41 (1979) 28–39.

Levi, L.: Quality of the working environment: promotion and protection of occupational mental health. Reports from the Laboratory for Clinical Stress Research Nr. 88. Karolinska Institute, Stockholm 1978.

Levi, L.: Stress and coronary heart disease – causes, mechanisms, and prevention. In: Rosenman, R. H., H. Ray (eds.): Psychosomatic Risk Factors and Coronary Heart Disease: Indications for Specific Preventive Therapy, pp. 15–21. Huber, Bern 1981.

Levi-Strauss, C.: Strukturale Anthropologie. Suhrkamp, Frankfurt 1967.

Levin, A. P., S. E. Hyler: DSM-III personality diagnosis in bulimia. Comprehens. Psychiat. 27 (1986) 47–53.

Levin, D., A. D. Bertelson, P. Lacks: MMPI differences among mild and severe insomniacs and good sleepers. J. Pers. Assessment 48 (1984) 126–129.

Levin, R. B., A. M. Gross: The role of relaxation in systematic desensitization. Beh. Res. Ther. 23 (1988) 187–196.

Levin, S.: Libido equilibrium. In: Zinberg, N. E., I. Kaufman (eds.): Normal Psychology of the Aging Process, pp. 160–168. Int. Univ. Press, New York 1963.

Levine, B., T. Roehrs, E. Stepanski, F. Zorick, T. Roth: Fragmenting sleep diminishes its recuperative value. Sleep 10 (1987) 590–599.

Levine, J., E. Zigler: Denial and self image in stroke, lung cancer and heart disease patients. J. consult. clin. Psychol. 43 (1975) 751.

Literatur

Levine, J. et al.: The role of denial in recovery from coronary heart disease. Psychosom. Med. 49 (1987) 109–117.

Levine, J. D., N. C. Gordon, R. T. Jones, H. L. Fields: The narcotic antagonist naloxone enhances clinical pain. Nature (Lond.) 272 (1978) 826–827.

Levine, J. D., N. C. Gordon, H. L. Fields: The mechanism of placebo analgesia. Lancet II (1981) 654–657.

Levine, S., N. A. Scotch (eds.): Social stress. Aldine, Chicago 1970.

Levinson, D. F., G. M. Simpson: Antipsychotic drug side effects. In: Hales, Frances (eds.): Psychopharmacology: Drug Side Effects and Interactions. APA Annual Review 1986.

Levitan, H. L.: Patterns of hostility revealed in the fantasies and dreams of women with rheumatoid arthritis. Psychother. and Psychosom. 35 (1981) 34–43.

Levitan, S. J., O. S. Kornfeld: Clinical and costs benefits of liaison psychiatry. Amer. J. Psychiat. 138 (1981) 790–793.

Levy, L. (ed.): Society, stress, and disease, Vol. 1: The psychosocial environment and psychosomatic diseases. London 1971.

Levy, N. B.: Sexual adjustment to maintenance hemodialysis and renal transplantation. In: Levy, N. B. (ed.): Living or Dying, pp. 127–141. Thomas, Springfield/Ill. 1974.

Levy, N. B.: What's new on cause and treatment of sexual dysfunction in end-stage renal disease. In: Levy, N. B. (ed.): Psychonephrology 1, pp. 43–48. Plenum, New York 1981.

Levy, N. B. (ed.): Psychonephrology 1. Plenum, New York 1981.

Levy, N. B. (ed.): Psychonephrology 2. Plenum, New York 1983.

Levy, N. B.: Sexuelle Probleme von Dialysepatienten. In: Balck, F. B., U. Koch, H. Speidel (Hrsg.): Psychonephrologie. Springer, Heidelberg 1984.

Levy, N. B., G. D. Wynbrandt: The quality of life on maintenance hemodialysis. Lancet 1 (1975) 1328–1330.

Levy, P. E.: Die Rolle der Psychotherapie in der Behandlung der Ischias. Über die Notwendigkeit einer Kombination der Physio- und Psychotherapie. Zbl. Psychoanal. 4 (1913) 1–9.

Levy, R.: The immobilized patient and his psychologic well-being. Postgrad. Med. 40 (1966) 73–77.

Levy, S., R. Herberman, M. Lippman, T. d'Angelo: Correlation of stress factors with sustained depression of natural killer cell activity and predicted prognosis in patients with breast cancer. J. clin. Oncol. 5 (1987) 348–353.

Levy, S. M.: Behavior and cancer: life-style and psychosocial factors in the initiation and progression of cancer. Jossey Bass, San Francisco 1985.

Lewin, B.: The body as phallus. Psychoanal. Quart. 2 (1933) 24–47.

Lewin, B. D.: The psychoanalysis of elation. Norton, New York 1950.

Lewis, A. J.: Diagnosenschlüssel und Glossar psychiatrischer Krankheiten. 5. Aufl, korrigiert nach der 9. Revision der ICD. Springer, Berlin 1980. Vorwort der englischen Ausgabe (1974).

Lewis, B. I.: Hyperventilation syndromes: clinical and physiologic observations. Postgrad. Med. 21 (1957) 259–271.

Lewis, J. W., J. T. Cannon, J. C. Liebeskind: Opioid and non-opioid mechanisms of stress analgesia. Science 208 (1980) 623–625.

Lewis, W. C., M. Berman: Studies of conversion hysteria, I. Operational study of diagnosis. Arch. gen. Psychiat. 13 (1965) 275–282.

Lewis-Faning, E.: Report on an inquiry into the aetiological factors associated with rheumatoid arthritis. Ann. rheum. Dis. (Suppl.) Vol. 9 (1950) 1–94.

Lewy, A. J., R. L. Sack, C. M. Singer: Treating phase-typed chronobiological sleep and mood disorders using appropriately timed artificial light. Psychopharmacol. Bull. 21 (1985) 368–372.

Ley, P.: Psychological studies of doctor-patient communication. In: Rachman, S. (ed.): Contributions to Medical Psychology, Vol. I, pp. 9–42. Pergamon, Oxford-New York 1977.

Leyendecker, G.: The pathophysiology of hypothalamic ovarian failure – diagnostic and therapeutic considerations. Eur. Obstet. Gynec. reprod. Biol. 9 (1979) 175.

Leyendecker, G., L. Wildt, E. J. Plotz: Die hypothalamische Ovarialinsuffizienz. Gynäk. 14 (1981) 84.

Leyendecker, G., H. Stock, L.Wildt (eds.): Brain and pituitary peptides II. Karger, Basel 1983.

Liang, B., R. L. Verrier, J. Melman, B. Lown: Correlation between circulating catecholamine levels and ventricular vulnerability during psychological stress in conscious dogs. Proc. soc. exp. Biol. 161 (1979) 266–269.

Liang, M. H., M. Rogers, M. Larson, H. M. Eaton, B. J. Murawski, J. E. Taylor, J. Swafford, P. H. Schur: The psychosocial impact of systemic lupus erythematosus and rheumatoid arthritis. Arthr. and Rheum. 27 (1984) 13–19.

Liberson, W. T., K.Akert: Hippocampal seizure states in guinea pig. Electroenceph. clin. Neurophysiol. 7 (1955) 211.

Liberthson, R. R., E. L. Nagel, J. C. Hirschman et al.: Pathophysiologic observations in prehospital ventricular fibrillation and sudden cardiac death. Circulation 49 (1974) 790.

Liberthson, R. R., D. V. Sheehan, M. E. King, A. E. Weyman: The prevalence of mitral valve prolaps in patients with panic disorders. Amer. J. Psychiat. 134 (1986) 511–515.

Libman, E.: Observations on individual sensitiveness to pain. J. Amer. med. Ass. 102 (1934) 335–341.

Lichtenberg, I.: Psychoanalysis and infant research. Psychoanalytic inquiry book series 2. Analytic Press, New Jersey 1983.

Lichtenberg, I.: Einige Analogien zwischen Befunden der Säuglings- und Kleinkindforschung und klinischen Beobachtungen an Erwachsenen, insbesondere bei Patienten mit borderline- und narzißtischen Persönlichkeitsstörungen. Vortrag vor der René-Spitz-Gesellschaft, München 1984.

Lichtenberg, J.: The testing of reality from the standpoint of the body-self. J. Amer. psychoanal. Ass. 26 (1978) 357.

Lichtenberg, J. D., J. W. Slap: On the defensive organization. Int. J. Psychoanal. 52 (1971) 451–461.

Lichtwitz, L.: Pathologie der Funktionen und Regulationen. Sijthoff's, Leiden 1936.

Lieberz, K.: Zur Psychologie des Rentenbegehrens. In: Willert, H.-G., G. Wetzel-Willert (Hrsg.): Psychosomatik in der Orthopädie. Huber, Bern-Stuttgart-Wien 1989.

Liebeskind, J. C., D. J. Mayer, H. Akil: Central mechanisms of pain inhibition: Studies of analgesia from focal brain stimulation. In: Bonica, J. J. (ed.): International Symposium on Pain. Advances in Neurology 4, pp. 261–273. Raven, New York 1974.

Liebman, R., S. Minuchin, L. Baker: An integrated treatment program for anorexia nervosa. Amer. J. Psychiat. 131 (1974) 432–436.

Liebman, R., S. Minuchin, L. Baker: The use of structural family therapy in the treatment of intractable asthma. Amer. J. Psychiat 131 (1974) 535.

Liedtke, R.: Familiäre Sozialisation und psychosomatische Krankheit. Springer, Berlin-Heidelberg-New York 1987.

Liedtke, R., K. Schemmel, S. Zepf: Behandlungsmodalität der Colitis ulcerosa und symptomfreies Intervall. Med. Klinik 67 (1972) 1666–1671.

Liedtke, R., H. Freyberger, S. Zepf: Personality features of patients with ulcerative colitis. Psychother. and Psychosom. 28 (1977) 187–192.

Lienert, G. A.: Testaufbau und Testanalyse, 3. Aufl. Beltz, Weinheim 1969.

Lifton, R. J.: Observations on Hiroshima survivors. In: Krystal, H. (ed.): Massive Psychic Trauma. Int. Univ. Press, New York 1968.

Light, K., P. Obrist: Cardiovascular reactivity to behavioral stress in young males with and without marginally elevated casual systolic pressures. Comparison of clinic, home and laboratory measures. Hypertension 2 (1980) 802–808.

Light, K. C., P. A. Obrist: Cardiovascular response to stress: effects of opportunity to avoid shock experience and performance feedback. Psychophysiol. 17 (1980) 243–252.

Light, K. C., J. P. Koepke, P. A. Obrist, P. W. Willis: Psychological stress induces sodium retention and fluid restriction in men at high risk for hypertension. Science 220 (1983) 429–431.

Like, A. A., L. Butler, R. M. Williams, C. Appel, E. J. Weringer, A. A. Rossine: Spontaneous autoimmune diabetes mellitus in the BB rat. Diabetes 31 (Suppl.1) (1982) 7–13.

Lim, V. S., V. S. Fang: Gonadal dysfunction in uremic men. Amer. J. Med. 58 (1975) 655.

Lima, B. R., A. Vanneman: Propranolol, benzotropine, fluphenazine decanoate, and delirium. Amer. J. Psychiat. 140 (1983) 659–660.

Lincoln, G. A.: Luteinizing hormone and testosterone in man. Nature 252 (1974) 232–233.

Lindemann, E.: Symptomatology and management of acute grief. Amer. J. Psychiat. 101 (1944) 141–148.

Linden, G.: The influence of social class in the survival of cancer patients. Amer. J. publ. Health 59 (1969) 267–274.

Linden, W.: Psychologische Perspektiven des Bluthochdrucks. Karger, Basel 1983.

Linder, M.: Psychiatrische Probleme in der Therapie mit Benzodiazepinen. Schweiz. med. Wschr. 113 (1983) 654–658.

Lindholm, E., G. Shumway, C. U. Grijalva, T. Schallert, M. Ruppel: Gastric pathology produced by hypothalamic lesions in rats. Physiol. Behav. 14 (1975) 165–169.

Lindner, J.: Biologie des Alterns. In: Lang, E. (Hrsg.): Praktische Geriatrie. Enke, Stuttgart 1988.

Lindsay, P. H., D. A. Norman: Human information processing. Acad. Press, New York 1972.

Lindsay, R. M., D. G. Oreopoulos, H. Burton, J. Conley, G. Wells, S. S. A. Fenton: Adaptation to home dialysis: a comparison of continuous ambulatory peritoneal dialysis and hemodialysis. In: Legrain, M. (ed.): Continuous Ambulatory Peritoneal Dialysis, p. 120. Excerpta Medica, Amsterdam-Oxford-Princeton 1980.

Lindsley, D. B.: Psychophysiology and motivation. In: Jones, M. R. (ed.): Nebraska Symposium on Motivation. Univ. of Nebraska Press, Nebraska 1957.

Lindsley, O. R.: Direct measurement and prosthesis of retarded behavior. J. Educ. 147 (1964) 62–81.

Lindström, L. H., G. Besev, L. M. Gunne, L. Terenius: CSF levels of receptor-active endorphins in schizophrenic patients: correlations with symptomatology and monoamine metabolites. J. psychiat. Res. 19 (1986) 93–100.

Linn, M. W., B. S. Linn, J. Jensen: Stressful event, dysphoric mood and immune responsiveness. Psychol. Rep. 54 (1984) 219.

Linn, R., R. L. Kahn: Patterns of behavior disturbance following cataract extraction. Amer. J. Psychiat. 110 (1956) 251–259.

Linneweh, F.: Die Prognose chronischer Erkrankungen. Springer, Berlin-Göttingen-Heidelberg 1960.

Linnoila, M., F. Karoum, H. M. Calil, I. J. Kopin, W. Z. Potter: Alteration of norepinephrine metabolism with desipramine and zimelidine in depressed patients. Arch. gen. Psychiat. 39 (1982) 1025–1028.

Lipowski, Z. J.: Review of consultation psychiatry and psychosomatic medicine I: General Principles. Psychosom. Med. 29 (1967a) 153–171.

Lipowski, Z. J.: Review of consultation psychiatry and psychosomatic medicine II: Clinical aspects. Psychosom. Med. 29 (1967b) 201–224.

Lipowski, Z. J.: Review of consultation psychiatry and psychosomatic medicine III: Theoretical issues. Psychosom. Med. 30 (1968) 395–422.

Lipowski, Z. J.: Physical illness, the individual and the coping process. Psychiat. Med. 1 (1970) 91–102.

Lipowski, Z. J. (ed.): Psychosocial aspects of physical illness. Advanc. Psychosom. Med. Vol. 8. Basel 1972.

Lipowski, Z. J.: Psychiatry of somatic diseases: epidemiology, pathogenesis, classification. Compr. Psychiat. 16 (1975) 105–124.

Lipowski, Z. J.: Current trends in psychosomatic medicine II. Psychiat. Med. 6 (1975) 3–311.

Lipowski, Z. J.: Psychosomatic medicine in the seventies: an overview. Amer. J. Psychiat. 134 (1977) 233–244.

Lipowski, Z. J.: Psychosocial reactions to physical illness. Canad. med. Ass. J. 128 (1983) 1069–1072.

Lipowski, Z. J.: Somatization: the concept and its clinical application. Amer. J. Psychiat. 145 (1988) 1358–1368.

Lipowski, Z. J.: Delirium in the elderly patient. New Engl. J. Med. 320 (1989) 578–582.

Lipowski, Z. L.: Organic mental disorders: introduction and review of syndroms. In: Kaplan, H. I. (ed.): Comprehensive Psychiatry, pp. 1359–1392.

Lipowski, Z. L.: Delirium updated. Comprehens. Psychiat. 21 (1980) 190–196.

Lipowski, Z. L.: Delirium. Acute brain failure in man. Thomas, Springfield 1980.

Lipowski, Z. L.: The need to integrate liaison psychiatry and geropsychiatry. Amer. J. Psychiat. 140 (1983) 1003–1005.

Lippman, M., R. Haltermann, S. Perry, L. Leventhal, E. B. Thompson: Glucocorticoid proteins in human leukaemic lymphoblasts. Nature (New Biol.) 242 (1973) 157–158.

Lipton, J., S. Dworkin, J. Marbach, S. Gold, B. Gurion: Psychosocial considerations in the MPD syndrome. J. dent. Res. 53 (spec. issue) (1974) 127.

Lipworth, L., T. Abelin, R. R. Connelly: Socioeconomic factors in the prognosis of cancer patients. J. chron. Dis. 23 (1970) 105–116.

Litin, E. M.: Should the cancer patient be told? Postgraduate Med. 28 (1960) 470–475.

Little, B. C., P. Benson, R. W. Beard, J. Hayworth, F. Hall, J. Dewhurst, R. G. Priest: Treatment of hypertension in pregnancy by relaxation and biofeedback. Lancet 1 (1984) 865–867.

Littler, W. A., A. J. Honour, P. Sleight, F. D. Stott: Continuous recording of direct arterial pressure and electrocardiogram in unrestricted man. Brit. med. J. 3 (1972) 76–78.

Liu, S. J., R. J. H. Wang: Increased analgesia and alterations in distribution and metabolism of methadone by desipramine in rat. J. Pharmacol. exp. Ther. 195 (1975) 94–104.

Liungberg, L.: Hysteria: Clinical, prognostic and genetic study. Acta psychiat. Scand. 32 (1957) 1–162.

Livesley, W. J.: Psychiatric disturbance and chronic hemodialysis. Brit. med. J. 2 (1979) 306–308.

Livesley, W. J.: Symptoms of anxiety and depression in patients undergoing chronic hemodialysis. J. psychosom. Res. 26 (1982) 581–584.

Loader, P. J., W. Kinston, J. Stratford: Is there a „psychosomatogenic" family? J. Fam Therapy 2 (1980) 311.

Lobitz, W., H. Brammel: Anxiety management training vs.

aerobic conditioning for cardiac stress management. Annual meeting APA, Los Angeles 1981.

Loch, W.: Voraussetzungen, Mechanismen und Grenzen des psychoanalytischen Prozesses. Huber, Bern-Stuttgart-Wien – Klett-Cotta, Stuttgart 1965.

Loch, W.: Über die theoretischen Voraussetzungen einer psychoanalytischen Kurztherapie. Jahrbuch der Psychoanalyse, Band IV. Huber, Bern-Stuttgart-Wien 1967.

Loch, W.: Seelische Ursachen psychischer Störungen. Praxis der Psychotherapie, Band XV (1970) 49.

Loch, W.: Sprechstunde-Psychotherapie. In: Loch, W. (Hrsg.): Theorie, Technik und Therapie der Psychoanalyse, S. 283. Frankfurt 1972.

Loch, W.: Der Analytiker – Gesetzgeber und Lehrer. Psyche 28 (1974) 431–460.

Loch, W.: Die Balintgruppe. In: Loch, W. (Hrsg.): Über Begriffe und Methoden der Psychoanalyse, S. 159. Bern-Stuttgart-Wien 1975.

Loch, W.: Grundriß der psychoanalytischen Therapie. In: Loch, W. (Hrsg.): Die Krankheitslehre der Psychoanalyse. 4. Aufl. Stuttgart 1983.

Lochs, H., M. Egger-Schödl, R. Pötzi, Ch. Kappel, R. Schuh: Enterale Ernährung – eine Alternative zur parenteralen Ernährung in der Behandlung des Morbus Crohn? Leber Magen Darm 14 (1984) 64–67.

Locke, S. E.: Stress, adaptation and immunity: studies in humans. Gen. Hosp. Psychiat. 4 (1982) 49–58.

Locke, S. E.: Psychosocial and behavioral treatments for disorders associated with the immune system: an annotated bibliography. Institute for the Advancement of Health, New York 1986.

Locke, S. E., M. Hornig-Rohan: Mind and immunity: Behavioral immunology: an annotated bibliography 1976–1982. Institute for the Advancement of Health, New York 1983.

Locke, S. E., R. Ader, H. Besedovsky, N. Hall, G. Solomon, T. Strom: Foundations of psychoneuroimmunology. Aldine, New York 1985.

Lodewick, L.: Die körperliche Untersuchung. Ein Atlas für die allgemeine Praxis und Ausbildung. Fischer, Stuttgart-New York 1981.

Loebell, H.: Seelentaubheit. Arch. Ohren-, Nas.- u. Kehlk.-Heilk. 154 (1944) 157.

Loebert, L.: Kenntnisse des Diabetikers über seine Krankheit. Dtsch. med. Wschr. 97 (1972) 1055–1057.

Loew, Th. (Hrsg.): Anamnesegruppen – Patientenzentrierte Medizin erleben. Centaurus, Pfaffenweiler 1989.

Loewald, H.: Psychoanalyse. Aufsätze aus den Jahren 1951–1979. Kap. 9: Primärprozeß, Sekundärprozeß und Sprache. Klett-Cotta, Stuttgart 1986.

Loewit, K.: The communicative function of human sexuality: A neglected dimension. In: Forleo, R., W. Pasini (eds.): Medical Sexology, p. 234–237. PSG, Littleton 1980.

Loewit, K.: Sexualität und Partnerschaft. In: Viefhues, H. (Hrsg.): Lehrbuch Sozialmedizin, S. 79–89. Kohlhammer, Stuttgart 1981 .

Loewit, K.: Sessualita Umana. In: Cappelletti, V. (ed.): Enciclopedia del novecento. Istituto dell'enciclopedia italiana. Vol. VI Roma (1982) 516–525.

Loewit, K.: Geheimsprache Sexualität. Tyrolia, Innsbruck-Wien 1988.

Lohmann, R.: Autogenes Training. Dtsch. Ärztekalender 1981, S. 316–323. Urban & Schwarzenberg, München-Berlin-Wien 1980.

Lolas, F., H. Kordy, M.v. Rad: Affective content of speech as predictor of psychotherapy outcome. (1984, unpublished).

Lonergarn-Cooper, E.: Group intervention. How to begin and maintain groups in medical and psychiatric settings. Jason Aronson, North Vale NJ 1982.

Loosen, P. T.: The TRH-induced TSH response in psychiatric patients: a possible neuroendocrine marker. Psychoneuroendocrinology 10 (1985) 237–260.

Loosen, P. T., A. J. Prange jr.: Serum thyrotropin response to thyrotropin-releasing hormone in psychiatric patients. Amer. J. Psychiat. 139 (1982) 405–416.

Lo Piccolo, J.: Direct treatment of sexual dysfunction in the couple. In: Money, J., H. Musaph (eds.): Handbook of Sexology. pp. 1227–1244. Elsevier, New York-Oxford 1978.

Lord, F. M., M. R. Novick: Statistical theories of mental test scores. Addison Wesley, London 1968.

Lorenzer, A.: Zur Begründung einer materialistischen Sozialisationstheorie. Suhrkamp, Frankfurt 1972.

Lorenzer, A.: Über den Gegenstand der Psychoanalyse oder: Sprache und Interaktion. Frankfurt 1973.

Lorenzer, A.: Die Wahrheit der psychoanalytischen Erkenntnis. Frankfurt 1974.

Lorenzer, A.: Zur Dialektik von Individuum und Gesellschaft. In: Leithäuser, T., W. R. Heinz (Hrsg.): Produktion, Arbeit, Sozialisation, S. 13–47. Suhrkamp, Frankfurt 1976.

Lorig, K. R., T. Cox, Y. Cuevas, R. G. Kraines, M. C. Britton: Converging and diverging beliefs about arthritis: caucasian patients, spanish speaking patients, and physicians. J. Rheum. 11 (1984) 76–79.

Louhivuori, K., T. Huupponen, T. Riita, E. Sormunen: Leukemic children and their families. Psychiatria Fennica Helsinki 8 (1976) 113.

Lourens, P. J. D.: Crohn' s disease, ulcerative colitis and psychology. Ala. J. med. Sci. 10 (1973) 285–293.

Lovallo, W. R., V. Pishkin: A psychophysiological comparison of Type A and B men exposed to failure and uncontrollable noise. Psychophysiology 17 (1980) 29–36.

Lovallo, W. R., G. A. Pincomb, M. F. Wilson: Heart rate reactivity and Type A behavior as modifiers of physiological response to active and passive coping. Psychophysiol. 23 (1986) 105–112.

Lovallo, W. R., G. A. Pincomb, M. F. Wilson: Predicting response to a reaction time task: heart rate reactivity compared with Type A behavior. Psychophysiol. 23 (1986) 648–656.

Lowen, A.: Bio-Energetik. Bern-München 1975.

Lown, B., M. Wolf: Approaches to sudden death from coronary heart disease. Circulation 44 (1971) 130.

Lown, B., R. L. Verrier: Neural activity and ventricular fibrillation. New Engl. J. Med. 294 (1976) 1165–1170.

Lown, B., R. A. DeSilva: Roles of psychologic stress and autonomic nervous system changes in provocation of ventricular premature complexes. Amer. J. Cardiol. 41 (1978) 979–985.

Lown, B., R. L. Verrier, R. Corbalan: Psychologic stress and threshold for repetitive ventricular response. Science 182 (1973) 834.

Lown, B., M. Tykocinski, A. Garfein et al.: Sleep and ventricular premature beats. Circulation 48 (1973) 691–701.

Lown, B., J. V. Temte, P. Reich et al.: Basis for recurring ventricular fibrillation in the absence of coronary heart disease and its management. New Engl. J. Med. 294 (1976) 623–629.

Lown, B., R. L. Verrier, S. H. Rabinowitz: Neural and psychological mechanisms and the problems of sudden cardiac death. Amer. J. Cardiol. 39 (1977) 890–902.

Lowrie, M. R.: Frequence of depressive disorder in patients entering hemodialysis. J. nerv. ment. Dis. 167 (1979) 199–204.

Lowy, M. T., A. T. Reder, J. P. Antel, H. Y. Meltzer: Glucocorticoid resistance in depression: the dexamethasone suppression test and lymphocyte sensitivity to dexamethasone. Amer. J. Psychiat. 141 (1984) 1365–1370.

Luban-Plozza, B.: Zehn Jahre Balint-Gruppen mit Studenten. Dtsch. Ärztebl. 9 (1979) 585–590.

Lubar, J. F., A. A. Peracchio: One-way and two-way learning and transfer of an active avoidance response in normal and cingulectomized cats. J. comp. physiol. Psychol. 60 (1965) .

Lubin, B., A. K. Lubin, R. Taylor: The group psychotherapy. Literature 1978. Int. J. Group Psychother. Oct. 1979.

Luborsky, L.: Einführung in die analytische Psychotherapie – Ein Lehrbuch. Springer, Heidelberg 1988.

Luborsky, L., B. Singer, L. Luborsky: Comparative studies of psychotherapies. Arch. gen. Psychiat. 32 (1975) 995–1008.

Luborsky, L., J. Mintz, P. Christoph: Are psychotherapeutic changes predictable? Comparison of a Chicago counseling center project with a Penn psychotherapy project. J. consult. clin. Psychol. 3 (1979) 469–473.

Luborsky, L., M. Chandler, A. Auerbach, J. Cohen, H. Bachrach: Factors influencing the outcome of psychotherapy: a review of quantitative research. Psychol. Bull. 75 (1971) 145–185.

Luborsky, L., J. Mintz, A. Auerbach, P. Christoph, H. Bachrach, Th. Todd, M. Johnson, M. Cohen, C. P. O'Brien: Predicting the outcome of psychotherapy. Arch. gen. Psychiat. 37 (1980) 471–481.

Lucas, A. R., C. M. Beard, J. S. Kranz, L. T. Kurland: Epidemiology of anorexia nervosa and bulimia. Int. J. Eat. Dis. 2 (1983).

Ludin, H. P.: Das Parkinson-Syndrom. In: Baumgartner, G., R. Cohen, O. J. Grüsser, H. Helmchen (Hrsg.): Psychiatrie, Neurologie, klinische Psychologie. Grundlagen – Methoden – Ergebnisse. Kohlhammer, Stuttgart 1988.

Ludwig, A.: Psychiatric considerations in rheumatoid arthritis. Med. Clin. N. Amer. 39 (1955) 447–458.

Lue, T. F., T. Takamura, M. Umraiya, R. A. Schmidt, E. A. Tanagho: Hemodynamics of canine corpora cavernosa during erection. Urology 24 (1984) 347.

Lühring, H., H. D. Seibel: Beanspruchung durch die Arbeit und psychische Gesundheit: Auswirkungen von Diskrepanzen zwischen Arbeitserfahrungen und Arbeitserwartungen bei Industriearbeitern. Zeitschrift für Soziologie 10 (1981) 395–412.

Lüscher, E.: Psychische Faktoren bei Hals-, Nasen- und Ohrenkrankheiten. Arch. Ohren-, Nas.- u. Kehlk.-Heilk. 175 (1959) 69.

Lützenkirchen, J., K. Lamprecht, J. Walter, A. Dietz: The sociomedical situation and personality after heart surgery. In: Speidel, H., G. Rodewald (eds.): Psychic and Neurological Dysfunctions after Open-Heart Surgery. Thieme, Stuttgart 1980.

Lukas, K. H.: Die psychologische Geburtserleichterung. Schattauer, Stuttgart 1968.

Lukas, K. H.: Psychologische Aspekte der Geburtshilfe. Dtsch. Ärztebl. 10 (1972) 555.

Lukesch, H.: Der Einfluß sozialer Beziehungen auf das Schwangerschaftserleben. In: Prill, H. J., D. Langen (Hrsg.): Der psychosomatische Weg zur gynäkologischen Praxis. Schattauer, Stuttgart 1983.

Lum, L. C.: The syndrome of chronic habitual hyperventilation. In: Hill, O. W. (ed.): Modern Trends in Psychosomatic Medicine. Butterworth, London 1976.

Lumpkin, M. D.: The regulation of ACTH secretion by IL-1. Science 238 (1987) 452–454.

Lund, R., E. Rüther: Schlafstörungen und ihre psychosomatische Bedeutung bei alten Menschen. In: Bergener, M., B. Kark (Hrsg.): Psychosomatik in der Geriatrie. Steinkopff, Darmstadt 1985.

Lund-Johansen, P.: Hemodynamics in early essential hypertension. Acta med. scand. (Suppl.) 482 (1967) 1–15.

Lund-Johansen, P.: Hemodynamic trends in untreated essential hypertension. Acta med. scand. (Suppl.) 602 (1976) 68–81.

Lundberg, U., M. Frankenhaeuser: Pituitary-adrenal and sympathetic-adrenal correlates of distress and effort. J. psychosom. Res. 24 (1980) 125–130.

Lundberg, U., T. Theorell, E. Lind: Life changes and myocardial infarction: individual differences in life change scaling. J. psychosom. Res. 19 (1975) 27–32.

Lundin, P. M., U. Schelin: The effect of steroids on the histology and ultrastructure of lymphoid tissue. III. Thymus in prolonged steroid-induced involution. Pathol. Eur. 4 (1969) 58–68.

Luparello, T. J., N. Leist, C. H. Lourie, P. Sweet: The interaction of psychologic stimuli and pharmacologic agents on airway reactivity in asthmatic subjects. Psychosom. Med. 32 (1970) 509–513.

Luparello, T. J., H. A. Lyons, E. R. Bleecker, E. R. McFadden jr.: Influences of suggestion on airway reactivity in asthmatic subjects. Psychosom. Med. 30 (1968) 819–825.

Lupton, D.: Psychosocial aspects of temporomandibular joint pain dysfunction. J. Amer. dent. Ass. 79 (1969) 131.

Lussier, A.: The physical handicap and the body ego. Int. J. Psychoanal. 61 (1980) 179–185.

Luthe, W. (ed.): Autogenic therapy, Vol. I–IV. Grune & Stratton, New York-London 1969–1973.

Lyell, A.: Dermatitis artefacta in relation to the syndrome of contrived disease. Clin. exp. Derm. 1 (1976) 109–126.

Lynch, J. J.: The broken heart. Medical consequences of loneliners. Basic Books, New York 1977.

Lynch, J. J., S. A. Thomas, D. A. Paskewitz et al.: Human contact and cardiac arrhythmia in a coronary care unit. Psychosom. Med. 39/3 (1977) 188.

Lyndon, R. W., J. D. Russell: Benzodiazepine use in a rural general practice population. Aust. N. Z. J. Psychiat. 22 (1988) 293–298.

Lysle, D. T., M. Lyte, H. Fowler, B. S. Rabin: Shock-induced modulation of lymphocyte reactivity: suppression, habituation, and recovery. Life Sci. 41 (1987) 1805–1814.

Lysle, D. T., J. E. Cunnick, H. Fowler, B. S. Rabin: Pavlovian conditioning of shock-induced suppression of lymphocyte reactivity: acquisition, extinction, and preexposure effects. Life Sci. 42 (1988) 2185–2194.

Maalseed, R. T., F. J. Goldstein: Enhancement of morphine analgesia by tricyclic antidepressants. Neuropharmacol. 18 (1979) 827–829.

Maas, G.: Praktisches Vorgehen bei Herzneurose. Med. Welt 26 (1975) 592.

Maas, J. W., S. A. Contreras, E. Seleshi, C. L. Bowden: Dopamine metabolism and disposition in schizophrenic patients: studies using debrisoquin. Arch. gen. Psychiat. 45 (1988) 553–559.

Maas, J. W., S. H. Koslow, J. Davis, M. Katz, A. Frazer, C. L. Bowden, N. Berman, R. Gibbons, P. E. Stokes, H. Landis: Catecholamine metabolism and disposition in healthy and depressed subjects. Arch. gen. Psychiat. 44 (1987) 334–337.

Maas, S., J. Kuypers: From thirty to seventy. Jossey-Bass, San Francisco 1974.

MacDougall, J. M., T. M. Dembroski, L. Musante: The structured interview and questionnaire methods of assessing coronary-prone behavior in male and female college students. J. behav. Med. 2 (1979) 71–83.

MacDougall, J. M., T. M. Dembroski, J. Dimsdale, T. Hakkett: Components of Type A, hostility and anger – further relationship to angiographic findings. Health Psychol. 4 (1985) 137–152.

Mackay, A. V. P., L. L. Iversen, M. Rossor, E. Spokes, E. Bird, A. Arregui, I. Creese, S. H. Snyder: Increased brain dopamine receptors in schizophrenia. Arch. gen. Psychiat. 39 (1982) 991–997.

MacKenzie, J. N.: The production of the so-called „rose cold" by means of an artificial rose. Amer. J. med. Sci. 91 (1886) 45–57.

Mackintosh, N. J.: Conditioning and associative learning. Oxford Univ. Press, New York.

Mackintosh, N. J.: A theory of attention: variations in the associability of stimuli with reinforcement. Psychol. Rev. 82 (1975) 276–298.

Mackintosh, N. J.: Kognitive Lerntheorien. In: Zeier, H. (Hrsg.): Psychologie des 20. Jahrhunderts, Bd. IV. Kindler, Zürich 1977.

Mackintosh, N. J.: Cognitive or associative theories of conditioning: implications of an analysis of blocking. In: Hulse, S. S., H. Fowler, W. K. Honig (eds.): Cognitive Processes in Animal Behavior. Erlbaum, Hillsdale 1978.

MacQueen, G. M., S. Siegel: Conditioned immunomodulation following training with cyclophosphamide. Behav. Neurosci. 103 (1989) 638–647.

Macris, N. T., R. C. Schiavi, M. S. Camerino, M. Stein: Effect of hypothalamic lesions on immune processes in the guinea pig. Amer. J. Physiol. 222 (1970) 1054–1057.

Macy and Allen, zit. nach: Weiss, E., O. S. English: Psychosomatic Medicine, p. 9. Saunders, Philadelphia-London 1949.

Madanes, C.: Strategic family therapy. Jossey-Bass, San Francisco 1981.

Madden, J., H. Akil, R. L. Patrick, J. Backsas: Stress-induced parallel changes in central opioid levels and pain responsiveness in the rat. Nature 265 (1977) 358–360.

Madeya, S., G. Börsch: Zur Differentialdiagnose des Morbus Crohn: Segmentale intestinale Metastasierungen beim Mamma- und Magenkarzinom. Leber, Magen, Darm 19 (1989) 140–152.

Magarian, G. J.: Hyperventilation syndromes: infrequently recognized common expressions of anxiety and stress. Medicine 61 (1982) 219–236.

Magnus, G., M. Cavallini, F. Halberg, G. Cornelissen, D. E. R. Sutherland, J. A. Najarian, W. J. M. Hrushesky: Circadian toxicology of cyclosporin. Toxicol. appl. Pharmacol. 77 (1985) 181–185.

Magnusson, D., N. S. Endler (eds.): Personality at the crossroads: current issues in interactional psychology. Erlbaum, Hillsdale NJ 1977.

Magora, A., A. Schwartz: Relation between the low back pain syndrome and X-ray findings. 4. Lysis and olisthesis. Scand. J. Rehab. Med. 12 (1980) 47–52.

Maguire, G. P., E. G. Lee, D. J. Bevington, C. S. Küchemann, R. J. Crabtree, C. E. Cornell: Psychiatric problems in the first year after mastectomy. Brit. med. J. 1 (1978) 963–965.

Mahl, B. F.: Anxiety, HCl-secretion and peptic ulcer etiology. Psychosom. Med. 12 (1950) 158.

Mahler, M. S.: Über Psychose und Schizophrenie im Kindesalter, autistische und symbiotische frühkindliche Psychosen. Psyche 21 (1952) 895.

Mahler, M. S.: On human symbiosis and the vicissitudes of individuation. New York 1968. Deutsch: Symbiose und Individuation. Klett, Stuttgart 1972.

Mahler, M. S.: Symbiose und Individuation. Die psychische Geburt des Menschenkindes. Psyche 7 (1975) 609.

Mahler, M. S., L. Rangell: A psychosomatic study of 'maladie des tics' (Gilles de la Tourette disease). Psychiat. Quart. 17 (1943) 579–603.

Mahler, M. S., F. Pine, A. Bergman: The psychologic birth of the human infant. New York 1975. Deutsch: Die psychische Geburt des Menschen – Symbiose und Individuation. Fischer, Frankfurt 1980.

Mahoney, M. J.: Cognition and behavior modification. Ballinger, Cambridge/Mass. 1974.

Mahoney, M. J., E. Thoresen: Self-control. Power to the person. Brooks & Cole, Monterey 1974.

Mahoney, M. J., K. Mahoney: Permanent weight control: a total solution to the dieter's dilemma. Norton, New York 1976.

Mai, F.: Graft and donor denial in heart recipients. Amer. J. Psychiat. 143 (1986b) 1159–1161.

Mai, F., J. Burley: The psychosocial aspect of heart transplantation. Transplant. Today 2 (1985) 16–21.

Mai, F., N. McKenzie, W. Kostuk: Psychiatric aspects of cardiac transplantation: preoperative evaluation and postoperative sequelae. Brit. med. J. 292 (1986a) 311–313.

Maier, S. F., M. L. Laudenslager: Inescapable shock, shock controllability, and mitogen stimulated lymphocyte proliferation. Brain, Behavior, and Immunity 2 (1988) 87–91.

Maier, W., R. Buller, H. Rieger, O. Benkert: The cardiac anxiety syndrome – a subtype of panic attacks. Europ. Arch. Psychiat. Neurol. Sci. 235 (1985) 146–152.

Main, T. F.: The hospital as a therapeutic institution. Bull. Menninger Clin. 10 (1946) 66.

Mains, R. E., B. A. Eipper, N. Ling: Common precursor to corticotropins and endorphins. Proc. Nat. Acad. Sci. 74 (1977) 3014–3018.

Mains, R. E., B. A. Eipper, C. C. Glembotski, R. M. Dores: Strategies for the biosynthesis of bioactive peptides. Trends in Neurosci. 6 (1983) 229–235.

Malan, D. H.: Psychoanalytische Kurztherapie. Eine kritische Untersuchung. Klett, Stuttgart 1962.

Malan, D. H.: A study of brief psychotherapy. London 1963.

Malan, D. H.: The outcome problem in psychotherapy research. Arch. gen. Psychiat. 29 (1973) 719–729.

Malan, D. H.: Toward the validation of dynamic psychotherapy: a replication. Plenum, New York 1976.

Malchow, H.: Morbus Crohn. In: Caspary, W. F. (Hrsg.): Handbuch der inneren Medizin, Bd. III/3: Dünndarm. 5. Aufl. Springer, Berlin 1983.

Malchow, H., W. Daiss: Diagnostik des Morbus Crohn. Dtsch. med. Wschr. 109 (1984a) 1770–1775.

Malchow, H., W. Daiss: Therapie des Morbus Crohn. Dtsch. med. Wschr. 109 (1984b) 1811–1816.

Malchow, H., U. Riker, K. Dietz: Lebenserwartung bei Morbus Crohn. Lebensversicherungsmedizin 33 (1981) 27–30.

Malchow, H., H. Jenss, H.-J. Steinhardt, R. Schütze, F. Hartmann: Welche Wertigkeit kommt dem coloskopischen Befund bei der Differenzierung zwischen Colitis ulcerosa und Morbus Crohn zu? Verh. Dtsch. Ges. Inn. Med. 84 (1978) 1019–1022.

Malchow, H., K. Ewe, J. W. Brandes, H. Goebell, H. Ehms, H. Sommer, H. Jesdinsky: European Cooperative Crohn's Disease Study (ECCDS): Results of drug treatment. Gastroenterology 86 (1984c) 249–266.

Malcuit, G.: Cardiac responses in aversive situations with and without avoidance possibility. Psychophysiol. 10 (1973) 295–306.

Maler, T.: Musiktherapie. In: Feiereis, H. (Hrsg.): Diagnostik und Therapie der Magersucht und Bulimie. Marseille, München 1989.

Mall, H.: Katamnesen eines integrierten internistisch-psychosomatischen Behandlungskonzeptes zur Anorexia nervosa. Med. Diss., Ulm 1983.

Mall-Haefeli, M.: Die hormonale Antikonzeption und ihre

Auswirkungen auf die Psyche. Schweiz. med. Wschr. 104 (1974) 878.

Mallett, S. J., J. E. Lennard-Jones, J. Bingley, E. Gilon: Colitis. Lancet II (1978) 619–621.

Malmo, R. B.: Activation. In: Bachrach, A. J. (ed.): Experimental Foundations of Clinical Psychology. Basic Books, New York 1962.

Malmo, R. B., C. Shagass: Physiologic studies of symptom mechanisms in psychiatric patients under stress. Psychosom. Med. 11 (1959) 25–29.

Malmquist, A. et al.: Factors in psychiatric prediction of patients beginning hemodialysis: a follow-up of 13 patients. J. psychosom. Res. 16 (1972) 19–23.

Malmquist, A., B. Hagberg: A prospective study of patients in chronic hemodialysis: V. A follow-up of 13 patients in hemodialysis. J. psychosom. Res. 18 (1984) 321–326.

Malony, M. J., M. K. Farell: Treatment of severe weight loss in anorexia nervosa with hyperalimentation and psychotherapy. Amer. J. Psychiat. 137 (1980) 310–314.

Malzman, J.: Theoretical conceptions of semantic conditioning and generalization. In: Dixon, T. R., D. L. Horton (eds.): Verbal behavior and general behavior theory. Prentice Hall, Englewood Cliffs 1968.

Mancini, B., A. Carbonara, J. F. Heremons: Immunochemical quantification of antigens by single radial immunodiffusion. Immunochem. 2 (1965) 235–254.

Mandel, A., K.-H. Mandel, E. Stadter, D. Zimmer: Einübung in Partnerschaft durch Kommunikationstherapie und Verhaltenstherapie. Pfeiffer, München 1971.

Mandel, K.-H., A. Mandel, H. Rosenthal: Einübung der Liebesfähigkeit. Pfeiffer, München 1975.

Mandelbrote, B. M., E. D. Wittkower: Emotional factors in Graves' disease. Psychosom. Med. 17 (1955) 109.

Mandell, A. J.: Psychological management of coronary artery disease. In: Wahl, H. (ed.): New Dimensions in Psychosomatic Medicine. Little, Brown & Co., Boston 1964.

Mandler, G., I. Kremen: Autonomic feedback: a correlational study. J. Pers. 26 (1958) 388–399.

Maneros, A., A. Rohde, K. M. Otto: Infektionsbedingte psychische Störungen. Dtsch. med. Wschr. 112 (1987) 796–800.

Manhold, J. H.: Report of a study on the relationship of personality variables to periodontal conditions. J. Periodont. 24 (1953) 248.

Mann, Th.: Der Zauberberg. Fischer, Berlin 1924.

Mann, Th.: Buddenbrooks, S. 277. Fischer, Frankfurt 1960.

Mansour, A., H. Khachaturian, M. E. Lewis, H. Akil, S. J. Watson: Anatomy of CNS opioid receptors. Trends in Neurosci. 11 (1988) 308–314.

Mantovani, A.: Effect of glucocorticoid hormones on the immune system. In: Mihich, E., Y. Sakurai (eds.): Biological Responses in Cancer, Vol. 3, pp. 155–180. Plenum, New York 1985.

Manuck, S. B., B. Giordani: Heart rate reactivity, blood pressure and report of parental hypertension. Soc. psychophysiol. Res., Vancouver 1980.

Manuck, S. B., J. M. Proietti: Parental hypertension and cardiovascular response to cognitive and isometric challenge. Psychophysiol. 19 (1982) 481–489.

Manuck, S. B., D. S. Krantz: Psychophysiologic reactivity in coronary heart disease. Behav. Med. Update 6 (1984) 11–15.

Manuck, S. B., J. R. Kaplan, B. Thomas, D. V. Clarkson: Behaviorally induced heart rate reactivity and atherosclerosis in cynomolgus monkeys. Psychosom. Med. 45 (1983) 95–108.

Manuck, S. B., J. R. Kaplan, T. B. Clarkson: Atherosclerosis, social dominance and cardiovascular reactivity. In: Schmidt, T., T. Dembroski, G. Blümchen (eds.): Biological

and Psychological Factors in Cardiovascular Disease, pp. 459–475. Springer, Berlin-Heidelberg-New York 1986.

Manz, R.: Gütekriterien der Instrumente zur Fallidentifikation. In: Schepank, H. (Hrsg.): Psychogene Erkrankungen der Stadtbevölkerung, S. 235–238. Springer, Berlin-Heidelberg-New York 1987.

Mappes, G.: Zur Klinik und Therapie des Morbus Crohn. Therapiewoche 29 (1979) 1028–1034.

Marbach, J. J.: Phantom bite. J. Amer. Orthodont. 70 (1976) 190.

Marble, A.: The natural history of diabetes. Horm. metab. Res. (Suppl.) 4 (1974) 153–158.

March, H.: Menschenschicksale in Gutachten. Psyche 7 (1954) 711–720.

Marchand, H.: Die Suggestion der Wärme im Oberbauch und ihr Einfluß auf Blutzucker und Leukozyten. Psychotherapie 1 (1956) 154–164.

Marcovic, B. M., V. J. Djuric, M. Lazarevic, B. D. Jankovic: Anaphylactic shock-induced conditioned taste aversion. I. Demonstration of the phenomenon by means of three modes of CS-US presentations. Brain, Behavior, and Immunity 2 (1988) 11–23.

Marcuse, H.: Triebstruktur und Gesellschaft. Frankfurt 1965.

Margolin, S. G.: The behavior of the stomach during psychoanalysis. A contribution to a method of verifying psychoanalytic data. Psychoanal. Quart. 20 (1951) 349.

Margolin, S. G.: Genetic and dynamic psychophysiological determinants of pathophysiological processes. In: Deutsch, F. (ed.): The Psychosomatic Concept in Psychoanalysis. New York 1953.

Mariotti, S., E. Martino, C. Cupini, R. Lari, C. Giani, L. Baschieri, A. Pinchera: Low serum thyroglobulin as a clue to the diagnosis of thyreotoxicosis factitia. New Engl. J. Med. 307/7 (1982) 410–412.

Marks, R. M., E. J. Sachar: Undertreatment of medical inpatients with narcotic analgetics. Ann. intern. Med. 78 (1973) 173–182.

Markson, E. W.: Patient semiology of a chronic disease. Rheumatoid arthritis. Soc. Sci. Med. 5 (1971) 159–167.

Marmot, M. G.: Socio-economic and cultural factors in ischaemic heart disease. Advanc. Cardiol. 29 (1982) 68–75.

Marmot, M. G., W. Winkelstein: Epidemiological observations on intervention trials for prevention of coronary heart disease. Amer. J. Epidemiol. 101 (1975) 177–181.

Marmot, M. G., S. L. Syme: Acculturation and coronary heart disease in Japanese-Americans. Amer. J. Epidemiol. 104 (1976) 225–247.

Marmot, M. G., A. Adelstein, N. Robinson, G. Rose: Changing social distributions of heart disease. Brit. med. J. 76 (1978a) 1109–1112.

Marmot, M. G., G. Rose, M. Shiple, P. J. Hamilton: Employment grade and coronary heart disease in British civil servants. J. Epidemiol. Comm. Health 32 (1978b) 244–249.

Marris, P.: Widows and their families. Routledge & Kegan, London 1958.

Marsden, C. D.: The mysterious function of the basal ganglia: the Robert Wartenberg Lecture. Neurology 32 (1982) 514–539.

Marsh, J. T., J. F. Lavender, S.-S. Chang, A. F. Rasmussen: Poliomyelitis in monkeys: decreased susceptibility after avoidance stress. Science 140 (1963) 1414.

Marsh, L. C.: zit. n. Schulte-Herbrüggen: Historische Aspekte der Gruppenpsychotherapie. In: Heigl-Evers, A. (Hrsg.): Sozialpsychologie. Beltz, Weinheim-Basel 1979.

Marshall, G. D., P. G. Zimbardo: Affective consequences of inadequately explained physiological arousal. J. Pers. soc. Psychol. 37 (1979) 970–988.

Literatur

Marshall, H. E. S.: The incidence of physical disorders among psychiatric in-patients. Brit. med. J. 2 (1949) 468–470.

Marshall, H. K., R. Jerauld, N. C. Kreger, W. McAlpine, M. Steffa, J. Kenell: Maternal attachment. Importance of the first post-partum days. New Engl. J. Med. 286 (1972) 460.

Marshall, L. S. A.: Soldaten im Feuer. Frauenfeld, Zürich 1951.

Martin, F.: Subgroups in anorexia nervosa: a family systems study. In: Darby, P. L., P. E. Garfinkel, D. M. Garner, D. V. Coscina (eds.): Anorexia nervosa. Recent Developments in Research, pp. 57–63. Liss, New York 1983.

Martin, F. E.: The treatment and outcome of anorexia nervosa in adolescents: a prospective study and 5-year follow-up. J. psychiat. Res. 19 (1985) 509–514.

Martin, J. L.: The impact of AIDS on gay male behavior patterns in New York City. Amer. J. publ. Health 77 (1987) 578–581.

Martin, P. R., A. M. Mathews: Tension headache: psychophysiological investigation and treatment. J. psychosom. Res. 22 (1978) 389–399.

Martin, P. R., P. R. Nathan, D. Milech, M. Van Keppel: The relationship between headache and mood. Behav. Res. Ther. 26 (1988) 353–356.

Martin, R. L., W. V. Roberts, P. J. Clayton: Psychiatric status after hysterectomy. J. Amer. med. Ass. 244 (1980) 350.

Martini, G. A., J. W. Brandes: Increasased consumption of refined carbohydrates in patients with Crohn's disease. Klin. Wschr. 54 (1976) 367–371.

Martini, G. A.: Morbus Crohn. Dtsch. Ärztebl. 85 (1988) 1796–1801.

Martini, P.: Eröffnungsansprache des Internistenkongresses. In: Lasch, H. G., B. Schlegel (Hrsg.): Hundert Jahre Deutsche Gesellschaft für Innere Medizin. Bergmann, München 1982.

Marty, P.: La relation objectale allergique. Rev. franç. Psychanal. 22 (1958) 5–35. (Dt. in: Brede, K. (Hrsg.): Einführung in die Psychosomatische Medizin, S. 420–445. Frankfurt 1974).

Marty, P.: Notes cliniques et hypothèses à propos de l'économie de l'allergie. Rev. franç. Psychanal. 33 (1969) 243–253. (Dt. in: Brede, K. (Hrsg.): Einführung in die Psychosomatische Medizin, S.446–455. Frankfurt 1974).

Marty, P.: Zit. nach Schneider, P.: Zum Verhältnis von Psychoanalyse und Psychosomatischer Medizin. Psyche 27 (1973) 21–49.

Marty, P.: Les mouvements individuels de vie et de mort. Paris 1976.

Marty, P., M. de M'Uzan: La pensée opératoire. Rev. franç. Psychanal. 27 (1963) 345–356 (Suppl.). (Dt. in: Psyche 32 (1978) 974–984).

Maruta, T., D. W. Swanson, W. M. Swenson: Pain as a psychiatric symptom: comparison between low back pain and depression. Psychosomatics 17 (1976) 123–127.

Maruta, T., D. W. Swanson, W. M. Swenson: Chronic pain: which patients may a pain-management program help? Pain 7 (1979) 321–329.

Marx, J. L.: Brain peptides: Is substance P a transmitter of pain signals? Science 205 (1979) 886–889.

Marxkors, R., H. Müller-Fahlbusch: Psychogene Prothesenunverträglichkeit. Hanser, München-Wien 1976.

Maschewsky, W.: Zwischenauswertung der schriftlichen Befragung des Herzinfarktprojekts am Wissenschaftszentrum Berlin. In: Friczewski, F., W. Maschewsky, F. Naschold, P. Wotschack, W. Wotschack (Hrsg.): Arbeitsbelastung und Krankheit bei Industriearbeitern, S. 85–126. Campus, Frankfurt-New York 1982.

Maschewsky, W., U. Schneider: Soziale Ursachen des Herzinfarkts. Campus, Frankfurt-New York 1981.

Mascia, A. V. et al.: Manual on the standardization of care of the severely asthmatic child. J. Asthma Res. 13 (1976) 115–127.

Måseide, P.: Sincerity may frighten the patient: medical dilemmas in patient care. J. Pragmatics 5 (1981) 145–167.

Maser, J. D., S. J. Keith: CT scans and schizophrenia: report on a workshop. Schizophrenia Bull. 9 (1983) 265–283.

Mash, E. J., L. G. Terdal: Kompendium der verhaltenstherapeutischen Diagnostik. Fachbuchhandlung f. Psychologie, Frankfurt 1980.

Masica, D. N., J. Money, A. A. Ehrhardt: Fetal feminization and female gender identity in the testicular feminizing syndrome of androgen insensitivity. Arch. sex. Behav. 1 (1971) 131–142.

Maslach, C.: Negative emotional biasing of unexplained arousal. J. Pers. soc. Psychol. 37 (1979) 953–969.

Mason, J. H., J. J. Anderson, R. F. Meenan: A model of health status for rheumatoid arthritis. Arthr. and Rheum. 31 (1988) 714–720.

Mason, J. W.: A review of psychoendocrine research on the pituitary-thyroid system. Psychosom. Med. 30 (1968) 30.

Mason, J. W.: A review of psychoendocrine research on the pituitary-adrenal cortical axis. Psychosom. Med. (Suppl.) 30 (1968) 576–607.

Mason, J. W.: Organization of psychoendocrine mechanisms: a review. In: Greenfield, N. S., R. A. Sternbach (eds.): Handbook of Psychophysiology. Holt, Rinehart and Winston, New York 1972.

Mason, J. W.: Emotions as reflected in patterns of endocrine integration. In: Levi, L. (ed.): Emotions: Their Parameters and Measurement. Raven, New York 1975.

Mason, J. W., E. N. Mongey: Thyroid (plasma BEI) response to chair restraint in the monkey. Psychosom. Med. 34 (1972) 441.

Mason, J. W., J. V. Brady, W. W. Tolson: Behavioral adaptations and endocrine activity. In: Levine, R. (ed.): Proc. of the Association for Research in Nervous and Mental Diseases. Williams & Wilkins, Baltimore 1966.

Maß, R., H. Harden, M. Ramm, R. Simeit, R. Richter, B. Dahme: Evaluation of an airway resistance biofeedback training for asthmatics. J. Psychophysiol. 2 (1988) 144–145.

Massarrat, S., E. Heuser, L. Hausmann, R. Schubotz: Langzeitprophylaxe des Ulcus duodeni mit Cimetidin. Dtsch. med. Wschr. 107 (1982) 1085–1088.

Massie, M. J., J. Holland, E. Glass: Delirium in terminally ill cancer patients. Amer. J. Psychiat. 140 (1983) 1048–1050.

Masters, W. H., V. E. Johnson: Human sexual response. Little, Brown and Co, Boston 1966. Dtsch. Ausg.: Die Sexuelle Reaktion. Rowohlt, Reinbek (1970).

Masters, W. H., V. E. Johnson: Human sexual inadequacy. Little, Brown and Co, Boston 1970.

Masters, W. H., V. E. Johnson: Homosexuality in perspective. Little, Brown and Co, Boston 1979.

Mastrovito, R. C., K. S. Deguire, J. Clarkin, T. Thaler, J. L. Lewis, E. Cooper: Personality characteristics of women with gynecological cancer. Cancer Detect. Prev. 2 (1979) 281–287.

Matta, R. J., J. E. Lawler, B. Lown: Ventricular electrical instability in the conscious dog. Effects of psychologic stress and beta-adrenergic blockade. Amer. J. Cardiol. 34 (1974) 692.

Matta, R. J., R. L. Verrier, B. Lown: The repetitive extrasystole threshold as an index of vulnerability to ventricular fibrillation. Amer. J. Physiol. 230 (1976) 1469–1473.

Mattern, H.: Ich klage an ! Im Namen des Patienten. Ärztl. Praxis 106 (1988) 3255.

Matthes, A.: Über das Suchtmoment bei einem Fall optisch induzierbarer Epilepsie. Z. Kinderheilk. 75 (1954) 162–169.

Matthes, K.: Prinzipien und Probleme der klinischen Forschung. Mitt. Ges. Dtsch. Naturforscher u. Ärzte 1 (1963) 5–10.

Matthews, J., H. Akil, J. Greden, D. Charney, V. Weinberg, A. Rosenbaum, S. J. Watson: β-endorphin/β-lipoprotein immunoreactivity in endogenous depression: effect of dexamethasone. Arch. gen. Psychiat. 43 (1986) 374–381.

Matthews, K. A.: What is the type A (coronary-prone) behavior pattern from a psychological perspective? Psychol. Bull. 91 (1982) 293–323.

Matthews, K. A.: Assessment issues in coronary-prone behavior. In: Dembrowski, T. M., T. H. Schmidt, G. Blümchen (eds.): Biobehavioral Basis of Coronary Heart Diseases. Karger, Basel 1983.

Matthews, K. A.: Coronary heart disease and Type A behaviors: update on and alternative to the Booth-Kewley and Friedman (1987) quantitative reviews. Psychol. Bull. 104 (1988) 373–380.

Matthews, K. A., C. J. Rakaczky: Familial aspects of the Type A behavior pattern and physiologic reactivity to stress. In: Schmidt, T., T. Dembroski, G. Blümchen (eds.): Biological and Psychological Factors in Cardiovascular Disease, pp. 228–245. Springer, Berlin-Heidelberg-New York 1986.

Matthews, K. A., S. G. Haynes: Type A behavior and coronary disease risk – update and critical evaluation. Amer. J. Epidemiol. 123 (1986) 923–960.

Matthews, K. A., C. M. Stoney: Familial influences on cardiovascular response to behavioral stress. Psychophysiol. 25 (1988) 423.

Matthews, K. A., R. H. Rosenman, T. M. Dembroski et al.: Familial reassemblance in components of the Type A behavior pattern. A reanalysis of the California Type A twin study. Psychosom. Med. 46 (1984) 512–522.

Matthews, K. A., D. C. Glass, R. H. Rosenman, R. W. Bortner: Competitive drive, pattern A, and coronary heart disease: a further analysis of some data from the Western Collaborative Group Study. J. chron. Dis. 30 (1977) 489–498.

Matthews, K. A., S. M. Weiss, T. Detre, T. M. Dembroski, B. Falkner, S. B. Manuck, R. B. Williams jr. (eds.): Handbook of Stress, Reactivity, and Cardiovascular Disease. Wiley, New York 1986.

Matthysse, S., P. S. Holzman, K. Lange: The genetic transmission of schizophrenia: application of mendelian latent structure analysis to eye tracking dysfunctions in schizophrenia and affective disorders. J. psychiat. Res. 20 (1986) 57–67.

Mattussek, S., H.-H. Raspe: Psychometrische Untersuchungen zur Aggressivität von Patienten mit chronischer Polyarthritis. Akt. Rheum. 13 (1988) 18–24.

Maturana, H. R.: The neurophysiology of cognition. In: Garvin, P. (ed.): Cognition: A Multiple View. Spartan Books, New York 1969.

Maturana, H. R.: Die Organisation des Lebendigen. In: Maturana, H. R.: Erkennen: Die Organisation und Verkörperung von Wirklichkeit. Vieweg, Braunschweig 1982.

Maturana, H. R.: Cognition. In: Hejl, P. M., W. K. Köck, G. Roth (Hrsg.): Wahrnehmung und Kommunikation. Lang, Frankfurt 1978.

Maturana, H. R., F. Varela: Autopoietic systems. In: Biol. Computer Laboratory, Report 9, 4. Urbana, Illinois 1975.

Maturana, H. R., F. Varela: Autopoiesis and cognition. Boston Studies in the Philosophy of Sciences. Reidel, Boston 1980.

Maturana, H. R., F. J. Varela: Der Baum der Erkenntnis. Wie wir die Welt durch unsere Wahrnehmung erschaffen – die biologischen Wurzeln des menschlichen Erkennens. Scherz, Bern-München-Wien 1987.

Matussek, P.: Zum Problem der Psychogenese in der modernen Medizin. In: Matussek, P., J. P. Lotz (Hrsg.): Studien und Berichte der Katholischen Akademie in Bayern, Heft 2. München 1958.

Matussek, P.: Funktionelle Sexualstörungen. In: Giese, H. (Hrsg.): Die Sexualität des Menschen, S. 786–828. Enke, Stuttgart 1971.

Maurer, Y.: Der Körper in der psychiatrischen und psychotherapeutischen Behandlung. Schweizer Arch. Neurologie und Psychiatrie 138 (1987a) 49.

Maurer, Y.: Körperzentrierte Psychotherapie. Hippokrates, Stuttgart 1987b.

Maurin, J., J. Schenkel: A study of the family unit's response to hemodialysis. J. psychosom. Res. 20 (1976) 163–168.

Mavissakalian, M., J. M. Perel: Imipramine dose-response relationship in panic disorder with agoraphobia. Preliminary findings. Arch. gen. Psychiat. 46 (1989) 127–131.

Mavissakalian, M., L. Michelson: Tricyclic antidepressants in obsessive-compulsive disorder. Antiobsessional or antidepressant agents? J. nerv. ment. Dis. 171 (1983) 301–306.

May, P. R. A.: Psychotherapy and ataraxic drugs. In: Bergin, A. E., S. L. Garfield (eds.): Handbook of Psychotherapy and Behavioral Change: An Empirical Analysis. Wiley, New York 1971.

Mayer, A.: Über seelisch bedingte Menstruationsstörungen. Geburtshilfe Frauenheilk. 6 (1944) 178.

Mayer, B., F. J. Marx, T. Spiro: Sexualverhalten nach Prostataoperation. Sexualmedizin 12 (1983) 366.

Mayer, D. J., D. D. Price: Central nervous system mechanisms of analgesia. Pain 2 (1976) 379–404.

Mayer, D. J., D. D. Price, A. Rafii: Antagonism of acupuncture analgesia in man by the narcotic antagonist naloxone. Brain Res. 121 (1977) 368.

Mayer, H., B. Stanek, P. Hahn: Biometric findings on cardiac neurosis: II. ECG and circulation findings of cardiophobic patients during standardized examination of the circulatory system. In: Freyberger, H. (ed.): Topics of Psychosomatic Research, pp. 283–288. Karger, Basel 1973.

Mayer, J., D. W. Thomas: Regulation of food intake and obesity. Science 156 (1967) 328–337.

Mayr, E.: The growth of biological thought. Diversity, evolution and inheritance. Harvard Univ. Press, Cambridge/Mass. 1982.

Mayring, P.: Qualitative Inhaltsanalyse. In: Jüttemann, G. (Hrsg.): Qualitative Forschung in der Psychologie. Grundfragen, Verfahrensweisen, Anwendungsfehler. Beltz, Weinheim-Basel 1985.

Mazur, A.: Effects of testosterone on status in small groups. Folia primatol. 26 (1976) 214–226.

Mazur, A., T. A. Lamb: Testosterone, status and mood in human males. Horm. Behav. 14 (1980) 236–246.

McCain, G. A., R. A. Scudds: The concept of primary fibromyalgia and significance to other chronic musculoskeletal pain syndromes. Pain 33 (1988) 273–287.

McCall, W. A.: Measurement. New York 1939.

McCann, B., K. A. Matthews: Influences of potential for hostility, Type A behavior, and parental history of hypertension on adolescents' cardiovascular responses during stress. Psychophysiol. 25 (1988) 503–511.

McCaul, K. D., J. M. Malott: Distraction and coping with pain. Psychol. Bull. 95 (1984) 516.

McClelland, D. C., D. Burnham: Power is the great motivator. Harvard Business Rev. 25 (1975) 159–166.

McClelland, D. C., C. Alexander, E. Marks: The need for power, stress, immune function, and illness among male prisoners. J. abn. Psychol. 91 (1982) 61–70.

McClelland, D. C., G. Ross, V. Patel: The effect of an academic examination on salivary norepinephrine and immunoglobulin levels. J. hum. Stress 11 (1985) 52–59.

McClelland, D. C., E. Floor, R. J. Davidson, C. Saron: Stressed power motivation, sympathetic activation, immune function, and illness. J. hum. Stress 6 (1980) 11–19.

McClintock, M. K.: Menstrual synchrony and suppression. Nature (Lond.) 229 (1971) 244–245.

McCombs, R. P., F. C. Lowell, J. L. Ohmann: Myths, morbidity, and mortality in asthma. J. Amer. med. Ass. 242 (1979) 1521–1524.

McConnell, R. B.: Genetics in Crohn's disease. Z. Gastroenterol. 17 (Suppl.) (1979) 61–65.

McCoy, D. F., T. L. Roszman, J. S. Miller, K. S. Kelly, M. J. Titus: Some parameters of conditioned immunosuppression: species difference and CS-US delay. Physiol. Behav. 36 (1986) 731–736.

McCoy, G. C., S. Fein, E. B. Blanchard, D. A. Wittrock, R. J. McCaffrey, L. Pangburn: End-organ changes associated with the self-regulatory treatment of mild hypertension? Biofeedback & Selfregul. 13 (1988) 39–46.

McCranie, E. W., L. O. Watkins, J. M. Brandsma, B. D. Sisson: Hostility, coronary heart disease (CHD) incidence, and total mortality: lack of association in a 25-year follow-up study of 478 physicians. J. behav. Med. 9/2 (1986) 119–225.

McCraw, R., J. Tuma: Rorschach content categories of juvenile diabetics. Psychol. Rep. 40 (1977) 818.

McCreary, C., J. Turner, E. Dawson: Differences between functional versus organic low back pain patients. Pain 4 (1977) 73–78.

McDaniel, S. H., L. C. Wynne, T. T. Weber: The territory of systems consultation.In: Wynne, L. C., S. McDaniel, T. T. Weber (eds.): Systems Consultation, pp. 16–28. Guilford, New York – London 1986.

McDonald, C.: The pattern of neurotic illness in the elderly. Aust. N. Z. J. Pychiat. 1 (1966) 203–210.

McDonald, C.: An age-specific analysis of the neuroses. Brit. J. Psychiat. 122 (1973) 477–480.

McDougall, J.: The psychosoma and the psychoanalytic process. Int. Rev. Psychoanal. 1 (1974) 437–459.

McDougall, J.: A child is being eaten. Contemp. Psychoanal. 16 (1980) 417–459.

McDougall, J.: Ein Körper für zwei. Forum Psychoanal. 3 (1987) 265–287.

McEwen, B. S.: Neural gonadal steroid actions. Science 211 (1981) 1303–1312.

McFarlane, A. C., P. M. Brooks: An analysis of the relationship between psychological morbidity and disease activity in rheumatoid arthritis. J. Rheum. 15 (1988 a) 926–931.

McFarlane, A. C., P. M. Brooks: Determinants of disability in rheumatoid arthritis. Brit. J. Rheum. 27 (1988 b) 7–14.

McFarlane, A. C., R. S. Kalucy, P. M. Brooks: Psychobiological predictors of disease course in rheumatoid arthritis. J. psychosom. Res. 31 (1987) 757–764.

McGivney, J. Q., B. R. Cleveland: The levator syndrome and its treatment. Sth. med. J. 58 (1965) 505.

McGlashan, T. H., F. J. Evans, M. T. Orne: The nature of hypnotic analgesia and the placebo response to experimental pain. Psychosom. Med. 31 (1969) 227–246.

McGrath, J.: Settings, measures and themes. An integrative review of some research on social and psychological factors in stress. In: McGrath, J. (ed.): Social and Psychological Factors in Stress. Holt, Rinehart & Winston, New York 1970.

McGue, M., I. I. Gottesman, D. C. Rao: The transmission of schizophrenia under a multifactorial threshold model. Amer. J. hum. Genet. 35 (1983) 1161–1178.

McGuffin, P., R. Katz, P. Bebbington: The Camberwell Collaborative Depression Study III. Depression and adversity in the relatives of depressed probands. Brit. J. Psychiat. 152 (1988) 775–782.

McHugh, P. R., P. R. Slavney: The perspectives of psychiatry. Johns Hopkins Univ. Press, Baltimore 1983.

McIntosh, J.: Processes of communication, information seeking and control associated with cancer. A selective review of the literature. Soc. Sci. Med. 8 (1974) 167–187.

McIntosh, J.: Patients' awareness and desire for information about diagnozed but undisclosed malignant disease. Lancet 20 (1976) 300–303.

McIntosh, J.: Communication and awareness in a cancer ward. Watson, London-New York 1977.

McKegney, F. P.: The incidence and characteristics of patients with conversion reactions. A general hospital consultation service sample. Amer. J. Psychiat. 124 (1967) 542–545.

McKegney, F. P., R. O. Gordon, S. M. Levine: A psychosomatic comparison of patients with ulcerative colitis and Crohn's disease. Psychosom. Med. 32 (1970) 153–166.

McKegney, F. P., C. Runge, R. Bernstein et al.: Severe psychiatric disorder in dialysis-transplant patients. In: Levy, N. B. (ed.): Psychonephrology 1, pp. 49–60. Plenum, New York 1981.

McKegney, P. F.: The intensive care syndrom: the definition, treatment and prevention of a new „disease of medical progress". Connect. Med. 30 (1966) 633.

McKegney, P. F.: Consultation-liaison teaching of psychosomatic medicine: opportunities and obstacles. J. nerv. ment. Dis. 154 (1972) 198–205.

McKeown, T.: Die Bedeutung der Medizin. Traum, Trugbild oder Nemesis. Suhrkamp, Frankfurt 1982.

McKinlay, J.: Who is really ignorant – physician or patient? J. Health Soc. Behav. 16 (1975) 3–11.

McKusick, L., W. Horstman, T. J. Coates: AIDS and sexual behavior reported by gay men in San Francisco. Amer. J. publ. Health 74 (1985) 493–496.

McLean, P. D.: Psychosomatic disease and the visceral brain. Recent developments bearing on the Papez theory of emotion. Psychosom. Med. 11 (1949) 338.

McLean, P. D.: New findings on brain function and sociosexual behavior. In: Zubin, J., J. Money (eds.): Contemporary Sexual Behavior, pp. 53–76. Johns Hopkins Univ. Press, Baltimore 1973.

McLusky, N. J., F. Naftolin: Sexual differentiation of the central nervous system. Science 211 (1981) 1294–1303.

McMahon, A. W., P. Schmitt, J. F. Patterson, E. Rothman: Personality differences between inflammatory bowel disease patients and their healthy siblings. Psychosom. Med. 35 (1973) 91–103.

McNally, R. J.: Preparedness and phobias: a review. Psychol. Bull. 101 (1987) 283–303.

McSweeney, A. J.: Life quality of patients with chronic obstructive pulmonary diseases. Arch. intern. Med. 142 (1982) 473–478.

McWilliam, J. A.: Cardiac failure and sudden death. Brit. med. J. 1 (1889) 6.

McWilliam, J. A.: Ventricular fibrillation and sudden death. Brit. med. J. 2 (1923) 215.

McWilliam, J. A.: Blood pressure and heart action in sleep and dreams: their relation to hemorrhages, angina and sudden death. Brit. med. J. 2 (1923) 1196.

Meadow, R.: Factitious epilepsy. Lancet II (1984) 25–28.

Meares, A.: A regression of osteogenic sarcoma metastases associated with intensive medication. Med. J. Aust. 2 (1978) 433.

Meares, R.: Features which distinguish groups of spasmodic torticollis. J. psychosom. Res. 15 (1971) 1–11.

Mechanic, C.: zit. n.: Coelho, G. V., D. A. Hamburg, J. E.

Adams: Coping and Adaptation. Basic Books, New York 1974.

Mechanic, D.: The concept of illness behavior. J. chron. Dis. 15 (1962) 189.

Medalie, J. H., H. A. Kahn, H. N. Neufeld et al.: Five-year myocardial infarction incidence. Association of single variables to age and birthplace. J. chron. Dis. 26 (1973) 329.

Medalie, J. H., M. Snijder, J. J. Groen et al.: Angina pectoris among 100 men. Five-year incidence and univariate analysis. Amer. J. Med. 55 (1973) 583.

Medansky, R.: Dermatopsychosomatics: an overview. Psychosomatics 21 (1980) 195–200.

Medawar, P. B., J. S. Medawar: The life science. Harper & Row, New York-Hagerstown-San Francisco-London 1977.

Mednick, S. A., F. Schulsinger: Some premorbid characteristics related to breakdown in children with schizophrenic mothers. In: Rosenthal, D., S. S. Kety (eds.): Transmission of Schizophrenia. Pergamon, New York 1968.

Meenan, R. F., P. M. Gertman, J. H. Mason: Measuring health status in arthritis. Arthr. and Rheum. 23 (1980) 146–152.

Meermann, R. (Hrsg.): Anorexia nervosa. Enke, Stuttgart 1981.

Meermann, R., M. Fichter: Störungen des Körperschemas (body image) bei psychischen Krankheiten: Methodik und experimentelle Ergebnisse bei Anorexia nervosa. Psychother. med. Psychol. 32 (1982) 162–169.

Meermann, R., W. Vandereycken: Therapie der Magersucht und Bulimia nervosa. De Gruyter, Berlin 1987.

Meerwein, F.: Über die Führung des ersten Gespräches. Schweiz. med. Wschr. 18 (1960) 497.

Meerwein, F.: Tiefenpsychologische Aspekte der zahnärztlichen Tätigkeit. Schweiz. Mschr. Zahnheilk. 77 (1967) 776.

Meerwein, F.: Der Krebspatient und sein Arzt im 19. Jahrhundert. Ursprünge der Psychoonkologie? Züricher Medizingeschichtliche Abhandlungen. Neue Reihe Nr. 189. Junis, Zürich 1980.

Meerwein, F.: Die Arzt-Patient-Beziehung des Krebskranken. In: Meerwein, F. (Hrsg.): Einführung in die Psycho-Onkologie, S. 84–165. 2. Aufl. Huber, Bern-Stuttgart-Wien 1981.

Meerwein, F.: Probleme und Konflikte des Onkologen und seiner Mitarbeiter. Münch. med. Wschr. 126 (1984) 219–222.

Meerwein, F., S. Kauf, G. Schneider: Bemerkungen zur Arzt-Patienten-Beziehung bei Krebskranken. Z. psychosom. Med. Psychoanal. 22 (1976) 278–300.

Meffert, H. J., A. Boll, B. Dahme et al.: Der relative Anteil somatischer und psychischer Befunde an der Vorhersage psychopathologischer Auffälligkeiten nach Herzoperationen. In: Studt, H. H. (Hrsg.): Psychosomatik – Forschung und Praxis. Urban & Schwarzenberg, München 1983.

Mehan, H.: Learning lessons. Harvard Univ. Press, Cambridge 1979.

Mei-Tal, V., S. Meyerowitz, G. L. Engel: The role of psychological process in a somatic disorder: multiple sclerosis. 1. The emotional setting of illness onset and exacerbation. Psychosom. Med. 32 (1970) 67–86.

Meichenbaum, D.: Cognitive behavior modification. Plenum, New York 1977. Deutsch: Kognitive Verhaltensmodifikation. Urban & Schwarzenberg, München 1979.

Meichenbaum, D., R. Cameron: Stress inoculation training – toward a general paradigm for training coping skills. In: Meichenbaum, D., M. E. Jaremko (eds.): Stress reduction and prevention. Plenum, New York 1983.

Meikle, M., S. Griesst: The perceived localization of tinnitus. In: Feldmann, H. (ed.): Proceedings of the International Tinnitus Seminar Münster 1987, p. 183. Harsch, Karlsruhe 1987.

Meissner, W. W.: Family dynamics and psychosomatic processes. Family Process 5 (1966) 142. Deutsch: Familiendynamik und psychosomatische Prozesse.In: Brede, K. (Hrsg.): Einführung in die psychosomatische Medizin, S. 193. Fischer, Frankfurt 1974.

Mekhjian, H. S., D. M. Switz, C. S. Melnyk, G. B. Rankin, R. K. Brooks: Clinical features and natural history of Crohn's disease. Gastroenterology 77 (1979) 898–906.

Mellinger, G. D., M. B. Balter, E. Uhlenhut: Prevalence and correlates of the long-term regular use of anxiolytics. J. Amer. med. Ass. 251 (1984) 375–379.

Mellinger, G. D., M. B. Balter, H. J. Parry, D. J. Mannheimer, J. H. Cisin: An overview of psychotropic drug use in the USA. In: Josephson, E., E. E. Carroll (eds.): Drug Use: Epidemiological and Sociological Approaches. Wiley, New York 1974.

Meltzer, H. Y., S. M. Stahl: The dopamine hypothesis of schizophrenia: a review. Schizophrenia Bull. 2 (1976) 19–76.

Meltzer, H. Y., T. Kolakowska, V. S. Fang, L. Fogg, A. Robertson, R. Lewine, M. Strahilewitz, D. Busch: Growth hormone and prolactin response to apomorphine in schizophrenia and major affective disorders: relation to duration of illness and depressive symptoms. Arch. gen. Psychiat. 41 (1984) 512–519.

Meltzoff, J., M. Kornreich: Research in psychotherapy. Atherton, New York 1970.

Melzack, R.: Pain perception. Res. Publ. Ass. nerv. ment. Dis. 48 (1970) 272–285.

Melzack, R.: The puzzle of pain. Basic Books, New York 1973.

Melzack, R., T. Scott: The effects of early experience on the response of pain. J. comp. physiol. Psychol. 50 (1957) 155.

Melzack, R., P. D. Wall: Pain mechanisms: a new theory. Science 150 (1965) 971–979.

Melzack, R., W. S. Torgerson: On the language of pain. Anaesthesiology 34 (1971) 50–59.

Melzer, E.: Psyche und Tuberkulose. Hippokrates 28 (1957) 1; 35.

Mendell, L. M., P. D. Wall: Presynaptic hyperpolarisation: A role for fine afferent fibers. J. Physiol. 172 (1964) 274–294.

Mendeloff, A. I.: Epidemiology of Crohn's disease. Z. Gastroenterol. 17 (Suppl.) (1979) 66–69.

Mendeloff, A. I.: The epidemiology of inflammatory bowel disease. Clin. Gastroenterol. 9 (1980) 259–270.

Mendelson, J., P. Solomon, E. Lindenmann: Hallucinations of poliomyelitis patients during treatment in a respirator. J. nerv. ment. Dis. 126 (1958) 421–428.

Mendlewicz, J., C. C. Schulmann, B. de Schutter, J. Wilmotte: Chronic prostatitis. Psychosomatic incidence. Psychother. and Psychosom. 19 (1971) 118.

Mendlewicz, J., P. Linkowski, M. Kerkhofs, D. Desmedt, J. Golstein, G. Copinschi, E. Van Ceuter: Diurnal hypersecretion of growth hormone in depression. J. clin. Endocr. 60 (1985) 505–512.

Meng, H.: Das Problem der Organpsychose bei seelischer Behandlung organisch Kranker. Int. Z. Psychoanal. 20 (1934) 400–410.

Menninger, K.: Polysurgery and polysurgical addiction. Psychoanal. Quart. 3 (1934) 173–199.

Menninger, K. A.: Psychoanalytic study of a case of organic epilepsy. Psychoanal. Rev. 13 (1926) 166–187.

Menninger, K. A.: Some observations on the psychological factors in urination and genitourinary afflictions. Psychoanal. Rev. 28 (1941) 117.

Menninger, W.: Psychoanalytic principles applied to the treatment of hospitalized patients. Bull. Menninger Clin. 1 (1936) 35.

Mense, S.: Nervous outflow from skeletal muscle following chemical noxious stimulation. J. Physiol. 267 (1977) 75–88.

Mensen, H.: ABC des autogenen Trainings. Goldmann, München 1988.

Mensen, H.: Autogenes Training – ein physiologisch-rationales Naturheilverfahren für Prävention und Rehabilitation. Perimed, Erlangen 1988.

Menzel, W.: Zur Physiologie und Pathologie des Nacht- und Schichtarbeiters. Arbeitsphysiologie 14 (1950) 304–318.

Menzel, W.: Über labile und paroxysmale Hypertonie. Med. Welt 12 (1961) 560–565.

Mergner, U., M. Osterland, K. Pelte: Arbeitsbedingungen im Wandel. Schwartz & Co., Göttingen 1975.

Merion, R. M., D. J. G. White, S. Thiru, D. B. Evans, R. Y. Calne: Cyclosporine: five years' experience in cadaveric renal transplantation. New Engl. J. Med. 310 (1984) 148–154.

Merl, H.: Das Problem der Indikationsstellung in der Familientherapie. Voraussetzungen und methodische Überlegungen. Material. Psychoanal. 9 (1983) 167–241.

Mersereau, B. S.: Regional ileitis in depressed patients. Amer. J. Psychiat. 119 (1963) 1099–1100.

Merskey, H.: The role of the psychiatrist in the investigation and treatment of pain. In: Bonica, J. J. (ed.): Pain. Raven, New York 1980.

Merskey, H. (ed.): Chronic pain syndromes and definition of pain terms. International Association for the Study of Pain. Subcommittee on Taxonomy. Pain Suppl. 3 (1986) 1–225.

Merskey, H., F. G. Spear: Psychological and psychiatric aspects. Ballière, Tindall and Cassell, London 1967.

Merskey, H., N. A. Buknih: Hysteria and organic brain disease. Brit. J. med. Psychol. 48 (1975) 359–366.

Merskey, H., D. Bryd: Emotional adjustment and chronic pain. Pain 5 (1978) 173–178.

Mertens, W.: Psychoanalyse. Stuttgart-Berlin-Köln-Mainz 1981.

Merton, R. K., G. G. Reader, P. L. Kendall (eds.): The student physician. Harvard Univ. Press, Cambridge/Mass. 1957.

Mesmer, F. A.: zit. in: Schultz, J. H.: Psychotherapie. Hippokrates, Stuttgart 1952.

Mester, H.: Der Dermatozoenwahn: Ein hautärztliches und psychiatrisches Problem. Extracta dermatologica 4 (1980) 205–222.

Mester, H.: Die Anorexia nervosa. Springer, Berlin-Heidelberg-New York 1981.

Mester, H.: Dysmorphophobie: Klinische Bilder und die ihnen gemeinsame Psychodynamik. Extracta dermatologica 7 (1983) 113–135.

Metal'nikov, S., V. Chorine: Rôle des reflexes conditionnels dans l'immunité. Ann. Inst. Pasteur 40 (1926) 893–900.

Mettlin, C.: Occupational careers and the prevention of coronary-prone behavior. Soc. Sci. Med. 10 (1976) 367–372.

Meves, C.: Die Verantwortung des Arztes im Hinblick auf die frühe Kindheit. Berl. Ärztebl. 89/11 (1976) 550.

Meves, C.: Der Weg in die neurotische Verwahrlosung. Vorschläge zu Heilung und Vorbeugung. Berl. Ärztebl. 19 (1977) 874.

Mewes, J.: Testpsychologische Untersuchung der Persönlichkeitsstruktur Colitis-ulcerosa-Kranker. Diss., Lübeck 1973.

Meyer, A. E.: Das Syndrom der Anorexia nervosa. Arch. Psychiat. Nervenkr. 202 (1961) 31–59.

Meyer, A. E.: Die Anorexia nervosa und ihre für die Allgemeinmedizin wichtigen Aspekte. Z. Allgemeinmed. 46 (1970) 1782–1786.

Meyer, A. E.: Beitrag im Workshop Anorexia nervosa. Congr. Int. Coll. Psychosom. Med., Rom 1975.

Meyer, A. E.: The Hamburg Short Psychotherapy Comparison Experiment. Karger, Basel-München-Paris 1981.

Meyer, A. E.: Die Psychosomatik der Kranken mit Störungen des oberen Verdauungstraktes – mit funktionellen Oberbauchbeschwerden (FOB). In: Jores, A. (Hrsg.): Praktische Psychosomatik, 2. Aufl. Huber, Bern 1981.

Meyer, A. E.: Taxonomic subgroups within psychosomatic disease entities: an alternative strategy to the specificity approach. Psychother. and Psychosom. 42 (1984a) 26–36.

Meyer, A. E.: Mündliche Mitteilung. 1984b.

Meyer, A. E., W. Weitemeyer: Zur Frage krankheitsdependenter Neurotisierung. Psychometrisch-varianzanalytische Untersuchungen an Männern mit Asthma bronchiale, Lungentuberkulose oder mit Herzvitien. Arch. Psychiat. u. Zschr. f.d. ges. Neurologie 209 (1967) 21.

Meyer, A. E., M. v. Kerekjarto: Umgang mit zum Tode Kranken. In: Jores, A. (Hrsg.): Praktische Psychosomatik. Huber, Bern-Stuttgart-Wien 1980.

Meyer, B. C., R. S. Blacher, F. Brown: A clinical study of psychiatric and psychological aspects of mitral surgery. Psychosom. Med. 23 (1961) 194–218.

Meyer, E., M. Mendelson: Psychiatric consultations with patients on medical and surgical wards: patterns and processes. Psychiatry 24 (1961) 197–220.

Meyer, E. A. 1988. (Unveröffentlichtes Vortragsmanuskript).

Meyer III, E., L. R. Degoratis, M. J. Miller, A. J. Reading, J. H. Cohen, L. C. Park, G. A. Whitmarsh: Addition of time-limited psychotherapy to medical treatment in a general medical clinic. J. nerv. ment. Dis. 169 (1981) 780–790.

Meyer, J. E.: Tod und Neurose. Vandenhoeck und Ruprecht, Göttingen 1973.

Meyer, J. E.: Einstellungen zu Tod und Sterben in der Gegenwart. In: Bitter, W. (Hrsg.): Alter und Tod – annehmen oder verdrängen? S. 49–58. Klett, Stuttgart 1974.

Meyer, J. E.: Über abnorme Trauerreaktionen. Z. psychosom. Med. Psychoanal. 23/4 (1977) 303–309.

Meyer, J. E.: Todesangst und das Todesbewußtsein der Gegenwart. Springer, Heidelberg 1979.

Meyer, J. E.: Die Aufklärung des unheilbar Kranken. In: Bönisch, E., J. E. Meyer (Hrsg.): Psychosomatik in der klinischen Medizin, S. 140–148. Springer, Berlin 1983.

Meyer, R., D. Beck: Zur Frage des psychogenen Fiebers. Schweiz. Rundsch. Med. 64/50 (1975) 1599.

Meyer zum Gottesberge, A.: Über Ohrgeräusche. Arch. Ohren-, Nas.- u. Kehlk.-Heilk. 169 (1956) 344.

Meyer-Bahlburg, H. F. L.: Aggression, androgens, and XYY syndrome. In: Friedman, R. C., R. M. Richart, R. L. van de Wiele (eds.): Sex Differences in Behavior, pp. 433–453. Wiley, New York 1974.

Meyerowitz, S.: The continuing investigation of psychosocial variables in rheumatoid arthritis. In: Hill, A. (ed.): Modern Trends in Rheumatology, Vol. 2, pp. 92–105. Butterworth, London 1971.

Meyerowitz, S., R. F. Jacox, D. W. Hess: Monozygotic twins discordant for rheumatoid arthritis: a genetic, clinical and psychological study of 8 sets. Arthr. and Rheum. 11 (1968) 1–21.

Meyers, O. L., A. G. Hall: Talking to patients with arthritis. S. Afr. med. J. 52 (1977) 673–676.

Meyers, S., H. D. Janowitz: „Natural history" of Crohn's disease: an analytical review of the placebo lesson. Gastroenterology 87 (1984) 1189–1192.

Meyers, S., J. S. Walfish, D. B. Sachar, A. J. Greenstein, A. G. Hill, H. D. Janowitz: Quality of life after surgery for Crohn's disease: a psychosocial survey. Gastroenterology 78 (1980) 1–6.

Mezey, E., T. Reisine, M. Palkovits, M. J. Brownstein, J. Axelrod: Direct stimulation of β2-adrenergic receptors in rat anterior pituitary induces the release of adrenocorticotropin in vivo. Proc. Natl. Acad. Sci. (Wash.) 80 (1983) 6728–6731.

Miall, W. E., P. D. Oldham: The hereditary factor in arterial blood pressure. Brit. med. J. 1 (1963) 75.

Miall, W. E., H. G. Lovell: Relation between change of blood pressure and age. Brit. med. J. 2 (1967) 660.

Michaelis, R.: Beitrag zur Kenntnis ätiologisch-pathogenetischer Faktoren der essentiellen juvenilen Hypertonie. Z. psychosom. Med. 12 (1966) 1.

Michallik-Herbein, U., E. Frieling, W. Langosch: Psychologische Arbeitsanalysen bei Herzinfarktpatienten. Z. Arb. Wiss. 35 (7 NF) (1981) 156–161.

Michel, L.: Allgemeine Grundlagen psychometrischer Tests. In: Heiss, R., K. J. Groffmann, L. Michel (Hrsg.): Psychologische Diagnostik. Handbuch der Psychologie, Bd. 6, S.19. 3. Aufl. Hogrefe, Göttingen 1971.

Michel, S.: HIV-Antikörpertest und Verhaltensänderungen. Literaturstudie. Wissenschaftszentrum, Berlin 1988.

Miehke, K.: Der Weichteilrheumatismus unter besonderer Berücksichtigung des sogenannten Muskelrheumatismus. Therapiewoche 8 (1973) 598–608.

Miehke, K.: Zur Ätiologie und Pathogenese rheumatischer Erkrankungen. Therapiewoche 26 (1976) 2855.

Miethe, E.: Das Kontaktgranulom. Vortrag, II. Kommunikationsmedizinische Tage, 15.-17. 4.1988, Bad Rappenau.

Mikami, D. B.: A review of psychogenic aspects and treatment of bruxism. J. prosth. Dent. 37 (1977) 411.

Mikhail, A. A.: Stress and ulceration in the glandular and nonglandular portions of the rat's stomach. J. comp. Physiol. Psychol. 85 (1973) 636–642.

Miles, L. E., W. C. Dement: Sleep and aging. Sleep 3 (1980) 119–220.

Milgram, P., M. Marder, B. Williams, R. Beaton, P. Weinstein: Stress and gingivitis. J. dent. Res. 62 (1983) 187.

Miller, A.: Das Drama des begabten Kindes. Suhrkamp, Frankfurt 1979.

Miller, B., F. Fervers, R. Rohbeck, G. Strohmeyer: Zuckerkonsum bei Patienten mit Morbus Crohn. Verh. Dtsch. Ges. Inn. Med. 82 (1976) 922–924.

Miller, D., J. Green, D. J. Jeffries, A. J. Pinching, J. R. W. Harris: HTLV-III: Should testing ever be routine? Brit. med. J. 292 (1986) 941.

Miller, G. A., E. Galanter, K. H. Pribram: Plans and the structure of behavior. Holt, Rinehart and Winston, New York 1960.

Miller, G. E., T. Fülöp: Educational strategies for the health professions. WHO, Geneva No. 52. 1974.

Miller, L.: Is alexithymia a disconnection syndrom (a neuropsychological perspective)? Int. J. Psychiat. Med. 16 (1986) 199–209.

Miller, N. E.: Learning of visceral and glandular responses. Science 163 (1969) 434–445.

Miller, N. E.: Biofeedback and visceral learning. Ann. Rev. Psychol. 29 (1978) 373–404.

Miller, N. E., L. V. Dicara: Instrumental learning of heart rate changes in curarized rats: shaping and specifity to discriminative stimulus. J. compar. physiol. Psychol. 63 (1967) 12–19.

Miller, N. E., B. J. Cohler: Psychodynamic research perspectives on development, psychopathology and treatment in later life. Psychoanal. Psychol. 1 (1984) 77–82.

Miller, N. E., L. V. Dicara, H. Solomon, J. M. Weiss, B. Dworkin: Learned modifications of autonomic functions: a review and some new data. Circul. Res. 26/27 (1970) 3–11.

Miller, R. E.: Experimental approaches to the physiological and behavioral concomitants of affective communication in rhesus monkeys. In: Altmann, S. A. (ed.): Social communication among primates. Univ. of Chicago Press, Chicago 1967.

Miller, R. R.: Analgetics. In: Miller, R. R., D. J. Greenblatt (eds.): Drug Effects in Hospitalized Patients, pp. 133–164. Wiley, New York 1976.

Miller, W. B., R. Rosenfeld: A psychophysiological study of denial following acute myocardial infarction. J. psychosom. Res. 19 (1975) 43–54

Mills, J. E., J. G. Widdicombe: Role of the vagus nerves in anaphylaxis and histamine-induced bronchoconstriction in guinea pigs. Brit. J. Pharmacol. 39 (1970) 724–731.

Milman, L., T. C. Todd: Families of children with psychosomatic problems. Amer. J. Orthopsychiat. 43 (1973) 243.

Milne, J. F.: Psychosocial aspects of renal transplantation. Urology (Suppl.) 9 (1977) 82–88.

Milne, J. F., J. S. Golden, L. Fibus: Sexual dysfunction in renal failure. Int. J. Psychiat. Med. 8 (1978) 335–345.

Milner, B.: Memory and the medial temporal regions of the brain. In: Pribram, K. H., D. E. Broadbent (eds.): Biology of Memory. Acad. Press, New York 1970.

Miltner, W.: Verhaltensanalyse in der Verhaltensmedizin. In: Miltner, W., N. Birbaumer, W. D. Gerber: Verhaltensmedizin. Springer, Berlin-Heidelberg 1986.

Miltner, W., N. Birbaumer, W. D. Gerber: Verhaltensmedizin. Springer, Berlin-Heidelberg 1986.

Mindham, R. H. S., A. Bagshaw, S. A. James, A. J. Swannell: Factors associated with the appearance of psychiatric symptoms in rheumatoid arthritis. J. psychosom. Res. 25 (1981) 429–435.

Minuchin, S.: Families and family therapy. Harvard Univ. Press. Deutsch: Familie und Familientherapie. Lambertus, Freiburg 1974.

Minuchin, S., H. C. Fishman: Family therapy techniques. Harvard Univ. Press, Cambridge/Mass. 1981.

Minuchin, S., L. Baker, B. Rosman, L. Liebman, L. Milman, T. Todd: A conceptual model of psychosomatic illness in children. Arch. gen. Psychiat. 32 (1975) 1031.

Minuchin, S., L. Baker, B. L. Rosman: Psychosomatic families. Anorexia nervosa in context. Harvard Univ. Press, Cambridge/Mass. 1978. Deutsch: Psychosomatische Familie. Klett, Stuttgart 1982.

Minuchin, S., B. L. Rosman, L. Baker: Psychosomatische Krankheiten in der Familie. Klett-Cotta, Stuttgart 1986.

Minuck, M., R. Perkins: Long-term study of patients successfully resuscitated following cardiac arrest. Anesth. Analg. 49 (1970) 115–118.

Mirsky, I. A.: Physiologic, psychologic, and social determinants in the etiology of duodenal ulcer. Amer. J. digest. Dis. 3 (1958) 285–314.

Mirsky, I. A., P. Futterman, S. Kaplan: Blood plasma pepsinogen: II. The activity of the plasma from „normal" subjects, patients with duodenal ulcer, and patients with pernicious anemia. J. Lab. clin. Med. 40 (1952) 188.

Mischel, W.: Personality and assessment. Wiley, New York 1968.

Mischel, W., E. Straub: Effects of expectancy on working and waiting for larger rewards. J. Pers. soc. Psychol. 2 (1965) 625–633.

Mishler, E. G.: Viewpoint: critical perspectives on the biomedical model. In: Mishler, E. G., L. R. AmaraSingham, S. T. Hauser, R. Liem, S. Osherson, N. E.Waxler (eds.): Social contexts of health, illness, and patient care. Cambridge Univ. Press, Cambridge 1981.

Mishler, E. G.: The discourse of medicine. Dialectics of medical interviews. Ablex, Norwood 1984.

Mitchell, D. N., R. J. W. Rees: Possible role of infectious agents in Crohn's disease. Z. Gastroenterol. 17 (Suppl.) (1979) 98–100.

Mitchell, J. E., J. P. Bantle: Metabolic and endocrine investigations in women of normal weight with the bulimia syndrome. Biol. Psychiat. 18 (1983) 355–365.

Mitchell, J. E., G. Goff: Bulimia in male patients. Psychosomatics 25 (1984) 909–913.

Mitchell, J. E., D. C. Laine: Monitored binge-eating behavior in patients with bulimia nervosa. Int. J. Eat. Dis. 4 (1985) 177.

Mitchell, J. E., C. Pomeroy: Medizinische Komplikationen der Bulimia nervosa. In: Fichter, M. M. (Hrsg.): Bulimia nervosa. Enke, Stuttgart 1989.

Mitchell, J. E., D. Hatsukami, E. D. Eckert, R. L. Pyle: Characteristics of 275 patients with bulimia. Amer. J. Psychiat. 142 (1985) 482–485.

Mitchell, J. E., L. Davis, G. Goff, R. Pyle: A follow-up study of patients with bulimia. Int. J. Eat. Dis. 5 (1986) 441–450.

Mitchell, J. E., D. Hatsukami, G. Goff, R. L. Pyle, E. D. Eckert, L. E. Davis: Intensive outpatient group treatment for bulimia. In: Garner, D. M., P. E. Garfinkel (eds.): Handbook of Psychotherapy for Anorexia Nervosa and Bulimia. Guilford, New York 1985.

Mitler, M. M., W. F. Seidel, J. van der Hoet, D. J. Grennblatt, W. C. Dement: Comparative hypnotic effect of flurazepam, triazolam and placebo: a long-term simultaneous nighttime and daytime study. J. clin. Psychopharmacol. 4 (1984) 2–13.

Mitscherlich, A.: Freiheit und Unfreiheit in der Krankheit. Hamburg 1948.

Mitscherlich, A.: In: Die Umschau 50 (1950) 29 (zit. nach Brede, K.: Sozioanalyse psychosomatischer Störungen, S. 114. Frankfurt 1972).

Mitscherlich, A.: Krankheit als Konflikt. Studien zur psychosomatischen Medizin. Suhrkamp, Frankfurt 1966.

Mitscherlich, A.: Zusätzliche Gedanken über die Chronifizierung psychosomatischen Geschehens. In: Mitscherlich, A.: Krankheit als Konflikt. Studien zur psychosomatischen Medizin 2, S. 94–124. Suhrkamp, Frankfurt 1967.

Mitscherlich, M.: The psychic state of patients suffering from parkinsonism. Advanc. psychosom. Med. 1 (1960) 203–207.

Mitscherlich, M.: Beitrag zur Psychologie des Tic und des Torticollis spasticus. Advanc. psychosom. Med. 3 (1963) 203–207.

Mitscherlich, M.: Zur Psychoanalyse des Torticollis spasticus. Nervenarzt 42 (1971) 420–426.

Mitscherlich, M.: Analytische Behandlung von Hyperkinesen. Med. Welt 24 (1973) 1058–1062.

Mitscherlich, M.: Ein Beitrag zur Frage der Alexithymie. Therapiewoche 26 (1976) 909–915.

Mitscherlich, M. : The significance of the transitional object for psychosomatic thinking. Psychother. and Psychosom. 28 (1977) 272–277.

Mitscherlich, M.: Zur Theorie und Therapie des Torticollis. In: Studt, H. H. (Hrsg.): Psychosomatik in Forschung und Praxis, S. 401–410. Urban & Schwarzenberg, München-Wien-Baltimore 1983.

Mittelmann, B.: Psychosomatic medicine and the older patient. In: Kaplan, O. J. (ed.): Mental Disorders in Later Life, 2nd ed. Univ. of Stanford Press, Stanford CA 1956.

Mixter, W. J., J. S. Barr: Rupture of the intervertebral disc with involvement of the spinal canal. New Engl. J. Med. 211 (1934) 210–215.

Modell, A. H.: Denial and the sense of separateness. J. Amer. psychoanal. Ass. 9 (1961) 533–547.

Modestin, J.: Psychiatrische Morbidität bei intern-medizinisch hospitalisierten Patienten. Schweiz. med. Wschr. 107 (1977) 1354–1361.

Möhlen, K., S. Davies-Osterkamp: Psychische und körperliche Reaktion bei Patienten der offenen Herzchirurgie in Abhängigkeit von präoperativen psychischen Befunden. Zschr. Psychosom. Med. Psychoanal. 25 (1979) 128.

Möhlen, K., E. Brähler, H. Rohde, G. Overbeck: Zur Psychosomatik des operierten Ulkus-Kranken – eine 4-Jahres-Katamnese. Psychother. med. Psychol. 32 (1982) 19–26.

Möhler, H., T. Okada: Benzodiazepine receptor: demonstration in the central nervous system. Science 198 (1977) 849–851.

Möhring, P.: Mit Krebs leben – maligne Erkrankungen aus therapeutischer und persönlicher Perspektive. Springer Berlin 1988.

Möller, H. C.: Treatment of the irritable colon. Modern Treatment 2 (1965) 988–1002.

Moeller, M.L: Selbsthilfegruppen. Selbstbehandlung und Selbsterkenntnis in eigenverantwortlichen Gesprächsgruppen. Rowohlt, Reinbek 1978.

Moeller, M.L: Möglichkeiten, Grenzen und Gefahren psychotherapeutisch arbeitender Selbsthilfegruppen. Psychother. med. Psychol. 33 (1983) 69.

Moeller, M. L.: Die Gruppe kann mehr als der Einzelne. Psoriasis 39 (1983) 617–625.

Moeller, M. L.: Psychosomatic aspects of psoriasis. In: Current Research Problems in Psoriasis. Grosse, Berlin 1984.

Mörl, F. K., P. Matis: Die Colitis ulcerosa aus der Sicht der Chirurgen. Med. Welt 18 (1967) 2844–2852.

Mörl, M., H. Koch, W. Rösch, P. Frühmorgen, J. Zeus: Familiäre Enterocolitis regionalis Crohn. Dtsch. med. Wschr. 101 (1976) 493–496.

Moertel, C. G., D. L. Ahmann, W. F. Taylor, N. Schwartau: A comparative evaluation of marketed analgesic drugs. New Engl. J. Med. 286 (1972) 813–815.

Moffaert, M. van: Psychosomatics for the practising dermatologist. Dermatologica 165 (1982) 73–87.

Mohr, F.: Psychotherapie bei organischen Erkrankungen. Thieme, Leipzig 1930.

Mohr, G. J., I. M. Josselyn, J. Spurlock, S. H. Barron: Studies in ulcerative colitis. Amer. J. Psychiat. 114 (1958) 1067–1076.

Moldofsky, H.: Stress, disordered sleep and fibrositis syndrome. In: Weiner, H., D. Hellhammer, I. Florin, R. C. Murison (eds.): Neuronal Control of Bodily Function: Basic and Clinical Aspects: Vol IV: Frontiers of Stress Research. Huber, Toronto 1989.

Moldofsky, H., W. J. Chester: Pain and mood patterns in patients with rheumatoid arthritis. Psychosom. Med. 32 (1970) 309–318.

Moldofsky, H., A. I. Rothman: Personality, disease parameter and medication in rheumatoid arthritis. J. chron. Dis. 24 (1971) 363–372.

Molinski, H.: Bilder der eigenen Weiblichkeit, Ärger während der Geburt und Rigidität des Muttermundes. Z. psychosom. Med. Psychoanal. 14/2 (1968) 90.

Molinski, H.: Archaische Mütterlichkeit, Grundlage psychogener Störungen von Schwangerschaft und Geburt. Sexualmed. 3 (1972) 140.

Molinski, H.: Die unbewußte Angst vor dem Kind. Kindler, München 1972.

Molinski, H.: Gesprächsführung bei Schwangerschaftskonflikten. Dtsch. Ärztebl. 46 (1975) 3183.

Molinski, H.: Die fokussierende Deskription. Praktische Hinweise für die Behandlung funktioneller Sexualstörungen aus analytischer Sicht. Sexualmed. 5 (1976) 712.

Molinski, H.: Das psychosomatisch orientierte Sprechstundengespräch in der Gynäkologie und Geburtshilfe. Therapiewoche 28 (1978a) 9486.

Molinski, H.: Larvierte Depression. Geburtsh. Frauenheilk. 38 (1978b) 199.

Molinski, H.: Unterleibsschmerzen ohne Organbefund und eine Bemerkung zum pseudoinfektiösen Syndrom der Scheide. Gynäk. 15 (1982) 207.

Molinski, H.: Gegenwärtiger Stand der Psychosomatik in der Frauenheilkunde. In: Richter, D., M. Stauber (Hrsg.): Psychosomatische Probleme in Geburtshilfe und Gynäkologie. Kehrer, Freiburg 1983.

Molinski, H.. Sexualstörungen der Frau. Sexualmedizin 12 (1983) 182.

Molinski, H., D. G. Hertz: Zielsetzung der Psychosomatik in der Frauenheilkunde. In: Prill, H. J., D. Langen (Hrsg.): Der psychosomatische Weg zur gynäkologischen Praxis. Schattauer, Stuttgart 1983.

Molinski, H., J. Rechenberger, D. Richter: Psychosomatik in der Sprechstunde des niedergelassenen Arztes – eine Utopie? Dtsch. Ärztebl. 76 (1979) 3307.

Monat, A., R. S. Lazarus: Stress and coping. Columbia Univ. Press, New York 1977.

Money, J.: Determinanten der geschlechtsspezifischen Identität und des Sexualverhaltens. In: Heigl-Evers, A. (Hrsg.): Handbuch der Ehe-, Familien- und Gruppentherapie, Bd. 2, S. 718–747. Kindler, München 1973.

Money, J., A. A. Ehrhardt: Man and woman, boy and girl. Johns Hopkins Univ. Press, Baltimore 1972.

Money, J., H. Musaph (eds.): Handbook of Sexology. Elsevier, New York-Oxford 1978.

Monjan, A. A.: Stress and immunologic competence: studies in animals. In: Ader, R. (ed.): Psychoneuroimmunology, pp. 185–228. Acad. Press, New York 1981.

Monjan, A. A.: Effects of acute and chronic stress upon lymphocyte blastogenesis in mice and humans: „Of mice and men." In: Cooper, E. L. (ed.): Stress, Immunity, and Aging, pp. 81–108. Dekker, New York 1984.

Monjan, A. A., M. I. Collector: Stress-induced modulation of the immune response. Science 196 (1977) 307–308.

Monk, M., A. I. Mendeloff, C. I. Siegel, A. Lilienfeld: An epidemiological study of ulcerative colitis and regional enteritis among adults in Baltimore. III. Psychological and possible stress-precipitating factors. J. chron. Dis. 22 (1970) 565–578.

Montagu, A.: Touching: the human significance of the skin. Columbia Univ. Press, New York 1971.

Montagu, A.: Körperkontakt. Klett, Stuttgart 1974.

Monteleone, P., M. Maj, M. Iovino, L. Steardo: Growth hormone response to sodium valproate in chronic schizophrenia. Biol. Psychiat. 21 (1986) 588–594.

Moody, R. A.: Leben nach dem Tod. Rowohlt, Hamburg 1977.

Moore, B. E., D. L. Rubinfine: The mechanism of denial. In: Fine, F. D. et al. (eds.): The Mechanism of Denial. Monograph III. Int. Univ. Press, New York 1969.

Moore, C., L. Ocko: Night walking in early infancy. Arch. Dis. Childhood 32 (1957) 333–342.

Moore, N.: Behavior therapy in bronchial asthma: a controlled study. J. psychosom. Res. 9 (1965) 257–276.

Moore-Ede, M. C., F. M. Suzlman, C. A. Fuller: The clocks that time us. Harvard Univ. Press, Cambridge/Mass. 1982.

Moos, R. H.: Personality factors associated with rheumatoid arthritis: a review. J. chron. Dis. 17 (1964) 41–55.

Moos, R. H.: Psychological techniques in the assessment of adaptative behavior. In: Coelho, G. V., D. A. Hamburg, J. E. Adams: Coping and Adaptation. Basic Books, New York 1974.

Moos, R. H.: Coping with physical illness. Plenum, New York 1977.

Moos, R. H., G. F. Solomon: Minnesota Multiphasic Personality Inventory response patterns in patients with rheumatoid arthritis. J. psychosom. Res. 8 (1964a) 17–28.

Moos, R. H., G. F. Solomon: Personality correlates of the rapidity of progression of rheumatoid arthritis. Ann. rheum. Dis. 23 (1964b) 145–151.

Moos, R. H., G. F. Solomon: Psychologic comparisons between women with rheumatoid arthritis and their non-arthritic sisters, Parts I, II. Psychosom. Med. 27 (1965) 135–149, 150–164.

Moreno, J. L.: Gruppenpsychotherapie und Psychodrama. Thieme, Stuttgart 1959.

Morgan, D. H.: Neuro-psychiatric problems of surgery. J. psychosom. Res. 15 (1971) 41–46.

Morgan, H. C., G. M. F. Russel: Value of family background and clinical feature as predictors of long-term outcome in anorexia nervosa: four-year follow-up of 41 patients. Psychol. Med. 5 (1975) 355.

Morgan, H. G., D. G. H. Sylvester (1976). Zit. n. Hill, O. W.: Anorexia nervosa. Modern Trends in Psychosomatic Medicine 3 (1976) 382–403.

Morgan, L., H. Dallosso, S. Ebrahim, T. Arie, P. H. Fenten: Prevalence, frequency, and duration of hypnotic drug use among the elderly living at home. Brit. med. J. 296 (1988) 601–602.

Morgan, M., D. L. Patrick, J. R. Charlton: Social network and psychosocial support among disabled people. Soc. Sci. Med. 19 (1984) 489–497.

Morgan, R., L. Luborsky, P. Crist-Christoph, H. Curtis, J. Solomon: Predicting the outcomes of psychotherapy by the Penn Helping Alliance Rating Method. Arch. gen. Psychiat. 39 (1982) 397–402.

Morgan, W. L., G. L. Engel: Der klinische Zugang zum Patienten. Huber, Bern-Stuttgart-Wien 1977.

Morgenstern, H., G. A. Gellert, S. D. Walter, A. M. Ostfeld, B. S. Siegel: The impact of a social support program on survival with breast cancer: the importance of selection bias in program evaluation. J. chron. Dis. 37 (1984) 273–282.

Morin, C. L., M. Roulet, C. C. Roy, A. Weber: Continuous elemental enteral alimentation in children with Crohn's disease and growth failure. Gastroenterology 79 (1980) 1205–1210.

Morley, J. E.: The neuroendocrine control of appetite: the role of the endogenous opiates, cholecystokinin, TRH, gamma aminobutyric acid and the diazepam receptor. Life Sci. 27 (1980) 355–368.

Morris, D.: Intimate Behaviour. Cape, London 1971.

Morris, D.: Der Mensch, mit dem wir leben. Droemer Knaur, München-Zürich 1978.

Morris, E. W., M. Di Paola, R. Vallance, G. Waddell: Diagnosis and decision making in lumbar disc prolapse and nerve entrapment. Spine 11 (1986) 436–439.

Morris, T., H. S. Greer: A „type C" for cancer? Low trait anxiety in the pathogenesis of breast cancer. Cancer Detect. Prev. 3 (1980) . Abstract 102.

Morrison, F. R., P. A. Paffenbarger: Epidemiological aspects of bio-behavior in the etiology of cancer. A critical review. In: Weiss, S. M., J. A. Herd, B. H. Fox (eds.): Perspectives on Behavioral Medicine. Acad. Press, New York 1981.

Morse, R. M., E. M. Litin: Postoperative delirium: a study of etiologic factors. Amer. J. Psychiat. 126 (1969) 388–395.

Morse, S. J.: Structure and reconstruction: a critical comparison of Michael Balint and D. W. Winnicott. Int. J. Psychoanal. 53 (1972) 487–500.

Morson, B. C.: Rectal and colonic biopsy in inflammatory bowel disease. Amer. J. Gastroent. 67 (1977) 417–426.

Morson, B. C., I. M. P. Dawson: Gastrointestinal pathology, 2nd ed. Blackwell, Oxford 1979.

Literatur

Mortola, J. F., J. H. Liu, J. C. Gillin, D. D. Rasmussen, S. S. C. Yen: Pulsatile rhythms of adrenocorticotropin (ACTH) and cortisol in women with endogenous depression: evidence for increased ACTH pulse frequency. J. clin. Endocr. 65 (1987) 962–968.

Morton, R.: Phtisiologia sen exencitationes de phtisi, p. 5. Frankfurt-Leipzig 1691. (Erstdruck London 1689).

Moser, T.: Körpertherapie innerhalb der Psychoanalyse. In: Rechenberger, H.-G., H.-V. Werthmann (Hrsg.): Psychotherapie und Innere Medizin. Grundlagen und Anwendungen, S. 126. Pfeiffer, München 1988.

Moser, U.: Zur Abwehrlehre: Das Verhältnis von Verdrängung und Projektion. Jahrbuch Psychoanal. 3. Huber, Bern 1964.

Moses, P. H.: Die Stimme der Neurose. Thieme, Stuttgart 1956.

Moss, A. J., B. Wynar, S. Goldstein: Delay in hospitalization during the acute coronary period. Amer. J. Cardiol. 24 (1969) 659–665.

Moss, A. J., J. J. Camilla, H. P. Dairs et al.: Clinical significance of ventricular ectopic beats in the early post-hospital phase of myocardial infarction. Amer. J. Cardiol. 39 (1977) 635.

Mosse, M., G. Tugendreich (Hrsg.): Krankheit und soziale Lage. Lehmann, München 1913.

Motsch, A.: Das sogenannte Costen-Syndrom – Neue Erkenntnisse. In: Drücke, W., B. Klemt (Hrsg.): Kiefergelenk und Okklusion, S. 99. Quintessenz, Berlin 1980.

Mowrer, O. H.: Two-factor learning theory reconsidered with special reference to secondary reinforcement and the concept of habit. Psychol. Rev. 63 (1956) 114.

Mowrer, O. H.: Learning theory and behavior. Wiley, New York 1960.

Mueldner, H.: Die Behandlung von Zwangsphänomenen mit Clomipramin. Therapiewoche 30 (1980) 5614–5616.

Müller, C.: Alterspsychiatrie. Thieme, Stuttgart 1967.

Müller, C. (Hrsg.): Lexikon der Psychiatrie. Springer, Berlin 1973.

Müller, E.: Die multiple Sklerose des Gehirns und des Rückenmarks. Ihre Pathologie und Behandlung. Fischer, Jena 1904.

Müller, J.: Handbuch der Physiologie des Menschen, Bd. 2. Koblenz 1840.

Müller, N., M. Ackenheil, R. Eckstein, E. Hofschuster, W. Mempel: Reduced suppressor cell function in psychiatric patients. Ann. N. Y. Acad. Sci. 496 (1987) 686–690.

Müller, P.: Organisation des Wochenbettes aus psychosomatischer Sicht. In: Richter, D., M. Stauber (Hrsg.): Psychosomatische Probleme in Geburtshilfe und Gynäkologie. Kehrer, Freiburg 1983.

Müller, W.: Der Weichteilrheumatismus. In: Kaganas, G. et al. (Hrsg.): Der Weichteilrheumatismus. Karger, Basel 1971.

Müller-Braunschweig, H.: Die Wirkung der frühen Erfahrung. Das erste Lebensjahr und seine Bedeutung für die psychische Entwicklung. Ergebnisse und Probleme. Klett, Stuttgart 1975.

Müller-Braunschweig, H.: Gedanken zum Einfluß der frühen Mutter-Kind-Beziehung auf die Disposition zur psychosomatischen Erkrankung. Psychother. med. Psychol. 130 (1980) 48.

Müller-Braunschweig, H.: Rezension zu Becker, H.: Konzentrative Bewegungstherapie. Integationsversuch von Körperlichkeit und Handeln in den psychoanalytischen Prozessen. Psyche 40 (1986a) 1038.

Müller-Braunschweig, H.: Psychoanalyse und Körper. In: Brähler, E. (Hrsg.): Körpererleben, S.19. Springer, Berlin-Heidelberg-New York 1986b.

Müller-Braunschweig, H.: Bild, Körperbild und Psychoanalyse. In: Janssen, P. L. (Hrsg.): Reichweite der psychoanalytischen Therapie. Springer, Berlin-Heidelberg-New York (in Vorbereitung).

Müller-Ehrenberg, K.-H.: Der Einsatz von Tranquilizern in der Urologie. Therapiewoche 31 (1981) 3811.

Müller-Seitz, P.: Multifaktorielle Belastungen am Arbeitsplatz aus arbeitswissenschaftlicher Sicht. Zbl. für Arbeitsmedizin, Arbeitsschutz und Prophylaxe 29 (1979) 94–103.

Müller-Wieland, K.: Das Beschwerdebild des Pankreaskranken. Dtsch. med. Wschr. 32 (1968) 391–394.

Müller-Wieland, K. (Hrsg.): Handbuch der inneren Medizin, Bd. III/4: Dickdarm. 5. Aufl. Springer, Berlin 1982.

Münstermann, J., K. Preiser: Schichtarbeit in der Bundesrepublik Deutschland. Humanisierung des Arbeitslebens (hrsg. vom Bundesminister für Arbeit und Sozialordnung), Bd. 8, Bonn 1978.

Mulcahy, R., N. Hickey: The rehabilitation of patients with coronary heart disease. Scand. J. rehab. Med. 2 (1970) 108.

Mullen, P. E., C. R. Linsell, D. Parker: Influence of sleep disruption and caloric restriction on biological markers of depression. Lancet II (1986) 1051–1055.

Multicenter International Study: Improvement in prognosis by long-term beta-adrenoreceptor blockade using practolol. Brit. med. J. 2 (1975) 735.

Mumenthaler, M.: Neurologie, 7. Aufl. Thieme, Stuttgart-New York 1982.

Mumenthaler, M., F. Regli: Kopfschmerzen. Sandoz, Nürnberg 1981.

Munck, A., P. M. Guyre, N. J. Holbrook: Physiological functions of glucocorticoids in stress and their relation to pharmacological actions. Endocr. Rev. 5 (1984) 25–44.

Mundinger, F., T. Riechert: Die stereotaktischen Hirnoperationen zur Behandlung extrapyramidaler Bewegungsstörungen (Parkinsonismus und Hyperkinesen) und ihre Resultate. Fortschr. Neurol. Psychiat. 31 (1963) 1–65; 69–110.

Munkwitz, W., G. Neulandt: Zur Psychologie der Tätowierungen bei jugendlichen Straffälligen. Mschr. Krim. 40 (1957) 227–233.

Munson, S.: Family-oriented consultation in pediatrics. In: Wynne, L. C., S. McDaniel, T. T. Weber (eds.): Systems consultation, pp. 219–239. Guilford, New York – London 1986.

Muralt, L. v.: Zur Frage der epileptischen Amnesie. Z. Hypnotismus, Suggestionsther., S.-Lehre, verw. Forsch. 10 (1900) 75–89.

Murphy, H. B. M.: Comparative psychiatry. The international and intercultural distribution of mental illness. Springer, Berlin-Heidelberg-New York 1982.

Murray, C. D.: A brief psychological analysis of a patient with ulcerative colitis. J. nerv. ment. Dis. 72 (1930) 617–627.

Murray, H. A.: Thematic Apperception Test Manual. Harvard Univ. Press, Cambridge/Mass. 1943.

Murray, J., G. Dunn, P. Williams, A. Tarnopolsky: Factors affecting the consumption of psychotropic drugs. Psychol. Med. 11 (1981) 551–560.

Murray, J. B.: Psychological aspects of migraine headaches. Psychol. Rep. 48 (1981) 139–162.

Murrell, St.A., S. Himmelfarb, K.Wright: Prevalence of depression and its correlates in older adults. Amer. J. Epidem. 117 (1983) 173–185.

Musante, L., J. M. MacDougall, T. M. Dembroski, A. E. van Horn: Component analysis of the Type A coronary-prone behavior pattern in male and female college students. J. Pers. soc. Psychol. 45 (1983) 1104–1117.

Musaph, H.: Psychogenic pruritus. Seminars in Dermatology 2 (1983) 217–222.

Musher, D. M., V. Fainstein, E. J. Young, T. L. Pruett: Fever patterns. Their lack of clinical significance. Arch. intern. Med. 139 (1979) 1225–1228.

Muslin, H. L., K. Gyarfas, W. J. Pieper: Separation experience and cancer of the breast. Ann. N. Y. Acad. Sci. 164 (1969) 802–806.

Muthny, F. A.: Postoperative course of patients during hospitalization following kidney transplantation. VII. World Congr. Int. Coll. Psychosom. Med., Hamburg 1983.

Muthny, F. A.: Zur Erkrankungsspezifität der Krankheitsverarbeitung ein empirischer Vergleich mit Dialyse- und Herzinfarktpatienten. Psychosom. Med. Psychoanal. 34 (1988a) 259–273.

Muthny, F. A.: Einschätzung der Krankheitsverarbeitung durch Patienten, Ärzte und Personal. Klin. Psychol. 17 (1988b) 319–333.

Muthny, F. A., U. Koch: Psychosoziale Situation und Reaktion auf die chronische Erkrankung – ein Vergleich von Brustkrebs- und Dialysepatientinnen. Psychother., Psychosom., med. Psychol. 34 (1984) 261–304.

Muthny, F. A., B. Häusler, U. Koch: Psychosoziale Aspekte der CAPD im Urteil von Dialyse-Ärzten. Nieren- und Hochdruckkr. 17 (1988) 306–314.

Muthny, F. A., M. Broda, A. Dinger, U. Koch, B. Stein: Aspekte der Lebensqualität bei verschiedenen Behandlungsverfahren der chronischen Niereninsuffizienz – ein empirischer Vergleich. In: Franz, H. E. (Hrsg.): Blutreinigungsverfahren – Technik und Klinik. Thieme, Stuttgart-New York 1989.

M'Uzan, M. de: Zur Psychologie des psychosomatisch Kranken. Psyche 31 (1977) 318–332.

M'Uzan, M. de, C. David: Préliminaires critiques à la recherche psychosomatique. Rev. franç. Psychanal. 24 (1960) 19–39.

M'Uzan, M. de, S. Bonfils, A. Lambling: Étude psychosomatique de 18 cas de recto-colite hémorrhagique. Sem. Hop. Paris (1958) 922–928.

Myers, A., H. A. Dewar: Circumstances attending 100 sudden deaths from coronary artery disease with coroners' necropsies. Brit. Heart J. 37 (1975) 1133.

Myers, D. E., W. D. McCall: Head pain as a result of experimental ischemic exercise of the temporalis muscle. Headache 23 (1983) 43–46.

Myers, J., J. Lindenthal, M. Pepper: Life events, social integration and psychiatric symptomatology. J. Health hum. Behav. 16 (1975) 421–427.

Myers, J. K., M. M. Weissman, C. L. Tischler, C. E. Holzer III, P. J. Leaf, H. Orvaschel, J. C. Anthony, J. H. Boyd, J. D. Burke, M. Kramer, R. Stolzmann: Six-month prevalence of psychiatric disorders in three communities. Arch. gen. Psychiat. 41 (1984) 959–967.

Myrtek, M.: Psychophysiologische Konstitutionsforschung – Ein Beitrag zur Psychosomatik. Hogrefe, Göttingen 1980.

Myrtek, M.: Typ-A-Verhalten. Untersuchungen und Literaturanalysen unter besonderer Berücksichtigung der psychophysiologischen Grundlagen. Minerva, München 1983.

Myrtek, M.: Type A behavior and myocardial infarction. Letter to the editor. Amer. Heart J. 111 (1986) 1215–1216.

Myrtek, M.: Erwiderung auf den Leserbrief von M. J. Halhuber (zum Beitrag: Myrtek, M.: Streß und Typ-A-Verhalten, Risikofaktoren der koronaren Herzkrankheit? Eine kritische Bestandsaufnahme. Psychother. med. Psychol. 35 (1985) 54–61). Psychother. med. Psychol. (1985).

Myrtek, M., M. W. Greenlee: Psychophysiology of Type A behavior pattern: a critical analysis. J. psychosom. Res. 28 (1984) 455–466.

Myrtek, M., P. Walschburger, G. Kruse: Psychophysiologie der orthostatischen Kreislaufreaktionen. Z. Kardiol. 63 (1974) 1034–1050.

Myrtek, M., F. Foerster, W. Wittmann: Das Ausgangswertproblem. Theoretische Überlegungen und empirische Untersuchungen. Z. exp. angew. Psychol. 24 (1977) 463–491.

Naber, D., C. Perro, U. Schick et al.: Psychiatrische Symptome und neuropsychologische Auffälligkeiten beim HIV-Infizierten. Nervenarzt 60 (1989) 80–85.

Nachemson, A. L., B. J. Andersson: Classification of low back pain. Scand. J. Work Environm Health 8 (1982) 134–136.

Nadelson, C., M. T. Notman, D. W. Preven: Medical student stress, adaptation, and mental health. In: Scheiber, S., B. Doyle (eds.): The Impaired Physician. Plenum, New York 1983.

Nadelson, C. C., D. B. Marcotte (eds.): Treatment interventions in human sexuality. Plenum Press, New York-London 1983.

Nadelson, T.: Emotional interactions of patient and staff: a focus of psychiatric consultation. Psychiat. Med. 2 (1971) 240–246.

Naftolin, F.: Understanding the bases of sex differences. Science 211 (1981) 1263–1265.

Nagera, H.: Sleep and its disturbances. Psychoanal. Study Child 21 (1966) 393–447.

Nagy, J. I.: Capsicain action on the nervous system. Trends in Neurosci. 5 (1982) 382.

Nahum, L. H.: Madness in the recovery room from open-heart surgery or „they kept waking me up". Connect. Med. 29 (1965) 771.

Nakagawa, T. et al.: Psychosomatic studies of japanese youth under social changes. Psychother. and Psychosom. 30 (1978) 216–228.

Nakai, Y., M. Sugita, T. Nakagawa, I. Araki, Y. Ikemi: Alexithymic features of the patients with chronic pancreatitis. Psychother. and Psychosom. 31 (1979) 205–217.

Nalven, F. B., J. F. O'Brien: Personality patterns of rheumatoid arthritic patients. Arthr. and Rheum. 7 (1964) 18–28.

Naschold, F., P. Novak: Bedingung für eine Systemanalyse des Gesundheitswesens: Integrale Erklärung von Krankheit in der heutigen Gesellschaft. In: Schönbäck, W. (Hrsg.): Gesundheit im gesellschaftlichen Konflikt, S. 3–27. Urban & Schwarzenberg, München-Wien-Baltimore 1980.

Nasrallah, H. A., M. McCalley-Whitters, C. G. Jacoby: Cortical atrophy in schizophrenia and mania: a comparative CT study. J. clin. Psychiat. 43 (1982) 439–441.

Nathamson, C. A.: Learning the doctor's role: a study of first and forth year medical students. M. A. (unpublished), Chicago 1958.

National Center for Health Statistics: Basic data on hearing levels of adults 25–74 years. United States 1925–1971. Vital and Health Statistics Publication Series 11, No. 215. U. S. Department of Health, Education, and Welfare, Hyattsville/Md. 1980.

National Diabetes Data Group: Classification and diagnosis of diabetes mellitus and other categories of glucose intolerance. Diabetes 28 (1979) 1039.

Naujoks, C., H. Lieb, D. Schwarz: Ergebnisse und Probleme einer katamnestischen Untersuchung bei anorektischen Patienten. In: Meermann, R. (Hrsg.): Anorexia nervosa, S. 205–237. Enke, Stuttgart 1982.

Naujoks, W., H. Köhle, K. Köhle: Zum Umgang von Krankenschwestern mit Schwerkranken, I. Die Entwicklung von Teilnehmerinnen einer Balint-Gruppe. Eine quantitative Verlaufsuntersuchung. Abschlußbericht Teilprojekt B 5, SFB 129, Universität Ulm 1985.

Nedbal, J., Z. Mařatka: Ulcerative proctocolitis in Czechoslovakia. Amer. J. Proctol. 19 (1968) 106–114.

Neftel, K. A., R. H. Adler, L. Käppeli, M. Rossi, M. Dolder, H. E. Käser, H. H. Brügesser, H. Vorkauf: Stage fright in musicians: a model illustrating the effect of beta blockers. Psychosom. Med. 44 (1982) 461–469.

Neisser, U.: Cognitive psychology. Appleton Century Crofts, New York 1967.

Nelson, A.: Orgone (reichian) therapy in tension headache. Amer. J. Psychother. 30 (1976) 103–111.

Nelson, J. B.: Embodiment: An approach to sexuality and christian theology. Augsburg, Minneapolis 1978.

Nemeroff, C. B., E. Widerlöv, G. Bissette, H. Walleus, I. Karlsson, K. Eklund, C. D. Kilts, P. T. Loosen, W. Vale: Elevated concentrations of CSF corticotropin-releasing factor-like immunoreactivity in depressed patients. Science 226 (1984) 1342–1344.

Nemiah, J. C.: Anorexia nervosa. Fact and theory. Amer. J. digest. Dis. 3 (1958) 249–274.

Nemiah, J. C.: The psychological management and treatment of patients with peptic ulcer. Arch. psychosom. Med. 6 (1971) 169–185.

Nemiah, J. C.: The psychosomatic nature of anorexia nervosa. In: Reichsman, F. (ed.): Advances in Psychosomatic Medicine, Vol. 7, pp. 316–321. Karger, Basel 1972.

Nemiah, J. C.: Psychology and psychosomatic illness: reflections on theory and research methodology. Psychother. and Psychosom. 22 (1973) 106–111.

Nemiah, J. C.: Conversion, fact or chimera. Int. J. Psychiat. Med. 5/4 (1974) 443–448.

Nemiah, J. C.: Theoretical considerations. In: Bräutigam, W., M.v.Rad (eds.): Toward a theory of psychosomatic disorders, pp. 199–206. Basel 1977.

Nemiah, J. C.: The varieties of human experience. Amer. J. Psychiat. (in press).

Nemiah, J. C., P. E. Sifneos: Affect and fantasy in patients with psychosomatic disorders. In: Hill, O. W. (ed.): Modern Trends in Psychosomatic Medicine. Butterworth, London 1970.

Nemiah, J. C., H. Freyberger, P. H. Sifneos: Alexithymia: a view of the psychosomatic process. In: Hill, O. W. (ed.): Modern Trends in Psychosomatic Medicine, pp. 430–439. London 1976.

Neraal, A.: Das Asthma bronchiale aus familiendynamischer Sicht, dargestellt an einem exemplarischen Fall. Monatsschr. Kinderheilk. 128 (1980) 476–479.

Nestel, P. J., N. H. Fidge, M. H. Tau: Increased lipoprotein-remnant formation in chronical renal failure. New Engl. J. Med. 307 (1982) 329–333.

Neuchterlein, K. H., M. E. Dawson: Information processing and attentional functioning in the developmental course of schizophrenic disorders. Schizophrenia Bull. 10 (1984) 160–203.

Neugarten, B. L.: Middle age and aging. Univ. of Chicago Press, Chicago 1968.

Neugarten, B. L.: Adaptation and the life cycle. J. geriat. Psychiat. 4 (1970) 71–87.

Neugebauer, R., B. P. Dohrenwend, B. S. Dohrenwend: Formulation of hypotheses about the true prevalence of functional psychiatric disorders among adults in the United States. In: Dohrenwend, B. P., B. S. Dohrenwend, M. Schwartz-Gould, B. Link, R. Neugebauer, R. Wunsch-Hitzig (eds.): Mental Illness in the United States, pp. 45–94. Praeger, New York 1980.

Neumärker, K. J., U. Dudeck, P. Plaza: Borrelien-Enzephalitis und Katatonie im Jugendalter. Nervenarzt 60 (1989) 115–119.

Neumann, F., H. Steinbeck: Hormonale Beeinflussung des Verhaltens. Klin. Wschr. 49 (1971) 790–806.

Neumann, P. B., H. Henriksen, N. Grosman, C. B. Christensen: Plasma morphine concentrations during chronic oral administration of patients with cancer pain. Pain 13 (1982) 247–252.

Neumayer, E.: Wirbelsäule, Nervensystem und Psyche. Wien. med. Wschr. 45 (1974) 61–67.

Neun, H. (Hrsg.): Psychosomatische Einrichtungen, Bd. 5: Was sie (anders) machen und wie man sie finden kann. I. A. d. DKPM; Vandenhoeck & Ruprecht, Göttingen-Zürich 1987.

Neundörfer, B.: Die Parkinsonsche Krankheit. Fischer, Stuttgart-New York 1978.

Neuser, J., U. W. Schäfer, K. H. Stäcker: Psychological stress under bone marrow transplantation: prevalence of mood disturbance. J. Canc. Res. clin. Oncol. 114 (Suppl.) (1988) 115.

Neveling, R.: Die akute Ertaubung. Universitäts-Verlag, Köln 1967.

Neveu, P. J., F. Crestani, M. LeMoal: Conditioned immunosuppression: a new methodological approach. Ann. N. Y. Acad. Sci. 496 (1987) 595–601.

New Engl. J. Med. 300 (1979) 1269–1270: Sex hormones and sexual behavior (editorial).

Newland, C. A., L. S. Illis, P. K. Robinson, W. E. Waters: A survey of headache in an English city. Res. clin. Stud. Headache 5 (1978) 1–20.

Newman, J.: A patient's perspective on patient-staff-interaction. In: Levy, N. B. (ed.): Psychonephrology 2, pp. 43–52. Plenum, New York 1983.

Newson, S., M. Darrach: Effect of corticotropin and corticosterone on hemolytic antibody production in the mouse. Canad. J. Biochem. 33 (1954) 372–374.

Nicassio, P., R. Bootzin: A comparison of progressive relaxation and autogenic training as treatments for insomnia. J. abn. Psychol. 83 (1974) 253–260.

Nicassio, P. M., K. A. Wallston, L. F. Callahan, M. Herbert, T. Pincus: The measurement of helplessness in rheumatoid arthritis. The development of the Arthritis Helplessness Index. J. Rheum. 12 (1985) 426–467.

Niederhoff, H., B. Wiesler, W. Künzer: Somatisch orientierte Behandlung der Anorexia nervosa. Mschr. Kinderheilk. 123 (1975) 343–344.

Nielsen, A., T. Williams: Depression in ambulatory medical patients. Arch. gen. Psychiat. 37 (1980) 999–1004.

Nieschlag, E.: The endocrine function of the human testis in regard to sexuality. In: Porter, R., J. Whelan (eds.): Sex, Hormones and Behavior. Ciba Foundation Symposium 62, pp. 183–208. Excerpta Medica, Amsterdam 1979.

Nieuwenhuys, R., J. Voogd, C. van Huijzen: Das Zentralnervensystem des Menschen. Springer, Berlin 1980.

Nijs, P.: Psychosomatische Aspekte der oralen Antikonzeption. Enke, Stuttgart 1972.

Nijs, P.: Psychological aspects of the pain experience. In: Renaer, M. (ed.): Chronic Pelvic Pain in Women. Springer, Berlin 1981.

Nijs, P.: Sexualmedizin im ärztlichen Alltag. Sexualmed. 11 (1982a) 86.

Nijs, P.: Sexualität nach einer Brustoperation. notabene medici 12 (1982b) 1022.

Nijs, P.: Psychological aspects of gynecological pain experience. In: Prill, H. J., M. Stauber (eds.): Advances in Psychosomatic Obstetrics and Gynecology. Springer, Berlin 1982c.

Nijs, P.: Unterleibsschmerzen ohne Organbefund sind Klagen/Anklagen bei psychosozialen, beruflichen, familiären oder sexuellen Schwierigkeiten. Gynäk. 6 (1983) 12.

Niklewski, G.: Psychosomatische Erkrankungen des Auges – eine Übersicht. Psychosom. Med. 28 (1982) 300–316.

Nilsson, I.-K., S. Colleen, P.-A. Märdh: Relationship between psychological and laboratory findings in patients with symptoms of nonacute prostatitis. In: Danielson, D., L. Juttlin, P. A. Märdh (eds.): Genital Infections and Their Complications, p. 133. Almquist & Wiksell, Stockholm 1975.

Nirkko, O., M. Lauroma, P. Siltanen, H. Tuominen, K. Vanhaler: Psychological risk factors related to coronary heart disease. Prospective studies among policemen in Helsinki. Acta med. scand. (Suppl.) 660 (1982) 137–146.

Niskanen, P., J. Jääskeläinen, K. Achté: Anorexia nervosa. Treatment, results and prognosis. Psychiat. fenn. (1974) 257–263.

Nitsch, K.: Babys haben ein Recht auf die Mutterbrust. Pirmasenser Z., 22. 3. 1975.

Nitsch, K.: Die Bedeutung der Deprivation für Entwicklung und Leben des Menschen. Berl. Ärztebl. 19 (1977) 88.

Nixon, B. G. F., H. J. N. Bethell: Preinfarction ill health. Amer. J. Cardiol. 33 (1974) 446–449.

Noback, C. R., R. J. Demarest (eds.): The human nervous system. McGraw-Hill, Guatemala-Lisbon-San Juan 1981.

Nögel, R.: Die Bulimia nervosa bei Zwillingen. Med. Diss., München 1988.

Noel, G. L., H. K. Suh, J. G. Stone, A. G. Frantz: Human prolactin and growth hormone during surgery and other conditions of stress. J. clin. Endocr. 35 (1972) 840–851.

Nolan, L., K. O'Malley: Patients, prescribing, and benzodiazepines. Eur. J. clin. Pharmacol. 35 (1988) 225–229.

Nolte, D., V. Korn: Oszillatorische Messungen des Atemwiderstandes. Dustri, München-Deisenhofen 1979.

Noonan, A. S.: Gonorrhea screening in an urban hospital family planning program. Amer. J. publ. Health 64 (1974) 700.

Nordmeyer, J.: Arzt-Patient-Beziehung während der Visite unter besonderer Berücksichtigung von Problempatienten. Phil. Diss., Universität Heidelberg 1978.

Nordmeyer, J.: Formal-quantitative Aspekte der Arzt-Patient-Beziehung während der Visite. In: Köhle, K., H.-H. Raspe (Hrsg.): Das Gespräch während der ärztlichen Visite, S. 58–69. Urban & Schwarzenberg, München 1982.

Nordmeyer, J., G. Steinmann, F.-W. Deneke, M.v. Kerekjarto: Dimensionen des ärztlichen Visitenverhaltens und ihr Zusammenhang mit ausgewählten Merkmalen von Arzt und Patient. Medizinische Psychologie 5 (1979) 208–228.

Nordmeyer, J., J.-P. Nordmeyer, F.-W. Deneke, M.v. Kerekjarto: Formalquantitative Aspekte des Sprachverhaltens von Arzt und Patient während der Visite. Zschr. f. Klinische Psychologie 10 (1981) 220–231.

Nordmeyer, J., G. Steinmann, F.-W. Deneke, J.-P. Nordmeyer, M.v. Kerekjarto: Verbale und nonverbale Kommunikation zwischen Problempatienten und Ärzten während der Visite. Medizinische Psychologie 8 (1982) 20–39.

Nordmeyer, J., H. J. Avenarius, R. Zick, H. Mielke, J. Anagnou, H. J. Mitzkatz, H. Freyberger: Psychosomatik der artifiziell erzeugten Erkrankung (factitious disease). Therapiewoche 33 (1983) 4725–4730.

Norman, D. K., D. B. Herzog: Persistent social maladjustment in bulimia: a 1-year follow-up. Amer. J. Psychiat. 141 (1984) 444–446.

Norton, C.: Attitudes toward living and dying in patients on chronic hemodialysis. Ann. N. Y. Acad. Sci. 164 (1969) 720–732.

Norwegian Multicenter Study Group: Timolol-induced reduction in mortality and reinfarction in patients surviving acute myocardial infarction. New Engl. J. Med. 304 (1981) 801–807.

Nothdurft, W.: Aspekte der Undurchlässigkeit in Visiten. Eine Untersuchung zu den kommunikativen Schwierigkeiten von Patienten, in Krankenhausvisiten einzugreifen. Psychol. Diplomarbeit, Universität Bonn 1978.

Nothdurft, W.: „Ich komme nicht zu Wort". Austausch-Eigenschaften als Ausschluß-Mechanismus des Patienten in Krankenhaus-Visiten. In: Frier, W. (Hrsg.): Pragmatik/Theorie und Praxis. Amsterdamer Beiträge zur Neueren Germanistik, Bd. 13, S. 321–342. Rodopi, Amsterdam 1981.

Nothdurft, W.: Zur Undurchlässigkeit von Krankenhaus-Visiten. In: Köhle, K., H.-H. Raspe (Hrsg.). Das Gespräch während der ärztlichen Visite, S. 23–35. Urban & Schwarzenberg, München 1982.

Novak, P.: Approbationsordnung und Reform ärztlicher Ausbildung. Med. Soziologie Jahrbuch 3. Campus, Frankfurt 1983.

Novak, P.: Patient, Arzt und Krankenhaus. In: Jork, K., W. Schüffel: Ärztliche Erkenntnis – Entscheidungsfindung mit Patienten. Springer, Heidelberg 1987.

Nowlin, J., E. Busse: Psychosomatic problems in the older person. In: Wittkower, E., H. Warnes (eds.): Psychosomatic Medicine. Its Clinical Applications. 1977.

Nüchtern, M.: Der Arzt, der Kranke, der Tod und der Teufel (Bild von Emil Nolde). In: Böhme, W. (Hrsg.): Der Arzt und das Sterben. Herrenalber Texte 37 (1981) 29–30.

Nunberg, H.: Allgemeine Neurosenlehre. Bern 1959.

Nunes, E. V., K. A. Frank, D. S. Kornfeld: Psychologic treatment for the Type A behavior pattern and for coronary heart disease: a meta-analysis of the literature. Psychosom. Med. 48 (1987) 159–173.

Nutzinger, D. O., M. de Zwaan: Verhaltenstherapie bei Bulimia: Rückblick und Ausblick anhand der bisherigen Forschung. In: Fichter, M. M. (Hrsg.): Bulimia nervosa. Enke, Stuttgart 1989.

Nutzinger, D. O., H. G. Zapotoczky, S. Cayiroglu, G. Gatterer: Panikattacken und Herzphobie. Wien. klin. Wschr. 99 (1987) 545–560.

Nylander, I.: The feeling of being fat and dieting in a school population. An epidemiologic interview investigation. Acta sociomed. scand. 3 (1971) 17–26.

Oberdisse, K.: Die Schilddrüse. Thieme, Stuttgart 1964.

Oberhoff-Looden, I.: Psychopathologie der multiplen Sklerose. Müller, Salzburg 1978.

Oberhummer, I., J. Grünberger, H. Tilscher, G. Zapotoczky: Somatisch bedingte Beschwerden beim Herzangst-Syndrom. Fortschr. Med. 97 (1979) 709–713.

Obrist, P. A.: The cardiovascular behavioral interaction – as it appears today. Psychophysiol. 13 (1976) 95–107.

Obrist, P. A.: Cardiovascular psychophysiology: a perspective. Plenum, New York 1981.

Obrist, P. A.: Blood pressure control and stress: a necessary dimension in understanding the etiology of coronary heart disease. In: Schmidt, T., T. Dembroski, G. Blümchen (eds.): Biobehavioral Factors in Coronary Heart Disease. Springer, Berlin-Heidelberg-New York 1984.

Obrist, P. A., D. M. Wood, M. Perez-Reyes: Heart rate during conditioning in humans: effects of UCS intensity, vagal blockade and adrenergic block of vasomotor activity. J. exp. Psychol. 70 (1965) 32–42.

Obrist, P. A., R. A. Webb, J. R. Sutterer, J. L. Howard: The cardiac-somatic relationship: some reformulations. Psychophysiol. 6 (1970) 569–587.

Obrist, P. A., J. E. Lawler, J. L. Howard, K. W. Smithson.

P. I. Martin, J. Manning: Sympathetic influences on cardiac rate and contractility during acute stress in humans. Psychophysiol. 11 (1974) 405–427.

O'Connell, R. A.: Psychopharmacology in the care of the critically ill. J. Thanatol. 2 (1972) 592.

O'Connor, J. F., G. Daniels, A. Karush, L. Moses, C. Flood, L. O. Stern: The effects of psychotherapy on the course of ulcerative colitis. Amer. J. Psychiat. 120 (1964) 738.

O'Connor, M., S. Touyz, P. Beumont: Nutritional management and dietary counselling in bulimia nervosa: some preliminary observations. Int. J. Eat. Dis. 7 (1988) 657–662.

Odegaard, C. E.: Dear Doctor. A personal letter to a physician. Henry J. Kaiser Family Foundation, Menlo Park CA 1986.

Oehler, G.: Ernährungstherapie bei chronisch-entzündlichen Darmerkrankungen. Med. Welt 35 (1984) 1547–1551.

Oehler, J. W.: An exploratory study of psychological reactions to visual loss and blindness in patients with diabetic retinopathy. PhD thesis. Boston, MA, Univ. Microfilms 1980.

Oehler, J. W., R. G. Fitzgerald: Group therapy with blind diabetics. Arch. gen. Psychiat. 37 (1980) 463–467.

Oehler, K.: Vorwort. In: Zeichen und Realität, Akten des 3. semiotischen Kolloquiums, Hamburg. Stauffenberg, Tübingen 1984.

Oehlert, W.: Klinische Pathologie des Magen-Darm-Traktes. Schattauer, Stuttgart 1978.

Öhman, A.: Electrodermal activity in schizophrenia. Biol. Psychol. 12 (1981) 87–145.

Oesterreich, K.: Psychiatrie des Alterns, 2. Aufl. Quelle & Meyer, Heidelberg 1981.

Ogden, T. H.: On projective identification. Int. J. Psychoanal. 60 (1979) 357–374.

Ogden, T. H.: Projective identification and psychotherapeutic technique. Aronson, New York 1982.

Ohlmeier, D., R. Karstens, K. Köhle: Psychoanalytic group interview and short-term group psychotherapy with postmyocardial infarction patients. Psychiat. Clin. 6 (1973) 240–249.

Ohly, A.: Gedanken zum Phänomen der Diagnose. In: Begemann, H., P. Voswinckel (Hrsg.): Identifikationen. Arzt und Patient unter Erfolgszwang. Urban & Schwarzenberg, München 1988.

Okamoto, K., K. Oaki: Development of a strain of spontaneously hypertensive rats. Jap. Circulat. J. 27 (1963) 282–293.

Oken, D.: What to tell cancer patients: a study in medical attitudes. J. Amer. med. Ass. 175 (1961) 1120–1128.

Olbrisch, M. E., S. W. Ziegler: Psychological adjustment and patient information in inflammatory bowel disease: development of two assessment instruments. J. chron. Dis. 35 (1982a) 649–658.

Olbrisch, M. E., S. W. Ziegler: Psychological adjustment to inflammatory bowel disease: informational control and private self-consciousness. J. chron. Dis. 35 (1982b) 573–580.

Olds, J.: Self-stimulation experiments and differentiated reward systems. In: Jasper, H. H. (ed.): Reticular Formation of the Brain. Little, Brown & Co., Boston 1958.

Olds, J.: Differential effects of drive and drugs on self-stimulation of different brain sites. In: Sheer, D. E. (ed.): Electrical Stimulation of the Brain. Univ. of Texas Press, Austin 1961.

Olds, J.: Pleasure centers in the brain. Sci. Amer. Oct. 1956. Auch in: McGaugh, J. L., N. M. Weinberger, R. E. Whalen (eds.): Psychobiology. Freeman, San Francisco 1966.

Olds, J., P. Milner: Positive reinforcement produced by electrical stimulation of septal area and other regions of rat brain. J. comp. physiol. Psychol. 47 (1954) 419–427.

Olds, M. E., J. Olds: Approach-escape interaction in rat brain. Amer. J. Physiol. 203 (1962) 803–810.

Olds, M. E., J. Olds: Drives, rewards and the brain. In: Barron, F. et al. (eds.): New Directions in Psychology. Holt, Rinehart and Winston, New York 1965.

Olefsky, J. M., O. G. Kolterman: Mechanisms of insulin resistance in obesity and noninsulin-dependent (type II) diabetes. Amer. J. Med. 70 (1981) 151–168.

Olson, D. H., H. I. McCubbin, H. L. Barnes, A. S. Larson, M. J. Muxen, M. A. Wilson: Families – what makes them work. Sage Publ., Beverly Hills-London-New Delhi 1983.

O'Morain, C. A., A. W. Segal, A. J. Levi: Elemental diets in the treatment of acute Crohn's disease: a controlled study. Gut 23 (1983) 891.

O'Neal, P., L. N. Robins: Childhood patterns predictive of schizophrenia. Amer. J. Psychiat. 114 (1958) 961–969.

Onel, Y., A. P. Friedman, J. Grossman: Muscle blood flow studies in muscle-contraction headaches. Neurology 11 (1961) 935–939.

Oosterhuis, H., G. Wilde: Psychiatric aspects of myasthenia gravis. Psychiat. Neurol. Neurochir. 67 (1964) 484–496.

Opitz, H. J.: Tinnitusentstehung und Beeinflussung. Z. Laryng. Rhinol. 59 (1980) 522.

Oppenheim, H.: Die myasthenische Paralyse (Bulbärparalyse ohne anatomischen Befund). Karger, Berlin 1901.

Oppenheim, H.: Lehrbuch der Nervenkrankheiten, 7. Aufl. Karger, Berlin 1923.

Oreopoulos, D. G.: An update on the Continuous Ambulatory Peritoneal Dialysis (CAPD). Int. J. artif. Org. 3 (1980) 231–234.

Oreopoulos, D. G.: Kontinuierliche ambulante Peritonealdialyse (CAPD). In: Franz, H. E. (Hrsg.): Blutreinigungsverfahren – Technik und Klinik. Thieme, Stuttgart-New York 1981.

Orgel, S. Z.: Effect of psychoanalysis on the course of peptic ulcer. Psychosom. Med. 20 (1958) 117.

Orlinsky, D. E., L. J. Howard: Process and outcome in psychotherapy and behavior change. Wiley, New York 1986.

Ornish, D.: Stress diet and your heart. Holt, Rinehart & Winston, New York 1982.

Ornish, D., A. M. Gotto, R. R. Miller et al.: Effects of a vegetarian diet and selected yoga techniques in the treatment of coronary heart disease (Abstr.). Clin. Res. 27 (1979) 720 A.

Ornish, D., S. Brown, L. Scherwitz, J. Billings, W. Armstrong, T. Ports, S. McLanahan, R. Kirkeeide, R. Brand, K. Gould: Can life style changes reverse atherosclerosis? Circulation 78 (Suppl. 2) (1988) 11.

Ornish, D., S. Brown, L. Scherwitz, J. Billings, W. Armstrong, T. Ports, S. McLanahan, R. Kirkeeide, R. Brand, K. Gould: Can life style changes reverse coronary heart disease? An interim report. Unpublished, 1989.

Ornish, D., L. W. Scherwitz, R. S. Doody, D. Kesten, S. M. McLanahan, S. E. Brown, E. G. DePuey, R. Sonnemaker, C. Haynes, J. Lester, G. K. McAllister, R. J. Hall, J. A. Burdine, H. M. Gotto: Effects of stress management training and dietary changes in treating ischemic heart disease. J. Amer. med. Ass. 249/1 (1983) 54–59.

Orpen, C.: Type A personality as a moderator of the effects of role conflict, role ambiguity, and role overload on individual strain. J. hum. Stress (1982) 8–14.

Orth-Gomer, K., M. E. Edwards, M. E. Erhardt et al.: Relation between ventricular arrhythmias and psychologic profile. Acta med. scand. 207 (1980) 31–36.

O'Shaughnesse, E.-J., P. S. Parrino: Chronic prostatitis – fact or fiction? J. Amer. med. Ass. 160 (1956) 540.

Osler, W.: The lumleian lectures on angina pectoris. Lancet 1 (1910) 697–702.

Oster, M. W., M. Vizel, L. R. Turgeon: Pain of terminal cancer patients. Ann. intern. Med. 138 (1978) 1801–1802.

Ostfeld, A. M.: What's the pay off in hypertension research? Psychosom. Med. 35 (1973) 1.

Ostfeld, A. M., D. A. d'Atri: Rapid sociocultural change and high blood pressure. Advanc. psychosom. Med. 9 (1977) 20–37. Karger, Basel.

Ostfeld, A. M., B. Z. Lebovits, R. B. Shekelle: A prospective study of the relationship between personality and coronary heart disease. J. chron. Dis. 17 (1964) 265–276.

Ostow, W.: Psychopharmaka in der Psychotherapie. Stuber, Stuttgart 1962.

Ostow, W. (ed.): The psychodynamic approach to drug therapy. Psychoanalytic Research and Development Found, New York 1979.

Ostwald, P. F.: The semiotics of human sound. Paris 1973.

Ott, K., F. Wöhr: Klinische Untersuchungen über funktionelle Störungen bei Patienten mit marginaler Parodontitis. Dtsch. zahnärztl. Z. 37 (1982) 643.

Otte, M., W. Stöcker, W. G. Wood: Chronisch-entzündliche Darmerkrankungen: Vitamin D-Mangel und immunologische Aspekte. Nordwestdtsch. Ges. Inn. Med., 101. Tagung, Lübeck 1983.

Otte, M., J. Herhahn, A. Hölsbeck, G. Baretton, U. Löhrs, H. Feiereis: Wertigkeit klinischer, serologischer, immunologischer, makroskopischer und histologischer Befunde für die Differentialdiagnostik entzündlicher Darmerkrankungen. Leber, Magen, Darm (im Druck).

Otte, M., D. Normann, H. Bellinger, H. J. Friedrich, G. Jantschek, J. Studt, D. Waller, W. G. Wood: 25 (OH)-Vitamin D3 und Parathormon im Serum bei M. Crohn oder Colitis ulcerosa. Verh. Dtsch. Ges. Inn. Med. 89 (1983) 883–885.

Ottenjann, R.: Atlas der Koloileoskopie. Enke, Stuttgart 1980.

Otto, H. F., J.-O. Gebbers: Pathomorphologie des Morbus Crohn. In: Müller-Wieland, K. (Hrsg.): Handbuch der inneren Medizin, Bd. III/4: Dickdarm. 5. Aufl. Springer, Berlin 1982.

Otto, H. F., J.-O. Gebbers, S. Kügler: „Miliarer" Morbus Crohn. Dtsch. med. Wschr. 100 (1975) 505–507.

Otto, H. F., I. Bettmann, J.-O. Gebbers: Immunhistologische und elektronenmikroskopische Untersuchungen bei Morbus Crohn. Verh. Dtsch. Ges. Pathol. 64 (1980) 256–261.

Otto, H. F., I. Bettmann, J. v. Weltzien: Morbus Crohn-assoziierte Karzinome des Intestinaltraktes. Z. Gastroenterol. 18 (1980) 583–593.

Overbeck, A., G. Overbeck: Das Asthma bronchiale im Zusammenhang familiendynamischer Vorgänge. Psyche 32 (1978) 929–955.

Overbeck, G.: Psychosomatische Aspekte bei unklaren Fieberzuständen. Z. psychosom. Med. 19 (1973) 145.

Overbeck, G.: Objektivierende Beiträge zur Pensée opératoire der französischen Psychosomatik. Habilitationsschrift, Gießen 1975.

Overbeck, G.: Das psychosomatische Symptom – Psychische Defizienzerscheinung oder generative Ich-Leistung? Psyche 31 (1977) 333–354.

Overbeck, G.: Familien mit psychosomatisch kranken Kindern. Vandenhoeck und Ruprecht, Göttingen 1985.

Overbeck, G., K. Möhlen, E. Brähler: Die Ulcus-Krankheit – Psychodiagnostik, kontrasoziales Arrangement und Prognose bei Ulcus duodeni. Kritische Retrospektive. Vandenhoeck & Ruprecht, Göttingen 1990.

Overs, R. P., E. L. Belknap: Educating stroke patient families. J. chron. Dis. 20 (1967) 45–51.

Owen, R. T., P. Tyrer: Benzodiazepine dependence: a review of the evidence. New Ethicals (August 1983) 115–141.

Owerbach, D., J. Nerup: Restriction-fragment-length-polymorphism of the insulin gene in diabetes mellitus. Diabetes 31 (1982) 275–277.

Paar, G.: Psychosomatische Aspekte der Nierensteinerkrankung. In: Sitzungsbericht: Psychosomatische Aspekte in der Urologie, Würzburg 1986.

Paar, G. H., A. Schaefer, W. Drexler: Über das Mitwirken psychosozialer Faktoren bei Ausbruch und Verlauf der akuten Virushepatitis – Bericht über eine Pilotstudie. Psychother. med. Psychol. 37 (1987) 23–30.

Paar, G. H., U. Bezzenberger, H. Lorenz-Meyer: Über den Zusammenhang von psychosozialem Streß und Krankheitsaktivität bei Patienten mit Morbus Crohn und Colitis ulcerosa. Z. Gastroenterol. 26 (1988) 648–657.

Packard, R. C.: Conversion headache. Headache 5 (1980) 266–268.

Packard, R. C.: What is psychogenic headache? Headache 16 (1976) 20–23.

Pahn, J., K. Friemert: Differentialdiagnostische und terminologische Erwägungen bei sogenannten funktionellen Störungen im neuropsychiatrischen und phoniatrischen Fachgebiet. Fol. phoniat. 40 (1988) 162, 168.

Paioni, R., P. Waldmeier, Delini-Stua, G. Langer, G. Schönbeck, H. Beckmann: Antidepressiva: Grundlagen und Therapie. In: Langer, E., H. Heimann (Hrsg.): Psychopharmaka – Grundlagen und Therapie. Springer, Wien 1983.

Paley, A., T. Luparello: Understanding the psychologic factors in asthma. Geriatrics 28 (1973) 54–62.

Pallenberg, C.: Dokumentation der universitären gerontologischen Lehrangebote, Bd. 45. Dtsch. Zentrum für Altersfragen, Berlin 1983.

Palmblad, J., K. Cantell, H. Strander, J. Fröberg, C. G. Karlsson, L. Levi, M. Ganström, P. Unger: Stressor exposure and immunological response in man: interferon-producing capacity and phagocytosis. J. psychosom. Res. 20 (1976) 193–199.

Palmblad, J., B. Petrini, J. Wassermann, T. Akerstedt: Lymphocyte and granulocyte reactions during sleep deprivation. Psychosom. Med. 41 (1979 a) 273–278.

Palmblad, J., C. G. Karlsson, L. Levi, L. Lidberg: The erythrocyte sedimentation rate and stress. Acta med. scand. 205 (1979 b) 517–520.

Palmer, R. L., E. Stonehill, A. H. Crisp, L. Waller, J. J. Misiewicz: Psychological characteristics of patients with the irritable bowel syndrome. Postgrad. med. J. 50 (1974) 416–419.

Palmer, S., L. Canzona, J. Conley, G. Wells: Vocational adaptation of patients on home dialysis: its relationship to personality, activities and support received. J. psychosom. Res. 27 (1983) 201–207.

Palmer, W. L.: Chronic ulcerative colitis. Gastroenterology 10 (1948) 767–781.

Palmtag, H.: Neurophysiologie und Pharmakologie der Blaseninnervation. Pharmakother. 4 (1981) 52.

Palos, E., J. Ring: Vortrag. Jahrestagung der AG Dermatologische Forschung 1983.

Pancheri, P., S. Teodori, U. L. Aparo: Psychological aspects of rheumatoid arthritis vis-à-vis osteoarthritis. Scand. J. Rheum. 7 (1978) 42–48.

Panse, F.: Pathopsychologie der Entstellung durch Hautkrankheiten. In: Gottron, H. A., W. Schönfeld (Hrsg.): Dermatologie und Venerologie. Thieme, Stuttgart 1970.

Papastamou, P. A.: Psychiatric consultations in a general hospital. Psychosomatics 11 (1970) 57–62.

Papciak, A. S., M. Feuerstein: Alexithymia and pain in an outpatient behavioral medicine clinic. Int. J. Psychiat. Med. 16 (1986–87) 347–357.

Papez, J. W.: A proposed mechanism of emotion. Arch. Neurol. Psychiat. (Chic.) 38 (1937) 725.

Papousek, H.: Soziale Interaktion als Grundlage der kognitiven Frühentwicklung. In: Fortschritte der Sozialpädiatrie. Urban & Schwarzenberg, München 1975.

Paracelsus, T.: Von der Bergsucht (1534). In: Peuckert, W.-E. (Hrsg.): Theophrastus Paracelsus Werke Bd. II, S. 284–361. Wissenschaftliche Buchgesellschaft, Darmstadt 1965.

Parekh, H., R. Manz, H. Schepank: Life events, coping, social support. Versuch einer Integration aus psychoanalytischer Sicht. Psychosom. Med. Psychoanal. 34 (1988) 226–246.

Pariser, S. F., E. R. Pinta, B. A. Jones: Mitral valve prolaps syndrome, anxiety neurosis and panic disorder. Amer. J. Psychiat. 13 (1978) 246–247.

Park, R.: Thanatology, a questionnaire and a plea for a neglected study. J. Amer. med. Ass. 58 (1912) 1243.

Parker, C. W.: Adrenergic responsiveness in asthma. In: Austen, K. F., L. M. Lichtenstein (eds.): Asthma: Physiology, Immunopharmacology, and Treatment, pp. 185–210. Acad. Press, New York 1973.

Parker, D. L., J. R. Hodge: Delirium in a coronary care unit. J. Amer. med. Ass. 201 (1967) 132–133.

Parker, J., C. McRae, K. Smarr, N. Beck, R. Frank, S. Anderson, S. Walker: Coping strategies in rheumatoid arthritis. J. Rheum. 15 (1988) 1376–1383.

Parker, R. T., C. P. Jones: Anaerobic pelvic infections and development in hyperbaric oxygen therapy. Amer. J. Obstet. Gynec. 96 (1966) 645.

Parkes, C. M.: Bereavement. Studies of grief in adult life. Tavistock, London 1972. Deutsch: Vereinsamung. Die Lebenskrise bei Partnerverlust. Psychologisch-soziologische Untersuchung des Trauerverhaltens. Rowohlt, Reinbek 1974.

Parkes, C. M.: The effects of bereavement on physical and mental health: a study of the case records of widows. Brit. med. J. 2 (1964) 274.

Parkes, C. M.: Bereavement: studies of grief in adult life. Penguin, Harmondsworth 1980.

Parkes, C. M., B. Benjamin, R. G. Fitzgerald: A broken heart: a statistical study of increased mortality among widowers. Brit. med. J. 1 (1969) 740–743.

Parkinson, J.: An essay on the shaking palsy. Whittingham Rowland, London 1817.

Parks, T. G.: Ischämische Kolonerkrankungen. Colo-Proctol. 2 (1980) 213–218.

Parlow, J., A. I. Rothman: Personality traits of first year medical students: trends over the 4 year period, 1967 1970. AAMC-RIME Conference 1971, p. 68.

Parlow, M. B.: Contributions and limits of psychotherapy research. In: Minsel, W.-R., W. Herff (eds.): Methodology in Psychotherapy Research. Proc. 1st Europ. Conf. Psychother. Res., Trier, Vol. I. Lang, Frankfurt-Bern 1981.

Parmley, L. F.: The heart, the psyche, and neurosis. Psychiat. Forum 3 (1972) 16–20.

Parnas, J.: Risk factors in the development of schizophrenia: contributions from a study of schizophrenic mothers. Dan. med. Bull. 33 (1986) 127–133.

Parnas, J., H. Schulsinger: Continuity of formal thought disorder from childhood to adulthood in a high-risk sample. Acta psychiat. scand. 74 (1986) 246–251.

Parry, C. H.: Collected works, Vol. I, p. 478. London, 1825.

Parry, H. J., M. B. Balter, G. D. Mellinger, I. H. Cisin, D. J. Mannheimer: National patterns of psychotherapeutic drug use. Arch. gen. Psychiat. 28 (1973) 769–783.

Parsons, T.: Struktur und Funktion der modernen Medizin, eine soziologische Analyse. In: König, R., M. Tönnesmann (Hrsg.): Probleme der Medizinsoziologie. Kölner Z. f. Soziol. u. Sozialpsychol., Sonderheft 3. Köln-Opladen 1958.

Parsons, T.: An approach to psychological theory in terms of the theory of action. In: Koch, S. (ed.): Psychology. A

Study of Science, Vol. 3: Formulations of the Person and the Social Context, pp. 612–711. New York 1959.

Parsons, T.: The contribution of psychoanalysis to social science. Science and Psychoanalysis 4 (1961) 28–38. Deutsch in: Wehler, H. U. (Hrsg.): Soziologie und Psychoanalyse, S. 96–106. Stuttgart 1972).

Parsons, T.: The structure of social action, 3rd ed. New York 1964.

Parsons, T.: Sozialstruktur und Persönlichkeit. Frankfurt 1968a.

Parsons, T.: Einige Reflexionen über das Problem psychosomatischer Beziehungen in Gesundheit und Krankheit. In: Parsons, T.: Sozialstruktur und Persönlichkeit, S. 140–158. Frankfurt 1968b.

Partinen, M., P. Putkonen, J. Kaprio, M. Koskenvuo, J. Hilakivi: Sleep disorders in relation to coronary heart disease. Acta med. scand. (Suppl.) 660 (1982) 69–83.

Pasamanik, B., W. R. Dean et al.: Publ. Health 923 (1957).

Pascher, W.: Funktionelle Krankheiten der Stimme. In: Berendes, J., R. Link, F. Zöllner (Hrsg.): Hals-Nasen-Ohrenheilkunde in Praxis und Klinik, S. 7, Bd. IV/1. Thieme, Stuttgart-New York 1980.

Pascher, W., H. Bauer: Funktionelle Stimmstörungen. In: Pascher, W., H. Bauer (Hrsg.): Differentialdiagnose von Sprach-, Stimm- und Hörstörungen, S. 8. Thieme, Stuttgart 1984.

Paskuda, P., M. Birk, St. v. Sommoggy, G. Henckel-Donnersmark, G. W. Prokscha, G. Blümel: Frequenzanalyse der Darmgeräusche. Med. Welt 30 (1979) 687–688.

Pasnau, R. O. (ed.): Consultation-liaison psychiatry. Grune & Stratton, New York 1975.

Passchier, J., H. v.d. Helm-Hylkema, J. F. Orlebeke: Personality and headache type: a controlled study. Headache 24 (1984) 140–146.

Pateisky, K.: Die electroencephalographische Aktivierung bei Epilepsie unter Berücksichtigung von Mechanismen des Erregungsfanges. Wien. klin. Wschr. 69 (1957) 713–715.

Patel, C.: Randomized controlled trial of yoga and biofeedback in management of hypertension. Lancet II (1973) 93–95.

Patel, C.: A new dimension in the prevention of coronary heart disease. In: Schmidt, T., T. Dembroski, G. Blümchen (eds.): Biobehavioral Bases of Coronary Heart Disease. Karger, Basel 1983.

Patel, C., K. K. Datey: 12-month follow-up of yoga and biofeedback in the management of hypertension. Lancet I (1975) 62–65.

Patel, C., K. K. Datey: Relaxation and biofeedback technique in the management of hypertension. Proc. Biofeedback Res. Soc., Colorado 1974.

Patel, C., W. R. S. North: Randomized controlled trial of yoga and biofeedback in the management of hypertension. Lancet I (1975) 93–99.

Patel, C., M. G. Marmot: Stress management, blood pressure and quality of life. J. Hypertension 5 (Suppl. 1) (1987) 521–528.

Patel, C., M. G. Marmot, D. J. Terry, M. Carruthers, B. Hunt, M. Patel: Trial of relaxation in reducing coronary risk: 4-years follow-up. Brit. med. J. 290 (1985) 1103–1106.

Patrick, V., D. L. Dunner, R. R. Fieve: Life events and primary affective illness. Acta psychiat. scand. 58 (1978) 48–55.

Patterson, P. H.: Environmental determination of neurotransmitter function. Trends in Neurosci. 1 (1978) 126–128.

Patterson, R. M., S. C. Little: Spasmodic torticollis. J. nerv. ment. Dis. 98 (1943) 571–599.

Pattison, E. M.: The fatal myth of death in the family. Amer. J. Psychiat. 133 (1976) 674.

Pattison, E. M.: Detached compassion and its detortions in thanatology. In: Schoenberg, B. et al. (eds.): Education of the Medical Student in Thanatology. Arno Press, New York 1981.

Patton, J. et al. (eds.): Introduction to basic neurology. Saunders, Philadelphia 1976.

Paul, G. L.: Strategy of outcome research in psychotherapy. J. consult. Psychol. 31 (1967) 109–118.

Paul, L.: Psychosomatic aspects of low back pain. Psychosom. Med. 12 (1950) 116–124.

Pauli, H. G. (Hrsg.): Das klinische Studium an der Universität Bern seit der Studienreform. Inst. f. Ausbildungs- und Examensforschung. Med. Fakultät der Univ. Bern, Bern 1977.

Pauli, H. G.: Konflikte zwischen psychosozial und somatisch orientierter Medizin. Standpunkte und Verständigungswege. Überlegungen zur medizinischen Aus- und Weiterbildung. Vortrag, 21.4.1978. Inst. f. Ausbildungs- und Examensforschung. Med. Fakultät der Univ. Bern, Bern 1978.

Pauli, H. G.: Approaches and constraints to health manpower development research. In: Health for All – A Challenge to Research in Health Manpower Development, pp. 95–102. Proc. XVIth CIOMS Round Table Conf., Ibadan, Nigeria, 24.-26.11.82.

Pauli, H. G.: Begriffe von Gesundheit und Krankheit als Grundlagen der ärztlichen Versorgung und Ausbildung sowie der medizinischen Wissenschaft und Forschung. Medizin Mensch Gesellschaft 8 (1983) 223–233.

Pauli, H. G.: Models of medicine: from a biomechanical to a biopsychosocial view. In: Shea, W. R., B. Sitter (eds.): Scientists and Their Responsibility. Watson.

Pauli, H. G.: Versuch einer Systemsicht von Krankensituationen und Krankheitsverläufen. Symposium „Zur Wissenschaftstheorie Medizin-Realität für Patient und Arzt". Univ. Frankfurt 1984.

Pauli, H. G.: Wissenschaftstheorie und Allgemeinmedizin. 10. bundesdeutsches medizinisches Dekan-Symposion, München 1984.

Pauli, H. G.: Von der Bekämpfung der Krankheit zur Erhaltung der Gesundheit – Paradigmenwechsel? In: Schüffel, W. (Hrsg.): Sich gesund fühlen im Jahre 2000, S. 34–48. Springer, Berlin 1988.

Paulley, J. W.: Ulcerative colitis. Gastroenterology 16 (1950) 566–575.

Paulley, J. W.: Psychotherapy in ulcerative colitis. Lancet II (1956) 215–218.

Paulley, J. W.: Crohn's disease. Lancet II (1958) 959–960.

Paulley, J. W.: Crohn's disease. Psychother. and Psychosom. 19 (1971) 111–117.

Paulley, J. W.: Psychosomatic and other aspects of ulcerative colitis in the aged. Modern Geriatrics 2 (1972) 30.

Paulley, J. W.: Psychological management of Crohn's disease. Practitioner 213 (1974) 59–64.

Paulley, J. W.: Cultural influences in the incidence and pattern of disease. Psychother. and Psychosom. 26 (1975) 2–11.

Paulley, J. W.: Psychological management of multiple sclerosis. Psychother. and Psychosom. 27 (1976/77) 26–40.

Paulley, J. W.: Psychological management of spastic (irritable) colon. Abstracts 7th World Congr. Gastroenterol., Stockholm. Abstract No. 1206 (1982) 303.

Paulley, J. W.: The psychological management of the irritable colon. Hepatogastroenterol. 30 (1984) 53–54.

Paulson, G. W.: Inhibition of seizures. Dis. nerv. Syst. 24 (1963) 657–664.

Pavlov, I. P.: Conditioned reflexes. Oxford Univ. Press, Oxford 1927.

Pavlovic, M.: Katamnestische Untersuchungen von Patientinnen mit Anorexia nervosa. Med. Diss., Düsseldorf 1981.

Pawlik, K.: Dimensionen des Verhaltens. Eine Einführung in die Methodik und Ergebnisse faktorenanalytischer psychologischer Forschung. Huber, Bern 1968.

Pawlik, K.: Dimensionen des Verhaltens. Eine Einführung in Methodik und Ergebnisse faktorenanalytischer Forschung, 2. Aufl. Huber, Bern-Stuttgart-Wien 1971.

Pawlow, I.: Sämtliche Werke. Akademie Verlag, Berlin 1953.

Payk, R.: Psychopathologische Besonderheiten bei Kranken mit Encephalomyelitis disseminata („Multiple Sklerose"). Nervenarzt 44 (1973) 378–380.

Payk, T. R.: Möglichkeiten der psychiatrischen Schmerzbehandlung. Med. Klinik 78 (1983) 331–333.

Paykel, E. S.: Life stress, depression and attempted suicide. J. hum. Stress 2 (1976) 3–12.

Pearce, J., J. M. H. Moll: Conservative treatment and natural history of acute lumbar disc lesions. J. Neurol., Neurosurg., Psychiat. 30 (1967) 13–17.

Pearlin, L., C. Schooler: The structure of coping. J. Health soc. Behav. 19 (1978) 2–21.

Pearson, J., L. Brandeis, C. Cuello: Depletion of substance P containing axons in substantia gelatinosa of patients with diminished pain sensitivity. Nature 295 (1982) 61.

Pecknold, J. C., R. P. Swinson, K. Kuck, C. P. Lewis: Alprazolam in panic disorder and agoraphobia: results from a multicenter trial. III. Discontinuation effects. Arch. gen. Psychiat. 45 (1988) 429–436.

Peeke, H. V. S., K. Dark, G. Ellman, C. McCurry, M. Salfi: Prior stress and behaviorally conditioned histamine release. Physiol. Behav. 39 (1987) 89–93.

Pelner, L.: The determination of sensitivity to pain. A simple clinical method. J. Lab. clin. Med. 27 (1941) 248–251.

Pelser, H. E.: Psychological aspects of the treatment of patients with myocardial infarction. J. psychosom. Res. 11 (1967) 47–49.

Pender, N.: Patient identification of health information received during hospitalization. Nursing Research 23 (1974) 262–267.

Penfield, W., B. Milner: Memory deficit produced by bilateral lesions in the hippocampal zone. Arch. Neurol. Psychiat. (Chic.) 79 (1958) 475–497.

Penin, H.: Prävention und Behandlung von Epilepsie. Nervenheilk. 6 (1987) 274–278.

Perez-Gay, B.: Fluor genitalis et cervicis aus psychosomatischer Sicht. In: Prill, H. J., D. Langen (Hrsg.): Der psychosomatische Weg zur gynäkologischen Praxis. Schattauer, Stuttgart 1983.

Perez-Gay, B.: Was bedeutet Schwangerschaftsbetreuung aus psychosomatischer Sicht? In: Richter, D., M. Stauber (Hrsg.): Psychosomatische Probleme in Geburtshilfe und Gynäkologie. Kehrer, Freiburg 1983.

Perinelli, K., C. Günther: Unverarbeitete Trauer in Familien mit einem psychosomatisch kranken Kind. Praxis der Kinderpsychologie 32 (1983) 89.

Perini, C., F. R. Bühler: Psychophysiological responses to mental stress in type A and type B subjects with and without a history of hypertension. Paper presented at the conference „Biobehavioral factors in coronary heart disease", Winterscheid 1984.

Perini, C., F. W. Amann, P. Bolli, F. R. Bühler: Personality and adrenergic factors in essential hypertension. Contr. Nephrol., Vol. 30, pp. 64–69. Karger, Basel 1982.

Perley, M. J., S. B. Guze: Hysteria – the stability and usefulness of clinical criteria. A quantitative study based on a follow-up period of 6–8 years in 39 patients. New Engl. J. Med. 266 (1962) 421–426.

Perls, F. S.: Grundlagen der Gestalttherapie. Einführung und Sitzungsprotokolle. Pfeiffer, München 1976.

Permutt, M. A., P. Rotwein: Analysis of the insulin gene in noninsulin-dependent diabetes. Amer. J. Med. 72 (1983) 1–7.

Perrez, M., H. Schenkel, M. Stauber: Eine experimentelle Untersuchung zur psychologischen Geburtsvorbereitung. Z. Geburtsh. Perinatol. 182 (1978) 149.

Persky, V. W., J. Kempthorne-Rawson, R. B. Shekelle: Personality and risk of cancer: 20-year follow-up of the Western Electric Study. Psychosom. Med. 49 (1987) 435–439.

Pert, C. B., S. H. Snyder: Opiate receptor: demonstration in nervous tissue. Science 179 (1973) 1011–1014.

Pertschuk, M.: Behavior therapy: extended follow-up. In: Vigersky, R. A. (ed.): Anorexia nervosa, pp. 305–313. Raven, New York 1977.

Pertschuk, M., L. Crosby, J. Mullen: Nonlinearity of weight gain and nutrition intake in anorexia nervosa. In: Darby, P. L., P. E. Garfinkel, D. M. Garner, D. V. Coscina (eds.): Anorexia Nervosa. Recent Developments in Research, pp. 301–310. Liss, New York 1983.

Peshkin, M. M., I. Friedman: Residential asthma treatment centers in the United States and problems in relation to them. J. Asthma Res. 12 (1975) 129–175.

Pesso, A.: Dramaturgie des Unbewußten. Klett-Cotta, Stuttgart 1986.

Peter, B.: Milton H. Ericksons Weg der Hypnose. Hypnose und Kognition 5 (1988) 46–53.

Peter, J. H., E. Becker, E. Fuchs, K. Meinzer, P. v. Wichert: Ambulante transkutane Langzeitregistrierung von arterieller Sauerstoffspannung und Herzrhythmusstörungen bei Patienten mit Schlafapnoesyndrom. Verh. Dtsch. Ges. Inn. Med. 88 (1982) 390–393.

Petermann, F.: Veränderungsmessung. Kohlhammer, Stuttgart 1978.

Peters, F., D. Richter, M. Breckwoldt: Sekundäre Amenorrhoe. Zusammenhänge zwischen endokrinologischen und psychosomatischen Befunden. Dtsch. med. Wschr. 103 (1978) 898.

Peters, U. H.: Die erfolgreiche Therapie des chronischen Kopfschmerzes. Perimed, Erlangen 1983.

Peters, U. H.: Vorwort des deutschen Herausgebers. In: Freedman, A. M., H. I. Kaplan, B. J. Sadock, U. H. Peters (Hrsg.): Psychiatrie in Praxis und Klinik, Bd. 1. Thieme, Stuttgart 1984.

Petersen, P.: Psychiatric disorders in primary hyperparathyroidism. J. clin. Endocr. 28 (1968) 1491.

Petersen, P.: Psychiatrische und psychologische Aspekte der Familienplanung bei oraler Kontrazeption. Thieme, Stuttgart 1969.

Petersen, P.: Chirurgische Kontrazeption der Frau und die seelischen Folgen/Erkenntnisse aus der Analyse von 40 Reihenkatamnesen freiwillig Sterilisierter. Sexualmed. 6 (1977) 13; 100; 204; 295.

Petersen, P.: Kontrazeption und Ursprung. Trennung von Fruchtbarkeit und Geschlechtlichkeit – Risiko und Chance. Sexualmed. 9 (1980a) 407.

Petersen, P.: Endgültige Fruchtbarkeitsverhütung in der Familienplanung. Zur Psychologie der freiwilligen Sterilisation. Münch. med. Wschr. 15 (1980b) 122.

Petersen, P.: Seelische Folgen der Sterilisation. Kritische Anmerkungen zur Methodik psychologischer Studien. Geburtsh. Frauenheilk. 4 (1983) 253.

Petersen, P.: Pers. Mitteilung 1984.

Peterson, W. L., R. A. L. Sturdevant, H. D. Frankl, C. T. Richardson, J. I. Isenberg, J. D. Elashoff, J. Q. Sones, R. A. Gross, R. W. McCallum, J. S. Fordtran: Healing of duodenal ulcer with an antacid regimen. New Engl. J. Med. 297 (1977) 341–345.

Petrides, P.: Sozialmedizinische Probleme. In: Oberdisse, K. (Hrsg.): Handbuch für Innere Medizin. Diabetes mellitus, S. 1147–1177. Springer, Berlin-Heidelberg-New York 1977.

Petrie, A.: Individuality in pain and suffering. Univ. of Chicago Press, Chicago-London 1967.

Pette, H.: Kritische Bemerkungen zum Kapitel des Bandscheibenprolapses. Münch. med. Wschr. 95 (1953) 1145–1148.

Petterson, F., H. Fries, S. J. Nillius: Epidemiology of secondary amenorrhea. Amer. J. Obstet. Gynec. 80 (1973) 117.

Pettingale, K. W.: Coping and cancer prognosis. J. psychosom. Res. 28 (1984) 363–364.

Pettingale, K. W., T. Morris, S. Greer, J. L. Haybittle: Mental attitudes to cancer: an additional prognostic factor. Lancet 8431 (1985) 750.

Petzold, E.: Anorexia nervosa. Habilitationsschrift, Heidelberg 1977.

Petzold, E.: Familienkonfrontationstherapie bei Patienten mit Anorexia nervosa. Vandenhoeck und Ruprecht, Göttingen 1979.

Petzold, E., A. Reindell: Psychosomatische Diagnostik und Therapie bei Herzinfarkt, Colitis ulcerosa und Morbus Crohn. Prax. Psychother. 22 (1977) 109–115.

Petzold, H.: Integrative Bewegungstherapie. In: Petzold, H. (Hrsg.):Psychotherapie und Körperdynamik, S. 289. Junfermann, Paderborn 1977.

Petzold, H., E. Bubolz (Hrsg.): Psychotherapie mit alten Menschen. Junfermann, Paderborn 1979.

Petzold, U., A. Frehen: Die Pflege im Spannungsfeld zwischen kurativem und palliativem Ansatz in der Onkologie. Deutsche Krankenpflegezeitschrift 40/12 (1987) 848–850.

Peyser, J. M., K. R. Edwards, C. M. Poser: Psychological profiles in patients with multiple sclerosis. Arch. Neurol. 37 (1980) 437–440.

Pfändler, U., E. Schnyder: Die rezessive Taubstummheit im Werdenberg (St. Gallen). Bull. Schweiz. med. Wiss. 15 (1959) 178.

Pfaff, D. W.: Estrogens and brain function. Springer, New York 1980.

Pfaffmann, C.: Taste preference and reinforcement. In: Tapp, J. T. (ed.): Reinforcement and Behavior. Acad. Press, New York 1969.

Pfaltz, C. R.: Sudden and fluctuant hearing loss. In: Alberti, P. W., R. J. Ruben (eds.): Otologic Medicine and Surgery, Vol. 2. Churchill Livingstone, New York-Edinburgh-London 1988.

Pfau, E. M.: Psychologische Untersuchungsergebnisse zur Ätiologie der psychogenen Stimmstörungen. Fol. phoniat. 27 (1975) 298.

Pfeiffer, E. F.: Rationale for sulfonylurea therapy. Excerpta Medica, Amsterdam 1983.

Pfeiffer, E. F., W. Kerner: Diabetestherapie: Künstliches endokrines Pankreas und tragbare Insulinpumpen. Dtsch. Ärztebl. 81 (1984) 3495–3503.

Pfeiffer, E. F., M. Pfeiffer, H. Ditschuneit, Chang-Su-Ahn: Über die Bestimmung von Insulin im Blute am epididymalen Fettanhang der Ratte mit Hilfe markierter Glukose. II: Experimentelle und klinische Erfahrungen. Klin. Wschr. 37 (1959) 1239.

Pfingsten, M., M. Bautz, D. Eggebrecht, J. Hildebrandt: Soziale Interaktion bei Patienten mit chronischen Rückenschmerzen. Psychother. med. Psychol. 38 (1988) 328–332.

Pflanz, M.: Sozialer Wandel und Krankheit. Ergebnisse und Probleme der medizinischen Soziologie. Enke, Stuttgart 1962.

Pflanz, M.: Sozialanthropologische Aspekte der Anorexia nervosa. In: Meyer, J. E., H. Feldmann (Hrsg.): Anorexia nervosa, S. 146–150. Thieme, Stuttgart 1965 .

Pflanz, M.: Psychosomatische Aspekte der essentiellen Hypertonie. In: Heintz, R., H. Losse (Hrsg.): Arterielle Hypertonie. Thieme, Stuttgart 1969.

Pflanz, M.: Psychologische und sozialmedizinische Aspekte der Hypertonie. Verh. Dtsch. Ges. Inn. Med. 80 (1974) 42–49.

Pflanz, M.: Allgemeine Epidemiologie. Aufgaben, Techniken, Methoden. Thieme, Stuttgart 1973.

Pflanz, M.: Epidemiologie des essentiellen Hochdrucks. Verh. Dtsch. Ges. Kreislaufforsch. 43 (1977).

Pflanz, M., Th. v. Uexküll: Psychosomatische Untersuchungen an Hochdruckkranken. Med. Klinik 57 (1962) 345–351.

Pflanz, M., E. Rosenstein, Th.v. Uexküll: Socio-psychological aspects of peptic ulcer. J. psychosom. Res. 1 (1956) 68.

Pflanz, M., H.-D. Basler, D. Schwoon: Use of tranquilizing drugs by a middle-aged population in a West German city. J. Health soc. Behav. 18 (1977) 194–205.

Pfleiderer, A., D. Richter, P. Thiessen, V. Kissel, B. Tibi, P. Nowara: Aktuelle Probleme bei der Nachsorge von Patientinnen mit Karzinomen der Zervix und des Corpus uteri. Onkologie 2 (1979) 62.

Phelan, P. W.: Cognitive correlates of bulimia: the bulimic thoughts questionnaire. Int. J. Eat. Dis. 6 (1987) 593–607.

Philip, A. E., E. L. Cay: Short-term fluctuations in anxiety in patients with MI. J. psychosom. Res. 23 (1979) 277–280.

Philippopoulos, G. S., E. D. Wittkower, A. Cousineau: The etiologic significance of emotional factors in onset and exacerbations of multiple sclerosis. Psychosom. Med. 20 (1958) 458–474.

Philips, C.: A psychological analysis of tension headache. In: Rachman, S. (ed.): Contributions to Medical Psychology, Vol. 2, pp. 113–130. Pergamon, Oxford-New York 1977.

Philips, C.: Recent developments in tension headache research: implications for understanding and management of the disorder. In: Philips, C., M. Hunter: The treatment of tension headache, II. EMG 'normality' and relaxation. Behav. Res. Ther. 19 (1981) 499–507.

Philips, C.: Headache and personality. J. psychosom. Res. 20 (1976) 535–542.

Piaget, J.: Six psychological studies. Random House, New York 1967.

Piaget, J.: Einführung in die genetische Erkenntnistheorie. Suhrkamp, Frankfurt 1973.

Piaget, J.: Der Aufbau der Wirklichkeit beim Kinde. Ges. Werke, Bd. 2. Klett, Stuttgart 1975.

Piaget, J.: Psychologie der Intelligenz. Kindler, München 1976.

Piazza, E. U.: Comprehensive therapy of chronic asthma on a psychosomatic unit. Adolescence XVI/61 (1981) 139–144.

Pick, A.: Störung der Orientierung am eigenen Körper. Psychol. Forsch. 1 (1922) 303–318.

Pickering, G. W.: Die Erblichkeit der Hypertonie. In: Bock, K. P., P. Cottier (Hrsg.): Essentielle Hypertonie. Springer, Berlin 1960.

Pickering, T. G.: Letter to the editor. New Engl. J. Med. 313 (1985) 450.

Pickering, T. G., N. E. Miller: Learned voluntary control of heart rate and rhythm in two subjects with premature ventricular contractions. Brit. Heart J. 39 (1977) 152.

Pickering, T. G., I. Goulding, B. A. Cobern: Diurnal variations in ventricular ectopic beats and heart rate. Cardiovascul. Med. 2 (1977) 1013.

Pickering, T. G., G. A. Harshfield, R. B. Devereux, J. H. Laragh: What is the role of ambulatory blood pressure monitoring in the management of hypertensive patients? Hypertension 7 (1985) 171–177.

Pieringer, W.: Psychosomatische und somatopsychische Aspekte der progressiv chronischen Polyarthritis. Wien. klin. Wschr. 90 (1978) 17–20.

Pieringer, W., H. Reisner: Psychopathologische Syndrome nach Herzoperationen. Wien. Z. Nervenheilk. 28 (1970) 246–254.

Pierloot, R. A.: Different models in the approach to the doctor-patient relationship. Psychother. and Psychosom. 39 (1983) 213–224.

Pierloot, R. A., W. Wellens, M. E. Houben: Elements of resistance to a combined medical and psychotherapeutic program in anorexia nervosa. An overview. Psychother. and Psychosom. 26 (1975) 101–117.

Pierloot, R. A., W. Vandereycken, S. Verhaest: An inpatient treatment program for anorexia nervosa patients. Acta psychiat. scand. 66 (1982) 1–8.

Pierpaoli, W., E. Sorkin: Relationship between thymus and hypophysis. Nature (Lond.) 246 (1973) 405–409.

Pimpl, W., M. Umlauft: Ischämische Kolitis nach kardiogenem Schock. Colo-Proctol. 5 (1983) 15–18.

Pincus, L.: ... bis daß der Tod euch scheidet. Zur Psychologie des Trauerns. Deutsche Verlagsanstalt, Stuttgart 1977.

Pinkerton, P.: The influence of sociopathology in childhood asthma. Psychother. and Psychosom. 18 (1970) 231.

Pinkus, T., L. F. Callahan, L. A. Bradley, W. K. Vaughn, F. Wolfe: Elevated MMPI scores for hypochondriasis, depression, and hysteria in patients with rheumatoid arthritis reflect disease rather than psychological status. Arthr. and Rheum. 29 (1986) 1456–1466.

Piper, H. C.: Gespräche mit Sterbenden. Vandenhoeck und Ruprecht, Göttingen 1977.

Pirella, A. (Hrsg.): Sozialisation der Ausgeschlossenen. Praxis einer neuen Psychiatrie. Rowohlt, Hamburg 1975.

Pirke, K. M.: Menstruationszyklus und neuroendokrine Störungen der Gonadenachse bei Bulimia. In: Fichter, M. M. (Hrsg.): Bulimia nervosa. Enke, Stuttgart 1989.

Pirke, K. M.: Störungen zentraler Neurotransmitter bei Bulimia. In: Fichter, M. M. (Hrsg.): Bulimia nervosa. Enke, Stuttgart 1989.

Pirke, K. M., D. Ploog: Psychobiology of anorexia nervosa. In: Wurtman, R. J., J. J. Wurtman (eds.): Nutrition and the Brain, Vol. 6, pp. 167–198. Raven, New York 1986.

Pirke, K. M., M. M. Fichter, M. Warnhoff, G. Dorsch, B. Spyra: Hypothalamic regulation of gonadotropin secretion in anorexia nervosa and in starvation. Int. J. Eat. Dis. 2 (1983) 151–158.

Plante, T. G., G. E. Schwartz: Defensive and repressive coping styles: physiological stress responsivity to psychosocial stressors. Psychophysiol. 25 (1988) 473.

Plassmann, R.: Der Arzt, der Artefakt-Patient und der Körper. Eine psychoanalytische Untersuchung des Mimikry-Phänomens. Psyche (1987) 883–899.

Plassmann, R., M. Teising, H. Freyberger: Ein ‚Mimikry‘-Patient: Bericht über den Behandlungsversuch einer selbstgemachten Krankheit. Praxis Kinderpsychol. Kinderpsychiat. 34 (1985) 133–141.

Plassmann, R., B. Wolff, H. Freyberger: Die heimliche Selbstmißhandlung, eine psychosomatische Krankheit. Z. psychosom. Med. 32 (1986) 316–336.

Platanov, K. J.: zit. nach: Molinski, H.: Bilder der eigenen Weiblichkeit, Ärger während der Geburt und Rigidität des Muttermundes. Z. psychosom. Med. Psychoanal. 14/2 (1968) 90.

Platt, D.: Biologie des Alterns. Quelle & Meyer, Heidelberg 1972.

Platt, R.: Das Wesen der essentiellen Hypertonie. In: Bock, K. P., P. Cottier (Hrsg.): Essentielle Hypertonie. Springer, Berlin 1960.

Platt, S., N. Kreitman: Trends in parasuicide and unemployment among men in Edinburgh 1968–82. Brit. med. J. 289 (1984) 1029–1032.

Plaut, S. M., S. B. Friedman: Psychosocial factors in infectious disease. In: Ader, R. (ed.): Psychoneuroimmunology, pp. 3–30. Acad. Press, New York 1981.

Plaut, S. M., R. Ader, S. B. Friedman, A. L. Ritterson: Social factors and resistance to malaria in the mouse: effects of group versus individual housing on resistance to Plasmodium berghei infection. Psychosom. Med. 31 (1969) 536–552.

Plenz, R.-J.: Psychogene Hörstörungen im Kindesalter. Fortschr. Med. 94 (1976) 134.

Plessner, H.: Die Frage nach der Conditio humana. Suhrkamp, Frankfurt 1976.

Plester, D.: Die einseitige Hörstörung. Arch. Oto-Rhino-Laryng. 219 (1978) 451.

Ploeger, A.: Die therapeutische Gemeinschaft in der Psychotherapie und Sozialpsychiatrie. Thieme, Stuttgart 1972.

Plotnikoff, N. P., R. E. Faith, A. J. Murgo, R. A. Good (eds.): Enkephalins and endorphins. Stress and the immune system. Plenum, New York 1986.

Plovnick, M. S.: Career orientations in the primary care specialties. J. med. Educ. 54 (1979) 655–657.

Plügge, H.: Anthropologische Beobachtungen bei primär chronischen Arthritikern. Z. Rheumaforsch. 12 (1953) 231–246.

Plügge, H.: Über das Befinden von Kranken nach Herzinfarkt. Ärztl. Wschr. 15 (1960) 61–66.

Plügge, H.: Über die Hoffnung. In: Plügge, H. (Hrsg.): Wohlbefinden und Mißbefinden. Beiträge zu einer Medizinische Anthropologie, S. 38–50. Niemeyer, Tübingen 1962.

Plügge, H., R. Mappes: Über das Leiden herzkranker Kinder. Internist 3 (1962) 49–56.

Plumb, M., J. Holland: Comparative studies of psychological functions in patients with advanced cancer, II. Interviewer-rated current and past psychological symptoms. Psychosom. Med. 43 (1981) 243–254.

Plutchik, R.: Cognition in the service of emotions: an evolutionary perspective. In: Candland, D. K., J. P. Fell, E. Keen, A. I. Leshner, R. Plutchik, R. M. Tarpy (eds.): Emotions. Brooks-Cole, Belmont 1977.

Plutchik, R., H. Kellerman (eds.): Emotion – Theory, Research and Experience, Vol. I: Theories of Emotion. Acad. Press, New York 1980.

Poddig, K.: Erstellung eines Fragebogens zur Messung veränderlicher Persönlichkeitsmerkmale bei Patienten mit Colitis ulcerosa und Morbus Crohn. Diss., Lübeck 1987.

Poeck, K.: Zur Psychophysiologie der Phantomerlebnisse. Nervenarzt 31 (1963) 241–256.

Poeck, K.: Phantome nach Amputationen und angeborenem Gliedmaßenmangel. Dtsch. med. Wschr. 46 (1969) 2367.

Poeck, K., B. Orgass: Über die Entwicklung des Körperschemas. Untersuchungen an gesunden, blinden und amputierten Kindern. Fortschr. Neurol. Psychiat. 32 (1964) 538–555.

Pöldinger, W.: Psychiatric problems in physically ill patients. In: Ansell, B. (ed.): Rheumatism and the Psyche. Huber, Bern 1976.

Pöldinger, W.: Psychosomatik der funktionellen sexuellen Störungen des Mannes. In: Jores, A. (Hrsg.): Praktische Psychosomatik, S. 292–297. Huber, Bern-Stuttgart-Wien 1981.

Poettgen, H.: Die Integration des autogenen Trainings in die geburtshilfliche Psychoprophylaxe. Geburtsh. Frauenheilk. 31/2 (1971) 150.

Poettgen, H.: Schwangerschaftskonfliktberatung bei der „Notlagen-Indikation". Dtsch. Ärztebl. 8 (1977) 515.

Poettgen, H.: Die Geburtsvorbereitung mit autogenem Training und analytisch orientierten Gruppendiskussionen. Vortr. gynäkol.-psychosom. Kongr., Gießen 1973.

Poettgen, H.: Larvierte Sexualstörungen unter der „Flagge" gynäkologischer Symptome. Gyne 8 (1983) 8.

Poettgen, H.: Paartherapie: Die Bedeutung des interaktionellen Aspekts. Gyne 11 (1983) 6.

Pötzl, O.: Über Störungen der Selbstwahrnehmung bei linksseitiger Hemiplegie. Z. ges. Neurol. Psychiat. 93 (1924) 117–168.

Poewe, W., F. Gerstenbrand, G. Ransmayr, S. Plörer: Premorbid personality of Parkinson patients. J. neural Transmission (Suppl.) 19 (1983) 215–224.

Pogge, R.: The toxic placebo. Med. Times 91 (1963) 773–778.

Pohlen, M., M. Bautz: Gruppenanalyse als Kurzpsychotherapie. Nervenarzt (1974) 514–533.

Pohlmann, H., W. Schramm: Erfahrungen in der psychosozialen Betreuung von Hämophilen. Stellungnahme vor der Enquête-Kommission AIDS des Deutschen Bundestages. Kommissionsarbeitsunterlage Nr. 270.

Poll, I. B., A. Kaplan De-Nour: Locus of control and adjustment to chronic hemodialysis. Psychol. Med. 10 (1980) 153–157.

Polley, H. F., W. M. Swenson, R. M. Steinhilber: Personality characteristics of patients with rheumatoid arthritis. Psychosomatics 11 (1970) 45–49.

Pollock, G. H.: Aging or aged: development of pathology. In: Greenspan, S. I., G. H. Pollock (eds.): The Course of Life, Vol. III. Adulthood and the Aging Process. Mental Health Study Center, Maryland 1981.

Pommer, W., M. Broda: Der Stand der psychosozialen Betreuung chronisch Nierenkranker in der BRD und Westberlin. Nieren- und Hochdruckkr. 14 (1985) 462–468.

Pomp, A. M.: Psychotherapy for the myofacial pain dysfunction syndrome: a study of factors coinciding with symptom remission. J. Amer. dent. Ass. 89 (1974) 629.

Pongratz, J.: Leitsymptom: Wirbelsäulenschmerz – Eine psychosomatische Studie. Z. psychosom. Med. 26 (1980) 12–39.

Pontalis, J. B.: Jeu et réalité, préface. Gallimard, Paris 1975.

Pontzen, W., G. Daudert, R. Dietz, H. Kappauf: Probleme und Möglichkeiten der Zusammenarbeit zwischen internistischen Onkologen und Psychosomatikern. In: Rechenberger, H.-G., H.-V. Werthmann (Hrsg.): Psychotherapie und Innere Medizin, S. 243–250. Pfeiffer, München 1988.

Pool, L.: Discussion of Miller, C. M. In: Fields, W. W. (ed.): Clinical Syndromes in Cerebral Hemorrhage. Pathogenesis and Treatment of Cerebrovascular Disease. Thomas, Springfield/Ill. 1961.

Pope, H. G., D. L. Katz: Affective psychotic symptoms associated with anabolic steroid use. Amer. J. Psychiat. 145 (1988) 487–490.

Pope, H. G. jr., J. I. Hudson, J. M. Jonas, D. Yurgelun-Todd: Bulimia treated with imipramine: a placebo-controlled, double-blind study. Amer. J. Psychiat. 140 (1983) 554–558.

Popper, K. R.: Objektive Erkenntnis. Ein evolutionärer Entwurf. Hoffmann und Campe, Hamburg 1973.

Popper, K. R.: Kritik des Materialismus. In: Popper, K. R., J. C. Eccles: Das Ich und sein Gehirn. Piper, München 1977.

Popper, K. R., J. C. Eccles: The self and its brain. Springer, Berlin-London-New York 1981.

Porot, M., B. Duclaux, A. J. Coudert: Les besoins psychiatriques des hôpitaux généraux. Ann. méd.-psychol. 130 (1972) 609–624.

Portal, A.: Beobachtungen über die Natur und Behandlung der Epilepsie. Bibliothek der ausländischen Literatur für praktische Medizin, Siebenter Band. Hartmann, Leipzig 1828.

Porter, J, H. Jick: Addiction rate in patients treated with narcotics. New Engl. J. Med. 302 (1980) 123.

Portis, S. A.: Idiopathic ulcerative colitis. Newer concepts concerning its cause and management. J. Amer. med. Ass. 139 (1949) 208–214.

Portmann, A.: Nachwort in Flanagan, G. L. (Hrsg.): Die ersten neun Monate des Lebens. Rowohlt, Reinbek 1963.

Portmann, A.: Biologische Fragmente einer Lehre vom Menschen. Schwabe, Basel-Stuttgart 1969.

Poser, C. M. (ed.): The diagnosis of multiple sclerosis. Thieme-Stratton, New York 1984.

Poser, S., G. Ritter: Multiple Sklerose in Forschung, Klinik und Praxis. Schattauer, Stuttgart 1980.

Poser, S., H. Friedrich: Psychiatrisch-psychotherapeutische Erfahrungen bei der multiplen Sklerose. In: Bönisch, E., J. E. Meyer (Hrsg.): Psychosomatik in der klinischen Medizin. Psychiatrisch-psychotherapeutische Erfahrungen bei schweren somatischen Krankheiten, S. 39–54. Springer, Berlin-Heidelberg-New York 1983.

Post, R. M., J. C. Ballenger, A. C. Roy, W. E. Bunney jr.: Slow and rapid onset of manic episodes: implications for underlying biology. J. psychiat. Res. 4 (1981) 229–237.

Postpischel, F.: Katamnestische Untersuchungen bei Patienten mit Anorexia nervosa. Psychol. Diplomarbeit, Tübingen 1981.

Potreck-Rose, F.: Anorexia nervosa und Bulimia. Deutscher Studienverlag, Weinheim 1987.

Pott, W., M. Junk, U. Pauli, M. Wirsching, W. Weidner: Psychosomatische Aspekte der chronischen Prostatitis. Praxis der klinischen Verhaltensmedizin und Rehabilitation 1 (1988) 45.

Potts, M., M. Weinberger, K. D. Brandt: Views of patients and providers regarding the importance of various aspects of an arthritis treatment program. J. Rheum. 11 (1984) 71–75.

Pow, J. M.: The role of psychological influences in rheumatoid arthritis. J. psychosom. Res. 31 (1987) 223–229.

Powell, G. F., J. A. Brasel, R. M. Blizzard: Emotional deprivation and growth retardation simulating idiopathic hypopituitarism I. New Engl. J. Med. 276 (1967) 1271.

Powell, L. H., M. Friedman, C. E. Thoresen, J. J. Gill, D. K. Ulmer: Can the Type A behavior pattern be altered after myocardial infarction? A second year report from the Recurrent Coronary Prevention Project. Psychosom. Med. 46 (1984) 293–313.

Powers, P. S., R. C. Fernandez (eds.): Current treatment of anorexia nervosa and bulimia. Karger, New York 1984.

Pratt, J. H.: The tuberculosis class: an experiment in home treatment. In: Rosenbaum, M., M. Berger (eds.): Group Psychotherapy and Group Function. Basic Books, New York 1963.

Pratt, L. W., H. T. Wayne, R. A. Gallagher: Globus hystericus – office evaluation by psychological testing with the MMPI. Laryngoscope (St. Louis) 86 (1976) 1540.

Premack, D.: On the abstractness of human concepts: why it would be difficult to talk to a pigeon. In: Hulse, S. S., H. Fowler, W. K. Honig (eds.): Cognitive Processes in Animal Behavior. Erlbaum, Hillsdale 1978.

Prest, A. P. L.: Die Sprache der Sterbenden. Göttingen 1970.

Presthus, J.: Psychiatric side effects occuring in parkinsonism during longterm treatment with levodopa alone and in combination with other drugs. In: Rinne, U. K., M. Klingler, G. Stamm (eds.): Parkinson's Disease – Current Progress, Problems and Management. Elsevier, Amsterdam 1980.

Pribram, K. H.: Languages of the brain. Prentice Hall, Englewood Cliffs/New Jersey 1971.

Pribram, K. H.: The biology of emotions and other feelings. In: Plutchik, R., H. Kellerman (eds.): Emotion – Theory, Research and Experience, Vol. I: Theories of Emotion. Acad. Press, New York 1980.

Pribram, K. H., D. McGuiness: Arousal, activation and effort in the control of attention. Psychol. Rev. 82 (1975) 116–149.

Pries, K., W. Wellmann, H. Freyberger: Kombinierter gastroenterologisch-psychosomatischer Therapieansatz bei M. Crohn-Patienten mit Anorexia nervosa Symptomatik. In: Meermann, R. (Hrsg.): Anorexia nervosa. Enke, Stuttgart 1981.

Prigogine, I.: Time, structure and fluctuation. Science 201 (1978) 43–58, 777–785.

Prigogine, I.: From being to becoming. Freeman, San Francisco 1980. Deutsch: Vom Sein zum Werden. Piper, München 1982.

Prill, H. J.: Zur Psychosomatik der funktionellen gynäkologischen Blutungsstörungen. Z. Psychother. med. Psychol. 10/1 (1960) 15.

Prill, H. J.: Psychosomatische Gynäkologie. Urban & Schwarzenberg, München-Berlin 1964.

Prill, H. J.: Geburtsstörungen sind oft die Folge von diversen psychischen Fehlhaltungen. Medical Tribune, Sondernummer Gynäkologie, Okt. 1968.

Prill, H. J.: Psychologie der Schwangeren, Gebärenden und Wöchnerin. In: Gynäkologie und Geburtshilfe, Bd. II. Thieme, Stuttgart 1976.

Prill, H. J.: Fortschritte der Psychosomatik in der Gynäkologie. In: Schwalm, H., Döderlein (Hrsg.): Klinik der Frauenheilkunde und Geburtshilfe IV. Urban & Schwarzenberg, München 1977 (Ergänzung).

Prill, H. J.: Psychosomatik und Psychopathologie der Schwangeren, Gebärenden und Mutter. In: Martius, G. (Hrsg.): Hebammenlehrbuch. Thieme, Stuttgart 1979.

Prill, H. J.: Psychosomatik der vorzeitigen Wehentätigkeit. In: Grospietsch, G., W. Kuhn (Hrsg.): Tokolyse mit Betastimulatoren. Thieme, Stuttgart 1983.

Prill, H. J.: Therapie der Pelipathia vegetativa in ätiologisch-diagnostischer Sicht. Internist. Praxis 4 (1984) 588.

Prill, H. J., C. Lauritzen: Das Klimakterium. In: Schwalm, H., Döderlein (Hrsg.): Klinik der Frauenheilkunde und Geburtshilfe. Urban & Schwarzenberg, München 1970.

Prior, I.: Civilization and cardiovascular changes – a pacific viewpoint. In: Doc. Geigy. Ciba-Geigy, Basel 1976.

Pritchard, P. B., V. L. Holstrom, J. Giacinto: Self-abatement of complex partial seizures. Ann. Neurol. 18 (1985) 265–267.

Probst, B.: Soziale Integration von Morbus Crohn- und Colitis ulcerosa-Patienten. Studie zur Wechselwirkung somatischer, psychischer und sozialer Faktoren. Med. Diss., Lübeck 1989.

Probst, B., J. v. Wietersheim, E. Wilke, H. Feiereis: Soziale Integration von Morbus Crohn- und Colitis ulcerosa-Patienten. Z. psychosom. Med. Psychoanal. (im Druck).

Procci, W.: Persistent sexual dysfunction following renal transplantation. Dial. Transplant. 7 (1978) 897–970.

Procci, W.: Psychological factors associated with severe abuse of the hemodialysis diet. Gen. Hosp. Psychiat. 3 (1981) 111–118.

Procci, W., D. A. Goldstein, O. A. Kletzky: Impotence in uremia. In: Levy, N. B. (ed.): Psychonephrology 2, pp. 235–246. Plenum, New York 1983.

Prokop, H.: Autogenes Training – seine Bedeutung in Gegenwart und Zukunft. Perlinger, Wörgl 1979.

Proust, M.: Die Suche nach der verlorenen Zeit (7 Bände). Bd. 1 Swanns Welt. Suhrkamp, Frankfurt 1953–1957.

Prugh, D. G.: The influence of emotional factors on the clinical course of ulcerative colitis in children. Gastroenterology 18 (1951) 339–354.

Prugh, D. G.: Toward an understanding of psychosomatic

concepts in relation to illness in children. In: Solmit, A. J., S. A. Provence (eds.): Modern Perspectives in Child Development. Int. Univ. Press, New York 1963.

Prugh, D. G., C. K. Tagluri: Emotional aspects of the respirator care of patients with poliomyelitis. Psychosom. Med. 16 (1954) 104–128.

Pszywij, A.: Die imaginative Anwendung des Wassers im Katathymen Bilderleben. In: Leuner, H. (Hrsg.): Katathymes Bilderleben, S. 216–223. Huber, Bern 1980.

Puchalski, Z., L. Szlendak: Angst als Zustand und Angst als Persönlichkeitseigenschaft bei Patienten mit Alopecia areata, Rosacea und Lichen ruber planus. Z. Hautkr. 58 (1983) 1038–1048.

Pudel, V.: Zur Wirksamkeit einer Fernsehkampagne auf das Gewichtsverhalten der deutschen Bevölkerung. Ern.-Umschau 10 (1979) 360–364.

Pudel, V.: Zur Psychogenese und Therapie der Adipositas. Springer, Berlin-Heidelberg-New York 1982.

Pullar, T., H. A. Capell, A. Millar, R. G. Brooks: Alternative medicine: costs and subjective benefit in rheumatoid arthritis. Brit. Med. J. 285 (1982) 1629–1631.

Puntis, J., A. S. McNeish, R. N. Allan: Long term prognosis of Crohn's disease with onset in childhood and adolescence. Gut 25 (1984) 329–336.

Purcell, K.: Critical appraisal of psychosomatic studies of asthma. N. Y. St. J. Med. (1965) 2103.

Purrmann, J., B. Miller, G. Strohmeyer: Zur Ätiologie chronisch-entzündlicher Darmerkrankungen. Z. Gastroenterol. 24 (1986) 357–363.

Pyke, D. A.: Concordance of diabetes mellitus in MZ twins. In: Mimura, G., S. Baba, Y. Goto, J. Köbberling (eds.): Clinico-Genetic Genesis of Diabetes Mellitus. Excerpta Medica, Internat. Congress Series 597 (1982) 309–313.

Pyle, R. L., J. E. Mitchell, E. D. Eckert: Bulimia: a report of 34 cases. J. clin. Psychiat. 42 (1981) 60–64.

Quasthoff, U. M.: Eine interaktive Funktion von Erzählungen. In: Soeffner, H. G. (Hrsg.): Interpretative Verfahren in den Sozial- und Textwissenschaften, S. 105–126. Metzler, Stuttgart 1979.

Quasthoff-Hartmann, U. M.: Frageaktivitäten von Patienten in Visitengesprächen: Konversationstechnische und diskursstrukturelle Bedingungen. In: Köhle, K., H.-H. Raspe (Hrsg.): Das Gespräch während der ärztlichen Visite, S. 70–101. Urban & Schwarzenberg, München 1982.

Querido, A.: Forecast and follow-up: an investigation into the clinical, social, and mental factors determining the results of hospital treatment. Brit. J. prevent. soc. Med. 13 (1959) 33–49.

Quindlen, E. A.: Neurosurgical approaches. In: De Vita jr., V. T., W. Hellman, S. A. Rosenberg (eds.): Cancer. Principles and Practice of Oncology, pp. 1666–1670. Lippincott, Philadelphia 1982.

Quinlan, D. M., C. P. Kimball, F. Osborne: The experience of open-heart surgery. Arch. gen. Psychiat. 31 (1974) 241.

Quint, H.: Die Hypertoniker in psychodynamischer Sicht. In: Eiff, A. W. v. (Hrsg.): Essentielle Hypertonie – Klinik, Psychophysiologie und Psychopathologie. Thieme, Stuttgart 1967.

Quinton, D., M. Rutter: Parenting behavior of children raised „in care". In: Nicol, A. R. (ed.): Longitudinal Studies in Child Psychology and Psychiatry. Wiley, Chichester 1984.

Raab, W.: Emotional and sensory stress factors in myocardial pathology. Amer. Heart J. 72 (1966) 538–564.

Rabavilas, A. D.: Electrodermal activity in low and high alexithymia neurotic patients. Psychother. and Psychosom. 47 (1987) 101–104.

Rabavilas, A. D., G. N. Christodoulou, J. Lappas, C. Perissaki, C. Stefanis: Relation of obsessional traits to anxiety in patients with ulcerative colitis. Psychother. and Psychosom. 33 (1980) 155–159.

Rabe, F.: Hysterische Anfälle bei Epilepsie. Nervenarzt 37 (1966) 141–150.

Rabe, F.: Die Kombination hysterischer und epileptischer Anfälle. Das Problem „Hysteroepilepsie" in neuer Sicht. Springer, Berlin-Heidelberg-New York 1970.

Rabe, F.: Hysterische Dämmerzustände. Differentialdiagnose gegenüber Status psychomotorius. In: Karbowski, K. (Hrsg.): Status psychomotoricus und seine Differentialdiagnose, S. 103–116. Huber, Bern-Stuttgart-Wien 1980.

Rabending, G., W. Röse: Besonderheiten der myasthenischen Krisen. In: Kuhn, E. (Hrsg.): Progressive Muskeldystrophie, Myotonie, Myasthenie, S. 383–387. Springer, Berlin-Heidelberg-New York 1966.

Rabiner, J. C., H. E. Wilner: Differential psychopathological and organic mental disorder at follow-up five years after coronary bypass and cardiac valvular surgery. In: Speidel, H., G. Rodewald (eds.): Psychic and Neurological Dysfunctions after Open-Heart Surgery. Thieme, Stuttgart 1980.

Rabiner, J. C., H. E. Wilner, G. Fishman: Psychiatric complications following bypass surgery. J. nerv. ment. Dis. 160 (1975) 342–348.

Rabinowitz, S. H., R. L. Verrier, B. Lown: Muscarinic effect of vagosympathetic trunk stimulation on the repetitive extrasystole threshold. Circulation 53 (1976) 622–672.

Rabinowitz, S. H., B. Lown: Central neurochemical factors related to serotonine metabolism and cardiac ventricular vulnerability for repetitive electrical activity. Amer. J. Cardiol. 41 (1978) 516–522.

Rabkin, S. W., F. Mathewson, R. B. Tate: Chronobiology of cardiac sudden death in men. J. Amer. med. Ass. 244 (1980) 1357–1358.

Racamier, P. C.: Le terrain psychologique des tuberculeux pulmonaires. Chantenay, Paris 1950.

Rad, G. v.: Das Erste Buch Mose, Genesis. In: Das Alte Testament, deutsch. Teilband 2/4, 10. Aufl. Vandenhoeck & Ruprecht, Göttingen 1976.

Rad, M. v.: Alexithymie: Empirische Untersuchungen zur Diagnostik und Therapie psychosomatisch Kranker. Springer, Berlin-Heidelberg-New York 1983.

Rad, M. v., F. Lolas: Psychosomatische und psychoneurotische Patienten im Vergleich. Psyche 32 (1978) 956–973.

Rad, M. v., S. Zepf: Psychoanalytische Konzepte psychosomatischer Symptom- und Strukturbildung. In: Uexküll, Th. v. (Hrsg.): Lehrbuch der Psychosomatischen Medizin, 3. Aufl. Urban & Schwarzenberg, München-Wien-Baltimore 1986.

Rad, M. v., M. Drücke, W. Knauss, F. Lolas: Alexithymia: Anxiety and hostility in psychosomatic and psychoneurotic patients. Psychother. and Psychosom. 31 (1979a) 223–234.

Rad, M. v., M. Drücke, W. Knauss, F. Lolas: A comparative study of verbal behavior in psychosomatic and psychoneurotic patients. In: Gottschalk, L. A. (ed.): The Content Analysis of Verbal Behavior, pp. 643–674. New York 1979b.

Radanov, B.: Schwindel mit dem Schwindel? Psychiatrische Aspekte. Ther. Umschau 41 (1984) 75.

Radebold, H.: Der psychotherapeutische Zugang zu Patienten mit einer cerebralen Gefäßinsuffizienz. Z. präklin. Geriat. 2 (1972) 195–200.

Radebold, H.: Der psychoanalytische Zugang zu dem älteren und alten Menschen. In: Petzold, H., E. Bubolz (Hrsg.): Psychotherapie mit alten Menschen. Junfermann, Paderborn 1979.

Radebold, H.: Möglichkeiten und Einschränkungen von Behandlungsverfahren in den Versorgungssystemen Psychotherapie/Psychosomatik und Soziale Therapie. Z. Geront. 12 (1979) 149–155.

Radebold, H.: Psychotherapeutische Möglichkeiten im höheren und hohen Lebensalter. In: Mester, H., R. Tölle (Hrsg.): Neurosen, S. 146–152. Springer, Berlin-Heidelberg-New York 1981.

Radebold, H.: Neurotische, reaktive und psychosomatische Erkrankungen. In: Platt, D., K. Oesterreich (Hrsg.): Handbuch der Gerontologie, Bd. 5: Neurologie, Psychiatrie. Fischer, Stuttgart-New York 1989a.

Radebold, H.: Psychotherapie. In: Kisker, K. P., H. Lauter, J. E. Meyer, C. Müller, E. Strömgren (Hrsg.): Psychiatrie der Gegenwart, 3. Aufl., Bd. 8: Alterspsychiatrie. Springer, Berlin-Heidelberg-New York 1989b.

Radebold, H.: Gibt es die „neuen Alten"? Eine zusammenfassende Einschätzung. In: Karl, F., W. Tokarski (Hrsg.): Die „neuen Alten". Beiträge der XVII. Jahrestagung der Dtsch. Ges. f. Gerontologie 1988. Kasseler Gerontol. Schriften 6, Gesamthochschulbibliothek Kassel 1989c.

Radebold, H.: Altern und Alter – psychosomatischer Forschungsstand. In: Speidel, H., B. Strauss (Hrsg.): Zukunftsaufgaben der psychosomatischen Medizin. Springer, Berlin-Heidelberg 1989d.

Radebold, H.: Psycho- und soziotherapeutische Behandlungsverfahren. In: Platt, D., K. Oesterreich (Hrsg.): Handbuch der Gerontologie, Bd. 5: Neurologie, Psychiatrie. Fischer, Stuttgart-New York 1989e.

Radebold, H., G. Schlesinger-Kipp: Gruppenpsychotherapie und Gruppenarbeit im Alter. Ein Literaturbericht. In: Radebold, H. (Hrsg.): Gruppenpsychotherapie im Alter. Vandenhoeck & Ruprecht, Göttingen 1983.

Radebold, H., M. Rassek: Zur Psychotherapie. Psychosomatische Syndrome bei alten Menschen. In: Bergener, M., B. Kark (Hrsg.): Psychosomatik in der Geriatrie. Steinkopff, Darmstadt 1985.

Radebold, H., H. Bechtler, I. Pina: Psychosoziale Arbeit mit älteren Menschen. Lambertus, Freiburg 1973.

Radebold, H., H. Bechtler, I. Pina: Therapeutische Arbeit mit älteren Menschen. Lambertus, Freiburg 1981.

Radvila, A.: Das Hyperventilationssyndrom. Schweiz. med. Wschr. 114 (1984) 562–565.

Radvila, A., H. P. Bruggisser, D. Hess: Kutan gemessenes pCO_2 beim Hyperventilationssyndrom. Schweiz. med. Wschr. 113 (1983) 1943.

Radvila, A., K. H. Adler, R. L. Galeazzi, H. Vorkauf: The development of a german language (Berne). Pain questionnaire and its application in a situation causing acute pain. Pain 28 (1987) 185–195.

Radvila, A., M. M. Reich, U. Wipraechtiger, R. Zenhaeusern, H. Vorkauf, P. Huber, R. H. Adler: Prospektive Evaluation von Rückenschmerzen bei pathologischem und normalem CT-Befund. Jahreskongreß Ges. zum Studium des Schmerzes, Basel 6.–8. 10. 1988. (Abstract).

Raedler, A., W.-H. Schmiegel, H.-G. Thiele: Sind Morbus Crohn und Colitis ulcerosa Folgen einer immunregulatorischen Störung? Immun. Infekt. 10 (1982) 175–179.

Raether, M.: Über psychogene „Ischias-Rheumatismus"- und „Wirbelsäulenerkrankungen". Arch. Psychiat. 57 (1917) 772–791.

Raetzke, P.: Einfluß von Stress auf die oralen Strukturen. In: Schneller, T., A. Fleischer-Peters (Hrsg.): Anwendung psychologischer Methoden in der Zahnmedizin, S. 173. Fachbuchhandlung für Psychologie, Frankfurt 1985.

Ragland, D. R., R. J. Brand: Type A behavior and mortality from coronary heart disease. New Engl. J. Med. 318 (1988a) 65–69.

Ragland, D. R., R. J. Brand: Coronary heart disease mortality in the Western Collaborative Group Study. Follow-up experiences of 22 years. Amer. J. Epidemiol. 127 (1988b) 462–475.

Rahe, R. H.: Group therapy in the outpatient management of postmyocardial infarction patients. Int. J. Psychiat. Med. 4 (1973) 77–88.

Rahe, R. H.: A liaison psychiatrist on the coronary care unit. In: Pasnau, R. O. (ed.): Consultation Liaison Psychiatry, pp. 115–122. Grune & Stratton, New York 1975.

Rahe, R. H.: Life change and illness studies: past history and future directions. J. hum. Stress 4 (1978) 3–14.

Rahe, R. H.: Life change measurement clarification. Psychosom. Med. 40 (1978) 95–98.

Rahe, R. H., R. J. Arthur: Life-change patterns surrounding illness experience. J. psychosom. Res. 11 (1968) 341–345.

Rahe, R. H., J. Paasiviki: Psychosocial factors and myocardial infarction. J. psychosom. Res. 15 (1971) 33–39.

Rahe, R. H., J. L. Mahan, R. J. Arthur: Prediction of near-future health change from subjects preceding life change. J. psychosom. Res. 14 (1970) 401–406.

Rahe, R. H., C. Scalzi, K. Shine: A teaching evaluation questionnaire for postmyocardial infarction patients. Heart and Lung 4 (1975b) 759–766.

Rahe, R. H., H. W. Ward, V. Hayes: Brief group therapy in myocardial infarction rehabilitation. Three to four year follow-up of a controlled trial. Psychosom. Med. 41 (1979) 229–241.

Rahe, R. H., M. Romo, L. Bennet, P. Siltanen: Recent life changes, myocardial infarction and abrupt coronary death. Arch. intern. Med. 133 (1974) 221–228.

Rahe, R. H., T. O. O'Neill, A. Hagan, R. J. Arthur: Brief group therapy following myocardial infarction: 18 months follow-up of a controlled trial. Int. J. Psychiat. Med. 6 (1975a) 349–358.

Rajka, G.: Atopic dermatitis. Saunders, London 1975.

Rakic, P.: Mode of cell migration of the superficial layers of fetal monkey neocortex. J. comp. Neurol. 145 (1972) 61–84.

Rakoff, V.: Multiple determinants of family dynamics in anorexia nervosa. In: Darby, P. L., P. E. Garfinkel, D. M. Garner, D. V. Coscina (eds.): Anorexia nervosa. Recent Developments in Research, pp. 29–40. Liss, New York 1983.

Ramazzini, B.: De morbis artificium diatriba (1700). Die Krankheiten der Künstler und Handwerker und die Mittel, sich vor denselben zu beschützen. Neubearbeitung von P. Patissier; Übersetzung aus dem Französischen von G. Schlegel, Ilmenau 1823.

Rambaugh, D. M.: The cardiac adjustment scale. Educational and Industrial Testing Service, San Diego/Calif. 1964a.

Rambaugh, D. M.: Prediction of work potential in heart patients through the use of the cardiac adjustment scale. J. cons. Psychol. 29 (1964b) 597.

Rambaugh, D. M.: The psychological aspects. J. Rehab. 32 (1966) 56–58.

Rammstedt, O.: Gewalt. In: Lexikon zur Soziologie, S. 247f. Westdeutscher Verlag, Opladen 1973.

Rangell, L.: Die Konversion. Psyche 23 (1969) 121–147.

Rangell, L.: The nature of conversion. J. Amer. psychoanal. Ass. 7 (1959). Deutsch in: Overbeck, G., A. Overbeck (Hrsg.): Seelischer Konflikt – körperliches Leiden, S. 17–45. Hamburg 1978.

Ransom, D. C.: Research on the family in health, illness and care – state of the art. Family Systems Medicine 4 (1986) 329–336.

Ranson, S. W.: The hypothalamus: its significance for visceral innervation and emotional expression. Trans. Coll. Physicians Philad. Ser. 4 (1934) 222–242.

Literatur

Rapp, J. P., K. Knudsen, J. Iwai, L. K. Dahl: Genetic control of blood pressure and corticosteroid production in rats. Circulat. Res. 32 and 22 (Suppl. 1) (1973) 139.

Rappaport, J.: Organisation and pathology of thought (1951). 2nd ed. New York 1956.

Raskin, M., J. A. Talbott, A. T. Meyerson: Diagnosis of conversion reactions. Predictive value of psychiatric criteria. J. Amer. med. Ass. 197 (1966) 530–534.

Raskin, N. H., O. Appenzeller: Headache. Saunders, Philadelphia 1980.

Raskind, M. A., S. C. Risse, T. H. Lampe: Dementia and antipsychotic drugs. J. clin. Psychiat. 48 (1987) 16–18.

Rasmussen, A. F., J. T. Marsh, N. Q. Brill: Increased susceptibility to Herpes simplex in mice subjected to avoidance-learning stress of restraint. Proc. Soc. exp. Biol. (N. Y.) 96 (1957) 183.

Rasmussen, O. C., F. Bonde-Petersen, L. V. Christensen, E. Moller: Blood flow in human mandibular levators at rest and during controlled biting. Arch. oral. Biol. 22 (1977) 539–543.

Raspe, H.-H.: Informationsbedürfnisse und faktische Informiertheit bei Krankenhauspatienten. In: Begemann, H. (Hrsg.): Patient und Krankenhaus, S. 40–70. Urban & Schwarzenberg, München 1976.

Raspe, H.-H.: Informationsbedürfnisse und faktische Informiertheit bei Krankenhauspatienten. Med. Klin. 71 (1976) 1016–1020.

Raspe, H.-H.: Informationsbedürfnisse von Patienten. Aufklärungsintentionen von Ärzten im Akutkrankenhaus. Med. Welt 28/49 (1977) 1990–1993.

Raspe, H.-H.: Warum fragen Krankenhauspatienten so wenig? Eine medizinsoziologische Untersuchung der Stationsarztvisite. Therapiewoche 3 (1980) 560–573.

Raspe, H.-H.: Zur Aufklärung des Patienten – zwischen Aufklärungspflicht und Pflicht zur Schonung und Barmherzigkeit. Vortragsmanuskript, Ulm 1982.

Raspe, H.-H.: Psychosoziale Probleme im Verlauf einer chronischen Polyarthritis. Int. Welt 9 (1982) 193–203.

Raspe, H.-H.: Visitenforschung in der Bundesrepublik: Historische Reminiszenzen und Ergebnisse formal-quantitativer Analysen. In: Köhle, K., H.-H. Raspe (Hrsg.): Das Gespräch während der ärztlichen Visite, S. 1–15. Urban & Schwarzenberg, München 1982.

Raspe, H.-H.: Aufklärung und Information im Krankenhaus. Vandenhoeck & Ruprecht, Göttingen 1983.

Raspe, H.-H.: Social and emotional problems in early rheumatoid arthritis. 75 patients followed-up for two years. Clin. Rheum. (Suppl.) 2 (1987) 20–26.

Raspe, H.-H.: Die chronische Polyarthritis aus psychosomatischer Sicht unter besonderer Berücksichtigung epidemiologischer und soziologischer Zusammenhänge. In: Klußmann, R., M. Schattenkirchner (Hrsg.): Der Schmerz- und Rheumakranke, S. 36–47. Springer, Heidelberg 1989.

Raspe, H.-H.: Psychosoziologische Diagnostik bei chronisch-rheumatischen Erkrankungen. In: Zeidler, H. (Hrsg.): Innere Medizin der Gegenwart. Rheumatologie. Urban & Schwarzenberg, München (im Druck).

Raspe, H.-H., S. Mattussek: Magische Vorstellungen zwischen Arzt und Patient in der Rheumatologie. Fortbildungskurse Rheumatologie 7 (1985) 41–64.

Raspe, H.-H., N. Ritter: Laientheorien, paramedizinische Behandlung und subjektive Medikamentencompliance bei Patienten mit einer chronischen Polyarthritis. Verh. Dt. Ges. Inn. Med. 88 (1982) 1200–1204.

Raspe, H.-H., J. Siegrist: Zur Gestaltung der Arzt-Patient-Beziehung im stationären Bereich. In: Siegrist, J. (Hrsg.): Wege zum Arzt, S. 113–138. Urban & Schwarzenberg, München 1979.

Raspe, H.-H., S. Mattussek, R. Scheiblich: Zur sozialen Isolation von Patienten mit einer chronischen Polyarthritis. Med. Klin. 78 (1983) 60–67.

Raspe, H.-H., A. Vorbeck, M. Robin-Winn: Sonderstellung der seronegativen chronischen Polyarthritis? Dtsch. med. Wschr. 111 (1986) 1474–1478.

Rathmann-Kessel, J., K. F. Masuhr: Gruppenpsychotherapie mit anfallskranken Jugendlichen während der beruflichen Erstausbildung. Rehabilitation 23 (1984) 19–21.

Rauchfleisch, U., R. Schuppli, T. Haenel: Zur Persönlichkeit von Patienten mit dermatologischen Artefakten. Z. psychosom. Med. Psychoanal. 29 (1983) 76–84.

Rausch, F.: Die Rheumatismusdiagnose in der ärztlichen Praxis. Europaeum Medicum Collegium, München 1967.

Ravenscroft, K.: Psychoanalytic family therapy approaches to the adolescent bulimic. In: Schwartz, H. J. (ed.): Bulimia: Psychoanalytic Treatment and Theory. Int. Univ. Press, Madison 1988.

Rayfield, E. J., J.-W. Yoon: The viral etiology of diabetes. In: Brodoff, B. N., S. J. Bleicher (eds.): Diabetes Mellitus and Obesity, pp. 381–386. Williams & Wilkins, Baltimore-London 1982.

Razin, A. M.: Psychosocial intervention in coronary artery disease: a review. Psychosom. Med. 44 (1982) 363–387.

Razran, G.: Mind in evolution: an east-west synthesis of learned behavior and cognition. Houghton Mifflin, Boston 1971.

Razran, G.: Semantic and phonetographic generalizations of salivary conditioning to verbal stimuli. J. exp. Psychol. 39 (1949) 642–652.

Rea, M. P., S. Greenspoon, B. Spilka: Physicians and the terminal patient: some selected attitudes and behavior. Omega 7 (1976) 291–302.

Reader, G. G., L. Pratt, M. C. Mudd: What patients expect from their doctors. Modern Hospital 89 (1957) 88–94.

Reading, A. E., J. R. Newton: Psychological factors in IUD use. A review. J. biosoc. Sci. 9 (1977) 317.

Rechardt, E.: An investigation in the psychosomatic aspects of prurigo Besnier. Kunallispaino, Helsinki 1970.

Rechenberger, I.: Der Übergang vom körperlichen zum psychischen Selbstverständnis des psychosomatisch Kranken. Prax. Psychother. Psychosom. 25 (1980) 173–178.

Rechenberger, I.: Körperlich und seelisch bedingte Hautveränderungen in der kosmetischen Sprechstunde. Ärztl. Kosmetologie 8 (1978) 239–241.

Rechenberger, I.: Tiefenpsychologisch ausgerichtete Diagnostik und Behandlung von Hautkrankheiten. Vandenhoeck & Ruprecht, Göttingen 1976.

Rechenberger, I.: Zur Psychodynamik des Psoriasiskranken. Akt. Dermatol. 8 (1982) 157–159.

Rechlin, E.: Psychologische Aspekte der Enteritis regionalis Crohn. Hausarbeit Dipl. Psych., Kiel 1977.

Rees, L.: An appraisal of the concept of cardiac neuroses. Acta psychother. 15 (1963) 673.

Rees, L.: Physical and emotional factors in bronchial asthma. J. psychosom. Res. 1 (1956) 1.

Rees, L.: Psychosomatic aspects of asthma in elderly patients. Psychosom. Res. 1 (1956) 212–218.

Rees, L.: The development of psychosomatic medicine during the past 25 years. J. psychosom. Res. 27 (1983) 157–164.

Rees, W. D., S. G. Lutkins: Mortality of bereavement. Brit. med. J. 4 (1967) 13.

Regestein, Q.: Sleep and insomnia in the elderly. J. geriat. Psychiat. 13 (1980) 153–171.

Regier, D. A., J. H. Boyd, J. D. Burke jr., D. S. Rae, J. K. Myers, M. Kramer, L. N. Robins, L. K. George, M. Karno,

B. Z. Locke: One-month prevalence of mental disorders in the United States. Based on five epidemiological catchment area sites. Arch. gen. Psychiat. 45 (1988) 977–986.

Regier, D. A., J. K. Myers, M. Kramer, L. N. Robins, D. G. Blazer, R. L. Hough, W. W. Eaton, B. Z. Locke: The NIMH epidemiologic catchments area program. Arch. gen. Psychiat. 41 (1984) 934–941.

Reich, P., M. J. Kelly: Suicide attempts by hospitalized medical and surgical patients. New Engl. J. Med. 294 (1976) 298–301.

Reich, P., P. W. Gold: Interruption of recurrent ventricular fibrillation by psychiatric intervention. Gen. Hosp. Psychiat. 5 (1983) 255–257.

Reich, P., Q. Regestein, B. Murawski, R. DeSilva, B. Lown: Unrecognized organic mental disorders in survivors of cardiac arrest. Amer. J. Psychiat. 140 (1983a) 1194–1197.

Reich, P., R. A. DeSilva, B. Lown, B. J. Murawski: Acute psychological disturbances preceding life-threatening ventricular arrhythmias. J. Amer. med. Ass. 246 (1981) 233–235.

Reich, W. (1933): Charakteranalyse. Kiepenheuer & Witsch, Köln 1970.

Reich, W.: Der psychogene Tic als Onanieäquivalent. Z. Sexualwiss. 11 (1925) 302–313.

Reich, W.: Die Funktion des Orgasmus. Kiepenheuer & Witsch, Köln 1969.

Reich, W.: Über den epileptischen Anfall. Int. Z. Psychoanal. 17 (1931) 263–275.

Reiche, R.: AIDS im individuellen und kollektiven Unbewußten. Z. Sexualforsch. 1 (1988) 113–122.

Reichenbak, D. D., N. S. Moss, E. Meyer: Pathology of the heart in sudden cardiac death. Amer. J. Cardiol. 39 (1977) 865.

Reichlin, S., R. J. Baldessarini, J. B. Martin (eds.): The hypothalamus. Res. publ. Ass. Res. nerv. ment. Dis. 56 (1978) 1–14.

Reichsman, F., N. B. Levy: Problems in adaptation to maintenance hemodialysis. In: Moos, R. H. (ed.): Coping with Physical Illness, pp. 311–328. Plenum, New York 1977.

Reichsman, F., N. B. Levy: Problems in adaptation to maintenance hemodialysis: a 4-year study of 25 patients. In: Levy, N. B. (ed.): Living or Dying. Thomas, Springfield/Ill. 1974.

Reid, D. D., W. Holland, S. Humerfeldt, G. A. Rose: Cardiovascular survey of British postal workers. Lancet 1 (1966) 614–618.

Reilly, F. D., P. A. McCuskey, M. L. Miller, R. S. McCuskey, H. A. Meineke: Innervation of the periarteriolar lymphatic sheath of spleen. Tissue and Cell 11 (1979) 121–126.

Reimann, H. A.: Periodic fever, an entity. A collection of 52 cases. Amer. J. med. Sci. 243 (1962) 162.

Reimer, C., L. Hempfing, B. Dahme: Iatrogene Chronifizierung in der Vorbehandlung psychogener Erkrankungen. Praxis Psychother. Psychosom. 24 (1979) 123–133.

Reindell, A., H. Ferner: Psychosomatische Aspekte zur Langzeitbehandlung der Colitis ulcerosa und granulomatosa. Therapiewoche 29 (1979) 6307–6313.

Reindell, A., H. Ferner, K. Gmelin: Zur psychosomatischen Differenzierung zwischen Colitis ulcerosa und Ileitis terminalis (M. Crohn). Z. psychosom. Med. 27 (1981) 358–371.

Reindell, H., H. Klepzig, H. Roskamp: Krankheiten des Herzens und des Kreislaufs. In: Kühn, A. (Hrsg.): Innere Medizin, Bd I. Springer, Berlin 1971.

Reinhardt and Gray, 1972: zit. n. Rezler, A. G.: Medical student's attitudes, 1970–1980; AMEE Pre-conference workshop on attitudinal aspects of medical education, London 1982 (unpublished).

Reinhart, J., R. Succop: Regional enteritis in pediatric patients. J. Amer. Acad. Child Psychiat. 71 (1968) 252–281.

Reis, D. J., J. E. LeDoux: Some central neural mechanisms governing resting and behaviorally coupled control of blood pressure. Circulation 76 (1987) 1–2.

Reischauer, F.: Wirbelsäulen- und Bandscheibenschäden. Röntgenbild und Wirklichkeit in der Therapie. Therapiewoche 8 (1957) 130–139.

Reiser, M. F.: Are psychiatric educators „losing the mind"? Amer. J. Psychiat. 145 (1988) 148–153.

Reiser, M. F.: Mind, brain, body. Towards a convergence of psychoanalysis and neurobiology. New York 1984.

Reiser, M. F., A. A. Brust, A. P. Shapiro, H. M. Baker, W. Rauschoff, E. B. Ferris: Life situations, emotions and the course of patients with arterial hypertension. Proc. Ass. Res. nerv. ment. Dis. 29 (1950) 870.

Reisine, D. T.: Molecular mechanisms controlling ACTH release. In: Weiner, H., D. Hellhammer, I. Florin, R. C. Murison (eds.): Neuronal Control of Bodily Function: Basic and Clinical Aspects. Vol. IV: Frontiers of Stress Research. Huber, Toronto 1989.

Reister, G. et al.: Zum Verlauf psychischer Gesundheit und psychogener Störungen. Weitere Ergebnisse des Mannheimer Kohortenprojekts. Z. psychosom. Med. (in Vorbereitung).

Rekola, J. K.: Rheumatoid arthritis and the family. Suppl. 3. Keskuskirjapaino, Helsinki 1973.

Relinger, H., P. H. Bornstein, D. M. Mungas: Treatment of insomnia by paradoxical intention: a time-series analysis. Behav. Ther. 9 (1978) 955–959.

Remschmidt, H., B. Herpertz-Dahlmann: Bulimia und Bulimarexie im Jugendalter. In: Fichter, M. M. (Hrsg.): Bulimia nervosa. Enke, Stuttgart 1989.

Remschmidt, H., F. Wienand, C. Wewetzer: Der Langzeitverlauf der Anorexia nervosa. Monatsschr. Kinderheilk. 136 (1988) 726–731.

Renaer, M. (ed.): Chronic pelvic pain in women. Springer, Berlin 1981.

Renaer, M.: Chronic pelvic pain without obvious pathology in women. Eur. J. Obstet. Gynec. reprod. Biol. 10 (1980) 415.

Renggli, F.: Angst und Geborgenheit. Soziokulturelle Folgen der Mutter-Kind-Beziehung im ersten Lebensjahr. Rowohlt, Reinbek 1974.

Rentrop, E.: Die Bedeutung der Penisprothese im Paarkonflikt. Materialien Psychoanalyse 9 (1983) 70.

Rentrop, E., M. Straschill: Der Einfluß emotionaler Faktoren beim Auftreten des idiopathischen Torticollis spasmodicus. Z. psychosom. Med. 32 (1986) 44–59.

Rentrop, E., M. Straschill: Über die Wirkung emotionaler Einflüsse auf den Verlauf des idiopathischen spasmodischen Torticollis. Z. psychosom. Med. 33 (1987) 42–51.

Rescorla, R. A.: Pavlovian conditioning: it's not what you think it is. Amer. Psychol. 43 (1988) 151–160.

Rescorla, R. A.: Pavlovian second order conditioning: studies in associative learning. Erlbaum, Hillsdale 1980.

Rescorla, R. A.: Some implications of a cognitive perspective on Pavlovian conditioning. In: Hulse, S. S., H. Fowler, W. K. Honig (eds.): Cognitive Processes in Animal Behavior. Erlbaum, Hillsdale 1978.

Rescorla, R. A., D. R. Furrow: Stimulus similarity as a determinant of Pavlovian conditioning. J exp. Psychol. Anim. Beh. 3 (1977) 203–215.

Rescorla, R. A., P. C. Holland: Behavioral studies of associative learning in animals. In: Rosenzweig, M. R., L. W. Porter (eds.): Annual Review of Psychology, Vol. 33. Annual Reviews, Palo Alto 1982.

Rescorla, R. A., R. L. Solomon: Two-process learning theory: relationship between pavlovian conditioning and instrumental learning. Psychol. Rev. 74 (1967) 151–182.

Rettori, V., J. Jurcovicova, S. M. McCann: Central action of interleukin-1 in altering the release of TSH, growth hormone, and prolactin in the male rat. J. Neurosci. Res. 18 (1987) 179–183.

Reubi, J. C.: Biochemie des Schmerzes. Über einige Neurotransmitter des Schmerzsystems: Biochemische Aspekte und Interaktionen. In: Kocher, R., D. Gross, H. E. Kaeser (Hrsg.): Nacken-Schulter-Arm-Syndrom. S. 294–297. Fischer, Stuttgart 1980.

Reuler, J. B., D. E. Girard, D. A. Nardone: The chronic pain syndrom: misconceptions and management. Ann. intern. Med 93 (1980) 588–596.

Reveley, A. M., R. M. Murray: Cerebral ventricular enlargement in nongenetic schizophrenia: a controlled twin study. Brit. J. Psychiat. 144 (1984) 89–93.

Revusky, S. H.: The role of interference in association over a delay. In: Honig, W. K., P. H. R. James (eds.): Animal memory. Acad. Press, New York 1971.

Revusky, S. H.: Drug interactions measured through taste aversion procedures with an emphasis on medical implications. Ann. N. Y. Acad. Sci. 443 (1985) 250–271.

Revusky, S. H., E. W. Bedarf: Association of illness with prior ingestion of novel foods. In: Seligman, M. E. P., J. L. Hager (eds.): Biological Boundaries of Learning. Appleton Century Crofts, New York 1972.

Rey, E. R.: Allgemeine Probleme psychologischer Tests. In: Strube, G. (Hrsg.): Die Psychologie des 20. Jahrhunderts, Bd. V: Binet und die Folgen. Testverfahren, Differentielle Psychologie, Persönlichkeitsforschung. Kindler, Zürich 1977.

Reynolds, C. W., J. R. Ortaldo: Natural killer activity: the definition of a function rather than a cell type. Immunol. Today 8 (1987) 172–174.

Reynolds, D. V.: Surgery in the rat during electrical analgesia induced by focal brain stimulation. Science 164 (1969) 444–445.

Rezler, A. G.: Medical student's attitudes, 1970–1980; AMEE Pre-conference workshop on attitudinal aspects of medical education, London 1982 (unpublished).

Rezler, A. G.: The assessment of attitudes. In: WHO: Public Health Paper 52 (Development of Educational Programmes for the Health Professions, Geneva 1973).

Ribble, M. A.: Ego dangers and epilepsy. Psychoanal. Quart. 5 (1936) 71–86.

Ribble, M. A.: Disorganizing factors in infant personality. Amer. J. Psychiat. 98 (1941) 459.

Ribble, M. A.: Infantile experience in relation to personality development. In: Hunt, J. M. V. (ed.): Personality and the Behavior Disorders. Ronald, New York 1944.

Rich, A. R., T. H. Cochran, D. C. McGoon: Marked lipemia resulting from the administration of cortisone. Johns Hopkins med. J. 88 (1951) 101–109.

Richards, D. H.: Depression after hysterectomy. Lancet 2 (1973) 430.

Richardson, H. B.: Patients have families. Common Wealth Found, New York 1948.

Richman, D. P., B. G. Arnason: Nicotinic acetylcholine receptor: evidence for a functionally distinct receptor on human lymphocytes. Proc. Natl. Acad. Sci. (Wash.) 76 (1979) 4632–4635.

Richmond, J. B.: Relationship of the psychiatric unit to other departments of the hospital. In: Kaufman, M. R. (ed.): The Psychiatric Unit in a General Hospital. Int. Univ. Press, New York 1965.

Richter, C. P.: On the phenomenon of sudden death in animals and man. Psychosom. Med. 19 (1957) 91.

Richter, D.: Diagnostik und Psychodynamik beim Pelipathiesyndrom. Vortr. 1. Seminar Univ. Frauenklinik ü. Psychosomatik in Geburtshilfe und Gynäkologie. Düsseldorf 1979.

Richter, D.: Die Adnexitis aus psychosomatischer Sicht. In: Prill, H. J., D. Langen (Hrsg.): Der psychosomatische Weg zur gynäkologischen Praxis. Schattauer, Stuttgart 1983.

Richter, D.: Die Adnexitis aus psychosomatischer Sicht. Therapiewoche 28 (1978) 9508.

Richter, D.: Geburtsvorbereitung – eine präventiv-psychologische Aufgabe familienorientierter Geburtshilfe. Therapiewoche 30 (1980) 612.

Richter, D.: Geburtsvorbereitung – präventiv-psychohygienische Aufgabe familienorientierter Geburtshilfe. In: Hillemanns, H. G., H. Steiner, D. Richter (Hrsg.): Die humane, familienorientierte und sichere Geburt. Thieme, Stuttgart 1983.

Richter, D.: Psychoanalytic differential diagnosis of the different neurotic disturbances in patients with pelvic pain and adnexitis. In: Carenza, L., L. Zichella (eds.): Emotion and Reproduction. 5th Int. Congr. Psychosomatic Obstet. Gynec. Acad. Press, London 1979.

Richter, D.: Psychologische Geburtserleichterung. In: Beck, L., H. Albrecht (Hrsg.): Analgesie und Anästhesie in der Geburtshilfe. Thieme, Stuttgart 1982.

Richter, D.: Psychosomatisch und endokrinologisch orientierte Diagnostik und Therapie des Sekundäre-Amenorrhoe-Syndroms. Behandlungsergebnisse von 100 Amenorrhoe-Patientinnen. Gynäk. 15 (1982) 173.

Richter, D.: Psychosomatisch und endokrinologisch orientierte Diagnostik und Therapie des Sekundäre-Amenorrhoe-Syndroms. Behandlungsergebnisse von 100 Amenorrhoe-Patientinnen. Habilitationsschrift, Freiburg 1980.

Richter, D.: Psychosomatische Aspekte während der Behandlung gynäkologischer Malignom-Patientinnen. In: Wannenmacher, M. (Hrsg.): Kombinierte chirurgische und radiologische Therapie maligner Tumoren. Urban & Schwarzenberg, München 1981.

Richter, D.: Psychosomatische Differentialdiagnose des Pelipathie-Syndroms und der Adnexitis. In: Oeter, K., M. Wilken (Hrsg.): Frau und Medizin. Hippokrates, Stuttgart 1979.

Richter, D.: Recurrence of adnexitis-specific conflict. In: Dennerstein, L., M. de Senarclens (eds.): The Young Woman. Psychosomatic Aspects of Obstetrics and Gynecology. Excerpta Medica 618 (1983) 65.

Richter, D.: Schwangeren- und Elternberatung aus der Sicht des ungeborenen Kindes. In: Hau, T. F., S. Schindler (Hrsg.): Pränatale und perinatale Psychosomatik. Hippokrates, Stuttgart 1982.

Richter, D.: Schwangerschaft und Sexualität. Diagnostik 11 (1978) 423; 487.

Richter, D.: Was bedeutet Geburtsvorbereitung aus psychosomatischer Sicht? In: Richter, D., M. Stauber (Hrsg.): Psychosomatische Probleme in Geburtshilfe und Gynäkologie. Kehrer, Freiburg 1983.

Richter, D.: Was bedeutet umfassende Geburtsvorbereitung? In: Prill, H. J., D. Langen (Hrsg.): Der psychosomatische Weg zur gynäkologischen Praxis. Schattauer, Stuttgart 1983.

Richter, D., F. Peters, M. Breckwoldt: Psychosomatische und endokrinologische Befunde bei sekundärer Amenorrhoe. Arch. Gynäk. 224 (1977) 489.

Richter, H. E.: Der Krebs als psychisches Problem. Med. Welt 32 (1981) 177–184.

Richter, H. E.: Die Psychosoziale Arbeitsgemeinschaft Lahn-Dill. Erfahrungen mit einem selbstorganisierten Kooperationsmodell. In: Richter, H. E.: Engagierte Analysen, S. 201–241. Rowohlt, Hamburg 1978b.

Richter, H. E.: Eltern, Kind und Neurose. Klett, Stuttgart 1963. 2. Aufl. Rowohlt, Reinbek 1967.

Richter, H. E.: Ist psychosomatische Medizin überhaupt zu

verwirklichen? In: Richter, H. E.: Engagierte Analysen, S. 89–111. Rowohlt, Hamburg 1978 a.

Richter, H. E.: Lernziel Solidarität. Rowohlt, Reinbek 1979.

Richter, H. E.: Patient Familie. Entstehung, Struktur und Therapie von Konflikten in Ehe und Familie. Rowohlt, Reinbek 1970.

Richter, H. E.: Statement zur Sachverständigenanhörung der AIDS-Enquête-Kommission des Deutschen Bundestages am 29. 9. 87.

Richter, H. E.: Zur Psychodynamik der Herzneurose. Z. psychosom. Med. 10 (1964) 253.

Richter, H. E., D. Beckmann: Herzneurose, 2. Aufl. Thieme, Stuttgart 1973.

Richter, K., W. Pieringer, H. Mayer: Psychische Aspekte bei der Hysterektomie. Wien. klin. Wschr. 88 (1976) 733.

Richter, R.: Auslösung und Aufrechterhaltung des Asthmas durch psychologische Faktoren. In: Schultze-Werninghaus, G., M. Debelic (Hrsg.): Asthma. Grundlagen – Diagnostik – Therapie, S. 190–201. Springer, Berlin-Heidelberg 1988 b.

Richter, R.: Erfahrungen mit Asthmapatienten auf einer internistischen Station. In: Rechenberger, H.-G., H.-V. Werthmann (Hrsg.): Psychotherapie und Innere Medizin, S. 145–159. Pfeiffer, München 1988 a.

Richter, R.: Zur Psychophysiologie der obstruktiven Atemnot – Untersuchungen der Atemmuskelaktivität unter fluß-resistiver Atmung bei Gesunden und Asthmatikern. Habilitationsschrift Medizin, Universität Hamburg 1985.

Richter, R., B. Dahme, A. Kohlhaas: Bemühungen zu einer clusteranalytischen Taxonomie des Asthma bronchiale. Psychother. med. Psychol. 35 (1985) 320–328.

Richter, R., B. Dahme: Bronchial asthma in adults: there is little evidence for the effectiveness of behavioral therapy and relaxation. J. psychosom. Res. 26 (1982) 533–540.

Richter, R. et al.: Changes in total respiratory resistance during a noise-avoidance task. Biol. Psychol. 11 (1980) 280.

Richter, R., H. J. Freyberger, M. Bührig, H.-J. Schwedler, B. Dahme: Psychophysiologische Funktionsdiagnostik – Methodische Überlegungen am Beispiel der Herzneurose und des Asthma bronchiale. In: Tewes, U. (Hrsg.): Angewandte Medizinische Psychologie. Klotz, Frankfurt 1984.

Richter, R., H. J. Freyberger, M. Bührig, H.-J. Schwedler, B. Richter-Heinrich, V. Homuth, B. Heinrich, U. Knust, K. H. Schmidt, R. Wiedemann, H. R. Gohlke: Behavioral therapies in essential hypertensives: a controlled study. In: Elbert, T., W. Langosch, A. Steptoe, D. Vaitl (eds.): Behavioral medicine in cardiovascular disorders. Wiley, New York 1988.

Richter, R., S. Ahrens: Psychosomatische Aspekte der Allergie. In: Fuchs, E., K.-H. Schulz: Manuale allergologicum, Kap. VIII. Dustri, Frankfurt 1988.

Richter-Heinrich, E., U. Knust, W. Müller, K. H. Schmidt, H. Sprung: Psychophysiological investigations in essential hypertensives. J. psychosom. Res. 19 (1975) 251–258.

Richter-Heinrich, E., V. Homuth, B. Heinrich, U. Knust, K. H. Schmidt, R. Wiedemann, H. R. Gohlke: Behavioral therapies in hypertensives: a controlled study. In: Elbert, T., W. Langosch, A. Steptoe, D. Vaitl (eds.): Behavioral medicine in cardiovascular disorders. Wiley, Chichester-New York 1988.

Richter-Heinrich, E., V. Homuth, H. R. Gohlke, B. Heinrich, K. H. Schmidt, R. Wiedemann, H. Heine: Effectiveness of behavioral treatment methods compared to pharmacological therapy and self-recordings of blood pressure in essential hypertensives. Act. Nerv. Super. Suppl. 3 (1982) 422–427.

Rickels, K.: Benzodiazepines: use and misuse. In: Klein, D. F., F. G. Rabkin: Anxiety: New Research and Changing Concepts. Raven, New York 1981.

Rickels, K., E. E. Schweizer: Current pharmacotherapy of anxiety and panic. In: Meltzer, H. J. (ed.): Psychopharmacology: The Third Generation of Progress. Raven, New York 1987.

Rickels, K., G. Case, R. W. Downing, A. Winokur: Longterm diazepam therapy and clinical outcome. J. Amer. med. Ass. 250 (1983) 767–771.

Riedell, H., E. Brähler: Prostatitis und Ehepaarbeziehung. In: Brunner, H., W. Krause, C. F. Rothauge, W. Weidner (Hrsg.): Chronische Prostatitis, S. 273. Schattauer, Stuttgart 1983.

Riehl-Emde, A., C. Buddeberg, F. A. Muthny, E. Landolt-Ritter et al.: Ursachenattribution und Krankheitsbewältigung bei Patientinnen mit Mammacarcinom. Psychother., Psychosom., med. Psychol. 39 (1989) 232–238.

Riemann, J. F.: Further electron microscopic evidence of virus-like particles in Crohn's disease. Acta hepato-gastroent. 24 (1977 b) 116–118.

Riemann, J. F.: Ultrastrukturelle Befunde bei Ileocolitis Crohn. Z. Gastroenterol. Verh. Bd. 12 (1977 a) 238–239.

Riemann, J. F., L. Demling: Enteritis regionalis Crohn. Dtsch. med. Wschr. 104 (1979) 787–789.

Riemer, M. D.: Ileitis-underlying aggressive conflicts. N. Y. St. J. Med. 60 (1960) 552–557.

Riesman, D.: Freud und die Psychoanalyse. Frankfurt 1963.

Riess, B. F.: Genetic changes in semantic conditioning. J. exp. Psychol. 36 (1946) 143–152.

Rifkin, G. B.: The treatment of cardiac neurosis using systematic desensitization. Behav. Res. Ther. 6 (1968) 239–241.

Riklin, F.: Hebung epileptischer Amnesien durch Hypnose. J. Psychol. Neurol. 1 (1902) 200–225.

Riklin, F.: Zur Anwendung der Hypnose bei epileptischen Amnesien. J. Psychol. Neurol. 2 (1903/4) 28–30.

Riley, T. L., A. Roy: Pseudoseizures. Williams & Wilkins, Baltimore-London 1982.

Riley, V.: Psychoneuroendocrine influences on immunocompetence and neoplasia. Science 212 (1981) 1100–1109.

Riley, V., M. A. Fitzmaurice, D. H. Spackman: Psychoneuroimmunologic factors in neoplasia: studies in animals. In: Ader, R. (ed.): Psychoneuroimmunology, pp. 31–102. Acad. Press, New York 1981.

Rimón, R.: A psychosomatic approach to rheumatoid arthritis. Acta rheum. scand. 13 (Suppl.) (1969) 1–54.

Rimón, R.: Depression in rheumatoid arthritis. Ann. clin. Res. 6 (1974) 171–175.

Rimón, R.: Rheumatoid factor and aggression dynamics in female patients with rheumatoid arthritis. Scand. J. Rheum. 2 (1973) 119–122.

Rin, H., T.-Y. Lin: Mental illness among Formosan aborigines as compared with the Chinese in Taiwan. J. ment. Sci. 108 (1962) 134–146.

Rinaldo, J. A., P. Scheinok, C. E. Rupe.: Symptom diagnosis. A mathematical analysis of epigastric pain. Ann. intern. Med. 59 (1963) 145–154.

Ringel, E., W. Kropiunigg: Der fehlgeleitete Patient. Facultas, Wien 1984.

Rinsley, D. B.: Successful treatment of a case of ocular tic utilizing brief, intensive psychoanalytic psychotherapy. Bull. Menninger. Clin. 50 (1986) 447–455.

Risch, S. C., G. P. Groom, D. S. Janowsky: Interfaces of psychopharmacology and cardiology – part two. J. clin. Psychol. 42 (1981) 47–59.

Rish, B. L.: A critique of the surgical management of lumbar disc disease in a private neurosurgical practice. Spine 9 (1984) 500–504.

Rissanen, V., M. Romo, P. Siltanen: Premonitory symptoms and stress factors preceding sudden death from ischemic heart disease. Acta med. scand. 204 (1978) 389–396.

Ritchie, J. K., J. Powell-Tuck, J. E. Lennard-Jones: Clinical outcome of the first ten years of ulcerative colitis and proctitis. Lancet I (1978) 1140–1143.

Ritter, J. (Hrsg.): Historisches Wörterbuch der Philosophie. Schwabe, Basel 1971.

Roberts, A. H., L. Reinhard: The behavioural management of chronic pain: Longterm follow-up with comparison groups. Pain 8 (1980) 151–162.

Roberts, N., S. Bennett, R. Smith: Psychological factors associated with disability in arthritis. J. psychosom. Res. 30 (1986) 223–231.

Robertson, E. K., R. M. Suinn: The determination of rate of progress of stroke patients through empathy measures of patient and family. J. psychosom. Res. 12 (1968) 189–191.

Robins, L. N.: Discussions of chapters by Clausen, Wilder, and Srole. In: Erlenmeyer-Kinnling, L., N. E. Miller (eds.): Life-span Research on the Prediction of Psychopathology, pp. 239–246. Erlbaum, Hillsdale NJ 1986.

Robins, L. N.: Longitudinal methods in the study of normal and pathological development. In: Kisker, K. P., J. E. Meyer, C. Müller, E. Stroemgren (Hrsg.): Psychiatrie der Gegenwart I/1, S. 627–684. Springer, Berlin-Heidelberg-New York 1979.

Robinson, E. T., L. A. Hernandez, W. C. Dick, W. W. Buchanan: Depression in rheumatoid arthritis. J. roy. College Gen. Practitioners 27 (1977) 423–427.

Robinson, J. S.: Psychologische Auswirkungen der Intensivpflege (pers. Erfahrungsbericht). Anaesthesist 24 (1975) 416–418.

Robinson, P.: Störungen gastrointestinaler Funktionen bei bulimischen Eßstörungen. In: Fichter, M. M. (Hrsg.): Bulimia nervosa. Enke, Stuttgart 1989.

Robinson, P. H., N. L. Holden: Bulimia nervosa in the male: a report of nine cases. Psychol. Med. 16 (1986) 795–803.

Robison, J. C., N. Gitlin, H. F. Morrelli, L. J. Mann: Factitious hyperamylasuria: A trap in the diagnosis of pancreatitis. New Engl. J. Med. 306 (1982) 1211–1212.

Rodenberg, L.v.: Psychische Faktoren bei einigen motorischen Störungen (Tic, Torticollis, Schreibkrampf, Tremor, allgemeine motorische Unruhe, Gangstörungen). Z. psychosom. Med. 8 (1962) 1–11; 77–94.

Rodin, G. M., D. Daneman, L. E. Johnson, A. Kenshole, P. Garfinkel: Anorexia nervosa and bulimia in female adolescents with insulin-dependent diabetes mellitus: a systematic study. J. psychiat. Res. 19 (1985) 381–384.

Roehrs, T., G. Vogel, F. Vogel, R. Wittig, F. Zorick, C. Paxton, J. Lamphere, T. Roth: Eligibility requirements in hypnotic trials. Sleep 8 (1985) 34–39.

Römcke, O.: Die Resultate der konservativen Therapie. In: Boller, R. (Hrsg.): Der Magen und seine Krankheiten. Urban & Schwarzenberg, Wien-Innsbruck 1954.

Römer, H.: Das Sexualleben der Frau und seine Störungen. In: Käser, O. et al. (Hrsg.): Gynäkologie und Geburtshilfe I. Thieme, Stuttgart 1969.

Römer, H.: zit. nach: Ruppin, E. et al. (Hrsg.): Deutscher Kongreß für Perinatale Medizin 1977.

Rösch, W.: Ulcusrezidiv-Prophylaxe – konservativ oder operativ? Dtsch. Ärztebl. 81 (1984) 26–27.

Röttgers, K.: Gewalt. In: Ritter, J. (Hrsg.): Historisches Wörterbuch der Philosophie, Bd. 3, S. 562 ff. Wissenschaftliche Buchgesellschaft, Darmstadt 1974.

Rogentine, G. N., D. P. van Kommen, B. H. Fox et al.: Psychological factors in the prognosis of malignant melanoma: a prospective study. Psychosom. Med. 41 (1979) 647–655.

Rogers, C. R.: Client-centered therapy. Houghton-Mifflin, Boston 1951.

Rogers, C. R.: Die klientbezogene Gesprächstherapie. Kindler, München 1973.

Rogers, C. R.: Therapeut und Klient. Kindler, München 1977.

Rogers, C. R.: Die Klient-bezogene Gesprächstherapie. Fischer, Frankfurt 1978.

Rogers, M. P., D. Dubey, P. Reich: The influence of the psyche and the brain on immunity and disease susceptibility: a critical review. Psychosom. Med. 41 (1979) 147–164.

Rogers, M. P., D. E. Trentham, R. Dynesius-Trentham, K. Daffner, P. Reich: Exacerbation of collagen arthritis by noise stress. J. Rheum. 10 (1983) 651–654.

Rogers, M. P., D. E. Trentham, W. J. McCune, B. I. Ginsberg, M. G. Rennke, P. Reich, J. R. David: Effect of psychological stress on the induction of arthritis in rats. Arthr. and Rheum. 23 (1980) 1337–1342.

Rogers, M. P., P. Reich, T. B. Strom, C. B. Carpenter: Behaviorally conditioned immunosuppression: replication of a recent study. Psychosom. Med. 38 (1976) 447–451.

Rogers, M. P., P. Reich: On the health consequences of bereavement. New Engl. J. Med. 319 (1988) 510–512.

Rohde, J. J.: Veranstaltete Depressivität: Über strukturelle Effekte von Hospitalisierung auf dier psychische Situation des Patienten. Internist 15 (1974) 277–282.

Rohde-Dachser, Ch.: Das Borderline-Syndrom. Bern-Stuttgart-Wien 1979.

Rohrmeier, F.: Langzeiterfolge psychosomatischer Therapien. In: Albert, D., K. Pawlik, K.-H. Stapf, W. Stroebe (Hrsg.): Lehr- und Forschungstexte Psychologie 3. Springer, Berlin-Heidelberg-New York 1982.

Rohrmoser, H. G.: Zur Psychogenese und Psychotherapie der Colitis ulcerosa. Psychotherapie 1 (1956) 105–114.

Roitt, I. M.: Essential immunology. Blackwell, London 1977.

Roitt, I. M., J. Brostoff, D. K. Male: Kurzes Lehrbuch der Immunologie. Thieme, Stuttgart 1987.

Romberg, M. H.: Lehrbuch der Nervenkrankheiten des Menschen. Duncker, Berlin 1851.

Romeis, J. C., R. W. Hamilton, C. A. Snavely: Modelling home dialysis success. In: Levy, N. B. (ed.): Psychonephrology 2. Plenum, New York 1983.

Rorschach, H.: Psychodiagnostik, 8. Aufl. Huber, Bern-Stuttgart-Wien 1962.

Rosa, K. R.: Das ist autogenes Training. Kindler, München 1973.

Rosa, K. R.: Das ist die Oberstufe des autogenen Trainings. Kindler, München 1975.

Rosahl, W.: Fehldiagnosen bei Morbus Crohn. Dtsch. Gesundh.-Wes. 37 (1982) 2108–2110.

Rose, F. C., M. Gawel: Migraine. The facts. Oxford Univ. Press, New York 1979.

Rose, G., M. G. Marmot: Social class and coronary heart disease. Brit. Heart J. 45 (1981) 13–19.

Rose, R. J.: Familial influences on cardiovascular reactivity to stress. In: Matthews, K. A., S. M. Weiss, T. Detre, T. M. Dembroski, B. Falkner, S. B. Manuck, R. B. Williams jr. (eds.): Handbook of Stress, Reactivity, and Cardiovascular Disease, pp. 259–272. Wiley, New York 1986.

Rose, R. J., C. D. Jenkins, M. W. Hurst: Health change in air traffic controllers: a prospective study. Psychosom. Med. 40 (1978) 142–165.

Rose, R. M.: Testosterone, aggression and homosexuality: a review of the literature and implications for future research. In: Sachar, E. J. (ed.): Topics in Psychoendocrinology. Grune & Stratton, New York 1975.

Rosen, J. L., G. L. Bibring: Psychological reactions of hospitalized male patients to a heart attack. Psychosom. Med. 28 (1966) 808–821.

Rosen, S.: Philosophie und Wertesystem Milton H. Ericksons. In: Peter, B. (Hrsg.): Hypnose und Hypnotherapie nach Milton H. Erickson, S. 98–110. Pfeiffer, München 1985.

Rosenberg, M., L. Pearlin: Social class and self-esteem among children and adults. Amer. J. Soc. 84 (1978) 53–77.

Rosenblum, L. A.: Management of spastic, irritable colon, with a note on ulcerative colitis. Amer. J. Gastroent. 29 (1958) 407–411.

Rosenbrock, R.: AIDS kann schneller besiegt werden. Gesundheitspolitik am Beispiel einer Infektionskrankheit. VSA-Verlag, Hamburg 1986.

Rosenbrock, R.: HIV-Positivismus. Plädoyer für die Einhaltung der Kunstregeln. Kursbuch 94 (1988) 21–42.

Rosenfeld, J., M. R. Rosen, B. F. Puffman: Pharmacologic and behavioral effects of arrhythmias that immediately follow abrupt coronary occlusion: a canine model of sudden coronary death. Amer. J. Cardiol. 41 (1978) 1075.

Rosenkötter, L., C. de Boor, Z. Erdely, J. Matthes. Psychoanalytische Untersuchungen von Patientinnen mit funktioneller Amenorrhoe. Psyche 22 (1968) 838.

Rosenman, R. H.: Current status of risk factors and Type A behavior pattern in the pathogenesis of ischemic heart disease. In: Dembrowski, T. M., T. H. Schmidt, G. Blümchen (eds.): Biobehavioral Basis of Coronary Heart Disease. Karger, Basel 1983.

Rosenman, R. H.: Einleitende Anmerkungen zur Bedeutung des Typ-A-Verhaltens bei der koronaren Herzkrankheit. In: Dembroski, T. M., M. J. Halhuber (Hrsg.): Psychosozialer „Stress" und koronare Herzkrankheit: 3. Verhalten und koronare Herzkrankheit, S. 31–42. Springer, Berlin-Heidelberg-New York 1981.

Rosenman, R. H.: The interview method of assessment of the coronary-prone behavior pattern. In: Dembroski, T. M., S. M. Weiss, J. L. Shields, S. G. Haynes, M. Feinleib (eds.): Coronary-Prone Behavior. Springer, New York 1978.

Rosenman, R. H.: Type A behavior pattern: a personal overview. Stress Med. (in print).

Rosenman, R. H., M. Friedman: Association of a specific overt behavior pattern in females with blood and cardiovascular findings. Circulation 24 (1961) 1173–1184.

Rosenman, R. H., M. A. Chesney: The relationship of Type A behavior to coronary heart disease. Act. nerv. sup. 22 (1980) 1–45.

Rosenman, R. H., M. Friedman et al.: Coronary heart disease in the Western Collaborative Group Study. A follow-up experiment of 4.5 years. J. chron. Dis. 23 (1970) 173.

Rosenman, R. H., M. Friedman: Modifying Type A behavior pattern. J. psychosom. Res. 21 (1977) 323.

Rosenman, R. H., M. Friedman: Neurogenic factors in pathogenesis of coronary heart disease. Med. Clin. N. Amer. 58 (1975) 259–269.

Rosenman, R. H., M. A. Chesney: Psychological profiles and coronary heart disease. In: Kielholz, P., W. Siegenthaler, P. Taggart, A. Zanchetti (eds.): Psychosomatic Cardiovascular Disorders – When and How to Treat. Huber, Bern-Stuttgart-Wien 1981.

Rosenman, R. H., R. J. Brand, C. D. Jenkins, M. Friedman, R. B. Straus, M. Wurm: Coronary heart disease in the Western Collaborative Group Study: final follow-up experience of 8,5 years. J. Amer. med. Ass. 233 (1975) 872–877.

Rosenman, R. H., R. J. Brand, R. J. Sholtz, M. Friedman: Multivariate prediction of coronary heart disease during 8,5 year-follow-up in the Western Collaborative Group Study. Amer. J. Cardiol. 37 (1976) 903–910.

Rosenthal, D., P. H. Wender, S. S. Kety, F. Schulsinger, J. Weiner, L. Östergaard: Schizophrenics offspring reared in adoptive homes. In: Rosenthal, D., S. S. Kety (eds.): The Transmission of Schizophrenia. Pergamon, Oxford 1972.

Rosenthal, F., P. Holzer: Über die nervöse Beeinflussung des Agglutininspiegels, zugleich ein Beitrag zum Mechanismus der leistungssteigernden parenteralen Reiztherapie. Berlin. klin. Wschr. 25 (1921) 675–679.

Rosenthal, R., D. B. Rubin: A simple, general purpose display of magnitude of experimental effect. J. educ. Psychol. 74 (1982) 166–169.

Roskamp, H.: Grundzüge der Neurosenlehre. In: Loch, W.: Die Krankheitslehre der Psychoanalyse. Hirzel, Stuttgart 1967.

Roskies, E., H. Kearney, M. Spevack: Generalizability and durability of the treatment effects in an intervention program for coronary-prone (Type A) managers. J. behav. Med. 2 (1979) 195–207.

Roskies, E., M. Spevack, A. Surkis: Changing the Type A coronary-prone behavior pattern in a nonclinical population. J. behav. Med. 1 (1978) 202–216.

Rosman, B. L., S. Minuchin, L. Baker, R. Liebman: A family approach to anorexia nervosa: study, treatment and outcome. In: Vigersky, R. A. (ed.): Anorexia nervosa, pp. 341–348. Raven, New York 1977.

Rosman, B. L., S. Minuchin, L. Baker, R. Liebman: Input and outcome of family therapy in anorexia nervosa. In: Brunner, M. (ed.): Successful Psychotherapy, pp. 129–139. New York 1976.

Rosman, B. L., S. Minuchin, R. Liebman: Der Familien-Lunch. Familiendynamik 1 (1976) 334.

Ross, A. O.: Das Sonderkind. Hippokrates, Stuttgart 1967.

Ross, R.: The pathogenesis of atherosclerosis – an update. New Engl. J. Med. 314 (1986) 488–500.

Ross, R., J. A. Glomset: The pathogenesis of atherosclerosis. New Engl. J. Med. 295 (1976) 369–377; 420–425.

Rost, D.: Objektpsychologische Modellvorstellungen zur Theorie, Erforschung und Behandlung psychosomatischer („alexithymer") Störungen. Fallstudie einer analytischen Gruppe. Med. Diss., Frankfurt 1981.

Rost, W., M. Neuhaus, I. Florin: Bulimia nervosa: sex role attitude, sex role behavior, and sex role related locus of control in bulimarexic women. J. psychosom. Res. 26 (1982) 403–408.

Rostenberg, A.: Psychosomatic concepts in atopic dermatitis – a critique. Arch. Derm. 79 (1959) 692–699.

Roth, G., H. Schwegler (eds.): Self-organizing systems. An interdisciplinary approach. Campus, Frankfurt-New York 1981.

Roth, H. P.: The peptic ulcer personality. Science 56 (1955) 32.

Roth, J. W.: Katathymes Bilderleben als Kurzpsychotherapie in der psychosomatischen Gynäkologie. Schweiz. Rundsch. Med. 65 (1976) 252–256.

Roth, J. W.: Über die Bedeutung der introspektiven Imagination des Katathymen Bilderlebens, dargestellt am Beispiel des Spannungskopfschmerzes. In: Leuner, H. (Hrsg.): Katathymes Bilderleben, S. 224–234. Huber, Bern 1980.

Roth, K. H.: Filmpropaganda für die Vernichtung der Geisteskranken und Behinderten im „Dritten Reich". In: Beiträge zur nationalsozialistischen Gesundheits- und Sozialpolitik 2 (1985) 125–193.

Roth, W. F., F. H. Luton: The mental health program in Tennessee. Amer. J. Psychiat. 99 (1943) 662.

Rothenberg, R. et al.: Survival with the Acquired Immundeficiency Syndrome. New Engl. J. Med. 317 (1987) 1297–1302.

Rothenberger, A.: Therapie der Tic-Störungen. Zschr. Kinder-, Jugendpsychiat. 12 (1984) 284–301.

Rothenberger, A.: Tic-Störungen erkennen – verstehen – behandeln. Kinderarzt 19 (1988) 1196.

Rothman, A. I., N. Bryne, J. Parlow: Longitudinal study of personality traits in medical students from application to graduation. AAMC-RIME Conference 1973, 18.

Rotmann, M., R. Karstens: Psychosomatische Beratung im Krankenhaus. Psyche 28 (1974) 669–683.

Rotter, J. B.: An introduction to social learning theory. In: Rotter, J. B., J. E. Chance, E. J. Phares (eds.): Applications of a social learning theory of personality. Holt, Rinehart and Winston, New York 1972.

Rotter, J. B.: Social learning theory and clinical psychology. Prentice Hall, Englewood Cliffs 1954.

Rotter, J. I., C. E. Anderson, D. L. Rimoin: Genetics of diabetes mellitus. In: Ellenberg, M., H. Rifkin (eds.): Diabetes Mellitus, Theory and Practice, pp. 481–503. Medical Examination Publishing, New York 1983.

Rotter, J. I., D. L. Rimoin, J. M. Gursky, P. Terasaki, R. Sturdevant: HLA-B 5 associated with duodenal ulcer. Gastroenterol. 73 (1977) 438–440.

Rotter, J. I., D. L. Rimoin: Peptic ulcer disease – a heterogenous group of disorders? Gastroenterol. 73 (1977) 604–607.

Rotter, J. I., J. M. Diamond: What maintains the frequencies of genetic diseases? Nature (Lond.) 329 (1987) 289–290.

Rotter, J. I., J. Q. Sones, I. M. Samloff, C. T. Richardson, J. M. Gursky, J. H. Walsh, D. L. Rimoin: Duodenal ulcer disease associated with elevated serum pepsinogen, I. An inherited autosomal dominant disorder. New Engl. J. Med. 300 (1979) 63.

Rousseau, A., B. Herrmann, S. Whitman: Effects of progressive relaxation on epilepsy: analysis of a series of cases. Psychol. Rep. 57 (1985) 1203–1212.

Routtenberg, A.: The two-arousal hypothesis: reticular formation and limbic system. Psychol. Rev. 75 (1968) 51–80.

Rowe, A., A. Rowe: Bronchial asthma in patients over the age of 55 years. Ann. Allergy 5 (1947) 509.

Rowland, C.: A survey of the literature and review of 30 cases. In: Rowland, C. (ed.): Anorexia nervosa and Obesity. Int. psychiat. Clin. 7 (1970) 37–137.

Rowlands, D. B., M. A. Ireland, D. R. Glover, R. A. B. McLeay, T. J. Stallard, W. A. Littler: The relationship between ambulatory blood pressure and echocardiographically assessed left ventricular hypertrophy. Clin. Sci. (Suppl.) 61 (1981) 101 s–103 s.

Roy, A.: Five risk factors for depression. Brit. J. Psychiat. 150 (1987) 536–541.

Roy, A., D. Pickar, A. Doran, O. Wolkowitz, W. Gallucci, G. Chrousos, P. W. Gold: The corticotropin-releasing hormone stimulation test in chronic schizophrenia. Amer. J. Psychiat. 143 (1986) 1393–1397.

Roy, A., D. Pickar, M. Linnoila, W. Z. Potter: Plasma norepinephrine level in affective disorders: relationship to melancholia. Arch. gen. Psychiat. 42 (1985) 1181–1185.

Roy, R.: Marital and family issues in patients with chronic pain. Psychother. and Psychosom. 37 (1982) 1.

Rozin, P., J. Kalat: Specific hunger and poison avoidance as adaptive specializations of learning. Psychol. Rev. 78 (1971) 459–486.

Rubenstein, H. S.: Behavior in a medical clinic of patients with well-controlled bronchial asthma. Lancet I (1976) 1011–1012.

Ruberman, W., E. Weinblatt, J. D. Goldberg, B. S. Chaudhary: Psychosocial influences on mortality after myocardial infarction. New Engl. J. Med. 311 (1984) 552–559.

Ruberman, W., E. Weinblatt, J. D. Goldberg et al.: Ventricular premature beats and mortality after myocardial infarction. New Engl. J. Med. 297 (1977) 750.

Rubin, R. T., J. M. Reinisch, R. F. Haskett: Postnatal gonadal steroid effects on human behavior. Science 211 (1981) 1318–1324.

Rubin, R. T., R. E. Poland, I. M. Lesser, R. M. Winston, A. L. Nelson Blodgett: Neuroendocrine aspects of primary endogenous depression. I. Cortisol secretory dynamics in patients and matched controls. Arch. gen. Psychiat. 44 (1987) 328–336.

Rubin, W.: Evaluation and treatment of dizziness. Modern Treatment 6 (1976) 504.

Rubinow, D. R., P. W. Gold, R. M. Post, J. C. Ballenger, R. Cowdry, J. Bollinger, S. Reichlin: CSF-somatostatin in affective illness. Arch. gen. Psychiat. 40 (1983) 409–412.

Rubinstein, D., J. K. Thomas: Psychiatric findings in cardiotomy patients. Amer. J. Psychiat. 126 (1969) 360–369.

Rubinstein, S.: Grundlagen der Allgemeinen Psychologie (1957), 6. Aufl. Berlin 1968.

Ruderman, A. J.: Bulimia and irrational beliefs. Behav. Res. Ther. 24 (1986) 193–197.

Rückert, W.: Soziographische Daten. In: Lang, E. (Hrsg.): Praktische Geriatrie. Enke, Stuttgart 1988.

Ruegamer, W. R., L. Bernstein, J. D. Benjamin: Growth food utilization and thyroid activity in the albino rat as a function of extra handling. Science 120 (1954) 184.

Rüger, U.: Diagnostische Probleme in der psychosomatischen Medizin. Internistische Praxis 24 (1984) 723–730.

Rühmann, F.: AIDS – Eine Krankheit und ihre Folgen. Qumran, Frankfurt 1985.

Ruesch, J.: The infantile personality. Psychosom. Med. 10 (1948) 134–144.

Ruesch, J.: Therapeutic communication (1961). Norton, New York 1973.

Ruesch, J., G. Bateson: Communication. The social matrix of psychiatry (1951). Norton, New York 1968.

Ruesch, J., R. E. Harris, C. Christiansen, M. B. Loeb, S. Dewees, A. Jacobson: Duodenal ulcer – a socio-psychological study of naval enlisted personnel and civilians. Univ. of California Press, Berkeley 1948.

Ruffler, G.: Grundsätzliches zur psychoanalytischen Behandlung körperlich Kranker. Psyche 7 (1953/54) 521–560.

Rugh, J. D.: Psychological stress in orofacial neuromuscular problems. Int. dent. J. 31 (1981) 202.

Rugh, J. D., W. K. Solberg: Psychological implications in temporomandibular pain and dysfunction. Oral Sci. Rev. 1 (1976) 3.

Ruppin, E., S. Bäßmann, C. Dreessen, J. Ruppin, H. H. Chelius, H. Meier: Testpsychologische Untersuchungen über den Effekt der Psychoprophylaxe nach Read. In: Ruppin, E. et al. (Hrsg.): Deutscher Kongreß für Perinatale Medizin 1977.

Russek, H. I.: Role of heredity, diet and emotional stress in heart disease. J. Amer. med. Ass. 171 (1959) 503–508.

Russel, R. T., J. Sipich: Cue-controlled relaxation in the treatment of test anxiety. J. behav. Ther. exp. Psychiat. 4 (1973) 47–49.

Russell, G. F. M.: General management of anorexia nervosa and difficulties in assessing the efficacy of treatment. In: Vigersky, R. A. (ed.): Anorexia nervosa, pp. 277–289. Raven, New York 1977.

Russell, G. F. M.: Bulimia nervosa: an ominous variant of anorexia nervosa. Psychol. Med. 9 (1979) 429–448.

Russell, G. F. M.: Bulimia revisited. Int. J. Eat. Dis. 4 (1985a) 681–692.

Russell, G. F. M.: Diagnostik und klinische Meßverfahren bei Bulimia nervosa. In: Fichter, M. M. (Hrsg.): Bulimia nervosa. Enke, Stuttgart 1989.

Russell, G. F. M.: The changing nature of anorexia nervosa: an introduction to the conference. J. psychiat. Res. 19 (1985b) 101–109.

Russell, G. F. M.: The present status of anorexia nervosa (editorial). Psychol. Med. 7 (1977) 363–367.

Russell, G. F. M., G. I. Szmukler, C. Dare, I. Eisler: An evaluation of family therapy in anorexia nervosa and bulimia nervosa. Arch. gen. Psychiat. 44 (1987) 1047–1056.

Russell, M., K. A. Dark, R. W. Cummins, G. Ellman, E. Callaway, H. V. S. Peeke: Learned histamine release. Science 225 (1984) 733–734.

Russell, R. P., A. T. Masi: Significant associations of adrenal cortical abnormalities with „essential" hypertension. Amer. J. Med. 54 (1973) 44–51.

Rutenfranz, J.: Arbeitsphysiologische Aspekte der Nacht- und Schichtarbeit. Arbeitsmedizin, Sozialmedizin, Arbeitshygiene 2 (1967) 17–23.

Rutenfranz, J., E. Werner: Schichtarbeit bei kontinuierlicher Produktion. Forschungsbericht Nr. 141 (hrsg. von der Bundesanstalt für Arbeitsschutz und Unfallforschung), Dortmund 1975.

Rutenfranz, J., H. Luczak, G. Lehnert, W. Rohmert, D. Szadkowski: Denkschrift zur Lage der Arbeitsmedizin und der Ergonomie in der Bundesrepublik Deutschland. Im Auftrag der Senatskommission der Deutschen Forschungsgemeinschaft. Boldt, Boppard 1980.

Rutherford, G. W.: The natural history of Human Immunodeficiency Virus infection in a cohort of homosexual and bisexual men: a 7-year prospective study. III. International Conference on AIDS (abstract M.3.1.), June 1–5, Washington, D. C. 1987.

Rutter, B. M.: Emotional factors in arthritis. Curr. med. Res. Opin. 6 (Suppl. 2) (1979) 33–41.

Rutter, M.: Helping troubled children. Plenum, New York 1975.

Rutter, M.: Resilience in the face of adversity: protective factors and resistence to psychiatric disorder. Brit. J. Psychiat. 147 (1985) 598–611.

Rutter, M., P. Graham, F. D. Chadwick, W. Yule: Adolescent turmoil: fact or fiction? J. Child Psychol. Psychiat. 17 (1976) 35–56.

Rutter, M., D. Quinton: Parental psychiatric disorder: effects on children. Psychol. Med. 14 (1984) 853–880.

Rybakowski, J. K., W. L. Dyson, J. D. Amsterdam: Alexithymia and somatic conditions in patients with affective illness during lithium prophylaxis. Psychother. and Psychosom. 49 (1988) 1–5.

Rydevik, B., M. D. Brown, G. Lundborg: Pathoanatomy and pathophysiology of nerve root compression. Spine 9 (1984) 7–15.

Sachar, E. J.: Neuroendocrine abnormalities in depressive illness. In: Sachar, E. J. (ed.): Topics in Psychoendocrinology, p. 135. Grune & Stratton, New York 1975.

Sachar, E. J., J. W. Mason, H. S. Kilmer jr., K. L. Artiss: Psychoendocrine aspects of acute schizophrenic reactions. Psychosom. Med. 25 (1963) 510–537.

Sachar, E. J., L. Hellman, D. K. Fukushima, T. F. Gallagher: Cortisol production in depressive illness: a clinical and biochemical clarification. Arch. gen. Psychiat. 23 (1970) 289–298.

Sachsse, U.: Symbolgestalten in der Gruppenimagination. In: Roth, J. W. (Hrsg.): Konkrete Phantasie, S. 81–87. Huber, Bern 1984.

Sack, W. T.: Die Haut als Ausdrucksorgan. Arch. Derm. Syph. 151 (1926) 200–206.

Sackett, D. L., R. B. Haynes, E. S. Bibson et al.: Randomized clinical trial of strategies for improving medication compliance in primary hypertension. Lancet 1 (1975) 1205–1207.

Sacks, M., E. A. Schegloff, G. Jefferson: A simplest systematics for the organization of turn taking for conversation. In: Schenkein, J. (ed.): Studies in the organization of conversational interaction, pp. 7–55. Academic Press, New York 1978.

Sacks, O. W.: Awakenings. Duckworth, London 1973.

Sacks, O. W.: Der Mann, der seine Frau mit einem Hut verwechselte. Rowohlt, Reinbek 1987.

Sacks, O. W.: Der Tag, an dem mein Bein fortging. Rowohlt, Reinbek 1989.

Sacks, O. W.: Neuropsychiatry and Tourette's syndrome. In: Mueller, J. (ed.): Neurology and Psychiatry. A Meeting of Minds, pp. 156–174. Karger, Basel-München-Paris 1989.

Sadger, J.: Über Urethralerotik. Jb. Psychoanal. Psychopath. Forsch. 2 (1910) 409.

Säring, W., M. Prosiegel, D. v. Cramon: Zum Problem der Anosognosie und Anosodiaphorie bei hirngeschädigten Patienten. Nervenarzt 59 (1988) 129–137.

Safian, P., J. Fauler, U. Koch, C. Jährig, K. Köhle: Inhaltliche und methodische Analyse von Visitengesprächen zweier klinischer Populationen mittels unterschiedlicher Rating-Verfahren. In: Köhle, K., H.-H. Raspe (Hrsg.): Das Gespräch während der ärztlichen Visite, S. 140–177. Urban & Schwarzenberg, München 1982.

Sakai, F., J. S. Meyer: Abnormal cerebrovascular reactivity in patients with migraine and cluster headache. Headache 19 (1979) 257–266.

Sakai, F., J. S. Meyer: Regional cerebral hemodynamics during migraine and cluster headaches measured by the 133Xe inhalation method. Headache 18 (1978) 122–133.

Salfield, D. J.: The placebo. Lancet 265 (1953) 940.

Salih, H., W. Brander, H. Flax, J. R. Hoobs: Prolactin dependence in human breast cancers. Lancet II (1972) 1103–1105.

Salo, O. P., J. Karvonen, K. K. Mustakallio: Nachweis von Nagelkratzspuren mittels eines fluoreszierenden Stoffes. Hautarzt 21 (1970) 424–425.

Saltz, E.: Higher mental processes as the bases for the laws of conditioning. In: McGuigan, F. J., D. B. Lumbsden (eds.): Contemporary approaches to conditioning and learning. Wiley, New York 1973.

Salvisberg, H.: Therapie von Zwangsneurosen mit dem Katathymen Bilderleben – ein Beitrag zu Kasuistik und Theorie. In: Leuner, H., O. Lang (Hrsg.): Psychotherapie mit dem Tagtraum, S. 94–111. Huber, Bern 1982.

Samora, J., L. Saunders, R. F. Larson: Medical vocabulary knowledge among hospital patients. J. Med. Educ. 51 (1976) 83–92.

Sampson, T. F.: Use of fantasy for conflict resolution in the pediatric hemodialysis patient. In: Levy, N. B. (ed.): Psychonephrology 1, pp. 177–184. Plenum, New York 1981.

Samuel-Lajeunesse, B.: Le pronostic de l'anorexie mentale. Rev. Neuropsychiat. infant. 15 (1967) 447–470.

Sanders, W. M., A. E. Munson: Norepinephrine and the antibody response. Pharmacol. Rev. 37 (1985) 229–248.

Sandler, J. (ed.): Projection, identification, projective identification. IUP, Conn. 1987.

Sandler, J.: Das Konzept der projektiven Identifizierung. Zschr. psychoanal. Theorie Praxis 3/2 (1988) 147–164.

Sandler, J., C. Dare, A. Holder: The patient and the analyst. The basis of the psychoanalytic process. London 1973. Deutsch: Die Grundbegriffe der psychoanalytischen Therapie. Klett-Cotta, Stuttgart 1973.

Sandler, J., W. G. Joffe: Die Persistenz in der psychischen Funktion und Entwicklung mit besonderem Bezug auf die Prozesse der Fixierung und Regression. Psyche 21 (1967) 138–151.

Sandler, S. A.: Camptocormia, or the functional bent back. Psychosom. Med. 9 (1947) 197–204.

Sandweg, R.: Zur Psychodynamik und Therapie chronisch-entzündlicher Darmerkrankungen. Prax. Psychother. Psychosom. 34 (1986) 73–81.

Sanes, S.: A physician faces cancer in himself. State Univ. of New York Press, Albany 1979.

Sannerstedt, R., R. Sivertson, Y. Lundgren: Hemodynamic studies in young men with mild blood pressure elevation. Acta med. scand. 200 (Suppl. 602) (1976) 61.

Sapira, J. D., S. Eileent, B. A. Heib, R. Moriarty, A. P. Shapiro: Differences in perception between hypertensive and normotensive populations. Psychosom. Med. 33 (1973) 3.

Sapolsky, R. M., L. C. Krey, B. S. McEwen: The neuroendo-crinology of stress and aging: the glucocorticoid cascade hypothesis. Endocrine Rev. 7 (1966) 284–301.

Sargent, J. D., E. E. Green, E. D. Walters: The use of auto-genic feedback training in a pilot study of migraine and tension headaches. Headache 12 (1972) 120–124.

Sarre, H. J.: Arterielle Hypertonie. Sandoz AG, Nürnberg 1971.

Sarwer-Foner, G. J. (ed.): The dynamics of psychiatric drug therapy. Thomas, Springfield 1960.

Sarwer-Foner, G. J.: Psychiatric symptomatology: its mean-ing and function in relation to the psychodynamic actions of drugs. In: Deuber, H. C. B. (ed.): Psychopharmacologi-cal Treatment: Theory and Practice. New York 1975.

Sarwer-Foner, G. J.: Psychodynamics of psychotropic medi-cation – an overview. In: DiMascio, A., R. J. Shader (eds.): Clinical Handbook of Psychopharmacology. Aronson, New York 1970.

Sashin, J., S. Eldered, S. Amerongen: A search for predictive factors in institute supervised cases: a retrospective study of 183 cases from 1959–1966 of the Boston Psychoanaly-tic Institute. Int. J. Psychoanal. 56 (1975) 343–359.

Satinsky, J., B. Kosowsky, B. Lown: Ventricular fibrillation induced by hypothalamic stimulation during coronary oc-clusion (Abstr.). Circulation 44 (Suppl. 2) (1971) 2.

Sattler, G.: Colitis ulcerosa, Colitis mucosa, Ileitis termina-lis. Diss., Freiburg 1960.

Saul, L. J.: A clinical note on a mechanism of psychogenic back pain. Psychosom. Med. 3 (1941) 190–191.

Saunders, J. B.: Alcohol: an important cause of hyperten-sion. Brit. med. J. 294 (1987) 1045–1046.

Sauvant, J. D., Ch. Hürny, R. Hemmeler, R. H. Adler: Vali-dität der Diagnose „Psychogener Schmerz" bei der Kon-trolle nach fünf Jahren. Schweiz. Rundsch. Med. 77 (1988) 1379–1382.

Sawchenko, P. E.: Functional anatomy of peptidergic neu-rons. In: Weiner, H., D. Hellhammer, I. Florin, R. C. Muri-son (eds.): Neuronal Control of Bodily Function: Basic and Clinical Aspects. Vol. IV: Frontiers of Stress Re-search. Huber, Toronto 1989.

Sawrey, W. L.: Conditioned responses of fear in relation to ulceration. J. comp. Physiol. Psychol. 54 (1961) 347–348.

Saxena, B. M., K. Bhaskaran, J. V. Anath: Social class and schizophrenia: a study based on the caste system in India. Transcultural Psychiatry Res. Rev. 9 (1972) 130–133.

Scadding, J. G.: Mechanism of bronchial asthma – definition and clinical categorization. In: Weiss, E. B., M. S. Segal (eds.): Bronchial Asthma. Mechanisms and Therapeutics, pp. 19–31. Little, Brown & Co., Boston 1979.

Scammon (1922) : zit. nach: Flanagan, G. L. (Hrsg.): Die ersten neun Monate des Lebens. Rowohlt, Reinbek 1963.

Scanzoni: zit. nach: Artner, J. (1870).

Schacht, L.: Subjekt gebraucht Subjekt. Psyche 27/2 (1973) 151–168.

Schachter, S.: Some extraordinary facts about obese humans and rats. American Psychologist 26 (1971) 129–144.

Schachter, S.: The psychology of affiliation. Stanford Univ. Press, Stanford/Ca. 1959.

Schachter, S., J. E. Singer: Cognitive, social and psychologi-cal determinants of emotional state. Psychol. Rev. 69 (1962) 379–399.

Schachter, S., J. E. Singer: Comments on the Maslach and Marshall-Zimbardo experiments. J. Pers. soc. Psychol. 37 (1979) 989–995.

Schadewaldt, H.: Der Arzt vor der Frage von Leben und Tod. Klin. Wschr. 47 (1969) 557–568.

Schadewaldt, H.: Medizingeschichtliche Betrachtungen zum Anorexia nervosa-Problem. In: Meyer, J. E., H. Feld-mann (Hrsg.): Anorexia nervosa. Thieme, Stuttgart 1965.

Schäfer, H.: Lebenserwartung und Lebensführung. Medizin, Mensch, Gesellschaft 1 (1976) 27–32.

Schäfer, H.: Plädoyer für eine neue Medizin. Warnung und Appell, S. 729. München-Zürich 1979.

Schäfer, H., Ch. Voss, H.-J. Heuschel, N. Hartmann: Jodthy-roxin-Dejodasen im Stoffwechsel der Schilddrüsenhor-mone. Hoppe-Seylers Z. physiol. Chem. 341 (1965) 268.

Schäfer, H., H. Heinemann: Modelle sozialer Einwirkungen auf den Menschen. Soziophysiologie. In: Blohmke, M. (Hrsg.): Hdb. Sozialmedizin, Bd. I, S. 92–131. Enke, Stutt-gart 1975.

Schäfer, H., M. Blohmke: Epidemiologie der koronaren Herzkrankheiten. In: Blohmke, M. et al. (Hrsg.): Hand-buch der Sozialmedizin, Bd. II, S. 1–67. Enke, Stuttgart 1977.

Schäfer, H., M. Blohmke: Herzkrank durch psychosozialen Streß. Hüthig, Heidelberg 1977 a.

Schäfer, H., M. Blohmke: Sozialmedizin. Einführung in die Ergebnisse und Probleme der Medizin-Soziologie und So-zialmedizin. Thieme, Stuttgart 1972; 2. Aufl. 1978.

Schaefer, K., D. Schwartz: Verhaltenstherapeutische Ansät-ze für die Anorexia nervosa. Z. klin. Psychol. Psychother. 22 (1974) 267–284.

Schäfer, N.: Umwelt und Blutdruck. Das Konzept des gro-ßen Regelkreises. Habilitationsschrift, Ulm 1976.

Schaeffer, F.: Pathologische Treue als pathogenetisches Prinzip bei schweren körperlichen Erkrankungen. Ein ka-suistischer Beitrag zur Dermatomyositis. Nervenarzt 32 (1963) 315–332.

Schaeffer, M. A., W. McKinnon, A. Baum, C. P. Reynolds, P. Rikli, L. M. Davidson, I. Fleming: Immune status of chronic stress at Three Miles Island. Psychosom. Med. 47 (1985) 85.

Schalch, D. S., L. Lee: „Unpublished observations reported in Schalch, D. S., S. Reichlin: Stress and growth hormone release". In: Pecile, A., E. E. Müller (eds.): Growth Hor-mone, pp. 211–215. Exc. Med. Found., Amsterdam 1968.

Schanberg, S., G. Evoniuk, C. Kuhn: Tactile and nutritional aspects of maternal care: specific regulators of neuroendo-crine function and cellular development. Proc. Soc. exp. Biol. (N. Y.) 175 (1984) 135–146.

Schandry, R.: Psychophysiologie. Körperliche Indikatoren menschlichen Verhaltens. Urban & Schwarzenberg, Mün-chen 1981.

Schara, J.: Aufforderung zu einer humanen Intensivtherapie. Schwester/Pfleger 20 (1981) 638–643.

Scharfetter, C.: Das AMP-System. Manual zur Dokumenta-tion psychiatrischer Befunde. Springer, Berlin 1972.

Scharrer, B. : Über sekretorisch tätige Nervenzellen bei wir-bellosen Tieren. Naturwissenschaften 25 (1937) 131–138.

Schaumann, H.-J., B. Stegarn, H. Neuss, H. Scheurlen: Eine kontrollierte klinische Studie über die Frühmobilisation von Infarktpatienten. Med. Klinik 72 (1977) 465–470.

Scheiber, S. C.: Emotional problems of physicians. In: Scheiber, S., B. Doyle (eds.): The Impaired Physician. Plenum, New York 1983 .

Scheingold, L. D., N. N. Wagner: Herz, Alter, Sexualität. Medical Tribune, Wiesbaden 1976.

Schellack, D.: Neurosenpsychologische Faktoren in der Aetiologie und Pathogenese der afebrilen Colitis ulcerosa chronica. Z. psychosom. Med. 1 (1954) 28–38.

Schellack, D.: Neurosenpsychologische Faktoren in der Ätiologie und Pathogenese der Tonsillitis. Z. psychosom. Med. 4 (1957/58) 15.

Schellack, D.: Psychische Faktoren bei Muskel- und Gelenk-erkrankungen. Z. psychosom. Med. 3 (1955) 161–172.

Scheller, R. N., J. D. Barchas: Molecular neurobiology – a conference sponsored by the NIMH. Science 242 (1988) 13–14.

Schenk-Danzinger, L.: Entwicklungsquotient und Schulleistung. Diagnostica 3 (1957) 30–32, 43–48.

Schepank, H.: Anorexia nervosa. Zwillingskasuistik über ein seltenes Krankheitsbild. In: Heigl-Evers, A., H. Schepank (Hrsg.): Ursprünge seelisch bedingter Krankheiten, Bd. 2, S. 705–719. Vandenhoeck & Ruprecht, Göttingen 1982.

Schepank, H.: Die stationäre Psychotherapie in der Bundesrepublik Deutschland. Zschr. Psychosom. Med. 33 (1987) 363.

Schepank, H.: Epidemiologie psychogener Störungen. In: Kisker, K. P., H. Lauter, J. E. Meyer, C. Müller, E. Stroemgren (Hrsg.): Psychiatrie der Gegenwart I, S. 1–27. Springer, Berlin-Heidelberg-New York 1986.

Schepank, H.: Psychogene Erkrankungen der Stadtbevölkerung. Eine epidemiologisch-tiefenpsychologische Untersuchung in Mannheim. Springer, Berlin-Heidelberg-New York 1987.

Schepank, H.: Seelische Gesundheit und psychogene Erkrankungen. Stabilität im Verlauf. (In Vorbereitung).

Schepank, H.: zit. nach: Mattern, H.: Ich klage an! Im Namen des Patienten. Ärztl. Praxis 106 (1988) 3255.

Schepank, H., W. Tress: Häufigkeit und Bedingungen psychogener Erkrankungen in der Stadtbevölkerung. Nervenheilkunde 6 (1987) 23–26.

Scherg, H.: Zur Kausalitätsfrage in der psychosozialen Krebsforschung. Psychother. med. Psychol. 36 (1986) 98–109.

Scherl, A.: Klinik, Therapie und Prognose der Colitis ulcerosa. Diss., Erlangen 1978.

Scherwitz, L., K. Berton, H. Leventhal: Type A assessment and interaction in the behavior pattern interview. Psychosom. Med. 39 (1977) 229–240.

Scherwitz, L., K. Berton, H. Leventhal: Type A behavior, self-involvement, and cardiovascular response. Psychosom. Med. 40 (1978) 593–609.

Scherwitz, L., L. E. Graham, G. Grandits, J. Buehler, J. Billings: Self-involvement and coronary heart disease incidence in the Multiple Risk Factor Intervention Trial. Psychosom. Med. 48 (1986) 187–199.

Scherwitz, L., R. McKelvain, C. Laman et al.: Type A behavior, self-involvement, and coronary atherosclerosis. Psychosom. Med. 45 (1983) 47–57.

Schettler, G.: Arzt und Patient in der Leistungsgesellschaft. Dtsch. Ärztebl. (1975) 1138–1141; 1218–1224.

Schettler, G., E. Nüssel: Neuere Resultate aus der epidemiologischen Herzinfarktforschung in Heidelberg. Dtsch. med. Wschr. 99 (1974) 2003–2008.

Scheytt, Ch.: Seelsorge an Sterbenden im Krankenhaus. In: Spiegel-Rösing, J., H. Petzold (Hrsg.): Psychotherapie mit Sterbenden, S. 409–430. Junfermann, Paderborn 1984.

Schied, H. W.: Sterben in der Klinik oder zu Hause. Bestehende und fehlende Daten vom Sterbeort in der Bundesrepublik Deutschland. Zt. Allg. Med 55 (1979) 1270–1274.

Schiefer-Hofmann, E., H. Jäger: Psychoimmunologische Aspekte der HIV-Infektion. In: Jäger, H. (Hrsg.): AIDS und HIV-Infektionen. Diagnostik – Klinik – Behandlung. ecomed, Landsberg 1989.

Schienstock, G., W. Karmaus, V. Müller: Arbeitsbelastungen, Streß und Bewältigungsmöglichkeiten. In: Karmaus, W., V. Müller, G. Schienstock (Hrsg.): Streß in der Arbeitswelt, S. 191–222. Bund-Verlag, Köln 1979.

Schiffer, R. B., H. M. Babigan: Behavioral disorders in multiple sclerosis, temporal lobe epilepsy and amyotrophic lateral sclerosis. An epidemiological study. Arch. Neurol. 41 (1984) 1067–1069.

Schiffter, R.: Neurologie des vegetativen Systems. Springer, Berlin-Heidelberg-New York 1985.

Schild, R.: Psychiatrische Aspekte des Muskelschmerzes. Therapiewoche 33 (1973) 2665–2672.

Schilder, P.: Das Körperschema. Ein Beitrag zur Lehre vom Bewußtsein des eigenen Körpers. Springer, Berlin 1923.

Schilder, P.: Introduction to a psychoanalytic psychiatry. Int. Univ. Press, New York 1951.

Schildkraut, J. J.: Current status of the catecholamine hypothesis of affective disorders. In: Lipton, M. A., A. DiMascio, K. F. Killam (eds.): Psychopharmacology: A Generation of Progress. Raven, New York 1978.

Schildkraut, J. J.: The catecholamine hypothesis of affective disorders: a review of supporting evidence. Amer. J. Psychiat. 122 (1965) 509–522.

Schippperges, H.: Medizin und Umwelt, S.131. Verlag für Medizin, Heidelberg 1978.

Schirren, C.: Praktische Andrologie. Hartmann, Berlin 1971.

Schlag, P., S. Frohmüller, J. Ophof, R. Leucht, J. Doersam, G. Ruoff: Mobile, ambulante Nachbehandlung (MAN). Erfahrungen mit einem Modellprojekt zur Verbesserung der Situation des Tumorkranken. Münch. med. Wschr. 130 (1988) 284–288.

Schlegel, H. J.: Der Arzt als Patient – Gespräch mit einem Professor der inneren Medizin einer schweizerischen Universitätsklinik, der seit 50 Jahren an einem Ulcus duodeni leidet. Swiss Med. 4 (1982) 23.

Schleifer, S. J., S. E. Keller, M. Camerino, J. C. Thornton, M. Stein: Suppression of lymphocyte stimulation following bereavement. J. Amer. med. Ass. 250 (1983) 374–377.

Schleiffer, R.: Zur Indikation einer Psychotherapie bei atopischer Dermatitis im Kindesalter. Akt. Derm. 14 (1988) 17–20.

Schlesinger-Kipp, G., H. Radebold: Familien- und Paartherapie im höheren und hohen Lebensalter. In: Radebold, H., G. Schlesinger-Kipp (Hrsg): Familien- und paartherapeutische Hilfen bei älteren und alten Menschen. Vandenhoeck & Ruprecht, Göttingen 1982.

Schlesser, M. A., G. Winokur, B. M. Sherman: Hypothalamic-pituitary-adrenal axis activity in depressive illness. Arch. gen. Psychiat. 37 (1980) 737–743.

Schlewinski, E.: Untersuchungen über den Einfluß psychischer Faktoren auf das Immunsystem. Z. psychosom. Med. Psychoanal. 22 (1976) 370–377.

Schlewinski, E.: Untersuchungen über den Einfluß psychischer Faktoren auf das Immunsystem: Die Wirkung infantiler Stimulation auf den Krankheitsverlauf von bakteriell infizierten Babymäusen. Z. psychosom. Med. Psychoanal. 21 (1975) 390–399.

Schlewinski, E.: Veränderung der Immunreaktion in Abhängigkeit von psychischen Reizen – Eine tierexperimentelle Studie. Z. psychosom. Med. Psychoanal. 26 (1980) 336–346.

Schliack, H., R. Schiffter: Klinik der sogenannten vegetativen Schmerzen. In: Sturm, A., W. Birkmayer (Hrsg.): Klinische Pathologie des vegetativen Nervensystems I, S. 498–537. Fischer, Stuttgart 1976.

Schlorhaufer, W.: Das hörgeschädigte Kind (Pädaudiologie). In: Berendes, J., R. Link, F. Zöllner (Hrsg.): Hals-Nasen-Ohrenheilkunde in Praxis und Klinik, Bd. 6. Thieme, Stuttgart-New York 1980.

Schlote, B. M.: Long-term registration of muscle tension among office workers suffering from tension headache. In: Bischoff, C., H. C. Traue, H. Zenz (eds.): Clinical Perspectives on Headache and Low Back Pain, pp. 46–63. Hogrefe & Huber, Toronto 1989.

Schmädel, D.: Nichtbefolgen ärztlicher Anordnungen. Med. Klinik 71 (1976) 1460–1466.

Schmale, A. H.: Importance of life setting for disease onset. Modern Treatment 6 (1969) 643–654.

Schmale, A. H.: Normal grief is not a disease. In: Goldberg, J. (ed.): Psychopharmacologic Agents for the Terminally Ill and Bereaved. Columbia Univ. Press, New York 1973.

Schmale, A. H.: Relationship of separation and depression to disease. A report on a hospitalized medical population. Psychosom. Med. 20 (1958) 259–272.

Schmale, A. H.: Giving up as a final common pathway to deranges in health. Advanc. psychosom. Med. 8 (1972) 20–40.

Schmale, A. H., G. L. Engel: The giving-up-given-up complex, illustrated in film. Arch. gen. Psychiat. 17 (1967) 135–145.

Schmale, A. H., H. Iker: The psychological setting of uterine cervical cancer. Ann. N. Y. Acad. Sci. 125 (1966) 807–813.

Schmale, A. H., H. Iker: Hopelessness as a predictor of cervical cancer. Soc. Sci. & Med. 5 (1971) 95–100.

Schmale, A. H., H. Iker: The affect of hopelessness and the development of cancer. Psychosom. Med. 28 (1966) 714–721.

Schmeling, C., U. Koch: Betreuung von Schwer- und Todkranken – Möglichkeiten und Grenzen der Ausbildung von Krankenhauspersonal. In: Howe, J., R. Ochsmann (Hrsg.): Tod – Sterben – Trauer, S. 101–106. Fachbuchhandlung für Psychologie, Frankfurt 1984.

Schmidbauer, W.: Die hilflosen Helfer. Über die seelische Problematik der helfenden Berufe. Rowohlt, Reinbek 1977.

Schmidbauer, W.: Helfen als Beruf. Die Ware Nächstenliebe. Rowohlt, Reinbek 1983.

Schmidt, D., E. Wölke: Morbus Crohn – eine Übersicht. Dtsch. Gesundh.-Wes. 36 (1981) 721–731.

Schmidt, D. D.: Family determinants of disease: depressed lymphocyte function following the loss of a spouse. Family Systems Medicine 1 (1983) 33.

Schmidt, D. D., S. Zysanski, J. Ellner, M. L. Kumar, J. Arno: Stress as precipitating factor in subjects with recurrent herpes labialis. J. Fam. Pract. 20 (1985) 359–366.

Schmidt, G.: Gedanken zum Ericksonschen Ansatz aus einer systemorientierten Perspektive. In: Peter, B. (Hrsg.): Hypnose und Hypnotherapie nach Milton H. Erickson, S. 31–57. Pfeiffer, München 1985.

Schmidt, J.: Behandlung des Colon irritabile. pharma-kritik 23 (1983) 89–92.

Schmidt, R. F.: Schmerz und Motorik. „Muskelverspannung und Kopfschmerz", W. Reimers-Stiftung, Bad Homburg, 23.–25. 2. 01.

Schmidt, R. F., A. Struppler: Der Schmerz. Ursachen, Diagnose, Therapie. Piper, München 1982.

Schmidt, T. H.: Cardiovascular reactions and cardiovascular risk. In: Dembroski, T. M., T. H. Schmidt, G. Blümchen (eds.): Biobehavioral Bases of Coronary Heart Disease, pp. 130–174. Karger, Basel-München-Paris 1983.

Schmidt, T. H.: Die Situationshypertonie als Risikofaktor. In: Vaitl, D. (Hrsg.): Essentielle Hypertonie. Springer, Berlin-Heidelberg 1982.

Schmidt, T. H.: Koronares Risiko und Typ-A-Verhalten. In: Köhle, K. (Hrsg.): Forum Galenus Mannheim 8: Zur Psychosomatik von Herz-Kreislauf-Erkrankungen. Springer, Berlin-Heidelberg 1982.

Schmidt, T. H.: Tagesperiodische und situative Veränderungen des Kreislaufverhaltens. In: Palm, D., W. Rudolph (Hrsg.): Symposion über den Beta-Rezeptorenblocker Carazolol, S. 112–135. Excerpta Medica, Amsterdam 1980.

Schmidt, T. H.: Verhaltenskorrelate kardiovaskulärer Reaktionen. Ein Beitrag zu den psychosozialen Aspekten kardiovaskulärer Risikofaktoren mit besonderer Berücksichtigung des Typ-A-Verhaltensmusters. Habilitationsschrift, Hannover 1988.

Schmidt, T. H., H. Rüddel, W. Langosch, K. Undeutsch, H.

Neus, R. Hahn, T. M. Dembroski, J. M. MacDougall: Psychophysiologische Untersuchung zum Typ-A-Verhalten und seine Beziehung zu traditionellen kardiovaskulären Risikofaktoren bei Polizeibeamten aus dem Raum Köln. In: Langosch, W. (Hrsg.): Psychische Bewältigung der chronischen Herzkrankheit, S. 79–113. Springer, Berlin-Heidelberg-New York 1985.

Schmidt, T. H., H. Rüddel, W. Langosch, K. Undeutsch, R. Hahn, T. M. Dembroski: Kardiovaskuläre Risikofaktoren und Typ-A-Verhalten. Verh. Dtsch. Ges. Inn. Med. 88 (1982) 1204–1209.

Schmidt, T. H., H. Thierse, G. Blümchen: Coronary-prone behavior pattern, myocardial infarction and angiographic findings in a German sample of 400 cardiac patients. Abstr. Int. Symp. Cardiac Rehab. 1. 11.–3. 11. 1984.

Schmidt, T. H., H. Thierse, S. Schug, G. Blümchen: Koronargefährdendes Typ-A-Verhalten bei Herzinfarktpatienten im Anschlußheilverfahren und bei gesunden Vergleichspersonen. Herzmed. 7 (1984) 12–19.

Schmidt, T. H., J. Eschweiler, H. Thierse: Behavioral correlates of cardiovascular reactions in school children. In: Schmidt, T., T. Dembroski, G. Blümchen (eds.): Biological and Psychological Factors in Cardiovascular Disease, pp. 187–227. Springer, Berlin-Heidelberg-New York 1986.

Schmidt, T. H., J. Eschweiler, H. Thierse: Verhaltenskorrelate kardiovaskulärer Reaktionen bei Schulkindern. Abschlußbericht an die DFG, 1986.

Schmidt, T. H., T. M. Dembroski, J. M. MacDougall, P. Leidig, J. Eschweiler, H. Thierse, S. Schug: Various perspectives on cardiovascular reactivity and the Type A behavior pattern. In: Orlebeke, J. F., W. Mulder, L. van Doornen (eds.): Psychophysiology of Cardiovascular Control. Models, Methods and Data, pp. 745–766. Plenum, London 1985.

Schmidt, T. H., W. Thomas, H. Thierse, E. Scharf-Bornhofen, G. Blümchen: Koronarangiographische Befunde, Myokardinfarkt und koronargefährdende Verhaltensweisen. Klin. Wschr. (Suppl. 4) (1985) 9–10.

Schmidt, W., E. Lehnhardt, R.-D. Battmer: Die Bedeutung der ERA für die Differenzierung von Hörstörungen unterschiedlicher Genese. HNO 31 (1983) 109.

Schmit, G.: L'obésité infantile. In: Lebovici, S., R. Diatkine, M. Soulé (eds.): Traité de psychiatrie de l'enfant et de l'adolescent, Vol. II., pp. 487–506. PUF, Paris 1985.

Schmit, G., M. Soulé: L'énurésie. In: Lebovici, S., R. Diatkine, M. Soulé (eds.): Traité de psychiatrie de l'enfant et de l'adolescent, Vol. II., pp. 507–526. PUF, Paris 1985.

Schmitz-Moormann, P., H. Malchow, B. Miller, J.-W. Brandes: Häufigkeit und Vorkommen der epitheloidzelligen Granulome im Rektum und Kolonbiopsien bei Morbus Crohn. Ein Beitrag zur formalen Pathogenese. Z. Gastroenterol. 17 (1979) 287–295.

Schmorl, G., H. Junghanns: Die gesunde und die kranke Wirbelsäule in Röntgenbild und Klinik, 5. Aufl. Thieme, Stuttgart 1968.

Schmucki, O., R. Asper, C. Zortea: Streß und Urolithiasis. Urol. int. 39 (1984) 159.

Schmucki, O., R. Asper, W. H. Weihe: Streßinduzierte Veränderungen der Elektrolytkonzentrationen im Urin bei der Ratte. In: Gasser, G., W. Vahlensieck (Hrsg.): Pathogenese der Harnsteine VII, S. 126. Steinkopff, Darmstadt 1979.

Schmucki, O., R. Asper: Qualitative und quantitative Urin- und Serumuntersuchungen unter Extrembedingungen. In: Gasser, G., W. Vahlensieck (Hrsg.): Pathogenese der Harnsteine V, S. 35. Steinkopff, Darmstadt 1977.

Schnabl, S.: Correlations and determinants of functional sexual disturbances. In: Forleo, R., W. Pasini (eds.): Medical Sexology, pp. 153–161. PSG, Littleton 1980.

Schneble, H. J.: Krankheit der ungezählten Namen. Ein Beitrag zur Sozial-, Kultur- und Medizingeschichte der Epilepsie anhand ihrer Benennungen vom Altertum bis zur Gegenwart. Huber, Bern-Stuttgart-Toronto 1987.

Schneemann, N.: Der Krisenbegriff V. v. Weizsäckers. Zeitschrift für Psychotherapie und medizinische Psychologie 17 (1967) 144–153.

Schneider, A.: Untersuchungen zur Retest-Reliabilität eines Einstellungsfragebogens für Medizinstudenten in Anamnesegruppen. Dissertation, Marburg 1981.

Schneider, A., F. Wenzel: Forschungsstrategie und emotionale Entwicklung der Projektgruppe. In: Schüffel, W. (Hrsg.): Sprechen mit Kranken – Erfahrungen studentischer Anamnesegruppen. Urban & Schwarzenberg, München 1983

Schneider, D. E.: The image of the heart. Int. Univ. Press, New York 1956.

Schneider, H.-J.: Zur psychischen Situation der Harnsteinpatienten. In: Sitzungsbericht: Psychosomatische Aspekte in der Urologie, Würzburg 1986.

Schneider, H.-J., S. Heyne, G. Dittrich, E. Riedel: Untersuchungen zur soziologischen Struktur der Harnsteinpatienten. Z. ärztl. Fortbild. 67 (1973) 735.

Schneider, K. (Hrsg.): Klinische Psychopathologie, 12. Aufl. Thieme, Stuttgart 1980.

Schneider, P. B.: Zum Verhältnis von Psychoanalyse und Psychosomatischer Medizin. Psyche 27 (1973) 21–49.

Schneider-Helmert, D.: 24-hour sleep wake function and personality patterns in chronic insomniacs and healthy controls. Sleep 10 (1987) 452–462.

Schneider-Helmert, D.: DSIP: Clinical application of the programming effect. In: Inoué, S., D. Schneider-Helmert (eds.): Sleep Peptides: Basic and Clinical Approaches, pp. 175–198. Japan Sci. Soc. Press, Tokyo/ Springer, Berlin 1988 a.

Schneider-Helmert, D.: Schlafmittel: Symptomatisch oder kurativ? Schweiz. Ärztezeitung 67 (1986) 1477–1483.

Schneider-Helmert, D.: Schlafstörungen – ein Problem unserer Zeit. Schweiz. Ärztezeitung 66 (1985) 1102–1108.

Schneider-Helmert, D.: Towards a concept for chronic insomnia. In: Koella, W. P., F. Obal, H. Schulz, P. Visser (eds.): Sleep '86, pp. 422–424. Fischer, Stuttgart-New York 1988 b.

Schneider-Helmert, D.: Why low-dose benzodiazepine-dependent insomniacs can't escape their sleeping pills. Acta psychiat. scand. 78 (1988 c) 706–711.

Schneider-Helmert, D., A. Kumar: Subjective and objective measures of wakings among normals and insomniacs. Sleep Res. 15 (1986) 166.

Schneiderman, N.: Anger, aerobics and autonomic reactivity. In: Schmidt, T., T. Dembroski, G. Blümchen (eds.): Biological and Psychological Factors in Cardiovascular Disease, pp. 304–364. Springer, Berlin-Heidelberg-New York 1986.

Schneiderman, N.: Animal behavior models of coronary heart disease. In: Krantz, D. S., A. Baum, J. E. Singer (eds.): Handbook of Psychology and Health 3: Cardiovascular Disorders and Behavior, pp. 19–56. Erlbaum, Hillsdale 1985.

Schneiderman, N.: Behavior, autonomic function and animal models of cardiovascular pathology. In: Dembroski, T. M., T. H. Schmidt, G. Blümchen (eds.): Biobehavioral Bases of Coronary Heart Disease. Karger, Basel-München-Paris 1983.

Schneierson, S. S., J. H. Garlock, B. Shore, W. D. Stuart, M. Steinglass, B. Aronson: Studies on the viral etiology of regional enteritis and ulcerative colitis: a negative report. Amer. J. digest. Dis. 7 (1962) 839–843.

Schnieder, E. A.: Funktionelle Syndrome in der HNO-Heilkunde. In: Uexküll, T.v., R. Adler, J. M. Herrmann, K. Köhle, O. W. Schonecke, W. Wesiack (Hrsg.): Psychosomatische Medizin, 3. Aufl. Urban & Schwarzenberg, München 1986.

Schnyder, U. W.: Neurodermitis – Asthma – Rhinitis. Karger, Basel 1960.

Schömig, A., B. Lüth, R. Dietz, F. Gross: Changes in vascular smooth muscle sensitivity to vasoconstrictor agents induced by corticosteroids, adrenalectomy, and differing salt intake in rats. Clin. Sci. Mol. Med. 51 (1976) 51–63.

Schoenbach, V. J., C. Z. Garrison, B. H. Kaplan: Epidemiology of adolescent depression. Publ. Health Rev. 12 (1984) 159–189.

Schoenhals, H., M. Bernatz: Asthma als Symptom Darstellung einer Familientherapie. Fachbuchhandlung f. Psychologie, Eschborn 1984.

Schöpf, J.: Withdrawal phenomena after long-term administration of benzodiazepines. A review of recent investigation. Pharmacopsychiatry 16 (1983) 1–8.

Schoeppe, W., K. M. Koch, F. Oppermann: Heimdialyse. Dtsch. Ärztebl. 68 (1971) 2167–2172.

Schöttler, C.: Zur Behandlungstechnik bei psychosomatisch schwer gestörten Patienten. Psyche 35 (1981) 111–141.

Schonecke, O. W.: Mündliche Mitteilung 1984.

Schonecke, O. W.: Psychosomatik funktioneller Herz-Kreislaufstörungen. Springer, Berlin-Heidelberg 1987.

Schonecke, O. W.: Wissenschaftstheoretische und methodologische Probleme der Psychosomatischen Forschung und Theoriebildung. Z. Psychosom. Med. 18 (1972) 352–368.

Schonecke, O. W.: Zur Bedeutung der Psychophysiologie für die Psychosomatische Medizin. Münch. med. Wschr. 43 (1988) 305–308.

Schonecke, O. W., W. Schüffel, N. Schäfer, K. Winter: Assessment of hostility in patients with functional cardiac complaints. Psychother. and Psychosom. 20 (1972) 272.

Schonfield, J.: Psychological and life experience differences between Israeli women with benign and cancerous breast lesions. J. psychosom. Res. 19 (1975) 229–234.

Schors, R.: Zur Problematik von Intensivstationen. Psyche 33 (1979) 343–363.

Schramm, A.: Polypathie und Multimorbidität. In: Lang, E. (Hrsg.): Praktische Geriatrie. Enke, Stuttgart 1988.

Schramm, A., H. Franke, W. Chowanetz: Multimorbidität und Polypathie im Alter. Z. Allgemeinmed. 58 (1982) 234.

Schramm, J., F. Oppel, W. Umbach, R. Wüllenweber: Komplizierte Verläufe nach lumbalen Bandscheibenoperationen. Nervenarzt 49 (1978) 26–33.

Schrappe, O.: Psychosen bei Endokrinopathien. Internist 16 (1975) 10.

Schreiner, G. E., J. F. Maher: Hemodialyses for chronic renal failure. Medical, moral, ethical and socioeconomic problems. Ann. intern. Med. 62 (1965) 509–518; 551–557.

Schreml, W.: Supportive Maßnahmen bei der internistischen Tumorbehandlung: Schmerztherapie. In: Drings, P., W. Schreml (Hrsg.): Supportive Maßnahmen bei der internistischen Tumorbehandlung. Aktuelle Onkologie 7. Zuckschwerdt, München 1983.

Schreml, W., W. Hugl, B. Kossmann, H. Heimpel: Stufenplan der medikamentösen analgetischen Therapie für Tumorpatienten – eine prospektive Studie. Tumordiagnostik und Therapie 4 (1983) 189–196.

Schreml, W., W. Merkle, H. Heimpel: Medikamentöse Schmerztherapie bei Krebspatienten. Med. Klin. 76 (1981) 43–47.

Schrenk, M.: Der Oralbereich als psychosomatisches Forum. Unveröffentlichtes Manuskript des 1. Symp. Int. di Psychodontia, Como 1962.

Schroder, H. M., M. J. Driver, S. Streufert: Human information processing. Wiley, New York 1967.

Schroeder, J. S., J. Motta, G. Guilleminault: Hemodynamic studies in sleep apnea. In: Guilleminault, G., W. Dement (eds.): Sleep Apnea Syndromes, pp. 177–196. New York 1978.

Schröpl, F.: Das nichtallergische Urticaria-Syndrom. Münch. med. Wschr. 123 (1981) 1007–1012.

Schröter, E.: Langzeitverlauf und Prognose der Colitis ulcerosa unter kombinierter konservativer Therapie. Diss., Lübeck 1977.

Schucker, B., D. R. Jacobs: Assessment of behavioral risk of coronary disease by voice characteristics. Psychosom. Med. 39 (1977) 219–228.

Schucman, H., W. N. Thetford: A comparison of personality traits in ulcerative colitis and migraine patients. J. abn. Psychol. 76 (1970) 443–452.

Schüffel, W.: Can medical students acquire patient-centered attitudes at medical schools? Psychother. and Psychosom. 40 (1983) 22–32.

Schüffel, W.: Definition: Funktionelle Entspannung. Praxis Psychother. Psychosom. 33 (1988) 130.

Schüffel, W.: (Hrsg.): Sprechen mit Kranken – Erfahrungen studentischer Anamnesegruppen. Urban & Schwarzenberg, München 1983.

Schüffel, W.: Inhalte des Lernens in der Anamnesegruppe und das Ziel einer integrationsorientierten Psychosomatik: Die Entwicklung eines individuellen und eines professionellen Situationskreises. In: Schüffel, W. (Hrsg.): Sprechen mit Kranken – Erfahrungen studentischer Anamnesegruppen. Urban & Schwarzenberg, München 1983.

Schüffel, W.: Patienten mit funktionellen Abdominalbeschwerden. Unveröffentlichte Habilitationsschrift, Ulm 1976.

Schüffel, W.: Psychosomatic medicine III: Patients of the psychosomatic consultant. Psychother. and Psychosom. 22 (1973) 192–195.

Schüffel, W.: Selbsthilfe der Patienten – eine Herausforderung für den Arzt und den Staatsbürger. In: Ditfurth, H.v. (Hrsg.): mannheimer forum. Boehringer Mannheim 1983/84.

Schüffel, W.: Sprechen in der Medizin. Das Konzept der Anamnesegruppen. Münch. med. Wschr. 130 (1988) 315–318.

Schüffel, W.: Training in anger: how not to communicate with one's medical seniors. Bibl. psychiat., Vol.159, pp. 39–47. Karger, Basel 1979.

Schüffel, W., O. Schonecke, W. Wolfert: Patienten mit funktionellen Beschwerden im Abdominalbereich – Psychologische Charakteristik und Konsequenzen für Behandlung und Umgang mit diesen Patienten. Verh. Dtsch. Ges. Inn. Med. 77 (1971) 118–119.

Schüffel, W., P. Hahn, J. Hehl: Zum Begriff der Herzneurose. Unveröffentlichtes Manuskript (1972).

Schüffel, W., S. Bregulla, U. Egle, C. Räber, A. Schneider, A. Steinert: Asconauten und Ackerbauern. In: Pöldinger, H. (Hrsg.): Beziehungdiagnostik und Beziehungtherapie. Springer, Heidelberg 1983.

Schüffel, W., U. Egle, A. Schneider: Studenten sprechen mit Kranken. Anamnesegruppen als Ausbildungsform. Münch. med. Wschr. 39 (1983) 845–848.

Schüffel, W., U. Egle, U. Schairer, A. Schneider: Does history-taking effect learning of attitudes? Psychother. and Psychosom. 31 (1979) 81–92.

Schürenberg, B., G. Kittel: Zur Pathogenese von Kontaktgranulomen. Arch. Ohr.-Nas.-Kehlk.-hk. (Suppl.) 1988/II, S. 76.

Schüßler, G., E. Leibing: Coping und Abwehr – erste empirische Befunde einer multidimensionalen Erfassung. In: Muthny, F. A. (Hrsg.): Krankheitsverarbeitung: Hintergrundtheorien und klinische Erfassung. Springer, Berlin 1989.

Schüßler, G., K. Spiess, U. Rüger: Krankheitsbewältigung bei der rheumatischen Arthritis. Psychosom. Med. Psychoanal. 34 (1988) 291–305.

Schütz, R.: Körperbezogene Therapien in einer psychosomatischen Klinik. In: Studt, H. H. (Hrsg.): Psychosomatik in Forschung und Praxis. Urban & Schwarzenberg, München-Wien-Baltimore 1983.

Schütz, R., F. Besuden, S. Mang: Körperbildtheorie und körperbezogene Psychotherapie analytischer Orientierung. Vortrag auf der 28. Arbeitstagung des DKPM, Innsbruck 1988.

Schütz, R., F. Besuden, S. Mang: Das körperliche Selbsterleben als Differenzierungshilfe bei der Colitis ulcerosa vs. Morbus Crohn. In: Rechenberger, H.-G., H.-V. Werthmann (Hrsg.): Psychotherapie und Innere Medizin. Pfeiffer, München 1988.

Schütze, G.: Anorexia nervosa. Huber, Bern 1980.

Schulte, D.: Der diagnostisch-therapeutische Prozeß in der Verhaltenstherapie. In: Schulte, D. (Hrsg.): Diagnostik in der Verhaltenstherapie. Urban & Schwarzenberg, München 1976.

Schulte, D.: Diagnostik in der Verhaltenstherapie. In: DGVT (Hrsg.): Theorien und Methoden der Verhaltenstherapie. Mitteilungen der DGVT 1980, Sonderheft II.

Schulte, D.: Ein Schema für Diagnose und Therapieplanung in der Verhaltenstherapie. In: Schulte, D. (Hrsg.): Diagnostik in der Verhaltenstherapie. Urban & Schwarzenberg, München 1976.

Schulte, D.: Verhaltenstherapeutische Diagnostik. In: DGVT (Hrsg.): Verhaltenstherapie – Theorien und Methoden. Forum für Verhaltenstherapie und psychosoziale Praxis 1986, Bd. 11.

Schulte, W.: Neurologie des Praktikers. Med. Klinik 44 (1954) 1760–1762.

Schulte, W., H. Neus, A. W. v. Eiff: Blutdruckreaktivität unter emotionalem Streß bei unkomplizierten Formen des Bluthochdrucks. Klin. Wschr. 59 (1981) 1243–1249.

Schulte, W., H. Neus, H. Rüddel: Zum Blutdruckverhalten unter emotionalem Streß bei Normotonikern mit familiärer Hypertonieanamnese. Med. Welt 29 (1981) 2–4.

Schulte, W., R. Tölle (Hrsg.): Psychiatrie. Springer, Berlin-Heidelberg-New York 1975; 5. Aufl. 1979.

Schulte Holthausen, G.: Das klinische Bild der psychogenen Dysphonie im Wandel der Zeit. Sprache-Stimme-Gehör 5 (1981) 108.

Schultheis, K.-H., Th. v. Uexküll: Psychosomatische Aspekte des Morbus Crohn. In: Uexküll, Th.v. (Hrsg.): Lehrbuch der Psychosomatischen Medizin. Urban & Schwarzenberg, München-Wien-Baltimore 1979.

Schultz, J. H.: Arzt und Neurose, 2. Aufl. Thieme, Stuttgart, 1953.

Schultz, J. H.: Das Autogene Training. Thieme, Stuttgart 1966, 1970, 1976.

Schultz, J. H.: Die konstitutionelle Nervosität. In: Bumke, O. (Hrsg.): Handbuch der Geisteskrankheiten, Bd. V, S. 28. Springer, Berlin 1928.

Schultz, J. H.: Hypnose-Technik, 8. Aufl. Fischer, Stuttgart-New York 1983.

Schultz, J. H.: Psyche und Parafunktionen. Dtsch. zahnärztl. Z. 16 (1961) 1459.

Schultz, J. H.: Psychotherapie. Fischer, Stuttgart 1963.

Schultz, J. H.: Psychotherapie. Hippokrates, Stuttgart 1952.

Schultz, J. H.: Übungsheft für das autogene Training, 15. Aufl. Thieme, Stuttgart 1972.

Schultz, U.: Status pseudoepilepticus. In: Studt, H. H. (Hrsg.): Psychosomatik in der inneren Medizin 1, S. 109–119. Springer, Berlin-Heidelberg-New York 1986.

Schultz, U., A. Stäbler, T. Weiss: CT-controlled history of conservatively treated acute lumbar disc lesions. In: Poeck, K., W. Hacke, R. Schneider (Hrsg.): Verh. Dtsch. Ges. Neurologie, Bd. IV, S. 598–600. Springer, Berlin-Heidelberg-New York 1987.

Schultz, U., D. Köhler, M. Kütemeyer, A. Stäbler-Lehr: Zum Spontanverlauf des lumbalen Discusvorfalls. Eine CT-kontrollierte prospektive Studie. Nervenarzt 59 (1988) 661–668.

Schultz, W.: Dtsch. med. Wschr. 87 (1962) 1568.

Schultz-Coulon, H. J.: Die Diagnostik der gestörten Stimmfunktion. Arch. Ohr.-Nas.-Kehlk.-khk. 227 (1980) 1.

Schultz-Hencke, H.: Einführung in die Psychoanalyse. Thieme, Jena 1927.

Schultz-Ruthenberg, C.: Untersuchung über die Auswirkung und Verarbeitung eines nicht erfüllten Kinderwunsches. Diss., Berlin 1980.

Schultz-Zehden, W.: Eine psychosomatisch ausgerichtete Augenpraxis – ein Erfahrungsbericht. In: Studt, H. H. (Hrsg.): Psychosomatik in Forschung und Praxis, S. 536–545. Urban & Schwarzenberg, München-Wien-Baltimore 1983.

Schultz-Zehden, W.: Psychosomatik in der Augenheilkunde. In: Jores, A. (Hrsg.): Praktische Psychosomatik. 2. Aufl. Huber, Bern-Stuttgart-Wien 1981.

Schultz-Zehden, W.: Sehen – Ganzheitliches Augentraining. Gräfe & Unzer, München 1989.

Schultz-Zehden, W., F. Bischof: Auge und Psychosomatik. Deutscher Ärzte-Verlag, Köln 1986.

Schultze, H., H. Strauss, B. Pitt: Sudden death in the year following myocardial infarction – relation to ventricular premature contractions in late hospital phase and left ventricular ejection fraction. Amer. J. Med. 62 (1977) 192.

Schultze-Werninghaus, G.: Prävalenz des Asthmas. In: Schultze-Werninghaus, G., M. Debelic (Hrsg.): Asthma. Grundlagen – Diagnostik – Therapie, S. 3–9. Springer, Berlin-Heidelberg 1988.

Schulz, K.-H., A. Raedler: Tumorimmunologie und Psychoimmunologie als Grundlagen für die Psychoonkologie. Psychother. med. Psychol. 36 (1986) 114–129.

Schulz, K.-H., R. Ferstl: Psychoimmunologische Forschung in der Bundesrepublik Deutschland 1987. In: Florin, I., K. Hahlweg, G. Haag, U. B. Brack, E.-M. Fahrner (Hrsg.): Perspektive Verhaltensmedizin, S. 20–32. Springer, Berlin 1989.

Schulz, R., D. Aderman: Clinical research and the stages of dying. Omega 5/2 (1974) 137–143.

Schulz, R., D. Aderman: How the medical staff copes with dying patients: a critical review. Omega 7/1 (1976) 16–17.

Schulz van Treeck, A.: Neuro-vegetative Hörstörungen und ihre Behandlung. Wiss. Z. Univ. Halle 5 (1956) 1059.

Schulze, A., K. Gaebler: Verlaufstypen des Torticollis spasmodicus – eine Evaluationsstudie. Psychother. med. Psychol. 38 (1988) 401–404.

Schumacher, W.: Das hirnorganische Psychosyndrom in Klinik und Praxis. Therapiewoche 33 (1983) 4159.

Schumacher, W.: Psychotherapeutische Aspekte zur Situation des Herzoperierten. Z. Psychother. med. Psychol. 15 (1965) 12–23.

Schumann, H.-J. v.: Erotik und Sexualität in der zweiten Lebenshälfte. Hippokrates, Stuttgart 1980.

Schumm, F.: Myasthenia gravis und Myastheniesyndrom. In: Brandt, Th., J. Dichgans, H. C. Diener (Hrsg.): Therapie und Verlauf neurologischer Erkrankungen, S. 840–863. Kohlhammer, Stuttgart 1988.

Schunk, J.: Emotionale Faktoren in der Pathogenese der essentiellen Hypertonie. Klin. Med. 152 (1954) 251.

Schur, M.: Comments on the metapsychology of somatization. Psychoanal. Stud. Child 10 (1955) 119–164. (Dt. in: Brede, K. (Hrsg.): Einführung in die Psychosomatische Medizin, S.335–395. Frankfurt 1974.)

Schur, M.: The ego in anxiety. In: Loewenstein, R. M. (ed.): Drives, Affects, Behavior, pp. 67–103. New York 1953.

Schur, M.: Zur Metapsychologie der Somatisierung (1955). In: Brede, K. (Hrsg.): Einführung in die Psychosomatische Medizin. Fischer, Frankfurt 1974.

Schuster, H. P., L. S. Weilemann: Behandlungsergebnisse internistischer Intensiveinheiten. Med. Klinik 76 (1981) 419–422.

Schuster, M.: Kunsttherapie. Die heilende Kraft des Gestaltens. DuMont, Köln 1986.

Schuster, P., H. Strotzka: Diskussionsbeiträge zu Harvey Bluestones Aufsatz. Forum Psychoanal. 1 (1985) 318–320.

Schuster-Erfmann, J.: Die Hypertoniepatienten in psychologischer Sicht. In: Eiff, A. W. v., G. Kloska, H. Quint: Essentielle Hypertonie. Klinik, Psychophysiologie und Psychopathologie. Thieme, Stuttgart 1967.

Schwab, J. J.: Differential characteristics of medical inpatients referred for psychiatric consultation: a controlled study. Psychosom. Med. 27 (1965) 112.

Schwab, J. J.: Problems in psychosomatic diagnosis I: A controlled study of medical inpatients. Psychosomatics 6 (1964) 369.

Schwab, J. J.: Psychiatric illness in medical patients: why it goes undiagnosed. Psychosomatics 23 (1982) 225–229.

Schwab, J. J., R. S. Clemons, M. J. Valder, J. D. Raulerson: Medical patients' reactions to referring physicians after psychiatric consultation. J. Amer. med. Ass. 195 (1966) 1120–1122.

Schwartz, D., M. Thompson, C. Johnson: Eating disorders and the culture. In: Darby, P. L., P. E. Garfinkel, D. M. Garner, D. V. Coscina (eds.): Anorexia nervosa. Recent Developments in Research, pp. 83–94. Liss, New York 1983.

Schwartz, D., M. Thompson, C. Johnson: Anorexia nervosa and bulimia: the socio-cultural context. Int. J. Eat. Dis. 3 (1982) 20–36.

Schwartz, E., L. Fair, S. Greenberg, M. R. Mandel, G. L. Klermann: Facial expression and depression. An electromyographical study. Amer. Psychosom. Soc., Annual Meeting, Philadelphia 1974.

Schwartz, F. W.: Annahmen und Wissen zum Gesundheitszustand alter Menschen. In: Karl, F., W. Tokarski (Hrsg.): Die „neuen Alten". Beiträge der XVII. Jahrestagung der Dtsch. Ges. f. Gerontologie 1988. Kasseler Gerontol. Schriften 6, Gesamthochschulbibliothek Kassel 1989.

Schwartz, G. E.: Toward a theory of voluntary control of response patterns in the cardiovascular system. In: Obrist, P. A., A. H. Black, J. Brener, L. V. Dicara (eds.): Cardiovascular Psychophysiology. Aldine, Chicago 1974.

Schwartz, G. E., P. L. Fair, P. S. Greenberg, M. R. Mandel, G. L. Klermann: Facial expression and depression. An electromyographic study. Ann. Meeting Amer. Psychosom. Soc., Philadelphia, March 29–31, 1974.

Schwartz, H. J.: Bulimia: psychoanalytic perspectives. In: Schwartz, H. J. (ed.): Bulimia: Psychoanalytic Treatment and Theory. Int. Univ. Press, Madison 1988.

Schwartz, H. J.: Bulimia: psychoanalytic perspectives. J. Amer. psychoanal. Ass. 34 (1986) 439–461.

Schwartz, K.: Über penetrierende Magen- und Jejunalgeschwüre. Beitr. klin. Chir. 67 (1910) 96–128.

Schwartz, P. J.: Cardiac sympathetic innervation and the sudden infant death syndrome. Amer. J. Med. 60 (1976) 167–172.

Schwartz, R. C.: Bulimia and family therapy: a case study. Int. J. Eat. Dis. 2 (1982) 75–82.

Schwarz, D.: Verhaltensanalyse und Verhaltenstherapie bei Patienten mit funktionellen Herzbeschwerden. In: Köhle, K. (Hrsg.): Zur Psychosomatik von Herz-Kreislauferkrankungen. Springer, Berlin 1982.

Schwarz, O.: Psychogenese und Psychotherapie körperlicher Symptome. Springer, Wien 1925.

Schwarz, O.: Über psychogene Nierenschmerzen. Allg. ärztl. Z. Psychother.1 (1928) 28.

Schwarz, R., S. Geyer: Social and psychological differences between cancer and noncancer patients: cause or consequence of the disease? Psychother. and Psychosom. 41 (1984) 195–199.

Schwöbel, A.: Analyse einer 60jährigen Migräne-Kranken. Z. psychosom. Med. 11 (1965) 164.

Schwöbel, G.: Der Blinzeltic und seine phänomenologische Bedeutung. Z. psychosom. Med. 12 (1966) 264–275.

Schyve, P. M., F. Smithline, H. Y. Meltzer: Neuroleptic-induced prolactin level elevation and breast cancer. Arch. gen. Psychiat. 35 (1978) 1291–1301.

Sclare, A. B., B. Lubin, R. Dopson: The group psychotherapy. Literature 1979. Int. J. Group Psychother. Oct. 1980.

Scotch, N. A., H. J. Geiger: The epidemiology of rheumatoid arthritis. J. chron. Dis. 15 (1962) 1037–1067.

Scotch, N. A., J. H. Geiger: Epidemiology of essential hypertension: psychology and sociocultural factors in etiology. J. chron. Dis. 16 (1963) 1183–1213.

Scott, B., P. Lindberg, L. Lyttkens, L. Melin: Psychological treatment of tinnitus. Scand. Audiol. 14 (1985) 223.

Scott, D. L.: Psychiatric problems of hemodialysis: their treatment by hypnosis. Brit. J. Psychiat. 122 (1973) 91–92.

Scott, N. A., R. A. DeSilva, B. Lown, R. J. Wurtman: Tyrosine administration decreases vulnerability to ventricular fibrillation in the normal canine heart. Science 211 (1981) 727–729.

Searle, J.: Minds, brains and science. Harvard Univ. Press, Cambridge/Mass. 1984.

Searles, H. F.: Collected papers on schizophrenia and related subjects. J. & P., New York 1965. Deutsch: Kindler, München 1974.

Sebastian, U.: Bioenergetische Analysen nach Lowen und Reich. In: Toman, W., R. Egg (Hrsg.): Psychotherapie. Ein Handbuch. Bd. I, S. 152. Kohlhammer, Stuttgart-Berlin-Köln-Mainz 1985.

Sebastian, U.: Psychoanalytische Theorie und Bioenergetische Analyse. MAKS Publikationen, Münster 1983.

Sedvall, G., L. Farde, A. Persson, F.-A. Wiesel: Imaging of neurotransmitter receptors in the living human brain. Arch. gen. Psychiat. 43 (1986) 995–1005.

Seeman, P., C. Ulpian, C. Bergeron, P. Riederer, K. Jellinger, E. Gabriel, G. P. Reynolds, W. W. Tourtellotte: Bimodal distribution of dopamine receptor densities in brains of schizophrenics. Science 225 (1984) 728–731.

Seer, P.: Psychological control of essential hypertension: review of the literature and methodological critique. Psychol. Bull. 5 (1979) 1015–1043.

Seer, P., J. M. Raeburn: Meditation training and essential hypertension: a methodological study. J. Behav. Med. 3 (1980) 59–71.

Segura, J.-W., J.-L. Opitz, F. Greene: Prostatosis, prostatitis or pelvic floor tension myalgia. J. Urol. 192 (1979) 168.

Seidel, A.: Beschwerden am Bewegungsapparat ohne ausgeprägten organischen Befund. Z. Allgemeinmed. 51 (1975) 1356–1359.

Seidensticker, J., M. Tzagournis: Anorexia nervosa – clinical features and long-term follow-up. J. chron. Dis. 21 (1968) 361–367.

Seidler, H.: Die psychosoziale Verarbeitung chronischer Krankheit bei Multiple Sklerose-Kranken und Querschnittsgelähmten. Med. Diss., Göttingen 1978.

Seifert, K.: Funktionelle Störungen der vorderen Halsorgane im ursächlichen Zusammenhang mit Funktionsstörungen des zervikokranialen Überganges. Neuroorthopädie 4 (1988) 211.

Seiler, T. B.: Kognitive Strukturiertheit – Theorien, Analysen, Befunde. Kohlhammer, Stuttgart 1973.

Selby, G., J. W. Lance: Observations on 500 cases of migraine and allied vascular headache. J. Neurol., Neurosurg., Psychiat. 23 (1960) 23–32.

Seligman, M. E. P.: Helplessness. On depression, development, and death. Freeman, San Francisco 1975.

Seligman, M. E. P.: Phobias and preparedness. Beh. Ther. 2 (1971) 307–320.

Seligman, M. E. P., D. Groves: Non-transient learned helplessness. Psychosom. Sci. 19 (1970) 191–192.

Seligman, M. E. P., J. L. Hager (eds.): Biological Boundaries of Learning. Appleton Century Crofts, New York 1972.

Seligman, M. E. P., R. A. Rossellini, U. M. Kozak: Learned helplessness in the rat: reversibility, time course, and immunization. J. comp. physiol. Psychol. 88 (1975) 542–547.

Sellheim: zit. nach Artner, J. (1929).

Sellink, J. L.: Radiological atlas of common diseases of the small bowel. Stenfert Kroese, Leiden 1976.

Sellschopp, A.: Psychosoziale Probleme bei der Rehabilitation kurativ behandelter Patienten. Münch. med. Wschr. 126 (1984) 227–230.

Sellschopp, A., H. Luedeke, G. Hartel: Structure and functions of the Heidelberg University Organization for aftercare of cancer patients. Psychother. and Psychosom. 36 (1981) 17–23.

Selvin, S. E.: Termination of hemodialysis treatment – staff reactions. In: Levy, N. B. (ed.): Psychonephrology 2, pp. 117–129. Plenum, New York 1983.

Selvini-Palazzoli, M.: Die Familie des Anorektikers und die Familie des Schizophrenen. Ehe, Zentralblatt für Ehe und Familienkunde 12 (1976) 107.

Selvini-Palazzoli, M.: Self-Starvation. Human Context, London 1974. Deutsch: Magersucht. Klett-Cotta, Stuttgart 1982.

Selvini-Palazzoli, M., L. Boscolo, G. Cecchin, G. Prata: Paradosso e controparadosso. Feltrinelli, Mailand 1975. Deutsch: Paradoxon und Gegenparadoxon. Klett, Stuttgart 1977.

Selvini-Palazzoli, M., L. Boscolo, G. Cecchin, G. Prata: Hypothesizing – circularity – neutrality: three guidelines for the conductor of the session. Family process 19 (1980) 3. Deutsch: Hypothetisieren – Zirkularität – Neutralität: Drei Richtlinien für den Leiter der Sitzung. Familiendynamik 6 (1981) 123.

Selye, H.: A syndrome produced by diverse noxious agents. Nature (Lond.) 32 (1936) 138.

Selye, H.: The evolution of the stress concept. Amer. Scientist 61 (1973) 692–699.

Selye, H.: The general adaptation syndrome and the diseases of adaptation. J. clin. Endocr. 6 (1946) 117–230.

Selye, H.: The stress of life. McGraw-Hill, New York 1956.

Selye, H.: Thymus and adrenals in the response of the organism to injuries and intoxications. Brit. J. exp. Path. 17 (1936) 234.

Selye, H.: The evolution of the stress concept – stress and cardiovascular disease. In: Levi, L. (ed.): Society, Stress, and Disease, Vol. I. Oxford Univ. Press, London 1971.

Selye, H.: The stress concept today. In: Kutash, I. L., L. B. Schlesinger et al. (ed.): Handbook on Stress and Anxiety. Jossey Bass, San Francisco 1981.

Selye, H., S. Szabo: Experimental model for production of perforating duodenal ulcers by cystamine in the rat. Nature (Lond.) 244 (1973) 458–459.

Senarclens, M. de, W. Fischer: Aménorrhée: féminité impossible? Etude socio-psychosomatique. Masson, Paris 1978.

Senf, W.: Problempatienten. In: Senf, W., H. Becker (Hrsg.): Praxis der stationären Psychotherapie. Thieme, Stuttgart 1988 b.

Senf, W.: Anthropologische Gesichtspunkte der Stimme. Vortrag, II. Kommunikationsmedizinische Tage, 15.–17. 4. 1988, Bad Rappenau.

Senf, W.: Was hilft in der Psychotherapie? Sicht der Therapeuten – Rückblick der Patienten. Prax. Psychother. Psychosom. 1988 a.

Senf, W., G. Schneider-Gramann: Was hilft in der Psychotherapie im Rückblick der Patienten? In: Tschuschke, V., D. Czogalik (Hrsg.): Wirkfaktoren in der Psychotherapie. Springer, Berlin-Heidelberg-New York (im Druck).

Senf, W., H. Becker (Hrsg.): Praxis der stationären Psychotherapie. Thieme, Stuttgart 1988.

Senf, W., H. Kordy, M.v. Rad, W. Bräutigam: Indications in psychotherapy on the basis of a follow-up study. Psychother. and Psychosom. 42 (1984) 37–47.

Senn, H. J., A. Glaus: Schmerzen und Schmerzbekämpfung bei Tumorkrankheiten. Schweiz. med. Wschr. 112 (1982) 1158–1164.

Seres, J. L., R. I. Newman: Results of treatment of chronic low-back pain at the Portland Pain Center. J. Neurosurg. 45 (1976) 437–441.

Sergl, H. G.: Zusammenhang zwischen psychischen Schwierigkeiten und dem Auftreten von Kieferanomalien. Fortschr. Kieferorthop. 28 (1967) 57.

Serratrice, G.: Aspekte der Migräne für die Praxis. Huber, Bern 1976.

Serratrice, G., S. Serbanesco, R. Sambuc: Epidemiology of headache in the elderly – correlations with life conditions and socio-professional environment. Headache 25 (1985) 85–89.

Seward, G. H. et al.: Personality structure in a common form of colitis. Psychol. Monogr. 65 (1965) 1–26.

Sewitch, D. E.: The perceptual uncertainty of having slept: the inability to discriminate electroencephalographic sleep from wakefulness. Psychophysiol. 21 (1984) 243–259.

Shader, R. J., D. R. Weinberger, D. J. Greenblatt: Psychopharmacologic approaches to the medically ill patient. In: Karasu, T. B., R. J. Steinmüller (eds.): Psychotherapeutics in Medicine. Grune & Stratton, New York 1978.

Shaffer, J. W., P. L. Graves, R. T. Swank, T. A. Pearson: Clustering of personality traits in youth and the subsequent development of cancer among physicians. J. behav. Med. 10 (1987) 441–447.

Shagaas, C., R. B. Malmo: Psychodynamic themes and localized muscular tension during psychotherapy. Psychosom. Med. 16 (1954) 295.

Shambaugh, P. W., S. Kanter: Spouses under stress: group meetings with spouses of patients on hemodialysis. Amer. J. Psychiat. 125 (1969) 928–936.

Shanan, J., A. Kaplan De-Nour, J. Garty: Effects of prolonged stress on coping style in terminal renal failure patients. J. hum. Stress 2 (1976) 19–27.

Shands, H. C.: An approach to the measurement of suitability for psychotherapy. Psychiat. Quart. 32 (1958) 1–22.

Shands, H. C.: How are psychosomatic patients different from psychoneurotic patients? Psychother. and Psychosom. 26 (1975) 270–285.

Shands, H. C. et al.: Psychological mechanisms in patients with cancer. Cancer 4 (1951) 1159–1170.

Shanfield, S. B.: Predicting bereavement outcome: marital factors. Family Systems Medicine 1 (1983) 40.

Shapiro, A. K.: A historic and heuristic definition of the placebo. Psychiatry 27 (1964) 52–58.

Shapiro, A. K.: Placebo effects in medical and psychological therapies. In: Bergin, A. E., S. L. Garfield (eds.): Handbook of Psychotherapy and Behavioral Change: An Empirical Analysis. Wiley, New York 1978.

Shapiro, A. K., E. S. Shapiro, R. D. Brown, R. D. Sweet: Gilles de la Tourette syndrome. Raven, New York 1978.

Shapiro, D. A., H. L. Caplan, P. D. Rhode, J. P. Watson: Personal questionnaire changes and their correlates in a psychotherapeutic group. Brit. J. med. Psychol. 48 (1975) 207–215.

Shapiro, M. F., A. F. Lehman, S. Greenfield: Biases in the laboratory diagnosis of depression in medical practice. Arch. intern. Med. 143 (1983) 2085–2088.

Share, L.: Family communication in the crisis of a child's fatal illness. Omega 3 (1972) 187.

Sharpe, E. F.: Psychophysiological problems revealed in language. An examination of metaphor. Int.J. Psychoanal. 21 (1940) 201–213.

Shavit, Y., J. W. Lewis, G. W. Terman, R. P. Gale, J. C. Liebeskind: Opioid peptides mediate the suppressive effect of stress on natural killer cell cytotoxicity. Science 223 (1984) 188–190.

Shaw, E. B.: The outbreak of meningitis (editorial). Calif. Med. 102 (1965) 234.

Shaye, R.: Bewältigungsstrategien bulimischer Frauen. In: Speidel, H., B. Strauß (Hrsg.): Zukunftsaufgaben der psychosomatischen Medizin. Springer, Berlin 1989.

Shea, B. et al.: Hemodialysis for chronic renal failure. Ann. intern. Med. 62 (1965) 558–563.

Shealy, C. N., J. T. Mortimer, J. B. Reswick: Electrical inhibition of pain by stimulation of dorsal columns: Preliminary clinical report. Anaesth. Analg. Curr. Res. 46 (1967) 489–491.

Shear, M. K.: Pathophysiology of panic: a review of pharmacology provocative tests and naturalistic monitoring data. J. clin. Psychiat. 47 (1986) 18–26 (Suppl.).

Shearn, M. A., B. H. Fireman: Stress management and mutual support groups in rheumatoid arthritis. Amer. J. Med. 78 (1985) 771–775.

Sheehan, D. V.: Current concepts in psychiatry: panic attacks and phobias. New Engl. J. Med. 307 (1982) 156–158.

Sheehan, D. V., J. H. Coleman, D. J. Greenblatt, K. J. Jones, P. H. Levine, P. J. Orsulak, M. Peterson, J. J. Schildkraut, E. Uzorgara, D. Watkins: Some biochemical correlates of panic attacks with agoraphobia and their response to a new treatment. J. clin. Psychopharm. 4 (1984) 66–75.

Sheehy, M. P., C. D. Marsden: Writer's cramp – a focal dystonia. Brain 105 (1982) 461–480.

Sheffield, B. F., M. P. Carney: Crohn's disease: a psychosomatic illness? Brit. J. Psychiat. 128 (1976) 446–450.

Shekelle, R. B.: Educational status and risk of coronary heart disease. Science 163 (1969) 97–98.

Shekelle, R. B., S. Hulley, J. Neaton, J. Billigs, N. Borhani, T. Gerace, D. Jacobs: Type A behavior and risk of coronary heart disease in MRFIT. In: Schmidt, T., T. Dembroski, G. Blümchen (eds.): Biological and Psychological Factors in Cardiovascular Disease, pp. 41–55. Springer, Berlin-Heidelberg-New York 1986.

Shekelle, R. B., A. M. Ostfeld, O. Paul: Social status and incidence of coronary heart disease. J. chron. Dis. 22 (1969) 381.

Shekelle, R. B., J. A. Schoenberger, J. Stamler: Correlates of the Type A behavior pattern score. J. chron. Dis. 29 (1976) 381–394.

Shekelle, R. B., M. Gale, A. M. Ostfeld, O. Paul: Hostility, risk of coronary heart disease, and mortality. Psychosom. Med. 45 (1983 a) 109–114.

Shekelle, R. B., S. Hulley, J. Neaton et al. for the MRFIT Research Group: The MRFIT behavior pattern study: II. Type A behavior pattern and incidence of coronary heart disease. CVD Epidemiol. Newsletter Jan. (1983 b) 34.

Shekelle, R. B., S. B. Hulley, J. D. Neaton: Type A behavior and risk of coronary heart disease in MRFIT. Paper presented at the conference „Biobehavioral factors in coronary heart disease", Winterscheid 1984.

Shekelle, R. B., W. J. Raynor, A. M. Ostfeld, D. C. Garron, L. A. Bieliauskas, C. Liu, C. Maliza, O. Paul: Psychological depression and 17-year risk of death from cancer. Psychosom. Med. 43 (1981) 117–125.

Sheldon, A., C. Ryser, P. Krant, J. Melvin: An integrated family orientated cancer care program. J. chron. Dis. 22 (1970) 743.

Shem, S.: The house of god. Dell, New York 1980.

Shenkin, H. A.: The effect of pain on the diurnal pattern of plasma corticoid levels. Neurology 14 (1964) 1112–1117.

Sheppard, G., J. Gruzelier, R. Manchanda, S. R. Hirsch, R. Wise, R. Frackowiak, T. Johnson: 15O-positron emission tomographic scanning in predominantly never-treated acute schizophrenic patients. Lancet II (1983) 1448–1452.

Sherman, E. D.: Sensitivity to pain (with an analysis of 450 cases). Canad. med. Ass. J. 48 (1943) 437–441.

Sherman, N. A., R. S. Smith, E. Middleton jr.: Effect of adrenergic compounds, aminophylline and hydrocortisone, on in vitro immunoglobulin synthesis by normal human peripheral lymphocytes. J. Allergy clin. Immunol. 52 (1973) 13–22.

Sherrington, R., J. Brynjolfsson, H. Petursson, M. Potter, K. Dudleston, B. Barraclough, J. Wasmuth, M. Dobbs, H. Gurling: Localization of a susceptibility locus for schizophrenia on chromosome 5. Nature (Lond.) 336 (1988) 164–167.

Shevach, E. M.: The effects of cyclosporin A on the immune system. Ann. Rev. Immunol. 3 (1985) 397–423.

Shewell, J., D. A. Long: A species differences with regard to the effect of cortisone acetate on body weight, globulin and circulating antitoxin levels. J. Hyg. (Lond.) 54 (1956) 452–460.

Shimizu, A., M. S. Richardson: Spouse relief program for home dialysis patients. Dial. Transplant. 10 (1981) 428–433.

Shimm, D. S., G. L. Logue, A. A. Maltbie, S. Dugan: Medical management of chronic cancer pain. J. Amer. med. Ass. 241 (1979) 2408–2412.

Shmerling, D. H.: Ileocolitis granulomatosa Crohn. pädiat. prax. 20 (1978) 197–205.

Shneidman, E. S.: Orientations toward death. In: White, R. W. (ed.): The Study of Lives, pp. 200–227. Atherton, New York 1966.

Shochet, B. R., E. T. Lisansky, A. F. Schubart, V. Fiocco, Sh. Kurland, M. Pope: A medical-psychiatric study of patients with rheumatoid arthritis. Psychosomatics 10 (1969) 271–279.

Shoelson, S., M. Haneda, P. Blix, A. Nanjo, T. Sanke, K. Inouye, D. Steiner, A. Rubenstein, H. Tager: Three mutant insulins in man. Nature 302 (1983) 540–543.

Short, M. J., W. P. Wilson: Roles of denial in chronic hemodialysis. Arch. gen. Psychiat. 20 (1969) 433–437.

Shorter, E.: Bedside manners: the troubled history of doctors and patients. Simon & Schuster, New York 1985.

Shorter, R. G., M. Chiba, W. R. Thayer, W. Bartnik, S. G. Remine: Further studies of cell-mediated immunity in inflammatory bowel disease – a preliminary report. Z. Gastroenterol. 17 (Suppl.) (1979) 72–76.

Shuval, J.: Professional socialization and medical care. In: Noack, H. (ed.): Medical Education and Primary Health Care. Croom Helm, London 1980.

Sibertin-Blanc, D.: Le nanisme psychogène. In: Lebovici, S., R. Diatkine, M. Soulé (eds.): Traité de psychiatrie de l'enfant et de l'adolescent, Vol. II., pp. 561–569. PUF, Paris 1985.

Sich, D.: Naeng: Begegnung mit einer Volkskrankheit in der modernen frauenärztlichen Praxis in Korea. Curare 2 (1979) 87–96.

Sicuteri, F.: Emotional vulnerability of the antinociceptive system: relevance in psychosomatic headache. Headache 21 (1981) 113–115.

Sicuteri, F., B. Anselmi, P. L. Del Bianco: 5-hydroxy-tryptamine supersensitivity as a new theory of headache and central pain: A clinical pharmacological approach with p-chlorophenylalanine. Psychopharmacologia 29 (1973) 347–356.

Sidman, M.: Avoidance behavior. In: Honig, W. H. (ed.): Operant Behavior, Areas of Research and Application. Appleton Century Crofts, New York 1966.

Sidman, M.: Normal sources of pathological behavior. Science 132 (1960) 61–68.

Siebert, G.: Kopfschmerz in der Zahnheilkunde. In: Gross, D., R. Frey (Hrsg.): Kopfschmerz, S. 68. Fischer, Stuttgart 1981.

Siegel, E. V.: Tanztherapie. Klett-Cotta, Stuttgart 1986.

Siegel, S.: Morphine analgesic tolerance: its situational specificity supports a pavlovian conditioning model. Science 193 (1976) 323–325.

Siegenthaler, W.: Die Medizin im Spannungsfeld der Umwelt. Verh. Dtsch. Ges. Inn. Med. 90, S. 47–56. Bergmann, München 1984.

Siegenthaler, W.: Eröffnungsansprache zum deutschen Internistenkongreß, Wiesbaden 1984.

Siegenthaler, W.: Klinische Pathophysiologie. Thieme, Stuttgart 1973.

Siegman, A., T. M. Dembroski, N. Ringel: Components of hostility and the severity of coronary artery disease. Psychosom. Med. 49 (1987) 127–135.

Siegrist, J.: Arbeit und Interaktion im Krankenhaus. Enke, Stuttgart 1978.

Siegrist, J.: Asymmetrische Kommunikation bei der klinischen Visite. Med. Klinik 45 (1976) 1962–1966.

Siegrist, J.: Asymmetrische Kommunikation bei klinischen Visiten. In: Köhle, K., H.-H. Raspe (Hrsg.): Das Gespräch während der ärztlichen Visite, S. 16–22. Urban & Schwarzenberg, München 1982.

Siegrist, J.: Die Bedeutung von Lebensereignissen für die Entstehung körperlicher und psychosomatischer Erkrankungen. Nervenarzt 51 (1980) 313–320.

Siegrist, J.: Erfahrungsstruktur und Konflikte bei stationären Patienten. Zschr. f. Soziologie 1 (1972) 271–280.

Siegrist, J.: Fragebogen zur Schlaf- und Belastungsanamnese bei Patienten mit koronarer Herzkrankheit bzw. koronaren Risikofaktoren (SBA): Information für den Arzt. Pharma-Schwarz GmbH, Monheim 1985.

Siegrist, J.: Lebensereignisse und Krankheitsausbruch – Ergebnisse und Probleme aus medizinsoziologischer Sicht. Sozialwissenschaftliche Annalen 1 B (1977) 57–69.

Siegrist, J.: Lehrbuch der medizinischen Soziologie. Urban & Schwarzenberg, München 1975.

Siegrist, J.: Soziale Belastung und psychische Dispositionen bei Herzinfarktpatienten – Grenzen individuenzentrierter Intervention. In: Langosch, W. (Hrsg.): Psychosoziale Probleme und psychosoziale Interventionsmöglichkeiten bei Herzinfarktpatienten, S. 47–62. Minerva, München 1980.

Siegrist, J.: Threat to social status and cardiovascular risk. Psychother. and Psychosom. 42 (1984) 90–96.

Siegrist, J., H. Matschinger, I. Weber, K. Siegrist, K. Dittmann, P. Brockmeier, D. Klein: Der Einfluß sozialer Belastungen und ihrer Verarbeitung auf die Entwicklung kardiovaskulärer Risiken – eine Längsschnittstudie an berufstätigen Männern. Arbeitsbericht zum DFG-Projekt Si 236/5–2, Marburg 1984 (unveröffentlicht).

Siegrist, J., H. Matschinger, P. Cremer, D. Seidel: Atherogenic risk in men suffering from occupational stress. Atherosclerosis 69 (1988) 211.

Siegrist, J., J. H. Peter: Schlafstörungen und kardiovaskuläres Risiko. Med. Klinik 81 (1986) 429–432.

Siegrist, J., K. Dittmann, K. Rittner, I. Weber: Psychosocial risk constellations and first myocardial infarction. In: Siegrist, J., M. J. Halhuber (eds.): Myocardial Infarction and Psychosocial Risks, pp. 41–57. Springer, Berlin-Heidelberg-New York 1981.

Siegrist, J., K. Dittmann, K. Rittner, I. Weber: Soziale Belastungen und Herzinfarkt. Enke, Stuttgart 1980.

Siegrist, J., K. Dittmann, K. Rittner, I. Weber: The social context of active distress in patients with early myocardial infarction. Soc. Sci. Med. 16 (1982) 443–453.

Siegrist, J., K. Dittmann: Lebensveränderungen und Krankheitsausbruch: Methodik und Ergebnisse einer medizinsoziologischen Studie. Kölner Z. f. Soziol. u. Sozialpsychol. 1 (1981) 132–147.

Siever, L. J., R. D. Coursey, J. S. Alterman, M. S. Buchsbaum, D. L. Murphy: Impaired smooth pursuit eye movement: vulnerability marker for schizotypal personality disorder in a normal volunteer population. Amer. J. Psychiat. 141 (1984) 1560–1565.

Siewert, J. R., F. E. Isemer: Prinzipien operativer Behandlung von Colitis ulcerosa und Morbus Crohn. In: Ottenjann, R., H. Fahrländer (Hrsg.): Entzündliche Erkrankungen des Dickdarms. Springer, Berlin 1983.

Sifneos, P. E.: The motivational process: a selection and prognostic criterias for psychotherapy of short duration. Psychiat. Quart. 42 (1968) 271–284.

Sifneos, P. E.: Problems of psychotherapy of patients with alexithymic characteristics and physical disease. Psychother. and Psychosom. 26 (1975) 65–70.

Sifneos, P. E.: The prevalence of „alexithymic" characteristics in psychosomatic patients. Psychother. and Psychosom. 22 (1973) 255–263.

Sigler, L. H.: Abnormalities in electrocardiogram induced by emotional strain. Amer. J. Cardiol. 8 (1961) 807.

Sigusch, V.: Aids als Risiko. In: Sigusch, V. (Hrsg.): Aids als Risiko. Konkret Literatur Verlag, Hamburg 1987.

Sigusch, V.: Sexualität und Medizin. Kiepenheuer und Witsch, Köln 1979.

Sigusch, V.: Therapie sexueller Störungen. Thieme, Stuttgart 1975.

Sigusch, V.: Über die Vergesellschaftung der Krankheit AIDS. In: Sigusch, V., S. Fliegel (Hrsg.): AIDS. DGVT, Tübingen 1988.

Silomon, H.: Darstellung der Aufgaben des Vertrauensärztlichen Dienstes aus sozialmedizinischer Sicht. Interne Beratungsunterlage der Arbeitsgemeinschaft für Gemeinschaftsaufgaben der Krankenversicherung, März 1983.

Silverman, J.: Anorexia nervosa: clinical and metabolic observations. Int. J. Eat. Dis. 2 (1983) 156–166.

Silverman, J.: Medical consequences of starvation. The malnutrition of anorexia nervosa: caveat medicus. In: Darby, P. L., P. E. Garfinkel, D. M. Garner, D. V. Coscina (eds.): Anorexia nervosa. Recent Developments in Research, pp. 293–299. Liss, New York 1983.

Silvers, I. J., M. F. Hovell, M. H. Weisman, M. R. Mueller: Assessing physician/patient perceptions in rheumatoid arthritis. Arthr. and Rheum. 28 (1985) 300–307.

Silverstein, M. D., D. E. Singer, A. G. Muky, G. E. Thibault, G. O. Barnett: Patienten mit Synkopen auf internistischen Intensivstationen. J. Amer. med. Ass. D 2/7 (1983) 239–243.

Silverstone, J. T., R. P. Gordon, A. J. Stunkard: Social factors in obesity in London. Practitioner 202 (1969) 682–688.

Sime, A. M.: Relationship of preoperative fear, type of coping, and information received about surgery to recovery from surgery. J. Pers. soc. Psychol. 34 (1976) 716–724.

Simmel, E.: Die psychoanalytische Behandlung in der Klinik. Int. Z. Psychoanal. 14 (1928) 352–370.

Simmel, E.: The psychoanalytic sanatorium and the psychoanalytic movement. Bull. Menninger Clin. 1 (1937) 133.

Simmet, H.: Wechselwirkung von Katathymem Bilderleben und kreativem Prozeß bei einem Fall von Enteritis regionalis (Morbus Crohn). In: Leuner, H. (Hrsg.): Katathymes Bilderleben, S. 312–322. Huber, Bern 1980.

Simmons, R. G.: Long-term reactions of renal recipients and donors. In: Levy, N. B. (ed.): Psychonephrology 2, pp. 275–287. Plenum, New York 1983.

Simon, F. B., H. Stierlin: Die Sprache der Familientherapie. Ein Vokabular. Klett, Stuttgart 1984.

Simon, M.: Ovarektomie muß keine Katastrophe auslösen. Med. Klinik 75 (1980) 842.

Simons, D. H., E. Day, H. Goodell, H. G. Wolff: Experimental studies on headache: muscles of the scalp and neck as sources of pain. Ass. Res. nerv. Dis. Proc. 23 (1943) 228.

Simpson, M. A.: Therapeutic use of truth. In: Wilkes, E. (ed.): The Dying Patient, pp. 255–262. MTP, London 1982.

Sims, A.: Neurosis and mortality: Investigation on association. J. psychosom. Res. 28 (1984) 353–362.

Sims, J.: Job stress: an aetiological factor in hypertension development? J. Psychophysiol. 2 (1988) 81.

Sims, J., D. Carroll, R. Turner, J. K. Hewitt: Cardiac and metabolic activity in mild hypertensive and normotensive subjects. Psychophysiol. 25 (1988) 172–178.

Sinaki, M., J. L. Merrit, G. K. Stillwell: Tension myalgia of the pelvic floor. Mayo Clin. Proc. 52 (1977) 717.

Singe, C. C., K. Ewe: Medikamentöse Therapie chronischentzündlicher Darmerkrankungen. Arzneimitteltherapie 3 (1985) 10–21.

Singer, H. C., J. G. D. Anderson, H. Frischer, J. B. Kirsner: Familial aspect of inflammatory bowel disease. Gastroenterology 61 (1971) 423–430.

Singer, M. T., L. C. Wynne: Thought disorder and family relations of schizophrenics: IV. Results and implications. Arch. gen. Psychiat. 12 (1965) 201–212.

Singh, U., D. S. Millson, P. A. Smith, I. I. T. Owen: Identification of β-adrenoreceptors during thymocyte ontogeny in mice. Eur. J. Immunol. 9 (1979) 31–35.

Singh, V., C. Saiphoo, D. Oreopoulos: Psychosocial and sexual aspects of diabetes on chronic peritoneal dialysis. Peritoneal Dialysis Bull., 1980.

Singleton, J. W.: Results of the national cooperative Crohn's disease study – USA. Z. Gastroenterol. 17 (Suppl.) (1979) 37–50.

Siris, S. G., A. Rifkin: The problem of psychopharmacotherapy in the medically ill. Psychiat. Clin. N. Amer. 4 (1981) 379–390.

Sirol, F.: Les céphalées récurrentes de la grande enfance et de l'adolescence. In: Lebovici, S., R. Diatkine, M. Soulé (eds.): Traité de psychiatrie de l'enfant et de l'adolescent, Vol. II., pp. 537–548. PUF, Paris 1985.

Sirol, F.: Les douleurs abdominales récurrentes de la grande enfance et de l'adolescence. In: Lebovici, S., R. Diatkine, M. Soulé (eds.): Traité de psychiatrie de l'enfant et de l'adolescent, Vol. II., pp. 551–560. PUF, Paris 1985.

Sitruk-Ware, R., B. Seradour, C. Lafaye: Treatment of benign breast diseases by progesterone applied topically. In: Maucais-Jarvis, Vickers, Wepierre (eds.): Percutaneous Absorption of Steroids. Acad. Press, New York 1980.

Sittaro, N.: Selbstwertgefühl und Verhaltensnormalität bei Patienten mit Colitis ulcerosa. Diss., Hannover 1980.

Skanse, B., W. v. Studnitz, N. Skoog: The effect of corticotropin and cortisone on serumlipids and lipoproteins. Acta endocr. (Copenh.) 31 (1959) 442–450.

Skelton, M., J. Dominian: Psychological stress on wives of patients with myocardial infarction. Brit. med. J. 2 (1973) 101–103.

Skinhøj, E.: Hemodynamic studies within the brain during migraine. Arch. Neurol. 29 (1973) 95–98.

Skinner, B. F.: Are theories of learning necessary? Psychol. Rev. 57 (1950) 193–216.

Skinner, B. F.: Science and human behavior. Free Press, New York 1953.

Skinner, B. F.: The behavior of organisms. Appleton Century Crofts, New York 1938.

Skinner, J. E., J. T. Lie, M. L. Entman: Modification of ventricular fibrillation latency following coronary artery occlusion in the conscious pig: the effect of psychological stress and beta-adrenergic blockade. Circulation 51 (1975) 656–667.

Sklar, L. S., H. Anisman: Stress and coping factors influence tumor growth. Science 205 (1979) 513–515.

Sklar, M.: Functional bowel distress and constipation in the aged. Geriatrics 27 (1972) 79–85.

Sklar, M.: Functional gastrointestinal disease in the aged. Amer. J. Gastroenterol. 53 (1979) 570.

Skrabal, F., J. Auböck, H. Hörtnagl: Low sodium/high potassium diet for prevention of hypertension: probable mechanisms of action. Lancet 1 (1981) 895–900.

Slade, P. D.: A short anorexic behavior scale. Brit. J. Psychiat. 122 (1973) 83–85.

Slade, P. D.: Towards a functional analysis of anorexia nervosa and bulimia nervosa. Brit. J. clin. Psychol. 21 (1982) 167–179.

Slade, P. D., G. F. M. Russel: Awareness of body dimensions in anorexia nervosa: cross-sectional and longitudinal studies. Psychol. Med. 3 (1973) 188–199.

Slade, P. D., G. F. M. Russel: Experimental investigations of bodily perception in anorexia nervosa and obesity. Psychother. and Psychosom. 22 (1973) 359–363.

Slany, E.: Kinderpsychotherapie im Rahmen der Dermatologie. Hautarzt 26 (1975) 419–422.

Slater, E., E. Glitherto: A follow-up of patients diagnosed as suffering from hysteria. J. Psychosom. Res. 9 (1965) 9–13.

Slater, P. (ed.): The measurement of interpersonal space by grid technique, Vol.I/II. London-New York 1977.

Slavicek, R.: Okklusionskonzept. Inf. aus Orthod. u. Kieferorth. 14 (1982) 171.

Slavson, S. R.: A textbook in analytic group psychotherapy. Int. Univ. Press, New York 1969.

Slawson, P. F., W. Flynn, E. J. Kollar: Psychological factors associated with the onset of diabetes mellitus. J. Amer. med. Ass. 185 (1963) 166–170.

Sloan, W. P., J. A. Bargen, R. P. Gage: Life histories of patients with chronic ulcerative colitis: a review of 2000 cases. Gastroenterology 16 (1950) 25–38.

Sloane, R. B., F. R. Staples, A. H. Cristol, N. J. Jorkston, K. Whipple: Psychotherapy vs. behavior therapy. Harvard Univ. Press, Harvard/Mass. 1975.

Sluzki, C. E.: Family consultation in family medicine. In: Wynne, L. C., S. McDaniel, T. T. Weber (eds.): Systems Consultation, pp. 168–180. Guilford, New York-London 1986.

Smalley, S. L., R. F. Asarnow, M. A. Spence: Autism and genetics: a decade of research. Arch. gen. Psychiat. 45 (1988) 953–961.

Smith, C. K., J. Barish, J. Correa, R. H. Williams: Psychiatric disturbance in endocrinologic disease. Psychosom. Med. 34 (1972) 69.

Smith, D. E., J. L. Glaser, R. H. Schneider, M. C. Dillbeck: Erythrocyte sedimentation rate (ESR) and the transcendental meditation (TM) program. Psychosom. Med. 51 (1989) 259.

Smith, G. R., C. Conger, D. F. O'Rourke, R. W. Steele, R.

Charlton, S. S. Smith: Modulation of cellular immunity by meditation. Psychosom. Med. 51 (1989) 246.

Smith, G. R., S. M. McDaniel: Psychologically mediated effect on the delayed hypersensitivity reaction to tuberculin in humans. Psychosom. Med. 45 (1983) 65–70.

Smith, M. L., G. U. Glass, Th. Miller: The benefits of psychotherapy. Johns Hopkins Univ. Press, Baltimore-London 1980.

Smith, P., R. Coles: Epidemiology of tinnitus. In: Feldmann, H. (ed.): Proceedings of the International Tinnitus Seminar Münster 1987, p. 183. Harsch, Karlsruhe 1987.

Smith, S. R., T. Bledsoe, M. K. Chnetri: Cortisol metabolism and the pituitary-adrenal axis in adults with protein-calorie malnutrition. J. clin. Endocr. 40 (1975) 43–52.

Smithers, D. W.: Maturation in human tumors. Lancet II (1969) 949–952.

Smits, A.: Familie und Krankheit: Eine theoretische Übersicht. Psychosozial 3 (1981) 66.

Smokler, L. A., H. Shevrin: Cerebral lateralization and personality style. Arch. gen. Psychiat. 36 (1979) 949–954.

Smotherman, W. P.: Glucocorticoid and other hormonal substrates of conditioned taste aversion. Ann. N. Y. Acad. Sci. 443 (1985) 126–144.

Smythe, H. A.: Fibrositis as a disorder of pain modulation. Clin. rheum. Dis. 5 (1979) 823–832.

Snell, L., S. Graham: Social trauma as related to cancer of the breast. Brit. J. Cancer 25 (1971) 721–734.

Snijders, J. T., N. Snijders-Oomen: Nichtverbale Intelligenzuntersuchungen für Hörende und Taube, 3. Aufl. Wolters, Groningen 1967.

Snow, H. (1893): Zit. in: Baltrusch, H. J. F.: Psyche – Nervensystem – neoplastischer Prozeß. Z. psychosom. Med. 9 (1963) 229; 10 (1964) 157.

Snyder, S. H.: Brain peptides as neurotransmitters. Science 209 (1980) 976.

Snyder, S. H.: Opiate receptors and internal opiates. Sci. Amer. 236 (1977) 44–67.

Sodemann, U., J. Toerkott, K. Köhle: Affekt-Themen in Visiten bei Patienten mit ungünstiger Prognose auf einer internistisch-psychosomatischen Krankenstation. In: Köhle, K., H.-H. Raspe (Hrsg.): Das Gespräch während der ärztlichen Visite, S. 210–231. Urban & Schwarzenberg, München 1982.

Söderberg, L., S. Sjöberg: On operated herniated lumbar discs. Acta orthop. scand. 31 (1961) 146–152.

Söllner, W., W. Wesiack: Zur Effizienz koordinierter Selbsthilfegruppen in der Behandlung psychosomatischer Störungen. In: Lamprecht, F. (Hrsg.): Spezialisierung und Integration in Psychosomatik und Psychotherapie. Springer, Berlin-Heidelberg 1987.

Sofia, R. D.: The effect of overcrowding stress on the development of adjuvant-induced polyarthritis in the rat. J. Pharm. Pharmacol. 32 (1980) 874–875.

Sokoloff, N. et al.: zit. nach: Montagu, A.: Körperkontakt. Klett, Stuttgart 1974.

Sokolow, M., D. Wedegar, H. K. Kain, A. T. Hinman: Relationship between level of blood pressure measured casually and by portable recorder, and severity of complications in essential hypertension. Circulation 34 (1966) 279–298.

Solberg, W. K.: Neuromuscular problems in the orofacial region: diagnosis – classification, signs and symptoms. Int. dent. J. 31 (1981) 206.

Solms-Wildenfels, I.: Psychosomatische Probleme im Altenheim. In: Bergener, M., B. Kark (Hrsg.): Psychosomatik in der Geriatrie. Steinkopff, Darmstadt 1985.

Solomon, C. M., P. S. Holzman, S. Levin, H. J. Gale: The association between eye-tracking dysfunctions and thought disorder in psychosis. Arch. gen. Psychiat. 44 (1987) 31–35.

Solomon, G. F.: Emotional and personality factors in the onset and course of autoimmune disease, particularly rheumatoid arthritis. In: Ader, R. (ed.): Psychoneuroimmunology, pp. 159–182. Acad. Press, New York 1981.

Solomon, G. F.: Immunologic abnormalities in mental illness. In: Ader, R. (ed.): Psychoneuroimmunology, pp. 259–278. Acad. Press, New York 1981 b.

Solomon, G. F.: Stress and antibody response in rats. Int. Arch. Allergy 35 (1969) 97–104.

Solomon, G. F., A. A. Amkraut, P. Kasper: Immunity, emotions and stress. Ann. clin. Res. 6 (1974) 313.

Solomon, G. F., A. A. Amkraut: Emotions, stress, and immunity. In: Insel, P. M., R. H. Moos (eds.): Health and the Social Environment, pp. 193–209. Heath, Lexington/Mass. – Toronto – London 1974.

Solomon, G. F., L. Temoshok: A psychoneuroimmunologic perspective on AIDS research: questions, preliminary findings and suggestions. J. appl. soc. Psychol. 17 (1987) 286–308.

Solomon, G. F., R. H. Moos: Emotions, immunity and disease. Arch. gen. Psychiat. 11 (1964) 657.

Solomon, G. F., S. Levine, J. K. Kraft: Early experience and immunity. Nature (Lond.) 220 (1968) 821–822.

Solomon, M. A., L. B. Hersch: Death in the family. J. Marital and Fam. Therapy 5 (1979) 43.

Solomon, S., D. S. Holmes, K. D. McCaul: Behavioral control over aversive events: does control that requires effort reduce anxiety and physiological arousal? J. Pers. soc. Psychol. 39 (1980) 729–736.

Sommer, M., G. Overbeck: Zur Psychodynamik der Kopfschmerzen. Prax. Psychother. 22 (1977) 117–127.

Sonnenberg, A., R. Arnold, A. Fritsch: Epidemiologie und Genetik der Ulcuskrankheit. In: Blum, A. L., J. R. Siewert (Hrsg.): Ulcus-Therapie, S. 3–22. 2. Aufl. Springer, Heidelberg 1982.

Sontag, S.: Krankheit als Metapher. Hanser, München 1978.

Sopko, J.: Die Welt des Klanges und der Stille. Schweiz. Ärztezeitung 67 (1986) 8.

Sopko, J.: Klinische Laryngologie 1. Inpharzam, Cadempino 1987.

Sopp, H.: Soziologie des Krankenstandes. Münch. med. Wschr. 100 (1958) 489.

Sopp, H.: Was der Mensch braucht … Econ, Düsseldorf 1958.

Sorensen, E. T.: Group therapy in a community hospital dialysis unit. J. Amer. med. Ass. 221 (1972) 899–901.

Sorotzkin, B.: Nocturnal enuresis: current perspectives. Child clin. Res. 4 (1984) 293–316.

Soulé, M.: Das Kind im Kopf. In: Storck, J. (Hrsg.): Neue Wege im Verständnis der allerfrühesten Entwicklung des Kindes. Frommann & Holzboog, Stuttgart 1989.

Soulé, M., K. Lauzanne: Les troubles de la défécation: encoprésie, mégacolon fonctionnel de l'enfant. In: Lebovici, S., R. Diatkine, M. Soulé (eds.): Traité de psychiatrie de l'enfant et de l'adolescent, Vol. II., pp. 527–535. PUF, Paris 1985.

Souques, M. A., P. Rosanoff-Saloff: La camptocormie. Incurvation du tronc, consécutive aux traumatismes du dos et des lombes. Considérations morphologiques. Rev. neurol. 27 (1915) 937–939.

Sours, J. A.: The anorexia nervosa syndrome. Int. J. Psychoanal. 55 (1974) 567–579.

Sovak, M., D. J. Dalessio, M. Künzel, R. A. Sternbach: Current investigations in headache. In: Bonica, J. J. (ed.): Pain, pp. 261–280. Raven, New York 1980.

Spain, D. M., V. A. Brades: Sudden death from coronary heart disease. Chest 58 (1970) 107.

Spangfort, E. V.: The lumbar disc herniation. A computer-aided analysis of 2504 operations. Acta orthop. scand. (Suppl.) 142 (1972) 1–95.

Sparacino, J.: Type A behavior pattern: a critical assessment. J. hum. Stress 5 (1979) 37–51.

Spearman, C.: Correlations calculated from faulty data. Brit. J. Psychol. 3 (1910) 281.

Specht, K. G.: Arbeit, Umwelt und Lebensgewohnheiten bei Früh- und Altersrentnern. Hrsg. vom Institut für empirische Soziologie, Nürnberg, der Landesversicherungsanstalt Baden, Karlsruhe und der Landesversicherungsanstalt Württemberg, Stuttgart 1977.

Speck, R. V., C. L. Attneave: A family networks – a new approach to family problems. Pantheon Books, New York 1973. Deutsch: Die Familie im Netz sozialer Beziehungen. Lambertus, Freiburg 1976.

Spector, N. H., K. Bulloch, B. H. Fox, B. D. Jankovic, A. P. Kerza-Kwiatecki, A. A. Monjan, W. Pierpaoli (eds.): Neuroimmunomodulation: Proc. of the First Int. Workshop on Neuroimmunomodulation. IWGN, Bethesda, MD 1985.

Speidel, H., A. Boll, B. Dahme, P. Götze et al.: Der herzchirurgische Patient und seine Familie. In: Angermeyer, M. C., H. Freyberger (Hrsg.): Chronisch kranke Erwachsene in der Familie, S. 63–75. Enke, Stuttgart 1982.

Speidel, H., B. Dahme, B. Flemming et al.: Psychosomatische Probleme in der Herzchirurgie. Therapiewoche 28 (1978) 8191–8210.

Speidel, H., B. Dahme, B. Flemming, P. Götze, G. Huse-Kleinstoll, H. J. Meffert, G. Rodewald, W. Spehr: Open-heart surgery unit. In: Freyberger, H. (ed.): Psychotherapeutic Intervention in Life Threatening Illness. Advanc. psychosom. Med. 20, pp. 30–56. Karger, Basel 1980.

Speidel, H., B. Dahme, B. Flemming, P. Götze, G. Huse-Kleinstoll, H. J. Meffert, G. Rodewald: Probleme der Klassifizierung psychopathologischer Auffälligkeiten nach Herzoperationen mit extrakorporaler Zirkulation. Psychiat. Clinics 12 (1979b) 57–59.

Speidel, H., B. Dahme, B. Flemming, P. Götze, G. Huse-Kleinstoll, H. J. Meffert, G. Rodewald: Psychische Störungen nach offenen Herzoperationen. Nervenarzt 50 (1979a) 85–91.

Speidel, H. et al.: Analyse von Behandlungsfaktoren der postoperativen psychopathologischen und neurologischen Auffälligkeiten bei Herzoperierten mit extrakorporaler Zirkulation. Bericht an die DFG, Hamburg 1981.

Speidel, H., F. Balck, M. Dvorak: Die Sexualität unter chronischer Hämodialyse und nach Nierentransplantation. Sexualmed. 12 (1983).

Speidel, H., F. Balck, U. Koch, J. Kniess: Problems in interaction between patients undergoing long-term hemodialysis and their partners. Psychother. and Psychosom. 31 (1979) 235–242.

Speidel, H., F. Balck, U. Koch: Empirical questionnaire survey of the situation of hemodialysis patients and their partners in various dialysis settings. In: Levy, N. B. (ed.): Psychonephrology 1, pp. 147–168. Plenum, New York 1981.

Speidel, H., F. Balck, U. Koch: Psychische und psychosoziale Probleme der chronischen Hämodialyse. Therapiewoche 28 (1978) 8262–8279.

Speidel, H., P. Kalmar, M. v. Kerekjarto et al.: Untersuchungen zur Psychopathologie der Schrittmacherpatienten. Verh. Dtsch. Ges. Inn. Med. (1969) 746.

Speidel, H., W. Bauditz, P. Bürger et al.: Beitrag zur Psychopathologie der Dauerdialysepatienten. Verh. Dtsch. Ges. Inn. Med. 76 (1970) 1040–1042.

Speirer, T. W., I. Weidelt: Wie Medizinstudenten sich selbst und ihre Kommilitonen sehen. Münch. med. Wschr. 40 (1984) 4–6.

Speizer, F. B.: Epidemiology, prevalence, and mortality in asthma. In: Weiss, E. B., M. S. Segal (eds.): Bronchial Asthma. Mechanisms and Therapeutics, pp. 43–53. Little, Brown & Co., Boston 1979.

Spence, J., W. S. Walton, F. J. M. Miller, S. D. M. Court: A thousand families in Newcastle-upon-Tyne. An approach to the study of health and illness in children. Oxford Univ. Press, London 1954.

Spence, K. W.: The differential response in animals to stimuli varying within a single dimension. Psychol. Rev. 44 (1937) 430–444.

Spence, K. W., J. A. Taylor: Anxiety and strength of the UCS as determiners of amount of eyelid conditioning. J. exp. Psychol. 42 (1951) 183–188.

Spencer, K. M., B. M. Dean, A. Tarn, J. Lister, G. F. Bottazzo: Fluctuating islet cell autoimmunity in unaffected relatives of patients with insulin-dependent diabetes. Lancet I (1984) 764–766.

Spencer, P. S. J., R. L. Lee, R. D. E. Sewell: Centrally-acting analgesics. A brief review of opiates and psychotropic drugs. Pharmaceut. Med. (1979) 6–20.

Spencer, R. F.: Medical patients: consultation and psychotherapy. Arch. gen. Psychiat. 10 (1964) 270.

Spencer-Gardner, C., L. Dennerstein, G. D. Burrows: Premenstrual tension and female role. J. Psychosom. Obstet. Gynec. 2/1 (1983) 27.

Spengler, A.: Psychische und sexuelle Störungen nach Genitaloperationen in der Urologie. Wien. med. Wschr. 138 (1988) 81.

Spergel, P., G. E. Ehrlich, D. Glass: The rheumatoid arthritis personality: a psychodiagnostic myth. Psychosomatics 19 (1978) 79–86.

Sperling, E.: Die Magersucht-Familie und ihre Behandlung. In: Meyer, J. E., H. Feldmann (Hrsg.): Anorexia nervosa, S. 156. Thieme, Stuttgart 1965.

Sperling, E., A. Massing, H. Georgi, G. Reich, E. Wöbbe-Mönks: Die Mehrgenerationen-Familientherapie. Vandenhoeck und Ruprecht, Göttingen 1982.

Sperling, E., A. Massing: Besonderheiten in der Behandlung der Magersuchtsfamilie. Psyche 26 (1972) 357–369.

Sperling, E., A. Massing: Der familiäre Hintergrund der Anorexia nervosa und die sich daraus ergebenden therapeutischen Schwierigkeiten. Z. psychosom. Med. Psychoanal. 16 (1970) 130–141.

Sperling, M.: Etiology and treatment of sleep disturbances in children. Psychoanal. Quart. 24 (1955) 358.

Sperling, M.: Psychoanalytic study of ulcerative colitis in children. Psychoanal. Quart. 15 (1946) 302–329.

Sperling, M.: The psycho-analytic treatment of a case of chronic regional ileitis. Int. J. Psychoanal. 41 (1960) 612–618.

Sperling, M.: The role of the mother in psychosomatic disorders in children. Psychosom. Med. 11 (1949) 377–385.

Spevack, M.: Behavior therapy treatment in bronchial asthma: a critical review. Canad. psychol. Rev. 19 (1978) 321–327.

Spiecker-Henke, M.: Logopädische Behandlung von Stimmstörungen. In: Biesalski, P., F. Frank (Hrsg.): Phoniatrie-Pädaudiologie, S. 302. Thieme, Stuttgart 1982.

Spiegel 21 (1983) 86: Wie gefährlich sind junge Ärzte? Ärztlicher Nachwuchs schlecht gerüstet.

Spiegel, J. S., B. Leake, T. M. Spiegel, H. E. Paulus, R. L. Kane, N. B. Ward, J. E. Ward: What are we measuring? An examination of self-report functional status measures. Arthr. and Rheum. 31 (1988) 721–728.

Spiegel, Y.: Der Prozeß des Trauerns. Analyse und Beratung. Kaiser-Grünewald, München 1973.

Spiegel-Rösing, I.: Ethik der Thanatologie. In: Howe, J., R. Ochsmann (Hrsg.): Tod – Sterben – Trauer, S. 43–50. Fachbuchhandlung für Psychologie, Frankfurt 1984.

Spiegel-Rösing, I.: Psychotherapie mit Sterbenden: Ein kritischer Überblick vorliegender Ansätze. In: Spiegel-Rösing, I., H. Petzold (Hrsg.): Die Begleitung Sterbender, S. 31–83. Junfermann, Paderborn 1984.

Spiegel-Rösing, I.: Thanato-Therapie. Anmerkungen zum Thema. Integrative Therapie 6 (1980).

Spiegel-Rösing, I.: Ziele psycho-sozialer Intervention beim Sterben. In: Spiegel-Rösing, J., H. Petzold (Hrsg.): Die Begleitung Sterbender, S. 141–182. Junfermann, Paderborn 1984.

Spiegelberg, U.: Colitis ulcerosa in psychiatrisch-neurologischer Sicht. Enke, Stuttgart 1965.

Spiegelberg, U.: Hypnose und autogenes Training in der gegenwärtigen inneren Medizin. Therapiewoche 18 (1968) 1061–1064.

Spiegelberg, U.: Pharmacotherapy and psychosomatics. Pharmacopsychiat. 1 (1968) 87–111.

Spielberger, C. D., G. Jacobs, S. Russel, R. Crane: Assessment of anger: the state-trait anger scale. In: Butcher, J., C. D. Spielberger (eds.): Advances in Personality Assessment 2. Erlbaum, Hillsdale 1983.

Spielberger, C. D., L. D. DeNike: Descriptive behaviorism versus cognitive theory in verbal operant conditioning. Psychol. Rev. 73 (1966) 306–326.

Spiro, H. R.: Chronic facticious illness. Munchhausen syndrome. Arch. gen. Psychiat. 18 (1968) 569–579.

Spitz, R. A.: Die Entstehung der ersten Objektbeziehungen. Klett, Stuttgart 1957.

Spitz, R. A.: Eine genetische Feldtheorie der Ich-Bildung. Fischer, Frankfurt 1972.

Spitz, R. A.: Nein und Ja. Klett, Stuttgart 1967.

Spitz, R. A.: The first year of life. New York 1965.

Spitz, R. A.: Vom Säugling zum Kleinkind. Naturgeschichte der Mutter-Kind-Beziehung im ersten Lebensjahr. Klett, Stuttgart 1965, 1967, 1969, 1983, 1985.

Spitzer, R. L., J. Endicott, E. Robins: Forschungs-Diagnosekriterien (RDC). Beltz, Weinheim 1982.

Spoendlin, H.: Akustisches Trauma. In: Berendes, J., R. Link, F. Zöllner (Hrsg.): Hals-Nasen-Ohrenheilkunde in Praxis und Klinik, Bd. 6. Thieme, Stuttgart-New York 1980.

Spoendlin, H., W. Lichtensteiger: The sympathetic nerve supply to the inner ear. Arch. klin. exp. Ohr.-, Nas.- und Kehlk.-Heilk. 189 (1967) 346.

Spring, A., R. Wittek, R. Wörz: The influence of lumbar disc disease on psychiatric symptomatology. Advanc. Neurochir. 4 (1976) 59–62.

Springer-Kremser, M., A. Springer: Sexuelle Funktionsstörungen – Diagnose und Therapie. In: Kongreßband Van Swieten Tagung, ÖÄK, Wien 32 (1978) 121–127.

Spry, C. J. F.: Inhibition of lymphocyte recirculation by stress and corticotropin. Cell Immunol. 4 (1972) 86.

Squires, R. F., C. Braestrup: Benzodiazepine receptors in rat brain. Nature (Lond.) 266 (1977) 732–734.

Squires, R. F., D. J. Benson, C. Braestrup, J. Coupet, C. A. Klepner, V. Myers, B. Beer: Some properties of brain specific benzodiazepine receptors: new evidence for multiple receptors. Pharmacology, Biochemistry and Behavior 10 (1979) 825–829.

Srole, L.: Measurement and classification in socio-psychiatric epidemiology: the Midtown Manhattan Study (1954) and Midtown Manhattan Restudy (1974). J. Health soc. Behav. 16 (1975) 347–364.

Srole, L., A. K. Fischer: Gender, generations, and well-being: the Midtown Manhattan Longitudinal Study. In: Erlenmeyer-Kinnling, L., N. E. Miller (eds.): Life-span

Research on the Prediction of Psychopathology, pp. 239–246. Erlbaum, Hillsdale NJ 1986.

Srole, L., T. S. Langner, S. T. Michael, M. K. Opler, T. A. C. Rennie: Mental health in the metropolis. The Midtown Manhattan Study. McGraw-Hill, New York-Toronto-London 1962.

Staats, J., G. Graber: Myoarthropathien unter individual-psychologischen Aspekten. Schweiz. Mschr. Zahnheilk. 92 (1982) 921.

Stabile, B. E., M. Borzatta, R. S. Stubbs, H. T. Debas: Intravenous mixed amino acids and fats do not stimulate exocrine pancreas secretion. Amer. J. Physiol. 246 G (1984) 274–280.

Stafford-Clark, D.: Anorexia nervosa. Brit. med. J. 2 (1958) 446.

Staines, N., J. Brostoff, K. James: Immunologisches Grundwissen. Fischer, Stuttgart 1987.

Stall, R., L. McKusick, J. Wiley, T. J. Coates, D. G. Ostrow: Alcohol and drug use during sexual activity and compliance with safe sex guidelines for AIDS: the AIDS behavioral research project. Health Educ. Quart. 13 (1986) 359–371.

Stallard, R. E.: Relation of occlusion to temporomandibular joint dysfunction: the periodontic viewpoint. J. Amer. dent. Ass. 79 (1969) 142.

Stalmann, H., L. Hartl, P. Pauli, F. Strian, R. Hölzl: Perception of arrhythmias in mitral valve prolaps, cardiac phobia and panic attacks. J. Psychophysiol. 2 (1988) 160.

Stamey, T. A.: Pathogenese und Behandlung rezidivierender Harnwegsinfekte bei Frauen. In: Verhandlungsbericht der Deutschen Gesellschaft für Urologie, S. 263. Springer, Berlin 1985.

Stamler, J., D. M. Berkson, A. Dyer et al.: Relationship of multiple variables to blood pressure. In: Oglesby, P. (ed.): Epidemiology and Control of Hypertension, pp. 41–48. Stratton, New York 1975.

Stange, G.: Behandlungsergebnisse bei Hörstürzen. Arch. klin. exp. Ohr.-, Nas.- und Kehlk.-Heilk. 194 (1969) 538.

Stange, G., R. Neveling: Hörsturz. In: Berendes, J., R. Link, F. Zöllner (Hrsg.): Hals-Nasen-Ohrenheilkunde in Praxis und Klinik, Bd. 6. Thieme, Stuttgart-New York 1980.

Stanton, M. D.: Strategic approaches to family therapy. In: Gurman, A. S., D. P. Kniskern (eds.): Handbook of Family Therapy. Brunner/Mazel, New York 1981.

Stappen, B.: Formen der Auseinandersetzung mit Verwitwung im höheren Alter. Roederer, Regensburg 1988.

Starkman, M. N., D. W. Schteingart, M. A. Schork: Depressed mood and other psychiatric manifestations of Cushing's syndrome: relationship to hormone levels. Psychosom. Med. 43 (1981) 3.

Statistisches Bundesamt (Hrsg.): Statistisches Jahrbuch 1988 für die Bundesrepublik Deutschland. Kohlhammer, Stuttgart-Mainz 1988.

Stauber, M.: In: Prill, H. J. (Hrsg.): Das Sprechstundengespräch. Ein schriftliches Symposion. Geburtshilfe Frauenheilk. 36 (1976) 461.

Stauber, M.: Psychohygienische Forderungen an die heutige Geburtshilfe. In: Hillemanns, H. G., H. Steiner, D. Richter (Hrsg.): Die humane, familienorientierte und sichere Geburt. Thieme, Stuttgart 1983.

Stauber, M.: Psychosomatik der sterilen Ehe. Grosse, Berlin 1988.

Stauber, M.: Psychosomatische Aspekte in der Geburtshilfe. Dtsch. Ärztebl. (1979) 797.

Stauber, M.: Psychosomatische Forderungen an das Geburtsgeschehen. In: Richter, D., M. Stauber (Hrsg.): Psychosomatische Probleme in Geburtshilfe und Gynäkologie. Kehrer, Freiburg 1983.

Stauber, M.: Untersuchungen zur sterilen Ehe unter besonderer Berücksichtigung psychosomatischer Aspekte. Habilitationsschrift, FU Berlin 1977.

Stauber, M., C. Haupt: Pruritus vulvae als psychosomatische gynäkologische Erkrankung. Therapiewoche 30 (1980) 599.

Stauder, K. H.: Über den Pensionsbankrott. Psyche 9 (1955) 481.

Stavraky, K., C. Buck, J. Lott, N. J. Wanklin: Psychological factors in the outcome of human cancer. J. psychosom. Res. 12 (1968) 251–259.

Stead jr., E. A., J. V. Warren: Clinical significance of hyperventilation. The role of hyperventilation in the production, diagnosis and treatment of certain anxiety symptoms. Amer. J. med. Sci. 206 (1943) 183–190.

Steele, T. E., S. H. Finkelstein, F. O. Finkelstein: Hemodialysis patients and spouses – marital discord, sexual problems and depression. J. nerv. ment. Dis. 162 (1976) 225–237.

Stefansson, J. G., J. A. Messina, S. Meyerowitz: Hysterical neurosis, conversion type: Clinical and epidemiological considerations. Acta psychiat. scand. 53 (1976) 119–138.

Steffen, H. et al.: Psychische Reaktionen bei niereninsuffizienten Kindern auf intermittierende Hämodialyse und Nierentransplantationen. Zschr. Kinderheilk. 116 (1974) 115–126.

Steffens, W., H. Kächele: Abwehr und Bewältigung – Mechanismen und Strategien. In: Kächele, H., W. Steffens (Hrsg.): Bewältigung und Abwehr. Springer, Berlin 1988.

Steffens, W., H. Kächele: Abwehr und Bewältigung – Vorschläge zu einer integrativen Sichtweise. Psychother. med. Psychol. 38 (1988a) 3–7.

Stegmann, M.: Darstellung epileptischer Anfälle im Traum. Int. Z. Psychoanal. 1 (1913) 560–561.

Stein, E.: Die Strahlenproktitis. Colo-Proktol. 5 (1983) 32–37.

Stein, E. H., J. Murdaugh, J. A. McLeod: Brief psychotherapy of psychiatric reactions to physical illness. Amer. J. Psychiat. 125 (1969) 1040–1047.

Stein, M., S. E. Keller, S. J. Schleifer: Stress and immunomodulation: the role of depression and neuroendocrine function. J. Immunol. 135 (1985) 827s–833s.

Stein, S. P., E. S. Charles: Emotional factors in juvenile diabetes mellitus: a study of early life experiences of adolescent diabetics. Amer. J. Psychiat. 128 (1971) 700–704.

Stein, S. P., E. S. Charles: Emotional factors in juvenile diabetes mellitus: a study of the early life experiences of eight diabetic children. Psychosom. Med. 37 (1975) 237–244.

Steiner, E.: Pragmatische Kurzzeittherapie von zwei Herzneurosen (Herzphobien) mit dem Katathymen Bilderleben. In: Leuner, H., O. Lang (Hrsg.): Psychotherapie mit dem Tagtraum, S. 190–202. Huber, Bern 1982.

Steinert, R.: Hörsturz heute – eine Übersicht. HNO 34 (1986) 453.

Steinglass, P.: The conceptualization of marriage from a systems theory perspective. In: Paolino, T. J., B. S. McCrady (eds.): Marriage and Marital Therapy. Psychoanalytic, Behavioral and Systems Theory Perspectives, p. 298. Brunner/Mazel, New York 1978.

Steinglass, P., S. Gonzales, I. Dosovitz et al.: Discussion groups for chronic hemodialysis patients and their families. Gen. Hosp. Psychiat. 4 (1982) 7–14.

Steinhausen, H. C.: Die Persönlichkeit von Eltern chronisch kranker Kinder. Prax. Kinderpsycholog. Kinderpsychiat. 26 (1977).

Steinhausen, H. C.: Psychosomatische Störungen und Krankheiten bei Kindern und Jugendlichen. Kohlhammer, Stuttgart 1981.

Steinhausen, H. C., H. Kies: Comparative studies of ulcerative colitis and Crohn's disease in children and adolescents. J. Child Psychol. Psychiat. 23 (1982) 33–42.

Steinhausen, H. C., K. Glanville: Retrospective and prospective follow-up studies in anorexia nervosa. Int. J. Eat. Dis. 2 (1983) 221–236.

Steinhauser, H., S. Borner, P. Koepp: The personality of juvenile diabetics. In: Laron, Z. (ed.): Psychological Aspects of Balance of Diabetes in Juveniles. Pediatr. Adolesc. Endocrinol. 3 (1977).

Stekel, W.: Der epileptische Symptomenkomplex und seine analytische Behandlung. Fortschr. Sexualwiss. Psychoanal. 1 (1924) 17–57.

Stekel, W.: Technik der analytischen Psychotherapie. Bern 1938.

Steltzer, J., F. Finkelstein, H. Feigenbaum: The spouse's role in home hemodialysis. Arch. gen. Psychiat. 33 (1976) 55–58.

Stemmler, G.: Psychophysiologische Emotionsmuster: Ein empirischer und methodologischer Beitrag zur inter- und intraindividuellen Begründbarkeit spezifischer Profile bei Angst, Ärger und Freude. Lang, Frankfurt-Bern 1984.

Stenback, A.: Psychosomatic states. In: Howee, J. (ed.): Modern Perspectives in the Psychiatry of Old Age. Brunner/Mazel, New York 1975.

Stengel, E., G. D. F. Steele: Unawareness of physical disability (anosognosia). J. ment. Sci. 92 (1946) 379–388.

Stepanski, E., J. Lamphere, P. Badia, F. Zorick, T. Roth: Sleep fragmentation and daytime sleepiness. Sleep 7 (1984) 18–26.

Stephanos, S.: Pathological primary identifications and their effects on the psychosomatic economy of the individual. Psychother. and Psychosom. 30 (1978) 56–67.

Stephanos, S., U. Auhagen: Objektpsychologisches Modell auf der Basis der Französischen Psychosomatischen Schule. In: Balmer, H. (Hrsg.): Psychologie des 20. Jahrhunderts, Bd. IX, S. 162–181. Zürich 1979.

Steptoe, A., A. Ross: Psychophysiological reactivity and the prediction of cardiovascular disorders. J. psychosom. Res. 25 (1981) 23–31.

Steptoe, A., A. Ross: Voluntary control of cardiovascular reactions to demanding tasks. Biofeedback & Self-Regul. 7 (1982) 149–166.

Sterman, M. B., L. R. MacDonald, R. K. Stone: Biofeedback training of the sensorimotor electroencephalogram rhythm in man: effects on epilepsy. Epilepsia 15 (1974) 395–416.

Sterman, M. B., L. R. MacDonald: Effects of central cortical EEG feedback training on incidence of poorly controlled seizures. Epilepsia 19 (1978) 207–222.

Stern, D.: Affect attunement. In: Call, J. D. et al. (eds.): Frontiers of Infant Psychiatry, Vol. II, pp. 3–13. Basic Books, New York 1984.

Stern, D.: Mutter und Kind. Die erste Beziehung. Klett, Stuttgart 1979.

Stern, D.: The interpersonal world of the infant. Basic Books, New York 1985.

Stern, E.: Zum Problem der Spezifität der Persönlichkeitstypen und der Konflikte in der psychosomatischen Medizin. Z. psychosom. Med. 4 (1957/58) 153.

Stern, K., M. Gwendolyn, B. A. Williams, M. Prados: Grief reactions in later life. Amer. J. Psychiat. 108 (1951/52) 289–294.

Stern, M. P., O. G. Kolterman, J. F. Fries, H. O. McDevitt, G. M. Reaven: Adrenocortical steroid treatment of rheumatic diseases. Effects on lipid metabolism. Arch. intern. Med. 132 (1973) 97–101.

Stern, W.: Die psychologischen Methoden der Intelligenzprüfung und deren Anwendung an Schulkindern. 5. Kongreß für Experimentelle Psychologie, Berlin 1912.

Sternbach, R. A.: Pain patients. Traits and treatment. Acad. Press, New York-San Francisco-London 1974.

Sternbach, R. A., D. S. Janowsky, L. Y. Huey: Effects of altering brain serotonine activity on human chronic pain. I. World Congress on Pain, pp. 249. Florence 1975.

Sternbach, R. A., G. Zimmermans: Personality changes associated with reduction of pain. Pain 1 (1975) 177–181.

Sternbach, R. A., R. W. Murphy, W. H. Akeson, S. R. Wolf: Chronic low-back pain. Postgrad. Med. 53 (1973) 135.

Sternbach, R. A., S. R. Wolf, R. W. Murphy, W. H. Akeson: Aspects of chronic low back pain. Psychosomatics 14 (1973a) 52–56.

Sternbach, R. A., S. R. Wolf, R. W. Murphy, W. H. Akeson: Traits of pain patients: the low-back „loser". Psychosomatics 14 (1973b) 226–229.

Stettler, C.: Zur Indikation des Katathymen Bilderlebens bei Suchtpatienten. In: Roth, J. W. (Hrsg.): Konkrete Phantasie, S. 47–52. Huber, Bern 1984.

Stevens, J. H., C. V. Turner, F. Rhodewalt, S. Talbot: The Type A behavior pattern and carotid artery sclerosis. Psychosom. Med. 46 (1984) 105–113.

Stevens, L. A., P. R. Muskin: Techniques for reversing the failure of empathy towards AIDS patients. J. Amer. Acad. Psychoanal. 15 (1987) 539–551.

Stevenson, I., H. S. Ripley: Variations in respiration and in respiratory symptoms during changes in emotion. Psychosom. Med. 14 (1952) 476–490.

Stewart, M. A. et al.: A study of psychiatric consultations in a general hospital. J. chron. Dis. 15 (1962) 331.

Stewart, S., R. Johansen: A family systems approach to home dialysis. Psychother. and Psychosom. 27 (1976/77) 86–92.

Stewart, W. A.: Psychosomatic aspects of regional ileitis. N. Y. St. J. Med. 49 (1949) 2820–2824.

Stief, C. G., W. Bähren, H. Gall, W. Scherb, A. Gallwitz, J. E. Altwein: Schwellkörper-Autoinjektions-Therapie (SKAT). Urologe A 25 (1986) 63.

Stief, C. G., W. Bähren, H. Gall, W. Scherb, A. Gallwitz, J. E. Altwein: Erektile Dysfunktion. Dtsch. Ärztebl. – Ärztl. Mitteilungen 18 (1987) 41.

Stieglitz, R. D., J. Albrecht, A. Lundt et al.: Psychopathometrie bei HIV-infizierten Patienten. Nervenarzt 59 (1988) 330–336.

Stierlin, H.: Die Anpassung an die Realität der „stärkeren Persönlichkeit". Einige Aspekte der symbiotischen Beziehung der Schizophrenen. In: Stierlin, H.: Von der Psychoanalyse zur Familientherapie, S. 50–64. Klett-Cotta, Stuttgart 1980.

Stierlin, H.: Die Familie des Krebskranken. Münch. Med. Wschr. 126 (1984) 231–233.

Stierlin, H.: Eltern und Kinder im Prozeß der Ablösung. Suhrkamp, Frankfurt 1975.

Stierlin, H., I. Rücker-Embden, N. Wetzel, M. Wirsching: Das erste Familiengespräch. Klett, Stuttgart 1977. 2. Aufl. 1980.

Stierlin, H., M. Wirsching, B. Haas, F. Hoffmann, G. Schmidt, G. Weber, B. Wirsching: Familienmedizin mit Krebskranken. Familiendynamik 8 (1983) 46.

Stober, B., W. Nützenadel, F. Ullrich: Elementardiät bei Morbus Crohn. Monatsschr. Kinderheilk. 131 (1983) 721–724.

Stocksmeier, U. (ed.): Psychological approach to the rehabilitation of coronary patients. Springer, Heidelberg 1976.

Stocksmeier, U.: Lehrbuch der Hypnose. Karger, Basel-München-Paris 1984.

Stöcker, W., M. Otte, P. C. Scriba: Zur Immunpathogenese des Morbus Crohn. Dtsch. med. Wschr. 109 (1984a) 1984–1986.

Stöcker, W., M. Otte, S. Ulrich, D. Normann, K. Stöcker, G. Jantschek: Autoantikörper gegen exokrines Pankreas und gegen intestinale Becherzellen in der Diagnostik des Morbus Crohn und der Colitis ulcerosa. Dtsch. med. Wschr. 109 (1984b) 1963–1969.

Stöhr, W.: Heart rate of the tree shrews and its persistent modification by social contact. In: Schmidt, T., T. Dembroski, G. Blümchen (eds.): Biological and Psychological Factors in Cardiovascular Disease, pp. 491–499. Springer, Berlin-Heidelberg-New York 1986.

Störmer, A.: Geriatrie in der täglichen Praxis. In: Platt, D. (Hrsg.): Handbuch der Gerontologie, Bd 1 Innere Medizin. Fischer, Stuttgart 1983.

Stokvis, B.: Allgemeine Überlegungen zur Hypnose. In: Frankl, V. E., V.v. Gebsattel, J. H. Schultz (Hrsg.): Handbuch der Neurosenlehre und Psychotherapie. Bd. IV, S. 71–121. Urban & Schwarzenberg, München-Berlin 1959.

Stokvis, B.: Hypnose in der ärztlichen Praxis. Karger, Basel-New York 1955.

Stokvis, B.: Hypnose, Suggestion, Entspannungstherapie. In: Stern, E. (Hrsg.): Die Psychotherapie der Gegenwart, Bd. II, S. 143–184. Rascher, Zürich 1958.

Stokvis, B.: Psychosomatik. In: Frankl, V. E., V.v. Gebsattel, J. H. Schultz (Hrsg.): Handbuch der Neurosenlehre und Psychotherapie. Bd. III. Urban & Schwarzenberg, München-Berlin 1959.

Stokvis, B.: Suggestion. In: Frankl, V. E., V.v. Gebsattel, J. H. Schultz (Hrsg.): Handbuch der Neurosenlehre und Psychotherapie. Bd. IV, S. 1–59. Urban & Schwarzenberg, München-Berlin 1959.

Stokvis, B., D. Langen: Lehrbuch der Hypnose, 2. Aufl. Karger, Basel-New York 1965.

Stokvis, B., E. Wiesenhütter: Der Mensch in der Entspannung, 4. Aufl. Hippokrates, Stuttgart 1979.

Stokvis, B., M. Pflanz: Die Psychologie der Suggestion. Karger, Basel-New York 1961.

Stolorow, R. D., F. M. Lachmann: Psychoanalysis of developmental arrests. Int. Univ. Press, New York 1980.

Stolze, H. (1979): Über die Erweiterung des therapeutischen Raums durch die Konzentrative Bewegungstherapie. In: Stolze, H. (Hrsg.): Konzentrative Bewegungstherapie, S. 466. 2. Aufl. Springer, Berlin-Heidelberg-New York 1989.

Stolze, H. (1981): Zur Geschichte der Konzentrativen Bewegungstherapie. In: Stolze, H. (Hrsg.): Konzentrative Bewegungstherapie, S. 278. 2. Aufl. Springer, Berlin-Heidelberg-New York 1989.

Stolze, H. (1984): „Agieren" und „Erinnern" in der Konzentrativen Bewegungstherapie. In: Stolze, H. (Hrsg.): Konzentrative Bewegungstherapie, S. 121. 2. Aufl. Springer, Berlin-Heidelberg-New York 1989.

Stolze, H.: Das obere Kreuz. Psychotherapie bei Erkrankungen der Halsregion. Lehmanns, München 1953.

Stone, A. A., D. S. Cox, H. Valdimarsdottir, L. Jandorf, J. M. Neale: Evidence that secretory IgA antibody is associated with daily mood. J. Pers. soc. Psychol. 52 (1987) 988–995.

Stone, L.: Die psychoanalytische Situation. Fischer, Frankfurt 1973.

Stonehill, E., A. H. Crisp: Psychoneurotic characteristics of patients with anorexia nervosa before and after treatment and at follow-up 4–7 years later. J. psychosom. Res. 21 (1977) 187–193.

Storch, H., P. Steck: Begleitende thymoleptische Therapie im Rahmen einer kontrollierten Studie mit Maprotilin (Ludiomil®) bei der Behandlung von Kreuzschmerzen. Nervenarzt 53 (1982) 445–450.

Storey, P. B.: The precipitation of subarachnoid hemorrhage. J. psychosom. Res. 13 (1969) 175–182.

Storms, M. D., R. E. Nisbett: Insomnia and the attribution process. J. Pers. soc. Psychol. 16 (1970) 319–328.

Stoudemire, A., T. L. Thompson: Medication noncompliance: systematic approaches to evaluation and intervention. Gen. Hosp. Psychiat. 5 (1983) 233–239.

Stoudemire, A., T. L. Thompson: The borderline personality in the medical setting. Ann. intern. Med. 96 (1982) 76–79.

Strain, J. J.: Impediments to psychological care for the chronic renal patient. In: Levy, N. B. (ed.): Psycho nephrology 1, pp. 19–34. Plenum, New York 1981.

Straub, C. D. K.: Morbus Crohn. Diss., Tübingen 1980.

Strauss, A., S. Fagerhaugh, B. Suczek, C. Wiener: Gefühlsarbeit – Ein Beitrag zur Arbeits- und Berufssoziologie. Kölner Zschr. Soz. Sozialpsychol. 32 (1980) 630–651.

Strauss, G. D., J. S. Spiegel, M. Daniels, T. Spiegel, J. Landsverk, P. Roy-Byrne, C. Edelstein, J. Ehlhardt, R. Falke, L. Hindin, L. Zackler: Group therapies for rheumatoid arthritis – a controlled study of 2 approaches. Arthr. and Rheum. 29 (1986) 1203–1209.

Strauss, J., M. Wohlschläger: Die Natur des Schlafes (Referat). Med. Tribune (1971).

Streltzer, J., R. A. Markoff, B. Yano: Maintenance hemodialysis in patients with severe pre-existing psychiatric disorders. J. nerv. ment. Dis. 164 (1977) 414–418.

Strian, F., R. Maurach, C. Klicpera: Das Mitralklappenprolapssyndrom als ätiologischer Faktor bei Herzphobie und juvenilem Insult. Fortschr. Neurol. Psychiat. 49 (1981) 200–203.

Strian, F., R. Maurach: Zytostatische Therapie: Neurologische und psychiatrische Syndrome. Med. Klin. 75 (1980) 478–484.

Strickland, B. R.: Internal-external expectancies and health-related behaviors. J. consult. clin. Psychol. 46 (1978) 1192–1211.

Striegel-Moore, R. H., L. R. Silberstein, J. Rodin: Towards an understanding of risk factors for bulimia. Amer. Psychol. 41 (1986) 246–263.

Strobel, W., G. Huppmann: Musiktherapie: Grundlagen – Formen – Möglichkeiten. Verlag für Psychologie, Göttingen 1978.

Strober, M.: An empirically derived typology of anorexia nervosa. In: Darby, P. L., P. E. Garfinkel, D. M. Garner, D. V. Coscina (eds.): Anorexia nervosa. Recent Developments in Research, pp. 185–196. Liss, New York 1983.

Strober, M.: Familial aspects of depressive disorder in early adolescence. In: Weller, E. B., R. A. Weller (eds.): Current Perspectives on Major Depressive Disorders. American Psychiatric Press, Washington 1984.

Strober, M., B. Salkin, J. Burroughs, W. Morrell: Validity of the bulimia-restricter distinction in anorexia nervosa. Parental personality characteristics and family psychiatric morbidity. J. nerv. ment. Dis. 170 (1982) 345–351.

Strober, M., J. L. Katz: Do eating disorders and affective disorders share a common etiology? A dissenting opinion. Int. J. Eat. Dis. 6 (1987) 171–180.

Strober, M., L. L. Humphrey: Familial contributions to the etiology and course of anorexia nervosa and bulimia. J. consult. clin. Psychol. 55 (1987) 654–659.

Stroebe, W., M. S. Stroebe: Bereavement and health. The psychological and physical consequences of partner loss. Cambridge Univ. Press, New York 1987.

Strohl, K. P., N. A. Saunders, C. E. Sullivan: Sleep apnea syndromes. In: Saunders, N. A., C. E. Sullivan (eds.): Sleep and Breathing, pp. 365–402. Dekker, New York 1984.

Strom, T. B., A. J. Sytkowski, C. B. Carpenter, J. P. Merrill: Cholinergic augmentation of lymphocyte-mediated cytotoxicity. A study of the cholinergic receptor of cytotoxic T lymphocytes. Proc. Natl. Acad. Sci. (Wash.) 71 (1974) 1330–1333.

Strotzka, H.: Einführung in die Sozialpsychiatrie. Reinbek 1965.

Strotzka, H.: Psychotherapie und soziale Sicherheit. Bern 1969.

Strotzka, H.: Psychotherapie und Tiefenpsychologie. 2. Aufl. Springer, Wien 1984.

Strotzka, H., I. Grunmiller: Krankheit als soziales Phänomen. Internist 13 (1972) 403.

Strotzka, H., I. Leitner, G. Czerwenka-Wenkstetten, S. R. Graupe, M. D. Simon: Kleinburg. Eine sozialpsychiatrische Feldstudie. Österreichischer Bundesverlag für Unterricht, Wissenschaft und Kunst, Wien-München 1969.

Strotzka, H., Mai: Ökonomische Aspekte Psychosomatischer und Psychosozialer Erkrankungen. Inst. f. Gesellschaftspolitik, Wien 1979.

Strunin, L., R. Hingson: Acquired Immundeficiency Syndrome and adolescents: knowledge, beliefs, attitudes and behaviors. Pediatrics 79 (1987) 825–828.

Strunk, C.: Die Pelipathie. Therapiewoche 28 (1978) 9523.

Strupp, H. H.: Die wirksamen Bestandteile in der Psychotherapie. Verh. Dtsch. Ges. Inn. Med. 90, S. 422–427. Bergmann, München 1984.

Strupp, H. H.: On failing one's patient. Psychotherapy: Theory, Research and Practice 12 (1975) 39–41.

Strupp, H. H.: Psychotherapie – eine nie endende Kontroverse. Psychother., Psychosom., med. Psychol. 33 (1983) 1–2.

Strupp, H. H.: Psychotherapie: Einige Bemerkungen zu Forschung, Ausbildung und Praxis. Vortrag Univ. Ulm 1986.

Strupp, H. H.: Psychotherapy research and practice: an overview. In: Garfield, S. L., A. E. Bergin (eds.): Handbook of Psychotherapy and Behavior Change. Wiley, New York 1978.

Strupp, H. H., A. E. Bergin: Some empirical and conceptual bases for coordinated research in psychotherapy: a critical review of issues, trends, and evidence. Int. J. Psychiat. 7 (1969) 18–90.

Strupp, H. H., J. L. Binder: Psychotherapy in a new key: a guide to time-limited dynamic psychotherapy. Basic Books, New York 1984.

Strupp, H. H., R. W. Levenson, S. B. Mauck: Effects of suggestion on total respiratory resistance in mild asthmatics. J. psychosom. Res. 18 (1974) 337.

Strupp, H. H., S. W. Hadley: A tripartite model of mental health and therapeutic outcomes. With special reference to negative effects in psychotherapy. Amer. Psychol. 32 (1977) 187–196.

Struthers, G. R., D. L. Scott, D. G. I. Scott: The use of „alternative treatments" by patients with rheumatoid arthritis. Rheum. International 3 (1983) 151–152.

Stuart, R. B.: Behavioral control of overeating. Behav. Res. Ther. 5 (1967) 257–265.

Stucke, W.: Psychosomatische Grundversorgung: Definition – Ziele – Abgrenzung. Prax. Psychother. Psychosom. 34 (1989) 22–26.

Studt, H. H.: Die Psychosomatik der Infektionskrankheiten. In: Jores, A. (Hrsg.): Praktische Psychosomatik. Huber, Bern 1976.

Studt, H. H.: Herzneurose. Med. Klin. 74 (1979) 1302–1305.

Studt, H. H.: Zur Problematik psychischer Faktoren bei der Lungentuberkulose, I–IV. Z. psychosom. Med. 19 (1973) 101; 201.

Studt, H. H., H. Mast: Zur Ätiopathogenese der Colitis ulcerosa und des Morbus Crohn. In: Studt, H. H. (Hrsg.): Psy-

chosomatik in der inneren Medizin. Springer, Berlin 1986.

Stübinger, D. K.: Psychotherapeutische Selbsthilfe-Gruppen in der BRD. Med. Diss., Univ. Giessen 1977.

Stuhr, U., A. Haag: Eine Prävalenzstudie zum Bedarf an psychosomatischer Versorgung in den Allgemeinen Krankenhäusern Hamburgs. Psychother. med. Psychol. 39 (1989) 273–281.

Stumpfe, K.-D.: Der Psychogene Tod. Hippokrates, Stuttgart 1973.

Stunkard, A. J.: New therapies for the eating disorders: behavior modification of obesity and anorexia nervosa. Arch. gen. Psychiat. 26 (1972) 391–398.

Stunkard, A. J.: From explanation to action in psychosomatic medicine: the case of obesity. Psychosom. Med. 37 (1975) 195–236.

Stunkard, A. J.: The pain of obesity. Bull, Palo Alto 1976.

Stunkard, A. J., A. J. Rush: Dieting and depression reexamined: a critical review of reports of untoward responses during weight reduction of obesity. Ann. intern. Med. 81 (1974) 526–533.

Stunkard, A. J., E. d'Aquili, S. Fox, R. Filion: The influence of social class on obesity and thinness in children. J. Amer. med. Ass. 221 (1972) 579–584.

Stunkard, A. J., L. W. Craighead, R. O'Brien: The treatment of obesity: a controlled trial of behavior therapy, pharmacotherapy and their combination. Lancet I (1980) 1045–1047.

Stunkard, A. J., M. McLaren-Hume: The results of treatment for obesity: a review of the literature and report of a series. Arch. intern. Med. 103 (1959) 79–85.

Sturm, J., M. Zielke: „Chronisches Krankheitsverhalten": Die klinische Entwicklung eines neuen Krankheitparadigmas. Praxis klin. Verhaltensmed. Reha. 1 (1988) 17–27.

Sturzenberger, S.: A follow-up of anorexia nervosa in adolescent females: the prediction of outcome. Dissertation Abstr. Int. 37 (1976) 992–993.

Sturzenberger, S., D. P. Cantwell, J. Burroughs, B. Sulkin, J. K. Green: A follow-up of adolescent psychiatric inpatients with anorexia nervosa. J. Amer. Child Psychiat. 16 (1977) 703–715.

Suarez, E. S., A. McRae, R. B. Williams: High scores on the Cook-Medley Hostility (Ho) Scale predict increased cardiovascular responses to harassment. Ann. meeting Amer. psychosom. Soc., Toronto 1988.

Sudnow, D.: Passing on: the social organization of dying. Prentice-Hall, Englewood Cliffs 1967. Deutsch: Organisiertes Sterben. Fischer, Frankfurt 1973.

Süddeutsche Zeitung Nr. 122 (1982) 10.

Suess, W. M., B. Alexander, D. D. Smith, H. W. Sweeney, R. J. Marion: The effects of psychological stress on respiration: a preliminary study of anxiety and hyperventilation. Psychophysiol. 17 (1980) 535–540.

Suinn, R. M.: Intervention with Type A behaviors. J. consult. clin. Psychol. 50 (1982) 933–949.

Suinn, R. M.: The cardiac stress management program for Type A patients. Cardiac Rehab. 5 (1975) 13–15.

Suinn, R. M.: The coronary-prone behavior pattern: a behavioral approach to intervention. In: Dembroski, T. M., S. M. Weiss, J. L. Shields, S. G. Haynes, M. Feinleib (eds.): Coronary-Prone Behavior. Springer, New York 1978.

Suinn, R. M., L. J. Bloom: Anxiety management training for pattern A behavior. J. behav. Med. 1 (1978) 25–36.

Sullivan, A. J., C. A. Chandler: Ulcerative colitis of psychogenic origin: a report of six cases. Yale J. Biol. Med. 4 (1932) 779–796.

Sullivan, H. S.: Conceptions of modern psychiatry. Washington 1947.

Sullivan, M. A., S. Cohen, W. J. Snape jr.: Colonic myoelectrical activity in irritable bowel syndrome. Effect of eating and anticholinergics. New Engl. J. Med. 298 (1978) 878–883.

Sullivan, P., T. P. Hackett: Denial of illness in patients with myocardial infarction. Rhode Island med. J. 46 (1963) 648–650.

Sullivan, P. R. C., J. A. Bollinger, S. Reichlin: Selective deficiency of tissue triiodothyronine: a proposed mechanism of elevated free thyroxine in the euthyroid sick. J. clin. Invest. 52 (1973) 83.

Suls, J. M., J. W. Gastorf, S. H. Witenberg: Life events, psychological distress and the Type A coronary-prone behavior pattern. J. psychosom. Res. 23 (1979) 315–319.

Summers, R. W., D. M. Switz, J. T. Sessions jr., J. M. Becktel, W. R. Best, F. Kern jr., J. W. Singleton: National Cooperative Crohn's Disease Study: results of drug treatment. Gastroenterology 77 (1979) 847–869.

Summerskill, J., C. G. Darling: Group differences in the incidence of upper respiratory complaints among college students. Psychosom. Med. 19 (1957) 315.

Sundsvold, M. Ø., P. Vaglum, B. Østberg: Movements, lumbar and temporomandibular pain and psychopathology. Psychother. and Psychosom. 35 (1981) 1.

Suppes, P., H. Warren: On the generation and classification of defense mechanisms. Int. J. Psychoanal. 56 (1975) 405–414.

Suppes, V.: Stimulus-response theory of finite automata. J. math. Psychol. 6 (1969) 327–355.

Surman, O. S., N. Tolkoff-Rubin: Use of hypnosis in patients receiving hemodialysis for end-stage renal disease. Gen. Hosp. Psychiat. 6 (1984) 31–35.

Surridge, D.: An investigation into some psychiatric aspects of multiple sclerosis. Brit. J. Psychiat. 115 (1969) 749–764.

Surtees, P. G., C. Dean, J. G. Ingham, N. B. Kreitman, P. McC. Miller, S. P. Sashidharan: Psychiatric disorder in women in an Edinburgh community: association with demographic factors. Brit. J. Psychiat. 142 (1982) 238–246.

Susser, M.: Causes of peptic ulcer. A selective epidemiologic review. J. chron. Dis. 20 (1976) 435–456.

Svarstad (1974): zit. nach: Podell, R. N., L. R. Gary (eds.): Hypertension and Compliance: Implications for the Primary Physician. New Engl. J. Med. 294 (1976) 1120–1121.

Svartz, N.: The treatment of ulcerative colitis. Gastroenterologia 86 (1956) 683–688.

Svedlund, J.: Psychotherapy in irritable bowel syndrome. Acta psychiat. scand. (Suppl.) 306 (1983) 67.

Svedlund, J., I. Sjödin, J. Ottoson, G. Doteval: Controlled study of psychotherapy in irritable bowel syndrome. Lancet II (1983) 589–592.

Swanson, D., J. D. Affliti: Group couples treatment. Dial. Transplant. 3/2 (1974) 23.

Swanson, O. W., T. Maruta, W. M. Swenson: Results of behaviour modification in the treatment of chronic pain. Psychosom. Med. 41 (1979) 55–61.

Sweeney, D. R., A. H. Schmale, D. C. Tincing: Differentiation of the giving-up affects, helplessness and hopelessness. Arch. gen. Psychiat. 33 (1970) 378–382.

Sweeney, D. R., D. J. Fine: Pain reactivity and field dependence. Percept. Motor Skills 21 (1965) 757.

Sweeney, D. R., J. W. Maas, G. R. Heninger: State anxiety, physical activity and urinary 3-methoxy-4-hydroxy-phenylglycol excretion. Arch. gen. Psychiat. 35 (1978) 1418–1423.

Swett, C. P., R. I. Shader: Cardiac side effects and sudden death in hospitalized psychiatric patients. Dis. nerv. Syst. 38 (1977) 69–72.

Swift, C. R., F. Seidman, H. Stein: Adjustment problems in juvenile diabetes. Psychosom. Med. 29 (1967) 555–571.

Swift, W. J.: The long-term outcome of early onset anorexia nervosa: a critical review. J. Amer. Acad. Child Psychiat. 21 (1982) 38–46.

Swift, W. J., M. Ritholz, N. H. Kalin, N. Kaslow: A follow-up study of thirty hospitalized bulimics. Psychosom. Med. 49 (1987) 45–55.

Syme, S., L. Berkman: Social class, susceptibility and sickness. Amer. J. Epidemiol. 104 (1976) 1–8.

Symonds, C.: Excitation and inhibition in epilepsy. Brain 82 (1959) 133–146.

Symons, D.: The evolution of human sexuality. Oxford Univ. Press, New York 1979.

Szasz, T. S.: The psychology of bodily feelings in schizophrenia. Psychosom. Med. 19 (1957) 11–16.

Szentivanyi, A.: The beta adrenergic theory of the atopic abnormality in bronchial asthma. J. Allergy 42 (1968) 203.

Szentivanyi, A., G. Filipp: Anaphylaxis and the nervous system, Part II. Ann. Allergy 16 (1958) 143–151.

Szmukler, G. I.: A study of family therapy in anorexia nervosa: some methodological issues. In: Darby, P. L., P. E. Garfinkel, D. M. Garner, D. V. Coscina (eds.): Anorexia nervosa. Recent Developments in Research, pp. 417–425. Liss, New York 1983 b.

Szmukler, G. I.: The epidemiology of anorexia nervosa and bulimia. J. psychiat. Res. 19 (1985) 143–153.

Szmukler, G. I.: Weight and food preoccupation in a population of English schoolgirls. In: Understanding Anorexia Nervosa and Bulimia. Reprint of the 4th Ross Conf. on Medical Research. Ross Laboratories, Columbus OH 1983 a.

Szmukler, G. I., G. F. M. Russell: Diabetes mellitus, anorexia nervosa and bulimia. Brit. J. Psychiat. 142 (1983) 305–308.

Szonn, G.: Trauerarbeit mit dem Katathymen Bilderleben. In: Leuner, H. (Hrsg.): Katathymes Bilderleben, S. 263–271. Huber, Bern 1980.

Szurszewski, J. H.: In: Johnson, L. R. (ed.): Physiology of the Gastrointestinal Tract, p. 401. Raven, New York 1987.

Sørensen, A., E. Strömgren: Frequency of depressive states within geographically delimited population groups. 2. Prevalence. Acta psychiat. scand. 37 (1961) Suppl. 162.

Taggart, P., D. Gibbons, W. Somerville: Some effects of motor-car driving on the normal and abnormal heart. Brit. med. J. 4 (1969) 130.

Taggart, P., M. Carruthers, W. Somerville: Electrocardiogram, plasma catecholamines and lipids, and their modification by oxprenolol when speaking before an audience. Lancet 2 (1973) 341.

Talbot, E., S. C. Miller: The struggle to create a sane society in the psychiatric hospital. Psychiatry 29 (1968) 165–171.

Tammen, A. T., G. Bierck, T. Fentrop, G. Blümchen: Prognostische Bedeutung des Belastungsblutdrucks bei Herzinfarktpatienten. In: Anlauf, M., K. D. Bock (Hrsg.): Blutdruck unter körperlicher Belastung. Steinkopff, Darmstadt 1984.

Tammen, H.: Katamnestische Untersuchungen von stationär oder ambulant behandelten Patienten mit Ulcus duodeni und/oder ventriculi in der Psychosomatischen Klinik Heidelberg. Unveröffentl. Diss. 1988.

Tapp, W. N., B. Lewin, B. H. Natelson: Stress induced heart failure. Psychosom. Med. 45 (1983) 171–176.

Targum, S. D.: Neuroendocrine dysfunction in schizophreniform disorder: correlation with six-month clinical outcome. Amer. J. Psychiat. 140 (1983) 309–313.

Tattersall, R. B.: Mild familial diabetes with dominant inheritance. Quart. J. Med. 43 (1974) 339–357.

Taub, J. M., R. J. Berger: Performance and mood following variations in the length and timing of sleep. Psychophysiol. 10 (1973) 559–570.

Tausch, A.-M.: Gespräche gegen die Angst. Rowohlt, Reinbek 1981.

Tausch, R.: Gesprächspsychotherapie. S. 134. Hogrefe, Göttingen, 2. Aufl.1968.

Tavormina, J. B., L. S. Kastner, P. M. Slater, S. L. Watt: Chronically ill children: a psychologically and emotionally deviant population. J. abnorm. Child Psychol. 4 (1976) 95–110.

Taylor, A. L., L. M. Fishman: Corticotropin-releasing hormone. New Engl. J. Med. 319 (1988) 213–222.

Taylor, G. J.: Alexithymia and the countertransference. Psychother. and Psychosom. 28 (1977) 141–147.

Taylor, G. J.: Alexithymia. Concept, measurement and implications for treatment. Amer. J. Psych. 141 (1984) 725–732.

Taylor, G. J.: Psychosomatic medicine and contemporary psychoanalysis. Int. Univ. Press, Madison 1987.

Taylor, G. J., K. Doody, A. Newman: Alexithymic characteristics in patients with inflammatory bowel disease. Canad. J. Psychiat. 26 (1981) 470–474.

Taylor, G. J., R. M. Bagby, D. P. Ryan, J. D. A. Parker, K. F. Doody: Criterion validity of the Toronto Alexithymia Scale. Psychosom. Med. 50 (1988) 500–509.

Taylor, H. C.: Vascular congestion and hyperemia. Amer. J. Obstet. Gynec. 57 (1949) 222.

Taylor, J. L., J. R. Tinklenberg: Cognitive impairment and benzodiazepines. In: Meltzer, H. J. (ed.): Psychopharmacology: The Third Generation of Progress. Raven, New York 1987.

Tecker, G. (Hrsg.): Morbus Crohn – Colitis ulcerosa, 2. Aufl. Thieme-Hippokrates-Enke, Stuttgart 1989.

Teichmann, A.: Neuere Erkenntnisse zu den psychosomatischen Bedingungen vorzeitiger Wehentätigkeit. Vortr. 45. Tag. Dtsch. Ges. Gynäkologie Geburtshilfe, Frankfurt 1984.

Teichmann, A. T., K. Bosse, F. Ahrens: Acne excoriée und Artefact. Über die Bedeutung der Selbstbeschädigung. Hautarzt 25 (1974) 494–497.

Teichmann, A. T., K. Bosse: Hautkrankheiten und Kommunikation. Hautarzt 25 (1974) 427–429.

Teitelbaum, P.: The encephalization of hunger. In: Stellar, E., J. M. Sprague (eds.): Progress in Physiological Psychology, Vol. 4. Acad. Press, New York 1971.

Telch, M. J., W. S. Agras, C. B. Taylor: Combined pharmacological and behavioral treatment for agoraphobia. Behav. Res. Ther. 23 (1985) 325–335.

Tellenbach, H.: Melancholie. Springer, Berlin Göttingen Heidelberg 1961.

Tellenbach, H.: Zur Phänomenologie der Verschränkung von Anfallsleiden und Wesensänderung beim Epileptiker (Versuch einer Wesensbestimmung des epileptischen Fürsten Myschkin). Jahrb. Psychol. Psychother. med. Anthropol. 14 (1966) 57–68.

Temkin, O.: The falling sickness. A history of epilepsy from the Greeks to the beginnings of modern neurology, 2nd ed. Hopkins, Baltimore-London 1971.

Temoshok, L.: Emotion adaptation and disease: a multidimensional theory. In: Temoshok, L., C. van Dyke, L. S. Zegans: Emotions in Health and Illness: Theoretical and Research Foundations. Grune & Stratton, New York 1983.

Temoshok, L.: Personality, coping style, emotion and cancer: towards an integrative model. Cancer Surv. 6 (1987) 545–567.

Temoshok, L., B. H. Fox: Coping styles and other psychosocial factors related to medical status and to prognosis in patients with cutaneous malignant melanoma. In: Fox, B. H., B. H. Newberry (eds.): Impact of Psychoneuroendocrine Systems in Cancer and Immunity, pp. 86–146. Hogrefe, Toronto 1984.

Temoshok, L., B. W. Heller: On comparing apples, oranges and fruit salad: a methodological overview of medical outcome studies in psychosocial oncology. In: Cooper, C. L. (ed.): Psychosocial Stress and Cancer, pp. 231–260. Wiley, Chichester 1984.

Temoshok, L., C. van Dyke, L. S. Zegans: Emotions in health and illness: theoretical and research foundations. Grune & Stratton, New York 1983.

Temoshok, L., G. F. Solomon, D. M. Sweet, S. Jenkins, J. Zich, K. Straits, I. Pivar, J. M. Moulton, D. P. Stites: Psychoimmunologic studies of men with AIDS and ARC. Presented at the IV. International Conference on the Acquired Immune Deficiency Syndrome (AIDS). Stockholm, June 12–16, 1988.

Ten Houten, W. D., K. D. Hoppe, J. E. Bogen, D. O. Walter: Alexithymia and the split brain, I: Lexical analysis. Psychother. and Psychosom. 43 (1985 a) 202–208.

Ten Houten, W. D., K. D. Hoppe, J. E. Bogen, D. O. Walter: Alexithymia and the split brain, II: Sentential content analysis. Psychother. and Psychosom. 44 (1985 b) 1–5.

Ten Houten, W. D., K. D. Hoppe, J. E. Bogen, D. O. Walter: Alexithymia and the split brain, III: Global-level content analysis, an overview. Psychother. and Psychosom. 44 (1985 c) 89–94.

Ten Houten, W. D., K. D. Hoppe, J. E. Bogen, D. O. Walter: Alexithymia and the split brain, IV: Gottschalk-Gleser content analysis, an overview. Psychother. and Psychosom. 44 (1985 d) 113–121.

Terenius, L.: Endogenous peptides and analgesia. Ann. Rev. Pharmacol. 18 (1978) 189.

Terenius, L.: Stereospecific interaction between a narcotic analgesic and a synaptic plasma membrane fraction in rat brain. Acta pharmacol. (Kbh.) 33 (1974) 377–384.

Terenius, L.: The effect of peptides and amino-acids on dihydro-morphine binding to the opiate receptor. J. Pharm. Pharmacol. 27 (1975) 540–541.

Terman, C. M., M. A. Merrill: Stanford-Binet Intelligence Scale. Manual for the 3 rd revision. Houghton Mifflin, Boston 1937.

Terrahe, K.: Hyperergische und hyperreflektorische Rhinopathie. Therapiewoche 35 (1985) 447.

Teufel, M., H. Meyer-Hohnloser, E. M. Mörcke, U. Stubig, K. H. Niessen: Nachuntersuchungen bei 60 Kindern mit Colitis ulcerosa und Morbus Crohn. Monatsschr. Kinderheilk. 136 (1988) 378–383.

Tews, H.: Soziologie des Alterns. Quelle & Meyer, Heidelberg 1979.

Thailer, S. A., R. Friedman, G. A. Harshfield, T. G. Pickering: Psychologic differences between high-, normal-, low-renin hypertensives. Psychosom. Med. 47 (1985) 294–297.

The Concise Oxford Dictionary, 6th ed. Oxford 1976.

The Coronary Drug Project Research Group: Prognostic importance of premature beats following myocardial infarction. J. Amer. med. Ass. 223 (1973) 1116.

Theander, S.: Anorexia nervosa. Monograph. Acta psychiat. scand. (Suppl.) 214 (1970).

Theander, S.: Long-term prognosis of anorexia nervosa: a preliminary report. In: Darby, P. L., P. E. Garfinkel, D. M. Garner, D. V. Coscina (eds.): Anorexia nervosa. Recent Developments in Research, pp. 441–442. Liss, New York 1983.

Theander, S.: Research on outcome and prognosis of anorexia nervosa and some results from a Swedish long-term study. Int. J. Eat. Dis. 2 (1983) 167–174.

Thefeld, W., H. Hoffmeister: Sozialmedizinische Aspekte beim Diabetes mellitus. Z. allg. Med. 58 (1982) 1396–1403.

Theorell, T.: Life events, job stress and coronary heart disease. In: Siegrist, J., M. J. Halhuber (eds.): Myocardial Infarction and Psychosocial Risks, pp. 1–17. Springer, Berlin-Heidelberg-New York 1981.

Theorell, T.: On risk factors for premature myocardial infarction in middle-aged building-construction workers. A comparison with other selected illness. In: Dembroski, T. M., M. J. Halhuber (Hrsg.): Psychosozialer „Stress" und koronare Herzkrankheit: 3. Verhalten und koronare Herzkrankheit. Springer, Berlin-Heidelberg-New York 1981.

Theorell, T.: Selected illnesses and somatic factors in relation to two psychosocial indices – a prospective study on middle-aged construction building workers. Psychosom. Res. 20 (1976) 7–20.

Theorell, T., B. Floderus-Myrhed: Workload and risk of myocardial infarction. Int. J. Epidemiol. 6 (1977) 17–21.

Theorell, T., E. Lind, J. Fröberg: A longitudinal study of 21 subjects with coronary heart disease: life changes, catecholamine excretion and related biochemical reactions. Psychosom. Med. 34 (1972) 505–516.

Theorell, T., P. Wester: The significance of psychological events in a coronary care unit. Acta med. scand. 193 (1973) 207–210.

Theorell, T., R. H. Rahe: Life change events, ballestocardiography and coronary death. J. hum. Stress 1 (1975) 18–24.

Theorell, T., R. H. Rahe: Psychosocial factors and myocardial infarction, I. An inpatient study in Sweden. J. psychosom. Res. 15 (1971) 25–31.

Thetford, W. N., H. Schucmann: Weitere Persönlichkeitstheorien. In: Freedman, A. M., H. I. Kaplan, B. J. Sadock, U. H. Peters (Hrsg.): Psychiatrie in Klinik und Praxis. Bd. 4. Psychosomatische Störungen, S. 26. Thieme, Stuttgart-New York 1988.

Thibault, G. E., A. G. Mulley, G. O. Barnett, R. L. Goldstein, V. A. Reder, E. L. Sherman, E. K. Skinner: Medical intensive care: indications, interventions and outcomes. New Engl. J. Med. 302 (1980) 938–942.

Thiel, H. G., D. Parker, T. A. Bruce: Stress factors and the risk of myocardial infarction. J. psychosom. Res. 17 (1973) 43–57.

Thiele, H. G.: Zum heutigen Stand der Knochenmarktransplantation. Immun. Infekt. 13 (1985) 237–244.

Thiele, H. G., W.-H. Schmiegel, F. Bläker: Immunophänomene bei chronisch entzündlichen Darmerkrankungen. In: Müller-Wieland, K. (Hrsg.): Handbuch der inneren Medizin, Bd. III/4: Dickdarm. 5. Aufl. Springer, Berlin 1982.

Thielemann-Jonen, I., H. Pichlmaier: Terminale Pflege Krebskranker. Münch. med. Wschr. 130 (1988) 279–283.

Thoma, P.: Arbeit und Krankheit. In: Geissler, B., P. Thoma (Hrsg.): Medizinsoziologie. Eine Einführung für medizinische und soziale Berufe, S. 93–115. 2. Aufl. Campus, Frankfurt-New York 1979.

Thomä, H.: Anorexia nervosa. Geschichte, Klinik und Theorien der Pubertätsmagersucht. Huber-Klett, Stuttgart 1961.

Thomä, H.: Anorexia nervosa: treatment. In: Reichsman, F. (ed.): Advances in Psychosomatic Medicine, Vol. 7, pp. 300–315. Karger, Basel 1972.

Thomä, H.: Erleben und Einsicht im Stammbaum psychoanalytischer Techniken und der „Neubeginn" als Synthese im „Hier und Jetzt". In: Hoffmann, S. O. (Hrsg.): Deutung und Beziehung. Frankfurt 1983.

Thomä, H.: Über die Unspezifität psychosomatischer Erkrankungen am Beispiel einer Neurodermitis mit zwanzigjähriger Katamnese. Psyche 7 (1980) 589–624.

Thomä, H., H. Kächele: Lehrbuch der psychoanalytischen Therapie. I. Grundlagen. Springer, Berlin-Heidelberg-New York 1985.

Thomä, H., H. Kächele: Lehrbuch der psychoanalytischen Therapie. Springer, Berlin-Heidelberg-New York 1988.

Thomas, C. B.: Precursors of premature disease and death: the predictive power of habits and family attributes. Ann. intern. Med. 85 (1976) 653–658.

Thomas, C. B.: Psychological dimensions of hypertension (1965). In: Stamler, J., R. Stamler, T. Pullman: The Epidemiology of Hypertension. Grune & Stratton, New York 1967.

Thomas, C. B., K. R. Duszynski, R. Karen: Closeness to parents and the family constellation in a prospective study of five disease states: suicide, mental illness, malignant tumor, hypertension and coronary heart disease. Johns Hopkins Med. J. 134 (1974) 251.

Thomas, C. B., D. C. Ross, K. R. Duszynski: Useful hypercholesteremia: its associated characteristics and role in premature myocardial infarction. Johns Hopkins med. J. 136 (1975) 193–208.

Thomas, C. B., K. R. Duszynski: Blood pressure levels in young adulthood as predictors of hypertension and the fate of the cold pressor test. Johns Hopkins med. J. 151 (1982) 93–100.

Thomas, G. W.: Psychic factors in rheumatoid arthritis. Amer. J. Psychiat. 93 (1936) 693–710.

Thomas, K.: Praxis der Selbsthypnose und des autogenen Trainings (nach J. H. Schultz): Formelhafte Vorsatzbildung und Oberstufe, 3. Aufl. Thieme, Stuttgart 1972.

Thompson, J. K., H. E. Adams: Psychophysiological characteristics of headache patients. Pain 18 (1984) 41–52.

Thompson, M. G., D. M. Schwartz: Life adjustment of women with bulimia nervosa and anorexic-like behavior. Int. J. Eat. Dis. 1 (1982) 47–60.

Thompson, S. C.: Will it hurt less if I can control it? A complex answer to a simple question. Psychol. Bull. 90 (1981) 89–101.

Thompson, T. L., M. G. Moran, A. S. Nies: Psychotropic drug use in the elderly. New Engl. J. Med. 308 (1983) 134–138; 194–198.

Thompson, W. G.: Clinical features and diagnosis of the irritable bowel. In: Chey, W. Y. (ed.): Functional Disorders of the Digestive Tract, pp. 299–310. Raven, New York 1983.

Thorbecke, R.: Untersuchung über die Befolgung ärztlicher Anordnungen bei Epilepsie-Patienten, die die Behandlung abbrechen. In: Meier-Ewert, K. (Hrsg.): Therapieresistenz bei Anfallsleiden, S. 110–117. Zuckschwerdt, München 1984.

Thorndike, E. L.: Educational psychology. Teachers College, Columbia Univ., New York 1913.

Thornton, W. E.: Dementia induced by methyldopa with haloperidol. New Engl. J. Med. 294 (1976) 1222.

Thorén, P., M. Åsberg, B. Cronholm, L. Jörnestedt, L. Träskman: Clomipramine treatment of obsessive-compulsive disorder. Arch. gen. Psychiat. 37 (1980) 1281–1285.

Thurn, A.: Die psychogenen Aspekte der perioralen Dermatitis. In: Bosse, K., P. Hünecke (Hrsg.): Psychodynamik und Soziodynamik bei Hautkranken. Vandenhoeck & Ruprecht, Göttingen 1976.

Tibblin, G., B. Lindström, S. Ander: Emotions and heart disease. In: Physiology, Emotion and Psychosomatic Illness. Excerpta Medica, Amsterdam 1972.

Timsit, M., E. Urbain, J. Sabatier, M. Timsit-Berthier: Statistische Untersuchung der psycho-somatischen Kopfschmerzen. Münch. med. Wschr. 117 (1975) 1515–1520.

Tissot, S. A. D.: Traité de l'épilepsie. Paris, 1770.

Titchener, J. L., J. Riskin, R. Emerson: The family in psychosomatic process. In: Hendel, G.: The Psychosocial Interior of the Family, p. 461. Aldine, Chicago 1967. Deutsch: Die Familie im Psychosomatischen Prozeß. In: Brede, K. (Hrsg.): Einführung in die psychosomatische Medizin, S. 214. Fischer, Frankfurt 1974.

Todd, A. D.: The prescription of contraception: negotiations between doctors and patients. Discourse Processes 7 (1984) 171–200.

Todd, P. B., C. J. Magarey: Ego defenses and affects in women with breast symptoms. Brit. J. med. Psychol. 51 (1978) 177–189.

Todes, C. J.: Idiopathic Parkinson's disease and depression: a psychosomatic view. J. Neurol. Neurosurg. Psychiat. 47 (1984) 298–301.

Todes, C. J., A. J. Lees: The pre-morbid personality of patients with Parkinson's disease. J. Neurol. Neurosurg. Psychiat. 48 (1985) 97–100.

Tölle, R.: Wahnentwicklung bei körperlich Behinderten. Nervenarzt 58 (1987) 759–763.

Tokarski, W.: Zur Situation von Lehre und Studium in der Gerontologie an den Hochschulen der Bundesrepublik. In: Karl, F., W. Tokarski (Hrsg.): Die „neuen Alten". Beiträge der XVII. Jahrestagung der Dtsch. Ges. f. Gerontologie 1988. Kasseler Gerontol. Schriften 6, Gesamthochschulbibliothek Kassel 1989.

Tolman, E. C.: Purposive behavior in animal and men. Appleton, New York 1932.

Tonkoff, W.: Zur Kenntnis der Nerven der Lymphdrüsen. Anat. Anz. 16 (1899) 456–459.

Tonnesen, E., J. Tonnesen, N. J. Christensen: Augmentation of cytotoxicity by natural killer cells after adrenalin administration in man. Acta path. microbiol. scand. Sect. C 92 (1984) 81–83.

Toone, B. K., J. Roberts: Status epilepticus. An uncommon hysterical conversion syndrome. J. nerv. ment. Dis. 167 (1979) 548–552.

Torrey, E. F.: Prevalence studies in schizophrenia. Brit. J. Psychiat. 150 (1987) 598–608.

Torrey, E. F., M. McGuire, A. O'Hare, D. Walsh, M. P. Spellman: Endemic psychosis in Western Ireland. Amer. J. Psychiat. 141 (1984) 966–969.

Toth, J. C.: Effect of structured preparation for transfer on patients' anxiety on leaving coronary care unit. Nurs. Res. 29 (1980) 28–34.

Totman, R. G., J. Kiff, S. E. Reed, J. W. Craig: Predicting experimental colds in volunteers from different measures of recent life stress. J. psychosom. Res. 24 (1980) 155–163.

Totman, R. G., J. Kiff: Life stress and susceptibility to colds. In: Oborne, D. J., M. M. Gruneberg, J. R. Eiser (eds.): Research in Psychology and Medicine, Vol. 1, pp. 141–149. Acad. Press, New York 1979.

Traue, H. C.: Gefühlsausdruck, Hemmung und Muskelspannung unter sozialem Streß. Hogrefe, Göttingen 1989.

Traue, H. C., A. Gottwald, P. R. Henderson, D. A. Bakal. Nonverbal expressiveness and activity in tension headache sufferers and controls. J. psychosom. Res. 29 (1985) 375–381.

Traue, H. C., A. M. Mahoney, C. Bischoff: Toward a new understanding of tension headache. In: Papakostopoulos, D. et al. (eds.): Clinical and Experimental Neuropsychophysiology. Croom Helm, Kent 1984.

Traue, H. C., C. Bischoff, H. Zenz: EMG-Unterschiede bei Personen mit und ohne Kopfschmerzen in einer sozialen Belastungssituation. In: Michaelis, W. (Hrsg.): Bericht 32. Kongr. der DGFPs, Zürich 1980, S. 742–745.

Traue, H. C., C. Bischoff: Training emotionaler Expressivität (in Vorbereitung).

Traue, H. C., C. Lösch: Visual stress and dysfunctional muscle activity in tension headache. Annual Meeting AASH, New York 1985.

Traue, H. C., W. Kraus: Ausdruckshemmung als Risikofaktor: Eine verhaltensmedizinische Analyse. Prax. klin. Verhaltensmed. Reh. 2 (1988) 89–95.

Travell, J. G.: Myofascial trigger points: clinical review. In: Bonica, J. J., D. Albe-Fessard (eds.): Advances in Pain Research, Vol. 1, pp. 919–926. Raven, New York 1976.

Travell, J. G., D. G. Simons: Myofascial pain and dysfunction. The trigger point manual. Williams & Wilkins, Baltimore 1983.

Trends in Neurosciences: Neurotransmitters and their actions (special issue). Vol. 6/8 (1983).

Tress, W.: Die diagnostische Bedeutung der „Alexithymie". Med. Psychol. 5 (1979) 95–106.

Tress, W.: Forschung zu psychogenen Erkrankungen zwischen klinisch-hermeneutischer und geisteswissenschaftlicher Empirie: sozialempirische Marker als Vermittler. Psychother. med. Psychol. 38 (1988) 269–275.

Tress, W.: Prävalenz und Behandlung psychischer Erkrankungen in der Allgemeinbevölkerung – Ergebnisse einer Feldstudie in 3 Gemeinden Oberbayerns. Bemerkungen zur Arbeit von S. Weyerer und H. Dilling. Nervenarzt 56 (1985) 50–51.

Tress, W.: Schweregradkriterien für psychogene Erkrankungen. In: Schepank, H. (Hrsg.): Psychogene Erkrankungen der Stadtbevölkerung. Eine epidemiologisch-tiefenpsychologische Untersuchung in Mannheim, S. 79–82. Springer, Berlin-Heidelberg-New York 1987.

Tress, W.: Zur Ätiologie psychogener Erkrankungen. In: Böcker, F., W. Weig (Hrsg.): Aktuelle Kernfragen in der Psychiatrie, S. 167–178. Springer, Berlin-Heidelberg-New York 1988.

Trichtel, F.: Das Licht und die Pathologie des Auges. Maudrich, Wien-München-Bern 1983.

Trichtel, F.: Überbelichtungssyndrom, Dyskinesie der Pupille und Retinopathia centralis serosa. Vortrag anläßlich der Jahreshauptversammlung d. Österr. Ophthalm. Gesellschaft in Baden 1979.

Trichtel, F.: Zur Ätiologie der Retinopathia centralis serosa. Wien. med. Wschr. 15/16 (1980) 533–538.

Trimble, M. R.: Serum prolactin in epilepsy and hysteria. Brit. med. J. 4 (1978) 1682–1683.

Trimble, M. R.: The relationship between psychiatry and neurology. A British perspective with particular reference to neuropsychiatry. In: Mueller, J. (ed.): Neurology and Psychiatry. A Meeting of Minds, pp. 14–30. Karger, Basel-München-Paris 1989.

Trimborn, W.: Probleme bei der stationären Behandlung von Borderline-Patienten. Psyche 37 (1983) 204–236.

Trinkl, W.: Das primäre Fibromyalgie-Syndrom. Dtsch. Ärztebl. 84 (1987) 2343–2348.

Troll, L., J. Turner: Overcoming age-sex discrimination. In: Cahn, A. F. (ed.): Women in Midlife – Security and Fullfillment. GPO, Washington D. C. 1978.

Troschke, J. v., J. Siegrist: Das Krankenhaus-Patientenheft in der Erprobung. Befragungsergebnisse über die Anwendungsmöglichkeiten eines Patientenleitfadens. Dtsch. Ärztebl. 40 (1977) 2393–2396.

Trousseau, A.: Clinical medicine. Philadelphia 1882.

Troxler, R. G., E. A. Sprague, R. A. Albanese, R. Fuchs, A. J. Thompson: The association of elevated plasma cortisol and early atherosclerosis as demonstrated by coronary angiography. Atherosclerosis 26 (1977) 151–162.

Trygstad, O.: Bulimie. Tidsskr. norske Laegeforen. 23 (1985) 1511–1516.

Tsaltas, M. O.: Children of hemodialysis patients. J. Amer. med. Ass. 236 (1976) 2764–2766.

Tsouyopulos, N.: Schelling, seine Bedeutung für eine Philosophie der Natur und der Geschichte. Frommann-Holzboog, Stuttgart 1979.

Tuchinda, M., R. W. Newcomb, B. L. DeVald: Effect of prednisone treatment on the human immune response to keyhole limpet hemocyanin. Int. Arch. Allergy 42 (1972) 533–544.

Tufo, H. M., H. M. Ostfeld, R. Shekelle: Central nervous system dysfunction following open-heart surgery. J. Amer. med. Ass. 212 (1970) 1333–1340.

Tuke, D. H.: Influence of the mind upon the body. Lindsay & Blakeston, Philadelphia 1872.

Tunis, M. M., H. G. Wolff: Studies on headache. Cranial artery constriction and muscle contraction headache. Arch. Neurol. Psychiat. 71 (1954) 425–434.

Tunis, M. M., H. G. Wolff: Studies on headache. Long-term observations of the reactivity of the cranial arteries in subjects with vascular headache of the migraine type. Arch. Neurol. Psychiat. 70 (1953) 551–557.

Turk, D. C., D. Meichenbaum: A cognitive-behavioral approach to pain management. In: Wall, P. D., R. Melzack (eds.): Textbook of pain. Livingstone, New York 1984.

Turk, D. C., D. H. Meichenbaum, W. H. Berman: Die Anwendung von Biofeedback bei der Schmerzkontrolle. Ein kritischer Überblick. In: Keeser, W., E. Pöppel, P. Mitterhusen (Hrsg)· Schmerz. Urban & Schwarzenberg, München 1982.

Turk, J. L., D. Parker: Effect of cyclophosphamide on immunological control. Immunol. Rev. 65 (1982) 99–113.

Turner, J. A., C. R. Chapman: Psychological interventions for chronic pain: a critical review. Pain 12 (1982) 1.

Turner, S. M., M. Hersen, A. S. Bellack, K. C. Wells: Behavioral treatment of obsessive-compulsive neurosis. Behav. Res. Ther. 17 (1979) 95–106.

Turpin, G., M. Lader: Life events and mental disorders: biological theories of their mode of action. In: Katschnig, H. (ed.): Life Events and Psychiatric Disorders: Controversial Issues. Cambridge Univ. Press, Cambridge 1986.

Twycross, R. G.: Overview of analgesia. In: Bonica, J. J., V. Ventafridda (eds.): Advances in Pain Research and Therapy, Vol. 2, pp. 617–633. Raven, New York 1979.

Twycross, R. G.: Choice of strong analgesics in terminal cancer. Diamorphine or morphine? Pain 3 (1977) 93–104.

Tyson, R. L., J. Sandler: Probleme der Auswahl von Patienten für eine Psychoanalyse. Psyche 28 (1974) 530–559.

Udelman, D. L.: Stress and immunity. Psychother. and Psychosom. 37 (1982) 176–184.

Udris, I. (Hrsg.): Arbeit und Gesundheit. Streß und seine Auswirkungen bei verschiedenen Berufen. Huber, Bern-Stuttgart-Wien 1982.

Uehara, A., S. Gillis, A. Arimura: Effects of interleukin-1 on hormone release from normal rat pituitary cells in primary culture. Neuroendocr. 45 (1987) 343–347.

Uexküll, J. v.: Nie geschaute Welten. Die Umwelten meiner Freunde. Fischer, Berlin 1936.

Uexküll, J. v.: Theoretische Biologie, 1. Aufl. Springer, Berlin 1920.

Uexküll, J. v.: Bedeutungslehre. J. A. Barth, Leipzig 1940, S. 107/108. Neudruck Fischer, Frankfurt 1970.

Uexküll, J. v.: Kompositionslehre der Natur. Biologie als undogmatische Naturwissenschaft. In: Uexküll, Th. v. (Hrsg.): Ausgewählte Schriften. Propyläen-Ullstein, Frankfurt-Berlin-Wien 1980.

Uexküll, J. v., G. Kriszat: Streifzüge durch die Umwelten von Tieren und Menschen. Springer, Berlin 1936. Neudruck Fischer, Frankfurt 1970.

Uexküll, Th. v. (Hrsg.): Lehrbuch der Psychosomatischen Medizin, 2. Aufl. Urban & Schwarzenberg, München-Wien-Baltimore 1981.

Uexküll, Th. v.: Historische Überlegungen zu dem Problem einer Medizinsemiotik. Zeitschr. f. Semiotik, Bd. 6, 1–2 (1984) 53–58.

Uexküll, Th. v.: Modelle eröffnen oder verschließen uns den Zugang zum Patienten. (Unveröffentlichter Vortrag).

Uexküll, Th. v.: Are functional syndromes culture-bound? Referat anläßlich der Konferenz „Anthropologies in Medicine", Hamburg Dez. 1988 (Manuskript).

Uexküll, Th. v.: Bilanz der Curriculum-Entwicklung in der BRD – Rückschau und Ausblick eines Beteiligten. Referat anläßlich des 20. Jahrestages des Bestehens der Anamnesegruppen. Student. Reader, Marburg 1989.

Uexküll, Th. v.: Mündliche Mitteilung.

Uexküll, Th. v.: Psychophysiologie – Historische und wissenschaftstheoretische Probleme. In: Uexküll, Th. v. et al. (Hrsg.): Lehrbuch der Psychosomatischen Medizin, 1. Aufl. Urban & Schwarzenberg, München 1979.

Uexküll, Th. v.: Was weiß die Medizin vom Menschen? In: Rössner, H. (Hrsg.): Der ganze Mensch, S. 146–168. Deutscher Taschenbuch Verlag, München 1986.

Uexküll, Th. v. (Hrsg.): Integrierte Psychosomatische Medizin. Modelle in Praxis und Klinik. Schattauer, Stuttgart 1981.

Uexküll, Th. v.: Die Chefarztvisite als Problem. Med. Klinik 72 (1977) 269–276.

Uexküll, Th. v.: Geleitwort. In: Köhle, K., H.-H. Raspe (Hrsg.): Das Gespräch während der ärztlichen Visite, S. X–XII. Urban & Schwarzenberg, München 1982.

Uexküll, Th. v.: Geschichte der deutschen Psychosomatik. Psychother. med. Psychol. 36 (1986) 18.

Uexküll, Th. v.: Untersuchungen über das Phänomen der „Stimmung" mit einer Analyse der Nausea nach Apomorphingaben verschiedener Größe. Z. klin. Med. 149 (1952) 132–210.

Uexküll, Th. v.: Unveröffentlichtes Manuskript. 1986.

Uexküll, Th. v.: 40 Jahre Psychosomatische Medizin. Münch. med. Wschr. 119 (1977) 795–800.

Uexküll, Th. v.: Confused consent: Organische Krankheit, Antwort des Internisten, Verlorenheit der Kranken. Psychosomatische Medizin 11 (1982) 24–34.

Uexküll, Th. v.: Das Deutsche Kollegium für psychosomatische Medizin. Versuch einer Standortbestimmung. Mitteilungen an die Mitglieder des DKPM, Nr. 7 (1984) 20–31.

Uexküll, Th. v.: Das Verhältnis der Heilkunde zum Tode. In: Sudnow, D. (Hrsg.): Organisiertes Sterben, S. 11–20. Fischer, Frankfurt 1973.

Uexküll, Th. v.: Die Religion und die Naturwissenschaften. Die Erziehung 12 (1936) 328–379.

Uexküll, Th. v.: Funktionelle Herz- und Kreislaufstörungen, II. Internistentagung Jena-Halle-Leipzig. VEB Thieme, Leipzig 1962.

Uexküll, Th. v.: Funktionelle Syndrome in der Praxis. Psyche 31/4 (1958) 481 f.

Uexküll, Th. v.: Funktionelle Syndrome in psychosomatischer Sicht. Klinik der Gegenwart IX (1960) 303.

Uexküll, Th. v.: Gescheiterte Reform der Medizinischen Ausbildung. Ein Rückblick auf das Schicksal der Nachbaruniversität Ulm. In: Galle, R. (Hrsg.): Gebremste Reform, ein Kapitel deutscher Hochschulgeschichte. Universitäts-Verlag, Konstanz 1977.

Uexküll, Th. v.: Grundfragen der Psychosomatischen Medizin. Rowohlt, Reinbek 1963 und 1965.

Uexküll, Th. v.: Responses of the health care system to maintain and restore health: Psychological considerations: The problem of bio-psycho-social models. Workshop on scientific analysis of health and health care: Paradigms, methodologies and organisation, WHO, Ulm 1983.

Uexküll, Th. v.: Zehn Jahre Deutsches Kollegium für psychosomatische Medizin. Versuch einer Standortbestimmung. Praxis d. Psychother. und Psychosom. 29 (1984) 157–162.

Uexküll, Th. v.: Zeichen und Realität als anthroposemiotisches Problem. In: Oehler, K. (Hrsg.): Zeichen und Realität. Akten des 3. semiotischen Kolloquiums, Hamburg. Stauffenberg, Tübingen 1984.

Uexküll, Th. v., E. Wick: Die Situationshypertonie. Arch. Kreislaufforsch. 39 (1962) 236–271.

Uexküll, Th. v., W. Wesiack: Theorie der Humanmedizin. Urban & Schwarzenberg, München-Wien-Baltimore 1988.

Literatur

Uexküll, Th.v., H. G. Pauli: The mind-body problem in medicine. Advances 3 (1986) 158–174.

Uexküll, Th.v., M. Pflanz: „Entlastung" als pathogenetischer Faktor. Klin. Wschr. 414 (1952).

Uexküll, Th.v., R. Adler, J. M. Herrmann, K. Köhle, O. W. Schonecke, W. Wesiack (Hrsg.): Psychosomatische Medizin, 3. Aufl. Urban & Schwarzenberg, München 1986.

Uexküll, Th.v., W. Wesiack: Wissenschaftstheorie und psychosomatische Medizin: ein bio-psycho-soziales Modell. In: Uexküll, Th.v. (Hrsg.): Psychosomatische Medizin. Urban & Schwarzenberg, München 1986.

Uhlenberg, P.: Death and the family. J. Family History 3 (1980).

Ulich, E., C. Baitsch: Schicht- und Nachtarbeit im Betrieb. 2. Aufl. gdi, Rüschlikon 1979.

Ullman, M.: Behavioral changes in patients following strokes. Thomas, Springfield 1962.

Ullmann, H.: Der Morbus Crohn aus psychosomatischer Sicht. Med. Klinik 77 (1982) 782–788.

Ullrich, G.: Projektive Reduplikation bei psychosomatisch Kranken II. Z. Psychosom. Med. Psychoanal. 34 (1988) 361–372.

Ulreich, A., F. Rainer, K. P. Pfeiffer: Compliance und chronische Polyarthritis. Therapiewoche 32 (1982) 5915–5919.

Ulrich, R. E.: Pain-aggression. In: Kimble, G. A. (ed.): Foundations of conditioning and learning. Appleton Century Crofts, New York 1967.

Ulshöfer, B., G. Paar, B. Cramer: Harnsteinmanifestationen und Streß. In: Gasser, G., W. Vahlensieck (Hrsg.): Pathogenese der Harnsteine IX, S. 46. Steinkopff, Darmstadt 1982.

United States National Center for Health Statistics: Vital and health statistics. Heart disease in adults 1960–62. Blood pressure of adults by race and area. Public Health Service Publ. No. 1000, Ser. 11 No 5. US Government Printing Office, Washington 1964.

Ursin, H.: Personality, activation and somatic health. A new psychosomatic theory. In: Levine, S., H. Ursin (eds.): Coping and Health. NATO Conf. Series III/12, pp. 259–279. Plenum, New York 1980.

Ursin, H., E. Baade, S. Levine: Psychobiology of stress. A study of coping men. Acad. Press, New York 1978.

Ursin, H., R. Mykletun, O. Tonder, R. Vaernes, G. Relling, R. Isaksen, R. Murison: Psychological stress-factors and concentrations of immunoglobulins and complement components in humans. Scand. J. Psychol. 25 (1984) 340–347.

Vachon, L., E. S. Rich: Visceral learning in asthma. Psychosom. Med. 38 (1976) 122–129.

Vägerö, D., G. Persson: Cancer survival and social class in Sweden. J. Epidemiol. Community Health 41 (1987) 204–209.

Vahlensieck, W.: Die Prostatakongestion. In: Helpap, B., Th. Senge, W. Vahlensieck (Hrsg.): Prostatakongestion und Prostatitis. Die Prostata, Bd. 3, S. 1. pmi-Verlag, Frankfurt 1985.

Vahlensieck, W.: Histologische Differenzierung von Prostatakongestion und Prostatitis. Niere, Blase, Prostata – aktuell 13 (1988) 4.

Vaillant, G. E.: Natural history of male psychological health, II. Some antecedents of healthy adult adjustment. Arch. gen. Psychiat. 31 (1974) 15–22.

Vaillant, G. E.: Natural history of male psychological health, III. Empirical dimensions of mental health. Arch. gen. Psychiat. 32 (1975) 420–426.

Vaillant, G. E.: Natural history of male psychological health: effects of mental health on physical health. New Engl. J. Med. 301 (1979) 1249–1254.

Vaitl, D.: Kontrolle der essentiellen Hypertonie durch Entspannungstechniken. In: Vaitl, D. (Hrsg.): Essentielle Hypertonie. Springer, Berlin-Heidelberg 1982.

Valenstein, E. S.: Brain stimulation and motivation. Scott, Foresman & Co., Chicago 1973.

Valentin, H., A. Zober: Über den permanenten Dialog zwischen Politik und Wissenschaft in der Arbeitsmedizin. Die Berufsgenossenschaft 1 (1982) 63–67.

Valentin, H., W. Klosterkötter, G. Lehnert, H. Petry, J. Rutenfranz, G. Weber, H. G. Wenzel, H. Wittgens: Arbeitsmedizin. Ein kurzgefaßtes Lehrbuch für Ärzte und Studenten, Bd. 1. 2. Aufl. Thieme, Stuttgart 1979.

Valins, S., A. A. Ray: Effects of cognitive desensitization on avoidance behavior. J. Pers. soc. Psychol. 7 (1967) 354–350.

Valkonen, T.: Psychosocial stress and sociodemographic differentials in mortality from ischemic heart disease in Finland. Acta med. scand. (Suppl.) 660 (1982) 152–164.

Van der Valk, J. M., J. J. Groen: Personality structure and conflict situation in patients with myocardial infarction. J. psychosom. Res. 1 (1967) 41–46.

Van Doorn, P., H. Folgering, P. Colla: Control of the end-tidal pCO_2 in the hyperventilation syndrome: effects of biofeedback and breathing instructions compared. Bull. europ. Physiopath. respir. 18 (1982) 829–836.

Van Kammen, D. P., W. B. van Kammen, L. S. Mann, T. Seppala, M. Linnoila: Dopamine metabolism in the cerebrospinal fluid of drug-free schizophrenic patients with and without cortical atrophy. Arch. gen. Psychiat. 43 (1986) 978–983.

Van Keep, P. A., M. Humphrey: Psychosocial aspects of climacteric. In: Van Keep, P. A. et al. (eds.): Consensus on Menopause Research. MTP, London 1976.

Vandereycken, W.: Körperschemastörungen und ihre Relevanz für die Behandlung der Bulimia. In: Fichter, M. M. (Hrsg.): Bulimia nervosa. Enke, Stuttgart 1989.

Vandereycken, W., J. Vanderlinden: Denial of illness and the use of self-reporting measures in anorexia. Int. J. Eat. Dis. 2 (1983) 101–107.

Vandereycken, W., R. Meermann: Anorexia nervosa. A clinician's guide to treatment. De Gruyter, Berlin-New York 1984.

Vandereycken, W., R. Pierloot: Combining drugs and behavior in anorexia nervosa: a double-blind placebo/pimozide study. In: Darby, P. L., P. E. Garfinkel, D. M. Garner, D. V. Coscina (eds.): Anorexia nervosa. Recent Developments in Research, pp. 365–375. Liss, New York 1983.

Vandereycken, W., R. Pierloot: Long-term outcome research in anorexia nervosa: the problem of patient selection and follow-up duration. Int. J. Eat. Dis. 2 (1983) 237–242.

Varela, F. G.: Autonomy and autopoiesis. In: Roth, G., H. Schwegler (eds.): Self-organising Systems. An Interdisciplinary Approach. Campus, Frankfurt-New York 1981.

Varela, F. G.: Principles of biological autonomy. Elsevier, Amsterdam-New York 1979.

Varela, F. G., H. R. Maturana, R. Uribe: Autopoiesis: the organisation of living systems, its characterisation and a model. Biosystems 5 (1974) 187–196.

Varney, N. R., B. Alexander, J. H. MacIndoe: Reversible steroid dementia in patients without steroid psychosis. Amer. J. Psychiat. 141 (1984) 369–372.

Vaughn, K. B., J. T. Lanzetta: Vicarious investigation and conditioning of facial expressive and autonomic responses to a model's display of pain. J. Pers. soc. Psychol. 38 (1980) 909–923.

Veale, W. C., N. W. Kasting, K. E. Cooper: Arginine vasopressin and endogenous antipyresis: evidence and significance. Fed. Proc. 40 (1981) 2750–2753.

Veith, R. C., A. Murray, M. D. Raskind, J. H. Caldwell, R. F. Barnes, G. Gumbrecht, J. L. Ritchie: Cardiovascular effects of tricyclic antidepressants in depressed patient with chronic heart disease. New Engl. J. Med. 306 (1982) 954–959.

Veldhuis, H. D., D. DeWied: Is conditioned immunosuppression an adequate research strategy? Behav. Brain Sci. 8 (1985) 411–412.

Velvolvski, I. S.: Erfahrungen mit der psychoprophylaktischen Methode zur Schmerzausschaltung bei der Geburt auf der Grundlage der Lehre I. P. Pawlows. In: Schmerzausschaltung bei der Geburt. Volk und Ges., 1953.

Veniar, F. A., R. S. Salston: An approach to the treatment of pseudohypocusis in children. Amer. J. Dis. Child 137 (1983) 34.

Verhagen, F., C. Nass, A. Appels: Cross-validation of the A/B typology in the Netherlands. Psychother. and Psychosom. 34 (1980) 178–186.

Verres, R: Krebs und Angst. Springer, Berlin 1986.

Verrier, R. L., B. Lown: Influence of neural activity on ventricular electrical stability during acute myocardial ischemia and infarction. Excerpta Medica, Amsterdam 1978.

Verrier, R. L., B. Lown: Behavocral stress and cardiac arrhythmias. Ann. Rev. Physiol. 46 (1984) 155.

Verrier, R. L., P. L. Thompson, B. Lown: Ventricular vulnerability during sympathetic stimulation: role of heart rate and blood pressure. Cardiovascul. Res. 8 (1974) 602.

Verwoerdt, A.: Communication with the fatally ill. Charles C. Thomas, Springfield 1966.

Verwoerdt, A.: Psychopathological responses to the stress of physical illness. In: Lipowski, Z. J. (ed.): Psychosocial Aspects of Physical Illness, pp. 119–141. Advanc. Psychosom. Med. Vol. 8. Basel 1972.

Verwoerdt, A., J. L. Elmore: Psychological reactions in fatal illness. The prospect of impending death. J. Amer. geriat. Soc. 15 (1967) 9–19.

Verzàr, F.: Biologische Veränderungen im Alter. In: Martin, E., J.-P. Junod (Hrsg.): Ein kurzes Lehrbuch der Geriatrie. Huber, Bern 1975.

Vetter, N. J., E. L. Cay, A. E. Philipp, R. C. Strange: Anxiety on admission to a coronary care unit. J. psychosom. Res. 21 (1977) 73–78.

Viederman, M.: Adaptive and maladaptive regression in hemodialysis. Psychiatry 37 (1974) 283–290.

Villard, H. P.: Consultation psychiatrique dans un service de dialyse et de greffe rénale. Union méd. Canada 98 (1969) 233–238.

Villard, R. de et al.: Le somnambulisme de l'enfant. Neuropsychiat. de l'Enfance 28 (1980) 222–224.

Vincent, S., H. Kaczkowski: Bulimia: sign, symptom, or entity. Int. J. Eat. Dis. 3 (1984) 81–95.

Violon, A.: The onset of facial pain. A psychological study. Psychother. and Psychosom. 34 (1980) 11.

Virchow, R.: Die Not im Spessart. Mitteilungen über die in Oberschlesien herrschende Typhus-Epidemie. Hildesheim 1868.

Visintainer, M. A., J. R. Volpicelli, M. E. P. Seligman: Tumor rejection in rats after inescapable or escapable shock. Science 216 (1982) 437–439.

Vismara, L., Z. Vera, J. M. Foerster et al.: Identification of sudden death risk in acute and chronic coronary artery disease. Amer. J. Cardiol. 39 (1977) 821.

Vliegen, J., T. Vogel: Kopfschmerzen beim depressiven Syndrom. In: Barolin, G. S., D. Sanrugg, W. Hemmer (Hrsg.): Kopfschmerz. München 1975.

Völgyesi, F.: Menschen- und Tierhypnose. Orell Füssli, Zürich-Leipzig 1938.

Vogel, A. V., J. S. Goodwin, J. M. Goodwin: The therapeutics of placebo. Amer. Fam. Physician 22 (1980) 105–109.

Vogel, P.: Grundfragen der klinischen Neurologie. In: Vogel, P. (Hrsg.): Victor von Weizsäcker: Arzt im Irrsal der Zeit. Eine Freundesgabe zum 70. Geburtstag am 21. 4. 1956, S. 179–189. Vandenhoeck & Ruprecht, Göttingen 1956.

Vogel, P.: Von der Eigenart der Neurologie. Dtsch. med. Wschr. 78 (1953) 527–530.

Vogel, P.: Von der Selbstwahrnehmung der Epilepsie – Der Fall Dostojewskij. Nervenarzt 32 (1961) 438–441.

Vogel, P. G.: Zur Acne excoriée des jeunes filles. Hautarzt 25 (1974) 333–336.

Vogt, H.-J.: Andrologie. In: Eicher, W. (Hrsg.): Sexualmedizin in der Praxis, S. 117–201. Fischer, Stuttgart-New York 1980.

Vogt, H., M.Blohmke: Häufigkeit psychischer und sozialer Problemfälle in einer Allgemeinpraxis. Der praktische Arzt 22 (1974).

Vogt, J.: Der Einfluß der Eltern auf den Drogenkonsum ihrer Kinder. In: Lukesch, H., M. Perrez, K. A. Schneewind (Hrsg.): Sozialisation und Intervention in der Familie, 1977.

Vogt, O.: Die direkte psychologische Experimentalmethode in hypnotischen Bewußtseinszuständen. Barth, Leipzig 1897.

Voigt, K. H., H. L. Fehm: Hormone und Emotionen. In: Euler, H. A., H. Mandl (Hrsg.): Emotionspsychologie. Urban & Schwarzenberg, München 1983 c.

Voigt, K. H., H. L. Fehm: Neuropeptides and pain. In: Holroyd, H. A., B. Schlote, H. Zenz (eds.): Perspectives in Research on Headache. Hogrefe, Toronto 1983 b.

Voigt, K. H., H. L. Fehm: Neurotransmitter/Neuromodulatoren: Spezifische Signalsubstanzen für psychische Prozesse? In: Amelang, M. (Hrsg.): Bericht des 35. Kongr. Dt. Ges. Psychologie, 1986, S. 47–57. Hogrefe, Göttingen 1987.

Voigt, K. H., H. L. Fehm: Psychoendokrinologie. Int. Welt 6 (1983a) 130–136.

Voigt, K. H., M. Ziegler, M. Grünert-Fuchs, U. Bickel, G. Fehm-Wolfsdorf: Hormonal responses to exhausting physical exercise: the role of predictability and controllability of the situation. Psychoneuroendocrinology (in press).

Voigt, K. H., S. Bretschneider, S. Bossert, A. Bliestle, H. L. Fehm: Disturbed cortisol secretion in man: contrasting Cushing's disease and endogenous depression. J. psychiat. Res. 15 (1985) 341–350.

Volger, J.: Indications for client-centered psychotherapy with headache patients. In: Holroyd, K. A., B. Schlote, H. Zenz (eds.): Perspectives in Research on Headache, pp. 198–203. Hogrefe, Lewiston-New York 1983.

Volhard, F.: Eröffnungsrede. In: Lasch, H. G., B. Schlegel (Hrsg.): Hundert Jahre Deutsche Gesellschaft für Innere Medizin. Bergmann, München 1982.

Volkholz, V.: Belastungsschwerpunkte und Praxis der Arbeitssicherheit. Hrsg. vom Bundesminister für Arbeit und Sozialordnung, Bonn 1977.

Vollhardt, B. R., S. H. Ackerman, A. I. Grayzel, P. Barland: Psychologically distinguishable groups of rheumatoid arthritis patients: a controlled, single blind study. Psychosom. Med. 44 (1982) 353–362.

Vollmer, W. M., P. W. Wahl, C. R. Blagg: Survival with dialysis and transplantation in patients with end-stage renal disease. New Engl. J. Med. 308 (1983) 1553–1558.

Vollrath, P., H. Ferner, P. Vetter et al.: Sexualverhalten hämodialysierter Patienten. Inn. Med. 3 (1976) 349.

Volmerg, B.: Die Vergesellschaftung psychopathologischer Strukturen im Produktionsprozeß. In: Leithäuser, T., W. R. Heinz (Hrsg.): Produktion, Arbeit, Sozialisation, S. 128–145. Suhrkamp, Frankfurt 1976.

Volmerg, U.: Identität und Arbeitserfahrung. Eine theoretische Konzeption zu einer Sozialpsychologie der Arbeit. Suhrkamp, Frankfurt 1978.

Volpe, R., J. Vale, M. W. Jonston: The effect of certain physical and emotional tensions and strains on fluctuations in the level of serum protein-bound jodine. J. clin. Endocr. 20 (1960) 415.

Volpert, W.: Der Zusammenhang zwischen Arbeit und Persönlichkeit aus handlungstheoretischer Sicht. In: Groskurth, P. (Hrsg.): Arbeit und Persönlichkeit, S. 21–46. Rowohlt, Reinbek 1979.

Von der Hardt, H., D. Hoffman: Das Asthmasyndrom. In: Fenner, A., H. von der Hardt (Hrsg.): Pädiatrische Pneumonologie, S. 297–346. Springer, Berlin 1985.

Voss, G.: Klinische Beiträge zur Lehre von der Hysterie. Fischer, Jena 1909.

Voukydis, P., S. Forwand: The effect of elicitation of the relaxation response in patients with intractable ventricular arrhythmias. Circulation 55, 56 (Suppl. 3) (1977) 111–157.

Vreeland, R., G. Ellis: Stresses on the nurse in an intensive care unit. J. Amer. med. Ass. 208 (1969) 332–334.

Wacker, A.: Arbeitslosigkeit als Sozialisationserfahrung – Skizze eines Interpretationsansatzes. In: Leithäuser, T., W. R. Heinz (Hrsg.): Produktion, Arbeit, Sozialisation, S. 171–187. Suhrkamp, Frankfurt 1976.

Wacker, A.: Arbeitslosigkeit. Soziale und psychische Voraussetzungen und Folgen. 2. Aufl. Europäische Verlagsanstalt, Frankfurt-Köln 1977.

Wacker, A.: Unfreiwillige Arbeitslosigkeit und psychosomatische Störungen. In: Deppe, H.-U., U. Gerhardt, P. Novak (Hrsg.): Medizinische Soziologie, Jahrbuch 1, S. 175–183. Campus, Frankfurt-New York 1981.

Waddell, G.: A new clinical model for the treatment of low back pain. Spine 12 (1987) 632–644.

Wächter, H. M.: Psychosomatische Aspekte des Streßulcus. In: Becker, H. D. (Hrsg.): Streßulcus. Thieme, Stuttgart 1981.

Waelder, R.: The structure of paranoid ideas: a critical survey of various theories. Int. J. Psychoanal. 32 (1951) 167–177.

Wagenhäuser, F. J.: Die Arthrosen. Therapiewoche 8 (1973) 577.

Wagner, A. R.: Expectancies and the priming of STM. In: Hulse, S. S., H. Fowler, W. K. Honig (eds.): Cognitive Processes in Animal Behavior. Erlbaum, Hillsdale 1978.

Wagner, A. R., F. A. Logan, K. Haberland, T. Price: Stimulus selection in animal discrimination learning. J. exp. Psychol. 76 (1968) 171–180.

Wagner, G., R. Green: Impotence. Plenum Press, New York-London 1981.

Waismann, M.: Pickers, pluckers and impostors. A panorama of cutaneous self-mutilation. Postgrad. Med. (1965) 620–630.

Waitzkin, H., J. D. Stöckle: Information control and the micropolitics of health care: summary of an ongoing research project. Soc. Sci. Med. 10/6 (1976) 263–276.

Waitzkin, H., J. D. Stöckle: The communication of information about illness. Clinical, sociological, and methodological considerations. Advanc. Psychosom. Med. 8 (1972) 180–215.

Wakeling, A., V. de Souza: Differential endocrine and menstrual response to weight change in anorexia nervosa. In: Darby, P. L., P. E. Garfinkel, D. M. Garner, D. V. Coscina (eds.): Anorexia nervosa. Recent Developments in Research, pp. 271–277. Liss, New York 1983.

Waldeyer, A.: Anatomie des Menschen, Bd. 2, 13. Aufl., S. 195–225. De Gruyter, Berlin-New York 1975.

Waldron, J.: The coronary-prone behavior pattern, blood pressure, employment and socio-economic status in women. J. psychosom. Res. 22 (1978) 79–87.

Waldron, J., A. Hickey, C. McPherson: Type A behavior pattern: relationship to variation in blood pressure, parental characteristics and academic and social activities of students. J. hum. Stress 6 (1980) 16–27.

Walker, E. L.: Action decrement and its relation to learning. Psychol. Rev. 65 (1968) 129–142.

Walker, H. E.: How to manage the hyperventilation syndrome. Behav. Med. 5 (1978) 30–37.

Walker, P., J. Luther, I. M. Samloff, M. Feldman: Life event stress and psychological factors in men with peptic ulcer disease. Gastroenterol. 94 (1988) 323–330.

Wall, S., N. Kaltreider: Changing social-sexual patterns in gynecologic practice. J. Amer. med. Ass. 237 (1977) 565–568.

Wallace, D. J.: The role of stress and trauma in rheumatoid arthritis and systemic lupus erythematosus. Semin. Arthr. Rheum. 16 (1987) 153–157.

Waller, S. L., J. J. Misiewicz: Prognosis in the irritable bowel syndrome. Lancet II (1969) 753–756.

Waller, V., M. Kaufman, F. Deutsch: Anorexia nervosa: a psychosomatic entity. Psychosom. Med. 2 (1940) 3–16.

Wallnöfer, H.: Gesund mit autogenem Training. Umschau, Frankfurt 1979.

Wallnöfer, H.: Seele ohne Angst – Hypnose, Autogenes Training, Entspannung. Hoffmann & Campe, Hamburg 1972.

Walschburger, P.: Zur Beschreibung von Aktivierungsprozessen. Dissertation, Freiburg 1976.

Walsh, B. T., S. P. Roose, A. H. Glassman, M. Gladis, C. Sadik: Bulimia and depression. Psychosom. Med. 47 (1985) 123–131.

Walsh, P., A. Dale, D. E. Anderson: Comparison of biofeedback pulse wave velocity and progressive relaxation on essential hypertensives. Percept. Mot. Skills 44 (1977) 839–843.

Walter, B., H. Rüddel, A. W. v. Eiff: Efficiency of behavioral intervention in hypertension. In: Elbert, T., W. Langosch, A. Steptoe, D. Vaitl: Behavioral Medicine in Cardiovascular Disorders. Wiley, New York 1988.

Walton, H., R. Kalucy: The extent of agreement about the nature of anorexia nervosa. Unpublished, 1975.

Waltz, G. M.: Soziale Faktoren bei der Entstehung und Bewältigung von Krankheit – Ein Überblick über die empirische Literatur. In: Badura, B. (Hrsg.): Soziale Unterstützung und chronische Erkrankung. Zum Stand epidemiologischer Forschung, S. 40–119. Suhrkamp, Frankfurt 1981.

Wanick, W., J. Bach, H. G. Hartmann et al.: Hirnleistungsstörungen bei Patienten eines Dialyse-Transplantationsprogramms. Schweiz. med. Wschr. 107 (1977) 832–835.

Ward, D. J.: Rheumatoid arthritis and personality: a controlled study. Brit. med. J. 1 (1971) 297–299.

Ward, M.: The pathogenesis of Crohn's disease. Lancet II (1977) 903–905.

Ward, N. G., V. L. Bloom, R. O. Friedel: The effectiveness of tricyclic antidepressants in the treatment of coexisting pain and depression. Pain 7 (1979) 331–341.

Wardrop, J.: Diseases of the heart. London 1851.

Wardwell, W. J., C. B. Bahnson: Behavioral variables and myocardial infarction in the Southeastern Connecticut Heart Study. J. chron. Dis. 26 (1973) 447–461.

Waring, E. M.: Family therapy and psychosomatic illness. Int. J. Fam. Therapy 2 (1980) 243.

Warner, G., J. Lance: Relaxation therapy in migraine and chronic tension headache. Med. J. Aust. 1 (1975) 293–301.

Warnes, H.: Twenty years lines of advances in psychosomatic medicine. Psychiat. J. Univ. Ottawa 4 (1979) 85–92.

Warren, S., S. Greenhill, K. G. Warren: Emotional stress and the development of multiple sclerosis: case control evidence of a relationship. J. chron. Dis. 35 (1982) 821–831.

Warren, W.: A study of anorexia nervosa in young girls. J. Child Psychol. 9 (1968) 27–40.

Wartegg, E.: Schichtdiagnostik. Hogrefe, Göttingen 1953.

Wasmus, A., H.-H. Raspe: Analyse der Verschlüsselung von Arbeitsunfähigkeitsdiagnosen mit der ICD-Nr. 714 anhand einer rheumatologischen Nachuntersuchung. Öff. Gesundh.-Wes. 50 (1988) 2–8.

Waters, W. E.: Migraine: intelligence, social class, and familial prevalence. Brit. med. J. (1971) 77–81.

Waters, W. E., P. J. O'Connor: Prevalence of migraine. J. Neurol., Neurosurg., Psychiat. 38 (1975) 613–616.

Watkins, L. R., D. J. Mayer: Organization of endogenous opiate and nonopiate pain control systems. Science 216 (1982) 1185.

Watson, C. G., C. Buranen: The frequency and identification of false positive conversion reactions. J. nerv. ment. Dis. 167 (1979) 243–247.

Watts, J. M., F. T. de Dombal, G. Watkinson, J. C. Goligher: Long-term prognosis of ulcerative colitis. Brit. med. J. I (1966) 1447–1453.

Watts, J. W., J. F. Fulton: The effects of lesions of the hypothalamus upon the gastrointestinal tract and heart in monkeys. Ann. Surg. 101 (1935) 363.

Watts, R. J.: The patient on renal dialysis. Strategies for sexual counseling. In: Levy, N. B. (ed.): Psychonephrology 2, pp. 107–115. Plenum, New York 1983.

Watzlawick, P., J. H. Beavin, D. D. Jackson: Pragmatics of human communication. Nortin, New York 1967. Deutsch: Menschliche Kommunikation. Huber, Bern-Stuttgart-Wien 1969.

Watzlawick, P., J. H. Weakland, R. Fisch: Lösungen – zur Theorie und Praxis menschlichen Wandels. Huber, Bern-Stuttgart-Wien 1974.

Waxenberg, S. H., M. G. Drellich, A. M. Sutherland: The role of hormones in human behavior. Changes in female sexuality after adrenalectomy. J. clin. Endocr. 19 (1959) 193.

Way, L. W.: Abdominal pain. In: Sleisenger, M. H., J. S. Fordtran (eds.): Gastrointestinal Disease, pp. 326–337. Saunders, Philadelphia-London-Toronto 1973.

Wayner, E. A., G. R. Flannery, G. Singer: Effects of taste aversion conditioning on the primary antibody response to sheep red blood cells and Brucella abortus in the albino rat. Physiol. Behav. 21 (1978) 995–1000.

Waysfeld, B., M. LeBarzic, P. Aimez, B. Guy-Grand: „Pensée opératoire" in obesity. In: Bräutigam, W., M.v.Rad (eds.): Toward a Theory of Psychosomatic Disorders, pp. 127–132. Basel 1977.

Weakland, J. H.: Family somatics – a neglected edge. Family Process 16 (1977) 263.

Webb, H. E., R. G. Lascelles: Treatment of facial and head pain associated with depression. Lancet I (1962) 355–356.

Webb, W. B., C. M. Levy: Effects of spaced and repeated total sleep deprivation. Ergonomics 27 (1984) 45–58.

Webb, W. B., D. Schneider-Helmert: A categorical approach to changes in latency, awakenings, and sleep length in older subjects. J. nerv. ment. Dis. 172 (1984a) 63–70.

Webb, W. B., D. Schneider-Helmert: Awakenings: subjective and objective relationships. Percept. Mot. Skills 59 (1984b) 63–70.

Weber, E.: Grundriß der biologischen Statistik, 5. Aufl. Fischer, Jena 1964.

Weber, G., H. Stierlin: In Liebe entzweit. Rowohlt, Reinbek 1989.

Weber, H.: Lumbar disc herniation. A controlled, prospective study with 10 years of observation. Spine 8 (1983) 131–140.

Weber, H.: The effect of delayed disc surgery on muscle paresis. Acta orthop. scand. 46 (1975) 631–642.

Weber, M.: Wirtschaft und Gesellschaft. Grundriß der verstehenden Soziologie. Mohr, Tübingen 1972.

Weber, R. G.: The effects of curriculum change on the „new" medical student. AAMC-RIME Conference 1972, 55.

Wechsler, D.: Die Messung der Intelligenz Erwachsener. Textband zum HAWIE, 3. Aufl. Bern-Stuttgart 1964.

Weder, A. B., S. Julius: Behavior, blood pressure variability and hypertension. Psychosom. Med. 47 (1985) 406–414.

Weed, L. L.: Medical records, medical education and patient care. The Press of the Western Reserve University, Cleveland 1969.

Weeda-Mannak, W., M. Drop, F. Smith, L. Strijbosch, I. Bremer: Toward an early recognition of anorexia nervosa. Int. J. Eat. Dis. 2 (1983) 27–37.

Wehr, T. A., F. K. Goodwin: Biological rhythms and psychiatry. In: Ariety, S. H., H. K. H. Brodie (eds.): The American Handbook of Psychiatry. Vol. 2. 2nd ed. Basic Books, New York 1981.

Weidner, G.: Self-handicapping following learned helplessness treatment and the Type A coronary-prone behavior pattern. J. psychosom. Res. 24 (1980) 319–325.

Weidner, G., J. Andrews: Attributions for undesirable life events. Type A behavior and depression. Psychol. Res. 53 (1983) 167–170.

Weidner, G., K. A. Matthews: Reported physical symptoms elicited by unpredictable events and the Type A coronary-prone behavior pattern. J. Pers. soc. Psychol. 38 (1978) 213–220.

Weidner, W.: Moderne Prostatadiagnostik. Habilitationsschrift. In: Klinische und experimentelle Urologie, Bd. 7. Zuckschwerdt, München 1984.

Weidner, W., W. Krause, H. Brunner, C. F. Rothauge, H. G. Schiefer: Zur Differentialdiagnose von chronischer Prostatitis und vegetativem Urogenitalsyndrom – Langzeitbeobachtungen an 267 Männern. In: Verhandlungsbericht der Deutschen Gesellschaft für Urologie, S. 25. Springer, Berlin 1979.

Weimann, G.: Das Hyperventilationssyndrom. Urban & Schwarzenberg, München 1968.

Weinberg, L. A.: An evaluation on stress in temporomandibular joint dysfunction-pain syndrome. J. prosthet. Dent. 38 (1977) 192.

Weinberg, S. J., J. M. Foster: Electrocardiographic changes produced by localized hypothalamic stimulation. Ann. intern. Med. 53 (1960) 332.

Weinberger, D. R.: Implications of normal brain development for the pathogenesis of schizophrenia. Arch. gen. Psychiat. 44 (1987) 660–669.

Weinblatt, E., W. Ruberman, J. D. Goldberg et al.: Relation of education to sudden death after myocardial infarction. New Engl. J. Med. 299 (1978) 60.

Weinel, E.: Übertragungs- und Gegenübertragungsprobleme bei der Behandlung von AIDS-Patienten. Psyche 43 (1989) 710–719.

Weiner, H.: Brain, behaviour and bodily disease: a summary. In: Weiner, H., M. A. Hofer, A. J. Stunkard (eds.): Brain, Behaviour and Bodily Disease. Raven, New York 1981.

Weiner, H.: Are psychosomatic diseases diseases of regulation? Psychosom. Med. 37 (1975) 289–291.

Weiner, H.: Bronchial asthma. In: Weiner, H. (ed.): Psychobiology and Human Disease. Elsevier, New York 1977.

Weiner, H.: Die Geschichte der Psychosomatischen Medizin und das Leib-Seele-Problem in der Medizin. Psychother. med. Psychol. 36 (1986) 361–391.

Weiner, H.: Eine Medizin der menschlichen Beziehungen. Psychother. med. Psychol. 39 (1989) 96–102.

Weiner, H.: Essential hypertension. In: Weiner, H. (ed.): Psychobiology and Human Disease. Elsevier, New York 1977.

Weiner, H.: Gesundheit, Krankheitsgefühl und Krankheit – Ansätze zu einem integrativen Verständnis. Psychother. med. Psychol. 33 (1983) 15–34.

Weiner, H.: The concept of „stress" in the light of studies on disasters, unemployment and loss. A critical analysis. Ann. Meeting of the American College of Psychiatrists, San Diego 1984.

Weiner, H.: Human relationships in health, illness and disease. In: Magnusson, D., A. Öhman (eds.): Psychopathology: An Interactional Perspective. Acad. Press, Orlando FL 1987.

Weiner, H.: Life events, depressive symptoms and immune function. Amer. J. Psychiatry 4 (1987) 144.

Weiner, H.: Psychobiology and human disease. Elsevier, New York 1977.

Weiner, H.: Psychosocial factors in autoimmune disease: In: Ader, R., D. L. Felten, N. Cohen (eds.): Psychoneuroimmunology II. Acad. Press, San Diego 1989 a.

Weiner, H.: Schizophrenia: Etiology. In: Kaplan, H. I., B. J. Sadock (eds.): Comprehensive Textbook of Psychiatry IV. Chapt. 15.3. Williams & Wilkins, Baltimore 1985.

Weiner, H.: Some unexplored regions of psychosomatic medicine. 9th. Congress of the International College of Psychosomatic Medicine. Sydney, 1987.

Weiner, H.: Stress, relaxation and asthma. Int. J. Psychosom. 34 (1987) 21–24.

Weiner, H.: The concept of stress and its role in disease onset. In: Lolas, F., H. Mayer (eds.): Perspectives on Stress and Stress-related Topics. Springer, Berlin-Heidelberg-New-York-Tokyo 1987.

Weiner, H.: The functional bowel disorders. In: Weiner, H., A. Baum (eds.): Perspectives in Behavioral Medicine: Eating Regulation and Discontrol. Erlbaum, Hillsdale 1988.

Weiner, H.: The hypothalamic-pituitary-ovarian axis in anorexia and bulimia nervosa. Int. J. Eat. Dis. 2 (1983) 109–116.

Weiner, H.: The illusion of simplicity: the medical model revisited. Amer. J. Psychiat. 135 (Suppl.) (1978) 27–33.

Weiner, H.: The organism in health and disease: contributions of modern biology to an integrated psychosomatic theory. Psychosom. Med. (in press).

Weiner, H.: The prospects for psychosomatic medicine: selected topics. Psychosom. Med. 44 (1982) 488–517.

Weiner, H.: Untersuchungen über den Zusammenhang zwischen Pepsinogenspiegel, psychischen Auffälligkeiten und Ulcus duodeni. Unveröffentlichter Vortrag anläßlich des 25 jährigen Bestehens der Psychosomatischen Poliklinik München, 1981.

Weiner, H.: Vortrag. Europ. Conf. Psychosom. Res., Heidelberg 1976.

Weiner, H.: Zentralnervöse Kontrollmechanismen und Krankheitsentwicklung – ihre Bedeutung für die psychosomatische Medizin. Psychother. med. Psychol. 35 (1985) 310–314.

Weiner, H., J. L. Katz: The hypothalamic-pituitary-adrenal axis in anorexia nervosa: a reassessment. In: Darby, P. L., P. E. Garfinkel, D. M. Garner, D. V. Coscina (eds.): Anorexia nervosa. Recent Developments in Research, pp. 249–270. Liss, New York 1983.

Weiner, H., M. Thaler, M. F. Reiser, I. A. Mirsky: Etiology of duodenal ulcer: I. Relation of specific psychological characteristics to rate of gastric secretion (serum pepsinogen). Psychosom. Med. 19 (1957) 1.

Weiner, M. F., T. Caldwell: Stresses and coping in ICU-nursing, II. Nurse support groups on intensive care units. Gen. Hosp. Psychiat. 3 (1981) 129–134.

Weingart, B.: Schwangerschaft und Geburt bei inhaftierten Frauen in Berlin (West). Diss., FU Berlin 1983.

Weinstein, E. A.: Symbolic neurology and psychoanalysis. In: Marmor, J. (ed.): Modern Psychoanalysis. New Directions and Perspectives. Basic Books, New York-London 1968.

Weinstein, E. A., R. L. Kahn: Denial of illness. Thomas, Springfield/Ill. 1955.

Weinstein, E. A., R. L. Kahn: Personality factors in denial of illness. Arch. Neurol. Psychiat. 69 (1953) 355–367.

Weinstein, E. A., S. Sersen: Phantoms in cases of congenital absence of limbs. Neurology 11 (1961) 905–911.

Weinstein, J., M. Pope, R. Schmidt, R. Seroussi: Neuropharmacologic effects of vibration on the dorsal root ganglion. An animal model. Spine 13 (1988) 521–525.

Weinstock, H. J.: Successful treatment of ulcerative colitis by psychoanalysis. J. psychosom. Res. 6 (1962) 243.

Weintraub, A.: Beitrag zur Psychosomatik der progredient chronischen Polyarthritis. Ther. Umschau 24 (1967) 368–372.

Weintraub, A.: Psychorheumatologie. Karger, Basel 1983.

Weintraub, A.: Psychosomatischer Beitrag zur Diagnose und Therapie des Kreuzschmerzes. Psychosom. Med. 7 (1977) 109–118.

Weintraub, A., R. Battegay, D. Beck, G. Kaganas, F. Labhardt, W. Müller (Hrsg.): Psychosomatische Schmerzsyndrome des Bewegungsapparates. Schwabe, Basel 1975.

Weisenberg, M.: Schmerz und Schmerzkontrolle. In: Keeser, W., E. Pöppel, P. Mitterhusen (Hrsg.): Schmerz. Urban & Schwarzenberg, München 1982.

Weisman, A. D.: Discussion: Modell: aging and psychoanalytic theories of regression. J. geriat. Psychiat. 3 (1970) 147–152.

Weisman, A. D.: Appropriate and appropriated death. In: Shneidman, E. S. (ed.): Death: Current Perspectives, pp. 502–506. Mayfield, Palo Alto 1976.

Weisman, A. D.: Coping behavior and suicide in cancer. In: Cullen, J. E. et al. (eds.): Cancer: The Behavioral Dimensions. Raven, New York 1976.

Weisman, A. D.: Coping with cancer. McGraw-Hill, New York 1979.

Weisman, A. D.: On dying and denying. Behavioral Publications, New York 1972.

Weisman, A. D.: Thanatology. In: Freedman, A. M., H. I. Kaplan, B. J. Sadock (eds.): Comprehensive Textbook of Psychiatry, pp. 1748–1759. Williams & Wilkins, Baltimore 1975.

Weisman, A. D.: The realization of death. Jason Aronson, New York-London 1974.

Weisman, A. D., J. W. Worden: The existential plight in cancer: significance of the first 100 days. Int. J. Psychiat. Med. 7 (1976/77) 1–16.

Weisman, A. D., J. W. Worden: Psychological analysis of cancer deaths. Omega 6 (1975) 61–75.

Weisman, A. D., R. Kastenbaum: The psychological autopsy. A study of the terminal phase of life. Human Sciences Press, New York 1976.

Weisman, A. D., T. P. Hackett: Denial as a social act. In: Lewin, S., K. J. Kahana (eds.): Psychodynamic Studies on Aging. Int. Univ. Press, New York 1967.

Weisman, A. D., T. P. Hackett: Organization and function of a psychiatric consultation service. Int. Rec. Med. 173 (1960) 306–311.

Weiss, D. A.: The pubertal change of the human voice. Fol. phoniat. 2 (1950) 126.

Weiss, E.: Cardiovascular lesions of probable psychosomatic origin in arterial hypertension. Psychosom. Med. 2 (1940) 249–264.

Weiss, E., O. S. English, H. K. Fisher, M. Kleinbart, J. Zatuchni: The emotional problems of high blood pressure. Arch. intern. Med. 37 (1952) 677.

Weiss, E., O. S. English: Psychosomatic medicine, the clinical application of psychopathology to general medicine, 3rd ed. Saunders, Philadelphia-London 1949 .

Weiß, H., A. Zacher: Konfliktstrukturen und Biographie bei Morbus Crohn-Kranken. II. Konflikte in den Bereichen Abhängigkeit/Unabhängigkeit, Nähe/Distanz. Z. klin. Psychol. Psychopath. Psychother. 34 (1986) 69–82.

Weiss, J. M.: Effects of coping responses on stress. J. comp. Physiol. Psychol. 65 (1968) 251–260.

Weiss, J. M.: Influences of psychological variables on stress-induced pathology. In: Ciba Foundation Symposium 8: Physiology, Emotion and Psychosomatic Illness. Elsevier, North Holland-London- Amsterdam 1972.

Weiss, J M · Psychological factors in stress and disease. Sci. Amer. 222 (1972) 104–113.

Weiss, J. M.: Somatic effects of predictable and unpredictable shock. Psychosom. Med. 32 (1970) 208–397.

Weiss, S.: Instantaneous physiological death. New Engl. J. Med. 223 (1940) 793.

Weiss, S. R., M. H. Ebert: Psychological and behavioral characteristics of normal-weight bulimias and normal-weight controls. Psychosom. Med. 45 (1983) 293–303.

Weiss, T., B. T. Engel: Operant conditioning of heart rate in patients with premature ventricular contractions. Psychosom. Med. 33 (1971) 301.

Weiss, W.: Persönliche Mitteilung, 1952.

Weissman, M. M., E. S. Gershon, K. K. Kidd, B. A. Prusoff, J. F. Leckman, E. Dibble, J. Hamovit, W. D. Thompson, D. L. Pauls, J. J. Guroff: Psychiatric disorder in relatives of probands with affective disorders. Arch. gen. Psychiat. 41 (1984) 13–21.

Weissman, M. M., E. S. Paykel, R. French, H. Mark, K. Fox, B. A. Prusoff: Suicide attempts in an urban community: 1955–1970. Soc. Psychiat. 8 (1973) 82–89.

Weissman, M. M., E. S. Paykel: The depressed woman: a study of social relationships. Univ. of Chicago Press, Chicago 1974.

Weissman, M. M., J. K. Myers: Rates and risks of depressive symptoms in an United States urban community. Acta psychiat. scand. 57 (1978) 219–231.

Weissman, M. M., K. K. Kidd, B. A. Prusoff: Variability in rates of affective disorders in relatives of depressed and normal probands. Arch. gen. Psychiat. 39 (1982) 1397–1403.

Weissman, M. M., P. Wickramantne, K. R. Merikangas, J. F. Leckman, B. A. Prusoff, K. A. Karuso, K. K. Kidd, G. D. Gammon: Onset of major depression in early adulthood: increased familial loading and specifity. Arch. gen. Psychiat. 41 (1984) 1136–1143.

Weissman, M. M., R. B. Jarrett, J. A. Rush: Psychotherapy and its relevance to the pharmacotherapy of major depression: a decade later (1976–1985). In: Meltzer, H. J. (ed.): Psychopharmacology: The Third Generation of Progress. Raven, New York 1987.

Weitz, J. D.: The etiology of experimental gastric ulceration. Psychosom. Med. 19 (1957) 61–73.

Weitzman, E. D., R. M. Boyar, S. Kapen, L. Hellman: The relationship of sleep stages to neuroendocrine secretion and biological rhythms in man. Recent Progr. Hormone Res. 31 (1975) 399.

Weizenbaum, J.: Die Macht der Computer und die Ohnmacht der Vernunft. Suhrkamp, Frankfurt 1977.

Weizman, R., A. Weizman, J. Levi et al.: Sexual dysfunction associated with hyperprolactinemia in males and females undergoing hemodialysis. Psychosom. Med. 45 (1983) 259–269.

Weizsäcker, V. v.: Dtsch. Z. Nervenheilkunde 88 (1925) 264.

Weizsäcker, V. v.: Ärztliche Fragen. Vorlesungen über allgemeine Therapie, 2. Aufl. Leipzig 1935.

Weizsäcker, V. v.: Arzt und Kranker. Koehler, Stuttgart 1949.

Weizsäcker, V. v.: Begegnungen und Entscheidungen (1949). 2. Aufl. Stuttgart 1951.

Weizsäcker, V. v.: Das Problem des Menschen in der Medizin. In: Kraft und Innigkeit. Festschrift für Hans Ehrenberg. Heidelberg 1953.

Weizsäcker, V. v.: Der Begriff der allgemeinen Medizin, Heft 1, S. 1–44. Stuttgart 1947.

Weizsäcker, V. v.: Der Begriff sittlicher Wissenschaft. In: Diesseits und Jenseits der Medizin, S. 188. 2. Aufl. Koehler, Stuttgart 1950.

Weizsäcker, V. v.: Der Gestaltkreis. Theorie der Einheit von Wahrnehmen und Bewegen (1940). 4. Aufl. Thieme, Stuttgart 1950. Auch: suhrkamp taschenbuch wissenschaft. Frankfurt 1973.

Weizsäcker, V. v.: Epileptische Erkrankungen, Organneurosen des Nervensystems und allgemeine Neurosenlehre. In: Krehl, L., J. v. Mering (Hrsg.): Lehrbuch der Inneren Medizin, II, S. 354–392. 16. Aufl. Fischer, Jena 1929.

Weizsäcker, V. v.: Fälle und Probleme. Enke, Stuttgart 1947. In: Gesammelte Schriften, Bd. 6. Suhrkamp, Frankfurt 1986.

Weizsäcker, V. v.: Gesammelte Schriften, Bd. 6, S. 508. Suhrkamp, Frankfurt 1986.

Weizsäcker, V. v.: Grundfragen medizinischer Anthropologie (1948). In: Weizsäcker, V.v.: Dieseits und Jenseits der Medizin. Stuttgart 1950.

Weizsäcker, V. v.: Klinische Vorstellungen, S. 5–6. 3. Aufl. Hippokrates, Stuttgart 1947.

Weizsäcker, V. v.: Kranker und Arzt. Koehler, Stuttgart 1928.

Weizsäcker, V. v.: Menschenführung, nach ihren biologischen und metaphysischen Grundlagen betrachtet. Kleine Vandenhoeck-Reihe Nr.8. Göttingen 1955, 6. Aufl. 1970.

Weizsäcker, V. v.: Nach Freud. In: Diesseits und Jenseits der Medizin, S. 251–262. 2. Aufl. Koehler, Stuttgart 1950.

Weizsäcker, V. v.: Pathosophie. Göttingen 1956.

Weizsäcker, V. v.: Psychosomatische Medizin. Psyche 3 (1949/50) 331–341.

Weizsäcker, V. v.: Psychotherapie und Klinik. Ther. d. Gegenwart 65 (1926) 241–248.

Weizsäcker, V. v.: Randbemerkungen über Aufgabe und Begriff der Nervenheilkunde. Dtsch. Z. Nervenheilk. 87 (1925) 1–22.

Weizsäcker, V. v.: Soziale Krankheit und soziale Gesundung. Vandenhoeck und Ruprecht, Göttingen 1955.

Weizsäcker, V. v.: Studien zur Pathogenese, S. 11. 2. Aufl. Thieme, Wiesbaden 1946.

Weizsäcker, V. v.: Studien zur Pathogenese. Thieme, Leipzig 1935.

Weizsäcker, V. v.: Über Medizinische Anthropologie (1927). In: Arzt und Kranker, S. 35–61. Koehler, Stuttgart 1949.

Weizsäcker, V. v.: Über Träume bei sogenannter endogener Magersucht. Dtsch. med. Wschr. 63 (1937) 253–257.

Weizsäcker, V. v., D. Wyss: Zwischen Medizin und Philosophie. Göttingen 1957.

Wekerle, H., U. P. Ketelsen: Intrathymic pathogenesis and dual genetic control of myasthenia gravis. Lancet I (1977) 678–680.

Welin, S., G. Welin: Die Doppelkontrastuntersuchung des Dickdarms. Thieme, Stuttgart 1980.

Wellens, H., A. Vermeulen, D. Durren: Ventricular fibrillation occurring on arousal from sleep by auditory stimuli. Circulation 46 (1972) 661–665.

Wellens, W., R. Pierloot: Clinical treatment of anorexia nervosa. Unpublished, 1975.

Wellhöner, H. H.: Allgemeine und systematische Pharmakologie und Toxikologie. Springer, Berlin 1982.

Wellish, D. K., M. B. Mosher, C. van Scoy: Management of family emotion stress: family group therapy in a private oncology practice. Int. J. Group Psychotherapy 28 (1978) 225.

Wellish, D. K., M. M. Cohen: The family therapist as systems consultant to medical oncology. In: Wynne, L. C., S. McDaniel, T. T. Weber (eds.): Systems Consultation, pp. 199–218. Guilford, New York – London 1986.

Wellmann, W., K. Pries, H. Freyberger: Die Kombination von Morbus Crohn und Anorexia-nervosa-Symptomatik. Dtsch. med. Wschr. 106 (1981) 1499–1502.

Weltgesundheitsorganisation: Gesundheitsdienste in Europa, Bd. 1 und 2. WHO, Copenhagen 1983.

Wenderlein, J. M.: Einstellung der Frau zur Sterilisation durch Unterbrechung der Tubenwege. Geburtsh. Frauenheilk. 34 (1974) 940.

Wenderlein, J. M.: Psychologische Aspekte bei der Hormonsubstitution im Klimakterium. In: Zander, J., R. Goebel (Hrsg.): Psychologie und Sozialmedizin in der Frauenheilkunde. Springer, Berlin-Heidelberg 1977.

Wendler, J., W. Seidner (Hrsg.): Lehrbuch der Phoniatrie. VEB Thieme, Leipzig 1987.

Wendt, H.: Das Liebesleben in der Tierwelt. Rowohlt, Reinbek 1962.

Wener, J., A. Polonsky: The reaction of the human colon to naturally occurring and experimentally induced emotional states; observations through a transverse colostomy on a patient with ulcerative colitis. Gastroenterology 15 (1950) 84–93.

Wenerowicz, W. J., J. H. Riskind, P. G. Jenkins: Locus of control and degree of compliance in hemodialysis patients. J. Dial. 2 (1978) 495–505.

Wenger, M. A.: Studies of autonomic balance in Army Air Forces personnel. Comp. Psychol. Monogr. 19 (1948) 4.

Wenger, M. A.: The measurement of individual differences in autonomic balance. Psychosom. Med. 3 (1941) 427–434.

Wenzl, H.: Zur biographischen Situation zu Beginn und im Verlauf epileptischer Erkrankungen. Med. Diss., Heidelberg 1965.

Werb, Z., R. Foley, A. Munck: Interaction of glucocorticoids with macrophages. Identification of glucocorticoid receptors in monocytes and macrophages. J. exp. Med. 147 (1978) 1684–1694.

Werkö, L.: Risk factors and coronary heart disease – fact or fancy? Amer. Heart J. 91 (1976) 87–98.

Werkstattgespräche zum Thema Körperbild mit Beiträgen von R. Schütz, E. Bay, F. Besuden, H. Freyberger, O. W. Schulte-Herbrüggen, I. Rechenberger, Th. v. Uexküll und P. Fürstenau. In: Mat. z. Psychoanal. und analyt. orient. Psychother. IX, H. 1 (1983).

Werner, E., N. Borchardt, R. Frielingsdorf, H. Romahn: Schichtarbeit als Langzeiteinfluß auf betriebliche, private und soziale Bezüge. Westdeutscher Verlag, Opladen 1980.

Werner, H.: Antibiotika-induzierte pseudomembranöse Kolitis durch Clostridium difficile. Fortschr. Med. 97 (1979) 1411–1414.

Werry, H., C. Arends: Untersuchung zur Objektivierung von Persönlichkeitsmerkmalen bei Patienten mit Retinopathia centralis serosa. Klin. Mbl. Augenheilk. 172 (1978) 363–370.

Wershub, L. P.: Sexual impotence. Springfield/Ill. 1957.

Wesiack, W.: Der ganze Mensch ist krank. Der prakt. Arzt 16 (1984) 1245–1251.

Wesiack, W.: Grundzüge der Psychosomatischen Medizin. Springer, Berlin 1984; 2. Aufl. 1985.

Wesiack, W.: Grundzüge der psychosomatischen Medizin. Beck, München 1974.

Wesiack, W.: Ist der Hiatus scientificus nur eine Berufskrankheit des praktizierenden Arztes? Dtsch. med. Wschr. 9534 (1970) 1724.

Wesiack, W.: Möglichkeiten der Früherkennung und Prophylaxe psychoneurotischer und psychosomatischer Erkrankungen in der internistischen Sprechstunde. Psychother. med. Psychol. 27 (1977) 31–34.

Wesiack, W.: Psychoanalyse und praktische Medizin. Klett-Cotta, Stuttgart 1980.

Wesiack, W.: Realitäten der psychotherapeutischen Versorgung. Praxis der Psychotherapie, Bd. XX H.4 (1975) 194.

Wesiack, W.: Psychosomatische Medizin in der ärztlichen Praxis. Urban & Schwarzenberg, München-Wien-Baltimore 1984.

West, K. M. (ed.): Epidemiology of diabetes and its vascular lesions. Elsevier, New York 1978.

West, L. J.: Liaison psychiatry and medical education. In: Pasnau, R. O. (ed.): Consultation-Liaison Psychiatry. Grune & Stratton, New York 1975.

West, M., M. O'Donnel: Personality type and curriculum preference in primary care. Med. Educ. 16 (1982) 94–96.

Westmeyer, H.: Allgemeine methodologische Probleme der Indikation in der Psychotherapie. In: Baumann, U. (Hrsg.): Indikation zur Psychotherapie – Perspektiven für Praxis und Forschung. Urban & Schwarzenberg, München-Wien-Baltimore 1981.

Westmeyer, H.: Die rationale Rekonstruktion einiger Aspekte psychologischer Praxis. In: Albert, H., K. Stapf (Hrsg.): Theorie und Erfahrung. Beiträge zur Grundlagenproblematik der Sozialwissenschaften. Klett-Cotta, Stuttgart 1979.

Westphale, C., K. Köhle: Gesprächssituation und Informationsaustausch während der Visite auf einer internistisch-psychosomatischen Krankenstation. In: Köhle, K., H.-H. Raspe (Hrsg.): Das Gespräch während der ärztlichen Visite, S. 120–139. Urban & Schwarzenberg, München 1982 a.

Westphale, C., K. Köhle: Visitengespräche: Gesprächssituationen und Informationsaustausch. Abschlußbericht 1, SFB 129, Teilprojekt B 5, Universität Ulm 1982 b.

Wetterhus, S., E. Aubert, C. E. Berg: The effect of trimipramine (Surmontil®) on symptoms and healing of peptic ulcer: a double-blind study. Scand. J. Gastroenterol. 12 (Suppl. 43) (1976) 33–38.

Wever, R. A.: Man in temporal isolation: basic principles of the circadian system. In: Folkard, S., T. H. Mark (eds.): Hours of Work: Temporal Factors in Work-Scheduling. Wiley, New York 1985.

Wever, R. A.: Order and disorder in human circadian rhythmicity: possible relations to mental disorders. In: Kupfer, D. J., T. Mark, J. D. Barchas (eds.): Biological Rhythms and Mental Disorders. Guilford, New York 1989.

Wey, W.: Schilddrüsenkrankheiten. In: Berendes, J., R. Link, F. Zöllner (Hrsg.): Hals-Nasen-Ohrenheilkunde in Praxis und Klinik, 2. Aufl. Bd. 2. Thieme, Stuttgart-New York 1977.

Weyerer, S., H. Dilling: Prävalenz und Behandlung psychischer Erkrankungen. Nervenarzt 55 (1984) 30–42.

Whalley, L. J., N. Borthwick, D. Copolov, H. Dick, J. E. Christie, G. Fink: Glucocorticoid receptors and depression. Brit. med. J. 292 (1986) 859–861.

Wheatley, D. (ed.): Psychopharmacology of old age. Oxford Univ. Press, Oxford 1982.

Wheatley, D.: The use of anti-anxiety drugs in the medically ill. Psychother. and Psychosom. 49 (1988) 63–80.

Wheaton, B.: The sociogenesis of psychological disorder. J. Health soc. Behav. 20 (1980) 100–124.

Wheeler, E. O., P. D. White, E. W. Reed, M. E. Cohen: Neurocirculatory asthenia (anxiety neurosis, effort syndrome, neurasthenia). J. Amer. med. Ass. 143 (1950) 878–889.

White, B. V., S. Cobb, C. M. Jones: Mucous colitis. A psychological medical study of 60 cases. Psychosom. Med., Monograph I. NRC 1939.

White, D. J. G. (ed.): Cyclosporin A. Elsevier, Amsterdam 1982.

White, P. D.: Delusional depression after infectious mononucleosis. Brit. med. J. 295 (1987) 97.

White, P. D., F. D. Jones: Heart disease and disorders in New England. Amer. Heart J. 33 (1978) 302.

White, R. B., R. M. Gilliland: Elements of psychopathology – the mechanisms of defense. Grune & Stratton, New York 1975.

White, R. L., S. C. Liddon: The survivors of cardiac arrest. Int. J. Psychiat. Med. 3 (1972) 219–225.

White, R. W.: Strategies of adaptation: an attempt at systematic description. In: Coelho, G. V., D. A. Hamburg, J. E. Adams: Coping and Adaptation. Basic Books, New York 1974.

Whitehead, W. E., M. M. Schuster: Gastrointestinal disorders; behavioral and physiological basis for treatment, pp. 155–177. Acad. Press, Orlando 1985.

Whitlock, F. A.: Psychophysiological aspects of skin diseases. Saunders, London 1976. Deutsch: Psychophysiologische Aspekte bei Hautkrankheiten. Perimed, Erlangen 1980.

Whitten, C. F., M. G. Pettit, J. Fischhoff: Evidence that growth failure from maternal deprivation is secondary to undereating. J. Amer. med. Ass. 209 (1969) 1675.

WHO-Expert Committee on Diabetes Mellitus. Second Report. WHO Technical Report Series 646, Geneva 1980.

WHO Public Health Papers: Health system support for primary health care. Geneva 1984.

WHO Technical Report Series No. 168 (1959): Hypertension and coronary heart disease: classification and criteria for epidemiological studies.

WHO Technical Report Series No. 231 (1962): Arterial hypertension and ischaemic heart disease: preventive aspects.

WHO: International Classification of Diseases. Geneva 1977 (Vol. I), 1978 (Vol. II).

WHO: International Classification of Impairments, Disabilities, and Handicaps. Geneva 1980.

Whorwell, P. J.: Infectious agents in Crohn's disease – fact or artefact? Scand. J. Gastroent. 16 (1981) 161–166.

Whorwell, P. J., C. A. Phillips, W. L. Beeken, P. K. Little, K. D. Roessner: Isolation of reovirus-like agents from patients with Crohn's disease. Lancet I (1977) 1169–1171.

Whybrow, P. C., A. J. Prange jr.: A hypothesis of thyroid-catecholamine-receptor interaction. Its relevance to affective illness. Arch. gen. Psychiat. 38 (1981) 106–113.

Whybrow, P. C., F. J. Kane, M. A. Lipton: Regional ileitis and psychiatric disorder. Psychosom. Med. 30 (1968) 209–221.

Whybrow, P. C., R. B. Ferrell: Psychic factors and Crohn's disease. An overview. In: Lindner, A. E. (ed.): Emotional Factors in Gastrointestinal Illness. Excerpta Medica, Amsterdam 1973.

Wickler, W.: Sind wir Sünder? – Naturgesetze der Ehe. Knaur, München-Zürich 1969.

Widdicombe, J. G.: Regulation of tracheobronchial smooth muscle. Physiol. Rev. 43 (1963) 1–37.

Widen, L., G. Blomqvist, T. Greitz, J. E. Litton, M. Bergström, E. Ehrin, K. Ericson, L. Eriksson, D. H. Ingvar, L. Johansson, L. G. Nilsson, S. Stone-Elander, G. Sedvall, F. Wiesel, G. Wük: PET studies of glucose metabolism in patients with schizophrenia. Amer. J. Neuroradiol. 4 (1983) 550–552.

Widok, W.: Die Colitis ulcerosa – ein Beitrag aus psychoanalytischer Sicht. Prax. Psychother. 21 (1976) 274–277.

Wieck, H. H.: (Hrsg.): Lehrbuch der Psychiatrie, 2. Aufl. Schattauer, Stuttgart 1977.

Wieck, H. H.: Psychovegetative Allgemeinstörungen in der Sprechstunde. Med. Welt 25 (1974) 291–294.

Wieck, H. H.: Zur Klinik der sogenannten symptomatischen Psychosen. Dtsch. med. Wschr. 81 (1956) 1343–1345.

Wieck, H. H.: Zur Lokalisation zyklothymer Mißempfindungen. Med. Welt (1965) 2452–2454.

Wiegelmann, E., H. G. Solbach: Effect of LH-RH on plasma levels of LH and FSH in anorexia nervosa. Horm. metab. Res. 4 (1972) 404.

Wiener, C.: The burden of rheumatoid arthritis: tolerating the uncertainty. Soc. Sci. Med. 9 (1975) 97–104.

Wiesel, S. W., N. Tsourmas, H. L. Feffer, C. M. Citrin, N. Patronas: A study of computer-assisted tomography. The incidence of positive CAT scans in an asymptomatic group of patients. Spine 9 (1984) 549–551.

Wiesel, T. N., D. H. Hubel: Comparison of the effects of unilateral and bilateral eye closure on cortical unit response in kittens. J. Neurophysiol. 28 (1965) 1029–1040.

Wiesenhütter, E.: Vorwort zu: Gräff, Ch.: Konzentrative Bewegungstherapie in der Praxis. Hippokrates, Stuttgart 1983.

Wietersheim, J.v.: Morbus Crohn-Persönlichkeit: Zusammenfassende Ergebnisse eines hypothesengeleiteten Interviews. Symposion Morbus Crohn und Colitis ulcerosa. 31. Arbeitstagung des DKPM, 9.-11. 11. 1989, Gießen (im Druck).

Wietersheim, J.v., A. Ennulat, B. Probst, E. Wilke, H. Feiereis: Konstruktion und erste Evaluation eines Fragebogens zur sozialen Integration. Diagnostica 35 (1989) 359–363.

Wijsenbeck, H., H. Munitz: Group treatment in a hemodialysis center. Psychiat. Neurol. Neurochir. 7 (1970) 213–220.

Wijsenbeek, H., B. Maoz, I. Nitzan, R. Gill: Ulcerative colitis. Psychiat. Neurol. Neurochir. 71 (1968) 409–420.

Wilde, K.: Über die Zuverlässigkeit psychologischer Untersuchungsmethoden. Psychol. Rundschau 2 (1951) 187–193.

Wildegans, H.: Die Krankheiten und Verletzungen des Dickdarms und Mastdarms. Enke, Stuttgart 1959.

Wilder, J.: Das „Ausgangswert-Gesetz" – ein unbeachtetes biologisches Gesetz; seine Bedeutung für Forschung und Praxis. Kli. Wschr. 10 (1931) 1889–1893.

Wilhelm, E.: Die Beckenbodenmyalgie, keine Prostatitis. In: Verhandlungsbericht der Deutschen Gesellschaft für Urologie, S. 494. Springer, Berlin 1985.

Wilhelm, R. (1961): Elsa Gindler. Eine große Pädagogin besonderer Art. In: Stolze, H. (Hrsg.): Konzentrative Bewegungstherapie, S. 234. 2. Aufl. Springer, Berlin-Heidelberg-New York 1989.

Wilhelmsson, C., J. Vedin, L. Wilhelmsen et al.: Reduction of sudden deaths after myocardial infarction by treatment with alprenolol. Lancet 2 (1974) 1157.

Wilke, E.: Das Katathyme Bilderleben bei der konservativen Behandlung der Colitis ulcerosa. In: Leuner, H. (Hrsg.): Katathymes Bilderleben, S. 186–208. Huber, Bern 1980.

Wilke, E.: Das Katathyme Bilderleben in der Therapie des Asthma bronchiale. In: Roth, J. W. (Hrsg.): Konkrete Phantasie, S. 103–116. Huber, Bern 1984.

Wilke, E.: Diagnostische und therapeutische Aspekte der Arbeit mit dem Katathymen Bilderleben bei Patienten mit Colitis ulcerosa und Morbus Crohn. In: Studt, H. H. (Hrsg.): Psychosomatik in Forschung und Praxis. Urban & Schwarzenberg, München 1983 a.

Wilke, E.: Die spezifische Wirkung der KB-Therapie bei

psychosomatisch Kranken. In: Wilke, E., H. Leuner (Hrsg.): Psychosomatische Medizin und Katathymes Bilderleben. Huber, Bern-Stuttgart 1990.

Wilke, E.: Die Wertigkeit des Katathymen Bilderlebens innerhalb der kombinierten konservativen Behandlung der Colitis ulcerosa. Diss., Lübeck 1978.

Wilke, E.: Möglichkeiten und Grenzen der Therapie mit dem Katathymen Bilderleben bei chronisch-entzündlichen Darmerkrankungen – Colitis ulcerosa und Morbus Crohn – unter Berücksichtigung einer 5-Jahres-Katamnese. 3. Wiss. Tagung Int. Ges. Katathymes Bilderleben, München 1983b.

Wilke, E.: Therapieverlauf bei einer 60jährigen Patientin mit Colitis ulcerosa. In: Schütz, R. M. (Hrsg.): Praktische Geriatrie 5. Lübeck 1985.

Wilke, E.: Tiefenpsychologisch fundierte (analytisch orientierte) Therapie und Katathymes Bilderleben. In: Feiereis, H. (Hrsg.): Diagnostik und Therapie der Magersucht und Bulimie, 2. Aufl. Marseille, München 1989.

Wilke, E., H. Leuner (Hrsg.): Psychosomatische Medizin und Katathymes Bilderleben. Huber, Bern-Stuttgart 1989.

Wilkinson, J. B.: Hypnotherapy in the psychosomatic approach to illness: a review. J. Res. soc. Med. 74 (1981) 525–530.

Wilkinson, M.: Ergotamin-Kopfschmerz. Münch. med. Wschr. 126 (1984) 14–15.

Wille, A.: Die Enkopresis im Kindes- und Jugendalter. Springer, Berlin 1984.

Willi, J.: Die Zweierbeziehung. Rowohlt, Reinbek 1975.

Willi, J.: Therapie der Zweierbeziehung. Rowohlt, Reinbek 1978.

Willi, J.: Zur Pathogenese des exogenen akuten Reaktionstypus bei körperlicher Krankheit. In: Bleuler, M., J. Willi, H. P. Bühler: Akute psychische Begleiterscheinungen körperlicher Krankheiten. Thieme, Stuttgart 1966.

Willi, J., R. Hagemann: Langzeitverläufe von Anorexia nervosa. Schweiz. med. Wschr. 106 (1976) 1459–1465.

Willi, J., S. Grossmann: Epidemiology of anorexia nervosa in a defined region of Switzerland. Amer. J. Psychiat. 140 (1983) 564–567.

William, J. G., B. Williams, J. R. Jones: The chemical control of preoperative anxiety. Psychophysiol. 12 (1975) 40–49.

Williams, C. B.: Koloskopie – Diagnostische Bedeutung bei entzündlichen Erkrankungen des Kolons. In: Ottenjann, R., M. Classen (Hrsg.): Gastroenterologische Endoskopie. Enke, Stuttgart 1979.

Williams, E.: Anorexia nervosa – a somatic disorder. Brit. med. J. 2 (1958) 190–195.

Williams, J. M., D. L. Felten: Sympathetic innervation of murine thymus and spleen: a comparative histofluorescence study. Anat. Rec. 199 (1981) 531–542.

Williams, R., I. Karacan, C. Hursch: EEG and human sleep. Wiley, New York 1974.

Williams, R. B.: The trusting heart. Times Books, Random House, Inc., New York 1989.

Williams, R. B., J. C. Barefoot, F. E. Haney, J. A. Blumenthal, D. B. Pryor, B. Peterson: Type A behavior and angiographically documented coronary atherosclerosis in a sample of 2289 patients. Psychosom. Med. 50 (1988) 139–152.

Williams, R. B., J. D. Lane, C. M. Kuhn, W. Melosh, A. D. White, S. M. Schanberg: Type A behavior and elevated physiological and neuroendocrine responses to cognitive tasks. Science 218 (1982) 483–485.

Williams, R. B., T. L. Hanes, K. L. Lee, Y. Kong, J. A. Blumenthal, R. E. Whalen: Type A behavior, hostility and coronary atherosclerosis. Psychosom. Med. 42 (1980) 539–549.

Williams, R. B., T. W. Bittker, M. S. Buchsbaum, L. C. Wynne: Cardiovascular and neurophysiologic correlates of sensory intake and rejection. Effects of cognitive tasks. Psychophysiol. 12 (1975) 427–433.

Williams, S. L., G. Zane: Guided mastery and stimulus exposure treatments for severe performance anxiety in agoraphobics. Behav. Res. Ther. 27 (1989) 237–245.

Williams, T., J. P. Zorley: The management of chronic illness. In: Conn: Family Practice, pp. 103–117. Saunders, Philadelphia 1973.

Willis, Th.: The London practice of physics. Basset & Crook, London 1685.

Wilms, H.: Nierentransplantation. In: Balck, F. B., U. Koch, H. Speidel (Hrsg.): Psychonephrologie. Springer, Heidelberg 1984.

Wilsch, L., O. P. Hornstein: Statistische Untersuchungen und Behandlungsergebnisse bei der perioralen Dermatitis. In: Bosse, K., P. Hünecke (Hrsg.): Psychodynamik und Soziodynamik bei Hautkranken. Vandenhoeck & Ruprecht, Göttingen 1976.

Wilson, C. J., S. A. Schneps, L. H. Muzekari, D. M. Wilson: Time-limited group counseling for chronic home hemodialysis patients. J. counsel. Psychol. 21 (1974) 376–379.

Wilson, G. T., E. Rossiter, E. I. Kleifield, L. Lindholm: Cognitive-behavioral treatment of bulimia nervosa: a controlled evaluation. Behav. Res. Ther. 24 (1986) 277–288.

Wilson, G. T., G. C. Davison: Processes of fear reduction in systematic desensitization. Psychol. Bull. 76 (1971) 1–14.

Wilson, J. D.: Sex hormones and sexual behavior. New Engl. J. Med. 300 (1979) 1269–1270.

Wilson, J. M. G., G. Jungner: Principles and practice of screening of disease. WHO, Geneva 1968.

Wilson, M. S., E. Meyer: Diagnostic consistency in a psychiatric liaison service. Amer. J. Psychiat. 119 (1962) 207.

Winckelmann, G., A. Lütke, J. Löhner: Über sechs Monate bestehendes rezidivierendes Fieber ungeklärter Ursache. Bericht über 85 Patienten. Dtsch. med. Wschr. 107 (1982) 1003–1007.

Winckelmann, N. W.: The use of chlorpromazine and prochlorperazine as adjuncts to psychoanalytic psychotherapy – general principles for combined therapy. In: Sarwer-Foner, G. J. (ed.): The Dynamics of Psychiatric Drug Therapy. Thomas, Springfield 1960.

Wing, J. K. (ed.): What is a case? The problem of definition in psychiatric community surveys. McIntire, London 1981.

Wing, J. K., S. A. Mann, J. P. Leff, J. M. Nixon: The concept of a „case" in psychiatric population studies. Psychol. Med. 8 (1978) 203–217.

Winkelstein, W., M. Samuel, N. S. Padian, J. A. Wiley: Selected sexual practices of San Francisco heterosexual men and risk of infection by the Human Immunodeficiency Virus. J. Amer. med. Ass. 257 (1987) 1470–1471.

Winnberg, G.: Psykodonti. Scandinavian University Books, Stockholm 1969.

Winnicott, D. W. (1931): A note on normality and anxiety. In: Collected Papers: Through Paediatrics to Psycho-Analysis. Tavistock, London 1958.

Winnicott, D. W. (1941): The observation of infants in a set situation. Ibid.

Winnicott, D. W. (1945): Primitive emotional development. Ibid.

Winnicott, D. W. (1948): Paediatrics and psychiatry. Ibid.

Winnicott, D. W. (1949): Birth memories, birth trauma, and anxiety. Ibid.

Winnicott, D. W. (1951): Transitional objects and transitional phenomena. In: Playing and Reality. Tavistock, London 1971.

Winnicott, D. W. (1952a): Anxiety associated with insecurity. Ibid.

Winnicott, D. W. (1952b): Psychosis and child care. Ibid.

Winnicott, D. W.(1954): Metapsychological and clinical aspects of regression within the psycho-analytical set-up. Ibid.

Winnicott, D. W.(1954): Primäre Mütterlichkeit. In: Von der Kinderheilkunde zur Psychoanalyse, S. 270–293. Kindler, München 1976.

Winnicott, D. W.(1956): Primary maternal preoccupation. Ibid.

Winnicott, D. W.: Primäre Mütterlichkeit. Psyche 14 (1960) 393–399.

Winnicott, D. W.(1960a): The theory of the parent-infant-relationship. In: The Maturational Processes and the Facilitating Environment. Hogarth Press and the Institute of Psycho-Analysis, London 1965.

Winnicott, D. W.(1960b): Ego distortion in terms of the true and false self. Ibid.

Winnicott, D. W.(1960c): Counter-Transference. Ibid.

Winnicott, D. W.(1962): Ich-Integration in der Entwicklung des Kindes. In: Reifungsprozesse und fördernde Umwelt, S. 72–81. Kindler, München 1974.

Winnicott, D. W.(1962a): Ego integration in child development. Ibid.

Winnicott, D. W.(1962b): Providing for the child in health and crisis. Ibid.

Winnicott, D. W.(1963a): Communicating and not communicating leading to a study of certain opposites. Ibid.

Winnicott, D. W.(1963b): Reifungsprozesse und fördernde Umwelt. Kindler, München 1974.

Winnicott, D. W.(1967): Mirror-role of mother and family in child development. In: Playing and Reality. Tavistock, London 1971.

Winnicott, D. W.(1968): Communication between infant and mother, and mother and infant, compared and contrasted. In: What is Psychoanalysis. Institute of Psycho-analysis, published by Ballière, Tindall & Cassel, London 1968.

Winnicott, D. W.: Übergangsobjekte und Übergangsphänomene. Psyche 23 (1969) 666–682.

Winnicott, D. W.(1969): The use of an object and relating through identifications. In: Playing and Reality. Tavistock, London 1971.

Winnicott, D. W.(1970): The mother-infant experience of mutuality. In: Benedek, E. J. A., T. Benedek (eds.): Parenthood: Its Psychology and Psychopathology. Little, Brown & Co., Boston 1970.

Winnicott, D. W.(1971a): Letter to Mme. Jeannine Kalmanovitch. In: Nouvelle Revue de Psychanalyse, Vol. 3, pp. 47–48.

Winnicott, D. W.(1971b): Playing and Reality. Tavistock, London 1971.

Winnicott, D. W.(1972): The basis for self in body. Int.J. Child Psychotherapy, Vol.1, No.1.

Winnicott, D. W.(1973a): Die therapeutische Arbeit mit Kindern. Kindler (Studienausgabe), München 1973.

Winnicott, D. W.(1973b): Vom Spiel zur Kreativität. Klett, Stuttgart 1973.

Winnicott, D. W.: Reifungsprozesse und fördernde Umwelt. Kindler, München 1974.

Winnicott, D. W.(1976): Von der Kinderheilkunde zur Psychoanalyse. Kindler, München 1976.

Winnicott, D. W.(1977): Le concept d'individu sain. In: L'Arc, 2e trimestre 1977, No. 1496, pp.13–26. Mistral, Cavaillon 1977.

Winnicott, D. W.: Collected papers. London 1958.

Winnik, H., V. Bental: Psychoanalytische Aspekte der Parkinsonkrankheit. Psyche 18 (1964/65) 89–106.

Wint, G.: The third killer. Meditations on a stroke. Abelard-Schulman, New York 1965.

Winter, I., J. J. Kellerman: Psychological factors involved in cardiac rehabilitation. In: Stocksmeier, U. (ed.): Psychological Approach to the Rehabilitation of Coronary Patients, pp. 156–172. Springer, Berlin-Heidelberg-New York 1976.

Winter: zit. nach: Prill, H. J.: Neuere Erkenntnisse der Mutter-Kind-Beziehung nach der Geburt. Vortr. 41. Tag. Dtsch. Ges. Gynäkologie Geburtshilfe, Hamburg 1976.

Wintersberger, H.: Arbeitswissenschaften in Italien. Österreichische Zeitschrift für Soziologie 2/3 (1976) 41–49.

Wirsching, M.: Familiendynamische Aspekte bei Colitis ulcerosa und Morbus Crohn. Ztsch. Psychosom. Med. Psychoanal. 3 (1984) 238.

Wirsching, M.: Familiendynamische Aspekte im psychosomatischen Konsiliardienst. Prax. Psychother. Psychosom. 28 (1983 b) 209.

Wirsching, M.: Krebs im Kontext. Patient, Familie und Behandlungssystem. Klett-Cotta, Stuttgart 1988.

Wirsching, M.: Unmöglicher Auftrag – Psychosomatische Konsiliararbeit aus analytisch-systemischer Sicht. Familiendynamik 8 (1983 a) 3.

Wirsching, M., H. Stierlin: Krankheit und Familie. Klett-Cotta, Stuttgart 1982.

Wirsching, M., W. Georg, F. Hoffmann, J. Riehl, P. Schmidt: Psychosocial factors influencing health development in breast cancer and mastopathia: a systemic study. In: Cooper, C. L.: Stress and breast cancer, pp. 97–107. Wiley, Chichester-New York-Brisbane 1988.

Wirth, G.: Stimmstörungen. Dtsch. Ärzte-Verlag, Köln 1987.

Wise, T. N., D. Feldheim, L. Mann: Patients' reactions to house staff ward rounds. Psychosomatics 26 (1985) 669–672.

Wise, T. N., J. N. Couper, S. Ahmed: The efficacy of group therapy for patients with irritable bowel syndrome. Psychosomatics 23 (1982) 465–469.

Witkin, H. A., R. B. Dyk, H. F. Faterson, D. R. Goodenough, S. A. Karp: Psychological differentiation studies of development. Wiley, New York-London 1962.

Witt, C.: Problem with cardiac pacemakers in the aged. Geriatrics 27 (1972) 92–95.

Wittchen, H. U.: A biobehavioral treatment program (SEP) for chronic migraine patients. In: Holroyd, K. A., B. Schlote, H. Zenz (eds.): Perspectives in Research on Headache, pp. 183–197. Hogrefe, Lewiston-New York 1983.

Wittgenstein, L.: Philosophische Untersuchungen. Suhrkamp, Frankfurt 1967.

Wittig, M. A., A. C. Petersen: Sex related differences in cognitive functioning. Acad. Press, New York 1979.

Wittkower, E.: A psychiatrist looks at tuberculosis. London 1949.

Wittkower, E.: Ulcerative colitis: personality studies. Brit. med. J. II (1938) 1356–1360.

Wittkowski, J.: Tod und Sterben. Ergebnisse der Thanatopsychologie. UTB, Quelle und Meyer, Heidelberg 1978.

Wittkowski, K. M.: Wann ist ein HIV-Test indiziert? Dtsch. Ärztebl. 85 (1988) 1510–1511.

Wittmann, W. W.: Evaluationsforschung. Aufgaben, Probleme und Anwendungen. Springer, Berlin 1985.

Witzel, L.: Beobachtungen an Sterbenden. Psychol. Heute 2/8 (1975) 76–77.

Witzel, L.: Das Verhalten von sterbenden Patienten. Med. Klinik 66 (1971) 577–578.

Witzel, L.: Der Sterbende als Patient. Med. Klinik 68 (1973) 1373–1375.

Wörz, R., R. Lendle: Schmerz. Fischer, Stuttgart 1980.

Wolcott, D. L., D. K. Wellish, C. R. Robertson, J. A. Ransom: Serum gastrin and the family environment in duodenal ulcer disease. Psychosom. Med. 6 (1981) 501–507.

Wolf, E., K. M. Spencer, A. G. Cudworth: The genetic susceptibility to type 1 (insulin-dependent) diabetes – analysis of the HLA-DR association. Diabetologia 24 (1983) 224–230.

Wolf, P.: Neue Aspekte zur Pathogenese und Therapie der hyperreflektorischen Rhinopathie. Z. Laryng. Rhinol. 67 (1988) 138.

Wolf, P., G. Wagner, F. Amelung (Hrsg.): Anfallskrankheiten. Nomenklatur und Klassifikation der Epilepsien, der epileptischen Anfälle und anderer Anfallssyndrome. Springer, Berlin-Heidelberg-New York 1987.

Wolf, S.: Cardiovascular reactions to symbolic stimuli. Circulation 18 (1958) 287–292.

Wolf, S.: Psychosocial influences in gastrointestinal function. In: Levi, L. (ed.): Society, Stress and Disease. London-New York-Toronto 1971.

Wolf, S.: The bradycardia of the dive reflex – a possible mechanism of sudden death. Cond. Reflex 2 (1967) 89.

Wolf, S.: The central nervous system regulation of the colon. Gastroenterology 51 (1966) 810–821.

Wolf, S., H. Goodell: Causes and mechanisms in psychosomatic phenomena. J. hum. Stress (March) (1979) 9–18.

Wolf, S., H. G. Wolff: Human gastric function. Oxford Univ. Press, New York 1947.

Wolf, S., T. P. Lamy, W. H. Bachrach, H. M. Spiro, R. Sturdevant, H. Weiner: The role of stress in peptic ulcer disease. J. hum. Stress (June) (1979) 27–37.

Wolfe, F., M. A. Cathey, S. M. Kleinheksel, S. P. Amos, R. G. Hoffmann, D. Y. Young, D. J. Hawley: Psychological status in primary fibrositis and fibrositis associated with rheumatoid arthritis. J. Rheum. 11 (1984) 500–506.

Wolfe, F., S. M. Kleinheksel, M. A. Cathey, D. J. Hawley, P. W. Spitz, J. F. Fries: The clinical value of the Stanford Health Assessment Questionnaire Functional Disability Index in patients with rheumatoid arthritis. J. Rheum. 15 (1988) 1480–1488.

Wolff, B. B.: Current psychosocial concepts in rheumatoid arthritis. Bull. rheum. Dis. 22 (1971/72) 656–661.

Wolff, C. T., S. B. Friedman, N. A. Hofer, J. W. Mason: Relationship between psychological defenses and mean urinary 17-hydrocorticosteroid excretion rates: I, II. A predictive study of parents of children with leukemia. Psychosom. Med. 26 (1964) 576–591.

Wolff, G., J. Brix, E. Ostermann: Psychosoziale Probleme bei der Hämodialyse und Transplantation von Kindern und Jugendlichen. In: Balck, F. B., U. Koch, H. Speidel (Hrsg.): Psychonephrologie. Springer, Heidelberg 1984.

Wolff, H. G.: Headache and other head pain, pp. 18–48. Oxford Univ. Press, New York 1963.

Wolff, H. G.: Stress and bodily disease. Baltimore 1950.

Wolff, H. G.: Stress and disease. Illinois 1953.

Wolff, H. G.: The contribution of the interview situation to the restriction of phantasy life and emotional experience in psychosomatic patients. Psychother. and Psychosom. 28 (1977) 58–67.

Wolff, H. G.: The contribution of the interview situation to the restriction of phantasy life and emotional experience in psychosomatic patients. In: Bräutigam, W., M.v.Rad (eds.): Toward a Theory of Psychosomatic Disorders, pp. 58–67. Basel 1977.

Wolff, H. G., S. Wolf: The management of hypertensive patients. Observations on the pertinence of life situations, attitudes and emotions to variations in the course of essential hypertension and to the occurrence of associated symptoms. In: Bell, E. (ed.): Hypertension. Univ. of Minn., Minneapolis 1951.

Wolff, W. H.: Multiple Sklerose und Beruf. Ein Beitrag zur anthropologischen Medizin. Z. ges. Inn. Med. 4 (1949) 434–437.

Wolinsky, H.: Long-term effect of hypertension on the rat aortic wall and their relation to concurrent aging changes. Circulat. Res. 30 (1972) 301.

Wolter, M., J. Zimmermann: Katamnestische Untersuchungen zum Verlauf der multiplen Sklerose. In: Seitz, D., P. Vogel (Hrsg.): Hämoblastosen, Zentrale Motorik, Iatrogene Schäden, Myositiden, S. 696–698. Springer, Berlin-Heidelberg-New York 1983.

Wong, D. F., H. N. Wagner, L. E. Tune, R. F. Dannals, G. D. Pearson, J. M. Links, C. A. Tamminga, E. P. Broussolle, H. T. Ravert, A. A. Wilson, T. K. Young, J. Malat, J. A. Williams, L. A. O'Tuama, S. H. Snyder, M. J. Kuhar, A. Gedde: Positron emission tomography reveals elevated D2 dopamine receptors in drug naive schizophrenics. Science 234 (1986) 1558–1563.

Wood, D., S. Sheps, L. Elveback, A. Schirger: Cold pressor test as a predictor of hypertension. Hypertension 6 (1984) 301–305.

Woodmansey, A. C.: Psychosomatic education: teaching or treatment? Psychother. and Psychosom. 40 (1983) 16–21.

Woodruff, R. A.: An evaluation of objective diagnostic criteria by the study of women with chronic medical illness. Brit. J. Psychiat. 114 (1968) 1115–1119.

Woodruff, R. A., P. J. Clyton, S. B. Guze: Hysteria. An evaluation of specific diagnostic criteria by the study of randomly selected psychiatric clinic patients. Brit. J. Psychiat. 115 (1969) 1243–1248.

Woods, P. A., P. J. Higson, M. M. Tannahill: Token-economy programmes with chronic psychotic patients: the importance of direct measurement and objective evaluation for long-term maintenance. Behav. Res. Ther. 22 (1984) 41–52.

Woodworth, R. S.: Personal Data Sheet. Stoelting, Chicago 1917.

Wooley, S., O. Wooley: Eating disorders: obesity and anorexia. In: Brodsky, A., R. Hare-Mustin (eds.): Women and Psychotherapy: An Assessment of Research and Practice. Guilford, New York 1980.

Wooley, S. C., B. Blackwell, C. Winget: A learning theory model of chronic illness behavior. Theory, treatment and research. Psychosom. Med. 40 (1978) 379–401.

Worden, J. W.: Beratung und Therapie in Trauerfällen. Huber, Bern-Toronto 1987.

Worden, J. W., A. D. Weisman: The fallacy in postmastectomy depression. Amer. J. Med. Sci. 273 (1977) 169–175.

Wright, E. T. et al.: Some psychological effects of cosmetics. Percept. Motor Skills 30 (1970) 12–14.

Wright, R. A., R. J. Contrada, D. C. Glass: Psychophysiologic correlates of Type A behavior. In: Katkin, E. S., S. B. Manuck (eds.): Advances in Behavioral Medicine. JAL, Greenwich/CT 1984.

Wrong, N. M.: Excoriated acne of young females. Arch. Derm. 70 (1954) 576–582.

Wrzesniewski, K.: Anxiety and rehabilitation after myocardial infarction. Psychother. and Psychosom. 27 (1976/77) 41–46.

Wulsin, L. R., A. M. Jacobson, L. I. Rand: Psychosocial aspects of diabetic retinopathy. Diabetes Care 10 (1987) 367–373.

Wuttke, W., A. Weindl, K. H. Voigt, R.-R. Dries (eds.): Brain and pituitary peptides. Karger, Basel 1980.

Wuttke, W., R. Horowski (eds.):Gonadal steroids and brain function. Exp. Brain Res., Suppl. 3. Springer, Berlin 1981.

Wybran, J.: Enkephalins, endorphins, substance P, and the immune system. In: Guillemin, R., M. Cohn, T. Melnechuk (eds.): Neural Modulation of Immunity. Raven, New York 1985.

Wyke, R. J., F. C. Edwards, R. N. Allan: Employment problems and prospects for patients with inflammatory bowel disease. Gut 29 (1988) 1229–1235.

Wyler, A. R., C. A. Robbins, C. B. Dodrill: EEG operant conditioning for control of epilepsy. Epilepsia 20 (1979) 279–286.

Wyllie, E., H. Lüders, J. MacMillan, M. Gupta: Serum prolactin levels after epileptic seizures. Neurology 34 (1984) 1601–1604.

Wyss, D.: Die Psychotherapie der juvenilen Hypertonie – katamnestische Beobachtungen. Dtsch. med. Wschr. 80 (1955) 822.

Wyss, D.: Psychosomatische Aspekte der juvenilen Hypertonie. Nervenarzt 26 (1955).

Yager, J., J. Landsverk, C. K. Edelstein: A 20-month follow-up study of 628 women with eating disorders, I: Course and severity. Amer. J. Psychiat. 144 (1987) 1172–1177.

Yalom, I.: Existential psychotherapy. Basic Books, New York 1980.

Yamada, A., M. M. Jensen, A. F. Rasmussen: Stress and susceptibility to viral infections, III. Antibody response and viral retention during avoidance-learning stress. Proc. Soc. exp. Biol. (N. Y.) 116 (1964) 677.

Yanagida, E. H., J. Streltzer, A. Siemsen: Denial in dialysis patients. Relationship to compliance and other variables. Psychosom. Med. 43 (1981) 271–280.

Yehuda, S., A. J. Kastin: Peptides and thermoregulation. Neurosci. biobehav. Rev. 4 (1980) 459–471.

Yodfat, Y., H. Silvian: A prospective study of acute respiratory infections among children in a Kibbuz. J. infect. Dis. 136 (1977) 26–30.

Yorke, C.: Some comments on the psychoanalytic treatment of patients with physical disabilities. Int. J. Psychoanal. 61 (1980) 187.

Yorkston, N. J., E. Eckert, R. B. McHugh, D. A. Philader, M. N. Blumenthal: Bronchial asthma: improved lung function after behavior modification. Psychosomatics 20 (1979) 325–331.

Youcha, I.: Short-term in-patient group: formation and beginnings. In: Rabin, H., E. Rosenbaum (ed.): How to Begin a Psychotherapy Group. Brosch, New York 1976.

Young, E.: Reading DSM-III on PTSD: an anthropological account of a core text in American psychiatry. Vortrag: Anthropologies of Medicine: A Colloquium on West European and North American Perspectives. Hamburg 1988.

Young, J.: Lower abdominal pains of cervical origin. Their genesis and treatment. Brit. med. J. 1 (1938) 104.

Young, M., B. Benjamin, C. Wallis: Mortality of widowers. Lancet 2 (1963) 454.

Young, M. A., W. A. Scheftner, G. L. Klerman, N. A. Andreasen, R. M. Hirschfeld: The endogenous sub-type of depression: a study of its internal construct validity. Brit. J. Psychiat. 148 (1986) 257–267.

Yu, D. T. Y., P. J. Clements, H. E. Paulus, J. B. Peter, J. Levy, E. V. Barnett: Human lymphocyte subpopulations. Effects of corticosteroids. J. clin. Invest. 53 (1974) 565–571.

Yunus, M. B.: Fibromyalgia syndrome: a need for uniform classification. J. Rheum. 10 (1983) 841–844.

Yunus, M. B., A. T. Masi, J. J. Calabro, K. A. Miller, S. L. Feigenbaum: Primary fibromyalgia (fibrositis): clinical study of 50 patients with matched normal controls. Semin. Arthr. Rheum. 11 (1981) 151–171.

Yunus, M. B., A. T. Masi, J. C. Aldag: Criteria studies of primary fibromyalgia syndrome (PFS). Arthr. and Rheum. 30 (1987) 27.

Zach, J., S. H. Ackerman: Thyroid function, metabolic regulation, and depression. Psychosom. Med. 50 (1988) 454–468.

Zacher, A.: Der Schreibkrampf – fokale Dystonie oder psychogene Bewegungsstörung? Eine kritische Literaturstudie. Fortschr. Neurol. Psychiat. 57 (1989) 328–336.

Zacher, A., H. Weiß: Konfliktstrukturen und Biographie bei Morbus Crohn-Kranken. I. Einleitung. Z. klin. Psychol. Psychopath. Psychother. 33 (1985) 259–269.

Zacher, A., U. Becker: Zur Psychosomatik des Morbus Crohn aus allgemein-ärztlicher, internistischer und chirurgischer Sicht. In: Rechenberger, H.-G., H.-V. Werthmann (Hrsg.): Psychotherapie und Innere Medizin. Pfeiffer, München 1988.

Zahradnik, H. P., E. Stengele, E. Kraut, G. Scharpf, M. Breckwoldt: Neue Aspekte zur Pathogenese und Therapie der Dysmenorrhoe. Prostaglandin F 2α im Menstrualblut. Dtsch. med. Wschr. 103 (1978) 1270.

Zanchetti, A., A. Stella: Neural control of renin release. Clin. Sci. Mol. Med. 48 (1975).

Zander, E., R. R. Engel, M. Kitscha, G. Wiedemann: Psychophysiologische Korrelationsuntersuchungen während eines halbstandardisierten Interviews bei Patienten mit Ulcus duodeni und Hypertonie. In: Zander, W. (Hrsg.): Experimentelle Forschungsergebnisse in der psychosomatischen Medizin, S.120–128. Vandenhoeck & Ruprecht, Göttingen 1981.

Zander, E., W. Zander: Das psychosomatische Konzept von Harald Schultz-Hencke. In: Rudolf, G., U. Rüger (Hrsg.): Die Psychoanalyse Schultz-Henckes. Thieme Stuttgart 1988.

Zander, W.: Psychosomatische Forschungsergebnisse bei Ulcus duodeni. Vandenhoeck & Ruprecht, Göttingen 1977.

Zander, W.: Streß und Strain. In: (Hrsg.): Soma und Psyche. Ciba-Geigy, Basel 1978.

Zander, W., F. Lehner, M. Birk, G. Blümel: Colitis ulcerosa und Morbus Crohn aus psychosomatischer Sicht. Med. Welt 33 (1982) 948–950.

Zander, W., F. Lehner, M. Birk, G. Blümel: Experimentelle Untersuchungen zur Psychodynamik der Colitis ulcerosa und des Morbus Crohn. Prax. Psychother. Psychosom. 27 (1982) 161–172.

Zaphiropoulos, G., H. C. Burry: Depression in rheumatoid disease. Ann. rheum. Dis. 33 (1974) 132–135.

Zarcone jr., V. P., K. L. Benson, P. A. Berger: Abnormal rapid eye movement latencies in schizophrenia. Arch. gen. Psychiat. 44 (1987) 45–48.

Zauner, J.: Das Krankheitsbild der Dysmorphophobie. Med. et Hyg. 37 (1979) 329–330.

Zauner, J.: Grundsätzliche Möglichkeiten der Entstehung psychogener Herzsymptome mit Indikation zur Psychotherapie. Psychosom. Med. 13 (1967).

Zauner, J.: Psychopharmaka und klinische Psychotherapie 2. Psychosom. Med. Psychoanal. 29 (1972) 138–148.

Zauner, J.: Psychosomatische Aspekte der Adoleszenz. Med. Psychoanal. 24 (1978) 17.

Zavalova, N. D.: Decision making models under stress. Biocybern. of the CNS (1969) 55–60.

Zawadski, B., P. Lazarsfeld: The psychological consequences of unemployment. J. soc. Psychol. 6 (1935) 224–251.

Zeidler, H., G. Krüskemper, M. Toeroek: Erhöhte Depressionsskala im MMPI in ihrer Beziehung zu somatischen Daten bei Patienten mit chronischer Polyarthritis und Spondylitis ankylopoetica (M. Bechterew). Verh. Dtsch. Ges. Inn. Med. 84 (1978) 1505–1508.

Zeig, J. K. (Hrsg.): Meine Stimme begleitet Sie überallhin. Ein Lehrseminar mit Milton H. Erickson. Klett-Cotta, Stuttgart 1985.

Literatur

Zeig, J. K.: Therapeutische Muster der Ericksonschen Kommunikation der Beeinflussung. Hypnose und Kognition 5 (1988) 5–18.

Zeitlin, D. J.: Psychological issues in the management of rheumatoid arthritis. Psychosomatics 10 (1977) 7–14.

Zeitz, M., K. Hartmann, C. Emde, J. Karow, H. Menge, E. O. Riecken: Untersuchungen zur Thrombozytose als Aktivitätsparameter bei Morbus Crohn. Verh. Dtsch. Ges. Inn. Med. 90 (1984) 565–567.

Zenner, H. P.: Diagnostik und Therapie allergischer Erkrankungen der Schleimhaut des oberen Respirationstraktes. Arch. Ohr.-, Nas.-, Kehlk.-Heilk. Suppl. 1 (1987) 85.

Zepf, S. (Hrsg.): Tatort Körper – Spurensicherung. Eine Kritik der psychoanalytischen Psychosomatik. Heidelberg 1986.

Zepf, S.: Das Katathyme Bilderleben in der Erforschung der Psychodynamik des Asthma bronchiale. In: Leuner, H. (Hrsg.): Katathymes Bilderleben, S. 105–123. Huber, Bern 1980.

Zepf, S.: Die psychosomatische Erkrankung in der „Theorie der Interaktionsformen" (Lorenzer) – Metatheorie statt Metasemantik. In: Tatort Körper – Spurensicherung, S. 129–151. Heidelberg 1986.

Zepf, S.: Die Sozialisation des psychosomatisch Kranken. Frankfurt 1976 a.

Zepf, S.: Grundlinien einer materialistischen Theorie psychosomatischer Erkrankungen. Frankfurt 1976.

Zepf, S.: Klinik der psychosomatischen Erkrankungen. In: Kisker, K. P., H. Lauter, J.-E. Meyer, C. Müller, E. Strömgren (Hrsg.): Psychiatrie der Gegenwart I, S. 63–102. Springer, Heidelberg 1986.

Zepf, S.: Psychosomatische Medizin auf dem Weg zur Wissenschaft. Frankfurt 1981.

Zepf, S., E. Gattig: Die Wiederbelebung der Todestriebhypothese – Das theoretische Konzept der französischen psychosomatischen Schule. In: Tatort Körper – Spurensicherung, S. 75–87. Heidelberg 1986.

Zepf, S., H.-W. Künsebeck, N. Sittaro: Körperbeschwerden und narzißtische Objektbeziehung bei Patienten mit Colitis ulcerosa. Z. psychosom. Med. 27 (1981 b) 59–72.

Zepf, S., H.-W. Künsebeck, N. Sittaro: Untersuchungen zum Selbstwertgefühl von Patienten mit Colitis ulcerosa. Psyche 35 (1981 a) 142–156.

Zepf, S., L. Tschirch: Zur empirischen Überprüfung der Alexithymie mit dem semantischen Differential. Psychother. Med. Psychol 37 (1987) 15–22.

Zepf, S., M.v. Rad: Aktuelle Probleme psychoanalytischer Psychosomatik. Psyche 39 (1985) 738–749.

Zerssen, D.v.: Klinische Selbstbeurteilungs-Skalen (KSb-S) aus dem Münchner Psychiatrischen Informations-System. Beltz, Weinheim 1976.

Zerssen, D. v.: Mood and behavioral changes under corticosteroid therapy. In: Itil, T. M., G. Laudahn, M. W. Herrmann (eds.): Psychotropic Action of Hormones, p. 195. Spectrum, New York 1976.

Zerssen, D.v., H.-J. Moeller, U. Baumann, G. Bühringer: Evaluative Psychotherapieforschung in der BRD und West-Berlin. Psychother. med. Psychol. 36 (1986) 8–11.

Ziegler, F. J., J. B. Imboden, E. Meyer: Contemporary conversion reactions: a clinical study. Amer. J. Psychiat. 116 (1959) 901–910.

Ziegler, G., R. Pulwer, D. Koloczek: Psychische Reaktionen und Krankheitsverarbeitung bei Tumorpatienten – erste Ergebnisse einer empirischen Untersuchung. Psychother. med. Psychol. 34 (1984) 44–49.

Ziegler, R.: Krisenhafte Zustände bei Störungen des Kalziumstoffwechsels. Der Krhs.-Arzt 49 (1976) 2–8.

Ziegler, R., J. A. Sours: A naturalistic study of patients with anorexia nervosa admitted to a university medical center. Comprehens. Psychiat. 9 (1968) 644–651.

Zilboorg, G.: Anxiety without affect. Psychoanal. Quart. 2 (1933) 48–67.

Zilkens, K. W., H. Peters: Die Ileitis terminalis als krebsbegünstigende Vorerkrankung. Med. Klinik 74 (1979) 1713–1715.

Zimmermann, M.: Dorsal root potentials after C-fiber stimulation. Science 160 (1968) 896–898.

Zimmet, P.: Epidemiology of diabetes mellitus. In: Ellenberg, M., H. Rifkin (eds.): Diabetes mellitus, pp. 451–468. Medical Examination Publishing, New York 1983.

Zinberg, N. E.: Psychiatric rounds on the private medical service of a general hospital. In: Zinberg, N. E. (ed.): Psychiatry and Medical Practice in a General Hospital, pp. 124–134. Int. Univ. Press, New York 1964.

Zinberg, N. E.: Psychiatry and medical practice in a general hospital. Int. Univ. Press, New York 1964.

Zinberg, N. E.: The psychiatrist as group observer: notes on training procedure in individual and group psychotherapy. In: Zinberg, N. E. (ed.): Psychiatry and Medical Practice in a General Hospital, pp. 322–336. Int. Univ. Press, New York 1964.

Zinberg, N. E.: The relationship of regressive phenomena to the aging process. In: Zinberg, N. E., I. Kaufman (eds.): Normal Psychology of the Aging Process, pp. 143–159. Int. Univ. Press, New York 1963.

Zinberg, N. E., D. Shapiro, W. Gruen: A group approach to nursing education.In: Zinberg, N. E. (ed.): Psychiatry and Medical Practice in a General Hospital, pp. 283–289. Int. Univ. Press, New York 1964.

Zinberg, N. E., D. Shapiro, W. Gruen: Some vicissitudes of nursing education. In: Zinberg, N. E. (ed.): Psychiatry and Medical Practice in a General Hospital, pp. 290–300. Int. Univ. Press, New York 1964.

Zintl-Wiegand, A., B. Cooper, B. Krumm: Psychisch Kranke in der ärztlichen Allgemeinpraxis. Beltz, Weinheim 1980.

Zintl-Wiegand, A., C. Schmidt-Maushardt, R. Leisner, B. Cooper: Psychische Erkrankungen in Mannheimer Allgemeinpraxen. Eine klinische und epidemiologische Untersuchung. In: Häfner, H. (Hrsg.): Psychiatrische Epidemiologie. Springer, Berlin-Heidelberg-New York 1978.

Zintl-Wiegand, A., et al.: Psychische Erkrankungen in Mannheimer Allgemeinpraxen. In: Häfner, H. (Hrsg.): Psychiatrische Epidemiologie. Springer, Berlin 1979.

Ziolko, H. U.: Anorexia nervosa. Fortschr. Neurol. Psychiat. 34 (1956) 353.

Ziolko, H. U.: Pers. Mitteilung. Berlin 1978.

Ziskind, E.: Isolation stress in medical and mental illness. J. Amer. med. Ass. 1968 (1958) 1427–1431.

Zitrin, C. M.: Differential treatment of phobias: use of imipramine for panic attacks. J. Behav. Ther. exp. Psychiat. 14 (1983) 11–18.

Zitrin, C. M., D. F. Klein, M. G. Woerner, D. C. Ross: Treatment of phobias. Arch. gen. Psychiat. 40 (1983) 125–145.

Zitrin, C. M., D. F. Klein, M. G. Woerner: Behavior therapy, supportive psychotherapy, imipramine, and phobias. Arch. gen. Psychiat. 35 (1978) 307–316.

Zitrin, C. M., D. F. Klein, M. G. Woerner: Treatment of agoraphobia with group exposure in vivo and imipramine. Arch. gen. Psychiat. 37 (1980) 63–72.

Zorick, F., N. Kribbs, T. Roehrs, T. Roth: Polysomnographic and MMPI characteristics of patients with insomnia. In: Hindmarch, I., H. Ott, T. Roth (eds.): Sleep, Benzodiazepines and Performance. Psychopharmacol. (Suppl. I) (1984) 2–10.

Zubek, J. P. (ed.): Sensory deprivation – 15 years of research. Appleton, New York 1969.

Zubin, J., B. Spring: Vulnerability: a new view of schizophrenia. J. abnorm. Psychol. 86 (1977) 103–126.

Zubin, J., J. Magaziner, S. Steinhausel: The metamorphosis of schizophrenia. Psychol. Med. 13 (1983) 551–571.

Zuckerman, M.: Hallucinations, reported sensations and images. In: Zubek, J. P. (ed.): Sensory Deprivation – 15 years of research, pp. 85–125. Appleton, New York 1969.

Zuckerman, M.: Variables affecting deprivation results. In: Zubek, J. P. (ed.): Sensory Deprivation – 15 years of research, pp. 47–84. Appleton, New York 1969.

Zurawski, R. M., T. W. Smith, B. K. Houston: Stress management for essential hypertension: comparison with a minimally effective treatment, predictors of response to treatment, and effects on reactivity. J. psychosom. Res. 31 (1987) 453–462.

Zussmann, L., S. Zussmann, R. Sunley, E. Bjornson: Sexual response after hysterectomy-oophorectomy: recent studies and consideration of psychogenesis. Amer. J. Obstet. Gynec. 140 (1981) 725.

Zweig, S.: Die Heilung durch den Geist. Insel, Leipzig 1932.

Zwicker, E.: Objective otoacoustic emissions and their correlation to tinnitus. In: Feldmann, H. (ed.): Proceedings of the International Tinnitus Seminar Münster 1987, p. 75. Harsch, Karlsruhe 1987.

Zwiebel, R.: Einige klinische Anmerkungen zur Theorie der projektiven Identifizierung. Zschr. psychoanal. Theorie Praxis 3/2 (1988) 165–186.

Zysanski, S. J., J. Wrzesniewski, C. D. Jenkins: Cross-cultural validation of the coronary-prone behavior pattern. Soc. Sci. Med. 13 (1979) 405–412.

Fallberichte

Personenverzeichnis

Sachverzeichnis

Die Zahlen verweisen auf Seiten, die Zusätze A auf Abbildungen, F auf Fallbeispiele und T auf Tabellen.
Halbfette Zahlen verweisen auf Hauptfundstellen.

Sachverzeichnis

Sachverzeichnis

Sachverzeichnis